19th DAAAM International Symposium on Intelligent Manufacturing and Automation (DAAAM 2008)

Focus on Next Generation of Intelligent Systems and Solutions

Trnava, Slovakia
22-25 October 2008

Volume 1 of 2

Editor:

B. Katalinic

ISBN: 978-1-5108-4000-3

Printed from e-media with permission by:

Curran Associates, Inc.
57 Morehouse Lane
Red Hook, NY 12571

CURRAN ASSOCIATES INC. proceedings.com

Some format issues inherent in the e-media version may also appear in this print version.

Printed by Curran Associates, Inc. (2017)

For permission requests, please contact DAAAM International Vienna
at the address below.

DAAAM International Vienna
Karlsplatz 13/311, A-1040
Wien, Austria, European Union

Phone: 0043-1-58801-31121
Fax: 0043-1-58801-31199

secretary@daaam.info

Additional copies of this publication are available from:

Curran Associates, Inc.
57 Morehouse Lane
Red Hook, NY 12571 USA
Phone: 845-758-0400
Fax: 845-758-2633
Email: curran@proceedings.com
Web: www.proceedings.com

ISSN 1726-9679

DANUBE ADRIA ASSOCIATION FOR AUTOMATION & MANUFACTURING

DAAAM INTERNATIONAL VIENNA

www.daaam.com

ANNALS OF DAAAM FOR 2008 & PROCEEDINGS

OF THE 19TH INTERNATIONAL DAAAM SYMPOSIUM
"INTELLIGENT MANUFACTURING & AUTOMATION:
FOCUS ON NEXT GENERATION OF INTELLIGENT SYSTEMS AND SOLUTIONS"
22-25TH OCTOBER 2008, TRNAVA, SLOVAKIA
(SECOND EDITION)

ORGANIZED BY:
DAAAM INTERNATIONAL VIENNA
SLOVAK UNIVERSITY OF TECHNOLOGY
FACULTY OF MATERIALS SCIENCE AND TECHNOLOGY IN TRNAVA
VIENNA UNIVERSITY OF TECHNOLOGY, UNIVERSITY OF APPLIED SCIENCES
TECHNIKUM, AND AUSTRIAN SOCIETY OF ENGINEERS AND ARCHITECTS - ÖIAV 1848

UNDER THE AUSPICES OF: THE DANUBE RECTORS' CONFERENCE &
RECTORS' HONOR COMMITTEE OF DAAAM INTERNATIONAL

SUPPORTED BY FESTO

EDITOR: B.[RANKO] KATALINIC

DAAAM INTERNATIONAL TABULA GRATULATORIA

Happy Birthday to:

Mr Manuel Del Vechio (25 Year) / Dr Radek Matusu (30 Years) / Mr Daniel Novoselovic (30 Years) / Dr Jun Su Ha (35 Years) / Dr Andreas Kandelbauer (35 Years) / Mr Kan Wongkasem (35 Years) / Professor Tiberiu Dobrescu (40 Years) / Dr Snjezana First Rogale (40 Years) / Dr Gabriela Patrascu (40 Years) / Dipl.-Ing. Jana Stefankova (40 Years) / Professor Miran Brezocnik (45 Years) / Mr Sergey Vasilevich Koukhta (45 Years) / Professor Neculai Eugen Seghedin (45 Years) / Professor Anton Hadar (50 Years) / Dr Miroslav Matysek (50 Years) / Professor George Constantin (50 Years) / Dr Kvetoslava Resetova (50 Years) / Professor Predrag Cosic (55 Years) / Professor Franc Cus (55 Years) / Professor Ivan Budic (60 Years) / Professor Ilija Cosic (60 Years) / Professor Malik Kulenovic (60 Years) / Miran Milfelner (60 Years)

University of Rijeka (35 Years) / Middle East Technical Universiy (45 Years) / Sevastopol National Technical University (45 Years)

Faculty of Mechnical –Engineering – Univesity of Sarajevo (50 Years) / Tallinn University of Technology (90 Years) / Austrian Society of Engineers and Architects (160 Years)

Congratulations to:
Professor Evgenij Valentinovich Pashkov on his election to the post of the Rector of SevNTU, Sevastopol
Professor Ivan Budic on his election to the post of the Dean of Polytechnic of Pozega, Pozega

Welcome to the 4808th DAAAM Author:
Since 1990 **4808** authors and co-authors published their scientific papers in the
Annals & Proceedings of DAAAM International Symposium

Special thanks to concern **Festo** for establishing of **"Festo Young Researchers and Scientist Support Scholarship"** for financial support of next generation of Researchers and Scientist at DAAAM international conferences

With love and appreciations
World-wide DAAAM International Community

Who is in the field, and not in the DAAAM he/she does not exist.
(Jyri Papstel 1996)

EDITOR'S NOTE This publication was reproduced by the photo process, using the manuscripts and soft copies supplied by their authors. The layout, the figures and tables of some papers did not conform exactly to the standard requirements. In some cases was the layout of the manuscripts rebuild. All mistakes in manuscripts there could not been changed, nor could the English be checked completely. The readers are therefore asked to excuse any deficiencies in this publication which may have arisen from the above causes. The editor and the DAAAM International Vienna are not responsible either for the statements made or for the opinion expressed in this publication.

DAAAM International is a member of CrossRef. Contributions in DAAAM Scientific Book 2006 and DAAAM International Journal of Simulation Modelling are cross-referenced and available on Internet. DAAAM International publications are cited and referenced in the: Inspec Database, Cambridge Scientific Abstracts, and ISI Scientific Proceedings Thomson Reuters. Papers are available at Thomson Gale.
ANNALS OF DAAAM FOR 2008 & PROCEEDINGS OF THE 19TH INTERNATIONAL DAAAM SYMPOSIUM, VOL 19, NO1,
Abbreviated key title: ***Ann. DAAAM proc. int. DAAAM Symp.***
ISSN 1726-9679, ISBN 978-3-901509-68-1, Editor B.[ranko] Katalinic
Published Annually by DAAAM International Vienna, Vienna 2008

Layout & Design: B.[ranko] Katalinic / Technical Editors: Cesarec P.[aulina], Kettler R.[oman], Alispahic, A.[smir], Brötzenberger R.[ene], Fuchs M.[artin], Richter M.[artin], Rupar M.[iljan], Saliger A.[lex]; Trauzettel, U.[lrich]
DAAAM Home Page Administrators: Malisa V.[iktorio] / Stopper M.[arkus] Graphics used in this Publication: M.C. Escher
Publisher: DAAAM International Vienna, TU Wien, Karlsplatz 13/311, A-1040 Vienna, Austria
E-mails: president@daaam.com katalinic@mail.ift.tuwien.ac.at / DAAAM homepage: http://www.daaam.com http://www.daaam.org

THE 19TH INTERNATIONAL DAAAM SYMPOSIUM

"INTELLIGENT MANUFACTURING & AUTOMATION:

FOCUS ON NEXT GENERATION OF INTELLIGENT SYSTEMS AND SOLUTIONS"

22-25TH OCTOBER 2008, TRNAVA, SLOVAKIA

ORGANISED BY:

Vienna University of Technology	DAAAM International Vienna	Slovak University of Technology MTF Trnava	University of Applied Sciences Technikum Vienna	Austrian Society of Engineers and Architect

UNDER THE AUSPICES OF
DANUBE RECTORS' CONFERENCE &
RECTORS' & PRESIDENTS' HONOR COMMITTEE OF DAAAM INTERNATIONAL FOR 2008

ECATERINA ANDRONESCU	RECTOR, UNIVERSITY POLITEHNICA OF BUCHAREST
JAVIER UCEDA ANTOLÍN	RECTOR, UNIVERSIDAD POLITÉCNICA DE MADRID
VLADIMÍR BÁLEŠ	RECTOR, SLOVAK UNIVERSITY OF TECHNOLOGY BRATISLAVA
GABRIELE BEIBST	RECTOR, JENA UNIVERSITY OF APPLIED SCIENCE
FLOELLA BENJAMIN	CHANCELLOR, THE UNIVERSITY OF EXETER
LAJOS BESENYEI	RECTOR, UNIVERSITY OF MISKOLC
ALEKSA BJELIS	RECTOR, UNIVERSITY OF ZAGREB
VLADIMIR BOBAL	VICE-RECTOR, TOMAS BATA UNIVERSITY UNIVERSITY ZLIN
ION BOSTAN	RECTOR, TECHNICAL UNIVERSITY OF MOLDOVA
ANTONIO CASTORANI	RECTOR, POLYTECHNIC OF BARI
TOMÁŠ ČERMAK	RECTOR, TECHNICAL UNIVERSITY OF OSTRAVA
ANTON CIZMAR	RECTOR, TECHNICAL UNIVERSITY OF KOSICE
FRANCE A. CÓRDOVA	PRESIDENT, PURDUE UNIVERSITY
GIOVANNI DEL TIN	RECTOR, POLITECNICO DI TORINO
JERZY DEMBCZYŃSKI	RECTOR, POZNAN UNIVERSITY OF TECHNOLOGY
MIKHAIL I. DEMCHUK	RECTOR, BYELORUSSIAN STATE POLYTECHNIC ACADEMY
AKOS DETREKŐI	RECTOR, BUDAPEST UNIV. OF TECHNOLOGY AND ECONOMICS
JOZEF GAWLIK	RECTOR, CRACOW UNIVERSITY OF TECHNOLOGY
SERGEY N. GRIGORYEV	RECTOR, MOSCOW STATE UNIVERSITY OF TECHNOLOGY
IAN GOULD	CHANCELLOR, UNIVERSITY OF SOUTH AUSTRALIA
BORIS V. GUSEV	PRESIDENT, RUSSIAN ACADEMY OF ENGINEERING, MOSCOW
VÁCLAV HAVLÍČEK	RECTOR, CZECH TECHNICAL UNIVERSITY IN PRAGUE
SAID IRANDOUST	PRESIDENT, ASIAN INSTITUTE OF TECHNOLOGY
ZEJNEL KELMENDI	RECTOR, UNIVERSITY OF PRISHTINA
MICHAEL L. KHEIFETZ	VICE-RECTOR, POLOTSK STATE UNIVERSITY
DO-YEON KIM	PRESIDENT, UNIVERSITY OF ULSAN
MARKKU KIVIKOSKI	RECTOR, TAMPERE UNIVERSITY OF TECHNOLOGY
GORDANA KRALIK	RECTOR, STROSSMAYER UNIVERSITY OF OSIJEK
VLADO MAJSTOROVIC	RECTOR, UNIVERSITY OF MOSTAR
SALVATORE MARZANO	RECTOR, POLYTECHNIC OF BARI
RADU MUNTEANU	RECTOR, TECHNICAL UNIVERSITY OF CLUJ-NAPOCA
HASAN MURATOVIĆ	RECTOR, UNIVERSITY OF SARAJEVO
CONSTANTIN OPREAN	RECTOR, LUCIAN BLAGA UNIVERSITY OF SIBIU
IVAN PAVIC	RECTOR, UNIVERSITY OF SPLIT
KAREL RAIS	RECTOR, BRNO UNIVERSITY OF TECHNOLOGY
ÁKOS RÉDEY	RECTOR, UNIVERSITY OF PANNONIA
NICOLAE ROBU	RECTOR, THE "POLITEHNICA" UNIVERSITY OF TIMISOARA
PAVEL V. ROSLYAKOV	VICE-RECTOR, MOSCOW POWER ENGINEERING INSTITUTE - TECHNICAL UNIVERSITY
IVAN ROZMAN	RECTOR, UNIVERSITY OF MARIBOR
DANIEL RUKAVINA	RECTOR, UNIVERSITY OF RIJEKA
PETR SAHA	RECTOR, TOMAS BATA UNIVERSITY UNIVERSITY ZLIN
TAKESHI SASAKI	PRESIDENT, UNIVERSITY OF TOKYO
SUPACHAI SAMAPPITO	PRESIDENT, MAHASARAKHAM UNIVERSITY
FRITZ SCHMÖLLEBECK	RECTOR, UNIV. OF APPLIED SCIENCES TECHNIKUM, VIENNA
SERGEY V. SEREBRYANNIKOV	RECTOR, MOSCOW POWER ENGINEERING INSTITUTE - TECHNICAL UNIVERSITY
VIKTOR V.SHALAI	RECTOR, OMSK STATE TECHNICAL UNIVERSITY
PETER SKALICKY	RECTOR, VIENNA UNIVERSITY OF TECHNOLOGY
SERGEY M. STAZHKOV	VICE-RECTOR, BALTIC STATE TECHNICAL UNIVERSITY
WILFRIED STOLL	PRESIDENT, CHAIRMAN OF SUPERVISORY BOARD - FESTO
NAM PJO SUH	PRESIDENT, KOREA ADVANCED INSTITUTE FOR SCIENCE AND TECHNOLOGY
PEEP SÜRJE	RECTOR, TALLINN TECHNICAL UNIVERSITY
WIESLAW TRAMPCZYNSKI	RECTOR, TECHNICAL UNIVERSITY OF KIELCE
DZEMO TUFEKCIC	RECTOR, UNIVERSITY OF TUZLA
ANTE UGLESIC	RECTOR, UNIVERSITY OF ZADAR
ION VISA	RECTOR, TRANSILVANIA UNIVERSITY OF BRASOV
BORIS A. YAKIMOVICH	RECTOR, IZHEVSK STATE TECHNICAL UNIVERSITY, IZHEVSK
WOJCIECH ZIELIŃSKI	RECTOR, SILESIAN UNIVERSITY OF TECHNOLOGY, GLIWICE

Annals of DAAAM for 2008 & Proceedings of the 19th International DAAAM Symposium, ISSN 1726-9679
Editor B. Katalinic, Published by DAAAM International, Vienna, Austria 2008
Make Harmony between Nature and Technology and your Mind will fly free as a Bird

DEANS' AND ACADEMIC DIRECTORS´ HONOR COMMITTEE FOR 2008

REVIEWING COMMITTEE FOR 2008

ORGANIZING COMMITTEE FOR 2008

Annals of DAAAM for 2008 & Proceedings of the 19th International DAAAM Symposium, ISSN 1726-9679
Editor B. Katalinic, Published by DAAAM International, Vienna, Austria 2008
Make Harmony between Nature and Technology and your Mind will fly free as a Bird

SCIENTIFIC DAAAM INTERNATIONAL COMMITTEE FOR 2008

SPONSORS & SUPPORTERS

CITY OF TRNAVA / FESTO / SELF-GOVERNMENTAL REGION OF TRNAVA

BEKAERT, A.S. HLOHOVEC
DENONA D.D.
DRILL, S.R.O. BRATISLAVA
HSTEC D.D., ZADAR
ING. MÁRIA ČUNDERLÍKOVÁ - KESA –TRNAVA
IBOK
INVEX, S.R.O. TRNAVA
JAVYS, A.S. JASLOVSKÉ BOHUNICE
KHS AG - FILLING AND PACKING - WORLDWIDE
NADÁCIA ARX - JASLOVSKÉ BOHUNICE
NADÁCIA JÁNA KORCA TRNAVA
PRVÁ ZVÁRAČSKÁ, A.S. BRATISLAVA
SLOVENSKÁ ELEKTRIZAČNÁ PRENOSOVÁ
SÚSTAVA, A. S. BRATISLAVA
SLOVENSKÉ ELEKTRÁRNE, A.S., ČLEN SKUPINY ENEL
BRATISLAVA
SLOVNAFT, A.S. BRATISLAVA
TECHNODAT, CAE - SYSTÉMY, S.R.O. TRENČÍN
VÚJE, A.S. TRNAVA
VUZ PI SR
ZLIEVAREN TRNAVA
ŽOS, A.S. TRNAVA
ZVS HOLDING, A.S. DUBNICA NAD VÁHOM

Annals of DAAAM for 2008 & Proceedings of the 19th International DAAAM Symposium, ISSN 1726-9679
Editor B. Katalinic, Published by DAAAM International, Vienna, Austria 2008
Make Harmony between Nature and Technology and your Mind will fly free as a Bird

International Program Committee for 2008

Annals of DAAAM for 2008 & Proceedings of the 19th International DAAAM Symposium, ISSN 1726-9679
Editor B. Katalinic, Published by DAAAM International, Vienna, Austria 2008
Make Harmony between Nature and Technology and your Mind will fly free as a Bird

Young Researchers' and Scientists' International Committee for 2008

The main aim of this International Committee is international promotion of talented young researchers and scientists. Target group are: young engineers, master students, young master, doctoral students, young doctors, talented graduate students and other tolerant and talented young persons from one of the fields of interests of DAAAM International. The young winners of Best Paper Award of DAAAM International are also the member of this committee. The candidate has to be under age 35 years and has to be author or co-author of paper published in one of DAAAM publications. The candidate has to be nominated by the supervisor/professor/team leader. The nomination (short CV, research: status, field, course, targets, results) has to be send to the President of DAAAM International.

Annals of DAAAM for 2008 & Proceedings of the 19th International DAAAM Symposium, ISSN 1726-9679
Editor B. Katalinic, Published by DAAAM International, Vienna, Austria 2008
Make Harmony between Nature and Technology and your Mind will fly free as a Bird

DAAAM International Industry, Business and Administration Committee for 2008

The Members of Permanent Committee of the University Network of DAAAM International for 2008 are:

Member organisation	Delegate
DAAAM International Vienna	*President Univ. Prof. Dipl.-Ing. Dr. techn. Dr.mult. h.c. Branko Katalinic*
Danube Rectors´ Conference	*Professor Dr. sc. Franci Cus*
Budapest University of Technology and Economics, Budapest	*The members of Permanent Committee have to be nominated by Budapest University of Technology and Economics*
Josip Juraj Strossmayer University of Osijek	*Professor Dr. sc. Dragomir Krumes*
Kobe University, Kobe	*Director, Chair and Professor Kanji Ueda*
Moscow State University of Technology "Stankin", Moscow	*Professor Dr. Sci. (tech.) Member of Russian Academy of Science Juri M. Solomentsev / Univ. Prof. Dipl.-Ing. Dr. Sergey Egorov*
Romanian-American University of Bucharest, Bucharest	*Univ.Sen.Lect. Dipl.-Ing. Dr. techn. George Carutasu* *Univ.Sen.Lect. Dipl.-Math. Dr. techn. Cezar Botezatu*
Slovak University of Technology Bratislava	*Professor Dr. Marian Vesely* *Professor Ladislav Javorcik*
Tallinn Technical University	*Professor Rein Kyttner*
Technical University of Brno, Brno	*Senator Prof. Ing. Jaroslav Balate, DrSc. Doc. Ing. Vladimir Vasek, CSc.*
Technical University of Cluj-Napoca, Cluj-Napoca	*Dean Prof. Dr.Eng. Petru Berce,* *Prof.dr.Eng. Csaba Gyenge*
Technical University of Kosice, Kosice,	*Prof. Ing. Karol Florian, DrSc., / Prof. Dr. Emil Wessely*
University Politehnica of Bucharest, Bucharest	*Professor Dr. Eng. Horia Gheorghiu* *Lecturer Dr. Eng. Costel Emil Cotet*
University of Maribor, Maribor	*Professor Dr. sc. Joze Balic*
University of Miskolc, Miskolc	*Professor Dr Gabor Terstyanszky*
University of Mostar, Mostar, B & H	*Rector Prof. Dr. Vlado Majstorovic, / Prof. Dr. sc. Vojo Visekruna*
University of Prishtina, Prishtina, Kosovo	*Vice-Rector Prof. Dr. Ahmet Geca / Prof.-Dr. Bajrush Bytyci*
University of Rijeka, Rijeka	*Professor Josip Brnic, Ph.D. / Professor Zoran Mrsa, Ph.D.*
University of Split, Split	*Rector Professor Dr. Ivan Pavic*
University Ovidius of Constanta, Constanta	*Professor Dr. Eng. Mihael Chircor*
University of Zagreb, Zagreb	*Rector Professor Dr. Aleksa Bjelis*

Annals of DAAAM for 2008 & Proceedings of the 19th International DAAAM Symposium, ISSN 1726-9679
Editor B. Katalinic, Published by DAAAM International, Vienna, Austria 2008
Make Harmony between Nature and Technology and your Mind will fly free as a Bird

DAAAM INTERNATIONAL NETWORK:
BACKGROUND, HISTORY, AIMS, STRUCTURE, ACTIVITIES & FUTURE

Branko KATALINIC
President of DAAAM International
www.daaam.com / www.daaam.org

0.0 Contents:

1.0 BACKGROUND

The DAAAM International was founded and based on the idea and initiative of Professor B. Katalinic, on November 5th, 1990 at the Vienna Symposium, which was organised in order to mark the 175th anniversary of the Vienna University of Technology.
The purpose of DAAAM International is to provide a world forum for engineers, scientists and industrial managers to present and discuss the current status and impact of advanced manufacturing and automation.
During the last seventeen years, the activities of DAAAM International have been supported by scientists and experts from over fifty countries all over the world. In the course of the last five years, DAAAM International has focused activities on the regional and global long-term co-operation between (technical) universities. The result of this initiative is the founding of the University Network of DAAAM International Vienna and DAAAM International Network for Advanced Technologies which was founded on 2002-10-25 in Vienna.

2.0 AIMS

The main aims of DAAAM are:

- To organise an annual DAAAM International Symposium each year.
- To found and organise the University Network of DAAAM International Vienna as a strategic alliance and permanent international open platform for long-term co-operation and partnership in all fields of technology, research and education.
- To organise long-term projects, initially known as Inter-university Doctoral Studies
- To organise a pilot project, initially known as the Virtual University.
- To enable and promote the exchange of people, scientific, and technical information.
- To enable and promote international co-operation among scientists and engineers.
- To enable and promote R & D, education and industrial application.
- To co-operate with other national and international organisations in matters of common interest.

3.0 ORGANISATION OF DAAAM INTERNATIONAL VIENNA

DAAAM International Vienna

President:
Branko **Katalinic**

Vice presidents for 2008	Joze **Balic** / Andrew **Kusiak** / Oliver **Moravcik** / Toma **Udiljak** / Kanji **Ueda** / Koloman **Ulrich** / Karol **Velisek**
Honour Vice presidents for 2008	Jaroslav **Balate** / Klaus **Meissner** / P. C. **Pandey** / Janez **Peklenik** / Emil **Wessely**
International Senior Advicer for 2008	Franci **Cus** / Vladimir **Filaretov**/ Malik **Kulenovic**
Secretary General for 2008	Paulina **Cesarec**
Team 2008	Branko **Katalinic** / Paulina **Cesarec** / Liane **Höller** / Roman **Kettler** / Markus **Stopper**

Presidents and Vice Presidents* of National DAAAM Organisations for 2008

Jaroslav Balate	*Czech Republic*
Arne Bilberg	*Denmark*
Jens Bliedtner	*Germany*
Iosif Breido	*Kazakhstan*
Glen Bright	*Republic of South Africa*
Costel Emil Cotet*	*Romania*
Fernando M. M. Cruz	*Portugal*
Mahmud Dikic	*Canada*
Jianhua Dong	*P.R. China*
Igor Drstvensek	*Slovenia*
Joost Duflou	*Belgium*
Vladimir I. Dynnikov*	*Brazil*
Sergey Egorov	*Russia*
Jan Goossenaerts	*Holland*
Johann Grabenweger	*Brazil*
Csaba Gyenge	*Romania*
Qing-Yao Han	*P.R. China*
Peter Holmstedt	*Sweden*
Velko Iltschev	*Bulgaria*
Mohammad A. Irfan	*Pakistan*
Ashutosh Khanna*	*India*
Yurij Kondratenko	*Ukraine*
Petri Kuosmanen	*Finland*
Andrew Kusiak	*USA*
Rein Kyttner	*Estonia*
Marcin A.Krolikowski	*Poland*
Luis Norberto Lopez de Lacalle	*Spain*
Vlado Majstorovic	*Bosnia and Herzegovina*
Ali Meghdari	*Iran*
Karim Mustafizul	*Malaysia*
P. C. Pandey	*India*
Hong-Seok Park	*Korea*
Gilberto Hererra Ruiz	*Mexico*
Kemajl Stuja	*Kosovo*
Dmitry Svirsky	*Belarus*
Kok Kiong Tan	*Singapore*
Mario T. Tabucanon	*Thailand*
Victor Timoschenko	*Moldavia*
Kanji Ueda	*Japan*
Toma Udiljak	*Croatia*
Erol Uyarr	*Turkey*
Ernesto Lo Valvo	*Italy*
Karol Velisek*	*Slovakia*
George Christopher Vosniakos	*Greece*
Emil Wessely	*Slovakia*

DAAAM International Coordinators for 2008

Miroslav **Badida**	*Technical Environmentalistics*
Ivan **Baranek**	*Production Technologies*
Joze **Balic**	*Intelligent Manufacturing*
Ivo **Cala**	*Maintenance*
Dragan **Covic**	*Management*
Costel Emil **Cotet**	*Discrete material flow management*
Franci **Cus**	*High Education Quality Management*
Mahmud **Dikic**	*Assembly Automation*
Zvonko **Dragcevic**	*Textile and Garment*
Alexey **Eliseev**	*Internet Lab Sharing and Networking*
Sergey **Egorov**	*Securities & Financial Management*
Vladimir F. **Filaretov**	*Teleoperators*
Johann **Grabenweger**	*Material Sciences*
Manfred **Gronalt**	*Logistics*
Branko **Katalinic**	*Intelligent Manufacturing Systems*
Viktorio **Malisa**	*Mechatronics / Robotics*
Klaus **Meissner**	*Precision Engineering*
Zoran **Mrsa**	*Optimal Shape Design*
Jozef **Nanasi**	*Automotive Industry*
Markus **Stopper**	*Production Information Technology*
*Ante **Uglesic***	*History of Artefacts*
Kanji **Ueda**	*Biological Manufacturing Systems*
Ante **Uglesic**	*History of Artefacts*
Koloman **Ulrich**	*Production Technologies*
Heinz **Weigl**	*Simulation*

NB: The further International Coordinators for different topics/fields are needed and welcome, if you want to be one of them please contact the president of DAAAM International

4.0 DAAAM Symposiums Clubs

Since founding of DAAAM International Symposium in 1990 **4808** authors and co-authors published at least once their scientific papers in the Annals & Proceedings of DAAAM International Symposium. The members of the club are the persons which presented and/or published their papers at all / not les than 15 / not les than 10 / not less than 5 DAAAM International Symposiums:

All DAAAM Symposiums Authors´ Club

Balic, Joze / Cus, Franci / Katalinic, Branko / Polajnar, Andrej / Veza, Ivica / Wessely, Emil

Members of 15+ DAAAM Symposiums Authors´ Club

Balate, Jaroslav / Boyadjiev, Iliya Kirilov / Buchmeister, Borut / Budic, Ivan / Gyenge, Csaba / Novakovic, Branko / Udiljak, Toma / Velisek, Karol

2009 Candidates for 15+ DAAAM Symposiums Authors´ Club

Following DAAAM authors are candidates and have a chance to join the 15+ Club in 2009
Alpek, Ferenc / Kulenovic, Malik / Micieta, Branislav / Vasek, Vladimir

Members of 10+ DAAAM Symposiums Authors´ Club

Adamczak, Stanislaw / Agic, Darko / Badida, Miroslav / Bobal, Vladimir / Brezocnik, Miran / Buble, Marin / Capek, Jan / Cohodar, Maida / Cyklis, Jerzy / Dragcevic, Zvonko / Drstvensek, Igor / Dulcic, Zelimir / Gojo, Miroslav / Grizelj, Branko / Hodolic, Janko / Janac, Alexander / Janacova, Dagmar / Jerbic, Bojan / Kheifetz, Michael / Kljajin, Milan / Knego, Nikola / Kolomaznik, Karel / Kostal, Peter / Koukhta, Sergei Vasilievich / Krolikowski, Marcin Andrzej / Krumes, Dragomir / Kubalcik, Marek / Kundrak, Janos / Kunica, Zoran / Kuric, Ivan / Leber, Marjan / Mahovic Poljacek, Sanja / Majetic, Dubravko / Majstorovic, Vlado / Malakov, Ivo / Math, Miljenko / Matysek, Miroslav / Meissner, Klaus / Milfelner, Matjaz / Mrsa, Zoran / Mursec, Bogomir / Nanasi, Jozef / Neumann, Petr / Nikolic, Gojko / Pahole, Ivo / Papstel, Jyri / Park, Hong Seok / Pavic, Ivan / Rogale, Dubravko / Siemiatkowski, Mieczyslaw / Stopper, Markus / Szalay, Tibor / Visekruna, Vojo / Voloder, Avdo / Vranjes, Bozo / Weiss, Zenobia / Zajac, Jerzy

2009 Candidates for 10+ DAAAM Symposiums Authors´ Club

Following DAAAM authors are candidates and have a chance to join the 10+ Club in 2009

Acko, Bojan / Ancau, Mircea / Anna, Vojtech / Brnic, Josip / Canadija, Marko / Cosic, Predrag / Cotet, Costel Emil / Dukovski, Vladimir / Karaulova, Tatyana / Kittl, Burkhard / Kralikova, Ruzena / Markos, Sandor / Mikac, Tonci / Palcic, Iztok / Pandilov, Zoran / Peterka, Jozef / Popescu, Diana / Poppeova, Viera / Seghedin, Neculai / Sysala, Tomas / Ueda, Kanji / Zorc, Davor / Zuperl, Uros

Members of 5+ DAAAM Symposiums Authors´ Club

Abhary, Kazem / Abram Zver, Marta / Abrudan, Ioan / Adamek, Milan / Andrlik, Vladimir / Anisic, Zoran / Apostol, Bogdan / Aracic, Stjepan / Arz, Gusztav / Aurite, Traian / Babic, Darko / Bachvarov, Angel / Badjgholi, Farsad / Balc, Nicolae / Barisic, Branimir / Baumann, Christian / Bekes, Jan / Belansky, Jan / Belohoubek, Pavel / Belous, Vitalie / Berce, Petru / Bibu, Marius / Bijelonja, Izet / Bleicher, Fritz / Bliedtner, Jens / Bohez, Erik L J / Bolanca, Stanislav / Bolanca, I / Bolanca, Zdenka / Bolcu, Dumitru / Brezak, Danko / Budic, Hrvoje / Bunjaku, Avdyl / Cala, Ivo / Car, Zlatan / Carija, Zoran / Carutasu, Nicoleta Luminita / Carutasu, George / Casalino, Giuseppe / Catrina, Dumitru / Cekic, Ahmet / Celar, Stipe / Chramcov, Bronislav / Cieloszyk, Janusz / Ciglar, Damir / Ciucescu, Doru / Constantin, George / Cosic, Ilija / Costea, Aurel / Covic, Dragan / Cristea, Luciana / Crnekovic, Mladen / Croitoru, Christian / Cselenyi, Jozsef / Cubonova, Nadezda / Dabacan, Mircea / De Filippis, L A C / Deacu, Liviu / Dimitrovski, Krste / Dolga, Lia / Dolga, Valer / Dostal, Petr / Dostalek, Petr / Dudas, Illes / Dumitras, Catalin Gabriel / Durakbasa, M Numan / Dvorak, Jiri / Dvorakova, Jana / Egorov, Sergey / Emert, Rudolf / Erdeljan, Aleksandar / Farcas, Cristian / Feld, Mieczyslaw / Ficko, Mirko / Filaretov, Vladimir F / First Rogale, Snjezana / Fratila, Domnita / Gatalo, Ratko / Gavlas, Josef / Gawlik, Jozef / Gecevska, Valentina / Ghionea, Adrian / Grabenweger, Johannes / Grgona, Jadranko / Grman, Jan / Gronalt, Manfred / Gudac, Goran / Guenther, Ulrich / Guran, Marius / Gyani, Karoly / Hajduk, Mikulas / Halbritter, Erno / Hernandez, Oscar / Hesselbach, Juergen / Hlava, Jaroslav / Hlavenka, Bohumil / Hofreiter, Milan / Honczarenko, Jerzy / Hradocky, Ladislav / Hrman, Miljenko / Hrubina, Kamil / Hruska, Frantisek / Hruskova, Erika / Ivakovic, Cedomir / Ivancic, Ivan / Jablonski, Wojciech / Jaehn, Hendrik / Janecki, Darius / Jardzioch, Andrzej / Jurecic, Denis / Jurum Kipke, Jasna / Kachanak, Anton / Kasac, Josip / Kerekes, Laszlo / Knezevic, Petar / Knosala, Ryczard / Kolakovic, Marko / Kopac, Janez / Krys, Nikolai / Krzyzanowski, Jozef / Kubinger, Wilfried / Kurtovic, Mate / Lalic, Bojan / Lange, Dirk / Lazea, Gheorghe / Lazibat, Tonci / Lee, Gyu Bong / Lendak, Imre / Lerch, Zoran / Lipa, Zdenko / Ljubetic, Juraj / Lopez De Lacalle, Luis Norberto / Ludovico, Antonio Domenico / Luong, Lee H S / Maglic, Leon / Majdandzic, Niko / Majernik, Milan / Malopolski, Waldemar / Mandic, Lidija / Marchelek, Krzysztof / Marusic, Vlatko / Matjasic Fris, Mateja / Matusu, Radek / Mehta, Narinder Kumar / Meszaros, Imre / Mihoci, Kristijan / Monostori, Laszlo / Morar, Liviu / Moraru, Sorin Aurel / Moskovenko, Igor B / Motika, Roman / Mrocheck, J / Muransky, Juraj / Nagy, Zoltan / Navratil, Zdenek / Navratil, Pavel / Nedelcu, Dumitru / Nekrassov, Grigori / Novoselovic, Daniel / Nowicki, Bogdan / Obsieger, Boris / Olaru, Adrian / Olteanu, Ciprian / Olteanu, Felicia / Oluic, Cedomir / Opruta, Dan / Oravec, Ladislav / Osanna, Peter Herbert / Osmani, Hysni / Otto, Tauno / Pajor, Miroslaw / Pandza, Krsto / Parpala, Lidia Florentina / Pasic, Mugdim / Patrascu, Gabriela / Pavletic, Dusko / Pavlic, Mile / Perinic, Mladen / Perutka, Karel / Peshkoun, Sergei Vasilievich / Petkovic, Darko / Petreus, Dorin / Pierzchala, Wieslaw / Pinteric, Andreja / Pisla, Arian / Pitica, Dan / Pivonka, Petr / Pokorny, Pavel / Poljak, Krunoslav / Pomenic, Loreta / Popa, Marcel Sabin / Popa, Cicerone Laurentiu / Popescu, Sorin G / Powalka, Bartosz / Prodan, Dan / Prokop, Roman / Prostrednik, Daniel / Pruteanu, Octavian Vasile / Przybylski, Wlodzimierz / Rachieru, Nicoleta / Rados, Jerko / Radu, Mihaela / Ravas, Rudolf / Repanovici, Angela / Rezic, Snjezana / Rizescu, Sabin / Robotin, Radu / Rogic, Kristian / Ros, Olimpia / Rovan, Mario / Rugescu, Radu Dan / Ryniewicz, Andrzej / Safran, Mario / Saks, Alo / Samardzic, Ivan / Samek, David / Sanchez, J A / Saric, Tomislav / Sebastijanovic, Slavko / Seda, Milos / Seminsky, Jaroslav / Seucek, Ivan / Sihn, Wilfried / Simunic, Juraj / Sinay, Juraj / Sisek, Boris / Skenderi, Zenun / Skoczynski, Waclaw / Slota, Adam / Sluga, Alojzij / Sokansky, Karel / Somodji, Zeljko / Sopta, Luka / Sostar, Adolf / Srdjak, Miroslav / Staicu, Stefan / Stan, Gheorghe / Stefanic, Nedeljko / Stoic, Antun / Stuja, Kemajl / Sugar, Peter / Syrova, Livia / Sysel, M / Szabo, Otto / Szadkowski, Jan / Szelig, Karoly / Tabucanon, Mario T / Teich, Tobias / Tkalic, Mladen / Tomas, Andrija / Tudjarov, Boris / Turkalj, Goran / Ujevic, Darko / Uricek, Juraj / Varga, Gyula / Vavrik, Ivan / Viducic, Ljiljana / Vincze, Markus / Virkus, Uwe / Vitecek, Antonin / Vitez, Ivan / Vrljicak, Zlatko / Wantuch, Edward / Weigl, Karl Heinz / Zamfira, C Sorin / Zeljkovic, Milan / Zhirabok, Alexey / Ziberna Sujica, Milena / Zichil, Valentin / Zivic Djurovic, Marijana / Zmijewski, Wladyslav / Zvirinsky, Vladimir

2009 Candidates for 5+ DAAAM Symposiums Authors´ Club

Following DAAAM authors are candidates and have a chance to join the 5+ Club in 2009

Adamczyk, Zbigniew / Axinte, Eugen / Bajic, Drazen / Baksa, Sarajko / Bana, Viktoria / Banabic, Dorel / Banyai, T / Benes, Petr / Benic, Djuro / Bliznak, Michal / Borza, Alexandru Ioan / Brajlih, Tomaz / Brandes, Arndt / Budin, Rajka / Buerger, Wolfgang / Bulgaru, Marius / Bytyci, Bajrush / Carp Ciocardia, Daniela Craita / Chodzko, Marcin / Comsa, Dan Sorin / Corluka, Venco / Cuesta, Eduardo Gonzales / Cukor, Goran / Curtu, Ioan / Curtu, Djurdjica / Dlapa, Marek / Dobras, Branka / Dolecek, Vlatko / Dolinay, Jan / Dragoi, George / Duerr, Holger / Dumitriu, Adrian / Dvorackova, Marie / Dzaferovic, Ejub / Eftimie, Lucian / Egorova, Tatyana / Ekinovic, Sabahudin / Evin, Emil / Gabrijelcic, Helena / Gallo, Mikulas / Gallova, Stefania / Gersak, Jelka / Ghionea, Ionut / Giannoccaro, Rosanna / Gil, Stanislaw / Gotovac,

Sven / Greguss, Pal / Grizelj, Dejan / Grubisic, Dragana / Grzesiak, Dariusz / Hadar, Anton / Haertig, Guenter / Hait, Alexander L / Hajrizi, Edmond / Hebbal, Shivanand / Hrzic, Dragana / Iltschev, Velko / Jadlovska, Anna / Jain, Pramod K / Janzekovic, Marjan / Jermak, Czeslaw Janusz / Johanyak, Zsolt Csaba / Jovic, Franjo / Jurda, Pavel / Kavran, Zvonko / Keran, Zdenka / Kirchfeld, Maria / Kiss, Akos / Klapka, Jindrich / Kodacsy, Janos / Kolibal, Zdenek / Kosmol, Jan / Kovacevic, Branko / Kovacevic, Drazen / Kovacic, Miha / Kowalski, Marek / Kozak, Drazan / Krajci, Vesna / Kuenanz, Klaus / Kuzle, Igor / Kuzman, Karl / Lache, Simona / Lamikiz, Aitzol / Ljubic, Franjo / Ljuhar, Davul / Macku, Lubomir / Manriquez, Jose / Marek, Bartosz / Marincic, Valentina / Markic, Brano / Marunic, Gordana / Matusova, Miriam / Mera, Mircea / Mihelic Bogdanic, Alka / Mlinaric, Tomislav J / Moeller, Klaus / Mohora, Cristina / Mrcela, Tomislav / Mudronja, Vedran / Mudura, Pavel / Nagla, Kuldeep Singh / Nardin, Blaz / Neage, Irina / Nedeff, Valentin / Nemes, Toderita / Neslusan, Miroslav / Oberlaender, Rudolf / Osmera, Pavel / Palaghita, Niculaie / Paralic, Jan / Parpala, Radu Constantin / Pavlickova, Jirina / Pawlukowicz, Piotr / Pechacek, Frantisek / Perez, Loseba M / Peric, Nedjeljko / Petrescu, Valentin / Piculin, Srecko / Pidany, Jan / Pintilie, Gheorghe / Pisla, Doina / Pizmoht, Petja / Plancak, Miroslav / Ploj, Tone / Popescu, Ion / Radic, Jure / Rajkovic, Borivoje / Rubio, Eva M / Rucki, Miroslaw / Rusu, Lazar / Sabol, Tomas / Salihu, Avdi / Salopek, Ivana / Samanovic, Josip / Sariguil, Erdal / Sebastian, Miguel Angel / Sebastijanovic, Nebojsa / Sebo, Dusan / Senchilo, Igor Arkadievich / Severincu, Miha / Siminiati, Dubravka / Simionescu, Gheorghe / Situm, Zeljko / Skoko, Miroslav / Skolud, Bozena / Skunca, Marko / Sladek, Jerzy / Sliwka, Janusz / Smutny, Lubomir / Sogorovic, Danijel / Sokhan, Serg / Solc, Frantisek / Solenicki, Goran / Starcevic, Gordan / Stefanek, Michal / Stefanuta, Ioan / Stjepanovic, Zoran / Strgar Kurecic, Maja / Sunje, Aziz / Surace, Rossella / Sutlovic, Igor / Svirsky, Dmitry / Tache, Florin / Takakuwa, Soemon / Tarca, Ioan / Tarca, Radu Catalin / Tierean, Mircea / Tomic, Drazana / Topliceanu, Liliana / Toth, Andras / Uhlik, Alexander / Ungur, Petru / Ungureanu, Catalin Sorin / Vaclavek, Pavel / Vaida, Liviu / Valter, Zdravko / Vaupotic, Bostjan / Viteckova, Miluse / Vrdoljak Raguz, Ivona / Vujasinovic, Edita / Vujica Herzog, Natasa / Vukmirovic, Srdan / Vukovic, Ivan / Weiland, Kristian / Weinert, Klaus / Wong, Hung Kun / Zebala, Wojciech / Zizka, Jan / Zjakic, Igor / Zurek, Witold

Authors of DAAAM 2008 - NewComer Club

Abrudan, Adrian / Abrudan, Gheorghe / Adamcikova, Andrea / Adamescu, Dorina / Agrianidis, Anastasios / Aisman, David / Alecu, Aurel / Alexa, Florin / Alexandrescu, Marius / Alexei, Adrian / Alionte, Cristian Gabriel / Almeida, Henrique / Alutei, Adrian / Alvarez, Roberto / Amariei, Olga Ioana / Amberg, Michael / Anania, Florea Dorel / Andersson, Paul / Andrei, Gabriel / Andrejcak, Imrich / Andres, Solomia / Anghel, Alina / Anghel, Cornelia / Anthymidis, Konstantinos / Antonescu, Ion / Ardelean, Felician / Argesanu, Veronica / Arion, Aurel / Aryassov, Gennady / Atanasiu, Costica / Atanasiu, Virgil / Avram, Calin / Avram, Eleonora Laura / Avram, Laura / Ayyaz, Ahmad / Azzouzi, Messaouda / Baba, Marius Nicolae / Badarau, Rodica / Badea, Adrian / Badea, Cristinel / Badescu, Loredana Anne / Baeckstroem, Mikael / Bagu, Constantin / Bajcicak, Martin / Bajramovic, Rasim / Bal, Carmen / Bal, Nicolaie / Balan, Corneliu / Balasoiu, Victor / Balcak, Stanislav / Balcu, Ion / Bancila, Daniel / Banciu, Felicia / Bandrabur, Diana Lilia / Banea, Mariana / Banerjee, Soumya / Banica, Mihai / Bara, Camelia / Bara, Lucian / Bara, Vasile / Barborak, Oto / Barbu, Teodora / Barbu, Valentin / Barglazan, Diana / Barglazan, Mircea / Barta, Jozef / Bartolo, Paulo / Bassoli, Elena / Baya, Alexandru / Beca, Paul / Bejan, Mihai / Beles, Horia / Bena, Irina / Benak, Michal / Bertrand, Philippe / Beznak, Matej / Beznila, Hortensia / Biclesanu, Cornelia / Biedunkiewicz, Witold / Bienert, Jan / Bilic, Bozenko / Bilik, Jozef / Birau, Nicolae / Birdeanu, Elena Daniela / Biriescu, Marius / Biris, Cristina / Birsan, Dan / Birsan, Iulian Gabriel / Bisu, Claudiu / Bizau, Viorel / Bizdoaca, Nicu / Bjoerkstrand, Roy / Blaga, Florin / Blaga, Vasile / Blidisel, Rodica / Bode, Florin / Bogoi, Alina / Bogovic, Tomislav / Boiangiu, Costin Anton / Boiangiu, Mihail / Boicea, Alexandru / Boieriu, Camelia / Bolanca Mirkovic, Ivana / Bolfa, Traian Eugen / Borangiu, Theodor / Borawska, Patrycja / Bordea, Nicolae / Borza, Ioana / Borza, Paul Nicolae / Bota, Josip / Botezatu, Elena / Botezatu, Mihai Alexandru / Boyadjiev, Ilia / Bozek, Pavol / Brabie, Gheorghe / Bratu, Anca / Braun, Barbu / Bretcu, Angela / Bria, Vasile / Brindasu, Paul Dan / Brodnianska, Zuzana / Brotac, Sorin / Bucur, Gabriela / Budimir, Verica / Budzik, Grzegorz / Buidos, Traian / Bulej, Vladimir / Bunjaku, Januz / Buracu, Victor / Buransky, Ivan / Burca, Ioan / Burger, Wolfgang / Butnariu, Silviu / Buzera, Marius / Caganova, Dagmar / Cahuc, Olivier / Calbureanu, Madalina / Calefariu, Gavrila / Camacho, Ana / Cambal, Milos / Cananau, Dan Cristian / Cananau, Sorin / Capko, Darko / Caplovic, Lubomir / Carausu, Constantin / Carbunar, Mihai / Carcea, Ioan / Carp, Iulia / Carutasu, Nicoleta / Catana, Dorina / Catana, Madalin / Catana, Razvan / Catas, Adriana / Cazacu, Maria / Cehajic, Arijana / Cernecky, Jozef / Cervenansky, Jaroslav / Cha, Sungdeok / Charbulova, Marcela / Charvatova, Hana / Chaus, Alexander / Chelariu, Romeu / Chelaru, Elena Silvia / Chelu, Gheorghe / Cherciu, Mirela / Cherlea, Ioan Valeriu / Chicos, Lucia Antoneta / Chioncel, Cristian Paul / Chioreanu, Nicolae / Chira, Flavia / Chircor, Lidia / Chiriacescu, Sergiu / Chirilus, Alina Alexandra / Chirita, Bogdan / Chiru, Anghel / Chita, Monica Anca / Choi, Hung Won / Chwajol, Grzegorz / Cibulka, Viliam / Cican, Grigore / Cioata, Vasile George / Cioban, Horia / Ciobanu, Bogdan / Ciobanu, Camelia / Ciobanu, Gheorghe / Ciobanu, Ioan / Ciobanu, Mircea / Ciocirlan, Doinita / Ciofu, Ciprian / Cionca, Dan / Ciotea, Margareta / Circiumaru, Adrian / Ciuca, Ion / Ciuca, Sorin / Ciupitu, Ion / Cociu, Nicolae / Cocorada, Sorin / Coculescu, Cristian Catalin / Coculescu, Cristina / Cohal, Viorel / Colic, Mirsad / Coman, Daniela / Coman, Liviu / Comanescu, Adriana / Comanescu, Brindus / Comanescu, Dinu / Constantin, Emil / Cooper, Edward / Copilus, Cristian / Cornescu, Viorel / Cosic, Kristina / Cotetiu, Adriana / Cotetiu, Radu / Cotrut, Cosmin Mihai / Covic, Daniela / Cozminca, Mircea / Craciun, Dan / Craciunescu, Puiu / Crainic, Nicolae / Cretu, Dan / Crisan, Aurel / Crisan, Gheorghe / Crisan, Lia / Croitoru, Bogdan / Croitoru, Catalin / Croitoru, Sorin Mihai / Cronskaer, Marie /

Crudu, Liviu / Cruz, Fernando / Cseki, Liviu Bogdan / Cuciumita, Cleopatra / Curak, Ivan / Curkovic, Milan / Cus, Franci / Dado, Miroslav / Dancsecz, Gabriella / Danihelova, Anna / Danisova, Nina / David, Ioan / David, Konstantinos / Davidescu, Arjana / De Agustina, Beatriz / Deaconescu, Andrea / Deaconescu, Tudor / Deaky, Bogdan / Dehelean, Nicolae / Del Vechio, Manuel / Delgado, Antonio / Demian, Gabriela / Demian, Mihai / Denzinger, Joachim / Dermanovic, Ivana / Diaconescu, Dorin / Diaconu, Ilie / Diaconu, Nicolae / Dichiu, Daniel / Dima, Bogdan / Dima, Dumitru / Dini, Gino / Dinita, Ion / Dinu, Doina / Dinu, Gabriela / Dobra, Andreea / Dobre, Daniel / Dobren, Flavius Andrei / Dodun, Oana / Dogaru, Florin / Doicin, Cristian / Domankova, Maria / Domuta, Cristian / Donevski, Davor / Dorian, Marin / Dormal, Thierry / Doroftei, Ioan / Dosoftei, Constantin Catalin / Dragan, Dinu / Draghia, Marius / Draghici, Sorin / Dragomir, George / Dragomir, Madalina / Dresvyannikov, Denis / Dreucean, Mircea / Drosescu, Radu / Druica, Elena / Duicu, Simona / Dujic, Lucijana / Dumitrascu, Adela / Dumitrescu, Constantin Dan / Dumitrescu, Corina / Dumitrescu, Diana Mariana / Dumitrescu, Lucia / Dumitriu, Adelina / Dumitriu, Dan / Dupta, Pavel Ernest / Duran, Dan / Duran, Irina / Duran, Vasile / Durica, Jan / Duris, Rastislav / Dusek, Daniel / Dvornic, Andrei Iulian / Dzimbeg Malcic, Vesna / Dziubek, Tomasz / Eberst, Otto / Eftimie, Adrian Catalin / Eftimie, Alina Luciana / Eftimie, Roxana / Elezi, Dervish / Enciu, George / Enescu, Ioan / Enescu, Monica / Esse, Martin / Ester, Jozef / Evans, Kenneth / Failli, Franco / Farcane, Nicoleta / Faur, Cosmin / Fedoryshyn, Roman / Fejzullahu, Xhemajl / Fetecau, Catalin / Fialka, Miloslav / Fidler, Branislav / Fidler, Martin / Fidlerova, Helena / Filip, Ioan / Filip, Iulian / Filipowicz, Krzysztof Z. / Florea, Claudia / Florescu, Daniela / Florescu, Iulian / Foeldi, Andrei / Folcut, Ovidiu / Forgaciu, Flavia Lelia / Fota, Adriana / Frietsch, Markus / Fucak, Sanjin / Fulconis, Farcane / Gabor, Gianina / Gaese, Thomas / Gajdos, Ivan / Galetic, Fran / Galic, Josip / Gapinski, Bartosz / Gatto, Andrea / Gavrilovic, S. Branislav / Gavrus, Cristina / Geaman, Virgil / Geca, Ahmet / Georgescu, Luminita / Gerard, Alain / Ghenghea, Dan / Gheorghe, Camelia Monica / Gheorghe, Gheorghe / Gheorghe, Ion Gheorghe / Gheorghiu, Liviu / Ghiban, Brandusa / Ghiban, Nicolae / Ghiculescu, Daniel / Ghinea, Mihai / Ghinea, Mihalache / Ghinea, Valentina Mihaela / Ghiorghe, Adrian / Ghita, Constantin / Ghitescu, Teodor / Gitea, Manuel / Godal, Daniel / Godec, Damir / Gogu, Grigore / Goh, Gyoung Tae / Goia, Ioan / Golubovic, Kristijan / Gombos, Mircea / Gombos, Radu / Gontean, Aurel / Gordan, Cornelia / Grebenisan, Dan / Grecu, Eugenia / Grecu, Luminita / Grigoras, Stefan / Grigore, Roxana / Grigorescu, Costin / Grigoriu, Aurelia / Grigoriu, Mircea / Guempel, Paul / Guiman, Maria Violeta / Guta, Carmen / Ha, Jun Su / Hajduchova, Luba / Hamat, Codruta Oana / Han, Changhyo / Harangus, Daniela / Haruta, Ovidiu / Hasan, Saad / Hazi, Gheorghe / Herceg, Tomislav / Hoffmann, Cristina / Hoghova, Katarina / Holm, Timo / Homa, Johannes / Horak, Karel / Horja, Ioana Monica / Horja, Ioana Monica / Horman, Izet / Horvath, Stefan / Hribernik, Ales / Husanu, Valerica / Iacovoiu, Viorela / Iancu, Anica / Iancu, Carmen / Ibraim, Cristina / Ignat, Andreea / Ignaton, Elemer / Ihan Hren, Natasa / Ihan Hren, Natasha / Ilcik, Stefan / Ileana, Ioan / Ilies, Paul / Iliescu, Mihaiela / Iliescu, Nicolae / Ilii, Sanda / Inaguma, Takanao / Ionescu, Alexandru / Ionescu, Cristian / Ionescu, Daniela Smaranda / Ionescu, Gabriela / Ionescu, Gheorghe Constantin / Ionescu, Mircea Dan / Ionescu, Nicolae / Ionescu, Octavian / Ionescu, Sorin / Ionescu, Stefan Alexandru / Iordachescu, Danut / Iordachescu, Mihaela / Iordan, Andrei / Iosub, Adrian / Istrat, Nicolae / Itu, Calin / Itul, Tiberiu / Iuras, Emilia / Ivan, Ioana / Ivan, Ioana Carmen / Ivan, Maria Cornelia / Ivan, Mihail Vincentiu / Ivanescu, Andrei Nick / Ivanescu, Nick / Ivanov, Marian / Ivascu, Costel / Ivetic, Dragan / Jadaneant, Mihai / Jancic, Zoran / Jedlicka, Martin / Jee, Eun Kyoung / Jelenc, Lara / Jeon, Changsu / Jeon, Seungjae / Jirkova, Hana / Jivanescu, Iulia / Jokanovic, Simo / Jones, Jason / Jula, Mihaela / Jurca, Lucian / Jurchela, Alin / Kachakov, Zharko / Kanerva, Jukka / Karabenciov, Adrian / Karakulov, Maksim / Karaoglanidis, George / Karapetrovic, Stanislav / Karuovic, Dijana / Kasala, Jozef / Kii, Konstantin / Kisicek, Tomislav / Knevez, Jean Yves / Kobrin, Alexandr / Koh, Kwang Yong / Kolenic, Frantisek / Kolosowski, Mariusz / Koniar, Jan / Konobritskiy, Dmitry / Koptioug, Andrei / Kostanjevec, Tomaz / Kostic, Aleksandra / Kramar, Vadim / Kramberger, Slavko / Krasniqi, Fehmi / Krizanova, Gabriela / Krupa, Marek / Krupka, Lothar / Kucher, Dimitry / Kwak, Hyun Soo / Laatsit, Toomas / Laget, Bernard / Lai, Kim Teng / Lajic, Branka / Lampart, Marek / Lancea, Camil / Langova, Nada / Latifi, Ahmet / Latinovic, Tiho / Laureyns, Isabelle / Lavrov, Valentin / Lazar, Liviu / Lazarescu, Cezara / Lee, Kyungkeun / Leemet, Tonu / Letinic, Svjetlana / Letunic, Stijepo / Lewandowski, Jakob / Lica, Dumitru / Lim, Mei / Lisjak, Dragutin / Liska, Jan / Liskova, Jana / Loeser, Carsten / Loncar, Damir / Lorenzen, Leif Erik / Lovasz, Erwin Christian / Luca, Gheorghe / Luchin, Milenco / Lungu, Andy / Lungu, Bogdan / Lungu, Ion / Lungu, Mihai / Lungu, Romulus / Luong, Lee / Lupeanu, Mihaela / Lupescu, Ovidiu / Lupinca, Cinca Ionel / Lupu, Ciprian / Lupulescu, Nouras / Luschmann, Christian / Machedon Pisu, Mihai / Maekitie, Antti / Magheti, Ioan / Majdik, Andras / Majerik, Marian / Majnaric, Igor / Maksarov, Viacheslav / Maksuti, Rrahim / Malciu, Raluca / Maleshkov, Stoyan / Malina, Jiri / Malos, Raul / Mamulea, Adina Ana / Mancia, Aurora / Manciulea, Ileana / Mandru, Dan / Manea, Adriana Sida / Maniu, Laura / Marascu Klein, Vladimir / Marc, Gheorghe / Marcu, Cosmin / Marcu, Florin / Margineanu, Dan / Marin, Alexandru / Marin, Cornel / Marin, Dumitru / Marin, Gheorghe / Marin, Marta / Marin Pantelescu, Andreea / Marinac, Antun / Marincel, Petra / Marinescu, Mirel / Marinescu, Niculae / Marinescu, Sorin Alin / Markowska, Olimpia / Markowski, Tadeusz / Maronek, Milan / Marta, Constantin / Martinescu, Ionel / Martinovic, Dunja / Maruf, Hasan / Masar, Alojz / Masarova, Renata / Masek, Bohuslav / Masiar, Harold / Matiko, Fedir / Matsi, Birthe / Matulja, Dunja / Mazilescu, Crisanta Alina / Mazuru, Sergiu / Meciarova, Julia / Meckl, Peter / Menyhardt, Karoly / Merticaru, Vasile Jr. / Meudtner, Andreas / Mezentsev, Mikhail / Micu, Ioan / Miechowicz, Stawomir / Mihai, Valia Maria / Mihail, Laurentiu / Mihaila, Stefan / Mihailidis, Athanassios / Mihartescu, Ana Andreea / Mihet Popa, Lucian / Mihic, Srdan / Mihljevic, Viktor / Mihok, Jozef / Miklosevic, Kresimir / Mikota, Miroslav / Mikulic, Zelimir / Milcic, Diana / Militaru, Gheorghe / Milos, Teodor / Milosan, Ioan / Minica, Mirela / Mircea, Octavian / Mironeasa, Costel / Mironeasa, Silvia / Misik, Ladislav / Misulic, Marko / Mjaku, Malush / Modog, Traian / Modrea, Arina / Molcho, Gila / Moldovan, Cristian / Moldoveanu,

Alin / Moldoveanu, Florica / Morariu, Zaharie / Moroianu, Corneliu / Morovic, Mira / Mrcelic, Zeljka / Mrenes, Mircea / Mrvac, Nikola / Mudrik, Jozef / Muhaxheri, Mustafe / Munguia, Javier / Muntean, Angela / Muntean, Sebastian / Munteanu, Adriana / Munteanu, Daniel / Munteanu, Eliza / Munteanu, Gabriel / Munteanu, Irena / Munteanu, Mihaela Violeta / Munteanu, Sorin Ion / Muriqi, Ali / Musca, Elena / Musca, Gavril / Mustea Serban, Razvan / Mutka, Alan / Nad, Milan / Nagi, Mihai / Nagy, Ramona / Nakatani, Hiromasa / Nanasi, Tibor / Nanu, Neculai / Nascak, Lubomir / Nastac, Silviu / Neagu, Dumitru / Nedelcu, Adrian Valentin / Nedelescu, Mihai / Negoitescu, Arina / Negrea, Virgiliu Dan / Negru, Mihai / Negru Strauti, Gabriela / Neidenbach, Norbert / Nerantzis, Ioannis / Nesvadba, Petr / Ngoc Anh, Tran / Nicodim, Mariana / Nishiyama, Daisuke / Nizetic, Josip / Nizic, Berislav / Nutu, Emil / Ocenasova, Lenka / Ocnarescu, Constantin / Olah, Iosif / Olarescu, Alin / Olt, Jueri / Olteanu, Vlad Gabriel / Opelsz, Helen / Opetuk, Tihomir / Oroian, Maria / Ostoia, Daniel / Otel, Calin / Paclisan, Dana / Pacurar, Razvan / Paecurar, Razvan / Palm, Stjepan / Paloheimo, Kaija Stiina / Paloheimo, Markku / Panduru, Lucian / Pantea, Ioan / Paraschiv, Marius Daniel / Pardanjac, Marjana / Park, Jin Woo / Parv, Luminita / Parvu, Felix / Pastrama, Stefan Dan / Patachia, Silvia / Pavlic, Daniel / Pavlovic, Ivana / Pavlovic, Tomislav / Pelau, Corina / Perminova, Anastasia / Pernevan, Ioan / Pernevan, Maria / Pernis, Rudolf / Petcu, Romulus / Petre, Mihai / Petrescu, Ligia / Petrlic, Dario / Petropoulos, Georgios / Pikna, Lukas / Pilipovic, Ana / Pirtea, Marilen / Pistun, Yevhen / Pleniceanu, Aristica / Podhorsky, Stefan / Poenaru, Violeta Domnica / Polakovic, Milos / Pop, Alin / Pop, Nicolae / Pop, Nicolae Alexandru / Pop Lupu, Anca / Popa, Adina / Popa, Elena / Popa, Ionut / Popa, Ramona / Popa, Viorel / Popentiu Vladicesu, Florin / Popescu, Cristina / Popescu, Daniel / Popescu, Jeni / Popescu, Mihaela / Popescu, Radu / Popescu, Rodica / Popescu Mitroi, Maria Monica / Popovici, Gheorghe / Prajova, Vanesa / Preda, Ion / Preda, Oana / Predoiu, Constantin / Prejmerean, Mihaela / Prem, Florica / Pribytkova, Marina / Prichici, Mariana Adriana / Pricope, Cornel / Prisacaru, Gheorghe / Prostean, Gabriela / Prostean, Octavian / Pruna, Liviu / Purcarea, Ramona / Racic, Ifigenija / Radakovic, Nikola / Radman Pesa, Anita / Radosav, Dragica / Radu, Adrian / Radu, Ciprian / Radu, Mioara / Raducanu, Bogdan / Radulescu, Alexandru / Radulescu, Bruno / Radulescu, Florin / Radulescu, Irina / Radulescu, Mara / Raennar, Lars Erik / Rahau, Loredana / Raicu, Lucian / Ramsay, Hans / Rat, Nadia Ramona / Recio Perero, Pablo / Rejda, Rudolf / Rennie, Allan / Ridzon, Martin / Rindasu, Venera / Risovic, Stjepan / Risteiu, Mircea / Rittidech, Sampan / Robu, Andreea / Robu, Raul / Rohan, Rodica / Roman, Costel / Romanescu, Maria Camelia / Roosimoelder, Lembit / Rosca, Ileana / Rosca, Ioan / Rosu, Dorin / Rotaru, Horatiu / Rubio, Eva M. / Rugescu, Dragos Ronald / Runkas, Martin / Rus, Alexandru / Rusnac, Lucian / Rusu, Lucian / Ruzarovsky, Roman / Ruzicka, Jan / Rybansky, Rudolf / Ryu, Kwangyeol / Saenz De Pipaon, Jose / Salminen, Kai / Samak, Samoil / Samakoski, Blagoja / Sandru, Diana / Sandu, Ion / Santos Martin, Francisco / Sarbu, Flavius / Saulescu, Radu / Savin, Adriana / Savu, Lucian / Scarneciu, Ioan / Schor, Vladimir / Schuermeyer, Jan / Schumann, Christian / Scukanec, Danijel / Scurtu, Dan / Scutaru, Maria Luminita / Scutelnicu, Elena / Seong, Poong Hyun / Serb, Adrian / Serban, Nicolae / Sercer, Mladen / Sergiu, Nicolae / Sewell, Neil Thomas / Shin, Seung Ki / Shirobokov, Konstantin / Sica, Mihaela / Siketova, Katarina / Silivestru, Valentin / Sima, Ion / Simeunovic, Nenad / Simion, Ionel / Sindila, Gheorghe / Sintea, Sorin / Sinyavskiy, Oleg / Skala, Tibor / Skalova, Ludmila / Slota, Jan / Smurov, Igor / Soare, Vasile / Sobolak, Mariusz / Sobota, Robert / Soceanu, Alexandru / Sokoloski, Zlatko / Spataru, Andrei Cristian / Spisak, Emil / Staicu, Florentiu / Stamate, Valentin / Stamatescu, Grigore / Stampfl, Juergen / Stan, Andrei / Stanasel, Iulian / Stanasel, Oana / Stanciu, Anka / Stanciu, Mariana / Stanciu, Mihai / Stanciu, Valerian / Stanescu, Cristina / Stanescu, Marius / Stanescu, Titu / Stanic, Nikolina / Stankovic, Ana / Starko, Kirill / Starshev, Denis / Stefan, Constantin / Stibritz, Guido / Sticlaru, Carmen / Stimac, Igor / Stipancic, Tomislav / Stirbu, Cosmin / Stirbu, Cristel / Stoerzner, Falko / Stoia, Dan Ioan / Stoica, Gina Florica / Stoica Guzun, Anicuta / Stoicanescu, Maria / Stollmann, Vladimir / Stroita, Daniel Catalin / Stuparu, Adrian / Suba, Roland / Suchomel, Jozef / Suciu, Lenuta / Surcel, Ioan / Suriansky, Jozef / Szeidert, Iosif / Szekely, Iuliu / Szentes, Balazs / Tabara, Iulian / Tallis, Andrea / Talpos, Ioan / Tamas, Levente / Tanasa, Raluca / Tanasa, Valentin / Tane, Nicolae / Taratinsky, Ilya / Tarcolea, Mihai / Tarnita, Dan / Tarnita, Daniela / Tatar, Olimpiu / Tatarciuc, Monica / Tataru, Mircea Bogdan / Tatucu, Iancu / Teodorescu, Horatiu / Tera, Melania / Teusdea, Alin / Thierheimer, Diana / Tiba, Zsolt / Tilina, Dana / Tilina, Dragos / Timar, Maria Cristina / Tion, Monica / Tittel, Viktor / Toca, Alexei / Todericiu, Ramona / Tofan, Mihai / Tokar, Adriana / Toma, Victoria / Tomasegovic, Tamara / Tomic, Mladen / Tomoiaga, Tiberius / Tomuletiu, Elena Adriana / Tonoiu, Sergiu / Toth Tascau, Mirela / Tothova, Maria / Trso, Martin / Tulbure, Adrian / Tuomi, Jukka / Turna, Milan / Turnova, Zuzana / Udrea, Andreea / Udroiu, Razvan / Ulrich, Koloman / Ungur, Patricia / Ungureanu, Dan / Ungureanu, Delia / Ungureanu, Miorita / Ungureanu, Nicolae / Ursu, Ioan / Vacarescu, Cella Flavia / Vacarescu, Valeria / Vacek, Vladimir / Valcuha, Stefan / Van Voorthuysen, Erik Jan / Vancea, Smaranda / Vartolomei, Mihaela / Vartolomei M, Mihael / Vasar, Cristian / Vasicek, Vesna / Vasilache, Simona / Vaxevanidis, Nikolaos / Veres, E Mircea / Viglasky, Jozef / Vilag, Valeriu / Villeta, Maria / Vintila, Georgeta / Visan, Dana / Visan, Daniel / Vitalariu, Anca / Vlad, Angelica / Vlad, Dorin / Vlase, Auremia / Vlase, Sorin / Voicu, Andreea Raluca / Voicu, Vasilica / Voicu, Vlad / Voisan, Emil / Volf, Irina / Vologdin, Sergey / Vujic Ziljak, Jana / Vujicic Tomic, Brankica / Vulpe, Anca Diana / Vulpe, Marius Cristian / Walekova, Gabriela / Weiss, Edmund / Wimpenny, David / Winkler, Sebastian / Wolf, Edwin / Wolny, Ryszard / Wongkasem, Kan / Yadroitsev, Igor / Yassin, Adrian / Yiangkamolsing, Chana / Zaharia, Sebastian Marian / Zaheu, Ioana / Zamfira, C. Sorin / Zelenay, Miroslav / Zeqiri, Fitim / Zezelj, Dragan / Zgardea, Emanuel / Zimmermann, Matthias / Zitinski Elias, Paula

5.0 IN MEMORIAM OF NICE DAAAM COLLEAGUES AND DEAR FRIENDS

Professor Dr Jyri Papstel (1940- 2004), Tallinn Technical University, Tallinn Estonia, Vice President of DAAAM International, Founder and President of DAAAM Estonia, Founding member of the pilot projects Interuniversity Doctoral Studies and Virtual University. Member of: International Organizing / Program / Scientific / Review Committees. President of Award Committee. Author. Honored with Medal of DAAAM International Vienna.

M. Sc. Dipl.-Ing. Ivan Ivancic (1952-2004), Ericsson - Tesla, Zagreb, Croatia, International Coordinator for the field Maintenance, Member of International Organizing / Program / Review Committees.

Professor Dr Zdenko Kordic (1949-2004), University of Mostar and University of Split, Founder and First Rector of the University of Mostar, Member of Rectors´ Honor Committee, Founding member of the pilot projects Interuniversity Doctoral Studies and Virtual University. Member of International Program and Scientific Committees. Author. Honored with Medal of DAAAM International Vienna.

6.0 DAAAM INTERNATIONAL MEDAL

On the occasion of 10th DAAAM International Symposium, DAAAM International introduced the Medal of DAAAM International "for excellence in the science, technology, and international scientific co-operation in the framework of DAAAM International Vienna" as a highest recognition for individual persons and institutions. The medal is designed by B. Katalinic, produced by Moscow State Mint Institute and sponsored by Moscow State University of Technology "Stankin" and Cassandra Trust – Ireland. Maximal one person and/or one institution are honoured by medal in one year.

7.0 UNIVERSITY CHARTER OF DUBROVNIK

The text of the University Charter of Dubrovnik: We, the representatives of different Universities and Institutions, after two years preparation at the meeting initiated and organized by The Danube Adria Association for Automation & Manufacturing, DAAAM International under the auspices of The Danube Rectors' Conference on the occasion of the Eighth International DAAAM Symposium "Intelligent Manufacturing & Automation" held at the International Center of Croatian Universities (University of Zagreb) in Dubrovnik, Croatia, decided to found the University Network of DAAAM International Vienna as a strategic alliance and permanent open platform for long-term co-operation and partnership in all fields of technology, research and education.

The co-operation has to be continuously organized and is based on the partnership, friendship, tolerance, appreciation and positive experience of international scientific co-operation within the framework of DAAAM International during the last eight years. The co-operation can use different flexible forms, structures and aspects. The practical co-operation will be project-based organized. The highest priority is given to the organization of two long - term projects, initially known as: Interuniversitary Doctoral Studies and a pilot project Virtual University. Signed by the founding members in Dubrovnik, Croatia, on the twenty-third of October 1997

8.0 DAAAM INTERNATIONAL NETWORK FOR ADVANCED TECHNOLOGIES

The text of the Network founding document: We, the representatives of different Universities, Institutions, and Firms at the meeting initiated and organized by The Danube Adria Association for Automation & Manufacturing, DAAAM International Vienna under the auspices of The Danube Rectors' Conference on the occasion of the Thirteenth International DAAAM Symposium "Intelligent Manufacturing & Automation: Learning from Nature" held at the Vienna University of Technology and Austrian Society of Engineers and Architects in Vienna, Austria, decided to found the DAAAM International Network for Advanced Technologies as a strategic alliance and permanent open platform within European Research Area for long-term co-operation and partnership in all fields and aspects of research, developing, transfer, education and use of advanced technologies.

The co-operation has to be continuously organized and had to be based on the partnership, friendship, tolerance, appreciation and positive experience of international scientific co-operation within the framework of DAAAM International during the last fourteen years. The co-operation can use individual flexible project oriented forms, structures and aspects and support the young researchers' mobility as well. The practical co-operation will be project-based organized. The highest priority is given to the organization of long - term international cooperation projects focused on advanced technologies for balanced development and harmony between men, technology and nature. The network has to be platform for dissemination of relevant information and ideas, supporting international partnership and encouraging of cooperation initiatives. Signed by the founding members in Vienna, Austria, European Union, on the twenty-fifth of October 2002

Activity No. 1: The conference bags and flags for DAAAM 2003 in Sarajevo are sponsored by the Faculty of Textile Technology, University of Zagreb, and organised by Professor Z. Dragcevic as the first activity in the frame of DAAAM International Network for Advanced Technologies.

9.0 DAAAM INTERNATIONAL NETWORK FOR MECHATRONICS AND ROBOTICS

The text of the Network founding document: We, the representatives of different Universities, Institutions, and Firms at the meeting initiated and organized by The Danube Adria Association for Automation & Manufacturing, DAAAM International Vienna under the auspices of The Danube Rectors' Conference on the occasion of the Fifteenth International DAAAM Symposium "Intelligent Manufacturing & Automation: Globalisation - Technology - Men - Nature" held at the Vienna University of Technology and Austrian Society of Engineers and Architects in Vienna, Austria, decided to found the DAAAM International Network for Mechatronics and Robotics as a strategic alliance and permanent open platform within European Research Area for long-term co-operation and partnership in all fields and aspects of research, developing, transfer, education and use of advanced technologies.

The co-operation has to be continuously organized and had to be based on the partnership, friendship, tolerance, appreciation and positive experience of international scientific co-operation within the framework of DAAAM International during the last fifteen years. The co-operation can use individual flexible project oriented forms, structures and aspects and support the young researchers' mobility as well. The practical co-operation will be project-based organized. The highest priority is given to the organization of long - term international cooperation projects focused on advanced technologies for balanced development and harmony between men, technology and nature. The network has to be platform for dissemination of relevant information and ideas, supporting international partnership and encouraging of cooperation initiatives. Signed by the founding members in Vienna, Austria, European Union, on the fifth of November 2004

10.0 DAAAM INTERNATIONAL SYMPOSIUMS (1990-2008)

One important activity is organising the DAAAM International Symposium. The main objective of the Symposium is to provide a world forum, which takes place annually in Central Europe to exchange knowledge, experience, results and information related to various aspects of the advanced manufacturing and modern automation.

The scope of the Symposium covers scientific, technological and practical concepts concerning research, development and realisation of advanced manufacturing and automation concepts, offering a unique opportunity for experts to meet and exchange ideas. The Symposium is concerned with the current status of automation and manufacturing, methods and results of research and development, as well as application problems and publishes proceedings accordingly.

2008	**19th International DAAAM Symposium "Intelligent Manufacturing & Automation: Focus on Next Generation of Intelligent Systems and Solutions"**
Date & Place	22-25th October 2008, Trnava, Slovakia
Organized by	DAAAM International Vienna, Slovak University of Technology Faculty of Materials Science and Technology in Trnava, University of Applied Sciences Technikum Vienna, Vienna University of Technology - Department of Production Engineering / Intelligent Manufacturing Systems (IFT-IMS), ÖIAV 1848 Austrian Society of Engineers and Architects
Under the auspices of	Danube Rectors' Conference & Rectors' Honor Committee of DAAAM International
Organizer	Katalinic, Branko (Chair) / Moravcik, Oliver (Co-Chair) / Cesarec, Paulina / Dragcevic, Zvonko / Höller, Liane / Kettler, Roman / Knezevic, Sime / Resetova, Kvetoslava / Stefankova, Jana / Stopper, Markus / Velisek, Karol
Publication	Annals of DAAAM for 2008 and Proceeding of the 19th International DAAAM Symposium, ISSN 1726-9679, ISBN 978-3-901509-68-1
Referred/indexed in	ISI Scientific Proceedings Thomson Reuters
Editor	B.[ranko] Katalinic
2007	**18th International DAAAM Symposium "Intelligent Manufacturing & Automation: Focus on Creativity, Responsibility, and Ethics of Engineers"**
On the occasion of	Ogranised in Order to Celebrate 70th Birthday of Dr Wilfried Stoll
Date & Place	24-27th October 2007, Zadar, Croatia
Organized by	DAAAM International Vienna, University of Zadar, University of Applied Sciences Technikum Vienna, Vienna University of Technology - Department of Production Engineering / Intelligent Manufacturing Systems (IFT-IMS), ÖIAV 1848 Austrian Society of Engineers and Architects
Under the auspices of	Danube Rectors' Conference & Rectors' Honor Committee of DAAAM International
In cooperation with	Croatian National Tourist Board
Organizer	Katalinic, Branko (Chair) / Uglesic, Ante (Co-Chair) / Cesarec, Paulina / Dragcevic, Zvonko / Höller, Liane / Kettler, Roman / Knezevic, Sime / Stopper, Markus / Stuja, Kemajl / Wenger, Monika
Publication	Annals of DAAAM for 2007 and Proceeding of the 18th International DAAAM Symposium,

	ISSN 1726-9679
Referred/indexed in	ISI Scientific Proceedings Thomson Reuters
Editor	B.[ranko] Katalinic
2006	**17th International DAAAM Symposium "Intelligent Manufacturing & Automation: Focus on Mechtronics and Robotics"**
On the occasion of	150th Birthday of Nikola Tesla (1856-1943) Inventor of Mobile Robot Teleoperation
Date & Place	8-11th November 2006, Vienna, Austria
Organized by	DAAAM International Vienna, Vienna University of Technology - Department of Production Engineering / Intelligent Manufacturing Systems (IFT-IMS), ÖIAV 1848 Austrian Society of Engineers and Architects , University of Applied Sciences - Technikum
Under the auspices of	Danube Rectors' Conference & Rectors' Honor Committee of DAAAM International
In cooperation with	Vienna Convention Bureau
Organizer	Buchmeister, Borut / Dragcevic, Zvonko / Egorov, Sergey / Höller, Liane / Katalinic, Branko (Chair) / Lazinica, Aleksandar / Malisa, Viktorio / Sagmeister, Nicole / Stuja, Kemajl / Viktorik, Hans / Viktorik, Silvia
Publication	Annals of DAAAM for 2006 and Proceeding of the 17th International DAAAM Symposium, ISSN 1726-9679, ISBN 3-901509-57-7
Referred/indexed in	ISI Scientific Proceedings Thomson Reuters
Editor	B.[ranko] Katalinic
2005	**16th International DAAAM Symposium "Intelligent Manufacturing & Automation: Focus on Young Scientists and Researchers"**
On the occasion of	15 Years of DAAAM International Vienna, 1990-2005
Date & Place	19-22nd October 2005, Opatija, Croatia
Organized by	DAAAM International Vienna, Vienna University of Technology - Department of Production Engineering / Intelligent Manufacturing Systems (IFT-IMS), ÖIAV 1848 Austrian Society of Engineers and Architects
Under the auspices of	Danube Rectors' Conference & Rectors' Honor Committee of DAAAM International
In cooperation with	Vienna Convention Bureau
Organizer	Anisic, Bojan / Barisic, B. / Blazevic, Dalibor / Buchmeister, Borut / Car, Zlatan / Dragcevic, Zvonko / Egorov, Sergey / Grabenweger, Johannes / Gronalt, Manfred / Höller, Liane / Huber, Ingrid / Jurkovic, Z. / Katalinic, Branko (Chair) / Lazinica, Aleksandar / Malisa, Viktorio / Merdan, Munir / Mikac, Tonci (Co-Chair) / Mirjanic, Valentina / Nanasi, Jozef / Schaumberger, Werner / Stopper, Markus / Stuja, Kemajl / Udiljak, Toma / Widtmann, Georg
Publication	Annals of DAAAM for 2005 and Proceeding of the 16th International DAAAM Symposium, ISSN 1726-9679, ISBN 3-901509-46-1
Referred/indexed in	ISI Scientific Proceedings Thomson Reuters
Editor	B.[ranko] Katalinic
2004	**15th International DAAAM Symposium "Intelligent Manufacturing & Automation: Globalisation - Technology - Men - Nature"**
On the occasion of	**Organised in Memoriam on Professor Dr Jyri Papstel (1940- 2004)**
Date & Place	3-6th November 2004, Vienna, Austria, European Union
Organized by	DAAAM International Vienna, Vienna University of Technology - Department of Production Engineering / Intelligent Manufacturing Systems (IFT-IMS), ÖIAV 1848 Austrian Society of Engineers and Architects
Under the auspices of	Danube Rectors' Conference & Rectors' Honor Committee of DAAAM International
In cooperation with	Vienna Convention Bureau
Organizer	Bartak, Harald; Buchmeister, Borut; Dragcevic, Zvonko; Egorov, Sergei; Frank, Willy; Grabenweger, Johann; Gronalt, Manfred; Höller, Liane; Hrelja, Igor; Katalinic, Branko (Chair); Kettler, Roman; Kordic, Vedran; Lazinica, Aleksandar; Löschner, Adam; Magyar, Teresa; Malisa, Viktorio; Merdan, Munir; Müller, Simon; Nanasi, Jozef; Piller, Johannes; Reinisch, Roland; Roth, Michael; Saliger, Florian; Schaumberger, Werner; Schmidt, Thomas; Stuja, Kemajl; Trauzettel, Ulrich; Udiljak, Toma; Widtmann, Georg; Wirths, Thomas.
Publication	Annals of DAAAM for 2004 and Proceeding of the 15th International DAAAM Symposium, ISSN 1726-9679, ISBN 3-901509-42-9
Referred/indexed in	ISI Scientific Proceedings Thomson Reuters
Editor	B.[ranko] Katalinic
2003	**14th International DAAAM Symposium "Intelligent Manufacturing & Automation: Focus on Reconstruction and Development"**
On the occasion of	155 Years of Austrian Society of Engineers and Architects ÖIAV 1848 / 45 Years of the Faculty of Mechanical Engineering University of Sarajevo
Date & Place	22-25th October 2003, Sarajevo, Bosnia and Herzegovina

Organized by	DAAAM International Vienna, University of Sarajevo - Faculty of Mechanical Engineering, Vienna University of Technology - Department of Production Engineering / Intelligent Manufacturing Systems (IFT-IMS), ÖIAV 1848 Austrian Society of Engineers and Architects
Under the auspices of	Danube Rectors' Conference & Rectors' Honor Committee of DAAAM International
In cooperation with	Federal Ministry of Education, Sciences and Culture - Vienna, ÖIAV 1848 Austrian Society of Engineers and Architects
Organizer	Katalinic, Branko (Chair), Kulenovic, Malik (Co-chair), Widtmann, G., et al.
Publication	Annals of DAAAM for 2003 and Proceeding of the 14th International DAAAM Symposium, ISSN 1726-9679, ISBN 3-901509-34-8
Referred/indexed in	ISI Scientific Proceedings Thomson Reuters
Editor	B.[ranko] Katalinic
Information	Contact: DAAAM International Vienna (Katalinic B.) president@daaam.com www.daaam.com
2002	**13th International DAAAM Symposium "Intelligent Manufacturing & Automation: Learning from Nature"**
On the occasion of	150 Years of Austrian Society of Engineers and Architects ÖIAV 1848
Date & Place	23-26th October 2002, Vienna, Austria
Organized by	DAAAM International Vienna, Vienna University of Technology - Department of Production Engineering / Intelligent Manufacturing Systems (IFT-IMS), ÖIAV 1848 Austrian Society of Engineers and Architects
Under the auspices of	Danube Rectors' Conference & Rectors' Honor Committee of DAAAM International
In cooperation with	Federal Ministry of Education, Sciences and Culture - Vienna, ÖIAV 1848 Austrian Society of Engineers and Architects
Organizer	Katalinic B. (Chair), Buchmeister B., Doganac A., Dragcevic Z., Gersak J., Grabenweger J., Kordic V., Ljoljic B., Nanasi J., Stopper M., Stuja K., Widtmann G., et al.
Publication	Annals of DAAAM for 2002 and Proceeding of 13th International DAAAM Symposium, ISBN 3-901509-29-1
Referred/indexed in	ISI Scientific Proceedings Thomson Reuters
Editor	B.[ranko] Katalinic
2001	**12th International DAAAM Symposium "Intelligent Manufacturing & Automation: Focus on Precision Engineering"**
On the occasion of	10 Years of Jena University of Applied Science / 150 Years of ÖIAV 1848 Austrian Society of Engineers and Architects
Date & Place	24-27th October 2001, Jena, Germany
Organized by	DAAAM International Vienna, Jena University of Applied Science, Vienna University of Technology - Department of Production Engineering / Intelligent Manufacturing Systems (IFT-IMS), Austrian Society of Engineers and Architects ÖIAV 1848
Under the auspices of	Danube Rectors' Conference & Rectors' Honor Committee of DAAAM International
In cooperation with	Thuringian Ministry of Sciences, Research and Culture - Erfurt,
Organizer	Katalinic B. (Chair), Meissner K. (Co-chair), Beez S., Buchmeister B., Doganac A., Dragcevic Z., Fidi L., Gersak J., Grabenweger J., Jacob M., Kordic V., Ljoljic B., Nanasi J., Pachmann A., Schoele H., Selke K., Stopper M., Stuja K., Walther J., Widtmann G.
Publication	Annals of DAAAM for 2001 and Proceeding of 12th International DAAAM Symposium, ISBN 3-901509-19-4
Editor	B.[ranko] Katalinic
2000	**11th International DAAAM Symposium "Intelligent Manufacturing & Automation: Men – Machine - Nature"**
On the occasion of	150 Years of ÖIAV 1848 Austrian Society of Engineers and Architects / 40 Years of Faculty of Engineering – University of Rijeka / 10 Years of DAAAM International
Date & Place	19-21st October 2000, Opatija, Croatia
Organized by	DAAAM International Vienna, Vienna University of Technology - Department of Production Engineering / Intelligent Manufacturing Systems (IFT-IMS)
Under the auspices of	Danube Rectors' Conference & Rectors' Honor Committee of DAAAM International
In cooperation with	Federal Ministry of Education, Sciences and Culture - Vienna, ÖIAV 1848 Austrian Society of Engineers and Architects, Union of Croatian Universities
Organizer	Katalinic B. et al.
Publication	Annals of DAAAM for 2000 and Proceeding of 10th International DAAAM Symposium, ISBN 3-901509-13-5
Editor	B.[ranko] Katalinic
1999	**10th International DAAAM Symposium "Intelligent Manufacturing & Automation: Past – Present - Future"**
On the occasion of	10th International DAAAM Symposium, Last International DAAAM Symposium in this Millennium

Date & Place	21-23rd October 1999, Vienna University of Technology - Department of Production Engineering / Intelligent Manufacturing Systems (IFT-IMS), Austrian Society of Engineers and Architects ÖIAV 1848, Vienna, Austria
Organized by	DAAAM International Vienna, Vienna University of Technology, Department of Production Engineering / Intelligent Manufacturing Systems (IFT-IMS)
Under the auspices of	Danube Rectors' Conference & Rectors' Honor Committee of DAAAM International
In cooperation with	Federal Ministry of Sciences and Traffic - Vienna, ÖIAV 1848 Austrian Society of Engineers and Architects
Organizer	Katalinic B. (Chair), Adamczak S., Alpek F., Balate J., Balic J., Beer M., Buchmeister B., Celar S., Cus F., Dinnyes C., Dragcevic. Z., Duflou J. R., Egorov S., Fidi L., Grabenweger J., Gyenge C., Hajrizi E., Kerekes L., Kordic V., Kordic Z., Ljoljic B., Llorente J. I., Meissner K., Nanasi J., Papstel J., Park H.-S., Stuja K., Tabucanon M. T., Udiljak T., Ueda K., Vranjes B., Wessely E.
Publication	Annals of DAAAM for 1999 and Proceeding of 10th International DAAAM Symposium, ISBN 3-901509-10-0
Editor	B.[ranko] Katalinic
1998	**9th International DAAAM Symposium "Intelligent Manufacturing, Automation & Networking"**
On the occasion of	50 Years of the Technical University of Cluj - Napoca 150 Years of ÖIAV 1848 Austrian Society of Engineers and Architects Founding of Permanent Committee of University Network of DAAAM International Vienna
Date & Place	22-24th October 1998, Technical University of Cluj - Napoca, Cluj - Napoca, Romania
Organized by	DAAAM International Vienna, Technical University of Cluj - Napoca, Vienna University of Technology - Department of Production Engineering / Intelligent Manufacturing Systems (IFT-IMS)
Under the auspices of	Danube Rectors' Conference & Rectors' Honor Committee of DAAAM International
In cooperation with	Ministry of Research and Technology - Bucuresti, Federal Ministry of Sciences and Traffic - Vienna, Journal of Intelligent Manufacturing, ÖIAV 1848 Austrian Society of Engineers and Architects
Organizer	Arsin M., Beer M., Bejan M., Berce P., Borzan M., Buchmeister B., Calinescu O., Car Z., Celar S., Dusa P., Fidi L., Gheorghe I., Gramescu T., Gyenge C., Hajrizi E., Ivan N.V., Katalinic B. (Chair), Kerekes L., Lazea G., Ljoljic B., Lörincz S., Nanasi J., Nemes S., Popescu S., Pusca M., Szava T., Udiljak T., Vranjes B.
Proceedings	ISBN 3-901509-08-9
Editor	B.[ranko] Katalinic
1997	**8th International DAAAM Symposium "Intelligent Manufacturing & Automation"**
On the occasion of	Founding of the University Network of DAAAM International Vienna
Date & Place	23-25th October 1997, ICCU - International Centre of Croatian Universities (University of Zagreb), Dubrovnik, Croatia
Organized by	DAAAM International Vienna, University of Zagreb, Vienna University of Technology, Department of Production Engineering / Intelligent Manufacturing Systems (IFT-IMS)
Under the auspices of	Danube Rectors' Conference & Rectors' Honor Committee of DAAAM International
In cooperation with	Ministry of Science and Technology - Zagreb, Federal Ministry of Sciences and Traffic - Vienna, Journal of Intelligent Manufacturing, ÖIAV 1848 Austrian Society of Engineers and Architects, Union of Croatian Universities, Croatian Society - KoREMA - Zagreb.
Organizer	Benic D., Brunsko V., Budic I., Burgjelez I., Celar S., Gojo M., Hajrizi E., Ivancic I., Jerbic B., Katalinic B. (Chair), Kovacic B., Majstorovic V., Mencer H.J., Nanasi J., Pavic I., Polic-Bobic M., Udiljak T. (Co-Chair), Veza I., Visekruna V.
Proceedings	ISBN 3-901509-04-6
Editor	B.[ranko] Katalinic
1996	**7th International DAAAM Symposium "Product & Manufacturing: Flexibility, Integration, Intelligence"**
On the occasion of	1,000 Years of Austria
Date & Place	17-19th October 1996, Vienna, Austria
Organized by	DAAAM International Vienna, Vienna University of Technology, Department of Production Engineering (IFT-IMS)
Under the auspices of	Danube Rectors' Conference
Organizer	Katalinic B., Celar S., Fabian E., Nanasi J.
Proceedings	ISBN 3-901509-02-X
Editor	B.[ranko] Katalinic
1995	**6th International DAAAM Symposium "Intelligent Manufacturing Systems"**
On the occasion of	50 Years of Technical University of Krakow
Date & Place	26 - 28th October 1995, Poland

Organized by	DAAAM International Vienna; Technical University of Krakow Technical University of Kielce; Vienna University of Technology - Department of Production Engineering / Intelligent Manufacturing Systems (IFT-IMS)
In cooperation with	IMECO; Ministry of Education, Warsaw; Ministry of Sciencees and Technology, Warsaw; Federal Ministry of Sciences, Technology and Art, Vienna; BIT, Bureau for International Research and Technology Cooperation, Vienna
Organizer	Adamczak S., Duch A., Gawlik J., Katalinic B., Osanna H. P., Sladek J., Wantuch E., Rusiniak K., Zebala W.
Proceedings	ISBN 3-901509-01-1
Editor	B.[ranko] Katalinic
Co-Editors	Osanna H.P., Gawlik J.
1994	**5th International DAAAM Symposium "Automation and Metrology: Challenge and Chance"**
On the occasion of	700 Years of City of Maribor, 35 Years of Faculty of Technical Sciences
Date & Place	27-28th October 1994, Maribor, Slovenia
Organized by	DAAAM International Vienna, University of Maribor, Faculty of Technical Sciences
In cooperation with	IMECO; Ministry of Sciencees and Technologies, LjubljanaFederal Ministry of Sciences, Technology and Art, Vienna
Organizer	Balic J., Cus F., Katalinic B., Novak D., Osanna H. P., Sujica M. Z.
Proceedings	ISBN 3-901509-00-3, ISBN 86-435-0084-4
Editors	Katalinic B., Osanna H.P., Balic J., Cus F.
1993	**4th International DAAAM Symposium "Automation and Metrology: Man, Technology, Environment"**
On the occasion of	750 Years of City of Brno
Date & Place	16-18th September 1993, Brno, Czechia
Organized by	DAAAM International Vienna; Technical University of Brno, Faculty of Mechanical Engineering
In cooperation with	IMECO, IUTAM
Organizer	Balate J., Dvorak J., Kadrnozka J., Katalinic B., Lacko B., Osanna H. P., Skopal M., Vavrin P.
Editors	Katalinic B., Osanna H.P:
1992	**3rd International DAAAM Symposium "Flexible and Intelligent Automation"**
Date & Place	4-6th November 1992, Budapest, Hungary
Organized by	DAAAM International Vienna; Technische Universität Budapest, Department of Production Engineering
In cooperation with	IMECO
Organizer	Alpek F., Arz G., Katalinic B., Osanna H. P., Szelig T., Jerzsabek P.
Editor	B.[ranko] Katalinic
1991	**2nd International DAAAM Symposium "Flexible Automation"**
Date & Place	9-13th December 1991, Strbske Pleso, Slovakia, Czechoslovakia
Organized by	DAAAM International Vienna; House of Technology, Bratislava; Technical University of Kosice, Faculty of Mechanical Engineering, Department of Automated Production Systems
Organizer	Sebo D., Wessely E., Kovacik J., Katalinic B.
Proceedings	ISBN 80-233-0225-6
1990	**1st International DAAAM Symposium „Automatization and Measurement Technique“**
On the occasion of	175 Years of Vienna University of Technology
Date & Place	5th November 1990, Vienna, Austria
Organized by	DAAAM International Vienna, Vienna University of Technology - Department of Production Engineering / Intelligent Manufacturing Systems (IFT-IMS), ÖIAV, Sections: Measurement and Automation, ASESA, Robotics
Organizer	Beer R. J., Osanna H. P., Katalinic B.
Editor	B.[ranko] Katalinic

11.0 BEST PAPER / PRESENTATION DAAAM AWARD

Since 1999 the DAAAM International Vienna recognizes excellence of contributions in a number of ways. There are three categories for papers:

- Best Paper Award of DAAAM International for the best paper presented at Symposium,
- DAAAM International Award for Best Oral Presentation of Paper,
- DAAAM International Award for Best Visual Presentation of Paper.

The Awards are made at the Closing Section and Best Paper Award Ceremony of the Symposium takes part on Friday at 17:00 hours. The Award Committee of DAAAM International makes the evaluation of presented papers during the Symposium. The final decisions are made by the President of the Award Committee of DAAAM International. Former Presidents of the Paper Award Committee of DAAAM International are: 2008 Professor, Master of Sport and Cosmonaut Alexey Eliseev, 2007 Professor Kanji Ueda & Professor Vladimir Filaretov, 2006 Professor Vladimir

Filaretov, 2005 Professor Ueda, 2004 Professor Franc Cus, 2003 Professor Jyri Papstel, 2002 Professor Kanji Ueda, 2001 Professor Toma Udiljak, 2000 Professor Jyri Papstel, and 1999 Professor Kanji Ueda. The President and the final list of the members of the Committee will be announced at Opening Ceremony of the Symposium. All (co)authors of awarded papers are getting the certificate, the paper (main author) additionally getting Best Paper Medal. The layout of the Best Paper Medal you can see at the cover of this publication. The papers which authors are member of the Award committee and the papers which are not presented at Symposium are excluded from competition and can not get the Award.

12.0 DAAAM INTERNATIONAL YOUNG RESEARCHER AND SCIENTIST AWARD

DAAAM International Vienna wants to recognize excellence of contributions of young researchers and scientists, and their supervisors and mentors. Starting with the 1st European DAAAM International Young Researchers´ and Scientists´ Conference, and 18th DAAAM International Symposium "Intelligent Manufacturing and Automation: Focus on Creativity, Responsibility and Ethics of Engineers", which are held 24-27th October 2007 in Zadar, Croatia, the best young researchers and scientists will be honoured by "DAAAM International Young Researchers and Scientist Award" Awards are possible in following categories:

Related to:	Categories:
Paper:	Best Paper / Best Oral Presentation / Best Graphic Presentation / Best Poster Presentation / Best Demonstration
Supervisor / Mentor:	Best Supervisor/Mentor / Best Team Manager / Best Project Manager / Best Leader / Best Female Supervisor
Quality of Project / Research:	Best Industrial Project / Best Research Project / Best Application / Best Innovation / Best Theory
Contest:	Interdisciplinarity / European Relevance / Future Oriented / Ethics / Harmony with Nature
Thesis:	Best Bachelors Thesis / Best Master Thesis / Best Diploma Thesis / Best MBA Thesis / Best PhD/Doctor Thesis
Team:	Best Team / Best University Team / Best Firm / Best International Team / Youngest Team

In the case of extraordinary contributions which can not be expressed by above mentioned recognitions is possible to make a Special Award. The Awards are made at the Closing Section of the Symposium. The final decision is made by President of the Award Committee of DAAAM International for actual year. Former Presidents of the Award Committee are recognised international leading experts and professors. All awarded authors and co-authors of paper are getting attractive **DAAAM International Best Award Certificate.** The Award Certificates are personal, very nice professional designed, framed and ready to hang in your room. Authors' team is getting one medal named: **"DAAAM International Young Researcher and Scientist Award".** This Medal is designed by: Branko Katalinic; Sabine Wulffert, and Igor Drstvensek who made the prototype and first edition of medals in 2007. This medal is sponsored by concern Festo.

Very Important: The papers which authors are member of the Award Committee and the papers which are not presented at Symposium are excluded from competition and can not get the Award.

13.0 FESTO YOUNG RESEARCHERS AND SCIENTISTS SUPPORT SCHOLARSHIP

Since 2007 concern Festo gives, by courtesy of Dr. Wilfried Stoll, support scholarship to excellent young researchers and scientists to present own research results at European DAAAM International Young Researchers´ and Scientists´ Conferences, and DAAAM International Symposiums "Intelligent Manufacturing and Automation". This scholarship covers conference fee, accommodation and (partly) travel expenses. Conditions for wining this scholarship are: candidate must come to conference and must present own research results, candidate must have high quality research and presentation, and candidates have to come from different countries. The winners are getting special "Certificate of Attendance" with special text " ...participated as winner of Festo Young Researcher and Scientist Support Scholarship". The awards are made at the closing award ceremony, and given by high representative of concern Festo. The first representative for 2007 was Ing. Wolfgang Keiner, general manager Festo Austria.

In 2007 thirty four young researchers and scientists from 11 countries won Festo Young Researcher and Scientist Support Scholarship. The 2007-winners are: Babic, Anita / Bagimov, Igor / Bartos, Stanislav / Beatrice, Nica / Belosevic, Ivan / Bodini, Ileana / Dokupil, Radek / Florina, Anghel / Hong, Dongphyo / Hursa, Anica / Jeong, Donghwa / Kim, Insu / Klevakin, Vladimir / Kovács, Zoltán / Lancini, Matteo / Lasunon, OnUma / Lorenc, Vlastimil / Loun, Kaia / Mączka, Janusz Józef / Mikhailov, Maxim / Mogilnikov, Pavel / Moise, Cristian / Noskovicova, Lucia / Ovari, Nora / Parpala, Radu / Pelikan, Radek / Stjepanovic, Goran / Sysala, Tomas / Tache, Florin / Vesely, Libor / Vesely, Milos / Vidović, Nataša / Zbranek, Pavel / Zuev, Alexandr.

14.0 BEST DIGITAL PHOTO DAAAM AWARD

Importance of digital photography for research, publishing and social relations is rapidly increased during last time. For that reason we are going to establish DAAAM International Photo Gallery with photos which are related to the symposium and our activities (place, people, situation, art, special effects). The gallery will be free accessed and searchable. For this project we are looking for amateur photographs among the participants to be official photographs of DAAAM International and member of DAAAM International Photo Team. Digital photos can be given during the symposium to the organizer, or send after symposium by e-mail. All photos which are technically good made will be included in the gallery. On the next symposium the up to three authors of nicest photos will get Best Digital Photo Award.

15.0 BEST TASTE AND SMELL OF DAAAM AWARD

"Best Taste and Smell of DAAAM International": Many of colleagues in DAAAM International network are excellent hobby cooks, liquor maker or winemaker. Each year we are making small tasting of their specialities under the motto "Best taste and smell of DAAAM International". Authors are encouraged to bring some self-made specialities to DAAAM symposium. The best will be honoured with "Best Taste and Smell DAAAM International Award".

16.0 DAAAM INTERNATIONAL SPECIALIZED CONFERENCES

2008	**2nd International Conference on Additive Technologies, DAAAM International Specialized Conference**
Founder and Chair	Drstvensek, I.
Date & Place	September, 17th till 18th, Hotel Primus, Ptuj, Slovenia
Organized by	DAAAM International Vienna/Interesansa (RAPIMAN – Rapid Prototyping and Innovative Manufacturing Network)/ University of Maribor - Faculty of Mechanical Engineering/University of Ljubljana, Institute for Innovation and Development
Organizers	Drstvensek, Igor (Chair) / Katalinic, Branko (Co-Chair) / Dolinsek, Slavko (Co-Chair) / Valentan, Bogdan / Brajlih, Tomaz / Bauer Fuzir, Gabriela / Reinhardt, Andrea / Ficko, Mirko / Pogacar, Vojko /
Proceedings	ISBN 3-901509-72-0, ISSN1992-5085
Editors	Drstvensek I.[gor] & Katalinic B.[ranko] & Dolinsek S.[lavko]
Published by	DAAAM International Vienna
2007	**1st DAAAM International Specialized Conference on Additive Technologies**
Founder and Chair	Katalinic B.
Date & Place	20th April 2007, Celje, Slovenia
Organized by	DAAAM International Vienna / RAPIMAN – Rapid Prototyping and Innovative Manufacturing Network / Vienna University of Technology - Department of Production Engineering / Intelligent Manufacturing Systems (IFT-IMS), / University of Maribor , Faculty of Mechanical Engineering - Department of Production Engineering / ÖIAV 1848 Austrian Society of Engineers and Architects / University of Applied Sciences - Technikum
Organizers	Drstvensek, Igor (Chair) / Katalinic Branko (Co-Chair) / Valentan, Bogdan / Brajlih, Tomaz / Tercek, Tomaz / Reinhardt, Andrea / Ficko, Mirko / Pogacar, Vojko / Dolinsek, Slavko
Proceedings	ISBN 3-901509-61-5, ISSN1992-5085
Editors	Drstvensek I.[gor] & Katalinic B.[ranko]
Published by	DAAAM International Vienna

17.0 DAAAM INTERNATIONAL CONFERENCES "ADVANCED TECHNOLOGIES FOR DEVELOPING COUNTRIES"

2006	**5th DAAAM International Conference on Advanced Technologies for Developing Countries, ATDC-2006**
Founder and Chair	Katalinic B.
Date & Place	28-30 June 2006, Rijeka, Croatia
Organized by	University of Rijeka, Faculty of Enineering, DAAAM International Vienna
Organizers	Mikac T.[onci] (Chair), Car Z.[latan] (Co-chair), Barisic B.[ranimir], Cukor G.[oran], Ikonic M.[ilan], Pavletic D.[usko], Perinic M.[laden], Blazevic D.[alibor], Jurkovic Z.[oran]
Proceedings	ISBN 3-901509-54-2 (Book), ISBN3-901509-55-0 (CD), ISBN 953-6326-45-0
Editors	Mikac T.[onci] & Katalinic B.[ranko]
Published by	University of Rijeka, Faculty of Engineering, DAAAM International Vienna
2005	**4th DAAAM International Conference on Advanced Technologies for Developing**

	Countries, ATDC-2005
Founder and Chair	Katalinic B.
Date & Place	September, 21-24th, 2005, Slavonski Brod, Croatia
Organized by	Faculty of Mechanical Engineering, Josip Juraj Strossmayer University of Osijek; DAAAM International Vienna
On the occasion of	30th Anniversary of Josip Juraj Strossmayer University of Osijek
Organizers	Budic, Ivan (Chair); Kljain Milan (Co-Chair); Kozak, Drazen (Secretary); Katalinic, Branko (Chair of Program Committee), Raos, Pero; Saric, Tomislav; Stojic, Antun; Simunovic, Goran; Novoselovic, Daniel; Karakasic, Mirko
Proceedings	ISBN 3-901509-49-6 ISBN 953-6048-29-9
Editors	Kljain M.[ilan], Katalinic B.[ranko] & Budic I.[van]
Published by	DAAAM International Vienna; Faculty of Mechanical Engineering Slavonski Brod Josip Juraj Strossmayer University of Osijek
2004	**3rd DAAAM International Conference on Advanced Technologies for Developing Countries, ATDC-2004**
Founder and Chair	Katalinic B.
Date & Place	June 23-26, 2004, Split, Croatia
Organized by	Faculty of Electrical Engineering, Mechanical Engineering and Naval Architecture, University of Split; DAAAM International Vienna
Organizers	Ivica Veža (Chairman); Željko Domazet (Co-Chairman); Branko Katalinic (Co-Chairman); Tonći Grubić (Secretary), et al.
Proceedings	ISBN 3-901509-41-0
Editors	Katalinic B.[ranko] & Veza, I.[vica]
Published by	DAAAM International Vienna; Faculty of Electrical Engineering, Mechanical Engineering and Naval Architecture, University of Split
2003	**2nd DAAAM International Conference on Advanced Technologies for Developing Countries, ATDC-2003**
Founder and Chair	Katalinic B.
Date & Place	25-28th June 2003, Tuzla, Bosnia and Herzegovina
Organized by	Faculty of Mechanical Engineering, University of Tuzla; DAAAM International Vienna
On the occasion of	40th Anniversary of Mechanical Engineering Study in Tuzla
Organizers	Tufekcic, Dzemo (Chair); Katalinic, Branko (Co-Chairman); Widtmann, Georg (Co-Chairman); Topcic, Alen (Secretary); Kapetanovic, Izrudin; Alagic, Sandira; Alic, Izet; Buljubasic, Indira; Hadzic, Seniha; Mehmedovic, Muhamed; Mujic, Elvedin; Nezirevic, Salahudin; Saric, Emir; Kapetanovic, Irena
Proceedings	ISBN 3-901509-35-6
Editors	Katalinic B.[ranko] & Tufekcic, Dz.[emo]
Published by	DAAAM International Vienna; University of Tuzla - Faculty of Mechanical Engineering
2002	**1st DAAAM International Conference on Advanced Technologies for Developing Countries, ATDC-2002**
Founder and Chair	Katalinic B.
Date & Place	12-14the September 2002, Slavonski Brod, Croatia
Organized by	Faculty of Mechanical Engineering, Strossmayer University of Osijek; DAAAM International Vienna
On the occasion of	40th Anniversary of Mechanical Engineering Study in Slavonski Brod
Organizers	Majdandzic, Niko (Chair), Kljajin M. (Co-Chair), Kozak D. (Secretary), Cicak M., Galeta T., Hnatko E., Ivandic Z., Lujic R., Maglic L., Maticevic G., Raos P., Stojic A., Saric T., Simunovic K., Simunovic G., Vujcic D.
Proceedings	ISBN 3-901509-32-1; ISBN 953-7611-64-X
Published by	DAAAM International Vienna; Strossmayer University of Osijek - Faculty of Mechanical Engineering; Grafika Osijek
Editors	Katalinic B.[ranko] & Kljajin M.[ilan]

18.0 DAAAM INTERNATIONAL WORKSHOPS "COMPUTER AIDED SYSTEMS AND TECHNOLOGIES"

This workshop is based on the bilateral co-operation between two similar departments of the Faculties of Mechanical Engineering of the University of Zilina, Slovakia and Cracow University of Technology - CUT, Cracow, Poland. Those activities are supported by DAAAM National Organisations of Poland and Slovakia, and DAAAM International. In order to install this workshop as stabile long-term regional platform for the *CA Systems and Technologies* fields it was decided to found:

Permanent Committee of Workshop:

Permanent Members:

I. Kuric *(University of Zilina, Zilina, Slovakia),*
B. Katalinic *(Vienna University of Technology and DAAAM International, Vienna, Austria),*
J. Duda *(CUT, Cracow, Poland).*

Academic Year	Members	Institutions
2008/2009	B. Barisic	*University of Rijeka, Croatia*
	J. Audy	*Edith Cowan University, Australia*
	L. Goldhahn	*FH Mittweida, Germany*
	J. Marcinčin	*TU Kosice, Slovakia*
	J. Matuszek	*ATH - University of Bielsko Biala, Poland*
	S. Legutko	*PUT Poznan, Poland*
	N. Ungureanu	*University of Baia Mare, Romania*
2007/2008	B. Barisic	*University of Rijeka, Croatia*
	R. Cotetiu	*University of Baia Mare, Romania*
	L. Goldhahn	*FH Mittweida, Germany*
	J. Marcinčin	*TU Kosice, Slovakia*
	J. Matuszek	*ATH - University of Bielsko Biala, Poland*
	J. Peterka	*TU Bratislava, Slovakia*
	N. Ungureanu	*University of Baia Mare, Romania*
2006/2007	B. Barisic	*University of Rijeka, Croatia*
	R. Cotetiu	*University of Baia Mare, Romania*
	L. Goldhahn	*FH Mittweida, Germany*
	J. Marcinčin	*TU Kosice, Slovakia*
	H. Sjolander	*University of Kristianstad, Sweden*
	P. Šugár	*TU Zvolen, Slovakia*
	N. Ungureanu	*University of Baia Mare, Romania*
2005/2006	R. Cotetiu	*University of Baia Mare, Romania*
	J. Marcinčin	*TU Kosice, Slovakia*
	J. Matuszek	*University of Bielsko Biala, Poland*
	S. Legutko	*PUT Poznan, Poland*
	P. Šugár	*TU Zvolen, Slovakia*
	N. Ungureanu	*University of Baia Mare, Romania*
2004/2005	L. Goldhahn	*FH Mittweida, Germany*
	J. Madl	*CVUT Prague, Czech Republic*
	J. Marcincin	*TU Kosice, Slovakia*
	E. Pay	*University of Baia Mare, Romania*
	D. Plinta	*University of Bielsko Biala, Poland*
	G. Varga	*University of Miskolc, Hungary*
2003/2004	D. Georgiev	*TU Varna, Bulgaria*
	L. Goldhahn	*FH Mittweida, Germany*
	J. Manlig	*TU Liberec, Czech Republic*
	J. Marcincin	*TU Kosice, Slovakia*
	G. Varga	*University of Miskolc, Hungary*
2002/2003	D. Kumicakova	*University of Zilina, Slovakia*
	S. Legutko	*PUT Poznan, Poland*
	J. Marcincin	*TU Kosice, Slovakia*
	I. Neaga	*University of Loughborough, U.K.*
	A. Pendrey	*TU Birmingham, U.K.*
	J. Peterka	*STU Trnava, Slovakia*
	L. Soos	*STU Bratislava, Slovakia*
2001/2002	N. Cubonova	*University of Zilina, Slovakia*
	S. Legutko	*PUT Poznan, Poland*
	J. Marcincin	*TU Kosice, Slovakia*
	I. Neaga	*TU Iasi, Romania*
	J. Peterka	*STU Trnava, Slovakia*
	L. Soos	*STU Bratislava, Slovakia*
2000/2001	N. Cubonova	*University of Zilina, Slovakia*
	S. Legutko	*PUT Poznan, Poland*

	J. Marcincin I. Neaga J. Peterka L. Soos	*TU Kosice, Slovakia* *TU Iasi, Romania* *STU Trnava, Slovakia* *STU Bratislava, Slovakia*
1999/2000	S. Legutko J. Marcincin I. Neaga	*PUT Poznan, Poland* *TU Kosice, Slovakia* *TU Iasi, Romania*

The main aim of the Permanent Committee is to prepare the DAAAM International workshops oriented on CA systems and technologies and propagate all accompanying activities during academic year (projects, WEB journal, common publications, PhD seminars, exchange of students and lectures, etc.). The list of workshops organised during the past.

2008	**13th DAAAM International Workshop "CA Systems and Technologies"**
Date & Place	6 – 8th October 2008, Cracow, Poland
Organized by	CUT Cracow, University of Zilina, DAAAM International Vienna
Organisers	Duda J., Kuric I., Katalinic B.
Proceedings	ISBN 978-83-7242-481-5
Editors	Duda J., Kuric I., Katalinic B.
2007	**12th DAAAM International Workshop "CA Systems and Technologies"**
Date & Place	17 – 19th October 2007, Žilina, Slovakia
Organized by	University of Zilina, CUT Cracow, DAAAM International Vienna
Organisers	Kuric I., Duda J., Katalinic B.
Proceedings	ISBN 3-901509-56-9
Editors	Kuric I., Duda J., Katalinic B.
2006	**11th DAAAM International Workshop "CA Systems and Technologies"**
Date & Place	2 – 4th October 2006, Cracow, Poland
Organized by	CUT Cracow, University of Zilina, DAAAM International Vienna
Organisers	Duda J., Kuric I., Katalinic B.
Proceedings	ISBN 3-901509-56-9
Editors	Duda J., Kuric I., Katalinic B.
2005	**10th DAAAM International Workshop "CA Systems and Technologies"**
Date & Place	12 – 14th December 2005, Strecno - Zilina, Slovak Republic
Organized by	University of Zilina, CUT Cracow, DAAAM International Vienna
Organisers	Kuric, I., Duda J., Katalinic B.
Proceedings	ISBN 3-901509-51-8
Editors	Kuric I., Katalinic B.
2004	**9th DAAAM International Workshop "CA Systems and Technologies"**
Date & Place	5 - 6th October 2004, Cracow, Poland
Organized by	CUT Cracow, University of Zilina, DAAAM International Vienna
Organisers	Duda J., Kuric I., Katalinic B.
Proceedings	ISBN 83-7242-320-2
Editors	Duda J., Katalinic B.
2003	**8th DAAAM International Workshop "CA Systems and Technologies"**
Date & Place	29th September - 1st October 2003, Strecno - Zilina, Slovak Republic
Organized by	University of Zilina, CUT Cracow, DAAAM International Vienna
Organisers	Duda J., Kuric I., Katalinic B.
Proceedings	ISBN 3-901509-39-9
Editors	Kuric I., Katalinic B.
2002	**7th DAAAM International Workshop "CA Systems and Technologies"**
Date & Place	8-9th October 2002, CUT Cracow, Poland
Organized by	CUT Cracow, University of Zilina, DAAAM International Vienna
Organisers	Duda J., Kuric I., Katalinic B.
Proceedings	ISBN 3-901509-33-X
Editors	Duda J., Katalinic B.
2001	**6th DAAAM International Workshop "CA Systems and Technologies"**
Date & Place	25-26th September 2001, Sulov, Slovak Republic
Organized by	University of Zilina, CUT Cracow, DAAAM International Vienna
Organisers	Duda J., Kuric I., Katalinic B.
Proceedings	ISBN 3-901509-27-5
Editors	Kuric I., Katalinic B.
2000	**5th DAAAM International Workshop "CA Systems and Technologies"**
Date & Place	28-29th September 2000, Cracow, Poland

Organized by	CUT Cracow, University of Zilina, DAAAM International Vienna
Organisers	Duda J., Kuric I., Katalinic B.
Proceedings	ISBN 3-901509-16-X
Editors	Duda J., Kuric I., Katalinic B.
1999	**4th DAAAM International Workshop "CA Systems and Technologies: Advanced Methodology of Process Planning"**
Date & Place	12-13th October 1999, Zilina, Slovak Republic
Organized by	University of Zilina, VUT Vienna, CUT Cracow, DAAAM International Vienna, DAAAM Slovakia, DAAAM Poland
Organisers	Kuric I., Katalinic B., Duda J., Poppeova V., Marcincin J.
Proceedings	ISBN 3-901509-12-7
Editors	Kuric I., Katalinic B.
1998	**3rd DAAAM International Workshop "CA Systems and Technologies"**
Date & Place	29-30th September 1998, Cracow, Poland
Organized by	CUT Cracow, University of Zilina, DAAAM Poland, DAAAM Slovakia, DAAAM International Vienna
Organisers	Kuric I., Katalinic B., Duda J., Poppeova V., Marcincin J.
Editor	Kuric I.
1997	**2nd International DAAAM Workshop "CA Systems and Technologies"**
Date & Place	19th March 1997, Zilina, Slovakia
Organized by	University of Zilina, Faculty of Mechanical Engineering, DAAAM International Vienna; DAAAM Slovakia
Organizer	Cubanova N., Janac A., Kuric I., Micieta B., Wessely E.
Editor	Kuric I.

19.0 DAAAM INTERNATIONAL WORKSHOPS "INTELLIGENT MANUFACTURING SYSTEMS"

Founder and chair of DAAAM International workshop on Intelligent Manufacturing Systems is Professor Emil Wessely, Technical University of Kosice.

2002	**4th DAAAM Workshop "Human Factor and Environmentalist"**
Date & Place	5th December 2002, Kosice, Slovakia
Organized by	Department of Environmental Studies and Operations Management of Mechnaical Engineering Faculty of Technical University of Kosice, DAAAM International Vienna
Organized in cooperation with	Slovak Technical Museum of Kosice Oesterreichische Sued-Ost Institut - Vienna
Organisers	Badida M., Wessely E., Bosak M., Majernik M., Lumnitzer E., Kralikova R., Evin E., Rovnak M., Hricova B., Karaszova I., Liberkova L., Verebova H.
Supported by	Action Austria - Slovakia
Proceedings	ISBN 3-901509-37-2
Editors	Katalinic B., Wessely E.
2001	**3rd DAAAM Workshop "Intelligent Manufacturing Systems"**
Date & Place	29 November 2001, Kosice, Slovakia
Organized by	Technical University of Kosice, DAAAM International Vienna, DAAAM Slovakia
Organized in cooperation with	Action Austria-Slovakia, Faculty of Mechanical Engineering - Technical University of Kosice, Slovak Community for System Integration, Slovak Mechanical Engineering Association
Organisers	Wessely E.(Chair), Badida M., Seminsky J., Evin E., Ljoljic B., Kordic V., Pecha C., Rovnak M., Taticek F.
Supported by	Action Austria - Slovakia
Proceedings	ISBN 3-901509-31-3
Editors	Katalinic B., Wessely E.
2000	**2nd DAAAM Workshop "Intelligent Manufacturing Systems"**
Date & Place	30 November 2000, Kosice, Slovakia
Organized by	Technical University of Kosice, DAAAM International Vienna, DAAAM Slovakia
Organized in cooperation with	Action Austria-Slovakia, Faculty of Mechanical Engineering - Technical University of Kosice, Slovak Committee for Management - ZVTS Bratislava, Slovak Mechanical Engineering Association
Organizer	Wessely E.(Chair), Seminsky J., Badida M., Evin E., Paralic J., Knoflicek R., Hradocky L., Bosak M., Kozak L., Vargova J.
Supported by	Action Austria - Slovakia
Proceedings	ISBN 3-901509-18-6
Editors	Katalinic B., Wessely E.

1997	**1st DAAAM Workshop "Intelligent Manufacturing Systems"**
Date & Place	28th November 1997, Kosice, Slovakia
Organized by	Technical University of Kosice, DAAAM International Vienna, DAAAM Slovakia
Organized in cooperation with	Action Austria-Slovakia, Faculty of Mechanical Engineering - Technical University of Kosice, Slovak Committee for Management - ZVTS Bratislava, Slovak Mechanical Engineering Association
Organisers	Wessely E. (Chair), Seminsky J., Evin E., Kovacik J.
Proceedings	ISBN 3-901509-09-7
Editor	Seminsky J.[aroslav]

20.0 DAAAM INTERNATIONAL WORKSHOPS

As an extension of the Symposiums, International Workshops focussing on key issues are arranged. A number of these are detailed below.

1999	**1st International Regional DAAAM-CEEPUS Workshop on Intelligent Machines and Technologies in the 21st Century**
Date & Place	27-29th May 1999, Miskolc, Hungary
Organized by	Faculty of Mechanical Engineering, University of Miskolc in co-operation with DAAAM International Vienna
Proceedings	ISBN 3-901509-14-3
Organizer	Cselenyi J., Alpek F., Katalinic B., Kalmar L., Terstyanszky G., Kerekes I., Siposs I., Peter J.
1996	**4th International DAAAM Workshop "CAD & Intelligent Manufacturing Systems"**
Date & Place	22nd February 1996, Kosice, Slovakia
Organized by	DAAAM International Vienna; DAAAM Slovakia; Technische Universität Kosice, House of Technology, Informatech
Proceedings	ISBN 80-967223-8-7
Organizer	Katalinic B., Seminsky J., Wessely E.
1995	**3rd International DAAAM Workshop "Intelligent Manufacturing Systems"**
Date & Place	31st May 1995, Kosice, Slovakia
Organized by	DAAAM International Vienna; DAAAM Slovakia; Technische Universität Kosice, KAVS; Slovak-Austrian Society Kosice
Organizer	Katalinic B., Seminsky J., Wessely E.
1995	**2nd International DAAAM Workshop "Integration of Applications"**
Date & Place	15th May 1995, IBM Austria, Vienna, Austria
Organized by	DAAAM International Vienna; IBM Österreich; IBM Central Europe and Russia; AKS
Organizer	Beer M., Hansjakob K., Katalinic B., Kaufmann A., Lerch Z., Steinbichler J., Zucker R.
1994	**1st International DAAAM Workshop "Automatic Identification in Theory and Praxis",**
Date & Place	28th April 1994, Vienna University of Technology, Vienna, Austria
Organized by	DAAAM International Vienna; DATALOGIC; Vienna University of Technology, Department of Production Engineering, IFT-IMS
Organizer	Bruck P., Katalinic B., Lerch Z.
1991	**1st International DAAAM Seminar "Concurrent Engineering"**
Date & Place	17-18th April 1991, Kosice, Czechoslovakia
Organized by	Schlumberger Technologies, CAD/CAM Division – Wien; Department for Automatic Production Systems, Faculty of Mechanical Engineering, Technical University of Kosice
Organizer	Wessely E., Katalinic B., Ljuhar D.

21.0 INTERNATIONAL CONFERENCES ORGANIZED IN COOPERATION WITH DAAAM INTERNATIONAL AND/OR DAAAM NATIONAL

2008	**4th International Textile, Clothing and Design Conference, ITC&DC 2008**
Date & Place	5-8th October 2008, Dubrovnik, Croatia
Organised by	Faculty of Textile Technology University of Zagreb
Under the auspices of	Croatian Ministry of Science, Education and Sports, University of Zagreb, Croatian Chamber of Economy
In cooperation with	DAAAM International Vienna; AUTEX Ghent, Belgium; The Textile Institute, Manchester, Great Britain; The International Textile Academy, New Delhi, India; Croatian Academy of Engineering; Croatian Society of Chemical Engineers; Society of Dyers and Colourist, Bradford, Great Britain; INA Oil Industry, Zagreb, Croatia; CARNet, Zagreb, Croatia
Organisers	Dragcevic Z. (Chair), Vuljanic N. (Co-Chair), Bogovic S., Hursa A., Mauser B., Runkas M., Simoncic K.N., Vujasinovic E.

Proceedings	978-953-7105-26-6
Editors	Dragcevic, Z.[vonko]
Published by	Faculty of Textile Technology University of Zagreb
2008	**1st International Conference "Vallis Aurea":Focus on Regional Development**
Date & Place	19th September 2008, Pozega, Croatia
Organized by	Polytechnic of Pozega, DAAAM International Vienna
Organizers	Budic, I. (Chair) / Katalinic, B. (Co-Chair) / Andrlic, V. / Cesarec, P. / Culo, I. / Del Vechio, M. / Fajdic, J. / Svitlica, B. / Zima, D.
Proceedings	ISBN 978-3-901509-60-5, ISBN 978-953-98762-7-0
Editors	Katalinic, B.[ranko]
Published by	Polytechnic of Pozega, Croatia and DAAAM International Vienna
2008	**6th International Conference of DAAAM Baltic "Industrial Engineering"**
Date & Place	24-26th April 2008, Tallinn Technical University, Estonia
Organized by	Tallinn University of Technology, Faculty of Mechanical Engineering, DAAAM Estonia
Under the auspices of	DAAAM International Vienna
In cooperation with	DAAAM International Vienna , BALTECH Consortium, Estonian Academy of Science, Federation of Estonian Engineering Industries, Association of Estonian Mechanical Engineers
Proceedings	ISBN 978-9985-59-783-5
Referred/indexed in	ISI Proceedings
Editor	Kyttner R.
2006	**3rd International Textile Clothing & Design Conference, ITC&DC 2006**
Date & Place	8-11th October 2006, Dubrovnik, Croatia
Organized by	Faculty of Textile Technology, University of Zagreb
Under the auspices of	Croatian Ministry of Science, Education and Sport; University of Zagreb; Croatian Ministry of Economy, Labour and Enterpreneurship; Croatian Chamber of Economy
In cooperation with	DAAAM International Vienna; Association of Universities for Textile, Ghent, Belgium; The Textile Institute, Manchester, Great Britain; The International Textile Academia New Delhi; Croatian Academy of Engineering; Croatian Societs of Chemical Engineers; AATCC, United States of America; IFKT, Reutlingen, Germany; INA Oil Industry, Zagreb, Croatia; City of Dubrovnik, Croatia
Organizers	Dragcevic Z. (Chair), Vuljanic N. (Co-Chair), Bobovcan M., Bogovic S., Geratovic I., Hursa A., Jandric Dj., Martek M., Matusic K., Mauser B., Petrunic I., Vujasinovic E.
Proceedings	ISBN 953-7105-12-1
Editors	Dragcevic, Z.[vonko]
Published by	Faculty of Textile Technology, University of Zagreb
2006	**11th International FMTÜ Conference for Young Researchers: Focus on IMS and Environmentally friendly technologies**
Date & Place	24-25 March 2006, Cluj -Napoca, Romania
Organized by	Technical Section of Transylvanian Museum Society (EME)
Under the auspices of	DAAAM International Viena
In cooperation with	Hungarian Academy of Sciences
Organizers	Gyenge Csaba - president of Scientific Committee Bitay Eniko - president of Organizing Committee
Proceedings	ISBN 973-82331-50-7
Editors	Gyenge C.[saba] & Katalinic B.[ranko]
Published by	Transylvanian Museum Society (EME) - external section of Hungarian Academy of Science
2006	**5th International Conference of DAAAM Baltic "Industrial Engineering – Adding Innovation Capacity of Labour Froce and Entrepreneur"**
Date & Place	20-22nd April 2006, Tallinn Technical University, Estonia
Organized by	Tallinn University of Technology, Faculty of Mechanical Engineering, DAAAM Estonia
Under Auspices of	DAAAM International Vienna
In cooperation with	DAAAM International Vienna , BALTECH Consortium, Estonian Academy of Science, Federation of Estonian Engineering Industries, Associations of Estanian Mechanical Engineers
Proceedings	ISBN 9985-894-92-8
Referred/indexed in	ISI Proceedings
Editor	Kyttner R.
2004	**3rd International Conference Business Systems Management UPS 2004**
Date & Place	23-24th September 2004, Mostar, Bosna and Herzegovina

Organized by	University of Mostar Faculty of Mechanical Engineering and Computing, DAAAM International Vienna, Austrian Society of Engineers and Architects OEIAV-1848, Scientific and Technical Forum of Mostar
Under Auspices of	Presidency of Bosnia and Herzegovina
Organizers	Majstorovic V. (Chair), Katalinic B. (Chair), Covic D., Visekruna, V.
Proceedings	ISBN 3-901509-24-0
Editors	Majstorovic V., Katalinic B., Covic D., Visekruna, V.
2004	**The 5th International Workshop on Emergent Synthesis**
Date & Place	May 24-25, 2004, Budapest, Hungary
Organized by	Computer and Automation Research Institute, Hungarian Academy of Sciences
Chair	L. Monostori, B. Katalinic, K, Ueda
2004	**4th International DAAAM Conference "Industrial Engineering – Innovations as Competitive Edge for "**
Date & Place	29-30th April 2004, Tallinn Technical University, Estonia
Organized by	Tallinn Technical University - Department of Machinery, DAAAM Estonia, DAAAM International Vienna
In cooperation with	Baltic Rector's Council; Association of Estonian; Mechanical Engineers, Federation of Estonian Engineering Industries
Proceedings	ISBN 9985-894-59-6
Editors	Papstel J., Katalinic, B.
2003	**The 1st European-Pacific Congress Globalization: Interaction of Economies, Cultures, Technologies, and Nature,**
Date & Place	23-26th September, 2003, Vladivostok, Russia
Organised by	Association "International Assembly of Capitals and Cities"; DAAAM International Vienna; Far Eastern State Technical University; Far Eastern National University; Maritime State University; International Engineering Academy Primorsk Chamber of Commerce & Industry; Far Eastern Branch of the Russian Academy of Sciences; Russian PEN Center; Vladivostok City Administration; Vladivostok Program Committee; Vladivostok State University of Economics
Organizers	Yu.M. Kopylov (Chairman); (Co-Chairman); G.S. Vishnyakova (Co-Chairman); I.V. Abrosjkin (Co-Chairman); Branko Katalinic (Co-Chairman); M.M. Sokolov, (Co-Chairman); A.A. Akhonin (Coordinator); R.G. Kazmiruk (Secretary)
Proceedings	in Print
2003	**The 8th Scientific Conference of Young Technical Researches, FMTU-2003**
Date & Place	21-22nd March 2003, Cluj-Napoca
Organised by	Technical Section of The Transilvanian Museum Association
Organizers	President of Scientific Committee: Professor Dr Csaba Gyenge
Proceedings	953-96408-8-1
Editors	Gyenge C.[saba]
2002	**1st International Textile Clothing & Design Conference, ITC&DC**
Date & Place	6-9th October 2002, Dubrovnik, Croatia
Organized by	Faculty of Textile Technology, University of Zagreb
Under the auspices of	DAAAM International Vienna; Association of Universities for Textile, Ghent, Belgium; The Textile Institute, Manchester, Great Britan; International Textile Academia; University of Zagreb; Croatian Academy of Engineering
On the occasion of	40th Anniversary of University Education and Scientific Research in the Field of Textiles and Textile Technology in Croatia
Organizers	Dragcevic Z. (Chair), Vuljanic N. (Co-Chair), Grundler D., Hursa A., Krmpotic N., Rolich T., Rogale D., Vujasinovic E.
Proceedings	953-96408-8-1
Editors	Dragcevic, Z.[vonko]
Published by	Faculty of Textile Technology, University of Zagreb
2002	**8th International Conference on Soft Computing: MENDEL 2002**
Date & Place	5-7th June 2001, Brno, Czech Republic
Organized by	Brno University of Technology, Faculty of Mechanical Engineering, Institute of Automation and Computer Science; DAAAM International Vienna, Austria; Foundation for Advanced Information Technology GENOPS, Ltd
Organizers	Ošmera P., Šeda M., Dvořák J., Roupec J., Matousek R.
2001	**5th International Conference Modern Technologies in Manufacturing MTeM 2001:** Assembly and disassembly, CIM, Cold Metal forming technologies, Concurrent Engineering, Ecological Design, Flexible Manufacturing Systems,Industrial Automation,,

	Nonconventional technologies, Rapid Prototyping, Quality Assurance, Robotics
Date & Place	4-6th October 2001, Cluj-Napoca, Romania
Organized by	Technical University of Cluj Napoca, Faculty of Machine Building, DAAAM International Vienna, S.C. Fortpres Cluj-Napoca, S.C. Plastor Oradea, AGIR- General Association of Engineers from Romania
Organizers	Gyenge C. (Chair) et all.
Proceedings	ISBN 973-85354-1-7
Editors	Gyenge C.
2001	**7th International Conference Maintenance 2001**
Date & Place	25-26th June 2001, Opatija, Croatia
Organized by	Croatian Maintenance Society
In cooperation with	DAAAM International Vienna, Union of Croatian Engineers
Organizers	Ivančić I., Cala I., Frković D., Sacevic T., Podgorčić D., Madžarević P., Kristovic D.,Poljak K., Kudelić, R., Ugarković Ž., Susanj B., Pavlovic A.
2001	**7th International Conference on Soft Computing: MENDEL 2001:** Evolutionary Computation, Genetic Programming, Fuzzy Logic, Rough Sets, Neural Networks, Fractals, Bayesian Methods
Date & Place	6--8th June 2001, Brno, Czech Republic
Organized by	Brno University of Technology, Faculty of Mechanical Engineering, Institute of Automation and Computer Science; DAAAM International Vienna, Austria; Foundation for Advanced Information Technology GENOPS, Ltd
Organizers	Ošmera P., Šeda M., Dvořák J., Roupec J., Matousek R.
2001	**2nd International Conference Business Systems Management UPS 2001**
Date & Place	31st May - 2nd June 2001, Mostar, Bosna and Herzegovina
Organized by	University of Mostar Faculty of Mechanical Engineering, DAAAM International Vienna, Austrian Society of Engineers and Architects OEIAV-1848
Organizers	Majstorovic V. (Chair), Katalinic B. (Chair), Covic D., Visekruna, V.
Proceedings	ISBN 3-901509-26-7
Editors	Majstorovic V., Katalinic B., Covic D., Visekruna, V.
2000	**6th International Conference Maintenance 2000**
Date & Place	19-20th June 2000, Opatija, Croatia
Organized by	Croatian Maintenance Society
In cooperation with	EFNMS - European Federation of National Maintenance Societies, DAAAM International Vienna, Union of Croatian Engineers
Organizers	Ivančić I., Cala I., et al.
2000	**5th International Conference on Soft Computing: MENDEL 2000:** Evolutionary Computation, Genetic Programming, Fuzzy Logic, Rough Sets, Neural Networks, Fractals, Bayesian Methods
Date & Place	7-9th June 2000, Brno, Czech Republic
Organized by	Brno University of Technology, Faculty of Mechanical Engineering, Institute of Automation and Computer Science; DAAAM International, Vienna, Austria; Foundation for Advanced Information Technology GENOPS, Ltd
Organizers	Ošmera P., Šeda M., Dvořák J.
Proceedings	ISBN 80-214-1609-2
2000	**2nd International Conference on Industrial Engineering – Activities Targeted to SME**
Date & Place	27 - 29th April 2000, Tallinn Technical University, Estonia
Organized by	Tallinn Technical University - Department of Machinery, DAAAM Estonia, DAAAM International Vienna
Under the auspices of	DAAAM International Vienna
In cooperation with	Baltic Rector's Council; Association of Estonian; Mechanical Engineers, Federation of Estonian Engineering Industries
Organizers	Kyttner R. (chair) Katalinic B., Kulu P., Papstel J., Reedik V., Riives J.
Proceedings	ISBN 9985-894-21-9
Editors	J. Papstel / Katalinic B.
1999	**District Heating Control ′99 – Control of Extensive Heating Systems**
Date & Place	26-27th October 1999, Zlin, Czech Republic
Organized by	Brno University of Technology, Faculty of Technology in Zlin; Czech Power Plants – Ltd.; Association for District Heating of Czech Republic; EUROHEAT & Power, Unichal – Brussels; DAAAM International, Vienna, Austria; DAAAM - Czech Republic
Organizers	Balate J., Vasek V., Vachutova I., Sysala T.
Proceedings	ISBN 80-214-1461-8

1999	**5th International Conference on Soft Computing: MENDEL ´99:** Evolutionary Computation, Genetic Programming, Fuzzy Logic, Rough Sets, Neural Networks, Fractals
Date & Place	9-12th June 1999, Brno, Czech Republic
Organized by	Technical University Brno, Faculty of Mechanical Engineering, Institute of Automation and Computer Science; DAAAM International, Vienna, Austria; Foundation for Advanced Information Technology GENOPS, Ltd
Organizers	Ošmera P., Šeda M., Dvořák J.
Proceedings	ISBN 80-214-1131-7
1999	**5th International Conference Maintenance 99**
Date & Place	7- 8th June 1999, Opatija, Croatia
Organized by	Croatian Maintenance Society
In cooperation with	DAAAM International Vienna, EFNMS - European Federation of National Maintenance Societies
Organizers	Ivančić I., Poljak K., Ugarković Ž., Frković D., Podgorčić D., Madžarević P., Kuzmanić A. Kudelić, R.
Proceedings	ISBN 953-97615-1-4
Editors	Ivančić I., Čala I.
1999	**XXIVth Seminary ASR 2000: Instruments and Control**
Date & Place	28th April 1999, Ostrava, Czech Republic
Organized by	VŠB - Technical University Ostrava, Faculty of Mechanical Engineering, Institute of Automation Equipment and Control; DAAAM National - Czech Republic, Zlín, Czech Republic; IMECO - Czech National Committee; Committee of Applied Cybernetics and Information Technology Ostrava
Organizers	Farana R., Smutny J., Szlachta D., Fojtik D., Piechota H.
Proceedings	ISBN 80-7078-774-0
Editors	Farana R., Smutny J., Szlachta D., Fojtik D., Piechota H.
1999	**XXIIth Seminary ASR ´99 : Computers on Measurements, Diagnostics and Control**
Date & Place	22nd May 1999, Ostrava, Czech Republic
Organized by	VŠB - Technical University Ostrava, Faculty of Mechanical Engineering, Institute of Automation Equipment and Control; DAAAM National - Czech Republic, Zlín, Czech Republic; IMECO - Czech National Committee
Organizers	Smutný L., Farana R., Vítecek A., Baláte J.
Proceedings	ISBN 80-7078-666-3
1998	**3rd International Conference „ New Trends in Automation of Energetic Processes ´98“**
Date & Place	19 - 20th May 1998, Zlín, Czech Republic
Organized by	Technical University Brno, Faculty of Technology in Zlín; Czech Power Company, Ltd.- CEZ, a.s.; Association for District Heating, Czech Republic; STU Bratislava, Faculty of Electrical Engineering and Information Technology; Slovak Power Plants, Ltd. - SE, a.s., Bratislava; DAAAM International Vienna, Austria; DAAAM National - Czech Republic
Organizer	Balate J., Vasek V., Sysala T. & Co.
Proceedings	ISBN 80-214-1094-9
Editor	Balate J., Sysala T.
1998	**4th International Mendel Conference: MENDEL ´98** on Genetic Algorithms, Optimization Problems, Fuzzy Logic, Neural Networks, Rough Sets
Date & Place	24 - 26th June 1998, Brno, Czech Republic
Organized by	Technical University Brno, Faculty of Mechanical Engineering, Institute of Automation and Information Technology; DAAAM International, Vienna, Austria; Foundation for Advanced Information Technology GENOPS, Ltd.
Organizer	Osmera P., Seda M., Simoník I.
Proceedings	ISBN 80 – 214 - 1199 - 6
1998	**XXIth Seminary ASR ´98 : Computers on Measurements, Diagnostics and Control**
Date & Place	22nd May 1998, Ostrava, Czech Republic
Organized by	VSB - Technical University Ostrava, Faculty of Mechanical Engineering, Institute of Automation Equipment and Control; DAAAM National - Czech Republic; IMECO - Czech National Committee
Organizer	Smutny L., Farana R., Vítecek A., Balate J.
Proceedings	ISBN 80-7078-559-4
1998	**9th International Conference on Information and Intelligent Systems IIS´98**
Date & Place	23-25th September 1998, University of Zagreb, Faculty of Organization and Informatics Varazdin, Croatia
Organized by	University of Zagreb, Faculty of Organization and Informatics Varazdin & DAAAM

	International Vienna
Organizers	Aurer B., Katalinic B., Kero K., et. al.
Proceedings	UDC: 007:681.3:651(082)
Editors	Aurer B., Logozar, R.
1998	**3rd National DAAAM Conference "Science ´98"**
On the occasion of	80 Years of Tallinn Technical University
Date & Place	September 1998, Tallinn Technical University, Estonia
Organized by	DAAAM Estonia, Tallinn Technical University, Department of Machinery
Organizers	Papstel J.
Proceedings	ISBN 9985-894-02-2
Editor	Papstel J.
1997	**4th International Conference Maintenance 97**
Date & Place	16-17th October 1997, Opatija, Croatia
Organized by	Croatian Maintenance Society
In cooperation with	DAAAM International Vienna, Croatian Academy of Science and Art, Zagreb EFNMS - European Federation of National Maintenance Societies
Organizers	Ivančić I., Čala I., Vidić J., Udiljak T., Ugarković Ž., Širola D., Podgorčić D., Keser A.
Proceedings	ISSN-1331-4114
Editors	I. Ivančić
1997	**1st International Conference "Industrial Engineering - Actual Activities"**
Date & Place	25-27th September 1997, Tallinn Technical University, Tallinn, Estonia
Organized by	Tallinn Technical University - Institute of Machinery, DAAAM Estonia Tallinn, DAAAM International Vienna
Under the auspices of	DAAAM International Vienna
In cooperation with	Baltic Rectors´ Council, Association of Baltic Material Societies, Federation of Estonian Engineering Industries, Association of Estonian Mechanical Engineers
Organizer	Kulu P. (vice-chair), Kyttner R. (general chair), Papstel J.(vice-chair), Katalinic B., Kübarsepp J., Reedik V., Riives J.
Proceedings	ISBN 3-901509-07-0
Editors	Papstel J., Katalinic B.
1997	**2nd International Conference "Enterprise in Transition"**
On the occasion of	1700 Years of the City of Split
Date & Place	22-24th May 1997, Split - Brela, Croatia
Organized by	University of Split, Faculty of Economics, Split, Croatia; DAAAM International Vienna
Under the auspices of	The Government of the Republic of Croatia, The Croatian Chamber of Economy, Bank of Zagreb
Organizer	Alfirevic N., Buble M., Derado D., Goic S., Katalinic B., Mijacika D., Pavic I. (Chair), Reic Z. (Secretary)
Proceedings	ISBN 953-6024-14-4
Publisher & Editor	Faculty of Economics, University of Split, Split
Co-Publisher	DAAAM International Vienna
1997	**3rd International Mendel Conference on Genetic Algorithms MENDEL '97**
Date & Place:	25-27th June 1997, Brno, Czech Republic
Organisation:	Technical University of Brno, Faculty of Mechanical Engineering, Institute of Automation and Computer Science; DAAAM International Vienna; Foundation Advanced Information Technology; GENOPS, Ltd.
Organizer:	Osmera P., Seda M., Simonik I.
1996	**The Second National DAAAM Conference SCIENCE '96**
Date & Place:	1996, Tallinn, Estonia
Organisation:	Tallinn Technical University - Estonia, DAAAM International Vienna
Organizer:	Papstel J. & Co
Proceedings:	ISBN 9985-59-041-4
Editor:	Papstel J.
1996	**2nd International Mendel Conference on Genetic Algorithms MENDEL '96**
Date & Place:	26-28th June 1996, Brno, Czech Republic
Organisation:	Technical University of Brno, Faculty of Mechanical Engineering, Institute of Automation and Computer Science; DAAAM Czech Republic & DAAAM International Vienna; Foundation Advanced Information Technology; AutoCont Brno Ltd. Computer Systems
Organizer:	Osmera P. & Co.

Proceedings:	ISBN 80-214-0769-7
1996	**13th Conference BIAM '96**
Date & Place:	18-20th June 1996, Zagreb, Croatia
Organisation:	KoREMA, Croatian Society for Communications, Computing, Electronics, Measurement and Control, DAAAM International Vienna
Organizer:	Surina T., Vranjes B., Katalinic B., Jerbic B., Cosic P., Husic S., Kovacic Z., Kunica Z., Majetic D., Momcilovic D., Schauperl Z., Stambuk-Borsic N., Stefanic N., Udiljak T.
Proceedings:	ISBN 953-6037-12-2
1996	**Nanotechnologie - Nanometrologie**
Date & Place:	26th April 1996, Brno, Czech Republic
Organisation:	Technical University of Brno, Faculty of Mechanical Engineering, Vienna University of Technology; DAAAM Czech Republic
Organizer:	P. H.Osanna, J. Svejcar, J. Vackar, M. Skopal
1996	**Rechenmaschinen in Messungen, Diagnostik und Steuerung**
Date & Place:	18th April 1996, Ostrava, Czech Republic
Organisation:	Technical University of Ostrava TU-VSB, KAKI Ostrava, DAAAM Czech Republic, IMECO Czech Republic
Organizer:	Smutny L., Farana R., Vitecek A.
1995	**„New Trends in the Automation of Energetical Processes"**
Date & Place:	17-18th May 1995, Zlin, Czech Republic
Organisation:	TU-Brno / FT-Zlin, KART, DAAAM Czech Republic & DAAAM International Vienna, and others
Organizer:	Balate J., Janacova J., Kubalcik M., Sysala T., Neumann P., Vasek V.
Proceedings:	ISBN 80-214-0654-2

22.0 DAAAM & CEEPUS JEP

Actual Project: HR-108 CEEPUS Project "Concurrent Product and Technology Development - Teaching, Research and Implementation of Joint Programs Oriented in Production and Industrial Engineering"
Network Co-ordinator: *Barisic Branimir - University of Rijeka*

Project Code:	**HR-108**
Project Duration:	***2006-***
Project Coordinator:	***Barisic Branimir - University of Rijeka***
2006/2007 Name:	**Concurrent Product and Technology Development**
Project Partner 06/07	Department of Production Engineering, Faculty of Engineering, University of Rijeka, Rijeka, Croatia / Department of Production Engineering, Faculty of Mechanical Engineering and Naval Architecture, University of Zagreb, Zagreb, Croatia / Institute of Production Engineering, Vienna University of Technology, Vienna, Austria / Department of Manufacturing Engineering, Budapest University of Technology and Economics, Budapest, Hungary / Department of Production Engineering, Faculty of Mechanical Engineering, Czech Technical University in Prague, Prague, Czech Republic / Department of Production Engineering, Faculty of Mechanical Engineering, University of Miskolc, Miskolc, Hungary /Institute of Mechanical Technology, Poznan University of Technology, Poznan, Poland / Department for Material Forming Technologies and Surface Engineering, University of Novi Sad, Novi Sad, Serbia / Department of Machining and Automation, Faculty of Mechanical Engineering, University of Žilina, Žilina, Slovakia / Department of Production Technologies and Systems, Faculty of Mechanical Engineering, University of Ljubljana, Ljubljana, Slovenia.
Organizer 06/07	**Barisic B.[ranimir],** Math M.[iljenko], Katalinic B.[ranko], Szalay T.[ibor], Kreibich V.[iktor], Varga G.[yula], Legutko S.[tanislaw], Plancak M.[iroslav], Kuric I.[van], Kuzman K.[arl]
2007/2008 Name:	**Concurrent Product and Technology Development**
Project Partner 07/08	Department of Production Engineering, Faculty of Engineering, University of Rijeka, Rijeka, Croatia / Department of Production Engineering, Faculty of Mechanical Engineering and Naval Architecture, University of Zagreb, Zagreb, Croatia / Institute of Production Engineering, Vienna University of Technology, Vienna, Austria / Department of Manufacturing Engineering, Budapest University of Technology and Economics, Budapest, Hungary / Department of Production Engineering, Faculty of Mechanical Engineering, Czech Technical University in Prague, Prague, Czech Republic / Department of Production Engineering, Faculty of Technology, Tomas Bata University in Zlín, Zlín, Czech Republic / Department of Production Engineering, Faculty of Mechanical Engineering, University of Miskolc, Miskolc, Hungary /Institute of Mechanical Technology, Poznan University of Technology, Poznan, Poland / Department for Material Forming Technologies and Surface Engineering, University of Novi

	Sad, Novi Sad, Serbia / Department of Machining and Automation, Faculty of Mechanical Engineering, University of Žilina, Žilina, Slovakia / Department of Production Technologies and Systems, Faculty of Mechanical Engineering, University of Ljubljana, Ljubljana, Slovenia.
Organizer 07/08	**Barisic B.[ranimir],** Math M.[iljenko], Katalinic B.[ranko], Szalay T.[ibor], Kreibich V.[iktor], Lukovics I.[mrich],Varga G .[yula], Legutko S.[tanislaw], Plancak M.[iroslav], Kuric I.[van], Kuzman K.[arl].
2008/2009 Name:	**Concurrent Product and Technology Development**
Project Partner 08/09	Department of Production Engineering, Faculty of Engineering, University of Rijeka, Rijeka, Croatia / Department of Production Engineering, Faculty of Mechanical Engineering and Naval Architecture, University of Zagreb, Zagreb, Croatia / Institute of Production Engineering, Vienna University of Technology, Vienna, Austria / Department of Manufacturing Engineering, Budapest University of Technology and Economics, Budapest, Hungary / Department of Machining and Assemmbly, Faculty of Mechanical Engineering, Technical University Ostrava, Ostrava, Czech Republic /Department of Production Engineering, Faculty of Mechanical Engineering, Czech Technical University in Prague, Prague, Czech Republic / Department of Production Engineering, Faculty of Technology, Tomas Bata University in Zlín, Zlín, Czech Republic / Department of Production Engineering, Faculty of Mechanical Engineering, University of Miskolc, Miskolc, Hungary / Institute of Mechanical Technology, Poznan University of Technology, Poznan, Poland / Department of Mechanical Technology and Metrology, Faculty of Mechatronics and Machinery Design, Kielce University of Technology, Kielce, Poland / Department for Material Forming Technologies and Surface Engineering, University of Novi Sad, Novi Sad, Serbia / Department of Machining and Automation, Faculty of Mechanical Engineering, University of Žilina, Žilina, Slovakia / Department of Production Technologies and Systems, Faculty of Mechanical Engineering, University of Ljubljana, Ljubljana, Slovenia.
Organizer 08/09	**Barisic B.[ranimir],** Math M.[iljenko], Katalinic B.[ranko], Szalay T.[ibor], Cep R.[obert], Kreibich V.[iktor], Lukovics I.[mrich],Varga G .[yula], Legutko S.[tanislaw], Adamczak S.[tanislaw], Plancak M.[iroslav], Kuric I.[van], Kuzman K.[arl]

PAST DAAAM – CEEPUS Joint European Annual Projects

Project Code:	***A-104***
Project Duration:	***1999-2004***
Project Coordinator:	***Branko Katalinic - Vienna University of Technology***
2003/2004 Name:	**Intelligent Manufacturing and Automation**
Project Partner 03/04	Institute of Production Engineering, Vienna University of Technology, Vienna, Austria / Department of Manufacturing Engineering, Budapest University of Technology and Economics, Budapest, Hungary / Department of Manufacturing Engineering, Technical University of Cluj-Napoca, Cluj-Napoca, Romania / Department of Technology, Faculty of Mechanical Engineering and Naval Architecture, University of Zagreb, Zagreb, Croatia / Department of Technological Devices and Systems, Faculty of Material Science and Technology Trnava, Slovak University of Technology Bratislava, Trnava, Slovakia / Institute of Production Engineering, University of Maribor, Maribor, Slovenia / Department of Environmental Studies and Process Control, Technical University of Kosice, Kosice, Slovakia / Department of Machine Technology, The Silesian University of Technology, Gliwice, Poland / Department of Production Engineering, Faculty of Engineering, University of Rijeka, Rijeka, Croatia / Institute of Mechanical Technology, Poznan University of Technology, Poznan, Poland
Organizer 03/04	**Katalinic B.[ranko],** Alpek F.[erenc], Balic J.[oze], Gyenge C.[saba], Kosmol J.[an], Legutko S.[tanislaw], Mikac T.[onci], Velisek K.[arol], Wessely E.[mil], Udiljak T.[oma]
2002/2003 Name:	**Intelligent Manufacturing and Automation**
Project Partner 02/03	Institute of Production Engineering, Vienna University of Technology, Vienna, Austria / Department of Manufacturing Engineering, Budapest University of Technology and Economics, Budapest, Hungary / Department of Manufacturing Engineering, Technical University of Cluj-Napoca, Cluj-Napoca, Romania / Department of Technology, Faculty of Mechanical Engineering and Naval Architecture, University of Zagreb, Zagreb, Croatia / Department of Technological Devices and Systems, Faculty of Material Science and Technology Trnava, Slovak University of Technology Bratislava, Trnava, Slovakia / Institute of Production Engineering, University of Maribor, Maribor, Slovenia / Department of Environmental Studies and Process Control, Technical University of Kosice, Kosice, Slovakia /

	Department of Machine Technology, The Silesian University of Technology, Gliwice, Poland / Department of Production Engineering, Faculty of Engineering, University of Rijeka, Rijeka, Croatia / Department of Industrial Robotics, University of Applied Sciences - Fachhochschule Technikum - Wien, Vienna, Austria / Institute of Mechanical Technology, Poznan University of Technology, Poznan, Poland
Organizer 02/03	**Katalinic B.[ranko],** Alpek F.[erenc], Balic J.[oze], Gyenge C.[saba], Kosmol J.[an], Legutko S.[tanislaw], Mikac T.[onci], Stopper M.[arkus], Velisek K.[arol], Wessely E.[mil], Udiljak T.[oma]
2001/2002 Name:	**Assembly Automation in Manufacturing Engineering**
Project Partner 01/02	Vienna University of Technology, Budapest University of Technology and Economics, Silesian University of Technology, Slovak Technical University Bratislava / Trnava, Technical University Cluj-Napoca, Technical University of Kosice, University of Maribor, University of Zagreb
Organizer 01/02	**Katalinic B.,** Alpek F., Kosmol J., Velisek K., Gyenge C., Wessely E., Balic, J., Jerbic B.,
2000/2001 Name:	**Assembly Automation in Manufacturing Engineering**
Project Partner 00/01	Vienna University of Technology, Slovak Technical University Bratislava / Trnava, Technical University Cluj-Napoca, Technical University of Budapest, Technical University of Kosice, University of Maribor, University of Zagreb
Organizer 00/01	Katalinic B., Velisek K., Gyenge C., Alpek F., Wessely E., Balic J., Jerbic B.
1999/2000 Name:	**Assembly Automation in Manufacturing Engineering**
Project Partner 99/00	Vienna University of Technology, Slovak Technical University Bratislava / Trnava, Technical University Cluj-Napoca, Technical University of Budapest, University of Maribor, University of Zagreb
Organizer 99/00	**Katalinic B.,** Velisek K., Gyenge C., Alpek F., Trzan F., Jerbic B.

Project Code:	***RO-01***
Project Duration:	***2002/2003***
Project Coordinator:	***Csaba Gyenge, Technical University of Cluj-Napoca***
2002/2003 Name:	**Computer-Aided Systems for Manufacture and Measurement of Machine Elements**
Project Partner	Department of Manufacturing Engineering / Technical University of Kosice, Department of Manufacturing Systems / University of Maribor, Faculty of Mechanical Engineering, Institute for Production Engineering / Poznan University of Technology, Faculty of Mechanical Engineering, Institute of Mechanical Technology / Kecskemet College, Faculty of Technology

Project Code:	***PL-1***
Project Duration:	***1999-2002***
Project Coordinator:	***Stanislaw Adamczak - Technical University of Kielce***
2001/2002 Name:	**Computer-Aided Systems for Manufacture and Measurement of Machine Elements**
Project Partner	Technical University of Kielce, Vienna University of Technology, University of Zilina, Faculty of Mechanical Engineering of The Technical University of Kosice, Faculty of Production Technologies of The Technical University of Kosice in Presov, University of Maribor, Technical University of Ostrava, University of Zagreb, Technical University of Bratislava, Technical University of Prague, Technical University of Brno
Organizer	**Adamczak S.,** Osanna P. H., Obmascik M., Wessely E., Vasilko K., Cus F., Vrba V., Mahovic S., Janac J., Dvorak R., Vackar J.
2000/2001 Name:	**Geometrical Surface Structure of Machine Parts, Constitution, Methods for Measurement and Assessment, Influence on the Operational Capacity of Mating Machine Parts**
Project Partner	Technical University of Kielce, Vienna University of Technology, University of Zilina, Technical University of Kosice, University of Maribor, Technical University of Ostrava, University of Zagreb, Technical University of Bratislava, Technical University of Prague, Technical University of Brno
Organizer	**Adamczak S.,** Osanna P. H., Obmascik M., Wessely E., Cus F., Vrba V., Mahovic S., Janac J., Dvorak R., Vackar J.
1999/2000 Name:	**Geometrical Surface Structure of Machine Parts, Constitution, Methods for Measurement and Assessment, Influence on the Operational Capacity of Mating Machine Parts**
Project Partner	Technical University of Kielce, Vienna University of Technology, University of Zilina, Technical University of Kosice, University of Maribor, Technical University of Ostrava, University of Zagreb, Technical University of Bratislava, Technical University of Prague, Technical University of Brno

Organizer	**Adamczak S.,** Osanna P. H., Obmascik M., Wessely E., Cus F., Vrba V., Mahovic S., Janac J., Dvorak R., Vackar J.

Project Code:	***RO-01***
Project Duration:	***1998/1999***
Project Coordinator:	***Csaba Gyenge, Technical University of Cluj-Napoca***
1998/1999 Name:	**Development and Application of Environment-Friendly Technologies in Manufacturing Engineering**
Project Partner	Technical University of Cluj-Napoca, Technical University of Budapest, University of Maribor, Technical University of Kosice, Technical University of Bucharest
Organizer	**Gyenge C.,** Alpek F., Balic J., Anna V., Iatan F.

Project Code:	***PL-1***
Project Duration:	***1998/1999***
Project Coordinator:	***Stanislaw Adamczak - Technical University of Kielce***
1998/1999 Name:	**Metrology in Quality Assurance Systems**
Project Partner	Technical University of Kielce, Vienna University of Technology, University of Zilina, Technical University of Kosice, University of Maribor, Technical University of Ostrava, University of Zagreb
Organizer	**Adamczak S.,** Osanna P. H., Obmascik M., Wessely E., Cus F., Vrba V., Mahovic S.

Project Code:	***H-78***
Project Duration:	***1998/1999***
Project Coordinator:	***Alpek,F.***
1998/1999 Name:	**Process Integrated Measurement Techniques and Methods for Quality Management**
Project Partner	Technical University of Budapest, Vienna University of Technology, Cracow Institute of Technology, Slovak Technical University Bratislava, University of Maribor
Organizer	**Alpek F.,** Osanna H. P., Sostar A., Gawlik J., Kralik,M.

23.0 DAAAM INTERNATIONAL BOOKS

The DAAAM International initiated publishing of books in the field of advanced manufacturing and automation. There are three series: "Manufacturing technology", "DAAAM International University Book" and “Management Sciences”. Till now the following books are published:

DAAAM Publishing series - Manufacturing technology:

1. **Joze Balic:** ***"Contribution to Integrated Manufacturing"***, DAAAM Publishing series - Manufacturing technology, Editor: B Katalinic, ISBN 3-901509-03-8, DAAAM International Vienna, Vienna 1999 (Promotion at DAAAM 1999 - Vienna)
2. **Borut Buchmeister:** ***"English-Slovene Glossary of Terms used in Production Engineering and Production Management"***, DAAAM Publishing series - Manufacturing technology, Editor: B Katalinic, ISBN 3-901509-17-8, DAAAM International Vienna, Vienna 2000 (Promotion at DAAAM 2000 - Opatija)
3. **Joze Balic:** ***"FMS: Development Structure, Operation, Handling, Tooling"***, DAAAM Publishing series - Manufacturing technology, Editor: B Katalinic, ISBN 3-901509-20-8, DAAAM International Vienna, Vienna 1999 (Promotion at DAAAM 2001- Jena)
4. **Joze Balic & Niko Majdandzic:** ***"Digit@l F@ctory"***, DAAAM Publishing series - Manufacturing technology, Editor: B Katalinic, ISBN 3-901509-67-4, DAAAM International Vienna, Vienna 2008 (Promotion at DAAAM 2008-Trnava)

DAAAM Publishing series - University Book:

5. **Vlado Majstorovic:** ***"Upravljanje proizvodnjom i projektima – Production and Project Management"***, (in Croatian) DAAAM International University Book, Editor: B Katalinic, ISBN 3-901509-23-2, DAAAM International Vienna and University of Mostar, Mostar – Vienna, 2001 (Promotion at International Conference in Mostar, May 31st, 2001)

6. **Dragan Covic, Vlado Majstorovic, Vojo Visekruna:** ***"Poslovni sustavi – Business Systems"***, (in Croatian) DAAAM International University Book, Editor: B Katalinic, ISBN 3-901509-22-4, DAAAM International Vienna and University of Mostar, Mostar – Vienna, 2001 (Promotion at International Conference in Mostar, May 31st, 2001)

DAAAM Publishing series – Scientific Monograph:

7. **Iztok Palcic,** Industrial Clusters, DAAAM International, Editor: B. Katalinic, ISBN-10 3-901509-80-1; ISBN-13: 978-3-901509-80-3, DAAAM International Publishing, Vienna, 2007

DAAAM Publishing Series – Management Science:

8. **Dragan Covic:** ***"Inventivnost Tima – Vodic za menedzere"***, (in Croatian) DAAAM International University Book, Editor: B Katalinic, ISBN 3-901509-25-4, DAAAM International Vienna and University of Mostar, Mostar – Vienna, 2003. (Promotion October 2003, Sarajevo – Mostar)

9. **Zvonko Kremljak:** ***"Decision making under risk"***, DAAAM International University Book, Editor: B Katalinic, ISBN 3-901509-44-5, DAAAM International Vienna, Vienna 2004 (Promotion at DAAAM 2004 – Vienna)

10. **Zvonko Kremljak & Borut Buchmeister:** ***"Uncertainty and Development of Capabilities"***, DAAAM Publishing Series – Management Science Editor: Branko KATALINIC, ISBN 3-901509-55-0, DAAAM International Vienna, Vienna, 2004 (Promotion at DAAAM 2006 – Vienna)

DAAAM International Scientific Book

11. ***DAAAM International Scientific Book 2002,*** ISBN 3-901509-30-5, hard cover, 713 pages, Publisher DAAAM International Vienna, Editor: B.[ranko] Katalinic, Vienna, 2002. (Promotion of the book at DAAAM 2002 Vienna, 2002-10-23/26)
12. ***DAAAM International Scientific Book 2003,*** ISSN 1726-9687, ISBN 3-901509-36-4, 654 pages, hard cover, Publisher DAAAM International Vienna, Editor: B.[ranko] Katalinic, Vienna 2003. *Abbreviated key title: DAAAM int. sci. book* (Promotion of the book at DAAAM 2003 Sarajevo 2003-10-22/25)
13. ***DAAAM International Scientific Book 2004,*** ISSN 1726-9687, ISBN 3-901509-38-0, 730 pages, hard cover, Publisher DAAAM International Vienna, Editor: B.[ranko] Katalinic, Vienna 2004. *Abbreviated key title: DAAAM int. sci. book* (Promotion of the book at DAAAM 2004 Vienna 2004-11-03/06)
14. ***DAAAM International Scientific Book 2005,*** ISSN 1726-9687, ISBN 3-901509-43-7, 728 pages, hard cover, Publisher DAAAM International Vienna, Editor: B.[ranko] Katalinic, Vienna 2005. *Abbreviated key title: DAAAM int. sci. book*
15. ***DAAAM International Scientific Book 2006,*** ISSN 1726-9687, ISBN 3-901509-47-X, 692 pages, hard cover, Publisher DAAAM International Vienna, Editor: B.[ranko] Katalinic, Vienna 2006. *Abbreviated key title: DAAAM int. sci. book*
16. ***DAAAM International Scientific Book 2007,*** ISSN 1726-9687, ISBN 3-901509-60-7-X, 694 pages, hard cover, Publisher DAAAM International Vienna, Editor: B.[ranko] Katalinic, Vienna 2007. *Abbreviated key title: DAAAM int. sci. book*

Book published under the auspices of DAAAM International

17. **Balc, N.; Gyenge, Cs. & Berce, P: *"Design for Competitive Manufacturing",*** Published under the auspices of DAAAM International, ISBN 973-7898-31-1, Alma mater Publishing, Cluj-Napoca, 2006.

Several other book projects are in process, if you want to join one of them or propose to write a new book please contact the Editor.

24.0 INTERNATIONAL JOURNAL OF SIMULATION MODELLING

Title	**International Journal of Simulation Modelling**
Abbreviated key title	**IJSIMM Int. j. simul. model.**
ISSN	ISSN 1726-4529
Editors	B.[orut] Buchmeister (Editor-In-Chief); B.[ranko] Katalinic (Editor)
Founded	2002
Published	Quarterly by DAAAM International Vienna
Citation and reference	Article published in Journal are crossreferenced and online available on Internet. Additionali they are cited and referenced in the: Inspec Database and Cambridge Scientific Abstracts
Info and Link	http://www.ijsimm.com, http://www.daaam.com,

25.0 20 YEARS DAAAM INTERNATIONAL 1990-2010 PROJECT

Aims: The main aims for future development of our organisation are to increase meaning, importance, quality, relevance and prestige of DAAAM International Network and members in international academic and scientific community. To achieve those aims it is necessary to improve quality and quantity of existing activities and to organize new creative future oriented activities using enormous potential of knowledge, abilities, prestige, relevance and creativity of our members.

Who: Each person in DAAAM International community is welcome to give his/her own contribution to achieving defined aims and can be responsible for one initiative or activity on the level of DAAAM International.

Focus: Focused on upper aims and the fact that we are going soon to celebrate first 20 years of our organisation we are starting an open platform under the name "20 Years of DAAAM International Project". The project is open platform the main parts and topics of the project are:

- To complete and publish Cumulative Bibliography of DAAAM International including all publications: proceedings of DAAAM Symposiums; proceedings of DAAAM Workshops; proceedings of ATDC conferences; DAAAM International Scientific Books; International Journal on Simulation Modelling; Books.
- Founding and organisation of web DAAAM International Archive
- Founding and organisation of web DAAAM International Photo Gallery
- Completing and publishing exclusive monograph "20 Years of DAAAM International"
- Founding, completing and publishing "Who is who in DAAAM International" (Founding members, members of committees since 1990, authors, honours, awards, personal contributions, …)
- improve quality and quantity of existing activities of DAAAM
- to motivate old and animate new people to take a part in our activities
- better use of human resources
- including of younger researchers in our activities
- sponsoring and partnerships
- to increase dissemination of information about our activities
- use of synergies of DAAAM Network in science, research, education and international scientific and academic cooperation
- awards and recognition for DAAAM International on the occasion of celebrating of anniversary of your institution
- to increase quality of our publications
- increase citing and indexing of our publications
- publishing of journals, monographs and student books
- international scientific and academic cooperation with joint application for co-operation of institutions and mobility of teachers and students
- cooperation in industrial and research projects
- DAAAM International certifications of Labs
- Marketing

- DAAAM International Experts data bank
- Cyber space and DAAAM
- Spectrum of Activities to celebrate 20 Years of DAAAM International in 2009
- and many of yours ideas

Call for cooperation: Please think about your wishes, possibilities, and field of interest and propose me one to three initiatives you or your team want to start, organize, manage and be responsible in the frame of DAAAM International. I will pick out of your proposal one initiative which is at the moment the most important from the point of view of DAAAM International. I will give you additional information and suggestions to support your initiative and make it relevant and usable for DAAAM International and synchronised with other activities.

Contact: I am looking forward to see soon your ideas, creativity and initiatives. Please be free to contact me at president@daaam.com I will be happy to hear from you.

26.0 DAAAM - INTERNET & CONTACT ADDRESS

Additional information about DAAAM activities your can obtain from DAAAM homepage: http://www.daaam.com and http://www.daaam.org If you are interested to take a part in DAAAM activities please contact:

✉ PROF. DR. BRANKO KATALINIC, DAAAM INTERNATIONAL PRESIDENT,
TU-WIEN, KARLSPLATZ 13/311, A-1040 WIEN, AUSTRIA, EUROPE
Phone: 0043 1 58801 31121, *Fax#* 0043 1 58801 31199,
E-mails: president@daaam.com, katalinic@mail.ift.tuwien.ac.at

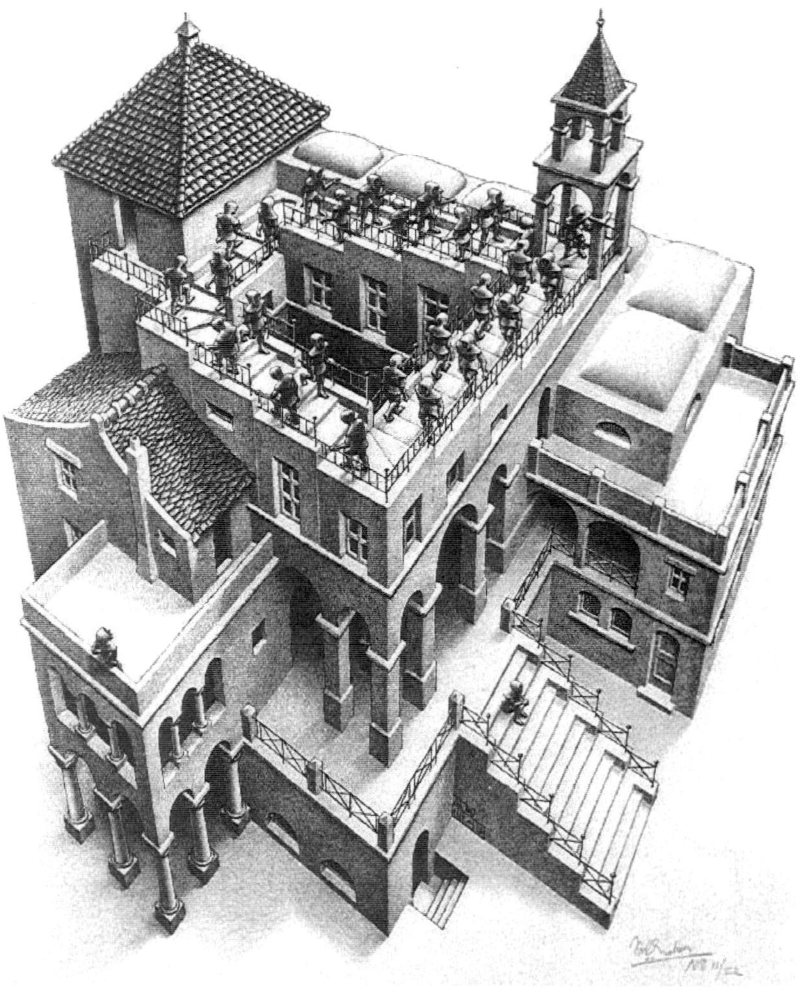

Annals of DAAAM for 2008 & Proceedings of the 19th International DAAAM Symposium, ISSN 1726-9679
Editor B. Katalinic, Published by DAAAM International, Vienna, Austria 2008
Make Harmony Between Technology and Nature, and Your Mind will Fly Free as a Bird

CONTENTS

D

H

M

N

O

P

S

W

Y

Annals of DAAAM for 2008 & Proceedings of the 19th International DAAAM Symposium, ISSN 1726-9679
Editor B. Katalinic, Published by DAAAM International, Vienna, Austria 2008
Make Harmony Between Technology and Nature, and Your Mind will Fly Free as a Bird

ANALYSIS OF OSCILLATORY CONVEYOR SEPARATOR

ABRUDAN, G[heorghe]; RUS, A[lexandru]; VESSELENYI, T[iberiu]; ABRUDAN, A[drian] & TATARU, M[IRCEA] B[OGDAN]

Abstract: *The purpose of this study is to find out which one of the operating states produces critical vibrations and which one of the operating states is best for the sorting.*
There may be operating states that provide a good sorting of the material but, would generate an increased level of vibration of the machine, as well as operating states that only generate a low level of vibration of the machine, but they can not ensure a good sorting of the final product.
Therefore, the problem addressed here consists in finding operating states that can satisfy both criteria, avoiding the spectrum of critical vibrations, meanwhile ensuring a good sorting of the product.
Key words: *Simulation, Vibrations, Oscillations, Sorting conveyors*

1. INTRODUCTION

Pea harvesting can be quite difficult and time-consuming. Two key components the producer must consider are: knowing when to harvest, and taking steps to ensure that the peas are the best quality possible to obtain a high grade. The first component is beyond the purpose of this article. The second point, which consists of many steps itself, incorporates the goal of this paper: to provide a better understanding of the sorting process and provide technical solutions for the development of the optimal separation process to minimize the damage and percent of impurities in the final product.

A good understanding of the way the product is processed and transported through a combine harvester is not so easy. Every section of the machine has its own impact on the flow depending on geometrical dimensions and crop properties (Maertens et al., 2001). Some work has already been performed on these individual machine parts (Dănilă & Neculăiasă, 1987; Neculăiasă & Dănilă, 1995). This study will look at all the parameters that affect the sorting process, determine the pea bean movement on the conveyer belt and optimize the drums geometry, kinematic and dynamic parameters. The mathematical and simulation models developed in this paper could be easily modified to suite other type of products, not just peas. Finally, the analysis of the effect of the optimal sorting process, from a structural point of view, will be a topic of future study.

2. DEVELOPING THE MATHEMATICAL MODEL

2.1 Purpose of the functional forming operation of the peas sorting machine

In order to analyze the operation of the peas sorting machine, considering the vibrations generated by the belt conveyor, it would be useful to develop its mathematical model. The mathematical design is the basis for the generation of the program simulating the operation of the machine. This way, with the help of simulations, different operating states and vibrations generated for different values of the working parameters can be studied (Maertens et al., 2001). The operation of the belt conveyor implies a dynamic and not geometrical analysis (as the drums of the conveyor are not cylindrical), that is necessary for the process of peas sorting. The complete elimination of vibrations and shocks is not possible, because of the fact that the sorting process itself is based on this kind of operation (Brîndeu, 2001).

2.2 Rotation speed of conveyor idle drum.

The pea sorting machine's transporting belt is different from a conventional transporting belt because it carries out an oscillating movement and, at the same time, it moves forward in the direction given by the rotation of the drums (Abrudan, 2007).

The oscillating movement is due to the fact that the drums do not have a cylindrical shape, but consist of bars that make up the edges of prisms with polygonal bases. Currently, in the practical construction, three sided polygons are used as bases for the driving drum, and four sided ones for the idle drum. The difference between the numbers of sides for the two drums generates a specific oscillating movement that allows the sorting of pea beams from the pods and other impurities. The necessary elements for calculating the speed of the idle drum is given in Fig 1. The formula of the angular speed of the idle drum is:

$$\omega_b = \frac{r_a}{r_b}\frac{\sin(\alpha+\gamma)}{\sin(\beta+\gamma)}\,\omega_a \qquad (1)$$

The above formula (1) is the basis for simulating the movement of point B with respect to the movement of point A. The formulas are defined on time intervals with a continuous variation. Meanwhile, the simulation programs can be conceived only for a discrete variation of the parameters. This is the reason why the calculation formulas defined above have to be converted.

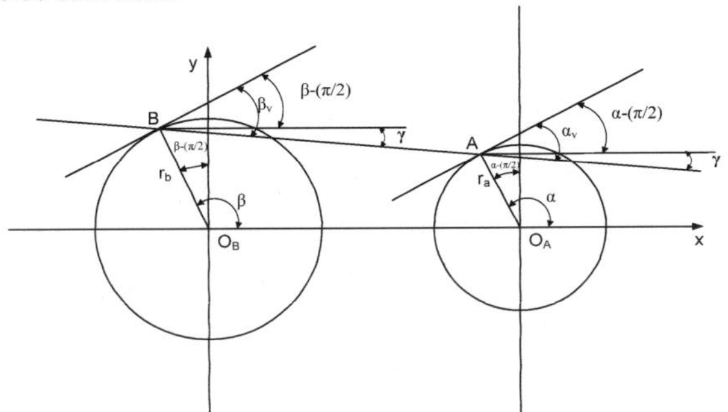

Fig.1. Diagram of the velocity components.

2.3 Determination of forces that influence the bearings of the idle drum

The forces, generated by the oscillating movement of the sorting belt, are transmitted to the frame of the machine through the bearings of the drums in Fig. 1, at points O_A and O_B. During the simulation program, the force acting on the bearing of the driving drum (point O_A) is calculated; this has an oscillating character and can play the part of an excitation force exerted on the frame of the machine generating vibrations, Fig.2, (Abrudan, 2007).

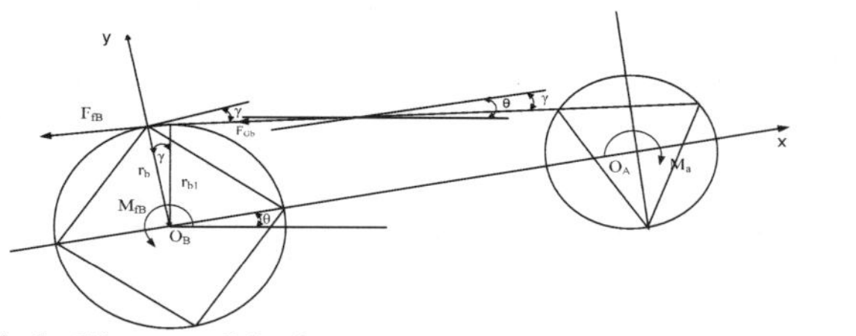

Fig.2. Diagram of the forces.

The resistance force (F_r) has two major components, the stretching force of the belt which is due to its weight (F_{Gb}) and the stretching force that is due to the torque of friction in the idle bearing (F_{fB}). After a few substitutions, the resistance force becomes:

$$F_r = \frac{M_{fB}}{r_b \cos\gamma} + G \sin(\theta - \gamma) \quad (2)$$

The simulation results are compared with experimental data to validate the simulation program.

2.4 Determination of the magnitude and direction of the pea's speed after the impact with the sorting belt

The simulation program accomplishes the calculation of the trajectory of the pea after the impact with the transporting belt. It is considered that the pea starts from a point above the belt, with the coordinates X_{ma0} and Y_{ma0} and has a uniformly accelerated movement until the moment of its first impact with the transporting belt. The position of the pea (X_{ma0}, Y_{ma0}) and its speed have to be calculated for each cycle of the simulation program (Abrudan, 2007). Also, the distance between the pea and the points *A* and B (Fig.1), is being tested in every cycle. If the sum of the distances to the points A and B is close to the distance between the points A and B that means that the pea is near the contact point. Next, the proximity of the pea to the sorting belt is tested, so that the following impact can be detected, using the same formulas as for the first impact. Previous to every impact, the components of the pea's velocity are updated and the iteration is continued.

2.5. Determination of the relationship between the movements of the sorting belt and the pea

In order to achieve the results of the real time working of the sorting belt, a correlation has to be established between the step size used in the simulation program and the real time. The size through which this correlation takes place is given by the number of simulation cycles that correspond to a second of real time working of the belt.

$$N = \frac{2\pi \frac{n}{60}}{\Delta\alpha} \quad (3)$$

Where n is the revolution of the driving drum in rotations/minute, and $\Delta\alpha$ is the angle that this drum turns for one of the simulating program's cycle. The real time is calculated (Δt), in seconds, which corresponds to a cycle in the simulating program, is inverse proportional to N. Using this correlation, the speeds and accelerations that characterize the movement of the belt and the pea can be calculated.

3. THE SIMULATION RESULTS

After running the simulation program, different diagrams were obtained that were used to analyze the mechanics of the sorting belt and detect the critical operating states.

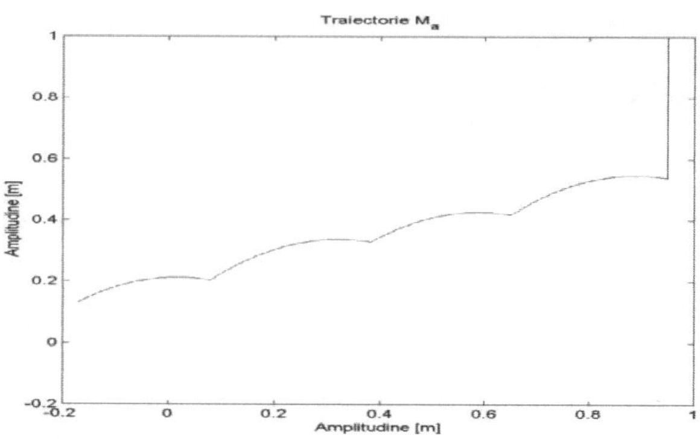

Fig.3. The pea's trajectory for a driving drum's speed of 28 rot/min and a return coefficient of 0.05.

Analyzing the pea trajectory for different parameters of the simulation (the revolution of the idle drum, the drums' number of bars, different coefficients of return), the mathematical values of these parameters, for which the sorting reaches its optimal level, can be found. A result of the pea's trajectory simulation is presented, in the case of a 3-bar driving drum and 4-bar idle drum, Fig 3. Simulations have been carried out for revolutions of the driving drum from 16 to 38 rotations/minute and for return coefficients of 0.05; 0.15, respectively 0.25.

4. CONCLUSIONS

As a result of this study, with the evolution of the kinetic and dynamic parameters of the separating belt on the pea's trajectory in mind, it has been concluded that: generally, once the number of revolutions increases, the number of collisions between the pea and the belt decrease slightly; concerning the influence of the restoration coefficient on the trajectory, it has been established that for higher values of the restoration coefficients, the number of collisions between the pea and the belt decreases. The pods, impurities, vegetal remainders, which have a low restoration coefficient and implicitly a high number of collisions, will be forwarded by the belt in its upper side; the number of bars of the drum is directly proportional to the acceleration of the separating belt and implicit to the belt's oscillations' amplitudes. Even though this study offers several important findings, there are some limitations to it. First, the transporting belt has not been realistically modelled. Also, a greater number of sensors used, at more locations, would have provided a more accurate description of the process when matched with the simulation program. Finally, the analysis of the effect of the optimal sorting process, from a structural point of view, will all be a topic of future study.

5. REFERENCES

Abrudan, G., (2007). The Dynamics of the Oscillating Belt Conveyer, Issue"Polytechnic", Timisoara, pages 60-65.

Brîndeu L(2001). *Vibrations and Vibro-percussions. Basic Mechanical Vibrations and Vibro-Percussions*, Issue Polytechnic, Timisoara

Dănilă, I.; Neculăiasă, V. (1987). *Agricultural Harvesting Machinery*, vol. III, Timisoara Polytechnic Institute.

Maertens, K.; De Baerdemaeker J.; Ramon H.; De Keyser R. (2001): An Analytical Grain Flow Model for a Combine Harvester, Part II: *Analysis and Application of the Model Journal of Agricultural Engineering Research*, Volume 79, Issue 2, June 2001, Page. 187-193.

Neculăiasă, V.; Dănilă, I. (1995). The Working Process and Agricultural Harvesting Machinery, Issue A92, Iasi.

Annals of DAAAM for 2008 & Proceedings of the 19th International DAAAM Symposium, ISSN 1726-9679
Editor B. Katalinic, Published by DAAAM International, Vienna, Austria 2008
Make Harmony Between Technology and Nature, and Your Mind will Fly Free as a Bird

APPLICATION OF THE C-TEST METHODOLOGY FOR THE VALIDITATION OF BOUNDARY CONDITION FOR OIL QUENCHING PROCESS

ADAMCIKOVA, A[ndrea] & TARABA, B[ohumil]

***Abstract**: The article is aimed at the heat treatment cooling process. In the first part, the quantification of the combined heat transfer coefficient (CHTC) by cooling in quenching oil ISOMAX 166 is presented. The cooling curve was obtained by the methodology of the Wolfson's test (DIN 9950). Using the statistical approach, inverse heat conduction method and computer modeling, the boundary condition (CHTC) as the function of the probe surface temperature was obtained. The validity check of the obtained CHTC is the objective of the second part. The methodology of data validation was based on the C-test. The tested part (tube) with a defined slot was cooled in the oil ISOMAX 166 at the same conditions as the Wolfson's probe. Computer modeling of temperature and stress-strain fields in the tested part and evaluation of the results proved that the obtained CHTC of the Wolfson's test is acceptable. The task was solved by finite element method, using the engineering-scientific program code ANSYS.*

Key words: *quenching, modeling, Wolfson's test, C-test, ANSYS*

1. INTRODUCTION

The aim of this paper is the presentation of partial results of the heat transfer phenomena research by cooling of parts in choosen coolants used in industrial production. Knowledge of parameters of cooling medium, physical, mechanical material properties and geometry of a quenched part allows to predict behavior of a part during cooling process. The quantification of the combined heat transfer coefficient (CHTC) by the cooling process in quenching oil and verification of its validity are presented. The obtained boundary condition for convection heat transfer was based on the Wolfson's test. Combining the experimental temperature measurement and numerical analysis, the cooling condition usable for vertical wall was determined. The validity of CHTC obtained was verified at the same experimental conditions as in Wolfson's test. Isomax 166 quenching oil in unagitated state at the temperature of 60 °C was used as a cooling medium for the C-shaped tested part. The Wolfson's probe and the tested part were made of DIN 1.4541material. ANSYS interpretation computer code was used and solution procedures were transient and nonlinear.

2. EXPERIMENTAL

The oil Isomax 166 belongs to intensive quenching oils of low viscosity ($\nu = 12.5\ 10^{-6}\ m^2.s^{-1}$ at 40 °C) and is generally applied for quenching non-alloyed, low-alloyed, alloyed and carbonized steels. It is resistant to evaporation. The recommended working temperatures of Isomax 166 range from 40 °C to 70 °C (www.petrofer.com.ua 2008). The experimental equipment consisted of an electrical resistance furnace of LM 212.10 type, cylinder-shaped experimental Wolfson's probe (www.extra.ivf.se 2008), Isomax 166 oil and NI USB 9211 for digital record of the temperatures measured. Before quenching, the probe was heated up to the initial temperature of 850 °C. The temperatures were measured by the encapsulated 304 SS thermocouple of K type by the diameter of 1.53 mm situated in the center of the probe. The tested "C" part (tube) had a longitudinal slot in the tube body by the initial width of 1.72 mm. In the process of cooling the probe and the tested part, the oil temperature was kept constant at 60 °C, where the cooling ability of Isomax 166 is maximum (Taraba & Laščcek 2006).

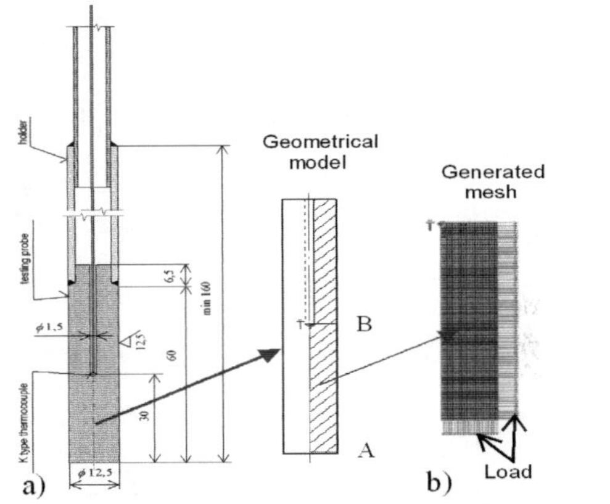

Fig. 1 Wolfson's probe, a) geometry, b) generated FEM mesh.

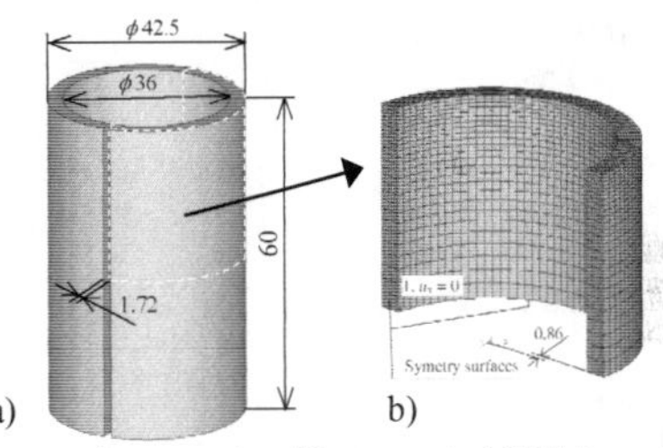

Fig. 2. Tested part, a) geometry, b) generated FEM mesh.

3. THEORETICAL BACKGROUND

Thermal problem. For the presented problem, temperature fields are transient and they can be described by the Fourier-Kirchoff differential equation of heat conduction in cylindrical coordinate system (probe) and the Cartesian coordinate system (tube), respectively. Material is considered to be isotropic. Thermal and mechanical properties were given by functions of temperature for the temperature interval from 20 °C to 900 °C (Tab. 1). Thermal load represented the boundary condition of the 3rd type, i. e. the heat transfer by convection (Incropera). Structural problem. The temperature field, tube shape, thermal expansion and mechanical properties (elastic modulus, Poisson's ratio and yield stress) generate thermal stress-strain states. Stress fields are described by the equation for temperature stress state (Trebuňa at al. 2002). An elastic-plastic material model with bilinear isotropic hardening (Fig. 3) was used. Generation of plastic strains was evaluated by the Huber-Mises-Hencky's hypothesis (Trebuňa at al. 2002). The reference temperature for structural task was 60 °C. The geometric models and generated meshes were generated for one half of the probe (Fig. 1b) and a quarter of the tube (Fig. 2b).

For probe meshing, 2D elements PLANE77 with the option axisymmetric were used.

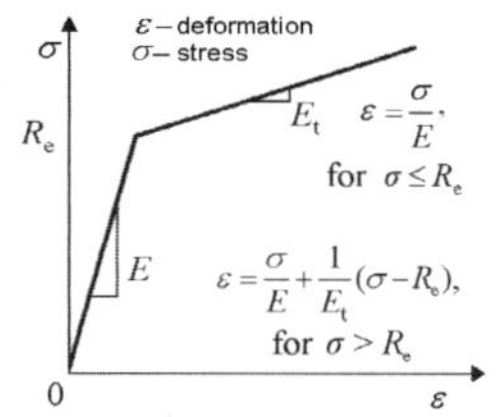

Fig. 3 Elastic-plastic material model.

The tube model was meshed by the element of BRICK90 (thermal problem) and BRICK 186 (structural problem) types (ANSYS 10.0 2005). For structural analysis, symmetry conditions according to the Fig. 2b were used. The "C" model was gripped in the point 1 considering the displacement in the direction of x axis.

Temperature T [°C]	Thermal conductivity λ [W.m^{-1}.K^{-1}]	Specific heat c [J.kg^{-1}.K^{-1}]	Density ρ [kg.m^{-3}]	Elasticity modulus E [GPa]	Thermal expansion coefficient $\alpha_t.10^6$ [K^{-1}]	Yield stress R_e [MPa]	Poison's ratio ν [-]	Tangent modulus E_t [MPa]
0	14.8	455	7940	200	16.8	235	0.3	1185
100	15.8	475	7911	195	17.2	233	0.3	1175
200	17.0	495	7871	188	17.6	230	0.3	1160
300	18.4	508	7830	181	17.8	222	0.3	1080
400	20.0	525	7787	172	18.0	206	0.3	950
500	22.0	550	7745	165	18.3	174	0.3	812
600	24.0	572	7703	157	18.5	137	0.3	660
700	25.7	602	7662	147	18.8	94	0.3	470
800	27.5	620	7620	135	19.0	55	0.3	250
900	29.4	630	7578	100	19.2	36	0.3	185

Tab. 1. Material DIN 1.4541, thermal & mechanical properties.

4. RESULTS

Fig. 4 shows the measured cooling curve from the Wolfson's test. The solution was searched via fitting the curves (measured vs. computed temperatures) and applying the iterative approach to the CHTC, as the function of surface temperature. The correlation coefficient by value of 0.995 was achieved by the fitting of curves. The CHTC curve is shown in the Fig. 5 and it can be considered the boundary condition for the vertical walls cooling.

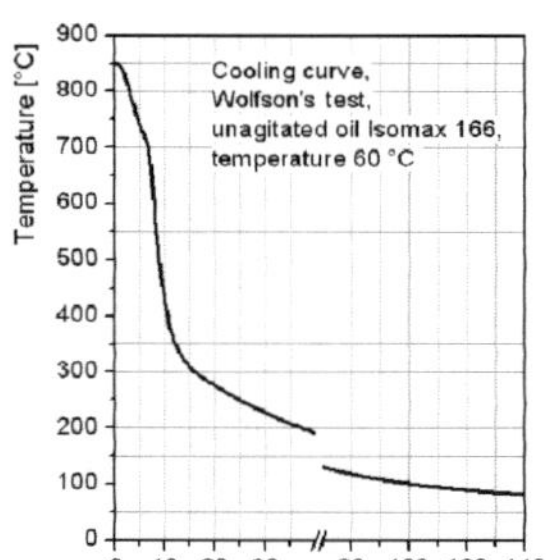

Fig. 4. Measured cooling curve from Wolfson's test.

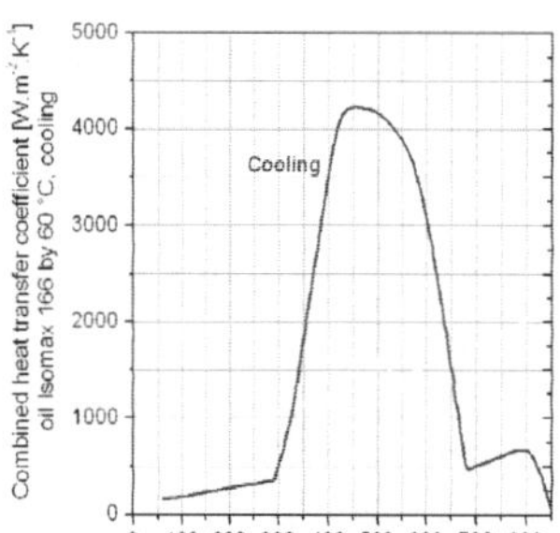

Fig. 5. CHTC determined from Wolfson's test.

The tested part. Fig. 6 illustrates the comparison of the temperature field in the time of 3.2 s after immersing the tube into the oil, as well as the real photo taken at the same time. Each of the three types of cooling process is evident: vapor blanked (A), boiling oil zone (B), convection after ending of boiling (C). There is a reference temperature field from numerical simulation in Fig. 6b. Distribution of residual stresses after cooling the part to the temperature of 60 °C is shown in Fig. 7b. It is evident from the Fig. 7b that in the process of cooling, the stresses reach and exceed the yield stress of material. The existing residual stresses resulted in the contraction of the slot width. Time history of the slot width can be seen in Fig. 7a. The maximum change of the slot width was observed during the cooling in the phase of oil boiling. The final slot width after cooling was 1.43 mm.

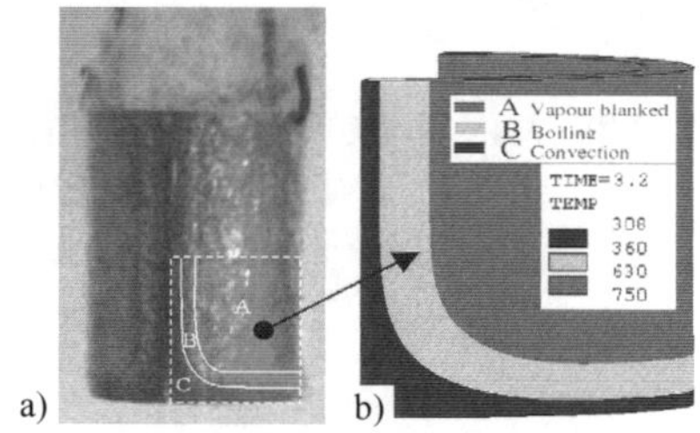

Fig. 6. Comparison of temperature fields, a) photo, b) computed.

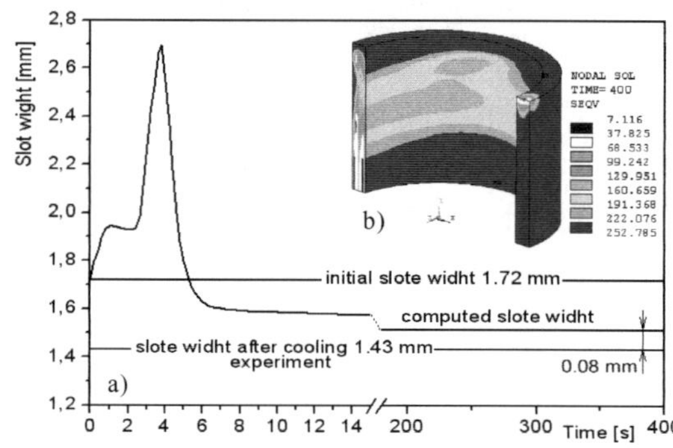

Fig. 7. a) Time history of slot width, b) residual equivalent Mises stress field.

5. CONCLUSIONS

The relation between the experiment and numerical simulation allows us to obtain the knowledge and better understanding of the relationships of cooling parameters in the process of oil quenching.

Loading boundary condition of cooling, obtained by combination of the Wolfson's test and numerical analysis, was proved acceptable for vertical walls cooling.

The influence of particular parameters on the stress-strain state of a tested part during cooling can be determined by the indirect measurement of the change in the slot width. The slot width after experimental cooling on one hand, and the computed slot width on the other hand, exhibited the difference of 0.08 mm.

The research was supported by the projects No. 1/0721/08, 1/0837/08 and 2/7167/27 within VEGA Ministry of Education and the Slovak Academy of Science, Slovak Republic.

6. REFERENCES

Ansys Theoretical Manual, Release 10.0. (2005). *Available from*: http://www.tsne.co.kr/intra/data_center/ansys/theory.pdf *Accessed* 2008-06-23

Available from: www.extra.ivf.se/smartquench/dokument/download.asp?id=21 *Accessed* 2008-06-23

*Available from:*www.petrofer.com.ua/content/hardening_compound/2_1.htm *Accessed* 2008-06-23

Incropera, F., P. (1996). *Fundamentals of Heat and Mass Transfer*, John Wiley Sons, ISBN 0-471-30460-3, New York

Taraba, B. & Laščak, M. (2006). The influence of Isomax 166 quenching oil temperature on its cooling properties. *Acta Mechanica Slovaca*, 10, 01, (01 2006) ISSN 1335-2393, 567-572

Trebuňa, F.; Šimčák, F. & Jurica, V. (2002). *Elasticity and strenght II*, Vienala, ISBN 80-7165-364-0, Prešov

Annals of DAAAM for 2008 & Proceedings of the 19th International DAAAM Symposium, ISSN 1726-9679
Editor B. Katalinic, Published by DAAAM International, Vienna, Austria 2008
Make Harmony Between Technology and Nature, and Your Mind will Fly Free as a Bird

THE IMPLEMENTATION OF ACROMATIC PRINCIPLES IN REPRODUCTION AND BLACK PRINTER CHANGES

AGIC, D[arko]; MANDIC, L[idija]; STRGAR-KURECIC, M[aja] & STANIC, N[ikolina]

Abstract: *The aim of this research was to determine some non-standard behavior of creating the separation by applying achromatic UCR and GCR principles. For selected tertiary colour patches various assets of replacements of common achromatic content was exchanged with black, by preserving colorimetric values as much as possible. In the area from visible to 1000 nm differences in response of patches containing various tone values of black are perceived.*
Key words: *achromatic reproduction, tertiary colours, colorimetric values, IR response*

1. INTRODUCTION

The black printer is developed in the same time as colour C M Y (cyan, magenta, yellow) separations. The intention of broadening the basic colours with black was to improve the reproduction meaning contrast, density range, some details etc. As the speed of printing machines grew higher, the total coverage amount was unable to dry in necessary time, what caused various technical and procedure difficulties. Implementation of UCR (under colour removal) or CCR (complementary colour removal) principles, meaning substitution of the common achromatic part in colours with black tint coverage (K), solved numbered problems including charge of chromatic dyes/tints in printing process (Kipphan, 2001). In some other situations caused decreasing in high densities (increasing of lightness L), deviations in chroma or hue for tertiary colours reproduction. Colorimetric measurements, determining Lab values and density measurements were carried out.

2. BACKGROUND

UCR was still possible to be used in photomechanical operations as a separate mask made, but also was implemented as a hardware function in older analogue scanners. It was practised for grey scale or neutral tones defined as C M Y coverage combination to be substituted with black (K) coverage in desired amount (Yule, 2000). This is quite important as reproductions are often judged on their neutral or grey regions. Digital scanners were able to distinguish reduction for grey scale (UCR) and other tertiary colours (GCR), also applied in various graphic arts computer programs (Johnson, 1984). Contemporary approach primary considers that as an achromatic substitution process for both principles.

Theoretically it is a simple process of substitution, but due to non ideal inks and substrates, various physical and optical deviations such as additivity failure, light scatter, boundary effects, have influence on the final result. One combination corresponds only to the target printing profile. In various manuals for processing in graphic arts (scanners or operating programs) the achromatic possibility is offered, but the amount of substitution is gave up to the technologist, or eventually some value for concerned process is recommended. In today's practice when ISO standardisation (ISO 1996) recommends achromatic methods, it is of significant interest to get more information about these issues.

Currently the most important color space based on the opponent-color theory is known as CIELAB (Berns, 2000). The L*, a* and b* coordinates can be calculated from the tristimulus values XYZ normalized to the white by equations (1)-(3):

$$L^* = 116\left(\frac{Y}{Y_n}\right)^{1/3} - 16 \qquad (1)$$

$$a^* = 500\left[(X/X_n)^{1/3} - (Y/Y_n)^{1/3}\right] \qquad (2)$$

$$b^* = 200\left[(Y/Y_n)^{1/3} - (Z/Z_n)^{1/3}\right] \qquad (3)$$

Where X_n, Y_n, Z_n are tristimulus values of light source.

3. EXPERIMENT

Two patches, grey and tertiary, were generated on monitor for this experiment. The prints were printed on Xeikon 32D digital printer. The grey patch has following values: 90 c, 80 m, 80 y and 0 k. The corresponding RGB values are 76, 80, 70 and Lab coordinates 33, -4, and 6. Various combinations of substitutions were applied. The first combination was traditional substitution with black. The substitution was performed in several steps. In the second combination, the tone value of black was constant and the other values were calculated to adjust colorimetric values constant. In the third combination, the cyan was holding constant and the other values were calculated to keep colorimetric values constant. The CMYK values for three combination and programmed Lab values are shown in Table 1.

Programmed patches were printed and measured by spectrophotometer and densitometer. The tone values of grey patch are: 90 c, 80m, 80 y and k 0) and Lab values: 42, -14 0.3. The results are shown in table 2.

It is evident that reduction principles in recalculated profile achieve more acceptable results than standard ones. The more black substitution than in standard procedure is achieved, that result in more stable better black is performed.

For further investigation a tertiary coloured patch was generated. Its purpose was to check on that proposed principles could be adopted more common. The basic tone values recalculated values for tertiary colour patch are C 80, M 60, Y 60, and K 0. Colorimetric Lab values are 47, -5, and -1. Related R G B values are 104, 112, and 112. The patches are recalculated for K coverage 10, 30, 50 and 70.

As can be seen colorimetric lightness by recalculated values do not considerably change, and chromatic coordinates a and b slightly shift to achromatic with increasing amount of black tint coverage amount coverage, even more than implied.

Comb		1. step	2.step	3.step	4.step	5.step
1	C	70	50	30	10	
	M	60	40	20	0	
	Y	60	40	20	0	
	K	20	40	60	80	
	L	45	51	50	40	
	A	0	1	0	-1	
	b	4	2	0	-1	
2	C	86	79	67	37	23
	M	73	64	48	10	0
	Y	77	71	62	41	30
	K	20	40	60	80	84
	L	33	33	33	33	33
	A	-5	-4	-4	-4	-4
	b	6	6	6	5	5
3	C	70	50	30	10	
	M	66	40	20	0	
	Y	60	40	20	0	
	K	55	73	81	86	
	L	31	31	32	33	
	A	0	1	0	-1	
	b	3	1	0	-1	

Tab. 1. Achromatic substitution for a grey patch and programmed Lab values

Comb		1. step	2.step	3.step	4.step
2	C	77	70	58	32
	M	58	46	29	0
	Y	56	49	37	16
	K	10	30	50	70
	L	46	46	46	45
	A	-4	-2	-5	-5
	b	-1	-2	-2	-1

Table 2: Recalculated coverage c m y k values for a grey patch

As can be seen colorimetric lightness by recalculated values do not considerably change, and chromatic coordinates a and b slightly shift to achromatic with increasing amount of black tint coverage amount coverage, even more than implied.

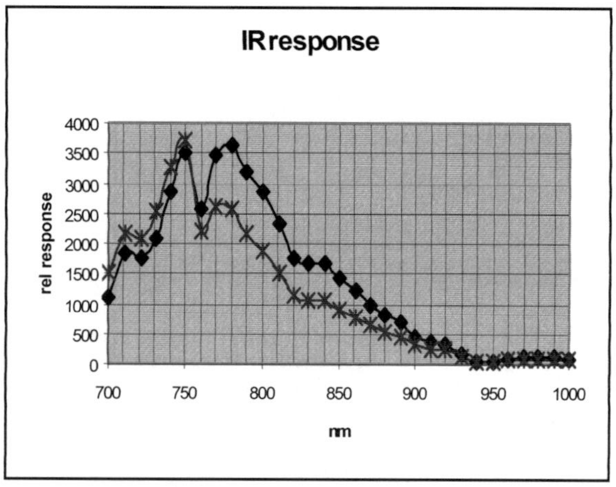

Fig. 1. IR response of recalculated patches. Patches are marked on their K tone values □ no reduction, × 40 K reduction

The overall visual response also does not change considerably also, so for possible discrimination the behaviour of patches containing black ink combination determination in the region from 700 to 1000 nm. is performed, fig 1.

4. RESULTS AND DISCUSSION

It is obvious that achromatic model of reproducing has various interesting features. In pure reproduction process the total coverage amount is reduced, but increase in lightness is occurred, what means lower density, as shown in table 2 and 3. Higher lightness (L) in standard graphic communications means lower density and reduced density range. The reason is in optical and physical properties of dyes and substrates, such as additivity failure, unwanted light dispersion and adsorption and boundary effects. If some complex recalculation is used, meaning keeping the colorimetric values as possible constant, lightness mainly stays unchanged. Not only for reasons claimed by ISO standardisation for chromatic reduction meaning total coverage decrease and lowering technical and technological conditions, such a reduction means more for practical purposes as other colorimetric values are more stabile (mainly unchanged).

Similar situation occurred by tertiary colour, when complex recalculation was used. The very interesting situation is that by preserving colorimetric values, what includes a visual response, higher black ink amount can be implemented for chosen reproduction profile. In such combination some coloured patches optically can appear near similar, but vary with the black (K) coverage amount. For now that that can be practised in generated (tertiary) bars and stripes, for a real picture a separate model algorithm should be made. From visual/IR response it can be concluded that reproduced colours containing less black coverage have lower reflection (higher density) in visual part and higher in IR part. Colours with more black coverage have higher reflection (lower density) in visible part, and lower reflection (higher density) in IR part of spectrum, so in certain conditions they can be distinguished. This behaviour is particularly closely linked with structure of dyes.

5. CONCLUSION

A grey and a tertiary colour were generated in this experiment. Various combinations of substitutions were applied. The aim was to preserve colorimetric conditions as much as possible the same as origin. The results show that colours with more black coverage have higher reflection (lower density) in visible part, and lower reflection (higher density) in IR part of spectrum. The more black substitution than in standard procedure is achieved, that result in more stable better black is performed. It leads to conclusion that in certain conditions they can be distinguished.

6. REFERENCES

Berns, R.; (2000). *Handbook of Billmeyer and Saltzman's Principles of Color Technology*, Wiley, ISBN 978-0471194590

ISO 12647-2: Graphic technology-Process control for the manufacture of half-tone 1996-01-10

Johnson, A. (1984), Practical Implementation of optimum Colour reproduction, *The Journal of Photographic Society*, Vol.32, No.4, (July, 1984) pp.146

Kipphan, H.; (2001). *Handbook of Printed Media*, Springer-Verlag, ISBN 3540673261, Boston

Yule, J.A.C. ;(2000), *Principles of Color Reproduction: Four color printing and the black printer*, GATF Press, *Pittsburg*

Annals of DAAAM for 2008 & Proceedings of the 19th International DAAAM Symposium, ISSN 1726-9679
Editor B. Katalinic, Published by DAAAM International, Vienna, Austria 2008
Make Harmony Between Technology and Nature, and Your Mind will Fly Free as a Bird

TESTING OF THE PARAMETRES OF THE Q-P PROCESS IN HIGH STRENGTH LOW-ALLOYED STEEL

AISMAN, D[avid]; JIRKOVA, H[ana]; SKALOVA, L[udmila] & MASEK, B[ohuslav]

Abstract: *The mechanical properties of new types of materials are dependent on their composition and the distribution of their microstructure. These microstructures are obtained innovative heat and thermo mechanical treatments. One of these is the Q-P process. During the Q-P process untransformed austenite is stabilized in a martensite matrix and carbides are suppressed in nucleation (Edmonds et al., 2006). The process was optimized in new low-alloy steel 42SiCr with the aim of creating a steel with ultimate strength 2000MPa and ductility over 10%. The resulting structure was analyzed using light and laser confocal microscopy. The quantity of the residual austenite was measured with X-ray diffraction analysis. The mechanical properties were validated with a tensile test.*
Key words: *Q-P process, low – alloyed high strength steel.*

1. INTRODUCTION

Industry is increasingly demanding materials with a good combination of mechanical properties, i.e. ultimate strength and ductility, at low cost. New trends are to reduce costs by reducing the content of high-alloy and creating the final microstructure using new heat and thermo mechanical treatments. One of these is the Q-P process. Achieving a good combination of strength and ductility can only be achieved by altering parameters such as length of heating, cooling rate, time and temperature quenching. All these parameters depend on the dimensions of the specimen, the chemical composition and other parameters of the components.

2. Q-P PROCESS

The quenching and partitioning process (Q-P process) consists of quenching at intervals between martensite start and martensite finish temperatures. After quenching, the temperature is held to stabilize untransformed austenite partitioning carbon. The carbon diffuses from the saturated martensite. This process is different from quenching and tempering (Fig.1).

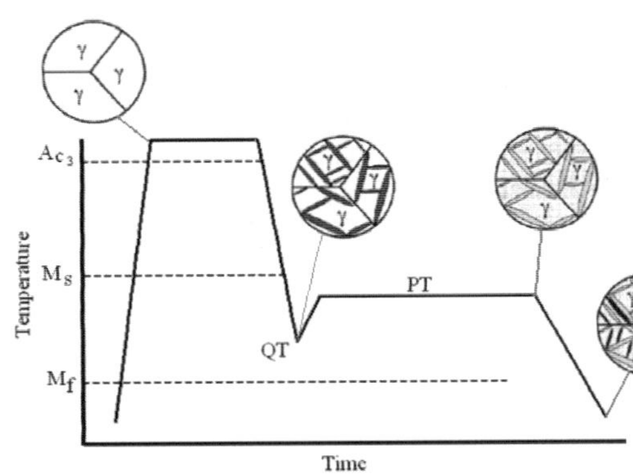

Fig. 1 Diagram of Q-P process showing microstructures (Edmonds et al., 2006)

During normal quenching and tempering, carbide is formed, but the Q-P process suppresses the creation of carbide using a suitable combination of heat treatment and alloying to stabilize the untransformed austenite.

The final structure consists of martensite and stabilized retained austenite. This structure ensures high ultimate strength and at the same time keeps very good ductility values (Gerdemann , 2004)(Speer et al. 2005).

3. EXPERIMENT

The aim of experiment was to propose parameters for a new heat treatment method with strength over 2000MPa and ductility over 10%. The first step towards optimization was to define a suitable austenization temperature, cooling rate and the stabilizing salt bath temperature. 42SiCr steel was used for the experiment. The main components this steel influencing stabilization of untransformed austenite and hardening solid solution are silicium, chromium and manganese. The quantity of all the components is less than 5%, which makes it economically viable. The basic microstructure was ferrite-pearlite with a higher proportion of pearlite. The hardness was 295HV in its base state (Fig.2).

Six heat treatment regimes were designed for testing the influence of Q-P process (Table. 1). The time of heating was between 20 - 30 min, the length of holding in salt bath was 5 – 10 min. The time in salt bath was established on the basis of dilatometric analysis so the temperature laid between martensite start and martensite finish temperatures. Dilatometer analysis showed the temperature of martensite start to be 305°C. To find the influence of cooling rate, two specimens were cooled in the water for 2s and then put in the salt bath . The specimens tested in this regime were 55 x 18 x 25 mm.

The final structure was evaluated with light and confocal laser microscopy. The sample was etched using 3% Nital. The structure of the heat treated specimens consisted of acicular martensite and retained austenite foil a few tens of μm thick

Fig. 2. The base state – ferrite and pearlite structure which were not visible with light microscopy.

Strategy	heating 900°C [min]	Cooling in water	Salt bath 250°C [min]
1	25	no	5
2	20	no	10
3	20	no	20
4	30	no	10
5	20	2s	10
6	30	2s	10

Tab. 1. Regime parameters

Different sizes of martensite were observed under light and confocal microscopy which were consistent with heat time. (Fig. 3. a, b)

3.1 The results of mechanical testing

The mechanical properties were evaluated in a tensile test on the mini specimens whose active part was 5 mm long. (Table. 2).

The ultimate strength of the material was 980MPa and its ductility was A_{5mm} = 31 %. The ultimate strength was over 2000MPa and ductility more than 14% higher after heat treatment. The specimen with parameters 900°C/20min – 250°C/10min had the highest ductility; A_{5mm} = 18%.

The resulting strength and ductility values correspond closely with the proportion of retained austenite in the microstructure (Tab. 2), which were measured by x-ray phase diffraction analysis. As the amount of retained stabilized austenite increases, the tensile strength decreases and the ductility increases.

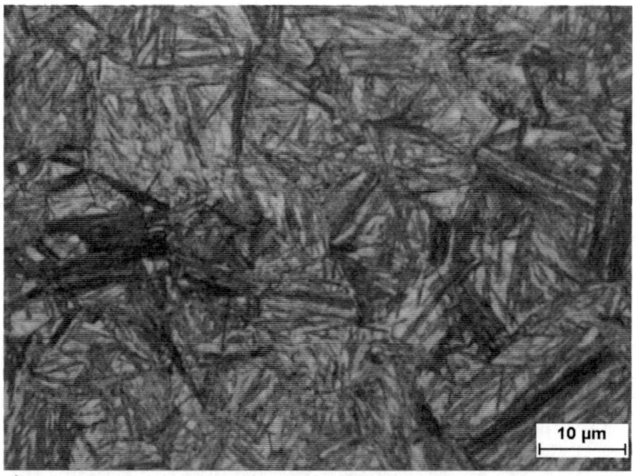

a)

b)

Fig. 3. Examples of final structure
a) heating 900°C/30min, time in salt bath 10min
b) heating 900°C/20 min, time in salt bath 10m

	$R_{p0,2}$ [MPa]	R_m [MPa]	A_{5mm} [%]	Proportion of RA [%]
Basic state	592	981	31	-
(900°C/25min – 250°C/5min)	1657	2157	14	5.1
(900°C/20min – 250°C/10min)	1714	2120	16	4.1
(900°C/20min – 250°C/20min)	1663	2038	18	9.8
(900°C/30min – 250°C/10min)	1728	2054	15	6.9
(900°C/20min – 2s water - 250°C/10min)	1765	2102	14	7.1
(900°C/30min – 2s water - 250°C/10min)	1852	2107	14	7.5

Tab. 2. Results of the tensile test and proportion of residual austenite

This is evident in the regime with parameters 900°C/20min – 250°C/20min where there is over 9% retained austenite. The ductility of this specimen was 18%, but ultimate strength was slightly decreased to 2030 MPa.

4. CONCLUSION

The optimization of the parameters of the Q-P process in a new experimental steel showed us the possibility of creating materials with an attractive combination of mechanical properties such as strength and ductility. The ultimate strength of the heat treated specimens was over 2000MPa and ductility around 15%. X-ray analysis confirmed the presence of retained austenite which was probably laid down as a thin film on the martensite plate. Cooling the sample in a medium with a high cooling intensity (water), and prolonged holding at austenization temperatures were found to be unsuitable treatments because of the resulting coarseness of the acicular martensite and deterioration of the samples' mechanical properties.

Influence and type of quenching and partitioning bath will be tested in the following step.

5. ACKNOWLEDGEMENTES

This paper includes results obtained within the project 1M06032 Research Centre of Forming Technology.

6. REFERENCES

Brokate,M. ; Sperkles, J. (1996). *Hysteresis and phase transitions,* Springer, *pp 59-60*

Edmonds D.V. et al. (2006). Quenching and partitioning martensite – A novel steel heat treatment, *Materials Science and Engineering,* A 438–440 25–34

Gerdemann F.L.H. (2004), *Microstructure and hardness of 9260 steel heat-treated by the quenching and partitioning process,* Aachen University of Technology, Germany.

Speer J. G.; Assunção F. C. R.; Matlock D. K.; Edmonds D. V; (2005), The "quenching and partitioning" process: background and recent progress, *Materials Research,* ISSN 1516-1439

Speer J. G.; Matlock D. K,; Cooman B.C.; Schroth J.G.; (2003). Acta Mater. 51, 2611–2622.

Annals of DAAAM for 2008 & Proceedings of the 19th International DAAAM Symposium, ISSN 1726-9679
Editor B. Katalinic, Published by DAAAM International, Vienna, Austria 2008
Make Harmony Between Technology and Nature, and Your Mind will Fly Free as a Bird

THE SHEET THICKNESS EFFECT ON SPRINGBACK AMOUNT OF A PART MADE FROM TAILOR WELDED STRIPES

ALBUT, A[urelian]

Abstract: *This paper presents the results obtained by numerical simulation regarding springback phenomenon of a U-shaped part manufactured from tailor welded stripes. The final shape of the formed part is seriously affected by springback phenomenon. This paper work is trying to prove the important role the metal sheet thickness on the springback effect. The U-shaped part has different springback values for each material from the welded assembly structure. The influence of the sheet thickness on the tailor welded stripes springback is examined by finite element method using Abaqus Standard for forming process and Abaqus Explicit for springback of the obtained part.*

Key words: *welded stripes, sheet thickness, springback*

1. INTRODUCTION

The automobile industries have been trying to develop various types of model and high-quality low-cost cars to meet the customer's requirements and to find new ways of establishing this goal effectively (Papeleux & Ponthot, 2002). For the purpose of achieving the above presented objectives, different methods using various welding processes (such as laser-welding, mash seam-welding processes, etc.) were developed (Samuel, 2000).

A tailor welded blank consists of two or more sheets that have been welded together in a single plane prior to forming (Saunders, 1995). The sheets can be identical, or can have different thickness, mechanical properties or surface coatings. Thickness of tailor welded sheets plays an important role in sheet metal forming since fracture, wrinkling and weak spots are strongly influenced by material behaviour (Liu et al., 2002).

In this study, the tailor welded stripes with two types of material having the same thickness, are used to investigate springback characteristics in U-shape bending.

2. MATERIAL PROPERTIES

Simulation of the forming processes based on finite element method using ABAQUS software requires as input data the mechanical properties of the used materials (Chirita & Brabie, 2003). Determination accuracy of the mechanical properties has an important influence on simulation results (Zhao et al., 2001).

To obtain the mechanical properties of the base materials and of the welding line, tensile tests were performed on a universal mechanical testing machine, equipped with Hottinger force cells of 5 tf and a Hottinger – Baldwin electronic measurement system for PCs – type Spider 8. The data acquisition, processing and visualisation were performed using Catman Express software.

To determine the mechanical properties of the welding line, from the original TWB, a 4 mm wide stripe which includes the welding line, has been removed using EDM wire cutting (Fig. 1).

Fig. 1. Sample used to determine the mechanical properties of the welding line

The mechanical properties of FEPO steel and E220 steel determined for 0° material rolling direction are presented in Table 1. In the table below are presented also, the mechanical properties of the welding line. In Fig. 2 are presented the stress – strain curves for FEPO ans E220 steels and for the welding line material.

Property	FEPO	E220	weld
Yield strength $\mathbf{R_{p0,2}}$ [MPa]	203	268	252
Tensile strength **Rm** [MPa]	381	458	417
Percentage elongation after fracture $\mathbf{A_{80}}$ [%]	–	35,3	–
Elongation for max. load $\mathbf{A_{gt}}$ [%]	31,8	20,4	17,3
Strain-hardening coefficient **n**	0,222	0,190	-
Plastic strain ratio **r**	1,860	1,420	1,29
Poisson's ratio **ν**	0,286	0,297	0,278
Young modulus **E** [MPa]	200825	204271	203253

Tab. 1. Mechanical properties

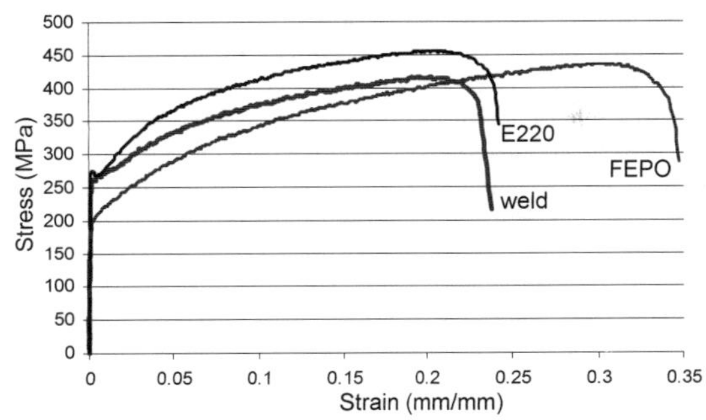

Fig. 2. Stress – strain curves for 0°

3. SIMULATION OF THE FORMING PROCESS

The objective is to create a model that allows an accurate prediction of springback intensity, stress and strain state at the end of the forming process. The analyzed geometrical parameters are sidewall radius ρ and springback angles θ_1 and θ_2.

The tailor welded stripes used in the simulation tests were made from FEPO and E220 steel. Stripes of 350 mm length and 30 mm wide, with thickness variation from 0,8 till 1,3 mm (Fig. 3).

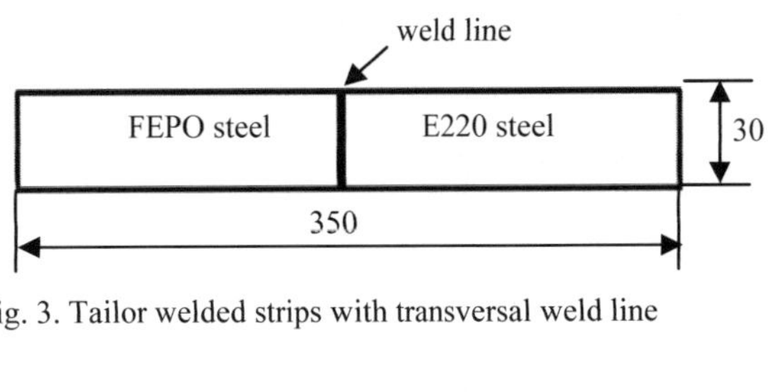

Fig. 3. Tailor welded strips with transversal weld line

The material was modelled as elastic-plastic, where elasticity is considered isotropic and plasticity is modelled as anisotropic using Hill quadratic anisotropic yield criterion.

In this simulation, fine mesh technology was applied to interested regions such as the punch profile radius and the tailor-welded strip model to improve the accuracy of the analysis for the springback (Lee et al., 2001).

The geometrical model is presented in Fig. 4. The sheet was modelled as deformable body with 400 shell elements (S4R) on one row with 5 integration points through the thickness. The tools (punch, die and blankholder) were modelled as analytical rigid because they have the advantage of reduced calculus efforts and a good contact behaviour (Uemori et al., 1998). Rigid body movements are controlled by reference points.

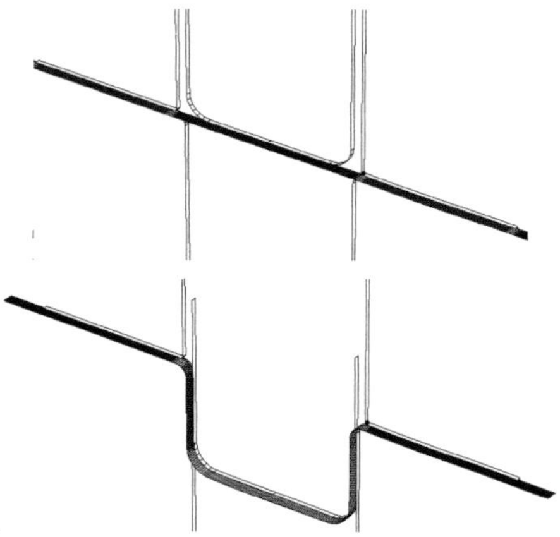

Fig. 4. Geometrical model

The boundary conditions imposed to the tools were intended to describe the experimental conditions as accurate as possible. For contact conditions a modified Coulomb friction law combined with penalty method was used (Mustafa et al., 1955).

For determination of the sheet thickness influence on springback phenomenon, the simulations have been done under the following conditions: the sample model is loaded with the material characteristics corresponding to 0° rolling direction; blank holder force F = 15 kN; the friction coefficient between the specimen and tools surfaces is 0.1, the thickness of the sample model are 0.8, 0.9, 1.0, 1.1, 1.2 and 1.3 mm.

Springback parameters that were observed during the simulation tests are presented in Fig. 5:

- θ_1 – sidewall angle between real profile and theoretical profile;
- θ_2 – flange angle between real profile and theoretical profile;
- ρ – curvature radius of the sidewall.
-

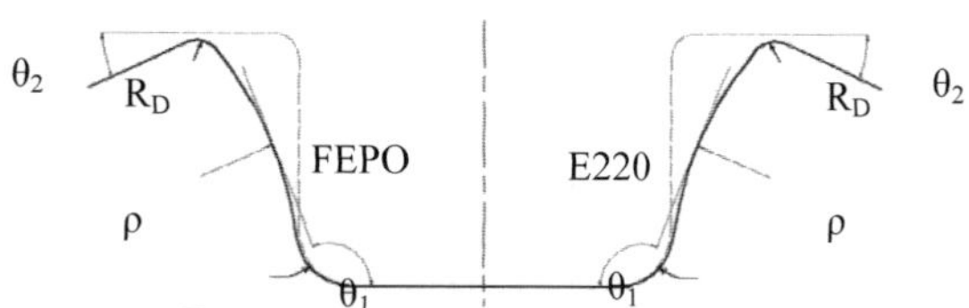

Fig. 5. SpringBack parameters

4. CONCLUSION

From the analysis of Table 2 data, the following observations can be presented concerning the influence of the sheet thickness on springback parameters:

- the modification of sheet thickness leads to important variations of springback parameters;
- increasing of the sheet thickness results in reduction of the springback effect, the final geometry of the formed part is closer to the ideal part shape;
- variation of the springback phenomenon proportional with sheet thickness is observed for both areas of the part;
- for sheet thickness over 1.0 mm the sidewall radius increases so much, that it can be considered a straight line;
- the springback parameters from the part zone made from FEPO steel record smaller values for both angles θ_1 and θ_2 with respect to the E220 part area;
- the part area made by E220 present a springback intensity higher that the part area made from FEPO steel.

Thickness [mm]	FEPO steel					
	Angle θ_1 [degrees]		Angle θ_2 [degrees]		Sidewall radius [mm]	
	Nominal value	Obtained value	Nominal value	Obtained value	Nominal value	Obtained value
0.8		100.6		14.9		205.0
0.9		99.7		14.1		307.6
1.0	90	99.0	0	13.4	:	385.8
1.1		96.8		10.3		1524.0
1.2		95.1		7.6		2875.5
1.3		93.6		5.5		3347.3
	E220 steel					
Thickness [mm]	Angle θ_1 [degrees]		Angle θ_2 [degrees]		Sidewall radius [mm]	
	Nominal value	Obtained value	Nominal value	Obtained value	Nominal value	Obtained value
0.8		102.8		19.3		188.4
0.9		101.0		17.5		265.7
1.0	90	100.1	0	15.6	:	322.9
1.1		98.2		11.8		818.5
1.2		96.7		9.5		2106.6
1.3		95.6		8.3		3071.5

Tab. 2. Springback parameters

5. REFERENCES

Chirita B. & Brabie G., (2003), *Experimental analysis of different influences on springback of parts formed by U-bending*, 7th International Research/Expert Conference Trends in the Development of Machinery and Asociated Technology, pp. 174-179.

Lee, J.K., Chun, B.K. & Kim, H.Y., (2001), *Numerical investigation of tailor welded blank forming and springback*, Simulation of Material Processing, Lisse, pp. 729–734.

Liu G., Lin Z., Xu W. & Bao Y., (2002), *Variable blankholder force in u-shaped part forming for eliminating springback error*, J. of Mat. Proc. Tech., vol. 120, pp. 531–537.

Mustafa, A.A., Brouwers, D. & Shulkin, L., (1995), *Deep drawing of round cups from tailor-welded blanks*, J. Mater. Process. Technol. 53, pp. 684–694.

Papeleux L. & Ponthot J.P., (2002), *Finite element simulation of springback in sheet metal forming*, J. of Mat. Proc. Tech., 125-126, pp. 210–215.

Samuel M., (2000), *Experimental and numerical prediction of springback and side wall in u-bending of anisotropic sheet metals*, J. of Mat. Proc. Tech. pp. 382-393.

Saunders, F.I., (1995), *Forming of tailor-welded blanks*, Ph.D. Dissertation, Ohio State University, Columbus, OH,.

Uemori, T., Okdas, T. & Yoshida, F., (1998), *Simulation of springback in V bending process by elasto-plastic finite element method with consideration of Bauschinger effect*, Met. Mater. Vol. 4, pp. 311–314.

Zhao, K.M., Chun, B.K. & Lee, J.K., (2001), *Finite element analysis of tailor welded blanks*, J. Mater. Process. Technol., Vol. 37, pp. 117–130.

Annals of DAAAM for 2008 & Proceedings of the 19th International DAAAM Symposium, ISSN 1726-9679
Editor B. Katalinic, Published by DAAAM International, Vienna, Austria 2008
Make Harmony Between Technology and Nature, and Your Mind will Fly Free as a Bird

DYNAMIC CHARACTERISTICS OF SPATIAL PIPE SYSTEM BY SEMIANALYTICAL METHOD

ALECU, A[urel]; BOIANGIU, M[ihail] & BURACU, V[ictor]

Abstract: *This paper is focused on the issue of calculating the frequencies and mode shape vibrations of a spatial pipe system by a semianayitical method combined with the transfer matrix method. The semianalytical method is based on approximating the unknown functions on a straight pipe segment by integer power series. The proposed method can be combined with other calculus methods and can be applied to other types of problems: integration of differential equations systems with variable coefficients, dynamical response. The method can be applied to the study of any systems that can be approximated, piece by piece, with straight bars (can be variable or unsymmetrical section). Generally speaking, our method can be applied to the study of any systems, whose motion is described by a system of differential equations whit variabe coefficients, coupled linearly.*
Key words: *semianalytic, vibration analyzis, pipe system.*

1. INTRODUCTION

The paper presents a semianalytical method combined with the transfer matrix method for the study of a spatial pipe structure filled with liquid. All possible motions of the pipe cross section are taken into consideration (extension-compression, torsion, bending) and their interdependence due to the space attributes of the system. This approach of the study combines analytical and numerical calculus. The semianalytical method is based on approximating the unknown functions on a straight pipe segment by integer power series. The differential equations system thus approximated leads us to a recurring matrix relation between the coefficients of all power series. Within the recurring relation only a limited number of coefficients are independent. The outstanding part of the method is the numerical evaluation of the influence of each independent coefficient on the approximated function. The independent coefficients and the unknown functions result from the boundary conditions. The proposed method can be combined with other calculus methods and can be applied to other types of problems: integration of differential equations systems with variable coefficients, problems of eigenvectors and eigenvalues, dynamical response. Generally speaking, our method can be applied to the study of any systems, whose motion is described by a system of differential equations coupled linearly or can be approximated well by linear differential systems. Until now the method has been developed and applied only to the behaviour of one-dimensional mechanical systems described by only one differential equation (Giurgiuţiu 1977), (Giurgiutiu V. & Stafford R.O. 1977) or by several coupled differential equations (Alecu et al., 2002). This method is an alternative to the FEM, for the study of mechanical systems that can be approximated as bar systems with concentrated masses (Yu J. Y. & Craggs A. 1995).

2. PRESENTATION OF THE SEMIANALITICAL METIIOD FOR THE STUDY OF VIBRATIONS

Consider a system of 2 maximum order $k = 4$ linear differential equations (without loosing the generality) with variable coefficients. This system describes the general vibration problem of a mechanical system with 2 unknown functions.

$$\sum_{k=0}^{4}[Q]_k\{v\}^{(k)} + \omega^2[P]_k\{v\}^{(0)} = \{F\} \ , \ \ \{v\} = \begin{Bmatrix} u(x) \\ v(x) \end{Bmatrix} \qquad (1)$$

with boundary conditions given through the zeros, first and second derivatives. The unknown functions are approximated by integer power series.

$$[u(x) \quad v(x)] = \sum_{i=1}^{\infty}[A_i \quad B_i]\, x^i \qquad (2)$$

The following recurring relation results from (1) and (2)

$$\begin{Bmatrix} A_{i+4} \\ B_{i+4} \end{Bmatrix} = [R]_4^{-1}\left(\{g\} + \sum_{j=3}^{j=-3}[R]_j \begin{Bmatrix} A_{i+j} \\ B_{i+j} \end{Bmatrix}\right) \qquad (3)$$

where $[A_{i+j} \quad B_{i+j}] = 0 \ , \ i + j < 0$ and only the coefficients identified in the components of the vector $\{D\} = [A_0\ A_1\ A_2\ A_3\ B_0\ B_1\ B_2\ B_3\ ]$ are independent. The influence of each independent coefficient on each approximated function can be written as

$$\begin{Bmatrix} u(x) \\ w(x) \end{Bmatrix} = \begin{bmatrix} U_1(x) & \cdots & U_8(x) \\ \vdots & \ddots & \vdots \\ W_1(x) & \cdots & W_8(x) \end{bmatrix}\{D\}^T \qquad (4)$$

The independent functions $U_n(x)$, W_n , $n = \overline{1,8}$ are defined by (5) and are evaluated numerically.

$$\begin{Bmatrix} U_n(x) \\ W_n(x) \end{Bmatrix} = \begin{Bmatrix} q_U \\ q_W \end{Bmatrix} x^{n-1} + \sum_{i=4}^{\infty}\begin{Bmatrix} {}^{n}A_i \\ {}^{n}B_i \end{Bmatrix} x^i \qquad (5)$$

$$q_U = \begin{cases} 1 & if \ n = 1,2,3,4 \\ 0 & if \ n = 5,6,7,8 \end{cases}, \quad q_W = \begin{cases} 0 & if \ n = 1,2,3,4 \\ 1 & if \ n = 5,6,7,8 \end{cases}$$

Coefficients ${}^{n}A_i$, ${}^{n}B_i$ result from the recurring formula (3) for 8 sets of initial values of the components of vector ${}^{i}\{D\} = [\ {}^{i}A_0\ {}^{i}A_1\ {}^{i}A_2\ {}^{i}A_3\ {}^{i}B_0\ {}^{i}B_1\ {}^{i}B_2\ {}^{i}B_3]$. The 8 values of the components of vector ${}^{i}\{D\}$ correspond to line „i" in the 8th order unit matrix. Relation (4) allows the numerical evaluation of the unknown functions and their derivatives. The semianalytical method can be combined with other methods of study of the vibration problem: proper frequencies, mode shapes, and dynamical response. One can easily associate this method with the transfer matrixes method.

3. STUDY FOR A SPACE PIPE SYSTEM

This paper is focused on the issue of calculating the frequencies and mode shape vibrations of a space pipe system. A space pipe system is modelled as a system of beams whose geometry is identical. The elastic characteristics of systems are localized in the ring shaped section of the pipe. The fluid inside the pipe is considered a distributed additional mass to the pipe system. The difficulty of the study stems from the three dimensional characteristic of the system. This feature leads us to the necessity of considering simultaneously all possible displacements of the cross section of the pipe: axial, bending in two planes and twisting. This paper suggests the combined use of the transfer matrix method and of the Frobenius power series for the approximation of the displacements of the cross section of the pipe. The pipe system is made up of different segments which are modelled as straight bars.

Let us consider a portion of a straight pipe of length l whose cross sectional dimensionless displacements are: u–axial displacement along axis Ox, v–displacement along axis Oy in the two planes bending, w–displacement along axis Oz in the two plane bending, φ-rotation around axis Ox due to torsion. The following differential equations for the vibrations of the pipe result taking into consideration the twisting of its cross section:

$$\frac{\partial^2}{\partial\xi^2}\begin{Bmatrix} u(\xi) \\ \varphi(\xi) \end{Bmatrix} + \omega^2 \begin{Bmatrix} a\cdot u(\xi) \\ e\cdot\varphi(\xi) \end{Bmatrix} = 0$$
$$\frac{\partial^4}{\partial\xi^4}\begin{Bmatrix} v(\xi) \\ w(\xi) \end{Bmatrix} + b\cdot\omega^2\frac{\partial^2}{\partial\xi^2}\begin{Bmatrix} v(\xi) \\ w(\xi) \end{Bmatrix} - c\cdot\omega^2\begin{Bmatrix} v(\xi) \\ w(\xi) \end{Bmatrix} = 0 \qquad (6)$$

where a , b , c , e are constants which depend on cross sectional characteristics of the pipe, the material the pipe is made of and the fluid inside it. For a cross section of the pipe a twelve component state vector is defined which include the displacements, the forces and the moments related to the four displacements. The matrix connection (transfer matrix with dimension 12x12) between the global state vectors in the cross sections $\xi = 0$ and ξ of the same uniform segment is in the form $\{z(\xi)\} = [M_{tr}(0,\xi)]\{z(\xi = 0)\}$. Transfer matrix $[M_{tr}(0,\xi)]$ results numerically in 2 steps:

- according to basic relations (4) one can obtain the connection between state vector $\{z(\xi)\}$ and the independent coefficients of power series $\{D\}$; in this case there are 12 independent coefficients.
- substitute the independent coefficients $\{D\}$; it results

$$\{z(\xi)\} = [T(\xi)]\{D\} = [T(\xi)][T(\xi = 0)]^{-1}\{z(\xi = 0)\} \qquad (7)$$

Where matrix $[T(\xi)]$ elements corespond to the components of the state vector calculated with relation (4). Relations (7) allow us to determinate the transfer matrixes between state vectors at the ends of a pipe segment. According to the classical methodology we define the jump matrixes over the inertial unevenness's, over the supports and the matrixes of the direction changes of the pipe segments. Finally, we will have the transfer matrix for the whole mechanical system. The transfer matrix method will be applied in a classical manner for the determination of the frequencies and mode shape vibrations.

4. NUMERICAL APPLICATION

The method presented in the previous paragraph will be applied for a space pipe specifically to a nuclear power station whose form and dimensions can be seen in the pictures showing the mode shape vibration. Raw material and sectional characteristics of the pipe are: elasticity modulus $E = 2{,}1\cdot 10^{11}\ N/m^2$, $G = 8\cdot 10^{10}\ N/m^2$, density of the pipe material $\rho_p = 7830\ Kg/m^3$, density of the fluid $\rho_f = 1100\ Kg/m^3$, radiuses of the pipe $R = 0{,}0254\ m$ and $r = 0{,}01665\ m$. The pipe is considered embedded at both ends and supported in 5 points on three rectangular directions. The elastic constants of all supports on each direction have been taken $k = 10^6\ N/m$.

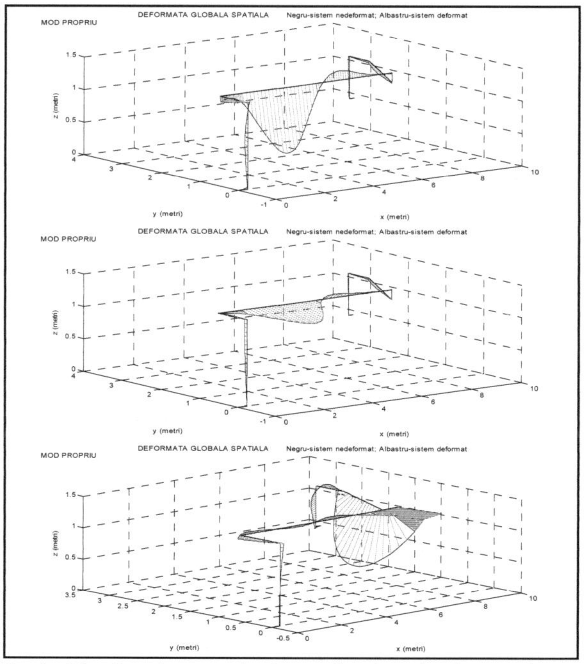

Fig. 1. The first three mode shape vibrations for frequencies.

No inertial discontinuities have been considered. The twelve proper frequencies and the corresponding modes shape of vibrations have been determined. This proper frequencies are (Hz): 17.86, 18.28, 27.42, 32.24, 33.06, 33.56, 37.72, 38.17, 41.33, 41.76, 42.80, 46.92. The mode shape is presented in a *space representation* along the pipe. A *global displacement* for each section of the pipe results adding vectors for each displacement $\vec{d} = \vec{u} + \vec{v} + \vec{w}$. Since the volume of information is large in figure 1 only first three modes shape vibrations have been given.

5. CONCLUSION

The proposed method can be applied to the study of any systems that can be approximated, piece by piece, with straight bars (can be variable or unsymmetrical section (Alecu et al., 2002)). For these systems motions are described by a system of differential equations coupled linearly or these can be approximated well with linear differential systems - bars with an initial twist (Aurel Alecu & Mihail Boiangiu 1999). For each analyzed system the analytic part of the method is specific

6. REFERENCES

Alecu A., Boiangiu M. & Anghel V.,(2002), Semi-analytical Method for the Dynamic Analysis. *Computer Assisted Mechanics and Engineering Sciences*, 9, 2002, pp 423-438, ISSN 1232-308X

Aurel Alecu & Mihail Boiangiu (1999) *Dynamic of Helicopter blade – Semianalytical method of study* BREN Press, ISBN 973-9493-36-X, Bucharest.

Giurgutiu V. (1977), *Vibration and Dynamic Stability of Rotor Blade*, Ph.D. Thesis, University of London.

Giurgiutiu V. & Stafford R.O., (1977), Semi analytic Method for Frequencies and Mode Shape of Rotor Blades, *Vertica*, Vol. I, nr. 4, 1977, pp 291-306, ISSN 0360-5450.

Yu, J., Y. & Craggs, A. (1995). A Transfer Matrix Method for Finite Element Models of a Chaine-like Structure Under Harmonic Excitations. *Journal of Sound and Vibration,* Vol.187, No.4, Nov. 1995

Annals of DAAAM for 2008 & Proceedings of the 19th International DAAAM Symposium, ISSN 1726-9679
Editor B. Katalinic, Published by DAAAM International, Vienna, Austria 2008
Make Harmony Between Technology and Nature, and Your Mind will Fly Free as a Bird

IMPROVING THE ENERGETIC EFFICIENCY OF THE PHOTOVOLTAIC ARRAYS USING MECHATRONIC SOLAR TRACKERS

ALEXANDRU, C[atalin] & POZNA, C[laudiu]

Abstract: *The paper presents researches on increasing the energetic efficiency of the PV arrays by designing and optimizing a solar tracker that simultaneously changes the daily position of the panels with a single motor source (there are panels with individual supports). The motion is transmitted from the driving source, which is a DC rotary motor, with a parallelogram mechanism, the revolute axes of the panels being parallel with the polar axis. The main task in optimizing the mechanism is to maximize the energetic gain by increasing the solar input, and minimizing the energy consumption for tracking. The analysis is made by developing the virtual prototype of the system, in mechatronic concept.*
Key words: *PV array, tracking, virtual prototype, optimization.*

1. INTRODUCTION

The paper is approaching a theme that belongs to a very important field: renewable sources for energy production - increasing the efficiency of the photovoltaic (PV) conversion. *"The PV systems can deliver energy on large-scale to a competitive price"* is the conclusion of the European Commission for Energy in the report "A Vision for Photovoltaic Technology for 2030 and Beyond" (2004). In this frame, the realization of the PV arrays (system of PV panels that function as a single electricity-producing unit) appeared as a necessity for the development of large systems for producing electric energy. The energetic efficiency of the PV arrays depends on the degree of use of the solar radiation, which can be maximized by use of tracking systems for the orientation of the panels in accordance with the paths of the Sun.

In practice, there are two solutions for developing the PV arrays with tracking, depending on the existence (or not) of a common frame: (a) PV platforms, where the panels are mounted on a common frame, the orientation being simultaneously realized by moving the entire platform; (b) array with individual panels, which are separately mounted on individual sustaining structures. In the second case, the orientation can be realized in two ways: (b.1) independent orientation for each panel (panel with own tracking system - self motor source); (b.2) simultaneous orientation of all panels of the array by using single driving source and mechanisms that transmit the motion to the panels (this solution, even is more complex by constructive aspects, ensures a greater energetic & economic efficiency because of the minimization of the consumers - motor sources). The simultaneous orientation, with the predicted advantages and the characteristic problems, opens a research area insufficiently explored since now, fact sustained by the literature and practical developments in the field that refer almost entirely to the individual orientation of the panels.

2. ACTUAL STAGE AND CONTRIBUTIONS

The bibliography research reveals a series of aspects and hardships in the literature concerning the tracking systems for PV arrays. In the literature there are no unitary models for the tracking mechanisms of the PV arrays referring to the structural, kinematical and dynamical issues. In the same time, there is no general approach for conceptual design and structural synthesis of these mechanisms. Thereby becomes obviously the necessity a method for the unitary modeling of the mechanisms and according to our strategy this method is based on the Multi-Body Systems (MBS) theory, which may facilitates the self-formulating algorithms for making possible the real-time simulation (Schiehlen, 1993).

The issue concerning the control of the tracking systems is approached mostly for the tracking systems of the individual panels, using different techniques (closed loop systems with photo sensors, opened loop systems based on astronomical computerized systems, or hybrid combinations), and types of controllers - PID, FNC, FNLC (Abdallah & Nijmeh, 2004; Chojnacki, 2005). The research is focused mostly on the quantity of the energy achieved by tracking, but there is no evaluation on the energy consumption for orientation. A possible cause of this case is due to the fact that the issue is not approached as an integrated assembly.

The disposing of the panels in the array configuration is approached in the literature mainly from the point of view of the self-shading avoiding (Perez et al., 2007; Kaushika et al., 2005). Complementary, our research approaches the issue regarding also the simplification of the solution for transmitting the motion from the driving source to the panels of the array, in order to obtain efficient and feasible solutions. Following this trend, we also investigate a modular approaching of the PV array by designing modules of panels with tracking which can transmit (or receive) the motion to (from) other modules.

Therefore, the paper is approaching the improvement of the energetic efficiency of the PV arrays by designing a tracking mechanism that simultaneously changes the daily position of the panels with a single driving source. The main task in optimizing the mechanism is to maximize the energetic gain by increasing the solar input and minimizing the energy consumption for tracking. The paper proposes the integration of the mechanic and control components at the virtual prototype level (modeling in mechatronic concept), which allows performing the energy balance: energy gain by orientation versus energy consumption for realizing the motion. Thus, the physical testing is greatly simplified, and the risk of the control law being poorly matched to the real system is eliminated.

3. RESULTS AND CONCLUSIONS

The application is made for a single-axis tracking system, which simultaneously change the daily position of the panels. The mechanism is designed for a string configuration array (panels disposed in line), but it supports the adaptation for other configurations (in the general case, the matrix configuration). The motion is transmitted from the driving source (i.e. the DC rotary motor), which directly drives the first panel, with a multi-parallelogram mechanism, the revolute axes of the panels being parallel with the polar axis. The geometric constraints between parts are revolute joints. Each panel rotates around its own support, which is mounted on the sustaining pillar.

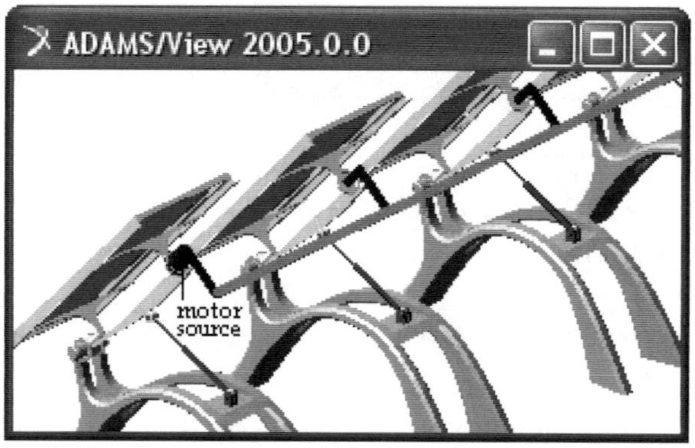

Fig. 1. The virtual prototype of the PV string with tracking.

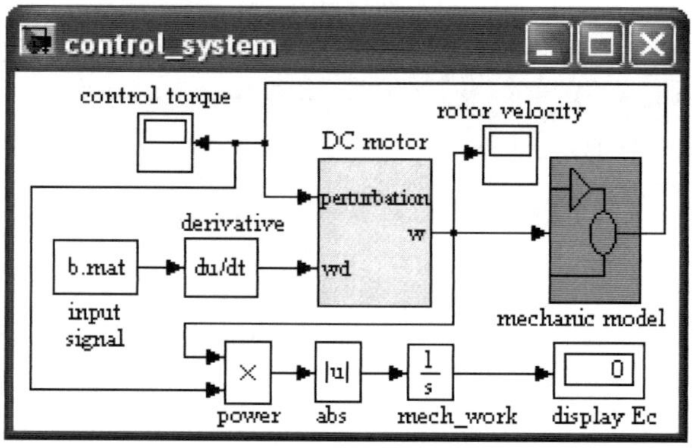

Fig. 2. The control system block.

For establishing the solution, the structural synthesis was made using a method based-on the MBS theory. In this way, a collection of possible schemes have been obtained. The solution for system used in the study was selected from the multitude of the structural solutions by using of the Multi Criteria Analysis. The evaluation criteria were referring to the tracking precision, the amplitude of the motion, the complexity and the reability of the system.

For simulating the dynamic behavior of the PV system, we have developed the virtual prototype (fig. 1), by using a digital prototyping platform that includes the following software solutions: CAD (CATIA) - to create the solid model, which contains information about the mass & inertia properties of the parts, MBS (ADAMS/View) - for analyzing the tracking system, and C&C (ADAMS/Controls & MATLAB/Simulink) - for modeling the control system.

For connecting the mechanical model and the electronic control system, the input & output parameters have been defined. The angular velocity of the rotor, which determines the daily position of the panels (there is the same angle for all panels because the motion is transmitted with a parallelogram mechanism, at which the input and output motions are identical) represents the input parameter in the mechanical model. The output transmitted to the controller is the control torque generated by the DC motor. The philosophy is to control the angular velocity of the rotor, which is perturbed with the control torque (the mechanism is considered as a perturbation for the DC motor). The computation of the control torque is based on the dynamic model of the tracking mechanism, including parts, constraints, internal & external forces.

The ADAMS and MATLAB models communicate by passing state variables back and forth. ADAMS/Controls save the input and output information in a specific file for MATLAB (*.m); it also generates a command file (*.cmd) and a dataset file (*.adm), which are used during the simulation. With these files, the control diagram has been developed in order to complete the connection between the mechanical device and the actuating & control system (fig. 2). The input signal block represents the database of the daily angle (the position angle of the rotor), which defines the motion law of the tracking system.

The electric energy produced by the PV array depends on the quantity of incident solar radiation, the panel efficiency, and the number of panels. The direct radiation is empirical established depending on the extraterrestrial radiation, the medium solar constant, the day number during a year, the distortion factor, the solar altitude angle, the solar declination, the latitude angle, the solar hour angle, and the local solar time. The incidence angle is determined from the scalar product of the sunray vector and the normal vector on panel. Thus, we are able to estimate the incident radiation in every day during a year, for different locations, and tracking strategies. In paper, the results correspond to the summer solstice day, which is a relevant situation for evaluating the energetic efficiency.

The panels can be rotated without brakes during the day-light, or can be discontinuously driven (step-by-step motion). The key idea for optimizing the motion law is to reduce the angular field of the motion, and the number of actuating operations, without significantly affecting the incoming energy, and with minimum energy consumption. Thus, we obtained the optimum angular field for the daily motion, [-60°, +60°], with 6 motion steps (after the sunset, the system returns in the initial position). In this way, the energy (mechanical work) balance can be performed, considering the energy produced by the PV array with tracking - E_T, the energy produced by the same array without tracking (fixed) - E_F, and the energy consumption for orientation - E_C. For example, considering a string with 6 panels, there are the following values: E_T = 10358.658 J, E_F = 7384.356 J, E_C = 52.697 J; this means an energy contribution of 39.57 % by orienting the panels, relative to the fixed array case.

Concluding, the optimization strategy leads to an efficient PV system, without developing expensive hardware prototypes. In this way, the behavioral performance predictions are obtained much earlier in the design cycle of the tracking systems, thereby allowing more effective and cost efficient design changes and reducing overall risk substantially.

The tracking system will be manufactured and tested in the Centre Product Design for Sustainable Development from Transilvania University, creating a real perspective for the research in the field. This allows a relevant comparison between the virtual prototype analysis and the data achieved by measurements. At the same time, we intend to develop new types of tracking mechanisms for different topologies of PV array, and new models for controlling the tracking systems.

4. REFERENCES

Abdallah, S. & Nijmeh, S. (2004). Two-Axis Sun Tracking with PLC Control. *Energy Conversion and Management*, Vol. 45, No. 11-12, 31-39, ISSN 01968904

Chojnacki, J.A. (2005). Application of Fuzzy Logic Neural Network Controllers in PV Systems, *Proceedings of the 20-th European Photovoltaic Solar Energy Conference*, WIP-Renewable Energies, pp. 2269-2272, ISBN 3936338191, June 2005, Published by WIP Munich, Barcelona

Kaushika, N.D.; Gautam, N.K. & Kaushik, K. (2005). Simulation Model for Sizing of Stand-Alone Solar PV System with Interconnected Array. *Solar Energy Materials and Solar Cells*, Vol. 85, No. 4, 499-519, ISSN 09270248

Perez, P.J.; Almonacid, G. & Vidal, P.G. (2007). Estimation of Shading Losses in Multi-Trackers PV Systems, *Proceedings of the 22-nd European Photovoltaic Solar Energy Conference*, WIP-Renewable Energies, pp. 2295-2298, ISBN 3936338221, September 2007, Published by WIP Munich, Milan

Schiehlen, W.O. (1993). *Advanced Multibody Systems Dynamics*, Springer-Verlag, ISBN 0792321928, London

Annals of DAAAM for 2008 & Proceedings of the 19th International DAAAM Symposium, ISSN 1726-9679
Editor B. Katalinic, Published by DAAAM International, Vienna, Austria 2008
Make Harmony Between Technology and Nature, and Your Mind will Fly Free as a Bird

ALGORITHMS FOR AUTOMATIC/SEMIAUTOMATIC GEOGRAPHIC DATA ACQUIRING AND ERROR ASSESSMENT

ALEXEI, A[drian]; TOMOIAGA, T[iberius]; MARINESCU, M[irel] & BADEA, A[lexandru]

Abstract: *In this paper are presented a few algorithms for automatic and semiautomatic vectorization of punctual, linear and polygonal symbols from topographic maps. All algorithms developed and tested are classified by Wenyin and Dov Dori indicators, based on performance.*
Key words: *GIS, vectorization, algorithms, geographic data.*

1. INTRODUCTION

With the development of the first commercial GIS at the beginning of 80's new methods of data acquiring developed too: **vector digitization** of paper maps, and later, after hardware improvement, **raster digitization** (Nitu, 1992).

Vectorial digitization uses paper maps and digitizers.

Raster vectorization uses scanned maps and is the most used method in present. It uses large formats scanners, plotters, powerful computers, DBMS's and raster – vector conversion software. Raster – vector conversion software's are not fully automated for data acquiring from scanned maps (Liu et al., 1997).

Because of the great importance and the difficulties involved by automatic raster – vector conversion, this process has been one of the major topics in the last decade researches.

These algorithms for automatic acquiring of polygonal, linear and punctual elements from scanned maps will have a big contribution to geodatabase creation from 1:25.000, 1:50.000 and 1:100.000 maps.

By implementing these algorithms, users will be allowed to produce their own geodatabases very easily and at a very low cost.

2. ALGORITHMS DESCRIPTION AND ERROR ASSESSMENT

Based on scheletisation algorithms there are 2 classes of vectorization algorithms (Alexei, 1997):

- One pass algorithms, applied directly on primary images;
- Two passes algorithms, which have scheletisation as preliminary phase.

2.1. Linear elements vectorization

First step involves determination of one end of the linear element. Second step involves linear element tracking using circle method.

After a complete exploration around the moving point (x_0, y_0), we can have the following situations:

1. One pixel set on 1 – means we have a new point of the linear element with coordinates (x, y). The current point becomes $(x_0 = x, y_0 = y)$ and this is followed by the continuation of the circle method.

2. Two or more pixels set on 1 – means we have an intersection of linear elements fig. 1. In this case the software requires human intervention or the point can be memorized for returning after all other linear element are analyzed (backtracking type algorithm).

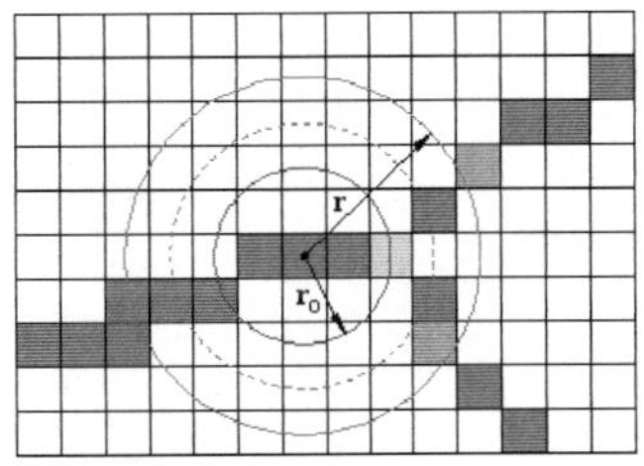

Fig. 1. Intersection point determination.

3. No pixel set on 1 – means the end of linear element and the end of analysis.

When we have two or more pixels set on 1 the way to evaluate and eliminate the ambiguity is presented in the fig. 2.

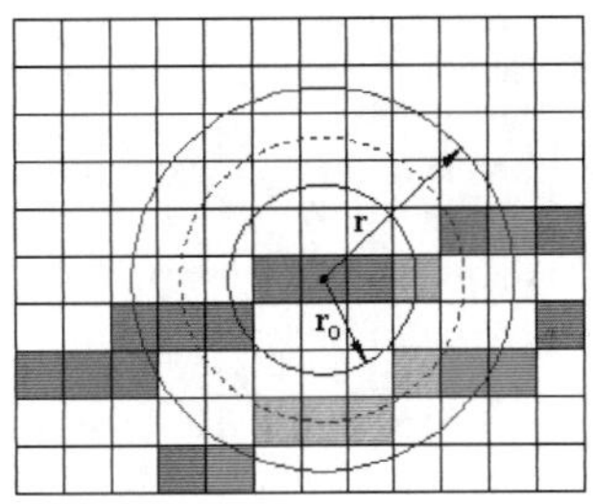

Fig. 2. Close linear elements.

If linear elements are not tangential, in order to eliminate the ambiguity, the analysis radius is decreased until a unique pixel set on 1 is found.

2.2. Polygonal elements vectorization.

First stage involves one pixel determination of the contour to be vectorized.

Stage two involves the following of the contour. This end when the current pixel is the same with the first one or when the current pixel leaves the map.

2.3. Punctual elements vectorization.

Shape signature is one variable function associated to a plane form. Shape signature is reversible – having a signature we can rebuild a shape.

Signatures are invariant characteristics by translation due the direct reference to the mass center (Maguire et al., 1991). The invariance by the scale factor can be obtained by determining the average scale factor between the distance on the reference form and the distance on the analyzed form. The invariance by the rotation factor can be obtained by a cyclic permutation of the signature.

Fourier Descriptors characterize the shape. This method is a supervised one and needs the sets of Fourier descriptors.

If we have the Fourier Descriptors $U^1(k)$ and $U^2(k)$ corresponding to some different shapes $u^1(n)$ respective $u^2(n)$, then the form of these shapes is similar if

$$d(u_0,\alpha,\phi_0,n_0) = \min_{u_0,\alpha,\phi_0,n_0} \left\{ \sum_{n=0}^{N-1} \left| u^1(n) - \alpha \cdot u^2(n-n_0) \cdot e^{i\cdot\phi_0} - u_0 \right|^2 \right\} \to 0 \quad (1)$$

Parameters u_0, α, ϕ_0 and n_0 results from minimizing condition of the effects of translation, scaling, rotation.

Mask based punctual symbols recognition. This algorithm is a simple recognition method of conventional symbols based on correlation between theoretical and real image of the symbol.

If *f(x,y)* is the theoretical image and *g(x,y)* scanned image we should have the following relation:

$$f\ AND\ g = f \quad (2)$$

Above relation is an ideal one, but because of the scanning errors we have:

$$f\ AND\ g = f' \text{ where } f' \neq f. \quad (3)$$

To see if the *f* and *g* images are in correlation we use the following algorithm:

1) Determination of differences between theoretical and scanned image using a logical or exclusive

2) Determination of plus pixels (*pP*) or minus pixels (*pM*).

If the *pP* or *pM* is lower than a certain thresholdε, then *g* is influenced by the noise of *f*. ε is determined experimental.

2.4. Geodatabases error assessment.

The elements acquired in a geodatabase cannot be more precise then the original information. For error assessment we will develop algorithms for the following accuracy indicators:

Accuracy indicators for linear elements. Linear element accuracy can be split in two components (Hauvelink, 2000):

- Positioning accuracy;
- Shape fidelity.

Accuracy indicators for polygonal elements. It will be used the indicators from linear elements.over the border.

Accuracy indicators for attributes. This algorithm will be based on Cohen's *Kappa (K)* coefficient.

Accuracy indicators for conceptual accuracy. Conceptual accuracy verifies the selection criterias and the quantity of information. It has two components:

- The missing information;
- Supplemental information.

3. TEST RESULTS

For testing the algorithms a software package was developed in Delphi 7.0. As input data scanned images in .bmp and .jpg format was used and as output format was used ESRI shape format.

The figure 3 shows how a church symbol is recognized and marked with the correspondent vectorial symbol from a symbol database.

All punctual elements identified on map are memorized in one file of ESRI SHAPE type, based on unic codes. The codes are used to associate the symbols from the symbol database for representation.

The figure 4 shows a comparation between the algorithm implemented for linear automatic vectorization and a similar solution offered by one of the most powerful GIS solution on the market, ESRI ArcGIS. The results are comparable both as accuracy and speed.

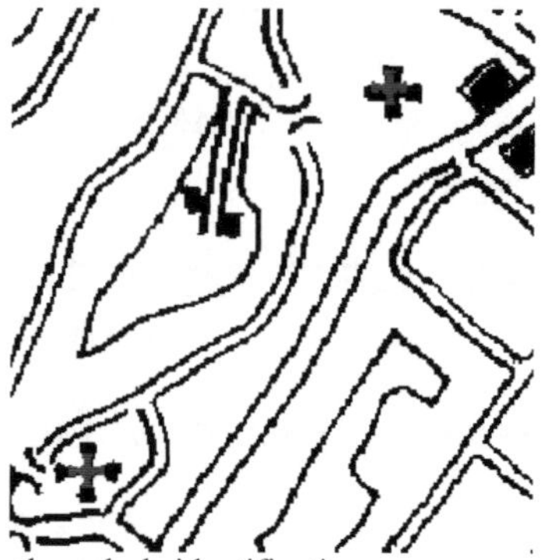

Fig. 3. Punctual symbols identification.

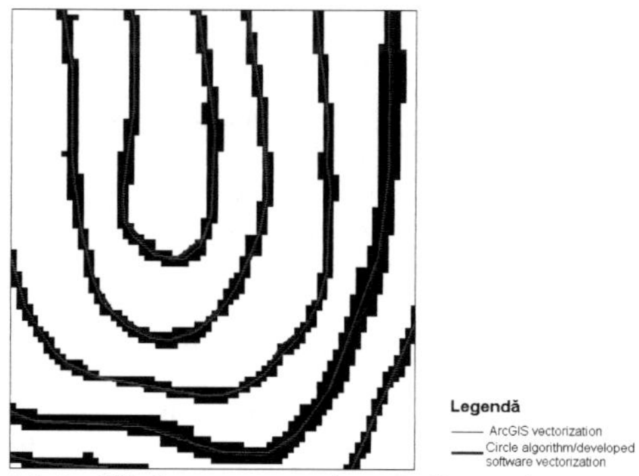

Fig. 4. Method/Software comparison for linear elements automatic vectorization

4. CONCLUSION

The developed algorithms for linear and poligonal elements vectorization was tested with good results and they can be used succesfuly for interactive vectorization of contour lines and hydrography. The false pixel detection rate can be lowered by decreasing the analisys radius, increasing the number of detected points.

The algorithm for automatic recognition of the shape of punctual symbols can be easily implemented and the speed of analisys for a class of objects is independent of the image. For increasing the speed, the approximative coordinates of the element can be pointed out using a locator, the algorithm finding teh precise coordinates and the associated code.

Having a dual character the propose solutions can be implemented both in military and civil environment. Military Topographical Department will be one of the major users of the proposed solutions.

5. REFERENCES

Alexei, A. (1997). Raster – vector conversion algorithms, *MTA Review*, Military Technical Academy, Bucharest, May 1997, pp. 59-62.

Hauvelink, G., (2000) – *Error propagation in environmental modeling with GIS*, Taylor & Francis, London.

Liu, W., Dov, D., (1997) - A Protocol for Performance Evaluation of Line Detection Algorithms, *Machine Vision and Applications*, Springer-Verlag GmbH Volume 9, Numbers 5-6, March 1997, p240 – 250.

Maguire, D. J., Goodchild, M. F., (1991) - Section I. Introduction. In: Maguire D J, Goodchild M F, Rhind D W (eds.) *Geographical Information Systems: principles and applications*. Longman, London, pp. 3-7, Vol 1

Niţu, C., (1992) – Developing trends in automated cartography – *Geodesy, Cartography and Cadastral Survey Review*, no. 1, 1992, pp 27.

Annals of DAAAM for 2008 & Proceedings of the 19th International DAAAM Symposium, ISSN 1726-9679
Editor B. Katalinic, Published by DAAAM International, Vienna, Austria 2008
Make Harmony Between Technology and Nature, and Your Mind will Fly Free as a Bird
2nd DAAAM-ICAT

BIOMANUFACTURING TECNOLOGIES AND BIOMATERIALS USED TO PRODUCE SCAFFOLDS FOR TISSUE ENGINEERING

ALMEIDA, H[enrique A.] & BARTOLO, P[aulo J.]

Abstract: *Tissue engineering involves the development of functional substitutes to replace missing or malfunctioning tissues and organs. Some strategies in tissue engineering have focused on using tri-dimensional units, scaffolds, fabricated of biopolymers and/or bioceramics. Scaffolds provide temporary mechanical support for tissue regeneration while shaping in-growing tissues. With the aid of advanced technologies, it is possible to design the most appropriate scaffold for a desired situation. Additive biomanufacturing processes are increasingly recognised as ideal techniques to produce scaffolds. This paper proposes a description of all existing additive biomanufacturing technologies and biomaterials. A novel computational tool combining modelling and several engineering tools, like structural, fluid dynamics and topological optimisation schemes to predict and optimise the scaffolds performance for medical applications is also presented.*
Key words: *Scaffolds, Biomanufacturing, Tissue Engineering, Biomaterials, Computer Aided Design of Scaffolds*

1. INTRODUCTION

Tissue engineering is a biomedical engineering discipline integrating biology with engineering to create tissue or cellular products outside the body (*ex vivo*) or to make use of gained knowledge to better manage the repair of tissues within the body (*in vivo*). This discipline requires the combined effort of cell biologists, engineers, material scientists, mathematicians, geneticists, and clinicians toward the development of biological substitutes that restore, maintain, or improve tissue function. The first definition of tissue engineering was provided by Skalak and Fox (Skalak and Fox, 1988) who stated it to be "the application of principles and methods of engineering and life sciences' toward the fundamental understanding of structure-function relationships in normal and pathological mammalian tissues and the development of biological substitutes to restore, maintain, or improve tissue function". An historical overview of this field can be found in a recent report published by National Science Foundation (NSF, 2003).

Therapeutic strategies in tissue engineering involve cellular implantation, where cells derived from an endogenous source in the patient or from a donor are either injected into the damaged tissue or are combined *in vitro* with a degradable scaffold and then implanted, and tissue regeneration where a scaffold implanted directly into the damaged tissue stimulates the cells to promote local tissue repair. These scaffolds are often critical, both ex vivo as well as in vivo, as they serve some of the following purposes (Leong *et al*, 2003): Allow cell attachment, proliferation and differentiation; Deliver and retain cells and growth factors; Enable diffusion of cell nutrients and oxygen and Enable an appropriate mechanical and biological environment for tissue regeneration in an organised way. To achieve these goals an ideal scaffold must satisfy some biological and mechanical requirements (Table 1 and 2):

Biological Requirements	
Biocompatibility	The scaffold material must be non-toxic and allow cell attachment, proliferation and differentiation
Biodegradability	The scaffold material must degrade into non-toxic products
Controlled degradation rate	The degradation rate of the scaffold must be adjustable in order to match the rate of tissue regeneration
Porosity	Appropriate porosity macro and microstructure of the pores and shape to allow tissue in-growth and vascularisation
Mechanical and Physical Requirements	
Strength and stiffness	Sufficient strength and stiffness to withstand stresses in the host tissue environment
Surface Finish	Adequate surface finish to guarantee that a good biomechanical coupling is achieved between the scaffold and the tissue
Sterilised	Easily sterilised either by exposure to high temperatures or by immersing in a sterilisation agent remaining unaffected by either of these processes

Tab. 1. Biological, mechanical and physical requirements

Scaffold characteristics	**Biological effect**
Biocompatibility	Cell viability and tissue response
Biodegradability	Aids tissue remodelling
Porosity	Cell migration inside the scaffold
	Vascularisation
Chemical properties of the material	Aids in cell attachment and signalling in cell environment
	Allows release of bioactive substances
Mechanical properties	Affects cell growth and proliferation response
	In-vivo load bearing capacity

Tab. 2. Relationship between scaffold characteristics and the corresponding biological effect (Mahajan, 2005)

A variety of biodegradable materials have already been used for tissue scaffolds, including ceramics and polymers. The primary use of ceramics has been in bone tissue engineering, presenting long degradation times often on the order of years. Polymeric scaffolds are used in the form of fibrous meshes, porous sponges, or hydrogels showing short degradation times, often on the order of days or months. The most common materials used for scaffolds are indicated in Table 3.

Biopolymers			
Natural	Alginate	**Synthetic**	Poly(glycolic acid), Poly(lactic acid) and their co-polymers
	Collagen		Poly(ε-caprolactone)
	Chitosan		Poly(dioxanone)
	Hyaluronic acid		Polyethylene oxide/polybutylene teraphthalate co-polymers
	Poly(hydroxybutyrate)		Poly(propylene fumarate)
Bioceramics			
Hydroxyapatite and other types of calcium phosphate like fluorapatite or tricalcium phosphate			
Biphasic hydroxyapatite/ tricalcium phosphate ceramics			
Bioactive glass ceramics			

Tab. 3. Biomaterials commonly used in tissue engineering

2. RAPID PROTOTYPING AND MANUFACTURING TECHNIQUES FOR TISSUE ENGINEERING

Rapid prototyping and manufacturing represents a new group of non-conventional fabrication techniques recently introduced in the medical field. The main advantages of rapid prototyping and manufacturing are both the capacity to rapidly produce very complex 3D models and the ability to use various raw materials. In the tissue engineering field, rapid prototyping and manufacturing have been used to produce scaffolds with customised external shape and predefined internal morphology, allowing good control of pore size and pore distribution (Bártolo, 2006).

Figure 1 provides a general overview of the necessary steps to produce rapid prototyping scaffolds for tissue engineering. The first step is the generation of the corresponding computer solid model through one of the currently available medical imaging techniques such as computer tomography, magnetic resonance imaging, etc. These imaging methods produce continuous volumetric data (voxel-based data), which provide the input data for the digital model generation (Bártolo, 2006). The model is then tessellated as an STL file, which is currently the standard file for facetted models in rapid prototyping and manufacturing. Finally, the STL model is mathematically sliced into thin layers (sliced model). Rapid prototyping and manufacturing technologies are similar to 2D printing and plotting technologies using both vector-based and raster-based imaging techniques.

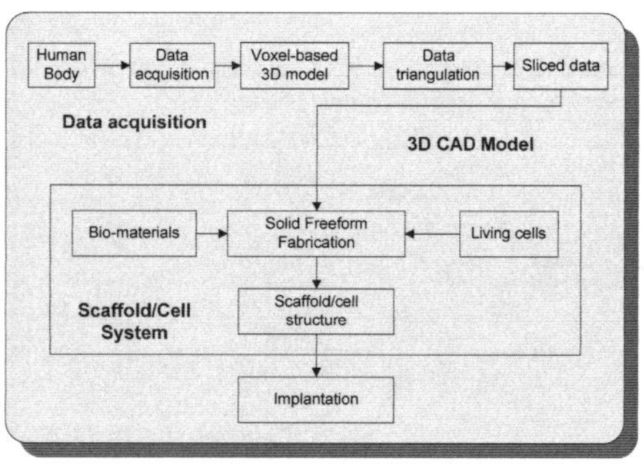

Fig. 1. Steps of rapid prototyping in tissue engineering

Rapid prototyping and manufacturing technologies are similar to 2D printing and plotting technologies using both vector-based and raster-based imaging techniques. The various rapid prototyping and manufacturing technologies used for fabricating scaffolds for tissue engineering applications include stereolithographic processes, laser sintering, extrusion and three dimensional printing processes.

Scaffold porosity, pore and filament/rod size that can be achieved using RP techniques are listed in table 4.The main advantages and limitations for the fabrication of tissue engineering scaffolds are listed below in Table 5, while Table 6 presents the main mechanical properties of scaffolds produced by one of each technology.

Process	Achieved Porosity	Pore Size	Rod Size
3DF	70 % - 87 %		
3DP	40 % - 90 %	45-150 μm	
FDM	21 % - 77 %	160-700 μm	250-370 μm
SLA	26 % - 90 %	366 μm	
SLS	37.5 % - 74 %		

Tab. 4. Scaffold porosity values [30]

Process	Advantages
SLA	Relatively easy to achieve small feature
SLS	Relatively higher mechanical strength and solvent free
FDM	No material trapped in the scaffold and solvent free
3DP	Low heat effect on raw powder; easy process and low cost
Bio-plotting	Large variety of materials for both soft and hard tissues
Process	**Limitations**
SLA	Limited by the development of photo-polymerisable, biocompatible and biodegradable materials; currently limited to reactive and mostly toxic resins
SLS	Powder material trapped in small inner holes can be difficult to remove; high temperatures in the chamber
FDM	High heat effect on raw material; limited geometrical complexity
3DP	Materials trapped in small inner holes; low mechanical properties
Bio-plotting	Low geometrical complexity

Tab. 5. Characteristics of rapid prototyping scaffolds.

Process	Biomaterials	Mechanical Properties
3DF	PEGT–PBT	Dynamic compressive modulus: 4.33 MPa
3DP	Cornstarch, dextran and gelatine	Compressive modulus: 0.059-0.102 MPa
	HA (followed by sintering)	Compressive strength: up to 22 MPa
	Poly(urethane)	Young's modulus: 580 MPa
FDM	PCL	Compressive modulus: 4-77 MPa
		Compressive strength: 2.4-20.2 MPa
	Alumina (indirect fabrication)	Compressive strength: 50 MPa
SLS	PCL	Compressive modulus: 52-67 MPa
	PCL – HA	Compressive modulus: 33-102 MPa

Tab. 6. List of achieved mechanical properties (Leong *et al*, 2008).

3. CONCLUSION

Rapid prototyping and manufacturing technologies have a great potential for tissue engineering. These technologies offer a high degree of freedom for tissue engineering either for the design of scaffolds (pore size, pore geometry, orientation, interconnectivity, etc.) or for its fabrication. Several materials can also be used enabling the production of both soft and hard scaffolds. These characteristics can enhance the fabrication of biomimetic scaffolds and scaffolds for complex biomechanical applications. Additive technologies are also the ideal processes to fabricate functional graded scaffolds.

4. REFERENCES

Skalak, R. and Fox, C.F. (1988) *Tissue Engineering*, New York: Alan R. Liss.

NSF (2003) 'The emergence of tissue engineering as a research field', Report, *National Science Foundation*, USA.

Leong, K.F., Cheah, C.M. and Chua, C.K. (2003) 'Solid freeform fabrication of the three-dimensional scaffolds for engineering replacement tissues and organs', *Biomaterials*, Vol. 24, pp.2363–2378.

Mahajan, H. P., Evaluation of chitosan gelatine complex scaffolds for articular cartilage tissue engineering, *MSc Thesis*, Mississipi State University, USA, 2005.

Bártolo, P.J. (2006) 'State of the art of solid freeform fabrication for soft and hard tissue engineering', *Design and Nature III: Comparing Design in Nature with Science and Engineering* (pp.233–243). Southampton, UK: WIT Press.

Leong, K.F., Chua, C.K., Sudarmadjia, N., Yeong, W.Y. (2008) 'Engineering functionally graded tissue engineering scaffolds', *Journal of Mechanical Behaviour of Biomedical Materials*, Vol. 1, pp. 140-152.

Annals of DAAAM for 2008 & Proceedings of the 19th International DAAAM Symposium, ISSN 1726-9679
Editor B. Katalinic, Published by DAAAM International, Vienna, Austria 2008
Make Harmony Between Technology and Nature, and Your Mind will Fly Free as a Bird

MODELING OF THE PC MILL100 MACHINE TOOL AND MILLING PROCESS USING DELMIA V5R17

ANANIA, F[lorea] D[orel]; ZAPCIU, M[iron] & MOHORA, C[ristina]

***Abstract**: In this paper there are presented some tehnics of simulation regarding the milling process on a PCMILL100 machine tool. The first step of the simulation of the cutting process strats with a CAM model, witch allows to generate the NC programm. The second step is to implement this NC program on a virtual machine tool in order to make a real simulation. If the result of the simulation is satisfying the NC program can be uploaded on the real machine.*
***Key words:** Design, simulation, virtual machine tool, manufacturing.*

1. INTRODUCTION

The 3D modeling, simulation and optimization using the computer of the real systems based on the mathematical models, became a very useful research tool in many fields of human activities.

In the field of the mechanical engineering the Computer Aided Design – CAD has become an indispensable tool in the conception phase of the product. (Ghionea & Anania 2002)

Since the age of the Industrial Revolution, the manufacturing process has undergone many dramatic changes. One of the most dramatic of these changes is the introduction of Computer Aided Manufacturing (CAM), a system of using computer technology to assist the manufacturing process.

Through the use of CAM, a factory can become highly automated, through systems such as real-time control and robotics. (Ispas et al., 2006) A CAM system usually seeks to control the production process through varying degrees of automation. Because each of the many manufacturing processes in a CAM system is computer controlled, a high degree of precision can be achieved that is not possible with a human interface.

The CAM system, for example, sets the toolpath and executes precision machine operations based on the imported design.(Ispas et al., 2008)

Computer Aided Manufacturing is commonly linked to Computer Aided Design (CAD) systems. The resulting integrated CAD/CAM system then takes the computer-generated design, and feeds it directly into the manufacturing system;(Ispas et al., 2004) the design is then converted into multiple computer-controlled processes, such as drilling or milling.

2. MILLING PROCESS SIMULATION AND NC FILE GENERATION

The complex CAD/CAM software allow to generate an NC file based on the simulation of the cutting process up to 5 axis CNC.

In Delmia V5R17 the CAM moduls offers some important advantage in order to obtain an optimum NC program. It could operate with tools and tools library (Ispas et al., 2007).

A flcxiblc administration of the programs, based on a simple and intuitive interface, is also one of its main characteristics.

The most important feature is that DELMIA allowed a powerful interaction between tool trajectory, verification and optimization.

We generate a 3 axis milling CAM process in order to generate the NC program for a piece shown in figure 1.

To create a CAM model, the 3D design of the pieces has to be modified according with the manufacturing process demands. (Ispas et al., 2007) There must be created some additional contours, plane and point in order to define all the CAM parameters.

In DELMIA – prismatic machining it has to define the workpieces, the pieces, the references, the tool parameters, the type of the machine tool etc (fig.2).

For each type of opertaion there have to be selected some surfaces or conturs in order to define the machining area and all the machining parameters (ex. Speed, cutting depth, etc).

From the optimisation point of view the main advantage in Delmia V5R17 is the freedom to choose and modifies the tool trajectory (fig.3depending on the cutting process (roughing, finishing etc), the material, and the quality of the surfaces. After all the parameters have been defined the NC file can be generated. For this it has to be assigned in DELMIA the NC postprocessor, specific to the machine tool command (ex. for PC MILL 100 SINUMERIK 810).

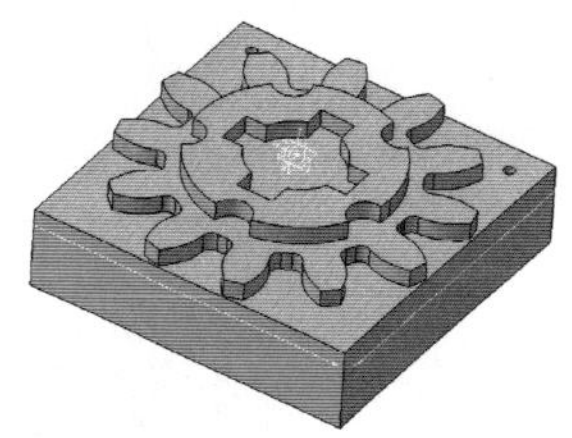

Fig.1. 3D model of the pieces.

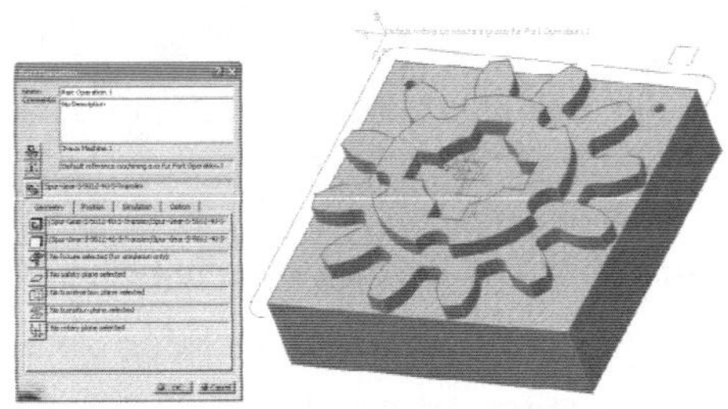

Fig.2. Define the machining parameters in Delmia

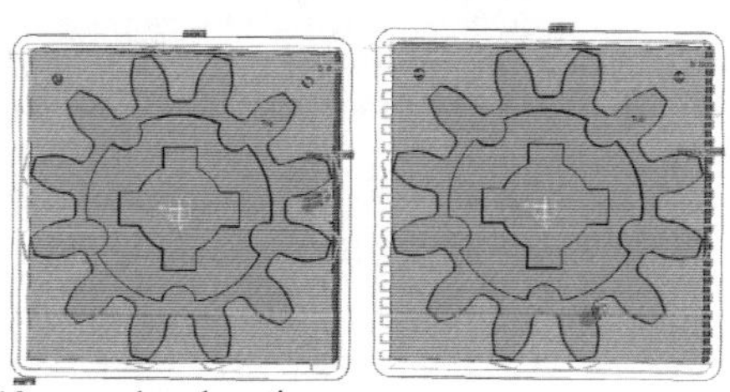

Fig.3 Difrent tool path option

3 MODELING THE PART AND KINEMTICS OF THE PCMILL100 MACHINE TOOL

The next step into complex milling simulation is to define the virtual machine tool with all the characteristics (dimensional and functional).

In DELMIA you must have all the parts of the machine defined as 3D model. It is not necessary to define all the machine tool components (as kinematic chaine or bearings), but these have to be define all the parts (as forms and dimensions) from the workspace and their relative position as accurate as possible. The machine tool functional parameters (domaine of speeds curses on each axis) also have to be highly defined.

The specifications of the PC MILL100 are: dimension of the machine tool: Lxlxh - 1730x875x1892 m; the workspace: X xYx Z -185x100x100; the main spindel speed – 60-5000 rot/min; linear speed 0-3000 mm/min.

Function of this specification the PC MILL100 machine tool has been defined in DELMIA (fig.4) All the structural parts as column, turret head, machine bed and machine table were highly designed. Also the workpiece mount point and the tool mount point were accurately defined.

The kinematics of the linear axis was defined as translation joints of the structural parts, according to the machine workspace.

The clamping device can be modeled as a part of the machine tool or separatly. In practice, the clamping devices are specific to the workpiece and/or pieces which have to be machined. For PC MILL 100 it was designed separately (fig.5).

3D CAD Models	 PC MILL100 parts	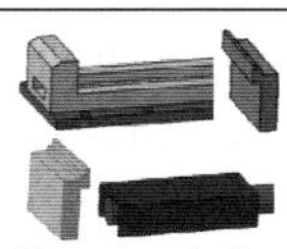 Clamping device parts	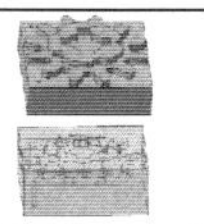 Piece and workpiece
Simulation Kinematics And CAM	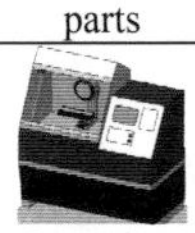 PC MILL100	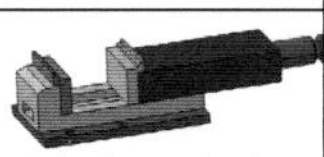 Clamping device part	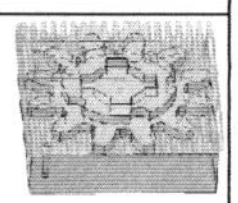 CAM
Machining Simulation	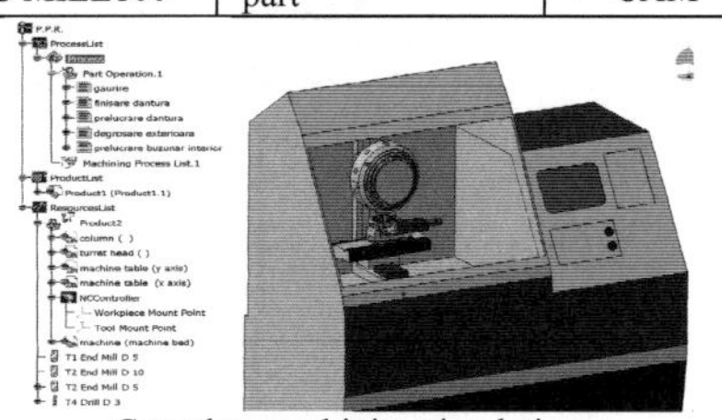 Complex machining simulation		

Fig.4 Level of simulation

4 MILLING PROCESS ON THE VIRTUAL AND REAL PCMILL 100 MACHINE TOOL

In order to obtain a simulation of the real conditions which appear on the machine tools it is necessary to go trough the following steps: upload the necessary elements for machining (machine tool, tool, the clamping device, the workpiece and the part); positioning of the parts on the machine table - the setting in position – figure 6; create or uploading the software already modified by using Post processor software; making necessary adjustments for simulation; setting up the collision detection.

This setting in position has a special importance, if the setting in position is realized with different errors, the machining simulation will be realized incorrectly.

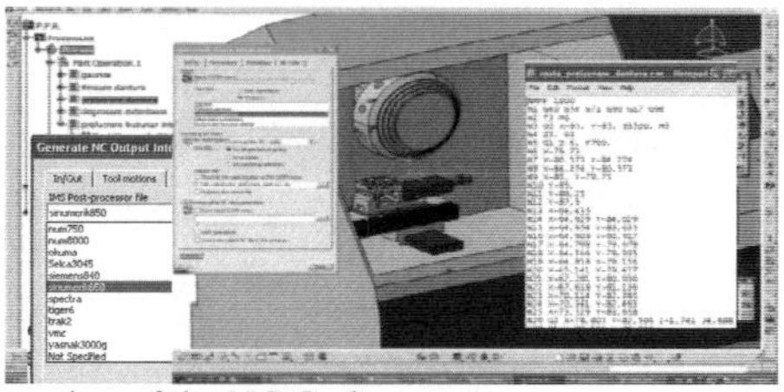

Fig.6 Generation of the NC Code

Fig.7. Simulation and real machining on PC MILL100

If the simulation results are satisfying, the CNC program is generated based on the choosen post procesor (fig.7).

The NC code was uploaded on the PCMILL100 and the pieces was machined.(fig.8)

To machine these polymer pieces it was used a two theeth mill with 5mm in diameter. The spindle speed was about 4000rot/min and the linear speed was 800 mm/min.

5. CONCLUSION

By simulation there could be obtained the optimum machining parameters in order to assure the surface quality and decrease the machining time;

Using CAD-CAM techniques an optimum NC file can be obtained for a specific machine tool.

By using a PCMILL100 virtual machine tool in machining simulation there were identified and eliminated all the possible clashes between machine components.

Before start using a virtual machine tool for simulation of the process, the model has to be calibrated in accordance with the real machine tool.

New calibration methods are required.

6. REFERENCES

Ghionea, I., Anania, D.(2002) – *„The management of the computer aided design using virtual prototype"*, Proceedings of the International Conference on Manufacturing Systems ICMaS 2002, Ed. Academiei Române, p.529…532, ISSN 0035-4074, ISBN 973-27-0932-4.

Ispas C-tin, Paraschiv, M.D.,Laboreau L, Anania F,D.(2008) *Research concerning the numerical errors correction for a NACA profile surface on a 5 axis machine tool*, Academic Journal of Manufacturing Engineering, p75-81, Ed. Politehnica, volume 6, ISSUE 3/2008, ISSN 1583-7904,

Ispas C-tin, Mohora C., Anania F,D, Paraschiv M,D.(2007) *Virtual design methods for the new products.*, Proceedings of the Inetrnational Conference on Manufacturing Systems, ICMaS, 22 Noiembrie, ISSN 1842-3183

Ispas C-tin, Zapciu M., Mohora C., Anania F,D.(2006), *Product development using cad-cam-cae software and internet facilities* - The International Conference Of The Carpathian Euro-Region Specialists In Industrial Systems – Baia Mare – Romania ISSN 1224-3264

Ispas C-tin, Ghionea I, Anania F.D.,(2004) *Computer aided design and manufacturing optimization process for a set of spur gears of a gearbox*..International Meeting of Carpatian Specialist in the Field of Gears, Baia-Mare 2004 ISSN – 1224 – 3264

Annals of DAAAM for 2008 & Proceedings of the 19th International DAAAM Symposium, ISSN 1726-9679
Editor B. Katalinic, Published by DAAAM International, Vienna, Austria 2008
Make Harmony Between Technology and Nature, and Your Mind will Fly Free as a Bird

VIRTUAL INSTRUMENTATION IN DATA ACQUISITION AND ANALYSIS OF TOOL WEAR MONITORING

ANGHEL, A[lina]; SARBU, I[onel] & SCURTU, D[an]

Abstract: *This paper presents the authors' researches on tool wear monitoring with turning applications, a recent domain with major economic impact. The chosen monitoring method uses detected vibrations as information carriers on the tool wear and extracts the useful information by processing the signal. Signal processing is made in the amplitude, time and frequency domains using virtual instruments designed by the authors in the LabVIEW environment. Following the tool wear signal processing with the presented virtual instruments, the authors have obtained concluding results on the monitoring process. We believe that these researches can be continued by adapting the presented virtual instruments or creating new ones to monitor tool wear for other machining operations.*
Key words: *Digital Signal Processing, Virtual Instrumentation.*

1. INTRODUCTION

Controlling and estimating machining tool wear is one of the aspects of monitoring machining tool state and conditions. This domain is grounds for research to increase the autonomy, productivity and quality of the manufacturing and reduce the economic loss caused by unexpected failures. Among the elements that make up the machine-cutting and device-work piece tools system, the machine cutting tool has the lowest reliability and can cause major defects.

Among the cutting tool wear monitoring methods, vibration analysis plays an important part and the main reason to consider vibration signals for wear monitoring lies in their capability to respond to excitations occurring at sources that are practically inaccessible, like the tool-work piece interface. The ease of incorporating sensors into the machining tool structure, its low price and the better performances comparing with other methods, are among the other reasons. This method is viable in spite of the inherent difficulties.

The methods that process vibration signals to evaluate the tool state are based on the comparison of the average amplitude signal with a reference signal.

The analysis of the signal picked up by the vibration sensor is made in the time domain, as well as in the frequency and amplitude domains to investigate which of these methods is most efficient for the desired monitoring. Estimating the wear state of the turning tool is made by correlating the information provided by the analysis functions.

By designing an entire package of virtual instruments dedicated to the monitoring of the wear of the turning tool, the authors provide a modern and efficient method that can easily replace the traditional monitoring equipment.

This paper's goal is to carry out experimental research using virtual instrumentation on diagnosing tool wear state for turning machining operation by vibration analysis.

The goal of these experiments is to determine a relationship between the tool's condition and the characteristics of the signal picked up by the vibration sensors.

2. CHOOSING THE PROCESSING SCHEME

From the analysis of the tool wear monitoring equipment by vibration analysis, it is made clear that the signal analysis is required in the amplitude domain as well as in the time and frequency domains. There are a variety of methods used, depending on the school and researcher that used them, but many times these methods aren't systematic and cannot be reproduced. The conclusions drawn are often contradictory.

The turning operation was chosen since it is the most widespread cutting operation. The installation of the vibration sensor can be made in the proximity of the tool-work piece contact, so that the detected signal is not distorted.

Monitoring systems that use vibrations as tool-wear information carriers operate using the following general scheme: the complex signal generated in the machining process is captured by a sensor, transformed into an electrical signal and sent to a preprocessing block (containing hardware and software elements) that increases the signal/noise ratio and performs data reduction. The next block extracts the features and, after an analog-digital conversion, carries out signal analysis in the time, amplitude and frequency domains (Dongfeng & Axinte, 2006). Through comparison between the analyzed signal and the reference signal, we can obtain information on tool-wear and identify the trends in its evolution.

In this article, the general scheme for monitoring the wear process is applied using a virtual instrument for acquiring and saving to files the signals received from the sensor. Data acquisition and transducers signal processing were realized using a NI-DAQ board and LabVIEW software (Wu & Liu, 2008).

The file data is processed by a main modulated program, containing four levels: time domain analysis, frequency domain analysis, amplitude domain analysis and dynamic analysis. After the file data is loaded, the analysis domain can be chosen (Fig.1).

In time domain analysis, the cross-correlation function is used. This function describes joint properties of the reference signal and the test signal. By normalizing this function, its calculation is obtained for the two sequences acquired in different manufacturing conditions and envelope detection.

In our program, the analysis in the frequency domain is based on the estimation of the spectra for two signals, the reference and the test signals, with dedicated virtual instruments. Another virtual instrument was designed to display two graphs: the first for the differences between spectra and the second for comparison with the warning and alarm levels. The virtual instrument for frequency domain analysis allows direct observation of the differences between the power spectra of the reference vector and the test vector. Spectra are displayed simultaneously and with contrasting colors. The calculation and the display of the spectra differences allow reading the difference of the spectra amplitude on the vertical axis.

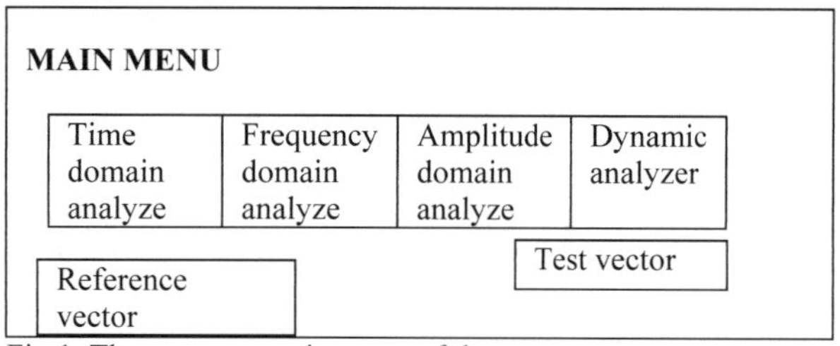

Fig.1. The program main menu of the program

The two vectors' spectra and the warning and alarm levels are displayed together on the virtual instrument's front panel and eventual equalities or overloads are made visible.

One of the necessary conditions for using these virtual instruments with good results is the correct setting of the warning and alarming levels and the widening number. The high sensitivity of the instruments is such that any warning or alarm situation will be signaled. The risk of false alarms is high if these levels are chosen incorrectly. However, the high sensitivity allows the advantage of being able to observe the tool wear from its very first appearance in the power spectrum. Based on his experience, the operator chooses the multiplying values for the reference spectra to obtain the warning and alarm levels. If these levels are too high then all the negative effects of tool wear will occur and the higher level of maximum wear is observed too late. If the predefined levels are too low, there is the risk of false alarms (Bajic & Lela, 2008).

The amplitude domain analysis displays the evolution of the peak-factor and the trend index, where the peak-factor is the ratio between the maximum and the effective values of the vectors and the trend index is the product of the mean intensity and the mean frequency of occurrence.

The analyzer is a modified complex virtual instrument from the LabView Library. Its adaptation for processing the vectors from the experiments plan enabled the access to different functions and the verification of some of the concluding remarks.

We consider that the possibility of accessing these functions is useful for using the program in other applications.

3. RESULTS AND PERSPECTIVES

Theoretically, it can be considered that the complex signal produced during the machining operation can be described by a function, depending on multiple variables: cutting conditions, dimensions, form and homogeneity of the work piece, tool geometry and so on. Practically, solving this function is very difficult, from mathematical reasons and requires a very large quantity of experimental data. This function is replaced by other functions that partially characterize the process (Bradley & Wong, 2006). For the theme of this paper, five independent variables were considered: revolution, n [rev/min]; feeds, s [mm/rev]; depth, t [mm]; hardness, HB; wear along the clearance face, VB [mm]. For each variable, five values are set up: medium, minimum, maximum and two intermediate values (Table 1). In order to reduce the number of experiments, an experimental project was used with factorial programming, second order, with central rotation. This experimental program reduces the number of experiments and obtains good results because it doesn't lose experiment significance (Dasic & Natsis, 2008). The trend index values seem well correlated with the tool-wear. Increasing these values indicates an active wear mechanism. The practical implications of this remark are the introducing of the trend index in monitoring schemes for other tool wear. For tool wear monitoring, the peaks that trend to zero of the Hilbert transform for the normalized cross-correlation function can be used. The presences of certain tight areas where the mentioned peaks are more wear sensitive, increase.

n [rev/min]	s [mm/rev]	t [mm]	HB	VB [mm]
260	0.025	1	165	0
425	0.05	1.25	180	0.15
660	0.1	1.6	200	0.3
850	0.12	2	210	0.5
1040	0.25	2.5	230	0.7

Tab. 1. Independent variables for tool wear monitoring

The modification of a complex virtual instrument from the LabVIEW Library and its adaptation to process the vectors from the experiments plan enabled access to different functions and the verification of some of the concluding remarks.

4. CONCLUSIONS

This paper presents a method of monitoring tool wear using virtual instrumentation that estimates the wear, constructed by wiring together objects that send and receive data, perform specific functions and control the execution flow.

For the research that led to this paper, the authors have leveraged theoretical and experimental methods for tool wear monitoring, in order to apply them to the turning operation

The factorial centered experimental program is also briefly described in this paper, in order to highlight the influence of the tool wear, cutting parameters and hardness of the piece work on the signal that is picked up by the vibration sensor during the turning operation.

A package of virtual instruments operating in the amplitude, time and frequency domains has been conceived to process the vectors containing the experimental data.

The use of this program allows complex processing by selecting different functions and comparative graphical display of the variables. The modular and flexible structure allows usage in other programs and makes it useful for digital signal processing.

The virtual instrumentation that was designed around this topic was utilized in experimental research. The results obtained recommend introducing virtual instrumentation for vibration analysis monitoring and diagnosing.

Other tool state monitoring systems for cutting operations can be based on the research method presented in this paper.

Moreover, future research could associate a greater number of parameters of different nature to establish a technical diagnosis of the machining tool and the machining process.

5. REFERENCES

Bajic, D. & Lela, B. (2008). Examination and Modeling of the Influence of Cutting Parameters in Longitudinal Turning. *Journal of Mechanical Engineering,* Vol. 54, No. 5/2008, pp 322-333, ISSN 0398-8673.

Bradley, C.& Wong, Y.S. (2006). Surface Texture Indicators of Tool Wear. *The International Journal of Advanced Manufacturing Technology,* Vol. 17, No. 6/2006, pp. 435-443, ISSN 0268-3768.

Dasic, P. & Natsis, A. (2008). Models of Rehabilitee for Cutting Tools. Examples in Manufacturing Engineering. *Journal of Mechanical Engineering,* Vol. 54, No. 2/2008, pp 122-130, ISSN 0318-6572.

Dongfeng, S.& Axinte, D.A. (2006). Online Machining Process Monitoring. *Instrumentation and Measurement Technology,* Vol. 23, No. 8/2006, pp 281-286, ISSN 1091-5281.

Wu, C.Y. & Liu, X.L. (2008). A Study of Cutting State Monitoring System Based on Virtual instrument. *Applied Mechanics and Materials,* Vol. 10, No 3/2008, pp. 568-572, ISSN 165-6424.

Annals of DAAAM for 2008 & Proceedings of the 19th International DAAAM Symposium, ISSN 1726-9679
Editor B. Katalinic, Published by DAAAM International, Vienna, Austria 2008
Make Harmony Between Technology and Nature, and Your Mind will Fly Free as a Bird

STATISTICAL MODELING OF BASIC MACHINABILITY PARAMETERS IN DRILLING OF METALS

ANGHEL, C[ornelia]; PETROPOULOS, G[eorgios]; VAXEVANIDIS, N[ikolaos] & DASIC, P[redrag]

Abstract: *The present study is aimed at the stochastic modeling of basic machinability parameters in the drilling of metals. With the application of the design of experiments methodology, probabilistic models are developed for cutting forces and torque, surface roughness (evaluated by the Ra parameter) and dimensional accuracy (hole oversize). The interpretation of the findings and the representation of the models are achieved through response surface graphs that show the main effects and the interactions of the factors that have physical significance.*
Keywords: *machinability parameters, drilling, modeling, surface roughness measurements.*

1. INTRODUCTION

Cutting processes are widely used in industry to manufacture both metallic and non metallic products, providing high form and dimensional accuracy, and surface quality with high degree of flexibility. Machinability of an engineering material is a crucial technological property that denotes its adaptability to machining processes in view of factors such as cutting forces, tool wear, and surface roughness. Effective and reliable machinability databases for cutting of metals are of paramount importance to assist manufacturers to apply proper machining conditions and relevant decision making. When carrying out machinability investigations data mining is necessary instead of a one-at-a-time factor approach. Such data mining techniques are Taguchi, response surface methodology and others accompanied by statistical design of experiments (Montgomery, 1997). Drilling is one of the most common cutting operations being necessary for machine building and assembling in a variety of applications (Bakkal et al., 2005). But it should be borne in mind that drilling especially when opening new holes is essentially a rough machining method and cannot meet satisfactorily the aforementioned requirements (Basavarajappa, 2007). The experimental investigation on basic machinability parameters in conventional drilling, which will be described in the following, attempts to give insight into the crucial parameters that rule the drilling performance and to model this complicated behavior (Petropoulos et al., 2007).
Parameters that do not contribute to safe conclusions are rejected during the statistical design of the models (Petropoulos & Vaxevanidis, 2005).

2. EXPERIMENTAL

2.1. Measuring force system set-up – Surface roughness measurements

The cutting force and torque measurements were undertaken by a 4-axis 9272 KISTLER dynamometer.
The *DynoWare* software was used for processing of the measurements. The evaluation of surface roughness was realized by a Sutronic 3+ Rank Taylor Hobson profile-meter using the *Talyprof* software.
The average values of three measurements of the parameter $\mathbf{R_a}$ are considered in each case.

2.2. Drills

The drills used in the experiments are twist drills of cylindrical shank made of high speed steel with 5% in cobalt according to DIN 338 HSS-Co (high speed steel cobalt alloyed).

Diameter (mm)	Total length (mm)	Twist length (mm)
8	117	75
10	133	87
12.5	151	101

Table 1. Dimensional characteristics of the drills used in the experiments

2.3 Work-piece materials – Specimens

Specimens made of St37 steel and Aluminum alloy 5005, according to DIN 17100 and DIN AlMg, respectively.
St37 steel is a carbon steel of wide use, relatively soft and possessing good machinability.
Al 5005 is an aluminum alloy of medium hardness, also widely used in structural applications.
Principal mechanical and physical properties of both materials used for processing are listed in Table 2. The great difference in hardness between the two alloys is evident, a fact that will be considered in the statistical analysis of the experiments.

	St 37	Al 5005
Density (Mg/m³)	7.91	2.70
Hardness (HB)	≤ 135	≤ 51
Maximum tensile strength (MPa)	415	145
Yield point (MPa)	205	35
Special heat (J/kgr·K)	447	900
Thermal conductivity (W/m·K)	360	205
Melt point (°C)	1510	652

Table 2. Mechanical and physical properties of the work-piece materials.

3. RESULTS AND DISCUSSION

3.1. Dry drilling.

In the tests 27 holes were drilled, corresponded to the combinations of 3 feed rates and 9 rotational frequencies. The results are illustrated in the corresponding diagrams in Fig. 1.

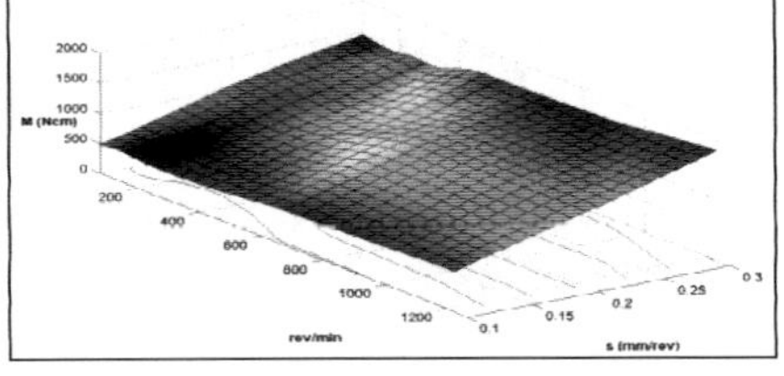

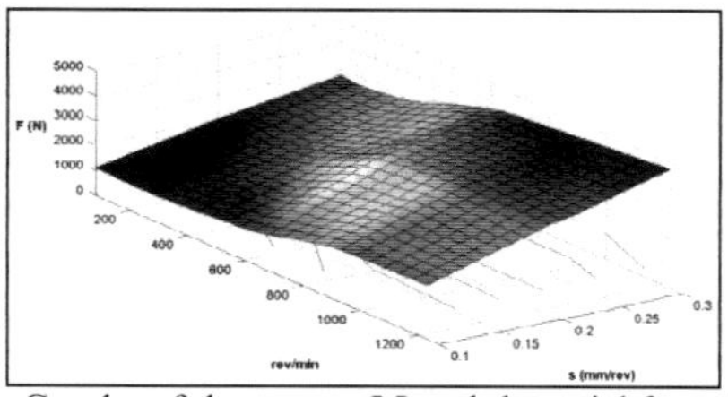

Fig.1. Graphs of the torque **M** and the axial force **F** in regard to rotational frequency and feed.

In view of the foregoing diagrams it is clear that two stages are noticed on the relevant curves. At low values of cutting speed both **M** and **F** tend to reduce, whilst they further increase with increase in rotational frequency. The effect of the latter is more pronounced on the axial force; the torque increases less significantly.
This behavior could be attributed to trapping of the chip and the difficulty to be removed from the opening hole due to the increased cutting speed. Concerning the influence of feed on **B** and **F** it is obvious that is more significant compared to that of cutting speed. Besides some exceptions surface roughness Ra increases when feed increases. The significant fluctuations observed are related to random phenomena like chip generation and removal under non repetitive conditions.

3.2. Wet drilling

The drilling tests were performed with the same cutting factors using three different cutting oils (neat oils) with viscosity varied from 4,80 cS to 11.30 cS at 80 °C .
Considering the experimental findings it was observed that when using the oil of intermediate viscosity the values of torque and force were higher compared to drilling performance with the rest oils; the use of the oil of the lowest viscosity gives slightly lower values of M and F than the oil of the highest viscosity.
As expected, the values of both magnitudes are pronounced in the case of dry drilling but the trends followed in association with the cutting conditions are similar.

4. STOCHASTIC MODELLING

4.1 Statistical experimental design

In Fig. 2 some response surface graphs illustrate the formed models. Summing up, in Table 3 comparative results from ANOVA are shown regarding the statistical validity and the most significant factors of the developed models.

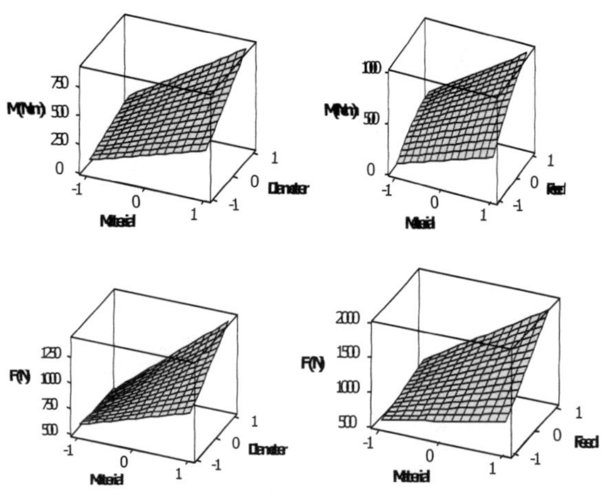

Fig. 2. Response surface graphs (indicative) of M and F against the controlling experimental factors

	F (N)	M (N cm)	R_a (μm)	ε_D (mm)
R^2	0.959	0.979	0.412	0.403
Conformity to hypotheses	full	full	partial	partial
Statistically more significant factor	Feed	Dia	Mat	Feed
Statistically most significant interaction	Dia*Feed	Dia*Feed	Mal*Dia	Mat*Dia

Table 3. Validity assumptions and statistically most significant factors of the stochastic models

Axial force

$\mathbf{F = 1371.5 + 491.7X_1 + 453.0X_2 + 641.2X_3 + 106.6X_4 - 38.7 X_1^2 + 145.1 X_4^2 + 133.5X_1X_2 + 187.7X_1X_3 + 238.9X_2X_3 + 11.9 X_1X_4 + 108.2X_2X_4 + 115.5X_3X_4 \quad R^2 = 0.959}$ (X_1:work-piece material, X_2: drill diameter, X_3: feed, X_4 : rotational frequency)

Torque

$M = 804.9 + 252.5X_1 + 294.0X_2 + 337.6X_3 - 8.0X_4 + 81.5X_1X_2 + 82.54X_1X_3 - 63.7X_1X_4 + 133.1X_2X_3 - 63.9X_3^2 \quad R^2 = 0.979$

Roughness

$\mathbf{Ra = 5.9 + 0.5X_1 - 0.3X_2 - 0.2X_3 - 0.2X_4 + 0.3X_1X_2 - 0.2X_2X_4 - 0.2X_3X_4 - 0.8X_4^2 \quad R^2 = 0.412}$

Hole oversize

$\mathbf{\varepsilon_D = 0.3 + 0.03X_1 + 0.02X_2 + 0.0X_3 + 0.02X_4 + 0.035X_1X_2 + 0.03X_1X_4 + 0.02X_2X_4 + 0.06X_3^2 + 0.04X_4^2 \quad R^2 = 0.403}$

5. CONCLUSIONS

From the initial tests conducted dry and replicated using cutting oils, the most physically significant factors leading to safe conclusions were detected that could be introduced to the stochastic models. In the lubricated tests the values of torque, cutting force and surface roughness were higher compared to dry drilling, as expected.
No clear connection of the viscosity of each oil type with the results was found.
The range of cutting speed was limited to medium values, as high cutting speed values increase the cutting forces and are unfavorable for chip removal.

6. REFERENCES

Bakkal M., A.J. Shih, S.B. Mc Spadden, R.O. Scattergood, *Thrust force, torque, and tool wear in drilling the bulk Metallic glass*, International Journal of Machine Tools and Manufacture (ISSN: 0890-6955), 45(7-8), pp. 863-872, 2005.

Basavarajappa S., Chandramohan , M. Prabu, K. Mukund, M. Ashwin, *Drilling of hybrid metal matrix composites-Work piece surface integrity*, International Journal of Machine Tools and Manufacture (ISSN: 0890-6955), 47(1), pp 92-96, 2007.

Montgomery D.C, *Design and analysis of experiments*, John Wiley & Sons, 4th ed (ISBN: 978-0-471-48735-7), 1997.

Petropoulos G, I, Ntziantzias, P. Reis, J. P. Davim, *Predicting machinability parameters on drilling Glass Fibber Reinforced Plastics using Response Surface Methodology*, International Journal of Materials and Product Technology (in press) (ISSN 0268-1900), Greece, 2007.

Petropoulos, G.P., Vaxevanidis N. *A topographic description of the bearing properties of electro-discharge machined surfaces*, Proc. 2nd Int. Conf. on Manufacturing Engineering – ICMEN (ISBN 960-243-615-8), Kassandra-Chalkidiki, 7 Oct, 2005, pp. 159-166, Greece.

Annals of DAAAM for 2008 & Proceedings of the 19th International DAAAM Symposium, ISSN 1726-9679
Editor B. Katalinic, Published by DAAAM International, Vienna, Austria 2008
Make Harmony Between Technology and Nature, and Your Mind will Fly Free as a Bird
*Young Researcher and Scientist Paper / * Supervisor, Mentor*

VIRTUAL MACHINE TOOLS MODELS IN ASSEMBLING SYSTEMS MATERIAL FLOW MANAGEMENT

ANGHEL, F[lorina]; CAZACU, D[ragos] A[lexandru] & DOBRESCU, T[iberiu]*

Abstract: *This paper presents the results of our late research in defining an algorithm for integrating virtual machine tools models in manufacturing systems material flow management. In order to illustrate this algorithm we are going to use the material flow simulation for an assembly made out of a milling machine main shaft and a pulley wheel. Our algorithm is based on the description of the modelling process in Witness, of the machines used and of the manufacturing parametrical values. The role of this assembly is to transmit movement the other parts of the milling machine.*

Key words: *simulation, manufacturing systems, architecture, model, production*

1. INTRODUCTION

Before starting a manufacturing process it is advisable to run a simulation for the production cycle. In this way one can have valuable information regarding productivity, the necessary number of machines, the personal involved in the production cycle and both the manufacturer and the beneficiary will know an approximate production cost.

The main purpose of this paper is to determine the necessary time for assembling the main shaft with the pulley wheel. In order to obtain the data necessary for simulating this assembly the first step was to use the material flow simulation for the two parts, separately. Thus we determined the production time and productivity for the shaft and the pulley wheel.

We define diffused manufacturing systems as architectures based at least 2 work points surrounded & assisted by transport, transfer & deposit facilities (Sanchez, 2001).

When taking in consideration concentrated systems we can define them as being a single assembling point, receiving parts from different working points.

We agree here with the thesis that within the class of stochastic simulation models, one further distinction is necessary: simulations can be either terminating (sometimes called finite) or non-terminating in nature, with specific algorithms for each category.

In our case the simulation is finite; we considered a finite period of time, a month, with 22 working days, 16 working hours a day.

2. VIRTUAL MACHINE TOOLS MODELS IN PARTS PRODUCTION

In figure 1 we can see how the optimized virtual model for the main shaft production looks (Anghel et al., 2007).

The working points symbolized in the model represent the machines used in the production cycle: debtor machine (D1), lathe machine (S1), milling machine (F1), boring and grinding machine (MAR1), each of them having an operator supervising the production cycle. After doing a number of optimizations in order to eliminate the material flow concentrators that appeared we obtained a productivity of 322 pieces per month (these means a necessary production time of 65.4 minutes).

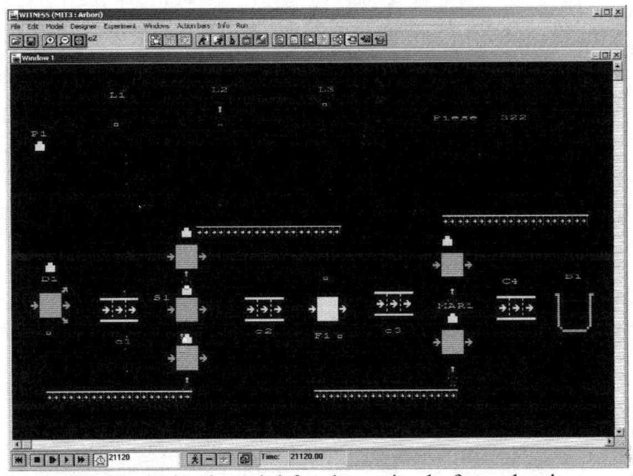

Fig. 1. The optimized model for the main shaft production.

In the same way we will build a virtual model describing the manufacturing process for the pulley wheel. Taken in consideration the operations required for manufacturing the pulley wheel we could establish the machine-tools used. These are, as symbolised in figure 2: a lathe machine (S2), milling machine (F2), boring and grinding machine (MAR2). For each machine there is an operator present with the role of supervising the operation. All parts are collected in a buffer.

For transporting blanks among working points we used conveyor belts.

After running a material flow simulation in the preliminary architecture we could see that the milling machine is a material flow concentrator do to the big manufacturing time that the blank spends on this working point. These is also do to the fact that the blank passes on quickly from S2 (the manufacturing time on this machine is 7.6 minutes).

Because of these we started an optimization that consisted in increasing the number of the working points. We decided to add another conveyor belt and another milling machine.

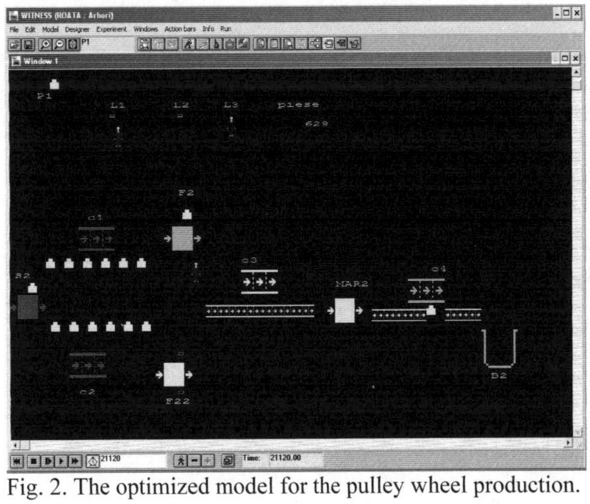

Fig. 2. The optimized model for the pulley wheel production.

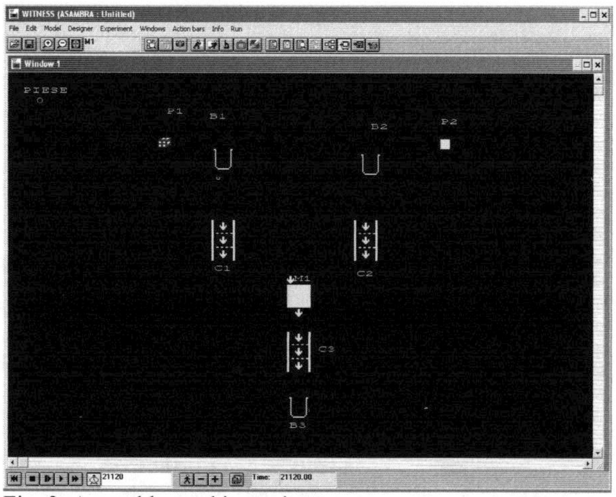

Fig. 3. Assembly working point.

By adding another machine and transport belt the total cost will increase and it is advisable to do a NPV analyses to see if the investment is justified.

After doing a number of optimizations we obtained a productivity of 629 pulley wheels per month (these means a necessary production time of 33.6 minutes).

3. VIRTUAL WORKING POINT ASSEMBLY

In this chapter we will build the model for the assembly between the two manufactured parts, the main shift and the pulley wheel.

In figure 3 we can see the preliminary architecture for the assembly unit, symbolized M1. The two parts ready to be assembled are symbolized P1 (shaft) and P2 (pulley wheel) and are deposited in B1 and B2 buffers. From buffer the parts are transported by conveyor belts to the assembly unit and from there to another buffer.

For the assembly process it is also possible to take the parts directly from the buffers in which they are deposit after the manufacturing process is finished. We choose not to do that, but to ship them in different buffers from which conveyor belts will take one P1 part and P2 part in the same time. The reason for this is given by the big difference between the manufacturing times of the two parts. The main shaft take almost twice as much time compared to the pulley wheel to be manufactured, so if we tried to assembly them directly from the buffers that they were initially deposited we wouldn't obtain a very good production site.

In this situation the assembly unit will be always waiting for shafts (P1) and the conveyor transporting pulley wheels (P2) will be continuously blocked with parts.

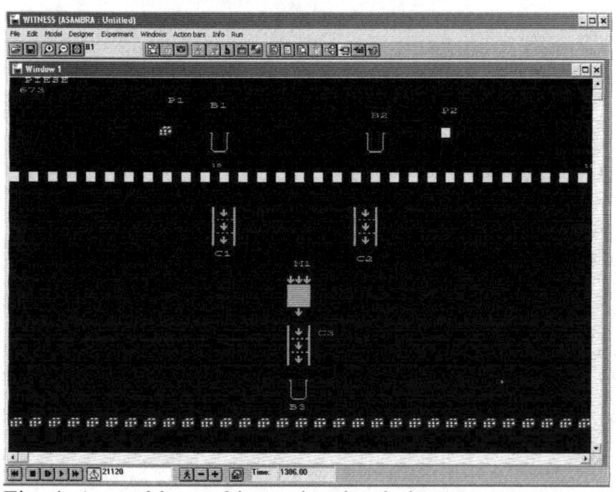

Fig. 4. Assembly working point simulation

In figure 4 we run a material flow simulation for the assembly site.

4. CONCLUSION

Three manufacturing systems interconnected in order to obtain a main shaft & pulley wheel assembly are acting like a virtual enterprise nodes (Tichkiewitch et al., 2006). The first manufacturing system is a diffused one and produces a main shaft from a milling machine. The second manufacturing system is also a diffused one and produces pulley wheel (Cotet, et al., 2007). The manufactured parts provided by those production sites are assembled on a third concentrated manufacturing system (Cotet et al., 2007).

The entire virtual manufacturing architecture where the 3 production sites are acting like partners in the same production network represents our case study used to illustrate an algorithm for integrating virtual machine tools models in manufacturing systems material flow management.

The model integrates material flow and process simulation in order to increase manufacturing system productivity. This kind of simulation shows important aspect regarding the manufacturing process: eventual material flow concentrators, how the presence of the operator affects the production cycle (for example: if we have two milling machines we will need two operators in order to operate these machines if we want to obtain an increasing of the production).

Another important aspect of this simulation is that we can determine the necessary number of machines for an optimum production site.

After doing a number of optimizations in both manufacturing sites, this consisted in adding machines in order to eliminate material flow concentrators from we managed to obtain a productivity of 322 pieces per month for the shaft's execution and 629 pieces for the pulley wheel. This means 65.4 minutes for the shaft and 33.6 minutes for obtaining the pulley wheel.

5. REFERENCES

Anghel, F., Cazacu, D. A., Bucur, C. C. & Coteţ, C. E. (2007). Classical versus numerical control machines in discrete material flow simulation based manufacturing architectures optimization, 18th International DAAAM Symposium – DAAAM 2007, pag. 025-026, ISSN 1726-9679, Zadar, Croaţia.

Cotet, C. E., Dragoi, G. & Carutasu, G. (2007). Material Flow & Process Synchronous Simulation In Concentrate Manufacturing Systems, Annals of DAAAM for 2007 & Proceedings of The 18th International DAAAM SYMPOSIUM, „Intelligent Manufacturing & Automation: Focus on Creativity, Responsibility and Ethics of Engineers", pag. 180-181, ISSN 1726-9679, ISBN 3-901509-58-5, Zadar, Croaţia

Cotet, C. E., Dragoi, G. & Abaza, B. F. (2007). Multipolar Synchronous Material Flow & Process Simulation In Difussed Manufacturing Systems, Annals of DAAAM for 2007 & Proceedings of The 18th International DAAAM SYMPOSIUM, „Intelligent Manufacturing & Automation: Focus on Creativity, Responsibility and Ethics of Engineers", 2007, pag. 179-180, ISSN 1726-9679, ISBN 3-901509-58-5, Zadar, Croaţia

Sanchez, S. M. (2001). ABC's of output analysis, *Proceedings of the 2001 Winter Simulation 2001*, Peters, B.A., Smith, J.S., Medeiros D.J. (Ed.), CD-ROM, Presses Association for Computing Machinery (ACM), New York

Tichkiewitch, S.; Radulescu, B. & Dragoi, G. (2006). *Knowledge management for a cooperative design system,* Advances in Design, ElMaraghy & Hoda A. Eds., pp. 97-110, Springer Verlag, ISBN 1-84628-004-4

Annals of DAAAM for 2008 & Proceedings of the 19th International DAAAM Symposium, ISSN 1726-9679
Editor B. Katalinic, Published by DAAAM International, Vienna, Austria 2008
Make Harmony Between Technology and Nature, and Your Mind will Fly Free as a Bird
*Young Researcher and Scientist Paper / * Supervisor, Mentor*

PRODUCTION OF ALUMINUM-STEEL ALLOY MATRIX COMPOSITES BY THE STIR-CASTING METHOD

ANTHYMIDIS, K[onstantinos]; AGRIANIDIS, P[avlos]; DAVID, K[onstantinos] & AGRIANIDIS, A[nastasios]

Abstract: *Interactions between solid materials and liquid aluminum lead to a dissolution of solid elements into aluminum, which in turn results in a subsequent growth of intermetallic and intermediate phases. It was established that the growth of the intermetallic phases could be governed by chemical reactions at the interfaces and by interdiffusion of the reacting elements through the different phases. The main objective of this study was to investigate both the formation of intermetallic and intermediate phases in the reaction zone and to examine the development of the diffusion structures of Aluminum-Steel alloy matrix reinforced with TiB and TiB_2 particles and to investigate the mechanical properties of the as-produced composite materials.*

Key words: *metal matrix composites, aluminium-steel reinforcement, mechanical properties, stir-casting*

1. INTRODUCTION

It is well known that the addition of ceramic phases in a metal or alloy e.g. aluminium, in form of fibers or particles infects its mechanical properties (Aylor 1987). This leads to a new generation of materials, which are called metal matrix composites. They have found a lot of application during the last twenty-five years due to their low density, high strength and toughness, good fatigue and wear resistance (Kiourtsidis & Skolianos 2002; Deuis et al., 1996; Song et al., 1995; Venkataraman & Sundararajan 1996; Zhang et al., 1995 ; Bindumadhavan et al., 2001). Automobiles industry use aluminum alloy matrix composites reinforced with SiC or Al_2O_3 particles for the production of pistons, brake rotors, calipers and liners (Prewo 1989). However, no reference could be cited in the international literature concerning aluminium or aluminium alloy matrix reinforced with TiB and TiB_2 particles although these composites are very promising for improving the mechanical properties of this metal or alloy without significantly alter it's corrosion behavior. Several processing techniques have been developed for the production of reinforced aluminum alloys. This paper is concerned with the study of TiB and TiB_2 reinforced aluminium-steel alloy produced by the stir-casting method.

2. EXPERIMENTAL PROCEDURE

The Aluminum - 34CrNiMo6 steel alloy was heated and thawed at about 1060°C. After fifteen minutes of isothermal holding and vigorous stirring of the alloy, with a tungsten impeller at 800 rpm, a satisfactory vortex was created and then equal quantities of K_2TiF_6 and KBF_4 were gradually added in order to produce TiB and TiB_2 particles. When this was completed the mix was maintained under stirring for further fifteen minutes and then the alloy was left to rest for about eight minutes and afterwards was poured in a preheated mold and left to be cooled in the atmosphere. The morphology of the produced composite materials was examined using optical and SEM microscopy. The composition of microstructural features was determined by EDX spectroscopy. The tribological properties of the as-produced composite material were evaluated using a pin on disc type equipment under dry wear conditions. Test conditions: pressure 154 KPa, SiC paper 220 grit, testing time 5 min, velocity 47.1 m/min. The experimental set up for tribological testing has been described in a previous publication (Skolianos et al., 1993).

3. RESULTS AND DISCUSSION

Figure 1 shows the typical morphology of the produced Aluminum-Steel alloy matrix reinforced with TiB and TiB_2 particles. In the present experiment synthesis of composite materials was carried out with the method described above, by mixing of 90% w.t. Al metal, 5% w.t. 34CrNiMo6 steel, 2,5% w.t. K_2TiF_6 and 2,5% w.t. KBF_4 at temperature of 1060°C. The Vickers microhardness of the produced composite material was measured 110 Hv (Fig.2).

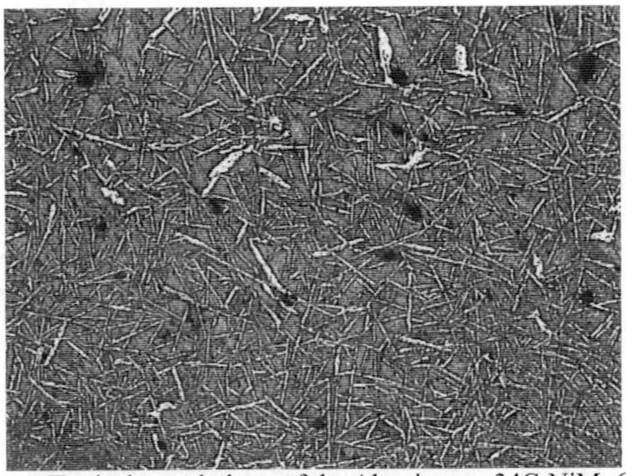

Fig. 1. Typical morphology of the Aluminum - 34CrNiMo6 steel alloy reinforced with TiB and TiB_2 particles (black spots) produced by the stir-casting method.

Fig. 2: Microhardness measurements of the Aluminum - 34CrNiMo6 steel alloy reinforced with TiB and TiB_2 particles produced by the stir-casting method.

From the EDX spectroscopy it was concluded that the as prepared material consists of three different phases: white phase comprises 80%w.t. Al and 20%w.t. Fe, dark color phase comprises 99%w.t. Al and 1%w.t. Fe and gray color phase which is almost pure Aluminum metal, TiB and TiB_2 particles (black spots) are scattered all over the material (Fig.3). The SEM mapping of the produced composite material is presented in Figure 4 and it is in good agreement with the above analysis of the experimental results.

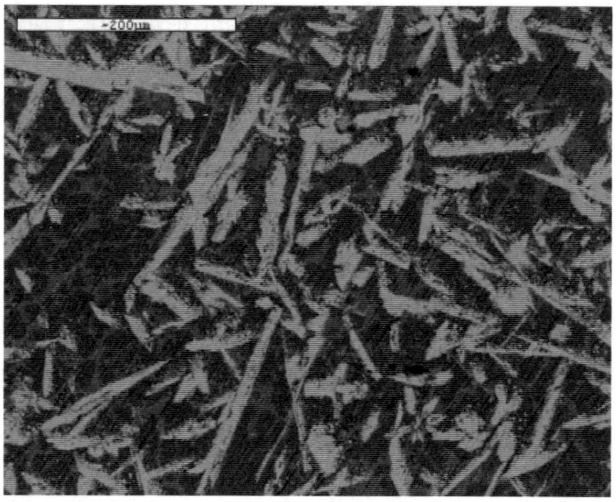

Fig. 3: SEM photos of the Aluminum - 34CrNiMo6 steel alloy reinforced with TiB and TiB_2 particles produced by the stir-casting method.

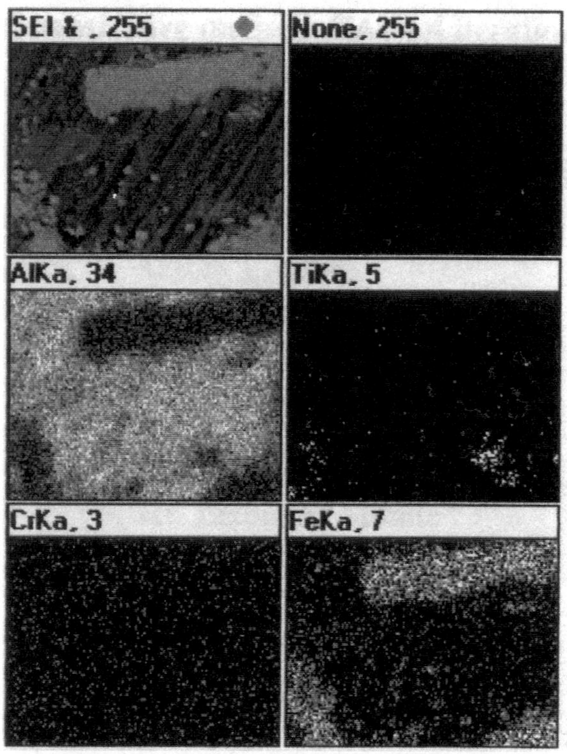

Fig. 4: SEM mapping of the Aluminum - 34CrNiMo6 steel alloy reinforced with TiB and TiB2 particles produced by the stir-casting method

In the tribological tests under dry wear conditions the as-produced composite material showed over 50% increased resistance to wear compared to pure Al metal (Fig. 5). Test conditions: pressure 154 KPa, SiC paper 220 grit, testing time 5 min, velocity 47.1 m/min.

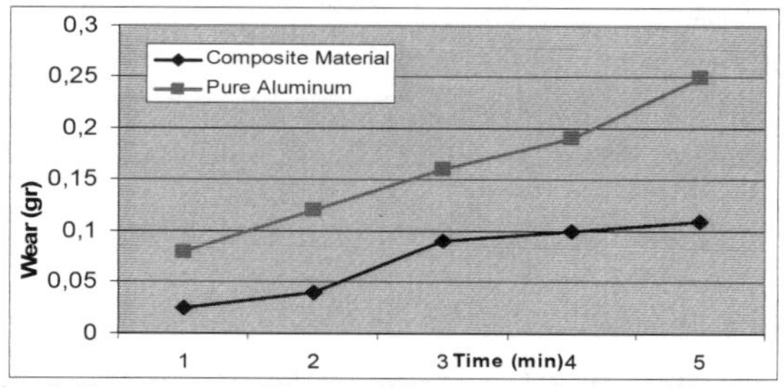

Fig. 5: Dry wear resistance of the composite material.

4. CONCLUSIONS

An Aluminum – 34CrNiMo6 steel alloy matrix reinforced with TiB and TiB_2 particles was produced using the stir-casting method. It was consisted of three different phases comprised 80%w.t. Al and 20%w.t. Fe, 99%w.t. Al and 1%w.t. Fe and almost pure Aluminum metal. A non-uniform distribution of the solute elements may causes a local concentration of these elements into the liquid alloy, which practically results in a non-uniform distribution of the produced TiB and TiB_2 particles in the composite material. Thus it is crucial to control the dissolution conditions such as thermodynamic conditions, temperature, stirring time so that the instabilities induced at the solid materials/aluminum-steel alloy interface are limited. The produced composite material showed over 50% increased dry wear resistance compared to pure Aluminum metal.

5. ACKNOWLEDGMENTS

The authors are indebted to the Research Committee of the Technical University of Serres for funding provided.

6. REFERENCES

Aylor, D. (1987). *Metals Handbook,* Vol.13, 9th ed., edited by ASM, Publisher: Butterworth-Heinemann , ISBN: 0872011542, Metals Park, OH, p. 859

Bindumadhavan, P.; Wah, H. & Prabhakar, O. (2001). Dual particle size (DPS) composites: effect on wear and mechanical properties of particulate metal matrix composites, *Wear,* Vol. 248, No.1-2, March 2001, p. 112-120, ISSN 0043-1648

Deuis, R.; Subramanian, C. & Yellup, J. (1996). Abrasive wear of aluminum composites: A review, *Wear,* Vol. 201, No.1, December 1996, p. 132-144, ISSN 0043-1648

Kiourtsidis, G. & Skolianos, S. (2002). Wear behavior of artificially aged AA2024/40 μm SiC_p composites in comparison with conventionally wear resistant ferrous materials, *Wear,* Vol. 253, No.9, November 2002, p. 946-956, ISSN: 0043-1648

Prewo, K. (1989). Fiber-Reinforced Ceramics: New Opportunities for Composite Materials, *Am. Ceram. Soc. Bull.,* Vol. 68, No.2, p. 395-400

Skolianos, S.; Hatziaugoustis, A. & Papaefthimiou, A. (1993). Tribological properties of the squeeze cast Al-6001, *Proceedings of the First Balkan Conference of Tribology Balkantrib*, Assenova, E (Ed.), pp.239-246, Sofia Bulgaria, October 1993, Technical University of Sofia, Sofia

Song, W.; Krauklis, P.; Mouritz, A. & Bandyopadhyay, S. (1995). The effect of thermal ageing on the abrasive wear behaviour of age-hardening 2014 Al/SiC and 6061 Al/SiC composites, *Wear*, Vol. 185, No.1, June 1995, p.125-130

Venkataraman, B. & Sundararajan, G. (1996). The sliding wear behaviour of Al-SiC particulate composites, *Acta Materialia*, Vol. 44, Issue 2, February 1996, p. 451-460, ISSN: 1359-6454

Zhang, Z.; Zhang, L. & Mai, Y. (1995). Particle effects on friction and wear of aluminium matrix composites, *Journal of Materials Science,* Vol. 30,No.23, December 1995, p.5999-6004

Annals of DAAAM for 2008 & Proceedings of the 19th International DAAAM Symposium, ISSN 1726-9679
Editor B. Katalinic, Published by DAAAM International, Vienna, Austria 2008
Make Harmony Between Technology and Nature, and Your Mind will Fly Free as a Bird

3D NUMERICAL SIMULATION OF THE TURBULENT FLOW IN AN INDUCER

ANTON, L[iviu]; STUPARU, A[drian]; MUNTEAN, S[ebastian] & BAYA, A[lexandru]

Abstract: *This paper presents the3D numerical investigation of the turbulent flow in an inducer by using commercial code FLUENT 6.3. First the investigated inducer is described and then the equations that govern the flow and the boundary conditions imposed on the computational domain. The results of the turbulent flow simulation are compared with the experimental results, and then the stream trace distribution and a set of conclusions are presented.*
Key words: *numerical simulation, turbulent flow, pump inducer, flow analyse.*

1. INTRODUCTION

The inducer is an axial impeller which is mounted on the same shaft with the impeller of a pump. By producing a supplementary specific energy on the inlet of the impeller of the pump, because of the rising of the pressure over the vaporising pressure, it prevents or reduces the cavitation. The inducer is used for the pumps with severe suction condition and it has a "sacrificial role", (Garay, 1996). It will be replaced after a certain operating time and that is why the technology for its execution has to be simple and economical. To use straight hydrofoil cascade for the construction of the blades of the inducer satisfies these requirements.

The aim of this paper is to study the structure of the flow given by the use of a straight hydrofoil cascade for the manufacturing of the inducer blades. This study is performed by analysing the stream trace distribution inside the inducer.

We numerically investigated the flow in an inducer, which has two blades manufactured from a straight hydrofoil cascade.

In order to validate the numerical investigations of the flow, the pressure distribution along the blade is compared with the experimental pressure distribution obtained from the investigations performed in a test rig upon the inducer.

2. COMPUTATIONAL DOMAIN, FLOW EQUATIONS AND BOUNDARY CONDITIONS

The computational domain, figure 1, was generated using the pre-processor GAMBIT from FLUENT, based on the existing geometry.

The geometrical characteristic of the blades of the inducer and the investigated operating condition are given in table 1, where we have informations about the significant diameters of the inducer, the interior diameter, *d*, and the exterior diameter, *D*, and the thickness of the hydrofoil, *s*, used for the blades.

The investigated inducer has two blades and it is numerical investigated while it is operating in air with a given rotational speed, *n*, and at a certain flow rate, *Q*.

The generated mesh for the computational domain is structured and has 460,000 cells, (Mejri et al., 2006).

Hydrofoil	Q [m^3/s]	D [mm]	d [mm]	s [mm]	n [rot/min]
Straight	0.335	300	100	3	1450

Table 1. Geometrical characteristics and operating conditions of the investigated inducer

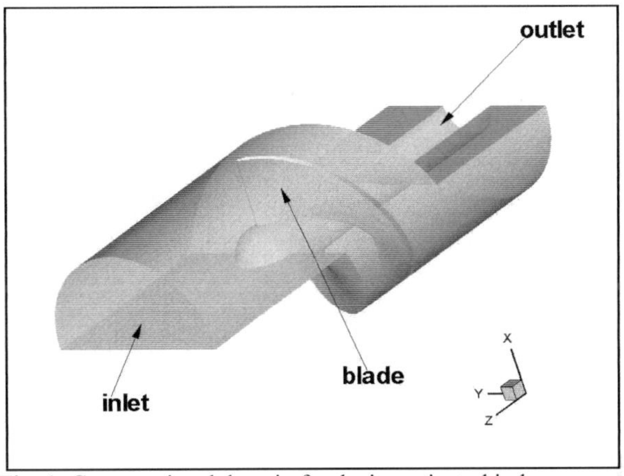

Fig. 1. Computational domain for the investigated inducer

For the flow analysis presented in this paper we consider a 3D turbulent flow model. A steady relative 3D flow is computed:

$$\nabla \cdot \vec{v} = 0 \qquad (1)$$

$$\rho \frac{d\vec{v}}{dt} = \rho g - \nabla p + \mu \Delta \vec{v} \qquad (2)$$

The numerical solution of flow equations (1) and (2) is obtained with the expert code FLUENT 6.3, (Fluent, 2001), using a Reynolds-averaged Navier-Stokes (RANS) solver. As a result, the viscosity coefficient is written as a sum of molecular viscosity μ and turbulent viscosity μ_T, and the last term in the right-hand-side of (2) becomes $\nabla \cdot \left[\left(\mu + \mu_T \right) \nabla \vec{v} \right]$.

We solve a relative flow, in a rotating frame of reference with angular speed $\vec{\omega} = \omega \vec{k}$ ($\vec{k}$ being the unit vector of the pump axis direction).

By introducing the relative velocity

$$\vec{w} = \vec{v} - \vec{\omega} \times \vec{r} \qquad (3)$$

With $\vec{r}$ the position vector, the left hand side of (2) becomes

$$\begin{gathered} \frac{\partial}{\partial t}\left(\rho \vec{w}\right) + \nabla \cdot \left(\rho \vec{w}\vec{w}\right) + 2\rho\vec{\omega} \times \vec{w} + \\ + \rho\vec{\omega} \times \left(\vec{\omega} \times \vec{r}\right) + \rho \frac{\partial \vec{\omega}}{\partial t} \times r \end{gathered} \qquad (4)$$

An important assumption used in the present computation is that ***the relative flow is steady***. This simplifies (3) by removing the first and last terms, and also allows the computation of flow on a single inter-blade channel.

The turbulent viscosity is computed using the RSM model.

We imposed on the inlet section the velocity, corresponding to the prescribed flow rate together with the turbulence parameters, (Susan-Resiga, 1999):

$$v = \frac{Q}{S_{IN}} \tag{5}$$

On the outlet section of the computational domain a radial equilibrium condition is chosen.

On the periodic surfaces of the domain the periodicity of the velocity, pressure and turbulence parameters were imposed:

$$\vec{v}(r,\theta,z) = \vec{v}\left(r,\theta+\frac{2\pi}{n_b},z\right) \tag{6}$$

$$p(r,\theta,z) = p\left(r,\theta+\frac{2\pi}{n_b},z\right) \tag{7}$$

The remaining boundary conditions for the domain correspond to zero relative velocity on the blade.

3. NUMERICAL RESULTS

The pressure coefficient is defined by the following equation:

$$c_p = \frac{p - p_{IN}}{\frac{\rho}{2} w_{IN}^2} \tag{8}$$

The pressure distribution is plotted on a surface placed at the radius r_m given by the following equation:

$$r_m = \frac{1}{2}\sqrt{\frac{D^2 + d^2}{2}} \tag{9}$$

From the comparison of the numerical and experimental (Anton, 1994) pressure coefficient distribution along the pressure side, *PS*, and suction side, *SS*, of the blade, figure 2, it results a very good agreement. This proves that the turbulent model and boundary conditions were adequate chosen.

The 3D effects due to the loading of the inducer blade lead to the deviation of the stream trace towards the hub on the pressure side and towards the shroud on the suction side, figure 3. The flow on the pressure side towards the blade exit is positioned at the same radius as on the inlet of the blade. Moreover, the stream trace near to the pressure side moves near to the blade, while the stream trace on the suction side is pushed away from the blade, figure 4.

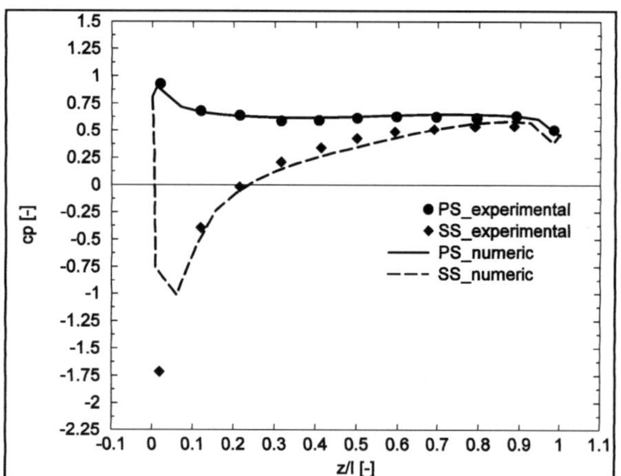

Fig. 2. Pressure coefficient distribution along the inducer blade at radius $r_m = 0.111$ m

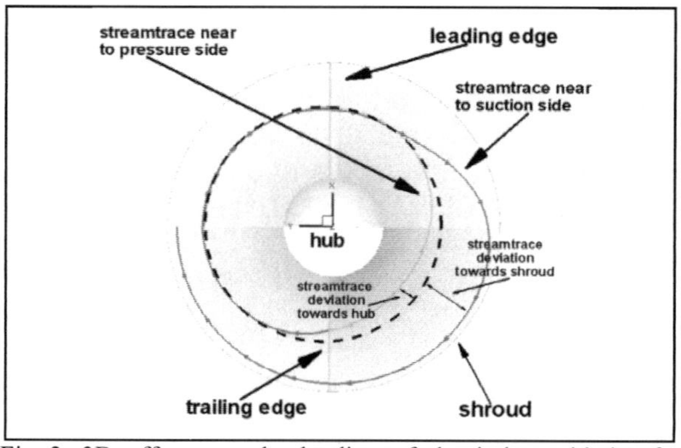

Fig. 3. 3D effects on the loading of the inducer blades for stream trace positioned at $r_m = 0.111$ m

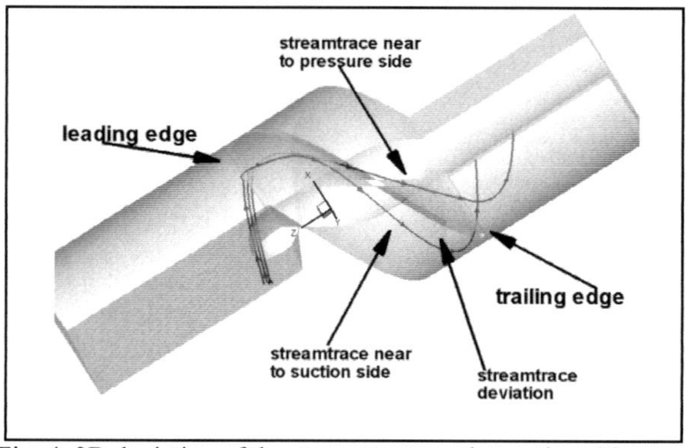

Fig. 4. 3D deviation of the stream trace on the suction side

4. CONCLUSIONS

The paper presents the 3D full turbulent numerical investigation and analyse of the flow in an inducer. The numerical results are in very good agreement with the experimental data at the section situated at $r_m = 0.111$ m.

The 3D effects are presented due to the loading of the blade. Consequently the stream traces on the suction side are deviated towards the shroud and the stream traces on the pressure side are deviated towards the hub.

Because these 3D effects could not be taken into account in the design process of the inducer, it results a poor operation of this type of inducer.

The numerical investigation of the structure of the internal flow for this type of inducer leads to the conclusion that it is recommended the use of another type of hydrofoil cascade for manufacturing the blades of the inducer.

5. REFERENCES

Anton, L. E. (1994). Determination of pressure distribution on the blades of an inducer, *Proceedings of XVII IAHR Symposium*, pp. 321-328, China, September, 1994, Beijing

Fluent Inc., (2001). *FLUENT 6.3 User's Guide*, Fluent Incorporated, Lebanon

Garay, P. N. (1996). *Pump Application Desk Book,* The Fairmont Press Inc., ISBN 0881732311

Mejri, I.; Bakir, F.; Rey, R. & Belamri, T. (2006). Comparison of Computational Results Obtained From a Homogeneous Cavitation Model With Experimental Investigations of Three Inducers. *Journals of Fluid Engineering*, Vol. 128, No. November 2006, 1308-1323

Susan-Resiga, R.; (1999). Periodic boundary conditions implementation for the Finite Element Analysis of cascade flows, *Scientific bulletin of Politehnica University Timisoara*, Vol. 44(58), pp. 151-160

Annals of DAAAM for 2008 & Proceedings of the 19th International DAAAM Symposium, ISSN 1726-9679
Editor B. Katalinic, Published by DAAAM International, Vienna, Austria 2008
Make Harmony Between Technology and Nature, and Your Mind will Fly Free as a Bird

LIP SEALS AND MECHANICAL FACE SEALS – PERFORMANCE CRITERIA

ARGESANU, V[eronica]; JULA, M[ihaela]; LUCHIN, M[ilenco] & RADU, A[drian]

Abstract: *A great diversity of contact-seals designs, materials, operating conditions, and factor that affect their performance have not yet allowed the general conclusions on friction and wear of these seals to be drawn. At the same time, the results of studies of particular cases may often lead the design engineer to an erroneous decision if the seal he develops is different in some way from that he has taken as a prototype. To a face seal design, choice of materials and type of seal arrangement, individual requirements such as minimum leakage, maximum life or minimum friction can be met.*
The paper presents the FEM (finite element method) analyze of the friction rates of mechanical face seal and lip seals.
Key words: *performance; lip seals; mechanical face seal, FEM.*

1. INTRODUCTION

Labyrinths, stuffing boxes, lip seals, bushings, spiral-groove seals, and mechanical seals made out of a very large number of materials, are the systems that are commonly used for sealing rotating shafts. The behavior of a seal is determined by the complex interaction of a number of factors. Advantages are usually attained at the price of disadvantages in the order directions. For example if the roughness is constant, an increase of the contact pressure reduces leakage, but the wear and frictional heat increase. As against this, increasing leakage losses can reduce the friction and the heat production, but the effectiveness of the unit as a seal is reduced. Again, a high friction may not only lead to increased wear but also, due a thermal distortion, to considerable leakage losses, or it may cause the seal to break down because of a thermal stress cracks.

Depending on application, sealing rubber should be strong, heat resistant, cold resistant, or resistant to chemical attack. The characteristics must often be combined; some of them are mutually incompatible. Anyhow, all sealing rubber applications require good friction properties: high wear resistance and low coefficient of friction.

2 FRICTION AND WEAR IN DYNAMIC CONTACT SEALS

The performance of seals is characterized by the degree of tightness, service life, power losses, by the extent of damage to the contacting surface in operation, etc. the degree of tightness, wear life t_w, and performance factor I are the most important characteristics of seal performance.(Argesanu, Madaras, 2003) In addition to the above factors, temperature, whose level is determined by their joints action, also affects the performance of dynamic seals. Whereas temperature has the major influence on the frictional effects in the contact area, the leakage is caused by reduction in the contact area pressure and distortions in the geometry of the rubbing surface due to wear, increased thermal deformations, etc. In some instances, these factors are interdependent. The service conditions of sliding contact seals in machinery, determined by combinations of the above factors, are very diverse. In a face seal (fig2), an axial force pressed a rotating floating ring 5 against a fixed counterface 6. The axial

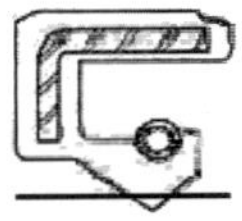

Fig. 1.a lip seal

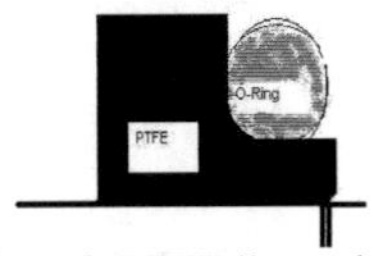

fig. 1.b PTFE lip seal

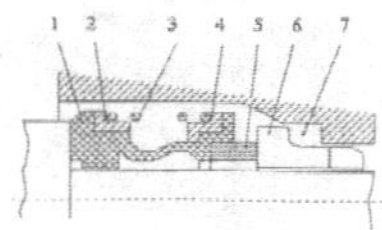

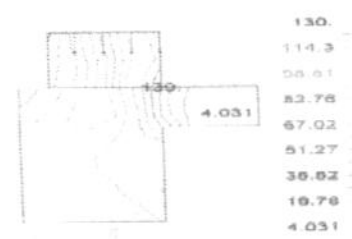

Fig. 2.mechanical face seal fig. 3.FEM-model mechanical seal

leakage path between the floating ring and the shaft is closed by a static seal such as an O-ring 7.(Mayer, 1987)

The static and sliding surface of the traditional stuffing box are effectively interchanged, with the advantage than the geometry if the sliding sealing surface can now be produced more accurately and less expansively and there is no longer any wear on the shaft or shaft sleeve. (Gheorghiu, Argesanu, 1996). To compensate for any lack of alignment of the seal faces and for longitudinal thermal expansion of machine and seal, as well as wear of the seal faces, the face seal must contain at least one flexible member such as diaphragm, bellows, elastometric seal, or springs 1,3.(fig2).(Argesanu, Madaras, 2003) In selecting sliding materials, consideration should be given to operating conditions, ease of manufacture and material costs. The chemical activities as well as the physical and mechanical properties have to be considered. By selecting materials with appropriate thermal conductivity coefficients, by additional cooling, lubrication and load "balancing". The sealing medium also has a considerable influence on the life of a mechanical seal. Mounting seals on elastomeric rings has a very beneficial effect on wear because of damping actions of the elastomer. Often the durability of a seal is determined not by the wear of the seal alone but by the resistance of ageing of any elastomers used. Intermittent operation as well as increases of contact pressure, friction coefficient, sliding speed and temperature will reduce the life. Since the effects of adhesive wear, abrasive wear, corrosive wear and erosive wear, let alone vibration, temperature and material effects, can be cumulative. If it defines the intensity of the power lost by friction in the area of contact as the ratio of the power lost by friction and land sliding (Gheorghiu, Argesanu, 1996):

$$P_{fr}/A_{al} = (\mu * p_d * v)_a \quad (1)$$

where: μ - friction coefficient in the contact area;
p_d - pressure in the area of contact;
v - relative speed.

This may give an appreciation of the value limit of operating of the sealing. If we consider for example μ=0.1…0.3 for PTFE lip seal(fig.1.b.) or μ=0.005…0.1 for the elastomer lip seal(fig.1.a.), the pressures of work of 3MPa sliding speeds of 12m/s: ${}^{P_{fr}}/_{A_{al}}$=3.6 W/mm^2 for PTFE and ${}^{P_{fr}}/_{A_{al}}$=7.2W/mm^2 for elastomers.

3 FRICTION CONTACT PROBLEMS BY FEM

The simulation consists in solving the moving equations of the floating ring associated with the interaction between the two rings due to the determination of the functional factors of a face seal. When friction is considered, the tangential displacement in the interface implies energy dissipation. The problem is solved by incremental computation.(Knothe, Wells, 1992) Portions of the structure can have areas of gaps which can open and close or slide in relation to each other. Similarly, boundary conditions can change dooring a nonlinear analysis. The distribution configuration of the equivalent stresses under the form insosurfaces (fig.3) exprimed in [Mpa] that reveals the most solicitated zones of the mechanical face seal. The configuration of those quantities is an expected one with a maximum to the inner radius of the rings of the mechanical seal. The FEM analysis is the only way to reestablish, by mathematical means, the rubbing contact pressure distribution from a face seal interface. This allows the fizico-mechanical and functional influence factors evaluation on the sealing performance of a seal.

4 PERFORMANCE CRITERIA

A comparative situation between the face seals and lip seals based on technological, operational and cost is presented in tab.1. When comparing the values it must be borne in mind that the sealed pressure p_1 for lip seals are lower that the corresponding mechanical seals. Despite this, it is evident that on account of smaller leakages of the buffer fluid and sealed product, and greater operational safety and reduced maintenance, mechanical seals are much superior to lip seals. Whereas the initial prices of the lip seals are lower than those of the face seals, the position is reverse upon installation and putting into operation. The labor-intensive maintenance costs are about 20 times higher with leap seals. Despite the more expensive spare parts for mechanical seals their total costs comes lower than for lip seals, on account of the long life of the former.(Mayer, 1987, (Muller, Waschle, 1990)

It also is taken into account that due to labor savings, smaller leakage losses, greater operational safety, and reduced-down time, face seals are even more economical than would appear. By means of a metallic support jacket in the carbon ring which is certainly deformed much more than the tungsten carbide seal under high pressure loads, because of its low modulus of elasticity, the previous very good operating characteristics of conventional thermohydrodynamic seals could be further improved at higher pressure. In seals, with the stabilized seal gap, the hydrodynamics of the circulations grooves are a better design. Seals gap are much less sensitive to pressure changes and have even longer service lives.

5. CONCLUSION

The behavior of a seal is determined by the complex interaction of a number of factors. Advantages are usually attained at the price of disadvantages in the order directions.

If we compare specific losses through friction: in the usual pressure for lip seals "uncharged" or classical "charged" with

$$\theta = {}^{F_{cr}}/_{\pi * d} \qquad (2)$$

Specification parameters /functional conditions	Seal tipe	
	Face seal	Lip seal
Materials and technology operation	High maintenance, involves polishing the active surfaces of rings	Easy to maintain, for the sealing edge, E class of precision
Space assembly		At d=idem
	maximum	Minimum
Condition processing of axle	correcting with advance	
	radial+axial	Only radial
can compensate for irregularities shape and position of axle	very good	medium/good
Axle wear	zero	existing
Power lost due to friction	reduced	loaded – big
		unloaded - reduced
maintenance expenses	reduced	
Lifetime expenses	comparable	
Initial expenses	high	low
Lubrications and cooling measures	necessary	
Type of fluid	no restrictions	limited – based on compatibility with the lip material

Tab. 1.Performance parameters of seals

face seals we can notice the net advantage of he lip seals, these having as a plus size and lower costs. However, in the field of higher speed and pressure and for hard working environments, the face seals are still irreplaceable. (fig. 4)

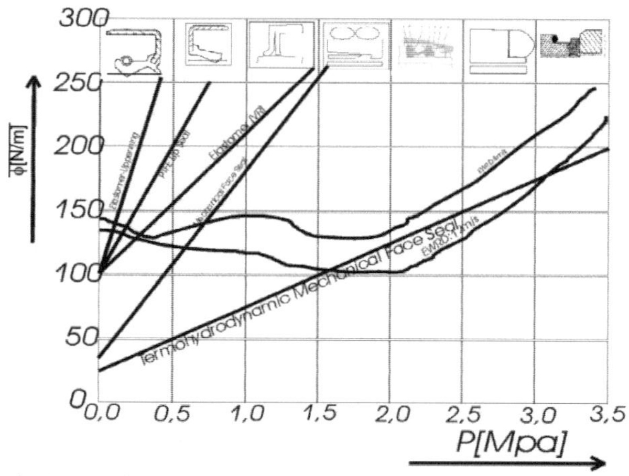

Fig. 3.Specific friction losses function pressure dependency for different types of seals.

6. REFERENCES

Argesanu, V; Madaras, L,(2003). *Leakage, wear and friction in the mechanical face analyzed by FEM, ROTRIP* Galati, Romania

Gheorghiu, N; Argesanu,V,(1996). *Comparison between the performances of the lip and face seals,* Arad

Knothe, K.; Welles, H, (1992).*Finite Element, Introduction for Engineers, Springer-Verlag,* Berlin, Heidelberg, New York, London, Paris, Tokyo, Hong Kong, Barcelona, Budapesta

Mayer, E, (1987). *Mechanical Seals, Hewnes* – Butterworth, London, Boston

Muller, H-K; Waschle, P,(1990). *EWDR–A new seal type for pressured shafts*, Anbetriebstechnik 29 Nr.10

Annals of DAAAM for 2008 & Proceedings of the 19th International DAAAM Symposium, ISSN 1726-9679
Editor B. Katalinic, Published by DAAAM International, Vienna, Austria 2008
Make Harmony Between Technology and Nature, and Your Mind will Fly Free as a Bird

CONSIDERATIONS UPON THE CIRCULAR SECTION CIRCLIPS/RETAINING RINGS AXIAL LOAD-CARRYING CAPACITY

ARGESANU, V[eronica]; LUCHIN, M[ilenco]; JULA, M[ihaela] & MARGINEANU D[an]

Abstract: *The reference standards as well as the specialty literature offer very many useful data to the designer. But, in specific applications, requiring small axial limits, they are insufficient to provide a satisfying safety without additional calculation. By means of FEM (finite element method), the authors are trying to point out the real behavior of the assembly in such cases*

Key words: *circlips, axial load-carrying capacity,FEM*

1. INTRODUCTION

Circlips / retaining rings are designed to position and secure component in bores and houses. Simultaneously they provide rigid end – play take – up in the assembly to compensate for manufacturing tolerances or wear in the retained parts. For the technical designer, who uses standardized and/or in a list manufactured rings on shafts or in housings with nominal diameter, a computation is not necessary. It is of crucial importance however with special applications of the normal rings and particularly with the construction of special rings.
The reasoning for the fundamentals of the bending calculus is presented in detail (Mesaroş-Anghel et al., 2006) for circlips with rectangular section, and FEM analysis has been done.
The authors propose to analyze circular section circlips. The conclusions of the paper can be directly applied in technical design. In the future, if experimental research results are added, also the reconsideration of the present standards regarding shape, dimensions, and materials can be made

2. FUNDAMENTALS

The strength calculation is based on the consideration that a circlip -for the shaft or for the housing – is a curved bended bar.(Argeşanu, 1999); (***. 1973); (Voinea, 1989) The ideal solution is a curved bar of same firmness (Hübener, 1970).

The circular section is particularized in fig. 1 and the following equations:

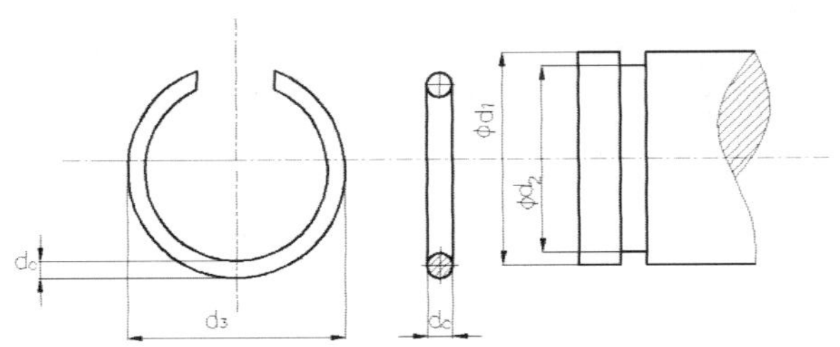

Fig.1. Circular sectionof the circlip

$$\frac{1}{r}=\frac{M_b}{EI} \quad (1)$$

If a bar with a neutral radius of curvature r, already curved, is deformed on a radius ρ, the equation is:

$$\frac{1}{r}-\frac{1}{\rho}=\pm\frac{M_b}{EI} \quad (2)$$

Using the names for the neutral diameters, usual with circlips

$$r=\frac{1}{2}\cdot D_3\ ;\quad \rho=\frac{1}{2}\cdot Dl\ ;\quad \frac{2}{D_3}-\frac{2}{D_l}=\pm\frac{M_b}{EI} \quad (3)$$

$$\frac{1}{D_l}-\frac{1}{D_3}=-\frac{\sigma_b}{E\cdot d_c} \qquad \frac{1}{D_3}-\frac{1}{D_l}=-\frac{\sigma_b}{E\cdot d_c} \quad (4)$$

$$\sigma_b=\frac{(D_l-D_3)\cdot E\cdot d_c}{D_l\cdot D_3} \qquad \sigma_b=\frac{(D_3-D_l)\cdot E\cdot d_c}{D_l\cdot D_3} \quad (5)$$

$$\sigma_b=\frac{1{,}15(d_l-d_3)\cdot E\cdot d_c}{(d_l+d_c)(d_3+d_c)} \qquad \sigma_b=\frac{1{,}15(d_3-d_l)\cdot E\cdot d_c}{(d_l-d_c)(d_3-d_c)} \quad (6)$$

$$\sigma_b=\frac{M_b}{W}=\frac{P\cdot l\cdot 6}{d_c^2\cdot s}\,becomes\,P=\frac{\sigma_b\cdot d_c^2\cdot s}{6\cdot l} \quad (7)$$

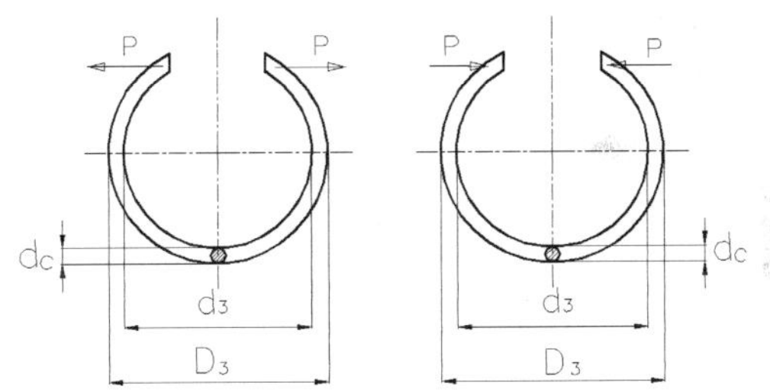

Fig.2. Forces acting on a circlip

3. COMPUTATION OF THE CIRCLIP

For the special cases mentioned above it is of interest the axial loading behavior (and axial load-carrying capacity) of the ring as well as its stability in the reserved groove (in the shift or housing)

Fig.2 presents the situation in which a machine part presses a circlip with an axial force. At first sight, shearing seems to condition for the drawing out of use of the ring, so that, at the beginning of the use of these machine elements it was very much insisted on this kind of calculation. It was observed that, because of the relationship between the depth of the groove and the thickness of the ring, the shear never takes place because at loadings under the maximum stress there takes place a loss of stability by deformation (as in fig.3a,b,c)

It is said that the ring is “inverted” The deformation is determined by the occurrence of a lever arm that modifies by a bending moment the shape of the ring that becomes conical. For a better understanding of the situation, the deformation (the characteristic angle ψ) is exaggeratedly enlarged. As to be seen, between the chamfer g and the level of the elastic deformation i of the groove edge, there appear the lever arm. It results in a displacement of the machine element adjacent on the distance f.

Similarly, the phenomenon appears in the case of rounded edges at adjacent machine elements. For sharp edges, plastic deformation of the groove edge (together with elastic deformation) contributes also to the apparition of the lever arm. For loads that determine a too high value of the angle ψ there appear permanent conical deformations or even the failure of the ring.

If the ring is considered an axial elastically element, the formula (8) is applied

$$P_R = C \cdot f \tag{8}$$

$$\frac{\pi \cdot E \cdot d_c^3}{6} \ln\left(1 + \frac{2 \cdot d_{mc}}{d_2}\right) = K \tag{9}$$

$$C = \frac{K}{h^2} \tag{10}$$

$$\psi = \frac{f}{h^2} \tag{11}$$

$$P_R = \frac{\psi \cdot K}{h \cdot S} \tag{12}$$

$$K^I = K \cdot \frac{E^I}{21000} \tag{13}$$

4. FEM RESULTS

The analyze was made for a ring of3.4 mm(***STAS 8436-69).
The applied load was scaled in 10 steps from 15 to 1500 N.
In figures . 4, 5 and 6. is presented the displacement of the ring for 150 and 300 N

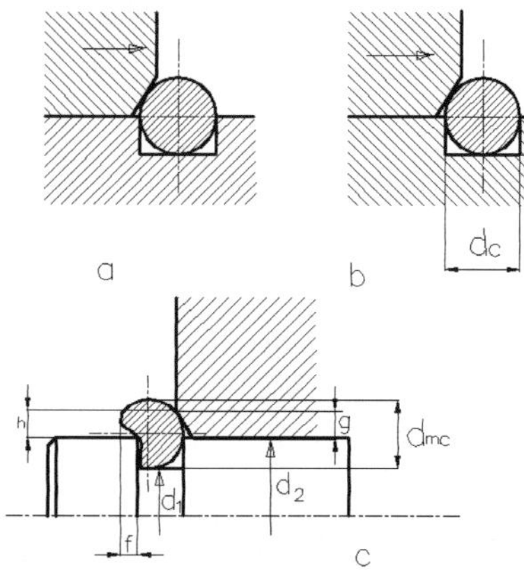

Fig.3. Loss of stability of the circlip

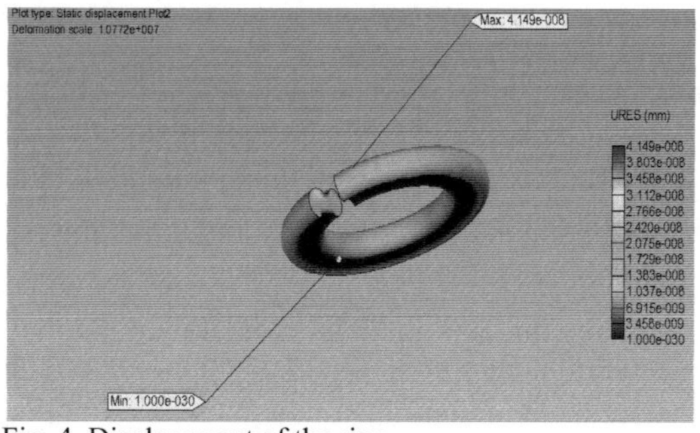

Fig. 4. Displacement of the ring

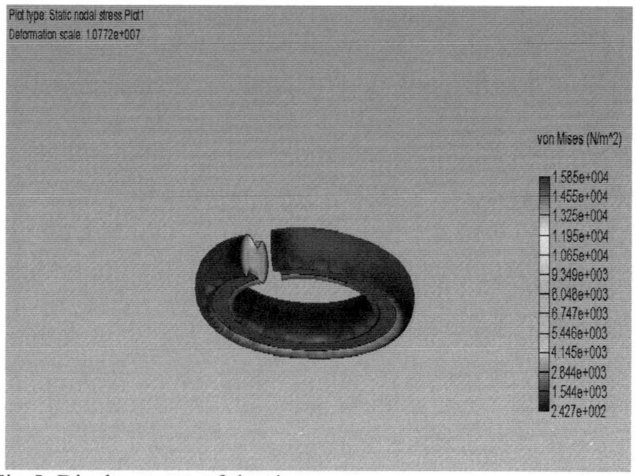

Fig.5. Displacement of the ring

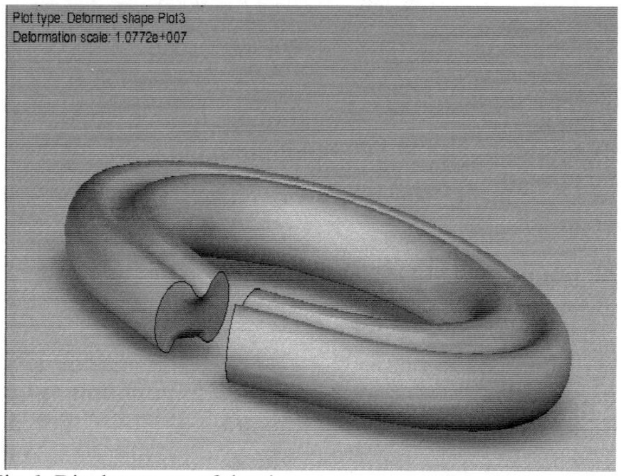

Fig.6. Displacement of the ring

5. CONCLUSIONS

For the reaching of high load-carrying capacities it is to be thus always aimed at that the effective lever arm is as small as possible h.

The conclusions of the paper can be directly applied in technical design and in the future if experimental research results are added, the reconsideration of the present standards regarding shape, dimensions, and materials can be made.

6. REFERENCES

Argeşanu, V., ş.a. (1999). *Element of mechanical engineering*, Ed. Eurostampa, Timişoara.

Hübener R.. (1970). *Seeger rings. A manual for the Constructor*, Seeger-Orbis Gmbh, Schneidhain/ Taunus.

Mesaroş-Anghel, V., Argeşanu, V., Mădăras, L., Cuc, A. (2006). *Rectangular section circlips/retaining ringsaxial load-carrying capacity considerations,* COMEFIM'8 The 8-th International Conference on Mechatronics and Precision Engineering, in Acta Technica Napocensis, series: Applied Mathematics and Mecahanics, 49 vol. IV, pp.843-850, ISSN 1221-5872, Technical University of Cluj-Napoca.

Voinea, R., ş.a. (1989). *Introduction in solid's mechanics with aplications in engineering*, Ed. Academiei Române, Bucureşti.

***. (1973) *Mechanical engineer's Manual, Materials, Material's Strength,Elastically Stability* , Ed. Tehnică, Bucureşti.

***STAS 8436 -69 *Wire Retaining Snap Rings for shafts and hole*

Annals of DAAAM for 2008 & Proceedings of the 19th International DAAAM Symposium, ISSN 1726-9679
Editor B. Katalinic, Published by DAAAM International, Vienna, Austria 2008
Make Harmony Between Technology and Nature, and Your Mind will Fly Free as a Bird
*Young Researcher and Scientist Paper / * Supervisor, Mentor*

ANALYSIS OF STRESS DISTRIBUTION IN ROOTS OF BOLT THREADS

ARYASSOV, G[ennady]* & PETRITSHENKO, A[ndres]

Abstract: *It is well known that axial load acting on a bolt is distributed between threads of threaded joints extremely nonuniformly. Most of existing calculation schemes of stress distribution in threaded joints is idealized and don't taking into account thread fit or tolerances in thread connections, which directly affect to precision of calculations. Due to thread fit or tolerances between the bolt and nut threads, solution of real stress distribution and stress concentration in thread connections is much more complicated and must be considered in reliable threaded connections, working within dynamic loads or made of sensitive materials to stress concentration. In this case, stresses in roots of first working threads can exceed nominal tensile stresses calculated by idealized calculation schemes and so, considerably reduce fatigue life of threaded connections.*
In this paper more exact mathematical models for analysis of stress distribution in threaded joints with thread fit, based on finite element method are proposed.
Key words: *threaded joints, finite element method, thread fit*

1. INTRODUCTION

Solution of nonuniform distribution of axial load between threads of threaded connections was first proposed by Russian scientist N. Zhukovski in 1902. He idealized that roots of threads are independent to each other and nonuniform distribution of axial load between independent roots of thread occur due to shear only.

More accurate solution of nonuniform distribution of axial load between roots of real threaded connections was reported by (Birger & Iosilevich, 1973). Roots of threads were considered as continuous and actual stress distribution in thread roots was defined by force flows, which cause tension of the bolt body, as well as bending and shearing of the threads. Solution of nonuniform distribution of axial load in some idealized threaded connections by finite element method was introduced and compared with experimental results by (Marujama, 1976). Later, (Strizhak et., 1999) and (Aryassov et al., 2000) in their studies showed, that stress distribution and stress concentration in threaded connections by existing calculation schemes depends upon loading schemes mainly and so cannot be extended to every kind of thread connections.

To overcome fore aft mentioned limits, more general calculation scheme of stress distribution and stress concentration is proposed. Proposed calculation scheme takes into account thread fit in thread connections and is based on more exact models of finite element method. Theory to use extended approach in calculations is given as well.

2. BASIS FOR CALCULATIONS

As far as threads in threads connections can be considered as a group of small recesses, in which circular stresses can be ignored in comparison with the axial and radial stresses, a task of calculation of threaded joints can be reduced to axisymmetric task of the elasticity theory.

Relation between displacements of nodal points and acting forces by finite element method is described as

$$[K]\{u\}=\{f\} \tag{1}$$

[K] – stiffness matrix,

$\{u\}$ and $\{f\}$ – vector-columns of nodal displacements and forces (external and contact forces).

Thread connection is meshed by square element with 16 degrees of freedom. Every nodal point has two linear and two angular generalized displacements. Stiffness matrix of square element is obtained by variation of potential energy of body deformation, which is described by

$$U=\int_S G\left\{\beta_1\left[\left(\frac{\partial v}{\partial y}\right)^2+\left(\frac{\partial w}{\partial z}\right)^2\right]+\beta_2\left(\frac{\partial v}{\partial y}\cdot\frac{\partial w}{\partial z}\right)+\left(\frac{\partial v}{\partial z}\cdot\frac{\partial w}{\partial y}\right)+\frac{1}{2}\left[\left(\frac{\partial v}{\partial z}\right)^2+\left(\frac{\partial w}{\partial y}\right)^2\right]\right\}dS \tag{2}$$

$\beta_1=\dfrac{1-\mu}{1-2\mu}$ *and* $\beta_2=\dfrac{2\mu}{1-2\mu}$ -nondimensional coefficients

v *and* w - displacements in direction of y and z axes,

μ – Poisson's ratio, G – shear modulus.

Displacements inside the element are to be approximated by interpolation polynomials as follows

$$\begin{aligned} v(r,z)&=\left(\alpha_1+\alpha_2 z+\alpha_3 z^2+\alpha_4 z^3\right)\left(\alpha_5+\alpha_6 r\right),\\ w(r,z)&=\left(\alpha_7+\alpha_8 r+\alpha_9 r^2+\alpha_{10} r^3\right)\left(\alpha_{11}+\alpha_{12} z\right) \end{aligned} \tag{3}$$

where

α - unknown coefficients of interpolation polynomials

z *and* r - local coordinate axes.

Displacements inside the elements with respect of generalized coordinates q_{ik} are

$$\begin{aligned} v(r,z)&=\sum N_i^v q_i^v;\quad i=1,4,5,8,9,12,13,16\\ w(r,z)&=\sum N_k^w q_k^w;\quad k=2,3,6,7,10,11,14,15 \end{aligned} \tag{4}$$

where

N_i^v *and* N_k^w - interpolation shape functions, according to kinematic relations by theory of elasticity.

Elements of stiffness matrix can be found by substituting equation (4) to expression of potential energy (2) and following variation (2) with respect of generalized coordinates q_{ik}.

Mathematical model for determination of load distribution in contact zone between threads is represented in Fig. 1.

For determination of load distribution, compatibility equations of displacements and equilibrium are used.

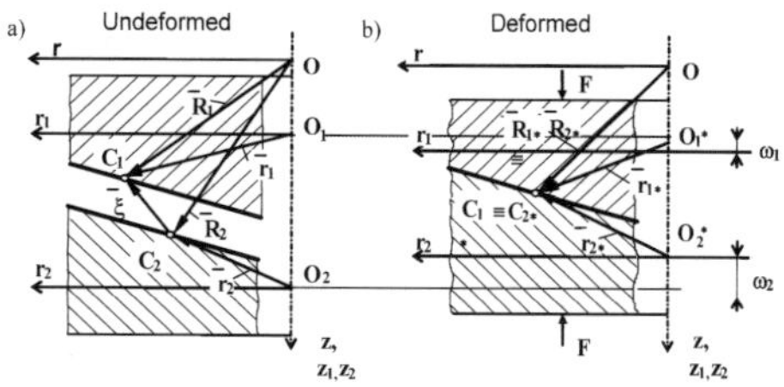

Fig. 1. Interaction between bolt and nut threads

Equation of equilibrium for one thread component can be expressed as

$$\bar{F} = \sum_{i=1}^{n} \int_{A_k} \bar{p}_i \cdot dA_k \tag{5}$$

where

p_i – contact pressure; i - number of thread root in contact,

A_k - area of contact zone.

The compatibility equation of displacement can be expressed by contact points C_1 and C_2 between bolt and nut thread as represented in Fig. 1 and are

$$\bar{\omega}_1 - \bar{\omega}_2 + \bar{\xi} = \bar{\delta}_2 - \bar{\delta}_1 \tag{6}$$

where

$\bar{\omega}_1 = \overline{OO_{1*}} - \overline{OO_1}$ and $\bar{\omega}_2 = \overline{OO_{2*}} - \overline{OO_2}$ - kinematical displacements in global coordinate system, $\bar{\delta}_1 = \bar{r}_{1*} - \bar{r}_1$ and $\bar{\delta}_2 = \bar{r}_{2*} - \bar{r}_2$ - displacements in local coordinate systems.

For determination of contact pressure p_i in finite element model, delta function p_{ik} is used.

Kinematical displacements are projected to the normal of the working plane of roots of thread using delta function and can be expressed as

$$\sum_{k=j=1}^{N} p_{ik} t_k \left[\left(K_{1jk}^{w} \cos\frac{\alpha}{2} + K_{1jk}^{u} \sin\frac{\alpha}{2}\right) - \left(K_{2jk}^{w} \cos\frac{\alpha}{2} + K_{2jk}^{u} \sin\frac{\alpha}{2}\right)\right] + \\ + F\left[\left(G_{1j}^{w} \cos\frac{\alpha}{2} + G_{1j}^{u} \sin\frac{\alpha}{2}\right) - \left(G_{2j}^{w} \cos\frac{\alpha}{2} + G_{2j}^{u} \sin\frac{\alpha}{2}\right)\right] = \omega_1^{n} - \omega_2^{n} \tag{7}$$

where

$K_{1jk}^{w}, K_{1jk}^{u}, K_{2jk}^{w}, K_{2jk}^{u}$ - influence function of contact pressure for bolt and nut threads,

$G_{1j}^{w}, G_{1j}^{u}, G_{2j}^{w}, G_{2j}^{u}$ - influence function of external forces for bolt and nut threads

t_k - width of k's block of shape function of contact pressure, α - angle between roots of thread

3. CALCULATION RESULTS

Thread M10 was used as calculation model and it was subjected to uniformly distributed stress σ_1=10 MPa on the bolt cross-section.

No of root	Load on the thread in per set of F			
	Present calcu-lation	Experimental (Marujama, 1976)	Analytical (Birger, 1973)	ANSYS program
1	40.93	37.35	36.50	36.48
2	23.12	22.83	23.75	21.97
3	15.07	16.78	16.50	15.64
4	11.56	13.40	12.75	13.05
5	9.32	9.64	10.50	10.82

Tab. 1. Contact load (pressure) in roots of thread

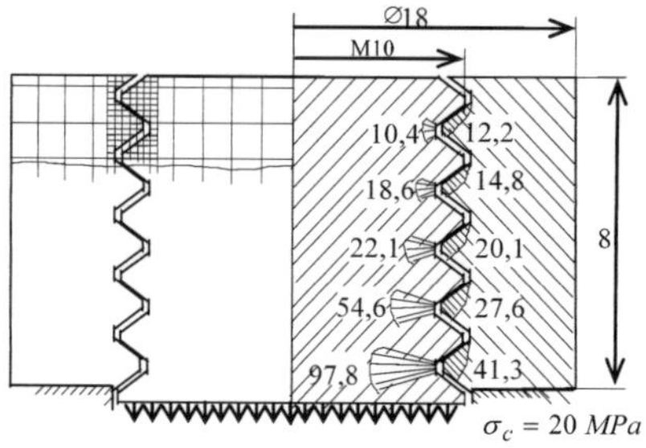

Fig. 2. Stress distribution in roots of thread

Modulus of elasticity E = 200 GPa, Poisson's ratio μ = 0.3. Influence of temperature and gravity were not considered in model. Model assumes unmovable hinged nodes on an axis r, passing within of free bolt edge, which corresponds to constrained deformation conditions appearing, when there are free threads of the bolt over the nut. Results of calculation of load distribution on threads are given in Table 1 and are compared with calculation results according (Birger & Iosilevich, 1973) and (Marujama, 1976) and with ANSYS program, where a standard triangular element with six nodal points were used.

Results of present calculation of stress distribution in roots of thread are represented in Fig. 2.

3-D model of the finite element with 20 degrees of freedom (every nodal point has 5 general displacements – two linear in-plane, one transverse linear and two angular displacement) was used as well. Fore aft mentioned 3-D model of finite element has not been used in calculations of threaded joints up to day. So calculation results, using 3-D model, have preliminary character and will be introduced hereafter in works of authors.

4. CONCLUSION

Calculations results of load distribution coincide with results which were made by (Marujama, 1976) and (Strizhak et al., 1999) and with results of experimental researches (Marujama 1974; 1976). Analysis of results shows, that loads on the first roots of thread calculated by the more exact model are higher as compared to the simplified ones by 10-15%.

Calculation results of stress distribution according to calculated load distribution between roots of thread show differences in stress state rate in comparison to stress state rate determined by theoretical coefficients of stress concentration.

In future calculated theoretical results will be compared to theoretical results calculated by more complex 3D finite element models. Theoretical results will be proved by experimental researches of threaded joints of ladder frames.

5. REFERENCES

Aryassov, G; Strizhak, V. & Zverkov, A. (2000). Influence of Scheme of Loading of Threaded Joints on Stress Concentration, pp. 38-40, *Proceeding of the 2th International DAAAM Conference*, Tallinn.

Birger, I. & Iosilevich, G. Threaded connections. (1973). Mashinostroenie, Moskva (in Russian)

Marujama ,K. (1976). Stress Analysis of a Bolt-Nut Joint by the Finite Element Method and the Copper-electroplating Method, *Bull. of ASME*, 1973 Vol. 16, No. 94, pp. 671-678; 1974 Vol. 17, No. 106, pp.442-450; 1976 Vol. 19, No. 130, pp. 360-368.

Strizhak, V; Aryassov, G & Penkov, I. (1999). Making More Precise of Calculation of Critical Threaded Joints. *Proc. of 10th World Congress on the Theory of Machine and Mechanism,* Vol. 6, pp. 2482-2487, ISBN 951-42-5293-4, Oulu, Finland, June 1999, Oulun Yliopisto

Annals of DAAAM for 2008 & Proceedings of the 19th International DAAAM Symposium, ISSN 1726-9679
Editor B. Katalinic, Published by DAAAM International, Vienna, Austria 2008
Make Harmony Between Technology and Nature, and Your Mind will Fly Free as a Bird

DESIGN AND TESTING OF A TRANSDUCER FOR MEASURING THE LOADS ON THE ROLLERS OF A CRANE CARRIAGE

ATANASIU, C[ostica]; ILIESCU, N[icolae] & PASTRAMA, S[tefan], D[an]

Abstract: *In this paper, a transducer designed for measurement of the loads acting on the rollers of a crane carriage is presented. The design process, the influence of the position of the applied load and calibration of the transducer are described. The load cell was tested on a 25 kN bridge crane for which the loads on the carriage rollers were determined in order to verify the strength structure.*
Key words: *Transducer, elastic element, strain gauge, stress.*

1. INTRODUCTION

The paper presents the design, testing and practical use of a transducer used to measure the load on the rollers of a crane carriage. The elastic element of the transducer is described extensively by Iliescu and Atanasiu (2006). It was conceived to measure loads up to 200 kN with accurate sensitivity. For the measurement range, the design of the elastic element was made to obtain a variation of the strain with the applied load as close to a linear one. Also, the maximum stress was limited to a value of 0.75 of the yield limit σ_y.

2. THE LOAD CELL

The chosen material was a Romanian chromium alloy steel 13CN35 with a yield limit of 670 *MPa*.

The elastic element which is a simply supported beam is presented in Fig. 1 and has the following geometry:

- Span: $L = 150$ *mm*;
- Height of the cross section: $h = 40$ *mm* in the 40 mm length middle part and $h_1 = 35$ *mm* in the rest of the length;
- width of the cross section: $W = 60$ *mm* (it was chosen taking into account the width of the track rail of the roller).

As one can see in Fig. 1, the upper face of the middle part was machined with a cylindrical surface and has two arms on which the roller is supported. In this way, a fixed position on the surface of the element is ensured.

Four strain gauges were glued on the elastic element (Fig. 1) at the distance $c = 40$ *mm* from the supports. Two gauges were placed on the upper face (denoted as T_1 and T_3) and the other two (T_2 and T_4) – on the lower face with their axis in the longitudinal central plane (Fig. 1).

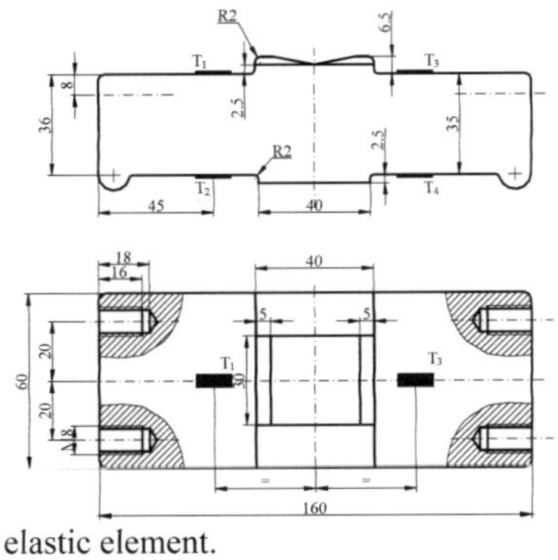

Fig. 1. The elastic element.

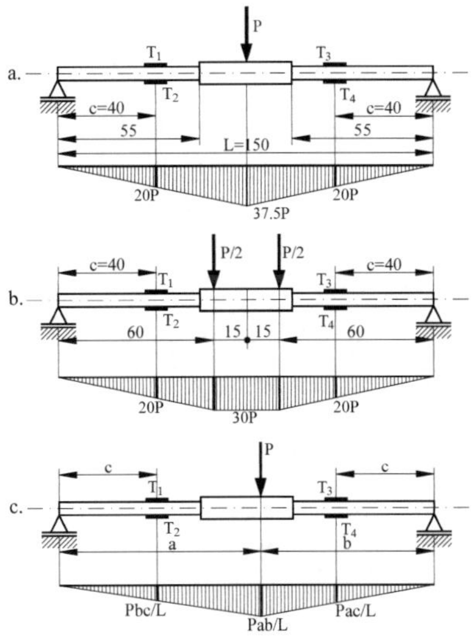

Fig. 2. The load cases.

The load cases and the corresponding bending moment diagrams are presented in Fig. 2. The measured load may be applied either in the middle of the beam, equally distributed on the two arms or eccentrically with respect to the middle.

The normal stress due to the bending of the elastic element under load is determined with the well known Navier formula (Timoshenko & Goodier, 1970):

$$\sigma = \frac{M}{I_y} \cdot z \tag{1}$$

where M is the bending moment, I_y is the second moment of area of the cross section and z is the coordinate of a point in the cross section with respect to the system of axes that passes through the centroid.

In order to verify the limitation of the stress imposed by design, the maximum stress values were calculated for the three load cases.

In the central section of the elastic element, the stress has the following values:

- For the case from Fig. 2, a:

$$\sigma_{max} = \frac{6 \cdot 37.5P}{Wh^2} = 2.34 \cdot 10^{-3}P \tag{2}$$

- for the case from Fig. 2, b:

$$\sigma_{max} = \frac{6 \cdot 30P}{Wh^2} = 1.875 \cdot 10^{-3}P \tag{3}$$

Since the diameter of the roller is bigger than the diameter of the cylindrical zone of the elastic element, the load is applied on the two arms, as in load case b.

In the section where the height of the cross section is changed, the maximum stress from load case a. has a lower value than in the middle:

$$\sigma_{max} = \frac{6 \cdot 27.5P}{Wh^2} = 2.25 \cdot 10^{-3}P \quad (4)$$

A maximum value of the stress σ_{max} = 450 *MPa* is obtained for a measured load of 200 *kN*, meaning that the condition imposed by design ($\sigma_{max} < 0.75\ \sigma_y$) is fulfilled.

In order to measure the loads using the designed transducer, the stress must be calculated in the points where the strain gauges are placed.

From the bending moment diagram from Fig. 2, b and using equation (1), one can calculate the value of the stress as:

$$\sigma_{max} = \frac{6 \cdot 20P}{Wh^2} = 1.63 \cdot 10^{-3}P \quad (5)$$

while the strain is:

$$\varepsilon = \frac{\sigma}{E} = 7.76 \cdot 10^{-9}P \quad (6)$$

If the strain gauges are connected in full bridge (Theocaris et al., 1976), the strain read by the measurement equipment is:

$$\varepsilon_{measured} = 4\varepsilon = 31.04 \cdot 10^{-9}P \quad (7)$$

When the two arms take over different parts of the load, and the overall load is eccentric to the center of the beam (Fig. 2, c) the strains indicated by the gauges are:
- For gauges T_1 and T_2:

$$\varepsilon_{1,2} = \pm\frac{6Pbc}{LEWh^2} \quad (8)$$

- For gauges T_3 and T_4:

$$\varepsilon_{3,4} = \pm\frac{6Pac}{LEWh^2} \quad (9)$$

For the connection of the gauges in full bridge, the strain measured in this case is:

$$\varepsilon_{measured} = \varepsilon_1 - \varepsilon_2 + \varepsilon_3 - \varepsilon_4 = \frac{12Pc}{EWh^2} = kP \quad (10)$$

where $k = 31.04 \cdot 10^{-9}$.

From eq. (10) it can be noticed that the obtained strain is proportional with the load and does not depend on the eccentricity. In this way, deviations from the load symmetry of the elastic element do not influence the values of the strain and eq. (6) is recovered in the case of eccentric load.

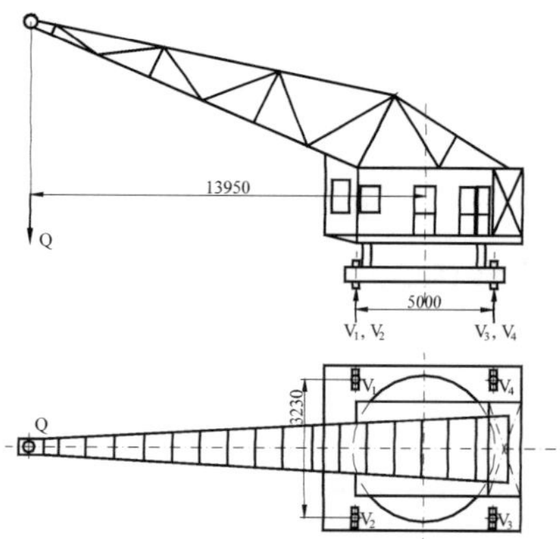

Fig. 3. The studied crane.

Loading case	Obtained loads on each roller [N]			
	V_1	V_2	V_3	V_4
I	80000	36700	49200	91200
II	33300	92000	106200	51400

Tab. 1. The obtained results

2. CALIBRATION AND TESTING OF THE TRANSDUCER

For the four rollers of the crane carriage (V_1, ..., V_4), four transducers were manufactured and placed on the track rails, under the rollers. Since the accuracy of machining is not the same for each transducer, they should be calibrated in order to determine the relationship between the strain ε and the applied load P (Sandu & Sandu, 1999).

For calibration, the transducers were loaded with known forces and the strains were measured with the strain gauge device. Before the calibration process, each transducer was loaded and unloaded in order to check the feed-back of the strain gauges and to avoid unwanted local permanent strains of the material. After the final measurements, the transducers were again calibrated. The differences between the initial and final calibrations were very small.

Finally, tests were performed in order to verify the behavior of the transducers under load. They were used to measure the load on the rollers of a crane carriage (Fig. 3). The maximum load that can be lifted by the crane is 25 *kN*. The experimental tests were necessary in order to perform a verification of strength structure of the crane.

The measurements were performed for two loading cases:
Case I: own weight of the carriage and the arm of the crane at 90° with respect to the axis of the track rail;
Case II: a load of 25*kN* at the hook of the crane (applied using another crane) and the arm of the crane at 90° with respect to the axis of the track rail.

3. RESULTS AND CONCLUSIONS

The loads on each roller of the carriage were determined using the measured strains from eq. (10) and the calibration curves for each transducer. The obtained results are presented in Table 1.

From Table 1, the following data can be obtained by adding the loads on the rollers for each loading case separately:
- For loading case I, the own weight of the carriage is calculated as: G_{own} = 257100 *N*.
- For loading case II, the own weight of the carriage and the weight in the hook is G_{total} = 282900 *N*.

The obtained results can be verified by determining the weight in the hook as $G_{hook} = G_{total} - G_{own}$ = 25800 *N*.

The difference between the obtained weight and the real one (25*kN*) is 1.2%, showing thus that the accuracy of the designed transducers is very good and they can be used further to measure elevated loads with precision.

4. REFERENCES

Iliescu, N. & Atanasiu, C. (2006), *Metode tensometrice în inginerie (Strain gauge measurements in engineering)*, AGIR Publishing House, Bucharest (in Romanian).

Sandu, M. & Sandu, A. (1999), *Captoare cu traductoare rezistive. Proiectare. Aplicatii (Strain gauge transducers. Design. Applications)*, Printech Publishing House, Bucharest, (in Romanian).

Theocaris, P., et al. (1976), *Analiza experimentală a tensiunilor (Experimental stress analysis)*, Vol. 1, Technical Publishing House, Bucharest (in Romanian).

Timoshenko, S.P. & Goodier, J.N. (1970) – *Theory of Elasticity*, McGraw-Hill Kogakusha, Ltd, New York. [1].

Annals of DAAAM for 2008 & Proceedings of the 19th International DAAAM Symposium, ISSN 1726-9679
Editor B. Katalinic, Published by DAAAM International, Vienna, Austria 2008
Make Harmony Between Technology and Nature, and Your Mind will Fly Free as a Bird

AN ANALYTICAL INVESTIGATION OF THE SHARED FRICTION LOADS OF SPUR GEAR PAIRS

ATANASIU, V[irgil]

Abstract: *The objective of this paper is to outline an alternative methodology to evaluate friction forces as a function of the shared normal load and instantaneous friction coefficient of the contacting teeth. The shared normal load between tooth pairs in mesh is computed in relation to the time-varying mesh stiffness. The calculus procedure of the instantaneous friction coefficient is developed according to the experimental results. The comparisons of the variation of the friction forces for a mesh cycle are presented for different calculus methodologies.*

Key words: *friction, gear, tooth, load, stiffness.*

1. INTRODUCTION

The friction force is considered as a dynamic excitation source in the models of gear systems (Velex & Sainsot, 2002). Advanced dynamic models are based on the assumption that the load is equally shared among the tooth pairs in contact (Vaishya & Singh, 2003). At the same time, the Coulumb friction is assumed with a constant coefficient of friction μ. These assumptions don't lead to a realistic dynamic model of gear pairs. The instantaneous friction coefficient is difficult to calculate accurately. (Martin, 1978) has provided an extensive review of the existing equations for the coefficient of friction in gearing, but the numerical results are not in good agrement with the experimental ones.

The objective of this paper is to propose an alternative calculus methodology of the frictional forces F_f on the gear tooth surfaces, by considering instantaneous friction coefficient and the sharing normal load of the meshing cycle in relation to the variable mesh stiffness. An analytical formulation of the variable friction coefficient is included in the analysis of the shared friction loads on the path of contact.

2. INSTANTANEOUS COEFFICIENT OF FRICTION

It is generally agreed that the coefficient of friction does not remain constant during a meshing cycle. The experimental values of the instantaneous coefficient of friction at the meshing teeth were obtained by (Radzimovsky & Mirarefi, 1976) and (Rebbechi et. al., 1996). On this basis, the curve drawn in Figure 1 is considered in the analytical model for the evaluation of the instantaneous coefficient of friction. In this figure, A, E, C denote the beginning of engagement, end of engagement and pitch point of the pinion, respectively. This model satisfies the boundary conditions of the sliding friction.

The pitch point C is taken as origin and the curve corresponding to the recess path can be approximately represented by the following expression proposed by (Rao, 1979)

$$\mu = a_1 \frac{y}{l} + a_2 \frac{y^2}{l} \tag{1}$$

where the coordinates y and l are shown in Figure 1.

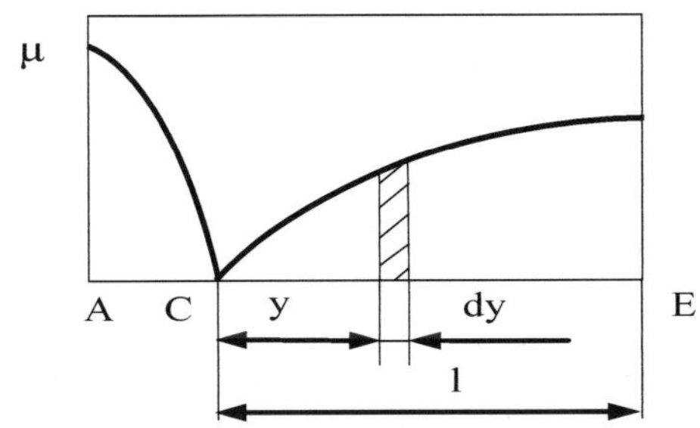

Fig. 1. Variation of friction coefficient of the analytical model.

The constants a_1 and a_2 can be established by considering Eq. (1) and the boundary conditions for the equation. From these conditions results $a_1 = 3\mu_m$ and $a_2 = -1.5\mu_m$. This methodology is also fully valid for the aproach length, by considering the specific geometrical parameters of this meshing zone. The analytical relation to determinate the average coefficient of friction μ_m in gear teeth are derived from roller and gear tests. Thus, in (ISO/TR 13989, 2000), the calculation formula of the friction coefficient can be used especially for spur gear pair with the pitch point in the middle of the path of contact. When $l_a \neq l_g$ it is necessary to consider the specific geometry parameters of the gear pair which is analyzed, where $l_a = AC$ and $l_g = CD$. In this way, the following expression is proposed

$$\mu_m = \frac{\mu_{ma} l_a + \mu_{mg} l_g}{l_a + l_g} \tag{2}$$

where μ_{ma}, μ_{mg} are calculated in the middle length of the approach path and of the recess length, respectively.

3. SHARED FRICTION FORCES

Usually, the shared factor of the friction force corresponding to a tooth pair is defined as

$$c_{f_I} = \mu_m F_n^I / F_n \tag{3}$$

In Eq.(3), F_n^I represents the normal force corresponded to the single gear pair I, and F_n is the total normal force, where $F_n^I = 0.5F_n$ is for double-tooth contact and $F_n^I = F_n$ for single-tooth contact. The instantaneous distribution factor c_{fi_I} is proposed as follows

$$c_{fiI} = \mu c_I \tag{4}$$

where μ represents the instantaneous coefficient of friction and

c_I is the shared factor of the normal load, where

$$c_I = k_s^I / (k_s^I + k_s^{II}) \quad (5)$$

and k_s represents the individual variable mesh stiffness of a gear pair (Atanasiu & Leohchi, 2005).
During the engagement cycle, the normal contact load does not remain constant. The load variation is mainly caused by the following factors: (i) the alternating engagement of single and double pairs of teeth; (ii) the variation of the mesh stiffness along the line of action. In the paper, the tooth stiffness along the path of contact is calculated by using an exact analytical procedure (Atanasiu & Leohchi, 2005).

4. CASE ILLUSTRATIONS

The specifications of the gear pairs which are selected in these illustrations are shown in Table 1, where x_1, x_2 are the addendum modification coefficients and ε_α is the contact ratio. These are for spur gear pairs having the number of teeth $z_1 = 18$ and $z_2 = 42$; face-width of gears, $b = 20$ [mm], tooth module $m = 3$ [mm] and center distance, $a = 90$ [mm]. Additionally, the operational parameters are chosen as: nominal torque, $T_1 = 40$ [Nm] and pinion speed, $n_1 = 3000$ [rpm].
A computer program is developed to compute the variable mesh stiffnes, shared normal load, instantaneous coefficient of friction and shared friction forces along the line of action.
An example of the variation of the friction coefficient is shown in Figure 2 as a function of tooth contact position. The values of the average coefficient of friction μ_m in gear teeth are shown in Table 1 in relation to the geometrical characteristics

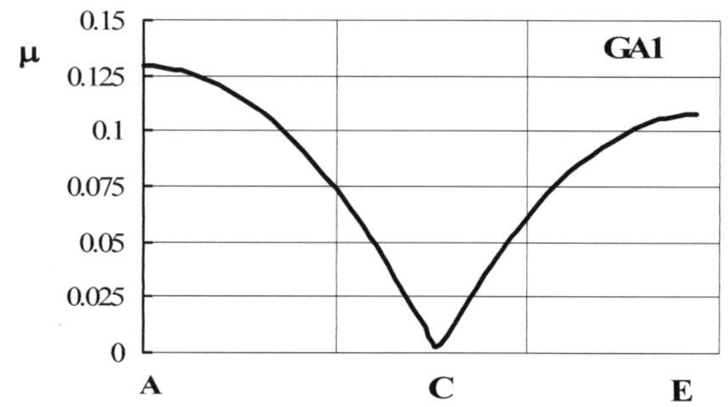

Fig. 2. Variation of the predicted friction coefficient.

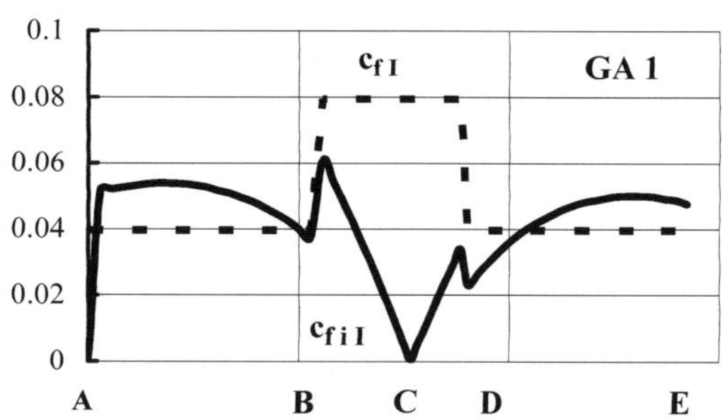

Fig. 3. Variation of the friction load factors c_{f_I} and c_{fi_I} .

Gear pairs	x_1	x_2	ε_α	AB [mm]	AC [mm]	μ_m
GA1	0	0	1.63	5.55	7.63	0.79
GA2	+0.6	-0.6	1.49	4.49	3.29	0.73

Tab. 1. Specifications of the gear pairs

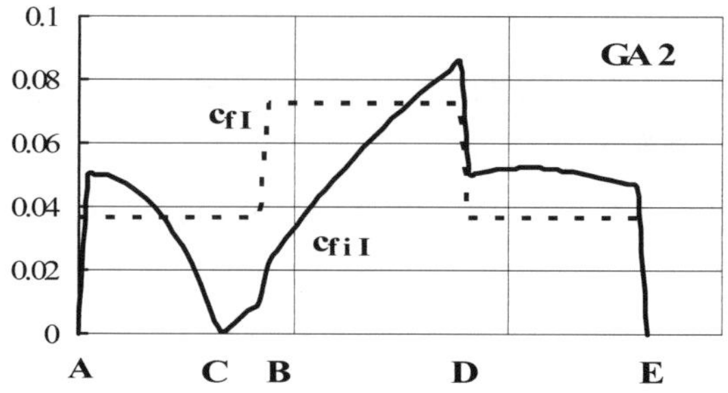

Fig. 4. Variation of the friction load factors c_{f_I} and c_{fi_I}

of gear pairs. These numerical results are in good agreement with the experimental results of the coefficient of friction obtained by (Rebbechi et al., 1996).
Figures 3 and 4 present the variation of the factors c_{fI} and c_{fi_I} corresponding to the friction forces along the engagement line, where BD represents the single tooth contact and AB and CD are the segments of double tooth contact. The analysis of these results permits to underline the effects of instantaneous coefficient of friction and the variable mesh stiffness on the shared friction load between pairs in mesh in relation to the geometrical parameters of gears.

5. CONCLUSIONS

This paper proposes an alternative calculus methodology for analysis of the friction load of contacting teeth as a function of the instantaneous friction coefficient and shared normal load. The time-varying mesh stiffness is included in this investigation. These formulations permits to obtain accurate numerical results in relation to the geometrical parameters of gear pairs and operation conditions. The further research should include the friction load into the dynamic model of spur gears in order to quantify their effect on dynamic characteristics.

ACKNOWLEDGEMETS

This paper is based on the financial support of the National University Research Council of Romania, grant ID_296.

6. REFERENCES

Atanasiu, V. & Leohchi, D. (2005). *Dynamic Analysis of Spur Gears with Tooth Profile Modifications*. Proc. of the Ninth IFToMM International Symposium on Theory of Machines and Mechanisms, Bucharest, Romania, 313-318.
Martin, K. F., (1978). *A Review of Friction Predictions in Gear Tooth.* Wear, 49, 2, 201 – 238.
Radzimovsky E. I. & Mirarefi A. (1975). *Dynamic Behaviour of Gear Systems and Variation of Coefficient of Friction and Efficiency During the Engagement Cycle.* ASME Journal of Engineering Industry, 97, 4, 1247-1281.
Rao, A. C. (1979). *Gear Friction Coefficient and Forces.* Wear, 53, 87-93.
Rebbechi, B., Oswald, F. B. & Townsend, D. P (1996). *Measurement of Gear Tooth Dynamic Friction.* DE-Vol.88, Power Transmission and Gearing Conf., ASME, 355-363.
Vaishya, M. & Singh, R. (2003). *Strategies for Modeling Friction in Gear Dynamics.* Journal of Mechanical Design, Transaction of the ASME, Vol.125, 383-393.
Velex, P. & Sainsot, P. (2002). *An Analytical Study of Tooth Friction Excitations in Spur and Helical Gears.* Mechanism and Machine Theory, 37, 641-658.
ISO/TR 13989 (2000). *Calculation of Scuffing Load Capacity of Cylindrical, Bevel and Hypoid Gears.*

Annals of DAAAM for 2008 & Proceedings of the 19th International DAAAM Symposium, ISSN 1726-9679
Editor B. Katalinic, Published by DAAAM International, Vienna, Austria 2008
Make Harmony Between Technology and Nature, and Your Mind will Fly Free as a Bird

TAX BURDEN IN BUSINESS CLIMATE

AVRAM, L[aura]; SAVU, L[ucian]; VANCEA, S[maranda]; IGNAT, A[ndreea]; HORJA, M[onica] & AVRAM, C[alin]

Abstract: *Due to economic expansion and social role of the state, but also because of continuous growth of public expenditure, Romanian tax system tended to increase the rates of existing taxes and fees and to set new taxes which means increasing the tax burden. An increase of fiscal pressure will cause taxpayers to seek solutions to reduce tax burden resorting to tax evasion*
In this paper we intend to analyze the way in which the modification of tax burden can influence the dimensions of tax evasion. Can we consider that the flat tax of 16% for earnings in Romania is an optimal tax?
Key words: *tax burden, tax evasion, The Laffer curve.*

1. INTRODUCTION

The intensity with which revenues are taken from individuals and legal entities, or the whole society through taxation, is tax burden and indicates the extent to which the nominal incomes of the population are adjusted through taxation, and show the extent or degree to which the state budget procure their income through taxes. We established that fiscal pressure represents a part of the taxpayers' income that they compulsorily give up in favour of the sate. These mandatory taxes established and collected through legal coercion by the state finance the state budget, social insurance, local budgets and public funds.

2. BODY OF PAPER

Tax burden, made visible by the **fiscal rate**, is calculated as a "ratio between the fiscal returns, including social insurance contribution, over a certain period, usually a year, AND the value of gross domestic product (GDP), achieved in the same period by a national economy" (Hoanta, 2000), and shows the share of GDP taken from the public budget to cover public expenditure.

For legal persons, the tax burden will be determined as a ratio between the total of tax payments supported by the firm and the value added obtained by the legal person' (Lepadatu, 2007).

Each tax can modify the taxpayers' economic and social behavior because the tax will have an impact on his/her expenditures or savings. Reducing the available income as a result of paying the tax the taxpayers' ability to obtain savings will be automatically reduced, and implicitly its consumption. However, this influence manifests itself differently according to the social category involved.

Apart form the fiscal rate at the state or legal person level, specialists in the field see as relevant in explaining the growth of tax evasion as a result of tax burden the **psychological tax burden** which represents the psychological tolerance threshold acceptable to the taxpayer. This tolerance to payment represents the **individual tax burden**, calculated as ratio between total levy and the total income obtained before tax payments (Buziernescu, 2001). This ratio expresses the "sacrifice" approved by law which has to be made by the taxpayers from the income over a period, usually a year. Thus, this indicator allows for the subjective elements of social life to enter economic calculus. The higher this pressure the higher the tendency toward tax evasion. However, it is difficult to determine which is the ideal level of individual tax burden to which taxpayers can be submitted to because it is influenced by random elements which depend on the taxpayers' education, their mentality, but also by the diversity of the levy, the volume of state supported social services, taxpayer's fiscal subjectivity (direct tax payments are visible and their amount can be easily determined, while the indirect tax payments included in the price of products and services are not that obvious to the taxpayers).

Given the fact that analysts dealing with problems (Basiat, Keynes) related to tax burden assessed the influence of the taxes on the taxpayers' behaviour these have always tried to establish an optimal tax system to minimalize financial pressure. In our opinion, of all the researches in the field, the American Arthur Laffer's work is the most relevant.

The Laffer curve (Laffer, 2004) illustrates the basic idea that changes applied to tax rates have two effects on taxes: the arithmetic effect and the economic effect. **The arithmetic effect** refers to the fact that if tax rates are lowered, taxes will be lowered by the amount of the decrease in the rate. The reverse is true for an increase in tax rates. **The economic effect**, however, recognizes the positive impact that lower tax rates have on work, output, and employment—and thereby the tax base—by providing incentives to increase these activities. Raising tax rates has the opposite economic effect by penalizing participation in the taxed activities. The arithmetic effect always works in the opposite direction from the economic effect.

According to Laffer's theory, a rise of tax burden does not necessarily determine the rise of the taxes, but on the other hand, a lowering of fiscal pressure creates favorable conditions to increase fiscal incomes. This increase of fiscal incomes as a consequence of tax rate cut, even by only one percent, determines, at least psychologically, a lowering tendency of tax evasion from the taxpayer's side. But a reduction of tax burden can be difficult because, a sudden tax cut by a significant percent can lead to a budgetary deadlock, while a timely moderate reduction of tax burden is sensed later and the tax payers' behaviour remains relatively unaffected.

If the existing tax rate is too high, then a tax-rate cut would result in increased taxes. The economic effect of the tax cut would outweigh the arithmetic effect of the tax cut.

A moderate tax burden will maintain the taxpayers' interest in working or extending the business and the attraction to tax evasion is diminished. At the other end an unbearable high tax burden will discourage supplementary work, will discourage the legal person and will encourage tax evasion.

In this paper, we want to use Laffer's curve as a tool to analyze the economic effect of tax rates cut, because this theory shows the complexity of the economic effect of taxation by analyzing the ways in which different factors interact.

To try to demonstrate the usefulness of Arthur Laffer's theory, we'll analyze on the tax due for individual and legal entities income in Romania. Thus, in the year 2004, profits made by a company is taxable with a fixed percentage rate of 25% and earnings of employees are taxable by progressive tax rate, respectively from 18% to maximum 40% for revenue higher than 13 million ROL, which represented approximately 325 Euros. Also wages generate employer contributions of 33.25% and employee contributions of 17%. In the year 2005 a single tax rate of 16% was introduced both for legal entities profits and for individuals, without affecting other contributions generated by the salaries. What was the impact of this tax diminishing?

In order to answer the question we determined the net salary obtained from an average salary of 10 million ROL (approx. 250 euros) in 2005 compared with 2004.

In 2004 when the progressive tax rate was applied the tax on salary was of 21.92%, which exceeds by 5.92% the unique tax rate. As a result of introducing the unique tax rate, the tax burden of the employees lowered from 30.15% to 26.28%, which represents a decrease of 3.87%.In all, the tax burden generated by the salary has decreased by almost 3% in 2005 as compared to 2004. In 2004, in the case of a 20 million ROL salary the tax on salary reached a maximum of 40%. In 2005, this tax rate cut to the unique tax rate of 16% determined a reduction of the tax on salary by 13.1% and the tax burden decreased by 7.53% for the employee. Finally, the entire tax burden from a 20 million salary decreased by 5.64%. It can be noticed that in the case of a double salary the tax burden decreases by almost double. Thus, the higher the incomes the lower the tax burden is. In other words, a cut of the tax on salaries from the progressive tax rates - that reached the maximum tax rate of 40% - to the unique tax rate of 16% favored the ones with higher incomes.

On the basis of the official reports issued by the Ministry of Finance we noticed that with the setting of the unit tax rate of16% on salaries had as a first effect the decreasing of the income receipts by the state budget from these taxes. Thus, the collected incomes receipts from the tax on salary, decreased by 5.27% in the first 8 months of 2005 as compared to the same period of the previous year. This can be seen as a result due to the fact that at the employer's level there were no cuts in taxation, the social contributions levied on them being 32.5%. If we add social contributions levied on the employees, a contribution of 17% we get 49.5% total contributions for social security only. Therefore, taxation of wages in Romania is among the highest in Europe. As a result, the legal entities still resort to the use of the "black labor", especially since the fines charged in the case of discovering the use of the undocumented labor are much smaller than "earnings" obtained from undeclared incomes from salaries.

Regarding the tax on profit, the budgetary income from the profit increased by 7.6% in 2005 as compared to 2004. Nevertheless, the 9% profit tax rate cut was not incentive enough for the legal entities. The fiscal relaxation for the economic entities is an illusion because the reduction of tax rate on profit did not counter-balance the raising of indirect taxes which determined an increase in raw materials, materials, goods and services and also a decrease in the gross profit. At the same time, the employer continues to pay the same high taxes to social insurances.

Nevertheless, after only 8 months since the introduction of the unique tax rate on income, the collected budgetary income increased by 15.2% as compared to the same period of the previous year.

The Romanian state, seeing that the tax burden for the legal entities regarding the social security contributions is too high, establishes a fiscal policy of reducing it. A first reduction of 2.75% is achieved in 2006, but the Romanian state establishes a new tax of 0.75% for leaves and medical allowance. Even in these conditions, in 2006 the collected fiscal revenues increased as compared to 2005 by 44.57% regarding the income tax and an increase of 21.71% in the case of profit tax. At the moment, the social contributions levied on the employer represent 27.6% a 5.63% decrease as compared to 2004. Social contributions levied on the employees represent 16.5% a 2.5% decrease as compared to 2004. Ideally, on the basis of the above presented pattern, we can state that the fiscal relaxation regarding social contributions will lead to an increase of revenues for the budget. We believe that this raising of the budgetary revenues from the income tax may also be a result of the newly gained visibility of a part of the underground economy, namely from the "black labour". This positive effect caused by the tax cut may be also noticed by the increasing number of documented work contracts which are registered at the territorial labour bureau.

These results demonstrate Laffer's theory according to which a reduction in tax rates may cause an increase in collected revenues. Although there were some opinions against the desirability of introducing the unique tax rate, we consider that this reform of the tax system was a positive one which did not affect the state budget

3. CONCLUSION

In conclusion, the introduction of the unique rate of taxation had a favorable effect; the fiscal reduction generated an economic effect and not an arithmetic one, as showed by Laffer's theory on collected budgetary revenues. The increasing of the collected budgetary revenues with 20-40% compared to the period previous to introducing the single tax rate, shows that this relaxation of fiscal policy was stimulated the taxpeyers voluntary compliance and we can consider that the flat tax of 16% is approaching an optimal tax. Also it is proper for the Romanians mentality.

Even the unique tax rate to 16 % had as an effect the raising of fiscal revenues to the state budget we consider that tax burden levied on the natural and legal entities are still very high in Romania as compare to the tax burden in other EU countries Today in Romania is more than 300 duties, taxes and contributions which must pay by the legal and natural persons in order to undertake an economic activity. Of these, only five categories of taxes make more than 80% of state revenues. These are taxes on income and profit, social contributions, VAT, excise and customs duties. Because of these, we believe that the government must review the policy and tax system, reduce the number of duties, taxes and contributions because, as we saw from the undertaken analysis, tax cuts can lead to an increase of the state budget, by encouraging taxpayers to declare all taxable incomes and by not resorting to tax evasion.

In the future, this study can be developed through the creation of a mathematical model that determines the impact of the reduction with another one or two percent of this tax, currently 16%, on revenues collected to the budget. The outcome of such reductions will increase the revenue collected to the budget or, on the contrary, a decrease?

4. REFERENCES

Balaban, C. (2003). *Tax evasion, Theoretical and juridical practice controversial aspects*, Rosetti Publising House, ISBN 973-8378-47-8, Bucuresti

Buziernescu, R. (2001). *Tax evasion in Romania*, Universitaria Publising House, ISBN 973-8043-50-6, Craiova

Hoanta, N. (2000). *Economie and public finance,* Polirom Publising House, ISBN 973-683-459-X, Bucharest

Laffer, A. (2004).The Laffer curve: Past, Present and Future, *Availablefrom:*http://www.heritage.org/Research/Taxes/bg1765 *Accessed:* 2008-07-15

Lepadatu, G. (2007) Tax burden and their quantification indicators, *The Economic Tribune,* No.42, 5 pages (33-37), ISSN 1018-0451

Annals of DAAAM for 2008 & Proceedings of the 19th International DAAAM Symposium, ISSN 1726-9679
Editor B. Katalinic, Published by DAAAM International, Vienna, Austria 2008
Make Harmony Between Technology and Nature, and Your Mind will Fly Free as a Bird
2nd DAAAM-ICAT

RP TECHNOLOGY: WHY LIMITED UPTAKE IN SME'S?

AYYAZ, A[hmad]; MARUF, H[asan] & VAN VOORTHUYSEN, E[rik] J[an]

Abstract: *Global competition for manufactured products places great emphasis on, short lead time, high quality and mass customization and it have become essential attributes for companies to stay competitive in world markets. The availability of a growing number of Time Compression Technologies (TCTs) has enabled companies to reduce the time to market as well as to reduce the development costs associated with their products.*

An anomaly that this paper discusses is that in spite of all its benefits, only a small proportion of the design and development community uses rapid prototyping technologies. If these technologies are so powerful than why is their use so limited? This applies specifically to small to medium enterprises in developing countries. This paper highlights the critical factors which can be considered as root cause for the slow uptake of these technologies. This paper also provides future directions to increase the use of these technologies, to assist the SME sector especially, to make them able to remain in the mainstream of global competition.

A case study of automotive cluster in Pakistan is also presented to elaborate the case.

Key words: *Time Compression Technology, Rapid Prototyping, Rapid Manufacturing*

1. INTRODUCTION

Product Development process has evolved significantly. With more powerful softwares and proliferation of traditional prototyping processes to smart 3D prototyping, the meaning of the product development process have been totally changed. Competition for manufactured products has been intensified in recent years. Short lead time and high quality has become essential competitive dimension to stay competitive in world markets. Heavy penalties are resulting for the companies failing to take decision in the early product development phase in the form of retooling and extra labour and loss of market share due to late entry into the markets.

In this scenario of increased competitiveness, one solution is the adoption of Time Compression Technologies. Bhattacharya & Jina (1997) report that TCTs has received increasing attention from academics and practitioners both in the analysis of the mode in which business operates and redefined the process to make them more effective. Time Compression Technology can mean any technology that can improve a design and manufacturing process to achieve better quality in minimum time (Kengpol *et, al* 2001). One example of time compression technology is rapid prototyping (RP) which can be applied to shorten design and development time. Other examples are rapid manufacturing (RM) and the use of CNC.

Boeing has made extensive use of RP machines to produce parts, tooling and manufacturing aids for the F18 and the other military aircraft (Ogando, 2007).Stratysys a supplier of additive manufacturing system revealed that 32 of the parts on its latest large format fused deposition modelling system (FDM) were produced via FDM, with saving of an estimated $100,000 of tooling costs and at least six weeks of tooling lead time (Design News, 2007).

Now the ultimate question which rise, that if the technology is so wonderful and also have the capability to meet the current and future challenges, why its growth is slow particularly at SME level? On one side the companies are exploiting the striking benefits of the technology creating new fronts of competition but on the other side the companies are facing tremendous pressure and challenges to stay and survive in world markets. This is particularly true for small to medium enterprises SMEs and especially in developing countries. The following questions need to be investigated in this scenario. What are the major obstacles which are preventing the growth in this sector and how such barriers can be tackled? There is no simple answer for this complex question. But to answer this to some extent, we need to break it down and then categorical analysis can guide us towards the major contributing factors involved in slow growth.

2. PREVIOUS RESEARCH

Wohler [2001] in his article "Obstacles to RP growth" highlighted certain issues by pointing out that RP is still used by few but not for the masses. He also raised issues like cost, resistance to change, outdated information and lack of awareness. We believe that these are valid issues but the only difference is the change in chemistry, as developing countries have their own problems which are different from that of developed countries and control mechanism to address these problems are also markedly different, so it is very important to take into account these differences when formulating solutions for such countries.

Auto part cluster in Lahore Pakistan was selected to find the major contributing factors responsible for slow growth.

3. AUTO PART INDUSTRY OF PAKISTAN

There are 400 organized units in the auto parts industry, most of which are registered vendors to assemblers. Along with the organized sector, a good number of small and large units (approx 1200) are operating in un-organized sector. In fact 90% of automotive parts industry constitutes of small to medium enterprises. These units produce a wide range of parts for the replacement market and have a total investment of more than 1 billion$(US).The problems which they are facing: **1**.Currently, the auto parts industry is not able to cater the increasing demand of automobiles in the country.**2**.Only 10% of the companies are capable to export only because their products comply with the International standards.**3**.The import of auto parts, which cannot be produced locally due to lack of technology, is projected to increase dramatically (Asif, 2006).

Interviews were conducted with the academics , chamber of commerce, representatives of Govt agencies responsible for SMEs affairs and member of the clusters to understand that why RP could not gain attention in spite of its capability to solve certain problems (Up to our knowledge only one RP system is available in cluster). Based on findings a model was developed and results are explained according to Fig1.

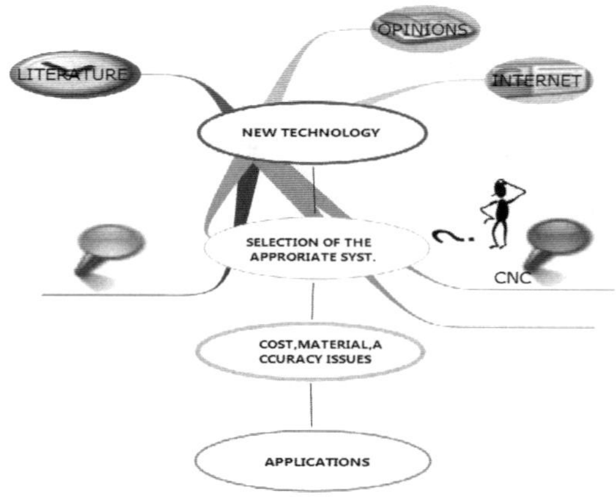

Fig.1 Technology Selection Process

4. RESULTS

4.1 Introduction of New Technology

Interviews showed that majority of the organizations were not able to cross even the first stage. Some of them were totally unaware about the technology and the other were confused and were not able to find answer which match their requirement. There are number of reasons for that 1. Education level at SME is relatively low 2. Most people acquire managerial positions based on their experience and they lack professional qualification. 3. R&D culture is absent and professional development is very limited as it is considered as financial burden.

4.2 Process Selection

Process selection involves the production capabilities and technological constraints for each process. Selection of the parts most suitable for TCT is an important part of the product development and process selection procedure.

Survey showed that large proportion, which entered in this phase end up with confusion, as there are no tools exist which can assist them in the selection of right TCT process. The abundance of CNC systems in the market at very cheap price force them to continue with the old process to avoid risk as new system is normally considered as a threat for their expertise developed over years.

We believe that this factor very critical for the slow growth of the prototyping technology.

5. FUTURE DIRECTION

5.1 Development of Common Resource Centres (CRCs)

The basic purpose of the CRCs is to improve the way industry works by expanding their vision and decision making ability by providing cutting edge technology & resources at one place, which otherwise is difficult to afford by individuals due to many reasons which includes, lack of awareness about the latest technology, lack of funds. Collaboration is the basic theme of such centres and is shared by the partners.

The idea of CRC differ in scope with service baueras as the later is for the sole purpose of revenue generation, while the former is normally developed by the funding either directly from the Government or from other sources and the basic purpose is to promote technology and assist the particular segment of the industry, which are not able to solve their problems either due to lack of expertise or their inability to excess the latest technology. It assists organizations to stay competitive and to cope with the changing requirements of the external environment to meet customer demands.

The development of such CRCs to promote RP education is an excellent starting point, where stakeholders can manage to solve their problems and also get hand on experience on technology. The involvement of academia is very critical for the continuous success and for the sake of research. Closure of such centres was observed due to lack of interest or expertise, where academia was not involved or partially involved.

5.2 RP as Engineering Curriculum

Rapid prototyping is an integral part of the early design stage. It helps to expose the design problems early in the design stage before it go to expensive tooling and production stage.

Engineering curriculum in universities does not cover the importance of such technology. The courses lack hand on experience and projects on rapid prototyping. Absence of equipments and trained manpower are the typical issues which the developing countries are facing today. It is need of the time for the academia to play its vital role and help this sector to become the part of competition .This is only possible when academic institutes realize the importance of this technology and make it an essential and comprehensive part of their curriculum. Limited buyers will end up with limited growth, and research. Awareness programmes, short courses designed for particular industry, exposure of the industry to latest technology and demonstration that how it can solve their problem are some powerful tools, which not only can assist the SMEs but also will be helpful in the rapid growth of technology.

6. CONCLUSION

Several issues responsible for the slow growth of TCT with a special focus of SMEs in developing countries were presented and discussed. It shows that some of them are critically responsible for the slow growth and need urgent attention of Govt. agencies and academics. Joint efforts should me made to achieve fundamental understanding of the potential benefits of the RP technology.

Major responsibility lies on the academia to develop strategies to overcome such growth barriers. By doing this it can be hoped that technology will find its way in this sector.

7. REFRENCES

Bhattacharya A, & Jina. J. (1997). Manufacturing Time Compression: Not a technique but a philosophy. *Production Planning & Control, 8*(1), 88-91.

Kengpol, A., & O'Brien, C. (2001). The development of a decision support tool for the selection of advanced technology to achieve rapid product development. *Int.J.Production Economics, 69*, 177-191.

M.Asif (2006). Cluster diagnostic study, auto part cluster Lahore.Retrieved 20May, 2008, from http://www.smeda.org.pk

Ogando, J. (2007). Rapid Manufacturing's role in the factory of the future. *Design News*.August 13, 2007

Wohlers, T. (2001). Obstacles to RP's growth *Time Compression Technology, October 2001.*

Stratasys' Latest Direct Digital Move, *Design News*, December 6, 2007.

Annals of DAAAM for 2008 & Proceedings of the 19th International DAAAM Symposium, ISSN 1726-9679
Editor B. Katalinic, Published by DAAAM International, Vienna, Austria 2008
Make Harmony Between Technology and Nature, and Your Mind will Fly Free as a Bird

EFFICIENT VIRTUAL REALITY PROJECTION SCHEME FOR PRODUCT DESIGN ENGINEERING

BACHVAROV, A[ngel G.]; MALESHKOV, S[toyan] & BOYADJIEV, I[lia]

Abstract: *This paper presents some considerations and concepts of Virtual Reality (VR) for the engineering practice in order to assist the design engineers in getting a better understanding about VR technology and developing their own low-cost VR systems. It introduces and discusses hardware implementation of a simple low-cost projection scheme for CAD-models exploration supporting the engineering decision on early stages of product development based on VR. It is based on author's work within newly established VR Lab at the Technical University of Sofia and supported by cooperation activities with the IMI at the University of Karlsruhe, Germany.*
Key words: *Low Cost Virtual Reality, Product Design, CAD-models exploration*

1. INTRODUCTION

The decisions made in the initial design stages predetermine up to 80% of the cost of product development and manufacture. 92% of communications within the design process are graphically based [Bertoline, 1997]. Information visualization allows designers to understand their developing designs and to present their ideas to others. Thus, efficient graphics communication tools can improve significantly design and decision-making processes. Presently most common practice is the use of conventional CAD tools such as 3D-modelling and publishing software packages (e.g. CATIA, ProEngineer, UG, SolidWorks). However, this practice enables users to examine 3D models only on flat computer monitors. The models and the observers are in different realms [Seth, 2004].

2. VIRTUAL REALITY IN PRODUCT DESIGN

Virtual Reality (VR) is an exciting and contemporary emerging area of computer graphics, supporting communication and collaboration within product design stage. There are a lot of definitions of VR. Commonly it could be referred to "*a high-end user-computer interface that involves real-time simulation and interactions through multiple sensorial channels*" [Burdea, 2003], based on the so called 3 "**I's**": *(i)* **Immersion**, *(ii)* **Interaction**, and *(iii)* **Imagination**. VR applications give users a sense of immersion into some sort of virtual space. This space can model physically existing reality, or be something entirely fictitious. VR has proven to be an effective tool for helping engineers to evaluate product designs (Seth, 2004). Using VR enables everyone on a design team to understand designs better, leading to more informed and meaningful decision making. The VR technology offers a lot of potential applications ranging from product development in diverse industries to architecture and medicine to event planning and art. Today VR applications are already being integrated in the production process itself. The reason for this is as simple as obvious: the possibility of a 1:1 representation of complex 3D structures allows the realistic viewing and evaluation of products or objects which do not yet actually exist. Several major industrial companies, coming in particularly from the automotive industry, have integrated VR technology into their design or production processes (e.g. BMW, Daimler-Chryser, GE etc.) [BMW, 2003]. In this way following advantages were achieved: *(i)* reduced number of physical product mockups needed; *(ii)* improvement of the overall product quality; *(iii)* obtaining quick answers, in an intuitive way, during the concept phase; *(iv)* enhancing communication flow and improving collaboration; *(v)* streaming up the design process; and *(vi)* significantly reducing cost and time to market.

3. HARDWARE IMPLEMENTATION

Commonly the VR effects immersion through stereographic visualization, providing separate images to the left and right eyes that correspond to the different views both human eyes perceive of the real world. When each eye is allowed only to view its intended image, the brain construct from both displays a 3D-view of the world, giving a sense of depth. There are two general approaches to accomplish such a stereo viewing: trough *(i)* head mounted displays (HMDs) and *(ii)* projection systems.

With head mounted displays, users wear a helmet-like device that physically places a different display screen in front of each eye. The advantage here is the portability. The drawbacks are that they are normally very expensive (several thousands of EUR) and only support a single user at a time.

The preferred approach for collaborative VR applications nowadays is the use a projection based system, which projects images onto a large display screen or combination of screens.

Presented approaches can be divided into two categories: (i) active stereo and (ii) passive stereo. With active stereo, users wear special LCD shutter glasses that alternate quickly between opening and closing of the left and right shutters. The displays on the screen also alternate between left and right eyes, in time with the glasses. Thus, the left shutter is open only while the left eye image is on the screen, and likewise for the right, allowing each eye to see only its intended view. Available LCD shutter glasses use various image switching techniques.

With passive stereo, images are shown simultaneously on the same screen but, once again, users wear special glasses (made from colour or polarized filters) that allow the eyes to see only their intended views. There are two types of passive stereo: *(i)* anaglyph based and *(ii)* polarized.

Anaglyphs are based on the principle that if you view an image through a certain colour filter, that filter's colour will disappear in the image. The main drawback with anaglyphs is that the colour spectrum is reduced; each eye will be unable to see one of the primary colours of the image.

The polarized passive stereo uses light polarization. Images for the left and right eyes are projected through opposing polarizing light filters. When users wear special glasses with the same filters, the eyes are only able to see the image projected through the same polarizer. If light is polarized in a single direction (north/south, east/west, or even diagonally), the light is linearly polarized. If a viewer changes the orientation of linearly polarized glasses by tilting his or her head, the resulting polarization orientation of the viewer's glasses will not match that of polarization filters mounted on the projectors, and there will be a loss of stereo information.

The polarized passive stereo setup relies on separate projectors for left and right eye images. Although there have been tremendous advances in data projector technology in the last few years, good projectors are still relatively expensive. One potential problem in selecting cost effective solution is that the vast majority of current data projectors use LCD technology, which itself relies on polarization properties of light. As a result, LCD projectors emit light that is already polarized. The alternative is to use high-resolution DLP projectors, which do not emit polarized light. Projector placement involves some special considerations. For best results, the two projectors should either be stacked or placed next to each other with parallel lines of projection to avoid any differential keystoning that would interfere with the proper alignment of the images. Typically, the projectors are stacked vertically using some sort of rigid custom designed framing. Some higher-end projectors have lens-shifting ability that allows the two images to be perfectly aligned while maintaining parallel projection. Another possibility for projectors with built-in keystone adjustment is to align the projection centres of the two images and then use the vertical keystone adjustment to align the edges of the images. A third approach is to simply keep the projectors parallel and only use the overlapping region for a cinemascope effect. In this case, viewports of the images need to be adjusted in the software so that they align properly.

The only special consideration for the graphics card is that it must be a dual output channels card to provide output to the two projectors. The main consideration in choosing a graphics card is driver support. Best suited for passive stereo visualization are quad-buffer graphics card.

Other important aspect for VR is the interaction and the user input, which normally comes from the conventional mice and trackballs, used for one handed "drag & drop" and "click & rotate" interaction in 2D desktop environments, where the second hand can be used for activating additional functions through the keyboard. Joysticks are also common input devices for navigation in 2D desktop environments. While the primary hand moves the stick in 2D and presses buttons on top of the stick, the secondary hand holds the device and manipulates additional buttons and throttles on the base. Two out of three translation or rotation directions (x, y, and z) can be selected and mapped to the joystick movement. Some joystick models also allow turning the stick, thereby adding an additional rotational axis. Additional degrees of freedom can be activated by pressing a button on the joystick or on the keyboard. However, these devices are not very "natural" for an immersive VR setting. There are several other types of interface devices which allow more than 3-DoF for better access to the virtual space (simple stylus, data glove, magic wand, flying stick etc.). Recently a lot of research interest is directed toward the video game controllers that can provide more flexibility than the above mentioned devices providing several axes for numerous degrees of freedom under drastically reduced cost (up to 10 times). One of such most promising devices enabling implementation of low cost VR solutions is Wiimote, the primary remote controller of Nintendo Wii play station. A main feature of the Wii is its motion sensing capability, which allows the user to interact with and manipulate items on screen via movement and pointing through the use of accelerometer and optical sensor technology. Another feature is its expandability.

4. BUILDING A SIMPLE VR SCHEME FOR CAD MODELS EXPLORATION

For testing and verification of the proposed approach and workflow for CAD model exploration a simple VR system with passive polarization technique was designed and implemented. (Fig.1). A computer cluster of 3 quad core CPU PCs (one master and two slaves) was built under 64 bit Windows 2003 server environment. The master computer calculates controls and synchronises the virtual world projection and data stream from the special input devices and the tracking system and "feeds" the slave PCs with data for the stereo projection. They have a dual-pipe graphic cards connected to vertically-stacked within a metal frame LCD projectors with resolution of 1024x768 which provide the images for the right and left eye. The projectors have a built-in test pattern that can be turned on for adjustment of vertical keystoning. An external linear polarizer is put in front of each projector. The image is rear projected on a semi-transparent power wall, the result is viewed through a standard (polarized) 3D glasses. As an interface device a conventional mouse, space-mouse, flying-stick and keyboard were used. The CAD models have been prepared in CATIA V5 (R17 and R18) and have been imported in Virtools 4 scene graph editor and VDP:Explore projection module. The experience stage of the experiment was carried out in the newly developed Laboratory for Virtual Engineering at the Technical University of Sofia. In both cases fully sufficient results were obtained. The VR projection of the CAD mode was of high quality and level of details providing more visual information than the onscreen flat image.

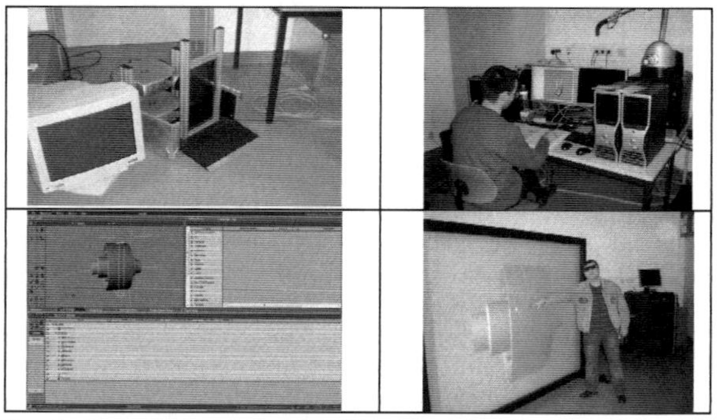

Fig. 1. Experimental simple VR-scheme for CAD-models view

5. CONCLUSION

In this work we discussed an efficient hardware set-up for building a simple, low cost VR scheme for use in the product engineering practice. An experimental implementation of such a scheme is presented. The experiments carried out within validation process of a method for CAD-models exploration in VR environment proved the efficiency of the proposed scheme for supporting the engineering decision on early stages of product development though improving and increasing the design communication and reducing design time and efforts.

The work is partially funded within project BY-TH-110-10 by the Bulgarian Ministry of Science and Education.

6. REFERENCES

Bennett, D., Farrell, P., Lee, M., Ruttan, A. (2000), *A Low Cost Commodity Based System for Group Viewing of 3D Images.* VDE, Berlin.

Bertoline, G. R., Wiebe, E. N., Miller, C. L., & Mohler, J. (1997). *Technical graphics communication (2nd ed.).* Burr Ridge, IL: McGraw-Hill

BMW (2003). Informationen der BMW Group Mai 2003. Virtuelle Realität im Automobilbau. *Available from* http://www.bmwgroup.com/scienceclub

Burdea, G., Coiffet, P. (2003). *Virtual Reality* Technology Hoboken, New Jersey: John Wiley and Sons.

Seth, A. & S-F Smith, S. (2004). PC-Based Virtual Reality for CAD Model Viewing. *The Journal of Technology Studies*, NY.

Annals of DAAAM for 2008 & Proceedings of the 19th International DAAAM Symposium, ISSN 1726-9679
Editor B. Katalinic, Published by DAAAM International, Vienna, Austria 2008
Make Harmony Between Technology and Nature, and Your Mind will Fly Free as a Bird

LAND COVER GIS DATABASES, SUPPORT FOR THE IMPLEMENTATION OF THE EUROPEAN AGRICULTURE AND ENVIRONMENT PROGRAMS

BADEA, A[lexandru]; DANA, I[ulia] F[lorentina]; MOISE, C[ristian]; MAMULEA, A[dina] A[na]; OLTEANU, V[lad] G[abriel] & POENARU, V[ioleta] D[omnica]

Abstract: *Land cover information is extremely useful for monitoring and identifying the landscape dynamics and the condition of vegetation. The complex analysis of the multi-temporal land cover databases provides important statistics for agriculture, natural hazard management and environment protection. At the moment, satellite images are the main source for the creation of the land cover databases. Together with other types of data (topographic maps, aerial images, ground truth information, etc) the land cover databases are usually stored and managed in a Geographic Information System (GIS). The project has two major objectives: the creation of the complex land cover database (based on satellite images acquired in 2000, 2003 and 2007) and the elaboration of the optimal methodology for data processing.*
Key words: *land cover, land use, remote sensing, GIS, LCCS*

1. INTRODUCTION

Land cover is the observed (bio) physical cover of the Earth's surface. Land use is characterized by the arrangements, activities and inputs people undertake in a certain land cover type to produce, change and maintain it (Di Gregorio & Jansen, 2000). Land Cover Classification System (LCCS) was initiated by the United Nations Environment Program (UNEP) and the Food and Agriculture Organization (FAO) in 1993. Its main objective was represented by the elaboration of a reference and standardized classification of land cover that may be applicable anywhere in the world.

In Romania, the FAO-LCCS project consists of two geo-databases created upon the coverage of Landsat TM satellite images corresponding to the years of 2000 and 2003. The project was able to provide to the Ministry of Agriculture an objective and accurate database of the land cover for the whole country and also detailed land cover/land use maps for areas of particular agricultural interest. Also, the output of the project was used in case of emergency situations, for example the floods that affected Romania in 2005 and 2006. Now, a third complete coverage of Romania, containing high resolution multispectral satellite images acquired in 2007, is added to this impressive dataset. The project is intended to develop a set of methods and rules for data processing and validation that can be applicable to images acquired once in three or four years.

Recently, European Space Agency (ESA) established an international partnership with UN FAO and others research institutes for the creation of a global and detailed portrait of Earth's land cover with a resolution never obtained before: GlobCover. GlobCover is based on the imagery acquired by the Medium Resolution Imaging Spectrometer (MERIS) – Envisat between 2005 and 2006 with the resolution of 300m. GlobCover is extremely useful for climate change study (modeling of the climate change extent and impacts), ecosystem management (study of the natural and managed ecosystems) and worldwide land-cover trends identification. GlobCover's thematic legend is compatible with LCCS (Arino et al., 2007).

2. METHODOLOGY

2.1 LCCS concept

In the frame of LCCS, the classification of the land cover classes is made in two main phases:

- Dichotomous Phase – in this phase there are identified 8 major land cover types: cultivated and managed areas, (semi) natural vegetation, cultivated aquatic areas, (semi) natural aquatic vegetation, artificial surfaces, bare areas, artificial water-bodies, natural water-bodies, snow and ice.
- Modular-Hierarchical Phase – in this phase a set of diagnostic criteria are defined for each major cover types.

A team of experts from different fields of activity was involved in the elaboration of the LCCS legend that comprises of land cover classes specific to Romania. During the realization of the three LCCS projects (the years of 2000, 2003 & 2007) the legend has been continuously updated.

The land cover classes that appear in the figure below represent: LDU- low density urban area, ORD – orchard, F – forest, SSH – small sized herbaceous field, LSH – large sized herbaceous field, GRL – grassland, GRL/SHR/TRS – grassland, shrubs, trees (mixed class). These are only a few of the complex legend which comprises of more than 50 land cover classes defined for the territory of Romania.

Fig. 1. Land cover map of Romania (LCCS-2003) - overview

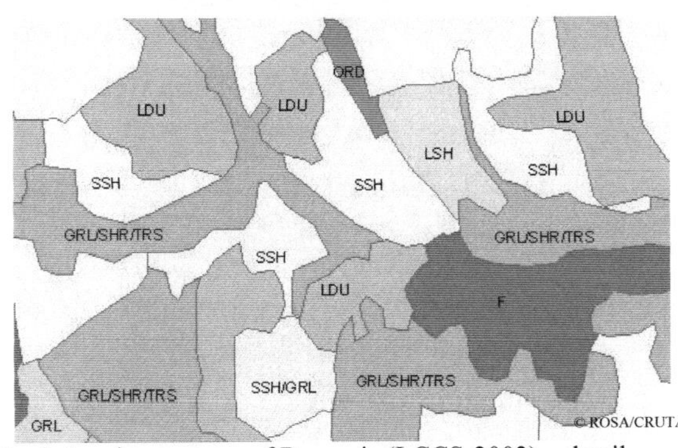

Fig. 2. Land cover map of Romania (LCCS-2003) – detail

2.2 Input data

The satellite images which represent the basis of the first two Romanian FAO-LCCS projects were acquired by Landsat TM in the years of 2000 and 2003. The identification of the land cover classes was made by direct photo interpretation.

Automatic image classification (Lillesand et al., 2004) was also performed for several test areas. The GIS interactive database (Arctur & Zeiler, 2004), which is used to guide the photo interpretation process, is land use/land cover focused and contains a description of the interpretation procedures according to the adopted methodology and also a more detailed classification of the identified classes. The geographic interface contains field data, pictures, satellite images, aerial photos, vegetation index images and other geo-referenced information (geology, soils, etc.).

In the framework of the RO-LCCS-2007 project, SPOT 5 multispectral images with the spatial resolution of 10m were acquired; these images allow a multi-temporal and detailed analysis adapted to specific characteristics of landscape.

2.3 Objectives

The data regarding the land use represent an important feature for the environment monitoring and for political decisional initiative. The land cover (Lambin & Geist, 2006) is the most important characteristic of nature and it is used for the following processes: diagnosis and forecast for sustainable management at the level of regions and countries; monitoring and use of natural resources (agriculture, water and forest), etc.

2.4 Work flow

The results are exported in vector format, compatible with the most known GIS systems. For each database, approximately 730 land cover maps are produced, at the scale of 1:50.000.

This research project takes into account the evolution of data acquired by the new Earth Observation satellites, as well as advanced methods for semi-automatic and automatic data processing. The automation limit that can be achieved without jeopardizing the quality of relevant information will be evaluated. A synthetic chart containing the processing steps of the LCCS project is presented next.

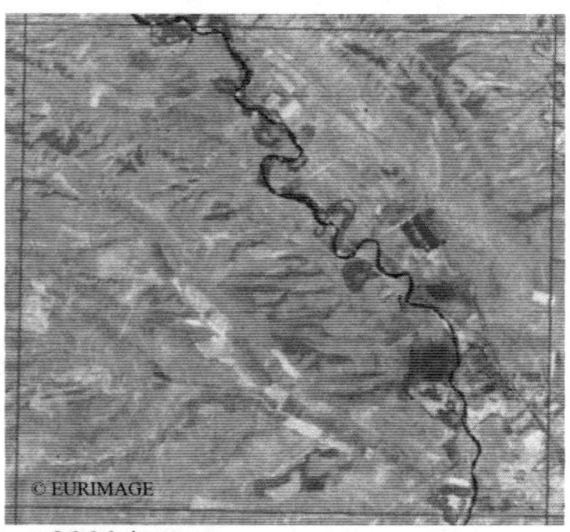

Fig. 3. Landsat 2000 image

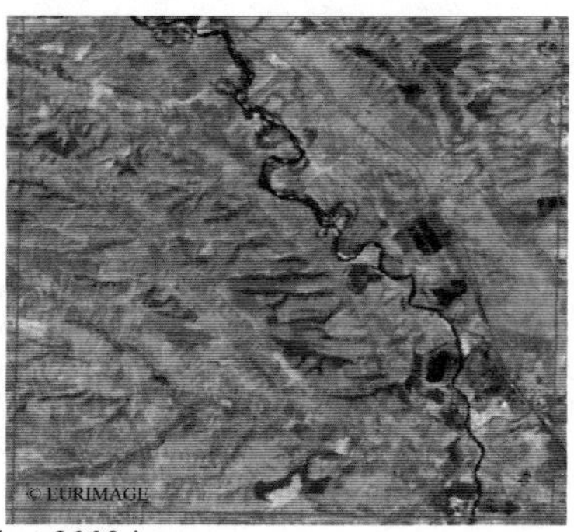

Fig. 4. Landsat 2003 image

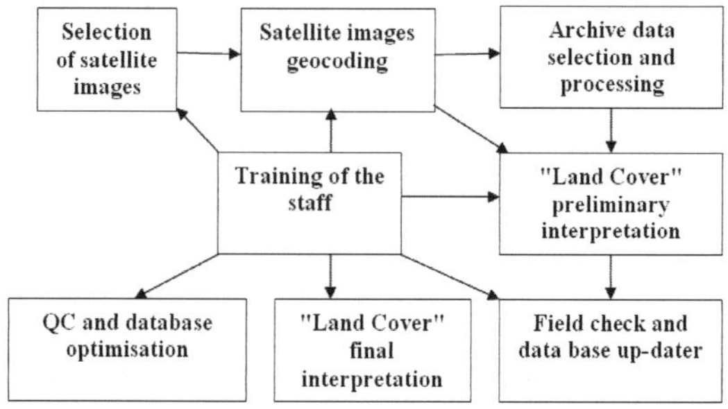

Fig. 5. Processing steps for RO-LCCS-2007

3. CONCLUSION

The RO-LCCS-2007 project approach follows the GMES (Global Monitoring for Environment and Security) actions in Romania, according to the development and use of remote sensing techniques for national environment protection and rehabilitation.

The resulting database can be included in the Integrated Administration and Control System/Land Parcel Identification System (IACS/LPIS) database and other already functional databases from the Ministry of Environment.

The analysis of this dataset provides support for the development plans of the territorial infrastructure, sustainable development in agriculture and implementation of "Natura 2000" directives.

Indirectly, correlated with statistical data produced by the National Institute for Statistics, this unique database will offer the possibility to analyze the socio-economical relations from more objective points of view.

RO-LCCS-2007 will be part of the "Technical Assistance to Develop the Environmental Related GIS Maps - Romania" Phare Project elaborated by a consortium led by IGN France International. The beneficiary of the project is the Romanian Ministry for Environment and Sustainable Development.

In the future, the LCCS geo-database will be permanently updated and completed with different types of data. The results and the elaborated methodology will be published in a scientific guideline.

RO-LCCS-2007 offered the possibility to obtain the 3rd updated series of the FAO-LCCS database by using recent advanced technologies, adapted to the Romanian conditions. This way, Romania possesses a dataset unique at global level.

4. REFERENCES

Arctur, D. & Zeiler, M. (2004). *Designing Geodatabases: Case Studies in GIS Data Modeling,* ESRI Press, ISBN 9781589480216, Redlands, California

Arino, O.; Gross, D.; Ranera, F.; Bourg, L.; Leroy, M.; Bicheron, P.; Latham, J.; Di Gregorio, A.; Brockman, C., Witt, R.; Defourny, P.; Vancutsem, C., Herold, M.; Sambale, J.; Achard, F.; Durieux, L.; Plummer, S. & Weber, J.-L. (2007). GlobCover: ESA Service for Global Land Cover from MERIS, *Proceedings of IEEE International Geoscience and Remote Sensing Symposium (IGARSS 2007),* IEEE, pp. 2412-2415, ISBN 978-1-4244-1211-2, Barcelona, July 2007

Di Gregorio, A. & Jansen, L. (2000). *Land Cover Classification System (LCCS): Classification Concepts and User Manual,* FAO, ISBN 92-5-104216-0, Italy

Lambin, E. & Geist, H. (2006). *Land Use and Land Cover Change – Local Processes and Global Impact,* Springer, ISBN: 978-3-540-32201-6, Berlin, Heidelberg

Lillesand, T., Kiefer, R. & Chipman, J. (2004). *Remote Sensing and Image Interpretation – Fifth Edition,* John Wiley & Sons, Inc., ISBN 0-471-45152-5, USA

Annals of DAAAM for 2008 & Proceedings of the 19th International DAAAM Symposium, ISSN 1726-9679
Editor B. Katalinic, Published by DAAAM International, Vienna, Austria 2008
Make Harmony Between Technology and Nature, and Your Mind will Fly Free as a Bird

COMPLEX METHOD FOR DEVELOPING THE DIGITAL GEOMORPHOLOGIC MAP OF ROMANIA AT 1:200.000 SCALE

BADEA, A[lexandru]; MOISE, C[ristian]; DANA, I[ulia] F[lorentina]; OLTEANU, V[lad] G[abriel]; MAMULEA, A[dina] A[na] & BADEA, R[uxandra]

Abstract: *The development of GIS, the satellite imagery and the analysis of the digital terrain model (DTM) made possible the up-dating and the improvement of the geomorphologic maps content. This paper aims to reveal the practical importance and the necessity of the digital geomorphologic maps in the field of environment, biodiversity protection and preservation and to present steps in creating the digital geomorphologic map. The general geomorphologic map of Romania at scale 1:200.000 is used as support for creating the digital map and the digital library of map symbols. Also, based on different DTMs, a series of algorithms interactively validate the map symbols position, dimension and shape.*

Key words: *geomorphology, DTM, GIS*

1. INTRODUCTION

The first complete geomorphologic maps were achieved at the beginning of the 20th century and they included particular representations of the studied areas. In 1965, at the congress *International Geographical Union* (IGU) from Rio de Janeiro, there were presented two new ideas in approaching the geomorphologic maps: the introduction of certain specific elaboration methods in geomorphology and the adoption of a standard plotting system.

When Remote Sensing and Geographical information Systems developed, geomorphologic plotting was decisively influenced, as the interpretation and representation of the relief elements was eased by comparing the field data with the satellite ones or by the development of diverse algorithms of the *map algebra* type or of automatic calculus and integration of the morphometry, geodeclivity, insolation, or slopes' exposure data.

In spite of the fact that geomorphologic maps continued to evolve during the last century, there was not reached a universally accepted mapping system and there still are differences with regard to the form, content, and cartographic symbols (legend). Presently, a very powerful organism involved in unifying the geomorphologic approaches is the International Association of the Geomorphologists (IAG/AIG), where, under the coordination of EC, activates a work group that is working at a project the main objectives of which are the following:

- development and deepening of the theoretical base knowledge in the field of applied geomorphologic mapping
- development of standards, mapping procedures, and types of legend for different applications and scales
- dissemination of the importance and efficiency of using geomorphologic mapping as a base instrument for those who study the environmental problems the purpose of which is to establish links among different scientific and professional communities

In Romania, a major achievement of geomorphologic mapping was the elaboration of the general geomorphologic map of Romania at a mean scale of 1:200,000 by the Geography Institute of the Romanian Academy (Figure 1).

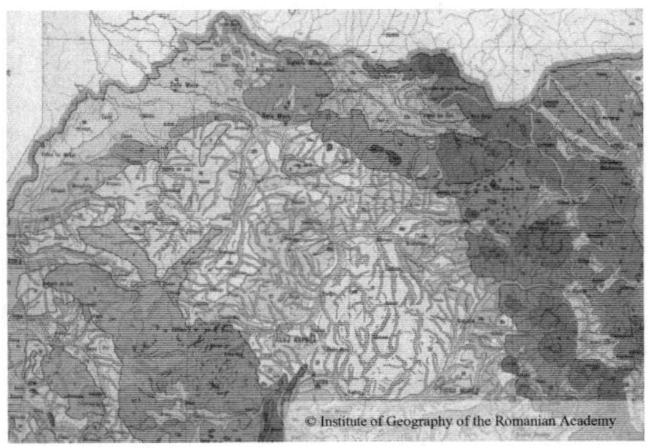

Figure 1- General geomorphologic map of Romania, 1:200.000 scale

The drawing of the map at a scale of 1:200,000 started in 1977 and ended in 1990. The legend was elaborated according to the recommendations and the principles adopted by the Research and geomorphologic mapping commission of U.I.G. The unique and universally valid legend cannot be applied at all the scales, which determined the replacement of generalized aspects, used in small-scale representations, with more and more detailed features of the forms, specific to large-scale representations. This is why legends were differentiated on categories of scales. Thus, the attention focused on establishing the principles for the elaboration of legends (Posea, Popescu and Badea, Niculescu – 1972) and for their application at different scales. The considerations presented in this paper are based on the above-mentioned general geomorphologic map of Romania at a scale of 1:200.000.

2. METHODOLOGY

The elaboration of a general geomorphologic map supposes the territory's investigation and knowledge display a uniform and detailed character in agreement with the map scale; moreover, it is necessary to make a geomorphologic analysis of the map, so that the fundamental morphometric elements (relief intensity and mean slopes) could be emphasized and adequately rendered. Relief intensity is a defining element in determining the classes of the relief forms, the main morphologic types – plains, hills, plateaus, and mountains, which represent, in fact, the association (assembly) of certain forms in their specific stages of evolution (Badea et al., 1983). Consequently, each trapeze has to be accompanied by a map rendering the classes of the relief forms or the relief units generically defined and hierarchically displayed by classes (as relief intensity and altitude), over which the lithological structure can be overlapped (Nastase, 1995).

The drawing of the general geomorphologic map was based on a set of original principles:

- the groups or types of characteristic elements have to be rendered by symbols, hachure, or colors, according to the importance of the phenomenon.
- the legend has to correspond to a certain scale or, more precisely, to a certain group of scales: small (1:1,500,000 up to 1:1,000,000 or even 1:500,000), mean (1:400,000 up to 1:100,000) and large (more than 1:100,000 up to 1:20,000) and to the features of the Romanian territory, which underlines the idea that the legend must not be universal.
- the possibility of including certain elements resulted from the elaboration of some special geomorphologic maps in the legend of the general geomorphologic map. The legend has to display an open character.

The elaboration of the legend respected the recommendations made by the IGU and the suggestions given for the legend of the Europe geomorphologic map at a scale of 1: 2.500.000 (1968,1971) and of the geomorphologic map at a mean scale (1973), obviously, by adapting it in order to better render the relief forms of Romania.

The legend of the general geomorphologic map (scale 1:200.000) contains five parts:

- the map of classes of landform
- the structural-lithological types and the cover deposits
- simple landforms and elements represented by symbols
- the genetic types of landforms (the most important section of the legend)
- declivity

This involves the classification of the map elements, and their separation to different types of vectors: polygon, line or point. Once these are set, they are integrated into a digital library of map symbols containing all necessary elements for the geomorphologic map.

The validation of the present geomorphologic maps is to be achieved by their correlation with the digital elevation models (DEM) at different resolutions and with high and very high-resolution satellite images.

From the geomorphologic point of view, a complete description of any relief forms can be achieved by using three elements: altitude, slope, and curvature (Posea et al., 1976). The slope is the most important element for describing the terrain surface. The determination of the slopes and of their exposure can be achieved by analyzing a digital terrain model.

There are used many types of digital terrain models, obtained from different data sources (topographical maps, satellite images undertaken by active and passive sensors, other data sources) (Figure 2).

For certain test zones, a digital terrain model will be generated by digitizing the contour lines from the topographical map at the scale of 1:25.000. The precision (accuracy) the relief forms are rendered within a topographical map is mainly

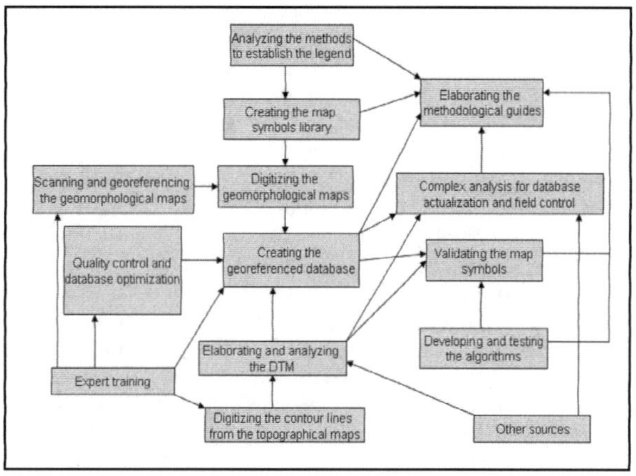

Figure 2- Development of the operations

determined by the density (equidistance) of the contour lines. The large- and mean-scale maps render, at high fidelity, the field elements, while the general topographical maps display an increased generalization level (Posea & Cioaca, 2003). Generally, the precision in altitude of a certain point determined by the interpolation of the contour lines is of 1/2-1/3 of the equidistance. In order to obtain certain results with a superior altimetric precision, a digital elevation model will be also generated on the basis of the topographical map at the scale of 1:5.000.

The processing of the satellite images overtaken by SPOT 5 HRS allowed obtaining a highly precision digital terrain model. This model contains up-dated information referring to the relief elements. As supplementary data, other digital terrain models are used: SRTM (generated through interferometry techniques) and DTED2 (informational content equivalent to the model generated by the digitization of the contour lines 1:50.000).

The precision of the digital terrain models are evaluated by means of statistical methods that include GPS measurements made in the field in the control points.

Starting from these numerical models, is to be achieved a comparative study, the purpose of which is the identification of the most adequate model for the up-dating and editing of the geomorphologic map at a scale of 1:200.000.

An extremely important scientific activity is represented by the development and testing of the correlated algorithms for validating the map symbols' position, dimension, orientation and shape.

3. CONCLUSIONS

The approaches of this project follow the line of implementing the European directives (Nature 2000) and the application and development of the INSPIRE (Infrastructure for Spatial Information in Europe) and GMES (Global Monitoring for Environment and Security) actions related to the environment protection and rehabilitation through the development and utilization of GIS/Remote Sensing techniques.

The output will be represented by the elaboration of the first geomorphological digital map of Romania at a scale of 1:200,000 and the elaboration of the first digital map symbols library specific to the Romanian territory.

The complexity of the project achievement solutions is represented by the study and testing of the methods of correlation between the different digital terrain models and the digital geomorphological layer, materialized through the development of software algorithms for the interactive validation of the map symbols characteristics.

We plan to use the results of this project in facilitating the drawing of the risk maps, of environment impact studies and optimizing the determination of the geomorphologic risk factors, all these being essential elements for most of the national, regional, and local economic targets.

4. REFERENCES

Badea, L.; Gastescu, P. &Velcea, V. (1983): *Geografia fizica (Physical geography)*, Editura Academiei Republicii Socialiste România, Bucharest

Nastase, A. (1995): *Cartografie (Cartography)*, Editura Fundatiei "Romania de maine", Bucharest

Posea, Gr. & Cioaca, A. (2003): *Cartografierea geomorfologica (Geomorphologic mapping)*. Editura Fundatiei "Romania de maine", Bucharest

Posea, Gr.; Grigore, M.; Popescu, N. & Ielenicz, M. (1976): *Geomorfologie (Geomorphology)*, Editura Didactica si Pedagogica, Bucuresti

****Atlas R.S.R.* (1979), Editura Academiei R.S.R., Bucharest

Annals of DAAAM for 2008 & Proceedings of the 19th International DAAAM Symposium, ISSN 1726-9679
Editor B. Katalinic, Published by DAAAM International, Vienna, Austria 2008
Make Harmony Between Technology and Nature, and Your Mind will Fly Free as a Bird

INTRODUCING GIS IN MARKETING ANALYSIS AND SALES MANAGEMENT

BADEA, R[uxandra]; BAGU, C[onstantin]; VOICU, V[lad]; LUNGU, A[ndy] & MOISE, C[ristian]

Abstract: *Modern commerce stimulates retail expansion. In Romania this market finds itself in a continuous expansion as can be seen in the growth registered every year. This phenomenon is sustained, most of all, by the increasing number of superstores and supermarkets. In the same time mentalities change and people become more and more receptive to what the market has to offer. Therefore it is necessary to answer the needs of the market with innovative measures, in order to sustain a new approach of the selling process, for all companies acting in retail. Combining classical methods and new advanced tools, we intend to show non conventional ways to improve sales management for Romanian retail companies, having as a starting point the Western-European Experience.*
Key words: *retail, GIS, sales management, decision making*

1. INTRODUCTION

In the present context of the globalization of the economic activity and the expansion of knowledge management, the structured information a company owns becomes a vital resource for maintaining the level of competitiveness. (Donaldson, 2005). The quality and the opportunity of information held influence the strategic decision process and have a major impact on the future development of the organization. In this context managers need to use powerful instruments for decision making, instruments that can capture, process and display a true image of the reality through a spatial representation of its characteristics.

Taking in consideration all these aspects, the paper aims to draw attention on the impact of using Geographical Information Systems (GIS) in the economic activity in Romania. Through these tools users can perform complete and complex market analysis, analysis that brings together geographical, economical and census datasets. This means using complex applications for market analysis, based on GIS technologies. (Badea et al, 2008)

Recent studies show that 80% of the total data handled by decision makers in any organization, public or private, has a geographic component. Key activities depend on knowing where the clients and the potential clients are. As a consequence, in order to efficiently manage the physical assets and the human resources necessary in forecasting and satisfying the demand, using the geographical component becomes compulsory. (Badea et al, 2007)

Big providers of GIS software started to pay attention to the business analysis component. For example the biggest provider of Geographic Information Systems, ESRI, developed an extension specialized in Business Analysis. In the USA the extension is provided together with a database that consists of business, household and demographic data. Seeing all that data on a map gives a better insight and helps in making better decisions. Modeling processes included in GIS help in obtaining a spatial representation of the market's characteristics. (Zeiler, 1999). Using these tools has a great relevance on elaborating the development strategy of different companies, as they give them the possibility to use these datasets in achieving their own objectives.

2. BACKGROUND

The use of spatial information in the economic activity was not documented in the specialty literature in Romania. However, even though bibliographic resources are limited, the international literature provides many articles written by western specialists which have studied the field.

David Bennison shows in his articles the importance of using the geographic information in management and marketing, especially for much more rational base-setting of the enterprise strategic decisions. (Bennison, 2005)

In countries like France or Sweden, applications which embed georeferenced data have been developed. Following the American model, adapted software instruments have been conceived to fit the national conditions and to satisfy the needs of local users. Such applications exist in England, but the datasets are not georeferenced. Only local delineations exist for demographic and other data. This is the cause of high degree of uncertainty of the analysis performed. (Badea et al, 2008)

3. THE CASE OF ROMANIA. RESULTS ACHIEVED

One project is going on in the present and focuses on the following objectives: 1) The establishment of a geo-referenced database for Bucharest and the parametric validation of the geographic information. The database will contain economic, geographic and census data offering the possibility to make complex analysis. 2) The elaboration of a modular software application; each module will be customized taking into account the field of interest and the structure of the database. 3) The elaboration of an analytical model which will allow the analysis of the obtained results and the establishment of synthetic indicators for economic prognosis. The project will be finalized in a Ph. D. thesis.

The project aims to implement and test the application on a complex database belonging to Bucharest, which incorporates geographic, economic and demographic data. The structure model for this data was defined in the first part of the project. As a punctual connection between elements contained in the thematic information layers, the following data was introduced in the system: the address, the postal number and the stereographic (x, y) or geographic coordinates.

Until now the datasets were gathered and the geographical information was attached. This is an important aspect of the project as it is the necessary condition to display all the information owned on an interactive map. The validation of the datasets on the field is the next logical step.

4. RESULTS TO BE ACHIEVED

Any commercial entity or public institution will be able to obtain with this application the necessary data in order to make its activity more efficient.

In practice, through a computer, any organization can gain access to a relational database to find all the necessary information for better knowing and understanding the market. Moreover, besides the valuable information about the market, the users have the possibility to exploit the information sets about their own activity and will be able to take real-time decisions. Even though they contain specific thematic information, these datasets must have a geographic component which, in correlation with GIS, will allow the localization of the features of interest on the digital map. The GIS application provides an efficient and ergonomic user friendly interface, which generates "thematic maps" through interrogations addressed to the databases administration system. The maps are being structured in such a manner that allows the economic entity to build an exact picture about a certain area (sector, neighborhood, street, address, objective etc.). In this manner, the user will be capable to analyze the area of interest using diversified interrogation criteria such as: demography (labor force availability, age group, gender group, income level, education level, marital status, etc.), infrastructure (communication and public utility network, education and health network, etc.). Because this information is accessible, the market on which they activate becomes more transparent, and the time period between analysis and decision decreases considerably.

This type of informatics solutions, based on predefined datasets, is successfully applied by U.S. and western-European companies in economic analysis domains such as: consumer profile quantification, complex marketing analysis, strategic segmentation, optimal placement determination for new outlets, transportation/supply or logistics. Also, even public institutions often use similar tools, adapted to census campaigns, population evidence updating, taxes assessment, etc.

5. CONCLUSIONS

Using these informatics platforms is, for sure, a tendency and a necessity of the near future. In the western European countries and in the U.S. this is a common practice. Romania must line up to this reality. Taking into account that our country must recover as soon as possible an important technological gap, this project aims to develop a complex marketing GIS-based instrument and a composite database which will allow the augmentation of Romanian economic entities' competitiveness. In order to be competitive, Romanian organizations must embrace the latest technologies and use up-to-date instruments which, once implemented, become a real competitive advantage. Adjusting these solutions to the Romanian market context develops into a necessity. Actually, the application developed by this project will be a powerful software tool for modeling, analysis and decision. Among the facilities offered, we can list the integration with the GIS technology and the possibility of storing a great amount of data organized into a minimum 12-month data archive, making possible a high level of statistical processing. Besides that, analysis can be performed using a large number of variables for setting interest values for unknown elements which are, most of the times, hidden for classic analysis of economic prognosis. In this context and according with global tendencies, the project proposes to elaborate a very useful application for decision-makers in Romania that will be adapted to the structure of Romanian economy, which finds itself in profound transformation.

The economic organizations must administer and use a large amount of thematic information. This information includes: a geographic localization, an address, a specific service's area coverage, a sales territory, a delivery route that may be visualized and analyzed on a map. The profitableness maximization is realized through a suite of GIS applications.

Figure 1 Example of the GIS application

These applications include selecting of the best emplacements, defining the consumer's profile, market polls and the supplying of services based on optimal localization.

These aspects can be controlled by extrapolating the known values, with the help of GIS techniques, to zones where relevant information is not available. Practically, when the necessary data for a certain customer group is known and it is arranged within a vector structure, an estimation of the stratified component of similar groups becomes available using computer assisted comparative analysis of satellite images (by taking in consideration the density of built areas and building types) (Figure 1)

Through GIS analysis, area with commercial potential that is similar or almost similar as structure with the known model may be established, without using subjective declarative information that are the basis for classic analysis approaches.

The improvement of the decisional results comes from the fact that understanding the factors that influence the demand correlated to the physical localization of the demand eases the manage process and thus:

- ◊ The decisions may be applied faster
- ◊ The opportunities and risks are easier identified
- ◊ Allocating finances and human resources are more rationally made

Efforts for mobilizing the organization's resources to satisfy the client's needs are more efficient

6. REFFERENCES

Badea, A.; Dana, I.F.; Moise, C.; Mamulea A.A. & Badea, R.(2007). ***Method for Multi-Source Data Exploitation, Support for Crisis Situations Management.***, *Proceedings of the* 18th International DAAAM Symposium, Katalinic, B. (Ed), pp 45-46., ISSN 1726-9679, Zadar, Croatia, 2007,, Viena

Badea,R.,Voicu, V.& Moise, C.(2008). *Ways to Improve Sales Management for Romanian Retailers Using Geographical Information Systems,* Proceedings Of 9th WSEAS International Conference On Mathematics & Computers În Business &Economics (MCBE '08), Bucharest, Romania, June 24-26, 2008,

Bennison, D.(2005). *Retailing and the Marketing of Urban Places: a UK perspective*, International Review of Retail, Distribution and Consumer Research, 15 (2), p. 191-215, ISSN 0959-3969, UK

Donaldson, B. (2005). *Sales Management: Theory and Practice 3rd edition*, Codecs Publishing House, ISBN CDCS-12, Bucharest

Zeiller, M., (1999), *Modeling our world, The ESRI Guide to Geodatabase design,* ESRI Press, ISBN 1-879102-62-5, California, USA

Annals of DAAAM for 2008 & Proceedings of the 19th International DAAAM Symposium, ISSN 1726-9679
Editor B. Katalinic, Published by DAAAM International, Vienna, Austria 2008
Make Harmony Between Technology and Nature, and Your Mind will Fly Free as a Bird
*Young Researcher and Scientist Paper / * Supervisor, Mentor*

SOFTWARE PRODUCT "DISTEDU" FOR DISTANCE EDUCATION LESSONS

BAGIMOV, I[gor]; KONOBRITSKIY, D[mitry] & KRAMAR, V[adim]*

Abstract: *In this paper presents developed by authors software product DistEdu, describes its structure and application. This program helps students of technical specialties to learn distantly systems of pneumatic automation, to get skills of assembling of technical systems and communicate with teachers during learning process.*
Key words: *Distance education, software, SYNERGY project*

1. INTRODUCTION

At present time distance education has being developed active. There are more and more people who want to study distantly and independently from place of living.

There are some problems in process of teaching of future technical specialists. The main difficulty is impossibility to get practical skills of operating with real equipment. Correspondence education student can study how equipment works in theory, but can't see or control it. Practically a student understands all symbols of elements in drawing but has no possibility to get skills of assembling of technical systems. It is very important to know how to interpret technical drawing into real device, using inputs/outputs of components and understanding rules of assembling.

There is a wellknown software product named FluidSIM (by FESTO) for creating and simulation pneumatic schemes on level of symbols of components.

Another product is LogicLab – Pneumatic workplace developed by Globula NetStudio for FESTO (http://www.logiclab.hu). Quite realistic animation of working pneumatic systems is realized in this application. Students have possibility to get some lessons about direct and indirect control, simple logic. Images of real components and corresponding symbolic view are presented.

Authors offer to develop the concept of e-learning software for online studuing and practicing. The aims of research are: to combine workplace simulation, remote control with real stands and real visual control of assembled schemes and interactive contact students with teachers.

Thus authors improve wellknown projects by possibilities of feedback communication between student and teacher (for example for receiving comments about mistakes from teachers and joint analyses of assembled schemes) and by possibility to assemble schemes for different purposes and more levels of complexity.

Software DistEdu helps to develop pneumatic scheme, send its image to laboratory, observe process of assembly and communicate with teacher. DistEdu provides audio and video connection between a student and engineer in laboratory, works on any computer via Internet with any operating system.

This article was prepared in context with SYNERGY project realized by universities of Russia, Kazakhstan, Ukraine and FESTO.

2. INFORMATION

Modern telecommunication instruments should be used to solve difficulties with distance access to didactic equipment. We have to use broadband Internet channels to provide video and audio connections and to show real functioning of pneumatic equipment to students.

2.1 Purpose of software

Main purpose of application is to provide proper learning process via distance access to FESTO didactic equipment.

There are several subtasks of DistEdu software:

2.1.1 To assemble pneumatic schemes using library of components (with description of components).
2.1.2 To transmit designed pneumatic scheme to computer in laboratory via Internet or local net of university.
2.1.3 To provide video and audio connection between a student and a teacher in laboratory with equipment using web cameras and microphones.
2.1.4 Virtual components should be looked as real ones.
2.1.5 Software should have reference information about components. This information could be added by users.
2.1.6 A student is able to save schemes on server; an engineer or a teacher in laboratory should have a possibility to open projects just after saving.

Students will have a possibility to work with equipment via Internet from any place on Earth. Thus it is necessary to install Flash-player on computer (small free application) to work with DistEdu and to have connection with server with DistEdu.

2.2 Tools for DistEdu development

ActionScript 3.0 language and Flash CS3 environment were used for development of software. Thus DistEdu was developed without necessity of its installation on computer of user. It is enough to load web page with application and start to work. DistEdu will function on any operational system with installed Flash Player.

PHP language, database MySQL and Flash Media Server were used for developing of server part of application. Software Zend Studio was used for developing of PHP scripts.

2.3 Describing of working principles of software

Software is placed on web server and it starts on computer of final user in Flash player application after a web page is opened. Database MySQL and PHP scripts placed on web server are used for storage and loading of component base. Flash Media Server is used for providing of video and voice connections. It allows to see video from laboratory and to control video flow for several students simultaneously. FMS3 uses H.264 format for data communications, thus Internet bandwidth is used economically but with high image quality.

It is possibly to record video into file on hard drive for further demonstration.

2.4 Interaction of client and server parts

DistEdu uses client-server architecture (Please see Figure 1). Server part is used for storage of application, component library and organization of video communication.

After loading of web page with application user browser loads DistEdu from web server and starts it in Flash player which is installed on computer earlier. During loading of application PHP script on server returns data about component library into XML. Application loads icons and images of

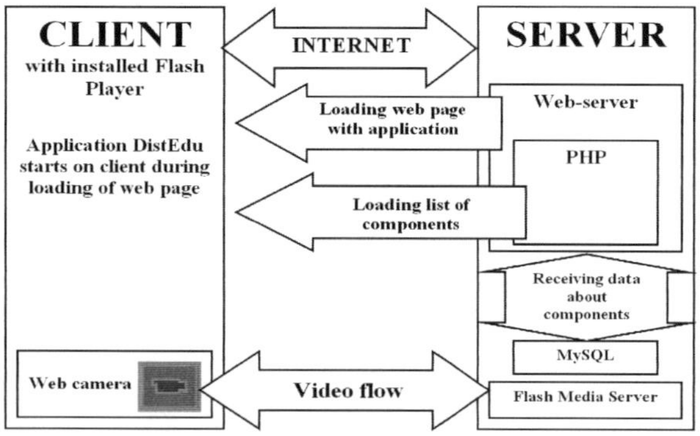

Fig.1. Interaction of program with server

components. PHP script turns to MySQL using query language SQL to receive data from database.

Application connects with FMS and receives video flow from laboratory to provide video connection. Also application directs video flow from client web camera to FMS to realize two-side video connection. The same operations are realized with audio flow to make audio connection. Interaction of data flows is realized on FMS.

Thus we have the following data flows: data flow from web server on HTTP protocol and data flow from FMS on RTMP protocol.

2.5 Review of program interface

Program interface is presented on Figure 2. It is quite simple and allows to work with application right away. There are three areas:

a) Working area for allocation of scheme
b) Component library consisted of list of component images and inscriptions
c) Video window for video from web cameras mounted in laboratories.

Student has to place components on working area and to connect them with pipes. Images of components are photos of real elements, so students can select required elements on appearance. Scale of presented elements corresponds to assembled schemes on a real stand.

It is necessary to mark start and final connection points in order to connect elements with pipes. It is possible to turn and delete elements from working area.

User may save scheme on server after virtual assembling.

Places of areas might be changed if it's necessary. Control panel of video area allows to connect different video cameras in different laboratories.

Users with proper rights may add components. New component will be accessible for all users after reset of program.

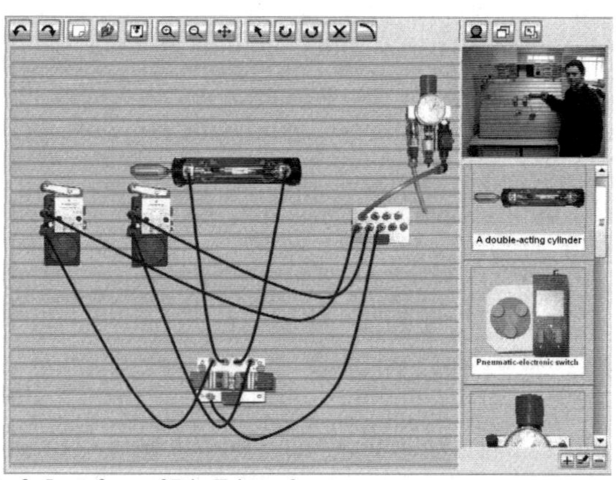

Fig.2. Interface of DistEdu software

2.6 Content of a lesson

Students receive a task to learn (or investigate) pneumatic scheme. Theoretic material and methodical instructions are accessible on server of university.

It is strongly recommended to use FluidSim software from FESTO to learn principles of operation of components and to make virtual model of pneumatic scheme during executing of the task.

After that students have to learn theory and to assemble scheme using DistEdu in compliance with the task. After preparing virtual scheme student have to save it on server and switch on video connection with laboratory.

An engineer in laboratory opens image with scheme and assemble it "as is". A student observes this process and check if scheme is assembled as it is shown on image. Then scheme is ready an engineer switchs on air pressure and demonstrates if it really works. Also it is possible to investigate scheme by measuring of times, pressure, length of movement etc.

After that an engineer (or a teacher) and a student discuss working scheme, look for mistakes if scheme does not work and talk over theoretic material.

Student disconnects, scheme is disassembled in offline mode, teacher and engineer are ready to work online with next student.

2.7 Plans for future

Directions of future development of DistEdu software are:

- Replenishment of component library with elements
- Testing of software and creation of methodical instructions of usage DistEdu as a part of learning process.
- Teaching of correspondence education students, organizing of distance trainings for engineers, assembling schemes as a demonstration of designed projects for industrial customers.
- Developing of applications for other types of equipment: hydraulics, sensors, electro drives ets.

3. CONCLUSION

Interactive software DistEdu for teaching technical specialists was created. Students received a possibility to study a theory, to get practice distantly with equipment placed in university and to discuss technical questions with a teacher.

Correspondence educated students received a possibility to spend more time for practice, thus they'll be better prepared to "face time" communication with teachers in university.

The created structure may be added with new components and expanded for new equipment.

4. REFERENCES

Cove, A.; Eide-Goodman, H.; Lecky-Thompson, E.; Nowicki, S. *Pfofessional PHP5*, Dialictica, ISBN 0-7645-7282-2, Moscow

Bagimov, I.., Klevakin, V., Kramar, V., Mikhailov, M., Mogilnikov, P., Roslyakov, P.,Shalai, V., Stazhkov, S. (2007). International Educational Project "SYNERGY", *Annals of DAAAM for 2007 & Proceedings of the 18th International DAAAM Symposium,* Katalinic, B. (Ed.), pp. 473-474, ISSN 1726-9679, Zadar, Croatia, 24-27th of October 2007, DAAAM International Vienna, 2007

Burlakov, M. (2007). *Adobe Flash CS3 manual for self-tuition*, Dialictica, ISBN 978-5-8459-1319-7, Moscow

Konobritskiy, D. (2007). *Development of software for distance using of FESTO didactic stands*, Qualification work for bachelor's degree, Sevastopol State Technical University, Sevastopol, Ukraine, 2007

LogicLab e-learning software. *Available from http://www.logiclab.hu, Accessed: 2008-07-11*

Annals of DAAAM for 2008 & Proceedings of the 19th International DAAAM Symposium, ISSN 1726-9679
Editor B. Katalinic, Published by DAAAM International, Vienna, Austria 2008
Make Harmony Between Technology and Nature, and Your Mind will Fly Free as a Bird

THE FRICTION EFECT IN ONE SERIAL SONIC CIRCUIT FORMED BY ONE SMALL CONDENSER AND A FRICTION RESISTANCE

BAL, C[armen] & BAL, N[icolaie]

Abstract: *In this paper we present the practical application of the thermal effect in the theory of sonicity. The sonic actions permit the best combination of facilities offered by the processing of electrical signals (reduces energy) with sonic actions of high power and efficiency and gives the possibility of eliminating the biggest parts of a classical hydraulically system (hydraulic reservoir, valve of regulating the debit). The result of this action which combine the opportunities offered by the processing the signals of low energy and the compact sonic actions with high efficiency are a reduced volume and a very economical solution. In the paper we present the effect of the friction used the sonic pressure and the sonic flow in one sonic installation. This installation is formed by one serial cylinders and one frictions resistance. The effects produced in this case are the caloric effects by the increase of the temperature in the friction resistance.*

Key-words: *sonic serial circuit, sonic pressure, sonic flow, friction resistance, temperature.*

1. INTRODUCTION

One of the fundamental problems of mechanical engineering is energy transmission (presented in different forms) at distance, where at a certain point it can be transformed into a useful mechanical work (Constantinescu, 1985).The methods of moving the energy into liquids, used in present paper, are generally based on continuous transmission of pressure and flow, thus these produced at one end of a line can be taken at the other end, the liquid being considered incompressible. The science which is based on the application of elastic proprieties of matter at energy transmission carries the name of *sonic science* or *sonicity*. (Constantinescu , 1985) The energy in the new system is transmitted from one point to other point, at a distance which could be considered considerable, by inducing some periodical variations of compression which produce longitudinal vibrations in the solid, liquid or gas columns. The energy which is transmitting through this periodical pressure and volume vibrations in the longitudinal direction could be characterized as transmission power through *sonic waves*. (Constantinescu , 1985)
The aim of the experimental study is to obtain the caloric effect as succession of the transmission of the caloric effect at distance by vibration (through the sonic waves in liquids).

2. THE SONIC INSTALATION WITH ONE SMALL CONDENSER AND A FRICTION RESISTANCE

This experimental study used the installation presented in figure 1, starting from different frequencies of the drive motor acting the piston of the sonic generator. For each frequency were made three measures corresponding to a static pressure in the installation of 0,25 bar, 0,5 bar.

In the figure 1 is presented the installation were the small condenser are connected in series with the friction resistance. The installation is formed by the sonic generator who is connecting by a pipe with the friction resistance R_f., this resistance is connected also by a pipe to the capacity cylinder (condenser).
The volume of the small condenser is $V = 1462{,}411 \cdot 10^3\ mm^3$.

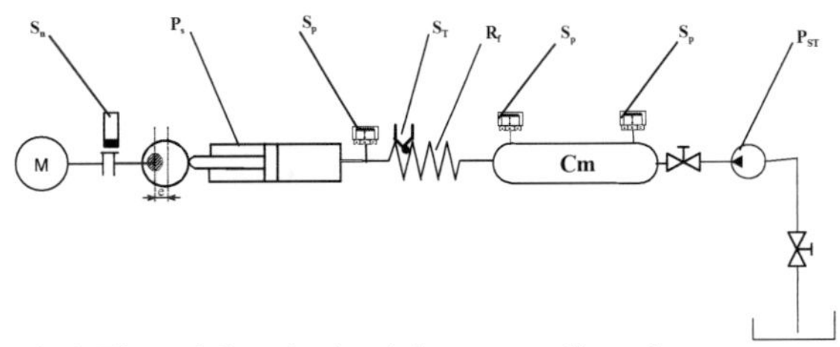

Fig.1 The serial sonic circuit by one small condenser

After processing of the experimental dates obtained from the three sensors placed in the system, results the primary histograms represented in the figure 3. This figure shows the evolution of the generator pressure and also the pressure at the extremity of the capacity cylinder. Also we can see the revolution speed of the generator. (Bal, 2006)
The evolution of the pressure curve shows the existence the phase difference between the generator pressure and the pressure of the capacity cylinder (figure 2).

To put in evidence the effects of the friction we have studied the effects of the sonic pressure in the system. For this we have charged the system with static pressure. For the same static pressure we obtained the diagrams corresponding to these charges. The evolution of the pressure curves shows the existence of a phase shift between the pressure of the generator and the condenser pressure.

In the figure 3 and 4 is presented the variation in time of the diagrams of the pressure and the temperature and also the revolution speed for a the static pressure of 0,25 bar (0,25E+05 Pa). The situation was registered at 50 seconds after the installation start. In this situation we see that the pressure in the pressure sensor, situated on the condenser, is constant and equal with 40E+05 Pa, the generator pressure has the value of 160E+05 Pa, moment when the electrical motor is stopped. (Bal, 2006)

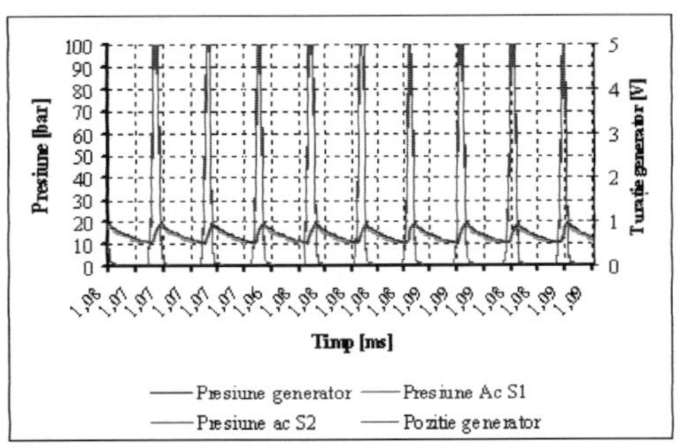

Fig. 2 The evolution in time of the pressure in the sonic circuit with a small condenser

n = 600 rot/min p$_s$ = 0,25E+05 Pa

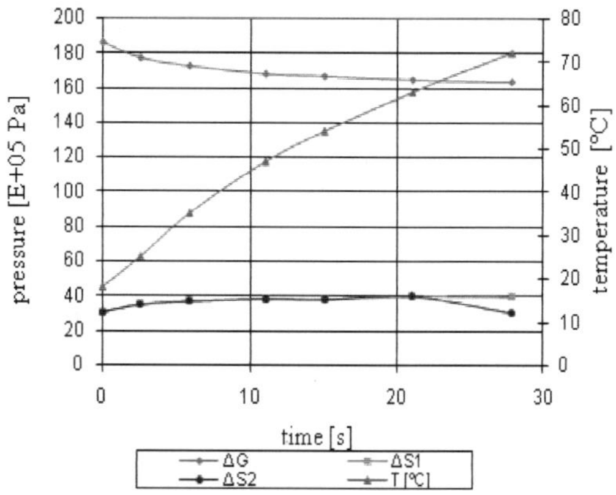

Fig. 3 The variation of the pressure and the temperature as function of revolution speed at 0,25E +05Pa static pressure

The start revolution speed of the electrical motor was 600 rot/min, this speed decreases in time at 350 rot/min and during this time the temperature at the surface of the friction resistance

p$_s$ = 0,25E+05 Pa

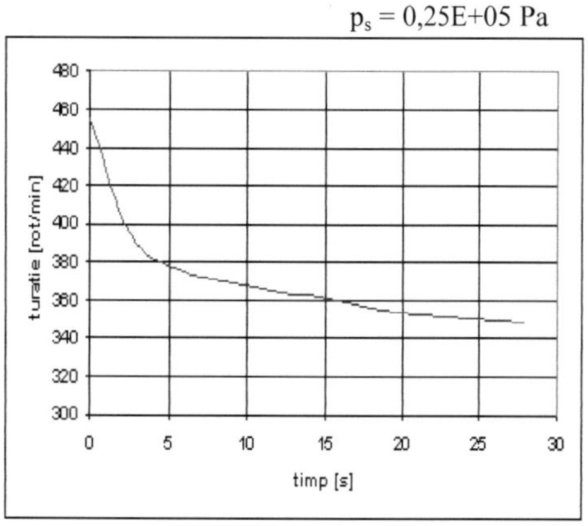

Fig. 4 The variation in time of the speed for the static pressure 0,25E+05 Pa

increases to 70°C after 1 minute and 30 seconds from start of the installation. (Bal, 2007)

In the presented diagrams are following notations:

ΔG – variation of the generator pressure;

n = 800 rot/min p$_s$ = 0,5E+05 Pa

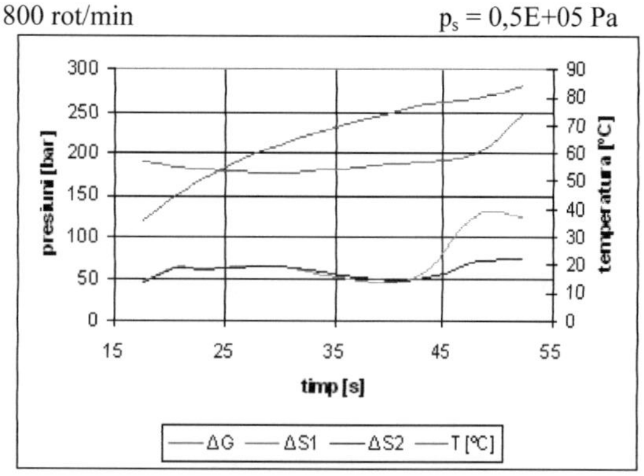

Δ

Fig.5 The variation of the pressure and the temperature as function of speed at the static pressure of 0,5 E+05 Pa

S$_1$ –variation of the pressure obtained with the first sensor of pressure place to the la left of the capacity cylinder;

ΔS$_2$ – the variation of the pressure obtained with the right sensor of pressure placed to the capacity cylinder;

T – temperature. (Bal, 2006)

In the figure 6 are presented the diagrams representing the variation in time of the pressure and temperature and also the revolution speed for the static pressure of 0,5 E+05 Pa. In this case the experiment starts at a small speed 200 rot/min, this speed continuously changes up to 800 rot/min (Bal, 2006). We observed the silence function of the installation, but the big pressure created in the generator (who reached 250E+05 Pa) and in the condenser sensor (125E+05 Pa) produced the electrical motor brake. (Bal, 2006)

Because the big decrease of the pressure into the friction resistance 125E+05 Pa the temperature increased to 85°C.

3. CONCLUSION

By the analyses of the diagrams we observed:

1. The influence of the big pressure and the temperature in the installation has the revolution speed and when the speed at start is high the pressure and the temperature rise quickly;
2. The static pressure in the installation has small influence in the evolution of the temperature in the friction resistance;
3. Because the electrical motor is stopped after short time it is not recommended to use the installation with a small condenser and one friction resistance.

p$_s$ = 0,5E+05 Pa

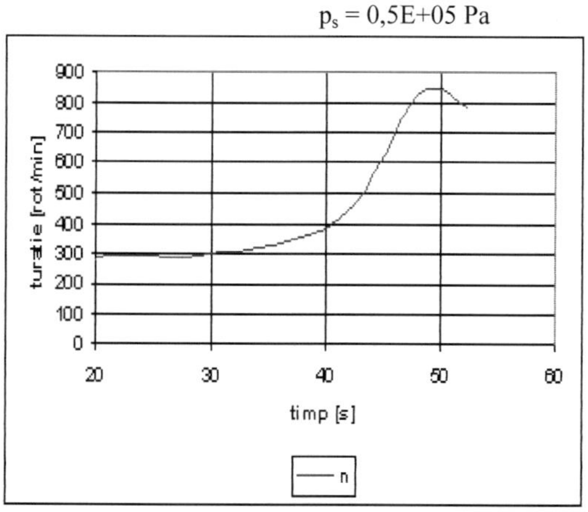

Fig. 6 The variation in time of the speed for the static pressure 0,5E+05 Pa

4. REFERENCES

Constantinescu, G., (1985), *The theory of the sonicity*, Bucureşti, Editura Academiei.

Carmen Bal, (2007), *The caloric effect in the circuits by harmonic flow*, Cluj Napoca, Ed. ALMA MATER, ISBN 978-973-7898-75-3.

Carmen Bal, (2006), *Research and contributions about the drive systems with the harmonic flow*, the doctoral thesis Technical University of Cluj Napoca.

Pop I. Ioan, Carmen Bal, Marcu Lucian ş.a., (2007), *The sonicity applications. Experimental results*, Iaşi, Ed. Performantica, ISBN 978-973-730-391-2.

Pop I. I., (2006), *The treaty of the sonicity theory*, Iaşi, Ed. Perfomantica.

Annals of DAAAM for 2008 & Proceedings of the 19th International DAAAM Symposium, ISSN 1726-9679
Editor B. Katalinic, Published by DAAAM International, Vienna, Austria 2008
Make Harmony Between Technology and Nature, and Your Mind will Fly Free as a Bird

THE FRICTION EFFECTS IN THE SONIC SERIAL CIRCUIT (SMALL CONDENSER AND ONE FRICTION RESISTANCE)

BAL, C[armen] & BAL, N[icolae]

Abstract: *In this paper we propose to analyses the alternative flow in the serial sonic installation including one small capacity and one friction resistanc.e In this paper we show the caloric effect due to the displacement of the low in the friction resistance calculate. The effects makes in this case are the effects caloric show by the upper of the temperature in the friction resistance.*
Key words: *sonic serial circuit, sonic pressure, sonic flow, friction resistance, temperature.*

¶

1. INTRODUCTION

¶

The science which is based on the application of elastic proprieties of matter at energy transmission carries the name of *sonic science* or *sonicity*. The sonicity is different by the hydraulic in the sneeze of this last one, in practical applications considered fluids as being practically incomprehensible. (Constantinescu, 1985)

The energy in the new system is transmission from one point to other point, at a distance which could be considered considerable, with the help of imparting some periodical variations of compression which produced longitudinal vibrations in solid, liquid or gaze columns. The energy with is transmitting through this periodical pressure and volume vibrations in longitudinal direction could be characterized as power transmission through *sonic waves*. (Constantinescu, 1985)

For the transmission of the power through mecanic waves, it is necessary to realize means through which vibrations from the transmitting line could be received and converted for use.

We can suppose that in the case on closed pipe, with an equal length with the multiple of wave length, it will be set in comunication with it, and in the neighboring of piston, a vessel (C) with rigid walls, with a considerable volume from the moved volum at piston movement (P), figure 1.

At each advance course of piston, it will be created a flow through the orifice of vessel, which wills comprime the liquid

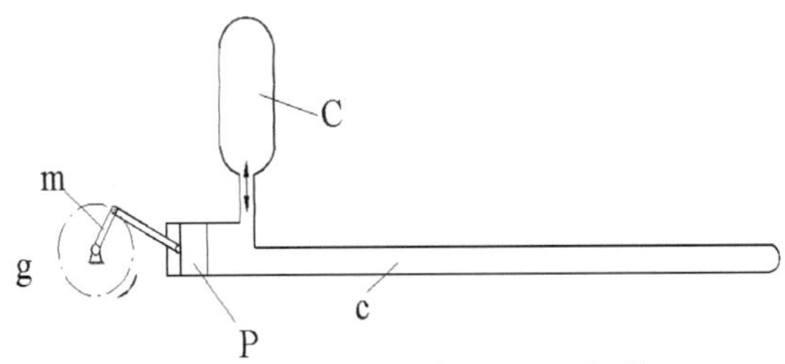

Fig.1 The installation to create the alternant sonic flows

from precincts and at each course of with drawing of it, the liquid will relax. Taking into account the volume of vessel, a certain smaller or bigger quantity of liquid will flow in or from the vessell, at each rotation of crank (m). The capacity (C) will work as an arch, which will absorb the energy of direct or reflected waves, when the pressure is high and given it back when the pressure is decreasing. The medium pressure (C) in pipe will be the same, though after the waves reflected successive in the pipewill reach a certain amplitude equivalent with medium pressure, the piston will develop energy in order to comprime the liquid from (C) at moving forward course, and in this case the liquid working as a resort, will restitute to the piston thee nergy needed for the back race (Constantinescu, 1985). The effect consists in that that the producing of reflected waves will give birth in the piston of the pipe a serie of stationary waves, none raising energy in liquid will results, pressures won't outrun fixed limit.

2. THE FRICTION EFFECTS IN THE SONIC SERIAL

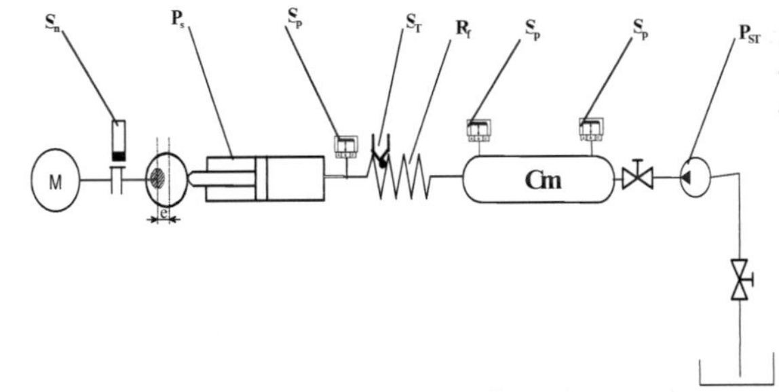

Fig.2 The serial sonic circuit by one small condenser

3.CIRCUIT FORMED BY SMALL CONDENSER AND ONE FRICTION RESISTANCE

This experimental study used the installation present by the figure 1, to start from different frequency of the drive motor ram of the sonic generator. For same frequency are use three measures of the static pressure in the installation with the value 0,75 bar and 1 bar.

In the figure 3 are present the installation were the small condenser are connected in serial with the friction resistance.

The evolution of the pressure curve to notice the existence

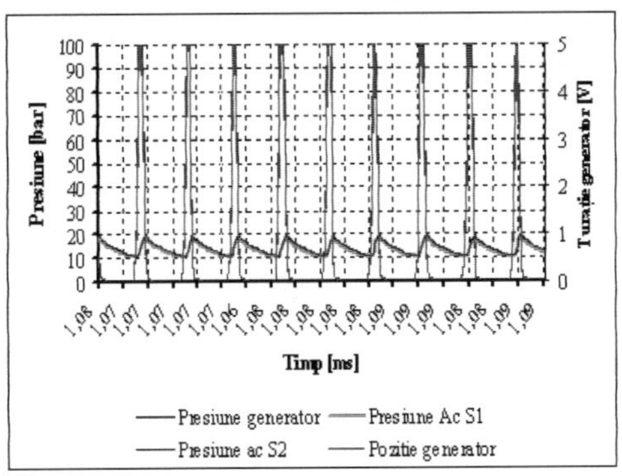

Fig. 3 The evolution in time of the pressure for the sonic circuit with the small condenser

the phases difference by the generator pressure and the pressure of the capacity cylinder (fig. 3) (Pop et al., 2007).

For make in evidence the effects of the friction we can study the effects of the sonic pressure in the system. For this we can have charge the system with static pressure. For same static pressure we obtained the diagrams for this charges. The evolution of the pressure curves shows the existence emphases by the pressure of the generator and by the condenser pressure.

In the figure 4 and 5 we represent the variation in time of the diagrams of the pressure and the temperature and also the speed for the static pressure by 0,75 bar (0,75E+05 Pa). (Bal, 2006)

n = 1400 rot/min p_s = 0,75E+05 Pa

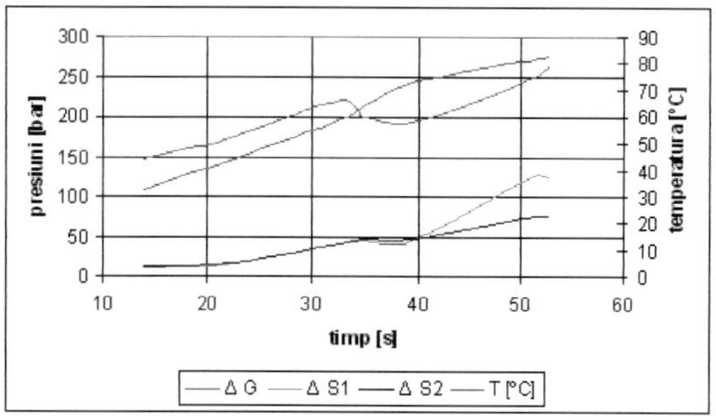

Fig.4 The variation of the pressure and temperature function by speed, to the static pressure 0,75 E+05 Pa

The begin speed of the electrical motor was 1400 rot/min.

p_s = 0,75E+05 Pa

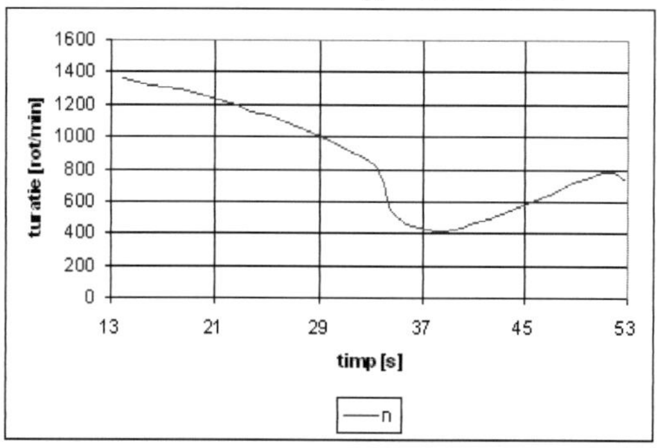

Fig. 5 The variation in time of the speed for the static pressure 0,75E+05 Pa

We are study with the sensor pressure situeted by the condenserare equal to 40E+05 Pa, after these the pressure by the firt senso up to the 130 E+05 Pa and the pressure to the generator up quickly to the 260E+05 Pa, time moment are stopped the electrical motor. The temperature of the friction surface resistance up to 90ºC after 1 munite to move of the instalation.

In the figure 6 and 7 we represent the variation in time of the diagrams of pressure and temperature and also the speed for the static pressure by 1 bar (1E+05 Pa). (Bal, 2006) We see the pressure by the sensors situated by the condenser are equal and he stabisesed to the values 50E+05 Pa, the generator pressure are fluctuation because the big variation of the speed who begin to the value la 834 rot/min and down to 275 rot/min.

That result the pressure 160E+05 Pa to generator, moment that produce the stop the electrical motor. After 1 minute to move the temperature touch to 80ºC. (Bal, 2007)

The test to the static pressure up to 0,75E+05 Pa dont give the results favorable, instalation stopped quickly, but to the static pressure up to 1,25E+05 Pa, the electrical motor cant to move the hydraulic generator.

n = 1667 rot/min p_s = 1E+05Pa

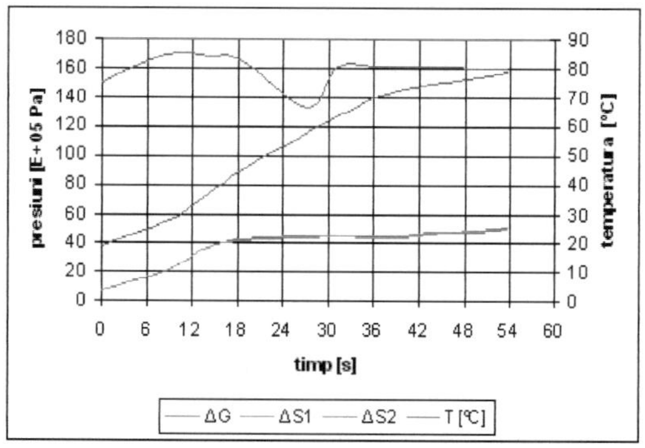

Fig. 6 The variation of the pressure and temperature function by speed, to the static pressure 1E+05 Pa

p_s = 1E+05Pa

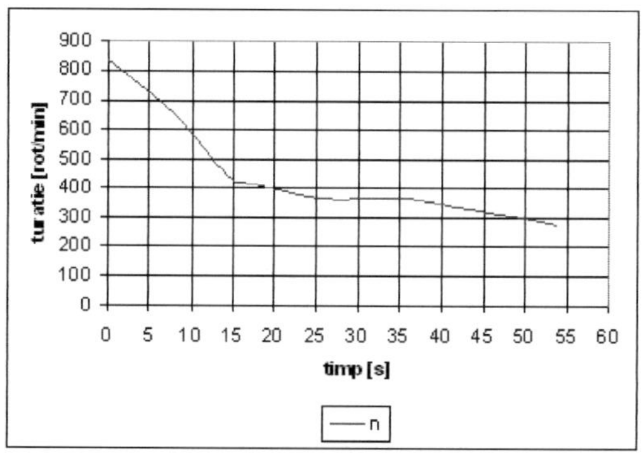

Fig. 7 The variation in time of the speed for the static pressure 1E+05 Pa

4. CONCLUSION

By the analyses of the diagrams we observed:

1. The influence of the big pressure and the temperature in the instalation has the speed, when the speed to begin it is up the pressure and the temperature up quickly;
2. The static pressure in the installation is small influence in the evolution of the temperature in the friction resistance;
3. Because the electrical motor is stopped after short time it is not recommended used the installation with one small condenser and one friction resistance.

5. REFERENCES

Bal Carmen, (2007), *The caloric effect in the circuits by harmonic flow*, Cluj Napoca, Ed. ALMA MATER, ISBN 978-973-7898-75-3.

Bal Carmen, (2006), *Research and contributions about the drive systems with the harmonic flow*, the doctoral thesis Technical University of Cluj Napoca.

Constantinescu, G., (1985), *The theory of the sonicity*, Bucureşti, Editura Academiei.

Pop I. Ioan, Carmen Bal, Marcu Lucian ş.a., (2007), *The sonicity applications. Experimenthal results*. Iaşi, Ed. Performantica, ISBN 978-973-730-391-2.

Pop I. I., (2006), *The theory of the sonicity*, Iaşi, Ed. Perfomantica.

Annals of DAAAM for 2008 & Proceedings of the 19th International DAAAM Symposium, ISSN 1726-9679
Editor B. Katalinic, Published by DAAAM International, Vienna, Austria 2008
Make Harmony Between Technology and Nature, and Your Mind will Fly Free as a Bird

THE CAVITATION PHENOMENON IN THE FLOW ZONE DISTRIBUTOR BODY-SPOOL VALVE

BALASOIU, V[ictor]; BORDEASU, I[lare]; POPOVICIU, M[ircea] O[ctavian]; GOMBOS, R[adu] & GOMBOS, M[ircea]

Abstract:*The cavitation phenomena occuring in the hydraulic machinery and equipments, is for many years a reserch topic, wich led to great running improvements of the hydraulic systems. Cavitation occurs, in some elements of hydraulic drive devices, as a resukt of great increases of the velocity. In distributors, is represents the major limiting factor of the pressure difference of the hydraulic resistence, and as a resukt, it limits the possibility to reduce the flow capacity beyond some values. The cavitation erosion is general not a very stressing factor. On the hand, as a results of important modifications in the characteristics of the working fluid, when the pressure decreases the running of the systems is heavily disturbed. Having as a starst point the general definition of the cavitation coefficients and the criteria for the cavitation inception there were defined the cavitation coefficients for the directional valves having cylindrical spools.*
Key words:*cavitation erosion, distributors, hydraulic systems,*

1.INTRODUCTION

The hydraulic resistance describe hydraulic system elements with diverse functional role, and the diversity of type construction assures the purpose for their role. The majority of hydraulic resistance works by strangle the flow vein . From this reason the cavitation phenomena can occurs during with increasing speed and the different pressure values in different points of the hydraulic resistance.

For studying the cavitation regime in hydraulic resistance distributor body-spool valve , the research had focus to determine energetic spectrum of analyse cavitation effect among the hydraulic distribution apparatus (Anton 1985).

In a series of anatomization Numachi broach the problem of cavitation inside Venturi tubes and inside the diaphragm to determine the cavitation start moment ,by using the relation:

$$\sigma_{inst} = \frac{p_i - p_d}{\Delta p} = \frac{p_i - p_d}{\rho \frac{v_0^2}{2}} \qquad (1.a)$$

Where: p_i pressure on the strangled section , p_d the vaporization pressure of the liquid, v_0 the reference medium speed to input. The initiation and the development of the cavitation inside the hydraulic servo valve with cylindrical spool valve is possible due to local flowing conditions, which can occurs by increasing the speed and lowering the pressure and it is characterized through the cavitation coefficient (Raszga 1998, Gerald 1994, Toshiuki et al. 1993,)

Numachi, starting from the definition relation of the flow rate coefficient, he determine by experimental the coefficient :

$$C_k = \frac{Q_k \sqrt{1 - m^2}}{\frac{\pi}{4} d_2^2 \sqrt{2g(p_{1k} - p_{2k})/\gamma}} \qquad (2)$$

Where: C_k –flow rate coeficient in cavitation regime, Q_k – flow rate in cavitation conditions, p_{1k}, p_{2k} input and output pressure in cavitations. In figure 1, is presented the variation of the flow coefficient versus the cavitation coeficient.

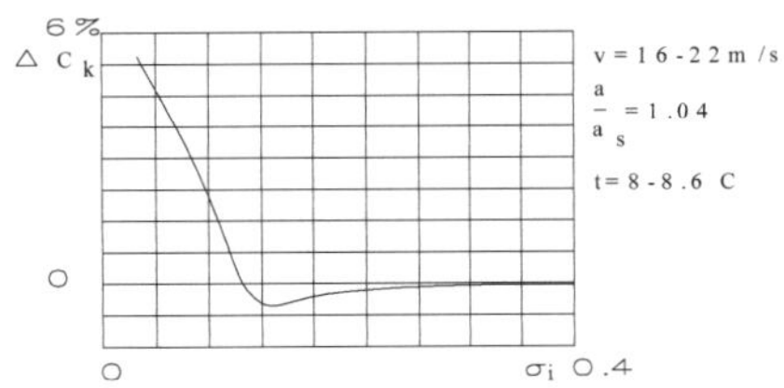

Fig.1. The variation of the flow coefficient versus the cavitation coeficient.

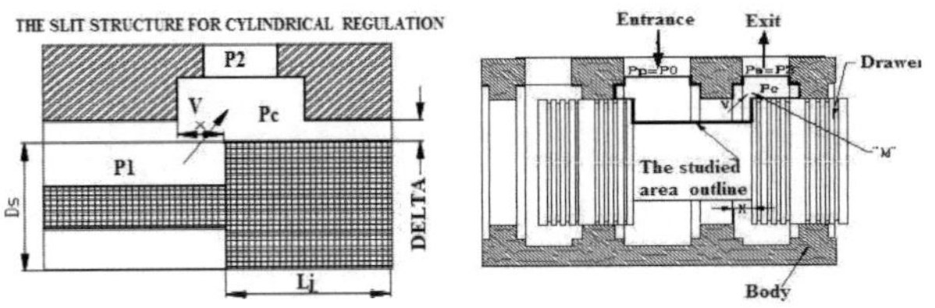

Fig.2. Exit circuit of apparatus

The spool-valve distributor its characterized by Martin and Wiggert (Anton 1985) as being made of several hydraulic resistance, which works at high pressure and increasing speed. The most important is high speed jet with circular form, which appears at the exit of adjusting slit through cylindrical room and the exit circuit of apparatus.(Fig 2). If the jet is free for small apertures and two-dimensional flow model, it will reach to a value for flow coefficient Cd of 0.673, and a 69° in ratio with spool-valve axis, calculated by McCloy and Martin in 1973 (Raszga 1998).

In distributors with cylinder spool-valves can be notice two types of incipient cavitation correlated to free or attached jet. In case of attached jet can be notice a cavitation development along wall and the cavitation bubbles appears due to increased friction tension near to wall, that leads to a liquid break and appearance of new liquid-gas interface.

The development of cavitation in distributors with cylindrical spool-valve can be identified by recording dynamical values, which are pulsating pressure in the exit room of circular jet, which represent a different noise from the turbulent jet, to be distinguished in energetic spectrum of measured values. (Martin et al., 1981) due to development conditions of phenomenal and assumed measurements for experimental appliance. The type of cavitation that appear is developed cavitation, not a gas cavitation that appears in other author's works incriminated. Point out the presence of dissolved air in 10% has a reduced effect due to short time of existence of cavity in case of cavitation implosion. But they are responsible whit forming of air bubble with high dimensions which can affect system dynamics.

2. GENERAL EXPRESSION OF CAVITATION COEFFICIENT FOR CONTROL VALVES

If we consider the structure of hydraulic trace for a distributor with cylindric spool-valve, in maximum speed assumption, respective lower pressure, appears near to minimal section of trace from the adjusting slit area. (Point M from fig.3) Making use by cavitation coefficient of distribution section, ratio to input section we obtain:

$$\sigma_{corp0} = \sigma_{D0} = \left(\frac{V_{max}^2}{V_0^2} - 1\right) + \frac{h_{p0M}}{\frac{V_0^2}{2g}} \tag{3}$$

The energetic loses therm can be write starting from the equation of energy transfer, to take into account that the hydraulic loses are preponderant due to perturbation of speed field. Reference to entrance section we obtain cavitation coefficient for distributor:

$$\sigma_{D0} = \left(\frac{V_{max}^2}{V_0^2} - 1\right) + \frac{h_{p0M}}{\frac{V_0^2}{2g}} = \left(\frac{V_{max}^2}{V_0^2} - 1\right) + \varsigma_{0M} \tag{4}$$

Reporting the difference of pressure between in and out (1.a), we take input section as reference:

$$\sigma_{inst0}^{*} = \frac{p_0 - p_v}{\Delta p_{02}} \tag{1.b}$$

$$\sigma_{D0}^{*} = \frac{\rho V_0^2}{2\Delta p_{02}}\left[\left(\frac{V_{max}}{V_0}\right) - 1 + \zeta_{0M}\right] \tag{5}$$

for output section considered as reference, relation (5) becam:

$$\sigma_{D2}^{*} = \frac{\rho V_2^2}{2\Delta p_{02}}\left[\left(\frac{V_{max}}{V_2}\right) - 1 - \zeta_{M2}\right] \tag{6}$$

Reporting to section "x", relation becam: :

$$\sigma_{Dx}^{*} = \frac{\rho V_x^2}{2\Delta p_{02}}\left[\left(\frac{V_{max}}{V_x}\right) - 1 + \zeta_{xM}\right] \tag{7}$$

Similar to cavitation coefficient (Raszga 1998), (Toshiuki et all. 1993) the local loosing coefficient signify the energetic looses which emerge on trace of the hydraulic distributor with cylindrical spool valve, due to strong perturbation geometrical structure of the speed field of the inner hydraulic trace. The major loose it take place in adjusting slit area with x opening. At the slit output appears a drown annular jet in the exit room ("A" or "2") .In fig. 4 is represented the spectrum of current lines on the all considered domain, and in fig. 5 we have represented a detailed flow structure in adjusting split area. Inquiring the energy transfer equation between section "0" and "f" the local energetic losses coefficient ζ_{0M} that appear on hydraulic trace with cylindrical spool valve becomes:

$$\zeta_{M2/0} = \frac{p_f - p_2}{\rho\frac{V_0^2}{2}} - \frac{K_0^2}{K_2^2} + \frac{1}{C_c^2}\frac{K_0^2}{x^2} \tag{8}$$

The reserved cavitation coefficient in all matters of connote, marking out the difference between cavitation coefficient of installation and the cavitation coefficient of device (apparatus) – hydraulic distributor. Long time as the value of reserved cavitation coefficient is positive , the regime of work remain cavitation free.

In the moment when the value of reserved cavitation coefficient is null, the working cavitation regime is incipient , and in one point from the hydraulic trace the pressure drops enough to make possible the air bubble emerge ,and the liquid break and the gas vapour development is high enough . In this working regime and only now ,the value of cavitation coefficient of device is equally numeric with value of cavitation coefficient of the distributor , although they differ as analytic relations. From analytic point of view , the characterization of free cavitation regime of incipient, developed and industrial cavitation and super cavitation it is realised. Practical knowledge of coefficients values is difficult by using local values characteristic flowing field through distributor spool valve.The all definition for cavitation coefficient ,in particular for distributor with cylindrical spool valve ,was started from rigorous definition of cavitation regime and cavitation coefficient.

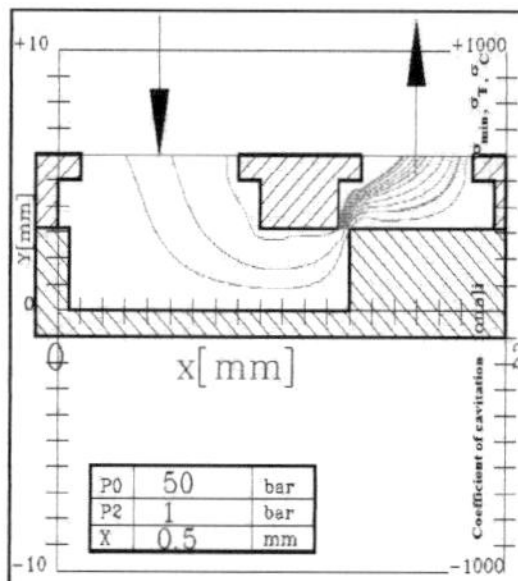

Fig.4. The spectrum of current lines

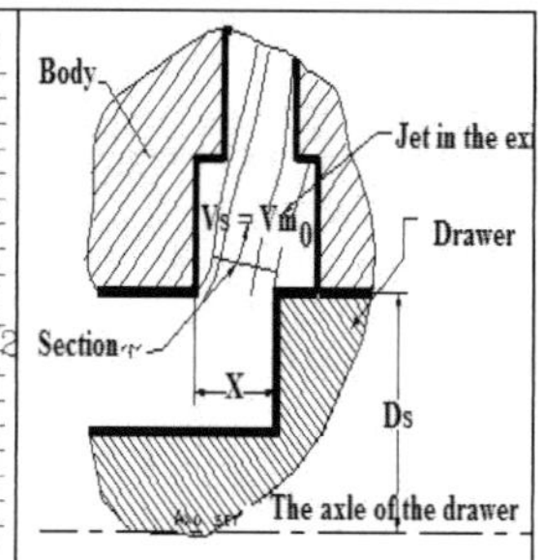

Fig.5. Detailed flow structure

3. CONCLUSION

Elaborated documentation about studied phenomenon – the cavitation phenomenon in the hydraulic distributor – spool valve, allow knowledge from theoretical research and experimental, and offer the based elements for extend knowledge of phenomenon. To accentuate the existence of cavitation phenomenon inside the acting hydraulic installation, which works usually with mineral oil, and especially in case of distributor with spool valve, it was accomplished by experimental way and deduced with numerical way by using calculus software with method of finite element.

Correlation of experimental result by Martin and Wiggert (1981) with indicated numerical results by (Gerald C, 1994), (Toshiyuki H. et al. 1993), (Anton, 1985) . To determine an analytic numerical relation even approximated it is necessary if we want to describe the initiation and cavitation phenomenon effect in operation of the distributors with spool valve.

ACKNOWLEGMENTS

The present work has been supported from the National Univ. Research Council Grant CNCSIS - IDEI nr.35/ 68 / 2007.

4. REFERENCES

Anton, I. (1985), *Cavitation* , Vol II, Ed. Academiei Romaniei, Bucuresti,

Gerald,C.F.,Wheatley.P.O.,(1994), *Applied numerical Analysis,* Edition Hardcover, ISBN:13.9780201565539, pp.536

Raszga, C., (1998)..*The Cavitation of phenomenal in the distribution of cilynder sertar,* Doctoral Degree Thesis,

Toshiuki, H., Cheng, P., Hayashi, S. (1993) *Numerical Analysis of Transient Flow Through a Spool Valve,* Reports. Inst. Fluid Sciences, Tohoku Univ, pp.123-133, ISSN:0916287

Martin, C.S., Medlarz, S., Wiggert,D.C., Brennen, C. (1981), *Cavitation Inception in Spool Valves,* Journal Fluids Engineering, ASME, Vol 103 (4), pp564-576, ISSN: 0098-2202.

Annals of DAAAM for 2008 & Proceedings of the 19th International DAAAM Symposium, ISSN 1726-9679
Editor B. Katalinic, Published by DAAAM International, Vienna, Austria 2008
Make Harmony Between Technology and Nature, and Your Mind will Fly Free as a Bird

COMPUTER AIDED EDUCATION IN AREA OF AUTOMATIC CONTROL: CAAC INFORMATION SYSTEM

BALATE, J[aroslav]; NAVRATIL, P[avel] & CHRAMCOV, B[ronislav]

Abstract: *This paper is deals with a presentation of the CAAC (Computer Aided Automatic Control) information system. The CAAC information system has been arranged for the time being into 15 problem areas, the so called subsystems which include the automatic control theory. In all subsystems, the so-called "problem specification" is formulated by means of tree structure. The tree structure serves as a basis for the creation of structure of given subsystem on webpages. In this paper is shown the CAAC information system tree structure of concrete subsystem - analysis. The use of the CAAC information system is anticipated especially in pedagogical processes in area of automatic control.*

Key words: *automatic control, dynamic webpages, education, Internet*

1. INTRODUCTION

When solving the particular design of automatic control of the technological process, it is necessary to use some of the methods of analysis of controlled systems, or their identification, synthesis of control loops, solution of robustness of the designed control systems, or adaptive algorithms, design of optimum algorithms for control, simulation of behaviour of the controlled system, including control processes, or predication of time series and predictive control, with an aim at improving quality of the course of the control processes, simultaneously providing safety at technological processes, their economy and influence on the environment. The described CAAC information system represents a continuously created information system of partial problem areas which include the automatic control (Navrátil, 2004). This information system is a certain form of eLearning.

2. CAAC INFORMATION SYSTEM

2.1 Conception of the CAAC information system

The CAAC information system structure is open, with a possibility to formulate further problem areas in the area automatic control. The CAAC information system has been arranged for the time being into 15 subsystems (see Fig. 1) (Navrátil, 2004).

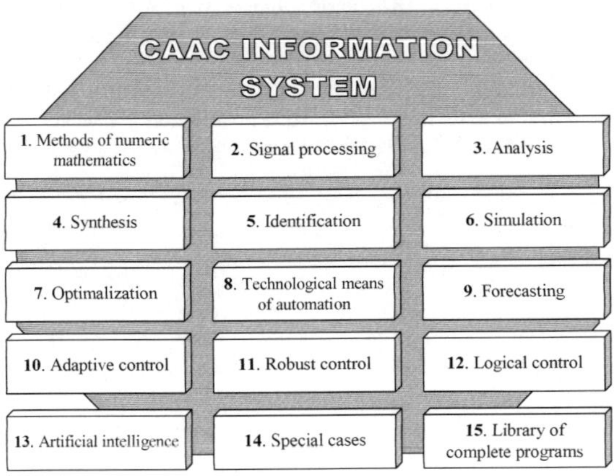

Fig. 1. Problem areas of the CAAC information system

2.2 Structure of the CAAC information system

In all subsystems, the so-called "problem specification" is formulated. Each subsystem is split up into the modules, which are further split up into the sub-modules up into the basic sub-modules, which is the lowest level of hierarchic arrangement of the CAAC information system (see Fig. 3). The basic sub-module solves the concrete problem (e.g. system stability according to the Lyapunov theory) of the given subsystem (e.g. Analysis) and should include description of solving of the problem, example, references and computer program. An exception is the subsystem „Library of complete programs", where the basic sub-module should include a computer program created for a chosen parts of the concrete subsystem or subsystems and also description of this program (Navrátil, 2004).

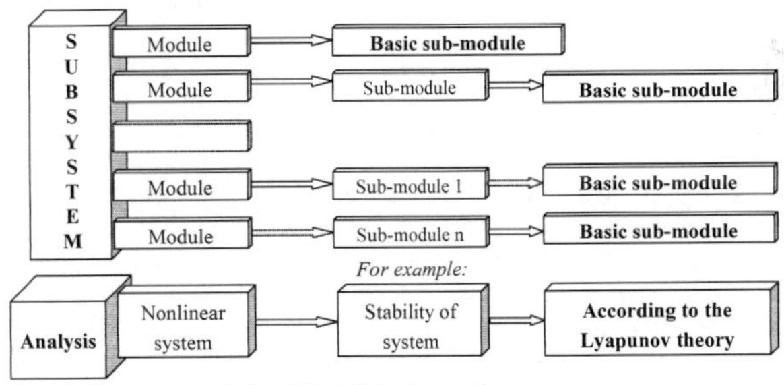

Fig. 2. Structure of the CAAC information system

2.3 Tree structures of subsystems of the CAAC information system

Tree structure of a chosen CAAC information system subsystem is shown as an example in Fig. 3. You can also find other examples of structures in (Navrátil, 2004).

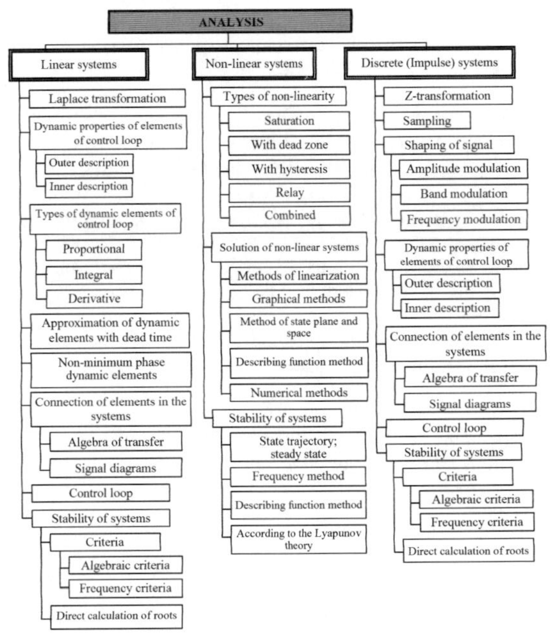

Fig. 3. Tree structure of analysis subsystem

3. CAAC INFORMATION SYSTEM ON WEBPAGES

The CAAC information system is very extensive it consists of a large quantity of the individual basic sub-modules which determine due to their contents the quality of the whole CAAC information system. Therefore it was very important to determine such claims in order that it may be possible to complete the CAAC information system with the new basic sub-modules, or as the case may arise, to modify the existing basic sub-modules easily and quickly.
Therefore the following claims are posed on the CAAC information system on WWW pages:
- hierarchical arrangement and intuitive names of individual directories and files
- unambiguous structure of a particular webpages of the CAAC information system
- safety at updating and adding of individual parts of this information system
- fast searching of files for their updating

One of possible approaches is the creation of the directories structures and a proposal of the files location in the CAAC information system, the creation of the contents of webpages for individual parts of CAAC information system and the creation of the information lists about the present state of the particular parts of the CAAC information system solution (Navrátil, 2004). To creation of the CAAC information system have used possibilities of PHP language, structured query language (SQL), HTML and cascade style sheets (CSS) (Veselý, 2006).

3.1 Structure of directories and files location in the CAAC information system

For correct and safe adjustment or replacement of one of the files of the CAAC information system, it is necessary to keep a certain location and names of the directories and files. Therefore, they were created 3 types of the directories structures and proposed the files location in these directories. These 3 types cover the whole problem of layout of directories and files in the CAAC information system on webpages. The first structure deals with the most fundamental arrangement of the directories and files in the CAAC information system. The other two structures refer to location of the files inside individual subsystems of the CAAC information system.

3.2 Contents of webpages

Each contents of webpage includes a certain part of the structure of the CAAC information system, which was shown in Fig. 3, where each subsystem is split up into the modules, which are further split up into the sub-modules up into the basic sub-modules. There have been created 3 contents of webpages, i.e. contents of webpages of the subsystem, contents of the information webpage of the subsystem "Library of complete programs" and contents of webpage of the basic sub-module.

3.3 Information lists about the situation of the particular parts of the CAAC information system solution

For better orientation in items of information on status of individual solved parts of the CAAC information system, it was necessary to create a structure of lists of this information system. With help of these lists it could be better to search for information on actual status of particular problems, and on the basis of information gained like this, modifications in this information system might be done. Two principal types of the information lists have been created. The first type of the list will comprise information on basic sub-modules of the particular subsystem. This list will be used for subsystems 1 to 14. For subsystem 15, i.e. the subsystem „Library of complete programs", the second type of the list has been created, as not so much information is required for this one.

4. USAGE

The use of the CAAC information system is anticipated mainly for didactic purposes whereas utilisation of webpages on the Internet is supposed. Concrete problems of given problem areas - subsystems 1-14 (see Fig. 1) will be described in basic sub-modules in detail. The finished programs that will solve problems in the given problem area or areas will be loaded in subsystem 15 - "Library of complete programs" (see Fig. 1). Last figure (see Fig. 4) shows the way of loading of information into the CAAC information system. The solved task will pass through the evaluation on basis of which the solved task will be recommended or not recommended to be included into the CAAC information system. In case of positive assessment of the solved task, this will be made available to the users by means of the service of webpages. The users will be able to give their comments, based on their experience, to the solution through the author (Navrátil, 2004).

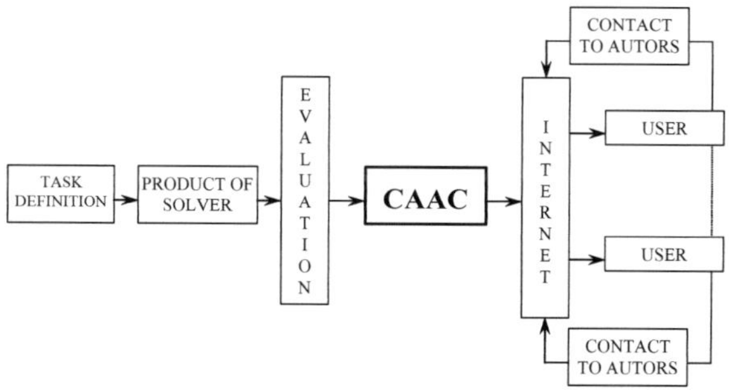

Fig. 4. Information flows in the CAAC information system

Other electronic utilities for didactic purposes in area of the automatic control theory have been created especially at universities, i.e. TU in Ostrava, TU in Liberec, etc. (Wagnerová & Minár, 2000), (Wagnerová & Minář, 2000), (Vítečková, 1999), (Linka, 2007).

5. PRESENT SITUATION

At present, the CAAC information system has on webpages partially loaded the subsystems "Analysis", "Synthesis", "Logic control" and „Library of complete programs", i.e. in the subsystem "Analysis" there are 10 basic sub-modules, in the subsystem "Synthesis" 20 basic sub-modules and in the subsystem "Logic control" 10 basic sub-modules. The subsystem "Library of complete programs" comprises 6 finished programs. At present time, chosen webpages are accessible on the Internet (***http://www.caac.zde.cz***). For the time being, these webpages have been created in the Czech language.

6. CONCLUSION

The aim of this paper was to demonstrate one of the possible approaches to education realization for the area of automatic control by utilization of webpages on the Internet. The CAAC information system is proposed as a tool for realization of an electronic support of education that is called eLearning.

7. REFERENCES

Linka, A. (2007). *TUL E-learning system*, *Available from*: http://e-learning.vslib.cz *Accessed:* 2007-05-02 (in Czech)

Navrátil P. (2004). *CAAC information system - Computer aided automatic control* (Dissertation), TBU in Zlin, Czech Republic. (in Czech)

Veselý, L. (2006). Usage of the modern tools to modification of CAAC information system on webpages (Bachelor's work), TBU in Zlín.

Vítečková, M. (1999). *The mathematic methods used in the area automatization and control*, *Available from*: http://www.fs.vsb.cz/books/matmet/ATR.htm *Accessed:* 1999-06-18 (in Czech)

Wagnerová, R. & Minár, K. (2000). Tutorial for the analysis of control loop, *Available from*: http://www.fs.vsb.cz/books/Analyza/Index.htm *Accessed:* 2000-07-21 (in Czech)

Wagnerová, R. & Minář, M. (2000). Control loop synthesis, *Available from*: http://www.fs.vsb.cz/fakulta/kat/352/uc_texty/Synteza/ index.htm *Accessed:* 2000-08-22 (in Czech)

Acknowledgements: This work was supported in part by the Grant agency of Czech Republic under grant No: 101/06/0920 and in part by the Ministry of Education of the Czech Republic under grant No. MSM 7088352102.

Annals of DAAAM for 2008 & Proceedings of the 19th International DAAAM Symposium, ISSN 1726-9679
Editor B. Katalinic, Published by DAAAM International, Vienna, Austria 2008
Make Harmony Between Technology and Nature, and Your Mind will Fly Free as a Bird
2nd DAAAM-ICAT

CAD FOR OPTIMAL SCALING OF THE 3D MODEL, TO COMPENSATE THE SLS POST-PROCESSING ERRORS

BALC, N[icolae]; BERCE, P[etru] & PACURAR, R[azvan]

Abstract: *The paper presents a new CAD method and a new software package to calculate better scaling factors for the Selective Laser Sintering (SLS) technology. The errors during the SLS process and during post processing are analysed. FEA was used to estimate the 3D deformations during post processing in the oven the SLS metal parts, in order to infiltrate them to decrease their porosity and improve their mechanical properties.*
Key words: *CAD for Rapid Prototyping, Selective Laser Sintering, Post Processing, Finite Element Analysis*

1. INTRODUCTION

It is well known the Selective Laser Sintering (SLS) process and technology, which can be used to produce complex parts in different type of materials. Besides the advantages of the SLS process, as compare to the classical manufacturing technologies, there are some disadvantages, as well.
One of the major disadvantages of all the Rapid Prototyping (RP) technologies is the accuracy to produce metal parts, which is bellow the accuracy provided by some classical technologies, such as grinding.
The use of the RP technologies has become more and more frequent in the manufacturing industry, because they offer the quick solution to produce prototypes of the very complex shapes, without being necessary any special tools. Anyhow, there is a lot of work to be done to increase the accuracy of the metal parts produced by RP.
The latest advances in the recent developed processes, such as selective laser melting and electron beam melting, offer the possibility to quickly produce a complex fully dense metal part, but research is steel necessary to be done, in order to improve the accuracy and the surface roughness of the metal parts made using these technologies.
This paper presents a new CAD method to find the necessary scaling factors, in order to compensate the SLS manufacturing errors.
New software has been developed by the authors, in order to compensate the shrinkage which might occur during different post processing methods of the SLS metal parts.

2. SLS ERRORS AND EXISTING METHODS TO COMPENSATE SOME ERRORS

It is very complex the SLS process and there are different types of errors which might be involved [Yan, et al., 1998] and [Tang et al., 2003], caused by: the scanning system, material contractions, layers scanning, etc.
On top of that, there are random errors and systematic errors in connection to a particular SLS machine. The Sinterstation 2000 – SLS machine drives the laser beam using the two mirrors galvanometric system, illustrated in figure 1.
F-Theta lens is used to focus the laser beam onto a horizontal working plane. The software package provided by the SLS machine manufacturer does not have the possibility to compensate the distortions caused by the scanning system.

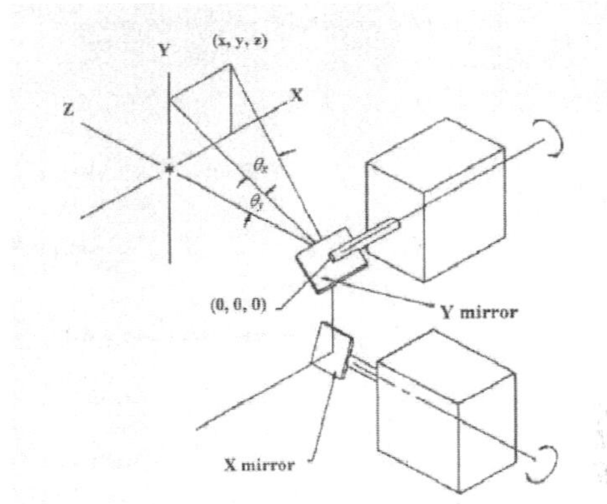

Fig. 1 The scanning system with 2 mirrors

There is a calibration procedure for the SLS machine, in order to estimate some compensation factors, just for the SLS process onto the Sinterstation 2000 machine, but not for the post processing. For plastic parts made by SLS, the accuracy improvements were discussed by Raghunath [2007] and Yang [2002]. For the PolyJet TM process the scale factors were optimized [Brajlih et al., 2006].

3. NEW METHOD TO CALCULATE THE X, Y, Z COMPENSATION FACTORS

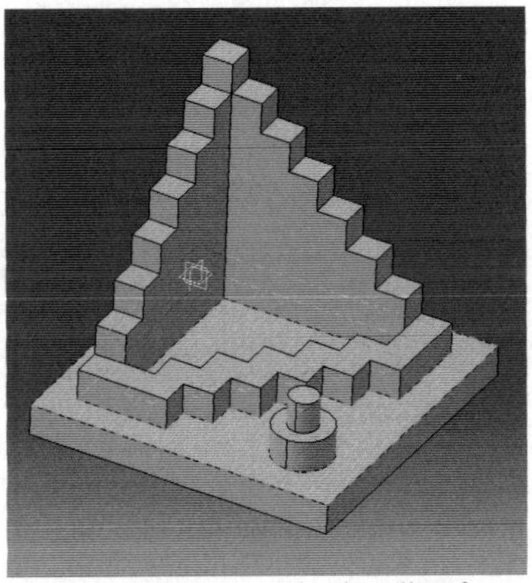

Fig. 2 New test part to find the optimal scaling factors

New original software package has been developed by the authors, to calculate better scaling factors for the SLS process, which take into account the post processing deformations.
The test part, illustrated in figure 2 was designed in order to test and validate the new software, called FOS (Optimal Scaling Factors). The FOS software calculates the scaling factors by

taking into account both the systematic errors of the Sinterstation machine measured onto the green parts and the post processing shrinkage. During post processing the SLS metal parts, 3D deformations occur, depending of the type and temperature of the post processing cycle.
It was used the Finite Element Analysis method, to estimate the deformations during post processing the sintered metal parts.
The FEA has been done in two stages. First stage consisted in a transient thermal analysis.

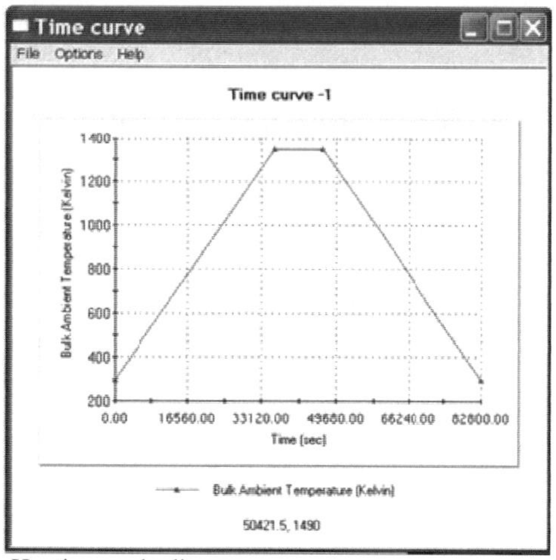

Fig. 3. Heating cycle diagram

For the undertaken case study, the warming up time was 12 hours (up to 1070 ° C), the maintaining period was 3 hours and the cooling period was 12 hours.
The purpose of this stage was to find out the temperature distribution in the part, at different moments, during the heating cycle illustrated in figure 3.
The second stage consisted in a thermoelastic analysis. Two cases have been studied:

- Thermoplastic expansion during the heating process from room temperature to 1070 °C. In this case, the thermal expansion coefficient has been set to a higher value, due to the fact the sintered material is in a powder state.
- Thermal contraction during the cooling process in the oven, from 1070 °C to the room temperature. In this case, the thermal expansion coefficient has been set to a lower value, due to the fact the part is already in a compact state.

As conclusion, the effects of the infiltration process can be modeled by modifying the value of the thermal expansion coefficient (α - coefficient). The other values for the parameters used within the FEA are presented in figure 4.

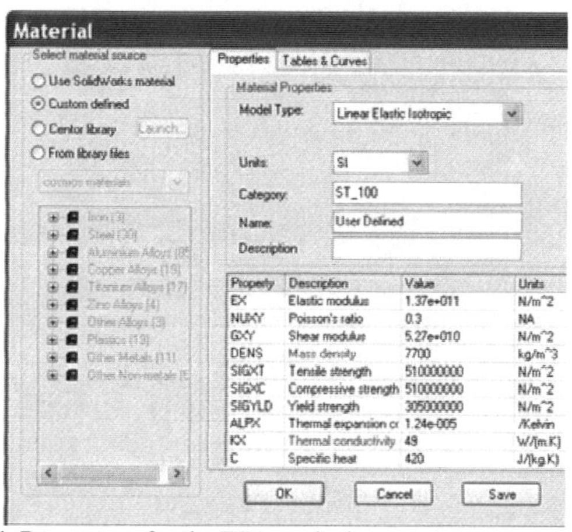

Fig. 4. Parameters for the FEA analysis

The thermoelastic analysis has been performed imposing the minimum amount of kinematical restraints. More precisely, just the translation degrees of freedom corresponding to three perpendicular surfaces were locked: the basis and two lateral surfaces of the part. In this way, the thermal deformation could take place along all three axes.

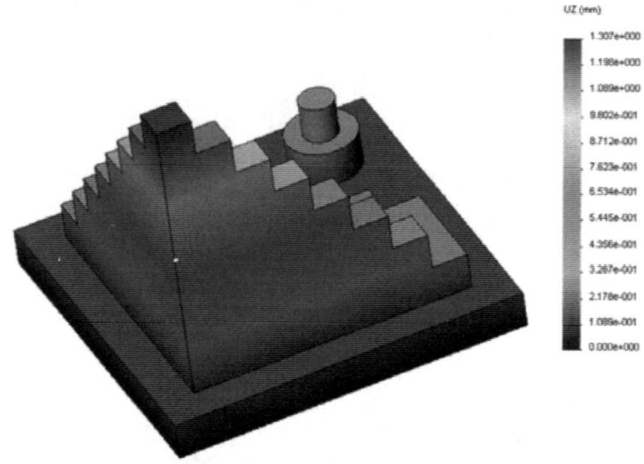

Fig. 5. Thermoplastic expansion, when heating up to 1070 °C

For the 100 mm high part, the thermoplastic expansion was 1.307 mm and the thermal contraction was 1.158 mm, so the errors on Z axis would have been 0.149 mm.
All deformations on x,y,z axis are presented in table 1.

Deformation	X	Y	Z
Warming up	1,451	1,451	1,307
Cooling down	1,285	1,285	1,158
Residual deformations	0,166	0,166	0,149

Tab. 1. Estimated deformations, using the FEA

4. RESULTS AND CONCLUSIONS

The new software developed in C++ was tested and validated onto the test parts and some case studies of producing complex metal parts by SLS. The FEA was used to estimate the 3D deformations during post processing by infiltrating the SLS steel parts with epoxy resins. Both experimental data measured onto green parts and post processing data estimated using FEA were used as input data for the new FOS software, which calculates better scaling factors to alter the dimensions of the 3D virtual model, in order to obtain good dimensions of the SLS part, as close as possible to the theoretical dimensions.

5. REFERENCES

T. Brajlih et al., Optimizing scale factors of the PolyJet TM rapid prototyping procedure by genetic programming. *J. Achiev. Mater. Manuf. Eng.*, May-Jun. 2006, vol. 16, iss. 1/2, str. 101-106

N. Raghunath, Pulak M. Pandey, Improving Accuracy Through Shrinkage Modeling by using Taguchi method in Selective Laser Sintering, *International Journal of machine Tools & Manufacture* 47 (2007), 985-995.

Y. Tang et al, Accuracy Analysis and Improvement for Direct Laser Sintering, *International Journal of machine Tools & Manufacture* 43 (2003), 985-995.

M. Yan, et al., Analysis of machine accuracy for rapid prototyping of quality components, *Proceedings of SPIE*, The International Society for Optical Engineering, V 3517 (1998), 91-101.

H.J. Yang et al., A study on shrinkage compensation of the SLS¬process by using Taguchi method, *International Journal of machine Tools & Manufacture* 42 (2002), 1203-1212.

Annals of DAAAM for 2008 & Proceedings of the 19th International DAAAM Symposium, ISSN 1726-9679
Editor B. Katalinic, Published by DAAAM International, Vienna, Austria 2008
Make Harmony Between Technology and Nature, and Your Mind will Fly Free as a Bird

ADVANCED DESIGN OF HIGH PRESSURE DIE CASTING MOULDS

BALTES, L[iana]; TIEREAN, M[ircea] & EFTIMIE, L[ucian]

Abstract: *The paper presents some modern design strategies in high pressure die casting moulds field, using different mould profiles with complete sets of interchangeable inserts. There are presented the rules of design and the measures introduced to develop a modular design, to increase the maintainability of the high pressure die casting moulds.*
Key words: *design, mould, interchangeability, casting.*

1. INTRODUCTION

The fast going changes in the economy of the last years has determined several strategic changes in the industrial manufacturing. Aspects as:

- decreasing the designing time of the products due to the development without precedent of software and computers; in automotive industry for example, which is one of main factors in the development of casting industry, the design time for a new car decreased to six months;
- the decreasing of replacement period of parts with a new one due to the same development of IT industry and market demands;
- the increasing pressure on manufacturing price, effect due greatly to the globalization of world economy;
- the demands to obtained the final parts without other further machining, which means time and money;
- the necessity to modulate the tools due to the decreasing of manufacturing cycle, the design and processing costs;

asses the necessity to rethink the strategies in high pressure die casting execution (Bralla, 1999).

2. ACTIVE PARTS DESIGN

The classical design system is based, for the most parts, on massive constructions, strategy due in the main part to the large-scale series production with quite rarely changes of manufacturing type, in which the massively mould (single casting construction) was the guarantee of a high number of casting pieces (Madan et al., 2007).

The reduction of the manufacturing exchanging rate, the decreasing of mould casting number of pieces until the product replacement in addition with the new equipments and manufacturing techniques, warrants the new directions in designing.

It is possible to see in figure 1 the simply positioning system of the sliders (1) in relation to the mould body (2) inclusive one face to other, that allow their changing with other sliders, having only the overall size and sealing system identically, but other active parts, so pieces with different profile.

Assembling, disassembling, maintenance and correction working are very simple through the replacement or maintenance of part symbols 3 and 4.

As well it is possible to remark in figure 2 how the active part was divided in critical surfaces end/or interchangeable.

Thus, central part (2) for reason of ventilation, to prevent the complete filling of the shape, has a simple positioning due only to the constructive shape which prevent the wrong positioning (rotation, translation without iteration) not only plane but also in elevation. The same principles are applied to the fixed part of the mould, figure 3.

Positions 3, 4, 5 and 6 (fig. 2) are interchangeable inserts which assure good ventilation, easier manufacturing as component parts, facile displacement without high qualified personnel, directly in the casting house. Their replacement could allow the execution of other profiles of cast moulding.

The total costs of a mould obtained through this designed system, for a family composed by different three-pieces, are appreciatively with 45% lower than the execution of tree

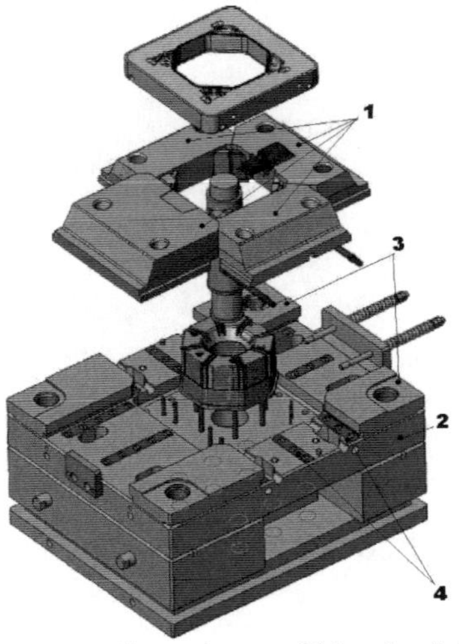

Fig. 1. High pressures die casting mould for aluminium cover: 1-sliders; 2-mold body; 3-guiding inserts; 4-sliding inserts; 5-active parts.

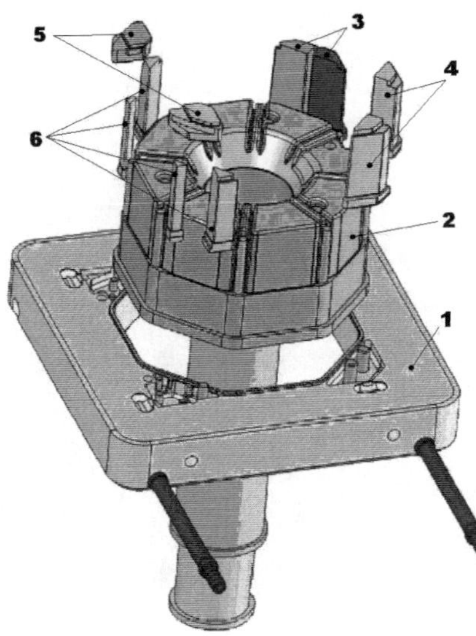

Fig. 2. Active parts assembly for mould presented in figure 1 1-active part; 2-core; 3, 4, 5, 6-interchangeable inserts.

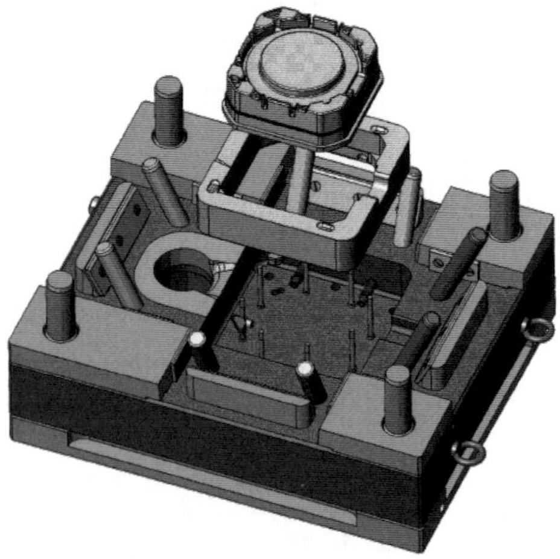

Fig. 3. Fixed part assembly of the mould.

different moulds. The maintenance costs decrease with 20-25%. Statistically, the down-time, the maintenance work, and reconditioning of the mould due to the accidental lock-out, decrease with 20-30% depending on the mould complexity, dimensions, cavity number, etc (Driscoll, 2002).

High reliability and the decreasing of the maintenance time are assured by sliding guides with high hardness, rigorous unstrained as position, only as thickness, in order to assure the free distance gliding of sliders. It is also considered the specific heat expansion which appears for different type of materials, dimensions and shapes. These, together with gliding slides for sliders displacement in the work direction, are spare parts easier to dismantling, to replace and to repair.

In this way it is possible to realise gliding systems for sliders, considering only the keeping of geometry in the guiding area, inclusive the strict bordering in tolerance zone provided for all interchangeable sliders.

The interchangeable inserts which can be delivered in assemblage sets, allow the different shapes parts execution. Thus from design we realised different mould shapes within the same project through the activation/visualisation of a certain configuration of the project (Zhao et al., 2005).

These insertion pieces can be assemble either through the positioning of pins and screws compression, or through the direct positioning on "key" type profiles, which lock up at assemblage the wrong assembling or the assembling of other interchangeable component instead of the correct one.

Such a mould exceed the tools cost and the time for completion in comparison with a simple mould for one part realising. The design time and costs increase but, in opposition, the overall cost with the casting tools decreases dramatically, due to incorporation of several moulds in one.

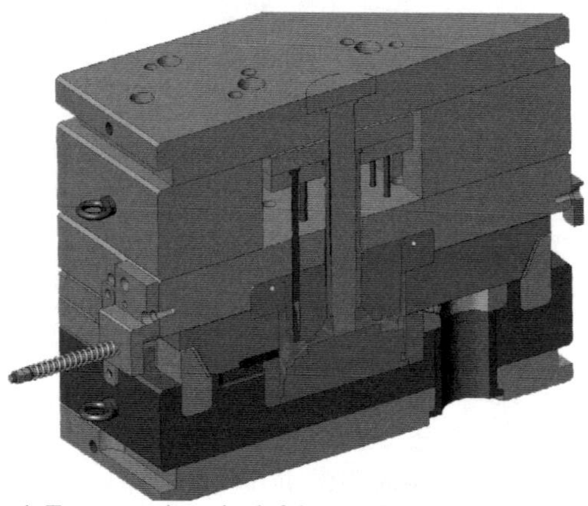

Fig. 4. Transversal section of the mould presented in figure 1.

Fig. 5. Mould for aluminium covers manufactured at Aeromold Ltd.

3. ADVANTAGES OF THE NEW DESIGN

The advantage of this design and manufacture type consist in a better controlling of the uncertain collisions and interferences on each mould shape, the whole project having different profiles attached to each casting part.

Other real advantage of this type of design consist in fact that designer being forced to use insertion pieces due to the multiple shapes of the pieces rank from the demand pieces family, the active tools instead of the manufacturing from the massively ingot are realised from insertion pieces, configuring in this way the piece geometry. This fact increase the lifetime of the mould and the down-time of the tools, witch is very costly. At each accidentally damage or due to the wear, the mould is get out from the pressing machine table and with nonqualified personnel the defective parts are replaced with new ones.

As it is observed in figure 4, the different shapes of the mould are grouped almost in totality in the centre of the mould, in the active part.

4. CONCLUSION

Last years changes concerning the demand of casting aluminum alloys parts forced major changes in design and manufacturing strategies of this type of moulds. The present paper joins the new trend of the market.

5. ACKNOWLEDGMENTS

The authors thank to Aeromold Ltd. Brasov, Romania for credit and support in the design and experimental research.

6. REFERENCES

Bralla, J.G. (1999) *Design for Manufacturability Handbook*, McGraw Hill, New York

Driscoll, M. O. (2002) Design for manufacture, *Journal of Materials Processing Technology*, 22, pag. 318–321

Madan, J.; Rao, P. V. M. & Kundra, T. K. (2007) Computer Aided Manufacturability Analysis of Die-cast Parts, *Computer-Aided Design & Applications*, Vol. 4, No. 1-4, pag. 147-158

Zhao, Z. & Shah, J. J. (2005) Domain independent shell for DfM and its application to sheet metal forming and injection molding, *Computer-Aided Design*, 37, pag. 881–898

SolidWorks Co. (2007) Solid Works documentation

Annals of DAAAM for 2008 & Proceedings of the 19th International DAAAM Symposium, ISSN 1726-9679
Editor B. Katalinic, Published by DAAAM International, Vienna, Austria 2008
Make Harmony Between Technology and Nature, and Your Mind will Fly Free as a Bird

HIGH SPEED MACHINING WITH INTERNET MONITORING

BALTES, L[iana]; TIEREAN, M[ircea] & EFTIMIE, L[ucian]

Abstract: *The paper presents a combined strategy for using much more efficient the production time, increasing the capabilities of the workshops and reducing the manufacturing costs. Having these goals it is proposed a method to increase the manufacturing availability on the night and weekend by remote controlling of the CNC machine through Internet connection.*
Key words: *remote controlling, Internet monitoring, manufacturing, machining.*

1. INTRODUCTION

High speed machining among the intensive removing of metal means also the integral using of machining time to increase the productivity. To solve this problem it was realised an Internet application for the monitoring of the machining process. The presented solution refers to a combination of techniques designed to decrease production costs while shortening the work time through a set of combined solutions.

2. CHOOSING OF THE TOOL

The choosing of the tool according the type of application, in any cutting tool process, is crucial in obtaining of the good results. In this paper we propose to realise an high speed milling, remote controlled by the operator for a piece of hardened Duralumin 6061 T651 used in a polyurethane injection mould, by casting technology Reaction Injection Moulding. The piece is presented in figure 1.

The chosen tool (fig. 2) is a milling head, similar to "bull nose" mill, having inserts with high productivity geometry, which ensures high speed for the metal removing. There are assured such high operating conditions, as in table 1.

S [rot/min]	F [mm/min]	V [m/min]	F_z [mm]	A_p [mm]
7000	12000	703.36	0.571	3

Table 1. Milling process parameters.

Where:
S = speed spindle,
F = feeding rate,
V = cutting speed.
F_z = tooth loading,
A_p = cutting depth,

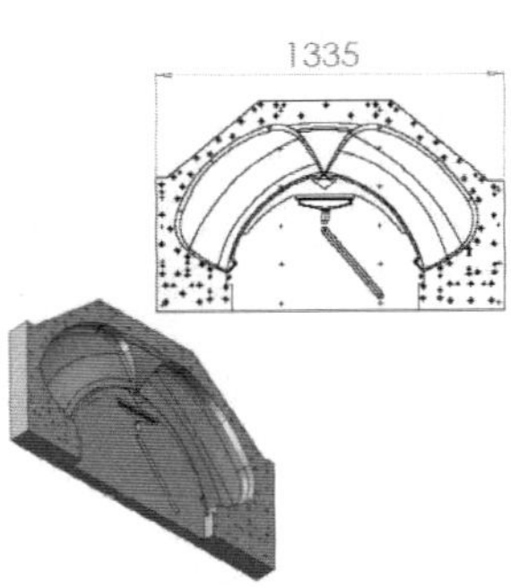

Fig.1 Duralumin manufactured part.

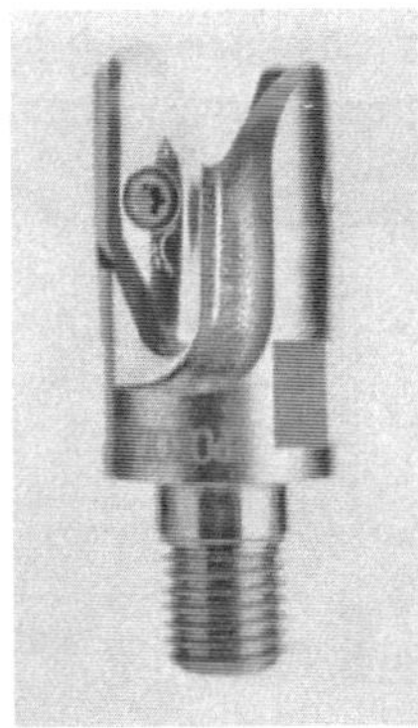

Fig. 2. Milling head used in the processing of figure 1 part.

For the roughing milling, it was used an high degree of tool coverage, with adjustment of cutting speeds depending on the tool path and differentially feeding rate on tooth insertion into material. As result, a big volume of removed material approximate 38 dm^3, can be processed in an hour. The equipment used for machining was a vertical processing centres HAAS.

3. OPTIMIZATION OF MILLING TOOL PATHS

Using of adequate software allows through attached routines the process optimisation, the most important being the computing of inactive positioning movement G_0 and the choosing of the shortest paths of positioning and movements between milling areas. Shortness of cutting time is approximate 10-15% over the total process time. The software chosen for this application is ESPRIT CAM (fig. 3).

After compiling and optimizing of the milling tool paths (fig. 4), in order to prevent collisions and to find the stock that must be re-processed in previous operations, the simulation process was done (fig. 5).

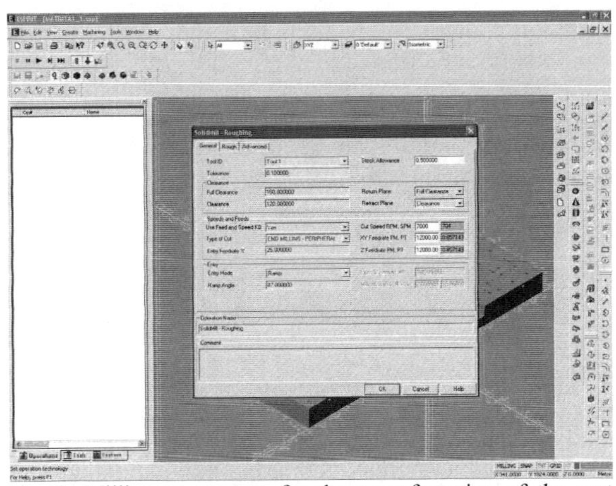

Fig. 3. Milling parameters for the manufacturing of the part from figure 1.

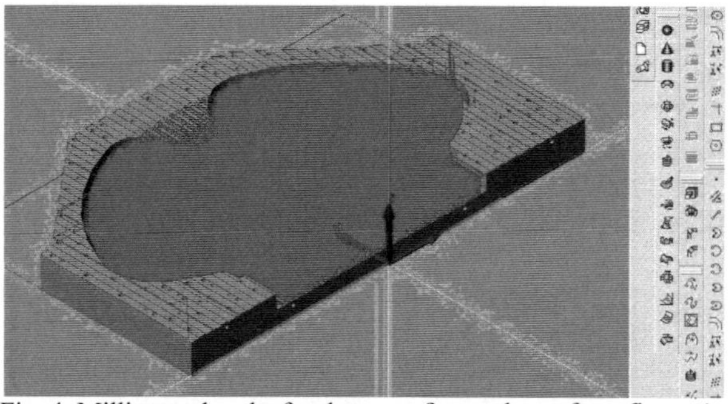

Fig. 4. Milling tool paths for the manufactured part from figure 1.

Fig. 5. Manufacturing process simulation for the part from figure 1.

The final result is post-processed and is estimated the working time for the production planning. The estimation process is extremely important in planning and organizing of the technological production flow (fig.6).

4. REMOTE MANUFACTURING

Increasing of the processing cost, especially due of working time, can be reduced using the third measure, among the proper tool choosing and path optimisation, the remote controlling of the machine. This measure can be applied for "non-hazardous" processes considering the process precision, but big time consumers, as roughing and semi finishing processes (Cilliers, 2002).

It was developed a software application, compatible as data transfer and monitoring system with HAAS VF5 processors, which allows on-line working with the equipment, using the Internet and/or GSM link.

Such systems have been developed only on high-cost CNC equipments, but which, by their amortization extremely high (prices of equipment is 50-100% higher) rise the processing time cost, thus increasing of the product cost. The on-line

Fig. 6. Simulation process in order to estimate the manufacturing time for the part from figure 1.

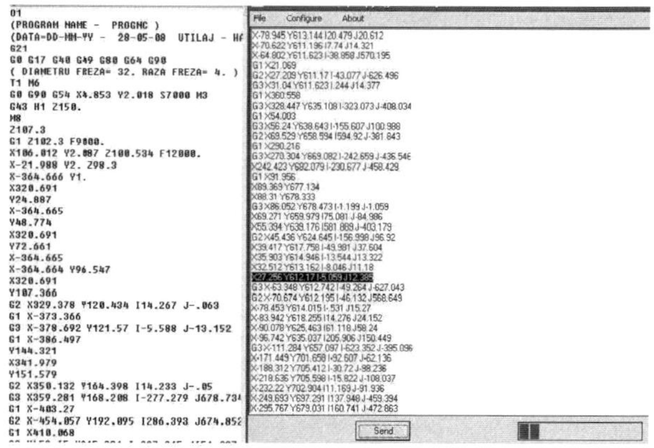

Fig. 7. On-line transmission of the program from PC to CNC machine tool for the manufacturing of the figure 1 part.

system allows the loading of some programs much bigger than the equipments memory, thus assuring a long operating time without stopping. All these stops cause additional costs.

In the tools' storehouse are introduced the adequate tools for roughing cut and semi finishing, that can be reached without the presence of operator. Being a more critical operation, the finishing phase requires the operator assistance.

For each equipment the monitoring is done in real time and it is possible to connect several equipment to a single PC (fig.7) (Kao et al., 1996).

The system warns through SMS message (mobile phone) only in the event of an accidental problem and betimes announces the stop of the machining process.

In this way it is possible to ensure the simultaneous working of several processing centres with a single operator or use the whole weekend's period. The small resources involved, as well as costs, make such an application to have extremely low monitoring prices and practically required only one network and Internet access.

The application has proved extremely useful in the case of parts which consume a lot of time. Practically, the only minimal technical condition required is to remove efficiently the chips in the case of concave parts, using an efficient multi-liquid jet cooling systems cooling or air (Song et al., 2005).

Another utility is the processing of a series of parts, with moderate tolerance, at which the tools wearing is at the range of 0.05 mm.

5. CONCLUSION

By cumulating manufacturing strategies, choosing the cutting conditions, tools and monitoring, the delivery time for products decreased significantly by increasing the number of working hours per week in combination with the decrease of industrial time costs per hour by reducing the labour costs per unit of product.

6. REFERENCES

Cilliers, C. (2002) *Remote monitoring and diagnosis for control of EDM parameters*, PhD thesis, Rand Afrikaans University

Kao Y.C. & Lin G.C.I. (1996) CAD/CAM collaboration and remote machining, *Computer-integrated manufacturing systems*, vol. 9, no3, pag. 149-160

Song, I.H.; Park, J.M. & Chung, S.C (2005) Web based interference verification system for injection mold design, *Computer-Aided Design and Applications*, 3 (1-4), pag. 129-138

DP Technology Corp. *ESPRIT CAD/CAM Software*, Camarillo, CA

HAAS Automation Inc.; *Programming and operation Manual VF-series*, Oxnard, CA

Annals of DAAAM for 2008 & Proceedings of the 19th International DAAAM Symposium, ISSN 1726-9679
Editor B. Katalinic, Published by DAAAM International, Vienna, Austria 2008
Make Harmony Between Technology and Nature, and Your Mind will Fly Free as a Bird

A HYBRID ALGORITHM TO MINIMIZE THE NUMBER OF TARDY JOBS IN SINGLE MACHINE SCHEDULING

BANCILA, D[aniel] & BUZATU, C[onstantin]

Abstract: *This paper studies the problem of scheduling n jobs on a single machine in order to minimize the number of late jobs. Each job has a release time, a certain processing time and a due time. Taking into consideration all the aspects involved in scheduling this specific case of problem is one of the most important because of its applicability on the engineering field, many solutions being proposed for its solving. By combining priority rules with genetic algorithms, this paper offers a hybrid solution for solving this problem.*

Key words: *scheduling, genetic, algorithm, optimization, priority.*

1. INTRODUCTION

This study regards the general problem of scheduling on a single machine, denoted by $1|r_j|\sum U_j$. A set of n jobs needs to be scheduled on one machine, the processing cannot be interrupted, there are no precedence relations among jobs and only one job can be processed at a time. Each job J_j has a release time r_j, needs p_j time units to be processed on the machine and must be completed until the due time d_j. The starting time of a job is denoted by t_j $(t_j \geq r_j)$ and the completion time by $C_j(C_j = t_j + p_j)$. If the J_j job is completed after its due date $(C_j > d_j => U_j = 1)$ it is said to be *late*. Otherwise, if $C_j \leq d_j$, the job is *on time*. Thus, our objective is to minimize the number of late jobs$(\sum U_j)$.

In literature, this problem is considered to be NP-hard, series of solving methods being proposed over time.

In (Dauzere-Peres & Sevaux, 2004) the authors solve this problem by developing a branch-and-bound method. This algorithm defines a "master sequence" by using dominant rules.

In (Crauwels et al., 2005) the same problem is also treated by branch-and-bound techniques, but the jobs are grouped into families. A two alternative branching schemes is proposed.

The two methods presented above have a highly degree of efficiency in the case of small problems. However, in the case of more complex problems the computational time increases, making them impossible to be solved.

In (Gelinas & Soumis, 1997) is presented the scheduling problem on a single machine having as objective the minimization of completion time of all jobs. In this case the authors used the dynamic programming in order to solve the problem.

(Vakhania, 2004) solved the same problem by the means of an algorithm called "balancing procedure", which used the GT heuristic.

In the last years, more and more interest gained the suboptimal methods, among these the genetic algorithms being frequently used.

In (Sevaux & Sorensen, 2004), is presented a genetic algorithm applied on a Just In Time scheduling, while in (Miller et al., 1999) the genetic algorithm is combined with an Wagner-Whitin algorithm and with TSP modeling.

Taking into consideration the fact that genetic algorithms have better results in shorter time we developed a hybrid genetic algorithm in order to solve our problem. In the next section the used hybrid algorithm will be described.

2. THE HYBRID ALGORITHM

The proposed hybrid algorithm (H.A.) (Fig.1) is based on the fundamentals of genetic algorithms, being modified according to the following steps:

1. Generate an initial population;
2. Produce a new generation P;
3. Invert the new obtained generation to produce InvP generation;
4. Evaluate the objective function for both P and InvP generations;
5. Choose from P and InvP the best n individuals to form the next generation (n – number of individuals from a generation);
6. Repeat steps 2 – 5 until the prescribed stop criterion is satisfied.

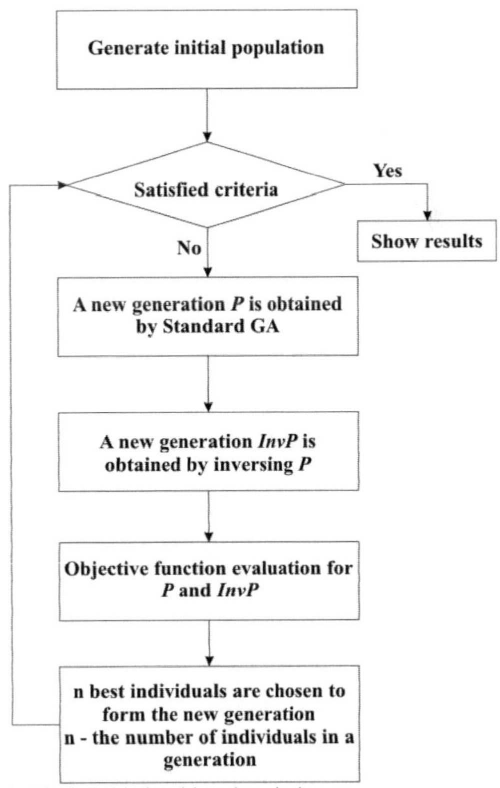

Fig. 1. The hybrid algorithm description

3. COMPUTATIONAL RESULTS AND COMPARISON

In this section we will discuss the computational results of the hybrid algorithm in the case of single machine scheduling.

In order to properly compare the performances of a standard GA and the HA, we applied both algorithms on several scheduling problems, a simple case being outlined in the following section.

The problem consists of seven jobs which have to be scheduled on a single machine. The release, processing and due time for each job are given in Table 1.

J	1	2	3	4	5	6	7
r_j	0	2	4	5	7	11	18
p_j	3	5	2	1	7	6	3
d_j	4	7	9	8	18	22	27

Table 1. Release, processing and due time for each job

When comparing their performance, the HA and GA have the same population size, consisting of six individuals, the genetic operators are the same in both methods, and the number of generations is set to 60. The results of one sample are illustrated in fig. 2 and fig.3.

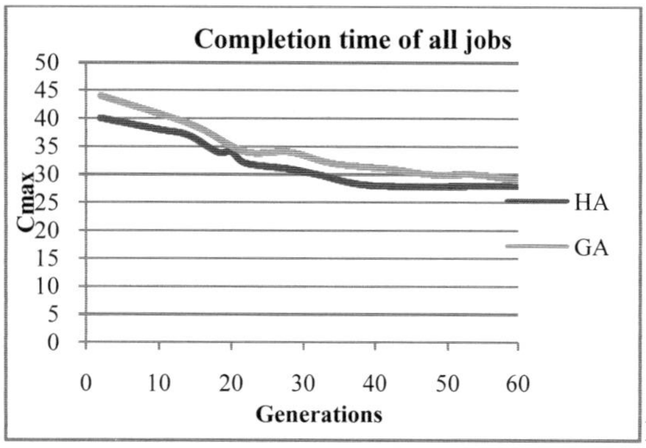

Fig. 2. Comparison between HA and GA – completion time of all jobs

From fig. 2 it is shown that the HA improves much faster than the GA, which is an advantageous feature of the HA. In addition, the execution time for the HA is almost the same as the GA. The time spent on inverting procedure in negligible.

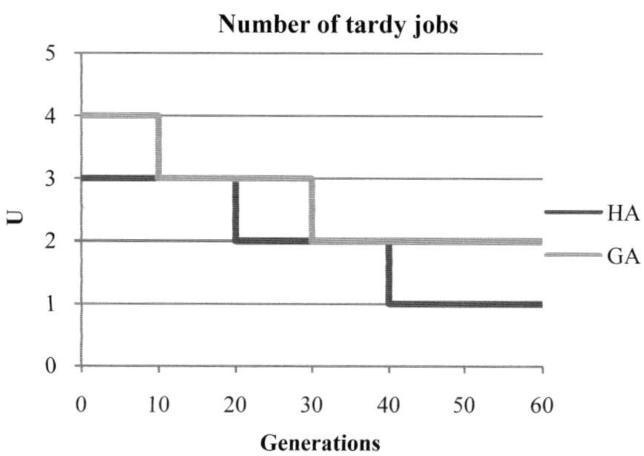

Fig. 3. Comparison between HA and GA – number of tardy jobs

As seen in fig. 3 the number of tardy jobs obtained by applying HA to our problem is smaller than in the case of GA. The two optimal solutions generated by applying HA and GA on our problem become:

1→3→4→5→6→7→2 (generated with HA);

1→3→5→6→7→2→4 (generated with GA).

By applying active-decoding procedure on these two solutions we obtained the two Gantt diagrams in fig. 4 and fig. 5.

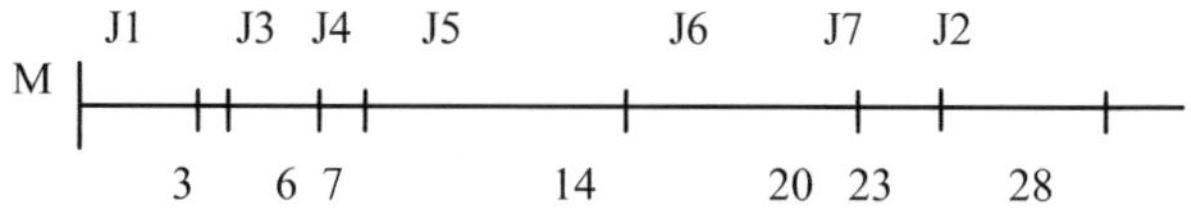

Fig. 4. Gantt diagram obtained by active-decoding the solution generated with HA

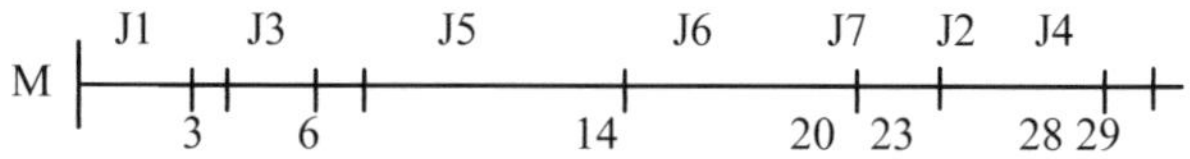

Fig. 5. Gantt diagram obtained by active-decoding the solution generated with GA

4. CONCLUSIONS

A hybrid algorithm based on an integration of a genetic algorithm and heuristic rules is proposed in this paper in order to solve the problem of minimizing the number of tardy jobs in single machine scheduling. The most prominent feature of our algorithm is a combination of heuristic rules (used to generate the initial individuals) with genetic operators to generate new individuals, which can effectively guide and speed up the genetic search. Another characteristic of the algorithm is consisted in the way the next generation is formed, by using the inverted generation, greatly increasing the flexibility of the algorithm. We have compared the hybrid algorithm HA to a standard genetic algorithm GA and the computational results showed that our algorithm is superior to the GA.

As a future research, by combining the presented algorithm with other scheduling methods like: linear programming, simple and complex heuristic rules, branch & bound techniques, other space search methods or even artificial intelligence, new algorithms for solving different problems like scheduling on parallel machines or the generalized job shop scheduling, can be obtained.

5. REFERENCES

Buzatu, C.; Bancila, D. (2008). A Hybrid Algorithm for Job Shop Scheduling, *Proceedings of the 6th International Conference of DAAAM Baltic Industrial Engineering*, Kyttner, R., pp. 309-314, ISBN 978-9985-59-783-5, Tallinn, April 2008, Tallinn University of Technology, Tallinn

Crauwels, H.A.J.; Potts, C.N.; Van Oudheusden, D.; Van Wassenhove, L.N. (2005). Branch and Bound Algorithms for Single Machine Scheduling with Batching to Minimize the Number of Late Jobs. *Journal of Scheduling,* Vol. 8, (2005) pp. 161-177

Dauzere-Peres, S.; Sevaux, M. (2004). An Exact Method to Minimize the Number of Tardy Jobs in Single Machine Scheduling. *Journal of Scheduling,* Vol. 7, (2004) pp. 405-420

Gelinas, S.; Soumis, F. (1997). A Dynamic Programming Algorithm for Single Machine Scheduling with Ready Times. *Annals of operations research,* Vol. 69, (1997) pp. 135-156

Miller, D.; Chen, H.; Matson, J.; Liu, Q. (1999). A Hybrid Genetic Algorithm for the Single Machine Scheduling Problem. *Journal of Heuristics,* Vol. 5, (1999) pp. 437-454

Sevaux, M.; Sorensen, K. (2004). A Genetic Algorithm for Robust Schedules in a One-Machine Environment with Ready Times and Due Dates. *Quarterly Journal of the Belgian, French and Italian Operations Research Societies,* Vol. 2, (2004) pp. 129-147

Vakhania, N. (2004). Single-Machine Scheduling with Release Times and Tails. *Annals of Operations Research,* Vol. 129, (2004) pp. 253-271.

Annals of DAAAM for 2008 & Proceedings of the 19th International DAAAM Symposium, ISSN 1726-9679
Editor B. Katalinic, Published by DAAAM International, Vienna, Austria 2008
Make Harmony Between Technology and Nature, and Your Mind will Fly Free as a Bird

IMPROVING STRUCTURAL BEHAVIOR OF A MEDICAL ROD

BANDRABUR, D[iana] L[ilia]; BEJAN, M[ihai]; MUNTEAN, A[ngela]; Moroianu, C[orneliu] & SAMOILESCU, G[heorghe]

¶

Abstract: *After fracture reduction, osteo synthesis is the second aim in a bone fracture treatment. Metal implants used in osteo synthesis must allow the reconstruction of the fractured bone. First requirement is to ensure a robust structural bone – metal implants assemble and the second one is to be able to induce no human body rejecting reaction. Many times, in organism versus metal implant the last one achieves reaches failure point. The simplest approach, in order to increase the structural behavior of a medical rod, is by finding the most favorable rod position related to fracture. This allows us putting forward some recommendations for this particular medical rod studied and some general patterns for other rod types.*

Key words: *Rod, implant, behavior, position, fracture.*

1. INTRODUCTION

After fracture reduction, osteo synthesis is the second aim in a bone fracture treatment. The shape and function of a human limb are restored after this approach. Metal implants, rods, brooches, screws or shells, used in osteo synthesis must have such physical and chemical properties which will allow reconstruction of the fractured bone (Gheorghiu & Cotet 1999). There are two important requirements for this. The first one is to ensure a robust structural bone – metal implants assemble and the second one is to be able to induce no human body rejecting reaction. Many times, in organism versus metal implant the last one reaches the failure point. The simplest approach in order to improve structural behavior of a medical rod is by finding the most favorable rod position related to bone fracture – figure 1. This allows us putting forward some simple recommendations for this particular medical rod studied and may be some patterns for other rod types (Hadar & Gheorghiu 2005); (Gheorghiu et al., 2000 a); (Gheorghiu et al., 2000 b).

2. FEM ANALYSIS OF A STRAIGHT MEDICAL ROD

Structural analysis was performed using finite element method. Side rod surface is 740 mm^2. Material is chrom.Young modulus is E = 210000 MPa and Poisson coefficient $\nu = 0.3$.

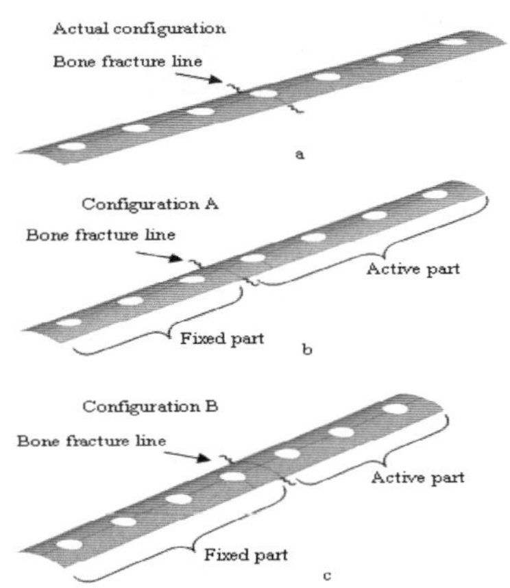

Fig. 1. Reconstructive configurations

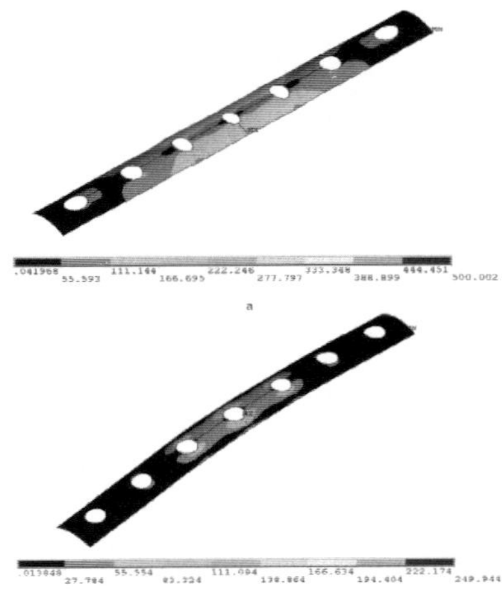

Fig. 2. Actual configuration structural behaviour

We set the limit stress at $\sigma_{axial\ lim}=500$ MPa for axial loading and $\sigma_{bending\ lim}= 250$ MPa for bending.

There are two mechanisms producing the rod failure: fatigue and overloading. This study will consider the overloading case.

It is important to quantify the effects of relative positioning of medical rod related to bone fracture. Von Misses equivalent stresses [MPa] and result displacements [mm] reveals structural responses for all load cases.

2.1 Actual configuration

The most frequenly used configuration is the bone fracture at the middle point of the rod – fig. 1a.

Compression loading take into consideration distributed force on three holes. The 500 MPa limit stress is produced by a $F_{axial\ case\ A}$ = 1252 N resultant force – fig. 2a. The maximum resultant displacement for this load case is $\delta_{res\ max\ case\ A}=0.02$ mm. All important results are syntetic revealed in table 1.

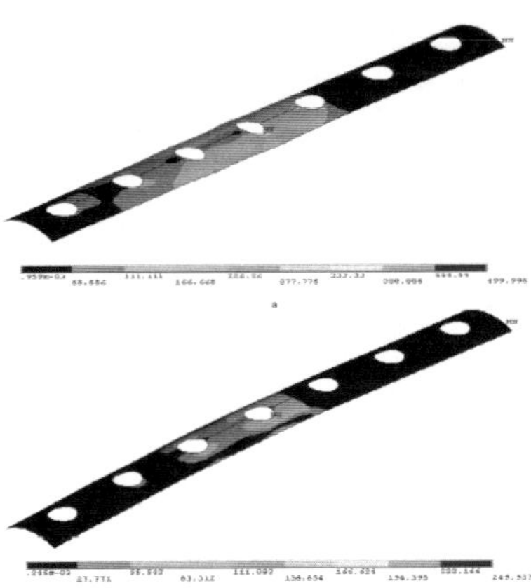

Fig. 3. Alternative A configuration structural behaviour

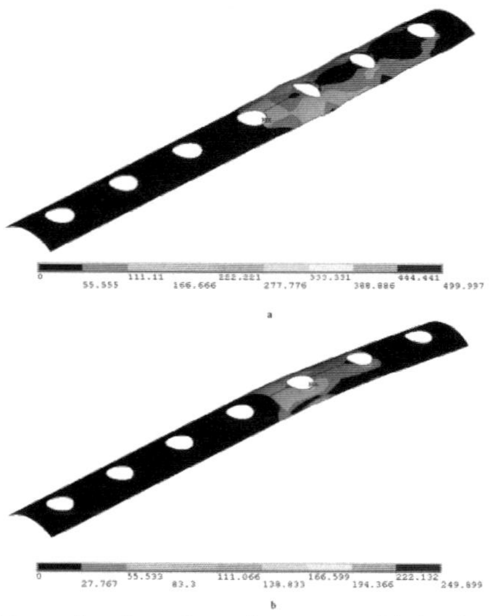

Fig. 4. Alternative B configuration structural behaviour

The bending load is applied on half of the rod. The 250 MPa limit stress is produced by a $p_{bending\ case\ B}$ = 0.03131 MPa which correspond to a $F_{bending\ case\ B}$ = 11.6 N resultant force – fig. 2b. The maximum resultant displacement for this load case is $\delta_{res\ max\ case\ B}$ =0.23 mm.

2.2 Alternative configurations

Alternative A configuration has the bone fracture between the third and the fourth holes and load on the biggest side surface of the medical rod. This will be considered the active part of the rod and the smaller side will be the fixed one – fig. 1b.

Compression loading take into consideration distributed force on four holes. The 500 MPa limit stress is produced by a $F_{axial\ case\ C}$ = 1815 N resultant force – fig. 3a. The maximum resultant displacement for this load case is $\delta_{res\ max\ case\ C}$=0.03 mm.

The bending load is applied on 410 mm^2. The 250 MPa limit stress is produced by a $p_{bending\ case\ D}$ = 0.03111 MPa which correspond to a $F_{bending\ case\ D}$ = 12.76 N resultant force – fig. 3b. The maximum resultant displacement for this load case is $\delta_{res\ max\ case\ D}$ =0.42 mm.

Alternative B configuration has the bone fracture between the forth and the fifth holes and the load on the smallest side surface of the medical rod. This is the active part of the rod and the bigger side is the fixed one – fig. 1c.

Load type/result	Actual configuration	Alternative A configuration	Alternative B configuration
Axial/maximum equivalent force [N]	1252	1815	917
Axial/maximum result displacement [mm]	0.02	0.03	0.018
Bending/maximum equivalent force [N]	11.6	12.76	20.02
Bending/maximum result displacement [mm]	0.23	0.42	0.26

Tab. 1. Absolute results

Load type/result	Actual	Alternative A	Alternative B
Axial/force	1	1.45	0.73
Axial/displacement	1	1.5	0.9
Bending/force	1	1.1	1.72
Bending/displacement	1	1.82	1.13

Tab. 2. Normed results

Compression loading is applied distributed on three holes. The 500 MPa limit stress is produced by a $F_{axial\ case\ E}$ = 917 N resultant force – fig. 4a. The maximum resultant displacement for this load case is $\delta_{res\ max\ case\ E}$=0.019 mm.

The bending load is applied on 330 mm^2. The 250 MPa limit stress is produced by a $p_{bending\ case\ E}$ = 0.0607 MPa which correspond to a $F_{bending\ case\ E}$ = 20.02 N resultant force – fig. 4b. The maximum resultant displacement for this load case is $\delta_{res\ max\ case\ F}$ =0.26 mm. Values of alternative configurations are normed regarding actual configuration –table 2.

3. CONCLUSIONS

Comparing A configuration with actual one the maximum equivalent axial force is 563 N bigger which reveals a 45 % increase of load carring capacity. For alternative B configuration 335 N smaller reveals a 26.7 % decrease. Bending the rod the A alternative configuration brings a 1.1 N bigger force which is a 10 % increase. More significant load carring capacity increase is for alternative B configuration where 8.42 N more represents 72.6 %. Results are consequences of structural design, load type and stress limits.

For this particular medical rod, both alternative configurations bring an increase in load carring capacity when bending is the applied load. When axial load is implied the alternative A configuration increases the load carring capacity while the alternative B configuration decreases it. For all medical rods the recomandation is to place the bone fracture between screw holes which are stress concentrators.

Further research can be orientated to create some general patterns for related types of medical rods (Hadar et al., 2004).

4. REFERNCES

Gheorghiu, H. & Cotet, C. (1999). *Preliminary Selection of Biomaterials for Orthopaedic Implants*, Technical Publishing House, Bucharest

Gheorghiu, H., Hadar, A., Jiga, G. & Cotet, C.E. (2000). Mechanical Behavior Study of Rod Implant For Cypho-Scoliosis Correction, *The 8-th International Conference of Stress Measurement, Material Testing*, pp. 526-531, 973-31-1492-8 and 973-31-1494-4, June 2000, Constanta

Gheorghiu, H., Dumitrescu, E., Hadar, A. & Cotet, E. C. (2000). Preliminary Three-dimensional Model Regarding Improvement of Harrington Medical Devices Used for Some Spine Malformation Correction, *The 8-th International Conference of Stress Measurement*, pp. 512 – 517, 973-31-1492-8 and 973-31-1494-4, June 2000, Constanta

Hadar, A., Gheorghiu, H., Cotet, C. E. & Ciobanu, L. F (2004). *3d Model For a New Type of Implant in the Treatment of Cypho-Scoliosis*, International Conference "Biomaterials & Medical Devices" – BiomMedD' 2004, p. 132, November 2004, Bucharest

Hadar, A. & Gheorghiu, H. (2005). An Optimized Solution for the Implant used in the Treatment of Cypho-Scoliosis, *22nd DANUBIA-ADRIA Symposium on Experimental Methods in Solid Mechanics*, Terme, M., pp. 38-39, September 2005, Parma

Annals of DAAAM for 2008 & Proceedings of the 19th International DAAAM Symposium, ISSN 1726-9679
Editor B. Katalinic, Published by DAAAM International, Vienna, Austria 2008
Make Harmony Between Technology and Nature, and Your Mind will Fly Free as a Bird
2nd DAAAM-ICAT

LASER PRINTING OF SOLUBLE TONER FOR RAPID MANUFACTURING

BANERJEE, S[oumya] & WIMPENNY, D[avid]

Abstract: *The Rapid prototyping & Manufacturing Group (RPMG) at De Montfort University is actively investigating a novel laser printing based RP process where toner formulated thermoplastic powder is deposited and therafter fused to form consolidated layers. Earlier published research focused on the development of tough toners using a range of engineering polymers, including polyethylene and polypropylene and thermoplastic elastomer to provide a build material for components made with the new process. This paper describes trials to establish the feasibility of developing a soluble toner as a support material*

Key words: *Rapid Manufacturing, electrophotography,laser printing*

1. INTRODUCTION

Standard laser printers work on the principle of electrophotography, where fine polymeric toner powder is picked up and precisely deposited on a substrate, usually paper, using electrostatic forces. Laser printing offers several key advantages for layer manufacturing applications; it can process dry powders at high speed, offers excellent resolution and it is capable of forming functionally graded structures. Once the toner has been deposited it is fused to form a solid layer, typically by contact with heated rollers or via infrared radiant heating. Laser printing is, in principle, highly adaptable to a range of toner materials based on thermoplastics, cermics and metallic powders. The potential application of laser printing in layer manufacturing has been recognized by several researchers (Bynum.K.D 1992, Grenada 2001, Kumar 2000, 2003 ,2004, Cormier 2002 and Banerjee 2006,2007).

At least two fully functional toners are required to enable complex 3D objects with overhanging features to be formed produce. One should be tough toner for the build material and the other toner should be support material which can be easily removed from the finished object. This paper, aims to study the feasibility of developing soluble toner support material.

2. TONER MATERIAL DEVELOPMENT

2.1 Selection of Toner Materials

Toner can be of two types; either mono-component or dual component. Mono-component toner is very popular for desktop applications whereas dual component toner is more common for industrial printers. Dual component toner is generally comprised of magnetic carrier particles (30 to 300 μm) and fine toner particles (5 to 15 μm) see Figure 1. Dual component systems have an advantage that the carrier particles "do most of the work" in terms of electrostatic transfer and this makes them more flexible in terms of toner formulation. In addition dual component printers offer much high printing speeds, increased layer thickness and better resolution than mono-component machines. Research was carried out at DeMontfort University to check the feasibility of developing a soluble dual component toner material using
P400 support material, an acrylic based copolymer employed as a soluble support material in the FDM process.

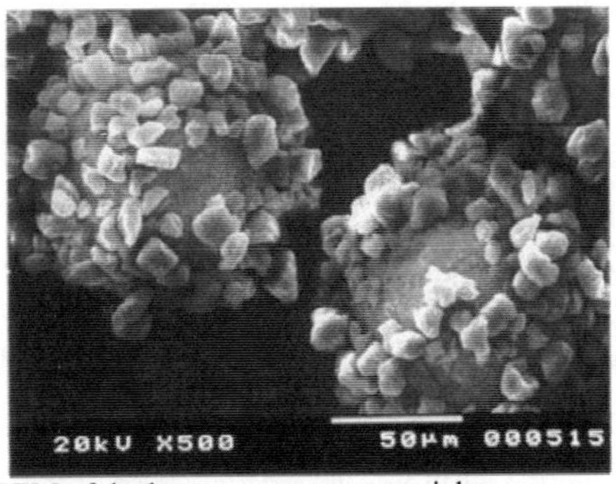

Fig. 1. SEM of dual component toner particles

2.2 Powder Grinding Trial

Initial trials were carried out to check if P400 support material can be readily processed to produce fine powder. Short lengths (6mm long) of P400 filament were loaded into a Retsch PM100 impact mill and milled for 2 hours. The powder produced was then sieved using a Retsch AS200 sieve shaker (with mesh size of 20, 36, 50 and 100 micron).It was found that and 30% of the material was below 20 micron. See Figures 2 and 3 below showing the P400 before and after milling respectively.

Fig. 2. P400 before milling

Fig. 3. P400 after milling

To reduce the grinding time, further samples were processed using a more efficient industrial scale Hosokawa 100 AFG mill. After grinding the particle size was checked by laser diffraction and it was found that approximately 60 % of 'P400 support material' was within 20 micron particle size after grinding (see Figure 4 below).

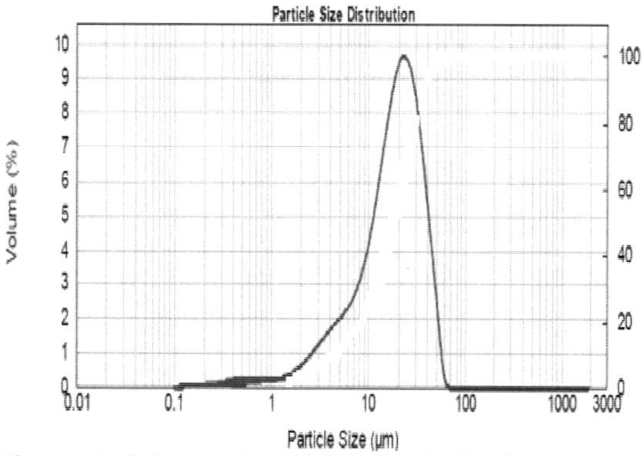

Fig. 4. Particle size distribution for P400 after grinding with 100 AFG mill.

2.3 Infrared Sintering trial

The ability to sinter the powder material using an infrared radiant source was then evaluated. Loose P400 powder was deposited manually into a mould (tensile test specimen 100mm long x 2mm thick) bone. The mould was then placed under infrared radiated heater (average wavelength of 0.78 to 2µm) at a distance of 100 mm from the source of radiation. Particles are joined together and formed continuous layers with 2mm thickness approximately (see Figure 5. A little shrinkage was observed after sintering.

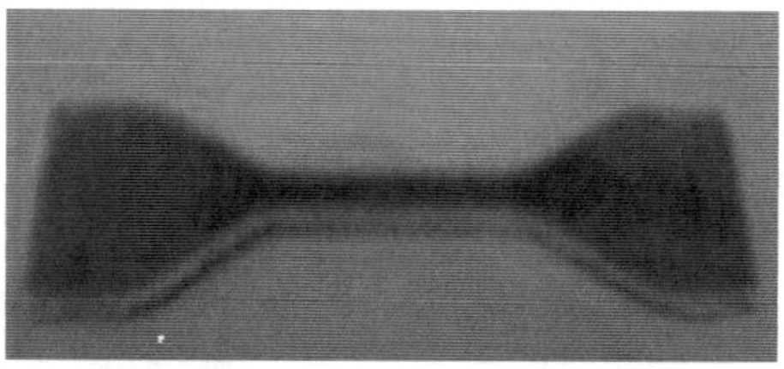

Fig. 5. Sintered sample of P400 material

2.4 Printing trials

An office dual toner laser printer, Ricoh 7000, was selected for the printing trial (see Figure 6 below). P400 powder was surface coated by mechanical mixing using fumed silica (HDK20TX) with a particle size between 0.17 to 0.25µm. This flow/charge control agent is normally used for developing negatively charged toner. The surface coated P400 support powder was mixed with ferrite carrier to give a developer with a 10% toner concentration.

Fig. 6. Ricoh 7000 printer

The standard black cartridge of the printer was filled with the toner (developer) formulation and test prints produced (see Figure 7 below). It was noted that deposition of powder particle was not homogeneous throughout the printed areas , with patches of greater and less deposition.

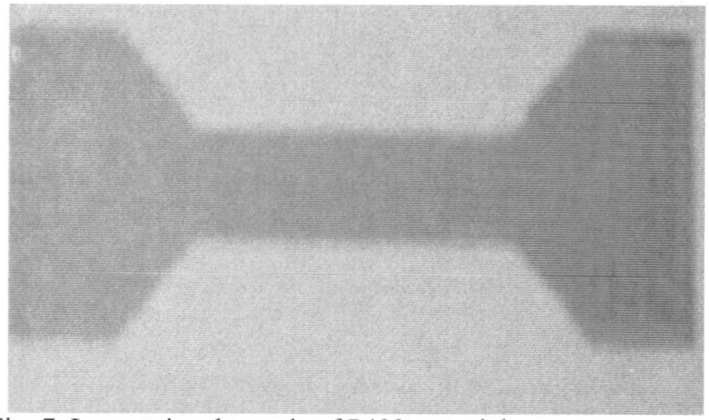

Fig. 7. Laser printed sample of P400 material

3. DISCUSSION

These initial trials indicate that P400 soluble acrylic material can be processed to produce a suitable dual component toner formulation which can be successfully printed and fused. However, there are clearly aspects which require further development. During printing toner particles were prematurely dislodged from the surface of the carrier. This was primarily due to weak adhesive force between P400 support and carrier particles. This adhesive force needs to be strong enough to hold toner particles until they are deposited selectively on OPC (Organophotoconductor) drum of the printer. It should be possible to reduce this problem by fine tuning of the carrier particle size, shape, surface texture as well as applying the most appropriate polymeric coating onto its surface.

The dielectric constant of powdered materials is strongly influenced by moisture content in the powder and the humidity level in the surrounding environment. This problem is more acute for materials which are hygroscopic in nature, such as water soluble material like P400. Care must be taken while making the full functional toner to reduce the absorption of moisture. This could be achieved by using silica gel/dehumidifier or similar device to control the humidity and moisture content of the operating environment.

4. REFERENCES

Bynum.K.D. (1992) Automated manufacturing system using thin sections, US Patent 5088047.

Grenda E.P. (2001) Apparatus of fabricating 3 dimensional objects by means of electrophotography, ionography or similar process, US patent 6206672.

Kumar A.V. (2000). Solid free form fabrication using powder deposition, US patent 6066285.

Kumar A. V. and Dutta A. (2003) Investigation of an electrophotography based rapid prototyping technology. *Rapid Prototyping Journal*, Vol 9,Nos 2, pp 95-103. Emerald, ISSN1 355-2546

Kumar A. V. and Dutta A. (2004) Electrophotographic printing of part and binder powders, *Rapid Prototyping Journal*, Vol 10,Nos 1, pp 7-13.

Cormier D. (2002). An investigation of selective colouring with 3-D Laser printing. *Journal of manufacturing Processes*, , vol 4, No2, ASME, ISSN: 0022-1817

Banerjee S. and Wimpenny D. (2006) Laser printing of polymeric materials, 17th Solid Freeform Fabrication Conference, University of Texas.

Banerjee S. and Wimpenny D. (2007) Rapid Manufacturing of Thermoplastic Parts by Laser Printing, International Conference on Polymers & Mould Innovations, Belgium.

Annals of DAAAM for 2008 & Proceedings of the 19th International DAAAM Symposium, ISSN 1726-9679
Editor B. Katalinic, Published by DAAAM International, Vienna, Austria 2008
Make Harmony Between Technology and Nature, and Your Mind will Fly Free as a Bird

NUMERIC EXPRESSION OF THE QUALITY OF PRODUCTION TECHNOLOGY

BARBORAK, O[to]; CERVENANSKY, J[aroslav]; BOZEK, P[avol]; MIHOK, J[ozef]; RYBANSKY, R[udolf] & TOTHOVA, M[aria]

***Abstract**: Contribution presents numeric expressions of the car components new testing method. The testing method is related to the product quality in its final phase. Authors also concern complex information system of the quality, operation, automatic testing of vehicle subsystem. Numeric analysis is carried out on the basis of automatic collection and systematic recording of commercial car operation. Critical components verified in laboratory conditions are further detected by numeric analysis of reliability (Černecký et al., 2003).*
***Key words:** measuring information, trial automation, new methodology*

1. INTRODUCTION

Tendency to ensure car's high technical and operational parameters, and consequently successful realisation of any difficult technical system and mechanism comes out from perfect knowledge of operational conditions, knowledge of operation - technical characteristics and parameters of own system. This fact conditions to increase requests for technical level of measurement, actual time of measure results, precision of measurement results, possibility of measurement results processing, level of measurement automation i.e. measurement as method of objective quantification of physical values, or about relation between two or higher number of physical values. The analysis of results achieved in solving and car security area in the whole scale of connectivity and relations confirms that the role of laboratory verification is not replaceable in the new quality management system. A complex approach to the questions of testing in a new quality management system is getting inevitable.
The logic part of the car production process and, as the analysis of trends in car industry shows, the effect of solution in testing automation has an objective need (Štollmann, 2001). It is necessary to concern the automation data gathering about component operation mode in real conditions. It is possible to define their reliability based on gathered data about operating and component failure accepting. It is possible to design device for test tracing of components and test method after defining critical elements.

2. PROBLEMS WITH QUALITY MANAGEMENT

Today brings significant changes in product area, in all industry forwarded countries. Understanding of changes in quality is possible to characterise as a movement from clearly technical soundness to integral product quality. Pro-duct quality became significant criteria of science and research effect and big productivity reserve of common work.
This general definition of development in quality field and objective need to continually increase quality in application for cars, which reports to increasing requirements for high technical and operating parameters, requires:

- to direct systematically all related actions, i.e. arrange questions of quality management to whole creating process of gearbox,
- to complex evaluate quality, which requires to use quantitative evaluation,
- to identify quality, i.e. to evaluate reached quality level, which has basic importance to control whole creating process and innovation gearbox process,
- to compare objective gearbox parameters with comparable world level types directly by measuring of chosen parameters, or indirectly by data from progressive technical information.

Quality management is concerned as complex, if it is expanded to all fields, which can have influence the final gearbox quality, including user's operation. This system, which prevails today in all advanced countries, leads to create conditions in flow pre-production, production and after production activities by using objective methods for apply testing need (Štollmann, 2006).

2.1 Analyse of reliability influence to profitability and economic effect

Before analysis of gearbox reliability influence to economic effect, it is important to realise basic ideas of reliability theory and to check relation of coefficients quantitative reliability to economic coefficient and efficiency coefficient. Related basic ideas:
Quality of product is properties summary, expressing ability to fulfil functions, for which it is dedicated. We take into account also product economic coefficients, its outfit by accessories, spare parts etc., as well as premises, which producer creates to provide services connected to product use.
Reliability is general object attribute, reposing in ability to fulfil requested functions with keeping of values for specified operational coefficients in specified limits and in time according to technical conditions. Reliability is one of the most important groups of product quality marks.

3. RELIABILITY TRIALS METHODICS PROPOSE

Automated reliability verifying and gear mechanism life cycle is the goal of test.
It is necessary to change dimensions, eventually construction nodes of least reliable elements before test. Test methodology is deduced from the way of speed degree gearing during car running. Speed degrees gearing flow is introduced in previous part. The flow repeats always the same way with aspect to chosen speed degree.
It is necessary to assure handle control power, path length of default and path length of speed degree gear. Gear mechanism lines effort by power intensity, which is prescript in tolerances during car running and followed construction nodes, ensure fulfil these conditions (Šurianský & Naščák 2000).
Whole system is specified to follow up gear mechanism in aspect of car reliability during car running. Test commerce car measuring results are base to propose the test method about control mechanism use in aspect of number of individual speed gear. The selected ratio must be kept during testing and gear mechanism test in number, what is used and record in tables. Proposed ratio method seems to be the most suitable.

3.1 Ratio method

Ratio method comes from distribution of percentage ratio elements use of gear mechanism during speed degrees gearing. Values assigned in tables are the base of distribution. Maximum level is reached when each component of gear mechanism will be running by distribution in table one during follow-up. Mathematic interpretation of number of gear z_u in one level and trial process is:

$$z_u = \frac{1}{u} \times \left[\sum_{i=1}^{n} n_{1i} + \sum_{i=1}^{n} n_{2i} + \sum_{i=1}^{n} n_{3i} + \sum_{i=1}^{n} n_{4i} + \sum_{i=1}^{n} n_{5i} \right] \quad (1)$$

Where:
n_1-n_5 - number of gear in u-level
u - dividing level of max. number of gear,
z_{1i}-z_{5i}. -number of gear of individual speed degrees,
n - maximum number of speed degrees gear of related speed.
Example: if we want to follow-up gear mechanism element reliability in speed degree 5, then number of individual speed degrees gears will be:

n_1= 97555 / 5 = 19511 times
n_2=114100 / 5 = 22820 times
n_3=154050 / 5 = 30810 times
n_4 = 91300 / 5 = 18260 times
n_5 = 18805 / 5 = 3761 times

Use coefficient estimate exactness or gear mechanism element reliability is higher, when dividing level is higher. It is important to know, same number of gears ratio in every level is kept in ratio method.
The graph of the testing method would show the ratio number of gears of speed degrees in one level. Level 5 has 5 times smaller number of gears in same ratio as a basic level. It is possible to fill this area by trial in 2 ways:
1) vertical classification,
2) horizontal classification.

3.2 Vertical classification

The level of test according to vertical classification is divided in sequence according to speed percentage use. It means, that control program give command to gear speed degree 5 and then 4, 1, 2, and 3 as the last one. It is because all gear mechanism elements would be follow-up from the speed degree which use is the smallest. We have the biggest chance, that we verify biggest number of gear mechanism elements. Mathematic total number of gear z_{uV} in vertical classification, is expressed by relation:

$$z_{uV} = u \times \left[\sum_{j=1}^{n_5} z_{5j} + \sum_{j=1}^{n_4} z_{4j} + \sum_{j=1}^{n_1} z_{1j} + \sum_{j=1}^{n_2} z_{2j} + \sum_{j=1}^{n_3} z_{3j} \right] \quad (2)$$

Where:
n_1-n_5 - number of gear in u-level,
u - dividing level of max. number of gear,
z_{1i}-z_{5i} - number of gear of individual speed degrees.

Vertical classification is risky, because rest parts of gear mechanism will stay untested for example by testing 4 speed degree by failure.This negative solves trial according horizontal classification.

3.3 Horizontal classification

The test of gear mechanism elements by horizontal classification is possible to describe by a graph. All speed degrees gearing are in phase one. After that, control system gives command to test control mechanism from 4 speed degree. Test till 3 speed degrees is done the same way. It is gear separately from neutral position. We can mathematically describe the number of gears Z_{uH} by horizontal classification test like this:

$$z_{uH} = u \times \left[z_u - \sum_{j=1}^{n_5} z_{5j} - \sum_{j=1}^{n_4} z_{4j} - \sum_{j=1}^{n_1} z_{1j} - \sum_{j=1}^{n_2} z_{2j} - \sum_{j=1}^{n_3} z_{3j} \right] \quad (3)$$

Where:
n_1-n_5 - number of gear in u-level,
u - dividing level of max. number of gear,
z_{1i}-z_{5i} - number of gear of individual speed degrees.
Cycle is repeating after one level of test is finished according to proposed method.
Slovak technical norms do not contain gear mechanism test method of car. That's why a space is given to this part, which will be needed to improve according to reached knowledge by test realisation.

4. CONCLUSION

To know the technical, running parameters and running conditions present the basis for evaluation, formulation of the parameters assurance of the technical level, running quality and car reliability or its components. It is important to know them in the operating mode, dynamic stress, the way of operating and use during running.
Quality requirements are increasing also requirements to function models laboratory testing, their parts, prototypes and their new components (Javorek et al., 2000). This comes out of facts. Practise in foreign countries showed, and it should be relevant, that it is profitable to subject the car problem parts to further development and research. The contribution is aimed at increasing the quality level not only of the final product, but also of the time development of the car and its accessories automation testing.

Acknowledgements: The contribution was elaborated within VEGA No.1/0421/08 "Research of possible applications of non-metal and composite materials by the action members production in the production technology" carried out at Trenčín Univerzity of A. Dubček in Trenčín together with related university workplaces and VEGA No. 1/0582/08 "Generalized convergence types in fuzzy environment".

5. REFERENCES

Černecký, J. ; Pivarčiová, . ; Dubovská, R. (2003). *Holografi and her technikal aplications*, Radom – Poľsko, ISBN 80-2564-36-07, pp. 283 – 289

Javorek, Ľ. ; Naščák, Ľ. ; Oswald, J. (2000). Monitoring proceeding digging, Starý Smokovec, ISBN 458 – 235 – 36 – 5 page 87 – 92

Štollmann,V. ; (2001). *Automatization / Robotics in theory and practice*, TU SjF Košice, ISBN 673 – 569 – 364 – 5, page183-186.

Štollmann, V. ; (2006). *Progress forest techniques*, In: OPTIROB 2006 : first international conference Optimization of the robots and manipulators / ed. Adrian Olaru, Mihai Stanciu, Editura Bren, Bucuresti, ISBN 973-648-572-2. - page 249-255.

Šuriansky, J. ; Naščák, L. (2000). *Measures systems with computer*, TU Zvolen, ISBN 80-228-1017-7, pages 79

Annals of DAAAM for 2008 & Proceedings of the 19th International DAAAM Symposium, ISSN 1726-9679
Editor B. Katalinic, Published by DAAAM International, Vienna, Austria 2008
Make Harmony Between Technology and Nature, and Your Mind will Fly Free as a Bird

TRACKING CAPTURE IN VIDEO MOTION

BARITZ, M[ihaela] & CRISTEA, L[uciana]

Abstract: *In this paper, we will explore how can we are able to obtain some information's about video motion and tracking capture and why the demands of animation lead to a particularly difficult challenge for capture techniques. In the first part of paper are presented some considerations about the problems develop in motion capture and the possibilities to avoid them. In the final part we present some recordings of the human body movements in gait process and the conclusions.*

Key words: *computer vision, tracking, image capture, video motion*

1. INTRODUCTION

Motion capture of human body can be an effective method of creating realistic human motion for animation or for different technical or medical studies. Unfortunately, the quality demands for animation place challenging demands on a capture and visualization system. The capture solutions that need these demands have required specialized hardware that is invasive and expensive solutions. For that computer vision could make animation data much easier that another method and in plus it can obtain information's to use them in different fields. In this paper it is presented a methodology for analyzing the 2D motion given image observations; use this as a tool for understanding the problem and for survey the evolution of a human body in different moments of walking process.

Synthetic experiments confirm that these situations would arise in practice. The experiments show how even simple visual tracking information can be used to understand the sort of gait for different people or to create a standard human motion but even with video tracking. (Bregler C., Malik, J. 2002)

Motion capture is an integrate method for creating the movement for computer animation or for different technical or medical studies, but motion capture also has its share of weaknesses.

The development of new and improved methods of editing and processing motion capture data has made great strides in making motion capture a more viable tool for animation production or for understanding biological process, technical and medical structures.

Another problem of motion capture has been the practical challenges of acquiring data. While research has made progress on using the data, capture techniques have evolved slowly, having problems with inertial time of acquisition and visualization. In special tracking technologies, based either on mechanical or magnetic sensors, or specially designed cameras viewing light markers, are required to create the observations that are processed into motion data. While these systems have improved in their dimensions, reliability, precision, and range, they are still generally expensive and some times hard to be used. These activities of motion capture must be performed by dedicated places providing specific environments for the recordings. Any "performer" could be captured in any setting that is desired. Using standard, high speed or thermal video cameras and special conditions for recording is an integrated structure as it could meet these goals.

The use of a single camera is a particularly situation only for static position and small or medium movements.

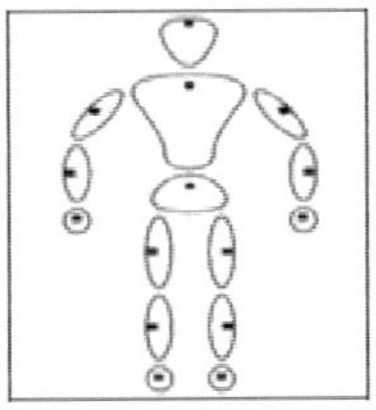

Fig.1.Main humn body parts

It offers the lowest cost, simplified setup, the potential use of natural or artificial light sources and a space without any modifications. The creation of motion capture data from a single video stream seems like a plausible idea but only for small and medium movements (short distance).

2. THE MOTION CAPTURE PROBLEM

The goal of motion capture is to record the movement of a "performer" (typically, but not always, human) in a compact, usable, repetitive manner and to be possible to rebuild these movements.

For this paper, we are concerned with the gross motion of the body, on the short distance (3 m), because the specific capture of facial or hands motion poses a different set of problems like skin color, 3D images and vibration stability. In computer graphics/computer vision studies usually the human body is divided into a small number of rigid segments that rotate relative to one another. This approximation is simply because the human knees, elbows and ankles do not have a single pivot point. The true motions of more complex joints are sometimes examples of kinematics approximations. The motion capture problem we consider, therefore, must have the following form: given a single stream of video observations of a people with normal/disabilities gait, analyze the tracking of each point, and compute a 3D skeletal representation of the motion using sufficient quality to be useful for animation. But for medical studies of the movement of the human body, like in Parkinson disease, are sufficient to use the rigid form of the human body parts. (Gleicher M., Ferrier, N 2003)

3. EXPERIMENTAL SETUP

Virtual humans are articulated figures modeled with multiple layers: a virtual *skin* is usually attached to an underlying *skeleton,* which animates the whole body. The skeleton is a hierarchically organized set of *joints*, and this set depends on the animation requirements and the fields of applications. To create morphologically correct skeletons it will be better but this can turn out to be quite costly. Real humans have so many degrees of freedom that virtual models frequently are necessary to omit some of them or to minimize the importance into movement actions.

Fig.2. Experimental setup to record the human body movement

The capture problem is sometimes difficult: the articulated model does not accurately reflect the real people, articulations lead to self-occlusions, even the articulated models contains many degrees of freedom, the skeleton is internal and therefore cannot be observed directly, also the clothes are an important part into the observations and recordings the movement's images. Our information sources are 2D-bidimensional and some occlusion is possible in time of recordings (like hands passing in front of hip joints). In addition, the optical medium for recordings provides a finite resolution (spatially and temporally), and the parameters of real cameras are difficult to obtain precisely (light, vibration, humidity). Having these limitations, it is not a problem that the practical approach to motion capture for moving tracking of the human body involves can be used in this method without these limitations. For example, if we observe a point on the people in an image created by a camera, we cannot determine the position of the point; only constrain its location to lie along a ray. For that, we assume an idealized pinhole camera model such that the ray is defined by the camera's focal point and the point on the image plane. In practice, a video camera has a finite resolution so observations are only localized to a region of the image plane and the space of human body movement must be in this region. (Monzani, J.S., 2002)

Additional information is needed to determine the position of a point in space, like axis system and calibration. A variety of sources can be utilized in various computer vision techniques, and a few can be applied to motion reconstruction. Most methods assume strong models to place further restrictions on possible poses. There are a wide range of visual tracking techniques in the practice ranging from edge feature based to region based tracking, and brute-force search methods to differential approaches. Edge feature based tracking techniques usually require clean and good data with high contrast object boundaries. But on human bodies such features are very difficult to obtain; clothes have many folds, environmental light is sometimes inconstant and also the images background has different colors in different directions.

Also, if the left and right leg has the same color and they overlap, they are separated only by low contrast boundaries.

Region based techniques can track objects with arbitrary texture and attempt to match areas between consecutive frames.

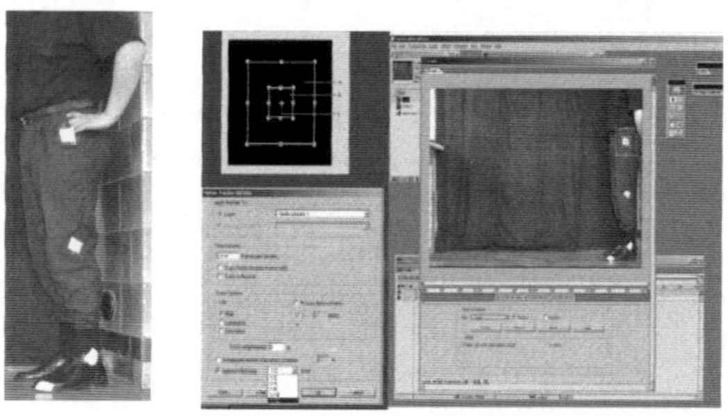

Fig.3.Experimental recordings of human body gait for tracking

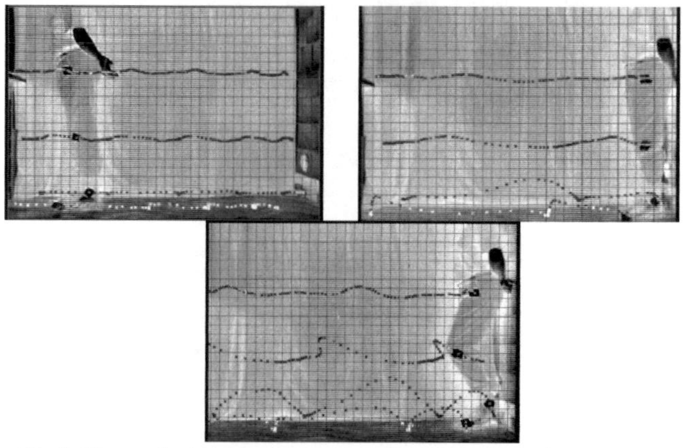

Fig.4. Human body image recordings in gait and running process

This requires the estimation of 6 free parameters that describe this deformation (x/y translation, x/y scaling, rotation, and shear).

4. RESULTS AND CONCLUSIONS

In our researches we used two video cameras, one of them withhigh speed and other with normally speed, to record the human body movements between initial point and end point situated at 3 meters distance. Initial the people wearing black cloth walks with a normal speed (also measuring the ground forces develop in gait process by a force plate) on this distance and we're recording the image of all points attached on the legs joints. Using these trajectories of the leg joints we can transfer to a human virtual model to analyze the movements, the joints positions in different movement moments and also to establish a data base for each people. (www.adobe.com)

Also, these recordings are possible to be used to simulate a movement and to establish the limits or to calculate the forces develop in muscles or joints during gait process.

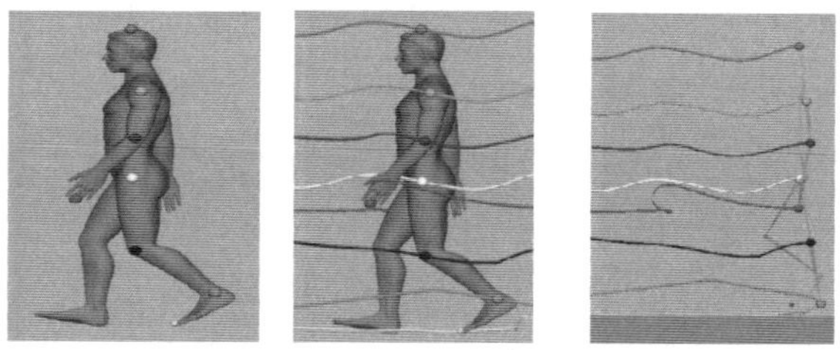

Fig.5. Virtual human body with simulated trajectories

5. ACKNOWLEDGMENT

Researches are part of Grant A1088 with CNCSIS Romania and we've developed the investigations with apparatus from Research Platform "SAVAT", University Transylvania Brasov.

6. REFERENCES

Bregler C., Malik, J. 2002, *Video Motion Capture*, UCB/CSD 97-973;

Gleicher M., Ferrier, N 2003, *Evaluating Video-Based Motion Capture*;

Monzani, J.S., 2002, *An architecture for behavioral animation of virtual humans* Ecole Politehnique Federale de Laussanne, Suisse;

www.adobe.com, accessed 2008-06-15.

Annals of DAAAM for 2008 & Proceedings of the 19th International DAAAM Symposium, ISSN 1726-9679
Editor B. Katalinic, Published by DAAAM International, Vienna, Austria 2008
Make Harmony Between Technology and Nature, and Your Mind will Fly Free as a Bird

TRANSIENT EVAPORATIVE LASER CUTTING WITH FINITE VOLUME METHOD

BEGIC, D[erzija]; BIJELONJA, I[zet]; KULENOVIC, M[alik] & CEKIC, A[hmet]

Abstract: *A computational model for the simulation of a transient laser cutting process has been developed using a finite volume method (FVM). Mathematical model based on the energy balance equation is used. A numerical method consists of transforming the energy balance equation into a system of non-linear algebraic equations. In order to achieve this the time, the space and the equation has to be discretized. For a numerical modelling of the laser cutting process a continuous laser beam is considered. The quantative analysis in respect of groove depth and shape is performed.*

Key words: *evaporative, laser, cutting, model, finite, volume.*

1. INTRODUCTION

Laser cutting is a very important application of laser technologies in the manufacturing industry. There have been various investigations about the material removal processes by laser beam from the early days (Schuöcker, 1986). Thermal models for investigation of laser cutting may be divided into two main categories: models with detailed treatment of thermal conduction and models where details of phase transition (melting, vaporization) are considered. Further, the methods of investigations can also be divided into three categories: analytical, numerical and experimental methods. Major numerical methods used so far are finite element method (FEM) and boundary element methods (BEM). (Junke & Wang, 2008) have used FEM software Ansys to analyse the temperature and thermal stress distributions in laser cutting of glass. The glass is machined by a single CO_2 laser beam and by dual CO_2 laser beams. (Kim, 2004) developed an unsteady convective heat transfer model using BEM considering moving continuous Gaussian laser beam for the prediction of groove shape, groove depth, temperature and flux distribution. A FEM based unsteady heat transfer model for the prediction of amount of material removal and groove smoothness during evaporative cutting with a Gaussian wave pulsed laser has also developed (Kim, 2001).

The objective of this paper is to develop a numerical model based on the finite volume method in order to predict the groove shapes in evaporative laser cutting. Considering the heat input from laser beam as a fixed heat source, an unsteady heat transfer equation is considered that deals with the material cutting process using a continuous laser beam. The convergence analyses are performed for the groove shapes with various meshes and time steps.

2. MATHEMATICAL FORMULATION

A diagram of the laser processing is illustrated in Fig.1. Dimension of aluminium sample is 10 mm x 1 mm and the sample is subjected by a continuous CO_2 laser beam.

In order to build a mathematical model, some assumptions should be made: (1) material is isotropic and physical parameters of aluminium are temperature-independent, (2) material removal is a surface phenomenon and phase change from solid to vapour occurs in one step, (3) evaporated material is transparent and does not interfere with incident laser beam, and (4) heat loss by radiation is neglected.

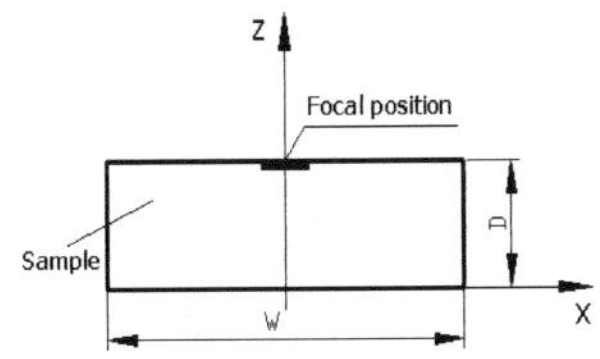

Fig. 1. The diagram of the laser processing

Based on the above assumptions, the mathematical heat transfer model can be written as follows:

$$\frac{\partial}{\partial t}\int_V \rho c T dV = \int_S \mathbf{q}\cdot\mathbf{n}\, dS + \int_V h_s\, dV \quad (1)$$

$$T(t) = T_o \text{ at } t = 0, \quad (2)$$

where V is volume of sample bounded by the surface S, $\mathbf{q}$ is the heat flux vector, $\mathbf{n}$ is the unit outward surface normal, h_s is the sink of heat; c and ρ are the specific heat and the density, respectively. T_o, denotes the initial temperature of sample.

The relation between the heat flux and temperature gradient is given by Fourier's law:

$$\mathbf{q} = k\, grad\, T, \quad (3)$$

where k is the thermal conductivity.

The boundary condition on the surface which is subjected to the laser beam is obtained from the balance of energy on the surface as:

$$-k\frac{\partial T}{\partial z} + h(T_s - T_\infty) = \alpha I, \quad (4)$$

and boundary conditions on the other boundaries are:

$$-k\frac{\partial T}{\partial \mathbf{n}} = h(T_n - T_\infty) \quad (5)$$

where, h is the convection heat transfer coefficient, T_s denotes the temperature of the heated zone surface and T_n denotes the temperature of the sample area without laser heating, α is the laser absorptivity, I is the intensity of the laser beam, and T_∞ is the environment temperature.

3. NUMERICAL METHOD

The governing equation, together with non-linear boundary conditions, applied to the two-dimensional cutting geometry is solved using finite volume method. In order to obtain the discrete counterparts of equations (1), (4) and (5), the time is discretized into an arbitrary number of time steps of the size δt and the solution domain is subdivided into a finite number of contiguous quadrilateral controls volumes (CV) or cells of volume V bounded by the surface S. The computational points (nodes) are placed at the centre of each CV, while boundary nodes, needed for the specification of boundary conditions, reside at the centre of boundary cell-faces. The inertial term is approximated using the backward differencing, and volume integrals are estimated using midpoint rule. A detailed description of the finite volume method applied in this work is given in (Demirdzic & Muzaferija, 1994). The set of linearized equations for each variable is solved at each time step by iteration. A fully implicit scheme is used for the time dependent terms.

At each time step, the energy equation is solved to obtain the temperature in each computational point of control volumes. When the temperature at any cell centre achieves the evaporation temperature, the last term in equation (1) is activated. Via this last term, the sink of heat is activated so it simulates the latent heat. For a time step, a size of sink of heat is determined so that the temperature of cell remains at a constant value, in other words, the same as the evaporation temperature. After a convergent solution is obtained for the given time step, it is checked whether the sink of heat is less than the latent heat. If the sink of heat is less than the latent heat the next time step will start. This procedure is repeated until the remains latent heat is spent. In the case that in a time step the sink of heat is greater than the latent heat, which is a physical non realistic solution, the size of time step decreases so that the sink is equal the latent heat. In this case after convergent solution is obtained, the cells by this status are considered as the evaporated cells and the new boundaries of the solution domain must be generated.

4. RESULTS AND DISCUSSION

For numerical analyses an aluminimum sample of size 10 x 1 mm is subjected by a continuous laser beam. The top side of the sample is subjected to the laser beam as shown in Fig.1. The initial temperature of sample is the same as the environment temperature. Due to the symmetry only a half of the aluminium plate is analysed.

The physical parameters of aluminium, air and laser are shown in Table 1. The computations are performed using various meshes and time steps in order to get a converged solution. The computations are performed for three different the time step $2 \cdot 10^{-4}$, $1 \cdot 10^{-4}$ and $0.5 \cdot 10^{-4}$ s. There were not considerable differences among the results with different time steps. Also, the computations are carried out for four various meshes. For all cases uniform meshes are used. Mesh refinement is stopped, when a difference between results obtained by two consecutive mesh refinement were less than 1%.

	Properties	Values
Aluminium	T_{evap}	933 K
	H_L	3.88 x 10^5 J kg^{-1}
	c	921 J kg^{-1} K^{-1}
	α	0.15
	ρ	2700 kg m^{-3}
	k	238 W m^{-1} K^{-1}
Air	h	30 W m^{-2} K^{-1}
Laser	P	1000 W
	r	0.1 mm

Table 1. Physical parameters of material (Al), air and laser

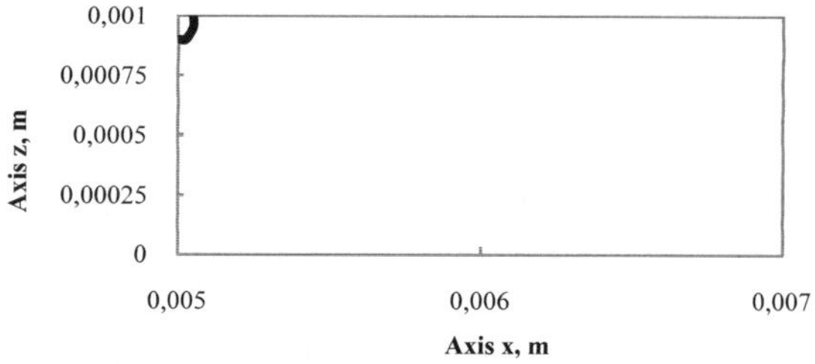

Fig.2. Transient surface groove shape at the time $6.10965 \cdot 10^{-5}$ s

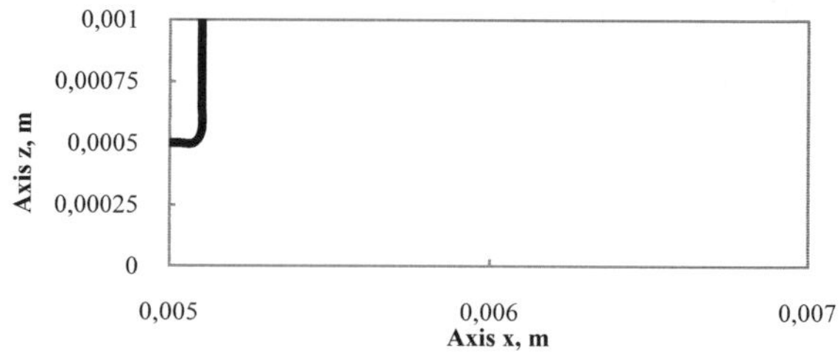

Fig. 3. Transient surface groove shape at the time $3.06792 \cdot 10^{-4}$ s

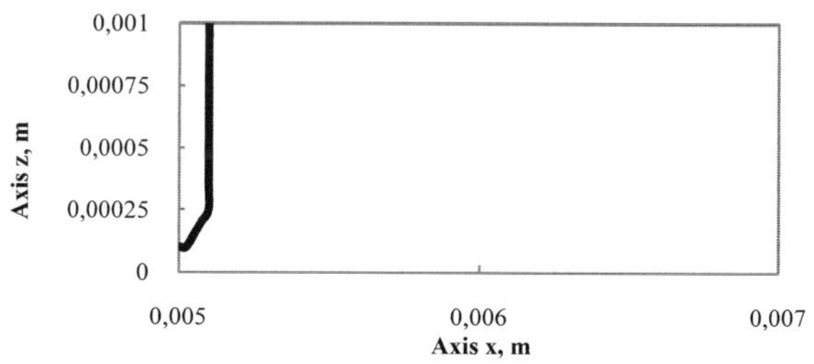

Fig. 4. Transient surface groove shape at the time $5 \cdot 10^{-4}$ s

In Figs. 2, 3 and 4, the groove shapes are shown for the finest mesh 200 x 10 CVs at the times $6.1096 \cdot 10^{-5}$ s, $3.0679 \cdot 10^{-4}$ s and $5 \cdot 10^{-4}$ s, respectively. The obtained numerical results yield the qualitative good results. A quantitative estimate of results including the shape and the depth of the groove and temperature fields will be performed by the experiment.

5. CONCLUSION

A finite volume method to study the time-dependent material removal of a sample subjected to a continuous laser beam of constant power is developed. For this analysis, the sample and the laser are fixed at a position to simulate the transient drilling process.

Studies show that the method yields convergent results with both mesh refinement and time step. These results showed a qualitative analysis in respect of depth and shape of a groove. A quantitative estimate of numerical results will be performed by experiments.

Also, in the future work, numerical analysis of transient evaporative laser cutting with a moving laser beam by finite volume method will be considered.

6. REFERENCES

Demirdzic, I. & Muzaferija, S. (1994). Finite volume method for stress analysis in complex domains. *International Journal for Numerical Methods in Engineering*, 37, 21, (January 1994) page numbers (3751-3766), ISSN: 0029-5981

Junke, J. & Wang, X. (2008). A numerical simulation of machining glass by dual CO2-laser beams. *Optics & Laser Technology,* 40, 2, (March 2008) page numbers (297-301), ISSN 0030-3992

Kim, M.J. (2004). Transient evaporative laser cutting with moving laser by boundary element method. *Applied Mathematical Modelling*, 28, 10, (October 2004) page numbers (891-910), ISSN 0307-904X

Kim, M.J. & Zhang, J. (2001). Finite element analysis of evaporative cutting with a moving high energy pulsed laser. *Applied Mathematical Modelling*, 25, 3, (January-February 2001) page numbers (203-220), ISSN 0307-904

Schuöcker, D. (1986). Dynamic Phenomena in Laser Cutting and Cut Quality. *Applied Physics B: Lasers and Optics*, 40, 1, (May 1986) page numbers (9-14), ISSN 1432-0649

Annals of DAAAM for 2008 & Proceedings of the 19th International DAAAM Symposium, ISSN 1726-9679
Editor B. Katalinic, Published by DAAAM International, Vienna, Austria 2008
Make Harmony Between Technology and Nature, and Your Mind will Fly Free as a Bird

COMPARISON OF DIFFERENTLY SHAPED TESTING SPECIMENS FOR SIMULATION OF THERMOMECHANICAL CYCLES

BEHULOVA, M[aria]

Abstract: *New processes of thermomechanical treatment (TMT) are designed and experimentally tested using laboratory simulators. For the following examination of the influence of thermomechanical cycles on the microstructure, mechanical and utility properties of materials, testing specimens for TMT are required to have a proper shape in order to assure sufficiently uniform temperature and stress-strain fields in their active zones. Moreover, they must be suitable from the point of view of stability by applied deformation loading. In this paper, five differently shaped testing specimens for a simulator of thermomechanical cycles are compared using electric, thermal and buckling analyses in the program code ANSYS.*
Key words: *thermomechanical treatment, electrical heating, temperature fields, buckling analysis, ANSYS.*

1. INTRODUCTION

In the field of production and treatment of materials for automotive industry, the great attention is paid to relatively cheap low alloyed steels (Staňková *et al*, 2005, Mašek *et al*, 2006). A special combination of microstructural, mechanical and utility material properties of these steels can be achieved by proper alloying and following suitable heat or thermo-mechanical treatment.

Newly designed processes of thermomechanical treatment (TMT) are tested using laboratory simulators. In this paper, five testing specimens for simulator of thermomechanical cycles with different shapes are analysed using FEM simulation to evaluate and compare the homogenity of temperature fields in their active parts and stability by applied clamping force.

2. EXPERIMENTAL MEASUREMENT

Laboratory simulator of TMT SMITWELD with module TTU 2002 installed at the TU Chemnitz (Fig. 1a) is designated for experimental investigation of material behavior during TMT with the maximum deformation rate of 2 $mm.s^{-1}$ and clamping force of 20 kN.

The specimen with the length of 81 mm and active cylindrical part with the diameter of 6 mm and the high of 11 mm is at this time used for experimental measurements (Fig. 1b). Before experiment, the Ni-CrNi thermocouple is welded on the surface in the central part of active cylindrical zone for very accurate temperature measurement (Fig. 1c). The surface in the active zone is covered by a protective antioxidant coating. Testing sample is then screwed in jaws of simulator and subjected to TMT. Specimen heating is realised by direct electrical heating, cooling can be assured by air, nitrogen or water. Process of TMT is controlled by computer using prescribed thermal, deformation and loading regimes.

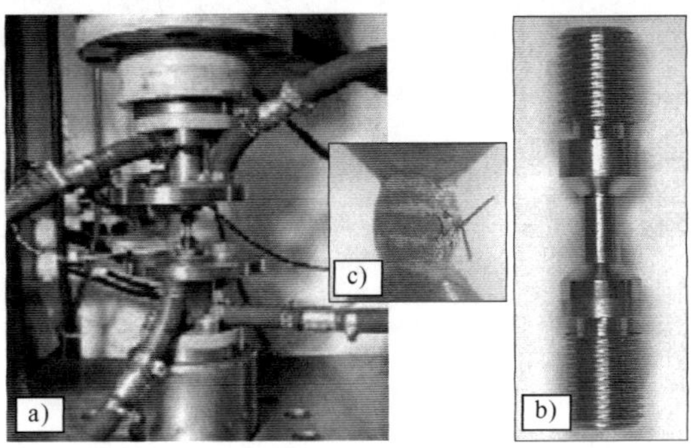

Fig. 1. Laboratory simulator of TMT SMITWELD with a detail of testing sample (a) before TMT (b) and after TMT (b).

3. SIMULATION MODEL

Five specimens with different shapes (Fig. 2) were analysed using the finite element code ANSYS 10.0. The length of all specimens is 81 mm. The active part of specimens has cylindrical shape with the diameter of 6 mm and the length of 11 mm. The middle carrying part of the specimen 1 (Fig. 2a) is comprised of conical and cylindrical parts finished by a notch. The specimen 2 (Fig. 2b) has extended middle part with the successive change in diameter from the active to the clamping part. The middle part of the specimen 3 (Fig. 2c) is cylindrical with the diameter larger then in the active part. The specimen 4 (Fig. 2d) is equipped in the middle cylindrical part by five notches and the specimen 5 (Fig. 2e) by five threads.

For the analysis of temperature fields in specimens during direct electrical heating as well as stability analysis, simulation models were prepared including development of geometrical model, FEM model, definition of nonlinear material properties, initial conditions, boundary conditions and loading for electric, thermal and static problems.

Analysis of electric fields was based on the solution of Laplace equation in the form (Novák, 2001)

$$div(\sigma\, grad V) = 0 \quad (1)$$

where σ is electric conductivity and V electric potential.

Transient temperature field in solids can be described by Fourier-Kirchhoff partial differential equation (Incropera & DeWitt, 1996)

$$c\rho \frac{\partial T}{\partial t} = div(\lambda\, grad T) + q_v , \quad (2)$$

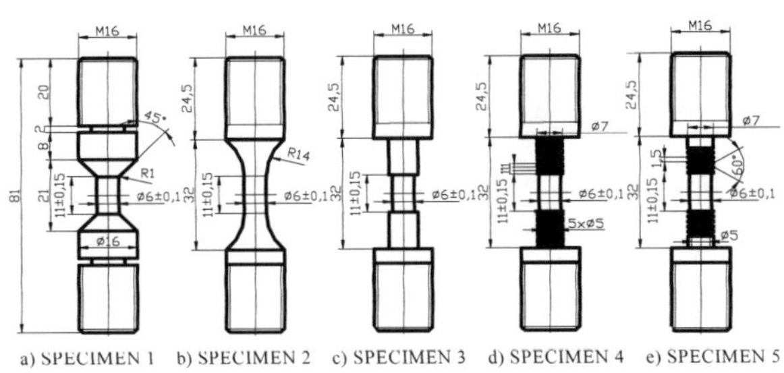

Fig. 2. Geometrical characteristics of analysed specimens.

in which ρ is the density, c - the specific heat capacity, λ - thermal conductivity and q_v - volume density of internal heat sources, i. e. the heat generated in unit volume per second. By resistance heating this term corresponds to the Joule heat computed by electric analysis.

Nonlinear buckling analysis with gradually increasing loads was applied to find the load level at which a specimen became unstable (Ansys, 2005).

Specimens with suggested shapes were numerically tested for the steels with the chemical composition of 0.1C-0.25Si-1.25Mn-0.3Ni-0.01N-0.03Al [wt. %]. Electric, thermal and mechanical properties were supposed to be temperature dependent. Specimens with the initial temperature of 20 °C were loaded by the time variable voltage during heating period of 60 s. During the first 10 seconds, the voltage was 0.5 V with its following decrease to the value 0.47 V. Specimen cooling by mechanisms of free convection and radiation to the air with the temperature of 20 °C was considered. For buckling analysis, specimens were constrained on the bottom surface and loaded by buckling axis force on the top surface. Developed simulation model and solution procedure were verified using experimental measurements at TU Chemnitz (Behulova *et al*, 2006).

4. RESULTS

During direct electric heating, specimens are heated due to the Joule heat generated in specimens by passing of electric current. For illustration, the distribution of current density j and internal heat sources (Joule heat) q_v in the specimen 1 at the end of heating in the time of 60 second is shown in Fig. 3. Maximal values of current density and Joule heat were computed for the specimen 4.

Fig. 4 illustrates temperature fields in analysed specimens and their active zones in the time of 60 seconds. Temperatures in the clamping parts of specimens are approximately the same for all specimens attaining the values from 20 °C to 200 °C. Maximal temperature at the end of heating of 1105 °C was reached in the specimen 1. Maximal temperatures of specimens 3, 4 and 5 are lower than maximal temperatures reached by heating of specimens 1 and 2 but the temperature differences in the active cylindrical parts are considerably lower. The most homogeneous temperature field in active part is in the specimen 4 with five notches in middle part.

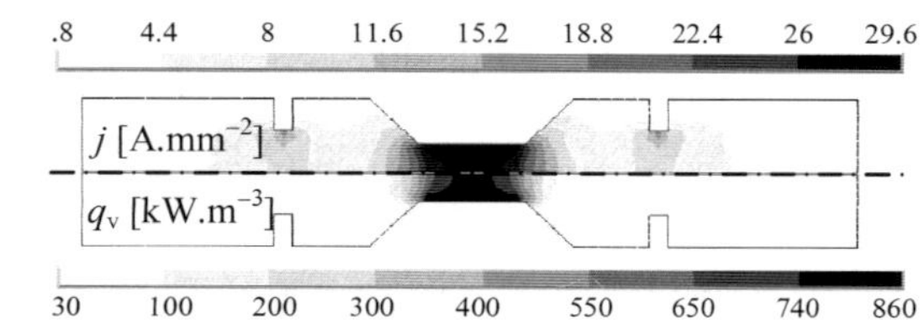

Fig. 3. Distribution of current density j and Joule heat q_v in the specimens 1 at the end of heating.

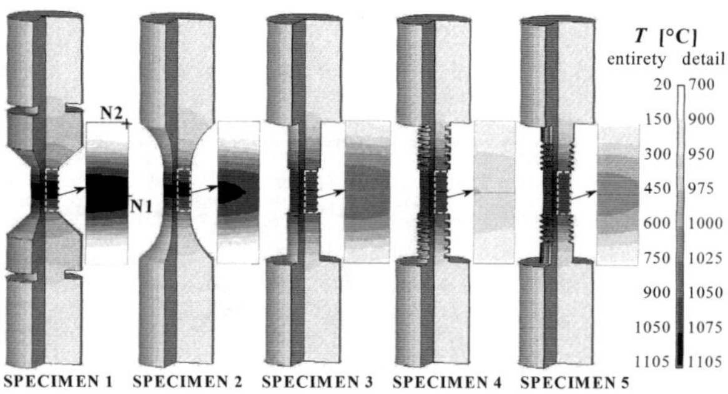

Fig. 4. Temperature distribution in the analysed specimens and their active parts at the end of heating in the time of 60 seconds.

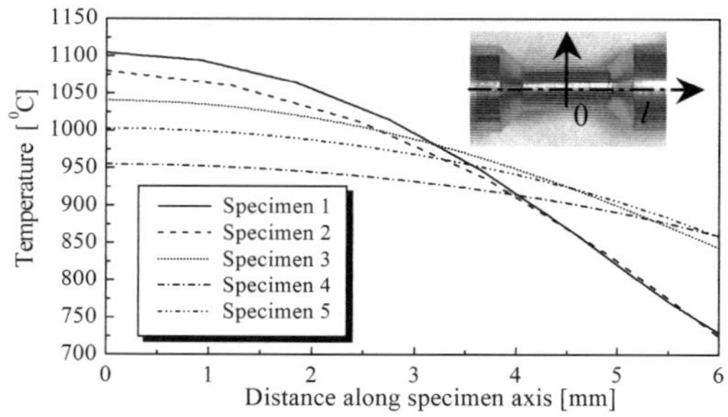

Fig. 5. Dependence of temperatures along rotation axis of specimens in the active parts at the end of heating.

Specimen	Buckling force [kN]	
1	20.1	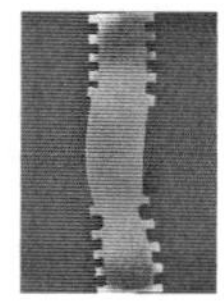
2	16.3	
3	15.9	
4	8.4	
5	9.8	

Tab. 1. Computed buckling forces for analysed specimens.

This result can be confirmed also by the temperature distribution along the rotation axis of specimen active parts at the end of heating (Fig. 5). Maximal temperature differences approximately 330 °C were computed for the specimen 1. In term of homogenity of temperature distribution, the specimen 4 seems to be the most favorable.

On the other hand, the results of buckling analysis (Table 1) revealed that this specimen can be loaded during experimental TMT only by smaller clamping force. The real behavior of the specimen 4 under higher deformation loading can be seen from figure in the Table 1.

5. CONCLUSIONS

Based on the numerical analysis of temperature fields by direct electric heating, the specimen 4 can be recommended for TMT testing of steels using the laboratory simulator of thermomechanical cycles. However, according results of buckling analysis, the specimen 4 can be loaded by smaller clamping force that currently used specimen 1.

6. REFERENCES

Ansys Theoretical Manual, Release 10.0, SAS IP, Inc., (2005).

Behúlová, M., Staňková, H. & Mašek, B. (2006). Analysis of temperature distribution in samples by direct electrical heating. *Materials Science and Technology* [online]. Vol. 7, No. 1. ISSN 1335-9053.

Incropera, F., P. &DeWitt, D. P. (1996). *Fundamentals of Heat and Mass Transfer*. New York, J. Wiley&Sons, ISBN 0-471-30460-3.

Mašek, B.; Staňková, H.; Nový, Z. & Meyer, L. W. (2005). Development of new incremental forming strategies for low-alloyed TRIP steels, *8th Int. Conference on Technology of Plasticity*, Verona, ISBN 88-87331-74-X

Novák, P. (2001). *Základy elektrotepelnej techniky*. (*Fundamentals of electrothermal Technique)*. Košice, Mercury-Smekal, 2001.

Staňková, H.; Mašek, B. & Meyer, L. W. (2006). The Influence of the Incremental Deformation Intensity on the Microstructure Development, *7th Int. Conference on Production Engineering and Design for Development*, PEDD 7, ISBN 12-7340-49-8, 2006, Cairo, Egypt.

7. ACKNOWLEDGEMENT

The research has been supported by the project VEGA MS and SAV of the Slovak Republic No. 1/0837/08 and DAAD A/07/01388.

Annals of DAAAM for 2008 & Proceedings of the 19th International DAAAM Symposium, ISSN 1726-9679
Editor B. Katalinic, Published by DAAAM International, Vienna, Austria 2008
Make Harmony Between Technology and Nature, and Your Mind will Fly Free as a Bird

METHODOLOGY FOR ORGANIZING E-CREATIVE RESEARCH AND E-CREATIVE DESIGN SESSIONS

BEJU, L[ivia Dana] & BRINDASU, P[aul] D[an]

Abstract: *The paper aims at developing a creative research and designing methodology for internet that can be used in the engineering area. Such a methodology is useful in the context of globalization and computer science facilities (hardware and software. E-creative design and E-creative research terms were introduced by the authors. The needed steps for organizing the E-sessions for research and design activities was elaborated in accordance with the general creativity theories, research and design principles, project management methodology and computer science facilities. The methodology has been used for research and design activities in Romanian INPRO national network for the last months.*

Key words: *creativity, research, design, brainstorming*

1. INTRODUCTION

Research and Design activities are a very important component in the life and development of an organization which otherwise cannot cope with the huge pressure of the global market. Research and design activities are carried out generally in teams and require specialists from different areas. The research equipments are expensive and can belong to research centers, organizations, companies or research networks. An example of such a research network is the VRL-KCiP network created in Europe. The network offers joint research, consulting and training for manufacturing techniques, requirements engineering, risk management, collaborative engineering, knowledge capture and management. In the same way the INPRO national network was created in Romania. The network's mission is to bring together the specialists in production process research in order to share their knowledge in a common structure.

Creativity (or "creativeness") is a mental process involving the generation of new ideas or concepts, or new associations of the creative mind between existing ideas or concepts. At this moment there are structured three main categories of creative methods in the area of engineering research and design. The first category has the aim of elaborating as many ideas, solutions, methods etc. as possible. In this category, brainstorming is the main group creative method. There were elaborated also some other methods connected with the mentioned one. There are sinectics, Delphy method, Panel discussion, 6-3-5 method, Philips 66 method, Frisco method etc. These group of methods use and stimulate the whole potential of the human thinking including the subconscious too. The second category includes the morphological analysis method developed by Fritz Zwicky. Its aim is to explore all the possible solutions of a problem. The approach is rational, very well structured and needs computer support in order to be efficiently tackled. The third category is a complete algorithmic approach and contains TRIZ (Theory of Inventive Problem-Solving) and ARIZ (Algorithm for inventive problem solving) both developed by the Russian scientist Genrich Altshuller (Orloff, 2005). Today, TRIZ is a methodology, tool set, knowledge base, and model-based technology for generating innovative ideas and solutions for problem solving. TRIZ provides tools and methods for use in problem formulation, system analysis, failure analysis, and patterns of system evolution. TRIZ, in contrast to techniques such as brainstorming (which is based on random idea generation), aims at creating an algorithmic approach to the invention of new systems, and the refinement of old systems.

2. E- CREATIVE DESIGN, E-CREATIVE RESEARCH

Computer Science and the new software allow an easier and efficient communication between persons located at more distances (Anderson, 2007). We enumerate the facilities of E-mail, Messenger, Skype, the possibility of viewing and intervention on documents (Team viewer) and more complex software which allows the access to database, specialists of great value and accredited laboratories contained in the research networks (Team Center). In this context of better communication between specialists and easier access to information, the creative activities in research and design areas can be done using network facilities.

We elaborated the E-Creative Design and E-Creative Research concepts which mean the use of creative methods in the research and design on the Internet network (Brindasu 2006). Analyzing the existing creative methodologies and taking into account the fact that each research activity has a unique character and can be planned and tracked through Project Management methodology, we underline the fact that for each research project three main steps must be covered. These are: problem definition, elaboration of new solutions and selection of the best (optimum) solution.

2.1. Problem definition

In management and engineering, a problem is the difference between the actual situation and the desired one. Generally, the true causes are not known and an attentive analysis of the root causes is necessary in order to establish the true nature of the problem. The correct and complete understanding of the problem is very important and necessary in each situation, more in the situation of the network collaboration where the information exchange is made in a synchronous or asynchronous way (McWay Lynch, 2002). The problem must be explained in a textual and/or graphical way and communicated in the network.

The main steps that must be followed for a problem definition in the situation of network communication are the following:

- Establishing a virtual team for the problem definition;
- Individual studies or E-sessions in order to determine the analyzed areas or perspectives (technical, managerial, economic, ecological, legislative, etc.) to identify the key parameters or indicators, behaviors for short and long periods, possible feedback loops caused by different parameters, contradictions, risks and uncertainty.
- E-session for establishing the problem and the research-design topic.

In order to communicate information in the network in the best way, the methods of root cause analysis are very useful. These are 5Whys, Failure mode and effects analysis, Pareto Analysis, Fault tree analysis, Bayesian inference and Ishikawa diagram. Some other useful methods for problem definition are specified by TRIZ methodology. These are: 9 box solution map (used to accurately capture the history of each problem and the problem context, to set the system context, to define its environment (super-system) and all the details (subsystems)) and Eight Trends of Technical Evolution (show how to predict new markets and new products, how all products and industries develop).

Taking into account the importance of a correct and complete problem definition is useful to organize more E-sessions with this aim.

TRIZ methodology underlines the characteristics of a clear problem definition (Orloff, 2005). These are: the end point, the purpose, the main function we are seeking, all functionality / benefits (everything we want and everything we don't want), where and when we want our functionality, conflicts identification in time and space, real functionality checking (Why do we want this?), everyone's ideal outcome.

2.2 Elaboration of new solutions

In order to elaborate new solutions, it is necessary to establish a virtual team. The methodology of choosing the team members was presented in a previous research (Brindasu, 2007). The next steps can be done in a synchronous or asynchronous way.

For synchronous sessions the steps can be:

- *ESTABLISH the E-design virtual group members;*
- *COMMUNICATE the research-design topic to the E-design group members- mail;*
- *ORGANIZE an E-session for establishing the analyzing criteria. Each member can propose one or more criteria. Each member will note all criteria. A specific soft will determine the most important criteria and their weight;*
- *ORGANIZE an E- creative session in order to develop new constructive solutions from the design topic and in concordance with the criteria established before – use a specific E-creative method;*
- *ORGANIZE an E-session in order to analyze the existing constructive solutions from the design topic and in concordance with the criteria established before; analyze all combination variants – Morphological Analysis Method;*
- *New creativity E-session.*

Of course, these E-sessions can have different characteristics depending on hardware equipments and software. The synchronous E-sessions can be done on Messenger, Skype or more sophisticated video systems. Software packages such as Teamviewer allow visualization, modifications and additions to the drawings made by all participants at a network meeting. When Teamcenter Engineering is used, the participants have accesses to more information. Team Center is a *Product Data Management (PDM)* system that helps the managing of all the processes, applications, and information required to research, design, manufacture, and support a product throughout its life cycle. The goal of a PDM system is to provide a single, common interface for managing and accessing all data within an organization (this can be a research network).

For asynchronous E-sessions, the problem definition, research topic and the other necessary documents are sent to the team members after an algorithm presented next.

- *Each participant will receive a code (P1, P2, P3, ...);*
- *Knowing the problem, each participant will write three proposals for the problem solving on the computer;*
- *Send the file to the participant with the following code;*

Participant	Proposal1	Proposal2	Proposal3
Pi			
Pi+1			
Pi+2			
Pi+3			
...			

Tab. 1. Structure of the list that must be completed by each participant to asynchronous E-sessions

- *Complete, improve, establish details, change or write the opinion about the ideas elaborated before (by the other participants) until all the ideas are discussed by all the participants (table 1).*

The E-brainstorming sessions can be organized in a classical way or using some Creativity Methods for breaking Psychological Inertia. These are creativity, intuitive and heuristic techniques: consonants association, analogy and extrapolation, inversion, empathy, combination, modifying, improving, developing, input-output techniques, interrogative lists. TRIZ methodology uses a list of 40 Inventive Principles which must be followed step by step or/and the 9 box solution map. These creativity E-sessions are organized depending on the research topic, importance, team skills etc.

2.3. Selection of the best (optimum) solution

The selection of the best solution can be done only using Value Analysis principles. The analyzing criteria were established before. Each solution is marked by each team member for each analyzing criteria. The method is more useful if the members have no information about how the other members mark. Specific software for mark addition is very useful. The optimum solution can be improved if necessary.

3. CONCLUSION

The paper develops a methodology for E-research and E-design activities in engineering area with special application for the research networks created in Europe (VRL-KCiP) and in Romania (INPRO). First, a rewiev of the main creative methodologies uses by individuals and groups in engineering area was done. Problem definition, elaborating of the new solutions and selection of the best solution was analyzed and the logic steps were elaborated in accordance with the general creativity theories, research and design principles, project management methodology and computer science facilities. The methodology establishes a large frame and offers the possibility of specific application depending on the characteristics of the problem, team members, resources etc.

4. REFERENCES

Anderson, T., Kanuka, H. (2007) *E-Research-Methods, Strategies and Issues.* Available from: http://www.e-research.ca/ Accessed: 2008-5-15.

Brindasu, P.D., Beju, L.D. (2006). Contributions to Cutting Tools Design using Web Facilities, *Proceedings of the 15th International Conference on Manufacturing Systems – ICMaS*, , pp. 29-34, ISSN 1842-3183, Bucharest, Romania, October, 2006, Published by Editura Academiei Romane.

Brindasu, P.D., Beju, L.D. (2007). E-Creative Design for Cutting Tools Area, *Annals of the MTeM for 2007 and Procedings of the 8th International Conference Modern Technologies in Manufacturing,* Cs Gyenge (Ed.), pp. 83-86, ISBN 973-9087-83-3, Cluj Napoca, Romania, 2007.

McVay Lynch, M. (2002). The Online Educator. *A Guide to Creating the Virtual Classroom*, RoutledgeFalmer, ISBN 978-0-415-26369-6, New York.

Orloff, M. (2005) *Inventive Thinking through TRIZ. A practical Guide*, Springer, ISBN 10-3-540-33222-7, Berlin.

Annals of DAAAM for 2008 & Proceedings of the 19th International DAAAM Symposium, ISSN 1726-9679
Editor B. Katalinic, Published by DAAAM International, Vienna, Austria 2008
Make Harmony Between Technology and Nature, and Your Mind will Fly Free as a Bird

STYLOMETRY – DEFINITION AND DEVELOPMENT

BELAK, S[tipe]; RADMAN PEŠA, A[nita] & BELAK, B[ranko]

Abstract: *The aim of this paper is to define stylometry and describe the main methods used by stylometrists in resolving the problem of author anonymity. The inductive method is used to define stylometry, its development, history, methods and implications of its future development and application. Identifying the authorship of a text, as the main task of stylometry, is presented on the basis of several key methods developed over the last fifteen years, including mathematical and statistical methods, as well as modern artificial intelligence systems. Besides identifying the authorship of a text as the stylometry's main task, the method of intentional anonymization of a text for the protection of author's privacy and security is also described. It is equally interesting to look at stylometry from the point of view of diplomacy, political and economic espionage, security, public relations and academic communities.*
Key words: *stylometry, authorship identification, text, document, identity*

1. INTRODUCTION

Written texts of uncertain authorship attracted the attention of numerous experts from different fields, mostly linguists and literary theorists whose research focused on many famous literary works of uncertain authorship, with the aim of establishing the authenticity of the author and his oeuvre. Relentless studies and searches of texts for the hallmarks of an author's linguistic style, expressions and usage of certain words or metaphors, did not fail to yield results. However, fast technological development in the last fifteen years enabled stylometrists to develop different methods of authorship identification on the basis of a given text. Those methods include mathematical tools, statistical methods and artificial intelligence methods, the result of which is specialized software for text analysis and authorship identification, but also for intentional concealment of a document's authorship. Initially, stylometry was mostly applied in literature and other fine arts. Most stylometrists began their research on certain widely known texts, such as Catholic epistles, ancient Greek and Latin texts, the plays of Shakespeare and Marlowe, and the Federalist Papers (Mosteller & Wallace, 1964), a series of 85 articles published anonymously in 1787 and 1788, advocating the ratification of the new Constitution of the United States of America (Klarreich, 2003). The hypothesis examined in the paper is that stylometry will become more important over time due to the frequency and significance of communication in the realm of the economy and politics. Analyses of anonymous texts from various sources will be crucial for gaining any kind of advantage, be it political or economic, all within business intelligence as a modern form of protection. The paper is based on the project entitled "031-2/2008 Research Into Matters Warranting, Economically and Situation-wise, Adaptive Restructuring of an Organization in a Dynamic Environment" University of Zadar, Department of Economics, Centre for Economic Research, MER-Evrocentar Slovenia, 2008, and is the continuation of earlier research into the concept of terotechnology within the study of organization and business intelligence protection (Belak & Čičin-Šain, 2005).

2. STYLOMETRY - DEFINITION

Stylometry (Greek *stylos* (style) + *metron* (measure)) is a scientific method of studying linguistic style, which has been successfully applied to other areas in the last several years. In addition to art (literature, music, painting), stylometry in applied in forensic science, security affairs, the economy, politics, but also at universities, i.e. whenever it is necessary to identify the author of an anonymous text or a text penned by more potential authors, whether they all claim to be the author or the author of the text in question is simply not known. This is usually the case with complex groups of texts on which several authors worked together. The linguistic style of an author can more readily be identified by analyzing only the *form* of writing, rather than by studying only the *content,* because the content can easily be copied while it is very difficult or almost impossible to imitate the form of writing of an author, it being strongly influenced by the author's subconscious. A text sample used in stylometric analysis should contain a minimum of 1000 words. It is also important to use the pure, original text which was not changed in the course of the years.

3. HYSTORY AND DEVELOPMENT

In 1439 the Italian humanist Lorenzo Valla proved that the Donation of Constantine was a forgery dating from the period between 750 and 850. His conclusion was based on the analysis of the written text and the comparison of Latin with that used in authentic 4th century documents. Some Latin words and phrases used in the Donation did not exist in the 4th century, and came into use much later. This is one of the first examples of stylometric analysis. In modern times stylometry came to be widely used for the study of authorship issues in English Renaissance drama. The development of computers and computer languages for analyzing large quantities of data greatly improved stylometric efforts. In the early 1960s, Andrew Q. Morton developed a computer programme for the analysis of the fourteen Epistles of the New Testament attributed to St. Paul. The analysis showed that the epistles had been written by six different authors.

4. METHODS

The main aspect of authorship identification is the process of selecting features to be analyzed in a text. The most immediate idea that comes to mind at the mention of linguistic analysis of a text could be misleading – that an author's identity can be revealed through complicated or specific words since they mark the author's unique style and set him apart from others. Stylometrists have proven that it is the exact opposite which matters. Authors differ in their usage of the most frequently used simple words such as *with* and *in,* because one's subconscious uses words in daily parlance, such as these prepositions, automatically and without reflection. Words such as these are in fact an authorial "fingerprint" enabling experts to identify the author. Rare and specific words have a strong impact on readers but can easily be consciously inserted into a text to imitate a specific author. It is much more difficult to imitate the usage of simple prepositions and thus it is deemed a safer technique, according to Holmes (1998). Ideally the two techniques should be combined, says Holmes, analyzing first the frequency of simple widespread words and then of the rare and author-specific ones.

4.1. Frequency in Text Segments

In the early 1960s Frederick Mosteller and David Wallace first demonstrated the stylometric method for authorship attribution of the aforementioned Federalist Papers of the United States of America. Mosteller and Wallace found that Hamilton used the word *upon* ten times as often as Madison did. They conducted analyses assuming both men were authors and examined the frequency of thirty words, word upon word, to establish differences in the frequency of repetition. Using the frequency method they concluded that the author of all twelve disputed articles was Madison, which was confirmed by subsequent stylometric analyses.

4.2. Writer Invariant

The writer invariant is a specific distinctive value of a text that can be attributed to a certain author. An example of a writer invariant is the frequency of function words used by the writer. In one such method (Binongo, 2003), the text is analyzed to find the 50 most common words. The text is then broken into 5,000 word chunks and each of the chunks is analyzed again to find the frequency of those 50 words in that chunk. This generates a unique 50-number identifier for each chunk of text. The identifier is then displayed as a point in a 50-dimensional space, flattened into a plane using the PCA technique. Principal Components Analysis is a new technique used to represent texts in multidimensional space and to reduce multidimensional data sets to lower dimensions. The latest similar technique of sample recognition and representation of chunks of text in multi-dimensional space called *support-vector machines* was used by Fung (2003).

4.3. Neural Networks

A neural network is an artificial intelligence method for identifying the author from the text which imitates the neural network of a human brain. One such network is built with random links and works by trial and error using the method *learning by doing.* The network is presented with a prepared text of a known author. Any time the network guesses the authorship in a text segment, the correct assumption strengthens its links until the network can properly identify known texts. Once the training period is complete, the neural network can determine the authorship of texts by authors that it had been trained on previously. Matthews & Merriam (1993) were the first to create the neural network method for the purpose of separating Shakespeare's plays from the plays of his contemporary Christopher Marlowe. Once applied to all Shakespeare's works, the programme only recognized Part Three of Henry VI as Marlowe's.

4.4. Genetic Algorithms

The genetic algorithm is another author identification method which uses a set of rules. It is based on Darwin's principle of natural selection. For example, a rule might be, "If the conjunction *but* appears more than 1.7 times in every thousand words, then the text is by author X". The program is presented with a text and uses the rules to determine authorship. The rules are tested against a set of known texts and each rule gets a score depending on correct or incorrect results. The 50 rules with the lowest scores are thrown out. The remaining 50 rules are slightly changed and reintroduced into the algorithm. This is repeated until, after 256 generations, the evolved rules correctly attribute the text. In the end the finalized rules included only 8 words. Using this method on the Federalist Papers, Holmes and Forsyth (1995) proved once again that the author of all 12 disputed articles was Madison.

4.5. Rare Pairs

This method of authorship identification relies upon specific individual habits of collocation, i.e. using certain words with one another. The use of certain words may, for a particular author, entail the use of other words, which that author usually combines with the first word. By itself, one word means nothing, but combined with another word it often says a lot about the author, according to Craig (2004), who developed this method. Still, stylometrists caution that the thorough analysis of author's style and metaphors he uses render this method open to manipulation and document falsification.

4.6. Cryptography – Stylometry Reversed

There are many reasons to identify authorship. However there are also many reasons for staying anonymous. A department or company manager may wish to stay anonymous when sending out a memo with unpleasant corporate news or some other important document. Kacmrick and Gamon (2006), working on the famous Federalist Papers, tested the possibility of creating a tool for intentional anonymization of a text, believing that there are authors who wish to stay anonymous. Their technique presupposes that an author A wishes to preserved anonymity for a particular document D. Using cryptographic tools, the author A creates a new document D which moves away from A. Combining stylometric and cryptographic methods, they successfully showed that an author's identity can be intentionally concealed in a text.

5. CONCLUSION

This paper offers a definition of stylometry, describes its development and presents basic authorship identification methods applied in stylometry today. The initial research of the project "031-2/2008" points to the future development of stylometry marked by the discovery of a unique method which will, with the support of high technology and artificial intelligence as well as mathematical and statistical methods, combine the most important and efficient stylometric methods. That unique method would have to be applicable to any text, in any language, on any subject-matter and from any source, and provide reliable authorship identification or deliberate authorship obfuscation.

6. REFERENCES

Belak, S., Čičin-Šain, D., (2005) The Development of The Concept of Terotechnology, *Journal of Maritime Studies*, 19, ISSN 1332-0718

Binongo JNG; Smith MWA, (1999). The Application of Principal Component Analysis to Stylometry, *Literary and Linguistic Computing* 1999 14(4):445-466, ISSN 1477-4615

Halteren H. Oostdijk N.H.J., (2004). Linguistic Profiling of Texts for the Purpose of Language Verification, *Annual Meeting of the ACL,* July, 2004, Barcelona

Holmes D., (1998). The Evolution of Stylometry in Humanities Scholarship, *Literary and Linguistic Computing* 13(3):111-117, ISSN 1477-4615

Kacmarcik G. i Gamon M., (2006). Obfuscation Document Stylometry to Preserve Author Anonymity, *Natural Language Processing Group,* ACL, 2006, Sidney.

Mosteller F. i Wallace D. (1964). Inference and Disputed Authorship: The Federalist. *Reading,* MA: Addison-Wesley, 366pp.

Annals of DAAAM for 2008 & Proceedings of the 19th International DAAAM Symposium, ISSN 1726-9679
Editor B. Katalinic, Published by DAAAM International, Vienna, Austria 2008
Make Harmony Between Technology and Nature, and Your Mind will Fly Free as a Bird

THEORETICAL RESEARCH REGARDING THE ACTIVE SAFETY SYSTEMS

BELES, H[oria]; PREDA, I[on]; DRAGOMIR, G[eorge]; BLAGA, V[asile] & ARDELEAN, F[elician]

Abstract: *In this study were developed theoretical and experimental researches regarding the active safety systems for passenger cars. In order to validate the ABS/ASR system working algorithm and to accentuate the influences over the braking or starting processes with and without ABS/ASR, a planar dynamic model was designed for a single wheel of the passenger car. This simplified model has two inertial elements: a wheel in rotation and a mass in translational movement (the "quarter" vehicle).*
Key words: *ABS, dynamic, Simulink, model*

1. INTRODUCTION

The passenger car safety represents its capacity to prevent accidents and to diminish the consequences of an accident produced by other causes than technical break downs. (Bosch 2004) This safety comprises two major categories:

- Active safety.
- Passive safety.

The active safety consists in prevention or avoidance of some accidents that take place as a consequence of an improper operation of the passenger car, and the passive safety consists in reducing the effects of an accident on passengers.

The main characteristic of active safety systems consists in maintaining the directional stability and manoeuvrability of the passenger car in critical rolling conditions as: wheels locking on a slippery surface, skidding in curves, the unexpected appearance of some obstacles.

2. THE MODEL FOR ABS (ANTI-LOCK BRAKING SYSTEM) SIMULATION

2.1 The dynamic model of ABS

To check the ABS working algorithm and to emphasize the different influences on the braking process, with and without ABS, a planar dynamic model was used for a single wheel of the vehicle. The selected model has two inertial elements: a wheel and a mass in translational movement (the vehicle).

The influences of the suspension and direction mechanisms weren't explicitly considerate. (Preda 2005) Also were neglected the jerk and jumping movements of the wheel, respectively the lift up and pitch movements of the translational mass. These simplifications were realised deliberately to highlight the anti-lock braking system parameters that influence the braking process. In these conditions, the dynamic model has only two degrees of freedom: the wheel rotation around the spindle and the translation of the mass sustained by this wheel (including the wheel mass) on the travelling direction (longitudinal) (Preda 1997).

2.2 The Simulink model of ABS

Simulink is a work module of the general mathematical calculus program Matlab (Mathworks Inc.). It is suitable for the simulation of technical systems because it has some very valuable qualities (Simulink 2007):

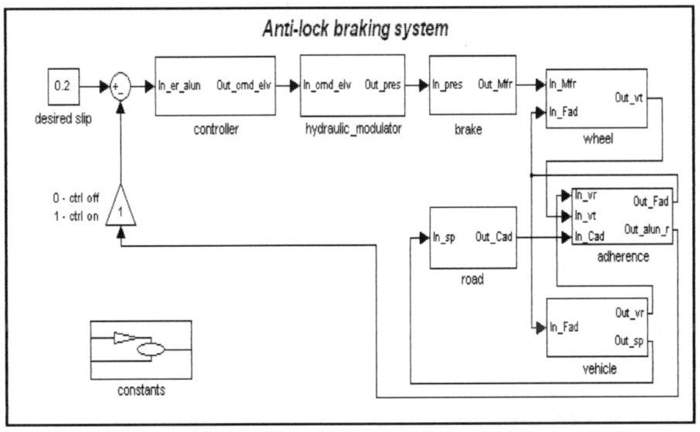

Fig. 1. Diagram of the Simulink model for ABS simulation

- It allow the description of the modelled system using block diagrams interconnected with signal lines; from here results the ease in creation, understanding and use.
- The block diagrams can be hierarchically grouped, which permits the description of some assemblies and subassemblies of great complexity.
- The writing of the algebraic and differential equations with initial values that constitute the mathematical model of the system is done automatically.
- For solving the mathematical model more advanced methods of numerical integration are available.

In figure 1 it is presented the main diagram of the Simulink model realised for the simulation of the ABS operating mode. The blocks that compose it are defined so that it permit the identification and understanding of the ABS functioning.

The "Wheel" and "Vehicle" blocks correspond to these two degrees of freedom of the dynamic model – wheel rotation and "quarter" vehicle translation. These two movements are connected through the grip between the wheel and the road ("Adherence" block) that depends on the road properties defined in the "Road" block.

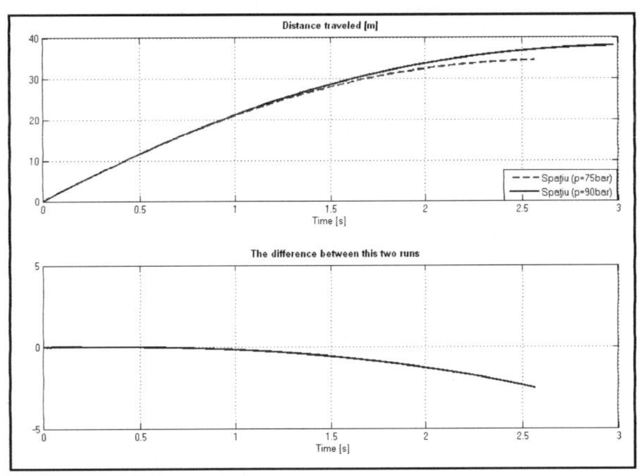

Fig. 2. Braking simulation with the pressure in the hydraulic circuit equal with 90/75 bar: a – braking distance; b – the difference between the two simulations

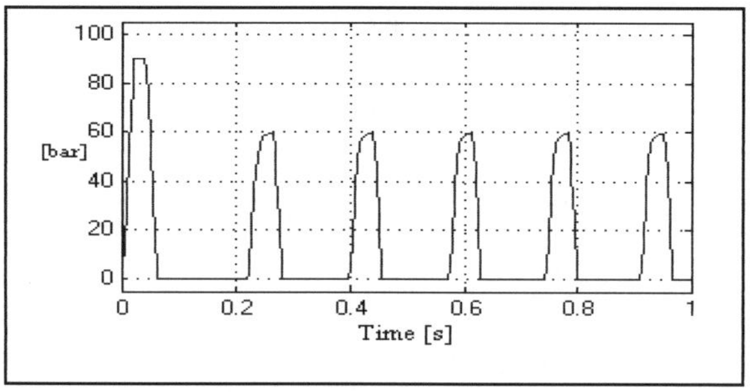

Fig. 3. Simulation of the braking pressure evolution in the hydraulic circuit – braking on snow

The comparator (the round block from the left side of the figure) calculates the difference between the desired values of the slip (the "Desired Slip" block) and the real slip (calculated in the "Adherence" block). Using this difference, the "Controller" block that simulates the operation mode of the adjustment algorithm implemented in the ABS computer will establish what kind of command must be executed by the ABS hydraulic assembly (modelled by "Hydraulic Modulator" block). The pressure generated by this is finally applied to the brake (modelled by the "Brake" block).

The "Ctrl off / Ctrl on" block was introduced in the inverse connection path (to enable or disable the feedback loop) in order to simulate with the same model both an anti-lock braking system (ABS) and a classic braking system (without ABS).

Even if it is very simple, the model was realized in a systemic vision, chasing the accentuation of the functional links between the different components of the ABS.

2.3 Simulation of braking with a pressure of 90/75 bar

The simulation parameters for braking with the pressure inside the hydraulic circuit of 90/75 bar, figure 2 and 3 (Beles 2007):

- Three states controller (able to increase, maintain or decrease the braking pressure).
- The pressure in the hydraulic circuit is 90/75 bar.
- The vehicle mass is 300 kg per wheel.
- The inertia moment of the wheel is 0.75 kgm^2.
- The velocity from which the braking starts is 25 m/s.
- The road surface used for the simulation begins with a portion of 5 m with dry asphalt with an adherence coefficient equal with 1, then a portion of 2 m with snow with adherence coefficient equal with 0.2, then again dry asphalt.

3. THE COMPARATIVE ANALYSIS OF THE THEORETICAL AND EXPERIMENTAL RESULTS

Experimental data was acquired on an experimental ABS system fitted on a Daewoo Nubira passenger car.

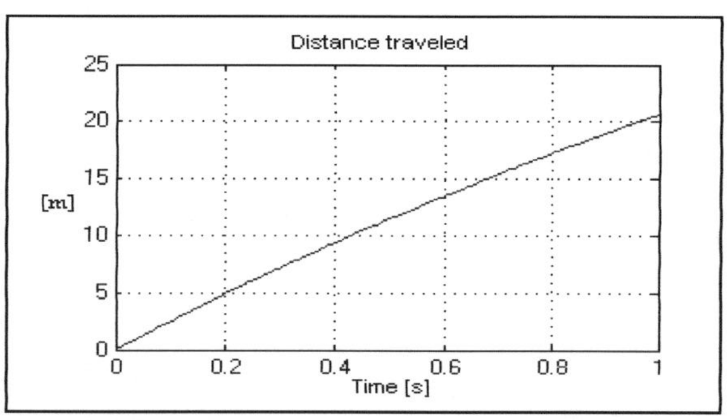

Fig. 4. Braking distance vs. time; braking on dry asphalt, simulation

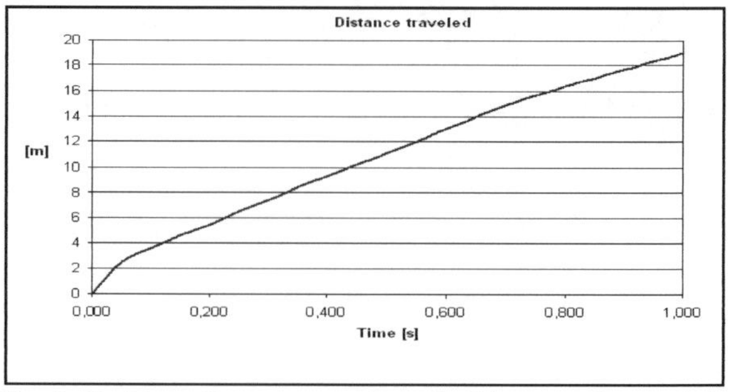

Fig. 5. Braking distance vs. time; braking on dry asphalt, experimental results

Comparing the experimental results with the theoretical results obtained from simulation, it can be seen that the results obtained with the mathematical model have similar evolutions with those determined experimentally. This proves that the simulation model realized and presented in this paper surprise accurately enough the phenomena related to the passenger car braking dynamics. Thus, the main objective of this work was achieved: the one to realize a simplified mathematical model of the vehicle that permits the simulation of the braking process, able to reproduce with enough accuracy the ABS behaviour.

The simulation parameters on a road with dry asphalt figure 4 and 5:

- The pressure in the hydraulic circuit is 90 bars.
- Three states controller (able to increase, maintain or decrease the braking pressure).
- The pressure in the hydraulic circuit is 90 bars.
- The vehicle mass is 300 kg per wheel.
- The inertia moment of the wheel is 0.75 kgm^2.
- The velocity from which the braking starts is 25 m/s.
- The road surface used for the simulation is with a grip coefficient equal with 1 corresponding to a road with dry asphalt.

The experimental conditions for braking:

- Rectilinear braking (with ABS), with warm brakes on dry asphalt.
- Constant grip coefficient, equal with 1 (dry asphalt) for all the wheels.
- The velocity from which the braking starts is 22.8 m/s.

4. CONCLUSION

The paper presented a dynamic and mathematic model realised with the idea of simplicity, but still able to generate results with enough accuracy, that permit to put in evidence the main influences of each ABS subsystems. Experimental results acquired during road tests on a car ABS permitted to tune and validate the simulation model. The model can be easily adapted for different subsystems or for an entire vehicle.

5. REFERENCES

Beles, H. (2007). *The research regarding the development of active safety for passenger cars in order to reduce the accidents with serious consequences*, Doctoral thesis (in Romanian), "Transilvania" University, Brasov.

Bosch R. GmbH (2004). *Automotive handbook – 6th Edition*, Plochingen

Matlab Simulink tutorial (2007). *Modelling an Anti-Lock Braking System*

Preda, I.; Todor, I.; Ciolan, Gh. (2005). *Contributions to the simulation of vehicle longitudinal dynamics*. In: Bulletin of the "Transilvania" University of Brasov, vol.12 (47), Seria A1, p.177-186, ISSN 1223-9631, Brasov

Preda, I. si Ciolan, Gh. – Wheel-road interaction modelling. CAR Pitesti, 1997, pg. 85-90.

Annals of DAAAM for 2008 & Proceedings of the 19th International DAAAM Symposium, ISSN 1726-9679
Editor B. Katalinic, Published by DAAAM International, Vienna, Austria 2008
Make Harmony Between Technology and Nature, and Your Mind will Fly Free as a Bird
*Young Researcher and Scientist Paper / * Supervisor, Mentor*

DETERMINATION OF STRAIN VIA 3D SCANNING IN A BIMETAL FABRICATED BY EXPLOSION WELDING

BENAK, M[ichal]; BURANSKY, I[van] & TURNA, M[ilan]*

Abstract: *The contribution is aimed at explosion welding of aluminium with austenitic CrNi steel. The objective was to fabricate the above-mentioned bimetal and to assess its quality. S 30 Semtex explosive was used for welding. The accelerated material was aluminium. The welded materials were situated parallelly. Bimetal was fabricated in co-operation of the Welding Department, Faculty of Materials Science and Technology (FMST), STU in Trnava with Explosia Inc., Pardubice. Its quality was assessed in co-operation with the Department of Materials Engineering, and the Department of Machining and Assembly of FMST, STU, Trnava.*

To determine the formed strain in Al-CrNi bimetal, the optical 3D scanner ATOS I 350 from GOM company was applied. Bimetal was scanned and individual scans were processed by GOM ATOS software. The scanned model of bimetal was compared with the reference model of bimetal which was designed by PowerSHAPE software. A colour map of deviations was attained from this comparison.

Key words: *Bimetal, Explosion welding, 3D scanning, Semtex, ATOS I 350*

1. INTRODUCTION

Formation of welded joints at collision of two different materials was observed at the end of forties in the study of cumulative phenomenon. Since that time the explosion welding (EW) technology underwent a dynamic progress both in theoretical and application fields. In technological processing of metals by explosion the substrate and accelerated materials are bonded by acting of compression formed by detonation of a suitable explosive located over entire area of accelerated material (Referativnyj žurnal „SVARKA“. (2005)). Compressive effect of explosive acts normally on accelerated metal, which is colliding on the stable metal at v_0 velocity. Due to dynamic effect of collision not only bonding but also deformation of both metals occurs (Turňa, 1989). An optical scanner was used for its measurement which measures with accuracy of 0.02 mm (Peterka et al., 2007).

2. PRINCIPLE OF EW TECHNOLOGY

Explosion welding is considered for a specific kind of cold pressure welding. Welded joints fabricated by explosion have a typical undulated boundary, characterizing both, welding process and bond quality. This methods allows to weld relatively broad assortment of metals and their alloys. Principal and kinematic schemes of explosion welding are show in Figs. 1. and 2 (Yuheng et al., 2005).

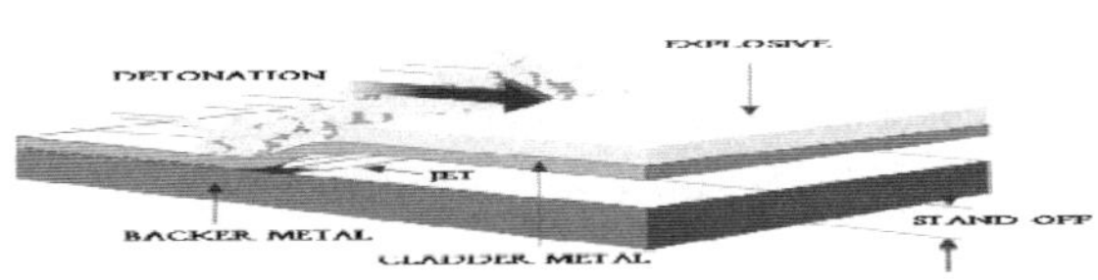

Fig. 1. Principle scheme of explosion welding.

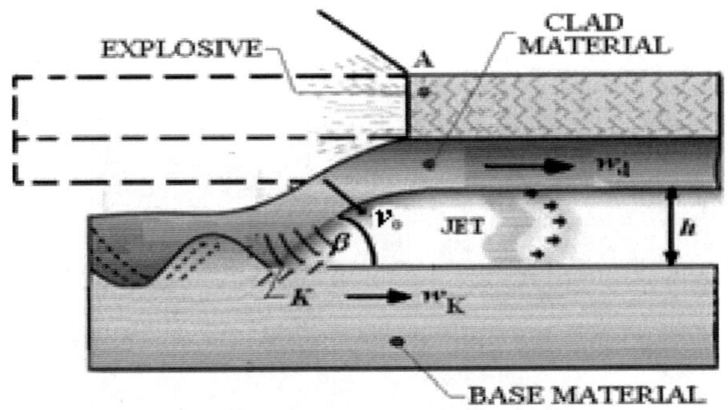

Fig. 2. Kinematic scheme of explosion welding.

3. CHARACTERISTIC OF WELDED MATERIALS

In explosion welding process a stable material Al 99.5%, delivered in soft condition according to STN 42 4005 standard. Semi product was an aluminium plate of rectangular shape 15 mm in thickness.

CrNi austenitic steel type18/8 (304, 1.4310) was used as accelerated material, defined by STN EN 10 088 standard. This material was heat treated by solution annealing and by stress relief heat treatment. The delivered material was a sheet of rectangular shape 1 mm in thickness.

4. WELDING PARAMETERS

The input data for calculation of parameters and conditions of welding process are given in Table 1.

Name	Desighnation	Value
Accelerated metal thickness (austenitic steel type 18/8)	H_D [mm]	1
Material density (austenitic steel type 18/8)	ρ [kg. m^{-3}]	7850
Thickness of stable metal (Al 99.5)	H_D [mm]	13
Material density (Al 99.5)	ρ [kg. m^{-3}]	2700
Explosive thickness (Semtex S 35)	H_E [mm]	10.30
Distance spacing	h [mm]	2
Initial angle of setup	α [°]	0
Collision speed of sheet	v_0 [m.s^{-1}]	606.34
Detonation pressure	p_{cj} [GPa]	1.87
Time constant	τ [μs]	3.15
Dynamic collision angle	β [°]	19.9
Resultant detonation speed	v_d [m.s^{-1}]	2164.89

Tab. 1. Parameters and conditions of welding process.

5. FABRICATION OF Al-CrNi BIMETAL

The mechanically machined weld edges were cleaned from surface impurities and degreased closely prior to welding. Aluminium plate was backed with chipboard, which created also relational plane for the ance spacing of stainless steel plate. A wooden frame was placed on accelerated material with dimensions (279 x 185 x 10.3 mm), to which a Semtex S 35 explosive was poured. The point of ignition and ignition explosive Semtex was selected in such a manner that the detonation wave would be directed approximately in parallel with material length. The charge was ignited with an electronic detonator inserted into ignition explosive. The fabricated bimetal Al-CrNi austenitic steel (see Fig. 3) was trimmed to final size 146 x 112 x 16 mm.

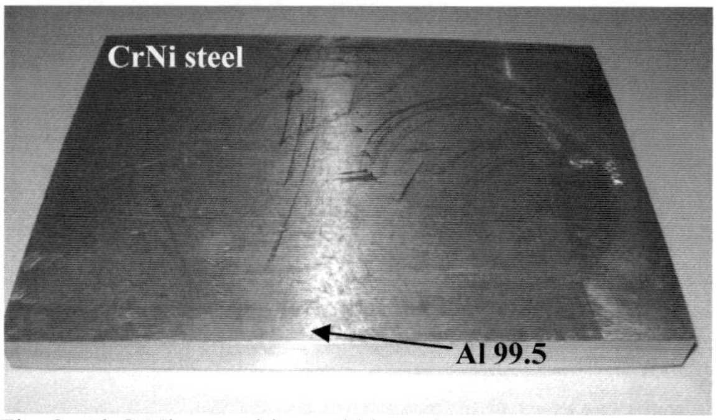

Fig. 3. Al-CrNi austenitic steel bimetal.

6. MEASUREMENT OF BIMETAL DEFORMATION

An optical 3D scanner type ATOS I 350 from GOM company was used for determination of formed deformation of Al-CrNi steel bimetal. The optical 3D scanner consists of two CCD cameras, projection lamp (projector), stand and a powerful computer with pre-installed GOM ATOS software.
Uncoded reference points were stuck on Al-CrNi steel bimetal. A chalk spray was sprayed on material surface, since the glossy surface of bimetal is not suitable for optical scanning. The uncoded reference points were then cleaned form chalk spray, in order that optical scanner could sense them. The uncoded points are used for determination of 3D coordinates and are automatically detected with GOM ATOS software (ATOS v.6 User Manual.(2006)). The result of bimetal scanning is a cloud of points. For creation of a digitalised model it was necessary to perform 12 scans form different sides. After scanning the bimetal, the „align" function was used, which serves for reducing the deviation from individual scans. The function „polygonized" was used, which created triangles from the cloud of points.

The digitalised model created (modelled) in DELCAM PowerSHAPE software was then compared with a CAD model in dimensions (146 x 112 x 16 mm). CAD model was subsequently imported to GOM ATOS software. The reference CAD model and the digitalised model of bimetal were placed into one coordinate system by use of „best-fit registration" function. For comparison of above-mentioned points the „deviation to mesh" function was used, which resulted in the colour map of deviations, (Fig. 4).

The colour map of deviation makes visible the greatest deformation along the sides of fabricated bimetal. The measured deviation is (0.8 až 1)mm (Fig. 5). The deviation varies on the surface and it attains values from -1.6 to 0.9 mm. Fig. 5 also shows Al-CrNi bimetal deflection. Deflection occurs mainly at smaller dimensions of bimetal, while it is rare at greater dimensions.

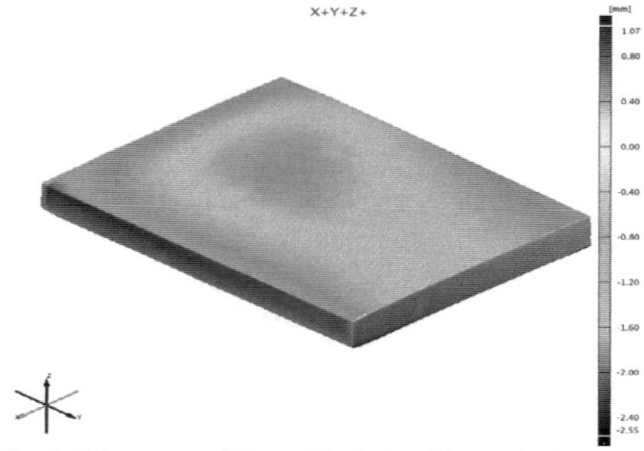

Fig. 4. Colour map of bimetal deviation (above view).

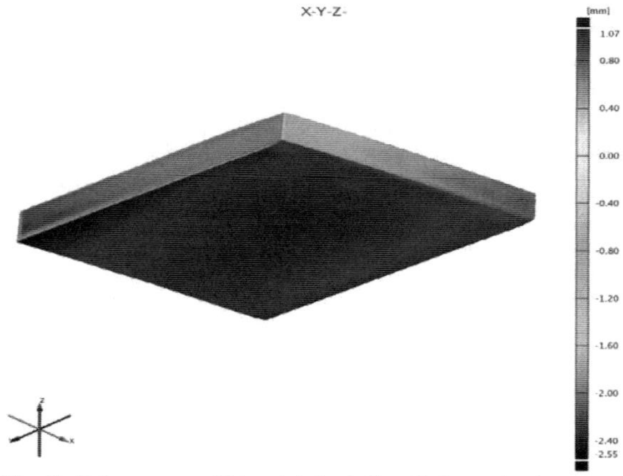

Fig. 5. Colour map of bimetal deviation (below view).

7. CONCLUSIONS

Development of combined materials Al-CrNi steel (aimed at utilisation of priority mechanical and physical properties of these metals) is at present relatively discussed topic also from the viewpoint of searching for optimum welding technology. Experimental results have shown that the quality of joints is acceptable and deformation of bimetal is not extreme. The greatest change in shape is observed on the fringe parts of bimetal, which may be removed by machining. Material deflection should be then solved, which was the subject of measurements in the mentioned work.

8. REFERENCES

ATOS v.6 User Manual, GOM mbH, Germany 2006.

Peterka, J.; Janac, A.; Morovič L. (2007) *Optical methods of capturing the metrological quantities of mechanicalparts,* ICPM 2007, IV International Congress of Precision Machining, ISBN 978-83-88906-91-6

Referativnyj žurnal „SVARKA". (2005). Solid state welding. Explosion welding. Moscow.

Turňa, M. (1989).; *Special welding processes*, Alfa, ISBN 80-05-00097-9, Bratislava.

Yuheng Li.; Deming Shu.; Tuncer M. K.(2005). Explosion Bonding of Dissimilar Materials for Fabricating APS Front End Components, *Available from*: http://www.aps.anl.gov/Facility 2005 , Accessed: 2005-06-12.

ACKNOWLEDGEMENT

This work is a part of VEGA no. 1/3191/06 project.

Annals of DAAAM for 2008 & Proceedings of the 19th International DAAAM Symposium, ISSN 1726-9679
Editor B. Katalinic, Published by DAAAM International, Vienna, Austria 2008
Make Harmony Between Technology and Nature, and Your Mind will Fly Free as a Bird

THE INFLUENCE OF COATING AND TOOL GEOMETRY ON THE TOOL LIFE IN A THREAD CUTTING PROCESS

BENGA, G[abriel] & CIUPITU, I[on]

Abstract: *The paper deals with the influence of coatings on the tool life of the different shapes of taps in a thread cutting process. The influence of the tap geometry was very important and affected significantly the tool life measured in the number of holes tapped. The wear pattern for each type of tap was analyzed as well.*
The TiAlN coatings promote the formation of protective alumina films composed of amorphous-crystalline structures improving the tribological behavior of the coated tools.
Keywords*: tapping, tap geometry, PVD coatings, tool wear*

1. INTRODUCTION

Internal thread cutting is a very demanding operation taking into account the possibility of damaging both the tool and the workpiece material. Usually the thread cutting is performed in the late stage of the manufacturing process and therefore a major investment in labor cost, energy and material is already done. Thus a breakage of the tap can significantly decrease the productivity of the process (Veldhuis et al, 2007). Recent improvements regarding the tool life of cutting tools have been achieved by the development of titanium aluminum nitride TiAIN coatings. Coatings as TiAIN display a very interesting combination of properties, including high hardness at elevated temperature together with thermal and chemical stability, as well as low thermal conductivity (Fox-Rabinovich et al, 2004; Benga, 2008). TiAlN offers superior performance for a range of metal machining and fabrication applications. The reason for this lies in the addition of aluminium to TiN or specifically the formation of aluminium oxide on the surface of the tool which increases the operational temperature range of the coating to 800^0 C compared with 500^0 C for TiN. As the TiAlN coating is heated in air a thin layer of amorphous aluminium oxide forms on the surface of the coating, which protects the coating from further oxidation. This results in better hot hardness than most other coatings (Iwai et al, 2001).

2. EXPERIMENTAL PROCEDURE

The work piece material used was AISI P20 hardened mold steel with a chemical composition presented in table 1. The dimension of the work piece material was 150mm × 150mm × 30mm and the hardness of each block was 35±2HRC.

Chemical element wt.%	C	Si	Mn	Cr	Ni	Mo	S
Composition	0.31	0.4	0.75	1.2	0.8	0.41	0.008

Table1. Chemical composition of the AISI P20 mold steel

The tests were performed on an Okuma Cadet Mate CNC vertical milling center and the tool wear was measured using a Mitutoyo optical microscope. As a wear criterion was chosen the V_B flank wear V_B=0.3 mm.

The cutting regime employed was as follows:
- cutting speed: 8 m/min approximately 260rpm;
- feed rate: 1.587 mm/rev;
- length of thread: 10 mm.

The holes tapped were through holes in order to facilitate the chip evacuation.

Three different tap designs were employed for the tests as follows: normal taps, spiral taps, pointed spiral taps.

The taps are presented in figure 1.

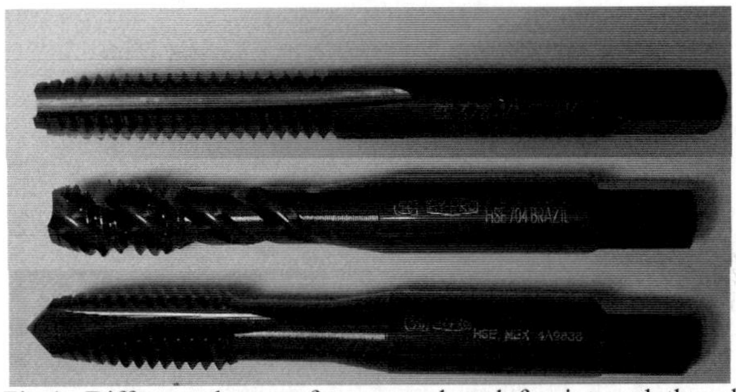

Fig.1. Different shapes of taps employed for internal thread cutting

The cutting tools were made of HSS and then coated using PVD technique with TiAlN. The diameter of the taps was 3/8" which approximately means 9.5 mm.

3. RESULTS AND DISCUSSIONS

The tests were performed initially using both coated and uncoated taps of each type and the wear was measured after 4,5 holes tapped. At the beginning the tool was unclamped after each hole in order to measure the wear but it was observed that the wear does not advance so fast. The following figures present the wear patterns for each tap.

Fig.2. Wear patterns for normal taps without coating V_B=0.385 mm after 20 holes

Fig. 3. Wear patterns for normal taps with coating V_B=0.35 mm after 27 holes

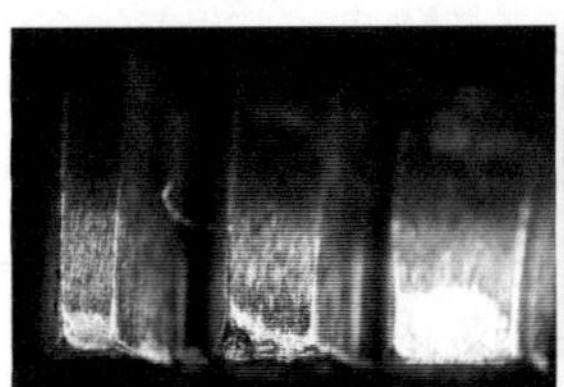

Fig.4. Wear patterns for pointed spiral taps without coating, V_B=0.3 mm after 15 holes

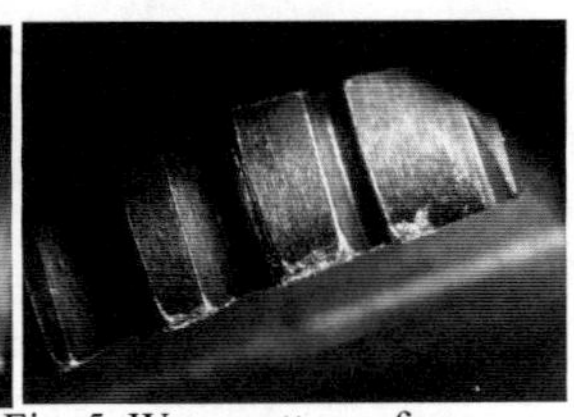

Fig. 5. Wear patterns for pointed spiral taps with coating V_B=0.31 mm after 32 holes

Fig.6. Wear patterns for spiral taps without coating V_B=0.27mm after 194 holes

Fig. 7. Wear patterns for spiral taps with coating V_B=0.28 mm after 380 holes

Figures 2 and 3 present the wear patterns when normal tapes were used. In the case of the HSS tap it is obvious that breakage occurred on one flute of the tap after tapping 20 holes. A possible explanation in this case is that the wear did not occur mainly due to the friction between the tool and the work piece material but due to the cutting load developed during the internal thread process which caused breakage of the tool. The wear pattern when the TiAlN coated tap was employed is different and seems to be the result of the abrasive wear.
This happens due to the TiAlN coating which does not exhibit edge brittleness and can be used for interrupted cuts without chipping. The use of TiAlN has increased the tool life by 1.35 times taking into considerations the number of holes tapped.

Figures 4 and 5 show the wear pattern for the spiral pointed taps. In the case of the uncoated tap (fig.4) the dominant wear mechanism seems to be chipping of the edge. There is no breakage like happened with the uncoated normal tap but chipping of the edge is present. This can be due to the vibrations occurred during the thread cutting process combined with the cutting load.

The TiAlN spiral pointed coated tap develops a regular wear on the clearance face. Again the presence of TiAlN coating improves the toughness of the tool and the wear is produced mainly because of the abrasion between the tool and the work piece material and the delaminating of the coating layer. The use of TiAlN in this case results in a tool life improving of about 100%. Anyway comparing with the performance of the normal tap, the tool life foe spiral pointed tap is not very good, being even lower in the case of the uncoated tap. A possible explanation could be the fact that the number of flutes is lower (3 flutes for pointed spiral tap) than for normal tap (4 flutes for normal tap) and therefore the number of channels for chip evacuation is lower as well. As a consequence the chip evacuation is worse and this may cause chips to be blocked and even tool breakage. There is no significant difference between the tool life of the TiAlN normal tap coated and the TiAlN spiral pointed tap coated. This can be attributed to its nano layers and the high hardness of the TiAlN coating (Santos et al, 2007).

In figures 6 and 7 the tool wear of the spiral taps is presented. This taps were by far the best in terms of tool life. Regarding the TiAlN coated spiral tap, the tool life exhibited was 14 times higher than for coated normal tap and almost 12 times higher than for coated pointed spiral tap. This tool life can be attributed on one hand on the improving of chip evacuation and on the other hand on the very good tribological behavior offered by the TiAlN coating. The spiral flute taps are especially designed for machine tapping of blind holes. The right hand spiral flutes direct the chips back out of the drilled hole, minimizing clogging and tap damage at the cutting chamfer. In the following figures are presented several tool life graphics for the cutting tools employed.

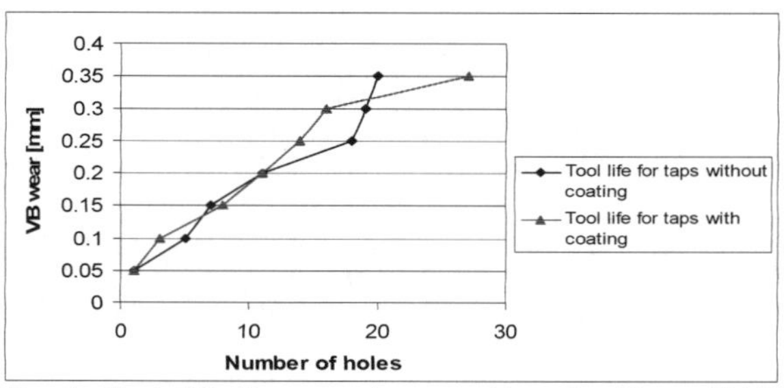

Fig. 8. Tool life for normal taps with and without TiAlN coating

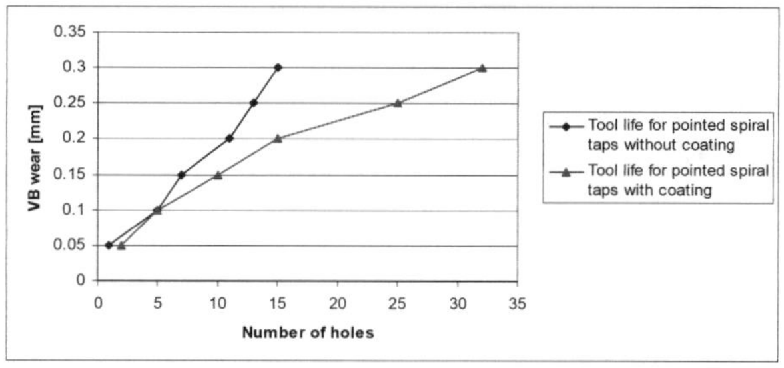

Fig.9. Tool life for pointed spiral taps with and without TiAlN coating

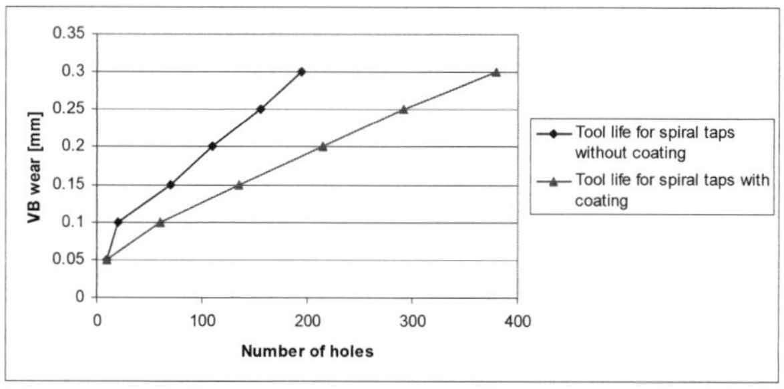

Fig.10. Tool life for spiral taps with and without TiAlN coating.

4. CONCLUSIONS

I. Irrespective to the tool geometry it was obvious that cutting tools coated with TiAlN have showed higher tool life than the HSS tools. The high hardness of this coating does not allow the appearance of an irregular wear. The abrasion wear and the delaminating of the coating layer appear to be the dominant wear mechanism of these cutting tools. Abrasion wear is caused by the friction generated between the tool flank and work piece material.

II. The tool geometry has a significant influence on the tool life. While there is no big difference of using normal taps or pointed spiral taps, the use of spiral fluted taps has dramatically improved the tool life. The spiral taps allow the chips to be directed out of the hole much more easily minimizing clogging and tap damage at the cutting edge.

5. REFERENCES

Benga, G. (2008). The influence of different PVD coating techniques on cutting forces in a drilling process, *Proceedings of the 6th International Conference DAAAM Baltic Industrial Engineering*, R Kyttner (Ed), pp. 407-412, ISBN 978-9985-59-783-5, Tallinn, Estonia, April 2008

Fox-Rabinovich, G.; Weatherly, G.C. Dodonov, et al (2004) Nano-crystalline filtered arc deposited (FAD) TiAlN PVD coatings for high-speed machining applications, *Surface and Coatings Technology,* Volumes 177-178, January 2004, pp. 800-811, ISSN 0257-8972

Iwai, Y.; Honda, T.; Yamada, et al. (2001) Evaluation of wear resistance of thin hard coatings by a new solid particle test, *Wear,* Volume 251, 2001, pp. 861-867, ISSN 0043-1648

Santos, J.A.B.O, Sales, V.F. et al (2007) Tribological evaluation of TiN and TiAlN coated PM-HSS gear cutter when machining 19MnCr5 steel, *The International Journal of Advanced Manufacturing Technology*, Volume 31, Numbers 7-8/ January, 2007, pp. 629-637, ISSN 0268-3768

Veldhuis, S.C., Dosbaeva, G.K., Benga, G. (2007) Application of ultra-thin fluorine-content lubricating films to reduce tool/workpiece adhesive interaction during thread-cutting operations *Int. Journal of Mach. & Manufact.*, vol. 47, 2007, pp. 521-528, ISSN 0890-6955

Annals of DAAAM for 2008 & Proceedings of the 19th International DAAAM Symposium, ISSN 1726-9679
Editor B. Katalinic, Published by DAAAM International, Vienna, Austria 2008
Make Harmony Between Technology and Nature, and Your Mind will Fly Free as a Bird
2nd DAAAM-ICAT

SLS PARAMETERS OPTIMIZATION USING THE TAGUCHI METHOD

BERCE, P[etru]; PACURAR, R[azvan]; BALC, N[icolae] & PACLISAN, D[ana]

Abstract: *In present work, the relationship between the shrinkage and various process parameters is developed. Two test parts are proposed and manufactured from Laserform St-100 powder by Selective Laser Sintering (SLS) technology. As expected, the shrinkages along X, Y and Z-axes are not independent. Taguchi method for the design of experiments and analysis of experimental data has been used in order to find the optimal parameters for manufacturing the SLS metal parts.*

Key words: *Rapid Prototyping, Selective Laser Sintering, Taguchi method, S/N ratio*

1. INTRODUCTION

Taguchi approach uses three major steps namely, system design, parameters design and tolerance design for optimizing a process or product. In system design, scientific and engineering knowledge is applied to produce a basic functional prototype design. It contains selection of materials, components and production equipment, process parameter design, which is used to optimize the settings of process parameter values for improving quality characteristics (Pham et al., 1999). Final step of the optimization is tolerance design, used to determine and analyze tolerances around the optimal settings recommended by the parameter design (Bagchi, 1993).

Yang et al. propose compensation test pieces for the X, Y and Z axes to compensate the shape distortions caused by phase changes during the sintering process and the shrinkage rates have been measured experimentally. The scale factors obtained from the proposed building compensation test pieces of X, Y and Z axes satisfy the required dimensional accuracy, even if there are changes in the build positions and in the size of the SLS parts (Yang et al., 1998).

Figure 1 shows the simplified steps of applying the Taguchi method to the SLS process through the selection of the optimum scale factors and the build orientation. With this method it is possible to maintain the robustness to the noise factors generated by the temperature deviation in the build chamber (Yang et al., 2002).

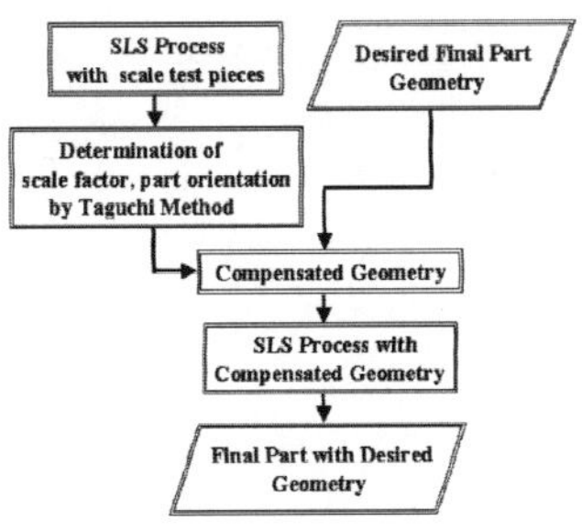

Fig.1. SLS process distortion using Taguchi method

N. Raghunath and Pulak M. Pandey propose in their paper an optimization method in order to find a relationship between the shrinkage and various process parameters of the SLS system. One case study is successfully presented to show the effectiveness of developed models for improving the accuracy of SLS plastic prototypes (Raghunath & Pandey, 2006).

The current paper presents a research developed at Technical University of Cluj-Napoca (TUCN) focused mainly on the manufacturing of the metallic parts from a dedicated material (Laserform St-100 powder) by using the Sinterstation 2000 equipment. Taguchi method has been applied successfully in this case in order to find the optimum manufacturing parameters for the SLS metal parts.

2. TAGUCHI METHOD APPLIED

Two test parts were proposed, as illustrated in Figure 2. Several pairs were manufactured on the SLS system and then measured by an optical Werth Video Check IP 250/400 machine (TUCN), on "green-stage" level and by Zeiss Eclipse CMM-550 (TUCN) after the parts were post-processed in the oven. Each dimension is measured and compared to the designed one, in order to calculate the shrinkage percentage, using the following formula:

$$S = \frac{L_c - L_m}{L_c} \times 100 \quad (1)$$

where S is the shrinkage (%), L_c is the CAD dimension and L_m is the measured one

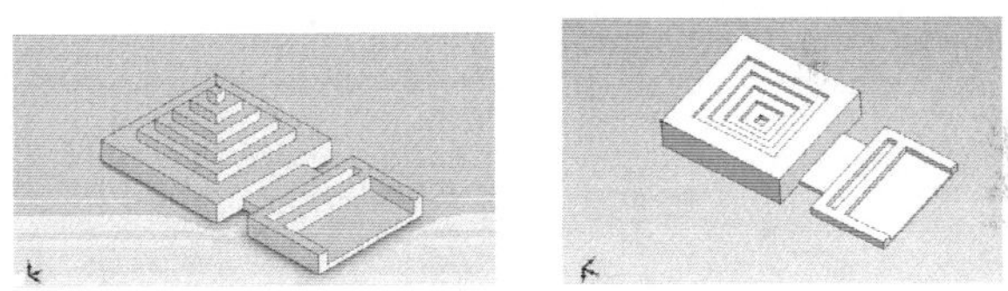

Fig.2. Scale test parts manufactured by SLS

In order to optimize the SLS manufacturing parameters Taguchi method has been applied. At the beginning, the most significant SLS parameters were chosen as presented in Table 1. There are so called levels of importance given for each parameter as could be observed in the same Table. In this case Level 1 is the most insignificant level and Level 5 is the maximum level.

No	Parameter	Level 1	Level 2	Level 3	Level 4	Level 5
1	Fill laser power [W]	24	25	26	27	28
2	Fill scan speed [mm/s]	1200	1300	1400	1500	1600
3	Outline scan speed [mm/s]	50	100	200	300	400
4	Slicer fill scan spacing [mm]	0.074	0.076	0.078	0.08	0.082
5	Part bed temperature [° C]	90	92	94	96	98
6	Powder layer thickness [mm]	0.074	0.076	0.078	0.08	0.082

Tab. 1 Parameters levels of importance

For this experiment an L_{25} orthogonal array with 6 columns and 25 rows is used as presented in Table 2. To select an appropriate orthogonal array, total degrees of freedom need to be computed. The degrees of freedom are the number of comparisons to be made between designed parameters. Each parameter is assigned to each column of the orthogonal array, so 25-parameter combinations are available using L_{25} array.

Exp No	Columns					
	Factor 1	Factor 2	Factor 3	Factor 4	Factor 5	Factor 6
1	1	1	1	1	1	1
2	1	2	2	2	2	2
3	1	3	3	3	3	3
4	1	4	4	4	4	4
5	1	5	5	5	5	5
6	2	2	1	3	4	5
7	2	1	2	4	5	3
8	2	4	5	2	3	1
9	2	5	3	1	2	4
10	2	3	4	5	1	2
11	3	3	2	4	5	1
12	3	2	3	5	1	4
13	3	5	1	3	4	2
14	3	1	4	2	3	5
15	3	4	5	1	2	3
16	4	4	3	5	1	2
17	4	3	4	1	2	5
18	4	1	2	4	5	3
19	4	2	5	3	4	1
20	4	5	1	2	3	4
21	5	5	4	1	2	3
22	5	4	5	2	3	1
23	5	2	3	5	1	4
24	5	3	1	4	5	2
25	5	1	2	3	4	5

Tab. 2 Taguchi L_{25} orthogonal array

3. EXPERIMENTAL DATA ANALYSIS

Experimental data were analyzed using the S/N ratio by measuring the quality characteristics and meantime the deviation from the desired value. The term signal (S) represents the desirable value (mean) and the noise (N) represents the undesirable value (Standard Deviation from Mean - MSD).
The S/N ratio (η) and the MSD can be defined as:

$$\eta = -10 \cdot \log(MSD) \quad (2)$$

$$MSD = \frac{1}{n}\sum_{i=1}^{n} Y_i^2 \quad (3)$$

where n is the total number of the experiments in the orthogonal array and Y_i is the mean percentage shrinkage for the i-th experiment (measured dimension). S-N ratio for each dimension is calculated and presented in Table 3. The effect of a factor level is defined as the deviation it causes from the overall mean.
The overall mean S/N ratio can be calculated using the formula:

$$m = \frac{1}{n}\sum_{i=1}^{n} \eta_i \quad (4)$$

where η_i is the mean S/N i-th experiment ratio.

All five levels of every factor are equally represented in 25 experiments. Thus, m is the mean of the entire experiments.
Mean response is the average of quality characteristic for each parameter at different level. S/N ratio and shrinkage (%) for each parameter at each level can be calculated from mean S/N ratio and shrinkage (%) value of each of the experiment. For example, the mean percentage shrinkage for fill scanning speed at Level 1 can be calculated by averaging shrinkage (%) from the experiments 1, 7, 14, 18 and 25.

Ex No	Measured dimensions			MSD			η		
	X	Y	Z	X	Y	Z	X	Y	Z
1	5,07	5,06	25,18	0,068	0,044	0,009	-8,336	-6,515	0,279
2	15,13	15,10	20,20	0,023	0,011	0,023	-3,775	-0,428	-3,684
...	...	...	...	...	...	...	...	...	...
25	45,07	45,01	4,99	0,001	1,21 E-04	1,01 E-04	17,497	13,76	3,457
						m	2,115	3,328	5,443

Tab. 3. S-N ratio

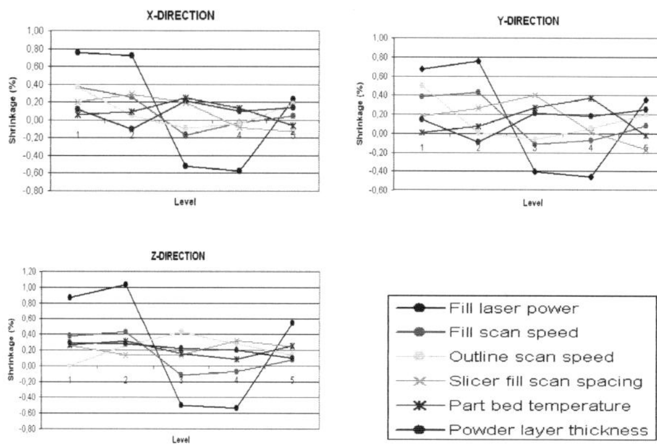

Fig. 3. Effect of process parameters on shrinkage

4. CONCLUSION

Further on, by following the same algorithm as presented above, the effect of process parameters in X, Y and Z direction is obtained as illustrated in Figure 3. Finally, the optimum SLS manufacturing parameters for minimum shrinkage were obtained, as the ones presented in Table 4.

No.	Parameter	Measuring unit	Value
1	Fill laser power	W	28
2	Fill scan speed	mm/s	1500
3	Outline scan speed	mm/s	300
4	Slicer fill scan spacing	mm	0.08
5	Part bed temperature	° C	90
6	Powder layer thickness	mm	0.08

Tab. 4. Optimum parameters obtained for the SLS process

5. REFERENCES

Bagchi, T. P. (1993). *Taguchi Methods Explained,* Prentice-Hall, ISBN 0-87692-808-4, India.

Pham, D. T. et al., (1999). Selective laser sintering: applications and technological capabilities. *Journal of Engineering Manufacture,* Vol. 213, No. 5 (1999), 435-449, ISSN 0954-4054.

Raghunath, N. & Pandey, P. M., (2006). Improving accuracy through shrinkage modeling by using Taguchi method in selective laser sintering. *International Journal of Machine Tools and Manufacture,* Vol. 47, No. 6 (2006), 985-995, ISSN 0890-6955.

Yang, H. -J. et al. (2002), A study on shrinkage compensation of the SLS process by using the Taguchi method. *International Journal of Machine Tools and Manufacture,* Vol. 42, No.11 (2002), 1203-1212, ISSN 0890-6955.

Yang, W. H. et al., (1998). Design optimization of cutting parameters for turning operations based on the Taguchi method. *Journal of Materials Processing Technology,* Vol. 84, No. 1 (1998), 122-129, ISSN 0924-0136.

Annals of DAAAM for 2008 & Proceedings of the 19th International DAAAM Symposium, ISSN 1726-9679
Editor B. Katalinic, Published by DAAAM International, Vienna, Austria 2008
Make Harmony Between Technology and Nature, and Your Mind will Fly Free as a Bird

ANALYSIS OF CAST TOOL STEEL OF 55NiCrMoV7 TYPE CAST IN PRECISION CERAMIC MOULD

BEZNAK, M[atej]; CHAUS, A[lexander] & BAJCICAK, M[artin]

Abstract: *The main aim of the present paper is to analysis structure and properties of cast tool steel of 55NiCrMoV7 type in the case of die insert produced in ceramic moulds. The investigations have been carried out after casting, annealing, quenching and tempering. The paper introduces the technological operations and materials for producing precision ceramic moulds using so-called Shaw process.*
Key words: *ceramic, mould, steel, structure*

1. INTRODUCTION

In the past castings, as a rule, had relatively large allowances for processing due to bad quality of their surface. Today the modern trend is to produce castings with the final dimensions and good surface quality. One of the methods, which provide needed accuracy and quality of castings, is producing composite ceramic moulds in accordance with Shaw process. The application of Shaw process makes it possible to produce castings from all known foundry alloys with no presently known size restriction that can be the main advantage of this technology. The very castings produced in ceramic moulds by Shaw process have high degree of size precision and outstanding metal soundness in term of density as well as surface quality, that benefits in improvements of both energy and materials savings, and processing time reduction. In this term importance of Shaw process increases dramatically when casting expensive tool steels, high alloy steels, and special alloys, which are the difficult-to-machine materials (Beznák, 2003).

2. SHAW PROCESS IN PRODUCTION OF CERAMIC MOULDS

For casting die inserts the composite ceramic mould, which is comprised of a backing layer and a facing layer, is produced as shown in Fig. 1. Pouring a mixture of a chamotte refractory and a water glass binder about an oversized pattern forms the backing layer. After the backing layer hardens, the facing layer is then formed integrally with the backing layer by pouring a refractory slurry-like mixture between the oversized backing layer and a dimensionally correct pattern. When this mixture forms a somewhat flexible gel, the mould can be stripped off the pattern. After it facing layer is fired and then baked (Beznák, 2004).

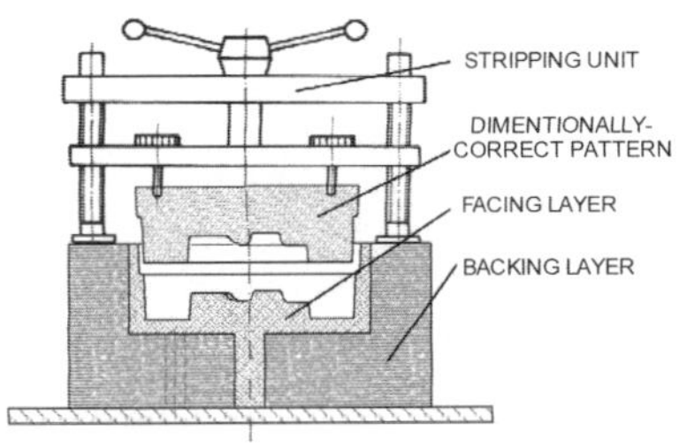

Fig. 1. Scheme of producing composite ceramic mould in accordance with Shaw process.

In general, mixture for producing facing layer of a ceramic mould is composed of a highly refractory material, a gelling agent, and a binder. As a refractory material may be used sillimanit, mullite, zircon. We used two types of the refractory materials, which differ in particle size. The first refractory material was of a finer grade with the particle size less than 0.5 mm. The second refractory material was of a coarser particle size in the range from 1 till 3 mm. The slurry-like mixture hardens when gelling agent, which is 15 % sodium hydroxide solution, is added in amount of 10-12 g per 1 kg of the mixture to ethyl silicate binder. Due to hydrolysis of ethyl silicate, induced by NaOH, silica gel is formed. Rate of the gel formation or gelling rate depends on PH of a sodium hydroxide solution.

When mixing all components, the slurry-like mixture is formed and its PH is about 10 just before pouring. For castings of common sizes gelling time of the mixture is 30 – 180 seconds. This time depends on the both pattern sizes and amount of the slurry-like mixture used (Beznák, 2005).

The slurry-like mixture has very good contact with the pattern that results in lower surface roughness of castings. Due to high elasticity of the mould served in the first stage it can be easy stripped off the pattern without risk of destruction.

In the Shaw process, after setting and stripping the pattern, the mould is immediately subjected to a rapid, uniform and intense flame firing with the aim to remove the alcohol generated during hydrolysis. By flame firing water is also removed from the mould. Dehydratation is accompanied by increase in amount of the retained SiO_2 that results in higher hardness and lower plasticity of the ceramic mould material. Last but not list, when the moulds are fired the rapid burning and intense heat cause micro-cracks to develop, which renders a dimensional freezing, so that the moulds are immune to subsequent severe thermal shocks. Beside this micro-cracks provide better gas permeability of the ceramic mould.

After flame firing ceramic moulds are subjected to a baking at 1000 °C. After the baking, moulds are assembled and then cooled for 24 hours. After it the moulds are prepared for pouring.

Fig. 2a shows the composite ceramic mould produced for die insert casting and comprised of the facing layer and the chamotte-basis backing layer. Fig. 2b shows the die insert cast in the composite ceramic mould produced in accordance with Shaw process.

(a) (b)

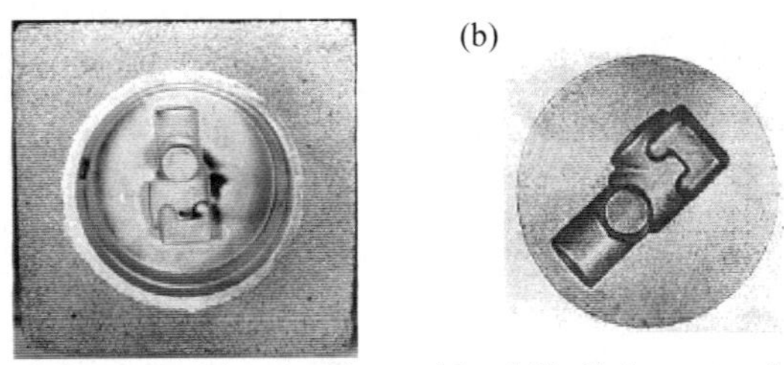

Fig. 2. (a) Composite ceramic mould and (b) die insert cast in this composite ceramic mould

3. STUDY OF THE TOOL STEEL AFTER CASTING AND HEAT TREATMENT

3.1 Experimental Procedure

Chemical composition of cast tool steel of 55NiCrMoV7 type used in this study is presented in Table 1. The steel was melted in an electric middle-frequency induction furnace and after additional alloying and deoxidising was poured into ceramic moulds shown in Fig. 2a. Four specimens from this steel for different structural states (after casting, annealing, quenching, and tempering) were prepared from the cast die insert shown in Fig. 2b. The specimen's sizes were 7x7x17 mm.

The experimental castings and specimens were isothermally annealed at 880°C for no less than 2 h. The cooling down to 500°C was performed in a furnace and, then, in air. To prevent decarburisation, when annealing, the castings and specimens were covered with crashed cast-iron chips. The final heat treatment of the materials included quenching and tempering. When quenching, the specimens were cooled from the temperature of 850°C in oil. The specimens were held at austenitising temperature in a molten salt (44% NaCl + 56% KCl) for 1 min per 1 mm of the specimen cross section. Cooling of the specimens during quenching was carried out in the oil. The specimens were tempered at 550°C for 2 h in a molten salt KNO_3.

3.2 Metalographic analysis of the cast tool steel studied

Fig. 3 shows microstructure of the steel of 55NiCrMoV7 type after casting.

Microstructure is composed of fine ferritic and perlitic constituents without defects like non-metallic inclusions, gas and shrinkage porosity. In some local areas the martensite as well as the primary carbides can be seen as shown in Fig. 3b. According to the typical shape of these carbides they can be classified as MC-type carbides, namely vanadium-reach VC carbides. The microhardness of these carbides, which is 1800 – 1850 HV0.1, is in line with such prediction. The VC carbides seem to be formed due to strong segregation of vanadium in the remaining melt during primary solidification of the steel.

Fig. 4a shows the structure of 55NiCrMoV7 tool steel after annealing. It is seen from this figure that the steel matrix has typical dendritic structure after annealing that inherited from cast state.

Martensitic and troostitic structure is typical for this tool steel after quenching that shows Fig. 4b. The structure is formed in the frame of the primary dendrites, which have the evident steps of segregation of alloying elements that leads to the different level of etching of the dendrites in their cross sections as shown in Fig. 4b.

Steel	Chemical composition (wt.%)												
	C	Si	Mn	P	S	Cr	Mo	V	Ni	Ti	Nb	N	Al
55NiCrMoV7	0.39	1.03	0.35	0.02	0.016	5.22	1.32	1.10	1.12	0.04	0.03	0.02	0.014

Table 1. Chemical composition of cast tool steel of 55NiCrMoV7 type.

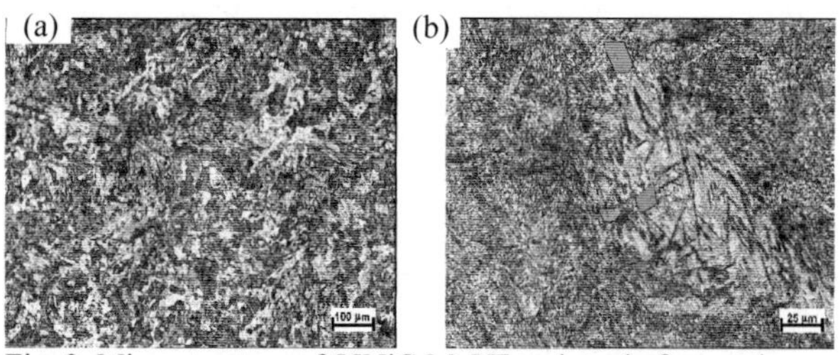

Fig. 3. Microstructure of 55NiCrMoV7 tool steel after casting (specimen 1)

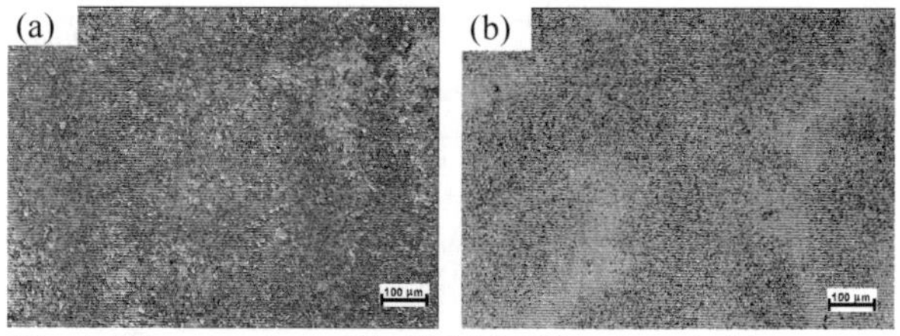

Fig. 4. Microstructure of 55NiCrMoV7 tool steel after annealing (specimen 2) and after quenching (specimen 3)

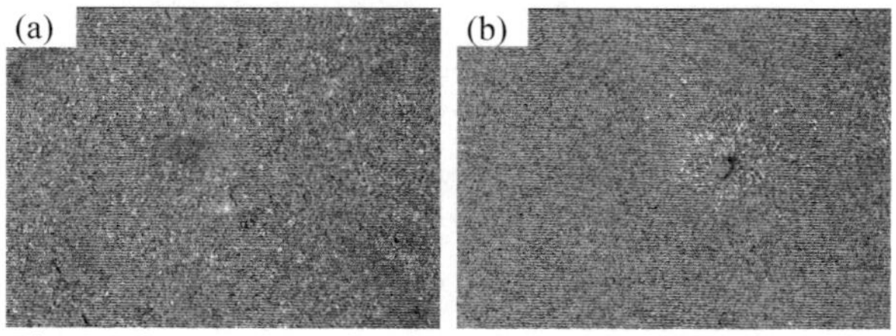

Fig. 5. Microstructure of 55NiCrMoV7 tool steel after tempering (specimen 4)

Fig. 5a shows that the structure of the steel after tempering is composed of fine troostite. Despite this the structural heterogeneity is also seen in Fig. 5b that resulted from the presence of the random primary carbides of VC type in the tempered structure.

4. CONCLUSION

Solidification sequence of 55NiCrMoV7 tool steel cast in the ceramic mould provides formation of the fine solidified structure, which is composed of the fine ferrite and perlite without defects like non-metallic inclusions, gas and shrinkage porosity. Heat treatment does not change the typical signs of cast structure in general but leads to formation of the fine troostitic structure after tempering. Such structure seemed to be suitable in terms of mechanical and working properties of cast die inserts cast in the ceramic moulds in accordance with Shaw process (Beznák, 2007/a, b). Employment of the tool steel specially designed for cast die tools is very effective from viewpoint of the impact toughness and durability of the cast die tools.

5. ACKNOWLEDGMENT

The financial support of grants from the Ministry of Education of the Slovak Republic VEGA 1/4109/07 and VEGA 1/3191/06 is gratefully acknowledged.

6. REFERENCES

Beznák, M. (2007/a). Production of the precision castings in ceramic moulds using permanent pattern. *International science conference of materials science and manufacturing technology*, 157-161, ISBN 978-80-213-1650-8, Prague, 2007

Beznák, M. (2007/b). *Production of die tools by precision casting into ceramic moulds.* Alumni press, ISBN 978-80-VM/2007, Trnava

Beznak, M. (2005). Production of forging tools by precision casting into ceramic moulds. *Masinostroenie.* Vol. 21, No. 1, (2005) 281-285 ISBN 985-479-322-2, Minsk

Beznák, *M. (2004).* Possibilities of forging die inserts production by precision casting using permanent pattern. *Cooperation international conference. (10,2004),* 97-102

Beznák, M. (2003). Main aspects influencing of the technology of precision ceramic moulds. *CO-MAT-TECH 2003,* 60-63, ISBN 80-227-1949-8, Bratislava

Annals of DAAAM for 2008 & Proceedings of the 19th International DAAAM Symposium, ISSN 1726-9679
Editor B. Katalinic, Published by DAAAM International, Vienna, Austria 2008
Make Harmony Between Technology and Nature, and Your Mind will Fly Free as a Bird

EXPERT SYSTEMS FOR THE DENTAL AESTHETIC

BICLESANU, C[ornelia]; CHERLEA, I[oan] V[aleriu] & YASSIN, A[drian]

***Abstract:** The purpose of this study consists in implementing a new software program that integrates simple and time efficient methods into the aesthetic techniques of the clinicians and into the technical expert daily routine. The application was developed using Microsoft Visual C++ 2005, by a software developer. The basis of the program is girding virtually over the existing pixels and recognizing them, after which the user will point out the points (pixels) and name them using A,B,C...*
The next step is to determine which segments would be calculated and comparing the outcome (ratio) to the golden cut. The initial results made clear that the natural development of the aesthetic frontal group is more close to ø Phi (golden proportion of 1 to1.6), than any other medical intervention made by doctors and technicians alike.
In conclusion, this software is still at the beginning of its journey in application, but from primary results, it might have an impact on the way we understand, use and deliver aesthetics to our patients.
***Key words:** Dental Proportion, Microsoft Visual C++ 2005, Dental Aesthetic.*

1. INTRODUCTION

The phi relationship, 1.618:1, known as the Golden Proportion or Divine Proportion, has long been used by artists, architects, designers, engineers, musicians, photographers, sculptors, surgeons and stock analysts alike to achieve excellence in their results. The Golden Proportion results from the division of a straight line in such a way that the shorter part is to the longer part as the longer part is to the whole. This theory states that for any two objects to be in esthetical harmony, they should exist in the ratio of 0.618 to 1.0 (Ricketts, 1982 a, b).

An arrangement of teeth in Golden Proportion would yield relative widths of 1.618 : 1.0 : 0.618 for the central incisor, lateral incisor, and canine respectively.

We have applied this relationship in aesthetic dentistry. The proposed application is a graphic analyses and design tool that allow to see and apply golden proportion to any digital image.

Additionally the before and after patient's dental status could be easily compared. At the actual stage the software is used within the dental aesthetics area aiming to explore new ways for achieving a better cosmetic of dental crowns.

The goal of this software is to integrate simplicity and time efficient methods into the esthetical techniques of the clinicians and the technical expert daily routine.

2. MATERIALS AND METHODS

The study was performed on 280 patients aged between 18 and 25 years. Part of the measurements were performed on natural teeth, on patients carefully selected to fulfil certain criteria (no prior orthodontic treatments, gaps), in order to determine statistically the accordance with the golden proportion. Other measurements were made on patients that had suffered aesthetic restoration treatments using comparative images prior and after the treatment for measuring the M-D dimensions of teeth.

A standardized frontal image of each subject's teeth was taken using a common digital camera (i.e. Canon Power Shot A 550; 7.1 Mp).

The images were downloaded on to a personal computer. To improve upon image accuracy, a zoom function of the program was used to allow more precise measurement The application was developed using Microsoft Visual C++ 2005, by a software developer.

The basis of the program is girding virtually over the existing pixels and recognizing them, after which the user will point out the points (pixels) and name them using A,B,C...

The next step is to determine which segments would be calculated and comparing the outcome (ratio) to the golden cut. Simple and easy to use this software will aid us later on in the completion of a study which will test the aesthetic awareness, the demand for it and last but not least the efficiency of the GOLDEN CUT software.

3. APPLICATION DESCRIPTION

The main functions of the application are selected from the buttons presented and explained in the fig.1

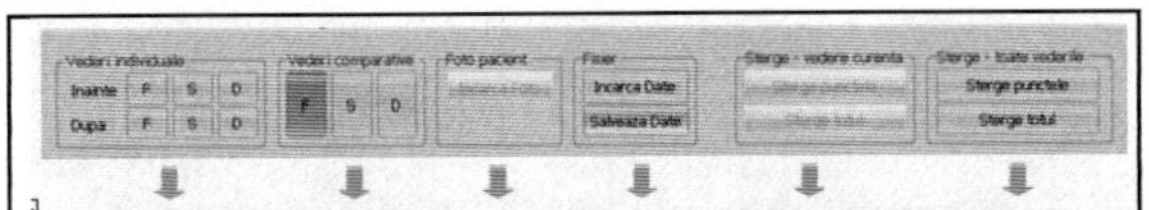

Input/output data buttons			Data management buttons		
Individual views	Comparative views	Patient's photo	File	Erase - current view	Erase – all views
Displaying one single captured image	Displaying two comparative images (before and after treatment)	Button for loading pictures of the patient (one at a time)	Load/save data buttons	Button for corrections of the current view (erasing the selected point(s))	Button for corrections of the all loaded views (erasing the selected point(s))

Fig. 1. Toolbox of the application

The application allows the upload and display of the teeth in three different views (frontal and lateral), and permit the storage of a big number of photos. The measurements are done automatically based on the chosen points by clicking on the selected photo. (Fig. 2-3)

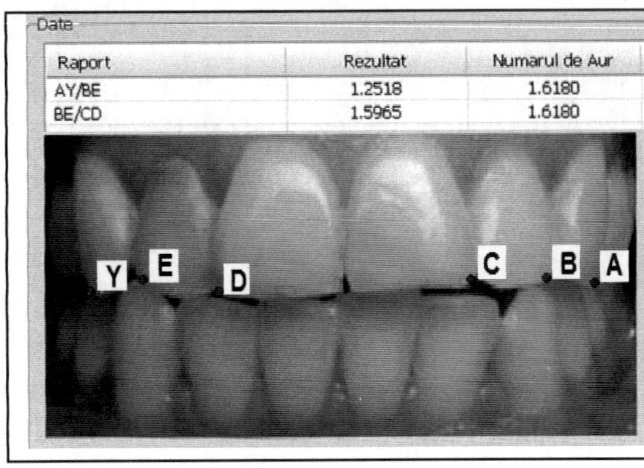

Fig. 2. Selection of measurement points

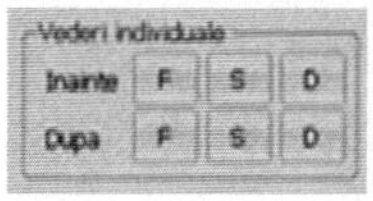

Individual views			
Before	F (frontal view)	S (left view)	D (right view)
After	F (frontal view)	S (left view)	D (right view)

Fig. 3. Individual views function

4. RESULTS AND DISCUSSIONS

It is important to determine a mathematical or geometrical relationship between teeth in order to achieve an aesthetic restorative result. These findings are also supported by other investigations (Mahshid M, et al., 2004 & Gillen RJ, et al., 1994) .

In our study based on experimental results the golden proportion (1.618 : 1.0) was found as report of the widths between the central right incisor and the lateral incisor in 37.5% of women and 25.7% of men.

Among them 14.6% of male subjects and 18.9% of female subjects have the width between their right lateral incisor in golden proportion with the width between their right canine. Within other studies (Preston JD, 1993) the data show that 46.2% of the male subjects and 49.5% of the female subjects have the width between their right central incisor in golden proportion with the width of the right lateral incisor, while 24.7% of males and 35.2% of females have the width between the right lateral incisor in golden proportion with the widths between their right canine. The difference of results might be explained by several specific factors such as the differences of the research methodology used as well as the differences of the ethnicity of the selected subjects.

5. CLINICAL AND TECHNICAL APPLICATION OF THE SOFTWARE TOOL

The measurements have been done both on natural teeth and natural teeth compared with direct or indirect restorations. It has been carried out restorations with composite materials and porcelain veneer, together with acrylic prosthesis. In order to obtain suitable results (the golden proportion) the better way is to use iterative alternation of measurement and adjustment procedures (fig. 4, 5). Dentists need to exercise extreme caution when considering restorative treatment alone to correct anterior width proportion deficiencies.

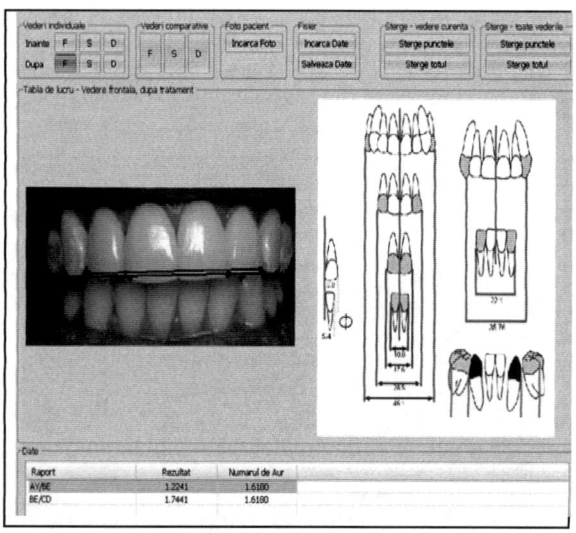

Fig 4. Practical application

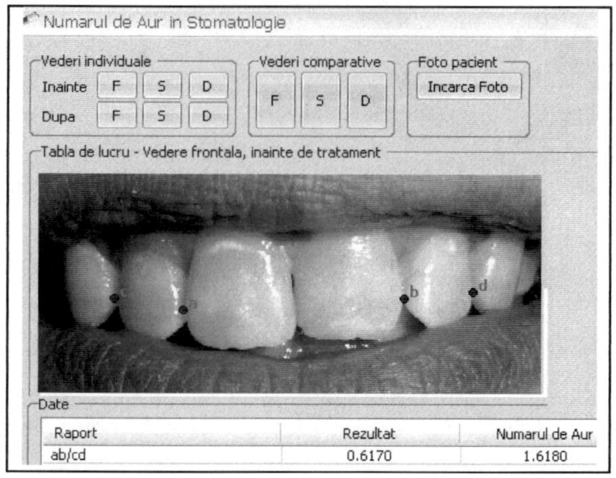

Fig. 5 Clinical case: direct restoration

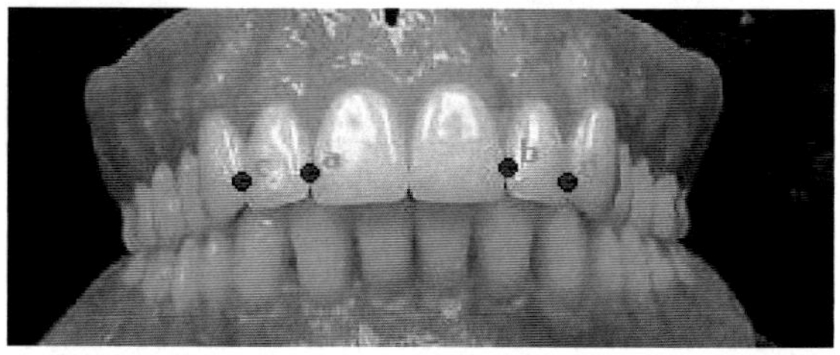

Raport	Rezultat	Numarul de Aur
ab/cd	0.6089	1.6180

Fig. 6. Technical case: acrylic prosthesis

6. CONCLUSION

The initial results made clear that the natural development of the aesthetic frontal group is more close to ø Phi (golden proportion of 1 to 1.6), than any other medical intervention made by doctors and technicians alike.

This software is still at the beginning of its journey in application, but from primary results, it might have an impact on the way we understand, use and deliver aesthetics to our patients.

The area of application is also much larger than frontal teeth aesthetics, it could be used in different fields of expertise such as plastic surgery and many other fields the require an extra Eye for aesthetics.

7. REFERENCES

Ricketts, R.E. (1982 a): The biological significance of the divine proportion. *American Journal of orthodontic* **81:** 351-70.

Ricketts, R.E. (1982 b) : The divine proportion in facial esthetical. *Clinical & Plastic Surgery.* **9 :** 401-22.

Mahshid M, et al., (2004). Evaluation of "golden proportion" in individuals with esthetical smile. *J Esthet Restor Dent*; **16**: 185-192.

Gillen RJ, et al., (1994) Schwartz RS, Hilton TJ, Evans DB. An analysis of selected normative tooth proportions, *Int J Prosthodont*; **7**: 410–417.

Preston, JD (1993). The golden proportion Revisited. *J Esthet Dent;* **5**: 247–251

Annals of DAAAM for 2008 & Proceedings of the 19th International DAAAM Symposium, ISSN 1726-9679
Editor B. Katalinic, Published by DAAAM International, Vienna, Austria 2008
Make Harmony Between Technology and Nature, and Your Mind will Fly Free as a Bird
*Young Researcher and Scientist Paper / * Supervisor, Mentor*

THE TOOL STEEL SURFACE LAYER STRENGTHENING BY SHOT PEENING

BILIK, J[ozef]; RIDZON, M[artin] & SUBA, R[oland]

Abstract:*The paper is focused on research area of tool steels surface layers mechanical strengthening, based on surface plastic deformation. The surface plastic deformation is made by dynamical shot peening on single purpose machine with one shooting wheel after heat treatment. The paper mentions also the results of some experiments with STN 419552 tool steel for forging dies.*
Keywords: peening, strengthening, surface, deformation

1. INTRODUCTION

The lifetime of dynamically stressed forming tools and other parts depends in especially on surface layer. Surface layer is critical for wear and fatigue fracture. So the paper is focused on surface layers strengthening based on surface plastic deformation by the method of shot peening.

2. THE EFFECTS OF DYNAMICAL IMPACT DURING SURFACE LAYER STRENGTHENING BY DYNAMICAL SHOT PEENING

The high speed impact of bodies causes the physical – mechanical processes of elastic – plastic waves and it affects the properties of metals during and after stress (Bača, 2005), (Bača, 1995). The accompanying effects of impact are thermal effect, localization of plastic deformations in surface layers, effects in structure (Bača, 2005), (Polák & Höbemägi 1988). The important demonstrations of surface layers strengthening at dynamical shot are the density and configuration of thermally stable dislocations, dislocations nets, dislocation barriers, density of slip lines, slips and twins, fragmentations of carbides or impurities joined with equiaxed grains properties homogenization levelling, very fine austenitic and martensitic structures with two or even three times higher wear resistence (Bača, 1995), the creation of beneficient pressure stresses system in surface layer, increasing of hardness and strength in surface layer, the creation of texture in strengthened surface layer, the change of surface quality.

3. CHOSEN RESULTS OF EXPERIMENTS FOCUSED ON FORGING DIE LIFETIME INCREASING BY APPLICATION OF DYNAMICAL SHOT PEENING

The experiments of surface layers strenghtening by dynamical shot peening included: a) the testing of samples cyclic stress resistance (fig.1). Samples were quenched and tempered on required hardness and so called strengthening effectivnes was measured by number of cycles to failure at impact energy 160 J, b) the measurement of hardness distribution from surface by gradual grinding of thin layers and measurement of microhardness distribution of grinded and polished cross section, c) the measurement of surface roughness and its character after shot peening, d) the measurement of samples size change after shot peening, e) the estimation of residual austenite content before and after strenghtening by x-ray difraction analysis, f) the estimation of character and morphology of fracture surfaces after cyclic stress resistance test. (Pfeiffer, 2005), (Kirk, 2005)

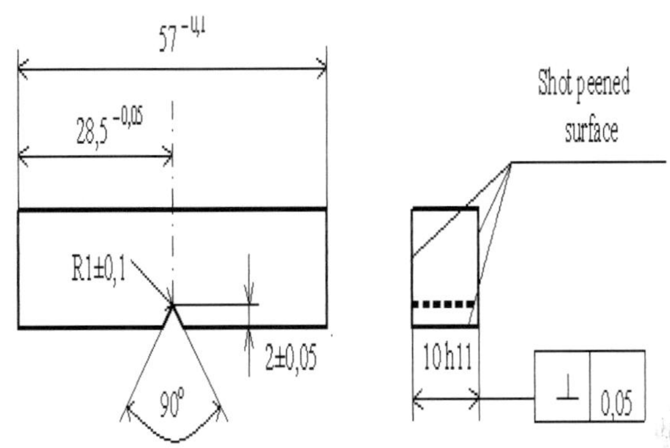

Fig.1 Shape and size of test specimen for measurement of cyclic stress resistance and some other parameters of strenghtened surface layer

There would be mentioned only some of obtained results, due to restricted number of papers page, for Cr-Mo-Si-V steel STN 419552, surface layers strenghtening. The chemical composition of this tool steel is 0,32 ÷ 0,42 % C, 0,2 ÷ 0,5 % Mn, 0,8 ÷ 1,2 % Si, 4,5 ÷ 5,5 % Cr, 1,1 ÷ 1,6 % Mo, 0,35 ÷ 0,6 % V, max. 0,03 % P, max. 0,03 % S. For experiments single purpose machine with one shooting wheel with tip speed approximately 32 ms-1 was used. The steel balls with diameter 1 up to 1,5 mm and weight 85 ±10 kg.min-1 were used for shot peening.

4. THE MEASUREMENT OF RESIDUAL AUSTENITE CONTENT BY X-RAY DIFFRACTION ANALYSIS

The residual austenite content was measured by y-ray diffraction analysis in surface layers of shot peened specimens andspecimens without treatment from STN 419552 steel. The analysis was made on PHILIPS PW 1710 x-ray diffractometer. There was established that the content of residual austenite decreased by 4 to 5 % after shot peening due to plastic deformation and residual pressure stresses in surface layers. It was transformed to martenzite.

5. THE MEASUREMENT HARDNESS DISTRIBUTION AFTER SHOT PEENING

The measurement of hardness at gradual grinding of layers with 0,05 mm thickness showed that all measured specimens with initial hardness from 48 to 50 HRC and shot peening time 4, 6, 8, 10 minutes has depth of strengthened surface layer from 0,35 to 0,4 mm. The beneficial distribution of hardness is proved also by mocrohardness measurement on specimens from STN 419552 steel with initial hardness before stzrengthening 40 up to 48 HRC. The measured values of microhardness are in table 1.

Hardness of specimen before shot peening	Distance of indention from surface [mm]	Measurement No 1	Measurement No 2	Average hardness HV 0,1
		Hardness HV	Hardness HV	
40 HRC	0,04	478	478	478
	0,06	478	464	471
	0,15	444	396	420
	0,24	409	422	415
	0,36	411	394	402
	0,45	398	398	398
	0,55	404	394	399
48 HRC	0,04	530	540	535
	0,08	527	511	519
	0,14	499	527	513
	0,23	508	496	502
	0,32	502	493	498
	0,44	476	511	494

Tab.1 Measured values of microhardness on various distances from surface for STN 419552 steel (shot peening time 8 minutes)

6. THE OBSERVATION OF INITIAL HARDNESS INFLUENCE ON SURFACE ROUGHNESS AFTER SHOT PEENING

The dependence of roughness after shot peening on initial hardness before shot peening for STN 419552 steel is on Fig.2. The surface roughness before shot peening was Ra = 0,46 μm.,

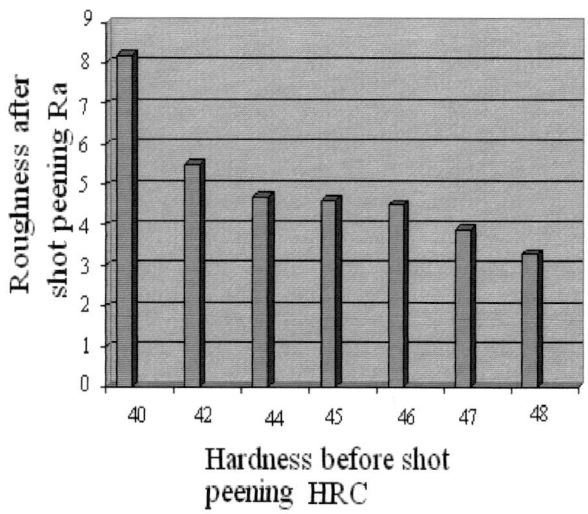

Fig.2 The dependence of surface roughness after shot peening on initial hardness (shot peening time 8 minutes) for STN 419552 steel

7. THE FRACTURE SURFACES OBSERVATION AFTER STRENGTHENING EFECTIVNESS TEST

The character of fracture surfaces changed at specimens with strengthened surface layers and crack initiation and their development is suppressed. It is visible in different morphologies of fracture surface near notch and in the middle of specimen. The morphology of fracture surface of specimen from STN 419552 steel shot peened 8 minutes on the surface near notch (Fig.3a) and in the middle of specimen is shown (Fig.3b).

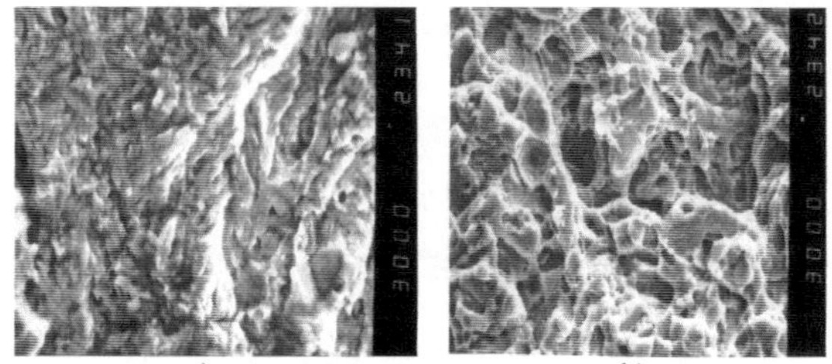

a) b)

Fig.3 The morphology of fracture surface of specimen from STN 419552 steel shot peened 8 minutes. a) on the surface near notch, b) in the middle of specimen, magnification 3000x

It is visible from Fig.3a and 3b that fracture surface of strengthened specimen have different morphologies on surface near notch and in the middle. It is connected with plastic deformation of surface layer during shot peening. The split facettes are finer in strengthened surface layer and thus the energy required for their creation and development is higher than at non-strengthened specimen.

8. CONCLUSION

The experimental measurements proved beneficial effects of surface layer strengthening of STN 419552 tool steel, used for forging die and forging die inserts production, by dynamical shot peening. The analysis of deformation size and measurements of hardness and microhardness allowed to establish the depth of plastic deformation penetration, i.e. the thickness of surface layer, which in dependence on initial hardness was from 0,3 to 0,5 mm. The observed specimens proved the minimal increase of surface layer hardness from 2 to 3 HRC at initial hardness 40 up to 50 HRC. The observation of surface roughness after shot peening of heat treated and grinded surface established that its roughness was increased in dependence of initial state of specimens. The specimens with hardness 40 up to 48 HRC, which corresponds to usual forging die hardness, has surface roughness from Ra = 3 to 8 μm. But the shot peening significantly changed the roughness profile character in comparison with grinded surface. The observation of fracture surfaces on strengthened and non-strengthened specimens showed different character of fracture surfaces in their surface layer. The mechanical strengthening of surface layers is one of very effective fatigue strength increasing methods for dynamically stressed tools and parts. This is evident from international conferences focused on this problem.

9. ACKNOWLEDGEMENTS

The paper was realised with VEGA 1/3192/06, VEGA 1/0060/08 support.

10. REFERENCES

Bača, J. (2005). Volumen forming – Drop forging Bratislava, STU

Bača, J. (1995). Contribution to problematics of cavity tools production by forming (collection of papers). Habilitation thesis. Trnava, MTF STU

Polák, K & Höbemägi, A.(1988). Production of cavity tools by forming. Bratislava, Alfa

Pfeiffer, W. (2005). Characterization of peened components by x-ray diffraction. ICSP-9, Paris

Kirk, D. (2005). Evolution of shot peening experimental technigues. ICSP-9, Paris

Annals of DAAAM for 2008 & Proceedings of the 19th International DAAAM Symposium, ISSN 1726-9679
Editor B. Katalinic, Published by DAAAM International, Vienna, Austria 2008
Make Harmony Between Technology and Nature, and Your Mind will Fly Free as a Bird

ASPECTS OF OPTIMISING THE POSITIONING AND CONTOURING PRECISION OF CNC LASER CUTTING MACHINES

BIRIS, C[ristina]; DEAC, C[ristian] & TERA, M[elania]

Abstract: *The paper analyses the conditions of realising high-precision cutting operations of steel sheets by means of numerical-controlled laser cutting machines, studying the various factors that influence the positioning and contouring precision under certain conditions. The findings of the experimental researches presented in this paper will be used for improving mathematical models employed for the optimisation of the NC equipment's functioning.*

Key words: *laser cutting, precision, machine-tool, numerical control, contouring error.*

1. INTRODUCTION

Nowadays, the means and techniques employed for cutting various materials, and especially metallic materials, are very diversified, so both producers and users of cut parts are often faced with a difficult choice with regard to the right technique and machine to use for a certain application, the more so when the part has a complex shape.

The precision of parts realized by cutting can be improved significantly by using equipments for numerical control (NC) on the machine-tools used for this purpose. However, there are a variety of factors that can negatively affect precision even under these conditions (Biris, 2008), so special precautions have to be taken when programming a machine-tool for an operation that demands a high accuracy.

One of the most critical components for achieving a high precision are the feed systems, assimilated with electromechanical motion control structures that can be realised in a modular, reconfigurable structure. The possibility of using modular cutting systems also allows a significant cut of the costs of such an equipment, without implying a loss of performances (Tu&Xie, 2001).

The experimental testing of performances in the case of modular reconfigurable cutting systems implies the study of the dynamic behaviour of motion control systems on each axis as well as of the kinematic precision (viewed as positioning precision on each axis and as contouring precision in the case of multiaxial movement), thus determining the main causes for errors and the possibilities of compensating them (Wang et al., 2005).

When it comes to selecting a precise type of cutting system to analyse, there are nowadays several non-conventional choices: laser cutting, plasma cutting, abrasive jet cutting etc. Even if the laser cutting equipment is much more expensive than other equipment types, it still offers the best precision.

The present paper therefore aims to present some considerations and experimental results concerning the factors influencing the precision obtained by laser cutting steel sheet parts on NC machine tools.

2. THE LASER CUTTING TECHNOLOGY

Laser cutting uses a focused beam of high energy laser light to cut materials by selectively burning or melting a precise spot, while an assist gas is used to remove the molten material from the resulting cut. It is one of the fastest and most accurate methods for cutting a variety of metals and non-metals.

Nowadays, both gaseous (CO_2) and solid-state (Nd:YAG) lasers are employed for cutting various materials. Since, in the case of CO_2 lasers, the type of gas flow can affect the cutting performance, several laser subvariants can be identified, such as fast axial flow, slow axial flow, transverse flow, and slab lasers.

Compared to other cutting technologies, there are several disadvantages to be considered (Olsen, 2008): high cost of the equipment; the thickness of materials that can be cut is very limited (10-20 mm at best); the cutting of metals with highly reflective surfaces can cause problems for the equipment (especially for the focusing lens); it can cause micro-fracturing in some materials; variations in the material's quality can affect the cutting results etc.

However, the major advantages of laser cutting over other cutting methods recommend its employment once the equipment is available and the thickness range is an appropriate one: it can cut a variety of metals and non-metals; it can produce part accuracies better than 0.08 mm; it can cut thinner metals at over 170 mm/s; it produces a narrower heat affected zone than plasma etc.

3. PRECISION OF CNC LASER CUTTING SYSTEMS

In the case of CNC machine-tools, the feed movements are realized in an automated regulation regime. The information on displacement and speed, needed for the part's processing, is generated by the machine's control unit and then transmitted to the numerical axes.

The precision of parts processed by laser cutting, on CNC machine tools, is significantly influenced by following two main factor categories:

- the configuration parameters of the position control system on each movement axis;
- the feed rate on each movement axis.

Previous researches presented in the speciality literature (Breaz&Bogdan, 2003; Bohez, 2002) indicate that errors introduced by the functioning of the numerical axes represent the main part of errors in the processing on CNC machine-tools.

The errors that appear in the functioning of the movement control systems are directly translated into contour generating errors. This is especially true for trajectories which require variable feed rates on the various axes, such as circles or curved lines.

The main factors influencing the precision of the CNC cutting machine-tool's machining feed can be grouped into dynamic constraints, uncertainty factors and non-linearities.

Among the dynamic constraints, there is the system's dynamic disturbance, which leads either to different speed amplification factors on the axes, leading to contouring errors in stationary regime, or to different time constants on each axis, which negatively influences the transitory regime. In the case of cutting a circular contour, for example, this contour will become rather an ellipse, with a "thinning" along the axis with a smaller amplification factor. Errors may appear especially during changes of direction or stops, which influences negatively the cutting of corners.

This shows that by precisely controlling the elements of the feed kinematic chain, and especially the parameters of the control system of the numerical axes, we can significantly improve the CNC laser cutting machine-tool's performance.

However, even when using an optimal feed rate, recommended by the speciality literature, there can still appear contouring errors due to a dynamically imbalanced control system. This underlines the importance of using mathematical models and experimental researches for an optimal tuning of the working parameters of laser cutting machine-tools.

4. EXPERIMENTAL RESEARCHES

The experimental researches regarding the cutting precision were carried out on a NTX-48 Champion laser cutting machine manufactured by Mazak (Japan) using a CO_2 laser and having following characteristic working parameters:

- maximal power output: 1.5 kW;
- impulse frequency: from 0 to 2000 Hz;
- maximal machining feed rate: 10 m/min;
- positioning precision on X and Y axes: ±0.01/100 mm;
- positioning precision on Z axis: ±0.01/500 mm;
- NC control unit: Mazak L-32B.

The blanks used for these researches were sheets of OL 52 2k steel, STAS 500/2-88, with thicknesses of 3 mm, 6 mm and 8 mm, respectively. The aim was to obtain, on the one hand discs with a diameter of 60 mm, and on the other hand rectangular shapes with sides measuring 40 and 60 mm, respectively, to assess the dimensional errors and thus the machine's precision, for different working regimes and NC setups.

Figure 1.a shows the result of cutting a disc from an 8 mm thick OL 52 steel sheet, with a feed rate v = 700 mm/min, while figure 1.b presents the contouring errors in the case of cutting a disc from the same type and thickness of steel sheet, but with a feed rate v = 1500 mm/min.

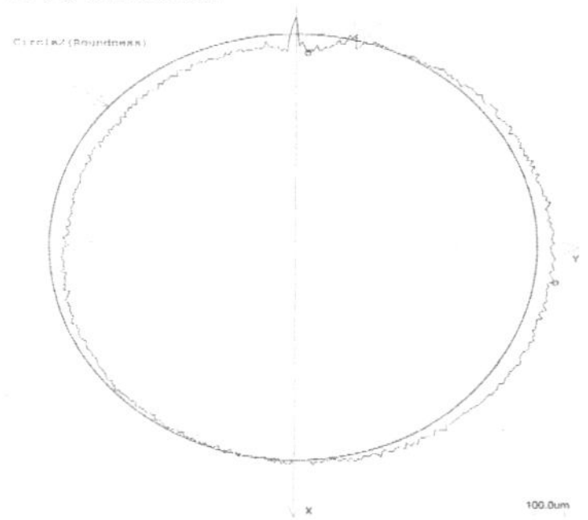

a)

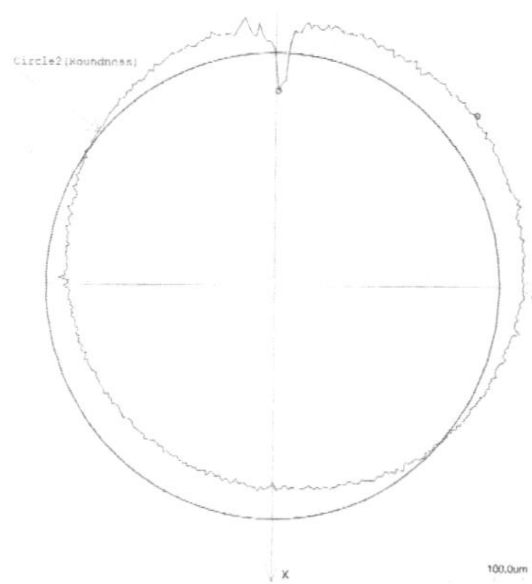

b)

Fig. 1. Contouring errors at the cutting of a disc from a steel sheet of 6 mm thickness with: a) v = 700 mm/min b) v = 1500 mm/min

Figure 2.a presents the contouring errors in the case of cutting a rectangular shape from a metal sheet of 3 mm thickness with a feed rate v = 800 mm/min, while figure 2.b shows the result of cutting the same part from the same steel type and thickness, but with a feed rate v = 2000 mm/min.

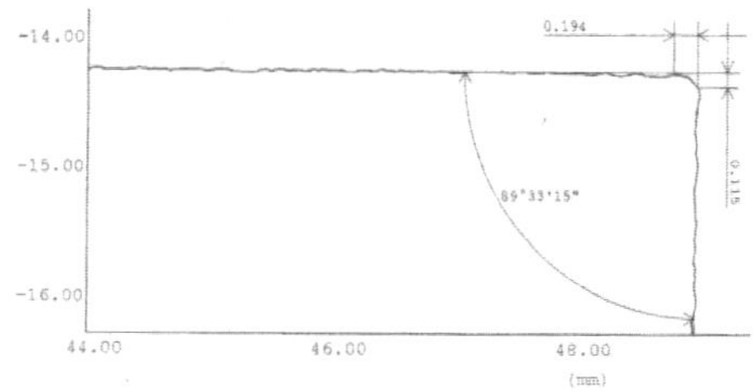

a)

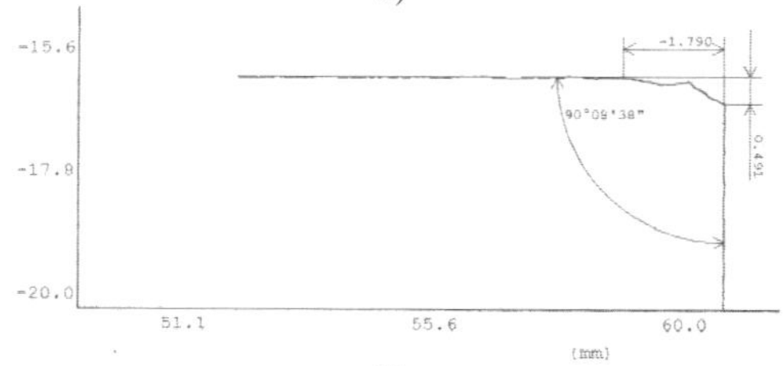

b)

Fig. 2. Contouring errors at the cutting of a rectangle from a steel sheet of 3 mm thickness with: a) v = 800 mm/min; b) v = 2000 mm/min

5. CONCLUSIONS

The researches presented in this paper have underlined, on the one hand, the benefits of laser cutting systems with regard to precision, even if the high costs of the laser cutting equipment may pose some start-up problems. On the other hand, the importance of a precise control of the feed rate with the help of rotation transducers has been proven and an optimal feed rate has been found for the studied case.

As a conclusion, for both part types, at lower feed rates the contouring errors are relatively small compared to the case of higher feed rates, a phenomenon that aggravates with decreasing sheet metal thickness. This indicates that the optimal setting for this laser cutting machine is achieved when cutting alloyed steel types or harder materials in general, whereas for softer and/or thinner materials, which can be processed at higher feed rates, a modifying of the machine's dynamics is needed for ensuring an acceptable precision.

In future, it is sought to use the outcome of the experimental researches presented here for generating and optimising mathematical models that could take into account the above-mentioned error-inducing factors and thus lead to a more precise cutting technology.

REFERENCES

Biris, C. (2008). *Proiectarea unui sistem de comanda si control aferent echipamentelor de debitare* (Design of a command and control system for cutting equipments), Ph.D. paper no.2, "Lucian Blaga" University of Sibiu.

Bohez, E.L.J., (2002), *Compensating for systematic errors in 5-axis NC machining*, Computer Aided Design no. 34, pp. 391-403.

Breaz R., Bogdan L. (2003). *Automatizari in sisteme de productie* (Automations in production systems), Publishing House of the "Lucian Blaga" University of Sibiu, Sibiu.

Olsen, C. (2008). *Waterjets.org: Waterjet Web Reference*, http://www.waterjets.org/index.html

Tu, Y. L., Xie, S. Q. (2001), *An Information Modelling Framework to Support Intelligent Concurrent Design and Manufacturing of Sheet Metal Parts*, Intl. J of Advanced Manufacturing Technology, 18 (12), pp. 873–883;

Wang, S.-M., Yu, H.-J., Liao, H.-W., (2005) *A new high-efficiency error compensation system for CNC multi-axis machine tools*, Intl. J of Advanced Manufacturing Technology 16, pp. 623-631.

Annals of DAAAM for 2008 & Proceedings of the 19th International DAAAM Symposium, ISSN 1726-9679
Editor B. Katalinic, Published by DAAAM International, Vienna, Austria 2008
Make Harmony Between Technology and Nature, and Your Mind will Fly Free as a Bird

THE USE OF FUZZY COLOURED PETRI NET IN MODELING AND SIMULATION OF FLEXIBLE MANUFACTURING CELL

BLAGA, F[lorin]-S[andu]; POP, A[lin]; STANASEL, I[ulian] & PANTEA, I[oan]

Abstract: *The paper presents an application of Fuzzy Coloured Petri Net model in flexible manufacturing cell of University of Oradea. The functioning in automatic state of the flexible manufacturing cell requires the use of effective control methods. The control systems based on fuzzy set bring beside the control attribute, the modeling and simulation facilities.*

The paper presents a Fuzzy Coloured Petri Nets used in modelling flexible manufacturing cell from University of Oradea.

Key words: *modeling, simulation, Coloured Petri Net, Fuzzy Sets*

1. INTRODUCTION

The functioning optimization of the flexible manufacturing system requires them modeling and simulation. One of the tools used in this is Coloured Petri Net (David & Alla 1992).

The controlling systems witch are serving the components of the flexible manufacturing systems are using more often automatic fuzzy regulators (Preitl & Precup 1997).

It may be combined Coloured Petri Net and Fuzzy sets so to result a new tool for modeling, simulation and control of flexible manufacturing system components. This possibility is materializing through Fuzzy Coloured Petri Net (Yeung et al., 1996).

2. THE MODEL OF THE CELL

We consider a cell (figure 1) with a robot 1 which is able to pick up an electronic component from a storage room 3 and this robot moves around a Printed Circuit Board (PCB) in order to assemble component in PCB. After mounting the component, robot returns to pick up a new electronic component from storage room. Therefore, the process will keep on operating until PCB is complete and able to abandon the system.

The application problem is how to use Fuzzy Coloured Petri Nets in modeling robotic assembly. In order to create the model, we will use specific Coloured Petri Nets concepts, (Jensen, 1995) (Blaga et al, 2008).

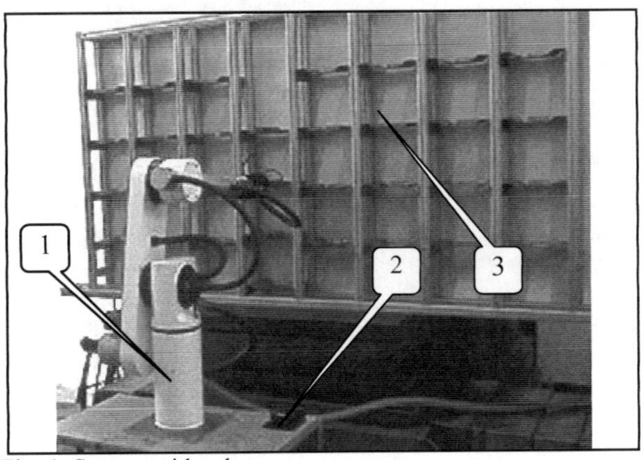

Fig. 1. System with robot

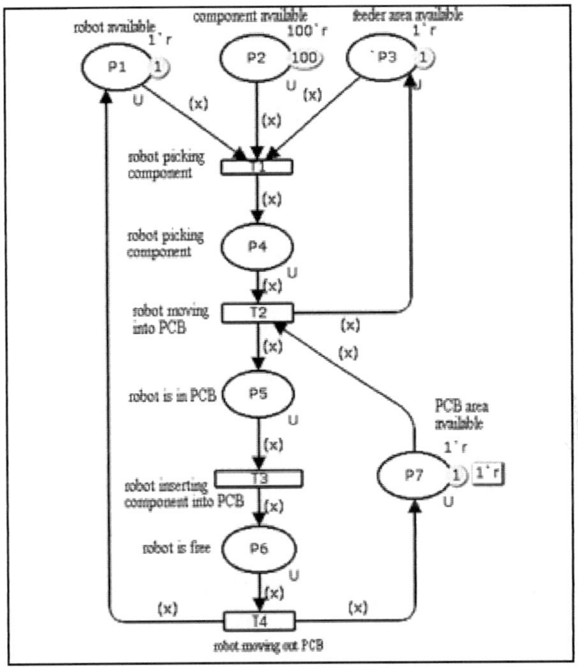

Fig 2 Model structure

Model structure is presents in figure 2. We will begin from initial marking with T1 transition so r marking moves out from P1 position, c marking moves out from P2 position and f marking from P3 position. In the same time, in the P2 position is stored a complex color {r,c}. This complex color will appear in the model until robot moves out from the PCB area (after execution of T4 transition). During the insertion process of an electronic component into PCB, the robot must rotate in an angle such that the legs of the components are perpendicular to the PCB area.

The rotation of the robot is ordered and controlled by a system witch can be modeled with a Fuzzy Coloured Petri Net (figure 3).

The adjustment procedure follows these sequences (steps):

- **Sequence 1.** At first, we have to read the values of the displacements on tree angles of the component witch will be inserted into the PCB. The three measurements are used as inputs and these are associated with a finite set of colors d={x_d, y_d, z_d}.
- **Sequence 2.** Every color can be considered as a fuzzy variable and all these are fuzzified as the followings:

$$x\_d:\begin{cases} x\_s - \text{displacement small on x angle;} \\ x\_m - \text{displacement medium on x angle;} \end{cases} \quad (1)$$

$$y\_d:\begin{cases} y\_s - \text{displacement small on y angle;} \\ y\_m - \text{displacement medium on y angle;} \end{cases} \quad (2)$$

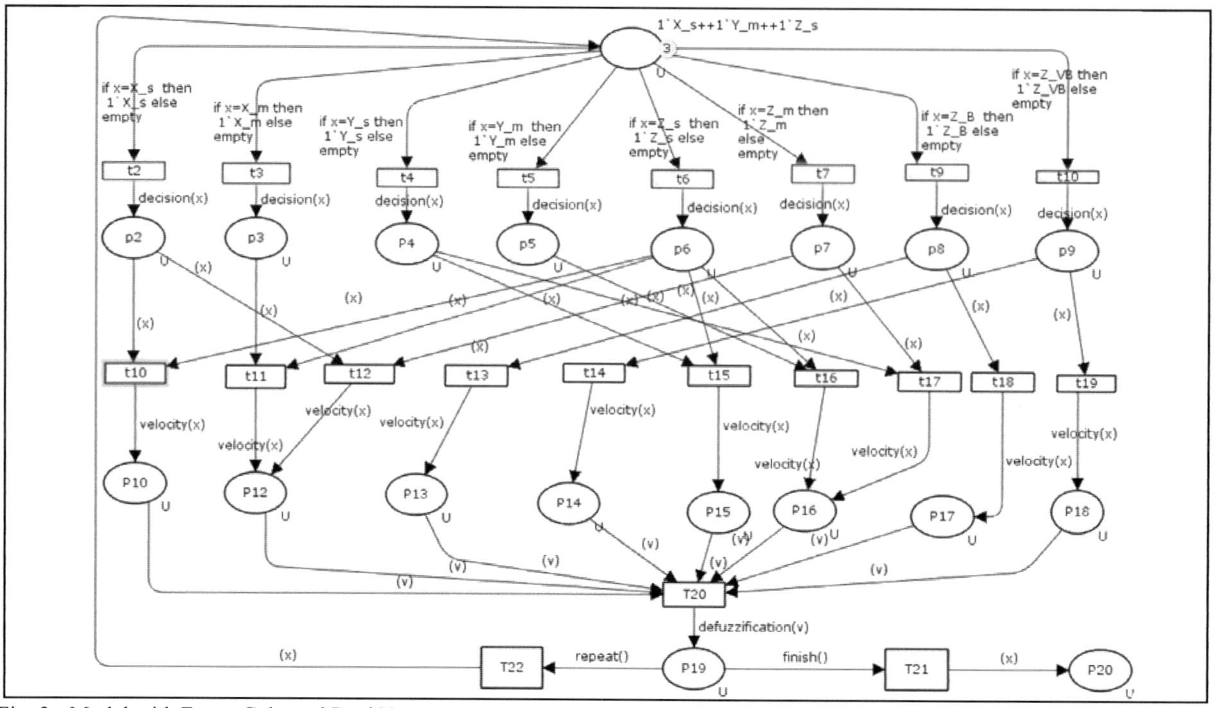

Fig. 3. Model with Fuzzy Coloured Petri Nets

$$z\_d:\begin{cases} z\_s-\text{displacement small on z angle;}\\ z\_m-\text{displacement medium on z angle;}\\ z\_B-\text{displacement big on z angle;}\\ z\_VB-\text{displacement very big on z angle;}\end{cases} \quad (3)$$

- **Sequence 3.** These eight fuzzy variables are used as inputs to so called fuzzy production rules. The rules outputs sets the velocity needed to rotate electronic component until it will be in the right position for inserting into PCB.
  Fuzzy variables are fuzzified as the followings:

$$v\_x:\begin{cases} v\_xn\text{ - velocity zero on x angle;}\\ v\_xs\text{ - velocity small on x angle;}\\ v\_xmd\text{ - velocity medium on x angle;}\\ v\_xB\text{ - velocity big on x angle;}\end{cases} \quad (4)$$

For the velocity component v_y we have the same linguistic degree (v_yn, v_ys, v_ymd, v_yB).

This set of ten fuzzy production rules are modeled by transitions: T10, T11, … , T19. The T10 transaction models the following rule:

$$\text{IF } x\_s = x\_m \text{ AND } z\_s = z\_m \text{ THEN } v\_x = v\_xn \quad (5)$$

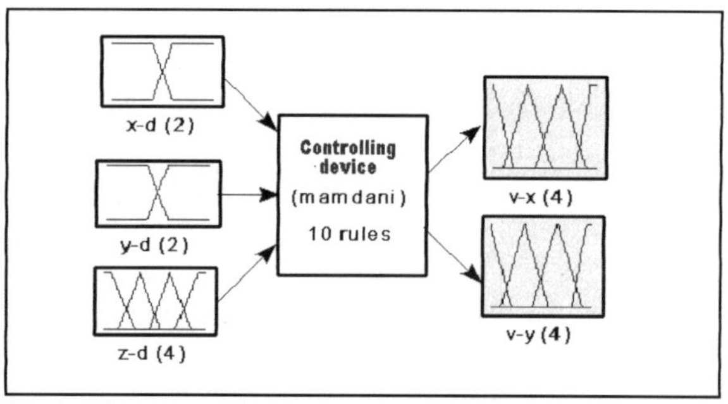

Fig. 4. The controller structure made with Fuzzy Logic Toolbox

The other 9 rules are defined in the same way as the first one.

The adjustment of the displacement can be implemented in Fuzzy Logic Toolbox from Matlab. In figure 4 presents the system based on fuzzy sets with tree inputs (the arm position in comparison with x,y,z axes) and two outputs: displacement on x axis (v_x) and displacement on y axis (v_y).

The inference engine is compound by ten roles. These are the same like the roles defined in Fuzzy Colored Petri Nets.

3. CONCLUSION

Future researches will pursue modeling with Coloured Petri Net of same system in witch will find more components (robots, machine and others). By using the features of Fuzzy Sets it will pursue the accuracy improvement of the model, so that this to be closer to the real system.

The combine use of Coloured Petri Nets and Fuzzy Sets in modeling flexible manufacturing systems succeeds in mixing advantages of both methods. The main advantage is that such a complex system can be relatively simple modeled.

4. REFERENCES

Blaga, F., Pop, A., Stanasel, I., Pele, A-V, (2008), *Coloured Petri nets modeling using CPN Tools*, ANNALS of the ORADEA UNIVERSITY, Fascicle of Management and Technological Engineering, Volume VII (XVII), pp. 1224-1228, ISSN 1583-0691

David, R. ,Alla, H. (1992), *Grafcet on Petri Nets,* Ed. Hermes, (in French), Paris, ISBN: 2866013255

Jensen, K. (1995) *Coloured Petri Nets: Basic Concepts, Analysis Methods and Practical Use (Volume 2),* Springer-Verlag, Berlin, ISBN: 3-540-58276-2

Preitl,Ş., Precup,E. (1997), *Introduction to fuzzy control*, Ed. Tehnică, (in Romanian), Bucharest, ISBN 973-31-1081-1

Yeung, D.S.; Liu, J.N.K.; Shiu, S.C.K.; Fung, G.S.K. (1996) Fuzzy coloured petri nets in modelling flexible manufacturing systems, *ISAI/IFIS*, pp. 100-107, ISBN: 968-29-9437-3

Annals of DAAAM for 2008 & Proceedings of the 19th International DAAAM Symposium, ISSN 1726-9679
Editor B. Katalinic, Published by DAAAM International, Vienna, Austria 2008
Make Harmony Between Technology and Nature, and Your Mind will Fly Free as a Bird

THE MATHEMATICAL MODELLING OF THE ELECTRONIC INJECTION GASOLINE

BLAGA, V[asile]; BARA, C[Amelia]; BARA, V[asile]; BARA, L[ucian] & GITEA, M[anuel]

Abstract: *The author proposes a personal model for the calculation of pressure regulator and electromagnetic injection, the volume of fuel injected in cycle and the duration of the injection with the number of rotation at the total full charge load. This kind of mode can be used in the modeling system of electronic gasoline injection moonlit or multiunit. To carry out this model it is necessary modeling engines with spark lighting cycle with gasoline injection with a model helping cycle proposed by the author. The used method is mathematical modelling.*
Key words: pressure regulator, valve, pulverization, duration of injection.

1. INTRODUCTION

The working of the curled engine with lambda transmitter and catalyst, makes coefficient λ to be mentioned as close as possible to λ=1 (ideal dosage) (Băţagă, 2000).

On the base of a personal model the author realized analytic calculation of pressure in the manifold p_{ga} and the admission pressure p_a, the calculation of engine pressure regulator and the duration t_i of the electromagnetic injector.

For the modeling cycle engines with spark lighting with engine injection are noticed the initial dates. All these expression were analytic calculated and correlated among them in order to be introduced in the calculator. The calculator gets the command to repeat this operation till the getting of the imposed error of the engines with spark lighting parameters (Turcoiu et al, 1987).

It is calculated the mechanical theoretic work proposed L_{tp}, the average pressure proposed p_{tp}, theoretical proposed efficiency and the mechanical loosing of the cycle.

The calculation of algorithm is made up of the following steps:

1) The initial parameter relation T_{ao}, T_{zo}, T_{uo} şi k_{co}, k_{vo}, k_{uo}, k_{do}, k_{eo} for one revolution.
2) The cycle is going through, obtaining new values for k coefficients and T_a, T_z, T_u temperatures.
3) It is made differences between old and new calculated values (Δk, ΔT).
4) If the differences between (Δk, ΔT) are lower than the imposed error, then the new calculated values are valid and they are going to the next revolution value and 1-4 points are repeated.
5) If the differences (Δk, ΔT) are above the imposed errors then the cycle is going through by the recalculation of the adiabatic coefficient values and temperature, considered new value as initial parameter. This re-going through is realized until the descend of the differences (Δk, ΔT) under the imposed errors.

2. THE OPTIMUM CALCULATION OF THE PRESSURE REGULATOR

Pressure regulator maintains constant pressure injection in supplying installation (Blaga, 2005). Calculation scheme of the pressure regulator is presented in figure 1.

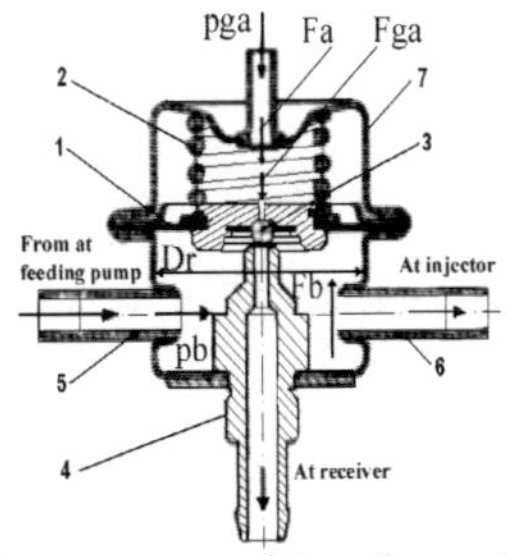

Fig.1. Scheme of pressure regulator: 1-membrane; 2- springs; 3- valve; 4- connection to the tank; 5- entrance connection; 6- exit connection; 7- housing.

Static balance equation of regulator membrane is given by the following relation (Negrea & Sandu 2000):

$$F_a = F_{ga} + F_b, \tag{1}$$

where: $F_a = K_a \cdot f$

$$F_{ga} = \frac{\pi \cdot D_r^2}{4} \cdot p_{ga} \; ; \; F_b = \frac{\pi \cdot D_r^2}{4} \cdot p_b ;$$

and K_a is elastic constant springs; f- spring arrow; D-diameter of the regulator membrane; F_a-power pressure of spring; F_{ga} – pressure power from manifold; p_{ga} –pressure from the manifold; p_b –gasoline pressure in masterly of the injection; F_b –the gasoline injection power.

$$\frac{\pi \cdot D_r^2}{4} \cdot p_b = K_{a \cdot f +} \frac{\pi \cdot D_r^2}{4} \cdot p_{ga} \; ,$$

After the changing, relation (1) becomes:

$$p_b = \frac{4}{\pi \cdot D_r^2} \cdot K_{a \cdot f + p_{ga}} \; ;$$

or:

$$p_b = K_r + p_{ga} ; \tag{2}$$

where:

$$K_r = \frac{4 \cdot K_a \cdot f}{\pi \cdot D_r^2} \; ; \; K_r = 1..4$$

K_r – is the constant regulator.

3. THE OPTIMUM CALCULATION OF THE ELECTROMAGNETIC INJECTION

The section of the passing pulverization hole is determined

by the following relation (Băţagă, 2000):

$$A_a = \pi \cdot \overline{A_c}\left[\frac{d}{2}+\left(\frac{d}{2}-\overline{A_c}\right)\right] = \pi \cdot s_a \sin\left(\frac{\beta}{2}\right)\left(\frac{d-1}{s_a \cdot \sin\beta}\right); \quad (3)$$

where: A_I-passing section offered by conic top needle; s_a-raising up needle; d_v-diameter of the needle in top zone; β-cone angle tight; d_p-sack diameter.

Rising up needle s_a is considered to be constant.

It was marked with A_I the area of the leaking section near the needle top of injector with conic top (Grünwald, 1980).

$$A_i = f(s_a, \beta, d) = ct; \; A_i = \pi(d_p - 0{,}5\, s_a \sin\beta) s_a \sin \beta/2;$$

$$s_a = 0{,}15 \text{ mm}; \; d_p = 1 \ldots 1{,}2 \text{ mm}; \; \beta = 60^0.$$

Discharge of gasoline which passes the injector leaking section is calculated by the relation:

$$Q_b = \mu_i \cdot A_i \sqrt{\frac{2(p_b - p_{ga})}{\rho_b}}; \quad (4)$$

where μ_I coefficient of discharge in the section offered by the needle;

$$\mu_i = 0{,}8 - 0{,}93;$$

Taking in consideration relation (2):

$$Q_b = \mu_i \cdot A_i \sqrt{\frac{2K_r}{\rho_b}} \quad (5)$$

where: A_i is the leaking section at the injector; p_b – gasoline pressure at the entrance of injector; p_{ga} – air pressure from the manifold; ρ_b – gasoline density; K_r - constant regulator pressure; Q_b- discharge of the gasoline.

In the other hands it is LAN own from the relation of the discharge definition that is the leaking fuel volume during a time unit.

Maybe written:

$$Q_b = \frac{V_b}{t_i} = \frac{m_{cb}}{\rho_b \cdot t_i} \quad (6)$$

From the equality of relation (5) and (6) results un duration of the injection, t_i:

$$t_i = \frac{m_{cb}}{\rho_b} \cdot \frac{1}{\mu_i \cdot A_i \cdot \sqrt{\frac{2K_r}{\rho_b}}} = \frac{m_{cb}}{\mu_i \cdot A_i \cdot \sqrt{2K_r \cdot \rho_b}}; \; d = \frac{1}{\lambda \cdot L_o} = \frac{m_{cb}}{m_{aer}};$$

$$m_{cb} = \frac{m_{aer}}{\lambda \cdot L_o} = \frac{m_{aer}}{m_{ad}} \cdot \frac{m_{ad}}{\lambda \cdot L_o} = \xi \cdot d \cdot m_{ad}; \; \xi = \frac{m_{aer}}{m_{ad}};$$

Where ξ is the coefficient which represents the ratio between necessary air quantity for burning moor and the mixed quantity fuel accepted m_{ad}.

Results:

$$t_i = \frac{\xi \cdot d \cdot m_{ad}}{\mu_i \cdot A_i \cdot \sqrt{2K_r \cdot \rho_b}} \text{ [s]}, \quad (7)$$

Where: m_{cb} is the volume of the fuel; m_{aer} – the volume of the air aspirated by the engine; m_{ad} – quantity of mixed fuel admitted; d – proportioning. In figure 2 is represented the variation of the duration of injection t_i with the revolution and the temperature of the environment surroundings.

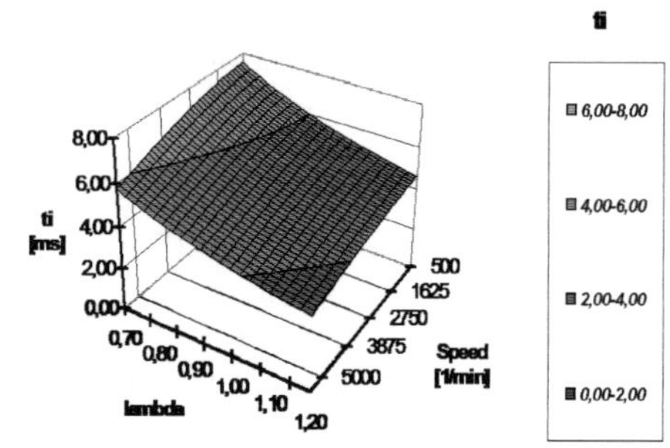

Fig. 2. The variation of the duration of injection t_i with the revolution and the temperature of the environment surroundings.

4. CONCLUSIONS

The engines with spark lighting modeling cycle with the gasoline injection proposed by the author consist of the presentation of the initial dates of the calculation program, calculation and the correlation between the engines with spark lighting parameter expression with the gasoline injection for the achievement of the program calculation. The engines with spark lighting cycle calculation proposed, with gasoline injection is an helping cycle for the simulation on the calculator of the gasoline injection. The simulation on the calculator permit the determination of the theoretical economical-technical parameters proposed; mechanical work theoretical proposed corresponding to the diagram, theoretical pressure proposed, theoretical efficiency, theoretical specific consumption proposed. The loosing mechanical pressure and the pumping are calculated, the theoretical economical-technical parameters of the engine. Sing the proposed model by the author the volume of the fuel is calculated which reset of the engine cycle in function with the revolution of the total load, the debit of the gasoline which goes through the electromagnetic section of the injection, reaching the aim in view, the determination of the injection duration with the revolution in total load. It is conceived the calculation of the pressure regulator and of the electromagnetic injector. It was made the calculation of the injection duration t_i with the revolution and load which represents the model proposed by the author. It was effectuated a engines with spark lighting program for the calculation of parameters with the gasoline injection with the under- program: the calculation program of engines with spark lighting parameters (depending on n and λ at t_o=-35...+45°C and $p_o=1 \cdot 10^2$ kPa), the calculation program of engines with spark lighting parameters (depending on n and t_o at λ=1 and $p_o=1 \cdot 10^2$ kPa) (Blaga, 2005).

5. REFERENCES

Băţagă, N. (2000). *Driving with inward alight,* The Didactic and Pedagogical Publishing House, 973-9471-20-X, Bucharest.

Blaga, V. (2005). *Engine with gasoline injection,* The Publishing House of University Edition Oradea, ISBN 973- 613-981-6, Oradea.

Grünwald, B. (1980). *The Theory, calculation and construction engine's for road motor vehicle,* The Didactic and Pedagogical Publishing house, Bucharest.

Negrea, V.D. & Veneţia Sandu (2000). *The combating of medium pollution in motor vehicle,* The Technical Publishing House, ISBN 973-31-1455-3 Bucharest.

Turcoiu,T., Boncoi, J. & Time, Al., (1987). *Equipment's from injection for engine with internal burning,* The Technical Publishing House, Bucharest.

Annals of DAAAM for 2008 & Proceedings of the 19th International DAAAM Symposium, ISSN 1726-9679
Editor B. Katalinic, Published by DAAAM International, Vienna, Austria 2008
Make Harmony Between Technology and Nature, and Your Mind will Fly Free as a Bird

THE DETERMINATION OF THE INJECTIONS DURATION FOR THE MODEL SUGGESTED BY THE AUTHORS

BLAGA, V[asile]; BARA, L[ucian]; BARA, C[amelia]; BARA, V[asile] & CARBUNAR, M[ihai]

Abstract: *In this paperwork is presented the fuel charging scheme for both the multipoint injection system design by authors. The software outputs are the 3-dimensional variations between angular speed and outside temperature respectively angular speed. It was effectuated a engines with spark lighting program for the calculation of parameters with the gasoline injection with the under- program: the calculation program of engines with spark lighting parameters (depending on n and* λ *at* t_o*=-35...+45°C and* $p_o=1\cdot10^2$ *kPa);- the calculation program of engines with spark lighting parameters (depending on n and* t_o *at* $\lambda=1$ *and* $p_o=1\cdot10^2$ *kPa). It is used mathematical modelling method. The experimental research was performed at CESAR S.A. Colibasi Pitesti, Romania (Research Center for Motor Vehicles Subassemblies).*

Key words: *debit meter, electronic injection, lambda, electric pump, revolution.*

1. INTRODUCTION

The system suggested by author's main task is the correlation establishment between the mass of the engines inhaled air and the mass of the injected combustibles on cycle forming a mixture of maximum economically for the engines each workings conditions. For model of system proposed from author was proposed one system from injection of gasoline Bosch Motronic who power for air with ultrasonor wave. It be used plate from acquisition from trial, program from acquisition from date,electrical sounder from temperature, electrical sounder from pressure, electrical sounder from revolution and position angular, analyzer from gas. It is be make one study for select excess coefficient from air λ=1 (electronic dosage) (Băţagă, 2000); (Apostolescu & Chiriac, 1998). The electric injection systems the dependence between the injected gasoline quantity on cycles at engines each work conditions and the injectors opening time establishes in adherence to the evidence stall, with control units manually operated according to the effective engine moment and the minimum polluted emissions (carbon monoxide, hydrocarbon, nitrogen, oxide) after wards this is motorized in the injection. The adjustment of the injected petrol quantity on cycle for the different engines working conditions depend on the inhaled air quantity thanks to the air debit meter which is equipped with a sounder which transmits information in regard of the airs debit to the calculator. In figure 1 is presented the combustibles feeding scheme of the Renix system, used to the molding of the system suggested by the authors. The air debit meter with palette and a sounder element are with an air debit meter of Karman Vartex type that improves the engines performance. A study of the choosing of the airs excess coefficients. It is made the pressure regulator; it is molded; the pressure regulator constant it is determinate. The electromagnetic injectors molding. It is made and it is calculated the injections duration.The molding of engine with spark lighting cycles with gasoline injection supposed by the authors is carried out by the 3-D dimensional and by dimensional parameters in two cases in figure 2 and 3.

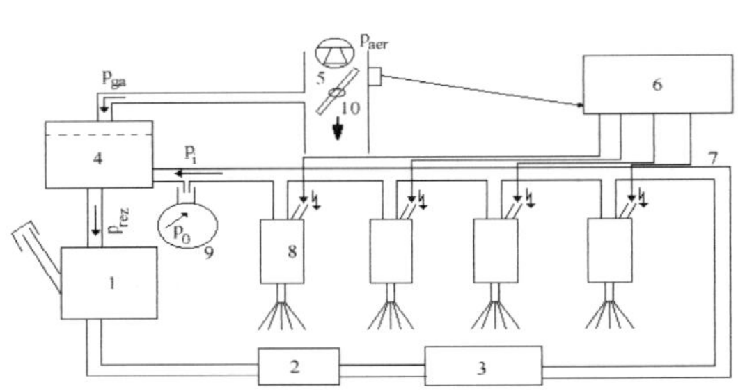

Fig. 1 The estimation scheme of combustibles feeding suggested by the authors: a – Renix system; 1– petrol tank; 2. – electric pump; 3. – electric filter; 4 – pressure regulator; 5 – debit meter (traducer for the measure of the air quantity); 6 – calculator; 7 – injection main line; 8 – electromagnetic injection; 9 – little lid valve for the measure of the pressure in the system; 10 – acceleration flap.

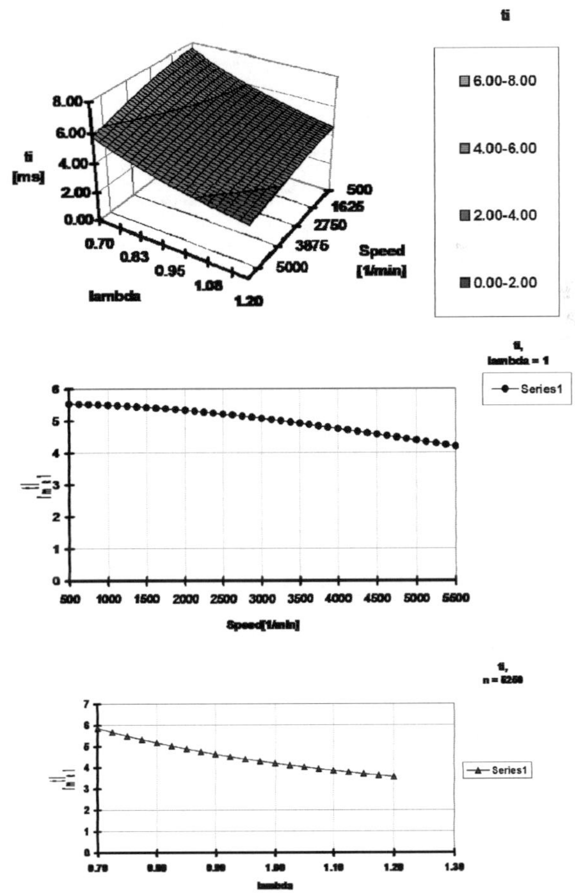

Fig.2 The dependence on the engines rotation and the airs excess co-efficient of the injections duration for the model suggested by the authors.

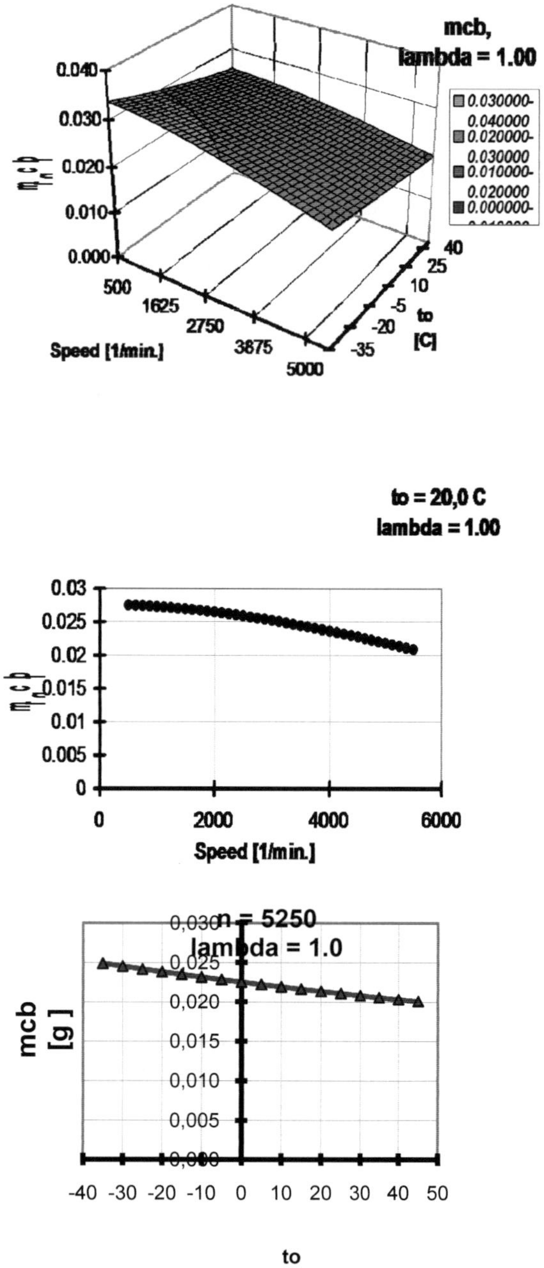

Fig.3 The dependence of the engines rotation and the environment temperature of the combustibles mass for the authors suggested model.

2. THE THEORY FROM CALCULATION

A delicate problem from point of program on the computer, adiabatic coefficient is it starting calculation, temperature in different point is unknown and the adiabatic exponents not can be determinable. For solution of this problem has been used at impose to some value, found initial more correct, who permit a cycle crossing and determination more precise temperature, (Blaga, 2000). The temperature at the finished admission T_a was to consider known as initial date and she's expression calculation it be used at the finished as equation from verify at finished the cycle. Through calculate again of adiabatic exponent on amount foundation temperatures to get from the new cycle (Grünwald, 1980). In this mode, through some cycle the error can be reduce upside a calculate estimate admit in a thermal calculus very pretentious. Must be introductory simplify hypothesis that one with starting burning at finished corresponding thermal agent composition it is instantaneous beckoning that finished corresponding the excess coefficient from air λ with who are place burning combustible. It be to achieve a experimental equipment for try the propose model by the author, gifted with device and electrical sounders necessary of modeling.

3. THE PROPOSE CONTRIBUTIONS

It be realize a program from calculate parameter spark ignition engines with electronic injection from gasoline with subrogates:

- the program from calculate to parameter spark ignition engines with electronic injection from gasoline (dependence after revolution *n* and excess coefficient of air λ at temperature ambient medium t_o=-35...+45°C and pressure ambient medium $p_o=1\cdot10^2$ kPa);
- the program from calculate to parameter spark ignition engines with electronic injection from gasoline (dependence after revolution *n* and temperature ambient medium t_o at excess coefficient of air λ=1 and pressure ambient medium $p_o=1\cdot10^2$ kPa) (Blaga, 2000).
- debit meter point of cycle, coefficient from fill, dosage, measure thermal combustion of unity cylinder engine, rapport from grow of pressure in isochors combustion, rapport growth volume in post burning, the technical-economical of driving and duration injection (Blaga, 2005).

It be representation 3-D variation of parameter technical-economic of model proposed with revolution engine and temperature ambient medium. It be to achieve a comparation between the calculate and measured of duration injection and angle injection with revolution growth from at 500 at 6500 rot/min, of engine with electronic injection gasoline type Bosch Motronic. It be realization one comparation between diagram theoretic from cycle proposed by the author and diagram of engine with electronic injection of gasoline, that was obtain one the experimental installation.

It is to effect calculation error and the correction factor face atmospheric condition from reference, on the engine electronic injection of gasoline system. For engines fitting out with propose from author it well be raise on engine stall, from foundation diagram 3-D for duration injection, angle from kindling at total load and partial load and enrichment coefficients at total load.

Investigation realized to be continue for determine mechanical loss at electronic injection of gasoline, in domain little revolution (Negrea & Sandu 2000).

4. REFERENCES

Apostolescu, N. & Chiriac, R., (1998). *The Process of burning in driving with inward alight,* The Technical Publishing House, ISBN 973-31-0646-1, Bucharest.

Băţagă, N. (2000). *Driving with inward aligh,.* The Didactic and Pedagogical Publishing House, 973-9471-20-X, Bucharest.

Blaga, V. (2000). *The modeling injection from gasoline at engines with spark lightning through sparking*, The Publishing House of University from Oradea, ISBN 973-8083-41-9, Oradea.

Blaga, V. (2005). *Engine with gasoline injection,* The Publishing House of University Edition Oradea, ISBN 973- 613-981-6, Oradea.

Grünwald, B. (1980). *The Theory, calculation and construction engine's for road motor vehicle*, The Didactic and Pedagogical Publishing house, Bucharest.

Negrea, V.D. & Veneţia Sandu (2000). *The combating of medium pollution in motor vehicle,* The Technical Publishing House, ISBN 973-31-1455-3 Bucharest.

Annals of DAAAM for 2008 & Proceedings of the 19th International DAAAM Symposium, ISSN 1726-9679
Editor B. Katalinic, Published by DAAAM International, Vienna, Austria 2008
Make Harmony Between Technology and Nature, and Your Mind will Fly Free as a Bird

THE EXPERIMENTAL STAND OF ENGINES WITH GASOLINE INJECTION

BLAGA, V[asile]; BARA, L[ucian]; BARA, C[amelia]; BARA, V[asile] & DOMUTA, C[ristian]

Abstract: *With help to present parameters in this working are realize on the stand engine: the characteristic from speed at all load, the characteristic from speed at partial load, the combustible consumption's at totally task, the combustible consumption's at partially task In the writing "The characteristic's for revolution at partial's of engine load type 106-10 with injection for gasoline ", the experimental test of settle proposal was determine the performance dynamics and combustible consumption. The equipment for to try is more feet that is existing now. Was using equipment for test of Engines with Spark lightning with help injection of petrol that is present in number one imagines. It was used mathematical modelling method. The experimental research was performed at CESAR S.A. Colibasi Pitesti, Romania (Research Center for Motor Vehicles Subassemblies).*

Key words: *gasoline, injection, power, task.*

1. INTRODUCTION

From on the object to take proceeding against principals trial are: from adjust, from taking over, from inspection testing and control, specials trials, standard and from investigation.

The trials from investigation is make with object from has to work out different problems referring to thermal process, changing from gas, to working mechanism, user materials from construction, and from exploitation or from strength indication stabilization, in seeing engines perfecting existing and new engines realization. The equipment for to try are constituted by: sample stall equipped with apparatus and the feeding devices with fuel and coping water, the pipeline of the gases evacuation, apparatus and devices for the measurement's effectuation. The equipment for to try of the engine with fuel injection is made by: foundation with a plate and a frame, on wick it is fixed the equipment for testing; the engine supply system with consumption and the devices wick to allow -the determination of the consummated combustible quantity; the engine's starting system; the evacuation system of the burned gases and of the damping of the sound; the system for of the oil, air's delft and water's delft consumption; apparatus for the measure of temperature of pressure etc.; the equipment for to broke, wick can be either hydraulic or electric; (Negrea, & Sandu, 2000).

2. THE TESTING PRESENTATION

At the engines stall have been brought in order to try a Mono-Motronic (Grünwald, 1980), injection system engine type 106-20 equipped with carburetor 28/30 DCI. The injection's and lighting map drawings used at the engine with fuel injection expert meted by the author constitute some networks of characteristics with are put in the memory under the numerical form into an electronic module. They do not alter net much on the whole duration of the engine's work-ink. The electronic variation of the advance at the lighting has two important advantages the relative information to the revolution speed is directly taken from the shaft engine thank to in inductive sound with a highly precision of measure and the using of a map drawing allows to obtain adjustment of the advances angle at the lighting. The testing have carried on retiring to the followings the engine's running in; the determination of the dynamic and of consumption's .At the engines stall have been brought in order to try a Mono-Motronic injection system engine type 106-10 equipped with carburetor 28/30 DCI.

The injection's and lighting map drawings used at the engine with fuel injection expert meted by the author constitute some networks of characteristics with are put in the memory under the numerical form into an electronic module. They do not alter net much on the whole duration of the engine's work-ink. The electronic variation of the advance at the lighting has two important advantages the relative information to the revolution speed is directly taken from the shaft engine thank to in inductive sound with a highly precision of measure and the using of a map drawing allows to obtain adjustment of the advances angle at the lighting (Bățagă, 2000). The testing have carried on retiring to the followings the engine's running in; the determination of the dynamic and of consumption's performances according to SR ISO 1585/1998 and the carburetor variant; the determination of the STAS 6635/87 and of the combustible consumption's performance according to; in the fuel injection's variant; the comparison of the obtained results in the two equipment variant presented in the figures 1,2, and 3 (Blaga, 2000), (Blaga, 2005).

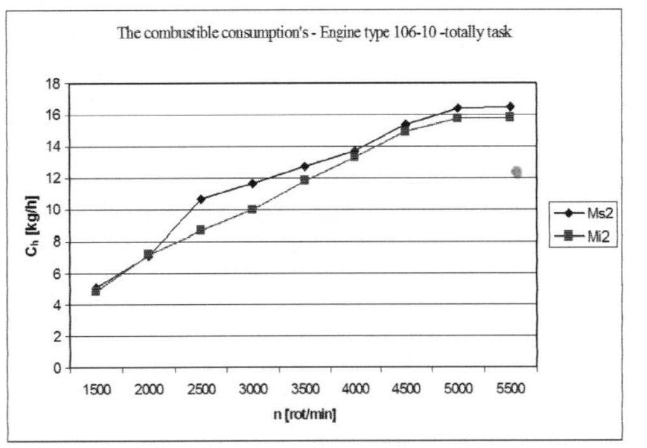

Fig. 1. The combustible consumption's - Engine type 106-10 - totally task

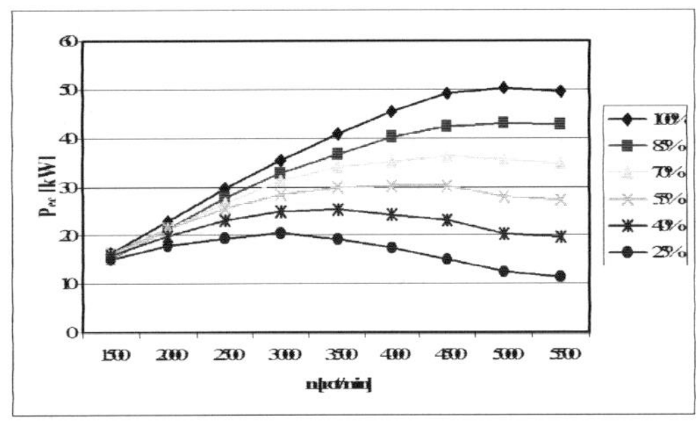

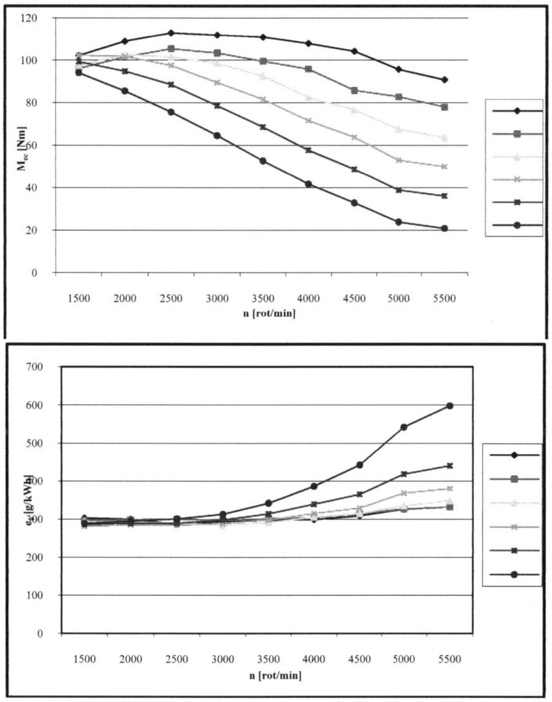

Fig. 2. The rotation characteristic's at partially task. Engine's type 106-10 with carburetor 28/30DCI.

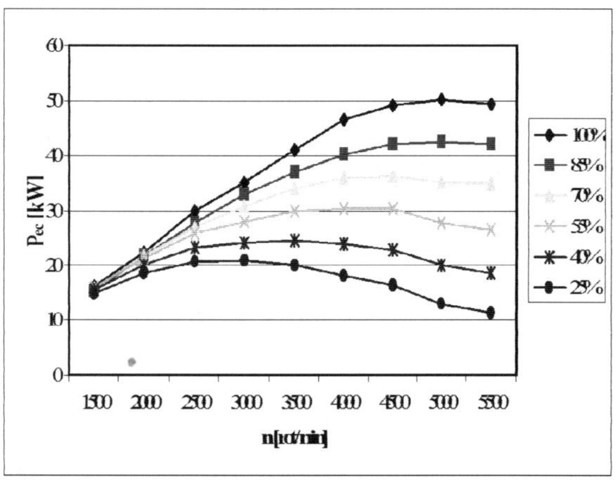

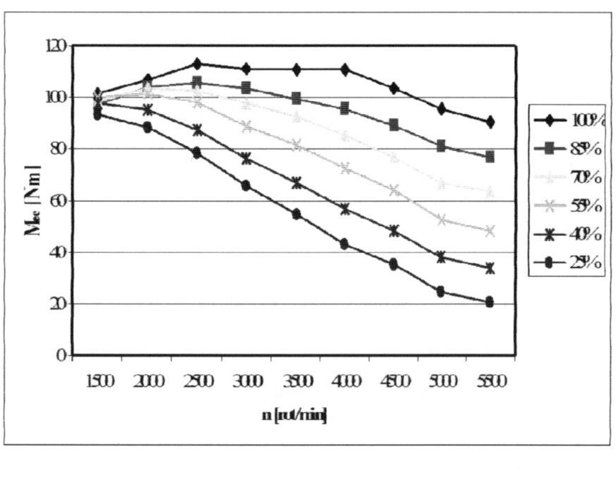

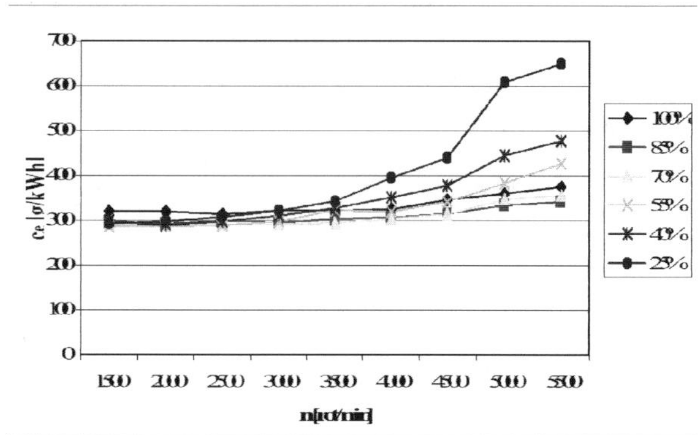

Fig. 3. The rotation characteristic's at partially task. Engine's type 106-10 with engine injection.

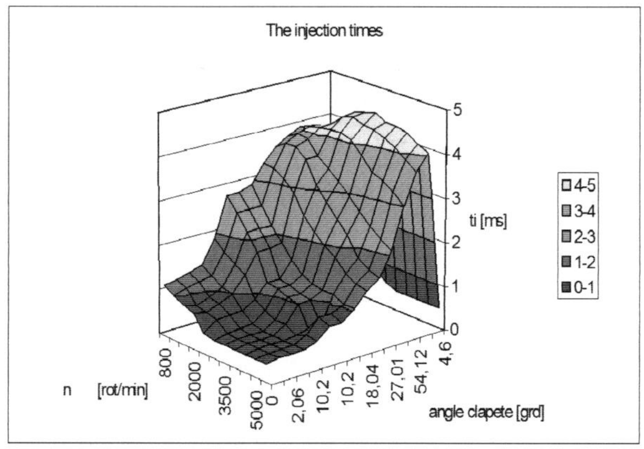

Fig. 4. The characteristic for injection's times

3. CONCLUSION

For the equipped engine's with the model supposed by the author, have been raised up on the engine stall, the main cartograms for the injection's times, at totally and partially task (Grünwald, 1980) and the enrichment's coefficient at the totally task which are presented in the figure no.4 (Blaga, 2000).

To make a comparison betwen characteristic's at partially task. Engine's type 106-10 with carburetor 28/30DCI and characteristic's at partially task. Engine's type 106-10 with engine injection is obvious the superiority of last type, technic, economic and environmental protection.

4. REFERENCES

Băţagă, N. (2000). *Driving with inward alight,* The Didactic Publishing house and Pedagogical, ISBN 973-9471-20-X, Bucharest.

Blaga, V. (2000). *The modeling injection from gasoline at engines with spark lightning through sparking*, The University Publishing house from Oradea, ISBN 973-8083-41-9, Oradea.

Blaga, V. (2005). *Engine with gasoline injection,* The University Publishing house from Oradea, ISBN 973- 613-981-6, Oradea.

Grünwald, B. (1980). *The Theory, calculation and construction engine's for road motor vehicle*, The Didactic and Pedagogical Publishing house, Bucharest.

Negrea, V. D. & Sandu Veneţia. (2000). *The combating of medium pollutant in motor vehicle,* The Technical Publishing house, ISBN 973-31-1455-3, Bucharest.

Annals of DAAAM for 2008 & Proceedings of the 19th International DAAAM Symposium, ISSN 1726-9679
Editor B. Katalinic, Published by DAAAM International, Vienna, Austria 2008
Make Harmony Between Technology and Nature, and Your Mind will Fly Free as a Bird

THE VARIATION OF THE PARAMETERS TECHNIC AND ECONOMIC WITH THE SPEED AND THE AIRS EXCESS COEFFICIENT

BLAGA, V[asile]; BARA, V[asile]; BARA, C[amelia]; BARA, L[ucian] & BORZA, I[oana]

Abstract: *Plane from realization of writing:elaboration programmes from calculate for modeling of the electronic injection gasoline from engine; programme from calculation; publication source code; experimental stall for control and to draw a parallel between the date to gate with the help's of programme. Through writing we must to pressent: a-growth of the dynamical parameters of motor vehicle;b-growth economicals parameters; c-reduction of pollution gase of escapement. It was used mathematical modelling method. The experimental research was performed at CESAR S.A. Colibasi Pitesti, Romania (Research Center for Motor Vehicles Subassemblies).*

Key words*: valve, couple, duration of injection*

1. INTRODUCTION

In this presence working itself contribute at modeling injection from gasoline at spark ignition engines; the model of fuel pump from gasoline, the model of regulator from pressure and injector (Turcoiu et al, 1987). For model of system proposed from author was proposed one system from injection of gasoline Bosch Motronic who power for air with ultrasonor wave.It be used plate from acquisition from trial, program from acquisition from date,electrical sounder from temperature, electrical sounder from pressure, electrical sounder from revolution and position angular, analyzer from gas. It is be make one study for select excess coefficient from air λ=1(electronic dosage). It is calculating the parameters for optimum for to run cycle at spark ignition engines with electronic injection from gasoline and to simulate on the computer (Bățagă, 2000).

2. THE THEORY FROM CALCULATION

A delicate problem from point of program on the computer, adiabatic coefficient is it starting calculation, temperature in different point is unknown and the adiabatic exponents not can be determinable. For solution of this problem has been used at impose to some value, found initial more correct, who permit a cycle crossing and determination more precise temperature. The temperature at the finished admission T_a was to consider known as initial date and she's expression calculation it be used at the finished as equation from verify at finished the cycle It is to determine the expression temperature's in succession from route of cycle $T_c,T_z,T_u,T_d,T_{d1},T_r$ and T_{r1}, as well as expression of gas temperature from evacuation T_e and the expression temperature thermodynamics medium to process from burning T_{med}. It is calculating the expression coefficient gas burnt residual γ_r. (Turcoiu et al, 1987). Must to remark that all the expression that was used at mathematical modeling cycle spark ignition engines with electronic injection from gasoline proposed from the author it is correlation betweens to can be introduce on the computer. This expression is correlation only in function from temperatures from characteristic point of cycle; pressuring sluices this point it is substitute with pressure's exponents and coefficient from initial date. Through calculate again of adiabatic exponent on amount foundation temperatures to get from the new cycle. In this mode, through some cycle the error can be reduce upside a calculate estimate admit in a thermal calculus very pretentious. It is make calculus mechanical this theoretically proposed L_{tp}, theoretical medium pressure proposed p_{tp} and of theoretical output propose η_{tp}. It be calculate effective pressure p_e, effective output η_e and mechanical output engine η_m, theoretic power propose P_{tp}, effective moment M_e, from combustible mass that have been on cycle engine m_{cb}, the stole theoretic Ch_t, the stole theoretic specific from combustible c_e and drawing cycle spark ignition engines with electronic injection from gasoline for propose model by the author; the models by calculus for pressure-volume and pressure-angle motor shaft diagram (Grünwald, 1980).

It be to achieve a experimental equipment for testing the propose model by the author, gifted with device and electrical sounders necessary of modeling. It by raise experimental diagram indication a engines equipment with system from electronic injection from gasoline type Bosch Motronic and will be to compare with indicated diagram theoretic for propose model by the author. In figure 1 is represented the variation of the effective power with the speed and the airs excess co-efficient.

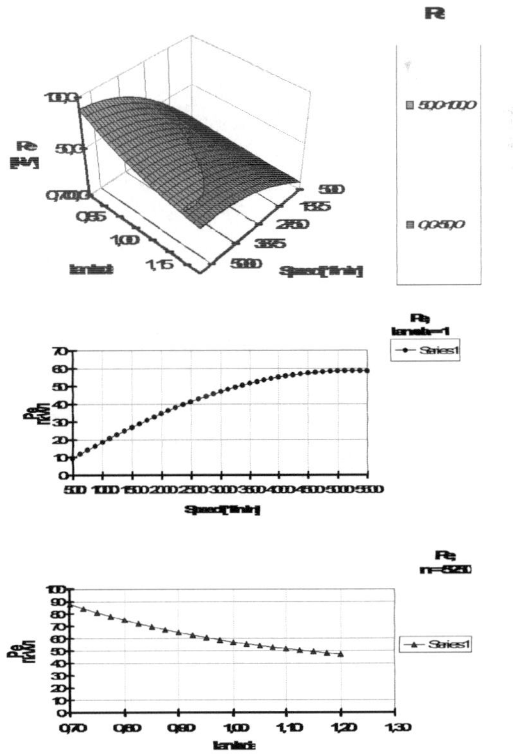

Fig. 1. The variation of the effective power with the speed and the airs excess coefficient.

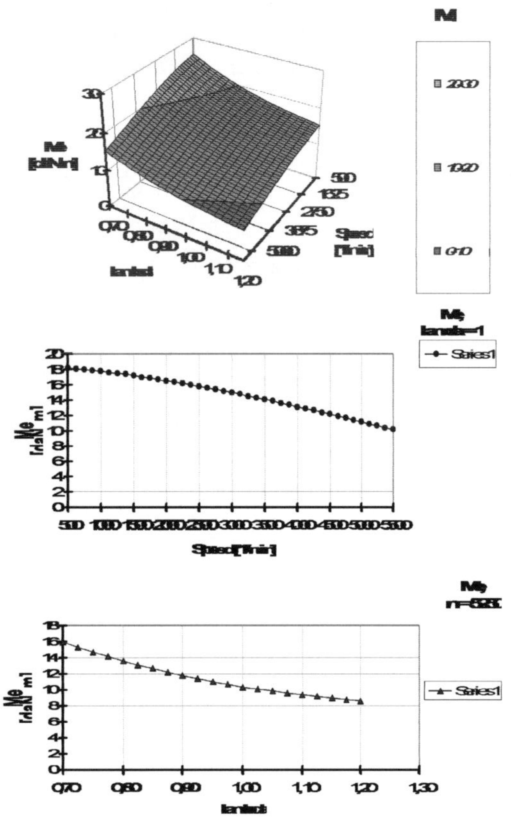

Fig. 2. The variation of the moment of the couple with the speed and the airs excess co-efficient.

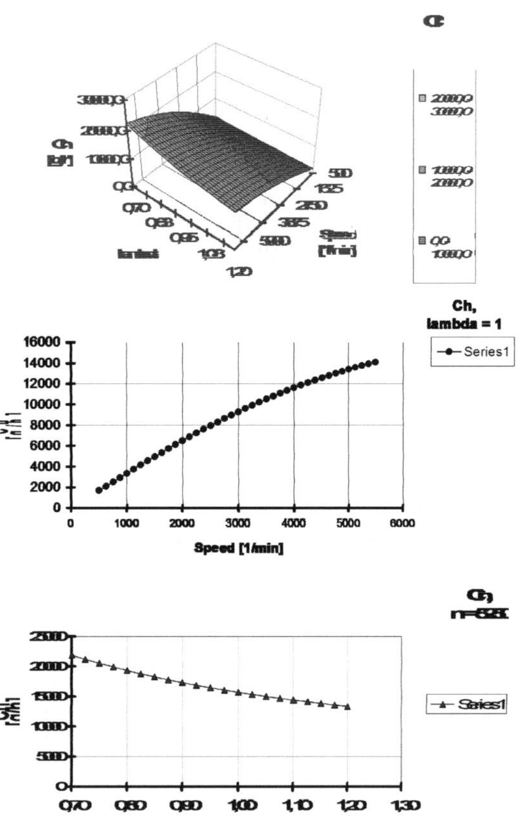

Fig. 3. The variation of the hour consumption with the speed and the airs excess co-efficient.

In figure 2 is represented the variation of the moment of the couple with the speed and the airs excess co-efficient. In figure 3 is represented the variation of the hour consumption with the speed and the airs excess co-efficient.

3. THE PROPOSE CONTRIBUTIONS

a. It is to achieve from modeling principle to some equipment from gasoline, modeling regulator from pressure of gasoline, the same general model spark ignition engines with electronic injection from gasoline.

b. It be to conceive a personal model for calculation a pressure on the received gallery p_{ga} and a pressure at finished admission.

c. It be conceive the model from calculus of regulator from pressure and a electromagnetic injector. It be effect the model of calculation duration injection t_i with revolution and load who represent even model suggestion by the author. (Blaga, 2005)

d. It be realize a program from calculate parameter spark ignition engines with electronic injection from gasoline.

e. With helping from calculate it be calculating and represent variation 3-D adiabatic and polytrophic coefficient, temperature and pressure in the characteristic point of cycle, coefficient from fill, dosage, measure thermal combustion of unity cylinder engine, rapport from grow of pressure in isochors combustion, rapport growth volume in post burning, the technical-economical of driving and duration injection. (Negrea, ,& Sandu, 2000).

f. It be raise the indicate diagram for the little two types from engine that has equipping car with the model has been proposed by the author. It is raise the diagram pressure –volume and pressure-angle from rotate tortuous arbor for these engines.

g. In the frame of experimental try of propose installation would be determine the dynamic performance and the combustible conformable in vigor according. The equipment from to test used it is that more performance that is now. It be used from try installation of one spark ignition engines with electronic injection from gasoline proposed on car.

h. It is realization one comparation between diagram theoretic from cycle proposed by the author and diagram of engine with electronic injection of gasoline, that was obtain one the experimental installation.

i. For engines fitting out with propose from author it well be raise on engine stall, from foundation diagram 3-D or duration injection, angle from kindling at total load and partial load and enrichment coefficients at total load. Investigation realized to be continue for determine mechanical loss at electronic injection of gasoline, in domain little revolution.

4. REFERENCES

Bățagă, N. (2000). *Driving with inward alight,* The Didactic and Pedagogical Publishing House, 973-9471-20-X, Bucharest.

Blaga, V. (2005). *Engine with gasoline injection,* The Publishing House of University Edition Oradea, ISBN 973- 613-981-6, Oradea.

Grünwald, B. (1980). *The Theory, calculation and construction engine's for road motor vehicle*, The Didactic and Pedagogical Publishing house, Bucharest.

Negrea, V.D. & Veneția Sandu (2000). *The combating of medium pollution in motor vehicle,* The Technical Publishing House, ISBN 973-31-1455-3 Bucharest.

Turcoiu,T., Boncoi, J. & Time, Al., (1987). *Equipment's from injection for engine with internal burning,* The Technical Publishing House, Bucharest.

Annals of DAAAM for 2008 & Proceedings of the 19th International DAAAM Symposium, ISSN 1726-9679
Editor B. Katalinic, Published by DAAAM International, Vienna, Austria 2008
Make Harmony Between Technology and Nature, and Your Mind will Fly Free as a Bird

THE DIFFERENTIAL EQUATION OF THE ELECTROMAGNETIC INJECTION PERIOD

BLAGA, V[asile]; TEUSDEA, A[lin]; CHIOREANU, N[icolae]; CATAS, A[driana] & HARUTA, O[vidiu]

Abstract: *The object of this paper is to derive the differential equation of the injection period and the time constant T. Moreover, the fuel discharge through the electro-injector, plan area (section area) and the lifting period of the needle, dependent on lifting height of the needle are determined.*
Key words: *differential equation, electromagnetic injection, linearization.*

1. INTRODUCTION

One writes the Bernoulli relation between sections 1-1 and 2-2 as shown in the figure 1. We obtain a Riccatti-type differential equation of the duration of the injection. The flow passes through the electro-injector and the flow area is also computed.

2. DYNAMIC ANALYSIS OF THE DISCHARGE

We deduce the Bernoulli relation between the sections 1-1 and 2-2 as is shown in figure 1, (Blaga, 2000). We have

$$\frac{\alpha_1 v_1^2}{2g}+\frac{p_1}{\gamma}+z_1=\frac{\alpha_2 v_2^2}{2g}+\frac{p_2}{\gamma}+z_2+\frac{1}{g}\int_{S_1}^{S_2}\beta\frac{\partial v}{\partial t}ds+h_{r1-2} \qquad (1)$$

Where we consider the section $S_1=S_2=S$ (constant section), $v_1=v_2=v$ the speed through the section, $\beta\cong 1$ coefficient of the mean speed, $z_1=z_2$ position heights, γ specific weight of the gasoline, g gravity acceleration,

$$h_{r1-2}=(\lambda\frac{l}{D}k^2+\Sigma\xi)\frac{v^2}{2g}$$

drop in energy due to resistances, λ coefficient of the carrier loss, l length of the pipe, D diameter of the pipe, $k=\frac{v_c}{v}$ speed ratio, $\Sigma\xi$ sum of coefficients of local loads losses, p_1, p_2 pressures from the sections 1 and 2 respectively.
The relation (1) becomes

$$\frac{p_1-p_2}{\gamma}=\frac{1}{g}\cdot\frac{dv}{dt}\int_{S1}^{S2}ds+(\lambda\frac{1}{D}k^2+\Sigma\xi)\frac{v^2}{2g}. \qquad (2)$$

Taking in account the relations

$$\gamma=\frac{G}{V}=\frac{mg}{V},\ \int_{S1}^{S2}ds\cong l,\ V=l\cdot S$$

$$\Delta p=p_1-p_2=l\cdot\rho\frac{dv}{dt}+(\lambda\frac{l}{D}k_2+\Sigma\xi)\frac{\rho}{2}\frac{v^2}{S}$$

where $m=l\cdot S\cdot\rho$ is the gasoline mass accelerated through the pipe and $\rho\in[0.682;0.767]$ is the density of the gasoline.

One obtains the Riccatti-type (Micula & Pavel 1992) differential equation

$$m\frac{dv}{dt}+cv^2=\Delta p\cdot S \qquad (3)$$

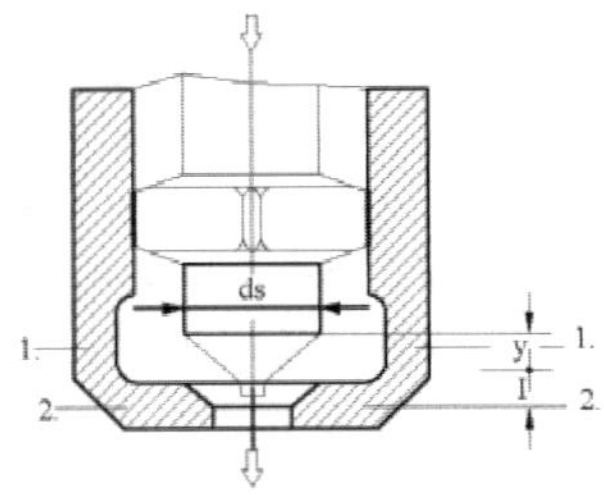

Fig. 1. Electro-injector with cone point

where

$$c=(\lambda\frac{l}{D}k^2+\Sigma\xi)\frac{\rho\cdot S}{2}.$$

The global resistance coefficient is denoted by $\lambda\frac{l}{D}k^2+\Sigma\xi=\xi_g$. Since $\xi_g=\frac{1}{\alpha^2}$ where α is the discharge coefficient (α = 0,7…0,8) we obtain

$$c=\frac{\rho\cdot S}{2}\cdot\frac{1}{\alpha^2}=\frac{\rho\cdot S}{2\alpha^2} \qquad dt=\frac{mdv}{\Delta p\cdot S-cv^2}$$

Let v_0 be the speed from the section y after the stabilization and let be T the time constant (natural time). When the speed is stabilized, from the equation (3) one obtains

$\frac{dv}{dt}=0$. Hence $v_0^2=\frac{\Delta p\cdot S}{c}$ or $v_0=\sqrt{\frac{\Delta p\cdot S}{c}}$ and $c=\frac{\Delta p\cdot S}{v_0^2}$.

We have after some computations

$$dt=\frac{m\cdot vo}{\Delta p\cdot S}\frac{d(\frac{v}{vo})}{1-(\frac{v}{vo})^2},\ \frac{1}{vo}dv=d(\frac{v}{vo}).$$

The time function is (O'Neill, 1991)

$$t=\frac{mv_0}{\Delta p\cdot S}arcth\frac{v}{v_0}+C=T\cdot arcth\frac{v}{v_0}+C \qquad (4)$$

Where $T = \frac{m \cdot v_0}{\Delta p \cdot S} = \frac{m \cdot \sqrt{\frac{\Delta p \cdot S}{C}}}{\Delta p \cdot S} = \frac{m}{\sqrt{\Delta p \cdot S \cdot C}}$.

From the relation (4) we deduce $arcth\frac{v}{v_0} = \frac{t}{T} - C, v = v_0 th(\frac{t}{T} - C)$ .For $t=0$ one obtains $v=0$ hence $C=0$ and $v = vo \cdot th\frac{t}{T}$. The stationary regime state $v = v_0$ is achieved for $t = \infty$ therefore

$$t_0 = 2{,}647T = 2{,}647\frac{m}{\sqrt{\Delta p \cdot S \cdot C}}.$$

3. THE DISCHARGE THROUGH THE INJECTOR

$Q = v \cdot S \cdot y = v \cdot S_t$, where: $y = \frac{a_c \cdot t^2}{2}$, $t \in (0, t_0)$

a_c is the lifting acceleration of the needle S_t is the plan area (section area) of the needle of the injector, which has a frustum of a cone profile as it can be seen in figure 2. The base of this profile formed by the needle point has the diameter d_2, and the big base has the diameter of the needle $d_1 = d_2 + 2e$.

The length of the generating line is $h = y\cos\alpha$ and the area of the passing section is $S_t = \pi\frac{d_1 + d_2}{2}h = \pi\frac{d_1 + d_2}{2}y\cos\alpha$, hence $Q = \pi(\frac{d_1 + d_2}{2})\cos\alpha \cdot \frac{a_c \cdot t^2}{2} \cdot v_0 th\frac{t}{T}$.

We denote by $A = \pi(\frac{d_1 + d_2}{2})\cos\alpha \cdot \frac{a_c}{2} \cdot v_0$, $t \geq t_0$, $t \in (t_0, t_1)$

Where t_1 is the time up to the end of the stroke of needle and t_0 is the stabilization time of the liquid and denote by $Q_1 = v_0 \cdot S \cdot y$. With the notation $A' = \pi(\frac{d_1 + d_2}{2})\cos\alpha \cdot \frac{a_c}{2} v_0 = A$ we obtain $Q_1 = A' \cdot t^2$ or $Q_1 = A \cdot t^2$, $t \geq t_0$; $t \in (t_0, t_1)$. The gasoline volume which passes over the injector for $t \in (0, t_0)$ is

$$V = \int_0^{t_0} A \cdot t^2 \cdot th\frac{t}{T}dt = A\int_0^{t_0} t^2 \cdot th\frac{t}{T}dt = A\int_0^{t_0} t^2 \cdot \frac{v}{v_0}dt \qquad (5)$$

where $th\frac{t}{T} = \frac{v}{v_0}$. In order to integrate in (5) we will proceed to the linearization of the function (Chioreanu, 2006)

$$tg\alpha = \frac{0{,}99v}{t_0} = \frac{v}{t} \text{ with } t_0 = T \cdot arcth0{,}99 = 2{,}647T.$$

After linearization we obtain :

$$v = \frac{0{,}99v_0}{2{,}647T} \cdot t = 0{,}374\frac{v_0}{T} \cdot t.$$

Finally, one obtains

$$Q = \pi \cdot (\frac{d_1 + d_2}{2})\cos\frac{a_c \cdot t^2}{2} \cdot 0{,}375\frac{v_0}{T}t = A\frac{0{,}375}{T}t^3 = Bt^3,$$

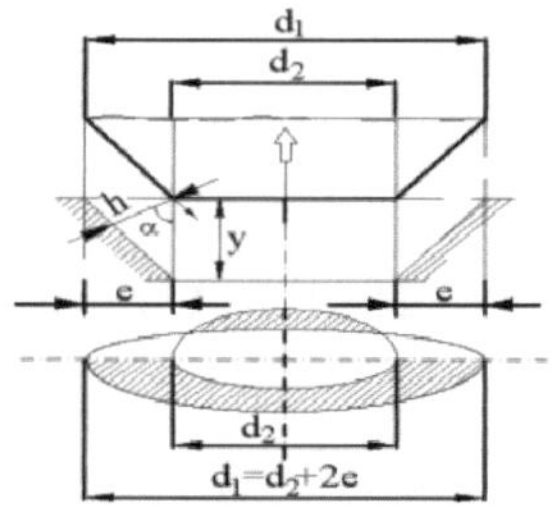

Fig. 2. Passage section through the orifice of the needle where of the injector

$B = 0{,}375\frac{A}{T}$. After a short computation, the gasoline volume which passes over the injector it can be determined from the formula $dV = Bt^3dt$. One obtains

$$V_0 = \int_0^{t_1} Bt^3dt = B\frac{t^4}{4}\Big|_0^{t_0} = \frac{B}{4}t_0^{\,4}; V_1 = \int_{t_0}^{t_1} At^2dt = A\frac{t^3}{3}\Big|_{t_0}^{t_1} = \frac{A}{3}(t_1^{\,3} - t_0^{\,3})$$

where $t_1 = \sqrt{\frac{2y_{max}}{a_c}}$. The gasoline overall volume V_t is

$$V_t = V_0 + V_1 = \frac{B}{4}t^4 + \frac{A}{3}(t_1^{\,3} - t_0^{\,3}).$$

We are presented the gasoline discharge dependent on lifting time t and the gasoline discharge dependent on the lifting height of the needle of the electro-injector, respectively, for the following values: $d_1 = 3mm$; $d_2 = 1mm$; $y = 0.15mm$; $\rho = 0.7kg/dm^3$; $\Delta p = 1.1daN/cm^2$; $t = 9\mu s$.

4. CONCLUSIONS

The quality of the pulverization of the fuel depends on a several factors related to the construction conditions (e.g. injection system type and also injector system type, the number, shape and the sizes of structural dimensions of pulverization orifices) and working conditions (e.g. injection pressure, backpressure and air density from the combustion chamber, flow rate of the jet). Very important factors are also the physic-chemical properties of the fuel: combustion value, fractional composition, octane number, gasoline purity and the running conditions (Blaga, 2007).

5. REFERENCES

Blaga, V. (2000). *The modeling injection from gasoline at engines with spark lightning through sparking,* The University from Oradea Publishing House, ISBN 973-8083-41-9, Oradea.

Blaga, V. (2007). *Engines for motor vehicles and tractors,* The University from Oradea Publishing House, ISBN 978-973-759 408-2, Oradea.

Chioreanu, N. & Chioreanu, S. (2006). *Heat engine for nonconventional motor vehicles*, The University from Oradea Publishing House, ISBN 973-759—065-1, Oradea.

Micula, Gh. & Pavel, P. (1992). *Differential and Integral Equations through Practical Problems and Exercises*, Kluwer Academic Publishers, ISBN 0-7923-1890-0, U.K.

O'Neill, P.V. (1991). *Advanced Engineering Mathematics*, Wadsworth Eds, ISBN-10: 0534552080, U.K.

Annals of DAAAM for 2008 & Proceedings of the 19th International DAAAM Symposium, ISSN 1726-9679
Editor B. Katalinic, Published by DAAAM International, Vienna, Austria 2008
Make Harmony Between Technology and Nature, and Your Mind will Fly Free as a Bird

DISCLOSURE LEVEL OF RISK MANAGEMENT INFORMATION: THE CASE OF ROMANIAN COMPANIES

BLIDISEL, R[odica]; POPA, A[dina]; FARCANE, N[icoleta] & FULCONIS, F[rancois]

Abstract: *The paper emphasizes aspects regarding risk and uncertainty. Nowadays many risk management methods are implemented in organizations. This perspective reveals that managers are linked in different forms to the activities they are managing, depending on the conditions and levels of uncertainty they are in. Actually, these multiple levels of uncertainty lead to the conclusion that any situation in an organizational system can be classified in two different models of organizational phenomena: risk and uncertainty. In this paper we survey the disclosure level of risk management information in the annual report of top Romanian listed companies.*
Key words: *decision, disclosure, risk management, uncertainty.*

1. INTRODUCTION

Mass media and academic literature are full of examples of organizations recently face "uncertainty". The term "risk" is recurrent (Pennings et al., 2008; Pesaran et al., 2008). Uncertainty and risk are now the usual vocabulary of the managers; stakeholders are also interested about the risk of the business they are connected to. If the first term "uncertainty" is still seen today in a largely negative sense, around the second one "risk" a mythology has been formed, which often leads us away from its first meaning given in management or in political science. Sometimes, these two terms appear to be used interchangeably, especially when we talk about a situation "risky" or "uncertainty". Sometimes, on the contrary, it seems that they can not be confused. In interviews with corporate executives, staff members of the same organizations, we have seen the same ambiguity in their words.

2. THEORETICAL FRAMEWORK AND METHODOLOGY

When one hears the words risk or uncertainty, it usually thinks about negative events that could occur. In some cases, there is advanced knowledge and information of the likely occurrence of some negative event. Economists use the terms "risk" and "uncertainty" to define the level of knowledge and information about an event or occurrence. The importance of this knowledge is that it affects the manager plan, the decisions making, and the tools used (Juniper, 2000; Alcaras et al., 2004). Knight (1935) highlights the difference between uncertainty and risk.

Risk management is becoming an increasingly important activity within firms and organizations. As a part of the management activities, risk management helps an organization to meet its objectives through the allocation of resources to undertake planning, decision making, and carry out productive activities. Risk management is focused on uncertainties that an organization faces: uncertainties in the probability of occurrence of events, uncertainties in the value to the organization of consequences of events, and other uncertainties that fall outside the "normally expected" range of variation. Generally risks are low probability, but high consequence events that can cause major disturbance to the organization.

The criterion for real judging of a risk management policy is whether it increases shareholder value. A systematic and dynamic use of risk management tools significantly increases the value. This is true not only for tools to identify and assess risks, but also for those who provide information to management to decide what it can cover or reduce risk, what risk transfer or sell, and which risks keep.

It should be a clear call for caution for the leaders and regulators are often inclined to prescribe, codify and standardize risk management techniques.

A rigorously analytical approach to risk measurement is only the imperative one of a good internal political risk management. The other elements were necessary in our view, the degree of risk transparency, timeliness and quality of information, the effectiveness of policies and internal controls, the extent of supervision made by management and independent bodies, the importance of diversification and spreading risk and, finally, judgment and experience of staff who knows the limits of models.

There is a discrepancy between the techniques used to measure and aggregate risks in circumstances "normal" and the methods used to assess the risks induced by extreme events.

The days when organizations were content to measure their market risk by the VAR (Value at Risk) are gone. It is admitted to supplement VAR by various forms of "stress tests", professionals are aware and in agreement on how to conduct these tests rather than on how to use them. It is irrational to evaluate the compromise between risk and performance when extreme events and highly improbable (such as those assumed in "stress tests") receive a probability almost equal to that of ordinary events.

This leads to the thought: we are now significantly more sensitive to the value of liquidity and its fragility, but there are no simple way to integrate this value methods of risk management.

Experience shows that deal with shocks in the markets (or the sudden bankruptcy of a significant contribution), the market prices start to be developed in an unusual way, and the liquidity is reduced, and will even up to disappear.

Starting from the presumption that risk management, like other management activities, must be practical, cost effective, and help the organization survive and prosper we survey in this section of our paper risk management disclosure of the Top 15 Bucharest Stock Exchange (BSE) Romanian listed companies. A good corporate governance practice faced company with a better disclosure and become more transparent also in the field of risk management.

The data was collected from Annual Reports of the Top 15 Romanian companies for the years 2006-2007. Also, during January-March 2008 numerous discussions with representatives, corporate executives, staff members of the studied companies took place in order to sustain our hypotheses. The empirical research is based on a comparative case study through content analysis, using a qualitative research methodology.

3. RISK MANAGEMENT FRAMEWORK AND DISCLOSURE: THE CASE OF ROMANIAN LISTED COMPANIES

Starting from the presumption that risk management, like other management activities, must be practical, cost effective, and help the organization survive and prosper we survey in this section of our paper risk management disclosure of the Top 15 Bucharest Stock Exchange (BSE) Romanian listed companies. A good corporate governance practice faced company with a better disclosure and become more transparent also in the field of risk management.

The growth in risk management is directly linked to the increasing number of risks an organization faces due to more complexity and interactions in the world, greater scrutiny by stakeholders and the media, and so forth.

Risk management is an integral part of the corporate governance arrangements and has been built into the management processes as part of the corporate overall framework to deliver continuous improvement.

Risks are usually described by a list of risks, arranged in priority order with the largest risks first.

A risk management framework is a description of an organizational specific set of functional activities and associated definitions that define the risk management system in an organization and the relationship of the risk management organizational system. A risk management framework defines the processes and the order and timing of processes that will be used to manage risks. A good risk management framework should enhance and improve risk management by:

1. Making it more transparent and understandable to stakeholders,

2. Making its processes more efficient,

3. Allowing for cross fertilization of risk controls, risk estimation, risk assessment from others because of standardization of terms, processes, tools etc.

Finally, risk management must produce a net value for the organization. This value is estimated and reviewed and consists of three basic elements: costs, financial benefits, and trust and respect of stakeholders and the public.

Risk is inherent in any decision, at any level in the organization.

Stakeholders need to be better informed regarding the risk and opportunities company is faced with, that's make managers to voluntary or mandatory disclose information regarding risk management, usually as a part of the Annual Report. In this part of our paper we present the results of a survey we conducted on the top 15 BSE listed companies.

Our findings are:

- Only one Romanian company from Top 15 BVB are not disclosing information regarding risk management,
- 8 companies present a synthesis of the risk categories that affects their activity during the reported year,
- 6 companies' presents more detailed information regarding their risk management activity.

4. CONCLUSIONS

Looking back a moment to the presentation of the problem which this article is intended to provide some answers, it seems that the analysis of literature developed here goes beyond the issue of the distinction between uncertainty and risk. Indeed, the preparation of a research on a specific field sometimes raises questions outnumber those originally asked. If the uncertainty, in generic terms, is the ignorance or total or partial failure of events, one can, according to the body of assumptions that we adopt, be seen as irreducible or reduced as a given intangible, or as the result of a more or less off time.

In trying to give advice the enterprise policy makers we can show that the statistical uncertainty is "the stochastic nature or error from various sources, as described by statistical methodology". The technical consultation defines risk as "the probability of the occurrence of something unfortunate". It is noteworthy that in terms of decision-making theory, the risk is defined as the average loss or forecasted loss when something unfortunate happens.

The process of communication the risk to policy makers is in its infancy and represents major challenges for both technicians and managers. In return, managers and participants must find a way to objectively evaluate the potential costs of adverse events, define acceptable levels of risk and quantify the short-term production to waive that to reduce these risks.

There are no standard methods to reflect the uncertainty and risk to policy makers. The statistics provide a basic set of means to account for variability, which can be used to indicate the uncertainty associated with a specific estimate, or the likelihood that an adverse event occurs.

The method choused by managers for measure the probability of uncertainty and risks will depend on their level of technical knowledge. In most developing countries, it will be important to link the uncertainty of environmental characteristics that are well known. For many economically important stocks, it is justified to try to quantify the uncertainty and risk.

Regarding our sample companies we may conclude that the risk management in the majority of Top 15 BSE Romanian listed companies 60% are in an incipient stage of implementation, only 40% of this Top 15 companies have already defined their objectives regarding risk management or already have an integrated system of risk management.

Because our empirical study is limited to a descriptive examination of the level of risk management disclosure in annual reports of the Top 15 BSE companies, some aspects can be mentioned as reason for future research:

- to extent the study to all companies listed on BSE;
- to investigate the level of implementation of risk management by the Romanian listed companies;
- to investigate the association between the risk management disclosure, the investors' behavior and the companies' performance indicators.

5. REFERENCES

Alcaras J.R.; Gianfaldoni P. & Paché G. (2004)., *Decision in organizations - Dialogue between economy and administration*, L'Harmattan, ISBN 2-7475-6066-X, Paris – Budapest – Torino.

Cooper J.C.& Selto F.H. (1993). Is risk preference induction a reliable method of controlling risk preferences?, *Journal of Management Accounting Research*, Vol. 5, ISSN 1443-9905, 109-123.

Juniper J. (2000). *Uncertainty, Risk, and Chaos*, Cambridge University Press, ISBN 0-521-62030-9, Cambridge, New York and Melbourne, 37-60.

Pennings J.M.E.; Isengildina-Massa O.; Irwin, Sc.H.; Garcia, Ph.& Good, D.L. (2008). Producers' Complex Risk Management Choices, *Agribusiness*, Vol. 24, No. 1, 2008, 31-54, *Available from*: http://www3.interscience.wiley. com/cgibin/fulltext/117878732/pdfstart *Accessed*: 2008-05-11

Pesaran M.H.; Schleicher Ch.& Zaffaroni P. (2008). Model Averaging in Risk Management with an Application to Futures Markets, Faculty of Economics, University of Cambridge, *Cambridge Working Papers in Economics*, ISSN (printed): 0951-0079, 24 p.

Annals of DAAAM for 2008 & Proceedings of the 19th International DAAAM Symposium, ISSN 1726-9679
Editor B. Katalinic, Published by DAAAM International, Vienna, Austria 2008
Make Harmony Between Technology and Nature, and Your Mind will Fly Free as a Bird

ULTRA-PRECISION MACHINING OF OPTICAL SURFACES WITH SLOW-TOOL SERVO

BLIEDTNER, J[ens]; BUERGER, W[olfgang] & FROEHLICH, M[aik]

Abstract: *The technological preparation for the machining of optical surfaces of different materials by means of Slow-Tool-Servo was carried out during the research work.*
Key words: *ultra-precision turning, optical surfaces, Slow-Tool-Servo, measuring technology*

1. INTRODUCTION

The continuous technical development in the field of ultra-precision machining enables a high-quality, ultra-precise manufacturing of moulds for the injection moulding of plastic optics in accordance with the users' requirements. The high requirements in terms of roughness demand the application of high-quality surface measuring technology for assuring the quality of the component ranges.

By means of examples the article shows results of experimental investigations of precision-machined toric surfaces, which were realised in co-operation with Jenoptik Polymer Systems GmbH.

2. EXPERIMENTAL PROCEDURE, RESULTS

The investigated component range (see fig. 1) consists of tree metallic moulds and two plastic prototypes with toric surfaces (metallic moulds: radiuses R 800 and R 600; plastic moulds: radiuses R 800 and R 100).

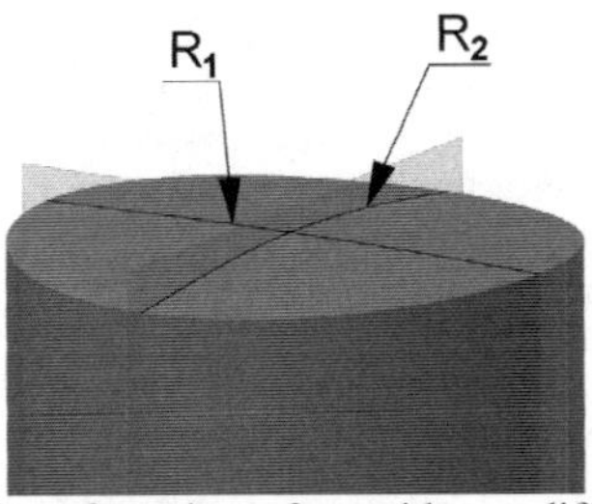

Fig. 1: Illustration of a toric surface with two different radiuses R_1 and R_2

There were three specimens of each material available, so that fifteen specimens had to be measured in five positions each. The components with toric surfaces were manufactured on the ultra-precision turning lathe Nanoform 350 of the company Precitech (see fig. 1). It has three servo-controlled axes (XZC) and provides different machining options.

The component range under investigation was exclusively manufactured with the Slow-Tool-Servo (STS) option. This machining option is one of three Servo-Tool options which are currently used in ultra-precision manufacturing of optical surfaces (Osm2007).

The advantage of the Servo-Tool options is the economical manufacturing of non-rotation-symmetric geometries, as they can be manufactured at significantly lower costs in comparison with a conventional manufacturing. Furthermore, the Servo-Tool options enable an on-axis manufacturing of e.g. aspheres compared to previous off-axis manufacturing. During the STS machining the work piece is mounted on the air-bedded spindle (C-axis). The C-axis is mounted above the hydrostatically bedded linear guiding of the X-axis and rotates around the Z-axis (see fig. 2).

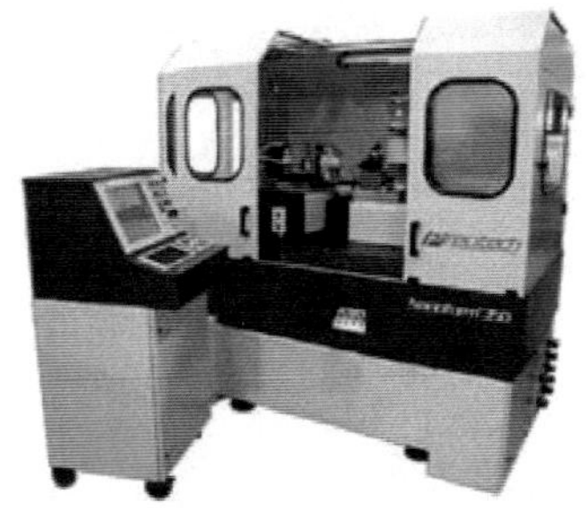

a) Nanoform 350

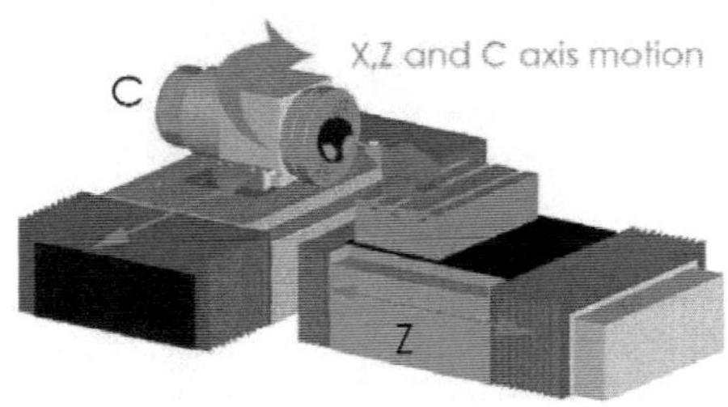

b) Slow-Tool-Servo (STS)/CNC-axes

Fig. 2: Ultra-precision turning lathe Nanoform 350 and machining with Slow-Tool-Servo/(CNC) - axes

The STS procedure is used for manufacturing of non-rotation-symmetric connected surfaces such as toric surfaces. Simple arithmetic relations between excitation frequency f_{an} [Hz] and rotation speed [min^{-1}] are used for creating the optical structures.

If a surface is to be machined at a rotation speed of $n = 400\ min^{-1}$, a tool excitation frequency of $f_{an} = 400/60$ Hz = 6.66 Hz is necessary. For the creation of a toric surface where two radiuses are staggered by 90°, the tool has to make two moves for one rotation. The optical surfaces of the component range are toric surfaces with radiuses of R = 600 mm and R = 800 mm (metallic moulds) or radiuses of R = 800 and R = 100 (plastic moulds) which are staggered by 90°.

All turning work pieces were scrubbed at a constant rotation speed of $n = 149\ min^{-1}$ with a cutting edge radius of $r_\varepsilon = 0.5$ mm and finished at a rotation speed of $n = 133\ min^{-1}$ with a cutting edge radius of $r_\varepsilon = 1.5$ mm. The following materials were machined with MKD tools in the investigations:

1. CuproNickel: good machinability in cold state, good thermo formability
2. Aluminium RSA-905 AE: universally applicable, corrosion resistant, high stiffness

3. Aluminium AlMgSi1: versatilely applicable, good corrosion resistance
4. Zeonex E48R (Plastics/Cyclo Olefin Polymer (COP)): excellent heat resistance, low water absorption, high transparency
5. PMMA gs (Polymethylmethacrylat (PMMA)): good transparency, UV-resistance, low weight compared to glass.

The mechanical precision machining was carried out with selected technological parameters. For determining the micro topography (roughness) in the nanometer range an AFM (Atomic Force Microscope) was used in the investigations. The scanning range for determining the roughness was 50 x 50 µm². Furthermore, the Power Spectral Density–Function (PSD) was determined as a useful instrument for analysing the surface roughness. This function determines the amplitude for a roughness value in the roughness profile of the technical surface as a function of the spatial frequency or the wavelength. Fig. 3 exemplarily shows the PSD-function of all surfaces manufactured by means of STS.

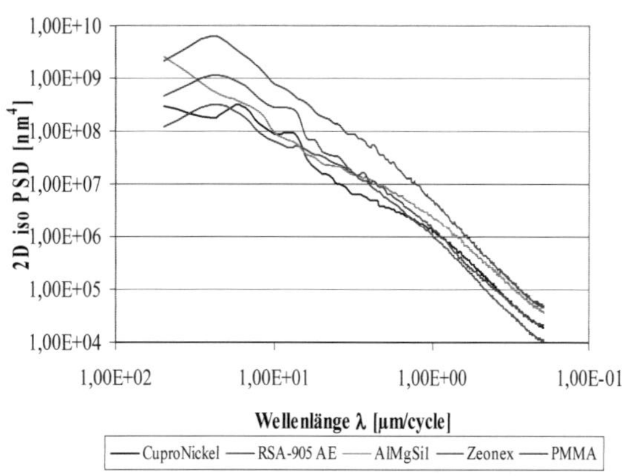

Fig. 3: PSD-function of the machined technical surfaces

An elevation image, an image of the three-dimensional surface, the PSD-function for each specimen and the PSD-functions for all specimens of a material (compilation of all PSD-functions of the investigated material) were realised in the investigations. From the individual PSD-functions of a material an averaged PSD-function can be derived, which was also illustrated. From the results of the investigations important technical expertise can be gained, which can only briefly be introduced in this article (Dupa2002); (Weck2000).

Concerning the roughness of the technical surfaces the following values (average values) could be achieved with Slow-Tool-Servo (STS) and the same technological machining parameters:

Material	roughness R_q [nm]
CuproNickel	5.4
Aluminium RSA-905	5.2
Aluminium AlMgSi1	7.4
Plastics Zeonex E48R	14.1
PMMA gs	7.7

Except for the material Zeonex E48R, with all other materials the restriction $R_q \leq 10$ nm was kept. From the illustrations of the surfaces (e.g. see fig. 4) it can be realised very well that the specific micro topography of the machined technical surfaces mainly results from the different machinability of the used materials (material influence/machinability). Further investigations on the problem specific roughness differences in relation to the machining position on toric surfaces have to be carried out.

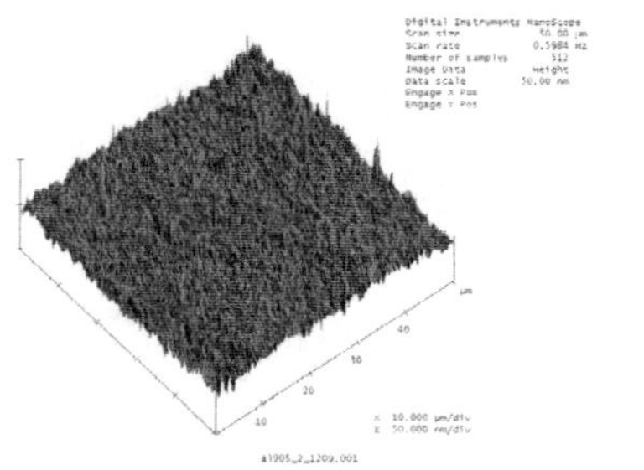

Fig. 4: Micro topography creation material RSA-905 AE No. 2 (R_q = 5.0 nm) (3D-surface AFM)

A comparison of the individual surfaces clearly shows the different micro topography creation with the different materials, although all components were manufactured with the same procedure (STS) and with the same technological parameters. The specific micro topography of the machined technical surfaces results mainly from the different machinability of the used materials (material influence). It is a consequence of the overlay of the roughness, which is due to the geometry of the diamond tool and the technological parameters with the micro roughness, due to the machinability of the material.

The comparison of the PSD-functions (see fig. 4) of the STS-machined surfaces shows that the PSD-function of all finished components has nearly the same gradient. The deviations result from the differing RMS-values, which again result from the specific machinability of the individual materials (micro roughness).

3. CONCLUSIONS

The evaluation of the determined power spectral density distributions (PSD-functions) of selected surfaces of the tooling inserts and their plastic mouldings aims at increasing the statistical certainty of the quality evaluation of these technical surfaces on the one hand and detecting the differences between the technical surfaces on the other hand. The evaluation of these specific technical surfaces can be qualitatively increased by the power spectral density distribution.

By means of Slow-Tool-Servo sophisticated technical surfaces can be manufactured from the different materials (moulds for injection moulding). Furthermore the extensive experimental investigations have proved valuable for:

- objectification of production issues by components with optical surfaces
- further optimization of the manufacturing process
- the quality assurance of such components.

4. REFERENCES

Osmer, J.; Weingärtner, S.; Riemer, O.; Brinksmeier, E.; Bliedtner. J.; Bürger, W.; Fröhlich, M.; Müller, W.

Diamond Machining of Free-Form Surfaces: A Comparison of Raster Milling and Slow Tool Servo Machining. In: Proceedings of the 7th International Conference euspen Volume I – Volume II, May 20 th – May 24th 2007, Bremen, Germany, S. 189 - 192, Published by euspen, Bedford, 2007

Duparre´, A.; Notni, G.: Charakterisierung nanorauher Oberflächen. DAKOM 2002: Charakterisierung von optischen und technischen Oberflächen. Darmstadt, 27. Februar 2002

Weckenmann, A.; Ernst, R.: Anforderungen und Randbedingungen für den Einsatz von Meßsystemen in der Mikro- und Nanotechnik. VDI-Berichte Nr. 1530 „Sensoren und Meßsysteme 2000“, pp. 297-307, VDI Verlag GmbH, Düsseldorf, 2003

Annals of DAAAM for 2008 & Proceedings of the 19th International DAAAM Symposium, ISSN 1726-9679
Editor B. Katalinic, Published by DAAAM International, Vienna, Austria 2008
Make Harmony Between Technology and Nature, and Your Mind will Fly Free as a Bird

EXPERIMENTAL RESEARCH ABOUT DYNAMICS BEHAVIOUR OF LANDING GEAR

BOGDAN, C[onstantin]; BOGDAN, M[ihaela] L[iana]; POPA, D[ragos] & GHERGHINA, G[eorge]

Abstract: *The paper presents experimental investigation for testing of scale model main landing gear. It is made a special soft for data acquisition of the measured parameters. Also, it is elaborated another soft for experimental data processing for the obtaining diagrams of variation for many parameters which describe the dynamical behavior of the landing gear. The values of the reference parameters can be determined in various function conditions. It can estimate the service life for separate elements of landing gear structure.*
Key words: *experimental investigation, testing, landing gear.*

1. INTRODUCTION

The society development proposes an eficient and sure transport system not only for persons, but also for goods. The air transport had imposed in last years by security and comfort. The security implies the realization of reliable, robust and safely aircrafts able to coresspond at the normal and function limit conditions. This objective must be taken into account from design procedure. Many aircrafts events are caused by the landing gear. One of the main elements which influence the realibility and security of aircraft is the resistance of the landing gear. This is determined by technico-constructive solutions adopted from the conceptual design procedure, volume of information obtained by tests made on prototype or scale model, and, also, service and technical maintenance politique.

In this paper is presented the experimental equipment and tests of main landing gear.

Having in view the complexity and function conditions of the landing gear, it is made landing gear scale model (Bogdan C. et al., 2007; Bogdan M.L. et al., 2007; Oţăt et al., 2006) for testing.

Next, it is presented the experimental testing comprises: tests, testing conditions, results obtained after data processing by a special soft.

2. EXPERIMENTAL EQUIPMENT USED FOR TESTING OF MAIN LANDING GEAR SCALE MODEL

Experimental testing equipment for measurement and processing of all parameters which describe main landing gear behaviour includes: portable equipment for data acquisition Multi Tester MT01, accelerometers, transducers of linear and angular course, charge amplifiers and computer Notebook with special software for acquisition and processing data under TestPoint-Keithley.

2.1. Portable Equipment for Data Acquisition MULTI TESTER MT01

MultiTester MT01 is portable diagnosis equipment having the following elements: acquisition interface µDaq – USB - 30A16; backplane for 16 conditioners with galvanic isolation type MB; impulses programmer.

3. TESTS MADE FOR DETERMINATION OF MECHANICAL CHARACTERISTIC AT THE IMPACT BY DROP AT LANDING AND AT RETRACTION/TAKE OFF OF THE MAIN LANDING GEAR WITH DOUBLE TWIN TANDEM WHEELS

3.1 Measured or Determined By Calculating Mechanical Characteristics at Impact /Shock by Drop at Landing of Main Landing Gear with Double Twin Tandem Wheels

The measured parameters are the following:

1. Linear or drop course of landing gear at landing Crs (mm);
2. Linear or drop velocity of landing gear at landing V (m/s);
3. Linear (drop) accel of landing gear at landing Acc (m/s^2);
4. Vibration accel of base plate of the stand, Acc 1(m/s^2);
5. Vibration accel of landing gear, Acc 2 (m/s^2);
6. Vibration accel of the landing gear superior part, Acc 3 (m/s^2);
7. Impact force, F Impact (N).

The mechanical tests were made on the main landing gear scale model with twin dual in tandem wheels (Bogdan C. et al., 2007) in the following conditions:

1. - drop from a height approx. 132 mm, no additional load (0) with initial mass of landing gear model scale of 8,360 kg;
2. - drop from a height approx. 132 mm, with additional load (1) of 5,580 kg with total mass of 13,940 kg ;
3. - drop from a height approx. 132 mm,, no additional load (2) of 2,235 kg with total mass of 16,175 kg
4. - drop from a height approx. 97 mm, no additional load (0), with initial mass of landing gear model scale of 8,360 kg;
5. - drop from a height approx. 97 mm, with additional load (1) de 5,580 kg with total mass of 13,940 kg;
6. - drop from a height approx. 97 mm, with additional load (2) de 2,235 kg with total mass of 16,175 kg.

3.2. Results obtained after data processing

In the tables 1 and 2 are centralized the characteristic (the maxim values) mechanical parameters (defined above) determined during tests. (Bogdan M.L. & Bogdan C.,2005 a,b; Oţăt et al., 2006)

It is determined the course taking as reference the superior drop point. The transducer course is linearized by soft using an interpolation table. The drop velocity is determined by numerical derivation of drop course. Drop acceleration is determined by numerical derivation of drop velocity.

No. test	Height (mm)	Load (kg)	Crs Max (mm)	V Max (m/s)	Acc Max (m/s^2)	F Impact (N)
1.	132	8,360	132,57	1,1549	81,499	681,33
2.	132	13,940	133,29	1,2597	84,007	702,3
3.	132	16,175	133,65	1,2818	86,445	1398,2
4.	97	8,360	95,618	0,9853	75,307	629,57
5.	97	13,940	98,173	1.0454	67,825	945,48
6.	97	16,175	97,087	1,028	64,35	1040,9

Tab. 1. Values of height, load and parameters Crs, V, Acc, F Impact for different six tests.

No. test	Crs Max (mm)	Acc 1 (m/s^2)	Acc 2 (m/s^2)	Acc 3 (m/s^2)
1.	132,57	37,253	252,53	90,847
2.	133,29	36,692	284,75	60,483
3.	133,65	34,592	247,55	91,514
4	95,618	30,729	307,38	81,94
5.	98,173	31,725	208,69	92,778
6.	97,087	23,069	330,13	85,58

Tab.2.Values of the course (maxim values), vibration acceleration of base plate of the stand, Acc 1, vibration acceleration of landing gear, Acc 2, vibration acceleration of the landing gear superior part, Acc 3.

For example, in the following figures (fig.1, 2) are represented the variation diagrams of mechanical parameters, diagrams obtained in the mechanical testing conditions mentioned above for case of test no.1. Others diagrams obtained in the other five cases (no.2 - no.7) are similary as shape as the variation diagrams from figure 1.

In each diagram on the horizontal axis appear the time in seconds and in the table under the diagram are given the values for the moments indicated by markers. Markers are vertical point lines from the diagram plane. The marker positions show the important moments during tests.

Each set of values corresponds to the values of the points from intersection between diagram and markers. Notations from tables (under the diagrams) are the same meaning with parameters mentioned above. In the field of each diagram appear different vertical scales for the various parameters which are tested. The diagrams are made at the scale sometimes different for the "catching" of the phenomenon in the drawing plane. The values from diagram tables are directly calculated by TESTPOINT soft, being correlated with the above mentioned scales. For the processing of the signals, it is achieved a calculus program under the programming medium TESTPOINT which can made the following sequence of the operations:

- the effective value determination of the vibration acceleration for all length of the recording;
- the graphical representation of the experimentally determined specific features.

These variation diagrams of the parameters describe the dynamical behavior of the main landing gear. From statistical analysis of these diagrams, it can estimate the service life for separate elements of landing gear structure.(Bogdan M.L.. et al., 2007)

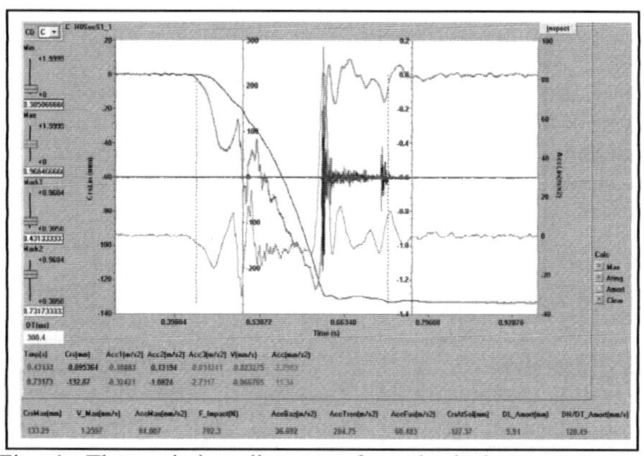

Fig. 1. The variation diagram of mechanical parameters at impact / shock test by drop from a height approx. 132 mm, no additional load (0) that is a drop with initial mass of main landing gear model scale of 8,360 kg

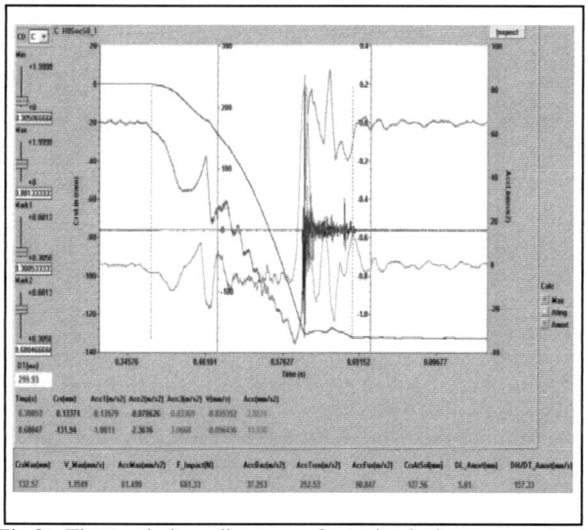

Fig.2. The variation diagram of mechanical parameters at impact / shock test by drop from a height approx. 132 mm, no additional load (0) that is a drop with initial mass of main landing gear model scale of 8,360 kg. Detail of impact

4. CONCLUSION

The testing of main landing gear is made usually with equipment for testing at ground from reason of cost and low risk comparatively with testing on aircraft.

Because of this reason, the testing of main landing gear on scale model presents many advantages such as: the service life estimation for different elements from landing gear structure; the determination of stresses appeared in function which should lead to the appearance of yields; an additional experimental information which should characterize the landing gear behaviour in various working conditions; short periods for the asimilation of new landing gear design; the increased safety and service life of landing gear.

5. REFERENCES

Bogdan, C., Bogdan, M.L., Simniceanu, L. & Dincă, L.(2007) Experimental Equipment for Study of Landing Gear Mechanisms, *Annals of DAAAM for 2007 & Proceedings of the 18th Theory of INTERNATIONAL DAAAM SYMPOSIUM "Intelligent Manufacturing & Automation: Focus on Creativity, Responsibility and Ethics of Engineers"*, 24-27th October 2007, ISBN 3-901509-58-5, ISSN 1726-9679, pp.51, Editor B. Katalinic, Published by DAAAM International, Vienna, Austria

Bogdan, M.L. & Bogdan, C. (2005). About Kinematics of the Landing Gear Assembly, *Annals of the University of Craiova*, "Mechanics" series, vol. I, nr.1, 2005, pp.5-10, ISSN 1223-5296

Bogdan, M.L. & Bogdan, C. (2005). *Mechanics. Dynamics and Mechanical Vibrations*, "Universitaria" Publishing House, ISBN 973-742-329-1, ISBN 978-973-742-329-0, Craiova, Romania

Bogdan, M.L.; Gherghina, G. & Popa, D. (2007). Aspects concerning Virtual Prototyping of Some Mechanisms from Landing Gear, *11th European Automotive Congress*, 30 May-1 June 2007, Budapest, Hungary, Book of Abstracts, pp.228

Oţăt, V.; Dincă, L. & Simniceanu, L. (2006). Aircraft landing run at the Asymmetric Braking, *The 5th International Conference on Advanced Engineering Design*, Prague, Czech Republic, June, 2006 (CD-ROM)

Annals of DAAAM for 2008 & Proceedings of the 19th International DAAAM Symposium, ISSN 1726-9679
Editor B. Katalinic, Published by DAAAM International, Vienna, Austria 2008
Make Harmony Between Technology and Nature, and Your Mind will Fly Free as a Bird

COMBINED APPROACHES IN AUTOMATIC PAGE CLUSTERING FOR CONTENT CONVERSION

BOIANGIU, C[ostin] A[nton] & CANANAU, D[an] C[ristian]

Abstract: *The purpose of this paper is to present a model for separator detection using the Delaunay triangulation and a common sampling algorithm based on a triangular filter. By using a sampling algorithm, the blur effect is created and together with the output triangles, obtained from the modified Delaunay algorithm, the separators can be determined after processing.*
Key words: *Delaunay triangulation, filter, contour, resampling, separators.*

1. INTRODUCTION

The first step to determine the areas on a page (text areas, pictures, tables) is to emphasize what separates them from each other. And so, the problem that arises when attempting to detect separators is that most algorithms are dependent on the geometry of the separators.

A large number of such algorithms detect the line separators by using simple facts, such as a line has either the width or the height greater as a value with respect to the other. Other algorithms use the fact that lines are oriented on an axis, either vertical or horizontal or even rotated with a certain angle.

There are also approaches that use more advanced mathematical knowledge. An example can be seen in (Guru et al., 2004; Fortune, 1997), where small eigenvalues analysis stands at the basis of the idea.

Delaunay triangulation for various text areas detection has been presented in a number of papers, an example of such being (Xiao & Yan, 2004), where it has been used for author and title regions detection.

However, in this paper the solution to the already mentioned issue of separators is an algorithm that is independent on geometry or shape and uses only characteristics common to all separators.

2. THE SOLUTION

2.1 How to find the separators

The idea is the following: if we see each character, image or collection of pixels as an entity we can connect each of them with the neighboring entity by using the Delaunay triangulation. So, the entity is defined as the collection of neighboring pixels of same color, different from the background.

The purpose of this paper is not to present the Delaunay triangulation algorithm, but to point out how the results from such a mathematical approach can be used to further determine various output for an entirely different domain. For further information on the algorithm please check (Shewchuck, 2002).

First a set of vertexes has to be created for the algorithm to work. In order to obtain this input the contour of each entity is created. Every entity can be seen as a collection of horizontal segments of pixels. The contour is actually made out of the pixels on the outside border of the entity and the ones on the inside which can be seen from the exterior. To sum up, the contour is given by the extremities of the entity's segments.

Fig. 1. Constrained Delaunay triangulation applied on a section of a newspaper document.

Every entity can be seen as a collection of horizontal segments of pixels. The contour is actually made out of the pixels on the outside border of the entity and the ones on the inside which can be seen from the exterior. To sum up, the contour is given by some of the extremities of the segments contained in the entity.

Next, the constrained Delaunay algorithm is applied and some triangles are obtained between entities as in the example in Fig. 1.

However, a filtering is required, because the purpose of this first stage is to create connections between two entities.

The triangles that connect one or three entities are eliminated (Fig. 2 shows an example of applied constrained Delaunay filtered as mentioned above) and so, the input for the next step is created.

Most of the characters will be connected to up to eight entities which are the two entities on the same row and three above and three bellow, and so, if we compute the ratio between the connected points on the current entity (further referred as *current points*) and the connected points on all other entities connected by triangles (further referred as *target points*) it will give us a relatively high result, around one.

However, the separators do not generate such a value. The fact that they are a special case of entities with respect to the letters or symbols also influences the outcome of different approaches applied on them. And so, their unique properties separate them from the symbols.

Fig. 2. Constrained Delaunay transformation of the initial image in which the triangles are created only between two entities.

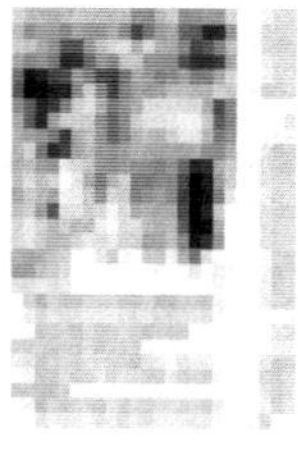

Fig. 3a. The page after resampling and resizing it with the triangle filter of fixed window.

Fig. 3b. The page resampled with the same filter to the initial size. The blur effect is visible (the reconstructed image).

Fig. 3c. A simple black and white transformation applied to the reconstructed image.

Because they extend to a few rows, are connected to far more entities and have far more connectable Delaunay points, the number of connected points on the separator will be significantly greater than the number of connected points on the other entities connected by triangles with the line.

2.2 How to find the zones

Another part of the solution is to find the zones based on a simple resampling method. The initial image is taken and downsampled using a triangle filter with a fixed very small window and the obtained result is a blurred version of the initial image at a very small dimension (Fig. 3a).

Next, the image is upsampled by using the same filter and brought back to the initial size. By doing this operation, the size of the picture was maintained, but the content inside has been changed, the new image being a blurred version of the initial one (Fig. 3b).

The difference between the small version of the image after applying the downsampling and the reconstructed image created by upsampling is noticeable. Even though both of them seem to be a blurred image of the initial, the first one is actually a compact version. At a closer look on Fig. 3a the pixel can be seen entirely and their colors can be noticed without any zooming effects (see Fig. 4).

The fact that the second image, Fig. 3b, is blurred is a result of successively applying the filter. A simple transformation of the initial image (see: Pratt, 2002; Mann, 2002; for image processing) would not have given such a good effect and zones that should have been united would have been left singular.

On the blurred image, a black and white transformation based on a simple centre-thresholding technique is applied and so, the result from Fig. 3c is obtained. The goal is to obtain white pixels in almost white areas.

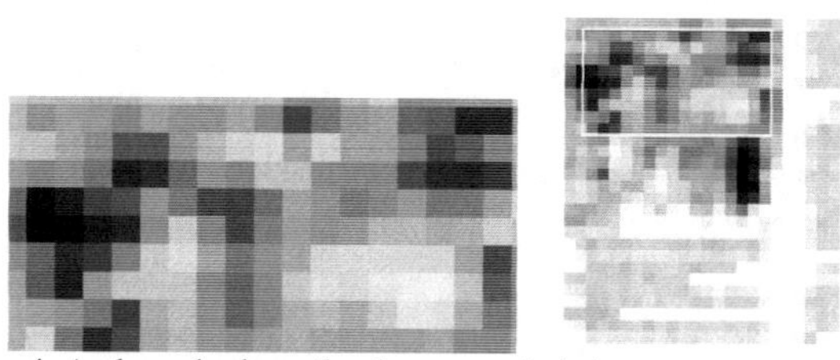

Fig. 4. A closer look on the downsampled picture.

2.3. Correct zone detection

The black and white image provides an overview of the possible join probabilities of different zones.

The effect of the resampling before the actual conversion is to emphasize the white spaces with respect to the text or other elements, from which they can be clearly differentiated. This approach favors the large spaces that have a continuous distribution in the page.

Fig. 5a. The zones are the parts which have all the elements connected by Delaunay triangles.

Fig. 5b. After detecting the separators and removing the connections to them, the zones are correctly found.

This last concept of distribution does not refer to the spaces that are bounded nor have very large dimensions on both axes between the bounds.

As an example, in the obtained black and white image it can be clearly observed that text from the right is not in the same zone as the text from the middle or the picture and so, a difference is determined (Fig. 5a). However, the picture is also clearly different from the text in the middle. And here is where the Delaunay separator detection algorithm appears.

If all the separators on a page are known then all the connections to the separator are deleted and so the zones are differentiated. And in this case the Delaunay algorithm determines the image as a separator, because of the high ratio between *source* and *destination points*. The final and correct result is displayed in Fig. 5b.

3. CONCLUSION

There have been many approaches used for page clustering, but the presented algorithm brings a new view on this topic by starting with the detection of the basic and also most important elements in a document page: the separators. By using a combination of mathematical algorithms and common knowledge on scanned document pages, this approach yields good results with a high degree of certainty.

Because of its innovative character further improvements will be made and it can open new research paths for the future.

4. REFERENCES

Fortune, S. (1997). Voronoi Diagrams and Delaunay triangulations, In: *Handbook of discrete and computational geometry,* pp. 377-388, CRC Press , ISBN:0-8493-8524-5, Boca Raton, Florida

Guru, D. S.; Shekar, B. H. & Nagabhushan P. (2004). A simple and robust line detection algorithm based on small eigenvalues analysis, *Pattern recognition Letters,* Elsevier Science, Issue 1, Vol. 25, (January 2004) pp. 1-13, ISSN: 0167-8655

Mann, S. (2002). *Intelligent Image Processing,* John Wiley & Sons, ISBN: 9780471221630, United States of America

Pratt, W. K. (2002). *Digital Image Processing,* John Wiley & Sons, ISBN: 0-471-85766-1, United States of America

Shewchuck, J. R. (2002). Delaunay refinement algorithms for triangular mesh generation, *Computational Geometry: Theory and Applications,* Vol. 22, No. 1, pp. 21-74, May 2002

Xiao Y. & Yan H. (2004). Location of title and author regions in document images based on the Delaunay triangulation, *Image and Vision Computing,* Vol. 2, No. 4, April 2004

Annals of DAAAM for 2008 & Proceedings of the 19th International DAAAM Symposium, ISSN 1726-9679
Editor B. Katalinic, Published by DAAAM International, Vienna, Austria 2008
Make Harmony Between Technology and Nature, and Your Mind will Fly Free as a Bird

MODERN PREPROCESSING TECHNIQUES FOR AUTOMATIC CONTENT CONVERSION SYSTEMS

BOIANGIU, C[ostin] A[nton] & DVORNIC, A[ndrei] I[ulian]

Abstract: *In this paper we address the preprocessing problem of both scanned and electronic documents as an initial stage before document content analysis. Different approaches are presented for each type of input images. For scanned books and newspapers a new multi-threshold technique is proposed as a solution for noise interferences. Apart from that, for black and white conversion, we suggest a new method consisting of a contrast enhancement step, followed by a conversion based on adaptive levelling of the document. Finally, three algorithms aimed at bitonal transformation of electronic documents are presented. Experimental results are given in order to verify the effectiveness of the proposed techniques.*

Key words: *content analysis, conversion, bitonal, denoising*

1. INTRODUCTION

Document image analysis has increasingly become a very active research area due to the need of both converting old documents to digital media and acquiring useful and on-time information in electronic and searchable format. To this goal, faster and better methods for automatic content conversion need to be developed as variations in layout, orientation, size, quality and characteristics of printed documents (both in real and electronic form) made this task a very complicated one. In order to cope with all these aspects, latest developments in this domain require image binarization before the actual features extraction in order to reduce the computational load and simplify the analysis methods (Chang, 2001). As a result, in this paper we focus on proposing a series of conversion techniques for different types of documents, as well as some preprocessing algorithms which can increase the quality of bitonal transformations.

Global thresholding and adaptive binarization (Sheikh et al., 2005) have proven to be the most popular binarization methods so far. Unfortunately, apart from having their limitations (it is very hard to find a global threshold or a predefined window size of analysis for all types of documents), this techniques are not addressing critical issues specific to electronic documents (detecting text inside images or uneven background, converting areas where both the text and background is multi-coloured, etc.) and hence are not a viable solution for this type of conversions.

In this view, we propose three methods which are aimed at solving some of the problems areas on which actual methods fail.

2. CONTRAST PREPROCESSING TECHNIQUE

Bitonal conversion using contrast preprocessing is a two-stage method which tries to solve problems like large brightness variations and low contrast ratios in scanned documents. These issues, which are the direct result of both document degradation and poor calibration of automatic scanners, are requiring an adaptive local approach because any global thresholding attempt will most probably fail on some parts of the input document. The first stage of the algorithm is an auto-stretch contrast step which is performed using two succesive iterations through the input document: on the first one the contrast stretch bounds are computed and then the actual transformation is performed.

The stretch bounds are computed by appling a horizontal threshold on the intensity histogram(s) of the document. For grayscale images, the histogram is computed by taking into consideration the frequency of each pixel value (in the case of color documents individual histograms are considered for each color channel).The bounds of the longest horizontal segment cutting the histogram(s) at the threshold level are used as references for the contrast stretch. If some color indexes are missing from the histogram, a triangular filter is applied repeatedly before the longest segment estimation, until all the color values have a representation in the histogram.

From the efectiveness point of view, experimental results performed on scanned books and newspapers from British Library have shown that conversion of documents with good contrast ratio might not need this preprocessing step. The increase in precision during the conversion does not compensate for the time cost of such preliminary transformation. Apart from that, this technique increases the contrast ratio between the background and the foreground pixels without performing a contrast equalization. Hence if an effective black and white conversion is desired, the second stage of the binarization process must be a local one.

Taking all of the above into consideration, the second stage of the proposed technique is an actual binarization process, based on adaptive leveling of the preprocessed document. By this, we mean applying a threshold to the difference between the image obtained after the first stage (involving the stretch of the document's contrast) and a leveled version of it, obtained using one of the following three methods: Gaussian unsharp (Dobrescu et al., 2006), downsampling/upsampling (Fischer, 2000) or local contrast stretch. The basic idea behind this is that areas in the image which belong to the background will have intensity values below the leveled ones, whereas the pixels belonging to the foreground will have intensity values above them.

In the following example we show the results of all three leveling methods. In the case of downsampling/upsampling the following interpolation modes have been tested: Lanczos, Hermite, Triangle, Mitchell, Bell and B-Spline. The last one yield the best results.

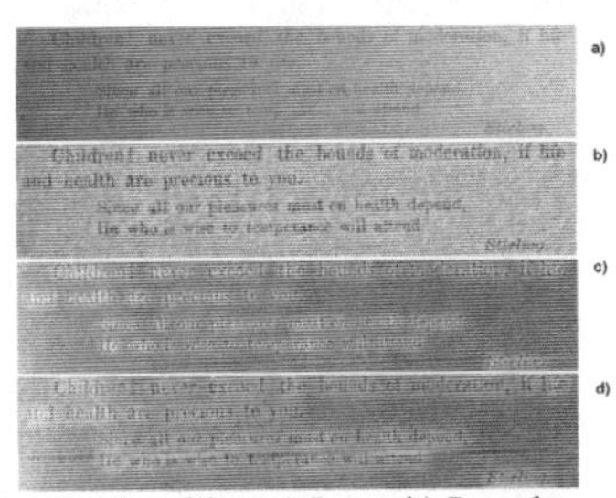

Fig. 1. Adaptive levelling a) Input b) Local contrast stretch c) Gaussian unsharp d) Donwsampling/upsampling

3. MULTI-THRESHOLD CONVERSION

Multi-threshold conversion is a binarization technique proper for documents in which the noise aspect is very sensitive. This includes both documents which have large background variations and documents where OCR detection of text needs to be very precise (Wenzel & Grigat 2005).

Four different intermediary threshold-based conversions are used in order to obtain the final black and white document: threshold applied to the grayscale transformation of the image; threshold applied to the K component of the CMYK color model; thresholding each component of the RGB color model with the minimum average intensity of channels and a variable thresholding technique based on the hue component. The least accurate intermediary conversion is considered to be that conversion which outputs the least number of black pixels. Starting from this, the proposed technique adds recursively undetected foreground pixels, provided that they are detected in one of the other intermediary images and are neighboring a foreground pixel which has been already detected. The final conversion is obtained when no more pixels can be classified as belonging to the foreground based on the proposed algorithm.

Experiments have been performed in order to compare multi-threshold method with both adaptive and global thresholding. It was noticed that the noise level was obviously lower when using the proposed method due to the threshold levels of the intermediary conversions which were considered such as to be "safe" (this means that an intermediary conversion would contain the least level of noise possible, even if some relevant information would be lost – this missing information would be recovered from the other intermediary versions). In this way the final black and white documents was obtained similar to a puzzle, using pieces from the previously computed intermediary images in which undesired pixels level was very low.

4. ELECTRONIC DOCUMENTS CONVERSION

In the case of electronic documents (like PDFs and online newspapers) the focus of attention during the binarization process is shifted from aspects like contrast or noise to translating the colour combinations of texts, images and backgrounds into a meaningful black and white version. To this goal three different approaches for black and white transformation of electronic documents have been tried.

The first method is the "outside-in" local technique. This algorithm starts from the outer white background and decides to convert everything which neighbours it black; then everything that neighbours the previously detected area is set to white and so on until the whole document is processed. This approach works better than the threshold-based global ones, avoiding the disappearance of letters printed with colours that are scarce in the picture. Apart from that, using this technique cancels the risk of errors generated by the same colour playing both the role of background and foreground in different parts of the document.

The "alternate colours" approach is also a local conversion method which tries to solve the binarization decision using horizontal scan lines. This iterative process takes each pixel at a time (on each individual horizontal line in the input document) and decides its final value based on the previous made decisions. Conversion to white or black is alternated each time a colour shift occurs. Whenever a white or black pixel is encountered in the original document, the output colour is automatically set to that value, no matter the alternation rule. This algorithm manages to solve some cases in which the "outside-in" method would fail, like bi-coloured areas in which the background and foreground are combinations of two distinct colours. A threshold is used in order decide when a colour shift has occurred and output colour must be changed.

Fig. 2. Conversion Results: a) Input; b) "Alternate-colour" method; c) "Outside-in" method; d) "Edge-detection" method

The third and final proposed method is an edge-detection based technique. This algorithm uses a threshold in order to decide if a pixel is part of an edge between foreground and background or not. In this way, the problems resulting from background variations can be reduced by adjusting the threshold. Research is still carried out in order to find a way to decide which of the detected contours are belonging to foreground and how they could be filled up with black pixels in order to generate a more relevant conversion.

5. CONCLUSIONS

Due to the increasing interest in document content conversion there is the need for new image binarization methods that can cope with problem areas in a large variety of documents. As the best technique to be chosen depends on the desired output and the set of images to be converted, we have tried to present in this paper a series of alternatives to current algorithms.

Further research must still be conducted in order to improve current methods, as well as develop new techniques that can cope with the new challenges of modern document analysis.

6. REFERENCES

Chang, F. (2001). Retrieving information from document images: problems and solutions, *International Journal on Document Analysis and Recognition*, Vol. 4, No.1, (August 2001) pp. 46-55, ISSN 1433-2833

Dobrescu, R.; Dobrescu, M.; Mocanu, S. & Taralunga, S. (2006). Development platform for parallel image processing, *Proceedings of the 6th WSEAS International Conference on Signal, Speech and Image Processing*, pp. 31-36, ISBN 123-4567-99-3, Portugal, September 2006, WSEAS Press, Lisbon

Fischer, S. (2000). *Digital Image Processing: Skewing and Thresholding*, Master of Science thesis, University of New South Wales, Sydney, Australia

Sheikh, L.M.; Hassan, I.; Sheikh, N.Z.; Bashir, R.A.; Khan, S.A. & Khan, S.S. (2005). An Adaptive Multi-Thresholding Technique for Binarization of Color Images, *Proceedings of the 9th WSEAS International Conference on Computers*, Article No. 104, ISBN 960-8457-29-7, Greece, July 2005, WSEAS Press, Athens

Wenzel, F. & Grigat, R.R. (2005). A Framework for Developing Image Processing Algorithms with Minimal Overhead, *Proceedings of the 5th WSEAS International Conference on Signal, Speech and Image Processing*, pp. 185-190, ISBN 940-4271-11-5, Greece, August 2005, WSEAS Press, Corfu Island

Annals of DAAAM for 2008 & Proceedings of the 19th International DAAAM Symposium, ISSN 1726-9679
Editor B. Katalinic, Published by DAAAM International, Vienna, Austria 2008
Make Harmony Between Technology and Nature, and Your Mind will Fly Free as a Bird

EFFECTS OF DATA FILTERING TECHNIQUES IN LINE DETECTION

BOIANGIU, C[ostin] A[nton] & RADUCANU, B[ogdan]

Abstract: *Line detection has a key role in any Automatic Content Conversion system. Such systems generate digital models and documents from scanned paper copies of newspapers, books, etc. During this process, one of the subtasks is the classification of regions from the input image as different document entities. For example groups of characters should be classified as paragraphs, whereas groups of aligned non-character elements should be classified as lines. The detection of lines is very helpful in the understanding of the structure of a parsed document: lines can separate columns in a newspaper page, or they can form a table whose structure is otherwise impossible to extract.*

Key words: *automatic content conversion, line detection, Hough Transform, document digitization, data filters*

1. INTRODUCTION

The Line Detection module is typically implemented at the software application level of an Automatic Content Conversion System. This module is being fed a black and white image which holds enough information to solve the considered task. The desired output is a collection of lines represented either by their analytical parameters or by a list of pixel collections (each made of a series of connected black pixels).

This task proves to be less straightforward than it might seem as lines are often imperfect, containing noises or deformations. The line detection module should be developed such that to be able accept a set of calibrating parameters (in order to help it cover the broad range of possible line types).

There are a number of approaches to the line detection problem. Among the most popular and researched are the ones based on voting, firstly introduced along with the Hough Transform method. Their efficiency lies in their ability to deal with noises (Duda & Hart, 1972).

2. THE HOUGH TRANSFORM

When using the Hough Transform, lines are viewed as a pair of polar parameters, (ρ,θ). A line equation in terms of these parameters would be (Princen et al., 1992):

$$\rho = y*\sin\theta + x*\cos\theta \qquad (1)$$

Every point in the input image counts as a vote for each line it may lay on. In the generated parameter space (or accumulator) lines are identified either as a maximum or by applying a threshold (Sklanksy, 1978).

3. DISCRETE LINE PARAMETERIZATION

Following the approach of the Hough Transform, we have developed a system which employs a different parameterization in order to achieve better results for our particular targets. The parameterization of a line is set in terms of the pair (x_0, dx) and called Discrete Parameterization - it was designed to resemble a digital line, as drawn onto a discrete memory space.

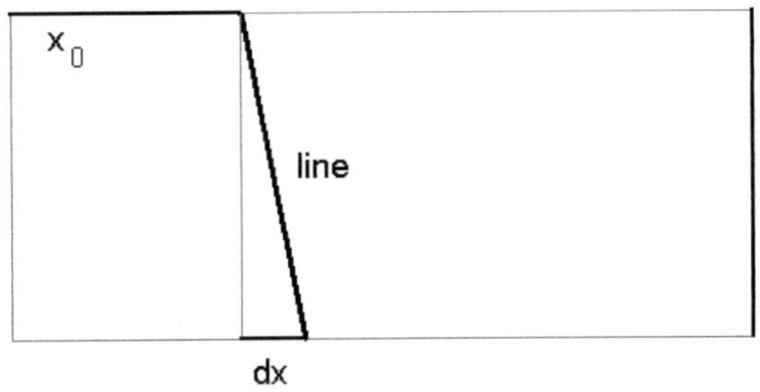

Fig. 1. Discrete Line Parameterization

In an image with height *h*, a line described by (x_0, dx) passes through the points $(x_0, 0)$ and $(x_0 + dx, h)$.

Retrieving the lines from the voting space is a fundamental problem and requires a well thought approach (Niblack & Petkovic, 1990).

When working with scanned newspaper or book pages, the accumulator is flooded with votes from the large contents of characters. It is important to find a stable threshold mechanism so characters do not yield many false lines.

4. PARAMETER SPACES

Looking at the voting space as a greyscale image (brighter pixels correspond to higher voting counts) we obtain a visual of the voting process.

One way to extract information from the two spaces is to find points with values above a threshold. Lines imply a high number of votes, thus they will pass the threshold. The problem is that characters are responsible for the majority of votes. The difference between a false line point and a correct line point, in terms of votes, could be very small (5%) and it is impossible to establish a threshold in this case.

A second approach is to analyze local maxima points. Characters participate with high probability with the same number of votes to a set of lines so they will not influence local maxima. The problem with this approach is that the parameter space is constructed point by point and this heavily affects the continuous nature of local maxima.

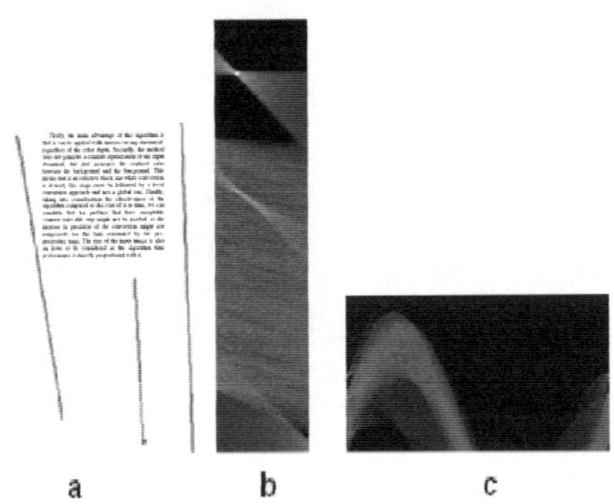

Fig. 2. a) Input Image, b) Discrete Space, c) Hough space

5. FILTERING THE PARAMETER SPACES

Both problems presented above can be solved by using resampling filters. Such a filter modifies the points in the parameter space and assigns them new values based on a weighted sum their neighbours' values. The filters have a macro scale effect of blurring or smoothing the space, which is the exact solution for the local maxima sweep problem. Particularly useful for this is the Triangle filter (Umbaugh, 2005).

$$T(t)=\begin{cases}1-|t|,|t|<1\\0,otherwise\end{cases} \quad (2)$$

Equation (2) shows a one dimensional Triangle filter. To apply this to the two dimensional parameter space, we first apply it for every row and obtain a preliminary array. Finally we apply the filter again for each column, thus obtaining a 2D effect. This approach is only possible for symmetrical filters such as this one.

The Triangle Filter gives higher weights to points closer to the origin point and lower values as points are further apart, thus real local maxima are enhanced while false maxima are decimated.

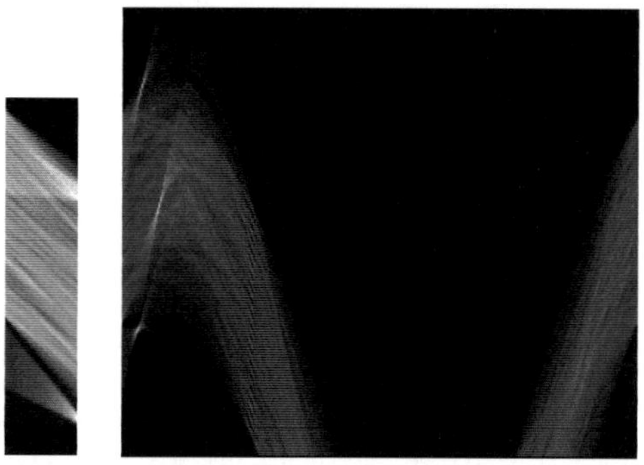

a b

Fig. 3. Triangle Filtering, a) Discrete Space, b) Hough Space

The parameter spaces in Fig. 3. have been filtered with a Triangle Filter. Correct local maxima (represented by bright pixels) are more obvious now, while noise generated maxima disappeared.

To solve the thresholding problem, a different, more complex, filter is adopted. This is the Lanczos scaling filter.

$$L(t)=\begin{cases}\sin c(t)*\sin c(\frac{t}{a}),-a<t<a\\0,otherwise\end{cases} \quad (3)$$

Equation (3) represents a one dimensional Lanczos filter. It is applied to the two dimensional parameter space first row-wise then column-wise.

The Lanczos filter is typically used for scaling images because it has the property that it retains protruding aspects of the original image, producing results of higher quality.

To diminish the effects of the large number of characters on the voting space, we first apply the Lanczos filter on the space and then we subtract the original space from it. This is called a Negative filter effect and the results are shown in Fig. 4.

Before applying this filter, the parameter space was heavily composed of the votes generated by characters. Lines were difficult to detect because their vote count was negligible higher than the average.

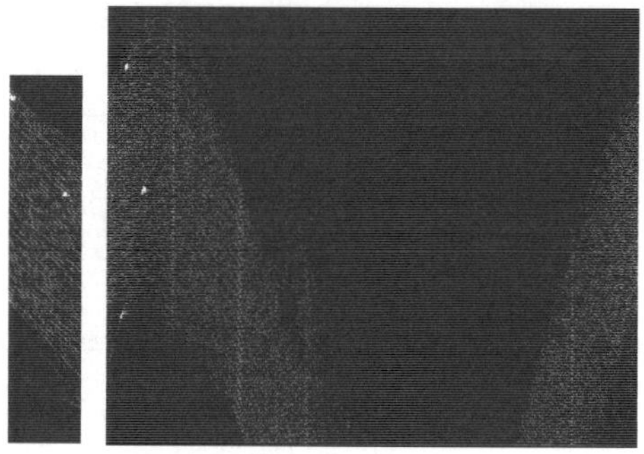

a b

Fig. 4. Negative Lanczos Filtering, a) Discrete Space, b) Hough Space

The filter has overcome this problem and in the resulting space, the lines have become more obvious. A thresholding scheme can now be applied with greater stability as the range of possible values has been greatly decreased.

Points with high vote counts are now more visible, in the sense that they strongly disperse from the vote average. This effect is caused by the special quality of the Lanczos filter to normalize high frequency values while enhancing peak values.

These examples show how filtering the parameter spaces can enhance particular characteristics and turn out helpful in determining the intended points.

Depending on the desired characteristics, other filters can be used. A median filter can be used first to diminish noises propagated in the parameter space.

To further enhance local maxima, a Gauss convolution filter may be applied. For larger kernel dimensions, a Gauss filter can prove more accurate than the Triangle filter.

6. CONCLUSIONS

When working with digital signals, such as images or sounds, there is no stable way of extracting certain characteristics like maximums, frequencies or gradients. Hence, an important preprocessing is the naturalization of the signal which leads to resampling of the discrete data in order to conform to a continuous domain.

In this paper we have shown how resampling an input image, considered as a two dimension array of data, can enhance its characteristics and improve the computation of useful information about it.

Statistical methods, like the ones proposed in this paper, are a powerful method of solving difficult tasks without a deterministic solution. Hence, these filtering techniques may prove useful in determining a good approximating solution for a problem in any domain.

7. REFERENCES

Duda, R. & Hart, P. (1972). Use of the Hough Transformation to Detect Lines and Curves in Pictures, *Comm. ACM*, Vol. 15, Issue 1, (January 1972) pp. 11–15, ISSN: 0001-0782

Niblack, W. & Petkovic, D. (1990). On improving the accuracy of the Hough Transform, *Mach. Vision Appl,*. Vol. 3, Issue 2, (March 1990) pp. 87–106, ISSN: 0932-8092

Princen, J.; Illingworth, J. & Kittler, J. (1992). A Formal Definition of the Hough Transform: Properties and Relationships, *JMIV*, Vol. 1, No. 2, (July 1992) pp. 153-168, ISSN: 0924-9907

Sklanksy, J. (1978). On the Hough Technique for Curve Detection, *IEEE Transactions on Computers,* Vol. 27, Issue 10, (October 1978) pp. 923-926, ISSN: 0018-9340

Umbaugh, S. E. (2005). *Digital Image Analysis and Processing*, CRC, ISBN: 0849329191

Annals of DAAAM for 2008 & Proceedings of the 19th International DAAAM Symposium, ISSN 1726-9679
Editor B. Katalinic, Published by DAAAM International, Vienna, Austria 2008
Make Harmony Between Technology and Nature, and Your Mind will Fly Free as a Bird

THE "BETA-STAR SHAPE" ALGORITHM FOR DOCUMENT CLIPPING

BOIANGIU, C[ostin] A[nton]; BUCUR, I[on] & DVORNIC, A[ndrei] I[ulian]

Abstract: *An important problem in many content conversion applications is reconstruction (as precise as possible) of a polygonal contour bounding a given set of input points. The basic scenario is the following: considering a set of planar points placed in a certain area, there is the need to construct a polygon that encapsulates all the input points and reproduces the area's shape as fairly as possible. The main usage of this polygon is to detach/clip the inside polygon area from the plane, or to separate multiple areas by non-overlapping polygons.*
Key words: *alpha shape, convex hull, tight fitting, contour*

1. INTRODUCTION

There are several theories and techniques used to solve the tight-fitting polygonal contour generation for a set of points defined in a certain space (Edelsbrunner, 1987). One of the most efficient, and probably the most used, is the theory of "Alpha-Shapes". For a good understanding of what this concept means, let us consider a very simple and intuitive example: imagine that all input points from a certain plane are the heads of some nails fitted into a wooden panel. Consider now a ball of radius *R* that is rolling over the exterior of the "nailed surface" and must keep the same rolling direction at all times. Suppose the direction is clockwise-oriented. It is obvious that the ball will enter between nails that are positioned at a distance $d >= R$ and will keep rolling over the ones situated at a distance $d < R$. After the ball has performed a full rotation and reached the initial position, the nails heads "touched" by the ball in the full-rotation process will represent a non-self-intersected polygon, clockwise oriented (as previously supposed). *This polygon is the R -Radius "Alpha-Shape" for the given input set of points.* Most of the algorithms that compute "alpha-shapes" for a set of input points are based on Delaunay (Aurenhammer, 1991) triangulation process, an elegant but not very economical approach. Among the disadvantages of this theory is the fact that points in the plane cannot be controlled with respect to the distances between them. For this reason, generating the "alpha-shape" using a small radius can yield a polygon that does not contain all the input points. On the other hand, if the radius is big enough to ensure only one output polygon, the "alpha-shape" may not reproduce the contour very accurate, a coarse polygon being practically unusable in a demanding multimedia application.

2. MOTIVATION

Research aimed at constructing polygonal contours that approximate sets of points started with the development of an application that was separating various items belonging to an input written page (or scanned document). The goal was to detect elements like normal texts, headings, pictures, etc. and separate them using tight-fitting contours which don't overlap (in order to successfully execute selective clipping). Since the distance between adjacent items or elements that compose these items (like text characters) is not controllable and only the shape that bounds each item is needed, the usage of "alpha-shapes" was not an option. In this paper we focus on a new technique that has been developed: the "Beta-Star Shape" - a robust algorithm that has a controllable behaviour and creates only one shape for an input set of points.

3. ALGORITHM OUTLINE

An arbitrary collection of planar points, defined by their coordinates (x, y), is considered as the input for the algorithm. The proposed technique is based on refining the convex hull of the considered points in successive iterations. Hence, the computation of the convex hull is the first step of the proposed algorithm and can be performed using any well-known robust method (Avis & Bremner, 1995). At this stage the convex hull is the starting "Beta-Star Shape" polygon that will be successively refined. The points that are not part of the convex hull are called *candidate points* or simple *candidates*. The algorithm specifies the way to insert candidates into the "Beta-Star Shape" in order to refine it (preserving the precondition that it contains all the input points). The algorithm is controllable both as tightness of the shape and execution time (it can be stopped whenever the result is refined enough for the application purposes). Choosing candidates to be inserted in the "Beta-Star Shape" is done such that it allows a uniform smoothing of the contour and avoids self-intersections. For this reasons every iteration step is in fact a usable output polygonal contour that verifies all the conditions needed by an application.

For constructing the "Beta-Star Shape", a distance parameter (named d) is needed; this is somehow equivalent with the R parameter used in the description of the "Alpha-Shape". If two adjacent points from the constructed shape in a certain moment are placed at a distance lower than *d*, then the segment will be finally integrated into the "Beta-Star Shape" and will suffer no further refinements. Otherwise, the process of successive refinements will continue to try splitting the segment by finding the best candidate for it and inserting it into the chain of points.

For a good search process the algorithm uses square search areas, because they do not encourage the usage of potentially better candidates for other edges and they restrict the search to a much lower subset of the input dataset. If one candidate is closer to an edge than any other candidate and the insertion of that candidate in the polygonal chain does not generate self-intersections then it is considered the best candidate for that edge. Having the lowest distance ensures the fact that all the other points in the input dataset remain inside the polygonal

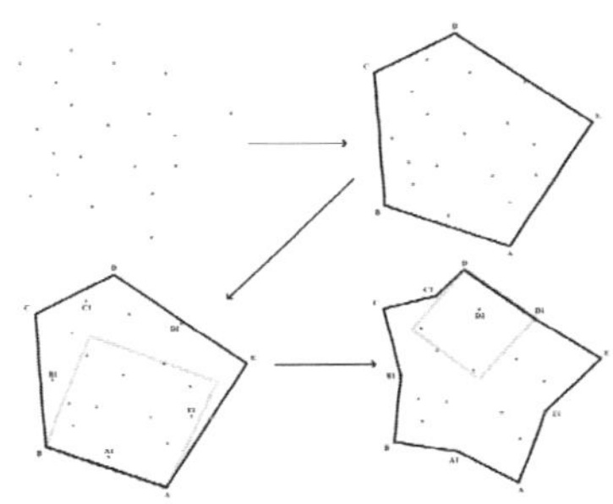

Fig. 1. "Beta-Star Shape" process outline

"Beta-Star Shape" contour. If no successful candidate is to be found in a certain area then all of the other remaining candidates are taken into account.

In order to check if a point is inside a certain search area, the *(a, b, c)* coefficients that define the support line of the edge to be refined are computed based on the coordinates of the two endpoints of the line $P_1(x_1, y_1)$ and $P_2(x_2, y_2)$:

$$a = y_2 - y_1;\quad b = x_1 - x_2;\quad c = -a \cdot x_1 - b \cdot y_1 \tag{1}$$

The parameters for the edges that are perpendiculars onto P_1P_2 are:

$$\begin{aligned} a_1 &= a_2 = b;\quad b_1 = b_2 = -a \\ c_1 &= -b \cdot x_1 + a \cdot y_1;\quad c_2 = -b \cdot x_2 + a \cdot y_2 \end{aligned} \tag{2}$$

Taking into account the results from equations (1) and (2) and assuming a clockwise orientation of the convex hull resulted polygon (the orientation of the polygon cannot change after any refinement that has been applied) then the conditions for a valid candidate are:

$$\begin{aligned} &b \cdot x - a \cdot y + c_1 >= 0;\quad b \cdot x - a \cdot y + c_2 <= 0 \\ &a \cdot x + b \cdot y + c <= EdgeLength;\quad a \cdot x + b \cdot y + c >= 0 \end{aligned} \tag{3}$$

While the first two conditions regard the position of the point *(x, y)* inside the two perpendicular axes, the third condition means that the "depth" of the point is lower than the edge length (square search area); the forth inequality is a precondition that must be verified for all points in the search dataset in order to ensure that the polygon has the correct orientation and that points are inside the "Beta-Star Shape" at any moment. This specific search mode inside a square area defined by the current edge allows a good trade between speed and quality of the output polygon. Apart from that, it makes local use of the input data, thus enabling fast-recovery of the search point domain by means of a simple two-dimensional hash-table. The algorithm searches the best candidate for every edge, until all edges have potential candidates or there are no more candidates to be considered. Using the previously computed distances, the proposed method takes them one by one in ascending order and merges the candidates (denoted A1, B1, C1, D1 and E1 in Fig. 1.) into the "Beta-Star Shape" chain. None of the chosen candidates generate self-intersections; hence they can be safely inserted.

The algorithm continues by considering new search areas until all remaining candidates are generating self-intersection (internal status of "deadlock") or all the edges have lengths lower than *d*. The refinement of the shape is not continued if there are no "obstacle" points contained in the current shape as opposite to the *standard "Beta-Shape"* (Boiangiu, 2003). This allows the time consumed for the refinement process to be reduced to the maximum acceptable. We call "obstacle" points all the points that should not be part of the current points set collection (that are not desired in the resulted shape).

Fig. 2. Results of the "Beta-Star Shape"

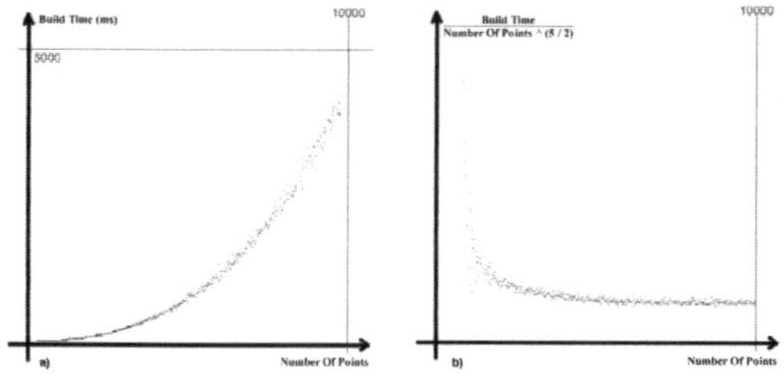

Fig. 3. "Beta-Star Shape" performance a) time vs. dataset input size b) complexity for random input data set

4. CONCLUSIONS

The proposed algorithm has been implemented and tested in order to prove its robustness and make various performance measurements.

The "Beta-Star Shape" algorithm has a variable complexity of about $O(n^{5/2})$ for random input data and about $O(n^{3/2})$ for data sets generated by real images. Fig. 3. shows some statistics about how the input time is affected by the number of points presented in the input dataset. The results are plotted without performing an average in order to enable measuring of time deviations (1000 time-measure tests have been conducted on random point datasets).

To sum up, the "Beta-Star Shape" algorithm:

- can be controlled as time vs. refinement quality using the *d* parameter.
- is robust and generates a single output polygon for any input dataset.
- is fully valid at every step (the intermediate result); hence it is possible to trade between execution time and refinement quality and also stop whenever the output polygon satisfies application needs and/or the allocated time for refinement is consumed.
- allows adaptive refinement since only "obstacle" points are avoided during shape update comparing with the *standard "Beta-Shape"* (Boiangiu, 2003).
- does not need much auxiliary memory; practically all the needed memory space is *O(n)*, since the final "Beta-Star Shape" is a subset of the input dataset.
- has a clear end since every step performed consumes at least one candidate from the input dataset.
- always generates an output polygon containing the convex hull points.
- can be easily adapted for the 3-dimensional (and even for the n-dimensional) space since the only concepts used are: the convex-hull (de Berg et al., 1987) and space separation equations, which in the 2-dimensional case are the line equations.

5. REFERENCES

Aurenhammer, F. (1991). Voronoi diagrams - A survey of a fundamental geometric data structure, *ACM Computing Survey*, Volume 23, Issue 3, September 1991, pp. 345-405

Avis, D. & Bremner, D. (1995). How good are convex hull algorithms? *Proceedings 11th A.C.M. Symposium on Computational Geometry*, June 1995, pp. 20-28, Vancouver, Canada

Boiangiu, C.A. (2003) The "Beta-Shape" Algorithm for polygonal contour reconstruction, *The 14th International Conference on Control System and Computer Science*, Vol. II, July 2003, pp. 29-33

De Berg, M.; Otfried, C.; van Kreveld, M. & Overmars, M. (2008). *Computational Geometry: Algorithms and Applications*, Springer-Verlag

Edelsbrunner, H. (1987). *Algorithms in Combinatorial Geometry*, Springer-Verlag, Berlin

Annals of DAAAM for 2008 & Proceedings of the 19th International DAAAM Symposium, ISSN 1726-9679
Editor B. Katalinic, Published by DAAAM International, Vienna, Austria 2008
Make Harmony Between Technology and Nature, and Your Mind will Fly Free as a Bird

STATISTICAL APPROACHES USED IN AUTOMATIC MERGE OF SCANNED IMAGES

BOIANGIU, C[ostin] A[nton]; BUCUR, I[on] & SPATARU, A[ndrei] C[ristian]

Abstract: *This paper describes methods used for the purpose of image merging, required when scanning a page that is bigger in size than the scanner. The image merge scenarios differ significantly, so a number of algorithms are presented, taking into consideration the geometrical attributes of detected page elements, either independently or locally.*
Key words: *image merge, image conversion, image analysis, image scan, text characteristics, page overlap*

1. INTRODUCTION

As digital libraries worldwide tend to expand significantly in size, and at a high rate, the need for document digitization and content extraction has been ever more present in the last years. The emphasis is placed on the speed and quality of the image digitization process, however satisfactory end-results cannot be obtained without a thorough pre-processing phase of the input (Baird, 2003), as in the case of large newspaper page scans that do not completely fit the scanner, or magazines that have a headline or paragraph composition spread logically over two consecutive pages. When working with such documents, a given number of logically connected page elements will appear on different scanned images, making the content extraction and conversion increasingly difficult. Hence, a solution must be found in order to perform a connection between these images.

2. INPUT SCENARIOS

While performing image analysis on a wide array of scanned images, two main scenarios were isolated. The first situation can be described as a page scanned in two parts, these parts having common elements, as exemplified in Fig. 1 and Fig. 2.

As it can be observed, the scanned documents in Fig. 1 and Fig. 2 contain common elements, and in following, this scenario will be referred to as "overlapping" input images. This case of overlapping images can be encountered in pages having either a Top-Bottom or a Left-Right displacement.

The second situation is encountered when the two parts of the document have no common elements. This situation will be referred to in following as "non-overlapping" input images, and can be seen in Fig. 3.

Because the most common scenario of non-overlapping input images is found in magazine articles, these are generally left-right separated. Also, in Fig. 3, the scanned images are skewed, an issue that will have to be addressed in order to accomplish the image merging task.

Eine Symbolfigur der Luftfahrt
Vor 150 Jahren wurde Otto Lilienthal geboren – Gebrüder Wright bauten auf seinen Erkenntnissen auf

Fig. 1. Top half of a newspaper page (top-bottom separated)

Fig. 2. Bottom half of the same newspaper page

CDU beschwört
„Jobmaschine"
Türkische Polizisten oh

Fig. 3. Non-overlapping sections of a newspaper page (left-right separated)

3. IMAGE MERGING SOLUTIONS

The image merging algorithm will be composed of several steps: first, the input images will undergo a pre-processing stage, comprising a transformation from a given colour space into Black & White, and an extraction of connected foreground (black) pixels, which will be referred to in following as "Entities".

After this step, a specific routine will be run, according to the input scenario (overlapping or non-overlapping). The steps are presented in detail in the following sub-sections.

3.1 Pre-processing stage:

The actual Black & White conversion algorithms are beyond the scope of this paper, so further attention will not be given to this subject. However, a simple threshold method can be easily implemented in order to obtain a binary image. This is also the case of Entities extraction, where a run-length algorithm can accomplish the task. For further reading, algorithms for Black & White image conversion can be found in (Di Zenzo et al., 1996), while run-length connected pixels extraction algorithms are available in (Bovik, 2000).

3.2 Overlapping input images:

In the case of overlapping input images, a "comparison band" is chosen on each image, at a safety ratio of half of the scanned document. For example, for the Top-Bottom case, the lower half of the Top part and the upper half of the Bottom part are chosen. Inside these bands, random entities are selected, and a geometrical comparison is made between entities in the upper and the lower comparison bands, until a positive match is found. When a sufficient number of entity matches is achieved, the algorithm considers segments between the matching entities in both bands, and searches for the best rotation-translation function that can superimpose corresponding segments from the bands. This resulting function represents the transformation that has to be made in order to correctly merge overlapping images, and has the output parameters: x-axis translation (t_x), y-axis translation (t_y) and rotation angle (α).

The criteria by which the entities are compared are of geometrical nature: area of the bounding rectangle, bounding rectangle fill ratio, number of black pixels.

The values returned by the function described above are placed in three histograms (t_x, t_y, α). These histograms are then filtered, using a triangle window filter of size 7, where the peak is 100% of the initial value, decreasing by 25% for each neighbouring value. The algorithm then tries to match the two images according to all plausible height differences of peaks on the histograms, choosing the transformation that obtains the smallest difference in height of the peaks. Once these measurements are done on all three types of histograms, the rotation-translation function is computed.

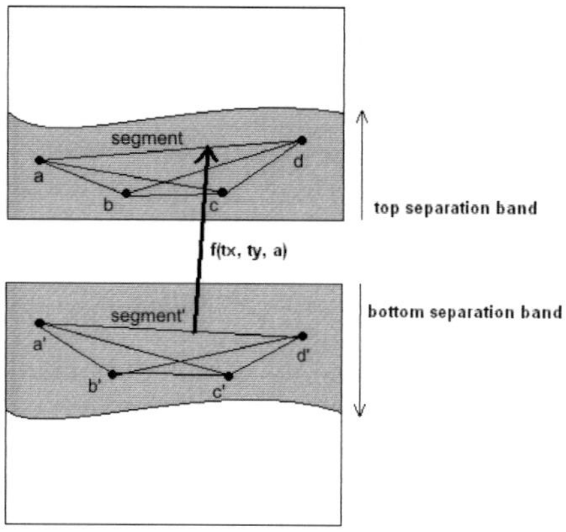

Fig. 4. Graphical representation of the algorithm, where (a, b, c, d) are entities matching (a', b', c', d') respectively, with the segments that connect them.

3.3 Non-overlapping input images:

The first step of the algorithm for merging non-overlapping images is the detection of text lines (Zheng et al., 2003) from an input array of Entities, and comparing the elements of the two sections at the level of text lines.

As the input images may be skewed when scanned, a de-skewing algorithm (Yuan & Tan, 2003) has to be applied independently on the input images, as to have the text lines horizontal in the output. This will lead to a better "fit" of the two page portions after applying the third and final step.

The third step is obtaining a "Text Characteristic" measurement for each detected text line, and, based on this comparison, the algorithm finds the translation function (only on the y axis) that best fits the lines in the two input images.

The "Text Characteristic" measurement is composed of a number of routines that detect parameters related to the geometry of the text. The main stages of the "Text Characteristics" extraction are:

- Filtering of entities: in order to take into consideration only relevant entities (excluding noise, punctuation marks, etc.), the input entity array is filtered. These filters join broken parts of a letter, eliminate separator lines and punctuation marks.
- Font Size measurement: the heights of the bounding boxes of the entities are placed in a histogram upon which a triangle filter is applied. The histogram will present three distinct peaks, the first (lowest heights) representing noise or various punctuation marks that have not been eliminated by the filters, the second being the small caps of letters, and the third – the high caps of letters. The last two peaks will be considered, representing the font size of the text.
- Boldness measurement: the boldness, or pen width, of the text is measured by finding, for each pixel inside a letter, the longest vertical and horizontal (black) segments it belongs to. The smaller of these two values is placed in a histogram, and the highest peak on this histogram represents the boldness.
- Italics measurement: in order to check if a letter is italic, the width of the bounding rectangle of the letter is compared to the width of the rotated letter's bounding rectangle. The rotation is done by -16 and the +16 degrees (considered a common value for the italic characters' slant angle). The decision is taken by applying the rule: if the width of the bounding rectangle increases after the rotation with +16 degrees but decreases when rotating by -16 degrees, that letter is italic.

By applying all the above measurements, "Text Characteristics" values are assigned to each text line in both images. The images are compared line by line in the following way: costs are computed for the differences in characteristics between a line in the first image and a line in the second one, and a match is found between lines with the lowest cost between them. When lines are matched from the characteristics point of view, the algorithm finds the y-translation value necessary to match these lines geometrically.

4. CONCLUSIONS

The approach presented in this paper is based on a geometrical statistic, and, due to its' nature, is of high confidence. In the case of overlapping input images, geometrical measurements are obtained and used in order to search for corresponding elements in the separation bands of the two images. When the input images are non-overlapping, the algorithm searches for matching characteristics of the text, relying on higher-order geometrical computations. By using the methods presented in this paper, the task of image merging can be accomplished in most scenarios, as tests have shown. In both cases, a singular output image is obtained, serving as a better starting point for content extraction procedures. As a consequence, the digital content resulted from these images is closer to the physical documents, thus reducing or eliminating altogether the need for manual corrections.

5. REFERENCES

Baird, H.S. (2003). Digital Libraries and Document Image Analysis, *Proceedings of the Seventh International Conference on Document Analysis and Recognition,* Vol. 1, pp. 2-15, ISBN 0-7695-1960-1, Scotland, August 2003, Edinburgh

Bovik, A. (2000). *Handbook of Video and Image Processing*, Academic Press, ISBN-10: 0-12-119792-1

Di Zenzo, S; Cinque, L. & Levialdi, S. (1996). Run-Based Algorithms for Binary Image Analysis and Processing, *IEEE Transactions on Pattern Analysis and Machine Intelligence,* Vol. 18, No. 1, January 1996, pp. 83-89, ISSN: 0162-8828.

Yuan, B. & Tan, C.L. (2003). Skewscope: The Textual Document Skew Detector, *Proceedings of the Seventh International Conference on Document Analysis and Recognition,* Vol. 1, pp. 49-53, ISBN 0-7695-1960-1, Scotland, August 2003, Edinburgh

Zheng, Y; Li, H. & Doermann, D (2003). A Model-based Line Detection Algorithm in Documents, *Proceedings of the Seventh International Conference on Document Analysis and Recognition,* Vol. 1, pp. 44-48, ISBN 0-7695-1960-1, Scotland, August 2003, Edinburgh

Annals of DAAAM for 2008 & Proceedings of the 19th International DAAAM Symposium, ISSN 1726-9679
Editor B. Katalinic, Published by DAAAM International, Vienna, Austria 2008
Make Harmony Between Technology and Nature, and Your Mind will Fly Free as a Bird

WHITE-SPACE DETECTION TECHNIQUES BASED ON NEIGHBOURHOOD DISTANCE MEASUREMENTS

BOIANGIU, C[ostin] A[nton]; CANANAU, D[an] C[ristian] & DVORNIC, A[ndrei] I[ulian]

Abstract: *The purpose of this paper is to present a model for white space separators detection by using simple mathematical algorithms based on distances or font measure. The algorithms present new features as they are independent of the position in page, but they start from common ideas and improve them in order to obtain a satisfactory output.*
Key words: *white-spaces, neighbourhood distance, crosshair, text areas, text rows.*

1. INTRODUCTION

The idea of processing an electronic document layout is to determine the areas of interest, such as text areas, pictures, tables or many other collections of entities gathered as one based on simple mathematical conditions or characteristics (Di Zenzo et al., 1996).

In this paper four algorithms are provided for solving the detection problem. All four are based on measurements of various distances (Breuel, 2003) in the document. The most important thing provided by the algorithm is the fact that is independent of the position in page.

2. THE SOLUTION

2.1 How to find the separators – A font size approach

First of all the white space separators surround the elements of the page such as text areas or pictures. These contain white spaces inside them, but they cannot be regarded as separators and a selection must be made (Saitoh et al., 1994).

This is the starting point for the first algorithm. It centres each pixel inside a square with a given size and checks for the percentage of white pixels inside the selected geometric figure. Then, if the percentage is close to 100% it means that the pixel is inside a white space, which will be surely a separator for letters, lines or areas. This approach will determine the spaces between lines and areas, but the most important thing is to determine the correct size for the square in order to remove undesired output. An example of the algorithm applied on a newspaper page can be seen in figure 2. As it can be seen, all the white spaces were found, using a search square of a fixed size.

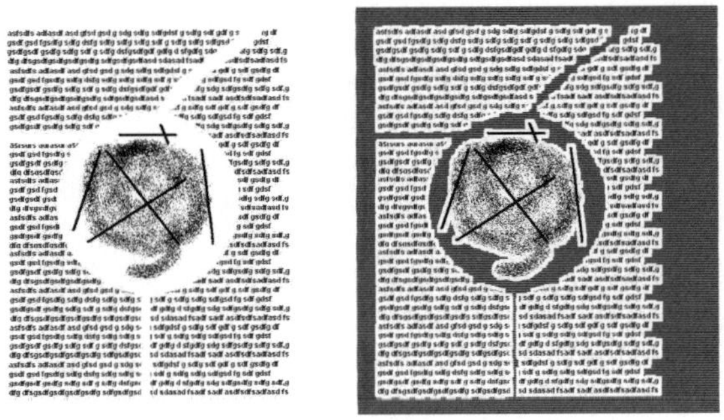

Fig. 1. Correct detection using font size

So, it is clear that a better measure has to be found. First of all, the selection of the output has to satisfy a number of criteria. It needs to eliminate the space between letters and text lines and the most appropriate measure for this task is the font size (Mann, 2002).

In order to have a good result, the algorithm has to be applied either on a document with the same font size on the whole page, or to apply it on areas. The results can be seen in Fig. 1.

2.2 How to find the separators – The distances approach

Distance map:

- A distance value for each pixel;
- If p is a white pixel, the distance value equals the number of pixels between the closest two black pixels on a specified direction;
- If p is a black pixel, the distance value is 0;

Types of distance maps:

- Horizontal distance map (the horizontal direction is considered when searching for the closest two black pixels);
- Vertical distance map (the vertical direction is considered when searching for the closest two black pixels);
- Arbitrary direction distance map;

Combining two or more distance maps:

- Very restrictive:
  - Final map = min(map0, map1, ..., mapN);
- Less restrictive:
  - Final map = max(map0, map1, ..., mapN);
- Advanced:
  - Final map = max(alpha0 * map0, alpha1 * map1, ..., alphaN * mapN);

For what the final map is used:

- Identify a page main fragments
- Progressively lowering a threshold value, identify for each main fragment its sub-fragments
- Recursively continue this procedure until the smallest convenient fragment has been identified

Fig. 2. Erroneous detection using a threshold independent of font size

How the final map is used:

- Determine a binary image with separators pixels and non-separator pixels (separator pixels will be those pixels with values above a certain threshold in the distance map; non-separator pixels will be those pixels with values below a certain threshold in the distance map).
- A set of connected pixels surrounded by page edges or separator pixels will form a fragment.

2.3. How to find the separators – Using a crosshair to improve the distance approach

As it has been presented above, the distance algorithm has a high rate of success, but it can be improved. A better approach towards the determination of white space separators has to take into consideration the orientation together with the horizontal and vertical distance between the centred white pixel and the first new black one (*crosshair)*, found on the corresponding direction.

The algorithm computes the distances on all axes from the Cartesian system, rotated with 5 degrees to the left and 5 degrees to the right. In this manner, the white spaces will be found even if the page is not perfectly centred and has a small deflection from the normal orientation.

After computing the distances for each pixel, the maximum corresponding to both axes in an orientation is selected and assigned to the corresponding pixel. In the next step a threshold is applied successively to eliminate the white spaces found between letters and text rows. The maximum stored for each pixel is matched with the threshold and the selection is made. The final result and the steps towards achieving this result can be seen in Fig. 3.

2.4. How to find the separators – Another distance approach

The distance approach to white separator detection refers to the usage of a run-length routine that checks distances between white and black pixels on the input image (Bovik, 2000).

Step 1: The algorithm finds, for every white pixel in the image, the distances to the first black pixel on four directions, top, bottom, left and right, and inserts these values into a matrix.

Then, three threshold values are computed, adjusted to the dimensions of the input page and the distances to the nearest black pixels. These thresholds will decide if pixels will be united horizontally and/or vertically.

$$h = 1 - \frac{top\_distance + bottom\_distance}{height + width} \quad (1)$$

$$w = 1 - \frac{left\_distance + right\_distance}{height + width} \quad (2)$$

$$sum = w + h \quad (3)$$

Fig. 3. The output obtained by successively increasing the threshold for the crosshair approach.

Fig. 4a. Input image

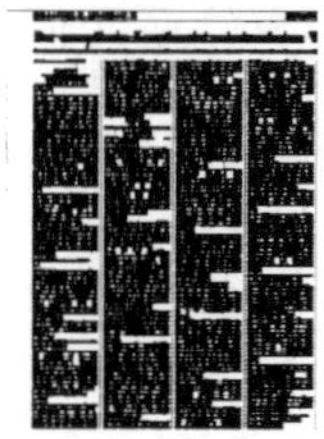

Fig. 4b. The partial result on the input image, after the first step of the algorithm

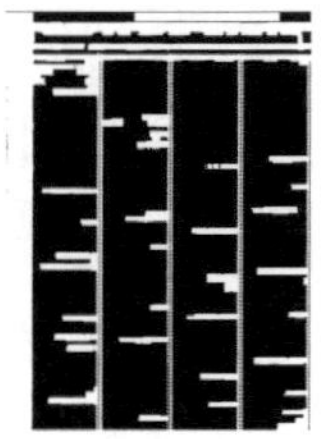

Fig. 4c. The resulting image after step two of the algorithm

In relations (1) and (2), the formulas for the horizontal and vertical thresholds are presented, where *top_distance* and *bottom_distance*, *left_distance* and *right_distance* are the distances to the nearest black pixel in the respective direction, while *height* and *width* are the dimensions of the input image. Relation (3) presents the third threshold, as being the sum of the horizontal and vertical components. Based on these thresholds, the algorithm connects the black pixels in the image, thus completing the first step.

Step 2: As it can be observed in Fig. 4b, the first step of the algorithm results in an image that connects homogeneous text zones, but the zones are not totally covered (white patches within the paragraph areas may still appear). In the second step, these white patches are detected and covered (integrated into their respective paragraphs). This is done by finding the number of passes from white to black, for every white pixel, on all four directions. It is obvious that if a white pixel is surrounded by black ones in a paragraph of text, the number of passes from white to black has to be even.

Step 3: At this point, the text and white space separators are clearly distinguished, and all that remains to be done is to obtain the coordinates of these areas. This is done by applying any contour detection algorithm, and so, the end-result will be the coordinates of the paragraphs and text areas.

3. CONCLUSIONS

All of these algorithms have in a common a very useful characteristic of white spaces: the size of the space on a given axis. By taking into account this value and combining it with simple mathematical concepts, the results are a starting point for further improvements in document processing.

4. REFERENCES

Bovik, A. (2000). *Handbook of Video and Image Processing*, Academic Press, ISBN-10: 0-12-119792-1, New York, United States of America

Breuel, T.M. (2003). An Algorithm for Finding Maximal Whitespace Rectangles at Arbitrary Orientations for Document Layout Analysis, *Proceedings of the Seventh International Conference on Document Analysis and Recognition,* Vol. 1, pp. 66, ISBN 0-7695-1960-1, Scotland, August 2003, Edinburgh

Di Zenzo, S; Cinque, L. & Levialdi, S. (1996). Run-Based Algorithms for Binary Image Analysis and Processing, *IEEE Transactions on Pattern Analysis and Machine Intelligence,* Vol. 18, No. 1, (January 1996) pp. 83-89, ISSN: 0162-8828.

Mann, S. (2002). *Intelligent Image Processing,* John Wiley & Sons, ISBN: 9780471221630, United States of America

Saitoh, T; Yamaai, T & Tachikawa, M. (1994). Document Image Segmentation and Layout Analysis, *IEICE Trans. Information and Systems*, Vol. 77, No. 7, (March 1994) pp. 778-784, ISSN: 0916-8532

Annals of DAAAM for 2008 & Proceedings of the 19th International DAAAM Symposium, ISSN 1726-9679
Editor B. Katalinic, Published by DAAAM International, Vienna, Austria 2008
Make Harmony Between Technology and Nature, and Your Mind will Fly Free as a Bird

BENDING VIBRATIONS OF BARS WITH AXIAL FORCES

BOIANGIU, M[ihail] & ALECU, A[urel]

Abstract: *In this paper we study the influence of the axial force on the natural frequencies of the bars that execute bending vibrations. Firstly is done a theoretical study that is applied in the concrete case of a bar. Is built the transfer matrix and the saltus matrix for the study of the bending vibrations of the bars systems with axial forces.*

Key words: *Vibrations, Bars, Forces, Frequencies, Matrices*

1. INTRODUCTION

A system of bars can execute bending vibrations in the conditions which the bars are and are not stressed of axial forces. For example for the system of bars from figure 1, if is neglected the weight of the bars, the stresses in bars are equal with zero. If about the system acts the forces F_1, F_2, then only a part from the efforts from bars is non-zero.

We want to know if an axial force influences or not the natural frequencies of the bending vibrations of the bars.

Also we build the state vectors and specific matrices for the study of bending vibrations of the bars systems with axial forces by using the transfer matrix method.

2. THEORETIC BACKGROUND

We consider an element of right bar, of constant section, that is stressed to bending in the *xOy* plan, as in figure 2. The bar is stressed of the distributed load $\vec{p}$ [N/m] and a constant axial tensile force. From the equations of equilibrium

$$T - pdx - (T + dT) = 0, \qquad (1)$$

$$Tdx + M - Ndy - (M + dM) - pdx\frac{dx}{2} = 0, \qquad (2)$$

neglecting $(dx)^2/2 \approx 0$, we obtain (Buzdugan 1986)

$$\frac{dT}{dx} = -p \ ; \quad T = \frac{dM}{dx} + N\frac{dy}{dx} \qquad (3)$$

or

$$\frac{d^2M}{dx^2} + N\frac{d^2y}{dx^2} = -p. \qquad (4)$$

We consider a right bar of constant section, stressed of constant axial force *N*. It executes bending vibrations in the *xOy* plan (figure 3) in the case Euler Bernoulli (Voinea, et al.,1989).

The distributed load is the force of inertia

$$p = F^{in} = -\rho A\frac{\partial^2 y}{\partial x^2}. \qquad (5)$$

Knowing the relation $M = -EI_z\frac{d^2y}{dx^2}$, from (4) and (5) is obtained the partial differential equation for the vibrations of right bars of constant section, stressed of constant axial forces:

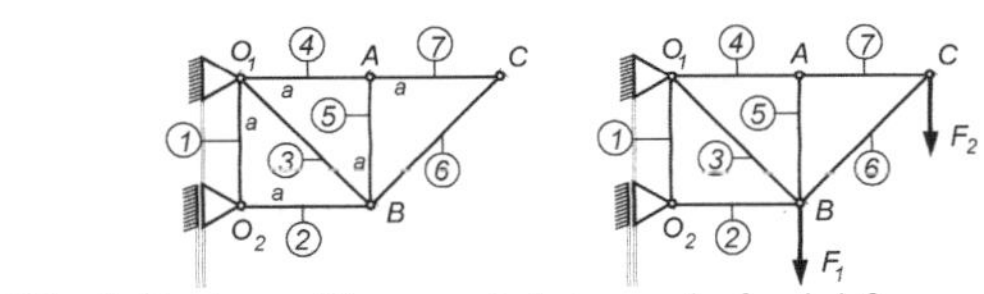

Fig. 1. System of bars partial stressed of axial forces

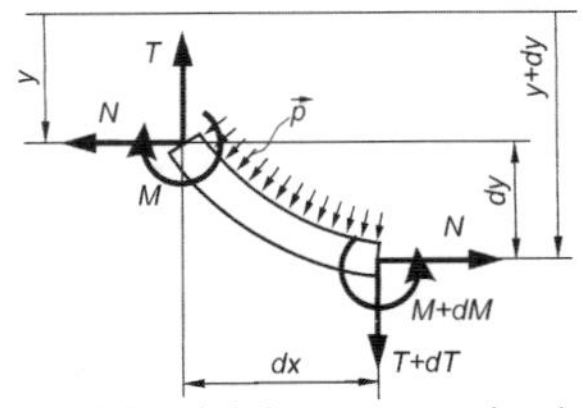

Fig. 2. Bar element with axial force stressed at bending

$$\frac{\partial^4 y}{\partial x^4} - \frac{N}{EI_z}\frac{\partial^2 y}{\partial x^2} + \frac{\rho A}{EI_z}\frac{\partial^2 y}{\partial t^2} = 0, \qquad (6)$$

where: ρ is the material density, E is the modulus of Young, A is the area of transversal section, I_z is the momentum of inertia of the transversal section.

We search solutions of type $y = Y(x)\cos(\omega t - \psi)$. It results:

$$Y^{IV}(x) - \frac{N}{EI_z}Y^{II}(x) - \frac{\rho A}{EI_z}\omega^2 Y(x) = 0. \qquad (7)$$

Is noted $\alpha^4 = \frac{\rho A}{EI_z}\omega^2$; $2\beta = \frac{N}{EI_z}$. It results (Ceausu,et al.,2006):

$$Y(x) = C_1 ch(\gamma x) + C_2 sh(\gamma x) + C_3\cos(\delta x) + C_4\sin(\delta x) \quad (8)$$

where: $\gamma = \sqrt{\beta + \sqrt{\beta^2 + \alpha^4}}$; $\delta = \sqrt{-\beta + \sqrt{\beta^2 + \alpha^4}} = \frac{\alpha^2}{\gamma}$.

3. CAS OF A BAR EMBEDDED AT HEADS

We consider a bar of length *l* embedded at both heads. The boundary conditions:

$$x = 0; \quad y = 0 \ and \ \varphi = \frac{\partial y}{\partial x} = 0 \Leftrightarrow Y = 0 \ and \ Y^I = 0$$

$$x = l; \quad y = 0 \ and \ \varphi = \frac{\partial y}{\partial x} = 0 \Leftrightarrow Y = 0 \ and \ Y^I = 0$$

leads to the algebraic linear and homogeneous system

$$\begin{cases} [ch(\gamma l) - cos(\delta l)]C_1 + \left[sh(\gamma l) - \frac{\gamma}{\delta}sin(\delta l)\right]C_2 = 0 \\ [\gamma sh(\gamma l) + \delta sin(\delta l)]C_1 + \gamma[ch(\gamma l) - cos(\delta l)]C_2 = 0 \end{cases}. \qquad (9)$$

From the condition as the system (10) to admit non-zero solutions ($\Delta = 0$) is obtained the natural frequencies equation:

$$\alpha^2 - \gamma^2 ch(\gamma l)cos(\delta l) + \beta sh(\gamma l)sin(\delta l) = 0\ . \qquad (10)$$

For the concrete case of a bar with length 1m and diameter 0.03m, the variation of the natural frequencies, depending of axial force, is presented in the figure 4. Axial force of tensile is positive and axial force of compression is negative.

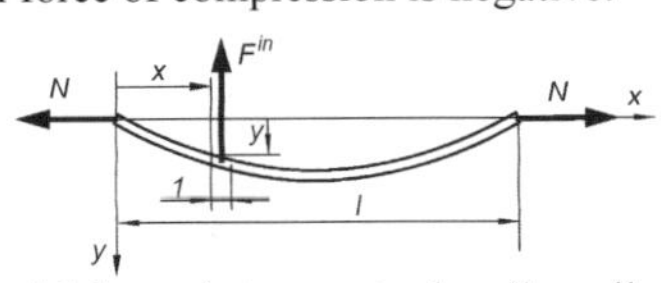

Fig. 3. Bar with axial force that executes bending vibrations

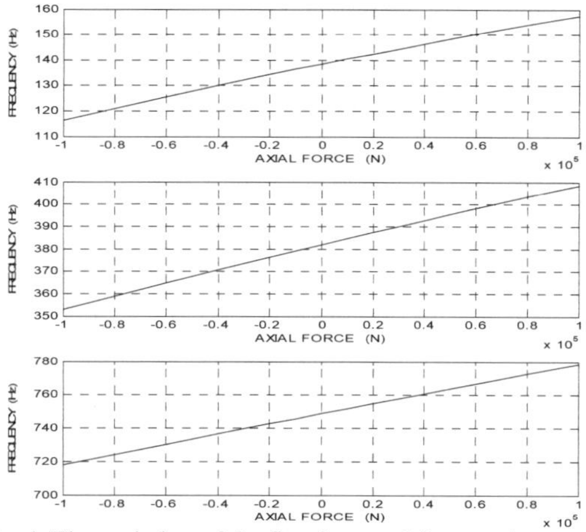

Fig. 4. The variation of the first 3 natural frequencies

If the bar is warmed from the temperature of *10°* to *40°* then it appears an axial force of compression of *53438 N*. In this case the first natural frequency decrease with 9.7%.

4. TRANSFER MATRIX

From the relations

$$\begin{cases} y = Y(x)\cos(\omega t - \psi) \\ \varphi = \frac{\partial y}{\partial x} = Y^{I}(x)\cos(\omega t - \psi) \\ M = -EI_z\frac{d^2y}{dx^2} = -EI_zY^{II}(x)\cos(\omega t - \psi) \\ F = \frac{\partial M}{\partial x} + N\frac{\partial y}{\partial x} = -EI_z[Y^{III}(x) - 2\beta Y^{I}(x)]\cos(\omega t - \psi) \end{cases} \quad (11$$

it results if $x = 0$

$$\begin{cases} y_0 = C_1\cos(\omega t - \psi) + C_3\cos(\omega t - \psi) \\ \varphi_0 = \gamma C_2\cos(\omega t - \psi) + \delta C_4\cos(\omega t - \psi) \\ M_0 = -EI_z\gamma^2C_1\cos(\omega t - \psi) + EI_z\delta^2C_3\cos(\omega t - \psi) \\ F_0 = -EI_z\alpha^2\delta C_2\cos(\omega t - \psi) + EI_z\alpha^2\gamma C_4\cos(\omega t - \psi) \end{cases} \quad (12$$

From the algebraic system (12) we find $C_1\cos(\omega t - \psi)$, $C_2\cos(\omega t - \psi)$, $C_3\cos(\omega t - \psi)$, $C_4\cos(\omega t - \psi)$, that, together with $Y(x), Y^{I}(x), Y^{II}(x), Y^{III}(x)$ from (8) are replaced in (11). We obtain thus the expressions for movement, rotation, momentum and force at the distance *x* from the head of the bar.

Generalizing for an element of bar, found out between the points (the nodes) *k-1* and *k* (figure 5), between the state vectors from the nodes *k-1* and *k* we can write the matrices relation:

$$\{q\}_k^l = \frac{1}{\sqrt{\beta^2+\alpha^4}}[A]_k\{q\}_{k-1}^r \quad (13)$$

where $\{q\}_{k-1}^r$ and $\{q\}_k^l$ are the state vectors in the nodes *k-1* (left) and *k* (right). $[A]_k$ is the transfer matrix for the element *k* (Boiangiu & Alecu 1999):

$$[A]_k = \begin{bmatrix} S_1 & \alpha T_1 & \alpha^2 U_1 & \frac{\alpha^3}{\gamma\delta}V_1 \\ \frac{\gamma\delta}{\alpha}V_1 & S_2 & \alpha T_1 & \alpha^2 U_1 \\ \frac{\gamma^2\delta^2}{\alpha^2}U_1 & \frac{1}{\alpha}V_2 & S_2 & \alpha T_1 \\ \frac{\gamma\delta}{\alpha^3}T_2 & \frac{\gamma^2\delta^2}{\alpha^2}U_1 & \frac{\gamma\delta}{\alpha}V_1 & S_1 \end{bmatrix}_k .$$

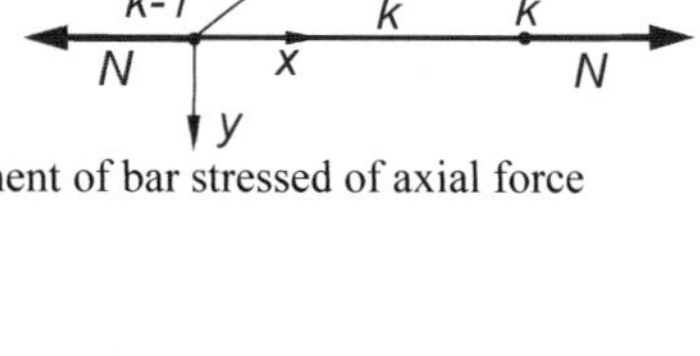

Fig.5. Element of bar stressed of axial force

$$\{q\}_k^{l,T} = \left\{y_k \quad \frac{\varphi_k}{\alpha} \quad -\frac{M_k}{\alpha^2 EI_z} \quad -\frac{F_k}{\alpha^3 EI_z}\right\}_k^l ;$$

We used the notations:

$$S_1 = \frac{\delta^2 ch(\gamma x) + \gamma^2 cos(\delta x)}{2}; \quad T_1 = \frac{\gamma sh(\gamma x) + \delta sin(\delta x)}{2};$$

$$U_1 = \frac{ch(\gamma x) - cos(\delta x)}{2}; \quad V_1 = \frac{\delta sh(\gamma x) - \gamma sin(\delta x)}{2};$$

$$S_2 = \frac{\gamma^2 ch(\gamma x) + \delta^2 cos(\delta x)}{2}; \quad V_2 = \frac{\gamma^3 sh(\gamma x) - \delta^3 sin(\delta x)}{2};$$

$$T_2 = \frac{\delta^3 sh(\gamma x) + \gamma^3 sin(\delta x)}{2}.$$

5. SALTUS MATRIX

If in a node *k* appears a saltus of section, then, between the state vectors from left and right of the node *k* , we can write the matrices relation (Boiangiu & Alecu 1999):

$$\{q\}_k^r = [B]_k\{q\}_k^l \quad (14)$$

where $\{q\}_k^r$ and $\{q\}_k^l$ are the state vectors at the right and left of the node *k*. $[B]_k$ is the saltus matrix 4x4 in the node *k*. It is a rare matrix. The non-zero elements are established using the relations of continuity: $y_k^r = y_k^l$; $\varphi_k^r = \varphi_k^l$; $M_k^r = M_k^l$; $F_k^r = F_k^l$:

$$B_{1,1} = 1; \quad B_{2,2} = \sqrt[4]{\frac{I_{z,k+1}}{I_{z,k}}\frac{A_k}{A_{k+1}}}; \quad B_{3,3} = \sqrt{\frac{I_{z,k}}{I_{z,k+1}}\frac{A_k}{A_{k+1}}};$$

$$B_{4,4} = \sqrt[4]{\frac{I_{z,k}}{I_{z,k+1}}\frac{A_k^3}{A_{k+1}^3}}.$$

The elements of the saltus matrix don't depend of axial force.

6. CONCLUSIONS

The axial force influences the natural frequencies of a bar for bending vibrations. The results differ if the axial forces are changed.

In future, the study will be applied at plane and spatial bars systems by using the transfer matrix method, because the axial forces can influence the dynamic behavior of the bars system.

It is possible to intervene about the system with forces so that the natural frequencies to be modified and keep them in a desirable interval (Alecu et al., 2008).

7. REFERENCES

Alecu, A.; Boiangiu, M.; Ursu, I. & Munteanu, E. (2008). Equivalent mechanical system for the study upon the problem of vibration control, *Bulletin of the Politechnic Institute of Jassy,* Tom LIV(LVIII), Fasc.1, Section Machine Construction, pp.265-270, ISSN1011-2855

Boiangiu, M. & Alecu, A. (1999). *Coupled vibrations of crankshaft,* Printech, ISBN 973-652-021-8, Bucharest

Buzdugan, Gh. (1986). *Strength of materials*, Romanien Academy, Bucharest

Ceausu, V.; Tiron, A.; David, M.; Stanila, R. & Costache, A. (2006). Transversal vibrations of bars with axial loadings. Transfer matrix method, Proceedings of SIMEC 2006, pp.45-49, ISBN 973-7797-83-3, Bucharest

Voinea, R.; Voiculescu, D. & Simion, F.-P. (1989). *Solid-State Mechanics with applications in Engineering*, Romanien Academy, ISBN 973-27-0000-9, Bucharest.

Annals of DAAAM for 2008 & Proceedings of the 19th International DAAAM Symposium, ISSN 1726-9679
Editor B. Katalinic, Published by DAAAM International, Vienna, Austria 2008
Make Harmony Between Technology and Nature, and Your Mind will Fly Free as a Bird

STUDY OF VIBRATIONS OF PLANE BARS SYSTEMS BY USING THE TRANSFER MATRIX METHOD

BOIANGIU, M[ihail] & ALECU, A[urel]

Abstract: *In this paper we study the simultaneous vibrations of bars systems, considering, as been continuous medium. We divide the system in cycles and we apply the transfer matrix method associated with transit matrix method for each cycle. We develop the free vibrations cases. We apply the purposed method to the concrete cases of a system statically determinate and of two systems statically indeterminate.*
Key words: *Vibrations, Transfer matrix method, Natural angular frequencies, State vectors, Cycles.*

1. INTRODUCTION

The precision in manufacturing can be influenced of vibrations. The machine tools, robots, equipments can contain bars structures. To control the vibrations it means to know the dynamic behaviour of these structures.

In literature it studies the coupled vibrations by utilization the finite elements method, but this method introduces specific feature. In few papers (Ohga et al., 1993) (Yu & Craggs, 1995), it has been tried to improve the finite elements method by combination with the transfer matrix method.

In this paper we propose a method of study based on transfer matrix method, applied first time at arborescent bars network. The bars system is divided in cycles and nodes and the transfer matrix method is applied for each cycle.

We introduce specific matrices for bars systems and special matrices for ramification and meeting nodes. We apply the purposed method to concrete cases.

2. DESCRIPTION OF METHOD AND SIMPLIFYING HYPOTHESIS

2.1 Hypothesis

We consider that the system is formed from straight elements (bars) of constant section (figure 1) which are effecting bending vibrations in two perpendicular plans which are containing longitudinal axis of a element xOy and xOz), torsion vibrations and axial vibrations around and respectively in length of this axis (Voinea, et al.,1989).

2.2 Transfer matrix for one element

For one element of the system, choosing the reference system like in figure 1, between state vectors of ends of the k element, $\{q\}_{k-1}^{r}$ and $\{q\}_{k}^{l}$, we can write matrix relation (1):

$$\{q\}_{k}^{l} = [A]_{k}\{q\}_{k-1}^{r} \qquad (1)$$

The matrix (12x12), named $[A]_{k,}$ is the transfer matrix for the element k. The elements of this matrix are the Krilov – Rayleigh functions. The elements of the state vectors are quantities in proportion to displacements, rotating, forces and momentums (Boiangiu & Alecu 1999).

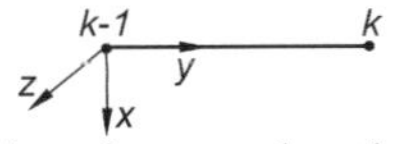

Fig. 1. Application of proposed method to a bar

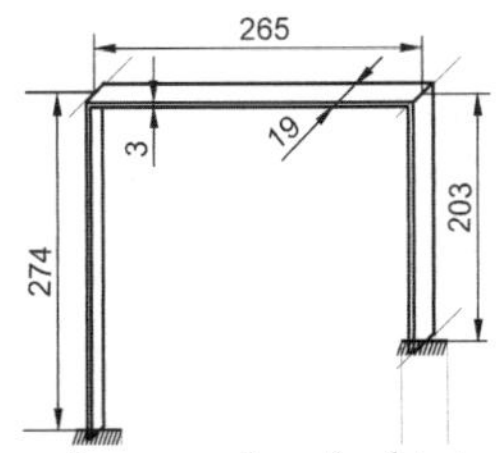

Fig. 2. Application of proposed method to tested bars system

2.3 Special matrices

In some nodes is possible to appear special matrices (Boiangiu & Alecu 2001). These are rare matrices and can be:
- a transition matrix, $[C]_{k}$, when we move from an element to another and the coordinate system is modified;
-a saltus matrix, $[B]_{k}$, when in a node k appears a saltus of section;
- a transfer matrix over a concentrated mass in node k, $[D]_{k}$.

Between the left state vectors and right state vectors of section k, $\{q\}_{k}^{l}$ and $\{q\}_{k}^{r}$, we can write relation:

$$\{q\}_{k}^{r} = [S]_{k}\{q\}_{k}^{l}, \qquad (2)$$

The matrix $[S]_{k}$ can be one from the matrices $[C]_{k}$, $[B]_{k}$ or $[D]_{k}$.

2.4 Determination of the natural angular frequencies

We begin from the left terminal point, 0, and we traverse the system from an element to another till the right terminal point, n. Depending of configuration of the bars system, on route, intervene the transfer matrix, [A], and the matrices [C], [B], [D]. We obtain the matrix relation between the state vectors of ends of the n element, $\{q\}_{n}^{l}$ and the first (point of start) $\{q\}_{0}^{r}$ (Boiangiu & Alecu 2001):

$$\{q\}_{n}^{l} = [Q]_{k}\{q\}_{0}^{r} \qquad (3)$$

The matrix (12x12), [Q], is called transfer matrix for system.

Depending on the type of the leaning we write the boundary conditions and we obtain an algebraic system. We put the condition that the system to admit non-zero solution and we obtain the natural angular frequencies.

3. EXAMPLE

The method presented here is applied in concrete case of a steel frame in a "U" shape with unequal sides and rigid fixed

Mode number	Theoretical results(Hz)	Experimental results (Hz)
Mode 1	40,87	39,2
Mode 2	138,66	132
Mode 3	139,14	138
Mode 4	212,21	230
Mode 5	255,77	253

Tab. 1. The first five natural angular frequencies

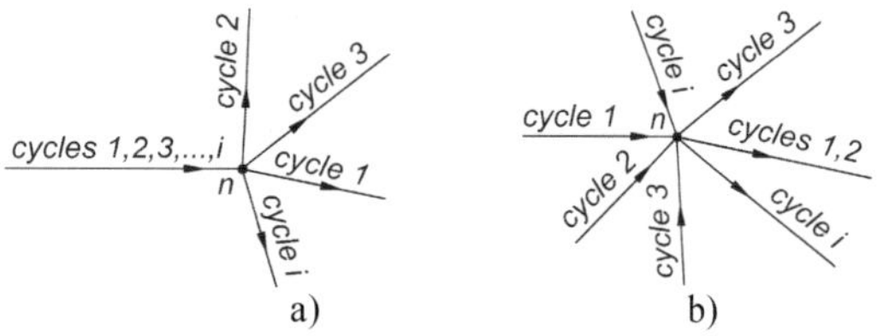

Fig. 3. Types of nodes

in the terminal points (figure 2).

We have effectuated experiments on the same steel frame. As shown in Table 1, the theoretical results approximately correspond to the experimental results.

4. CASE OF SYSTEM STATICALLY INDETERMINATE

In case of the bars systems statically indeterminate we divide the system in more cycles and we introduce special matrix and equations in ramification nodes. All the cycles must start from the same node (point), named "0". The start and the end points of the cycles must be in a connection point of the system with the outside or in a free end point of the bars. We can find two types of nodes: nodes with one input and more output, named ramification nodes, (figure 3a), and nods with more input and more output, named meeting nodes, (figure 3b)

In a node we consider a main cycle, for example the cycle number 1. The other cycles are secondary cycles.

In a ramification node we keep the bars from the main cycle and we replace the bars from the secondary cycles with the forces and momentums, unknown, from the heads in this node. From this reason, in all state vectors we introduce new lines corresponding to these forces and momentums. Also in all transfer and special matrix we introduce new lines and columns. The number of the new lines and columns (in state vectors and matrices) is equal with the total number of the unknowns from the secondary cycles in the ramification nodes. In these nodes we introduce a special matrix named ramification matrix and written $[R]_2$.

For example we consider a plane bars system as in figure 4a. The node 2 is a ramification node.

We divide the bars system in 2 cycles. The cycle 1 (0-1-2-4-5) and the cycle 2 (0-1-2-3). The node 2 is a ramification node. In this node the main cycle is the cycle 1.

The ramification introduces 6 unknowns: F_x^2, F_y^2, F_z^2, M_x^2, M_y^2, M_z^2, that is why should be introduced in the state vectors 6 new lines.

For the cycle 2, in ramification node 2, intervenes also the transition matrix [C]. We obtain for cycle 1 the matrix relation:

$$\{q\}_5^l = [A]_5[C]_4[A]_4[R]_2^1[A]_2[C]_1[A]_1\{q\}_0^r = [Q]^1\{q\}_0^r \quad (4)$$

We obtain for cycle 2 the matrix relation:

$$\{q\}_3^l = [A]_3[C]_2^2[R]_2^2[A]_2[C]_1[A]_1\{q\}_0^r = [Q]^2\{q\}_0^r \quad (5)$$

where $[Q]^1$ and $[Q]^2$ are the transfer matrices for cycle 1 respectively 2. The superscript number shows the cycle.

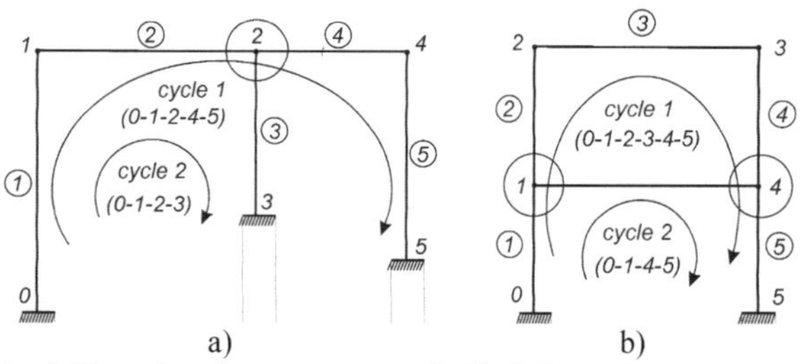

Fig. 4. Two plane bars systems statically indeterminate

From the relations (4) and (5) we obtain the system

$$\begin{Bmatrix}\{q\}_5^l\\ \{q\}_3^l\end{Bmatrix} = \begin{bmatrix}[Q]^1\\ [Q]^2\end{bmatrix}\{q\}_0^r . \quad (6)$$

In a meeting node we keep also the bars from the main cycles. We replace the bars from the secondary cycles, with the forces and momentums from the heads in this node, but only for the bars that leave the node. These forces and momentums are unknowns. The bars from the secondary cycles that come in node are replaced with the forces and momentums at the head of the cycle in this node. If the output is common for a main cycle and a secondary cycle, we keep only the output for main cycle. In a meeting node are written special equations.

For example we consider a plane bars system as in figure 4b. The node 1 is a ramification node and the node 4 is a meeting node. In node 1 we introduce the ramification matrix $[R]_1$ as in the previous example. In node 4 we introduce a special ramification matrix $[R^{F,M}]_4$ written only for forces and momentums. We can write for cycle 1:

$$\{q\}_4^r = [A]_5([Q]_{0-4}^1 + [R^{F,M}]_2^2[Q]_{0-4}^2)\{q\}_0^r = [Q]\{q\}_0^r \quad (7)$$

where, the square matrix [Q] is called transfer matrix for system. The superscript number shows the cycle.

In node 4 displacements on the cycle 1 are equal with the displacements on the cycle 2. In this case we introduce a special ramification matrix $[R^D]_4$ written only for displacements (displacements, rotations). We can write for the node 4:

$$\{0\} = ([R^D]_4^1[Q]_{0-4}^1 - [R^D]_4^2[Q]_{0-4}^2)\{q\}_0^r = [R^D]_4^{1-2}\{q\}_0^r \quad (8)$$

$[R^D]_4^{1-2}$ is the ramification matrix for displacements in the node 4. The superscript number shows the cycles.

From the relations (7) and (8) we obtain the system

$$\begin{Bmatrix}\{q\}_5^l\\ \{0\}\end{Bmatrix} = \begin{bmatrix}[Q]\\ [R^D]_4^{1-2}\end{bmatrix}\{q\}_0^r . \quad (9)$$

5. CONCLUSIONS

The proposed method gives good results, is flexible and can be easily applied on different types of plane bars systems.

This method is indicated to be used in computational engineering, especially for the complicated bars systems.

The next step will be to apply the method for spatial bars systems.

For complicated plane and spatial bars systems, in future, this method will be applied associated with the graph theory, because is difficult to identify the cycles and the types of nodes.

6. REFERENCES

Boiangiu, M. & Alecu, A. (1999). *Coupled vibrations of crankshaft,* Printech, ISBN 973-652-021-8, Bucharest

Boiangiu, M. & Alecu, A. (2001). The simultaneous vibrations analysis of bars systems, *Proceedings of 8th International Symposium on Theory of Machines and Mechanisms*, pp. 77-82, ISBN 973-8143-38-1, Bucharest, Sept 2001

Ohga, M. et al. (1993). A Finite Element-Transfer Matrix Method for Dynamic Analysis of Frame Structure. *Journal of Sound and Vibration,* Vol.167, No.3, Nov. 1993

Voinea, R.; Voiculescu, D. & Simion, F.-P. (1989). *Solid-State Mechanics with applications in Engineering*, Romanien Academy, ISBN 973-27-0000-9, Bucharest

Yu, J., Y. & Craggs, A. (1995). A Transfer Matrix Method for Finite Element Models of a Chaine-like Structure Under Harmonic Excitations. *Journal of Sound and Vibration,* Vol.187, No.4, Nov. 1995

Annals of DAAAM for 2008 & Proceedings of the 19th International DAAAM Symposium, ISSN 1726-9679
Editor B. Katalinic, Published by DAAAM International, Vienna, Austria 2008
Make Harmony Between Technology and Nature, and Your Mind will Fly Free as a Bird

QUALITY OF FINISHED SURFACES FOR LIGNIN-CELLULOSE BASED MATERIALS

BOIERIU, C[amelia]; CURTU, I[oan]; TIMAR, M[aria] C[ristina] & LICA, D[umitru]

Abstract: *The paper presents the influence of the wood anatomical structure and of the machining on the surface quality of some coated lignin-cellulose based composite panels made of three species of hardwood lamellas glued together. A quantitative evaluation of the absorption power of the three species of wood (cherry, maple and walnut wood) and investigations on the evolution of the surface quality after machining and finishing, through the quantitative evaluation of the roughness, brings new data about the final quality of the finished panels.*

Key words: *surface quality, roughness, finishing.*

1. INTRODUCTION

Sanding is an important phase in wood processing. Sanding smoothens and homogenizes the surface reducing the influence of the anatomical structure on the roughness profile, though the largest anatomical elements (e.g. vessels) are scarcely visible, even on the smoothest surfaces obtained with high grits (De Moura & Hernández 2006). Various methods of measuring the roughness of the wooden surfaces (with stylus, optical or laser beams), are accepted to be applied in the conditions of the lack of a unitary technical specification for measuring this surface quality parameter. Because wood surface contains irregularities due to both the sanding process and wood anatomy, any quantitative evaluation of the roughness of a sanded surface requires data filtering (Gurau et al., 2006), as in the case of oak wood. The anatomical roughness can distort the measuring results (Krisch & Csiha, 1999). Sanding precedes wood surface coating with lacquers or paints directly influencing the quality of the finished surfaces and it is a time-consuming process requiring optimisation (Sinn et al., 2004).

What happens when different species of wood (as lamellas) are glued together in composite panels (Boieriu, 2007), (Boieriu et al., 2007), being further machined and finished in the same conditions? Is the quality of the resulted finished surface being influenced by the anatomical structure of the wood? These are the questions this paper intends to answer to.

For this purpose, investigations on the quality of the surface of composite panels made of glued lamellas of three hardwood species with different micro- and macro-structure characteristics (number and size of pores, annual rings): cherry, maple and walnut were made. These investigations included water absorption tests and roughness measurements. The evolution of surface roughness after sanding with grit sizes of P60 and P120, P150 (final sanding) and further finishing was determined and correlated to the anatomical structure of wood.

2. APPLIED METHODS

The three species of wood studied in the present paper are hardwood species, with uniformly distributed pores structure, having certain particularities: maple wood has very numerous (61%), but small pores and no visible differences between the late and early wood; cherry wood has small, numerous (62%), and sometimes visible pores (in the early wood) and visible differences between the early and the latewood; walnut wood has visible but fewer pores than the previous species (57%), situated in the early wood and containing parenchyma. The microstructure of the three species of wood is shown in Fig. 1.

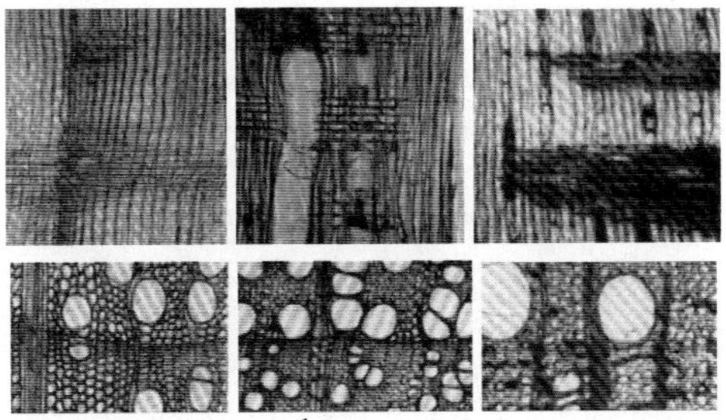

Fig. 1. The microstructure on radial (up) and crosscut (down) section for maple (a), cherry (b) and walnut (c) wood.

The composite panels were sanded successively with grit sizes of P60, P120 and P150 and further successively coated with UV curable finishing materials: UK 1168 (0,035 kg/m^2) and UM 1178-0025 lacquer type (0,004 kg/m^2). The surface roughness of the panel on different areas corresponding to the three wood species was quantitatively measured after each of these steps and the measurement was done perpendicularly to the wood grain direction. A MicroProf equipment with optical sensor has been used for determining the roughness after sanding, before finishing, whilst a stylus instrument, Mar PS1 was employed for measuring the surface quality after finishing.

The quality of the surface after the final sanding, just before finishing, in terms of absorption and wetting capacity was appreciated through two additional tests. The first one was a quantitative one. A constant volume of 0,05 ml of water was placed on the panel surface and the absorption time (T) alongside the area of absorption (L×l) was measured. The second one, that could indicate an inactivation of surface reducing its wetting capacity, used a colour indicator: a drop of 3 mm diameter was placed on the wood surface. Total absorption within 1 minute and appearance of a green round spot with orange-blue margins are indications for a corresponding surface (Timar, 2006). On a non-corresponding surface the drop remains on the same blue colour. The indicator consists of: methylene blue crystallised – 0,07 weight parts, aniline sulphate – 0,5 weight parts, water – 100 weight parts.

3. RESULTS AND DISCUSSION

The results in Table 1 indicate the best absorption time (maple wood) and the worst one (nut wood).

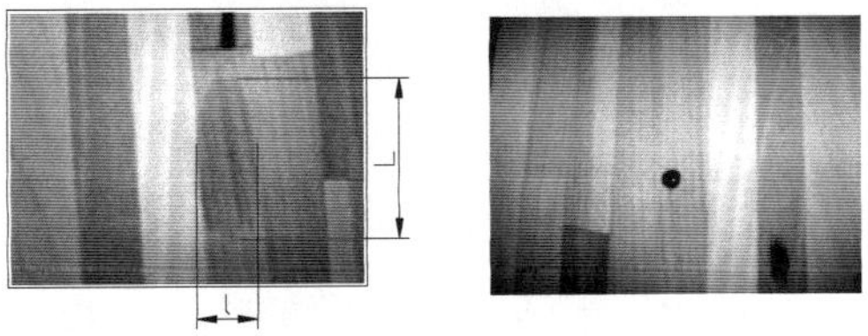

F ig. 2. The results of the absorption test.

The sizes of the measured water spots (L, l) are shown in Fig. 2 and Table 1 and the non-corresponding surface for the walnut wood, where the drop of indicator remained on the blue colour is shown in Fig.2.

The roughness measurements after sanding with grit size P60 and P150 showed that the anatomic structure of the wood is apparent in the diagram of the roughness profile for cherry (Fig. 3b, Fig. 4) and walnut wood (Fig. 3c, Fig. 6) as minimum and maximum peaks, corresponding to the early wood, where the pores are bigger and more numerous than in the late wood.

Wood species	Absorption test		Rz, µm after P60 sanding	Rz, µm after P150 sanding	Rz, µm after coating
	Time of absorption T, in s	Water spot size Lxl, mm			
Cherry	620	50x20	45.3	37.6	8.2
Maple	218	40x25	32	27.6	8.6
Nut	4004	40x15	45	34.7	8.1

Tab. 1. Results of the applied methods for the three species of wood.

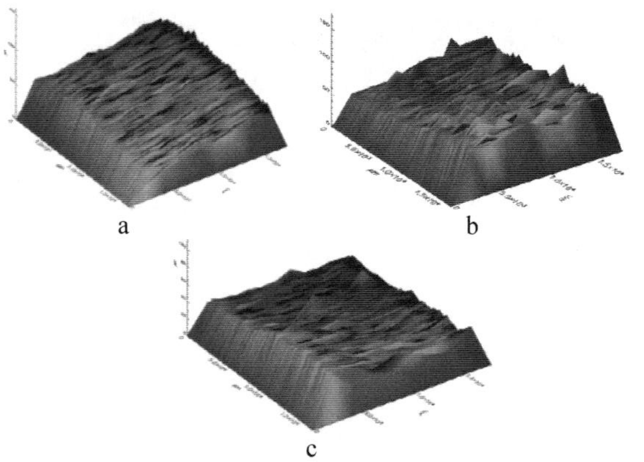

Fig. 3. The 3D roughness profile for maple (a), cherry (b) and nut wood (c), after sanding with P60 grit size.

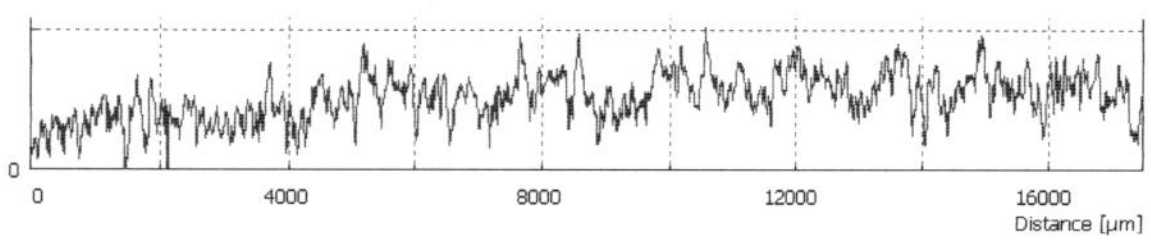

Fig. 4. Roughness of the cherry wood after sanding (P150).

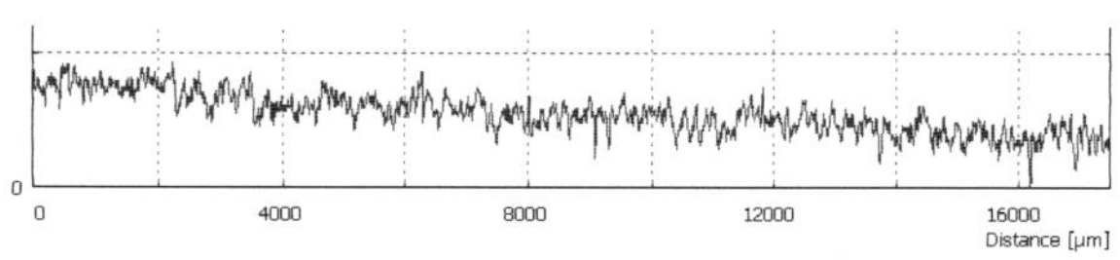

Fig. 5. Roughness of the maple wood after sanding (P150).

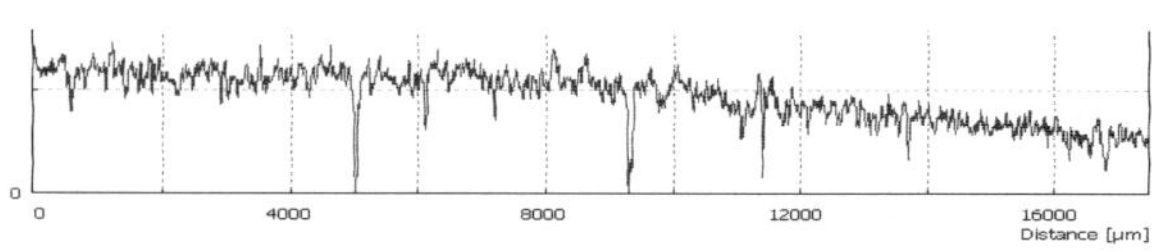

Fig. 6. Roughness of the nut wood after sanding (P150).

The maple, apart to the other two species, has a uniform distribution of the roughness profile, which shows its homogeneous structure with small pores and poor differences between the latewood and the early wood (Fig. 3a, Fig. 5).

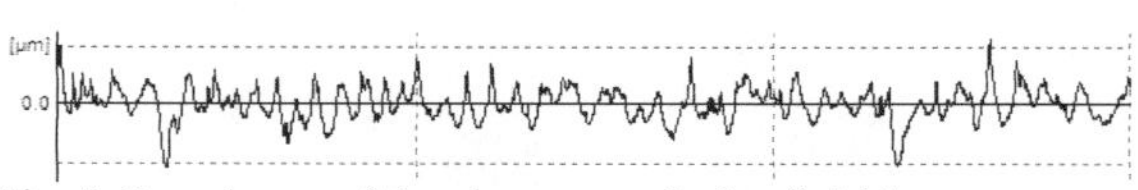

Fig. 7. Roughness of the cherry wood after finishing process.

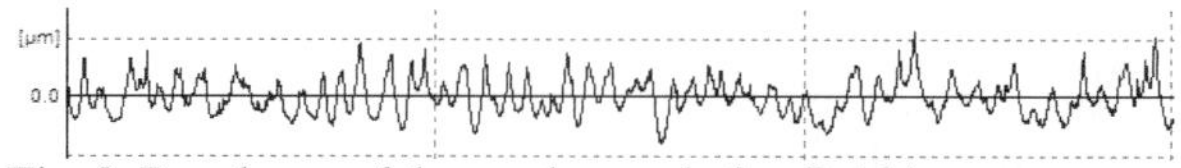

Fig. 8. Roughness of the maple wood after finishing process.

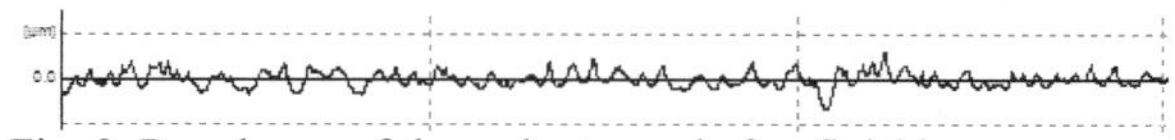

Fig. 9. Roughness of the walnut wood after finishing process.

The roughness profiles after finishing (Fig. 7, 8 and 9) show that the walnut finished surface is smoother than for the other two species. The evolution of the roughness of the surface after machining and finishing is shown in Fig. 10.

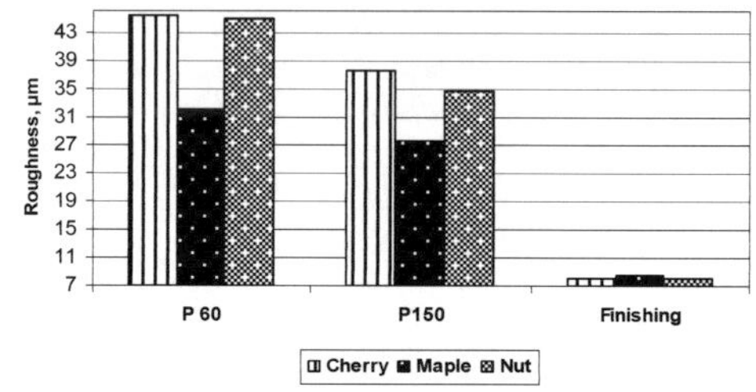

Fig. 10. Evolution of the roughness including finishing.

4. CONCLUSIONS

The anatomical structure of wood is visible on the roughness profile (for walnut and cherry wood), and the surface appears rougher than for the maple wood. For the wood species with a lower absorption capacity (walnut) the final roughness of the finished surface is better than for the others. Optimising the sanding process is not enough for the quality of the finished surface. The different capacity of the wood species to absorb the finishing material is also important and it depends on the wood anatomy. Further research can be done on measuring the roughness in different moments of the sanding process, to determine the right moment to apply the finishing material.

5. REFERENCES

Boieriu, C. (2007). *Composite Panels Made of Glued Hardwood Lamellas* (in Romanian language), Editura Universităţii Transylvania din Brasov, ISBN (10) 973-635-843-7, Braşov

Boieriu, C.; Curtu, I.; Lica, D. & Brenci, L. (2006). New composite materials for furniture design, *Materialy XX sesji naukowej Badania dla Meblarstwa,* pod redacja Jerzego Smardzewskiego, pp. 7-16, ISBN 978-83-89887-89-4, Poznan, August 2006, Poland

De Moura, L. F. & Hernández, R. E. (2006). Effects of abrasive mineral, grit size and feed speed on the quality of sanded surfaces of sugar maple wood, *Wood Science and Technology*, 40, March 2006, 517-530, ISSN 0043-7719

Gurau, L.; Mansfield-Williams, H. & Irle, M. (2006). Filtering the roughness of a sanded wood surface, *Holz als Roh – und Werkstoff*, 64(5), January 2006, 363-371, ISSN 0018-3768

Krisch, J. & Csiha, C. (1999). Analysing Wood Surface Roughness Using an S3P Perthometer and Computer Based Data Processing, *Proceedings of XIII Sesja Naukowa Badania dla Meblarstwa*, pp. 145-154, ISBN 83-88518-04-6, Poland, 1999

Sinn, G.; Gindi, M.; Reiter, A & Stanzi-Tschegg, S. (2004). Changes in the surface properties of wood due to sanding, *Holzforschung*, 58, 2004, 246-251, ISSN 0018-3830

Timar, M. C. (2006). *Wood adhesives,* Editura Universităţii Transilvania din Brasov, ISBN (10) 973-635-760-0, Braşov

Annals of DAAAM for 2008 & Proceedings of the 19th International DAAAM Symposium, ISSN 1726-9679
Editor B. Katalinic, Published by DAAAM International, Vienna, Austria 2008
Make Harmony Between Technology and Nature, and Your Mind will Fly Free as a Bird

RECYCLING OF THE AGED OFFSET PRINTS

BOLANCA MIRKOVIC, I[vana] & BOLANCA, Z[denka]

***Abstract**: The investigation results of the influence of accelerated ageing conditions (dynamics of the ageing process, exposition of prints to different temperatures and air humidity) on gamut of prints and efficiency of the recycling process are presented in the paper. The characteristics of fibers in different phases of the recycling process are attended with the chromatic values a* and b* and the number and size of the remained ink particles. The investigation results are the contribution to the explanation of the influence of print ageing on detaching the ink from fibers and removing it from the suspension with the aim of obtaining the secondary raw material of adequate quality for the production of fine graphic papers.*
***Key words:** offset prints, accelerated ageing, recycling*

1. INTRODUCTION

Wastepaper recycling is of growing importance both for better utilization of the natural resources and for reduction of the amount of solid waste. Today, in the paper industry, flotation deinking is the dominating process worldwide for the removal of printing inks from recovered paper (Kumar Pati et al., 2008).

Recent studies are focused on the type and amount of chemicals used in processing stages: pH, pulp consistency, time of disintegration, slushing method and other chemical and physical operating conditions which influence the efficiency of deinking operation (Liu et al., 2007; Labidi et al. 2007). One of the problems here is the detachment efficiency and removal of ink particles. Wang J.P. and others study the quality of the wastewater and the technological purification as well as evaluation of deinking by-products (Wang et al. 2007; Ait-Benichou et al. 2008).

Other consideration in deinkability of waste paper includes pulp ageing. The durability of paper depends mainly on the physical and mechanical characteristics of the raw materials, impact of microclimatic factors such as heat, humidity or emanation and on contamination by ions and gas from the environment and action of micro-organisms (Dupont et al. 2007; Lattuati-Derieux at al. 2006).

The influence of the ageing process on recycling efficiency of wastepaper is generally less studied than other unmentioned factors of the deinking flotation process. In the frame of our investigations in this area the process of natural and accelerated ageing of different formulations of the printing substrates and ink on stability of prints observed over relevant optical values as well as FT-IR spectroscopy has been studied. Characteristic properties of the recycled fibers dependent on the exposed ones of the input material in earlier described system have also been studied.

In this work, the investigation results of the prints expositions in controlled conditions (temperature and air humidity, dynamics of the ageing) before deinking flotation depending on characteristics of the recycled fibers have been presented. The specific characteristics of fragmentation in relation to the ageing conditions have been discussed and the optical properties of handsheets as the indicator and proof of the mentioned mechanism of the mentioned process have been presented as well as.

In scientific sense the investigations are the contribution to explanation of the detaching mechanism of the ink particles from the cellulose fibers and the influence of that process on the efficiency of their removing from the suspension in the flotation phase. The expected investigation results could be used in the production of the recycled papers for obtaining the printing papers of better quality.

2. EXPERIMENTAL

The samples of colour prints were obtained by the conventional offset printing. The printing form contained the standard CMYK wedges in the range from 10-100 screen value, standard ISO illustration, textual positive and negative microelements and standard wedge for the production of ICC profile and 3D gamut.

The prints were made on the fine art mat paper 115 g/m^2. One series of prints was accelerated aged in the Käterman chamber in the period from 10, 20 and 30 days at 80° C and relative humidity of 65% and the other was aged at 100 ^{0}C and relative humidity of 80%.

Samples prepared in such a way were used in the process of alkaline deinking flotation (Bolanca & Bolanca, 2005). The handsheets were made using a laboratory sheet former, according to standard method T 205.

The optical parameters were performed by X-Rite spectrophotometer with the support of ColorShop program. The measuring results were processed by means of Data Analysis program and technical Graphic Origin Professional.

Residual ink particles size and number were assessed with image analysis-based software systems: Spec*Scan (Apogee System). Size intervals were defined according to TAPPI methods T 213 and T 437. Threshold value (100), white level (75) and black level (65) were chosen after comparing computer images to handsheets.

3. RESULTS AND DISCUSSION

Gamut represents the whole information range about colour, including the hue, saturation and brightness, which is possible to reproduce on the given medium. The two dimensional and three dimensional gamuts of the non aged and accelerated aged prints for 30 days at 80^0 C and 60 % relative humidity have been presented.

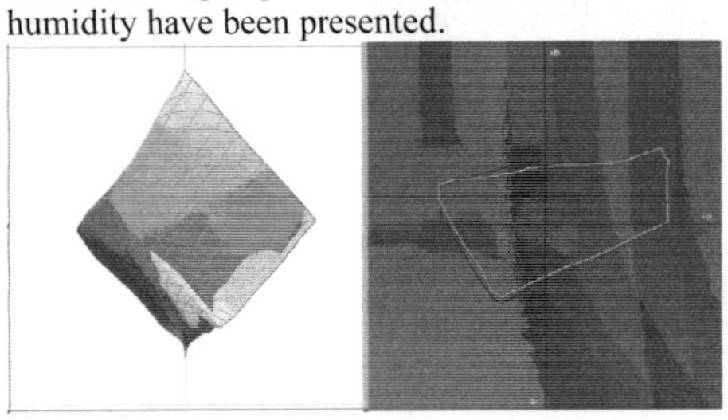

Fig.1. Comparative two dimensional and three dimensional gamut presentation of the non aged and for 30 days accelerated aged print at 80^{0}C and 65% humidity.

How the described changes, which appear during the ageing of prints, can have influence during the recycling process are presented in the investigation results which follow.

As the reflection of the print ageing somewhat weaker intensity of inking in blue and red area can be noticed (Fig.1.). In accordance with this the weaker intensity of inking, i.e. gamut ranges uniformly through the whole purple area. Smaller aberration on prints with smaller inking density, ranges in the area of cyan to green. The remaining part of the spectral area does not almost differ on aged and non aged print.

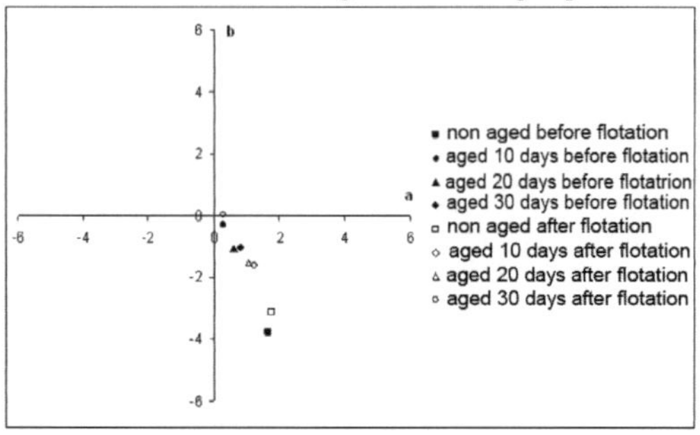

Fig.2. Chromatic values a* and b* of handsheet made from the fibers before and after the flotation of the non aged and aged prints at 80⁰ C and 65% humidity.

The investigation results show almost linear shift towards the area of achromaticity for handsheets made from the fibers before and after flotation with the increasing time of print ageing.

The influence of prints ageing on the efficiency of the flotation process and on the removal of the ink particles is presented in figure 3.

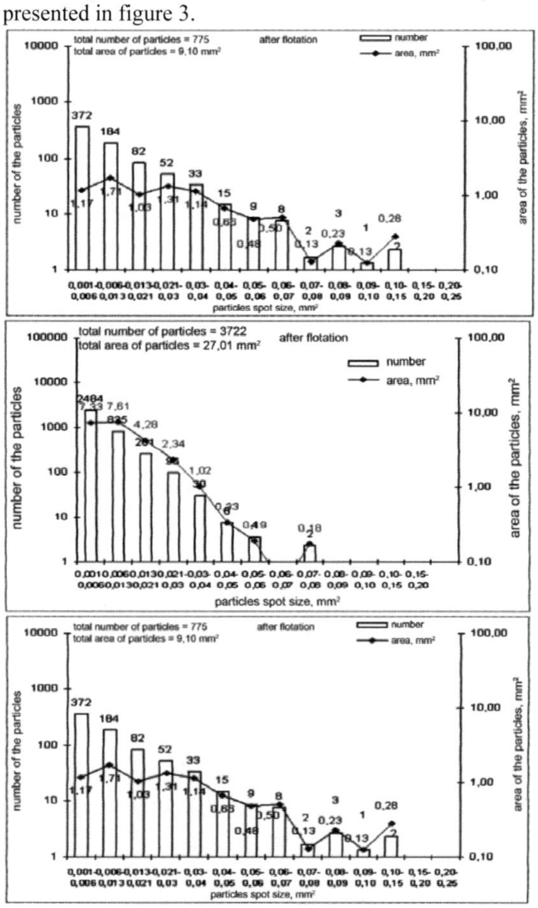

Fig.3. The distribution of the ink particles and their surface on handsheet made from the recycled fibers of the non aged and aged prints at different conditions

The investigation results show that the recycling efficiency is decreased by print ageing dependent on the conditions of their exposition. The increase of fragmentation by the increase of temperature and humidity in the process of the print ageing is observed. The greatest majority of particles after disintegration is in the first size class from 0,001-0,006 mm^2 in the shares as follows: non aged print 62,0%, aged print at 80^{0}C and 65,0% humidity 46.4% and in the extreme ageing conditions 50,0%. The efficiency of the particle removal with flotation, for the same class and in the same series is: 83,8%, 47,1% and 80,9%. From the basis of the results, the share of the influence on the removal of the particles concerning their size, the surface properties as well as the form can be deducted. Connected with the ageing process the specific characteristic of the particle distribution is determined with the final size class in which the particles are distributed, which, for handsheet after disintegration in the same series of samples, as earlier presented, are the following ones: 0.15-0,20 mm^2, 0,08-0,09 mm^2 and 0,25-0,30 mm^2. The removal of the whole number of ink particles by flotation is on the non aged print 86,0%, while on the aged print at the temperature of 80^0 C and the humidity of 65% it decreases at 53,5% and on handsheet made from the recycled fibers of the prints aged in more extreme conditions, it is 83,6%.

4. CONCLUSION

Only one segment of the broad investigations has been presented in this article. It comprises the study of endogenous and exogenous factors in the process of print ageing. The results of these investigations show weaker recycling efficiency of the aged prints dependent on the ageing conditions. The obtained results could be explained by the oxidation processes in which the chemical interactions between ink and paper are increased dependent on the variables of the investigated systems. In the context of the presented discussion the further investigations in combination with the experimental design and contemporary statistical methods will be continued.

4. REFERENCES

Ait-Benichou,S.;Cabral,A.R.& PanarottoC.T.(2008).Evoluation of biodegradation of deinkingby-products as alternative cover material. *Waste Management,*28,1,85-96

Bolanca I. & Bolanca Z.; (2005). The optical properties of deinked pulp. *Journal of Imaging Science and Technology* 49(3) 284-292

Dupont, A.L.; Egasse C.; Morin, A.; Vasseur F.; (2007). Comprensive characterisation of cellulose and lingocellulos -degradation products in aged papers: *Carbohydrate Polymers,*68,1, 1-16

Kumar Pati, R.; Vrat, P. & Kumar, P. (2008). A goal programm ing for paper recycling system. *Omega* 36,3,405-417

Labidi, J.; Pelach, M.A.; Turon, X. & Mutje, P. (2007). Predicting flotation efficiency using neural networks. *Chemical Engineering and Processing,*46,4,913-920

Lattuati-Derieux, A.; Bonnassies-Termes, S. & Lavedrine, B. (2006). Characterisation of compounds emitted during natural and artificial ageing of a book. Use of headspace-solid-phase microextraction/gas chromatography/mass spectrometry. *Journal of Cultural Heritage,*7,2,123-133

Liu, J.; Vandenberghe, J.; Masliyah, J.; Xu, Z. & Yorda, J. (2007). Fundamental study on talc-ink adhesion for talc-assisted flotation deinking of wastepaper. *Minerals Engineering,* 20,6,566-573

Wang J.P.; Chen, Y.Z.; Ge, X.W.& Yu, (2007).Optimization of coagulation-flocculation process for paper-recycling wastewater treatment using repouse surface methodology. *Colloids and Surfaces A: Physicochemical and Engineering Aspects,*302,1-3,204-210

Annals of DAAAM for 2008 & Proceedings of the 19th International DAAAM Symposium, ISSN 1726-9679
Editor B. Katalinic, Published by DAAAM International, Vienna, Austria 2008
Make Harmony Between Technology and Nature, and Your Mind will Fly Free as a Bird

PARTICULARITIES OF THE EHD LUBRICATION SYSTEM IN THE CASE OF POINT CONTACT

BOLFA, T[raian] E[ugen]

Abstract: *At the level of the contacts between the rolling bodies and the ring races of the bearings are, generally, accomplished the conditions of EHD lubrication working conditions. At the level of the contacts between the cage and the rolling frames, the lubrication systems vary among large limits (limited, mixed, EHD or HD conditions), while at the contacts between the cage and the guiding level prevail the mixed or HD lubrication conditions. The paper presents the theoretical and experimental importance of studying the thickness of the lubricant film and proposes methods of determining it. The diminished quantity of lubricant generates the phenomenon which determines the reduction of the film's thickness.*
Key words: *radial bearings, high speeds bearings, thickness of lubricant film*

1. INTRODUCTION

In the case of the bearings, the accidental contact voltage reach high values (1000-3000)Mpa , this way developing lubrication systems dependent on factors such as: load, springiness of the interlocking gear's materials, variation of the lubricant's viscosity with the pressure, peripheral speed and also geometry Dowson D.& Hamrock B.(1976),.

2. THEORETICAL AND EXPERIMENTAL ASPECTS

In the case of the point contact, solving the Reynolds equation becomes difficult because of the sideway leaks that can not be ignored. So, in this case, the starting point is Reynolds equation for the bidirectional leak:

$$\frac{\partial}{\partial x}\left(\rho\frac{h^3}{\eta}\cdot\frac{\partial p}{\partial x}\right)+\frac{\partial}{\partial y}\left(\rho\frac{h^3}{\eta}\cdot\frac{\partial p}{\partial y}\right) = 12\frac{\partial}{\partial x}(\rho\cdot h\cdot v_x)+12\frac{\partial}{\partial y}(\rho\cdot h\cdot v_y)$$

and

$$v_x=\frac{v_{1x}+v_{2x}}{2};\qquad v_y=\frac{v_{1y}+v_{2y}}{2} \tag{1}$$

where V_x and V_y represent the medium peripheral speeds on the rolling Ox direction and on the lateral Oy direction, according to figure 1

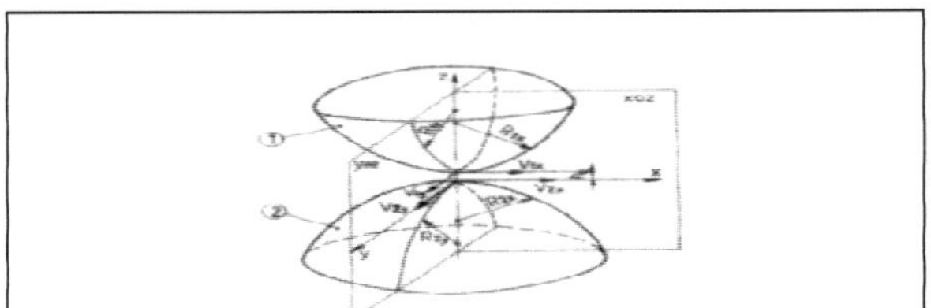

Fig.1. The geometry of the Hertzian rolling point contact

The contact between the two bodies (1) and (2) may be equated with a contact between a plane and an equivalent revolution bodies which has in the contact point the equivalent curved shaped Rx and Ry beams, where:

$$\frac{1}{R_x}=\frac{1}{R_{1x}}+\frac{1}{R_{2x}}\ \text{și}\ \frac{1}{R_y}=\frac{1}{R_{1y}}+\frac{1}{R_{2y}}$$

In the case when the two bodies are distorted elastically under the action of the normal load Q, the vertical distance between 2 points situated on the distorted surfaces is identical with the similar distance between the equivalent body and the plane, like in figure 2, where:

$$h(x,y)=h_0+s(x,y)+w(x,y) \tag{2}$$

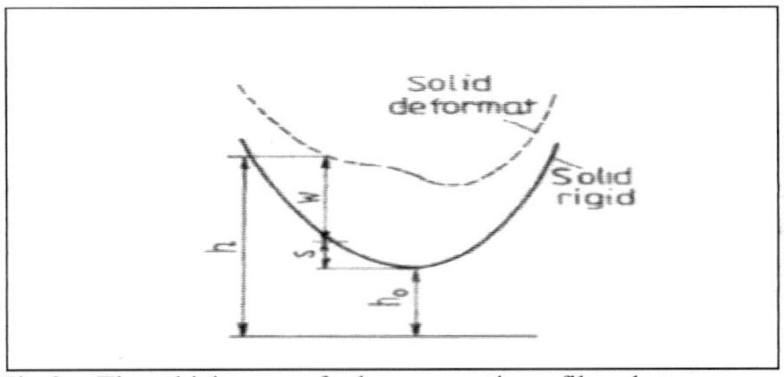

Fig.2. The thickness of the separation film between an equivalent ellipsoid and a flat surface

In equation number 2, h_0 is the thickness of the lubricant film in the central area, $s(x,y)$ represents the geometrical separation of the two bodies, and $w(x,y)$ is the elastic distortion.
The geometrical separation $s(x,y)$ is given by the formula:

$$S(x,y)=\frac{x^2}{2R_x}+\frac{y^2}{2R_y} \tag{3}$$

The elastic distortion, in the case of perfect elastic, homogenous and isotropic bodies, is given by the relation:

$$W(x,y)=\frac{2}{\pi E'}\int_{-\infty}^{+\infty}\int_{-\infty}^{+\infty}\frac{p(\xi,\delta)d\xi d\delta}{\sqrt{(x-\xi)^2+(y-\delta)^2}} \tag{4}$$

In formula (4), $p(\xi,\delta)$ represents the pressure under a point (ξ,δ) from the contact area.
But there have to be taken into consideration both the changes of viscosity, but also of the lubricant's density, as a consequence of the high pressures from the contact area. Thus, it can be used Barus's relation for viscosity, where:

$$\eta=\eta_0 e^{\alpha p} \tag{5}$$

For the variation of the lubricant's density with the pressure, it can be used Dowson's relation:

$$\rho=\rho_0\left(1+\frac{0{,}6p}{1+1{,}7p}\right) \tag{6}$$

Solving the first equation (1) is very difficult, but as a consequence of experimental determinations, was emphasized the presence of a central plate, with h_0=the thickness of the film and of a horseshoe-shaped notch, situated in the exit area of the contact.

The practical importance presents h_{min} (the minimum thickness of the lubricant) and h_0 (the central thickness). Essentially, these depend on U, the speed parameter, and w the loading parameter (figure 3).

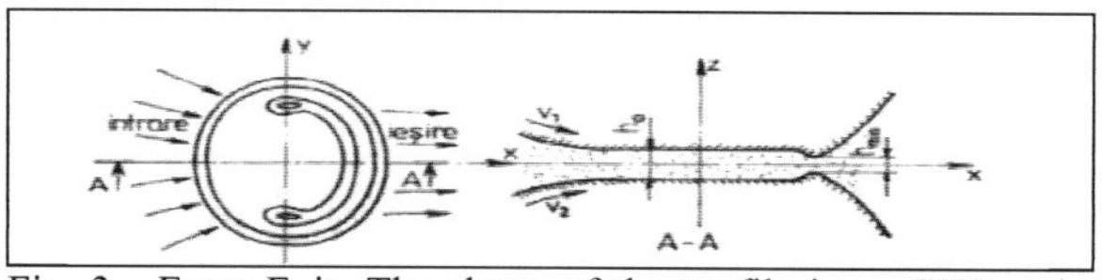

Fig. 3 - Entry-Exit- The shape of the profile in an EHD point contact

Using the principle of resistive method and considering the contacts between balls and bearing races, as a consequence of their separation through the lubricant film, realizing electrical resistances when the flow passes from a level to another, the electrical resistance of a contact varies exponentially with the thickness of the film:

$$R = Ke^h \tag{7}$$

where proportionality factor K, depends on the pressure, temperature, lubricant type, contact's dimension, rigor, etc. The resistive method allows qualitative, comparative measurement, of the lubrication system for diverse lubrication mediums. If it is considered a cylindrical roller bearing and the equivalent electrical scheme, it can be written that the global resistance of the bearing is:

$$R = \sum_{n=1}^{2} \frac{1}{(R_{c,i}+R_{c,e})_n} \tag{8}$$

where $R_{c,e}$ and $R_{c,i}$ represent the electrical resistances at the level of interior and exterior contacts and "2" is the number of balls. The use of capacitive method means using a radio frequency flow of low voltage applicable to the frames among which is found lubricant's film whose thickness has to be determined, the lubricant's film functioning like a dielectric. The method requires that while testing, the dielectric constant of the oil should not vary. The method is also practicable because, when the thickness of the film is small, the measured capacity is higher and so, the measurement's precision is correct. The difficulty was caused by the frequency of the piercement of the film by the contacts of low resistance, among the surfaces' asperities or conductive particles'.

3. CONCLUSIONS

Due to the radial loadings, in relation number (8) takes part only the resistances from the loaded balls (balls 1 and 2), in the case of other contacts the electrical resistance is ∞.

Due to the small thicknesses of lubricant film (<1 μm), the flow that passes through the contacts is low, varying among the limits (1÷50) μA, the unique value of the flow is dictated by avoiding the appearance of the electrical loadings at the level of the contacts.

In figure 4 is presented the electrical schematic diagram, the supply voltage is of 1 V, and the additional resistance Ra has the same value as the quantity order of the electric bearing resistance.

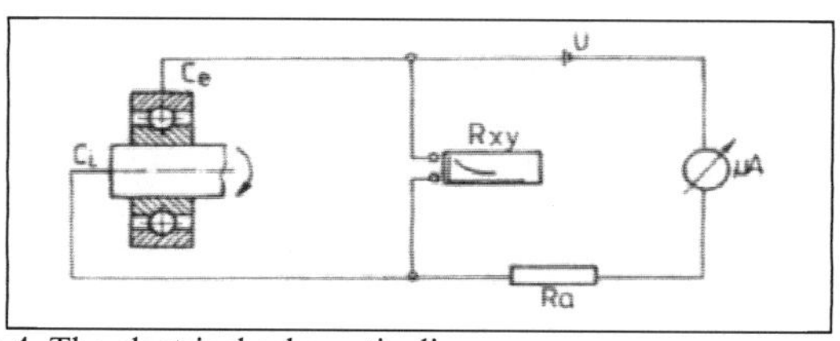

Fig.4. The electrical schematic diagram

With the help of an acquisition plank have been registered the variations of potential dropping on the bearing and also the variation of electrical resistance of the bearing (the collector from the external ring that is immobile, while the collector from the internal ring is mobile) (Bolfa T. (2006).

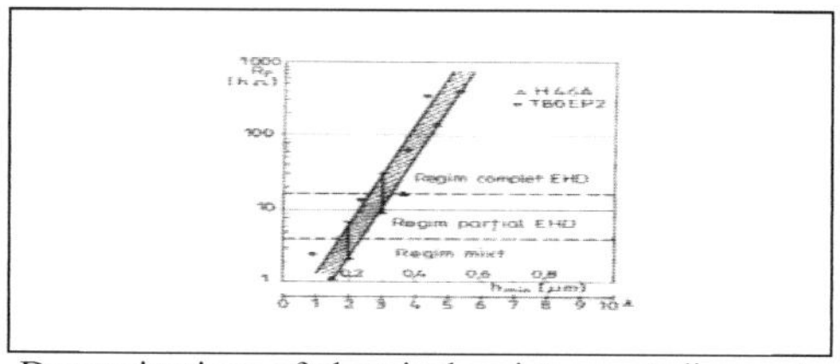

Fig.5. Determinations of electrical resistance at diverse types of oil

In figure 5 are presented electrical resistance determinations at 2 tested oils. At the same time, on the diagram from the picture, according the λ parameter's values, defined as the proportion between the minimum thickness of the lubricant and the compounded rigor of the 2 surfaces in contact, have been separated the lubrication working conditions, according to global electrical resistance of the bearing. The diminishing of the electrical resistance under 3-4 kΩ leads to the existence of mixed or limited lubrication systems. This reduction is due to the lubricant's deterioration and precedes always its out of use.

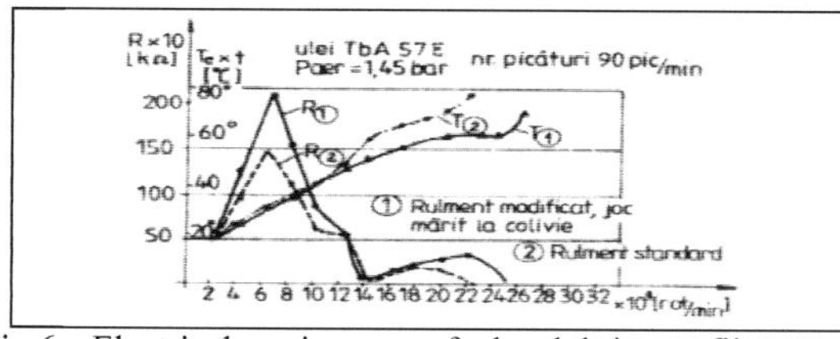

Fig.6. Electrical resistance of the lubricant film and the temperatures in the case of the particular bearing 6306 MAUP.

In figure 6 are presented the evolutions of the electrical resistance of the lubricant film, and also of the temperatures of 6306 MAUP bearing, in two variants of the cage's play.

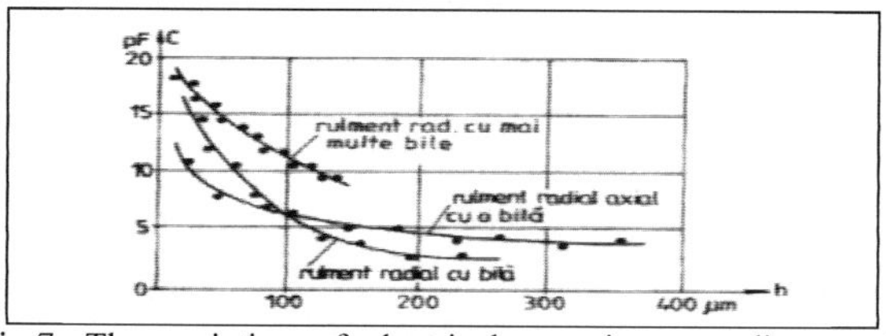

Fig.7. The variation of electrical capacity according to the thickness of the lubricant film

Using the initial principle of capacitive method for experiments (figure 7), was used a sectioned angular contact ball bearing, the pressure being taken by a single ball, and in the third phase was used a cylindrical roller bearing with many balls.The lubrication is essential to assure the bearings' friability and the durability is directly influenced by the parameter λ of the lubrication film.

4. REFERENCES

Blaga A.&Robu C. (1991),*The technology of Organic Coverings*, Bucharest Technical Publishing House

Bolfa T. (2006) *Mechanics Contact and Tribology*, Lux libris Publishing House

Bolfa T. Phd. Thesis, *Contributii privind imbunatatirea perfomantelor calitative ale rulmentilor de turatie ridicata*

Dowson D.& Hamrock B.(1976), *Isothermal EHD Lubrication of Point Contact.*, Transf. of ASME, vol 98

The Contract No.47/87, *Researches Regarding New Materials*

Annals of DAAAM for 2008 & Proceedings of the 19th International DAAAM Symposium, ISSN 1726-9679
Editor B. Katalinic, Published by DAAAM International, Vienna, Austria 2008
Make Harmony Between Technology and Nature, and Your Mind will Fly Free as a Bird

SERIAL DATA COMMUNICATION CONNECTION OVER GSM DATA CHANNEL

BORDEA, N[icolae]; SINTEA, S[orin]; MORARU, A[urel] S[orin] & KRISTALY, D[ominic] M[ircea]

Abstract: *This paper describes data communication device over GSM Network, used for delivery of serial data packages (in words or package of words data format). The device is able to connect simultaneous two RS232 lines and one RS485. On RS232 lines the device ensures full data synchronisation protocol and data buffering for each line. Programming of the device is possible through internal web site.*

Key words: *Serial data package communication over GSM network*

1. INTRODUCTION

In many occasions we need to transfer data between serial isolated devices (controllers, programmable logic controllers or computers), located at long distance or in inaccessible locations. In these locations the only way to communicate is over GSM network (Tisal, J., 1999; Hammuda H., 1999). Also, all these devices are produced by third party producers and we do not have possibility to upgrade their software or add drivers (Bajenarescu T, 2000). In this case we need new equipment which adapt the signals and connect our isolated devices through GSM network. The equipment will ensure end to end communication connection protocol between three independent serial lines (two RS232 lines and one RS485 line).

Until now, many manufacturers offer dedicated devices for communicating over GSM using modems connected through USB cables (Bajenarescu T, 2000). But we have a lot of old devices that not support USB connections. Also the modems request software driver or additional hardware modules implemented into connected device. Presented solution solves these requests by introducing a hardware protocol level between two communicating end points equipments.

2. GENERAL DESCRIPTION

The system proposed is composed by two communication devices. Each device is able to communicate with the pair device, and create a direct channel between each corresponding port (Fig. 1.). The device creates three parallel communication channels for each port. Attached to each port the device creates an internal 1024 Bytes buffer for data synchronization. For communication, each port status is show using 3 status LEDs (for data transmission, data receiving and buffer status/busy).

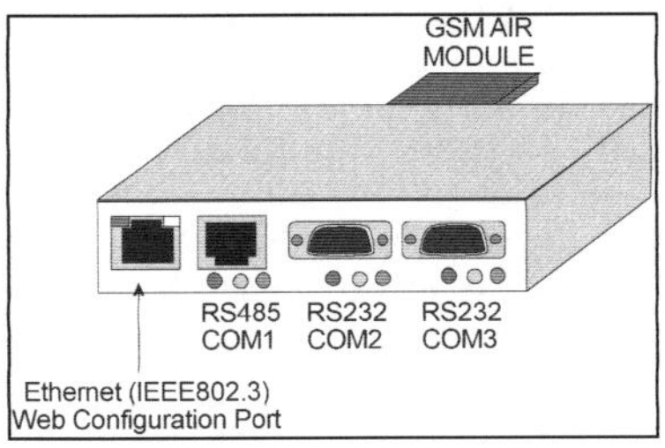

Fig. 1. GSM Communication device.

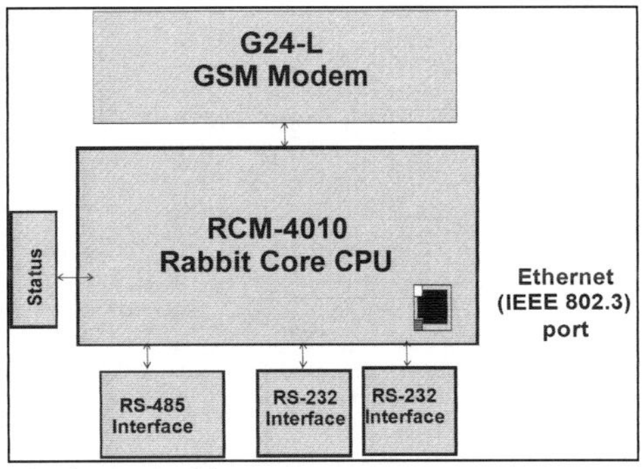

Fig. 2. Hardware Architecture.

Additional the communication device have attached an Ethernet port. Over this port the operator can connect, using a notebook, with internal CPU unit data and configure the communication device parameters.

To ensure link with GSM network the communication device contains the GSM air connection modem.

The activity of all modules is synchronized by micro-controller unit. The microcontroller run dedicated software for implementation of communication protocol and link connection between unit and GSM modem.

2.1 Hardware Architecture

To solve all this requests (see Fig. 2) we are using an RCM-4010 Rabbit 8 bits CPU unit @ 59 MHz (Rabbit Semiconductor Inc, part of Digi International). This CPU is offering us possibility for connection of four serial USART connection ports (Rabbit Semiconductor Inc., 2006). Also the CPU has one Ethernet port on board for TCP/IP connection and implemented of one Web Server in system driver library. For saving data the CPU contain a data flash on board. Two USARTs from CPU board are connected with RS232 external devices (COM2 and COM3) using RS232 interface line adapters (see Fig. 2.) and one USART is connects RS485 devices (over COM1 connector) with our controller through RS485 interface line adapter. The connection with GSM modem is ensuring using low voltage serial port COM4 from CPU board (Rabbit Semiconductor Inc., 2006).

The Rabbit 4000 series CPU offer us possibility for parallel multitasking software programming through Dynamic-C compiler (Rabbit Semiconductor Inc., 2006).

Modem used for communication with GSM Network is Motorola G24-Lite (G24-L) modem. This modem implements Motorola G24 AT commands for communication and control. On GSM network the modem supports GSM/CSD protocol for data transfer on quad band GSM networks (850/900/1800/1900 MHz) and GPRS multislot class 10 (Motorola Inc, 2008).

Modem, micro-controller and interfaces are integrated in the same compact enclosure (see Fig. 1). The module can be powered from 8-30V power supply. The external ports (communication ports and Ethernet port) are placed one the front of enclosure for easy access.

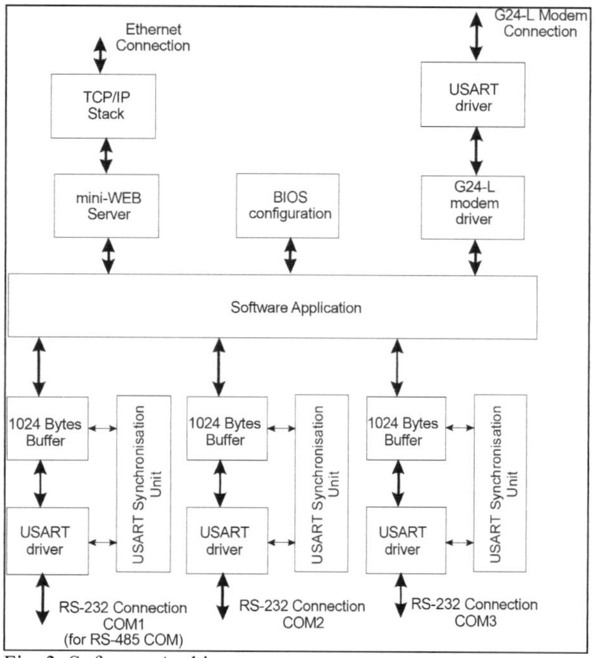

Fig. 3. Software Architecture

2.2 Software Architecture

All software is created based on parallel programming mechanism, ensured by Dynamic-C compiler language for Rabbit 4000 processors (Rabbit Semiconductor Inc, 2006). The software is divided in five threads, started by main application (see Fig. 3):

- Three of them are responsible with serial data processing incoming over RS232 connections with local communication devices (named **serialProcess1**, **serialProcess2**, **serialProcess3**);
- One thread is responsible with processing of web interface (named **webProcess**);
- One thread is monitor communication with GSM modem, and monitor GSM communication line (for testing GSM signal strength, verifying GSM incoming connection), named **gsmProcess**;
- One thread is responsible with data delivery over GSM communication line (named **gsmDataProcess**).

serialProcess thread, is the base thread which processing incoming data from RS232 line (COM1, COM2 or COM3). This process is buffering incoming data from RS232 controller's device driver. The process also drive synchronisation signals between controller and serial attached device. Also, the same process analyse one of the data receiving from GSM remote unit from data queues. If some data are found in attached queue, the data is extract from that queue and send through corresponding serial device driver, byte by byte.

gsmProcess thread is responsible with monitoring of GSM line. This thread monitors if is signalling a GSM connection from other threads. The process will force a connection with remote GSM unit, if GSM signal is able to ensure clear connection with the remote unit. When the connection with remote unit is established the process will send a signal in system. In case of incoming request from remote GSM unit, the process is responsible with ensure of correct answer to establish the connection with remote unit. After this, the process will signal in system that connection is ensured. All protocol syntax is described into G24-L modem device driver (Motorola Inc., 2008).

gsmDataProcess thread is responsible with processing of stream of data transferred through GSM communication line with remote unit. The activity of this process is divided into two main tasks. One task is involved with testing of serial receiving queue; in case of detecting at least one character, the thread will request a GSM connection with the remote unit over G24L modem. The other task of this thread is monitoring of incoming data stream from GSM network. When data are receiving, the task get data extract from data package the serial information data and transfer it to destination data buffer.

And at last is the **webProcess** thread. This thread is responsible with presentation of the status of the unit and configure of the unit on a web page. This web page can be accessed through TCP/IP layer over Ethernet port. Using this thread we can configure the following parameters:

- The communication USART parameters (speed, parity, even bit, serial word width, hardware protocol levels etc)
- The phone number of remote unit
- Buffer size for each channel (1 for direct connection)

All settings are saved into a local flash memory.

3. RESULTS

The link channel realised with this devices ensures a serial connection at 600-19200 bps speed between end equipments.

In domain 600-4800 bps speed, the connection protocol layer does not request data buffering. The data delay transmission (between data send and data recreated at end point) is about $2\,ms + \frac{1}{Speed}$ (where *Speed* is the data transmission speed).

When we are using the higher speed than 4800 bps, the data buffer is essential, and also the terminal must use hardware handshake signals to synchronize data transfer. The transmission delay is about $2\,ms$.

The connection time is depending by the calling time in GSM network. If you want to obtain good results (low time connection response) we recommend using an "auto connection" mode; in this case the communication devices will try to keep the data line always open.

4. CONCLUSIONS

Current equipment is created for an easy way of connection between two PLCs or controllers at long distance range using GSM network. The connection does not request any other equipments or software drivers.

This solution is very efficient when we do not have possibility of upgrading of equipments or software implemented into the controllers or PLCs. The adopted solution does not interpret the transferred information; it is presented as a protocol layer between the end points of communications.

In future the solution can be developed by sending data from PLC or custom controller to PC central point using GSM/CDMA or GSM/GPRS network infrastructure (Geier, J., 2005).

5. REFERENCES

Bajenarescu, T. (2000). *Personal Communication Systems,* Teora, ISBN 973-20-0236-0, Bucharest, Romania

Geier, J. (2005). *Wireless Networks first-step,* Corint, ISBN 973-653-709-9, Bucharest, Romania

Hammuda, H. (1999) *Celular Mobile Radio Systems,* Teora, ISBN 973-20-0004-X, Bucharest, Romania

Motorola Inc. (2008). *Motorola G24-L and G24-LC Developer's Guide. Module Hardware Description.,* Motorola, 6802984C05-C, USA

Rabbit Semiconductor Inc. (2006-2007). *RabbitCore RCM4000 User's Manual,* Rabbit Semiconductors Inc, 019-0157, 070928-F, USA

Tisal, J. (1999). *GSM: rèseau et services,* Teora, ISBN 973-20-0192-5, Bucharest, Romania

Annals of DAAAM for 2008 & Proceedings of the 19th International DAAAM Symposium, ISSN 1726-9679
Editor B. Katalinic, Published by DAAAM International, Vienna, Austria 2008
Make Harmony Between Technology and Nature, and Your Mind will Fly Free as a Bird

THE NECESITY OF CONSIDERING CAVITATION EROSION A MECHANIC PHENOMENA, AGAINST CHEMICAL COROSION

BORDEAŞU, I[lare]; ANTON, L[iviu] E[ugen]; BAYA, A[lexandru] & JURCHELA, A[lin]

Abstract: The paper is an attempt to establish the necessity of considering cavitations erosion like a distinct process of corrosion. It is analyzed cavitation erosion from mechanic point of view, considering results of cavitation erosion upon different materials in distillated water and in other liquids, chemical aggressively. Numerical results of the paper are obtained both in Hydraulic Laboratory of „Politehnica" University of Timisoara and other world wide authorized laboratories. Eroded mass during a cavitation process is considered reference parameter.
Key words: cavitation erosion, eroded mass, aggressive chemical environment, material, vibratory apparatus,

1. INTRODUCTION

In last 30 years (Franc & Michel 2004, Mitelea et al. 2005), cavitations erosion is considered, by specialists, a typical local fatigue phenomena of the material, which suffered deformations or/and breakings under repeating action of micro jets and shock waves generated by cavitation bubbles implosion. Scientific researches, done in special cavitation laboratories, shows that damage level produced by cavitation erosion depends, both by cavitaion intensity, specific to hydrodynamics of the cavitating flow (which includes nature and physical properties of the liquid) and, by the nature of the material (Bordeasu 1997). Because of these, large theoretical and experimental studies are realized in order to correlate resistance to cavitation erosion with mechanical characteristics of the material and with hydrodynamic parameters of cavitaion phenomenon (Garcia et al. 1967, Bordeasu 1997, Franc &Michel 2004).
In the paper there are presented a number of caviattion erosion results tests, made in various liquids, and using materials with different cavitation erosion resistance, in order to reveal mechanical aspects of cavitation erosion and to motivate the cavitation process approach by hydraulic specialists, against general corrosion. To reach this goal, mass loss at various cavitation regimes is used like analyze parameter. In the paper are used only results of cavitation erosion produced in vibratory apparatus, because of the high intensity of damages process.

2. MECHANICAL ASPECT OF CAVITATION EROSION

If cavitation erosion was observed long time ago, the firs technical cases (ship propellers and hydraulic turbine runners) eroded by cavitation phenomena, were noticed about 1890 (Bordeasu 1997). General aspect of the erosion areas, present following characteristics:

- cavitation erosion of the material is selective;
- individual damages have the aspect of roughness grooves, often parallel disposed ;

The opinion that cavitation erosion is mostly a mechanic phenomena is sustained also by following experimental results (Franc & Michel 2004):

a) it have been obtained rapid erosions in situations in which in non cavitating flow (monophasic flow), for the same couple of liquid – solid, chemical reaction were unrelevant (for example oil product/glass, mercury/plexiglas);
b) microscopic observation of some caverns produced in few imploded bubble areas leads to the conclusion that the caverns have the same configuration to moon's craters, that is symmetric holes with often singular border ring

3. EXPERIMENTAL FACILITY

The paper results are obtained on cavitation eroded materials in vibratory apparatus characterized by the highest damaging intensity (Bordeasu et al.2007).
It have been analyzed cavitation eroded materials in vibratory apparatus from Hydraulic Machinery Laboratory of Timisoara (named THML) (Bordeasu et al.2007) and from Michigan University Laboratory (named MUL) (considered the best one)) (Bordeasu 1997, Garcia et al. 1967) .
Operating parameters of Michigan (Garcia et al. 1967) vibratory apparatus, which results have been used in present paper considerations, are as follows:

- Power consumption 200W
- vibrations aplitude 12.5 -100 μm
- vibrations frequency = 20000 Hz
- probe diameter = 14.3mm

4. RESULTS AND DISCUTIONS

For a pertinent analyze in order to justify the necessity of approaching cavitation erosion as a mechanical process, in figure1 and 2 were shown Bordeasu results on various materials probes tested in TMHL on magnetostrictive vibratory apparatus with nickel tube (Bordeasu et al.2007), also in various liquids. For the same goal, in figure 4 are presented (Garcia et al. 1967 results, and in figure 3 are presented Sisak (Sisak et al. 1980) results.

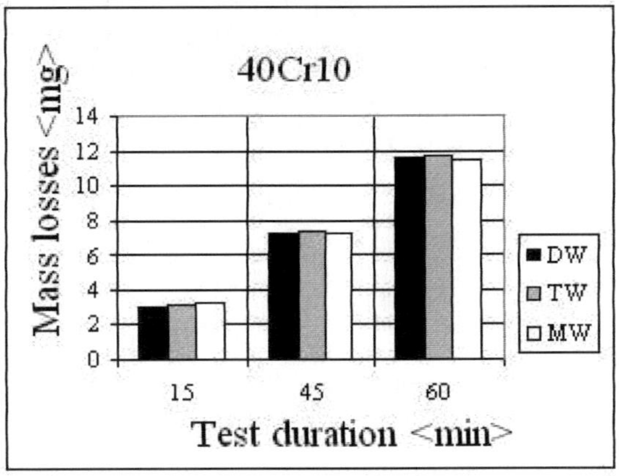

Fig.1. Cavitation erosion of alloyed steel 40Cr10

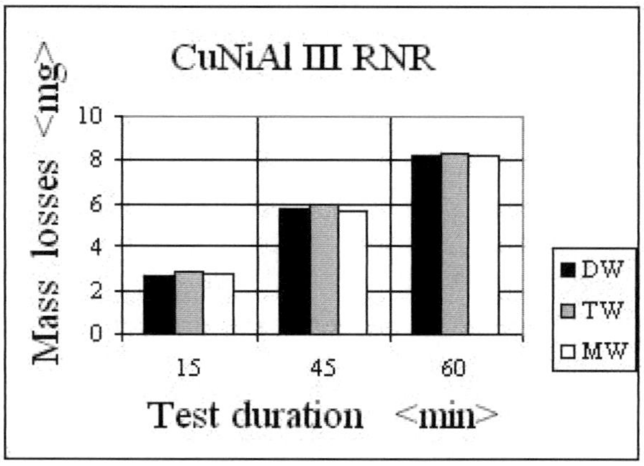

Fig.2. Cavitation erosion of bronze CuNiAl IIIRNR

Note: *DW- distillated water; TW – drinking water; MW- standard sea water; Em – emulsion; Ra-average roughness of probe surface before cavitation erosion. Here, Ra=0,2μm.*

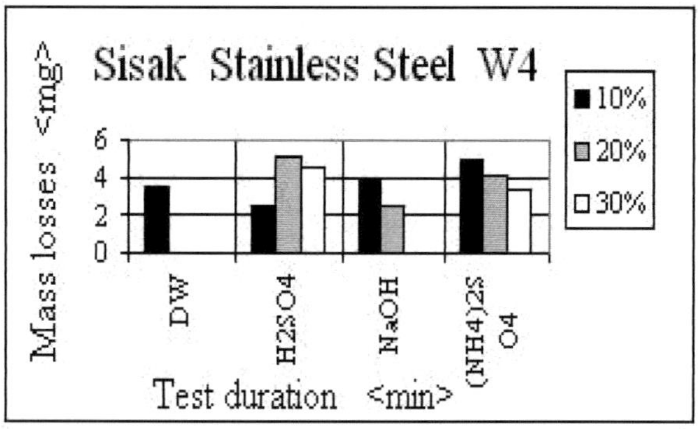

Fig.3. Processing after Sisak e.a.

Note: *DW- distillated water;* H2SO4 – H_2SO_4; (NH4)2SO4- $(NH_4)_2SO_4$; 10%, 20% 30% - substance percentage in water).

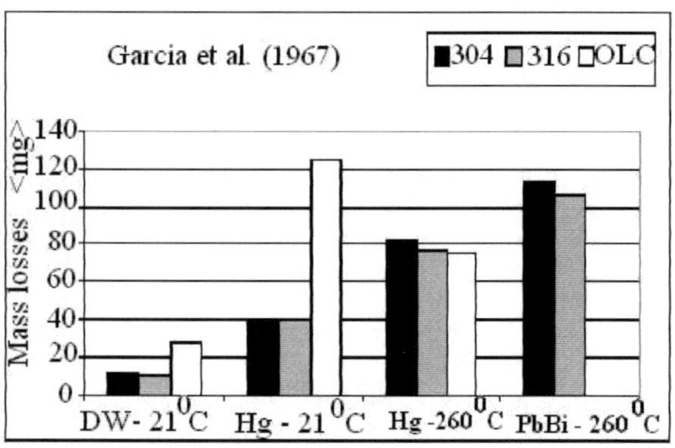

Fig.4. Processing after Garcia e.a.

Note: *DW- distillated water at 21C; Hg –mercury at 21C;PbBi-lead - bismuth at 260 C; 304 and 316 stainless steel; OLC – carbon steel.*

The results presented in fig. 1 and 2, show for the same material (alloyed steel 40Cr10, bronze CuAl III RNR) independent by working fluid (distillated water, drinking water and sea water), mass losses are produced only by mechanical effect of cavitation erosion, without substantial contribution of surface or intergranular corrosion.

Cavitation liquids and probes were chosen so, to reveal both cavitation erosion independence by chemical corrosion, through effect of upper mentioned factors on cavitation damages, as well as the plurality of cavitation and chemical corrosion effect.

These mass losses are mostly the same, the differences between them are included in deviation limits according to ASTM standard, under ±15%, specific to complx phenomena, such as cavitation erosion.

Datas showh in fig.3,4 reveal the effect of substance percentage in chemical aggressively solution upon material damage during cavitation erosion..

4. CONCLSIONS

1. It have been demonstrated why cavitation erosion must be considered like a mechanical process of materials damaging and a distinct part of corrosion. In ordre to reach this goalit have been analyzed the results of cavitation erosion tests upon various metals tested on vibratory apparatus in TMHL and MUL. Basic of mechanical aspect were done by cavitation erosion tests performed in distillated water drinking water standard sea water and emulsion

2. It was clarified the effect of cumulated action of chemical corrosion – cavitation erosion uopm damage level of materials during cavitation phenomenon. It was possible because various liquids chemical aggressively were used.

ACKNOWLEDGMENTS

The present work has been supported from the National University Research Council Grant (CNCSIS) PNII, ID 34/77/2007 (Models Development for the Evaluation of Materials Behavior to Cavitation)

5. REFERENCES

Bordeaşu, I. (1997). *Eroziunea cavitaţională asupra materialelor utilizate în construcţia maşinilor hidraulice şi elicelor navale. Efecte de scară*, Teză de doctorat, Timişoara. (*Cavitational erosion upon materials used in manufacturing hydraulic machinery and ship propellers.* Doctoral degree thesis)

Bordeaşu, I., Popoviciu, M.O, Mitelea, I., Anton, L.E., Bayer, M., Funar, S.P., (2007).. *Cavitation Eroded Zones Analysis For G-X 5CrNi13.4 Stainless Steel* Proceedings of The 18th International DAAAM Symposium "Intelligent Manufacturing & Automation: Focus on Creativity, Responsibility and Ethics of Engineers", Zadar Croatia, 24-27.10.2007, pp.105-106, ISSN 1726-9679, ISBN 3-901509-58-5

Franc, J., P., Michel, J., M. (2004). Fundamentals of cavitation, Kluwer Academic Publishers-Dordrecht/Boston/London, 2004, ISBN 1-4020-2232-8 (HB)

Garcia, R., Hammitt, F.G., Nystrom R.E. (1967). *Correlation of Cavitation Damage, with Other Material and Fluid Properties*, Erosion by Cavitation or Impingement, ASTM STP 408, American Society Testing Materials, p.239-279

Mitelea I., Bordeasu I., Hadar A., (2005). *The Effect of Nickel Content Upon Cavitation Erosion for Stainless Steels with 13% Chromium and less than 0,1% Carbon*, Chem.Abs. RCBUAU 56(11) Revista de chimie Vol.56 Nr.11, vol.56, pp.1169-1174, ISSN:0034-7752,

Sişak A, Kuzman F.F., Kuzman-Anton R., Potencz I. (1980). *Eroziunea cavitaţională în soluţii apoase de diferite concentraţii ale unor substanţe chimice*, Memoriile secţiilor Ştiinţifice ale Academiei RSR, Seria IV, Tomul III, nr.2, Iaşi, 1980, p.197-221, ISSN 1224-6077.

*** *Standard method of vibratory cavitation erosion test*, ASTM, Standard G32-8

Annals of DAAAM for 2008 & Proceedings of the 19th International DAAAM Symposium, ISSN 1726-9679
Editor B. Katalinic, Published by DAAAM International, Vienna, Austria 2008
Make Harmony Between Technology and Nature, and Your Mind will Fly Free as a Bird

THE DAMAGE OF AUSTENITE - FERRITE STAINLESS STEELS BY CAVITATION EROSION

BORDEASU, I[lare]; GHIBAN, B[randusa]; POPOVICIU, M[ircea] O[ctavian]; BALASOIU, V[ictor]; BIRAU, N[icolae] & KARABENCIOV, A[drian]

Abstract: *The present paper presents some aspects concerning cavitation erosion damages of stainless steels with austenitic ferrite structure used for casting the blades of hydraulic turbines. The steel composition was elaborated in Resita Siderurgic Plant. The structure type was established with a Schäffler diagram (E_{Cr}= 21.37 %, E_{Ni}=8.62 %). Chemical composition was locally determined in the "Special Materials Research Center, University Politehnica Bucharest". The investigations were carried out by scanning electron microscopy and after cavitation tests.*
Key words: *erosion, cavitations, stainless steel, electronic microscope, microstructure austenito-ferite.*

1. INTRODUCTION

Martensitic stainless steels are the most used materials for manufacturing hydraulic turbine components, such as runners and blades (Bordeaşu 2006). But, different pieces made of stainless steels, with two phases austenite and ferrite, is compulsory analyzed in order to establish the cavitation erosion behavior. The present paper aim is to study some aspects of austenitic-ferritic microstructure, damaged by intense cavitation attack produced after testing them in the vibrator magnetostrictive device of Timisoara Hydraulic Machinery Laboratory.

2. TESTED MATERIAL

The experimental steel was elaborated in Resita Siderurgic Plant, after a receipt not standardized. The local composition determined by EDAX analysis in Laboratory of Special Materials Research Center from University Politehnica Bucharest is: C = 0.1%, Si = 0.76%, Mn =3.28%, Cr = 20.23%, Ni =3.98%, Al =0.63%, Fe=rest. In according with Schäffler diagram (E_{Cr}= 21.37 %, E_{Ni}=8.62 %), the steel structure is formed by austenite ≅50 %, ferrite ≅45 % and martensite ≅5 %.

The mechanical characteristics were obtained in The Material Strength Laboratory of the Timişoara "Politehnica" University, and are: ultimate strength σ_m= 610 MPa, yielding strength $\sigma_{0,2}$ = 338 MPa, and Brinell hardness 185 HB.

3. TEST FACILITIES AND TESTING METHOD

The cavitation erosion tests were carried out in the magnetostrictive facility, of Timisoara Hydraulic Machinery Laboratory, in accordance with ASTM G32-85 (Bordeaşu et.al., 2007), using as cavitation liquid drinking water at 20 ±1^{0}C.

The microstructure, analyzed by a scanning electron microscope, figure 1-3, put in evidence the following aspects of the cavitation damages after 165 minutes:

- Fractures by cavitation, with brittle grain aspect. The cause of brittle behavior is due to the high amount of manganese, which makes the austenite grain and the needles of martensite more roughness (with a more fibrous structure).
- Mixed surface with brittle aspect, by cleavages having transcrystalline and intergrain propagation. The explanation is connected with the fact that microstructure damages may begin at the grain boundary between ferrite and austenite, respectively austenite and martensite needles. By continuing the cavitation attack, the damage is propagated in the ferrite matrix till massive breakdown. Because the bonds between austenite and martensite needles became softer, even the complete expulsions of austenitic grains and martensite needles occur during the attack. Also, almost complete damages of ferrite islands were locally put in evidence by scanning electron microscopy.
- Cavitations with intergrain propagation. These cavitations are specific to ferrite mass damage, which from the three constituents of the steel structure, is the phase with the lowest resistance to cavitation erosion (Mitelea et al., 2005)

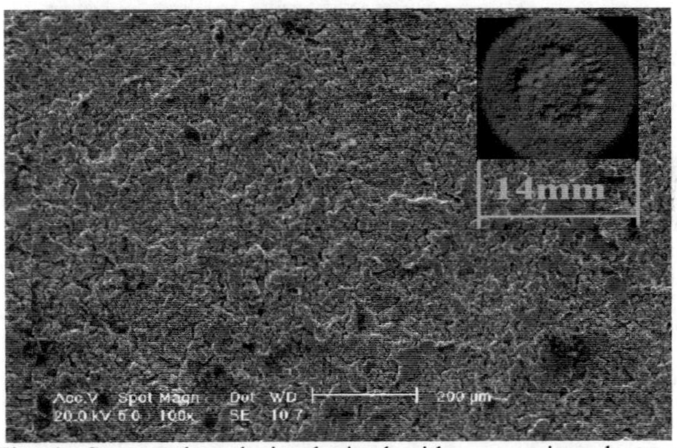

Fig. 1- Structural analysis obtained with a scanning electron microscope (SEM), after 165 minutes of cavitation (x100)

For a correct analysis of the steel behavior at cavitation erosion, in figures 4 and 5, there are presented the characteristics curves by cavitation erosion (cumulative mass and erosion rate) compared with the martensitic stainless steel OH12NDL (C = 0.1%, Si = 0.3%, Mn =0.04%, Cr = 12.75%, Ni =1.25%, P =0.085%, S =0.025%, Cu =0.9%, Fe=rest, σ_m = 610 MPa, $\sigma p_{0,2}$ = 225 MPa, Brinell hardness 400 HB, martensite ≅74 %, ferrite ≅26%) generally used for hydraulic turbine blades in our country and being considered with a good resistance to cavitation erosion.

The evolution of eroded volume, respectively eroded rate versus attack time, from figures 4 and 5 show clearly that the studied austenite- ferrite steel (A-F) has a better behavior at cavitation attack than the OH12NDL steel, usually used for hydraulic turbines blades in Romania.

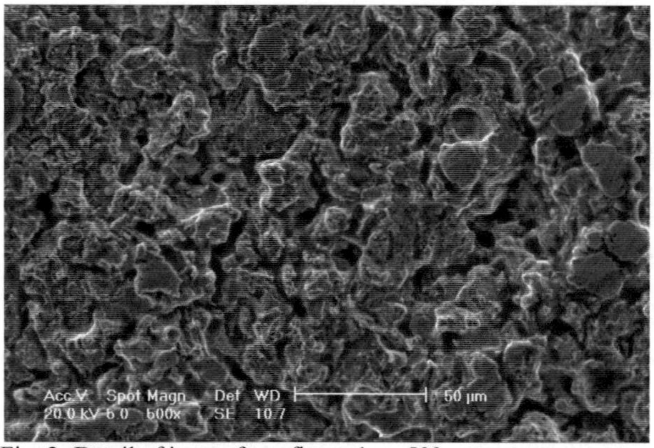

Fig. 2- Detail of image from figure 1, x 500

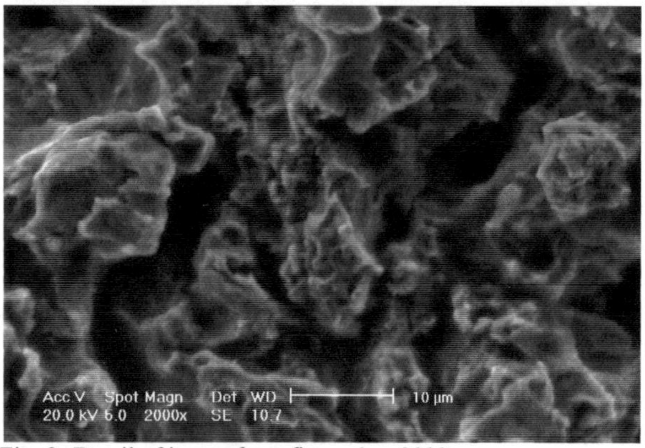

Fig. 3- Detail of image from figure 1, x 2000

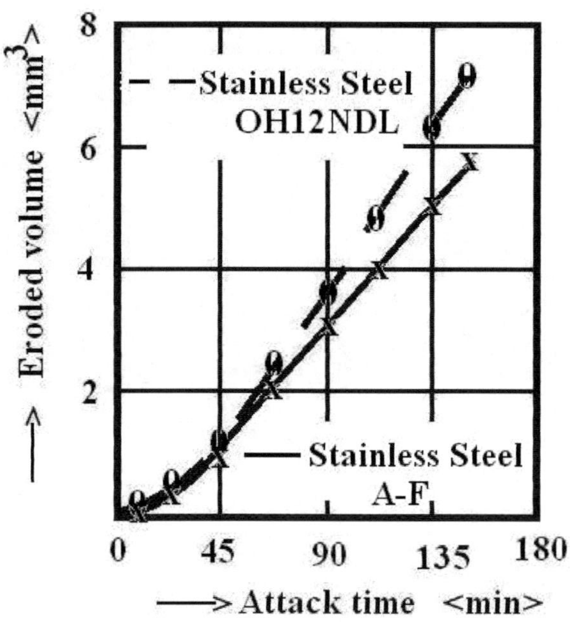

Fig. 4 – Eroded volume versus attack time (o- experimental points of OH12NDL steel; x- experimental points of austenitic-ferritic steel, A-F)

The evolution of eroded volumes, respectively eroded rate versus attack time, in figures 4 and 5 clearly show that the studied austenite - ferrite steel (A-F) has a better behavior at cavitation attack than OH12NDL steel, usually used for casting hydraulic turbines blades in our country.

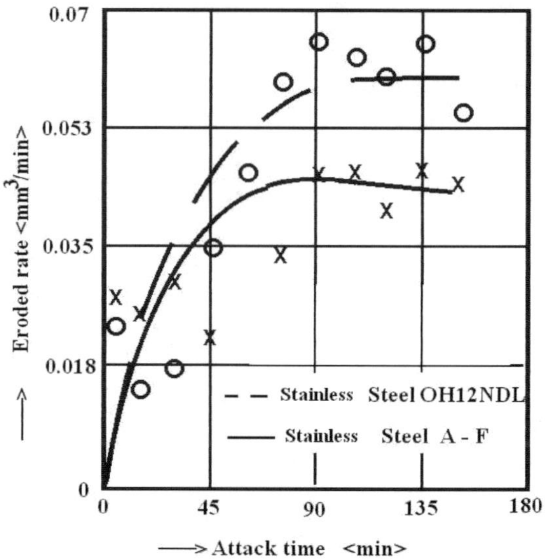

Fig. 5 – Eroded rate versus attack time (o- experimental points of OH12NDL steel; x- experimental points of austenite-ferrite steel, A-F)

The dispersion of the experimental points in figure 5, in comparison with the average curve, may confirm the previous affirmations concerning the manner of ferrite breakdown, with expelling of ferrite, austenite grains and martensite needles.

4. CONCLUSIONS

The stainless steel having austenite and ferrite as predominant constituents is appreciated as a material whose structure may resist very well to cavitation erosion. The comparison with the usually steel OH12NDL, based on erosion characteristics, recommend both of them as materials, which can be used to cast different hydraulic devices (turbine and pumps) and also as a repair material for cavitation damaged zones.

ACKNOWLEDGMENTS

The present work has been supported from the National University Research Council Grant (CNCSIS) PNII, ID 34/77/2007 (Models Development for the Evaluation of Materials Behavior to Cavitation)

5. REFERENCES

Bordeaşu I. (2006). *Eroziunea cavitaţională a materialelor (Cavitation Erosion of Materials).*Editura Politehnica Timişoara, ISBN: (10)973-625-278-7; (13)978-973-625-278-5

Bordeaşu, I., Popoviciu, M.O., Mitelea, I., Anton, L.E., Bayer, M., Funar, S.P. (2007). *Cavitation Eroded Zones Analysis For G-X 5CrNi13.4 Stainless Steel* Proceedings of The 18th International DAAAM Symposium "Intelegant Manufacturing & Automation: Focus on Creativity, Responsibility and Ethics of Engineers", Zadar Croatia, 24-27.10.2007, pp.105-106, ISSN 1726-9679, ISBN 3-901509-58-5.

Mitelea, I., Bordeaşu, I., Hadăr, A. (2005). *The Effect of Nickel Content Upon Cavitation Erosion for Stainless Steels with 13% Chromium and less than 0,1% Carbon*, Revista de Chimie, Chemical Abstracts RCBUAU 56(11), pp.1169-1174, ISSN: 0034-7752.

*** (1985). *Standard method of vibratory cavitation erosion test*, ASTM, Standard G32-85.

Annals of DAAAM for 2008 & Proceedings of the 19th International DAAAM Symposium, ISSN 1726-9679
Editor B. Katalinic, Published by DAAAM International, Vienna, Austria 2008
Make Harmony Between Technology and Nature, and Your Mind will Fly Free as a Bird

ANALYSIS OF THE SUBJECTIVE QUALITY ESTIMATION OF PHOTO PRINTS WITH REDUCED JPEG QULITY

BOTA, J[osip]; MILCIC, D[iana] & DONEVSKI, D[avor]

Abstract: *The growing use of digital photo camera has influenced the development of different type of computer files. Digital cameras and the users mostly use JPEG format. In this case quality stands for the best reproduction image of the photographed element.*
In former times the quality of photography depended on the quality of the negative film and the photographic paper. Today, for most users, it depends on the sensor of the photo camera, type of file used and the print quality.
JPEG format uses certain amount of compression which influences the file size and the image quality. This paper investigates the relation between the subjective estimation and the objective results of different JPEG compression levels.
Key words: *JPEG, JPEG compression, spectrophotometer, quality control, subjective estimation*

1. INTRODUCTION

The accelerated development of digital electronic devices influences our daily routine. Today, a personal computer and a digital camera make a part of the inventory of almost every household. Digital photography became a standard, replacing the conventional photography. Due to its simplicity, using it does not require a great deal of technical knowledge. Differences in camera quality (optics, sensors, file format, etc.) and computer processing determine the final image quality.
The growing demands for transferring greater amounts of data push hardware solutions to their limits. That is the point where software solutions come into place, attempting to compress the date so the same information could be transferred in less time. Transfer of large image files is the bottleneck of distribution systems, and therefore every reduction of file size represents a success for saving both time and storage space.
Image compression is a vital instance in multimedia communications (Singh et al., 2007). Deviations in color reproduction appear in every color reproduction process. The two important instances about compression are its platform dependence and performance. Today's compressions are mostly platform independent because they conform to international standards. Human evaluation of color and quality is highly individual, and calls for the determination of the right quality level satisfying most users' needs.

2. THEORETICAL

JPEG is a standard procedure for image compression. The Joint Photographic Experts Group is the standard author. JPEG was created for compression of both monochrome and color images, and is used for photographs, natural painting and similar applications. It is not suited for text, simple drawings and technical drawings. JPEG compression uses the flaws of the human eye, i.e. the fact that human perception is more sensitive to small differences in lightness than the differences in color (Kayand & Levine, 1994). JPEG compression uses the DCT (Discrete Cosine Transform), and divides its coefficients by their corresponding quantum. Figure 1 shows the block diagram of JPEG compression. It should be noted that different image processing applications use different quantization tables. Therefore, the results achieved at same compression levels differ, and we cannot express any recommendations as certain compression level values. Approximate quality factors are better indicator and should be considered for this purpose.

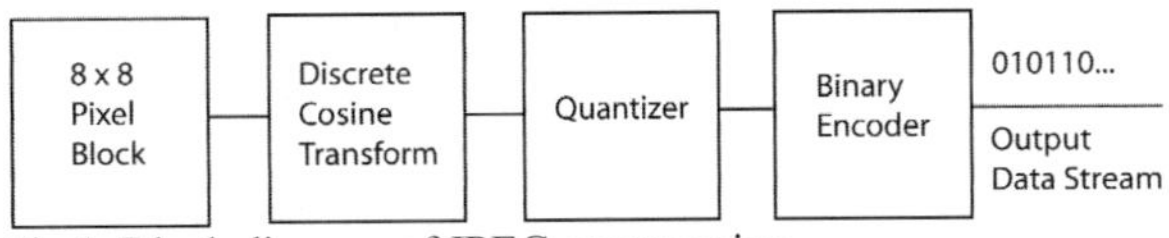

Fig.1. Block diagram of JPEG compression

3. EXPERIMENTAL

3.1 Hypothesis

The subjective human estimation of the compressed images quality is inferior to the objective estimation. This brings the importance of optimizing the reproduction process. It is known that digital images using greater JPEG compression suffer greater amounts of loss in details. This particularly relates to small differences in color on more or less uniformly colored parts of the image. Based on the experience as a photographer and a graphic designer, I expect that one part of the examinees will not notice any differences in quality of the images.
It is assumed that the reproduction quality does not differ significantly among different photo studios due to the use of standard substrates and printing machines of similar technical characteristics. The assumption is based on the equable market price and opens the possibility of determining the needs and the cost effectiveness on our market and most likely on the global market as well.

3.2 Materials and methods

The determination of subjective estimations requires a template containing familiar objects with as wider variety of chromaticity and luminance points.
One of the color consistency control tools is GretagMacbeth ColorChecker Chart. It allows the control of color reproduction by measuring the values of the 24 reference patches. The colors of the patches are chosen to allow the tracking of the most common colors which appear in photography (flowers, light and dark human skin, etc.).
The examination template and the ColorChecker Chart are combined in the same file. The file was saved using the image processing software in compression levels 7 to 12. The reference level is 12, and the comparison with levels 10, 8, 6, 4, 2, and 0 was made. For the purpose of collecting the data, every level was assigned a letter as an indicator (Table 2). Those photographs were printed in photo studios, two samples for each compression level. The overall template format was 1772 x 2488 pixels, 300 dpi or 15 x 21 cm in size.
On the delivery of files, the photo studio personnel were instructed to print the photographs on a glossy photographic

paper in a given format without any corrections. Such corrections, commonly made in photo studios, would result in uneven conditions of producing the photographs, and incomparable results for the analysis.

The photographs were later detached from the ColorChecher Chart so it wouldn't affect the evaluation. 50 examinees of different age and sex analyzed the templates in different conditions, and marked those on which they noticed difference in quality compared to the reference. The chart patches were measured using a spectrophotometer with standard illuminant D50, and measured values compared to the referent values. The differences are expressed as ΔE_{00}.

3.3 Results

Table 1 shows the ΔE_{00} difference between the reference print (A) and for the rest for each patch. In addition to ΔE_{00} for each patch, mean values of ΔE_{00} for each print are presented in Table 2. Table 3 summarizes the analysis results. The number of examinees which notices reduction in compressed images quality compared to the reference is expressed as percentage.

Patch	ΔE_{00} B	ΔE_{00} C	ΔE_{00} D	ΔE_{00} E	ΔE_{00} F	ΔE_{00} G
A1	0,41	0,45	0,32	10,90	0,31	0,54
A2	0,49	2,79	0,67	11,27	0,53	2,02
A3	0,55	1,84	0,17	11,03	0,41	2,06
A4	0,29	0,27	0,75	10,48	0,42	0,85
A5	0,36	1,16	0,30	7,54	0,60	1,29
A6	0,83	0,67	0,92	5,15	1,00	0,87
B1	0,31	0,40	0,40	6,61	0,13	0,48
B2	0,63	0,30	0,72	10,46	0,63	1,23
B3	0,17	0,34	0,43	7,89	0,25	0,71
B4	0,20	0,32	0,21	11,09	0,24	0,46
B5	0,30	0,38	0,21	10,53	0,15	0,64
B6	0,27	0,19	0,17	10,12	0,52	0,28
C1	0,49	0,51	0,70	10,43	0,37	1,59
C2	0,13	0,58	0,34	8,87	0,23	0,67
C3	0,12	0,95	0,35	10,62	0,47	1,36
C4	0,33	0,68	0,56	5,57	0,39	1,05
C5	0,44	0,93	0,42	10,49	0,64	0,46
C6	0,30	0,23	0,32	10,26	0,57	0,56
D1	0,84	1,95	1,04	7,92	0,60	1,67
D2	0,45	1,18	0,61	10,31	0,50	1,15
D3	0,44	0,73	0,66	10,92	0,43	0,73
D4	0,53	0,55	0,49	9,35	0,79	1,48
D5	0,66	0,64	0,88	11,13	0,63	1,97
D6	0,94	2,03	0,95	10,61	0,95	2,38

Table 1. ΔE of the prints compared to the reference

	B	C	D	E	F	G
Mean ΔE_{00}	0,43	0,83	0,52	9,56	0,49	1,10

Table 2. Mean ΔE_{00}

	B	C	D	E	F	G
Noticed difference	0%	6%	0%	54%	0%	26%

Table 3. Subjective estimation

3.4 Discussion

The analysis of the measurement results shows that images with greater compression levels suffer greater deviations. There is no specific patch that suffers greater deviation than the other patches. Mean value of ΔE_{00} and relatively small data dispersion indicate that greater compression levels result in greater values of ΔE_{00}. These results are especially noticeable in prints "E" and "G".

50 examinees evaluated 6 prints comparing them to the reference in different conditions. The examinees are not graphic technology experts, nor did they evaluate the prints under standard lighting conditions. The examination resulted with the following answers: 6% for print "C", 26% for print "G", 54% for print "E" i.e. that was the percentage of the examinees who noticed the difference in the print quality compared to the reference, while 46% did not notice any difference. Non expert examinees and non standard viewing conditions were used to inspect what results can be expected with the majority of population and most common viewing conditions. Comparing the file sizes (Table 2) reveals that they differ significantly. Comparing the files "A" and "B" reveals that file "B" is approximately 1/3 the size of file "A". Comparing "B" to "F" reveals that file "F" is approximately ½ the size of file "B". Other files, "D", "C", "G", "E", are more similar in size.

4. CONCLUSION

The investigation determined that using JPEG compression which is adapted to the flaws of the human eye, and using the premium quality digital printing, 46% of the examinees did not notice any difference between the samples. That indicates that the examinees either were not able to recognize the difference, or that the difference was not in significant factors which would draw their attention. The 54% noticed the difference, but only 6% noticed the difference at the compression level 4. Considering that the examinees compared the samples directly, there is still an open question if the results would differ if they had compared the samples by memorizing them. The results can be used as guidelines for users to use the greater compression levels safely, which most of them usually avoid. The results show that the viewers estimated that the best ratio between quality and compression level is achieved at compression level 8, which corresponds to approximate quality factors of 88,28 for luminance, and 90,19 for chrominance. At that compression level, the sample image file was ¼ the size of the original, and was not on the borderline where viewer starts noticing reduction in quality.

This paper calls for further investigations of factors influencing the subjective estimation of quality. The viewing conditions, age and sex are all factors which influence the subjective estimation to higher or lower extent.

5. REFERENCES

Aldaba A., M.; Linhares M.M., J.; Pinto D., P.; Nascimento M.C., S.; (2006). *Visual sensitivity to color errors in images of natural scenes*, Visual Neuroscience, United Kingdom, printed in the USA

Alkholidi, A.; Alfalou, A.; Hamam, H.; (2007) *A new approach for optical colored image compression using the JPEG standards*, Signal Processing, Elsevier North-Holland, Inc. Amsterdam, The Netherlands

G.K, W.; (1991). *The JPEG Still Picture Compression Standard*, Communications of the ACM, ACM New York, NY, USA

K.M. Au, N.F. Law, W.C. Siu, (2007). *Unified feature analysis in JPEG and JPEG 2000-compressed domains*, Pattern Recognition, Elsevier Science Inc. New York, NY, USA

Kayand C., D.; Levine R., J.; (1994). *Graphics File Formats*, Windcrest, McGraw-Hill

Singh, S.; Kumar, V.; H.K. Verma, (2007). *Reduction of blocking artifacts in JPEG compressed images*, Digital Signal Processing, Academic Press, Inc. Orlando, FL, USA

Annals of DAAAM for 2008 & Proceedings of the 19th International DAAAM Symposium, ISSN 1726-9679
Editor B. Katalinic, Published by DAAAM International, Vienna, Austria 2008
Make Harmony Between Technology and Nature, and Your Mind will Fly Free as a Bird

CONFORMITY ISSUES FOR HEALTH AND SAFETY AT WORK – PART OF INTEGRATED MANAGEMENT SYSTEMS

BOTEZATU, C[ezar]; BOTEZATU, C[ornelia] P[aulina] & CARUTASU, G[eorge]

Abstract. *Establishing integrated management systems is a necessary precondition without which some efficient activities cannot exist, in the context of an increasing competition, in an increasingly dynamic multi-layered development of information technologies and communication, of their integration in specific processes, in other words of globalization. The authors present the results of a project which began in September 2007, on "The Development and Implementation of the Integrated Management Systems (IMS) in the Field Of Energy (IMS-FE)" within the „4th Framework Programme (Partnerships)", a project during which they participated in integrating the work safety management subsystem in IMS.*
Key words: *information system, health, management, risk*

1. INTRODUCTION

Health and security at work is constantly gaining importance for the enterprises' management teams, under social or humanitarian pressures, these issues gaining significant implications in the economic and legal fields, being an important part of the integrated management systems, irrespective of their level of operation. The principle of risk evaluation is also widely accepted, and this process engages two stages: the first – identifying the risks and the second – analyzing and evaluating the risks.

Developing a management system for health and safety at work may contribute in an efficient manner to anticipating events that have a high-risk of affecting the health and/or safety of workers at workplace, to reducing the negative effects when they occur, while maintaining the productivity of the company at a decent level (Botezatu et al., 2006).

Workplace accidents and professional illnesses mustn't be perceived only as a fatality, as it is more lucrative to consider them as a dysfunctionality of the processes taking place in an enterprise.

Three different ways of action may be identified:

- Prevention or precaution regarding the events and/or situations through which dysfunctions in the work processes lead to damaging the heath of the workers or to the technological deterioration of the environment (named risks, in this context);
- Intervention through actions in the cases described above for the elimination and/or reduction of the negative effects or of their intensity;
- Analysis of the situations that determined the negative effects on humans, environment or on the technological security and post-event actions to rehabilitate or compensate them.

As a result of appliance of EU Directives and national laws and regulations (more the 300 documents), regarding Health and security at work, one of the implementation activity, as a prerequisite for certification, is evaluation of accident emerging risk, being dependent (as method of cuantification) of industry, work complexity, number of employees, type of product manufactures etc.

2. RISK EVALUATION

The considerable impact of the precautionary principle is based on a lot of opinions. Associating the concept of sustainable development with environment damage at planetary level is translated ad a reaction caused by the flaws in prevention, observed during several recent sanitary crises, so the aspirations towards a better life are associated to the management of technological development.

The precautionary principle tends to gain weight in Europe in the fields of environment, food and health (Green, 2004). It is compatible with the "a priori" control practice, widely spread in the European area. Precaution aims at limiting hypothetic, potential risks, while prevention regards the control of identified risks (Burt et al. 2008).

The distinction between a potential and a verified risk fundaments the distinction between precaution and prevention: precaution for potential risks and prevention for verified risks.

For risk evaluation a very popular classification introduces three levels of tolerance: intolerable risks, risks tolerable only with adequate surveillance and intervention methods and insignificant risks that only necessitate actions to ensure they maintain this level.

Tolerable risks are considered all risks between intolerable and insignificant. It considered that an activity can be tolerated with these risks when they cannot be reduced, while there are certain benefits of the respective activity, under special coordination and surveillance, remaining that they are eliminated or reduces when possible.

At the same level are the risks for which the actions for their reduction are too costly compared to the advantages obtained.

The tolerable risks area has two subdivisions. The first comprises the risks that cannot be reduced or whose reduction costs are disproportionately big compared to the improvement. The second part of tolerable risks is represented by lower risks, without being insignificant.

An example of how a risk can be estimated, according to two characteristics: frequency (Frequent, Probable, Occasional, Seldom, Unlikely, and Incredible) and consequences (Catastrophic, Critical, Marginal, and Negligible) is presented in the table 1 (Botezatu, 2001).

Considering the level of tolerable risks as being made up of two classes (II and III), we can make an interpretation (table 2).

Frequency	Consequences			
	Catastrophic	Critical	Marginal	Negligible
Frequent	I	I	I	II
Probable	I	I	II	III
Occasional	I	II	III	III
Seldom	II	III	III	IV
Unlikely	III	III	IV	IV
Incredible	IV	IV	IV	IV

Tab. 1. Censequences vs. Frequency

3. RISK ELIMINATION – REDUCTION

In the countries of the European Community (including Romania) and not only there is a long experience and a lot of good practices both in classifying, identifying and evaluating risks and in eliminating and reducing them.

Among others we can mention that at enterprise level, a lot of attention is paid to: air quality (ventilation systems), pressure equipments, elevators, lifting equipments and accessories, noise, construction sites, shipyards, illumination, electricity, individual protection equipments, explosives, extraction of fat using flammable substances, fires, fixed refrigeration systems, heat systems, machinery, fruit and vegetable maturation, divers and deep-sea activities, flammable liquids or liquefied gasses, gates and doorways, ionized areas, chemical hazards, depots, air cooling towers (according to INRS) (Yule et. al.2007).

Each European country (including Romania) has, through a system of regulations, standards and good practices guides, its own set of procedures, more or less compulsory for all economic agents.

4. STUDY CASE IMS-FE

The authors are currently participating in a research project which began in September 2007, named *"The Development and Implementation of the Integrated Management Systems in the Field Of Energy (SIM-FE)" within the „4th Framework Programme (Partnerships)" which* aims through its objectives to contribute to the development of the socio – economic research to enhance the Romanian firms' competitiveness.

Within this project we elaborate procedures to integrate the field of health and safety at work within the integrated management system which is created, projected and implemented in a company, after developing an adequate theory which takes into account the results on international (ILO for the subject matter) European and national levels, both for regulations (including accepted standards) and for institutional, labour force, efficient methods or accredited quality assessment procedures, efficient technologies (including IT&C) as well as for good practice guides corresponding to specific activities.

The estimated results are: generating an "integrated systems" theory, elaborating a number of procedures conforming to the aim of the integrated system and not the subsystems (a global optimization), a methodology to implement the integrated systems, a technological transfer web site and programs to assist the managerial decision.

The novelty aspect claimed is the integration of work and safety procedures into a larger concept and it degree of generality. Furthermore, we will be able to offer, at the end of the project a set of tools (knowledge system, database with reference documents, evaluation algorithm, step-by-step procedure to follow by enterprises to inform and obtain health and safety national certification, mandatory in Romania and EU. At this stage, we collect a large amount of unstructured information, related to this subject, a classifying process being on course.

Risk class	Interpretation
Class I	Intolerable risks
Class II	Undesirable risks, tolerable only is their reduction is impossible or if the costs of the reduction are disproportionately high compared to the results
Class III	Risks that are tolerable only if the costs of the reduction are lower than the achieved results
Class IV	Negligible risks

Tab. 2. Risk classes

5. CONCLUSIONS

As conclusion, presently it exist many reference documents regarding Health and Safety at work, information unstructured, all effort of the companies being focused to obtain the mandatory certification imposed by national or EU regulations. In this matter, an important activity consists in risk evaluation, a very complex process, with many depending parameters. So, as first step in integration of Health and Safety into an integrated management system, we emphasize the activities regarding labour health and safety at work together with the proposed instruments, inside of SIM-FE project.

Firs was realised the documentation regarding the current situation and the need for conformity assessment in the field of labour security and health, at European and international level, using studies and analyses regarding the current situation of the methods used for conformity assessment at European and international level. Also, will be documentated activities regarding the current regulations in the field of labour security and health in Romania and European and international level for conformity assessment methods, from point of view of inclusion of the conformity assessment within the enterprise general management system.

Now, will develop the experimental models of the conformity assessment methodology and of the support - software product, by consultation with partner economic agent regarding the general requirements of the methodology and the corresponding work instruments and the requirements of the software product. Next step is the analysis of the data collected during documentation and of the specific requirements regarding the methodology of conformity assessment. Is followed by development and testing of an experimental model of methodology (procedures and work instruments) and software product. Preparing documentation for the methodology and software product models and proving the functionality and utility of the developed models together with logistical and information support for identifying the general requirements of the methodology and of the software product and for testing the resulting models is the last step.

Next two years we will develop of the conformity assessment methodology and the software product based on the experimental model. We hope, at the project end (2010), to have a clear, integrated methodology for conformity in Health and Safety, using an integrated system. The overview of this problem, the method, transposed in software and integration in larger system is the main contribution of the authors, having a relevant experience in the field.

6. REFERENCES

Botezatu, C. (2001). *Professional lesions by economic activity, 1991-2000,* The directory of the statistics of work" 2001; Table 8B - ISIC-Rev3 /ST125-56 (Left 8), Romania

Botezatu, C.P., Botezatu, C. Căruţaşu, G. (2006),The development and the implementation of the integrated management system, *Annals of DAAAM for 2006 & Proceedings of the 17th International DAAAM Symposium*, pp. 053-054, ISSN 1726-9679, ISBN 3-901509-57-7

Burt, C.D.B., Sepie, B., McFadden, G. (2008) The development of a considerate and responsible safety attitude in work teams, *Safety Science*, v 46, n 1, January, 2008, p 79-91, ISSN: 0925-7535 CODEN: SSCIEO, Elsevier

Green, S. (2004) Safety is the priority, *Materials Handling News*, n 558, June 2004, 25-6 ISSN: 0025-5351 CODEN: MAHNA4 Highbury Nexus, UK

Yule, S., Flin, R.; Murdy, A (2007). The role of management and safety climate in preventing risk-taking at work, *International Journal of Risk Assessment & Management*, v 7, n 2, 137-51, ISSN: 1466-8297 CODEN: IJRABZ, Inderscience Enterprises, Switzerland

Annals of DAAAM for 2008 & Proceedings of the 19th International DAAAM Symposium, ISSN 1726-9679
Editor B. Katalinic, Published by DAAAM International, Vienna, Austria 2008
Make Harmony Between Technology and Nature, and Your Mind will Fly Free as a Bird

TECHNICAL EDUCATION AND ECONOMIC DEVELOPMENT IN THE NEW MEMBER STATES: ROMANIA CASE STUDY

BOTEZATU, E[lena] & BOTEZATU, M[ihai] A[lexandru]

Abstract: *Economic competitiveness nowadays stands under the umbrella of knowledge and information. Rapid technological progress, associated with the development of information technologies is what the third industrial revolution is all about, and the new panacea for economic development.*
The key to capitalizing the advantages of the knowledge society stands not only in possessing certain assets, such as: a good research and innovation infrastructure, a highly-skilled labour force, a competitive industrial base, but also in creating the necessary synergies so as to be able to have a good transfer of knowledge from education and research towards the business environment.
Key words: *education, development, knowledge, management*

1. INTRODUCTION

An extremely important indicator in identifying industries with good development potential is the one which quantifies the number of newly created jobs. Analysis of the relevant fields shows that the firms activating in construction, transport, storage and communications fields are the most dynamic, they are in the process of expansion and have a much higher capacity to generate new jobs.

These are in fact the sectors that have a deficit of labor force, and the request for qualified personnel is high; on the other hand, there are multiple opportunities for work in other countries for those with such qualifications.

Another problematic issue refers to the incapacity of the superior education system to integrate into the Romanian research and innovation system. Thus, even if the quality of the students would meet the requirements of the new economy, the technical education system would still lack a vital component: the connection to the business environment, the capacity to transfer knowledge and to capitalize the potential synergies between a good human resources capital, innovation and the market.

According to this, the success of the technical education system is not only defined mainly by a high level of knowledge and the effective possession of cognition techniques, but also the total preparation for professional activity (Laporte, 2007).

The paper starts from the following basic premises:

1. Among the key factors for development, the most important were identified as a good human resources capital and innovation.
2. The existence of these factors cannot be achieved unless the country has a good education system, to prepare the future labour force.
3. Among the key qualifications in today's knowledge economy (or aspiring to become a knowledge economy), technical skills are crucial; a good pool of qualified technicians will most certainly provide the leverage for competitiveness;
4. The economy must have a functional transfer of knowledge between universities and RDI units towards the market

Authors' contribution is related to determine the key elements for re-designing the education system on long terms.

2. REGIONAL CONTEXT

Economic competitiveness nowadays stands under the umbrella of knowledge and information. Rapid technological progress, associated with the development of information technologies is what the third industrial revolution is all about, and the new panacea for economic development.

The key to capitalizing the advantages of the knowledge society stands not only in possessing certain assets, such as: a good research and innovation infrastructure, a highly-skilled labour force, a competitive industrial base, but also in creating the necessary synergies so as to be able to have a good transfer of knowledge from education and research towards the business environment (Botezatu, 2007).

The Lisbon Agenda has made these issues a priority for the European Union, which was found lagging behind the US and Japan in terms of technical education and RDI – business transfers. Repeatedly, it has been pointed out that qualified engineers of core industries do not meet the demand of the market, whereas the number of students who chose to specialize in a technical field is decreasing (Pokholkov, 2001).

The European Commission has identified something called the *European paradox.* This refers to "*... the European seeming inability to turn excellent research results into globally competitive products ... It does seem that Europe's advanced science is too often taken elsewhere to be exploited commercially -usually to the United States.* (EU performs excellent science, but lacks transfer capacity). "

A few significant characteristics emerge when looking at the current European superior education systems:

- universities often receive less funding for research
- university staff is usually paid less than in the US
- general mentality is reluctant to the cooperation between universities and businesses.

Apart from these features, that are common throughout the European Union, a few othrs emerge in SEE countries:

- severe underpayment for staff ;
- lack of facilities;
- outdated, theory-oriented curricula;
- little or no research activities;
- no link between to the business environment.

On the other hand, businesses regard RDI institutions, including universities, as completely outside market needs and thus do not take into account research results; moreover, universities are regarded not as sources of highly-trained, qualified people, but instead of people with basic specialization that need extensive further training in order to become productive, increasing costs (Johnson, 2008), (Mastromatteo, 2007).

This, along with the economic developments that have unfolded in the region, the transion and passage to capitalism has lead to a reorientation of students towards legal and economic studies, which proved more lucrative and with better changes of becoming employed, whereas technical qualifications, such as engineering, were left behind, along with the depreciation of the economy.

3. THE ROMANIAN SITUATION

Romania is currently facing the challenge of getting its education system, particularly its higher education system, recognized as part of the European education. Although Romania is aligned to the Bologna process and has noticeble results in international competitions or technical fairs, it constantly fails to provide a pool of well trained, highly skilled technical professionals. This in turn translates into several effects on the economy:

- an increasing pressure on the national and regional labour market, as companies encounter difficulties in finding qualified labour force;
- low productivity, low capacity of internalizing new techniques, methods or strategies, in other words a vicious circle of poor economic results;
- brain drain;

Recently, a new effect is increasingly visible: an artificial upward pressure on wages, as a consequence of lack of labour force, without correspondence in productivity or skill, which is leading to constant unbalances in the economy, causing inflation.

As one can observe, while the total number of graduates increased almost four times, the share of technical students decreased severly. The year/year developments presented in the figure 1 below provide an even more interesting picture.

While the total number of university graduates in Romania increased almost four times, technical university education went through a severe decline, reaching a record 14.935 graduates in 1998/1999, in the context of the economic recession and restructuring of all industrial sectors.

At present, the country benefits from a steady growth, fuelled partly by the end of the transition, by significant foreign direct investments and by a revival of the economy. In this context, the demand for technical skills has picked up slightly and the trend is expected to continue near future, and the share will increase (Turowski, 2002).

As a positive thing, employment in trade (Figure 2), a sector with huge potential in Romania, continued to grow both in relative terms and in absolute terms, reaching almost 10% of the total labour force occupied, although the creation of jobs and economic growth in this sector is possible and even desirable (weighted average trade in employment is 21.2% in southern Europe, respectively 18.6% in the north) (Voinea et al. 2005).

4. CONCLUSIONS

In order to prepare highly qualified technicians, it is universally recognized that the traditional understanding of professional education as mastering a certain sum of knowledge based on teaching fixed subjects is insufficient and hinders the forming of the new style of thinking. The basis of education must not be taught subjects but rather ways of thinking and activities.

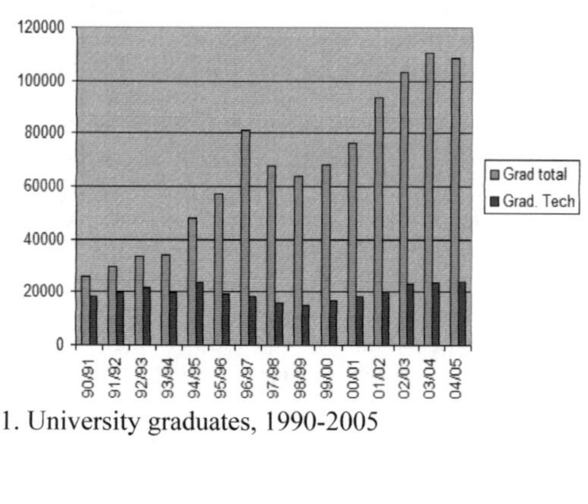

Fig. 1. University graduates, 1990-2005

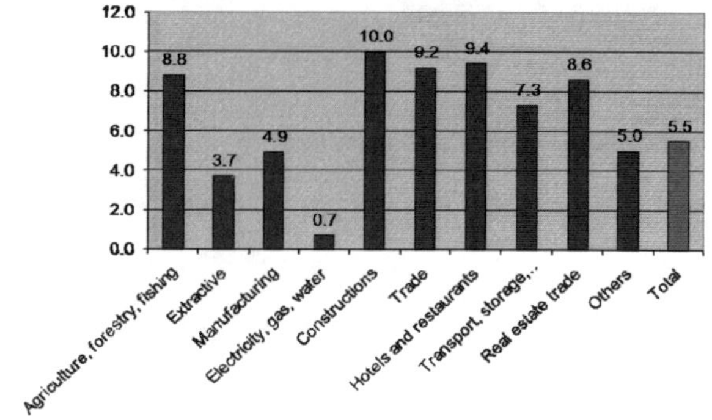

Fig.2. The rate of creating new jobs per economic activity

Since the 1990's the number of Romanian graduate students has increased from 25.927 (of which 18.344 in technical education) to 108.475 in 2004/2005 ((of which 24.054 in technical education).

Fueled by the European integration, Romania has the unique opportunity to make a shift in its development path and to redesign its education system, so as to make technical studies not only more attractive, but also more market oriented, so that they can provide a solid knowledge base and an incentive to the future labour force.

However other problems arise, which arerelated more to quality, than to quantity. The Romanian technical university education is not prepared to make the leap into the knowledge economy: it lacks the means, the people and the mentality.

In redesigning the education system, some key issues must be considered:

- improving university endowment and staff motivation and qualification;
- encouraging research and knowledge transfer;
- encouraging collaborations between the universities and businesses, by promotig clusters, industrial parks etc.
- preventing brain drain by incentivating companies to retain graduates.

5. REFERENCES

Botezatu, E. (2007). Romania Between the Challenges of Competitiveness and Regional Cohesion. *Available from:* http://ssrn.com/abstract=1015206 *Accessed:* 2008.06.23

Johnson, W.C. (2008). Innovation and Student Achievement: Why Employers are Worried and How Educators Can Respond, paper presented at the *Association of American Colleges and Universities Annual Meeting*, January 2008, Washington, D.C.

Laporte, B.(2007). Building Knowledge Economies: Education in a Global and Competitive World, paper presented at *IFEES Global Engineering Education Summit*, Istanbul, September 2007

Mastromatteo, M (2007). *Opening debate on engineering education reform,* Engineering Dimensions, vol. July/August 2007, pages 19-21, Professional Engineers Ontario, Toronto

Pokholkov, Yu.P. (2001). Problems of engineering education in the field of high technologies, paper presented at the *4th UICEE Annual Conference on Engineering Education,* February 2001, Bangkok, Thailand

Turowski, J. (2002). Technology and technical innovation transfer in engineering education, *World Transactions on Engineering and Technology Education,* Vol.1, No.2, 2002, pages 159-162, UICEE

Voinea, L.; Simionescu, L. (2005). *Annual Survey Report on research, development, innovation and competitiveness in the Romanian industry*, Group of Applied Economics, Bucharest

Annals of DAAAM for 2008 & Proceedings of the 19th International DAAAM Symposium, ISSN 1726-9679
Editor B. Katalinic, Published by DAAAM International, Vienna, Austria 2008
Make Harmony Between Technology and Nature, and Your Mind will Fly Free as a Bird

MIGRATION OF LABOUR FORCE IN ROMANIA

BOTEZATU, M[ihai] A[lexandru] & BOTEZATU, E[lena]

Abstract: *The study and evaluation of the processes within the labour market regarding the size, structure and trends have a special importance for defining the organizers' action plans on the medium term (2-4years) or long term (5-10years) perspectives. It should be taken into consideration that these processes have influence on active employment policies and social protection under the most varied forms, which determines policy constraints regarding organizers and / or managers of activities. It requires the development of a proper behavior of economic agents regarding employment and efficient use of labour.*
Key words: *occupation, employment, labour, management*

1. INTRODUCTION

After 1990, the restructuring process of the Romanian economy has been accompanied by a strong negative impact on the level of employment, causing chronic unemployment. Between 1990-2002 the occupied population decreased by about 2.5 million people, and the occupancy rate decreased by 19.6%, standing at 62.4%.

Romania faces a rapidly ageing population, due to negative natural increase and migration. In 2000, the aged population exceeded the young one, and this trend tends to continue in the future, with negative consequences for the labour market. This demographic pattern tends to follow the West European countries' pattern, but it's development in our country was and is strongly influenced by the dynamics and effectiveness of economic and social reforms (MLFEO, 2006).

The importance and complexity of the structural analysis of labour force and of the market where this is traded comes from the fact that work, unlike other factors, cannot be dealt with strictly from an economic perspective, but only together with its social side (Ciobanu et al. 1999).

The process of economic restructuring has generated significant changes in the occupational structure in Romania: new occupations occurred, and on the other hand, the content of many others suffered changes. In some branches of activity, some occupations began to group together increasing numbers of occupied population, and others have undergone a process of disappearance (Vasile, 2004).

The main factors that generate changes in activities were: adapting to the needs of customers, increased competition and technological reabilitation. Development and expansion of the private sector, the need of companies to stabilize certain segments of the market, led to the development of "aimed at the customer/client" policies and therefore the increase of the role of factors such as "customer needs adaptation" and "increasing competition in the field.

In this paper will be emphasized the main influences that causes labour force migration on national level and abroad having as support MLFEO records for recently past years togheter with governamental measures to balance that migration, depending on national and local needs.

The link between causes and effects, clearly show a national picture about labour force migration in context of integration of Romania in EU, having in mind also the liberalization of labour force market in EU.

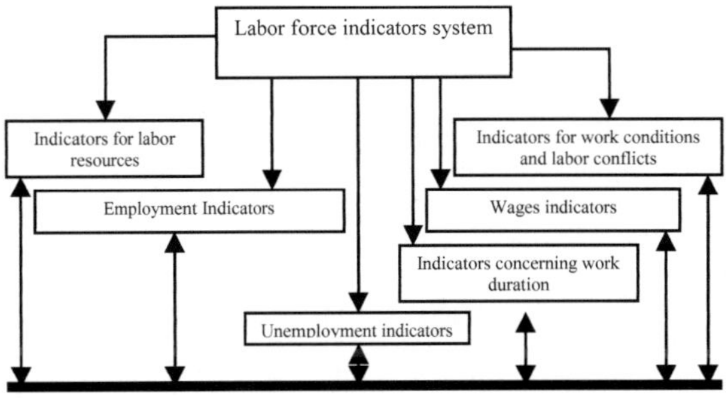

Fig.1. Labor force indicators system

2. LABOUR FORCE INDICATORS

The system of employment indicators have several categories of indicators, which aims to characterize the labor force and the labour market taking into consideration the economic conditions of globalization and the free movement of citizens.

The ability to generate new jobs is the lowest for micro-enterprises. When you look at the capacity of the economy to generate new jobs, from the perspective of the number of jobs newly created, and not from the perspective of the share of operators who have created new jobs in total sample, things are more detailed, and the micro sector, sector that gathers together a low overall share of the population employed (figure 1).

The rate of creating new jobs, defined as the ratio between the number of new jobs created per year and the number of existing employees at the end of the year recorded in 2005 the highest values in the South (11.7% vs. the 5.5 national average), followed at a distance by South - East (7.1% vs. 5.5 national average) and North - East (6.3 vs. 5.5 average). The lowest rates of creating new jobs were registered in Bucharest (3.4% vs. 5.5 national average) and Center (3.8%) (figure 2).

On a sectoral basis, the highest rate of creating new jobs was recorded in constructions (10% - almost double compared to an average of 5.5% nationally), hotels and restaurants (9.4% versus the average national 5.5%) - both sectors using largely seasonal activities - and trade (9.2% versus an average of 5.5%).

This dimension of restructuring process makes any deficiencies in the labour market, such as obstacles to mobility, reduced access to information.

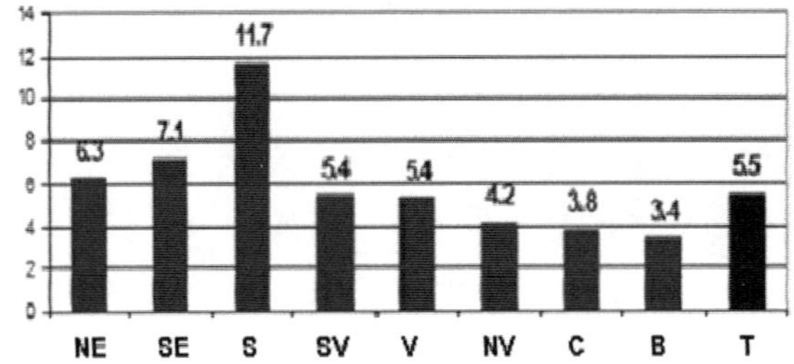

Fig.2. The rate of creating new jobs in regions

3. REALOCATION OF THE LABOUR FORCE

The reallocation of labour is closely related to the reform process and therefore is practically unavoidable. Using widely a large number of employees to compensate for low productivity makes the migration of labour from agriculture to industry and services, a process called disindustrialization, to be one of the main objectives of economic restructuring during the transition periods (Dostie et al. 2006).

Secondly, the reallocation of labour is performed by outside labour market, meaning that it can be achieved only through labour mobility, which leaves a business, probably from a declining sector, and finds a new workplace elsewhere, probably in a growing sector.

The skills associated with changes in the work are computer skills, problem solving, resources management and communication. Targeting profit and "client-oriented" services have made certain skills such as "solving problems", "communication" and "resource management" (especially time management) to become increasingly valued at all levels.

Specific excessive qualification is needed for the previous job that now becomes useless, or specific industry rents for the previous job generate harsh adverse effects (Botezatu, 2006).

At least in the context of the transition labour market, restructuring is an issue that regards the industrial composition of an economy. A change in industrial composition will be generally associated with declining state sector and an increase of the number of jobs in the private sector, but in the context of reallocating labour the most important subject is modifying the structure of employment in the sectors of industry.

According to the latest forecasts developed by the National Commission for Prognosis, until 2013-2015 the labour market will experience shortage of labour, of wage earners in particular, which is considered to be covered from two main sources - those who work outside working age and from a second increasing category, manpower from other countries outside EU borders.

4. LABOUR FORCE MIGRATION

Work abroad was a novelty, an innovation in the context of Romanian realities after 1989 and followed the course of any social innovation, with an off period, followed by a maximum social contagion (table 1).

The dominant migration path has changed over time. The changes between the phases are not only regarding the field of migration, but in the volume of departures as well. During pre-Schengen period of time, the intensity of the phenomenon is doubled compared to the stage 1990-1995, and after 2001 the intensity of the phenomenon of migration for work abroad increases three times compared with the previous timeframe.

Departures in order to work abroad are strongly localised. Thus, Moldova is by far the most important exporter of workers abroad, followed by Wallachia and Transylvania, showing small differences regarding intensity of the temporary emigration. West and south-west of the country, through Crişana Maramures, Banat and Oltenia regions.

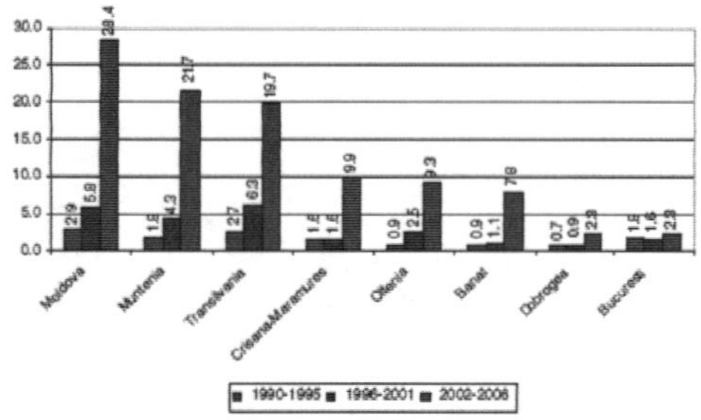

Fig. 3. Rate of migration

		Migration timeframe			Total
		1990-1995	1996-2001	2002-2006	
Italy		8	22	50	40
Spain		2	9	24	18
Germany		7	6	5	5
Hungary		8	9	4	5
Israel		18	17	0	6
Turkey		10	7	1	3
Greece		2	4	2	2
Canada		0	7	1	2
Belgium		5	1	0	1
Russia		5	0	0	1
Others		21	10	9	11
Total departures	%	100	100	100	100
	Nr.	61	107	360	528
Total left		33	81	203	317
Departures/person		1,84	1,32	1,77	1,66

Tab. 1. Countries of destination for migration for employment

5. CONCLUSIONS

Research carried out by the authors during the Doctoral studies have topics in the fields "workforce", "competitiveness", "modern management" and the specific interconnections. Significant results will be included in doctoral studies; this paper presents only a few aspects of employment, considered relevant by the authors primarily for management of activities, authors contribution underlying the link between causes and effects and determining measures taken by MLFEO.

In the new context, a series of programs developed by the Ministry of Labour, Family and Equal Opportunities, financed by the European Union, are aimed at preparing the workforce for new types of occupations, new requirements for the jobs, taking into consideration the economic restructuring, the mutations caused in the structure, organization and technology, future research aiming to discover if that measures have the predicet effect the extensive use of information and modern communication technology.

Some of these programs, very suggestively entitled: "Promoting Life Long Learning for qualification and Requalification of Labour Force" and "Active Measures for Employment" are in progress try to meet both employers demands and the demands of those who are searching for a suitable job, being in progress, try to meet both employers demands and the demands of those who are searching for a suitable job.

6. REFERENCES

Botezatu, M.A. (2006). Labour Migration Its Impact On Domestic Economy – The Case Of Romania, *Proceedings 2006 International Conference on Commerce*, AES, Bucharest, Romania, March 27-29

Ciobanu, A.; Parciog, S. (1999). Background study on employment and labour market in Romania, research report, *Availble from:* http://www.etf.eu.int/etfweb.nsf/pages/downloadromania *Accessed:* 2008.05.23

Dostie, B.; Sahn, D. E. (2006). Labor Market Dynamics in Romania during a Period of Economic Liberalization, Cahiers de recherche 06-17, HEC Montréal, Institut d'économie appliqué

Ministry of Labour, Family and Equal Opportunities (2006) – Employment tendencies for Romanian labour market, for 2010

Vasile, V. (2004). Demographic Changes and Labour Market in Romania, *Availble from:*www.ier.hit-u.ac.jp/pieJapanese/seminar/workshop040220/Vasile.pdf *Accessed:* 2008.05.23

Annals of DAAAM for 2008 & Proceedings of the 19th International DAAAM Symposium, ISSN 1726-9679
Editor B. Katalinic, Published by DAAAM International, Vienna, Austria 2008
Make Harmony Between Technology and Nature, and Your Mind will Fly Free as a Bird

VIRTUAL PROGRAM IMPORTED INTO THE REAL TECHNOLOGICAL WORKPLACE

BOZEK, P[avol]; MORAVCIK, O[liver]; STOLLMANN, V[ladimir]; SURIANSKY, J[ozef]; PRAJOVA, V[anesa] & WALEKOVA, G[abriela]

Abstract: *The virtual development process presents an essential condition for a successful origin of a new product or a workplace. The scientific project is a logical continuation of the research activities of the interdisciplinary investigators group as well as it represents the follow-on contractual scientific cooperation of university workplaces on an international basis. The project concept is based on the sustainability of the HCS model 3E designed by us and verified by practice. It is built on the „National Strategy of Sustainable Development of the Slovak Republic". Generating the user program from the virtual into real technological workplace is a necessary strategic condition for the implementation of the mentioned strategy of sustainable (Smrček & Hajtinger 1996).*
Key words: *virtual, imported, program, workplace .*

1. INTRODUCTION

Economic, Environmental and Ergonomic pillars have found their way to the list of political priorities of national economics. The concept of virtual programming of the real technological workplace being solved is significant also from the point of teaching development strategy at STU MTF. It allows to design the technology as well as it permits the definition of the basic principles of the technological process control system. The virtual methods imported in the beginning phase of the development is highly effective due to more 3E vital and acceptable aspects.

2. STRATEGIC ASPECTS OF HCS 3E MODEL IN A VIRTUAL TECHNOLOGICAL WORKPLACE

The *concept* virtual model of a technological workplace is designed in the phase of the project. It allows to design the technology in real as well as it permits the definition of the basic principles of the technological process control system. The implementation of virtual methods leads to immediate decisions on effectivity from various aspects. There are three acceptable aspects (3E) of sustainable development:

1.) Environmental aspects - each of the technologies carry an environmental burden which is also reflected to the economic aspects because of the operational measures that have to be followed. The advantage of the proposed virtual technology application means that it is not necessary to produce the model of a technological device, nor it is necessary to create the real technological workplace. By the production of the aforementioned devices, the environmental burden is minimized (Černecký, 2005).

2.) Economic aspects – there are few of them. They can be divided into partial groups and subgroups. The most important aspects that have to be in the optimization of such a technology that are taken into consideration are the investment costs(the amount of finances related to the specific virtual equipment purchase) and operation costs (technological devices, electricity consumption and other related costs) which can influence also the decision making in a new technology purchase.

3.) Ergonomic aspects of virtual technologies – working environment, lighting, noise, protective clothing necessary for the operation of such a technology, etc. (Barborák, 2007)

At present the optimization is used and applied in different information systems oriented on conventional technologies whereas the possibilities of virtual technologies are neglected. The necessity of computer aided tool that will take these aspects into account is vital in national as well as in European research. The issue is vital also for the top European companies involved in the subsupply of reputated car factories.

3. SUPERIORITY IN VIRTUAL PROGRAMMING

By the project investigation and implementation the following aspects from the view of basic principles of a new paradigm will be fullfilled:

1.) Certain deviation from the principles of traditional rational pradigm followed by the majority of organizations and schools in the past and still followed by several traditional ones.

2.) In contrast with the rational opinion, the supporters of the current paradigm of technological workplaces think, that the organization effectiveness is needed not only internally but externally as well.

3.) All faculty structures (organizational structure, study programmes, organizational control, university control, etc.) are the organization reflections to the external changes.

4.) Current trend of teaching technical subjects has to be researched as a socioeconomic and environmental system (HCS 3E model).

Designed virtual/real workplace reflects as well as accepts the aforementioned three most important aspects of sustainable development/environmental, economic and ergonomic aspects.

4. ESSENTIAL INVESTIGATION PROCESS

In the first phase the project concept will be elaborated. The whole scope of works will be defined, by what method will be investigated.

1.) To carry out collective investigators´ brainstorming to identify the key terms and the hierarchy of partial tasks/goals and their intrefaces.

2.) To carry out analysis for partial tasks/goals in the selected universities and companies in the Slovak Republic.

3.) To generalize partial analysis conclusions.

4.) To propose methods for partial tasks/goals.

- a general model of robotic workplace,
- method to investigate the control system of a laboratory robot,
- method of the effectivity evaluation of the environment-friendly oriented workplace at universities and companies,
- possibilities of using the exact methods for teaching at universities,

- interface for the user program transfer and control.

In the second phase of the project investigation:

- to verify related methods of individual tasks/goals at the Institute laboratory equipment,
- to carry out final corrections of the related methods,
- to generalize the knowledge into final proposals of individual systems of control programming and interface,
- to process the synthetic final research report.

Upon project completion, the programming itself will be carried out with the programming of all the needed functions so that there is no duplication and the application is capable to simulate the industrial robot as well as the entire robotic workplace.
The final phase will be devoted to testing. After testing the application the virtual application will be transformed into the real robotic workplace.

5. ANIMATION APPLICATION AND INTERFACE

Animation as a significant part of the application illusatrates the current state of all units and parts of the robotic workplace. It is impressive not only by the manual control but also by the data processing.
The animation will be carried out by the means of object oriented Microsoft visual C+ with the use of graphic library Open GL, both providing wide possibilities of the use of a large number of orders and functions .
Library Open GL is compatible with Linux operation system and represents a standard in 3D graphics.
Iterface has to be compatible with the data processing generated from the virtual scene, then transformed into the real environment of a robotic workplace (Ožvoldová et al., 2007).

6. PROPOSAL OF THE BOUNDARY APPLICATION

In the phase of the proposal of the boundary application, it is essential to define the boundary between the application itself and the application user. The ergonomics of the boundary application is an important point, i.e. the simpler the control the better. Another important condition of an application control overview is represented by the smallest possible number of control units for the user.

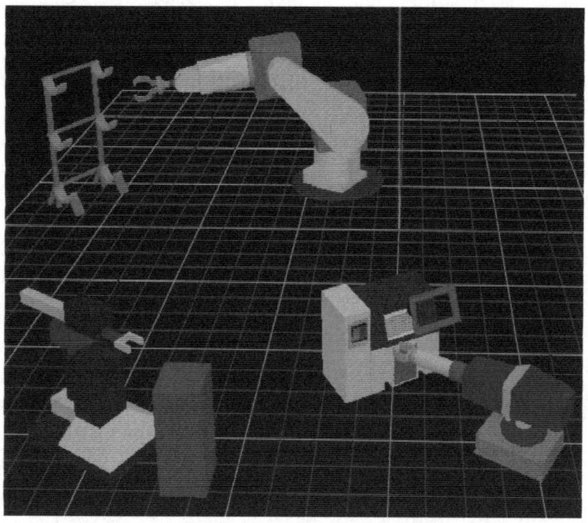

Fig. 1. An example of a virtual scene creation

In the simulation application of the virtual robotic workplace, there will be many control units but they will be arranged and implemented in such a way so that they are user friendly.
Application control. It is necessary so that the application control is unified as a whole and that there is one control unit per function. Individual control units will be called by names or abbreviations and the control will be assisted by a helper.

7. CONCLUSIONS

Results are considering the new trends related to the environmental aspects and operational economics of the workplace. The necessary training of new managers in the field is focused prevailingly on ergonomics and workplace control of a new type. The aforementioned results are succesfully being carried out in ŠKODA AUTO, a.s. Mladá Boleslav. The HCS Model 3E reflecting sustainable development (micro-solutions of macro-problems) of the virtual reality in economics, environmentalism and ergonomics as a vital environmental prerequisite is fully accepted. The abovementioned issues are highly topical and often also controversial once they carry both environmental and economic aspects (related to *„externalisation and internalisation costs“*) (Hatiar & Sakál 2007).
On one hand, the various ways of environmental protection influence the economic structure, on the other hand, they create conditions for future survival of man. At the same time they are closely related to the environmental impact caused by the technologies in mechanical engineering. Therefore, the gradual implementation of new virtual technologies is suitable as soon as in the phase of the production and technological processes begins. Now when the energy sources are limited, the mentioned technologies can win a reasonable share on the market.
The contribution was elaborated within the research project VEGA No. 1/0368/08 *The use of artificial intelligence in the control of flexible production systems* and the research project VEGA MŠ SR No. 1/3523/06 *Research of new technical and technological principles of wood concentration* together with the KEGA project No. 3/3111/05 *The creation of virtual robotic laboratory for supporing teaching the subject Robots and manipulators in a newly accredited study programme.*
The paper is also aimed at contributing to the vision of the *Agenda 21* and *Lisbon Strategy* in the field of industrial pillars of sustainable development strategy at the STU MTF research and pedagogical processes.

8. REFERENCES

Barborák, O. (2007). *Technical drive.* Dubnica nad Váhom, ISBN 978-80-969615-6-6, pp. 146

Černecký, J. (2005). *Technical resources measuring and monitoring.* Zvolen, TU Zvolen, ISBN 80-228-1439-3 pp - 95

Hatiar, K. & Sakál, P. (2007). *Transforming industry in Slovakia Trought participatory ergonomic,* II. Internet Journal, MtF Trnava, http://www.mtf.stuba.sk/docs//internetovy_casopis/2007/2/sakal.pdf *Accessed 2008-05-13*

Ožvoldová, M. ; Kvetan, K. & Krajčovič, J. (2001). *Designing a Basic Course in Physics Withing IDEP Programme,* In: Proceedings of the 2nd International Conference Virtual University 2001, STU Bratislava, ISBN 80- 227-1652-9, pp. 137-141

Smrček,J. & Hajtinger, K. (1996). *The analysis of the robots implementation possibilities in the non-standard environment.* In: Int. Cong. „MATAR 96 – Machine Tools, Automation and Robotics in Mechanical Engineering“, Praha, ISBN 80-364-2568-6, pp. 185-189

Annals of DAAAM for 2008 & Proceedings of the 19th International DAAAM Symposium, ISSN 1726-9679
Editor B. Katalinic, Published by DAAAM International, Vienna, Austria 2008
Make Harmony Between Technology and Nature, and Your Mind will Fly Free as a Bird

OPTIMUM DESIGN OF TOOL AND PROCESS IN THE CASE OF METAL SHEETS DRAWING BY APPLYING THE TAGUCHI EXPERIMENTAL DESIGN METHOD

BRABIE, G[heorghe] & NANU, N[eculai]

Abstract: *The present work performs the application of the Taguchi method in the design of tool and process parameters in order to increase the accuracy of drawn parts by reducing or eliminating the effects of springback phenomenon. Based on the optimal parameters obtained from the application of such technique, it was possible the accurate design of tools geometry and technological parameters just from the designing stage.*
Key words*: deep drawing, springback, tool accuracy*

1. INTRODUCTION

The most frequent defects generated by cold plastic forming in parts made from metal sheets are caused by the springback phenomenon; its effects mainly determine the decrease of the accuracy of drawn part by changing its shape and dimensions after the tool removing (Tang, 2000 and Arwidson, 2005). The classical methods and techniques applied to eliminate or reduce the effects of such phenomena were as follows: tools correction based on data obtained from their using in production; the use of tools and devices having a special construction in order to reduce or eliminate the springback effects; the optimization of the forming process by using mathematical models that establish a relation between the parameters of the springback phenomenon and the factors that influence its intensity (Brabie, 2002 and SSAB Tunnplat,2002). The utilization of the above mentioned methods and techniques in order to reduce or eliminate the springback effects is expensive and requires many experimental tests. As a consequence, new methods must be developed to avoid and reduce the springback effects in optimum technical and economical conditions. An optimal solution - that allows the optimization of tools and process parameters just from the designing stage - can be obtained by using the finite elements method combined with some statistical and mathematical models that permit the description of the influence of different process parameters on the geometrical and shape parameters of the worked part. The present paper analyses the conditions and steps needed and the results obtained from the application of the Taguchi Experimental Design method for the optimization of the drawing process parameters.

2. STEPS AND RESULTS OF TAGUCHI METHOD

The Taguchi's Experimental Design Method is a design optimization methodology that improves the quality of existing products and processes and simultaneously reduces their costs very rapidly, with minimum engineering resources (Taguchi, 1994). The method uses a special design of orthogonal arrays to study the entire parameter space with only a small number of experiments and is developed in the following main steps: *Step1 - Formulation of the problem.* The problem to be optimized can be formulated as follows: to optimize the accuracy of geometric parameters for hemispherical drawn parts made from metal sheets by minimizing the errors generated by the springback phenomenon. *Step 2 - Identification of the output performance characteristics most relevant to the problem.* The output performance characteristics are materialized by the following geometric parameters: R_p-part radius, R_f-flange connection radius, α- springback angle and ρ - radius of flange connection circle (Fig. 1a). *Step 3- Identification of control factors, noise factors and signal factors.* The control factors are the geometrical factors of the drawing tools (a) and the technological factors of the drawing process (b) that are shown in Fig. 1b. The optimization has been developed for the economic design of drawing process. The conventional optimum is determined without considering noise factors. *Step 4 - Selection of factor levels, possible interactions and interaction effects.* The influence factors on three levels of variation that were chosen are presented in Table 1. *Step 5 - Design of an appropriate orthogonal array.* The orthogonal array based on the Taguchi concept is used to arrange the discrete variables. In order to reduce the experimental trials and to increase the accuracy, the orthogonal array triangular tables were developed; for example, the L_9 orthogonal array (Table 2) has four columns corresponding to the factors and nine rows for the possible combination of factor levels; thus, Taguchi's experimental design offers 9 experiments instead of 3^4 (=81) trials needed for four factors and three level factorial design. *Step 6 - Preparation and running of the experiment/ simulations.* The drawing process has been simulated using the ABAQUS-Explicit software. A three dimensional model (Fig. 2) used for simulation was created in order to ensure the simulation of the quasi-static problem and to obtain the state of equilibrium after the forming operation; the part was considered deformable with a planar shell base. In simulation a quarter of part having two symmetry conditions (yz plane - A and xy plane -B) was used. The materials elastic properties used for simulation were as follows: Young's modulus 2.1×10^5 MPa, Poisson's ratio 0.3, density 7800 kg /m^3. The material was considered elastic-plastic with an isotropic hardening. The coefficient of friction used for the contact between blank, punch, die and blankholder was $\mu = 0.1$. The process parameters were as follows: drawing depth = 61.8 mm, drawing speed = 54 mm/s. The results of the simulation performed according to Taguchi's array are presented in Table 3 and Fig. 3. *Step 7 - Running of the experiment with appropriate data collection. Step 8 - Statistical analysis and experimental results processing.* The statistical analysis and interpretation of experimental results was performed using the Design Expert software. In optimization, the software can arrange that combination of influence parameters levels for witch the geometrical parameters of the part have optimal values.

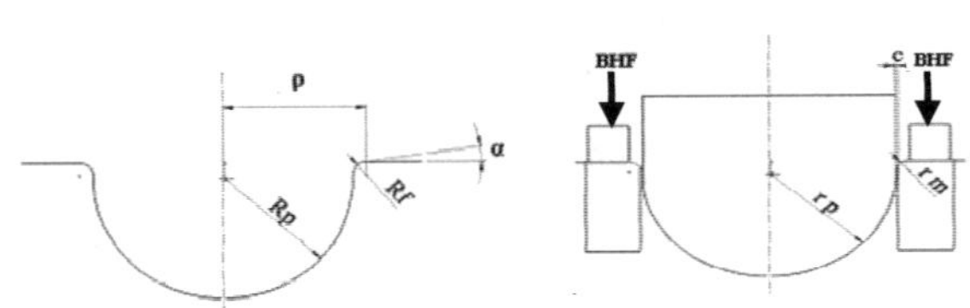

a. b.

Fig.1. Output performances and control factors

Factors	Levels of variations		
	1st level	2nd level	3rd level
Punch radius Rp[mm]	54	55	56
Die radius Rf[mm]	5	6	7
Blankholder force BHF[kN]	30	50	70
Clearance between punch and die c[mm]	1	1.2	1.4

Tab. 1. Influence factors

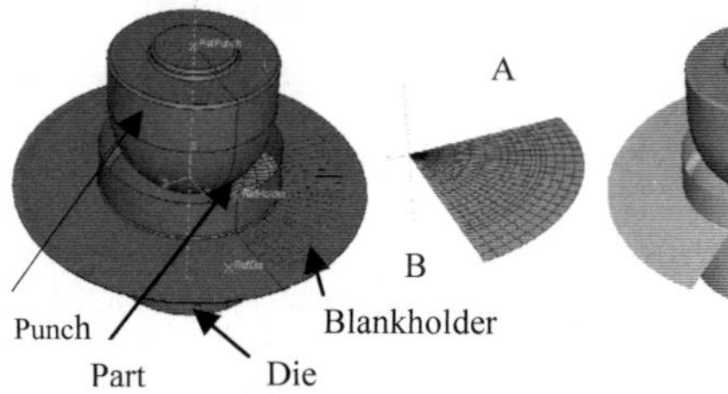

Fig. 2. Model used in simulation

No	Rp [mm]	Rf [mm]	BHF [kN]	c[mm]
1	1	3	3	3
2	3	3	2	1
3	2	1	2	3
4	3	1	3	2
5	1	2	2	2
6	2	2	3	1
7	2	3	1	2
8	3	2	1	3
9	1	1	1	1

Tab. 2. L_9 orthogonal array

Influence parameters				Part springback parameters			
Rp	Rf	BHF	c	R	r	α	ρ
mm	mm	kN	mm	mm	mm	grad	mm
1	3	3	3	54.14	7.46	0.34	62.46
3	3	2	1	56.48	7.86	0.95	64.31
2	1	2	3	55.22	5.39	1.10	61.41
3	1	3	2	56.65	5.85	0.41	62.92
1	2	2	2	54.68	6.53	0.47	61.74
2	2	3	1	55.48	6.69	0.32	62.25
2	3	1	2	55.92	7.01	1.37	59.12
3	2	1	3	56.93	6.54	1.53	63.63
1	1	1	1	54.20	6.4	1.28	61.28

Tab. 3. Results from simulations

Notes for MyDesign.de7
Design (Actual)
Summary
Graph Columns
Evaluation
Analysis
Part radius (Analyze
Flange radius (Anal
Angle (Analyzed)
Abscissa (Analyzed
Optimization
Numerical
Graphical
Point Prediction

Select	Std	Run	Factor 1 A:Punch radius	Factor 2 B:Die radius	Factor 3 C:BHF	Factor 4 D:Clearance	Response 1 Part radius mm	Response 2 Flange radius mm	Response 3 Angle grd	Response 4 Abscissa mm
	2	1	1	3	3	3	54.141	7.4609	0.34	62.467
	1	2	3	3	2	1	56.48	7.86	0.95	64.312
	5	3	2	1	2	3	55.22	5.395	1.1	61.418
	7	4	3	1	3	2	56.65	5.856	0.41	62.92
	6	5	1	2	2	2	54.68	6.53	0.47	61.74
	3	6	2	2	3	1	55.48	6.69	0.32	62.256
	4	7	2	3	1	2	55.92	7.01	1.37	59.121
	9	8	3	2	1	3	56.93	6.546	1.53	63.635
	8	9	1	1	1	1	54.206	6.4	1.28	61.28

Fig. 3. Taguchi's experimental array

The results of optimization are shown in Fig. 4. The following two solutions corresponding to the following arrangement of the influence parameters levels were found: 1st solution: R_p = 55 mm, R_m= 5 mm, BHF = 70 kN and c = 1 mm and 2nd solution: R_p = 55 mm, R_m= 5 mm, BHF = 50 kN and c = 1 mm. *Step 9 - Undertaking a confirmatory run of the experiment.*

The configuration of the optimized tools and the process parameters are shown in figure 5 a *(1st solution)* and b *(2nd solution)*. The geometrical parameters of the part obtained by using the optimized tool and process are given in Table 3.

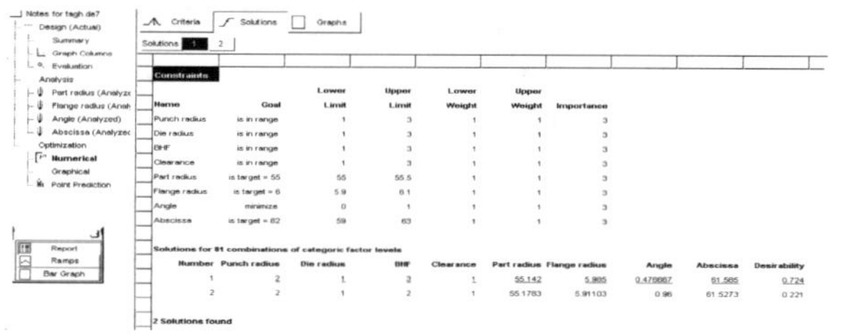

Fig. 4. Results of optimization *(report)*

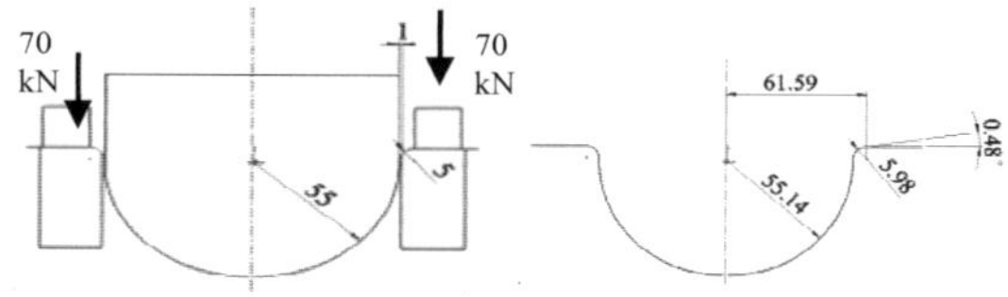

a. 1st solution

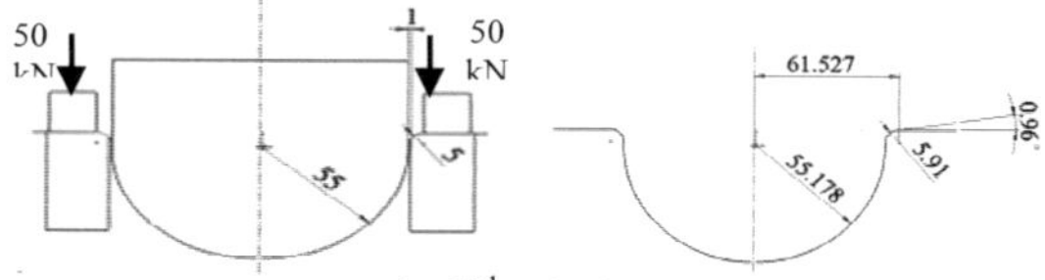

b. 2nd solution

Fig. 5. Optimized tools and parts obtained using the new tools

Initial/Nominal parameters		1st solution	2nd solution
R_p[mm]	55	55	55
R_f[mm]	6	5	5
BHF[kN]	35	70	50
c[mm]	1	1	1
R[mm]	55,33/55	55,14	55,18
r[mm]	6,12/6	5,96	5,91
α[°]	1,23/0	0,47	0,46
ρ[mm]	60,95/62	61,58	61,52

Tab. 4. Results from Taguchi method

3. CONCLUSIONS

The application of the Taguchi method for the design of deep drawing tool and process can be more advantageous by comparing to classical technical solutions or theoretical models. The geometric deviations from the nominal values of the finished part made by using the optimized tool and process are smaller with more than 50% than the deviations resulted by using initial tools. Thus, based on the optimal parameters obtained from the application of such technique, it is possible just from the designing stage the increase of accuracy of the tools geometry and technologic parameters of deep drawing process by minimizing the effects of springback phenomenon.

4. REFERENCES

Arwidson, C. (2005). *Numerical simulation of the sheet metal forming for high strength steels*, Lulea Univ. of Techn., Division of Manufacturing Systems Engineering, Finland

Brabie, G.; Chiriţă, B. (2002). Methods of reducing or eliminating effects of instability phenomena that occur in metal sheets forming processes, *Modelling and Optimization*, no. 8, p. 40-44, Romanian Acad. of Tech. Sciences, ISSN 1224-7480

Taguchi, G. & Yokoyama, Y. (1994). *Taguchi Methods. Design of Experiments*, American Supplier Inst., Deaborn MI, USA

Tang, S.C. (2000). Analysis of springback in sheet forming operation, *Advanced Technology of Plasticity*, vol. I, p.193-197, ISBN 0-387-17915-1

*** *SHEET STEEL Forming Handbook* (2002), Ed. SSAB Tunnplat, ISBN 91-971592-9-8, Sweden

Annals of DAAAM for 2008 & Proceedings of the 19th International DAAAM Symposium, ISSN 1726-9679
Editor B. Katalinic, Published by DAAAM International, Vienna, Austria 2008
Make Harmony Between Technology and Nature, and Your Mind will Fly Free as a Bird

FACTORS AND CAUSES THAT INFLUENCE THE SPRINGBACK INTENSITY IN THE CASE OF CONICAL DRAW PARTS

BRABIE, G[heorghe] & RADU, M[ioara]

Abstract: *The present paper performs an analysis by simulation using ABAQUS software of the factors that influence the springback intensity in the case of conical drawn parts made from steel sheets. The springback phenomenon and its effects on part geometry are caused by the differences resulted between the states of residual stresses on the outer and inner faces of part. The springback intensity can be influenced to a great extent by the variation and values of the major components of residual stresses that act along the part profile.*
Key words: *springback, deep drawing, conical part.*

1. INTRODUCTION

Springback is a phenomenon of elastic nature determined by the distribution of residual stresses on the section of the formed part. Springback is not only manifested by the modification of the state of stress/strains in the formed material but also its main effect is the modification of the geometric shape of the formed parts. (Pearce, 1991) In the case of deep drawing the springback problem is very complex because of the complex applied loading and complex geometry of formed part. In the case of a drawn part the springback parameters can be as follows: modification of the curvature radius and inclination angle of part walls and the difference in height. (Brabie, 2006) The main factors that influence springback phenomenon after drawing are as follows: punch and die radii, initial clearances, lubricating conditions, blankholder force and shape, material chemical composition and mechanical properties, sheet thickness (Han, 1999; Samuel, 2000 and Tang 2001). The general conclusions presented by different researchers concerning the influence of different factors on springback intensity indicate different senses of its variation. The present paper is a research concerning the influence of different factors on springback intensity in the case of conical drawn parts made from steel sheets by using conical and cylindrical punches.

2. CONDITIONS OF SIMULATION

The analysis concerning the behaviour of drawn parts was performed by simulation using the ABAQUS-Explicit software. The simulation was performed for parts made from FEPO 5MBH steel sheets by using a two dimensional model. The material was considered elastic-plastic with an isotropic hardening and the blanks were considered deformable with a planar shell base. The geometry and dimensions of the part and the model used in simulation are shown in figures 1 and 2. The material elastic properties used for simulation were as follows: Young's modulus $2x10^5$ MPa, Poisson's ratio 0.3, density: 7800 kg /m3.

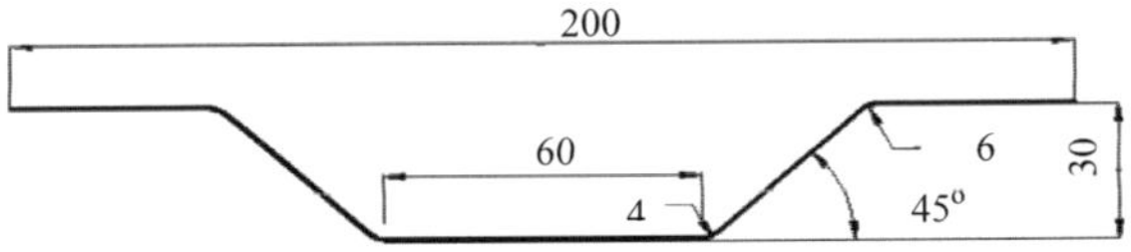

Fig. 1 Part Geometry

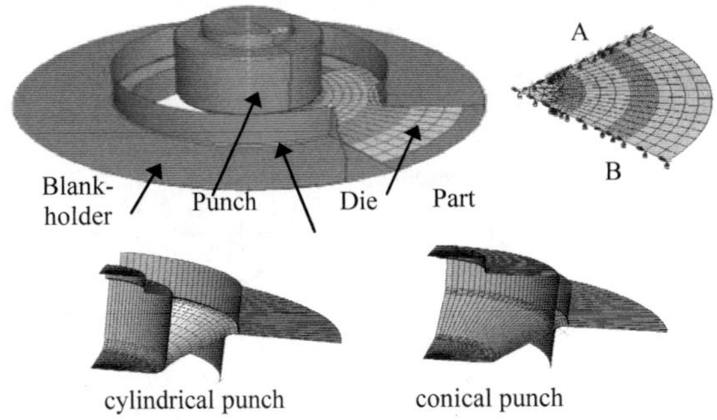

Fig. 2 Geometry of the model used in simulation

3. INVESTIGATION RESULTS

The investigations concerning springback were performed in the above presented conditions and in the following two drawing cases: using cylindrical punch and using conical punch. The springback parameters resulted for different values of blankholder force are presented in Fig. 3. The deviations of the parts from the theoretical profile - measured in the most important points along the part profile are graphical presented in Fig. 4 and 5. The variation of deviations from the circularity of part measured in the zone of connection between wall and flange is presented in Fig. 6.

BHF [kN]		20	35	50
cylindrical punch	R_2 [mm]	6.44	6.48	6.81
	R_1 [mm]	4,87	4.31	4.26
	α [°]	-0.95	-0.53	-0.25
conical punch	R_2 [mm]	6.33	6.37	6.412
	R_1 [mm]	4,425	4.419	4.416
	α [°]	-0.67	-0.38	-0.13

Fig. 3. Springback parameters

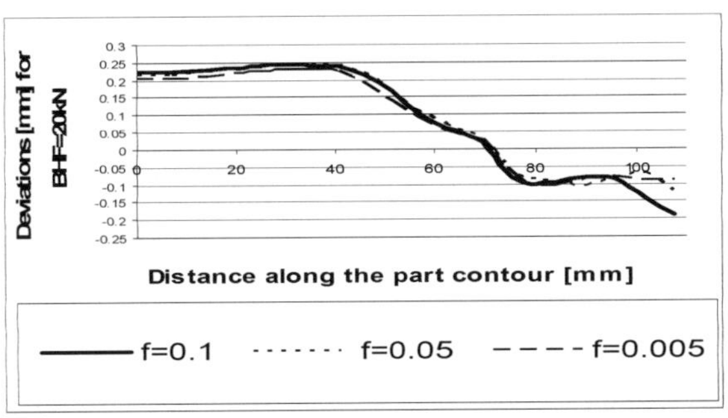

a. case of conical punch

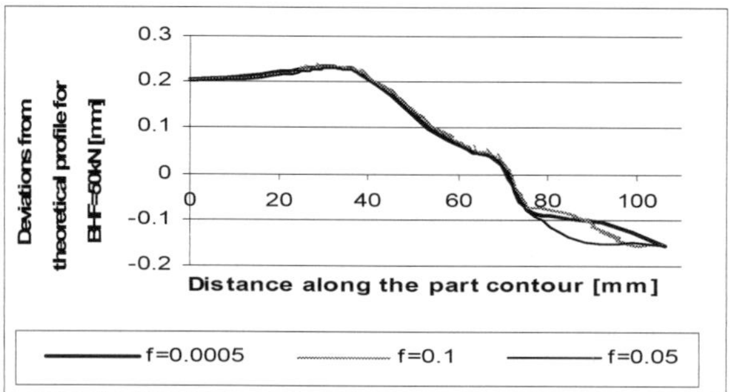

b. case of cylindrical punch

Fig. 4. Variation of deviations measured along the part profile - as a function of different friction coefficients

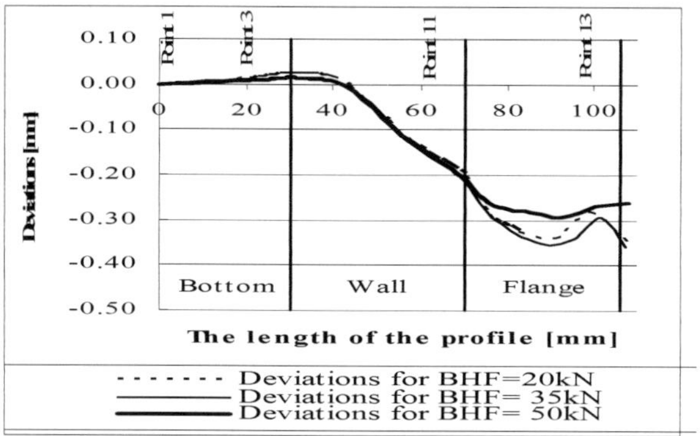

a.case of cylindrical punch

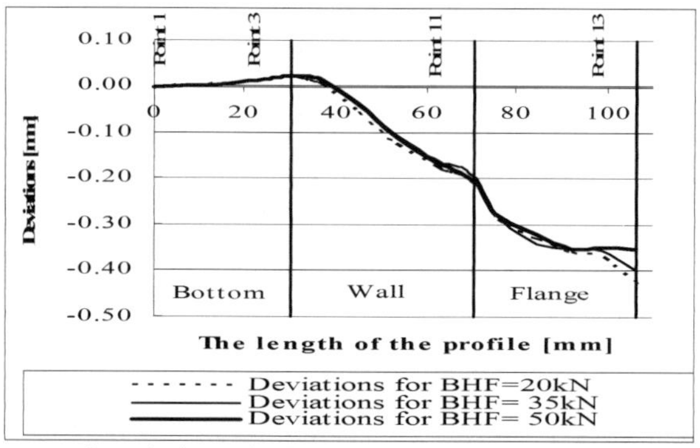

b. case of conical punch

Fig. 5. Variation of deviations measured along the part profile - as a function of different blankholder forces

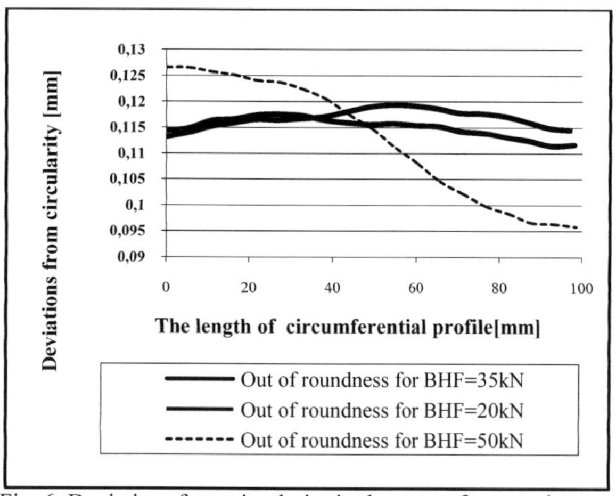

Fig. 6. Deviations from circularity in the zone of connection between wall and flange

The variation of springback parameters in the case of a conical drawn part as a function of blankholder force presents the following aspects: the increase of blankholder force leads to the increase of the radius of connection between part bottom and wall, to the decrease of the radius of connection between part flange and wall and of the flange springback angle; in the case of cylindrical punch the springback parameters have higher values than in the case of conical ones. The variation of deviations from the theoretical profile due to springback - measured in different points along the part profile - presents the following aspects: the deviations have approximately the same variation along the part profile for all the values of blankholder force; the deviations have approximately the same variation along the part profile for all the values of friction coefficient; on the part bottom the deviations are approximately equal to zero; in the zone of connection bottom - wall the deviations have small positive values but above of such zone the deviations become negative for all blankholder forces and will touch the highest values on the part flange; the deviations resulted in the case of conical punch have approximately the same values like in the case of cylindrical ones, an exception from such variation being registered in the flange zone and in the zones of connection wall - bottom and wall - flange where the deviations have higher values in the case of conical punch; the deviations from the circularity vary on the length of the part circumference as a function of blankholder force.

5. CONCLUSIONS

The springback phenomenon and its effects on the part geometry are caused by the differences resulted between the states of residual stresses on the outer and inner faces of part, especially in the zones of connection wall - bottom and wall – flange. The variation of springback parameters is caused by the modification of the stresses state generated in material depending on blankholder force values; thus, the utilization of high blankholder forces will block the flow of the material into flange and die cavity. The material flow blocking will lead to the elimination of differences between the states of stress on the outer and inner faces of part. The cancelling of the differences between the states of residual stresses will have positive consequences on the reduction of springback parameters.The springback intensity is also influenced by the variation and values of the major components of residual stresses that act along the part profile. The deviations from the part circularity are caused by the variation and values of the minor components of residual stresses that act along the part circumference.

Acknowledgments:
The present research was performed with the financial support of the Romanian Ministry of Education and Research in the frame of the CEEX Programme.

6. REFERENCES

Brabie, G. (2006). *Sheets metal forming. Instability Phenomena* (in Romanian), Ed. Junimea, ISBN 973-37-1098-9, Iassy

Han, S. & Park, K. (1999). An investigation of the factors influencing springback by empirical and simulative techniques, *Proc. of Numisheet'99*, Gelin, J. (Ed.), pp 53-57, ISBN 2-84449-001-8, sept. 1999, BURS, Besancon

Pearce, R. (1991). *Sheet metal forming*, The Adam Hilger Series - New Manufact. Processes and Materials, Bristol

Samuel, M. (2000). Experimental and numerical prediction of springback in U-bending of anisotropic sheet metals, *J. of Mat. Proc. Tech.*, pp. 382-393, ISSN 0924-0136

Tang, C. S. (2001). Analysis of springback in sheet forming operation, *Advanced Technology of Plasticity*, vol. I, pp 193-197, Springer, ISBN 0-387-17915-1, Stuttgart

Annals of DAAAM for 2008 & Proceedings of the 19th International DAAAM Symposium, ISSN 1726-9679
Editor B. Katalinic, Published by DAAAM International, Vienna, Austria 2008
Make Harmony Between Technology and Nature, and Your Mind will Fly Free as a Bird

ON THE POST PROCESS DIMENSIONAL CONTROL AIDED BY COMPUTER

BRAUN, B[arbu]; BECA, P[aul]; OLTEANU, C[iprian] & GHEORGHE, G[heorghe]

Abstract: *The paper describes the way in which our team implemented an automatic system for the post process dimensional control about the hydraulic pump cases. To increase the efficiency and quality of the process, we took into account the following aspects: 1. To insure manufacturing precision; 2. To protect the polished active surfaces of the pump cases; 3. To increase the efficiency of the process, we programmed two applications to insure the aided by PC post process dimensional control.*
Key words: *probe, sensors, control, user, LabVIEW*

1. INTRODUCTION

Nowadays, the work productivity and the costs reducing is impossible without the automatic systems and processes aided by PC (Cristea & Ionescu 1998)

Besides, in the present the paradox between the quality and productivity insuring and energy consumption seems to be a problem. An answer for this problem could be a continuous control of the manufacturing process, to avoid the scraps. That means to eliminate the superfluous times and energy consumption, in case of scraps (Cristea, 2006).

The machines with numerical command developed by different companies could be a solution to ensure the commanded manufacturing process in real time. The coordinate measuring machines could be a good solution for the dimensional control of the probes with complex geometry. But the main inconvenience for both is the high cost.

For this reason, for the manufacturing of some components with non-complex geometry, in our research we implemented different systems to insure the dimensional control aided by PC. The way in which we thought to reduce the costs was to dispose some not very expansive displacement transducers coupled to the computer and to program the data processing to have in real time the complete information about the process.

Until the present we made some experiments for two types of manufactured probes with non-complex geometry. It's about the entrained in rotation and translation probes. Making an analyse we concluded that the costs should represent about 25 ÷ 30 % to the costs in case of dedicated measuring gauges.

In our further research we propose to develop to command the active control via PC, by programming the interfacing of the sensors and actuators that command the process.

2. THE DEVELOPPED POST PROCESS CONTROL GAUGE

Our research was realised together with the **National Institute for Research and Development in Precision Mechanics, Bucharest, Romania** about the ***waterwheel pump metallic cases***, which active surfaces precision work must be inspected (Alexandrescu et al., 2008).

In the present our study is in progress, grace to a research contract between our Department and the Institute for Research and Development in Precision Mechanics, Bucharest. It contains two main steps: the developing of the manufacturing gauge and the dimensional control process improvement.

Fig. 1. The three axis robot machine for the transporting of the probes to the manufacturing and control gauges

To transport each probe to the manufacturing machine, a three linear axis robot was used. It was developed by YAMAHA (Japan), which precision is about 0.2 μm (Braun et al., 2008).

As measuring device we used two displacement transducers with non contact principle, both developed by OMRON Company (Japan). The first one is based by the optical principle, and its detection distance is 25 mm. The second transducer is equipped with a laser source and its detection distance is 38.5 mm. For both transducers the measuring resolution is 0.01 mm.

To eliminate the measuring errors caused by glitter of the tested surface, both transducers were disposed to one side and to other side of the tested probe, so that the distance between the treated surfaces could me measured for each position along the direction of its transportation. In this way, considering that the distance between the transducers is constant, the distances between the transducers and the surfaces could vary. If L represents the distance between the transducers and d_1 and d_2 the distances between the transducers and the controlled surfaces, the relation to determine the distance d between the surfaces is the following:

$$d = L - (d_1 - d_2) \qquad (1)$$

Regarding the interfacing with the PC, first of all, to obtain the measured data about the distance between the pump's surfaces, we used the Cscape software, which algorithm of calculus is presented as a Ladder diagram (Fig. 2).

The first row represents the data input providing from the first transducer. The second addresses the data from the other transducer. In the third row there are defined two functions to substract from the distance between the two transducers the sum of the two measured distances between the transducers and the surfaces.

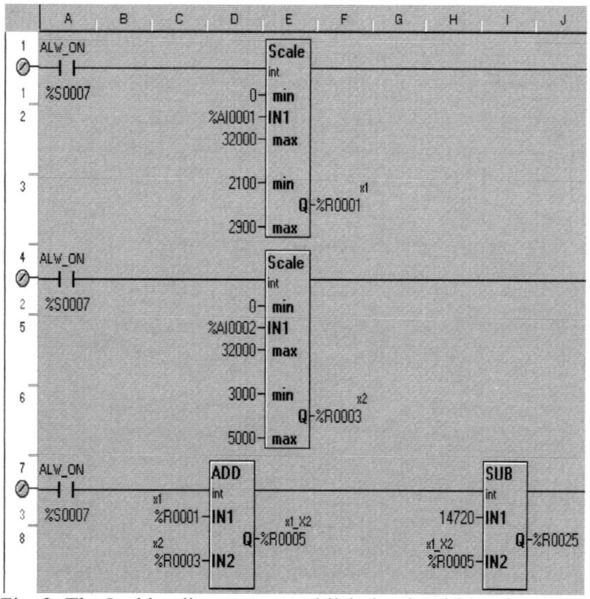

Fig. 2. The Ladder diagram to establish the algorithm of calculus of the distance between the polished surfaces of the probe

Grace to this interface there it was obtained an EXCEL file as a list of measured data.

To complete the information regarding the post process dimensional control, aided by PC, we developped an application in LabVIEW to interface with the user the manufacturing process. The interface with the user is displayed like a virtual dialog through, via some numerical controls the process parameters can be established. Each of the control adresses a Text Ring dialog to define a specific numeric parameter, as required. (Ciobanu, O.; Cottet, F., 1998).

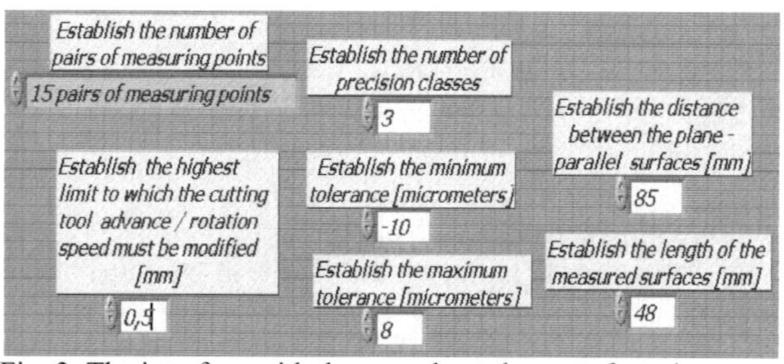

Fig. 3. The interface with the user about the manufacturing process parameters

The user can establish not only the measuring parameters for each dimension of the probe, but some other parameters, about the advance or speed of the cutting tool, during the manufacturing process.

Grace to the measured data providing from the sensors and the established parameters, the running of the program generates quickly and simultaneously all the information about the results of the manufacturing process. There are presented some of the most important aspects of the manufacturing process, which could be classified in two categories: information about the dimensional control process and information about the manufacturing parameters, as a result of the dimensional control. The first category refers to the result of the control process, the user being informed if the quality of the manufacturing of the probe corresponds (Fig. 4).

Some numerical data about the form deviation could also be useful as a statistics of the process (Fig. 5) (Braun et al., 2008). The second category refers to the user to allows to modify the cutting tools parameters if necessary.

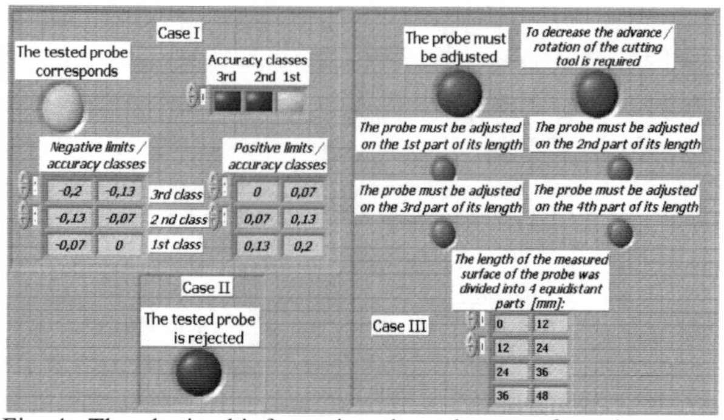

Fig. 4. The obtained information about the manufacturing parameters for a running cycle

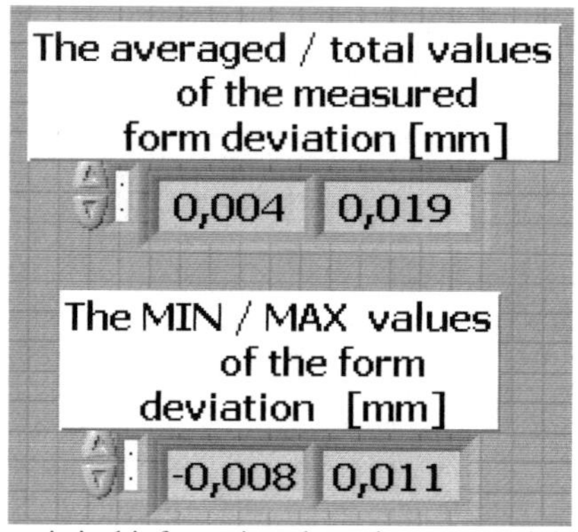

Fig. 5. The statistical information about the measured data

3. CONCLUSION

Our software application will serve to the Institute to evaluate very efficient and quickly low or high series of probes like pneumatic components, which geometry is not very complex. Another advantage is that using this principle, different, the manufacturing process for different dimension ranges will be possible.

Our further research will be concentrated to the increasing of the control process during the manufacturing. That means that we propose to develop an application to ensure the active control aided by PC. Another aspect refers to the aided by PC control of the surface quality of some manufactured components. First of all it's about the roughness of the polished active surfaces.

4. REFERENCES

Alexandrescu, N.;Besnea, D., Dontu, O., (2008) *Some problems regarding the design and execution of mechatronics parts using CNC centers,* Bulletn of the Politechnical Institute of Jassy, Romania, ISSN: 1011-2855

Braun, B.; Costas, C., Druga, C. *The virtaul instrumentation in the automatic dimensional control process optimisation*, The 6th International DAAAM Baltic Conference Tallinn, Estonia, ISBN: 978-9985-59-783-5

Ciobanu, O.; Cottet, F., (1998) *The fundaments of programming in LabVIEW*, MATRIX ROM publishing house, Bucharest, ISBN: 973-9390-56-0

Cristea, L.; Ionescu, E (1998). *Automatic control plants in industry*, Didactic and Pedagogic publishing house, Bucharest, Romania, ISBN: 973-30-5047-4

Cristea, L. (2006) *Considerations regarding the improvement of throust bearings automatic dimensional inspection*, Acta Technica, Cluj-Napoca, Romania, ISSN: 1221-5872.

Annals of DAAAM for 2008 & Proceedings of the 19th International DAAAM Symposium, ISSN 1726-9679
Editor B. Katalinic, Published by DAAAM International, Vienna, Austria 2008
Make Harmony Between Technology and Nature, and Your Mind will Fly Free as a Bird

APPLYING BENCHMARKING IN ROMANIAN ENTERPRISES

BRETCU, A[ngela]

Abstract: The *Benchmarking is a tool of continuous improvement, frequently used in the USA and Europe, through which the organisations constantly improve their performances. In Romania, this method is rarely met and has not brought very good results. The reasons of these happenings are searched for, as well as the possibilities for remedying the difficulties. The necessity for using this tool is expressed in Romania in view of diminishing the existent economic disparities.*
Key words: *continuous improvement, benchmarking, phase*

1. INTRODUCTION

The Benchmarking can be defined as the practice to improve the organisations performances by comparing it to the best ones on the market. This represents a systematic and continuous process of comparing own products, services, techniques, technologies, practices with similar ones, in order to learn the significant differences with a purpose of assimilating them. Practicing Benchmarking is comparable to seeking the best methods utilized within an activity, this method allowing the companies to improve their performances (Scurtu et al., 2006).
The Benchmarking is very popular in USA (Rank, Xerox, Rover, Alcoa), as well as in Europe (Great Britain, Denmark, Belgium, Austria, Germany, Sweden)(Magd et al.,2003). This concept bases itself on the fact that the world is in a continuous evolution and represents the comparison of the own enterprise with one or more identified organisations as being a point of reference in a certain domain. In Romania, the Benchmarking is poorly applied. The enterprises which have tried the implementation of the method have had good results, but not as good as they expected, although the phases and concrete modalities of approach have been precisely followed. Due to certain economic particularities, determined by the passing from the state economy to the market one, the implementing phases must subject to some modifications, by which should be kept into account the concrete conditions in which the economic activity takes place. We must take into account that in Romania, the principles of quality management are not often applied, reason for which the Benchmarking takes longer than in other countries, takes more costs and extra efforts. The external environment also intervenes in specific conditions in Romania. Nevertheless, in this framework, the application of Benchmarking is necessary for the Romanian enterprises to align to European environment, as to performances and visions. It is important that the research to be continued in the future, more profoundly, for optimizing the benchmarking implementing conditions. In this sense concrete research for applying the Benchmarking will be undertaken in large enterprises from Romania, in order to highlight the particularities of quality management.

2. INFORMATION

Therefore, the essence of Benchmarking is the process of borrowing ideas and adapting them in order to achieve important advantages, so it is a tool for continuous development(Juran,2002).Starting from the idea that benchmarking is a tool for continuous development; The European Quality Organisation proposes an example of approach for benchmarking which contains four phases of continuous improvement: planning, collection, analysis and adaptation (Scurtu et al., 2006).
These phases of benchmarking process development imply setting out and following some steps which differ depending on the stage of the benchmarking process. For example, in the planning phase, the building up of the team is essential, taking into account the performances of each member, their knowledge and training regarding the process and the details of the studied organisation.
The next step is the documentation related to the analysis process and contains activities such as: identification of the analysis area, identification of successfully critical factors, the development of the quantification factors. Also the decision regarding the setting out of the analysis goal for benchmarking must be taken. During the planning phase the drawing up of a declaration concerning the purpose of benchmarking takes place so as any deviation of the team's original goal will be avoided.
The selection of the partners is also a very important step and therefore the determination of the selection criteria for the partners must be connected to the successful critical factors. The last step of the planning phase consists in the determination of the collection for the data plan, establishing therefore the instruments and techniques for collecting of the information.
The collection phase implies the identification of the best practices and data collection. The developing of the tools for data collecting is the next step and after the selection of the target-organisations, the information must be revaluated, collected and processed.
Before putting into practice the selected tools for the analysis, they need to be tested within the own organisation, in other words, an internal evaluation of the selected tools for the analysis takes place. The selected partners for the analysis must be contacted and requested for collaboration in the framework of the projected branch marking. Within this phase, the selection of the partners may be done depending on the interest for the proposed analysis (Falniţă, 2000).
Within the analysis phase of the benchmarking process, comparison of own performances with the ones of the competition implies the sorting and the processing of the data, adapting the gathered up information to a specific form, favourable for comparison and identifying the differences. The next step is the identification of operations such as „the best practice" and the development of the implementation strategy, which implies the evaluation of solutions adaptability, the identifying of the opportunities for improvement and the analysis of the costs(Stanciu,2002). An important problem within this phase is the communication and the discussion of the implementing plan with the involved parties, making it possible for certain changes to occur, in this initial phases, if they are needed. The last phase proposed by The European Quality Organisation, the adopting phase, implies in the first place the implementation of the plan. At this level it is very important to apply all the solutions initially elaborated, without being interpreted or „dissolved".Furthermore, the superior leadership must periodically request reports regarding the

implementing of the improved plan, therefore achieving the *monitoring and reporting of the progress.* The information resulted from the study, as well as all of the developed methodology must be systematized and registered in order to be of service to other studies. This step is called the recalibration and recycling of the study. Finally, a continuous improvement plan is elaborated, by identifying new opportunities for benchmarking. Because the Benchmarking process involves more imitation and adaptation than pure invention, the time and money are saved. Nevertheless, by developing such a process one must take into account certain requests and the satisfaction of these requests contributes to achieving the proposed goals. In this respect, there are four fundamental principles of benchmarking. A first principle is reciprocity, namely a good service deserves a good answer. Another principle, the analogy, states that the products of the benchmarking partners must be comparable. Likewise, the correct choice for the measurement system is very important because the gathered information must be seen in a favourable form for comparison. More, the correctness of the information which is to be exchanged must be guaranteed. In conclusion, the benchmarking is not a strategy and does not intend to be a business philosophy; it is instead a developing tool and in order to be efficient, must be used as such. All of the economic agents of Romania must take interest in Benchmarking, regardless the form of property. It is important to underline the fact that Romania has lost two major events because of the old system: the technological and managerial revolution. In this framework, the private capital enterprises took over from the state capital enterprises, together with the means of production and the work force, a number of difficulties, connected to the technological depreciation of the equipment and the employers' mentality. At present time in Romania, there are a few enterprises with a private capital, equipped with the nowadays technological level, with an organisational culture and managerial practices proper for benchmarking implementation. In the planning phase, in order to achieve a benchmarking activity, best practices in certain precise conditions must be identified. Without a detailed knowledge of the conditions in which a practice can be successful, the benchmarking may offer incorrect conclusions, because similar practices cannot generate identical results in different contexts. First of all, there must be taken into consideration the fact that there are few enterprises which implement quality management. The superficiality of principles implementation manifests itself from the managers' side and from the employees' side. Besides that, it is very frequent that the continuous improvement is not applied and as far as the diagnosis analysis goes, if it is applied, it is only in a formal way. Under these conditions, the team that implements the Benchmarking process will have to put in time and additional resources for the correct identification of the successful critical factors. Second of all, there are, even in the present time a series of anomalies which could alter the Benchmarking effects. One of this is the employees' mentality, which can act as a brake within the economic activity, however performing the technology may be. The introduction of practices that depend on the environment, in which the economic agent operates, must be then delayed until the amelioration of those conditions.
Within the analysis phase, it is very important for the external environment to be understood by the leaders of the Benchmarking study. A first aspect would be the proportion in which the general environment may create barriers while introducing identified practices, on its different components. In the framework of economic segment, the identified barriers are the inflation rate and interest rate, in the political and legal context, the laws regarding taxes and incomes and in the technological context; the identified barriers are the levels of telecommunication and infrastructure compatibility.
Another aspect that needs to be studied is the proportion in which the conjuncture may diminish the effects of practice application. Therefore, in the political and legal context, the effects are diminished by the legislation regarding economic and rivalry contracts and the social and cultural context influence these effects through the employees' mentality and attitude. Also important is the proportion in which the positive effects of the Benchmarking process justify the efforts. Considering that by introducing the best practices through an economic agent in Romania, this meaning a significant financial effort, the indicators may be smaller than the ones that may be achieved within the same project but in developed countries.This is due to the conditions of obtaining credits, the high level of fiscality and other barriers. Because of the low wages level, the contributions to costs reduction are therefore less significant. Within the adaptation phase, the success of the project will nevertheless open, through the economic agent, the way to a process of continuous improvement. The monitoring of the process is difficult because of the lack of monitoring systems that measure the main performance indicators. In spite of these deficiencies, the Benchmarking represents an effective way for the producers to assimilate the best practices, reason for its continuance. By learning the best achievements of the time, some research stages, already covered by others, can be eliminated, saving therefore significant financial and time resources. The Benchmarking can also contribute to avoiding repeated mistakes, by analysing unsuccessful experiences. By using Benchmarking the organisations will be obligated to change their traditional behaviour and introduce new practices within their current activity, such as: permanent evaluations, implementing measurements regarding each sector, depending on strong points, week points, opportunities and revaluation of employees, through the process of continuous improvement. A favourable climate is essential for Benchmarking implementation, by eliminating wrong perceptions regarding this method, by sustaining the knowledge of the system through governmental programmes, that should keep into account not only the dissemination of information but also the organisation and development of research institutes, by establishing doctorates and master degrees containing this profile within the universities, by creating of an interactive website for the partners actively involved in Benchmarking implementing.

3. CONCLUSION

Efforts need to be made for a future Benchmarking study, because this is the only way to hurry the change of wrong mentalities regarding economic activity and to achieve the positive role of competitiveness for a profitable business, the practices regarding research and development, understanding the need for transparency. By implementing Benchmarking in reference enterprises some important contributions will be therefore brought to the coherent development of innovation, investment, in the regional and social domain.

4. REFERENCES

Falniţă E. (2000e). *Benchmarking in small and medium entreprises*, VEB, nr. 53:9,16

Juran, J.M. (2002). *Supremacy through quality*, Teora Publishing House, ISBN 973-601-965-9, Bucharest

Magd, H. & Curry, A. (2003). *Benchmarcking achieving best value in public-sector organisations*, Benchmarking, 10,3:261-273

Scurtu, V.; Russu, C. & Popescu I. (2006). *Benchmarking Theory and applications*, Economic Publishing House, ISBN973-709-165-5, Bucharest

Stanciu, I. (2002). *Calitology, The Science of merchandise quality,* Oscar Print Publishing House, ISBN 973-8338-03-4, Bucharest

Annals of DAAAM for 2008 & Proceedings of the 19th International DAAAM Symposium, ISSN 1726-9679
Editor B. Katalinic, Published by DAAAM International, Vienna, Austria 2008
Make Harmony Between Technology and Nature, and Your Mind will Fly Free as a Bird

NEURAL NETWORK CONTROL FOR WIG WELDING PROCESSES

BUCUR, G[abriela]; POPESCU, C[atalin] & POPESCU, C[ristina]

Abstract: *The improvement of the welding quality is achieved by the torch trajectory correction, based on following the welding joint, processing the acquired signals in the command system and modifying, if necessary, the welding parameters. He aim of the paper is to deal with the problem of using artificial neural networks in the welding process, respectively the use of adaptive linear filters in the command of the technological process, for controlling the position of the welding head, based on the analyze of the electric arc voltage.*

Key words: *neural networks, control, welding, measurement, acquisition*

1. INTRODUCTION

The final purpose of a welding operation is to achieve a welding seam that satisfies many imposed conditions. These conditions results from technological general system analysis, system witch is composed by power source – welding source – arc welding system, welding seam, manipulation system and control system.

The paper present the way to process the welding-arc voltage with neural network. That result can be applied in automated robotic welding based on the corection of the trajectory welding torch through the welding seam.

2. CONTROL POSITION SYSTEM

The values of welding parameters regime (welding current, arc voltage and speed penetration) depend on relative position between the components of welding joint. Some parameters are controlled by welding equipment and the others by the geometrical ratio between welding head and piece. For example, on WIG welding process, the arc voltage depends on electrode-piece distance (Stefanescu & Cupcea 2002) .

For welding head position control we use an acquisition system for arc voltage values from welding process. So, the arc voltage is a measure of welding head position.This signal will be the input for a neural network structure especially made by the authors for modifies the position of welding head. This neural network will be the controller for position closed loop, shown in figure 1, where $U_{arc}^{process}$ is process arc voltage, $U^*{}_{arc}^{ref}$ is reference calculate arc voltage, a is output of neural network, q is controlled position of welding torch.

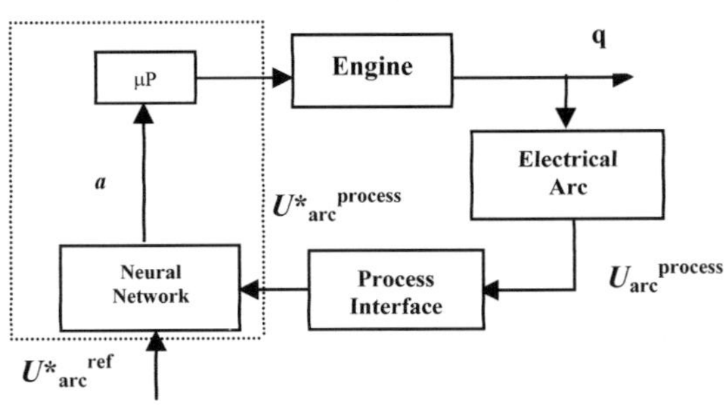

Fig.1. Position control with neural network controller.

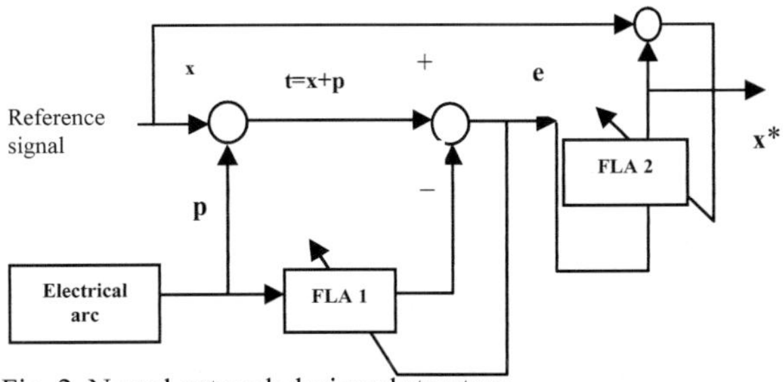

Fig. 2. Neural network designed structure.

3. NEURAL NETWORK STRUCTURE

The neural network structure is presented in figure 2, where x is the reference signal, p - electrical arc voltage, $U_{arc}^{process}$, e is the adaptive estimation of signal x, x^* is the neural network output, FLA1,2 are the linear adaptive filters. This „neural controller", is composed by two interconnected adaptive linear filters. FLA1 work on interferences compensation principle and that filter extract the useful signal from welding process signal. This useful signal contains the information about the deviation of the trajectory welding torch through the welding seam. FLA 2 work on inverse modeling principle and that filter eliminate the error between reference signal and FLA1 output signal (Haykin, 1999).

4. EXPERIMENTAL STALL

The process signal (figure 3) is obtained from a real WIG welding process realized with an experimental stall, especially builder for that (Bucur et al.,2002). Characteristics:

- Electrical engine: MCC 12V.
- Welding source: LUD 450W ESAB ARISTO.
- Welding machine: AGA DW 22: 220V, 0.4 A, 50W.
- Welding process: WIG without addition material; wolfram electrode 2.4 mm; protection gas: argon; protection gas pressure: 10 bar; welding current: 65 A; welding voltage: 10...30V; advance speed: 70 mm/min; oscillation frequency: 0.5 Hz; oscillation amplitude: ±1.5 mm.
- Process-computer interface: AX 5411.

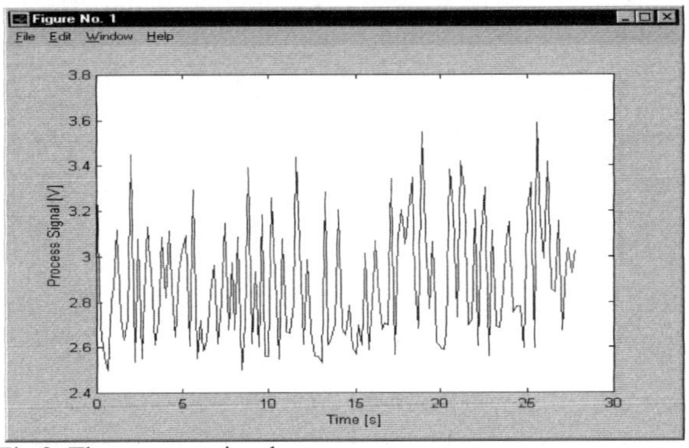

Fig.3. The process signal.

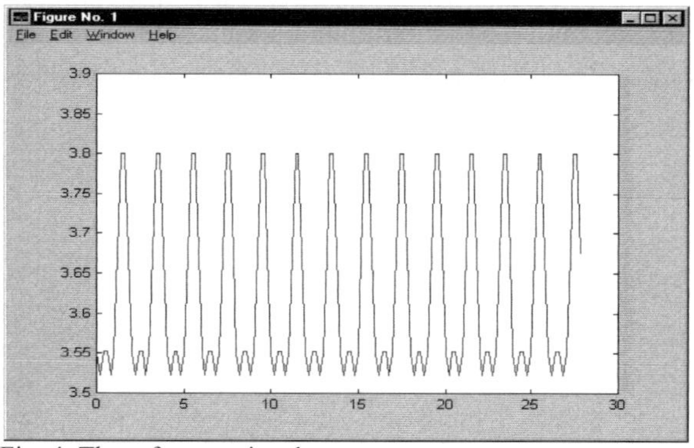

Fig. 4. The reference signal.

5. ANALYTICAL COMPUTATION OF THE REFERENCE SIGNAL

The reference signal (figure 4) is analytically calculated for a welding seam with the next dimensions: *L* – seam width – 1 mm; *d* – seam thickness = pieces high – 2.3 mm; *B* – distance from welding torch to the seam – 5.3 mm, δ – deviation of welding torch to the medium plan of the seam – 0.

We know that the correspondence between the arc welding parameters: welding current I_S, arc voltage U_{arc} and length of electrical arc l_{arc}, for WIG welding processes is done by:

$$U_{arc} = a_1 I_S + a_2 + a_3 / I_S + a_4 l_{arc} . \quad (1)$$

For welding current I_S = 65A (Bucur et al.,2002):

$$U_{arc} = 8.412 + 0.3 \cdot l_{arc} \quad (2)$$

A function that approximates the configuration of U_{arc} is:

$$U^*{}_{arc}{}^{ref} = -0.13\sin(\pi t) - 0.07\cos(2\pi t) + 3.62 \ \ [\text{V}]. \quad (3)$$

6. NEURAL NETWORK SIMULATION

Neural network simulation was realized with the MATLAB program (Bucur & Bucur 2007). The program is:

```
time=0:0.2:27.8;
x= - 0.13sin(πt) - 0.07cos(2πt) + 3.62 ;
plot(time,x) ; (figure 4)
p=[3.33984375
2.69775390625
2.55859375
2.49755859375
2.802734375
2.63427734375
3.4521484375
2.5341796875…];
plot(time,p); (figure 3)
t=x+p';
[w,b]=initlin(p',t);
[a,e]=adaptwh(w,b,p',t,0.00001);
[w,b]=initlin(e,x);
[a,e]=adaptwh(w,b,e,x,0.01);
plot(time,a,time,x) ; (figure 5)
plot(time,e) ; (figure 6)
```

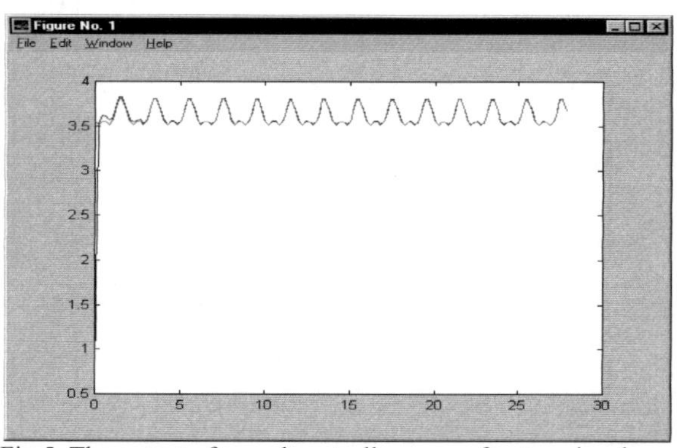

Fig.5. The output of neural controller over reference signal.

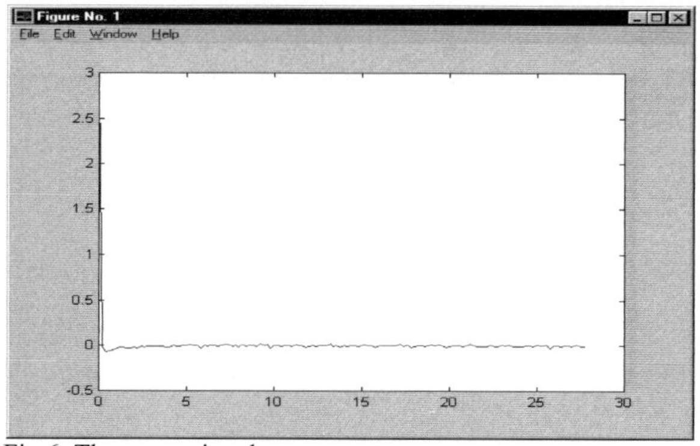

Fig.6. The error signal.

In this program x is the reference signal, p is the process signal, w and b is adjusted weights, signal a is the output of neural controller and e is the error signal. The time of acquisition for arc voltage values is 28 seconds (Kharab & Guenther 2002).

We observe that time of transition is under 1 second and the deviation of equilibrium position for welding torch is near to zero value (figure 6).

7. CONCLUSION

This paper show the neural controller with linear adaptive filters can move the welding torch in a good position near to reference position. Also, during the WIG welding process, if the oscillation frequency is modified over the reference frequency applied to the neural network, we can demonstrate that the designed network can make the correction of the oscillation frequency of welding torch (Bucur et al.,2002).

In the future, the solution of neural network control of trajectory welding torch to the welding seam is very interesting, especially for MIG-MAG welding process, because will appear the burned drop transfer phenomena thru the welding arc.

8. REFERENCES

Bucur, G. & Bucur, L. (2007). Automatic WIG Welding Control using Feed Forward Neural Network. *Annals of the University of Petrosani, Electrical Engineering*, Vol.9, pp.260-265, ISSN 1454-8518

Bucur, G., Dumitrescu, St. & Micloşi, V. (2002). Neural Network Control in Robotic Welding Processes. *Journal of Symposium "35 de ani de activitate a Universităţii Petrol-Gaze la Ploieşti"*, Vol.LIV, No.2/2002, pp. 38-43, ISSN 1221/9371

Haykin, S. (1999). *Neural Networks*, Prentice Hall, ISBN 0-13-273350-1, U.S.A.

Kharab, A. & Guenther, R. (2002). *An Introduction to Numerical Methods in MATLAB*, Chapman & Hall/CRC, ISBN 1-58488-281-6, U.S.A.

Stefanescu, C. & Cupcea, N. (2002). *Control and Measurement Intelligent Systems*, Blue Publishing House, ISBN 973-650-058-6, Romania

Annals of DAAAM for 2008 & Proceedings of the 19th International DAAAM Symposium, ISSN 1726-9679
Editor B. Katalinic, Published by DAAAM International, Vienna, Austria 2008
Make Harmony Between Technology and Nature, and Your Mind will Fly Free as a Bird

MODELING AND SIMULATION OF HYDRAULIC SYSTEMS WITH PRESSURE INTER-CONDITIONING CONSUMERS

BUCURESTEANU, A[nca]

Abstract: *The paperwork presents the mathematical modeling and the results of simulations made for studying the hydraulic systems with consumers that inter-condition by means of pressure. The studied system consists of two hydraulic cylinders which, while working, ensure a certain working cycle that is conditioned by reaching some pre-selected pressures.*

Key words: *hydraulic system, simulation, pressure inter-conditioning*

1. INTRODUCTION

Modeling and simulation of hydrostatic systems represent modern methods of research and engineering design that allow determining static and dynamic characteristics before making a prototype (Prodan, 2006). There are situations when the hydraulic consumers (cylinders) must perform a working cycle conditioned by reaching some pre-selected pressures (Prodan, 2004). Figure 1 presents such a kind of installation.

The pumps P_1 and P_2 feed the system with oil from the tank T. The flow Q_1 defines the pump P_1, while Q_2 defines P_2, and usually $Q_1 > Q_2$. Absorption is made through suction filters F_1 and F_2. Pressures p_{1max} and p_{2max} adjusted by means of the pressure valves PV_1 and PV_2 to the two pumps meet the condition $p_{1max} < p_{2max}$. The circuits of the two pumps are connected through the check valve CV_1. The directional control valve DCV helps starting on the system and reversing the directions. The two hydraulic cylinders C_1 and C_2 represent the consumers. Two sequencing valves (SV_1 and SV_2) equipped with check valves (CV_2 and CV_3) ensure the desired working cycle based on two pre-adjusted pressures (p_1 and p_2). The pilot operated check valve POCV prevents accidental charge failure of the cylinder C_2.

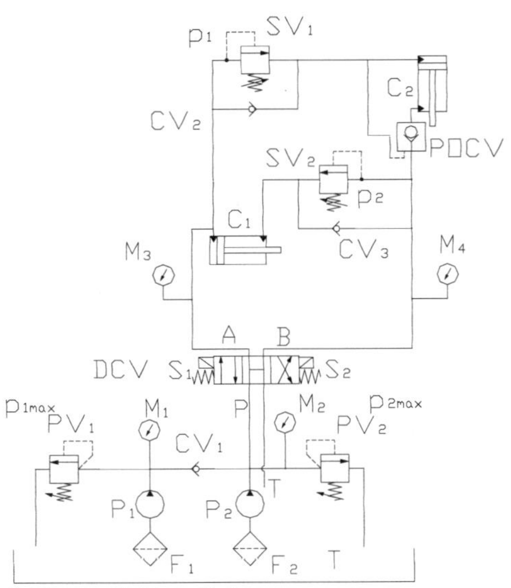

Fig. 1. The studied installation

The pressure gauges M_1 - M_4 display the four adjusted pressures.

If considering the start positions of the cylinders as being those in the figure, then the system works as follows:

When powering the electromagnet S_1, cylinders C_1 moves to the left-hand side with the high speed resulted by summing of the flows Q_1 and Q_2. The pressure required for the movement is lower then pressures p_{1max} and p_1. In this situation the sequencing valve SV_1 is closed and cylinder C_2 does not move. Moreover, the pilot operated check valve prevents its accidental failure (Bucuresteanu, 2003).

If cylinder C_1 reaches the end of the travel, then the pressure increases and exceeds the values p_{1max} and p_1. The valves PV_1, SV_1 and POCV get open and the cylinder C_2 starts descending. The flow supplied to this cylinder is Q_2 only, because the pump P_1 discharges to the tank through the pressure valve PV_1. Therefore, the cylinder C_2 starts only when the pressure p_1 adjusted on SV_1 is reached.

If powering the electromagnet S_2, the cylinder C_2 ascends due to the oil that gets trough at the lower part of the pilot operated check valve POCV. The ascending is accomplished with the flows summed from the two pumps. When the pressure p_2 adjusted on the check valve CV_3 is reached, the cylinder C_1 withdraws to the initial position. The pilot operated check valve POCV ensures the locking of the cylinder C_2 in case of accidental power failure.

2. MODELING AND SIMULATION OF THE SYSTEM

If the characteristics of the system components and the working parameters are known, then the modeling and simulation of the system can be done after having made the operating diagram. The cylinders have a significant influence from the dynamic point of view, among the components of the system. The mathematical model specific to the hydraulic cylinders working as those in Figure 1 is the one below (Oprean et al., 1998):

$$Q_{HC} = S_1 \cdot v + a_{HC} \cdot p + \frac{V_{MHC}}{E_O} \cdot \frac{dp}{dt} \qquad (1)$$

$$M \cdot \frac{dv}{dt} + b_{HC} \cdot v + \Sigma F = p \cdot S_1 \qquad (2)$$

where: Q_{HC} - input flow of the hydraulic cylinder; S_1 - active surface of the cylinder; v - instant speed; a_{HC} - linear coefficient of the flow loss in proportion with the pressure; p - pressure on the active surface of the cylinder; V_{MHC} - average volume of pressured oil inside the active chamber; E_0 - elasticity module of the oil; t – time; M – weight of the moved assembly; b_{HC} - linearised coefficient of force losses in proportion with the speed; ΣF - sum of resistant forces within the driving cylinder.

Figure 2 shows the speed characteristics for C_1 during a working cycle.

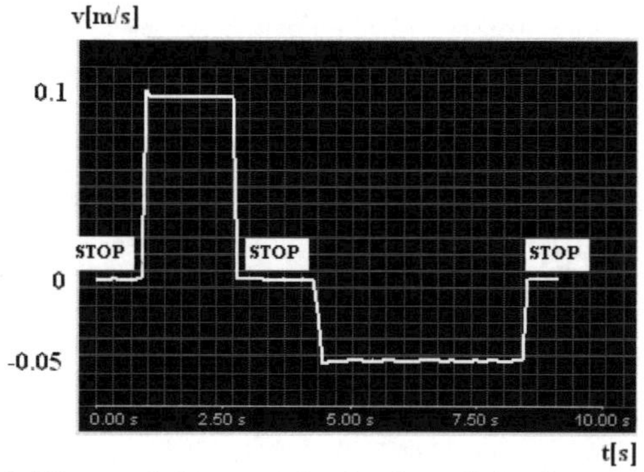

Fig. 2. The speed characteristics for the cylinder C_1

When moving from the left-hand side to the right-hand side, by summing the flows, a speed of 0.1 m/s is ensured. When withdrawing, only the pump P_2 supplies oil, and a speed of only 0.05 m/s is reached. When actuating to the left-hand side, the pressure on the active surface develops according to the curve in Figure 3. The cylinder C_2 performs a cycle with the speed developing as shown in Figure 4. When descending, the useful flow comes only from pump P_2. When returning, the speed is higher because of summing of the flows, at a pressure lower than the one adjusted on the pressure valve PV_1. Figure 5 shows the development of the pressure on the active surface.
The simulation allows establishing the influence that each component has over the system working. Due to specialized software, optimal components can be selected and the dynamic balance of the system can be achieved (Oprean et al., 1989). Thus, by using the curve in Figure 6, one can study the transit flow of the pilot operated check valve during the whole cycle.

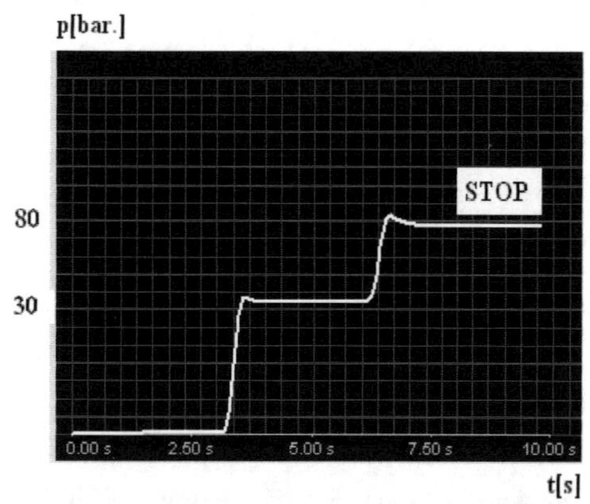

Fig. 3. The pressure on the active surface of the cylinder C_1

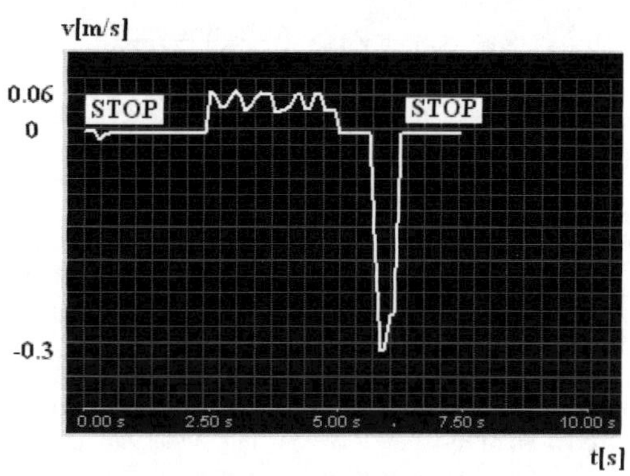

Fig. 4. The speed characteristics for the cylinder C_2

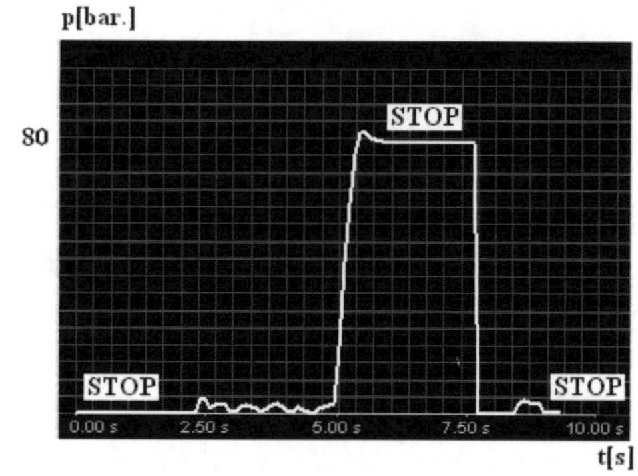

Fig. 5. The pressure on the active surface of the cylinder C_2

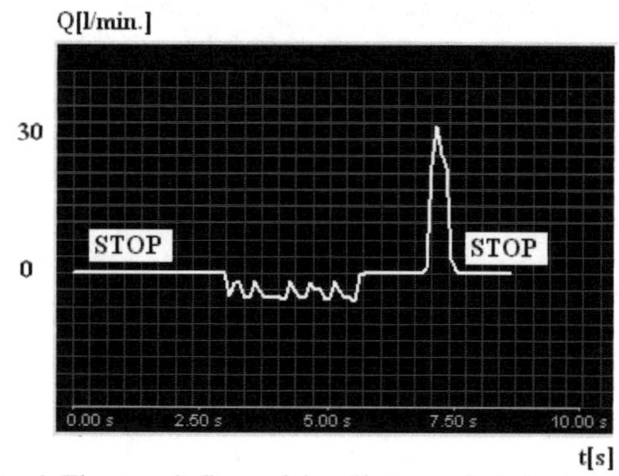

Fig. 6. The transit flow of the pilot operated check valve

These curves/characteristics can be compared with those provided by the components manufacturer and, if necessary, modifications can be made from the design stage, so as the costs generated when making the prototype will decreased.

3. CONCLUSION

The hydraulic systems with pressure inter-conditioning are recommended for devices and presses actuation and for machine-tools. They are simple and reliable systems that can operate even in hazardous environment. The inter-conditioning being purely hydraulically, the electric installation is very simple. There is specialized software for the modeling and stimulation of those systems, which allows the optimization of the design and the selection of the components based on both technical and cost criteria. Modeling and simulation reduce the risks of a faulty working of the prototype, the test time and the costs, as well.

4. REFERENCES

Bucuresteanu, A. (2003). *Hydraulic and Pneumatic Actuators,* Printech Publishing House, ISBN 973-652-819-9, Bucharest

Oprean, A.; Ispas, C.; Olaru, A. & Prodan, D. (1989). *Hydraulic Actuators and Automation,* Technical Publishing House, ISBN 973-98652-2-4, Bucharest

Oprean, A.; Dorin, A.; Olaru, A. & Prodan, D. (1998). *Hydraulic Actuators Equipments,* Bren Publishing House, ISBN 973-98652-2-4, Bucharest

Prodan, D. (2004). *Machine-Tools Hydraulics,* Printech Publishing House, ISBN 973-718-109-3, Bucharest

Prodan, D. (2006). *Elements and Hydrostatical Systems Modelling and Simulation,* Printech Publishing House, ISBN (10)973-718-572-2, Bucharest

Annals of DAAAM for 2008 & Proceedings of the 19th International DAAAM Symposium, ISSN 1726-9679
Editor B. Katalinic, Published by DAAAM International, Vienna, Austria 2008
Make Harmony Between Technology and Nature, and Your Mind will Fly Free as a Bird
2nd DAAAM-ICAT

STL FILES PARAMETERS OF THE SELECTIVE OBJECTS FOR RAPID PROTOTYPING

BUDZIK, G[rzegorz]; MARKOWSKA, O[limpia] & MARKOWSKI, T[adeusz]

Abstract: *The article presents STL files parameters analysis of the selective geometrical objects. The examinations included STL models received from reverse engineering (RE) process and CAD modeling. During the reverse engineering process were used methods like: computer tomography and laser scanning. The investigations were carried out on assortment virtual models of real objects: cranium, turbocompressor rotor. The paper presents possibilities of acquisition and processing data from RE process. The article also clarified the CAD model export of the machine elements with complex shapes: rotor blade of air engine, turbocompressor rotor, and gear-wheel. The programme data processing was as well prepared for virtual models and figures designed for entertainment industry, e.g. computer games. The parameters influence on files exporting from chosen CAD programs as well as STL files parameters influence on physical model accuracy received from rapid prototyping methods were also explained. Recommendations as to the accuracy of a STL model have been presented.*

Key words: *STL files parameters, export complex shapes, reverse engineering (RE), CAD modeling*

1. INTRODUCTION

Parameters of STL file have fundamental influence on accuracy of physical model made with Rapid Prototyping method (3D Ligtyear 2001). Accuracy of STL model created in CAD systems is dependent on export module in CAD programme (Budzik & Sobolak 2006). Accuracy of STL model obtained by 3D scanning process is dependent on scanning method and the way of data processing (Budzik 2007). Fundamental problem of STL file export is volume optimization and accuracy representation of STL data (Road & Tan 2000).

Generally publication of STL file processing can be divided on a few parts. First part is a literature concerning STL model creation on the basis of 3D scanning (Pieraccini et al 2001, Schemenauer et al. 2002). Next part are instructions of CAD systems. They describe mainly ways of STL data export without description of model accuracy (Skarka & Mazurek 2006). Next part of literature describes possibilities of STL data processing and repairing STL model (Lee & Woo 2000).

Some publications describe STL file preparation direct to RP apparatus, there are RP/RE systems instructions (Geomagic Studio 2006, ZScan 3D 2006, ZEdit 2006).

The article presents analysis of STL model from reverse engineering (computer tomography – CT, laser scanning LS) process and 3D-CAD modelling (CATIA and Mechanical Desktop systems).

Article describes STL file parameters from reverse engineering process of human skull and turbocharger rotor. There are STL data export from CAD models of: aircraft engine blade, bevel gear and computer game model of man.Future research will include comparison of STL data exported from a few different CAD systems and optimization of STL models parameters.

2. GENERATING AND PROCESSING OF STL FILE

Data file formats are developed by CAD and Rapid Prototyping producers. CAD system have possibilities of data export in different formats for RP systems. STL format is the application most often used in Rapid Prototyping and Reverse Engineering systems.

STL format describes surface of CAD model with the use of triangles. Every single triangle is described by normal vector with outside of model surface direction (Stroud & Xirouchakis 2000).

2.1 Reverse Engineering process

Non-contact methods used in a process of Reverse Engineering can be divided on four groups: non-contact scanning, scanning with laser beam, photogrammetry, Computed Tomography. A non-contact scanning is performed with portable or stationary spatial scanners. In case of stationary closed scanners the operation area of a scanner limits dimensions of examined element. Portable scanners allow to scan objects with theoretically unlimited dimensions. A major advantage of Computed Tomography is a possibility of obtaining external and internal geometry of examined object in non-destructive way. A processing of data acquisited from computer tomograph using a specialist software enables to receive files for standards typical for CAD/CAM/CAE.

Non-contact scanning can be executed with a hand scanner. Such type of device is a hand scanner type "ZScanner 700", which allows to convert geometric objects into three-dimensional digital forms (digitalizing).

Scanning precision in "ZScan" programme can be changed by virtual working space adjustment (a space with white markers) to the scanning space (virtual cube) (Fig. 1). During the scanning a scanner application presents a digital representation of the object.

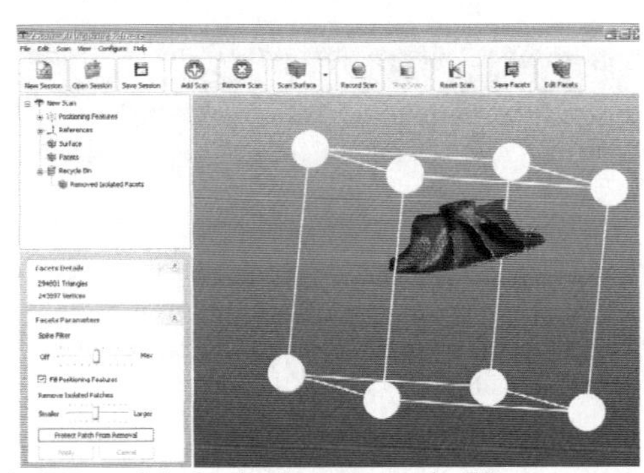

Fig. 1. STL model of turbocharger rotor in ZScan system

All scanning methods require using a specialist software to create a CAD model of examined blade. Point cloud or STL files processing is possible directly in CAD systems (e.g. CATIA, Mechanical Desktop) or with help of a specialist software compatible with RE or RP formats. A scanning method ought to be selected depending on a shape of examined object, also digitalizing process precision, scanning time, the service cost, and format of received data in the aspect of its further processing, should be taken into consideration. Figure 2 shows STL model from CT process (Miechowicz at al. 2005).

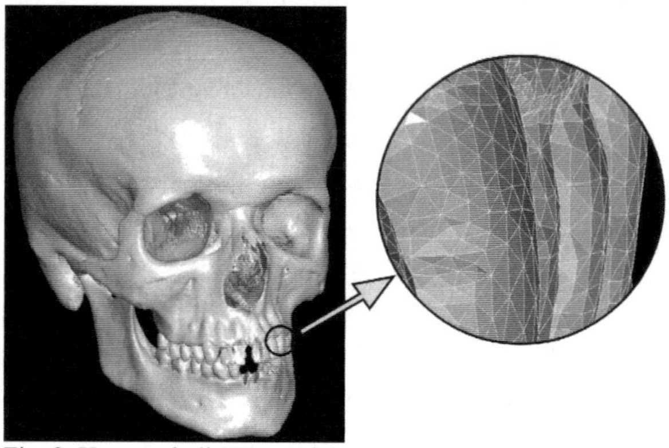

Fig. 2. Human skull STL model (1322123 triangles)

2.2 STL file generating from CAD systems

Most of CAD systems has possibilities of data export in STL format e.g. AutoCAD, Alibre, CATIA, I-DEAS, IronCAD, Mechanical Desktop, ProE, Rhino, SolidDesigner, SolidEdge, SolidWorks, Think3, Unigraphics, 3D Studio Max. Data export is mostly realized by "save as" command (CATIA V5 system), sometimes export to STL command and separate command (e.g. amstlout in Mechanical Desktop System). Some programs allow to define accuracy of STL model. A few CAD systems allow to define geometry STL model very precisely (Mechanical Desktop). Figure 3 shows STL model exported from MDT system depending on angular tolerance (Stasiak 2002).

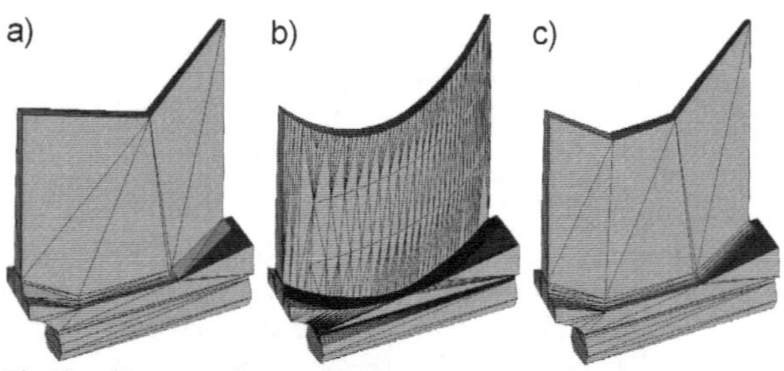

Fig. 3. STL model of aircraft engine blade, depending on angular tolerance (k): a) $k = 0°$, b) $k = 2°$, c) $k = 30°$

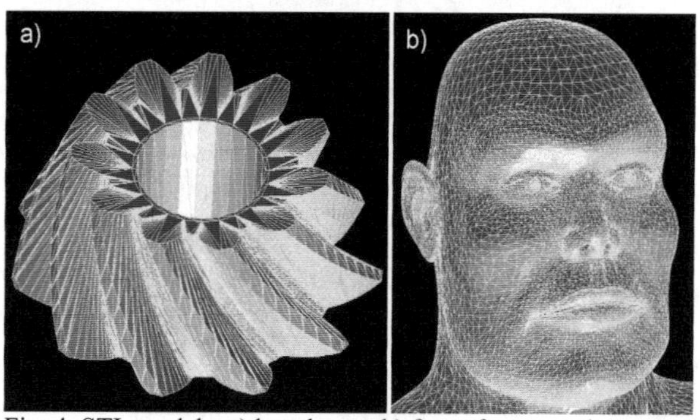

Fig. 4. STL models: a) bevel gear, b) face of a game character

Complex shape elements e.g. bevel gear or aircraft engine blade demand very precise STL models. Figure 4a shows STL model exported from CATIA system (24700 facets).

A computer game character also demands high accuracy of STL data export. Figure 4b shows STL model of a face of computer game character – all character consist of 150 024 facets.

3. CONCLUSION

Optimization of STL models in Rapid Prototyping process is very important for accuracy of physical prototype. Optimization process is depended on the origin of virtual model. Reverse Engineering process requires data processing in specialized system and STL data has large volume of file.

STL model exported straight from CAD systems is very precise. Export process allows to control the accuracy and volume of STL file. Influence of STL data parameter for physical prototype can be evaluated uniquely after physical prototype building.

As a result of research carried out by the authors found that for the presumed accuracy of ±0.05 mm best results obtained at a STL model surface accuracy of 0,005 mm. Less accuracy of the model STL aggravated by the accuracy of the model, while its bigger does not improve.

This research work is financed from the funds allotted for science in 2008-2010, as a research and development project (R03 0004 04).

4. REFERENCES

3D Lightyear SLA File Preparation Software – User's Guide 2001.

Budzik, G. (2007). The Selection Criteria of Scanning Method in Process of Reproducing an Aircraft Engine Blade Geometry, In: *Archives of Mechanical Technology and Automation*, Vol. 27 No 2/2007, pp. 63-70, WPP Poznan

Budzik, G.; Sobolak, M. (2006). Generating stereolitographic (STL) files from CAD systems. *Acta Mechanica Slovaca*, No 2B/2006, pp. 73-78

Geomagic Studio 9 – Shape Phase, Quick Start Guide, Raindrop Geomagic, Inc 2006.

Lee, K.; Woo, H. (2000). Direct integration of reverse engineering and rapid prototyping, *Computers & Industrial Engineering* No 38/2000, pp. 21-38, Pergamon, PII: S0360-8352(00)00017-6

Miechowicz, S.; Sobolak, M. Budzik, G. (2005) The complex rapid prototyping STL free surface generation, *Progressive Technologies and Materials,* No 1/05, Rzeszów, pp. 45-49.

Pieraccini, M.; Guidi, G.; Atzeni, C. (2001). 3D digitizing of cultural heritage, *Journal of Cultural Heritage,* No 2/2001, pp. 63-70, Elsevier, PII: S1296207401011086/FLA

Schemenauer, N.; Peng, X.; Zeng, Y.; Gu, P. (2002). A reverse engineering system for rapid manufacturing of complex objects, *Robotics and Computer Integrated Manufacturing,* No 18/2002, pp. 53-67, Pergamon

Skarka, W.; Mazurek, A. (2006). *CATIA – podstawy modelowania i zapisu konstrukcji*, Helion, ISBN: 83-7361-599-7, Gliwice

Stasiak, F. (2002). *Mechanical Desktop 4.0 PL*, Helion, Gliwice

Stroud, I.; Xirouchakis; P. (2000). STL and extensions, *Advances in Engineering Software*, No 31/2000, pp. 83-95 Elsevier, PII: S0965-9978(99)00046-0

Road, C.; Tan S. (2000). Incremental tessellation of trimmed parametric surfaces, *Computer Aided Design*, No 32/2000, pp. 279-294, Elsevier

ZScan 3D Digitizing Software Version 1.2 Build 1216/2006

ZEdit™ User Manual Version 2.15. (2006). Z Corporation

Annals of DAAAM for 2008 & Proceedings of the 19th International DAAAM Symposium, ISSN 1726-9679
Editor B. Katalinic, Published by DAAAM International, Vienna, Austria 2008
Make Harmony Between Technology and Nature, and Your Mind will Fly Free as a Bird
2nd DAAAM-ICAT

RAPID PROTOTYPING METHODS ADAPTED TO ACCURACY COMPATIBILITY ANALYSIS OF SKULL DECREASE

BUDZIK, G[rzegorz]; MIECHOWICZ, S[tawomir]; DZIUBEK, T[omasz] & MARKOWSKA, O[limpia]

Abstract: *The paper presents the accuracy compatibility analysis of skull decrease executed by selected rapid prototyping techniques. The traumatic decrease model was received from scanning by multi-slice spiral computer tomograph - Siemens Sensation 10. The TC data processing of the skull model was made by 3D-Doctor program Able Software Corp. The adopted data acquisition process enabled employing the worked out method not only for medical aims but also to manufacture technical apparatus and attachments. The virtual model of skull disease also was elaborated. The model data was configurated in STL format in aim to create the prototype using selected rapid prototyping techniques by RP machine (3DP).*
The principal element of the research were microscopic measurements of slot between the implant of traumatic scalp opening model and skull model. These measurements serve the purpose to control accuracy of models compatibility. The investigations were carried out in determinated circumferential points of disease.
Key words: *reverse engineering (RE), rapid prototyping (RP), cranial decrease*

1. INTRODUCTION

Rapid prototyping methods (RP) and reverse engineering (RE) allow creating physical models for medical requirement. Digital three-dimensional objects can be received from several methods: tactile and optical external surface measurements and no tactile measurements like magnetic resonance (MRI) or computer tomography (CT) for both external and internal surfaces (Miechowicz & Sobolak 2004). The CT measurement results give the opportunity to obtain the cloud of points, which describes the geometrical shape of investigated object (Bajaj et al., 1999). These data are basis to create physical model by chosen RP technique. On polish market are many specialised firms, which occupy of rapid prototyping. However due to a small absorption from polish medical community firms do not have much experience in manufacturing medical models. The shortage of orders from medical domain is caused by poor knowledge condition about capabilities of RP techniques application in medicine. Patients usually have a choice between three manners of reparation cranial decreases: acrylic plate, polypropylene-polyester plate or bony own plate (Andrzejak et al., 2005). These techniques are not efficient if considering the accuracy compatibility of the implant and the cranial disease and they are time-consuming especially in bony own plate case. Another disadvantage is poor aesthetic. Patients care about postoperative image that is why precise, individual matching of the plate is the best solution. It is possible to achieve by using mentioned before methodology of manufacturing reconstructions of scalp defects by RP methods.

2. TRAUMATIC DAMAGE MODELLING – STL MODEL

Through raster data (DICOM) received from computer tomography is creating 3D vector model. Tomography investigations were carried out in Chair of Radiology, Collegium Medicum of Jagiellonian University in Krakow by multi-slice spiral computer tomograph - Siemens Sensation 10.
Received 2D (DICOM) representations are subsequently segmented – separating only elements, which are the base to elaborate the medical model (e.g. scalp defect). Generated contour representations including informations about internal and external borders of isolated elements. Set of contour images serve to create three-dimensional representation – superficial model (Miechowicz & Urbanik 2004) (fig.1).

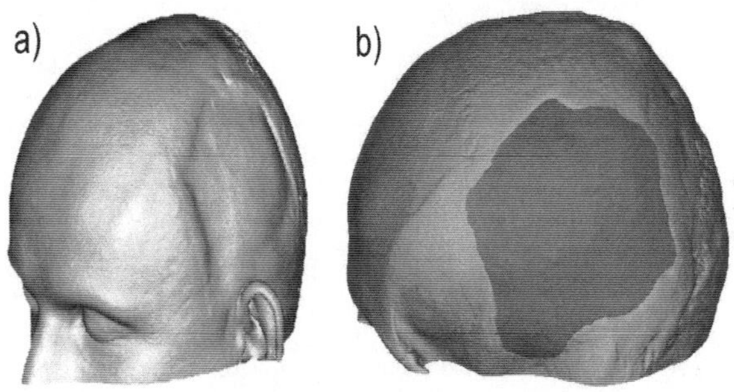

Fig. 1. Superficial STL model: a) shape representation of injured head b) implant model

Reconstruction of the defect was made by mirror reflection of patient intact hemisphere (Krishnan et al. 2003).

3. RAPID PROTOTYPING BY 3DP METHOD

Worked out digital model is a foundation to create physical 3DP model of implant (fig.2). During the process technology model is building via layers. CAD (STL) model is segmented and every segment contains detailed information about particular layer. Following layers of model are forming by spreading powder layers on working platform. 3DP technology is similar to ink printing, an adhesive is joining powder with previously layer only on the space where the model has to be formed. Next the platform goes down giving the possibility to spread subsequent powder layer, which by the adhesive is joined with the previous one. The process is repeated to obtain complete model. At the end the object is hardening by the infiltrator (Chumnanklang et al., 2007, Budzik, 2007).

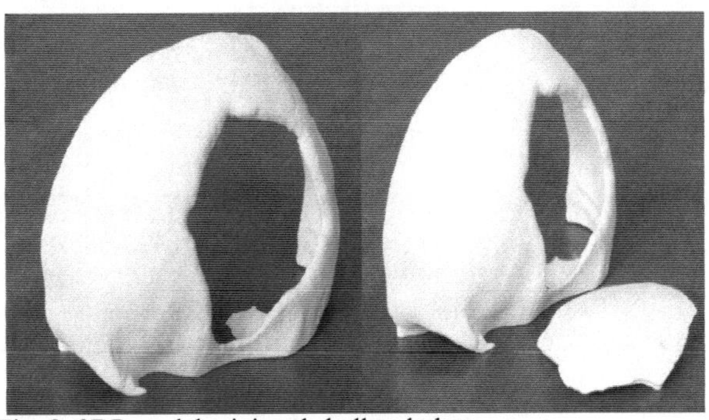

Fig. 2. 3DP model – injured skull and plate

On the polish market are many techniques of rapid prototyping however the choice of 3DP technique was made mainly by economical reasons. Three-dimensional printing is one of the lower-cost method of RP. Model was made in the Faculty of Mechanical Construction, Department of Mechanical Engineering and Aeronautics, Rzeszów University of Technology.

4. ACCURACY COMPATIBILITY ANALYSIS OF SKULL DECREASE

Model of injured cranium and model of it's complement were generated precisely fitted by 3D-doctor system (fig.1). The accuracy compatibility analysis of 3DP models – trauma and implant – was made in eighteen chosen peripheral points along the edge of disease. Measurements were done using reflected light optical microscope with 60x magnification (fig.3). Table 1 presents results of surveys.

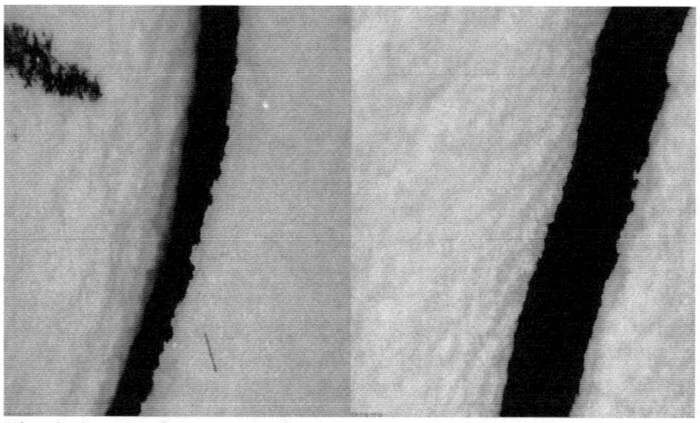

Fig. 3. Image from optical microscope during measurements

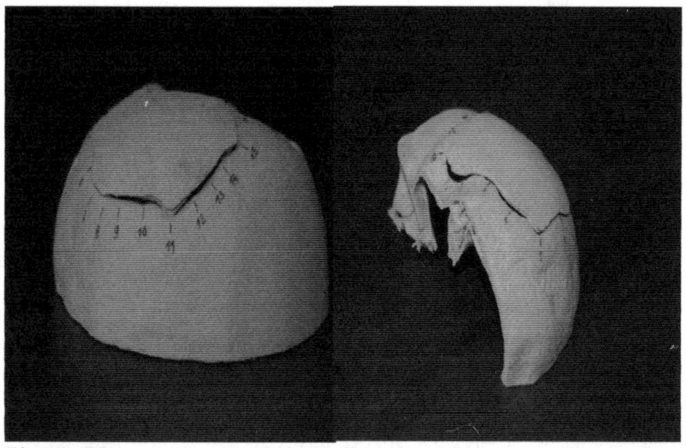

Fig. 4. The prototype of skull disease with complement before matching

MEASURE MENT	SLOT SIZE mm	MEASURE MENT	SLOT SIZE mm	MEASURE MENT	SLOT SIZE mm
1.	2.4	7.	0.2	13.	0.9
2.	1.8	8.	0.25	14.	0.65
3.	0.6	9.	0.4	15.	1.8
4.	0.6	10.	0.65	16.	0.15
5.	0.45	11.	0	17.	0.1
6.	0.2	12.	0.7	18.	1.0

Tab. 1. Measuring results from reflected light optical microscope of slot between cranial decrease and complement

Because of unfitting the implant to the trauma opening (fig.4) technological treatment of excess was made in points 4-5, 5-6, 11, 14-15, 15-16, 16-17. After excess treatment both elements fitted perfectly (fig.5).

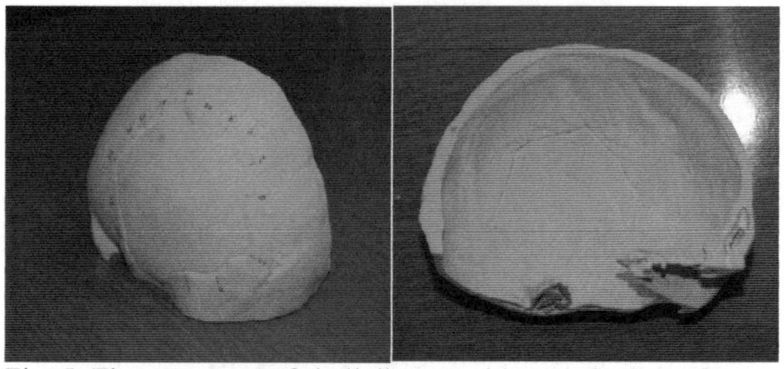

Fig. 5. The prototype of skull disease with complement after matching

5. CONCLUSION

Supporting medicine by reverse engineering (RE) and rapid prototyping (RP) have significant influence on productivity of surgeon's work. Applying mentioned above methods give possibility to reduce preoperative and operative time. Aesthetic and accuracy of individually fitted prototypes is also important especially for patients. Models carried out through RP methods (3DP) may be used afterwards to create complements of trauma openings e.g. by vacuum casting. Despite the fact that 3DP technique is not the most precisely it is accurate sufficiently for medical models. Prices of working out 3DP models are the lowest from among RP methods. During the analysis of adoption the implant to the decrease of skull noticed sharp edges around the model, which obstruct exact adhesion. Taking this into consideration sharp edges should be reduced to minimum in STL file. Consequently afterwards treatment would be redundant.

Next investigations predict exploit another RP techniques to manufacturing process of trauma defects, also by using biocompatible materials (FDM method). As well forming implants with the assistant of vacuum casting from materials exploited in cranial surgery.

This research work is financed from the funds allotted for science in 2008-2010, as a own project (N N501 268134).

6. REFERENCES

Andrzejak, S., Fortuniak, J., Jaskólski, D., Papierz, T., Shariati, B., Tybor, K.& Zawirski, M. (2005). Supplementation of cranial defects by an autologous bone flap stored in the abdominal wall. *Neurologia I Neurochirurgia,* Vol.39, No.3 (220-224)

Bajaj, C.L., Chandrajit, L., Coyle, J. & Kwun-Nan Lin (1996). Surface and 3D Triangular Meshes from Planar Cross Sections. *CSDTR-96-002, Technical Report*, Department of Computer Science, Purdue University

Budzik G., (2007). Possibilities of utilizing 3DP technology for foundry mould making, Archives of Foundry Engineering, Vol. 7, Issue 2/2007, s. 65-68

Chumnanklang, R., Panyathanmaporn, T., Sitthiseripratip, K. & Suwanprateeb, J. (2007). 3D printing of hydroxyapatite: effect of binder concentration in pre-coated particle on part strength. *Materials Science & Engineering C*, Vol.27, No.4 (May 16, 2007) (914-921)

Krishnan, K.G., Muller, A.& Uhl, E. (2003). The application of Rapid prototyping techniques in cranial reconstruction and preoperative planning in neurosurgery. *Craniofacial Surgery*, Vol.14, No.6, (899-914)

Miechowicz, S. & Sobolak, M. (2004). The efficiency of utilization of Siemens Sensation 10 CT scanner for Reverse Engineering, *Prace Naukowe Instytutu Technologii Maszyn i Automatyzacji Politechniki Wrocławskiej*, No. 85

Miechowicz, S. & Urbanik, A. (2004). Spatial analysis of CT accuracy. *Polish Congress of Radiology*, Vol. 69, suppl. 1,372(319) Mikołajki 2004, Polish Journal of Radiolog

Annals of DAAAM for 2008 & Proceedings of the 19th International DAAAM Symposium, ISSN 1726-9679
Editor B. Katalinic, Published by DAAAM International, Vienna, Austria 2008
Make Harmony Between Technology and Nature, and Your Mind will Fly Free as a Bird

USE WIRELESS SENSOR NETWORKS FOR PATIENTS IN HOME MONITORING – DATA FLOW

BUJDEI, C[atalin]; MORARU, S[orin] A[urel]; DAN, S[tefan]; SISAK, F[rancisc] & VULPE, A[nca] D[iana]

Abstract: *In this paper it will be described the data flow between different components of a system dedicated for monitoring the patients at home. The system use a wireless sensors network (wsn) for record the parameters of the patient (pulse, blood oxigen level) and transmit the recorded data for being stored into a database. Since it is a large amount of recorded data the speed of transmiting the data throught the system is a very important aspect. The consumed energy is proportional with the quantity of data transported. Because of this the consume of energy is taken in consideration especially into the wsn which should run properly as long as possible. The security of data is a secondary aspect which is taken in consideration.*
Key words: *wireless sensors network, patient, monitor, data flow*

1. INTRODUCTION

The wireless sensors network (WSN) is a new tehnology developed since few years ago which have impact over different areas of interest. Being a new technology different functionalities and characteristics need to be developed or improved. One of the major interests of using the WSN is into the medical domain, and especially for monitoring the patients which suffer of different ills. The patient could stay at home without being disturbed by the aspect of a hospital. He could walk from one room to another, to develop different activities, during the monitoring process. The only requirement is to keep the monitoring equipments attached to them. Nowadays, the monitoring equipments become smaller and smaller in size, but this also depends on their functionalities.

We have under developement a system which permits the monitoring of the level of oxygen from blood and the pulse for a patient using a small medical equipment, capable to monitor this parameters. The system records this data and stores all the data into a database. The data from the database could be accesed real-time for being vizualized and analyzed. Different supplementary functional modules could be attached to the system for let generating different alarms depending on the measured data variation, but this will be for future implementing. Into the system under development another requirement for the patient is to not exit the area of wireless communications, and which, in our system case, covers the entire living space.

Since the system record a large amout of data, which need to be stored into the database, it must to process and transmit the data as quikly as possible. It is now allowed to exist locks of the system or lose of the data. Also, the energy consumed by the *wsn* must be minimum possible. Since the monitoring have to run continuously the optimization methods of the system are different than the system which measure values from time to time. The most systems developed for monitoring the patients consider more the mobility of the patient or the analyze of the recorded data (Proulx et al., 2006; Chen et al., 2004) and less the possibilities of minimize the quantity the data transmited and in this way, the consumed energy. Since it is an important system for the medical domain we want get the best solution possible for optimization. In this way we want to ensure an stable and long life monitoring system.

2. SYSTEM ARCHITECTURE

The main functionalities of the system are: to record the necessary information (parameters), to transport the information securely and with minimum delay to a central database and to permit the visualization of the data, in different formats on different users computers. The system architecture is presented into the figure 1.

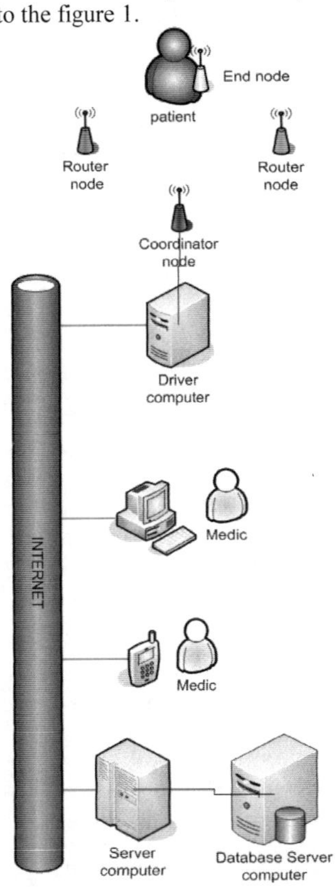

Fig. 1. System architecture

The first part of the system is represented by the monitoring network, represented by the top part of the figure 1. The monitoring network is formed by the end node, attached to the patient and which is connected to the medical equipment, router node, necessary for increase the area of wireless signal and the coordinator node which transfer the data from the wireless network to a Driver computer.

The Driver is the second important part of the system. Its role is to get the recorded information by the network and transfer it to the Database Server computer. The Server computer has the role of processing the data of the database and from the system. Different users could connect to the system, using a personal computer or a mobile phone, and vizualize the recorded data. They could see the instant values or history graphs with data.

Similar systems, for monitoring of the patients, have been developed by different researchers or centers of research. Some of them preferred to use instead the WSN, a PDA equipment which to transmit the data using Bluetooth or mobile phones which to transmit the data using the existing GPRS service (Proulx et al., 2006; Chen et al., 2004). The Bluetooth technology consumes more power energy than the ZigBee, which is used by the most wireless sensor networks. Also, the Bluetooth range is smaller than the ZigBee range. Using the GPRS service the mobile consumes its energy power very quikly and the system could not run for a long term. The system must run continuously without errors or interruptions.

3. THE FLOW OF DATA IN WSN

In our system we want to record only the heart-rate and the blood oxygen level. The analogic signals are transferred to the network end node for being proceesed. First the analog signal is transformed in digital and when this signal is transmited through the network to the coordinator node. This requires lot energy to be consumed for processing and transmiting of the data. The most part of energy is consumed at data transmission. For this reason the rate of transmission and the quantity of data transmited should be minimized.

For sending the data into the wireless network the data must be packed in telegrams which to respect a establish format.

The format of the telegram is presented into the table 1.

Begin mark	ID Patient	Date, hour	First data value	Offset value 1	...	Offset value *n*	End mark

Tab. 1. Telegram format

For minimize the quantity of data transmited, into the telegram are not inserted all real measured values but only the first value measured into a period of time, and then the differences of the measured values and the first value for that period of time (these represent the *offset* values). A measured value is a real value represented in memory by a number of octets. This number is determined by the precision which we want for storing data and the range of values for the monitored parameters. The offset values are smaller than the real values recorded for that moment of time, and this means that they can be represented on a smaller number of octets. In this way we can reduce the number of total octets for a telegram; the telegram become shorter and the quantity of data transmited into the network become smaller, but without losing precious information. For a further development we want to pack the information from the telegram using different archiving alghoritms (Fullford-Jones et al., 2004).

4. THE FLOW OF DATA IN INTERNET

The Driver computer sent the data, using the Internet network, to the Server computer. The data must be sent without delay and safety throught the Internet network. If the Internet network is not accessible for a period of time the data will be buffered on the Driver computer and send to the Server computer when the Internet network becomes available. For the transmission of data in this direction .Net software application have been implemented. The data is packed using a *zip* tool existent into the .Net platform. For security reason 2 types of encrypting the data could be used: using symmetric keys or using asymmetric keys. Both, implies a collaboration between the users of the application for exchange the public keys (Northrup et al., 2006).

5. THE DATABASE

The database is an important part of the system which could affect the performances of the entire system. It is necessary to have an optimization of the database for increase the performances of the system. No paper was found till yet which to discuss the optimization of a database inside such a system.

The system will have to be functional not for a year or less but for many years. The database has the role to store a lot of information (millions of record) and the processing of the data from the database must be done in the shortest time possible. If an alarm signal must be generated it should not be delayed.

We had to consider the optimization of the database especially for the *insert* and *select* operations. The optimization of the database is essential for such big number of records for allow the system to function in normal conditions. It is not desired that when a user want to view a graphic with the recorded information for a specific period of time he have to wait a while, about minutes, since the information is retrieved from the database, processed and then sent to the user's application interface for view. Also, since the record of the information is done continuously is necessary that the insert operation of the information into the database to be done in the shortest time possible. In this way there will not be system locks, by creating unlimited waiting queue of information which to be inserted.

The most important methods, which have been tested and which could lead at a database's performances improvement:

- Normalization;
- Using indexes; recreate indexes; indexes defragmentation;
- Statistics update;
- Manage concurrency;
- Table and index partitioning (Kimberly, 2005);
- SQL tuning;
- Disk defragmentation (Henderson, 2003).

6. CONCLUSION

The optimization of the data flow has a major impact over the system. In this way the system could execute rapidly different function without having disturbing delays. The success of the system depends on this optimization. The further tests will permit us to discover and other optimization solutions.

7. REFERENCES

Chen, S.; Gaur, A.; Mutthukrishnan, S. & Rosenbluth, D. (2004), *Wireless in loco Sensor Data Collection and Applications*, *Available from:* http://www.cs.rutgers.edu/~muthu/inloco.pdf, *Accesed:* 2008-07-02

Fullford-Jones, T.R.F.; Wei, G.Y. & Welsh, M. (2004), A Portable, Low-Poer, Wireless Two-Lead EKG System, *Proceedings of the 26th Annual International Conference of the IEEE EMBS*, pp 2141-2144, 0-7803-8439-3, San Francisco, CA, USA, 1-5 september 2004

Henderson, K. (2003), *Proceduri stocate in SQL Server. XML. HTML.(Stored Procedures in SQL Server. XML. HTML.)*, Teora, 973-20-0613-7, Bucharest

Kimberly, L. (2005), *Partitioned tables and indexes in SQL Server 2005*, *Available from:* http://msdn2.microsoft.com/en-us/library/ms345146.aspx, *Accessed* 2008-04-11

Northrup, T.; Wildermuth, S. & Ryan, B. (2006), *Microsoft .Net Framework 2.0 Application Development Foundation,* Microsoft Press, 0-7356-2277-9, Redmond, Washington

Proulx, J.; Clifford, R.; Sorensen, S.; Lee, D.J. & Archibald, J. (2006), Development and Evaluation of a Bluetooth EKG Monitoring Sensor, *Proceeding of the 19th IEEE Symposium on Computer-Based Medical Systems*, pp 507-511, Salt Lake City, Utah, 22-23 june 2006

Annals of DAAAM for 2008 & Proceedings of the 19th International DAAAM Symposium, ISSN 1726-9679
Editor B. Katalinic, Published by DAAAM International, Vienna, Austria 2008
Make Harmony Between Technology and Nature, and Your Mind will Fly Free as a Bird

DESIGN OF VENTILITATION DUCT WITH MINIMAL FLOW RESISTANCE

BUNJAKU, J[anuz]; KRASNIQI, F[ejzullah]; FEJZULLAHU, X[hemajl]; MUHAXHERI, M[ustafe] & MURIQI, A[li]

Abstract: *One of the important characteristics of the ventilation branch ducts, on which the level of pressure losses and the equipment price depends, is the duct cross section area. The optimum level value determination based on the dedicating aspect of aerodynamics, as well as of the equipment costs, is now imposed on us as a prior duty related to the engineering design. For the purpose of determination of the optimum model the losses during the flow from the main duct to the ventilation branch duct have been analyzed. The pressure losses in the duct length and the local losses are integral parts of total losses.*

Key words: duct, ventilation, resistance

1. INTRODUCTION

As a initially section of ventilation branch duct, towards the main ventilation duct, the transitional section from the main duct to the ventilation branching duct, as one of the essential sections against the local resistance of ventilation equipment, can be taken.

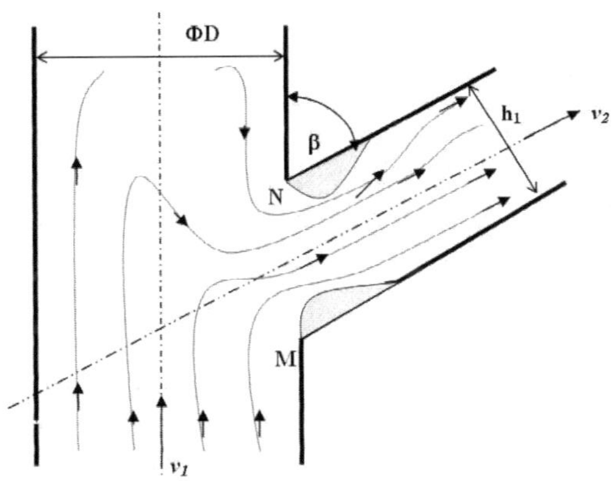

Fig.1. The flow scheme in the connection of main duct and the ventilation branch duct

In Fig. 1. the typical scheme of air flow into the respective air distribution element has been presented, where it can be seen that a visible disorder of air flow lines occurs in it, a vigorous air velocity increase in the ventilation duct entrance – as well as a detachment of the air flow from the duct walls allowing the creation of whirling zones. As a result we have significant reduction of the cross section versus the flow and an increasing of pressure losses in the relevant air distribution element.

2. SELECTION OF THE OPTIMAL SHAPE OF DUCTWORK ELEMENTS

Related to the selection of the optimal shape of air distribution system elements we have to start from the Navier-Stokes equation which describe the motion of whirling viscous fluids:

$$\frac{d\vec{v}}{d\tau} = \vec{F} - \frac{1}{\rho}.\nabla P + \nu.\Delta\vec{v} \tag{1}$$

where the first member describe the total unsteady acceleration of fluid particles, the second is the acceleration under the influence of mass forces, the third one – to the acceleration under the action of pressure forces.

Hence the required solution of the rational element form has to provide the fluid flow similar to the potential one, the equation (1), now gets simplified and receives the following form:

$$\frac{d\vec{v}}{d\tau} = \vec{F} - \frac{1}{\rho}.\nabla P \tag{2}$$

Providing that the β angles of the existing duct and the free flow significantly distinguishes, the disturbance of flow mass will appear not on just the very characteristic points M and N, but on the areas, a little more distanced. In this case the duct walls determine the form of the free flow limits.

Regarding the sizing of the optimal dimensions of the interconnecting ventilation duct to the main duct, characteristic types of respective pressure losses have to be determined. Furthermore, we shall use the determination methodology of cross section dimensions of ventilation duct, to achieve the minimum pressure losses.

Having in mind, that we refer to the duct with minimal aerodynamic resistance, then it is to be supposed that the duct has a square cross section and that the other optimum norms of ventilation system functioning have to be taken into account.

Thus the air flow local losses passing from the main duct to the branch ventilation duct are determined according to:

$$\Delta H_t = \frac{\rho.Q^2}{2.F_2^2}.\frac{(\pi+2)^2}{4.\pi}.\frac{F_1}{F_2}\left(1-\frac{F_2}{F_1}\right)^{-0,46.\frac{F_2}{F_1}} = \frac{\rho.Q^2}{2.F_2^2}.\frac{\pi}{4}.\frac{1}{\varepsilon^2}.\frac{F_1}{F_2}A \tag{3}$$

where ε - is a contraction(shrinking) coefficient of flow .

The pressure losses in the duct length (along the duct) are determined according to:

$$\Delta H_{gj} = \frac{\rho.Q^2}{2.F_2^2}.\frac{\lambda.l}{\sqrt{F_2}} \tag{4}$$

where λ - is a losses coefficient related on resistance of flowing inside the duct{ Locjanskii, 1987} :

$$\lambda = \left(\frac{1{,}15}{\ln\frac{3{,}7.\sqrt{F_2}}{\Delta}}\right)^2 \tag{5}$$

Where Δ– is the absolute ventilation duct wall roughness.
The total losses (Voronjec, 1985) in other elements connected to the ventilation duct (duct refraction parts, fittings, etc.), can be determined according to:

$$\Delta H_n = \frac{\rho . Q^2}{2 . F_2^2} . \sum_{i=1}^{m} \varsigma_i \quad (6)$$

Where ζ (i=1, 2, 3, …, m) – are the coefficients of local losses, whose size does not depend on the duct dimensions.
Thus, the total losses can be determined according to:

$$\sum_{j=1}^{3} \Delta H_j = \frac{\rho . Q^2}{2 . F_2^2} . \left[\frac{\pi}{4} . \frac{1}{\varepsilon^2} . \frac{F_1}{F_2} + \frac{\lambda . l}{\sqrt{F_2}} + \sum_{j=1}^{n} \varsigma_j \right] \quad (7)$$

From the condition of determination of extreme value $\frac{\partial}{\partial F_2}\left(\sum_{j=1}^{3} \Delta H_j \right) = 0$, we now can obtain:

$$(F_2)_{OPT} = \frac{1}{256} \frac{(-10\lambda L \varepsilon^2 + 2\sqrt{25\lambda^2 L^2 \varepsilon^4 - 24 \sum_{i=1}^{n} \zeta_i \varepsilon^2 \pi F_1})^2}{\zeta^2 \varepsilon^4} \quad (8)$$

If we substitute equation (8) in equation (7) we will have:

$$\left(\sum_{j=1}^{3} H_j \right)_{OPT} = \frac{\rho Q^2 \sum_{i=1}^{n} \varsigma_i \varepsilon^{10} \pi F_1}{4 \cdot 10^{-7} \cdot \psi^6} + \frac{\rho Q^2 \lambda L}{2 \cdot 10^{-6} \cdot \left(\frac{\psi^2}{\sum_{i=1}^{n} \xi_i \varepsilon^4} \right)^{5/2}} + \frac{\rho Q^2 \left(\sum_{i=1}^{n} \xi_i \right) \varepsilon^8}{3 \cdot 10^{-5} \cdot \psi^4} \quad (9)$$

Where $\psi = -10\lambda L \varepsilon^2 + 2\sqrt{25\lambda^2 L^2 \varepsilon^4 - 24 \sum_{i=1}^{n} \xi_i \varepsilon^2 \pi F_1}$

(10)

In fig. 2 the dependence $F_2 = F_2(F_1)$ for the designing parameters L=2.5 (m), λ = 0,075 , $\sum_{i=1}^{n} \varsigma_i = 2.5$

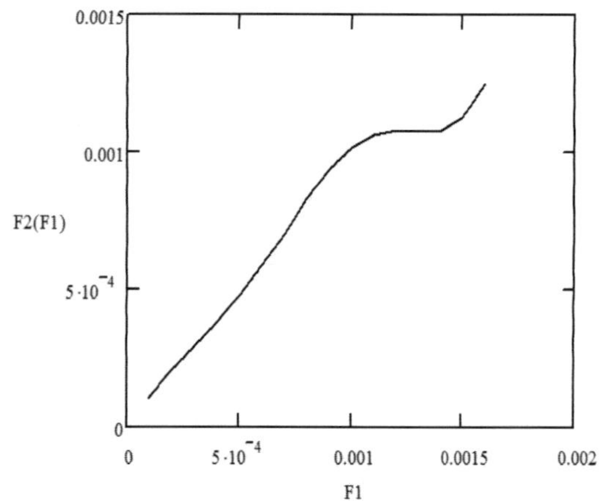

Fig.2. Dependence $F_2 = F_2(F_1)$

In the fig. 3. its shown dependence $\left(\sum_{j=1}^{3} H_j \right)_{OPT} = f(F_1)$, like a minimal resistance of ventilation duct surface against of air flows. The diagram presents total losses during the air flow calculated on base of the designing parameters mentioned above, according expression Q = $F_1 . v$, where is v the average air velocity of the flow with supposed value v=1.5 (m/s) The total losses have been calculated in Pascal's (Pa), for L = 2.5 (m) , λ = 0.075 and $\sum_{i=1}^{n} \varsigma_i = 2.5$

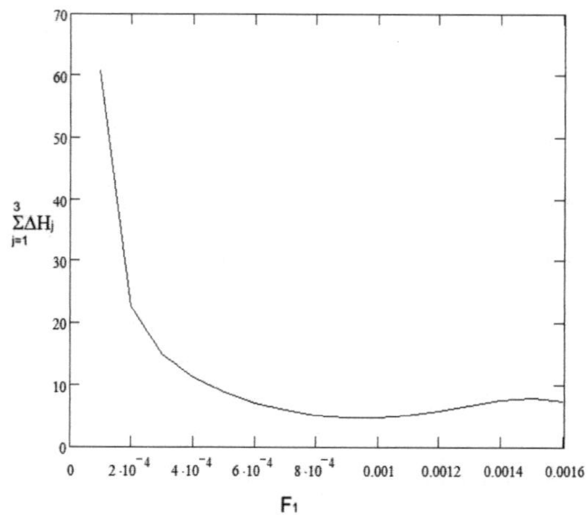

Fig.3 Minimal resistance in depend of section

3. CONCLUSIONS

In this article the issue of design of ventilation duct system with the minimum air flow resistance has been treated. In the beginning the extremely important role of the connection angles (β) of the branch ventilation duct versus the main ventilation duct has been specified. The presented model is very compatible regarding the designing of ventilation ducts with minimum air flow resistance.
By studying the given criteria dependencies, and the presented values in diagrams, we can conclude that optimum relation (F2/F1) varies within the limits from 96% to 98%.
The next step related the study of this topic would be the determination of optimum values of angle β, and the observation of a closer functional dependence of optimum relation in the function of angle β, because this is the essence of discussed optimization.

4. REFERENCES

Aleksandrov,V.(1986), *Technical hydromechanics ,* Moskva, 1986
Munson B, Young, D.& Okishi, Th. (2002), *Fundamentals of Fluid Mechanics*, USA, 2002
Carafoll, E., (1986), *High – Speed Aerodynamics*, Bucharest, 1986
Locjanskii, I.(1987 *), Fluid Mechanics*, Moskva, 1987
Voronjec, K & Obradovic, N.(1985), *Fluid Mechanics*, Beograd,1985

Annals of DAAAM for 2008 & Proceedings of the 19th International DAAAM Symposium, ISSN 1726-9679
Editor B. Katalinic, Published by DAAAM International, Vienna, Austria 2008
Make Harmony Between Technology and Nature, and Your Mind will Fly Free as a Bird

HUMAN MULTIBODY MODEL USED IN ATHLETICS

BURCA, I[oan]; TOFAN, M[ihai]; PURCAREA, R[amona]; CHIRU, A[nghel] & MICU, I[oan]

Abstract: *In this paper the authors wish to make an adequate mathematical model to study some special motion of an athlete. Such motion has some particularities as the fact that there are great forces and great accelerations acting on different parts of the human body. A cinematic model to study the motion of different parts of the body is made. The mathematical model used for dynamic analysis is strong non-linear and for this reason involves some difficulties in solving the problem. This need an adequate parameterization of human body and the identification of the best modality to introduces power in the system.*

Key words: *Multibody, human model, kinematics, athletics*

1. INTRODUCTION

The present paper aims to elaborate a mathematical model designed to clear up the technical perfecting in hurdle from its biomechanical point of view in order to improve the athletes' performances.

The biomechanical research of sports technique tends to become the most important way of improving performance (Burcă, 1979), (Burcă, 1981), opposing somehow to the biochemical methods used to serve this purpose, which are considered to be harmful to the human body.

The biomechanical research encourages the natural development of all athletic events' performances, including the hurdle (Burcă, 1970) ,(Burcă, 1985) .

Coaches must be able to use the latest knowledge available in order to assist the sportsmen in one of the most important fields of their activity, namely the improvement of the athletic events' technique.

2. THE RUNNER'S MULTIBODY MODEL

(Burcă et al., 2003) presents a mathematical model designed for the hurdle runner's movements kinematics. The model is made of 12 rigid elements, interconnected by spherical or cylindrical joints. For reporting of various segments, a non-inertial reference system tied to the athlete's trunk was considered. For the dynamical analysis of the system the multibody system will be dismembered in its constitutive parts. The previous kinematic analysis allows to determine the couplings between the kinematic elements. Then the movement equations for every constitutive element of the athlete's body must be written. The constitutive elements of the system are connected through cylindrical and spherical joints. The forces occuring inside the system are internal and external forces. The interior forces are performed by the athlete's muscles that move the segments of the body in order to accomplish the various phases of running. For the four main phases of the hurdle they activate different groups of muscles that move the active segments. The dynamical analysis identifies the groups of muscles having the crucial influence on the movement. The exterior forces are the weights of the segments and if this influence is considered as significant, the air resistance force.

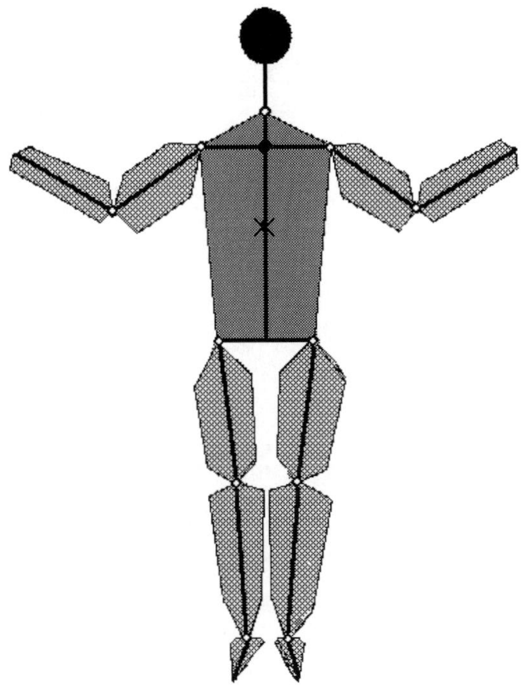

Fig. 1. Runner's multibody model

3. MOTION EQUATIONS

The moment he rises in order to pass the hurdle, the athlete will suffer only the influence of the weight, the position of his weight center will move in a determinate way within the gravitational field and the movement of the active segments during his passing the hurdle will generate a compensatory movement of the other segments in order to respect the mechanical theorems. Fig. 3 represents the cylindrical and spherical joints and the forces that appear within this type of couplings.

In order to write the movement equations the system has to be dismembered in its constitutive parts. Fig. 2 presents the constitutive parts of the athlete and the forces appearing within the couplings, without the representation of the forces given by the muscle groups and the weights. In each cylindrical joint appear two components of a reaction while in each spherical one, there are three reaction components.

These forces are joined by the ones due to various groups of muscles which depend on the segment relative position and the weights that operate within the segments weight centers.The connection betweeen two random segments of the body are made through spherical or cylindrical joints. The spherical joints suppress three movement possibilities of a rigid and equalizes the introduction of a reaction. The movement of the various segments will be related to the athlete's trunk movement. His position is unimportant, so it results that his position towards an inertial reference system can be defined using six scalar parameters. The number of freedom degrees defining the position of the entire system at a certain moment is 27.

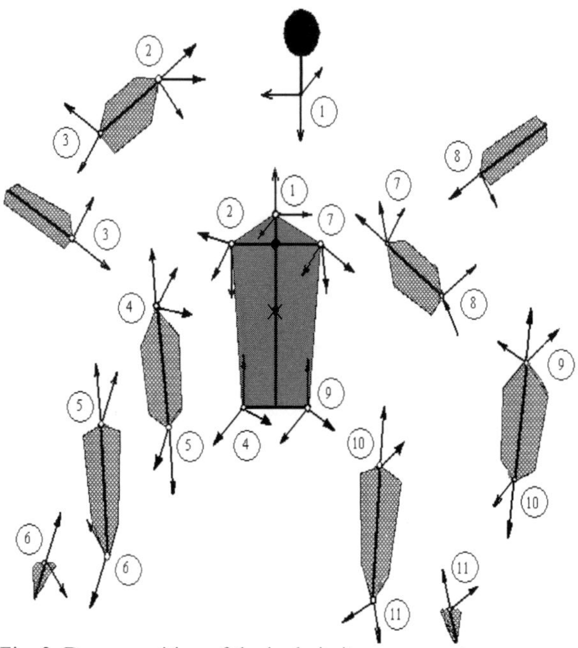

Fig. 2. Decomposition of the body in its components

$$[R]=[\{i\} \quad \{j\} \quad \{k\}] \tag{1}$$

Where:

$$\{i\}=\begin{Bmatrix}\cos\psi\cos\varphi-\sin\psi\sin\varphi\cos\theta\\ \sin\psi\cos\varphi+\cos\psi\sin\varphi\cos\theta\\ \sin\theta\sin\psi\end{Bmatrix}$$

$$\{j\}=\begin{Bmatrix}-\cos\psi\sin\varphi-\sin\psi\cos\varphi\cos\theta\\ -\sin\psi\sin\varphi+\cos\psi\cos\varphi\cos\theta\\ -\sin\theta\cos\psi\end{Bmatrix} \tag{2}$$

$$\{k\}=\begin{Bmatrix}\sin\varphi\sin\theta\\ -\sin\theta\cos\varphi\\ \cos\theta\end{Bmatrix}$$

$$[R]=\begin{bmatrix}\cos\psi\cos\varphi-\sin\psi\sin\varphi\cos\theta & -\cos\psi\sin\varphi-\sin\psi\cos\varphi\cos\theta & \sin\psi\sin\theta\\ \sin\psi\cos\varphi+\cos\psi\sin\varphi\cos\theta & -\sin\psi\sin\varphi+\cos\psi\cos\varphi\cos\theta & -\sin\theta\cos\varphi\\ \sin\theta\sin\psi & -\sin\theta\cos\psi & \cos\theta\end{bmatrix}$$

A limitative condition is the athlete's moving on a horizontal track which will decrease the number of system's freedom degrees.

The scalar parameters defining the position of a random segment will appear in second degree differential equation systems that, once solved, will reveal the trajectory and the movement rule.If φ,ψ,θ are Euler' angles that describe the movement of the reference system attached to a segment tied to a spherical joint, then the attached rotation matrix will have the following shape:

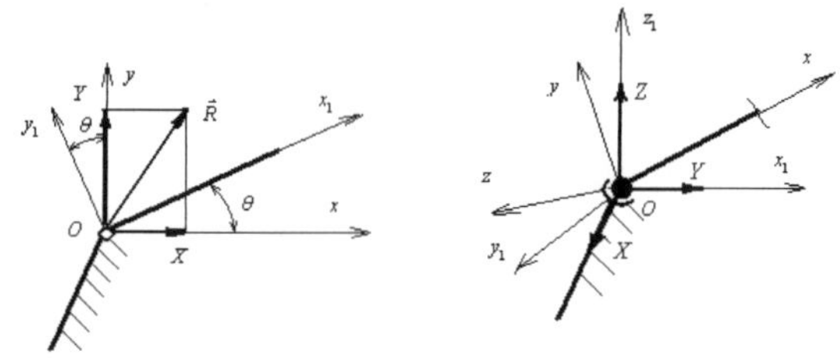

Fig. 3. Cylindrical and spherical joint

The angular speed, reported to the global reference system, will be obtained from:

$$\{\omega\}=[\dot{R}][R]^T=\begin{Bmatrix}\dot{\theta}\cos\psi+\dot{\varphi}\sin\theta\sin\psi\\ \dot{\theta}\sin\psi-\dot{\varphi}\sin\theta\cos\psi\\ \dot{\psi}+\dot{\varphi}\cos\theta\end{Bmatrix} \tag{3}$$

And the angular acceleration from:

$$\{\varepsilon\}=\{\dot{\omega}\}=$$

$$=\begin{Bmatrix}\ddot{\theta}\cos\psi-\dot{\theta}\dot{\psi}\sin\psi+\ddot{\varphi}\sin\theta\sin\psi+\dot{\varphi}\dot{\theta}\cos\theta\sin\psi+\dot{\varphi}\dot{\psi}\sin\theta\cos\psi\\ \ddot{\theta}\sin\psi+\dot{\theta}\dot{\psi}\cos\psi-\ddot{\varphi}\sin\theta\cos\psi-\dot{\varphi}\dot{\theta}\cos\theta\cos\psi+\dot{\varphi}\dot{\psi}\sin\theta\sin\psi\\ \ddot{\psi}+\ddot{\varphi}\cos\theta-\dot{\varphi}\dot{\theta}\sin\theta\end{Bmatrix}$$

4. CONCLUSIONS

In what the athelete's body model (Burcă, 1979) - (Burcă, 1985) using a multibody system is concerned, we obtain a set of differencial equations describing the movements of the various segments that form the system (Tofan, 1981), (Vlase, 1993). The unknown variables are the angular positions of the various segments of the body and the coupling forces appearing within the spherical and cylindrical joint type couplings. A mixed differential and algebrical system results from it, and after eliminating the unknown algebric variables we obtain a differential equation system in which the amount of unknown variables, generalized coordinates is equal to the amount of equations we get. Solving these differential equations for concrete situations in hurdle are the goals established for further research.

5. REFERENCES

Burcă, I.; Tofan, M.; Vlase, S. & Modrea, A. (2003) Model multicorp al alergătorului de garduri. *Conferința științifică internațională: COMPETIȚIA (The multibody model of the runner of the fences. International scientific conference: COMPETITION)*, pp 40-46, București (Bucharest), 24 octombrie 2003 (october 2003)

Burcă, I. (1979) *Atletism- Analiza tehnicii probelor de atletism* (*Athletics- The analysis of the athletics samples technique*). Institutul de învățământ superior Târgu-Mureș (The Institute for Higher Education Targu-Mures)

Burcă, I. (1981), *Atletism (Athletics)*. Curs opțional (Optional course). Institutul de învățământ superior Târgu- Mureș (The Institute for Higher Education Targu-Mures)

Burcă, I. (1985) *Relația dintre pregătirea tehnică și cea fizică la nivelul grupei de avansați, alergători de garduri, în cadrul Clubului Sportiv Scolar (The relationship between technical and physical training at the advanced group level, runner fences, inside the Scholastic Sports Club)*.Lucrare, grad didactic I (Paper, 1st teaching grade)

Tofan, M. (1981) *Mechanics. Kinematics*. Transilvania University of Brasov

Vlase,S. (2008) *Mechanics. Kinematics*. Ed. Infomarket

Annals of DAAAM for 2008 & Proceedings of the 19th International DAAAM Symposium, ISSN 1726-9679
Editor B. Katalinic, Published by DAAAM International, Vienna, Austria 2008
Make Harmony Between Technology and Nature, and Your Mind will Fly Free as a Bird

ASPECTS OF DEVELOPEMENT OF PLANETARY MECHANISM USING SYNCHRONOUS BELTS DRIVE

BUTNARIU, S[ilviu]

Abstract: *Drive belts of different type are some of the most widely used rubber-based structural elements in the automotive industry. This paper tries to present a new vision of planetary mechanisms, using synchronous belts drive and by excluding classical gear drive. It proposes a classification and a determination of the general characteristics of these transmissions from the constructive and cinematic.*
Key words: *synchronous, timing, belts, planetary, mechanism*

1. INTRODUCTION

The mechanisms built on the basis of planetary transmissions through synchronous belts are similar in terms of cinematic those with toothed wheels, according the formulas of transmission ratio (fig. 1). The transmission, when the driver wheel $\mathbf{Z_1}$ is central, is calculated as follows (Butnariu & Tanasescu, 2002):

$$i_{13}^{H} = \frac{\omega_1 - \omega_H}{\omega_3 - \omega_H} = \frac{Z_2}{Z_1}\cdot\frac{Z_3}{Z_2'}; \qquad i_{1H} = 1 - i_{13}^{H} = 1 - \frac{Z_2}{Z_1}\cdot\frac{Z_3}{Z_2'};$$
$$i_{H1} = \frac{1}{i_{1H}} = \frac{1}{1 - \frac{Z_2}{Z_1}\cdot\frac{Z_3}{Z_2'}} \qquad (1)$$

2. INFORMATION

Planetary mechanisms, in which the gear pair is replaced by a belt drive, are used in various branches of machinery. The main feature of these mechanisms is the presence of planets: links that perform complex planar motion, making it possible to obtain a great variety of trajectories of points belonging to these links (Abdullah & Suleiman, 2004). The loads are transferred between two parallel axles through a synchronous belt allowing theoretically the same range of transmission ratio as in the case of gear drive built in the same space (fig. 2, a,b) (Emri et al., 2006). In terms of constructive, more precisely - arrangement of the satellite wheels, differ two types of planetary mechanisms that use synchronous belts:

1) When $\mathbf{R_2 \le a \le R_1 + R_2}$ (fig. 2, a, fig. 3) – synchronous belt will be required large width, at least twice the width of wheels. On one half of the width will render the central wheel and the other half satellite wheels, which lies in a plan parallel to the central wheel. The disadvantage of this type comes from the fact that the belt width is not be all used simultaneously, inducing the need for an imbalance and build a new type of belt, with income metal cross.

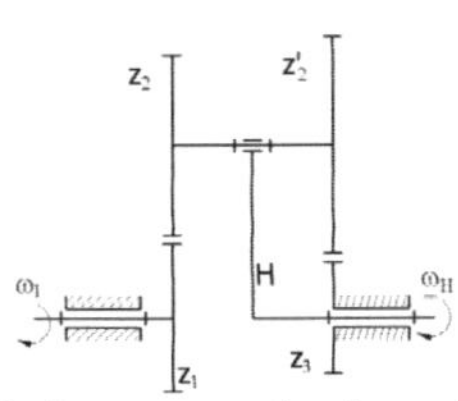

Fig. 1. The classical planetary mechanism with gear drive

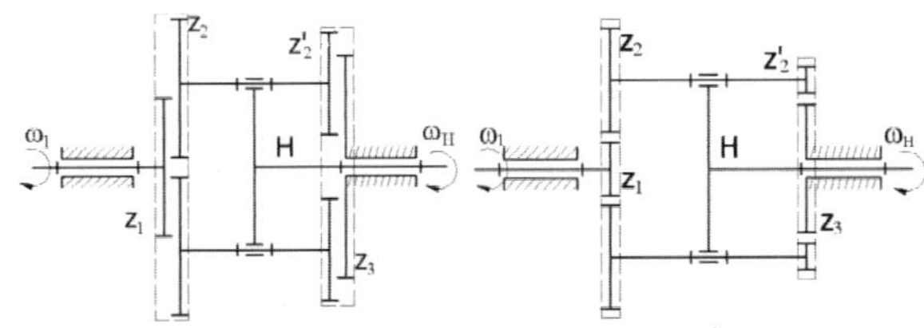

a. b.

Fig. 2. The planetary mechanism with synchronous belts

To prevent the belt fall on wheels, is required to mount flanges on the sides of wheels and use a larger number of wheels satellite. Another disadvantage arises from geometrical considerations, namely the satellite wheels should have the dimensions so that gearing could take, $\mathbf{R_1 \le a + R_2}$. These two inequalities create a geometrical condition to be used in the design process. It may introduce neighbourly conditions, depending on the number of satellite wheels. In this particular case, the maximum dimension of satellite wheel will be $\mathbf{R_2}$ $\mathbf{R_1/2}$, otherwise will be obtain the results of the case 2) and the maximum ratio value will be 2.

2) When $\mathbf{a > R_1 + R_2}$ (fig.2, b and fig.4) - in this case, the synchronous belt width will be equal to the width of the belt wheels, they being in the same plan. This case has the advantage that you can use standardized synchronous belts. The range of transmission ratio is much wider than in the first case, because it can be used to rotate belt very different sizes, and, if necessary, it is possible to mount tensioning wheels (fig. 5, a,b).

To design a mechanism that uses planetary synchronous belts, the only impediments will be: geometric rules, gearing conditions and conditions of use for belts. Belt manufacturers offer a limited range of belt types and belt wheels; ideally it would be that planetary mechanism to be built with standard elements. Analyzed the construction and operation of these mechanisms it may be distinguished several advantages:

- flexible transmission, less noise - direct contact will no longer be between two metal wheels, but between the wheel and belt;
- possibility to adjust the dimensional transmission if they have differences with the tensioning wheels;

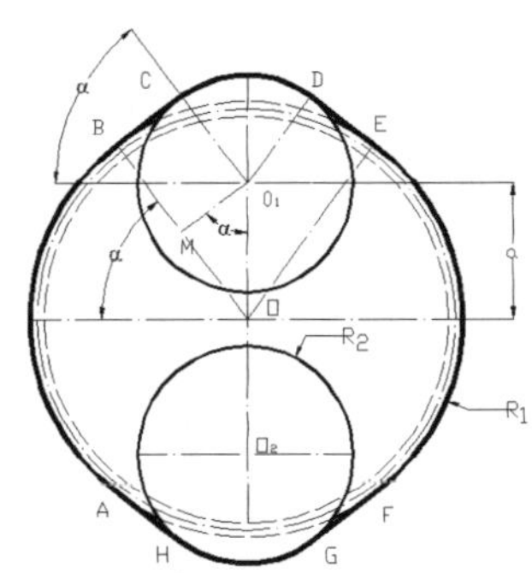

Fig. 3. The constructive scheme ver.1

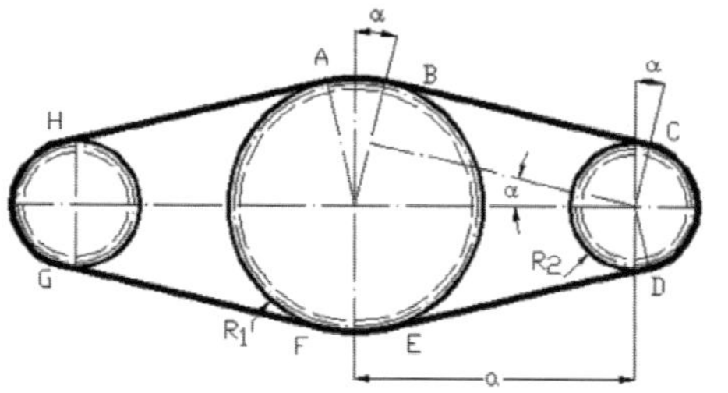

Fig. 4. The constructive scheme ver. 2

- transmission ratio with a range of higher values;
- the conditions of classical planetary transmission are eliminated: condition of coaxiality, condition assembly, condition of the neighbourhood (Callegari et al., 2003).

In the same context, transmissions through synchronous belts can generate certain disadvantages (Childs et al., 1998):

- gearing condition - a minimum of 6 belt teeth are concurrently in gearing on the same wheel;
- conditions of use (temperature, lubricants, air);
- it is preferable to use standardized wheels and belts (possible restrictions on the geometrical design).

3. ASPECTS OF DESIGN

According fig. 3 and fig. 4, the belt length **L** can be calculated as follows (Butnariu & Tanasescu, 2002):

$$L = 2l_1 + 4l_2 + 2l_3, \tag{2}$$

where:

$$l_1 = AB = \frac{2\alpha}{180^o}\pi R_1;\ l_2 = BC = \sqrt{a^2 - (R_1 - R_2)^2}\ ; \quad \text{and}$$

$$l_3 = CD = \frac{(180^o - 2\alpha)}{180^o}\pi R_2, \tag{3}$$

so,

$$L = \frac{2\pi}{90^o}(R_1 - R_2)\alpha^o + 2\pi R_2 + 4\sqrt{a^2 - (R_1 - R_2)^2}. \tag{4}$$

For given value of **L**, equation (4) can be written:

$$\alpha^o = \frac{90^o}{2\pi} \cdot \frac{L - 2\pi R_2 - 4\sqrt{a^2 - (R_1 - R_2)^2}}{R_1 - R_2}. \tag{5}$$

The distance between the axles of wheels can write as the product of a real number and value of belt pitch and the length **L** is a multiple of the belt pitch

$$a = k_0 \cdot p = \frac{k}{2\pi} \cdot p; \qquad L = Z \cdot p, \tag{6}$$

where **Z** is the number of teeth and **p** is belt pitch.

In the same way, the circumferences of the wheels are:

$$2\pi R_1 = Z_1 \cdot p \ \text{and}\ 2\pi R_2 = Z_2 \cdot p\ , \tag{7}$$

where $\mathbf{Z_{1,2}}$ are the teeth number of wheels 1 and 2.

Replacing $R_1 = \frac{Z_1 \cdot p}{2\pi}$ and $R_2 = \frac{Z_2 \cdot p}{2\pi}$ on the basis of equation (6), it can still write the equation (4) such:

$$Z = \frac{\alpha^o}{90^o}(Z_1 - Z_2) + Z_2 + \frac{2}{\pi}\sqrt{k^2 - (Z_1 - Z_2)^2}, \tag{8}$$

or, in other form, if the number of belt teeth it is known:

$$\alpha^o = \frac{90^o}{(Z_1 - Z_2)}(Z - Z_2 - \frac{2}{\pi}\sqrt{k^2 - (Z_1 - Z_2)^2}), \tag{9}$$

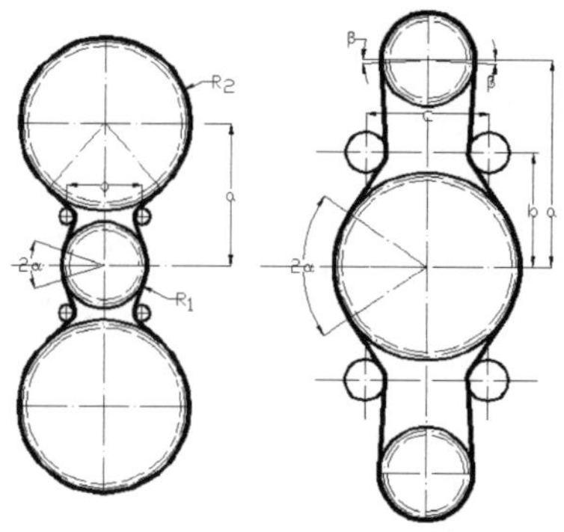

Fig. 5. Different variation of constructive scheme ver. 2

Designers and manufacturers limited belt in case of transmission through synchronous belts, the number of teeth in contact between the belt and the wheel at least 6. For this reason, construction of the planetary mechanism will suffer some restrictions dimensional. This is written so (v.fig.3, fig.4):

$$l_1 = \frac{2\alpha}{180^o} \cdot \pi R_1 \geq 6 \cdot p \quad l_3 = \frac{(180^o - 2\alpha)}{180^o} \cdot \pi R_2 \geq 6 \cdot p\ , \tag{10}$$

so

$$6 \cdot p \cdot \frac{90^o}{\pi R_1} \leq \alpha \leq (\pi R_2 - 6p)\frac{90^o}{\pi R_2}\ , \tag{11}$$

or

$$\frac{6}{Z_1} \cdot 180^o \leq \alpha \leq (\frac{1}{2} - \frac{6}{Z_2}) \cdot 180^o. \tag{12}$$

According equation (9), the angle **α** depends on the numbers of the wheels teeth and the belt, and the distance between the axles of the two wheels.

4. CONCLUSION

This work presented a model for a planetary mechanism with synchronous belt drive, with details about the basis concept of designing of this type of mechanism.

If in terms of design, execution is possible, this type of mechanism should be studied and the possibility of using it in practice, highlighting the applications where it could implement it.

5. REFERENCES

Abdullah, F. & Suleiman, E. (2004). Modelling and dynamic analysis of a planetary mechanism with an elastic belt, in: *Mechanism and Machine Theory*, Vol. 39, 2004, pp. 343-355, Elsevier, Amsterdam.

Butnariu, S. & Tanasescu, I. (2002). Aspects of using the synchronous belts in developing of planetary mechanisms, *Proceedings of PRASIC'02,* ISBN 973-635-076-2 pp.309-312, Brasov, Oct. 2002, Transilvania University, Brasov.

Callegari, M.; Cannella, F. & Ferri, G. (2003). Multi-body modelling of timing belt dynamics. *Proceedings of the Institution of Mechanical Engineers*, 2003, vol. 217, part K, pp. 63-75.

Childs, T.H.C.; Dalgarno, K.W.; Day, A.J. & Moore, R.B. (1998). Automotive timing belt life laws and a user design guide, *Proceedings of the Institution of Mechanical Engineers*, 1998, vol. 212, part D, pp. 409-420.

Emri I.; Kramar J.; Hribar A.; Nikonov, A. & Florjancic, U. (2006). Time-dependent constitutive modeling of drive belts – I. The effect of geometry and number of loading cycles, *Mech Time-Depend Mater,* Vol. 10, 2006, pp. 245–262, ISSN 11043-006-9021-2.

Annals of DAAAM for 2008 & Proceedings of the 19th International DAAAM Symposium, ISSN 1726-9679
Editor B. Katalinic, Published by DAAAM International, Vienna, Austria 2008
Make Harmony Between Technology and Nature, and Your Mind will Fly Free as a Bird

THE INFLUENCE OF THE CONSTRUCTIVE AND TECHNOLOGICAL PARAMETERS OF THE TOOL SYSTEM ON THE SURFACE FINISH AT GRINDING

BUZATU, C[onstantin] & BANCILA, D[aniel]

Abstract: *In this paper, the influence of the constructive and technological parameters on the roughness variation during grinding operation is shown, based on the mathematical model and graphic dependences. Starting from an original model of machining with abrasives, the paper takes into consideration together the constructive and technological parameters and the influence of the cooling liquid on the surface finish. Using a mathematical model of the maximum height of the irregularities (Ry), the influence of the next parameters is studied: feed, cutting speed, depth of cutting, tool grade, diameter of the abrasive wheel, quality of the cooling liquid by the friction coefficient between grain and work piece surface. As a result, recommendations regarding the optimization of machining process at grinding operation are given.*

Key words: *grinding, roughness, liquid, friction.*

1. INTRODUCTION

It is shown that to increase the reliability and the working life of the parts, which have an important functional role in the engineering equipment structure, they are subjected to surface treatments and grinding operation to improve their dimensional accuracy.

The optimization of grinding operation requires even from the design step, the knowledge of the quantitative and qualitative influence of the factors that determine the technological parameters which give the accuracy needed to the parts according with the technical requirements (Buzatu, 1981).

To improve the parts quality it is possible to modify the technological and constructive parameters of the grinding operation.

Besides the technological and constructive parameters of the grinding operation, the influence of the cooling liquid has been least researched under quantitative aspects, an applicative study being imposed.

Based on these considerations, in the paper are presented the theoretical and experimental studies and researches regarding the influence of these technological and constructive parameters on the surface finish.

2. THE THEORETICAL EQUATION OF THE ROUGHNESS SURFACE OBTAINED IN GRINDING OPERATION

Based on the theoretical modelling of machining process with abrasives, in the conditions of simplifying assumptions (Secara et al., 1989), according with the schematic illustration from fig.1 and after a series of theoretical calculations (Buzatu et al., 2007), the next equation for the maximum height of the profile Ry, at grinding cylindrical surfaces was found:

$$Ry = h = \frac{t \cdot s_l \cdot e}{\pi \cdot k1 \cdot D^2 \cdot \sqrt{D} \cdot v_s} \cdot \frac{2 \cdot \mu\left[\left(1-\mu^2\right)ctg\gamma - 1\right]}{1-\mu^2+2\cdot\mu\cdot ctg\gamma} \qquad (1)$$

where: t – is the depth of cut, in mm, S_l – is circular feed of work piece in mm/sec, V_s – is angular speed of tool in mm/min, e – Gmed = average dimension of the abrasive grain from the grinding wheel in μm, D – is the grinding wheel diameter in mm, k1 = 0.7 – 0.9 is a coefficient that depends of the number of abrasive grains that work at maximum capacity, μ – is the friction coefficient between abrasive grain and work piece surface in the presence of cooling liquid, γ – is the rake angle of the abrasive grain in radians(fig.1).

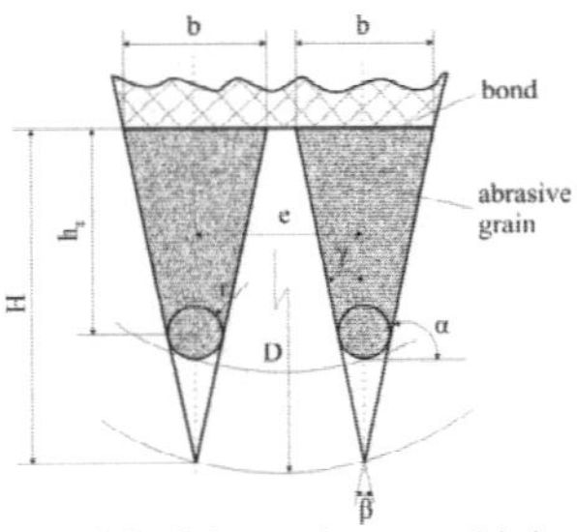

Fig. 1. Spherical model of the grain at machining with abrasives

3. STUDY REGARDING THE INFLUENCE OF TECHNOLOGICAL AND CONSTRUCTIVE PARAMETERS ON THE SURFACE ROUGHNESS AT INFEED GRINDING

The Ry equation (1) can be written synthetically as:

$$Ry = h = f\,(S_l, V_s, t, \mu, D, e) \qquad (2)$$

where, S_l, V_s, t and μ, represent the technological parameters, D and e, are the constructive parameters of the too, system. Taking into account the values recommended in the literature (Vlase et al., 1985), (Picos & Pruteanu, 1992), which are used frequently at grinding (S_l = 0.001 – 0.075 mm/rev; V_s = 0.0005 – 0.002 m/min; t = 0.1 – 0.4 mm; e = 100 – 300 μm; D = 200 – 500 mm; μ = 0.01 – 0.2), if fig. 2 – fig. 7 are shown the graphical dependences between roughness and variable parameters from equation (2) in certain working conditions.

Also, using a special software were deducted the mathematical models of these dependences, adopted based on the optimal correlation coefficients presented on the graphical dependences.

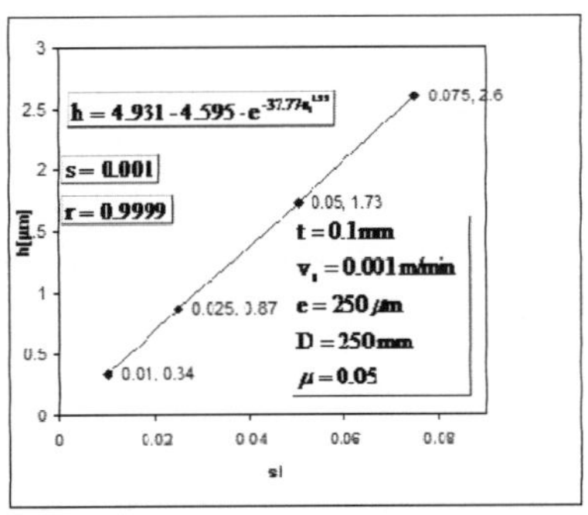

Fig. 2. Roughness variation Ry = h, function by feed S_l

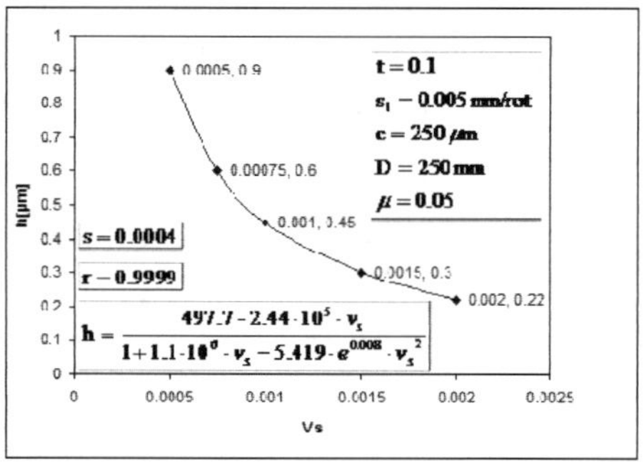

Fig. 3. Roughness variation Ry = h, function by the angular speed V_s

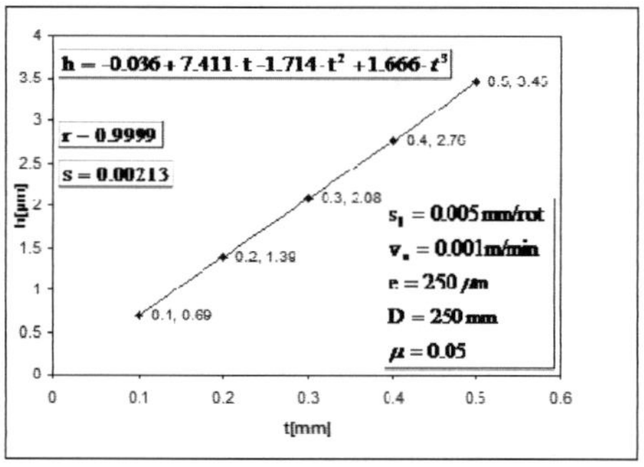

Fig. 4. Roughness variation Ry = h, function by the depth of cut, t

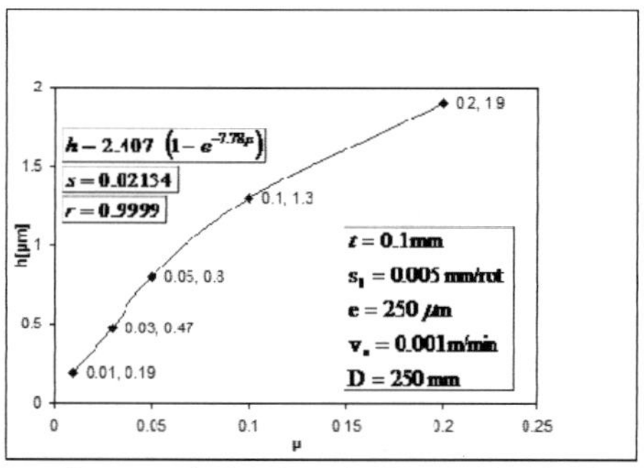

Fig. 5. Roughness variation Ry = h, function by the friction coefficient μ

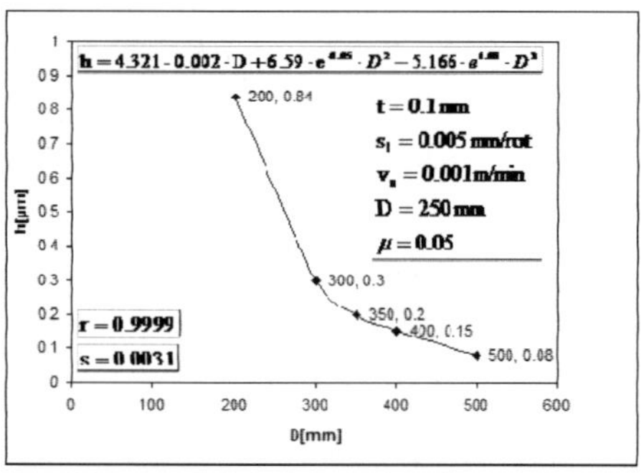

Fig. 6. Roughness variation Ry = h, function by the abrasive wheel diameter D

In fig. 8, the dependence of the roughness Ry = h, according to all considered parameters is shown.

Also, from fig.8 results the optimal domain of values to obtain the roughness at grinding between Ra = 0.8 – 1.6 μm.

By using the mathematical models and data of different parameters (technological and constructive), other bi or three dimensional graphic dependencies can be obtained.

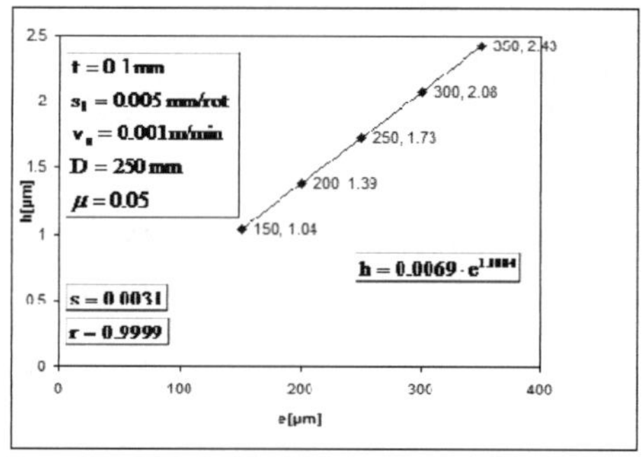

Fig. 7. Roughness variation Ry = h, function by the abrasive wheel grade e

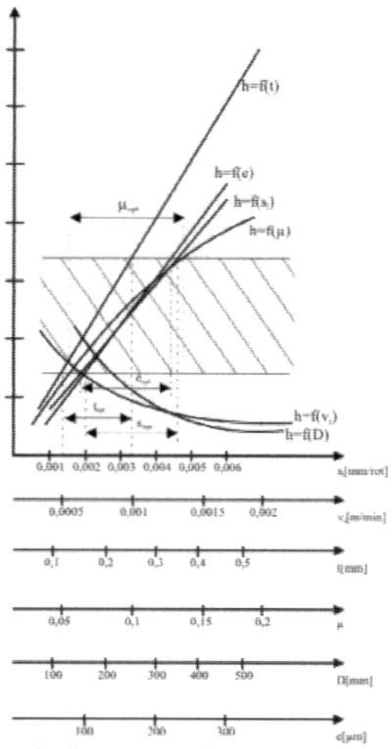

Fig. 8. Roughness variation Ry = h= f(S_l, V_s, t, μ, D, e)

4. OBSERVATIONS AND CONCLUSIONS

Based on the previous data the next observations arise:

- The graphical dependencies from fig.2 – fig.7 and their mathematical models present an optimal choice of the cutting and constructive parameters of tool according with the requirements of the part design;
- It can easily be observed (fig. 5), that the cooling liquid has an important influence on the roughness of the surface;
- The friction coefficient μ has a great influence on the roughness and it depends on the cooling liquid characteristics.

5. REFERENCES

Buzatu, C. (1981). Contributions on the factors study which affect increasing the accuracy of parts made by turning and external grinding. *Doctoral thesis,* Transilvania University of Brasov, Brasov, 1981

Buzatu, C.; Lepadatescu, B. & Dumitrascu, A.E. (2007). Studies regarding the influence of the process parameters on the surface roughness at manufacturing by grinding, *Annals of DAAAM for 2007& Proceedings*, pp. 127 - 128, ISBN 3-901509-58-5

Picos, C.; Pruteanu, O. (1992). *Design of manufacturing technologies. Design handbook,* Universitas Publishing House, ISBN 5-362-00971-0, Chisinau

Secara, Gh.; Rosca, D.; Mares, Gh.; Ditu, V.; Bodocan, S.; Lupulescu, N.B. & Deaconu, V. (1989). *Fundamentals of cutting and surface generating,* University of Brasov

Vlase, A.; Sturza, A.; Mihail, A. & Bercea, I. (1985). *NCutting parameters stock removal and cutting time,* Technica Publishing House, Bucharest.

Annals of DAAAM for 2008 & Proceedings of the 19th International DAAAM Symposium, ISSN 1726-9679
Editor B. Katalinic, Published by DAAAM International, Vienna, Austria 2008
Make Harmony Between Technology and Nature, and Your Mind will Fly Free as a Bird

ASSESMENT TECHNIQUES OF THE PRODUCTS ON THE BASIS OF SHAPE ANALYSIS

BUZERA, M[arius]; PROSTEAN, G[abriela] & STEFAN, C[onstantin]

Abstract: *Shape is one of the most important parameters that have to be considered when assessing the products, especially for vegetal products. The differences between the parameters in question may very broadly vary according to the species, variety and even the weather. Shape is an easy parameter to assess for human brain, but difficult to assess for the computer. Still, on the basis of shape analysis, one can assess the quality of products, their size, as well as their integrity. Considering all that several new algorithms assessing the shape of the products were to be carried out, in order to support the new research in the field of artificial intelligence.*
For this purpose have they carried out an experimental stand using the machine vision principles and which is to allow the assessment of the algorithms in question, and part of the conclusions of the tests are presented throughout this paper.
Key words: *machine vision, shape, neural network, classification, image processing*

1. INTRODUCTION

Shape is an essential parameter when analyzing the vegetal products, especially due to the fact that, by taking It into consideration on can appreciate the integrity of the products, their quality as well as their size. If in the case of industrial products analysis, it is but easy to make an assessment for vegetal products, having an endless number of possible shapes one can find it very hard to assess. (Baoping, 1999) Also the usage of classical algorithms under classification, irregardless of the calculus ability to use, is to became inefficient. (Tao & Wen, 1999)

Thus, the algorithms assessing the products throughout this article based on conexionist patterns – neural networks and they ensure the assessment of the quality of products according to the basic shape of the products under three classes: class A – good products, class B – average products and class C – under standard products, according to the integrity of the products: class A – good products, class B – damaged products and according to the size: class A – big products, class B – medium products and class C – small products.

The assessment of the quality of the products in question was carried out on an experimental stand, on the basis of machine vision principles.

2. THE EXPERIMENTAL DEVICE THE „MACHINE VISION" SYSTEM

The device of video-inspection has been carried-out around transporters with a belt activated by an electric engine, which ensures the moving of the product in face of the video-inspection system, made of two video cameras with manual focusing. The cameras are disposed in the same frame, realizing an angle of 60^0, towards the analyzed object (Laykin *et al.*, 2003). These allow the acquisition of 4 coloured RGB pictures of 512*512 pixels, which they transmit in a real time, to the analyzing mechanism represented by a PC of Pentium IV/3200 MHz type.

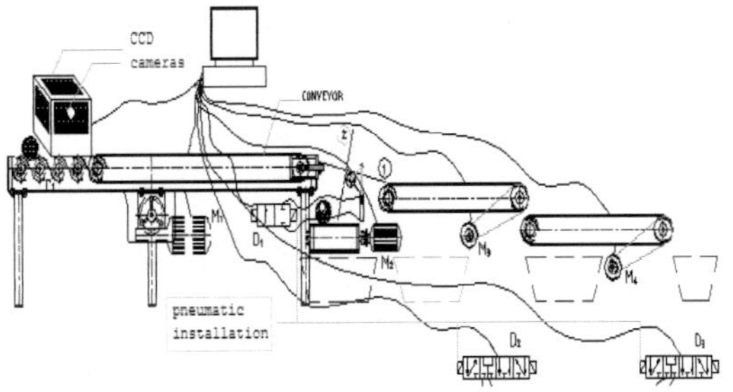

Fig. 1. Conveyor and device of video-inspection

For a better illumination of the product, the interior walls of the illuminating room were painted white, such that they can ensure a diffuse illumination of the object under inspection. (Laykin *et al.*,2002) Also, the illuminating room was implemented on a mobile metallic framework that allows to change the distance between electric bulbs and the object, and to reposition the electric bulbs on the prop, as shown in Fig. 1. (Buzera *et al.*, 2008)

3. THE SHAPE. THE ANALYSIS OF THE SHAPE

After the outline of the object had been carried-out, we must obtain the signature of the shape of the object. The signature of the shape is a feature that can be used to identify the objects. A main part in getting a right answer is that of the precision of the weigh centre. (Sudharkara, 2003)

For each image the shape of the object was automatically determined through an analytical way, due to the help of an application carried-out in the Matlab program. The polygonal line that defines the outline analytically and successively over crossed with a right line. The angle of orientation was considered of 5^0 (Buzera & Stefan, 2007).

In order to detect the shape of the object from the picture the most important information was represented by the position of the pixels that describe the outline. Other information can be ignored.

The signature of the shape of the object supposes that the weight-centre of the outline should be determined. These can be determined due to the moments of stillness.

$$x_c = \frac{\left[\iint_{x,y\in O} x\,dx\,dy\right]}{\iint\limits_{x,y\in O} dx\,dy}, \quad y_c = \frac{\left[\iint_{x,y\in O} y\,dx\,dy\right]}{\iint\limits_{x,y\in O} dx\,dy} \qquad (1)$$

We will continue by presenting some notions necessary to the introduction of these theorems using the basic formula called Green-Riemann.

If MCR^2 is an elementary compact with the FrM border and P and Q are two functions of class C^1, on a simple convex domain which contain M, then

$$\iint_M \left(\frac{\partial Q}{\partial x} - \frac{\partial P}{\partial y} \right) dxdy = \oint_{FrM} Pdx + Qdy \tag{2}$$

By using the fundamental Green-Riemann's formula with P and Q chosen, from formula (1) we get:

$$x_c = \frac{\left[\frac{1}{2}\oint_{x,y\in C}\left(yxdx - \frac{1}{2}x^2 dy\right)\right]}{\left[\frac{1}{2}\oint_{x,y\in C}(ydx - xdy)\right]} \quad y_c = \frac{\left[\frac{1}{2}\oint_{x,y\in C}\left(\frac{1}{2}y^2 dx - xydy\right)\right]}{\left[\frac{1}{2}\oint_{x,y\in C}(ydx - xdy)\right]} \tag{3}$$

By using the definition with Riemann's sums in (3) and then, for each point from the outline we can find out the ray by using the relation:

$$R_k = \sqrt{(x_k - x_k)^2 + (y_k - y_c)^2} \tag{4}$$

Talking into account the fact that fruits have usually unique shapes and thus, they can have sizes and shapes that are very different, the rays have been normalized with the medium ray and next by applying the Fourier transformed over the signature of the shape, we get:

$$F(u) = \frac{1}{N}\sum_{k=1}^{N} r(k)\exp[-j2\pi uk/N], u = 0,1,2....N \tag{5}$$

The histogram applied to the Fourier transformed shows how many times each of the values of the rays can appear.

4. SHAPE RECOGNITION BASED ON ARTIFICIAL NEURAL NETWORK

Consequently, as a result of tests and attempts carried out, a neural network was chosen, of the type back propagation having 16 neurons on the first stratum, 24 neurons on the second one, as for the third one, the number of neurons varies from 2 to 3 according the requirements.

Considering the fact that the algorithm is supposed to classify to be of hyperbolical tangent type for the first two strata and of sygnoidal type for the third stratum.

$$f(v) = \frac{1}{1+e^{-qv}}, \quad f(v) = \frac{1-e^{-2v}}{1+e^{-2v}}; \tag{6}$$

The exit stratum is formed by 2 or 3 neurons (Fig.2) that signify the two or three categories: according to the basic shape – good products (1,0,0), average products (0,1,0) and under standard products (0,0,1), according to the integrity of the products - good products (1,0), damaged products (0,1) and according to the size: big products (1,0,0), medium products (0,1,0) and small products (0,0,1).

The precision of the algorithms implemented was tested by using tomatoes and the network was trained up to the point when the average global square error reached the value of under 0,0001 during the training stage.

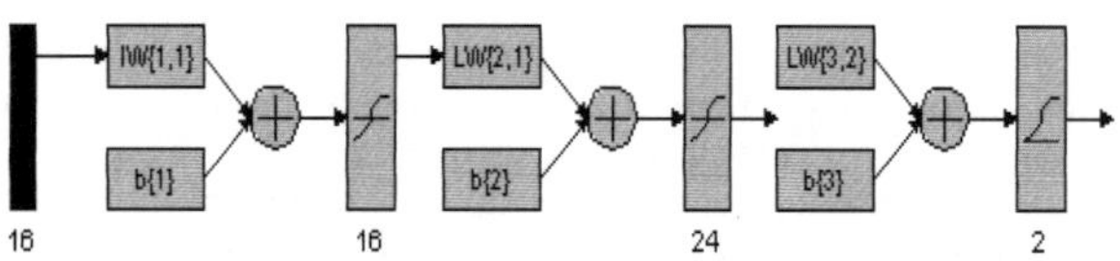

Fig. 2. The block scheme of the algorithm establishing the integrity of the products on the basis of the shape analysis.

5. RESULTS AND DISCUSSION

After the training of the network, tests were carried out upon it while using 50 tomatoes belonging to all classes according to the size, quality or the presence of fault, and the evaluation of tomatoes belonging to each class was carried out by a human operator.

Very good results of the tests were obtained, with an average of the classification precision of over 94 % in the case of setting the integrity of products, 99 % in the case of evaluating the size and 91 % for the evaluation of quality.

We consider that the lower percent than 91 % in the case of products quality assessment is mainly due to the small amount of this training stage, as well as the inappropriate choice of products from the training set.

Consequently, we can take it that the usage of neural networks in the determinating process is a fast, secure one, very easy to implement.

Also, it allows the assessment of a large number of features and it can be easily correlated to field of artificial intelligence such as the fuzzy logics reason.

6. CONCLUSION AND FUTURE WORK

After the test the following conclusions have been formulated:

The technique of the signature of rays is quite easy to apply, is quick and can be used both in the classification of the products according the shape and in the analysis of the products quality.

It can be easily correlated with the classification of neuronal type.

Taking into account the numberless amount of forms that the vegetal products can have, the utilization of the classificatory of neuronal type can be considered an important alternative (in comparison with the rest of the techniques of classification). The HIS pattern (used for image analysis) also proved efficient, saving a large amount of time and resources.

7. REFERENCES

Baoping, J., (1999). *Non-destructive Technology for Fruits Grading. International Conference on Agricultural Engineering Beijing*

Buzera, M.; Stefan C., (2007). *Detecting the integrity of the shape of vegetal products by using non-destructive techniques*, Proceedings of the International Conference, Bulgarian National Society of Agricultural Engineers, Lozenec, Vol.2 (2007) 252-256

Buzera, M.; Voicu, G.; Prostean, G. & Prostean, O. (2008) *Techniques of Analysing the Colour of Produces for Automatic Classification*, 12th IEEE International Conference on Intelligent Engineering Systems, Miami, Florida, USA, 209-214

Heinemann, P. H.; Pathare, N. P. & Morrow, C.T. (1996), *An Automated Inspection Station for machine vision grading of potatoes*. Machine Vision and Applications 9. 14-19

Laykin, S.; Alchanatis, V. ; Fallick, E. & Edan, Y. (2002). *Image Procesing Algorithms for Tomato Classification.* Transaction of the Assae. Vol. 45(3) 851-858

Laykin, S.; Edan, Y.; Alchanatis V. & Regev, R. (2003). *Development of a Quality Sorting Machine Using Machine Vision and Impact*. Transaction of the ASAE

Sudharkara, R. P.; Gopal, A.; Revathy, R. & Meenakshi K. (2003). *Colour Analysis of Fruits Using Machine Vision System for Automatic Sorting and Grading.* Journal Instrum. Soc. India. Vol.34(4) 284-291

Tao, Y. & Wen, Z. (1999). *An adaptive spherical Image Transform for High Speed Fruit Defect Detection.* Transaction of the ASSAE. Vol. 42(1) 241-246

Annals of DAAAM for 2008 & Proceedings of the 19th International DAAAM Symposium, ISSN 1726-9679
Editor B. Katalinic, Published by DAAAM International, Vienna, Austria 2008
Make Harmony Between Technology and Nature, and Your Mind will Fly Free as a Bird

SYLLABUS DESIGN

CAGANOVA, D[agmar]

Abstract: *The paper presents the outcomes of the language syllabus design developed at the Faculty of Materials Science and technology, the Slovak University of Technology. Consequently,the emphasis is put on the syllabus design process and methodology in foreign language teaching. However, the conditions of the research are briefly summarized,and the need for the consequent research is mentioned.*

Key words: *course design, syllabus, programme, process, English for Specific Purposes (ESP)*

1. INTRODUCTION

In the Twenty – First century political and social change is typical at the global level. There is a great need for foreign language instruction in the modern world. In teaching foreign languages there is a strong tendency towards a communicative approach and an expansion of information technologies and e-communication. Computerization influences all aspects of the teaching and learning process. The Faculty of Materials Science and Technology (MtF), as one of the seven faculties of the Slovak University of Technology of Bratislava (STU), provides complete training for experts in the production, surfacing, and testing of metallic and non-metallic materials. In addition to this training, students must learn a foreign language. Due to their language competence and training in core curriculum, our graduates have the ability to find a position in the labor market. The aim of our research was to illuminate the new demands on graduates and recent changes in our curriculum, and consequently the necessity to prepare a new syllabus corresponding with the demands of the future employers of our graduates.

In this paper the author tries to summarize their own process of course design.

2. COURSE DESIGN PROCESS

From the authors' point of view, course curriculum is primarily a topic/content form of education.

The author's approach to the course design process is learning centered one focusing on the needs of the individual student.

The objectives are as follows:

- To examine the practical needs of students' and graduate's in the academic environment
- -To create a a relevant questionnaire
- To process the questionnaires and evaluate the results
- - To design a course curriculum (syllabus)
- To choose materials
- To design appropriate materials
- To decide on a suitable methodology
- To sort materials into a logical (units)
- To rehearse
- To revise
- To prepare a final version of the course curriculum - syllabus
- To implement the course curriculum into the teaching-learning process.

2.1 Course curriculum content

General English (GE) combined with English for Specific Purposes (ESP)

- English for Specific Purposes (ESP)
- Flexible; based on skills, tasks, functions
- Proportional amount of relevant questions

To ensure that future engineers are properly prepared, the author believes it is appropriate to combine the two basic components of General English and English for Specific Purposes-into the syllabus. General English will be used on two levels, intermediate and advanced, and English for Specific Purposes will focus on technical and business English. General English will enable students to obtain skills for everyday communication, while technical and business English will be useful in their professional careers.

The flexible content of a syllabus, with its openness to systematic fine-tuning, is endorsed by author/educator Janice Yalden (1987). The syllabus is an instrument for synchronizing the various aspects of teaching languages. It shouldn't be rigid and static, but flexible and dynamic. Its themes should be reviewed systematically by means of class feedback.

2.2 Methodology

According to assoc. prof. Gadušová (1997, p. 481), the term methodology has different meanings for different users, and she cites several interpretations: Rives (1968) defines methodology as the universal technique of teaching various linguistic materials within a specific skill, e.g. presentation, practice and production.

Cohen (1990) defines methodology as the diversified teaching techniques of basic language skills such as reading, listening, speaking, and writing.

According to Richards and Rodgers (1986), methodology is the description of various certified methods and instructionssuch as Direct method, Suggestopedia, Silent method, etc.

Rivers (1981) characterizes methodology as an eclectic collage of techniques reflecting the ideologies of varied teachers.

Freeman (1992) sees methodology as intentionally modeling the practices of experienced teachers.

For teachers, the term methodology indicates how the teacher organizes the teaching-learning process in the classroom.

Holliday (1994) identifies three basic types of foreign language methodology:

- methodology of procedures and methods (as regards the teacher's activities in the classroom and how they were trained for their profession)

- syllabus design (its preparation and realization)
- acquiring information regarding the social context in which foreign languages are taught (the aim of this methodology is ethnographic action research)

However, Gadušová (1998) mentions that on the basis of various criticism in 90ś there is evident shift of methods from methodology. Brown (1994) declares the end of methods. Woodward (1996) persuades that "teacher researcher" is in accordance with contemporary post-method thinking. Nowadays it is favourite eclectic method i.e. contemporary element connection of different methods together.

The authorś intention is to find an inspiration in particular methods, to choose positive elements and suitable effective style of educational process.

Our preference:

- communicative methodology
- learner-centred approach
- problem task solving
- student's project work

Theoretically, according to Tandlichová (1995) communicative methodology emphasizes integration of all 4 (or more) skills in teaching English as a foreign language. The focus is put on receptive and productive skills. It is necessary to highlight the fact that aim of skills integration is significant due to language acts as a means of communication via all these channels.

As far as the learner centered approach Nunan, Lamb (1996) characterize learner – centred classrooms as classrooms in which are students dynamically implemented into their own learning process. It is highly appreciated if the students are directly and actively participating on their own learning process. In Nunanś, Lambś theory is evident strong connection with experimental teaching, humanistic psychology and problem task solving.
The problem task solving is also supported by Widdowsonś (1990) declaration that pedagogical movement in 70ś, 80ś shows the tendency to assume that meaning is not exclusively signaled by language themselves but it is also reached implicitly. Whatever necessary for the student is implemented in problem task solving.

3. RESEARCH

The research, which was a part of institutional research at MtF STU, was focused on analysis of the needs of students and graduates in practice. From this analysis we should be able to specify the amount and extent of relevant topics needed in foreign language teaching. Also, we should be able to suggest effective methods and approaches to teaching and create standard criteria for both teachers and students. The results of the needs analyses offered a reliable and persuasive argument for implementation of the new model into practice.

Within the research the author tested the following hypotheses:
Hypothesis One
We believe that the quality of the syllabus will be improved through the use of foreign language educational needs analysis.
Hypothesis Two
We suppose that suitable students´ project work inclusion will improve independence and autonomy of the university students in a foreign language.
This research was conducted at the Faculty of Materials Science and Technology of the Slovak University of Technology in 2004, and the research sample is comprised of:

110 students before taking ESP course
116 students after taking ESP course
(The questionnaire No. 1 about needs analysis)

107 graduates in practice
(The questionnaire No. 2 about needs analysis)

110 students
(The questionnaire No. 3 about project work)

100 students
(Participants of interview)

All of the data from the questionnaires and interviews have been analyzed and the results of these analyses were given in the form of graphs, pie charts, tables and statistics. Support was found for both of the hypotheses.

4. CONCLUSION

Hypothesis One was supported by three experts, ESP teachers, on the basis of Kotarbińskiś (in Švec) features of effective syllabus design. Hypothesis Two was verified by questionnaire in the form of individual and group interviews on the sample of 100 university students. 76% university students consider project work inclusion as supporting and improving independence and autonomy of the university students in a foreign language.
The results obtained from our research can be used in the education of future engineers. The author believes that the new syllabus and also questionnaires about needs analysis can be used, after some adjustments, at other technical faculties and also as an initial stimulus for further research or for modification of existing syllabi. It could be useful to suggest a transparent model of continuous assessment. An interesting fact could be to compare our syllabus with other syllabi existing at particular faculties of the Slovak University of Technology and for example to use content analysis for comparison of topics and subtopics and constantly continue in improving our programme and syllabi.

5. REFFERENCES

Caganova, D.; Ruskova, D.(2002*.) Step by step: From needs analysis to project preparation and presentation at the students´ scientific conference.* In: CO – MAT – TECH 2002. 10. International Scientific Conference, Bratislava

Gadusova, Z. (1997*). Methods in Methodology*. Zborník CO-MAT- TECH 97. 5. medzinárodná vedecká konfere ncia. Trnava

Nunan, D. (1993*). Syllabus Design*. Oxford University Press.

Nunan, D.;Lamb,C.(1996) *The self – directed teacher. Managing the learning process*. CUP

Rivers, W. M.(1981).*Interactive Language Teaching*

Rodgers,(1986) (in: Nunan, D. 1991. *Designing Tasks for the Communicative Classroom*. CUP

Švec,Š. (1989). *Theory of didactic planning*. In: Paedagogica 6., SPN Bratislava.

Tandlichova, E. (1999/2000). *Knowledge Changes*. In: Foreign Languages, number 43 .

Widdowson, H.G. (1990). *Aspects of Language Teaching.* OUP.

Woodward, T. 1992.*Ways of Training.* Longman Group UK Limited

Yalden, J. (1987).*Principles of Course Design for* Language Teaching. Cambridge. Cambridge University Press.

Annals of DAAAM for 2008 & Proceedings of the 19th International DAAAM Symposium, ISSN 1726-9679
Editor B. Katalinic, Published by DAAAM International, Vienna, Austria 2008
Make Harmony Between Technology and Nature, and Your Mind will Fly Free as a Bird

ADVANCED MACHINING TECHNOLOGIES FOR AIR-CRAFT COMPONENT PARTS WITH THIN WALLS

CALBUREANU, M[adalina]; MALCIU, R[aluca]; LUNGU, M[ihai] & LUNGU, R[omulus]

Abstract: *The paper presents the machining process with advanced NC programs (SURFCAM 2005) of a bracket for a cargo air-craft door, made of aluminium alloy (zycral). The part has thin walls and it is subjected to thermal and static transitorial loads and strong vibrations. As a result of a heat transfer, static and dynamic analyses, there were determined the critical sections that had to be taken into account by the manufacturing process.*
Key words: *bracket, thin, walls, zycral, loads*

1. INTRODUCTION

The problem is to realise an advanced conception for manufacturing thin walls parts subjected to static, thermal and dynamic tranzitorial loads, such as a bracket for a cargo air-craft door.

The bracket is a component of the air-craft resistence structure, a first class part, which supposes supplementary safety measures and, implicitely, supplementary quality assurance measures.

Likewise parts were manufactured on universal machine-tools or, recently, on 2&1/2 axes NC machine-tools. The part manufacturing process approach on universal machine tools involves a major manufacturing preparation, the bigger execution time and serious problems in assuring the quality of the product. The manufacturing process on 2&1/2 axes NC machine-tools leads to complex surfaces generation with stock to leave for the final operation of adjusting the boundaries, in order to eliminate the milling traces (Groover, 2006).
The processing on NC machine tools with 5 axes will lead to (Kief, 2000):
-shortening the time of the conception of the manufacturing preparing;
- enhancing the quality of execution especially by eliminating human intervention;
- process repeatability and part interchangeability are obtained;
- reduced rate of scrap;
- parts with technological difficulties such as thin walls, considerable depths, can be machined;
- elimination of adjustment operations and improved quality of the obtained surfaces;
- increasing the flexibility of the process.

2. THE PART MANUFACTURING CONCEPTION

In order to manufacture the part with the geometrical configuration as shown in the figure 1, 2 and 3, the following steps are realised:

- the 3D model part elaboration starts from the part drawing using ProEngineer program;

- the critical sections determination are found by using the finite element analysis in COSMOSM program, performing static, dynamic and heat transfer through importing the 3D model. The geometrical and shape optimization are performed as well reducing the width of the walls considerably (approx. 20%) and saving a lot of material;

- the ProE model is imported in SURFCAM 2005 in order to perform the NC programs;

- the tools trajectories are obtained for each operation using SURFCAM, taking into account the determined configuration of the critical sections and avoiding the appeareance of supplementary tensions during the process;

- the post-processing programs and the part execution on the NC machine (Mattson, 2001).

The final form of the part will be as shown in fig. 6.

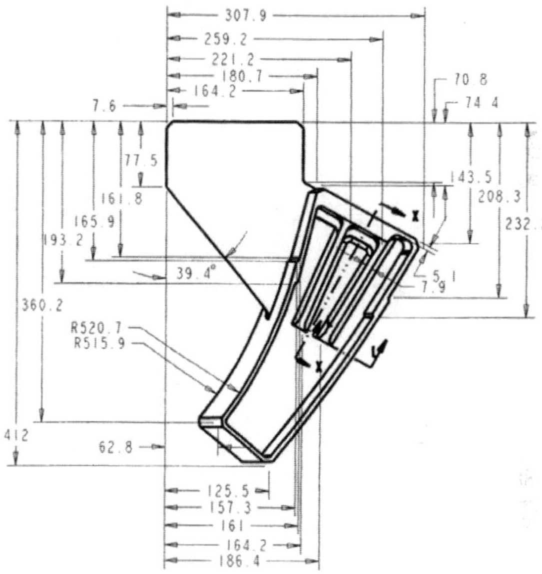

Fig. 1. View of Part face 1

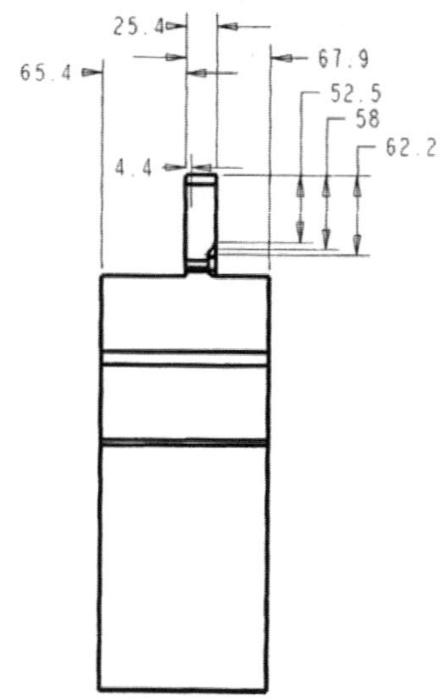

Fig. 2. Lateral view of the part

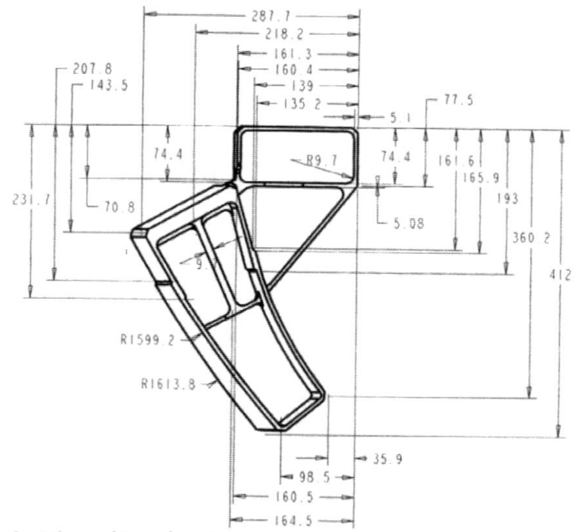

Fig. 3. View of Part face 2

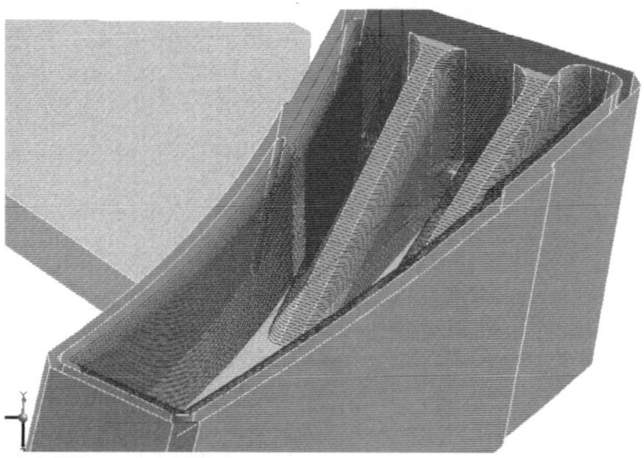
Fig. 4. Milling finishing pocket walls – tool routes

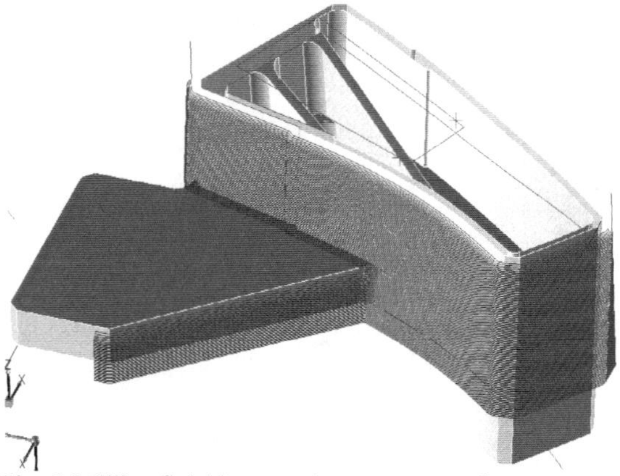
Fig. 5. Milling finishing exterior contour - tool routes

3. SIMULATION AND OPTIMIZATION OF COMPUTER PROGRAMS

After getting the routes of the tools, the computer simulation of all processing operations is done, respecting the established order by the technological route (Lancea, 2004). It aims to obtain the correct geometry and required surface quality, to eliminate unwanted interferences, material missing from surfaces and edges. After computer simulation analysis, the routes of tools are restored in order to obtain optimal joining of all processing operations, taking care so that the step of the finishing cuts set to be enough as not to cause an overheating of the part and the introduction of additional tensions (Bannister et al., 2001).

Fig. 6. The part close to the final form

4. CONCLUSION

It was realized the processing technology optimization not only under geometric criterion, but also under technological criterion. According the geometric criterion, the optimum is obtained primarily by minimizing the total length of trajectories (curves) with which the whole area is covered, in terms of maximum accuracy providing from the geometric point of view and in terms of a calculation effort as lower as possible.

Among the technological criteria there were considerred the maximization of the tool capacity of cutting. The great importance, as a condition for general optimum, is to establish the direction of processing trajectories so that to allow a comfortable manual finishing. It is about giving such a direction to the "rifles" which mark processing trajectories that can be conveniently attacked by the tool of the adjusting worker on a parallel direction to the direction of finishing.

In order to reduce the weight and the overall size of the air-crafts component parts with thin walls are frequently used. Their machining requires special attention for not introducing additional tensions in the critical sections determined by finite element analysis using previous experience of authors in optimizing and modeling the thin walls parts. Also, the finite element analysis and optimization saves 20% of material and providing the small widths for walls where the model is not overloaded. Thus, interdisciplinary approach of the construction and technology design, of the execution and the control makes the parts with thin walls to require advanced technologies for processing.

Using of lightweight alloys as titanium alloys and others more resistant in mechanical and thermal terms will be a future research direction for improving the processing technology of the parts with thin walls.

5. REFERENCES

Bannister, K.; Gabrel, S.; Polywka, J. (2001), *Programming of Computer Numerically Controlled Machines*, Industrial Press, Inc., 2-nd edition, ISBN 978-0831131296, New York.

Groover, M. P. (2006), *Fundamentals of Modern Manufacturing: Materials, Processes, and Systems*, 3rd Edition, John Wiley & Sons, ISBN: 978-0-471-74485-6, New York.

Kief, H. B., (2000), *CNC for Industry*, Hanser Gardner Pubns, ISBN 978-1569902967, Cincinnati, SUA.

Lancea, C.T.S.(2004), *A Computer Simulation Program for NC Milling of 3D Parts*, First International Conference “Mechanics and Machine Elements”, pp. 200-203, ISBN: 954-580-173-5, Technical University of Sofia, Bulgaria, 2004.

Mattson, M.(2001), *CNC Programming Principles and Applications*, Delmar Cengage Learning, 1st edition, ISBN 978-0766818880, Canada.

Annals of DAAAM for 2008 & Proceedings of the 19th International DAAAM Symposium, ISSN 1726-9679
Editor B. Katalinic, Published by DAAAM International, Vienna, Austria 2008
Make Harmony Between Technology and Nature, and Your Mind will Fly Free as a Bird

CREATING OF OPTIMAL CORPORATE CULTURE THROUGH COMPLEX TRAINING PROGRAM

CAMBAL, M[ilos]

Abstract: *Assurance of organization prosperity is influenced by creating of optimal corporate culture, because the effective organization operation depends on people, not on investment or technology. Creating the optimal corporate culture isn't one-shot campaign activity, but it is an integrated and systematic process, the base of which has to be the continual organization's education. The most effective approach to the organizational continual learning process and the development of human sources is a realization of the self-contained complex of training programs.*
Key words: *organization, culture, optimalization, learning, training*

1. INTRODUCTION

The main aim of each organization is to succeed on the market. This success is mostly fixed by the height of achieved profit in which can be also seen the attraction of the organization for business partners and customers. This is the reason why organizations try the dynamics of profit to have the increasing tendency, because the profit enables their further existence and development. Many organizations therefore invest a great amount of finances into machines or other technologies in order to increase their value. However the results are often not so conspicuous as originally expected. The way in which entries in the organization are appraised depends primarily on the people, on their attitude to the organization and to their work as well as on their attitude to themselves. These factors are greatly influenced by the „business presence phenomenon" – the corporate culture. The culture of organization supports, or to the contrary it defend the achievement of organizational aims, the increase of efficiency and therefore it obstructs the development of competitive position on the market.

2. THE CORPORATE CULTURE – THE MAIN POSTULATES

If a firm wants to be successful, it's needed to be "set up" on this kind of functioning. Nowadays the conception of corporate culture is being analyzed by many authors from the management field of theory and experience, but at the same time (maybe because of this) it is an idea very ambiguously defined. It is caused by many different points of view of these authors and by the reality that people create the basis of corporate culture. In a great portion the ideas, attitudes and values of individual workers of the organization originate it, while it's obvious that all what is connected with human thinking and behavior is very rarely possible to strictly determine. (Vaňová, 2005)

We can consider the corporate culture to be a summary or mixture of "inner – recognizable" values and ways of standard behavior, relationships in internal (colleagues, managers, subordinates, ...) and external (clients, business partners, distributors, competitors, ...) organizational environment, of thought patterns and attitudes, which are in firm considered to be a norm. It is important to realize that in any case it cannot be a summary or statistical average of individual value preferences, attitudes or norms of behavior, but a group phenomenon that is up individual and has a great social character. (Čambál, 2007)

Another important postulate is a reality that the culture of a firm itself is one of the discerning factors influencing a set of elements representing inner running forces of the activities of individual workers, which direct their acting and living. It means that even the corporate culture is created and influenced by the individual employees, on the other hand it greatly influences unwritten patterns of their behavior, and so it influences the employees motivationally or contrary, as a strong de-motivator.

If corporate culture acts like a motivator, it becomes a conspicuous energizing factor. Though if it acts like a de-motivator, the results of such a cultural influence is conscious or unconscious decrease of energy used for achieving the aim. This, of course, causes wasting the most precious source that a firm has – employees, their talents and creativity.

From stated upwards is obvious, that if organizations are trying to have a long-term success on a present turbulent market, the key presumption of fulfilling this aim is creating an optimal culture of the organization.

3. CONNECTION OF THE CORPORATE CULTURE AND THE ORGANIZATIONAL LEARNING PROCESS

When creating an optimal corporate culture in practical conditions of Slovak enterprises we often find a relevant problem. Many organizations approach creating their culture as a campaign (here needs to be said that in many cases it is more or less outward declared effort for change without any specific practical presentation), it means that although they intensively develop, their effort is short-term, evolving only in a particular way; this effort is temporary and in certain point it ceases. (Horňák, 2007) Building up an optimal (so in given conditions highly effective) culture has to be a long-term work, which influences the orientation of values and the development of firm for a longer time horizon.

More successful enterprises differ from the less successful ones in the investments to creating their culture according to the way of firm's development, but mostly they do it directly and programmatically. They do it with conscious of the fact that nowadays the internal settings and adjustments of people are very important and they need to be accomplished. The rate of investing the time into the training of people is from the means of creating the corporate culture in such enterprises a controlled component.

From the viewpoint of fulfilling the requirements of long-term firms' success on the market is a key "factor" the way, in which employees react in certain working conditions. While creating the corporate culture is therefore a priority to focus on the effort to influence the major kind of behavior, attitudes, relationships and values of employees, which cause that in a particular situation they always react in the same (required) way.

From the stated upwards we see that the corporate culture is mostly about the kind of way. The success and the ability of competition of enterprises is certainly the result of certain way in which they function. Sure, it is unable to avoid the entrance investments, operating costs and technological know-how, but all of this is just an "injection" on the beginning; while for the maintaining of a smooth run there is needed a particular way of functioning. If a businessman places the first and forgets to place the second, his investment was wrong. The effective functioning of an organization depends on people, not on the investments or technologies!

If we want to think about the providing of required way of firm's functioning, so about the creating (moving in the wished direction) its optimal culture, it's necessary to pay special attention to the basic building stone of this procedure – to the teaching the employees. But attention! It's needed to realize that it has to be a whole systematic and primarily the continuous process. The interventions into the corporate culture cannot be reduced on a series of trainings, because the "problem" usually isn't in employees (or not only in them), but mainly in the environment in which they work. And this environment isn't going to be changed by one seminar.

If a firm wants to invest into the development of employees and through this affect the building up its culture, and while doing this it doesn't want to waste its money, it can't use the way of "isolated" trainings and courses. These partial steps are not only misleading us away from our aim, but also can cause the contrary of what we want to achieve – they can frustrate the employees. By showed isolated (although thoughtfully planned and well-organized) courses it's possible to replace the deficit of employees' and whole firm's knowledge, but it's not possible to replace the area, which is distinctive from the viewpoint of creating the optimal corporate culture. And therefore individual trainings of particular area do not lead to the aim. If we want to work on a creating the optimal corporate culture, the only possible answer is an integrated complex training program "perfectly suitable".

Successful enterprises are conscious of this requirement, and therefore they pay a great attention to realization of the continual learning process, which is becoming a part of whole firm. The main characteristic of such enterprises is an effort for maximization the opportunities for education and usage of the obtained knowledge, while achieving the firm's aims. It means that this process of continuous improvement is being taken seriously on all directed levels and it actually happens to be a part of the corporate culture.

4. TRAINING FORM OF THE ORGANIZATIONAL LEARNING PROCESS

One of the progressive and nowadays the most effective approaches to the organizational continual learning process and to the development of human sources is a realization of the self-contained complex of training programs. The essence of training programs comes out of an educational philosophy focused not only on the achievement of new needed knowledge, but also oriented on achieving and improving the skills and ability of individuals as well as of whole working groups, and on forming the attitudes – this is exactly what we are "interested in" when creating the corporate culture. (Holková, 2006)

These tough goals are achievable through the stated form of education, because during the experimental trainings there are being used modern technologies of a participative learning. The techniques of participative learning are based on the social teachings through personal experiences, where attendants of such a learning program get the knowledge and skills based on lived experience in learning procedures. They analyze, evaluate, make conclusions out of given experience, as well as they plan other steps how to apply this experiences of learnt knowledge to the particular organizational conditions. (Čambál, 2001)

This is the only way how complex training programs enable realization of highly specialized "perfectly suitable" educational activities. Opposite to this, the firms get only "the standard product" – this is the last thing that can currently help to organizations in a competitive environment. If the enterprises want to have a long-term success and ability to compete, they have to respect the following trend – customer oriented organizations do not have to primarily outdo the competition, but above all they have to try to differ from it. The main position here has the optimal corporate culture, because in the market environment, where all the products and all the technologies are similar, the competition contest moves into the area of human sources. Effectiveness and success as a result of aimed influencing the attitudes of employees towards the organization and their work became a requirement for existence on the market.

5. CONCLUSION

Today, the firms have to be firstly "able to act", what means for them to be willing and able to react quickly, being in a closer relationship with a customer, supporting the internal business activity by offering the operative autonomy and continuously to innovate it. The key factor of accepting all given rules of organizational functioning is creating such corporate culture that will enable the subject not only to adapt to given conditions but also to use them as a competition asset.

The presumption of creating this kind of corporate culture is realization of a continual organization learning process in the form of complex training programs that represent intensive learning in a situation of aimed support and stimulation. This form of organization learning evokes in employees the need of further development by using the gained knowledge; the need of self-education and self-improvement.

Therefore our future research will be focused on creating process of corporate culture optimalization with application of complex training programs.

6. REFERENCES

Čambál, M. (2007). *Continuous Education of Employees as Necessary Condition in Forming Optimal Corporate Culture.* AlumniPress, ISBN 978-80-8096-013-1, Trnava

Čambál, M. (2001). Company Education – A Supposition of Building Optimum Company Culture. *Research Papers Faculty of Materials Science and Technology SUT in Trnava.* No.11, pp. 23 - 27, ISBN 80-227-1648-0

Holková, A. (2006) Managers Possibilities How to Involve Employees into Process of Effective Change Manage. *Forum of manager.* No.4, pp. 11 - 13, ISSN 1336-7773, Bratislava

Horňák, F. (2007). *Development of Creative Potencial of Managers as a Support for Innovation Processes in an Enterprise.* AlumniPress, ISBN 978-80-8096-018-6, Trnava

Vaňová, J. (2005). The Corporate Culture and the Relationship with other Incorporeal Phenomena of Enterprise. *Research Papers Faculty of Materials Science and Technology SUT in Trnava.* No.18, pp. 165 - 170, ISSN 1336-1589

Annals of DAAAM for 2008 & Proceedings of the 19th International DAAAM Symposium, ISSN 1726-9679
Editor B. Katalinic, Published by DAAAM International, Vienna, Austria 2008
Make Harmony Between Technology and Nature, and Your Mind will Fly Free as a Bird

CORPORATE CULTURE APPLICATION IN ENTERPRISES AS ASSUMPTION THEIR LONG-TERM PROSPERITY

CAMBAL, M[ilos] & HOGHOVA, K[atarina]

Abstract: *Corporate culture is the determining assumption to long-term organization prosperity. This article descript research results, that was oriented on the real corporate culture optimalization in the Slovak enterprises. The purposes of the survey was to see, whether companies operating in Slovakia consider corporate culture an important factor their ability to be competitive, what the critical impulses that would lead the company leaders into analyzing the culture their companies and what tools and techniques used to analyze their corporate culture*

Key words: *organization, culture, research, analysis, application, prosperity.*

1. INTRODUCTION

The corporate culture is a phenomenon, which is very complex, hard to define and hard to explain, but it has great influence on the long-term company success. On one side it may be the source of company power and the advantage in competition, on the other hand it may also be the source of slowing down the progress of the company, as well as it may be the source of destruction. (Jančíková, 2006)

The corporate culture is a complex of basic predictions, values, behavior standards and approaches, which are shared by the members of the company. It has a significant part in ensuring the competitive ability of the company. These days, when competition is growing at individual markets, and when the competitive environment is much more turbulent and less predictable, the role of corporate culture is being much more important. This phenomenon is referred to by many writers as a significant factor, which influences the management of company's high performance, and therefore the managers should pay close attention to the creation and application of corporate culture. (Lukášová, 2004)

It was a reason what for we (Institute of Industrial Engineering, Management and Quality Faculty of Material Science and Technology) realized research that was focused on the actual corporate culture application in companies operating in Slovakia.

2. THE SURVEY REALIZATION

The survey has taken place simultaneously in Slovakia as well as abroad from June 2007 to November 2007. Each company could fill out only one questionnaire. The survey focused on companies mainly in the automobile industry, chemistry, engineering, foodstuffs industry and building industry.

The questionnaire consisted of eight questions, of which many focused on the analysis of corporate culture, as this complex analysis is the key phase in creating a desirable corporate culture.

In the beginning, the survey provided the following aims: find out, whether and in what context do the leading employees in Slovak and foreign companies know the term corporate culture; find out the approach of leading employees to the creation and implementation of corporate culture; find out, whether and how often do Slovak and foreign firms conduct an analysis of corporate culture; identify the most common reasons and impulses of conducting the analysis of corporate culture; identify the methods and techniques used to analyze corporate culture.

2.1 Survey results

The purpose of the first question "How do you define corporate culture, in what context do you perceive the term corporate culture?" was to find out, whether the employees were familiar with this term, and whether their ideas comply with the actual definition of the term. The survey pointed out, that all of the answers outlined the general definitions of corporate culture. This draws a conclusion, that corporate culture is a well-known term in Slovak and foreign companies.

The second question „Do you consider it important to create and implement suitable corporate culture?" was a follow-up to the first, its purpose was to find out, whether leading employees are aware of the practical need of corporate culture, whether they determine it to be an important factor or not. The questioned would provide a positive answer to this question, and confirmed the importance of corporate culture.

When asked "Why?", the most common reasons would be: the corporate culture is a uniting point in the company, it motivates the people to work better, it has a significant influence on the overall performance of the company as well as the quality of its services. It helps in communication, applies particular rules and many facts are more predictable. It helps to achieve common goals, it improves competitiveness of the company, as it uniquely identifies it and sets it apart from others. These answers show, that all survey participants consider corporate culture to be an important and essential element on the road to success.

The purpose of third question "Do you conduct analysis of corporate culture in your company?" was to see the actual relation of companies to corporate culture, as the leading employees of all questioned companies claimed, that they consider the corporate culture to be an important element. This question divided firms into two categories – depending on whether the companies conduct analysis of corporate culture or not. 26% of questioned companies claimed to be conducting such analysis which means that 74% of questioned companies claimed that they do not conduct the analysis of corporate culture.

How the main reasons was mentioned: „we solve problems which are recognizable even without corporate culture" (38%), "such analysis of corporate culture would be very expensive for our company" (20%). Despite the fact that all questioned leading employees consider corporate culture to be important, 75% of them do not conduct the analysis of corporate culture in the firm that they run or own. When it comes to companies, which have claimed to solve problems identified without analysis, it is important to know, that the probability of identifying a problem without an analysis is very low. It is very common, that this way the leader will only identify a secondary aspect of the problem, but will not solve the whole problem, which may pop up later more intensively and unpredictably, as everyone assumes that it has already been solved. This process does not lead to solving the problem, but only to make notice of it. If we want to persuade companies to pay more attention to corporate culture, we have to know their needs and views about the most suitable process of corporate culture analysis. (Čambál, 2007). Because of this we will still refer to the companies in two ways – to those, which do conduct analysis on corporate culture and to those which do not conduct such analysis.

The fourth question then followed – "When do you conduct the analysis of corporate culture?". The first option was "operatively (in case of need)", and this option was preferred by 44% of those companies that really did conduct the analysis. The remaining 56% ticked the option "periodically" – and chose the time interval either twice a year, annually or once in every two years. Two thirds of the companies which really conduct the analysis of corporate culture, incline to conducting in periodically, which brings them a continuous view to the life of the company, it allows them to identify even the first indications of adverse influences to the company's progress. In operatively conducted analyses, there are generally two impulses. The first is a feedback reaction to the noticed problem, and the second impulse may be a radical change that the company has gone through. It is recommended to conduct the analysis on a somewhat regular basis. Only in cases when the management is preparing the company for a considerable internal or external change which influences the company, then it is useful to conduct the analysis operatively.

The purpose of the fifth question „What was the impulse for your company to conduct an analysis?" is to identify the reasons, to which the companies react most often by an analysis. 44% of the companies would choose two most common reasons – either the inconvenient or insufficient communication between particular company units, or the second most common reason, which was the moment of going from one development phase to another. 33% of the companies would consider making an analysis at the time when their company would change in size. 22% of companies would conduct analysis because of problematic interpersonal relationships in the company, or because of incompatibility of present corporate culture with the strategic plans. When it comes to the group of companies that do not conduct analysis of corporate culture, 57% would probably do so if they had experienced high staff turnover. This reason was not very significant to the companies who did analysis of corporate culture regularly or periodically.

The same situation was with the second impulse that showed different views of companies that have experience with analysis of corporate culture compared to those, who do not. The ones who do not conduct analysis, would probably do so if they had knowledge of wasting time in their company (46%), although the companies that do conduct analysis did not state the „waste of time" in any matter. The third most influencing impulse (34% of companies in the second group) would be the decrease of working productivity – this impulse was not marked by any of the companies in the first group.

Results show, that the impulses to conduct an analysis were inverse when it comes to comparing the companies in both groups. While companies that have experience with analyzing corporate culture, had their impulses in company size change, insufficient communication and the change from one development phase to another, the companies that did not analyze culture had their impulses focused on the in-work productivity of employees or staff turnover issues. This makes us assume that companies which do conduct corporate culture analysis have either solved these types of problems, or solve them continuously during the analysis, and so they do not see them as high priority impulses.

The seventh question „How did you manage to conduct the analysis of corporate culture?" focused more on the group of companies that do not conduct the analysis, and the expected answer was to be hypothetic. The purpose was to find out, which process they would choose. Even in this matter, the two groups of companies did not choose common answers. 55% of companies in the first group use their own manual, and process the results through their own employees. The remaining 45% outsource this matter to specialized companies. In the second group 61% of companies would outsource the conduct of the analysis and 31% would want to work with their own manual. The results show, that almost 70% of those companies, which do not analyze their corporate culture, would outsource this task to an external company.

With the last question „What tools where used in the analysis of corporate culture?" we wanted to find out, which techniques are most suitable for the companies. The most preferred tool to analyze corporate culture was the questionnaire with open-ended or close-ended questions (77%). The next order was: questionnaire with specific statements, group interview, structured interview, observation techniques and analysis of internal documents. (Hoghová, 2008)

3. CONCLUSION

The conducted survey shows, that there is a positive change in feeling the importance of corporate culture over the past. Today, even companies in Slovakia are aware that corporate culture is a very significant factor that influences the long-term prosperity of the company. However, only 26% of the questioned Slovak firms are active in implementing the analysis of corporate culture. In companies questioned abroad, the implementation of corporate culture is a standard. Even though the Slovak firms have very good theoretical knowledge of corporate culture, the implementation is insufficient. (Horňák, 2007) Therefore these companies should concentrate on applying the suitable corporate culture. In order to achieve this, the firms should conduct a complex analysis of corporate culture as a priority, and the information gained should be systematically implemented. An appropriate corporate culture is one of the main aspects that can positively influence the company's long-term prosperity. Therefore it is necessary to make an appeal to the companies operating in Slovakia, so that they practically start applying corporate culture, because appropriate corporate culture is considered to be a great advance in competition.

Therefore our future research will be focused on identification appropriate corporate culture optimalization process in conditions of Slovak organizations and definition settings on realization of analyze in this process.

4. REFERENCES

Čambál, M. (2007). *The Entire Company Continuous Staff Training as a Key Supposition for the Optimum Corporate Culture Development.* AlumniPress, ISBN 978-80-8096-013-1, Trnava

Hoghová, K. (2008) Slovak Companies are Familiar with Theory of Corporate Culture, Foreign Firms Apply It. In: *Proceedings of Internal Doctoral Seminar.* Vol. III, pp. 117 - 124, ISBN 978-80-8096-058-2, Trnava

Horňák, F. (2007). *Development of Creative Potencial of Managers as a Support for Innovation Processes in an Enterprise.* AlumniPress, ISBN 978-80-8096-018-6, Trnava

Jančíková, A. (2006). Organizational culture and performance: An overview of empirical results. In: *Proceedings of The development trends of organization II.* Vol. I, Masaryk university, Brno.

Lukášová, R.; Nový, I. (2004). *Corporate culture: from Common Values to Better Performance of Organization.* Grada, ISBN 8024706482, Praha

Annals of DAAAM for 2008 & Proceedings of the 19th International DAAAM Symposium, ISSN 1726-9679
Editor B. Katalinic, Published by DAAAM International, Vienna, Austria 2008
Make Harmony Between Technology and Nature, and Your Mind will Fly Free as a Bird

PITTING MODELING OF A SPUR GEAR IN MESH

CANANAU, S[orin] & STOICA, G[ina]

Abstract: *To improve the current generation of diagnosis techniques, the paper presents a new model for the initiation and growth of the pits and cracks in the contact area of spur gears. The main criterion to split the process into initiation phase and subsequent growth of cracks is the number of cycles. The process of crack initiation is based on the theory of initiation along inclusions (grains) interfaces. The second phase, the short crack growth is studied using finite element method for simulation of the crack propagation. The comparison with numerical results is discussed.*
Key words: *gear, pitting, contact, FEM*

1. INTRODUCTION

The principal mode of failure of gears systems is the surface pitting of gear teeth flanks. Pitting occurs within small cracks are initiated on the tooth surface or just below the surface (Johnson, 1989). The main problem which concern also this paper is to give the correct prediction of possible gear failure and is bounded by several requirements as high loads, high speed, minimal weight etc. The main cause of the initiation consists in metal-to-metal contact of asperities or defects due to low lubricant film thickness. The cracks may start at inclusion in the gear flank (or foreign particle contamination of lubricant), which act as stress concentrators and propagate due to normal and tangential loads (Glodež et al., 1998) This process is dependent of material and operating conditions. In any case the complete fatigue process of micro-pit formation consists of several stages: the crack initiation, the crack growth, the destruction of the surface layer at high rate and the destruction of the surface layer at low rate. In the paper we are going to study these stages using the tools of FEM analysis.

2. PHISICAL MODEL OF PITTING DESTRUCTION

Consequent with the physical model the process and the stages presented in introduction, the calculus model will be divided into two parts: the first model is associated with fatigue crack initiation under load and the second model is associated with fatigue crack development under contact loading.

The main criterion for separation of these two models is present also in classical standardized procedures DIN 3990 and AGMA 218.01for prediction of service life in gears operation and consists in defining two numbers of stress cycles (Glodež et al., 1998): N_i (the number required for the appearance of the initial crack in the material), N_j (the number required for growth of the crack in the material). The theoretical model presented in figure 2 represents the model with an inclusion (Glodež et al., 1998). The contact projection of a spherical inclusion is characterized by the diameter d_{in} and the shear modulus G_{in}. This inclusion is residing in a slip zone, an ellipse with semi-major axis L and semi-minor axis l.

The shear modulus of the material, G, is the shear modulus of the gear material (Glodež et al., 1999).

The model of crack initiation due to a grain presence assumes that the dislocation stress. This model assumes that

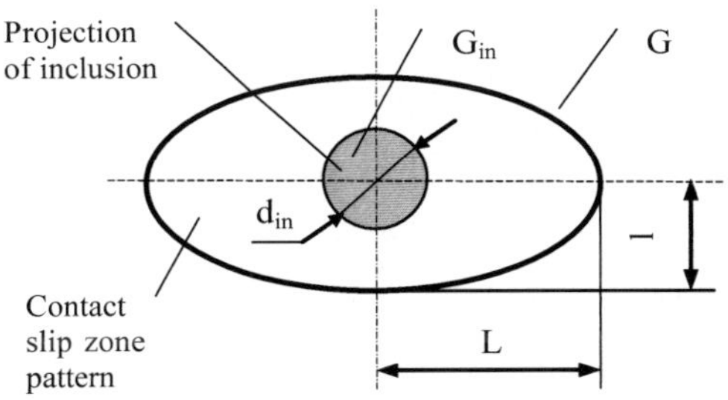

Fig. 1. Theoretical model for inclusion in the surface slip zone

the dislocation stress (consequent tangential load) τ_d is equal to (Tanaka & Mura, 1982):

$$\tau_d = \gamma_1 \frac{G \cdot G_{in}}{G + G_{in}} \tag{1}$$

where γ_1 is the plastic strain in the grain due to one cycle loading with applied shearing stress τ_{max}:

$$\gamma_1 = \frac{(l+L)}{l} \cdot \frac{\tau_{max} - \tau_f}{G} \tag{2}$$

and the meaning of τ_f is the tangential stress due to frictional forces in the meshing process.

We consider the process of grains dislocation is irreversible. Connected with this hypothesis we can say that the number of cycles N_i will be a function of plastic strain amplitude (Sun et al., 1991). The constant k will take into account the irreversibility factor and this factor takes values:

$$k = 10^{-1}..10^{-4} \tag{3}$$

In this case, the number of cycles N_i is expressed in equation (4):

$$N_i = k \frac{8(G + G_{in})}{d_{in} \cdot G} \cdot \frac{\tau_{max} - \tau_f}{(\Delta\tau - 2\tau_f)^2} \tag{4}$$

Concerning the crack development we can assume that the cracks are defined as being short because their length are small and are associated with the process of surface pitting (Wang, 1998)

The theory of short crack growth is used to model the crack development associated with pitting phenomenon. We can use the theory of short crack growth only considering the characteristic dimension of material microstructure (the grain dimension) of the same seize with the length of the crack development.

But the dimension of the diameter of the hole is from units to thousands of grain characteristic dimension. That means the cracks extend from a dimension of a small grain to a much larger scale. The model of crack growth must taken into account the transfer of the phenomenon from the grain of the initial crack to tens (or thousands) neighboring grains. This implies the discontinuity of this process. We have also to consider the decrease of crack growth rate as the crack approaches, grain by grain, to a grain-boundary. We can express a relation between stress intensity range ΔK and plastic displacements Δu, using the geometric characteristics as are shown in fig.2:

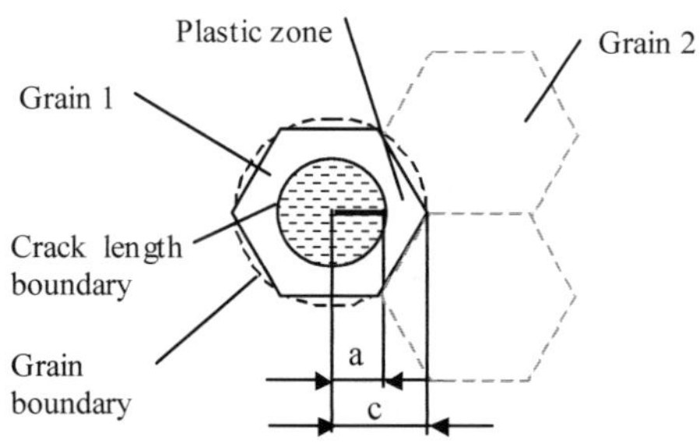

Fig. 2. Short crack development

3. FEM ANALISYS

Numerical analysis has the aim to establish the relationship ΔK=f(a) utilizing a virtual crack extension method. In the first step we will study the possibility of the crack initiation as consequence of normal and tangential load applied.

The finite element analysis was conducted to asses with large approximation grains-polycrystalline distribution at the surface, considering a contact zone. The analysis is a 2D-plain strain analysis, with 4-nodes quadrilateral elements. The number of total elements is 2176. The present model taken into account the load applied corresponding to single pair teeth engagement and the maximum value for the load applied in inner point of this load distribution (Cananau, 2005).

In accordance with our hypotheses we consider in the same time a frictional load at the contact surface as tangent vectors of profile geometry. The values of these forces are determinate using Amontons-Coulomb friction laws and a coefficient of friction μ=0.06 – an average value considered for lubricated gears. We suppose that the pitting phenomenon is the cause of the initiation of crack. In the FEM analysis we are going to simulate the presence of a short crack and we are going to investigate the effect of the crack in the gear function.

The diameter d_{in} of the inclusion (grain) we assume to be as the same size as the finite element size in the meshing contact area. For the geometric model presented and for the chosen scale, we can assume the average value d_{in} = 100μm.

In the second step of this analysis we develop on the base of previous results, the study of crack propagation. The local meshing around the initial crack is sown in Fig.3. For this step we used special fracture finite elements. With this type of element, we can calculate the stress intensity factor.

	Pinion	Gear
Module mm	5	
Standard pressure angle	20^0	
Number of teeth	15	21
Addendum coefficient modification	+0.571	0
Tip circle diameter mm	93.210	117.5
Center distance mm	90	
Face width mm	18	16
Contact ratio	1.246	

Table 1. Geometrical characteristics of spur gear model

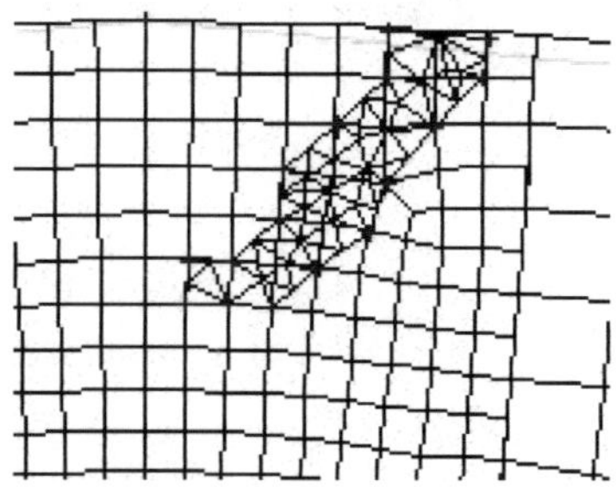

Fig. 3. Crack propagation study

In the next figure (Fig.4) is shown the torsional mesh stiffness result for a short crack presence (about 2 mm) at the root of the tooth of the driven gear.

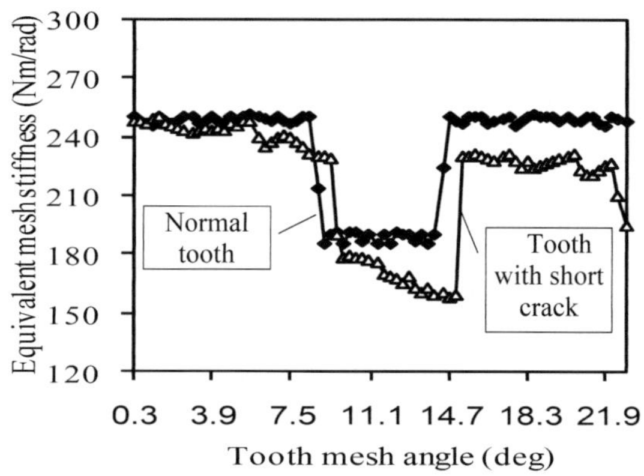

Fig. 4. Mesh stiffness result

4. DISCUSSION AND CONCLUSION

The paper presents a new model for the initiation and growth of the pits and cracks in the contact area of spur gears The FEA model showed that the tooth crack affects the stiffness of the tooth over the phase of engagement of the tooth. The stress intensity factor was investigated, as the crack propagates through material to the surface.

5. REFERENCES

Cananau, S., (2005) Pitting model in the contact area of spur gears, *Tribology, The Annals of University "Dunarea de Jos" of Galati,* Fasc.VIII, p. 89-92, ISSN 1221-4590

Glodež, S., Ren Z., Flăsker, J., (1998) Simulation of surface pitting due to contact loading, *Int. J. Num. Meth. In Eng.,* Vol. 43-23, p.33-50.

Glodež, S., Ren Z., Flăsker, J., (1999), Surface fatigue of gear teeth flanks, *J. Comp. and Struct*., Vol. 73, p.475-483.

Johnson, K.L. (1989) The strenght of surfaces in rolling contact, *J. Mech. Eng. Science,* Vol. 203, p. 151-163.

Sun, Z., Rios, E.R., Miller, K.J., (1991), Short and long fatigue cracks interacting with grain boundaries, *Fatigue and Eng. Mat.,* Vol. 14, p.277-291.

Tanaka, K., Mura, T., (1982), A theory of fatigue crack initiation at inclusion, *Met. Trans.,* Vol. 13A-23, p.117-123.

Wang, C.H., (1998), Effects of Stress Ratio on Short Fatigue Crack Growth, *ASME J. Eng. Mater. Techn*., Vol. 118, p.362-366

Annals of DAAAM for 2008 & Proceedings of the 19th International DAAAM Symposium, ISSN 1726-9679
Editor B. Katalinic, Published by DAAAM International, Vienna, Austria 2008
Make Harmony Between Technology and Nature, and Your Mind will Fly Free as a Bird

DEPENDENCE OF THE NUMBER OF DEVELOPERS AND THE NUMBER OF ERRORS IN TEAM SOFTWARE DEVELOPMENT

CANDRLIC, S[anja]; PAVLIC, M[ile] & PAVLIC, D[aniel]

Abstract: *This paper presents the analysis results of the influence of the number of developers on the total number of conflicts during software version merging. Conflicts in code merging can arise when more developers simultaneously and independently make changes of the same part of the programming code, and the goal is to make the number of conflicts as small as possible. The authors' starting hypothesis is that a larger number of developers in the development team will cause more conflicts. However, the research results show an opposite tendency.*
Key words: *version, source code, team, software, conflict.*

1. INTRODUCTION

The development of complex business applications does not rely on a sole developer, but on the whole development team. Having a larger number of developers concurrently developing or maintaining the same software can lead to errors (conflicts) when implementing a part of source code into the main software version (Thomas & Hunt 2004). In order to avoid code losses during parallel development (Bays 1999) and software maintenance, development companies usually choose a version control tool which supports team software development. Certain development tools have inbuilt version control mechanisms, and other tools can be found as independent applications (Candrlic et al., 2003). The aim of this paper is to present the development flow data of two chosen applications and check whether it is possible to find a connection between the number of developers in a team and the number of conflicts which arose during version merging.

This research used the version merging data of certain development teams in a software company. The observed development teams had 1 to 13 developers which use the following tools for software development: DB – Oracle and CASE – Clarion. Since the company could not find a version control tool on the market which would satisfy its needs, it approached the development of its own tool for version control, team development and software maintenance. The tool was named Builder (Candrlic et al., 2006) and it is being used since 2002. It works on the basis of optimistic locking or copy-modify-merge model (Bar & Fogel 2003). This research covered data from 2003 to 2007, and analyzed the development of two applications: AO (car responsibility) and IMIS (insurance sales).

2. DATA ANALYSIS

Table 1 shows data which are related to the AO application development. Column Number of developers shows how many developers were engaged in the application development. The number of conflicts which arose during version merging is shown for each month in the column Number of conflicts, while the column Number of merging shows the total number of merging in the observed period. The last column Share of conflicts shows the percentage of the number of conflicts in the total number of version merging.

AO					
Months		Nr of developers	Nr of conflicts	Nr of merging	Share of conflicts
2003	7	1	0	0	0
	8	1	0	0	0
	9	1	0	0	0
	10	1	0	0	0
	11	1	1	206	0,49
	12	1	0	12	0
2004	1	1	0	5	0
	2	1	4	22	18,18
	3	1	0	49	0
	4	1	0	14	0
	5	1	0	20	0
	6	1	0	0	0
	7	1	1	5	20,00
	8	2	1	41	2,44
	9	1	1	4	25,00
	10	2	1	52	1,92
	11	2	0	32	0
	12	2	0	19	0
2005	1	3	0	83	0
	2	2	2	53	3,77
	3	1	0	0	0
	4	1	0	0	0
	5	1	0	0	0
	6	1	14	23	60,87
	7	1	0	1	0
	8	1	0	3	0
	9	1	0	0	0
	10	1	0	0	0
	11	2	1	70	1,43
	12	3	2	161	1,24
2006	1	3	3	211	1,42
	2	3	0	74	0
	3	2	0	39	0
	4	2	8	56	14,29
	5	2	5	57	8,77
	6	2	0	27	0
	7	1	0	9	0
	8	2	0	2	0
	9	2	0	11	0
	10	2	0	15	0
	11	2	1	34	2,94
	12	2	1	11	9,09
2007	1	2	1	21	4,76
	2	2	0	18	0
	3	2	0	14	0
	4	4	2	72	2,78
	5	3	1	11	9,09
	6	3	0	8	0

Table 1. Version merging in AO project

A data that stands out from the rest is the value of 60,87% of conflicts, May 2005. When the data for the last few months is looked into, it is visible that in March, April and May not even a single merging took place. This was a period of intense development without version merging, and all changes were integrated into the software in May. It is visible that the number of conflicts in the total number of merging is bigger if continuous work on a problem lasts longer and version merging is rare. This hypothesis should be researched on more projects.

Year	Month	Nr of developers	Nr of conflicts	Nr of merging	Share of conflicts
IMIS					
2003	7	12	15	651	2,30
	8	9	7	239	2,93
	9	10	4	695	0,58
	10	12	13	1215	1,07
	11	11	1	1405	0,07
	12	12	10	1457	0,69
2004	1	12	11	813	1,35
	2	12	3	820	0,37
	3	11	15	1016	1,48
	4	13	3	562	0,53
	5	12	3	424	0,71
	6	11	3	485	0,62
	7	11	6	582	1,03
	8	10	0	143	0,00
	9	10	6	632	0,95
	10	11	12	446	2,69
	11	10	0	495	0,00
	12	10	4	375	1,07
2005	1	11	19	438	4,34
	2	11	1	598	0,17
	3	7	8	183	4,37
	4	6	5	129	3,88
	5	6	23	299	7,69
	6	7	7	116	6,03
	7	6	5	107	4,67
	8	3	1	38	2,63
	9	7	2	122	1,64
	10	5	2	72	2,78
	11	5	10	200	5,00
	12	6	3	197	1,52
2006	1	8	457	1254	36,44
	2	8	3	380	0,79
	3	8	1	119	0,84
	4	6	0	161	0
	5	6	0	207	0
	6	6	2	91	2,20
	7	7	6	193	3,11
	8	5	0	75	0
	9	5	1	144	0,69
	10	8	5	240	2,08
	11	7	0	212	0
	12	4	5	148	3,38
2007	1	6	0	183	0
	2	8	1	497	0,20
	3	7	5	324	1,54
	4	6	5	202	2,48
	5	8	0	284	0
	6	7	4	150	2,67

Table 2. Version merging in IMIS project

Table 2 shows data which is related to project IMIS. One data item is extreme – January 2006 when 8 developers working on the IMIS project created 457 conflicts in the total number of 1254 version merging. It is also presumed that this data value was caused by a longer development process.

Data from the tables are graphically shown in Figure 1. The x-axis shows the number of developers in a team, and the y-axis shows the share of conflicts. The decrease of the upper border of the number of conflicts with a larger number of developers can be observed. One can draw a conclusion that the number of conflicts is mostly less than 10%.

3. CONCLUSION

This paper presented the results of a research based on the analysis of the influence of the number of developers on the total number of conflicts which emerge during their work on same programming modules. The share of conflicts is mostly between 1% and 5%. It seems unusual that conflicts exist even in one-man teams, but they arise if the person is working on more than one software version simultaneously.

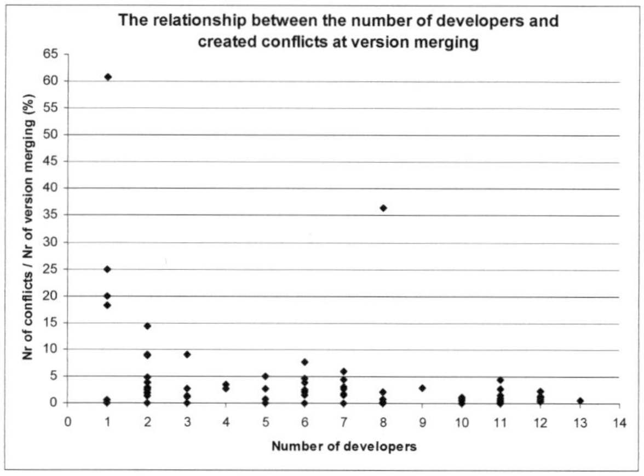

Fig. 1. Illustration of the relationship between the number of developers and created conflicts at version merging

Authors' starting hypothesis was that a larger number of developers in a team cause a larger share of conflicts. Research on actual projects has shown that a maximum percentage of conflicting merging in the total number of merging is on the decrease when the number of developers is getting larger.

A question emerges: why is this so? The authors cannot answer this question in this paper, but further research which would analyze this subject in detail is suggested.

Further research plans to increase the number of observed projects and define the difference between software in development and software in maintenance.

Also, the difference between modules for approaching tables in database and procedures will be defined. Besides that, it would be good to differentiate actions which caused a conflict: changing, creating new or deleting program code.

Further research can also encompass other parameters which can influence conflicts at software version merging. Some of these are: the size of the finished software (the number of programming modules, number of procedures, number of code lines), the size of the project (number of entity types, number of attributes, number of relationships, number of keys), software development time-frame, complexity of documents being modelled, development organization, etc. An expanded research which aims at answering these questions is in process.

Similar research of software development, conducted on teams which use different development tools, or which have no tools for team software development at all, could give different results.

4. REFERENCES

Bar, M & Fogel, K. (2003). *Open Source Development with CVS,* Paraglyph Press, ISBN 1-932111-81-6, Scottsdale Arizona

Bays, M. E. (1999). *Software release methodology,* Prentice Hall, ISBN 0-13-636564-7, Upper Saddle River

Candrlic, S.; Golem M.; Pavlic, M.; Zamlic, I. & Kaic, A. (2003). Network development of Applications, *Proceedings of the 14th International DAAAM Symposium*, Katalinic, B. (Ed.), pp. 079 – 080, ISBN 3-901509-34-8, Vienna, Austria, October 2003, DAAAM International, Vienna

Candrlic S.; Pavlic M. & Poscic P. (2006) Model of a System for Team Software Development, *Proceedings of the 28th International Conference on Information Technology Interfaces*, Kalpic D. & Hljuz Dobric V. (Ed.), pp. 117-122, ISBN 953-7138-05-4, Cavtat, Croatia, June 2006;. SRCE University Computing Centre, Zagreb

Thomas, D. & Hunt, A. (2004) *Pragmatic Version Control with CVS,* The Pragmatic Bookshelf, ISBN 0-9745140-0-4, Dallas, Texas

Annals of DAAAM for 2008 & Proceedings of the 19th International DAAAM Symposium, ISSN 1726-9679
Editor B. Katalinic, Published by DAAAM International, Vienna, Austria 2008
Make Harmony Between Technology and Nature, and Your Mind will Fly Free as a Bird

DEPENDABILITY AND SECURITY OF THE INFORMATION SYSTEMS

CAPEK, J[an]

Abstract: *The contribution deals with problem of obtaining trustworthy data from information sources for next processing in numerous systems. The trustworthy data will be preferably stored not only in the dependable information systems but also in the secure information systems. For example decision making systems, warehouses, control systems, management of business organizations, etc.*
Key words: *dependability, reliability, confidentiality, risk.*

1. INTRODUCTION

The importance of the information systems dependability and security is increasing. In most organizations the role of the information systems are vital for operations. Therefore, unplanned interruptions are not welcome as they can impact production, sales and numerous other processes and procedures within the organization. The dependability and robustness of the information systems are critical issues in today's networked world (Anderson 2008; Sirkemaa & Ylonen, 2007; Fabian 2007; Capek 2001).

2. SYSTEM - DEPENDABILITY AND RELIABILITY

A system is a set of members (systems elements), which are bound by some relations between these elements. Each system is defined by its behaviour and structure. The inputs boundary elements of system are for example information sources. The safety system is as safe as is each element of this system. The technological leaps of the past decades in computers, electronics, optics, and advanced, high-performance, complex control systems have created the need for extra reliability and safety. An underlying feature of all "safety critical" systems entails a reliable fault-detection (and of then isolation and reconfiguration) system." For the reason, we focus our future steps into these boundary elements of system – information sources. First of all is need to notice that the term reliability according the IEEE (Wikipeida 2008) defines the *Reliability* as ". . . the ability of a system or component to perform its required functions under stated conditions for a specified period of time.", began to be become overloaded and was being used outside of it's originally intended definition, as a measurement of failures in a system, to encompass more diverse measures which would now come under other classifications such as safety, integrity, etc. With term reliability is connected well known standby coefficient K, according given formula (1).

$$K = \frac{MTBF}{MTBF - MTTR} \qquad (1)$$

Where: MTBF is Mean Time Between Failure and MTTR is Mean Time To Restart
Jean-Claude Laprie thus first used the term *Dependability* to encompass these related disciplines in the early 1980 (Wikipeida 2008). Avizienis et all (Avizienis et al., 2001, 2004) showed that the general, qualitative, definition of *dependability*

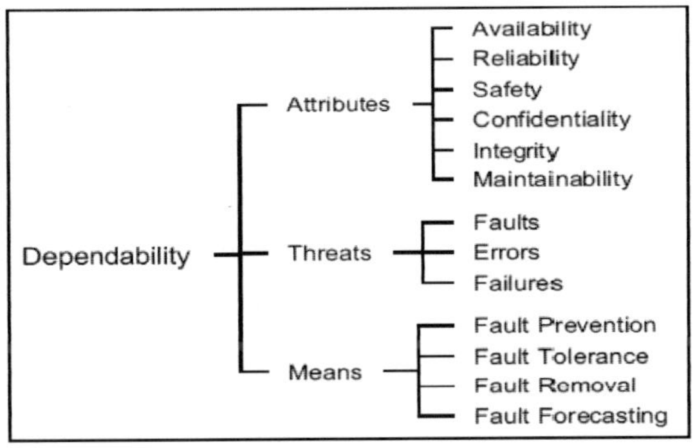

Fig.1 The dependability tree (Avizienis at al., 2001, 2004)

is: the ability to deliver service that can justifiably be trusted (Figure 1). This definition stresses the need for justification of trust. The alternate, quantitative, definition that provides the criterion for deciding if the service is dependable is: dependability of a system is the ability to avoid service failures that are more frequent and more severe than is acceptable to the user(s). As developed over the past three decades, dependability is an integrating concept that encompasses the following attributes:

Security is the concurrent existence of
a) availability for authorized users only,
b) confidentiality, and
c) integrity with 'improper' meaning 'unauthorized'.

The dependability specification of a system must include the requirements for the dependability attributes in terms of the acceptable frequency and severity of failures for the specified classes of faults and a given use environment.

Risk is a function of the ***likelihood*** of a given ***threat-source's*** exercising a particular potential ***vulnerability***, and the resulting ***impact*** of that adverse event on the organization (OISRMH 2006)
According Ryan (Ryan & Ryan,1995) we can previous opinion put into following formula with adding the Countermeasures:

$$\text{Risk} = \frac{\text{Threat} \times \text{Vulnerability} \times \text{Impact}}{\text{Countermeasures}} \qquad (2)$$

Threats are posed by organizations or individuals who both intend us harm and have the capability to accomplish their intentions. These types of threat and measures that may be taken to reduce or eliminate the risks they may take the form of enemy armed forces, spies, criminals, terrorists, psychotics, computer hackers, drug lords or saboteurs. Threats must be

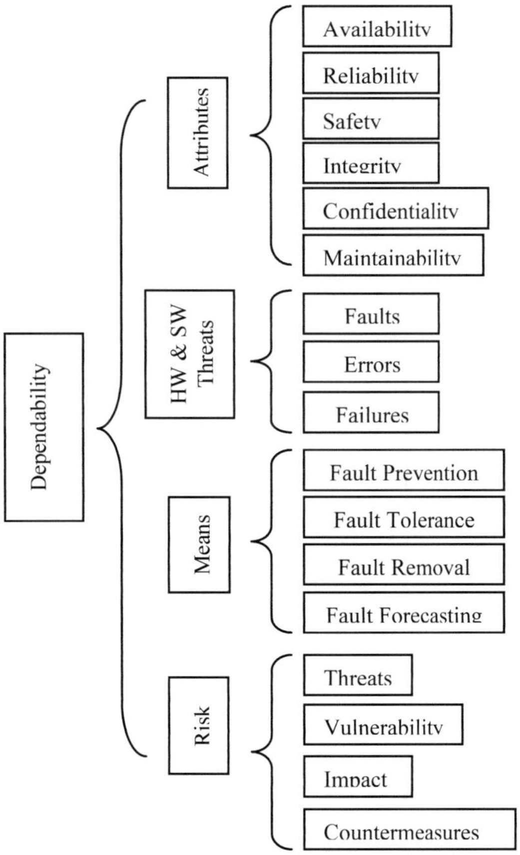

Fig.2 Modified dependability tree

coupled with threat-sources to become dangerous. So these types of threats are different then the threats described by Avizienis et all (Avizienis et al., 2001, 2004)

3. CONCLUSION

Recently, "the new economy" exposed the growing importance of non-tangible assets that are at the heart of business processes. Dependability of the information systems and and/or information sources with data certainly play a special role here.
The dependability of the information systems is more complex problem then Avizienis et all (Avizienis et all 2001, 2004) showed. It is need to include risk and security dimension into this scheme, because many damages of the information systems are due to hackers and intruders on the one side and own employee on the other side.
The modified dependability tree which risk include is on the figure 2.
It is often argued that Internet is the main source of risks as it is not controlled by anyone Blakley (Blakley,at al., 2001). Internet is the hideaway for hackers and intruders, who are ready to hack organizations connected to the global Internet. New ways to attack and damage are continuously being developed, and so there are many kinds of threads. Risks from the Internet can be divided into a) external attacks, b) intrusions and c) malicious software like viruses and worms. It is still important to notice that often security problems have their roots inside the company Anderson (Anderson 2001).
External attacks try to harm operations in some way. One of the most common types of external attacks is DoS (Denial of Service). Massive amount of requests consume system resources so that server crashes or users are unable to use the system because of the massive network traffic generated by the attacker Blakley (Blakley,at al., 2001), British Security Standard (BSS1999).

As it was mentioned above, taking into account the risk in the Avizienis' dependability tree, one can obtained more complex point of dependability view.
Future work will be focused to simulation more factors according figure 2 for modelling the dependability of the information system.

4. REFERENCES

Anderson R. (2008) : *Security Engineering.* John Willey 2008 ISBN: 978-0-470-06852-6, Indianopolis

Anderson, R., (2001) "Why Information Security is Hard- An Economic Perspective", 17th Annual Computer Security Applications Conference, Dec. 2001. New Orleans, Louisiana Dec. 10-14, 2001, ISBN: 0-7695-1405-7 *Available on*: http://*citeseer.ist.psu.edu/640465.html* *Accessed:* 2008-05-21

Avizienis at all (2001) Avizienis A, Randell B. Laprie J.C : Fundamental Concepts of Dependability. Technical report, LAAS - NewCastle University - UCLA, 2001. LAAS Report no. 01-145, NewCastle University Report no. CS-TR-739, UCLA CSD Report no. 010028, *Available on* www.cs.ncl.ac.uk/research/trs/papers/739.ps *Accessed:* 2008-05-21

Avizienis at all (2004) Avizienis,A,. Laprie J.-C,. Randell B, and Landwehr C,: Basic Concepts and Taxonomy of Dependable and Secure Computing, *IEEE Transactions on Dependable and Secure Computing,* vol. 1, pp. 11-33, 2004., ISSN 1545-5971

Blakley, B.(2002) , "The Measure of Information Security is Dollars", *Workshop of Economics and Information Security*, May 2002. Univ. of California, Berkeley, 16-17-May 2002, *Available on:* www.cl.cam.ac.uk/~rja14/econws/54.pdf, *Accessed:* 2008-05-21

Blakley,at all (2001) Blakley, B., McDermott, E., Geer, D., "Information Security is Information Risk Management", *Proceedings of the 2001 workshop on New security paradigms*, 2001, pp. 97-104, ISBN 1-58113-457-6, Cloudfort, New Mexico Sept. 10-13,

BSS(1999)British Security Standard, BS 7799, *British Standards*, 1999, *Available on*: www.riskserver.co.uk/bs7799 *Accessed:* 2008-05-21

Capek J (2001) Trustworthy Information from system of sensors., *The 12th INTERNATIONAL DAAAM SYMPOSIUM "Intelligent Manufacturing & Automation: Focus on Precision Engineering"* 24-27th October 2001, Vienna, 2001

Fabian. P., et al (2007): Preserving the Experts' Knowledge for the New Learners' Generation: A Web-Based Knowledge Map, In *Proceedings of 5th International Conference on Emerging e-Learning Technologies and Applications (ICETA 2007),* pp.317-322 Stara Lesna, 2007, ISBN 978-80-8086-061-5, Elfa Kosice, 2007

OISRMH (2006) *Open Information Security Risk Management Handbook Version 1.0,* September 2006, *Available on:* www.somap.org/handbook/somap_handbook.pdf *Accessed:* 2008-05-19

Ryan,D.J., Ryan, J.,C.,H.: Risk Management and Information Security. In.: *11th Computer Security Applications Conference*, New Orleans, Louisiana, USA.,December 19, 1995, *Available on:* www.julieryan.com/riskmgt.htm *Accessed:* 2008-05-19

Sirkemaa,S Ylanen,E.(2007): The Challenge of Information Systems Security *WSEAS TRANSACTIONS on COMPUTERS*
Volume 6, March 2007 , Issue 3, pp 605-612, ISSN 1109-2750

Wikipeida(2008) en.wikipedia.org/wiki/Dependability, *Accessed on:* 2008-05-21

Annals of DAAAM for 2008 & Proceedings of the 19th International DAAAM Symposium, ISSN 1726-9679
Editor B. Katalinic, Published by DAAAM International, Vienna, Austria 2008
Make Harmony Between Technology and Nature, and Your Mind will Fly Free as a Bird

THE INFLUENCE OF TECHNOLOGICAL FACTORS UPON THE ACCURACY AND QUALITY OF THE COLD PLASTIC DEFORMATION REVOLUTION SURFACES

CARAUSU, C[onstantin]; PRUTEANU, O[ctavian]; SARBU, I[onel]; NEDELCU, D[umitru] & PRICOPE, C[ornel]

¶
¶

Abstract: *This paper presents the results of experimental research carried out at the industrial unit SC Rulmenţi SA in the town of Bârlad. The research focused on the influence of the deformation force, the deformation feed and the tool's speed of rotation upon the roughness and ovality of the annular ball bearings'bed bearers. The values of working parameters were set out so as to obtain quality parameters within the admissible limits for the normal running of bearings.*
Key words: *bearing, plastic deformation, roughness, ovality.*

1. INTRODUCTION

The cold-working of the ball-bearing races'bed bearers, especially cyclical with balls and cyclical with cylindrical rollers, is actually an innovation for the bearing industry in Romania.

Experimental research has showed the way in which working parameters of special technological equipments have an influence upon the roughness and ovality, i.e. upon the quality parameters of surfaces processed by means of cold plastic deformation (Wagoner & Chenot, 2001), (Lupescu et all, 2004).

2. EXPERIMENTAL CONDITIONS

2.1 Material and half-finished

We have used for the experiment the steel 100Cr6, specially designed for ball-bearing races, which has a good cold-workability and hot-workability, as well as cutting workability. The blank was a ball-bearing race, which was hot-formed and processed by turning, boring and facing, until the dimensions of the blank corresponded to those established in the technical documentation of SC Rulmenţi SA Bârlad.

2.2 Working Equipment

For the cold-working of the outer races' bed bearers of the bearing 6207 we have used the Japanese equipment CRF-120 OR which allows the variation of the deformation force, P, of the deformation feed, A, and of the tool's rotation speed, V.

The working diagram is shown in fig. 1.

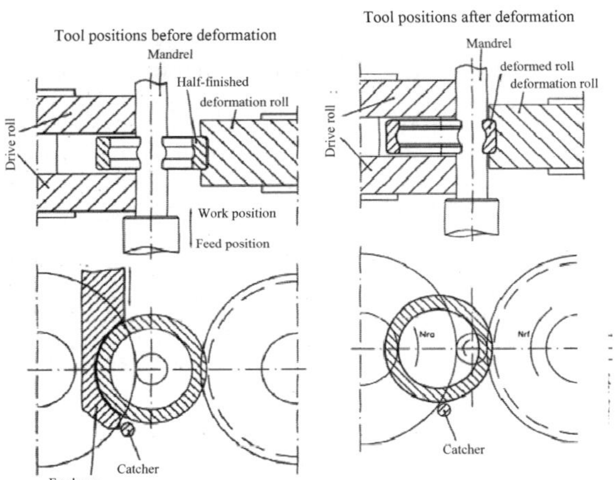

Fig. 1. Equipment CRF-120 OR - working diagram

2.3 Instrumentation

The roughness of surfaces was measured with the device Taylor-Hobson, and the shape deviations with the device Perthometer Marsurf CD 120.

3. EXPERIMENTAL RESULTS

3.1 The Influence of the deformation force, P

Table 1 shows the results of out tests. Each value represents the average of five measurements (Pruteanu et all, 2006).

We infer from the table that the prameters' minimal values have been obtained for the deformation force P= 6.5 Mpa. Fig. 2 shows the graphical representation of the variation of the bed bearers'roughness and ovality varying with the deformation force (Pruteanu et all, 2007).

3.2 The influence of the deformation feed, A

Table 2 shows the results of our tests. The parameter values represent the average of five measurements.

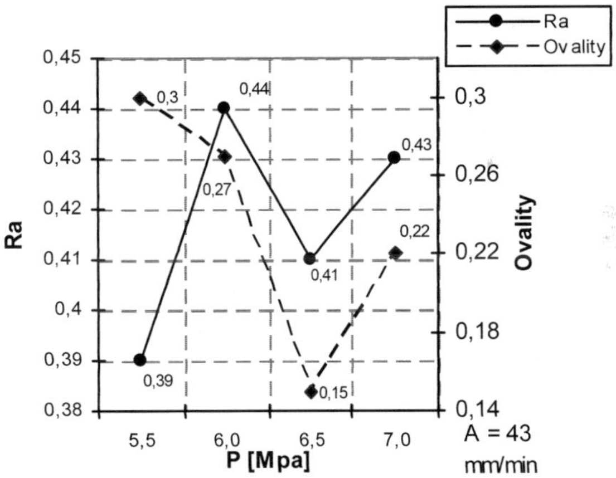

Fig. 2 The influence of the deformation force upon the quality parameters

P [MPa]	R_a [μm]	Ovality [mm]
5.5	0.39	0.30
6.0	0.44	0.27
6.5	0.41	0.15
7.0	0.43	0.22

Table 1. The Influence of the deformation force, P
A= 34 mm/ min, S= 95 rot/ min

A [mm/min]	R_a [µm]	Ovality [mm]
27.5	0.46	0.27
30	0.41	0.20
32	0.45	0.24
34	0.42	0.18

Table 2. The influence of the deformation feed
P= 6.5 Mpa, S= 95 rpm

We infer from the table that the minimal value for the shape deviation are obtained for A= 34 mm/min, while the minimal roughness will be obtained at an feed of 30 mm/min. Due to the fact that the roughness difference compared to that calculated for A= 34 mm/min is only of 0.01 µm, we have chosen this value for the deformation feed.

Fig. 3 shows the graphical representation of the variation of the bed bearers'roughness and ovality varying with the deformation feed.

We infer from the table that the minimal values for shape deviations will be obtained for the speed V= 95 rpm, while the minimal roughness of 0.41 µm will be obtained for V= 80 rpm. Due to the fact that the difference compared to that for V= 95 rpm is of only 0.03µm and that the value 0.44 µm is bellow the value stipulated in the technical documentation for a normal functioning of bearings, we have recommended for processing the value V= 95 rpm.

Fig. 4 shows the graphical representation of the bed bearers' roughness and ovality, varying with the tool's speed of rotation.

3.3 The influence of the tool's rotation speed, V

Table 3 shows the results of our tests. The parameters values represent the average of five measurements.

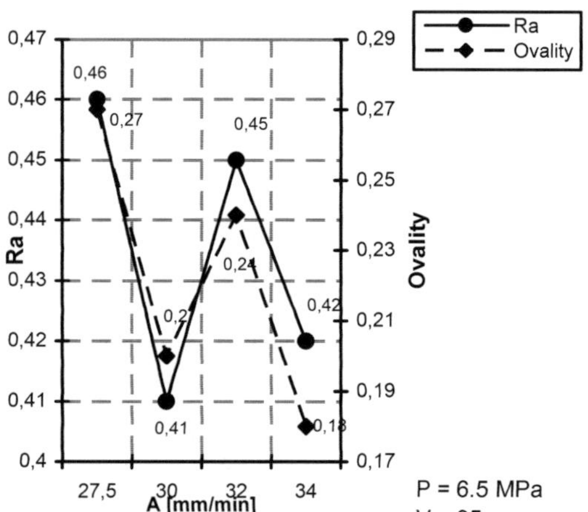

Fig. 3. The influence of the deformation advance upon the quality parameters

V [rpm]	R_a [µm]	Ovality [mm]
80	0,41	0,27
85	0,44	0,25
90	0,42	0,24
95	0,44	0,18

Table 3. The influence of the tool's rotation speed
P= 6,5 Mpa, A= 34 mm/ min

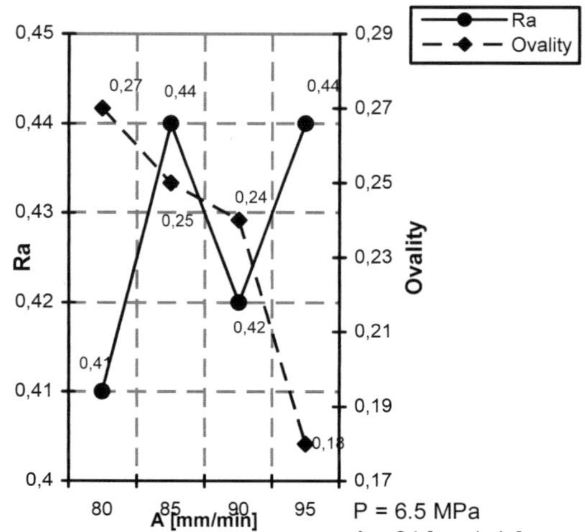

Fig. 4. The influence of the tool's rotation speed upon the quality parameters

4. CONCLUSION

From the parameters' variation range on the testing machine, we have determined the values for which we can obtain the best technical conditions for the ball-bearing races'bed bearers:

- The deformation force value P= 6,5 MPa ensures for all parameter values- roughness and ovality - the best values;
- The deformation feed value A= 34 mm/min ensures minimal values for shape deviation - ovality - as well as values acceptable within the framework of technical conditions imposed for the normal functioning of bearings to ensure the roughness of the bed bearers.
- The tool's rotation speed value V = 95 rpm ensures minimal values for the shape deviation and acceptable values for the bed bearers' roughness.

We will extend ours research to other bearings. Likewise, we will analyze the influence of the work parameters upon the circularity.

5. REFERENCES

Lupescu, O., Carausu, C., Lupescu Ov. (2004). *Structural modification of the OLl 37 steel to superficial plastic distorsion processing at cold through rolling*, Bul. Inst. Pol. Iaşi, Tomul L (LIV), Fasc V a, 213-216.

Pruteanu ,O.V., Carausu, C. Paraschiv, Dr., Sarbu, I., (2006). *The influence of the working parameters on the parameters of the superficial layer when procesing the steels through grinding in the rolling's technology*. Мировая зкономика и бизнес – администрирование, Minsk, Belarus, may, 186-197.

Pruteanu, O.V., Paraschiv, Dr., Pricope, C., Bunea, C.M., Sarbu, I., Carausu, C. (2007). *Considerations Regarding the 100Cr6 Material's Conduct through the cold Rolling of the Bearing Rings,* International Journal for Manufacturing Science & Technology, U.S. Library of Congress, vol. 9, no.1, p.51-57, ISSN 1524-1548.

Pruteanu, O.V., Paraschiv, Dr., Carausu, C., Sarbu, I., Bunea, M., Bardac, D. (2007). *Processing through plastic deformation of the bearing rings.* MicroCAD 2007, International Scientific Conference, Miskolc, 143-148.

Wagoner, R.H., Chenot, J.L. (2001). *Metal forming analysis*. Cambridge University Press. New York.

Annals of DAAAM for 2008 & Proceedings of the 19th International DAAAM Symposium, ISSN 1726-9679
Editor B. Katalinic, Published by DAAAM International, Vienna, Austria 2008
Make Harmony Between Technology and Nature, and Your Mind will Fly Free as a Bird

RESIDUAL TENSIONS IN COLD PLASTIC DEFORMED SURFACES

CARAUSU, C[onstantin]; SARBU, I[onel]; PRUTEANU, O[ctavian] ; GRAMESCU, T[raian]; NEDELCU, D[umitru] & HUSANU, V[alerica]

Abstract*: In the present paper there are presented the results of residual tensions' measurements in the surfaces of the rolling tracks of cold plastic deformed bearing rings. The method of residual tensions' measurements, the measurements' results and the variation graphics are presented.*

Key words*: bearing ring, plastic deformation, residual tensions, gauge strain.*

1. WORK CONDITIONS

1.1. Equipments and work conditions

For the rolling tracks' plastic deformation of the bearing's exterior ring type 6210, the Japanese machine, type CRF-120 OR was used, fig. 1, and for measuring the residual tensions, the Measurement System MTS 3000 produced by Hottinger Baldwin Messtechnik was used, fig. 2.

The residual tensions measurement was done by the method of piercing (Johnson et al., 1982), (Wagoner & Chenot, 2001).

The residual tensions' values determination was done by measuring the deformations according to three directions noted a, b and c on the gauge strain (1). The hole was produced with the help of a frontal milling machine set in the piercing device (2) which receives the main movement of splintering through a pneumatic turbine connected to the compressor (6) through a pressure regulator and a distributor (4).

Fig.1. Plastic deformation machine

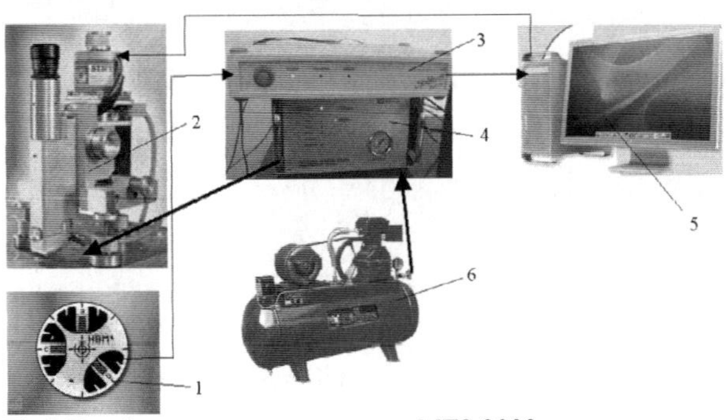

Fig.2 Tensions measurement system MTS 3000

The work progress is ensured by the programme installed on the computer (5) which, through the date acquisition board PC-LPM-16, commends step by step the engine situated on the device (2). The deformations measured by the gauge strain were transmitted to the tensometric bridge (3) type Spinder 8, which, being amplified, were transmitted to the computer where, with the help of the programme installed, were transformed in residual tensions (Pruteanu et al., 2006), (Pruteanu et al., 2007).

The processing was done in the following conditions:

- Material: 100Cr6 – specific to the bearing rings,
- Diamond milling machine with diameter d=1.7 mm for the plastic deformed ring and 2.0 mm for the semi-finished product,
- Work feed: 0.1 mm/min,
- Work rotation: 300,000 rot/min,
- Maximum hole depth: 2.5 mm.

2. EXPERIMENTAL RESULTS

The residual tensions were measured on two rings; semi-finished ring and cold plastic deformed ring.

Depth [mm]	Smin [MPa]	Smax [MPa]	Von Mises [MPa]	Tresca [MPa]
0.013	303.056	821.807	719.831	821.807
0.036	205.853	673.534	597.808	673.534
0.073	113.635	394.428	351.661	394.428
0.121	10.631	236.732	231.6	236.732
0.18	-47.793	189.051	216.933	236.844
0.252	-44.176	175.395	201.154	219.571
0.335	-32.788	161.976	180.616	194.764
0.431	-44.859	141.09	168.071	185.949
0.538	-70.859	119.421	166.567	190.28
0.657	-83.875	107.949	166.56	191.824
0.788	-73.514	106.815	157.054	180.329
0.931	-58.859	105.883	144.595	164.742
1.085	-60.784	89.55	130.985	150.334
1.252	-71.858	73.727	126.084	145.585

Depth [mm]	Strain (a) [1E 10^{-6}]	Strain (b) [1E 10^{-6}]	Strain (c) [1E 10^{-6}]	Alfa Angle [°]
0.013	-12.7	-5.143	-3.587	-16.68
0.036	-27.837	-7.712	-9.087	-24.454
0.073	-36.718	-13.805	-5.948	-13.035
0.121	-39.351	-23.561	4.79	7.942
0.18	-44.288	-35.039	14.836	17.247
0.252	-57.491	-45.492	19.209	17.247
0.335	-74.327	-52.691	22.623	14.486
0.431	-85.676	-57.512	35.102	14.043
0.538	-90.509	-63.052	58.502	16.136
0.657	-99.887	-71.236	81.017	17.171
0.788	-121.037	-78.752	90.668	15.493
0.931	-139.823	-77.847	92.706	12.515
1.085	-135.352	-67.218	104.286	11.667
1.252	-122.135	-66.454	120.013	14.187

Table 1. Experimental results - semi-finished product

Depth [mm]	Smin [MPa]	Smax [MPa]	Von Mises [MPa]	Tresca [MPa]
0.013	98.306	471.596	430.936	471.596
0.036	165.178	563.583	501.815	563.583
0.073	256.858	508.149	440.079	508.149
0.121	297.387	491.89	429.084	491.89
0.18	304.651	490.811	429.164	490.811
0.252	282.957	470.211	410.017	470.211
0.335	226.86	397.334	345.255	397.334
0.431	176.783	290.284	253.377	290.284
0.538	124.877	202.381	176.861	202.381
0.657	66.002	174.852	152.934	174.852
0.788	45.578	157.94	140.797	157.94
0.931	53.606	141.786	124.001	141.786
1.085	51.218	125.621	109.407	125.621
1.252	31.854	115.224	111	115.224

Depth [mm]	Strain (a) [1E 10-6]	Strain (b) [1E 10-6]	Strain (c) [1E 10-6]	Alfa Angle [°]
0.013	-5.654	-6.666	-1.726	28.287
0.036	-19.274	-23.199	-8.095	29.785
0.073	-39.52	-37.525	-17.94	19.591
0.121	-66.756	-53.475	-31.879	6.705
0.18	-99.152	-79.776	-49.187	6.325
0.252	-130.22	-114.933	-66.315	13.773
0.335	-148.062	-138.411	-76.618	18.062
0.431	-145.274	-127.25	-76.314	12.756
0.538	-126.111	-81.225	-69.106	-14.945
0.657	-105.503	-32.221	-65.708	-34.779
0.788	-94.185	-14.45	-74.521	-40.997
0.931	-86.788	-26.189	-92.044	-46.19
1.085	-69.333	-30.562	-105.492	-53.821
1.252	-44.203	-10.942	-108.464	-58.084

Table 2. Experimental results – plastic deformed ring

In table 1 there are the residual tensions'measurement results for the semi-finished product according to several criteria until the maximum depth of the hall of 1.252 mm, value until which there were noticed differences, and in table 2 the plastic deformed ring's measurement results are shown.

In figure 3, the graphic representations of the residual tensions for the finished (a) and semi finished (b) rings are shown.

3. CONCLUSIONS

From the graphs analysis we can draw the following conclusions:

- at the exterior surface of the semi finished ring, the residual tensions are very high and decrease rapidly with the depth (until 1.5 mm). These are due to the rough turning operation followed by the rectification with insufficient cooling,
- at the exterior surface of the plastically deformed ring at cold, the residual tensions are stretched throughout a much higher depth (0.4 mm) and tend to level out at a value of around 100 N/mm^2 (Von Mises method).

In future, we will research the difference between the residual tensions in ring processing through grinding (old technology) and in plastic deformed ring.

Likewise, we will compere the durability of the bearing processing through two technology.

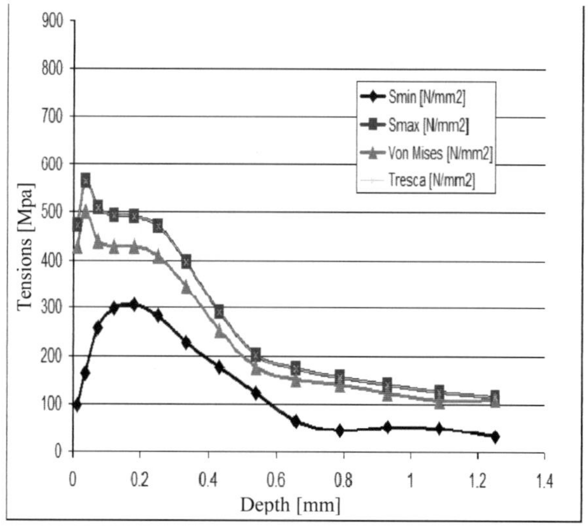

a

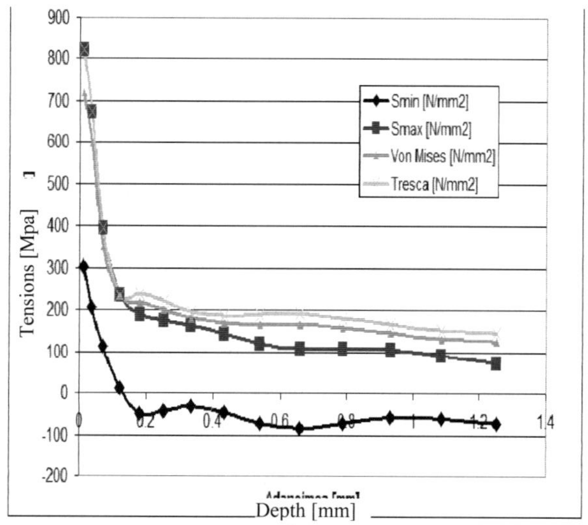

b

Fig.3 Comparative analysis regarding the residual tensions' distribution

4. REFERENCES

Johnson, W., Sowerby, R., Venter, R.D. (1982). *Plain strain slip for metal deformation processes*, Pergamon Press.

Pruteanu ,O.V., Carausu, C. Paraschiv, Dr., Sarbu, I., (2006). *The influence of the working parameters on the parameters of the superficial layer when procesing the steels through grinding in the rolling's technology*. Мировая зкономика и бизнес – aдминистрирование, Minsk, Belarus, may, 186-197.

Pruteanu, O.V., Paraschiv, Dr, Pricope, C, Bunea, C.M., Sarbu, I., Carausu, C. (2007). *Considerations Regarding the 100Cr6 Material's Conduct through the cold Rolling of the Bearing Rings,* International Journal for Manufacturing Science & Technology, U.S. Library of Congress, vol. 9, no.1, p.51-57, ISSN 1524-1548.

Pruteanu, O.V., Paraschiv, Dr., Carausu, C., Sarbu, I., Bunea, M., Bardac, D. (2007). *Processing through plastic deformation of the bearing rings.* MicroCAD 2007, International Scientific Conference, Miskolc, 143-148.

Wagoner, R.H., Chenot, J.L. (2001). *Metal forming analysis*. Cambridge University Press. New York.

Annals of DAAAM for 2008 & Proceedings of the 19th International DAAAM Symposium, ISSN 1726-9679
Editor B. Katalinic, Published by DAAAM International, Vienna, Austria 2008
Make Harmony Between Technology and Nature, and Your Mind will Fly Free as a Bird

KNOWLEDGE MANAGEMENT SYSTEM FOR IMPLEMENTENTING INTEGRATED MANAGEMENT IN ROMANIAN INDUSTRY

CARUTASU, G[eorge]; BOTEZATU, C[ezar] & BOTEZATU, C[ornelia] P[aulina]

Abstract: *Lack of integrated management approach, together with low performance of certificated systems (e.g. Quality Assurance, Labour Health and Safety, Environment Protection, Human Resources) conduct to need to design an integrated management systems that could sustain all subsystems and improve business performance of organization. The issue of such system is working with large amount of unstructured data, under various forms and tracking changes into the systems. Main concerning are related to structuring and querying necessary information, future work being related with integrated system implementation after the methodology is finalised, a study case being presented as result of R&D project finced by Romanian Agency for R&D.*
Key words: *KMS, DSS, integrated management, information system*

1. INTRODUCTION

In case of management systems, the necessity of integration derive from large efforts needed to be made in order to assure effective tools used to manage large organization. So, as history, in 60', in the management subsystems were considered decisional subsystem, financial subsystem, IT subsystem and human resources. In the next years, quality has gained more importance and as consequence a QA (Quality Assurance) has been added. In this case experts, procedures, equipments, and information used in order to guarantee that of products and services quality is achieved. These systems, named nowadays "quality management systems" appeared as a consequence of deep changes in science and technology, initially for the aerospace, nuclear and military industry. After 1990, the quality management systems have been extended to the machine tools industry, electronic, computers, motor vehicles, and after 2000 they were also designed for services, public services included.

After 1990, environmental management systems have appeared and these systems allow the enterprise to adopt environmental protective measures. In the meantime, that there are some procedures, which are similar to those of the quality system. The term of IMS is used only with this meaning, defining in ISO standards (from 2000) the audit procedures, corrective and preventive procedures used for quality and environment system (Breen, 1997, Abromeit, 2006).

In Romania, there are remarkable performances regarding the designing of the quality management system, numerous enterprises already having a certified system. In European Union, also the enterprises, including SMEs, have such systems. Regarding the other Romanian systems, there are some manufacturing systems which have environmental systems. Recently, food safety systems have been developed. Less than 20% of Romanian enterprises have a certified quality management system, and less than 5% have a certified environmental system (study of R&DITTMA). In Europe, many enterprises have environmental systems, food safety systems, but so far the health issue is concerned, only few German enterprises started to take an interest in it. Regarding the social responsibility, except for the USA, theoretical preoccupations of implementing such systems are present only in Italy.

The issue, regarding the lack of integration on top management level, is viewed as a source of loss, in general picture of enterprise activity. Supporting the integration of subsystem presented before create the opportunity of an increased overall profit. Scientifically speaking, in Romania, specific researches regarding the quality system have been developed only for a number of universities and research institutes.

As for researches on other types of subsystems, the number of published results is rather small; nevertheless, some papers exist. The experts involved in this project established a partnership made from organizations with remarkable performances in managerial subsystems' analysis and design, this also resulting from their list of published papers.

An analysis concerning the Romanian enterprises management shows a certain discrepancy between the methods used and the desired performances.

The project „Development and the Implementation of the Integrated Management System (DI - IMS)" aims at generating knowledge that leads to an efficient management, thus allowing the enterprises a smooth integration on the European market (Botezatu et al., 2006).

2. DECISION SUPPORT SYSTEM

As generic term *decision support system* is referring to a computerized program, which gathers various data from different resources on enterprise level, in order to assist the decision maker in decision making activity, by presenting a set of indicators results from aggregating collected data (see figure 1). This generic term could be used to describe any type of software used to support the decision making process. Most common categories in business application field, covering various enterprise management functions are Enterprise Resource Planning, Customer Relationship Management, Supply Chain Management, Product Lifecycle Management and Business Intelligence. This entire software categories offer dedicated tools to help the decision maker, by statistical models, simple management science models, spreadsheets, and graphics and so on (Naiman et al., 2006).

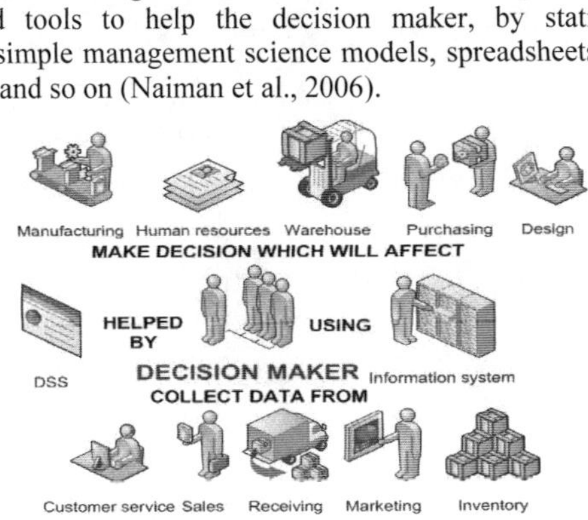

Fig. 1. Decision making process

3. KNOWLEDGE MANAGEMENT

First, we have to categorize the decision making process, using a framework which specifies the decision types and categories of problems solved.

So the structured process are based on repetitive and routine problems which have standardized solution, such as accounts, assets, receivable, order entry, data used to perform budget analysis, short-term forecasting, decision helped being investment management, warehouse location and distribution systems.

The decision making process, based on knowledge management is three cumulative step processes, having the components:

- *intelligence*, to determine the conditions that call for decision;
- *design*, where different possible course of action is invented, developed and analyzed;
- *choice*, from available courses of action one is selected.

The *unstructured process* does not use structured solution for those three step described above, most of them are based on human intuition. These types of processes are fuzzy, complex problems with no well-known and accepted solution, such as R&D planning, new technology development, quality assurance planning, integrated management policies.

Between those two types lies *semistructured process* where parts of its components are structured. The most common applications are referring to new product planning, mergers and acquisition, new production facilities.

We present a three-tier structure, using large implemented Operating Systems and Database Servers: MS W2003 Servers and MS SQL 2005. The concept of data' is referred to three different layers: Data layer, Information layer and Knowledge layer. Data layer, where primary data are inserted in the system, information as final stage (Niculita et al., 2007).

All data are stored in a data warehouse as main repository of the enterprise historical data. Information is data obtain after aggregation process, using aggregate functions depending on subject, interconnected by communication nods existent between organizations involved named data mart (Caron et al., 2004).

This concept supposes a repository of data gathered from operational data and other sources that is designed to obtain knowledge (figure 2). So, the information is obtained from data warehouse after suffering a data mining process. Also, an OLAP (On- Line Analytical Processing) system is provided, to assure a framework for the analysis of multidimensional data. Metadata are expressed in a condensed data mine-able format, or one that facilitates the practice of data mining (Hsu, 2007).

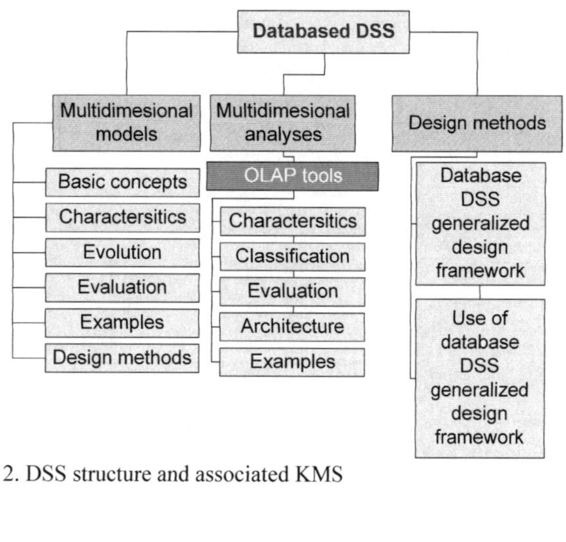

Fig. 2. DSS structure and associated KMS

4. CONCLUSION

The limited awareness of the integrated management systems advantages by organizational decision – making bodies (especially for SMEs), in this case a suitable DSS architecture is required.

So, in this paper we propose a database DSS, the project being on course having the deadline September 2010. Integrating all management subsystems could be a challenge because of lack of integrated information systems on enterprise level and, on second hand, by lack of procedures in particular cases. Preliminary results show that on EU and national level just several subsystem are defined, so another achievement of DI-IMS will be the complete defining of integrated management system, together with all associated procedures and methods. Next step in this research is to implement the integrated management system in large number of organization and to produce the software connected with local ERP systems.

As technical point of view, designing a DSS involves a various set of techniques and for such complexity more studies must be made in order to structure and formalize the problems needed to be assisted. More methods have to be used to study the same problem and the suitable form of DSS, in this case is the *cooperative* way

An approach in this direction was to envelop the files into a data structure, covering the subject, activity, owner, repository, type, volume etc. and other information needed for link the document with subject of search. Future work will explore the possibility to use XML structure (where all files are enveloped with a header and trailer containing above mentioned information) or to design an external database for managing the files.

5. REFERENCES

Abromeit, M. & Sassano L (2006). *Integrated Management System For Safety, Health And The Environment: A Continual Improvement Management System*, Available from: http://www.iapa.on.ca/documents/IMS_Presentation.pdf *Accessed:* 2006-03-20

Botezatu, C.P., Botezatu, C. Carutasu, G. (2006). *The development and the implementation of the integrated management system*, Annals of DAAAM for 2006 & Proceedings of the 17th International DAAAM Symposium, pp. 053-054, ISSN 1726-9679, ISBN 3-901509-57-7

Breen, K., Pecora, G., Gregerson, D. & McCabe, T (1997). *A new direction: Integrated Management Systems*, *Available from:* http://www.qualitydigest. com /aug97/ html/ims.html *Accessed:* 2006-04-07

Caron, E., Daniels, H. (2004). *Diagnosis in the OLAP Context*. ERIM Report Series Reference No. ERS-2004-063-LIS, 2004

Hsu, C.H., Wang, S.Y., Lin, L.T. (2007), *Using Data Mining to Identify Customer Needs in Quality Function Deployment for Software Design*, pp.170-174, Proceedings of the 6th WSEAS Int. Conf. on Artificial Intelligence, Knowledge Engineering and Data Bases, Corfu Island, Greece, ISSN: 1790-5117, ISBN: 978-960-8457-59-1

Naiman, K., Kopackova, H., Simonova, S., Bilkova, R. (2006). *Approaches of quality outputs from the business systems*, Proceedings of the 5th WSEAS Int. Conf. on COMPUTATIONAL INTELLIGENCE, MAN-MACHINE SYSTEMS AND CYBERNETICS, pp. 282-285, ISSN: 1790-5095, ISSN: 1790-5117, ISBN: 960-8457-56-4,

Niculita, L., Carutasu, N. L., Carutasu G. (2007). *Governance Intelligence for Romanian R&D Funding*, Proceedings of 7th WSEAS International Conference on APPLIED INFORMATICS AND COMMUNICATIONS (AIC'07), , ISSN: 1790-5117, ISBN: 978-960-6766-01-5

Annals of DAAAM for 2008 & Proceedings of the 19th International DAAAM Symposium, ISSN 1726-9679
Editor B. Katalinic, Published by DAAAM International, Vienna, Austria 2008
Make Harmony Between Technology and Nature, and Your Mind will Fly Free as a Bird

RESIDUAL STRESS CALCULATION FOR BUTT WELDING

CATANA, D[orin]; POPESCU, R[odica] & CATANA, D[orina]

Abstract: *Due to butt welding procedure internal stresses will appear inside the pipes and their behaviour will be influenced for next phases such as machining and operating. For diminishing the residual stresses of the welded products, the tension-releasing thermal treatment is being employed. Establishing the efficiency of the aimed thermic treatment also consists in calculating the residual tensions, which are left in products. Proposed method consists in dividing thermal tensions release process in preset time counting in order to establish the equation of residual stress and elastic modulus. This is possible because the values within the equations are known. The established equations are related with thermal treatment parameters and for this reason it is possible to adjust them in order to avoid residual stresses within welded seam. The values calculated through the method suggested in the work are close to the ones established through Runge-Kutta method.*

Key words: *steel, strain relief, welding, stresses.*

1. INTRODUCTION

The evolution of the world metal materials production enforces the implicit development of the thermal treatment technologies. Perfecting the heating methods plays an important role in both quantitative and qualitative development of the thermic treatment technologies. For welded products this evolution is critical for heating rate, process energetic threshold and for technological effects afterward. These desiderates can be achieved through improving the thermic treatment technology alongside assisting the thermic treatment by means of computer, which can be applied in the case of the butt head welding pipes. Taking into account the big quantity of steel for pipes, destined to performing the welded structures, conclusions can be drawn on the importance of the welded technologies and thermic treatment on the welded joints. The welding technologies have known a distinguished boom through the knowledge of the materials' behaviour during the welding process and afterwards, perfecting the welding procedures allowing the joints to be executed for different materials in a wide range of dimensions. The new joint technologies' economical implications are:

- reduced material expenses;
- reduced energy costs;
- isotropic and homogeneous joints.

The thermal treatment applied to butt welding pipes plays an important role because an inappropriate tension-release of the performed welding can result in consequences as to the subsequent functioning of the welded set. It is worth mentioning that most of these pipes are underground which makes it extremely difficult to locate and fix the cracks.

2. THEORETICAL CONSIDERATIONS

During the operations of welding the parts to be joined, within the products are internal tensions that equilibrate themselves within their volumes and stay integrally or partially within them, as residual tensions. The presence of the residual stresses has a sensitive bearing on the products' behaviour during the following operations of treatment and exploitation, too. As for the cause that generates them, the internal tensions are of two types (Balauca & Popescu, 2007):

- thermal tensions, due to the expansions' irregularities and contractions resulted from lack of simultaneity of the heating and cooling process in different micro-volumes of the metallic body;
- structural tensions, due to expansions' irregularities and contractions which come alongside the phase transformations that are caused by alternative heating and cooling in different micro-volumes of the metallic body. The thermal tensions turn up as a result of the cold steel hindering the deformation process; the steel is not heated by the thermic source. The thermal tensions can be assessed on the basis of Hook's law taking into account the elastic features of the steel:

$$\sigma(t) = \varepsilon \cdot E(t_1 - t_0), \quad (1)$$

where:

ε - specific linear deformation;
E - elasticity module;
t_1 - heated area temperature;
t_0 - cold area temperature.

The residual tensions that come during the welding process can even reach the material's flowing limit. The practical need for diminishing and removing these tensions is determined by their vicious influence in certain conditions on some exploitation properties, such as: by the tendency for fragile rupture, cracking by corrosion and exhaustion.

In view of diminishing the level of residual tensions in the welded joints, which can produce brittleness-cracking phenomena, there are some methods, but the most known is that of applying (after the welding process) the thermal tension-releasing treatment (TTD), which consists of heating to a certain temperature, keeping at that temperature and cooling. Among the metallurgical implications of this treatment there are the following:

- restoring the deformed crystalline networks;
- rearranging the dislocations;
- blocking the sliding and limits of the grains by a fine dispersion of the stabile particles.

From analyzing the mathematical methods of the tension-releasing process, a conclusion can be drawn that this is firstly dependent on the plastic flowing speed.

The present standards take into account the experimental results, which showed that the plastic flowing speed is proportional with the value of the residual tension, but it does not reflect another phenomenon encountered in practice, such as that the plastic flowing speed increases with approaching to the flowing limit of the material (Candea & Popescu, 2007).

Here is an equation for the elastic flowing speed:

$$\dot{\varepsilon} = \frac{K \cdot \sigma}{\sigma_c - \sigma}, \quad (2)$$

where:

K - flowing creep rate;

σ - initial residual tension;

σ_c - material's flowing tension.

During the process of tensions' relaxation, the specific extension can be determined with equation (3).

$$\varepsilon_0 = \varepsilon + \varepsilon_p \tag{3}$$

Deriving the equation (3) in relation with time and taking into account equation (2), we get the differential equation of the thermal tension-releasing process (equation 4).

$$\dot{\sigma} + \frac{K \cdot E \cdot \sigma}{\sigma_c - \sigma} - \frac{\dot{E}}{E} \cdot \sigma \tag{4}$$

All values in equation (4) are dependable on time. A possibility of resolving the equation (4) is adopting sizes E, σ and K as temperature functions which in its turn are dependable on time. The functions E (t), σ_c(t) can be found in the product norms or they can be experimentally determined through known methods (Popescu et al., 2005).

Even if there are particular values of sizes, regression analytical functions can be calculated. The values of coefficient K can be determined using the relation (2), if the plastic flowing speed is known.

The internal tensions in different moments of the thermal tension-releasing process are given by the solutions of the differential first-degree equations. Because an analytical solution cannot be found to this equation, a numerical solution was found by using Runge-Kutta method.

Another possibility consists in solving the equation (4) through other procedure than Runge-Kutta method, this being suggested from now on. The process is divided in time intervals by pitch p (Popescu et al., 2007), and the solutions σ(t) and E (t) can be written with equation (5).

$$\sigma(t) = \frac{\sigma(I) - \sigma(I-1)}{p}\ ;\ E(t) = \frac{E(I) - E(I-1)}{p} \tag{5}$$

By replacing equation (5) with equation (4) and performing the calculations, a second-degree equation results, such as:

$$A \cdot \sigma^2(I) - B \cdot \sigma(I) + C = 0 \tag{6}$$

where:

$$A = \frac{E(I)}{E(I-1)}$$
$$B = \sigma_c(I) \cdot \frac{E(I-1)}{E(I)} + \sigma(I-1) + K(I) \cdot E(I) \cdot p \tag{7}$$
$$C = \sigma_c(I) \cdot \sigma(I-1)$$

$E(t)=A_1*t^3+B_1*t^2+D_1$	A_1=-1.584559*10^{-4} B_1=3.042152*10^{-3} C_1=-32.70786 D_1=210654
$\sigma_c(t)= A_2*t^3+B_2*t^2+D_2$	A_2=9.949638*10^{-7} B_2=-1.112616*10^{-3} C_2=-6.420578*10^{-3} D_2=366,56
$K(t)=K_{min}+A_3[\exp(B_3*t)-1]+C_3*t$	K_{min}=10^{-7} A_3=2.895451*10^{-9} B_3=1.303368*10^{-2} C_3=6.363989*10^{-2}

Table 1. Material characteristics according to temperature.

t_0 [°C]	t_{max} [°C]	Time [h]	p	σ_0 [MPa]	Final stress [MPa] R-K	Final stress [MPa] Eq. (6)
20	300	3	1000	100	93.6733	93.6731
		10			93.5363	93.5361
	600	3	100	200	148.0739	147.9977
		3	1000		148.0740	148.0658
		10			143.3567	143.3380

Table 2. Stress values at the end of heating cycle.

All values from equation (6) and (7) are known. The material characteristics E, σ_c and K have known functions during the whole process, and the tensions only turn up with the value σ(I-1), which is known from the previous calculation step. On the basis of the pattern suggested, there have been determined the expressions of the material's characteristics according to temperature (see table 1), only for the heating phase during the thermal treatment process, applied to a steel brand OLC45 (0.45 %C). By comparing the accuracy of the two methods, table 2 shows the tension's values at the end of heating cycle (Balauca et al., 2005).

The flowing stress's values at different temperatures can be obtained from the function σ_c(t) presented in table 1. Following the calculation, the flowing stress's values are: σ_c(20)=365.99 MPa, σ_c (200)=329.73 MPa, σ_c(400)=249.05 MPa, and σ_c (600)=177.8 MPa.

3. CONCLUSIONS

As a result of pipe's butt welding, internal tension within the material will appear but critical tensions from butt seam are increased with 34.44% over steel yield limit. After applying the tension-releasing thermal treatment (TTD), the butt seam diminishes with 35. 6% in comparison to the flowing limit, the measurements performed on the internal tensions highlighting the efficiency of the local thermic treatment.

The theoretical and experimental researches performed establish the mathematical equations of the elasticity module and flowing limit after tension releasing. Also the mathematical equation for creep rate was established. Based on determined equations, thermal treatment parameters can be established in order to avoid cracks in welded joint while operating.

By using the same procedure these characteristics can be determined for any kind of material.

4. REFERENCES

Balauca, I.; Ploscariu, C. & Tont, F. (2005). Final heat treatment for welded joints in piping, *Proceedings of International Conference on Materials Science and Engineering*, pp. 135-138, ISBN 973-635-454-7, University Transylvania, 02.2005, University Transylvania, Brasov

Balauca, I. & Popescu, R. (2007). Contribution to establishing the reliability of ash delivery piping in steam power plants, *Proceedings of International Conference on Materials Science and Engineering,* Catana, D., pp. 279-282, ISSN 1223-9631, University Transylvania, 02-2007, Supplement of Bulletin of Transilvania University of Brasov, Brasov

Candea, V. & Popescu, R. (2007). The mechanism of fissuring in welded joints, *Bulletin of Polytechnic Institute of Jassy*, Vol. LIII, No. 4, 05.2007, pp. 111-114, ISSN 1453-1690

Popescu, R.; Candea, V. & Balauca, I. (2005). The influence of allied elements on steels used in steam power plants, *Bulletin of Polytechnic Institute of Jassy*, Vol. LIII, No. 4, 05.2007, pp. 341-344, ISSN 1453-1690

Popescu, R.; Balauca, I. & Medan, R. (2007). Methods for determining the reability of service pipes for transporting cinder in power plants, *Bulletin of Polytechnic Institute of Jassy*, Vol. LIII, No. 4, 05.2007, pp. 337-340, ISSN 1453-1690

Annals of DAAAM for 2008 & Proceedings of the 19th International DAAAM Symposium, ISSN 1726-9679
Editor B. Katalinic, Published by DAAAM International, Vienna, Austria 2008
Make Harmony Between Technology and Nature, and Your Mind will Fly Free as a Bird

THERMAL TREATMENT INFLUENCE ON MICRO HARDNESS TITANIUM ALLOY BIOMATERIALS

CATANA, D[orin]; SCARNECIU, I[oan] & POPESCU, R[odica]

Abstract: *Increased life expectation was due to the usage in well determined conditions of artificially made products implanted in human body. To be used in human body, the materials have to accomplish some features, one of these being the biocompatibility. Considering that human body is an aggressive environment for implants, these have to resist a long period of time until next surgical intervention. This paper presents the influence of buffer solution hardening treatment over micro hardness for 4 types of titanium alloys. The treatments have been differentiated by sample heating temperatures and the environment in which the cooling took place. As a confirmation of the results, coded treatment CT1 has given the best solution regarding micro hardness evolution.*
Key words: *micro hardness, titanium alloy, biomaterials, human-implant*

1. INTRODUCTION

Human desire for ameliorating his health state, affected by diseases or accidents has imposed the idea of the use in well determined conditions, of some artificially produced parts which are made of naturals or artificial materials. It appeared two alternatives even by the time of the first substitution assays:

- dead or alive tissues and organs transplantation;
- parts, devices or systems made of natural or artificial material implants.

Present implants advantages are: functionality, availability and the capacity of reproduction. Implants major disadvantages are determined by nature, way of synthesis and materials different features used in relation with alive tissues. Implant materials lack two important characteristics of the alive tissues:

- the capacity of modifying its structure and features depending on the solicitations that they support (for example: mechanical charging for the osseous tissue or the blood flow for the blood-vessels);
- auto-annealing capacity.

Hip articulation has as main physiological functions: the mechanical loads transmission between pelvis and femur, as well as the relative movement between the system's two parts, osteo articular. Hip articulation (cox-femoral) is a spherical type articulation, with tree degree of movement facility: flexion-extension, abduction-adduction and internal-external rotation. This articulation is greased with synovial fluid, synovial cartilage – fluid combination carrying out the diminution friction coefficients. Total replacement of the joint is made by using prosthesis with its main purpose of taking over the biomechanical function of natural joint.

2. THEORETICAL CONSIDERATIONS

The biomaterials are synthetic materials used for the parts, devices and artificial systems carrying out, with the object of replace and/or take over, totally or partially, the function of an alive tissue for a limited or unlimited time. For the manufacture of some articulation prothesis components, the metallic biomaterials must have:

Alloy type	Chemical composition [%]			
	Al	V	Fe	Nb
Ti6Al4V	5.5-6.5	3.5-4.5	0.25	-
Ti6Al7Nb	5.5-6.5	-	0.25	6.5-7.5
Ti5Al2.5Fe	4.5-5.5	-	2-3	-
Ti6Al2Nb1Ta1Mo	5.5-6.5	-	-	2

Tab 1. Main alloying elements for titanium implant alloys.

- an elasticity module as similar as possible, to that of the long bone;
- good fatigue resistance and lastingness;
- wear and electrochemical corrosion resistance as advanced as possible.

The metallic materials that can be used for the hip articulations replacement are: stainless steels, Co- Cr alloys and titan alloys (Catana, 2002).

From all these presented metallic materials, the titanium and its alloys are considered to be the most biocompatible, with the best corrosion resistance in the human organism average and with the most powerful relation between the implant and the surrounding osseous tissue. The alloys for the implants are described in table 1 (Wusinczky et al., 2007).

Titanium base alloys used for implants were subject of quenching thermal treatment followed by artificial ageing. The thermal treatments have been carried out in argon atmosphere. Thermal treatments differentiated by the heating temperature, cooling temperature and cooling average. Cooling time has been 1 hour for all the tests (Wusinczky & Popescu, 2005).

To accentuate the metallurgical processing influence on the micro hardness used as mechanic features modification indicator, there have been carried out measurements of this, for all the materials with titan base taken to be studied, respectively, for all the metallurgical materials applied to these.

In the case of the Ti-6Al-4V alloy (Popescu et al., 2005) the applied thermal treatment has been the placing into the solution with the following parameters (micro hardness evolution is described in figure 1):

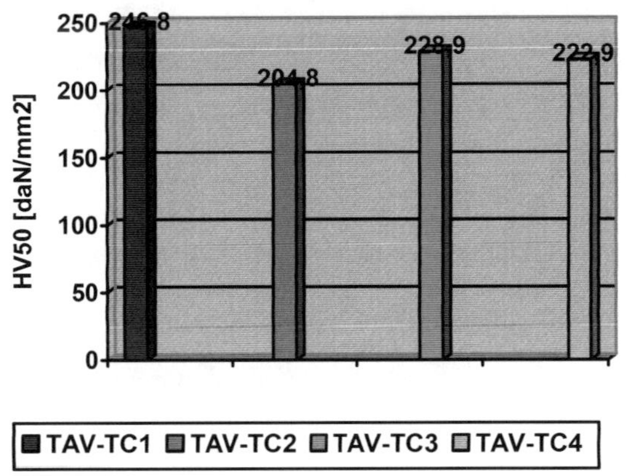

Fig. 1. Average micro hardness value for Ti-6Al-4V alloy.

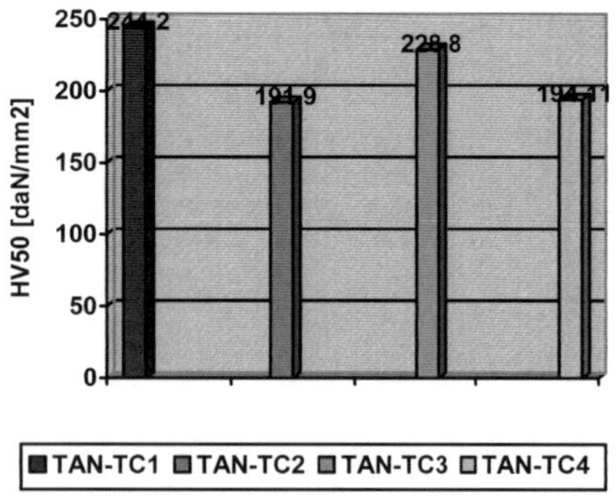

Fig. 2. Average micro hardness value for Ti-6Al-7Nb alloy.

TC1 – heating at 950 - 970°C water cooling;
TC2 – heating at 950 - 970°C furnace cooling;
TC3 – heating at 850 - 870°C water cooling;
TC4 – heating at 850 - 870°C furnace cooling.

Maintaining time has been for 1 hour and, after the hardening, followed an ageing, its parameters being: 2 hours of heating at 700-710°C followed by air cooling.

For Ti-6Al-7Nb alloy the placing in solution hardening has been performed in the following parameters:

TC1 – heating at 920 - 940°C water cooling;
TC2 – heating at 920 - 940°C furnace cooling;
TC3 – heating at 820 - 840°C water cooling;
TC4 – heating at 570 - 580°C furnace cooling.

The ageing parameters were: heating at 570 - 580°C, 2 hours maintenance, followed by cooling.

Alloy micro hardness evolution is described in figure 2.

For Ti-5Al-2.5Fe alloy the placing in hardening solution has been performed in the following parameters:

TC1 – heating at 1010 - 1030°C water cooling;
TC2 – heating at 1010 - 1030°C furnace cooling;
TC3 – heating at 900 - 920°C water cooling;
TC4 – heating at 900 - 920°C furnace cooling.

Ageing parameters were: heating at 555 - 545°C, 2 hours maintenance, followed by airy cooling. Alloy micro hardness evolution is described in figure 3 (Wusinczky et al. 2005).

The last alloy studied is Ti-6Al-2Nb-1Ta-1Mo which has in composition thallium and molybdenum, two essential chemical elements for implants. For this alloy the placing in hardening solution has been performed in the following parameters:

TC1 – heating at 1020 - 1040°C water cooling;
TC2 – heating at 1020 - 1040°C furnace cooling;
TC3 – heating at 920 - 940°C water cooling;
TC4 – heating at 920 - 940°C furnace cooling.

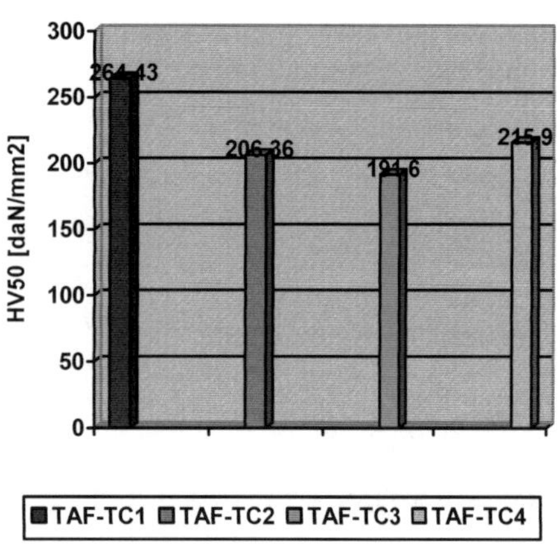

Fig. 3. Average micro hardness value for Ti-5Al-2.5Fe alloy.

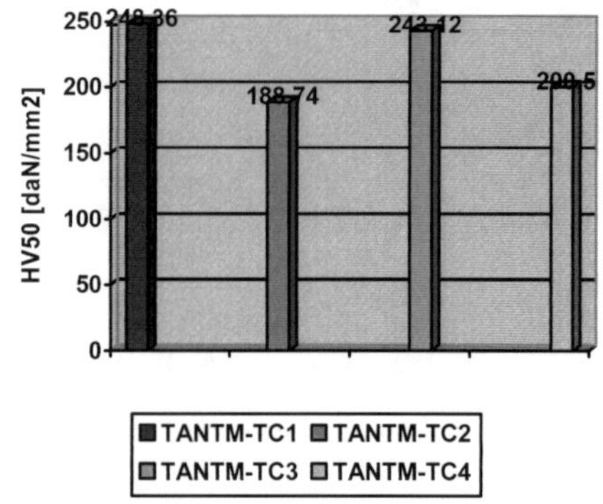

Fig. 4. Average micro hardness value for Ti-6Al-2Nb-1Ta-1Mo alloy.

Ageing parameters were: heating at 595 - 605°C, 2 hours maintenance, followed by airy cooling. Alloy micro hardness evolution is described in figure 4. To be used in medicine, the presented treatments have been completed with an anodization thermal treatment and a thermochemical in electrolytic plasma in order to make the alloys biocompatible.

3. CONCLUSIONS

For titanium base biocompatible materials the application of adequate hardening thermal treatments will determine the improvement of micro hardness which is essential when these materials are used for joint prosthesis. For Ti6Al14V alloy, the TC1 treatment permits the attainment of the highest hardness. For the Ti6Al7Nb alloy the entire TC1 treatment is that which determine the wearing resistance. Also, the hardness is increased, especially, due to the cooling increased speed values, more than due to maintenance at placing in solution hardening different temperatures. Ti5Al2.5Fe and Ti6Al2Nb1Ta1Mo alloys have a similar behavior with the prior presented alloys at the TC1 treatment. In order to have a significant increasing of titanium alloys micro hardness the recommended placing in solution thermal treatment is that one coded TC1. The researches will be completed with corrosion and biocompatibility tests in order to determine which one is the best solution for thermal treatment. When all test results will be available, finally will be established which thermal treatments will increase the micro hardness of studied titanium alloys and implicitly the best one to produce the prostheses.

4. REFERENCES

Catana, D. (2002). *Advanced Materials Processing*, Lux Libris, ISBN 973-9428-73-8, Brasov

Popescu, R.; Catana, D. & Wusinczky, D. (2005). Optimal parameters experimentally set for the temper heat treatment with laser beam applied to steel C120, *Proceedings of International Conference on Materials Science and Engineering*, pp. 125-128, ISBN 973-635-454-7, University Transylvania, 02.2005, University Transylvania, Brasov

Wusinczky, D. & Popescu, R. (2005). New material used in the medical area, *Proceedings of International Conference on Materials Science and Engineering*, pp. 181-184, ISBN 973-635-454-7, University Transylvania, 02.2005, University Transylvania, Brasov

Wusinczky, E.; Wusinczky, D. & Popescu, R. (2005). Hip joint implants, *Bulletin of Polytechnic Institute of Jassy*, Vol. LIII, No. 4, 05.2007, pp. 333-338, ISSN 1453-1690

Wusinczky, D.; Wusinczky, E. & Popescu, R. (2007). Titan based biomaterials behaviour used in joint replacements, *Bulletin of Polytechnic Institute of Jassy*, Vol. LIII, No. 4, 05.2007, pp. 339-344, ISSN 1453-1690

Annals of DAAAM for 2008 & Proceedings of the 19th International DAAAM Symposium, ISSN 1726-9679
Editor B. Katalinic, Published by DAAAM International, Vienna, Austria 2008
Make Harmony Between Technology and Nature, and Your Mind will Fly Free as a Bird

TEMPERATURE RANGE DETERMINATION FOR BUTT WELDING

CATANA, D[orin]

Abstract: *The process of butt welding is based on hot plastic deformation for the parts of which free ends were previously brought into the contact. In order to simulate this process the main requirement is to know the temperature of the parts to be heated for each cross section on the whole length of longitudinal axis. The paper presents the way of solving this requirement, furthermore was considered that the heated parts to be welded are to be assimilated with acicular ribs. Based on this hypothesis were developed the equations which determine the temperature along longitudinal axis. By comparing calculated results with simulated ones was noticed that the differences are not significant and confirms that physical-mathematical model proposed for this theme might be correct.*
Key words: *butt welding, heating, temperature, simulation.*

1. INTRODUCTION

Heat transfer is the process which transforms thermal energy from high value parameters to low value parameters. Temperature is the parameter which estimates the quality of heat and could be defined as a global measure of process intensity which gives the level of energy inside the body. Heat exchange complies with the second principle of the thermodynamics which establishes the natural sense of heat propagation which is always from the source with a high temperature to the source with a low temperature. Heat transfer penetrates more or less in all current areas of method and its importance is growing. Heat transfer is produced in three distinct ways: conduction, convection and radiation. Thermal conduction is the process of heat transfer from an area with a high temperature to an area with a low temperature. The transfer takes place inside of a solid, liquid or gaseous medium, or between different medium in direct contact. Conduction is the only heat transfer process through opaque bodies.

2. THEORETICAL CONSIDERATIONS

Butt pressure electric welding is presented in figure 1 (Iovanas, 2004). This method is differentiated by other welding processes such as spot welding, line welding etc. not only regarding the theoretically scheme but also considering that the welding seam is stretched on both welded section. From a dimensional point of view a single, continuously part is obtained and from a metallurgical point of view a homogeneous with uniform mechanic characteristics is the result of this process (Eftimie et al., 1998).

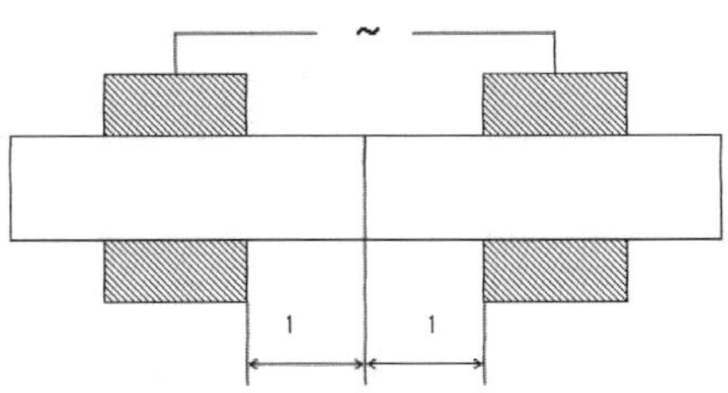

Fig. 1. Butt head welding theoretically scheme.

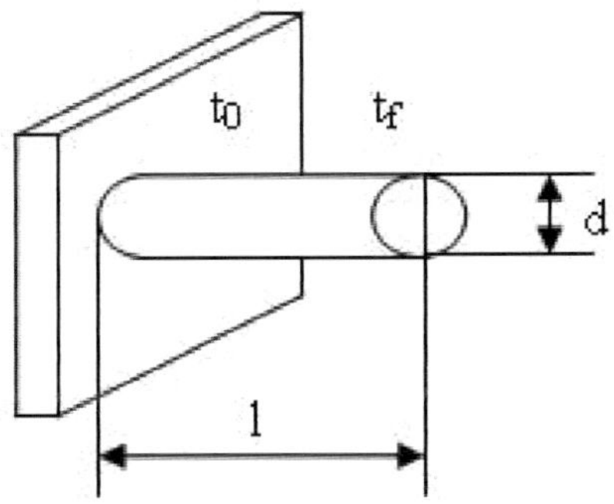

Fig. 2. Acicular rib.

The parts to be welded are heated using Joule-Lentz effect and by connecting them to electrical current. Maximum temperature is achieved on contact surfaces area between the parts and diminishes toward the fastening area. In order to determine the evolution of temperature (the establishment of temperature range) a model of a thermal system which is similar in behaviour should be found in order to see what is happening when butt welding is in progress.

For butt welding is to be considered as a model one that is closest to reality and that is a wide surface (conductive-convective systems). This system is made of an acicular rib (figure 2) with limited values of the cross section (d) which are smaller in comparison with the rib length (Catana, 2007).

For a rib with a cross section area (S = S(x)) and perimeter (P = P(x)), both variable, in contact with a fluid (air) with constant temperature and convection coefficient (α = ct), in some sections, including side perimeter, rib temperature is the same and is given by equation:

$$t = t(x) > t_f \tag{1}$$

The differential equation for the analyzed thermal transfer process is:

$$\frac{d^2\theta}{dx^2} + \frac{1}{S}\cdot\frac{dS}{dx}\cdot\frac{d\theta}{dx} - m^2\cdot\theta = 0 \tag{2}$$

where:

- $\theta = t - t_f$;
- $m^2 = \dfrac{\alpha \cdot P}{\lambda \cdot S}$.

Equation (2) can be used for different cross section shapes. For an acicular bar with length (l), diameter (d) and constant section (S = ct) the equation becomes:

$$\frac{d^2\theta}{dx^2} - m^2\cdot\theta = 0 \tag{3}$$

Distance from free rib end [mm]	Temperature [°C] through		Δt
	Calculus	FEM	[%]
0	1200	1200	0
2.5	945	980	-3.70
5	876	830	5.25
7.5	642	685	-6.70
10	631	608	3.56
12.5	583	559	4.12
15	486	497	-2.26

Tab 1. Temperature evolution with distance.

General solution of the equation is:

$$\theta = C_1 e^{mx} + C_2 e^{-mx} \quad (4)$$

By solving the equation and considering the limit conditions, temperature distribution on rib's length is:

$$\frac{\theta}{\theta_0} = \frac{t - t_f}{t_0 - t_f} = \frac{ch\, m \cdot (b - x)}{ch\, m \cdot b} \quad (5)$$

where:

- $ch\, x = \frac{e^x + e^{-x}}{2}$;
- b = l;
- t_0 temperature at the rib's toe (Catana, 2006).

In case of the butt welding, temperature value in contact area is known (the free rib end) and based on this value, temperature range can be determined during rib's length, from tip to toe.

For a hexagonal steel bar with a 15 mm free length, 12 mm key aperture and λ_{1200} = 21.17 W/(m°C) temperature values established with equation (5) are presented in the table 1 (Catana, 2005).

In table 1 difference between temperatures do not exceed -6.70% and that is why a mathematical method is recommended. Due to laborious calculus FEM is recommended.

When FEM software is not available the equations can be easily used by technical personnel in order to calculate the temperatures for butt welding.

It is necessary to know the temperature of the parts to be welded in order to use butt welding and to finalize the process. Knowing the temperatures range in piece makes possible to perfect the welding process.

In figures 3 and 4 is presented the evolution of established temperature through simulation for a steel bar with specific features presented before.

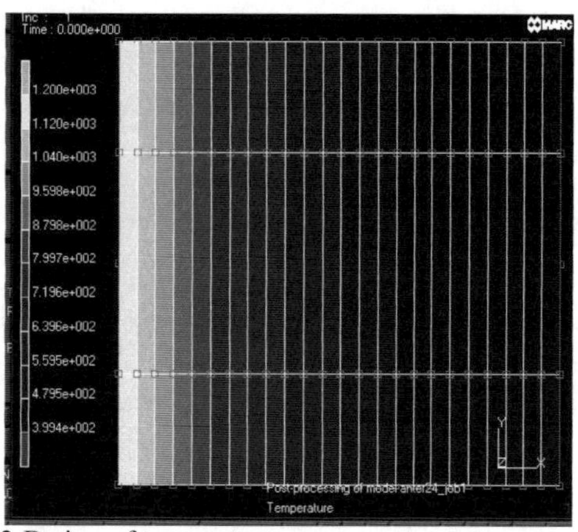

Fi. 3. 2-D view of temperature range.

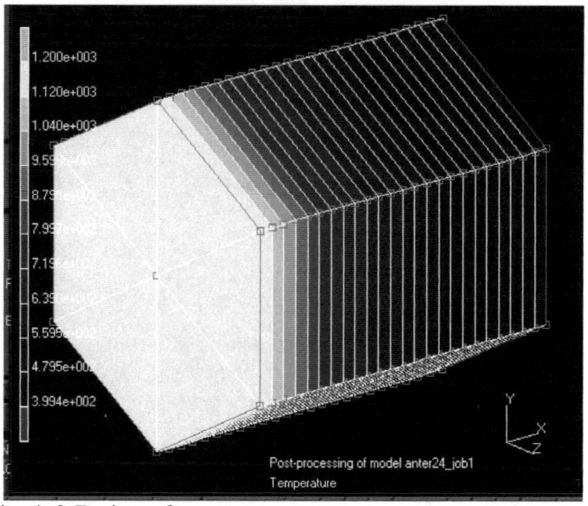

Fig. 4. 3-D view of temperature range.

3. CONCLUSIONS

Considering that is very laborious to determine thermal transfer value during welding processes using numerical method could be a real advantage. Besides the rapidity of getting the results, another benefit is given by the accuracy of the solutions when suggested model is closer to reality.

Obtained temperature range can be used when hot plastic deformation is studied for bars butt welded. Also if the simulation is made for more diameters, the diagram can be used in order to determine the temperature evolution. The bar type of diagram can be used in order to facilitate the access of specialists from production considering that mathematical relations are difficult and final results are similar.

The developed model offers close results to simulated ones (+5.25% … -6.70%), which confirms the initial analogism. As well the validity of developed model for different cross section and free lengths will be checked.

After improving the model, digital integration into dedicated software to determine these temperatures is expected. Calculated and simulated results will be verified by measuring the temperature of the parts while welding them.

This final test will determine a substantial improvement of the developed theoretical model for butt welding.

4. REFERENCES

Catana, D. (2005). Theoretic contributions for the plastic deformation simulation process, *Proceedings of International Scientific Conference „Modern Technologies, Quality, Restructuring" TMCR 2005*, pp. 333-337, ISBN 9975-9875-4-0, Technical University of Moldova, 05-2005, University of Moldova, Chisinau

Catana, D. (2006). Possibilities of simulation of the axial deformation for the butt head welding, *Proceedings of International Scientific Conference „Modern Technologies, Quality, Restructuring" TMCR 2006*, Musca, G., pp. 1135-1140, ISSN 1011-2855, Jassy, 05-2006, Bulletin of Polytechnic Institute of Jassy, Jassy

Catana, D. (2007). The evolution of the temperature field in the time of butt head welding, *Proceedings of International Conference on Materials Science and Engineering,* Catana, D., pp. 481-484, ISSN 1223-9631, Transylvania University, 02-2007, Supplement of Bulletin of Transylvania University of Brasov, Brasov

Eftimie, L.; Dinescu, I. & Catana, D. (1998). *Materials Engineering*, Lux Libris, ISBN 973-9240-55-0, Brasov

Iovanas, R., (2004). *Spots pressure welding*, Lux Libris, ISBN 973-9240-71-8, Brasov

Annals of DAAAM for 2008 & Proceedings of the 19th International DAAAM Symposium, ISSN 1726-9679
Editor B. Katalinic, Published by DAAAM International, Vienna, Austria 2008
Make Harmony Between Technology and Nature, and Your Mind will Fly Free as a Bird

TRACTION BEHAVIOUR SIMULATION OF SPOT WELDED JOINTS

CATANA, D[orin]

Abstract: ***For a*** *lot of production areas but particularly for automotive industry spot welding is one of the mostly used welding procedures. Most of these welding spots are arranged in line but this displacement is not always the best choice regarding traction behaviour. There are situations when no physical space is available to line up the welding spots in which case another arrangement should be found. In this paper are presented simulation and test results for spot welded joints under traction stress but for different spot arrangement. This study is based on tests on spot welded joints in which the spots were disposed in different patterns such as: square, parallelogram and rhombus. Based on tests, square arrangement turned out to be the most convenient solution. A better traction resistance for welded joint means less welded spots and means energy saving.*

Key words: *FEM, behaviour, welding, spots.*

1. INTRODUCTION

Finite element method is a powerful and efficient instrument in design and scientific research with multiple and various applications, but also a main component of computer research and design.

The expansion of using the finite element method is due, on the one hand, to its general character of its fundamental concept formulation, fact that allowed its utilization in numerous domains of science and technology, and on the other hand, its capacity to improve studied physics phenomena comparable with other numeric computation methods, fact that allowed to be preferred by many specialists.

A high fidelity reflection of materials and structures behaviour is a major exigency required by designers and used calculus methods being the primary condition for a rigorous dimensioning and accomplishing for important material savings, energy and labour. From this point of view, finite element method (FEM) is superior to all the other calculus methods.

The possibility of a precise investigation of materials behaviour is also encouraged by study and spreading of replacing materials which is another economic advantage.

Developing of this method is reflected by increasing number of publications and worldwide scientific sessions. Highly significant is the increasing number of FEM calculus software, the fact that software companies are preoccupied with supplying of these products together with calculus equipments as base soft, and as well, the development of dedicated computers for finite element calculus and its hardware components, with software included, which allow general use computers to be able to perform finite element calculus by a simple installation of the software.

2. THEORETICAL CONSIDERATIONS

FEM is an approximated solving procedure of a large variety of engineering issues by using the computer. Parts of these problems are those involving metallic materials welding.

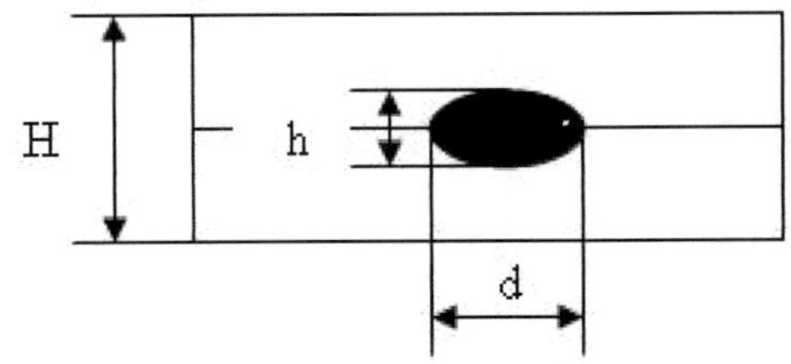

Fig. 1 Welding spot dimension for the hard welding regime.

In order to shape a geometric domain can be used dots, straight or curved lines, plane or curve surfaces. The geometrical modelling algorithm is generally based on sub-domains, with a convenient choice for passing from one sub-domain to another. For the finite elements with straight lines, digitised error can be reduced by increasing the number of nodes and therefore the number of finite elements. For finite elements with curve edges, digitised error can be reduced due to curve boundaries of finite elements (Catana, 2004).

For spot welding case (see figure 1) these dimensions and therefore the allowable stress depends on welding process parameters. Welding can be performed in a soft or hard regime. In the case of hard regime, only the welded spot area is brought to a high temperature, welding being finished before the surface of elements to be welded in contact with electrodes to reach a high temperature (Iovanas, 2004). In case of soft regime, a larger metal volume is heated which exceeds spot welded area and the electrodes in contact with the metal in plastic state will leave deep marks in the components to be welded. By comparison the spot dimensions obtained with these two welding regimes we have:

- $d_{hr} > d_{sr}$ which means that shear resistance of welded spot with hard regime is higher than that one obtained with soft regime welding (Novac, 1994);
- $h_{hr} < h_{sr}$ which means that at soft regime a bigger volume of melted metal is obtained and therefore more dendrite structure, which is more fragile and less resistant; for hard regime $h_{hr} \cong 0.3$ H, and for soft regime $h_{sr} \cong 0.9$ H.

Therefore the traction behaviour of a spot welded joint will be tested (see figure 2) (four different spots will be arranged).

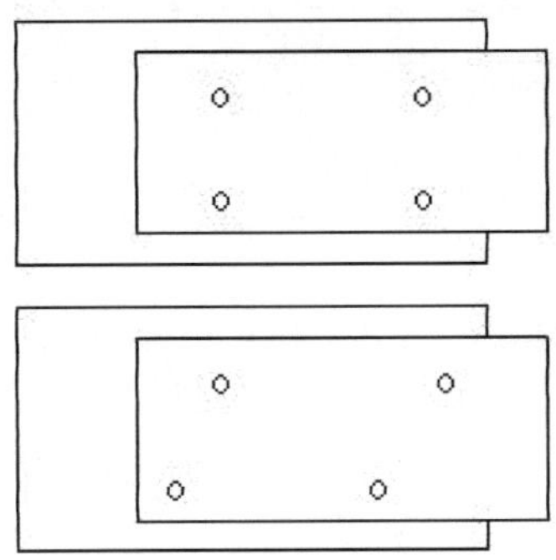

Fig. 2 Spots welding arrangement: in square and in parallelogram.

Spot arrangement	Regime I	Regime II	Regime III
o o o o	256	433	571
o o o o	258	497	596
o o o o	260	532	583
o o o o	229	458	546

Tab 1. Breaking strength of spot welded joints.

The welding points were arranged as follow:
- in line;
- in square;
- in parallelogram;
- in rhombus.

Due to forces that stress the joining, within the material stress will be present and when these exceed material breaking strain will determine the damage of welded joint.

The study of welded spot arrangement influence has been done on A_3 (STAS 9485-80) metal sheet samples with the next chemical composition:

C – max 0.08 %;
Mn – 0.2 … 0.4 %;
Si – 0.03 %.

Used parameters of the welding regime as well as the results are shown in table 1 (Catana & Machedon-Pisu, 2003). The regime parameters were:
- regime I: I=3 kA, t=3 [per];
- regime II: I= 5 kA, t=5 [per];
- regime III: I= 6 kA t=5 [per].

The parts are warmed (due to Joule-Lentz effect) by connecting them to the electrical power supply. The amount of necessary heat to perform the welding was calculated with equation 1:

$$Q = I_s^2 \cdot \left(R_c + 2 \cdot R_p\right) \cdot t \qquad (1)$$

where:

R_c – resistant of contact;
R_p – own resistant of the free heads.

For all regimes the traction speed was: v=6.5 mm/min and the dimensions of the electrodes were (10 mm diameter for body and 6 mm for peak). The thickness of metal sheets was of 0.4 mm, spot welding method being used to join two components of equal thickness and with the electrodes having the same constructive parameters. After having welding joint done, these have been tested at shearing traction according to the standard (Catana, 2005).

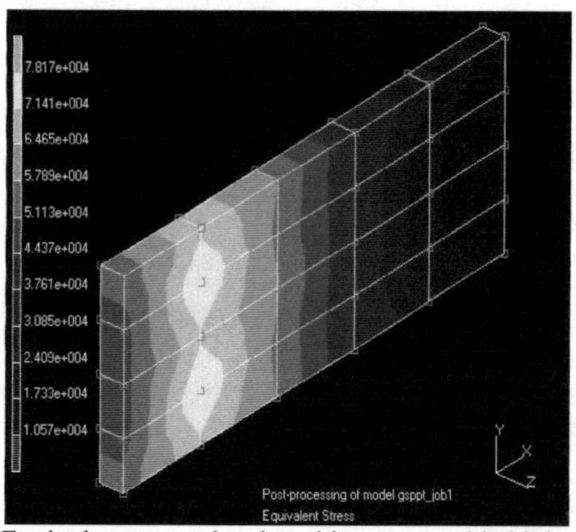

Fig. 3. Equivalent stress developed in a spot welded joint with points arranged in square.

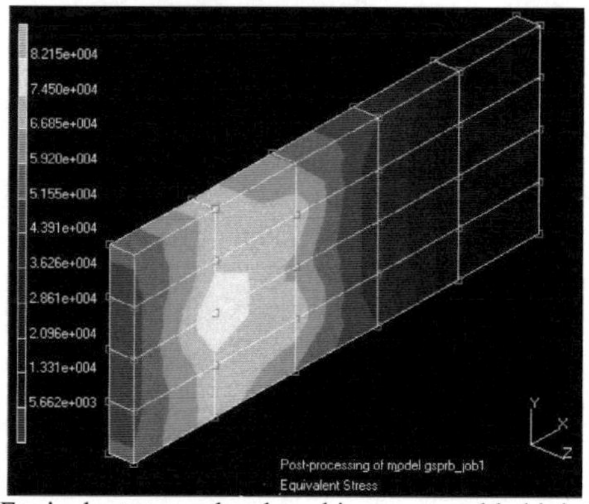

Fig. 4. Equivalent stress developed in a spot welded joint with points arranged in parallelogram.

Mechanical characteristics of used material are:
- yield limit $R_{p0,2}$ – Max 240 MPa;
- breaking strength – 270 … 370 MPa;
- breaking elongation – A_5 min 34 %.

In figure 4 is presented the unwanted case of parallelogram arrangement welding points. In this case the stress has maximum value and asymmetrical distribution. The tests were confirmed by simulation trials for all types of welding points' arrangement.

3. CONCLUSIONS

Spot welding is used to produce body cars of all types in car industry, electro technical and home appliances. This method fits most welding operation automation. From the analysis of results obtained through simulation a conclusion derives that the most convenient arrangement of welded spots is the square one where developed stresses are symmetrical and have lower values. Based on tests, line and rhomb arrangements are not to be preferred taking into consideration the value of tensions and the required space. Test results showed that parallelogram arrangement is unwanted and should be avoided when spot welding procedure is used. Research plan will continue with studying the influence over traction resistance for spot welded joints as well for the distance on longitudinal and across direction for different spot welded arrangements.

The simulation can be improved through a realistic modelling closer as possible of the way in which forces strain the joint, more precisely by establishing the volume of welded spot and its position within the joint.

4. REFERENCES

Catana, D. & Machedon-Pisu, T. (2003). The simulation of behaviour at traction of spots welded joints. *Sudura,* Vol. XIII, No. 2, 2003, 23-27, ISSN 1453-0384

Catana, D. (2004). The study through simulation of the stress developed in the spots welded joints solicit at traction, *Proceedings of International Scientific Conference „Modern Technologies, Quality, Restructuring" TMCR 2004*, Musca, G., pp. 461-464, ISSN 1011-2855, Jassy, 05-2004, Bulletin of Polytechnic Institute of Jassy, Jassy

Catana, D. (2005). Theoretic contributions for the plastic deformation simulation process, *Proceedings of International Scientific Conference „Modern Technologies, Quality, Restructuring" TMCR 2005*, pp. 333-337, ISBN 9975-9875-4-0, Technical University of Moldova, 05-2005, University of Moldova, Chisinau

Iovanas, R., (2004). *Spots pressure welding*, Lux Libris, ISBN 973-9240-71-8, Brasov

Novac, G., (1994). *Welded joints calculus*, Lux Libris, ISBN 973-96308-7-1, Brasov

Annals of DAAAM for 2008 & Proceedings of the 19th International DAAAM Symposium, ISSN 1726-9679
Editor B. Katalinic, Published by DAAAM International, Vienna, Austria 2008
Make Harmony Between Technology and Nature, and Your Mind will Fly Free as a Bird

NUMERICAL PROOF OF CALORIMETER CHEMICAL FREEZING

CATANA, R[azvan]; CICAN, G[rigore]; PREDOIU, C[onstantin]; SILIVESTRU, V[alentin] & RUGESCU, R[adu]

Abstract: *The observation that the chemical equilibrium between the combustion products of solid propellant samples within static calorimeters is unexpectedly freezing at high temperatures is proved through a general numerical simulation of the isochoric cooling with chemical reactions between the gaseous products. A proprietary, direct linearization method of thermochemical computation is used that enables following any chemical reaction in equilibrium with high convergence. The observed chemical freezing within calorimeters is proved.*

Key words: *calorimetry, thermochemistry, chemical equilibrium, propellants*

1. INTRODUCTION

The process of combustion of solid propellant samples within a small constant volume calorimeter is considered as a typical slow evolution. The motion of the developed gas mixture is fairly limited to the closed vessel space and no fast moving gasses are ordinarily considered to appear. After the inner volume is fast filled with the gas mixture at high temperature and pressure a very slow cooling of the content follows, almost down to the surrounding water, preserving at the room temperature. There is no apparent reason to consider that the chemical equilibrium is not fulfilled during this process (Kuhl *et al.* 2006; Beckstead, 2006; Kubota, 2007).

It was observed however, during repeated calorimetric experiments, followed by frequent chemical composition analyses, that the composition of the residual gas from the calorimeter does not obey the chemical equilibrium at room temperature, but at an unexpectedly high temperature instead (Rugescu, 2005). The observation was striking and led to numerical simulations of isochoric cooling that show that the composition of the residual gases in the vessel remains frozen to a temperature as high as 1650ºK after hours of standing at the final room temperature.

2. CHEMICAL FREEZING: STATE OF THE ART

Freezing in chemical composition of the air was early suspected and experimentally detected afterwards in supersonic wind tunnels and shock tubes. A similar behaviour was later attributed to the rocket nozzle gasses that flow with supersonic speed in the diverging extension of the nozzle. This phenomenon greatly complicates the computations of high-speed flow, requiring strong numerical schemes, involving differential equations of chemical kinetics. A simplifying treatment of this kind of flows was proposed during the 50's that led to approximate description of the high-speed flow with considerable computational time savings (Bray, 1959).

Many methods of simulating fast chemical interacting flows were developed (Westenberg 1962, Williams, 1965, Hill et al, 1967, Schuricht, 2001). They confirm the sudden chemical freezing in supersonic nozzles. However, no observation is reported on any non-equilibrium in calorimetric evolutions.

Chemical freezing, when identified in nozzles, produces a constant chemical composition to occur after a certain area of the fast expanding gas flow in the nozzle (Fig. 1).

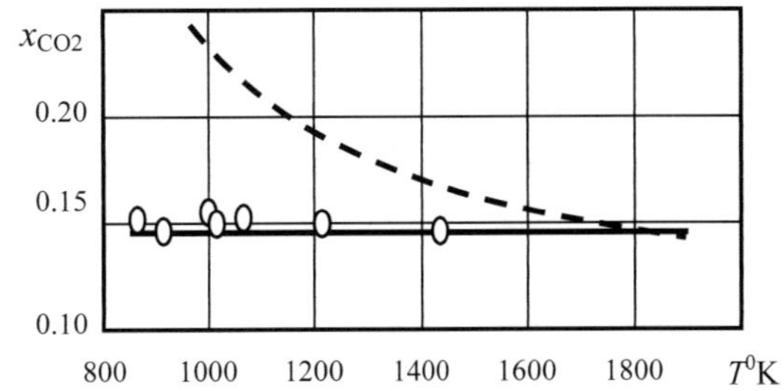

Fig. 1 Nozzle experimental and computed CO_2 concentration.
.... Chemical equilibrium computation.
—— Non-equilibrium flow computation.
O Experimental measurements (Hill, 1967).

Although in the diagram the temperature is a reference, the chemical freezing in nozzles is only related to high velocities of the flow, hindering the molecules from further interacting. It is important to note however, that the temperature range where the kinetical freezing was observed to occur is sensibly similar to the temperature range where a chemical departure from equilibrium within calorimeters was observed (Rugescu, 2005).

3. THERMOCHEMICAL NUMERICAL METHOD

In order to cover all possible dissociation and formation processes during the calorimeter cooling of the combustion products, a universal method is built, based on a direct linearization procedure. The equations of conservation are first written as the conservation of chemical elements,

$$A_i\eta = \sum_{j=1}^{n} {}^{G}\alpha_{ij} p_j + \sum_{j=1}^{n} {}^{C}\eta\alpha_{ij} n_{j*}, \quad i = \overline{1,l}\,, \tag{1}$$

chemical reactions with equilibrium constants K_r:

$$\sum_{i=1}^{n} {}^{E}\nu'_{ri}\mathsf{M}_i \xleftrightarrow{K_r} \sum_{i=1}^{n} {}^{E}\nu''_{ri}\mathsf{M}_i\,, \tag{2}$$

their gas-phase equilibrium (Q - the number of condensed spec)

$$\ln K_r(T) = \sum_{i=1}^{n} {}^{G}(\nu''_{ri} - \nu'_{ri}) \ln p_i, \quad r \in \{R \setminus Q\}. \tag{3}$$

mono-condensed species:

$$\ln K_q = \nu''_{qi} \sum_{i=1}^{n} {}^{G} p_i, \quad q \in \{Q\} \tag{4}$$

local chemical freezing:

$$p_i / p_\Sigma = (p_i / p_\Sigma)_{\,local}, \quad n_i^* = (n_i^*)_{\,local}, \quad i \in \{n\}, \tag{5}$$

equilibrium of phases:

$$\ln p_k^S = K_k(T),\ \ p_k = \delta_k p_k^S + (1-\delta_k) p_k,\ \ \overset{*}{n}_k = \delta_k \overset{*}{n}_k, \quad (6)$$

equation of state with η for the constant volume loading

$$\eta = p_\Sigma R_0 T \left(1 - \frac{\sum_{j=1}^{n} {}^{Q}\mu_j \overset{*}{n}_j + \sum_{j=1}^{n} {}^{\Xi}\delta_k^0 \mu_j \overset{*}{n}_j}{1000} \right) \quad (9)$$

Energy conservation governs the adiabat-isochoric combustion

$$\frac{1}{\eta}\sum_{j=1}^{n} {}^{G}\delta_k^0 p_j H_j^0(T) + \sum_{j=1}^{n} {}^{C}\delta_k^0 \left[\overset{*}{n}_j H_{j*}^0(T) + \frac{p_k^S(T)}{\eta} H_j^0(T) \right] - H_0 + R_0(n_0 T_1 - n_T^G T) = H_p \quad (10)$$

while the isochoric cooling through

$$\sum_{i=1}^{n} \left\{ \delta_k^0 \left[\frac{p_k^S(T)}{\eta} H_i^0(T) \right]^G + \left[\overset{*}{n}_i H_{i*}^0(T) \right]^Q + \delta_k^0 \left[\overset{*}{n}_i H_{i*}^0(T) + \frac{p_k^S(T)}{\eta} H_i^0(T) \right]^{\Xi} \right\} - n_G(T) R_0 T = H_p + H_0 - n_0 R_0 T_1 - Q_V. \quad (11)$$

The direct linearization with the following corrections applies,

$$p_j \leftarrow p_j^0 x_j,\ p_j \leftarrow p_j^0 x_j,\ \eta \leftarrow \eta^0 x_n. \quad (12)$$

Improvements in corrections are successively extracted, at each iteration, from the linearized system of equations

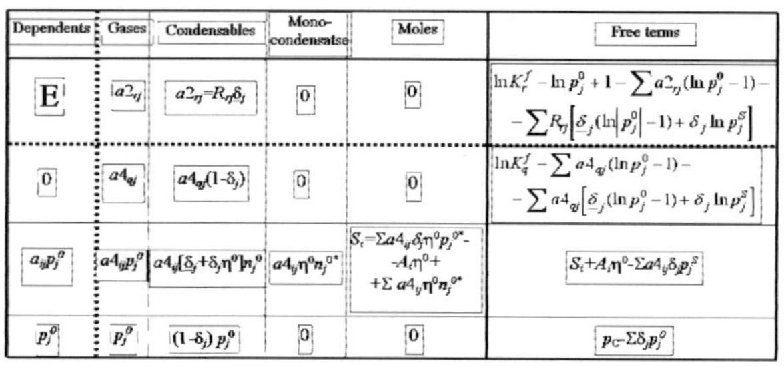

that give the composition of the gas at a given temperature T.

4. EXPERIMENTAL MEASUREMENTS

A standard combustor was used for the isochoric combustion and cooling of compositions A-100 and A-300:

A_o=33.7517 A_n=9.62048 A_c=22.66504 A_h=29.62565 A_{pb}=0.07938 A_{ca}=0.167173
A_o=35.697 A_n=10.24663 A_c=19.729682 A_h=28.14021 A_{pb}=0.0815 A_{ca}=0.077884

For the first propellant the chemical composition of the residual gas in Table 1 was found through gas chromatography, while for the second propellant the composition of the residual gas given in Table 2 was similarly found:

Species	H_2	N_2	CO	CO_2
Sample 1	22.927	13.657	51.467	12.669
Sample 2	22.641	13,646	51.283	13.668
Sample 3	22.829	12.898	51.513	12.766
Sample 4	22.772	12.861	51.463	12.584
Sample 5	22.783	12.923	51.582	12.892
Mean	22.847	13.197	51.481	12.665

Tab 1. Residual calorimeter gas of the A-100 propellant.

Species	Sample 1	Sample 2	Mean
H_2	16.793	16.519	16.6560
N_2	16,484	16.643	16.5640
CO	42.255	42.579	42.4170
CO_2	24.868	24.255	24.5615

Tab 2. Residual calorimeter gas of the A-300 propellant.

At room temperature the water is completely condensed so that only four gas concentrations were measured.

5. COMPARISON AND CONCLUSIONS

Only the four main gases are listed below, in order to compare the results. It was surprisingly found that the following high temperatures equilibrium values equal the composition, measured at *room* temperature:

Temperature	H_2	N_2	CO	CO_2
2000	21.816	13.679	53.445	10.963
1711	22.816	13.504	51.481	12.103
1704	22.847	13,499	51.420	12.138
1608	23.309	13.418	50.513	12.665
1403	24.571	13.197	48.035	14.103
700	22.783	12.923	51.582	12.892

The temperatures where the computed composition is identical with the measured one are well grouped. This shows that the phenomenon is not random. The adiabatic temperature of isochoric combustion (explosion) would have been around 2775 K, far above the values of consistency from the table. It is difficult to suppose that the temperature, where the experimental composition fits the computed one, is the actual temperature of combustion, as the far more energetic formulation A-300 provided almost equal or even a little lower temperature of freezing. This completely removes the idea that the observed temperatures are the actual temperatures of isochoric combustion. The above computations follow previous ones with very different codes and all give the same conclusion: the chemical composition greatly departs from normal equilibrium. A proof of the freezing temperatures is expected from pressure measurements during isochoric combustion.

The computations and measurements clearly show that the chemical composition of the residual gas after a long and slow cooling remains blocked at the composition that the gas went through at an early stage of cooling. The reasons for this freezing are now considered for a detailed investigation through a national CNCSIS grant, currently under evaluation.

6. REFERENCES

Beckstead, M. (2006), Recent progress in modeling solid propellant combustion, *Combustion, Explosion, and Shock Waves* (*CESW*), Vol. 42, No. 6, Nov., pp. 623-641 (19).

Bray, K. N. C. (1959), Atomic recombination in a hypersonic wind-tunnel nozzle, J. of Fluid Mech., **6**(1), p. 1-32.

Hill F.; Unger H. & Dickens W. (1967), *AIAA Journal*, 5(5) 54.

Kubota N. (2007), Propellants and Explosives-Thermochemical Aspects of Combustion, 2nd Ed, Wiley, N. York.

Kuhl, A., Neuwald, P., and Reichenbach, H. (2006), Effectiveness of combustion of shock-dispersed fuels in calorimeters of various volumes, *CESW*, **42**(6), p. 731-734

Rugescu, R.D. (2005), Chemical Freezing, Engineering Meridian, TUM Chisinau, Moldova

Schuricht, S. R. (2001), *Numerical simulation of high speed chemically reacting flows*, Purdue Univ., AAT 3075724.

Williams, J. C., III (1965), Correlation of the sudden freezing point in nonequilibrium nozzle flows, *AIAA J.*, **3**(6), p. 1169

Westenberg ,A. A.; Favin,S. (1962), *Nozzle Flow with Complex Chemical Reaction*, Johns Hopkins Univ., Appl. Phys. Lab, AD0275464.

Annals of DAAAM for 2008 & Proceedings of the 19th International DAAAM Symposium, ISSN 1726-9679
Editor B. Katalinic, Published by DAAAM International, Vienna, Austria 2008
Make Harmony Between Technology and Nature, and Your Mind will Fly Free as a Bird

THE MATHEMATICAL MODELLING OF THE WORK SPACES WITH VARIABLE VOLUME IN THE WANKEL ROTARY ENGINE CASE

CATAS, A[driana]; CHIOREANU, N[icolae] & BLAGA, V[asile]

Abstract: *Taking in consideration the work spaces with variable volume which are created during the engine running, the authors develop a theoretical approach of the determination of the volume function variation in the Wankel rotary engine case. The paper presents the parameter equations of the inner surface section of the carcass of the piston section and the determination of the volume function variation of one chamber. Due to this fact one can obtains an optimization of the operational factors in the case of this rotary engine.*
Key words*:* *Wankel engine, hypocycloid, Green's formula.*

1. GENERAL CONSIDERATIONS

The *Wankel* rotary engine is the most known rotary heat engine, patented in 1954 by Felix Wankel. The engine is built by a carcass which has an inside surface with an appropriate geometry. Inside the carcass a rotary piston is mounted over an eccentric haft. Because the rotation axes of the piston and stopper are not focused the work spaces with variable volume are created. The piston has an equilateral triangle shape with curved sides. During the rotating motion the peaks of the piston are in permanent contact with the inner surface of the carcass, forming three chambers. The volume of these chambers has a cyclical variation. For the achievement of the permanent contact between the peaks of the piston and the inner surface of the carcass, the paths of the peaks must be hypocycloids which are identical with the inner profile of the carcass. In this paper we will follow the notations used in (Chioreanu & Chioreanu, 2006). For more constructing details we can also consult (Marr, 2004 and Owen & Coley, 1990).

2. THE CARCASS PROFILE

The carcass has the bore in the form of hypocycloid (Aubin, 1976 and Libeskind, 2008), with 2 lobes. It is obtained by rolling a circle by the internal gear fixed on the rotor which is rolling on the external gearing. The engine executes three functioning cycles in four periods at one rotation of the rotor and three rotations of the engine shaft. The hypocycloid is the curve described by the point V_1, fixed relative to the mobile circle of r radius, which rolls without sliding on another fixed circle of r radius. The circles are internal. This curve closes if and only if the ratio r/R is rational. In the case of the *Wankel* engine the ratio of the radii is $r/R = 2/3$ (two lobes, the rotor rotates only once for each three rotations of the engine shaft), therefore the mobile circle runs along the fixed circle for three times. The center of the mobile circle describes a circle of e radius, $e = R - r$ called the eccentricity, and it has its center at the point M_0, the initial position. The center of the fixed circle is P. In order to determine the parameter equation of the hypocycloid we consider an orthogonal system of (x, y) coordinates with the origin at P. Let A be the contact point of the two circles at the initial position and let M be the center of the mobile circle after the angle rotation.

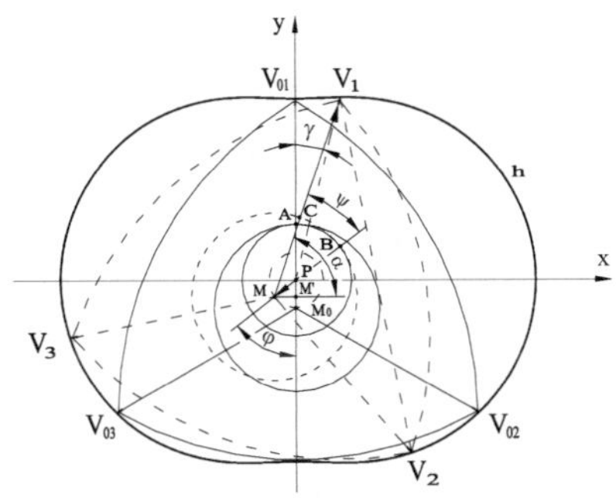

Fig. 1. Hypocycloidal path of the generated surface

Denote $c = d(M_0, V_{01})$. The contact point between the two circles is denoted by B and by $\psi = \measuredangle(BMV_1)$ (figure 1). From the rolling condition $\widehat{AB} = \widehat{CB}$ we obtain $r\varphi = R\psi$ or $\psi = (2/3)\varphi$ and from the vector equality $\overrightarrow{PV_1} = \overrightarrow{PM} + \overrightarrow{MV_1}$ we deduce the parameter equations which generate the sectional area of the carcass

$$x = c\sin(\varphi/3) - e\sin\varphi,\ \ y = c\cos(\varphi/3) - e\cos\varphi. \qquad (1)$$

Let us denote by M' the projection of point M on the Py axis and put $\alpha = \measuredangle(M'MV_1)$. One obtains $\alpha = (\pi/2) - (\varphi/3)$.

3. THE PISTON PROFILE

The peaks of the piston achieve a permanent contact with the inner surface of carcass. When the engine shaft rotates, the rotor will make a movement on a relative orbit compared to the engine shaft axis. The peaks of the piston are given by the intersection between a mobile circle, of c radius with the center at the point M and the hypocycloid of the carcass. The parameter equations of the circle are

$$x_c = c\ \sin\gamma_c - e\sin\varphi_c,\ y_c = c\ \cos\gamma_c - e\cos\varphi_c, \qquad (2)$$

where φ_c is the position of the center of the circle of c radius. The solution of the following system

$$\begin{cases} c\ \sin\gamma_c - e\sin\varphi_c = c\sin(\varphi/3) - e\sin\varphi, \\ c\ \cos\gamma_c - e\cos\varphi_c = c\cos(\varphi/3) - e\cos\varphi \end{cases} \qquad (3)$$

is $\gamma_c = (\varphi_c + 2k\pi)/3$. For $\varphi_c = \varphi$, the center of the circle coincides with the center of the piston P i.e. $\gamma_c = \gamma = (\varphi_c + 2k\pi)/3,\ k = 0,1,2$. We note that $\Delta\gamma_c = [\varphi + 2(k+1)\pi]/3 - (\varphi + 2k\pi)/3 = 2\pi/3$. For $\varphi = 0$, the positions of the peak piston: $\varphi_1 = 0, \varphi_2 = 2\pi/3, \varphi_3 = 4\pi/3$.

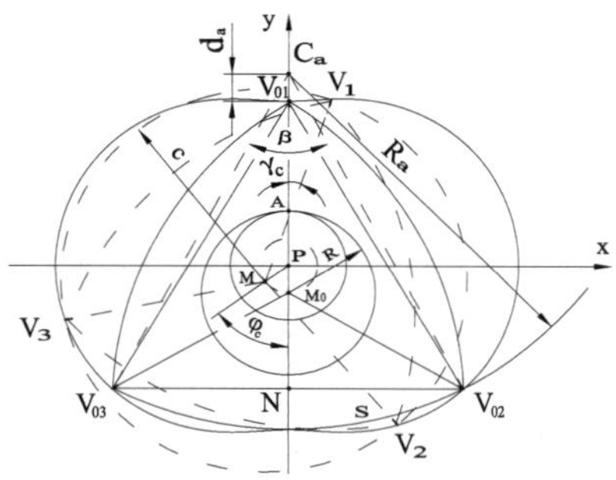

Fig. 2. The piston profile

Therefore, the peaks of the piston form an equilateral triangle with the radius of the circumscribed circle equal to c and the radius of the inscribed circle equal to $c/2$.

For construction reasons we have $R \le c/2$ so $\rho := c/e \ge 6$. The sides of the piston are considered curve arches with the center on the line determined by the M center and the opposite peak. The rotating of the piston is achieved if and only if $2(c-e) \ge R_a - d_a$, where R_a is the radius of the curve arch and $d_a = d(V_{01}, C_a)$ (figure 2). Let us denote by $k = d_a/e$. The triangle $\Delta C_a V_{02} V_{03}$ is isosceles having $C_a V_{02} = C_a V_{03} = R_a$; $V_{02}V_{03} = \sqrt{3}\cdot c$. From $\Delta C_a N V_{02}$ one obtains $R_a = e\cdot\sqrt{k^2 + 3\cdot\rho\cdot k + 3\cdot\rho^2}$. We denote by $k_1 = j/e$ where j is the free running between the carcass and piston, the peak V_1 being in initial position. We obtain $2\cdot(c-e) - j = R_a - d_a$ and

$k = -\left[\rho^2 - 4(k_1+2)\rho + (k_1+2)^2\right]/\left[\rho - 2(k_1+2)\right]$. For $k = 0$

$\rho = \left(2+\sqrt{3}\right)\cdot(k_1+2)$ with $\rho > 2\cdot\left(2+\sqrt{3}\right) = 7.4641$.

Writing the triangle $\Delta C_a V_{02} V_{03}$ area in two ways we get $\sin\beta = \left(\sqrt{3}/2\right)\left[\rho(3\rho+2k)\right]/\left[k^2+3\rho k+3\rho^2\right]$. For $k = 0$ we have $\beta = \pi/3$. The area of circle segment $V_{03}SV_{02}V_{03}$ is $A_s = (\beta - \sin\beta)R_a^2/2 = (\beta-\sin\beta)\left(k^2+3\rho k+3\rho^2\right)e^2/2$. For $k=0$ we get $A_s = 3\rho^2\left[(\pi/3) - \left(\sqrt{3}/2\right)\right]e^2/2$.

4. THE VOLUME FUNCTION VARIATION

The three sides of the triangular piston determine with the carcass three workspace, separated one from another. The volume of this space is modified during the rotation of the piston. The volume of one chamber formed between piston and carcass has a cyclical variation as a function of the φ angle given by the relation $V_p(\varphi) = l_p \cdot A_p(\varphi) + V_{ca}$, where: l_p is the breadth of the piston, $A_p(\varphi)$ is the plane area of one chamber and V_{ca} is the volume of the box fire. The $A_p(\varphi)$ is given by the Green's formula $\iint_D dxdy - A_s = \oint_{\Gamma^+} -ydx - A_s$ where D *is* the compact domain, simple with respect to the Py axis, bounded by peritrochoidal arch extended between the variable points V_1, V_2 and the segment V_1V_2. The curve $\Gamma = \partial D$ is formed by the concatenation of the curves Γ_1, Γ_2

$$\Gamma_1 : \begin{cases} x = c\sin(\alpha/3) - e\sin\alpha, \\ y = c\cos(\alpha/3) - e\cos\alpha \end{cases}, \Gamma_2 : \begin{cases} x = x_{V_1} + t\left(x_{V_2} - x_{V_1}\right), \\ y = y_{V_1} + t\left(y_{V_2} - y_{V_1}\right) \end{cases},$$

where $\alpha \in [\varphi, \varphi+2\pi]$, $t \in [0,1]$, $\oint_{\Gamma^+} -ydx = -\oint_{\Gamma_1^+} ydx - \oint_{\Gamma_2^+} ydx$,

$$\begin{cases} x_{V_1} = -e\sin\varphi + c\sin(\varphi/3), \\ y_{V_1} = -e\cos\varphi + c\cos(\varphi/3) \end{cases} \begin{cases} x_{V_2} = -e\sin\varphi + c\sin(\varphi+2\pi)/3, \\ y_{V_2} = -e\cos\varphi + c\cos(\varphi+2\pi)/3 \end{cases}$$

$$A_p(\varphi) = \pi e^2 + \left[\left(4\pi - 3\sqrt{3}\right)/12\right]\rho^2 e^2 + u - A_s \qquad (4)$$

where $u = \left[\left(3\sqrt{3}\right)/2\right]\rho e^2 \sin\left[(2\varphi/3) + \pi/6\right]$. The maximal and minimal values of the function $V_p(\varphi)$ are

$$V_{p\max} = V_p(\pi/2) = V_p(7\pi/2); V_{p\min} = V_p(2\pi) = V_p(5\pi)$$

The cylinder capacity V_s is $V_s = V_{p\max} - V_{p\min} = 3\sqrt{3}\cdot\rho\cdot l_p\cdot e^2$ and the compression ratio is $\varepsilon = V_{p\max}/V_{p\min}$, thus

$$V_{ca} = \left[\frac{3\sqrt{3}}{2}\frac{\varepsilon+1}{\varepsilon-1}\rho - \frac{4\pi - 3\sqrt{3}}{12}\rho^2 + \frac{(\beta-\sin\beta)\left(k^2+3\rho k+3\rho^2\right)}{2} - \pi\right]l_p e^2$$

For $k = 0$ and $\sin\beta = \sqrt{3/2}$, we have

$$V_p(\varphi) = \frac{3\sqrt{3}}{2}\rho\left[\frac{\varepsilon+1}{\varepsilon-1} + \sin\left(\frac{2\varphi}{3} + \frac{\pi}{6}\right)\right]l_p e^2 \qquad (5)$$

The maximum and the minimum volume of one chamber is

$$V_{p\max} = \left[\varepsilon/(\varepsilon-1)\right]V_s; \quad V_{p\min} = \left[1/(\varepsilon-1)\right]V_s.$$

In the figure 3 it is presented the variation of $V_p(\varphi)$ for one chamber of an engine with the following parameters: $V_s = 250\ cm^3$, the compression ratio $\varepsilon = 10.5$, eccentricity $e = 10.5\ mm$, $\rho = 7.5$, $l_p = 58\ mm$.

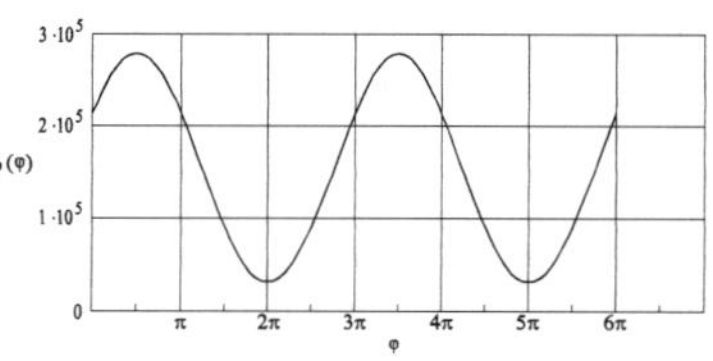

Fig. 3. Variation of $V_p(\varphi)$ for one chamber

If it is compared a *Wankel* rotary engine with an internal combustion four-cycle engine, having the same cylinder capacity, which at a certain number of rotations develops the same power, we conclude: a *Wankel* rotary engine is equivalent to an internal combustion four-cycle engine having two cylinders.

5. REFERENCES

Aubin, T. (2001). *A course in differential geometry*, Providence, R.I. : American Mathematical Society, ISBN 082182709X.

Chioreanu, N. & Chioreanu, S. (2006). *Heat engine for nonconventional motor vehicles*, Ed. Univ. of Oradea.

Libeskind, S. (2008). *Euclidian and Transformational Geometry*, ISBN-13: 9780763743666, Jones and Bartlett Publishers, Inc.

Marr, A. (2004). *Wankel Rotary Combustion Engines (WRCE) and Vehicles*. April 7, 2000. Accessed October 18.

Owen, K. & Coley, T. (1990) *Automotive Fuels Handbook.* Warrendale, PA: Society of Automotive Engineers, ISBN1-56091-064-X.

Annals of DAAAM for 2008 & Proceedings of the 19th International DAAAM Symposium, ISSN 1726-9679
Editor B. Katalinic, Published by DAAAM International, Vienna, Austria 2008
Make Harmony Between Technology and Nature, and Your Mind will Fly Free as a Bird

FORMALISM OF INDUSTRIAL ROBOTS KINEMATICS APPLIED TO NUMERICAL COMMAND LATHES

CATRINA, D[umitru]; GHIONEA, A[drian]; CONSTANTIN, G[eorge] & LUNGU, B[ogdan]

Abstract: *This paper presents a study concerning kinematics analysis method used in CNC machine-tools. Some lathe structures are analyzed having revolution (R) and translation (T) joints similar to industrial robots, that enable generation motions combined for generating trajectories that define the machined surfaces. This method is intended to be applied to different types of machine tools, and gives an instrument for easy choice of the machine for generating certain surfaces.*
Key words: *kinematic couples, kinematic structures, fictitious mechanism, generated trajectory.*

1. INTRODUCTION

For an industrial robot, the typical translational kinematic chain consists of an electric motor, reducer and ball screw-nut, which moves a translational module (Catrina et al., 1993). The same kinematic chain structure is used in several CNC machine tools types (Weck, 2001). Based on this observation, similarly the machine tool can be considered as the industrial robot equipped with at least two gripping systems, one for each component of the fictitious mechanism (*FM*) workpiece-tool.

The fictitious mechanism piece-tool, defined for the first time by Emil Botez (Botez, 1966) can be extended for all types of closed or opened kinematic chains. We accept that the fictive mechanism workpiece-tool is the one that appears during the generation of surfaces, as there is a real mechanism between the tool (*T*) and workpiece (*Wp*).

The kinematic chain produces at its end a relative translation *T* or rotation *R* motion between two elements of the robot or machine tool connected by a couple (joint).

The kinematic branch is constituted as an element assembly united by couples driven by kinematic chains, which start on the base (bed) and end at a gripping mechanism (*GM*).

A robot having a gripping mechanism has only a kinematic branch, and a machine-tool, having several clamping devices (*GM*), has in its structure more than one kinematic branches.

For the machines that at a moment of time cut one workpiece using only one tool, this *FM* has an input and an output; for the machines that cut a single workpiece with several tools, the *FM* has several inputs and outputs. Regarding these considerations, one proposes the extension of some methods used in kinematic study in industrial robots in machine tool kinematic analysis (in our case with application to lathes). Some studies in this field are known for up to 5 axes CNC machine tools (Bohez, 2002).

For an industrial robot, the structural formulae O - C_i^j (T_i - R_j) – GM is of open kinematic chain type, with one input O (bed) and an output GM (gripping mechanism), where C_i^j (T_i - R_j) is considered the opened kinematic chain consisting of combinations of translation T_i and rotation R_j joints.

For a CNC machine tool, the structural formula is:

$$GM_T - C_i^j\,(T_i - R_j) - O - C_m^{\,n}\,(T_m - R_n) - GM_{Wp}, \qquad (1)$$

where GM_T is the gripping mechanism for tools (one or more), and GM_{Wp} is the gripping mechanism for workpieces (W) (one or more), all mounted on the same bed.

2. CNC LATHE STRUCTURES

In Figs. 1 and 2 some principle schemes for horizontal and vertical CNC lathe types are presented. The notation are as follows: *MS* – main spindle, *LS*/*TS* – longitudinal/transversal slide, *Ts* - tailstock, *HS*/*VS* – horizontal/vertical slide, *RH* – revolver head, *C* – column, *B* -bed.

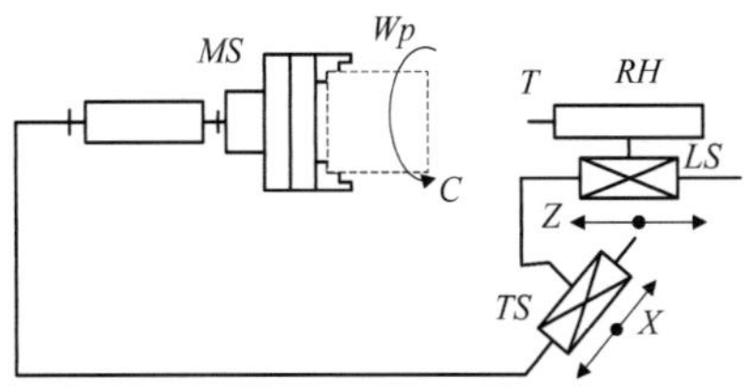

Fig. 1. Front lathe structure: *R* (*MS*) – *O* – *T* (*TS*) – *T* (*LS*).

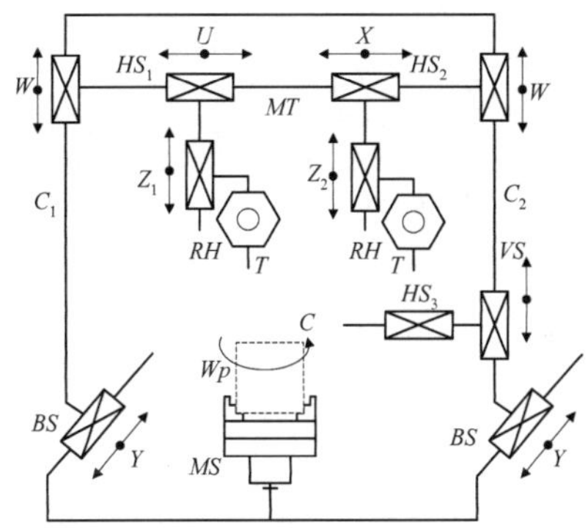

Fig. 2. Vertical lathe with two columns and movable bridge having the structure:

$$R(MS) - O - T(BS) - \left[\begin{array}{l} -T(VS_3) - T(HS_3) \\ -T(MT) - \left[\begin{array}{l} -T(HS_1) - T(VS_1) \\ -T(HS_2) - T(VS_2) \end{array}\right. \end{array}\right.$$

3. THEORETICAL ASPECTS. APLLPICATIONS

These kind of representations and structural formulae are based on formalisms used in kinematic structural analysis for industrial robots. For example, the Denavit-Hartenberg convention (Popescu et al., 1994) allow the transformation matrix calculation between two reference systems attached to two consecutive elements from robots structure. The positioning of an element relatively to an adjacent one is achieved by attaching a reference system to every element having its index number corresponding to the element number i.

For solving this model, one should know the meaning of the following operators:

- α_i – angle about x_{i-1} axis measured between z_{i-1} and z_i;
- a_i – distance along x_{i-1} from the origin O_{i-1} to z_i axis;
- θ_i – angle about z_i axis measured between x_{i-1} and x_i;
- d_i – distance along z_i axis measured between intersection point of x_{i-1} and z_i axes and O_i origin.

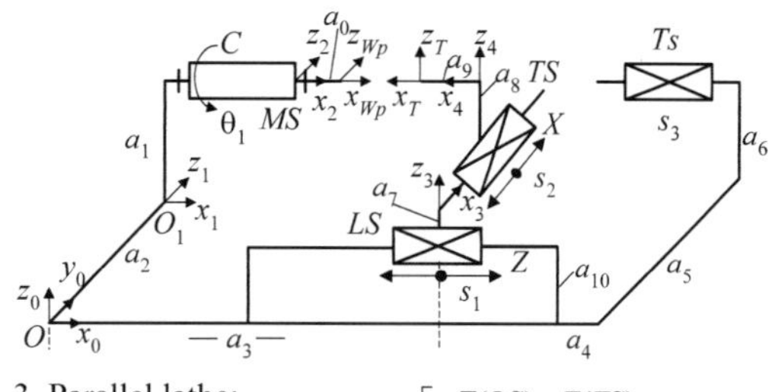

Fig. 3. Parallel lathe: $R(MS) - O - \left[\begin{array}{l} -T(LS) - T(TS) \cdot \\ -T(Ts) \end{array} \right.$

	Bed-workpiece branch			Bed-tool branch		
Operator	1	2	p	3	4	S
α_i	- 90°	90°	$-\theta_1$	0	0	0
a_i	0	0	a_0	$a_3 + s_1$	s_2	a_9
θ_i	0	0	0	90°	90°	0
d_i	a_2	a_1	0	$a_{10}+a_7$	a_8	0

Tab 1: Denavit-Hartenberg formalism for the parallel lathe.

The transformation matrix for an entire kinematic branch is the product of the transformation matrix of the reference systems attached to each element of the branch:

$$T_0^n = T_0^1 \cdot T_0^2 \cdot ... \, T_{i-1}^i \cdot T_{n-1}^n, \tag{2}$$

where

$$T_{i-1}^i = \begin{bmatrix} \cos\theta_i & -\sin\theta_i & 0 & a_i \\ \cos\alpha_i \cdot \sin\theta_i & \cos\alpha_i \cdot \cos\theta_i & -\sin\alpha_i & -d_i \cdot \sin\alpha_i \\ \sin\alpha_i \cdot \sin\theta_i & \sin\alpha_i \cdot \cos\theta_i & \cos\alpha_i & d_i \cdot \sin\alpha_i \\ 0 & 0 & 0 & 1 \end{bmatrix}. \tag{3}$$

As example, we present the structure of a parallel lathe having a main spindle *MS*, longitudinal slide *LS*, transversal slide *TS*, and tailstock *Ts* (Fig.3).

In Table 1, the Denavit-Hartenberg parameters for bed-workpiece and bed-tool kinematic branches are presented.

The transformation successive matrix for bed (0)-workpiece (*Wp*) branch is:

$$T_0^1 = \begin{bmatrix} 1 & 0 & 0 & 0 \\ 0 & 0 & 1 & a_2 \\ 0 & -1 & 0 & 0 \\ 0 & 0 & 0 & 1 \end{bmatrix}, \; T_1^2 = \begin{bmatrix} 1 & 0 & 0 & 0 \\ 0 & 0 & -1 & -a_2 \\ 0 & 1 & -0 & 0 \\ 0 & 0 & 0 & 1 \end{bmatrix},$$

$$T_2^{Wp} = \begin{bmatrix} 1 & 0 & 0 & a_0 \\ 0 & \cos\theta_1 & \sin\theta_1 & 0 \\ 0 & -\sin\theta_1 & \cos\theta_1 & 0 \\ 0 & 0 & 0 & 1 \end{bmatrix}. \tag{4}$$

The transformation matrix for the attached end elements of the branch 0-*Wp*:

$$T_0^{Wp} = T_0^1 T_1^2 T_2^{Wp} = \begin{bmatrix} 1 & 0 & 0 & a_0 \\ 0 & \cos\theta_1 & \sin\theta_1 & a_2 \\ 0 & -\sin\theta_1 & \cos\theta_1 & a_1 \\ 0 & 0 & 0 & 1 \end{bmatrix}. \tag{5}$$

In the same way the transformation matrix for the attached elements of the 0-*T* branch is calculated:

$$T_0^T = T_0^3 T_3^4 T_4^T = \begin{bmatrix} 1 & 0 & 0 & -a_9 + a_3 + s_1 \\ 0 & -1 & 0 & s_2 \\ 0 & 0 & 1 & a_7 + a_8 + a_{10} \\ 0 & 0 & 0 & 1 \end{bmatrix}. \tag{6}$$

The transformation matrix between the workpiece and tool (fictitious mechanism *Wp-T*) results in the form:

$$T_{Wp}^{\ T} = (T_0^{\ Wp})^{-1} \, T_0^{\ T} \tag{7}$$

On the basis of relation (5), the coordinates (x_{Wp}, y_{Wp}, z_{Wp}) of a point in the workpiece reference system result. As well, one can obtain the coordinates (x_T, y_T, z_T) for a point in the tool reference system. In both cases, one knows the point coordinates in the bed reference system (x_0, y_0, z_0). These have the form:

$$\begin{cases} x_{Wp} = x_0 + a_0 \\ y_{Wp} = y_0 \cos\theta_1 + z_0 \sin\theta_1 + a_2 \\ z_{Wp} = -y_0 \sin\theta_1 + z_0 \cos\theta_1 + a_1 \end{cases} ; \quad \begin{cases} x_T = x_0 - a_9 + a_3 + s_1 \\ y_T = -y_0 + s_2 \\ z_T = a_7 + a_8 + a_{10} + z_0 \end{cases}. \tag{8}$$

Subtracting the equations corresponding to the two equation systems the relations between the inputs and outputs of the workpiece-tool fictitious mechanism result:

$$\begin{cases} x_{Wp} = x_T + a_0 + a_9 - a_3 - s_1 \\ y_{Wp} = (s_2 - y_T)\cos\theta_1 + (z_T - a_7 - a_8 + a_{10})\sin\theta_1 + a_2 \\ z_{Wp} = (y_T - s_2)\sin\theta_1 + (z_T - a_7 - a_8 - a_{10})\cos\theta_1 + a_1 \end{cases}. \tag{9}$$

Equation (7) expresses the parametric equation of the generated path through the fictitious mechanism *Wp-T*.

Application. For simultaneous motions with constant speeds w_X, w_Z and ω corresponding to the CNC axes *X*, *Z* and *C* the equations of a conical helix are obtained based on eq. (9):

$$\begin{cases} x_T = w_Z t \\ y_T = w_X t \\ \theta_1 = \omega t \end{cases}, \quad \begin{cases} x_{Wp} = w_Z t + a_0 + a_9 - a_3 - s_1 \\ y_{Wp} = (s_2 - w_X t)\cos\omega t + (z_T - a_7 - a_8 - a_{10})\sin\omega t + a_2 \\ z_{Wp} = (w_X t - s_2)\sin\omega t + (z_T - a_7 - a_8 - a_{10})\cos\omega t + a_1 \end{cases}. \tag{10}$$

If $w_X = 0$ the equations of a cylindrical helix with constant pitch is obtained, and also for $w_Z = 0$ an Archimedean spiral.

For the studied case, The presented method has a general feature and is utile for interpolation methods (linear, circular, etc.) in case of generation of analytical or non-analytical generatrice *D* and directrice *G* curves.

4. CONCLUSIONS

In this paper some formalisms specific to industrial robots were applied to CNC machine tools. The kinematic structural formulas for two lathes were established The exemplification is made for a parallel lathe with two kinematic branches: bed-workpiece and bed-tool.

The Denavit-Hartenberg formalism was applied, resulting the transformation matrix between the reference systems attached to the bed and workpiece, and to the bed and tool.

The relations between fictitious mechanism *Wp-T* input and the output were established. For the studied case, the trajectory generated by the *Wp-T* contact point is a conical helix, a cylindrical helix, or an Archimedean spiral.

Deriving the equations of these helixes, the speeds and the accelerations are obtained for the *Wp-T* contact point. Differentiating these equations, the influences over the generated helix precision of some geometrical and constructive parameters are obtained for a cutting process.

5. REFERENCES

Bohez, E. L. J. (2002). *Five-axis milling machine tool kinematic chain design and analysis*. International Journal of Tools & Manufacture, 42, Pergamon, 2002, pp. 505-520.

Botez, E., (1966). *Bazele generării suprafeţelor pe maşini-unelte* (Fundamentals of surface generation on machine tools), Edit. Tehnică, Bucharest.

Catrina, D.; Moraru, V. & Dinu, G. (1993). *Maşini-unelte cu comandă numerică* (CNC Machine Tools), Vol. I and II, University "Politehnica" of Bucharest.

Popescu, P; Negrean, I; Vuscan, I & Haiduc, N. (1994). *Mecanica manipulatoarelor şi roboţilor* (Mechanics of manipulators and robots), Edit. Didactica si Pedagogica, ISBN 973-30-3580-7, Bucharest.

Weck, M (2001). *Werkzeugmaschinen. Automatisierung von Maschinen und Anlagen* (Machine tools. Automation of machine and equipment). VDI Springer, ISBN 3-540-67713-9, Berlin.

Annals of DAAAM for 2008 & Proceedings of the 19th International DAAAM Symposium, ISSN 1726-9679
Editor B. Katalinic, Published by DAAAM International, Vienna, Austria 2008
Make Harmony Between Technology and Nature, and Your Mind will Fly Free as a Bird

FOUNDRY TECHNOLOGIES AND MATERIALS FOR CAST CUTTING TOOLS

CAUS, A[lexander]; BEZNAK, M[atej] & CAPLOVIC, L[ubomir]

Abstract: *The paper deals with the foundry technologies of production of cutting tools. Effect of different additions, whose surface activity in the iron melt has been evaluated using such criteria as melting temperature, surface energy, specific heat of sublimation, entropy in the standard condition, statistic generalized moment, total electron potential barrier, on the structure and properties of standard HSS of AISI M2 and T30 types are studied. New findings on the structure and properties of HSS specially designed for cast tools are also introduced.*
Key words: *HSS, tool, casting, structure, properties.*

1. INTRODUCTION

Cutting tools that are made from wrought HSS are shaped to the required geometries by machining. When machining a great amount of tool material is removed in the form of chips to the wastes and the production time is long. Production of cutting tools by foundry technologies results in dramatic improvements in material saving and production efficiency (Gelin & Chaus, 2007). Since the application of precision casting of cutting tools allows casting near-net-shape work-pieces of tools this reduces the volume of roughing operation and large depth of cut, and consequently, the material consumption. Table 1 and 2 show the material consumption and processing time for the cutting tools produced from wrought HSS bars using machining, and by casting (Čaus, 2008).

Foundry technology provides more effective tool material utilisation because multiple using of a tool scrape is possible when melting charge as well as the use of both wrought wastes and chip briquettes of HSS. It is necessary to emphasise that utilisation of high alloy powder wastes produced upon grinding of hard metals is also possible at melting.

On the other hand, absence of special casting alloys for cutting tool production inhibits broad application of chipper cast tools in industry. As a rule, HSS of conventional chemical composition are usually used as a material for cast cutting tool but their casting and mechanical properties, primarily impact toughness do not suit perfectly. Therefore, in order to exhibit good all-round performance the impact toughness enhancement of as-cast HSS is obligatorily needed.

Tool	Mass of work-piece [kg]		Saving of HSS	
	Wrought	Cast	[kg]	[%]
Slotting tool	2.153	0.750	1.403	65.1
Set of inserted blades	2.748	1.682	1.066	65.3
Side milling cutters	1.040	0.440	0.600	57.7
Shell mill	1.360	0.410	0.950	68.7

Tab 1. Material consumption upon production of tools from wrought HSS bars by cutting, and by casting.

Tool	Processing time [min] upon production of tool by:		Time saving [%]
	Casting	Machining	[%]
Slotting tool	393.4	360.7	8.3
Set of inserted blades	47.2	38.07	19.3
Side milling cutters	135.0	90.1	33.9
Shell mill	105. 0	67.45	35.7

Tab 2. Processing time upon production of tools from wrought HSS bars by cutting, and by casting

2. FOUNDRY TECHNOLOGIES FOR CAST CUTTING TOOLS PRODUCTION

To date different foundry technologies can be used for metal cutting tools production. Among them the most attractive technology is casting in metal moulds. The reason is that due to high cooling rate of the melt upon solidification in a metal or graphite mould the high density of castings as well as the structure refinement can be achieved (Fig. 1) that leads to enhanced mechanical properties of HSS. In addition the casting into metal mould offers other advantages, which are as follows:

- Use of sand mixtures is not obligatory that simplify the very foundry technology and accelerates production cycle;
- Investments into both technology and equipment is low;
- Implementation of technology is quick and easy;
- Technology provides high stability and accuracy of both dimension and shape of castings;
- High surface quality allows reducing the allowance for machining;
- Technology improves hygienic and ecological conditions of casting process.

The main advantage of lost-wax casting is extremely high quality of casting surface, which allows producing near-net-shape work-pieces practically without allowance for rough machining (Fig. 2) that benefits to improvements in material saving and production efficiency.

Electro slag remelting (ESR) is also attractive for production of cast tools primarily with large cross sections. Dropping mechanism of remelting of consumable electrode, at which the material in small portions (drops) is conveyed from the consumable electrode through a slag pool to forming ingot, provides the high density and lower level of structural heterogeneity in HSS (Chaus et al., 1997) even though the diameter of ingot is larger than 100 mm. Beside this, ESR uses highly reactive slags thus reduces the amount of sulphides and other types of inclusions present in HSS.

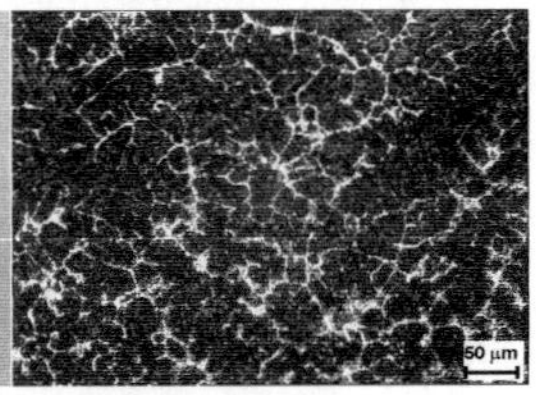

Fig. 1 Cast milling cutter (∅ 65 mm) and its structure

Fig. 2 Cast milling cutter (∅ 80 mm), lost wax technology

3. MODIFIED AS-CAST HIGH-SPEED STEELS

Different methods of structure and properties improvement of cast HSS can be commercially used. In terms of efficiency and the production cost of all methods used for as-cast HSS quality improvement, it appears that modification of the melt is more simple and effective one. However, the effect of modifying additions in HSS has been studied insufficiently. As a consequence, a restricted number of additions are used for as-cast HSS compared to common cast alloys.

In order to evaluate the surface activity of the elements in molten iron the calculations, using such criteria as melting temperature, surface energy, specific heat of sublimation, entropy in the standard condition, statistic generalized moment, total electron potential barrier for iron and addition have been carried out. According to the received activity series built up from the derived values on surface activities, Bi leads the list of the most surface-active elements followed by Ca, Sr, Sn, Sb, Cd, Mg, and so on. It was shown that Ti, Zr, Hf, Nb, Ta and B belong to so-called inoculating additions (Chaus, 2005).

Experimental verification of effects of different additions in as-cast M2 and T30 type HSS have been carried out. Calculated amounts of additions were as follows, mass%: for Zr and Bi 0.05-0.1-0.3-0.6; Nb, Ti, Ge, Ni, SiMM, Y, Cd 0.1-0.3-0.6; B 0.05-0.1-0.3-0.6-1.0; FeCe 0.1-0.3-0.6-1.2; Si and Al 0.4-0.8-1.2. Relationships between the structural parameters and mechanical properties have been established (Chaus & Rudnickii, 1989). The comparison of the structure of non-modified and modified as-cast HSS is shown in Fig. 3. On the basis of the received results a comprehensive range of new-patented as-cast HSS with the wide choice of improved structures and properties for cast cutting tool applications have been designed (Gelin & Chaus, 2007).

4. SPECIAL AS-CAST HSS

After casting the structure of the HSS of 11M5F type contains small volume fraction of the eutectic carbides (Fig. 4), which total amount does not exceed 1-5% (Chaus, 1998). As a consequence, the impact toughness of this steel is more than a factor of 1.8 higher than that of M2 HSS. Another feature of the 11M5F steel is the presence of carbides with the secondary origin, which dissolute at lower austenitising temperature and thus provide higher saturation of a solid solution with alloying elements. As a consequence, the 11M5F steel has hardness and a heat resistance no worse than those of the M2 steel.

For carburised cast cutting tools a new-patented M2-based HSS with ferritic matrix has been developed (Chaus & Latyshev, 1999). Composition of the steel is as follows, wt. %: 0.80-0.88 C; 3.8-4.4 Cr; 4.9-6.5 W; 4.5-5.5 Mo; 2.6-3.8 V; 0.4-1.8 Ti; 0.8-2.4 Nb. Varying the chemical composition of the steel it is possible to achieve different level of carbide dissolution during austenitising and, as consequent, different level of carbide heterogeneity in the diffusion layer after heat treatment (Fig. 4). The mechanical properties of the carburised HSS after heat treatment were as follows (Chaus et al., 2001): impact toughness 20 J/cm^2, hardness 68-69 HRC, heat resistance 675°C and wear resistance (rate of wear) 36.4 mg/h which is higher than that of the M2 HSS, 76.2 mg/h.

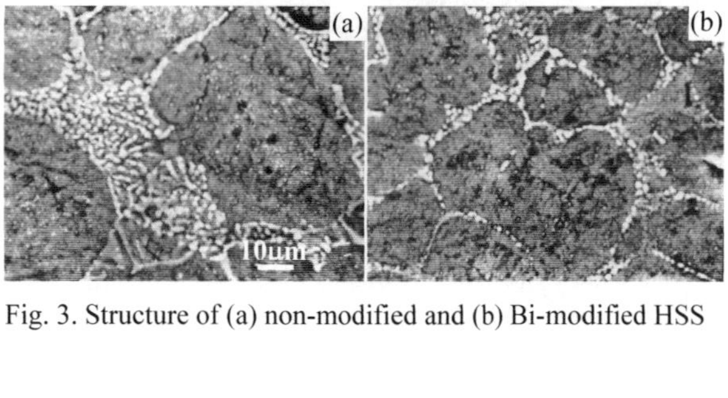

Fig. 3. Structure of (a) non-modified and (b) Bi-modified HSS

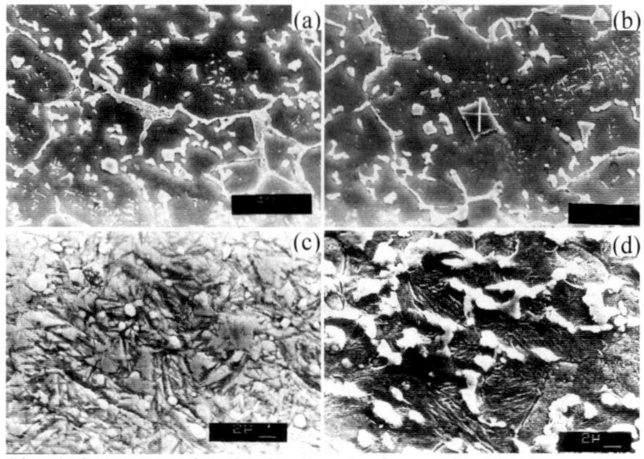

Fig. 4. Structure of carburised HSS after (a, b) casting and (c, d) carburising and heat treatment.

The results of durability tests performed in turning proved excellent durability of the carburised cast cutting tool in comparison with conventional one (Chaus & Latyshev, 1999).

5. CONCLUSION

Modification of as-cast HSS with inoculants and surface-active elements provides favourable structural changes that benefits in improved mechanical properties of as-cast HSS and finally in enhanced cutting performance of these steels. Employment of HSS with the chemical composition specially designed for cast tools is also very effective from viewpoint of the impact toughness and durability of cast tools.

6. ACKNOWLEDGMENT

The financial support of grants from the Ministry of Education of the Slovak Republic VEGA 1/4109/07 and VEGA 1/3191/06 is gratefully acknowledged.

7. REFERENCES

Čaus, A. S. (2008). *Advanced Materials and Technologies of Production of Cast Cutting Tools.* AlumniPress, ISBN 978-80-8096-060-5, Trnava

Chaus, A. S. (2005). Application of Bismuth for Solidification Structure Refinement and Properties Enhancement in As-cast High-Speed Steels. *ISIJ International.* Vol. 45, No. 9, (2005) 1297-1306, ISSN 0915-1559

Chaus, A. S. (2001). Heat Treatment of As-cast Carburised High-Speed Steel. *Metal Science and Heat Treatment.* Vol. 43, Nos. 5-6, (2001) 220-223, ISSN 0026-0673

Chaus, A. S. & Latyshev, I. V. (1999). Effect of V, Ti and Nb on the structure and properties of cast tungsten-molybdenum HSS. *Physics of Metals and Metallography.* Vol. 88, Iss. 5, (1999) 152-156, ISSN 0031-918X

Chaus, A. S. (1998). On the Prospects of the Use of Low-Alloy Tungsten-Free HSS 11M5F for Cast Tools. *Metal Science and Heat Treatment.* Vol. 40, Nos. 7-8, (1998) 319-325, ISSN 0026-0673

Chaus, A. S. et. al. (1997). Structural Inheritance and Special Features of Fracture of HSS. *Metal Science and Heat Treatment.* Vol. 39, Nos. 1-2, (1997) 53-56, ISSN 0026-0673

Chaus, A. S. & Rudnickii, F. I. (1989). Effect of Modification on the Structure and Properties of Cast W-Mo HSS. *Metal Science and Heat Treatment.* Vol. 31, Nos. 1-2, (1989) 121-128, ISSN 0026-0673

Gelin, F. D. &Chaus, A. S. (2007). *Metallic materials,* Vyshejshaja shkola, ISBN 978-985-06-1362-2, Minsk

Annals of DAAAM for 2008 & Proceedings of the 19th International DAAAM Symposium, ISSN 1726-9679
Editor B. Katalinic, Published by DAAAM International, Vienna, Austria 2008
Make Harmony Between Technology and Nature, and Your Mind will Fly Free as a Bird

INTER – VEHICLE AD – HOC COMMUNICATION PROTOCOL FOR ACQUIRING LOCAL TRAFFIC INFORMATION

CEHAJIC, A[rijana]; KOVACEVIC, D[razen] & STIMAC, I[gor]

Abstract: *In this paper, we propose an inter – vehicle ad- hoc communication protocol in order for each vehicle to acquire its local traffic information. Car drivers always want to access not only trunk road information but also back street road information. To solve this problems, we are doing research on disseminating and propagating road information using mobile inter-vehicle ad-hoc communication protocol. In this research, we have developed a mobile ad-hoc network simulator that produces inter-vehicle data propagation situation from the location information of all the cars and various kinds of statistics information such as packet collision ratios and reception message amount.This simulator uses traffic flow simulator NETSTREAM II that is developed by an automaker and produces actual traffic flow.*

Key words: *road information, ad – hoc communication protocol,network simulator, inter – vehicle communication*

1. INTRODUCTION

Car navigation systems are going to equip GPS receivers and wireless LAN cards to acquire the information about traffic jams, road surface conditions and free parking lots along the street in real time. On the other hand, car drivers always want to access not only trunk road information but also back street road information. To solve these problems, we are doing research on disseminating and propagating road information using mobile inter – vehicle ad-hoc communication protocol. For example, suppose that a preceding car – A holds a set of its own speed, location direction and surrounding facilities information for the fast several minutes, and disseminates it to surrounding cars. Car – B, which is driving on the opposite lane, receives the information, moves to another place and re – disseminates car-A′ information together with car - B′ s information. At that time another car – C may receives the car A′ s information. That means car C can know its preceding traffic conditions and road surface situations using inter – vehicle ad – hoc communication. (Ni at al.,1999.)

2. LOCATION DEPENDENT SERVICES AND INFORMATION EXCHANGE

By diffusion of GPS systems and progress of car navigation systems, we will be able to use road information services interlocked with map information as follows:

a) Global road information services: the service that provides traffic jams and road-repairing information of trunk roads and highways through radio broadcasting and/or inquiring by cellular phones
b) Local road information services: the service which is used within a narrow region such as the service notifying an approaching vehicle at a crossing of other vehicles and walkers, back road traffic jam information service and free parking lots along the street service (Xu at al., 2004.)

When many vehicles moving to the same direction are stopping and/or moving at very low speeds, we can predict the traffic jam. If this information can be propagated by opposite lane's vehicles' relay and if it is reached to the tail of the traffic jam, drivers just before the jammed area can recognize the length of the queue of the traffic jam, and they may be able to avoid it by re-routing. Recent vehicle sensors help other situations. For example, freeze road surfaces can be detected by slide conditions of tires, and rainy conditions can be detected by movement of wipers. To propagate such information, other vehicles' drivers can prepare danger situations such as advancing slowing down. So combining global and local road information services, we are able to have much safer driving environments.

2.1. Ad – hoc communication protocol

At first, vehicles within the circles of 100 m's radius can communicate each other, however, data communication success probability decreases linearly according to the distance. When the distance between two nodes becomes large, the receiving radio power becomes weaker, so that reception error ratios will be increased. When a vehicle can communicate with vehicles within the range of 100 m, it can communicate with the vehicles that are running at a speed of 60 km/h on its opposite lane only for about 6 seconds. On the other hand, in case of both of lanes are jammed and vehicles are waiting 10 m intervals, the vehicle at the center may receive the road information from more than 40 vehicles simultaneously. Here, we assume that the bandwidth is 100 Kbytes/s and exchange data size is 10 kbytes at the maximum. We also assume that one second is divided into 10 slots, and each data transmission occupies one slot (Kellerer at al., 2001.).For this reason, when two vehicles send data at the same slot, we assume that any vehicle in the over lapped area of 100 m radius circles cannot receives the both of data because of collision.

2.2. Road information dissemination protocol

In this paper, we use SDRP (Speed Dependent Random Protocol) for road information dissemination. In SDRP, according to the vehicle speed v, random transmission interval is calculated between the minimum value min (v) and the maximum value max (v). In case of traffic jams, since there are many vehicles around a vehicle, the transmission interval becomes large. In addition, in case of high speed driving, the interval becomes small to increase propagation probability. For example, we can define SDRP random interval as follows:

- v is less than 30 km/h, min (v) is 3 seconds and max (v) is 5 seconds
- v is more than 30 km/h, min (v) is 1 second and max (v) is 2 seconds

3. DEVELOPMENT OF THE MOBILE AD – HOC NETWORK SIMULATOR

3.1. The structure and the functions of the simulator

To evaluate road information propagation situations in an inter – vehicle mobile ad – hoc network setting, we have developed the network simulator. NETSTREAM II simulator is a traffic flow simulator for performing the effective prediction for traffic jams and prior evaluation of ITS introduction (Teramoto at al., 1998.) It defines traffic flow characteristics and the lengths of signal for all road links, and then calculates all vehicles' actions for every second. For each road, each vehicle calculates its speed V as $V=V_{max}$ ($1-K/K_{congestion}$). Here let the distance between a front vehicle and itself be S and vehicle density K be K=1/S.

V_{max} is the speed when the traffic density is 0, that means free running. $K_{congestion}$ is the vehicle density when vehicles are stopping because of traffic jams. Therefore, in wide areas, such as the whole city, it can calculate the traffic flows with considerably high accuracy. Based on those parameter values ,this simulator produces inter – vehicle data propagation situation from the local information of all the vehicles and records varoius kinds of statisticsc information such as packet collision ratios and reception message amount. The simulator's processing flow is as follows:

a) Setting the given network environment
b) Defining car navigation equipped vehicles by the given equipping ratios
c) Calculating the following NETSTREAM II vehicle log data for every second
    - Defining the vehicles to disseminate information by the given algorithm
    - Searching the vehicles within attainment distance and determing received vehicles based on their reception probabilities
    - Holding received vehicles data for next data transmission
d) Calculating data propagation situation

3.2. Simulation environment

Using our mobile Ad – hoc network simulator, we have evaluated the inter – vehicle road information propagation with the following input parameters: *Road Environment*: 20 km x 20 km, the number of signals is 198, *Simulation Time:* 60 minutes (the last 40 minutes data are used for evaluation), *Location Information of Vehicles:* Every second, *The number of Vehicles*: 8570 (The total number for 60 minutes), *Equipping Ratios of Car Navigation Systems*: 30 %, 60 %, 90 %, *Network Environment*: Attainment distance 100 m, Bandwidth 100 Kbytes/s, *Dissemination Algorithm*: SDRP (Speed Dependent Random Protocol) with two kinds of transmitting intervals, bordering on speed of 30 km/h, *Receiving Probability*: Linearly change based on the distance

3.3. Simulation result

At first we have measured the amount of transmitting data and network collision ratios for each equipping ratios and each SDRP protocol' s transmitting interval. (Nadeem at al., 2004.) In SDRP protocol, the transmitting interval for speed v is defined as [min (v), max (v)]. Here, we assume that min(v)=max (v)/2, that is, [max (v)/2, max (v)] is used as the interval. Hereafter, we call max (v) the interval for the speed v as an abbreviation of [max (v)/2, max (v)]. Fig. 1 and Fig. 2 show the results of fixed transmitting intervals. When the transmitting interval of SDRP are set as <A,B>, it means that A is interval of less than 30 km/h and B is interval of more than 30 km/h. Fig. 1 and Fig. 2 show the results by changing the value of < A, B > from <1,1> to <16,16>. Fig. 3 and Fig. 4 show the results for the cases that B is set to 1 second, and the value of A is varied from 1 to 16. show the results by changing the value of < A, B > from <1,1> to <16,16>. Fig. 3 and Fig. 4 show the results for the cases that B is set to 1 second, and the value of A is varied from 1 to 16.When the equipping ratios are 90 %, the total transmitting data amount is the maximum when a is 4 seconds, and it decreases if the value of A is far from 4 seconds. Similarly, when equipping ratios are 60 %, it is the maximum when A is 2 seconds. The collision ratios at that time are 52 % and 48 % respectively, and the total transmitting data amounts are 64 Gbyites and 26 Gbytes, respectively.

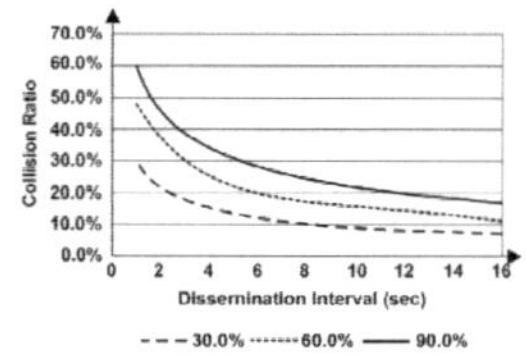

Fig. 1. Dissemination Interval and Collision Ratio (A=B=1~16)

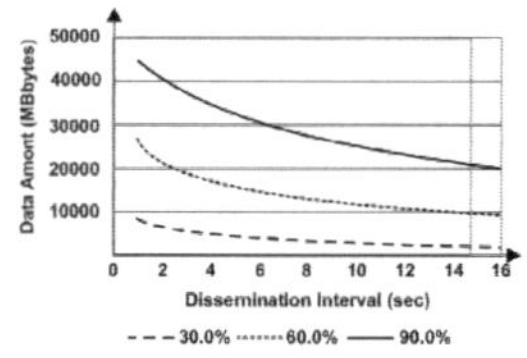

Fig. 2. Dissemination Interval and Total Amount (A=B=1~16)

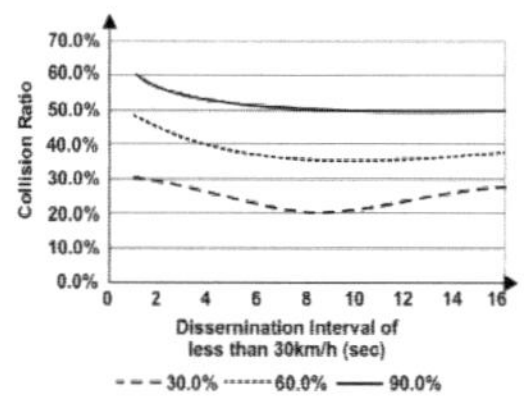

Fig. 3. Dissemination Interval and Collision Ratio (A=1~16,B=1)

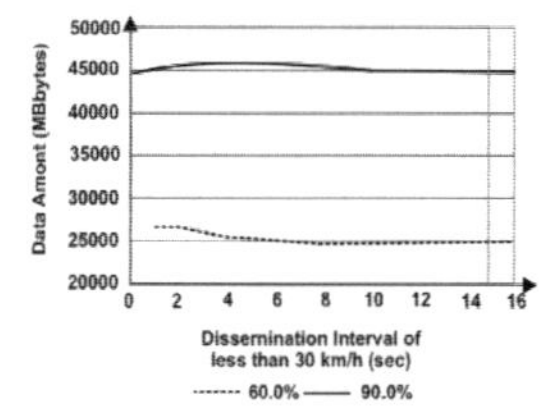

Fig. 4. Dissemination Interval and Total Data Amount (A=B=1~16)

4. CONCLUSION

When the equipping ratios are 90 %, the total transmitting data amount is the maximum when A is 4 seconds, and it decreases if the value of A is far from 4 seconds. Similarly, when equipping ratios are 60 %, it is the maximum when a is 2 seconds. In our setting, road information for each vehicle is 100 bytes so that about 4000 vehicles information can be acquired within a minute. We are sure that this data amount is enough to acquire local road information around several kilometers regions. As a future subject, we are studying more efficient protocol like RMDP, data aggregation technique, and more detailed analysis especially for each vehicle by defining actual exchanged information. We also anticipate the simualtion in the real world by implementing the protocol to car navigation systems.

5.REFERENCES

Kelleler, W.;Bettstetter, C. &Sties, P.(2001.): Mobile Communication in a Heterogeneous and Converged World, *IEEE Personal Communications*,Vol. 8, No.6, pp. 41 – 47, 2001.

Nadeem, T., Dastinezhad, S.,Liao, C.&Iftode, L (2004.).:A Scalable Traffic Monitoring System, *Proc. of 2004 IEEE Int. Conf.Mobile Data Management (MDM2004)* pp.13-26, ISBN 0-7695-2070-7, 24-08-2004.

Ni, S. Y., Tseng, Y.C., Chen, Y. S. & Sheu, J.P.(1999.): The Broadcast Storm Problem in a Mobile Ad-hoc Network, *Proc. of 5th Annual ACM/IEEE Int. Conf. on Mobile Computing and Networking,* pp.151-162, 1999.

Teramoto, E. & al.(1998.): Prediction of Traffic Conditions for the Nagano Olympic Winter Games Using Traffic Simulator:NETSTREAM, *Proceedings of 5th World Congress on Intelligent Transport Systems*, Vol. 4, pp.1801-1806, 1998.

Xu, B.; Ouksel, A. & Wolfson, O.(2004): Opportunistic Resource Exchange in Inter–vehicle Ad–hoc Networks, *Proc. of2004., IEEE Int. Conf. on Mobile Data Management* pp.4 – 12,ISBN 0-7695-2070-7, 24-08-2004.

Annals of DAAAM for 2008 & Proceedings of the 19th International DAAAM Symposium, ISSN 1726-9679
Editor B. Katalinic, Published by DAAAM International, Vienna, Austria 2008
Make Harmony Between Technology and Nature, and Your Mind will Fly Free as a Bird

ROUGHNESS AS PARAMETER OF CUT QUALITY DURING CO_2 LASER CUTTING OF HIGH ALLOY STEEL FOR THE SPECIAL PURPOSE

CEKIC, A[hmet]; KULENOVIC, M[alik] & BEGIC, D[erzija]

Abstract: *This article presents a preliminary study to evaluate the effect of the cutting speed and the thickness materials on the quality of cut during laser cutting of high alloy steels for a special purpose. The experiments were performed on two different materials using a CW CO2-laser with a nominal power output of 2800 W. The variable process parameters are taken: cutting speed from 0.05 m/min to 1.5 m/min, three different thicknesses 5 mm, 10 mm and 15 mm. The effect of these mentioned process parameters on the surface roughness is analyzed. Results are shown that necessary to take in consideration more process parameters in order to achieved improvement in cut quality. Given results also are shown very different behaviours during laser cutting thick and thin sheets, while the cutting speed is less relevant for both tested materials.*

Key words: *Cutting, CO_2 laser, speed, roughness, alloys steel*

1. INTRODUCTION

In the world today, there is rapidly introducing of new technologies into manufacturing systems from more reasons such as: saving materials, precision of operation, short processing time, high accuracy of products and high flexibility. Nowadays laser technologies are increasingly used in manufacturing and automotive industries. In most of the use of laser technologies is the laser beam cutting. The laser beam cutting can be successfully used for the cutting different materials such as: metals, polymers, ceramics, composites, high alloy steels and others (Avanish & Vinod, 2008). The use of high alloy steels in car bodywork has spread in recent years. The AISI institute is forecasting a dramatic increase in the use of high alloy steels in the car industry (Lamikiz et al., 2005). In the literature, there are no more experimental studies of the laser cutting of high alloy steels. The exothermal reactions of iron and other alloy elements are complex and high alloy steels have a high content of alloy elements, which can lead to differences with regard to the cutting of common steels.

Among various types of lasers used for machining in industries, CO_2 (wavelength 10.6 μm) and Nd: YAG (wavelength 1.06 μm) lasers are most established in the automotive industries. In order to use advantages of this technology, especially by aspect of saving materials is necessary to evaluate the optimum process parameters for each kind materials, thickness and others process conditions to the achieving desired cut quality with maximum productivity. The setting of process parameters mainly depends of the material characteristics (density, specific heat, absorptivity, thermal conductivity, chemical composite, surface quality and etc) and thickness of materials. The most important process parameters are, such as: laser power input, mode of operation (pulsed mode and continuous wave), pulse frequency, pulse duration, kind and pressure of assistant gas, focus position, diameter and kind of nozzle. These parameters are usually determined so those insure the desired cut quality, and the productivity is less relevant. The setting of process parameters which would satisfy both the desired cut quality and high productivity is specific problem (Rajaram et al., 2003). The quality of the CO2 laser cutting was evaluated by a standard process: measuring of surface roughness, kerf width and kerf deviation along the length of cut, kerf taper, size of heat affected zone, micro-hardness and metallographic and etc. Previous work in the literature elucidated the effects of oxidation dynamics and laser beam velocity on the quality of the laser cut (Avanish & Vinod, 2008). It suggested that there is an optimum range of cutting speeds for a given material, thickness and laser power. Surface roughness is an effective parameter representing the quality of machined surface. Some researchers (Nagels et al., 2007) shown that surface roughness value reduces on increasing cutting speed and frequency, and decreasing the laser power and gas pressure. Also nitrogen gives better surface finish than oxygen. In spite of an increase in available laser power during the last decade, the laser cutting process faces difficulties in cutting thicker parts. The main problems are losing the process stability and the deterioration of the cut quality.

2. EXPERIMENTAL PROCEDURE

The preliminary experiments were carried out with a laser system consisting of a 2800 W continuous CO_2 laser (Bystronic), a three axes CNC controlled table with work volume 2.0 m × 1.5 m × 0.5 m. Experiments were carried out on two different materials with three thickness of 5, 10, and 15 mm. The chemical compositions of the examined materials are given in Table 1.

Geometry of the sample shapes are given in Fig. 1. Sample I (EN 10083-3) was previous prepared for laser cutting by using the saw and milling machine. It was hot rolled square rod with dimensions of 140x140mm. Sample II (GX40CrNiSi22-10) is given from tube, produced by centrifugal casting.

The beam profile is nearly Gaussian (TEM_{00}) with beam quality k≈0.75. Others process parameters were kept constant throughout the experimentation: the laser power of 2520 W, the focal length of lens 127 mm, the assistant nitrogen gas pressure of 14 bar, nozzle diameter of 2.0 mm, focus position under bottom side of sample. Namely, previous experiments are shown that was not possibility to realize the cutting of tested materials when the oxygen used as assistant gas. In the previous experiments, the optimal value of assistant gas pressure and the focus position are obtained. In the experiments where the laser beam has been focused on the surface of the sheet and even above the sheet, the results show poor quality in cutting areas.

Chemical composition of GX40CrNiSi22–10 steel							
	C	Cr	Ni	Si	Mn	S	P
Min	0,40	18,0	12,0	1,80	....	0,20	....
Max	0,70	21,0	14,0	2,20	1,50	0,40	0,06
Chemical composition of EN 10083 - 3 steel							
	C	Cr	Ni	Si	Mn	S	P
Min	0,42	1,10	0,05	0,17	0,50	....	....
Max	0,49	1,40	0,25	0,24	0,80	0,05	0,05

Tab. 1. Chemical compositions of examined materials

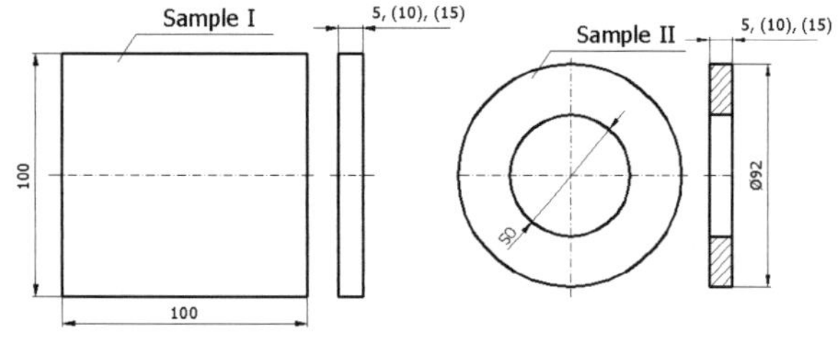

Fig.1.The samples for cutting

3. ROUGHNESS MEASUREMENT

Surface roughness as parameter of cut quality is analyzed. Roughness parameter R_a - mean deviation is measured. The surface roughness is measured by using the device Perthometer Concept. DIN 2310 standard prescribed the measuring of surface roughness on the exactly determined depth. It depends of the material thickness. Because of the specific process and the obtaining of valid results, the measurements are performed at the five parallel lines which are uniform sorted along the depth of cut, Fig.2. Namely, the high of unevenness increase from the top to the bottom of cut along the length of cut. The exchange of conditions of the intersection laser beam and material is reason for that, as and the exchange of the local rate between the thermal conductivity and the coefficient of absorption (it depends of the temeparture exchange). Fig. 2 shows the increasing of surface roughness along the depth of cut in the depending of the process parameters as well as the kind and the thickness of material. During executing of experiments, the maximum deviation of 6.6 μm is obtained.

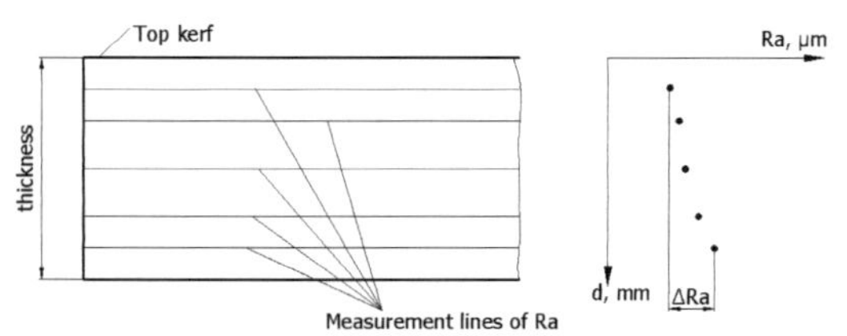

Fig. 2. The exchange of roughness along the depth of cut

In the aim of the obtaining of the valid results, the visual selections of samples are made. The samples with large dross were not taken in consediration.

4. RESULTS AND DISCUSSION

The experimental values of parameter roughness *Ra* for different cutting speeds during laser processing of GX40CrNiSi22 - 10 steel is given in Fig. 3 and in Fig.4 for EN 10083 -3 steel.

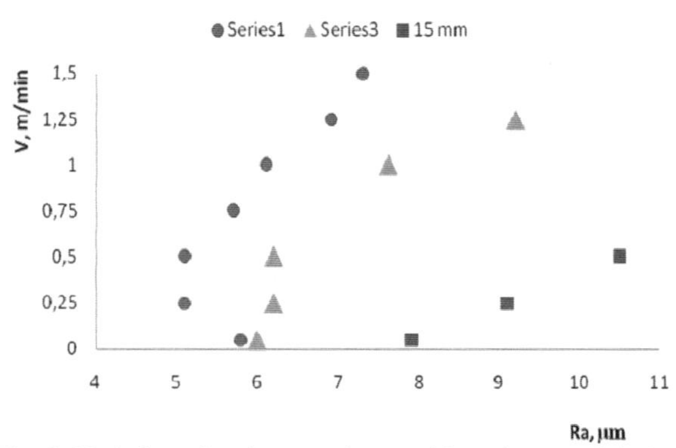

Fig. 3. Variation of surface roughness with cutting speed during CO_2 laser cutting of GX40CrNiSi22 – 10 steel

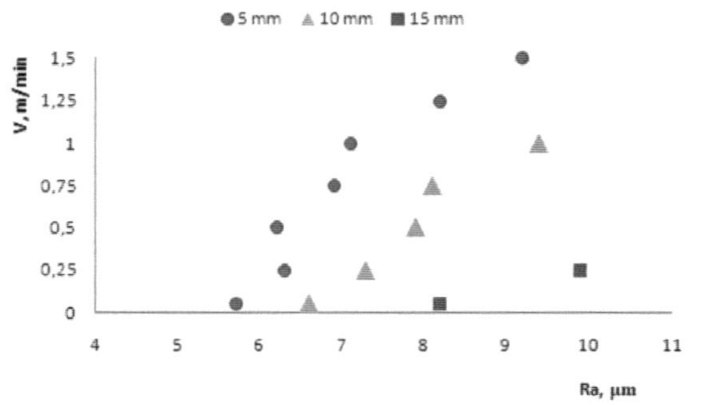

Fig. 4. Variation of surface roughness with cutting speed during CO_2 laser cutting of EN 10083 -3 steel

It is observed that, the surface roughness increases by increasing the material thickness at keeping constant others process parameters during CO_2 laser cutting of both examined high alloy steels.

5. CONCLUSION

In this paper are carried out the experimental investigations the effect of the cutting speed and the thickness of material as well as the kind and the pressure of the assistant gas on the surface roughness during the CO_2 laser cutting two different materials.

Following conclusions can be drawn on the basis of results obtained:

- Once the correct value has been established, small variations in pressure do not determine the quality of the cut. Thus, assisting gas pressure can be maintained for different values of power, cutting speed and focal position as well as kind and thickness of tested materials.
- Maximum of the possibility cutting speed is depends of the kind and thickness of material for the given laser system. Above this maximum there is no possibility to obtain the accepted level of cut quality.
- Better quality of cut and the large range of cutting speed is possibility during laser cutting of thinner materials.

In the future work, the influential process parameters during laser cutting of high alloy steels for the special purpose will be considered. These process parameters will be optimized with consideration of multi-performances characteristics of the cut.

6. REFERENCES

Al-Sulaiman, F.A.; Yilbas, B.S. & Ahsan M. (2006). CO2 laser cutting of a carbon/carbon multi-lamelled plain-weave structure. *Journal of Material Processing Technology,* 173, (April 2006) page numbers (345-351), ISSN: 0924-0136

Avanish, K. D. & Vinod, Y. (2008). Laser beam machining-A Review. *International Journal of Machine Tools and Manufacture,* 48, (May 2008) page numbers (608-628), ISSN: 0890-6955

Lamikiz, A.; Lopez de Lacalle, L. N.; Sanchez, J. A.; Pozo, D.; Etayo.J. M. & Lopez. J. M. (2005). CO2 laser cutting of advanced high strength steels (AHSS). *Applied Surface Science,* 242 (April 2005) page numbers (362-368), ISSN: 0169-4332

Nagels, E.; Dufloub, J. R. & Humbeeck Van, J. (2007). The influence of sulphur content on the quality of laser. *Journal of Materials Processing Technology, 194,* (November, 2007) page numbers (159-162), ISSN 0924-0136

Rajaram, N.; Sheikh-Ahmad, J. & Cheraghi, S. H. (2003). CO2 laser cut quality of 4130 steel. *International Journal of Machine Tools and Manufacture,* 43, (March 2003) page numbers (351-358), ISSN: 0890-6955

Annals of DAAAM for 2008 & Proceedings of the 19th International DAAAM Symposium, ISSN 1726-9679
Editor B. Katalinic, Published by DAAAM International, Vienna, Austria 2008
Make Harmony Between Technology and Nature, and Your Mind will Fly Free as a Bird

ERP SOLUTION IMPLEMENTATION FRAMEWORK FOR SMALL AND MEDIUM ENTERPRISES

CELAR, S[tipe] & GOTOVAC, S[ven]

Abstract: *ERP solution implementations are important and time consuming projects for small and medium enterprises (SME). One of the biggest risks is constant user requirements changing. Implementation framework described in this paper makes possible to manage these requirements and the whole implementation project successfully. This framework and its main characteristics (preconfigured settings, two projects) proved to be a success in practice.*
Key words: *SW Development Life Cycle (SDLC), Information Communication Technology (ICT) project results, ERP critical success factors, ERP implementation framework*

1. INTRODUCTION

1.1. SW development lifecycle and success of ICT projects

The SW development project begins with the planning phase in which one should try to plan the duration of the whole project. Thereby it is a hardly to predict how much time each phase of the software development life cycle (SDLC) is going to take. One of the frequently used approaches is shown in Table 1, which indicates that design and implementation claim a huge period of time.

Table 2 shows an improvement of project successfulness in the last 15 years, which is the result of a systematic implementation of software engineering and other managerial methods and techniques in Information Communication Technology (ICT) industry in the last 15 – 20 years.

1.2. Project of application solution implementation

Summing up the results from Table 1 and Table 2 it is possible to conclude that the implementation phase for application solution is time demanding and can greatly influence the whole SDLC project. Therefore, it is necessary to approach the planning of this phase very carefully, no matter whether it is the application solution of a small, medium or big enterprise.

This paper deals with the implementation method of ERP (Enterprise Resource Planning – ERP integrated application) solutions in small and medium enterprises (SME) in the Republic of Croatia.

2. ERP IMPLEMENTATION CRITICAL SUCCESS FACTORS

Success factors emerge as critical key factors to success in complex implementation projects. Some of the most common critical success factors (CSF) in these implementation projects are (Bancroft et al., 1998; Gross, 2004; Sternad et al., 2006):

Project Phase	*Project Time Using the Planning Phase Approach*
Planning	15 %
Analysis	20 %
Design	35 %
Implementation	30 %

Tab 1 Standard for business applications (Dennis et al., 2006)

ICT project outcomes	*CHAOS Report (1994)*	*2006 CHAOS Research (PM Network (2007b))*
Success	16,2 %	35 %
Challenged	52,7 %	46 %
Cancelled	31,1 %	19 %

Tab 2 ICT projects outcomes

- top management support and involvement in project
- clear goals, objectives and scope
- project team competence and organization
- user training and education
- business process reengineering
- change management
- effective communication
- user involvement
- data analysis and conversion
- architecture choice (package selection).

If CSSs are successfully controlled in SME such ERP implementation projects could end in three to four months (from GAP Analysis to Go-Live, s. Fig. 1). Otherwise, even small mistakes could generate enough problems for SME. On large ERP projects there are more time and resources to amortize such mistakes.

All these require precise and consistent implementation framework for SMEs which covers the whole implementation process and make possible to manage project goals effectively.

3. ERP IMPLEMENTATION FRAMEWORK AND PROCESS

3.1. Implementation framework

During this research the stress is put on the issues related to the changes of the *information system as a whole* (Enterprise Resource Planning – ERP), as the most complex case. However it is important to point out that the introduced framework presents a generic model, whether the change refers just to a part of the system or to the information system as a whole. If it refers just to a part of the system, then some of the processes in the model will be simpler (e.g. there might be no need for an extensive education, or no education at all if it is a simple, minor change of the system).

One of the biggest risks on such projects arises from constant user requirements changing (s. CSF *clear goals, objectives and scope*). If it is the case the constant development is needed and the project results according to Table 2 are to expect.

In order to overcome this risk good SME-framework for "roll out" projects must consider that ERP solutions should be ***preconfigured***. It includes two important factors of the framework:

- ***ERP functionality overlapping***. ERP functionality for concrete vertical industry segment should overlap with customers request for proposal (RFP) at least 80-90%.
- ***Implementation tools***. In order to avoid *'reinventing the wheel'* each time ERP should be implemented, framework defines some supporting tools for project team:

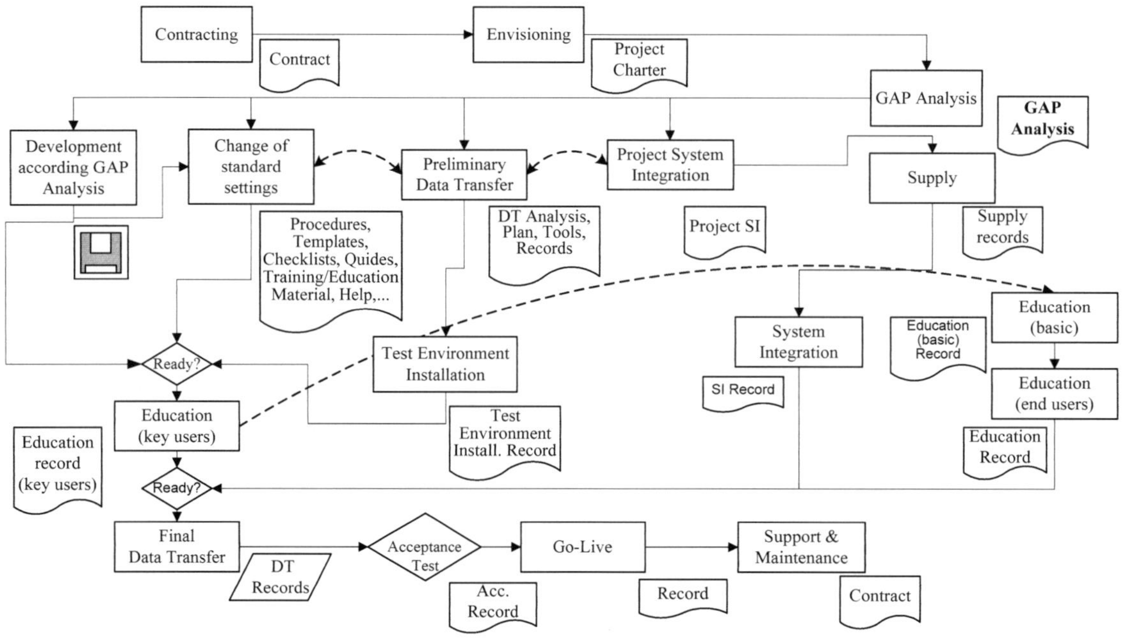

Fig. 1. ERP solution implementation process

- Predefined procedures
- Checklists
- Data transformation tools
- Templates
- Guides and
- Education/training materials.

By using these tools it makes possible to adopt existing ERP functionality and the best suitable practices and experience from all other customers who are using the same ERP solution.

3.2. Implementation process

Fig. 1 presents the ERP solution implementation process used by authors in the last ten years.

The most important step in the implementation process is good analysis of user requirements. Gap analysis is carried out for all business processes from the project scope.

After the gap analysis, a complete revision of the implementation preliminary plan of the whole information system is carried out.

It is to recommend to plan reasonable enough milestones with deliverables. On this way project manager could easier manage the whole project because good prepared milestone-meetings improve communication on project.

If the customer has too many additional requirements, implementer will take care for the most important processes in the **first project** and all additional requirements will be solved in a **second project**. It showed up in practice that some user requirements which were pushed for the second project became not important when ERP is implemented.

4. CONCLUSION

Typical successful example of framework applied in practice is the project of ERP implementation in BAKMAZ ltd, medium wholesale and retail enterprise from Zadar, Croatia.

First project lasted ten weeks from contracting to go-live and still three weeks for go-live. The project scope included following subsystems:

- Master Data
- Procurement and Distribution
- Wholesale and Retaile
- Financial Accounting and
- Reporting.

During the first project only few new user requirements were accepted – only those which were solved in days and improved warehouse management.

Scope of the second project, started a month after support and maintenance phase included following:

- OLAP and Data Warehouse Management
- Payroll and HRM
- Asset Management and
- New user requirements arising from the first project.

After go-live phase of the first project the client was ready to propagate his retail process to the whole Zadar county as a regular task of the maintenance process.

Success of the whole project is the proof of the efficient implementation framework which is developed, used and constantly improved by authors. The approach is the same for other type of solutions – whether the whole ERP or only a part of them (SCM, CRM, HRM, DMS, etc.).

In the next 12 months the approach will be:

- used in the local treasury implementation project in the Split Dalmatia County
- compared with some other methodologies (i.e. SAP's ASAP, Navision Implementation Methodology).

5. REFERENCES

Bancroft, N.; Seip, H. & Sprengel, A. (1998). *Implementing SAP R/3.* (2nd edition). Manning Publication Co., ISBN 0-13-889213-X, Greewinch

CHAOS Report (1994). The Standish Group International, *Available from:* http://standishgroup.com/sample_research/chaos_1994_1.php, *Accessed*: 2007-07-23

Dennis, A.; Wixom, B. & Roth R. (2006). Systems Analysis and Design, 3rd Edition, Copyright 2006© John Wiley & Sons, Inc. *Available from:* http://www.slideshare.net/timboeun/system-design-and-analysis-2, *Accessed*: 26.04.2008

Gross, P. (2004) How to Implement ERP Correctly, *Available from*: www.pemeco.com, *Accessed*: 2008-08-09

PM Network (2007). Here is How Software Projects fared in 2005-2006. *Project Management Network, Vol. 21, No. 6,* (June 2007), p. 10, PMI, ISSN 1040-8754

Sternad, S. & Bobek, S. (2006). Factors which have fatal influence on ERP implementation on Slovenian organizations, *JIOS*, Vol. 30, No. 2, pp. 279-293, Varazdin, Croatia, ISSN: 0351-1804

Annals of DAAAM for 2008 & Proceedings of the 19th International DAAAM Symposium, ISSN 1726-9679
Editor B. Katalinic, Published by DAAAM International, Vienna, Austria 2008
Make Harmony Between Technology and Nature, and Your Mind will Fly Free as a Bird

INTERFEROMETRIC RESEARCH OF THE FLOW WITHIN A TUBE BANK AT LOW REYNOLDS NUMBER VALUES

CERNECKY, J[ozef]; KONIAR, J[an]; BRODNIANSKA, Z[uzana] & PRAJOVA, V[anesa]

Abstract: *The present contribution deals with interferometric research of heat transfer during both the natural and forced air convections in the system of horizontal bank of tubes with constant surface temperatures (Taraba* et al., 1999*).The holographic variant of Mach-Zehnder interferometer was used in the experiment aimed at gaining the interferograms pictures of temperature fields needed for the definition of heat transfer coefficients. In the contribution the results of partial measurements are evaluated and compared with the values calculated in the FLUENT programme.*
Key words: *interferometric, tube bank, Reynolds number*

1. INTRODUCTION

During the flow across lateral boundaries of a cylinder a boundary layer is formed at its front surface, that always has its final thickness in the area between the leading and separation points. The separation point position depends on the Reynolds number. The shifting of the separation point influences the dependence of the mean heat transfer coefficient as well as the aerodynamic resistance coefficient on Re. The heat transfer intensity changes non-uniformly along the circumference of the cylinder (Barborák, 2007).
During the circumfluence of the tube bank the conditions for the first row of tubes are only a little different from those around one tube. In the following rows the heat transfer intensity increases under the influence of more intensive flow turbulence. As a result of the aerodynamic wake behind the first row of tubes the circumfluence character of the second and the following rows changes as well. The flow character stabilizes only in the fourth row of tubes at their alternating arrangement.
In the contribution we focused on the flow visualization at low Re values. For the experiment a model of a bank of 20 mm–diameter tubes heated by a heat–transfer medium (water) was used. Taking into consideration the viewing field area of the objective only the surrounding space of three tubes of alternating arrangement was studied at both the free and the forced convections (Fig. 1).

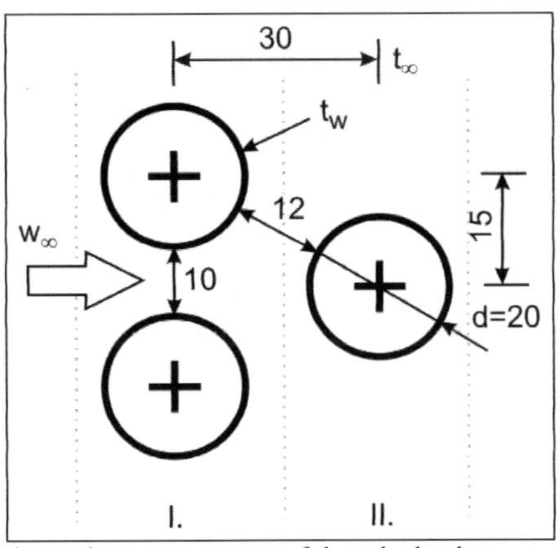

Fig. 1. Alternating arrangement of the tube bank

The interferometric method of research was chosen as it is an exact, objective and non-contact method that makes it possible to visualize non-homogeneities of transparent objects (changes of refractive index under the influence of a temperature field) in the whole studied space at the same time, which makes it possible to reveal the substance of the studied phenomena better. The results of interferometric visualization of thermal fields within transparent objects are represented by interferograms that are possible to consider qualitatively and evaluate quantitatively.
At quality consideration of the two–dimensional temperature field interferograms where the interferometer is adjusted to infinite fringe width, the interference fringes represent isotherms of a temperature field (at free convection or low Re values (Bird et al., 1960).
The quality consideration of interferograms makes it possible to reveal the shapes of temperature fields and their influence on local as well as mean heat transfer parameters.

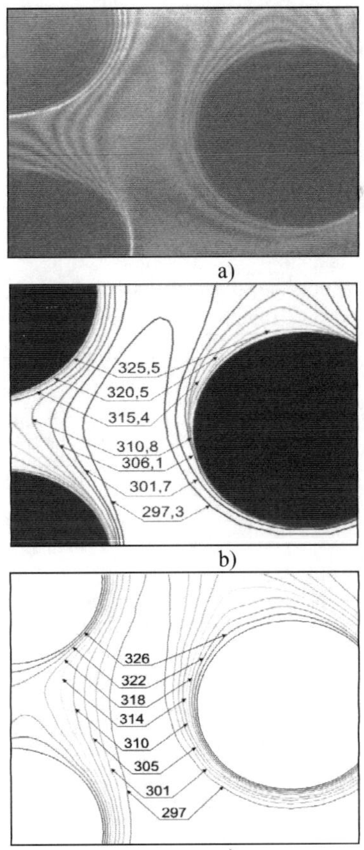

Fig. 2. Temperature field at free convection
a – holographic interferogram, b – quantitative analysis,
c – simulation
$v = 0$ m.s^{-1}, $t_w = 60°C$, $t_\infty = 20°C$, Re = 0

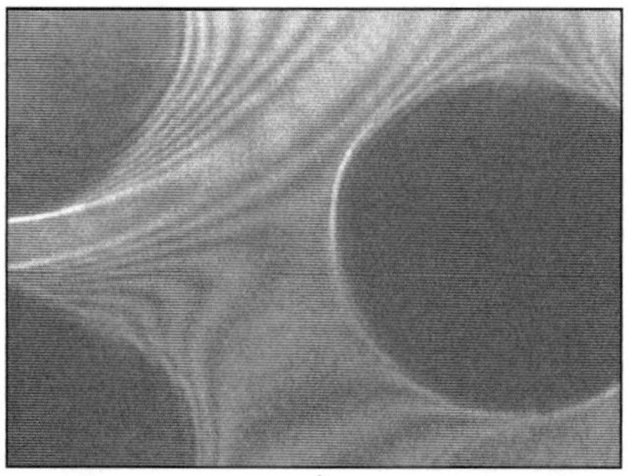

a)

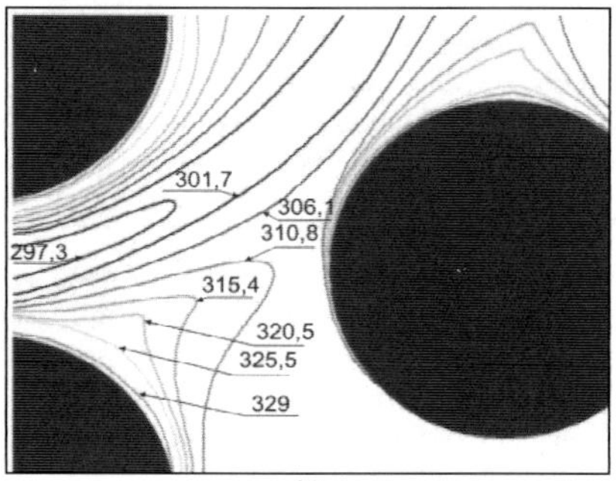

b)

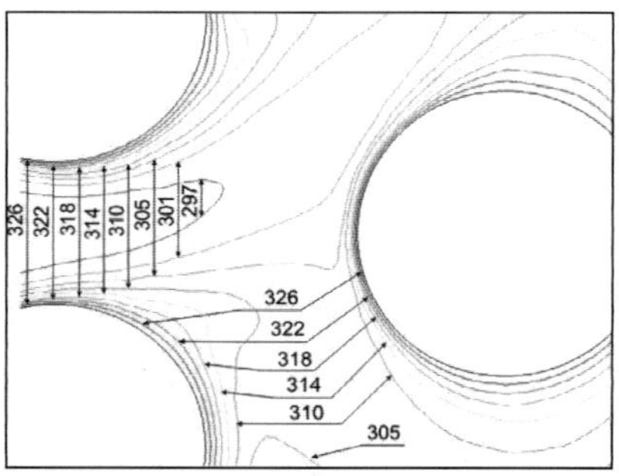

c)

Fig. 3. Temperature field at forced convection
a – holographic interferogram, b – quantitative analysis, c – simulation
$v = 0{,}2\ m.s^{-1}$, $t_w = 60°C$, $t_\infty = 20°C$, Re = 255

From the study of interferograms in real time or evaluation of dynamic records it is possible to make conclusions about the influence of parasite flow in the system, behaviour of dynamic temperature fields and dynamic development of heat transfer parameters.

2. DISCUSSION

With the aid of a holographic variant of the Mach-Zehnder interferometer interferograms of the temperature fields in the surroundings of the heated horizontal tube bank were obtained (Oosthuizen & Naylor 1999). The arrangement of the first three tubes and their diameters is displayed in Fig. 1. During the measurements the interferometer was adjusted to infinite fringe width. Considering the small viewing field area the tube bank must be shifted horizontally so that the interferograms from another row of tubes can be recorded as well.

From the obtained interferograms the interference order depending on state quantities of the measured environment and the wave length of the used light was determined (Taraba et al., 2004). From the interference images the positions of interference fringes were also determined and the respective temperature values for each fringe were calculated.

The temperature field in the surroundings of the tubes at free convection is illustrated in Fig. 2. The first picture (Fig. 2a) represents a holographic interferogram of a temperature field at the tube surface temperature of $t_w = 60°C$ and the temperature of surrounding environment of $t_\infty = 20°C$. In Fig. 2b its quantitative analysis is shown and in Fig. 2c the result of numerical simulation is displayed.

From the interferograms the thermal boundary layer at free convection is visible, in which the air flows upwards. Outside the boundary layer there is surrounding environment with constant air temperature. The expansion of the thermal boundary layer above the tube causes a decrease of the heat transfer gradient.

At forced convection the temperature conditions in the tube surroundings are changing. In Fig. 3 the temperature field in the surroundings of the tubes of surface temperature of $t_w = 60°C$ at the temperature of surroundings of $t_\infty = 20°C$ and the forced air flow with the value of Re = 255 is illustrated. As it follows from Fig. 3a, the thermal boundary layer is influenced by air flow, which influences the temperature field of the tubes in the second row.

Considering the fact, that the Re values are low, the forced convection influence is relatively small. In Fig. 3b quantitative evaluation of the holographic interferogram from Fig. 3a is displayed. The obtained results are possible to compare with the values calculated in the FLUENT programme (Fig. 3c).

3. CONCLUSION

In the present contribution it is indicated that interferometry has proved competent in the given research and, therefore, it can be applied to solve further research tasks. The effective utilization of interferometry, however, is conditioned by the development of image computer processing, which is also the subject of our contemporary research. By complementation of the interferometric method with computer image processing we will obtain a highly effective experimental tool for different research fields. Interferometry is a prospective method not only in the field of research, but also in education where it can complement the physical problems explanation with suitable and interesting image material. The obtained values can be used for further processing, i.e. calculation of heat transfer coefficients. The experimentally obtained results are also possible to compare with the simulations in the FLUENT programme use them for the CFD models verification. The results achieved can also be used in optimizing the contruction of tube heat exchangers in relation to their energetic effectiveness improvement.

The contribution was elaborated within the research project /KEGA No. 3/6431/08 Determination of the Characteristics of Emission and Indicator Quantification According to European Legislation

4. REFERENCES

Barborák, O. (2007). *Technical drive,* Dubnica nad Váhom, ISBN 978-80-969615-6-6, pp. 146

Bird, R.B. ; Steward, W.E. &Lighfood, E.N. (1960). *Transport phenomena, J. Wiley*, New York, ČSVTS, Praha

Oosthuizen, P.H. & Naylor, D. (1999). *An Introduction* to *Convective Heat Transfer Analysis,* WBC/McGraw-Hill, New York, ISBN 258-458-73-02

Taraba, B. ; Behúlová, M. & Kraváriková, H. (2004). *Liquid drive,* STU Bratislava, ISBN 80-227-2041-0, pp. 214

Taraba, B. ; Behúlová, M. & Kraváriková, H. (1999). *Hydromechanic Thermomechanic,* ES STU Bratislava, ISBN 80-227-2041-0, pp. 118

Annals of DAAAM for 2008 & Proceedings of the 19th International DAAAM Symposium, ISSN 1726-9679
Editor B. Katalinic, Published by DAAAM International, Vienna, Austria 2008
Make Harmony Between Technology and Nature, and Your Mind will Fly Free as a Bird

THE SOFTWARE APPLICATION FOR MATHEMATIC DESCRIPTION OF EXTRACTION PROCESS COURSE

CHARVATOVA, H[ana]; JANACOVA, D[agmar]; KOLOMAZNIK, K[arel] & FIALKA, M[iloslav]

Abstract: *Many technological operations are based on a process of undesirable components removing from the solid material. For propose of an optimal course of these components extraction, the physico-chemical formulation of the process is necessary to know. In the contribution we deal with creation of the software application of the removed component concentration field in the solid phase calculation during the extraction process. The calculation is based on analytical solution of the extraction process mathematical model. We visualized the formated dependence of the removing component concentration in the solid phase on the operating time in 2D and 3D projection for both real and dimensionless variables.*
Key words: *model, concentration field, extraction, software*

1. INTRODUCTION

During nature polymers processing extraction process of undesirable components from a solid material is very often used. For a suitable technological process proposal, it is necessary time course of the operation to know (Janáčová, 2002). The proposal can be determined by both experimentally (i.e. appropriate analytical method using) and appropriate mathematical model solving. The operation time course is often hardly experimental determined and in addition time-consuming. Therefore, it is advantageous to use the second above mentioned proposed procedure, i.e. the mathematical model based on physical mechanism of extraction process to suggest (Kolomazník, Janáčová, Charvátová, Fialka, 2007).

In this contribution we formulated mathematical model of the bound components extraction from the solid material to bath containing the extraction solvent in one cycle.

Next we derived the analytical solution of the model that describes a concentration field of the extracted component in the solid phase. The model was obtained on the basis of many experimental determinations and technological experiences.

We used the under mentioned solution for making of a software application which can calculate and visualize the time course of the extraction process for real or dimensionless values in a short time interval.

2. MATHEMATICAL MODEL OF THE EXTRACTION PROCESS

We supposed that the extracted component is strong bound to the solid material and the extracted component is very well soluble during extraction process.

Under given conditions the process of undesirable component removing from the solid phase to the bath containing extraction solvent can be described by the following equations (1) – (7) (Crank, 1968) and by figure 1 .

$$\frac{D}{1+A}\cdot\frac{\partial^2 c(x,t)}{\partial x^2}=\frac{\partial c(x,t)}{\partial t},\quad t>0,\quad 0\le x\le b \tag{1}$$

$$\frac{\partial c}{\partial x}(b,t) = -\frac{V_0}{D\cdot S}\cdot\frac{\mathrm{d}c_0}{\mathrm{d}t}(t) \tag{2}$$

$$c(x,0)=c_p \tag{3}$$

$$c_0(0)=0 \tag{4}$$

$$\frac{\partial c}{\partial x}(0,t)=0 \tag{5}$$

$$c(b,t)=\varepsilon\cdot c_0(t) \tag{6}$$

$$c_A = A\cdot c\,. \tag{7}$$

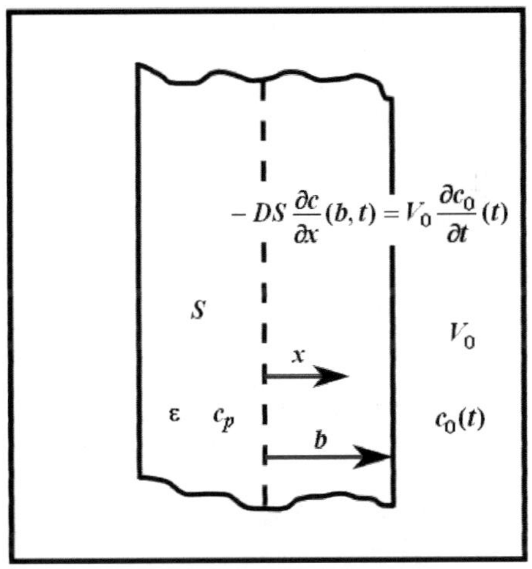

Fig. 1. Mathematical model of extraction process

Using the dimensionless variables

$$C=\frac{c}{c_p},\quad F_o=\frac{D\cdot t}{b^2(1+A)},\quad X=\frac{x}{b},\quad C_0=\frac{c_0}{c_p} \tag{8}$$

we obtained by means of a Laplace's transformation analytical solution of the concentration field in solid phase for dimensionless variables (Kolomazník, 1990)

$$C(X,F_o)=\frac{\varepsilon(1+A)}{\varepsilon(1+A)+Na}-2Na\sum_{n=1}^{\infty}\frac{\cos(Xq_n)\,\exp\left(-F_o\,q_n^{\,2}\right)}{\varepsilon(1+A)\cos(q_n)-\frac{\varepsilon(1+A)}{q_n}\sin(q_n)-Na\cdot q_n\sin(q_n)} \tag{9}$$

or for real variables

$$c(x,t) = c_p \left(\frac{\varepsilon(1+A)}{\varepsilon(1+A) + \frac{V_0}{V}} - 2\frac{V_0}{V} \sum_{n=1}^{\infty} \frac{\cos\left(\frac{x}{b} q_n\right) \exp\left(-\frac{D \cdot t}{b^2 (1+A)} {q_n}^2\right)}{\varepsilon(1+A)\cos(q_n) - \frac{\varepsilon(1+A)}{q_n}\sin(q_n) - \frac{V_0}{V} \cdot q_n \sin(q_n)} \right), \quad (10)$$

where q_n are positive roots of the following transcendent equation

$$\tan(q) = \frac{Na \cdot q}{\varepsilon(1+A)}. \quad (11)$$

Symbol	Sense	Measuring unit
V	volume of solid phase (i.e. material)	m^3
V_0	volume of extraction solvent	m^3
t	time	s
c	concentration of extracted component in solid phase	$kg \cdot m^{-3}$
c_0	concentration of extracted component in liquid phase (i.e. in bath with extraction solvent)	$kg \cdot m^{-3}$
c_A	concentration of bound component in solid phase	$kg \cdot m^{-3}$
c_p	initial concentration of extracted component in solid phase	$kg \cdot m^{-3}$
D	effective diffusion coefficient of extracted component from solid phase	$m^2 \cdot s^{-1}$
x	Cartesian coordinate (one-dimensional diffusion)	m
b	half thickness of solid phase	m
ε	porosity of solid phase	1
Na	dimensionless volume of extraction solvent	1
q_n	n^{-th} root of a certain transcendent equation	1
A	sorption coefficient (determined from Langmuir's sorption isotherm)	1
S	from one-side area of pelt	m^2
F_o	Fourier´s number (i.e. dimensionless time)	1
C	dimensionless concentration of extracted component in solid phase	1
X	dimensionless space coordinate	1

Tab 1. Sense of the used symbols.

3. THE SOFTWARE APPLICATION

We made up the software application for calculation and visualization of extracted bound component concentration field in solid phase in the computer algebraic system Maple (Charvátová, 2007). The calculation consists of three basic parts. Firstly the positive roots q have to be numerically solved for input parameters from equation (11).

Subsequently both dimensionless or real concentration fields can be calculated and displayed for one or two variables. We used equation (9) for computation of the dimensionless concentration field and equation (10) for computation of the real concentration field. The obtained graphics can be exported to postscript graphic format, when appropriate.

The software application can calculate the values of extracted component concentration in a given time and a place of the solid phase too (figure 2). Furthermore, it is possible to re-count the real and the dimensionless values each other.

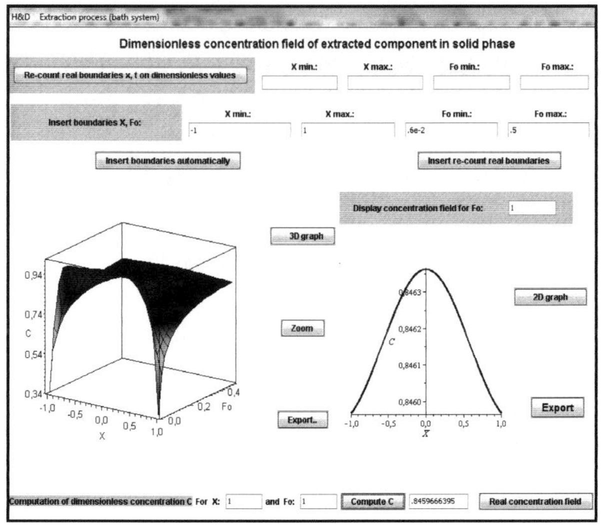

Fig. 2. Part of the window for solving and visualization of dimensionless concentration field in solid phase

4. CONCLUSION

The above mentioned analytical solution of mathematical model in case of the one-cycle bound component extraction from the solid material to the extraction solvent enabled us to make the software application for calculation and visualization of time-course of the extraction process for both real and dimensionless variables.

The software application speeds up very complicated computing and gives us the fast information about a surveyed technological process. The obtained informations can be used for optimization of many technological operations based on extraction process. It will lead to significant cost reduction of the operations.

Acknowledgement: *This paper was supported by the Ministry of Education, Youth and Physical Training of the Czech Republic in the range of the research project No. 7088352102: Modelling and controlling of natural and synthetic polymers processing.*

5. REFERENCES

Crank, J. & Park, G., S. (1968). *Diffusion in Polymers,* Academic Press, ISBN 0121970507, London

Charvátová, H. (2007). *Modelling of Pelt Chemical Deliming,* (Disertation work), Tomas Bata University in Zlín, Zlín

Janáčová, D. (2002). *Modelling of Extraction Procesess,* (Habilitation work), VŠB – Technical university in Ostrava, ISBN 80-248-0194-9, Ostrava

Kolomazník, K. (1990). *Modelling of Manufacturing Processes,* (Lecture notes), Faculty of Technology in Zlín, ISBN 8021401141, Zlín

Kolomazník, K.; Charvátová, H.; Fialka, M. & Janáčová, D. (2007). Chemical Engineering Experience for Tannery Optimization, [In CD] *Proceedings of XXIX Congress of the IULTCS 103rd Annual Meeting of the ALCA*, p. 6, ISBN 0002-9726, Washington, D.C., American Leather Chemists Association.Washington, D.C., June, 2007, Washington, D.C.

Annals of DAAAM for 2008 & Proceedings of the 19th International DAAAM Symposium, ISSN 1726-9679
Editor B. Katalinic, Published by DAAAM International, Vienna, Austria 2008
Make Harmony Between Technology and Nature, and Your Mind will Fly Free as a Bird

SCALAR CONTROL STRUCTURE OF AN ASYNCRONOUS MOTOR AT MAXIMUM TORQUE

CHIONCEL, C[ristian]; CHIONCEL, P[etru] & GILLICH, N[icoleta]

Abstract: *This paper presents a scalar control structure for a asynchronous motor at maximum torque. Voltage and frequency changes in steps until it reaches the final value based on a computing block. The final values for the request voltage and the frequency realised by the control scheme with one, two controller's ore just implementing the computing blocks, are obtained though solving algebraic systems and simulation in transient operating system, by solving differential equations, in an software based on the "Maple" module.*
Key words: *asynchronous motor – ASM, control algorithm*

1. INTRODUCTION

The speed control at asynchronous and synchronous machine is realised in presents through the modification of the frequency of the voltage supply, based on the fundamental principle:

$$n_1 = \frac{f_1}{p_1} \qquad (1)$$

where n_1 – speed of the stator rotating magnetic field, f_1 – frequency of stator voltage, p_1 – number of pair pools.

The number of pair pools can be changed, but difficult, through construction rates, but not enough precisely, so that the only solution is to change the frequency of the voltage supply.

The control of the asynchronous machine is studied in many papers and different publications (Garlasu et al., 2007), where complex control methods are proposed, with difficult practical implementation. The proposed method, with variable rotor pulsation during the speed up acceleration, ω_r, is simpler and presents a high efficiency, as the obtained results are showing.

In further research, the proposed method will be studied in the case of constant rotor pulsation ω_r during the speed up acceleration.

2. CONTROL SCHEME FOR MAXIMUM ELECTROMAGNETIC TORQUE

The control algorithm for ASM realize the speed [rpm] control by: constant stator flux ψ_s, constant useful flux ψ_h and constant rotor flux ψ_r. By constant stator flux, we obtain the highest values for the electromagnetic torque, by constant useful flux with 20% luster, and the lowest values are obtained by constant rotoric flux, case where the dynamic characteristics are linear and ideal for rapid controls.

The lowest currents at the same rotation speed are by rotoric constant flux and the highest, by stator constant flux (Biriescu , 1997).

Next we propose a control scheme by maximum electromagnetic torque, maxim through the entire control period. The maximum torque value is given by the value of the rotor flux ψ_r and maximum allowed rotor pulsation ω_r, taking in account the limitations dictated to the stator current, so that the maximum current protection should not act.

The maximum torque is achieved at the maximum values of the rotoric current and flux (Babescu et al, 2005):

$$M_{max} = 3p_1\psi_{r\,max}I_{r\,max} \qquad (2)$$

where $\psi_{rmax} = \psi_{rN}$ (nominal rotoric flux) and I_{rmax} – maximal allowed value for the rotor current.

The maxim rotor current (Babescu et al, 2005) is limited by the maxim allowed value for the statoric current, based on the relation (Biriescu, 1997)

$$I_{r\,max} = \frac{\omega_r M}{\sqrt{R_2^2 + (\omega_r L_2)^2}} I_{sadmx} \qquad (3)$$

where M – electromagnetic torque, R_2 - rotor resistance and L_2 - rotor inductance.

The allowed stator current, I_{sadm} is known from the catalogue data and can vary between the nominal value I_{sN} and *k* times higher as the nominal value, depending on the machines load (given by the number and length of control cycles during on hour).

From the value of the current I_{sadm} , using the relation (Quang et all, 2005)

$$I_{sadmx} = \frac{\psi_{rN}}{MR_2}\sqrt{R_2^2 + (\omega_r L_2)^2} \qquad (4)$$

the computing rotor pulsation ω^*_r will be calculated (for $\Delta\omega_{mec} > 0.5$, the case NO in figure 1):

$$\omega_r^* = \frac{1}{L_2}\sqrt{\left(\frac{MR_2 I_{sadm}}{\psi_{rN}}\right)^2 - R_2^2} \qquad (5)$$

In this way the computing statoric pulsation ω^* can be work out with the relation:

$$\omega^* = \omega_{mec} + \omega_r^* \qquad (6)$$

With ω^* and ω^*_r determinate, the stator voltage, that changes also, will be computed:

$$U = \frac{\psi_{rN}}{L_U}\sqrt{R_1^2 + (\omega L_1)^2 + \left(\frac{\omega_r L_2}{R_2}\right) 2\omega R_1 (1-\sigma) L_1 + \left(\frac{\omega_r L_2}{R_2}\right)^2 \left[R_1^2 + (\omega\sigma L_1)^2\right]} \qquad (7)$$

The described steps, included in a structural scheme for speed controlling (Chioncel, 2008), for maximum allowed acceleration, on reasons of maxim allowed rotor flux and maxim allowed stator current, is presented in figure 1.

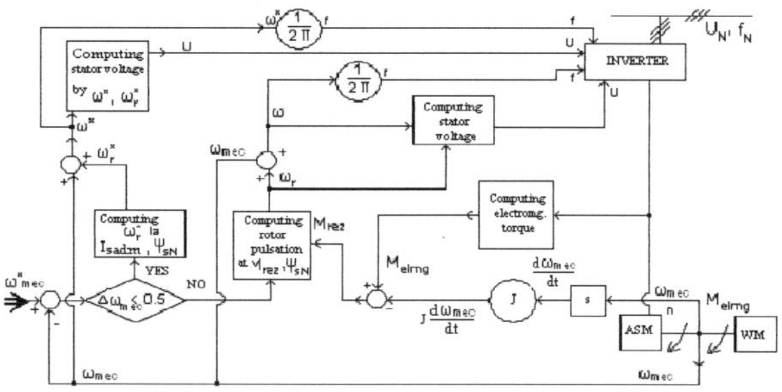

Fig. 1 Scalar speed control by maximum torque

During the acceleration period, the maximum torque value is reached and assured through maximal values of the voltage and frequency. When the prescribed value of the rotor speed ω_{mec} is achieved, voltage and frequency returns to values regarded by the stationary regime, adequate to the prescribed value.

The electromagnetic torque, in stationary regime and for $\omega_{mec} = \omega^*_{mec}$ has the same value with those of the resistant torque M_{rez} His value is computed from the movement equation:

$$M_{rez} = M_{elmg} - J\frac{d\omega_{mec}}{dt} \tag{8}$$

From this value of the resistant torque, by nominal rotor flux, the rotor pulsation will be computed:

$$\omega_r = \frac{R_2 M_{elmg}}{3p_1\psi_{rN}^2} \tag{9}$$

3. COMPUTING AND SIMULATION RESULTS OF THE PROPOSED ALGORITHM FOR AN ASM

The simulation results are obtained for an ASM, with the following nominal parameters (Chioncel et all, 2007):
$L_1 = L'_2$=0.1[H]; R_1=R'_2=5[Ω]; $L_{1\sigma}$=$L'_{2\sigma}$=0.02[H]; σ = 0.36, L_U=M=0.08H, U_N = 380V, p_1 =1, ψ_{sN}=1.2[Wb]

The initial conditions are: stator flux ψ_s=1.2[Wb], $\omega_{mec}(0)$=280[rad/s]. To get the maximum acceleration, we need following computing vales

- For I_{Sadm} = 30 [A], the rotor pulsation ω^*_r is:

$$\omega_r^* = \frac{1}{L_2}\sqrt{\left(\frac{MR_2 I_{sadm}}{\psi_{rN}}\right)^2 - R_2^2} = 115\left[\frac{rad}{s}\right] \tag{10}$$

- Computing stator pulsation: $\omega^* = \omega_{mec} + \omega^*_r$ = 280 +115 = 395 [rad/s]
- Stator voltage (7) Us = 696[V].

The accelerations subsist until the prescribed speed is reached, whereupon voltage and frequency get back to the prescribed values of torque and speed: M_{elmg} = 16.66[Nm], n = 2961 [rpm]. For the simulation, in the case where the inertial moment is J = 0,01[kgm^2], we get the following results, figure 2:

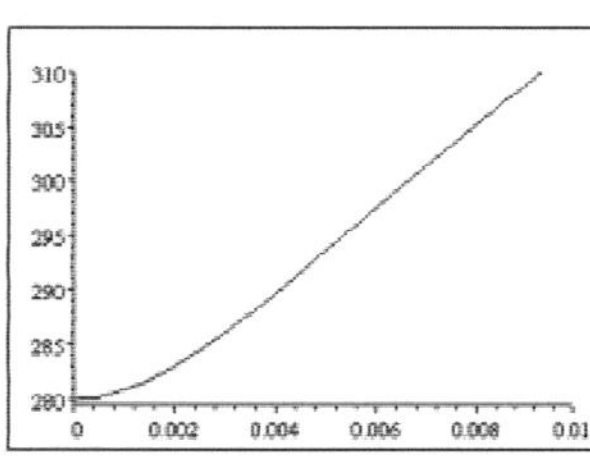

Fig. 2 Time evolution of the mechanical angular speed.

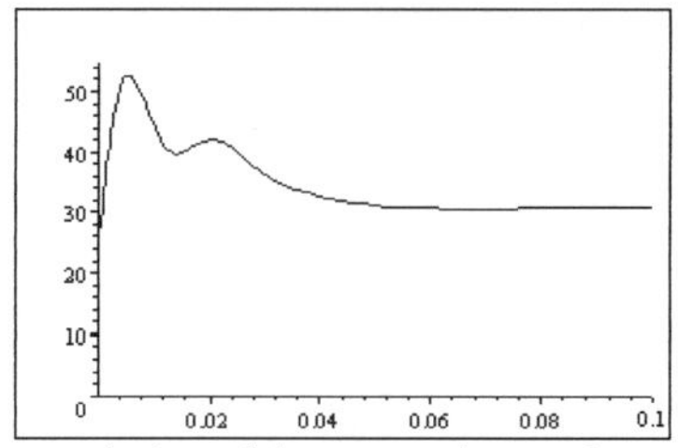

Fig. 3 Current evolution of the current by maximum torque

It can be observed that the response time during the speed reaches the prescribed value, is 0.01[s], much lower (ten times) than the control method when the maximum torque is not chose during the control period (Chioncel, 2008).

The initial value of the stator current I_{Sadm} = 30[A] is with 76% exceeded and from that reason, the maximum current protection has to be adjusted by I_{max} = 53[A], figure 3. Taking in consideration the very short time of the transient process, the thermal protection does not act.

During the transient control period, when the electromagnetic torque is maxim, his value is almost double comparing to the case when this value is not maximised (Chioncel, 2008).

4. CONCLUSIONS

The presented computing method for ASM speed control by maximising the electromagnetic torque, represents a modern solution for the automatic control for many electric drive systems meet in practice, where a very quick answer to the appeared perturbations is necessary.

The prescribed speed is achieved in an asymptotic way. As well, the same trend in time has the other values too: stator current, electromagnetic torque, rotor flux.

Until the prescribed speed is reached, the rotor flux exceeds the nominal value, that makes that the motor functions in saturation regime.

The presented control algorithm repeats after a period, so that each change that appears during the control process, are always corrected.

6. REFERENCES

Babescu, M & Paunescu, D. (2005) *Analiza matematica a dinamicii maşinilor electrice (Mathematical analyze of the dynamic behavior of electric machines)*, Ed. Politehnica, Timişoara

Biriescu M., *Maşini electrice rotative. Parametrii, caracteristici, încercări (Rotating electric machines. Parameters, test, characteristics)*, Ed de Vest Timisoara, 1997

Cristian P. Chioncel, Marius Babescu, Petru Chioncel, Nicoleta & Gilbert-Rainer Gillich, (2007) *Speed control method for asynchronous motor* Annals of DAAAM 2007 & Proceedings of the 18th International DAAAM Symposium, Zadar 2007, pp. 137-138

Cristian P. Chioncel (2008), *Contributii privind controlul turatiei la masina asincrona (Contribution regarding the speed control of asynchronous machine)*, Seria 1, Nr.9, Ed Politehnica, Timisoara

Gârlaşu Şt., Ruja I, Breabăn F., (2005) *Controlul mişcării (Movement control)*, Ed Orizonturi Universitare, Timişoara

Quang, Ph. & Schoenfeld, R. (2005) *Dynamische Stromregelung zur Drehmomenteinpregung in Drehstromantrieben mit Pulswechselrichter (Dynamic current control for pulsation in alternative machines with pulse inverter),* Journal Electrical Engineering, Springer, Berlin, page numbers 317 – 323

Annals of DAAAM for 2008 & Proceedings of the 19th International DAAAM Symposium, ISSN 1726-9679
Editor B. Katalinic, Published by DAAAM International, Vienna, Austria 2008
Make Harmony Between Technology and Nature, and Your Mind will Fly Free as a Bird

THE DETERMINATION OF THE VOLUME FUNCTION VARIATION OF PANU-STANESCU ROTARY ENGINE

CHIOREANU, N[icolae]; CATAS, A[driana] & BLAGA, V[asile]

Abstract: *The paper presents the parameter equation of the inner surface section of the carcass of the piston section and the determination of the volume function variation of one chamber. Due to this fact one can obtains an optimization of the operational factors in the case of this rotary engine.*
Key words: *Panu-Stanescu, rotary engine, volume.*

1. INTRODUCTION

The Panu-Stănescu rotary engine is a heat rotary piston engine, Romanian letters-patent of engineers Mihai Panu and Gheorghe Stănescu. As we seen in figure 1, (Chioreanu & Blaga, 2002 and Panu & Stănescu, 2000), the engine consists on a cylindrical carcass (1) with two different values and antipodal ovalizations. The rotary piston (rotor) (2) is mounted inside the carcass and includes six box-fires for each chamber. On the external surface of the piston there are six radial channels disposed at sixty degree, in which six obturation bars ($P_i, i=\overline{1,6}$) glides. During the running, the bars are in permanent contact with the surface of the carcass due to centrifugal forces. For other geometrical details see also (Chioreanu & Chioreanu, 2006; O'Neill, 1991 and Nice, 2004).

2. THE GEOMETRY OF THE INNER SURFACE OF THE CARCASS

Denote by (p_1,φ) the polar coordinates of the point M situated at the external extremity of the bar P_1 when this passes through the first ovalization and let (p_2,φ) be the polar coordinates of the point M when the bar passes through the second ovalization. From the triangles ΔOO_1P_1 and ΔOO_1M we get the relations

$$R_1^2 = R^2 + e_1^2 - Re_1;\ \rho_1^2 - 2e_1 \sin\varphi\, \rho_1 - \left(R_1^2 - e_1^2\right) = 0\,.$$

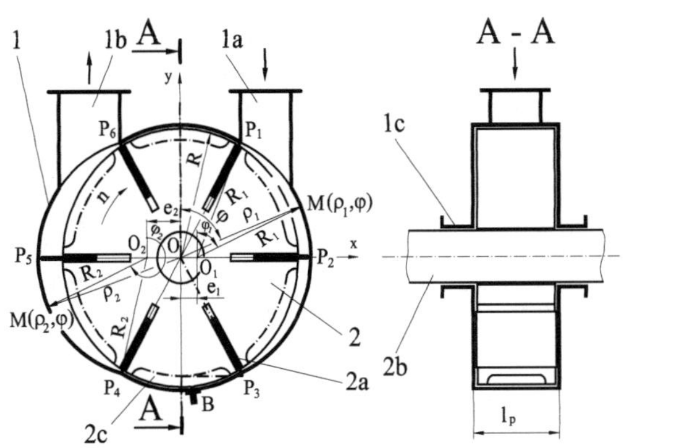

Fig. 1. Cross section of the engine

The equation yields the polar line of the first ovalization

$$\rho_1(\varphi) = e_1 \sin\varphi + \sqrt{R^2 - \mathrm{Re}_1 + e_1^2 \sin^2\varphi}\quad . \tag{1}$$

Similarly, from the triangles ΔOO_2P_4 and ΔOO_2M we get the polar line of the second ovalization

$$\rho_2(\varphi) = -e_2 \sin\varphi + \sqrt{R^2 - \mathrm{Re}_2 + e_2^2 \sin^2\varphi}\,. \tag{2}$$

3. THE VOLUME FUNCTION VARIATION

Between carcass, piston and the bars six chambers with variable volume are formed. When the bar P_1 passes through the first ovalization, inside the chamber $P_1 - P_6$ the admission-compression processes take place while when the bar passes through the second ovalization the expansion-evacuation processes take place.

The volume of a chamber formed between the piston and carcass has a cyclical variation as a function of φ angle given by the formula

$$V(\varphi) = l_p A(\varphi) + V_{ca} \tag{3}$$

where l_p is the breadth of the piston, $A(\varphi)$ is the plan area of one chamber and V_{ca} is the volume of the combustion chamber.

The φ position of the piston determines the following expressions of the $A(\varphi)$ area:

a) for $\varphi \in \left[0, \frac{\pi}{6}\right]$, we obtain

$$A_1(\varphi) = \iint_D dxdy = \iint_{D'} |J| dr d\varphi = \frac{1}{2}\int_{\frac{\pi}{6}}^{\varphi+\frac{\pi}{3}} \left(\rho_1^2(\varphi) - R^2\right) d\varphi\,,$$

b) for $\varphi \in \left[\frac{\pi}{6}, \frac{\pi}{2}\right]$,

$$A_2(\varphi) = \frac{1}{2}\int_{\varphi}^{\varphi+\frac{\pi}{3}} \left(\rho_1^2(\varphi) - R^2\right) d\varphi\,;$$

c) for $\varphi \in \left[\frac{\pi}{2}, \frac{5\pi}{6}\right]$,

$$A_3(\varphi) = \frac{1}{2}\int_{\varphi}^{\frac{5\pi}{6}} \left(\rho_1^2(\varphi) - R^2\right) d\varphi\,;$$

d) for $\varphi \in \left[\frac{5\pi}{6}, \frac{7\pi}{6}\right]$,

$$A_4(\varphi)=\frac{1}{2}\int_{\frac{7\pi}{6}}^{\frac{\pi}{3}+\varphi}\left(\rho_1^2(\varphi)-R^2\right)d\varphi\,;$$

e) for $\varphi\in\left[\frac{7\pi}{6},\frac{3\pi}{2}\right]$,

$$A_5(\varphi)=\frac{1}{2}\int_{\varphi}^{\varphi+\frac{\pi}{3}}\left(\rho_1^2(\varphi)-R^2\right)d\varphi\,;$$

f) for $\varphi\in\left[\frac{3\pi}{2},\frac{11\pi}{6}\right]$,

$$A_6(\varphi)=\frac{1}{2}\int_{\varphi}^{\frac{11\pi}{6}}\left(\rho_2^2(\varphi)-R^2\right)d\varphi\,;$$

g) for $\varphi\in\left[\frac{11\pi}{6},2\pi\right]$,

$$A_7(\varphi)=\frac{1}{2}\int_{\frac{\pi}{6}}^{\varphi-\frac{10\pi}{6}}\left(\rho_1^2(\varphi)-R^2\right)d\varphi\,.$$

The volume function variation $V(\varphi)$ depends continuously on φ and it has the branches $V_i(\varphi)=l_pA_i(\varphi)+V_{ca},\ i=\overline{1,7}$ with the angle φ in the suitable intervals. The maximal volume at the admission compression is obtained for $\varphi=\frac{\pi}{3}$ and the maximal volume at expansion-evacuation is obtained for $\varphi=\frac{4\pi}{3}$.

Denote by $\varepsilon=V_{\max}/V_{\min}$ the ratio of compression and by $\delta=V_{d\max}/V_{\min}$ the ratio of expansion, where $V_{d\max}$ is given by the formula

$$V_{d\max}=l_pA_5\left(\frac{4\pi}{3}\right)+V_{ca}\qquad(4)$$

The volume of combustion chamber $V_{ca}=V_{\min}$ is given by the relation

$$V_{ca}=\frac{1}{\varepsilon-1}l_pA_2\left(\frac{\pi}{3}\right).\qquad(5)$$

By means of (4) and (5) we get

$$\delta=1+(\varepsilon-1)\frac{A_5(4\pi/3)}{A_2(\pi/3)}\qquad(6)$$

At a single rotation of the piston, six working cycles take place. Figure 2 shows the variation of the volume of a rotary engine with the following specific parameters: the ratio of compression $\varepsilon=8.5$, the values of ovalizations $e_1=9.25\ mm$, $e_2=14.72\ mm$, the radius of the piston $R=130\ mm$, the breadth of the piston $l_p=125\ mm$.

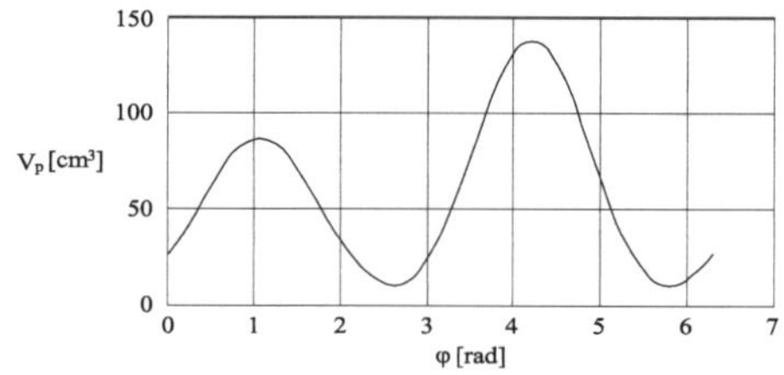

Fig. 2. Variation of $V(\varphi)$ of one chamber

4. ENGINE POWER

For a chamber with variable volume, the working cycle is performed at one rotation of the piston $(\varphi=2\pi)$. Let $\omega[rad/s]$ be the angular speed and $n[rot/\min]$ be the number of rotations of the motor shaft. The time $\tau_c[s]$ necessary to perform a working cycle is given in the following relation $\tau_c=\frac{2\pi}{\omega}=\frac{60}{n}[s]$. We also denote by $z_c=6$ the number of the chambers with variable volume, by i_r the number of the rotors and by $p_e\left[N/m^2\right]$ the averaged effective pressure at a one cycle. The effective power P_e of the engine is one obtains by the following relation

$$P_e=z_c\cdot i_r\cdot\frac{p_e\cdot V_s}{\tau_c}=\frac{z_c\cdot i_r}{60}\cdot n\cdot p_e\cdot V_s\ [W]\qquad(7)$$

where V_s is the cylinder of one chamber.

5. CONCLUSION

If it is compared a Panu-Stănescu rotary engine $(z_c=6, i_r=1)$ with an internal combustion four-cycle engine $(\upsilon=4)$, having the same cylinder capacity, which at a certain number of rotations develops the same power

$$\frac{i}{30\cdot\upsilon}\cdot n\cdot p_e\cdot V=\frac{z_c\cdot i_r}{60}\cdot n\cdot p_e\cdot V_s\ \Rightarrow i=12\cdot i_r\qquad(8)$$

From the implication (8) we conclude: a Panu-Stănescu rotary engine is equivalent to an internal combustion four-cycle engine having twelve cylinders.

6. REFERENCES

Chioreanu N. & Blaga V. (2002). *Panu Stănescu rotary engine*, Proceedings of the International Conference AMMA, vol.II, 227-230.

Chioreanu, N. & Chioreanu, S. (2006). *Heat engine for nonconventional motor vehicles*, Ed. Univ. of Oradea.

Nice, K. (2004). *How Rotary Engines Work,* 2002. Accessed October 18, 2004.

O'Neill, P.V. (1991). *Advanced Engineering Mathematics*, Wadsworth Eds.

Panu, M. & Stănescu, Ghe. (2000). *Heat engine with rotary pistons*, X-th National Conference on Heat Engineering, Sibiu, May, 25-27.

Annals of DAAAM for 2008 & Proceedings of the 19th International DAAAM Symposium, ISSN 1726-9679
Editor B. Katalinic, Published by DAAAM International, Vienna, Austria 2008
Make Harmony Between Technology and Nature, and Your Mind will Fly Free as a Bird

ON THE VARIATION OF THE FUNCTIONAL PARAMETERS OF THE ASYMMETRIC GEARS ON RELATION WITH THE DESIGNING VARIABLES

CHIRA, F[lavia]; BANICA, M[ihai] & LOBONTIU, M[ircea]

Abstract: *In order to improve the performances of the gear transmission with involutes profile gear teeth, in the last years have been developed method for studying and manufacturing asymmetric gears. These are characterised by asymmetrical involutes teeth profile, having different values for the base circles diameters. The research of the paper authors, on the asymmetric gears, have as result calculus algorithms and computational method for designing and analysing the performances in function of those special gears. The developed applications can be used by any designing engineer for better design of asymmetric gears transmission. This paper is focused on emphasize the variation of asymmetric gears performances in relation with the designing variables.*

Key words: *asymmetric, involutes, gears, stress, efficiency.*

1. INTRODUCTION

A few years ago the aim of the designing engineers in gears transmission has been to obtain large quantity of products at low coasts. Now it is more important to obtain by optimal design better performances in function. One of the methods to added supplementary advantages to those already known of the involutes gears, by improving the load capacity and efficiency (Kapelevich, A.,L., 2000), (Litvin F., L. et al 2000) and reducing the transmission error and vibrations (Karpat, F. et al 2006), is to use the asymmetric gears (Novikov, A.S., et al 2008).

By using the computational method of design the difficulties in manufacturing asymmetric gears can be solved without significant supplementary costs.

One of the most important parameters of asymmetric gears is the coefficient of asymmetry, noted with "k" and defined as the ratio between the base circle diameter of the inactive flank and the base circle diameter of the active flank (Kapelevich, A., L., 2000). There are authors that study and emphasize the advantage of gears with the coefficient of asymmetry bigger than the unit and others that emphasize the advantage of using gears having the coefficient of asymmetry smaller than the unit. The coefficient of asymmetry must be established in relation with the quality indicators imposed by the beneficiary.

2. THE VARIABLES OF THE DESIGN METHOD

In this paper author's approach it is considered that an asymmetric gear transmission, formed by two gears with asymmetric teeth profiles, with geometric parameters established, represent two gear transmissions with different performances, depending on the profile choose as active profile. Booth of them have specific advantages, reducing the contact stress or reducing the bending stress, so it is important to study in the same time the booth variant, comparing with the classical symmetrical gears and find the better solution. For this aim has been developed, as application in MATLAB, a packet of software for designing, modelling and analyzing the geometrical and functional parameters of asymmetric teeth spur gears. The profile with bigger pressure angle it has been called "direct profile" and the profile with smaller pressure angle it has been called "inverted profile" considering that the profile „d" belong to the positive side of an involute with the base circle diameter d_{bd} and the profile „i" belong to the negative side of an involute with the base circle diameter d_{bi}. If the direct profile is the active one, the gear is named "direct asymmetric gear", having $k = d_{bi} / d_{bd} > 1$. If the inverted profile is the active one the gear is named "inverted asymmetric gear", having $k = d_{bd} / d_{bi} < 1$ (Chira, F. & Banica, M., 2007).

A short presentation of the calculus method used for obtained the geometrical parameters is given in the following. For the centre distance (a), number of teeth (z_1, z_2) as initial data, the designing engineer chooses, as designing variables, the pressure angle for the direct profile (α_{wd}) and the pressure angle for the inverted profile (α_{wi}). Implicitly is also chosen the degree of asymmetry of the tooth and the coefficient of asymmetry of the direct gear "k".

Another designing variable is the coefficient „f", which is called coefficient of modification of the gear rack angle. That has been introduced, in the calculation program, to determine the angle of the rack profile for generating the direct profile of the tooth (α_{dc}):

$$\alpha_{dc} = \alpha_{wd} - f \cdot \left(\alpha_{dc\,max} - \alpha_{dc\,min}\right) / \alpha_{gwd}, \qquad (1)$$

where α_{gwd} is the pressure angle on the direct profile expressed in degrees. The limits between can be modified the gear rack angle are: $\alpha_{dc\ min} = \arccos(1/k)$, $\alpha_{dc\ max} = \alpha_{wd}$ (Kapelevich, A.,L., 2000).

The value of coefficient „f" shows how much the angle of the direct gear rack profile (α_{dc}) is different from the pressure angle of the direct profile of the gears (α_{wd}). The shift rack of the pinion X_1 and the shift rack of the gear X_2, and also the ratio of the pinion tooth thickness on the pitch circle and gear tooth thickness on the pitch circle (S_{w1} / S_{w2}) are determined by the α_{dc} angle of the rack, by choosing the value for „f":

$$(X_1 + X_2) = \frac{p_c(z_1 + z_2)}{2\pi \cdot (\tan\alpha_{dc} + \tan\alpha_{ic})}(\mathrm{inv}\alpha_{wd} + \mathrm{inv}\alpha_{wi} - \mathrm{inv}\alpha_{dc} - \mathrm{inv}\alpha_{ic}); \quad (2)$$

$$\frac{S_{w1}}{S_{w2}} = \frac{\left(\frac{p_c}{2} + X_1(\tan\alpha_{dc} + \tan\alpha_{ic}) + \frac{z_1 \cdot p_c}{2\pi}(\mathrm{inv}\alpha_{dc} + \mathrm{inv}\alpha_{ic} - \mathrm{inv}\alpha_{wd} - \mathrm{inv}\alpha_{wi})\right)}{\left(\frac{p_c}{2} + X_2(\tan\alpha_{dc} + \tan\alpha_{ic}) + \frac{z_2 \cdot p_c}{2\pi}(\mathrm{inv}\alpha_{dc} + \mathrm{inv}\alpha_{ic} - \mathrm{inv}\alpha_{wd} - \mathrm{inv}\alpha_{wi})\right)}, \quad (3)$$

where p_c, α_{dc}, α_{ic} are the pitch and the profiles angles of the asymmetric generation gear rack. In the calculus program first are determined the parameters of the involutes profiles and second the gear rack parameters which determine the joint profiles from the base of the direct and inverted involutes profiles of the tooth. X_1, X_2 are established on the base of correct meshing conditions.

3. THE FUNCTIONAL PARAMETERS

For offer the possibility of rapid evaluation of the behaviour in function of designed gears, the research course contains the following steps:

- Equations of the involutes and joint asymmetric profiles and realisation of the routine for mathematical modelling of the asymmetric gear geometry.
- Algorithm and routine for mathematical modelling of the meshing, for booth variant of using of asymmetric gear.
- Algorithms and realisation of the routines for calculating and determine the variation during the meshing cycle of some important functional parameters.
- Many routines used for studying the variation of the asymmetric gears performances in relation with different designing variables.

Some of the functional parameters that can be determined with the developed applications are: the elasticity, implicitly the rigidity, of the asymmetric tooth and of the pairs of teeth in contact; the normal force on the active profile; the transmission error; the deformations of the teeth; the relative sliding speed and the specific sliding between flanks; the instantaneous efficiency and then the medium efficiency for a meshing period; the bending stress at the bottom of the tooth for pinion and for the gear and the contact stress. All the parameters are determined for the "j" point of contact, which number, during the meshing cycle, can be established by the designing engineer. With the values obtained for all the points of contact are represented the variation diagrams during the meshing cycle. The significant values from these diagrams are used as quality indicators of the asymmetric gears, for compare different possible solutions for given initial date. Have been developed routines for the direct asymmetric gears and for the inverted asymmetric gears, taking in considerations the similarities and, most important, the differences that appear when the active profile is modified.

For verifying the method have been obtained the 3D models of asymmetric gears with different coefficients of asymmetry, and have been compared the values of the displacements and stress resulting from the MATLAB applications with those resulting by using finite element method analysis.

4. EXAMPLES OF RESULTS OBTAINED WITH THE DEVELOPED DESIGNING APPLICATIONS

Optimal design is facilitated by using the routines realised for studying the variation of the performances depending on the choosing designing variable. Any designing engineer can easy compare different possibilities, obtained for different values of f, k, α_{wd} or α_{wi} and find what is the better combination.

4.1 Gears with the same coefficient of asymmetry

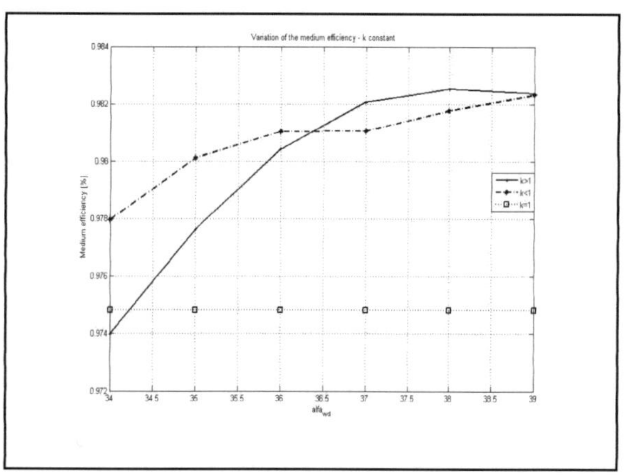

Fig. 1. Medium efficiency for gears with constant k

4.2 Gears with different coefficients of asymmetry

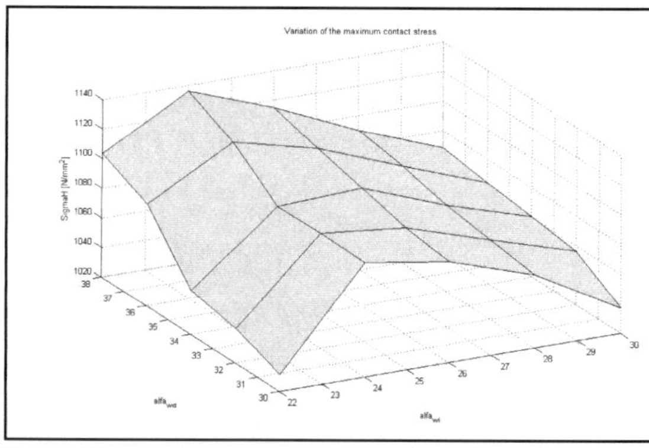

Fig. 2. The maximum contact stress for different k values

In the figure 1 is presented the diagram for comparison, using as evaluation function the medium efficiency, of gears designed for the initial date: $a = 100\,\text{mm}$, $z_1 = 13$, $z_2 = 60$. It have been used also k=1,15-for direct gears, k=0,87-for the inverted gears, $f = 3$ and $\alpha_{wd} = [34, 35, 36, 37, 38, 39]$. There are calculated the values for direct gears, for inverted gears and the horizontal line represent the value corresponding with the symmetric gears generated with the gear rack with 20 degrees profile angle.

In the figure 2 is presented the diagram for comparison, using as evaluation function the maximum contact stress, of a number of 25 inverted asymmetric gears designed with the initial date: $a = 120\,\text{mm}$, $z_1 = 16$; $z_2 = 57$. It has been used $f = 5$ and the values of the pressure angles are:

$\alpha_{wd} = [30, 32, 34, 36, 38]$ and $\alpha_{wi} = [22, 24, 26, 26, 30]$.

5. CONCLUSION

The computational method for designing and analysing the performances of the asymmetric gears on relation with the designing variable, having a short presentation in this paper, is very efficient, considering the time necessary for evaluation, and very easy to use. The method will be developed in the future by added routines for evaluation the dynamic behaviour.

6. REFERENCES

Chira, F. & Banica, M., (2007), *Studii asupra roților dințate cu dinți asimetrici (Studies on the gears with asymmetric teeth)*, Risoprint, ISBN 978-973-751-495-0, Cluj-Napoca, Romania.

Karpat, F., Cavdar, K. & Babalic, C.,F., (2006) *An investigation on analysis of involute spur gears with asymmetric teeth: Dynamic load and Transmission errors*, Proceedings of the 2th International Conference "Power Transmissions 2006", 69-74, ISBN 86-85211-78-6, 25-26 April 2006, Faculty of Technical Sciences, Novi-Sad, Serbia & Montenegro.

Kapelevich, A.,L., (2000), *Geometry and design of involute spur gears with asymmetric teeth*, Mechanism and Machine Theory, No.35, 117-130, ISSN 0094-114X.

Litvin F., L., Lian, Q. & Kapelevich A., L., (2000), *Asymmetric Modified Gear Drives: Reduction of Noise, Localization of Contact, Simulation of Meshing And Stress Analysis*, Computer Methods in Applied Mechanics and Engineering, No.188, 363-390, ISSN 0045-7825.

Novikov, A.S., Paikin, A.G., Dorofeyev, V.L., Ananiev, V.M., Kapelevich, A.L., (2008), *Application of Gears with Asymmetric Teeth in Turboprop Engine Gearbox*, Gear Technology, January/February 2008, 60-65, ISSN 0743-6858.

Annals of DAAAM for 2008 & Proceedings of the 19th International DAAAM Symposium, ISSN 1726-9679
Editor B. Katalinic, Published by DAAAM International, Vienna, Austria 2008
Make Harmony Between Technology and Nature, and Your Mind will Fly Free as a Bird

ON CHOOSING THE NONLINEAR DYNAMIC VIBRATION ABSORBER

CHIRIACESCU, T. S[ergiu]; ROSCA, I[oan]; BALCU, I[on]. & DINITA, I[on]

Abstract: *In this paper a method for choosing the nonlinear dynamic vibration absorbers. This can be used also for designing of such isolators. Taking into consideration different types of elastic and damping nonlinear characteristics, and using the harmonic linearization method, the authors present the equation which describes the dynamic behaviour of the isolators at the resonance regime.*
Key words: *vibration, absorber, harmonic linearization method, passive absorber.*

1. INTRODUCTION

As is known, many companies offer a large variety of vibration absorbers and give instructions of how they can be used (Chiriacescu, et al., 2008). The main application of isolators is to diminish the amplitude of the buildings vibration during an earthquake. In spite that, especially in the field of buildings, is very difficult to be designed or to be choose the right isolators because these have to work at the seismic waves low frequencies. The elastic and damping characteristics of these devices can be linear or nonlinear (Liu at al., 2005). In last period it turns that the nonlinear isolators are the most used.

As function of the oscillation centre position (the equilibrium position of the mounted mass on the isolators, in case of dynamic loads absence) the vibrations that are developed during the isolators run can be both symmetrically and asymmetrically (Chiriacescu at al., 2008). Referring to the nonlinear elastic characteristic from Figure 1, where f_e represents the elastic force developed by the isolator, and x_s is its static deformation, if the oscillation centre is situated in point O, then the vibrations are symmetrically, and if the centre is situated in point O_1 the vibrations are asymmetrically.

For simplicity it will be considered the dynamic model from Figure 2, where m is the mounted mass on the isolator, $F = F(t)$, $x = x(t)$. If $x_1 = x_1(t) = 0$ the isolation is called active and if $F(t) = 0$ and $x_1 \neq 0$ is defined as a passive one.

In this paper there will be took into consideration only the passive vibroisolation.

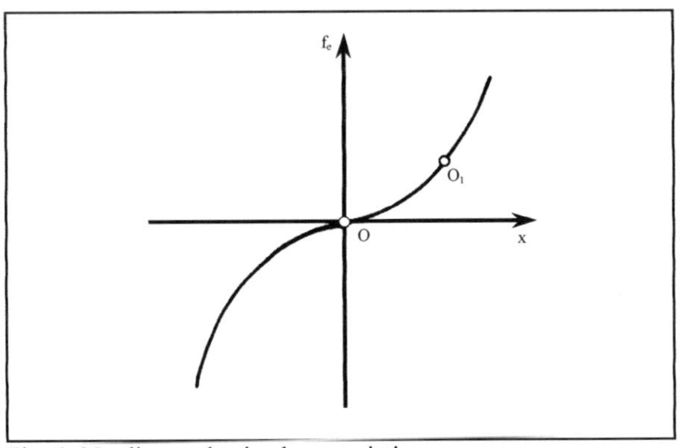

Fig. 1. Nonlinear elastic characteristic

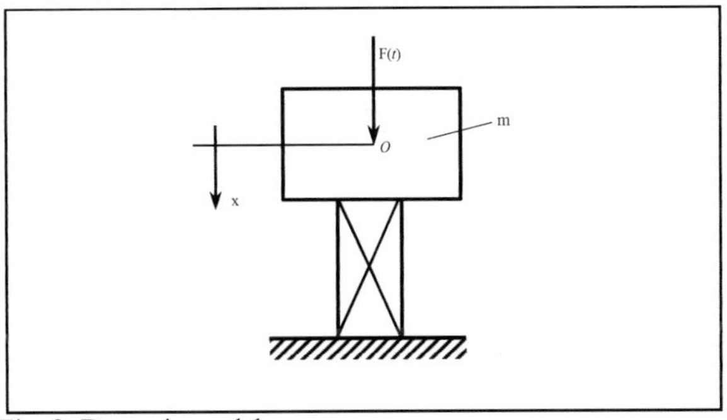

Fig. 2. Dynamic model

2. THE ISOLATOR WITH NONLINEAR ELASTICITY AND LINEAR DAMPING

Using the notations from Figure 2, for $x_1 = 0$ (active isolation) and $F(t) = F_0 \cos \omega t$, the motion equation of the mass m mounted on the nonlinear isolator with the characteristic $f(x, \dot{x})$ is:

$$mp^2 x + f(x, px) = F_0 \cos \omega t \tag{1}$$

where $p = d/dt$ represents derivatives operator or Laplace variable.

The equation (1) satisfies the low-pass filter property (Chiriacescu, 1990) that offers the possibility to consider:

$$x \cong x_0 \sin \omega t = x_0 \sin \psi \,. \tag{2}$$

In this case, the equation (1) can be rewritten as:

$$\left(mp^2 - \frac{F_0}{x_0} \cos \varphi \right) x + \frac{F_0 \sin \varphi}{y_0 \omega} px = f(x, px) \,. \tag{3}$$

Applying to the nonlinear function $f(x, px)$ the harmonic linearization method (Chiriacescu, 2007) (Popov, 1973) it is obtained:

$$f(x, px) = q_1 x + \frac{q_2}{\omega} px \,, \tag{4}$$

where

$$\begin{cases} q_1 = \dfrac{1}{\pi x_0} \displaystyle\int_0^{2\pi} f(x, px) \sin \psi \, d\psi \,, \\ q_2 = \dfrac{1}{\pi x_0} \displaystyle\int_0^{2\pi} f(x, px) \cos \psi \, d\psi \,. \end{cases} \tag{5}$$

To illustrate the way of using equation (3) to the isolator choose or design it considered the particular case:

$$f(x, px) = ax^3 + cpy \quad (6)$$

that has the quantities: $q_1 = (3/4)ax_0^2$ and $q_2 = c\omega$.

In this case, for $p = j\omega$, where $j = \sqrt{-1}$, the equation (3) leads to the system:

$$\begin{cases} -m\omega^2 - \dfrac{F_0}{y_0}\cos\varphi + \dfrac{3}{4}ax_0^2 = 0, \\ \dfrac{F_0}{x_0}\sin\varphi + q_2 = 0. \end{cases} \quad (7)$$

The dissipated energy on a cycle of vibration motion is:

$$E = \pi F_0 x_0 \sin\varphi \quad (8)$$

that has the maximum absolute value for $\varphi = 3\pi/2$.

Considering the illustrative case of a mass $m = 10kg$, $F_0 = 100N$, $x_0 = 0.01m$, and $\omega = 10 rad/s$, then for $\varphi = 3\pi/2$ results:

$$c = 100\,Ns/m, \; a = \frac{4}{3}\frac{m\omega^2}{y_0^2} = 1.33 \cdot 10^3\, kg/m^2s^2$$

3. THE VIBROISOLATOR WITH NONLINEAR ELASTICITY AND DAMPING CHARACTERISTICS

Considering the Figure 3, for:

$$f(x, px) = \left(a_1 - a_2x^2\right)x - bx_0\left(1 - \frac{x^2}{x_0^2}\right)\mathrm{sgn}(px) \quad (9)$$

and $|x| \le x_{\max}$, a_1, a_2, b are constants, one can obtain from relationships (5):

$$q_1 = a_1 - \frac{3}{4}a_2x_0^2\text{, and } q_2 = -0.68b. \quad (10)$$

In this case, from maximisation condition, the dissipated vibrating energy, we obtain:

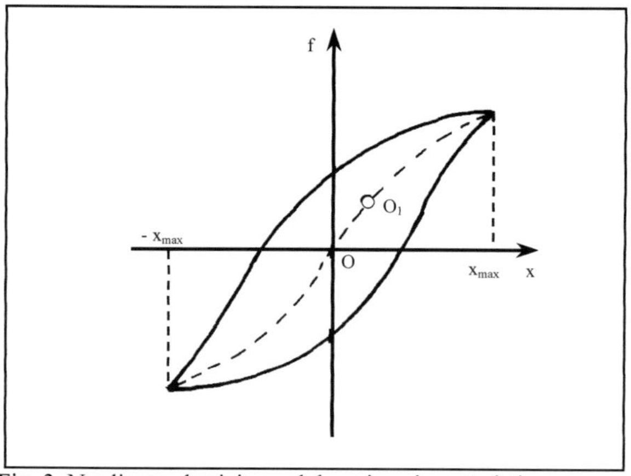

Fig. 3. Nonlinear elasticity and damping characteristic.

$$\begin{cases} a_1 = m\omega^2 + \dfrac{3}{4}a_2x_0^2, \\ b = -0.68\dfrac{F_0}{x_0}. \end{cases}$$

(11)

4. CASE OF ASSYMETRICAL VIBRATIONS

Let's consider the case of asymmetrical vibrations with O_1 as a centre of vibrations as in Figure 1, and the damping force is a linear one. In this case:

$$x = x^0 + x^*,\; x^* = x_0\sin(\omega t + \varphi) = x_0\sin\psi \quad (12)$$

Applying the harmonic linearization method we obtain:

$$f(x, px) = f^0 + \left(q_1 + \frac{q_2}{a}p\right)x, \quad (13)$$

where

$$f^0 = \frac{1}{2\pi}\int_0^{2\pi} f\left(x^0 + x_0\sin\psi, x_0\omega\cos\psi\right)d\psi. \quad (14)$$

Taking into consideration the procedure described in the above one can obtain the equation:

$$q_2^2 + \left(q_1 - m\omega^2\right)^2 = \frac{F_0}{x_0^2} \quad (15)$$

that permits to analyze the relationship between q_1 and q_2.

5. CONCLUSIONS

Using the method of harmonic linearization in the presented paper it was proposed a procedure that can be used to choose or to design the vibration absorbers.

As performance index was considered the dissipated energy on vibration cycle.

In a future paper there will be considered both cases of asymmetrically vibrations for active absorbers.

For the active control of vibration the equation of the form (4) and (9) are used to obtain the control law (Chiriacescu, 2007).

6. REFERENCES

Chiriacescu, T. S. (1990). *Stability in the Dynmics of Metal Cutting,* Elsevier, ISBN 0-444-98868-8, Amsterdam

Chiriacescu, T. S. (2007). *Linear Mechanical Systems. An Introduction, (in Romanian),* Romanian Academy Press House, ISBN -978-973-27-1492-8, Bucuresti

Chiriacescu, T. S.; Munteanu, R.; Dinita, I. (2008). A method for designing passive isolators, *Scientific Bulletin of Petrosani University, Vol.10, pp. 27-30, Petrosani, Romania.*

Harris, C. M.; Piersol, A., G., Harris (2002). *Shock and Vibration Handbook,* Fifth Edition, McGraw-Hill Handbook, New York

Liu,Y.; Matsuchisa, U.; Utsuno, H.; Park, J. (2005). Vibration isolation by variable stiffness and damping system. *JSME International Journal, Series C, vol. 48, No. 2, pp. 305-310.*

Popov, E., P. (1973). *Applied Theory of the Process Control for Nonlinear Systems,* (in Russian), Izd. Nauka, Moskow

Annals of DAAAM for 2008 & Proceedings of the 19th International DAAAM Symposium, ISSN 1726-9679
Editor B. Katalinic, Published by DAAAM International, Vienna, Austria 2008
Make Harmony Between Technology and Nature, and Your Mind will Fly Free as a Bird

ENCAPSULATION OF VANILIN IN AGAR BEDS FOR CONTROLLED RELEASE

CHIRILUS, A[lina] A[lexandra]; STOICA GUZUN, A[nicuta]; STROESCU, M[arta]; DOBRE, T[anase] & TACHE, F[lorin]

Abstract: *Controlled release of food ingredients at the right time and place can be provided by using encapsulation, method that improves the application range of food ingredients and ensure optimal dosage. Encapsulation of vanilin in agar beds is presented in this paper, using emulsion method. Release studies from the agar beds obtained were performed in vitro. The release data were analyzed using several release models.*
Key words: *agar, vanilin, encapsulation, micro-spheres, diffusion*

1. INTRODUCTION

Controlled release is a technology that can be used to increase the effectiveness of many ingredients. Despite the wide range of encapsulated products that have been developed, micro-encapsulation is still far from being fully developed in the food industry (Pothakamury et al., 1995). The active agent is released from the controlled release delivery systems by diffusion, biodegradation, swelling or osmotic pressure.

Two main types of microcapsules exist: reservoir systems and matrix systems. A reservoir system consists of an active agent contained within a rate-controlling barrier. The release rate from a reservoir system depends on the thickness, the area and the permeability of the barrier. In matrix systems, the active agent is homogeneously dissolved or dispersed throughout the polymer mass. The release pattern depends on the geometry of the system, the type of carrier material and the loading of the active agent (Arifin et al., 2006).

Agar is a gelatinous substance derived from red algae. Agar consists of a mixture of agarose and agaropectin.
Vanilin is the major component of natural vanilla, which is one of the most widely used and important flavouring substances worldwide. Vanilin displays antioxidant and antimicrobial properties and has the potential for use as a food preservative (Walton et al., 2003).

The aim of this paper is to present experimental data concerning vanilin release from agar capsules.

2. MATERIALS AND METHODS

2.1 Preparation of encapsulated vanilin

The agar granules were obtained starting from a solution of agar in hot water. The solution was prepared by mechanical stirring. Another solution of synthetic vanilin of known concentration was poured in the solution. The mixture was poured with a syringe as droplets which were solidified by cooling in sunflower oil under vigorous stirring for ten minutes. The resultant agar particles have a particle size ranging between 2.5-7 mm. Their diameters were measured using a BX 51 Olympus microscope.

2.2 Analysis of vanilin in agar particles

A certain quantity of agar granules was treated with distilled water in the same ratio solid/liquid as in the release experiments in an ultrasonic bath for 10 minutes. The granules were completely disintegrated in order to release the entire vanilin content.

The mixture was then filtrated and the liquid was analyzed for the vanilin content using UV-VIS spectroscopy.

2.3 In vitro release study

In vitro release was performed in a vessel under magnetic agitation. A certain quantity of agar particles having the same diameter was put in contact with a known volume of alcoholic solution (80% vol/vol alcohol). The vanilin content was analyzed in the liquid phase using UV-VIS spectroscopy.

2.4. Treatment of data from release studies

The release data were analyzed by applying several release models proposed in literature.

Active agent released from the studied systems can be diffusion-controlled, swelling-controlled and erosion-controlled. Combinations between diffusion, swelling and erosion can be encountered for one particular system.

For diffusion-controlled micro-spheres, drug release profile is obtained by solving Fick's second law of diffusion subject to appropriate boundary conditions.

The simplest way to model the release from a matrix system is to consider an infinite, flat matrix system of radius R and with the active agent dissolved in the polymer at or below the saturation concentration. The amount of active agent released when the device is dissolved in a well-agitated, infinite medium is given by:

$$\frac{M_t}{M_\infty} = 1 - \frac{6}{\pi^2}\sum_{n=1}^{\infty}\frac{1}{n^2}\exp\left(\frac{-D_{ef}n^2\pi^2 t}{R^2}\right) \qquad (1)$$

For Sh>1, Baker and Lonsdale (Baker & Lonsdale, 1974) have proposed two simplified solutions by using early time and late-time approximations to explain the drug release from dissolved matrix system of a sphere during early and late-time periods respectively. The approximation results are expressed as follows:

$$\frac{M_t}{M_\infty} = 6\cdot\left(\frac{D_{ef}t}{\pi\cdot r^2}\right)^{1/2} - \frac{3D_{ef}t}{r^2} \quad for \quad \frac{M_t}{M_\infty} < 0.4 \qquad (2a)$$

$$\frac{M_t}{M_\infty} = 1 - \frac{6}{\pi^2}\exp\left(-\frac{\pi^2 D_{ef}t}{r^2}\right) \quad for \quad \frac{M_t}{M_\infty} > 0.6 \qquad (2b)$$

The polymer used as matrix may also have swelling properties. For swelling-controlled system, the hydrophilic polymer is susceptible to swelling, as water tends to penetrate and relax the polymer matrix. In this case, the composition of the hydrophilic polymer will determine the extent of swelling.

In swelling controlled systems, since both diffusion and dissolution occur altogether, they are quite indistinguishable. A very simple equation proposed to describe the release from these systems is:

$$\frac{M_t}{M_\infty} = k \cdot t^n \quad (3)$$

The value of n indicates the type of active-agent transport. When n = 0.5, the active agent is released by simple Fickian diffusion. When n = 1.0, diffusion is described as 'case II diffusion'. In case II diffusion, the rate of the solvent uptaken by the polymer is largely determined by the rate of swelling and relaxation of the polymer chain. 'Super case II transport' occurs when $n > 1.0$. In the case $0.5 < n < 1$, diffusion is a combination between Fickian and non-Fickian diffusion, and is known as anomalous diffusion.

The release of an active agent from a matrix type delivery system may be controlled also by erosion or by a combination between diffusion and erosion. Erosion controlled processes may be heterogeneous or homogeneous and depend on the hydrophobicity and morphology of the polymer. Heterogeneous erosion is more common with hydrophobic polymers, whereas homogeneous erosion is common with hydrophilic polymers (Arifin et al., 2006). The amount of active agent released in heterogeneous erosion is given by:

$$\frac{M_t}{M_\infty} = 1 - \left[1 - \frac{k_{ero} t}{C_0 r}\right]^n \quad (4)$$

The values of n are different, depending on the matrix geometry: n = 3 for a spherical one, n = 2 for a cylindrical matrix and n = 1 for a slab shaped matrix.

3. RESULTS AND DISCUSSIONS

In order to understand the release kinetics the particles were selected with respect to their diameters. Figure 1 shows in vitro release profile of vanilin in alcoholic solutions for practically the same vanilin content in the particles 2.2 x10^{-4} g vanilin/g capsules.

For all experiments, a burst effect is observed for the first 5-10 minutes from the beginning of the experiments. For the calculation of the proposed models parameters, only vanilin release after 5 or 10 minutes was considered, because in the first period of release a "non-steady state" behaviour occurs.

In their article (Chattopadhyaya et al., 1998), the authors suggested that there exists an important quantity of surface vanilin that can be recovered by washing agar beds with absolute alcohol since vanilin is soluble in absolute alcohol.

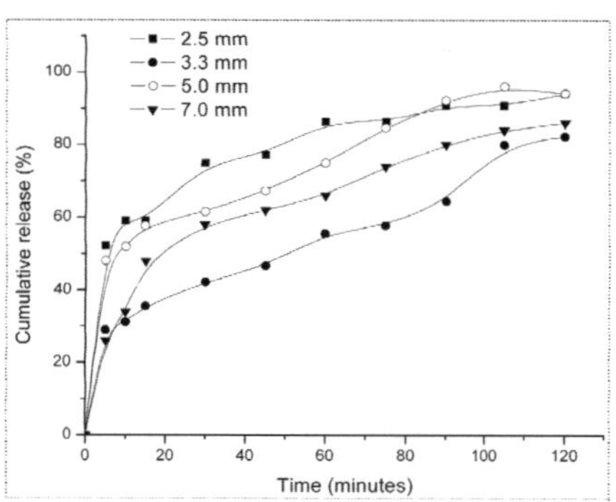

Fig. 1. Cumulative vanilin release in time for agar particles having different diameters, with the same loading for vanilin (2.2 x10^{-4} g vanilin/g capsules)

By our estimation, the surface vanilin varies between 30-55% from the total amount of the existing vanilin in the capsules. This surface vanilin is released in the first 5-10 minutes from the beginning of the experiment.

Table 1 summarizes the model parameters for the three models described by equations 2-4 and the correlation coefficients (R).

Particle diameter (mm)	**Model 1 Eq. 2a and 2b D_{ef} (m^2/s)**	**Model 2 Eq. 3**	**Model 3 Eq. 4 k_{ero}**
2.5	5.9x10^{-11} R=0.989	n=0.200 k=0.164 R=0.985	3.55x10^{-11} R=0.983
3.3	2.78x10^{-11} R=0.991	n=0.282 k=0.054 R=0.957	2.26x10^{-11} R=0.970
5.0	1.50x10^{-10} R=0.945	n=0.214 k=0.134 R=0.969	6.75x10^{-11} R=0.927
7.0	2.50x10^{-10} R=0.984	n=0.371 k=0.033 R=0.987	6.55x10^{-11} R=0.999

Tab 1. Model parameters for vanilin release.

The diffusion model describes best the cumulative release for the particles with small diameters. For the particles with larger diameters, the phenomena of swelling and erosion are prevailing. Diffusion and dissolution occur altogether for the larger diameter particles.

4. CONCLUSIONS

The purpose of this paper was to present a study of food flavour encapsulation processes and methods, employing agar as matrix for vanilin encapsulation. The results obtained from release experiments were analyzed using different models and the parameters determined from these models suggested that, for small particles, the diffusion prevails, but for those with larger diameters, the swelling and erosion cannot be neglected.

Encapsulation has the potential to be used in the food industry, helping to protect flavours against oxidation or other reactions that occur in the presence of light, as well as undesired interactions with the food itself. This process could eventually be a viable solution to the problem of food flavour degradation with time. The research was funded by research grant type PN II, no. 61045/2007.

5. REFERENCES

Arifin D.Y., Lee L.Y. & Wang, C-H (2006). *Mathematical modeling and simulation of drug release from microspheres: Implications to drug delivery systems*, Advanced Drug Delivery Reviews, 581274–1325

Baker R.W. & Lonsdale H.K. (1974). *Controlled release: mechanisms and rates*, in: Tanquarry, A.C. & Lacey, R.E. (Eds.), *Controlled Release of Biologically Active Agents*, Plenum Press, New York, NY, pp. 15–71

Chattopadhyaya S., Singhal R.S. & Kulkarni P.R. (1998). *Oxidised Starch as Gum Arabic Substitute for Encapsulation of Flavours*, Carbohydrate Polymers 37, 143–144

Pothakamury, U.R. & Barbosa-Canovas, G.V. (1995). *Review: Fundamental Aspects of Controlled Release in Foods*, Trends in Food Science & Technology 6, 397-406

Walton, N.J., Mayuer, M.J. & Narbad, A. (2003). *Molecules of Interest–Vanilin*, Phytochemistry 63, 505-515.

Annals of DAAAM for 2008 & Proceedings of the 19th International DAAAM Symposium, ISSN 1726-9679
Editor B. Katalinic, Published by DAAAM International, Vienna, Austria 2008
Make Harmony Between Technology and Nature, and Your Mind will Fly Free as a Bird

SHEET METAL FORMING OPTIMISATION FOR THE REDUCTION OF SPRINGBACK

CHIRITA, B[ogdan]

Abstract: *The use of modern materials in the forming processes has brought both advantages and disadvantages. High strengths steels are lighter and stronger but at the same time they yield more springback after forming than deep drawing steels. The present paper uses the response surface method (RSM) for the optimization of the forming process aiming to reduce the springback in the case of the U-bending test.*
Key words: *springback, response surface, sheet metal.*

1. INTRODUCTION

The non uniform distribution of the internal stresses throughout a formed part generates differences between the shapes of the part at the end of the forming stage and after unloading. This modification of the geometry is best known as springback. The phenomenon is influenced by a multitude of factors such as material properties, sheet thickness, tools geometry, forming conditions, lubrication etc. Springback is a major fault causing problems especially in assembling the parts and it has been the object of numerous studies lately.

Application of optimization techniques to metal forming problems (Chirita & Brabie, 2007, Naceur et al., 2005) often leads to high numbers of expensive function evaluations especially in the case when cost and constraint function resulted from finite element simulations fine meshes, nonlinear geometrical and material behavior etc. An alternative method is given by response surface methodology (Myers & Montgomery, 2002, Naceur et al., 2008) that replaces complex models with approximate ones based on the results calculated at various points in the design space.

2. FORMULATION OF THE OPTIMISATION PROBLEM

Process optimization represents the minimization of the cost function $F(x)$, $x \in R^n$, under some constraining conditions $g_j(x) \leq 0$, when the design variables x_i are defined over a region of interest $[l_i, u_i]$. The RSM approach consist in solving a problem where the cost function and the constraints are replaced with some approximations $\tilde{F}$ and $\tilde{g}$. The approximations are based on a set of experimental values of the function F.

Knowing the function values for a set of experimental points x_i distributed according to a certain design of experiments, the function $\tilde{F}$ may be defined in terms of basic functions p and some adjusting coefficients a as:

$$\tilde{F}(x) = p^T(x)a(x) \tag{1}$$

The coefficients a_i are determined by a weighted least squares method minimizing the error between the experimental and approximated values of the objective function:

$$J(a) = \sum_{i=1}^{N} w(\|x_i - x\|)\left(p^T(x_i - x)a - F(x_i)\right)^2 \tag{2}$$

where N is the number of performed experiments and x_i are the experimental designs (Bahloul et al., 2006)

The weights w_i insure the continuity and the locality of the approximation and are defined $w_i>0$, decreasing within a fixed region around the point i called the domain of influence of x_i and vanish outside. The weight function are determinant by influencing the way that the coefficients a_i depend on the location of the design point x.

3. NUMERICAL SIMULATION

The simulation of the springback process was carried in two steps: the forming step (U-bending) and the actual springback.

The U-bending was simulated considering a plain strain state, using an explicit formulation of the finite element code. Given the symmetric shape of the part, only half of the assembly was modeled in order to reduce the calculus efforts. The geometry of the model is illustrated in fig. 1. The sheet is 350 mm long, 30 mm width and 0.8 mm in thickness and was modeled using shell elements (S4R, ABAQUS® elements) on one row with 5 integration points through thickness.

The material properties are presented in table 1. The steel was modeled as elasto-plastic, considering an isotropic elasticity and anisotropic plasticity. Taking the anisotropy into account the yield locus was represented using Hill 48 quadratic criterion:

$$\sigma_f^2 = \sigma_1^2 - \left(\frac{2r_0}{1+r_0}\right)\sigma_1\sigma_2 + \left(\frac{1+r_{90}}{1+r_0}\right)\sigma_2^2 \tag{3}$$

where σ_f, σ_1, σ_2 are the equivalent stress and principal stresses and r_0, r_{90} are the Lankford parameters for the parallel and transverse to rolling direction.

For contact conditions a modified Coulomb friction law combined with penalty method was used.

In the springback step, the results of the forming step were imported and an implicit formulation of the finite element method was used.

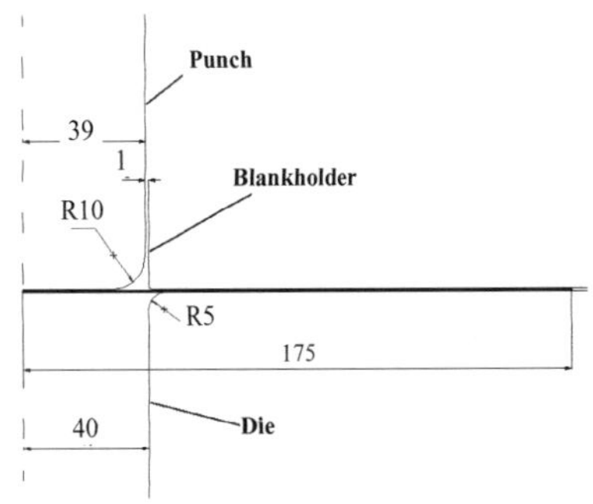

Fig. 1. Geometrical parameters of the model.

Material properties	
Young's modulus E	200 GPa
Poisson coefficient ϑ	0.3
Yield stress	306 MPa
Anisotropy coefficients	
r_0	0.82
r_{45}	0.77
r_{90}	0.81

Tab 1. Material parameters.

4. OPTIMISATION OF THE FORMING PROCESS

Three parameters were taken into account in this design of experiment: blankholder force (A), punch radius (B) and die radius (C). The domains of variations for these parameters are given in Table 2. The objective function is the maximum opening distance of the final part formulated as follows:

$$F = \max \|d_i\| \quad (4)$$

where d_i represents the difference between the positions of each node of the mesh before and after springback, as illustrated in fig. 2. The above relation may be written as:

$$F = \sqrt{\sum_{i=1}^{n_d} \left(x_i^2 + y_i^2 + z_i^2\right)} \quad (5)$$

where x_i, y_i and z_i are the components of d_i in a Cartesian reference system and n_d is the number of nodes of the mesh.

The optimization of the forming problem was carried using central composite design (CCD) formulation of the response surface method. CCD's are designed to estimate the coefficients of a quadratic model. The design requested a set of 20 simulations according to the table of experiments. The objective function was modeled as quadratic, and according to the analysis of variance (ANOVA) the equation of the model was:

$$\begin{aligned} F = &-398.009 - 3.979A - 170.385B + \\ &+ 583.803C - 82.268C^2 + 0.179AB + \\ &+ 0.356AC + 27.223BC \end{aligned} \quad (6)$$

The optimization is then carried out and the solution obtained had a desirability factor of 80.5%. The response surface for this model is illustrated in figure 4.

The optimum combination of values for the process factors was:

- blankholder force (A) 199.73 kN;
- punch radius (B) 10 mm;
- die radius (C) 5 mm.

The objective function for this set of parameters had an estimated value of 41.728.

Parameters	Minimum value	Maximum value
A: Blankholder force [kN]	40	200
B: Punch radius [mm]	10	12
C: Die radius [mm]	5	6

Tab 2. The domains of variation for process parameters.

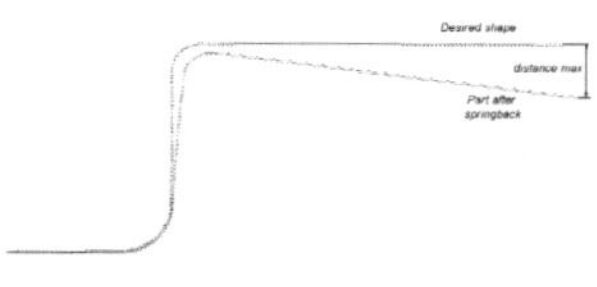

Fig. 2. The opening distance.

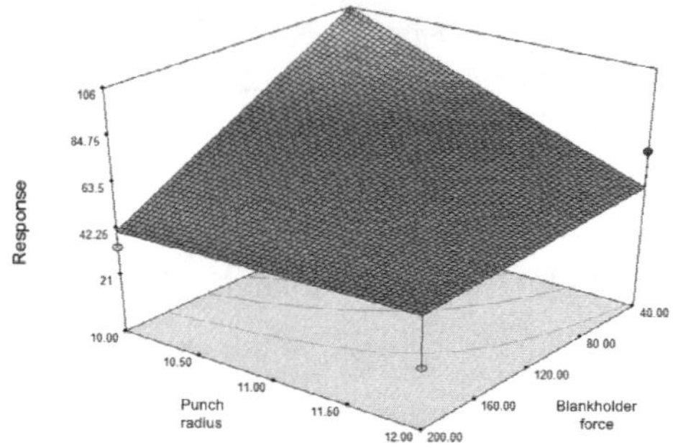

Fig. 3. Response surface plot.

For the verification of the proposed solutions a new simulation was carried in ABAQUS® under these conditions. The opening of the resulted part was 42.869. It means that the error of the estimation is only 2.734% so the response surface method has created very good results.

5. CONCLUSIONS

This paper is the result of an extended study concerning the springback phenomenon and the methods to reduce its impact on the final shape of the sheet metal formed parts.

The optimization method that was chosen used the response surface methodology applied for the sheet metal U-bending benchmark. Consequently a plan of experiment was created and a series of simulations were carried according to it. The results analysis allowed selecting the most significant parameters that were used to formulate the equation of the objective function.

The test proved that the proposed method has given promising results as the error estimation of springback is less than 3%.

Further improvements of the study will be searched by taking also into considerations other process parameters, a better material model for the simulations and nonlinear constraints.

6. REFERENCES

Bahloul, R. Ben-Elechi, S. & Potiron, A. (2006). Optimisation of springback predicted by experimental and numerical approach by using response surface methodology, *Journal of Material Processing Technology,* Vol. 173, No. 1, (march 2006) pp. 101-110, ISSN 0924-0136

Chirita, B. & Brabie, G. (2007). Control of springback in sheet metal U-bending through design experiment, *Proceedings of the 9th International Conference on Numerical Methods in Industrial Forming Processes NUMIFORM 07*, Cesar de Sa, J.M.A. & Santos, A.D. (Eds), pp. 829-834, ISBN 978-0-7354-0416-8, Porto, Portugal, June 2007, American Institute of Physics AIP, New York

Myers, R.H. & Montgomery, D.C. (2002). *Response surface methodology: process and product optimization using designed experiments*, John Willey & Sons, ISBN 0-471-41255-4, New York, USA

Naceur, H., Ben-Elechi, S., Batoz, J.L. & Knopf-Lenoir, C. (2008). Response surface methodology for the rapid design of aluminum sheet metal forming parameters, *Materials & Design,* Vol. 29, No. 4, (2008) pp. 781-790, ISSN 0261-3069

Naceur, H., Ben Elechi, S. & Batoz J.L. (2005). On the design of sheet metal forming parameters for springback compensation, *Proceedings of VIII International Conference on Computational Plasticity COMPLAS VIII*, Onate, E. & Owen, D.R.J. (Eds.), ISBN 84-95999-80-3, Barcelona, Spain, September 2005, CIMNE, Barcelona

Annals of DAAAM for 2008 & Proceedings of the 19th International DAAAM Symposium, ISSN 1726-9679
Editor B. Katalinic, Published by DAAAM International, Vienna, Austria 2008
Make Harmony Between Technology and Nature, and Your Mind will Fly Free as a Bird

EFFECT OF THE GEOMETRIC, STRUCTURAL AND DIMENSIONAL DIFFERENCES TO FIBER REINFORCED COMPOSITES

CHIRU, A[nghel]; GOIA, I[oan]; MODREA, A[rina]; PURCAREA, R[amona]; GUIMAN, M[aria] V[ioleta] & BURCA, I[oan]

Abstract: *During the manufacture process of a composite and after this, during the use of the material, there exists many ways in which the real material can have differences in comparison with the theoretical composite considered as a succession of a repeating cell. Sometime these differences can be great enough and they can have a major influence on the properties of the composite. In the part I of the paper we try to present haw these differences can influence the elastic constants of such material.*
Key words: *composite material, fibers, geometric differences, engineering constants.*

1. INTRODUCTION

In the manufacturing process of a fiber reinforced composite there are many causes that produce difference between the theoretical model and the real ones. These differences conduce to differences between the computed values of the elastic constants of the material and the real comportment of this. In the paper are analyzed these differenced in order to correct the formulas known for the engineering constants calculus.

2. GEOMETRIC, STRUCTURAL AND DIMENSIONAL DIFFERENCES

In the following are presented some of the current difference that there are between the theoretical model and the real composite material. This analysis is focus on the fiber reinforced composites.
The real modality to manufacturer the composite conduces to many points where the theoretical arrangement can not be respected. For these reasons can appear difference between the theoretical dimensions, structures or shape and the real characteristic obtained after the manufacturer process.

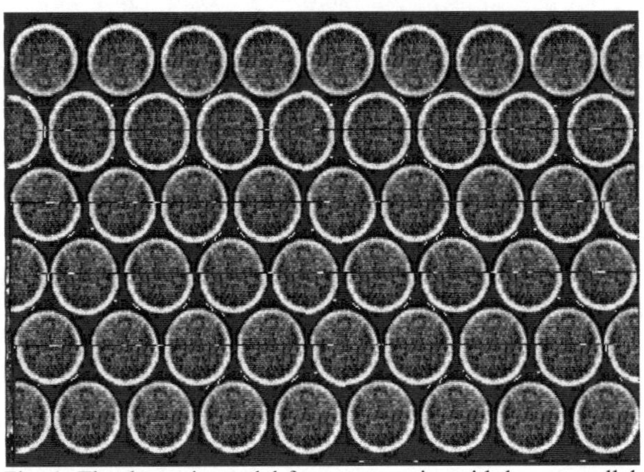

Fig. 1. The theoretic model for a composite with long, parallel, cylindrical fibers

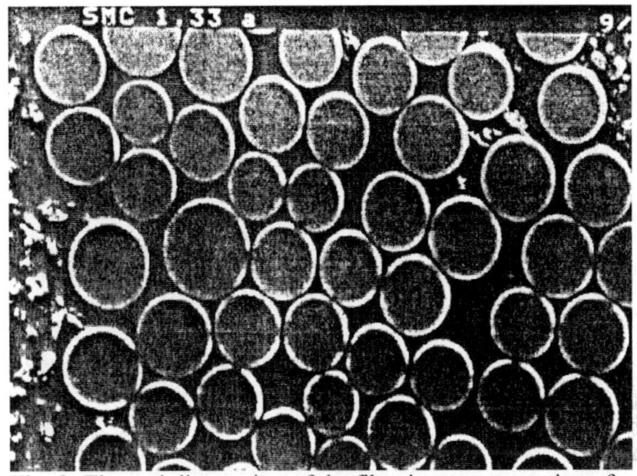

Fig. 2. The real disposation of the fiber in a cross section of a fiber reinforced composite

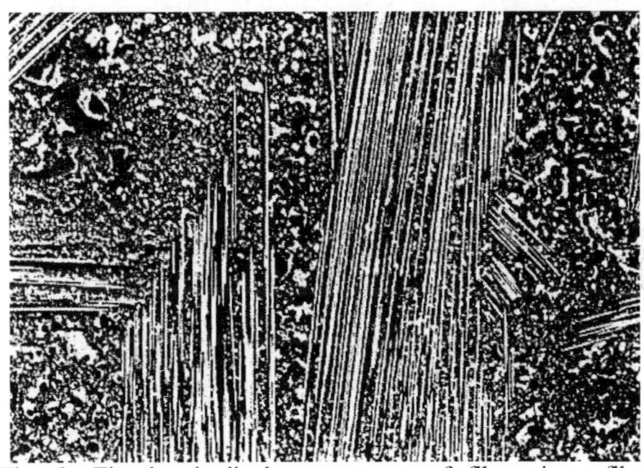

Fig. 3. The longitudinal arrangement of fibers in a fiber reinforced composite

In Fig. 2 and Fig. 3 are presented some of determined micrographs of a composite material. It is easy to observe the existing differences between the theoretical models and the real arrangements. These differences conduce to difference between the computed values of the engineering constants and the values obtained via experimentations. The aim of this section is to identify these differences for o composite with cylindrical, parallel and long fibers.

For the long fibers, cylindrical and parallel, it is considered that they are made by circular cylinders arrange into a regular network. The most common network considered is the hexagonal arrangement (for this case the resulting material is transversally isotropic) (Hashin, 1965); (Hill, 1965). The composite is considered obtained by the repetition of the basic cell (hexagonal in our case), (Goia & Modrea, 1999). The real modality to manufacturer the composite conduces to many points where the theoretical arrangement can not be respected. For these reasons can appear difference between the theoretical dimensions, structures or shape and the real characteristic obtained after the manufacturer process.

3. SOME CONSIDERATIONS

From the presented examples it can be observed that during the manufacturer process, but too after this, during the use of the composite, can appear many differences between the real and the theoretic model of the composite. The made analyze can identify the following differences between the real composite and the theoretical model:

- •Differences between the size of the mechanical characteristic of the matrix and of the fiber and the real values;
- •Differences between the geometric shape of the materials for reinforce and the real geometric shape. For example, is difficult to obtain circular cylindrical fibers and the real shape will be generally, elliptic, with variable semi-axes ratio.
- •Dimensional differences between the fibers of the same composite. In this case is necessary to know the distribution of these properties to realize a better calculus.
- •Difference in the geometrical arrangement of the fibers. In the mechanical models is considered that the fiber are disposed into a regular geometric network (hexagonal, quadratic, etc). In the reality the arrangement can be very different.

All these differences will have an influence on the computed values and it is necessary to do an analyse to determine which formulas obtained in prevoius papers is better to use. It is obvious that is necessary a great effort to analyse the existing relations and in the same time is necessary to make experience to compare the computed values with the values obtained by experience(Milton & Kohn,1988).

4. INFLUENCE OF THE DIMENSIONAL DIFFERENCE ON THE VALUES OF THE MECHANICAL CONSTANTS

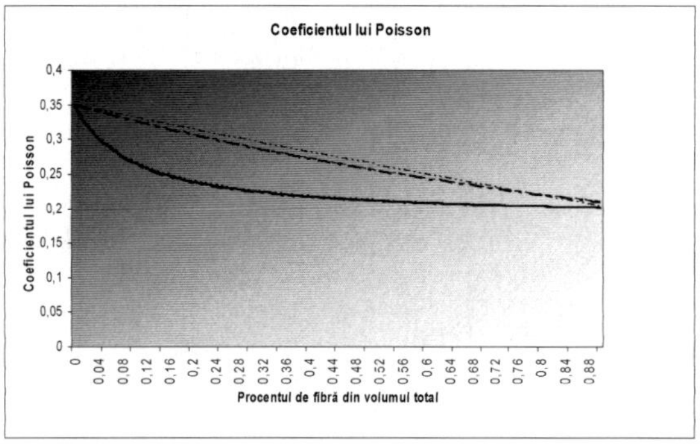

Fig. 4. The Poisson ratio for a variation of ± 10% of the Young modulus

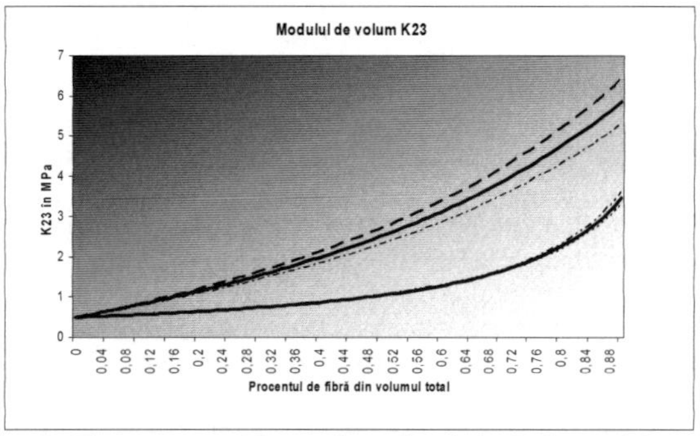

Fig. 5. The upper and lower bounds of the bulk modulus for variations of ± 10% of the Young's modulus, in the case nr.1.

In the following we will make a numerical analysis of the formulas propose in the literature to determine the sensitivity of these formulas to different variations of the parameters that describe the composite materials, presented in Fig. 4 and Fig. 5 (Modrea; 2001, a, b, c).

5. CONCLUSIONS

The analyse made for the relations presented to calculus of the engineering constants conduce to the conclusion that small variations of the properties of the fiber or of the matrix can produce variations, sometime very large, of the properties of the composite material. The properties associated to the longitudinal comportament of the fiber reinforced composite, that generally are described by formulas that respect, aproximativelly, the mixture law have a linear variation with the properties of the components. For example is the case of the bulk modulus and the Young's modulus. The Poisson ratio is not sensitive to different variations of the parameters describing the composite material. The values of the properties that describe the transversal comportment of the composite are very different in comparison with experimental determined values. This is determined by the fact that the models are less perfect to study the transversal properties and better for the longitudinal properties.

A general conclusion is that is necessary to consider the real dimension, shape and structure of a composite. The difference that exists between the theoretical model and the real composite can conduce to large variation of the computed values and the composite can have different properties.

6. REFERENCES

Goia,I., Modrea,A., ş.a. (1999). Calculus of the Mechanical Properties for the Composite Materials. *A IX-a Conferinţă internaţională CONAT, (IXth CONAT International Conference)*, Braşov

Hashin, Z. (1965). On the Elastic Behaviour of Fiber Reinforced Materials of Arbitrary Transverse Phase Geometry. *J. Mech.Phys. Solids*. Vol.13, pp. 119-134.

Hill, R. (1965). A Self-Consistent Mechanics of Composite Materials. *J. Mech. Phys. Solids*. Vol.13, pp. 213-232.

Milton, G.W., Kohn, R.V. (1988). Variational Bounds on the Effective Moduli of Anisotropic Composites. *J. Mech. Phys. Solids,* Vol. 36, No.6, pp. 597-629

Modrea, A. (1998). Evaluarea parametrilor mecanici omogenizaţi pentru un material alcătuit din mai multe componente. *A XXII-A Conferinţă Naţională De Mecanica Solidelor, (XXIIth National Conference of solids mechanics)*, Braşov

Modrea, A., ş.a. (2001, a). Evaluation of the elastic parameter for a composite when the strain/stress field is obtained via finite element method. *A III-a Conferinţă de dinamica maşinilor (IIIrd Dynamics cars Conference)*, pp. 365-370, Braşov

Modrea, A., ş.a. (2001, b). Evalution of homogenized coefficients for fiber reinforced plastic. *A III-a Conferinţă de dinamica maşinilor (IIIrd Dynamics cars Conference)*, pp. 371-374, Braşov

Modrea, A., ş.a. (2001, c). Mathematical Background for Evaluation of Homogenized Coefficients for Fiber Reinforced Plastic (FRP). *A III-a Conferinţă de dinamica maşinilor (IIIrd Dynamics cars Conference)*, pp. 375-378, Braşov

Annals of DAAAM for 2008 & Proceedings of the 19th International DAAAM Symposium, ISSN 1726-9679
Editor B. Katalinic, Published by DAAAM International, Vienna, Austria 2008
Make Harmony Between Technology and Nature, and Your Mind will Fly Free as a Bird
*Young Researcher and Scientist Paper / * Supervisor, Mentor*

OPTO-ELECTRONIC SYSTEM WITH OPTICAL FIBRES SENSOR FOR IONIC CONCENTRATION MEASUREMENT INTO SANGUINE SERUM

CHITA, M[onica] A[nca]; SIMA, I[on]; STIRBU, C[osmin] & GHITA, C[onstantin]

Abstract: *The paper presents an opto-electronic system with optical fibres sensor for ionic concentration measurement into sanguine serum. The performed study describes the behaviour of a fluorescence marker (fluorescein) into sanguine serum in the presence of hydrogen, calcium, magnesium and sodium ions. These studies can emphasise the marker selectivity under different conditions. Another important advantage of the measurement system is that it is very flexible, which allows very diverse studies in this domain, including the adaptation for sensors with selective-permeable membranes. Obviously, the study will be continued in order to observe also the influence of red cells from blood (the main suspect of the values dispersion), as others measurements. Also the performed system showed that it can be used for the subsequent studies concerning the behaviour of the others fluorescence markers, as in the tampon chemical solutions and in the complex organic solutions, like sanguine serum.*
Key words: *system, sensor, concentration, measurement, serum*

1. INTRODUCTION

The experimental part of this paper it refers at realisation of a chemical sensor, which uses the fluorescence phenomena of fluorescein.

The most used chemical indicators in the optical fibres technique (Stanciu & Stanciu, 2003) are they that use absorption and fluorescence phenomena, the last variant being preferred because it presents a higher sensitivity. One of the important class of fluorescence chemical indicators (Eggins, 2002) is the of ftaleines, the most important and used ftaleinels being the fluorescein: $C_{20}H_{12}O_5$ (the neutral form).

So the fluorescein (Wolfbeis, 1991) and his derivates are the very used indicators in the case of the chemical optrods wits optical fibres. The fluorescein may be founds in the neutral form; for low pH values, fluorescein appears under the colourless lactones form. Through molecular rearrangement it obtains the chinoid neutral form.

In the experiment it was used the sodium salt of fluorescein ($C_{20}H_{12}O_5Na_2$). It presents like a brick dust colour and in order to be used as indicator it is necessary to admix it with distillate water. The tampon solution obtained is yellow-green coloured and it can be used for the ionic analysis of sanguine serum.

2. THE BLOCK SCHEME OF THE EXPERIMENTAL OPTO-ELECTRONIC SYSTEM

The opto-electronic system used for the experimental determinations has the experimental scheme presented in the fig. 1 and contains:

- the optrod interfaces with two optical fibres PMMA;
- a light source, namely a blue "ultrabright" LED;
- an ORIEL-7180 photodiode;
- an optical filter;
- a multifunctional electro-optical source for the LED command;
- a received electronic block.

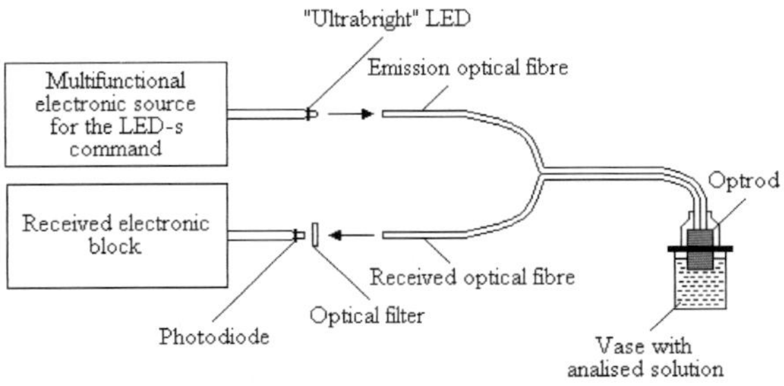

Fig. 1. The block scheme of the opto-electronic system.

Lowing from the spectral characteristics of fluorescein it was decided its use as fluorescent indicator (Stanciu et al., 1997) for concentration measurement of some ionics species presented generally into chemical solutions and particularly into sanguine serum.

3. EXPERIMENTAL DATA

In the sanguine serum case the following steps were followed:

- it was selected three serum samples (in three days);
- it was choosed twenty one valid probes;
- for the valid probes it was made the partial ionograme, respectively, it was determined the concentration of total calcium, of magnesium, sodium and potassium ions;
- for the valid probes it was measured pH with using a digital pH-meter CORNING-150 having electrode with lateral membrane and internal temperature compensation;
- after the presented determination, the serum probes was marked with fluorescein (fluorescein was initial dissolved in distillate water, fluorescein concentration being 10^{-3}; this "marked" substance was mixed in 1% proportion with sanguine serum, corresponding of different values of pH);
- the measures made with the electro-optical system with optical fibres were unfurled at the room temperature (23 ºC).

As about the normal values domain of the physiological pH, in literature is specified that it is light basic, the usual domain being, 7 ÷ 7.4 and normal average value being 7.2 (Lown & Prichard, 2003). For the sanguine serum from the effected measures, it was ascertained that this domain is 8.3 ÷ 8.8. The explanation of this phenomenon consists in the fact that the measures were effected in-vitro (after the blood harvesting): in contact with the air, the concentration of hydrogen ions decreases because a part of them tie of the oxygen ions from air, forming the hydroxilical groups (OH^-). From the definition relation of pH results that pH value increases. The equilibrium corresponding of a new state is extremely rapid reached. This was put in evidence through three measurements of pH (with indicator paper and with digital pH-meter CORNING – 150), five minutes after the harvesting of three blood samples. The following specification must be made: the measurements cannot be effected on the integral blood, because the existing "residues" (including the red cells) load the pH electrode membrane. The serum is obtained through the natural decantation at room temperature and the process build about one hour. The only solution was to work on plasma which can be obtained in few minutes through centrifugation. The measurements pointed out that pH value was the same for the serum and plasma.

Also it was ascertained that after three days (3x24 hours) the value of pH was modified slight enough (under 0.1 pH units), in the conditions in which in this period the probes was freeze. This thing do not happens however for the other ions, the prescriptions emphasized that the measurements must be made the same day in which it harvests the blood.

As about the voltage variation U_{0vv-f} as a function of pH, it ascertains an enough large dispersion of the values (fig. 2). The causes of this phenomenon are:

- the used optrod do not have a selective-permeable membrane and sanguine serum is a mixture of organic compounds, including proteins, which interactions with the fluorescein affecting its selectivity for hydrogen ions;
- the displacement of the pH values for the sanguine serum in-vitro in the range 8.3 ÷ 8.8, zone of reduced sensitivity of fluorescein (the maximum sensibility of the fluorescent marker is at pH = pKa, respectively at 6.4 ÷ 6.5 in this case);
- the measurements could not be effected at the human body temperature (the sanguine analysers, function of the measured measure, achieves the measures after bringing probes to 37 ºC temperature).

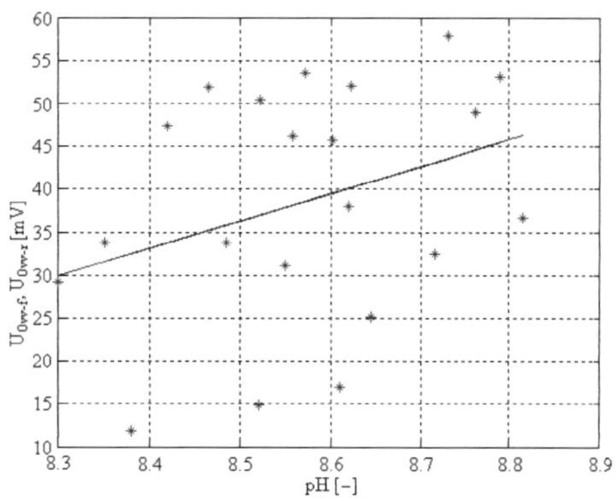

Fig. 2. The utile voltage variation V_{0vv-f} versus pH for sanguine serum probes.

For the values obtained by measure, the regression straight line, U_{0vv-r} = f(pH), was calculated and represented in the fig. 2 too.

Also it was represented the dependence between the useful voltage U_{0vv-f} and the concentration of the calcium, magnesium, sodium and potassium ions (fig. 3, fig. 4 and fig. 5).

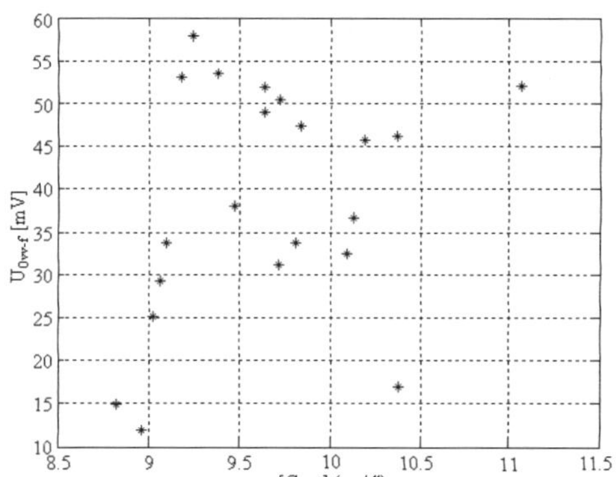

Fig. 3. The variation of utile voltage with concentration of calcium ions for the sanguine serum probes.

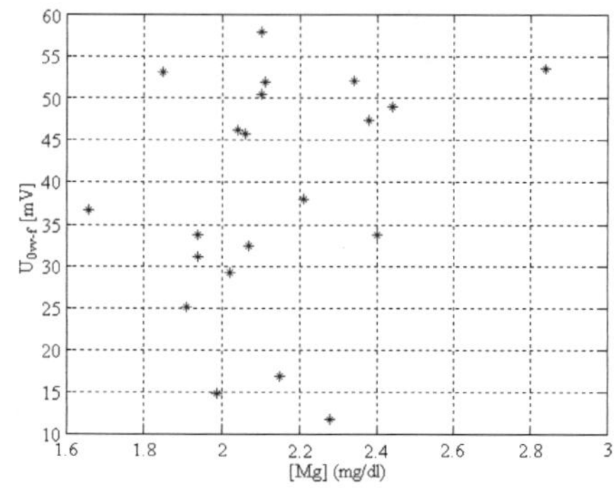

Fig. 4. The variation of utile voltage with concentration of magnesium ions for the sanguine serum probes.

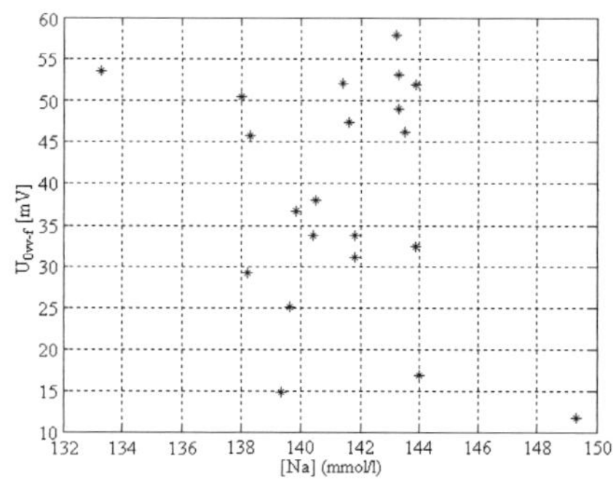

Fig. 5. The variation of utile voltage with concentration of sodium ions for the sanguine serum probes.

4. CONCLUSIONS

The performed study described the behaviour of a fluorescence marker (fluorescein) into sanguine serum, in the presence of hydrogen, calcium, magnesium and sodium ions. Also, the presented system can be used for subsequent studies concerning the behaviour of other fluorescence markers, as into tampon chemical solutions, or into complex organic solutions, such as sanguine serum. These studies can emphasise the marker selectivity under different conditions.

Another important advantage of the measurement system is that it is very flexible, allowing very diverse studies in this field, including the adaptation for sensors with selective-permeable membranes.

Obviously, the study will be continued in order to observe the influence of red blood cells (the main suspect of the values dispersion), and others measurements too.

5. REFERENCES

Eggins, B. (2002). *Chemical Sensors and Biosensors*, Wiley & Sons, Inc., ISBN 9780471899143, New York, USA.

Lawn, L. & Prichard, E. (2003). *Measurement of pH*, Royal Society of Chemistry Edition, ISBN13: 9780854044733, Cambridge, UK.

Stanciu, M & Stanciu, A.D. (2003). *Chemical and biochemical sensors with optical fibres*, Electra Edition, ISBN 973-8067-92-8, Bucharest, Romania.

Stanciu, M.; Iliescu, C.; Pantelimon, B. & Ilie, L (1997). The use of optical fibres sensors with fluorescence indicator for the pH measure. *The First international Conference of Electromechanical Systems*, pp.265-268, ISBN 9975-910-22-X, Chisinau, October 1997, University of Chisinau, Chisinau, Republica Moldova.

Wolfbeis, O. (1991). *Fiber Optic Chemical Sensors and Biosensors*, Boca Raton: CRC Press, ISBN 0849355087 9780849355080 0849355097, FL, USA.

Annals of DAAAM for 2008 & Proceedings of the 19th International DAAAM Symposium, ISSN 1726-9679
Editor B. Katalinic, Published by DAAAM International, Vienna, Austria 2008
Make Harmony Between Technology and Nature, and Your Mind will Fly Free as a Bird

ACTIVE PROCESS IMPROVEMENT IN A MONTAGE ENTERPRISE

CIBULKA, V[iliam]

Abstract: *In contribution there is characterized a progress of an actively managing in a montage enterprise on the base of system valuation efficiency of main processes of logistics chains of logistics system in installation enterprise, which consists of matrix model, then from relation of valuation of logistics process effective action and applying of simulation program Project Management Forecast for proposal of variants and valuation of logistics process efficiency, logistics chains and system with the aim of company competitive ability increasing and increasing of a customer productivity.*

Key words: *assembly, logistics, process, efficiency, productivity.*

1. INTRODUCTION

A need of making more efficient a processes (production, montage and logistics) within integrated supply chains is well known in our country for longer time. It is mentioned by more authors in their books and items. It is soundy recorded by this problemacy (Pernica, 2004). First stated author analyses an integrated supply chains and their total trash for company, allocate for necessarity of this problemacy solution, but they don't offer a complex solutions of problemacy given. There is no doubt about solution of this problemacy, because its successfully solution creates a premise for increase of company competitive ability in the market. Accordly took an author a decision because of his more years praxis from solution of such problems to bring his knowledge to problem stated for put on system access and progress list for solution of object problem.

2. SYSTEM ACCESS TO MAKING PROCESSES MORE EFFICIENT IN A MONTAGE COMPANY

System access (Horváth a, 2007) for making processes more efficient in a montage company comes out from index list and progress file, by which the complex valuation of process efficiency is realized, consequently their monitoring and improvement , with the aim of increase of company competitive ability and an increase of customer productivity. In my work I have mainly fixated to making logistics processes more efficient in a montage company, because they are the biggest potential force of added and wided goods value and because of this it is possible to increase an company competitive ability in the market and a customer productivity also.

2.1 A list of indexes

Main requirements for logistics indexes system (Horváth b, 2007):

- to ensure a whole company system efficiency valuation of whole company system efficiency,
- to cover all areas of logistics aims,
- to display a behaviour of an individual parts of logostics chain and behaviour of logistics system as total,
- to allow an observation of connection in a process structure, in a structure of a body hierarchy in a diagonal connection.

By fulfillment of logistics aims it is observing an efficient repression of plôace and time with the aim of assurance of customer requirements satisfaction after goods and services (logistics efficiency).

Logistics efficiency consists of two parts:
- logistics performances,
- logistics costs.

Logistics performances includes:
- logistics services level (logistics quality or an increase of customer productivity),
- productivity in a liogistics.

Logistics index system includes following index groups (Horváth c, 2007):
- logistics (supply) services level,
- logistics productivity,
- logistics costs,
- logistics systemu structure,
- logistics potential.

2.2 Body of action

Body of valuation actions of logistics system efficiency is very applicable for continuous managing of logistics system and also for ist improvement.

Objects, to which refers logistics indexes:
- logistics system as total,
- an individual variants of customer requirements,
- an individual processes and a place of performance.

2.2.1 Matrix model of valuation of logistics system efficiency

Complex progress of valuation of logistics system efficiency by logistics indexes, which makes provision for before stated principles, is represented by matrix model.

Matrix model allows to obtain:
- repertoire of detail logistics indexes at the horizontally level (index x process x variant of customers requirements),
- different final values of indexes for individual kinds of logistics chains at the vertical level,
- aggregate indexes for logistics system.

Applying of matrix model in a montage company allows following to performance many targeted analyses, (Václav et. al. a, 2007):
- Analysis of an influence on the individual processes to the value of system indexes,
- Analysis of different cost loading of an individual variants of customer requirements,
- Analysis of relationships among indexes from individual groups of indexes from system of given process indexes, it means f. e. analysis of relationship among indexes from group of logistics services indexes and from a logistics costs group of given process or comparison of together random groups from a system of given process indexes
- Comparison of processes of identical kind, but founded in diverse places of logistics chain (processes of manipulation in diverse parts of production).

Matrix model of valuation of logistics system efficiency is open and sectional model.

2.2.2Algorithm in valuation of logistics process changes efficiency

In a next part there are characterized a ground conditions of algorithm for valuation of logistics process changes efficiency. This allows by a simple way to value an efficiency of each change proposal till now ineffective performanced logistics process of integrated logistics chain.

A condition of logistics process change efficiency of suppliers chain is following. Modification proposed. in logistics process of suppliers chain have to show a higher intensity of profit formation, as primarily supply chain process.

Applied utilization of this condition is following.

If a cost increase ratio for logistics process change to a depression of an all continuous time of goods delivery to the customer elicited by changed logistics processes smaller than is a profit ratio of suppliers chain to the original general running time of goods delivery to the customer, then a realization of such change is very effective in term of efficiency valuation within integrated supply chain. .

Logistics processes in an industrial company don't create a value for a customer, but they take part on production a added value for customer. They create a conditions, which allow in a value – creating activities to reach a higher efficiency, it means that they allow to increase a productivity of value - creating activities.

Efficiency of logistics processes changes within an integrated supply chain consist in potential utilization that supports a company profit creation mainly in decreasing of all time of supply processes and not everytime in cost decreasing to the logistics pocesses.

First stated algorithm conditions is able to apply also for valuation of montage processes change efficiency by the same way.

2.2.3 Application simulation Project Management Forecast (PMF) at the proposal and valuation of logistics processes variants

Available for a user there are three following kinds of programs on the basis of simulation instrument Project Management Forecast (PMF):

- for valuation of change efficiency of logistics process of logistics chain - EFEKproc,
- for optimalisation of parameter changes of logistics chain processes - OPTIproc,
- for valuation of change efficiency of logistics chain processes of logistics system - LOGRET.

Programs EFEKproc, OPTIproc a LOGRET in a satisfactory area cover a valuation of a process changes efficiency of logistics chains of company logistics system on a defined level. Package of programs is open and it is possible to extend it at editional requirements, which are necessary for making a provision for valuation of process efficiency of logistics chains in logistics system.

Output datas from first stated programs on the base of PMF are processed values of variable in a form of protocols, graphs and histograms in a required area, which relieves a decisions mainly by solution of making more effective all complex logistics systems, which consist of bigger amount of logistics chains in company logistics system. With out applying of these programs is a problem solution of making more effective all process of logistics chains very hard resolvable.

An important is an aspect, that simulation offers a possibility to compare of bigger amount of logistics chain process change variants on the part of taken standards and with this, it augment a place of knowledge about simulated object, because performance by another way is much more harder and time - consuming. such information are very worth.

2.2.3Active improvement of processes of logistics system for purpose of a customer productivity increase

The base is to define an index from a system of indexes (matrix model) for individual logistics processes and variants of customer requirements of company logistics system. Consecutive applying of first stated progresses, which besides valuation of process efficiency allows non-stop to innovate a logistics processes in a company on the part of extended and added product values (value innovation), to reach a jump values for customers and company. To set up also a marginal values of indexes, necessary for checking a company logistics system.

A part of active access of continual and radical repair planting of logistics process efficiency by system of indexes in matrix model and system of progress, which creates conditions for company competitive ability increasing and an increase of customer productivity (Václav et. al. b, 2007).

Each set of changes claims for realization a different investment costs, time of realization and consist of different structure of logistics processes, which brings to the customer a different utility and they influence in different way a customers productability. All these factors are necessary to revalue as a complex and on the base o fit choose a final variant of logistics chain process changes. A priority in decision has all ways an aspect of maximal increasing of customer productivity.

3. CONCLUSION

Stated system access and a progress collection of logistics system efficiency valuation were applied in chosen montage companies. At its application has been achieving an important economic additions for companies and at the same time it has been achieving an important increase of customer productivity. An applied research was mainly centred on efficiency of information and material flow in a logistics. In a next , it is necessary to expand it to the financial flow also. In a next research is necessary to concentrate on connection to the financial flow (cash flow, net present value), which provide new, complex view to the problemacy. An auxilium is a part of a problem solution VEGA No. *1/3163/06.*

4. REFERENCES

Cibulka, V. (2007), An acces to the valuation and procedure of valuation of logistics System efficiency in an industrial company, In: 7. Vserossiskaja naučnaja konferencija s meždunarodnym učastijem, Ufa, 11/2007, UGATU Ufa, pp. 97-104, ISBN 978-5-86911-691-8.

Horváth, G. & Cibulka, V. (2007), *Logistics in a production company*, ZČU Plzeň, ISBN 978-80-7043-634-9, Plzeň.

Horváth, G. & Cibulka, V. (2007), *Logistics in a production company*, ZČU Plzeň, ISBN 978-80-7043-634-9, Plzeň

Horváth, G. & Cibulka, V. (2007), *Logistics activitie, logistics processes and a value creationy*, Quality Innovation Prosperity, ISSN 1335-1745, 11th Vol., No. 2, pp. 42-49, TNU Trenčín, Trenčín.

Pernica, P. (2004). *Logistics (Supply Chain management) for 21. century,* RADIX, ISBN 80-86031-59-4, Praha.

Václav, Š. ; Peterka, J.; Pokorný, P. & Cibulka, V. (2007), *The Assembly AS Systém*, 7. International Multidisciplinary Conference, North University oF Baia Mare, 05/2007, Baia Mare, ISSN-1224-3264, pp. 717-724.

Václav Š. ; Pokorný P. ; Áčová M. & Cibulka, V. (2007), *Principle of Mechanical Structures Assembly*, IV. International Congress on Precision Machining ICPM 2007, UT Kielce, 09/2007, Sandomierz-Kielce, pp. 225 - 229, ISBN 978-83-88906-91-6.

Annals of DAAAM for 2008 & Proceedings of the 19th International DAAAM Symposium, ISSN 1726-9679
Editor B. Katalinic, Published by DAAAM International, Vienna, Austria 2008
Make Harmony Between Technology and Nature, and Your Mind will Fly Free as a Bird

ANALYSIS OF THE SILVER HALIDE PRINTING PLATES AFTER PRINTING PROCESS

CIGULA, T[omislav]; MAHOVIC POLJACEK, S[anja] & GOJO, M[iroslav]

Abstract: *One of the most important features in lithography is the stability of the printing plates during the printing process. It is highly influenced by the consistency and the wear of the photoactive and the aluminium oxide layers. In this paper the optical properties of the silver halide photoactive layer were observed by measuring coverage value on the printing plates before and after the printing process. The results have shown that the wear of silver halide layer is not significant but is visible after 80000 prints made. This fact leads to conclusion that this kind of printing plate enables high quality prints through the long printing run.*

Key words: *lithography, printing plates, quality, stability.*

1. INTRODUCTION

The quality of the plate making process and the stability of the printing plates during the printing process have an essential role in the final product quality. Digital plate making process, Computer to Plate (CtP), which has been observed in this paper, enables greater control and standardization in the plate making process in a contrast to analogue process (Schmitt, 2005). As the Computer to Plate technology is introduced into the graphic reproduction in the last ten years is important to observe all the factors and aspects of this process which could cause instability of the printing process. The most important factors are definitely the consistency and quality of the photoactive and the aluminium oxide layers. According to this fact, the aim of this research includes defining and monitoring the stability of the photoactive layer during the printing process.

2. BACKGROUND

The lithography printing plates are mainly made of aluminium foils covered with a thin photoactive layer. It has a significant role in the printing process as it represents the printing areas and must attract oil based printing inks. The plate making process starts with the exposure of the photoactive layer with defined electromagnetic radiation that causes physical and chemical changes in photoactive layer. After exposure printing plate must be immersed in alkaline solution where photoactive layer become soluble and is removed from nonprinting areas (Urano et al., 2004; Fiebag & Savariar-Hauck, 2003).

Printing plates with silver halide coating, which have been observed in this study, are often used in newspaper reproduction. During the printing process, silver halide layer is chemically and mechanically distorted and therefore changed in its optical characteristics (Mahovic et al., 2003). As within the lithography it is possible to print only one tone of the colour, the halftones are made with screening and presented in different coverage values (MacPhee, 1998). Screening elements are generally small sized and greatly influenced by changes in the characteristics of the photoactive layer. Those changes can cause different degradations of prints quality and consequently, degradation of the coverage values on prints (Milos et al., 2008). In accordance with these facts, the aim of this paper was to observe the changes of the coverage values on the printing plates after the printing process is finished and to define what has happened with screening elements in a case that changes occur.

3. EXPERIMENTAL

In this research the printing plates samples were made in a digital plate making process and automatically developed in an alkaline solution (pH≈12). There were made two plates for four main colours (cyan, magenta, yellow and black). The first one plate was used for printing and the second one was used for measuring the starting values of coverage value before printing. To avoid possible differences in coverage values in plate making process, the exposure and developing parameters were equal for all samples (Cigula et al., 2007).

For precise measuring results a special control wedge was designed for this research. This wedge contained 26 control fields. The difference between two neighbour fields was 5% of the coverage. For detail observation in lower (0–20%) and higher (80–100%) coverage areas more control fields were defined.

The printing process was carried out on a web offset printing press on a light weighted paper (45 gm^{-2}) with a cold set printing ink. This kind of ink is melted on heated ink rollers and printing plate and it hardens in contact with cold paper.

The measuring was performed by a Gretag Macbeth iCPlate II Platereader. This kind of measuring unit enables coverage value measurement, dot radius measurement and gives a preview of the measured area.

4. RESULTS AND DISCUSSION

In Fig. 1. the coverage values measured on the printing plates of four main colours before printing process are presented. One can see that there is no significant difference in coverage values on the printing plates for observed colours. This result is implicating that the Platesetter unit and the developing unit are functioning without any fluctuations.

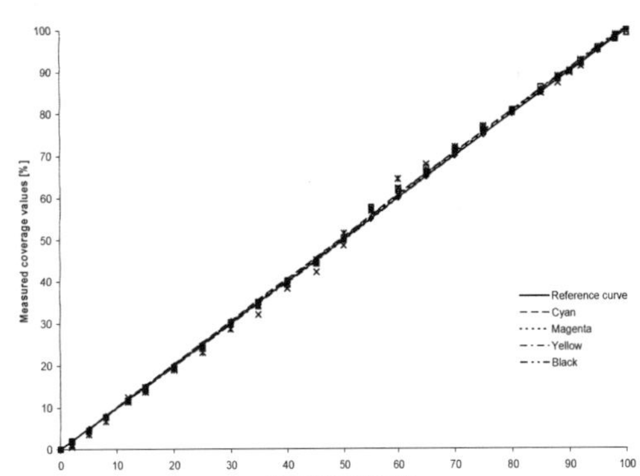

Fig. 1. Coverage values measured on the printing plates of four main colours before printing process.

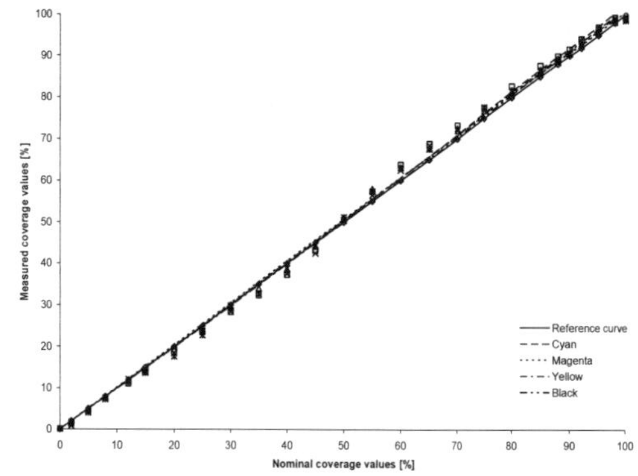

Fig. 2. Coverage values measured on the printing plates of four main colours after the printing process.

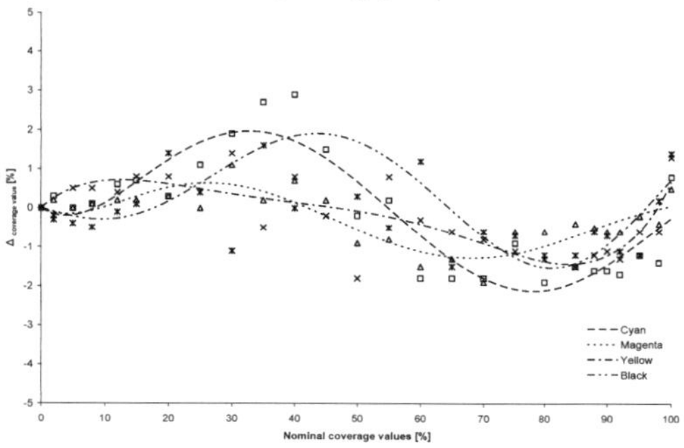

Fig. 3. Difference between coverage values on the printing plates before and after the printing process.

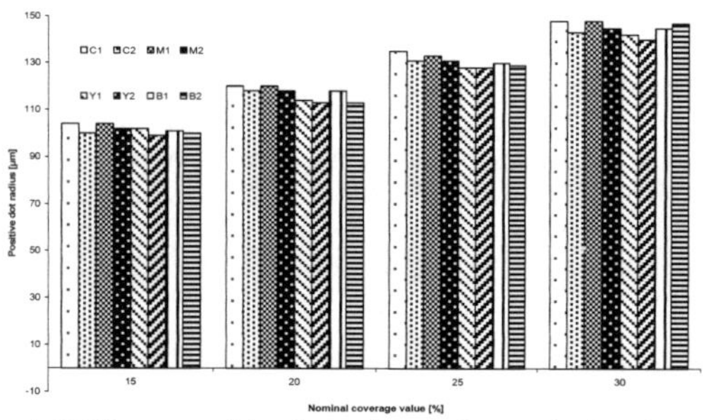

Fig. 4. Difference of the dot radius before and after the printing process in the low coverage areas.

The results of the coverage values measured on the printing plates after the printing process are shown in Fig. 2. One can see that there is not significant change of the coverage values at lower coverage areas on the printing plates after the printing process. In the medium and higher coverage area changes of the coverage values can be seen. On cyan printing plate it is visible that difference between coverage values and reference curve are increasing with increasing nominal coverage value. In opposition, the smallest difference is accomplished between coverage value on magenta printing plate and the reference curve.

In Fig. 3. the difference between coverage values on a printing plate before and after the printing process is shown. It is visible that the most significant differences are in the area from 30 to 50 percent coverage value and are decreased after the printing process. From 70 to 90 percent coverage value, values are increased after the printing process. The changes of the coverage values are inside 2%. It can also be seen that coverage values on cyan plate is most changeable.

In Figs. 4 and 5 the results of the screening elements – dots radius before and after the printing process are presented. C1, M1, Y1, B1 are dot radius values on the printing plate before the printing process. C2, M2, Y2, B2 are values of dot radius on the plate after the printing process. One can see in Fig. 4 that in the lower coverage area the dot radius is decreased which causes the decreasing of the coverage values. Decreased values are smallest on yellow. It can be seen in Fig. 5 that in the higher coverage area the greatest increase of dot is accomplished by cyan printing plate. These results indicate that yellow ink has smallest influence on wearing of photoactive layer and consequently causes minimal decrease of coverage value.

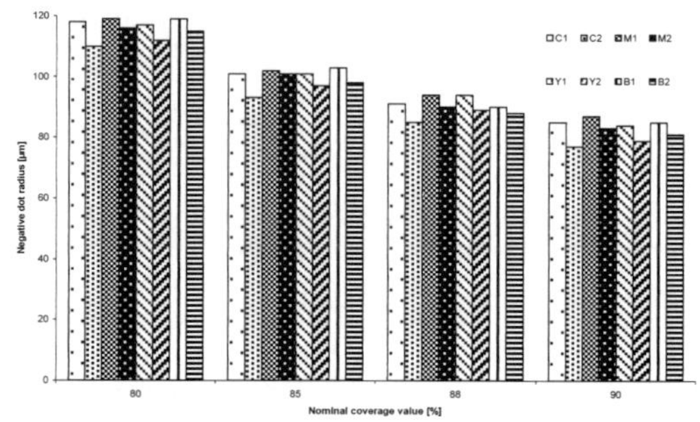

Fig. 5. Difference of the dot radius before and after printing.

5. CONCLUSION

Computer to plate process with automatic developing unit produces printing plates of the same quality with no significant changes in coverage values. In addition, results obtained by this research also declined presumption that there will be significant changes in coverage values caused by printing process. One could say that changes had occurred, but not causing decrease in print quality. The silver halide printing plates proved to have large endurance and could be used for longer printing run with no significant decrease in print quality. This paper has shown that with correct maintenance it is possible to standardize the silver halide plate making process. Our further research will be point at the thermal CtP plate making processes, endurance and possibility of standardization.

6. REFERENCES

Cigula, T.; Mahovic Poljacek, S. & Gojo, M. (2007). The Significance of Laser Exposition in CtP Plate Making Process, *Annals of DAAAM for 2007 & Proceedings of the 18th International DAAAM Symposium*, Katalinic, B. (Ed.), pp. 143-144, ISBN 3-901509-58-5, ISSN 1726-9679, Zadar, October 2007, DAAAM International Vienna, Vienna

Fiebag U. & Savariar-Hauck C., *Aqueous Developer for Lithographic Printing Plates*, Patent No.: US 6,649,324 B1, 2003.

MacPhee, J. (1998). *Fundamentals of Lithography Printing, Volume I, Mechanics of Printing,* GATFPress, ISBN 0-88362-214-9, Pittsburg

Mahovic, S.; Agic, D. & Gojo, M. (2003). Mechanical and Optical Differences in Long Run Printing in Conventional and CtP Offset Systems, *Proceedings of IARIGAI*, Lovrecek, M. (Ed.), pp. 213-221, ISBN 953-96276-8-0, Dubrovnik-Cavtat, September 2003, Acta Graphica Publishers, Croatia

Milos, S.; Cigula, T.; Mahovic Poljacek, S. & Gojo, M. (2008). Contribution to the Printing Ink – Printing Plate – Run Length Interactions, *Proceedings of MATRIB*, Grilec, K.; Maric, G.; Jakovljevic, S. (Ed.), pp. 205-210, ISBN 978-953-7040-14-7, Vela Luka, June 2008, Croatian Society for Materials and Tribology, Zagreb

Schmitt U. (2005). Korrekter Einsatz der FOGRA-CtP- und der FOGRA-Prozessor-Testform, *FOGRA Symposium: Computer to Plate*, München

Urano T., Kohori K. & Okamoto H., *Photosensitive Lithographic Printing Plate and Method for making a Printing Plate*, Patent No.: US 6,689,537 B2, 2004.

Annals of DAAAM for 2008 & Proceedings of the 19th International DAAAM Symposium, ISSN 1726-9679
Editor B. Katalinic, Published by DAAAM International, Vienna, Austria 2008
Make Harmony Between Technology and Nature, and Your Mind will Fly Free as a Bird

DETERMINING THE MACHINING ERROR DUE TO WORKPIECE - FIXTURE SYSTEM DEFORMATION USING THE FINITE ELEMENT METHOD

CIOATA, V[asile] G[eorge]

Abstract: *In this paper work there are described the error's sources due to workpiece-fixture compliance which appear while the workpieces are clamped and processed; and an example of using the finite element method in order to determine the machining error for a practical example. The differences between the results obtained using the finite element method and the results obtained in similar conditions but using other kind of methods from the technical literature are very small, which demonstrates that the finite element method can be used for determining the machining error due to workpiece-fixture compliance.*

Key words: *workpiece, fixture, deformation, machining error*

1. INTRODUCTION

In the workpieces machining process the accuracy conditions (dimensions, shape and position) for the machined features are obtained with few errors towards the ones provided in the workpiece's drawings: errors regarding dimensions, shape and position.

These errors appear due to the individual or simultaneous influence of various factors regarding the evolution of the machining process, also of the ones which operate inside the technological system: the machine tool, the fixture, the cutting tool, the workpiece and sometimes the control device.

In order for the workpiece to be considered adequate, the errors to the accuracy conditions obtained after machining must be framed in the limits of the prescribed tolerance.

In the machining process the fixtures are used for the orientation and clamping, for the workpiece to be machined through diverse methods, assembling and controlling. Once the workpiece is orientated and clamped with locators and clamping devices the workpiece can be machined in order to accomplish the imposed accuracy conditions.

The errors due to fixtures are major ones and influence the workpiece's machining accuracy; the errors can amount to 20-60% of the overall machining error. Therefore, performance evaluations of the workpiece-fixture system constitute a significant task for fixture design optimization and control of the machining error before manufacturing and the application in production (Qin et al., 2007).

The errors due to the workpiece-fixture system can be classified in two categories: errors due to orientation of the workpiece in the fixture and errors due to deformations of the workpiece-fixture system during clamping and machining.

The errors due to deformations that appear in the workpiece-fixture system under the action of the workpiece's gravity, the clamping forces, the cutting forces and torque are illustrated in Fig. 1 (Qin et al., 2007). During clamping and machining the workpiece, for a point P situated on the machined feature, the force assembly which acts on the workpiece-fixture system determines the appearance of three types of errors: errors due to contact deformation (Fig. 1, b), errors due to locators deformation (Fig. 1, c) and errors due to workpiece deformation (Fig. 1, d).

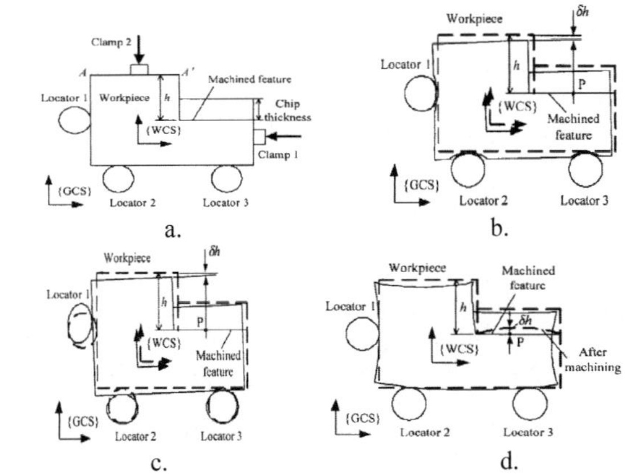

Fig. 1. Source error due to workpiece-fixture compliance

In technical literature the determination of the error due to workpiece-fixture system compliance is realized, generally, through the independent evaluation of the influence which the elements of the workpiece-fixture system have on it. Thus, (Li & Melkote, 2001) they estimated the error due to contact deformation between workpiece and fixture; (Raghu & Melkote, 2005) they estimated the error due to locators and contact deformation. In (Rong & Zhu, 1999) the authors are studying the influence that the locator deformation has on the workpiece machining precision. In (Siebenaler & Melkote, 2006), the authors are describing the modality where through the workpiece deformation is determined with the finite element method, used on thin-walled workpieces (diminished rigidity). A complete evaluation of the machining error due to workpiece-fixture system compliance and the description of some mathematical models for its determination are presented in (Qin et al., 2007).

2. THE EVALUATION OF THE MACHINING ERROR WITH THE FINITE ELEMENT METHOD

A quick and efficient evaluation of the machining errors due to workpiece-fixture system, taking in consideration the three types of deformations which appear during clamping and machining can be realized with the finite element method. Forwards it is presented an example of application of this method for a practical situation of machining.

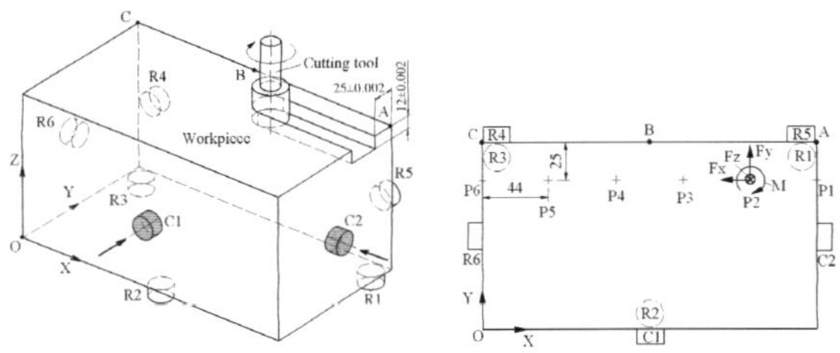

Fig. 2. Scheme of workpiece-fixture system

2.1. The workpiece-fixture system

Considered the machining of a prismatic workpiece into a device formed by 6 locators emplaced in three reciprocal perpendicular planes-which allow locking the workpiece's 6 DOFs- and two clamping systems C1 and C2 (Fig. 2). It is necessary milling a feature to 25±0.002 mm and 12±0.002 mm dimensions, measured from the AC edge. The workpiece is made of aluminum, with the sizes: 220x122x112 mm, with the density 2700 kg/m^3, Young's modulus E_P=70 GPa and Poisson's ratio υ_P=0.334. The locators and the clamping elements have plane surfaces, the diameter is 18 mm and are made of hardened steel with E=207GPa and υ_P=0.292. Clamping forces of 640 N and 690 N are applied simultaneously by clamp C1 and clamp C2. The static friction coefficient is assumed to be 0.25 between the workpiece and fixture elements.

The cutting force's constituents and torque used in milling the considered feature are estimated to be Fx=131 N, Fy=232 N, Fz=55 N and M=2.77 N·m as in Fig. 2.

2.2. Model development

The solid model of the system chosen to be analyzed was realized in Autodesk Inventor Professional and then transferred in Ansys Version 11.0 for the FEA analysis. All the elements of the system have been modeled as isotropic elastic bodies. The locators and the fixture tips have cylindrical shape, the contact surface is plane and has circular shape with the aria assumed to be 254 mm^2. The 10-node tetrahedral element Solid92 was used to mesh all solid bodies. Contact between the workpiece and fixture was simulated using the quadratic surface-to-surface contact elements Targe170 and Conta174. A constant static coefficient of friction was used to establish contact properties at the interfaces. To simulate the locators being rigidly fixed in place, the surface of each locator tip opposite to the contact was restrained in all three translational degrees of freedom. In order to simulate the clamping force, there have been applied single forces over the surface of both fixture elements, in opposite contact.

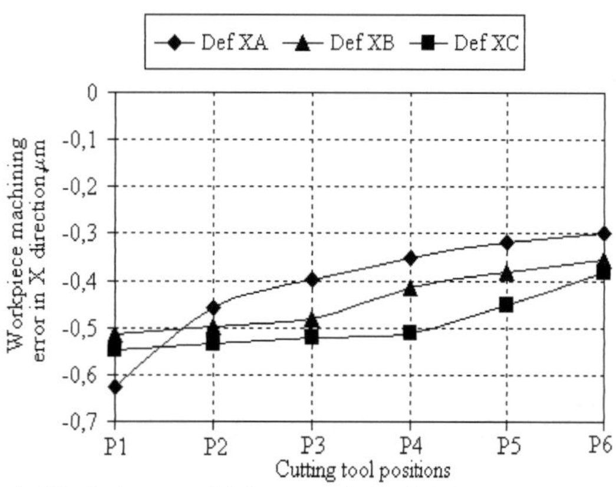

Fig. 3. Workpiece machining error in X direction

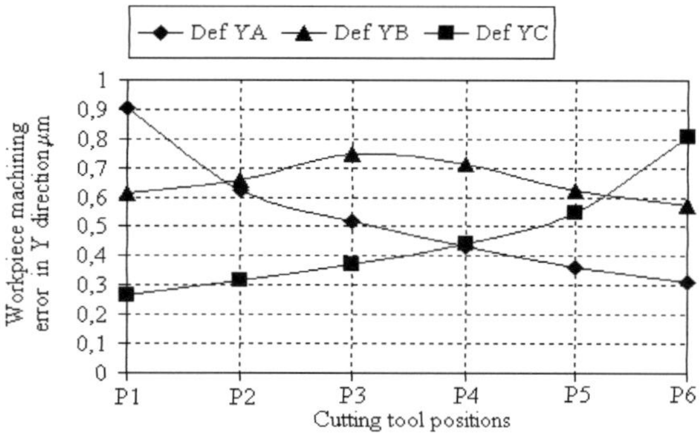

Fig. 4. Workpiece machining error in Y direction

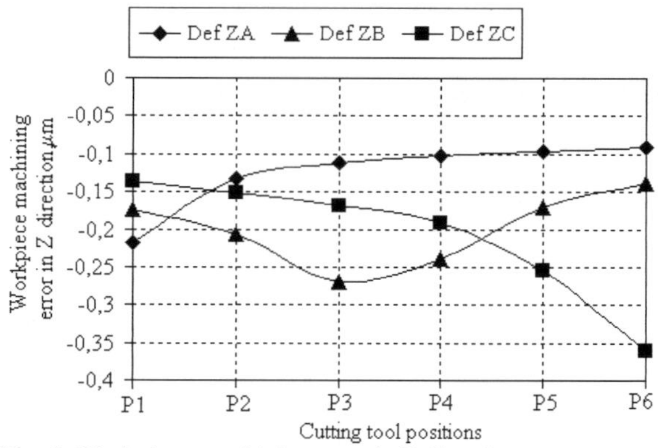

Fig. 5. Workpiece machining error in Z direction

The deformation of the workpiece, analyzed in the subsequent sections, was found at three points on the workpiece. The positive directions of these three points are shown in Fig. 2. The overall machining errors due to workpiece-fixture system compliance to the determinant conditions are caused by the maximum displacement of the datum points A, B, C in Y and Z directions. With this purpose in view, the cutting force and torque have been applied in six equidistant points (P1...P6), on the upper surface of the workpiece having a 25 mm distance to the AC edge. For each loading situation, there have been determined the displacement of the A, B, C points to the three directions of the coordinate system OXYZ. The results obtained are shown in Fig. 3, 4 and 5.

Analyzing the Fig. 3 it occurs that the machining error at dimension 25 is assumed to be 0.90 µm, value which is under the allowed tolerance for this size. Also, the machining error at dimension 12 is assumed to be 0.35 µm (Fig. 4), value under the allowed tolerance for it.

3. CONCLUSION

In (Qin et al., 2007), the authors calculated the error for a similar machining situation, but neglecting the contact deformations. Comparing the presented results with the ones obtained using the finite element method there are noticed relative small differences, these being explicable because the contact deformations between the workpiece and locators have been neglected. Also, the illustrated curves in Fig. 3, 4 and 5 have the same rate as the ones presented in (Qin et al., 2007).

These results demonstrate that the finite elements method permits a quick and efficient evaluation of the machining errors due to workpiece-fixture system, taking in consideration the three types of deformations which appear during clamping and machining.

4. REFERENCES

Li, B. & Melkote, S. N., (2001). Fixture clamping force optimization and its impact on workpiece location accuracy. *Int. J. Adv. Manuf. Technol.*, Vol. 17, No. 2, (January 2001), 104–113, ISSN 0268-3768

Qin, G. et al. (2007). Systematic modeling of workpiece-fixture geometric default and compliance for the prediction of workpiece machining error. *J. Manuf. Sci. Eng.*, Vol. 129, No. 4, (August 2007) 789-801, ISSN 1087-1357

Raghu, A. & Melkote, S. N., (2005). Modeling of workpiece location error due to fixture geometric error and fixture-workpiece compliance. *J. Manuf. Sci. Eng.*, Vol. 127, No. 1, (February 2005), 75–83, ISSN 1087-1357

Rong, Y. M. & Zhu, Y. X., (1999). *Computer Aided Fixture Design*, Marcel Dekker, ISBN 0-8247-9961-5, New York

Siebenaler, S., P. & Melkote, S., N, (2006). Prediction of workpiece deformation in a fixture system using the finite element method. *Int. J. Mach. Tools Manuf.*, Vol. 46, No. 1, (January 2006), 51-58, ISSN 0890-6955

Annals of DAAAM for 2008 & Proceedings of the 19th International DAAAM Symposium, ISSN 1726-9679
Editor B. Katalinic, Published by DAAAM International, Vienna, Austria 2008
Make Harmony Between Technology and Nature, and Your Mind will Fly Free as a Bird

SOME ASPECTS REGARDING THE INFORMATION AND THE QUALITY MANAGEMENT IN A CAD SERVICES COMPANY

CIOBAN, H[oria] & POP, N[icolae]

Abstract: *Having as model the IT structure of big companies, a CAD organization can manage projects and activities using a quality management system as a tool to orient business to the client requirements. This paper presents validated solutions for an information system and possibilities to improve the quality management system using in-house software applications.*
Key words: *CAD, PLM, ERP, Quality Management System.*

1. INTRODUCTION

Automotive industry is one of the major factors of CAD activity development. Dynamics of the automotive fields guide also the trend of the design activity, who becomes more and more a significant part in a PLM process. The increased volume of the outsourcing gives a chance to a services oriented company. As part of the product lifecycle, design can get up to 40% of the manufacturing costs. On the other side, taking into consideration only computer simulation, it reduces the time and efforts for producing the prototypes. To survive on the market, a CAD company must be able to react very quickly to the changing needs, offering at the same time a good quality products and services. It has to prepare a strong organization and infrastructure, an accurate information management, that means:

- flexible and well trained personnel;
- up to date hardware, network and communications;
- software platforms related to the client.

Avoiding terms like "knowledge management" this paper will use *managing information* instead, as a more appropriate meaning for the role of databases, projects, files and archives manipulated inside the organization or delivered outside as products (Miller, 2002). A quality management system is one of the first references asked by a company looking for a supplier.

2. NETWORK STRUCTURE IN A BIG COMPANY

The IT model given by the informational structure of a big company can be scaled and applied into an appropriate shape in small enterprises. It can takes as an example the network containing more than 50.000 computers of a corporation from automotive industry.

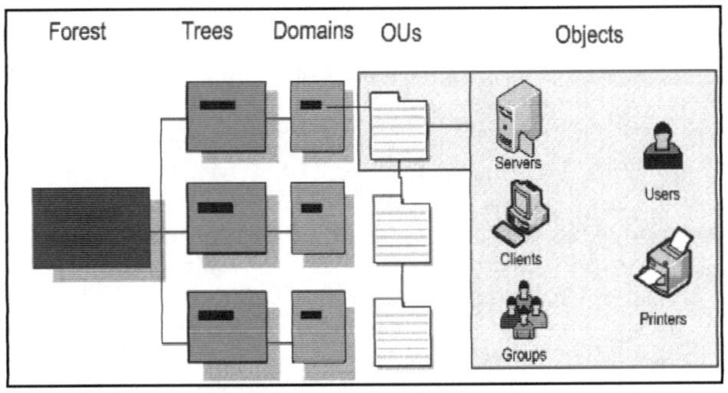

Fig. 1. A hierarchical framework of objects in a network.

This huge amount of equipment can be administrated only by organizing it into a *hierarchical framework* of objects. The new servers OS offer services like Active Directory – AD (Richards et al., 2006).

The framework that holds the objects is viewed as a number of levels, having at the top of the structure *the Forest*. Below, the hierarchy contains *trees, domains* and *OUs* (organizational units), as in figure 1.

These OUs are containers with groups of clients (PC's and laptops), servers, printers and users. The properties and the behavior of the groups from containers and guided through policies, using a console GPMC (Group Policy Management Console). The most important thing is that all applications running on server, including AD and GPMC can be accessed remotely by administrators from any client in the network.

Like Active Directory, there are several categories of applications used for managing information and activities in a company, like: Enterprise Resource Planning (ERP), Product Lifecycle Management (PLM), Supplier Relationship management (SRM), most of them having similar consoles as interfaces. All these solutions are integrated and change information through importing/exporting data procedures. As examples, antivirus and security solutions get all necessary inputs from Active Directory (Microsoft, 2008).

3. SOLUTIONS FOR INFORMATION MANAGEMENT

A CAD company can be a small to medium enterprise offering services in a niche of businesses. As a results of globalization, many automotive industry producers and suppliers have found outsourcing as a possibility to reduce costs. Design activity is one of externalized functions of a manufacturing company. Having several big clients, a supplier can have a comfortable life on the market.

Regarding the infrastructure, it is very important for the supplier to have the same hardware and software, including the working environment as the customer. For example, they have the same version of CAD solution, including service pack and going to the same hotfix. Figure 2 explains how a corporate network can be scaled, using only the necessary components and functions from the mother structure, starting from Organizational Units branch to the right of figure 1.

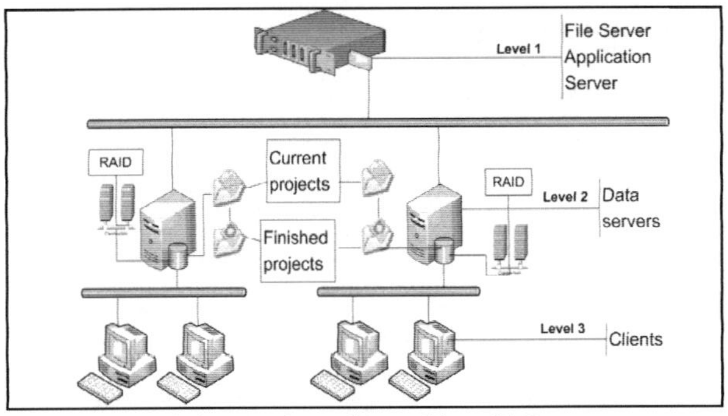

Fig. 2. A scaled network can be used in a CAD company.

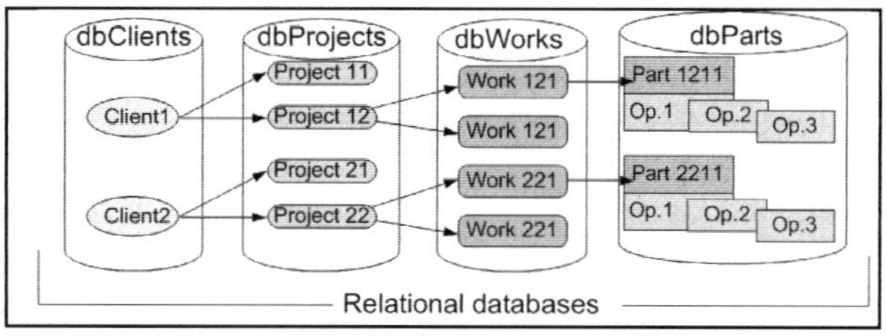

Fig. 3. Database application for company main process

Any change in the customer's software must to be replicated to the supplier and this is one of the most important condition for a good cooperation. Of course, not all components have to be replicated to the supplier. For example, if the customer uses a CAD application, the supplier have to install the same software, the same service pack and hotfix. Supplier have to setup also a compatible environment with the client.

The design activity in a CAD company is divided into projects. Each order obtained from the customer is one project administrated by a project manager. In figure 2, on the level 3 there are groups of clients/PCs and each team save the current projects data on servers from level 2. One of the properties of data servers is that they have the storage disks organized in mirrored or RAID schema. On each data server there are two folders: when a project is finished, the project manager moves it from the *current folder* to the *finished projects folder.*

On the top level, the server acts as file and application server and has also the storage disks in RAID. The backup utility of the OS server has a daily task to make backups from "Current projects" folder of every data servers (level 2). To reduce the space reserved on disks, the incremental backup method is used. It makes monthly backups from the "Finished projects" folder of every data servers. A second safety mechanism for data is "shadow copy", which is a feature that allows taking manual or automatic backup copies of a specific volume, folder or file at a specific point in time (Microsoft, 2008). These two layers of security offer the possibility to restore information from few months back.

Applications like licenses management servers and antivirus management solutions run also on the application and management server. To reduce costs, less then 50 users and clients can be managed without Active Directory service.

4. SOLUTIONS FOR QUALITY MANAGEMENT

Many processes of a CAD company can be integrated in a quality management system and this is one of the most important requirements for an automotive industry supplier. Because the price of an ERP solution is quite high, a small to medium enterprise can found an in-house solution, even though not all features will be covered (Godfrey, 1999).

Using a Database Management System (DBMS) it can develop a client/server application that can store information about: employees, customers, projects, works, parts, operations and estimated hours, like in figure 3.

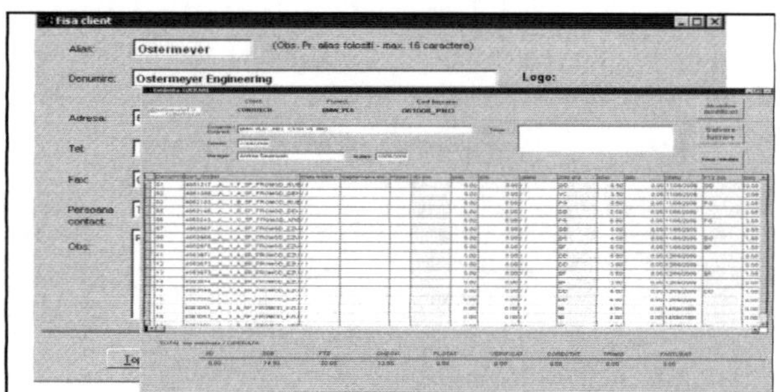

Fig. 4. Customer form and work fiche according to QMS

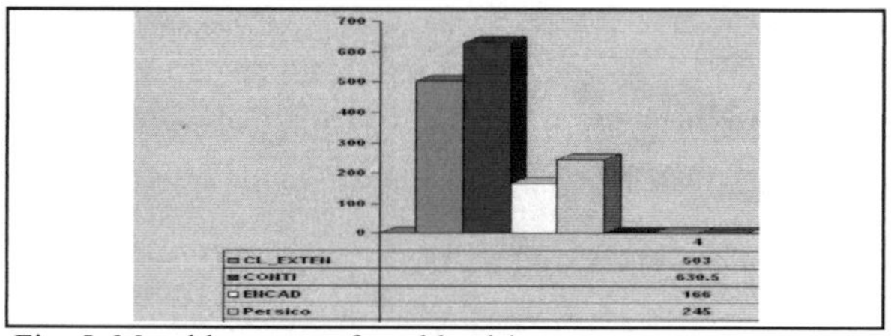

Fig. 5. Monthly report of workload / customer.

A client can order one or more projects, a project can contains one or more works and a work can have one or more parts or models.

For each part, the application administrator can define specific operations and 3 fields belong to each operation: designer's code (who works on this operation), estimation time (how long it takes) and delivery time. Input data are filled using forms, like in figure 4.

Sample models for customer's form, project fiche or work fiche are according to the quality manual requirements.

A work fiche contains all necessary information to keep the traceability of a project and can make arithmetic operations on lines and columns. It is also a useful tool to estimate the workload or offers.

Outputs are lists and reports like: monthly report of estimated hours/customer (figure 5), monthly report of estimated hour/employee, workload/customer. Application offers also information regarding quality indicators for design activity like deadline factor - days after deadline/no.of projects / month - or individual performance factor – estimated hours approved by the customer in a month / working hours in a month. The software is running on the file/application server and can be accessed by any user in the company.

5. CONCLUSIONS

Having as model the IT structure of big companies, a CAD organization can manage projects and activities using a quality management system as a tool to orient business to the client requirements. Using applications applied on QMS, managers can supervise the processes with respect to the system and its performance. These solutions reduce the cycle and the volume of papers and improves the decision process. Company becomes more flexible, reacting to the market inputs.

These managing systems for information and quality were designed, tested and implemented by authors in a medium size CAD company, during 3 months with satisfied results. Now they are in a dynamic and a continuous process of development.

Reacting to the positive impact, the company management intends to add more new functions to the QMS application and to integrate it with other software.

6. REFERENCES

Godfrey, A. B. (1999). *Juran's Quality Handbook*, McGraw-Hill, ISBN 978-0-07-034003-9, New York.

Miller, F.J. (2002). *I=0 (Information has no intrinsic meaning)*, Information reasearch, vol.8 No.1, (October 2002), ISSN 1368-1613. *Available from:* http://informationr.net/ir. *Accessed:* 2008-06-28.

Microsoft (2008). *Windows Server 2003: Active Directory Infrastructure*. Microsoft Press, 1-8 – 1-9. ISBN 0-7356-1438-5.

Microsoft (2008). *Volume Shadow Copy Service. MSDN.* Microsoft. *Available from:* http://en.wikipedia.org/wiki/Shadow_Copy, *Accessed:* 2008-06-28.

Richards J.; Allen R.; Lowe-Norris, A (2006): *Active Directory: design and deployment of Microsoft's Active Directory,* O'Reilly, ISBN:0596101732, Sebastopol, USA.

Annals of DAAAM for 2008 & Proceedings of the 19th International DAAAM Symposium, ISSN 1726-9679
Editor B. Katalinic, Published by DAAAM International, Vienna, Austria 2008
Make Harmony Between Technology and Nature, and Your Mind will Fly Free as a Bird

MAGNETIC FLUID GRINDING (MFG) OF INTERNAL ROTATING SURFACES

CIOBANU, M[ircea]

Abstract: *The paper describes a new finishing process that uses nonmagnetic grains suspended into a magnetic liquid. The main assumption behind the operating principle is that the process involves a system in which a medium for energy transfer exists and this acts as a working environment for finishing the surface of interest. The paper also presents the design principle of experimental equipment and the effect of the operating speed and working gap on the surface quality.*

Key words: *magnetic, grinding, quality, experimental device*

1. INTRODUCTION

In recent years there has been a considerable increase in use of magnetic technique for many purposes in the metalworking field.The four popular types of magnetic abrasive process are: Magnetic Abrasive Finishing (Baron et al., 2005); Magnetic Fluid Grinding (Ciobanu, 2007) also called Magnetorheological Finishing (Kordonsky, 2002), Magnetic Abrasive Jet Machining (Jha, 2004) and Magnetorheological Abrasive Flow Finishing (Sun & Yang 2007).

Magneto-abrasive finishing is based on the capacity of a magnetic field to rise and carve small grained particles that possess abrasive and magnetic properties. Between the poles of the magnetic system and the piece to be finished a "magnetic abrasive brush" characterized by orientation of the particles along the magnetic force lines occurs.

The dimensional manufacturing, which involves material removal, is done by transferring from exterior to the worked piece a quantity of energy. In this case, the energy is of magnetic type and it also acts as "glue" for magneto-abrasive particles. Meanwhile, the magnetic energy generates an important pressure upon these particles.

The energy transfer medium, that is the magnetic field, acts with a force or a pressure upon the working piece achieving a process of micro-abrasion. This process takes place entirely in the working zone.

2. A PHYSICAL MODEL FOR MFG

In the case of a magnetic abrasive finishing process, a ferrofluid mixed with abrasive grains is used as a polishing compound. The physical model presented in this paper is proposed by the author.

This process is based on the levitation property of non - ferromagnetic bodies sunken in a ferrofluid. The levitation occurs due to the force which a ferrofluid, placed into a magnetic field gradient, exerts upon these bodies. This force is oriented in an opposite direction to the gradient. The expression of electromagnetic levitation force depends on the shape of non - magnetic grains (spheres, revolution ellipsoids, cubes, cones, double cones, tetrahedrons) and on their orientation with respect to the coordinate axes.

The correct evaluations of the levitation force which acts upon a certain grain is quite a complex operation, because it must consider both the presence of the detached micro - chip, which can be magnetic or non -magnetic, and the properties of the involved ferrofluid. It is possible to determine experimental values for a correction coefficient that should be taken into account in the above mentioned situations.

A precise evaluation of the levitation force should use the complete system of equations from magneto-hydrodynamics as well as the distortions that occur among the force assembly when introducing the abrasive grains.

A fluid suspension containing 30% SiC grains was used as a polishing compound. The magneto - gravitational force acting upon grains immersed in ferrofluid is known to represent the force F resulting from combination of the magnetic force F_{MZ} with the gravitational force F_G, according to Figure 1:

$$F + F_G + F_{MZ} = 0 \qquad (1)$$

The force acting on the unit volume of a grain placed in a ferrofluid is:

$$F_{MZ} = \mu_o (M - M_g) dH/dz\ K \qquad (2)$$

where: M - ferrofluid magnetization; Mg - magnetization of the grain immersed in the ferrofluid; μ_o - magnetic permeability in vacuum.

The levitation force, acting on non-magnetic grains immersed in the ferrofluid is:

$$F_{lk} = \frac{\mu_0}{V_g} \int M \frac{dH}{dz} \vec{K}\, dv \qquad (3)$$

where: V_g - the volume of the immersed grains; dH/dz – the magnetic field gradient; $\vec{K}$ – the unit vector.

Gravitational force acting on the unit volume of the grain in ferrofluid is:

$$\bar{F}_G = (\rho_g - \rho_f) \cdot g \cdot \bar{J} \qquad (4)$$

where: J - vertical unit vector.

The resultant force is give by Equation (5) and depends on the direction of action of the levitation forces.

The magneto - gravitational force F acts upon the grains in the ferrofluid, lifting them or bringing them down. This force needs an adequate change of the shape of polar piece or even of the entire finishing equipment.

$$FK + (\rho_g - \rho_f) gJ + \mu_0 (M - M_g) \frac{dH}{dz} K = 0 \qquad (5)$$

3. EXPERIMENTAL METHOD AND DEVICE

Figure 2 shows, schematically, an original equipment used for experimental investigations. It comprises the main mechanical assembly, the block diagram and the apparatus used to measure the variable parameters of interest. In this figure, I denote the mechanical part of the device which achieves two working movements and II indicates the magnetic system. The ferrofluid 1 is introduced in the bowl 2, made of a non-

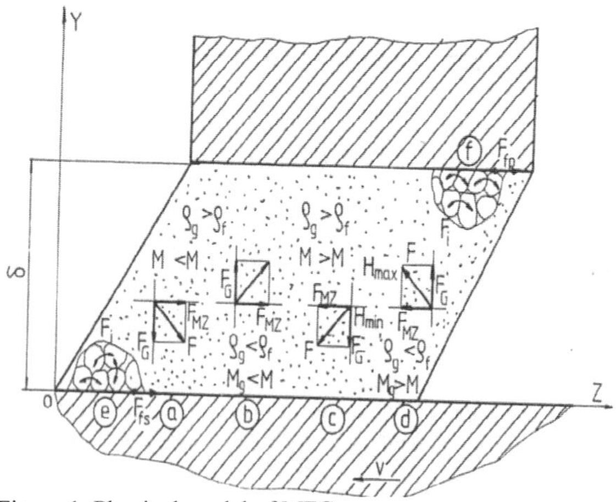

Figure 1. Physical model of MFG

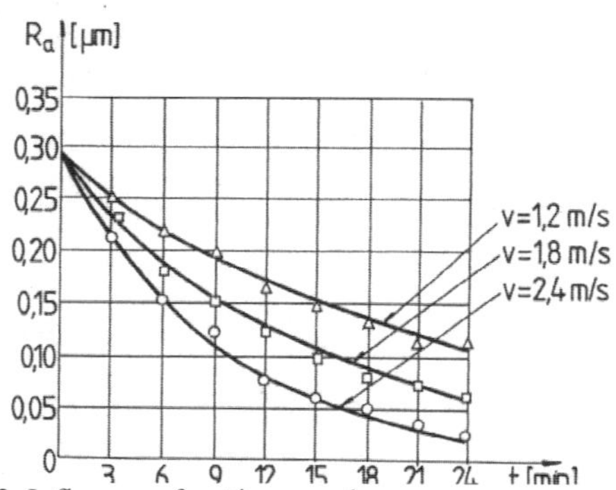

Figure 3. Influence of cutting speed

magnetic material. In the bowl are also introduced the polar pieces S, N, S. The polar piece N, having different profiles A, B, C, is positioned inside the processing piece 3.

The experiments were made under the following conditions: worked material: plain carbon steel OLC 45, containing 0.45% carbon; B- magnetic flux density: (0.8...1) Tt – finishing time: 0...24 minutes; abrasive particles: 30 % SiC, in ferrofluid.

4. EXPERIMENTAL RESULTS AND CONCLUSIONS

Experimental investigations were performed on the above tester to evidence the effect of operating parameters upon surface quality. The results shown in Figures 3 to 4 are useful to this end.

Magnetic fluid grinding has been developed as a new process which can remove materials more quickly than the lapping and honing which are conventionally used. It can be concluded that the experiments performed proved that ferrofluids, mixed with abrasive grains, can be successfully used in the process of internal rotating surface finishing.

It can be concluded that the experiments performed proved that ferrofluids, mixed with abrasive grains, can be successfully used in the process of internal rotating surface finishing.

They offer high precision and surface quality indices for the machined surfaces. Further research they will establish that the process can be simply controlled by means of the magnetic field, making the finishing compound to behave as a smart material.

The finishing efficiency and the surface roughness can be improved effectively when magnetic abrasives are mixed with

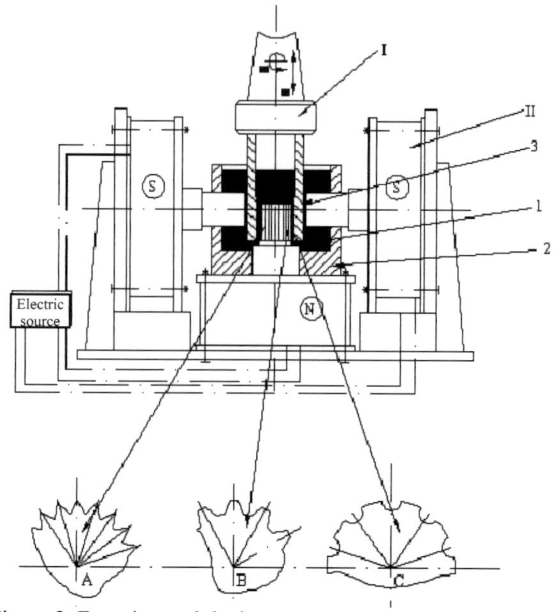

Figure 2. Experimental device

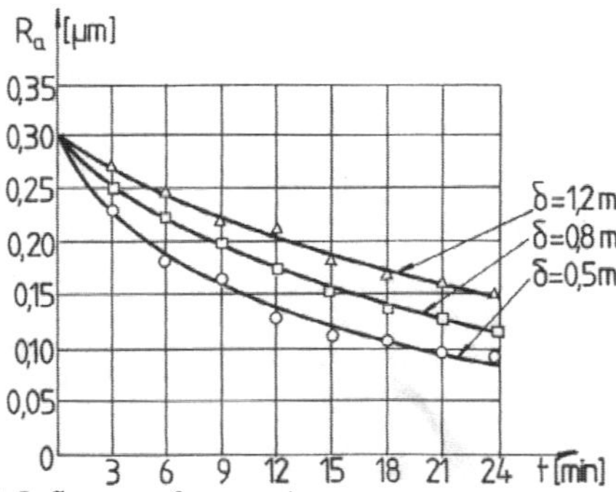

Figure 4. Influence of magnetic gap

reasonable mesh size iron powers in rational weight percent (Wang & Hu, 2005).

5. REFERENCES

Baron, Y.M.; Ko, S.L. & Park, J.I. (2005). *Characterization of the MAM and Its Applications to Deburing.* In: Key Engineering Materials, vol.2, pp 291-296, ISSN 1013-9826

Ciobanu, M. (2007). *Multiple aplications of magnetic abrasive finishing*. In: Annals of MteM for Proceedings of the 8-th International Conference Modern Technologies in Manufacturing. pp. 101-104, ISBN 973-9087-83-3, Cluj-Napoca, Romania

Jha, B.H. (2004). *Design and development of MAF process*. In: International Journal of Machine Tools and Manufacture, Vol. 1, pp. 101-109, ISSN 0890-6955

Kordonsky, W. (2002). *Multiple Application of MRF Effect in High Precision Finishing.* In: Journal of Intelligent Material System and Structure, Vol.13, pp. 401-404, ISSN 1045-389X

Sun, H.W. & Yang, S.C. (2007), *Study of the finishing mechanism of fluid magnetic abrasive*. In: International Journal of Computer Application in Technology, nr. 2-4, pp. 141-144, ISSN 0952-8091

Wang, Y. & Hu, D. (2005), *Study on the inner surface finishing of tubing by magnetic abrasive finishing*, In: International Journal of machine tools and manufacturing, ISSN 0890-6955, vol. 45, pp. 43 – 49.

Annals of DAAAM for 2008 & Proceedings of the 19th International DAAAM Symposium, ISSN 1726-9679
Editor B. Katalinic, Published by DAAAM International, Vienna, Austria 2008
Make Harmony Between Technology and Nature, and Your Mind will Fly Free as a Bird

ROM@IT PROJECT – THE DESIGN FRAMEWORK OF A ROBOTIC WELDING SYSTEM

CIOBANU, R[omeo]-M[ihai]; COHAL, V[iorel]; SARBU, I[onel] & IVASCU, C[ostel]

Abstract: *Many of the actually Romanian industrial structures are simply not adequate to respond to the new market requirements. The implementation of flexible manufacturing systems is a hard challenge which requires an up-down paradigm shift and it has to be based on the best practices, expertise, and models. ROM@IT (ROM@nian Intelligent Technology) is a proposed project regarding a robotic welding system. It aims to intensify the pro-active measures in order to develop the performance and competitiveness of the Romanian industrial companies.*

Key words: *flexibility, manufacturing, robotic, welding, systems.*

1. INTRODUCTION

The implementation of flexible manufacturing systems (FMS) is in the same time a necessity and actuality, constituting "one of the roofs of protection" for an industrial company, which may enable them to successfully resist the market economy requirements (Abrudan, 1996). The robotic welding systems represent one of the particularly FMS. The harmful work atmosphere, the physical solicitation, the static and sensorial fatigue are some of the decisive elements for the automation of the welding operations. The both technical and economical advantages of robotic welding systems have already been demonstrated.

Now, the implementation of such systems in the Romanian industrial companies requires particularly studies according the both market and industrial environment requirements. Thus, ROM@IT (ROM@nian Intelligent Technology) is a proposed project regarding a robotic welding system for the Romanian industrial companies. The aim of this paper is to present some of the author's opinions regarding the design framework and defining elements of this kind of systems.

2. PROBLEM STATEMENT

The representative products to be done in the ROM@IT system are a family of commonly tanks made by the 20 OLC iron sheet and 0.75-2 mm thickness with different dimensional characteristics. These tanks can be used to storage the fuel at normal pressure for the industrial and household needs. Each tank has two cavities. On the lower (inferior) cavity there are two profiles type angle with equal wings according to the Romanian STAS 424-86, which are made by hot rolled steel sections OL 37 (STAS 500/2-80). The profiles have to be assembled by welding points and they have the roles to assure the rigidity and positioning of the tank. The two cavities have to be assembled by continuous welding on the tank contour by using the WIG method (Wolfram Inert Gas).

3. STATE OF THE ART

Within a process of manufacturing, the human actions may be included into one of the following categories:

- *Category A*, which provides the continuous carrying out of the technological processes.
- *Category B*, which is specific for the operations with discontinuous character, such as start-stop, load-unload, change-handling tools, storage-extraction, etc.

The main functions performed by the welding robots are the following:

a) *Basic functions.* The actually robots accomplish the welding operation as a part of the A actions category. The specific applications are regarding to the point by point welding as well as the continuous linear or complex contour welding.

b) *Auxiliary functions.* The robots accomplish certain auxiliary operations in the welding processes as the manipulation of pieces for automatic machines of welding, as a part of the B actions category.

The world wide achievements of robotic welding systems are multiple, complex, very efficient and represent a lot of models for the ones who initiate the implementation of these systems.

4. RESEARCH COURSE, METHOD USED AND TECHNICAL SOLUTIONS

Our research is a development of the following methodology steps in order to conceive a robotic welding system (Cojocaru & Kovacs, 1986):

a) The study of pieces to be welded

This stage aims to group the pieces in order to emphasize the functional relationships which must be obtained by assemblage. The pieces of the tank (the two cavities and two profiles) can be grouped as the following:

- The first group contains the pieces without formed links which have to be assembled (the two cavities);
- The second group contains the ensemble made in previous assembly phases (the inferior cavity and welded profiles).

b) The study of assembly geometry

The study of assembly geometry aims to define the reference systems of pieces and assembles in order to establish the main spatial positions according the assemble geometry. In the present case we use the following reference systems:

- *The main system of reference (OXZY)* which is interrelated with the last element of the industrial robot.
- *The secondary system of reference (O'X'Z'Y')* having its origin in the centre of gravity of the assembly to be welded.

c) The study of points and contours of welding

This study has as objective the grouping of the welding points and contours, as well as the definition of the tools of welding. The welding will be done using the following procedures:

- *The welding by points.* Each profile will be temporarily welded by two points on the inferior cavity within a manually conformation station. Then, the first robot will execute all the basic points of welding. The number of these points will result for each specific length of profile according a constant step (for example a step of 20...50 mm).
- *The continuous welding.* The continuous welding on the tank contour will be done by the second robot by using the WIG

method in order to assemble the two cavities after the complete welding of the two profiles on the each lower cavity.

d) The study of flexibility

The study of flexibility is oriented on the following directions:

-*The technological flexibility,* which is the degree of adaptability of the system related to the different products to be assembled. From this point of view the proposed system is a polyvalent manufacturing line equipped in order to achieve the assembly of different variants of a basic tank, and others different products, too.

-*The flexibility of substitution,* which is the capacity to continue the work in a normal or reduced rhythm in the situations of some malfunctioning. From this point of view the continuity of functioning requires the implication of a human operator.

e) The definition of system configuration

We have analyzed some variants of layouts and different types of industrial robots. Based upon a certainty multi-criteria decision making process we have established the ROM@IT system layout shown in the fig. 1 with the following structure:

➢OU: the first manually station which is assisted by a human operator for the conformation welding of the profiles on the lower cavity of each tank.

➢P1: the first robotic welding station with a UNIMATION 6000 robot (UNIMATION INC., USA) placed on the ground for the achievement of the basic points of welding.

➢P2: the second manually station assisted by the same human operator for the conformation welding of the two cavities.

➢P3: the second robotic welding station with a ROMAT 320/350/410 robot (CLOOS, Germany) for the continuous welding by using the WIG method.

➢ P4: the delivery station of the final products (tanks).

The same human operator is presented in the both manually conformation stations in order to emphasize the necessity of his/her interventions. The system presented also includes the specific storage, positioning and transportation equipments (Trif & Joni, 1994; Mudrikova et al., 2007).

The system functioning planning is shown in the fig. 2 and it is established based upon some macro-structural and micro-structural analysis according the personal previous contributions (Ciobanu, 2002). The symbols used have the following significances:

t – the time coordinate; op_j – the operations coordinate; a – the conformation welding of the profiles on the lower cavity; b – the transportation to P1; c – the welding of the each two profiles basic points; d – the transportation to P2; e – the conformation of the two cavities; f – the transportation to P3; g – the continuous welding on the tank contour; g – the delivery of the tank; 1, 2… n: the tank number.

5. RESULTS AND FURTHER RESEARCH

The main results of the authors' research are the following:

a) The development of a general framework in order to design a robotic welding system according the both actually knowledge, and the Romanian market and industrial requirements.

b) The application of the multi-criteria decision methods in order to establish the appropriate welding robots and system configuration.

c) The achievement of certain models of layout and functioning planning which are useful for the both producers and users of robotic welding systems as a tool of decision making process.

d) The using of Computer Aided Design for the system layout.

e) The initiation of actions in order to identify the future robotic welding processes in certain Romanian industrial companies.

The further authors' researches will be oriented towards the following main directions:

a) To develop the actually design framework and collection of models for different flexible systems as welding, manufacturing, etc. for the potential industrial users and educational process, too.

b)To raise awareness and managerial support for the robotic welding systems implementation in the Romanian industry.

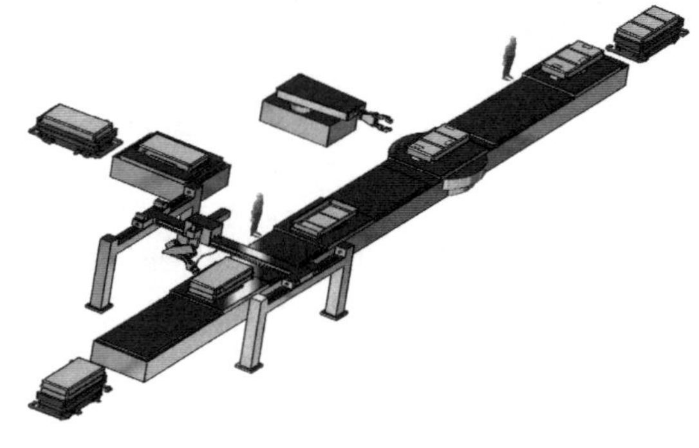

Fig. 1. ROM@IT Project: the system layout.

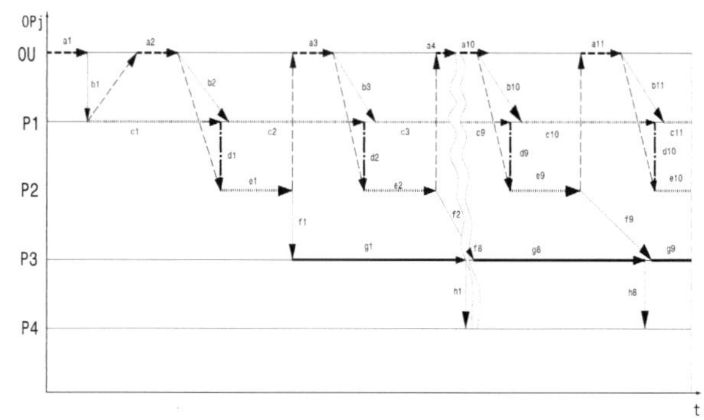

Fig. 2. ROM@IT Project: the system functioning planning.

6. CONCLUSION

The implementation and exploitation of the robotic welding systems is considered one of the most innovative, creative and risky actions by the technical, economical and managerial points of view. But, it is in the same time one of the key of success in the actually hard industrial environment. ROM@IT project is one of the pro-active measures in order to develop the competitiveness and performance of the Romanian industrial companies. It is very important for the both managers and employees to understand and accept the new technical systems as a fundamental way to survive an organization.

7. REFERENCES

Abrudan, I. (1996). *Sisteme flexibile de fabricatie.Concepte de proiectare si management (Flexible Manufacturing Systems. Design and Management Concepts),* Dacia Publishing House, ISBN 973-35-0568-4, Cluj-Napoca.

Ciobanu, R. M. (2002). *Fabricatia flexibila. Elemente de concepere si proiectare (Flexible Manufacturing. Elements of design),* ISBN 973-8075-36-X, Performantica Publishing House, Iasi.

Cojocaru, G & Kovacs, F. (1986). *Robotii in actiune. Probleme ale sintezei sistemelor de fabricatie flexibila (Robots in Action. Syntesis Problems of the Flexible Manufacturing Systems),* Facla Publishing House, Timisoara.

Mudrikova, A.; Kostal, P. & Velisek, K. (2007). Material and informational flow in flexible manufacturing cell, *Annals of DAAAM for 2007 & Proceedings of the 18th International DAAAM Symposium,* Katalinic, B. (Ed.), pp. 485-486, ISBN 3-901509-58-5, Zadar, Croatia, 24-27 October, DAAAM International, Vienna.

Trif, I. N. & Joni, N. (1994). *Robotizarea proceselor de sudare (Robotization of the Welding Processes),* Lux Libris Publishing House, ISBN 973-96308-9-8, Brasov.

Annals of DAAAM for 2008 & Proceedings of the 19th International DAAAM Symposium, ISSN 1726-9679
Editor B. Katalinic, Published by DAAAM International, Vienna, Austria 2008
Make Harmony Between Technology and Nature, and Your Mind will Fly Free as a Bird

NC GRINGING MACHINE – IMPLEMENTATION ONTO CLASSIC MACHINE

CIOFU, C[iprian]; CARAUSU, C[onstantin]; NEDELCU, D[umitru] & FILIP, I[ulian]

Abstract: *With the increased automation of manufacturing processes with CNC machining, considerable improvements in consistency and quality have been achieved. CNC automation reduced the frequency of errors and provided CNC operators with time to perform additional tasks. CNC automation also allows for more flexibility in the way parts are held in the manufacturing process and the time required to change the machine to produce different components.*
Key words: *CNC, steppers, machine tools conversion, grinding machine*

1. INTRODUCTION

Today's modern machinery are CNC (Computer Numeric Control) milling machines and lathes. A microprocesser in each machine reads the G-Code program that the user creates and performs the programmed operations. Personal Computers are used to design the parts and are also used to write programs by either manual typing of G-Code or using CAM (Computer Aided Manufacturing) software that outputs G-Code from the users input of cutters and toolpath (Vilan, 2002).

An entire manufacturing technology known as CAD/CAM has developed around the NC concept and, in addition, CNC with its powerful microprocessors and other enabling technologies proffered from the personal computing phenomenon has enabled the NC concept to branch into many variants, even a variant that is essentially record/playback. The latter of which are known in the industry as "teach lathes".

2. INFORMATION

The adopted constructive solution is refered to realise a conversion of a grinding equipment for profiled surfaces with manual control into a NC equipment.

In figure 1 is presented the general view of the transformed grinding equipment. This one is composed by a computer, NC drivers and steppers motors connected to profiled grinding machine through a positioning and clamping device (figure 3).

Mach 3 (http://www.machsupport.com/) is the software package used to translate 2D drawing and to send the G code date to the drivers. The operator uses a CAD and CAM software to execute the 2D drawing and NC program.

The stepper motor has a wide use through his abilities of digital functions. This abilities was proved by wide use of the integrated electronic in machine controls.

The stepper motor is used in numerical control of the machine because it has abilities of the digital-analog electromechanical transformer.

The stepper motor relised the direct transforming of the entrance signal, in digital form, to an by step angular or incremental movement. In this manner the movement of the machine is realized in discontinuous displacement concording through discret command signals. This solutions has the advantage of use as it is without reaction loops for movement control.

An other advantage of this technical solution is low cost of the implementation of this kind of control devices. This advantage is used to implement this kind of solutions into technological educations.

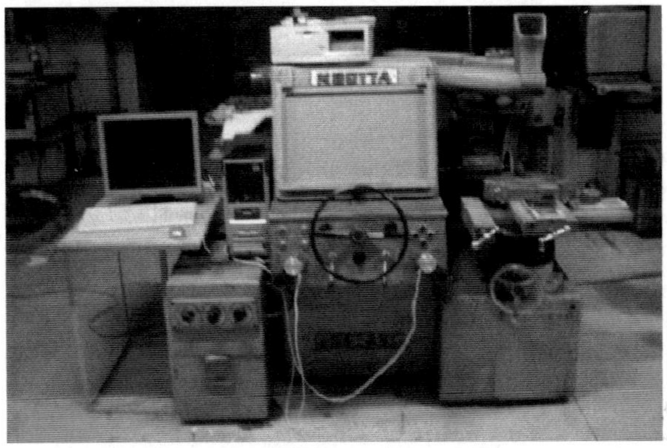

Fig. 1. Ansambling view of the machine tool

The main disadvantage of this system is low efficiencies and high vibrations through his oscillator behavior.

In figure 2 is presented the drivers unit photo.

Fig. 2. Drivers unit: 1 – controllers; 2 – power supply; 3 – motor controls; 4 – PC connections; 5 –AC 220V supply; 6 – stepper motors

An DC power supply is used to charge the stepper drivers.

The steppers motors (http://en.wikipedia.org/wiki/Stepper_motor) are protected by a mechanical safety system for the case of the stuck of the machine spindle (Picoş et. al, 1992).

The draw of the pieces which was machined was realized with AutoCad software. The draw in autocad was printed on a scale of 20:1 – machine optical system scale. This printed draw was used to check in real time the movements of the grinding machine.

To control the machine with Mach 3 Software we obtain the G-code for our draw using LazyCam software. This one convert the .dwg (.dxf) format of the AutoCad draw into G-code.

Mach 3 is a software package which runs on the normal computer, with Windows operating system, 1 GHz processor, 1 Gb RAM and 1024 x 768 pixels resolution. It is preferred to use a desktop system instead a notebook.

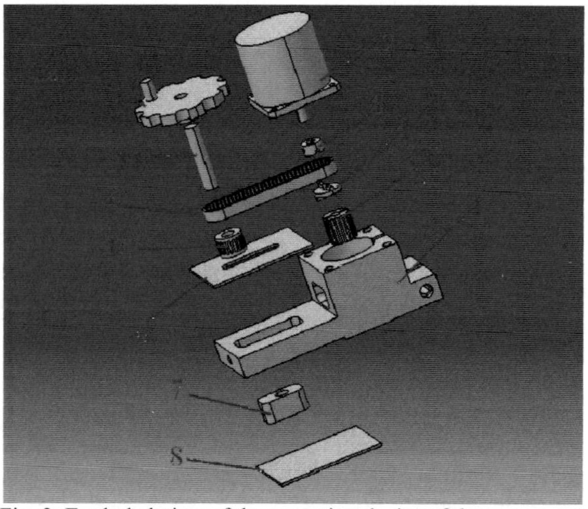

Fig. 3. Exploded view of the mounting device of the stepper motor: 1 – main support; 2 – stepper motor; 3 - machine conector; 4 – intermediary conector; 5 – engine conector; 6 – driving belt; 7 – intermediary part; 8 – lower clamping part; 9 – upper clamping part; 10 – gear; 11 – crank holder; 12 – crank.

This can be used for an other function i.e. to drawing into a CAD/CAM software when Mach 3 is not running.

The data are send from the PC to driver through parallel or serial port.

In figure 4 and figure 5 is presented the grinding process of the profiled grinding machine during the NC control.

To put in value the advantage of the NC machine versus initial control of the grinding machine we made parts on both system and compare them.

The results shown contignous surfaces on computer controlled.

In future, on this machine could be implemented some automatizations systems like: dimensions controls, controls in real time of the movments e.g.

To test the comportament of the machine with this new equipment mount on it, we execute a lot of pieces using manual control of the machine – the original system, and computer control of the machine – using stepper motors.

At the end, we observe two main advantages: a superior quality of the surfaces and a higher precision for samples obtained using computer numerical control.

Fig. 4. Working front panel of the grinding machine

Fig. 5. General view of the working grinding machine – NC controlled

3. CONCLUSION

The presented technical solution has follow advantages:
- working parameters ar controlled by computer
- dual working system
- PC control interface to guide the operator during working process
- axis positioning
- ISO, HPGL eg. programing

The main disadvantages of this adopted solution are:
- the clamping system for the piece is manual
- axis limit is manual
- is difficult to identify zero point
- this system does not have a real time wear compensation

This technical solution is the cheapest one possibility to transform an old classical machine in NC machine. This procedure is the most reliable solution for didactical purposes because show the advantages of the computer uses into manufacturing process.

In future, we intend to improve this machine by using a system for real time control of the dimension and shape the grinding wheel.

Also, another improvement it will be a system which should control the real movement of the grinding wheel along X and Y.

After all these improvement will be made, this machine could be implemented in industry with success due to the advantage of precision that could be obtained and also that one operator could control several machines.

Due to didactical point of view, this machine is extremely important to students because they have the acces to a classic machine with manual movement command and also a CNC machine, both in same time.

This technical solution is also usable into smale workshops.

4. REFERENCES

Picoş C., Pruteanu O., C.Bohosievici, V. Braha (1992) *Proiectarea tehnologiilor de prelucrare mecanică prin aşchiere, Mechanical Cutting Technology Design*, ISBN 5-632-00970-2 vol 1, Chişinău,

Vilan Gh. (2002) *Proiectarea unui Circuit de Comanda Pentru un Motor Pas cu Pas, Designing of an Electronic Control Circuit for a Stepper Motor, ISBN 973-32-0772-3* Bucuresti

http://www.grc.nasa.gov/www/RT19934567679/5000/5950yufmtz.html, Accessed: 2007-02-15

http://www.machsupport.com/, *Accessed:* 2008-01-15

http://en.wikipedia.org/wiki/Stepper_motor, *Accessed:* 2007-11-07

Annals of DAAAM for 2008 & Proceedings of the 19th International DAAAM Symposium, ISSN 1726-9679
Editor B. Katalinic, Published by DAAAM International, Vienna, Austria 2008
Make Harmony Between Technology and Nature, and Your Mind will Fly Free as a Bird

RELIABILITY OPTIMISATION FOR FLEXIBLE MANUFACTURING SYSTEM BY USING INTELLIGENT AGENTS

CIOTEA, V[alentin] F[lorin]; CIOTEA, M[argareta] & RAHAU, L[oredana]

Abstract: *This paper describes an optimisation method for flexible manufacturing systems reliability witch use the intelligent agents specialized in optimal temporary flows negotiation for described system. This method working with much type of agents: system supervisor agent, flow agents, working modules or machines agents. In case of failed or in case of structural changes in system, all this agents we negotiate, there by for this situation to obtain a benefit near by maximum benefit of system. In this paper we also describe concepts of commuting flows, redundant flows and reliability factor, required in reliability optimisation of economical-productive systems. The economical-productive systems can be assimilated by viewpoint of their reliability, with slow systems with commutation potential, for any predefined material, energetic or human flows. We define here economical-productive systems by clusters of manufacturing or production systems, witch can interaction in some condition, by forming some lines or networks with flexible dependencies.*

Key words: *reliability, optimisation, intelligent, agent.*

1. INTRODUCTION

Flexible Manufacturing System (FMS) represents a new manufacturing system, which determines a well manufacturing process economically. It is orientated in sense of real products from human society, means that diversify large goods which are produced in low quantity. In system manufacturing the principal conception problem are centred on reliability and automation (Abrudan, 1996). FMS are in general high level investment and in that condition, its reliability problem is very important. The particularity FMS's functioning will generate any particularity in determining our Flexible Manufacturing System. We will make some consideration about FMS Flexible Manufacturing System. Our contribution is determining SFM Flexible Manufacturing System use analytic method, a program from analytic method and simulation (Monte Carlo) method (Ciotea, V. 1997).

2. RELIABILITY OF FSM

FSM functioning can be see on to the many point of view:

- Possibility that one good may out from SFM on one moment (Ciotea, 2003)..

- Possibility that all goods out from SFM on a given moment;

- Possibility that a specified good to came out from SFM on a given moment. We consider here the first point within tree, because it implies a big volume of processing.

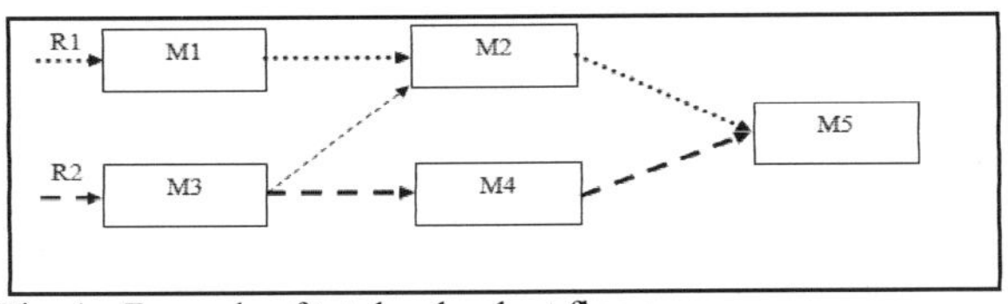

Fig. 1. Example of total redundant flows

3. THE EMPLOYMENT OF INTELLIGENT AGENTS IN THE OPTIMIZATION OF FLEXIBLE MANUFAC-TURING SYSTEMS RELIABILITY USING FLOW DYNAMICS

Flow dynamics represent a method witch identified any alternative fabrication flows. This flow are using in case of a flow is down and is necessarily its replacement.

I identify two alternative flows (Ciotea, 2006).:

1.Total redundant flows – it was introduced in system to increase the reliability of a fabrication flow. Total redundant flows can be beneficiate by a new working module, transportation module, specially introduced for increasing flow reliability.

Costs to involve total redundant flows are:

-Implementation cost – with introducing a new working module, transportation module

-Functional costs – with flexibility of manufacturing system, supplementary stock

Implementation of total redundant flows is justified when it was respected condition as sum of implementation costs and functional costs must be less than costs determine when flow is down.

2.Partial redundant flows – it was introduced in system to increase the reliability of a fabrication flow without introducing a new working or transportation modules. Partial redundant flows can be beneficiate only by flow witch exist in system. In this case implementation costs are null. This method increases the reliability of flexible manufacturing system less than first method, but also cost it is less than first method.

The number of factors that can influence flow dynamics in flexible manufacturing systems reliability is relatively high. Thus, one has to take into account all types of operations that can be performed by installations, the load of installations, the type of markers that are being processed in distinct installations, and the benefits that they bring to the system, the processing stages, module reliability, system flexibility costs, storage costs, loses determined by product failure, etc. the high number of factors that may influence flow dynamics and relationships among these makes the use of classic methods of optimization even more complicated. One of the most flexible and elegant method to obtain optimal reliability using flow dynamics, is the use of intelligent agents.

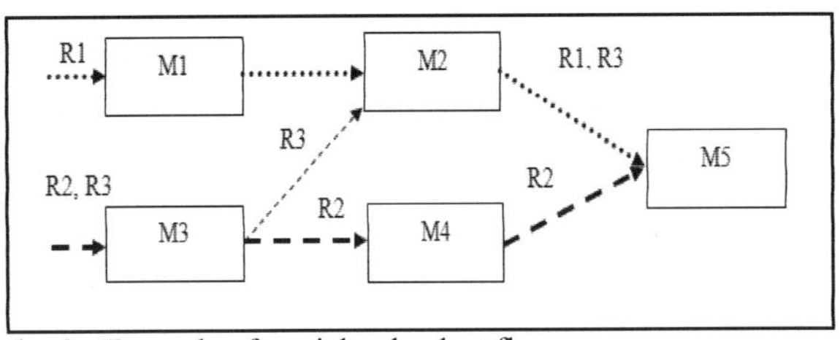

Fig. 2. Example of partial redundant flows

By using negotiation and a minimum of artificial intelligence-based on neuronal networks, fuzzy systems, genetic algorithms or hybrids – these can be satisfactorily used to determine the optimal reliability.

The main advantage of this method is that the introduction of new parameters and of transformations does not trigger alterations of the method, but only lets the agents know, teaches them, about the changes that have taken place.

In principle the method consists of defining the module for each marker that is processed. It is also necessary to define one agent for the entire manufacturing system. The roles of different agent types are as follows:

- The installation agent represents an installation or a group of installations (a module). He may possess knowledge about the operations that the installation can perform, the processing stages for different operations, processing costs, processing quality, degree of occupation, etc.

- The marker agent represents a marker that is being processed. It possesses knowledge about the technologic stages that have to be followed, the type of technological stages, the processing stages for different operations, processing quality, etc.

- The system or subsystem agent is the most important in the system. It must supervise the negotiation procedure between he marker agent and the module agent, so that negotiation is clear and optimal, and the result of the negotiation is as close as possible to the optimal result.

The main duty of the system agent is to trigger negotiation inside the system.

Stages of intelligent agent's method, used in flow dynamics:

1. It starts from a system initially designed as satisfactory in terms of reliability. Then the working modules, the markers that are being processed, the existent flows, the current costs and benefits are defined.

2. Supposing an installation or a module breaks down the system agent is warned. The system agent calculates potential losses determined by the event.

3. The system agent warns the agents of critical markers that are currently on the same flow with the defective installation or module.

4. Marker agents negotiate their distribution on working modules, going from priorities to the less vital. Negotiation with modules of the respective flows is done according to criteria settled for each marker: type of processing, processing quality, processing costs, time, etc. These criteria, as well as the order of negotiation are established by the system agent.

5. The markers' agent designated to negotiate with modules' agent creates a negotiation priority order according to its experience acquired in other negotiations, or to the distance to as certain installation, or simply at random (if none of the previous conditions can be applied).

6. The marker and the system agents establish whether the chosen module interrupts the technologic flow of the procedure, that is, if there is a way to return to the normal manufacturing flow, or whether the marker can be terminated using the new flow. If this condition is met, one can go to the next step.

7. The marker and the module agents establish whether the marker can technically be processed on the respective module. If it is possible one passes to the next step.

8. The marker agent inquires the module agent about the markers that are already being processed. If the marker cannot be taken over, then, with the system agent's approval, those markers that do not represent a priority will be temporarily eliminated.

9. The system agent is informed about the result of the negotiation. In case of a positive result the system agent compares the costs of system status changes. If the costs are higher than the losses determined by the breaking down of the module, then negotiation is considered to be a failure and one must go back to stage 6.

10. The marker agent that put an end to negotiation conveys to the system agent the new technologic flow, so that the later is able to direct marker agents of identical markers, on the new technologic flow.

11. The agents of markers that were suspended from processing will demand the system agent to put the respective markers on the markers-being-negotiated list, in a position corresponding to their priority. The algorithm is resumed from stage 4, until all markers on the list are dealt with.

12. The system agent calculates cost and benefits once the negotiation is over. If the new benefits are higher than those of the previous situation the new situation is considered to be the most propitious.

Agents' involvement is done in real time, as events occur in the system – defaults or changes in the system structure-, which is, called system status. The knowledge possessed by the agents can be ensured through:

- An initial loading of the system with information. In this case each type of agent must posses a certain amount of knowledge. Module agents must know the parameters of the modules they represent, such as the number of installations, types of processing that they can perform, processing costs, markers that are initially processed, initial occupation times, etc. Marker agents must know the stages of the technological flow of the markers they represent, processing costs, potential benefits, etc. The system agent must know the structure of the system, initial and potential manufacturing flows and posses a top of markers' importance.

- Previous answers of the system to different real changes of the system status.

- Previous answers of the system to different simulated changes of the system status.

4. CONCLUSION

After enterprise study of SFM reliability, as probability that one product to leave the system, we can conclusion:

- The last reliability module of system has a substantial influence in system's reliability than other module from system.
- Intelligent agents in the optimization of flexible manufacturing systems reliability method involve a maximum flexibility, low costs for implementation and dynamic structures of system.

In the next step of this project we want to extend area above economical systems by researching reliability optimisation for these systems by using intelligent agents. This method is possible to extend to social systems, where it can possible to simulate social group behaviour, by using one agent for any person or subgroup from studied social group (Ciotea, V. & Ciotea, M., 2007).

5. REFERENCES

Abrudan, I. (1996). *Flexible Manufacturing System,* Publisher Editura Dacia, ISBN 973-35-0568-4, Cluj-Napoca.

Ciotea, V. (1997). *About reliability of FSM,* Petru Maior University, Tg. Mures.

Ciotea, V. (2003). Intelligent agents and FSM, in *Revista Inginerului Economist*, no. 1, page 5-21, Cluj Napoca.

Ciotea, V. (2006). *Research by reliability of FSM – doctoral thesis,* Publisher Tehnical University of Cluj Napoca.

Ciotea, V. & Ciotea, M. (2007). Application for computations of the reliability flexible manufacturing system, In: Inter-Ing 2007, Petru Maior University, page I51-I54, Targu Mures.

Annals of DAAAM for 2008 & Proceedings of the 19th International DAAAM Symposium, ISSN 1726-9679
Editor B. Katalinic, Published by DAAAM International, Vienna, Austria 2008
Make Harmony Between Technology and Nature, and Your Mind will Fly Free as a Bird

CARBON FIBER BASED FILLED EPOXY COMPOSITES

CIRCIUMARU, A[drian]; ANDREI, G[abriel]; BIRSAN, I[ulian] G[abriel] & DIMA, D[umitru]

Abstract *The main goal of our studies is to identify the ways in which it is possible to intervene in order to improve the electromagnetic, thermal, thermo-mechanical and mechanical properties of composite materials. In the case of polymeric matrix composites this goal may be reached or designing the reinforcement by arranging the fibres in various 2D, 3D, 4D structures or by filling the polymer with various powders. This study's approach is a trail and error one because the fillers' influences are not very well known despite the multitude of world wide carried out studies.*
Keywords: *carbon fiber, filled epoxy, CNT, Ferrite, composites*

1. INTRODUCTION

The most important characteristic of a composite is given by the fact that such a material shows properties which are not belonging to its components. Many properties of polymeric composites are results of forming technique and, sometimes consequences of hazard. However the most important aim of nowadays researches is to find a way to change the basic properties of polymeric composites in order to make them more and more reliable in their competition against metals. Fact is that, generally, it is accepted that the properties of formed composites depend on nature and quality of interface (in the case of bi-component composites such as filled composites or fiber reinforced composites) (Schadler et al., 2007). In the case of multi-component composites the problem becomes more complicated (Vasiliev & Morozov 2001). It is obvious that using filled polymer instead of polymer as composite's matrix there will be necessary to take into account, at least, two interfaces polymer-filler one and polymer-fiber one.

Last studies show that an essential role in nature and quality of interface in the case of filled polymers is played by the dimensions of filler's particles (Imry 2002). The easiest way to describe the properties of a multi-component composite is to use twice the mixing rule; on one hand for the filled polymer matrix and on the other hand for the reinforced composite using first results instead of polymer properties (Cîrciumaru 2007). The mixing rule (Callister 1997) takes into account the parameter values of components and the reinforcement/polymer or filler/polymer volume ratio in order to describe the parameter value of formed composite.

Our study is a trail-and-error one designed in order to identify the influence of fillers' concentration and of forming conditions on the formed composite properties.

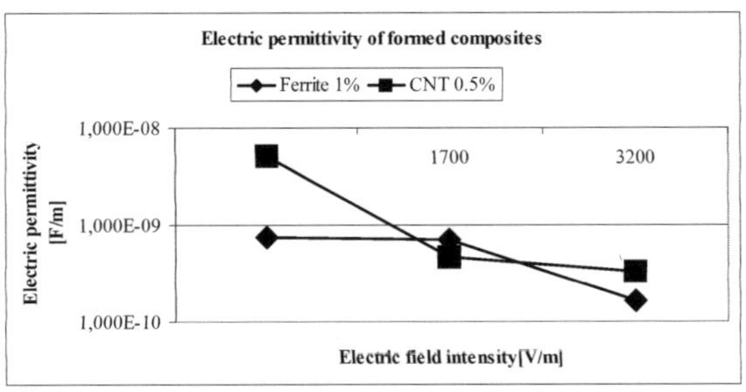

Fig. 1. Permittivity dependence on the electric fieldprints.

2. MATERIALS

Pseudo laminate plates 100 mm x 100 mm x 6 mm were formed using a hybrid technique. Simple type carbon fiber fabric was used as reinforcement and CNT and Ferrite were used as fillers for epoxy Epiphen 4020 RE resin. Taking into account the electric and magnetic properties of CNT and Ferrite, hybrid composites were formed in external electric or magnetic field in order to identify their effects on final properties of composites. The fabric was covered with a thin film of PNB in order to ensure the adhesion of epoxy to the carbon fiber. The reinforcement of each plate is realized from 13 sheets of carbon fiber fabric alternately arranged at 0 or 45 degrees.

Ferrite and CNT were used for fill the polymer in order to investigate their effects on electric permittivity, electrical conductivity and bonding mechanical resistance of formed materials. One interesting aspect is connected to the possibility of fillers particles arrangement in the polymer matrix and it was expected that the electric and magnetic external fields to generate a quasi-order in the filled matrix such as the properties to be modified.

3. MESUREMENTS AND RESULTS

The electric permittivity of each sample was evaluated using a measurement technique described in (Misra 1999). The measurement cell allows the four point measurement method and is respecting the imposed technical specifications.
The same measurement fixture was used for electric resistance measurement according to standard specification according to (Heaney 1999), in order to evaluate the electric conductivity across the reinforcement sheets, also using the van der Paw technique, the electric resistivity along reinforcement was evaluated and the results were presented in (Cîrciumaru et al. 2007). The initial study was carried out taking into account more concentrations for fillers but, the external fields effects were studied only on the most significant concentrations for the two fillers. For higher concentrations than the showed ones the results are not very good, especially because the filler particles clustarization. The electric parameters values were presented in (Cîrciumaru et al. 2007, 2) for three different concentrations of fillers. The high value of electrical conductivity showed by CNT filled matrix composite and, generally electrical conductivity values are effect both of reinforcement and filler conductivity.

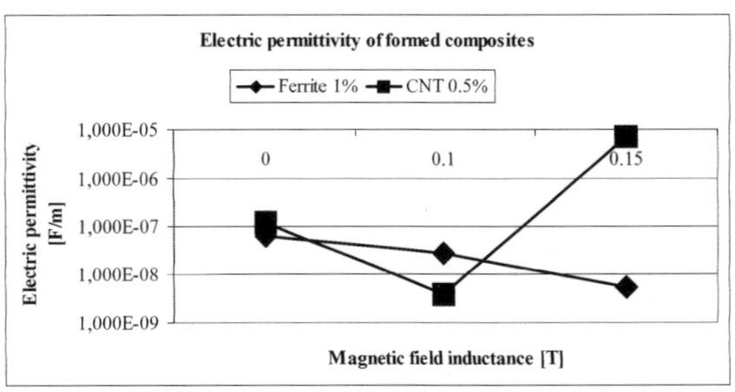

Fig. 2. Permittivity dependence on the magnetic field

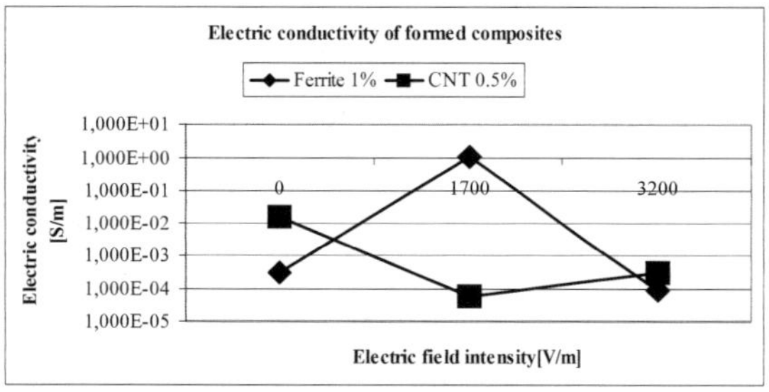

Fig. 3. Conductivity dependence on the electric field

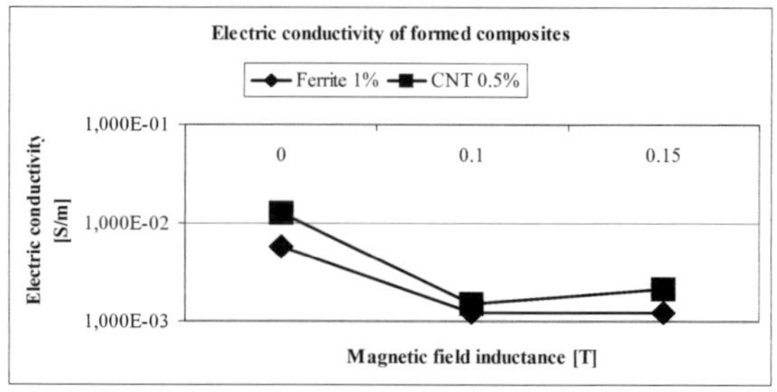

Fig. 4. Conductivity dependence on the magnetic field

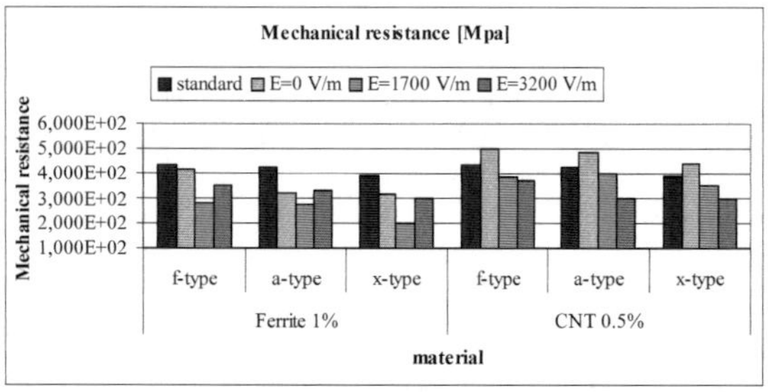

Fig. 5. Mechanical resistance in electric external field

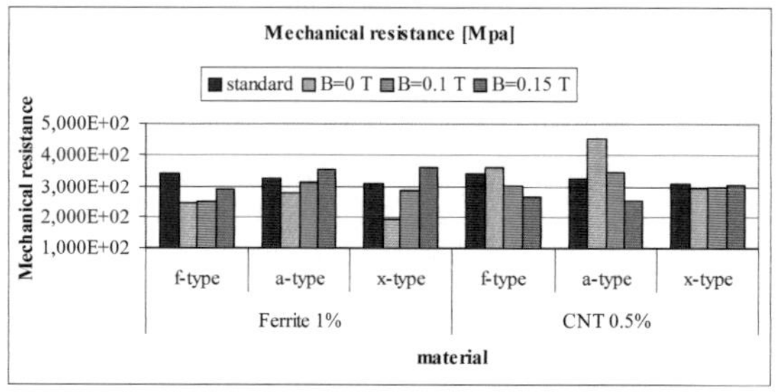

Fig. 6. Mechanical resistance in magnetic external field

Fig. 1 and Fig. 2 are showing the electric permittivity, evaluated at 1kHz and for CNT and Ferrite the electric permittivity are $-6{,}98\cdot10^{-11}$ F/m, respectively $-9{,}173\cdot10^{-11}$ F/m, measured with the same method. The permittivity and the conductivity (Fig. 3 and Fig. 4) had been evaluated on perpendicular direction both on reinforcement and external field direction (between large faces of samples).

Standard DIN EN ISO 10545-4 three point bending test was performed. Three types of samples were cut using a high pressure jet machine: f-type along the external field, a-type across the external field and x-type at 45 degrees relative to external field direction in order to test also the relative isotropy of the material.

The three types of samples were tested in order to point out the influence of reinforcement, a relative decrease of anisotropy because of the even sheets. In Fig. 5. and Fig. 6. the differences between values are not very high. In the both figure there are represented the values of mechanical resistance for a sample without filler (named standard). The differences between the two standard samples are due to the different fabric which were used (one without special treatment and one treated – unfortunately the producer did not agree to specify the nature of treatment). The small values in the case of x-type samples are due to the relatively short fibers and subsequently short interfaces while in other two cases there are long fibers (the length of the sample).

4. CONCLUSIONS

The above presented results allows the conclusion that it is possible to form a composite based on a certain geometry of reinforcement (fiber fabric) using various types of filled polymers. It is clear that these results are just a part of an exhaustive characterization of samples. Mechanical, thermal and thermo-mechanical analysis is necessary.

Another aspect to be taken into account is the fact that because of the fillers and of the reinforcement such a composite material is frequency sensitive *i.e.* the electric and electromagnetic properties depend on the frequency of measuring signal.

The study of mechanical and thermal properties will allow manufacturers to make the right decision about the best reinforcement, best geometry of the reinforcement, fillers concentrations for a given application. Based on these results it is possible to start verifying some models regarding electric and electromagnetic properties of complex composites.

The use of external electric or magnetic fields during the formation could represent an alternative to control the final properties of formed composites due their actions on fillers particles. Also the use of laser could be an alternative to induce particles arrangement.

One of the options for further studies might consist in the use of two ore more than two fillers in various layers of composites ensuring electrical conductive external layers with magnetic core, for example. Also, the use of alternate sheets of reinforcement might lead to improvement of some mechanical properties – such as impact resistance, or fire resistance of PMCs. Also, it would be interesting to identify the possibility to obtain metamaterials by using appropriate fillers in various polymeric matrix, in this case the dimensions and the shape of fillers particles are very important.

5. REFERENCES

Callister, W. D. (1994) *Materials Science and Engineering*, John Wiley & Sons, ISBN 0471736961.

Cîrciumaru, A., *Polymeric Composites Design*, Galaţi, 2007 (unpublished).

Cîrciumaru, A., Andrei, G., Bîrsan, I.-G., Dima, D., *Electric and electromagnetic Properties of Fiber Fabric Based Filled Epoxy Composites*, The Annals of "Dunărea de Jos" University of Galaţi, Fascicle IX, Faculty of Metallurgy and Materials Science, XXV (XXX), May 2007, no. 1, pp. 97-102, ISSN 1453083X.

Heaney, M. B. (1999) Electrical conductivity and resistivity in Webster, J. G. (ed), *Measurements, Instrumentations, and Sensors*, CRC Press, 49, ISBN 0849383471.

Imry, Y. (2002) *Introduction to Mesoscopic Physics*, Oxford University Press, ISBN 0198507380.

Misra, D. K. (1999) Permittivity measurement, in Webster, J. G. (ed), *Measurements, Instrumentations, and Sensors*, CRC Press, 46, ISBN 0849383471

Schadler, S. Linda, Kumar, K. S., Benicewicz, C. B., Lewis, L. Sarah, Arton, E. S. (2007) *Designed Interfaces in Polymer Nanocomposites: A Fundamental Viewpoint*, MRS Bulletin, 32.

Vasiliev, V. V., Morozov, E. V. (2001) *Mechanics and Analysis of Composite materials*, Elsevier, ISBN 0080427022.

Annals of DAAAM for 2008 & Proceedings of the 19th International DAAAM Symposium, ISSN 1726-9679
Editor B. Katalinic, Published by DAAAM International, Vienna, Austria 2008
Make Harmony Between Technology and Nature, and Your Mind will Fly Free as a Bird

STRATFIED HYBRID COMPOSITES

CIRCIUMARU, A[drian]; BIRSAN, I[ulian] G[abriel]; ANDREI, G[abriel] & DIACONU, N[icolae]

Abstract *Fillers are used in order to improve the properties of polymeric composites. It is possible to use the filled polymer in order to form reinforced composites. In both cases it is assumed that using some strategies the fillers' particles are uniform distributed in the volume of matrix. Clearly, using filled polymers the mechanical properties of formed composites will be changed. This study is about the investigation of filled epoxy layers' position effect on the final properties of formed material. Electrical conductivity, dielectric permittivity, bending mechanical resistance and bending apparent elastic modulus were evaluated.*
Keywords: *mixed fiber fabric, epoxy resin, carbon black, talc*

1. INTRODUCTION

Both for engineers and physicists a new challenge arose in last years: how to improve the electro-magnetic properties of a composite. The necessity of metal replacement in aircraft and spacecraft application requires not only excellent mechanical properties but also good electromagnetic properties. More, it seems to be possible to obtain composite materials having electromagnetic properties better than metals ones. There are many researches and also many results published in various journals covering a range of interests from physics to engineering. There are also many papers concerning the modeling of electric properties of composites in order to ensure the design of composites (Kim & Mai 1998). However it is considered that such an approach is just a partial one because the design has to take into account not only those properties but also the mechanical and thermal properties. Obviously this is a very difficult approach and it can be found an extensive study over its complexity (Scadler et al. 2007). One of the cheapest and most convenient methods is to fill the polymer with various powders (metal powders, CNT, carbon black, clay, ferrites etc.) (Gay 1991). The cited paper is an excellent review of actual orientations in the domain of filled polymer composites and it emphasizes the importance of filler dimensions and shapes bringing in attention the importance of interface. One of the most important assessments in last years is the highly contribution of interfaces at the macroscopic properties of composites (Hubca et al. 1999).

The response of a composite to an external change is not only the superposition of its parts responses. That is why it was considered that is important to identify the changes induced by the use of two fillers in the same material.

2. MATERIALS

This is just a "trail and error" approach viewed as a starting point both for further studies and decision making in forming a composite with certain properties. It is extremely difficult to mathematically describe a four-components composite even if there are various models for bi-component composites. The aim of this study, based on intuition, is to present some empirical results in order to help the manufacturers in decision making of forming a special composite.

A set of 8 different materials were realized in order to evaluate the influence of filler over the electric and electromagnetic properties of the composite materials. Excepting the standard sample (NULK) there were made other seven samples using talc and carbon black at the same concentration as fillers for epoxy resin. Two of these samples are formed, in the way above described, but using filled epoxy as matrix. One time the epoxy was filled with Talc (TALC) and the other time with carbon black (CARB). Also, a sample was formed using both fillers (TACA). Considering that fillers were well distributed in polymer these four samples might be considered of near homogenous matrix. The other four samples, formed in order to identify the influence of matrix structure, were realized alternating the matrix. So, there are two samples in which the external three sheets of reinforcement were imbued with Talc or Carbon Black filled epoxy while the internal seven were imbued with, respectively, Carbon Black (CARC) or Talc (TALN) filled epoxy. From the matrix point of view the structure of such a sample is 3T-7C-3T or 3C-7T-3C. The last two samples are formed by using alternatively Talc filled epoxy imbued sheets and Carbon Black filled epoxy imbued sheets. One has the external layers of Talc filled epoxy (TBON) the other one has the external layers of Carbon Black filled epoxy (CALC). From the matrix point of view the structure of these samples is T-C-T-... or C-T-C-...-T-C.

Sheets of kevlar and carbon fiber fabric were used as reinforcement. The fabric consists on alternate untwisted fascicles of kevlar and carbon filaments as warp and fills and is of simple type and was covered with a thin film of PNB in order to ensure the adhesion of epoxy to the fibers.

3. MESUREMENTS AND RESULTS

The electric permittivity and the electric conductivity of each sample were evaluated using standard measurement techniques described in (Misra 1999) and (Heaney 1999).

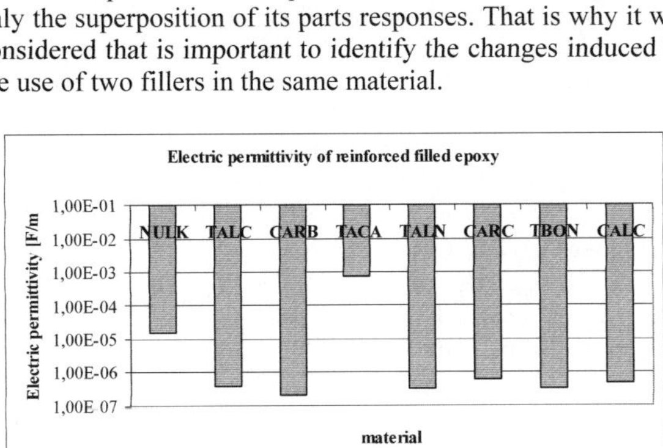

Fig. 1. Electric permittivity of formed composites

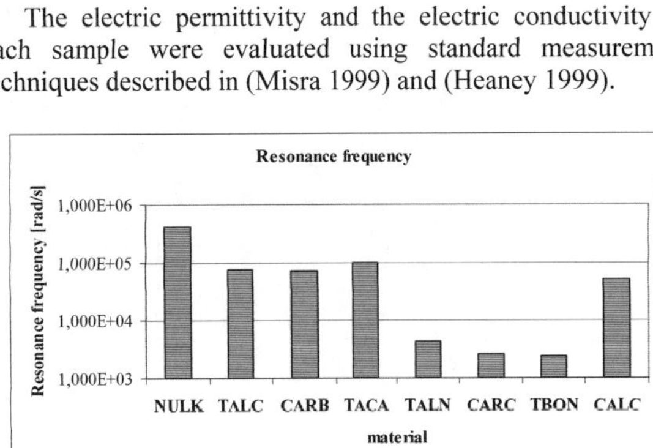

Fig. 2. The resonance frequencies for formed materials

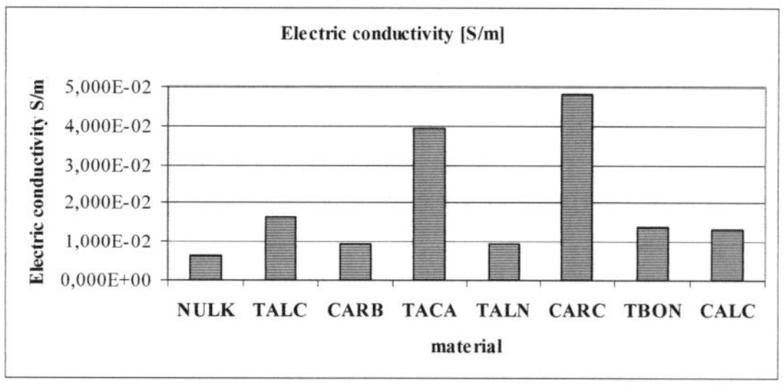

Fig. 3. The electric conductivity of formed materials

Also, one interesting parameter is the resonance frequency evaluated as $\omega = 1/\sqrt{LC}$ where L is the measured value of electric inductance and C is the measured value of electric capacitance.

Complete description of a material, from the electro-magnetic point of view, means not only the electric permittivity and electric conductivity but also the magnetic permeability and their frequency dependences. The measurement fixture allows the direct reading of electric inductance and capacitance and, as immediate consequence, the evaluation of resonance frequency.

Standard DIN EN ISO 10545-4 three point bending test was performed. Three types of samples were cut using a high pressure jet machine: f-type along the external field, a-type across the external field and x-type at 45 degrees relative to external field direction in order to test also the relative isotropy of the material. The main challenge in order to perform this test was to cut samples from the formed plates being known the low machinability of fibre reinforced composites. The problem was solved cutting the samples on a high pressure water jet machine.

As a major deficiency of the polymeric composites the electric conductivity is the one which has to be improved in order to use PMCs instead of metals.

All the samples were formed using the same concentration of fillers excepting the double filled one for which the weight ratio is double. It can be noticed that the last two samples in which the filled epoxy layers alternate have almost the same behaviour while the best result are obtained when both fillers are used.

Using the three point bending test the Bending mechanical resistance and the apparent elastic modulus were evaluated as is recommended in [3], $E = F[N]/y[mm]$ in any point of the linear zone of the curve $F(y)$ were F is the applied force and y is the travel. It must be underlined that all the curves $F(y)$ have almost the same profile presenting a linear zone and two peaks. The mechanical resistance are evaluated at the first peak while the elastic modulus is evaluated at the middle of linear zone.

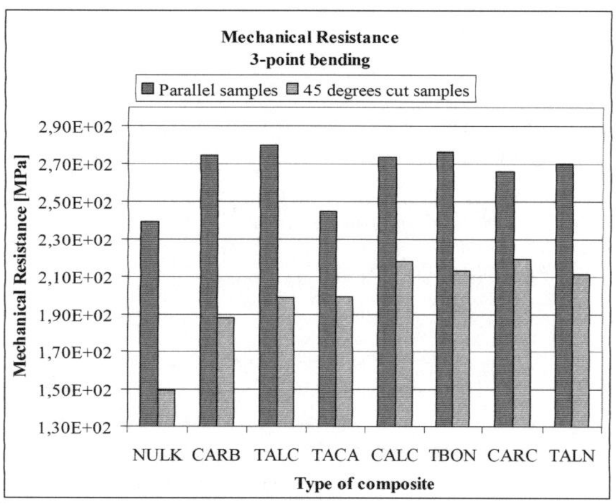

Fig. 4. Mechanical resistance

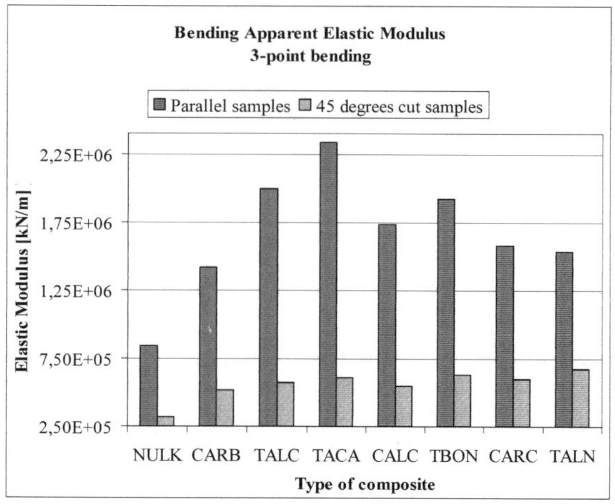

Fig. 5. The apparent elastic modulus

4. CONCLUSIONS

It is obvious that filling the epoxy resin the mechanical properties (at lest the ones regarding the bending) are improved. In the case of TACA materials the filler's concentration was double because we used both fillers in the same amount as in other samples. An interesting conclusion is that increasing the fillers' concentrations the bending apparent elastic modulus will be increased but the mechanical resistance is not very good. Regarding the bending mechanical resistance it can be noticed that the use of the talc leads to better results. Also, for the last two types of samples in which the resin layers are alternated we may say that the mechanical parameters are not changed. In this case important are the other properties (such as electromagnetic or thermal) in order to make the right decision.

It was our concern, based on known properties of kevlar and carbon fibers, to use the mixed fabric in order to evaluate the effects of fillers. The presence of distributed long kevlar fibers leads to a lower machining possibility. It is true that kevlar fibers ensure the impact resistance but, because of their machining resistance, the composite pieces or structures have to be directly formed, excluding post machining processes. Using fillers as carbon black or CNTs in small amounts it can be obtained surfaces with better electric conductivities and using talc it can be obtained materials with harder cores. The use of the nanopowders might be the solution of properties control of the composites.

Using various fillers for various sheets of resin in the same composite seems to be possible to stratified like materials having the same type of reinforcement-matrix interface and acting more like bulk materials than as laminate composite materials.

5. REFERENCES

Gay, D. (1991) *Matériaux composites*, Hermes, Paris.

Heaney, M. B. (1999) Electrical conductivity and resistivity in Webster, J. G. (ed), *Measurements, Instrumentations, and Sensors*, CRC Press, 49, ISBN 0849383471.

Hubca, G., Iovu, H. (1999) *Materiale compozite*, (*Composite Materials*), Editura Tehnică, Bucureşti.

Kim, J.-K., Mai, Y.-W. (1998) *Engineered Interfaces in Fiber Reinforced Composites*, Elsevier.

Misra, D. K. (1999) Permittivity measurement, in Webster, J. G. (ed), *Measurements, Instrumentations, and Sensors*, CRC Press, 46, ISBN 0849383471

Schadler, L. S., Kunar S. K., Benicewicz, B. C., Sarah L. Lewis, Harton, S. E. (2007) *Designed Interfaces in Polymer Nanocomposites: A Fundamental Viewpoint*, MRS Bulletin, V. 32, pp. 335 / 340.

Annals of DAAAM for 2008 & Proceedings of the 19th International DAAAM Symposium, ISSN 1726-9679
Editor B. Katalinic, Published by DAAAM International, Vienna, Austria 2008
Make Harmony Between Technology and Nature, and Your Mind will Fly Free as a Bird

INFLUENCE OF REINFORCEMENT ON THE PROPERTIES OF FILLED EPOXY COMPOSITES

CIRCIUMARU, A[drian]; BIRSAN, I[ulian] G[abriel]; ANDREI, G[abriel]; BRIA, V[asile] & CRUDU, L[iviu]

Abstract *Tailoring the electromagnetic properties of PMC is one of the most important aims in composites researches. One of the most common solutions is to fill the composite's matrix with various powders. In this case the dimensions of filler's particles are important while they determine the dimensions of interface changing not only the electro-magnetic properties but also the mechanical and thermal properties. When a PMC is designed the properties of reinforcement, fillers and polymer have to be taken into account. The current research is focused in showing the influence of filled polymeric layers over the properties of fabric reinforced composite. Both carbon fiber fabric and Kevlar and carbon fiber fabric were used as reinforcements. The symmetry of filled polymeric layers was changed in order to point out the influence of fillers over the composite's properties and to investigate the possibility of modifying the electro-magnetic properties form the surface to the middle plan of the sample.*
Keywords: *mixed fiber fabric, carbon fiber fabric, epoxy resin, nanosized fillers*

1. INTRODUCTION

While the major aim of composites is replacing the metals in aircraft, special applications the electromagnetic properties of such materials are very important. As it is well known metals show high electric conductivities while the composites (excepting metal matrix, metal reinforced or metal-metal ones) show poor electric conductivity. It is obvious that the polymeric composites are showing extremely small values of electric conductivity since their matrix are traditionally used as insulators but their major advantage is the low specific weight. Polymers are also relatively sensitive to the action of aggressive media and show high values of water absorption. The most common solution to improve the polymers' properties is to fill them with various solid systems (such as small metallic spheres) or powders obtaining filled composites, according to (Jones 1999). The mechanical properties of composites are improved in the case of reinforcements (reinforced composites or fibrous composites). There are many studies regarding the design of the reinforcement in various geometries (2D, 3D, 4D) in order to obtain dedicated composite structures (Vinson & Sierakowski 1986) and regarding the mechanical analysis of laminate composites (Constantinescu & Alămoreanu 1995, Vasiliev & Morozov 2001). In the case of laminate composites, the use of reinforcements induces a high level of anisotropy while the use of fillers has the opposite effect.

Combining both alternatives of polymers' properties improvement it is possible to obtain better materials but the problem is to predict the final properties which generally are unknown. The design approach is based on the knowledge of components' properties and on the researchers' abilities to preview the major effects of their utilisation. It is generally accepted that the most important feature of the composites is the reinforcement/matrix or filler/matrix ratio (weight or volume) and this parameter is used in order to describe the macroscopic properties of finite material (Callister 1994, Morozov 2004).

2. MATERIALS

A set of 10 different materials were realized in order to evaluate the influence of filler over the electric and electro-magnetic properties of the composite materials. The results are showed in (Cîrciumaru et al. 2007). In order to identify the effects of reinforcement over composite's electric and electromagnetic properties two types of samples were formed. Both types of samples have reinforcements of 13 sheets of simple type of fiber fabric. The mixed fabric is realized of alternate untwisted fascicles kevlar and carbon filaments. For each type of reinforcement four types of matrix were realized using filled epoxy resin in various setups. The samples were realized through layer-by-layer method. In this case the matrix was realized, each time, by using the same concentration of various fillers (CNT, Ferrite, Talc) but the filled resin was used in layers. In fact, the samples are named with four characters the first one denoting the type of reinforcement (K for Kevlar-carbon fiber fabric and C for carbon fiber fabric). The other three characters are denoting the epoxy's filler (C for CNT, F for ferrite, T for talc). Assuming the symmetry of reinforcement reported to medial plane there were used, for example, three layers of Ferrite filled epoxy (external layers), three layers of CNT filled epoxy (middle layers) and two of Talc filled epoxy (intermediate layers). So, the structure of the sample from the matrix point of view is 3F-2T-3C-2T-3F and for carbon fiber fabric reinforcement the sample is CFTC.

In the C-type samples there are alternate 0 degrees 45 degrees sheets of reinforcements wile in K-type of samples all sheets are placed such as fill and warp are parallel. In order to ensure the adhesion of epoxy to the two types of fibers the fabrics were covered with a thin film of PNB rubber.

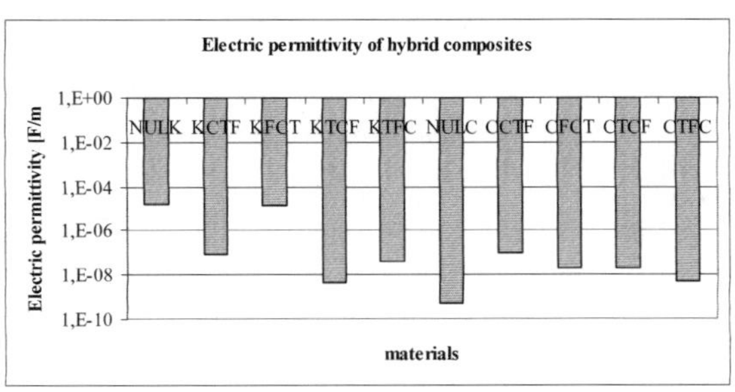

Fig. 1. Electric permittivity of hybrid composites

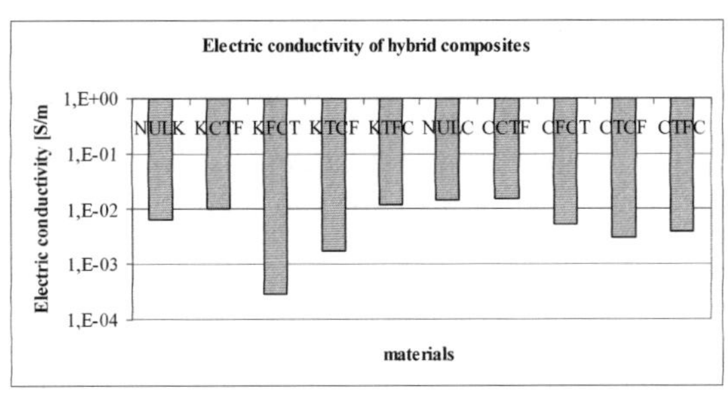

Fig. 2. Electric conductivity of hybrid composites

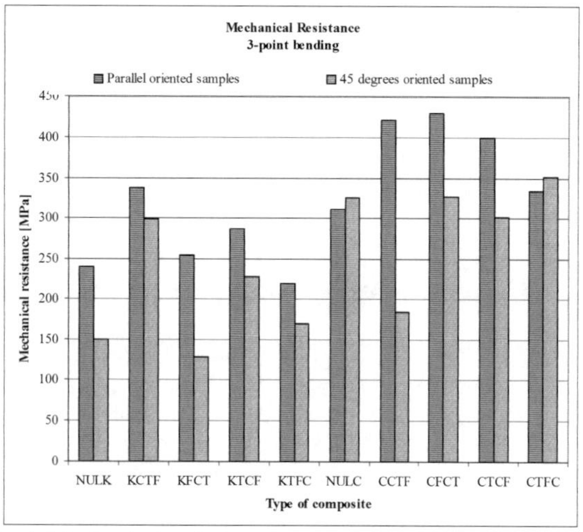

Fig. 4. Mechanical resistance

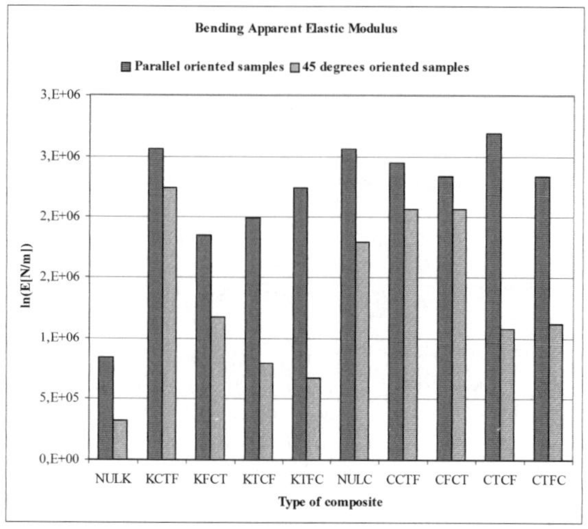

Fig. 5. Bending apparent elastic modulus

3. MESUREMENTS AND RESULTS

Standard DIN EN ISO 10545-4 three point bending test was performed. Three types of samples were cut using a high pressure jet machine: p-type along the warp, a-type across the warp and fill at an angle of 45 degrees. The electric permittivity of each sample was evaluated using a measurement technique described in (Misra 1999, Heaney 1999). The measurement cell allows the four point measurement method and is respecting the imposed technical specifications. The same measurement geometry is used in order to measure the electric resistance and evaluate the electric conductivity. The apparent elastic modulus is evaluated and it is a measure of composite properties in the domain of linear deformations.

4. CONCLUSIONS

We might say that in the C-type samples the degree of anisotropy is decreased. In both types of samples can be easily noticed the effect of external talc filled layers. Analyzing the graphs in above page the main conclusion is the known one about the improving of mechanical properties by using various oriented sheets as reinforcement. Also we may conclude that the solution with more than one layer of filled matrix leads to weak results than the alternate layers one.

The carbon fiber fabric seems to be a better reinforcement solution in order to form strong composites. An impact analysis has to be performed in order to identify chock resistance of such materials. Also we intend to develop our research by forming kevlar-carbon fiber fabric based composites but alternating the orientations of the sheets. It is our concern that it can be found the better concentrations combination of two or more fillers, in the same polymer, in order to reach the intended properties of the final material.

Thermal and thermo-mechanical analysis might be performed in order to refine the solution about composition and the architecture of pseudo-laminate plate. We use the term pseudo-laminate because in our case we cannot say that we used laminae in order to obtain the final laminate, in fact our reinforcement sheets cannot be treated as laminae.

Analyzing the graphs the main conclusion is the known one about the improving of mechanical properties by using various oriented sheets as reinforcement. Also we may conclude that the solution with more than one layer of filled matrix leads to weak results than the alternate layers one. The carbon fiber fabric seems to be a better reinforcement solution in order to form strong composites. An impact analysis has to be performed in order to identify chock resistance of such materials. Also we intend to develop our research by forming kevlar-carbon fiber fabric based composites but alternating the orientations of the sheets. It is our concern that it can be found the better concentrations combination of two or more fillers, in the same polymer, in order to reach the intended properties of the final material. Also it might be possible a design having as result a composite structure into which the reinforcement plays the role of electric circuit. Using small amounts of clay we succeed in improving the nanosized particles dispersion into the polymeric matrix and identifying other possibilities to control the electric properties (Cîrciumaru et al. 2008).

Thermal and thermo-mechanical analysis might be performed in order to refine the solution about composition and the architecture of pseudo-laminate plate. We use the term pseudo-laminate because in our case we cannot say that we used laminae in order to obtain the final laminate, in fact our reinforcement sheets cannot be treated as laminae.

5. REFERENCES

Callister, W. D. (1994) *Materials Science and Engineering*, John Wiley & Sons, ISBN 0471736961.

Cîrciumaru, A., Andrei, G., Bîrsan, I.-G., Dima, D., (2007) *Electric and electromagnetic Properties of Fiber Fabric Based Filled Epoxy Composites*, The Annals of "Dunărea de Jos" University of Galați, Fascicle IX, Faculty of Metallurgy and Materials Science, XXV (XXX), May 2007, no. 1, pp. 97-102, ISSN 1453083X.

Cîrciumaru, A., Bîrsan, I.-G., Andrei, G. (2008) *Electric Conductivity of Fabric Based Filled Epoxy Composites*, ECTP – 18th European Conferenece in Thermophysical Properties, Pau, France (Poster prezentation).

Constantinescu, D. M., Alămoreanu, Elena, (1995) *Proiectarea plăcilor compozite laminate* (*Design of composite laminate plates*), Editura Academiei Române, București.

Heaney, M. B. (1999) Electrical conductivity and resistivity in Webster, J. G. (ed), *Measurements, Instrumentations, and Sensors*, CRC Press, 49, ISBN 0849383471.

Jones, R. M., (1999) *Mechanics of composite materials*, Taylor & Francis.

Misra, D. K. (1999) Permittivity measurement, in Webster, J. G. (ed), *Measurements, Instrumentations, and Sensors*, CRC Press, 46, ISBN 0849383471

Morozov, E. V., *Mechanics and Analysis of Fabric Composites and Structures*, AUTEX, vol. 4, No. 2, june 2004, p. 60.

Vasiliev, V. V., Morozov, E. V. (2001) *Mechanics and Analysis of Composite materials*, Elsevier, ISBN 0080427022.

Vinson, J. R., Sierakowski, R. L., (1986) *The behavior of structures composed of composite materials*, Martinus Nijhoff Publishers, Dordrecht.

Annals of DAAAM for 2008 & Proceedings of the 19th International DAAAM Symposium, ISSN 1726-9679
Editor B. Katalinic, Published by DAAAM International, Vienna, Austria 2008
Make Harmony Between Technology and Nature, and Your Mind will Fly Free as a Bird

A RESTRICTING APPROACH OF THE RAPID PROTOTYPING BY MATERIAL ACCRETION

CIUCESCU, D[oru] & SIMIONESCU, G[heorghe]

Abstract: *There were developed many methods of rapid prototyping by material accretion. Despite the increasing rate of their diffusion and of their infusion, the rapid prototyping by material accretion methods present some limits, more or less important. This paper presents the limits of rapid prototyping by material accretion in order to obtain micro-electromechanical systems.*

Key words: *slice thickness, curl distortion.*

1. INTRODUCTION

Since 1982, when Chuck Hull invented stereolithography (figure1), have been developed many methods of rapid prototyping by material accretion, which involve building a part by defining its geometry layer by layer, but between the main methods may be mentioned as follows: stereolithography, selective laser-sintering, laminated object manufacturing, fused deposition modeling (Ciucescu & Bontas, 1998).

The rapid prototyping by material accretion permits to eliminate the aberration which consists for an engineer to conceive an object in three dimensions, to represent it in two dimensions, and then, in workshop, to interpret this representation for making a prototype in three dimensions (Askeland,1984).

These methods and other ones under development assures for the rapid prototyping by material accretion an increasing rate of its diffusion, the spread of these new methods into the broader market place, and of its infusion, the process of developing applications of these new methods to their maximum potential.

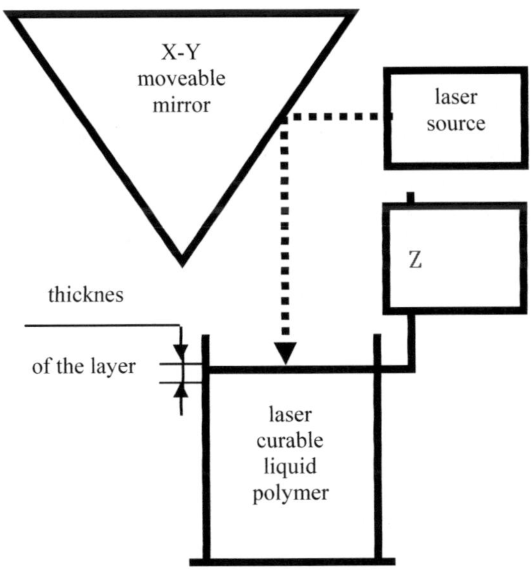

Fig. 1. A schematic overview of the stereolithography

The methods of rapid prototyping by material accretion	Vendor	The number of rapid prototyping by material incress installations
Stereolithography	3D SYSTEMS CMET/ Mitsubishi DEMEC/ Sony EOS GmbH Mitsui Engineering	469 43 28 19 1
Selective Laser Sintering	DTM EOS GmbH	43 4
Laminated Object Manufacturing	Helisys Sparx Kira	52 15 10
Fused Deposition Modeling	Stratasys Sanders	44 1

Tab. 1. The number of rapid prototyping by material accretion installations sold in the year 2004.

Despite these achievements, the rapid prototyping by material accretion methods present some technological limits, more or less important.

This paper presents the technological limits of rapid prototyping by material accretion in order to obtain micro-electromechanical systems.

2. THE MAIN METHODS OF RAPID PROTOTYPING BY MATERIAL ACCRETION

All methods of rapid prototyping by material accretion involve building a part by defining its geometry layer by layer.

The stereolithography method was developed by 3D Systems Corporation (Valencia, California), which presented at AUTOFAC fair in 1987 the first machine: SLA 190. All SLA units direct the laser beam downward onto the free surface of the liquid photopolymer, which passes from liquid state to solid state. The focused spot of the laser, those the diameter is about 200 μm, may be oriented by scanning some mirrors. The laser beam traces, firstly, the boundaries of the slice cross-section being drawn; afterwards accomplishes the solidification of the appropriate cross-section, where the part is full. Non-uniform resin shrinkage produces distortion of the part (Ciucescu, 2007).

The selective laser-sintering method appeared in the year 1994 at a branch of Zeiss Group, named Electro-Optical Systems (EOS GmbH) company, placed in Planegg, near Munich. This method is based on a selective fusing or sintering of small particles by means of high power (50-watt) CO2 laser. A quantity of powdered material is moved upward by a piston within a feed cylinder, and is spread over a over the working area by means of a counter-rotating roller to obtain a thin layer. A laser beam, generating infrared radiation at 10,6 μm, is then directed and focused with appropriate optics, onto the working surface. Some portion of the laser energy is absorbed and transformed into heat. In order to minimize the required laser output, the powder is maintained at an elevated temperature, just below the fusing point. Also, to avoid oxygen contamination of the bonding surfaces and the potential explosion or combustion hazard, the environment is made of nitrogen with maximum 2 % oxygen.

The laminated object manufacturing method appeared at Helisys company (Torrance, California) in the year 1991. The part is fabricated using laminated sheet material. Consecutive layers are joined using an adhesive that is both temperature and pressure sensitive. Each cross-section is cut using a 25-or-50 watt CO_2 laser, emitting in infrared, at a wavelength of 10,6 μm.

The fused deposition modeling (FDM) was developed by Stratatys company (Eden Prairie, Minnesota) in the early years of '90. The material is heated just beyond the melting point in a delivery head. The molten thermoplastic is then extruded through a nozzle in a form of a thin ribbon and deposited in zones, where the part is full.

Some usages of the rapid prototyping by material accretion are presented in figure 2.

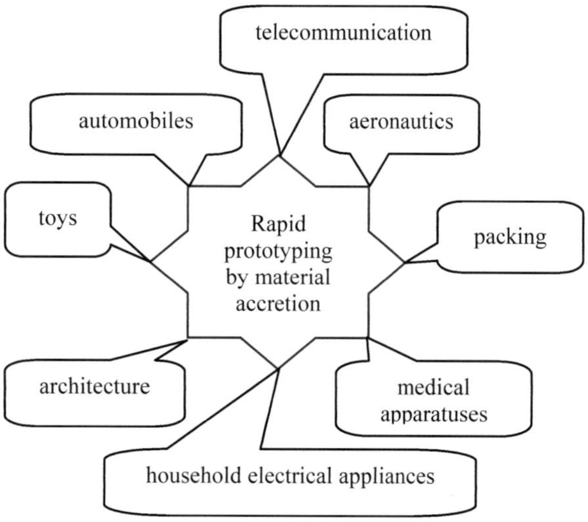

Fig. 2. Some usages of the rapid prototyping by material accretion

3. THE MICROELECTROMECHANICAL SYSTEMS

The micro-electromechanical systems are considered for maximum feature size of 50 μm. Firstly, the structure of micro-electromechanical systems was characterized by two aspects:

-geometric aspects by the fact that the structure was basically limited to planar geometry with a slight variation in the vertical direction;

-material aspect by the fact that the structure was limited to silicon-based materials.

The development of top technologies related to the computer manufacturing. demands a greater geometric and material flexibility.

Various and combined processes like bulk etching and wafer bonding permitted to increase partially the geometric flexibility.

By micro-machining methods and post-release assembly methods are able to make more complex 3D micro-electromechanical systems.

Later, another methods like deep X-ray lithography, laser induced photo deposition and etching permitted to micro-scale three dimensional patterns.

The appearance of rapid prototyping by material incress methods gave a great hope to enable new structure and applications to micro-electromechanical systems.

4. THE TECHNOLOGICAL LIMITS OF RAPID PROTOTYPING BY MATERIAL ACCRETION

In the case of stereolithography, the most important thing is the slice thickness, which now arrived very small, at 50 μm in order to reduce as much as possible stair-stepping errors in the Z axis of the object. Another important thing is the curl distortion of supports which hold the object in place during cladding. The release of Ciba-Geigy epoxy resins Cibatol SL 5170 and Cibatol SL 5180 appears like the best combination of mechanical properties. The advantages of epoxy objects made with these resins are:

-minimal mechanical anisotropy;
-high green strength;
-green dimensional stability;
-improved flatness;
-high curled strength;
-high curled dimensional stability.

The selective laser-sintering method produces a roughness surface due to:

-stair-stepping errors;
-horizontal aliasing errors;
-significant interstices between spherical particles which range in diameter from 50 to 125 μm.

Also, it must to be taken in account that the volumetric shrinkage of crystalline materials is quite large leading to a significant distortion.

The laminated object manufacturing method has the negative aspect that the surrounding volume must be diced to enable to be broken away from the part upon completion. If for large solid parts this operation is not a problem, for finely detailed geometries it becomes a difficult one.

In the fused deposition modeling method, the spooled filament has a diameter of 1,78 mm, which means that the fine parts are obtained with great difficulty.

5. CONCLUSIONS

The micro-electromechanical systems are considered for maximum feature size of 50 μm.

The development of top technologies related to the computer manufacturing. demands a greater geometric and material flexibility.

Even if the appearance of rapid prototyping by material accretion methods gave a great hope to enable new structure and applications to micro-electromechanical systems, up-to-now, the technological limits make them to be not considered for micro-electromechanical systems, exempting the streolithography.

As presented above, the slice thickness, which now arrived very small, at 50 μm , reduces as much as possible stair-stepping errors in the Z axis of the object and permits manufacturing of micro-electromechanical systems with feature size ranged at highest limit of 50 μm.

6. REFERENCES

Askeland, R.D., (1984), *The Science and Engineering of Materials,* PWS-KENT Publishing Company, ISBN 0-534- 029757-4, Boston, Massachussetts

Ciucescu, D., (2007), *The Science and Engineering of Materials,* ISBN 973-30-1465-6, E.D.P., Bucharest

Ciucescu, D. & Bontaş, D., (1998), *The Rapid Prototzping bi Material Accretion,* ISBN 973-30-5467-4, E.D.P., Bucharest

Annals of DAAAM for 2008 & Proceedings of the 19th International DAAAM Symposium, ISSN 1726-9679
Editor B. Katalinic, Published by DAAAM International, Vienna, Austria 2008
Make Harmony Between Technology and Nature, and Your Mind will Fly Free as a Bird

INNOVATIVE SYSTEM WITH ABRASIVE WATER JET

CIUPAN, C[ornel]; MORAR, L[iviu] & CIUPAN, E[milia]

Abstract: *The authors propose to achieve an experimental system for milling, cutting and drilling with abrasive water jet. The system will be developed on the model of a water jet cutting machine with five NC axis. The paper presents the structure of the system and aspects related to the mathematical modelling of the erosion process.*
Key words: *system, mechatronic, processing, energetic, jet.*

1. INTRODUCTION

Milling and lathing processes have been developed using the erosion effect of the abrasive particles in the AWJ cutting process. AWJ milling is in fact a complex erosion process of the manufactured material. This operation allows the manufacturing of parts with complex surfaces such as moulds, gears, etc. Research on international level in this domain aims at two aspects: perfecting the cutting equipment and optimizing the cutting parameters with the help of adequate control strategies; the development of processing procedures by water jet milling and turning.

Which is the problem? The authors propose to achieve an experimental system for ultra precise processing with energetic beam on a mechatronic support. This system uses an abrasive high speed water jet as the energetic beam.

How is it solved? We can notice a lack of machine tools that attain milling operations of complex surfaces, necessary for making molds, world wide. Even the lathes in water jet cutting machines have a drawback regarding the operations of angled cutting, tapped holes and the quality and the surface accuracy.

What is my plan? The innovation and the originality of the project are assured by the concentration of a single system of the following operations: cutting, drilling, milling. The mathematical model for determining the eroded section, the traverse rate f and the productivity is very important for the control of the system.

What is new? The novelty of the project consists of making a mecathronic product, flexible from the point of view of using the energy from the abrasive water jet for the entire range of possible operations (milling, drilling, cutting). Until now there have been, world wide, researches for separate operations, especially cutting, drilling and milling. For assuring the flexibility in the proposed range of operations, the system will use a numerical command processing machine, with 3 axes of movement (x,y,z) and two axes of orientation of the cutting head (A,B). The mathematical model for establishing the eroded material volume and the shape of the processed surface will be an element of high originality and will fill a gap in this domain.

What is next on the to do list? The authors wish to develop the analytical model so that this would allow control over the eroded surface during the milling process.

2. THE SYSTEM

The proposed system enables carrying out the following operations: cutting, drilling, milling (fig. 1). AWJ cutting allows obtaining parts with a surface which is perpendicular on the positioning surface or positioned under a certain angle.

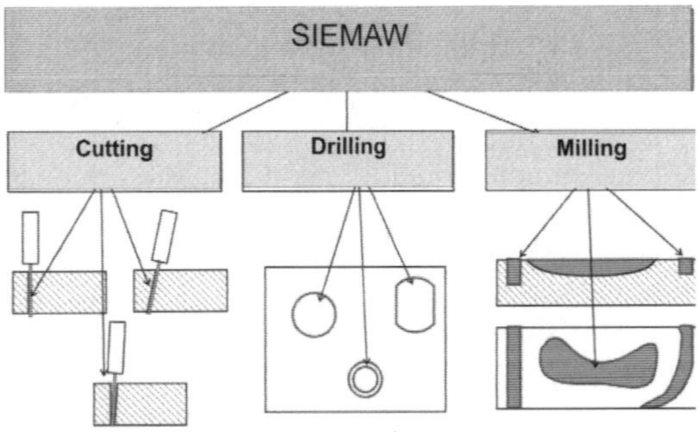

Fig. 1 Specific AWJ process operations

A correction of the kerf is also possible by tilting the cutting head.

Cylindrical, conical or profiled holes may be obtained similarly with the control over interior or exterior surfaces.

Milling operations will be used for obtaining 3D channels or complex surfaces.

The scheme of a waterjet milling machine is composed of parallel robot and a cutting head with a AWJ jet (Ciupan et al., 2007). The main parts of a parallel robot are a fixed platform , a mobile platform linked through six kinematics axis. The mobile platform along with the cutting head is able to perform 3 translation movements and 3 angular movements which allow obtaining the trajectory and the orientation of the AWJ jet.

A mathematical model for the control of the volume of eroded material and of the shape of the surface resulted will predict the processing depth. It will make it possible to process some 3D surfaces of high dimensions, with applications in the automotive field in conditions of precision and efficiency.

3. THE MATHEMATICAL MODEL

Creating a mathematical model that allows driving the system with the purpose of creating the proposed operation range is a very important objective. If for the cutting operations there are numerous models (Momber et al., 1998, Pop et al., 2004), for the milling operations there aren't any mathematical models that allow the control (prediction) over the resulted surfaces after AWJ erosion.

Fowler has developed a mathematical model that studies the abrasive water-jet controlled depth milling of Ti6Al4V alloy (Fowler et al., 2005, Shipway et al., 2005), but which doesn't allow the prediction of the resulted surface. Vikram and Ramesh have developed a model that enables the study of the topography of the resulted surface (Vikram & Ramesh, 2002).

The mathematical model of the eroded material volume control will form the base of calculating the modelling of the resulted shape of the surface. As a basic model for determining the eroded material volume we have the model (Ciupan & al., 2007).

The following steps were set for the development of the mathematical model for controlling the eroded material volume and the resulted surface:

– the mathematical model for the erosion process with abrasive jet;

– control of the eroded material volume and the resulted surface;
– optimization criteria;
– simulations on the virtual model;
– optimizing the virtual model.

The scheme in figure 2 is used to determine the abrasive particles trajectory. The abrasive particles trajectory is determined in relation to an uw reference system, which has the w axis on the previously manufactured surface and the u axis perpendicular on this.

Considering that the particle speed is slowed by the eroding strength coefficient of the material ε_M and that the same force will also have a deflection effect on the jet in the direction of the u axis, the following equations may be written:

$$\begin{cases} \int F \cdot dt = \int m \cdot dv \\ F_u = F \cdot \sin(\theta_t + \alpha) \\ F_w = F \cdot \cos(\theta_t + \alpha) \\ v^2 = v_u^2 + v_w^2 \end{cases}, \quad (1)$$

where:
– F, F_u, F_w are the resistant force that act on the particle and its projections on the axes u, w;
– m is the mass of the abrasive particle;
– v, v_u, v_w are the speed of the abrasive particles and its projections on the axes u, w.

The erosion force is determined by equation:

$$F = \frac{\varepsilon_M \pi d_a^2}{4}, \quad (2)$$

where d_a represents the diameter of an abrasive particle.

The abrasive particles may be considered spherical and equal to the diameter of the jet in order to simplify their study.

A MathCAD program based on relation (1) and (2) was developed for the study of the particles trajectory, the volume of eroded material and the abrasive mass flow.

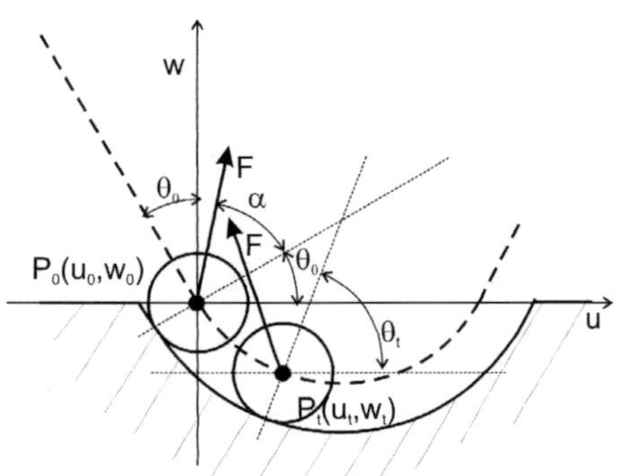

Fig. 2 The deflection of the AWJ jet

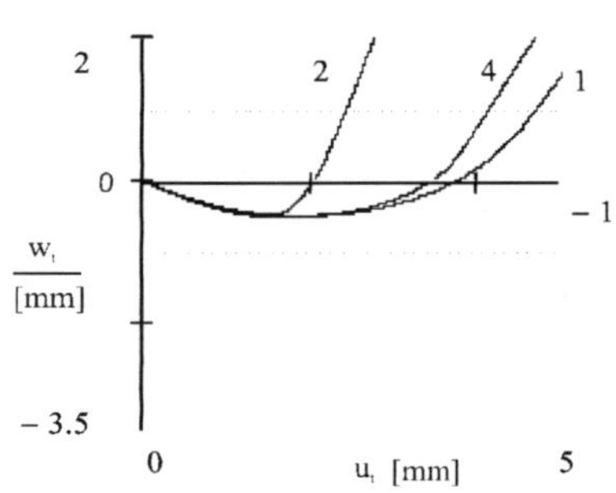

Fig. 3 The abrasive particles trajectory

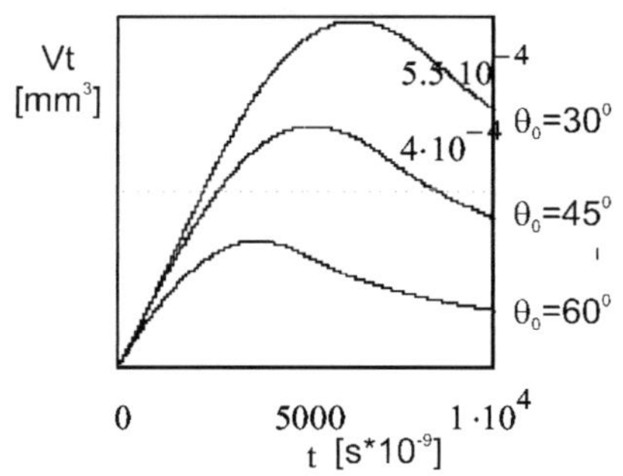

Fig. 4 The volume of eroded material

The eroded material surface is determined by the AWJ jet by taking into account the speed of the jet when exiting the focusing nozzle v_0 and the jet impact angle θ_0.

The u, w coordinates are determined based on the (1) and (2) equations. The abrasive particles trajectory is presented in figure 3 for θ_0=20^0 and v=(600; 800; 1000) m/s.

Figure 4 shows the volume of eroded materials in relation to the jet impact angle θ_0.

4. CONCLUSION

The proposed system has the following advantages:
– precise processing of parts with plane or profiled surfaces;
– the flexibility of the system with respect to milling, drilling, cutting operations;
– using an energetic beam such as the abrasive water jet allows processing of any material without any limitations.

The proposed model is just one of the first steps in building a working analytical model for the control of the eroded surface.

Optimizing the models, especially those for AWJ milling, simulating and optimizing the virtual model will lead to fulfilling the objectives by studying the technological parameters and command strategies in comparison with the AWJ operation range.

Another advantage of this procedure is the non-thermal influence of the processed area. Unconventional processes by which result similar surfaces with those obtained by milling and turning, can be realized by corrosion with abrasive water jet.

5. REFERENCES

Ciupan C., Pop A., Morar L. (2007). The Mathematical Model of Abrasive Water Jet Milling Process. *Advanced Material Research,* vol. 23, pp. 187-190.

Fowler, G. et al. (2005) Abrasive water-jet controlled depth milling of Ti6Al4V alloy – an investigation of the role of jet–workpiece traverse speed and abrasive grit size on the characteristics of the milled material. *Journal of Materials Processing Technology*, vol. 161, pp. 407–414.

Momber, A. W.& Kovacevic, R. (1998). *Principles of Abrasive Water Jet Machining,* Springer-Verlag, Berlin Heidelberg, 1998.

Pop A. (2004). *Study of Computer Control Strategy for Jet Cutting Integrated Systems*, Ph.D. Thesis, Technical University of Cluj-Napoca, 2004.

Shipway, P.H.; Fowler, G.& Pashby, I.R. Characteristics of the surface of a titanium alloy following milling with abrasive waterjet. *Wear*, Vol. 258, Issues 1-4, January 2005, pp. 123-132.

Vikram, G. & Ramesh, N. B. (2002). Modelling and analysis of abrasive water jet cut surface topography. *International Journal of Machine Tools & Manufacture*, Vol. 42, pp. 1345–1354.

Annals of DAAAM for 2008 & Proceedings of the 19th International DAAAM Symposium, ISSN 1726-9679
Editor B. Katalinic, Published by DAAAM International, Vienna, Austria 2008
Make Harmony Between Technology and Nature, and Your Mind will Fly Free as a Bird

SOFTWARE DESIGNED FOR MODELLING AND SIMULATING USING THREE-LAYER NEURAL NETWORKS

CIUPAN, E[milia]

Abstract: *The paper presents a software designed for processes modelling and simulating using neural networks. The used neural networks are perceptron type with three layers. The software describes the back propagation training algorithm of feed forward neural networks. There are two implementation methods: the descent gradient method and the Levenberg-Marquardt method. In addition, the software offers the possibility of processes simulating with the aid of neural models obtained during the training phase.*
Key words: *software, modelling, simulating, neural, network.*

1. INTRODUCTION

The paper presents original software called MLP developed by the author. The goal of this application is solving problems belonging to different domains such as engineering, economy etc. using methods of neural computing.

Process modelling and simulating using neural networks assumes more phases. First, the choice of an appropriate architecture of the neural network is necessary. Then, the network is trained so that it simulates the process as accurately as possible. The last phase is testing and using the network.

The software is divided into more modules which carry out different tasks such as: creating neural network, network training, testing and recall function.

The software was developed in Visual C++ language.

2. CONSIDERATIONS REGARDING THE IMPLEMENTED METHODS

The "MLP" software implements the back propagation algorithm as a training method by the means of two methods: the descent gradient method (GD) and the Levenberg-Marquardt method (LM).

Both methods are iterative and solve an optimum problem: determining the synaptic weights w of the network so that they would minimize the error function expressed as a sum of squares of the other functions (Hagan & Menhaj, 1994):

$$E(w) = \frac{1}{2}\sum_{i=1}^{m}(e_i(w))^2 \tag{1}$$

The GD method is sequential and related to gradient. This assumes a step by step sequential presentation of the training examples. The error is calculated as the difference between the desired output and the effective output for each example. This error is then used for the calculation of the weights modification Δw through back propagation.

The new weights are calculated using the following equation (Dumitrescu & Hariton, 1996; Zilouchian & Jamshidi, 2001):

$$w(k+1) = w(k) + \alpha \cdot \Delta w \tag{2}$$

where α is a constant that ranges from 0 to 1 called learning rate.

The LM method is based on Hessian matrix. This is a batch method. This means that the calculation of the weights modification assumes presenting all of the examples of the training set and determining the total error e(w). This is described by equation (3) (Hagan & Menhaj, 1994):

$$[J^T(w) \cdot J(w) + \mu I] \cdot \Delta w = -J^T(w) \cdot e(w) \tag{3}$$

where:

$e : R^n \rightarrow R^m, e = (e_1, e_2, ..., e_m)$ is an error function,
J is the Jacobian matrix of the function e
μ is a variable damping factor
I is the identity matrix.

The matrix J^TJ is the approximate Hessian of the error function E (Madsen et al., 2004).

An iteration consists of determining a Δw modification of the weights which will lead to a reduction of the error.

3. SOFTWARE TESTING

The software testing was done in two ways: by comparing the results obtained with MLP with the ones obtained with MATLAB 7.0 and by comparison with the mathematical model. Two mathematical functions described by examples 1 and 2 were modelled (Ciupan, 2008).

Example 1: Dampened wave function

A dampened wave function described by equation (4) was studied.

$$f(t) = e^{-\frac{1}{t}\sin t}, t > 0 \tag{4}$$

Because of the random values of the initial weights an average performance was calculated during 6 training sessions, 100 epochs per training. A 30 examples training set was used consisting of (t, f(t)) pairs obtained with the mathematical model (Eq. 4).

Both methods, GD and LM, were used. Other features of the neural models are presented in table 1.

The model with the best performance on the LM method, $6.3*10^{-4}$, was chosen in order to compare the neural model with the mathematical one. This comparison may be observed in figure 2, where f(t) are the theoretical values of the function f given by equation (4) and f(t)R are the values simulated by the neural model.

Method	Network architecture	Desired performance	Reached performance	
			MLP	MATLAB
LM	1-8-1	10^{-3}	$4.0*10^{-3}$	$1.9*10^{-3}$
GD	1-8-1	10^{-5}	$1.0*10^{-4}$	$38.3*10^{-4}$

Tab. 1. Characteristics of the neural models

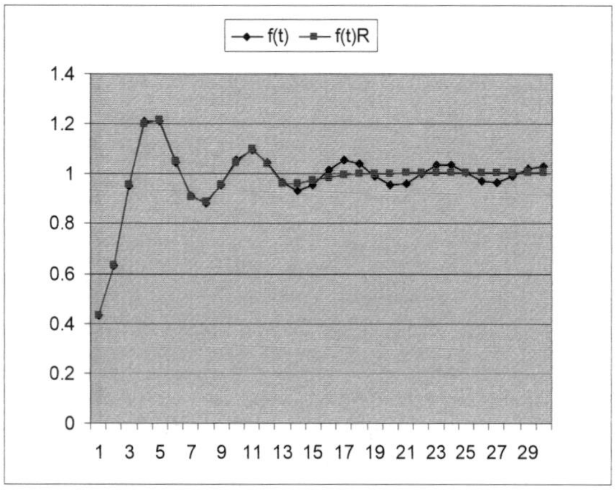

Fig. 2. The approximation of the dampened wave function

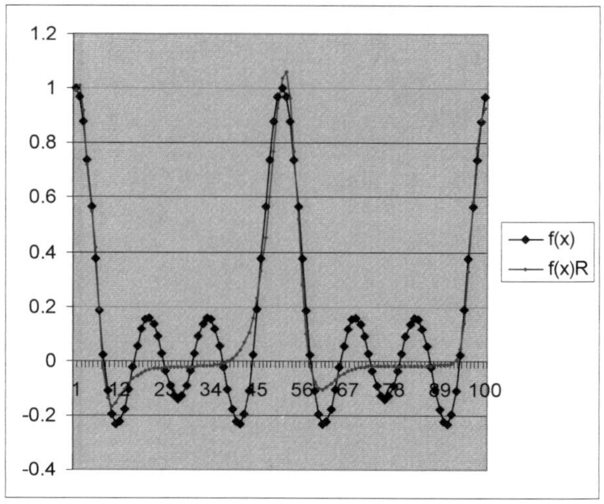

Fig. 3. The approximation of the Dirichlet function

Example 2: Dirichlet function

The general definition of the Dirichlet function is described by equation (5):

$$f(x)=\begin{cases}-1^{k(n-1)}, & \text{if } x=2k\pi,\quad k=0,\pm1,\pm2,\\ \dfrac{\sin(nx/2)}{n\sin(x/2)}, & \text{otherwise}\end{cases} \tag{5}$$

where n is a positive integer parameter.

Consider the case when n=7. The function is periodical in this case and the period is 2π. A 1-8-1 network was trained to approximate the function. The neural network was trained during 5 training sessions, each session consisting of 100 epochs. A training set containing 100 input/output pairs of type (x, f(x)), $x \in [0, 4\pi]$ was created using the mathematical model (5). The value 10^{-3} was chosen for desired performance.

Tables 2 and 3 present the performance reached over these 5 training sessions.

Further testing was done by choosing the best neural model obtained in the training process through the LM method (table 2, session #1, performance $6.0*10^{-3}$). The approximation of the function was carried out by the neural model mentioned above.

Session	Performance	
	MLP	MATLAB
1	0.00607977	0.00681455
2	0.08666790	0.00959747
3	0.08496530	0.01249476
4	0.07041760	0.01129421
5	0.05172580	0.00697923
Average	0.05997127	0.00943605

Tab. 2. The performance obtained with the LM method

Session	Performance	
	MLP	MATLAB
1	0.00314583	0.09752570
2	0.02416600	0.03229670
3	0.02190806	0.11534800
4	0.01805316	0.09075010
5	0.05172580	0.07358620
Average	0.02379977	0.08190134

Tab. 3. The performance obtained with GD method

Figure 3 illustrates the comparison between the simulated values obtained by the neural model and the theoretical values obtained by the mathematical one.

4. CONCLUSIONS

The results analysis shows that:

a) Both programs, MLP and MATLAB, lead to close results when modelling using three-layer neural networks in general.

b) It is impossible to conclude which program, MLP or MATLAB, leads to better results when training is done through the LM method or through the GD method. The results depend on more factors, including the initial random weights. In the case of each program the performance may be increased by more training sessions or by increasing the number of examples.

c) It is possible to notice that in the case of each programs the neural model approximates the theoretical values of the studied functions for those values of the argument for which the values of the function vary sufficiently. For those values of the argument for which the amplitude of the function values is small, the neural model tends to make an average of the theoretical values.

d) The neural model maintains certain features of the mathematical model such as oscillating, dampening or periodicity.

5. REFERENCES

Ciupan, E. (2008) *Integrated Management of the Systems Using Open Control Platforms*. Ph D Thesis, Technical University of Cluj-Napoca, pp. 90-100, 2008.

Dumitrescu, D. & Hariton, C. (1996) *Retele neuronale-teorie si aplicatii (Neural networks- theory and applications).* Ed. Teora, ISBN 973-601-461-4, Bucuresti.

Hagan, M. T. & Menhaj, M. B. (1994) *Training Feedforward Networks with the Marquardt Algorithm.* IEEE Transactions on Neural Networks, vol. 5, no. 6, November 1994, pp 989-993.

Madsen, K.; Nielsen, H.B. & Tingleff, O. (2004) *Methods for Non-Linear Least Squares Problems*. Second Edition. Informatics and Mathematical Modelling, Technical University of Denmark, 2004.

Zilouchian, A & Jamshidi, M. (2001) *Intelligent Control Systems using Soft Computing Methodologies*, ISBN 0-8493-1875-0, CRC Press LLC, 2001

Annals of DAAAM for 2008 & Proceedings of the 19th International DAAAM Symposium, ISSN 1726-9679
Editor B. Katalinic, Published by DAAAM International, Vienna, Austria 2008
Make Harmony Between Technology and Nature, and Your Mind will Fly Free as a Bird

A MODEL FOR THE KINEMATICAL ANALYSIS OF A THREE DEGREES OF FREEDOM MECHANISM

CIUPAN, E[milia]; ITUL, T[iberiu]; MORAR, L[iviu] & CIUPAN, C[ornel]

Abstract: *The paper presents a neural model for the kinematical analysis of a plane mechanism with tree degrees of movement. The modelling consists of two stages. The first stage is choosing a three-layer perceptron type neural network and training it so that it learns a set of training data well enough. The second stage means testing the model obtained through training, during the generalization phase. Both tasks were carried out using personal software.*
Key words: *neural, model, kinematical, analysis, mechanism.*

1. INTRODUCTION

The authors make a kinematical analysis of a three degrees of movement type BONEV mechanism using a neural model.

Figure 1 shows the schema of a mechanism of this type. This consists of two platforms: a rigid platform $B_1B_2B_3$ and a smaller, mobile one.

The movement of the mobile platform with the mobile reference system Oxy is done with the aid of three motor couplings q_1, q_2 and q_3. These produce the modification of the position angles θ_1, θ_2 and θ_3 of the arms. The couplings u_1, u_2 and u_3 are passive because their position is determined implicitly.

It will be possible to move the mobile reference system Oxy tied to the centre of the mobile platform in the workspace and related under an angle φ depending on the position of the motor couplings q_1, q_2 and q_3 (Itul, 2000).

2. THE MATHEMATICAL MODEL

The kinematical analysis used in the field of industrial robots assumes establishing certain equations between the input values (done by the motor drive) and the output values (the position, speed and acceleration of the final effecter).

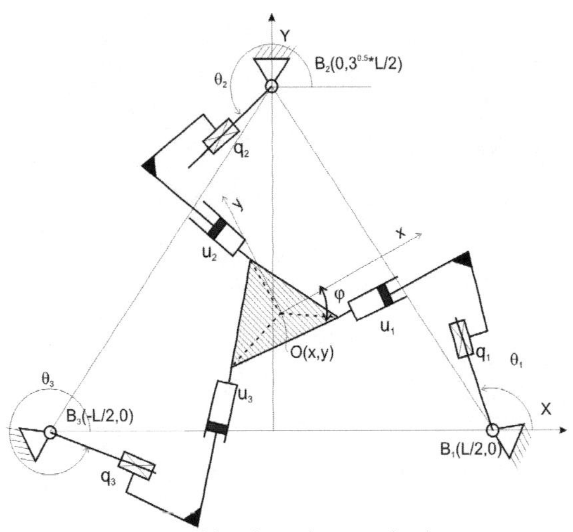

Fig. 1 Three degrees of freedom plane mechanism

The kinematical analysis is comprised of two methods (Ispas & Blebea, 2003):

1. The direct kinematics: the movement laws of the motors are considered to be known and the movement law of the effecter is determined;
2. The inverse kinematics: the movement laws of the effecter are considered to be known and the movement law of the motors is determined.

The mathematical model for the analysis of the inverse kinematics is briefly described by equation (1)-(3) (Itul, 2000):

$$q_1 = \frac{(X_{B1} - X)\cdot\sin(\delta_1 + \varphi) - (Y_{B1} - Y)\cdot\cos(\delta_1 + \varphi)}{\sin\alpha} \quad (1)$$

$$q_2 = \frac{(X_{B2} - X)\cdot\sin(\delta_2 + \varphi) - (Y_{B2} - Y)\cdot\cos(\delta_2 + \varphi)}{\sin\alpha} \quad (2)$$

$$q_3 = \frac{(X_{B3} - X)\cdot\sin(\delta_3 + \varphi) - (Y_{B3} - Y)\cdot\cos(\delta_3 + \varphi)}{\sin\alpha} \quad (3)$$

where

$$\delta_i = \frac{2\pi}{3}\cdot(i-1), i \in \{1,2,3\} \quad (4)$$

and the angles θ_i are calculated by the following equation:

$$\theta_i = \alpha + \varphi + \delta_i \,. \quad (5)$$

The mathematical model for the analysis of the direct kinematics is described by the following system of equations (Itul, 2000):

$$\begin{aligned}
&(X_{B1} - X)\cdot\sin(\delta_1 + \varphi) - (Y_{B1} - Y)\cdot\cos(\delta_1 + \varphi) - q_1\cdot\sin\alpha = 0\\
&(X_{B2} - X)\cdot\sin(\delta_2 + \varphi) - (Y_{B2} - Y)\cdot\cos(\delta_2 + \varphi) - q_2\cdot\sin\alpha = 0 \quad (6)\\
&(X_{B3} - X)\cdot\sin(\delta_3 + \varphi) - (Y_{B3} - Y)\cdot\cos(\delta_3 + \varphi) - q_3\cdot\sin\alpha = 0
\end{aligned}$$

The following constructive feature has been considered for the mechanism that is studied in the following part of the paper: the side of the fixed platform, meaning the side of the equilateral triangle $B_1B_2B_3$ is L=40 cm long.

3. THE NEURAL MODEL

A neural model was created for the inverse kinematical analysis of the mechanism described prior (Ciupan, 2008; Zilouchian & Jamshidi, 2001). The chosen neural network is three-layer type perceptron having architecture 3-50-3. The input layer has three neurons that correspond to coordinates X, Y and φ of the effectors' position. The three neurons belonging to the output layer correspond to motor couplings coordinates q_i, i=1, 2, 3. Good results for a hidden layer consisting of 50 neurons were obtained when the network was trained. The activation functions are sigmoid for the neurons in the intermediate layer and linear for the ones in the output layer.

The training was carried out with a training set having 121 input/output pairs. The training set was obtained using the mathematical model designed in MATHCAD. Back propagation was chosen training algorithm using the Levenberg-Marquardt method (Hagan & Menhaj, 1994). A mean square error of $3.5*10^{-5}$ was reached following training.

The model was tested conceiving some applications. The effecter's movement along certain trajectories in the plane describes by the positions (X, Y, φ) was taken into account. The coordinates of the motor couplings were calculated using the neural model. These were than used as input data for the mathematical model for the direct kinematical analysis and lead to the recalculation of the coordinates (XR, YR, φR).

The approximation error of the robot's trajectory through the neural model results from the comparison of the initial positions (X,Y,φ) with the recalculated positions (XR, YR, φR).

3.1 Application 1: a bunch of straight lines

The trajectories chosen for the testing of the model consist of a bunch of straight lines with different angles of the bow (30°, 45° and 90°). The performance obtained when simulating fit between 10^{-5} and 10^{-3}.

Figure 2 presents the approximation of the theoretical trajectories by the neural model. The approximation errors charts when the effecter moves on a straight trajectory having a 45° angle of bow are presented in figure 3.

3.2 Application 2: a circular trajectory

The robot's effecter movement was simulated on a circular trajectory in the workspace. The approximation error is around 10^{-4}.

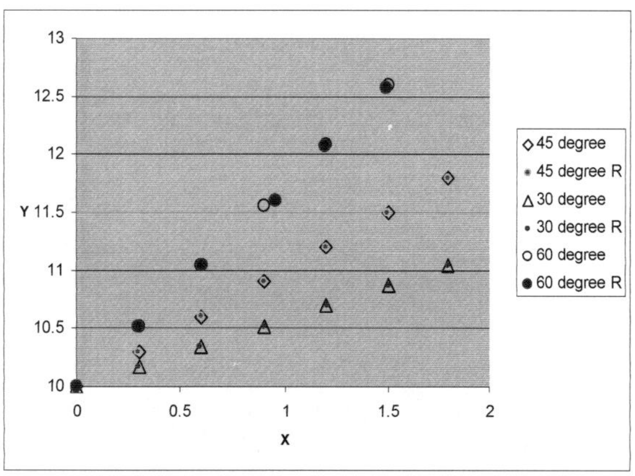

Fig. 2. The approximation of the bunch of straight lines

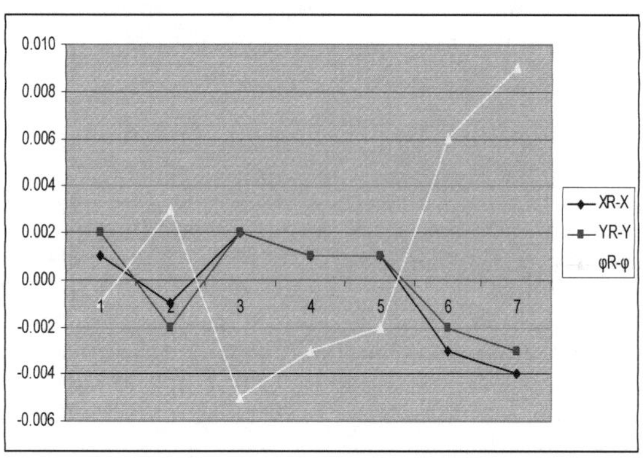

Fig. 3. The approximation errors-straight line at 45° angle

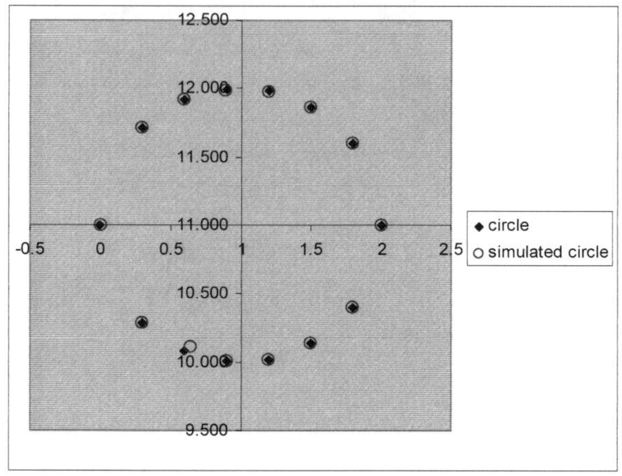

Fig. 4. The approximation of the circular trajectory

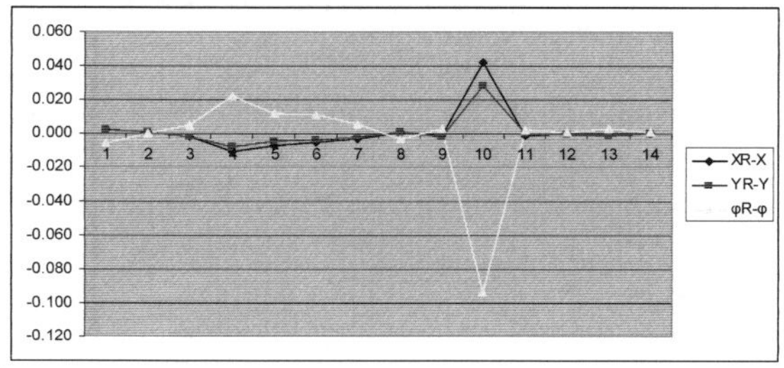

Fig. 5. The approximation errors-circular trajectory

The way in which the model approximates the movement of the robot on this trajectory is shown in figure 4. Figure 5 illustrates the charts of the approximation errors in this case.

4. CONCLUSIONS

Starting from the analysis of the above results as well as from the results of some research that is not presented in this paper, it is possible to affirm the following:

1. The neural model simulates the movement on different trajectories of the plane mechanism described at the beginning well enough;
2. Training the neural network on a small set of examples doesn't ensure a good performance in the generalization stage, even if the training performance is good;
3. Better result may be obtained through training on a larger set of examples that are equally distributed within the workspace;
4. A well built neural model can be used for the control of a mechanism the same type as the one presented in the paper successfully.

5. REFERENCES

Ciupan, E. (2008) *Integrated Management of the Systems Using Open Control Platforms*. Ph D Thesis, Technical University of Cluj-Napoca, pp. 115-128, 2008.

Hagan, M. T. & Menhaj, M. B. (1994) *Training Feedforward Networks with the Marquardt Algorithm*. IEEE Transactions on Neural Networks, vol. 5, no. 6, pp 989-993, November 1994.

Ispas, V. & Blebea, I. (2003) *Robotics. Parallel Robots. Service Robots*. UT Pres, Cluj-Napoca, 2003.

Itul, T. P. (2000) *Roboti paraleli de tip B. Geometrie si cinematica (Type B Parallel Robots. Geometry and Kinematics)*. Todesco, Cluj-Napoca, 2000.

Zilouchian, A & Jamshidi, M. (2001) *Intelligent Control Systems using Soft Computing Methodologies*, CRC Press LLC, ISBN 0-8493-1875-0.

Annals of DAAAM for 2008 & Proceedings of the 19th International DAAAM Symposium, ISSN 1726-9679
Editor B. Katalinic, Published by DAAAM International, Vienna, Austria 2008
Make Harmony Between Technology and Nature, and Your Mind will Fly Free as a Bird

DISTRIBUTED CONCENTRATOR FOR SOLAR POWER GENERATION

CIUPE, V[alentin]; DEHELEAN, N[icolae] & MANIU, I[nocentiu]

Abstract: *The paper presents a solar energy converter based on a distributed solar concentrator and a Stirling engine. The concentrator is made of a matrix of several relatively small Fresnel lenses that are able to automatically track the sun. Heat from the focus point of these lenses is collected and transported to the hot end of the engine.*

Key words: *concentrator, solar, lens, tracking, energy.*

1. INTRODUCTION

The use of solar generated thermal and electric power is an every increasing topic of interest worldwide. The two main areas of interest for electric power are photovoltaics and solar-thermal-electric. So far, the last method gives better results regarding efficiency and cost-effectivness, by using concentrator systems.

Unlike traditional power plants, concentrating solar power systems provide an environmentally benign source of energy, produce virtually no emissions, and consume no fuel other than sunlight. About the only impact concentrating solar power plants have on the environment is land use. Although the amount of land a concentrating solar power plant occupies is larger than that of a fossil fuel plant, both types of plants use about the same amount of land because fossil fuel plants use additional land for mining and exploration as well as road building to reach the mines (NREL, 2001).

In order to determine if a solar concentrator system is feasible in the region of interest, one must check the charts regarding the yearly average insolation and also the average number of clear-sky days, as concentrator systems can only use direct sunlight in order to function.

Figure 1 shows the yearly sum of global insolation, according to the EC's Joint Research Centre (PVGIS, 2008) for Europe and particulary for Romania (the region of interest). The map on the right shows a variation from 1700 kWh/m^2 to 1300 kWh/m^2.

Another factor to be taken into account beyond insolation is the availability of solar radiation in the area. This represents the average number of clear-sky days and the average number of diffuse light days (figure 2). The two maps show for the month of July about 75-80% clear-sky days, and less than 40% of diffuse light days.

Considering the paragraphs above, solar-thermal and solar-thermal-electric can be used in the region and allows for a fairly efficient systems

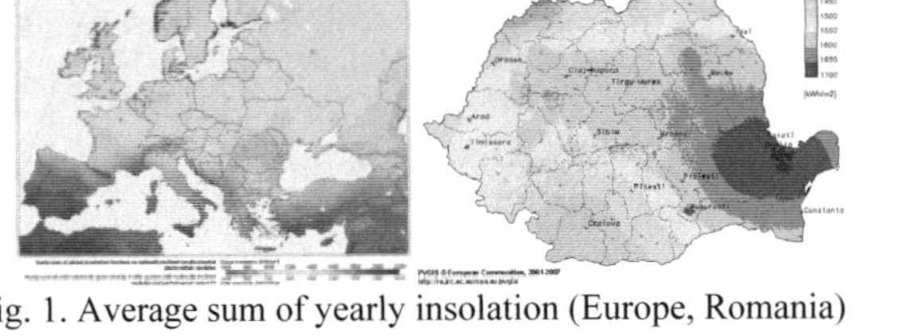

Fig. 1. Average sum of yearly insolation (Europe, Romania)

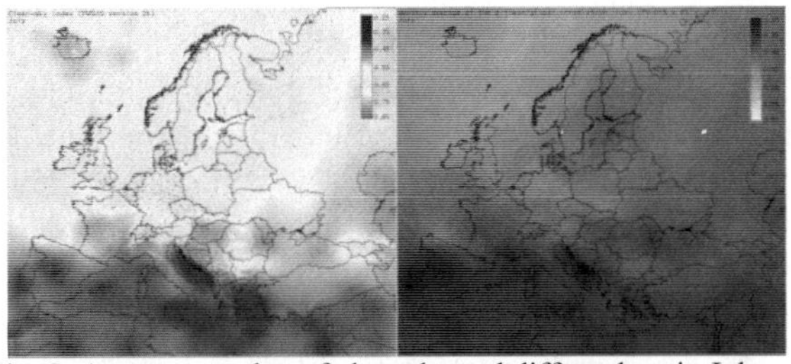

Fig. 2. Average number of clear-sky and diffuse days in July

2. CURRENT SOLAR CONCENTRATORS

There are four main types of concentrating solar power systems: parabolic troughs, dish/engine systems, central receiver systems and compact linear Fresnel reflectors. These technologies can be used to generate electricity for a variety of applications, ranging from remote power systems as small as a few kilowatts (kW) up to grid-connected applications of 200-350 megawatts (MW) or more. A concentrating solar power system that produces 350 MW of electricity displaces the energy equivalent of 2.3 million barrels of oil. (NREL, 2001)

Trough Systems solar collectors use mirrored parabolic troughs to focus the sun's energy to a fluid-carrying receiver tube located at the focal point of a parabolically curved trough reflector as shown in figure 3, left (Elrod, 2007). The energy from the sun sent to the tube heats oil flowing through the tube, and the heat energy is then used to generate electricity in a conventional steam generator.

Dish systems use dish-shaped parabolic mirrors as reflectors to concentrate and focus the sun's rays onto a receiver, which is mounted above the dish at the dish center. A dish/engine system is a standalone unit composed primarily of a collector, a receiver, and an engine (figure 3, right). It works by collecting and concentrating the sun's energy with a dishshaped surface onto a receiver that absorbs the energy and transfers it to the heat engine, usually a Stirling engine.

Central receivers (or power towers) use thousands of individual sun-tracking mirrors called "heliostats" to reflect solar energy onto a receiver located on top of a tall tower (figure 4, left). The receiver collects the sun's heat in a heat-transfer fluid (molten salt) that flows through the receiver. The salt's heat energy is then used to make steam to generate electricity in a conventional steam generator, located at the foot of the tower.

Fig. 3. Parabolic: trough concentrator and dish concentrator

Fig. 4. Central receiver concentrator and CLFR concentrator

Compact Linear Fresnel Reflector (CLFR) use flat reflectors (instead of parabolic mirrors), that move on a single axis and Fresnel lenses to concentrate the solar thermal energy in collectors (figure 4, right). The advantage of CLFR is that it allows for a greater density of reflectors in the array and flat mirrors are much cheaper than parabolic ones (CLFR, 2008).

3. THE DISTRIBUTED CONCENTRATOR

Because this level of efficiency is not currently possible or at least economically feasible (in the case of concentrated photo-voltaics) a new technical solution must be found.

The starting point is represented by a solution consisting in a large Fresnel acrylic lens, mounted on a pivoting frame which allows it to track the sun during the day while keeping the focused spot approximately in the same place. In the focal plane of the lens the hot cylinder of a modified alfa-type Stirling engine is placed. The engine remains fixed during lens movement.

The optical quality acrylic is the most widely applicable material for a concentrator Fresnel lens, and is a good general-purpose material in the visible. Its transmittance is nearly flat and almost 92% from the ultraviolet to the near infrared between 0.2 μm and 2.2 μm (Fresnel, 2003). This first solution seems to be the easy way from the constructor's point of view but it has some drawbacks: due to the large area of the lens, the focal distance is also long, so the resulting height of the system will be large rendering it less desirable. Another concern is represented by the wind problem, which, considering the surface of the lens becomes quite difficult to solve elegantly.

The second approach is to split the large collecting area into several smaller concentrators arranged either as a strip or as a matrix. Each individual concentrator consists of a smaller Fresnel lens with a lot shorter fcal length monted on a pivoting frame to allow for sun tracking. The smaller lenses are more available and less costly than big lenses (consider decomissioned overhead projector lenses). The focused spot from the lens falls onto the end of a copper pipe containing a thermal agent. The end of the pipe is hemispherical having the diameter larger than the focused spot. This receptor head is blackened in order to collect as much heat as possible and the rest of the pipe is insulated. The other end of the pipe is bolted onto the head of the hot cylinder of the Stirling engine, along with other collector pipes in the matrix (figure 5).

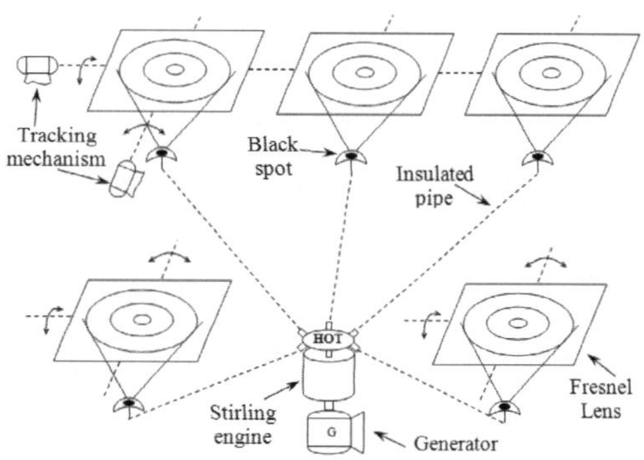

Fig. 5. The distributed collector matrix

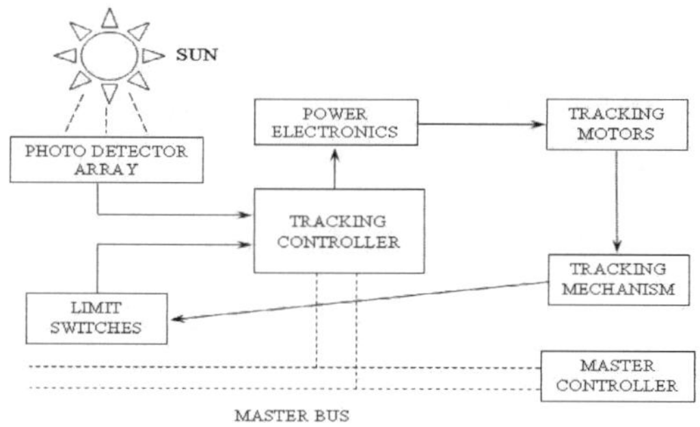

Fig. 6. Tracking system block diagram

The heated end of the pipe is hose coupled to the rest of the pipe to allow the movement of the head with the focus point of the lens durind tracking. Following this strategy, system scalability is provided by adding or removing lenses to the matrix while keeping the same engine.

The proposed method for the tracking is based on heliostat operation, or active sun tracking (Red Rock, 2008). The logic modules also present themselves as a descentralized system, each concentrator assembly having its own sensors, motors and control module. These can communicate over a bus and exchange information with a master controller which can also monitor the electrical energy generation and for strong winds or generally bad weather, it can park the concentrators (figure 6).

4. FUTURE RESEARCH

The proposed solution for the sun concentrator must be further optimized and tested in real life conditions.

The next step presumes building a working model of the concentrator and testing the ability to orient and to transport heat.

The implementing of these concentrators will be benefic for remote locations like holiday houses where they can be used as heat collectors for electricity generators or as water heaters.

The downside of the proposed solution could be reciprocal shading of the lenses at dusk and dawn. Another problem that needs attention is lens cleaning if dust or dirt represents a problem in the installation area.

5. CONCLUSION

The proposed system can be used in the area of interst mainly as a power generator for remote locations (with power output up to a few hundred watts) or as grid connected application in plain field or hill areas that are not usable for agricultural purposes. Providing the building and maintenance costs are kept low combined with a long lifespan through system optimisations, it can prove itself effective and gain popularity in the area.

6. REFERENCES

Compact Linear Fresnel Reflector. (2008), *Available from:* http://jcwinnie.biz/wordpress/?p=2470, *Accessed:* 2008-06-09

Elrod, S. (2007), Solar Concentrators. *Available from:* http://www.parc.com/research/publications/files/5706.pdf, *Accessed:* 2008-06-11

Fresneltech. (2003). Fresnele Lenses product guide, *Available from:* http://www.fresneltech.com, *Accessed:* 2008-05-05

National Renewable Energy Laboratory (NREL), US D.O.E. (2001). Concentrating Solar Power: Energy from Mirrors, *Available form:* http://www.nrel.gov/docs/fy01osti/28751.pdf, *Accessed:* 2008-05-05

PVGIS Project, (2008) EC's Joint Research Centre. *Available from:* http://re.jrc.ec.europa.eu/pvgis/, *Accessed:* 2008-05-02

Red Rock Energy. (2008) Solar Power Heliostat Arrays. *Available from:* http://www.redrok.com/main.htm, *Accessed:* 2008-03-21

Annals of DAAAM for 2008 & Proceedings of the 19th International DAAAM Symposium, ISSN 1726-9679
Editor B. Katalinic, Published by DAAAM International, Vienna, Austria 2008
Make Harmony Between Technology and Nature, and Your Mind will Fly Free as a Bird

OPTIMUM SYNTHESIS OF MOTION LAWS USED BY ROBOT CONTROLLERS

CIUPITU, L[iviu]; SIMIONESCU, I[on] & IVANESCU, A[ndrei] N[ick]

Abstract: *The robot controllers are implemented with only few motion laws: usually of trapezoidal shape and/or of parabolic shape, in term of speed. But transmission functions given numerically and resulted for example from an optimization process made off-line could be used as well for moving the characteristic point of robot to a specific trajectory. In fact this optimization problem should merge together with the problem of optimization the trajectories between 2 positions. It is also known that the robot controllers are using usually only 2 imposed trajectories: linear and circular.*

The proposed paper is dealing with the synthesis of the transmission functions so that an objective function representing the minimum energy consumption to be fulfilled.
Key words: *robot, controller, optimization, motion law.*

1. INTRODUCTION

Usual the robot controllers are implemented with only few motion laws of movement between 2 positions (configurations) from working space: of trapezoidal shape and of parabolic shape in term of speed (see for example Eshed Robotec Controllers series A (***, Eshed Robotec, 2001)). But transmission functions given numerically and resulted for example from an optimization process made off-line could be used as well for moving to a specific trajectory. In fact this optimization problem should merge together with the problem of optimization the trajectories between 2 positions (configurations) from an application. It is also known that the robot controllers are using usually only 2 imposed trajectories: linear and circular. Some other possible curves that the robot could use between 2 positions (configurations) from working space are presented in (Ciupitu, 1997).

The problem of optimization the trajectories between 2 positions (configurations) from an application was studied and solved in (Simionescu & Ciupitu 1997).

In paper (Antonelli et al., 2004) an industrial implementation of a minimum time path-following algorithm is presented but only common trajectories that exhibit a trapezoidal velocity profile; point-to-point linear and circular motions as well as fly movements were considered.

The discontinuity of transmission functions of second degree lead to shocks and vibrations (Simionescu et al., 1996), and because of elasticity of elements the errors between the ideal trajectory and real trajectory could be unacceptable. The inertial forces of manipulating object could be bigger than friction forces from gripper and the manipulating object could be dropped. Also, the stresses and wears of links depend to inertial forces, which imply the finding of new transmission functions.

The choosing of the optimal transmission function depends to:

- trajectory between the start point P_s and the target point P_t,
- maximum accepted values of speed and/or acceleration along this trajectory,
- continuity or discontinuity conditions for transmission function of second degree,
- weight of manipulating object.

2. SYNTHESIS OF TRANSMISSION FUNCTIONS

Generally, the transmission function is expressed non-dimensional (Artobolevskii et al., 1959) as function $y = y(x)$ where $x \in [0, 1]$, $y \in [0, 1]$, and has usually 3 sectors (Fig. 1):

1) acceleration sector: $y''(x) > 0, x \in [0, u_1]$;
2) uniform movement sector: $y''(x) = 0, x \in [u_1, 1 - u_2]$;
3) deceleration sector: $y''(x) < 0, x \in [1 - u_2, 1]$,

where: $u_1 > 0$, $u_2 > 0$, $u_1 + u_2 \le 1$. The uniform movement sector could miss if $u_1 + u_2 = 1$.

Real transmission function $S = S(t)$ which defines the distance from start point P_s along the trajectory to target point P_t is:

$$\begin{aligned} &S = yL\,; \\ &\frac{\mathrm{d}S}{\mathrm{d}t} = \frac{\mathrm{d}y}{\mathrm{d}x}\frac{L}{T}\,; \\ &\ldots \\ &\frac{\mathrm{d}^n S}{\mathrm{d}t^n} = \frac{\mathrm{d}^n y}{\mathrm{d}x^n}\frac{L}{T^n}\,; \end{aligned} \quad (1)$$

where L represents the length of trajectory between start point P_s and target point P_t and T represents the time of moving from start point P_s to target point P_t.

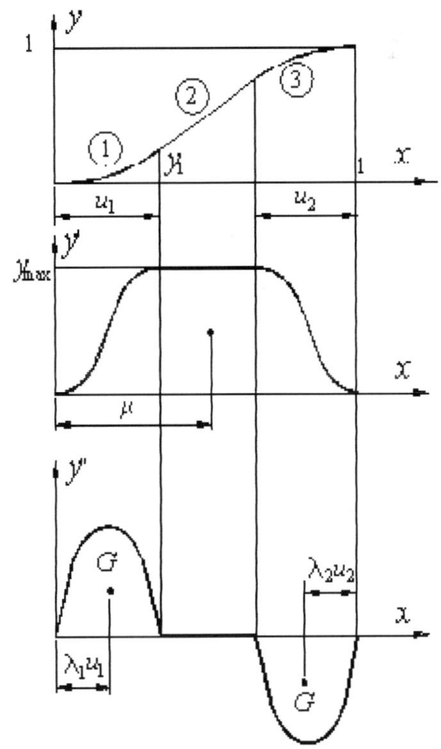

Fig. 1. Transmission functions with 3 sectors

If in acceleration and deceleration sectors are used the same transmission function (Fig. 1), then:

$$\begin{aligned} & k\, y_1(x) + y_2(1-k\,x) = 1; \\ & y_1'(x) - y_2'(1-k\,x) = 0; \\ & \dots \\ & y_1^{(n)}(x) + (-1)^n k^{n-1} y_2^{(n)}(1-kx) = 0, \end{aligned} \quad \text{where } k = \frac{u_2}{u_1} \qquad (2)$$

and: $y_1 = y_1(x)$, $x \in [0, u_1]$, and $y_2 = y_2(x)$, $x \in [1-u_2, 1]$, are non-dimensional transmission functions from acceleration and deceleration sectors respectively.

The maximum value of first derivate of transmission function depends to the shape of second derivative. It is known (Simionescu et al., 1996) that the value of a function at the end of an interval is equal with the product between the area of second derivative diagram and the distance from its mass center to the end of interval, if first derivative is monotony and null at the beginning of the interval.

For an asymmetrical transmission function with 3 sectors, results:

$$y_{1\max} = y_1(u_1) = (1 - \lambda_1)\, u_1\, y'_{\max} \qquad (3)$$

and

$$y_{2\min} = y_2(1 - u_2) = (1 - \lambda_2)\, u_2\, y'_{\max} \qquad (4)$$

so that:

$$y'_{\max} = \frac{y_{1\max} + y_{2\min}}{u_1 + u_2 - \lambda_1 u_1 - \lambda_2 u_2}. \qquad (5)$$

Observations:

a. Maximum value of first derivative for transmission function with 2 sectors ($u_1 + u_2 = 1$) is:

$$y'_{\max} = \frac{1}{1 - \lambda_1 u_1 - \lambda_2 u_2}. \qquad (6)$$

b. The variation of maximum value of first derivative is according with the variation of expression $\lambda_1 u_1 + \lambda_2 u_2$. So, the reduction of maximum value is done by moving the mass centers G_1 and G_2 (Fig. 1) to the extremities (λ_1 and $\lambda_2 < 0.5$).

If $\lambda_1 = \lambda_2 = \lambda$ and $u_1 = u_2 = 0.5$, then: $y'_{\max} = \frac{1}{1-\lambda}$; (7)

c. Minimum values for first derivative result in case of diagrams with 3 sectors ($u_1 + u_2 < 1$). Minimum minimorum of this value is $y'_{\min\,\min} = 1$ uniform movement ($u_1 = u_2 = 0$), but that leads to discontinuities in second derivative diagram.

d. The reduction of the acceleration and deceleration sectors increases the maximum values of second derivative:

$$y''_{\max} = c\,\frac{y'_{\max}}{u} \quad \text{where:} \quad c = \frac{y''_{\max}}{y''_{\text{med}}} \qquad (8)$$

For $u_1 = 0.5$ and $\lambda_1 = \lambda_2 = 0.5$ results the minimum values of second derivative: $y''_{\min} = 4$.

e. The so called time-section has the value:

$$\Sigma = \int_0^T S\, \mathrm{d}t = L\,T \int_0^1 y\, \mathrm{d}x, \text{ and } \quad \sigma = \int_0^1 y dx = 1 - \mu, \qquad (9)$$

where μ represents the x-coordinate of mass center of first derivative diagram (Fig. 1). The increasing of time-section by reducing the value of μ is limited by increasing of maximum acceleration.

Transmission functions could be expressed by polynomial functions or by trigonometric functions (Simionescu et al., 1996). The problem of synthesis is to find the coefficients of these functions with respect to specific criteria.

3. OPTIMIZATION PROBLEM

In order to formulate the optimization problem 2 models must to be known: the dynamic model of robot with maximum acceptable values for speed and acceleration for each driving motor and the application given by coordinates of positions (configurations) that the robot must reach, the trajectories and the weights of manipulating objects between these.

In selecting the optimum movement, one has to consider the following criteria:

- the movement time should be minim;
- robot acting power (or energy) should be minim;
- input moments and forces should be minim;
- the mechanical work for robot acting should be minim,

single or in combination.

The objective function of optimization problem is a of minimized parameter:

$$\boldsymbol{O} = E = \sum_{i=1}^{l} E_i \qquad (10)$$

where l represents the number of intervals of motion and E is the energy consumption of all driving motors correspondent to each interval (Ciupitu & Simionescu, 2007). In complex expression of energy consumption (in fact sum of integrals) the coefficients of polynomial transmission functions are unknowns, under constrains which represents the maximum accelerations and speeds for each motor and along the trajectory to each movement interval.

4. CONCLUSION

The solving of this optimization problem together with the problem of optimization the trajectories and the computing of optimum location of robot base with respect to an application (Ciupitu & Simionescu, 2007) is and important off-line task of controllers of new generation, leading to an important economy of energy.

5. REFERENCES

Antonelli,G.; Chiaverini,S.; Palladino,M.; Gerio,G.P. & Renga, G. (2004). Cartesian space motion planning for robots. An industrial implementation, *Proceedings of the 4-th Intl Workshop RoMoCo'04*, Kozłowski, K. (Ed.), pp. 279–284, ISBN 9788371432729, Puszczykowo, Poland, June 2004.

Artobolevskii, I. I.; Levitski, N. I. & Cercudinov, S. A. (1959). *Sintez ploskih mehanizmov* (*Synthesis of Planar Mechanisms*), Gosu. Izd. Fiz. Mat. Lit., Moskva.

Ciupitu, L. (1997). Complex Analytic Curves for Industrial Robot Trajectories, *Proceedings of the 3-rd Asian Conference on Robotics and Its Application*, pp. 211-218, Tokyo, Japan, October 1997.

Ciupitu, L. & Simionescu, I. (2007). Optimal Location of Robot Base With Respect to the Application Positions, *Proceedings of the 18-th International DAAAM Symposium*, Katalinic, B. (Ed.), pp. 161-162, ISBN 3-901509-58-5, Zadar, Croatia, October 2007, Published by DAAAM International, Vienna, Austria.

Simionescu, I. ; Moise, V. & Ciupitu, L. (1996). Sinteza funcţiilor de transmitere ale mecanismelor (Synthesis of transmission functions used in mechanisms), *Journal of Romanian Academy "Studii şi cercetări de mecanică aplicată"*, tome 55, no. 5-6, pp. 409-426.

Simionescu, I. & Ciupitu, L. (1997). Optimum Programming of Industrial Robot Trajectories, *Proceedings of the 3-rd Asian Conference on Robotics and Its Application*, pp. 219-224, Tokyo, Japan, October 1997.

*** (Eshed Robotec Ltd.). "ACL for Controller-A, Version 1.43, F.44", Reference Guide 9501, 4-th Edition.

Annals of DAAAM for 2008 & Proceedings of the 19th International DAAAM Symposium, ISSN 1726-9679
Editor B. Katalinic, Published by DAAAM International, Vienna, Austria 2008
Make Harmony Between Technology and Nature, and Your Mind will Fly Free as a Bird

SOME POSSIBILITIES OF OPTIMIZATION IN THE ROMANIAN FIRM MANAGEMENT

COCIU, N[icolae]; IGNATON, E[lemer] & TAROATA, A[nghel]

Abstract: *A few optimisation possibilities in the Romanian firm management are analysed, giving example the firm development, the optimal choice of the activity field, the firm location and the optimal selection of the CAD/CAM/CAE system. Some optimization possibilities in developing an existing firm and a new one using the multi-attribute decisions are shortly analyzed.*
Key words: *optimization, multi-attribute decisions, development, firm, placement.*

1. INTRODUCTION

Studying the firm management optimization aims to increase its competitiveness and to ensure a rational decisional and managerial process.

Choosing mathematical models and methods is made according to the actual conditions of the Romanian firm, the information that can be collected and processed in due time, its costs, the competition strengths, in order to obtain the competition advantage.

Using the mathematical models, the optimization methods and the information technology in management and the firm's activity leads to an increase of its efficiency, a competitive advantage of low costs and better quality for the products placed on market.

The firm's management system is defined as an assembly of elements with decisional, managerial, informational and motivational characteristics; it triggers all management processes and relations in order to obtain a better efficiency and effectiveness. When projecting the firm's management system, a few elements should be considered: the firm's size and goals, its profile and complexity, the capital structure and dimension, the human resources, the firm's position in its competition with other firms, the conception of the owners and managers.

The management system can be structured as follows: the organizational, informational, decisional, management methods and techniques subsystems. (b. Cociu, 1999)

2. OPTIMIZATION POSSIBILITY

Optimization represents the choice and enforcement of the best solution of many possible ones. A model is a simplified or abstracted representation of reality. Often in practice it is difficult to simultaneously obtain simplified and representative characteristics of the studied reality.

Some optimization possibilities are:

- optimization of the firm's management system: the informational, decisional, and management methods and techniques subsystems;
- optimization of management functions; foresight, management, decision, coordination and control;
- optimization in the firm's functioning and development;
- optimization of the company's activity.

Resorting to classical methods of optimization (linear, multi-objective and parametric programming, the transport problem, the multi-attribute decisions, the algorithms in the graph theory, dynamic programming, ' divide et impera ' algorithms, ' greedy ' algorithms, stochastic programming etc.), non-classical methods (neuronal calculation, evolutionary calculation, fuzzy algorithms, etc.) or very good calculation programs, intelligent systems (neuronal, decision support and expert systems etc.), according to the actual conditions of the Romanian local and global firms, will lead to an increase of the decisional process quality and to the firm's management optimization as well as to the increase of the firm's competitiveness.

There are at least three steps to solve a management problem using a mono-criterion or multi-criterion algorithm /an optimal or rational program:

- The optimal / rational choice of the algorithm class (determinist, stochastic, fuzzy, genetic algorithms etc.);
- The optimal / rational choice of the algorithm from the chosen class of algorithms;
- The optimal / rational choice of the programming language in order to program the chosen algorithm (the enforcement time of the program on the computer, the necessary computer memory, the portability of the realized program, etc., can be considered efficiency functions).

According to the probability that an element may or may not belong to a multitude and the degree to which an element belongs to a multitude, the mathematic models are: deterministic, stochastic and fuzzy (vague).

The deterministic models are characterized by variables and secure, achievable relations using the 1 probability and are exact.

When using stochastic (probable) models some variables that describe the studied process are random variables, thus variables have certain values with certain probabilities.

The fuzzy models (vague) use the fuzzy mathematics with variables that cannot be characterized neither exactly (deterministic) nor probably, one can only quality evaluations of their values.

The problem of distinguishing between random (stochastic) and fuzzy (vague) emerges. The random phenomenon results from the uncertainty that an element belongs or not to a multitude, with different probabilities. In case of a fuzzy (vague) phenomenon there are degrees for an element of belonging or non-belonging to a multitude and it is not based on the excluded third from the bivalent mathematic logic. The fuzzy phenomena are based on fuzzy logic, which is a continuous logic that includes the bivalent and n-valent discrete logics.

The decisions in multi-criteria decisional processes with a finite number of variants are called multi-attribute decisions. A multi-attribute decision problem is given, with m decisional variants and n evaluating criteria for variants , $m \in N$, $m \geq 2$, $n \in N$, $n \geq 2$. (b. Cociu, 1999)

A. The multi-criteria choice of firm activity field and placement

Choosing the activity field and the firm placement can be made in more than one ways: without considering any optimization criteria, using the intuition and experience, and decisions based on a single criterion, thus mono-criteria decisions, employing the multi-attribute decisions.

A1. Choosing the activity field can be looked at as a solvable multi-attribute decision problem in at least two stages. In the first stage the decisional options can be considered: industrial, agricultural, commercial, transportation, banking and service business etc., and the evaluation criteria may be: market size, competition, estimated profit, competence in a certain field, available capital etc. The optimal option determined in the first stage is detailed in the second stage at definite business fields and those will be the new decisional options in a new multi-attribute problem, and the criteria may be the same ones or others. By a multi-attribute algorithm the optimal business variant is determined (from this stage) from the considered n criteria point of view, other stages to detail the activity field remaining possible. (a. Cociu, 1999)

A2. The firm placement problem is very important, as it involves a great investment, and the misplacement cannot be rectified without important loss. The ideal location is the one that allows the lowest cost for the firm's activity. The firm placement can be made on a multi-criteria base in two stages: in the first stage the region from Romania is chosen: Alba, Arad, Bihor, Timis etc., that represents the decisional options, considering as criteria: placement cost, necessary capital, labor force availability a etc., possibly with certain importance coefficients.

In the next stage a few locations in the chosen geographic area are considered, and the options might be: a city, another town in the same area, different localities etc. The evaluation criteria for the options are: the placement cost, the utilities cost, the existence of roads, railroads, airport, electrical power, water and air quality, competition stage. It can be observed that there is a problem of multi-attribute decision, which is solvable if a specific algorithm is used.

If the firm's location has been decided, another optimization problem refers to the actual placement options: renting, buying a building, buying land and building a facility. The evaluation criteria for the decisional options can be considered: the placement cost, the period of time assigned for placement, the placement duration etc. A problem of multi-attribute decision is obtained, which is solvable with a multi-attribute algorithm.

B. The optimal selection of the CAD/CAM/CAE system is a very important problem. A company can make a mono-criteria or multi-criteria selection of the CAD (Computer Aided Design) system, CAM (Computer Aided Manufacturing) system and CAE (Computer Aided Engineering) system.

The mono-criteria selection of the CAD/CAM/CAE system is realized by choosing the optimal variant from among several possible by using a single criterion, for example the price of the system, that is mono-objective decision.

In the multi-criteria selection of the CAD/CAM/CAE system it is considered a multi-attribute decision problem, with m decisional variants and n assessment criteria of the variants. For example, the following decisional variants are considered:

V_1 = CATIA system;
V_2 = SIGRAPH system;
V_3 = EUCLID 3 system;
V_4 = Pro/ENGINEER system;
V_5 = MECHANICA system.

The assessment criteria of the variants can be:
C_1 - price of the system;
C_2 - language in which the system can be used;
C_3 - time of supply;
C_4–Information exchange with other CAD system;
C_5 – after-sale service.

It results the consequence matrix
$A = (a_{ij})$, $i = 1, 2, \ldots, m$; $j = 1, 2, \ldots, n$; $m = 5$, $n = 5$, and by a multi-attribute decision algorithm: maxmin, Onicescu, Topsis, Electre, Leader, Promethee II, Smrt (Simple Multi-attribute Rating Technique) etc. is determined the optimal CAD/CAM/CAE system variant for the company which wants to buy it. Obviously, other decisional variants can be taken into account (CAD/CAM/CAE systems) and other assessment variants. (a. Cociu, 1999)

C. The development of a firm, its expansion in the business, can be made by two complementary ways: internal and external growth. (a. Cociu, 1999)

Choosing the growth solution can be looked at as a multi-criteria optimization problem, in the form of a multi-attribute decision problem, with decision variants already mentioned, for example:

- V_1 – the internal growth of the firm;
- V_2 – the external growth of the firm that buys shares from another firms (the options are: one or more firms, and if there are more firms , for each firm whose stocks are bought, another decisional variant is considered);
- V_3 – the firm buys entirely another firm , thus absorption;
- V_4 – external growth by creating new firms.

Other firm development options may be considered.

The criteria for choosing the options may be: the firm growth cost, the estimated profit, the firm's independency, the period of time for the firm's growth etc., and each criterion may have attached an importance coefficient (Andreica et.al., 1998). The optimal solution and the options top are determined through a multi-attribute decision algorithm, for example: Electre, Topsis, Leader, Onicescu, maxmin, Promethee II, Smrt (Simple Multi-attribute Rating Technique) etc. (Filip, 2007), (Ionescu et al.,1999).

3. CONCLUSIONS

Using some optimization methods in conceiving, designing, exploiting and developing the Romanian firms, along with management methods, leads to increased performances and values for the firm, in a dynamic competition environment.

A Decision Support System (DSS) may be designed and implemented to solve such decision problems and other structured and unstructured problems that are specific to the Romanian economic environment.

The research results can be applied for the placement and development of a real Romanian firm.

4. REFERENCES

Andreica, M; Stoica, M & Luban, F. (1998) – Metode cantitative in management, Quantitative methods in management, Editura Economica, ISBN 973-590-027-0, Bucuresti

a. Cociu, N. (1999) – Optimizarea deciziilor in sistemele de productie,Decisions optimisation in the production systems, Editura Eurobit, 973-9441-65-3, Timisoara

b. Cociu, N. (1999) – Optimizari in conceperea si exploatarea sistemelor de productie, Optimisations in conceiving and exploitation of production systems, Editura Eurobit, 973-9441-65-15, Timisoara

Filip F. Gh.(2007) – Sisteme Suport pentru Decizii, Support systems for Decisions, Editura Tehnica, ISBN 978-973-31-2308-8, Bucuresti

Ionescu, Gh; Cazan, E & Negruta A. (1999). Modelarea si optimizarea deciziilor managerial, Modelling and optimisation of managerial decisions, Editura Dacia, ISBN 973-35-0950-7, Cluj – Napoca.

Annals of DAAAM for 2008 & Proceedings of the 19th International DAAAM Symposium, ISSN 1726-9679
Editor B. Katalinic, Published by DAAAM International, Vienna, Austria 2008
Make Harmony Between Technology and Nature, and Your Mind will Fly Free as a Bird
*Young Researcher and Scientist Paper / * Supervisor, Mentor*

IMPROVING THE RELIABILITY OF MULTICAST TRANSMISSIONS IN IEEE 802.11 WIRELESS LANS

COCORADA, S[orin] & SZEKELY, I[uliu]*

Abstract: *Multicasting is a more efficient method of supporting group communication than unicasting or broadcasting, as it allows transmission and routing of packets to multiple destinations using fewer network resources. Because the IEEE 802.11 standard handles multicast transmissions as simple broadcasts it can not provide a reliable multicast service. In this paper we propose and analyze an extension to the IEEE 802.11 multicast mechanism to improve the reliability. The proposed solution differs from previous work by providing a simple, low delay leader based solution using a cross-layer design approach. Calculations and simulation results show that, in worst case, the packet error rate of the group member's decreases with a value approximately equal to the packet error rate of the designated group leader.*

Key words: *reliable multicast, IEEE 802.11, wireless LAN, cross-layer design.*

1. INTRODUCTION

Multicasting is a more efficient method of supporting group communication than unicasting or broadcasting, as it allows transmission and routing of packets to multiple destinations using fewer network resources. Applications of wireless multicast support group-oriented multimedia conferences, command and control and video streaming of live events.

Because of their wide deployment wireless local area networks (WLANs) have become an important component of the Internet infrastructure. Multicast capability make WLANs suited for supporting group oriented applications but their behavior differs greatly from the wired networks. It is commonly accepted that there are three major factors contributing to errors on wireless channels: the signal energy degradation, interference and the random noise and fading. Packet losses up to 30% are assumed. According to the standard, there is no Medium Access Control (MAC) level recovery on multicast frames. As a result, the reliability of this traffic is reduced, relative to the directed traffic, due to the increased probability of lost frames from interference, collisions, or varying channel properties (IEEE 802.11 1999).

The rest of the paper is structured as follows: in the next section some aspects from related work are presented. In Section 3 we propose and analyze a leader based solution to improve the reliability of the multicast service. Some concluding remarks are provided in the final section.

2. RELATED WORK

Up to now several papers have been written especially on designing reliable muticast protocols which try to ensure that each frame was received by each multicast group member.

An extension to the IEEE 802.11 ad-hoc mode broadcast mechanism to improve the reliability is proposed in (Xie et al. 2005) by employing a broadcast acknowledgement (BrACK) mechanism. Receiving the BrACK the sender calculates and records the indices of lost frames and retransmits them.

Because the protocol tries to ensure the correct delivery of each data frame, the throughput decreases and delay increases if packet interarrival times are low because of the high overhead introduced by retransmissions.

Two types of reliable multicast protocols at the MAC layer are analyzed in (Bao & Liao 2005): one ACK based and one leader based. For the ACK based mechanism the access point (AP) monitors the frame reception progress for all receivers and retransmits frames whenever no ACK is received from any receiver; for the leader based mechanism, the AP retransmits the data only when no ACK frame is received from the selected leader. From the analysis it can be concluded that for both protocols the throughput decreases and delay increases with the increase of the group size.

The design of a cross-layer protocol to provide reliable multicast transmissions is provided in (Ge et al. 2007). First, the transmit rate optimization of MAC layer multicast is addressed to prevent losses, and then forward error correction at the transport layer is proposed to enhance the reliability of multicast transmissions.

Some open challenges and new opportunities for cross-layer design of wireless networks are discussed in (Srivastava & Motani 2005). The advantages offered by the broadcast nature of the wireless medium cannot be accommodated in layered architectures and hence inevitably require some degree of architecture violation, in other words, cross-layer design.

3. THE PROPOSED SOLUTION

We will propose a solution for improving the reliability of multicast transmissions which increases the real throughput by using retransmissions, but without ensuring that each group member has received all addressed frames. This approach is applicable in scenarios where the integrity of the received data is not crucial, but a low delay is required such as in voice conferencing. We propose that each multicast frame transmitted by the access point to be acknowledged by one of the group members, known as the group leader. The proposal is based on some experimental work done in (Tang & McKinley 2003) where it has been shown that there is some degree of spatial correlation between multicast packet losses. Considering the worst case, when packet losses at any station are independent from packet loses at the station which was designated to acknowledge the transmissions, the packet error rate (PER) can be calculated as follows:

$$\begin{aligned} PER_i^{'} &= 1 - \sum_{j=0}^{r} (1 - PER_i) PER^{j} PER^{*j} = \\ &= 1 - (1 - PER_i) \sum_{j=0}^{r} \left(PER_i \cdot PER^{*} \right)^{j} \end{aligned} \quad (1)$$

where PER_i is the packet error rate at node i, PER^* is the packet error rate at the leader node and r is maximum number of retransmissions. For the group leader the packet error rate is:

$$PER^{*'} = PER^{*(r+1)} \quad (2)$$

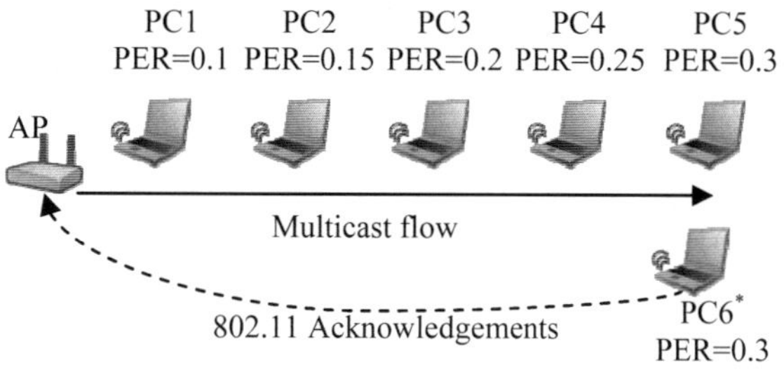

Fig. 1. Simulation scenario

To maximize the performance of the proposed algorithm the leader must be the station experiencing the lower reception probability. Assuming the same channel model among the whole multicast group the highest PER will be at the station where frames transmitted by the access point are received whit the lowest signal to noise ratio.

Because of the varying channel condition and of the mobility of the multicast group members it is desirable that the group leader will be dynamically designated for each transmitted frame; to accomplish this, a possible solution would be using IEEE 802.11 four address frames, with address four being the layer two address of the designated leader. In case the access point does not receive an acknowledgement the frame will be retransmitted using the same channel access mechanism as for unicast transmissions. To determine which of the group members has the highest PER we use a customized version of the Internet Group Membership Protocol (IGMP), modified so that it enables explicit tracking of mobile hosts and collects additional information from the physical and MAC layer (cross layer design approach).

4. SIMULATION RESULTS

We evaluate the performance of the proposed solution using Omnet++ (Omnet++ Simulator 2008). The network topology used in our simulation is presented in Fig 1. The packet error rates are also specified for each member station; the node designated to acknowledge the multicast frames is PC6 for the whole simulation. The multicast flow is generated from a fixed station (not figured) connected by cable to the access point. The multicast flow is a constant bit rate user datagram protocol (UDP) flow. Packets have a payload of 44bytes/packet at the application level and are transmitted with a period of 8ms (typical G.726 codec). The access point retransmits the frames using mode 7 (48Mbps); acknowledgements are transmitted using mode 1 (6Mbps). The contention windows was set up at 31 slot intervals and the number of retransmissions was limited at 3. To simplify the analysis we assume there is no other traffic in the network, there are no losses on the uplink and the PER is constant for all nodes. After running the simulation for 3 hours we obtain the values presented in Table 1.

The same values can be obtained using formulas (1) and (2). From Table 1 we can conclude that the number of lost packets at the nodes PC1-PC6 decreases with about 30%, this value being the PER at the leader node.

Station	PER	PER'	Improvement	
			Absolute (PER'-PER)	Relative PER'/PER
PC1	0.1	0.0722	-0.0278	0.722
PC2	0.15	0.11	-0.04	0.733
PC3	0.2	0.148	-0.052	0.74
PC4	0.25	0.189	-0.061	0.756
PC5	0.3	0.23	-0.07	0.76
PC6*	0.3	0.3^4=0,0081	-0.2919	0.027

Table 1. Values for the PER obtained by simulation

Station	Standard (no ACKs)		Proposed solution	
	Mean delay	Jitter	Mean delay	Jitter
PC1	109us	0.055us	123us	26us
PC2	110us	0.056us	130us	39us
PC3	110us	0.051us	138us	53us
PC4	110us	0.056us	147us	68us
PC5	110us	0.055us	168us	104us
PC6*	110us	0.052us	311us	287us

Table 2. Values for delay and jitter obtained by simulation

The same values can be obtained by calculation as follows:

$$\frac{PER_i^{'}}{PER_i} = \frac{1-(1-PER_i)\sum_{j=0}^{r}\left(PER_i \cdot PER^*\right)^j}{PER_i} \approx 1 - PER^* \quad (3)$$

considering $PER_i PER^*=0$.

The obtained values for the mean delay and jitter are presented in Table 2. As expected the values of the mean delay and jitter increase slightly after recovering lost frames by retransmission, because some of the frames with were previously lost are now received but with higher delays. Because the network is not saturated, ACK frames do not introduce additional delays, the time necessary to complete a transaction being much lower than the packet period.

5. CONCLUSION

In this paper we presented a simple leader based solution to improve the reliability multicast transmissions in IEEE 802.11 infrastructure networks. An analytical model to evaluate the performance of the proposed algorithm has also been proposed. We conducted simulations in order to validate the proposed analytical model.

Results show that, in worst case, the packet error rate of the group member's decreases with a value approximately equal to the packet error rate of the designated group leader.

6. REFERENCES

***, Omnet++ Simulator, http://www.omnetpp.org/

***, IEEE 802.11, Part 11: Wireless LAN Medium Access Control (MAC) and Physical Layer (PHY) Specifications, *Available from:* http://grouper.ieee.org/groups/802/11/ *Accessed:* 2008-07-20

Bao, C., Liao & W. (2005), Performance analysis of reliable MAC-layer multicast for IEEE 802.11 Wireless LANs, *IEEE International Conference on Communications- ICC 2005*, Vol. 2, pp.1378-1382 ISBN: 0-7803-8938-7, May 2005, Seoul

Ge, W., Zhang, J. & Shen, S. (2007). A Cross-Layer Design Approach to Multicast in Wireless Networks. *IEEE Transactions on Wireless Communications*, Vol.6, No.3, March 2007, pp. 1063-1071, 1536-1276

Srivastava V. & Motani, M. (2005). Cross-Layer Design: A Survey and the Road Ahead. *IEEE Communications Magazine*, Vol.43, No.12, Dec 2005, pp.112-119, 0163-6804

Tang, C. & McKinley, P. K. (2003). Modeling Multicast Packet Losses in Wireless LANs, *Proceedings of the 6th ACM international workshop on Modeling Analysis and simulation of wireless and mobile systems*, pp. 130-133, 1-58113-766-4, Sept 2003, San Diego

Xie, J., Das, A., Nandi & S. Gupta, A.K (2005). Improving the reliability of IEEE 802.11 broadcast scheme for multicasting în mobile ad hoc networks, *IEEE Wireless Communications and Networking Conference*, Vol. 1, pp. 126-131, 0-7803-8966-2, March 2005, New Orleans

Annals of DAAAM for 2008 & Proceedings of the 19th International DAAAM Symposium, ISSN 1726-9679
Editor B. Katalinic, Published by DAAAM International, Vienna, Austria 2008
Make Harmony Between Technology and Nature, and Your Mind will Fly Free as a Bird

APPROACHING OF BALANCING PROBLEM THROUGH SIMULATION

COCULESCU, C[ristina] & COCULESCU, C[ristian] C[atalin]

Abstract: *In this work we'll describe an algorithm of simulation of a technical process, relied on random generation of the main characteristics of working phases, which will be considered in solving balancing problem of a technical line whereon an only model of a product is assembled, operating times are established and line operating rhythm is fixed. Simulation algorithm makes the possibility of random generation of big number of instances of approached balancing problem, being known that their gaining from real technical processes is difficult.*

Key words: *simulation, balancing, manufacturing, management.*

1. INTRODUCTION

In many cases, experiences on real objects are difficult to make and need a big amount of funds and human resources for make then acceptably.

Nowadays, grace of computing facilities of electronic computers, that are big amount of memory and computing speed, it acts modeling methods improving, complex systems modeling with simulation techniques having a privileged place.

The problem of technical process balancing supposes the grouping of phases in workstations so the sum of operating times of the phases which are in a station to be less then or equal to a value *R* called rhythm. For approached model of balancing problem, *R* is fixed and is the same for all stations. The phases of technical process must be executed in a certain order in the meaning that certain phases can't have place before others. Order relations between phases reflect technologically, real development of the process and are called relation of precedence. Mathematically, they form an acyclic digraph (James, 2000). Balancing objective is the reaching of the smallest number of station having prior property.

Because in simulation programs, random number generation has a relatively big part of total computer operating time, it is necessary to find efficient ways to do this, to get random strings with big cyclicity.

Another important feature in technical line simulation is also the testing of acyclicity of the graph of precedence relations.

In this work we follow making an algorithm for simulate a technical process, relied on random generation of the main characteristics of working phases, which will be considered in solving balancing problem of a technical line whereon an only model of a product is assembled, operating times are established and line operating rhythm is fixed, relied also on solving the two essential features above mentioned.

2. SIMULATION OF A TECHNICAL LINE WHICH IS SUBJECT TO THE BALANCING PROCESS

2.1 Possibilities of random number generation

For realistic reproduce certain elements of simulated system, and also for solving some numeric problems, it appears the need of existence of some numbers "happily chosen".

In simulation programs, generation of random numbers has a relative big part of computer running time.

We'll mark $F = \{1,2,...,n\}$ the manifold of technical process' phases which follow to be simulated and $t = \{t(1), t(2),...,t(n)\}$, the vector of operating times associated to the phases.

Within this context, we purpose to build a function having property that it has equal likelihoods to generate its values. So, we consider the function **rnd** defined like this: for a natural number n, $rnd(n)$ is a natural number between 0 and *n*-1. If we want a real value, let's say a number between -2 and 3, having two exact digits, we use an expression like $rnd(500)/100-2$.

If maximum number of phases is marked *maxphases* the number of the phases from process will be $rnd(maxphases)+1$.

We consider operating time of each phase at most equal to *maxtime* time units, therefore for the phase *i*, $t(i) = rnd(100*\max time)/100+0.01$ (it is computed with at most two exact digits and cannot be zero).

As follows, we consider that arithmetic operations that will be used for generate such function (***rnd***), are with 32 bits that is, every result of an arithmetic operation is considered as the rest of the division of real result by 2^{32}. This value has chosen because it is enough for usual needs and is easy to be used in programming.

There is considered the string $\{x_n\}_{n\geq 0}$ defined like:

$$\begin{cases} x_0 = c \\ x_{n+1} = a * x_n + b \end{cases}, \quad for \quad n \geq 1 \qquad (1)$$

where *a* and *b* are odd numbers (*a* is a number of shape $a = 4*k+1$).

Such string, especially a and b are prime numbers, has an extreme long cyclicity and is good enough for usual needs.

If to its value there is added also that of the standard function which generates random numbers, from the libraries of several programming languages, and it is set by the time counter and computer's internal clock, the result can be of a big variety.

There is also the possibility of combination of two or three strings of mentioned type, their sum being a string having a very long cyclicity and in this case the result being very easy to compute too.

2.2. Acyclicity test of the graph of precedence relations

Let be *maxphases* maximum number of process' phases and *maxtime* maximum operating time of a phase.

At the beginning we'll establish the number of phases (*nrphases*) and operating times of phases.

For $i = 1$, *nrphases* $t(i) = rnd(\max time)+0.01$.

The number of phases is randomly generated.

After this, phases' names are read. For each phase, we generate then the operating time. Making a precedence relation follows.

A precedence relation (shortly, a precedence) can be considered as a couple (f_1, f_2) where $f_1 = rnd(nrphases)$ $f_2 = rnd(nrphases)$ and $f_1 \neq f_2$.

That means that f_2 operates before f_1. Clearly, the first care we have is not to have the relation already generated and $f_1 \neq f_2$. After that, total operating time and maximum operating time for a phase (the biggest operating time of generated phases) are computed. This computing is very important because if in balancing problem, a rhythm is smaller than maximum working time for a phase, the problem is wrong (Vernadat, 1996).

After that, we must decide whether precedence relation we put in is correct. It not, it will be cancelled in the sense of cancellation of its value in precedence matrix.

There are two ways to do the test. First, it is necessary to clarify what means that precedence (f_1, f_2) is correct, that is, if f_2 can precede f_1. The relation is correct if and only if no phase precede itself. If we have precedence of type (f_1, f_2) and (f_2, f_1), between indirect predecessors of f_1 is also f_1 as predecessor of f_2 and we reach up-mentioned contradiction.

Acyclicity test of precedence relations reduces to establish if no phase precede itself. A checking method consists of the establishing of all predecessors of each phase. For this, we purpose the following algorithm:

Algorithm 1

Step 1. We consider F_1 the manifold of f predecessors.

Let $k = 1$.

Step 2. $k = k + 1$. We consider F_k the manifold of direct predecessors of the phases from F_{k-1}.

If $F_k \neq F_{k-1}$ we repeat step 2.

Step 3. If $f \in F_k$, the precedence is not correct.

Otherwise it is correct.

Another method supposes weight computing of all technical process phases. The weight of a phase is the sum of its time with times of all its direct or indirect predecessors. Therefore, if a phase has no predecessor, its weight is equal to its working time. If the phase has predecessors, there is computed the sum of the times of precedent phases with computed weights. To have an acyclic graph of precedence relations, we must compute the weight of all phases and reciprocally, if there is a phase whose weight can't be computed, the graph is cyclic (Swamidass, 2002).

2.3 Simulation Algorithm

As follows, we'll describe an algorithm of simulation of a technical process, in the meaning of random generation of the main characteristics of working phases that will be considered for solving balancing problem of a technical line whereon an only model of a product is assembled, operating times are established and line operating rhythm is fixed.

The algorithm is the fruit of authors' research in the wide and very complex domain of the problematic of technical line balancing (Coculescu, 2005).

Algorithm 2

Step 1. There is randomly generated the number of phases and for each phase, working time is generated.

Step 2. Matrix *PRED* of precedence relations is initialized by zeros. This matrix has a number of lines equal to the number of phases and 20 columns.

Step 3. $nrfail = 0$;

Step 4. A triplet (f_1, f_2, p) is generated where $f_1 = rnd(nrphases)$, $f_2 = rnd(nrphases)$, $p = rnd(20)$. If $f_2 \neq f_1$ we make $PRED(f_1, p) = f_2$ in the meaning that the phase f_2 precedes f_1. If $f_1 = f_2$ or $PRED(f_1, f_2) \neq 0$, we repeat step 4. .

Step 5. There is checked with algorithm 1 if the matrix *PRED* corresponds to an acyclic graph of precedence relations,

Step 6. If the graph is cyclic, $PRED(f_1, p) = 0$ and if $nrfail < 10000$ go to step 4. Else STOP.

Step 7. If the graph is acyclic, go to step 3.

3. CONCLUSIONS

Balancing using in industrial practice, can be facilitated by making a program which additional to a more efficient implementation of selected algorithms must offer the possibility of a comparative analysis of the methods used for solving balancing problem approached concerning the quality of given solutions.

Because this action demands the using of a relative big number of problem instance, we considered necessary the making of possibility of random generation of these models, being known that their gaining from real technical processes is difficult.

The idea we started from, was that to build a function which to have equal likelihoods of generation its values. For this, we combined standard random number generator of C programming language, with the values of random string (1) of natural parameters *a*, *b*, *c*.

By helping of gained function (*rnd*) there was realized a simulation model of real technical process in the meaning of random generation of the main characteristics of working phases that will be considered in solving balancing problem of a technical line whereon an only model of a product is assembled, operating times are established and line operating rhythm is fixed.

Simulation algorithm and its informatics implementation have been then integrated in a program, proving very useful within the process of testing performances of used methods for solving the approached balancing problem (Coculescu, 2005).

General rules from technical line simulation process shown in this work, can be applied for simulation other models, from other domain, especially of economic ones.

This paper was developed within the research Project „*Development and the Implementation of the Integrated Management System* (DI-IMS)", no. 2387/2007, PNCDI_II, developed on National Research Plan, it course being set it up to 2010. The institutional partners cover different activity fields, from variuos industries.

4. REFERENCES

Coculescu, C.; (2005). *Modeling and Simulation of Economic Processes. Concepts, Models and Balancing Algorithms*, Juristic Universe Editing Press, ISBN 973-8446-44-9, Bucharest

Filip, F.G.; Bărbat, B. (1999). *Industrial Informatics – New Paradigms and Applications*, Technical Editing Press, ISBN 973-31-1324-7, Bucharest

James, A.J. (2000), *Next Generation Manufacturing. Methods and Techniques*, Wiley, ISBN 978-0471360063

Swamidass, P.M. (2002). *Innovations in Competitive Manufacturing*, AMACOM, ISBN 978-0814471401

Vernadat, F. (1996). *Enterprise Modeling and Integration: Principles and Applications*, Springer, ISBN 978-0412605505

Annals of DAAAM for 2008 & Proceedings of the 19th International DAAAM Symposium, ISSN 1726-9679
Editor B. Katalinic, Published by DAAAM International, Vienna, Austria 2008
Make Harmony Between Technology and Nature, and Your Mind will Fly Free as a Bird

THE HARDNESS OF THE SURFACE AFTER THE FLAT LAPPING

COHAL, V[iorel] & SERB, A[drian]

Abstract: *The paper presents some research results concerning the hardness value of the surface layer after the flat lapping operation for bearings hardened steel. Lapping consists in the final smoothing of previously grinding surfaces. The result of researches was analysis with* ***NCSS.*** *Regression techniques analyze the relationship between a dependent (Y) variable and one independent (X) variables.*

Key words: *lapping, flat, hardness, capacity, cutting*

1. INTRODUCTION

The lapping operation, being a complex process because of the numerous influence factors that appears, it is hard to be overall analyzed, consequently, both theoretical and experimental research are imposed An objective factor for evaluation of the lapping process is the hardness value of the surface layer. Because of the contact temperature the abrasive particle cut onto a modified surface layer, whose properties are different from those that characterize the material. This fact explains the absence of the steel hardness influence onto the cutting workability that characterize the lapping operation, the principal factors that determine the properties of the surface layer, are the chemical composition of the steel and the properties of the surface layer after the previous operation (usually, the grinding operation).

Lapping consists in the final smoothing of previously grinding surfaces. This is done by means of certain fine abrasive particles impressed on the lap or freely interposed between the part to be processed and the lap. Abrasive pastes applied on the lap can also be used. Through the relative motion of the lap as compared to the part, in the presence of abrasive grains, particles are being removed from the processed material.

Most of the interactions between the workpiece and the granule occur when a load is applied by the lap. A variety of abrasive jet cutting and bead blasting processes rely on granule kinetic energy either shearing work material or initiating and propagating fracture. Mechanical aspects of the interaction of a granule and a workpiece surface have been widely studied. Significant insights are provided by a body of work in indentation and scratch testing, although considerable care needs to be exercised when applying such physical insights to lapping. In lapping, penetration depths tend to be much smaller than in most indentation and scratch testing, although this is not true for some of the recent nano-indentation studies. Workpiece surfaces exhibit property variation (subsurface damage) over a depth which will be a larger fraction of the penetration depth in lapping than in scratch testing. For polycrystalline work materials, the relationship between grain size of the material and penetration depth must also be considered. Combining this with insights from classical cutting mechanics, we can describe granule interactions with materials as a function of load and the mechanics of the contact. Simply stated, if the workpiece yield stress is not exceeded, the granule will simply slide across the surface imposing elastic deformation and removing no material. Larger loads form dislocations, plastically deform material. The extensive literature in these fields provide ample descriptions of large scale single grit interactions; the challenge is to define the contact conditions of granules used in lapping processes such that predictive kinetic models (rather than parametric correlations) can be developed.

In conventional lapping, the applied load is born by granules embedded in the lap system. As more granules embed into the lap, there are more sites, for material removal. However, the load per granule decreases as the granule concentration is increased. These effects cancel each other leading to the observation that in some circumstances the conventional abrasive lapping rate is independent of (granule) concentration. However, since the lap .area is finite, abrasive concentration extremes give different behavior. At low abrasive concentration, the individual abrasive particles may be pushed too deeply into the lap or the workpiece and become ineffective. And at high abrasive concentrations the lap can become saturated with granules and form a sliding surface with the excess material (Konig, 1990).

The recent literature regarding the interactions of the workpiece-lap-fluid triplet in chemical mechanical polishing consists of studies carried out over a broad range of material combinations under a variety of process conditions. As a result, models attempting to capture the resulting physical behavior consist of an equally broad range of assumptions and approaches. The range includes consideration of complete wafer-pad separation and the hydrodynamic modeling of the workpiece-lap-fluid contact to the case where there is pad and wafer contact in the presence of a fluid (Evans & Paul, 2003).

In lapping, the tooling allowance is minimum and it over tops only a little the height of the roughness resulting from previous grinding. That is why mechanical lapping should be accompanied by the selfcentring of the parts or of the tool and it cannot correct the geometrical shape obtained as a result of the previous operation.

2. EXPERIMENTAL RESULTS OF RESEARCHES

In the experimental frame were machined materials bearings hardened steel (RUL 1). The abrasive material was the mechanical paste of chrome oxide, having the precision FFF, which is characteristic for finishing operations. The experiments utilized a plane-lapping machine with eccentric.

The values of the hardness, before (t=0) and after the flat lapping operation are presented in fig. 1. In order to obtain a better visualization the experimental data were presented into a graphical way, the working parameters being beside specified.

The optimal regression function was chosen from the five types of regression functions (which are linear, logarithmical, power, polynomial and exponential), using the R^2 criterion. The value of the R^2 criterion is between 0 and 1. As close the R^2 criterion is to value 1, as close the drawn curve pass nearest the experimental points. When $R^2=1$, the regression function pass through all the experimental points. The presented mathematical model is valuable only for the considered time interval. During the lapping process on to the material bearings hardened steel the optimal regression function ($R^2=1$) is:

Analysis of Variance Report

Page/Date/Time 1 4/7/2008 8:54:54 AM
Database
Response C2

Tests of Assumptions Section

Assumption	Test Value	Prob Level	Decision (0.05)
Skewness Normality of Residuals	0.0000		
Kurtosis Normality of Residuals		1.000000	Accept
Omnibus Normality of Residuals			
Modified-Levene Equal-Variance Test			

Box Plot Section

I = f(t)

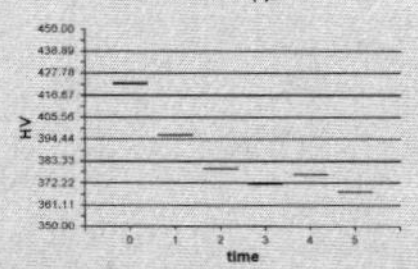

Expected Mean Squares Section

Source Term	DF	Term Fixed?	Denominator Term	Expected Mean Square
A: C1	5	Yes	S(A)	S+sA
S(A)	0	No		S(A)

Note: Expected Mean Squares are for the balanced cell-frequency case.

Analysis of Variance Table

Source Term	DF	Sum of Squares	Mean Square	F-Ratio	Prob Level	Power (Alpha=0.05)
A: C1	5	2126.833	425.3667			
S(A)	0	0				
Total (Adjusted)	5	2126.833				
Total	6					

* Term significant at alpha = 0.05

Analysis of Variance Report

Page/Date/Time 2 4/7/2008 8:54:54 AM
Database
Response C2

Kruskal-Wallis One-Way ANOVA on Ranks
Hypotheses
H0: All medians are equal.
Ha: At least two medians are different.

Test Results

Method	DF	Chi-Square (H)	Prob Level	Decision(0.05)
Not Corrected for Ties	5	5	0.415880	Accept H0
Corrected for Ties	5	5	0.415880	Accept H0

Number Sets of Ties 0
Multiplicity Factor 0

Group Detail

Group	Count	Sum of Ranks	Mean Rank	Z-Value	Median
0	1	6.00	6.00	1.4639	423
1	1	5.00	5.00	0.8783	397
2	1	4.00	4.00	0.2928	380
3	1	2.00	2.00	-0.8783	372
4	1	3.00	3.00	-0.2928	377
5	1	1.00	1.00	-1.4639	368

Means and Effects Section

Term	Count	Mean	Standard Error	Effect
All	6	386.1667		386.1667
A: C1				
0	1	423	0	36.83333
1	1	397	0	10.83333
2	1	380	0	-6.166667
3	1	372	0	-14.16667
4	1	377	0	-9.166667
5	1	368	0	-18.16667

Analysis of Variance Report

Page/Date/Time 3 4/7/2008 8:54:54 AM
Database
Response C2

Plots of Means Section

d = f(T)

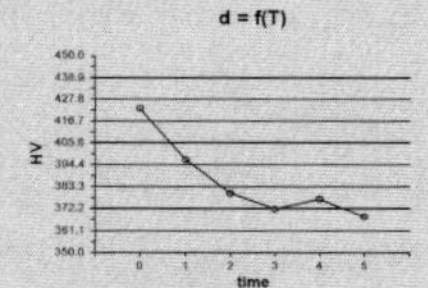

Kruskal-Wallis Multiple-Comparison Z-Value Test (Dunn's Test)

C2	0	1	2	3	4
0	0.0000	0.3780	0.7559	1.5119	1.1339
1	0.3780	0.0000	0.3780	1.1339	0.7559
2	0.7559	0.3780	0.0000	0.7559	0.3780
3	1.5119	1.1339	0.7559	0.0000	0.3780
4	1.1339	0.7559	0.3780	0.3780	0.0000
5	1.8898	1.5119	1.1339	0.3780	0.7559

Regular Test: Medians significantly different if z-value > 1.9600
Bonferroni Test: Medians significantly different if z-value > 2.9352

Kruskal-Wallis Multiple-Comparison Z-Value Test (Dunn's Test)

C2	5
0	1.8898
1	1.5119
2	1.1339
3	0.3780
4	0.7559
5	0.0000

Regular Test: Medians significantly different if z-value > 1.9600
Bonferroni Test: Medians significantly different if z-value > 2.9352

Fig.1. Analysis of Variance Report

$$HV0,2 = -0,475t^5+7,625t^4-45,808t^3+130,87t^2-198,32t+529 \quad (1.)$$

The rest of the working parameters that describe the plane lapping process were maintained constants, their values being settled on the basis of the previous experiments, into a manner that the roughness and the working efficiency to be as productive as possible. The hardness value of the surface layer during the lapping process decrease approximately with 60 hardness units of measure in the very first minute afterwards, the decrease trend continue less sharp until the value of working time is close to t=3 minutes, later the growth of the hardness is limited, having a maximum value for t=4,5 minutes. In this particular case the obtained regression function have a 5 degree polynomial form which on the considered interval of time is characterized by an ascending trend (Cohal, 1998).

This result of researches was analysis with **NCSS** (fig.1). **Number Cruncher Statistical System** (**NCSS**) is an advanced, easy-to-use statistical analysis software package. Regression techniques analyze the relationship between a dependent (Y) variable and one or more independent (X) variables. **NCSS** has regression procedures for many different situations.

The assumptions of the **one-way analysis of variance** (fig.1.) are:

1. The data are continuous (not discrete).
2. The data follow the normal probability distribution. Each group is normally distributed about the group mean.
3. The variances of the populations are equal.
4. The groups are independent. There is no relationship among the individuals in one group as compared to another.
5. Each group is a simple random sample from its population. Each individual in the population has an equal probability of being selected in the sample.

The assumptions of the **Kruskal-Wallis test** are:

1. The variable of interest is continuous (not discrete). The measurement scale is at least ordinal.
2. The probability distributions of the populations are identical, except for location. Hence, we still require that the population variances are equal.
3. The groups are independent.
4. All groups are simple random samples from their respective populations. Each individual in the population has an equal probability of being selected in the sample.

There are few limitations when using these tests. Sample sizes may range from a few to several hundred. If your data are discrete with at least five unique values, you can assume that you have met the continuous variable assumption. Perhaps the greatest restriction is that your data come from a random sample of the population. If you do not have a random sample, your significance levels will be incorrect (Hintze, 2007).

Given that the analysis of variance (ANOVA) test finds a significant difference among treatment means, the next task is to determine which treatments are different. Multiple comparison procedures (MCPs) are methods that pinpoint which treatments are different.

If the F-test from an ANOVA for this experiment is significant, we do not know which of the three possible pairs of groups are different. MCPs can help solve this dilemma.

3. CONCLUSIONS

Lapping is successfully applied in fine mechanics manufacturing, increasing the cutting tools lifetime and precision and also in finishing and superfinishing the ceramic workpieces.

The statement that during the lapping operation an improving of the material structure occurs, could be accepted only after a superfinishing of the surface and the removal of a part from the surface layer that is deformed and cold-harden.

4. REFERENCES

Cohal, V., *Contributions about surfaces lapping*. Doctor's degree thesis., Iaşi, Romania, 1998

Evans, C.J., Paul, E., Material Removal Mechanisms in Lapping, *Annals of the CIRP* Vol. 52/2/2003

Hintze, J., *NCSS Quick Start & Self Help Manual*, Kaysville, Utah, USA, 2007.

Konig, W., *Finish Processing, Ceramic Materials*, Zunch, 1990

Annals of DAAAM for 2008 & Proceedings of the 19th International DAAAM Symposium, ISSN 1726-9679
Editor B. Katalinic, Published by DAAAM International, Vienna, Austria 2008
Make Harmony Between Technology and Nature, and Your Mind will Fly Free as a Bird

EXPERIMENTAL RESEARCHES CONCERNING LAPPING OF FLAT SURFACES

COHAL, V[iorel]; SERB, A[drian] & CIOBANU, R[omeo]-M[ihai]

Abstract: *The paper presents some research results concerning roughness, cutting capacity on lapping of flat surfaces. Lapping pressure and friction lead to a decreasing of the abrasive grains dimensions. Consequence of this, the cutting capacity decreases and the surface roughness is better. The result of researches was analysis with* ***NCSS.*** *Regression techniques analyze the relationship between a dependent (Y) variable and one independent (X) variables.*

Key words: *lapping, roughness, abrasive, grain*

1. INTRODUCTION

Lapping consists in the final smoothing of previously grinding surfaces. This is done by means of certain fine abrasive particles impressed on the lap or freely interposed between the piece to be processed and the lap. Abrasive pastes applied on the tool may also be used. Through the relative motion of the lap compound to the piece, in the presence of abrasive grains, particles from the processed material are removed.

Lapping is necessary to obtain only special smoothness of the surface or to obtain extremely high dimensional precision together with special smoothness of the surface.

All lapping processes can be described as four component systems (a lap, a granule, a carrier fluid and a worckpiece) and that the mechanisms involved can be grasped by first understanding the interactions among those components. Lapping is differentiated technologically, but not in this mechanistic view; the relative size of the "grit" and the surface layer removed may change dramatically, but the processes all rely on interactions between these basic elements.

The object of lapping is to modify the workpiece. Workpieces vary in bulk chemical composition and may have property variations as functions of both lateral dimension and depth.

Depth variations may be intrinsic (due to different bonding for surface and bulk species) or may arise during the manufacture of the workpiece. They may arise from processing conditions (like plastic work of the near surface, or sub-surface cracking) or may be part of the lapping design process.

The fluid phase of the slurry may be characterized by its chemical composition and by its physical properties.

Chemical compositions of fluids include water and non-aqueous fluids like hydrocarbons and alcohols. The pH of the fluid may be controlled by addition of acids or bases, or by the use of a buffer system. In CMP, the fluid also contains a primary chemically active ingredient (like hydrogen peroxide or other oxidizers for metal) and may include secondary chemical ingredients (like inhibitors to control chemical interactions), or physically active additives (like surfactants).

Physical properties of the fluids affect both fluid dynamics and material transport in lapping. These properties include viscosity, density and thermal conductivity, all of which are pressure and temperature dependent. These properties can also be varied by changes in the chemical composition of the fluid.

Workpiece-fluid interactions involve both chemical and physical effects. The chemical effects include dissolution, etching and passivation, each of which change the workpiece surface in some way, and each of which has separate applications. Chemical changes in surface properties are sometimes described as "softening" of the workpiece, since they can make it more susceptible to lapping. Physical and mechanical effects involve transport of material and heat away from the workpiece surface during lapping; these processes are typically also affected by the nature of the lap and will be considered as three-way interactions below. All of these processes depend on the nature of the workpiece and the fluid.

Dissolution occurs when the chemical properties of the fluid and the workpiece surface match in specific ways. This match may hold for the bulk workpiece or only for its surface layer. In etching, a chemical reaction removes electrons from metals to produce soluble metal ions, which can then dissolve into the fluid (Evans & Paul, 2003).

Dissolution and etching are two-component methods of material removal. They are equivalent to lapping without either abrasive or lap, and are standard techniques in many applications which are not discussed here.

Passivation in particular, or coating in general, occurs when a chemical reaction between the workpiece and chemicals in the fluid results in a surface film that has different properties than the bulk workpiece. For metals, this film can reduce electrical or thermal conductivity, and is a desired or detested outcome depending on the particular application. For conventional lapping, the effects of this surface film are usually negligible. However, chemical mechanical lapping uses alternating cycles in which a surface film is chemically created and then mechanically removed. Thus chemical effects can be important. If the surface film is soluble, as is the case for most copper compounds, then this consideration must be included in describing any material removal process (Konig, 1990).

2. EXPERIMENTS AND RESULT OF RESEARCHES

The experiments have been carried out for cylindrical (Φ20 x 10) workpieces made by bronze (70 HV), brass (105 HV), annealed OLC45 steel (205 HV), 40Crl0 (220 HV) steel and bearings hardened steel RUL1 (700 HV). The lapping disk pressure was kept constant at 5 daN/cm^2, working time t = 3 min and cutting speeds values were 16, 32 and 44 m/min. A universal lapping paste was used composed of: white alundum —30%, vaseline oleum — 47 % and stearin — 23 %.

The experiments showed that together with the work hardness increasing to 200...220 HV, the material removal rate decreases significantly. For greater values of hardness, the decrease is slower (Cohal, 1998).

The surface roughness decreases when the hardness is smaller. When machining bronze and brass workpieces (with smaller hardness), the minimum roughness can be obtained for a cutting speed v=44 m/min. For the other materials, the optimum value for the cutting speed is v=*32* m/min.

One of the most important cutting parameter is the disk pressure upon the work. Its great values influence positively the cutting capacity but the influence is opposite to the surface roughness.

Analysis of Variance Report

Page/Date/Time 1 4/7/2008 8:46:41 AM
Database
Response C1

Tests of Assumptions Section

Assumption	Test Value	Prob Level	Decision (0.05)
Skewness Normality of Residuals	0.0000		
Kurtosis Normality of Residuals		1.000000	Accept
Omnibus Normality of Residuals			
Modified-Levene Equal-Variance Test			

Box Plot Section

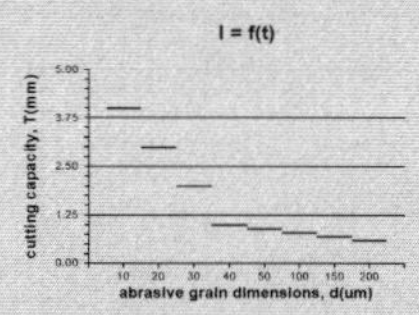

Expected Mean Squares Section

Source Term	DF	Term Fixed?	Denominator Term	Expected Mean Square
A: C2	7	Yes	S(A)	S+sA
S(A)	0	No		S(A)

Note: Expected Mean Squares are for the balanced cell-frequency case.

Analysis of Variance Table

Source Term	DF	Sum of Squares	Mean Square	F-Ratio	Prob Level	Power (Alpha=0.05)
A: C2	7	11.175	1.596429			
S(A)	0	0				
Total (Adjusted)	7	11.175				
Total	8					

* Term significant at alpha = 0.05

Analysis of Variance Report

Page/Date/Time 2 4/7/2008 8:46:41 AM
Database
Response C1

Kruskal-Wallis One-Way ANOVA on Ranks
Hypotheses
H0: All medians are equal.
Ha: At least two medians are different.

Test Results

Method	DF	Chi-Square (H)	Prob Level	Decision(0.05)
Not Corrected for Ties	7	7	0.428880	Accept H0
Corrected for Ties	7	7	0.428880	Accept H0

Number Sets of Ties 0
Multiplicity Factor 0

Group Detail

Group	Count	Sum of Ranks	Mean Rank	Z-Value	Median
10	1	8.00	8.00	1.5275	4
20	1	7.00	7.00	1.0911	3
30	1	6.00	6.00	0.6547	2
40	1	5.00	5.00	0.2182	1
50	1	4.00	4.00	-0.2182	0.9
100	1	3.00	3.00	-0.6547	0.8
150	1	2.00	2.00	-1.0911	0.7
200	1	1.00	1.00	-1.5275	0.6

Means and Effects Section

Term	Count	Mean	Standard Error	Effect
All	8	1.625		1.625
A: C2				
10	1	4	0	2.375
20	1	3	0	1.375
30	1	2	0	0.375
40	1	1	0	-0.625
50	1	0.9	0	-0.725
100	1	0.8	0	-0.825
150	1	0.7	0	-0.925
200	1	0.6	0	-1.025

Analysis of Variance Report

Page/Date/Time 3 4/7/2008 8:46:41 AM
Database
Response C1

Plots of Means Section

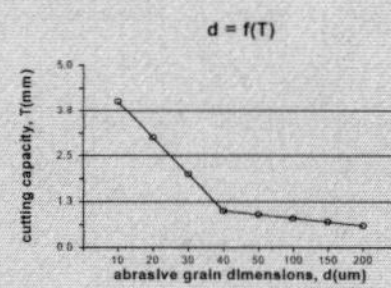

Kruskal-Wallis Multiple-Comparison Z-Value Test (Dunn's Test)

C1	10	20	30	40	50
10	0.0000	0.2887	0.5774	0.8660	1.1547
20	0.2887	0.0000	0.2887	0.5774	0.8660
30	0.5774	0.2887	0.0000	0.2887	0.5774
40	0.8660	0.5774	0.2887	0.0000	0.2887
50	1.1547	0.8660	0.5774	0.2887	0.0000
100	1.4434	1.1547	0.8660	0.5774	0.2887
150	1.7321	1.4434	1.1547	0.8660	0.5774
200	2.0207	1.7321	1.4434	1.1547	0.8660

Regular Test: Medians significantly different if z-value > 1.9600
Bonferroni Test: Medians significantly different if z-value > 3.1237

Kruskal-Wallis Multiple-Comparison Z-Value Test (Dunn's Test)

C1	100	150	200
10	1.4434	1.7321	2.0207
20	1.1547	1.4434	1.7321
30	0.8660	1.1547	1.4434
40	0.5774	0.8660	1.1547
50	0.2887	0.5774	0.8660
100	0.0000	0.2887	0.5774
150	0.2887	0.0000	0.2887
200	0.5774	0.2887	0.0000

Regular Test: Medians significantly different if z-value > 1.9600
Bonferroni Test: Medians significantly different if z-value > 3.1237

Fig.1. Analysis of Variance Report

Recommended values in case of lapping for this pressure are 0,4...5 daN/cm^2. The real machining pressure is much greater because of the machined surface which is much smaller than the workpiece surface and is continuously changing during machining. At the beginning, the disk-work contact is made only on the peaks of the surface roughness, the pressure is greater and the progress of machining is faster (approx. 3 μm/min in the case of steel). After a short time, the contact area increases and both pressure and cutting capacity are lower (approx. 1,5 μm/min).

This result of researches was analysis with **NCSS** (fig.1). **Number Cruncher Statistical System** (**NCSS**) is an advanced, easy-to-use statistical analysis software package. Regression techniques analyze the relationship between a dependent (Y) variable and one or more independent (X) variables. **NCSS** has regression procedures for many different situations

The assumptions of the **one-way analysis of variance** (fig.1.) are:

1. The data are continuous (not discrete).
2. The data follow the normal probability distribution. Each group is normally distributed about the group mean.
3. The variances of the populations are equal.
4. The groups are independent. There is no relationship among the individuals in one group as compared to another.
5. Each group is a simple random sample from its population. Each individual in the population has an equal probability of being selected in the sample.

The assumptions of the **Kruskal-Wallis test** are:

1. The variable of interest is continuous (not discrete). The measurement scale is at least ordinal.
2. The probability distributions of the populations are identical, except for location. Hence, we still require that the population variances are equal.
3. The groups are independent.
4. All groups are simple random samples from their respective populations. Each individual in the population has an equal probability of being selected in the sample.

There are few limitations when using these tests. Sample sizes may range from a few to several hundred. If your data are discrete with at least five unique values, you can assume that you have met the continuous variable assumption. Perhaps the greatest restriction is that your data come from a random sample of the population. If you do not have a random sample, your significance levels will be incorrect (Hintze, 2007).

Given that the analysis of variance (ANOVA) test finds a significant difference among treatment means, the next task is to determine which treatments are different. Multiple comparison procedures (MCPs) are methods that pinpoint which treatments are different.

If the F-test from an ANOVA for this experiment is significant, we do not know which of the three possible pairs of groups are different. MCPs can help solve this dilemma.

3. CONCLUSIONS

Lapping pressure and friction lead to a decreasing of the abrasive grains dimensions. Consequence of this, the cutting capacity decreases and the surface roughness is better. This is why there are recommendations to start the cutting process with sharp grains and to continue with worn ones.

Lapping is successfully applied in fine mechanics manufacturing, increasing the cutting tools lifetime and precision and also in finishing and superfinishing the ceramic workpieces.

4. REFERENCES

Cohal, V., *Contributions about surfaces lapping*. Doctor's degree thesis. Iaşi, Romania, 1998

Evans, C.J., Paul, E., Material Removal Mechanisms in Lapping , *Annals of the CIRP* Vol. 52/2/2003

Hintze, J., *NCSS Quick Start & Self Help Manual*, Kaysville, Utah, USA, 2007.

Konig, W., *Finish Processing, Ceramic Materials*, Zunch, 1990

Annals of DAAAM for 2008 & Proceedings of the 19th International DAAAM Symposium, ISSN 1726-9679
Editor B. Katalinic, Published by DAAAM International, Vienna, Austria 2008
Make Harmony Between Technology and Nature, and Your Mind will Fly Free as a Bird

CRANE'S DRIVING MECHANISM WARM UP TEMPERATURE CALCULATION FOR BRAKES WITH TWO SHOES

COLIC, M[irsad] & MUMINOVIC, A[dil]

Abstract: *Depending on their purpose and their build-in mechanism crane's brakes are exposed to the various working conditions. Characteristic of their work, like the work on all other brakes is the transformation of mechanical energy into heat, resulting in temperature increasement of the sliding elements. Increased temperature on the sliding surfaces effects on the lifetime of the friction linings and also on the brakes reliability. In this paper a method for predicting the temperature on the sliding surfaces is shown on a stopping brake example. Finite volume method is used for solving this problem.*
Key words: *brakes, temperature, finite volume method.*

1. INTRODUCTION

Brakes are important safety components of every crane on which they are build on. That is the place where, through friction, mechanical energy is transformed into heat, that is stored in the sliding elements, what leads to warming up and their temperature increasement. Greater change of temperature affects on the change in the coefficient of friction as does increase the wearing of the sliding surfaces. Because of that the temperature should be hold inside the given limits which are defined by the type of the friction material. Brakes are known for a long period and it is still researched and developed on them because of the increased number of devices on which they are build on and new demands that should be satisfied and also because of the appearance of new materials and calculations (Dufrenoy & Weichert 2003).

In this paper a numerical simulation of the braking process on the crane's driving mechanism is made, where the possibility of monitoring the temperature change on the sliding surface of the braking drum during crane's work is given. Elastic clamp's standard pulley made of steel EN: X27CrMoV51 is used as the braking drum.

To solve this problem a numerical method (finite volume method) is used.

2. MATHEMATICAL MODEL AND SOLUTION METHOD

With assumption that braking drum's exchange of temperature is not performed on the girth, at the temperature fields calculations, it is possible to pick one segment in radial direction with sufficient small angle and by doing that the problem is reduced on two-dimensional and the gained calculations results have the satisfying accuracy.

The equation that describes the temperature conduction inside the body, written in integral form is as following

$$\frac{\partial}{\partial t}\int_V \rho c T dV = \int_S \lambda_d \frac{\partial T}{\partial x_j} n_j dS + \int_V q_i dV \qquad (1)$$

Where: ρ-density, c-specific heat, T-temperature, t-time, x_j- coordinates, λ_d- coefficient of heat flow, q_i- specific amount of heat which for a unit of time appears or disappears, V –volume of a cell, S-total wrapping area, n_j-unit vector perpendicular to the outside area S.

For solving the equation (1) it is necessary to know the starting and boundary conditions (Colic et al., 2005). With assumption that the starting temperature is uniform it can be written

$$T(x_{j,0}) = T_p = const \qquad (2)$$

Boundary conditions are defined with heat exchange between the body surfaces and the environment, so the boundary heat flux is calculated by

$$q_B = -\lambda_d \left(\frac{\partial T}{\partial n}\right)_B = q_{zr} + q_{konv} \qquad (3)$$

Where: $(\partial T / \partial n)_B$- temperature gradient on the bordering body, q_{zr} i q_{konv} - heat taken from free body surfaces by radiation and convection.

Part of the temperature that is lost through radiation to the outside environment is calculated by

$$q_{zr} = \varepsilon\beta\left(T_S^{\ 4} - T_O^{\ 4}\right) \qquad (4)$$

Where: ε- emission coefficient, β-Stefan-Boltzmann constant, T_S- manifold surface temperature, T_O- temperature of air at a distance from the manifold.

Interchange of heat by convection according to Newton is defined by the coefficient of the transfer of heat

$$q_{konv} = \alpha(T_S - T_O) \qquad (5)$$

Where: α - heat transfer coefficient.

Equation (1) with the starting and limiting values described above (2) and (3) can be solved with finite volume method (FVM) (Demirdzic & Ivankovic, 1996/97) where process time has to discretize the given number of time intervals δt, space domain to the N control volumes (CV). Integrating the equation to the control volume (CV) using linear space change variable and implicit time differentiation we are able to derive equation (1) algebratically

$$a_P T_P = \sum_K a_K T_K + b, \quad (K = E, W, N, S, T, B) \qquad (6)$$

Which is equated iteratively (Demirdzic & Ivankovic, 1996/97).

Solving method for the given mathematical model has been tested on examples with already known results. Compared results for one of the examples are given on figure 1 (Milincic, 1979).

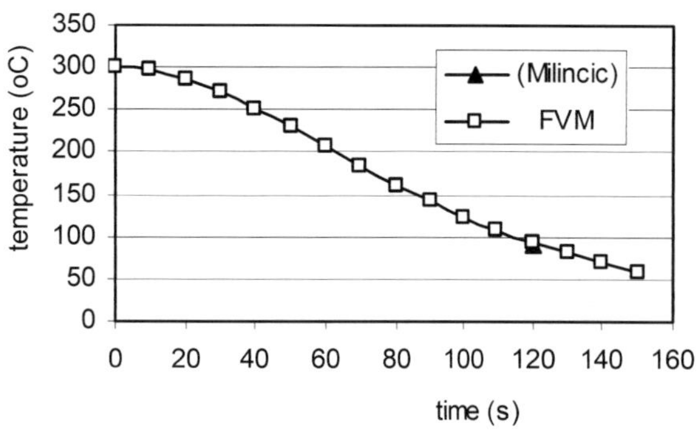

Fig. 1. Temperature change in the middle of the disc

3. TEMPERATURE ANALYSIS EXAMPLE

3.1. Braking modelling

Expression for the total kinetic energy of masses in motion at start of the braking is given by

$$E_C = \frac{J_R \cdot \omega_{K0}^2}{2} \qquad (7)$$

Where: J_R - reduced inertia momentum and ω_{K0} - angular velocity of the drum at the braking start.
Considering the fact that the part of the energy is spent on overcoming the driving resistance and the rest of the energy is transformed into heat during friction on the brake it is possible to write the next equation

$$E_C - F_{Wv}' s_z - Q_K = 0 \qquad (8)$$

Where: F_{Wv}' - constant driving resistance, s_z - crane's stopping path and Q_K - mechanical energy transformed into heat.
Stopping brakes stops the crane from the full velocity to the complete halt during which the velocity of the braking drum changes by the principle

$$\omega(t) = \omega_{K0} - \frac{\omega_{K0}}{t_z} t \qquad (9)$$

So the quantity of the mechanical energy transformed into the heat carried over to the brake in time and expressed in surface units (heat flux) is defined with the expression

$$q(t) = \left(E_C - F_{Wv}' s_z \cdot \eta\right) \frac{(t_z - t)}{A_p t_z^2} = (\mu p_{sr} v_0) \cdot \frac{t_z - t}{t_z} \qquad (10)$$

Where: $\mu p_{sr} v_0$ - specific braking power.
It is supposed, in calculation, that all heat is carried to the drum, by Ernst (Ernst, 1966)

$$q = q_b + q_p \cong q_b \qquad (11)$$

3.2. Calculation results

Brake with 200 mm and 960 min^{-1} braking drum diameter and 100 Nm available braking momentum is taken for the calculation example. Only temperatures on the sliding surface of the brake, immediately after removal of the brake shoe, during one braking process, are considered while displaying the calculation results.

Temperatures are fastly increasing and reach their maximum somewhere in the half of the braking cycle because the heat flux is maximal at the braking start.

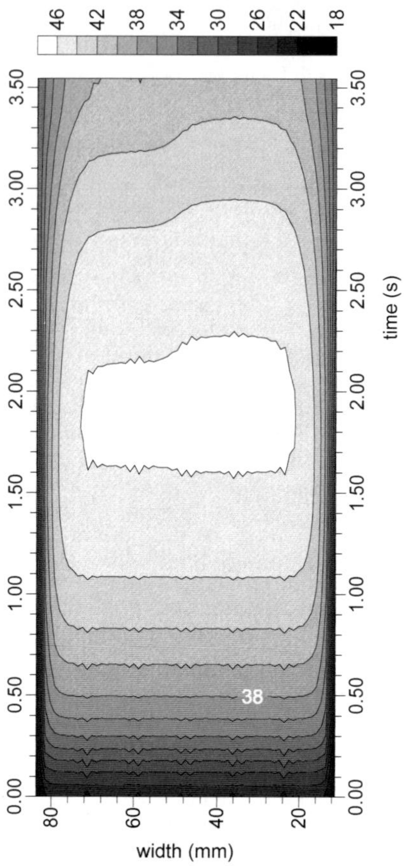

Fig. 2. The temperature change on the drums sliding surface during one braking process

4. CONCLUSION

The results show that it is possible to predict heating temperatures in the design phase or during the brake selection and this gives additional safety. It is possible to simulate the work of the brake during longer period of the crane's work and this positively influence on the lifetime and reliability of the brakes.

An experiment will be made to verify the calculation results.

5. REFERENCES

Colic, M.; Repcic, N. & Muminovic, A. (2005), A Thermal Analysis of the Crane Braking Drum for Regulation of Lowering Load Velocity, *The 9 th International Research/Ekspert Conference TMT,* Ekinovic, S. (Ed.), pp. 1013-1016, Antalya, September, 2005, Faculty of Mechanical Engineering in Zenica, Zenica

Demirdzic, I.; Ivankovic, A. (1996/97), *Finite Volume Stress Analysis,* Imperial College of Science, Technology and Medicine University of London

Dufrenoy; P.; Weichert, D. (2003), A termomehanical model for the analysis of disc brake fracture mechanisms. *Journal of Thermal Stresses,* 26, 2003, 815-828, ISSN: 0149-5739

Ernst, H. (1966), *Die Hebezeuge (eng. Cranes), Band I,* Braunschweig

Milincic, D. (1979), *Prostiranje toplote (eng. Heat Transmission),* Masinski fakultet u Beogradu, Beograd

Annals of DAAAM for 2008 & Proceedings of the 19th International DAAAM Symposium, ISSN 1726-9679
Editor B. Katalinic, Published by DAAAM International, Vienna, Austria 2008
Make Harmony Between Technology and Nature, and Your Mind will Fly Free as a Bird

USING PETRI NETS IN THE SOCCER ROBOT CONTROL ARCHITECTURE

COMAN, D[aniela]

Abstract: *Robot soccer is a challenging platform for multi-agent research, involving topics such as real-time image processing and control, robot path planning, obstacle avoidance and machine learning. A soccer robot has to take an appropriate decision based on environment situation. With the role of a robot fixed as goalkeeper, the supervisor, according to the game situation, assigns the role of attacking or defending to the other robots and then the respective controllers control the robots. The controller for the attacking robot is designed using Petri nets. The Petri net model are implemented in Petri Net Toolbox under MATLAB environment.*
Key words: *robot soccer, Petri net, MiroSoT.*

1. INTRODUCTION

Robot soccer games had been popular with educational institutions around the world since the inauguration of the FIRA MiroSoT competition in 1996 and the RoboCup competition in 1997. These initiatives provide a good platform for multi-agent domain research, dealing with issues such as co-operation by distributed control, effective and fault tolerant communication, real-time image processing, real time robot path planning and obstacle avoidance.

In robot soccer systems, images of objects on the field are processed by a vision system. Analysis of this raw data will yield information such as identification of objects including ball, team robots and opponents robots. Other information such as object identity (identity of robot), opponent, position, orientation and velocity can also be computed (Kim et al., 2004). Based on this information, each of the players carries out assigned roles including attacker, defender and goalkeeper. The simplest role selection strategy is to have a fixed role that does not change throughout the game. However, permanent role fixing causes undesirable behaviour such as a defensive robot not going for the ball even though the ball is near but outside its defence zone; or a forward robot giving up its possession of ball when it incidentally enters a defence zone (Shim et al., 1997).

In this present work it is presented a Petri-net state diagram approach for attacking robot for MiroSoT (Micro-Robot World Cup Soccer Tournament). In small league MiroSoT, a team has 3 team robots, inclusive of the goalkeeper. The robots are small in size (7.5cm x 7.5cm x 7.5cm), fully/semiautonomous.

Different types of schemes for soccer robot control are available (Shim et al., 1997). In a remote-brainless (visionbased) soccer robot scheme a host computer controls the robots by commanding the robot velocities like a radio-controlled car. In the brain on-board (vision-based) soccer robot system, the robots move as per the vision data, keeping away from obstacles autonomously. In a robot-based scheme, each autonomous robot can make a decision based on information it collects with its sensors and, if needed can communicate each other. Robots which can be operated autonomously are needed for the robot-based system. The control structure, behaviors and actions of robots will influence their performance in a big way in robot-soccer games. This paper deals with the remote brainless robot system.

In section 2, the basic theory of Petri nets it is presented. In section 3, it is illustrated how a attacking robot controller is designed using Petri net. In section 4, the Petri net model are implemented in MATLAB environment. Concluding remarks follow in section 5.

2. PETRI NET STRUCTURE AND GRAPH

Petri nets, developed by Carl Adam Petri in his Ph.D. thesis in 1962, are generally considered as a tool for studying and modelling of systems. A Petri net is foremostly a mathematical description, but it is also a visual or graphical representation of a system. A Petri net can be analysed in order to reveal important information about the structure and behavior of the modelled system. The information may for instance suggest improvements to the system.

A Petri Net consists of a number of places and transitions with tokens distributed over places. Arcs are used to connect transitions and places. When every input place of a transition contains a token, the transition is enabled and may fire. The result of firing a transition is that a token from every input place is consumed and a token is placed into every output place. The firing of transitions represents causality and inferencing relations. Petri nets have been generalized by allowing multiple token arcs, inhibitor arcs, place capacity, colored tokens etc. (Peterson, 1981).

Formally, the structure of a Petri net, defined by its places, transitions, input function and output function (Ziparo & Iocchi, 2006).

Definition 2.1 (Petri Net Structure): A Petri net structure, C, is a four-tuple,

$$C = (P, T, I, O) \tag{1}$$

where: $P = \{p_1, p_2, \ldots, p_n\}$ is a finite set of places, $n \geq 1$; $T = \{t_1, t_2, \ldots, t_m\}$ is a finite set of transitions, $m \geq 1$; $I : T \rightarrow P^\infty$ is the input function, a maping from transitions to bags of places, $O : T \rightarrow P^\infty$ is the output function, a mapping from transitions to bags of places. The set of places and the set of transitions are disjoint, i.e., $P \cap T = \varnothing$.

Definition 2.2 (Petri net graph): A Petri net graph G is a bipartite directed multigraph, $G = (V, A)$, where $V = \{v_1, v_2, \ldots, v_s\}$ is a set of vertices and $A = \{a_1, a_2, \ldots, a_r\}$ is a bag of directed arcs, $a_i = (v_j, v_k)$, with $v_j, v_k \in V$. The set V can be partitioned into two disjoint sets P and T such that $V = P \cup T$, $P \cap T = \varnothing$, and for each directed arc, $a_i \in A$, if $a_i = (v_j, v_k)$, then either $v_j \in P$ and $v_k \in T$ or $v_j \in T$ and $v_k \in P$.

3. ATTACKING ROBOT CONTROLLER

The principle to apply is simple: the robot assigned to attack should kick the ball to the opponent goal before any opponent robot can kick it.

Three simple states are assumed for the attacking controller:

- attacking robot behind the ball,
- attacking robot kicks the ball,

- attacking robot in contact with ball and behind.

The following places and transitions are defined for the modelling the attacking robot in Petri-net.

The key situations described above are represented by places p_i, $i \in \{1, 2, 3\}$.

Structure: (P, T, I, O).

Places: $P = \{p_1, p_2, p_3\}$.

Transitions: $T = \{t_1, t_2, t_3, t_4, t_5\}$.

Input function: $I(t_1) = \{p_2\}$, $I(t_2) = \{p_1\}$, $I(t_3) = \{p_1\}$, $I(t_4) = \{p_3\}$, $I(t_5) = \{p_4\}$.

Output function: $O(t_1) = \{p_1\}$, $O(t_2) = \{p_2\}$, $O(t_3) = \{p_3\}$, $O(t_4) = \{p_1\}$, $O(t_5) = \{p_3\}$.

The following places are defined to model the attack robot:

p_1 : moves behind the ball,

p_2 : kicks the ball,

p_3 : in contact with the ball and behind, so it has to move far away from the ball,

The transitions are defined as:

t_1 : tries to kick the ball, though it is not in a good position to kick,

t_2 : in front of the ball and at the following instant it is in a good position to kick,

t_3 : moves in front of the ball to the opponent goal,

t_4: in front of the ball, and escaping from that position,

t_5 : misses the ball, and it is in front of the ball.

With these defined places and transitions, the Petri net model for the attacking robot is arrived at (Fig. 1).

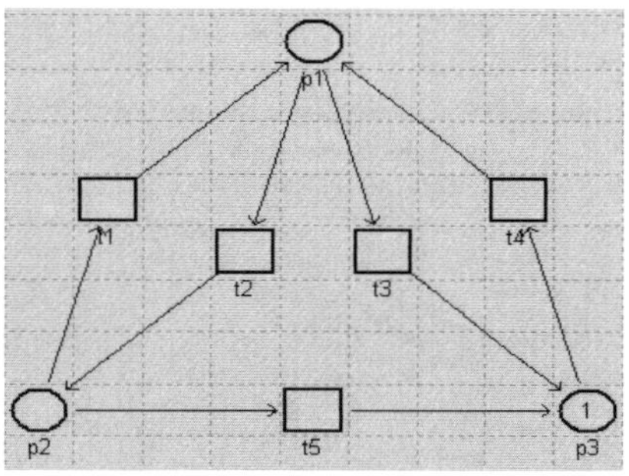

Fig. 1. Petri net model for attacking robot controller

4. SIMULATION STUDIES

The simulation studies are carried out for the Petri net giving statistics on events and resources for system (Pastravanu et al., 2003). The following two tables (Tables 1 and 2) present the complete lists of global indices associated with the places and the transitions considered in the architecture of Petri net that modelling the controller for attacking robot.

Place Name	Arrival Sum	Arrival Rate	Arrival Dist.	Throughput Sum	Throughput Rate	Throughput Dist.
p1	100	1	1	101	1.01	0.9901
p2	50	0,5	2	50	0.5	2
p3	76	0,76	1.3158	75	0.75	1.3333

Tab. 1. Global Statistics Places

Transition Name	Service Sum	Service Rate	Service Dist.
t1	25	0.25	4
t2	50	0.5	2
t3	51	0.51	1.9608
t4	75	0.75	1.3333
t5	25	0.25	4

Tab. 2. Global Statistics Transitions

A synthesis of this Petri net model is depicted in Fig. 2 and Fig. 3, which allows exploring the dependences of "Arrival Rate" and "Throughput Dist." indices on two "Design Parameters" (considered places p_1 and p_3).

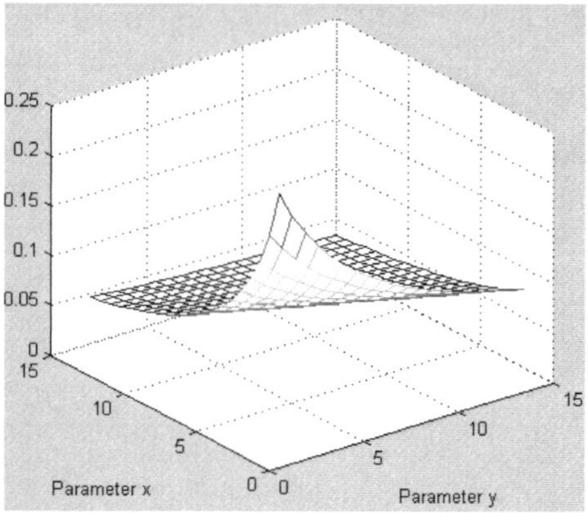

Fig. 2. Dependence of "Arrival Rate" on "Design Parameters"

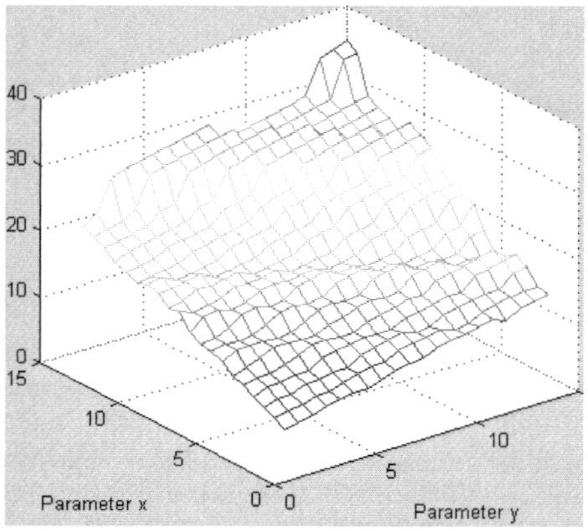

Fig. 3. Dependence of "Throughput Dist." on "Design Parameters"

5. CONCLUSION AND FUTURE WORKS

In this paper, a Petri net is used for designing a low level controller for the attack robot. The controller did not use the information of opponent team. But these will have to be studied to bring in more cooperation among robot agents. This Petri net model are implemented in MATLAB environment. The analysis of the acquired global performance indices confirm the accuracy of the proposed Petri net model.

6. REFERENCES

Kim J.H.; Kim D.H.; Kim Y. J. & Seow K.T. (2004). *Soccer Robotics*, Springer-Verlag, ISBN 3-540-21859-9, Germany

Pastravanu O.; Matcovschi M. & Mahulea C. (2003), Petri Net Toolbox – teaching discrete event systems under Matlab, In: *Advances in Automatic Control*, Voicu M. (Ed.), pp. 247-255, Kluwer Academic Publishers, ISBN 1-4020-7607-X, Iasi

Pastravanu O.; Matcovschi M. & Mahulea C. (2002), *Applications of Petri Net in Studying Discrete Event Systems*, Gh. Asachi, ISBN: 973-8292-86-7, Iasi

Peterson J. L. (1981). *Petri net theory and modelling of systems*, Prentice-Hall Inc., USA

Shim H.-S.; Kim H.-S.; Jung M.-J; Choi I.-H.; Kim J.-H. & Kim J.-O. (1997). Designing Distributed Control Architecture for Cooperative Multi-agent System and Its Real-Time Application to Soccer Robot, *Journal of Robotics and Autonomous System*, Vol. 21, No.2, September 1997, pp. 149-165

Ziparo V. & Iocchi L (2006). Petri Net Plans, *Available from:* http://www.dis.uniroma1.it/~iocchi/publications/iocchi-moca06.pdf, *Accessed*: 2007.11.12

Annals of DAAAM for 2008 & Proceedings of the 19th International DAAAM Symposium, ISSN 1726-9679
Editor B. Katalinic, Published by DAAAM International, Vienna, Austria 2008
Make Harmony Between Technology and Nature, and Your Mind will Fly Free as a Bird

WETTABILITY MONITORING BY SURFACE MODIFICATION WITH FEMTOSECOND LASER IRRADIATION

COMAN, D[iana]; VRINCEANU, N[arcisa] & GRIGORIU, A[urelia]

Abstract: *This paper presents the study of the surface modification of cellulose textiles in order to improve their hydrophilic properties as well as increase the absorption capacity of the material, as well. Surface treatment has been provided by laser femtosecond (fs) LPX 200 Excimer KrF, operating at 248 nm and surface properties have been characterized by measuring the contact angle of viscose fabrics with liquid. Textiles standard industrial permeability measurements have also been used.*

Key words: *surface, laser, viscose, textiles, permeability.*

1. INTRODUCTION

Laser irradiation of textiles can generate characteristic modifications of surface morfology. In many industrial applications there is a need to modify the polymer surface by keeping unchanged their bulk properties. Chemical activation of surfaces is the most commonly used method, however, the ecological requirements have forced the industry to search for alternative environmental safety methods. Surface modification of polymers by means of laser irradiation has also lead to a roughness alteration, achieving a ripple-like microstructure.

Due to low surface energy (approximately 20-25 mj/m^2), viscose has very weak hydrophilic properties. The aim of the present work is to investigate the surface modification of viscose fabrics by means of laser treatment so as to improve their hydrophilic properties. Many laser applications in the surface modification field were made at reduced fluences (Zorba et al, 2006). Excimer UV laser impact on textile properties as a result of surface modifications has been researched on (Kan 2008).

This work is focused on laser surface activation of viscose textiles and controlled observance of permeability properties. The novelty of this approach lies in the laser treatment application and the wettability monitoring on regenerated cellulose, so far such studies having been made mostly on synthetic fibres.

2. EXPERIMENTAL

2.1 Materials

In this study the laser activation of 100% viscose fabrics was investigated. The conditioned samples are employed before the preparation stage technology.

2.2. Laser treatment

The surface activation was provided by irradiation with a femtosecond (fs) LPX 200 Excimer KrF laser operating at 248 nm with the parameters shown in Table 1. In high-fluence laser irradiation, samples were irradiated directly from the laser beam without using either special photomask or focusing lens.

The laser energies in terms of applied fluence and number of pulses varied in different experiments so as to study their effects upon samples. Laser fluence was regulated in range from 29 to 43 mJ/cm^2 and the number of pulses varied between 0 and 4, with the pulse repetition constant at 1Hz to avoid any possible heat accumulation. At first, several attempts at optimizing the process of surface modification have been made.

Water contact angle was measured using a Sigma 700 computer-controlled research tensiometer. Water permeability of samples before and after laser treatment was determined by measuring the time necessary for the penetration of 5 ml of the testing liquid across the sample. The absorption properties of sample were measured as a difference of the weights of the sample before and after the dipping in the testing liquid. The contact angle was measured directly from the observation of the solid-liquid meniscus.

E (mJ)/0,4	S (cm2)	F (J/cm2)	Pulses
170.00	0.39	**0.435897**	4p
170.00	0.39	**0.435897**	2p
170.00	0.45	**0.377778**	**1p**
170.00	0.45	**0.377778**	**4p**
170.00	0.57	**0.298246**	2p
170.00	0.57	**0.298246**	1p

Tab. 1. Table with some parameters of laser irradiation treatment

3. RESULTS AND DISCUSSION

Laser treatment of viscose proceeds by a free-radical mechanism that introduces a wide variety of oxidized functional groups onto the surface of the treated polymer. These oxidized functional groups may include C-OH, C=O, COOH, C-O-C, or hydroperoxide and they are responsible for the change in the polymer surface properties. The surface properties of treated and untreated samples were characterized by means of the contact angle measurement and the water permeability measurement (Xiaodong et al, 2004). The contact angle of liquid on solid is closely related to surface free energy and this parameter is useful in the discussion of hydrophilicity, absorbency of sample and adhesivity. For the determination of total surface energy from contact angle measurement Wu's equation (Tammar et al, 2004) and Owens-Wendt-Kaeble equation (Lam et al, 2001) were used for plots from diagrams presented below. Figure 1 shows the dependence of the water contact angle and water permeability on the laser treatment number of pulses. Fast increase in permeability was observed during the first 2 pulses, followed by a slow increase with increasing the number of pulses. However, the water contact angle linearly decreases with increasing of pulses. The effect of the fluences on the water contact angle and water permeability was tested and the results are shown in Figure 2. Strong influence of the fluence of hydrophilic properties of viscose textiles was observed. The permeability exponentially increases with increasing fluence, however, the water contact angle decreases linearly. The contact angle decreased linearly with increasing laser treatment time. Laser irradiation treatment longer than 4 pulses can cause hydrophilisation of textile surface. As a result, the optimal laser treatment time at this condition is 4 pulses. After 4 pulses one can perform to the burning of fabric surface.

The aim of our work has been to increase the absorption capacity of the material, as well.

For the laser surface activation, it is possible to reach 500% absorption improvement of viscose textiles as is shown on Figure 4.

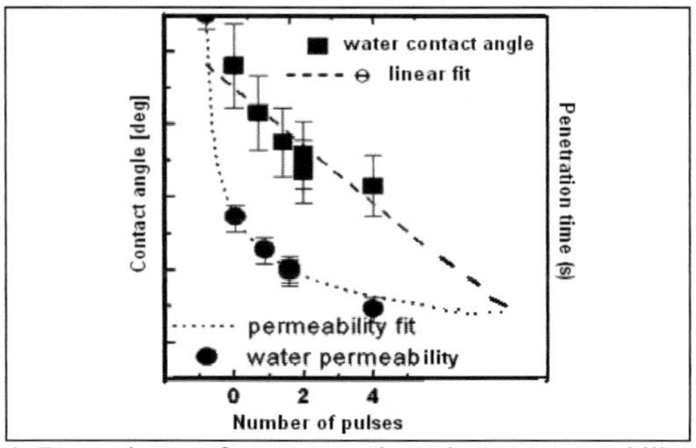

Fig. 1. Dependence of contact angle and water permeability as a function of surface treatment number of pulses

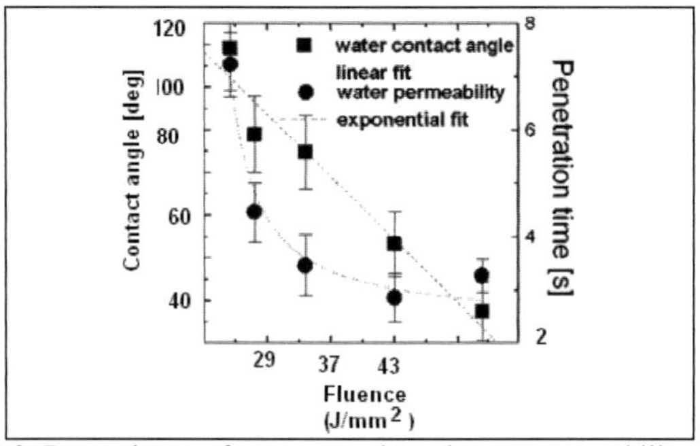

Fig. 2. Dependence of contact angle and water permeability as a function of fluence (J/cm^2)

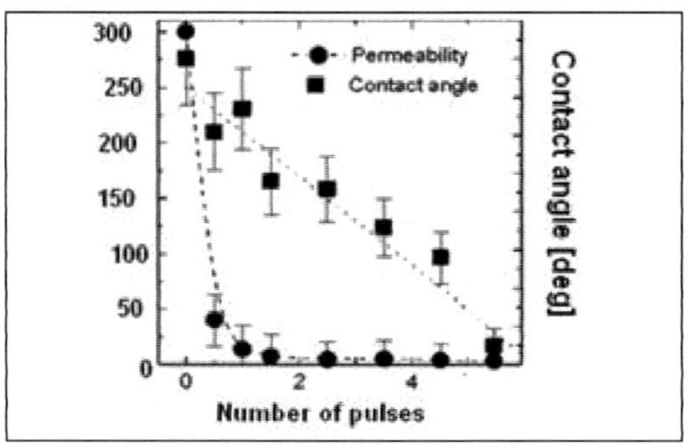

Fig. 3. Dependence of contact angle and water permeability as a function of laser treatment pulses and number

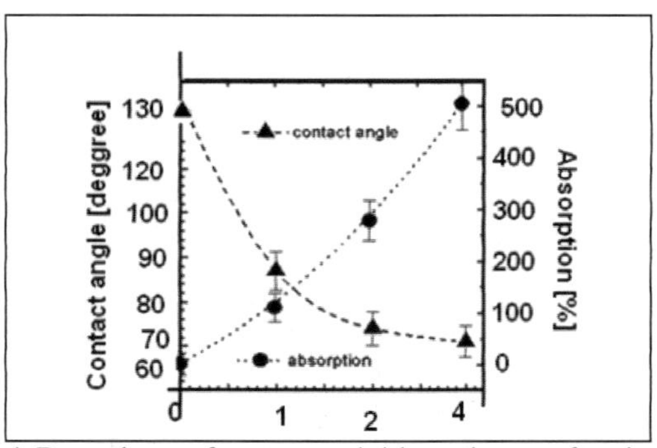

Fig. 4. Dependence of contact angle/absorption as a function of number of pulses

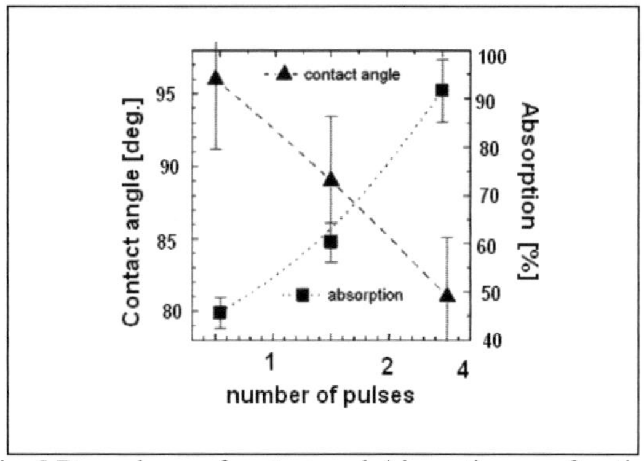

Fig. 5 Dependence of contact angle/absorption as a function of surface laser treatment pulses number

The water contact angle decreased to 70°. The laser treatment acted only on the surface of the viscose material i.e. the water contact angle decreased to 80º, however the absorption increased only up to 90 % (see Figure 5).

4. CONCLUSION

The experimental part of this study has shown the utility of laser treatment on viscose textiles as well as highlighted the following problems:

•It is obvious that femtosecond laser assisted-modification of viscose surface induces remarkable change in wettability of the textile material resulting from contact angle modifications and water permeability.

•High water absorption due to variation of irradiation parameters (fluence, pulse width, pulse number) will lead to a controlled improvement of comfort properties.

•Changes in the observed wettability may be attributed to a sinergy of surface chemistry and roughness effects.

•Laser surface modification will constitute the most effective tool for a non-contact, ecological technique for such textile types.

The aim of further experiments will be the optimization of laser treatment conditions with the purpose of reaching the lowest laser exposure time and increasing the stability of induced properties.

5. ACKNOWLEDGEMENTS

The authors would like to thank to Prof. FOTAKIS Costas, Director of Institute of Electronics Structure and Lasers (IESL), Foundation of Research Technology Hellas (FORTH), Heraklion, Greece, for having facilitated access to the above-mentioned institute's labooratories and also to the technical assistance offered by Dr. POULI and Ms. MELLESANAKI.

6. REFERENCES

Kan, C.W. (2008). Impact on textile properties of polyester with laser, *Optics & Laser Technology*, 40, 113–119

Lam, C.N.C.;Ko, R.H.Y.(2001). Dynamic cycling contact angle measurements: study of advancing and receding contact angles, *J.Colloid Interf.Sci*., 243, 208

Tammar, S.; Meiron, A.; Sam Saguy, I. (2004). Contact angle measurement on rough surfaces, *J.Colloid Interf.Sci.*,274, 6

Xiaodong W.; Xiaofenand, P.; Buxuan W. (2004). Contact angle hysteresis and hysteresis tension on rough solid surface, *Chinese J. Chem. Eng*. 12 (5) 615

Zorba, V.; Persano, L.; Pisignano, D.; Athanassiou, A.; Stratakis, E.; Cingolani, R.; Tzanetakis, P.; Fotakis, C. (2006). Making silicon hydrophobic: wettability control by two-lengthscale simultaneous patterning with femtosecond laser irradiation. *Nanotechnology,* Vol.17, (3234-3238).

Annals of DAAAM for 2008 & Proceedings of the 19th International DAAAM Symposium, ISSN 1726-9679
Editor B. Katalinic, Published by DAAAM International, Vienna, Austria 2008
Make Harmony Between Technology and Nature, and Your Mind will Fly Free as a Bird

ACHIEVING HIGH TEMPERATURE DEFORMATION FOR Al-Mg ALLOYS PROCESSED BY SEVERE PLASTIC DEFORMATION

COMANECI, R[adu]; ZAHARIA, L[uchian] & CHELARIU, R[omeu]

Abstract*. An Al-1.5%wt.Mg alloy was subjected to severe plastic deformation (SPD) by Equal Channel Angular Pressing (ECAP) at room temperature to a total strain of ≈ 6 that resulted in a submicronic grain size. Strain rate-change tests at temperatures between 463 and 523K and strain rates between $1.68x10^{-3}s^{-1}$ and $8.4x10^{-5}s^{-1}$ are carried out to determine the best elongation to failure that material shown. We are interested in a maximum elongation to failure as effect of SPD and thermal stability of the induced structure, without presence of secondary phase due to special alloying elements. The results show that ECAP is an effective procedure for achieving grain refinement and good ductility at elevated temperature.*
Keywords: *aluminum, deformation, plasticity.*

1. INTRODUCTION

There is now a great interest in developing highly formable Al-Mg based alloys for application leading to lightweight vehicles. Significant efforts have been made to improve the formability and strength of Al-Mg alloys. Wrought Al-Mg alloys (5xxx series) are non-heat-treatable. For a small amount of Mg, strength properties of these alloys are diminishing; but they have good corrosion resistance and weldability. Because of that, their commercial applications are restricted to very specific area. Scandium is the only alloying equilibrated, thermally stable, coherent L12 phase, i.e. Al_3Sc, in aluminum. Thus, Al_3Sc precipitate can be used as grain boundary resistant barrier to coarsening (Kyung-Tae et al., 2005, Musin et al., 2004).

The principle of developing (super)plasticity through SPD relies on the properties varying inversely with the submicronic grain size. On review of earlier works (Noda et al., 2004, Kyung-Tae et al., 2006) it is shown that to develop superplasticity, the material has to attain ultrafine grain sizes and retain the ultrafine sizes at elevated testing temperatures. If these ultrafine grains show reasonable stability at elevated temperatures, the alloy may exhibit a capability for achieving good ductility.

The present work thus initiated to explore the potential of achieving high temperature deformation for Al-1.5Mg alloy without the third alloying element, trough ECAP technique.

By improvement of mechanical properties and formability of Al-Mg alloys it is possible to introduce these non-heat-treatable but very good corrosion resistant alloys in near net shape forming processes.

2. EXPERIMENTAL PROCEDURE

Ingots of Al-1.35Mg with dimensions of 20x24x150mm were given a homogenization heat-treatment at 500°C for 24 hours (this resulted in coarsening of about 100μm) and then machined at final dimensions of 10x10x60mm.

These billets were pressed in an equal-channel angular die of 10 mm width channel, with a $\phi = 90°$ channel intersection angle and a $\psi = 28°$ curvature on the outer side of channel intersection, see Figure 1. The metal was subjected to a simple share strain in the bisect plan of the channels, under relative low pressure compared to the traditional extrusion process. The workpieces were pressed through the die 6 passes via route B_C (rotation of sample by 90° in same direction after every pass).

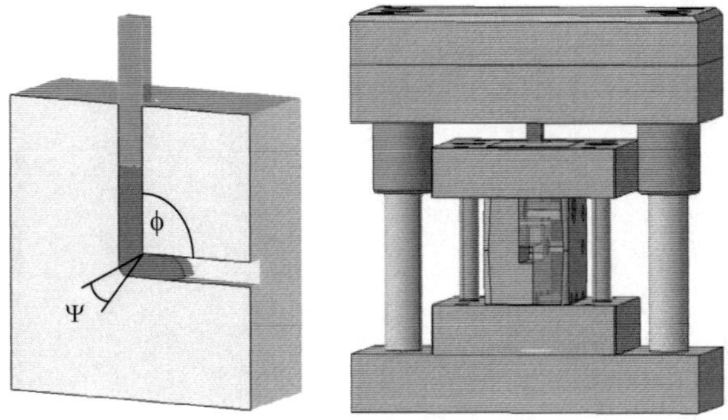

Fig.1. Die geometry and the ECAP assembly for SPD

The extrusions were carried out at 293 K at a speed of 8.75 mm/s. In order to reduce the friction, a solid lubricant was used. The total equivalent strain accumulated for N = 6 passes was ≈ 6 according to well-known Equation (1).

$$\varepsilon_N = \frac{N}{\sqrt{3}}\left[2\cot\left(\frac{\phi}{2}+\frac{\psi}{2}\right)+\psi\,\mathrm{csc}\left(\frac{\phi}{2}+\frac{\psi}{2}\right)\right] \quad (1)$$

where the significance of terms are revealed in Figure 1.

Samples of 24mm total length with 1.25mm gauge height, 10mm gauge length and 2.5mm gauge width were machined parallel to the longitudinal axis of the pressed rods by electro-discharge machine. The dimensions were established according to the rule of proportional rectangular tensile specimens:

$$L_g = k \cdot \sqrt{A_g} \quad (2)$$

where k = 5.65 (from ASTM Standard EN 10002-1) and L_g and A_g are the gauge length and gauge area respectively. The dimensions are imposed by the final dimensions of the work pieces after ECAP process which step down, Figure 2.

Fig.2. The billets after ECAP processing (0; 1; 2; 3; 4; 6 passes)

Tensile test is performed at 463, 493 and 523K at an initial strain rate of $1.68x10^{-3}s^{-1}$ and $8.4x10^{-5}s^{-}1$. The samples were subjected to tensile tests through a mounting device, Figure 3, under controlled temperature conditions using a universal INTRON 3382 testing machine and an environmental chamber INSTRON 3119-506. The samples were held at nominal temperatures for 120min.

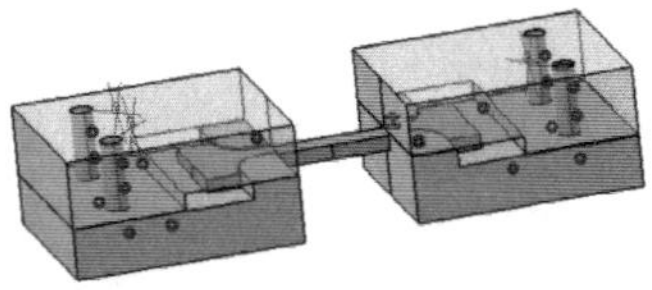

Fig.3. The sample-type and the mounting device for tensile test

3. RESULTS AND DISCUSSION

The Figure 4 shows the typical true stress - true strain curve for the ECAP processed Al-Mg alloy (six passes) at an initial strain rate of $8.4x10^{-5}s^{-1}$ for 220°C.

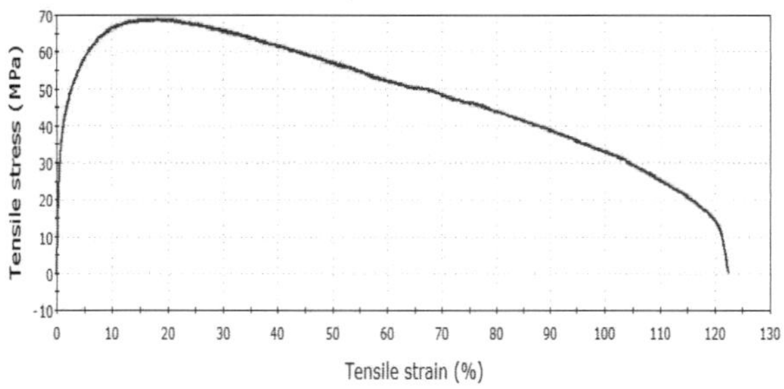

Fig.4. True stress vs. true strain for Al-1.35Mg ECAP 6 passes, (tensile test at 220°C and $8.4x10^{-5}s^{-1}$)

The complete results of performed tensile tests for a strain rate of $1.68x10^{-3}s^{-1}$ and $8.4x10^{-5}s^{-1}$ are presented in Fig. 5 and 6.

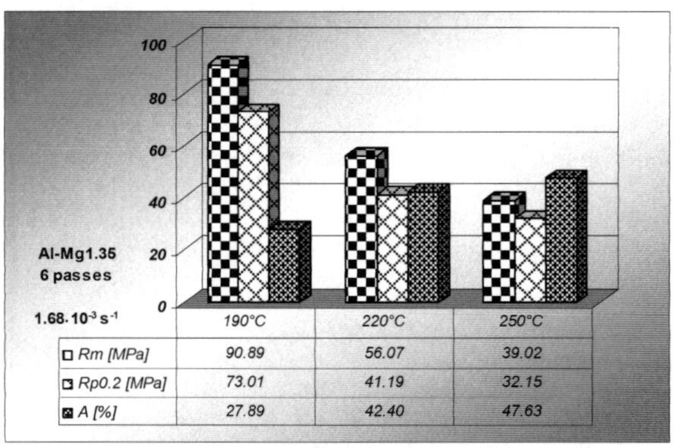

Fig.5. Evolution of mechanical properties for Al-1.35Mg after 6 passes ECAP (test performed at strain rate $1.68x10^{-3}s^{-1}$)

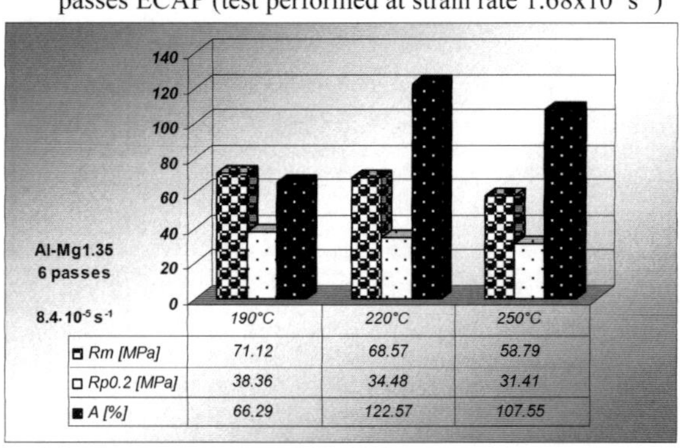

Fig.6. Evolution of mechanical properties for Al-1.35Mg after 6 passes ECAP (test performed at strain rate $8.4x10^{-5}s^{-1}$)

Extensive strain hardening takes place initially. After reaching a maximum, the flow stress continuously decreases until failure. The increase of temperature leads not to a shift of the peak stress to a higher strain as we expect. That because the lower strain rate sensitivity of alloy in the testing temperatures range. A steady-state flow was not found at all examined temperatures despite the fact that the value of elongation-to-failure is relatively high. No superplasticity was reported.

We found maximum tensile ductility (122%) at a middle of the testing temperatures range for the lower strain rate.

As is known from the Considère criterion (Wang & Ma, 2003) inhomogeneous deformation (necking) sets in when:

$$\left(\frac{\partial\sigma}{\partial\varepsilon}\right)_{\dot{\varepsilon}} \leq \sigma \tag{3}$$

where σ and ε are true stress and true strain, respectively.

For rate-sensitive materials, the presence of the strain rate sensitivity of the flow stress *(m)* helps to sustain the homogeneous deformation. The Hart's instability criterion is (Wang & Ma, 2003):

$$\frac{1}{\sigma}\left(\frac{\partial\sigma}{\partial\varepsilon}\right)_{\dot{\varepsilon}} - 1 + m \leq 0 \tag{4}$$

These equations indicate that for a high strength material (large σ), which is the case for UFG metals, sufficiently large strain hardening and/or strain rate hardening need to be present to sustain the uniform straining before the onset of localized deformation. A simple derivation based on Eq. 3 shows that the strain hardening exponent (*n*) corresponds to the true strain at necking under uniaxial tension. The strain hardening and strain rate hardening properties are therefore important for stable plastic deformation. It seems that without coherent Al_3Sc dispersoids which are highly effective in pinning of boundaries no exceptional elongations takes place because of the extensive static growth of structure takes place.

4. CONCLUSIONS

Due to ultra-fine grain structure achieved by ECAP, the Al-1.35Mg alloy exhibit low flow stress and large degree of ductility at high temperature and it becomes suitable for near net shape forming processes. The high formability expands the possibility of product design together with superior transferability of the die surface to the metal sheet.

5. REFERENCES

Kyung-Tae, P.; Chong, S. L. and Dong, H. S. (2005). Superplastic Behavior of as - Equal Channel Angular Pressed 5083 Al and 5083 Al - 0.2 Sc Alloys, *Materials Science Forum,* Vols. 475-479, 2005, pp. 2937-2940.

Kyung-Tae, P.; Chong, S. L.; Yong S. K. and Dong H. S. (2006). Superplastic Deformation of Ultrafine Grained Al Alloy Processed by ECAP and Post-Rolling, *Materials Science Forum,* Vols. 503-504,2006, pp. 119-124.

Musin, F.; Kaibyshev, R.; Motohashi, Y. and Itoh, G. (2004). Superplastic behaviour and microstructure evolution in a commercial ultra-fine grained Al-Mg-Sc alloy, *Materials Science Forum*, Vols. 447-448, 2004, pp. 417-422.

Noda, M.; Funami, K.; Hirohashi, M. and Kobayashi, M. (2004) Effect of Grain Size and Microstructure on Appearance of Low Temperature Superplasticity in Al-Mg Alloy,*Materials Science Forum*, Vols. 447-448, 2004, pp. 435-440.

Wang, Y.M. and Ma. E. (2003). Strain Hardening and Strain Rate Sensitivity of Ultrafine-Grained Metals, *Journal of Metastable and Nanocrystalline Materials,* Vol. 17, 2003, pp. 55-64.

Annals of DAAAM for 2008 & Proceedings of the 19th International DAAAM Symposium, ISSN 1726-9679
Editor B. Katalinic, Published by DAAAM International, Vienna, Austria 2008
Make Harmony Between Technology and Nature, and Your Mind will Fly Free as a Bird

ABOUT SPATIAL RIGID MEMORY CHARACTERISTICS

COMANESCU, A[driana]; COMANESCU, D[inu]; FILIPOIU, I[oan] D[an] & ALIONTE, C[ristian] G[abriel]

Abstract: *The paper deals with the computer aided synthesis of spatial cam used as kinematical elements of various mechanisms applied in industrial equipments. The follower has a planar translation displacement with different motion laws. The spatial cams profile parameters are determined for various laws of the follower. By using proper design software the profile parameters are used, for example, to draw the cylindrical, conical and hyperbolical cams and their profiles.*
Key words: *CAD, cams, analisys, synthesis.*

1. INTRODUCTION

The rigid memories materialized by planar or spatial cams are the main kinematical elements of various mechanisms applied in industrial equipments (food processing industry, textile industry, manufacturing processes, etc.) and robotics (Antonescu, 2005; Ivănescu, 2007).

The problem of cam planar mechanisms synthesis is comprehensively presented in the literature (Bishop, 2002; Hernandez et al., 2002). This problem is the only approach in some particular cases concerning cylindrical or truncated conic cams (Bloch 2000). This method presume that the spatial cams profile parameters can be determined for various motion laws of a follower which is in contact with the lateral or frontal spatial cam surface (Fig.1). After the calculation of the parameters, by using parametric CAD software, can be design cylindrical, conical and hyperbolical cams. The analytical synthesis of the spatial cam profile, presented in this paper, connected to adequate CAD software is firstly presented in the specialty literature.

2. CYLINDRICAL CAMS SYNTHESIS

In order to determine the line profile parameters of the cylindrical cam it is necessary to choose the following data: T – cam rotation period, r – cam cylinder radius, $LV = p$ – follower displacement amplitude and motion law for the follower.

By developing the cam cylinder a rectangle is obtain, with the length $LH = 2\pi r$ and the height equal to LV to which a xOy reference system is attached. The current point of the profile line is $A(x_{Ak}, y_{Ak})$. In this reference system its coordinates depends on the law for the follower, their graphic representations being given in Table 1.

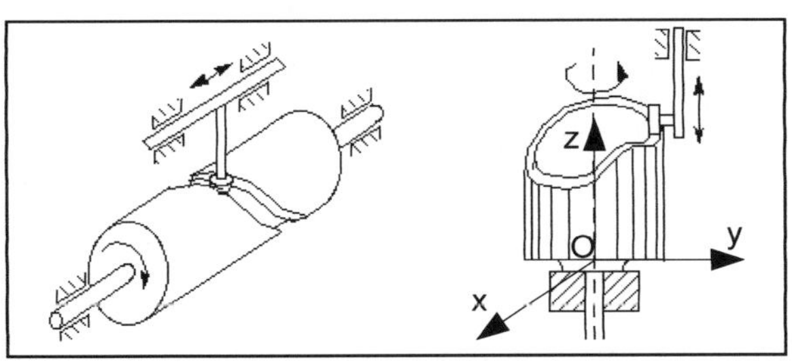

Fig. 1. Spatial cams

Law	Curve representation	Spatial curve profile line
Linear	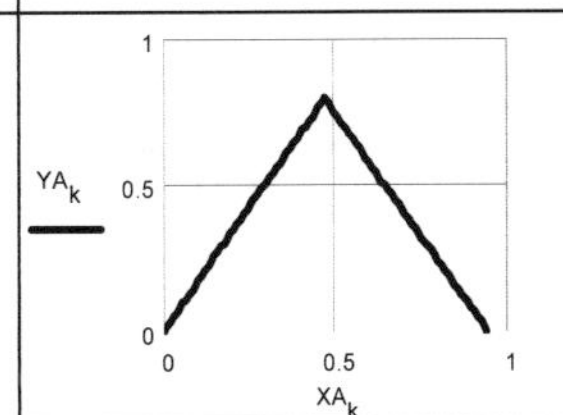	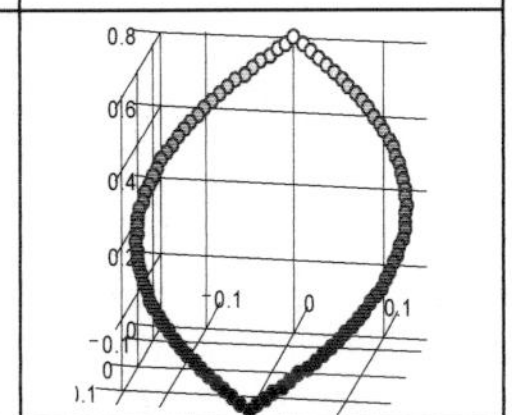
Parabolic	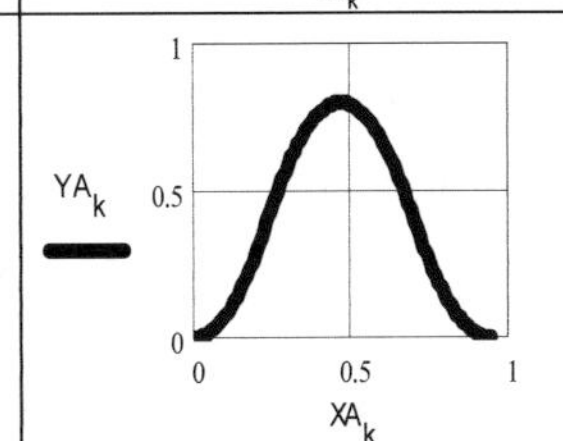	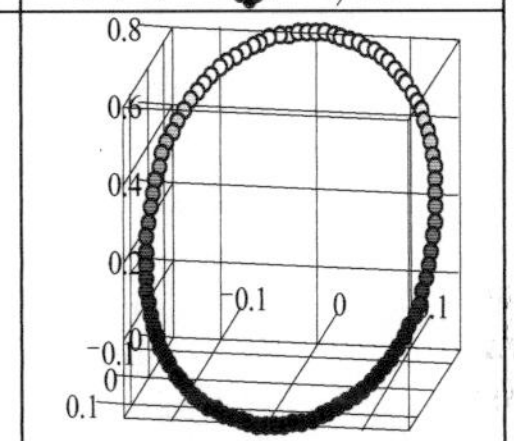
Sinus	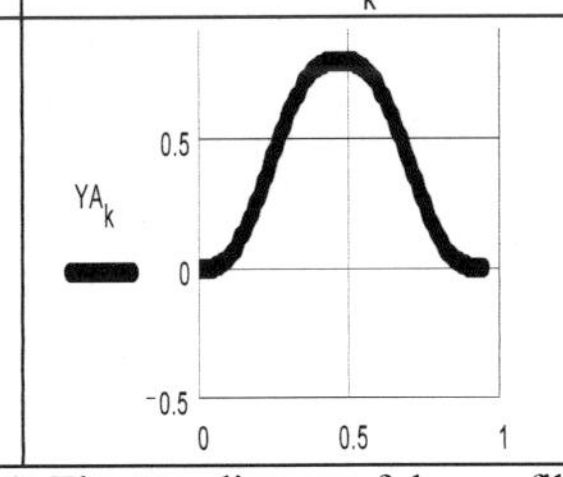	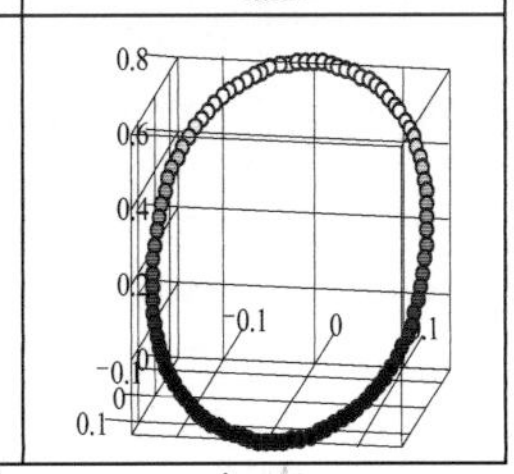

Tab. 1. The coordinates of the profile current point

The spatial curve has the following parameter equations

$$x_k = r\cos(\psi_k) \quad (1)$$

$$y_k = r\sin(\psi_k) \quad (2)$$

$$z_{Ak} = y_{Ak} \quad (3)$$

where $\psi_k = \frac{x_{Ak}}{r}$.

The cylindrical cams model may be achieved by using adequate design software such as AutoCad, Adams or SolidWorks. The parameters of the spatial curves profile are calculated. By applying a parametrization design, using CAD software, the design of such cams can be obtain, which are presented in Fig.2.

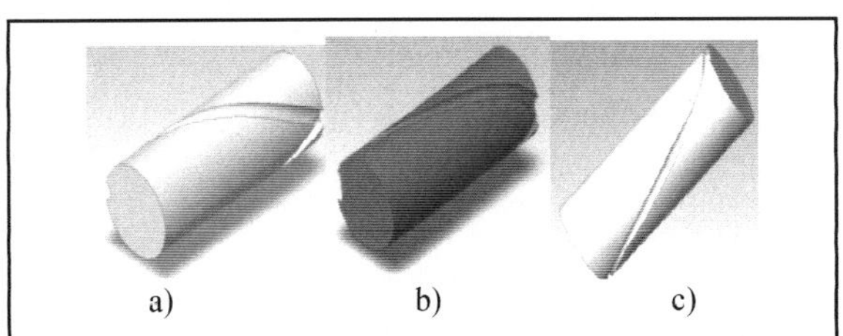

Fig. 2. Cylindrical cam: a) linear; b) parabolic; c) sincylindrical

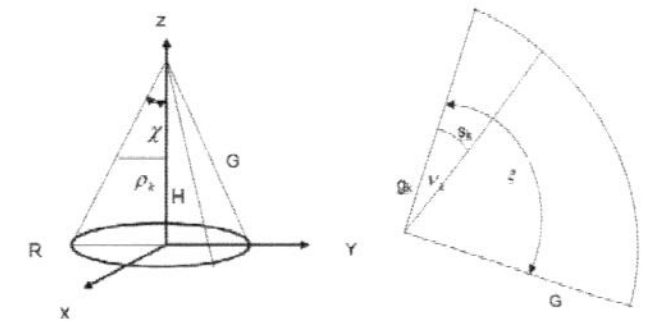

a) b)

Fig. 3. Characteristics to the truncated cone: a) The conic surface; b) the developed surface

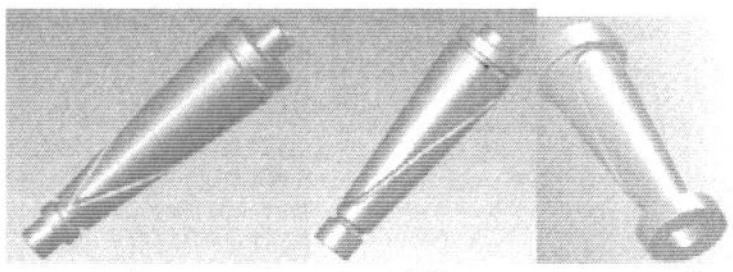

a) b) c)

Fig.4. Truncated conic cam: a) linear cam; b) parabolic cam; c) sin cam

3. TRUNCATED CONIC CAMS SYNTHESIS

The follower motion law on the Oz axis is imposed by the technological process. The characteristics to the truncated cone are presented in Fig.3

The truncated cone has the bases radius respectively R and r. The basis circle length is denoted by LH. The follower motion amplitude on Oz is LV. The current point coordinates of the curve placed on the truncated conical surface, $A(x, y, z_A)$ may be deduced using the Fig.3 and the following relations:

$$k_1 = \frac{LH}{T} \tag{4}$$

$$k_2 = 2\frac{LV}{T} \tag{5}$$

$$\chi = \operatorname{atan}\left(\frac{R-r}{LV}\right) \tag{6}$$

$$H = \frac{R}{\tan(\chi)} \tag{7}$$

For example, by adopting the following values for $T = 2\ s, R = 0{,}15\ m$, $r = 0.08m, LV = 0.8m$ there are determined the coordinates of the spatial curve current point for different motion laws.

These values were used in the modelling process using CAD software and the models are presented in Fig.4.

4. HYPERBOLOID CAMS SYNTHESIS

In the case of the hyperboloid cams the follower extremity is in contact with the curve described on a hyperboloid.

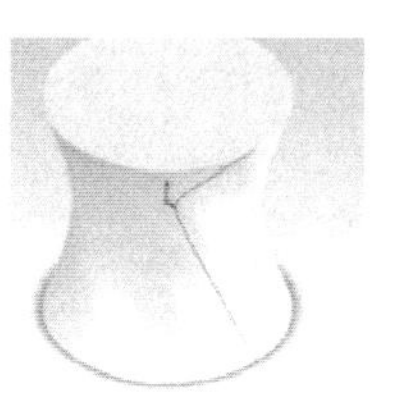

Fig.5. The hyperboloid

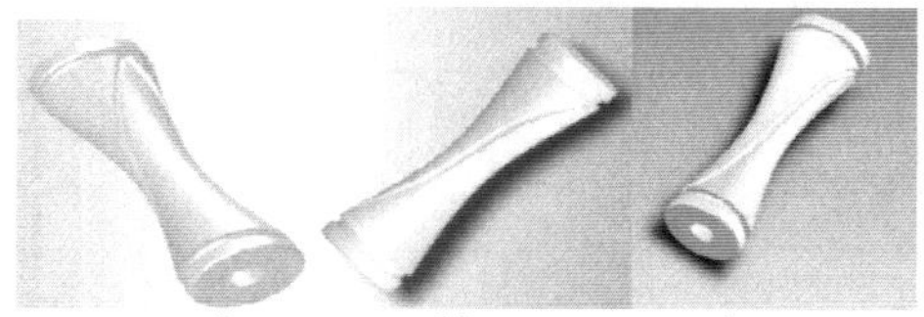

a) b) c)

Fig. 6. Hyperboloid cams: a) linear cam; b) parabolic cam; c) sin cam

The hyperboloid surface (Fig.5) may be obtained by rotating a hyperbole around the Oz axis.

The hyperbole equation in the yOz reference system is

$$\frac{y^2}{a^2} - \frac{z^2}{b^2} = 1 \tag{8}$$

where $b^2 = c^2 - a^2$ and $c > a$.

The current point of the spatial curve placed on the hyperboloid is $A(x_k, y_k, z_{Ak})$. z_{Ak} is determined in function of the selected motion law. The period of a complete rotation is T. R and r are respectively the maximum radius and the minimum one for the hyperboloid and $LV = p$ is the displacement amplitude on the Oz. For km time equal intervals can be determine $t_k = \frac{kT}{km}$, where $k \in [0, km]$. At a t_k time, the rotation angle is $\theta_k = t_k \cdot \omega$ (the cam angular velocity $\omega = \frac{2\pi}{T}$).

The ρ_k current radius is given by:

$$\rho_k = \sqrt{y_k^2}, \tag{9}$$

where $Y_k = \frac{a}{b}\sqrt{b^2 + z_{Ak}^2}$, thus

$$x_k = \rho_k \cos(\theta_k) \tag{10}$$

$$y_k = \rho_k \sin(\theta_k) \tag{11}$$

By using the advanced design techniques were obtained the hyperboloid cams presented in Fig. 6.

5. CONCLUSIONS

For rigid memories defined by spatial cams it is presented an adequate synthesis method.

The spatial parameters of the profile line may be determined by using calculus software and are introduced in the CAD algorithm for modelling. This will make the manufacturing easier because this can be done using CAM.

The future researches include other applications for spatial cams in function of the follower motion laws.

6. REFERENCES

Antonescu, P. (2005). *Mechanism and Machine Science*, Edit.PRINTECH. ISBN: 9737182916, Bucharest

Bishop, R.H. (2002). *The mechatronics handbook*, CRC Press. ISBN: 0849300665

Bloch, A.M. (2000). *Nonholonomic mechanics and control*, Springer. ISBN: 0387955356

Hernandez, A.; Pinto, C.; Petuya, V. & Agirrebeitia, J. (2002). *Mechanism theory*, Serv. Publicaciones de la E. T. S. Ingenieros de Bilbao, ISBN: 8495809117, Bilbao

Ivănescu, M. (2007). *From classical to modern mechanical engineering fundamentals*, Edit. Acad. Romane, ISBN: 9789732715611, Bucharest

Annals of DAAAM for 2008 & Proceedings of the 19th International DAAAM Symposium, ISSN 1726-9679
Editor B. Katalinic, Published by DAAAM International, Vienna, Austria 2008
Make Harmony Between Technology and Nature, and Your Mind will Fly Free as a Bird

A NEW HEAT PIPE COOLING DEVICE

COMANESCU, D[inu]; COMANESCU, A[driana]; FILIPOIU, I[oan] D[an] & ALIONTE, C[ristian] G[abriel]

Abstract: *The paper presents a specific application of fractals, which are used for a heat pipe cooling device design. A heat pipe is a simple device that can quickly transfer heat from one point to another. This new device was developed for some years in Politehnica University of Bucharest by a group of multi-disciplinary group researchers lead by this paper authors. To investigate the cooling characteristics it has been used three Peano fractals structures: Hilbert, Hutchinson and Koch curve. The design was validated through basic experimental tests.*
Key words: *heat-pipe, fractals, Peano, curve, cooling.*

1. INTRODUCTION

The fractals are well studied from mathematical point of view. Their simulation and their assimilation with various forms existing in the nature are also presented in the literature. By using proper design software the Peano, Hilbert and Koch profile parameters are used to obtain optimal constructions (Barnsley & Demko, 1986).

The paper presents a specific application of fractals, which are used in a heat pipe cooling device. This new device type was developed for some years in Politehnica University of Bucharest by a group of multi-disciplinary group researchers lead by this paper authors. The research found some original aspects unfound in the technical literature. Firstly mentioned on this occasion there are also specified new development research directions, which may be achieved on adequate projects.

2. ABOUT HEAT PIPES

The heat pipe is composed of three basic components (Zaghdoudi et al. 2004):

- the container;
- the working fluid;
- the wick or capillary structure which has a fractal form (two plates: superior and inferior) which can be placed in mobile computer structures. (Fig. 1).

Working fluid is vaporized in the evaporator and flows toward the condenser where it deposits its heat by condensation. Capillary forces in the porous wick return the condensed working fluid to the evaporator. The function of the container is to isolate the working fluid from the outside environment. It has to therefore be leak-proof, maintain the pressure differential across its walls, and enable transfer of heat to take place from and into the working fluid (Sobhan, 2005).

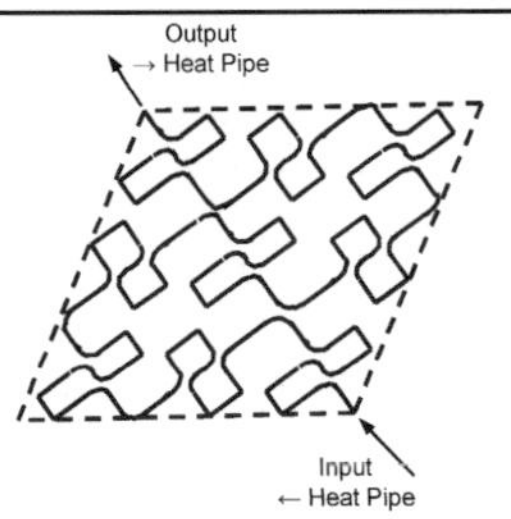

Fig. 1. The fractal capillary structure

Selection of the container material depends on many factors (Shankara Narayanan 2006). These are as follows:

- Compatibility (both with working fluid and external environment);
- Strength to weight ratio;
- Thermal conductivity;
- Ease of fabrication, including welding, machine ability and ductility;
- Porosity;
- Wet ability.

A high thermal conductivity ensures a minimum temperature drop between the heat source and the wick.

3. FRACTALS USED IN A HEAT PIPE COOLING DEVICE

Having in view the geometrical characteristics, the Peano curve fractal may be used for heat pipe device. In this situation the section of such a system are chosen in function of other requirements (heat transfer, flow, technological manufacturing, etc).

The main property of this curve, which its fractal dimension is two (Voinea & Stroe, 2000), is that a square may be filled with it (Fig. 2). For example a Peano curve is built by Hilbert (Fig. 2a). By using the property of its auto-similarity one may determine its fractal dimension, which it is also two as in the previous case (Voinea & Stroe, 2000).

The auto-similarity is a characteristic property of the mathematical fractals. This is the main difference between a mathematical fractal and a natural one, which it is governed by the dynamic phenomena, as such of growing.

By assumption the Hutchinson operator (Fig. 2b) as a succession of affine transformation operators (translation, rotation, reducing by similitude, affine reducing, mirror, cutting) a fractal appears as a fixed point of this operator.

In the case of the Koch curve (Fig. 2c) the Hutchinson operator (W) is composed successively by a 1/3 contraction for the OB segment (W1), followed by a 600 rotation and a 1/3 translation for the BC segment (W2), a similar transformation for the CD segment (W3) and a 1/3 contraction with a translation for the DK segment (W4), so that

$$W = W_1 \cup W_2 \cup W_3 \cup W_4 \quad (1)$$

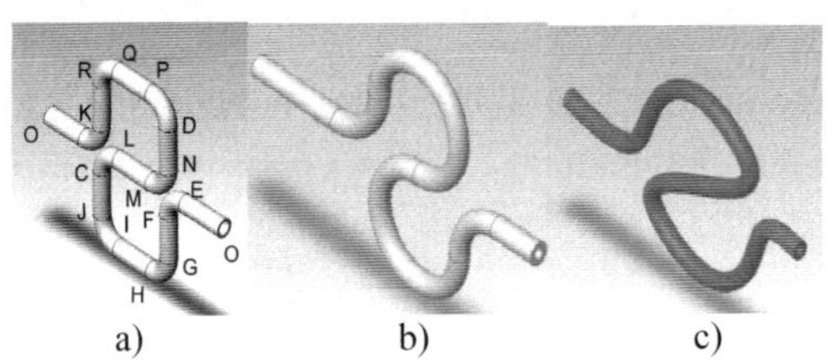

Fig. 2. Peano units with various forms: a) Hilbert curve ; b) Hutchinson curve; c) Koch curve

The W operator previously defined generating by successive iterations the Koch curve.

The previous units are inserted in Peano type networks, Fig. 3. The structures based on the Hilbert (Fig. 3a) and Hutchinson (Fig. 3b) curve and with different sections are possible to be used for units or heat transfer pipes and networks.

The last constructions (Fig. 3c) are based on the Koch curve. Even if they have rather strange image these are the first steps to attempt new field of applications for other fractals as Barnsley fern.

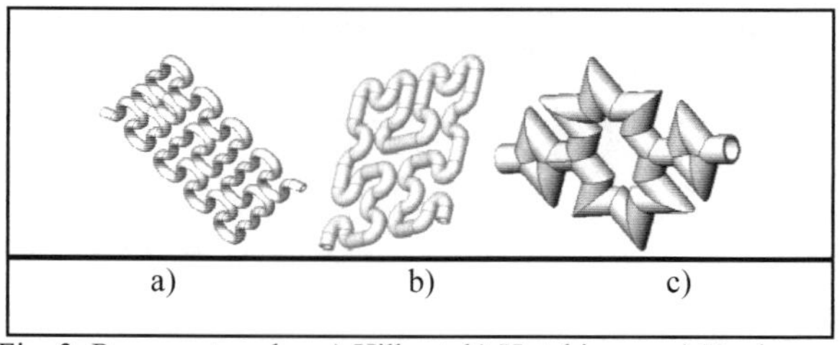

Fig. 3. Peano networks: a) Hilbert; b) Hutchinson; c) Koch

4. TESTING DATA RESULTS

For validation, basic tests have been conducted. In Fig. 4 is presented the experimental test equipment. Inside the container, placed on the heating system, is a liquid. Under its own pressure the liquid enters in the pores of the capillary material and is wetting all internal surfaces. Applying heat at any point along the surface of the heat pipe causes the liquid at that point to boil and enter a vapor state. When that happens, the liquid picks up the latent heat of vaporization. The gas has a higher pressure, moves inside the sealed container to a colder location where it condenses. Thus, the gas gives up the latent heat of vaporization and moves heat from the input to the output end of the heat pipe. This entire cycle usually happens with less than a 5°C differential from one end of the pipe to another. We used in the experiments the simplest type of wick structure, double-layer fractal mesh screen wick. The geometric and thermophysical properties of the wick have been selected are presented in the Table 1. Based on hereinbefore fractal capillary structure we obtained the following heat pipe specifications:

- Horizontal orientation
- Maximum heat transfer: 30 W;
- Nominal operating temperature: 40°C;
- Pipe diameter: 0.08 m;
- Heat Pipe length: Evaporator: 0.04 m; Adiabatic: 0.05 m; Condenser: 0.07 m.

5. CONCLUSIONS

The main purpose of this paper is that to find some new structures for the laptops and notebooks cooling systems. In order to increase its transfer efficiency and shape minimization the authors firstly use various fractals configuration as thin plates with large cooling surfaces. Because they are thin, the plates can be used in cooling device where the dimensions are critical.

Geometric properties	Value
wick diameter	2×10^{-5} m
wick width	1.2×10^{-4} m
working temperature	318.15 K
heat pipe radius of the cylinder	3.5×10^{-3} m
heat pipe inner radius of the cylinder	3.38×10^{-3} m
effective length of the heat pipe	0.115 m
total length of the heat pipe	0.16 m
cross-sectional area of the wick structure	2.971×10^{-6} m

Tab. 1. Geometric properties of the heat pipe device

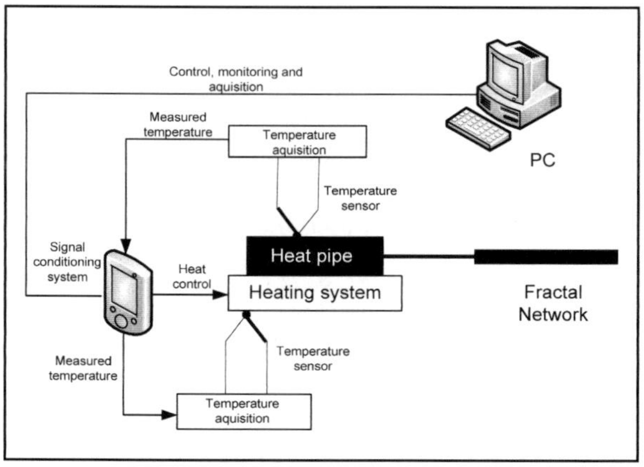

Fig. 4. Experimental test equipment

The design was validated through basic experimental tests which demonstrate the cooling capability of this structure (maximum heat transfer is 30 W). All of the heat transfer limitations, with the exception of the boiling limitation, exceed the specified heat transfer rate of 30W. The low value of 0.131W for the boiling limitation strongly suggests that the liquid will boil in the evaporator and possibly cause local dry spots to develop. The reason the liquid boils is because the effective thermal conductivity of the wick is equal to the conductivity of the liquid, which is very low in this case. Because the liquid is saturated at toe vapor-liquid interface, a low effective thermal conductivity requires a large amount of wall superheat which, in turn, causes the liquid to boil. This problem can be circumvented by using a high conductivity wire mesh or sintered metal wick, which greatly increases the effective conductivity. It should be noted, however, that because porous wicks have lower permeabilities, the capillary limitation should be lower as well.

The future researches include: optimizing the structure, researches regarding different cooling fluid using and investigation of full cooling characteristics of this device.

6. REFERENCES

Barnsley, M. F. & Demko, S. (1986). *Chaotic Dynamics and Fractals*, Academic Press ISBN 0120790602

Sobhan B. C. (2005). *Modeling of the Flow and Heat Transfer in Micro Heat Pipes. Available from:* http://www.rpi.edu/tphtl/research/mfht/mfht.html *Accessed:* 2006-07-20

Shankara Narayanan K.R. (2006). *What's A Heat Pipe? Available from:* http://www.cheresources.com/htpipes.shtml *Accessed:* 2006-06-16

Zaghdoudi M. C.; Tantolin C. & Godet C. (2004). *Use Of Heat Pipe Cooling Systems In The Electronics Industry. Available from:* http://www.electronics-cooling.com/html/2004_nov_a1.html *Accessed:* 2005-06-20

Voinea, R. & Stroe, I. (2000). *Introduction in theory of dynamical systems*, Ed.Academiei Romane, ISBN 973–27–0739 - 9, Bucharest

Annals of DAAAM for 2008 & Proceedings of the 19th International DAAAM Symposium, ISSN 1726-9679
Editor B. Katalinic, Published by DAAAM International, Vienna, Austria 2008
Make Harmony Between Technology and Nature, and Your Mind will Fly Free as a Bird

CREATING THE SERVICE PRODUCT BY APPLYING THE GENERAL WORK PROCEDURE MODEL

COSIC, I[lija]; RADAKOVIC, N[ikola]; SIMEUNOVIC, N[enad] & LALIC, B[ojan]

Abstract: *Even though the services have more and more important role in the society, the productivity of this is much smaller then the productivity of the production sector. The constant challenge represents the possibility of application of production methods on service systems.*

The work procedure of planning the service product by applying the general work procedure for different types of products, based on analyzed similarities and differences of the product itself and the process of the four generic groups of products classified in ISO 9000:2000, is shown in this work.
Keywords: *Work procedure, service, service system*

1. INTRODUCTION

The exceptional development of the service sector within the last two decades has focused the attention on the challenges of the efficient service organization and operation management, that are much different that the challenges in the classical production systems. However, even though the services have even more important role in the society as they grow, the productivity of this sector is much smaller than the productivity of the production sector. The devotion to the quality of the services and the productivity are the basic means of the struggle in the market, and, as such, they're directly connected to the need to improve the operative activities of the service systems.

The beginning of the market struggle represents the planning of the new product, I.E. service. The work procedure while planning the new product that has been presented in this work is the general model of the work procedure for the different types of products. This model is based on the analyzed similarities and differences of the product itself and the process of the four generic groups of products classified in ISO 9000:2000 and it consists of the five basic steps.

2. GENERAL WORK PROCEDURE MODEL

2.1 Defining and specifying the demands

Defining the demands, apropos specifying the demands is the beginning phase of every process. This is also one of the most important phases during the process development, the phase with the largest possibility for a big mistake. In his phase, the demands for the future system are being analyzed.

The functional demands describe the functions of the system, meaning what should it do, the outputs it gives for the given inputs, as well as its behavior in some of the situations. The nonfunctional demands express the limitations to the functioning of the system, the expected level of reliability, the esthetical demands, the economical demands, respecting the standards, the deadlines, the customer satisfaction, the influence to the environment etc.

The result is the document about demands, which describes what should the system be, if possible, without the assumptions how to accomplish that. A team, gathered mostly from planners, marketers, commissioners and users, is working on

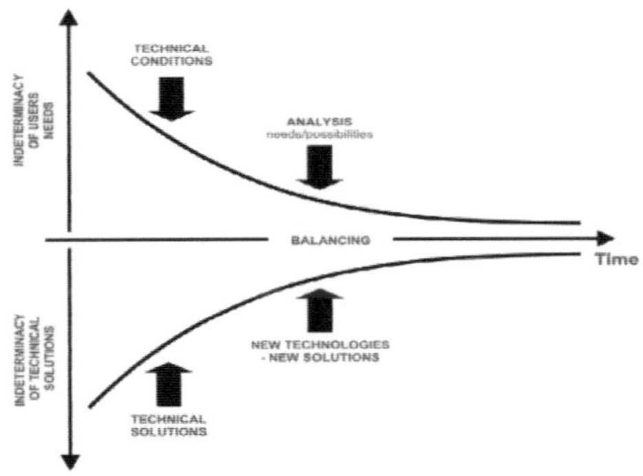

Figure 1: Balancing the Demands with Technical Possibilities

this specification. So, in this initial phase of this activity, the indeterminacy of the users needs, IE demands is on its maximum. In that phase the user knows what he doesn't want rather than what he wants, which implicates the maximum in the planners incomprehension. That phase is named "the indeterminacy of the technical solutions".

With the dialogue of the users and the planners, the level of the indeterminacy is lowering, because the user is getting to know the technical and the technological possibilities, and so, taking that in consideration, he profiles his demands. To the further balancing of the user's wishes and the technical/technological possibilities, the price of the given solutions has a great influence, viewed from the need/demand point of view, and so, it gives the determining impulse to the demand creation (Hawker 2001) (Figure 1). This is the iterative procedure and its length is determined by the quality of the communication, as well as the level of the users' criteria.

Establishing the demands. The information attained by the analysis is transformed into texts that determine the demands in two levels: "the users' demands" and "the systems' demands".

Demand Validation. Checks if the demands are objective (they can be reached with the given technology and budget), consistent (not being in mutual conflict), as well as if they're complete (involves all the functions and the limitations needed).

2.2 Generating the Concept

The transformation in the process itself adds up to the selection of the ideas, variables, concepts that come up as the possible answers to the demands given in the input (Slack et al., 2004).

For the idea to grow into the concept, it has to be shaped, meaning to acquire the certain features. In the first row, it has to have a form, meaning, the dimension that can be physical or sensitive (psychological) determined by the type of product. It has to have a function, meaning defined way of product/service functioning. The third agent is the purpose, meaning the need being satisfied by the product. And the last agent is the benefit, meaning the advantage that the product/service is giving to its customers.

2.3 The Evaluation And The Choice

Generated concepts run through the, so called, phase of scanning, meaning the check or the evaluation of the attained concepts (Slack et al., 2004). The main criteria are the marketing, technological and the financial evaluation, or, so called, feasibility studies. Based on these checks, the choice of the further developed concepts is being made.

2.4 The Usual Solutions

The performance specification is written for the product concept that passes the feasibility studies and then approves for the development. They describe the function of the product, meaning, what the product should do to satisfy the customers' needs. By developing the preliminary structure determined by the chosen concept, we reach the preliminary solution (Roberta 1999).

The preliminary solution is being developed, analyzed in detail, checked for the possible mistakes, possibility of realization, the costs are taken in consideration and the preliminary records are being arranged. This step is, actually, the mid-phase between the preliminary solution and the final solution. These three steps represent the mutually highly iterative procedures, so, often, they're almost impossible to split.

2.5 Verification, Validation and Documenting

Finally, the solution is verified, meaning checked if it fits the specification of demands and validated, meaning checked if it satisfies the demands of the user (if the nature of the product allows it, the making of the prototype and testing is made). The result is the specification of the product, and specification of the process, meaning the production instructions.

Watching the figure 1, we can tell that the foreseen process stream has two directions, meaning that it's iterative. This iterative process is especially accented in the phase of forming the solution.

The whole process, from making of the idea, over developing the concept and the choice of the solution, to the final solution, the process being run by using the knowledge base that transforms the given data into the available information that determine the nature of the planned system in the system of decision (Schroeder 1993).

This process flows through the usual phases: generating the concept, the evaluation and the choice of the adequate (so called scanning), Forming the solution and verification validation and documentation, as shown in figure 2.

3. CREATING THE SERVICE PRODUCT

The process described in applied in the creation of the service product of one educational facility, meaning the plan and the program of the specialist studies in the field of e-business. The applied process is described in the process flow chart (figure 3).

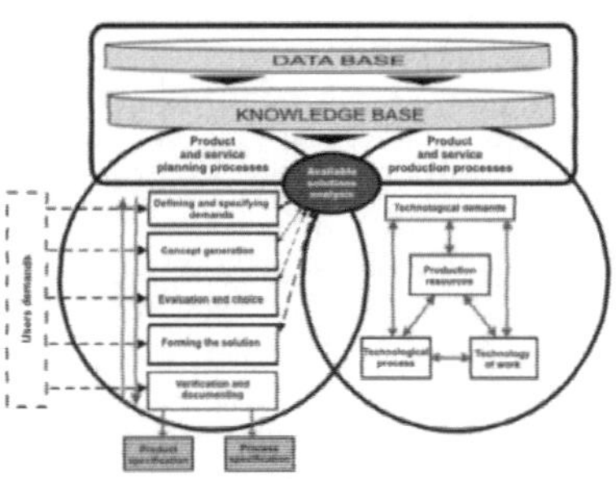

Figure 2: The Process Planning Phases in the Process Funnel

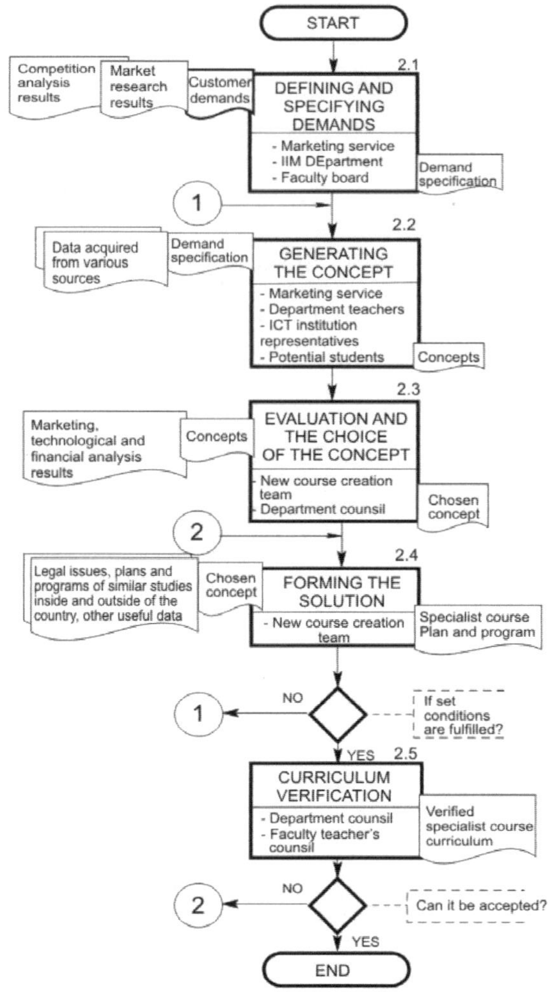

Figure 3: The Curriculum Creation Process Flow Chart

4. CONCLUSION

The multilayer architecture of the working process creation model is shown in this work, and we can clearly see the domains of the demands, form, and the process. That is not the assembly of the explicit rules for product creation, but a model that defines the mechanism of making the simultaneous project decisions, where the elements of the process are the always present factors. The model is applied in the realistic example of creating the product in one educational facility, and with that its applicability is shown.

5. REFERENCES

Hawker, S. (2001.) "*An Integrated Process/Product Component Model for Software Engineering*", University of Alabama

Roberta R. S. & Bernard, W. T., (1999) "*Operations Management*" Pearson Education,

Schroeder, G. Roger; (1993.) „*Operations Management – Decision making in the Operations Function*", McGraw-Hil

Slack, N.; Chambers, S.; & Johnston, R.; (2004). *Operation Management* McGraw-Hil

Annals of DAAAM for 2008 & Proceedings of the 19th International DAAAM Symposium, ISSN 1726-9679
Editor B. Katalinic, Published by DAAAM International, Vienna, Austria 2008
Make Harmony Between Technology and Nature, and Your Mind will Fly Free as a Bird

BUILD YOUR ANTIPHISHING TECHNOLOGY IN JUST 5 MINUTES

COSOI, A[lexandru] C[atalin]; SGARCIU, V[alentin] & VLAD, M[adalin] S[tefan]

Abstract: *Few countries still consider phishing as a new and emerging phenomenon. Fake websites impersonating both national and international financial institutions are now everywhere, trying to manipulate the users into giving away their credentials. This paper will try to deal with this problem at the browser level, by trying to combine both whitelisting and content based solutions into a single web-page forgery detector.*

Key words: *Bayesian Classifier, Antiphishing, Phishing, Forgery.*

1. INTRODUCTION

Phishing is a form of social engineering in which an attacker attempts to fraudulently acquire sensitive information from a victim by impersonating a trustworthy third party.

Nowadays, current AntiSpam technologies have obtained competitive detection rates on phishing emails, but since recently phishers are advertising their fake websites via a plurality of communication methods (e.g. email spam, instant messaging, social networks, blog posts and even sms (Cosoi & Petre, 2008; Hatlestad, 2006)), and having some starting information about their victims from social network profiles, (Jagatic et al., 2005) they can easily social engineer their way to the user's trust, which means that a browser level protection must be assured in order to prevent the user to access the website, even though he was persuaded to access the fake URL.

Current browser based technologies employ whitelists, blacklists, various heuristics to see if a URL is similar to a well-known URL, community ratings and content based heuristics (Cranor et al., 2006) and lately visual similarity (Wenyin et al., 2005). Blacklisting worked great so far, but the timeframe needed for a URL to become worldwide blacklisted is in most cases overlapping with the time in which the phishing attack is most successful. Also, current content based solutions, mostly blacklists and body heuristics (Cranor et al., 2006) do not always make use of whitelists, which sometimes might cause the filter to consider eBay's official website as a phishing website (SpamConference, 2008 and Wu et al., 2006).

2. PROPOSED METHOD

In developing our method we started from the following hypothesis: in a given language, the number of possible rephrases of a given text that transmits the same or similar information (e.g. "This is your online banking account. You must log in first in order to access your funds. Please be careful to phishing attempts") and not considering obfuscation purposes, is limited by the speaker's common sense (e.g. the information will be phrased in a simple readable and understandable form). In other words, we assume that all English log-in pages of financial institutions will have a large set of commons words, since they share common purposes and specialized financial vocabulary (Landauer et al., 1998; Kelleher, 2004; Shin & Choi, 2004; McConnell-Ginet, 1973; Merlo et al., 2003; Biemann & Quasthoff, 2007).

Considering two documents A and B (in our situation, websites of financial institutions like PayPal or BankOfAmerica), we can represent them as $A = C \bigcup N_1$ and $B = C \bigcup N_2$, where C represents the common words, and N_1 and N_2 the distinct words. This means, that the necessary number of words to construct a database with triples of the form (word, document, occurrences), is $|C| + |N_1| + |N_2| \leq 2|C| + |N_1| + |N_2|$, or in short $\left(A \bigcap B\right) \leq \left(A \bigcup B\right)$. If we consider the case of only two documents, this technique might not bring considerable improvements, but in the case of several documents which serve the same purpose (e.g. financial institutions websites), it is acceptable to presume that the outcome will consist in a large number of common words.

We will now define a similarity indicator between two documents, known as the Jaccard Distance for sets.

$$D = 1 - \frac{|A \cap B|}{|A \cup B|} \tag{1}$$

On identical documents, this distance will have a null value, while in case of almost similar documents, it will be close to 0. *Since these are not standard sets (e.g. in ordinary sets, identical elements appear just once, while in this set, we decided that each element (word) appears as many times as it is found in the document), the distance actually provides an acceptable similarity value, judging by the number of words.*

On a corpus of 101 financial institutions, from 3 different countries: top 5 phished banks from Romania, 7 websites from Germany and randomly chosen 89 phished US institutions which showed a high frequency on email phishing in our internal email corpus, with an average of 100 words per page, we obtained a database of just 4422 different words, instead of an expected minimum of 10000 words.

Considering a pool of WebPages (e.g. the ones presented above for example), we can construct a database in the format presented in Table 1.

If we consider that it is necessary to run the presented forgery filter on the target webpage, we then start computing the Jaccard distance for each institution on which the filter has trained on (e.g. the words from learned webpages are to be found in the database). The lowest distance obtained, indicates the highest similarity (judging by the specified distance) between the target webpage and one reference webpage from our database. If the computed distance is smaller then a predefined threshold, we will consider this website a forged page.

When dealing with this technology, using an up to date whitelist is a necesity, because after this filter has learned the original website, it will score a perfect match when visiting the target webpage. An up to date whitelist will inhibit runnig the forgery filter on original websites in order to avoid false positives.

Based on this initial background, our proposed method can be better understood from Figure 1. First, the presented webpage is verified against a blacklist and a whitelist. Afterwards, some simple heuristics are runned on the webpage's content, to check whether this page would actually try to mimic an official log-in page (e.g. contains a submit button or words like eBay, PayPal, etc). We introduced this step for speed optimization purposes (e.g. it would be pointless to check if a webpage without a submit form tries to duplicate a webpage which has such a form).

Words	OCCURENCES				
	Bank1	Bank2	Bank3	Bank4	Bank5
Word1	3	1	0	2	1
Word2	0	0	3	1	0

Tab. 1: Format of the filter dictionary

3. RESULTS

Usually, if the filter was trained on a certain webpage, we will have a similarity distance of a at least .01, and experimentally (e.g. 10 000 samples) we observed that on phishing websites, we never obtained a distance higher than 0.2. For training, we used a corpus of 101 pages (presented earlier) and a value of 0.25 for the similarity threshold.

We tested our filter on two different corpuses: one containing 10 000 forged websites of the exact pages on which the filter has trained on (randomly selected from real phishing pages) and the URLs published on PhishTank on a timeframe of 10 days.

We obtained a 99.8% detection on the first corpus, which means that we had 20 false negatives, mostly because they were generated as a screenshot from the original webpage and not showing enough text content for a discriminative decision, and a 42.8% detection on PhishTank URLs. Although it may seem low, our data indicates that we obtained these results due to two major reasons:

- According to Antiphishing Working Group, in December '07 there were 144 hijacked brands co-opted in phishing attacks (far more than our training corpus)
- PhishTank's database of fresh phishing submissions is sometimes polluted since anyone can submit a website (we even found BitDefender's website submitted as a possible phishing site).

This experiment can be easily reproduced if in a multicategorial Bayesian filter, we change the probability function with equation 1 and the probability for each word to belong to category, will actually represent the number of occurrences of that word in that specific category. Then, if instead of choosing the category with the highest probability obtained, we would choose the category with the smallest distance, we could obtain the same results as presented above.

As for false positives, on a corpus of 25 000 samples of WebPages containing login forms, or any other information that would activate the forgery filter, we obtained 10 false alarms. 8 of them were actually real financial institutions, which should have been in the whitelist if the filter would have been properly trained, while the other two were real false positives (2 online financial newspapers) and this problem can be easily solved by whitelisting them.

4. CONCLUSIONS

Since phishing websites are no longer advertised on just email spam, we believe that it is time for companies to invest more in research and development on browser level antiphishing protection.

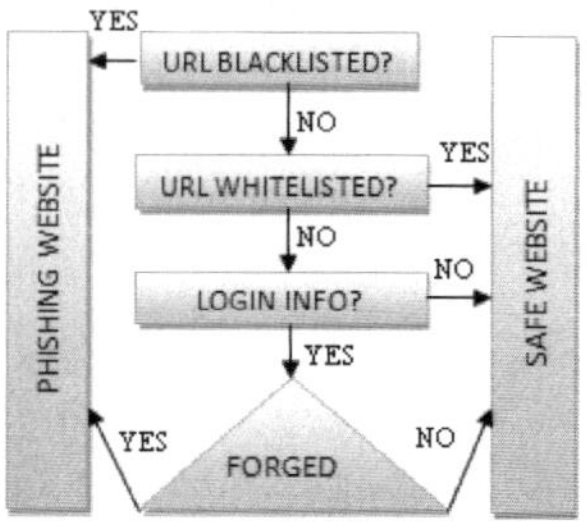

Figure 1 – Toolbar Algorithm

The proposed method comes as an add-on to current technologies, by providing the user with extra information about the visited webpages. Although, not a complete solution on its own, (it is ineffective on phishing websites that do not mimic the original website) used with other technologies (e.g. blacklists, content and URL heuristics) it increases the value of any antiphishing toolbar.

The obtained results show that this is a viable method to provide forgery detection to official financial institutions websites. Also, it is not necessary to run this system on all the pages visited by the user, focusing just on the ones that require information submission, and thereby, highly increasing the user's tolerance level by decreasing the time spent for analysis.

Acknowledgements

This work was entirely supported by BitDefender – AntiSpam Laboratory. Web: www.bitdefender.com

Also, a grateful thanks to Mr. Lucian Lupsescu and Mr. Razvan Visan for their precious help in developing this project.

5. REFERENCES

Biemann C., Quasthoff U.(2007). *Similarity of documents and document Collections using attributes with low noise*, Institute of Computer Science, NLP department, Universityof Leipzig, Johannisgasse 26, 04103 Leipzig, Germany

Cosoi A.C., Petre G. (2008). *Spam 2.0. Workshop on Digital Social Networks*, SpamConference 2008, Boston, MIT

Cranor L., Egelman S., Hong J., Zhang Y. (2006). *Phinding Phish: An Evaluation of Anti-Phishing Toolbars*, November 13, 2006, CMU-CyLab-06-018

Jagatic T., Johnson N., Jakobsson M., Menczer F. (2005). *Social Phishing*, School of Informatics, Indiana University, Bloomington, December 12, 2005

Kelleher D. (2004). *Spam Filtering Using Contextual Network Graphs, Available from* https://www.cs.tcd.ie/courses/csll/dkellehe0304.pdf, *Accessed:* 2006-06-18

Merlo P., Henderson J., Schneider G., Wehrli E. (2003). *Learning Document Similarity Using Natural Language Processing*, Geneva

Shin S., Choi K. (2004). *Automatic Word Sense Clustering Using Collocation for Sense Adaptation*, KORTERM, KAIST 373-1 Guseong-dong, Yuseong-gu, Daejeon, Republic of Korea

Wenyin L., Huang G., Xiaoyue L., Min Z., Deng X. (2005). *Detection of phishing webpages based on visual Similarity*, WWW2005, May 10-14, Chiba, Japan, ACM 1-59593-051-5/05/0005

Wu M., Miller R. C., Garfinkel S. L. (2006). *Do Security Toolbars Actually Prevent Phishing Attacks?*, MIT Computer Science and Artificial Intelligence Lab, CHI 2006, April 22-27, 2006, Montréal, Québec, Canada

Annals of DAAAM for 2008 & Proceedings of the 19th International DAAAM Symposium, ISSN 1726-9679
Editor B. Katalinic, Published by DAAAM International, Vienna, Austria 2008
Make Harmony Between Technology and Nature, and Your Mind will Fly Free as a Bird

INTRODUCING SYNTACTIC FEATURES INTO A BAYESIAN CLASSIFIER

COSOI, A[lexandru] C[atalin]; SGARCIU, V[alentin] & VLAD, M[adalin] S[tefan]

Abstract: *One of the most important problems of the bag-of-words classifiers today is that they don't take into consideration any syntactic information of the analyzed text. In this paper, we will try to experiment the effects on a Bayesian classifier when we add the syntactic function of each learned word. We will notice that this will increase accuracy by at least 0.5% on distinct categories and at least 10% on similar categories.*

Key words: *Bayes, Semantic, Syntactic, Statistic*

1. INTRODUCTION

Semantic classification can be considered the next level in what we like to call right now the future "Semantic Web", where information should be better classified and easier to access. Hence, the need for semantic processing technologies show an increasing importance in the future of internet. A key challenge in the semantic web is the mapping between different concepts. Many techniques for such mapping exist, but most of them induce a one-to-one mapping, which does not seem to correspond to real world problems. This project proposes a new approach, which tries to use the power of machine learning, and in particular classification algorithms, to solve the mapping task. Several techniques have been proposed in time, starting from Bayesian filters, KNN-algorithms, Neural Networks, CNG – (Contextual Network Graphs) (Ceglowski et al., 2003), LSI – (Latent Semantic Indexing) (Deerwester et al., 2001), TF-IDF – (Term Frequency – Inverse Document Frequency), NLP – (Natural Language Processing) and many others, yet few use other information in their decision process besides statistics, synonymy and other relations between documents and word frequencies.

Latent semantic indexing (LSI) is a vector-space technique that has shown great promise in addressing the problems of polysemy and synonymy in text collections, offering improved recall over standard keyword searches. One of the most striking features of LSI (Landauer et al., 2000) is the ability to return relevant results for queries even when there is no exact keyword match. Because the LSI vector model is a purely mathematical representation, the indexing technique is not limited to any particular language. (Jensen & Vikner, 1994)

A standard step in LSI is the creation of a term-document matrix (TDM), which is essentially a weighted lookup table of term frequency data for the entire document collection. In LSI, this matrix is interpreted as a high-dimensional vector space. An alternative interpretation of this matrix is possible, however, with the TDM representing a bipartite graph of term and document nodes where each non-zero value in the TDM corresponds to an edge connecting a term node to a document node. In this model, every term is connected to all of the documents in which the term appears, and every document has a link to each term contained in that document. The weighted frequency values in the TDM correspond to weights placed on the edges of the graph. We call this construct a *contextual network graph*. (Ceglowski et al., 2003)

Even though these two methods have shown great results so far, we believe that introducing syntactic information as a feature in the decision process will significantly improve their accuracy. In this paper we will experiment semantic classification using a multicategorial Bayesian filter and also including syntactic information about the analyzed documents.

2. PROPOSED METHOD

The Link Grammar Parser is a syntactic parser of English, based on link grammar, an original theory of English syntax. Given a sentence, the system assigns to it a syntactic structure, which consists of a set of labeled links connecting pairs of words. The parser also produces a "constituent" representation of a sentence (showing noun phrases, verb phrases, etc.).

If we think of words as blocks with connectors coming out, then there are different types of connectors; connectors may also point to the right or to the left. A left-pointing connector connects with a right-pointing connector of the same type on another word. The two connectors together form a "link". Right-pointing connectors are labeled "+", left-pointing connectors are labeled "-". Words have rules about how their connectors can be connected up, that is, rules about what would constitute a valid use of that word. A valid sentence is one in which all the words present are used in a way which is valid according to their rules, and which also satisfies certain global rules. (Sleator & Temperly, 1991)

The structure assigned to a sentence by a link grammar is rather unlike any other grammatical system that we know of (although it is certainly related to dependency grammar). Rather than thinking in terms of syntactic functions (like subject or object) or constituents (like "verb phrase"), one must think in terms of relationships between pairs of words. In the sentence below, for example, there is an "S" ("subject") relation between "dog" and "has"; a "PP" (past-participle) relationship between "has" and "gone"; and a "D" (determiner) relation between "the" and "dog". It may be seen, however, that parts of speech, syntactic functions, and constituents may be recovered from a link structure rather easily. For example, whatever word is on the left end of an "S" link is the subject of a clause (or the head word of the subject phrase); whatever is on the right end is the finite verb; whatever is on the left-end of a D link is a determiner; etc. (Grinberg et al., 1995).

The syntactic information representation can be represented as a prefix to each word included in a naïve Bayesian dictionary. As an example, we will use "sub_John" to represent that the word "John" is the subject of the phrase in which it appears. (Laerty et al., 1992). A naive Bayes classifier is a simple probabilistic classifier based on applying Bayes' theorem with strong (naive) independence assumptions. A more descriptive term for the underlying probability model would be "independent feature model" (Zhang, 2004)

```
+-----Ds-----+
|      +---A--+-Ss-+-PP-+
|      |      |    |    |
the black.a dog.n has gone
```

Fig. 1. Sample of parsed text using Link Grammar

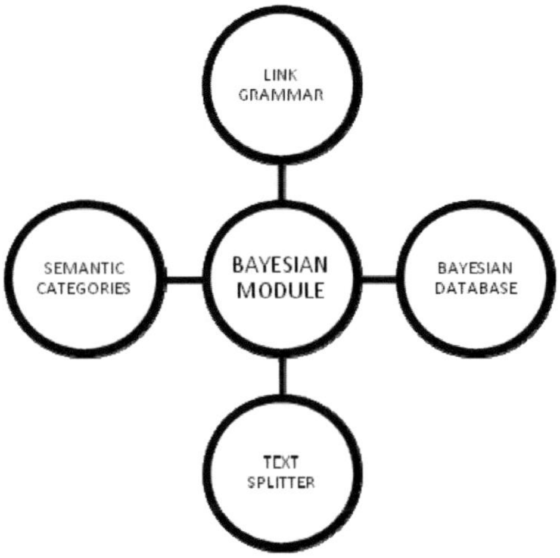

Fig 2: System Diagram

Depending on the precise nature of the probability model, naive Bayes classifiers can be trained very efficiently in a supervised learning setting. In many practical applications, parameter estimation for naive Bayes models uses the method of maximum likelihood; in other words, one can work with the naive Bayes model without believing in Bayesian probability or using any Bayesian methods.

Statistical classification is probably the most popular text classification method used by semantic classifiers. Compared with a rule based filter, it has the advantage of a learning algorithm. (e.g. the most it classifies, the better its results). Of course, being one of the pioneers of text classification methods, different methods against this technology have been found, but its mathematical grounds (e.g. the fact that the semantics of a word can be statistically established by the frequency of the documents in which it appears) still offers good results. (Ceglowski 2003). Our purpose was to add extra information to this classifier. Even though it is able to determine the semantics of a word statistically, we believe that this determination would be more accurate if we would also know the syntactic function of the analyzed word.

In spite of their naive design and apparently over-simplified assumptions, naive Bayes classifiers often work much better in many complex real-world situations than one might expect. Recently, careful analysis of the Bayesian classification problem has shown that there are some theoretical reasons for the apparently unreasonable efficacy of naive Bayes classifiers. An advantage of the naive Bayes classifier is that it requires a small amount of training data to estimate the parameters (means and variances of the variables) necessary for classification. Because independent variables are assumed, only the variances of the variables for each class need to be determined and not the entire covariance matrix. (Zhang 2004)

Also, it has been shown that on semantically distinct categories the Bayesian classifier performs quite well, but on similar ones, it lacks accuracy. Our results show that introducing syntactic features will increase accuracy and decrease misclassification on all documents. Basically, when using semantically distinct categories, a Bayesian module would obtain very good results, but when a large part of this categories have common parts (like physics, quantum physics and mechanics) there is a large risk of misclassifications. In other words, semantic equivalent classes will cause a Bayesian module to run inadequately.

3. RESULTS

Our experiment consisted in the following data:

- A naïve Bayesian module
- 25 scientific domains (among which some were branches some were sub-branches)
- A large corpus data consisting in publications and books used in the faculty

On an initial training and testing phase (without inserting the link grammar module), results showed a poor classification. Although on distinct categories, there was a 0.5% misclassification, on similar categories this percent raised to almost 25% - on the training corpus and 33% on a test corpus that was not presented in the training list of documents.

The next step was to repeat these tests after inserting syntactic information in the Bayesian module, using the canonical form presented in the proposed method section. This meant that new words had to be inserted in the dictionary (which led to approximately twice the size of the dictionary). Training also took longer since a pre-processing module was run before the Bayesian module. On distinct categories, the misclassification rate dramatically decreased to 0.01% while on similar categories it decreased to 15% on the training corpus and 27% on untrained corpus.

4. CONCLUSIONS

Although the presented results might seem low and inconcludent, we are confident that using syntactic information in a semantic classifier would increase its accuracy by at least 10% in the Bayesian case, and a lot more on better semantic classifiers like LSI, CNG and so on. Also, we had a mixed commission of people, consisting both professors and PhD students, which evaluated and performed a manual classification for the similar categories documents. Although it might seem incredible, the results provided by our classifier were higher than advised manual classification.

Acknowledgements

This work was entirely supported by BitDefender – AntiSpam Laboratory. Web: www.bitdefender.com

5. REFERENCES

Ceglowski M., Coburn A., Cuadrado J., (2003) *Semantic Search of Unstructured Data Using Contextual Network Graphs.*

Deerwester S., Dumais S., Furnas G., Landauer T., Harshman R. (2001), *Indexing by latent semantic analysis*, Journal of the American Society for Information Science

Grinberg D., Lafferty J., Sleator D. (1995) *A Robust Parsing Algorithm For Link Grammars,* Computer Science - Computation and Language

Jensen P., Vikner C. (1994) *Lexical knowledge and the semantic analysis of Danish genitive constructions,* Topics in Knowledge-based NLP Systems

Laerty J., Sleator D., Temperley D. (1992), *Grammatical trigrams: A probabilistic model of link grammar,* Proc. of the 1992 AAAI Fall Symp. on Probabilistic Approaches to Natural Language

Landauer T., Foltz P., Laham D., (2000) *An introduction to Latent Semantic Indexing,* Department of philosophy, University of Colorado at Boulder

Sleator D. K., B.; Temperley D. (1991). *Parsing English with a Link Grammar*, School of Computer Science Carnegie Mellon University Pittsburgh, PA 15213

Zhang H. (2004). *The Optimality of Naive Bays,* American Association for Artificial Intelligence (www.aaai.org)

Annals of DAAAM for 2008 & Proceedings of the 19th International DAAAM Symposium, ISSN 1726-9679
Editor B. Katalinic, Published by DAAAM International, Vienna, Austria 2008
Make Harmony Between Technology and Nature, and Your Mind will Fly Free as a Bird

HEURISTIC IMAGE RECOGNITION

COSOI, A[lexandru] C[atalin]; VLAD, M[adalin] S[tefan] & SGARCIU, V[alentin]

Abstract: *Using a distance specifically designed for AntiSpam filters, we developed an image retrieval system that can extract similar images ignoring most noise types. Sometimes, when taking a picture, random pixels change, or different objects appear or disappear from the image. When retrieval is performed, these minor modifications can cause serious problems. Using SID (short for spam image distance) our approach is insensitive to noise and combined to other image characteristics can boost up performance to any image retrieval algorithm.*
Key words: *SID, histogram, image retrieval, antispam.*

1. INTRODUCTION

Internet users have already become used to a new category of spam messages, one in which the content is actually delivered in the form of a picture attachment. Also, as they probably noticed, each image is slightly different from its predecessors, by having some noise added.

Two years ago, such "image spam" consisted of approximately 10% of the total amount of spam. Such message series typically consisted of about 10-15 spam images with some minor modifications. In the last 6 months, however, spammers noticed that many of the current AntiSpam solutions are almost ineffective against this new trick so they started attacking this niche in earnest. Image spam increased to 30-40% of the total amount of circulating spam, with random noise changing at almost every image sent. Detection rates dropped even further. (Graham 2006)

Usually noise consisted in:

- Adding random pixel in the image
- Animated GIF's with random bogus frames
- Modifying the size of the image
- A long line at the end of the image (some sort of border) with random part missing
- Splitting the image into sub-images and using the HTML tables to reconstruct it.
- Similar colors between different parts of text in the image.
- GIF's with transparent frames, with parts of the text in different frames to confuse OCR filters

As one may notice, part of this noise types are also met in an image retrieval system. Let's consider that a professional photographer has to take a picture of the Eiffel Tower. He will fix his camera on his tripod in the place where he can have the best view on the tower and take a few pictures. Even though he will take consecutive pictures during a small time interval, and for his eye the pictures are quite similar, computationally speaking, the pictures will be different. If a bird or a plane flies over the tower, the picture will be even more different.

2. PROPOSED METHOD

If in spam images this process is voluntary, in reality this process can happen by mistake, or simply because of the medium in which the pictures are taken. Another problem would be that the information contained in the image can be rotated with different angles in order to confuse OCR filters. The same problem would appear if the photographer would want to take a landscape picture, or a diagonal picture

BitDefender solved this problem by performing analysis only on the histograms of the spam images by using a special distance developed for this type of pictures, called SID (spam image distance). Using a frequently updated database of histograms of images met in spam emails, almost 98% of the spam images were correctly identified. (Cosoi, 2006)

Many histogram distances have been presented so far, (Stevens et al., 2002) among which we can enumerate: Histogram Euclidian distance (Sablak et al., 2006), Histogram intersection distance, Histogram quadratic (cross) distance, (Scheunders, 2003) where, if we consider h and g two color histograms, then Histogram Intersection Distance (the most appealing one for our purpose) represents

$$d(h,g) = \frac{\sum_A \sum_B \sum_C \min(h(a,b,c), g(a,b,c))}{\min(|h|,|g|)} \quad (1)$$

Where |h| and |g| give the magnitude of each histogram, which is equal to the number of samples. Colors not present in the user's query image do not contribute to the intersection distance. This reduces contribution of the background colors. The sum is normalized by the histogram with fewest samples. (Ling & Okada, 2006)

It is useful because it can ignore changes in background colors, but problematic when changes appear in foreground. Also, if the noise added by the spammers uses the same colors but in different quantity, these colors will be considered in the distance. The algorithm used was quite simple. There had to be a component that cleaned the image as much as possible from the noise added by spammers, and there had to be a distance which compared two histograms. Since this type of noise is particular for spam images, one way to avoid unnecessary problems was to ignore all colors that could be found less than x% in that image. Some noise persisted, but the major part is out. Of course, this eliminates good colors too, but only insignificant ones. The cleaning function *Ch* accepts a given histogram and returns a new shorter histogram in which unnecessary colors are eliminated. *Ch* can be defined like this:

$Ch(H) = H'$, where

$$H' = \left\{ c_i \in H \,\middle|\, \frac{|c_i|}{\sum_{j=1}^{|H|} |c_j|} \times 100 > \gamma \right\} \quad (2)$$

In other words, only colors that can be found in a large amount (normalized to the surface of the picture) are kept.

Experimentally, a good value for γ would be 0.01. SID is a modified intersection distance which can be defined like this:

$$SID(h,g) = \frac{\left(\sum_{A}\sum_{B}\sum_{C} \min\left(h(a,b,c), g(a \pm \delta, b \pm \delta, c \pm \delta)\right)\right)\Psi}{\min\left(|h|,|g|\right)} \quad (3)$$

where

$$\Psi = \frac{\left|h(a,b,c) - g(a \pm \delta, b \pm \delta, c \pm \delta)\right|}{\min\left(\left|h(a,b,c)\right|, \left|g(a \pm \delta, b \pm \delta, c \pm \delta)\right|\right)} \times 100 \leq \Delta$$

wherein h and g denote pixel counts of the target and reference histograms, respectively, a coordinate triplet (a, b, c) tags an individual color bin, A, B and C denote the basis colors of the color space, parameters δ and Δ denote a color neighborhood size and a color bin similarity threshold, respectively, while $|h|$ and $|g|$ denote the magnitude (i.e., total number of samples) of the target and reference histograms, respectively. The distance of eq. [3] is symmetrical with respect to h and g—i.e. the distance depends on the contents of the two images, but not on which of the two images is a target histogram and which is a reference histogram. In other words, we do not consider only identical colors, but also similar, which can differ from the original with no more than δ values, with the restriction that we consider this element in the sum only if the difference between the sizes of the bins is smaller than $\Delta\%$. These two parameters can be determined using ROC curves, trial and error or using any machine learning technique.

This way, SID is comparing a first bin of a first histogram representation to a range of second bins of a second histogram representation to determine a minimum distance between the first bin and the range of second bins, wherein the first histogram representation and the second histogram representation are selected from the target histogram representation and the reference histogram representation, and wherein the first histogram representation is distinct from the second histogram representation, and employing the minimum distance to determine the histogram distance between the target histogram representation and the reference histogram representation.A better representation of the SID algorithm can be seen in Figure 1.

Basically. this filter compares the query image with all the images contained in the database in order to find a match. The database contains only clean histograms.

The algorithm can be presented in the following steps

1. Extract histogram
2. q = Clean histogram (skip the unnecessary colors)
3. $y = \max(SID(q, h_i))$
4. If $y \geq \Omega$ return "We have a match"

It is obvious that this filter can be modified in order to develop an image retrieval system. The only difference is that the database has to contain the original images and not only the pre-computed histograms. All we have to do in the algorithm is to add another step and to modify the one where we extract only the maximum hit. We want all the hits.

3. RESULTS

On our test corpus, our image retrieval system has an accuracy of 97.9%, but since our query system permits the user to choose the degree of accepted noise, the degree of similarity between colors or the degree of similarity between pixel counts, it's accuracy can rise up even to 100%.

Also, working with complex pictures with a high number of colors, might cause some problems to SID, but if this method would be combined with other techniques for image similarity, we believe that there will be some amazing results.

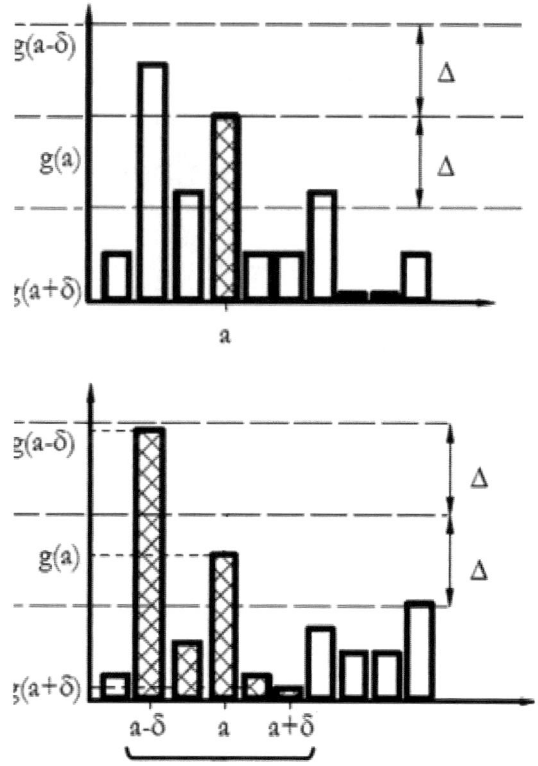

Fig. 1 – Description for SID functionality

4. CONCLUSIONS

By comparing each bin with a plurality of bins, SID is able to identify images that are slightly different from the original with high accuracy. Of course, since retrieval is performed only on the histograms of the images, we can't say that this process is flawless, but we are confident that this distance is one step further in image retrieval systems research.

Acknowledgements

This work was entirely supported by BitDefender – AntiSpam Laboratory. Web: www.bitdefender.com

5. REFERENCES

Cosoi A. C. (2006). *The medium or the message – Dealing with image spam,* Spam Bulletin

Graham-Cumming J. (2006). Title of conference paper, *The Rise and Rise of Image-Based Spam*, Spam Bulletin Ueda,

Ling H., Okada K. (2006). *Diffusion Distance for Histogram Comparison,* Center for Automatic Research, Computer Science Department, University of Maryland and Imaging and Visualization Department, Siemens Corporate Research Inc

Stevens, M.; Culberston B.; Malzbender T(2002) *A histogram-based Color Consistency Test for Voxel Coloring*, HP research labs

Scheunders P. (2003). *A comparison of clustering algorithms applied to color image,* Dept. of Physics, RUCA university of Antwerp

Sablak, S.; Boult T S. & Udiljak T. (2006). *Multilevel Color Histogram Representation of Color Images by Peaks for Omni-Camera,* Department of Electrical engineering & Computer Science (IASTED proceedings, October 18-21 1999, Nassau, Bahamas

Annals of DAAAM for 2008 & Proceedings of the 19th International DAAAM Symposium, ISSN 1726-9679
Editor B. Katalinic, Published by DAAAM International, Vienna, Austria 2008
Make Harmony Between Technology and Nature, and Your Mind will Fly Free as a Bird

QUALITY AND RISK MANAGEMENT ACTIVITIES IN VIRTUAL ORGANIZATION

COSTACHE, B[ogdan] M[ihai]; ADAMESCU, D[orina] & FUNAR, S[tefan] P[etru]

Abstract: *In the information society, the new economy is built on the main form of organization – the net. The fast change of the market and of the client's demands determines the apparition of virtual organizations. Thus, virtual organization is a market and client oriented organization. The market and client needs, the demands and their fulfilment, must be the decisive element for the success of virtual organization. By comparison to traditional organizations, the information and communication technology (ICT) is the distinctive characteristic of virtual organization. With multiple partners and a dynamic organizational structure of changes, coordination and cooperation between different partners must be the main content of the management for virtual organization¶*

Key words: *quality management, virtual organization, partnership.*

1. INTRODUCTION

1.1 The Components of the Quality Management System for Virtual Organization

Starting from the system analyses, it is being projected a quality management system for the virtual organization in three groups (Ticoll et al., 1998): client satisfaction group (CS), information technology group (IT), and coordination and audit group (CA). There are three modules of the quality management, one for each of the three groups of the quality management system (Durkin, 1994). The activities and the demands of the quality management are projected for each group as following:

1. The CS group is responsible for the client satisfaction quality management.

At this level, the quality management demands include:

- Investigation of client and market demands.
- Determination of plan quality according to the market and client demands.
- Investigation of client satisfaction and the dynamic market change.
- Propagation of response information on client and market back to the respective company or team, for facilitating the quality verification and improvements.

2. The IT group is responsible for the quality management of information system (including computers, network and software systems).

At this level, the quality management obligations include:

- Quality control of hardware and software projection systems, of partners operations and of the entire virtual organization.
- Quality control of information systems interface between different partners, specifying the standard for information and data exchange between partners in virtual organization.
- Quality control of data and information in virtual organization.

3. The CA group is responsible for the quality management of audit and partners coordination systems. In this direction, the quality management obligations include:

- Audit of the partners 'quality of guarantee validation, quality control systems and monitoring of their execution conditions.
- Coordination of quality activities between partners and conflict settling between them (Dragoi et al., 2005).

2. MODELS OF QUALITY FOR PRODUCTS AND INDUSTRIAL PROCESSES

For many organizations, risk and quality have become two essential and indispensable priorities (Rosu & Draghici, 2008). The reasoning determined in this step is mainly the following:

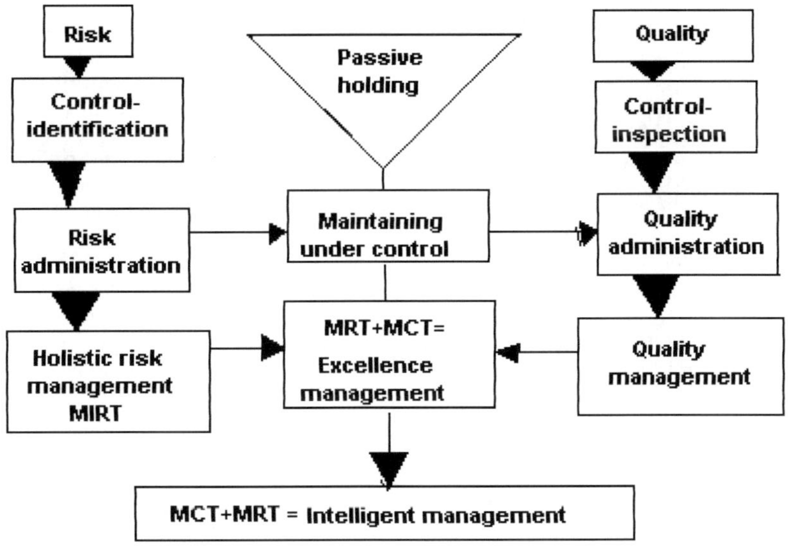

Fig.1. Parallel evolution of risk and quality management

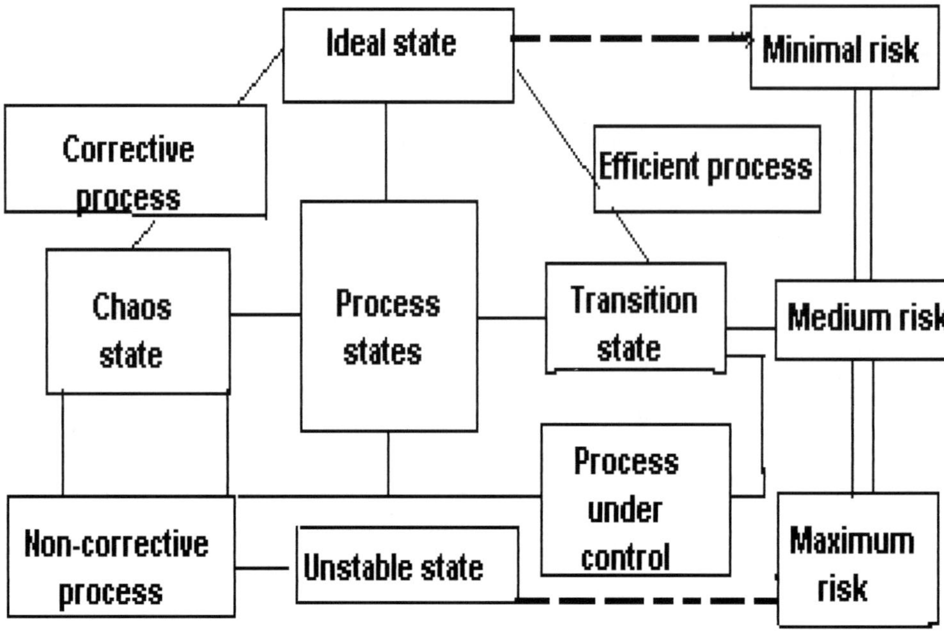

Fig.2. Process states

Critical examination of the processes shows important malfunctions. Security is an essential component of production quality in company processes.

Buyers become more exigent over the product and service quality which is being evaluated strictly related to life quality.

Complex processes impose the computer assisted quality approach.

Control evolution over preventive direction and risk administration.

Company space in the international context is being defined by the excellence coordinates: efficiency, environment, maintenance, fiability, security and quality (Schwarzer et al., 1997). The correlation risk-quality expresses the reciprocal relation by which one can identify the conditions and the technical-economic characteristics of avoiding the risk by projecting and monitoring the indicators of total flux quality and of industrial processes. A process can be expressed with the help of the four states (fig.2).

- Ideal state, when the process has capability and is under control, the risk being minimal, up to zero.
- Transition state at limit, when it is under control but it doesn't have capability and the risk is minimal.
- Unstable state, on the edge of chaos, when it has capability, but it is not under control and the risk is growing up to maximum.
- Chaos state, when it doesn't have capability and it is out of control.

3. CONCLUSION

For identifying the reasons for which virtual organization represents the apogee of work organizing in the information society, we must notice that similar to the industrial period and its bureaucratizing, the organizational form of the information era is the network.

If we add the major trends of the end of the second millennium that refer to the market globalization, to the individual consumer orientation and to the industrial sector towards services, to the orientation towards client's needs or to the shorter life cycles of the products, we can have an idea about the way in which the business today or in the future are being managed. Beyond all the mentioned trends, the main reasons for organizing activities in such organizations derive from two aspects:

- Flexibility need growing;
- Efficiency need by using the resources together with other partners.

The flexibility became necessary because of the growing of the environment and organizations exchange. Creating the adding value for clients became a complex process, implying the mingling of different types of knowledge and of a huge volume of knowledge. A group of organizations that, in order to make safe products/services, is using a common knowledge base for all the organizations which create a virtual organization.

Considering this, it is necessary the exact settlement of the quality assurance processes for the products delivered by the virtual organizations and the strict defining of the quality management system in virtual organization.

4. REFERENCES

Dragoi, G.; Cotet, C. E.; Patrascu, G.; Rosu, L. & Rosu, S. M. (2005). *Internet/Intranet/Extranet-based systems for virtual product development environment in the CESICED platform*, in the Proceedings of the 8th Conference on Management of Innovative Technologies (MIT'2005), 24th - 26th September 2005, pp. 67-72, ISBN 961-6238-96-5, Fiesa - Piran, Slovenia.

Durkin, John. Expert Systems. Design and Development. Pretice Hall, Englewood Cliffs, 1994;

Roşu, S. M. & Draghici, A. (2008). *Information system risk estimate as part of project development by virtual teams using expert systems,* Academic Journal of Manufacturing Engineering, no. 2/2008, pp. 135-142, ISSN 1583-7904.

Schwarzer, B., S., Zerbe, and H. Krcmar. An Eclectic Framework fat Understand – ing New Organizational Forms, in Social Sciences – Management and Network Technology. 1997. Brussels.

Ticoll, D., Lowy, A., Kalakota, R., Joined at the Bit – the Emergence of the E-Business Community, McGraw-Hill, New York, 1998.

Annals of DAAAM for 2008 & Proceedings of the 19th International DAAAM Symposium, ISSN 1726-9679
Editor B. Katalinic, Published by DAAAM International, Vienna, Austria 2008
Make Harmony Between Technology and Nature, and Your Mind will Fly Free as a Bird

GROUNDHOG DAY VERSUS BUTTERFLY EFFECT REVISITED IN DISCRETE MATERIAL FLOW MANAGEMENT

COTET, C[ostel] E[mil]; DRAGOI, G[eorge] & ABAZA, B[ogdan]

Abstract: *We present here an algorithm able to increase the productivity by synchronizing process and material flow simulation. We are modeling process and discrete material flow using two different theoretical models. The first theoretical model based on the Groundhog Day effect is used in process simulation. This model is reproducing every manufacturing cycle identically with the previous. The second theoretical model based on The Butterfly Effect is used in material flow simulation and allowed us to change some of the manufacturing cycles introducing stochastic distribution laws. In order to synchronize the material flow and process models we had to personalize the process simulation changing the Groundhog Day with the Butterfly effect approach.*

Key words: *Groundhog Day effect, discrete material flow management, process simulation, Butterfly effect.*

1. INTRODUCTION

In our increasing productivity research studies based on manufacturing systems architecture optimization we identified as one of the main difficulties in the area of synchronizing process and material flow simulation.

The main goal of our algorithm is to increase productivity by improving the discrete material flow management using the process simulation as a preliminary data. We analyze the results of the material flow simulation and we identify the flow concentrator for a preliminary manufacturing architecture based on process simulation results (Cotet & al., 2007). We propose a solution for eliminating flow concentrators. We perform a second simulation to validate the optimized manufacturing architecture by obtaining an increased productivity. Some of the necessary data for the material flow simulation like cycle times for the work points defined in our models are provided by process simulations describing each work point manufacturing cycles. The material flow simulator is integrating the process simulation results at the level at each work point in order to provide a complete model of the manufacturing system.

Analyzing the different behaviors in modeling for the process simulation and for discrete material flow simulation, we had defined two approaches starting from the plot of two films illustrating the models different characteristics.

The Groundhog Day film is about a man who finds himself living the same day over and over and over again. He is the only person in his world who knows this is happening, and even if he is free to change what he says and does from one Feb. 2 to the next, it will always be Feb. 2 for everyone else in the world, and he will always start from the same place. All the others will repeat themselves unless he changes the script for one day, but tomorrow they will have forgotten their new lines and be back to the first draft of Feb. 2. He is therefore trapped in a seemingly endless "time loop" to repeat the same day in the same small town (Ebert, 2005). We define the Groundhog Day effect the process simulation algorithm that reproduces every manufacturing cycle identically with the previous. According with this effect, by simulating one work point activity for multiple manufacturing cycles the same simulation will be reproduced multiple times with the same parameters. If we intend to change the script for one manufacturing cycle we can perform a new simulation and introduce this new model between the identical ones, but the previous and the next episodes in the manufacturing chain cycles will remain the same. It is impossible for us to have a script where a manufacturing cycle became different from the previous.

The Butterfly Effect film is based on the chaos theory teaching us that small events can have important consequences illustrated by butterfly flapping its wings in Asia could result in a hurricane halfway around the world. In the film, every time the main character changes his past, he goes to the exact moments when he blacked out. In the first timeline he simply blacks out traumatic moments; later, he is able to revisit these blackout moments by re-reading journal entries about them, which suggests that the blackouts could have been caused by his ability to revisit the past and he "blacks out" when his future self is revisiting his past self- a causality loop (Ebert, 2004).

We use the Butterfly Effect in material flow simulation because allowed us to define the script of a manufacturing cycles chain in which some of the manufacturing cycles are different by introducing stochastic distribution laws and not fixed values for MTBF (the mean time between failure) or MTTR (mean time to repair). In this case some of the manufacturing cycles will be different due to repair times who will personalize the multiple cycle chain.

As we already emphasize at the beginning of this chapter, one of the main difficulties in building an integrated model is that due to those two special effects characteristics process and material flow behavior are very different. In order to synchronize the material flow and process models of manufacturing cycles we had to modify the process simulation program in order to personalize the process simulation changing the Groundhog Day with the Butterfly effect approach.

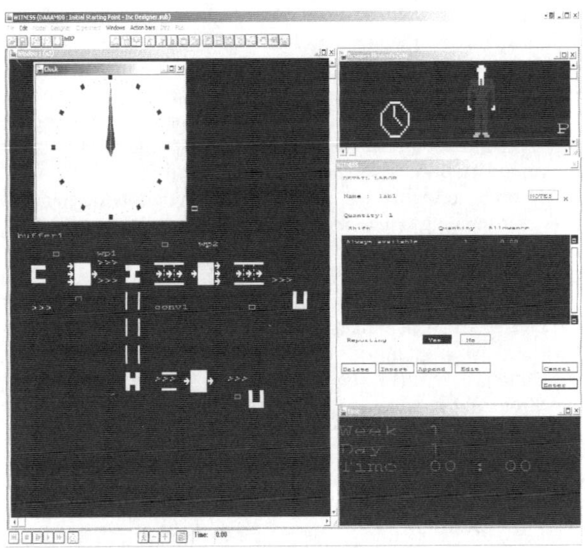

Fig. 1. Preliminary manufacturing architecture.

2. SIMULATING THE INTEGRATED MODEL

The first step of our algorithm is to realize the preliminary parametric model of the manufacturing system and to simulate the material flow in the system in order to identify the flow concentrator (fig.1). Different algorithms are used for the diffused and concentrate manufacturing systems, for terminating or non terminating simulations according with the specific constraints characteristic for each case. For a designed manufacturing architecture it is always useful to simulate the material flow conduct before applying our design into practice in order to avoid potential flow concentrators generating low productivity or even blockage. The leading actor able to manage this area will be the flow simulator.

We define diffused manufacturing systems as architectures with more than two work points connected by transport & transfer systems and using deposits at local or system level (Cotet & Dragoi, 2003). We agree here with the thesis that within the class of stochastic simulation models, one further distinction is necessary: simulations can be either terminating (sometimes called finite) or nonterminating in nature, with specific algorithms for each category (Sanchez, 2001). We consider a material flow and process synchronous simulation the simulation of a model where at the level at the work point the process simulation is concomitant with the material flow simulation. We consider that due to the complexity of the mean time between failure (MTBF) and mean time to repair (MTTR) modeling for various machines and manufacturing systems one can provide the best solutions for such productivity improvement based on stochastic distribution laws and not on fixed values, even if fixed values for these parameters would be used for some particular finite short simulations.

A running diffused manufacturing model for terminating simulations is presented in fig. 2. Each work point parameters were established based on process simulation results as described before.

Using the Witness reports one can identify the flow concentrators and according with manufacturing planning constraints can propose technological solutions or modifying manufacturing architecture based solutions to eliminate those concentrators. In order to validate the increased productivity obtained on the optimized manufacturing system architecture, a new simulation is necessary. If one weighs the productivity for the preliminary and optimized architecture the increasing productivity due to the optimization can be quantified. A very important issue of this algorithm is to evaluate if the optimized architecture supplementary costs are or not covered by the productivity gain. This financial analysis is based on a NPV algorithm and validates the financial profitability of the optimization. At the first level main parameters for the work points modeling in material flow simulation are provided form CAM simulations describing the manufacturing process (Lee, 1999). This way the material flow simulator is integrating the process simulation results at the level at the work point in order to provide a complete model of the manufacturing system.

In order to realize the synchronized simulation the material flow terminating simulation algorithm is applied for each manufacturing system of the enterprise. In this model the work points are the basic simulation model units describing the manufacturing system material flows. In the end at the level of the entire virtual manufacturing system the process simulation for each work point is synchronic with the material flow simulation describing the entire system activity. This local synchronizing process of the material flow simulation allowed us to change some of the manufacturing cycles introducing stochastic distribution laws values for MTBF, MTTR or failed parts adapted to each work point characteristics. If we use the integrated model versus the asynchrony classic one the accuracy of the simulation results for the presented case study increase with 10 %.

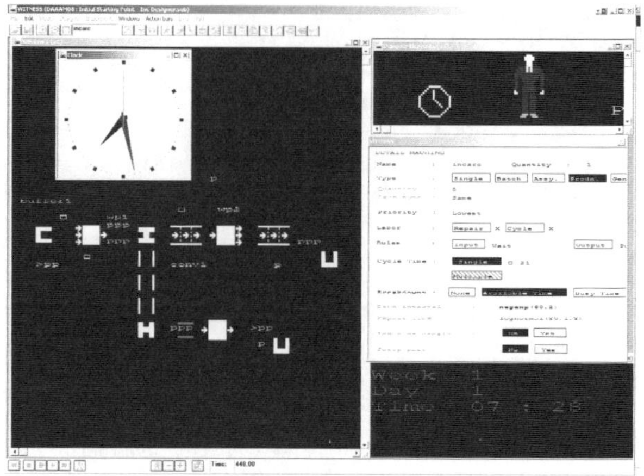

Fig. 2. Manufacturing architecture during material flow simulation.

3. CONCLUSION

Starting from the script of two films (The Butterfly effect & Groundhog Day) we describe here the theoretical models we defined in order to generate a manufacturing architecture optimization algorithm. The material flow and process simulation models based on specific software solutions are the main actors of this simulation project undertaken with the goals of demonstrating and confirming production rates of a manufacturing process based on a proposed design layout and operational data and of identifying ways of improving the design of the system in order to increase those production rates. According with this algorithm one can analyze the results of the material flow simulation and identify the flow concentrator for the manufacturing system. If an architecture modification is proposed as a solution for this problem a second simulation to validate the optimized architecture and the obtained increase of productivity is necessary.

4. REFERENCES

Cotet, C. E., Dragoi, G. & Carutasu, G. (2007). Material Flow & Process Synchronous Simulation In Concentrate Manufacturing Systems, *Annals of DAAAM for 2007 & Proceedings of The 18th International DAAAM SYMPOSIUM, „Intelligent Manufacturing & Automation: Focus on Creativity, Responsibility and Ethics of Engineers"*, Katalinic, B. (Ed.), pp. 180-181, ISSN 1726-9679, ISBN 3-901509-58-5, Zadar, Croaţia, October 2007, Publisher DAAAM International Vienna 2007.

Cotet, C.E., Dragoi, G.S. (2003). Material Flow Management in Validating Concentrate and Diffused FMS Architectures. In: *International Journal of Simulation Modelling IJSIMM, no. 4,* December 2003, pp.109-120, ISSN 1726-4529, Vienna.

Ebert, R. (2005). Groundhog Day, *Available from:* http://rogerebert.suntimes.com/apps/pbcs.dll/article?AID=/20050130/REVIEWS08/501300301/1023 *Accessed:* 2008-06-22

Ebert, R. (2004). The Butterfly Effect, *Available from:* http://rogerebert.suntimes.com/apps/pbcs.dll/article?AID=/20040123/REVIEWS/401230301/1023 *Accessed:* 2008-06-22

Lee, K. (1999). Principles of CAD/CAM/CAE Systems, *Addison Wesley Longman, Inc.*, ISBN 0-201-38036-6, USA.

Sanchez, S. M. (2001). ABC's of output analysis, *Proceedings of the 2001 Winter Simulation 2001*, Peters, B.A., Smith, J.S., Medeiros D.J. (Ed.), CD-ROM, Presses Association for Computing Machinery (ACM), New York.

Annals of DAAAM for 2008 & Proceedings of the 19th International DAAAM Symposium, ISSN 1726-9679
Editor B. Katalinic, Published by DAAAM International, Vienna, Austria 2008
Make Harmony Between Technology and Nature, and Your Mind will Fly Free as a Bird

TECHNICAL DEVELOPMENT PROGRAM WITH FRIENDLY MATERIALS

COTETIU, R[adu]; EBERST, O[tto]; ALEXANDRESCU, M[arius]; COTETIU, A[driana] & UNGUREANU, N[icolae]

Abstract: *During the last few years and for the future period in Romania there is a prognosis which shows an accelerated industrial development. The industry is increasing demands for up-to-date materials. As part of this process, the North University of Baia Mare gave currency to a technical developing program assisted by the Romanian Government on the basis of the results of the researches in using and replacing with friendly environmental materials. The paper shows the steps of the technical development project in correlation with the present status of the industries' preoccupation in developing the friendly technologies and materials like polymers and composites.*
Key words: *polymers, composites, friendly, materials.*

1. INTRODUCTION

Romania's economy has gone through a complete process of reorganization since the 1995s and the 1996s. The traditional industrial activities such as mining, the processing of minerals, mining machinery industry etc. have gradually lost ground so today they only make up less than 10% of total industrial output compared with the 1990s. As a result of the reorganization program since 1998 Romania's territory was divided into 8 regions, each one with their own goals and development strategies. Baia Mare which today is a part of the NW Region was a mining center connected with other industries: mining machine industries, mineral processing industries, nonferrous metallurgy etc. Now the areas of the development are: informatics and its subdivisions, aggregate machineries industry, agriculture, tourism, furniture manufacturing, research.

The action-plan put together here is matching with the above presented premium development areas.

The development regions in Romania are: 1. North-East; 2. South-East; 3. South Muntenia; 4. South-West Oltenia; 5. West; 6. North-West; 7. Middle; 8. Bucharest-Ilfov.

2. MATERIAL AND METODOLOGY

The major goal of "The technical development with composites" plan is to develop a research-projects and know-how implementing unit for the North-West region.

In Baia Mare, at the North University we are on the verge of opening a research-projects headquarter where we'll try to attract newly graduated students from the proposed organizations (University and partner companies) developing the skills of people from both university and economic fields and spreading the use of up-to-date, environment friendly plastics in the ranks of machinery engineering and aggregates industry.

As we take in consideration the proper university environment from the North University suitable for PhD dissertations about high performance plastics, we have to notice that the program presents solutions for developing research but also, provides a platform for training of the researchers and attracting highly skilled personal.

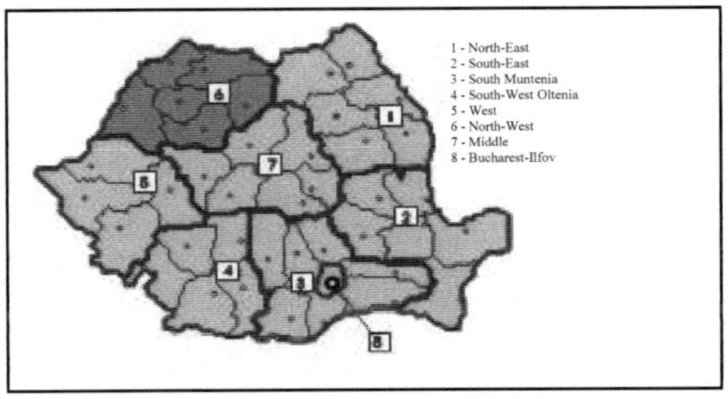

Fig. 1. The development regions in Romania.

3. RESULTS

3.1. Strategical Purposes

1. Implementing the "Technical development program with composites" in the interest of *technological development based on new, environment-friendly materials and technologies*, being integrated into a managerial research system, in harmony with the regional strategy called "The technology become resource" (material possibilities).
2. *Introducing research and development activities based on competitive technologies* for the most important areas of industries making possible the spread of using new composite plastics. That goal is reached by growing the material base of the University's research unit in the joint interests of university and industrial fields, efforts which have a great impact on regional and national development.
3. *Developing new or updated products using environment friendly materials*. We have set up a research-resulted and finished materials monitoring unit and we will study those materials maintainability and operational safety.
4. *Effectively using the commercial profitability concept* in the industries and inside those communities which need home made new products.

3.2. Presentation of Activities

The activities suggested in this plan are of two types: industrial research and experimental development activities.

1. Studying today used up-to-date materials and those applications on national and international level describing both the materials and their applications. A would be study will reveal the today used plastic materials and their applications in Romania and the perspectives of introducing high capability composite plastics (Cotetiu & Eberst 2003).

   Showing the today used environment friendly types of plastics. Studying and analyzing the national specific needs for this domain.
2. Studying the surrounding economic entities' machine equipments and finding ways to modernize those (Cotetiu & Eberst 2003, Eberst &Cotetiu 2003). We will establish a database for companies from the North-West Region's about level of machine equipments showing their maintainability and operational safety (Cotetiu et al., 2000). We will find new possibilities for using high capability plastics modernizing the

existing products, machines and aggregates (Cotetiu et al., 2003).

Finding out the effects of machineries and their maintenance and operational safety problems caused to the environment.

-Analyzing the possibilities of making up new products or by modernizing the existent ones due to the usage of new, environment friendly materials.

-Studying the production methodology of items which incorporate high performance plastics. (Kalácska et al. 2000). We present the world-wide novelty regarding researching-developing and modernization methodologies. We'll work out a research method on virtual models to be used on products and machineries which incorporate high capability plastics (Analyzing up-to-date European methodology tendencies about development and modernization. Learning about research methods based on virtual models).

3. Conducting researches about structures of products incorporating some high capability, environment friendly plastics. We'll present different and specific researching methods compared to metal works such as the following machinery (aggregate) elements made out of plastics: gears, slide bearings, rollers. We'll make the fez ability studies and the necessary technical documentation for materializing products fitted for new or modernized items to be found in this program.
4. Analyzing the specific designing methods for high performance, environment friendly plastics (Eberst, 2005). Detailed descriptions about products development. Identifying and learning about necessary new technologies for products incorporating high performance and environment friendly plastics.
5. Materializing prototypes and making up new prototypes for products and aggregates incorporating high performance plastics.
6. Research under factory, operational conditions, conducting operational test in labs, checking the already accepted technological solutions. Conducting studies about the obtained products' long term testing and compared operational safety. Conducting laboratory tests under overloaded, strained conditions for items including high performance plastics. (Operational test while working. Long term test and operational safety study. Speeded up operational tests (under strain).
7. Giving the research tests' results to the interested economic units, editing and spreading those results.

3.3. Partners

The project partners are: S.C. Adiss S.A., S.C. Proned-Control S.R.L., S.C. TechnoCAD S.A., Development of Small and Medium Enterprises Center from Maramures Foundation (CDIMM), "NORDTECH" - Research Center, Innovation and Technological Transfer Association (CCITT). Coordinator of the project is North University of Baia Mare (www.ubm.ro) and program coordinator: Prof. dr. ing. Radu Cotetiu.

All partners have experience and credibility in their domain of activity. Some of them are leaders in Romania in the matter of manufacturing of equipment. Also, the coordinator has the competence to lead this activity based on an experience and research activity in the domain of friendly materials in industrial and mechanical installations and equipments. A small part of this experience was published. Some papers are listed in title References.

4. CONCLUSIONS, SUGGESTIONS

The expected results, period of activity and quantity analyses are displayed in the Table 1.

Expected results	Quantity analyzes	Co-operating partners	Year
1. New or modernized items using environment friendly materials	At least four products	UNBM; ADISS; PRONED; TechnoCAD; CDIMM	2009
2. PhD students in this field of science	Two PhD students in Mechanical or Industrial Engineering Domains	UNBM	2011
3. Employees on the field of research	Four researchers	UNBM	2010
4. Economic units to use results of research	At least 100	UNBM; ADISS ; PRONED; TechnoCAD; CDIMM	2010
5. Developing work places in production, maintenance, servicing	At least 200	UNBM; ADISS; PRONED; TechnoCAD; CDIMM	2009 - 2014

Tab. 1. Expected results, period of activity and quantity analyses on partners

The training of the personal so much needed for materialization of this project will be guided through the necessary and interested area of activity promoting results which can be used to make competitive products and technologies. New jobs will appear at the producers and at the users of those products.

5. REFERENCES

Cotetiu, R., Cotetiu, A., Eberst, O. (2003). Experimental Investigations of the Influence of Lubrication on rolling Steel/Polyamide Structures, *Kutatasi es Fejlesztesi Tanacskozas* nr. 27, Magyar Tudomanyos Akademia, Godollo, pp. 72, ISBN 963 611 390 4.

Cotetiu, R., Pay, G., Cotetiu, A. (2000). Theoretical and experimental Studies About the Load Variation Influence on the Values of Friction Coefficients în Rolling Couples - Steel - Ploamidic Materials, *International Computer Science Conference, MicroCAD 2000*, pp. 39-45, Miskolc, ISBN 963 661 424 5.

Cotetiu R., Eberst, O. (2003). Research Analysis on the Replacement of Metallic by Non-Metallic Elements in the Mining Industry. *The 5th International Multidisciplinary Conference*, North University of Baia Mare, pp. 99-102, ISSN 1224-3264.

Eberst, O., Cotetiu, R. (2003). Studies About the Load and Material Influence on the Machine Elements from Ploamidic Materials. Rev. *Kutatasi es Fejlesztesi Tanacskozas* nr. 27, Magyar Tudomanyos Akademia, Gödöllő, pp. 68, ISBN 963 611 390 4.

Eberst, O. (2005). *Contributions Using Plastic Materials in Gears*, PhD Thesis, Universitatea de Nord Baia Mare.

Kalacska, G., Eberst, O., Cotetiu, R. (2000) Renewing of mining machines with the application of engineering plastics. *Magyar Tudomanyos Akademia, Kutatasi es fejlesztesi tanacskozas*, nr. 24, 3.kotet, pp. 83, ISBN 963 611 390 4.

Annals of DAAAM for 2008 & Proceedings of the 19th International DAAAM Symposium, ISSN 1726-9679
Editor B. Katalinic, Published by DAAAM International, Vienna, Austria 2008
Make Harmony Between Technology and Nature, and Your Mind will Fly Free as a Bird

CHANGE MANAGEMENT WITHIN A BUSINESS SYSTEM

COVIC, D[ragan] & COVIC, D[aniela]

Abstract: Change trend and their nature are difficult to be *controlled nowadays. In everyday business activities, changes are reaching one another even when we think we catch up with the one. Also, when we believe we have anticipated everything, we are often surprised with how much we actually fail to observe Numerous trends in thinking over a successful system can be added another one, based on change management within a business system., that combines a range of change interventions, and a diagnostic and prognostic approach with acceptable technologies. Change management model provides us with more confidence into continuance of life cycle of our business system on a daily basis.*
It is our wish to increase our immunity toward changes that happen due to our flexibility.
***Key words**: business system, management, changes, prognostic approach, diagnostic approach*

1. INTRODUCTION

While you are preoccupied with doing a daily, usual activity at your workplace, or reading a book and thinking about how to make your life better, time has passed unstoppably. During this relatively short time, incalculable number of activities in your environment changes their structure, position or attitude. Regardless of type or content of changes, whether they are of social, political, scientific or ethical nature, they all make you notice and respect them, no matter how strong or big you are. Today's business systems have a short life cycle unless they can follow change rituals in their environment. Due to a fast pace of changes, successful business systems are only those that can recognize and anticipate or predict a key change direction.
Trends of mega changes that happen under influence of men for justification of their dominant role, are recognizable and predictable. Trying to find a solution to observing changes within a business organizational structure has resulted even so far in a hardly calculable number of organizational forms and possible structures within a business system. Definitely there is not a single solution capable to meet requirements of a turbulent environment.
We are trying to find a way how to track changes in the environment, i.e. to manage changes within a business system by combining diagnosing and predicting techniques. If we rely on diagnostics we will run late in change trends, and if we accept exclusively a prognosis, we are statistically accepting a huge risk to choose a wrong side in trying to save a penalty throw. Management process is oriented toward reducing time needed for our positive reaction to key events in our environment.

2. CHANGES AND TRENDS

Intensity of a change wave and their diversity prove a necessity of their understanding. The very understanding leaves us a possibility to follow or even predict them. Change can be defined as a change of state and structure in a unit of time. Let's try to reveal a change process by directing it to everyday life we can easily understand.Change division proves that we have to presuppose a level of change understanding, keeping in mind achieved level of knowledge reached by a „modern“ man. If we are to supervise our sky on a daily basis and try to grasp different technological ideas as far as it's possible, we will find out very soon that changes, their temporal dynamics and intensity in the outer space represent changes that we cannot influence or this influence is minor. Heights of new questions that we pose turn us toward increased level of knowledge on what is really happening around us. We do understand majority of processes in our immediate environment, our country, our habitat, although we are only trying to investigate bigger part of it through knowing their history, rules and laws. Relying on evolutionary, unmanageable trend of changes, we are trying, to a large degree, to turn our dominance orientation into change of natural laws. We believe that we have conceived ourselves and that the gap between our knowledge and change dynamics is becoming increasingly smaller.
By studying changes in our business and social environment, we can recognize change trends that we need to respect and accept. Of course that we can oppose them too, but awareness that trends exist helps us a lot, since trends do not influence only our business systems. Our environment is also not exclusively affected but it's rather a global influence. Trends can be recognized as changes: from industrial toward information-based society, from forced technologies toward „high-tech“, from national toward super-region or world economy, from being short-term toward being long-term, from centralization toward decentralization, from institutional support toward self-help, from representative toward participative democracy, from hierarchic structures toward networks, from north toward south. All these trends have their time dimension that directs us, so it is inconceivable to manage sustainable business system if trends are neglected.

3. MANAGEMENT OF BUSINESS SYSTEM CHANGES

Current change process is intensive and comprehensive, and as such it is rather unconceivable and chaotic for those who cannot grasp it. Thus managing a business system today means managing changes. A key question a manger faces is how to react to changes. By analyzing behavior of sustainable business systems, the three strategies can be recognized: being assured that a change will happen and react to it – a strategy of late reaction, feeling change symptoms and trying to prevent them – change symptoms strategy, assuming changes and reacting to them while you are still stable and doing well – prognostic strategy. In line with diversity of these strategies, we can divide and categorize business systems, from those that survive to very successful ones. We need to understand that our business system is surrounded with uncertainty, unpredictability and chaos, in which dominates coincidence and statistical probability laws. Change implementation process within a business system has to be continuous. Managerial skills are used here to adjust organizational structure elements to the change pattern.
Change management phenomenon is to be understood as a permanent process of adjustment of changes that we don't

control in our environment. Complete process of changes that is to be controlled within a business process is reduced to selection of the most favorable information flows, documentation and materials in a business process. If we prevent floating within these flows, we will improve functioning of our business process. Change process in the environment could be simply observed if it didn't have multidimensionality that appears as synergy or different reaction of certain factors of organizational structure to an individual change. This phenomenon of multidimensionality of environment change influence on reactions in the environment and business system is extremely difficult to be fully observed.
Thus we look at the change management within a business system from the two points of view: in terms of size and breadth of the change intervention, and in terms of conception of change implementation.
From the point of view of size and breadth of an intervention, change process can be viewed as small, medium and large one, in a range of changes of an element within an organizational structure up to comprehensiveness of changes. Influence of a conception enables a division of change process on a diagnostic, prognostic and combined approach.

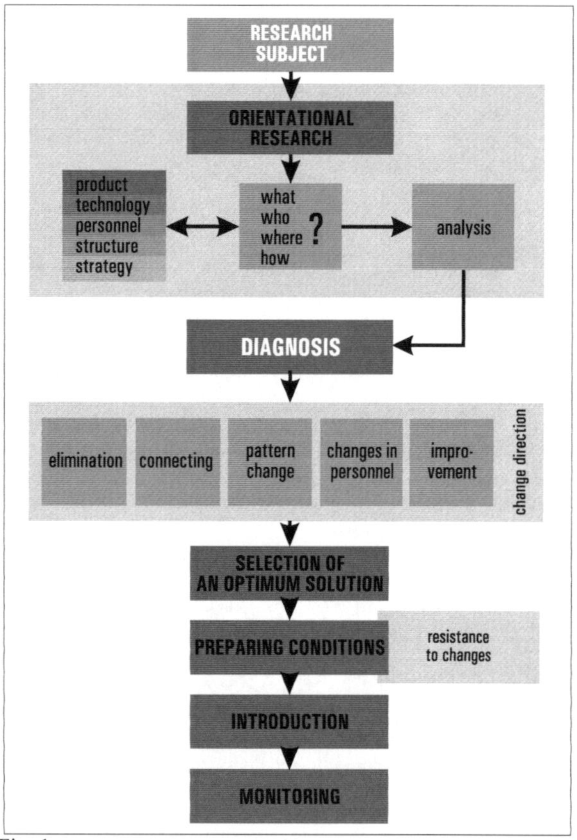

Fig. 1

Opposite to diagnostic approach that is regularly connected to tension, prognostic approach is extremely oriented toward future. It is regularly used in business systems that initiate change intervention prior to tension in a business system. Change in a business system is created as a wish to introduce some changes when things are still good. Complexity of change process management in a business system is only partially recognizable through these two concepts. By combining these two approaches, we can create a huge number of variants in modeling change process. Everything again depends on size and breadth of change implementation interventions. Which solution should be chosen and which way followed is to be decided by management. Whether we succeeded in choosing and implementing a right concept is found out very soon, but even in case we choose a wrong way, this mistake can be quickly corrected, since the process is continuous.

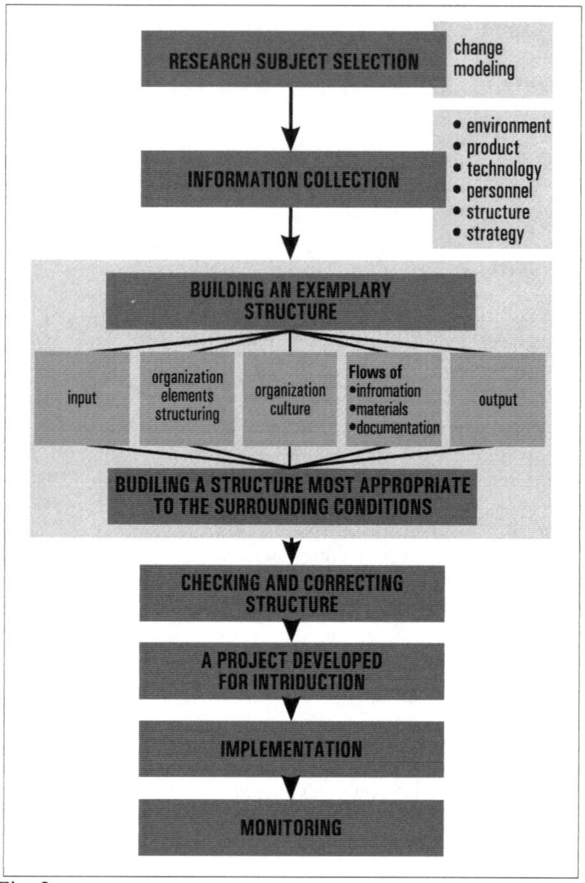

Fig. 2

4. CONCLUSION

Change trend within and around a business system is permanent and unstoppable. As changes influence a business system through every factor in business system organization, individually and synergistically, it is impossible to define a single change management process. It is necessary to develop at the same time prognostic and diagnostic model of change management. Our attention is to be primarily directed toward defining attributes of business systems that we will use for change management.

5. REFERENCES

Čović, D. (1996). *Modeliranje promjena organizacijske strukture u funkciji proizvoda i tehnologije*, Doktorska disertacija, Mostar

Jokić, B. (1994). *Poduzeće u promjeni*, ISBN 953-180-016-2, Alineja Zagreb

Whitmore, J. (2006). *Trening za postizanje učinkovitosti*, ISBN 953-246-002-0, „Mate“ Zagreb

Ferenčić, A. (1997). *Upravljanje promjenama*, ISBN 953-6157-54-3, „Izvori“ Zagreb

Naisbitt, J. (1985). *Megatrendovi*, „Globus“ Zagreb

Robbins, S. P. (1995). *Bitni elementi organizacijskog ponašanja*, ISBN 953-6070-30-8, UDK 159.94403(075.8), „Mate“ Zagreb

Adižes, I. (2004). *Upravljanje promjenama*, ISBN 86-7668-006-6, „Adižes“ Novi Sad

Annals of DAAAM for 2008 & Proceedings of the 19th International DAAAM Symposium, ISSN 1726-9679
Editor B. Katalinic, Published by DAAAM International, Vienna, Austria 2008
Make Harmony Between Technology and Nature, and Your Mind will Fly Free as a Bird

GEOMETRICAL ANALYSIS OF THE TOOL'S RAKE ANGLE AT CUTTING

COZMINCA, I[rina]; COZMINCA, M[ircea]; UNGUREANU, C[atalin] & MIHAILIDE, M[ircea]

Abstract: *The paper presents a theoretical approach regarding the values of the functional rake angle γ_f of the tool at cutting. Based on the results of the specialty literature, the authors determine a general expression for the real rake angle γ_r, according to the real chip's removal direction and the speed components v_x, v_y, v_z.*

Key words: *cutting, tool, rake angle, removal, speed.*

1. INTRODUCTION

Nowadays, the functional rake angle γ_f is defined as the angle between the plane normal to the instantaneous direction of the relative speed between tool and workpiece, known as pressure plane, and the plane tangent to the rake face of the tool. Based on this definition, the general equation of the functional rake angle, related to the geometrical parameters of the process, is (1), where: v_x, v_y, v_z are the components of the relative speed tool/ workpiece, γ_N – the orthogonal rake angle, λ - the edge inclination angle and K – cutting edge angle. These angles, also named constructive angles, are considered on a certain point of the cutting edge (Belous & Cozminca, 1971).

$$\cos\gamma_f = \frac{v_x(\tan\lambda\cos K - \tan\gamma_N\sin K) + v_y(\tan\gamma_N\cos K + \tan\lambda\sin K) + v_z}{\sqrt{(1+\tan^2\gamma_N+\tan^2\lambda)(v_x^2+v_y^2+v_z^2)}} \quad (1)$$

The relation (1) has a great disadvantage, namely it represents a particularly value for γ_f, the one measured along the direction of maximum gradient. The chip should be removed after this direction only if it's perfectly fluid, and that's unreal. Using the equation (1) to estimate γ_f for a possible practical situation, e.g. $\gamma_N = 0$, $v_x = 0$, $v_y = 0$ we obtain $\cos\gamma_f = \cos\lambda$, so $\gamma_f = \lambda$. It means that the chip, removing along the direction of maximum gradient, must flow along the cutting edge, and that's not possible (Trent & Wright, 2000).

As experimental tests have shown, the chip is removed along a direction at angle η, in the rake face plane, with the normal N_1 at the edge (Fig. 1). According to (Jawahir &von Lutterwelt, 1993) the angle η has the same value as the inclination angle λ.

However, (Rapier & Wilkinson, 1959) showed that angle η may reach larger values than λ, and (Russel & Brown, 1956) were concluded that the view θ of the chip's deviation angle η on a plane Q, normal to the cutting plane and parallel to the edge is practically equal to the inclination angle λ, $\theta = \lambda$.

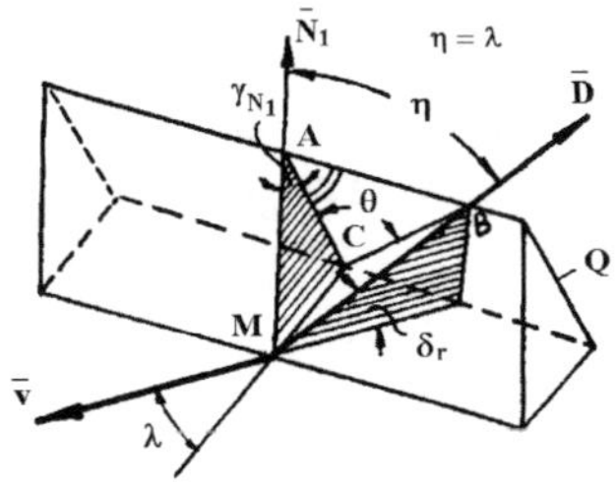

Fig. 1. The chip's removal direction, D, at cutting

As Figure 1 shows, $\tan\eta = \tan\theta\cdot\cos\gamma_{N1}$ so relation (2) result, where: γ_{N1} is the rake angle measured in a plane orthogonal to the cutting edge.

$$\eta = \arctan(\tan\theta\cos\gamma_{N_1}) = \arctan(\tan\lambda\cos\gamma_{N_1}) \quad (2)$$

Therefore, relation (2) shows that always $\eta < \lambda$. The fact that angle η is different than λ was relieved also by (Zorev & Granovski, 1967) which have been concluded that most of the times η exceeds λ, with maximum 5…7 degrees and it's increasing with λ and chip' width, at smaller values for tool rake angle γ.

On the other hand, the deviation angle η decreases the cutting speed v is increasing, as the experimental relation (3) shows:

$$\eta = \lambda\cdot v^{-0.08} \quad (3)$$

Equation (3) demonstrates that angle η is always smaller than the inclination angle λ , which is in accordance with the results of (Russel & Brown, 1956).

Based on the results presented above, we may conclude that for practical reasons we are allowed to approximate the deviation angle η by the inclination angle λ, in case of free cutting. But the chip deviation phenomenon becomes more complicate in case of complex cutting. In this situation come in at chips formation processes not only the major cutting edge, but also the minor or secondary edges (Gunay et.al., 2004).

Therefore the chip removal direction D, inclined with the angle η in the rake face plane regarding to the normal N_1 on the edge, cannot be determined only from geometrical reasons (as line with maximum gradient). We must consider also the aspects regarding the physics of chips formation.

That's why the real rake angle γ_r has to be measured in the chip's removal plane, which contains the direction v of the relative speed tool/ workpiece, and the direction D of chip's removal. This angle is defined as the one between a plane normal to the direction of speed v (pressure plane) and the tangent plane to the tool's rake face, in the considered M point on the edge.

According to this definition, the real cutting angle $\delta_r = \pi/2 - \gamma_r$ can be measured between the chip removal direction D and the speed direction v. See equation (4), where: $\eta \cong \lambda$, γ_{N1} – the constructive cutting angle, measured into a plane orthogonal to the edge direction (Zorev & Granovski, 1967).

$$\delta_r = \arccos(\cos\lambda\cos\eta\cos\gamma_{N_1} + \sin\lambda\sin\eta) \quad (4)$$

But this relation its lack in generality, due to the fact that the cutting speed v was considered included into the edge plane and that's valid only in the particularly situations, when $v_x = 0$, $v_y = 0$ and = v.

Therefore we try to determine a general expression for the real functional rake angle related to the geometrical parameters, when all the possible components of the cutting speed v_x, v_y, v_z occur.

2. THE GENERAL EXPRESSION OF THE REAL FUNCTIONAL RAKE ANGLE γ_r

To determine a general relation indicating the dependency between γ_r and the geometrical parameters λ, γ_{N1}, γ_N and η we use the schematization illustrated in Fig.2.

It illustrates views in the constructive base plane, in the edge plane, in the rake face plane, where appears the chip's removal direction D, the direction orthogonal to the edge N_1 and the vector T tangent to the edge and in the chip's removal plane (Belous & Cozminca, 1971).

If we take apart the vector D after the tangent to the edge and the one orthogonal to the edge, the vectors T and N_1 result with $\sin\eta$, respectively $\cos\eta$ modules. Projecting these vectors to the x, y, z tool's axis we find the rectangular projections of the vector D.

The vector N_1 has the expression (5) and the vector T the expression (6), so for D the relation (7) results:

$$\begin{aligned} N_1 &= \\ &= \left(-\cos\eta\cos\gamma_{N_1}\sin K - \cos\eta\sin\gamma_{N_1}\sin\lambda\cos K\right)\cdot i + \\ &+ \left(\cos\eta\cos\gamma_{N_1}\cos K - \cos\eta\sin\gamma_{N_1}\sin\lambda\sin K\right)\cdot j + \\ &+ \left(-\cos\eta\sin\gamma_{N_1}\cos\lambda\right)\cdot k \end{aligned} \quad (5)$$

$$T = \sin\lambda\cos\lambda\cos K\cdot i + \sin\eta\cos\lambda\sin K\cdot j - \sin\eta\sin\lambda\cdot k \quad (6)$$

$$\begin{aligned} D &= N_1 + T = \\ &= \left(-\cos\eta\cos\gamma_{N_1}\sin K - \cos\eta\sin\gamma_{N_1}\sin\lambda\cos K + \sin\lambda\cos\lambda\cos K\right)\cdot i + \\ &+ \left(\cos\eta\cos\gamma_{N_1}\cos K - \cos\eta\sin\gamma_{N_1}\sin\lambda\sin K + \sin\eta\cos\lambda\sin K\right)\cdot j - \\ &- \left(\cos\eta\sin\gamma_{N_1}\cos\lambda + \sin\eta\sin\lambda\right)\cdot k \end{aligned} \quad (7)$$

The functional real rake angle γ_r, measured in the removal plane and defined by v and D directions, is the angle between the pressure plane P, orthogonal to v, and the plane tangent to the rake face in M point (see Fig. 2). It is the complementary angle of the real functional cutting angle, between v and D, so the equation (8) occurs.

Equation (8) represents the general expression of the functional real rake angle γ_r and allow estimate it for any cutting process, according to the relative speed components v_x, v_y, v_z and the geometrical parameters of the tool.

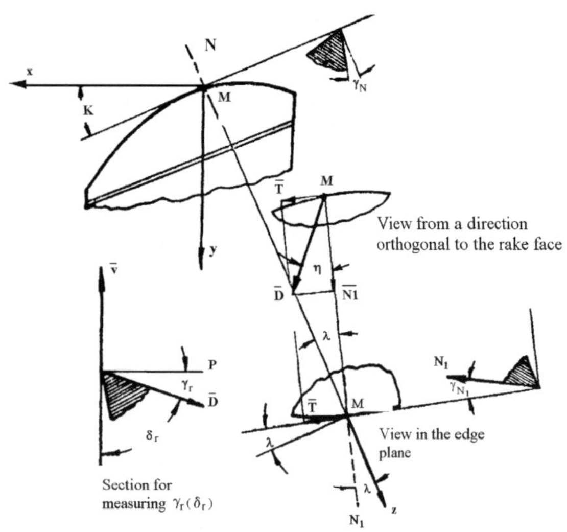

Fig. 2. The geometrical cutting parameters in different views

$$\begin{aligned} \sin\gamma_r &= \\ &= \frac{\left(\cos\eta\cos\gamma_{N_1}\sin K + \cos\eta\sin\gamma_{N_1}\sin\lambda\cos K - \sin\eta\cos\lambda\cos K\right)\cdot v_x}{\left(v_x^2+v_y^2+v_z^2\right)^{1/2}} + \\ &+ \frac{\left(\cos\eta\cos\gamma_{N_1}\cos K - \cos\eta\sin\gamma_{N_1}\sin\lambda\sin K + \sin\eta\cos\lambda\sin K\right)\cdot v_y}{\left(v_x^2+v_y^2+v_z^2\right)^{1/2}} + \\ &+ \frac{\left(\cos\eta\sin\gamma_{N_1}\cos\lambda + \sin\eta\sin\lambda\right)\cdot v_z}{\left(v_x^2+v_y^2+v_z^2\right)^{1/2}} \end{aligned} \quad (8)$$

But for most of the tools types the main constructive rake angle is γ_N, measured in the plane orthogonal to the apparent edge, therefore the simplified equation (9) result, based on the connection relations, i.e. $\tan\gamma_N = \tan\gamma_{N_1}/\cos\lambda$ or $\sin\gamma_{N_1} = \sin\gamma_N\left(\sin^2\gamma_N + \frac{\cos^2\gamma_N}{\cos^2\lambda}\right)^{-1/2}$ and neglecting the speed components v_x and v_y. For $\eta \cong \lambda$, the relation (10) results:

$$\sin\gamma_r = \cos\eta\cos\lambda\sin\gamma_N\left(\sin^2\gamma_N + \frac{\cos^2\gamma_N}{\cos^2\lambda}\right)^{-1/2} + \sin\eta\sin\lambda \quad (9)$$

$$\sin\gamma_r = \cos^2\lambda\sin\gamma_N\left(\sin^2\gamma_N + \frac{\cos^2\gamma_N}{\cos^2\lambda}\right)^{-1/2} + \sin^2\lambda \quad (10)$$

3. CONCLUSIONS

An analytical expression for estimating the real rake angle γ_r was established; it can be used when the values for the speed components v_x, v_y, v_z, the edge constructive angles γ_{N1}, λ, K and the deviation angle η are known.

If the speed components v_x and v_y may be neglected related to v, the simplified equations (9) and (10) are recommended. The influence of the normal constructive angle γ_N on angle γ_r is more pronounced as the absolute value of the inclination angle λ is lower; the λ sign also influences the γ_r values. The influence of λ on γ_r is higher at lower values for γ_N.

Applying the above relations for the real functional rake angle allow us to consider the real cutting plane in order to evaluate the cutting forces values needed in tools design or in cutting process simulation.

4. REFERENCES

Belous, V. & Cozminca, M. (1971). Contributions to the real tool's rake angle study in cutting processes, Bulletin of the Polytechnic Institute of Iaşi, XVII (XXI), Fasc. 3-4, pp. 145-154, ISSN 1011-2855.

Gunay, M.; Aslan, E.; Korkut, H. & Eker U. (2004). Investigation of the effect of rake angle on main cutting force, *Int. Journal of Machine Tools and Manufacture,* vol. 44, no. 9, pp. 953-959, ISSN 0890-6955.

Jawahir, J. S. & von Lutterwelt, C. A. (1993). Recent Developments in Chip Control Research and Application, *CIRP Annals*, vol. 42, no. 2, pp. 659-693, ISSN 1726-0604.

Rapier, A. C. & Wilkinson, P. T. (1959). *Plasticity Report No. 156*, Mechanical Engineers Research Laboratory, East Kilbride, Glasgow.

Russel, J. K. & Brown, R. H. (1956). The Measurement of Chip Flow Direction, *Int. J. of Machine Tool and Manufacture*, VI, Pergamon Press, London, pp. 129-138, ISSN 0890-6955.

Trent, E. M. & Wright, P. K. (2000). *Metal Cutting, Fourth Edition,* Publisher Butterworth – Heinemann, ISBN 0-7506-7069-X.

Zorev, N. N. & Granovski, G.I. (1967). *Development of Metal Cutting Science.* Publisher Mashinostroenia, Moskow.

Annals of DAAAM for 2008 & Proceedings of the 19th International DAAAM Symposium, ISSN 1726-9679
Editor B. Katalinic, Published by DAAAM International, Vienna, Austria 2008
Make Harmony Between Technology and Nature, and Your Mind will Fly Free as a Bird

THE INFLUENCE OF THE GEOMETRIC PARAMETERS AND CUTTING CONDITIONS ON THE THEORETIC DEVIATIONS OF WORMS BY WHIRLING THREAD DEVICE

CRETU, G[heorghe]; NEDELCU, D[umitru] & CRETU, D[an]

Abstract: *This paper established the link relations between theoretical deviation resulted at worm processing by the whirling thread cutting principle and the geometrical and processing parameters. The evolutions of these deviations amounts against module, diametrical coefficient and worm number of revolutions are represented graphically.*
Key words: *parameters, cutting, whiriling thread device.*

1. INTRODUCTION

The cutting of cylindrical worms by the whirling thread cutting principle is thought now as one of the most productive ways of processing. In this process the piece turns with slow speed and it is eccentric wrap up by the whirling thread cutting devices tools, which turn with high speed entering periodically into the splitter. Also for the correct achieving of the worm profile, the device's head inclines with an angle equal to that of the worm's helix reference.

The surfaces of flanks and bottom of the channels are cinematic generated, as successive wrap upping positions of the whirling thread cutting device tools. Because of this reason, the resulted surface presents successive deviations from the theoretical form of the profile (Nedelcu, D., et al, 2002). For the size determination of this deviation it is need to determine the surface equations described by two successive tools of the whirl head. Knowing the processed worm flanks equations it will be determined the relations that determine the theoretical deviation size resulted in cutting by this process.

2. TECHNOLOGY USE

For the determination of the surface equations described by the finishing tools it will be considered more reference systems (Bojan, L. 2006). First are written the surface equations described by the tools in the $O_0X_0Y_0Z_0$ reference system, in which O_0X_0 represents the symmetry axis of the disposal face of the whirling tools, and O_0Y_0 coincides with its own rotation axis :

$$(x_o^2 + z_o^2)\, tg^2\alpha_N = \left(y_0 - R_{0s} tg\alpha_N + \frac{\overline{e_{n1}}}{2} \right)^2 \quad (1)$$

In which D_{vs} represents the disposal diameter of the noses of tools from the rotation axis of the device, α_N the profiles angles ,e_{n1} is the normal reference cord of the worm tooth space, R_{os} is the circle radius described by the point form the tool edge blade which processes the reference diameter (Cretu, Gh., 1997).

A translation of the coordinate system is performed in the coordinates $O_0x_0y_0z_0$ in the length of the O_0X_0 axis with an amount e, so the axis O_0Y_0 should coincide with the axis of processed worm axis, resulting $O_1X_1Y_1Z_1$ reference system. Equation no. 1 becomes:

$$[(x_1 + e)^2 + z_1^2]\, tg^2\alpha_N = \left(y_1 - R_{0s} tg\alpha_N + \frac{\overline{e_{n1}}}{2} \right)^2 \quad (2)$$

If the system is turning round the O_1X_1 line with an γ angle (equal with the inclination angle of the worm helix on the reference diameter), it results the immobile OXYZ reference system. Knowing the link equation between the two reference systems, the equation of the conic becomes:

$$\left[(x + e)^2 + (-y\sin\gamma + z\cos\gamma)^2\right] tg^2\alpha_N = \left(y\cos\gamma + z\sin\gamma - R_{0s} tg\alpha_N + \frac{\overline{e_{n1}}}{2} \right)^2 \quad (3)$$

By geometrical considerations the value of ccentricity e it is determined:

$$e = R_{0s} - \sqrt{R_1^2 - \left(\frac{\overline{e_{n1}} \sin\gamma}{2} \right)^2} \quad (4)$$

In which R_1 represents the radius of the reference cylinder.

Then is determined the equation of the normal plane at the average helix of tooth space:

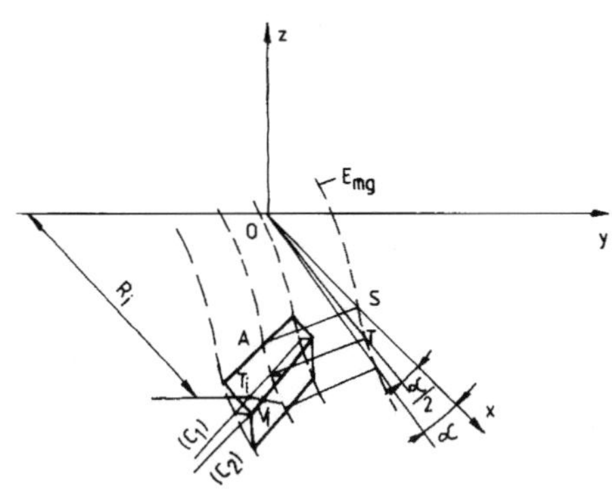

Fig. 1. The positions determination of the deviations from the theoretical surface.

$$xR_1\cos\varphi + yh - zR_1\sin\varphi - h^2\varphi + h\frac{\pi}{2} = 0 \quad (5)$$

For the determination of the theoretical deviation at worm flank processing with thread cutting devices, are considerate two successive positions of the finishing tools, offsite between them with an α angle, dependent on the parameters of the cutter conditions. The relation gives this angle:

$$\alpha = \frac{2\pi n_m}{n_s z_f} \tag{6}$$

In which N_m represents the worm rotation, N_s the rotation of the whirl head and Z_f is the number of finishing tools. The maxim deviations will be a half of this angle per each rotation. (figure 1). Their size is determined in the normal plan at the average helix of the tooth space, spine with a half of α angle around the axis Oy. For this the plan is intersected with the theoretic flank surface (resulting the curve C_1) and a surface generated by the tools (resulting the curve C_2).
The relation gives the C_1 curves equations:

$$\begin{cases} x\sin\frac{\alpha}{2} - \frac{h}{R_1}\left(y - \frac{h\alpha}{2}\right) + z\cos\frac{\alpha}{2} = 0 \\ x = R_d \sin\varphi + \mu\cos x_g \cos\varphi \\ y = h(\varphi - \varphi_0) + tgx_g\sqrt{R_1^2 - R_d^2} - \mu\sin x_g - \frac{p_x}{4} \\ z = R_d\cos\varphi - \mu\cos x_g \sin\varphi \end{cases} \tag{7}$$

To evaluate the amount of this deviation, on this curve we consider *m+1* equidistant points *Ti (i=0,m)*, between the head and bottom cylinders, situated on equal distance with R_i to the worm axis. In this points normal are layout to the helical surface of the worm and they are intersected with the surface described by the tools of the whirling thread cutting device, resulting V_i points. The resulting relations from the determinations of the coordination points T_i and V_i are transcendental and they are solved with the computer aid. Knowing the coordinates of the T_i and V_i points it can be determined the amount of the deviation with the relation:

$$\Delta h_i = \sqrt{(x_{V_i} - x_{T_i})^2 + (y_{V_i} - y_{T_i})^2 + (z_{V_i} - z_{T_i})^2} \tag{8}$$

The calculated deviation with this relation depends on many factors: the module, number of worm teeth, locating diameter of the whirling head teeth, diametric coefficient worm number revolutions, and the number of the finishing tools. For the determination of the deviation amount *Δhi (i=0,m)*

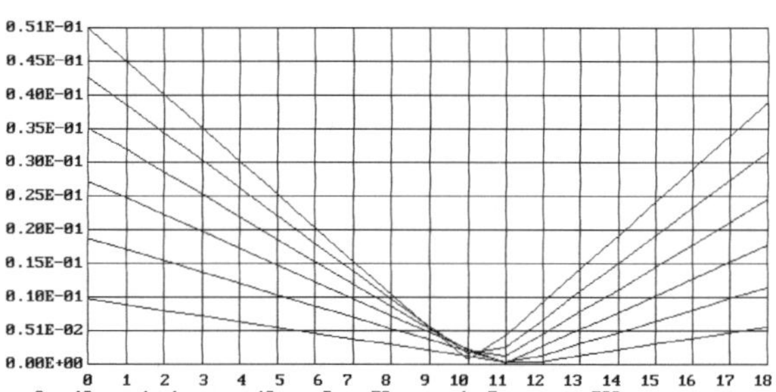

Fig. 2. The theoretical deviations values in the across flanks, from some module m_x

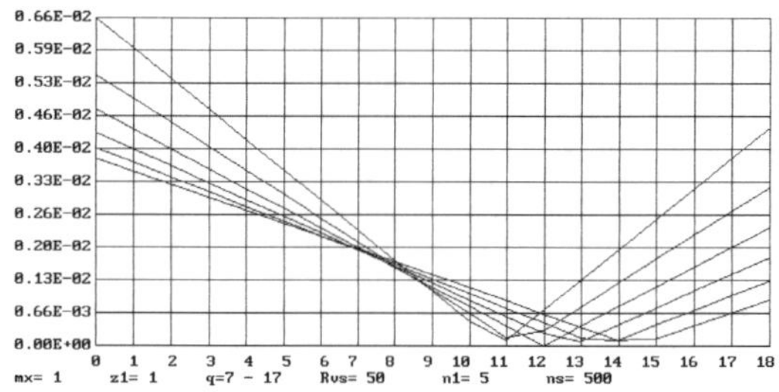

Fig. 3. The theoretical deviations values in the across flanks, from some diameter coefficient q.

and for the graphic representation of this dependence a special program was realized in QBASIC. The program is first receiving input data concerning the geometric parameters of the processed worm and of the cutting conditions. Solving the transcendent equation is achieved using the fractioned range method. After the coordinate determination of the T_i and V_i, points the amount of deviations *Δhi* is determined. The obtained data can be graphically represented depending on point position on flank. The graphics have in their abscissa the deviation position in a perpendicular plan on the medium helix of the flank (the point *0* corresponds the exterior diameter, and the point *18* on the bottom diameter), and in the ordinate the amount of the deviation. On the same graphic are represented the deviation results for *6* values of the variable specified parameter.

In the figures 2 and 3 are represented the deviation evolution on the worm flank for different values of module (figure 2) and diameter coefficient (figure 3).

3. CONCLUSIONS

Studying the evolution of the theoretical deviation to the cylinder worm processing with whirl processing, made with this program, a series of conclusions must be made:
-the deviations are having higher values at the nose of tooth and lower at the pitch diameter
-the deviations are higher with increase of the module and the piece rotation, with diminishing the tool number of revolutions and that of the diametric coefficient
-the disposal diameter of the tools influences very little the amount of the deviation.

4. REFERENCES

Creţu, Gh., (1997). *Contribuţii privind tehnologia de fabricare a melcilor folosind principiul filetării în vârtej*, *The contributions concerning the manufacturing technology of worm using the wriling process,* Technical University of Iasi Publishing, pp. 40-68, Ph.D. Thesis, Iaşi

Bojan, L., (2006). *Advanced Technologies*, Pro Literatur Verlag Publishing Germany, pp. 33-56, ISBN 3-86611-197-5, Mammendorf

Nedelcu, D., et al, (2002) *Aspecte comparative privind deformatiile termice ale cutitului de strung, The comparative aspects concerning of temperature strain for tool lathe,* Bulletin of Technical Institute of Iasi, pp. 165-171, ISSN 1011-2855, Iasi

Nedelcu, D., (2001) *The technological parameters influence upon over rolls dimension at forming exterior grooves using Taguchi method,* CITAF 2001, Technical Publishing Bucharest, pp.326-332, ISBN 973-31-2049-9, Bucharest

Annals of DAAAM for 2008 & Proceedings of the 19th International DAAAM Symposium, ISSN 1726-9679
Editor B. Katalinic, Published by DAAAM International, Vienna, Austria 2008
Make Harmony Between Technology and Nature, and Your Mind will Fly Free as a Bird

BUSINESS ORGANIZATIONS BASED ON KNOWLEDGE – AN ANSWER TO THE GLOBALIZATION OF INFORMATION AND OF TECHNICAL PROGRESS

CRISAN, G[heorghe]& CRISAN, L[ia]

Abstract: *The development of modern technologies generates multiple possibilities of information, and of course of knowledge. The technical progress becomes this way dependent not of the accessed information volume, but of the capacity that the organizations have, in order to select, process and transform them in the form of knowledge and explicit action. The organizations based on knowledge become this way, more than a necessity.*
Keywords: *information, knowledge, Internet, selection, processing.*

1. INTRODUCTION

The development of modern technologies, and especially of the Internet offers the possibility of storage, practically unlimited, by the business environment, of one of its essential resources, the information.

As a consequence, the firms may use o multitude of specific information, that allow them to develop in the direction of introduction of the new, of technical progress and implicitly of raising the competitivity.

The reality shows that this opportunities are limited by the existence of at least two factors, that they generate, such as the apparition and development of the Internet – the general access of any user, to the specific information, and the absence of a valorise filtration of them.

In this conditions, the technical progress will be directly dependent by the capacity of access and use by the business organizations of the truly valuable information, from the perspective of the specific of the activity that the organization develops.

2. THE ORGANISATIONS BASED ON KNOWLEDGE – AN ANSWER TO THE GLOBALIZATION OF INFORMATIONS AND THE PROMOTION OF TECHNICAL PROGRESS

The problematic of accessing and using in an efficient manner of the information, respectively, the capitalization of these, in the form of knowledge, brings into discussion the necessity of functioning of a new type of organization, specific the Information Society, respectively of a collectivity of employees with creation work, nested trough a computerized infrastructure.

This model, specific to the XXI century, it is characterized by a component part dominated by professionals, by a low number of intermediate management levels and also trough the coordination by means of non-authoritative nature.

This type of organization, known as "an organization centred on memory" (Le Moigne, 1990), "an intellectually intensive firm" (Nurmi, 1998), or "an intelligent organization", (Hendricks, 1998), may be appreciated as a viable alternative to the organizations based on control an authority. At the level of type of organizations, there is a problem that appears.

This problem is related to the validity and limits of knowledge, at least at the level of individual comportments of their members, that can identify, in a certain percent, the relations between the goals, means and results and also those between the organization and its environment.

In these conditions, the functioning of an organization based on knowledge should be institutionalized, at least on the next aspects:

- The transformation of knowledge stock in the main resource of the organization, vital for its performances
- Imposing the intellectual intensive processes as being determinant for reaching the organizations objectives
- The appearance of new roles and responsibilities connected with the management of knowledge and the processes connected to it
- The consensual introduction, at the level of organizational culture, of normative models for the permanence of the values, connected with creativity, competence, learning, communication
- Imposing of the aspects connected with knowledge as essential factors in building the identity of the organization and ensuring its integrity and coherence

Just in this type of organizational environment, those that have the information can transform it into knowledge, that later can be the basis, trough integrating, assimilation and understanding for an efficient capacity of action, followed by operationalization, in a concrete situation.

This way, the quality and accuracy of the information used, influence the level and quality of knowledge, adapted and operationalized to the objectives of the organization.

In front of the challenge to find ways of implementing what they have, the organizations build, in their functioning process, different forms of representation about their state of knowledge (and finding out, many times, that they are not fully aware of what they know or don't know).

But, in order to establish the route of the new informational and cognitive package, it is essential to know the milestone, respectively the initial level of accumulation of the organization and its members.

These desiderates are not sufficient to consolidate an organization, in its behaviour being necessary (Zack, 1999) new strategic and specific goals, that should determine it to represent, integrating and transparent, the accumulations of knowledge (at group or individual levels, or on artificial support), to continuously extend its knowledge basis (trough the stimulation of innovation and organizational learning processes) and to develop its capacity to transform, with intelligence and opportunity, the available knowledge into successful actions.

Practically, the capitalization process of intellectual actives it is implicitly associated with the concept of knowledge database (that include both human bearers and informatics intelligent systems), both as effect and in certain circumstances as a cause of the knowledge.

This database could have the attributes of an extended organizational memory, with the capacity to sustain cognitively specific autonomous projects (that benefit also from its results).

For this, the organizational actors employ themselves in a series of synergic behaviours, respectively of co-elaboration (interactive generation oh new knowledge), co-learning (reciprocal validation of new cognitive acquisitions) and co-managing of the capitalized knowledge, that implicitly generates systemic co-evolution effects in the cognitive plan.

On the other way, founding on knowledge has also an inter-organizational dimension because the contemporary society, and especially in the business environment, the organizations are interested to reciprocally evaluate themselves, to follow its leaders, to learn from each other, to compete between themselves, and if necessary to fusion, in order to create or use the new knowledge.

As a consequence, the organizations that play in this perimeter have more development alternatives and learning from external sources, including exigent etalons of performance (in a continuously evolution).

From the managerial perspective, the leading act it is concentrated on problems of elaborating a strategic vision and to facilitate a coordinated action of the main actors, that on competence basis and in a spirit of cooperation, become auto responsible, inclusively on decisional aspect.

On the basis of theoretical studies and own research upon a number of local firms, especially SMEs from the top areas, we can say that the manager becomes more and more a conceptual responsibility bearer (projecting of "architectures" of systems and processes, solution validations, proposal ratifications) that administrative power (Crisan, 2007).

In these conditions, the traditional roles of the managers, inter-personal, informational and decisional, are uploaded with those of facilitator, mentor, moderator or promoter.

In addition, the specific of organizations that were studied, generates more unusual managerial positions in the hierarchical environment, such as chief knowledge officer, chief information officer, project responsible for the development of competences, coordinator for organizational learning.

They are necessary because, from their nature, the organizations based on knowledge implies the management of collective competences, respectively of all it knows and it's capable of for a certain business community, in connection with their own objectives, environment conditions and based on the individual abilities of its members.

This way, on a market with fierce concurrence, the consequence will be that the winners will be those that will use in the most efficient way all their competences.

As an immediate reaction is the initiative of the Romanian Government, that fundaments its general framework of professional qualifications on the idea of competences, that will ensure clearer and stable reference points in organising and developing the process of learning.

This is a process where the universities will not be able to revendicate an absolute priority (that was traditionally theirs) in promoting a new generation of ideas, because more and more companies transform their activity in a research and learning space (the apparition of the concepts of "knowledge centre" or "corporate university").

3. THE OPERATIONALIZATION OF THE ORGANIZATIONS BASED ON KNOWLEDGE – A PREMISE FOR GENERATING THE TECHNICAL PROGRESS

From the perspective of operationalization of the organization based on knowledge, in the Romanian society, it is a possibility for a major risk for over solicitation of the technologic coordinate in comparison with those educational, cultural and managerial.

The main engine of transformation should be the promotion of the practices specific to the knowledge society and even the creation of new special organizations.

For these, there is a need for a coherent intervention, on multiple plans:

- Educational plan, that consists of the development of conception competences, of informatized work, and also of profesionalization of the roles involved in the functioning based on knowledge (virtual work platform administrators, knowledge engineers, competences trainers)
- The plan of organizational culture, that consists in assuming an ethics of responsible legitimacy and in centering of individual and collective behaviors on the values of community
- The plan of managerial intervention, that consists of assimilating and extending of the new generation managerial practices
- The plan for working methods and instruments:
  - Sustaining the production of knowledge at the organizational level, trough their own research and development effort, time for individual reflection, creativity reunions, intranet networks, electronic deliberation rooms, intelligent management systems for knowledge and competences, extended access to the data and knowledge bases, technological and concurential surveillance, cartography of organizational knowledge, stimulation to create thematically teams on the principle of competence centers, virtual mailboxes for new ideas and proposals, and support centers for expertise and assistance
  - Stimulating the organizational learning process, trough meetings for the dissemination of new ideas, facilitating of learning from experience, the adopting of the mentoring system, career planning, competences balances
  - The development of the organization as a professional community trough the interactive management, transparency of standards, rules and procedures, project-oriented work, the use of recommendations and procedures manuals, and stimulating the creation of professional communication networks.

4. CONCLUSIONS

The creation of this type of organizations must be an imperative at macro economic level, because just an economy oriented to knowledge can ensure the affirmation of the Romanian creativity as a main competitivity and identity source, in an environment more turbulent and dynamic.

At micro level, their function on long time level is justified, because the knowledge is continuously and on the other side, the accumulations are a process that is deliberately assumed as one of the essential needs governing the daily activities.

The technical progress is one the essential needs that conditions a more efficient daily activity and maximizes the profit.

My future research in this area is centered on projecting a building, organizing and functioning model of such a business organization, that will be able to selectively and profitably benefit from the specific information and from the knowledge generated by these.

5. REFERENCES

Crisan, G. (2007), *The Decalog of the Local Public Administration Reform*, Universitary Horizons Publishing House, Timişoara, pp. 122-143

Hendriks, P.H.J. (1999). *Do smarter systems make for smarter organizations ?* Decision Support Systems, 27, pp. 197-211

Le Moigne, J.L. (1993), *Sur l'ingénierie de la connaissance organisationnelle,* Note de Recherche 93-02, GRASCE, Université d'Aix-Marseille III, Aix-en-Provence

Nurmi, R. (1998), *Knowledge-intensive firms.* Business Horizons, 41(3), pp. 26-32

Zack, M. (1999), *Managing organizational ignorance,* Knowledge Directions, nr. 1 (summer issue), pp. 36-49.

Annals of DAAAM for 2008 & Proceedings of the 19th International DAAAM Symposium, ISSN 1726-9679
Editor B. Katalinic, Published by DAAAM International, Vienna, Austria 2008
Make Harmony Between Technology and Nature, and Your Mind will Fly Free as a Bird

RESEARCH CONCERNING THE IMPROVEMENT OF BEARINGS QUALITY AND PERFORMANCES USING AUTOMATIC SYSTEMS

CRISTEA, L[uciana]

Abstract: *The paper attempts to present a few results obtained during the research work sustained by the author within the domain of quality and automatic control systems used in bearing industry. The research objectives have been underlined within the frame of solving a specific problem raised in the field of machine manufacturing industry-automatic systems for dimensional inspection productivity improvement characteristics and performances.*

Key words: Bearings, dimensional inspection, flexibility, automatisation.

1. INTRODUCTION

The new idea of products' quality insurance as well the control effectiveness lay down the necessity of a strong interaction and integration among different branches of engineering. The bearing industry achieved in our country special results in the automatic inspection devices field.
The currently used automatic drive systems are manufactured for concrete service and are equipped with elementary components.
The author aimed the implementation of the newest measuring and analysis methods, essential in quality setting and insurance for bearings as main mechanic elements in the automotive industry.
The dimensional inspection automatisation is one of the present basic requirements concerning the upgrading and the retechnologizing of the manufacturing systems in all industrial fields.
The goal that the modern production system must achieve is the production with expected quality and characteristics, as well as small expenses. That tendency is visible on production systems that modified their own internal structure as well as their activities as a series of procedures aimed at defining new production systems with high performances. This task requires high flexibility and modularization in production planning and manufacturing.
The domain of the automat inspection can be considered in integral field including the classical domains of the mechanical, electronic, and computer science engineering. Bearings' production, due to the major economic implications, imposes a permanent improvement of the quality conditions, which leads to the necessity of assuring superior performances related to quality standards. As an integrated field, the automatic inspection devices includes the classic mechanic engineering, electronics and computer field in designing, control and evaluation for these systems (David 2005).
The design of improved control systems for the control automatic device has taken into account the following issues: the configuration of an automatic device with a certain number of mechanic modules - for supplying, distribution, sorting, etc, depending on the user's needs; it makes possible the checking of a larger product range from the same class or similar ones; it makes possible the checking according to one single parameter or a multi-parameter one, with many sorting possibilities of the codes, such as, the classification and the sorting in N dimensional classes according to the control criteria.
Special attention is dedicated to the analyses of dimensional control technology and product characteristics in order to segment the parts into homogenous groups of products with similarly technological characteristics based on concept of group technology.

2. EXPERIMENTAL ASPECTS

In designing the process, we emphasize that the modular control station could be organized as follows:

- All modularly and flexibly organized devices;
- Modular device designed in the most propitious way for each families of parts;
- Measuring and automation checked by an electric computer controlled device;
- High accuracy of operating;
- Interchangeable modulus;
- Different measuring and control transducers and sensors to be used;
- High productivity and efficiency.

All modular devices are defined in order to execute the control operations in correlation with all parts in group (Balachandra, 2002). In view of modulating and assuring a better flexibility and to estimate with better precision the results, the installation (fig. 1) has been designed under modular form.
The mechanical system and measurement unit of the test apparatus were designed to correlate the influential factors of parts assortment with measuring and control equipments. This system is based on client – server concept.
The user has the possibility to input data concerning his/her own application (part, measuring and process characteristics). After that, the main menu of the system works with data bases of parts, measuring and automation process, modular device projects and gives the process results for the most propitious dimensional control structure (Vetter 1994).
The automatic point allows both the measurements of heights and diameters and form deviations under both static and dynamic conditions.
The interactive work between user and server assures the possibility for varied analysis and changes of the inputs. In designing the process, we emphasize that the modular control station could be organized as follows: All modularly and flexibly organized devices;
Modular device designed in the most propitious way for each families of parts; Measuring and automation checked by an electric computer controlled device; High accuracy of operating; Interchangeable modulus; Different measuring and control transducers and sensors to be used; High productivity and efficiency (Alciatore & Histand 2005).
Figure 2 presents the block chart of this dimensional control automatic system. All modular devices are defined in order to execute the inspection operations in correlation with all parts in group.

The control point is made up of the following modules: base plate that is mounted on all the functional modules of the installation; test indicators support; interior attachment module; superior measurement module.

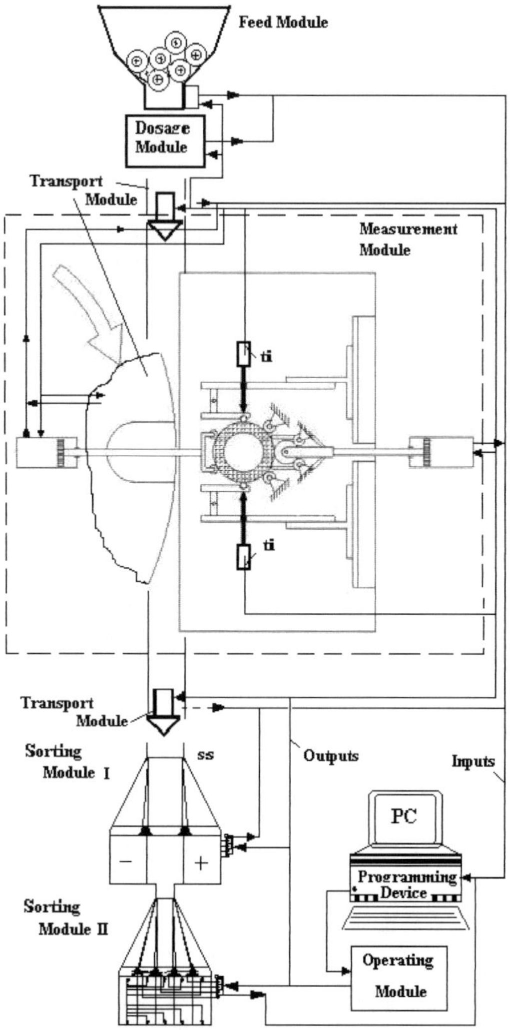

Fig. 1. Cinematic scheme of the modulated automatic inspection system.

The point is made up of modules disposed on a base plate provided with holes and milled keyway so as to be used for all the modules. Under the base plate there are mounted the dies chasers for fixing the pieces from the inside. The indicators are designed in order to be used for measuring a large range of diameters. In order to measure the pieces' height, the control point is provided with a mobile bar that supports the indicators; a pneumatic cylinder activates the bar (Fredic, 2007). The control point may determine the controlled parameters in fixed points imposed by the control procedure but it can also assure a continuous regulation by rotating the piece in the control point. The rotating movement is determined by the rotating manner at the moment the measurement device comes into contact with the presser roller. The electronic modular systems of decision are made up of association specific to the control theme within independent modules or deification integrates designated for feed, measurement, operating and dimensional classification. For an optimal selection and configuration of the modules in automatic dimensional control systems, our team generated a special soft. We consider that the interactive design process assures the most propitious structures and realize performing control systems allowing a suitable flexibility degree in the configuration of control automatic devices in modulated systems.

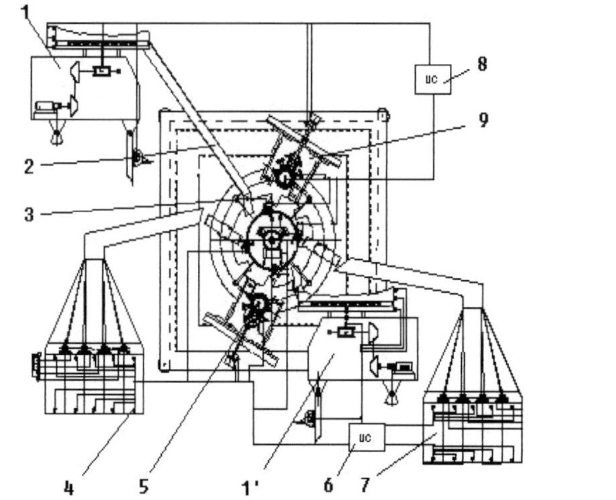

Fig.2. Block chart of automatic dimensional control system.

This solution is achieved and tested during this first step of the project.

3. CONCLUSION

In the bearing production, performing a complete control, of high precision and productivity, stands for an essential characteristic, constitutes an essential characteristic, an indispensable factor of the scientific management of the fabrication process, as well as in maintaining and raising the product quality at a level corresponding to the requirements.
Through the research carried out, there has been aimed at a particularly modern approach of the automatic control systems and at the implementation of the latest methods of measurement and analysis necessary for setting and ensuring the quality of bearings as essential mechanical elements in the framework of automotive industry. In the most quality problems during the manufacturing process, the solution has been set up accordingly to the multi-criteria analysis concerning: the necessary measuring accuracy for the tolerance size of the checked parameter, taking into account the increased uncertainty of a sufficient measuring accuracy on the acceptance-rejection decision with the respective involvement; the cost of the checking on a minimal reasonable level; the presentation way of the information raised from the measuring process, in order to reach the right shape at the right moment, necessary for different decisional factors which follow the evolution of the operating process to be stocked as an evidence of the achieved quality when necessary.

4. REFERENCES

Balachandra, R. (2002). Modular Design and Technological Innovation, *Report 2002-0,* Information Storage Industry Centre, Univ. of California.

Fredic, R. (2007). Towards a Methodology for Integrated Design of Mechatronic Servo Systems. In Doctoral thesis, Royal Institute of Technology, ISSN 1400-1179, Stockholm.

David G. (2005). Introduction to Mechatronics and Measurement Systems, McGraw-Hill Science/ Engineering Math; 3 edition, 2.

Vetter, G.(1994) The Dosing Handbook, Vulkan – Verlag, Essen.

Alciatore, G. & Histand, B. (2005), *Introduction to Mechatronics and Measurement Systems*, McGraw-Hill Science /Engineering /Math; 3 edition, 2005. http://www.tga.gov.au/devices/fs-sudman.htm, *Accessed on:* 2008-08-10

Annals of DAAAM for 2008 & Proceedings of the 19th International DAAAM Symposium, ISSN 1726-9679
Editor B. Katalinic, Published by DAAAM International, Vienna, Austria 2008
Make Harmony Between Technology and Nature, and Your Mind will Fly Free as a Bird

CUTTING FORCES SYSTEM FOR DEEP HOLES BORING

CROITORU, S[orin] M[ihai]; MINCIU, C[onstantin] & CONSTANTIN, G[eorge]

Abstract: *Firstly, some contributions to determining the cutting forces system in case of deep holes boring cutters are presented. The problem was to correctly and completely write the static equilibrium equations of the boring cutter for deep holes. In the second part experimental research in case of boring deep holes is presented.*
Key words: *Cutting Forces, Boring, Deep Holes, Stand.*

1. INTRODUCTION

The criteria used to establish the cutting capacity of the cutting tools are the followings: wear resistance or tool life, cutting efforts or specific cutting energy, roughness of the processed surface, dimensional accuracy and shape errors of the surface, dynamic stability during cutting process, reliability etc (Bălan, 2006).

Research made in the cutting laboratories revealed that for specific cutting operation, cutting tool, quality of the manufactured workpiece, material etc., an appropriate system of criteria to evaluate the cutting tool performances can be used in order to define and compare several tools of the same type (Croitoru et al., 2007).

For the design of the technological system elements (machine tool-cutting tool-workpiece-device) the main criterion of the cutting capacity becomes the cutting efforts or specific cutting energy. Consequently, the cutting forces system for deep hole boring was considered, as shown in Fig. 1.

In case of deep holes borer having fragmented edges not along its diameter, considered as the general case, the cutting forces system was represented in the constructive reference system *Oxyz* of the cutting tool.

Correspondingly, the equilibrium equations will be:

$$-F_{x1} - F_{x2} - \sum F_{xi} - \sum F_{xj} - F_{s1} - F_{s2} = R_x;$$

$$-F_{y1} + F_{y2} + \sum F_{yi}\cos\varphi - \sum F_{zi}\sin\varphi \sum F_{yi} + $$
$$+ N_1(\cos\beta_1 + \mu_v\sin\beta_1) - N_2(\sin\beta_2 + \mu_v\cos\beta_2) = R_y;$$

$$-F_{z1} - F_{z2} - \sum F_{zj} + \sum F_{zi}\cos\varphi + \sum F_{yi}\sin\varphi - \quad (1)$$
$$-N_1(\sin\beta_1 - \mu_v\cos\beta_1) + N_2(\cos\beta_2 - \mu_v\sin\beta_2) = R_z;$$

$$F_{z1}y_1 - F_{z2}y_2 + \sum F_{zj}r_i + \sum F_{zj}y_j + \mu_v N_1 r + \mu_v N_2 r = M_T.$$

where:

- μ_v and μ_s are tangential and axial friction coefficients;
- κ_1 and κ_2 are positioning angles of the edges, complementary to the angles ε_1 and ε_2;
- β_1 and β_2 are positioning angles of the guides;
- y_1 and y_2 are distances to the Oz axis of cutting forces F_y;
- $i = 4, 6, \ldots$ are the even and $j = 3, 5, \ldots$ are odd edges;
- φ is the angle between the edge directions;
- $r = d / 2$ is the radius of the borer.

This equations system must be considered together with the relationships between some terms of the equations system:
- friction forces and normal reactions:

$$F_{v1} = \mu_v N_1;\ F_{v2} = \mu_v N_2;\ F_{s1} = \mu_s N_1;\ F_{s2} = \mu_s N_2; \quad (2)$$

- cutting forces in *Oxy* plane:

$$F_{y1} = F_{x1}\tan\varepsilon_1;\ F_{y2} = F_{x2}\tan\varepsilon_2. \quad (3)$$

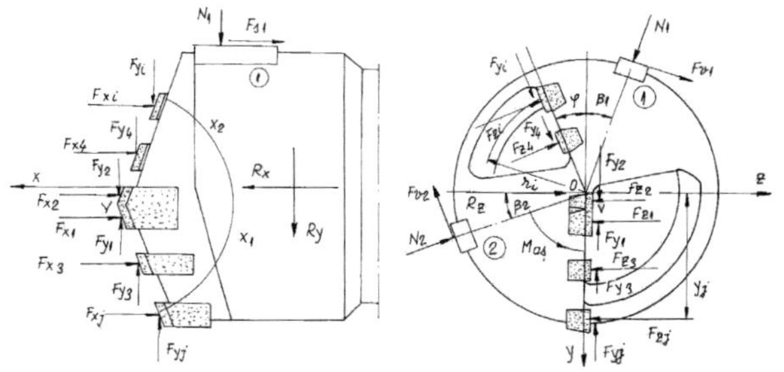

Fig. 1. Cutting forces system for deep holes boring.

This cutting efforts system could be considered as general case for the static loads of the deep holes cutting borer:

- if in the equilibrium equations should make $\varphi = 0$, would be found the equilibrium equations of the borer having fragmented edges along its diameter;
- if in the equilibrium equations should make $\varphi = 0$ and the cutting forces corresponding to the odd j and even i edges are considered equal to zero the obtained equilibrium system belongs to the "gun drill" (Wei Zhang, 2004).

After simplifications, we obtain the normal reactions:

$$N_1 = [(R_z + F_{z1} + F_{z2} + \Sigma F_{zj} - \Sigma F_{zi}\cos\varphi - \Sigma F_{yi}\sin\varphi)(\sin\beta_2\,\mu_v\cos\beta_2) + (R_y + F_{y1} - F_{y2} + \Sigma F_{yj} - \Sigma F_{yi}\cos\varphi + \Sigma F_{zi}\sin\varphi)(\cos\beta_2 - \mu_v\sin\beta_2)] / [\cos\beta_1\cos\beta_2 - \mu^2_v\sin\beta_1\sin\beta_2 - \mu_v\sin(\beta_2 - \beta_1)]; \quad (4)$$

$$N_2 = [(R_z + F_{z1} + F_{z2} + \Sigma F_{zj} - \Sigma F_{zi}\cos\varphi - \Sigma F_{yi}\sin\varphi)(\cos\beta_1 + \mu_v\sin\beta_1) + (R_y + F_{y1} - F_{y2} + \Sigma F_{yj} - \Sigma F_{yi}\cos\varphi + \Sigma F_{zi}\sin\varphi)(\sin\beta_1 - \mu_v\cos\beta_1)] / [\cos\beta_1\cos\beta_2 - \mu^2_v\sin\beta_1\sin\beta_2 - \mu_v\sin(\beta_2 - \beta_1)].$$

This equilibrium equation system is used to determine total cutting forces R_x, R_y, R_z and total cutting torque M_T.

The same equation system can be used to determine the normal reactions on the borer guides N_1 and N_2 when total cutting forces R_x, R_y, R_z and total cutting torque M_T are known.

In the experimental research total cutting forces R_x, R_y, R_z and total cutting torque M_T are measured and the normal reactions on the borer's guides N_1 and N_2 are calculated.

This calculation is used to establish the borer guidance conditions (Mironeasa, 1997).

2. RESEARCH ON CUTTING EFFORTS

Cutting efforts included: cutting forces, cutting torque and consumed power during the cutting process (Constantinescu, 1993). Cutting forces and cutting torque were measured on a stand mounted on a lathe.

The measuring system consists of the following elements: borer, Wheatstone bridge made of resistive stress gauges glued on the borer fastening part, digital amplifier and computer with data acquisition system. Measurements were done for deep holes borers CARMESIN-BOTEK type, having diameters d_A = 46 mm and d_B = 32 mm, and workpiece material OLC 45.

Considering the requirements in the technical literature and of the manufacturers, the following cutting regimes were used:

a) for the borer having $d_A = 46$ mm:

- spindle speed n_c = (230; 305; 380) rpm, which corresponds to the following linear speeds v_c = (33.23; 44; 55) m/min;
- feed s = (0.06; 0.1; 0.16; 0.2) mm/rev;

b) for the borer having $d_B = 32$ mm:

- spindle speed n_c = (305; 380; 460) rpm, which corresponds to the linear speeds v_c = (30.67; 38.20; 46.25) m/min;
- feed s = (0.06; 0.1; 0.16; 0.2) mm/rev.

For each experiment the diagrams showing the variation of the total feed force and total cutting torque were obtained. There are 24 diagrams for axial force and 24 diagrams for the cutting torque (12 + 12 diagrams for the borer with $d_A = 46$ mm and 12 + 12 diagrams for the borer with $d_B = 32$ mm).

For example, in Fig. 2 are presented 2 diagrams showing the cutting torque and axial force for the borer with $d_A = 46$ mm (experiment A3). After processing the diagrams as shown in Fig. 2, the axial force and total cutting torque were determined. The results were organized in tables. Using these experimental results the consumed power during the cutting process and the main cutting force were calculated.

Using all these results the variation diagrams of the measured or calculated cutting efforts versus the cutting speed and feed were drown. Two of such diagrams are presented in Fig. 3. The measured and calculated data and their variation diagrams underline the influence of the cutting speed and feed upon the

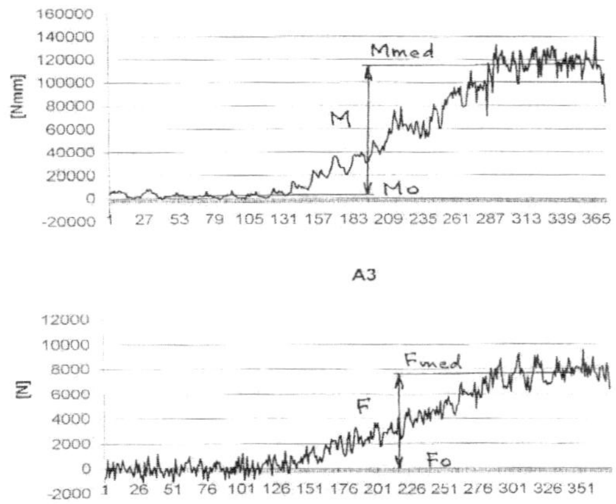

Fig. 2. Cutting torque and axial force diagrams (example).

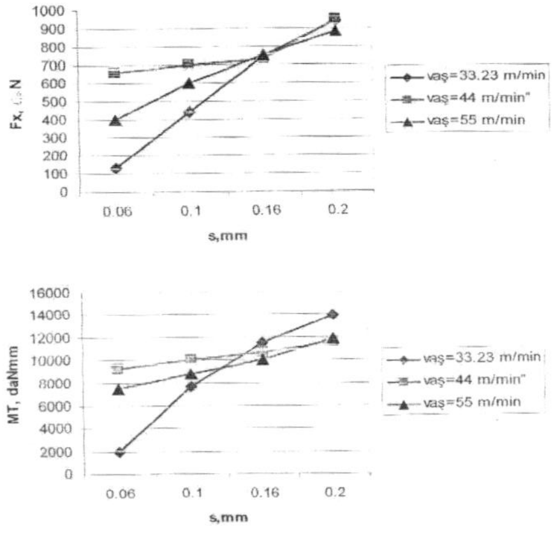

Fig. 3. Cutting efforts vs. cutting speed and feed diagrams.

cutting efforts. Using the measured and calculated cutting efforts the following Taylor functions were obtained:

In case $v_{aş}$ = constant,

$$F_x = C_{F_x} \cdot s^{y_x}; M_T = C_M \cdot s^{y_M}; F_{z,ec} = C_{F_z} \cdot s^{y_z} \quad (5)$$

In case s = constant,

$$F_x = C_{F_x} \cdot v_{as}^{z_x}; M_T = C_M \cdot v_{as}^{z_M}; F_{z,ec} = C_{F_z} \cdot v_{as}^{z_z} \quad (6)$$

Consumed power during the cutting process was measured on deep holes boring machine MGA 250 × 1000 using a watt-meter. The measured data were used to calculate the main cutting force and the total cutting torque.

Variation diagrams of the three cutting efforts were drown and Taylor functions depending on the cutting speed and feed were determined in case of workpiece material OLC 45 steel and a borer having the diameter $d = 46$ mm, as following:

$$F_z = 1.145.6\, s^{1.013} \text{ daN}; \; M_T = 26.368\, s^{1.013} \text{ daNmm},$$

$$P_c = 18.1\, s^{1.013} \text{ kW}. \quad (7)$$

3. CONCLUSIONS

Compared to other cutting operations (turning, milling, grinding) in case of boring deep holes the total cutting force includes components that are not produced by the cutting process itself. These components are contact forces between the drill and the workpiece and their values are comparable to the effective cutting forces. As a consequence, the cutting forces system in case of boring deep holes is very complex.

Measurements of the cutting efforts are useful for both designer and manufacturer of the machine tool. The cutting efforts decrease with the increase of the cutting speed. However, this conclusion is valid only in case of cutting without cutting fluids. In real cases, the coolant is a necessity.

The experimental stand is an original one, using a lathe instead of a deep hole boring machine. The experimental results confirmed the theoretical research and were confirmed by other experiments using a real deep holes boring machine.

4. REFERENCES

Bălan, M. (2006). *Contribuţii privind creşterea capacităţii de aşchiere a burghielor pentru alezaje lungi* (Contributions regarding the increase of cutting capacity of the long drills), PhD Thesis, University "Politehnica" of Bucharest, Romania.

Constantinescu, C. (1993). *Cercetări privind optimizarea constructivă şi geometrică a burghielor pentru prelucrarea alezajelor lungi* (*Researches regarding the geometric and constructive of the drills for long holes procesing*), PhD Thesis, Technical University of Iasi, Romania.

Croitoru, S.M.; Minciu C. & Bălan, M. (2007). Experimental Research on Measuring Forces and Moments at Deep Holes Machining, *The 5-th International Conference on Advanced Manufacturing Technologies ICAMaT 2007*, pp. 367-370, ISSN 1843-3162, AGIR Publishing House, Sibiu, Romania.

Mironeasa C., (1997). *Cercetări privind optimizarea autoghidării burghielor utilizate la aşchierea alezajelor lungi* (*Researches regarding the optimization of selfguiding of the drills used in long holes cutting*), PhD Thesis, Technical University of Iasi, Romania.

Wei Zhang, Fengbao He, Dilin Xiong (2004). Gundrill life improvement fordeep-hole drilling on manganesse steel, *International Journal of Machine Tools and Manufacturing*, No. 44.

Annals of DAAAM for 2008 & Proceedings of the 19th International DAAAM Symposium, ISSN 1726-9679
Editor B. Katalinic, Published by DAAAM International, Vienna, Austria 2008
Make Harmony Between Technology and Nature, and Your Mind will Fly Free as a Bird
2nd DAAAM-ICAT

APPLICATION OF ELECTRON BEAM MELTING TO TITANIUM HIP STEM IMPLANTS

CRONSKAER, M[arie]; RAENNAR, L[ars] E[rik]; KOPTIOUG, A[ndrei] & BAECKSTROEM, M[ikael]

Abstract: *The Free Form Fabrication Process (FFF) is nowadays an accepted technology widely used for prototyping and manufacturing. However, it is still in an expansive phase and new applications like direct manufacturing of implants are evolving continuously. Present work describes the possibilities provided by the electron beam melting (EBM) method for orthopedics; in particular hip stem implant manufacturing. The conventional machining used for individually adapted prostheses typically involves advanced milling with the drawback of removing up to 80% of the material. This paper addresses the economic feasibility of using an additive approach for the manufacturing of typical orthopedic implants. The studied implants were manufactured from biocompatible Ti-6Al-4V alloy using both EBM and conventional CNC technologies and compared according to material consumption, manufacturing time and cost.*

Key words: *orthopedic implants, hip stem, electron beam melting, free form fabrication, economic feasibility*

1. INTRODUCTION

During the last decade, FFF methods yielding 100% dense metal parts often referred to as Rapid Manufacturing (RM) methods have been developed. Selective Laser Melting (SLM) is one of these methods studied and used for the manufacturing of denture frameworks (Bibb, 2006) and production of complex biomedical structures (Wehmöller et al., 2005). Another RM method, Electron Beam Melting (EBM), has shown promising capabilities in the fabrication of custom orthopaedic implants with tailored material properties (Harrysson et al., 2008). It has also been used for generation of open-celled titanium components (Heinl et al., 2007). The EBM- processed Ti6Al4V ELI has been thoroughly investigated by Christensen et al. and proved to have "comparable and in some cases superior chemical and mechanical properties than those specified in ASTM F 136-02a and ASTM F 1108-04" (Christensen et al., 2007). The RM technique offers almost no design limitations and new possibilities for the customized hip stem prostheses design arise with this technique. For example modified surface structures for good bone ingrowths and open-celled net structures controlling the stiffness of the implant address the stress shielding problems in the implant-bone structure. The economic feasibility of using RM for implant manufacturers including issues as material consumption, manufacturing time and associated costs in comparison to the conventional manufacturing methods are however not yet fully investigated. Customized implants are commonly manufactured with CNC machining from a bar stock and for fabricating of a knee implant as much as 80% of the material is wasted (Harrysson & Cormier, 2006). The aim with present paper is to mimic a commercial, machined implant with the EBM technique to get a recent benchmark case and also an indication of the EBM's commercial potential in this area. The study is based upon experimental data for the EBM process.

2. METHODS

2.1 Comparison

The comparison has been done in cooperation with a Scandinavian Company developing and manufacturing customized hip stem prostheses using CNC methods ("comparing company", or CC). Seven solid customized hip stem prostheses designed by CC were fabricated in the ARCAM EBM machine at Mid Sweden University and using conventional CNC methods by CC at their site. Complete costs breakdown was available only in the case of EBM process due to the confidentiality policy of the CC. For the same reason the result is presented in relative numbers. In present study material, file preparation and manufacturing costs are taken into account, while all other costs are taken to be the same in the two cases.

2.2 File preparation

Files used for the EBM and CNC manufacturing were prepared from the same seven CAD files of customized hip stem prostheses provided by the CC. These files are produced from the DICOM image files using the SolidWorks software (SolidWorks Corp., Concord MA) and modules proprietary to the CC. Manufacturing files for the ARCAM EBM machine were prepared using the MagicsRP software (Materialize Leuwen, Belgium) and ARCAM's build assembler software. Support materials needed primarily for heat transfer from the parts are modelled in MagicsRP. For the conventional manufacturing the same starting CAD files were CNC-prepared in EDGEcam (www.edgecam.com) using the proprietary software modules for automations developed by the CC.

2.3 Electron Beam Melting (EBM)

The machine used for the EBM manufacturing is an ARCAM A2 (Arcam AB, Sweden) which operates in vacuum and uses an electron beam to melt consecutive layers of metal powder to form the solid object. When completed the parts are left in the machine to cool down and excess powder is removed in a special blast chamber. The process is further described in (Rannar et al., 2007). All seven implants were manufactured simultaneously using powdered Ti-6Al-4V ELI (grain size 50-100µm) with a layer thickness of 0.1 mm. The batch size varies from time to time and a batch of seven implants does not make use of the EBM machine's full capacity. When increasing the number of implants in the machine the manufacturing time increases but not proportionally to the number of implants. To determine the process duration at full machine capacity simulations were made using proprietary EBM Control software by ARCAM. The simulation showed a precision within ca 0.5% (35h 48 min vs 36 hrs in actual run).

2.4 Post processing of the EBM implants

To reduce the stress shielding effect in the body and longevity of the implant, having good bone ingrowths in the medial part of the stem is crucial. The CC is using titanium powder sputtering to create some rougher implant surface with better characteristics for bone ingrowths. Approaches to create a desirable surface structure within the EBM process are modifying build parameters or make a CAD design of the desired structure with optimized pore size. The possibilities to reach pore morphologies in the EBM process comparable with those that are obtained using the traditional bead sintering approach is currently being investigated (Harrysson & Cormier, 2006). The medial part of the stem is therefore left out of the post processing in this study. To reach the same tolerances as the CC implants, the cone (circular part at the top) was machined by a turning operation and the external parts of the stem were manually grinded and polished. Circular holders, Ø10mm with a total length of 40 mm were added to provide adequate mechanical grip in the post processing operations.

2.5 Conventional manufacturing

Traditionally implants are manufactured in a number of operations in a conventional machine tool shop. In this case a multi operational lathe is used to mill and sculpture the stem body. The cone part is then machined by a turning operation. Finally manual grinding and polishing operations are used to meet demands on high surface smoothness in chosen areas.

3. RESULTS

Except the excluded medial part of the stem, a mimic of the commercial customized hip stem prostheses was obtained with the same tolerances of the cone and the external parts by using EBM technique and post processing. The overall time required for the file preparation (including CAD file modification, support generation and compilation) was 40 minutes. The machine processing time was 36 hours and 8 hours were required for cooling the system. Total ARCAM machine operator time was 2 hours and 10 min, including preparation of the machine, supervising, removing excess powder and cleaning the implants and the machine. The milling of the cone took about two minutes per implant and the grinding and polishing of the external parts of the stem about 1.5 h per implant. The amount of wasted material in the EBM process for the seven implants was 24% of the total weight of the implants: 7% was used for the holders and 17% was support material and powder loss. The cost comparison with the conventional manufacturing indicated that the EBM-based manufacturing cost constitute about 65% of the conventional cost, see table 1. The comparison is based on the costs that CC has in the three comparing areas, which as all costs are influenced by surrounding factors not only concerning the manufacturing method. The comparing area number 3, (table 1), is the one most influenced by these factors. Assumptions were made concerning the profit margin and the different cost levels in the two countries to reduce the influence of these factors as much as possible.

Comparing areas	EBM costs in % of Conventional machining (seven implants)
1 Material	15%
2 File preparation	8%
3 Manufacturing	130%
Total cost	65%

Tab. 1. Cost comparison.

The cost calculation for EBM manufacturing is based on the experimental study and comprises costs of the machine time in EBM (including a 30% profit margin), operator time, rent for the premises, working materials (not including metal powder) and post processing operations. For the most efficient EBM manufacturing the machine should be running at full capacity. The simulation running time for 7 and 14 implants was about 36 and 38 hours correspondingly, as it is mainly determined by the number of melted layers in the build. The file preparation time is increasing in roughly the same proportion. This nonlinearity leads to the effective reduction of the cost per implant by about 20% (all other costs increase linearly with the number of implants).

4. DISCUSSION AND CONCLUSIONS

Present study strongly indicates that the EBM-manufacturing of the customized medical implants is commercially viable. Best commercial performance can be realized in the customized manufacturing, where the most savings are achieved in material costs and pre-processing (file preparation). The EBM process also facilitates some additional commercial advantages such as flexibility in the surface and bulk geometry selection. Important continuation in this application is finding a surface for the medial part of the stem that fulfills the requirments for ingrowth of bone and mecanical properties without any need of CNC-preparations. There is also significant room for the improvements in both the implant design and the post-EBM implant processing. Further work will be aimed at optimizing the mechanical and biomedical properties of the implants and their surfaces and at optimizing the EBM process and pre-processing stages.

5. REFERENCES

Bibb, R. (2006). Rapid manufacture of removable partial denture frameworks. *Rapid Prototyping Journal,* 12., 2., 95-99, ISSN 1355-2546

Christensen, A.; Lippincott, A. & Kircher, R. (2007). Qualification of electron beam melted (ebm) ti6al4v-eli for orthopaedic implant application. *Available from:* http://www.medicalmodeling.com/ebm/ Accessed: 2008-06-01

Harrysson, O.; Cansizoglu, O.; Marcellin-Little, D.; Cormier, D. & West Ii, H. (2008). Direct metal fabrication of titanium implants with tailored materials and mechanical properties using electron beam melting technology. *Materials Science and Engineering,* 28., 3., 366-373, ISSN 0928-4931

Harrysson, O. & Cormier, D. (2006). Direct fabrication of custom orthopedic implants using electron beam melting technology, In: *Advanced manufacturing technology for medical applications,* Gibson, I., (Ed.), 191-206, John Wiley & Sons, Ltd., ISBN-10: 0-470-01688-4, Chichester, England

Heinl, P.; Rottmair, A.; Körner, C. & Singer, R.F. (2007). Cellular titanium by selective electron beam melting. *Advanced Engineering Materials,* 9., 5., (may 2007) 360-364, ISSN 1438-1656

Rannar, L.E.; Glad, A. & Gustafson, C.G. (2007). Efficient cooling with tool inserts manufactured by electron beam melting. *Rapid Prototyping Journal,* 13., 3., 128-135, ISSN 1355-2546

Wehmöller, M.; Warnke, P.H.; Zilian, C. & Eufinger, H. (2005). Implant design and production-a new approach by selective laser melting. *International Congress Series,* 1281, (June 2005) 690-695, ISSN 0531-5131

Annals of DAAAM for 2008 & Proceedings of the 19th International DAAAM Symposium, ISSN 1726-9679
Editor B. Katalinic, Published by DAAAM International, Vienna, Austria 2008
Make Harmony Between Technology and Nature, and Your Mind will Fly Free as a Bird

MICRO NANO MANUFACTURING TECHNOLOGIES

CRUZ, F[ernando]

Abstract: *Micro Manufacturing is the production of small components and products, which have not been possible until the recent introduction of ground breaking technologies and techniques in the traditional manufacturing processes.*
This comprises the manufacture of parts in the micrometer and manometer ranges. This exciting emerging niche, allowing miniaturisation of components particularly in the consumer electronics, medical, pharmaceutical, automotive, aerospace, jewellery, micro optics and defence markets, is now available in all the major traditional groups of manufacturing processes (Macro Manufacturing), providing the production of micro-scale parts. Micro-and Nano Manufacturing Technologies (MNTs) make use of a variety of materials, components and knowledge-based technologies providing functionality and intelligence to highly miniaturised systems. This paper presents an updated overview of these emerging manufacturing technologies.
Key words: *Micro Manufacturing, Nano Manufacturing, Miniaturisation, Micro Machines.*

1. INTRODUCTION

Nowadays the technological advances have led to the ability to manufacture small parts or parts with extremely small features previously impossible to make.

These innovative techniques enable the manufacture of smaller and smaller components with incredible tolerances and accuracy.

MNTs has been initially developed and used mainly for the semiconductors industry, using has basic material the single crystal silicon. However, in the last few years has started to explore potential for other domains (e.g. medical and biotechnology applications), looking at same time for new materials for manufacture (micro-material micro manufacture).

Microstructures, in the context of this paper, are often described as structures down to about 1μm in dimensions, and nanostructures as having dimensions <1μm. The size of a component, however, is not important when discussing micro manufacturing, as relatively large components can require features less than ten microns (Dimov et al., 2007).

Development is required of a range of micromachining processes, e.g. micro-EDM, micro-milling, micro-injection, micro-prototyping, etc., in order to allow a viable alternative for serial production of high-aspect-ratio structures in metals, plastics and ceramics, as well as in the enhanced-forced micro actuation and high-precision aspects.

Figure 1 presents some examples of miniaturised components:

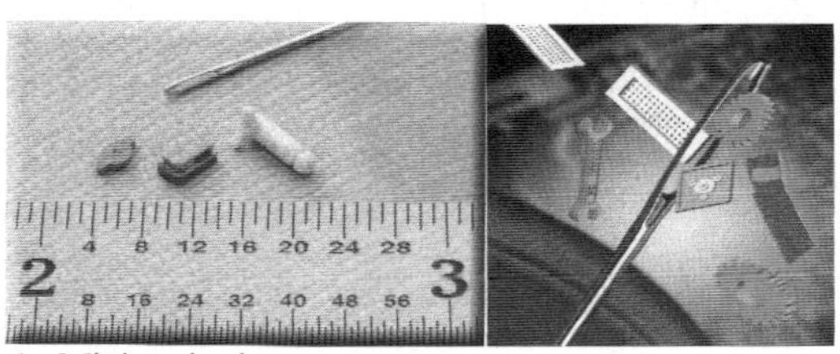

Fig. 1. Miniaturised components

2. MICRO MANUFACTURING TECHNOLOGIES

In the scope of this paper five of today's major micro-manufacturing technologies are presented, as follows:

- ✓ Micro Milling;
- ✓ Micro Injection Moulding;
- ✓ Micro Electro-Discharge Machining;
- ✓ Micro Laser Cutting and Machining;
- ✓ Micro Rapid Prototyping.

2.1 Micro Miling

Micro-milling is a technology that by employing the latest CNC systems is especially suitable for machining complex 3D geometries. In particular, important areas for the application of this technology are the manufacture of micro parts for watches, keyhole surgery, housings for micro engines, tooling inserts for micro injection moulding and hot embossing, and housings and packaging solutions for micro-optical and micro fluidics devices (Popov et al., 2008). A common challenge across all these application areas is the machining of micro features with dimensions smaller than 100 μm.

For that purpose micro machining centres, micro cutting tools and new machining strategies are required. Figure 2 presents images of micro milling machine and machined parts:

Fig. 2. Micro milling machine (left) and machined die inserts for micro injection moulding (right)

2.2 Micro Injection Moulding

Historically, machining has been the only feasible manufacturing alternative for manufacture low to moderate quantities of micro components. However, with advances in technology, micro moulding can now offer a range of cost-effective alternatives for components that are miniature, complex and require high precision tolerances.

Advances in material science and plastic injection mould equipment permit complex machined micro components to be inject or moulded in metal, plastic, or plastic with metal or ceramic filler.

Attempts to micro mould components using conventional moulding equipment often fail. Nowadays a large number of companies offer micro moulding engineering and manufacturing services capable of give response to market (Whynott, 2008). Figure 3 illustrates this micro technology:

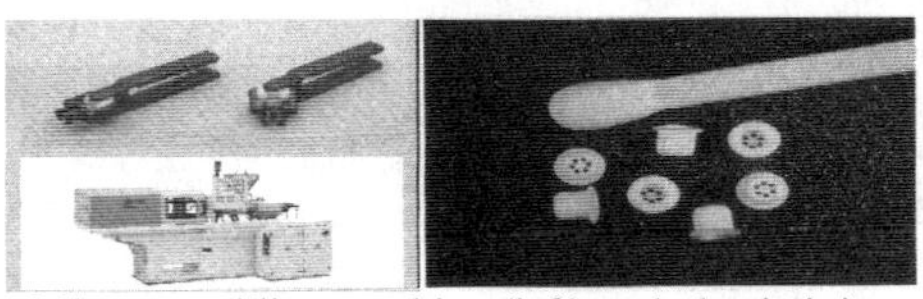

Fig. 3. Micro moulding machine (left) and plastic injected micro parts (right)

2.3 Micro Electro-Discharge Machining

EDM is a thermal process with the material removal mechanism being by spark erosion, and is ideal for metal and certain other conductive materials.

The micro wire EDM is a wire erosion process for profiling complex 2D structures and is a twin wire system, which incorporates a larger diameter wire (typically of 200 microns) for roughing work, and a smaller diameter wire (of diameters ranging from 20 to 70 microns) for the very fine finishing. This method is very much suitable to, for example, machine extrusion dies in tungsten carbide.

Micro EDM machining offers an excellent way of machining micro structures for many different applications from medical devices, drug delivery and diagnostic, micro injection mould tool making and many other applications (Hoyle, 2008). Figure 4 shows images of this micro technology:

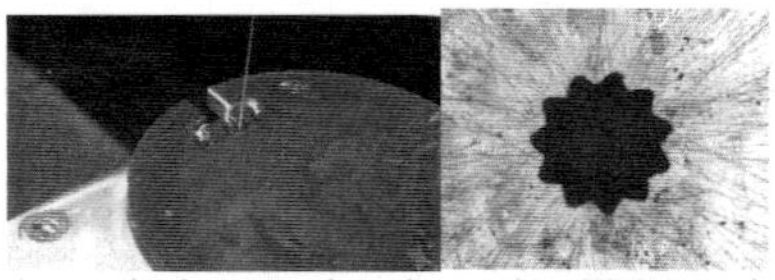

Fig. 4. Schematic image showing wire EDM technique used for profiling fine and micro structures (left) and Tungsten Carbide extrusion die with diameter 0.52 mm (right)

2.4 Micro Laser Cutting and Machining

The drive towards miniaturisation, particularly in the consumer electronics and medical devices market is resulting in a increase in applications for laser material processing to machine, permanently mark or join micro-scale components.

There is a natural synergy between the capabilities and benefits of lasers and the needs of modern manufacturing. The contribution of laser processes is simple, there is little maintenance required, running costs are low and reliability is extremely high.

Two examples of the application in micro manufacturing are given below (Rofin, 2008):

First, in the field of medical device manufacturing, there is a need to manufacture parts in bio-compatible material (e.g. stainless steel, titanium alloys, cobalt-chromium, platinum and gold) which are laser cut from fine diameter metal tubes. A second application where fine components are to be joined is in sensors used in electronic applications. Typically in this case, a component must be sealed hermetically with sensitive electronic components inside, which must be protected from excess heat and ingress of moisture. Figure 5 shows examples of those applications:

Fig. 5. Miniature laser cut medical devices in Nitinol (left) and Laser welded sensor in stainless steel (right)

2.5 Micro Rapid Prototyping

Some companies are investing resources to offer solutions for the additive fabrication of micro components, in materials such as high temperature ceramics, low temperature wax, industrial polymer types and bio medical grades, for medical, dental, jewellery and bio-tech applications.

One of these solutions uses a machine working on the principals of polymerisation to create a prototype layer by layer, by means of a DLP (Direct Light Projection) chip from Texas Instruments®. The process generates a layer mask to project a 1280*1024 pixel cross section image of the part being built onto a bath of photosensitive resin.

Thus, this technology allows the manufacture of high tolerance ultra fine detail and high tolerance parts in a short amount of time. Is capable of micro resolution at under 10μm in XY and under 15μm in Z (Envisiontec website, 2008).

A general view of this technology is given in Figure 6:

Fig. 6. Micro rapid prototyping machine and parts

3. CONCLUSIONS AND FUTURE TRENDS

Recent innovations in the area of micro/nanofabrication have created a unique opportunity for manufacturing structures in the manometer-milimeter range. This can be used to fabricate novel electronic, optical, magnetic, mechanical and chemical/biological devices with applications ranging from sensors, to computation and control.

While micro-manufacture is more established as a technology, nano-manufacturing is still in its infancy and very much based on laboratory-scale nano-science developments. The technologies and solutions for micro-manufacturing have developed rapidly over the recent years and are increasingly becoming a source of competitive advantage in major industrial sectors.

There is no one technology that will prevail. More probably the future trends comprise an innovative integration of complementary technologies and their implementation in the new manufacturing platforms. New developments will be made in reducing the features sizes towards nanoscale including continuous improvements of their accuracy and surface finish.

All current predictions indicate that micro-technologies will remain a critical enabling factor for the manufacture of high value products in a number of key sectors such as medical and pharmaceutical, communications, defence and aerospace.

4. REFERENCES

Dimov, S.S. et al. (2007). 4M (Multi-Material Micro Manufacturing) of Excellence: Progress Report 2004-2006, *Proceedings of International Conference 4M 2007,* Whittles Publishing, pp. xvii-xxxi, *Available from:* www.4m-net.org, *Accessed:* 2008-04-28.

Envisiontec website (2008). *Available from:* www.envisiontec.de, *Accessed:* 2008-05-13.

Hoyle, R., (2008). Angling for Success - an Application of Micro Wire EDM Machining, *MM: Micro Manufacturing & Commercial Nano Applications Magazine,* Volume 1, Issue 1, January/February, 2008, Published by Rapid News Publications plc., UK, pp. 46-49.

Popov, K., Dimov, S., Pham, D. T., and Ivanov, A., (2008). Micro Milling: Solutions for improving process reliability, *MM: Micro Manufacturing & Commercial Nano Applications Magazine,* Volume 1, Issue 2, March/April, 2008, Published by Rapid News Publications plc., UK, pp. 28-34.

Rofin website (2008). *Available from:* Www.rofin.co.uk, *Accessed:* 2008-05-15.

Whynott, J., (2008). Micro Moulding: A Cost Effective Alternative to Micro Machining, *MM: Micro Manufacturing & Commercial Nano Applications Magazine,* Volume 1, Issue 1, January/February, 2008, Published by Rapid News Publications plc., UK, pp. 18-22.

Annals of DAAAM for 2008 & Proceedings of the 19th International DAAAM Symposium, ISSN 1726-9679
Editor B. Katalinic, Published by DAAAM International, Vienna, Austria 2008
Make Harmony Between Technology and Nature, and Your Mind will Fly Free as a Bird
2nd DAAAM-ICAT

DEVELOPMENT OF A CONTINUOUS PASSIVE MOTION DEVICE FOR POST SURGICAL REHABILITATION OF THE KNEE

CSEKI, L[iviu] B[ogdan]; DRSTVENSEK, I[gor]; BALC, N[icolae] & KRAMBERGER, S[lavko]

Abstract: *The main purpose of this paper is to find a solution for speeding up the recovery of the approximate 500,000 people (only in Europe) who suffer some sort surgical procedure on the knee joint. This solution is embodied in a continuous passive motion device that is used post operational. This device consists of a leg support that forces the knee to bend gradually for a designated period of time while the muscles are electrically stimulated in order to accelerate the recovery time. The needs of the target market served as a key factor in the user oriented development of the mentioned rehabilitation apparatus. The existing devices have various drawbacks that make them less popular than it should. These drawbacks have been identified and guided the design of the apparatus towards eliminating them.*
Key words: *Knee, rehabilitation, device, design*

1. MEDICAL INSIGHT

The knee is essentially made up of four bones. The femur, which is the largest bone in your thigh, is attached by ligaments and a capsule to your tibia. Just below and next to the tibia is the fibula, which runs parallel to the tibia. The patella (the knee cap) slides up and down in a groove in the femur (the femoral groove) as the knee bends and straightens (Vaida *et al*., 2006).

The articular cartilage in the knee is called the menisci and it is located between the weight bearing joint surfaces of the femur and the tibia. The knee has two menisci, one located at the exterior side – lateral meniscus and one in the interior called medial meniscus. The menisci act as cushions and absorb the shocks that occur between the femur and the tibia.

The ligaments are very important in the knee because they stabilize the joint. There knee has four important ligaments: medial collateral ligament, lateral collateral ligament, anterior cruciate ligament and posterior cruciate ligament.

Knee problems can be the result of disease or injury. One of the most frequent injuries affects the ligaments and the meniscus and it can be corrected through reconstructive surgery. Diseases have degenerative causes that can even lead to the need for total knee reconstruction.

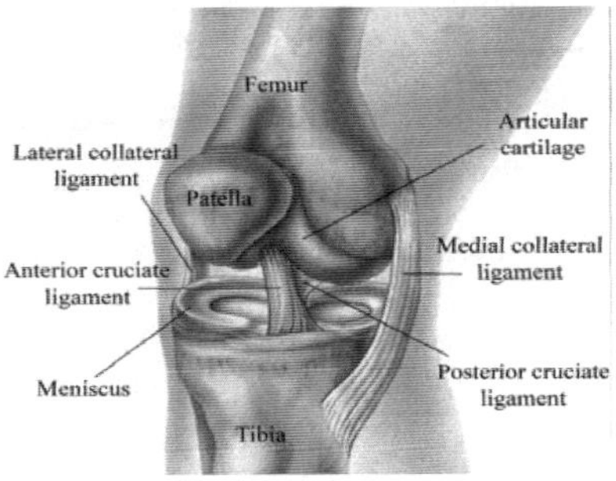

Fig. 1. Schematic drawing of the knee

2. REST VS. MOTION

One of the problems of orthopedics is that the effects of surgery on the soft tissue are more unfavorable to patient progress and recovery than the issue resolved by the procedure itself (Lenssen *et al*., 2003).

Rest and motion have varied as treatment of choice following surgery or injury for many centuries and the proper treatment is still in debate nowadays. Among the negative effects of immobilization there can be mentioned orthostatic hypotension, pneumonia and soft tissue contractures. Other unfavorable effects are edema, stiffness and pain at the affected site.

Studies proved that rest is actually more hazardous to proper healing than early motion (Prundeanu *et al*., 2004). At the beginning of the 20th century, researchers in the field of orthopedics started a trend toward manipulation and mobilization. This idea has been the subject of many controversies, but finally triumphed and has become a commonly practiced therapeutic technique following trauma.

Early motion can be achieved by physical therapy or with a continuous passive motion device. The benefits of early motion include decreased edema, decreased pain and shorter hospital stay.

3. CUSTOMER NEEDS

When developing a new product or service, exploratory market and customer research should be an essential and continuous component of the process. It provides the foundation and platform for effective idea generation and creativity management. This describes customer, or customer needs, wants, gripes, complaints, and problems that each have about the performance of a certain activity, function, process, life event, or product. This research prior to idea generation provides the basis for setting up problem-solving mindset towards idea generation.

The market for the continuous passive motion device was narrowed down to hospitals and rehabilitation centers. The main users of such a device are the patients and the hospital staff that handles the device. Discussions with hospital staff and patients have revealed a lot of useful information like the problems with our competitors' products and their view of the ideal product. This information was collected, sorted and prioritized using specialized software.

The needs of the customer with their importance served as a key component for a quality planning of the developing of the product. The House of quality was used for doing this and the result was the importance of each module of the product. This importance served as an indicator in the developing of the product, showing which part of the product needs the most attention and the higher price. This analysis was performed using the specialized quality control software Qualica QFD.

4. CONCEPT

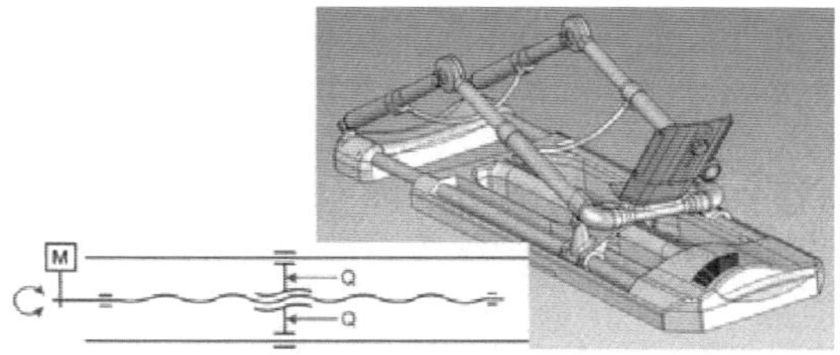

Fig. 2.

One of the biggest disadvantages of the competitors is the price of the device. In order to reduce the cost of the device the preferred option was a construction as simple as possible without affecting the performances of the device. The main focus was to eliminate as many parts as possible and design a safe, simple device that would have all the features necessary for a successful rehabilitation of the knee articulation.

Brainwriting (Blebea *et al.*, 2007) was used to generate a number of ideas which were analyzed at a later date. Using the data previously obtained in the brainwriting session we calculated the mechanisms that are part of the CPM device and built the virtual prototype based on the needs of the customers.

Calculi showed that the device needs only a 5W direct current motor and a belt transmission in order to overcome the resistance of the patient's leg. To control the motor there is the need for a regulator and also a current transformation unit from alternative current to direct current.

A very important feature of the CPM device is the speed of the exercise. This, along with the range of motion is designated by the physician based on clinical parameters. Discussions with several orthopedists revealed that the device should be capable of achieving a speed as low as 30 degrees / minute and as high as 150 degrees / minute. Anything lower than this would not have the desired effect and above this would cause pain to the patient and possibly harm the articulation.

5. DESCRIPTION OF THE DEVICE

In order to reduce weight as much as possible the metal structure of the device is fabricated from 6061 Aluminum, an alloy used for chassis and other similar structures. This type of aluminum offers very good rigidity and durability without adding to the weight of the device. The other components of the device can be fabricated through injection molding from plastic (Bâlc *et al.*, 2006).

The device is composed of 2 nickelled aluminum guiding cylinders that have role of guiding the footplate on a horizontal path. On the guiding cylinders slides the component that holds the weight of the leg and serves as a nut for the screw that ensures movement. This component is made up of three parts: the upper part, which is part of the joint corresponding to the ankle, the bottom part that holds the bearings in place and the part that holds the actual nut.

The foot of the patient rests onto the footrest. This part is connected to the lateral tubes of the device through an L-shaped part. This part has the role to hold the foot of the patient in place. It can be adjusted to fit all foot sizes with the help of an M5 screw that fixes in place the top sliding part to the lower part that can be adjusted to make the patient feel as comfortable as possible.

The CPM device is controlled by a remote control that translates the needs of the patient into electrical signals that are sent to the motor and controls the device.

A very important feature of the device is its ability of creating a gradual movement of the leg. In the rehabilitation of any member it is crucial that the articulation is not forced at all.

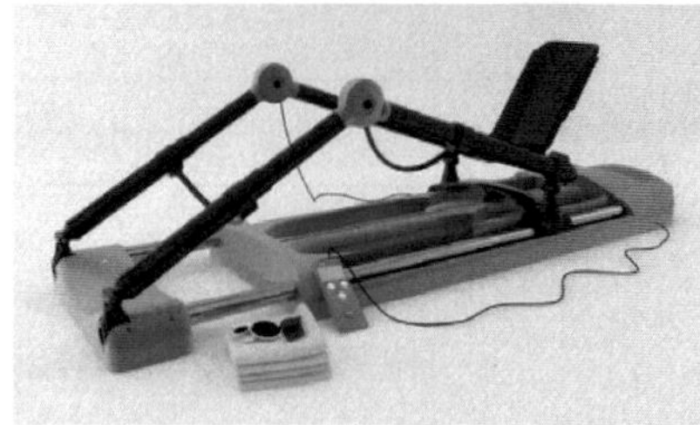

Fig. 3. Virtual prototype

Gradual movement is the only type of movement recommended. Because of this, our device bends the knee gradually. The muscles are stimulated using electrodes. Three types of electrodes were designed each of a different size to fit the desire of any user. This feature is very useful in eliminating the numbness of the muscles that occurs after the surgical intervention. The best type of movement in the rehabilitation is active movement, but that can not be achieved until the patient has control over his or hers muscles.

6. CONCLUSION

This project has started from an idea on how to help people who have suffered surgical interventions at the knee joint rehabilitate faster and safer. The premises of this study were to develop a device that could help patients in their rehabilitation after having had surgery performed on the knee articulation. A solution was needed for reducing costs without affecting the performance of the device and to reduce the weight in order to improve handling the device.

Using specialized software, the needs of the potential customers were analyzed and used as guides in the development of the virtual prototype which was used in the fabrication of the physical prototype. The final product is lighter than other continuous passive motion devices and because of its simple construction is cheaper.

Further research will focus on the gradual movement and the muscle stimulating feature. This implies creating the software and creating the link between the hardware and the software.

The result of this study is a product that is cheaper and better than the competition has to offer. With proper management this continuous passive motion device could become part of any hospital and bring significant benefits to all involved.

7. REFERENCES

Bâlc, N., Gyenge, C., Berce, P. (2006). *Proiectare pentru fabricatia competitiva: teorie, aplicatii si studii de caz,* U.T. Pres, 973-7893-31-1, Cluj-Napoca

Blebea, I, Dobocan, C. (2007). *Proiectarea produselor – de la teorie la practica*, U.T. Pres, 978-973-662-292-2, Cluj-Napoca

Lenssen, A. F., Koke, A. J. A., Bie de R. A., Geesink R. G. T. (2003). Continuous passive motion in rehabilitation following total knee arthroplasty; short and long term effects on range of motion. *Physical Therapy reviews*, 2003:113–121, doi: 10.1179/108331903225003028

Prundeanu H., Brad S., Dressler O., Prundeanu A., Epure V. (2004). *Notiuni de biomecanica,* Artpress, 973-653-12-4, Timisoara

Vaida, M., Bolintineanu, S. Niculescu, M. (2006). *Anatomie Humaine Compendium*, Eurostampa, 973-687-251-3, Timişoara

Annals of DAAAM for 2008 & Proceedings of the 19th International DAAAM Symposium, ISSN 1726-9679
Editor B. Katalinic, Published by DAAAM International, Vienna, Austria 2008
Make Harmony Between Technology and Nature, and Your Mind will Fly Free as a Bird
*Young Researcher and Scientist Paper / * Supervisor, Mentor*

THE TRANSFORMATION OF THE CFM 56-7 TURBOFAN INTO A TRIPLE FLOW JET ENGINE; A WAY OF REDUCING NOISE POLLUTION

CUCIUMITA, C[leopatra]; PETCU, R[omulus]; VILAG, V[aleriu] & STANCIU, V[irgil]*

Abstract: *Nowadays, air transportation becomes more and more restrictive, implying low chemical and noise pollution, low fuel consumption and high thrust force. The main objective of this paper is to transform the existing CFM 56-7 turbofan into a triple flow jet engine by separating the by-pass flow using a new fan stage. The calculus is based on the hypothesis that the turbine produces a maximum specific mechanical work at sea level static. The results show a decrease in noise pollution determined by an optimal distribution of the outlet velocity and also by the reduction of the velocity gradient between cold and hot gases from 45% (for the original engine) down to 21%.*
Key words: *noise, pollution, jet engine, optimal*

1. INTRODUCTION

Nowadays, air transportation becomes more and more restrictive, implying low chemical and noise pollution, low fuel consumption and high thrust force. In order to keep up with these restrictions, engineers must come up with new, improved solutions. The main objective of this paper is to investigate the characteristics and the performances of a triple flow jet engine in order to determine the eventual possibilities of improving the performances of jet engines to keep up with these restrictions. The idea is to transform the existing CFM 56-7 turbofan into a triple flow jet engine and to prove that this is a way of reduce noise. Experimental studies using sound absorbing materials on this type of engine resulted in a more silent engine (Ispas, 1991). But antipollution standards evolution implies further reduction of turbojet engines noise. The triple flow jet engine studied here implies the separation of the by-pass flow of a turbofan by adding a new fan stage down-stream of the existing one, as shown in figure no. 1. The calculus is based on the hypothesis that the turbine produces a maximum specific mechanical work at sea level static (SLS). Because the starting point is an existing engine, some of the engine's parameters remain the same. For this new type of engine are then calculated the performances and the outlet velocity for each flow which, combined, give the noise level of the engine, in order to compare them with the existing solution.

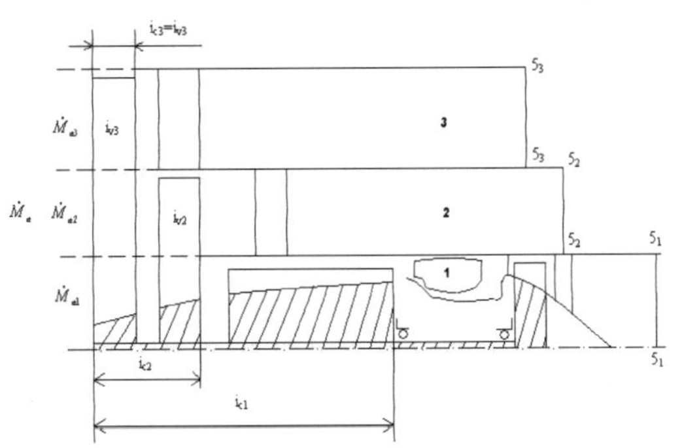

Fig. 1. Triple flow engine flow chart

2. CONTENT

The following conditions have to be satisfied for the engine to work properly and to assure the best performances for its configuration (Bejan, 1996; Pimsner, 1983):

- The operating at stationary regime at SLS:

$$P_{T1} = P'_{C1} \quad (1)$$

$$P_{T2} = P_{V2} \quad (2)$$

$$P_{T3} = P_{V3} \quad (3)$$

- The power produced by the turbine has a maximum known value:

$$P_{Tmax} = P_{T1} + P_{T2} + P_{T3} \quad (4)$$

- The compression specific work from the primary air flow is known:

$$l_{C1} = l'_{C1} + l_{V2} + l_{V3} \quad (5)$$

- According to the mass flow characteristics of the two fans:

$$K_2 \cdot l_{C2} \approx K_3 \cdot l_{C3} \quad (6)$$

- At these equations are added the expressions which connect the power to the related specific work
- The existence of the triple flow jet engine:

$$l_{C2} > l_{C3} \quad (7)$$

- Lowest level of noise possible obtained through the mixing of the three flows. Experimental noise studies made on triple flow coaxial cylindrical cells offer us the image of an optimal outlet velocity distribution for this type of engine compared with a double flow one, as shown in figure no. 2.

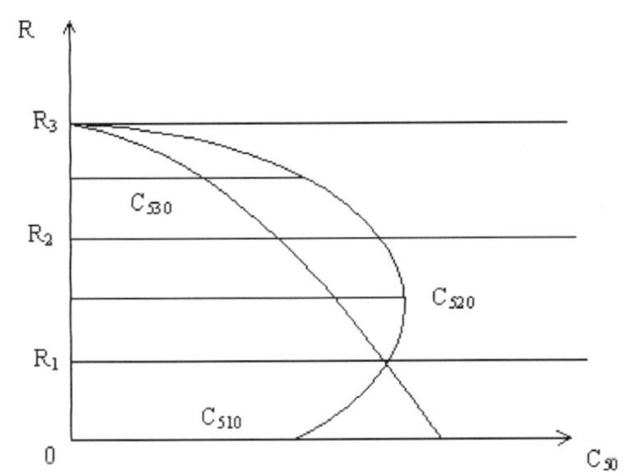

Fig. 2. Optimal outlet velocity distribution along the radius

Because in this study is intended a comparison with the existing solutions, the new engine also has to check the following hypotheses:

- The exterior fan must produce the same specific mechanical work

$$l_{C3}= l_{V3}=\text{const} \quad (8)$$

- The maximum diameter of the engine should remain the same:

$$K_2+K_3=K \quad (9)$$

- The transformed engine must provide the same pressure ration on the main flow.
- The new system must develop a maximum specific thrust force. Its optimization leads to:

$$K_3=4\cdot K_3 \quad (10)$$

According to these hypotheses, the parameters of the CFM 56-7 turbofan from table no. 1 remain constant after its transformation into a triple flow jet engine (Daly, 1997).

π_C=32.7
l_{C1}=700 kJ/kg
M_{a1}=59 kg/s
K=5.1
i_3=1725 kJ/kg
l_{V3}=29.93 kJ/kg

Tab. 1. Constant parameters for both types of engine

In table no.2 are given the results of the thermodynamic cycles' analysis for the main interest parameters in order to obtain a reasonable comparison between an existing double flow jet engine (CFM 56-7) and a theoretical triple flow one (Pimsner, 1983; Berbente et al., 1997)

Double flow jet engine	Triple flow jet engine
l_T=853 kJ/kg	l_T=944 kJ/kg
C_{51}=402.9 m/s C_{52}=222.6 m/s	C_{51}=301.6 m/s C_{52}=380.4 m/s C_{53}=222.6 m/s
F_{sp}=1538 m/s F=90737 N	F_{sp}=1597.6 m/s F=94260 N
C_{sp}=0.0435 kg/Nh	C_{sp}=0.0419 kg/Nh

Tab. 2. Comparative parameters results

Based on these results, it can be noticed in figure no. 3 and no.4 that the outlet velocity distribution is similar to the optimum experimentally established.

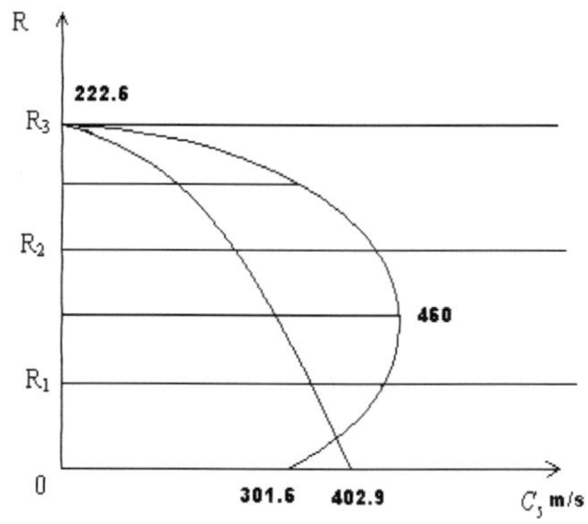

Fig. 3. The outlet velocity distribution obtained

Fig. 4. Results for outlet velocity and thrust force

3. CONCLUSION

By analysing the results obtained, some conclusions can be withdrawn with relation to the advantages and disadvantages implied by the new type of engine obtained by adding a new fan stage. The main advantage shown by this study is the reduction of the noise pollution by:

- An optimal distribution of the outlet velocity according to the experimental studies, as shown earlier.
- The decrease of the outlet velocity from the main flow with 5.6 %
- The decrease of outlet velocity medium gradient from 45 % to approximately 31 %
- The decrease of velocity gradient between the main flow, consisting of hot burnt gases and the second flow, which can be considered cold air, down to 21 %

Hence, it can be said that the triple flow jet engine is a way of reducing the noise as compared to the double flow existing solution. This research is, though, limited to a theoretical approach in which the efficiency of different components have been approximated based on other experimental data. For a complete image of the problem, the authors consider that experimental tests must be done in the future, to support this theory. The calculus also shows an increase in specific thrust and a decrease in specific fuel consumption with approximately 4%. Furthermore, these advantages are obtained through minimum constructive changes which assume adding a new, intermediate, fan stage. The small improvement given to the engine's performances is lost though, because the engine's weight also rises, which is a first disadvantage that occurs. But the biggest problem, yet to be solve, is that the new system requires a specific work developed by the turbine with approximately 90 kJ/kg more than the existing one. Thus, the real challenge is, as further research, to improve the turbine, by adding a new stage for example, so that it can provide the required specific work for the compressor and both fans.

4. REFERENCES

Bejan, A. (1996). *Termodinamica tehnica avansata (Advanced engineering thermodynamics),* Editura Tehnica, 73-31-0994-0, București

Berbente, C.; Mitran, S & Zancu, S. (1997). *Metode numerice (Numerical methods),* Editura Tehnica, 73-31-1135-X, București

Daly, M. (1997). *Aero-Engines,* Jane's, Coulsdon, UK

Pimsner, V. (1983). *Motoare aeroreactoare (Aero-jet engines),* Editura Didactica și Pedagogica, București

Ispas, S. (1991). *Motorul turboreactor (Turbojet engine),* Editura Tehnica, 973-31-0273-3, București

Annals of DAAAM for 2008 & Proceedings of the 19th International DAAAM Symposium, ISSN 1726-9679
Editor B. Katalinic, Published by DAAAM International, Vienna, Austria 2008
Make Harmony Between Technology and Nature, and Your Mind will Fly Free as a Bird

NEGOTIATIAN-BASED MULTI-AGENT SYSTEM FOR TIMETABLING

CURAK, I[van]

Abstract: *Timetabling problems are inherently distributed and timetabling applications function in distributed environment. In this paper, we present a negotiation based approach for creation of university timetable. In this distributed system, every actor is presented with an agent and the problem solution is found by using negotiation protocols between software agents. The process of finding solution is divided into two phases: in the first one agents are trying to find feasible solution and in the second phase agents are trying to optimize solution, minimizing the number of violated soft constraints. Regarding its flexibility this approach is a promising one as a solution for distributed timetabling problem.*

Key words: *multi-agent systems, timetabling, negotiation protocols, distributed systems.*

1. INTRODUCTION

Resource allocation problems such as timetable creation have usually been solved by constraint based or meta-heuristic methods. These methods implicate that all user's own data and private constraints have to be published to central authority. But, very often there is the case that the users don't want to reveal their private data and constraints. This problem can be resolved if we treat timetabling problem as a distributed problem, what is actually the real-life scenario. In such a system it's allowed to the users to keep their agenda private and the negotiations between actors are used for allocations of needed resources.

The potential of agent technology and multi-agent systems in the area of timetabling was recognized in the last decade. In the beginning, the usage of these systems was constrained to the solution of inter-departmental constraints at universities with many departments. Multi-agent based solution of this problem was proposed by some authors (Kaplansky & Meisels, 2004; Di Gaspero et al., 2004; Oprea, 2006).

In the next phase, multi-agents systems have been proposed as heuristic for solving course timetable problems. Verrons (Verrons & Mathieu, 2006) proposed the solution based on negotiation between software agents. The next example of agent based solution model for timetabling is the work of Yang (Yang et al., 2006). In this model, each agent represents a constraint in the course timetabling problem. Picard (Picard et al., 2005) proposed the system based on self-organization. As a mechanism of self-organization, they used cooperation.

In this paper we present the negotiation based distributed timetabling system. Our approach is the original one and it's designed in a way to be able to deal with timetabling problems from the praxis. It's realized as a multi-agent system (MAS) where agents represent teachers, students, courses and rooms. This system is flexible and able to adapt itself in response to dynamic changes in the environment. The changes in the environment occur when an agent is removed from the system, when a new agent is added or when some constraints are added or removed. Because of the system's flexibility, it's not needed to restart the system from the scratch after any change in the environment, but the system adapts itself.

In our negotiation-based approach for the solution of distributed university/school timetabling problem, the process of finding solution is split into two phases: in the first phase, when the process starts, the agents are trying to find any feasible solution (only hard constraints must be satisfied) and thereafter, in the second phase, the agents are optimizing the global solution, minimizing the number of violated soft constraints. We have implemented our multi-agent system by using JADE (Java Agent DEvelopment Framework).

In this paper we first present the architecture of the MAS. Then, we present the negotiation protocol, the results of this approach and give directions for future research.

2. SYSTEM ARCHITECTURE

Our starting point was to design the system that will be able to cope with real problems at universities and in schools. Because of that, in our distributed timetabling system all the known types of hard and soft constraints can be expressed or the system can be easily extended. Main agents in the system are *TeacherAgent*, *StudentAgent*, *RoomAgent*, *CourseAgent*. Their relations in a simple example configuration are presented in the Figure 1.

StudentAgent represents one student or one student group (students with common curricula). It has references to all the *CourseAgents* that are part of its curricula. *TeacherAgent* represents one teacher and has references to all the *CourseAgents* the teacher teaches. *CourseAgent* represents one course and has references to its *TeacherAgent* and to all the *StudentAgents* that are enrolled in this course. A course in timetabling problem can be regular (it belongs to only one student/student group) or it can be shared between many student groups. If the course is shared, it must be scheduled at the same time-slot in all the students' timetables that use this course. *CourseAgent* has the data about a number of lessons per week that should be scheduled. Soft constraints can be defined for this agent, too. For example, how the lessons are grouped across the week or the lessons should be scheduled in the starting time-slots of the day or the lessons should be scheduled right after the lessons of another course. *RoomAgent* represents

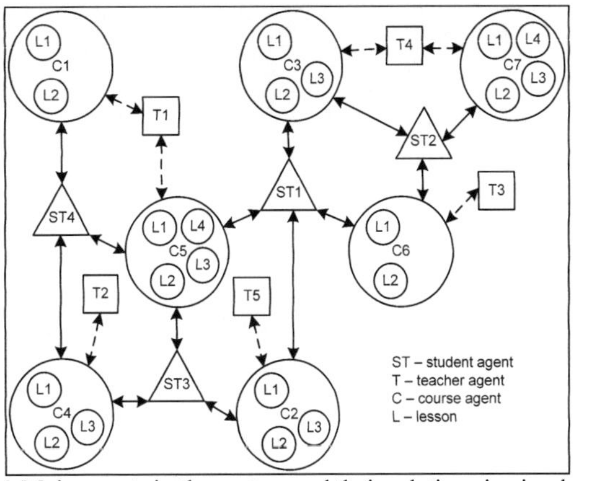

Fig.1 Main agents in the system and their relations in simple example configuration

one room and this agent holds resources that are subjects of negotiation, i.e. time-slots.

Using our system that is based on JADE framework, in extreme case, with the help of LEAP add-on, it is even possible that every actor (student, teacher) deploys its representing agent on his/her own handheld or mobile device and in that way participates in the process of timetable creation.

3. NEGOTIATION PROTOCOL

Multi-agent system we describe here has a goal to create timetables for teachers and students at universities or schools. For each student and teacher involved, the system has to allocate in time-slots all the lessons of every course the student must attend or the teacher must teach.

Every actor in the system has constraints to be satisfied. In order for the final solution to be feasible, all the hard constraints must be satisfied. We want to make one step further and make the optimisation of global solution, minimizing the number of violated soft constraints. Each soft constraint has its weight coefficient that is used for calculation of the global goal function.

Our negotiation protocol is an extension of Contract-Net protocol (Smith, 1980) and it's presented in Figure 2. In the first phase of allocation process, *ControlAgent* randomly chooses a *CourseAgent* and sends the message to it to start the allocation of its lessons. *CourseAgent* uses its local reasoning and finds the most suitable time-slots for its lessons. Then, it starts the negotiation protocol with the other participants (teacher, student(s) and rooms). At the end of negotiation process, if it's successful, the contract between all the participants is made.

The negotiation starts with a proposal for the contract from *CourseAgent* (initiator) to *TeacherAgent*, *StudentAgent* and *RoomAgent* (participants). They can accept or reject this proposal. If it's rejected, the initiator sends modification request to the participants and collects their propositions. Thereafter, the initiator synthesizes them and proposes a new contract. *ControlAgent* decides which *CourseAgents* can be in negotiations at the same time because only courses that belong to different students and teacher can do negotiation simultaneously.

When all the courses have done initial allocation of their lessons, the second phase starts. In this phase the system tries to minimize global goal function (Eq. 1).

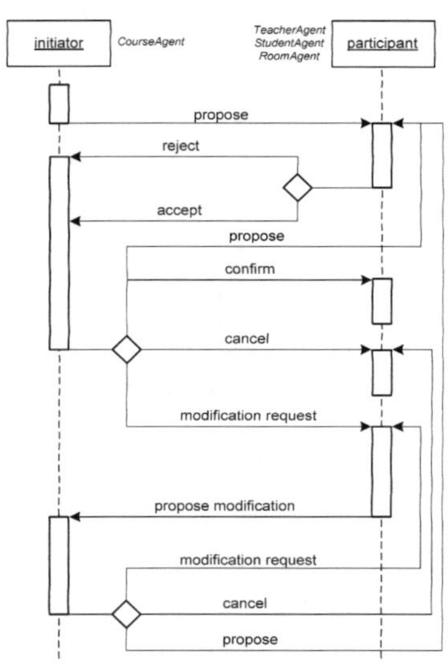

Fig. 2 Negotiation protocol

$$G=S+T+C \qquad (1)$$

In this equation, G represents global goal function and S, T and C represent the weighted number of violated soft constraints for all the students, teachers and courses, respectively. Using the same negotiation protocol, a *CourseAgent* tries to change the time-slots for its lessons, accepting the change only if the global goals function is decreasing. In order to check if the global function is decreased, it's necessary only to calculate and compare the values that are related to agents in negotiation.

4. RESULTS

Our system is still developing but the early results are encouraging. The system is tested only on small problems, with maximum 40 agents in the system deployed on three JADE containers on different hosts. Since the agents can't conclude when the optimization minimum is reached, the administrator user must stop optimization process or we can set the optimization time as a parameter. The system is able to adapt to dynamic changes: for each change in the environment it is automatically renegotiated. In the future work, in order to avoid the explicit use of negotiation protocol and to improve system's performances, we plan to use semantic agents with knowledge base.

5. CONCLUSION

Negotiation-based approach is a novel approach for solving of timetable creation problem and the promising one. By using it, teachers and students can find a timetable taking into account their constraints, without giving them to the other system participants. This approach is interesting because of its flexibility, and not for the results it currently gives. We think that this system gives contribution to the development of distributed timetabling systems and it can be used as a base for future research.

6. REFERENCES

Di Gaspero, L., Mizzaro S. & Schaerf, A. (2004). A MultiAgent Architecture for Distributed Course Timetabling, *Proceedings of 5th International Conference on the Practice and Theory of Automated Timetabling (PATAT)*, pp.471-473., Pittsburgh

Kaplansky, E.; Meisels, A. (2004). Negotiation among Scheduling Agents for Distributed Timetabling, *Proceedings of 5th International Conference on the Practice and Theory of Automated Timetabling (PATAT)* pp.517-520., Pittsburgh

Oprea, M. (2006), Multi-Agent System for University Course Timetable Scheduling, *ICVL Proceedings*, pp. 231-238., Bucharest

Picard, G., Bernon, C. & Gleizes, M.-P. (2005). ETTO: Emergent Timetabling by Cooperative Self-Organisation, *Third International Workshop on Engineering Self-Organising Applications (ESOA'05)*, Utrecht, pp. 31-45.

Smith, R. G. (1980). The Contract Net Protocol : high-level communication and control in a distributed problem solver, *IEEE Transactions on computers*, C-29(12), pp. 1104–1113

Verrons, M.-H., Mathieu, P. (2006), How to solve a timetabling problem by negotiation?, *Proceedings of 6th International Conference on the Practice and Theory of Automated Timetabling (PATAT)*, pp.502-505., Brno

Yang, Y., Paranjape, R. & Benedicenti, L. (2006). An Agent Based General Solution Model for the Course Timetabling Problem, *The Fifth International Joint conference on Autonomous Agents, (AAMOS'06)*, Hakodate, Kokkaido

Annals of DAAAM for 2008 & Proceedings of the 19th International DAAAM Symposium, ISSN 1726-9679
Editor B. Katalinic, Published by DAAAM International, Vienna, Austria 2008
Make Harmony Between Technology and Nature, and Your Mind will Fly Free as a Bird

COOPERATIVE COEVOLUTIONARY APPROACH TO DUAL ARM ROBOTS' OPERATION

CURKOVIC, P[etar]; JERBIC, B[ojan] & STIPANCIC, T[omislav]

Abstract: *This paper presents discussion on coevolutionary approach applied to operation of the dual arm robot. The arms have to pick an object and transport it coordinately to a predefined position. The dual arm configuration is a naturally distributed system. To be able to perform distributed work together, high level of coordination between arms is required. Two independent evolutionary algorithms optimize motions of each hand of the simulated robot. Mutual fitness function merges behaviours of collaborating robots to complete the task.*
Key words: *robotics, coevolution, coordination, optimization.*

1. INTRODUCTION

Dual arm robots exhibit behaviours similar to humans in terms of manipulating objects in highly coordinated fashion. To achieve such a coordination is a task difficult to solve for its inherent complexity (Fei et al., 2004). The task is naturally distributed – two arms, two robots, yet tightly coupled through high level of coordination required.

Coevolutionary algorithms draw their inspiration from the phenomena widespread in nature that involve the evolution of one species in close relation to another (Eiben & Smith, 2007). There are many examples of coevolution in nature, ranging from collaborational behaviours observable in relations of flowers and insects for pollination, competitive coevolution of predator and prey strategies, or relationships of host – parasite type. Coevolutionary algorithms are augmentation of traditional evolution algorithms in the sense that individuals fitness is assigned on the base of interaction of that particular individual with other individuals from the same population and/or other populations.

Most of the work in evolutionary robotics considers competitive approach to coevolution based on predator-prey relationship (Mermigikis & Petrou, 2006), (Cai & Peng, 2002). Work directed toward cooperative coevolution is mainly demonstrated in various robo-soccer applications (Uchibe et al., 1998).

The idea of this paper is to present some insight in possibilities of application of cooperative coevolutionary algorithm to a dual arm robot system as a cooperative coevolutionary architecture. The task of such a system is to explore the environment and find an object positioned in the environment. The hands have to coordinate at the location of the object since (i.e. the object is too heavy for one arm) and in coordinated fashion transport the object to the destination position. The paper will present simplified kinematics' structure adopted for initial research (Kasać & Novaković, 2001) definition of individuals representation in genotypic space, fitness function and collaboration strategy between individuals from different populations.

2. KINEMATIC STRUCTURE OF A DUAL-ARM ROBOT SYSTEM

The environment of the dual-arm robotic system is organized under following restrictions:

- The robots operate in an environment without obstacles
- Positions of the two robots are known in every time step
- Collisions of the robots' arms are not allowed
- Positions of the object and target position are known

The cinematic structures of the robots are for simplicity reasons chosen as structures with two degrees of freedom, with one rotational and one translational degree of freedom. In such manner, each point in the work space is reachable, if there are no obstacles present in the environment. Fig.1. presents simplified robotic arm with two degrees of freedom.

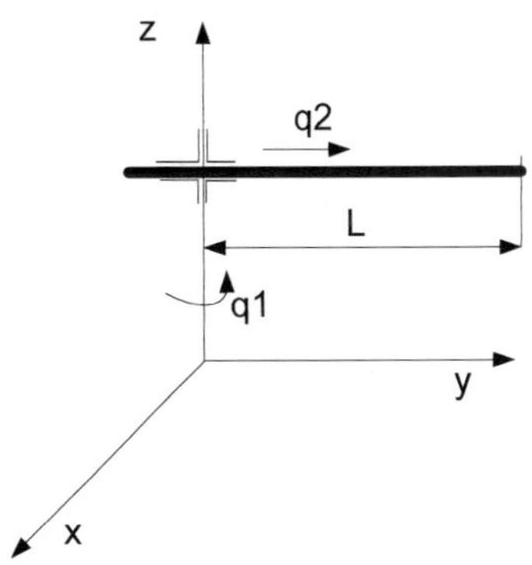

Fig. 1. Kinematic structure of the robot

Setup with two robots is presented in Fig. 2. The dimensions of the environment are known.

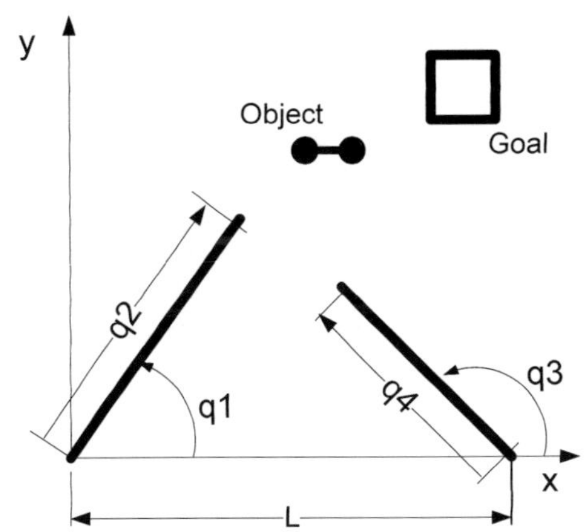

Fig.2. Environment with two robots working cooperatively to transport the object

3. REPRESENTATION

One of the crucial steps in design of an evolutionary-based algorithm is to choose proper representations of the individuals.

Complete evolutionary process is taking place in the genotypic space and it is important to enable finding (optimal) solution to the given problem by choosing proper representation. Phenotypic space is later revealed through decoding of individuals. In this paper, genotypic space will contain encoded information about moving of the robots in consecutive time-steps. There are two main possibilities of representation of the robots' movement: 1. occupancy grid representation, with environment divided in to finite number of elements and corresponding nodes in intersections. The movement of the robot would be indirectly encoded through intersections of the mesh. And 2. to encode movement of the robot directly as real-valued parameters of lengths and angles of robotic arms. Although both of the approaches were considered, it is decided to adopt direct real-valued encoding approach. The reason is in increased flexibility of the robot being able to continuously move through the space, as opposing the occupancy grid where the arm of the robot can move only discretely from one grid intersection to another.

$$\langle \underbrace{x_1, x_2, \ldots, x_{i-1}, xi}_{q_{21}}, \underbrace{y_1, y_2, \ldots, y_{i-1}, y_i}_{q_{11}} \rangle; x_i, y_i \in \mathbf{R} \quad (1)$$

The above expression presents one candidates' form, whereat first part of the string corresponds to lengths of arm of robot 1 and second part of the string corresponds to angles of the same robot. In such manner it is possible to reveal movement of the robot via corresponding pairs (x_i, y_i). The length of the chromosome is important parameter. It is environment dependant, and it should be incorporated into the algorithm through some sort of heuristic knowledge. For example, the environment presented in Fig. 2 is solvable with two strings of length l=4, because it is necessary only to determine two consecutive positions in the environment. It might be interesting to limit the amount of change for which the robot can move within one time step to achieve more realistic behaviors.

4. SELECTION

Selection of the candidates for the next generation is not based on the performance of particular candidate per se. It has rather to be evaluated according to collaboration with candidates from another population. In this paper; two independent populations of individuals competing to collaborate with individuals from other population are assumed.

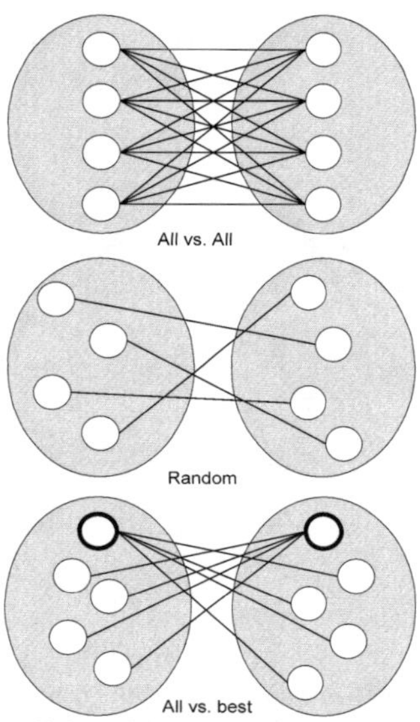

Fig. 3. Different collaboration patterns for two populations

Naturally, one population will represent the work of first robot, and another population the work of second robot.

It is important to mention that complete evaluation of these two populations can be done in parallel, on separate processors (Sims, 1994). The only point where these two populations interact with each other is the fitness evaluation.

The objective function has take in account following parameters: distance from the first robots' gripper to the object d_1, distance from the second robots' gripper to the object d_2, distance from the object to the goal d_3, number of collisions C. The objective function is thus of the following form:

$$min \left\{ \Theta = \Theta\left(d_1, d_2, d_{3,} C\right) | \left\langle d_1, d_2, d_3 \right\rangle \in \mathbf{R}, C \in \mathbf{Z} \right\} \quad (2)$$

Relative distance from one robots' gripper to the other ones must be constant after grasping the object. Otherwise, the object could be damaged. In the study presented here, the object is considered a point in the work space, and corresponding distance equals 0.

5. CONCLUSIONS

The paper presents possibilities of application of coevolutionary principles to an operation of a dual arm robot. Several critical points such as: representation, selection, fitness function, robot kinematics were discussed and solutions proposed. Representation chosen enables to easily add more degrees of freedom to the robot. This should be beneficial in the case of obstacles present in the work space or in the case of expanding the problem to 3D. At the same time, it would be able to evolve optimized architecture of the robot in terms of degrees of freedom of a robot for a specific environment. Future work will include more detail development of the proposed model with computer simulations of robots work.

6. ACKNOWLEDGEMENTS

The authors kindly acknowledge Croatian Ministry of Sciences Education and Sport for the support through grant No. 120-1201948-1941 *Autonomous multiagent automated assembly* and Croatian Institute of Technology for the grant TP-E-46.

7. REFERENCES

Cai, Z.; Peng, Z. (2002). Cooperative coevolutionary adaptive genetic algorithm in path planning of cooperative multi-mobile robot systems. *Journal of intelligent and robotic systems.* Vol 33, pp 61-67.

Eiben, A.E.; Smith J.E. (2007). *Introduction to evolutionary computing.* Springer, 978-3-540-40184-1, Berlin, Heidelberg, New York

Fei, Y.; Fuqiang, D. & Xifang, Z. (2004). Collision-free motion planning of dual-arm reconfigurable robots. *Robotics and computer-integrated manufacturing,* Vol.20. pp. 351-357

Kasać, J.; Novaković, B. (2001). Neural network application to optimal control of nonlinear systems. *Computer aided optimal design of structures,* Vol. 10.

Mermigikis, I.; Petrou, L. (2006). Exploring coevolutionary relations by alternations in fitness function: Experiments with simulated robots, *Journal of intelligent robotic systems,* Vol. 47, pp. 257-284.

Sims, K. (1994). Evolving 3D morphology and behaviour by competition. *Artificial life IV Proceedings,* MIT Press, pp.28-39.

Uchibe, E.; Asada, M.; & Hosoda, K. (1998). Cooperative behavior acquisition in multimobile robots environment by reinforcement learning based on state vector estimation. *In Proc. of IEEE International Conference on Robotics and Automation.*

Annals of DAAAM for 2008 & Proceedings of the 19th International DAAAM Symposium, ISSN 1726-9679
Editor B. Katalinic, Published by DAAAM International, Vienna, Austria 2008
Make Harmony Between Technology and Nature, and Your Mind will Fly Free as a Bird

RISK MANAGEMENT STRATEGIES BASED ON EXPERIENCE FROM NITRATE TOXICITY CASE STUDY

CURSEU, D[aniela]; POPA, M[onica] & SIRBU, D[ana]

Abstract: *In this paper we summarize findings from cases diagnosed with infant methemoglobinemia in Cluj County, Romania, over a 5-years period (2002-2006). Contaminated well water was the major source of nitrate in the cases reviewed, though early incorporation of vegetable juice and home-prepared foods into infant diet was also incriminated. High-nitrate drinking water was most often associated with improperly constructed and maintained wells. The study suggests some risk management strategies for preventing infant methemoglobinemia in this area.*

Key words: *case study, environment, management, prevention, pollution*

1. INTRODUCTION

The nitrate ion (NO_3^-) is the stable form of combined nitrogen for oxygenated systems. Although it is chemically unreactive, it can be microbially reduced to the reactive nitrite ion. Nitrite converts oxygen-carrying hemoglobin to methemoglobin, which cannot bind oxygen, by oxidizing the Fe^{2+} in heme to Fe^{3+}. The condition of methemoglobinemia is characterized by cyanosis, stupor, and cerebral anoxia. Naturally, 0.53-3.0% of the total hemoglobin (Hb) circulates as methemoglobin (MetHb). Generally, detectable clinical signs of methemoglobinemia appear at 10% MetHb (Brunning-Fann & Kaneene, 1993; Kross et al., 1992).

The aim of this study was to evaluate reported cases of infant methemoglobinemia in relation with environmental exposure to nitrate, in favor of appropriate intervention programs of risk management in Cluj County, Romania.

2. MATERIAL AND METHOD

To satisfy our objective, infants diagnosed with methemoglobinemia at the pediatric emergency departments from Cluj County, Romania, between 2002 and 2006 were identified. For each case, the following data were recorded: date of admission to the emergency department, age, sex, weight, place of residence, reason for consultation and associated symptoms, feeding characteristics in the last days, foods ingested in the previous 24 hours, results of laboratory investigations.

Water samples as well as fruit and vegetables cultivated in individual farms and used for infant's feeding, were collected 2 days after the infant was hospitalized. The nitrates/nitrites concentration was analyzed using the phenol-disulphonic acid method, respectively the sulphanilic acid and α-naphtylamine hydrochlorate method. In addition, information on household environmental conditions (including drinking water hygiene, sanitation and domestic hygiene) were provided.

3. RESULTS AND DISCUSSION

During a five years period (2002 – 2006), in Cluj county have been recorded 28 hospitalised cases of methemoglobinemia (42.9% boys and 57.1% girls).

Year	Cases	Live births	Incidence rate*
2002	12	5,021	239
2003	2	4,890	42
2004	3	4,795	63
2005	6	4,673	128
2006	5	4,334	115
Total	28	23,713	118

*Number of cases per 100,000 live births

Tab. 1. Incidence of methemoglobinemia in Cluj County.

The average age was 45 days, with a range of 12 days to 6.5 months. The majority of cases (79%) were under 4 month.

Annual data for live births in Cluj County were also available, so we were able to calculate the incidence rate of methemoglobinemia based on clinically reported cases (Table 1). The annual estimated incidence of methemoglobinemia ranged from 42 to 239 per 100000 live births during the study period, with a mean of 118/100000 over the five year period. Although these 28 cases are not representative for the entire population of Romania, the rates are in accordance with the incidence rates earlier reported by Ayebo et al. (1997) in contiguous areas from Transylvania.

A summary of selected clinical data is presented in Table 2. The average birth weight of infants was 2921±422 g, which is above the 2500 g, internationally acceptable definition of low birth weight, but statistically ($p<0.001$) lower than the Romanian national average of 3465±365g. The average weight of infants at the time of diagnosis with methemoglobinemia is 4020±450g, significantly lower than comparable age group infants (4020g vs. 4980g). Breastfeeding data indicate that about 74% of infants had been breast fed for at least one week postpartum. The duration of breastfeeding was generally less than one month. No infant was diagnosed with methemoglobinemia while being breast-fed.

Medical examination revealed a high incidence of anemia in infants reported with methemoglobinemia. The hemoglobin of infants averaged 11.2 g/dL for all ages, which is within the normal range of Romanian national hemoglobin values of 10.8-18.0, with a mean of 13.3 g/dL. However, when compared to national average data according to age distribution, 11 cases (39.2% of the infants) were classified as anemic (Table 3). This observation is important because the anemia emphasizes the anoxia status. Thirty-one percent of the infants had symptoms of diarrhea on initial diagnosis, and all cases presented cyanosis.

	Mean/±SD (range)[1]	Mean/±SD (range)[2]
Birthweight (g)	2921±422 (2200 – 4105)	3465±365
Weight at admission (g)	4020±450 (3300- 6750)	4,980±365
Breastfeeding (days)	28 (10-45)	NA[3]
Hemoglobin (g/dL)	11.2±1.5 (6.8-13.1)	13.3 (10.8-18.0)

[1]N=28 cases; [2]Romanian National Average; [3]not available

Tab. 2. Clinical data of hospitalized infants

	Anemia cases	Cases of methemoglobinemia Hb (g/dL)	National average Hb (g/dL)
Age (month)	N	Mean (range)	Mean (range)
under one month	2	11.3 (11.1-11.7)	16.6 (13.2-23.0)
1-4 month	8	11.0 (7.5-12.5)	13.3 (10.8-18.0)
over 4 month	1	11.2 (9.6 – 14.7)	12.4 (10.2-15.0)

Tab. 3. Comparison of infant hemoglobin levels

Geographically, all infants hospitalized with acute methemoglobinemia lived in rural and received their drinking water from private wells. Another important area of concern was well-water sources, well depth, the level of well-water nitrate (mg/L), and boiling of liquids. Findings in this study indicate that all communities where methemoglobinemia was reported had nitrate levels above of allowable limit of 50 mg/L. The highest recorded concentration was 265 mg/L. Well construction methods, placement, and general hygiene were primary causes of poor well water quality and high nitrate level. The most wells were dugouts of less than 8 meters, with some as shallow as 3 to 4 meters, without casing and no protective cover over the wellhead. Also, the wells were located very close to human traffic areas and livestock rearing facilities. Moreover, household laundry and washing of utensils is done around the open shallow well head. Nitrate contamination from agricultural use of fertilizers and animal manure may occur, but the most likely source of nitrate for the wells observed in this study is local human and animal waste.

The infants were weaned completely after a short breastfeeding period to cow milk diluted (1:1) with boiled water. All mothers interviewed routinely boiled water to kill pathogenic microorganisms in well water before use for infant feeding. However, this practice actually increases nitrate concentration.

No reported episodes of methemoglobinemia occurred within the rural communities with normal nitrate well water levels or in urban population supplied with treated tap water. According to the well water nitrates concentration, we categorized two groups at risk of methemoglobinemia:

-moderate risk – well water pollution between 50 - 100 mg/L;

-high risk - well water pollution above of 100 mg/L.

A significantly higher mean value of methemoglobin was found when nitrate concentration in well water exceeded 100 mg/L (Table 4).

Nitrites were detected in concentration up to 0.25 mg/L in all samples of well water.

In one multiple episode of methemoglobinemia reported the first episode occurred when mother offered some tea to her baby, and the repeat episode was when infant's feed included spinach, carrot and apple juice. They are given to supplement the infant's diet of diluted cow milk.

Frequently the fruits and the vegetables cultivated in individual farms constitute an additional source of nitrates. Table 5 shows the nitrates and nitrites average concentrations in vegetables comparing to maximum allowable level (MAL) according to Direction 975/1998 of Romanian Ministry of Health.

Nitrate level	50 - 100 mg/L			>100 mg/L		
Values of MetHb (%)	N	Mean	±SD	N	Mean	±SD
	10	12.1	±2.5	18	17.2	±2.9
t-test (p-value)	t=4.67 (p=0.0001)					

Tab. 4. The average values of methemoglobinemia in relation with nitrate levels in well water

Species	Nitrates concentration (mg/kg)		Nitrites concentration (mg/kg)
	Mean ± SD	MAL	Mean ± SD
carrots	395.9 ±23.3	400	0.45±0.09
potatoes	311.3±37.01	300	0.78±0.27
spinach	291.7±69.4	2000	0.58±0.12
squash	48.5±11	60	0.32±0.05

Tab. 5. The nitrates and nitrites average concentrations in vegetables

It has to be noted the over limit nitrates concentrations in potatoes and carrots samples. This observation is important because carrot soup is often recommended if the infant is suffering from diarrhea, so the high nitrate level in carrots and gastrointestinal illness go „hand in hand" for bacterial conversion of nitrate to nitrite in the stomach. For this reason many other studies associate methemoglobinemia with infantile cases of severe diarrhea (Hanukoglu & Danon, 1996). Also, others authors suggested that bacterial growth within the bottle or stored homemade soup / purees of mixed vegetables and exogenous conversion of nitrate to nitrite is possible the source of the problem (Echaniz et al., 2001; Fewtrell, 2004).

4. CONCLUSIONS AND RECOMMENDATIONS FOR RISK MANAGEMENT

1. The greatest risk of methemoglobinemia occurs in infants less than 4 months bottle-fed with diluted cow milk or formulas prepared with nitrate-contaminated water from wells.

2. Well construction methods, placement, and general hygiene are primary causes of poor well water quality and high nitrate levels. All prenatal and well-infant visits should include questions about the home water supply. If the source is a private well, the water should be tested for nitrate.

3. Since many vegetables are high in nitrate, consumption of home-prepared infant foods from vegetables (spinach, potatoes, carrots) should be limited until the infant is 4-6 months old.

4. An educational program for mothers to increase duration of breast feeding could be a very effective method of preventing infant methemoglobinemia.

5. REFERENCES

Ayebo, A.; Kross, B.; Vlad, M. & Sinca, A. (1997). Infant methemoglobinemia in the Transylvania region of Romania. *Int J Occup Environ Health.* 3(1): 20–29, ISSN 1077-3525.

Brunning-Fann, C.S. & Kaneene, J.B. (1993). The effects of nitrate, nitrite and *N*-nitroso compounds on human health: a review. *Vet Hum Toxicol.* 35(6): 521–538, ISSN 0145-6296.

Echaniz, J. S.; Fernández, J. B. & Raso S. M. (2001). Methemoglobinemia and Consumption of Vegetables in Infants. *Pediatrics.* 107(5): 1024-1028, ISSN 1098-4275.

Fewtrell, L. (2004). Drinking-Water Nitrate, Methemoglobinemia, and Global Burden of Disease: A Disscussion. *Environ. Health Perspect.* 112(14): 1371-1374, ISSN 0091-6765.

Hanukoglu, A. & Danon, PH. (1996). Endogenous methemoglobinemia associated with diarrheal disease in infancy. *J Pediatr Gastroenterol Nutr.* 23:1-7. *Available from:*http://pediatrics.aappublications.org/cgi/reprint/107/5/1024 *Accessed:* 2008-04-02.

Kross, B.C.; Ayebo, A.D. & Fuortes, L.J. (1992) Methemoglobinemia: nitrate toxicity in rural America. *Am Fam Physician.* 46(1): 183–188, ISSN: 0002-838X.

Annals of DAAAM for 2008 & Proceedings of the 19th International DAAAM Symposium, ISSN 1726-9679
Editor B. Katalinic, Published by DAAAM International, Vienna, Austria 2008
Make Harmony Between Technology and Nature, and Your Mind will Fly Free as a Bird

THE PROPAGATION OF FORCED VIBRATIONS IN COUPLED PLATES OF GUITARS

CURTU, I[oan]; STANCIU, M[ariana] & SAVIN, A[driana]

Abstract: *The paper analysed the frequency responses of the top, back and ribs of guitar box using finite element method. First it was modelled the resonance boxes of different types of guitar. Each structure was stressed virtually by unitary force and the frequency response was sampled from different nodes of structures. The results show the dynamical response of each part of acoustic box and the vibration amplitudes in different points of structures.*

Key words: *plates, amplitudes, FEM, resonance, dynamic, wood*

1. INTRODUCTION

Dynamical behaviour of guitar is linked by the kinetic energy of the strings.

The transfer of the vibration energy of the strings is done through the solid system (s_1 – string) - solid (s_2 – bridge) - solid (s_3 –sound board) – fluid (f_1 – the air from the cavity) - solid (s_4 - back) - solid (s_5 - side) - fluid (f_1 – the air from the cavity and the sound hole). The speed emission of acoustic waves through solid (string-bridge-plate-box) is higher than the one through air, therefore the mechanical structure of the instrument influences the acoustic quality (Curtu, 2008).

In this paper we analyzed which are the areas of plates with maximum and minimum amplitudes, how the vibrations are propagates in wooden structures and the radiation fields of plate.

2. LITERATURE OVERVIEW

Description of acoustical and structural guitar modes is far extensive but each reference depicts certain approaches in this field. Much of the researches in musical acoustic carried out at the University of Wales in Cardiff where the finite element analysis were used by Roberts (1986), Wright (1996). Recently studies were made up by Paul M. Shaheen (2004), Becache, Chaigne, Derveaux and Joly (2004), Bader (2005), Serghei Vladimirovici (2006), Elejabarrieta, Ezcurra and Santamaria (2007). Compared to the work of mentioned references, we modeled the classical guitar bodies according with the Romanian typical bracing patterns and with materials properties of Romanian resonance spruce.

3. DYNAMICAL BEHAVIOR ANALYSIS

3.1 The modelled structure

First, there was geometrically modeled a classical guitar's body incorporating the 3 top-braces plate taking into account the dimensions used in the musical instruments factory *S.C. Hora S.A. Reghin, Romania*. For dynamical analysis with FEM there were used shell type elements with 4 nodes (QUAD4), being very helpful the Patran Nastran 2004 package. The structures were clamped in the area of neck and body jointing. The program was used for the parameters: thicknesses $h=2.5$ *mm*, modulus of elasticity $E=14000 MPa$, density $\rho= 450\ kg/m^3$, Poisson's coefficient $\upsilon=0.36$, damping factor $\delta=0.02$ and shear modulus $G=5000\ MPa$. The values of these parameters were taken from the specialized literature in this domain.

3.2 The load case and frequency response

It was applied a unitary force ($F=1\ N$) in the node P1 placed on the bridge area of the soundboard. The force varied in the frequency range between 0 to 1000 Hz. The dynamic response was analyzed in the nodes placed on the top plate, on the back and on the sides.

3.3. Results and discussion

A series of quantitative and qualitative aspects regarding to the dynamical behaviour of acoustic box are presented in Fig. 1, 2, 3, 4, 5, and 6.

The amplitudes decrease in the modes further from the force's point of application on the cross axis of the plate, which, on the one hand, assumes that the vibration is transmitted longitudinally, being favored by the elastically characteristics and on the other hand it is necessary the assurance of a width big enough for the vibrations to radiate on a useful optimum surface (Fig. 1). With increasing the distance between the point of application force and a certain point of structures, the nodes of plate resonate at a higher frequency. Compare to the frequency response of the top, the vibration amplitude and the resonance frequency in the back is different. First, it can be noticed that the point P1 on the back which is corresponding with the point of application force on the top, has the highest amplitude at the frequency 280 Hz (Fig. 2). Secondly, the variation curves of amplitude versus frequency for the points of the back plate are similarly as shape (Fig. 2). Third, emplacement of the internal struts, the thickness and the pattern increase the mass and the stiffness of structures which conduct to the lowest frequency and modification of the longitudinal flexural modes.

Fig. 3 shows the comparison between the amplitudes of five points placed on the soundboard, back and sides. It can be noticed that the top plate vibrate in a large band of frequency. The common frequency for all parts of acoustic box is around 220 Hz, value obtained by Wright (1996) too.

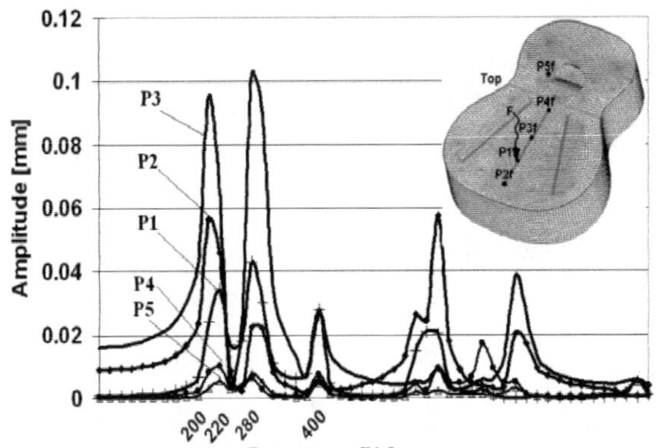

Fig.1. The amplitude variation of the nodes placed on longitudinal axis of the 3 braces pattern

The coupled plates in the acoustic box vibrate at the lower frequencies compared to the free plates (Fig. 4). The normal frequency has close value with the resonance frequency which means that the plate under the forced vibrations resonates in the same time with fundamental frequency. The eigenfrequencies occurring in the assembled instrument do not correspond to the eigenfrequency of the separate parts of the guitar, aspect noticed by Bader (2005) too. The study on vibration behaviour of the guitar box demonstrated the coupling between top and back plates through the air (Fig. 5).

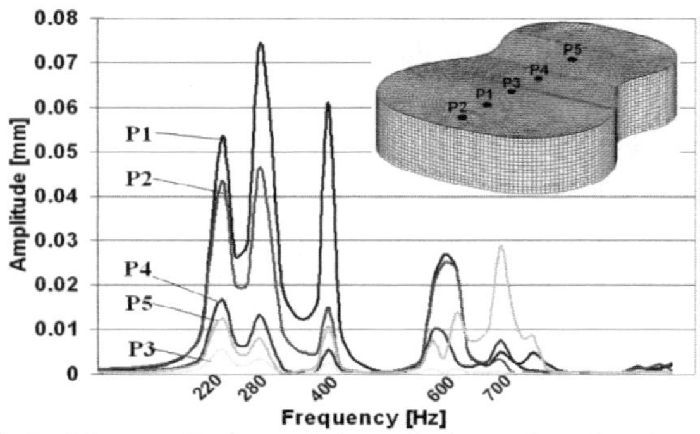

Fig.2. The amplitude variation of the nodes placed on longitudinal axis of the guitar's back plate

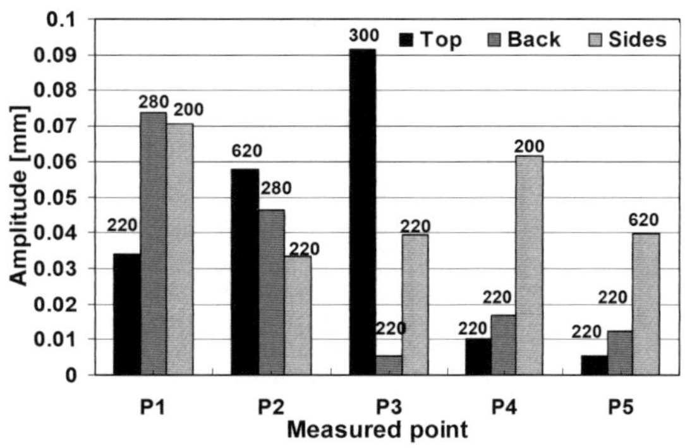

Fig.3. Comparison between amplitudes and resonance frequency of different point placed on the top, back and ribs

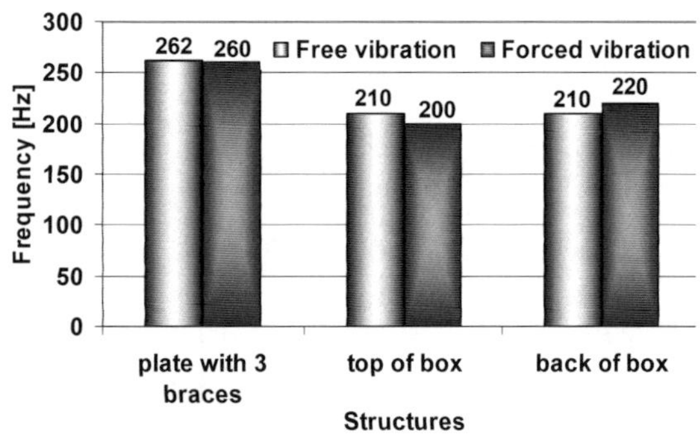

Fig.4. Comparison of guitar body response with and without coupling to the fundamental mode

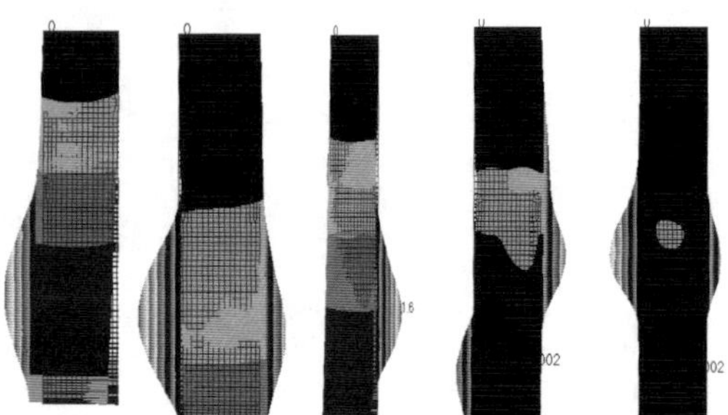

Fig.5. Contour plots of vibration amplitude and size views of modal shapes

Due to the initial boundary conditions, first the top and the plate is moving in phase; then they move out of phase to alternately increase and decrease the cavity volume as shown in Fig. 5 and finally the travelling of vibration in the plates is repeated at a higher frequency and with more nodal and anti –nodal areas.

4. CONCLUSION

The frequency response of the ligno-cellulose plates with different structure and material characteristics offers useful information about the way in which the frequency of the excitation force overlaps with the plate's normal frequency. The modal shapes show the elastically behavior of the plate, respectively the areas in which are formed the node and the amplitude loops for different vibration modes. One of the most important characteristics of the stringed instruments is the possibility of resonating at excitation frequency of the strings. On this line, the amplification box made up of the ligno-cellulose plates takes over this function, being the reason of the necessity of analyzing the frequency response of the plates with FEM. The resonating domain of the plates takes shape between the bridge (the force's application point) and the sound hole. From the analysis of the dynamical behavior of the plates it can be noticed that there are a series of factors which influence the frequency response and the propagation of waves in plates: the plates' structure, the material's features, the magnitude and the application point of the force. The dynamical analysis of the plates aimed just the structural analysis, the acoustical one being supposed to be researched in the future.

This study is a part of a wider research whose purpose is optimizing structures in order to improve the acoustic performance of classical guitar. The future work consists of experimental research on the frequency characteristics of the plate's structures.

5. REFFERENCES

Bader, R. (2005). *Computational Mechanics of the Classical Guitar,* Springer-Verlag Berlin Heidelberg N.Y. , ISBN 3-540-25136-7, in Netherland

Bécache, E., Chaigne, A., Derveaux, G., Joly, P. (2005). Numerical simulation of a guitar. In: *Computers and Structures* , **83**,. 107–126.

Curtu, I., Stanciu, Mariana, Itu, C., Grimberg, R., (2008) Numerical Modeling of the Acoustic Plates as Constituents of Stringed Instruments, in *Proceedings of the 6th International Conference of DAAAM Baltic Industrial Engineering, 24-26th April 2008,* ISBN 978-9985-59-783-5, p. 53-58, Tallinn, Estonia.

Haines, D. (2000). The essential mechanical properties of wood prepared for musical instruments. In: *Catgut Acoustic Society Journal*, **4(2),** pp. 20-32.

Rossing, T., Fletcher, N. (2004). *Principle of Vibration and Sound* – second edition. Springer Science, New York, ISBN 0-387-40556-9.

Vladimirovici, S. (2004). *The Computational Method of The Musical Instruments Body Parts.* PhD Thesis, The State Thechnical University of Marii, Russian Federation, in Russian.

Wright, H., (1996). *The Acoustics and psychoacoustics of the guitar.* PhD Thesis, University of Wales, 1996.

ACKNOWLEDGEMENT

This paper is supported by Romanian Ministry of Education and Research under Research of Excellence Program –Contract PNII 71-016/2007. This work was accomplished under the tehnical support of Transilvania University of Brasov, Romania.

Annals of DAAAM for 2008 & Proceedings of the 19th International DAAAM Symposium, ISSN 1726-9679
Editor B. Katalinic, Published by DAAAM International, Vienna, Austria 2008
Make Harmony Between Technology and Nature, and Your Mind will Fly Free as a Bird

CNC END MILLING OPTIMIZATION USING EVOLUTIONARY COMPUTATION

CUS, F[ranc]; ZUPERL, U[ros] & GECEVSKA, V[alentina]

Abstract: *In this paper, a new evolutionary computation technique, particle swarm optimization (PSO), is proposed and implemented to efficiently and robustly optimize multiple machining parameters simultaneously for the case of milling. An artificial neural networks (ANN) predictive model for critical process parameters is used to predict the cutting forces which in turn are used by the PSO developed algorithm to optimize the cutting conditions subject to a comprehensive set of constraints. Next, the algorithm is used to optimize both feed and speed for pocket-milling. Machining time reductions of up to 41% are observed. In addition, the new technique is found to be efficient and robust.*
Key words: *Machining, multi-objective optimization, PSO optimization.*

1. INTRODUCTION

NC programs generated today, experience a large variation in cutting forces due to non-uniformity in metal removal along the cutter path. This may be due to a variety of factors, surface nature (curvature), tool inclination, cornering etc. In order to increase productivity, process parameters should be assigned according to the NC tool path in addition to the conditions of the part, tools, setup, and the machine. The idea is to change these variables according to the current in-process part geometry and tool path so that the cutting force is in control.

Most optimization studies state one of two objectives: Minimum manufacturing cost (Cus & Balic 2000), maximum production rate (Milfelner et al., 2004).

It has also been realized that a combination of the minimum production cost and minimum production time (Cus & Balic 2000), is the most effective objective since neglecting either requirement alone does not do justice to the problem at hand. There are a variety of constraints that have been considered applicable by many researchers for different machining situations: 1. available feed and speeds (machine tool related), power, arbor rigidity, and arbor deflection. 2. Maximum available machine power and maximum permitted cutting edge load for roughing, and allowed maximum tool deflection for finishing (Liu & Wang 1999).

Cutting force is found to be one of the most important process parameters used as a constraint in the cutting operation, as it relates to a large number of abnormal occurrences such as tool breakage and excess tool wear as well as basic data for estimation of chatter vibration and machining error . In this paper, a PSO optimization method is proposed to obtain the optimal parameters in milling processes (Figure 1).

2. PARTICLE SWARM OPTIMIZATION

Particle Swarm Optimization (PSO) is a relatively new technique, for optimization of continuous non-linear functions. It was first presented by (Shi & Eberhart 1998).

PSO is a very simple concept, and paradigms are implemented in a few lines of computer code. It requires only primitive mathematical operators, so is computationally inexpensive in terms of both memory requirements and speed. PSO has been recognized as an evolutionary computation technique and has features of both genetic algorithms (GA) and evolution strategies (ES). Other evolutionary computation (EC) techniques such as genetic algorithm also utilize some searching points in the solution space. It is similar to a GA in that the system is initialized with a population of random solutions.

While GA can handle combinatorial optimization problems, PSO can handle continuous optimization problems. However, unlike a GA each population individual is also assigned a randomized velocity, in effect, flying them through the solution hyperspace. PSO has been expanded to handle also the combinatorial optimization problems. As is obvious, it is possible to simultaneously search for an optimum solution in multiple dimensions. Unlike other EC techniques, PSO can be realized with only small program. Natural creatures sometimes behave as a swarm. One of the main goals of artificial life researches is to examine how natural creatures behave as a swarm and reconfigure the swarm models inside a computer.

PSO has two simple concepts. Swarm behaviour can be modelled with a few simple rules. Even if the behaviour rules of each individual (particle) are simple, the behaviour of the swarm can be very complex. The behaviour of each agent inside the swarm can be modelled with simple vectors. This characteristic is the basic concept of PSO.

The applications of PSO are: Neural network learning algorithms (Boyd 2003), Rule extraction in fuzzy neural networks (He et al., 1998), computer controlled milling optimization , power and voltage control. Application of PSO to other fields is at the early stage. More applications can be expected. Most of papers are related to the method itself, and its modification and comparison with other EC methods (Eberhart & Shi 2003).

3. PSO ALGORITHM

is developed through simulation of bird flocking in two-dimension space. The position of each agent is represented by XY axis position and also the velocity is expressed by vx (the velocity of X axis) and vy (the velocity of Y axis). Modification of the agent position is realized by the position and velocity information.

Bird flocking optimizes a certain objective function. Each agent knows its best value so far (pbest) and its XY position. This information is analogy of personal experiences of each agent. Further, each agent knows the best value so far in the group (gbest) among (pbests).

This information is analogy of knowledge of how the other agents around them have performed. Each agent tries to modify its position using the following information: - the current positions (x, y), - the current velocities (vx, vy), - the distance between the current position and (pbest) - the distance between the current position and (gbest).

This modification can be represented by the concept of velocity. Figure 2 shows the general flow chart of PSO strategy.

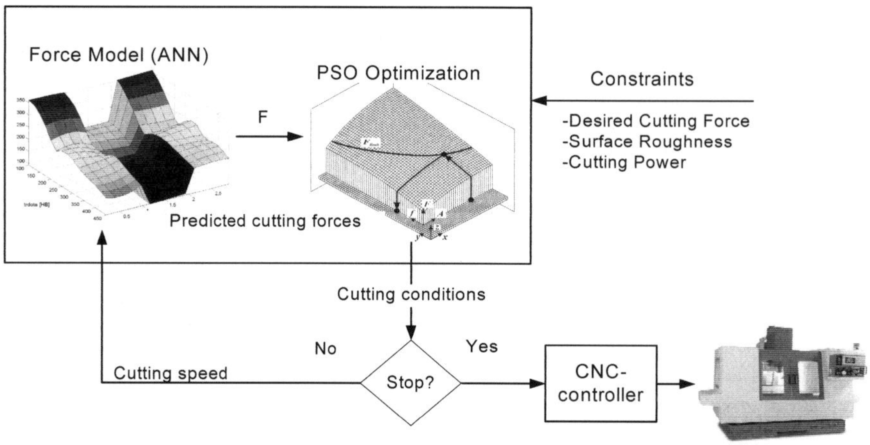

Fig. 1. The scheme of the proposed milling optimization approach.

4. MACHINING OPTIMIZATION PROBLEM

The optimization process executes in two phases. In first phase, the neural prediction model on the basis of recommended cutting conditions generates 3D surface of cutting forces, which represent the feasible solution space for the PSO algorithm. PSO algorithm generates a swarm of particles on the cutting force surface during the second phase. Swarm of particles flys over the cutting force surface and search for maximal cutting force. The coordinates of a particle which has found the maximal (but still allowable) cutting force represent the optimal cutting conditions.

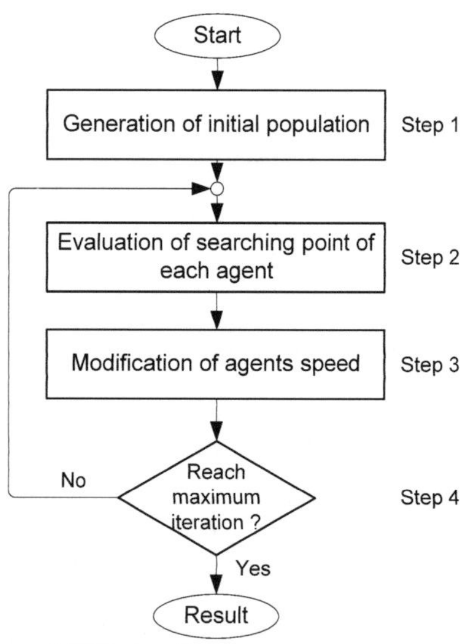

Fig. 2. A general PSO algorithm.

The optimization process is depicted by the following steps:

1. Generation and initialization of an array of 50 particles with random positions and velocities. Velocity vector has 2 dimensions, feed rate and spindle speed. This constitutes Generation 0.
2. Evaluation of objective (cutting force surface) function for each particle.
3. The cutting force values are calculated for new positions of each particle. If a better position is achieved by particle, the pbest value is replaced by the current value.
4. Determination if the particle has found the maximal force in the population. If the new gbest value is better than previous gbest value, the gbest value is replaced by the current gbest value and stored. The result of optimization is vector gbest (feedrate, spindle speed).
5. Computation of particles' new velocity
6. Update particle's position by moving towards maximal cutting force.
7. Steps 1 and 2 are repeated until the iteration number reaches a predetermined iteration

5. CONCLUSION

This work has presented a new approach to optimizing the cutting conditions in end milling subject to a near to comprehensive set of constraints. Next, a production cost objective function was used to define the parameter to optimize. An algorithm for PSO was then developed and used to robustly and efficiently find the optimum cutting conditions. Both feed and speed were considered during optimization. The new technique has several advantages and benefits and is suitable for use with ANN based models where no explicit relation between inputs and outputs is available. The research described in this paper will lead to the development of new intelligent optimization software. Next step will be implementation of on-line optimization of cutting parameters during machining.

6. REFERENCES

Cus, F. & Balic, J. (2000). Selection of cutting conditions and tool flow in flexible manufacturing system. *The international journal for manufacturing science & technology*, Vol. 2, pp. 101-106

Boyd, J. (2003). Thinking is Social: Experiments with the Adaptive Culture Model. *Journal of Conflict Resolution*, Vol. 42, pp. 56-76

Eberhart, R.C. & Shi Y. (2003). Comparison Between Genetic Algorithm and Particle Swarm Optimization, *Proceedings of the 7th ICEC*, pp. 611-616

He, Z.C.; Wei, L.; Yang, X.; Gao, S; Yao, R & Eberhart, (1998). Extracting Rules from Fuzzy Neural Network by Particle Swarm Optimization, *Proc. of IEEE International Conference on Evolutionary Computation (ICEC'98)*, pp. 66-71

Liu, Y. & Wang, C. (1999). Neural Network based Adaptive Control and Optimisation in the Milling Process. *International Journal of Advanced Manufacturing Technology*, Vol. 15, pp. 791-795

Milfelner, M.; Zuperl, U. & Cus, F. (2004). Optimisation of cutting parameters in high speed milling process by GA. *Int. j. simul. model.*, Vol. 3, pp. 121-131

Shi Y., & R. Eberhart (1998). Parameter selection in particle swarm optimization. *In Evolutionary Programming VII: Proc. EP98, New York: Springer-Verlag*, pp. 591-600.

Annals of DAAAM for 2008 & Proceedings of the 19th International DAAAM Symposium, ISSN 1726-9679
Editor B. Katalinic, Published by DAAAM International, Vienna, Austria 2008
Make Harmony Between Technology and Nature, and Your Mind will Fly Free as a Bird

COMPLEXITY THEORY IN THE CONTROL OF ASSEMBLY SYSTEM

CYKLIS, J[erzy]; SLOTA, A[dam] & ZAJAC, J[erzy]

Abstract: *New concepts of manufacturing systems are based on distributed, co-operative manufacturing control principle, so these systems can take advantage of the ideas of complexity theory. The paper presents the concept of assembly control system wich limits the application of complexity theory to the effect of collaboration of system components with the similar type of behavior. Object Observable Petri Net model is used for concept presentation.*
Key words: *complexity, assembly systems, distributed control*

1. INTRODUCTION

The basic concept of complexity is that all things tend to organize themselves into patterns, e.g., ant colonies, immune systems, and human cultures; further, they go through cycles of growth, mass extinction, regeneration, and evolution (The Columbia Encyclopedia, 2008). This concept is very promising in modern science and technology. Complexity theory can be also applied to address manufacturing control challenges. Manufacturing engineers are beginning to explore distributed architectures of relatively autonomous agents whose behaviors emerge as the system runs rather than being defined in advance and then rigidly executed (Parunak, 1995). This concerns most of the new generation manufacturing systems concepts like: holonic (Van Brussel et al., 1998), multi-agent (Shen & Norrie, 1998; Bussmann & Schild, 2000; Zajac&Chwajol, 2008) or reconfigurable (Koren et al., 1999), which are based on distributed co-operative manufacturing control principle.

The idea of Complex Manufacturing System (CMS) assumes that manufacturing system is an effect of collaboration of independent resources where each of them is represented by the same configurable elementary module. The control system of CMS should be based on integration of unified, easy to re-configure, intelligent and cooperative agents. Moreover, it ought to be dynamically configurable and reconfigurable in consideration of new products, introduction of structural changes in manufacturing equipment or even of unexpected breakdowns.

An abstract description of a manufacturing system property, which is called a model, is required to build a manufacturing control system. Model consists of finite set of symbols and mathematical relations that determine formal representation of the real manufacturing system. The most popular modeling tools are Petri Nets (PN) (Dicesare et al., 1993; MengChu Zhou & Zurawski R., 1995) and Unified Modeling Language (UML) (Booch et al., 1999). Modeling of manufacturing systems must take into consideration the applied manufacturing control concept. Petri Nets, as a graphical and mathematical tool, provide a uniform environment for modeling, formal analysis, and design of discrete event systems. Coloured Object Observable Petri Net (COPN) is suitable for modeling distributed manufacturing system, in particular CMS.

The introduction of new products usually requires to perform new manufacturing or assembly operations. Assembly systems are very common in a contemporary discrete manufacturing production processes. The following chapter shows that such systems can be treated as Complex Manufacturing System.

2. COPN MODEL OF ASSEMBLY SYSTEM

Assembly system consists of a set of assembly stations, storages, input/output buffers and transport devices. Components of an assembly system mentioned above are referred to as elementary modules. Elements which are components of products assembled in the system are referred to as elements/units. Each elementary module is represented by its model identified by a unique identifier, e.g. by its name. Such a module's model is a universal configurable entity. Its operation and interactions with other modules are defined by simple rules. Operation of the module is defined by COPN model. Graphical representation of a universal model of an elementary module built with the use of COPN is presented in figure 1.

Elementary module works in a cyclic way. The cycle consists of: (i) receiving a set of elements/units from proceeding module, this activity is represented in the model by *input transition* t_{IN}; (ii) processing a set of elements/units (e.g. assembly operation) represented in the model by *process transition* t_P marked by dashed line; (iii) transfer elements/units to the subsequent module, this activity is represented in the model by *output transition* t_{OUT}. States of elements/units are marked in the figure 1 by grey circles. States E_1, E_2, ..., E_n (marked by dashed circles) proceeding *input transition* t_{IN} are included in the model only if the module is the source of elements/units in the system. State U (marked by dashed circles) succeeding *output transition* t_{OUT} is included in the model only if the module is the final place of units/products in the system. As a result of processing activity for assembly station a set of elements/units is transformed into a unit/product. In this case the module's COPN model contains equivalence relation between a unit/product and its components – a set of elements/units (Cyklis et al., 2005).

Module is integrated with other modules which precede and succeed it according to the technological route of products. Input and output transitions are natural interfaces of the module. Thus modules' integration is achieved by definition pairs of transitions: an output transition in the module transferring elements/units and an input transition in the module receiving the same elements/units.

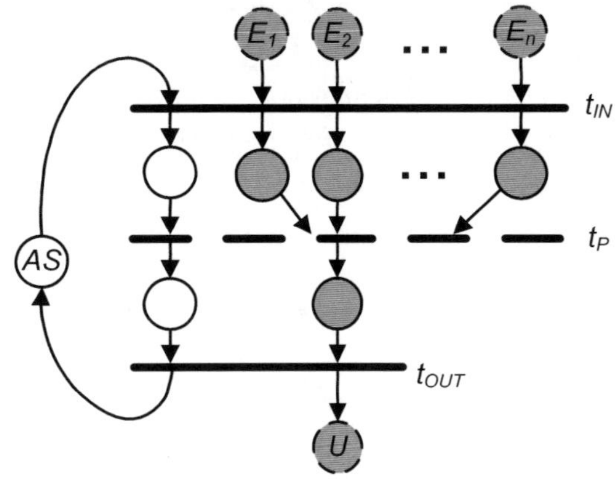

Fig. 1. COPN model of an elementary module.

The pairs of transitions are unique in the whole model – every input transition matches exactly one output transition in all other modules and every output transition matches exactly one input transition in all other modules.In figure 2 pairs of input - output tranisitions are marked by matching patterns.Formal definition and firing conditions of such pairs of transitions are presented in (Slota, 2006).

Module's model described above is a basic building block for the model of the whole system. It allows to model operation of basic modules of assembly system. For some modules, e.g. buffers, *process transition* t_P is not defined. In some cases receiving elements/units by a module can be modelled by a set of *input transitions* t_{IN} (activities) instead of a single *input transition*. In some cases of *process transition* t_P hierarchical model can be used (*process transition* t_P is unfolded into a subnet which defines assembly activity).

Structure of a simplified model based on complexity theory of assembly system for assembling a reduction gear is presented in figure 2. Elementary modules *Buffer1* and *Buffer2* store elements for input and output shaft units respectively (shafts, gear wheels, keys), elementary module *Buffer3* stores housing parts, bolts and bearings, elementary module *Buffer4* stores final product.

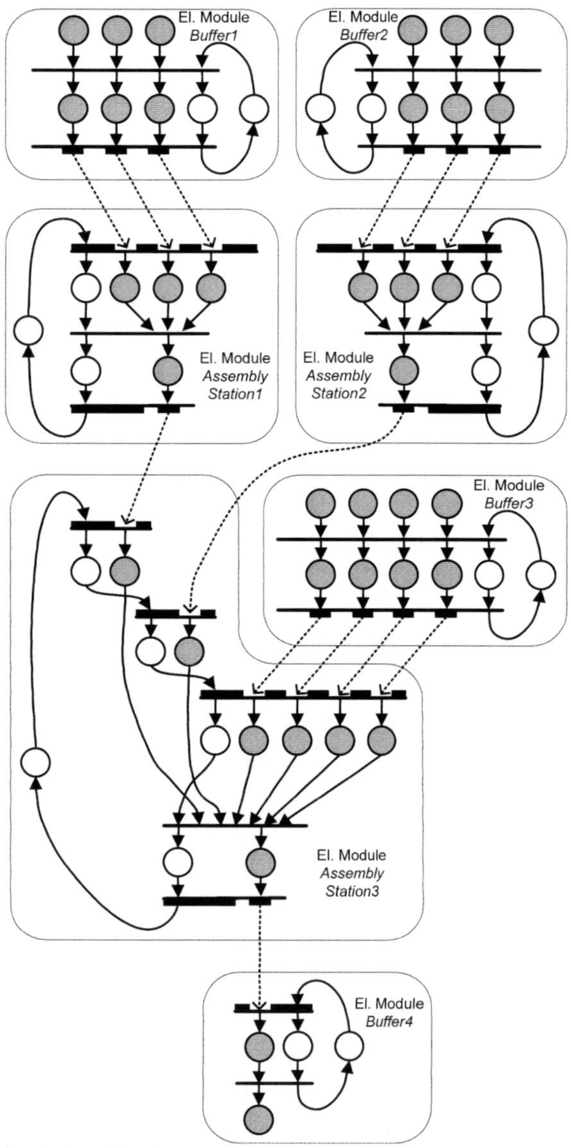

Fig. 2. Model of an assembly system based on complexity theory.

Elementary module *AssemblyStation1* assembles input shaft unit, elementary module *AssemblyStation2* assembles output shaft unit, elementary module *AssemblyStation3* assembles input and output shaft units and housing into the final product. Flow of elements/units between modules is marked by dotted line arrows. For the legibility purpose models of transport modules are not included in the model.

3. CONCLUSIONS

Factories are *de facto* complex systems (Parunak, 1995). Dynamical development of information technology allows to implement new theoretical ideas to the industrial environment. The presented example of building assembly control system shows only some of the aspects of this complicated task. The concept is much more general. It allows to apply hierarchy inside the elementary module that allows to perform a sequence of assembly operation on the same station, and moreover, flexibility that allows to add or modify assembly station. The presented approach can be also applied for the kanban technique in the control of assembly system.

4. REFERENCES

Booch, G.; Rumbaugh, J. & Jacobson, I. (1999). *The Unified Modeling Language User Guide*. Addison Wesley Longman.

Bussmann S. & Schild K.: Self-Organizing Manufacturing Control: An Industrial Application of Agent Technology. *Proc. of the Fourth Int. Conf. on Multi-Agent Systems*, Boston 2000, pp. 87-94.

Columbia Encyclopedia , Sixth Edition (2008). *Available from:* http://www.encyclopedia.com/topic/complexity.aspx *Accessed:* 2008-07-12.

Cyklis J.; Zajac J. & Slota A. (2005). Description of Flexible Assembly systems by means of Coloured Object Observable Petri Net. *Annals of DAAAM 2005 & Proceedings of the 16th International DAAAM Symposium*, Opatia, pp. 83-84.

Dicesare, F.; Harhalakis, G.; Proth, J.M.; Silva, M. & Vernadat, F.B. (1993). *Practice of Petri Nets in Manufacturing*. Chapman & Hall, London.

Koren, Y.; Heisel, U.; Jovane, F.; Moriwaki, T.; Pritchow, G.; Van Brussel, H. & Ulsoy, A.G. (1999). Reconfigurable Manufacturing Systems, *CIRP Annals*, Vol. 48, No. 2, (keynote paper).

MengChu Zhou & Zurawski R. (1995): Introduction to Petri Nets in flexible and agile automation. Petri nets in flexible and agile automation, Kluwer Academic Publishers, Boston/Dordrecht/London 1995, pp. 1-42.

Parunak, H.V.D. (1995). Complexity Theory in Manufacturing Engineering: Conceptual Roles and Research Opportunities, *Industrial Technology Institute Tech memo* 93-01, Ann Arbor, pp. 1-7.

Shen, W. & Nordie. D.H. (1998). Agent-Based Systems for Intelligent Manufacturing: a State-of-the-Art Survey. *Knowledge and Information Systems*. pp. 129-156.

Slota A. (2006). An environment for modular modeling and simulation of FMS. *International Scientific Conference MicroCad, Section M: Production Engineering and Manufacturing Systems*. Miskolc, pp. 133-138.

Van Brussel, H.; Wyns, J.; Valckenaers, P.; Bongaerts, L. & Peeters, P. (1998). Reference Architecture for Holonic Manufacturing Systems: PROSA. *Computers In Industry*, Special Issue on Intelligent Manufacturing Systems, Vol. 37 No. 3, 1998, pp. 255-276.

Zajac, J. & Chwajol, G. (2008). Towards Agent-Based Manufacturing Systems. *Annals of DAAAM for 2008 & Proc. of the 19th International DAAAM Symposium*, Trnava.

Annals of DAAAM for 2008 & Proceedings of the 19th International DAAAM Symposium, ISSN 1726-9679
Editor B. Katalinic, Published by DAAAM International, Vienna, Austria 2008
Make Harmony Between Technology and Nature, and Your Mind will Fly Free as a Bird
*Young Researcher and Scientist Paper / * Supervisor, Mentor*

SUCCESS FACTORS AND CRITERIA OF INTERNATIONAL SPORT EVENT PROJECTS

DANCSECZ, G[abriella] & SZABO, L[ajos]*

Abstract: *In recent years more and more attention has been devoted to international sport competitions. The positive effects of a successful event can be felt not only by the organizing sport association but also by the organizing towns, regions or even countries. The purpose of the study is to define success factors and criteria of international sport events as well as to analyze the intensity of relationships between the factors and criteria. We apply empirical analysis: questionnaire survey and statistical analyses as well as secondary research to validate the research model. The findings of the study show that not only technical factors play important role in organizing successful international sport events but also soft factors are essential.*
Key words: *Project Management, Sport Event Projects, Project Success Factors and Criteria, Task and Relation Orientation*

1. INTRODUCTION

National sport associations of West and Central Europe play more and more significant and year-by-year extending roles in organizing international sport events. The reason for this is, on one hand, providing financial background for sport organizations, on the other hand, the propagation of sports and national athletes. (Nyerges & Petróczi, 2007) In addition to this international sport competitions are more and more in the focus of interest each year, their preparation and implementation are constantly followed by media coverage and news reports.

2. THEORETICAL BACKGROUND

It is a wide-ranging and extremely complicated task to organize such large sport events, to manage the activities and to co-ordinate the participants' work. Therefore, nowadays it is ineluctable to apply the methods of project management and event organization for the sake of organizing successful and efficient events. Studying the success of sport event projects we need to distinguish the concept of project success factors and that of success criterion. Project success factors are facts and circumstances that influence the accomplishment of a project, whereas project success criteria are principles and requirements on the basis of which project results can be evaluated (Lim & Mohamed, 1999)

2.1 Project success criteria

Concerning the results of sport competitions arises the question when to consider a project successful. Among several approaches mentioned in literature towards project success, the most commonly accepted conception is the so-called project triangle, according to which a project can be regarded successful if it is carried out by an appointed deadline and under cost constraints, fulfilling specified quality parameters. (Turner & Cochrane, 1993) However, this approach does not satisfactorily evaluates project success, since examining the fulfilment of deadline, cost constraints and quality parameters indicates merely the direct contribution to profit, whereas it omits taking into account if the project was implemented properly, in the right way. (Kerzner, 2006) Theoretical studies and empirical researches of success criteria reveal that the analysis of project success is an extremely complicated task even if studied from diverse view-points and by different dimensions. Among the aspects of evaluation we can find quantitative, objectively measurable criteria, like cost and time constraints; and there are qualitative criteria rather measured in indirect, subjective ways, just as stakeholders' satisfaction or long term effects.

2.2 Project success factors

When analysing project success factors, two groups of critical factors can be defined, which date back to the task orientated and relation orientated categories established within the frame of researches of personal management at Ohio and Michigan Universities. Task orientation focuses on objectives, achievements and activity of staff, whereas relation orientation focuses on interest towards staff, teamwork of colleagues as well as on the relationship between different departments, customers and clients. (Blake & McCanse, 1992) This twofold objective also emerges in the area of project management, during the implementation of projects. Task orientation, in this respect, consists of business activities, systems and processes applied by the organization in order to accomplish the project, whereas relation orientation focuses on individual members of the organization, groups involved and communication. (Kendra & Taplin, 2004)

3. THE RESEARCH MODEL

The research model, elaborated based on analysis of professional literature and practical experience, presents success factors and criteria of sport event projects as well as relationships potentially existing between the elements of the model. In the model, factors contributing to the success of sport event projects are split into three main parts: task orientation, relation orientation and random impacts. The partition into task and relation orientation is based on the professional analysis, whereas the category of random impacts was formed during the analysis of concrete events, as we can deduce from experience that unexpected incidents (e.g. extremes of weather, riots, etc.) can impose significant influence on the efficiency of events. These three factors together affect the success of sport events, which we can measure on the basis of success criteria, that is, by analysing the implementation of objectives and the satisfaction of stakeholders.

4. APPLIED METHODOLOGY

We apply empirical analysis and secondary research to validate the elaborated research model. In the course of the empirical survey we made personal interviews with the organizers of sporting events in order to learn about their experience. In addition to this, we gathered information from organizers of European and World Championships, held between 2000 and 2008 in selected European countries, using questionnaires. In the present phase of the research, the questionnaire data collection, the interviews as well as the analysis of the documentation of selected sport events have been completed. In the course of statistical analyses so far we have done sorting of the model's indicators, applying Factor Analysis. The next steps are going to be studying of correlation between elements of the model with Regression Analysis.

5. RESULTS OF THE STUDY

5.1 Results of the Factor Analysis

As a result of a questionnaire data collection we have a basis of 102 questionnaires to make Factor Analysis supported by the SPSS program. In the sample organizers of European and World Championships organized in Germany, Austria, Switzerland, Hungary, Poland, Czech Republic, Slovakia and Slovenia evaluated the experiences of their latest sport events. In the course of statistical analysis we grouped both success factors and success criteria in order to create latent variables, which describe the behaviour of each variable in a suitable way. Six components can be separated during the Factor Analysis of **success factors**. The first factor is the project objectives. It indicates the detailed development of project aims, tasks and project plans. The next factor, contract strategy, includes the financial conditions of sub-contracts as well as the responsibility sharing among organizers and contractors. The third factor is project leadership, which contains variables connected to the competence, commitment, scope and responsibilities of project leader and project team. The fourth factor, organizational culture, includes cultural characteristics of the organizing committee, such as information sharing and communication within the committee, support of teamwork and organizational learning. The fifth factor is co-operation and communication. It means the co-operation and communication with direct stakeholders, such as contractors, sponsors and international and national associations. The sixth factor, partnership, contains partnership and conversation with indirect stakeholders, such as national and local authorities, the city and its inhabitants. Due to the content of the factors it is reasonable to assume, that not only hard, task oriented factors are essential in organizing successful international sport events but also soft, relation oriented factors are determining. In the area of **success criteria** four factors can be separated according to the analysis. The fulfilment of primary objectives factor indicates the adherence of budget as well as the satisfaction of association as client, and of project team. The next factor, fulfilment of further aims, includes the number of athletes and spectators as well as the boost of local economy and popularization of sport and home athletes. These two factors describe the implementation of project objectives, whereas the other two factors, direct and indirect stakeholders' satisfaction, include the content of these stakeholders as well as their approach to long term partnerships.

Fig. 1. The research model

5.2 Results of Interviews and Document Analysis

In the course of study supposed correlation between elements of the research model personal interviews and document analysis had been done so far. As far as **task orientation** it is essential to find appropriate suppliers, to co-operate with them effectively, to keep everything under permanent control in favour of providing adequate resources. This is what caused some problems in case of several events, especially because of misunderstandings and delays resulting from the lack of permanent control and discussions. Furthermore the proper defining of tasks, responsibilities and competences is crucial, as based on organizers' experience, failing this may result in a considerable overlap or the lack of responsibles may lead to the failure of implementation. In the scope of **relation orientation**, one of the important factors is employing participants who are proficient and competent in their specialties, can work effectively in a project team, and maximally contribute to the realization of the sport event. The essential factor of co-operations are establishing a good relationship with the relevant international association, gaining its support, as well as sharing a national and local commitment, which can strongly contribute to a successful event in case there is a positive attitude, however, in case of a negative attitude it may put difficulties in the way of carrying out the event, for instance in the way of constructions or renovations of buildings. The analysis of **random impacts** indicates interesting results, where extremes of weather and other accidents (strikes, riots, etc.) may exert a huge effect, as the example of European Swimming Championship 2006 in Hungary showed, where a mounsunlike storm interrupted the program and the broadcasting and timing equipment got soaked.

6. CONCLUSION

The findings of the factor analysis and the results of interviews and document analysis show that not only hard, technical factors play important role in organizing successful international sport events but also soft factors are essential. Regarding the evaluation of success of international sport events beside the implementation of project objectives the satisfaction of project stakeholders is also an important criteria. The next steps of the research will be the preparation and implementation of regression analysis to study correlation between elements of the research model. The statistical analysis of questionnaires and the research results based on experiences gained from case studies reveal the factors that influence the successful organization and implementation of sport events.

7. REFERENCES

Blake, R. R. & McCanse, A. A. (1992). *Das GRID-Führungsmodell (Leadership Dilemmas – Grid Solutions),* Econ Verlag, ISBN 3430113946, Düsseldorf

Kendra, K. & Taplin, L. J. (2004). Project Success: A Cultural Framework, *Project Management Journal,* April 2004, pp. 30-45, ISSN 87569728

Kerzner, H. (2006*). Project Management, A Systems Approach to Planning, Scheduling and Controlling,* Wiley & Sons, ISBN 0471741876, Hoboken

Lim, C. S. & Mohamed, M. Z. (1999). Criteria of project success: an exploratory re-examination, *International Journal of Project Management,* 17(4), pp. 243-248, ISSN 02637863

Nyerges, M. – Petróczi, A. (2007). *A sportmenedzsment alapjai (The basics of sport management),* Semmelweis Egyetem (Semmelweis University), ISBN 9637166432, Budapest

Turner, J. R. & Cochrane, R. A. (1993). Coals-and-Methods Matrix: Ill Defined Goals and/or Methods of Achieving Them, *International Journal of Project Management,* 11(2), pp. 93-102, ISSN 02637863

Annals of DAAAM for 2008 & Proceedings of the 19th International DAAAM Symposium, ISSN 1726-9679
Editor B. Katalinic, Published by DAAAM International, Vienna, Austria 2008
Make Harmony Between Technology and Nature, and Your Mind will Fly Free as a Bird

DESING OF ADDITIONAL CHECK STATION WITH INTELLIGENT CAMERA SYSTEM

DANISOVA, N[ina]; ZVOLENSKY, R[adovan] & VELISEK K[arol]

Abstract: *In this contribution is presented a complex design of intelligent camera system placed on intelligent manufacturing cell. Intelligent manufacturing cell is situated at Institute of manufacturing systems and applied mechanics. Complex design of check station with camera system is going out of intelligent manufacturing systems knowledge. Intelligent manufacturing systems as systems of new generation are gradually loaded in to the mechanical production, when they are removing human operation out of production process and they also short production times.*
Key words: *design, system, camera, cell, production*

1. INTRODUCTION

In these days is industrial manufacture still in progress. Industry is using except computer techniques and classical automation devices also flexible manufacturing frameworks. It is also effective to speak about possibility which will take manufacturing systems in the area of so called intelligent manufacturing systems. All subsystems of intelligent manufacturing systems are including machine intelligence elements. Using of such intelligent systems will lead to the labour elimination from the manufacturing system.

Basic in these kinds of intelligence in the manufacturing systems is so called monitoring, which allows scanning of own system stay and it also scan changing surrounding conditions.

Monitoring devices are using sensors which are places on the right places (machine frame, machine, tool, manipulating device), are scanning parameters, which are used are input data coming in the control system. Following to these data, can be realized some technological, manipulating of some helping process.

2. VISUAL SYSTEMS

One of many important changes placed in the automation research and mainly placed in the research of robotized systems of higher generations with higher adaptability and intelligence is using of visual systems. Sensors are used in the visualisation process. Sensors are realizing some functions of man senses such as vision or touch.

In real designs of manufacturing systems can be visualization mainly used for:

- objects sorting following to their typical marks
- stay monitoring of machines, tools and other devices
- automated industrial vehicle navigation
- automated control of surfaces, edges and dimensions

Visual systems working principle:

First working principle used in visual system working is scanning of optical information and ist modification for using with other electronic cameras. Picture frequency is nearly same as frequency used by ordinary televisions. In other one block is analog signal changed to the numerical signal. Format is organizing by controller and then it is sent in to the special memory placed in the computer. Digitalized picture, which is placed in the memory, is modificated following to the type and goals of various tasks. Intelligent and camera systems are mostly used in the visual systems (Liška 2002).

3. ADDITIONAL CHECK SENSORICAL STATION

Flexible manufacturing cell is placed at the department of manufacturing devices and systems.

Flexible manufacturing cell is realized from these subsystems.

1. **Industrial robot with Cartesian workplace**
2. **Shelf storage system**

Five working phases is integrated in the flexible manufacturing cell.

1. Storegeing (storageing of semi products, final products before its expedition)
2. Transport and manipulation (transport and manipulation with semi products and final products)
3. Manufacture (manufacturing of single semi products to the final product)
4. Assembly (assembly of single parts in to the one final product)
5. Expedition

Additional check station was designed, which will check all parts placed on the system pallet. Parts will be checked by rotary feeding unit placed on the shelf storage, which is putting pallets in to the working place of Cartesian robot (Velíšek, Javorová, Zvolenský, Danišová & Košťál 2007).

Additional check station if verifying, if the part which is coming in to the working place is really that one which was needed. One of designed solutions, which is proper for solving of this task, is using of proper camera system. Used camera is the camera of **DMK 21F04** type. (Fig.1).

Fig.1. Industrial camera DMK 21F04

Used camera is delivered without object glass. Each object glass is elected following to the working conditions, such as dimension of scanning chip, working distance between camera and scanned object, dimensions of scanned object. Leading parameter used for proper object glass designation is focal distance, which can be calculated following to the mentioned parameters.To reach full visual display of whole scanned object on the scanning chip, we have to choose focal distance which has smaller value as calculated focal distance. For camera of type DMK 21F04 and other given conditins is proper using of object glass with focal distance which has smaller value as 2,8mm. These conditions are correct for object glass of T 2314 FICS type.

Following to the more complicated way of camera control and parts database transcription by any new income of part in the working area of flexible manufacturing cell, is better when the camera will be controlled by main control computer. Designed camera is proper also for this condition (Whelan 2000).

By scanning with industrial camera is very important to ensure that all parts placed on the system pallet have to be placed in the same position. Other one condition is to ensure the same lights conditions. This can be realized by additional light, which is an optional part of industrial camera system, or this can be realized by external light, which need proper light inbound to the whight place.

After picture scanning is this picture send to the control computer. Picture will be worked and evaluated by help of algorithms. (Madarász 2006).

4. INDUSTRIAL CAMERA PLACEMENT

Placement of industrial camera was designed on the frame of industrial robot. Camera will be placed on the front side, above to the input rotary table (Fig.2).

Camera will be bound to the metal holder. System pallet with scanned object will come to the viewing zone of industrial camera every time as the palled come in to the system.

Holding to the metal holder of industrial camera is at the (Fig.3). At the (Fig.4) are parameters used for object glass selection and at the (Fig.5) is Object glass T 2314 FICS-3 made by firm Computar

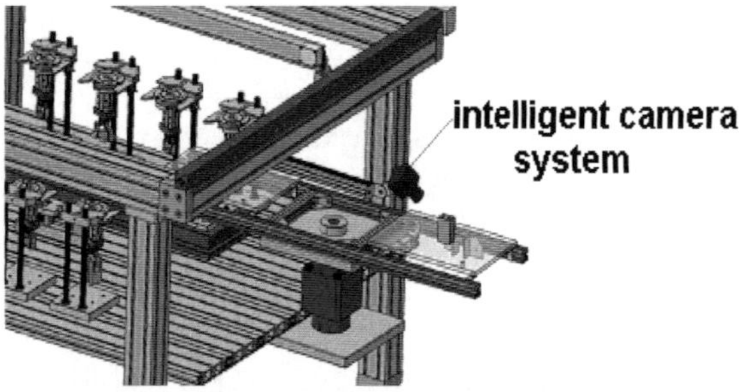

Fig.2 Camera placement on the industrial robot arm

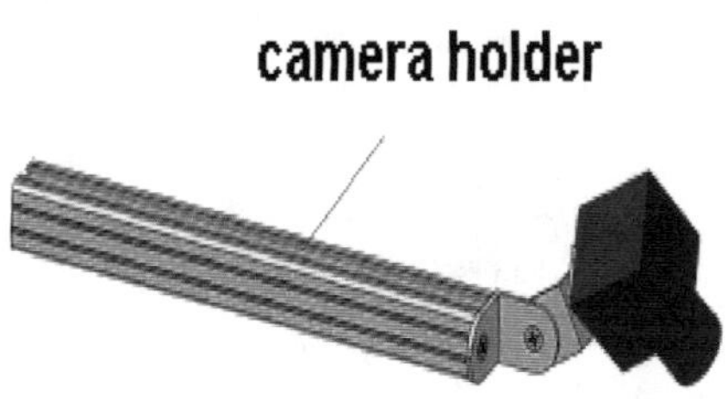

Fig.3 Camera holding on the metal holder

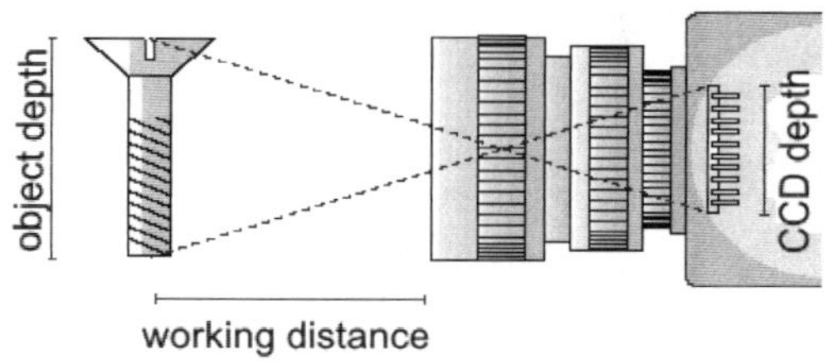

Fig.4 Parameters used for object glass selection

Fig.5 Object glass T 2314 FICS-3 made by firm Computar

Camera will be place on the stationary, solid place, where also proper light conditions can be realized. Also proper scanning background has to be realized. Control of whole camera system will be realized by control program implemented in to the control computer of whole flexible manufacturing cell (Košťál & Velíšek 2004).

5. CONCLUSION

Intelligent manufacturing systems can be considered as higher development degree of flexible manufacturing systems.

Each IVS is also composed of single subsystems such as technological, manipulating, transport and storageing subsystem. Each of these subsystems has to be equipped by elements of machine intelligence, which are giving some degree of intelligence to these subsystems.

This article was realized with help of grand tasks.
VEGA 1/3193/06 Multifunctional manufacturing and assembly cell.

6. REFERENCES

Košťál, P. & Velíšek, K. (2004). Flexible assembly systems, *Proceedings of Comatech 2004*,pp. 201- 206, ISBN 83-920972-4-6, Protection of Mechanical Systems in the Mining Industry Against the impact of high Energy conference location, October 2004, ZINT, Ustron

Liška, O. (2002). Funkcie a vývojové znaky monitorizačných systémov. *Transfér inovácií,* No5., (2002)

Madarász, L. (2006).*Intelligent technologys and applications in complex systems,* VTI SR, Bratislava, Košice

Velíšek, K.; Javorová, A.; Zvolenský, R.; Danišová, N & Košťál, P. (2007). Multifunctional manufacturing and assembly system, In: *Development of Mechanical Engineering as a Tool for Enterprise Logistics Progress:Science Report Project CII-PL-0033-01-0506,*

Whelan, P.(2000).*Vision Systems Laboratory*, Dublin city University, Dublin University of technologi, pp.(135-158), ISBN 83-89873-28-1, Poznaň

Annals of DAAAM for 2008 & Proceedings of the 19th International DAAAM Symposium, ISSN 1726-9679
Editor B. Katalinic, Published by DAAAM International, Vienna, Austria 2008
Make Harmony Between Technology and Nature, and Your Mind will Fly Free as a Bird

CREDITING STRATEGY FOR SMALL AND MEDIUM COMPANIES IN ROMANIA IN 2007

DAVID, O[ana]; OSICEANU, S[anda] & MIHAI, V[alia] M[aria]

Abstract: *This paper takes a trip into the activity of crediting the small and medium companies and will present in a nutshell several types of credits granted to small and medium companies by the Romanian banks. We also intend to display some trade practices of the banks and other suppliers of financial services that are considered unfair towards the consumers, population or companies, including the ones in the technical field.*

Key words: *small and medium companies, credit, cash-flow, interest.*

1. INTRODUCTION

One of the main activities of the commercial banks is the crediting of population and companies of either field, including the technical one. The analysis of such banking exercise has been performed since crediting entered economy, in order to make an assessment of this activity's evolution and profitableness.

In the field there are numerous statistics drafted for semesters and years, as well as for different credit categories for population and for small and medium companies (SMC).

This paper presents a statistics of the credit granted to the SMCs in Romania, in 2007 and attempts to show the situation of the Romanian banking market. In conclusion, the authors have tried to display the disadvantages on the Romanian banking market in the activity of crediting of both the population and companies, including the ones in the technical field.

2. CONTAIN PAPER

The credit relations generate the crediting as a method of financing the resources need. It is a modality of constituting the financial resources need as a completion of their own – but, most of the time, the credit represents the only financing source for a business. (Stroe & Armeanu, 2001) As still a sole financing source is the credit for formation of the abilities in order to be able to cover the financial needs during the year peak periods.

The essential requirement for calling and utilization of crediting, as means of completion your own financial resources, is the efficiency of using these agents, so that, upon the reimbursement time, all the loan amounts be returned and the relevant interest be paid accordingly.

The wide range of the crediting situations has imposed, from practical reasons, the classification of credits, as these are numerous: creditor, debtor, credit object, crediting source, method of available supply formation, crediting period, etc.

The Romanian banking market displays a multitude of credit forms, both for population and companies, including small and medium companies. Thus, various commercial banks grant the following credit categories to the small and medium companies:

- ❖ Transilvania Bank grants:
  - 'Credit for small and medium companies, 0% interest', meant to cover the short term payment points;
  - 'Credit for SMCs, 1 hour', designed to meet the expectations of the companies that are looking at a rapid and simplified access to the bank's financial support;
  - 'Credit for SMCs START UP', meant to start a business;
  - 'Credit for SMCs, Discounting 100%', for companies that need financing resources for the development and growth of their current activity;
- ❖ OTP Bank grants:
  - 'Short-term cash credit for small and medium companies' and 'Credit line for small and medium companies' destined for current activity financing;
  - 'Credit for small and medium companies from BERD and EU sources', utilized for financing both new projects and updating or extension of the existent business, with the following destinations: investments projects and operating capital;
- ❖ B.R.D. grants 'Export Credit' in foreign currency, representing a method of financing the small and medium companies needs derived from the current activity , which cannot be covered by the ordinary formulae of collecting the debts;
- ❖ C.E.C. grants 'Credit for small and medium companies from BERD and European Union sources', used to finance the investments credits, stocks, revolving credit line, etc, with the following destinations: investments projects, operating capital required for activity of production and services, new projects, the updating or extension of the existent businesses.

Romania's accession to European Union has favorably impacted the banking sector, as the investing activity in this field is extremely intense. More financial groups in the community region have considered that the local business environment is suitable for opening further units.

Besides the commercial banks in Romania before 2007, the accession lead to the opening a series of new branches of foreign banks focused on the crediting of the retail activity and companies'. (Draguta, 2004) Among these, Bank of Cyprus pledged to itself to focus on corporations that the mother-bank serves in Cyprus and Greece. At the same time, the bank is trying to attract clients from the medium companies sector.

Millennium Bank estimates that it will have total investments in Romania in the amount of 300 million euro until 2011. This bank wishes to set itself apart as a universal bank, the target clientele being represented by both small, medium and large companies and population.

The credits aimed at companies amount up to 70 billion lei, out of which more than half are foreign currency financing, namely 37.4 billion lei. The loans attracted from the banks only represent 8% of the companies' liabilities, when the financing from the mother-companies abroad keeps a high level, being a half of the corporate credit balance.

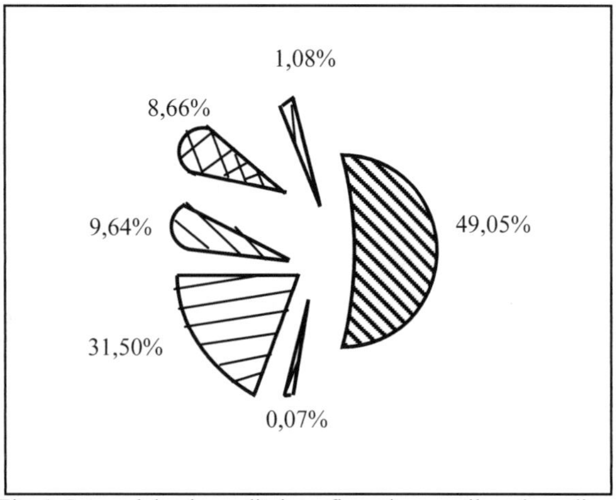

Fig. 1. Internal bank credit into financing small and medium companies (***, 2007)

Almost a half (49.05%) of the loans granted by the banks to the small and medium companies are treasury credits, having the company's cash flow as a guarantee, i.e. its income and liquid assets, and 31.5% of financing are meant to acquire equipment. Most of the treasury credits are current financing, but in the classification drafted by the Romanian commercial banks, the current credits are only 1.08%.

The data provided by the National Bank of Romania also show a high percentage for the credits granted to the small and medium companies for real estate, with 9.64% of balance, whilst the credits for export are only 0.07%.

Although the financing for the small and medium companies represent 63% of the volume of credits granted to the companies in Romania, only 14% of the small and medium companies have filed for bank loans. The small and medium companies sector has a great growth potential, but its evolution is dependent on the inclination of these banks towards these companies and the feasibility studies quality.

Along the implementation of Basel II capital agreement, which represents the capital adjusting regulation depending on the risks related to the bank activity in effect since 01.01.2008, the risk associated with the small and medium companies is lowering, a decision that is based on the large exposure variance.

In reality, both small and medium companies and population are facing a series of incorrect practices, from a legal point of view, of the commercial banks and the other financial institutions. (Jurvale, 2000)

A separate category of the incorrect commercial practices in Romania is constituted by the stipulation of abusive clauses in the pre-formulated standard contract concluded with the population and the companies, starting from the privileged position that the banks and the other financing institutions towards the consumers. Thus, we may enumerate:

- the right of the bank or of the financial institution to unilaterally modify the contract clauses without a proper reason;
- the interdiction or restriction of the consumer right to terminate the contract or unilaterally renounce, when the bank or the financial institution is entitled to it;
- the right of the bank or the financial institution to raise the service price, without granting to the consumer the right of terminating the contract, in case the final price is too high in comparison to the price convened upon the contract conclusion.

One of the widely used elements in the incorrect practices of the commercial banks or other financial institutions is the credit interest, which includes two aspects:

- former, the income of capital, i.e. the equivalent of the credit usage, and the interest in this case has been assimilated with the capital specific income;
- latter, is the one of penalty or reimbursement for the damage caused to the creditor by a late honoring the credit rate. In this case, it is about credit related penalty interests, which in the contract is similar with the delay penalties calculated by applying the value of the outstanding amount of a percentage between 0.5 and 2% per day.

As far as the interest in the role of a price for the granted credit is concerned, the sale of credits with promotional interests, fixed and initially low, but which become variable and rise suddenly or progressively, based on the contractual clauses – an incorrect practice by deceiving omission, in the sense that the sponsor omits an essential information required for the consumer to take a decision related to the credit or offers this information in an illegible way, in an ambiguous method or a rushed one, which makes the consumer request the credit, a decision which he would have not taken otherwise.

As far as the penalty interests applied to the consumers who are late with payments, parties may also determine, by contract, the extension of the damage that the creditor will bear as a consequence of late payments honoring, as well as the damage percentage that will cover it, before the damage occurrence. (Bistriceanu,, 2007)

These interests may be considered abusive when the paid amounts are out of proportionally high. In this sense, in the bank system entirety, the yearly average of the penalty interests is between 5 and 7% over the credit base interest (unlike European Union, where the average is 1%). Under conditions of adding these percentage points to the interest-cost of the credit of 7-9%, this issue may lead to a double penalty amount compares to what normally pays a client for a credit.

These penalties may not be considered disproportionate, since they are aligned to the international practices and because the banks have various costs, 'which may be quantified, at any moment', for a late rate payment, 'costs that may harm the image of a bank

3. CONCLUSION

Upon analysis of the credits for small and medium companies, there is ascertained a diversification of them compared to the previous years, the majority being held by the treasury credits and the credits meant for equipment acquisition, which represents 50% of the total credits granted to small and medium companies in 2007.

The penalty interests are way over the ones practiced in the European Union, and the banking officials should understand that they need to align themselves to the new codes of good practice in European Union and to self-regulate regarding the costs. A species of the credit cost is represented by the prepayment commission, whose quantum is also considered to be abusive in comparison to the one adopted by the EU (on average, 2-3% versus 1%).

4. REFERENCES

Bistriceanu, Gh. (2007). *Little encyclopedia in finance, money, ensuring*, Publisher Universitara, Bucharest

Draguta, V. (2004). *Practical landing in the firm finance*, Publisher IRECSON, Bucharest

Jurvale, V. (2000). *Fiscal methods and technics*, Publisher Rokris, Bucharest

Stroe, R.; Armeanu, D. (2001). *Finance*, Second Edition, Publisher ASE, Bucharest

***, (2007). Static Dates of the National Institute of Statistic from Romania

Annals of DAAAM for 2008 & Proceedings of the 19th International DAAAM Symposium, ISSN 1726-9679
Editor B. Katalinic, Published by DAAAM International, Vienna, Austria 2008
Make Harmony Between Technology and Nature, and Your Mind will Fly Free as a Bird

DIGITAL IMAGE PROCESSING IN SCANNING ELECTRON MICROSCOPY ANALYSIS OF PROTECTIVE COATINGS

DAVIDESCU, A[rjana]; SAVII, G[eorge] & STICLARU, C[armen]

Abstract: *The paper presents a graphical user interface (GUI) developed in Matlab for processing Scanning Electron Microscopy (SEM) images of HVOF-sprayed MCrAlY coatings used in high-temperature environments. The GUI allows an easy and more accurate way to evaluate the separation of the three existing layers: the oxide one, the diffusion one and the beta phase. The thicknesses of oxide and diffusion layers are calculated by means of statistical techniques. Also, the percentage of beta phase from the substrate is estimated.*
Key words: *Image processing, SEM, oxide layer, diffusion layer, coatings.*

1. INTRODUCTION

High temperature coatings are used to avoid surface degradation or to insulate the material against the hot environment. As the lifetime of structural components is frequently controlled by surface degradation, significant performance improvements are obtained by application of coatings. Coatings are applied to provide oxidation, corrosion or thermal protection depending on the nature of the operating environment and thermal loads to be endured. The coating should possess the required mechanical properties, adhesion and metallurgical stability in contact with the substrate.

MCrAlY's are a family of materials, which have a base metal (M) of cobalt, nickel and/or iron, plus chromium, aluminium, yttrium and sometimes other alloying elements. MCrAlY's are used as overlay coatings for turbine components improving their resistance and providing a longer lifetime for turbines, even under hard environmental conditions. High temperature oxidation resistance is achieved by one or more alloy components, which have to form a dense, stable, slow-growing, external oxide layer such as Al_2O_3 (Sloof & Jeurgens, 2004).

In the last few years, a new thermal-spraying process, high-velocity oxygen fuel (HVOF), was developed and used to produce MCrAlY coatings. The application of HVOF spraying has implications for the oxidation of powder particles, because of the free-oxygen content in the combustion gas and high temperatures required to melt the powder to assure a certain homogeneity level of the coating. There are two elements in the metallic powder, aluminium and yttrium, which have a high affinity for oxygen. These elements are oxidized during thermal spraying, so that the as-sprayed coatings contain oxides, such as Al_2O_3. After annealing, the oxides are uniformly distributed in the material. The HVOF coatings showed good oxidation behaviour at high temperature in various atmospheres (synthetic air, He–O_2 mixture or He–synthetic air), which suggested that the HVOF process can be used as a technological alternative to the more expensive VPS technology (Toma et al, 1999).

Aluminium content in the coating alloy is important, because selective oxidation of the Al occurs only on the surface of alloys with adequate Al contents. However, a high Al content in coatings leads to coating brittleness and a strong tendency to crack (Tang et al, 2004). Such cracks can propagate into the substrate material and lead to premature failure of the coated component. It has been proved that a fine grain size has a positive effect on the oxidation behaviour of alumina forming alloys (Neidel & Riesenbeck, 2004).

2. GRAPHICAL USER INTERFACE

The objective of the work presented in this paper was to process SEM images of HVOF MCrAlY coatings in order to study the uniformity of the coating, the thicknesses of the oxide and diffusion layer and the percentage of the β – phase from the substrate.

An instance of the GUI is presented in Figure 1.

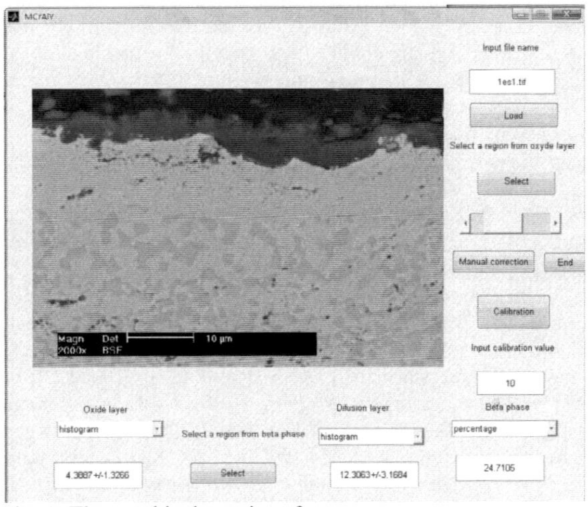

Fig. 1. The graphical user interface

The GUI allows loading a specific image file with tif extension (tagged image file). This goal is achieved by means of an edit box; where one has to type the name of the file, and a push button to update the image in the memory.

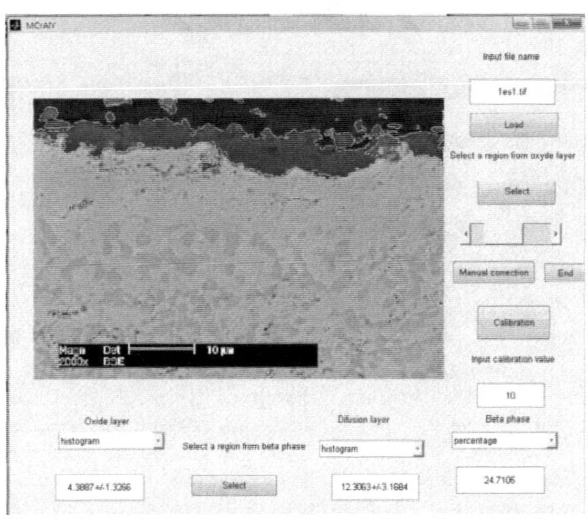

Fig. 2. The oxide layer perimeter overlaid on the original image

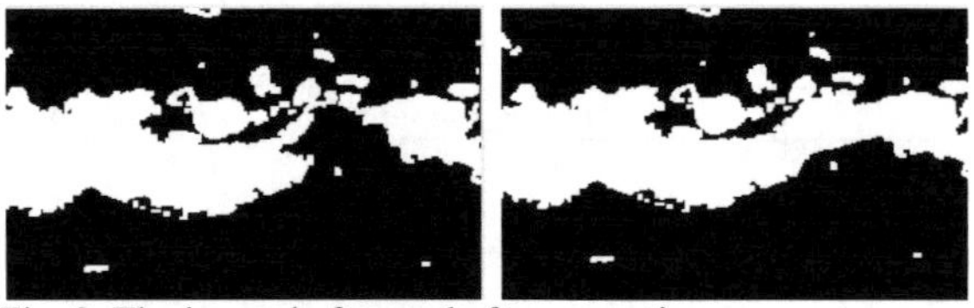

Fig. 3. The image before and after corrections

In order to depict the proper threshold level for the oxide layer, in the left side of the GUI a push button is presented which allows selection of a region from oxide layer. The threshold value is the mean of the pixel's grey levels extended by ±10. Some morphological operations are applied on binaries image: erosion, filling holes and removing isolated pixels. Finally, the perimeter of the binaries image is overlaid on the original image so the user can observe if any error occurs. The result can be seen in Fig. 2.

Two corrections can be made: an automated one – by means of the slider element which establishes the threshold level, and a manual one – by means of a push button. It has been introduced for the case the automated correction is not efficient enough. When this correction is selected from the push button, the cursor transforms into a cross-hair and the user can select two pixels from the image. A white line is drawn between the two pixels in order to attach a region to the interested feature. The results (a detail) can be seen in figure 3. Pushing End button indicates that all corrections are finished and only the largest feature of the image (the oxide layer) is maintained.

In order to obtain quantitative results the image must be calibrated by pushing Calibration button. The cursor transforms into a cross-hair; the user has to select the extremities of the image scale and to type the length value indicated in the text box below. When this operation takes place the GUI shows the original image for avoiding the influence of the applied morphological operations upon the scale.

At this point the image of the oxide layer is ready for measuring. At the bottom of the GUI there is a pop-up menu which allows selecting the image, thickness and the thickness histogram. The results are presented in Fig. 4. A linear grid (Russ & Dehoff, 2002) is overlaid on the oxide layer image consisting of 100 equally spaced vertices. The lengths between the two intersections are calculated for each line generating a series of 100 values for the thickness which allows statistical calculations. The mean value is associated with the confidence limit. Also, the histogram of the data is presented.

A region from the β phase must be selected in order to calculate the thickness of the diffusion layer. Due to the very small differences between grey levels of β phase and the substrate, an intensity adjustment was applied to the initial image with oxide layer removed. Also, some morphological operations (closing, removing isolated pixels) were applied.

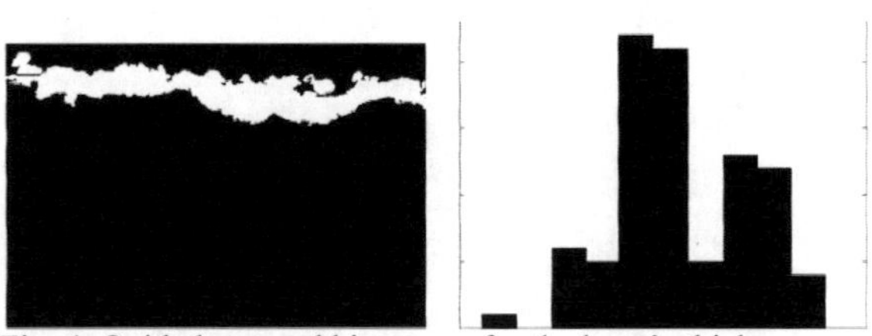

Fig. 4. Oxide layer and histogram for the layer's thickness

After cropping a region from β phase, the image is binaries by a threshold value equal to the mean of the pixel's grey levels extended by ±5. The inner contour of the oxide layer was overlaid in order to depict the thickness of the diffusion layer and the picture can be seen in Fig. 6. The diffusion layer's thickness was calculated in the same manner as the thickness of the oxide layer. The two intersections of the vertices were depicted by Boolean operations. A confidence interval is attached to the mean value of the thickness considering a probability of 95%. The histogram of the layer's thickness is shown. The results presented in Fig. 5 can be selected by a pop-up menu.

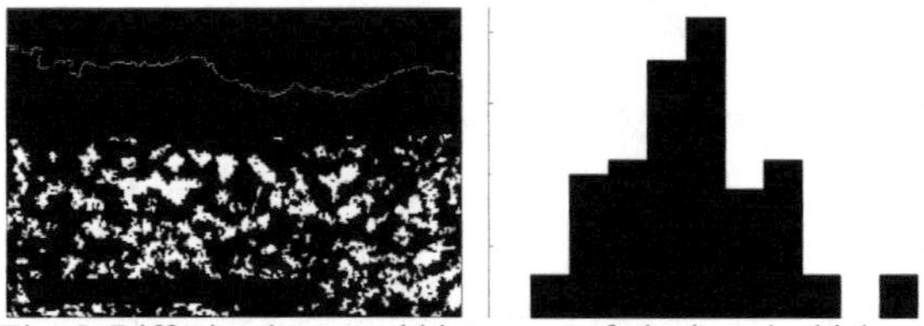

Fig. 5. Diffusion layer and histogram of the layer's thickness

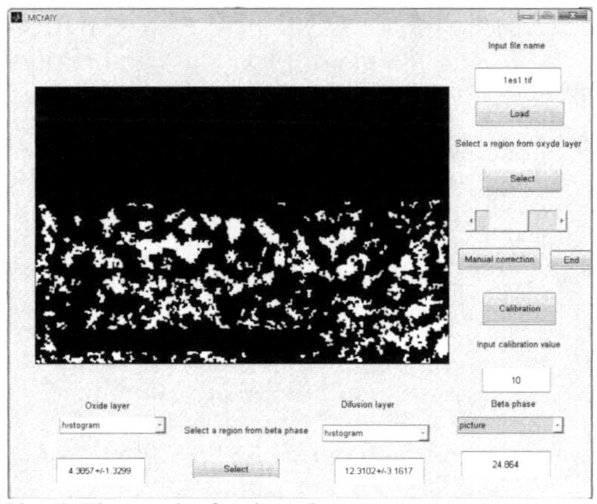

Fig. 6. The results for the substrate

The last objective was to depict some characteristics of the substrate. The pop-up menu has two options:

- Picture – showing the binary image of β – phase;
- Percentage – indicating the percentage of β – phase from the substrate.

For proper calculation the scale image is removed from the picture and the results can be seen in Fig. 6.

3. CONCLUSIONS

Digital image processing is a challenging field, considering that each application requires a specific algoritm, depending on the nature of the target parameters of the investigated process.

In this paper, we have presented the results in designing a GUI for objective, rapid, accurate analysis of SEM images for HVOF-sprayed MCrAlY coatings in order to obtain by statistic analysis the key parameters: the uniformity of the coating, the thicknesses of the oxide and diffusion layer and the percentage of the β – phase from the substrate. An advantage of the GUI is the fact that no expertise in digital image processing techniques is needed for the user.

The authors will further improve and extend the application capabilities in order to meet the specific requirements of other types of protective coating layers analysis.

4. REFERENCES

Neidel, A. et al. (2004). *Journal of Metalography*, Structure e-2, 14, Available from : http://www.struers.com/resources/elements/12/38856/e-Structure%202_EN.pdf. Accessed: 15.03.2008 .

Russ, J.C. & Dehoff, R.T. (2002). *Practical Stereology*, Plenum Publisher, ISBN 0-306-46476-4, New York.

Sloof, W.G. & Jeurgens, L.P.H, (2004). *Microchimica Acta*, vol.145, 1-4, p. 220, Wien, ISSN 0026-3672.

Tang, F.; Ajdelsztajn, L. & Schoenung, J.M. (2004). *Oxidation of Metals*, 61, 219-238, ISSN 0030-770X.

Toma, D.; Brandl, W. & Köster, U. (1999). *Oxidation of Metals*, 53, 125-137, ISSN 0030-770X.

Annals of DAAAM for 2008 & Proceedings of the 19th International DAAAM Symposium, ISSN 1726-9679
Editor B. Katalinic, Published by DAAAM International, Vienna, Austria 2008
Make Harmony Between Technology and Nature, and Your Mind will Fly Free as a Bird
*Young Researcher and Scientist Paper / * Supervisor, Mentor*

FIRST APPROACH TO THE ANALYSIS OF THE SURFACE ROUGHNESS OBTAINED DURING THE DRY TURNING OF THE UNS A97050-T7 ALUMINIUM ALLOY

DE AGUSTINA, B[eatriz]; MARIN, M[arta] & RUBIO, E[va]*

Abstract: *The main goal of this study is to analise the relationship between the surface roughness obtained during the dry turning of aluminium UNS A97050-T7 bars and the cutting parameters (cutting speed and feed) using tools with coating of TiN. The followed methodology consisted of a series of short tests (no longer than 10 seconds) for different values of the cutting parameters that mentioned above. The principal result was that the surface quality, improves mainly with the descent of the feed and, to a lesser extent, with the increase of the cutting speed. As a first conclusion, it would be convenient to carry out new tests using values of the nearer feed to the values that have lead to obtain the best surface quality in this study.*
Key words: *Dry turning, UNS A97050-T7, Cutting parameters Surface Roughness*

1. INTRODUCTION

The aluminium alloys are considered strategic materials in the aeronautical, motor and aerospace sectors due to their excellent weight to resistance ratio. Specifically, they are employed in the production of different elements that compose aircraft and aerospace vehicles. For such application an improved surface quality for the mechanized parts is required.

In spite of the important role these materials have from a competitive point of view, they can commonly show problems of machinability associated with the heat generated during the machining process. For this reason, although it is possible to evacuate the heat by means of the chip, the tool and the workpiece, cutting fluids are still widely used (Agustina *et al.*, 2008). However, the use of cutting fluids seriously degrades the environment quality and increases the cost of machining. As a result, dry machining has been widely studied in recent years (Nouari *et al.*, 2003).

In this study in order to analyse the evolution of the surface quality of an aluminium UNS A97050-T7 bar with respect to the cutting parameters employed (cutting speed and feed rate) a series of dry turning tests (no longer than 10 seconds) were carried out using tools with coating of TiN.

2. METHODOLOGY

The present work is framed within a serial of studies in which different materials, types of tools and cutting conditions are involved. The main steps of the methodology in which this work was developed are (Agustina *et al.*, 2007):

- Previous activities to the machining operations. These activities consist of the identification of the resources used and the preparation of the protocols both to calculate cutting parameters values and to register data and observations of the machining process.
- Turning tests. In each test a workpiece was mechanized during less than 10 seconds (short tests) under certain conditions of feed, cutting speed and depth of cut using both coated and uncoated cutting tools.
- Monitoring of the process. In order to have graphic documents that can be analysed after the process, all the turning tests described previously were recorded by video and both the chips obtained and the inserts used in each one of them was photographed with a high resolution camera.
- Roughness measurement. In order to systematize test-pieces roughness measurements, firstly, the measurement process has to be defined. This is, basically, to measure the roughness along four lines separated $\pi/2$ radians.
- Analysis of tools. From the results obtained, a tool preselection was made. Then, the selected tools were analysed employing both macroscopic and microscopic techniques. The former, using the macrographs taken and a profile projecter that allows measuring the quantity of the adhered material and the latter, by means of techniques of Scanning Electron Microscopy (SEM) and Energy Dispersive Spectrometer (EDS) in order to verify the alterations of the tools geometry.

3. APPLICATIONS

For this study, the workpiece used for the turning tests was a cylindrical bar with a diameter of 54 mm and length of 90 mm of UNS A97050-T7 aluminium alloy.

The cylindrical bar was horizontally dry turned on an EmcoTurn 120 CNC lathe equipped with an EMCO Turn 242 numerical control. The cutting conditions applied were cutting speeds from 40 m/min (0.66m/s) up to 170 m/min (2.83 m/s) and feeds from 0.05 mm/rev up to 0.30 mm/rev. Cutting depth was maintained at 1 mm in all the tests.

TiN coated tools (manufacturer reference SECO DCMT 11T308-F2-TP1000) were employed for the tests.

To observe the machining tests carried out, photographs and videos of the tools and the resulting chips were taken systematically during the tests using a Sony Cybershot DSC-P100 digital camera of high resolution.

To define the surface quality of the workpiece it was selected the parameter Ra (the arithmetical average roughness). The cut-off length was taken as 0.8 mm and the sampling length as 4 mm.

Fig.1. Roughnessmetre SJ-401 from Mitutoyo manufacturer with SurfTest SJ-401 software.

4. RESULTS

Table 1 represents the values of the rougness (Ra) measured along four lines separated π/2 radians: G1, G2, G3 and G4, for each cutting parameters applied.

To facilitate the analysis of the surface quality obtained on the workpiece under the different combinations of cutting parameters tested, two similar graphics were design to show the evolution of the roughness with respect to the feed (figure 2) and with respect to the cutting speed (figure 3). The values of the roughness Ra represented on both graphics correspond to the arithmetical average of the parameter roughness Ra measured along G1, G2, G3 and G4 for each test.

From figure 2, it can be seen that at cutting speeds of 40, 85 and 170 m/min, the higher the feed, the higher values of roughness was obtained, as was expected (Rubio, *et al.*, 2005), (Kulenovic, *et al.*, 2007).

On the other hand, analysing the evolution of the roughness with the cutting speed, in figure 3, it can be observed that the tendency of the values of the roughness measured was the opposite as it was with respect to the feed, though less accentuated, specially at 0.30 mm/rev for the higher cutting speeds applied 85, 125 and 170 m/min. Under such cutting conditions the values of roughness hardly varied.

	Ra (μm)			
v(m/min), *f*(min/rev)	G1	G2	G3	G4
170,0.30	2.796	3.689	2.875	3.616
170,0.20	1.569	1.537	1.522	1.537
170,0.10	0.592	0.576	0.593	0.578
170,0.05	0.316	0.403	0.396	0.424
125,0.30	3.386	3.326	3.313	3.313
125,0.10	0.488	1.383	0.564	1.780
125,0.05	0.482	0.537	0.458	0.627
85,0.30	3.351	3.496	3.398	3.215
85,0.20	1.556	1.633	1.651	1.823
85,0.10	0.812	1.011	1.025	1.058
85,0.05	0.881	0.799	0.905	1.082
65,0.30	3.754	3.669	3.780	3.624
65,0.10	1.385	1.573	1,690	1.730
65,0.05	1.135	1.369	1.127	1.507
40,0.30	4.310	3.935	4.509	4.617
40,0.20	3.225	3.799	2.929	3.628
40,0.10	1.603	1.814	1.583	2.115
40,0.05	1.900	2.215	1.785	2.579

Tab. 1. Ra (μm) measured under the different cutting parameters applied.

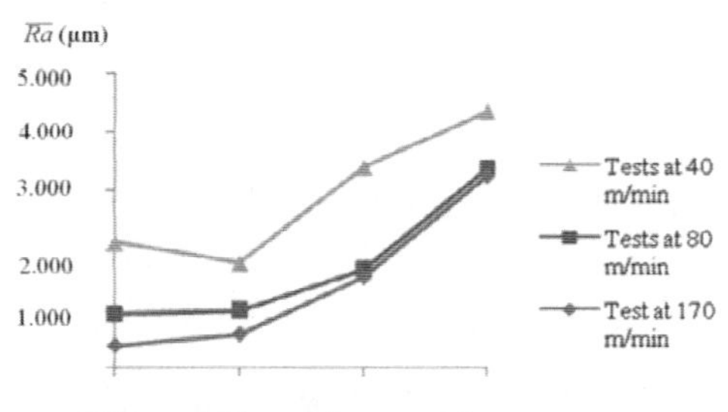

Fig 2. Evolution of the roughness versus the feed rate for different values of the cutting speed.

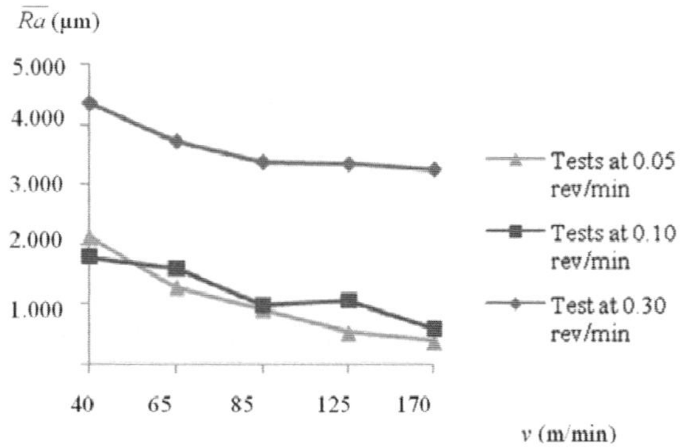

Fig 3. Evolution of the roughness versus the cutting speed for different values of the feed.

5. CONCLUSIONS

The surface quality of the aluminium UNS A97050-T7 bar which is obtained during dry turning short tests (no longer than 10 seconds) improves with the descent of the feed and with the increase of the cutting speed. From these two parameters, the feed is the parameter more influencial in the surface roughness.

Taking into account these results, it would be convenient to carry out new tests using values of the nearer feed to the values that have lead to obtain the best surface quality in this study as well as other values of the cutting depths in order to complete the work.

6. ACKNOWLEDGMENTS

Funding for this work was provided in part by the Spanish Ministry of Education and Science (Directorate General of Research), Project DPI2005-09325-C02-02.

7. REFERENCES

Agustina, B.; Rubio, E.M.; Sanz, A., Domingo R. (2007). A classification of the UNS A97050-T7 aluminium alloy chips in short duration tests under dry cutting conditions, *Proceedings of the MESIC-CISIF'2007*, Sebastián Pérez, M.A. (Ed) , pp 1-8, ISBN: ISBN(CD): 978-84-611-8001-1, July 2007, Madrid.

Agustina, B.; Rubio, E.M.; Marín, M.M.; Sebastián, M.A. (2008). Analysis of the material adhered on inserts with and without TiN coating during the dry turning of the aluminium alloy UNS A97050-T7, *Proceeding of the CIRP ICME'08*, Roberto Teti, R. (Ed), pp 1-6, July 2008, Naples.

Agustina, B.; Rubio E.M.; Marcos, M. (2007). Study of the adhered material to the cutting tools on dry turning of aluminium alloys, *Proceedings of the 18th International DAAAM Symposium, DAAAM International*, Katalinic, B. (Ed), pp 215-216, October 2007, Zadar.

Kulenovic, M.; Begic,; Cekic, A. (2007) Experimental Investigation of Carbon Steel in High Speed Cutting, *Proceedings of the 18th International DAAAM Symposium, DAAAM International*, Katalinic, B. (Ed), pp 411-412, October 2007, Zadar.

Nouari, M.; List, G.; Girot, F.; Coupard, D. (2003). Experimental analysis and optimisation of tool wear in dry machining of aluminium alloys. *Wear 255* (7-12) August – September 2003, 1359-1368, ISSN: 0043-1648.

Rubio, E.M.; Camacho A.M.; Sánchez-Sola, J.M.; Marcos, M. (2005) Surface roughness of AA7050 alloy turned bars. Analysis of the influence of the length of machining, *J. Mater. Process. Technol.* 162-163C, May 2005, 682-689, ISSN: 0924-0136.

Annals of DAAAM for 2008 & Proceedings of the 19th International DAAAM Symposium, ISSN 1726-9679
Editor B. Katalinic, Published by DAAAM International, Vienna, Austria 2008
Make Harmony Between Technology and Nature, and Your Mind will Fly Free as a Bird

ON THE ASSESSMENT OF THE QUALITY OF CASTING NONFERROUS SMALL-SIZE PARTS BY NUMERICAL SIMULATION

DEAC, C[ristian]; BIBU, M[arius]; PETRESCU, V[alentin]; TERA, M[elania] & BIRIS, C[ristina]

Abstract: *Quality is an important issue in casting, but it becomes essential in the case of small-size cast parts, such as dental prosthetic elements. As in most other areas of manufacturing, numerical simulation can be used for predicting potential failure areas in a cast part. The current paper analyses the various issues in assessing a cast part's quality and presents researches regarding the usage of a simulation software package designed for industrial castings for an accurate prediction of quality-related problems in the case of small-size cast parts.*

Key words: *casting, quality, numerical simulation, porosity, small-size parts.*

1. INTRODUCTION

The casting of small-size metallic parts, such as components of dental or bone prostheses, has become, in the last few years, a very important element in the activity of many specialized companies and even of many medical and technical laboratories.

This particular type of casting involves a high degree of precision and is very demanding in its execution, given the fact that parts often have wall thicknesses in the range of tenths of millimetres, the imposed tolerances being also very tight, sometimes in the range of ±0,05...±0,1 mm (Deac V. et al., 1995).

This makes any quality problem in a part, even if it would seem insignificant in terms of size, to still become a major issue and the cause for an immediate rejection of that part.

Therefore, it is of critical importance for the manufacturers of such parts to be able to predict the outcome of a casting operation in terms of quality. This is however complicated by three factors: the variety and complexity of the shapes that need to be cast, the variety of available commercial casting materials and not least the variety of available technologies and equipments (Deac C., 2003).

It is thus no longer possible to rely on one's experience or on simple techniques used for determining a material's castability that in turn would give a clue on whether problems would or not appear in a cast part. The development of numerical simulation and modelling techniques and software can be of great help here. However, currently there are very few casting simulation software packages designed specifically for small parts, and they are much more expensive than software designed for the industrial casting of large parts.

Therefore, starting from the analysis of the defect types that might occur, and from their importance in castings, the authors of the present paper tried to determine whether an industrial casting simulation software can be customised and used for the accurate prediction of quality issues in small-size cast parts, i.e. whether a system designed primarily for the relatively large tolerances of industrial cast parts can be used also for small-size applications..

2. THE QUALITY OF CAST PARTS

Given the small-batch or often even unique part production style employed for the casting of small-size parts, very often the producers use the traditional "lost wax" technique for realising these elements. This technique implies realising a model of the part in a wax-like material, covering it with a so-called investment material, heating the assembly until the wax melts and then casting the metallic alloy in the resulting mold inside the investment material (Deac V, 1995).

The notion "quality of the cast parts", which refers to the presence or absence in the cast parts of defects such as pores, shrinkage holes etc., is strongly related to the notion of "castability" of the material of which the parts are made. The latter can refer to both the completeness of the casting, in the sense of the cast part proportion which could be correctly reproduced during the casting process, and to the casting precision, i.e. to the minimal dimensions of the details which could be reproduced. This explains also the different categories of casting defects that need to be taken into consideration in quality assessments. These categories are (Marxkors, 1999):

- Investment defects, related to the components of the casting system. The main causes for this type of problems are:
    - secondary runners of inadequate length or diameter, or too many secondary runners connected to an under-dimensioned intermediate runner.
    - an oversized or undersized casting cone, or a casting cone that is not exactly aligned with the main runner;
    - strangled runners or strangled connections between runners;
    - uncontrolled forming and hygroscopic expansion rate of the investment material, and differences to the metal's expansion rate etc.;
- Pores, determined by the existence of gas bubbles, initially dissolved in the molten metal, that remain trapped in the metallic mass during its cooling and solidification. The main causes for the occurrence of pores in a cast part are:
    - inadequately dimensioned sprue and/or inadequately dimensioned gas removal runners;
    - overheating of the molten metal before the actual casting;
    - the presence of residual gases in the casting area;
    - a too large casting pressure;
    - a casting area that is too cold when casting starts.

  The area with the highest probability to generate pores is the junction between the casting pattern (model) al sprue, although pores could be found also in the casting part itself.

  Actually, as proven by several casting experiments and also by the speciality literature, porosity is the main problem in small-size cast parts.
- Incomplete castings, related also to the problem of under-dimensioning. Among its main causes, we can cite:
    - an insufficient amount of material used for melting and casting;
    - the metallic material was not heated to the optimal casting temperature, or there was a too long delay between the moment of reaching of this temperature and the moment of the actual casting;
    - the molten alloy starts flowing from the casting cone into the sprue before the actual casting force (e.g. centrifugal force) sets in;
    - the casting mould contains gases but there are no gas evacuation runners etc.

3. THE NUMERICAL SIMULATION OF CASTING SMALL-SIZE PARTS

All elements listed above have to be taken into account when manufacturing a small-size part by casting, but at the same time, if we want to be able to correctly predict the outcome of a casting operation from the point of view of quality, we need to use a proper simulation software which can take all these problems into account.

Although in the last few years there have been some attempts to use numerical modelling techniques for a better understanding or even solving of aspects related to the melting and casting of dental alloys, by using software packages like MAGMASOFT, FLUENT or JSCAST, there has been no systematic approach to the usage of computer-aided modelling and simulation software for the optimisation of the design of small-size casting moulds (Wu, 2002; Bernad et al., 2002).

The authors of this study have focused on the software package MAGMASOFT, developed by the company MAGMA GmbH from Aachen, Germany and meant for the use with relatively large parts from the machine-manufacturing industry.

The parts included in the simulation were simplified crowns, of cylindrical shape and with cylindrical cavities. From the software's materials database, the base metal alloy Fe-Cr-Ni type SA-6NM / ASTM A743 was chosen, with following characteristics:

- chemical composition: 0.07 % C; 1 % Si; 1.5 % Mn; 13 % Cr; 4.25 % Ni; 0.7 % Mo; remainder Fe;
- melting point: 1467ºC;
- casting temperature 1630ºC.

The discretisation process carried out with the simulation software has led to a mesh consisting of 998.944 elements.

In order to assess the influence of the various factors, several simulations were run, aiming at realising three identical parts at once, using different values for the casting pressure, protective atmosphere pressure and nature, mould temperature, as well as different design layouts for the sprue and patterns in the mould and various investment material types.

Several output types were obtained and analysed, including the part's porosity, the temperature distribution in the parts (which could indicate a problem in the dimensioning of the runners etc.).

Figures 1 and 2 present the assessment of porosity by simulations carried out for two different sprue designs.

It can be seen that, when using a sprue that allows the direct feeding with molten metal of each part, the porosity level is reduced and pores tend to form mostly in the sprue, rather than in the cast part itself. This is confirmed by experimental castings.

However, when comparing the simulation result with that of actual castings of the parts (figure 3), beneath an obvious similarity, several "fineness" differences were noticed, caused most likely by the lack of sufficient precision in the simulation software. This is understandable given the actual usage domain of the employed software, which is the casting of large parts, where a difference of a tenth of a millimetre, for example, does not play a significant role.

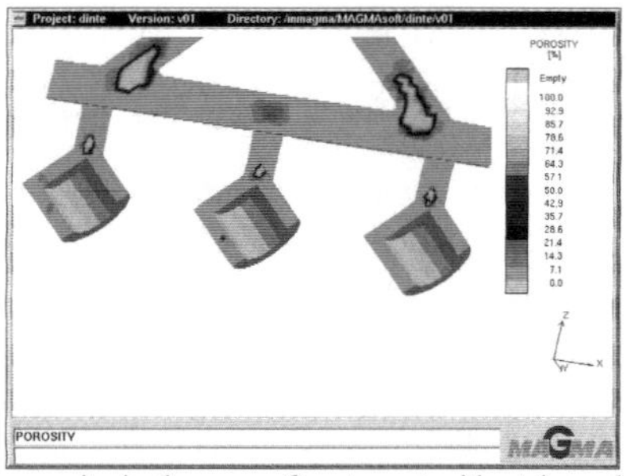

Fig. 1. The porosity in the case of parts cast with a triangular-shaped sprue (indirect feeding of the individual parts)

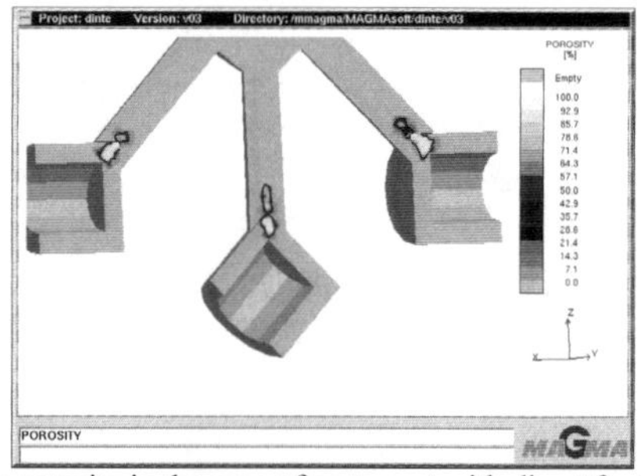

Fig. 2. The porosity in the case of parts cast with direct feeding.

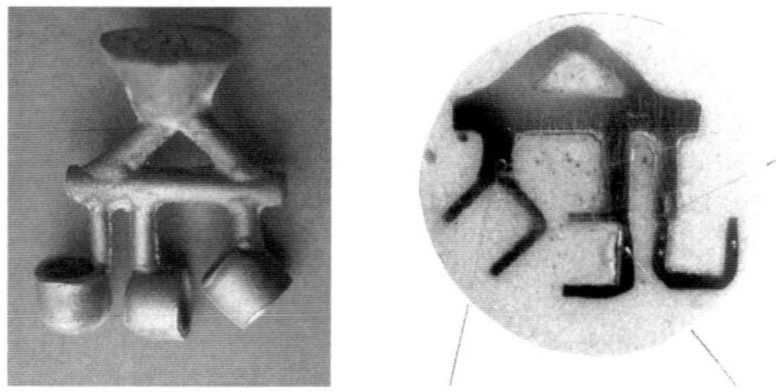

a) b)

Fig. 3. a) Parts cast in the indirect feeding variant; b) Cross-section through a casting system in the direct feeding variant

Therefore, it can be concluded that the usage of an unspecialised, unmodified simulation software cannot lead to a correct assessment of the quality-related problems in small-size cast parts. After modifying some of the software parameters, it was nevertheless possible to obtain a satisfactory result.

5. CONCLUSIONS

The researches described in this paper have proven that it is possible to predict the occurrence of quality-related problems in small-size parts obtained by casting. However, the appropriate program parameters have to be carefully tuned to the actual conditions, in order to obtain the desired precision.

The future researches will target a more detailed part modelling (e.g. more closely resembling a dental crown) and the realising of broader researches, aimed at developing a mathematical model that could be used for example to predict the position and precise size of pores inside a future cast part.

6. REFERENCES

Bernad, S.I, Susan-Resiga, R., Muntean, S. (2002) *Fluent Numerical Simulation of the Mold Filling Process for Titanium Dental Casting Applications*, Southeastern Europe FLUENT Users Group Meeting, Thessaloniki, Greece, 2002

Deac, C. (2003) *Contribuţii la dezvoltarea tehnologică a topirii şi turnării unor materiale metalice utilizate în tehnica dentară.* (Contributions to the technological development of melting and casting some metallic materials used in dental technics). Ph.D. thesis, "Lucian Blaga" University of Sibiu, 2003.

Deac, V. et al. (1995) *Turnarea titanului în protetica dentară"* (Casting of Titanium in Dental Prosthetics). Editura Universităţii din Sibiu, Sibiu.

Marxkors, R. (1999) *Ursachen, Auswirkungen und Behebung von Misserfolgen* (Causes, effects and repairing of problems). Deutsche Zahnärztliche Zeitschrift, vol. 54, nr. 7, p. 600-610.

Wu, M. et al. (2002) *Numerical simulation of the casting process of titanium removable partial denture frameworks.* Journal of Materials Science: Materials in Medicine, vol. 13, p. 301-306.

Annals of DAAAM for 2008 & Proceedings of the 19th International DAAAM Symposium, ISSN 1726-9679
Editor B. Katalinic, Published by DAAAM International, Vienna, Austria 2008
Make Harmony Between Technology and Nature, and Your Mind will Fly Free as a Bird

PNEUMATIC MUSCLE ACTUATED GRIPPER SYSTEMS

DEACONESCU, A[ndrea] & DEACONESCU, T[udor]

Abstract: *In view of the gradually growing number of industrial applications of pneumatic muscles, the motive for the research was the need for studying their performance as part of industrial robot gripper systems. The goal was to develop and study constructive solutions of non-anthropomorphic muscle actuated grippers. Performance parameters of the two presented models were measured by state of the art equipment. The results reveal that pneumatic muscles ensure accurate positioning at an improved eigen-weight to load ratio. The conclusions show that pneumatic muscles lend themselves to soft gripping operations of parts, particularly in mobile robotics requiring light actuators, generating large forces at moderate speeds.*
Keywords: *pneumatics, muscle, gripper, actuator, compliance*

1. INTRODUCTION

Automation applications in industry tend to implement a gradually increasing number of pneumatic actuation systems, the execution elements of which are typically cylinders.

One of the most attractive aspects of pneumatic actuation is the reduced weight of the utilized components, and implicitly, their favourable behaviour related to their response to commands. Favourable responding to commands, also called compliance is due to air compressibility, and can consequently be influenced by controlling the command pressure. Compliance can prove an important feature in relation to man – machine interaction or to performing highly sensitive operations like the manipulation of fragile objects. Compliance ensures soft contact as well as the safety of the man – manipulated object interaction.

At present industry makes current use of certain types of pneumatic actuation elements, like cylinders, bellows, rotating pneumatic motors and even step-by-step pneumatic motors. A less well-known type of actuation element is the so-called artificial pneumatic muscle. Pneumatic muscles are contracting linear motors actuated by the pressure of a gas. Their basic structural element is a flexible, closed and reinforced membrane, fixed at both ends to joining elements conveying the mechanical power to the load to be moved. Depending on the inflated or deflated state of the membrane, it expands or contracts, respectively. Once with its radial expansion or contraction the membrane length shortens or increases, in the former case generating a pulling force that acts upon the load to be moved (fig.1) (Hesse, 2003), (Neumann et al., 2008), (Pashkov et al., 2004).

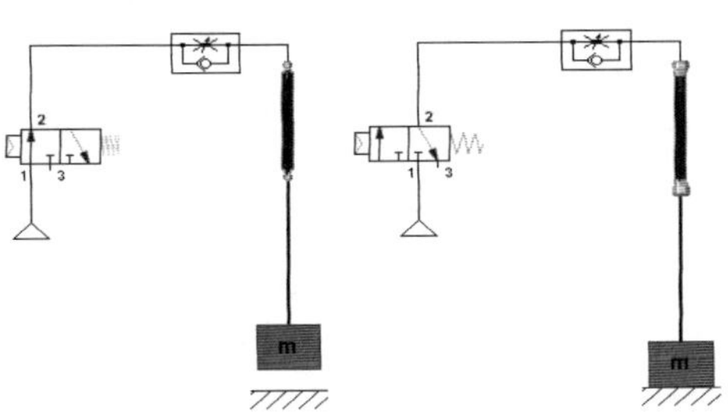

Fig. 1. Operating principle of pneumatic muscles

In literature pneumatic muscle actuated execution elements are known under various names, like pneumatic muscle actuator, fluidic actuator, fluid driven tensioning actuator, axial contraction actuator, tensioning actuator. Further on, this paper uses the term pneumatic muscle actuator (Murrenhoff, 2006).

The Fluidic Automation Laboratory of the Transilvania University of Brasov, Romania is engaged in research related to the utilization of pneumatic muscles in the construction of industrial robots, with a focus on gripper systems.

The paper presents two variants of pneumatic muscle actuated gripper systems, of parallel and radial type construction, respectively.

2. PARALLEL TYPE GRIPPER

The distinctive feature of parallel grippers - over their entire displacement range - is the parallel position of the jaw surfaces designed for the fitting of add-on fingers.

Figure 2 shows the cinematic diagram of the proposed variant of parallel gripper. The transversal position of the pneumatic muscle should be noted, as well as the fact that one of the jaws is fixed. The second jaw is able to glide along to guides, thus performing its designed functions of gripping and clamping of the part.

The construction of this gripper is aimed at achieving minimum overall dimensions, while also including the pneumatic valve necessary for controlling the two fingers. The construction of this gripper is presented in figure 3.

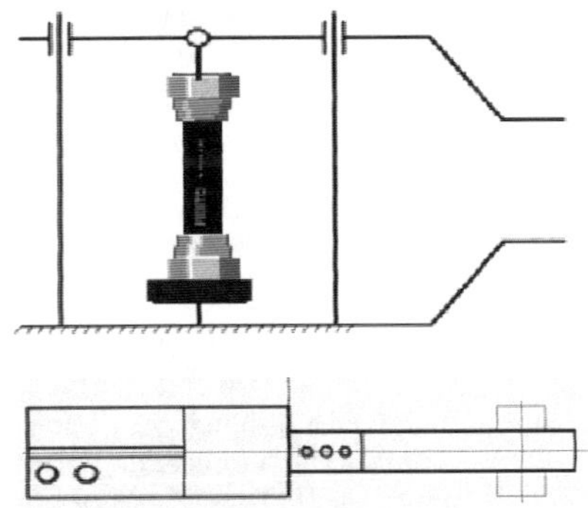

Fig. 2. Cinematic diagram of the parallel gripper

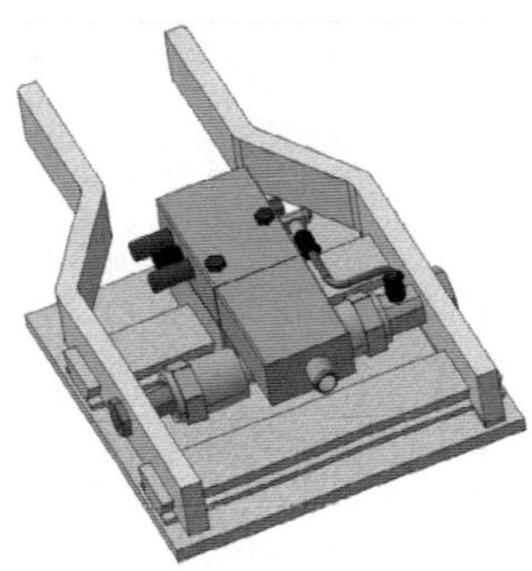

Fig. 3. Construction of the parallel type gripper

The parallel gripper was designed based on the following input data: maximum stroke: 5 mm; maximum charging pressure: 4 bar; maximum gripping force: 50 N. With these input data the computed length of the pneumatic muscle is of 40.3 mm.

3. RADIAL TYPE GRIPPER

The cinematic diagram of the radial gripper is presented in figure 4, its construction in figure 5.

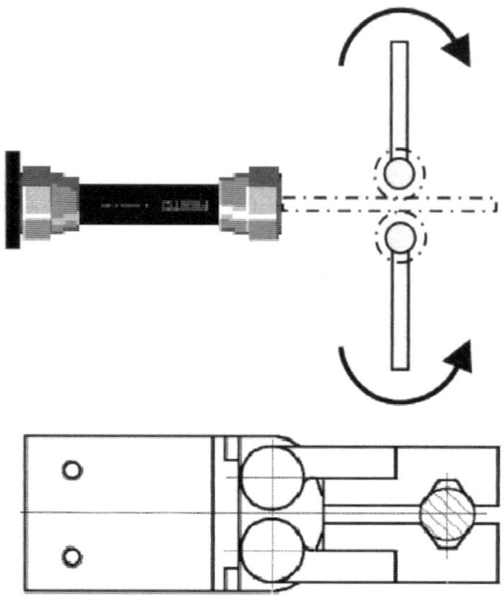

Fig. 4. Cinematic diagram of the radial gripper

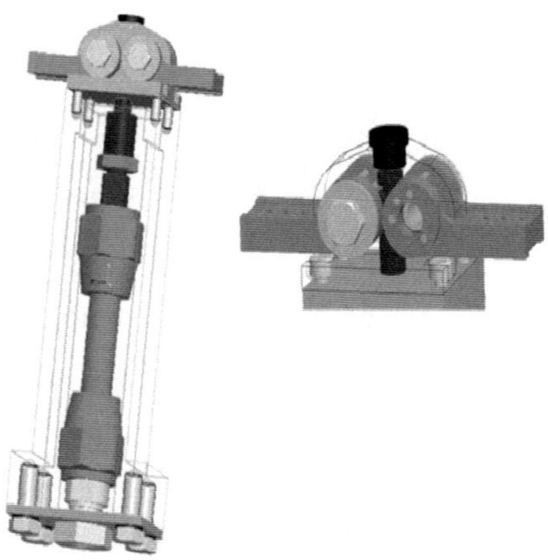

Fig. 5. Construction of the radial type gripper

In the case of the radial gripper the input data were: maximum stroke: 10 mm; maximum charging pressure: 6 bar; maximum gripping force: 40 N. The calculated gripper length in this case is of 51.8 mm.

The operational characteristics of the parallel and radial grippers are presented in figure 6. The curves plotted for the two grippers describe the evolution of the positioning speed of the muscles and the accuracy of maintaining the controlled position.

The analysis of the graphs allowed the extraction of the following results:

- For the parallel gripper the stroke time is of 0.04 s, while the return time is of 0.24 s. The maximum speed is if 0.43 m/s, the mean value being 0.11 m/s.
- For the radial gripper the stroke time is of 0.05 s, while the return time is of 0.15 s. The maximum speed achieved during the gripping stroke is of 0.82 m/s, the mean value being 0.21 m/s.

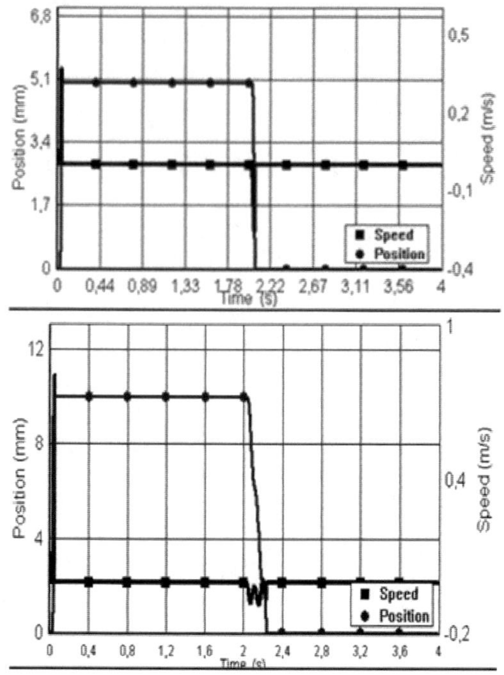

Fig. 6. Characteristics of the parallel and radial gripper

4. CONCLUSIONS

The research results expressed by the above graphs show that in both studied constructive variants the dynamic performance of the grippers, expressed by values for active and return stroke speeds and times, and the maintainability of the end position, as well as the compliance ensured by the very nature of pneumatic muscles best meet the requirements of soft gripping and clamping operations of parts, particularly in mobile robotics requiring light actuating systems, capable of generating large forces and torques at moderate velocities.

The inclusion of pneumatic muscles into the structure of industrial robots also contributes to an improved eigen-weight to load ratio.

The conclusion is yielded, that the issue of identifying actuating systems in the field of robotics that meet at the same time the conditions of strength and lightness has found an eligible solution in the utilization of pneumatic muscles.

While significant, the results and conclusions of the research described in this paper need to be seen as the basis for further in-depth research of pneumatic muscle applicability in the field of robotics.

5. REFERENCES

Daerden, F., Lefeber, D. (2001) The concept and design of pleated pneumatic artificial muscles. In *International Journal of Fluid Power*, vol. 2, no. 3, pp. 41–50

Hesse, S. (2003) *The Fluidic Muscle in Application.* Blue Digest on Automation, Esslingen

Murrenhoff, H. (2006). *Grundlagen der Fluidtechnik. Pneumatik. (Fundamentals of Fluid Technology. Pneumatics)* Shaker Verlag GmbH, ISBN 3-8322-4638-X, Aachen

Neumann, R., Hildebrandt, A., Knubben, E. (2008) Pneumatic Drive Systems for the Assistance Robotics. *Proceedings of the 6th International Fluid Power Conference* , pp. 257-266, Dresden, DVF Dresden

Pashkov, E., Osinskiy, Y. & Chetviorkin, A. (2004) *Electropneumatics in Manufacturing Processes.* Isdatelstvo SevNTU, ISBN 966-7473-60-0, Sevastopol

Annals of DAAAM for 2008 & Proceedings of the 19th International DAAAM Symposium, ISSN 1726-9679
Editor B. Katalinic, Published by DAAAM International, Vienna, Austria 2008
Make Harmony Between Technology and Nature, and Your Mind will Fly Free as a Bird

PNEUMATIC MUSCLE ACTUATED ROTATION MODULE FOR REHABILITATION ROBOTS

DEACONESCU, A[ndrea] & DEACONESCU, T[udor]

Abstract*: The discussed research was motivated by the necessity of an in depth study of the performance of pneumatic muscles as actuators in rehabilitation robotics. The consequent goal was to develop and study the dynamics of an alternative rotation module of light structure, actuated by pneumatic muscles, part of a rehabilitation robot mounted on a wheelchair. Laboratory studies were carried out by means of advanced data acquisition and processing equipment. The main outcomes include, in addition to the design and achievement of the new structure and its control, the relevant results of the processed experimental data, proving the superior dynamic performance of pneumatic muscles in relation to pneumatic cylinders, presented as the main conclusion of the paper.*
Keywords*: pneumatics, muscle, rehabilitation, robot, controller*

1. INTRODUCTION

Numerous international studies have revealed increasing number of physically incapacitated persons worldwide. The aggravation of their physical disabilities causes the gradual loss of their autonomy and dependence on the help of medical and caretaking staff in solving everyday tasks. Such a situation may lead to the financial overstraining of the respective social system, preventable by the implementation of new technology in the assistance process. Based also on the aspiration of the physically incapacitated to attain increased autonomy in everyday and professional life, the last three decades have witnessed intensive research aimed at the development, improvement and implementation of robots in physical rehabilitation, medical and occupational assistance. A series of so-called rehabilitation robots have emerged, available as finished products or laboratory prototypes. The wide range of applicability of these systems covers areas like "support in managing everyday tasks", "care", and "assistance in the professional environment" (Hesse, 2003).

Until recently robotized systems were actuated mostly by electric motors. Pneumatic solutions have been generally avoided, due to control and compliance issues. Nevertheless, their advantages, like compactness of the actuation elements, favourable power (force) to weight ratio, reduced cost, easy maintenance, and a clean work environment have facilitated the increased use of pneumatic actuation systems in robotics.

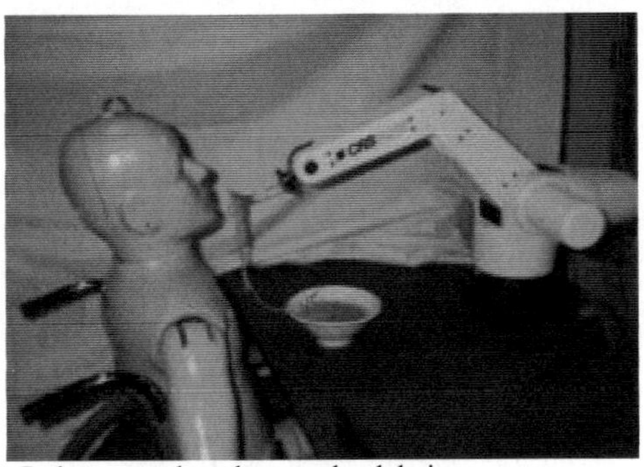

Fig.1. Robot arm placed on a wheelchair

Research carried out over the last few years at the Transilvania University of Braşov, Romania, has highlighted the advantages of the utilization of pneumatic muscles in robotics. The paper presents such a utilization example, concerning a robotized manipulation system, effective in rehabilitation activities of the physically handicapped. The construction of the arm is achieved by easily standardized rotation and translation modules, their combination allowing the development of robots with several degrees of freedom. Robots actuated by pneumatic muscles benefit from an extremely light structure, increased flexibility and meet the safety requirements for equipment operated in the immediate proximity of man or in narrow spaces.

The manipulators mounted on wheelchairs (Fig. 1) have the highest degree of flexibility in relation to their applicability. They allow the gripping of objects placed randomly in space, the opening of doors, drawers, and taps or can carry out simple manipulation tasks like pouring liquids into a glass and raising it to the mouth. The costs of such flexibility depend on the cognitive capacity of the users during accomplishing a certain task, namely on the degree of their physical handicap.

2. CONSTRUCTION OF THE ROBOT

Figure 2 presents the structure of the robotized system actuated by pneumatic muscles. The underling diagram allows the development of three degrees of freedom (rotation – translation – rotation), attained by three pairs of pneumatic muscles.

Further on the development of the first rotation is analyzed, achieved by the pair of muscles denoted by 1. The operational system of the two muscles is similar to the human one (agonist and antagonist), designed to generate by their counter-time action a ± 90° rotation, as well as to ensure the balance of a certain intermediary position (Fig. 3.). The free ends of the two muscles are connected by a flexible steel cable drawn over a pulley of 2r diameter (Daerden & Lefeber 2001).

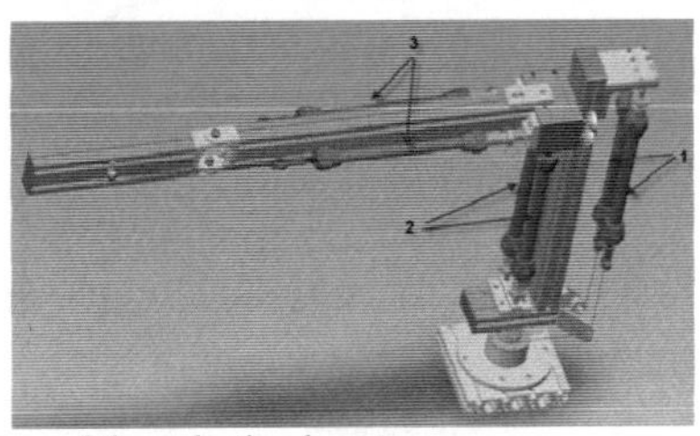

Fig.2. View of the robotized system

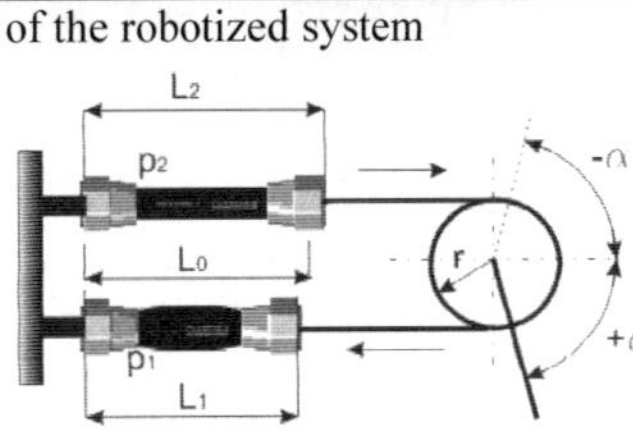

Fig.3. Kinematics of the rotation module

At rest the two muscles are inflated by the same pressure p_0 equal to ½ of the maximum working pressure, thus achieving length L_0. In order to carry out a rotation movement by a certain angle α, one of the muscles will be additionally fed air until its pressure reaches the value $p_1 = p_0 + \Delta p$, while the other muscle will be deflated until its pressure reaches value $p_2 = p_0 - \Delta p$. By feeding different pressures to the two muscles, their lengths will be modified in relation to their initial state as follows: the muscle inflated at pressure p_1 will shorten to a length of $L_1 = L_0 - \Delta L$, while the second muscle will be elongated to $L_2 = L_0 + \Delta L$.

The position of the two muscles in the robotized system and the generation of the rotation movement can be observed in figure 4; the actuation principle is shown in figure 5.

Control of the counter-time movement of the pneumatic muscles is achieved by means of a proportional 5/3-way directional valve, of a PID-controller and of a displacement encoder. The electronic testing equipment designed to create stimulus signals included a function generator capable of sine, square and triangle waveform generation. Following the feeding of a triangle signal of 2 Hz frequency and 5V amplitude into the system, the motion diagrams of the two muscles were obtained, as shown in figure 6 (Hering et al., 2005), (Pashkov et al., 2004).

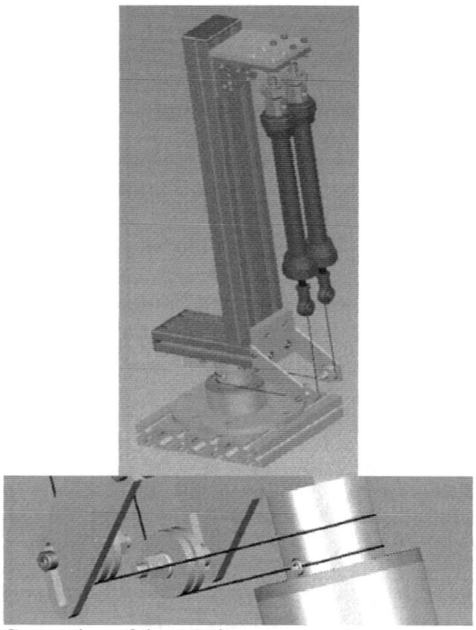
Fig. 4. Generation of the rotation movement

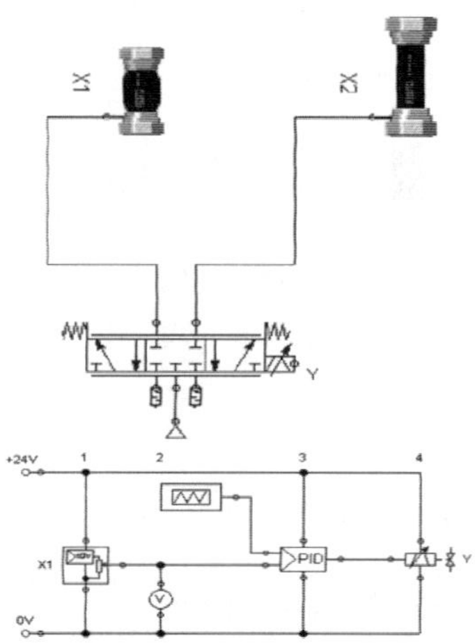

Fig. 5. Control diagram of the pneumatic muscles

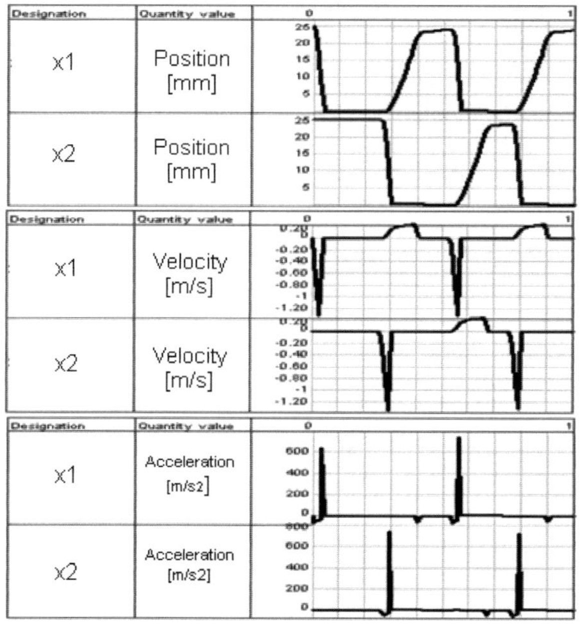

Fig. 6. Motion diagrams of the two pneumatic muscles

3. CONCLUSIONS

The processed data show that the use of pneumatic muscles for the rotation module (Fig. 4) determines a 50% increase of the velocity of the final element, compared to the use of a pneumatic cylinder. Also, the weight of the entire assembly is about 20% smaller. In addition further advantages of pneumatic muscles over pneumatic cylinders have been confirmed:

- exceptionally high power to weight ratios: >1kW/kg;
- being pneumatic, muscles are highly flexible, soft in contact and have excellent safety potential;
- force and position control by using antagonistic pairs of muscles for compliance regulation;
- high tolerance of rotational and translational deviation, reducing complexity and cost (Murrenhoff, 2006).

Due to its advantages the pneumatic muscle actuated rotation module represents an improved alternative to pneumatic cylinders in rehabilitation robotics. The research, while having yielded relevant conclusions, is not exhaustive. Pneumatic actuation systems, at present insufficiently applied in a wider context, require further detailed study of their performance, such as to successfully to replace pneumatic cylinders in an increasing number of applications.

Acknowledgement: This research project has been supported by a CNCSIS-IDEI Contract No. 220/2007, ID_184.

4. REFERENCES

Daerden, F. & Lefeber, D. (2001). The Concept and Design of Pleated Pneumatic Artificial Muscles. In *International Journal of Fluid Power*, vol. 2, no. 3, pp. 41–50.

Hering, E., Bressler, K. & Gutekunst, J. (2005). *Elektronik für Ingenieure und Naturwissenschaftler.* (*Electronics for Engineers and Natural Scientists).* Springer-Verlag.

Hesse, S. (2003). *The Fluidic Muscle in Application.* Blue Digest on Automation, Esslingen

Murrenhoff, H. (2006). *Grundlagen der Fluidtechnik. Teil 2: Pneumatik. (Fundamentals of Fluid Technology. Part 2: Pneumatics)* Shaker Verlag, ISBN 3-8322-4638-X, Aachen

Pashkov, E., Osinskiy, Y. & Chetviorkin, A. (2004). *Electropneumatics in Manufacturing Processes.* Isdatelstvo SevNTU, ISBN 966-7473-60-0, Sevastopol.

Annals of DAAAM for 2008 & Proceedings of the 19th International DAAAM Symposium, ISSN 1726-9679
Editor B. Katalinic, Published by DAAAM International, Vienna, Austria 2008
Make Harmony Between Technology and Nature, and Your Mind will Fly Free as a Bird

USING CHAINED RECORDS TO OPTIMIZE THE PREPARATION AND STORAGE INTO RAM FOR LARGE SETS OF STRUCTURAL DATA

DEAKY, B[ogdan]

Abstract: *Some of applications developed by the author (i.e. modules that implement the ISO calculus method for cylindrical gears) work with large sets of data and therefore require much time to compute the results. Code and algorithm optimizations had to be done. Research to determine the best methods used self-developed software which can track performance improvements. The paper refers to the data preparation and storage optimizations, an important stage because processing does frequently reference the data already loaded and organized in the RAM. Great improvements of performance were obtained using chained records and this paper presents a couple of existing and self-designed implementation methods, along with applications that benefit of these optimizations.*
Keywords: *software optimization, chained records, application.*

1. INTRODUCTION

The author was faced with the problem of having large amounts of data which induced very big calculation times, especially for the self-developed software to implement ISO calculus methods. As this is a problem encountered by more software developers, research was conducted to determine optimization methods to decrease the result delivery time.

This paper focuses on data preparation and storage into RAM, one of the main steps to get from the initial data to the needed results (fig.1). The usual method is the usage of static or dynamic (open) arrays of unique values or arrays of records (grouped variables of the same or different types). A less known approach, of using chained records, was tested and the result is a great performance improvement. This alternative was offered from inside the Bluemind Software Company (Bluemind Software 2005) which grants the author permission to publish the information in this article.

Interest was shown by other researchers including some working in astronomic research and therefore this paper briefly presents the research done, existing and designed implementation methods, some future research directions and applications that already benefited of these optimizations.

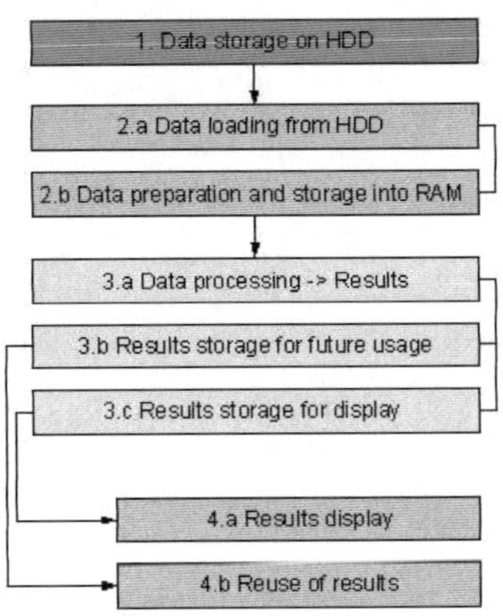

Fig.1. Steps from initial data to results

2. PERFORMANCE COMPARISON AND STRUCTURES

Fig.2. Representation of an array of records

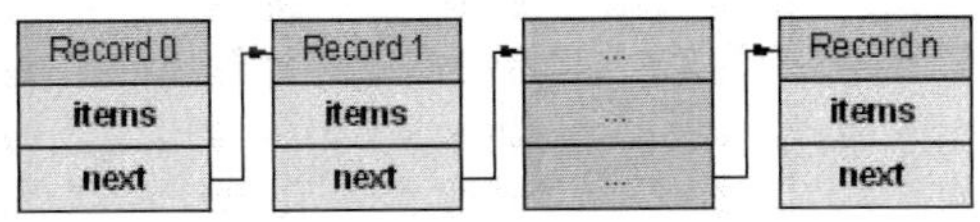

Fig.3. Representation of chained records

The array (fig. 2) actually contains pointers to the records, which makes the finding of an element by its index in an array very fast (Gajic 2000).

Usually chained records are records which have one additional field: next – this actually contains a pointer to the next record. The graphical representation is in fig.3. Borland Delphi implementation codes can be looked up in (Glaser 1998).

In theory this should make searches for values much quicker. To be able to make a decision, test software has been developed to compare the time needed to execute certain operations on both an array of records and the same number of chained records. The test system was a notebook with a 2.60 GHz Pentium 4 processor and 512 MB of RAM.

The results do speak for themselves (times are shown in milliseconds):

- Adding 10.000 new records :
  - CR Add: 280
  - Array Add: 12077
- Deleting 1000 records (average times):
  - CR Delete middle: 50
  - Array Delete middle: 2053
- Finding a record containing given values for 10000 times:
  - CR Find values: 50
  - Array Find values: 731
- Finding a record by index for 10000 times:
  - Array Find index: 0.0000001
  - CR Find index: 20

One can see that chained records have an enormous advantage over arrays of records in almost all situations: when adding, deleting or searching for records by values. The speed is **14 - 60 times faster**! The only exception is while finding a record by index. It was expected since arrays contain pointers to each record and therefore the access is instant. For our needs, chained records were the obvious choice for storing loaded data into RAM.

Meanwhile the author started to look for improvements to the classic chained records which might serve better for our needs. A couple were found or thought up. Most of them are presented in (Deaky 2006).

Using a base record with a "last" field – pointing to the last record (which makes adding even much faster) plus extra "previous" fields (fig.4) makes it possible to parse the chained records in the other direction as well. Rumor has it that reverse

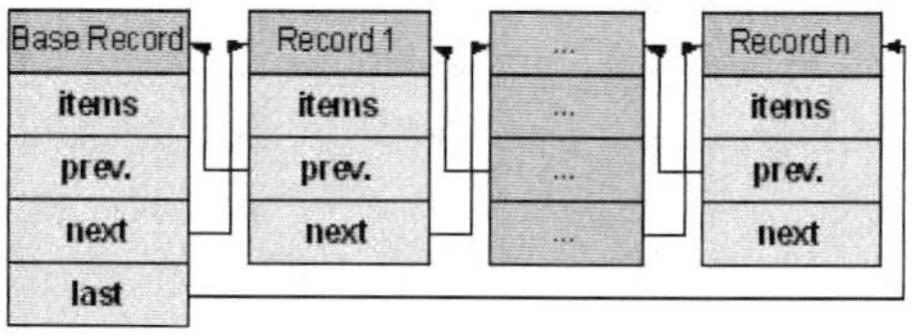

Fig.4. Base record with "last" + "previous" fields

parsing is actually faster due to RAM configuration, but the author did not test this yet.

Speeding up searches even more may be achieved by using pointers to indexes in the structure. Research also revealed a complex indexing and grouping structure (IBM 2005) developed by IBM and used in Airline Control System version 2. Due to lack of space the developed structure is not shown here, but usage of indexing can be noticed in this paper in the future research directions chapter.

3. APPLICATIONS

The CR structures were successfully implemented in more software projects, resulting in noticeable performance improvements. Some of them are presented bellow.

- ISO/DIN provides calculus methods for the factors which intervene in the **cylindrical gear calculus**. Yet, there are only a few hints for gear designers regarding the exact influence of the gear parameters on these factors. Several software modules have been developed to study the influences and **generate customizable diagrams** which can quickly show the researcher quantifiable modifications and **help in the decision making process**.
- Shafts D/M (fig.5) is an **original CAD/CAM system for shafts**. Being specialized software, it is more productive as generic drawing software, uses an **"engineer's way" of drawing** and features a couple of innovative features. A previous version won the ICAMES 2004 international engineering projects competition.
- 3DCar™ (fig.6) is probably the world's first interactive **3D car configuration software**. Software optimizations helped it work well even on 5-6 years old PCs. Interest was shown recently by Daimler AG for their Mercedes-Benz models. 3DCar™ is property of Bluemind Software.

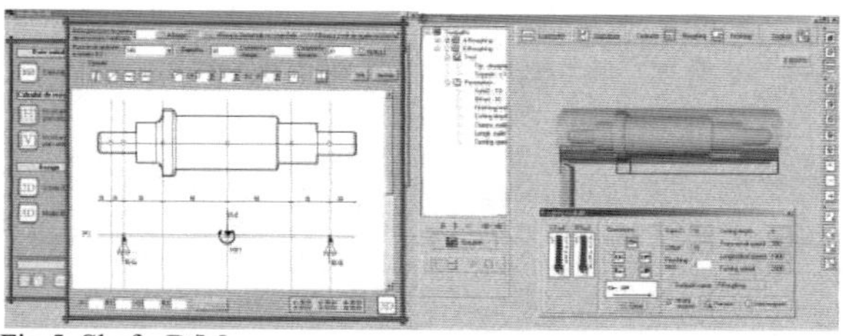

Fig.5. Shafts D/M

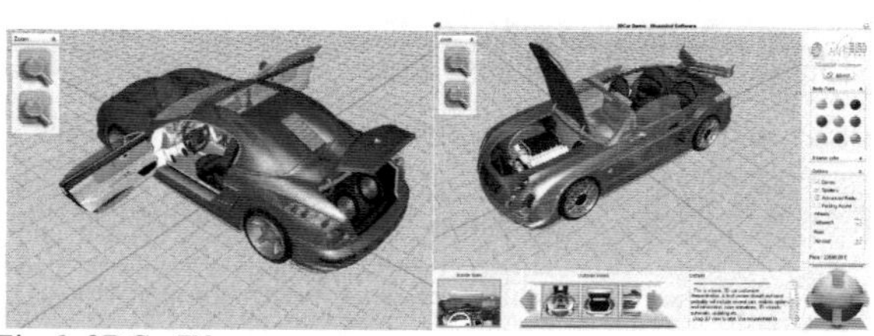

Fig.6. 3DCar™

4. FUTURE RESEARCH DIRECTIONS

In the diagram generation software, storage of the results is done in a matrix (array of array). To improve searching speed by means of CR models the following structure was designed (fig.7) and awaits implementation.

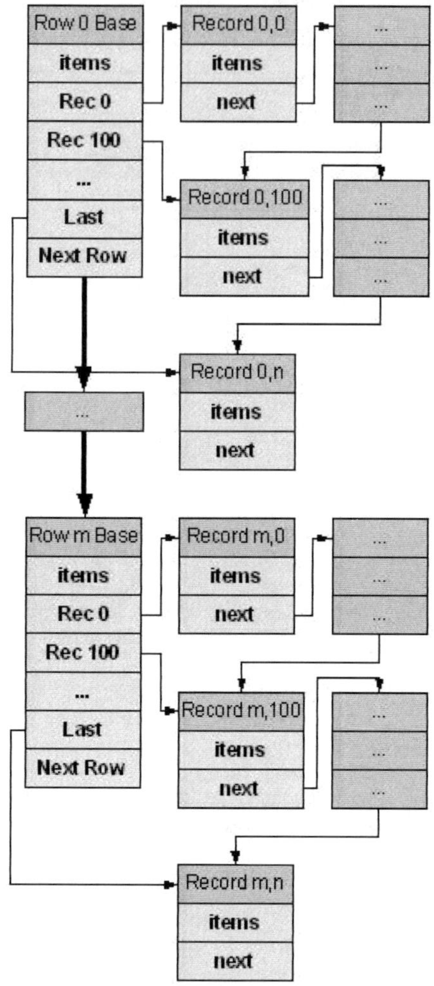

Fig.7. Chained records structure to replace matrix

5. CONCLUSION

As proved above, using the implementation methods presented, existing and future software result delivery times can be significantly decreased - making the difference between a successful piece of software and one that stays unused. We are looking forward to future research that most probably will result in even faster software.

6. REFERENCES

Bluemind Software (2005). Company website *Available from*: http://www.bluemind-software.ro *Accessed:* 2005-02-01

Deaky, B. (2006). Data preparation and storage methods for optimized data processing. *Proceedings of PRASIC' 06,* Vol. II, pp. 221-226, ISBN 978-973-635-825-9, Braşov, Romania, November 2006

Gajic Z. (2000). Understanding and Using Pointers in Delphi. *Available from*: http://delphi.about.com/od/objectpascalide/a/pointers.htm *Accessed:* 2006-09-20

Glaser F. (1998). Turbo Pascal Memory. New: Delphi pointer considerations, *Available from*: http://www.geocities.com/franzglaser/tpsrc/tpmem.html. *Accessed:* 2006-09-23

IBM (2005). *Airline Control System Version 2. Concepts and facilities.* IBM, SH19-6953-12, Denmark.

Annals of DAAAM for 2008 & Proceedings of the 19th International DAAAM Symposium, ISSN 1726-9679
Editor B. Katalinic, Published by DAAAM International, Vienna, Austria 2008
Make Harmony Between Technology and Nature, and Your Mind will Fly Free as a Bird

MODIFIED MECHANISM OF A STIRLING ENGINE FOR USE WITH SOLAR POWER GENERATORS

DEHELEAN, N[icolae]; CIUPE, V[alentin] & MANIU, I[nocentiu]

Abstract: *The paper presents a modified mechanism structure used to link the two pistons of the Stirling engine. The proposed structure is of crank and rod type and it is developed in such a way that to enhance the theortical pressure-volume cycle of the engine. The mechanical structure is described and performance estimations are made for the modified engine.*
Key words: *mechanism, Stirling, cycle, solar, energy.*

1. INTRODUCTION

Solar energy has to become a practical alternative to fossil fuels and humans must find out other efficient ways to convert solar radiation into electricity, other synthetic fuels and heat. The need for better conversion technologies is a driving force behind many recent developments in biology, materials, mechanical engineering, electronics and opto-electronics.

The Sun delivers to Earth 10^{17} joules of energy in one second. Earth's ultimate recoverable resource of oil, estimated at 3 trillion barrels, contains 1.7×10^{22} joules of energy, which the Sun supplies to Earth in 1.5 days. The amount of energy humans use annually, about 4.6×10^{20} joules, is delivered to Earth by the Sun in one hour.

The Sun continuously delivers to Earth, 1.2×10^{5} terawatts, so this issue surpasses every other energy source, renewable or nonrenewable. It dramatically exceeds the rate at which human civilization produces and uses energy, currently about 13 TW.

The main purpose of this paper is to find out an answer to the question: "How to convert more efficiently the energy that comes from the Sun?"

There is an enormous gap between the potential of solar energy and our conversion capacity. Solar electricity accounts for a minuscule 0.015% of world electricity production, and solar heat for 0.3% of global heating of space and water. Biomass produced by natural photosynthesis is by far the largest use of solar energy; its combustion or gasification accounts for about 11% of human energy needs. The use of biomass as fuel is limited by the production capacity of the available land and water.

When thinking about converting heat from the sun in either electricity or mechanical work, the most viable solution that comes in mind is to use a concentrator that collects sun radiation over a larger area and focuses it onto a small spot, where a conversion "machine" or "device" is placed. Here the choice is limited to steam generators for use with turbines or steam engines, and to Stirling type engines.

Fig. 1. Parabolic solar concentrator with Stirling engine.

Figure 1 presents such a concentrator dish having a Stirling engine placed in the focal plane (Wikipedia, 2008). This approach gives better performance over steam use but it is still relatively low on efficiency.

The paper proposes a technical solution meant to improve the heat transfer or p-V diagram for a modified version of an alfa-type Stirling engine.

2. THE CLASSICAL STIRLING ENGINE

Stirling engines are classified into three distinct types, shown in Figure 2. The Alpha type engine relies on interconnecting the power pistons of two cylinders to move the working gas, with the cylinders held at different temperatures. The Beta and Gamma type Stirling engines use a displacer piston to move the working gas back and forth between hot and cold heat exchangers in the same cylinder (Wikipedia, 2008).

The ideal cycle of any Stirling engine is the goal for every technical solution. This diagram is represented and marked in Figure 3. Overimposed is the real cycle of an alfa type engine. By looking at the two diagrams one can easily spot a problem in the design of these engines (Karlsruhe Uni, 2000).

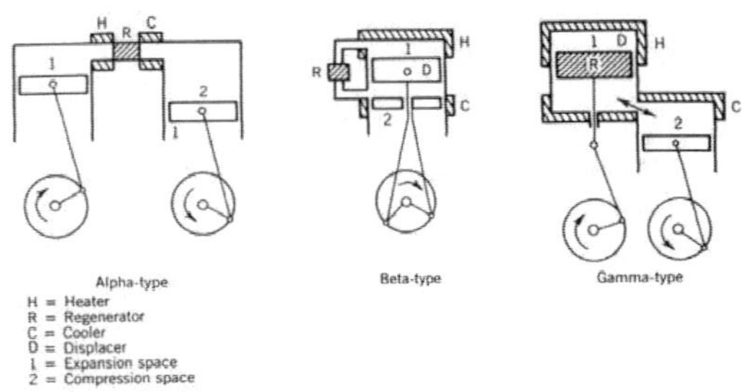

Fig. 2. The three main Stirling engine types.

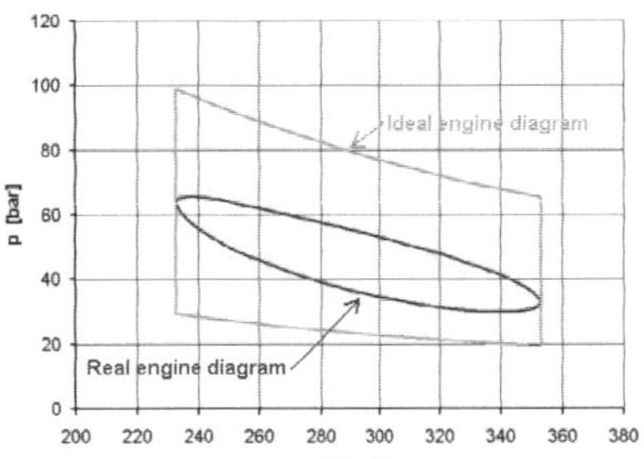

Fig. 3. Ideal vs. real p-V diagram for an usual Stirling engine.

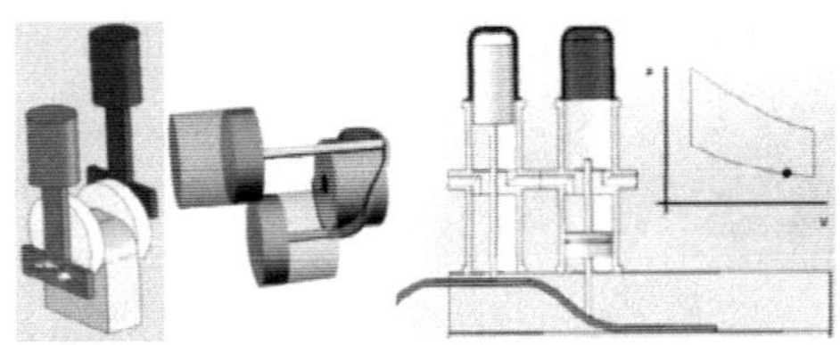

Fig. 4. Proposed mechanisms that could improve the diagram.

The solutions found for enhancing the behavior of the engine, from the mechanism point of view, usually rely on the modification of the piston driving mechanisms. A few methods for approaching the ideal cycle are presented in Figure 4 showing a modified scoth yoke mechanism, a spatial cam with a special profile and profiled slider mechanism (Youtube, 2008). These approaches have the disadvantage of sliding contacts and points of increased acceleration on the sliding profile, drastically reducing the reliability of the solution.

3. THE MODIFIED MECHANISM

In order to overcome the above stated problems, a new mechanical structure was proposed for an alfa-type Stirling engine and it works as follows:

The two cylinders are moved by two crank and connecting rod mechanisms. The space law movement is sinusoidal for both pistons. The expansion cylinder needs a movement law with two stop points to improve the p-V diagram of the engine. We consider being an improvement to obtain a new diagram that seems like the Carnot one as near as it is possible. In order to improve the efficiency of the engine we have to increase the area between the superior and inferior curve, considering the ideal diagram in in Figure 5 (Herzog, 2005). This is possible by increasing the enthalpy difference between admission and expansion into the expansion cylinder.

It starts in 1 point of the diagram by an isothermal compression stroke (process 1→2). This phase takes place by removing heat, Q12, from the working fluid in order to keep the temperature at a constant low value, TC. This is followed by an isochoric compression (process 2→3). The later takes place because heat of the amount Q23 is added to the working fluid. Expansion (process 3→4) takes place at constant temperature, TH. In order to keep the temperature constant, heat Q34 has to be added to the working fluid. Finally, the pressure is lowered at constant volume by removing heat, Q41, from the working fluid to close the cycle.

In order to achieve this, we need to replace the crank and connecting rod mechanism with another one that is able to perform a space function with two stop points.

This mechanism is presented in Figure 6 and is essentially a crank-rod mechanism coupled to a swivel mechanism for the secondary piston. The motion law is greatly enhanced (not over the entire cycle) and is constructed with 5th class joints only.

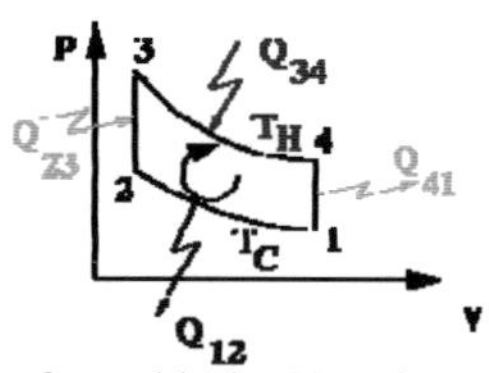

Fig. 5. Heat transfer for an ideal p-V cycle.

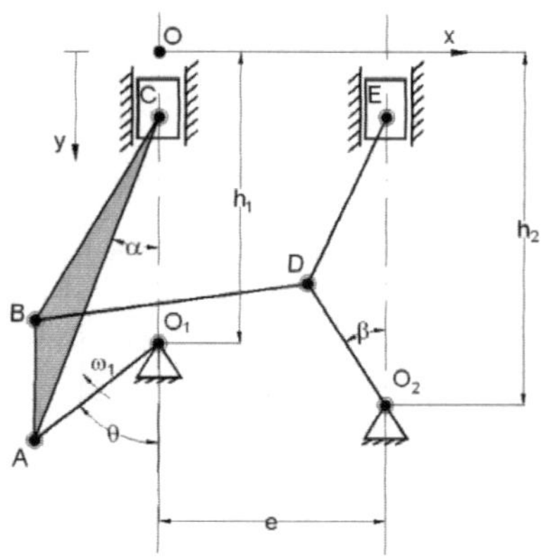

Fig. 6. The proposed mechanical structure.

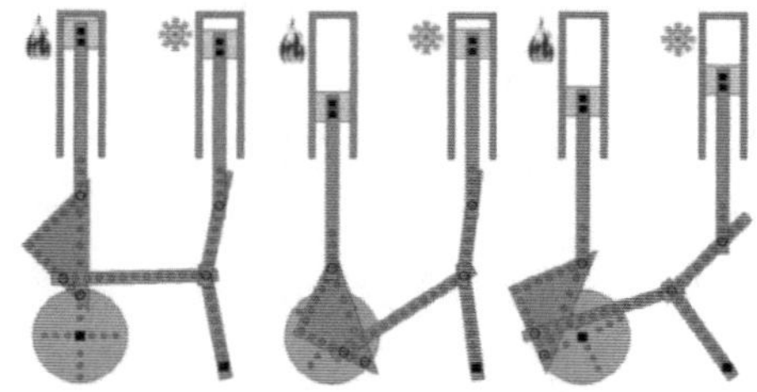

Fig. 7. Movement simulation of the proposed mechanism.

Figure 7 pesents three positions of the simulated mechanism (Kanagawa, 2005) in order to observe the functioning of the two pistons and their movement laws. The first position represents the starting point of the first stop point for the cold piston and this is held for about 120 degrees, until the torque generated by the hot piston drops with the cranck angle. The second stop point is on a smaller portion of the compression stroke (3rd position). This setup greately enhances the torque of the engine by allowing the gas in the hot cylinder to expand fully and give more energy to the engine shaft. The diameters of the two cylinders differ so that they have the same displacement despite their different strokes.

4. FUTURE RESEARCH

The proposed solution regarding the modified mechanism must be further optimized and tested in real life conditions. This presumes building a working model of the engine that can be run with different heat sources. The engine must be built to accept a wide variety of working gases (air, hydrogen, carbon dioxide or methane). Performance analisys must be conducted on the modified engine in order to the optimal solution. Also comparative analisys between simulated performance and real engine performance has to be conducted.

The practical implementation of such an engine should have positive impact over power generation and could easily replace portable electrical generators used in remote locations, running alternatively on solar power or a wide variety of fuels.

As a drawback of this engine is the difficult self-starting problem and, if higher efficiency is required (hydrogen as the working gas) maintaining quantity and pressure inside cylinders require additional costs and system complications.

5. CONCLUSION

Using clean, renewable energy is a very appealing thing. But the goal is to convert this energy as efficiently as possible, taking into account both the initial cost of the conversion unit and the reliability issues. The proposed mechanism is presumed to give a boost in performance while maintaining a simple and cost effective technical solution.

6. REFERENCES

Herzog, Z. (2005). The ideal Stirling Cycle and Heat Load on the Regenerator, *Available from:* http://mac6.ma.psu.edu/stirling/ideal_stirling_cycle/index.html *Accessed:* 2008-05-07

Kanagawa, P.E.C. (2005). Linkage mechanism simulator, *Available from:* http://www.edu-ctr.pref.kanagawa.jp/LinkWeb/index_e.htm *Accessed:* 2008-05-16

Karlsruhe University, Institut für Kolbenmaschinen (2000). Investigation of concepts for high power Stirling engines, *Available from:* http://www-ifkm.mach.uni-karlsruhe.de/Html-e/Project/Stirling/stirling.html *Accessed:* 2008-05-07

Wikipedia (2008). Stirling engine, *Available from:* http://en.wikipedia.org/wiki/Stirling_engine *Accessed:* 2008-04-23

Youtube (2008). Ideal cycle Stirling engines, *Available from:* http://www.youtube.com/results?search_query=ideal+cycle+stirling+engines&search_type=&aq=f *Accessed:* 2008-05-28

Annals of DAAAM for 2008 & Proceedings of the 19th International DAAAM Symposium, ISSN 1726-9679
Editor B. Katalinic, Published by DAAAM International, Vienna, Austria 2008
Make Harmony Between Technology and Nature, and Your Mind will Fly Free as a Bird

REALIZING A RULE-BASED CALCULATION SYSTEM FOR DIE CASTING MOULDS

DENKENA, B[erend]; LORENZEN, L[eif] E[rik] & SCHUERMEYER, J[an]

Abstract: *An inquiry for the price of a die casting mould will usually result in as many different quotations as companies inquired. The difference in price can amount to up to 300%. At least, that is the alarming result of a case study carried out by the Institute of Production Engineering and Machine Tools (IFW) in this domain. In addition, a discrepancy between the effort necessary for generating an offer and the amount of orders commissioned by the customer has been found. This article provides an overview about the essential issues and describes the development of a rule-based calculation approach facing this challenge.*

Key words: *design automation, decision making, product modelling, product planning and scheduling*

1. INTRODUCTION IN BID GENERATION

The amount of bids inquired from manufacturers of die casting moulds increased rapidly within the last few years. At the same time, the bid profitability measured in the amount of orders commissioned decreased. Only the traceability and informational value are of particular importance for the success of the bid (N.N, 2001). A convincing technical description of the mould with logically presented costs supports an easy internal and external evaluation. It enables the customer to judge whether the conditions offered are superior to those of other competitors regarding price and performance. An analysis among ten manufacturers of die cast moulds shows that this is not yet the status quo in practice and emphasizes the need to develop new calculation systems. The following report deals with this.

2. STATUS QUO OF BID GENERATION

The quantity of bids calculated per year is an indicator for the effort of the calculation a company has to face. An analysis among ten manufacturers of pressure moulds showed that most of them calculate an average of 1,000 bids per year, which means about 4 bids per day. In combination with an average creation time of about 5 hours for each offer, this little share quantifies the enormous financial and time effort spend on order acquisition. If the actual costs are higher than the calculated costs, the companies produce uneconomically. If the calculated costs are too high above the actual costs, the probability to get an order decreases.

The status quo can be summarized in two main statements: firstly, many offers are calculated but a high number of orders are lost every year due to a low success rate; and secondly, imprecise calculations imply the risk for the companies of producing under costs or of decreasing the order quantity.

3. THE RULE-BASED APPROACH

A promising approach is the integration of geometrical information about the die casting mould, the resulting die cast component, and the necessary production processes by means of a rule-based methodology. Current calculation models, as for example similar-based calculation models or calculation models using statistical cost functions, employ past data in order to make statements about future manufacturing costs (Fonseca, 2007), (Duverlie, 1999). In opposition to these models, analytic calculation models are based on detailed technological information of the future product. Interdependencies between the mould, the production resources and the resulting processing costs represent the basis of analytic functional coherences between design characteristics and costs. For the purpose of realizing an analytic calculation of costs, the functions are based on detailed parts lists and process plans. While the accuracy of expected costs increases with a more detailed description of the product within the model, the calculation effort also increases (Ort, 2000). Due to this growing effort especially for the manufacturing of complex geometry like die casting moulds, calculations based on analytic cost evaluations are not yet common (Nagahanumaiah, 2005). In order to meet the demands of the manufacturers of die casting moulds, the development of a calculation method based on an analytic calculation model seems to be successful. Thus the use of rule-based design and process knowledge for the bid calculation allows combining a fast calculation with an exact, meticulous cost structure. For this purpose the knowledge of the production process has to be used to define rules in order to support the calculation right from the start.

4. RULE-BASED CALCULATION

The IFW aims to develop and establish a novel quotation-costing system for manufacturers of diecasting moulds. For this purpose interdependencies between steps of the production process and steps of the calculation have been identified. In order to reproduce interdependencies systematically with rules the calculation process has been divided in the three processing steps: dimensioning, manufacturing structure and the calculation itself. The following section will introduce the developed calculation system by giving rule-based examples for each step.

4.1 Dimensioning of the pressure diecasting mould

The dimensioning provides the optimal positioning and order of one or several casting parts in a pressure die casting mould. This affects the function structure and the solution principle. The positioning of the casting part and the parting plane as well as the number and order of cavities are decisive for the amount of casting material the mould later requires.

In order to estimate the dimensions of the mould among others the dimensions of the contour element have to be calculated. In order to estimate the dimensions of the mould, among other things the dimensions of the contour element have to be calculated. This requires, for example, a specification of the arrangement of cavities within the contour element, the sprue angle and the position of the gaiting point. Using interdependencies between these parameters, each defined arrangement can be described in a rule-based way in order to estimate the dimensions of the contour element. The sealing area between a cavity and the boarder of the contour, for instance, can be approximated by a linear correlation between the sealing area and the size of the contour (Fig. 1).

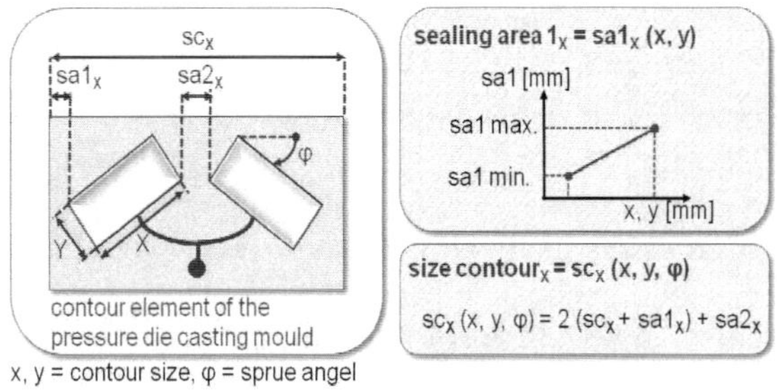

Fig 1: Rule-based dimensioning of the contour element

As result of the dimensioning, the pressure die cast mould is as a preliminary design at the disposal of the sales. This is a good starting point for developing the rules for a rule-based automation of the following calculation steps.

4.2 Generating die casts manufacturing structure

The specification of a manufacturing structure determines a technical solution which is suitable and meets the requirements of the customer based on a variety of feasible variants. This substantially influences the mould to be calculated. The manufacturing structure is represented by the design and the number of components and assemblies. In order to generate a parts list and process plans, the manufacturing structure has to be specified. For this purpose the developed calculation system represents interdependencies between different solutions and possible components and working operations in a rule-based way. These rules allow to automatically identify necessary components and operations from a complete parts list and a complete list of operations and to generate a parts list for the pressure die casting mould and process plans of the components (Fig. 2).

As a result of the rule-based generation of a manufacturing structure, a parts list of the pressure die casting mould and process plans for the components exist. For the purpose of calculation, the amount of work of the working operation has to be quantified analytically with regard to the geometry of the casting part.

4.3 Calculating the manufacturing costs

Rule-based interdependencies between the known geometry of the casting part and the calculated dimensions of the pressure die casting mould as well as the generated process plans of the components permit to analytically simulate the expenditure of production. Taking the production processes into account, the processing time of manufacturing operations can be calculated by dividing the amount of work by the specific performance of the machine tool. The volume of the pocket for the contour element in the frame of the pressure die casting mould, for example, changes with the size of the contour element. The size of the contour element again is determined by the geometry of the cast component and has been calculated during the dimensioning of the mould.

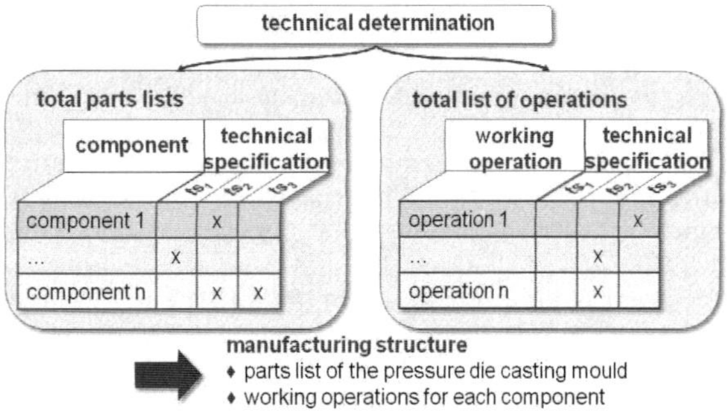

Fig 2: Rule-based generation of a parts list and process plans

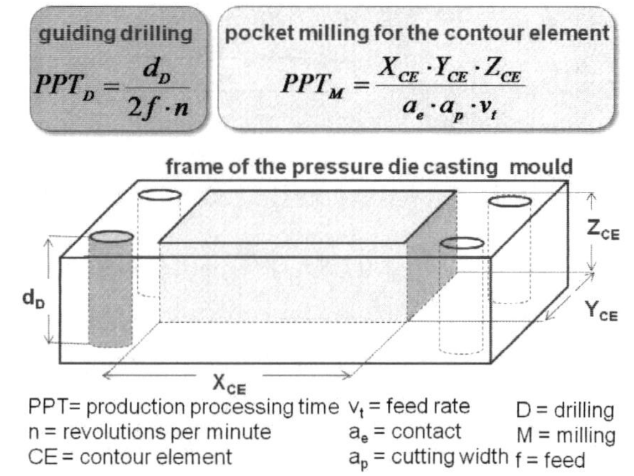

Fig 3: Rule-based calculation of the production processing time

Taking parameters of the specific performance of e.g. a milling machine into account, the production processing time of milling (PPT_M) for the specific working operation can be calculated with regard to the geometry of the cast component (Fig. 3).

In order to calculate the expenditures for the working operation, the calculated processing time has to be multiplied with the machine hourly rate of the machine tool. By summing up the manufacturing costs for each component and adding material cost and overhead costs, a calculation of the pressure die cast mould is realized.

5. CONCLUSION

The case study of the calculation process in the pressure die cast mould industry showed a laborious process with a low level of detail and a low calculation quality. The high effort results in high costs and the low level of detail hinders an internal and external traceability of the service offered. The low calculation quality increases the risk of not working on cost-effective orders. With the help of a rule-based mould dimensioning, an automated generation of a parts list as well as accompanying process plans, it is possible to analytically specify the amount of work of each operation and thus to charge costs according to source. Therefore, the presented model calculates costs with the help of technical principles instead of using past data. The experiences of the employees are documented in rules that permit to calculate a high level of detail at low effort.

6. REFERENCES

Crees, R.C.; Adithan, M.; Pabla, B.S. (1992): *Estimating and costing for the metal manufacturing industries*, Dekker, New York

Duverlie, PJ, Castelain, M. (1999): *Cost estimation during design step: parametric method versus case based reasoning method*, Int J Adv Manuf Technol 15: 895-906

Fonseca, M. et.al. (2007): *Assisting Mould Quotation Through retrieval of similar data*, In: Digital Enterprise Technology, Cunha, P; Marapanlos, P.G.;527-534

N.N., Verein Deutscher Gießereifachleute e.V. (VDG) (2001): *Rationellere Angebotserstellung durch Aufbereitung von Expertenwissen und Schaffung rechentechnischer Hilfsmittel,* VDG Fachbericht, Düsseldorf

Nagahanumaiah; Ravi, B; Mekherjee, N.P. (2005): *An integrated framework for die and mold cost estimation using features and tooling parameters*, Int J Adv Manuf Technol 26: 1138-1149

Ort, A. (2001): *Entwicklungsbegleitende Kalkulation mit Teilebibliotheken*, TU Clausthal

Annals of DAAAM for 2008 & Proceedings of the 19th International DAAAM Symposium, ISSN 1726-9679
Editor B. Katalinic, Published by DAAAM International, Vienna, Austria 2008
Make Harmony Between Technology and Nature, and Your Mind will Fly Free as a Bird
*Young Researcher and Scientist Paper / * Supervisor, Mentor*

A STUDY OF REWARD FUNCTIONS IN REINFORCEMENT LEARNING ON A DYNAMIC MODEL OF A TWO-LINK PLANAR ROBOT

DENZINGER, J[oachim]; LAUREYNS, I[sabelle]; FRIETSCH, M[arkus]; BURGER, W[olfgang] & MECKL, P[eter]*

Abstract: *Reinforcement Learning is one approach for executing a point-to-point movement using complex multi-link robots in an environment. The work presented in this paper, we: (1) use torque commands as input for a two-link manipulator; position commands do not take the dynamics of the system into consideration, (2) identify the optimal composition of a set of reward functions in fulfilling certain requirements. These requirements include: a smooth velocity profile while reaching the target, no overshooting and a minimum number of steps. We experimentally validate our approach on a two-link planar robot for reaching different target positions.*
Key words: *Machine Learning, Reinforcement Learning, Multi-Link Robot, Reward Function.*

1. INTRODUCTION

One of the many approaches to successfully control robot manipulators is Reinforcement Learning (Sutton & Barto, 1998). Reinforcement Learning is based on an interaction between an agent (decision-maker) and an environment. The agent chooses an action, which is performed by the manipulator within the environment. The action results in a reward and, mapping rewards to the states and actions, it is possible to supply the agent with enough information to successfully deal with an unknown system at the beginning of the learning process. This is a major advantage of Reinforcement Learning. Because the agent seeks high rewards, it is capable of finding good solutions without the need to supply the information of how the goal can be achieved. It is the job of the agent to find a policy, π, mapping states to actions that maximize the reward in the long run, to reach the goal.

In the work presented in this paper we use two methods in Reinforcement Learning, Q-Learning (Watkins, 1998), and SARSA (Rummery, 1994) to update the Q-table in order to learn the optimal policy for executing point-to-point movement in a two-link planar robot. The Q-table stores the expected return (Q-value or $Q^{\pi}(s,a)$), when the agent takes action a in state s and follows the policy π.

In the case of multi-linked robots, dimensions of the Q-table for states and actions would grow exponentially and result in extremely slow convergence (large number of steps). This is called the "*Curse of Dimensionality*" (Bellman, 1954). To address this issue, Martin & de Lope (2007) proposed a distributed approach to Reinforcement Learning; instead of dealing with a huge state-action matrix and one agent, they use several agents, one for each actuator so that the table for the states and actions is reduced significantly. However, in their approach they use position for the states (inverse kinematics) without explicity considering the dynamic aspects such as damping or inertia; the effects of different reward functions were also not studied.

In this paper we actually use torques as the action-vector. This is motivated by the fact that humans predominantly use torques for controlling joints. Therefore, considering the dynamical model is crucial in developing control methods for Humanoid Robots. The two-link planar robot is a simplification of a human arm.

In this work, we identify the optimal combination of several reward functions that would fulfill the following requirements:

- A smooth velocity profile while approaching the target position is necessary to mimic human motion.
- No overshooting after reaching the target.
- Reaching the target in a minimum number of steps.

This paper is organized as follows. Section 2 introduces the two-link planar robot and the corresponding dynamic model. Section 3 describes the reward functions that are being used. We present the experimental results in section 4.

2. TWO-LINK PLANAR ROBOT

Figure 1 shows the model of the two-link robot that was used in the current work. The equations of motion were derived using Lagrange's method. For the simulation, the shoulder and the elbow link are allowed to move $\pm \pi$ *rad*.

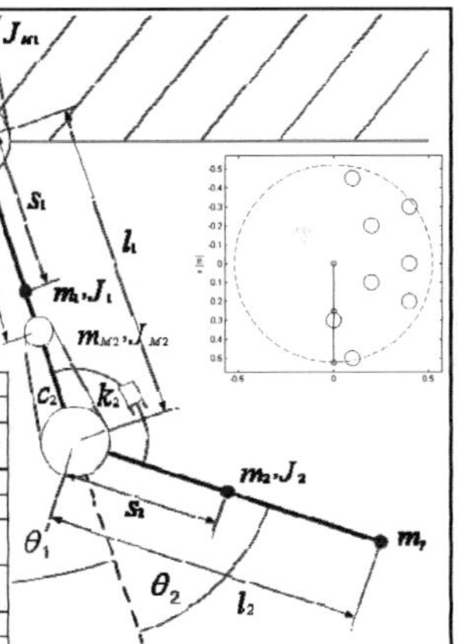

1/2. Link	Absolute angular position	Θ_i
	Length of 1st/2nd link	L_i
	Distance between 1st/2nd joint and COG of 1st / 2nd link	S_i
	Mass of 1st link	M_i
	Moment of inertia about COG	J_i
	Coefficient of Viscous friction in the bearing of the link	k_i
	Coefficient of coulombic friction	c_i
1/2. Motor	Gear ratio	i
	Distance between 1st/2nd joint and COG of motor	s_M
	Mass of 1st/2nd motor	m_{Mi}
	Moment of inertia about COG	J_M

Fig. 1. Schematic of the two-link planar robot model; Randomly generated target positions are also shown.

3. REWARD FUNCTIONS

In Reinforcement Learning, the reward that is used for updating the Q-table plays an important role in convergence and the learning rate. This behaviour is independent of the algorithm used (Q-Learning or SARSA). The reward given to the agent denotes the quality of the reached state as a result of the actions. All the following reward functions presented have one thing in common: the better the quality of the state, the higher the reward. The criteria for evaluating the states are based on the *distance*, the *velocity*, the *direction*, and an additional *penalty*. The reward based on velocity is calculated by taking the distance from the target into account. The penalty is applied when the state variable exceeds the maximal value (the robot moves beyond a prescribed boundary).

Distance

$$r_{dst,p} = 1 - \frac{d}{d_0} \tag{1}$$

$$r_{dst,e} = \frac{\beta}{1+(\frac{2d}{d_0})^n} \tag{2}$$

Velocity

$$r_{vel,p} = \left(1 - \frac{d}{d_0}\right)\left(1 - \frac{v}{v^*}\right) \tag{3}$$

$$r_{vel,g} = exp\left(-\frac{1}{2}\left(\frac{d/d_0}{\sigma_d}\right)^2\right) exp\left(-\frac{1}{2}\left(\frac{v/v^*}{\sigma_v}\right)^2\right) \tag{4}$$

$$r_{vel,e} = \left(\frac{1}{1+\left(\frac{d}{d_0}\right)^n}\right)\left(\frac{1}{1+\left(\frac{v}{v^*}\right)^n}\right) \tag{5}$$

Direction

$$r_{dir,p} = 2\left(0.5 - \left(\frac{\delta}{\pi}\right)\right) \tag{6}$$

$$r_{dir,e} = \frac{1}{1+\left(\frac{2\delta}{\pi}\right)^2} \tag{7}$$

Additional penalty

$$pen = \frac{-1}{1+\exp\left(\frac{50 s_i}{s_{i,j}} - 40\right)} \tag{8}$$

Each reward function was evaluated individually and then a combination that contains distance, velocity, direction and the penalty component was also evaluated. The composite reward function (r) is designed as a weighted sum of five of these components:

$$r = w_1 r_{vel,cc} + w_2\, r_{vel,g} + w_3\, r_{dst,e} + w_4 r_{dir,e} + w_5 pen \tag{9}$$

In order to find the suitable weighting, a fully factorial experiment is carried out. Each partial reward function is assigned a weighting between 5% and 45% ($w_{i,min} \approx 0.05$; $w_{i,max} \approx 0.45$; $i = 1, 2, \dots 5$).

The values of $w_1 \dots w_4$ are determined independently whereas the value of w_5 is calculated by enforcing the constraint $\sum_{i=1}^{5} w_i = 1$. The values for $w_{i,min}$ and $w_{i,max}$ are used to generate uniformly distributed weights for the factorial experiment. Each factor (w_i) is assigned 4 values between (and including) the minimal and maximal values.

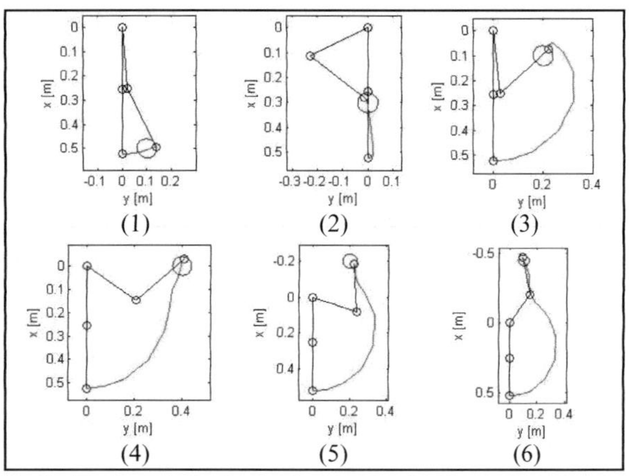

Fig. 2. Trajectories obtained with optimal combination of reward for velocity, distance, direction and penality.

4. EXPERIMENTAL RESULTS

A set of points within the workspace was randomly generated and provided as target positions for the RL algorithm. Figure 1 shows some of the randomly generated target positions. The performance (number of steps) of the RL algorithm in reaching these targets using each of the reward functions (eqns. 1-7) is shown in Table 1.

Using a four-level, four-factor fully factorial experiment and subsequent analysis of variance (ANOVA), the best reward function was obtained as:

$$r_{opt} = 0.29\, r_{vel,cc} + 0.07\, r_{vel,g} + 0.34\, r_{dst,e} + 0.14 r_{dir,e} + 0.16\, pen \tag{10}$$

The trajectory of the robot for six of the target positions using the optimal reward function (eq. 10) is shown in Figure 2. Each target was reached in a minimum number of steps, without overshoot and with a smooth velocity profile when reaching the target position (see Fig.2). The experiment details can be found in (Denzinger, 2008).

Target / Reward Function	Number of steps					
	1	2	3	4	5	6
$r_{dst,p}$	4	38	28	35	35	37
$r_{dst,e}$	4	38	28	35	35	37
$r_{vel,p}$	4	7	150	150	150	150
$r_{vel,g}$	4	8	10	9	25	34
$r_{vel,e}$	4	18	9	12	24	37
$r_{dir,p}$	4	28	14	19	20	26
r_{opt}	4	8	8	10	14	16

Table 1. The number of steps taken to reach each target position using each reward function.

5. CONCLUSION

In this paper, we empirically obtain the optimal composition of a set of reward functions for a dynamic model of a two-link planar robot, where torque commands are used as part of the action-vector. These reward functions were evaluated based on fulfilling certain requirements, and repeated for several target positions. Based on this study, we find that these reward functions are dependent on the dynamic model of the robot. Future work consists of identifying reward functions that are *independent* of the underlying robot model. More investigation is also needed to include velocity profiles within the framework of Reinforcement Learning through appropriate reward functions. This is especially important for mimicking human behaviour.

7. REFERENCES

Denzinger, J. (2008) Implementation of distinct Reinforcement Learning algorithms for the control of a 2-DOF manipulator model, Diplomarbeit, Institut für Produktentwicklung, Universität Karlsruhe (TH).

Martin-H., J. A. and De-Lope, J. (2007) A Distributed Reinforcement Learning Architecture for Multi-Link Robots. In *ICINCO 2007,* pp. 192-197.

Rummery, G. A. and Niranjan, M. (1994) On-line Q-Learning using connectionist systems. Technical Report CUED/F-INFENG/TR 166, Engineering Department, Cambridge University.

Sutton, R.S. and Barto, A.G. (1998). *Reinforcement Learning, an introduction*, The MIT press.

Watkins, C. J. and Dayan, P. (1992). Technical note Q-Learning. *Machine Learning,* 8:279.

Annals of DAAAM for 2008 & Proceedings of the 19th International DAAAM Symposium, ISSN 1726-9679
Editor B. Katalinic, Published by DAAAM International, Vienna, Austria 2008
Make Harmony Between Technology and Nature, and Your Mind will Fly Free as a Bird

SURFACTANTS INFLUENCE ON THE PAPER OPTICAL CHARACTERISTICS EXPOSED TO AGEING

DERMANOVIC, I[vana]; BARBARIC MIKOCEVIC, Z[eljka]; DZIMBEG MALCIC, V[esna] & COSIC, K[ristina]

Abstract: *In this work the influence of different surfactants and their mass fractions on the optical characteristics of paper was examined. Sodium lauryl ether sulfate, cocamidopropyl betaine and fatty alcohol etoksylate like surface active agent were used. The research was carried out with filter paper made from virgin fibres and photocopy paper made from recycled fibres. The reflective spectra of treated papers were measured before and after accelerated ageing. The differences in the optical properties caused by ageing and by surface active agent have been discussed. The results have shown that surfactants reduce filter paper reflectance and increase photocopy paper reflectance. Increasing of surfactant mass fractions inhibits the influence of visible and near UV radiation on fibres.*
Key words: *Surfactants, samples, reflectance, ageing*

1. INTRODUCTION

Almost all of the paper that is used today is made of wood fibers (Stenius, 2000). Besides fibers as main ingredients for paper manufacturing, pigments/dyes, adhesives, additives, fillers, starch and other chemicals are necessary. According to paper type these materials are added in bigger or smaller quantities (Neimo, 1999). Paper products are an important part of everyday life (households, offices and industry). Nearly all paper is eventually thrown away. The use of recycled fibers for paper manufacture is highly desirable.

The paper recycling process is a way to wash, remove, and bleach unwanted particles from collected printed paper so that it can be sold as an alternative to virgin paper (Behin & Vahed, 2007). Because of their physical and chemical characteristics surfactants are widely used in some subprocesses of recycling.

Apart from their primary function cleaning, surfactant choice and blending can influence product structure, visual appearance, foaming characteristics, sensory perception and mildness. The type of surfactant employed depends on requirements like removal of dirt from fibers and surfaces, conditioning of fibers or structuring of products.

A surfactant is a molecule with two distinct regions – a water soluble polar head and a non polar hydrocarbon tale which is sparingly soluble in water. There are three types of polar heads: ionic (cationic and anionic), zwitterionic and non-ionic. However, anionic and nonionic surfactants are more commonly used (Zhao et al., 2004). Surfactants will absorb at an air-water or oil-water interface and at the surface of a solid. The long hydrocarbon tail interacts with fats and oils whilst the water soluble head group bonds with water. This action removes and captures dirt particles, and prevents their re-deposition.

Paper optical characteristics in graphic profession are extremely important. Because of different terms of recycling process, as well as paper ageing (Jääskeläinen & Liitiä, 2007), optical characteristics are significantly changed. Surfactant as well plays an important role in the optical properties.

The purpose of this paper was to examine the surfactants influence on the optical characteristics of filter paper made only from virgin fibres and of photocopy paper made from recycled fibres. The research of that chosen samples was carried out at the same experimental conditions.

2. METHODS AND MATERIALS

Photocopy paper made from recycled fibre and filter papers were used like experimental samples.

The samples have been separately soaking (10 minutes) in three different surfactants solution: i) sodium lauryl ether sulfate, SLES, (anionic), ii) cocamidopropyl betaine, CAPB, (zwitterionic) and iii) fatty alcohol etoksylate, FAE , (nonionic) for two surfactant mass fractions (1,0%, 2,5%).

The samples were dried naturally at room temperature and after that accelerating ageing by electric lamp that emits visible and ultraviolet electromagnetic radiation was performed. Examined samples were exposed to radiation in time interval of 90 minutes.

Papers reflectance measurements were processed using X-rite Spectrophotometer, Digital Swatchbook in the interval of the wavelengths from 410 nm to 700 nm for every 10 nm. These measurements were supported by ColorShop 2.0 software and analysed by Technical Graphic Origin 6.0 Professional. Absolute reflectance was measured over white background. The reflectance is the basis for the evaluation of experimental samples optical properties.

In purpose of better evaluation, reflectance spectra of non soaked samples in surfactants solution, called original sample, were measured also.

3. RESULTS

The reflectance values of filter paper soaked in surfactants solution are increased for the wavelengths lower then 460 nm and decreased for the wavelengths higher then 460 nm in comparison with original filter paper (Fig. 1.).

Reflectance values of original filter paper exposed to electromagnetic radiation are significantly reduced. The highest reduction is noted for the wavelengths lower then 550 nm. Filter paper reflectance values, soaked in surfactant solution (w_{SAA} = 1,0 %), are reduced in comparison with original for the wavelengths higher then 450 nm, while the reflectance values of aged soaked filter paper are extra reduced for the same wavelengths. In a whole wavelengths area soaked filter paper reflectance differences before and after ageing are significantly lower than original reflectance differences (Mikac Dadić et al., 2003).

Surfactant mass fractions increase leads to soaked filter paper reflectance values increase for wavelength lower then 460 nm and to reduce reflectance values at wavelength higher then 460 nm in comparison with non aged original (Fig. 2.). With ageing reflectance of soaked filter paper is slightly changed. Reflectance differences soaked filter paper before and after ageing are reduced with increasing of surfactant mass fractions (Fig. 1., Fig. 2.).

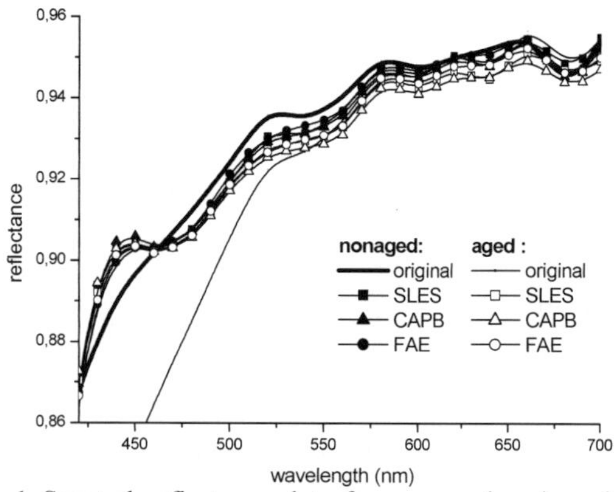

Fig. 1. Spectral reflectance data for non aged and aged filter paper soaked in different surface active agent; (w_{SAA} = 1,0 %).

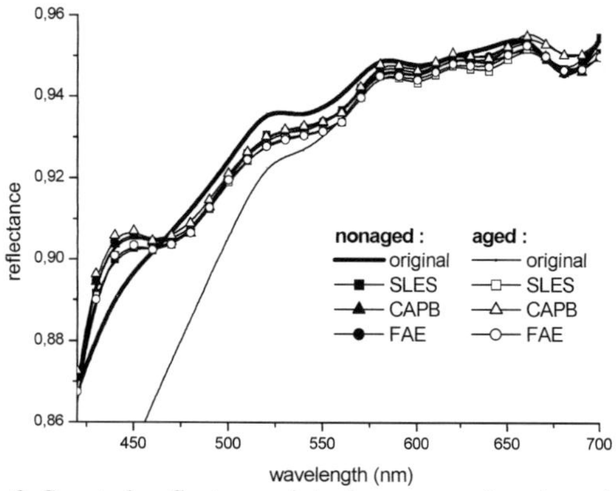

Fig. 2. Spectral reflectance data for non aged and aged filter paper soaked in different surface active agent; (w_{SAA} = 2,5 %).

Reflectance of aged original photocopy paper is slightly reduced for wavelength lower then 550 nm, and increased for wavelength higher then 550 nm (Fig. 3.) when we compare it with non aged photocopy paper.

Reflectance differences for aged and non aged original photocopy paper are significantly smaller then for original filter paper (Fig. 1., Fig. 3.). With soaking in all surfactant solution (w_{SAA} = 1,0 %) photocopy paper reflectance values are increased in comparison with original (non treated sample).

The surfactant mass fractions increase from 1,0% to 2,5% lead to slightly reflectance values increase of soaked non aged photocopy paper (Fig. 4.).

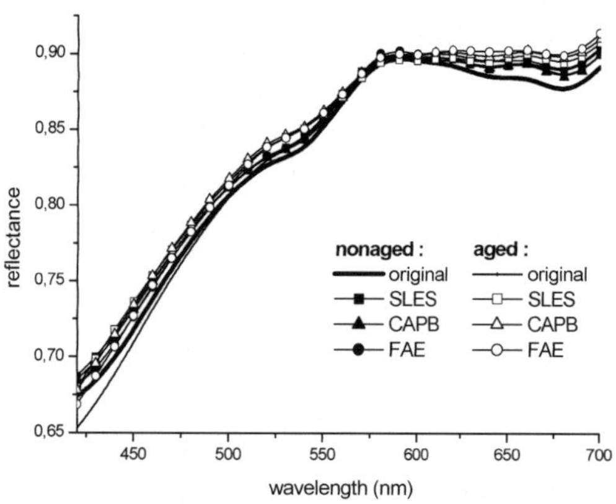

Fig. 3. Spectral reflectance data for non aged and aged photocopy paper soaked in different surface active agent; (w_{SAA} = 1,0 %).

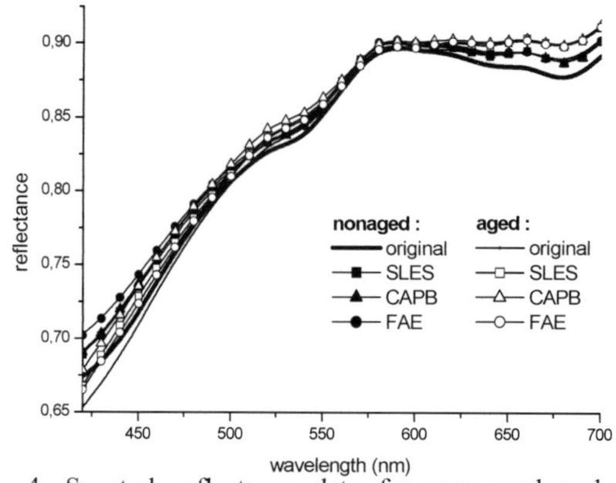

Fig. 4. Spectral reflectance data for non aged and aged photocopy paper soaked in different surface active agent; (w_{SAA} = 2,5 %).

With ageing those experimental samples reflectance is reducing for wavelength lower then 500 nm and increasing for wavelength higher then 500 nm. Reflectance differences of soaked photocopy paper before and after ageing are higher with increase of surfactant mass fractions (Fig. 3., Fig. 4.), especially at wavelength lower then 500 nm.

4. CONCLUSION

Paper structure and surfactant mass fraction have important influence on reflectance spectra of examined samples. Filter paper is made only from virgin fibers unlike photocopy paper which is made from recycled fibers.

Near UV and visible electromagnetic radiation much more decrease reflectance values of filter paper compare to photocopy paper. Previously manufacturing and recycling processes had caused exhausted hydrogen bonds in the photocopy paper. Those fibers were chemically saturated so the influence of UV radiation on paper is significantly reduced.

Molecules of ionic, anionic and zwitterionic surfactants reduce filter paper reflectance and increase photocopy paper reflectance. Increasing of surfactant mass fractions inhibit the influence of UV radiation on fibres.

Assumptions based on results are: i) surfactants support removing the impurities from photocopy paper or ii) surfactants molecules associate with fibres what increase reflectance in comparison with original photocopy paper.

Those assumptions would be interesting to verify and affirm in further researches.

5. REFERENCES

Behin J. & Vahed Sh. (2007). Effect of alkyl chain in alcohol deinking of recycled fibres in flotation process. *Colloids and Surfaces A,* 297(2007) (131.-141.)

Jääskeläinen, A-S. & Liitiä (2007). UV/vis reflectance spectroscopy reveals the changes in fibre chemistry during ageing. *SpectroscopyEurope,* Vol. 19, No. 5, (October / November 2007) (11.-13.)

Mikac Dadić V.; Džimbeg-Malčić V.; Bolanča I. & Bolanča Z. (2003). Optical characteristic of prints exposed to natural and accelerated ageing. *DAAAM International Scientific Book*, (July 2003) (401.-414.)

Neimo L. (1999.) *Papermaking Chemistry*, Fapet Oy, ISBN 952-5216-04-7, Helsinki, Finland.

Stenius, P. (2000). *Forest Products Chemistry*, Fapet Oy, ISBN 952-5216-03-9, Helsinki, Finland.

Zhao, Y.; Deng, Y. & Zhu, J.Y. (2004). Roles of Surfactants in Flotation Deinking. *Progress in Paper Recycling,* Vol. 14, No. 1, (November 2004) (41.-45.)

Annals of DAAAM for 2008 & Proceedings of the 19th International DAAAM Symposium, ISSN 1726-9679
Editor B. Katalinic, Published by DAAAM International, Vienna, Austria 2008
Make Harmony Between Technology and Nature, and Your Mind will Fly Free as a Bird

ON A GENERALIZED ALGORITHM OF THE TECHNICAL PRODUCT CONCEPTUAL DESIGN

DIACONESCU, D[orin]; JALIU, C[odruta]; NEAGOE, M[ircea] & SAULESCU, R[adu]

Abstract: *Based on a comparative analysis of the conceptual design algorithms proposed in literature, a new variant of the conceptual design modeling, whose algorithm generalizes the existing algorithms, is proposed in the paper. The structure of the generalized algorithm starts from the requirements list and contains: a) global function statement, b) global function description by structures of sub-functions, c) solving structures generation, and d) selection of the best solving structure (the concept) by evaluation.*

Key words: *Conceptual design, product concept, algorithm.*

1. INTRODUCTION

The conceptual design denomination is derived from the older notion of product concept, used for the result of this design. The new terminology promotes the denomination of the product principle solution instead of the concept notion; momentary both notions coexist in literature.

Based on a comparative analysis of the algorithms that are proposed in literature (see references), the authors propose a new modeling variant, for the conceptual design modeling, whose algorithm generalizes the existing algorithms.

2. COMPARATIVE ANALYSIS OF CONCEPTUAL DESIGN MODELS

Several models for conceptual design modeling are proposed in literature. The most representative of them belong to Cross, Ulrich & Eppinger, Dieter, Pahl & Beitz and VDI. From the comparative analysis of these models, the following useful conclusions come out:

2.1. On the requirements list elaboration

Unlike the first three mentioned models (Cross, 1994; Ulrich & Eppinger, 1995; Dieter, 2000), in which the requirements list is considered a step of the conceptual design, the requirements list is input entity in the German models (Pahl & Beitz, 1995; VDI, 1997); therefore, in the German view, the requirements list is a distinct activity, relatively autonomous.

2.2. On the strictness of terminology

From the analysis of the used terminology in the mentioned models, some interesting observations come out:

- Cross's model (Cross, 1994) uses a terminology of general type, like: objective, problem and sub-problem, function and sub-function, requirement, characteristic, alternative (variant), sub-solution, solution etc.
- The models of the American authors, Ulrich & Eppinger (Ulrich & Eppinger, 1995) and Dieter (Dieter, 2000), use a more elaborated terminology which has also specialized notions, like: target specification, product concept, concept development, the concept of a sub-problem, integrated solution, project development etc.
- The German models use the most rigorous terminology, and between them, the VDI model (VDI, 1997) has the most accurate form. Beside the general notions, there are used strictly specialized notions as: structure of functions, solving principle, physical effect, effect carriers, configurations of the effect carriers, solving structure, principle solution, effects plan, configuration plan etc.

2.3. On the models' common denominator

Though they have different formulations, all the mentioned design models can be reduced, through simplification, to the same common denominator: the algorithm for solving cycle of a technical problem. This algorithm can be found in the simplified variant of the type: global problem → sub-problems → sub-solutions → global solution in the structure of each design model that was previously mentioned.

3. GENERALIZED ALGORITHM

Starting from the product design variant modeled in Fig. 1, a new variant that models the product conceptual design is proposed in Fig. 2, derived from the previous mentioned models by generalization. The structure of the generalized algorithm from Fig. 2 is centered on an adequate base of information, starts from a requirements list and contains:
1°. Global function statement. Result: The global function,
2°. Global function description by structures of sub-functions. Result: The structure of the global function,
3°. Solving structures generation by: solving the sub-functions, sub-solutions composition and elimination of the inadequate solutions. Result: Solving structures,
4°. The best solving structure selection by evaluation.
Result: The product principle solution (the concept).

The inverse connections between steps (Fig. 2), which are necessary in the iterative optimization and in readapting the requirements list, can be made directly, through the peripheral information flow, or indirectly through the information base.

Among the four steps (Fig. 2), the solving structures generation (step **3**) represents the key step; for this step, illustrated shaded in Fig. 2, a general algorithm was detailed in Fig. 3. This algorithm (Diaconescu et al., 2008), derived from the homonymous algorithm proposed by Ulrich and Eppinger (Ulrich & Eppinger, 1995) considering the VDI terminology, is centered on an adequate information base, starts from a known structure of sub-functions and contains (see Fig. 3):
1°. Grouping of the sub-functions into: a) functions of **U.S.P.** type (**U**nknown **S**olving **P**rinciples) and b) functions of **K.S.P.** type (**K**nown **S**olving **P**rinciples).
Results: Functions of **U.S.P.** type and functions of **K.S.P.** type.
2°. Solving principles synthesis (for **U.S.P.** functions), i.e. synthesis of physical effects and configuration of effects carriers. Result: New solving principles.
3°. Identification of known solutions. Result: existent solutions.
4°. Generation of the solving structures, which contains: 4.1° solving structural variants generation and 4.2° pre- establishing and keeping the variants whose technical features satisfy quantitatively the requirements list. Result: solving structures.
In the case of the **K.S.P.** functions, the identification of the existent solutions is based on two main sources: the technical literature and the existent technical systems. For the **U.S.P.**

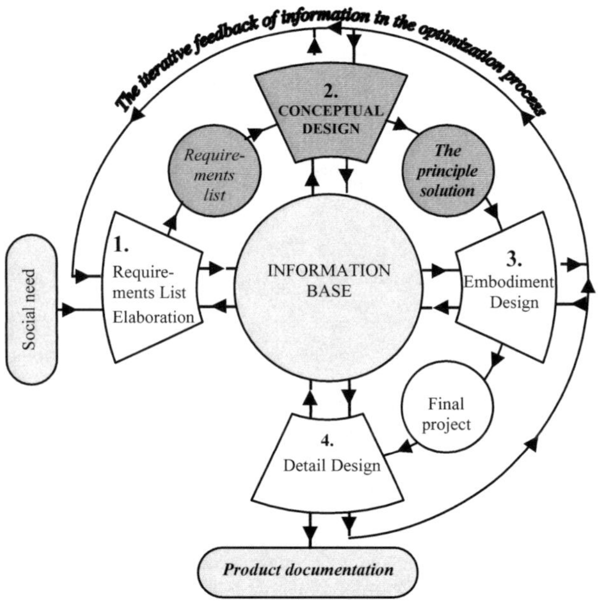

Fig. 1. Generalized algorithm of the Product Design process.

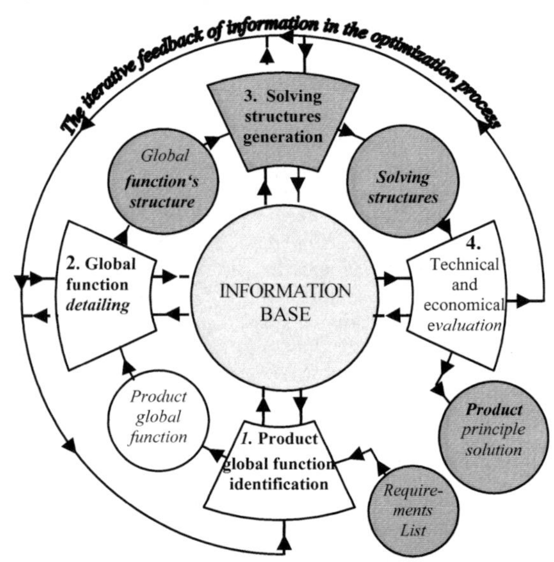

Fig. 2. Generalized algorithm of the conceptual design.

functions, the physical effects and/or the configuration of the effects' carriers can be found by: conventional methods (as: analysis of the natural systems, models analogy etc.), intuitive methods (as: Delphi method, Brainstorming, etc.) and logical methods (as: search by using catalogues of physical effects, the variation of the effects' carriers configuration for known solutions, etc.).

The sub-solutions are usually combined by using morphological matrixes (VDI, 1997).

4. CONCLUSIONS

The following conclusions can be formulated based on the algorithms from Fig. 1, 2 and 3:

a. Each step from Fig. 1 represents a distinct activity, relatively autonomous, with its own notions and methods.

b. According to Fig. 2 and Fig. 3, the global function of a product has elements of novelty if: $\mathbf{b_1}$) it has at least one **U.S.P.** function and/or if: $\mathbf{b_2}$) at least one original connection interferes in the global function structure.

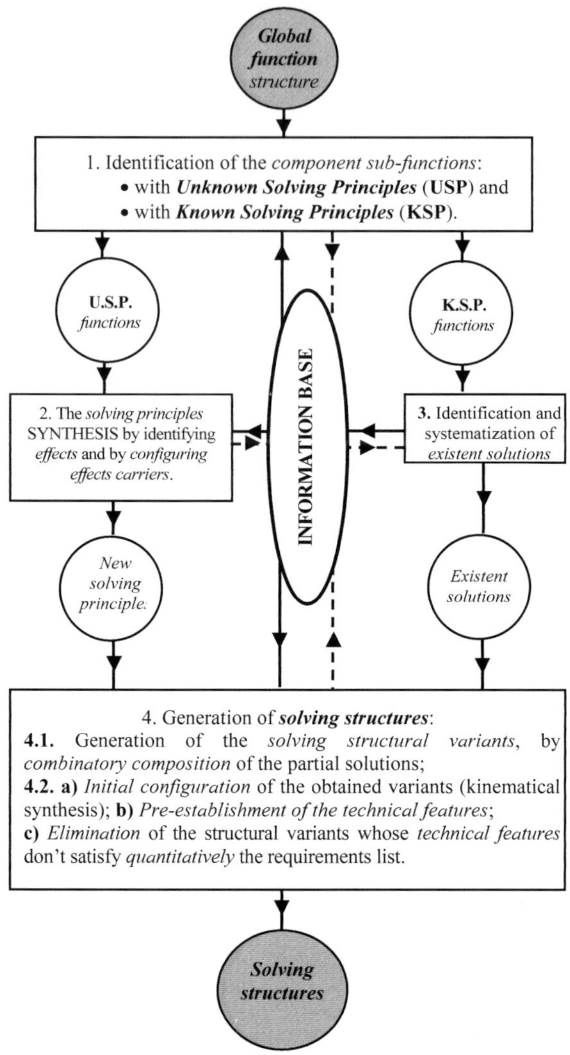

Fig. 3. Algorithm of solving structures generation.

c. In terms of novelty, more design types can come out from the global function of a product (Pahl & Beitz, 1995): 1) Original design, 2) Adaptive design and 3) Variant design; in each of them, the principle solution contains: 1) elements of originality at least in the plan of effects or at least in the plan of connections; 2) elements of novelty only regarding the configuration of the effects carriers; 3) known effects, known effects carriers configurations and known connections; therefore, the variant design is a form of design through similitude (by the „model").

5. REFERENCES

Cross, N. (1994). *Engineering Design Methods*, J. Wiley & Sons, ISBN 0-4719-4228-6, New York

Diaconescu, D. et al. (2008). *Products' Conceptual Design*, Transilvania University Publishing House, ISBN 978-973-598-230-0, Brasov

Dieter, G. (2000). *Engineering Design*, McGraw Hill, ISBN 0-07-116204-6, Boston

Pahl, G. & Beitz,W. (1995). *Engineering Design*, Springer, ISBN 3540504427, London

Ulrich, K. & Epinger, S. (1995). *Product Design and Development*, McGraw-Hill Inc. ISBN 0-07-113742-4, New York

VDI - Verein Deutscher Ingenieure (1997). *Richtlinien 2221 and 2222*

Annals of DAAAM for 2008 & Proceedings of the 19th International DAAAM Symposium, ISSN 1726-9679
Editor B. Katalinic, Published by DAAAM International, Vienna, Austria 2008
Make Harmony Between Technology and Nature, and Your Mind will Fly Free as a Bird

STUDY OF DEFORMATIONS IN RAM-BORING SPINDLE

DINU, G[abriela]; ATANASIU, C[ostica]; BACIU, F[lorin] & NUTU, E[mil]

Abstract: *In this paper we present a ram-boring spindle assembly of a boring and milling machine with spindle sleeve which is analyzed by methods of classical strength calculation, by calculation involving finite elements and by experimental tests.*
Using these two methods, we have established the assembly deformations values, which are used by the machine-tool operator to introduce the required corrections to obtain the prescribed precision.
Analyzing the three operation modes in which these two assemblies (ram and boring spindle) can work, the highest values of the deformations was obtained for the case with cumulated strokes of the ram and the boring spindle.
Key words: *contact stress, deformations, ram, spindle*

1. INTRODUCTION

The machine tools are designated to generate the parts surfaces, using the technological process of cutting, following some conditions for dimensional accuracy, for surface quality and for productivity. The technological accuracy of a machine tool is a sum of multiple precision types: geometric, static, cinematic and dynamic accuracy.

The ram type horizontal boring mills with displaceable column are used for the machining of the cast or welded parts from the field of nuclear, energetic, metallurgic or military industry by mean of milling, drilling, boring and threading operations. The machines can be used both for roughing, finishing operations and superfinishings(Gheorghiu et al., 2007, Hadar et al., 2007). The lay-out of the machine is presented in Fig. 1, where the main assemblies are shown.

A simplified section along the machine headstock is presented in Fig. 2.

The mechanical assemblies of different types represent weak parts in the operation of the machine-tools. The forces transmitted by assemblies may lead to concentrations of stress.

Being able to generate strains, this stress affects the accuracy of the machine and is also able to create supplementary contact loads that could produce the breakage of the surface, by fatigue (Hadar et al., 2007).

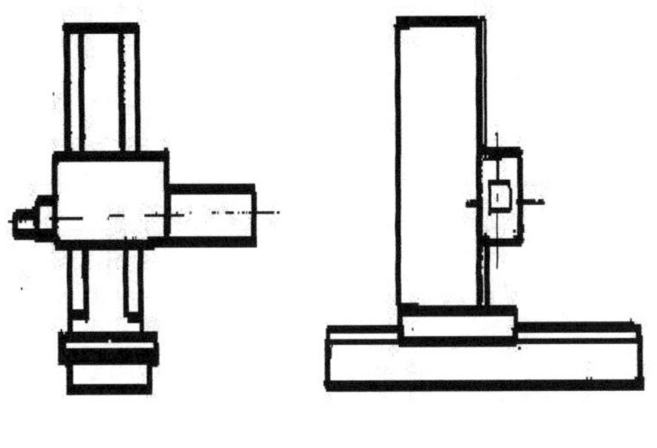

Fig. 1. The machine block diagram.

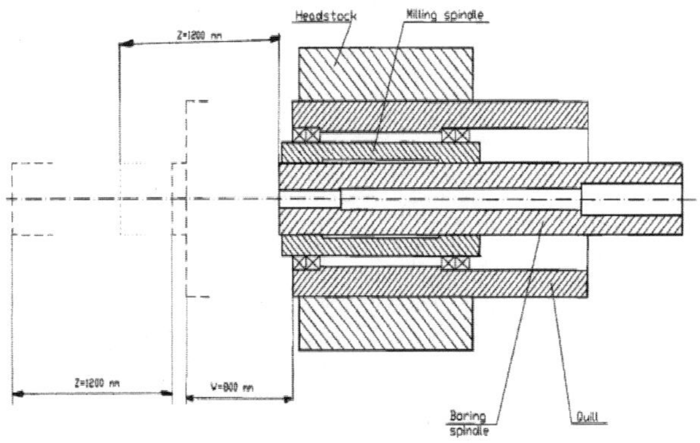

Fig. 2. The machine section.

The boring spindle is a pipe type part, guided inside the milling spindle by two bronze sleeves. The ram (box shape) is guided inside the headstock on hydrostatic guide-ways, with controlled pressure on each pocket.

2. NUMERICAL AND EXPERIMENTAL STUDY

The finite elements method study, performed by the COSMOS program, refers to the situation in which the assembly is loaded by its own weight and by the forces resulted in the cutting process, Fx, Fy and Fz, in three positioning cases (Gheorghiu et al., 2006):

CASE 1 - The ram slides out by 800mm stroke;
CASE 2 - The boring spindle slides out by 1200mm stroke;
CASE 3 - The ram is out by 800mm and the boring spindle is out by 1200mm.

The assignment of the deformations that were obtained is:
CASE 1: 0,0133mm;
CASE 2: 0,0288mm;
CASE 3: 0,038mm.

The distribution of deformations is presented in Fig. 3 (for CASE 1), Fig. 4 (CASE 2) and Fig. 5 (CASE 3).

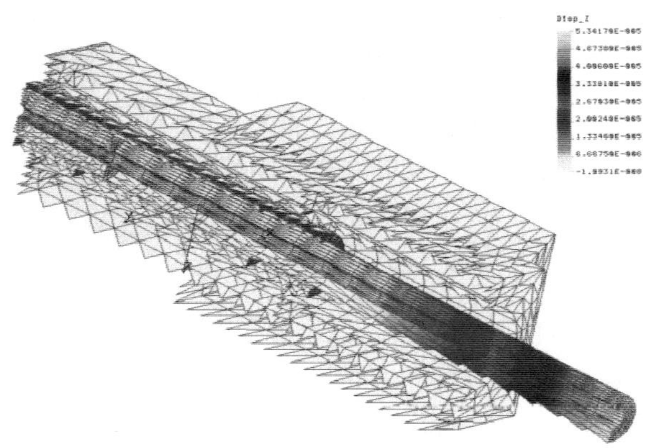

Fig. 3. The distribution of deformations for case 1

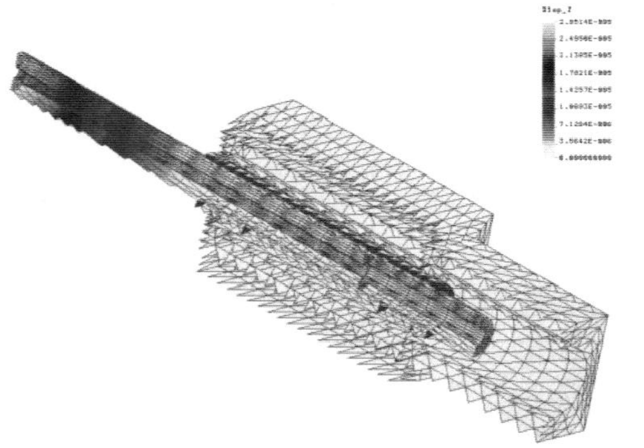

Fig. 4. The distribution of deformations for case 2

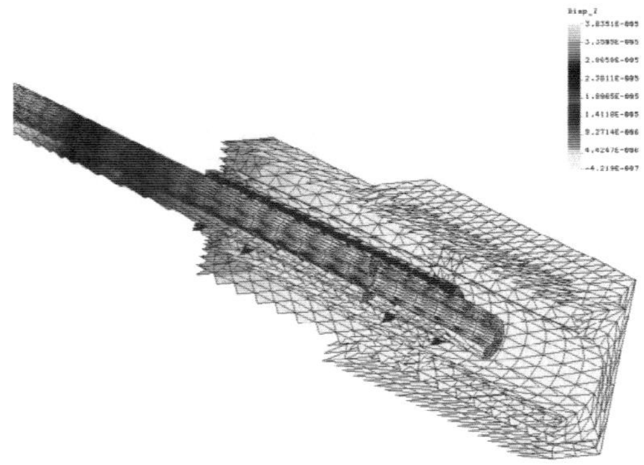

Fig. 5. The distribution of deformations for case 3

The experimental tests have been performed by a specialized company which made the measurements of the deformations by the help of micro-comparators and laser systems. The field point was in the same location for all situations, which is, at the free socket head of the boring spindle. The values of deformations obtained are given in:

CASE 1: 0,009mm;

CASE 2: 0,031mm;

CASE 3: 0,040mm.

As shown in Fig. 6 and Fig. 7, there are contact stresses along the contact surfaces of the assembly ram-headstock, especially for top and bottom surfaces.

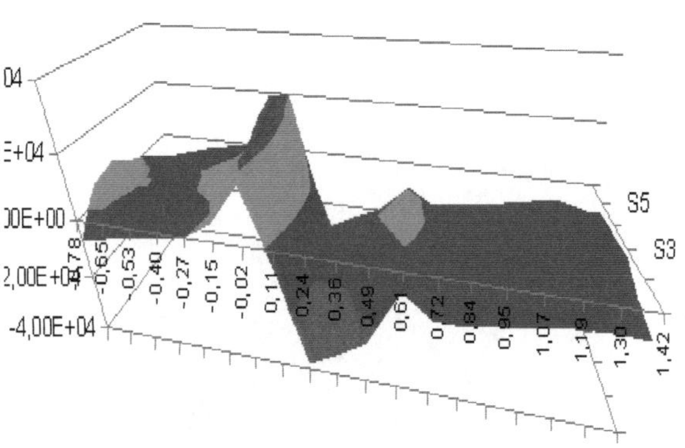

Fig. 6. The top are contact stress.

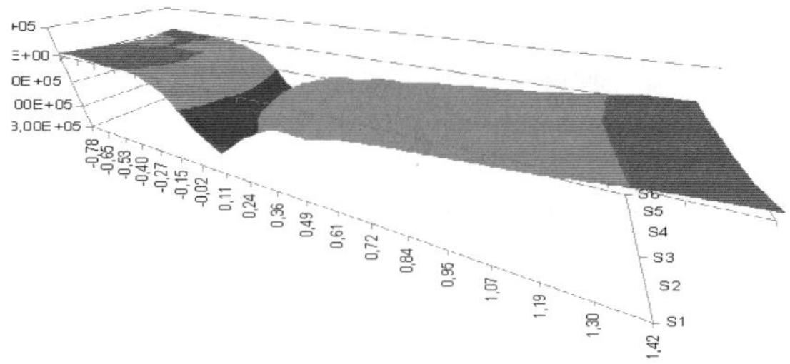

Fig. 7. The bottom are contact stress.

The calculated values range between the following limits: $2{\cdot}10^4$ - $3{\cdot}10^5$ Pa (Buzdugan, 1980). Even those relatively small values affect the precision of machine-tool. For the case when the ram is at maximum stroke out, one can see that the values of the contact stress on the bottom surface in front of the headstock are several times higher than the values on the top surface. The results can be used by the designer or by the machine operator (user) to set higher values of oil pressure for the first bottom pockets of the ram.

3. CONCLUSION

The conclusions are:

•It is to be noticed that deformations in normal limits have been obtained and, in the most unpropitious situation, as CASE 3, with deformations of 3,83·10-2mm;

•CASE 1 has the lowest deformations and CASE 2 has average ones;

•The results obtained by the two methods are comparable, the unwanted situations being noticed;

•The results are useful for the machine-tool's user, because the user can insert corrections so that the machine's accuracy will not be affected;

•The values of contact stress along the contact surfaces of the assembly ram-headstock, especially for top and bottom surfaces range between the following limits: $2{\cdot}10^4$ - $3{\cdot}10^5$ Pa. Even such relatively small values can affect the precision of machine-tool;

•For CNC machine tools, the value of contact deformations can be adjusted with some electronic corrections.

4. REFERENCES

Buzdugan, Gh. (1980). Strength of Materials, ed. XI-a, Ed. Tehnica, Bucuresti

Gheorghiu, H. & Hadăr, A. (2007). Determining Deformations and Existing Tensions in Structures Made from Bedded Composite Materials with Fibers, Proceedings of the 9th International Conference on Management of Innovative Technologies, 9-10 October, Fiesa, Slovenia, p. 191-195 and on C.D

Gheorghiu, H.; Constantinescu, I.N.; Hadar, A. & Petre, C. (2006). The numerical methods for calculation in strength structural , Ed. BREN, Bucuresti

Hadar, A.; Constantinescu, I.N.; Gheorghiu, H. & Coteţ, C.E (2007). Simulation and models for estimating in mechanics engineering, Ed. Printech, ISBN 978-973-718-759-8, Bucureşti

Hadar, A.; Szabo, A.; Bordeaşu, I.; Gheorghiu, H. & Szabo Şt. (2007). Contributions regarding force differences usage in weighting devices, The 18th INTERNATIONAL DAAAM SYMPOSIUM, "Intelligent Manufacturing & Automation: Globalisation – Technology – Men – Nature", p. 315-316, ISSN 1726-9679

Annals of DAAAM for 2008 & Proceedings of the 19th International DAAAM Symposium, ISSN 1726-9679
Editor B. Katalinic, Published by DAAAM International, Vienna, Austria 2008
Make Harmony Between Technology and Nature, and Your Mind will Fly Free as a Bird

STUDY THE STRENGTH AND STIFFNESS OF A BORING AND MILLING MACHINE BED

DINU, G[abriela]; BACIU, F[lorin]; VLASCEANU, D[aniel] & DRAGHICI, S[orin]

Abstract: *This paper's goal is to study the strength and stiffness for the bed of a boring and milling machine.*
Along the studied bed the slide of the machine column is moving. The headstock, with the main spindle has a vertical movement on the column guide-ways. So, during the working process, the bed is variably loaded (value, direction, application surfaces).
For that reason, is very important the construction of this structural part and it clamping and adjusting devices on the foundation, too.
The bed was studied by the finite elements method, using the performance program named "COSMOS/M". The range of values derived for the bed strains is comprised between 0.05-0.12 mm.
Key words: *contact deformations, foundation*

1. INTRODUCTION

The machine tools are designated to generate the parts surfaces, using the technological process of cutting, following some conditions for dimensional accuracy, for surface quality and for productivity. The technological accuracy of a machine tool is a sum of multiple precision types: geometric, static, cinematic and dynamic accuracy (Gheorghiu et al., 2006).

The requirements imposed to a machine-tool are the processing accuracy, the operating safety and the operating and maintenance manageability (Budynas, 1999). The processing accuracy is defined by the stiffness of different elements of the structure and by the strains and displacements that occur during the operating period.

2. STUDY ON THE STIFFNESS COMPUTATION

This paper's goal is to study the strength and stiffness of a boring and milling machine bed. The sliding guide of the bed is represented in Fig.1, effecting some approximations as regards to the real sections of the bed and using simple sections and parameters on the entire length.

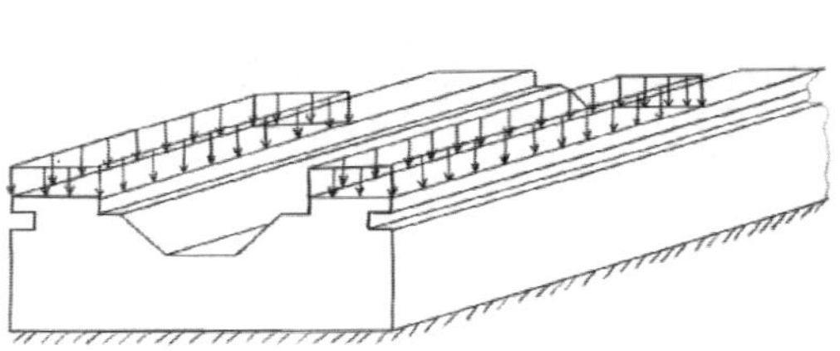

Fig. 1. The sliding guide of the bed.

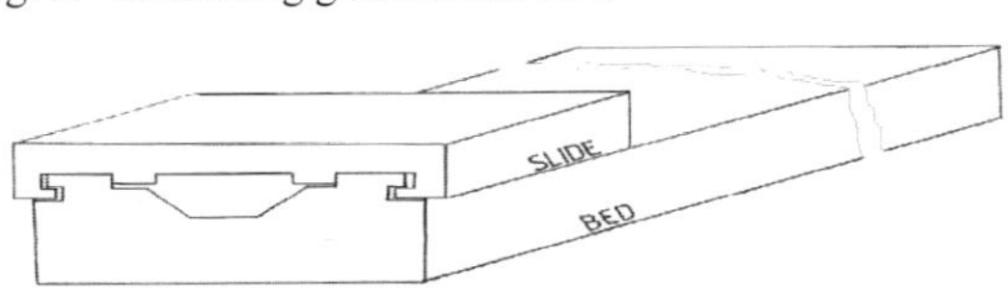

Fig. 2. The assambly slide and bed

The bed is put to flexion and torsion being loaded up according to the Fig.2. The uniformly distributed load reacts on the bearing faces of the transit path but only on the length of the pedestal base slide. The operating version of the pedestal base slide laid on the studied bed is the socket head position which represents the most unfavorable situation.

The bed was studied by the finite elements method, using the performance program named "COSMOS/M" (Jiga et al., 2007). Fig. 3 presents equivalent displacements values and the displacements values on the uniformly distributed load direction are presented in Fig. 4.

The displacements into the vertical plane on the bed longitudinal direction are analyzed first of all. The machine tool clamping system is more important than those into the cross plane of the bed, because the former ones have a bigger influence for the machine tool working accuracy (Iliescu et al., 2005).

The range of values derived for the bed displacements is between 0.02-0.12 mm.

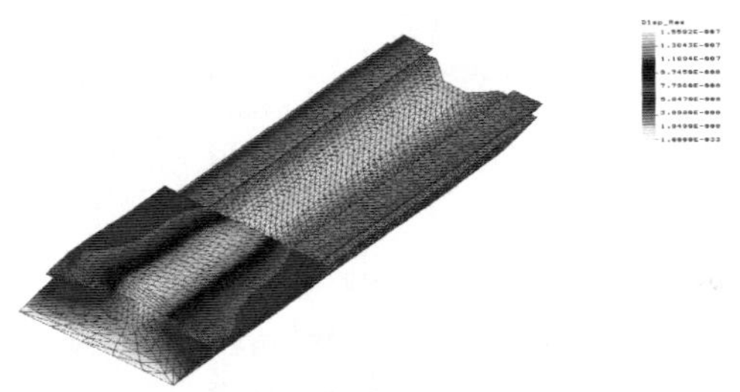

Fig. 3. The equivalent displacements.

Fig. 4. The displacements values on the load direction.

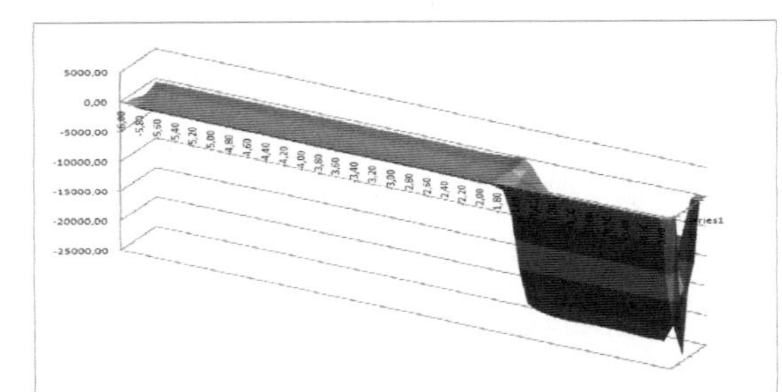

Fig. 5. The contact stress along the contact surfaces left guide-way.

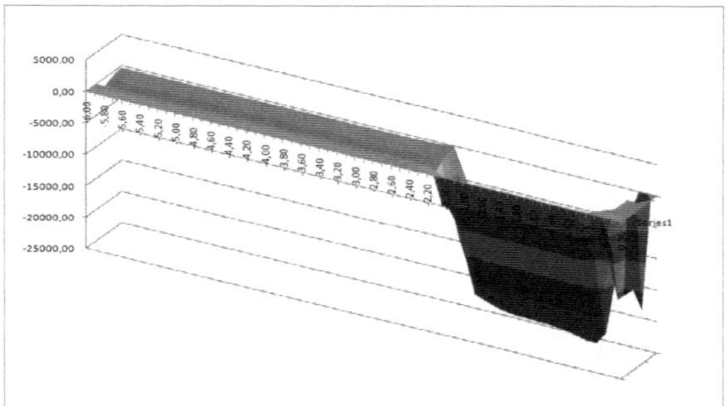

Fig. 6. The contact stress along the contact surfaces right guide-way.

Fig. 5 and Fig. 6 present the contact stress along the contact surfaces of the assembly slide-bed, especially for left and right guide-way(Atanasiu et al., 2007). As shown in the figures, the values of the contact stress on the bed guide-ways are in acceptable limits, but for a machine with a longer stroke on Y axis (that means a heaviest column), we suggested to use a bigger contact surface of the studied objects.

Using the same program, it is possible to observe in Fig.7, the equivalent Von Mises strains, which have acceptable values (Buzdugan et al., 1980).

On a plain model of the bed an experimental study has been performed, by tensiometric measurements, using an inductive transducer(Gheorghiu et al., 2007). The model was realized at a scale of 1:5 for the dimensions within the section and at a scale of 1:300 for the length. An L shaped plate was applied on the model bed by welding and sticking, exactly on the higher side, according to the Fig. 8.

For an equivalent force of F=200 kN a strain value of 0.069 mm in the measure has been obtained. A specialized company performed a technical research of the bed stiffness using high precision levels and a laser equipment and obtained real displacements during the operation between 0.01- 0.15 mm.

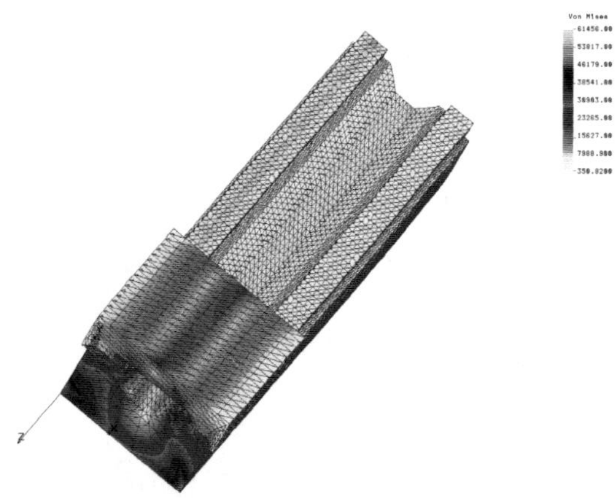

Fig. 7. The equivalent Von Mises strains.

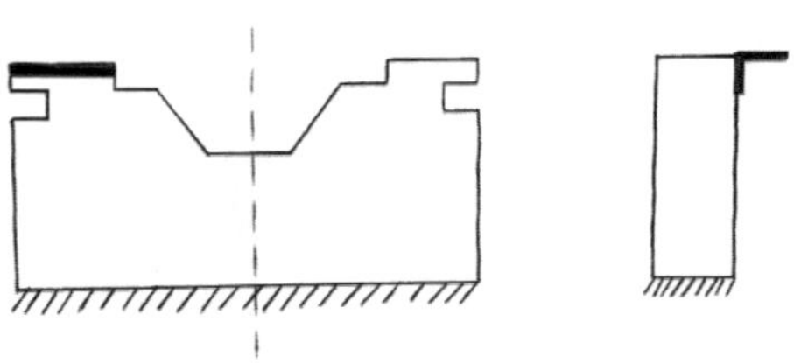

Fig. 8. The plain model of the bed there has been performed an experimental study.

3. CONCLUSION

•The displacements values obtained by the operating version presented range within the normal limits for a correct operating and are comparable to one another;

•The difference between the displacements values obtained by experiments and those obtained using the numerical method is determined by the amendment of the study. Namely, the bed modeling is plain because it was difficult to make a space model;

•Based on the results of the study, the designer and the producer of the horizontal boring machine decided to find some technical solutions to decrease the influence of the bed deformations up to the top end of the tool: a strong foundation, supplementary point to sustain the slide, special technology for the column guide-ways grinding etc;

• For CNC machine tools, the value of contact deformations can be adjusted with some electronic corrections;

• For high precision machine tools (grinding machines, gear cutting machines, jig boring machines, machining centers), the deformations and contact displacements, like those between the bed and the clamping keys, can have a major influence for the machine tool accuracy;

•The values of deformations obtained with these two methods are comparable;

•The values of contact stress along the contact surfaces of the assembly slide-bed, especially for left and right surfaces range between the following limits: $2 \cdot 10^4$ - $3 \cdot 10^5$ Pa. Even such relatively small values can affect the precision of machine-tool;

•The research (with numerical methods and an experimental model) will be continued for the assemblies slide-column-headstock and bed-table, up to find the influence of the structural parts deformations at the tool cutting edge;

•The results are useful for the machine-tool's user, because the user can insert corrections so that the machine's accuracy will not be affected.

4. REFERENCES

Atanasiu, C., Hadăr, A., Popescu, D. & Parpală, R. (2007). *Computational Analysis of Closed-Cells Cellular Structures*, The 18th International DAAAM Symposium "Intelligent Manufacturing & Automation: Focus on Creativity, Responsability and Ethics of Engineers", October, Zadar, Croatia, p. 313-314, ISSN 1726-9679

Buzdugan, Gh. (1980). *Strength of Materials,* ed. XI-a, Ed. Tehnica, Bucuresti

Budynas, R.C. (1999). *Advanced strength and applied stress analysis*, 2nd ed., McGraw-Hill International Editions, New York

Gheorghiu, H. & Hadăr, A., (2007). *Determining Deformations and Existing Tensions in Structures Made from Bedded Composite Materials with Fibers*, Proceedings of the 9th International Conference on Management of Innovative Technologies, 9-10 October, Fiesa, Slovenia, p. 191-195 and on C.D

Gheorghiu, H; Constantinescu, I.N.; Hadar, A. & Petre, C., (2006). *Methodes numeriques pour le calcul des structures de resistance*, Ed. BREN, Bucuresti

Iliescu, N.; Atanasiu, C. & Nastasescu, V. (2005). *Combined Researches on Mechanical Behavior of Guide Block*, 22nd DANUBIA-ADRIA Symposium on Experimental Methods in Solid Mechanics, Extended Abstracts, September 28-October 1, Monticelli Terme, Parma, Italia

Jiga, G., Vlasceanu, D.& Baciu, F. (2007). *Analysis of a roll over protective structure using the FEM,* Annals of DAAAM for 2007 & Proceedings of the 18th International DAAAM Symposium, October, Zadar Croatia, p.359-360

Annals of DAAAM for 2008 & Proceedings of the 19th International DAAAM Symposium, ISSN 1726-9679
Editor B. Katalinic, Published by DAAAM International, Vienna, Austria 2008
Make Harmony Between Technology and Nature, and Your Mind will Fly Free as a Bird

DESIGN OF MECHATRONICS APPLICATION

DJIKIC, M[ahmud]; COHODAR, M[aida]; KULENOVIC, M[alik] & VOLODER, A[vdo]

Abstract: *The main purpose of the paper is to point out an significancy automation of processes such as workpiece orientation, lifting and shifting parts, positioning, component handling - loading and unloading of conveyors and machine tools, etc. especially as it is known that 95% of the total time for production of the workpiece spends either waiting or moving workpiece. The paper gives an example of mechatronic system design - a 'pick and place' robot. Control of such system is realized by programmable logic controller (PLC) SIEMENS S7-200, for which programming STEP 7 MicroWin SP2 software package is used. Mechatronics approach has been resulted with application best suited for integration in automation of other processes in factory.*
Key words: *'pick and place' robot, ladder diagram, pneumatics, mechatronics*

1. INTRODUCTION

Analysis of the total time required for production of workpieces in the typical metal machining batch factory shows that only 5% of its time has spent on the machine tool (Groover, 2007). But of this 5%, 70% is the time required for feeding machine, moving tool in contact with workpiece, positioning, clamping and other auxiliary activities. In other words, 98,5% of the total time of a workpiece is nonprocessing time. On the other side, 40% metal workers are involved in processes of component placements and sub-assembly operations; machine tool, conveyors, die-casting machine, jigs and fixture and other processes loading and unloading; machine-to-machine transfer; handling dangerous or toxic materials; paletisation and package, assembly operations, etc.

Automation of such processes plays very important rule. Firstly, automation provides decreasing of human-worker participation in realization of processes, which finally leads to:

- Increasing of productivity,
- Improving of product quality (accuracy, repeatability and consistency),
- Integration with automation,
- Continuous availability for work,
- Decreasing of production costs and
- Humanization of working places which is especially important in the case of hazard environment.

Attention in the paper is devoted to design 'pick and place' robot. Its function is automation one of above-mentioned activities - transfer of component from one to the other storage. The paper presents the integrated so-called mechatronic approaches to design of robot system involving electronic, pneumatic component, sensors and computing. In mechatronic applications, a bulky mechanical system has replaced by a much more compact microprocessor system, which is readily adjustable to give a greater variety of programs (Bolton, 2007). This leads to cheaper, simpler, more reliable and grater flexible products than their predecessors. There are two approaches to design control system of products (Bradley at al.,1996):

- using embedded real-time microprocessor system and
- using data processing application.

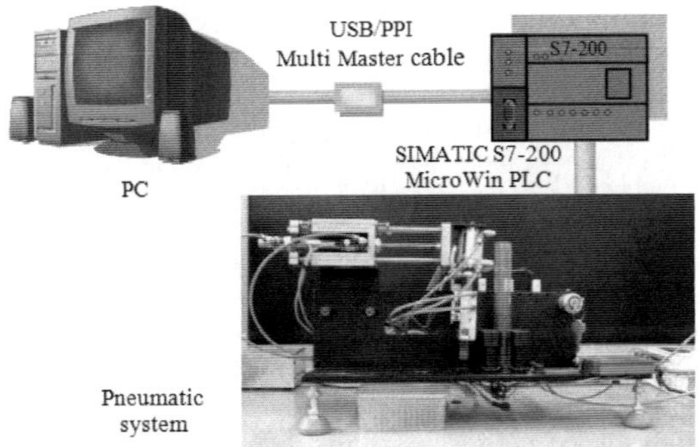

Fig.1. 'Pick and place' robot system

In the first one, control system can be realized by microprocessor system, which is subsystem of completed products. In the case of data processing application, microprocessor is built into an environmental designed for typical computer system such as personal computer or industrial microprocessor system – PLC, which is in most stand-alone devices. Control system of 'pick and place' robot is realized on the second proposed way using SIEMENS SIMATIC S7-200 Micro PLC. 'Pick and place' robot, described in the paper, is developed for education purpose, as well as research of different parameters – controlling pressure and position, activity and timing of control units, etc., and is placed in Laboratory of Automation and Robotics at Faculty of Mechanical Engineering in Sarajevo. Designer of system is the first author of the paper. In (Hesse, 2001) you can find a number of examples of pneumatic applications.

2. DESIGN OF 'PICK AND PLACE' ROBOT

Purpose of 'Pick and place' robot (Fig.1) is part transfer from one to the other storage. On this way, it bridges the gap between highly automated processes of machining and component handling. Transfer of components from one to the other place is repetitive operation, performed often under unpleasant and sometimes dangerous working conditions requiring little skill worker. Automation of such processes provides integration with other automated subsystems of the factory.

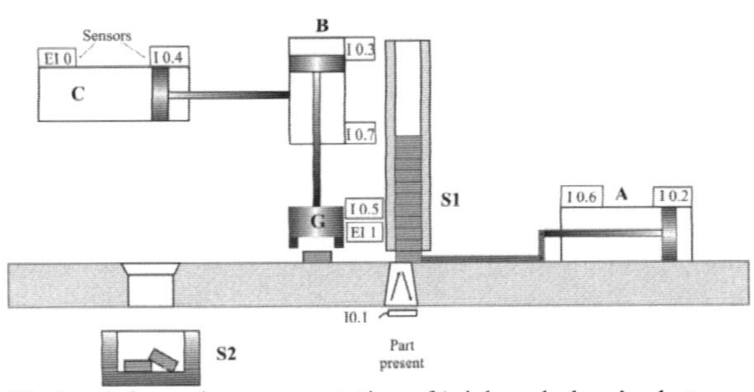

Fig.2. Schematics representation of 'pick and place' robot structure

Actuator	Symbol	Outputs	
		-	+
Cylinder 1	A	Q 0.2	Q 0.0
Cylinder 2	B	Q 0.4	Q 0.1
Cylinder 3	C	Q 0.5	EQ 1
Gripper	G	EQ 0	Q 0.3

Tab. 1. The outputs from the controller

'Pick and place' robot system consists of four cylinders, as shown in Fig.2. Cylinder A transfers parts from storage S1 to the place below gripper G. Cylinder B lets down gripper G, which grasps part. After that, cylinder B lifts workpiece in upper position and then cylinder C translates component to position over storage S2. When cylinder B set down component, gripper G releases it.

All cylinders are double acting, which piston motion is bounded with proximity sensors. Festo's proximity sensors SME-8-S-LED-24 are mounted on cylinder's slots. They are contactless sensors with resistors, which are sensitive to magnet-fields. Magnet mounted on the cylinder piston changes the switching signal status, as soon as the cylinder piston moves around of sensors. Diffuse sensor with background suppression (Baumer FHDK 10P1101/KS35) detects presence of workpiece in storage S1. To start 'pick and place' robot a pushbutton switch is included in system. Finally, system is also equipped with four solenoid valves.

3. CONTROL SYSTEM

The SIMATIC S7-200 Micro PLC, design as stand-alone CPU, coordinates all actions of transfer of components. It offers real-time control with Boolean processing speeds of 0,37 μs per instruction. (http://www.automation.siemens.com)

Between two possible programming options for connecting computer to S7-200, which SIEMENS provides, a direct connection with USB/PPI Multi Master cable RS485 is applied herein. The S7-200 is based on central processing unit CPU222 with 8 in/ 6 out and in order to cover 'pick and place' robot application which requires 10 inputs and 8 outputs, it is expanded by adding I/O expansion module EM223 with 4DI/4DO.

4. PROGRAMMING OF SYSTEM

All applied sensors are binary – 0 and 1 to indicate whether or not a cylinder piston has reached the end position. The inputs to the controller are signals from eight limit switches, the pushbutton and the diffuse sensor. The central processing unit (CPU) of PLC executes the various logic and sequencing functions depending on state of inputs as well as program instructions determining the appropriate output signals. They are on/off signals to operate solenoid valves which provide corresponding actions of cylinders (Table 1).

Sequences of actions of 'pick and place' robot with denoted activated sensors and produced outputs are shown in Fig.3.

The programming method, used in the paper, is based on the ladder diagram. The V. 4.0. STEP 7 MicroWIN SP2 programming package provides user-friendly environment to develop program sequences (System Manual, SIMATIC S7-200; Berger, 2006).

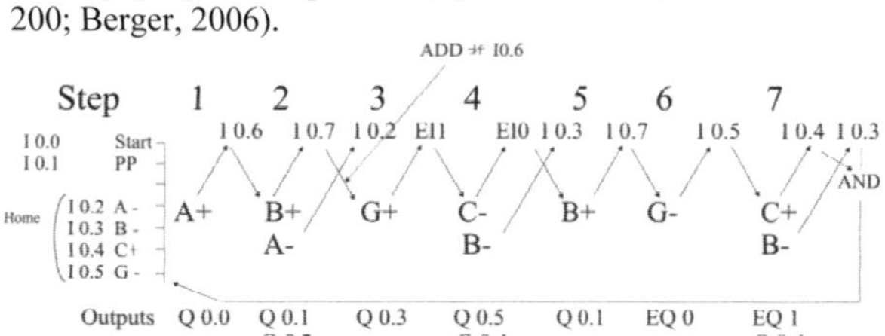

Fig.3. Operation steps of 'pick and place' robot system

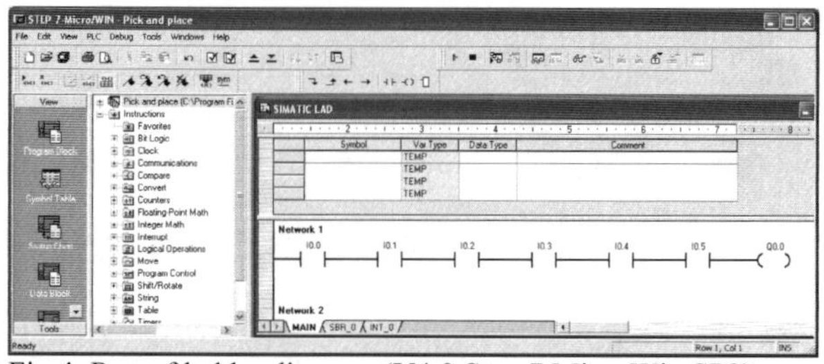

Fig.4. Part of ladder diagram (V4.0 Step 7 MicroWin SP2)

Start position of 'pick and place' robot is depicted in Fig.2. Sensors I0.2, I0.3, I0.4, I0.5 and I0.1 with condition that there are parts in the storage, are activated and produce 1 - signals. This start position is called home position and it is represented with network 1 on ladder diagram, Fig.4. In this start position when start pushbutton is activated for start operation of 'pick and place' robot system, six normally open contacts, arranged in series, will be closed and output will be energized.

Such designed system can be integrated very easily in complex automated production system. Also, if it is required, system can be simply reprogrammed for different sequences of activities.

5. CONCLUSION

The main purpose of the paper is to point out an importance of mechatronics approach to the design of technical system. Replacing a bulky mechanical control system with a microprocessor system gives an answer for mechanical problem with a little space.

Mechatronics approach based on integration of computer digital technique through electronic and electric interfaces with mechanical system leads to flexible, compact, modular, cheaper, simpler and more reliable solution.

These mentioned properties of mechatronic solution are considered on the example of 'pick and place' robot system with SIMATIC s7-200 MicroWin PLC as control system. It is developed (by M.Djikic) and placed in Laboratory for Automation and Robotics at Faculty of Mechanical Engineering in Sarajevo. On this way, students have a very nice example and opportunity that, through exercises, would be introduced with basis of mechatronic systems and their programming.

6. REFERENCES

Berger, H. (2006). *Automating with SIMATIC: Controllers, Software, Programming, Data Communication Operator Control and Process Monitoring,* Wiley-VCH; 3 Revised edition; ISBN-13: 978-3895782763

Bolton, W. (2004). *Mechatronics, Electronic Control Systems in Mechanical and Electrical Engineering,* Prentice Hall; 3 ed., ISBN-13: 978-0131216334

Bradley, D.A.; Dawson, D.; Burd, N.C. & Loader A.J. (1996). *Mechatronics, Electronics in products and processes,* Chapman & Hall; ISBN: 0 412 58290 2

Groover, M.P. (2007). *Automation, Production Systems, and Computer Integrated Manufacturing,* Prentice Hall; 3 ed., ISBN-13: 978-0132393218

Hesse, S. (2001). *99 Examples pf Pneumatic Applications,* FESTO AG & Co., D-73734 Esslingen, Germany, www.festo.com

http://www.automation.siemens.com/bilddb/index.asp?aktPrim=0&nodeID=5000646&lang=en&foldersopen=-1225-1224-1223-1222-1226-1228-1229-1231-&jumpto=1231 *Accessed:* 2008-07-2

System Manual Edition 08/2005, SIMATIC S7-200 Programmable Controller,

Annals of DAAAM for 2008 & Proceedings of the 19th International DAAAM Symposium, ISSN 1726-9679
Editor B. Katalinic, Published by DAAAM International, Vienna, Austria 2008
Make Harmony Between Technology and Nature, and Your Mind will Fly Free as a Bird

ROBUST CONTROL OF A TWO TANK SYSTEM

DLAPA, M[arek]

Abstract: *The paper deals with design of a robust controller via algebraic μ-synthesis for a two tank system, which is a well known benchmark problem. The controller is designed by decoupling two-input two-output system into two identical SISO (Single-Input Single-Output) plants. The task of robust controller design is then performed by finding a suitable pole placement for the SISO systems with the robustness assessed by structured singular value denoted μ. Final controller is verified by simulation for plant perturbed by worst case perturbation.*
Key words: *Robust control, decoupling, algebraic theory, structured singular value, evolutionary computation.*

1. INTRODUCTION

Algebraic methods (Kucera, 1972; Vidyasagar, 1985) are well known and easy to use for SISO (single-input single-output systems) systems described by continuous or discrete transfer functions. However, if applied to MIMO (multi-input multi-output) systems computational difficulties are increasing. In this paper the problem of MIMO system design is treated via decoupling the MIMO system into two identical SISO plants, which are then approximated by transfer functions with simple structure. This guarantees decoupled result control, and simplifies derivation of pole placement formulae.

In order to evaluate robust stability and performance structured singular value denoted μ is used (Doyle, 1982, 1985). The algebraic μ-synthesis (Dlapa & Prokop, 2003) overcomes some difficulties connected with *D-K* iteration, namely the fact that it does not guarantee convergence to a global or even local minimum, which leads to non-optimality of the resulting controller (Stein & Doyle, 1991). Moreover, controllers obtained via algebraic approach can have simpler structure as there is no need of absorbing the scaling matrices into generalized plant, and hence no need of further simplification causing deterioration of frequency properties of the controller.

In this paper functionality of algebraic μ-synthesis is verified by application to a model of a two tank system (Balas et al., 1998), which is a reference example for robust control design.

The following notation is used: $\mathbf{R}_{ps}$ denotes the ring of Hurwitz-stable and proper rational functions and $\mathbf{I}_n$ is the unit matrix of dimension n.

2. MODEL DESCRIPTION

Detailed description of two tank system is in Balas et al., 1998 and other papers (e.g. Smith et al., 1987; Smith & Doyle, 1988). Here only points important for the proposed method are presented.

The system consists of two water tanks in cascade depicted in Fig. 1. The upper tank (tank 1) is fed by hot and cold water via computer controllable valves. The lower tank (tank 2) is fed by water from an exit at the bottom of tank 1. A constant level is maintained in tank 2 by means of overflow. A cold water bias stream also feeds tank 2 and enables the tanks to have different steady-state temperatures.

It follows that the two tank system is a MIMO plant with two measured signals: t_1, t_2 and two inputs f_h and f_c. The quantities t_1, t_2 represent temperatures of tank 1 and 2 respectively. The input signals are commands to hot flow (f_{hc}) and cold flow (f_{cc}) actuators, which are transformed to hot water flow f_h and cold water flow f_c. Third measured signal is water level in tank 1 (h_1), which is not controlled. This quantity is, however, important for assessment of controller performance.

Due to linearization quantities h_1 and t_1 should be in a prescribed range:

$$0.25 \le h_1 \le 0.75 \quad (1)$$

$$0.25 \le t_1 \le 0.75 \quad (2)$$

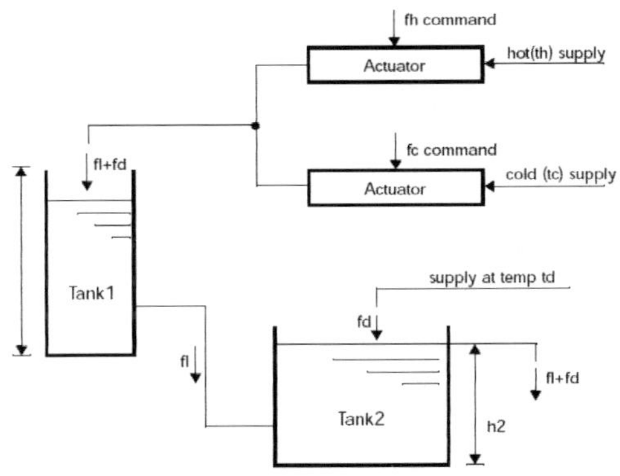

Fig. 1. Schematic diagram of the two tank system.

Nominal model from the inputs f_h, f_c to the outputs t_1, t_2 can be written in two dimensional transfer matrix

$$\mathbf{P}_{nom}(s) \equiv \begin{bmatrix} P_{11}(s) \, P_{12}(s) \\ P_{21}(s) \, P_{22}(s) \end{bmatrix} \equiv$$
$$\equiv \begin{bmatrix} \dfrac{0.0036s^2 + 0.0001s + 7.8157 \cdot 10^{-7}}{s^3 + 0.0491s^2 + 0.0007s + 3.0684 \cdot 10^{-6}} & \dfrac{0.0004s + 4.6643 \cdot 10^{-6}}{s^3 + 0.0491s^2 + 0.0007s + 3.0684 \cdot 10^{-6}} \\ \dfrac{-0.0109s^2 - 0.0004s - 2.3447 \cdot 10^{-6}}{s^3 + 0.0491s^2 + 0.0007s + 3.0684 \cdot 10^{-6}} & \dfrac{0.1562 \cdot 10^{-4}s + 0.0187 \cdot 10^{-4}}{s^3 + 0.0491s^2 + 0.0007s + 3.0684 \cdot 10^{-6}} \end{bmatrix} \quad (3)$$

For details on nominal models of tank 1 and 2 and uncertainty modelling see Balas et al., 1998.

3. ALGEBRAIC μ-SYNTHESIS

For the purposes of algebraic design the MIMO system is decoupled into two identical SISO plants. For decoupling of the nominal plant $\mathbf{P}_{nom}$ it is satisfactory to have the controller in the form

$$\mathbf{K}(s) = K(s)\mathbf{I}_2 \det[\mathbf{P}_{nom}(s)] \frac{1}{P_{11}(s)} [\mathbf{P}_{nom}(s)]^{-1} \quad (4)$$

The choice of decoupling matrix prevents controller from cancelling any poles or zeroes in the right half-plane so that internal stability of the nominal feedback loop is held. The MIMO problem is now transformed into finding a controller $K(s)$, which is tuned via setting the poles of the nominal feedback loop with the plant

$$\mathbf{P}_{dec}(s) = \frac{1}{P_{11}(s)} \det[\mathbf{P}_{nom}(s)][\mathbf{P}_{nom}(s)]^{-1}\mathbf{P}_{nom}(s) = \frac{1}{P_{11}(s)} \det[\mathbf{P}_{nom}(s)]\mathbf{I}_2 \quad (5)$$

Let

$$P_{dec} \equiv \frac{1}{P_{11}(s)} \det[\mathbf{P}_{nom}(s)] \tag{6}$$

Then the transfer function P_{dec} can be approximated by 2nd order system

$$P^{*}_{dec}(s) = \frac{b_{dec}(s)}{a_{dec}(s)} = \frac{0.0027s^2 - 0.0017s + 0.0001}{s^2 + 0.0164s + 0.0001} \tag{7}$$

and the controller $K = N/M$ is obtained by solving the Diophantine equation

$$A_{dec}M + B_{dec}N = 1 \tag{8}$$

with $A_{dec}, B_{dec}, M, N \in \mathbf{R}_{ps}$.

It can be shown that asymptotic tracking of reference signal is achieved if and only if $A_{dec}M$ is divisible by F_r and disturbance is suppressed if $A_{dec}M$ is divisible by F_d. Here F_r and F_d are Fourier transforms of the reference and disturbance respectively. By the analysis of the polynomial degrees of a_{dec} and b_{dec} the transfer functions A_{dec}, B_{dec}, M and N are chosen so that the number of closed loop poles is minimal and asymptotic tracking for step reference signal is achieved:

$$A_{dec} = \frac{a_{dec}}{\prod_{i=1}^{2}(s+\alpha_i)}, \quad B_{dec} = \frac{b_{dec}}{\prod_{i=1}^{2}(s+\alpha_i)} \tag{9}$$

$$M = \frac{sm}{\prod_{i=3}^{4}(s+\alpha_i)}, \quad N = \frac{n}{\prod_{i=3}^{4}(s+\alpha_i)} \tag{10}$$

Degrees of polynomials m and n are:

$$\partial m = 1, \partial n = 2 \tag{11}$$

Thus the characteristic polynomial of the nominal closed loop has 4 pairs of poles $-\alpha_i$, which represent the tuning parameters. Resulting controller has the structure:

$$K(s) = \frac{n}{sm} \tag{12}$$

The open-loop interconnection is in Fig. 2 with performance and reference weights:

$$W_{t1perf} = \frac{130s+1}{400s+1}, \quad W_{t2perf} = \frac{130s+0.5}{800s+1} \tag{13}$$

$$W_{t_{1cmd}} = 1, \quad W_{t_{diffcmd}} = 1 \tag{14}$$

$$\begin{bmatrix} t_{1err} \\ t_{2err} \end{bmatrix} = \begin{bmatrix} 1 & 0 \\ 1 & 1 \end{bmatrix} \begin{bmatrix} W_{t_{1cmd}} & 0 \\ 0 & W_{t_{diffcmd}} \end{bmatrix} \begin{bmatrix} w_{11} \\ w_{12} \end{bmatrix} - \begin{bmatrix} t_1 \\ t_2 \end{bmatrix} - \begin{bmatrix} W_{t1noise} & 0 \\ 0 & W_{t2noise} \end{bmatrix} \begin{bmatrix} w_{21} \\ w_{22} \end{bmatrix} \tag{15}$$

Input to the controller is tracking error of the measured temperatures t_{1err}, t_{2err}. The relaxed performance weight is justified by additional postulate of decoupled result control for the nominal loop, which is not present in standard design, and which makes the task of achieving the robust performance and stability more difficult. This modification of interconnection does not degrade the uncertainty model, and the resulting performance can be observed by simulation for nominal and perturbed plant.

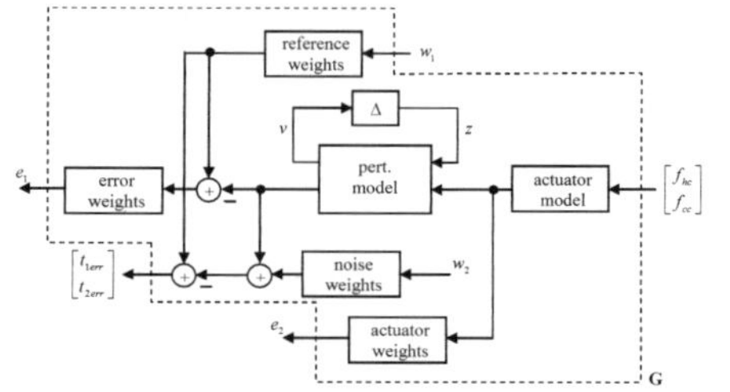

Fig. 2. Interconnection structure for algebraic approach.

The design of controller is then reduced to minimization of the peak of the μ-function. The cost function is then defined as

$$\sup_{\omega \in \mathbf{R}} \mu_{\Delta}[\mathbf{F}_l(\mathbf{G}, \mathbf{K})] \tag{16}$$

here Δ denotes augmented perturbation including performance weights in feedback loop. The optimal controller was obatined by optimization with Differential Migration (e.g. Dlapa & Prokop, 2006) and Nelder-Mead simplex method.

Simulation for worst case perturbation (Fig. 3) shows that no steady state error is present, and response to ramp reference signal is monotnous. This is the main difference from standard procedures, which, in most cases, do not achieve asymoptotic tracking and monotnous response.

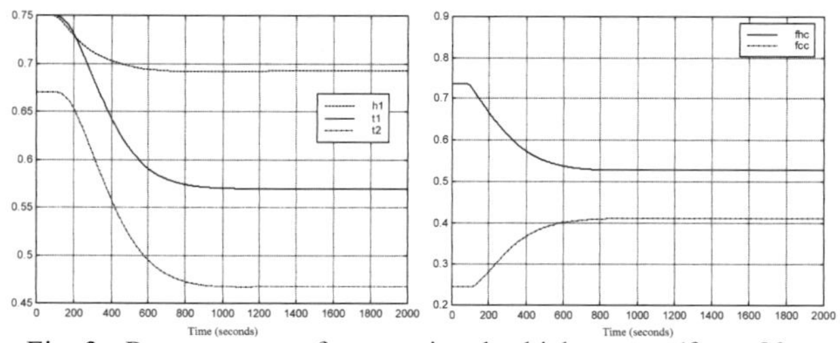

Fig. 3. Response to reference signal which ramps (from 80 to 100 seconds) t_1 from 0.75 to 0.57, and t_2 from 0.67 to 0.47 for algebraic approach and perturbed plant.

4. ACKNOWLEDGMENT

This work was supported by the project MSM7088352102.

5. REFERENCES

Balas, G. J.; Doyle, J. C.; Glover, K.; Packard, A. & Smith, R. (1998). *μ-Analysis and Synthesis Toolbox for Use with MATLAB*. The MathWorks, Inc., June 1998.

Dlapa, M. & Prokop, R. (2003). "μ-Synthesis: Simple Controllers for Time Delay Systems," *Proceedings of 11th Mediterranean Conference on Control and Automation*, Rhodes, Greece.

Dlapa, M. & Prokop, R. (2006). "Differential Migration: a new algorithm for global optimization," *Proceedings of the 6th Asian Control Conference*, Bali, Indonesia, pp. 618-626, ISBN 979-15017-0, July 18-21 2006.

Doyle, J. C. (1982). "Analysis of Feedback Systems with Structured Uncertainties," *Proceedings of IEEE, Part-D*, 129, pp. 242-250.

Doyle, J. C. (1985). "Structure Uncertainty in Control System Design," *Proceedings of 24th IEEE Conference on Decision and Control*, pp. 260-265.

Kučera, V. (1972). *Discrete Linear Control: The Polynomial Equation Approach*. Wiley, New York.

Smith, R. S.; Doyle, J.; Morari, M. & Skjellum, A. (1987). "A case study using μ: Laboratory process control problem," in *Proc. Int. Fed. Auto. Control*, vol. 8, pp. 403–415.

Smith, R. S. & Doyle, J. (1988). "The two tank experiment: A benchmark control problem," in *Proc. Amer. Control Conf.*, vol. 3, pp. 403–415.

Stein, G. & Doyle, J. (1991). "Beyond Singular Values and Loopshapes," *AIAA Journal of Guidance and Control*, Vol. 14, No. 1, pp. 5-16.

Vidyasagar, M. (1985). *Control Systems Synthesis: A Factorization Approach*. MIT Press. Cambridge, MA.

Annals of DAAAM for 2008 & Proceedings of the 19th International DAAAM Symposium, ISSN 1726-9679
Editor B. Katalinic, Published by DAAAM International, Vienna, Austria 2008
Make Harmony Between Technology and Nature, and Your Mind will Fly Free as a Bird

METHOD OF MEASURING PLANTAR PRESSURE WITH RESISTIVE SENSORS

DOBRA, A[ndreea]

***Abstract**: The paper presents a method of measuring the static plantar pressure with resistive sensors. Equipment designed and presented in this paper is intended to detection and prevention. It consists of a platform with resistive sensors and an acquisition board connected to the USB port on a PC. To view measured data has been made dedicated software in Visual Basic environment. The software allows recording measurements into a database-type MS ACCESS.*
***Key words**: measurement, plantar, pressure, sensor*

1. INTRODUCTION

The means of measurement are as diverse and feet as they measured, how each system offers a number of advantages, and comes with a number of limitations, so will not find a system that is suitable for all types of measurements.
Plantar pressure may be measure static and / or dynamic through several ways. All require expensive equipment and complex software. The methods of investigation of plantar pressure began in 1881 and evolved along with technology development.
The papers propose a measurement method for plantar pressure with resistive sensors. The device consists of a platform with 8 sensors, an acquisition board (Labjack U12 National Instrument) and a connection by USB with a PC.
Special software made in Visual Basic environment, allow viewing value measured, date & time of measurement and save this in a database.
For future development to this device it can multiply number of sensors and putting them in a specific area. The device is made for general use to preliminary trace or forewarns any problems of feet (heel, shank, etc.) in any place where is necessary.

2. GENERAL PROBLEMS OF DISEASES LEGS

Legs are very complex structures, composed of bones, joints, muscles, tendons, nerves, blood vessels and skin. They may develop some problems (with origins in other parts of the body) or may suffer injuries, traumas. The most obvious symptoms are pain at shank and dysfunction such as over a foot, inability to bear pressure on heel or fingers, etc. The pain can be caused by inappropriate footwear, injury or stretch the tendon and ligaments.
What are the most common problems?

2.1 General problems

-Fingered "hammer" (fig.1.1a)
-Bursitis hallux: (Hallux Abducto Valgus) (fig. 1.1b).
-Flat feet (fig. 1.1c) - Arch plantar develops between the ages of 3 and 10 years. At birth, and after a few years, all children have flat feet. Many people, although they have a small arc plantar not suffer of flat feet. Causes of this disease can be atrophying of muscles and ligaments submission by older, or years of standing in inappropriate footwear.
-Pes cavus (fig. 1.1d) is a medical term for a cavus deformity of the foot due to a fixed plantar flexion of the foot. There is a high arched foot - it is the opposite of flat feet. It is much less common than flat feet and the cause may be neurological, orthopedic or neuromuscular. Unlike most cases of flat feet, pes cavus can be painful because of metatarsal compression.

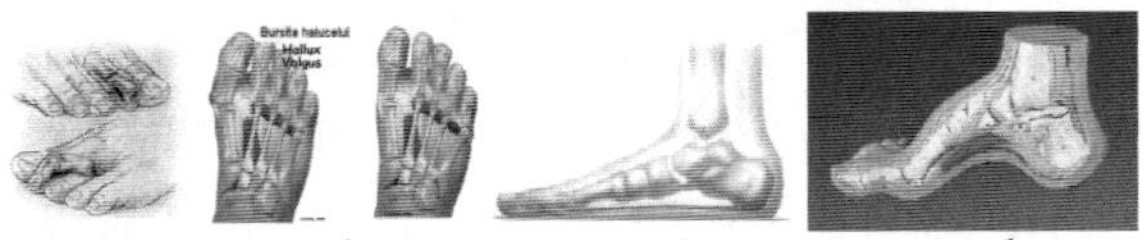

a. b. c. d.
Fig. 1. General foot problems

2.2 Problems due to sports activities

- Tendonitis Achilles
- Calcaneus's fracture is the most common cause of heels pain
- Dysfunction of posterior tibiae tendons

2.3 Problems caused by diabetes

High level of glucose in the blood of diabetes patients can cause major problems on two legs:
- Termination nerve damage: if nerves are deteriorating, it may not realize pain, heat or cold to the legs.
- Blood circulation deficit: this makes it very difficult wound healing and infections.

3. METHODS OF DETECTION OF DISEASES LEGS

3.1 Plantar Pressure measurement in footwear (static or dynamic)

There are complex systems that allow a very high degree of mobility patient, allowing play under normal pressure from the soles. They even allow recording of movies with patients, and real-time play map pressure from the soles. (Actis et al., 2005)
The system is composed of a receiver, fixed around the waist, which is connected to the inner soles that sit in footwear. The receiver stores all information relating force / pressure from the soles. Mobile Unit attached to the PC may transfer parameters and test data collected, allowing a view real-time data (Tekscan, 2006). This interface makes it possible to view 2-D and 3-D data recorded, showing curves of dependency pressure and force depending on time, allows comparison of data obtained pre-and post-treatment and isolation of specific areas of the soles.

3.2 Carpet sensory elements to measure the pressure / static forces static conditions

This system allows the evaluation, registration, documentation and data collected from the soles.

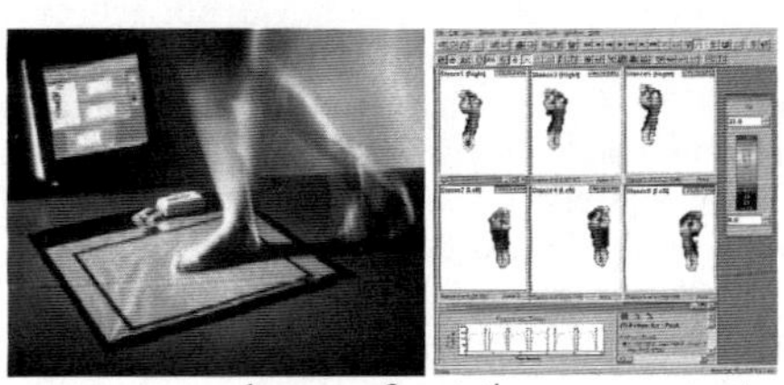

Fig.2. Carpet sensory elements for static measurements

It can be used to prescribe orthotic inner soles, and educate patients (Fig.2.)(Kirby, 2007).

3.3 Carpet sensory elements to measure the pressure / force in the dynamic conditions

This system allows measurements in both static and dynamic. It can do an analysis of normal or perturbed walking along several steps (Fig. 3.)(Urry, 1999)

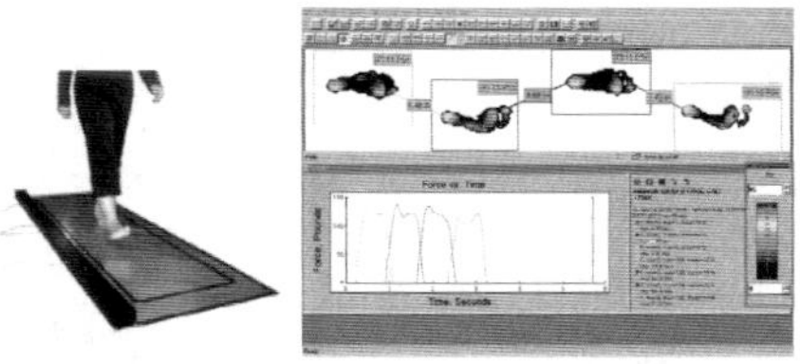

Fig.3. Carpet sensory elements for dynamic measurements

3.4 Assessing the plantar pressure by photographic methods

It is a method low cost viewing pressure plantar, which consists in a photocopier to support the weight of the human body. By this method can distinguish healthy feet, with flat feet or feet with plantar arch too high (Fig. 4).

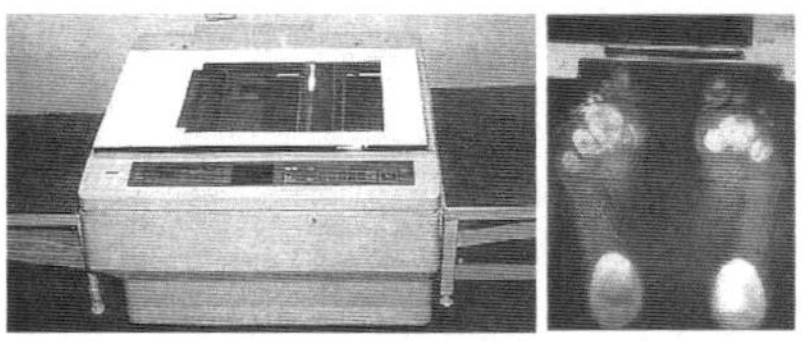

Fig.4. Photocopier adapted for plantar pressure measurements and published image of photocopier

3.5 Plantar pressure measurement through fiber-optic network

This method is based on the modification of light transmitted through an optical fiber network. By applying a force on it, is hereby-amended section fiber, which leads to diminishing light in this area (Fig. 5.).

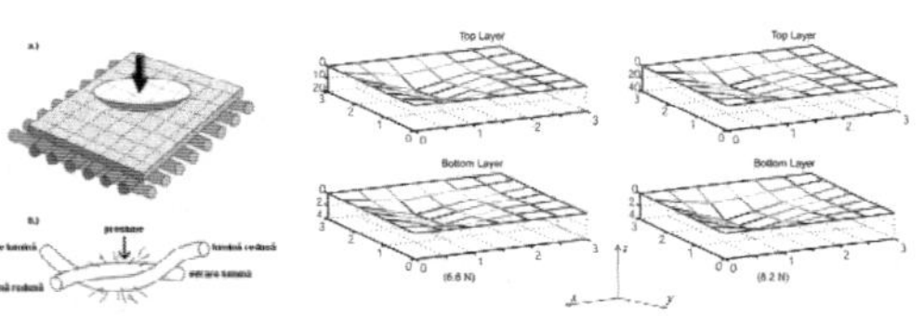

Fig. 5. The basic configuration of the sensorial system with optical fiber and the load distribution network on lower and the top layer of optical fiber

4. EQUIPMENT DESCRIPTION

The proposed system as a platform consists of acquisition board Labjack U12 and a PC (Fig.6.). The link between the platform and measuring acquisition board is achieved through two connectors for each plantar platform individual.

Multiplexing board is a printed circuit board, which takes over the role-multiplexing signal from the eight sensors. Multiplexer is an electronic switch, which carried out the operation of interruption, the ordering of input signals, namely the selection of an entry. The application is used two of the eight entries of the measuring, making it possible acquisition of six other types of signals, which can contribute to the diversification of the application, depending on the customer's requirements (Dobra & Albert, 2006). The used sensors are CZN - CP6 type (Membrane compression sensors). Software to measure a plantar pressure – is made in Visual Basic 6.0 environment. The user interface is shown in Fig. 7. is compose from 4 areas: patient's identification - name, gender, date of birth - (text boxes), pressure points visualization – sensors distribution picture (picture box), values tables and action area (command buttons: delete, read value, save & history, and end).In the picture box, the date on which the investigation is automatically generated based on the data system. Database can be view whenever with or without measuring step.

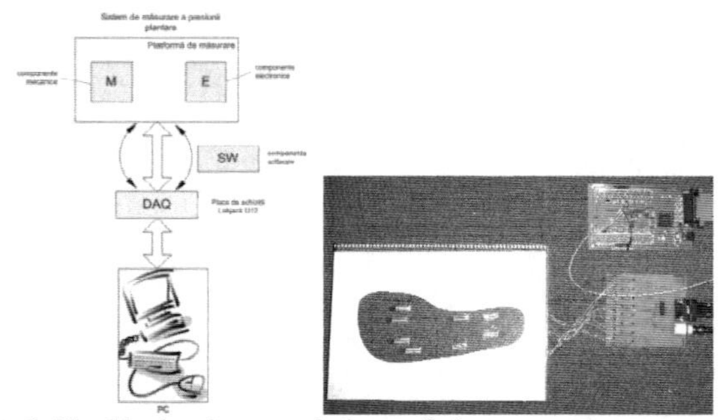

Fig.6. The block scheme and photo of plantar pressure measuring system

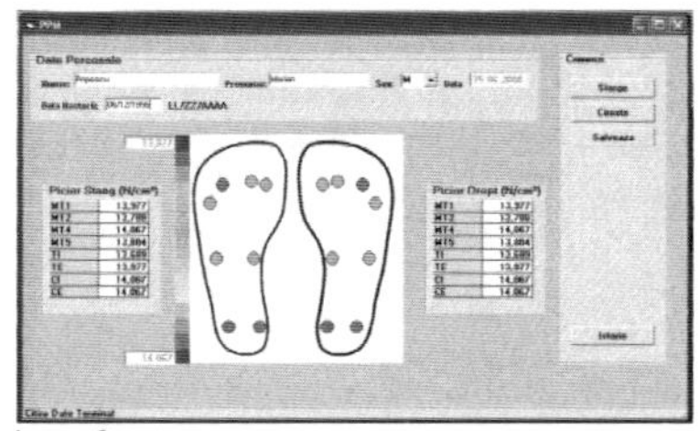

Fig.7. User interface

5. CONCLUSION

With equipment were made measurements (in different dates and times) and was created a database. By increasing the number of resistive sensors and putting, them in specific areas such increase measurement sensitivity and equipment performance. Improved equipment will allow to measure preventive and periodical, a static plantar pressure for everyone (children or adults), by placing the device in school or workplace.

6. REFERENCES

Actis, R. L.; Ventura, L. B.; Szabo, B. A. (2005). Plantar Pressure Distribution In The Diabetic Foot During Push-Off: Numerical Simulation Using The P-Version Of The Finite Element Method, *proceeding of XXth Congress of the International Society of Biomechanics And 29th Annual Meeting of the American Society of Biomechanics July 31 - August 5, 2005, Cleveland, Ohio*

Dobra, A. & Albert, E. (2006). *Measurement methods of plantar pressure*, Work License, POLITEHNICA University of Timişoara, Mechatronic Department (June 2006)

Kevin A., Kirby. (2007). Emerging Concepts in Podiatric Biomechanics - *Available from:* http://www.podiatrytoday.com/article/6436 *Accessed:* 2007-03-10

Tekscan, (2006), Medical page, *Available from:* http://www.tekscan.com, *Accessed:* 2006-01-27

Urry, S. (1999). Plantar pressure measurement sensor, *Measurement.Science and. Technology*, Volume 10, No.1, (1999) R16-R32, *Available from:* http://www.iop.org/EJ/abstract/0957-0233/10/1/017, *Accessed:* 2006-02-12

Annals of DAAAM for 2008 & Proceedings of the 19th International DAAAM Symposium, ISSN 1726-9679
Editor B. Katalinic, Published by DAAAM International, Vienna, Austria 2008
Make Harmony Between Technology and Nature, and Your Mind will Fly Free as a Bird

MULTI-CRITERIA ANALYSIS OF ELASTIC COUPLINGS WITH METALLIC FLEXIBLE MEMBRANES

DOBRE, D[aniel] & SIMION, I[onel]

Abstract: *The paper describes the analysis of stress and buckling state of flexible metallic membranes from the structure of the elastic couplings. These couplings assume significant misalignments and keep pace with the demand for higher powers and speeds. The finite element analysis (FEA) is developed in the paper for one membrane about three criterions: stress and buckling stability and dynamic behaviour. The FEA study and the corresponding results are given and discussed. This study is basic to find the solutions for improving membrane performance.*

Key words: *coupling, membrane, stress state, buckling analysis*

1. INTRODUCTION

These couplings use clamping packs of spoked membrane to combine an adequate level of torque transmission with a reasonable axial and angular misalignment capability.

In order to transmit the torque between two shafts of a kinematical chain, in conditions of compensation of some important misalignments, it is necessary for coupling producers to handle some multi-criteria requirements regarding size, safety in exploitation (by guaranteeing a superior mechanical strength), dynamic behaviour and constructive simplicity.

The paper task is analysis of stress and buckling state of the metallic membrane as component of the flexible couplings. This task is carried out by a Finite Element Analysis (FEA).

2. ON STRESS STATE FOR MEMBRANE

The couplings having spoke form membranes transmit the torque from the inner diameter to the outer diameter and reciprocally, by shear loading of the membrane. The flexible membranes are made from spring steel and have thin spokes by making radial indents in the central portion (figure 1), the deformation of the spokes giving to the coupling its flexibility and thus its ability to handle installation misalignments. The couplings use these membranes arranged in a packet and assembled rigidly at their inner and outer diameters to other components. An example of this coupling is given in the fig. 2.

The mechanical strength analysis is made only for a single membrane, for the critical situation of existing axial deviations, taking into consideration that the stress as greater as in the case of deviation absence.

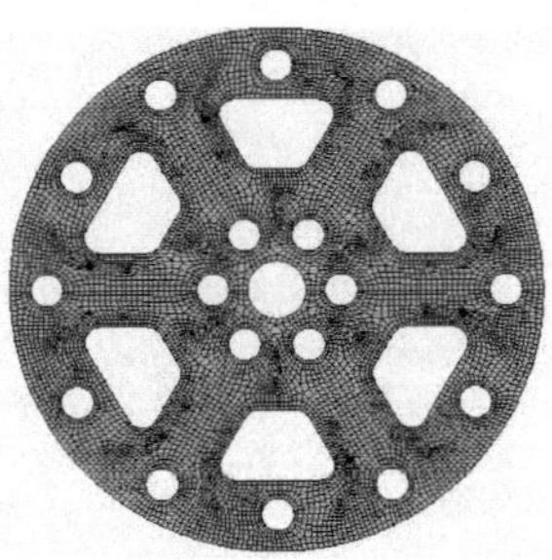

Fig. 1. The finite elements mesh of the membrane.

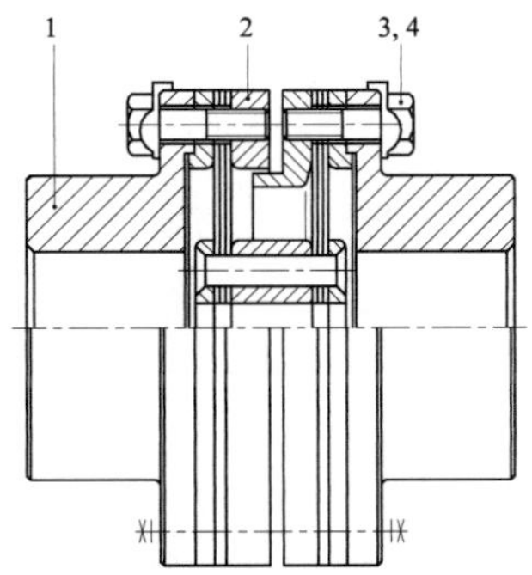

Fig. 2. Non spacer coupling.
1 – flanged hub; 2 – membrane unit; 3 – bolt; 4 – washer.

The stress state of the membrane spoke is determined by:

- torque transmission causing shear, bending and tensile stress;
- the angular and axial misalignments that causes bending and tensile stress;
- centrifugal loading.

For the membrane study by FEA the meshing is realised with finite elements of type SHELL, because the membrane depth is very small (0.3 mm), resulting 9054 nodes and 8316 elements. The meshing with the maximum size of 0.5 mm on the side resulted after a number of elements to ensure a sufficient convergence and accuracy of the FEA. The figure 1 shows the finite element mesh of the flexible membrane with trapezoidal holes using a uniform distribution of the elements.

The analysis was made for the nominal torque T=50 Nm and the rotation speed of 4500 rpm. A stress analysis was made for the cases without deviation and with axial deviation of 0.6 mm in the absence and with consideration of centrifugal load.

The tables 1 and 2 indicate the values for the principal stresses σ_1 and σ_3 and the maximum equivalent stress calculated using the von Mises theory $\sigma_{eq\ max}$.

One would see that the introduction of axial misalignment (Δa = 0.6 mm) increases almost three times the principal stress σ_1 and the maximal equivalent stress given the situation without axial deviation.

Parameter	
Principal stress σ_1 [N/mm^2]	71.8
Principal stress σ_3 [N/mm^2]	-65.2
Maximum equivalent stress $\sigma_{eq\ max}$ [N/mm^2]	69.7

Tab. 1. The results of stress analysis for the case without deviation (Δa=0), by considering the inertial loading.

Parameter	
Principal stress σ_1 [N/mm^2]	286
Maximum equivalent stress $\sigma_{eq\ max}$ [N/mm^2]	255

Tab. 2. The results of stress analysis for the case with axial deviation (Δa = 0.6 mm).

The equivalent (von Mises) stress distribution in the nodes of the structure is presented in figure 3 (with axial deviation of 0.6 mm) in grey colours. One would see that the maximum stress is located to the filleted root of the spokes. Other analysis at different torque (non-presented in this paper) shows that the maximal stress appears also on a large membrane area, that is the membrane solicitation is increasing with the load.

The maximum stress, $\sigma_{eq\ max}$, serves for the calculus of the safety factor, using the general expression:

$$c = \frac{\sigma_{0.2}}{\sigma_{eq\ max}}, \qquad (1)$$

where $\sigma_{0.2}$ represent the yield point of the material. The Romanian steel symbol used in calculus is OLC 65A, having $\sigma_{0.2}$ = 800 N/mm^2. The admissible safety factor indicated in literature for pieces subjected at fatigue (Niemann, 1986) is $c_a = 1.5 \ldots 3$. The values of safety factor for the two cases of stress analysis and the discussion regarding the checking at stress solicitation are indicated in the table 3.

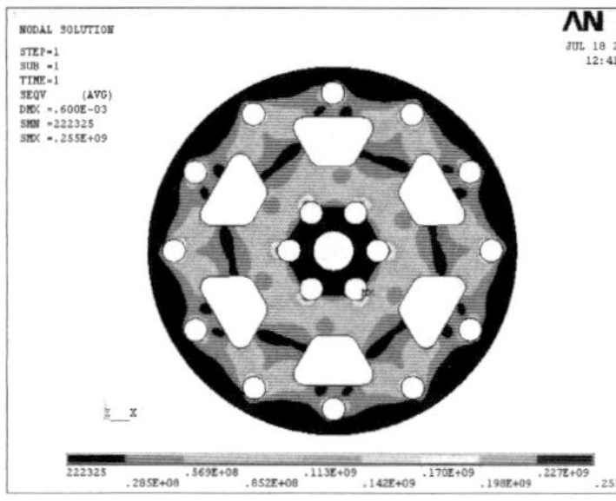

Fig. 3. The equivalent stress distribution with axial deviation.

Size	Analysis case	
	Without deviation	With axial deviation of 0.6 mm
Maximum equivalent stress $\sigma_{eq\ max}$ [N/mm^2]	69.7	255
Safety factor, c	11.48	3.14
Checking discussion	$c > c_a$	$c \cong c_a$

Tab. 3. The values of membrane stress safety factors.

3. BUCKLING ANALYSIS BY FEA

Buckling analysis determines the loading level at which the membrane becomes unstable. This analysis takes into account the effects resulting from the compressive stresses, which tend to lessen the capacity of the membrane to resist lateral loads. As the compressive stresses increase, the resistance to lateral forces decreases, so that, at some load level, this negative stress stiffening overcomes the linear structural stiffness, causing the structure to buckle (Sorohan & Sandu 1997).

The linear buckling analysis determines the scaling factors, c_f, for the stress stiffness matrix, which offset the structural stiffness matrix. The equation that describes linear buckling is:

$$([K] - c_f \cdot [S])\{u\} = 0 \qquad (2)$$

where: $[K]$ is the structural stiffness matrix; $[S]$-the stress stiffness matrix; c_f-the buckling load factor; $\{u\}$-the eigenvector representing the buckled shape.

The buckling load factor is a safety factor, defined as:

$$c_f = \frac{Limit\ buckling\ load}{Applied\ load} \qquad (3)$$

For the buckling analysis the membrane was blocked both on its inner and outer holes as displacement on the direction of the membrane axis with considering the centrifugal loading.

The buckling load factors for 10 shape modes has values in the interval $c_f \in [11.767,\ 17.625]$, being greater than the ones admissible conforming to Niemann (1986), $c_a = 3 \ldots 5$. This means that the membrane is stable at buckling.

The figure 4 shows an image of the second buckling shape. The buckling load factor, c_f, is noted FREQ, a denomination used in the ANSYS 6.1 program.

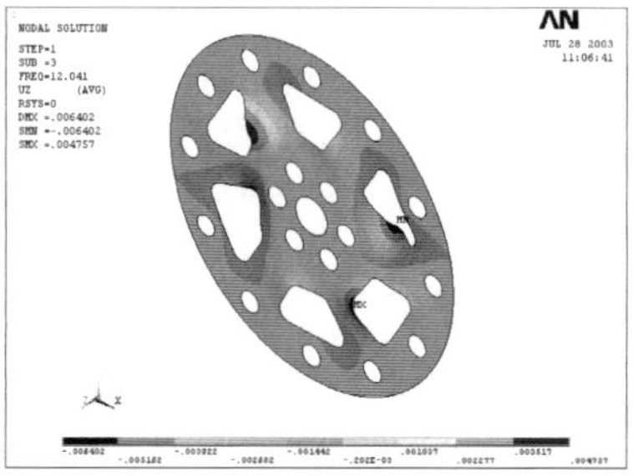

Fig. 4. Second buckling shape with centrifugal loading.

4. MODAL ANALYSIS OF THE MEMBRANE

The modal analysis is necessary to extract the natural frequencies and mode shapes of a flexible membrane. A membrane should be designed to produce natural frequencies that will prevent the coupling component from vibrating at one of its fundamental modes under operating conditions.

The dynamic torsional stiffness determines the torsional vibration behaviour of the drive system. This gives rice to very large natural frequencies and the operating speeds lie below the resonant speeds. The natural frequencies for 10 vibration modes have values in the interval $f \in [2013, 2577]$. This means that the membrane is stable at vibrations.

It is relevant to state that the increased of dynamic torsional stiffness causes an increase in the resonant speeds (Birsan & Jascanu 1998). This is important because the resonant speeds should be present above the operating speed range.

5. CONCLUSION

The finite element analysis (FEA) is basic to evaluate stress and buckling state of flexible membrane.

The areas of maximum equivalent stress are placed on fillet root of the membrane spokes. The maximum equivalent stress values increase with the axial deviation at the limit of the recommended admissible safety factor values.

The centrifugal loading has a significant stiffness effect which causes law increasing of the buckling safety factors, leading to the increased resistance to torsional deformations and to deformations caused by the misalignments of the two shafts.

The results are important for other researches for the case of the entire membrane packet and are in concordance with the experimental dates presented by Dobre (2004).

6. REFERENCES

Birsan, J.G. & Jascanu, M. (1998). *Elastic couplings' dynamic,* Tehnica Publishing House, ISBN 973-31-1232-1, Bucharest

Dobre, D. (2004). *Researches on multi-criteria optimization of elastic couplings with metallic flexible membranes,* Ph.D. Thesis, University "Politehnica" from Bucharest

Niemann, G. (1986) *Maschineelemente,* Springer Verlag, Berlin

Sorohan, S. & Sandu, M.A. (1997). Nonlinear FEM-Analysis of a Diaphragm Spring, ELFIN 4, Iasi, oct. 9-11, pp. 125-128.

Annals of DAAAM for 2008 & Proceedings of the 19th International DAAAM Symposium, ISSN 1726-9679
Editor B. Katalinic, Published by DAAAM International, Vienna, Austria 2008
Make Harmony Between Technology and Nature, and Your Mind will Fly Free as a Bird

INFLUENCE OF THE PHOTOSYNTHESIS ON THE POLLUTION WTIH CARBON DIOXIDE IN TIMISOARA

DOBREN, F[lavius] A[ndrei]; DUMITRESCU, C[onstantin] D[an]; ISTRAT, N[icolae] & MARINESCU, S[orin] A[lin]

***Abstract**: The paper presents the dependency report between the photosynthesis carbon dioxide absorption capacity and the carbon dioxide pollution. This report results, as shown in the biography, from the analysis achieved by using the green surface measured by Timisoara city hall's Environment Bureau and published in the green survey; this survey sums up the green surfaces of the city, by calculating the statistical crown volume and total plant foliar surface assessment by using coefficients shown in other studies. By calculating the foliar surface, we could estimate the carbon dioxide quantity absorbed by photosynthesis, and correlate it to the carbon dioxide emissions.*

***Keywords**: photosynthesis; process; adsorption; emissions; pollution*

1. INTRODUCTION

The development of industry and discovery of enhanced fossil fuels, allowed that, by the combustion process, numerous air pollutants such as: carbon mono and dioxide, nitrogen oxides, sulfured oxides, powder and carbon dioxide, are released into the atmosphere (Ionel et al., 2004).

Closely correlated with the enhanced emissions of carbon dioxide, a phenomenon indirectly proportional is represented by the plant photosynthesis process, a natural process by which plants use the energy from sun light to produce sugar, which cellular respiration converts into ATP (adenosine triphosphate), the "fuel" used by all living things. During the photosynthetic process, plants absorb carbon dioxide from the atmosphere and release oxygen. Technically speaking, this process is a reverse reaction of cellular respiration taking place only in the presence of the sun's rays (Coste et all., 2001).

2. CASE STUDY- THE CALCULUS OF THE LEAVES SURFACE IN TIMISOARA AND THE ESTIMATION OF THE CO_2 FROM PHOTOSYNTHESIS IN COMPARISION WITH THE EMISSIONS OF CO_2

Regarding the adsorption of CO_2, according to V.Ciupa and his co-authors in "*Timisoara verde (*Green Timisoara)", plant leaves assimilate CO_2 with an average intensity of 10-20mg $CO_2/dm^2/h$. According to this paper, a cubic meter of the crown has about 3 square meters of foliar surface (total leaves surface) (Ciupa et al., 2005).

In average, the carbon dioxide adsorption is 30mg/ dm2/ h* 300dm^2 (foliar surface of a cubic meter of crown), respectively 9g/h. Finally, the photosynthetic carbon dioxide adsorption capacity of a cubic meter of crown is 9g/h (Coste et al., 2001).

The assessment of the crown volume of all the trees in a whole city like Timisoara, inventorying all of the trees, other plants and green spaces, measuring the surface of all the green spaces and much more are described in the book "Timisoara Verde", also quoted in the bibliography. Using the data presented in this book, we can try to assess the quantity of carbon dioxide photosynthetically adsorbed by the town vegetation (Ciupa et al.,2005).

The total area occupied by grassy vegetation and flowers in Timisoara is 826503 m^2, plus the added 613.2 ha of the green forest. According to some studies (Ciupa et al.,2005), in 1 square meter area, the foliar surface is 4m^2, where as a result, the whole foliar area of the grassy vegetation is (826503 m^2 + 6132000 m^2) *4 = 27834012 m^2.

According to the same study, the volume of standard wood crown vegetation is 3548000 m^3, plus the 10650000 m^3 corresponding volume of the green forest's trees ((1), pg. 157) in the year 2004, a summing total 14198000 m^3. The growth factor is 5%/year, of which, in 2006 a crown volume of 15617800 m^3 results, which corresponds to a foliar area of 46853400 m^2. Summing up this surface with the foliar area of the grassy vegetation, we obtain 74687412 m^2. For an adsorption rate of 30mg CO_2/ 0.1m 2/h to 74687412 m^2 leaf surface, we obtain an adsorption capacity of 22406.22 kg CO_2/h, of which the adsorption rate in CO_2/h of the resinous plants is 2108.403kg/h - it represents 15% of total trees (foliar area of 7028010 m^2). Corresponding to the months of January, April, July and October, the periods of daily allowance are 282.72, 398.7, 469.03 and 341.62 hours respectively. As for the months, January photosynthesis is made only by resinous species, proper CO_2 adsorption of this month through the process of photosynthesis will be 282.72h (light of the month January) * 2108.403 kg/h meaning 596087.7 kg CO_2 corresponding to the month January. In April, the adsorption of CO_2 will be 398.7 h * 22406.22 kg CO_2/h meaning 8933360 kg; In July, CO_2 adsorption will be 69.03 h * 22406.22 kg CO_2/h, meaning 10509189 kg; In October adsorption of CO_2 will be 341.62 h * 22406.22 kg CO_2/h, meaning 7654413 kg.

The table bellow presents the adsorption rate of CO_2 through photosynthesis corresponding to the months of January, April, July and October, as well as the CO_2 emissions from road transport, Coltherm, and other combustion sources.

The next three charts present the status of the carbon dioxide emissions for the data of table1 and the carbon dioxide absorption by plants for the same time range.

Month	CO_2-emissions from Colerm[t]	CO_2-emissions road transportation[t]	CO_2 emissions from other burning instalations and incinera	Photosynthetic CO_2 adsorption[t]	Unabsorbed CO_2 [t]
January	82517	23381.17	79397.29	596.0877	184699.4
April	35239	23381.17	33906.72	8933.360	835953.53
July	3311	23381.17	3185.821	10509.189	19368.8
October	30362	23381.17	29214.11	7654.413	75302.86
Total	151429	23381.17	79397.29	27693.05	184699.4

Tab. 1: The status of the emissions of CO_2

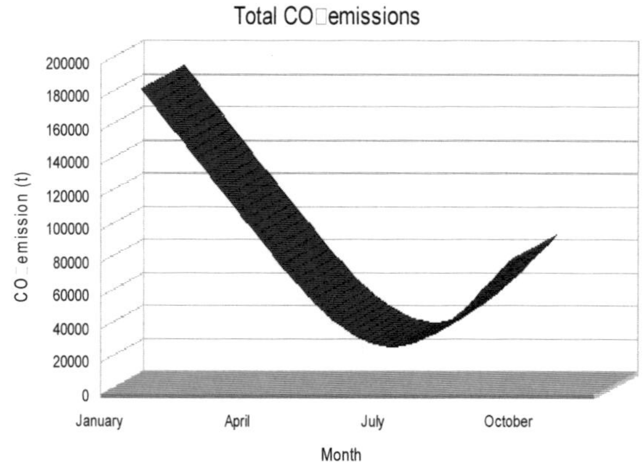

Fig.1 The evolution of CO_2 emissions according to the main sources

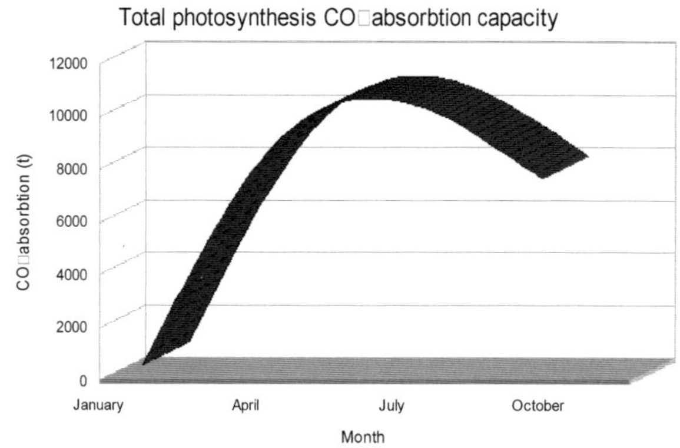

Fig.2 The evolution of the CO_2 adsorption capacity through the process of photosynthesis

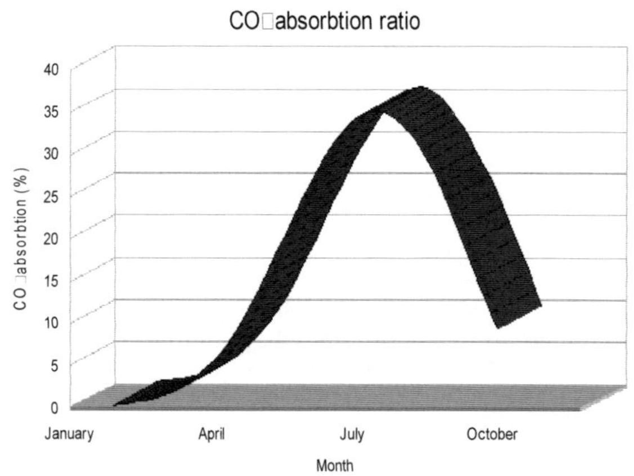

Fig.3. The CO_2 adsorption ratio through the process of photosynthesis depending on the total CO_2

The first chart represents, for the studied periods, the evolution of the CO_2 emissions according to the main sources (road transport, Colterm and other burning and incineration installations).

The second one presents the evolution of the CO_2 adsorption capacity through the process of photosynthesis:

The last chart presents the CO_2 adsorption ratio through the process of photosynthesis depending on the total CO_2.

3. CONCLUSIONS

The global heating potential (GHP) can be evaluated related to the carbon dioxide considered as standard: $CO_2 = 1$ (Ertl et all., 1999) and after analyzing the charts, we noticed that the adsorption of carbon dioxide by photosynthesis in comparison with the carbon dioxide emissions is extremely important because although we live in a city polluted with carbon dioxide coming from different sources, the photosynthesis attenuates the local summer temperature increase due to the local greenhouse effect and because from the total amount of emissions, the carbon dioxide adsorption through photosynthesis is 35%. While the number of cars, thermal power demands, housing and population are significantly and constantly increasing in the last 10 years (well above 100%), even the photosynthesis phenomenon will no longer face the increasing pollution if drastic measures will not be taken.

Due to such a high percentage of carbon dioxide adsorption, over 35% in July and quite low in the winter months in the absence of 85% of this phenomenon (vegetation outside the resinous, which represents 15% of all trees, does not have the leaves and does not synthesize substances, thus doesn't photosynthesize), we could draw the conclusion that during summer the greenhouse phenomenon is reduced with 35% which greatly prevents temperature increase and in winter in the absence of 85% of this phenomenon (photosynthesis), the greenhouse effect emphasizes and materializes in Timisoara by temperature increases during winter to little below 0 C^0 for a short period of time, low temperatures maintaining only when cold air fronts from Northern Europe are present.

At the same time, these issues bring along irreversible negative effects resulting in the disappearance of seasons and their homogenization (winter is heating and summer is cooling). These effects attract remarkable weather and climate phenomena, major climate changes, increasingly frequent occurrence of devastating weather phenomena (storm, hurricane, tornadoes, heavy rain, etc.), (Wehner et al., 2004), even in areas where there were no reported history of such phenomena (i.e. tornados in Dobrogea and along the Romanian coast) and which have an impact and repercussions on the population (floods, landslides, and so on). Although these "alarm signals" appear more and more often, carbon dioxide emissions are increasing more and more over the assimilation limit of plants direct proportionally to the increasingly wider areas of forest cleared due to human expansion.

4. REFERENCES

Ciupa, V. Radoslav, R.; Oarcea, C. & Oarcea, Z. (2005). *Timisoara verde. Sistemul de spatii verzial Timisoarei – Green Timisoara. The green spaces of Timisoara,* Marineasa Printing House, Timisoara

Coste, I.; Borza, I. & Arsene, G.G.(2001). *Ecologie Generală şi Agricolă - General and Agricultural Ecology,* Orizonturi Universitare Printing House, Timişoara

Ertl, G.; Knözinger, H. & Weitkamp, J. (1999). *Environmental Catalysis*, Weinheim

Ionel, I.; Popescu, Fr.; Oprisa-Stanescu, D.P.; Bisorca, D. & Gruescu, Cl. (2004). *Energoecologia combustibililor fosili. Teme experimental- Energoecology of fossil fuels. Experimental subjects,* pg. 130, ISBN 973 625 186 1, Politehnica Printing House, Timisoara

Wehner, B.; Philippin, S.; Wiedensohler, A.; Scheer, V. & Vogt, R. (2004). *Variability of non-volatile fractions of atmospheric aerosol particles with traffic influence. Atmospheric Environment* 38 (36), 6081–6090

Annals of DAAAM for 2008 & Proceedings of the 19th International DAAAM Symposium, ISSN 1726-9679
Editor B. Katalinic, Published by DAAAM International, Vienna, Austria 2008
Make Harmony Between Technology and Nature, and Your Mind will Fly Free as a Bird

OPTIMIZATION OF THE PRODUCTION STRUCTURES

DOBREN, F[lavius] A[ndrei]; DUMITRESCU, C[onstantin] D[an]; LAZARESCU, C[ezara]; ISTRAT, N[icolae] & AMARIEI, O[lga] I[oana]

***Abstract**: The objective of this paper is to research work carried out by the authors in order to achieve an instrument capable to evaluate/ monitor the production capacity, size, human resources and the investments required for the operation's achievement of a system in variable time horizons; the program is PcDHr / I.*

After the definition of the optimization concept and a short presentation of optimization methods, according to the applicability level (production, technology and technique) by the linear programming method, this paper presents a specific calculus program for the production capacity of a system, with applicability in SME, in all 4 cases of homogeneous / non-homogeneous conditions for production / operations.
***Keywords**: optimization, methods, resources, linear, calculus.*

1. INTRODUCTION

Optimization means to find, of the multitude of existing solutions, that precise solution that fulfills the maximum conditions, from one specific and established point of view. If that particular goal is oriented towards the economic domain, economic optimum represents that situation or state of economy that assures the highest efficiency. To choose from multiple solutions that particular one that assures the highest economic efficiency means to search for that solution which assures maximum results with minim effort and expenses, respecting certain restrictions that defined that particular domain.

Optimization of the processes in an industrial unit, considered as a complex system, is obtained based on three distinct levels (Nicolescu, 2001):

1. The level of the production system;
2. The level of technological production system;
3. The level of technical level;

The work presents the synthesis of the objectives through the optimization processes for each level of decision, the methods used for each level of optimization and the optimization criteria used for each level in part. During the optimization process, a specific objective, a certain optimization method, selected from a range of other methods, and a certain particular criteria of optimization correspond to each of these. A synthesis of these optimization processes pays attention to four factors: the level of optimization, the goals of the analysis processes, the maximization methods and the maximize criteria

Constantly, companies confront with the problem of assignation of the resources (Peel, 1993). For micro-units, small- and medium sized companies, the optimum solution of the problem can be reached through the usage of linear programming. A first phase is to transform the problem that conditions the existence of the restriction system into an equation and to establish the equation system; then the optimizing function is to be established, aiming the maximization/ minimization at the cost level of the whole of the distribution/ re-distribution process of the resources. (Andrei, 2005). The paper grants particular attention to the assessment of the human resources in companies because in small companies in Romania fluctuation employment is high, on the one hand and on the other hand the level of wages is negotiable from one person to another; these two elements compound a frequently used module.

To highlight this point, it is considered that an economic enterprise must produce P1, P2, …, Pn, benefiting at most of the b1, b2, …, bm quantities from the R1, R2, …, Rm resources. Knowing that the manufacturing of an unit from the Pj product needs the a quantity from the Ri resource, and that through the delivery of a quantity of the same product, the Cj, 1 ≤ i ≤ n profit is obtained, it is required to determine the x1, x2, …, xn quantities from the P1, P2, …, Pn products so that we obtain the highest benefit. So

$$\begin{cases} \max \sum_{j=1}^{n} c_j x_j \text{ in condition that} \\ \sum_{j=1}^{n} a_{ij} x_j \geq b, 1 \leq i \leq m \\ x_j \geq 0 \quad j = 1,2,\ldots, n \end{cases} \tag{1}$$

The restrictive conditions are also called the linear programming restrictions.If

$$\begin{aligned} &(x_1, x_2, \ldots, x_n)^T = X; \; (a_{ij})_{\substack{1\leq i\leq m \\ 1\leq i\leq n}} = A \\ &(b_1, b_2, \ldots, b_m)^T = L; \\ &(c_1, c_2, \ldots, c_n)^T = C \end{aligned} \tag{2}$$

Then PL – min is the matrix transcribed thus:

$$\begin{cases} \min f(x) = C^T \cdot X \\ A \cdot X = L \\ X \geq 0 \end{cases} \tag{3}$$

and PL – max:

$$\begin{cases} \max f(x) = C^T \cdot X \\ A \cdot X = L \\ X \geq 0 \end{cases} \tag{4}$$

Obviously that the A*X=L restriction system may be:

- incompatible
- exclusively compatible determined (only in the case m=n)
- undetermined compatible.

For our study, the last case is interesting because this is the situation where there is to choose from multiple solutions, the optimum one.

Let us assume that the range of A is m. If we mark the columns of the A matrix with aj, 1≤j≤n, then the restrictions of the problem are transcribed:

$$a1x1 + a2x2 + \ldots + anxn = L \tag{5}$$

A X Є Rn vector, whose components satisfy the restrictions of a linear problem, is called an admissible program (or a admissible solution, or a possible solution).An admissible X° program that minimizes (or maximizes, according to the problem) the linear function associated to that problem is called optimum program or optimum solution.

A X = (x1, x2, , xn)T program is called a base program if the vectors of the aj column, accordingly to the x components, non-null, are linear dependent. For the reason that A = m range a base program has at most m non-null components.

Optimum assignation of the resources

The moment when the resources allocation is solved, there is still the problem of their usage within the manufacturing process, so that the total costs would be minim, or the total profit would be maxim (Baron& Anghelache, 1996).

If it is considered that a plant/ department manufacturing system, manufactures Rj , j = 1,2,3,...,n guiding marks with the Mi , i = 1,2,3,..., m machines, with the operation sequence being random, the information about the process is:

the time standard for manufacturing a Rj guiding mark on the Mi machine is equal to tij (u.t./ piece)

the vacant time of a machine, for a time horizon (week, month, semester, year) is Ti .i = 1,2,3,...m

the assessed time for manufacturing a guiding mark is estimated considered as being cj j= 1,2,3,...n

it is thought that during manufacturing there is no vacant time, due to the fact that the operations are inter-changeable.

The quantity from the Rj guiding mark is named with Xj j=1,2,3,...,n, which is produced on the Mi machine.

With these elements, the mathematical model of the problem is:

$$\Sigma\ tij\ .\ Xj\ \leq Ti \qquad (6)$$

in which: i = 1,2,3,...,m, j = 1,2,3,...,n, Xj ≥ 0, şi Xj # N

The function for optimizing is:

$$\Sigma\ cj\ .\ Xj\ \rightarrow \min \qquad (7)$$

The problem's solution is obtained by using the Gomory solving algorithm for the case of whole numbers, or by using the specific programs for linear programming (QSB. LINDO). Solving this problem leads to linear programming in whole numbers with real numbers coefficients.

2. OPTIMIZATION OF THE MANUFACTURING CAPACITY BY USING A SPECIFIC CALCULUS PROGRAM

In order to dimension the production structure according to the agenda enforced by the market conditions (from Romania, and also from the European Community market), the user may access the condition (Dumitrescu et al., 1998).

$$Cpyear = \Sigma Qyearj \rightarrow (pcs/\ year) \qquad (8)$$

If we restrict the area of the manufacturing capacity notion to the company level, from the point of view of outputs of manufacturing systems, the production capacity is actually a transfer capacity of that system (which can be a work place, work team, department or organization). For this case, the production capacity actually represents the maximum quantity of a certain array and quality obtainable along a specified period, in fixed technical - organizing conditions, without considering the narrow places.

The Calculus Program for the Production Capacity and Dimensioning of the Humane Resources / Investments (The PcDHr/I program) has the following distinctive sections:

- Section 1 - The calculus of the production capacity of a production system, paying attention to the four cases he may be found in.
- Section 2 - The optimum dimensioning of the productions system (the calculus of the production areas needed to the production process)
- Section 3 - Human resources/ payment fund/ calculus for the necessary investment in order to realize the production system.

The program also offers the possibility to establish the human resources essentials, in order to draw up the flow chart (in the case of small/medium enterprises). Furthermore, the value of the necessary investment, the flow of furnishing process of equipment, tools and the endowment of the work place with specialized equipments can be established. The program detains the calculus possibility of the production capacity, considering the characteristics of the manufacturing processes (homogeneous/non-homogeneous) and the arrayal characteristics (homogeneous/ non-homogeneous arrays); in these four cases, the manufacturing capacity is evaluated distinctively.

3. CONCLUSIONS

1. Optimization means to determine the solution that in the conditions of pre-established restrictions assures maximum of results with minimum effort or expenses. Within a company the process optimization may be achieved on three different levels: the level of production system, the technological level, and the technical level; each level is characterized through its goals, optimization methods and proper methods.

2. Applying the linear programming method in order to establish the optimum allocation of resources in the case of SME, we can generate a general mathematical model:

$$\Sigma\ Xj[\alpha i,\ \beta ij]\#\ [\gamma i\ ,\ \delta i\ ]i= 1,2,3,\ldots m, \qquad (9)$$

where: Xj ≥ 0, şi Xj # N j = 1,2,3,...,n

3. The production capacity of a SME may be optimized by enforcing a specific calculus program, starting from the condition Cpyearj = ΣQyearj → pcs/ year, generally accepted within Romania and the European economic area. The program includes three distinct sections mentioned before: the calculus of the production capacity of a production system, the optimum dimensioning of the productions system and the human resources/ payment fund / calculus for the necessary investment in order to realize the production system.

This system may be implemented in homogeneous / non-homogeneous conditions, as much as in the production process and in the arrayal characteristics.

4. Research has aimed several specific aspects:

- the program is for small and medium sized companies' use; the information is to be introduced fast & will be used immediately;
- the cost of this program is low; it permits to use a high data volume, relevant to the company;
- the program may be used to dimension a manufacturing structure and the necessary staff to activate the developed structure and also for a first evaluation of the necessary investment to enable the dimensioned structure.

4. REFERENCES

Andrei, S. *Statistics, Theory and Applications* (2005) All Publishing, Bucharest

Baron, T. & Anghelache, C. *The Statistics* (1996), Economical Publishing, Bucharest

Dumitrescu, C.; Fantana, N. & Militaru, C., *General Management Factors* (1998), EUROBIT Publishing, Timisoara

Nicolescu, O. *Small and Medium Entreprises' Management* (2001), Economical Publishing, Bucharest

Peel, M. *Introduction to Management* (1993), Alternativ Publishing, Bucharest

Annals of DAAAM for 2008 & Proceedings of the 19th International DAAAM Symposium, ISSN 1726-9679
Editor B. Katalinic, Published by DAAAM International, Vienna, Austria 2008
Make Harmony Between Technology and Nature, and Your Mind will Fly Free as a Bird

SURFACE GRINDING METHOD OF SILICON WAFERS

DOBRESCU, T[iberiu] & ANGHEL, F[lorina]

Abstract: *Slicing technology is the key to either process, but uncontrolled deep damage (10-50 (µm in depth) and large warpage and thickness variation of the wafers are unavoidable. In the traditional process, lapping is used next for minimizing the depth of the damage layer and the thickness variation. Etching then eliminates the damage layer formed by lapping, but the thickness variation is increased. By mechanic-chemical polishing, the final macro- and micro-flatness is obtained. Using this process, silicon wafers have 1.0 µm of local thickness variation (LTV: 15x15 mm^2) and are supplied for 1 Mbit DRAM processes.*

Key words: *silicon wafers, rotation grinding method, diamond wheel, cup wheels*

1. INTRODUCTION

The flatness value may be achieved by improvement of the traditional method, but it is not the final target of Si-wafers. For wafers over 200 mm in diameter, two technological disadvantages of the traditional method are exposed. One is the thermal and mechanical inhomogenity of a huge lapping exposed (Dobrescu, T., 1998).

The second is the difficulty of improving the stability of the etching rate, both wafer to wafer and within a wafer.

Because of the high normal forces induced when grinding brittle materials, unavoidable elastic deformation of the tool and workpiece occur. This is due to the kinematics of the cutting process, i.e. that the relative trajectories and speeds of the abrasive grains are varied. The elastic deformations caused by high normal forces bring about faults of geometry and surface and the sub-surface damage.

The conventional grinding methods for wafer have the above mentioned characteristics and therefore the wafer quality is limited. With the new rotation method, which is integrated in the ultra precision process, it is possible to suppress this disadvantage (Dorin, A. & Dobrescu, T., 2007).

2. THE PRINCIPLE OF THE ROTATION GRINDING METHOD

Figure 1 shows the principle of the rotation grinding method. A wafer is centered on a porous ceramic vacuum chuck. The work piece rotates relative to a cup wheel, which also rotates on its own axis. In this way, every point of the planar wafer surface comes into contact with the grinding wheel and the diminution of size is executed on for every point of the wafer. The cup wheel is moved in axial direction until the wafer has reached its final thickness (Inasaki, I., 1987).

Economical reasons require a splitting of the grinding process into a rough grinding and a finishing part. For this purpose, cup wheels with different grit sizes have to be used. In the rough grinding with a D46-cup wheel it is possible to reach high stock removal rates. Finishing with the grit size D7 permits a good surface quality (Dobrescu, T., 1996).

There are two construction principles for the rotation grinding method with stepped grit size:

- Concentric cup wheels. Two annular grinding wheels are mounted concentrically on one rotating support disk. The inner grinding-ring has the large grit size for rough grinding whereas the outer ring is provided for fine grinding. The two grinding rings are arranged in such a way that either the rough-grinding-wheel or the fine-grinding-wheel can be put in contact with the work piece whereby the other grinding wheel is not activated;
- Serially joined cup wheel. The grinding process is divided into two or more steps of serially joined cup wheels. The wafer runs through the workstations in order of declining grit size. In wafer rotation grinding, accurate straightness can be obtained theoretically. But in reality, the rotary axis of the grinding wheel and the work piece are not absolutely parallel because the plane parallelism of the ground wafer face cannot be achieved exactly.

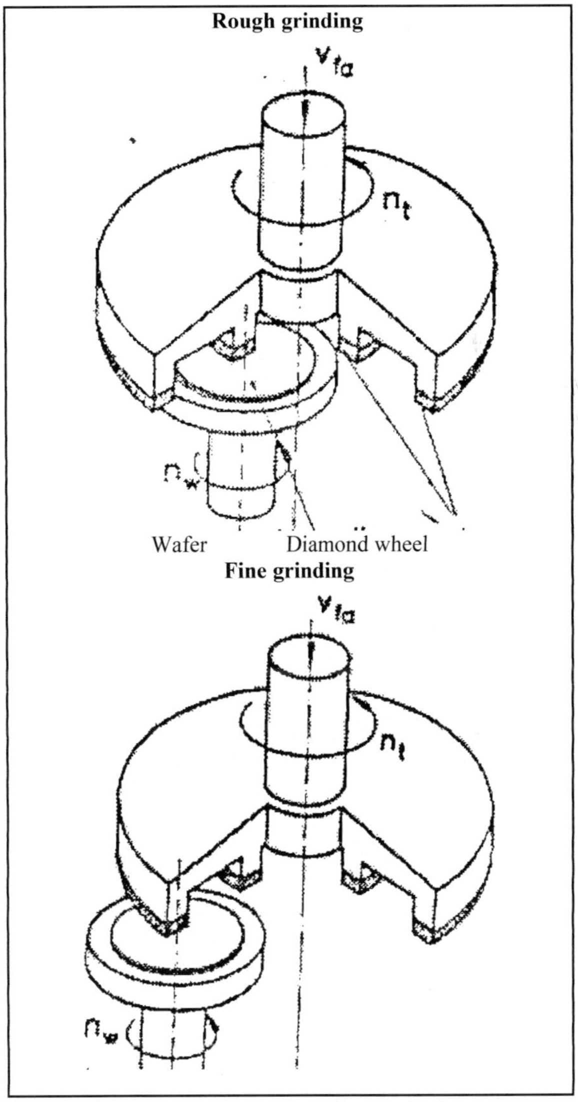

Fig. 1. Principle of the rotation grinding method.

The reasons for this fault are:

- machine deformation as a consequence of grinding forces;
- machine deformation due to thermal influence;
- insufficient adjustment;
- bearing clearance of tolerance.

Because of these reasons, a subsequent adjustment of the spindle is useful. The influence of the angular deviation on the shape of the ground work piece is quite simple.

The angular deviation can be divided up into a radial and tangential component of the angle between z - axis and the rotary axis of the grinding wheel.

If the spindle is inclined in positive α_r direction, the wafer has a convex profile on the wafer surface. In case of an inclination in negative α_r direction, the wafer would have a concave profile. Whether the inclination in a direction is positive or negative makes no difference. The wafer has a cone-like shape (Sun, W., Pei, Z. & Fisher, G., 2004).

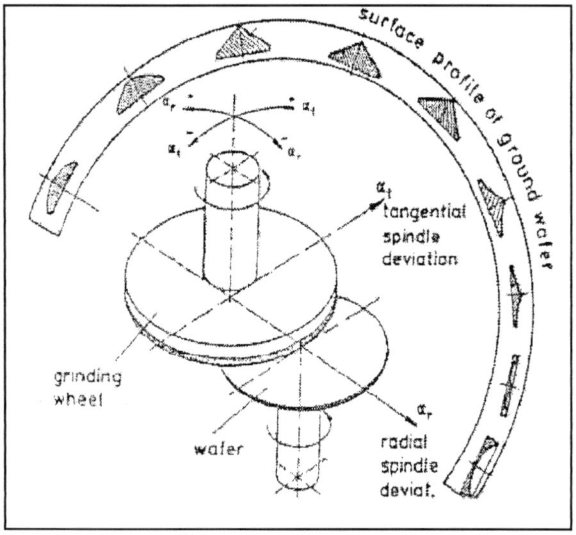

Fig. 2. Surface profile

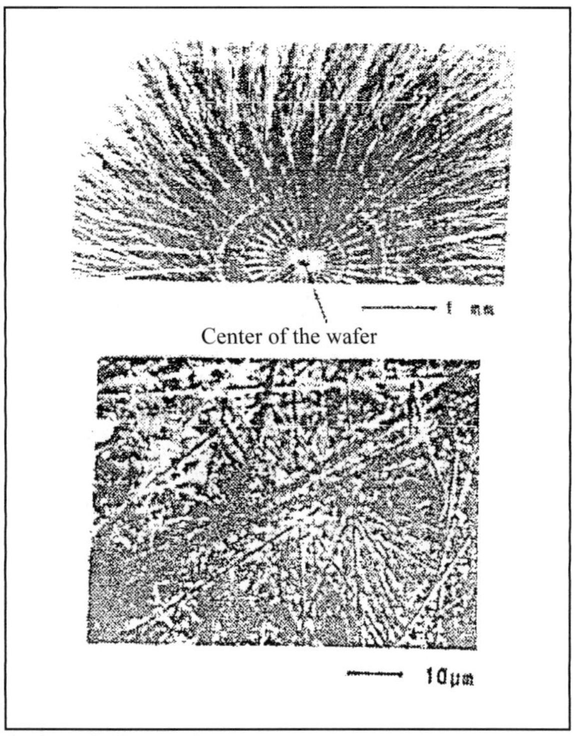

Fig. 3. Grinding marks

The spindles deviation is not only limited on a purely α_r or α_t direction, but all combinations between these two components are possible, so that all shapes between convex and concave on one side and "cone-like" on the other side are possible (Matsui, S., 1988).

A certain relation between α_r and α_t inclination leads to a typical surface-profile which is shown in Figure 2.

It is characteristic for grinding in general that the processed surface is not an absolutely homogenous but shows a certain orientation of grinding marks. The aim for every high quality grinding process is to reduce the visibility of the grinding marks although they always exist. The grinding marks depend on the kinematics of the process.

Figure 3 shows an example of the surface of a wafer ground by the rotation method.

For the best possible surface roughness the grit size of the grinding wheel should be very fin. However, there are certain limits below which grinding is not longer possible because the self-dressing effect decreases and finally disappears.

If smaller grit size than D7 is utilized to produce smoother surface, the diamond in resin bond tends to be buried by the silicon dust created during the processing.

3. CONCLUSION

The following conclusions can be drowning from this experiment:

- Especially for large wafers (200 mm and more), rotation grinding with a cup-wheel represents future technology, because the geometrical contact zone is small (thermal induction) and the tool is defined in its geometry;
- To reach values of less than 0.5 µm for the total thickness variation TTV, an automatic adjustment of the tool spindle will be necessary. The objective is to obtain a nearly perfect wafer quality after this step to reduce the polishing process expense;
- Lower wheel speed and higher feed rate produce rougher surface;
- In silicon wafer manufacturing, the removal amount of the subsequent polishing process to be large enough to eliminate all grinding marks generated in the simultaneous double side grinding or single side grinding operation. Further reduction of polishing amount necessitates optimization of the grinding process so that the grinding marks can be eliminated with minimum amount of polishing.

4. REFERENCES

Dobrescu, T. (1996), Rotation Grinding Method of Silicon Wafers, *Research Reports, LAPT,* University of Naples „FedericoII“, pp. 31-34, Italy

Dobrescu, T. (1998), Cercetari privind optimizarea masinilor de superfinisat materiale fragile, *PhD Thesess,*Universitatea „Politehnica“ din Bucuresti

Dorin, A & Dobrescu, T. (2004). A Study of Pad Dynamics in Wafer Polishing. *Proceedings of the International Conference of Manufacturing Systems*, Academy of Romania (Ed), pp. 513-516, Bucharest, October 2004, Romanian Journal of Technical Sciences

Inasaki, I. (1987). Grinding of Hard and Brittle Materials. *Annals of the CIRP*, no. 36, pp. 463-471

Matsui, S. (1988). An experimental study on the grinding of silicon wafers. *Bulletin of the Japan Society of Precision Engineering*, no. 22, pp. 295-300

Sun, W., Pei, Z. & Fisher, G. (2004). Fine grinding of silicon wafers: a mathematical model for the wafer shape. *International Journal of Machine Tools & Manufacturing*, no. 44, pp. 707-716

Annals of DAAAM for 2008 & Proceedings of the 19th International DAAAM Symposium, ISSN 1726-9679
Editor B. Katalinic, Published by DAAAM International, Vienna, Austria 2008
Make Harmony Between Technology and Nature, and Your Mind will Fly Free as a Bird

FLATTENING PROCESS OF SILICON WAFERS

DOBRESCU, T[iberiu] & DORIN, A[lexandru]

Abstract: *Three processes can be used to flatten the sliced wafers: lapping, single side grinding and simultaneous double side grinding. Lapping is a very complicated and random process resulting from the variation of abrasive grains by its size and shapes and from the numerous variables, which have an effect on the process quality. The advantages of surface grinding compared to lapping are the high material removal rates and the lower level of ceystal damage. This paper reviews manufacturing process flow for silicon wafers using lapping, face tangential grinding and face pluge grinding.*
Key words: *silicon wafers, lapping, face tangential grinding, face plunge grinding.*

1. INTRODUCTION

Most integrated circuit (IC) chips are built on single crystal silicon wafers. Wafer diameter has increased steadily from less than 50 mm in the 1970s to 200 mm today and possibly to 300 mm in the near future.

Manufacturing of silicon wafers starts with growth of silicon ingots. A sequence of processes is needed to turn an ingot into wafers. This typically consists of following processes (Dobrescu, 1996):

- Slicing, silicon ingot into wafers of this disk shape;
- Edge profiling (chamfering), to chamfer the peripheral edge portion of the wafer;
- Flattening (lapping or grinding), to achieve a high degree of parallelism and flatness of the wafer;
- Etching, to chemically remove the damage induced by slicing and flattering without introducing further mechanical damage;
- Rough polishing, to obtain a mirror surface on the wafer;
- Fine polishing, to obtain final mirror surface;
- Cleaning, to remove the polishing agent or dust particles from the wafer surface.

The lapping operation is shown in figure 1. A batch of wafers (for example, 20 pieces) is manually loaded into a lapping machine. The loaded wafers are then lapped by the abrasive slurry (typically a mixture of alumina and glycerin) injected between two lapping plates rotating in opposite directions (Dorin. & Dobrescu, 2007).

The lapping operation would generate sub surface damages in silicon wafers, which need to be removed by its sub sequent processes.

Doble-side lapping is an established process in modern wafer production.

A common feature of all lapping processes is that the material removal takes place by the action of loase abrasive grains, which are dispersed, in a watery or oily lapping fluid. The workpiece shape is produced by the relative motion between the lapping wheel pieces surfaces coupled with the action of the loose lapping compound located between these two active surfaces.

The engagement of the lapping grains takes place non-directionally and stochastically so that directional abrasion grooves are generally avoided. A lusterless, homogeneous surface result.

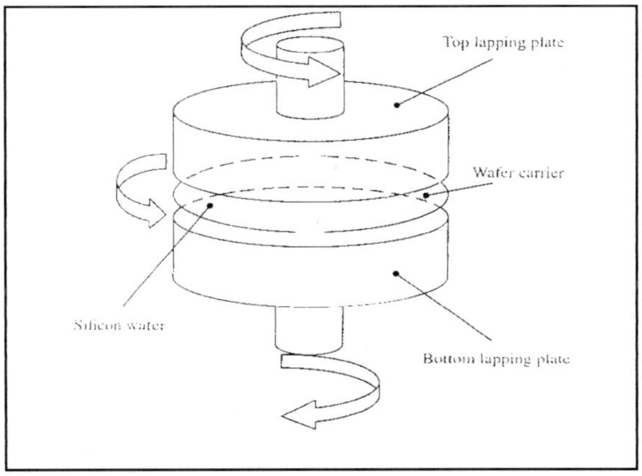

Fig. 1. Illustration of lapping process.

For large volume wafer production, mestly high powered special purpose doublesided lapping machines with planetary drive systems are use. The primary features of these modern lapping machines are (Dobrescu, 1998):

- Programme and gauging controls which allow pressure-controlled operation;
- The infinitely variable speed adjustment of the drive motors (soft-start control);
- The application of grooved, temperature-controlled lapping wheels.

The necessity to gradually increase the lapping pressure in parallel with ensuring a soft-start control may be attributed to the extreme sensitivity of the wafers to brittle fracture.

Single-sided surface grinding operations can be applied to removal diffusion layers or to reduce the thickness of completely processed semiconductor wafers (thinning). The wafer thickness is reduced from 450 - 600 μm down to approximately 150 - 380 μm. After the thinning process, the parallelism may not exceed a total thickness value of 5 μm, in certain cases of 3 μm. In pure thinning surface roughness's R_{max} of greater than 2 μm are generally not permitted. The advantages of surface grinding compared to lapping are the high material removal rates and the lower level of crystal damage. In addition, grinding is a cleaner process and has good potential for automatisation in which fully integrated cassette-to-cassette manufacturing can be achieved.

2. COMPARE BETWEEN FACE TANGENTIAL GRINDING AND FACE PLUGE GRINDING

All specially developed grinding machines for semiconductor manufacture have vertically oriented diamond cup wheels (Figure 2).

Grinding machines differ in process kinematics and the number of simultaneously machinable wafers, which are fixed on vacuum chucks. Depending on the depth of cut a_p and the table feed speed v_{ft} face tangential grinding processes can be further classified into "reciprocation grinding" and "creep feed grinding".

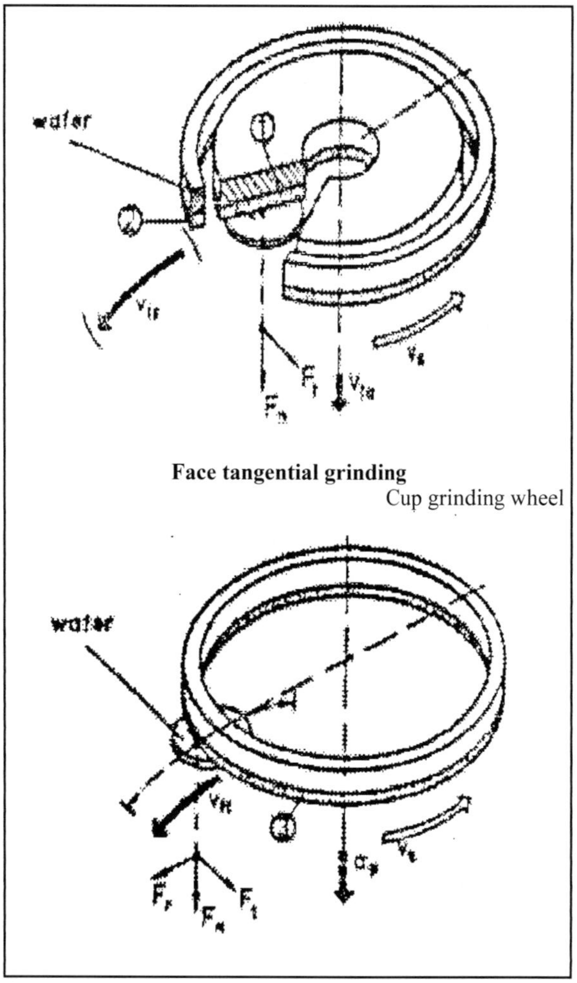

Fig. 2. Kinematics of face plunge grinding and face tangential grinding.

In the first case, the material removal takes place in several cuts each with low depths of cut (1 - 10 µm) and a high table feed speed (up to 8000 mm/min) whereas creep feed grinding is characterized by high depths of cut (100 - 200 µm) and low table feed speeds (100 - 300 mm/min). In face tangential grinding with a rotary table it is favorable to have the process conditions: contact length, contact area and grinding forces remain constant since this directly affects the form accuracy attainable.

In face plunge grinding, the rotating compound grinding, wheel, which covers the entire wafer surface, is fed continuously in axial direction (Figure 2). Shortly before reaching the final thickness, the axial feed is disengaged. Subsequently a tangential finishing process takes place in which the fine-grained diamond outer ring is engaged as the wafer leaves the grinding zone. In face pluge grinding, the engagement conditions concerning cutting speed v_c and cutting length l_{ck} vary within the wafer surface and the grinding layer. Furthermore, the grinding forces in pluge rough grinding are approximately an order of magnitude higher than those for tangential grinding due to the large wheel / wafer contact area A_k (Li et al., 2006)

Therefore, the final tangential finishing process diminishes form inaccuracies, which result from deflections of the machine tool. In this final operational step, very smooth surfaces can be obtained by using a fine grained outer ring. In tangential grinding, multi-step process as described above can only be realized on a multi-spindle machine tool system.

Face plunge grinding results in smaller depths of damage than that for tangential grinding. The normal component F''_n ranges between 2 and 36 N/cm^2 depending on the "process kinematics and the material removal rate. The highest values of area-related grinding forces have been obtained in tangential reciprocation grinding. The influence of the feed speed on the depth of damage and the forces per unit area is higher than that of the depth of cut for the same material removal rate (Sun et al., 2004).

Even at material removal rates of up to 1000 µm/min, considerably low depths of damage were found with face plunge grinding. This can be explained by the large number of diamond grains, which are engaged simultaneously. This causes low level process-induced stresses and damage in the crystal material although the material removal rate is remarkably high. The material removal process is governed by micro-plastic abrasion. Brittle chipping and cracking at the wafer surface usually associated with coarse abrasion occurred less often in face plunge grinding. Face plunge grinding is thus a very efficient abrasion process for roughing. In tangential grinding at material removal rate per unit area =1000 µm /min, more than 90 % of the wafers fractured during grinding or in subsequent processing for damage measurement. It was therefore not possible to evaluate depths of damage.

3. CONCLUSION

The following conclusions can be drowning from this experiment:

- When the two grinding wheels rotated in different directions the surface on one side of the wafer was different from the other side. This was a limitation for simultaneous double side grinding if the identical wafer surface on both sides were required;
- Lower wheel speed and higher feed rate produce rougher surface;
- In silicon wafer manufacturing, the removal amount of the subsequent polishing process to be large enough to eliminate all grinding marks generated in the simultaneous double side grinding or single side grinding operation. Further reduction of polishing amount necessitates optimization of the grinding process so that the grinding marks can be eliminated with minimum amount of polishing;
- The topography of the abrasives, in particular the larger ones, is reflected on the wafer surface in the form of arc-formed grinding grooves. Thus, a final tangential grinding process as a finishing operation is indispensable for high form accuracies and surface qualities.

4. REFERENCES

Dobrescu, T. (1996), Surface Grinding on a Rotary Table of Silicon Wafers, *Research Reports, LAPT,* University of Naples „FedericoII“, pp. 35-38, Italy

Dobrescu, T. (1998), Cercetari privind optimizarea masinilor de superfinisat materiale fragile, *PhD Thesess,* Universitatea „Politehnica“ din Bucuresti

Dorin, A & Dobrescu, T. (2007). The Manufacturing Process Flow for Silicon Wafers. *Proceedings of the 5th International Conference of Advanced Manufacturing Technologies*, AGIR (Ed), pp. 29-34, Sibiu, july 2007, Academy of Technical Sciences of Romania

Li. Z., Pei, Z. & Fisher, G. (2006). Simultaneous double side grinding of silicon wafers: a literature review. *International Journal of Machine Tools & Manufacturing*, no. 46, pp. 1449-1458

Sun, W., Pei, Z. & Fisher, G. (2004). Fine grinding of silicon wafers: a mathematical model for the wafer shape. *International Journal of Machine Tools & Manufacturing*, no. 44, pp. 707-716

Annals of DAAAM for 2008 & Proceedings of the 19th International DAAAM Symposium, ISSN 1726-9679
Editor B. Katalinic, Published by DAAAM International, Vienna, Austria 2008
Make Harmony Between Technology and Nature, and Your Mind will Fly Free as a Bird

MATLAB MODELING OF MATERIAL SURFACE ROUGHNESS PARAMETERS AT END MILLING PROCESS

DODUN, O[ana]; GHENGHEA, L[aurentiu]; SARBU, I[onel] & COTEATA, M[argareta]

Abstract: *The paper concerns a theoretical model proposed for the surface roughness parameters of a surface created by end-milling processes. A MATLAB function was created for calculate the theoretical values for R_a , R_z , R_y , R_q. This mathematical model can be used as prediction of the surface roughness, comparison with measured surface roughness value it is possible to investigate the effects and interdependences of tool radius corner, feed and radius of circle materialized by the peak tool on the machined material surface. The mathematical model permits us to observe that the roughness parameters are not influenced by the distance between the feed vector and the surface measurement profile.*
Key words: *milling, roughness, model, MATLAB.*

1. INTRODUCTION

It is known that the machined surfaces generated in ball-end or flat-end milling process have a cross-section consisting of elliptical or circular arcs, depending on the direction of the measurement line. This aspect of the machined material surface is due the perpendicularity deviation of the main shaft relative to the machined surface. In this case, the tool cutting edges, after a semi-rotation, did not reach the surface. If the deviation at perpendicularity is very small, the machined surface has curved tool paths on two directions.

In this paper we proposed a MATLAB function to establish, for a theoretical ideal profile, the following common surface roughness parameters: the arithmetic average roughness R_a , peak-to-valley roughness R_y , the average height of the asperities R_z and root-mean-square roughness R_q. The calculus establishes the coordinates of several points of the theoretic profile, determines the height of the least squares mean line and obtains the numeric values. To estimate the surface roughness parameters some approximation methods are commonly used. Since the 1960s, it is known the approximate solution to express the peak-to-valley roughness. The following mathematical relation was established assuming that R_y is small and neglecting the higher order term R_y^2 (Bohosievici, 1993, Boothroyd & Knight, 2006):

$$R_y \cong \frac{f^2}{8r} \ (\text{for } f < 2r, r \neq 0) \tag{1}$$

where r is the tool radius and f is the feed. The feed f is the distance between two adjacent peaks of the surface profile. The approximation of the arithmetic average roughness R_a can be expressed as (Bohosievici, 1993, Boothroyd & Knight, 2006):

$$R_a \cong 0.032\frac{f^2}{r} \tag{2}$$

An empiric relation, legitimate for a cutting speed between 18 … 44 m/min and feed of 14 … 28 mm/min, to estimate the arithmetic average roughness R_a at end milling process of a high-speed steel for bearings SH15 with a high-speed steel end mill was proposed in (Bohosievici, 1993):

$$R_a = 4.83\frac{f^{1.69} \cdot t^{0.15}}{v^{1.23} \cdot r^{0.14} \cdot \gamma^{0.46}} \ \mu\text{m}, \tag{3}$$

where t is the cutting depth and γ is the front rake angle.

Other empiric relation is defined in (Slatineanu, 1980):

$$R_z = \sqrt{\frac{Cr^u}{f \cdot t^x \cdot \chi^z \cdot \chi_1^z}} \ \mu\text{m} \tag{4}$$

where C, u, x, z are constants that are experimentally established.

Other researchers (Qu & Shih, 2003) proposed a comparison of close-form section profile solution for calculating the three roughness parameters and the approximation methods using a parabolic curve to match the elliptical or circular arcs.

2. MODELING PROPOSAL AND DISCUSSION

Figure 1 illustrates the grooved surface machined by an end mill with a tool radius r that rotates all around a central axis. Due to the perpendicularity deviation of the main shaft relative to the machined surface, after a semi-rotation, the tool cutting edges did not reach the surface. We supposed that the cutting tool motion is plotting a torus with the following parameters: R -the radius of circle materialized by peak tool and r -the radius of the circle that rotates all around an axis to obtain the torus. The plane of the surface measurement profile is parallel and at a distance d relative to the plan X-O-Z. The equation of the torus generated by the tool corner radius can be expressed as a biquadratic equation in the unknown z (Coxeter, 1989):

$$z^4 + A \cdot z^2 + B = 0 \tag{5}$$

where the coefficients can be expressed as:

$$A = 2(x^2 + y^2 + R^2 - r^2) \tag{6}$$

$$B = (x^2 + y^2 + R^2 - r^2)^2 - 4R^2(x^2 + y^2) \tag{7}$$

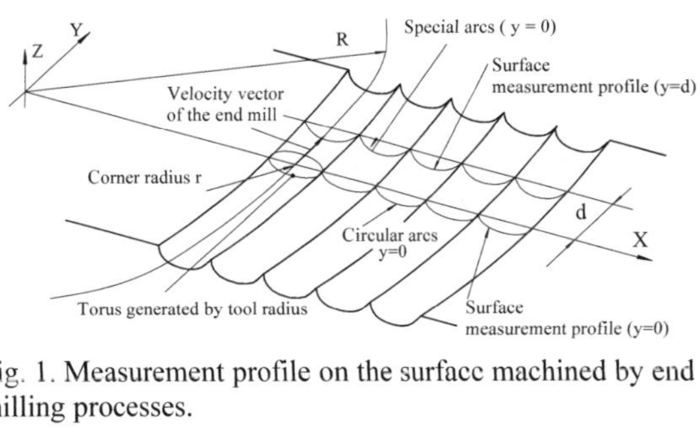

Fig. 1. Measurement profile on the surface machined by end milling processes.

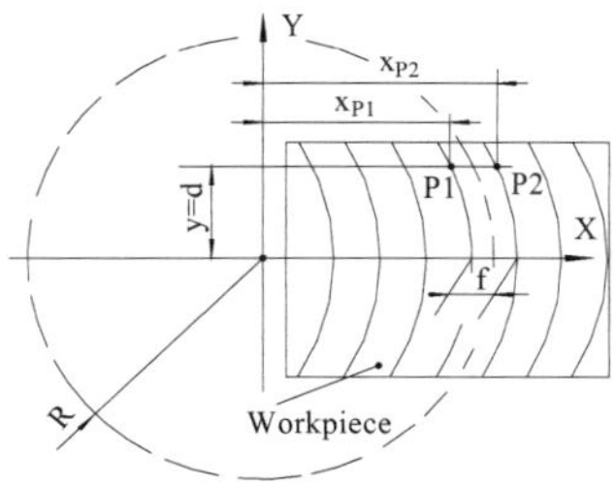

Fig. 2. Key parameters and coordinate system for modeling.

The first step to establish the surface roughness parameters is to resolve the equation (5). The coordinates of the points from theoretical ideal profile, x and y, shown in figure 2, can be represented by:

$$x_{P1} \leq x \leq x_{P2} \tag{8}$$

where
$$x_{P1} = \sqrt{(R - f/2)^2 - d^2} \tag{9}$$

$$x_{P2} = \sqrt{(R + f/2)^2 - d^2} \tag{10}$$

and
$$y = d \tag{11}$$

Afterwards, we determined zeros of the polynomial represented by equation (5), we made a translation of the coordinate system, for making the calculus easier. Figure 3 shows the theoretical surface profile and the new coordinate system. After coordinate translation, the first point of the theoretical profile has null abscissa and the point with minimum z-coordinate has null elevation. In a particular case, when y = 0, the cross-section view of theoretical surface profile consists in circular arcs.

The proposed MATLAB function follows these steps to obtain two vectors named: xR -vector containing the abscises and zR -vector containing the height of the points from theoretical profile. With the calculated points, we can draw the theoretical measurement profile. In program we can control n - the number of points of the profile by modifying the increment q between the abscissa of point $P1$ and point $P2$. The implicit value of q is 5 µm. We may obtain an increase of the calculus accuracy by growth the number of points that must diminish the value of q. The xR and zR vectors have the length equal with n.

According to STAS 5730/2-85, the reference mean line in surface roughness is the least squares mean line, a straight line parallel to the X-axis. Based on the definition of the least squares mean line, we calculate a parameter M defined as:

$$\sum_{1}^{n} z_i^2 = min \tag{12}$$

Some results obtain with the MATLAB functions are presented in the figure 3 and below:
-for $r = 1200\ \mu m,\ R = 60000\ \mu m,\ f = 400\ \mu m/rev,\ d = 0\ \mu m,$ we obtain n = 401 points and Ra = 4.2885 µm; Rz = 18.3371 µm; Ry = 16.7840 µm; Rq = 85.8762;
-for $r = 1200\ \mu m,\ R = 60000\ \mu m,\ f = 400\ \mu m/rev,\ d = 50000\ \mu m,$ we obtain n = 724 points and Ra = 4.2713 µm;
Rz = 18.2662 µm; Ry = 16.7840 µm; Rq = 114.9302 µm.

The proposed MATLAB function verifies the result by comparison the value of the first point form the theoretical profile with a parameter l that corresponds the geometrically establish equation:

$$\left(\frac{f}{2}\right)^2 = l \cdot (2R - l) \tag{13}$$

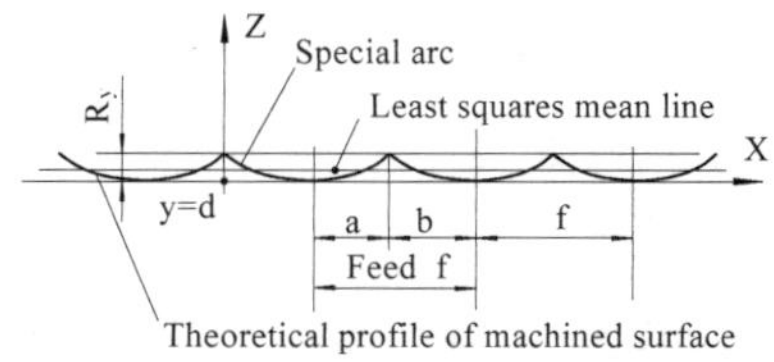

Fig. 3. Cross-section view of theoretical surface profile consisting of specials arcs (for y≠0).

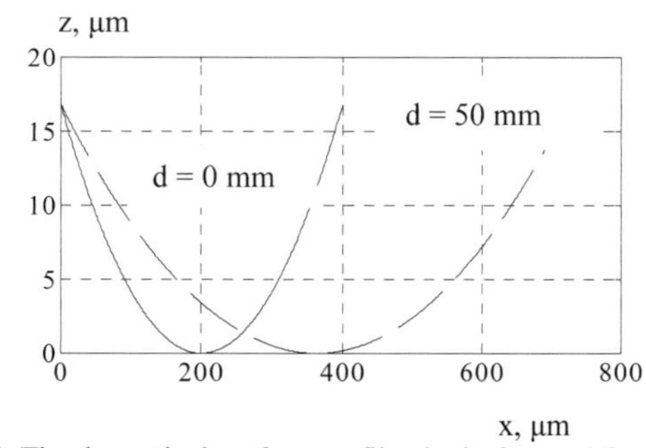

Fig. 4. The theoretical surface profile obtain for r = 1.2 mm, R = 60 mm, f = 0.4 mm/rev.

3. CONCLUSION

The proposed MATLAB function offers a solution to obtain theoretically four surface roughness parameters, R_a, R_z, R_y, R_q. We supposed that the trace for surface finish measurement is perpendicular relative to the velocity vector of the end mill. The mathematical model is based purely on geometric considerations. The real surface generated in machining processes will deviate from this theoretical profile owing to the tool wear, vibration of the tool during machining, elastic deformation and recovery of the tool and workpiece, build-up edge at tool tip, resolution of the measurement machine etc.

From the particular calculus presented above and using many times the proposed function one may observe an interesting fact that the roughness parameters are not influenced by the distance between the feed vector and the surface measurement profile d. This theoretical model can be used also as prediction of the surface roughness. By comparison with measured surface roughness value it is possible to investigate the effects and interdependences influences of the workpiece material, tool behavior, cutting parameters and other factors on the machined surface.

4. REFERENCES

Bohosievici, C. (1993). *Modern technologies for mechanical manufacturing (in Romanian)*, Polytechnic Institute, Iaşi

Boothroyd, G. & Knight, W. (2006). *Fundamentals of Machining and Machine Tools*, CRC Press, ISBN: 1574446592, Boca Raton

Coxeter, H. (1989). Introduction to Geometry, Wiley, ISBN: 978-0-471-50458-0, New York

Creţu, G. (1992). *Fundamentals of experimental research. Laboratory handbook (in Romanian).* Technical University "Gh. Asachi" of Iaşi, Romania

Ou, J. & Shih, A.J. (2003). Analytical Surface Roughness Parameters of a Theoretical Profile Consisting of Elliptical Arcs, In: *Machining Science and Technology,* Vol. 7, Issue 2, January, 281-294, ISSN 1532-2483

Slatineanu, L. (1980) *Contributions to the study of the some Romanian steels mahinability (in Romanian).* PhD thesis, Technical University of Iaşi, Romania, (in Romanian)

Annals of DAAAM for 2008 & Proceedings of the 19th International DAAAM Symposium, ISSN 1726-9679
Editor B. Katalinic, Published by DAAAM International, Vienna, Austria 2008
Make Harmony Between Technology and Nature, and Your Mind will Fly Free as a Bird

THE RESPONSE ANALYSIS OF THE CFRP LAMINATED PLATES DUE TO LOW VELOCITY IMPACT

DOGARU, F[lorin] & BABA, M[arius] N[icolae]

Abstract: *A comparative study, between the impact response of CFRP laminated plate and the force-displacement curve determined in static condition, is presented in this paper. For analytical studies, a simple solution using complete model for a rectangular plate subjected to low-velocity impact is used, based on Kirchhoff's plate theory. The results obtained analytically were compared with FEM results and experimental data. The study was conducted on composite laminates plates carbon/epoxy vinyl ester with dimensions 125x75mm², made of symmetrical laminate with 8 unidirectional laminae* $[0/-45/+45/90]_s$, *simply supported at the edges. It is to note the concordance between the results obtained theoretically, numerically and experimentally.*

Key words: *low-velocity impact, CFRP, FEM.*

1. INTRODUCTION

The CFRP material presents an increased susceptibility to damage due to impact. The weak behavior of the polymeric composite material reinforced with carbon fibers is due to the reduced absorption capacity of kinetic energy of the projectile.

There are many solutions in literature that were presented for different impact cases. Sun & Chattopadhyay (1975) used a complete model for studying a rectangular plate in the case of impact influenced by the wave propagation when the maximum displacement of the plate is out of phase with the projectile's displacement. Olsson (1992) gave an approximate solution for case of impact influenced by the wave propagation. Abrate (1998, 2005) studied deeper the impact response taking into account different models proposed by researchers in past years. Swanson (1997) showed that when the mass of the plate is much smaller than the projectile's mass the response is quasi-static.

In this paper the authors give a simple solution using complete model on the basis of Kirchhoff's plate theory concerning the impact response of CFRP laminated plate.

2. DESCRIPTION OF THE ANALYTICAL MODEL

A complete model takes into consideration the structure's dynamic, the projectile's dynamic and the contact force, (Sun 1975, Abrate 2005). In case of simply supported plate at the edges, Navier resolution can be used to obtain a closed-form solution to the transitory response. In accordance with the classic plate theory, for a symmetrically laminated plate, without bending-twisting coupling ($D_{16}=D_{26}=0$), the displacement equation of the plate loaded by a concentrated force may be written in the form:

$$D_{11}\frac{\partial^4 w}{\partial x^4}+2(D_{12}+2D_{66})\frac{\partial^4 w}{\partial x^2\partial y^2}+D_{22}\frac{\partial^4 w}{\partial y^4}++I_1\ddot{w}=F(t), \quad (1)$$

where D_{ij} stands for bending and torsion stiffness, I_1−ρh, ρ is density of the plate, h thickness of the plate, F(t) contact force and w stands for the transversal displacement of the plate. Considering a plate with axb dimensions we obtain the transversal plate's displacement of point (x,y), for the impact case in point (x_1, y_1):

$$w(x,y,t)=w_2=\frac{4}{\rho hab}\sum_{j=1,3,5}^{\infty}\sum_{k=1,3,5}^{\infty}\int_0^t F(\tau)\sin\left(\frac{j\pi x}{a}\right)\times \times\sin\left(\frac{k\pi y}{b}\right)\sin\left(\frac{j\pi x_1}{a}\right)\sin\left(\frac{k\pi y_1}{a}\right)\cdot\frac{\sin[\omega_{jk}(t-\tau)]}{\omega_{jk}}d\tau. \quad (2)$$

The projectile's displacement, w_1, is the sum of the local displacement due to contact, δ and plate's displacement, w_2. Considering the equilibrium projectile's equation with initial conditions $w_{1(t=0)}=0, \dot{w}_{1(t=0)}=V$, and so, through integration, the projectile's displacement becomes:

$$w_1=\delta+w_2,\ F(t)=-M_1\ddot{w}_1,\ w_1=V\cdot t-\frac{1}{M_1}\int_0^t F(\tau)(t-\tau)d\tau \quad (3)$$

where V and M_1 are the velocity and the mass of the projectile. In order to model the contact, Hertz's contact law was used. It was supposed that the projectile has spherical shape and the Hertz's contact law is valid for the dynamic case as well:

$$F(t)=\kappa_c\delta^{3/2},\ \kappa_c=4/3E_3R^{0.5},\ E_3=E_2, \quad (4)$$

where κ_c is contact stiffness' coefficient. Taking out δ from Eq.(3.1) and considering Eq.(3.3, 2) after substituting in Eq.(4.1), the non-linear integral equation, for central plate's impact becomes:

$$F(t)=\kappa_c\left(V\cdot t-\frac{1}{M_1}\int_0^t F(\tau)(t-\tau)d\tau- -\frac{4}{\rho\cdot a\cdot b\cdot h}\sum_{j=1,3,5}^{\infty}\sum_{k=1,3,5}^{\infty}\int_0^t\frac{1}{\omega_{jk}}F(\tau)\sin[\omega_{jk}(t-\tau)]d\tau\right)^{3/2} \quad (5)$$

3. NON-LINEAR EQUATION RESOLUTION

Trapezoid formula is used to solve Eq.(5). For particular case of equation (5), two kinds of functions under integral exist:

$$f_1(\tau)=F(\tau)\cdot(t-\tau),\ f_2(\tau)=F(\tau)\cdot\sin[\omega(t-\tau)] \quad (6)$$

Solving separately dividing (0, t) in i equidistant intervals, t_0, $t=t_0\cdot i$. The result is:

$$A(i)=\int_0^t f_1(\tau)d\tau=t_0^2\sum_{k=1}^{i-1}F(k)(i-k),\ i\geq 2, \quad (7)$$

where $F(i)=F(i\cdot t_0)$, and for the moment t=0 the two elements get into contact, $F(0)=0$. Quite in the same way, for the second function, taking in account $m\times n$ participating mode of the plate, the result is:

$$B(i)=\int_0^t f_2(\tau)d\tau=\sum_p c(p)t_0\sum_{k=1}^{i-1}F(k)\sin\omega_p\left[(i-k)\cdot t_0\right], \qquad (8)$$

$$i\geq 2\,,\omega_{jk}=\omega_p\,,1\leq p\leq m\times n\,,c(p)=\frac{4}{\rho\cdot a\cdot b\cdot h}\cdot\frac{1}{\omega_p}\,.$$

Using Eq.(7, 8) the Eq.(5) was solved iteratively with MATLAB program (Dogaru et al., 2005).

4. RESULTS AND CONCLUSIONS

The analytical and experimental analyses were conducted on a composite plate made of epoxy vinyl ester matrix (Derakane 470-30-S) reinforced with carbon fibers with dimensions 150x100mm^2 and 2.5mm thickness, 8 unidirectional laminae and symmetric orientation $[0/-45/+45/90]_s$. The characteristics of lamina's plate were: E_1=54GPa, υ_{12}=0.3, E_2=E_3=4.5GPa, G_{12}=G_{23}=1.65GPa. The plates used to cut off the specimens were manually manufactured and the resin was impregnated by brushing-on action and the fibers volumetric ratio was about 35%.

The impact test was done by the use of a device designed for this particular study, having the energy capacity of 1-50J obtained by adjusting the height and/or the weight. The specimen was simply supported at the edges against a metal plate (30mm thickness) with interior cutting-out of 125x75mm^2 by the intermediary of a wooden plate (6mm thickness) in order to avoid the specimen crushing at ends. The specimen was fixed at the edges during the impact at four points with screws having rubber disposed on the tip and manually screwed. The projectile had a 16-mm diameter semispherical head made of alloyed steel with increased hardness, 1.9kg weight and the impact was targeted at the plate's center. Behind the projectile, an accelerometer was attached (screwed), in order to measure the projectile's acceleration and then to calculate the contact force during the impact. The results were recorded by using an acquisition plate NI USB 6251 BNC. The velocity and the displacement of the projectile during the impact were calculated through integration of the measured acceleration curve using LabVIEW program (Dogaru et al., 2005).

A new static study was done because the impact is quasi static, it meant that the maximum force and the maximum displacement are reached simultaneously (Swanson 1997), using FEM (Ansys), in which the force was applied statically on the projectile which is considered in contact with the laminated plate. A 1000 elements were used, SHELL181 elements for plate's simulation, SOLID187 for projectile's simulation and CONTA174, TARGE170 for contact's simulation. The force applied on projectile was the maximum value recorded in dynamic experimental investigation, F=3750N. The analysis is done on quarter of the model due to the symmetry considering the large displacements and contact effects.

The analytical response of the composite plate was calculated using analytical solution, taking in account 5x5 modal parameters, for different cases of impact velocity.

Fig.(1) illustrates the variation of the maximum contact force due to impact related to the central maximum transversal displacement of the plate obtained analytically and experimentally. For comparison, the static solution obtained with FEM is presented, too. The solution due to Eq.(5) is accurate only for small displacements, generally for maximum transversal displacement smaller than the thickness of the plate fig.(1).

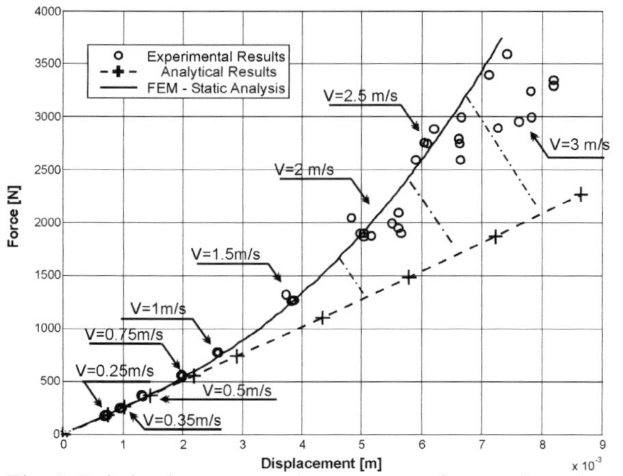

Fig. 1. Relation between maximum contact force and maximum central plate's displacement recorded for different initial impact velocity of the projectile.

Notice that for a projectile initial velocity higher than 1.5m/s, the results obtained using dynamical experiment are quite scattered and the mean value of the forces recorded, is between linear analytical solution and nonlinear numerical solution obtained with FEM. This occurred because at this loading level, the damages are introduced in the plate and they reduce the stiffness and also the contact force value. Also notice the concordance, for small displacement case, between the results obtained analytically, numerically using FEM static solution and experimentally. In the future the authors intend to investigate the damage introduced by impact, the level of the contact force that causes the damage and its effect on the residual properties. On the basis of these observations, the concept of a new equivalent model shall be also considered in the future in order to simulate the damage effects occurred in the composite plate, on the residual properties and impact response.

Acknowledgements. This research was done with financial support of MECT and ANCS, contract PN II – IDEI, ID_187, 110 / 1.10.2007.

REFERENCES

Abrate, S. (1998). *The Dynamics of Impact on Composite Structure,* In: *Impact Response and Dynamic Failure of Composites and Laminate Materials, Part 2,* editors J.K.Kim and T.,X.,Yu, Trans Tech Publications, ISBN-13: 978-0878497690, Switzerland.

Abrate, S. (2005). *Impact on Composite Structures,* Cambridge University Press, ISBN-13: 978-0521018326, United Kingdon.

Dogaru F. & Curtu I., et al., (2005). *Analytical And Experimental Investigation Concerning Response Of The CFRP Laminated Plates Due To Low Velocity Impact* - 9th International Research/Expert Conference "Trends in Development of Machinery and Associated Technology", ISBN 9958-617-28-5, TMT 2005, 26-30.09, Antalya, Turkey.

Olsson, R. (1992). *Impact response of orthotropic composite plates predicted from a one-parameter differential equation*, AIAA Journal, Vol.30, No.6, p.1587-1596.

Sun, C., T., Chattopadhyay, S. (1975). *Dynamic Response of Anisotropic Laminated Plates under Initial Stress to Impact of a Mass,* Journal of Applied Mechanics, p. 693-698.

Swanson, S. R. (1997). *Introduction to Design and Analysis with Advanced Composite Materials*, Pretice Hall, Upper Saddle River, ISBN-13: 978-0024185549, New Jersey.

Annals of DAAAM for 2008 & Proceedings of the 19th International DAAAM Symposium, ISSN 1726-9679
Editor B. Katalinic, Published by DAAAM International, Vienna, Austria 2008
Make Harmony Between Technology and Nature, and Your Mind will Fly Free as a Bird

MODELING OF CHOICE THROUGH THE TRADE COSTS ALGORITHM

DOGARU, V[asile]

Abstract: *The algorithm of the comparative advantage, using the modified prices related to the trade costs in terms of a real two product barter, brings us on a "previous" position towards the merchandise exchange using the currency. In this new algorithm, in the formal analytical plan, are included the prices increases, related to tariff and non-tariff measures and the reduction ones through subventions or other similar measures. The main remark is that now a higher consumption of resources takes place because of the products' transfer to and from far different economic areas and also in connection with some national custom's pass beyond which other legal and economic laws work.*

Key words: *comparative advantage, modeling, resources.*

1. INTRODUCTION

The constant task which we want to solve is to identify the context in which can be used the formulas of comparative advantages' measurement, identified by research of the defined hypotheses in typical cases of exchange. We rewriting the formulas from the deducted algorithm of the simple barter case (Dogaru, 2006) for a more exact approximation of the comparative advantages of the two economic entities, through taking in consideration the different clauses which involves external costs, excluded in the initial prices considered in this basic scheme. Observing in more details the buying-selling processes it appears obvious that each external goods exchange requires various tariff and non-tariff efforts from which the transportation ones which presently have a relatively increased role (Harrigan, 2003). Other international costs – regarding various tariff and non-tariff measures or, in general, all the trade costs – gives us a full picture of these efforts, sometimes called external costs.

2. RESEARCH COURSE

Modeling choice of goods, if the existence of several options exchange, can be done using comparative advantage scheme by removing successive the variants less effective. The most effective may be chosen in the last two remaining options.

The inclusion of various costs, which adds to the initial prices considers the necessity of formalization in the Evariste Galois meaning (Georgescu-Roegen, 1971), in order to be able to observe the change of the comparative advantage, according to Manoilescu (1937) generalized scheme. The identification of comparative advantage related to new efforts can be made using as instrument the new algorithm. Each from the initial prices of the two products will be changed with the unitary level of these costs (1).

$$p_e + c_{E\,ex} = p_{me}\ ;\ p_i + c_{E\,im} = p_{mi} \qquad (1)$$

where: p_e, p_i - internal initial prices from the E and I country; p_{me}, p_{mi} - internal changed prices from the E and I country; $c_{E\,ex}$, $c_{E\,im}$ - the external unitary costs of the E and I exporting entities, of the two products exchanged, Pr 1 and Pr 2.

Through the importation from I country to E country of Pr 2 product, the new price p_{me2} will *neutralize* the origin in our analytical scheme. Therefore the symbol refers to a product spatial found in E country although this is manufactured within I country. Initial price of Pr 1 product from E country, exported in I country, will be changed at new price, p_{mi1}, the exportation costs are included. The same suppositions are made also for the case regarding the change of the internal products' prices. In this analysis is alleged lack of internal costs.

The relative total advantage's formula, used in the new terms, will be (2).

$$Avmrt = \frac{p_{me2}}{p_{me1}} : \frac{p_{mi2}}{p_{mi1}} \qquad (2)$$

where:

$p_{me\,1/2}$, $p_{mi1/2}$ - the modified internal prices from E and I country of Pr1 and Pr2 products.

In the formula (2) the trade costs are proportionally distributed (see the difference versus (7)).

Correspondently, the relative modified advantage's formula for E economic entity will be:

$$A_{vmrE} = \frac{p_{me2}}{p_{me1}} : \frac{P_2}{P_1} \qquad (3)$$

where: P_1 and P_2 are international prices.

If these costs are proportionally distributed the relative advantage remains the same and the absolute one would increase, according to (1.5) formula (Dogaru, 2006), because the initial prices are increased with the trade costs. The resolution is not viable because in fact the comparative advantage – measured as gains from trade – will decrease. Therefore, *the reference point* is not well established.

The orienting point is necessary to become (temporary) a case in which each entity trades on its local national markets the two products obtained either through production or through buying. In order to get as closer possible to the real exchange terms we will suppose that the entities E and I can be simultaneously wholesale merchants, so that p_{e1}, p_{e2}, p_{i1} and p_{i2} will be accepted by the buyer in the en-detail net – or manufacturers, for the merchandises for the intermediate consumption (Dogaru, 2006).

The hypothesis from the previous case with all the proportionally modified prices is suspended but the respect guarantee of the hedonic price hypothesis must be fulfill. The necessary condition of the exchange goods is, according to interest principle, as each of the forward (modified) correspondent prices to the imported goods, to be smaller or at least equal to those existent on the internal market.

$$p_{me2(I)} \le p_{e2(E)}\ ;\ p_{mi1(E)} \le p_{i1(I)} \qquad (4)$$

The new prices are calculated in quality equivalent, as in forward the products are supposed, having this reference point, to be identical in the basis of the hedonic price's hypothesis (Eatwell, 1987). The letters between the brackets, E and I, from the above mentioned inequalities indicates us the country product origin and the small letters the country where they are sold. In the monetary competitive economy the prices of products are compared through the exchange rate.

In order to identify the comparative advantage the trade costs are going to be distributed in forward only for the imported products' prices. If the two economic entities distribute the costs over the imported products' prices – Pr2 for E entity and Pr1 for I entity – although the trade costs are made also for the exported price, but yet only in favor of the other product's importation, the relative comparative advantage changes towards the initial formula. In consequence we suppose the prices' change, p_{me2} for E country and p_{mi1} for I country. The relative advantage's formula, correspondently modified for E economic entity will be:

$$A_{vmrE} = \frac{p_{me2}}{p_{e1}} : \frac{P_2}{P_1} \quad (5)$$

Note: In order for a real exchange to be identified the respect of the condition, p_{me2} to be smaller or equal with p_{e2}, is necessary.

In the case of the simultaneous (4) and (5) conditions' fulfillment, economic entity E's relative advantage changes only in the limit of the initial relative comparative advantage, calculated without the trade costs. If (4) condition is not respected for maintaining or increase the absolute value advantage in country E, it is necessary that the influence in A_{vmrE}, because of p_{me2} internal price's increase, to be bigger than the influence of the absolute advantage's decrease through the probable decrease of the quantity for sale from Pr2 product. The twist point is case in which $p_{me2}=p_{e2}$. The analyzed case is similar as mechanism with the one of the profit's total maximization once with the simultaneous increase of the sold quantity and of the unitary profit's decrease across the quantity related to of minimum cost (the neoclassic curve of the total profit's maximization).

In forward we will observe the case of the trade cost algorithm in which only the exported prices will be affected. It has been indirectly deducted from the previous case that for Pr1 sample product, maintained for selling on E country's internal market, the inclusion of these costs in p_{e1} price of this internally sold product is not justified. The judgment is similar for p_{i2} price of Pr2 product from I country. The trade costs will add in this new case, changing correspondently the analytical border of the exchange process, only for the exported sample products. In conclusion c_{Eex2} and c_{Eim1} costs' size will be considered zero or included in c_{Eex1} and c_{Eim2}. In the new terms the formula of the changed total relative advantage, A_{vmrt}, will be (6):

$$Avmrt = \frac{p_{e2}}{p_{me1}} : \frac{p_{mi2}}{p_{i1}} \quad (6)$$

The case is different than the previous ones, of the proportional distribution of the costs over all prices or only over the imported ones. In this case it can be noticed the total relative advantage of the two economic entities reduces in comparison with the initial situation. The reversal of the trade costs between p_{me1} and p_{mi2} will not influence A_{vmrt} size according to a property from mathematics. The comparison of initial internal prices of the exported products, $p_{e1(E)}$ și $p_{i2(I)}$, with the changed prices, $p_{me1(E)}$ și $p_{mi2(I)}$, according to (4) condition, is not necessary in order to decide the opportunity of the internal exchange, separate on each of the two internal markets, as in the previous case.

$$Avmrt = \frac{p_{me2}}{p_{me1}} : \frac{p_{mi2}}{p_{mi1}} \quad (7)$$

Now another case requires be finished and which through generalizing includes the three previous cases and also other possible situations: the simultaneous change of all prices with condition (4) fulfillment.

	p_e	p_i	p_{me}	p_{mi}
Cloth	90	100	90	110
Wine	80	120	85	120

Tab. 1 Calculation of comparative advantage with the trade costs algorithm

In these terms a smaller comparative advantage is going to be achieved. In normal conditions the absolute comparative advantage, who in fact every economic entity is interested, cannot be bigger, because the condition (4) must be respected. The increase of the absolute advantage size is possible due to the quantities increase of imported (exported) products sold and if the value of comparative advantage increase can overcome prices decrease (smaller than the initial ones).

The identification of this analytical case with the change only of the exported product's price, p_{me1} and p_{mi2}, by each entity has as a purpose to make the comparison *in simplified terms possible*. Therefore, in case of such distribution of the trade costs, it has been quite easily remarked that each part's comparative advantage is smaller than the standard situation, the initial one, without the trade costs.

Economic entities calculate the initially comparative advantage and by taking into account the trade costs. Assume the following two exchange cases (table 1). In the new case, it is a reduction in the relative advantage compared to 1.35 at 1.155. Adequately the total profit, which is the monetary shape of comparative advantage, is decreased related to the reducing the quantities of goods traded.

3. FINDINGS

The scientists can use these formulas to measure the efforts of economic entities, according to analytical economy's principle. The trade costs algorithm, included in the Manoilescu generalized scheme, fills an analytical gap, of comparative advantages' measurement. The new algorithm is operable and could be used by the enterprisers for various cases occurring during negotiations.

The calculus with this algorithm will be finish in a quickly way. If the trade costs' size equals the initial comparative advantage's size, allegedly without these trade costs, the exchange cannot be made anymore, but for the case of an unequal exchange, in which at least one of the two partners will not achieve comparative advantage. The results obtained by algorithm of trade costs it is necessary to be developed further to identify other cases, which can provide a faster measuring of comparative advantage.

4. REFERENCES

Dogaru, V. (2006). *Manoilescu generalized scheme regarding the international goods exchange – a general presentation*, Economic Issues, nr. 226-227, National Institute for Economic Research, Romanian Academy, ISBN 973-7885-47-3, 978-973-7885-47-0, Bucharest.

Eatwell, J. & others, editors. (1987). *The New Palgrave. A Dictionary of Economics*, tome I-IV, The Macmillan Press Limited, ISBN 0-333-372352, London.

Georgescu-Roegen, N. (1971). *The Entropy Law and the Economic Process*, Harvard University Press, SBN 674–25780–4 , London.

Harrigan, James, Evans & Carolyn L. (2003). *Distance, Time, and Specialization*, National Bureau of Economic Research, Cambridge, MA 02138, 2003, Working Paper 9729.

Manoilescu, M. (1937). *The National Productive Forces and the External Trade*, Scientific and Encyclopedic Publishing House, 1986, Bucharest.

Annals of DAAAM for 2008 & Proceedings of the 19th International DAAAM Symposium, ISSN 1726-9679
Editor B. Katalinic, Published by DAAAM International, Vienna, Austria 2008
Make Harmony Between Technology and Nature, and Your Mind will Fly Free as a Bird

THE ELABORATION OF THE PROCESS DRAWINGS WITHIN THE GENERATIVE CAPP SYSTEMS

DOICIN, C[ristian]; IONESCU, N[icolae] & TONOIU, S[ergiu]

Abstract: *A generative CAPP system, TehnoCIN, based on an algorithm composed of four main stages was developed by the authors (Doicin, 2001; Doicin & Tonoiu, 2002). In accordance with this four stages structure, the generative system is composed of 3 main modules that generate the process plans for rotational parts (Doicin, 2001).*

The paper presents an important element of the system, those of automatically elaborating the process drawings for all operations within all feasible variants of process plans. The algorithms and the activities described by this stage are running in the frame of the third main module, and involve, mainly, an independent sub-module (block) for generating the process drawings.

Key words: *CAPP, generative system, process drawing.*

1. INTRODUCTION

All the software systems, developed in the field of generating the manufacturing process plans, use two main methods: a variant method and a generative one (Allada & Anand, 1995). A method to create a generative CAPP system, based on an algorithm composed of four main stages was developed by the authors, as follows: 1) Define the initial data, necessary to generate the variants of process plans, by creating the solid model of the part and after that, by filling in with technological data; 2) Define the constraints and preliminary establish the feasible variants of process plans, by preliminary establish the raw stock dimensions and intermediate dimensions, the machining parameters and the time required for machining, and by preliminary generating the set of variants of process plans satisfying specific criteria; 3) Establish the feasible variants of process plans and select the optimum one; 4) Design the jigs and fixtures used in the optimum variant of process plan.

In accordance with the above-mentioned stages, TehnoCIN, a generative CAPP system was created by the authors. The system is designed for revolution parts and is composed from 3 modules: The Module for Identifying the Parts' Shape - MGF (corresponding to the stage 1); The Module for Generating Variants of Process Plans - MGPT (corresponding to the stages 2 and 3); The Module for Designing the Jig and Fixtures - MGES (corresponding to the stage 4);

The paper presents the final step of the third stage of the method, those of elaborating the process drawings for all machining operations. The block which implements this step uses AutoLISP® and ARX® routines which automatically generate the process drawings, within an Autodesk Mechanical Desktop® environment.

2. CONCEPTS AND DEFINITIONS

In the process of developing the TehnoCIN system we've used some old and new concepts (with some particular definitions given by the authors) to define the constructive-technological entities (or part components) describing the machined part. We named these components Distinctive Features (Doicin, 2001; 2002).

Let us consider that a Simple Distinctive Feature (shortly, Distinctive Feature, abbreviated DF) is the simplest constructive piece of a part – made from a surface or a group of surfaces that can be machined together – in which a part can be divided (the atom of the part).

Related to this definition, other two important concepts can be described (Doicin, 2002): Complex Distinctive Feature (CDF), as the distinctive feature composed by several neighbouring distinctive features (a group of adjacent DF's), connected by precedence relations and having a certain role in the activities of defining the process plans variants, and Dominant Distinctive Feature (DDF), as those distinctive feature within a Complex Distinctive Feature having the biggest dimensions, and which gives the complex feature the main characteristics from constructive point of view.

3. ROUTINES AND FUNCTIONS

The routines for generating the process drawings are written by the system in a specific file, having a structure defined by two types of functions: a) Generative functions, defined before the system starts to run, as the core of the graphical representation kernel of the TehnoCIN, which are able to draw standard graphical components of the drawings (points, lines, squares, rectangles etc.) and b) Transfer functions, which depend on the currently analysed variant of process plan, and contain the lists with 3D point coordinates of the part, extracted from the database of the CAD model of the part.

The generative functions create the process drawings. The transfer functions are automatically generated by the CAPP system for each variant of process plan. They have the role of transferring the data – calculated or extracted from the database (Doicin, 2001; Doicin & Tonoiu, 2002) – to the generative functions. The routines contain technological information about machine-tools, tooling etc. and also create the list with 3D point coordinates required in order to draw the part having machined surfaces corresponding to the current operation. In this way, the generative functions will use the 3D lists of points made by the transfer functions in order to generate the 2D drawing of the part, as main component of the process drawing.

4. THE TRANSFER MATRIX

Prior to generate the transfer functions, a matrix (named Transfer Matrix) containing the intermediate dimensions during the machining phases is generated. The transfer matrix contains all intermediate dimensions important during the manufacturing process and is generated by passing the following steps: 1) Corresponding to each DDF contained by each CDF, the intermediate dimensions (the dimensions of the machined distinctive feature after each machining sequence) (Vlase, 1996), are calculated. The TehnoCIN system uses a database containing the types of machining phases able to generate all the technical possible types of DDFs. The calculus of the intermediate dimensions takes into account the longest succession of machining sequences of the same type identified in the above-mentioned database for each distinctive feature; 2) After that, the intermediate dimensions are extracted from the database and assigned to each distinctive feature within all the CDFs defining the analysed part; 3) The complete information regarding the intermediate dimensions for all feasible variants of process plans developed for the current analyse part are now available. All these data are stored in a matrix with intermediate dimensions (Transfer Matrix), built in accordance with the structure described in the table 1.

It is considered that each machining phase is defined by two important parameters (Doicin, 2001): a) the type of the machining phase (i.e. turning, milling, boring etc.) and b) the nature of the machining phase (i.e. roughing, finishing etc.).

In the transfer matrix, only those cells corresponding to the feasible machining sequences are storing data (i.e., the distinctive feature DF1 will be machined by a succession of only MP1 and MP2 machining sequences).

The implemented method is subject to improve because it is not able yet to offer precise values for simple distinctive features of type chamfer and fillet. This is the result of the fact that these kind of features are „borrowing" their tolerance and limits from the associate DDF, usually, the cylinder contained by the current CDF. It is considered that the entire part is composed by 3n simple distinctive features (DFs). For each machining sequence corresponding to a distinctive feature, the nominal value of the intermediate dimensions is retained, together with the upper and lower tolerance limits. This data will be used to generate the process drawings.

Machining phase (type, nature)	*DF1*	*Upper tol. limit, As*	*Lower tol. limit, Ai*	. . .	*DF_{3n}*	*As*	*Ai*
MP 1 (t_1, n_1)	Dim.1	As1	Ai1		Dim1	As1	Ai1
MP 2 (t_1, n_2)					Dim2	As2	Ai2
MP 3 (t_2, n_1)	Dim.2	As2	Ai2		...	...	...
. . .					Dimg	Asg	Aig
MP k (t_j, n_i)							

Tab. 1. The Transfer Matrix

5. THE PROCESS DRAWINGS ELABORATION

In order to elaborate one process drawing, it is necessarily to respect the following algorithm:

1. For each i operation in the current analysed variant of process plan, all the feasible machining sequences are skimmed through (each machining sequence is made over one simple distinctive feature, DF);
2. For each machining sequence, the DF to which the sequence is assigned has to be identified. If the shape of the same distinctive feature is repeatedly modified by the same type of machining sequences (having different natures - i.e. roughing turning and finishing turning) only the last machining sequence (finishing turning) is retained in a data-vector;
3. For each Pj machining sequence within the i operation, the values of the dimensions and of the tolerance limits corresponding to the assigned distinctive features are extracted from the Transfer Matrix;
4. The previously extracted values are assigned to the Pj machining sequence;
5. The complex distinctive feature containing the previously assigned DF (step 2) – machined during currently analysed machining sequence – is identified;
6. The succession of steps from 2 to 5 is repeated until all the machining sequences within the current operation are analysed;
7. All the simple distinctive feature compounding the machined part are grouped by their originate CDF;
8. All Complex Distinctive Features are ordered by the axial coordinates of a so called characteristic point (Doicin, 2001; 2002), unique defined for each CDF;
9. The list containing geometric information is generated;
10. The steps from 1 to 9 are repeated for all the operations within the current variant of process plan.
11. The steps from 1 to 10 are repeated for all the feasible variants of process plan generated by the TehnoCIN CAPP system.

The previously determined data are exported as lists. For each feasible variant of process plan, such a list is defined. The list contains point coordinates, used to draw all the process drawings for those variant.

For revolution parts, a variant of process plan composed from k operations was considered. First two of them, Cutting and Machining Planar surfaces are not analysed by the algorithm. Thus, by first operation the TehnoCIN will understand the third operation of the process plan.

By running the routines, the process documentation – containing all the description of the machining operations within all the feasible variants of process plans – will be automatically generated.

6. CONCLUSION

The paper brings in some important theoretical contribution related to the elaboration of the process drawings within generative CAPP systems.

Thus, the general algorithm of generate the process drawings is presented. This is based on a large amount of data and complex calculus for establish the dimensions and the tolerances of the surfaces machined in each machining sequence, and to find the coordinates of all the points defining the process drawings.

The algorithms are programmed using AutoLISP® and ARX® routines, automatically generated by the CAPP system, depending on the shape of the current analysed part.

The generative method assumes that a specific routine is written to draw each process drawing. All the routines related with the operations belonging to the same variant of process plan are written in a single file. This will be loaded and ran within an AutoCAD® or Autodesk Mechanical Desktop® working session.

The result is represented by the process specifications / process drawings, the documents required in order to machine the raw stock and to obtain the finished part.

7. REFERENCES

Allada, V.; Anand, S. (1995). Feature-based modeling approaches for integrated manufacturing: state-of-art survey and feature research directions, *International Journal of Computer Integrated Manufacturing*, vol. 8, no. 6, November 1995, pp. 411-440, ISSN 0951-192X.

Allada, V.; Anand, S. (1996). Machine Understanding of Manufacturing Features, *International Journal of Production Research*, Vol. 34, No. 7, July 1996, pp. 1791-1819, ISSN 0020-7543;

Doicin C.V. (2002). Concepts regarding the Description of the Revolution Parts through Constructive - Technological Entities, *Proceedings of the International Conference on Integrated Engineering*, G. Drăghici & S. Tichkiewitch (Ed.), pp. 94 - 98, ISBN 973-8247-92-6, Timisoara, April 2002, Ed. Politehnica, Timisoara;

Doicin C.V.; Tonoiu S. (2002). The codification and transfer of the data within the CAPP generative systems, *Proceedings of the 4th Workshop "Human Factor and Environmentalist"*, Branko Katalinic & Emil Wessely (Ed), pp. 27-28, Vienna-Kosice, December 2002, DAAAM International Vienna, Vienna-Kosice;

Vlase A. (1996). *Machine Building Technology (in Romanian)*, E.T., Bucharest, ISBN 973-648-228-7, 1996;

Annals of DAAAM for 2008 & Proceedings of the 19th International DAAAM Symposium, ISSN 1726-9679
Editor B. Katalinic, Published by DAAAM International, Vienna, Austria 2008
Make Harmony Between Technology and Nature, and Your Mind will Fly Free as a Bird

THE SELECTION OF THE MACHINE-TOOLS AND TOOLING WITHIN THE GENERATIVE CAPP SYSTEMS

DOICIN, C[ristian]-V[asile]; TONOIU, S[ergiu] & IONESCU, N[icolae]

Abstract: *The authors developed a method to create a generative CAPP system, based on an algorithm composed of four stages, as follows: (1) define the initial data, (2) define the constraints and establish in advance the variants of process plans, (3) establish the technical possible variants and select the optimum one and (4) design the jigs and fixtures.*
The method was used for developing an original generative CAPP system, TehnoCIN.
The paper presents an important step of the third phase, those of establishes the tools and machine-tools which could be used in a certain machining phase / operation.
Key words: *CAPP, machine-tool, process plan.*

1. INTRODUCTION

There are two main directions regarding the process of generate different variants of process plan in a computer aided environment: Variant Process Planning (VPP) and Generative Process Planning (GPP) (Allada & Anand, 1996). Both of them take into account specific tools and machine-tools, according with the constructive and technological parameters describing the part to be machined. First direction, the most used within the CAPP systems, is based on large databases storing the process plan variants for different groups of parts, developed in a previous stage. The second direction, more difficult to define, but much more flexible, is based on a method of defining the variants of process plan (Allada & Anand, 1996) in a generative way. Following the second direction, a method to create a generative CAPP system was developed by the authors.

The paper presents an important step of the third stage of the process of generating the variants of process plans, those of establishes the tools and machine-tools which could be used in a certain machining operation (Doicin & Gheorghe, 1999), (Doicin, 2001).

2. PROCESS STAGES WITHIN THE CAPP SYSTEM

The CAPP system needs detailed information in order to complete his work and to generate variants of process plans (Doicin & Gheorghe, 1999).

Thus, in the **first stage**, the 3D model of the part is defined, usually outside the TehnoCIN system, using a CAD package. Then, using specific routines specially developed, geometrical data are extracted from the CAD model and translated in a specific language, used by the system. The data are stored in a transfer file (Doicin, 2001), (Doicin & Tonoiu, 2002).

After that, in the **second stage**, a lot of technological data are added to the transfer file, by attaching a list – containing technological attributes – to the geometrical description of each geometric Distinctive Feature (Doicin, 2002). Thus, the constructive description (extracted from the CAD model) of the part is completed with technological data, all the information being stored in the transfer file. Based on these data, some preliminary variants of process plans are generated.

Then, during the **third stage,** the variants of process plans are generated. The algorithm which generates process plans uses a database containing information about part materials, machining parameters, cutting forces, tools, machine-tools etc.

The TehnoCIN system provides a minimal initial database (containing, in the machine-tools tables, data for 41 lathes and 5 milling machines), but user can add new data to the database in order to enlarge the set of parts that may be analysed.

3. ALGORITHM FOR SELECTING THE TOOLS AND THE MACHINE-TOOLS

The next stage is that of establishing the technical possible variants of process plans for the analysed part.

In order to detect the appropriate cutting tools and machine-tools for each variant of process plan generated by the system, a specific algorithm is used, composed from the following steps:

1) Based on the data extracted from the database and using the formulas presented in the literature (Vlase, 1996; Fabian et.al., 2007), for each machining phase within each machining operation, the depth of cut (t) is calculated;

2) Depending on the previously calculated value for the depth of cut, the axial feed rate values (mm/rev) are extracted from the database. There are a lot of restrictions which influences the value of the feed rate (Vlase, 1996), their influence being captured in the values contained in the database. Due to the fact that the database contains only certain discreet values – acquired from the literature – an interpolation (automatically made by the system) is usually required at this point, in order to calculate an exactly feed rate value corresponding to the cutting depth value.

According to the way the values are given in literature, an interval of feed rate values will be obtained during this activity. Let's call it recommended interval (I_s^r) of feed rate values (mm/rev),

3) The recommended interval is restrained by about 10% (value considered as safety coefficient: the experience shows that there are difficult to continue the calculus with the marginal value of the recommended feed rate values due to the lack of data required further on). The restrain is applied only to the maximum recommended values.

4) Based on the previous identified values of t and I_{sr}, for each machining phase, the values of cutting speed, cutting force and the power consumed by the machine-tool engine are extracted from the database. The system will use them in the next phases.

5) Taking into account the already known values for depth of cut and feed rate, two sets of values of machining parameters, an average and a maximum one, are calculated for each machining phase. These values are obtained by multiplying the recommended values of the cutting speed, cutting force and engine power with the average / maximum values of the correction coefficients described in the literature (Vlase, 1996), (Fabian et. Al, 2007). Let's name the obtained values as average / maximum machining parameters.

6) Let L_{UT}^r be the list of available machine-tools and Sij the set of part's surfaces machined during the Pj machining

phase, component of the i operation. For each machine-tool (UT) from L^r_{UT} list, some checks are made in order to verify if each available machine-tool is also an acceptable one (it may be used for machining the part, respecting all the prescribed dimensional and quality characteristics of the Sij part surfaces). As a first step, this is established by searching the feed rate tables in the database for some real values in the restrained recommended interval of feed rate (identified during the step 3). The search is made for each m machine-tool in the L^r_{UT} list and for each Pj machining phase in each i operation. Initially, all the available machine-tools (compounding the L^r_{UT} list) are considered to be also accepted ones.

After the above-mentioned verification, the machine-tools which do not contain the values of the feed rate within the restrained recommended interval (I_{sr}) are removed from the L^r_{UT} list. Let L^a_{UT-Pij} be the list of acceptable machine-tools assigned to the Pj machining phase within i operation. We may define this list as in the equation 1:

$$L^a_{UT-Pij} = L^r_{UT-Pij} - UT_e \qquad (1)$$

where L^r_{UT-Pij} is the list of available machine-tools for Pj machining phase within operation i and UT_e is the machine-tool current removed from the L^r_{UT-Pij} list);

7) For each machining phase and each acceptable machine-tool, the values of the determined feed rate are verified from the point of view of the feed mechanism resistance.

As a result, the machine-tools whose feed mechanisms don't accomplish the resistance criteria are removed from the list of acceptable machine-tools. Finally, for each machining phase within each operation a new (restrained) list of acceptable machine-tools is obtained;

8) For each machining phase and each acceptable machine-tool, the values of the engine consumed power are also verified. The machine-tools whose engine power is less than the power required for machining a certain surface is removed from the list of the acceptable machine-tools.

As a result, the final list of machine-tools, assigned to each Pj machining phase within each i operation will contain only those machine-tools capable to machine the Sij surfaces using the maximum recommended values of the machining parameters (depth of cut, feed rate, speed), by respecting the prescribed characteristics for Sij surfaces.

9) For each i operation, the L^a_{UT-Pij} lists assigned to each Pj machining phases are intersected. The result is a list of the acceptable machine-tools, not only for a machining phase, but for entire i operation (see equation 2).

$$L^a_{UT-Opi} = \bigcap_{j=1}^{nu} L^a_{UT-Pij} \qquad (2)$$

where nu is total number of acceptable machine-tools for machining Sij surfaces in operation i;

10) Depending on the type (turning, milling, drilling etc) and the nature (roughing, finishing etc.) of each Pj machining phase, on one hand, and on the dimensional and quality characteristics of the Simple Distinctive Feature (Doicin, 2001; 2002) to whom the Pij is assigned, on the other hand, the list of acceptable tools are extracted from the database.

11) For each Pij machining phase from each i operation and each acceptable machine-tool from the L^a_{UT-Opi} list, each tool from the acceptable tool list is verified from the point of view of tool body resistance. The calculus is made based on the assumption of using the maximum values of the machining parameters. The tools which don't fulfill the imposed condition are removed from the list of acceptable ones.

12) Based on the average values of the machining parameters, the unit values of the setup and machining time are extracted from the database;

13) The total time required for machining each Sij set of surfaces is calculated. The machining time (which is function of feed rate and speed – with no values at this moment) is estimated based on the maximum values of the machining parameters;

14) The system displays, in a specific window of the system interface, all the acceptable variants of process plans, ordered by the values of machining time, previously establish;

15) The user selects, from the list of acceptable machine-tools – assigned to each i operation, and from the list of acceptable tools – assigned to each Pj machining phase, one machine tool for each i operation and one or more tool(s) for each Pj machining phase.

As a result, lists of acceptable machine-tools and lists of acceptable tools, proposed by the CAPP system to the user, are obtained for each variant of process plan previously generated.

4. CONCLUSION

The paper presents an algorithm for automatically establishing the cutting tools and the machine-tools for each manufacturing phases within each operation of each variant of process plan made in a generative CAPP system. The algorithm proved his functionality within the TehnoCIN package, a generative CAPP system developed by the authors.

Starting from technical principles, the algorithm provides a list of acceptable tools, respectively, machine-tools and user action is required in order to select a single tool / machine-tool for the current machining phase or operation.

Efforts are being made in order to reduce at minimum the user intervention, and the system is continuously evolving. Thus, a friendlier interface it is planned for this module in the future.

5. REFERENCES

Allada, V.; Anand, S. (1996). Machine Understanding of Manufacturing Features, *International Journal of Production Research*, Vol. 34, No. 7, July 1996, pp. 1791-1819, ISSN 0020-7543;

Doicin, C.V.; (2001). Contributions to the CAPP of Shafts from Machine Building Industry (in Romanian), *PhD Thesis*, University POLITEHNICA of Bucharest, pp. 130 – 145, Bucharest, 2001.

Doicin C.V. (2002). Concepts regarding the Description of the Revolution Parts through Constructive - Technological Entities, *Proceedings of the International Conference on Integrated Engineering*, G. Drăghici & S. Tichkiewitch (Ed.), pp. 94 - 98, ISBN 973-8247-92-6, Timisoara, April 2002, Ed. Politehnica, Timisoara;

Doicin, C.V.; Gheorghe, M.; (1999), A Feature-Based CAPP Methodology for Rotational Parts, *Proceedings of the International Regional DAAAM-CEEPUS Workshop on INTELLIGENT MACHINES and TECHNOLOGIES in the 21-st CENTURY*, Kalmar L. (Ed.), pp. 97 – 102, ISBN 963 661 3710, Miskolc, May 1999, Miskolc.

Fabian, M; Spisak, E.; Dovica, M.; Draganovská, D.; Šeminský, J. (2007). The study of the effect of pre-set CAM parameters for cutting surface quality, *Ovidius University Annual Scientific Journal: Mechanical Engineering Series*, Vol. 9, No. 1, July 2007, pp. 33-38, ISSN 1224-1776.

Vlase A. (1996). *Machine Building Technology (in Romanian)*, E.T., Bucharest, ISBN 973-648-228-7, 1996;

Annals of DAAAM for 2008 & Proceedings of the 19th International DAAAM Symposium, ISSN 1726-9679
Editor B. Katalinic, Published by DAAAM International, Vienna, Austria 2008
Make Harmony Between Technology and Nature, and Your Mind will Fly Free as a Bird

EXPERIMENTAL DAMPING DETERMINATION IN BEAM BIFURCATION PROCESS

DOLECEK, V[latko]; ISIC, S[afet] & VOLODER, A[vdo]

Abstract: *This paper presents an experimental determination of dumping in beam vibration, caused by buckling under axial load, which magnitude depends on deformation of buckled beam. It is assumed viscous damping scheme, and damping coefficient is determined in function of beam length and second moment area of cross-section. For specific beam, damping coefficient is calculated from logarithmic decrement and vibration frequency, which are determined from an acquiesced accelerograms of a bifurcation process. It is presented an equipment for experimental measurement, which consists of supports for beam-like specimen, mechanism for force adjustment with dynamometer, accelerometer and amplifier.*
Key words: *damping, bifurcation, buckling, experiment.*

1. INTRODUCTION

A beam is the most used element in many engineering constructions. Because of its specific dimensions, stability loss most likely occurs than overcome of allowable stress. So, beam bifurcation of stability is still the problem of interests of many authors (Isić et al.,2007). Bifurcation of beam equilibrium position is mostly analyzed by using static approach, where postcritical equilibrium path is the subject to be determined (Thompson & Hunt,1973; Życzkowski, 2005). Because of that, dynamics of bifurcation process is not widely analyzed.

Load causing bifurcation is almost assumed as the constant, while in many practical problems load changes its magnitude according to deformation of buckled beam. Bifurcation of equilibrium of beam straight-line position under axial load, which changes magnitude with beam deformation, may produce moderate displacement and beam still remains in elastic range (Isić et al.,2007). As the result of the bifurcation, after departure from primary position, vibration around new secondary equilibrium position appears. While stiffness and stress stiffness change due to large displacement may be taken into account by using higher order terms of cross-section curvature and beam shortening (Isić et al.,2007; Thompson & Hunt,1973), damping remains unknown. Experimental modeling may only be used to its determination (Beards, 1996; Virgin, 2000).

In this paper experimental determination of dumping in beam vibration is presented, caused by bifurcation buckling under axial load, which is produced by compression of elastic body with known stiffness. The load magnitude depends on deformation of buckled beam. Equipment for experimental measurement is also presented, which consists of supports for beam-like specimens, mechanism for force adjustment with strain-gauge dynamometer, piezo-accelerometer and digital amplifier. It is assumed viscous damping scheme, and damping coefficient is determined in function of beam length and second moment area of cross-section. For specific specimen, viscous damping coefficient is calculated from logarithmic decrement of amplitude and vibration frequency, which are determined from time-domain analysis of acquiesced accelerograms of bifurcation process. Acceleration in imposed vibrations is measured by mounted piezo-accelerometer, connected to PC over digital amplifier, while applied force is controlled by mechanism with dynamometer.

2. EXPERIMENT SETUP AND MEASUREMENT

On the Fig. 1 scheme of the experiment setup is presented. Beam-like specimen 4 is connected to the testing platform 7 using two supports 3 and 5 (pinned or clamped). Axial force is applied over mechanism in the axially movable support 3 and measured with strain gage dynamometer 2. The dynamometer is hollow steel tube with bonded strain gages LY11 6/120, connected in the full bridge. The dynamometer is also main guide bush of axially movable support. An acceleration is measured on the middle point of the specimen by connected piezo-accelerometer KD37V. The specimen is restrained in straight-line initial position by auxiliary support 6, until force is regulated to the desired value, greater than calculated critical buckling force. After that, specimen is allowed to buckle and record of axial force and acceleration is acquired using digital amplifier SPIDER 8 and PC with CATMAN 5.0 software.

Using presented experiment setup the acceleration for given axial force and stiffness of dynamometer are measured. Characteristic accelerogram is given on the Fig. 2.

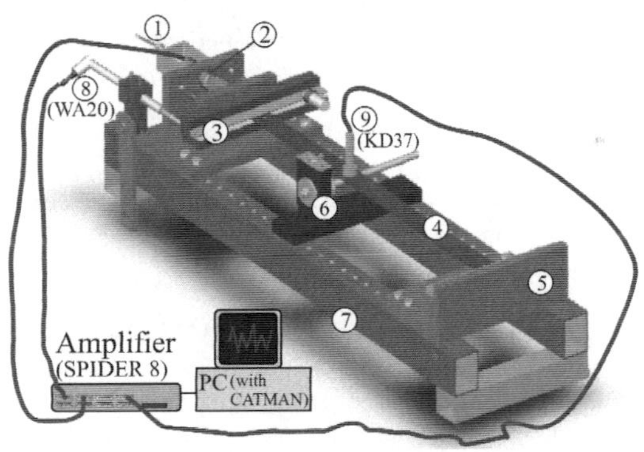

Fig. 1. Schematic view of experimental setup.

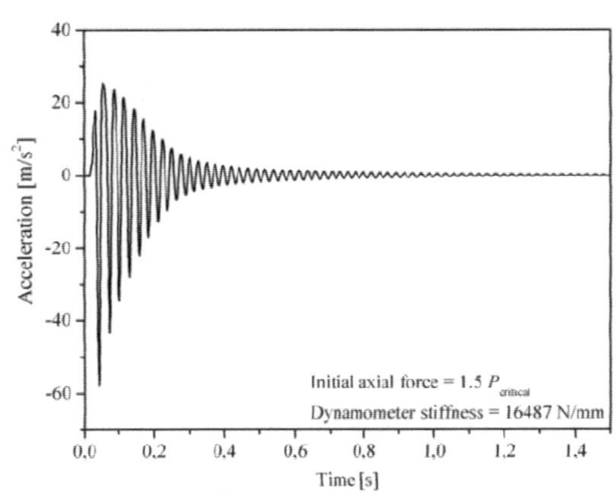

Fig. 2. Acceleration of beam middle point.

Measured accelerograms show two different rates of amplitude decay. In the beginning of the process large amplitudes appear, which are non-symmetric around time axes, what shows presence of departure of beam from initial straight-line position. After that process is finished in the vibration around new buckled configuration. The accelerograms of the process enable measurement of acceleration amplitudes, vibration period and frequency of vibration.

3. CALCULATION OF DAMPING COEFFICIENT

In case of a single d.o.f. system, logarithmic decrement of the amplitude may be calculated from the following expression

$$\Lambda = \frac{C\pi}{M\omega} \tag{1}$$

where C is viscous damping, M is mass of the system, ω is vibration frequency and Λ is logarithmic decrement.
Previous expression could be easily derived from corresponding differential equation of damped vibration. It is shown in [2] that beam deform according to buckling eigenmode. The vibration in the bifurcation process could then be described by using single nonlinear differential equation, which is derived by using finite elements method. Mass M and damping C in (1) is then replaced by

$$\begin{aligned} M &= \mathbf{D}_0^T \mathbf{M} \mathbf{D}_0 \\ C &= \mathbf{D}_0^T \mathbf{C} \mathbf{D}_0 \end{aligned} \tag{2}$$

where **M** is mass matrix, **C** is viscous damping matrix for damping coefficient κ = 1, and $\mathbf{D}_0$ is normalized buckling eigenvector in which characteristic displacement (in experiment - displacement of the specimen middle point) has unit value.
Using equation (2), coefficient of viscous damping κ, in case of considered vibration of buckled beam, could be calculated as

$$\kappa = \frac{M}{C\pi} \omega \Lambda \tag{3}$$

Equation (3) implies that damping factor for specific case could be calculated from measured vibration frequency and logarithmic decrement of a acceleration amplitude.

4. MATHEMATICAL MODEL OF DAMPING

It is assumed that viscous damping coefficient κ may be expressed in shape of polynomial function of length and second moment area of the beam-like specimen as

$$\kappa = b_0 + b_1 L + b_2 I_z + b_{12} L I_z \tag{4}$$

where b_0, b_1, b_2, b_{12} are constants to be determined, L is specimen length and I_z is second moment area.

Point of plan of experiment	Value of experimental factors		Viscous Damping Factor κ	
	L [m]	I_z [m^4]	Outcome ( )$_1$	Outcome ( )$_2$
1	0.300	$0.7840 \cdot 10^{-11}$	4.911	4.875
2	0.339	$0.7840 \cdot 10^{-11}$	2.921	2.764
3	0.300	$0.1065 \cdot 10^{-10}$	4.355	5.158
4	0.339	$0.1065 \cdot 10^{-10}$	9.033	9.328

Tab. 1. Results of experiment.

Unknown constants in the polynomial expression (4) are determined from 2^2 factor experiment with repetition in every point of experimental plan. Experimental measurements are done with four specimens with two different lengths and two different second moment areas, from which is possible to construct four points in plan of the experiment. Obtained results in eight provided measurements are given in the Table 1.

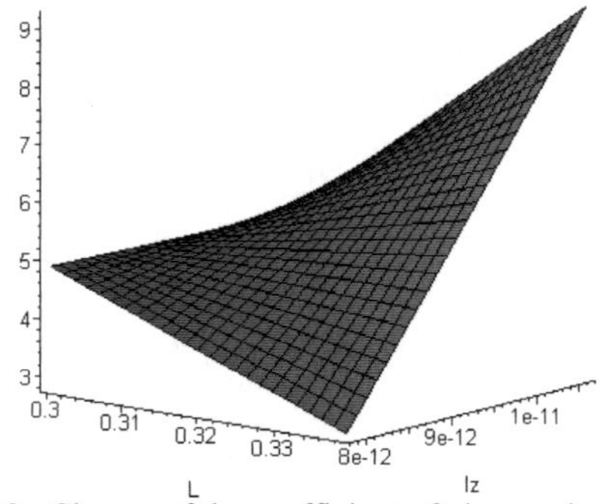

Fig. 3. Change of the coefficient of viscous damping.

The analysis of given results shows that all coefficients b_i, i = 1,2,12 are significant if Student t-test is used with confidence level of 95%. From presented experimental results following mathematical model for damping coefficient could be derived

$$\kappa = 159.98 - 515.70L - 17.77 \cdot 10^{12} I_z + 59.08 \cdot 10^{12} L I_z \tag{5}$$

which provides regression coefficient $R \geq 0.95$.
Fig. 3 shows dependency of damping coefficient on used factors of experiment: length and second moment area.

5. CONCLUSION

Experimental damping determination of vibration induced by bifurcation buckling under displacement dependent axial force is presented. Damping is assumed as viscous linear damping and damping coefficient is experimentally determined. Polynomial expression is given for damping coefficient, where appropriate coefficients are given. It is shown that damping could be modelled in function of beam geometry (length and second moment area). Given mathematical model for viscous damping calculation is chosen to give average amplitude decay in buckling process, while experimental results show different amplitude decay in the short beginning part than in the other part of measured accelerograms.

6. REFERENCES

Beards, C.F. (1996). *Structural Vibration – Analysis and Damping*, Arnold, ISBN 0-340-64580-6, London.

Isić S.; Doleček, V.; Karabegović, I. (2007). Numerical and Experimental Analysis of Postbuckling Behaviour of Prismatic Beam Under Displacement Dependent Loading, *Proceedings of First Serbian Congress on Theoretical and Applied Mechanics*, pp. 331-338, ISBN 978-86-909973-0-5, Sumarac, D., Kuzmanovic, D. (Ed.), Kopaonik, Serbia, April 2007.

Thompson, J.M.T.; Hunt, G.W. (1973). *A General Theory of Elastic Stability*, John Wiley & Sons, ISBN 0 471 85991 5, London.

Virgin, L.N. (2000). *Introduction to Experimental Nonlinear Dynamics: A Case Study in Mechanical Vibration*, Cambridge University Press, ISBN 0-521-66286-9.

Życzkowski, M. (2005). Post-buckling analysis of non-prismatic columns under general behaviour of loading, *International Journal of Non-Linear Mechanics*, Vol. 40, Issue 4, pp. 445-463, May 2005, ISSN 0020-7462.

Annals of DAAAM for 2008 & Proceedings of the 19th International DAAAM Symposium, ISSN 1726-9679
Editor B. Katalinic, Published by DAAAM International, Vienna, Austria 2008
Make Harmony Between Technology and Nature, and Your Mind will Fly Free as a Bird

THE TENSOMETRIC FORCE SENSOR. DESIGN, EVALUATION AND OPTIMAL SELECTION

DOLGA, L[ia] & DOLGA, V[aler]

Abstract: *Force sensors are essential components in mechatronic systems of various types. An efficient running service of the entire system requires optimal design. The paper proposes a unitary methodology to select the appropriate force sensor type and to optimize the design solution. The methodology was synthesized in an educational software tool for students, to train them in determining the best possible solution for a tensometric force sensor. Due to the modular structure, the computer application got further improvement becoming an assistant in designing and optimizing tensometric force sensors for industrial requirements or applied research.*
Key words: *force, sensor, model, selection, optimisation.*

1. INTRODUCTION

Effective monitoring of a manufacturing process is essential to guarantee the quality of the final product. Numerous industrial systems require data acquisition regarding a torque vector; in addition, the value of other external forces is often required. The measuring devices are specific; they are included in the multicomponent (multi-axes) force sensors. The existing references comprehensively present constructive solutions, criteria and design methods for this kind of sensors (Ahili, 1997), (Dickert, 1999), (Kang, 2005). A large range of sensorial elements for force measuring (in a generalized meaning) is available. The requirement of a stable, accredited and responsive method for evaluation and selection of appropriate force sensors is evident (Dolga, 2002), (Wang et al., 2003).

The paper outlines the authors' opinion concerning a unitary methodology about the analysis, the comparison, the assessment and the selection of tensometric sensorial elements. This concept was implemented through a software application, "PROTRA", developed in Visual Basic, which can be later upgraded for other types of sensors.

2. THE STRUCTURE OF "PROTRA"

Starting from the main request of the force sensors design, the authors modularly organized "PROTRA" in four sections; the user can retrieve each one either separately or sequentially.

The first design strategy, "Selecting the optimal sensor's variant", defines the "PROTRA" first module. This selection is applied to a set of "n" sensor variants V_i (i=1, 2... n), compared by "m" evaluation criteria, C_j (j = 1, 2... m). The "TOPSIS" method (Technical Ordered Preference by Similarity to Ideal Solution) ranks the variants (Dolga, 2002).

The module "Expert system" allows accessing a specific knowledge base, including heuristic knowledge of experienced designers. The module recommends potential decisions on the design alternative, proposes the sensor type and structure, etc.

The module "Sensors databases" comprises information about ready-made force sensors (commercial) or sensor components (Figure 1). These databases are relational, including "BLOB" fields (binary large objects). The user may consult either constructive databases or databases referring to the functional parameters and the potential applications.

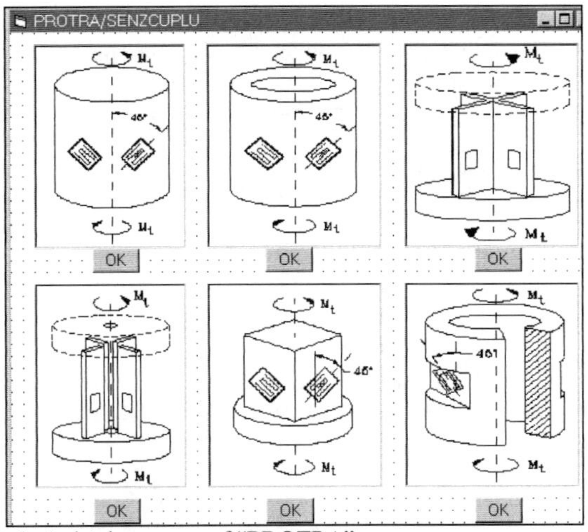

Fig.1. A database page of "PROTRA"

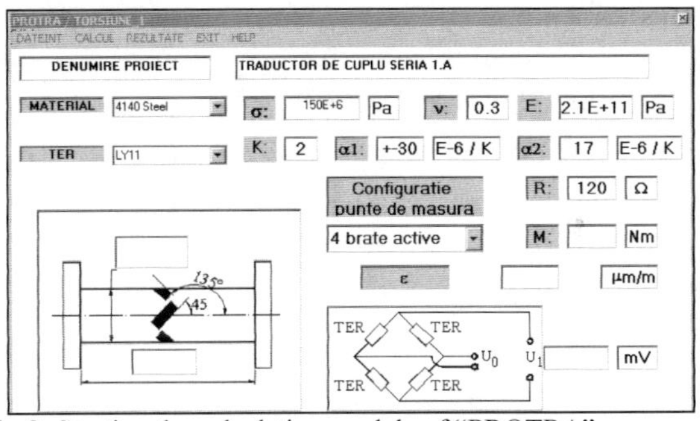

Fig.2. Starting the calculation module of "PROTRA"

The module "Calculus" ensures the conditions to apply electrical and mechanical analytical determinations for different components of the force sensors. Figure 2 shows a page in the module with recommended elastic elements for a single component force sensor. This module can call databases detailing the materials for the elastic element/ structure or strain gauges databases. The proof and confirmation of the designed solution should be completed using finite element programs.

The structure of the sensorial element often requires an analysis starting from the mathematical model of this element. An evaluation of the agreed solution follows, based on a set of reference criteria, availed by a proper mathematical support.

3. THE MATHEMATICAL MODEL OF THE SENSOR

3.1 Introduction

The input –output quantities dependence, specified in the relationship (1), gives the static characteristic of a force sensor:

$$[Y] = f([X],[P]) = [\Psi]\cdot[X] \qquad (1)$$

where $[X]=[[F][M]]^T_{6x1}$ is the vector of the input signals, including the components of the measured force-torque vector, [P] comprises the perturbation signals and perturbation effects, [Y] is the output quantity, represented by the output signals S_i (i=1,2...n) and $[\Psi]$ signifies a transfer function that is called "the measuring matrix of the sensor" (Bicchi, 1992).

Once correlating these considerations with the real input- and output quantities and with the applied notations, the equation system (1) becomes:

$$[S]_{nx1}=[C]_{nx6}\cdot[[F][M]]^T_{6x1} \tag{2}$$

The "coupling matrix" [C] outlines the influence of each force-torque vector component on the measuring channel S_i.

The equation system (2) relieves studying the system stability and the errors propagation during the design stage and the analysis stage. An important algebraic instrument in evaluating the perturbation sensitivity derived from the coupling matrix is the condition number (Bicchi, 1992):

$$K_{(C)}=\|C\|\cdot\|C^{-1}\| \tag{3}$$

$\|*\|$ signifies the norm of the matrix.

A minimum value for the condition number - the unit value being the reference rate – involves a higher stability of the sensor against gaussian perturbations, because a minimum ratio noise /signal is guaranteed.

In order to generalize the study and to remove the inconsistence of the relationships due to different unit measures for the force components and for the moment components, the normalized coupling matrix was introduced (Svinin, 1995):

$$[\overline{C}]=[N_\varepsilon]^{-1}\cdot[C]\cdot[N_n] \tag{4}$$

$[N_e]=diag[\varepsilon_{1m}\quad \varepsilon_{2m}\quad \ldots\quad \varepsilon_{nm}]$ is the diagonal matrix of the maximal stresses on the "n" measuring channels of the sensor; $[N_e]=diag\lfloor F_{xm}\quad F_{ym}\quad F_{zm}\quad M_{xm}\quad M_{ym}\quad M_{zm}\rfloor$ is the diagonal matrix containing the nominal values of each component in the force-torque vector.

Based on the normalized coupling matrix, the condition number is defined (Svinin, 1995) as below:

$$K_{(\overline{C})}=\lambda_{\max}/\lambda_{\min}\geq 1 \tag{5}$$

The numbers λ_{max} and λ_{min} are the largest singular value and the smallest singular value of the normalized coupling matrix. The ideal value $K_{(\overline{C})}=1$ points out a perfect isotropy of the measuring structure.

3.2 The dimensional optimisation

One can group the design parameters of the force sensors in three categories: geometric parameters (**X**), material parameters (**Y**) and loading parameters (**Z**). A set of design parameters, a set of constraints G_i and a set of possible optimisation criteria characterize the sensor during the design stage. The optimisation problem, expressed in (6), requires finding the design parameters U(X, Y, Z).

$$f_{(U)}\rightarrow\{\boldsymbol{min,\ max}\} \tag{6}$$

3.3 Application

Figure 3 presents a basis construction of a force sensor for two components whose measuring matrix is given in (7):

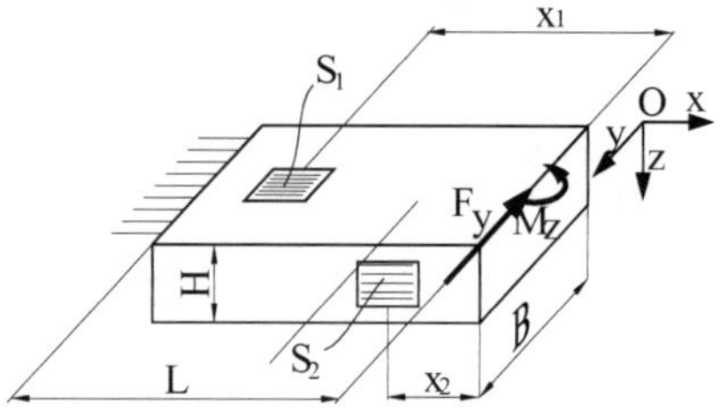

Fig.3. A force sensor for two components

$$[C]=\begin{bmatrix}\dfrac{6x_2}{EB^2H}\cdot F_{y\max} & 0\\ 0 & \dfrac{6x_2}{EBH^2}\cdot F_{y\max}\end{bmatrix} \tag{7}$$

If the clause of minimizing the condition number is applied, one obtains the solution (8) (Dolga, 2002):

$$x_2\cdot F_{y\max}/B=M_{y\max}/H \tag{8}$$

Besides the relationship (8), the geometrical constraints relationships and resistance relationships must be also satisfied for a comprehensive optimization of the force sensor.

4. CONCLUSION

The optimal design of a force sensor is an essential requirement in many types of mechatronic systems. The tensometric principle defines the most used force sensor type. A unitary methodology to design and evaluate these devices became necessary. To answer this requirement, the authors grouped appropriate design-, evaluation- and optimization procedures in a dedicated software application, "PROTRA".

In designing the elastic structure of the force sensors, helpful mathematical instruments are the pseudo-inverse matrix and the condition number. Appropriate selection- and optimization procedures were added to the shape calculation of the elastic element. Suitable databases make up the modular application. "PROTRA" evolved from a didactic tool to a design- and research assistant. New force sensor types stay in mind, but further progress focuses too on creating a simulator to select the sensorial elements for reconfigurable systems.

5. REFERENCES

Ahili, F. (1997). Dynamics and Control of direct-drive robots with positive joint torque feedback, *Proc. IEEE Int. Conference on Robotics and Automation,* pp. 2865 – 2870

Bicchi, A. (1992). A criterion for optimal design of multi-axis force sensors. *Journal of Robotics and Autonomous Systems,* Vol.10, No.4, pp. 269-286

Dolga, V. (2002). The optimal design of force-torque sensors (in Rom.) – moment, *Proc. of 6th Conf. on Fine Mechanic and Mechatronic,* pp.245 – 252, ISSN 1220-6830, Braşov, October, "Transilvania" University Publishing House

Kang, CG. (2005). Performance Improvement of a 6-axis force –torque sensor via novel electronics and cross-shaped double-hole structure, *International Journal of Control, Automation and Systems,* Vol.3, No.3, pp.469-475

Svinin, M.; Uchiyama, M. (1995). Optimal geometric structures of force / torque sensors. *Journal of Robotics Research,* Vol. 14, No.6, pp.560-573

Wang, L.; Kannatey-Asibu, E.; Mehrabi, M. (2003). A method for sensor selection in reconfigurable process monitoring. *Journal of Manufacturing Science and Engineering,* Vol.125, February 2003, p.95-99

Annals of DAAAM for 2008 & Proceedings of the 19th International DAAAM Symposium, ISSN 1726-9679
Editor B. Katalinic, Published by DAAAM International, Vienna, Austria 2008
Make Harmony Between Technology and Nature, and Your Mind will Fly Free as a Bird

ANALOGY IN MODELLING MECHATRONIC SYSTEMS

DOLGA, V[aler]; DOLGA, L[ia] & FILIPESCU, H[annelore]

Abstract: *The mechatronic systems integrate electronics and information technologies in constructing advanced mechanical systems capable of new functions. They include one or more of the following energy domains: electrical, mechanical, magnetic and thermo-fluidic. The systemic approach of each component requires a unitary methodology regardless of the component's physical nature. The authors outlined common features of these elements and studied several possibilities of homogeneously analyzing mechatronic systems and components: using the transfer function, the energy ports, the object-oriented paradigm and/ or the principles of the analytical mechanics. Examples and results obtained in modelling and simulation processes of mechatronic systems, detail the considered aspects. The main conclusion on finding analogies between different physical subsystems that compose a mechatronic system and on defining a broad approaching methodology leads to the inevitability of applying analytical mechanics. This result has a practical interest, for team working, and a theoretical relevance, for the mechatronic design philosophy.*

Key words: *mechatronics, model, simulation, physical system.*

1. INTRODUCTION

The mechatronic systems represent an integration of the electronic and information technologies in constructing advanced mechanical systems. In accordance to the authors' evaluation, based on experts' opinions, the essential characteristics of the mechatronic systems are high speed and accuracy, enhanced efficiency, robustness and miniaturization (Amerongen, 2007); (Dolga & Dolga, 2007).

A judicious design is essential, by balancing modelling, analysis, experiment validation and construction. Modelling, analysis and prediction ensure efficiency and quality, just from the first stages, after the launching of the design theme. A common working language plays an important role (Bishop, 2002). This language origins in the analogies between the physical systems defining the objective of mechatronics; it will be also useful in implementing efficient working methods and tools in mechatronics. The systemic study of the dynamic behaviour expects the modelling of various electromechanical systems (Krus, 2003); (Hostert, 2007). Extending the utility of several principles from within the analytical mechanics over the electromechanical systems is vital (Preumont, 2006).

2. THE MECHATRONIC SYSTEMS STRUCTURE

Mechatronics expresses the integration of mechanical, electromagnetic and computer elements to produce devices and systems that monitor and control machine and structural systems. Figure 1 shows a concise representation of a mechatronic system, omitting the interdependencies between components and distinguishing four sub-systems: mechanical-, informational-, electrical- and computing system. Within the mechanical process, which is a basic component of the mechatronic system, mainly three flows are present: substance-, power (energy)- and informational flow. Modelling and simulation of the mechatronic systems must consider analogies between the subsystems, in order to apply a unitary approach.

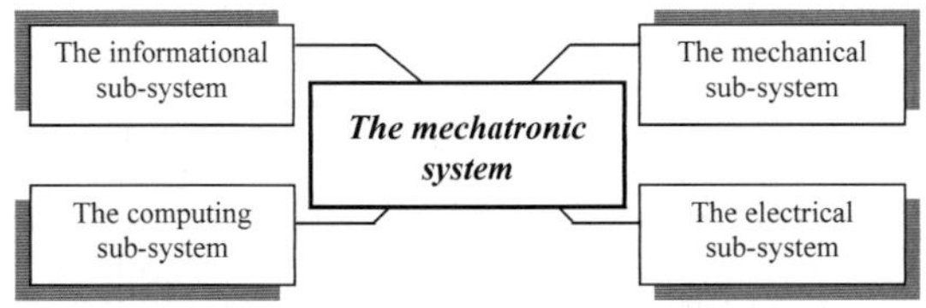

Fig.1. The main types of elements of a mechatronic system

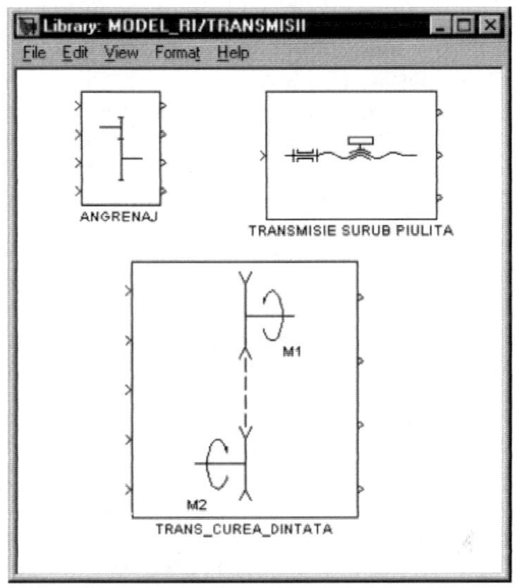

Fig. 2. Starting window of the "MODEL_RI" software application created in Matlab/ Simulink

3. ANALOGY AND MODELLING

The systems theory provides a fundamental "instrument" in modelling and analysis: the transfer function. The methodology based on the transfer function involves the advantage of an acknowledged theory for the systems analysis and synthesis as well as for a software package in which Matlab/ Simulink is the preferred working environment. Figure 2 shows the starting window of a Matlab/ Simulink library, called "MODEL_RI" created by the authors during this study.

However, the approach of representing a system by the transfer function, referring to an input quantity and an output quantity leaves aside the energetic aspects that are specific and important in physical systems and may not be neglected.

The theory of physical systems is based on the concept of energy (E), defined as power accumulated along the time. Starting from this viewpoint, one can define the notion of "generalized" power, Π, as being the product of two physical quantities, observable and complementary: quantities across two points (α) and through a point respectively (τ). These approaches are not singular. The theory of physical systems adds to the previous topics, named by some authors "pressurelike" and "flowlike", the equivalency between the effort e and the flow f. There is a very close similarity (almost an identity) between these two modes of definition.

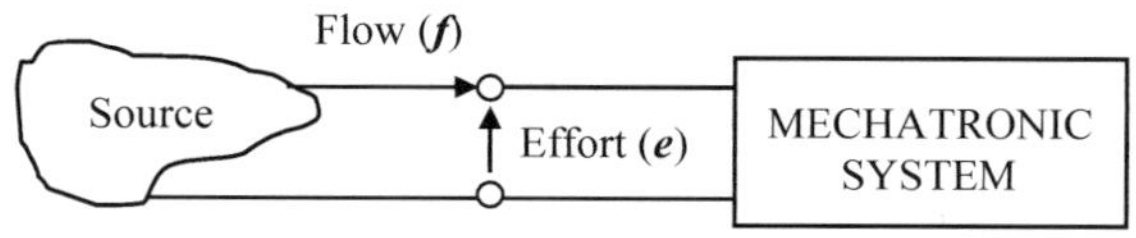

Fig.3. The generalized variables and the energy ports

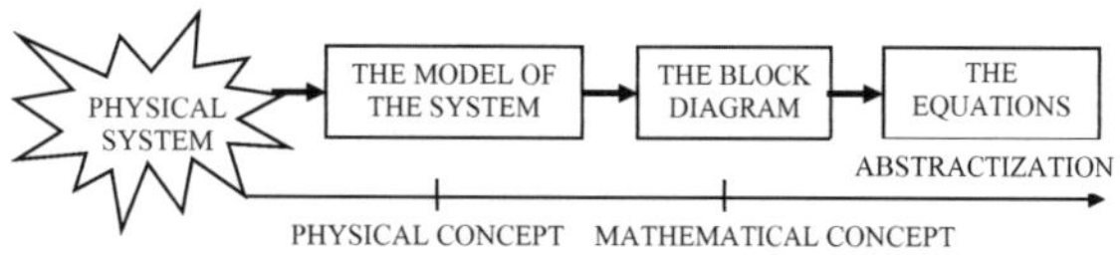

Fig. 4. From simple to complex in a simulation process

In accordance to the physical principles referring to systems operation, the topic of "energy ports" can be introduced, through which systems interact one with each other, by an energy exchange (multiport) (see Figure 3) (Amerongen, 2007).

Figure 4 proposes the stages in modelling physical systems starting from the physical system towards the corresponding equations. The design stage comes in a chronological order between the analysis stage and the implementation stage. The object-oriented development of the mechatronic product incorporates all these three stages.

The object-oriented approach is a perfect tactic for the teamwork, which is a fundamental characteristic of the mechatronic philosophy on design. The paradigm involves an efficient communication between the members of different groups that work together at the same project.

The authors worked in a dedicated modelling environment, Dymola (Dynamics modeling laboratory), to develop a mechatronic project, because it provides modelling- and simulation tools based on object-oriented paradigm (Figure 5). Using a principle similar to the bond-graph method, Dymola considers the energetic flow, a common characteristic for the majority of the physical systems (Elmqvist et al. 2003).

The principles of Newtonian mechanics are overall simple in their form, but sometimes difficult to apply. These aspects brought Lagrange to create the principles of the analytical mechanics "a mechanics of the systems – with a finite number of parameters which should not use any concrete figure or picture". The mechanics of Lagrange, the mechanics of the non-holonome systems, the Hamiltonian mechanics, and the variational principles use the analytical mechanics.

Variational methods and the associated extremum principles are fundamental scientific techniques of a notorious physical significance. Yet, this treatment will be restricted to variational methods of analysis for lumped physical systems, characterized by the energy variables "effort" and "flow".

Figure 6 shows a case of integrating a mechanical and an electrical system, a proper mathematical model being defined.

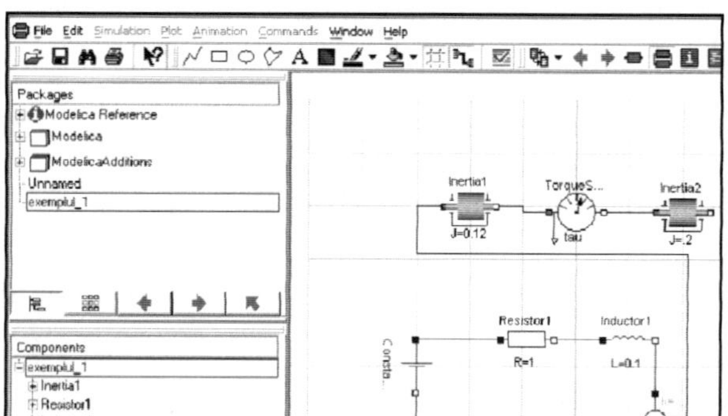

Fig. 5. A physical system modelled in Dymola

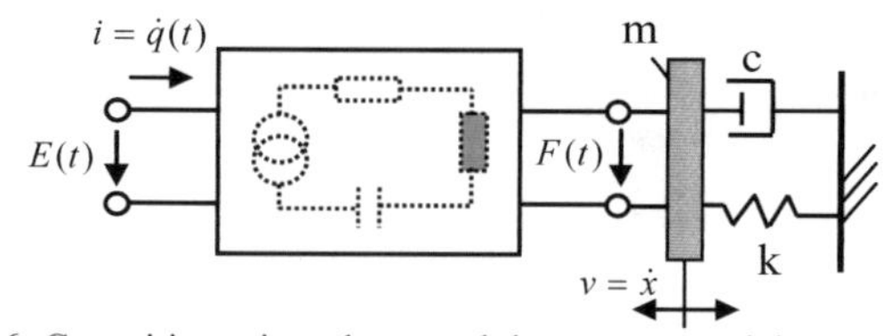

Fig. 6. Capacitive microphone and the system model

A condenser microphone consists of a movable plate mounted on a circular spring parallel to a fixed back plate (Preumont, 2006). The plates form a variable capacitor charged by a constant voltage source through an RL circuit. The dynamics of the moving plate is modelled by a spring-mass system of constants m, k, c. The equations (1), (2) govern the mathematical model developed from the analytical mechanics principles, where the symbols correspond to those in Figure 6:

$$L\ddot{q} + R\dot{q} + (q/C) = E(t) \tag{1}$$

$$m\ddot{x} + b\dot{x} + kx + (q^2/2\varepsilon_0\varepsilon_r A) = F(t) \tag{2}$$

4. CONCLUSION

Mechatronics is not only the sum of the implied engineering domains; consequently, the concepts, the language, the methods and tools in mechatronics are more advanced. The product design approach requires consistent concepts and methodologies based on an appropriate language. Analogies between physical systems and quantities of different nature can lead to a common language and various possibilities are available for a homogeneous approach, but the advantage of applying analytical mechanics principles is certain.

The authors outlined that a successful modelling and simulation is conditioned by an integrated approach and an appropriate experiment that should confirm/ infirm the developed model and the performed simulation.

The result of the study is useful in team working on a mechatronic product. The study will continue in view of improving the mechatronic design philosophy by specific concepts.

5. REFERENCES

Amerongen, J. (2007). Mechatronic design – a port-based approach, *Proc. Of the 4th Int. Symp. On Mechatronics and Applications*, Sharjah, March 26-29, *Available from:* http://www.ce.utwente.nl/rtweb/publications/2007/pdf-files/Sharjah.pdf. *Accessed:* 2008-05-12

Bishop, H. (2002). *The mechatronics handbook,* CRC Press, ISBN 0-849-300-665, London-New York-Whashington

Dolga,V.; Dolga,L. (2007). The principles of analytical mechanics applied to the dynamical analysis of the mechatronic systems, *Proc. of 2nd Intern. Conf. "Computational mechanics and virtual engineering"*, pp.221-226, ISBN 978-973-598-117-4, Brasov, 11-13 X 07

Elmqvist, H. et at. (2003). Real-time simulation od detailed automotive models, *Proc. of 3rd Intern. Modelica Conference,* pp.29-38, Linkoping, 3-4 November 2003

Hostert, C.; Maas, S. (2007). Dynamic simulation of mechatronic systems, *Available from:* http://www.aliai.lu / rt/rt20071/ rt20071d.pdf. *Accessed* 2008-05-10

Krus, P. (2003). Modelling and Simulation of heterogenous engineering systems, *Available from:* http://www.machine.ikp.liu.se/edu/post/ multidomain/ modelling_simulation.pdf *Accessed* 2008 -04-20

Preumont, A. (2006). *Mechatronics. Dynamics of Electromechanical and piezoelectric systems*, Springer, ISBN 1-4020-4695-2, Dordrecht (Netherlands)

Annals of DAAAM for 2008 & Proceedings of the 19th International DAAAM Symposium, ISSN 1726-9679
Editor B. Katalinic, Published by DAAAM International, Vienna, Austria 2008
Make Harmony Between Technology and Nature, and Your Mind will Fly Free as a Bird

ROBOTICS, RAPID CONTROL PROTOTYPING AND "dSPACE" HARDWARE

DOLGA, V[aler]; DOLGA, L[ia] & FILIPESCU, H[annelore]

Abstract: *Conventional methodologies for controllers' development become inadequate in robotic systems with uncertainties and nonlinear features. Experts' investigated controller design methodologies, dynamic fuzzy neural networks controllers, rapid prototyping of model-based robot controllers and rapid prototyping with dSPACE hardware. This paper presents the authors' realizations in developing applicative platforms with complex tasks of rapid control prototyping for robotic systems, outlining the wide usability of the rapid prototyping method in this area. The platforms are based on dSPACE equipment and Matlab/ Simulink software tools and offer high flexibility in applying the methodology. An integrated real-time hardware-in-the-loop simulation system becomes available and executable code for microcontrollers can be automatically generated. Next research will apply concepts of advanced control: "full path" and "bypass".*
Key words: *control, prototyping, robot, fuzzy logic, dSPACE*

1. INTRODUCTION

Intensive and extensive development of robotics involved various approaches for designing and improving controllers that are particularly destined to robot systems. Nonlinear features with many uncertainties characterize the comprehensive inquiries of the mechatronic system embodied by the robotic manipulator. Conventional methodologies for controller development became insufficient: either they hardly solved the problem, or they did not find a solution for dynamic modelling.

Designing a feedback control system typically requires several steps: identifying a dynamic model from experimental data, using computer aided design tools to construct the controller, verifying the design using computer simulations and finally, implementing the controller. Knowledgeable reference articles and books propose a solution for the specified question: a multivariable non-linear coupled dynamic system with uncertainties (Tagagi, 1992) – (Neo & Er, 1996). The experts' investigations regarded both controller design methodologies that should consider various uncertainties- dynamic fuzzy neural networks controller (Low, 2004) and rapid prototyping of model-based robot controllers (Bona, 2003) or rapid prototyping with dSPACE hardware (Ridley, 2004; Gattringer, 2006; Ionescu et al., 2008).

The subject of rapid control prototyping was one of the main tasks within the frame of the national excellence research grant "Simulation, Control and Testing Platform with Applications in Mechatronics «ConMec»" (Dolga et al., 2007).

The paper presents the authors' realizations in developing applicative platforms with comprehensive tasks of rapid control prototyping for robotic systems.

2. RAPID CONTROL PROTOTYPING APPROACH

The idiom of "rapid prototyping" is extensive and outlines remarkable facilities in creating new products; it provides the possibility to demonstrate the veracity of a concept or an idea. Previous studies outlined the superiority of hardware solutions, for which several variants were considered. The authors applied a multi-attribute decision making process with six evaluation criteria (the processor type, the memory, the inputs and the outputs, the signal conditioning, the operating mode and the platform sensitivity). The results revealed the dSPACE framework (Dolga et al., 2007) advantages, because dSPACE equipment perfectly matches the requisite of using the same software and hardware platform for controller design, simulation and implementation processes (Figure 1).

Figure 2 shows a typical software development process for an Electronics Control Unit (ECU).

Prototyping can help answer the following questions: will the design work properly, can the design be produced economically, which approach can be taken to get from concept to product, how can prototyping support product design specification. The main benefits of prototyping are: better communication with the user, improved design through feedback and iteration, provides training / learning medium.

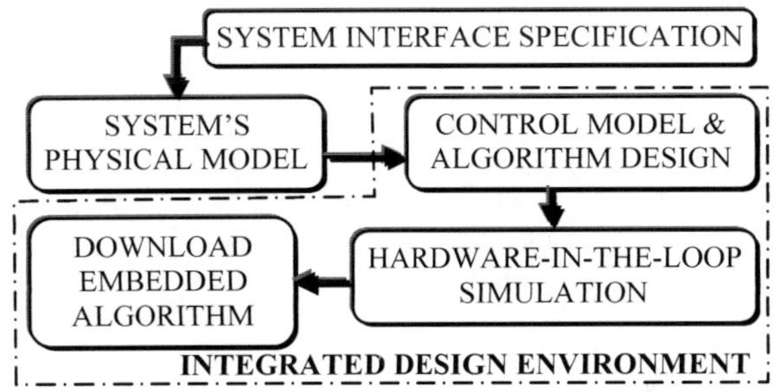

Fig. 1. The design principle for control project.

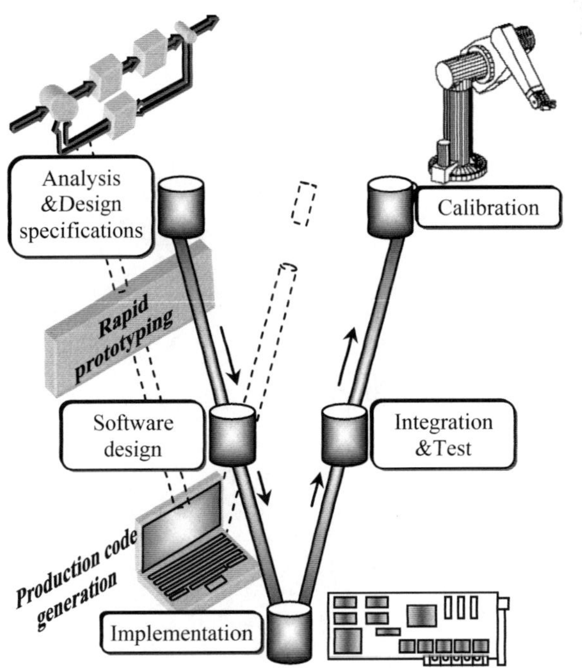

Fig. 2. A representative software development process in mechatronics

3. SOLUTIONS FOR ANALYSIS PLATFORMS

The first developed platform allows applying the concept of rapid control prototyping for a multi-axes robot (Figure 3).

While Matlab and the SIMULINK block diagram environment are useful for control design and analysis, the dSPACE-DS1005 provides the means for acquiring system identification data and implementing discrete-time controllers for analogue plants (a 5 axes Yamaha robot was bought). The dSPACE software ControlDesk is a graphical user interface that offers the functions to control monitor and automate experiments and make the controllers progress more efficient.

Another platform with a similar goal was focused on the analysis of an autonomous mobile robot which operates in obstructive environments (Figure 4). Therefore, the controller takes into consideration multiple uncertainty elements; for that reason, a fuzzy algorithm was provided with the aim of developing an appropriate controller (Figure 5).

The sensing elements supply the data required by the fuzzy controller. The CCD video sensor connected to a computer system together with the data processing in Matlab/ Image Processing Toolbox deliver the desired position Φi. The mobile robot position transducer gives the real navigation angle Φ.

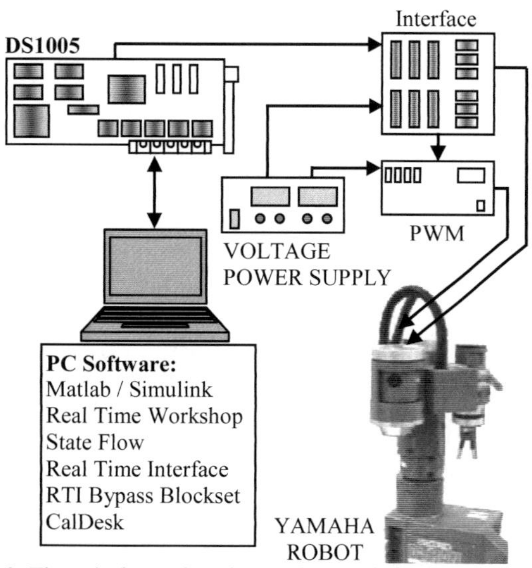

Fig. 3. The platform for the analysis of the rapid control prototyping in the case of the multi-axes Yamaha robot

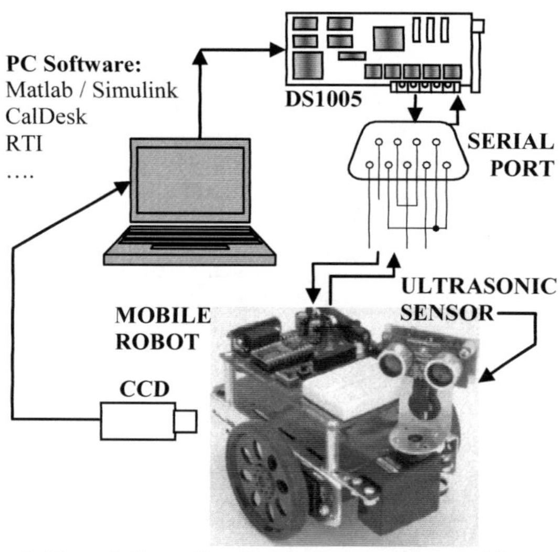

Fig. 4. The platform for the analysis of the rapid control prototyping in the case of the mobile robot

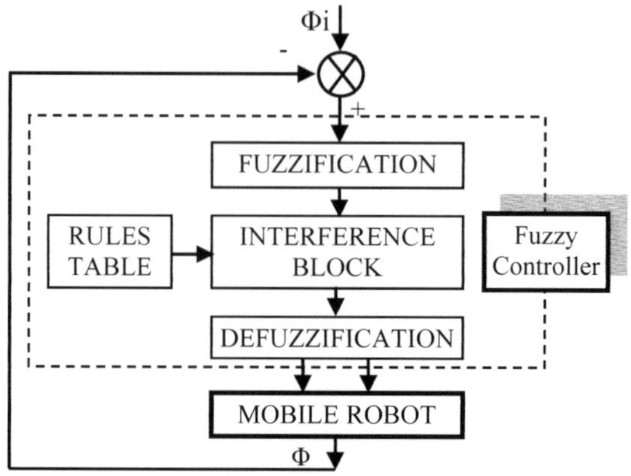

Fig. 5. The proposed fuzzy algorithm for the robot controller

The mobile robot is provided with a multiple use of the sensing elements (video, proximity, position). The interface RS232 aids the implementation of a control strategy compatible with the robot environment or the upload of the control code. The dSPACE 1005 board interfaces the computer and the robot.

4. CONCLUSION

The created platforms offer a concrete answer to a critical requirement in robotics control: rapid control prototyping.

The modular structures of the platforms make the developed applications to be supple and friendly and guarantee excellent results in rapid control prototyping for various mechatronic systems. An integrated real-time hardware-in-the-loop simulation system becomes available and executable code for microcontroller can be automatically generated.

To date only Matlab/ Simulink tools were used on the platforms; other software environments for dynamic simulations are also available (Dymola, 20SIM) and will be considered to increase the platforms applicability.

The intention is of further developing applications that suppose implementations of appropriate controllers. Analyses for levitation mechatronic systems and for shape memory alloys actuators are planned to be approached.

The described platforms for rapid control prototyping offer high flexibility; next research will make possible to apply concepts of advanced control like "full path" and "bypass".

5. REFERENCES

Bona, B.; Indri, M. & Smaldone, N. (2003). Architectures for rapid prototyping of model-based robot controllers, In: *Advances in control of articulated and mobile robots*, ISBN 978-3-540-20783-2, Springer Berlin, Heidelberg

Dolga, L.; Dolga, V. & Filipescu, H. (2007). Rapid prototyping within the simulation and control platform for mechatronics, *Proceedings of the 18th International DAAAM Symposium "Intelligent Manufacturing & Automation: Focus on Creativity, Responsibility and Ethics of Engineers"*, pp. 243-244, Zadar, Oct. 2007

Gattringer, H. (2006). The Bipedal Robot. *dSPACE News*, no.1, 2006, pp.12-13, Paderborn, Germany

Ionescu, F. (2008). Vlad, C. & Arotaritei, D.: Advanced Control of an Electrohydraulic Axis, In: *Mechatronic System Control, Logic and Data Acquisition*, edited by R.H. Bishop, CRC Press, ISBN 9260-0-8493-9260-8, 2008

Low, C. B.; Nah, K. H. & Er, M.J. (2004). Real –Time implementation of a dynamic fuzzy neural networks controller for a SCARA. *Journal of the Institution of Engineers*, vol.44 issue 3 pp. 43-56, Singapore

Ridley, P. (2004). The World's Largest Industrial Robot. *dSPACE News*, no. 2, 2004, pp.4 -5, Paderborn, Germany

Annals of DAAAM for 2008 & Proceedings of the 19th International DAAAM Symposium, ISSN 1726-9679
Editor B. Katalinic, Published by DAAAM International, Vienna, Austria 2008
Make Harmony Between Technology and Nature, and Your Mind will Fly Free as a Bird

SIMPLE LEARNING KIT FOR MICROCONTROLLER PROGRAMMING LESSONS

DOLINAY, J[an]; DOSTALEK, P[etr] & VASEK, V[ladimir]

Abstract: *This paper presents simple learning kit for students who learn microcontroller programming. The kit is very easy to make and cheap which allows students to build it at home and experiment with it. It is based on Atmel AVR microcontroller and besides this microcontroller it contains only LED and switches. Various options of microcontroller programming suitable for students are also analysed.*

Key words: *Microcontroller, Atmel, Learning kit, AVR.*

1. INTRODUCTION

Microcontrollers are found in virtually every area of life nowadays and their usage increases. As the usage increases so does the need for qualified experts able to program them. It is the task of universities and colleges to prepare such experts - software engineers able to program microcontrollers.

In this paper we will describe simple and cheap microcontroller tool which can be used in microcontroller programming courses. Our aim is to make it possible for a student to "play" with microcontroller (MCU) not only at the lessons but also at home, to obtain some real experience with the methods used for programming and so on.

The question why we strive for cheap solution may appear. It is usually accepted that schools are on tight budget but the budget is usually not as tight that it would not allow for buying proper equipment. Besides, microcontroller manufacturers are often glad to provide their tools for free or with discount as they are well aware that this investment will repay when the students start their career and make decisions in their work which microcontroller brand to use. So the motivation for cheap solution is not driven by the inability to obtain more sophisticated equipment for our lessons but by two other factors. First, it is the desire of some students to further investigate the topic. For such a student, buying the same equipment that is used in lessons may be too expensive and borrowing it from school is not always possible.

Second reason is simplicity. The devices used in lessons are professional products which more or less hide the basic principle in their sophisticated design. On a board with hundreds of parts it may be hard to even find the microcontroller itself. It is desirable, at least at the beginning, to allow the students to see how things are made. That is why simple board containing just the MCU and few other parts is helpful. Better yet, if the student can make the board himself. Such a board is described in this paper.

2. METHODS

Let us start with the description of our lessons. The students first learn programming in assembly and then in C. We use evaluation kit M68EVB908GB60 with Freescale (former Motorola) HCS08 microcontroller, which is 8 bit MCU with FLASH and RAM memory (Freescale, 2004). This kit is equipped with LCD display; several push switches, LED diodes and even a buzzer. Besides the built-in peripherals we have also developed several extension modules which can be attached to the kit. These modules include 7-segment, 4 digit multiplexed display and keyboard, DC motor drive and simple heating plant (Dolinay, 2007). The development kit is very well suited for the purpose of learning MCU programming but the advantage of many interesting peripherals brings also the disadvantage of complexity. This makes it hard to see the hardware principles and the cost of such a kit is beyond student's budget. Moreover, the student may soon want to create his own device (application) with microcontroller for which a surface mounted MCU with over 60 pins is not the right choice and the kit itself is too complicated to give much inspiration without good experience in electronics. For such a student and such a purpose the kit described here has been designed.

2.1 Choice of the microcontroller

There are many brands of microcontrollers available. We limited our choice to three brands we have personal experience with and which belong among the most used: Freescale, Atmel and Microchip. Luckily, the principles are very much the same for all microcontrollers. The choice of particular brand is affected by factors such as personal experience (and preference), price, features and development tools available.

Freescale microcontrollers are widely used in professional applications - in automotive industry, etc. For hobbyist use they are not so popular, perhaps because it is not easy to obtain them (situation in the Czech Republic). Atmel microcontrollers (Atmel 2008) are becoming very popular among the hobby users because of low price, high quality and availability of excellent free software including full featured C compiler (AVR studio, WinAVR). PIC microcontrollers (Microchip 2008) are very popular in hobby area also, but it seems they are loosing the battle to Atmel at present due to lack of quality free development tools - there is no full featured free C compiler for PIC available in contrast with WinAVR GNU for Atmel or Code Warrior for Freescale and the free assembly development IDE is also little behind current standard as to the ease of use and intuitiveness.

For microcontroller manufacturers the hobby-users are probably not the most important target but for our aim of providing simple and affordable kit for students which they could make at home, the popularity in hobby market segment is an important clue.

2.2 Learning kit design

In general we can come to these options for designing a learning kit for student:

1. Cheapest kit containing basically just the MCU itself - which cuts down the cost to about 5 EUR.
2. Kit containing some circuitry for communication between the MCU and PC but still very cheap, under 20 EUR.
3. Kit containing another MCU to facilitate the communication with PC and programming of our MCU.

Kit resulting from option 3 is the most comfortable to use. Its great advantage is that it may support USB port connection to PC which makes it easy to develop MCU programs or any new computer. However, the primary disadvantage is that it requires knowledge of microcontroller programming to build the kit. The solution could be to provide parts including programmed MCUs to students, but this introduces costs to school which is not in accordance with our aim. Also the kit will be too complex to understand for beginners and moreover home building of this kind of device has doubtful effectiveness. It may be easier, and in the end cheaper, to buy a professional programmer or development kit. For the above reasons we eliminate the option 3. The option-2 kit will most likely be connected with PC via serial line. It will have little higher cost and relatively complex design which requires some experience with electronics. For a student who lacks this experience this kit may still be too complex to show the basic principles. Option 1 means making learning kit actually just from the MCU and some parts to provide input/output (LED, a switch). The advantage is in the price and easiness of building such a kit. It may be put together on a breadboard without PCB and soldering, if desired, and the principle of MCU use is very clear here – there are no other parts to confuse the student. Disadvantage is in the connection to PC. This kind of programming device is limited to parallel or serial ports, which both are disappearing from modern computers. This may be problem especially with laptops, as will be discussed later.

The kit described here is based on the option 1. This does not mean that we would discard the other options as unsuitable for learning MCU programming. We just believe this is the best approach for the first contact with microcontrollers. At later stage a more sophisticated device can be used.

3. THE LEARNING KIT

This section will describe the MCU learning kit which we designed. For our kit we use Atmel AVR microcontroller Tiny2313 but the kit supports also other MCUs with compatible pin layout (Tiny25/45). The AVR MCUs can be programmed in-system, using 4 of its pins. The MCU may be clocked by internal oscillator, eliminating the need for a crystal. To put the MCU into the ISP programming mode, low level (0V) must be present at the Reset pin during power on or reset. This is a very simple way compared to methods used for similar programming mode by PIC and Freescale MCUs, which require higher voltage to be applied and/or special combination of logical levels at several pins for the MCU to switch to programming mode. The schematic of the kit is depicted in figure 1. Besides the socket for MCU there is just integrated voltage regulator 7805 to obtain 5V for powering the MCU with necessary capacitors, and then 2 LEDs and 2 push switches to allow some inputs/outputs for the MCU. A trim is used to generate variable voltage input for experiments with analogue comparer or A/D converter. It should be noted that the peripherals are not connected directly to the MCU, but to a DIP8 socket. The MCU pins are made available using sockets also. This way any peripheral can be connected to any pin of the MCU using a piece of wire. The kit is connected to PC via parallel or serial port. For parallel port it is possible to connect it directly, but to protect the port a simple circuit with 74HC is recommended (Kamal 2008). For serial line, simple circuit is also required which uses several Zener diodes and a transistor. Schema of the circuit can be seen in figure 2.

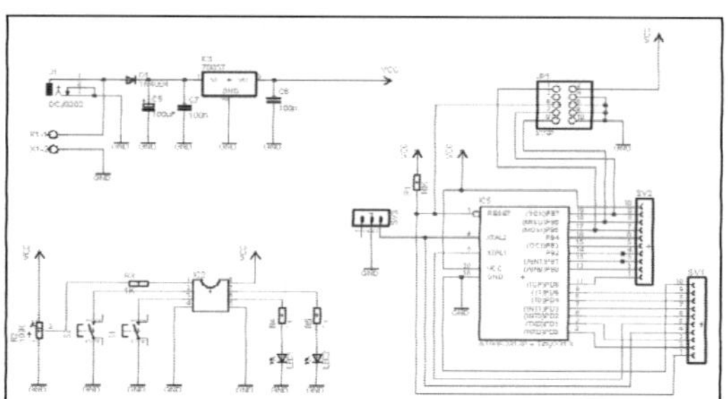

Fig. 1. Schematics of the learning kit.

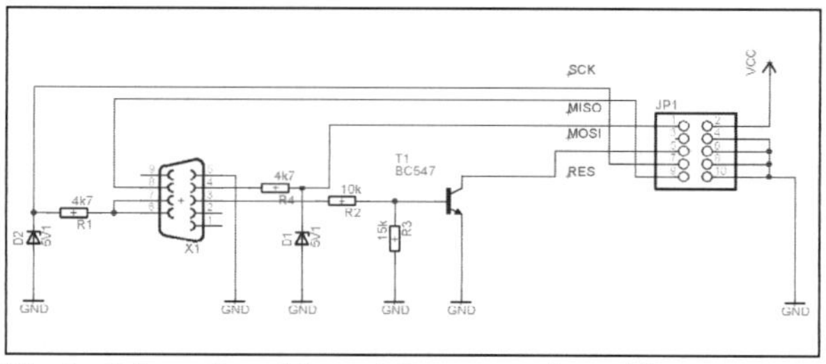

Fig. 2. Schematics of the serial programming cable.

The MCU is programmed by well-known free software PonyProg (Lanconelli, 2008) .

The connection to PC is main limiting factor for use of this kit, as it requires the development PC to have either serial or parallel port, but on many laptops only USB ports are found. The option to connect the kit via USB to serial programmer seems to be troublesome, as our tests with two different brands of such converters showed the same behavior – programming was very slow (about 10 minutes to read or program the MCU memory). The reason is probably that the programmer uses the serial port on hardware level which is not supported by the USB to serial converters (or their drivers). However, it is possible that there are drivers/converters working properly.

4. CONCLUSION

This paper presents very simple learning kit for students of microcontroller programming courses. The purpose of this kit is to allow students to build their own MCU application. It is not intended to be the main development platform for learning MCU programming - for this the school is equipped with professional learning kits. Described kit may be considered a tool for home experiments. It provides starting point which is guaranteed to work and opens the way for student's own designs. Its purpose it to illustrate how to put microcontroller into real use and how to program it. We also plan to make tutorial-like sample designs using this kit available soon.

This work has been supported by the Ministry of Education of the Czech Republic in the range of project No. MSM 281100001 and Grant Agency of the Czech Republic project No. 102/00/0526.

5. REFERENCES

Atmel corporation (2008). AVR 8-bit RISC. *Available from*: http://www.atmel.com/products/avr/default.asp *Accessed*: 2008-06-12.

Dolinay, J.; Dostalek, P. & Vasek, V. (2007). Educational models for lessons of microcontroller programming, Proceedings of 11th International research/expert conference TMT 2007, pp. 1447-1450, ISBN 978-9958-617-34-8, Tunisia, September 2007, Hammamet.

Freescale semiconductor, Inc. (2004). MC9S08GB60: Technical Data Sheet for MC9S08GB/GT Rev. 2.3, 2004. HC08 GB60, *Available from*: http://www.freescale.com/files/microcontrollers/doc/data_sheet/MC9S08GB60.pdf *Accessed*: 2008-06-16.

Kamal, I. (2008). In System Programming (ISP) for ATMEL chips, *Available from:* http://www.ikalogic.com/isp.php *Accessed:* 2008-06-10

Lanconelli, C. (2008). PonyProg serial device programmer, *Available from*: http://www.lancos.com/prog.html *Accessed:* 2008-06-16.

Microchip Technology, Inc. (2008). 8-bit PIC Microcontrollers, *Available from:* http://www.microchip.com/stellent/idcplg?IdcService=SS_GET_PAGE&nodeId=74 *Accessed:* 2008-06-11.

Annals of DAAAM for 2008 & Proceedings of the 19th International DAAAM Symposium, ISSN 1726-9679
Editor B. Katalinic, Published by DAAAM International, Vienna, Austria 2008
Make Harmony Between Technology and Nature, and Your Mind will Fly Free as a Bird

DATA FILTERING

DOLINAY, V[iliam]; LAMPART, M[arek] & VASEK, V[ladimir]

Abstract: *This contribution describes methods of processing data measured by special body belt which is used to monitor pregnant women. The task is to monitor movement of the belly with focus on catching the moment and length of contraction. Output data are values of belt tension measured by strain gages placed at the ends of belt. These raw data has to be filtered first and than analyzed to find the contractions. For this task three simple filter algorithms were proposed. After their testing it was found, that best results are obtained from their mutual combination. All above mentioned algorithms are implemented in an experimental program which also performs reading data from belt memory and simple visualization. This paper aimed to describe methods of data filtering algorithm.*

Key words: *filtering, data, contraction, belt, algorithm*

1. INTRODUCTION

The invented special body belt monitors pregnant women during risky section of pregnancy (Trojan 1999), (Guyton 2000). Stream of relative pressure values produced by belt must be filtered before final analyse and visualization (Svatos 1998). In this case quick motion and breathing of examined woman is typical source of noise. Naturally, there are such methods but we were looking for fast and powerful but also simple to be programmed in short code. Another requirement was possibility of real time processing with low memory needs. The invented algorithm was based on maths with simple idea of triangular center of gravity (Rektoris 2000). This idea can be advented in many ways when only certain area is presented in this paper (Adams 2003)

2. ALGORITHMS

There were arranged three easy methods which will construct from a group of initial values one new value – this value will approximate real situation of the up to now process. Each of three methods comes from the same kernel process.
The kernel of designed algorithm is the procedure of finding a centre of gravity of a triangle in the plane and comparing such a center of gravity with the following values.

Firstly, let us denote D as a set of initial data. One can imagine this set as a sequence

$$D := \{[t_i, v_i]\}_{i=1}^{k}, \tag{1}$$

where t_i are times in which we measure values v_i.

All the measuring process is passing in finite time interval, so we can assume that the sequence contains k elements. Moreover, the measuring process proceeds in constant time intervals, we have

$$\frac{t_{j+1} - t_j}{2} = c, \tag{2}$$

for every $j = 1,2,3...(k-1)$
Now, we will define algorithms (maps) T_1, T_2 and T_3, whose progressing application (compositions) we will receive the final algorithm.

2.1 Algorithm T$_1$

Algorithm is arranged in such a way that it assigned to a successive triple of points from D the value of gravity of the triangle arranged by this triple of vertexes. The output is a new set of data which contains one third of points with respect to the initial set of data D. Hence, is a map from to define by the formula:

$$T_1 = T_e \circ T_r, \tag{3}$$

where $T_r : D \rightarrow D^3$ is defined by the rule

$$T_e(\{[t_{i+j}, v_{i+j}]\}_{j=0}^{2}) = [[t_{i+j}, v_{i+j}]]_{j=0}^{2}, \tag{4}$$

where $i = 1,2,3... \frac{k - \mathrm{mod}_3(k)}{3}$; $[[t_{i+j}, v_{i+j}]]_{j=0}^{2}$ denotes ordered triple of points $[t_i, v_i]$, $[t_{i+1}, v_{i+1}]$, $[t_{i+2}, v_{i+2}]$, by analogy we will denote ordered senary of points; and $T_e : D^3 \rightarrow D_1$ is defined by:

$$T_e([[t_{i+j}, v_{i+j}]]_{j=0}^{2}) = \left[t_{i+1}, \frac{v_i + v_{i+1} + v_{i+2}}{3}\right] \tag{5}$$

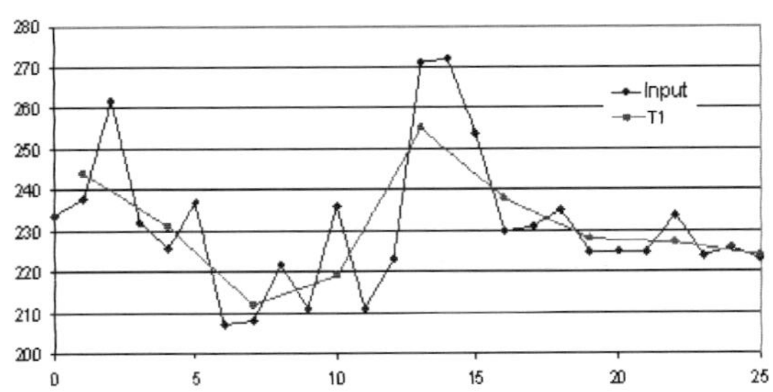

Fig. 1. Algorithm T1

2.2 Algorithm T$_2$

Algorithm T_2 is arranged in such a way that it assign to each triple of points from D_1 (resp. D) the value of gravity of the triangle arranged by this triple of vertices and then it compares this value of gravity with the following two values. If both this values are greater or smaller then the value of the value of gravity, then it replaces the second one with the value of gravity. If one of these values is greater and second one smaller, then it replaces the foregoing triple by new value of gravity. The output is a new set of data D_2. Consequently, T_2 is a map from D_1 to D_2 defined by:

$$T_1 = T_{es} \circ T_s \quad (6)$$

where $T_S : D_1 \rightarrow D^6$ is defined by

$$T_s\left(\left\{\left[t_{i+j}, v_{i+j}\right]\right\}_{j=0}^{5}\right) = \left[\left[t_{i+j}, v_{i+j}\right]\right]_{j=0}^{5}, \quad (7)$$

where i = 1,2,3…, $\frac{k - \mathrm{mod}_6(k)}{3}$ and $T_{es} : D^6 \rightarrow D_2$ is defined by

$$T_{es}\left(\left[\left[t_{i+j}, v_{i+j}\right]\right]_{j=0}^{5}\right) = \begin{cases} \left[t_{i+1}, te_i\right], \left[t_{i+4}, te_i\right] \text{if } v_{i+3}, v_{i+4} < te_i, \\ \qquad \text{or } v_{i+3}, v_{i+4} > te_i, \\ \left[t_{i+1}, te_i\right], \left[t_{i+4}, te_{i+3}\right] \text{otherwise}, \end{cases} \quad (8)$$

where $te_i = \frac{v_i + v_{i+1} + v_{i+2}}{3}$

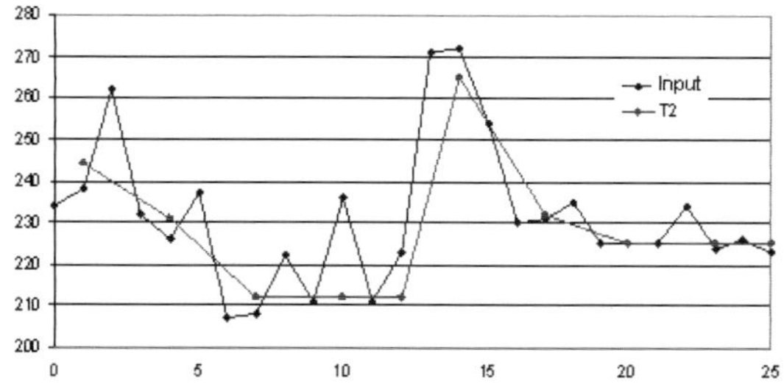

Fig. 2. Algorithm T2

2.3 Algorithm T_3

Algorithm T_3 is arranged in such a way that it assign to each triple of points from D_2 (resp. D_1 or D) the value of gravity of the triangle arranged by this triple of vertices and then it compares the foregoing triple of values with the value of gravity. If every point from the triple has the value greater or smaller then the value of gravity, then it replaces the third one with the value of gravity. If one value of the triple is smaller (resp. greater) and the last values of the triple are greater (resp. smaller), then it replaces this triple by a new value of gravity. The output is a new set of data D_3. Now, T_3 is a map from D_2 to D_3 defined by:

$$T_3 = T_{ee} \circ T_s \quad (9)$$

where $T_{ee} : D^6 \rightarrow D_3$ is defined by

$$T_{ee}\left(\left[\left[t_{i+j}, v_{i+j}\right]\right]_{j=0}^{5}\right) = \begin{cases} \left[t_{i+1}, te_i\right], \left[t_{i+5}, te_i\right] \text{if } v_{i+3}, v_{i+4}, v_{i+5} < te_i, \\ \qquad \text{or } v_{i+3}, v_{i+4}, v_{i+5} > te_i, \\ \left[t_{i+1}, te_i\right], \left[t_{i+4}, te_{i+3}\right] \text{otherwise}, \end{cases} \quad (10)$$

where $te_i = \frac{v_i + v_{i+1} + v_{i+2}}{3}$

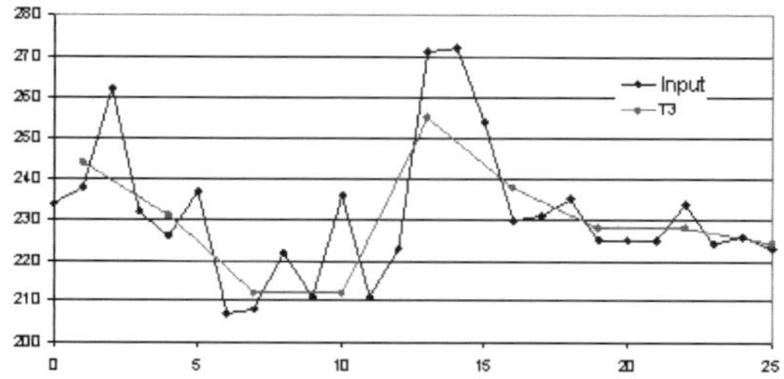

Fig. 3. Algorithm T_3

3. RESULTING ALGORITM

The resulting algorithm used for filtration of the initial data is the composition of the foregoing algorithms T_1, T_2 and T_3 exactly in this order. Denote it by A.
Finally, A is defined as a map $A : D \rightarrow D_3$,
by the rule: $A : T_3 \circ T_2 \circ T_1$.

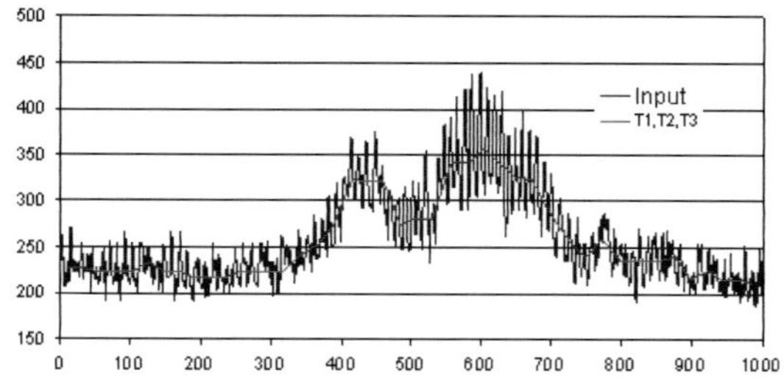

Fig. 4. Algorithm A

The output depiction is made in a primitive way (i.e., by linear joining of neighbour points from the final set of data).

4. CONCLUSION

The designed algorithm gives surprisingly first rate results. Its best advantage is its easy way to programme it and its unpretentiousness for the hardware. All the algorithm or its parts can be used for filtrating of similar data of other process.
Finally, it was made a faithful to the algorithm to be able to work on real time. It was achieved and it is possible to use this algorithm for a mobile hardware.
The output graph is a piecewise linear map, which is possibly make as a smooth curve by standard methods or by new one.

5. ACKNOWLEDGMENT

The work was performed with financial support of grant reg. No 102/03/0625 and research project 7088 352 102 MSM, by Grant Agency of the Czech Republic grant 201/07/P032 and the Ministry of Education of the Czech Republic No. MSM6198910027

6. REFERENCES

Adams, R. A. (2003). *Single Variable Calculus (paperback) (Edition: 5),* Pearson Education Canada, ISBN: 0-201-79805-0

Guyton, M. D. & Hall J. E, (2000). *Textbook of Medical Physiology*, W B Saunders, ISBN: 072168677X

Rektorys, K. (2000). *Summary of used maths (Edition: 7)*, Prometheus, ISBN 80-7196-180-9, Prague

Svatos, J. (1998). *Biological signals*, CVUT, ISBN: 8001018229, Prague

Trojan, S. (1999). *Medical physiology*, Grada Publishing, ISBN: 8071697885, Prague

Annals of DAAAM for 2008 & Proceedings of the 19th International DAAAM Symposium, ISSN 1726-9679
Editor B. Katalinic, Published by DAAAM International, Vienna, Austria 2008
Make Harmony Between Technology and Nature, and Your Mind will Fly Free as a Bird

AN EMPIRICAL RELATIONSHIP BETWEEN CUTTING FORCES AND LENGTH IN DRY DRILLING IN ALUMINIUM ALLOYS

DOMINGO, R[osario]; ALVAREZ, R[oberto] & SEBASTIAN, M[iguel] A[ngel]

Abstract: *This paper analyses the relationship between the cutting forces and length, according to experimental results obtained in aluminium alloys, in particular UNS A97050-T7, UNS A92024-T3, during the dry drilling. Box-Cox transformations have allowed finding a fitted model and determining a significant relationship between maximum cutting forces and cutting length, for each hole. Thus, it is possible to establish prediction limits for new observations of cutting forces.*
Key words: *Dry drilling, Aluminium alloys, Cutting forces, UNS A97050-T7, UNS A92024-T3.*

1. INTRODUCTION

Dry drilling continues being an important process in aeronautical industry, and also in the machining of aluminium alloys, in particular in UNS A97050-T7 and UNS A92024-T3.

The trust cutting forces can be calculated by dynamometers, some of them of high precision. However these forces are different respect to theoretical forces (Hamade, et al., 2006; Tansel, et al., 2000). For this reason it is possible to affirm that the prediction of forces is a question still does not resolve. Other studies have demonstrated that cutting forces for the first holes, being different, are similar statistically (Domingo et al., 2008a; Domingo et al, 2008b). This paper intends to clarify some relationships between cutting forces and length that allow predicting forces in dry drilling.

Cutting forces have been calculated by a piezoelectric dynamometer, and afterwards the results have been statistically analysed by Box-Cox transformations (Box & Cox, 1964) by SPSS software.

2. EMPIRICAL RESULTS

The tests have been realised in aluminium alloys, UNS A7050-T7 and UNS A2024-T3. Dry drilling has been processed to cutting speed of 83 m/min and 60 m/min for the first alloy, and for 50 m/min for the second one. Feed rate takes the value 0.175 mm/rev always. Drills used in the tests have the characteristics showed in Table 1.

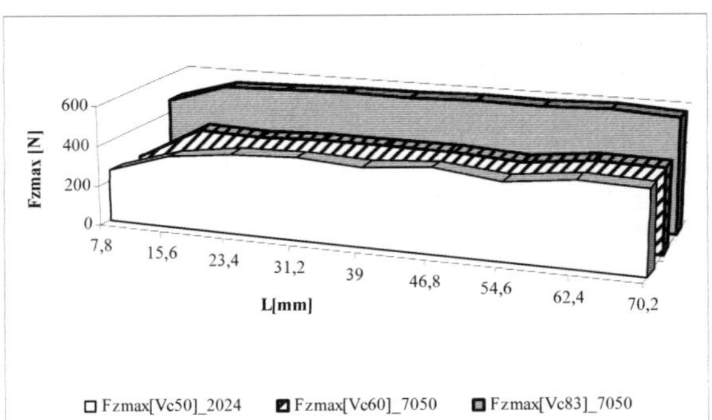

Fig. 1. Cutting forces.

Drill type	Tool material	Point angle	Coating
NH	VHM	140°	TiAlN

Tab. 1. Drill description.

Drilling process has been performed with three drills. Each drill has machined different holes, according to increment the cutting length. Cutting forces are expressed in Newton and cutting length expressed in mm. Data from UNS A97050-T7 and UNS A92024-T3 are shown in Fig. 1.

3. BOX-COX TRANSFORMATIONS

Box-Cox transformations have been used to determine whether, in this case, a significant relationship exists between maximum cutting forces, Fzmax (dependent variable) and cutting length (independent variable), *L*. This transformation with power allows minimizing the mean squared error.

Moreover, an analysis of variance (ANOVA) has been realised to determine the P-value. Thus the effect of various power transformations of Fzmax can be compared on the linear regression between *L* and Fzmax.

For Vc83 (Box-Cox Transformations - Power = 20.11 Shift = 0.0), Vc60 (Box-Cox Transformations - Power = 10.2112 Shift = 0.0) and Vc50 (Box-Cox Transformations - Power = 7.46592 Shift = 0.0), the equations of the fitted model, expressed in Fig. 2, Fig. 3 and Fig. 4 are the following,

$$\text{BoxCox}(F_{zmax\_Vc83}) = 11.23 + 0.82 \cdot L \quad (1)$$

$$\text{BoxCox}(F_{zmax\_Vc60}) = 29.79 + 0.94 \cdot L \quad (2)$$

$$\text{BoxCox}(F_{zmax\_Vc50}) = 21.36 + 1.37 \cdot L \quad (3)$$

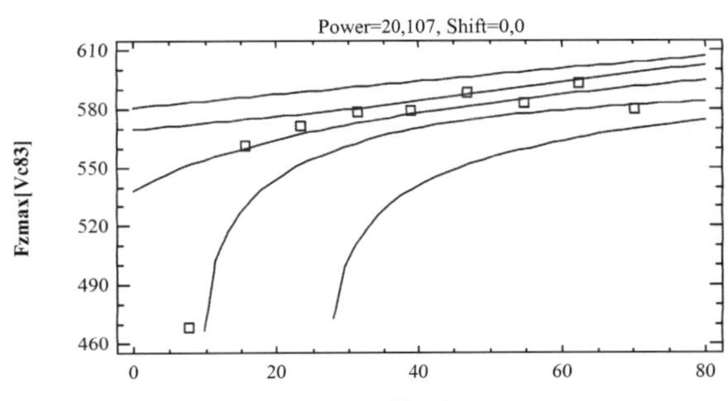

Fig. 2. Fitted model for UNS A97050 and Vc83.

	P-Value	Corr. Coeff.	R-squared	Standard Error Est.
V83	0.006	0.83	68.40%	12.73
V60	0.019	0.64	41.24%	24.35
V50	0.019	0.75	56.64%	24.89

Tab. 2. Data from ANOVA analysis.

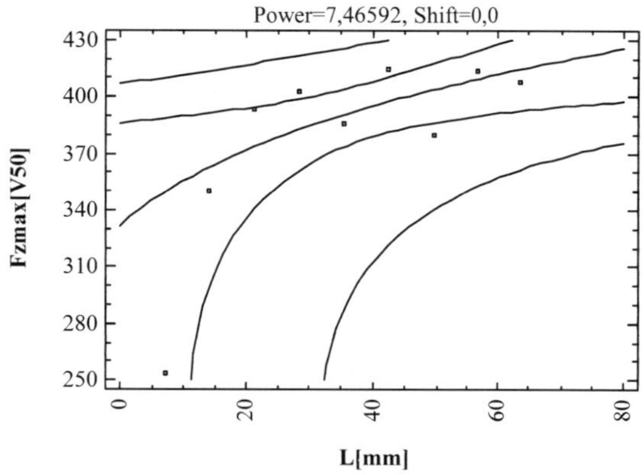

Fig. 3. Fitted model for UNS A97050 and Vc60.

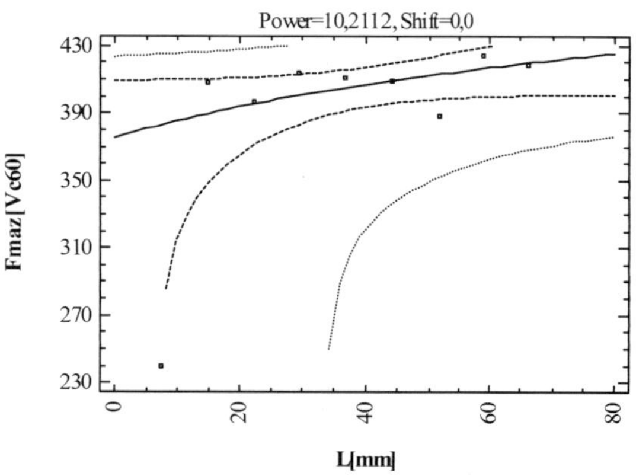

Fig. 4. Fitted model for UNS A92024 and Vc50.

In the ANOVA analysis (see Table 2), the P-value is less than 0.01, there is a statistically significant relationship between the transformed values of Fzmax_Vc83 and *L* at the 99% confidence level, for Fzmax_Vc60 and *L* at the 90% confidence level (P-value is less than 0.10), and for Fzmax_Vc50 and *L* at the 95% confidence level (P-value less than 0.05).

For Fzmax_Vc83, the R-squared statistic points to that the fitted model gives explanation of 68.40% of the variability in Fzmax_Vc83, of 41.24% of the variability in Fzmax_Vc60, and of 56.64% of the variability in Fzmax_Vc50.

Moreover, in Table 2 can be appreciated the correlation coefficient (0.83 for Fzmaz_V83) that it involves a moderately strong connection between the variables, and the standard error of the estimate that it shows the standard deviation of the residuals to be 12.73.

For Fzmax_V60, the correlation coefficient (0.64) describes a moderately strong relationship between the cutting forces and the length, and a standard deviation of the residuals of 24.35.

Finally, in the case of Fzmax_V50, the correlation coefficient (0.75) points to a moderately strong link between the cutting variables, and a standard deviation of the residuals of 24.89.

Thus, in the three cases, significant statistical relationships have been found, and the obtained value of cutting forces can be utilised to generate prediction limits for new observations.

4. PREDICTED VALUES

In accordance with Section 3, the Table 3 exposes the predicted values for Fzmax according the minimum and maximum cutting length (Lm and LM) that it has been drilled, at the 95% confidence level, the prediction limits and the limits of confidence. Lower (Low) and upper (Upp) limits are indicated.

	Predic-ted	Prediction Limits (95%)		Confidence Limits (95%)	
L	Fzmax _V83	Low	Upp	Low	Upp
Lm	551.12		583		572
LM	591.02	569	603	581	598
L	Fzmax _V60	Low	Upp	Low	Upp
Lm	383.02		425	245	409
LM	419.93	368	443	400	433
L	Fzmax _V50	Low	Upp	Low	Upp
Lm	349.59		410		388
LM	414.79	361	443	392	431

Tab. 3. Predicted values.

The Table 3 reveals the best forecasts, at 95% of confidence level, in the prediction intervals for new observations and for the mean of the observations (confidence intervals). These predictions can be distinguished in the Fig. 2, Fig. 3 and Fig. 4, by means of discontinuous curves on the fitted models.

5. CONCLUSIONS

This paper provides a study that it permits to categorise, by sectors, the prediction of cutting forces in aluminium alloys, considering the influence of the cutting length on the maximum force. Thus, a statistical relationship has been found between cutting forces and length, after to accomplish an ANOVA analysis and Box-Cox transformations. The results allow establishing a prediction limits at 95% confidence level.

These preliminary results can suppose an important element in the experimental tests due to different values of thrust forces that dynamometers provide.

Future researches could determine the connection between cutting forces and higher lengths, and also the influence of different cutting parameters and of the drill wear on the forces variability. Analogous advances will consent to complete and verify the model.

6. ACKNOWLEDGEMENTS

This work has received financial support from the MCYT (Spanish Government), by means of project DPI2005-09325-C02-02.

7. REFERENCES

Box, G.E.P. & Cox, D.R. (1964). An analysis of transformations. *Journal of the Royal Statistical Society*, Series B Vol. 26, 211–246, ISSN 1369-7412.

Domingo, R.; Álvarez, R.; Rubio, E.M. & Sebastián, M.A. (2008a). Experimental analysis of cutting forces in dry drilling of UNS A92024 alloy. *Journal of Machine Engineering*, Vol. 8, No. 2 (2008), 73-78, ISSN 1895-7595.

Domingo, R.; Arenas, J.M.; Rubio, E.M. & Marcos, M. (2008b). Experimental analysis of cutting forces in dry drilling of UNS A97050-T7 alloy, *Proceedings of Intelligent Computation in Manufacturing Engineering,* Teti, R. (Ed.), 3p Naples, July 2008, University of Naples Federico II, Naples.

Hamade, R.F.; Seif, C.Y. & Ismail, F. (2006). Extracting cutting force coefficients from drilling experiments. *International Journal of Machine Tools & Manufacture*, Vol. 46, No. 3-4, 387-396, ISSN 0890-6955.

Tansel, I.N.; Arkan, T.T.; Bao, W.Y.; Mahendrakar, N.; Shisler, B.; Smith, D. & McCool, M. (2000). Tool wear estimation in micro-machining. Part I: tool usage–cutting force relationship. *International Journal of Machine Tools & Manufacture*, Vol. 40, 599–608, ISSN 0890-6955.

Annals of DAAAM for 2008 & Proceedings of the 19th International DAAAM Symposium, ISSN 1726-9679
Editor B. Katalinic, Published by DAAAM International, Vienna, Austria 2008
Make Harmony Between Technology and Nature, and Your Mind will Fly Free as a Bird
2nd DAAAM-ICAT

RAPID MANUFACTURING OF MOULD INSERTS FOR TOOLING IMPROVEMENTS

DORMAL, T[hierry]

Abstract: *This paper will summarize some results of two EU projects in the Rapid Tooling field. The objectives are to demonstrate on real case studies the technological and economical advantages of layer manufacturing techniques for mould makers and injection moulders. Cost, lead time and quality enhancements will be involved here. The **Hipermoulding** project is dedicated to mould optimization with conformal cooling channels to reduce the cycle time and improve the filling of the mould. The only way to build inserts with such complex channels is to use layer manufacturing technologies (laser melting or sintering, 3DPrinting). Four moulds were built with both conventional and additive manufacturing.*
*In the **EuroTooling 21** project, a new process chain was experimented to find a low cost solution for the injection of small series of large parts.*
Key words: *Rapid Manufacturing, Layer based Mould Manufacturing, Hybrid Tooling, 3DPrinting, Conformal cooling channels*

1. INTRODUCTION

Two important EU projects ending in 2008 were dedicated to improve EU competitiveness of the injection moulding and mould making sectors.
Hipermoulding means Extreme Cycle Time Reduction for Injection Moulding Processes by using High Performance Injection Moulds and Moulding Processes.
The strategic objectives addressed by this collective research project are the cycle time reduction of injection moulding processes, the energy consumption reduction and the improvement of the efficiency of cooling.
So the partners focused on the use of conformal cooling channels and rapid manufacturing techniques to obtain a very efficient thermal management.
The basic objective of **EuroTooling 21** is to increase the European tooling industry competitiveness by increasing toolmakers' share in the total added value of the extended supply chain of final products, measurable in added value per employee per hour.
The real breakthrough is here a new paradigm of high added value tools and tool making business models based on technology and process innovation that will reposition toolmakers as key strategic partners in complex industrial value chains, through a bottom-up research approach based on Case Studies representative of major industry needs and challenges.
The project approach is based on three different Case Studies, involving SME tooling companies and their tools OEM end-users. Innovative technologies were developed in an industrial context and implemented, tested and validated. In this paper we will only focus on one new process chain based on the 3DPrinting of sand powder.

2. HIPERMOULDING CASE STUDIES

A large panel of metal rapid manufacturing techniques was validated as well as many powder materials and post finishing techniques to fit with the high level of quality required by production tool inserts for plastic injection. The following LBMM (Layer Based Mould Manufacturing) techniques were experimented: DMLS (EOS), 3DPrinting (Prometal), ConceptLaser, SLM (MCP).
The choices made during the project are also linked to the materials available on the market at the time of this choice. So for example, the maraging steel MS1 (EOS) and the stainless steel 17-4PH were not available in 2006 and the CoCr material was not the best solution for the post machining process, so most of the time the ConceptLaser was the only viable solution for some specific geometries.
As we built real production tools, we decided to apply an offset on the CAD file to allow systematically the post machining of the surface to guarantee a good accuracy around 0.05 mm even if some LBMM processes are not so far from this performance directly out of the machine.

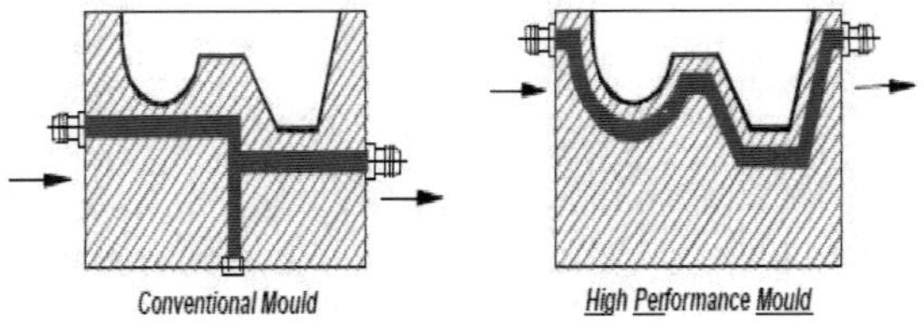

Fig. 1. Conformal cooling channels.

The four moulds were produced either by conventional techniques either by LBMM technologies to allow a comparison of the cycle time and other filling conditions and results.

Case study 1 (Poland)

This Polish case is a small part in PA6 with 25 % GF. The wall thickness varies from 2.6 to 4 mm and so it requires a long cycle time and results in rather big warping.

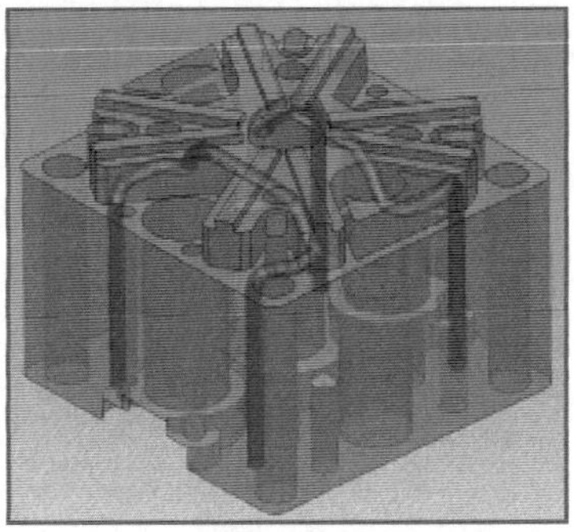

Fig. 2. 3DCAD tool design of case study 1.

A reduction of 31 % of the cycle time was observed but this saving didn't compensate the extra cost of the LBMM inserts.since the production is limited to 22.000 parts/year.

Fig. 3. DMLS inserts for case study 1

Case study 2 (Belgium)

The Belgian case is a small PP part for the medical sector with 2 mm wall thickness requiring a lot of competence to perform a good filling of the part. There was a very narrow space for the internal cooling but the result was efficient (34 % reduction) and the overall annual profit is 8500 €.

Case study 3 (Portugal)

The Portuguese case is a complex automotive part in PA6 with 30 % GF. The wall thickness varies from 0.7 to 1.9 mm. The objective is here to prevent warping and guarantee dimensional conformance.
The internal zone available for the cooling optimization is rather small and intricate. This is why the reduction of the cycle time is limited to 15 % but for such a large production (600.000 parts/year), the benefit is still 8600 euros/year.

Case study 4 (Spain)

The Spanish case is a small part for the food sector in HDPE. The wall thickness range is 1.2-1.7 mm and the conventional cycle time is very long (38 seconds). The cycle time reduction is rather small (15 %) but for such a large production (6.000.000 parts/year), the extra cost of the LBMM inserts becomes negligible and the overall benefit is 222.000 euros per year.

3. EUROTOOLING 21 CASE STUDIES

The objective of this case study is to find an efficient process chain to produce a tool for the injection of 50-100 large parts. One of the innovative solutions validated during ET21 is based on the fastest and largest rapid manufacturing technology. The 3DPrinting of the sand cores with the Prometal RCT technique allows the direct casting of the large aluminium inserts of the injection tool. The global lead time of this first step is only 5 days.

Fig. 4. Sand cores produced by 3DPrinting

The next step is more conventional since it is the post machining and the integration of the inserts in the mother mould. An additional and simultaneous rapid manufacturing step was performed for the small insert because of the complexity of this geometry.

Fig. 5. Cast aluminium insert

The final techno-economical summary shows that there is still some space for enhancements but a cost reduction up to 35 % is already realistic as well as a reduction of material up to 70 % compared to conventional aluminium milling.

4. CONCLUSION

The rapid tooling techniques developed during the last 15 years did not really reach the point where they are commercially viable. The Interest in using additive processes to solve tooling problems continues and still presents opportunities to reduce time and cost. The competition between layer manufacturing and high speed milling is always very active and the best way to reach a a real breakthrough seems to be, more than ever, an hybrid solution combining complex internal geometries for cooling optimization and fast post machining.
The EU projects described above bring many exploitable results to start some tooling services with conformal cooling channels as well as for large short-run production tools, but a real large scale exploitation is not yet for tomorrow.

5. REFERENCES

Dormal T. Capabilities and Limitations of actual Rapid Tooling solutions for the Plastic Injection Moulding of different Materials and Part Numbers. *µRapid 2006.*

Gravet D. & Dormal T. Rapid tooling case studies: success and limitations. *EuroMold SPE conference* (Francfort) 2004.

Voet A.; Mingneau J.;Van Pee B.;Cardon L. & Houtkier R. The effect of multiple material moulds and coolin strategy on cycle time and quality of injection and blow moulding products. *De Nayer Instituut & Hogeschool Gent 2006*

Kruth JP.; Vandenbroucke B., Van Vaerenbergh J.& Mercelis P.; Benchmarking of different SLS/SLM processes as Rapid Manufacturing techniques. JP Kruth, (KULeuven) *PMI 2005. Hogeschool Gent.* April 20-23 2005

Pelaingre C.; Velnom L. & Barlier C. (Cirtes) A cooling channels innovating design method for rapid tooling in thermoplastic injection moulding. *VRAP 2003 Leiria (Portugal)* October 1-4 2003

Wohlers T. *Wohlers Report 2006.* Rapid Prototyping & Manufacturing. State of the industry.

Annals of DAAAM for 2008 & Proceedings of the 19th International DAAAM Symposium, ISSN 1726-9679
Editor B. Katalinic, Published by DAAAM International, Vienna, Austria 2008
Make Harmony Between Technology and Nature, and Your Mind will Fly Free as a Bird

A HYBRID WHEEL-LEG MOBILE ROBOT

DOROFTEI, I[oan]; MARTA, C[onstantin]; HAMAT, C[odruta]; SUCIU, L[enuta] & PRISACARU, G[heorghe]

Abstract: *One of the merits of walking vehicles is their superior terrain adaptability and also stepping over obstacles without applying any forces on them. On the other hand, traditional wheeled platforms provide sufficient robustness, mechanical simplicity and energetic performance. They are fast, powerful in terms of load to weight ratio, stable, and easy to control. Hybrid systems were developed to exploit the terrain adaptability of legs in rough terrain and simpler control as well as high speed associated with wheels. In this paper the design of a small wheel-legged robot is presented.*
Key words: *hybrid locomotion, kinematics, design, control.*

1. INTRODUCTION

One of the merits of walking vehicles is their superior terrain adaptability and also stepping over obstacles without applying any forces on them. The use of legs is convenient for locomotion on soft ground or in unstructured environments, where the performance of wheels and tracks are considerable reduced. On the other hand, they are moving with smaller speed comparing to traditional wheeled platforms. Wheeled robots also provide sufficient robustness, mechanical simplicity and energetic performance. They are fast, powerful in terms of load to weight ratio, stable, and easy to control. One way of improving the drawback of walking robots may be to develop hybrid, in order to exploit merits of both walking and wheeled systems.

From an historical point of view, a vehicle using both legs and wheels was not new, because the wheel barrow, bullock cart, etc. have been used since the early days of civilization. But, legged animals were an integral part of these systems and the design of fully artificial hybrid systems started much later Kar, 2003). What can be observed from these systems is that the work traction is done by the legs that allow a better adhesion with the soil, while most of the weight is carried by the wheels. Based on this concept, there are several researches on wheel-leg hybrid robots. In Japan has been developed a hybrid locomotion vehicle for nuclear power plants. It was a vehicle with five locomotion devices, each device consisting of a wheel and a leg., for nuclear power plants. A hybrid robot called SAPPHYR, with two free rear wheels and two traction legs in the front, was designed in France (Guihard et al., 1995). (Muscato & Nunnari, 1999) developed in Italy a robot with two pneumatically actuated front legs, each with three degrees of freedom, and two rear wheels, actuated independently. At University of Pennsylvania has been developed a hybrid wheelchair with two d.o.f. planar legs, for use on uneven terrain, and four wheels. One of the original ideas has been proposed by Hirose, in Japan. He developed a hybrid mobile robot called Roller-Walker with a special foot mechanism in each leg, to change between foot sole for walking mode and passive wheel for skating mode. Researchers at Tohoku University used the idea of combination of four legs (two in the front and two rear legs) and two wheels in the middle, to develop a leg-wheel type robot for the exploration. Another example of the hybrid system, called Walk'nRoll, has been developed by Adachi (Adachi et al., 1999). The robot has two legs with 3 d.o.f. in the front and and two rear wheels. The system usually moves on the surface with wheels like an usual wheeled vehicle. Entering to rough surface area, it switches the moving mode from wheel to leg mode. One of the problems of such systems is how to switch the mode according to the change of surface conditions. To do it, the robot is using ultra-sonic sensors. A leg-wheel hybrid robot, with two active 2 d.o.f. legs in the front and two passive rear wheels, has been developed in Thailand (Suwannasit & Laksanacharoen, 2004). In this paper the design of a small wheel-legged robot will be presented. This report is the result of a research conducted at the Robotics Laboratory, Mechanical Engineering Faculty, "Gh. Asachi" Technical University of Iasi, Romania.

2. ROBOT HARDWARE

2.1 Mechanical design

Most of the similar robots developed before, using the same architecture - two legs and two wheels, can only move using hybrid locomotion mode (Muscato & Nunnari, 1999; Suwannasit & Laksanacharoen, 2004). Thanks to its two legs with 2 d.o.f. each and passive wheels as feet in the front, and two actuated rear wheels, respectively, the robot developed by our laboratory (see Fig. 1) can move using three locomotion modes:

- Wheeled locomotion mode, using the rear actuated wheels (*2* and *11*) and the two passive wheels in the feet (*6* and 10);
- The first hybrid locomotion mode, using active front legs (*5* and *7*) and passive rear wheels;
- The second hybrid locomotion mode, using the legs and active rear wheels..

The rear wheels (*2* and *11*) can be free or actuated, thanks to the two electromagnetic clutches that realize the connection between them and their servo (*12* and *13*). In order to offer a continuous rotational motion, the servos used for rear wheels actuation have been modified.

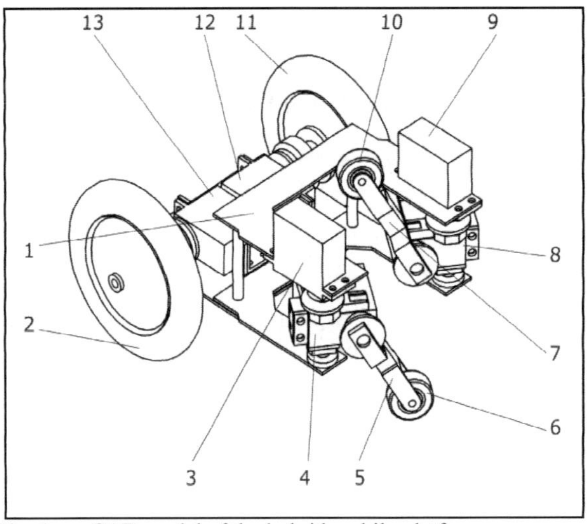

Fig. 1. CAD model of the hybrid mobile platform

The advantage of our robot configuration, comparing to other similar previous designs, is an improved stability, when only one leg is in support phase. In wheeled locomotion mode, the actuated wheels (the big ones) act in the front of the robot and the small wheels, as feet of the passive legs, are passive. One of the small passive wheels or both of them (depending on the instantaneously trajectory of the robot) will be in contact with the ground, only to keep the stability of the robot. We will have two small passive wheels on the ground for a straight or curved trajectory, and a single passive wheel in contact with the ground for turning on the spot. If we consider the wheeled locomotion mode, we can write:

$$\begin{cases} v_r = v + \Omega \cdot l_1 \\ v_l = v - \Omega \cdot l_1 \end{cases}, \tag{1}$$

$$\begin{cases} v = (v_r + v_l)/2 \\ \Omega = (v_r - v_l)/2 \cdot l_1 \end{cases}, \tag{2}$$

and

$$\begin{bmatrix} v_x \\ v_y \\ \Omega \end{bmatrix} = \begin{bmatrix} 1 & 0 \\ 0 & l_2 \\ 0 & 1 \end{bmatrix} \cdot \begin{bmatrix} v \\ \Omega \end{bmatrix} = \begin{bmatrix} \frac{R}{2} & \frac{R}{2} \\ \frac{R \cdot l_2}{2l_1} & -\frac{R \cdot l_2}{2l_1} \\ \frac{R}{2l_1} & -\frac{R}{2l_1} \end{bmatrix} \cdot \begin{bmatrix} \omega_r \\ \omega_l \end{bmatrix}, \tag{3}$$

where: v – the linear velocity of the robot body; Ω-its angular velocity; R – radius of the rear wheels; ω_r - angular velocity of the right rear wheel; ω_l - angular velocity of the left rear wheel; v_r - linear velocity of the right rear wheel; v_l - linear velocity of the left rear wheel. For hybrid locomotion mode next relation should be respected:

$$\omega_{med} \cdot r = \omega_{leg} \cdot l_3, \tag{4}$$

here

$$\omega_{med} = \frac{\omega_r + \omega_l}{2} \tag{5}$$

and l_3 is the horizontal projection of the leg. In order to simplify the control algorithm, the robot has been designed as well as

$$l_3 = R. \tag{6}$$

2.2 Control

The control board of the first hybrid robot is based on a PIC family micro-controller (Fig. 2). There are not sensors used for the first prototype that has been built only to test the architecture and locomotion modes. In order to save energy, for flat terrain only the rear wheels or the two front legs are actuated. For climbing obstacles or moving on soft or unstructured terrain, the rear wheels and the front legs are simultaneously actuated, in order to improve the traction force of the vehicle. Because the trajectories of the legs are crossing in the support phases, few precise rules should be established for the case when the robot is using hybrid locomotion mode:

- The legs could not be simultaneous in transfer phase (in that case the robot will fall down);
- The legs could not be simultaneously in support phase otherwise they will cross each other;
- When a leg is in support phase the other one should be in transfer phase, moving in opposite direction.

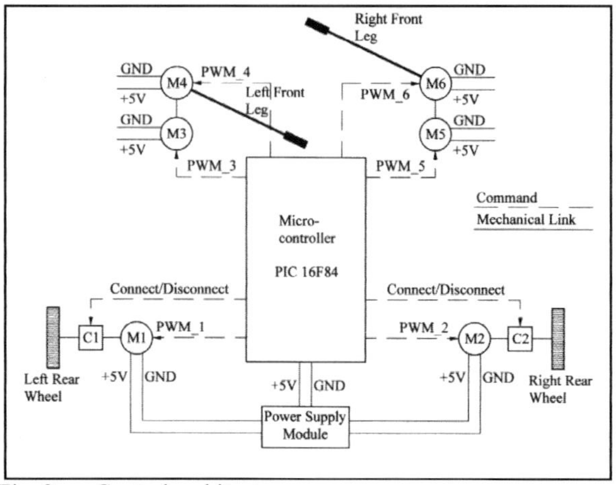

Fig. 2. Control architecture

The testing gave a qualitative view of the system's mobility performance. All the locomotion modes have been tested, using: forward, backward, turning right on the spot and turning left on the spot sequences.

We did not yet test a curved trajectory. All the tests have been done in the laboratory. We should also test the robot on a soft ground.

3. CONCLUSION

One of the merits of walking vehicles is their superior terrain adaptability and also stepping over obstacles without applying any forces on them.

On the other hand, they are moving with smaller speed comparing to traditional wheeled platforms, which also provide sufficient robustness, mechanical simplicity and energetic performance.

One way of improving the drawback of walking robots may be to develop hybrid, in order to exploit merits of both walking and wheeled systems.

Thanks to its two legs with 2 d.o.f. in the front and two passive wheels as feet and two actuated rear wheels, the robot proposed by the authors can move using three locomotion modes: wheeled mode and two hybrid locomotion modes.

4. REFERENCES

Adachi, H., Koyachi, N., Arai, T., Shimizu, A. & Nogami, Y. (1999). Mechanism and Control of a Leg-Wheel Hybrid Mobile Robot, *Proceedings of IEEE/RSJ International Conference on Robotics and Automation*, pp. 1792-1797.

Guihard, M., Gorce, P. & Fontaine, J.G. (1995). Sapphyr: Legs to pull a wheel structure, *Proceedings of IEEE International Conference on Robotics and Automation*, Nagoya, Japan, pp. 1303–1308.

Kar, D.C. (2003). Design of Statically Stable alking Robot: A Review. *Journal of Robotic Systems* 20(11), pp. 671-686.

Muscato, G. & Nunnari, G. (1999). Legs or wheels? Wheeleg - a hybrid solution. *Proceedings of the 1st International Conference on Climbing and Walking Robots - CLAWAR-99* (editors G.S. Virk, M. Randall, and D. Howard), Portsmouth, UK. Professional Engineering Publishing, pp. 173–180.

Suwannasit, K. & Laksanacharoen, S. (2004a). A BIO-Inspired Hybrid Leg-Wheel robot. *TENCON 2004, (2004 IEEE Region 10 Conference), Proceedings Analog and Digital Techniques in Electrical Engineering*, 21-24 Nov., Chiang Mai, Thailand.

Annals of DAAAM for 2008 & Proceedings of the 19th International DAAAM Symposium, ISSN 1726-9679
Editor B. Katalinic, Published by DAAAM International, Vienna, Austria 2008
Make Harmony Between Technology and Nature, and Your Mind will Fly Free as a Bird

SPEED CONTROL SYSTEM USING A FUZZY LOGIC CONTROLLER

DOSOFTEI, C[onstantin] C[atalin]; MASTACAN, L[ucian] & OLAH, I[osif]

Abstract: *The main purpose of this paper is to implement a fuzzy algorithm for control the velocity of the brushless motor. The algorithm was implemented and tested using a control structure MCK240 made of Technosoft Company. The experience of operators working with the process allows finding the fuzzy controller structure and the linguistic rule base. Using of lookup decision tables decrease the computation time, especially in the fuzzy inference reasoning and output defuzzifications. Implementing of the fuzzy logic on systems with microcontroller represent a good alternative for building of control systems for rapidly process.*
Key words: *fuzzy, look-up decision, dsp, velocity, algorithm*

1. INTRODUCTION

¶This year marks the 43th anniversary of fuzzy logic and fuzzy set theory (Zadeh – 1965), which has now evolved into a fruitful area containing various disciplines, such as calculus of fuzzy if-then rules, fuzzy graphs, fuzzy interpolations, fuzzy topology, fuzzy inferences systems and fuzzy modeling. The applications, which are multidisciplinary in nature, includes automatic control, consumer electronics, signal processing, time series prediction, information retrieval, data base management, computer vision and so on.
In this paper, focuses on a speed control system using a fuzzy logic controller for a Brushless AC Motor. In a digital controller of brushless AC Motor, the control accuracy is of a high level, and it has a fast response time. Recently, a brushless AC/DC Motor (BLAC/BLDC) has been rapidly demanded due to preciseness of industrial technology and increase of various kind of control device.

2. HARDWARE STRUCTURE

The MCK240 is a development kit for the Texas Instruments TMS320F240 DSP controller applied to digital motor control applications. The MCK240 provides to the designer a complete set of tools in order to quickly develop, implement and test digital motor control algorithms. It includes high level graphical programming tools and a DSP board equipped with a three phase's inverter.

A dynamic escape brushless motor combine with an incremental encoder, which can be directly driven in AC or DC brushless modes, is also included in the kit.

The MCK240 offers also an universal I/O interface for connection to various external power amplifiers, such as IGBT inverter bridges for controlling larger motors.

The MCK240 motion kit is combined with the extended software package MCWIN, which includes a basic monitor for serial communication with down/up load functions and debug facilities, and a Windows IDE for assistance on performance evaluation of motor control applications. All these tools offer a user friendly graphical interface, which facilitates the development, implementation and test of DSP motor control applications.

Ready-to-run examples of AC and DC brushless motor speed control are contained in the BLAC/BLDC packages. The dynamic behavior of the real time system can be easily analyzed during run-time through the data acquisition and graphical display of all the system variables, thus permitting a quick optimization of the control algorithms (MCK240 V1 1997).

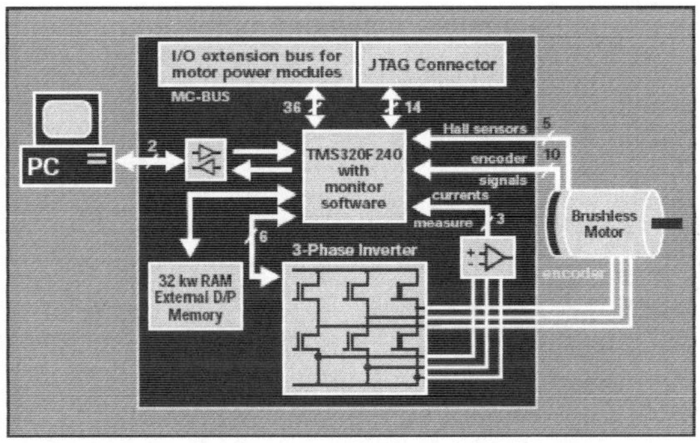

Fig. 1 MCK240 - A development platform for DSP motor control applications

3. IMPLEMENTING A FLC WITH LOOKUP TABLES

The design of a fuzzy logic controller (FLC) starts from a mathematical model of the process or from a linguistic model of the process, which emulate the behavior of a skilled operator. The experience of operators working with the process allows finding the fuzzy controller structure and the linguistic rule base.

For the rapidly process to implement a fuzzy logic controller the fuzzy algorithm computation time is an important criterion. Lookup decision tables are used in order to decrease the computation time, especially in the fuzzy inference reasoning and output defuzzifications. The computations are carried out offline and the results are stocked in memory in the form of lookup decision tables. (Mastacan 2000)

The figure on below shows the basic flow of information through the fuzzy logic decision mechanism. A fuzzy controller relies on fuzzy sets to represent the semantic properties of each input (sensor) and output (actuator). The variables are processed by a set of `IF-THEN` production rules that map the input values to the output space through a collection of fuzzy sets. These input sensor values may represent raw sensor metrics, synthesized sensor metrics processed by one or more sensor attached functions, qualitative classifications, or any combination of these.

A PI-like fuzzy logic controller is based on the "if-then" rules in the form:

if *error* is A_k and *error change* is B_k then *change in control* is ΔU_k (1)

The fuzzy control variable is computed as:

$$U_k = U_{k-1} + \Delta U_k \qquad (2)$$

In algorithm which was implemented for to control velocity of the brushless motor the *process output error*, *the process output change in error* and *change in control* was all quantized into an eleven number of values like in following table:

Q	e (rpm)	Δe (rpm)	Δu (Volţi)
-5	-100	-20	-2
-4	-80	-16	-1,2
-3	-60	-12	-0,9
-2	-40	-8	-0,6
-1	-10	-4	-0,3
0	0	0	0
+1	10	4	0,3
+2	40	8	0,6
+3	60	12	0,9
+4	80	16	1,2
+5	100	20	2

Tab. 1 Look-up decision e, Δe & Δu

In function of fuzzy rules from rules base of the FLC, through Mamdani inference and defuzzified to a crisp value using a strategy like the centroid are obtained the offline quantized values for control. These values are introduced into a lookup decision Mamdani table.(Mamdani 1974)

	Δe										
e	-5	-4	-3	-2	-1	0	1	2	3	4	5
-5	-5	-4	-3	-2	-1	-5	-1	-2	-3	-4	-5
-4	-4	-4	-3	-2	-1	-4	-1	-2	-3	-4	-4
-3	-3	-3	-3	-2	-1	-3	-1	-2	-3	-3	-3
-2	-2	-2	-2	-2	-1	-2	-1	-2	-2	-2	-2
-1	-1	-1	-1	-1	-1	-1	-1	-1	-1	-1	-1
0	-5	-5	-4	-3	-2	0	0	0	0	0	0
1	-5	-5	-4	-3	-1	0	1	1	1	1	1
2	-5	-5	-4	-3	0	2	1	2	2	2	2
3	-5	-4	-3	-2	0	3	0	2	3	3	3
4	-4	-3	-2	-1	1	4	1	2	3	4	4
5	5	5	5	5	5	5	1	2	3	4	5

Tab. 2 Look-up decision- offline quantized values for control

From this lookup decision table results the quant for command and then command on step k:

$$u_k = u_{k-1} + \Delta u_k \qquad (3)$$

4. EXPERIMENTAL RESULTS

Figure 4 presents the velocity fuzzy control of the motor without load and figure 5 presents the velocity fuzzy control of the motor with load.

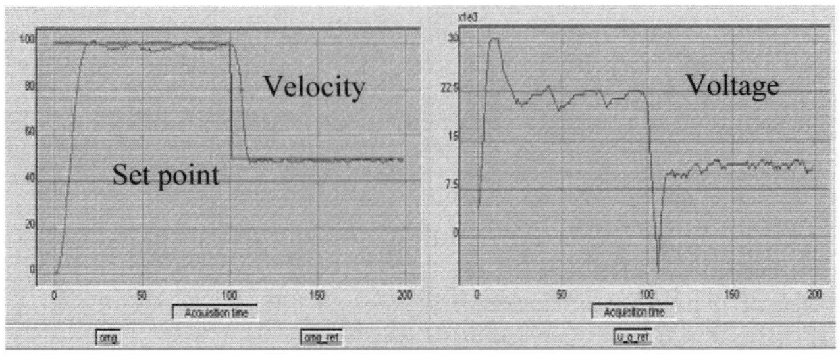

Fig. 2 Testing FLC (motor without load)

a) velocity; b) voltage for supply motor

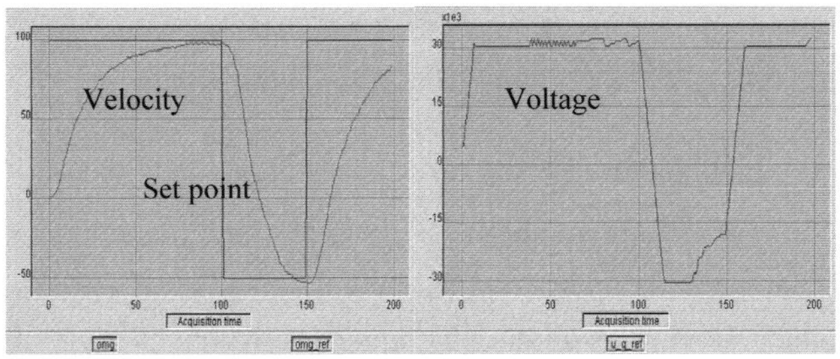

Fig. 3 Testing FLC (motor with load)

a) velocity; b) voltage for supply motor

5. CONCLUSION

The velocity of a brushless motor can be easily controlled in real time by the PI-like fuzzy logic as it shown in figures 4 and 5. The Mamdani method is an intuitive method and more competitive for applications were experience of operator is very high.

A well-performing and robust fuzzy logic controller to control the motor with/without load is implemented.

Implementing of the fuzzy logic on systems with microcontroller represent a good alternative for building of control systems for rapidly process. Explosion in developing of the DSP create new opportunities with control field. In the future we can develop other big system which can included this subsystem of controlling speed.

6. REFERENCES

Driankov, D.(1993); *An introduction to fuzzy control*, 1993, Springer-Verlag

Mamdani, E.H.(1974); Application of Fuzzy Algorithms for Control of a Simple Dynamic Plant, *Proc. IEE,* Vol. 121, No. 12, 1974, pp. 1585-1588

Mastacan, L.(2000); *Interpolation techniques for real time fuzzy control*, EURO XVII, 2000, Budapest, Hungary, pp.38-42

Zadeh, L. A.(1965): *Fuzzy sets. – Information and* control, 1965. p.8:338-353.

*** – MCK240 V1.0 User Manual, Tehnosoft, 1997

Annals of DAAAM for 2008 & Proceedings of the 19th International DAAAM Symposium, ISSN 1726-9679
Editor B. Katalinic, Published by DAAAM International, Vienna, Austria 2008
Make Harmony Between Technology and Nature, and Your Mind will Fly Free as a Bird

AN APPROACH TO DICOM EXTENSION FOR MEDICAL IMAGE STREAMING

DRAGAN, D[inu] & IVETIC, D[ragan]

Abstract: *Annual volume of medical images in PACS reaches 10 TB. This narrows the use of mobile devices in medicine. We present a transparent extension of the standard DICOM message by a parameter (JPIP) string. This enables streaming of medical images (mobile-health) and reduces storage demands over 10 times.*
Key words: *DICOM, medical image streaming, PACS, JPEG2000*

1. INTRODUCTION

The advance in technology enabled the use of handheld devices in medical environments. The mobile devices enable the physicians to access medical data wherever they are – in hospital buildings, outdoor or in their homes. Unfortunately, these devices are mainly used for receiving and displaying the medical record data without medical images. Even with the image compression, the size and resolution of medical images overcomes the capabilities of common mobile devices (Mirkovic et al., 2005). The mobile device limitations of processing power, memory capacity, display capabilities, and the mobile network band-with limitations increases the overall time needed to transmit, decompress and present the image. For example the calculated time that passes from requesting the JPEG2000 medical image until it is presented to the client on PDA is about one minute, which is unacceptable.

Picture Archiving and Communication System (PACS) is imaging system designed to steer medical image data and the corresponding information between various parts of healthcare system. The communication between various parts of PACS is usually achieved through Digital Image and Communication in Medicine (DICOM) protocol (Dragan & Ivetic, 2005). To enable the transmission of medical images to mobile devices, the image archive has to expand with the image copies suitable for different mobile devices. For example, to include mobile phone, PDA and Tablet PC into the PACS, the medical image archive has to contain images prepared for these devices, fig. 1. The PACS treats them as different medical images. Therefore, the DICOM client-side application has to know which image is adequate for its display and processing capabilities before requesting it. This is achieved only in querying the DICOM server prior to image request. There is no mechanism which will allow the client to request the image with the desired region of interest, resolution, and quality.

Medical image streaming is a more effective solution to enable the transmission of medical images to mobile devices (Skodras, 2006). Image streaming means that pixel data needed to represent a part of the image (or the whole image) in a certain resolution and quality are extracted from the one stored image codestream and transmitted to the client-side application, fig. 2. Therefore, the client only has to define the desired part, resolution, and quality of the image. The network communication is not overloaded with unnecessary data; the medical images are smaller in size; and the image resolution is suitable for display device resolution. The best results could be reached if image streaming is combined with a still image compression which achieves smaller spatial distortion.

Although it has been improved over the years, DICOM has not been modeled to support streaming of medical images. Therefore, DICOM restricts PACS to devices with higher power, storage, communication and display capacities. In this paper we describe the way in which DICOM can be extended to support image streaming. This will improve DICOM networks which will then support mobile client-side devices with limited processing and display capabilities. The proposed solution is transparent, simple to implement and easy to include into the existing DICOM based PACS.

The organization of the paper is as follows: the section 2 gives the overview of other's people researchers; the section 3 describes the proposed extension of DICOM; and the section 4 concludes the paper.

2. DICOM AND MEDICAL IMAGE STREAMING

DICOM defines the message format, the protocol for message interchange and the file structure for biomedical images and image-related information. Although it is not supported directly, the importance of medical image streaming is recognized by the DICOM standard. The standard defines the pixel data provider service which enables transmission of pixel data using a network protocol other than DICOM. Currently, the only service supported is the JPIP (JPEG2000 Interchange Protocol) pixel data provider which represents a mechanism for supporting the use of JPIP network protocol (Taubman & Prandolini, 2003). It enables interaction with JPEG2000 content and it is used for JPEG2000 image streaming. When this mechanism is used, pixel data are replaced with single DICOM attribute which contains URL string that represents the JPIP request, including the specific target information. The client-side application can expand the JPIP request with description of desired region of interest, resolution, and quality of the image.

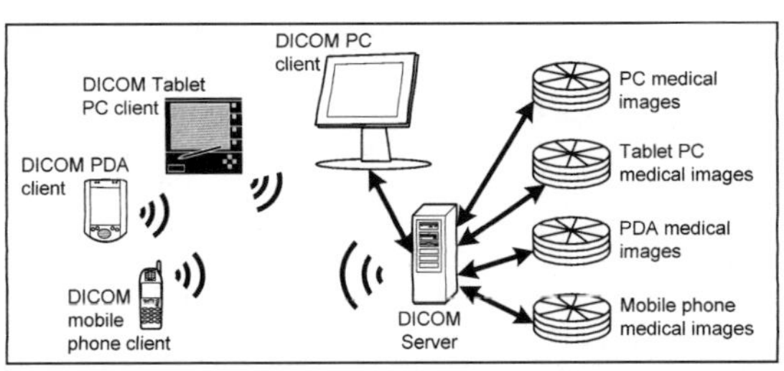

Fig. 1. PACS supporting 3 mobile devices – 4 image archives

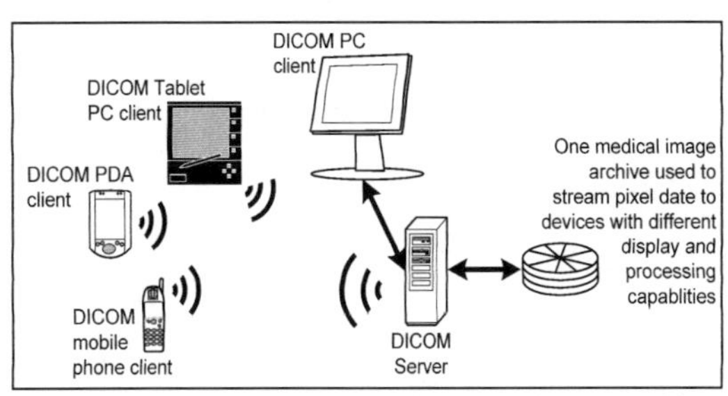

Fig. 2. The medical image streaming – one image archive

This service enables medical image streaming but not over DICOM network. Although the whole range of the JPEG2000 streaming capabilities is on disposal, this mechanism has several disadvantages. DICOM server has no control over the pixel data transmission. If, from some reason, image does not exists at JPIP server-side, or image can not be served, the transmission of the DICOM image has to be done from beginning, completely over DICOM protocol and the medical image has to be transmitted in its full size. Therefore, the JPEG2000 streaming can not be guaranteed. Also, the JPIP and the DICOM server archive have to be synchronized.

The streaming of DICOM images has been studied. The usual solution is based on two layer communication. One layer is used for communication with DICOM based PACS system and the second layer is used to stream images to the client-side devices. Tian proposed in his works (Tian et al., 2008) the two layer architecture in which the second part of the system is based on JPIP protocol. The medical image archive remained in original uncompressed DICOM format. The storage requirements are high, because the image streaming and compression are not combined. In this communication system, mobile devices are not internal part of the DICOM networks.

3. DICOM EXTENSION

It is possible to extend DICOM to support medical image streaming. Standard DICOM message should be extended with additional fields which will enable DICOM clients to request the desired region of interest, resolution, and quality of the medical image. Also, the extension should contain fields with the description of the DICOM server response. The additional fields should be transparently and easily incorporated inside standard DICOM message following DICOM rules for DICOM fields formatting and labeling.

The standard DICOM message can be extended in two ways: by a set of parameters or by a parameter string. The set of parameters would define a new DICOM field for each parameter of the requested medical image, fig. 3.A. For example, if DICOM client-side application requests part of a medical image of desired resolution and quality, DICOM request message should contain at least three attributes (one for each of the requested parameters). The parameter string would define only one new DICOM field, fig. 3.B. This DICOM field should contain the description of the requested image parameters, formatted according to some messaging format like HTTP, or JPIP. The choice for selecting one of the proposed extensions depends from the specific implementation.

The best result could be achieved when all the client requests are served from one basic image source. The basic image is processed according to the DICOM client request. The DICOM server would transmit only the data needed to represent the medical image in resolution and quality requested from the client, suitable for the client device. With this approach the DICOM medical image archive remains the same. It is even better to combine this mechanism with still image compression whit minimal spatial distortion. This combination leads to smaller medical image archive and high-quality resulting medical images.

The JPEG2000 compression supports image streaming from one JPEG2000 image with minimal spatial distortion in

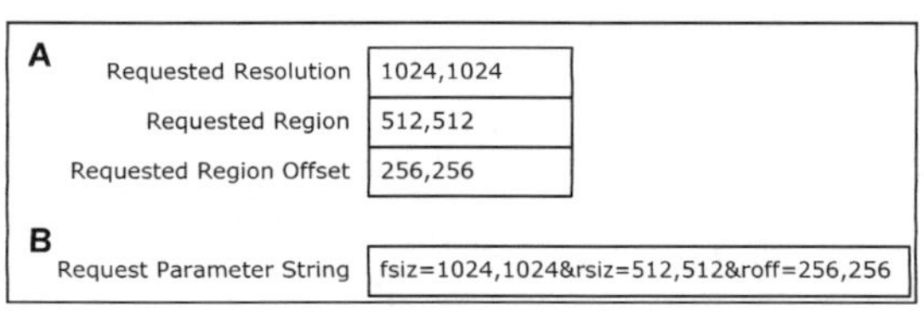

Fig. 3. DICOM extensions for medical image streaming; A –set of parameters; B –parameter string.

decompressed image (Taubman & Marcellin, 2001). The processed JPEG2000 images always yield the best quality for given resolution. Although DICOM does not support JPEG2000 streaming, it supports JPEG2000 image format and it is possible to replace native DICOM image format with JPEG2000 image format. It is possible to implement JPEG2000 streaming over DICOM network. The DICOM server should process JPEG2000 medical image according to client request, encapsulate it inside DICOM message and send it back to DICOM client. The JPEG2000 streaming has been tested in medical domain, but outside DICOM, and it achieved good results (Moshfeghi & Ta, 2004). Therefore, we propose the combination of the extended DICOM syntax with JPEG2000 image streaming.

4. CONCLUSON

We presented transparent extensions to standard DICOM syntax which enable streaming of medical images over DICOM networks. This enables fast browsing of medical images inside DICOM networks (even over low-band networks) and viewing of high-quality low-resolution medical images on limited size display devices. JPEG2000 compression combined with the DICOM extension represents the best solution for streaming of medical images. JPEG2000 reduces PACS storage demands more than 10 times and supports image streaming with minimal spatial distortion. Contrary to similar solutions, this extension offers transparent, efficient and simple way for achieving streaming of medical images over DICOM networks. Streaming mechanism is an integral part of DICOM network. Our future research is focused on an implementation of a hospital PACS based on the extended DICOM syntax and JPEG2000 compression, which will offer high quality medical images to physicians wherever they are, on devices they have.

Acknowledgements. This research was supported by IT Project No. 13013, financed by the government of Republic of Serbia.

5. REFERENCES

Dragan, D. & Ivetić, D. (2005). DICOM overview, *Proceedings of the VII International Symposium "Young People and Multidisciplinary Research"*; Zaberca, V. (Ed.), pp. 105-115, ISBN 1453-7394, Resita, Romania, Sep 2005.

Mirkovic, J.; Ivetic, D. & Dragan, D. (2007). Presentation of Medical Images Extracted From DICOM Objects on Mobile Devices, *Proceedings of the 9th International Symposium of Interdisciplinary Regional Research "ISIRR 2007" Hungary – Serbia – Romania*, Novi Sad, Serbia, Jun 2007. In Press 2008.

Moshfeghi, M. & Ta, J. (2004). Efficient Image Browsing with JPEG2000 Internet Protocol. *Medical Imaging 2004: PACS and Imaging Informatics, Proceedings of SPIE 2004*, Vol. 5371 No. 08, pp 31-42, ISSN 0277-786X.

Skodras A. N. (2006). The JPEG2000 Image Compression Standard in Mobile Health. In: *M-Health: Emerging Mobile Health Systems*, Istepanian R., Laxminarayan S., Pattichis C. (Ed.), pp. 313-327, Springer US, ISBN 978-0-387-26558-2.

Taubman, D. & Marcellin, M. (2001). *JPEG2000: Image compression fundamentals, standards and practice.* ISBN 978-0-7923-7519-7, Boston, Kluwer Academic Publishers, 2001.

Taubman, D. & Prandolini, R. (2003). Architecture, Philosophy and Performance of JPIP: Internet Protocol Standard for JPEG2000. *Available from:* www.ee.unsw.edu.au/~taubman/publications_files/JPIP-architecture-vcip03.pdf *Accessed:* 2008-04-23

Tian, Y.; Cai, W.; Sun, J. & Zhang, J. (2008). Accessing medical image databases on the go. *Available from:* http://spie.org/x19505.xml?highlight=x2416 *Accessed:* 2008-02-26.

Annals of DAAAM for 2008 & Proceedings of the 19th International DAAAM Symposium, ISSN 1726-9679
Editor B. Katalinic, Published by DAAAM International, Vienna, Austria 2008
Make Harmony Between Technology and Nature, and Your Mind will Fly Free as a Bird

ASPECTS OF GREEN BUSINESS DEVELOPMENT

DRAGHICI, A[nca]; IZVERCIANU, M[onica] & BARGLAZAN, D[iana]

Abstract: *The paper presents some aspects regarding the last corporate environmentalism evolution and its link with the companies' competitiveness. The green business practices can be support by trining and consultancy activities, as they are offer through the WEBT&C project in Romania. An integrated web platform will be designed to offer training and consultancy support activities, in the field of green business, for the SMS eco-efficinecy evaluation and improvement.*
Key words: *environmentalism, web platform, green business, company competitiveness.*

1. INTRODUCTION – THE CORPORATE ENVIRONMENTALISM EVOLUTION

As corporations move into the first decade of the new millennium, many expect them to work toward a full integration of economic and environmental goals by incorporating those goals in new products and resources as they are designed and built. The evolution of corporate environmentalism can be summarized as in the following:

1960 - Industrial environmentalism - internal problem solving;
1970 – Regulatory compliance - compliance with government regulations;
1980 – Social responsibility - impact reduction and protection of reputation;
1990 – Strategic environmentalism - recycling and waste reduction;
2000 - Sustainable environmentalism - integration of business goals and environmental goals.

Environmentalism is becoming a strategic issue and is being incorporated at the insurance liability, process and product design, and competitive strategy levels. Corporations seek managers and engineers who knows the environmental implications of the job, not an environmental specialist who knows the implications of environmental management marketing and human resources. These managers and engineers will be called on to integrate environmental issues with physical design and process factors (Lyon & Maxwell, 2004).

2. THE RELATION BETWEEN GREEN FIRMS AND COMPETITIVENESS

Companies operate in the world of dynamic competition. They are constantly creating innovative solutions to concerns raised by customers and regulators while responding creatively to competitive pressures and actions (Porter, 2001). Based on a reference research, we propose five attributes that reflect a company's commitment to invest in pollution prevention and green products: *green consumers - a* green firm invests in understanding the green demands of its customers and the market niches for green products; *technology investment - a* green firm invests in technology to prevent and clean up the pollution that its facilities generate; *materials choice* - a green firm is willing to increase product costs a small amount – approximately 5 % - to prevent toxic discharges; *redesign* - a green firm is willing to invest a portion of its product revenues - again, 5 % seems reasonable - toward a redesign that lowers pollution; *sustainability* - a green firm attempts to lower its use of resources to leave more for future generations.

In an increasingly competitive marketplace, firms must be able to benefit from their *green actions*. Often, companies can attain productivity increases from an innovation that simultaneously uses fewer resources and emits less waste. When environmental factors are internalized within a firm's business strategy, the firm can strive to generate positive organizational and technological benefits that accrue privately to the firm, thus, increasing its competitive advantage (Cunha & Putnik, 2005).

Companies are addressing the concept of pollution (waste) prevention and resource input reduction or materials substitution. Companies view as attractive the idea of moving toward high quality by means of reducing defects. In the same way defects can be reduced by quality product and process design. Many companies are using innovations to command price premiums for "green" products. Few companies have fully analyzed the true cost of wastes, toxicity, and resource use. As they trace the full cost up and down the supply chain they will find opportunities to reduce costs and environmental effects.

The first step is *pollution prevention* (Christmann, 2000). Pollution prevention focuses on reducing or eliminating waste before it is created. Pollution prevention pays off with reduced costs and better processes. *Product stewardship* focuses on all environmental impacts associated with a products life cycle. Step 3 is concerned with the *design for environment*, which is a method for making products easier to recover, reuse or recycle. Step 4 is concerned with *clean technology*, which focuses on creating technologies that are less impacting and more sustainable.

In addition, ten principles of sustainable business practices are essential for modern companies' green business: protect the biosphere and biodiversity through reducing pollution; promote sustainable use of natural resources; reduce waste; conserve energy; reduce environmental, health, and safety risks; design safe products and services; restore environmental damage; inform the public of environmental conditions; ensure management commitment; self-audit annually. This is a summary of the Coalition for Environmentally Responsible Economics (CERES) principles endorsed in 1992 by several leading global corporations (www.ceres.com).

3. THE GREEN BUSINESSES CONSULTANCY IN ROMANIA

According with statistics data, Europeans attach an overwhelming importance to protect the environment. During a research conducted at the end of 2007 by the European Commission in all EU member states (a special Eurobarometer on the environment), almost all the respondents (citizens of the EU) said that environment protection is *very* important to them (64%) or *fairly* important (32%). In total, 96% of European citizens believe environmental issues are important to them (European Commission, Special Eurobarometer, March 2008, p.11). Respondents who say that environmental protection is *very* important outnumber those who say it is *fairly* important in every country except Finland. There are very few people who belive that environmental protection is not important: the highest number of respondens were recorded in Austria and Romania (8% both), while in most European countries this percentage was less than 3%. Europeans tend to see environmental protection as a *joint responsibility*: 90% of them agree that the primary responsibility should lie with the biggest polluters, while 86% believe that they, as individuals, can play a role in protecting the environment in their countries. The most realistic interpretation that can be offer is that Europeans think that everybody can do something to protect the environment but they are still more inclined to support the "polluter pays principle", based on the idea that different *actors* should be held responsible in proportion to the environmental damage they cause.

In order to help companies transform their businesses into green businesses, and, by doing so, to reach eco-efficiency, innovative forms of training and/or consultancy systems were conceived and realized in the last year in Romania, started by a national initiative in the West part of the country. A good example in the field is the eco-design platform developed in Austria (www.ecodesign.at) – a web platform that offer: general information for product ecological design; courses; references; legislation and other links to service suppliers; examples of good practices; events; e-learning in ecological design; on-line questionnaire for a convenient optimization strategy identification for products, based on a checklist. The platform was developed by a consortium in which one of the main partners is the Technical University of Wien. Other examples of good practice can be found in: Holland (www.greenfacts.org), a web portal for non-specialists general information on health and environment; Australia (www.ecobusiness.com.au), web portal with information for sustainable business change, mainly information regarding capabilities, projects, examples of good practices; Columbia (www.tecnesoft.net), a company portal that offer training and consultancy in the field of information technology; Greece (Kritikou Y, et. al., User Profile Modeling in the context of web-based learning management systems, Journal of Network and Computer Applications, vol 12/6, 2007), with an e-learning platform development modality, based on a requirement and user behavior model; Romania (www.advancedelearning.com), where Siveco Romania has developed and implemented AeL, an universal e-learning platform for schools (480 beneficiaries), business environment and administration (general and specific elements for general qualification).

Started from these minimal, but representative reasons, Politehnica University of Timisoara, in Romania launched an initiative - the ***WEBT&C project*** developed since 2008 - aiming at developing multidisciplinary integrative researches that will offer an effectiveness, innovative system not only for training (based on web methods, tools and videoconferences), but also, for consultancy dedicated to the business environment with impact on competitiveness and sustainable development assurance. This objective can be achieved by building an integrated platform (web portal type) for managerial training and consultancy, adapted to the needs expressed by the companies (and identified by the project partners), for favorable strategies and policies elaboration, in the context of ecological modernization, as well as for the creation/implementation of the organizational culture focused on the environment preservation (in the terms of eco-efficiency and environment responsibility).

The ideas behind the project are: to develop frameworks to understand how organizations can strike a balance between business and environmental objectives while managing complex stakeholder relationships, and to build strategies to gain competitive advantage through environmentally sustainable practices, including product and process innovation and sustainable supply chain management.

The main objectives of the project are the following:

(1) The development of a partnership between different research entities (universities, a company and a research institute), to unroll multidisciplinary research activities and for the innovative mechanisms creation for the results implementation in the real business environment;

(2) The creation of an integrative, innovative and effectiveness system (virtual centre for formative and informative associated with a web portal, correlated with a collaborative learning system using (videoconference and an interactive tool for products/processes eco-efficiency on-line evaluation) for support processes with the business environment destination;

(3) National and international visibility growth for the project team and the research results, through publications (scientific papers published in national/ international conferences proceedings and national/ international journals) and through the implication in the European association EMIRAcle (www.vrl-kcip.org) activities for participation in the FP7 or other European projects.

4. CONCLUSION

There is a clear trend of adding an ecological dimension to any business that wants to be competitive. Especially in Europe, environment protection is seen as extremely important, and the main actors responsible for it are considered to be the companies. Eco-efficiency is a component of the companies' efficiency, whose importance is continuously growing. Soon, eco-efficiency will be just as important as the economic efficiency, for every company trying to stay competitive on a global market, in the long-run. There are several examples of successfully promoting eco-efficiency among companies, in different countries (Michelsen, 2006).

Since 2008, Politehnica University of Timisoara, together with 5 other partners, is creating a web platform meant to encourage Romanian companies to transform their business into a green one. We strongly believe such a platform is needed in Romania because while Romanians are thinking just as the other Europeans, that environment protection is extremely important, and the companies are the first actors who should assume responsibility for it, in our country the level of information in the field of environment is the lowest in all the European Union. Therefore the project is filling up (partially) a very large gap that currently exists between what Romanians know and what they want and need to know in order to act responsively in the field of environment policies (Draghici, 2008).

5. REFERRING

Christmann, P. (2000). Effects of "Best Practices" of Environmental Management on Cost Advantage: The Role of Complementary Assets, *The Academy of Management Journal*, Vol. 43, No. 4 (August, 2000), ISSN 00014273, 663-680.

Cunha, M., M. & Putnik, G., D. (Eds.) (2005). *Advanced Technology for Business Integration: Social, Managerial and Organizational Dimension*, Idea Group Publishing, ISBN 1-59904-048-4.

Kritikou Y, et. al. (2007). User Profile Modeling in the Context of Web-Based Learning Management Systems, *Journal of Network and Computer Applications*, vol 12/6, 35-42.

Lyon, Th. & Maxwell, J. (2004). *Corporate Environmentalism and Public Policy,* ISBN-13 9780521603768, Cambridge University Press.

Michelsen O, et. al (2006). Eco-efficiency in Extended Supply Chains, in *Journal of Environmental Management*, Volume 79, Issue 3, May 2006, 290-297

Porter, M. E. (Ed.) (2001). *Competitive Strategy. Techniques for Analyzing Industries and Competitors*. ISBN 9780684841489, New York: Free Press.

Annals of DAAAM for 2008 & Proceedings of the 19th International DAAAM Symposium, ISSN 1726-9679
Editor B. Katalinic, Published by DAAAM International, Vienna, Austria 2008
Make Harmony Between Technology and Nature, and Your Mind will Fly Free as a Bird

INTELLIGENT SYSTEM AND SOLUTIONS FOR KNOWLEDGE MANAGEMENT IN VIRTUAL RESEARCH TEAMS – ONTOLOGY AND EXPERTISE MAP

DRAGHICI, A[nca]; MOLCHO, G[ila] & DRAGHICI, G[eorge]

Abstract: *The paper presents the main steps in buiding the ontology in a virtual global team. This intelligent system and the associate solution for knowledge management was developed to enable knowledge sharing and reuse. The objectives of the presented approach are: (1) to ensure a common understanding of specific terms describing members' fields of expertise and relevant researches; (2) to provide the structure of the organization's knowledge map. The goal of the knowledge map was to enable explicit charting of member expertise to clearly define and locate experts within the network and to develop a concise core competency depiction.*

Key words: *ontology, virtual teams, competencie, knowledge management, knowledge map.*

1. INTRODUCTION

Misunderstandings between distributed team members and faulty translations of software applications contribute to the rising costs of interoperability in virtual, distributed organizations. Indeed, the growing implementation of distributed software agents necessitates developing and adopting a shared terminology and syntax for efficient and effective interoperability. Ontology offers a solution for solving the interoperability problems brought about by semantic obstacles, that is, obstacles related to definitions of business and scientific terms and software classes. Ontology is taxonomy of concepts and their definitions supported by a logical theory. It is often captured in the form of a semantic network - a graph whose nodes are concepts or individual objects and whose arcs represent relationships or associations among the concepts (Huhns & Singh, 1997). Ontologies may differ not only in their content but also in their structure and implementation. Various methodologies exist to guide the theoretical approach taken, and numerous ontology-building tools are available. The problem is that these procedures have not coalesced into popular development styles or protocols, and the tools have not yet matured as in other software practices. However, ontology is typically built in more or less the following manner (Denny, 2002): acquire domain knowledge; organize the ontology; check the work; commit to the ontology.

Based on these considerations and perspectives, the present paper will outline the detailed approach used to build the ontology and complementary knowledge map for a particular virtual organization, the Virtual Research Laboratory for a Knowledge Community in Production (VRL-KCiP), a Network of Excellence (NoE) established in the context of the 6th Framework Programme (www.vrl-kcip.org).

2. ONTOLOGY IN THE VRL-KCIP NoE

The central aim of the VRL-KCiP NoE is to create synergy by integrating the research expertise and capabilities of the different member teams to support product life cycle engineering in the modern manufacturing environment. Hence, knowledge sharing and collaborative research constitute the core competency and potential for the network's success and the essence of its existence. Among the central activities required to fulfill the VRL-KCiP vision are: (a) building an ontology with the purpose of generating a common reference language among the member teams that can overcome differences in culture, location, language, and fields of expertise; (b) implementing a central knowledge management system (KMS) that will allow expertise identification and knowledge-sharing capabilities; (c) implementing IT-enabled one-to-one or many-tomany communications capabilities to complement the face-to-face meetings of the distributed network. Because of the potential major impact of ontology development on the success of the network, significant effort was invested in completing this task efficiently and effectively.

Ontology-building focuses on what the ontology is required *for* (Gruber, 1993). The VRL-KCiP ontology was developed to enable knowledge sharing and reuse. Initially, the ontology had two objectives: (1) to ensure a common understanding of specific terms describing members' fields of expertise and research relevant to state-of-the-art life cycle engineering; (2) to provide the structure of the VRL-KCiP knowledge map. The goal of the knowledge map was to enable explicit charting of member expertise to clearly define and locate experts within the network and to develop a concise core competency depiction (Molcho, 2008).

The need for the VRL-KCiP ontology was magnified by the nature of the network—a virtual multilingual, multidisciplinary, multicultural dispersed research team, researching the state-of-the-art in the vast field of life cycle engineering that, contrary to most virtual enterprises, did not evolve gradually from a central core but rather emerged as a *fait accompli*.

During the process of developing the ontology it rapidly became evident that in addition to the points outlined above, the ontology would provide the structured context required to cultivate a high quality knowledge base for capturing, accessing, archiving, and validating knowledge objects in the VRL-KCiP knowledge management system (KMS). The following discussion focuses on the main stages in achieving the above goals.

2.1 Goal and methodology definition

Ontology definition is an art. Therefore, compromises had to be made. As a result, although the goals of building the ontology were clearly defined in the initial O stages of the network of excellence (NoE), opinions differed regarding which methods should be used to best achieve the goals, and many concerns were raised: (1) ontology construction is not yet well understood; (2) the size and complexity of the research domain is large; therefore, care must be taken to clearly define the scope; (3) there is no single correct methodology for ontology building.

In July 2004, a special knowledge management working group met in Troyes, France to establish vertical integration among tasks related to knowledge and to begin creating the network's ontology. A bottom-up approach was adopted, based on input from network members. An initial brainstorming session was held to establish the basic structure and instances of the VRL ontology.

2.2 Developing the ontology

A top-down approach was applied to define the ontology structure and determine its initial levels. At the highest level member expertise were divided into: (a) life cycle-related knowledge and (b) product-specific knowledge. At the next level, the life cycle-related knowledge was further detailed to specific life cycle stages (Design, Manufacture, Service, and End-of-Life (EOL)). These life cycle stages were then divided into substages. Next, emphasis was placed on collecting (1) approaches, (2) methods, and (3) tools. *A bottom-up approach* was then applied to explicate further levels of detail and gather instances and documents for each type of expertise. The ontology - both structure and content - was then further developed through iterative steps of collecting, analyzing, brainstorming, revising, and redistributing for further feedback. This process continued until a relatively stable ontology structure was formulated (Van Heijst, et. al., 1997).

2.3 Collecting expertise profile

Once the ontological structure was more or less defined and stabilized, the form was again distributed to all VRL-KCiP team members. To date, 250 responses have been received and entered into a collective knowledge base. Many new instances have been added to the basic structure, as members sought to define their personal expertise. The first stage of expertise collecting did not incorporate differential rating of personal expertise (Molcho, 2008).

2.4 Creating the VRL-KCiP knowledge maps

In accordance with the ontology structure and the responses from network members, all feedback was entered into a common database. This database made possible to map the expertise of the individual network members as well as to combine the input from individuals from each lab with the fields of expertise available in each lab. Four knowledge maps have been built: (a) individual expertise range, (b) individual expertise, (c) lab expertise strengths, (d) lab expertise. They were built by assigning the value 1 to all expertise fields relevant to each network member (Molcho, 2008).

2.5 Completing ontology instance profiles

To give the VRL ontology added value as well as added dimension and depth, a profile structure was defined for each instance. The profile for each field of expertise aids members in understanding the context of each instance. It also provides a basic introduction to the topic. Work is currently underway to complete and validate a profile for all instances in the ontology. Over 150 profiles have already been incorporated in the VRL-KCiP KMS. Each profile provides a definition of the instance, a short description, and good references to further information. It also provides a link to a more detailed description that outlines strengths, weaknesses, complementary tools, applications, and so forth.

2.6 Ongoing ontology updating

Since ultimately there is no correct ontological structure (each proposition has its benefits and drawbacks) and since a platform must be in place to initialize joint ventures and research, we have refrained from major changes in the structure. Nevertheless, the ontology continues to evolve for a number of reasons. The first reason for ontology evolution is determined by *the upper part of the tree* (product life cycle). The expertise maps indicate a lack of balance between the level of detail of the *design* phase, which is the most explicit, and the *service* and *EOL* phases, which lack detail. This imbalance appears to mirror the fact that the strength of the VRL-KCiP lies in the design phase (design approaches, methods, and tools), whereas the network lacks expertise in the service and EOL phases so that the structure is sparsely populated in these areas. More effort must be invested in further detailing the service and EOL product life cycle stages. One of the ways to further detail these areas is by applying bottom-up methodologies more common in ontology development (topic mapping or text mining methodologies). This work, currently underway, involves mining member CVs and current areas of research being collected on the central KMS (Draghici & Draghici, 2005). The second reason for ontology evolution is determined by *the lower part of the tree* (products section); the products section of the tree will be built from the bottom up, and branches are likely to be added as new members with expertise in specific product types join the network. New instances are added to the ontology as new members join the network, new fields of research evolve, and research projects begin. Hence, the bottom-up process of expanding the tree to include new fields of research relevant to the network and new tools or methodologies developed within the labs will continue. The ontological structure will expand further, both in depth (further detailing of existing branches) and in breadth (by introduction of additional fields of expertise).

2.7 Expertise level differentiation

On the first competence profile, some members filled in only those instances in which they were very highly knowledgeable, whereas others filled in all the instances in which they had basic knowledge. This discrepancy, apparently due to participant personality differences, resulted in an unbalanced picture for understanding lab capabilities. Hence, a differential rating was required. For each direct instance (leaf in the tree structure), the user can indicate his/her appropriate level of expertise (familiar, novice user, experienced user or teacher, and innovator or developer). Based on these results, the process of expertise differentiation will continue lab by lab until the level of expertise is mapped as well.

3. CONCLUSIONS

After almost two years of development, the ontology currently fulfills four major *purposes* in the network: (1) as a reference for a common understanding of terms in the fields of research relevant to the VRL-KCiP; (2) for collaboration definition and initiation; (3) as one of the indexes for the dual-index KMS; and (4) as the coordinates of the VRL-KCiP knowledge map describing the current expertise of each member in the network, thus representing its intellectual capability and core competence. Although the VRL-KCiP KMS is the core of the EMIRAcle association development (it will be continue up-date and enriche) as a fesable partner for industrial solutions in research and innovation (www.vrl-kcip.com).

4. REFERENCES

Denny, M. (2002). Ontology building: A survey of editing tools. Retrieved November 2, 2007, *Available from*: http://www.xml.com/pub/a/2002/11/06/ontologies.html, *Accessed*: 2008-05-13.

Draghici, A., & Draghici, G. (2005, November). *Building a knowledge share culture in VRL-KCiP NoE.* Presentation at the VRL-KCiP Knowledge Management Workshop, Nantes, France.

Gruber, T.R. (1993). A translation approach to portable ontologies. *Knowledge Acquisition, 5*(2), 199-220.

Huhns, M.N., & Singh, M.P. (1997). Ontologies for agents. *IEEE-Internet Computing, 1*(6), 81-83.

Molcho, G. (2008). Formulating an expertise map in the VRL-KCiP. In Bernard, A., Tichkiewitch, S. (Eds.), *Methods and Tools for Effective Knowledge Life-Cycle-Management*, Springer, pp. 169-184.

Van Heijst, G., Schreiber, A., & Wielinga B. (1997). Using explicit ontologies in KBS development. *International Journal of Human Computer Studies, 46*, 22-35.

Annals of DAAAM for 2008 & Proceedings of the 19th International DAAAM Symposium, ISSN 1726-9679
Editor B. Katalinic, Published by DAAAM International, Vienna, Austria 2008
Make Harmony Between Technology and Nature, and Your Mind will Fly Free as a Bird

VIRTUAL ENTERPRISE NETWORKS & VIRTUAL TEAMS: A BUSINESS MANAGEMENT APPROACH

DRAGOI, G[eorge]; COTET, C[ostel] E[mil]; DRAGOMIRESCU, C[ristian] & PARPALA, R[adu] C[onstantin]

Abstract: *This article will give the example of a specific market in which innovation in ICT technologies will allow to efficiently support new concepts of collaborative working. This system takes a top-down approach in the integration of Business System Support (BSS) and Operational System Support (OSS). The paper will feature application of service-oriented architecture applied to business-processed interaction across collaborative SME's in virtual enterprise environment. The core components of the architecture are described and evaluated in real organizational contexts.*

Key words: *Virtual enterprise, virtual networks, e-business, service-oriented architecture.*

1. INTRODUCTION

The ability of companies to form global networks on the Internet will transform the way that we do business in the future. Service-oriented computing is becoming the prominent paradigm for leveraging inter-enterprise information systems to complete higher-order business transactions at the heart of the modus operandi of the virtual enterprise (VE). This paper describes research aimed at supporting the formation and operations of virtual enterprises through the setting-up of service-oriented workspace environments. Today, the Challenge in the world is of the change agents in the countries and through its technological and social components, has among one of its objectives, to accelerate the digital economy's development process in businesses, especially the micro, small and medium ones, for increasing the competitiveness of economy; as well as develop a digitalization culture in society, particularly in the consumers (Dragoi, 2005).

As the e-economy evolves, we must reexamine our beliefs about what can be achieved. New ideas and consumer demand trigger new business opportunities, which proliferate so rapidly that three months in the e-economy is considered the equivalent of a full year in a traditional business development cycle. More than two thirds of businesses now have a web presence that offers product or catalog information for customer convenience. However, few organizations have really taken the trouble to find out what their channel partners need in terms of rapid information deployment to better manage the supply chain.

To grasp how the e-economy will affect your business, you need to think about how web technology is going to transform your relationship with your customers. During the next years, the supply chain model of the e-economy will bear little resemblance to today's model. Competitive forces and new technology will continuously alter the shape and flow of commerce (IBM, 2005). First, most ERP systems are, in the very nature of their design, focused on the manufacturing link of the supply chain. Yet, a fundamental shift in thinking is beginning to concentrate on the demand side of the supply chain. The critical difference: demand chain thinking starts with the customer and works in reverse. The new approach acknowledges that customers have individual needs and service requirements. Another problem with most ERP business models is their inability to work seamlessly within an organization to handle some of the critical links of the supply chain process. The net result of awkward integration into other business systems within the supply chain has created a new industry of Integration Service Providers (ISPs) that link ERP systems. New ways of sharing information are gaining acceptance. We propose to the above issues is the systematic study of nature's most successful living teams and the extraction of principles about their operational logistics, behavior patterns, command-structure, communication methods that can provide us with useful guidelines on how teams need to operate to be truly successful. It is our goal to analyze these principles and to see where and how some of the teaming principles can be transferred to the areas of organizational design, leadership, online collaboration, business networking, and human resources management. The solution is a virtual enterprise network (VEN):

- A way for businesses to achieve virtual scale enabling them to operate as if they had more resources and capacity than they actually have;
- Allowing them to operate with all the resources and reach of a large enterprise but without sacrificing their speed, agility and low overheads;
- Enabling them to compete for bigger more profitable contacts with higher innovation and design elements with bigger customers who are prepared to have strategic partnerships with their suppliers (Rosu & Draghici, 2008).

2. VIRTUAL TEAMS & VEN

Virtual collaboration for networked business teams is a complex and challenging activity in which there are major important components to be accounted for. Virtual business teams do not operate like traditional physical teams, as their requirements reflect a completely new way of communicating, working collaboratively, sharing information and mutually supporting other team members. A Virtual Enterprise Network needs its own Private Member Collaboration System to communicate and develop its projects and bids. It needs its distinctive Network Business Applications such as Capability Aggregation and Tender Matching to enable it to function effectively as a co-operative in both pre-sales and contract operations. It also needs a Public Web Site to manage its interactions with potential customers and new members(Rosu & al., 2007). In this context, the basic ingredients of VEN are presented in the figure 1.

The new technologies and approaches required to achieve this are completely alien to most of our present organizational culture. Moreover, this is why they fail. Cooperative processes are not the automatic results of implementing collaborative, real-time communication technologies, but the result of a carefully designed and systematically maintained virtual team development plan.

For those of you who have already exposed themselves to the positive advantages made available by the use of cutting-edge communication and collaboration technologies, this should sound as a familiar melody.

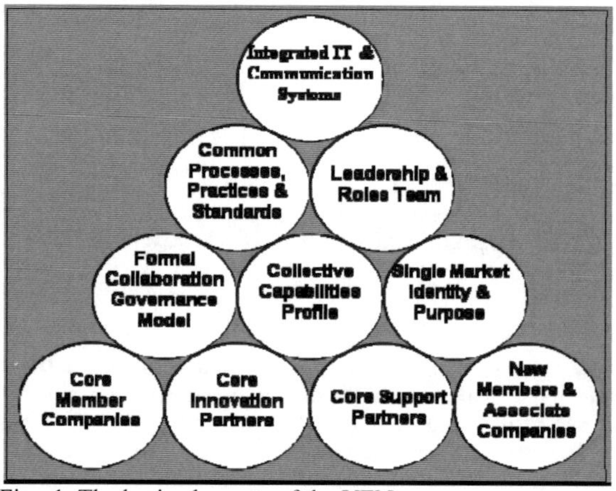

Fig. 1. The basic elements of the VEN

Why is there so much disjoint between technology potential and the productive use that business team members make of them? (Dragoi, 2005). If the solution is not in the technology enabling such networked business teams to easily interoperate, where is it then? The ideal solution framework suggests making a systematic pragmatically reference to the key interdependent components of a successful virtual business team - each of which must be set-up correctly and then kept in constant equilibrium as the team evolves and produces results.

3. A BUSINESS PROCESS MANAGEMENT APPROACH

The term Business Process Management (or BPM) is a set of activities that organizations can perform to either optimize their business processes or adapt them to new organizational needs (Botezatu & al., 2006). As software tools usually aid these activities, the term BPM is synonymously used to refer to the software tools themselves. BPM activities are process design, process execution, and process monitoring. Business driven development starts with modeling, then goes through implementation and testing, deployment to a server, and monitoring of the executing processes. BPM is about service assembly, where applications are assembled from components. WebSphere Business Modeler can generate them, or they can be developed using tools such as Rational Software Architect and Rational Application Developer (IBM, 2005).

We start with modeling the current system using WebSphere Business Modeler V6. In this phase, we create all the resources needed by our application like timetables, human resources, data arrays, etc. We also define simple business measures such as key performance indicators (KPIs) which we want to verify in the final application trough monitoring.

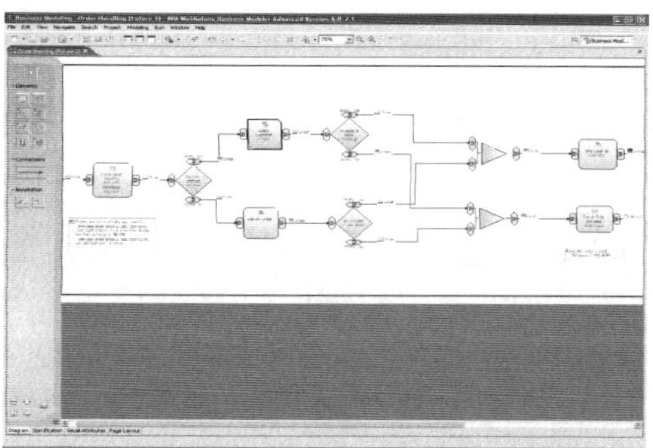
Fig. 2. Business modeler process diagram

We export the process model for implementation. In the second phase, we implement the process model as an enterprise application in WebSphere Integration Developer. This can be done even with minimal programming knowledge. We have to complete certain tasks by developing business rules, Java code, and a user interface for human tasks. The interface is very friendly and a basic GUI (Graphic User Interface) is created. Those default pages created for each human task can be further customized in order to facilitate the end user control.

We export the application for deployment into the Process Server. As part of development and testing we use the Monitor Development Toolkit, a plug-in in WID (WebSphere Integration Developer), to define the business measures (KPIs, alerts). We test the basic functionality of this monitor model using a Monitor Server test environment (see figure 2). In the third phase, we run the application in WebSphere Process Server. During the execution, events are captured and sent for monitoring.

In the fourth phase, we deploy the monitor model to the WebSphere Business Monitor, which then processes the events of the Process Server. The Monitor Dashboard is used to display the business measures and KPIs to a business analyst. In the fifth phase, we use the results of monitoring to compare the results with the model. Based on the analysis we modify the model and start a new cycle of implementation, deployment, and monitoring. This way even complex business models can be analyzed and optimized before they are implemented in real life situations. The process integration portfolio provides capabilities required for the delivery of the comprehensive enterprise wide business process management strategies and solutions. It offers a holistic approach to transform and manage a business by aligning strategic and operational objectives with business activities and supporting IT services.

4. CONCLUSIONS

The presented solution is based on ORGVIRT platforms that defines a standard for developing multi-tier, portable and platform neutral enterprise applications based on the technology of Web Services. Striving to discover the main problems of collaborative product data management, the theoretical backgrounds in the particular research area were analyzed, and interviews with the members of the design departments of SMS companies were carried out.

5. REFERENCES

Botezatu, C.P., Botezatu, C. & Căruţaşu G. (2006). The development and the implementation of the integrated management system, *Annals of DAAAM for 2006 & Proceedings of the 17th International DAAAM Symposium*, pp. 053-054, 2006, ISSN 1726-9679, ISBN 3-901509-57-7

Dragoi, G., (2005), Informational Infrastructure of modern enterprise (*Infrastructura informaţională şi de comunicaţii a îmtreprinderii moderne)*, Editura POLITEHNICAPRESS, ISBN 973-8449-73-1, Bucuresti, Romania.

IBM Software Development Platform, (2005), *IBM Rational Software solutions*, IBM Corporation, Bucharest, Romania

Roşu, S. M. & Draghici, A. (2008). *Information system risk estimate as part of project development by virtual teams using expert systems,* Academic Journal of Manufacturing Engineering, no. 2/2008, pp. 135-142, ISSN 1583-7904.

Rosu, S. M.; Draghici, A. & Guran, M. (2007). *Knowledge Transfer in the Enterprise Business Intelligence*, Annals of DAAAM for 2007 & Proceedings of the 18th International DAAAM Symposium, October 24th - 27th, B. Katalinic (Ed.), published by DAAAM International, Vienna, Austria, pp. 647-648, ISBN 3-901509-58-5.

Annals of DAAAM for 2008 & Proceedings of the 19th International DAAAM Symposium, ISSN 1726-9679
Editor B. Katalinic, Published by DAAAM International, Vienna, Austria 2008
Make Harmony Between Technology and Nature, and Your Mind will Fly Free as a Bird

DIESEL ENGINES POLLUTION AND FUNCTIONAL-CONSTRUCTIVE PERFORMANCES COMPROMISE OPTIMIZATION, USING FUZZY SETS

DRAGOMIR, G[eorge]; BELES, H[oria]; BLAGA, V[asile]; STANASEL, I[ulian]; NEGREA, V[irgiliu] D[an] & BLAGA, F[lorin] S[andu]

Abstract: *This paper presents a method for using fuzzy functions systems in order to research the improvement potential of the automobile Diesel engines, in correlation with the adaptation to the travelling conditions, reduction of fuel consumption and the control of noxious gas emissions.*
We study the influence of the admission / exhaust valves opening / closing times modifications and of their movement laws upon the engine traction performances, the effective specific fuel consumption and the nitric oxides NOx concentration.
Key words: *Diesel engines, valves, fuzzy, optimization.*

1. INTRODUCTION

1.1 Creating the fuzzy study model of the diesel engine thermo-gaseo-dynamic

By means of Diesel engines thermo-gaseo-dynamic model, we studied how the variations of the distribution phases and of the valves lift moving laws affect the traction, fuel consumption and NOx gases emission performances. The steps that we followed in order to elaborate the fuzzy-systems-based calculus model are described next.

Setting of the input quantities (criteria) according to which the determination of the technical, economic and NOx gases performances will be carried out. The input quantities are the following:

1. Opening advance of the admission valve (DSA);
2. Closing delay of the admission valve (ISA);
3. Opening advance of the exhaust valve (DSE);
4. Closing delay of the exhaust valve (ISE).

A variation range is associated to each input quantity, within which its specific values can be found. These variation ranges are:

$$\begin{aligned} &DSA: D_{DSA} = [L_{DSA}^{inf}, L_{DSA}^{sup}] \\ &ISA: D_{ISA} = [L_{ISA}^{inf}, L_{ISA}^{sup}] \\ &DSE: D_{DSE} = [L_{DSE}^{inf}, L_{DSE}^{sup}] \\ &ISE: D_{ISE} = [L_{ISE}^{inf}, L_{ISE}^{sup}] \end{aligned} \quad (1)$$

where: L^{inf}, L^{sup} are respectively the inferior and the superior limit of the variation range associated to the input quantity.

A linguistic variable is associated to each input quantity. Thus the input quantities become the input linguistic variables.

Setting of the linguistic degrees associated to each input linguistic variable

Linguistic degrees (Sofron et al., 1998) or linguistic terms (Preitl & Precup, 1997) are defined for each output linguistic variable. These will "vaguely" characterize the firm information. The sets of linguistic degrees associated to each input linguistic variable will be:

$$\begin{aligned} &DSA: \ GL^{DSA} = \{GL_1^{DSA}, GL_2^{DSA}, ..., GL_K^{DSA}\} \\ &ISA: \ GL^{ISA} = \{GL_1^{ISA}, GL_2^{ISA}, ..., GL_K^{ISA}\} \\ &DSE: \ GL_j^{DSE} = \{GL_1^{DSE}, GL_2^{DSE}, ..., GL_k^{DSE}\} \\ &ISE: \ GL_m^{ISE} = \{GL_1^{ISE}, GL_2^{ISE}, ..., GL_k^{ISE}\} \end{aligned} \quad (2)$$

Definition of the evaluation process output quantities
The decisional process output quantities are:
1. The effective power (Pe);
2. The effective fuel consumption (Ce);
3. The nitric oxides concentration (NOx).
Setting the value range of the output quantities
The value ranges of the output quantities are the following:

$$\begin{aligned} &Pe: D_{Pe} = [L_{Pe}^{inf}, L_{Pe}^{sup}] \\ &Ce: D_{Ce} = [L_{Ce}^{inf}, L_{Ce}^{sup}] \\ &NO_x: D_{NO_x} = [L_{NOx}^{inf}, L_{NOx}^{sup}] \end{aligned} \quad (3)$$

Definition of the linguistic variables corresponding to the output quantities

A linguistic value (Pe, Ce, NOx) is associated to each output quantity.

Setting the linguistic degrees associated to each output linguistic variable

The linguistic degrees or linguistic terms are defined for each linguistic variable associated to the output quantities. These will "vaguely" characterize the firm pieces of information that result from the inference procedures. The sets of linguistic degrees associated to each output linguistic variable are:

$$\begin{aligned} &Pe: \ GL^{Pe} = \{GL_1^{Pe}, GL_2^{Pe}, ..., GL_K^{Pe}\} \\ &Ce: \ GL^{Ce} = \{GL_1^{Ce}, GL_2^{Ce}, ..., GL_K^{Ce}\} \\ &NOe: \ GL^{NOx} = \{GL_1^{NOx}, GL_2^{NOx}, ..., GL_k^{NOx}\} \end{aligned} \quad (4)$$

An appurtenance function is associated to each linguistic degree that corresponds to an input linguistic variable.

The linguistic variables and linguistic degrees sets, to which appurtenance functions were associated, "vaguely" characterize the firm values of the input and output quantities. The inference machine is formed by a set of rules like the following:

$$\text{IF (premise) THEN (conclusion)} \quad (5)$$

By "de-fuzzy-ing" we mean the operation of getting a "crisp" value of the output quantity, based on the appurtenance function resulted from the fuzzy inferences. From the multitude of existing de-fuzzy-ing methods (Abate & Dosio, 1990), we will use the "gravity-centre" method, which is the most widely used in practice.

1.2 Implementing of the D118 diesel engine fuzzy model

In order to apply the fuzzy model, several motion laws for the valves were previously created, which differ by means of opening advances, closing delays, process durations and the speeds and accelerations of the valves motion. For this purpose, several cam profiles were designed by means of combined polynomial and sinusoidal laws, which assure shock-free and detachment-free motion of the tappet on the cam. After choosing three different moments for each valve opening and closing, nine different motion laws were obtained, which correspond to 81 versions for the engine distribution epures.

The study of the D118 Diesel engine performance variation was carried out for three speed operating conditions: 1400, 1700 and 2000 rpm, and then constructive and functional optimization criteria were established according to the match to the road condition and pollution control.

The determinist values corresponding to the fuzzy system input and output quantities (DSA, ISA, DSE, ISE, Pe, Ce, NOx) were extracted from the output files of the program that simulates the real functioning-cycle of the Diesel engine.

The values of the valves opening advances and closing delays were used in the fuzzy system for expressing the α ^{0}RA angle, which characterizes the piston position during the engine cycle, due to the clearer visualization of the distribution phases.

The value ranges for the input quantities were chosen taking into account their minimal and maximal values, corresponding to Diesel engines with similar characteristics:

DSA= [0, 20] °RA; ISA= [20, 60] °RA; DSE= [20, 60] °RA; ISE= [0, 40] °RA; Pe= [35, 61] kW, Ce= [245, 265] g/kWh NOx= [595, 900] ppm

The linguistic variables associated to the input and output quantities are DSA, ISA, DSE, ISE, respectively Pe, Ce, NOx. (Wenxiu et al., 1999)The appurtenance function used in the de-fuzzy-ing process is triangular-shaped. The linguistic degrees associated to the input linguistic variables are: SMALL, MEDIUM and BIG.

The linguistic degrees associated to the output linguistic variables are: VERY SMALL, SMALL, MEDIUM, BIG, VERY BIG (Wolkenhauer, 1999). These were given based on the values that resulted from the evaluation stage of the real engine functioning cycle by means of the calculus program.

The inference machine consists of 81 logic rules, based on the correspondence between the linguistic degrees associated to the input and output variables.

The determinist scalar value associated to the output variable is extracted by means of de fuzzy-ing.

The 3D variation diagrams of the effective power Pe, effective specific fuel consumption Ce and nitric oxides emission NOx in terms of two distribution phase moments, obtained by de-fuzzy-ing the functions by means of MATLAB® program, are shown in the next figure.

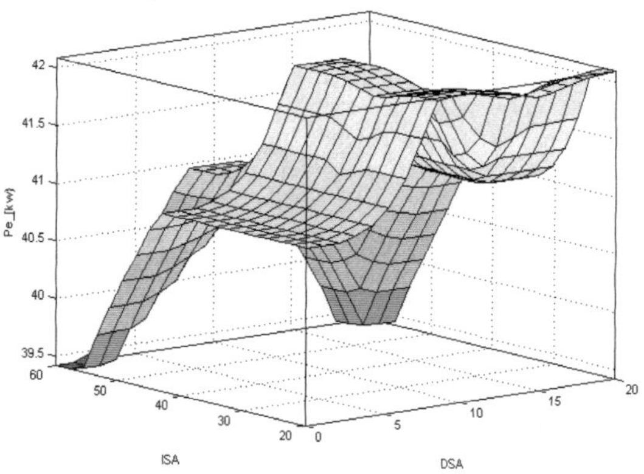

Fig. 1. Variation of the effective power Pe in terms of DSA and ISA at 1400 rpm.

2. THE OPTIMIZATION OF THE VALVES MOTION LAWS BY MEANS OF FUZZY SETS

Along with the 3D diagrams shown, other instructions can be written in MATLAB® in order to work with the files created by the fuzzy systems and return determinist values corresponding to the output quantities, for each set of input quantities from the value ranges. The values are determined by means of specific methods, using an intrinsic approximated logic and each interpolated numerical value agrees to the physical phenomena evolution logic.

The idea on which this optimization method was based consists of creating a calculus program that generates a very large number of value sets for the input quantities. Processing these value sets with fuzzy systems gives value sets for DSA, ÎSA, DSE, ÎSE, Pe, Ce and NOx, which are arranged in a value table or in a seven columns and "p^{4}" rows matrix, where "p" is a number given by the designer, that represents the evaluation points from the value range of an input quantity.

The present paper shows the attempts to set a method for determining the constructive parameters of the distribution epure which would assure the optimal engine functioning at different speed regimes. For this purpose, the following optimization criteria were adopted: getting the highest possible power, getting the lowest possible fuel consumption, getting the lowest nitric oxides NOx emission.

The optimization method took shape into a MATLAB® program which processes the data that is obtained by applying the fuzzy system based model.

3. CONCLUSION

The following conclusions can be drawn as a result of the studies concerning the engine performances by means of the above-mentioned calculus programs:

- at high speeds, in order to operate the engine with best power, fuel consumption and NOx emission performances, it is necessary that the cam profiles should assure the admission valve opening at 20°RA advance in regard to the superior dead point and its closing at 33÷40°RA delay in regard to the inferior dead point, the exhaust valve opening at 40°RA advance and closing at 20÷25°RA;
- at low speeds, in order to operate the engine with best power, fuel consumption and NOx emission performances, it is necessary that the cam profiles should assure the admission valve opening at 10°RA advance in regard to the superior dead point and its closing at 20°RA delay in regard to the inferior dead point, the exhaust valve opening at 20°RA advance and closing at 10÷20°RA.

The fuzzy functions system makes possible a complex analysis of the simultaneous influences of all the parameters of the distribution upon the engine functioning performances.

4. REFERENCES

Abate M., Dosio N. (1990), *Use of Fuzzy Logic for Engine Idle Speed Control*, SAE Paper 900594

Preitl, St. & Precup, E. (1997), *An introduction to fuzzy control process*, Editura Tehnica, ISBN 973-31-1081-1, Bucuresti

Sofron, E.; Bizon, N.; Ionita, S. & Raducu, R. (1998), *Fuzzy control systems-Modelling and computer aided design*, Editura All, ISBN: 973-9431-32-1, Bucuresti

Wenxiu C., Liguang L. et al. (1999), *Design and application of fuzzy control system of engine variable valve timing*, Vehicle Electronics Conference, (IVEC apos; 99) Proceedings of the IEEE International Volume, p. 26-29 vol.1

Wolkenhauer O., (1999), *Fuzzy Systems Toolbox for Use with Matlab*, Control Systems Centre, UMIST, Manchaster,

Annals of DAAAM for 2008 & Proceedings of the 19th International DAAAM Symposium, ISSN 1726-9679
Editor B. Katalinic, Published by DAAAM International, Vienna, Austria 2008
Make Harmony Between Technology and Nature, and Your Mind will Fly Free as a Bird
*Young Researcher and Scientist Paper / * Supervisor, Mentor*

INTENSIFICATION OF DRAWING PROCESS FROM SHEET BILLETS WITH USING OF LIQUID AND ELASTIC MEDIUM

DRESVYANNIKOV, D[enis]

Abstract: *Research of a key quality parameter – wall thinning of sheet billet – is carried out. Schemes of two processes (hydromecanical drawing and hydroelastic drawing) are presented. The schemes allow thinning to be reduced in the dangerous sections of sheet billet. Deformation at the initial stage of drawing is considered. A mathematical model is built defining dependence of thinning from main parameters of drawing.*

Key words: *hydromechanical drawing, hydroelastic drawing, sheet billet, thinning, equilibrium equation*

1. INTRODUCTION

Hydromecanical drawing and hydroelastic drawing are progressive methods of sheet-metal stamping allowing to decrease different thickness of produced parts. Thinning of the wall of a half-pnished product produced by drawing is one of the main indexes of its quality: the less deviation of the wall thickness of the initial sheet billet, the more qualitative it is.

This processes are widely used in producing of important parts from high alloy steels, aluminium alloys, high-temperature steel and others. High technical requirements to precision, equal strength, material structure and surface quality are claimed to such parts.

In hydromecanical drawing a billet and a matrix are separated by working fluid. It provides losses reduction in friction between the matrix and the billet and increase of useful friction forces along the punch which allows to produce more precise and qualitative parts of complex form in comparison with in rigid dies.

This scheme of drawing helps to lower tensile stresses in a dangerous section of a billet due to blocking of the side wall from deformation of tension; precision and quality of drawing parts are increased. Use of the given process for production of high cylindrical parts and parts of complex form allows to decrease the number of stamping operation 2-4 times in comparison with traditional drawing (Kazachenok V.I., 1978).

Hydroelastic drawing ("fluid-form") with high pressure requires only one rigid part of the stamp. A special diaphragm consisting of three layers simultaneously serves as a clamping of a sheet billet and a flexible element of the stamp. Under the influence of hydroelastic pressure of the fluid transferred by the diaphragm, the billet is deformed, taking the form of the rigid part of the die (Isachenkov E.I., 1976).

In hydroelastic drawing frictional forces acting on the surface of the membrane and billet contact have a positive influence, and frictional forces acting on the surface of the billet and clamping cantact have negative influence (Mihaylov Y.O., Dresvyannikov D.G., 2001).

The problem of drawing radius r_{cur} is very important in any methods of sheet stamping. In hydroelastic drawing the radius depends on pressure deformation. Thus, by means of pressure regulation it is possible to provide a large drawing radius in the beginning of the forming process and gradually to decrease this radius during the process.

2. WALL THINNING OF SHEET BILLET

Deformation at the initial stage of hydromechanical drawing and hydroelastic drawing is characterized by thickness change of the billet and the change of inner surface curvature. The process of deformation has a non-stationary character because relations between forces and moments are variable both in time and space. Fluid pressure increase in the stamping unit gives additional changes to the strained state of the billet.

For the analysis of stress-strained state of the billet at the initial phase of drawing was considered the stage when the billet contacts with the bending radius of the punch (Fig. 1).

At this stage when the force on the punch creates stresses in the billet flange equal to the resistance magnitude of the material to plastic deformation and contact stresses due to friction, drawing of the billet to the matrix opening and coverage of the punch radius begin. The experiments show that the greatest thinning (15%) is seen on the border of transfer of the radius punch edge into the cylindrical wall of the part. On achieving the relative depth of drawing H/d=0.2-0.3, which corresponds to full coverage of the punch radius by the billet, further thinning of this section is stopped.

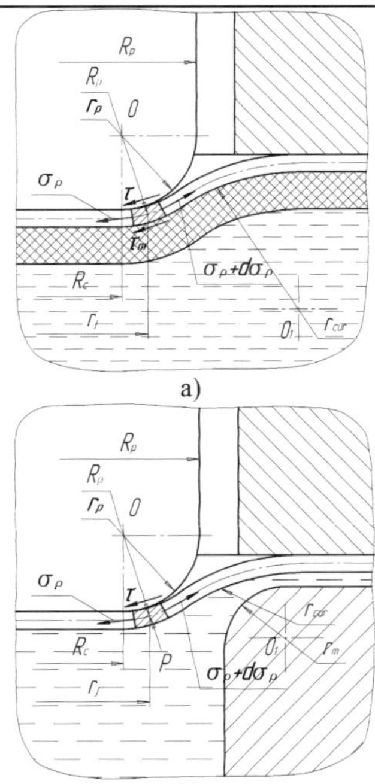

Fig. 1. Stressed-strained state of the billet at the initial stage: a) hydroelastic drawing; b) hydromechanical drawing.

Solving jointly the equilibrium equation for the initial stage of drawing, Laplas equation and coupling equation it is possible to receive the system of equations defining the change of the billet thickness and describing the process of metal deformation at the initial stage of hydroelastic drawing (equation 1-11). The given mathematical model allows to analyze peculiarities of deformation and find out the main technological parameters which influence the process of the billet thinning in hydroelastic drawing (Mihaylov Y.O., Dresvyannikov D.G., 2004):

$$h_j = (r_p + r_{cur})(1-\cos\theta_j); \qquad (1)$$

$$\varepsilon_{1j} = 1 - \frac{R_c + R_{\rho_j}\sin\theta_j}{R_c + R_{\rho_j}\Delta\theta_j}; \qquad (2)$$

$$R_{fj} = R_0\sqrt{1 - \frac{2\left(R_p + \frac{S_0}{2}\right)h_j}{R_0^2}}; \qquad (3)$$

$$\sigma_{s_j} = \frac{\sigma_t}{1-\psi}\left(\frac{\varepsilon_1}{\psi}\right)^{\frac{\psi}{1-\psi}}; \qquad (4)$$

$$R_{\rho_j} = r_p + \frac{S_0}{2};\quad R_c = R_p - r_p.; \qquad (5)$$

$$\sigma_\rho^{II} = A_i + P_0 \cdot B_i; \qquad (6)$$

$$A_i = \frac{\sigma_t}{1-\psi}\left(\frac{\varepsilon_\theta}{\psi}\right)^{\frac{\psi}{1-\psi}}\left[\ln\frac{R_{f_i}}{R_p + S_0/2} + \frac{S_0}{4r_{cur}}\right]; \qquad (7)$$

$$B_i = \frac{\mu-\mu_m}{S_0}\left(R_f - \left(R_p + \frac{S_0}{2}\right)\right) - r_{cur}\left(\frac{\mu_m + (\mu-\mu_m)\sin\alpha}{S_0}\right); \qquad (8)$$

$$\sigma_\rho^{III} = \sigma_\rho^{II}\left[1 - \frac{\mu+\mu_p}{r_{cur}}\left(\rho - \left(R_p + \frac{S_0}{2}\right)\right)\right] + \sigma_s \ln\left(\frac{R_p + \frac{S_0}{2}}{\rho}\right); \qquad (9)$$

$$\rho_j = R_0 + \left(R_p + \frac{S_0}{2}\right)\sin\theta; \qquad (10)$$

$$S_i = S_0\left(\frac{R_c}{\rho_i}\right)^{\left(\frac{2\sigma_{\rho i}^{III} - \sigma_s}{2\sigma_s - \sigma_{\rho i}^{III}}\right)}; \qquad (11)$$

where are: R_f – flange radius; S_0 – thickness of sheet billet; P_0 – pressure in matrix; μ and μ_m – coefficients of friction of metal over metal and metal over elastic membrane; σ_t – tensile strength of the material; σ_ρ – meridian stresses; σ_θ – tangential stresses.

Analogous system of equations is received for hydromecanical drawing as well (Mihaylov Y.O., 1988).

Analyze equation (11) we see that on the change of thick ness the stress influences, acting in the meridians direction that would be defined by the fluid pressure *P* in the matrix cavity equal to the clamping pressure *q*, fluid pressure *P* would be defined by the radius r_{cur} (Fig. 2).

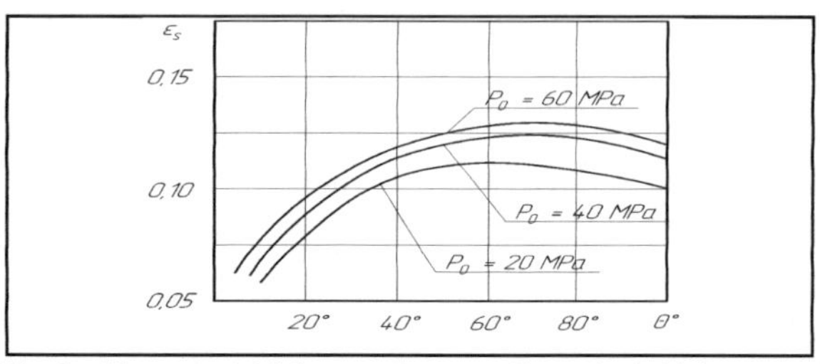

Fig. 2. Change of thickness of part walls at the initial stage of drawing.

Higher value of this pressure leads to undesirable increase of radial tensile stresses and thinning of the material in the dangerous section of the billet. With insufficient force of clamping the flange loses stability and obtained crimps could form folds that leads to decrease of the part quality or to unrecoverable reject. It means that for the decrease of thinning at the initial stage of drawing it is necessary to optimize the law of pressure change in the matrix. In hydromecanical drawing automatic regulation of lubrication pressure in the matrix is carried out depending on the drawing gap and clamping force; and in hydroelastic drawing it is possible to set the law of the change of liquid pressure by means hydroequipment (Mihaylov Y.O., Dresvyannikov D.G., 2003).

Experiments made to define thinning in hydromecanical drawing and hydroelastic drawing show that the optimum law of pressure change leads to more uniform distribution of stresses and as a consequence, to small thinning (Fig. 3).

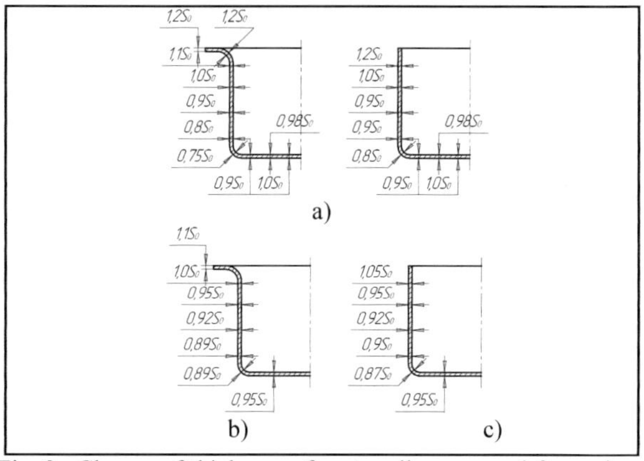

Fig. 3. Change of thickness of part walls, averaged for various materials (S_0 – billet thickness): a) in drawing in rigid die; b) in hydroelastic drawing; c) in hydromecanical drawing.

3. CONCLUSION

A system of equations defining the change of billet thickness and describing the process of metal deformation at the initial stage of hydroelastic drawing is obtained. It is found out, that thinning of sheet billet increases as the values of main parameters grow. The parameters are: drawing pressure, metal plasticity, extent of deformation. Maximum thinning is within 15 % of initial thickness of billet, as compared to 25-30 % in case of drawing in hard die.

4. REFERENCES

Isachenkov E.I. Development of technology of punching by elastic liquid and gas environments. *Fogging and stamping manufacture*. No.7, 1976, pp. 2-5.

Kazachenok V.I. *Stamping with liquid friction*. Machinebuilding, 1978, Moscow.

Mihaylov Y.O. *Theoretical principles of hydromecanical drawing*. Izhevsk, 1988, 354 p.

Mihaylov Y.O., Dresvyannikov D.G. Hydroelastic drawing of the parts of cylindrical form. *Vestnik of Izhevsk State Technical University*, No.2, 2001, pp. 60-63, Publisher ISTU, Izhevsk.

Mihaylov Y.O., Dresvyannikov D.G. Optimizing of pressure in the matrix in hydroelastic drawing. *Information technology in innovative projects*. pp. 94-95, Izhevsk, may 2003, Publisher ISTU, Izhevsk.

Mihaylov Y.O., Dresvyannikov D.G. Drawing by liquid and elastic medium as reduction thinning of the wall of parts. Higth technology. pp. 69-76, Izhevsk, may 2004, Publisher ISTU, Izhevsk.

Annals of DAAAM for 2008 & Proceedings of the 19th International DAAAM Symposium, ISSN 1726-9679
Editor B. Katalinic, Published by DAAAM International, Vienna, Austria 2008
Make Harmony Between Technology and Nature, and Your Mind will Fly Free as a Bird

CAD/CAM METHODS IN THE CUSTOMIZATION OF MEDICAL IMPLANTS

DREUCEAN, M[ircea]; TOTH TASCAU, M[irela] & RUSU, L[ucian]

Abstract: *The paper presents the case study of a patient with a severe skeleton defect at the mandible. The patient was taken in a pre-surgery planning procedure consisting of computer tomography scanning, 3D reconstruction with special software, design of the missing bone fragment with CAD software, and rapid prototyping (RP) of the part in Polyethylene layers. Based on the prototype, the surgeon prepared the auto implant for the patient in a separate intervention and implanted it in the mandible. The fixation of the implant during the osteosynthezis time was made with a special blade and screws inserted in the bone. For particular situations, the fixation blade was analyzed using Finite Element Analysis method (FEA).*
Key words: *tomography (CT), prototype (RP), CAD, FEA, implant.*

1. INTRODUCTION

The patient included in the study had a congenital skeleton defect at the level of the left mandible. The upper left limit of the mandible was missing. The pre-surgery CT reconstructed skeleton of the skull is presented in Fig. 1.

In the situation presented above, the normal procedure is to rebuild the mandible at the left condilys zone using a metal implant, a high density polyethylene implant [Drăgulescu D., Popescu M. (2006)] or an auto implant extracted from the long bones of the patient. In the study the authors tried to find a solution to pre-design the implant based on the real geometry of the patient's skeleton [Saringer W, Nobauer-Huhmann I, Knosp E. (2002)].

2. GENERATION OF THE MODEL

The 3D model of the mandible was generated in Mimics after the transfer of the CT scans from the computer tomograph. This software package offers the possibility of separation the bone from the rest of the tissues, according to the density of the material. The core idea of the paper is to use the reconstructed skull as a CAD object in order to determine the exact shape of an implant which will refill the missing part of the mandible [Tardieu PB, Vrielinck L, Escolano E. (2003)]. This activity is developed prior to the surgical intervention, so that the surgeon can shape the implant before the actual implantation. This is the way of obtaining the so-called "customized implant". In the lower right corner of Fig. 2 one can see the part extracted from the 3D model and mirrored in order to be used as a template.

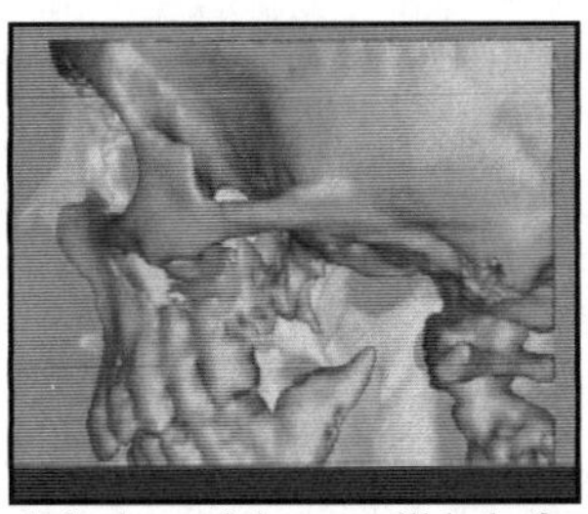

Fig. 1. Upper left view of the mandible before the operation for the implant.

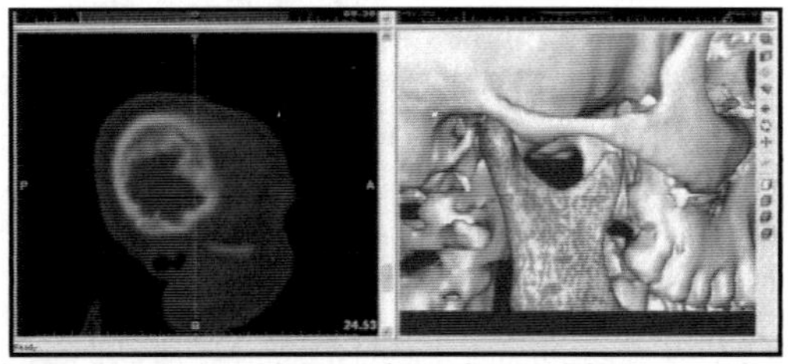

Fig. 2. The extraction of a part from the model as a template for the implant

After the detection of the shape of the implant on the right extremity of the mandible, this was separated using special Boolean operations and isolated as an individual part. In the next step the part was mirrored in order to fit in the congenital defect of the mandible at the left side of the face. With slight adjustments consisting of move and cut operations, the part filled the gap in the mandible and the conclusion drawn with the medical team was that the procedure of prototyping the implant as a model for the real auto implant can be started.

The process of production for the prototype of the implant was developed on a prototyping machine which works with polyethylene film. The part was produced in successive layers of thin film and compared with the shape and dimensions of the part in the model. The prototype had to be refined in several steps in order to satisfy the requests of the surgeon. In the end the part representing the model of the implant looked like in Fig. 3

3. GENERATION OF THE IMPLANT

The additive process which is the base of the rapid prototyping technology consists of the successive add of a new layer of material, with an individual contour [Yaxiong, Liu; et al. (2003)]. The layers are connected to each other in different ways, so that in the end the result is a solid part. One of the preparation stages for a model before starting the build process is the determination of the contour for each layer. The process goes automatically with the help of special software and the result is a closed contour for each section.

The Fig. 4 presents the mandible implant in the process of growth, at approximately one third of the height of the part. The generation of the prototype with the technology based on polyethylene film lasted one hour and 20 minutes.

With all the other preparation actions, the prototype was ready in about three hours from the moment when the patient left the CT room.

Fig. 3. The shape of the implant designed in symmetry with the right side of the mandible

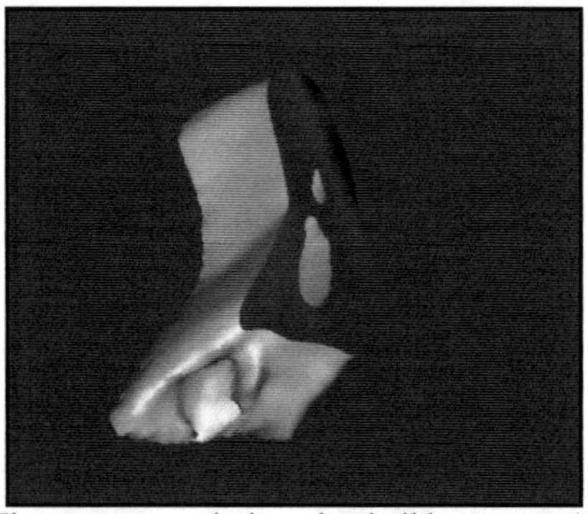

Fig. 4. The prototype during the build process (the waste material was removed)

Once the model of the implant was ready, the surgical process could start. The first step was the preparation of the implant from the femur bone of the patient, respecting the shape of the prototype. This was a separate surgical intervention. The implant was then prepared for implantation and stored for a short while in septic conditions until the patient was ready for a second surgical session. This second stage was much easier than a classical intervention of that type, because the implant had a proper shape and had only to be placed with very unimportant adjustments.

4. SOME ASPECTS OF THE FEA ON THE FIXATION BLADE

For the patient presented above the fixation of the implant on the rest of the mandible was achieved with a special blade made of Titanium alloy TiAlV6. The blade was taken from a set of different shapes of blades, especially designed and produced for aiding in maxilla-facial surgery.

The finite element analyze has two points of interest:

- The determination of stress and strain values in the least favorable load conditions;
- The determination of the influence of the number of fixation screws on the peak values of stress and strain.

The blade and the boundary conditions are presented in Fig. 5.

The load is considered to be developed by the pterygoid-masseter muscle [Gallas M., Fernandez R. (2004)], developing a force of 500 N in the plane of the plate. In other study there was also considered the occlusion force with a magnitude of 426.24 N. In the case of both loads acting in the same time, the value of the maximal equivalent stress is 197.62 MPa, very close to the yield limit of the material of the blade. The deformed shape of the blade is presented in Fig. 6.

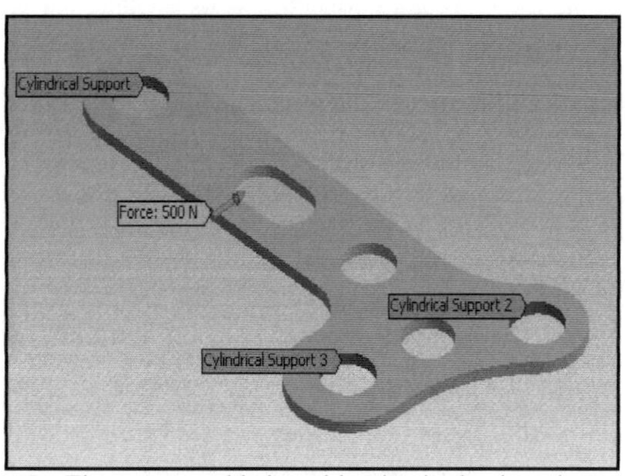

Fig. 5. The support blade with simple load and contact restrictions

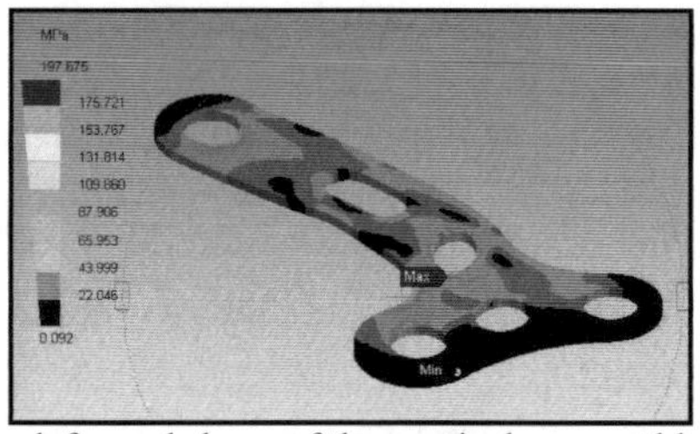

Fig. 6. The deformed shape of the surgical support blade

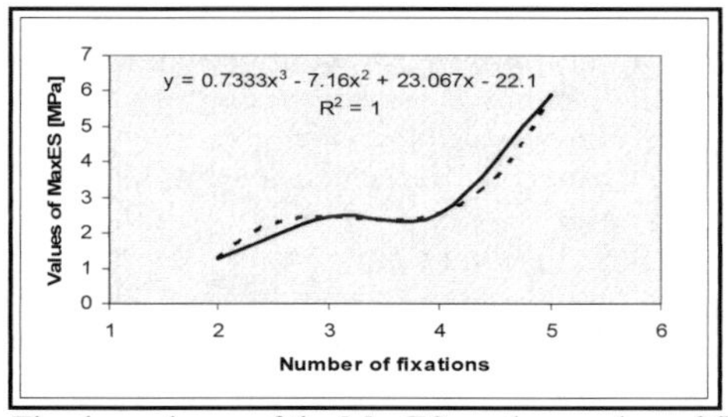

Fig. 7. The dependence of the MaxES on the number of fixation screws for a surgical fixation blade

The number of fixation screws has a great influence on the value of maximal stress. The study revealed o growth of the maximal equivalent stress (MaxES) of six times when the number of fixation screws goes from 2 to 5. The situation is presented in Fig. 7.

5. CONCLUSIONS

The case study presented in this paper is based on a real case included in the records of the research centre where the authors are involved. The contribution of the authors at the final solution consisted of the shortening of the surgery cycle for such an intervention and the enhancement of the resolution of the surgical operation by the use of an implant already prepared in a separate surgical stage.

The FEA analyze of the fixation blade revealed the level of stress and strain in the fixation element and recommends the fixation of the blade with a maximum possible number of screws in order to reduce the strain and stress in the bone.

For the future, the research team is oriented towards the production of customized implants direct in metal or reinforced HDPE.

6. REFERENCES

Drăgulescu D., Popescu M. (2006). *The Encyclopedia of Composites.* Ed. Politehnica, ISBN 973-625-272-8, Timişoara

Gallas M., Fernandez R. (2004). A three-dimensional computer model of the human mandible in two simulated standard trauma situations. *Journal of Crania- Maxillofacial Surgery* Volume 32, Issue 5, October 2004, page 303-307

Saringer W, Nobauer-Huhmann I, Knosp E. (2002). Cranioplasty with individual carbon fiber reinforced polymer (CFRP) medical grade implants based on CAD/CAM technique. *Acta Neurochir (Wien).* 2002 Nov; 144(11): page 1193-203

Tardieu PB, Vrielinck L, Escolano E. (2003). Computer-assisted implant placement. A case report: treatment of the mandible. *International Journal of Oro-Maxillofacial Implants.* 2003 Jul-Aug;18(4): page 599-604

Yaxiong, Liu; Dichen, Li; Bingheng, Lu; Sanhu, He; Gang, Li (2003). The customized mandible substitute based on rapid prototyping. *Rapid Prototyping Journal* v 9 Issue nr. 3 2003 page 167-174

Annals of DAAAM for 2008 & Proceedings of the 19th International DAAAM Symposium, ISSN 1726-9679
Editor B. Katalinic, Published by DAAAM International, Vienna, Austria 2008
Make Harmony Between Technology and Nature, and Your Mind will Fly Free as a Bird

THE OPPORTUNITY COST AND THE LAW OF DEMAND

DRUICA, E[lena]; CORNESCU, V[iorel]; BRATU, A[nca] & MUNTEANU, I[rena]

Abstract: *The present paper pursuits to emphasize a simple and elegant method that could help students understand, beyond intuition, the reasons why the law of demand is rather a general tendency than a generally valid truth, as well as to understand the notion of opportunity cost from a whole new perspective compared to the one usually provided by economics manuals. The Law of Demand's connections with several basic concepts in the economical theory, such as the opportunity cost, reveal new possibilities for the enterprisers of controlling the market.*

Key words: *Walras Law, downward sloping demand, opportunity cost, economic education.*

1. INTRODUCTION

In the field of economics a wide literature exists having undertaken the law of demand in view of numerous approaches: historical, psychological, intuitive or profoundly formal (Teira 2006), (Beattie & LaFrance 2006), (Zhang, 2005); many times our students do not have access to such articles or they regard them as being too difficult compared to their own purposes and background. The present article provides an approach based on Walras' Law, which is in the same time a rigorous and easy to understand method. As it is finding in any course note in Economics, Walras' Law states that a consumer always spends his entire budget, because the individual is tempted rather to consume more than to consume less when it comes to satisfying his own needs:

$$\sum_{i=1}^{N} p_i x_i(p,\omega) = \omega \qquad (1)$$

It is well known that very early on the hypotesis of Walras' Law have been discussed and that some specialists found them as being exaggerating claims and somehow unobservable (Walker, 1984). It is also true that further work improved the concept of economic rationality as being the basic of Walras' Law findings (Arrow, 1977) and lead to some other discussions. Hwever, for our purposes all these contradictions are not essential in our initiative to get the purpose of this paper.

In (1), p_i represents the price per unit of good i, ω is the consumer's budget, and $x_i(p,\omega)$ is the quantity of good i purchased upon prices $p = (p_1, p_2, ..., p_N)$ and budget ω. When modifications occur of price p_j of good j, for a fix j we obtain:

$$\sum_{i=1}^{N} p_i \frac{\partial x_i}{\partial p_j}(p,\omega) = -x_j(p,\omega) \qquad (2)$$

In order for equality (2) to take place, it is absolutely necessary for the sum in the left side to be negative, compensating by this the positive value of x_j. It is immediately deduced that in this sum there should exist at least one negative element, large enough in order to compensate for the potential positive nature of the other terms. Because p_k represents a price, it is understood that it can only be a positive measure, hence we finally deduce that:

$$(\exists) k \in \{1, ..., N\} \text{ such .that } \frac{\partial x_k}{\partial p_j}(p,\omega) < 0 .$$

The last formal writing can be interpreted by that there is at least one product, k for which the quantity purchased by the consumer decreases upon the increase of price p_j. Whereas the increase of a single price implies the decrease of the quantity for at least one of the products wanted by the consumer under the condition of unmodified budget, there can be expected that when the increase is recorded simultaneously for more prices, then the purchased quantities will decrease even more. What is essential in describing this case is the fact that nothing in the formal side of the phenomena can ensure for the time being that a decrease in the quantity is automatically reflected on the product the price of which has modified. The aspect must be kept in mind and we shall insist on it by further comments.

2. IMPLICATIONS OF WALRAS' LAW OVER THE LAW OF DEMAND

We consider that a certain consumer manifests a demand $x(p,\omega)$ which has N components, each position indicating the demand for the corresponding good or service from N existent and available in the economy. We are going to limit only to one arbitrary position $i \in \{1, ..., N\}$ and we'll know that $x_i(p,\omega)$ represents the demand for good i under the stated conditions. Accepting the context of *ceteris paribus*, we'll admit that only price p_i is variable, while the rest of the influence factors remain constant and hence the function $x_i(p,\omega)$ becomes of only one variable, $x_i(p_i)$. Further more, the economists say that a consumer will react promptly to the price conditions, meaning that along with the increase of p_i the demand function will acquire a decreasing behavior, which is actually a well known feature that we find stated in all specialized literature. Even if such attitude is somehow understandable, we must draw the attention on the fact that not a thing revealing out of the formalization of Walras' Law renders such with a mandatory nature.

In the first section we came to the conclusion that once the price p_i is increased, there definitely exists a quantity x_k of which we can assert that it decreases, but we have no valid arguments helping us know whether this is exactly the quantity x_i. The only context in which the decrease of x_i could be claimed is the one in which we are certain that there is no chance for this consumer to waive any unit from the other goods, although the situation described herein clearly exceeds the framework of *ceteris paribus*, which refers to prices and not to quantities. It is true that the choices made by the consumer are those which he prefers most, meaning it could become apprehensive to say that all the quantities from the other goods maximize the preferences and therefore their change is not intended. On the other hand, we know that preferences can change once the restrictions are modified. The increase of p_i represents exactly this sort of modification, thus the only argument which could help us remains invalid.

In conclusion, we have every reason to question why the economists consider that the "inverse" relation between price and the demanded quantity from the corresponding product is a general truth, usually stated under the name of Law of Demand and referring to the market demand. In order for us to be able to pertinently comment on this Law, which is somehow outside the common sense assertions, however not being definitely supported by what has been emphasized by formal tools, we need to go back at this point to identity (2) above and to the consequences of its validity. We have established and appropriately argued that an increase in price p_i determines a decrease in the quantity purchased by the consumer from at least one of the goods he wants, for instance k. We assume now that this good is not by any means the one for which the price increase was recorded, therefore $k \neq i$. This means that x_i, the purchased quantity from good i remained the same and that in the consumer's list the quantity of at least one product decreased. We define $\tilde{x}_k$ as the new quantity of product k, obviously smaller than x_k and $\tilde{p}_i$ as the new price of good i, obviously larger than p_i. If these two measures modify, then Walras' Law for the situations "before" and "after" the occurrence of the change is written as follows:

$$p_1 x_1 + ... + p_i x_i + ... + p_k x_k + ... + p_N x_N = \omega$$

$$p_1 x_1 + ... + \tilde{p}_i x_i + ... + p_k \tilde{x}_k + ... + p_N x_N = \omega$$

Diminishing each member of the two equalities, we obtain:

$$p_k \Delta x_k = -x_i \Delta p_i \qquad (3)$$

It is highly important to notice that the left side of the last equality represents the sum which the consumer should pay in addition for the quantitative difference from good k to which he is forced to give up consequent to the fact that he chooses to keep unchanged the quantity x_i from the good the price of which increased. If for a minute we don't take into account the minus sign, the right side represents the money difference which should be supported consequent to maintaining the consume of good i constant, although its price rises. Relation (3) actually illustrates a "cost" of the consumer's decision, which the economists call *opportunity cost*. Although the consumer does not actually spend more than his budget which remains unchanged, there is however a measurable effect of the potential waive, which is the opportunity price of the version of maintaining x_i constant.

In all probability relation (3) is one of the few manners in which this cost, considered as not-accountancy and somehow difficult to emphasize in the various situations we can encounter in practice, can be however formally expressed and represents one of the arguments in favor of Law of Demand's functionality. It is only natural to notice in this context that while $\tilde{p}_i$ increases, the difference Δp_i in its turn increases, so that the opportunity cost for keeping x_i constant increases as well, up to a critical point from which the consumer stops finding it reasonable to buy a quantity of the product which would compel him, grounded on relation (3) to reduce the quantity of one or more products he finds necessary. This perfectly real phenomenon cannot be viewed by simply applying the derivation operator in (3).

The opportunity cost is not the tool of a mechanism for reducing the demanded quantity which could be mandatory put into practice along with the increase in a product's price. The reason is a matter of the consumer's perception over the initial price p_i and over the new price $\tilde{p}_i$, whilst the economics theory refers to this context by means of the concept of "consumer surplus". This concept includes a side which is very difficult to measure and which has an impact which can hardly be foreseen over the quantity which the consumer is willing to purchase once the price increases. It can be argued that besides any subjective dimension the individual must pay from the same amount ω, and then the opportunity cost exists no matter how large or small is the consumer's surplus.Once more we must say that the formal relation (3) is not enough in order to understand the context. If the consumer's surplus for good i is clearly larger than for any other good on the list, then it is probably that this opportunity cost could function somehow in the opposite direction, generating a thinking such as: "the good k seemed kind of expensive, anyway; instead of maintaining the quantity I buy constantly, I'd rather reduce it and direct the money towards good i, which has a more reasonable price". Therefore, the opportunity cost manifests as an instrument for reducing the demand only when the consumers' surplus for the goods under discussion becomes a relatively uniform measure in the life of decisions made by the individuals. In order to agree with this last assertion, we must say that the consumer surplus implicitly includes the qualitative component, which is difficult to measure, namely the need of the individual for the respective product. Regardless whether it is a real need or if it describes a consume habit, keeping a tradition or an anticipation regarding the market situation, it is the first one of effective importance when it comes to interpreting the opportunity cost.

3. CONCLUSIONS AND IMPLICATIONS

We dare to assert that the Law of Demand is a truth which might not manifest itself as a prompt phenomenon consequent to the price modification, and that it is actually possible for the "curve" of the function modeling the price-quantity relation to be sometimes null and to ascertain by such a constant in the consume, until the economical entities decide to react to the impact of the opportunity cost. Far from being unimportant, this possible tendency of the demand of remaining constant for a period of time has a strong impact in the research of critical loss which might be registered consequent to increasing the price at a certain moment. The concept actually takes shape as "demand inertia", which once established is of help in deciding the level of increasing the price which allows the enterpriser to increase his profit throughout the period left until the consumers' reply.

4. REFERENCES

Arrow, K. J., (1977). *Social Choice and Individual Values (Paperback)*, Yale University Press; 2d Ed edition, ISBN-10: 0300013647, ISBN-13: 978-0300013641, USA

Beattie, B. & LaFrance, J. (2006). The Law of Demand versus Diminishing Marginal Utility. *Review of Agricultural Economics,* Vol. 28, No.2, March 2006, p. 263-271, ISSN 1058-7195

Teira, D. (2006). A Positivist Tradition in Early Demand Theory. *Journal of Economic Methodology,* Vol. 13, No.1, March 2006, p. 25-47, ISSN 1350-178X

Walker, D. A. (1984). William Jaffé's Essays on Walras. *Economica*, New Series, Vol. 51, No. 204, Nov., 1984, p. 480-481, ISSN 00130427

Zhang, Y. C. (2005). Supply and Demand under Limited Information, *Physica A,* Vol. 350, No.2-4, May 2005, p. 599-532, ISSN 0378-4371

Annals of DAAAM for 2008 & Proceedings of the 19th International DAAAM Symposium, ISSN 1726-9679
Editor B. Katalinic, Published by DAAAM International, Vienna, Austria 2008
Make Harmony Between Technology and Nature, and Your Mind will Fly Free as a Bird

COMPETITIVE ADVANTAGE OF INFORMATIONAL TECHNOLOGIES AND COMMUNICATION SECTOR FOR THE ROMANIAN WEST 5 REGION

DUDA DAIANU, D[ana] C[odruta]

***Abstract**: The present work paper aims to analyse and to point out the competitive advantage of the Western 5 Region in the field of information and communication technology (ICT). Applying the Michael Porter's model was determined that the competitive advantage is given by human resources. The results of this research will help the decision factors at local level to understand that it may enforce the combination of development strategies*

***Key words**: ICT, competitiveness, competitive advantage, human resources, strategy*

1. INTRODUCTION

The intensity of competition varies from industry or from a strategic group to another, where it appears a good strategic individualized segmentation. The intensity is determined by a number of factors that were be grouped by Michael Porter in a systemic model, called "Competitiveness's Diamond" (M. Porter, 1998).

The result of these forces will determine the potential performance potential of an industry, measured using the indicators of profitability. Depending on the resultant will be formulated certain strategies in an attempt to gain a more advantageous position in competition. Other economists have pointed out the modern concept of "knowledge" as the most recent new factor in competitiveness (N Negroponte, 1995).

2. THE METHODOLOGY OF THE RESEARCH

To analyse of external competitive environment of the ICT industry is resorting to the application of Michael Porter's model. It operates with the following concepts:

1) *External competitive environment* is defined by the immediate vicinity of the organization, sector, industry, consisting of items (individuals, groups or other organizations) with direct influence on business and on which may be exercised significant influence.

2) *The competitor* can be defined as that organization, sector, industry that is interested for the same type of resources like the considered organization, sector or industry.

The forces which represents Porter's model are shown in interaction in figure no.1. Applying the Diamond's model of Porter, analyse is carried out on national competitiveness for ICT domain.. It should be mentioned that were reconsidered the predefined forces and the model was adapted at situation of making the analyse. Thus, the analysis was based on:

a) The demands for ICT: Inside ICT's domain, the offer is greater than demand, the company has not fully aware of the advantages offered by ICT. Romanian companies that create exported software export "sporadically", on base of direct contracts with EU countries and the U.S.;

b) Local factors local: the graduates in ICT domain partially emigrate; native intelligence, innovative spirit, existing solutions are not yet integrated in a coherent planning.

c) Support industries and domains: education, research and development, production equipment.

d) Companies/competition in this field: 5 large external suppliers of hardware represented in Romanian West 5 Region; few Romanian manufacturers of computing equipment, few Romanian manufacturers for integrated solutions (systems' integrators); Romanian companies are not the force to compete with big international and external suppliers, due to conditions imposed; only few Romanian companies are ISO9000 certified.

3. THE RESULTS OF THE RESEARCH

According to analysis done, you can see that Romanian West 5 Region competitive advantage in this domain comes from the existence of human resources excellently prepared for this area. Romania's competitive advantage in ICT is closely related to resources and its potential in this field. The strategy to be adopted should ensure the optimal "exploitation" of existing potential in the context of present situation of Romanian opportunities' capitalisation. In figure no.1. is presented the system competitive advantage - critical factors of success - the resources of sectorial development strategy for the ICT sector or industry.

It should be highlighted the relation between national available resources, national achieved performances and the competitiveness, as shown in Figure no.2. These considerations point out the importance of domestic environmental analyse in which will be, in the final, implemented the strategy (Duda-Daianu, 2008).

The national internal medium contains all the elements on which industry or sector has a determining influence and, theoretically, a total control. The study of internal environment must establish what resources are available and if the sector is able to follow the chosen strategy. The understanding of internal environment is the first necessary step in establishing a strategy. The studies on resources generated a conceptual framework of analysis - the study of the sector or industry-based resources, used to explain the elements that build competitive advantages. The performance of an industry or sector is assigned to a configuration of own resources whose value is fixed by comparison in an external context.

The problem has two points of departure:

• The presumption of resources heterogeneity assumes that the ICT industry performs with an singular combination of its own resources;

• The presumption of resources immobility assumes that certain ICT own resources are difficult or costly to imitate or existing offer is inelastic.

Resources are attributes of the ICT industry which is presented as tangible and intangible forms and human resources (figure no.1). The items related to culture are analysed separately and are given special attention due to significant impact on the ICT industry. In the specialty literature (Amit R., 1993), the term of "resources" is associated with other concepts as skills or capabilities. Thus, basic or essential skills are collectively experience of industry shown particularly in combining the various skills associated with the production and technological integration of multiple streams.

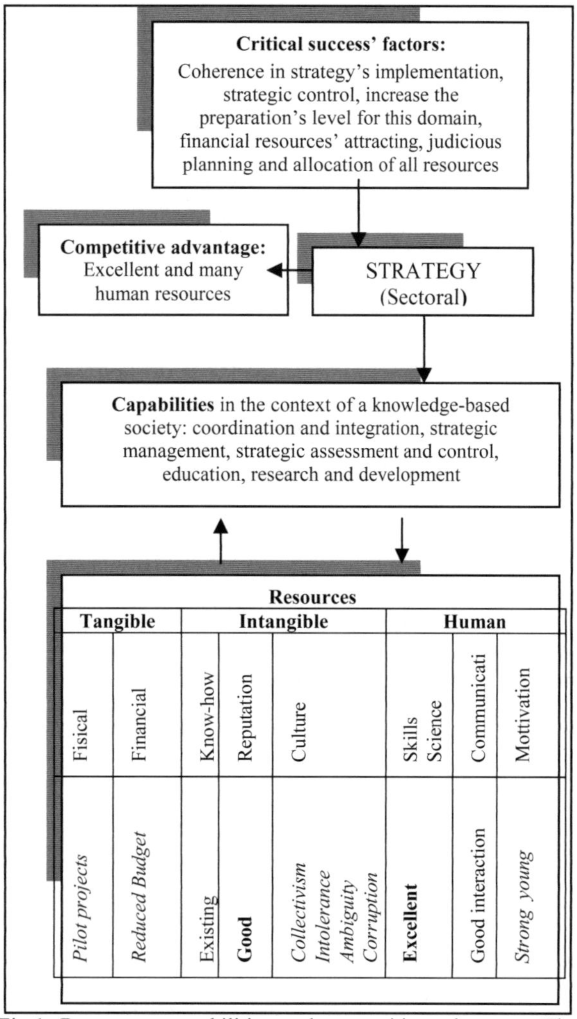

Fig.1. Resources, capabilities and competitive advantage. The methodology of strategy building

The strategic capability represents the industry's possibility of using value-generating resources at its disposal to create a privileged position in the market. Of principle, strategic capability implies the existence of resources and the possibility of its recovery.

Both skills and capability are associated to the industries, respectively to a group of people. The competence would represent the "theoretical" and the capability the "practice", in equal measure necessary to achieve a process of creative value. The relation between concepts is presented in figure no.2).

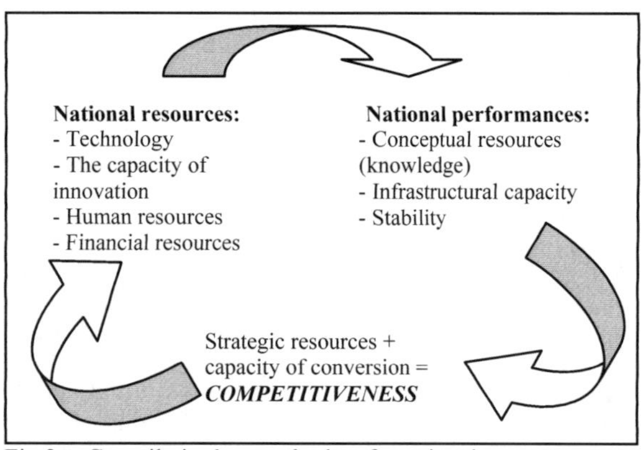

Fig.2. Compilation's method of national resources and performances for competitiveness

4. CONCLUSION

Terms	Competitiv advantage	Strategy
Short	Low cost	**Reduction of prices** (an appropriate option because some products or basic services offered are clearly defined)
Long	Low cost	**Establishing fixed prices** (wrong choice for this domain)
Short, medium and long	Low cost	**Penetration of new external** markets for ICT goods and services manufactured in Romania (right option)
Long	Lange range of services and products	The diversification of goods and products **(an adequate option)**
Short	Lange range of services and products	**The offer's diversification** (introduction of solutions that will increase demand for access to information- acceptable option)

Tab. 1. Strategic options for ICT sector

ICT sector is a sector marked by strong developments in technology. Stage of development of an industry is given by the existing market demand, by production and dissemination of "knowledge". ICT industry in the Region 5 Western Romania is at the beginning stage of growth. The "placing" stage of ITC products has been exceeded. The market's penetration has become increasingly aggressive, products become increasingly standardized, which, in future, will lead to a decrease in prices.

To increase the competitive advantage and the economic development of the Western Region 5 in this area may be adopted various strategic decisions (Cojocariu A., 2005) based on one or more options presented in the table no.1.).

The results of this research will help the decision factors at local level to understand that it may enforce a combination of these strategies which are integrated, closely linked one to another:

1.) Strategy of market development (start-ups companies, launching large projects, the integration of existing projects, encourage the export of ICT products and services);

2.) Strategy of penetration and market consolidation (for example, the introduction of IT solutions in all areas);

3.) Development offer new services and new products based innovation (electronic libraries, distance education for the disabled, public services provided in electronic format, etc.);

4.) Costs' reducing for services and products from the market (reduction of tariffs for communications, etc.).

5. REFERENCES

Amit, R.; Schomaker P.J.H. (1993). Strategic assets and organizational rent, *Strategic Management Journal*, Vol.14, No.1, pp.33-46, *Available from*: www.emeraldinsight.com

Cojocariu, A.; Munteanu, Alin & Sofran, Ofelia (2005). Verification, validation and Evaluation of expert Systems in Order to Develop a Safe Support in the Process of Decision Making, *Proceedings of the 7th International Conference on Informatics in Economy*, pp. 869-873, ISBN 978-973-594-921-1, Bucharest, Romania, May 2005, ASE Printing House, Bucharest

Duda-Daianu, D.C. (2008). Investments in the new technologies-factor of economic development, *Annals of the Oradea University, Fascicle of Management and Technological Engineering*, Vol.VII (XVII), pp.2153-2158, ISSN 1583-0691

Negroponte, N. (1995). *Being Digital*, Vintage Publisher, ISBN 0-679-76290-6, Hampshire

Porter, M.E (1998). *The Competitive Advantage of Nations*, Free Press Edition, ISBN 0-648-84147-9, New York

Annals of DAAAM for 2008 & Proceedings of the 19th International DAAAM Symposium, ISSN 1726-9679
Editor B. Katalinic, Published by DAAAM International, Vienna, Austria 2008
Make Harmony Between Technology and Nature, and Your Mind will Fly Free as a Bird

ASPECTS REGARDING THE CONTRIBUTION OF RESEARCH AND TECHNICAL PROGRESS AT GDP'S GROWTH

DUDA DAIANU, D[ana] C[odruta] & HARANGUS, D[aniela]

Abstract: *The work paper aims to point out the importance of the new technologies like factor of economic development. The new technologies represent the base of competitive advantage at micro and macro level. Those countries that support research could achieve higher GDP, which is an important issue. In the first part of the paper we would like to point out the importance of technical progress and innovations at GDP's growth, presenting the relation between investments and the aggregated function of production. The last part of the work paper is an analyse of Romania's situation in the context of European integration from the point of view of investments in the component of informational technology – transformation factor of an economic and social system.*
Key words: *labour force, research, investments, innovation, GDP*

1. INTRODUCTION

In the period of 2007-2013 more than 50 billion euros can be useable for R&D. the main goal of the state-support R&D is to improve sustainable growth throw the increasing economic competitiveness and number of employees.
Improvement of R&D is a common aim. According to the EU Commission Regulation, strengthen of "scientific and technical basic of common industry, urging improvement of the international competitiveness meanwhile support […] the sufficient research task" (Naresh K., Krishnan M., 2000). The Regulation declares the parameters and also the application fields of the funds and the whole system. By the paragraph 87 of the EC regulation, comparable with the common market: "those supports which improve the economic situation in regions where the level of living and the employment rate is quite low. These kinds of the supports, which effect the economic, do not harm the situation of the common market in different direction" (EC Regulation, 2006).

2. METHODOLOGY

The influence of investments against the aggregate production's function is represented by the relation between total quantity of labour force (L) and capital (K) involved and total production of the nation, GDP:

$$GDP = f(L,K)$$

The function "f" point out the relation between the input of capital and labour force and the result, respectively GDP. Thus, GDP can be interpreted like potential income.
The output Y, the income from economy or production, can be interpreted using the mentioned function and taking in consideration the technical level, as following:

$$Y = f(K,L)\cdot A \qquad (1)$$

where K= capital inputs; L= labour force input; A= technical level.

From the equation (1) results the increasing the output from the economy Y, that will be determined by the increasing of input factors, bur also by the improvement of technical level of production A. When the output rises in an proportional way with the determined factors, result the following relation:

$$\Delta Y = f(K,L)\cdot \Delta A + MPK\cdot \Delta K + MPL\cdot \Delta L \qquad (2)$$

where: ΔA= the level of production's technical spore; ΔK= the spore of capital production's factor; ΔN= the spore of labour factor; MPK= marginal product of capital factor; MPL= marginal product of labour factor

It's known that the marginal product of a factor represents the spore of production (income) realized on base of increasing with one unit of this factor, others remaining constants.
Simplifying the equation (2) by equation (1), it's obtained:

$$\Delta Y / Y = \Delta A / A + (MPK/Y)\cdot \Delta K + (MPL/Y)\cdot \Delta L \qquad (3)$$

Making an calculation artifice by multiplication and division with K and L of the second and respectively third terms from the second part of the equation (3), it's obtained:

$$\Delta Y / Y = \Delta A / A + \Delta K / K(MPK / Y\cdot K) + \Delta L / L(MPL / Y\cdot L) \qquad (4)$$

Analysing the terms from brackets, results: $(MPK/Y)\cdot K$ that represents the capital's percent in total income and $(MPL/Y)\cdot L$ represents the labour factor's weight in total income. The sum of these two terms will be 1 if only the two factors have contributed at income's achieving (Czarl, A.; Belovecz, M., 2007).

Noting: $(MPK/Y)\cdot K = v$ and $(MPL/Y)\cdot L = 1 - v$, the equation becomes:

$$\Delta Y / Y = \Delta A / A + \Delta K / K\cdot v + \Delta L / L\cdot (1-v) \qquad (5)$$

where : ΔY/Y represents the output's increase (production, income);ΔA/A represents the evolution of technical level of production; ΔK/K represents the income capital – production factor; ΔN/N represents the income of labour – production factor; *v* represents the weight of capital in total income; (v-1) represents the weight of labour in total income.

Equation 5) reflects the contribution of the progress and of other production's factors at result's increase, as following:
•The first term from right part reflect the contribution of production's technical and technological improvement, that determines the improvement of total productivity of the factors;
•The following two terms reflects the contribution of capital factor's increase and, respectively of labour factor, contribution equals with a proper increase's rate multiplied with the weight of the factor in income. (Duda-Daianu, D.C., 2008).

Analysing the equation (5) results that the output's increase in transition period will be determined by the increase of input's factors, measured by theirs weights in the income and by the technical progress, that determines differences regarding the increase of the output on hour. The contribution of technical progress and of innovation at output's increase is decisive in present global and national development's context. Many economists had demonstrated by researches and calculations that appreciatively 1/3 from this output's increase is determined by the technical progress and innovations and 2/3 of this output's increase is determined by the rise of production factors' total inputs. Regarding the factors' contribution at output rise, the most important weight is holed by labour factor (which can reach the level of 75% in the developed countries, while the capital contributions could represents only 25%), This aspect is suggested even by the equation (5), which demonstrates that, even in the condition of capital and labour's increase with the same rate, because their different weight in income, the labout factor will have a greater contribution to GDP's rise.

3. ROMANIA'S SITUATION

It's demanded a thoroughgoing study of the increase and development process's problems starting with the importance of technological innovation for investments. This affirmation is base on the actual situation of some countries from Europe and south-east Asia. These countries registered important successes regarding economic increase, infirming the old practices which sustained that only the powerful industrialized countries, with high incomes, possess the forces demanded for obtaining high growing rhythm. This way, can be demonstrated that the force which rule the increase in countries with medium and low development's level are not fundamentally different by those from developed countries.

Unfortunately, the economic experience of our country, after the revolution, proves that during the transition period through market economy has registered even a decline of economic increase. Mathematical expression of this decline is obtained by the separation of labour – production factor from the equation (5):

$$\Delta Y/Y - \Delta L/L = \Delta A/A + [(\Delta K/K) - (\Delta L/L)] \cdot v \qquad (6)$$

Indicators	*2004*	*2005*	*2006*	*2007*
GDP per capita in PPS, EU27=100	33.6	34.4	37.6	39.1
Labour productivity per person employed EU27=100	34.7	35.6	38.3	39.8 f
Employment rate	57.7	57.6	58.8	59.9
Gross R&D expenditure, %GDP	0.38	0.41	0.46	0.65
Business investments, %GDP	18.9	19.3	21.8	22.0

Tab. 1: The evolution of some Romania's macro-economic indicators

Source: Eurostat structural indicators, 2007, http://epp.eurostat.ec.europa.eu/portal/page?_pageid=1996,45323734&_dad=portal&_schema=PORTAL&screen=welcomeref&open=/&product=STRIND_ECOBAC&depth=2

f- represents Eurostat's forecasts

From equation (6) we can observe that the output (GDP/capita) increases on a hand because of total production factors productivity's increase like result of technical and technological improvement of production (ΔA/A) and, on the other hand, because of capital/person increase.

Resort, in Romania, the decline of economic growth produced because the decrease of total productivity of production factors, like a consequence of rapid disorganization of centralized system (without putting an other system in its place), and of capital accumulation's reduction, like result of disorganization by the bursting of relations of technical and supply, production's disorganization and financial blockade.

The errors of post-revolution period determined the start of a disinvestment process, a decline of GDP/capita. In the last few years was registered a positive economic growth (table no.1) and, for maintaining this positive tendency, it's needed a greater volition to invest and to work from the part of all persons.

4. CONCLUSION

The research and technical progress represents an important parts of the engine of the economic growth, which is getting slower in the last couple of years in Europe, thanks to the general conjuncture decline. Referring to the economical and social impact of investments, we think that an important accent must be put on its qualitative level, because this aspects can determines the market's enlargement or segments' losing.

For investments' relaunching, Romania should action for protect its industries, in view to meet the intern needs, the aggregate demand at national level should be assure from intern, because, as was shown in the paper, the demand of equipments and plants normally result from the demand of goods for which producing was utilized. If exists an increasing demand that will not be covered (satisfied) by production's rise in the context of existing industrial capacities, will be necessary new plants and equipments. The investments' expenditure appears in the period when the new capital is established and installed. If the wanted stock of capital goods increases, this will determine the apparition of investments' boom (that will create, in time, the new capitals). If nothing else does change, even if the economic conditions continue to be favorable, justifying the capital stock's increase, the investments for new plants and equipments will stopped when the capital's stock will reach a superior level.

The future research will analyse the relation between investments and the development level for West 5 Region of Romania.

5. REFERENCES

Czarl, A.& Belovecz, M. (2007). Role of Research and development in the 21st Century, *Proceedings of the Eight International Conference on Informatics in Economy*, pp.497-502, ISBN 978-973-594-921-1, Bucharest, Romania, May 2007, ASE Printing House, Bucharest

Duda-Daianu, D.C. (2006). Innovation process in Romania – essence of growth and economic competitiveness - *Annals of the Oradea University Fascicle of Management and Technological Engineering*, Volume V (XV), pp.1512-1518, ISSN 1583-0691

European Commision (2007). Eurostat - Structural Indicators, *Available from* http://epp.eurostat.ec.europa.eu/portal/page?_pageid=1996,45323734&_dad=portal&_schema=PORTAL&screen=welcomeref&open=/&product=STRIND_ECOBAC&depth=2 *Accessed on*: 2007-11-23

European Commision (2006). Science and technology in Europe, Office for Official publications of the European Comminities, ISSN 1725-5821, *Available from* http://europa.eu.int, Luxemburg, 2006-10-04

Naresh Kumar & K.S. Krishnan M. (2000). Foreign direct investment and technology capabilities in the developing countries: a review, *International Journal of Public Administration*, Vol.23, Issue 5-8, May-August 2000, pp. 1253-1268, ISSN 1532-4265

Annals of DAAAM for 2008 & Proceedings of the 19th International DAAAM Symposium, ISSN 1726-9679
Editor B. Katalinic, Published by DAAAM International, Vienna, Austria 2008
Make Harmony Between Technology and Nature, and Your Mind will Fly Free as a Bird

EXPERT SYSTEM FOR AUTOMATED OFFERING FOLLOWING A FEATURE BASED APPROACH

DUERR, H[olger]; TEICH, T[obias]; NGOC ANH, T[ran]; MILITZER, J[oerg]; UNGER, K[atja] & LOESER, C[arsten]

Abstract: *The phenomenon, that the requirements of a customer are more and more individualized, inevitably results in higher efforts required to fulfill customer's needs, especially to answer to him in an appropriate time and accuracy. In the future, order placement will not only depend on feasibility, but also and to high extent on the company's ability to respond to individual customer requests. Automated pre-calculation of production times and -costs becomes more and more decisive for the enterprises' profitability in order acquisition.*
The paper introduces a newly developed approach in the context of integrated feature modeling, which enables an automated pre-calculate of the production costs at the offering stage in an almost real-time manner.
Key words: *CAD, CAM, Feature, Offering, Pre-calculation.*

1. INTRODUCTION

The following procedures included in the three steps given below are focused in a research project, which lead to this paper:

- Feature based design (FBD) of components and manufacturing elements by means of a feature database – Computer Aided Design (CAD)
- Feature based main planning and specification of the manufacturing equipment – Computer Aided Manufacturing (CAM)
- Offering based on calculated machining times and pre-calculated production costs

This contribution is focused on making the offers in an integrated automated way. Thus, we could demonstrate that offering can be executed automatically and with sufficient accuracy of results. Within this context, the results obtained by main planning of the technological sequence are absolutely sufficient at the offering stage. When determining costs with this methodology, production time is calculated in a feature oriented way, and the characteristics of machine cost per hour are taken into consideration.

2. FEATURE BASED OFFERING

2.1 Overall concept

The general feasibility of the product is verified according to geometric, technological and physical properties. To do this, queries with corresponding information about individual technical elements of the product are sent to the database of the resource model. In this step, it can be determined whether the resources necessary to manufacture the individual geometrical shapes within particular constraints are existent. As a result, we obtain a set of manufacturing steps for the product. These steps are the basis for determining the manufacturing times and costs.

2.2 Feature-based design of the product's requirements

"Features" denotes the following: Features are regarded as objects used to represent work pieces, and these objects have functional, geometric and technological properties. They can also include specific application-oriented data, such as times and costs (Ehrenspiel & Kiewert 2007).

According to our methodology, the following steps must be followed for feature representation:

- Define a work piece coordinate system
- Represent global information
- Decompose work piece geometry by feature classes (base-, transition-, manufacturing- and machining shapes)
- Allocate dimensional relations to features
- Allocate technological requirements
- Generate feature catalogue as a means of communication among designing engineers and production planners.

The feature classes should be defined in a manner as closely related to the product as possible. Differentiating feature classes first of all makes it possible to simplify representation of location. The base shape features represent the work piece's initial shape. Manufacturing- and Machining features are understood as surface modifications to base shape features. Transition shape features are edge modifications of marginal contours. From the perspective of costs, they are irrelevant to the estimating process. For application of the feature description model, STEP (Standard for the exchange of product model data) is used. This ISO-Standard is segmented into different AP (Application Protocol). For our requirements AP 224 is suitable and is used for description of the demanded product. In Chapter 2.3, we introduce modeling of the manufacturing resources and rough process planning as another prerequisite for determining manufacturing times and costs.

2.3 Competency-based modeling of the manufacturing resources and rough process planning

The individual SMEs' competencies include human resources (engineering services, qualifications and levels of experience, know-how etc.), manufacturing resources (machinery, equipment and related items) and technologies (manufacturing methods, etc.) (Dürr et al. 2001; Hänel et al. 2006; Lässig et al. 2006). Competency-based descriptions of resources make it possible to balance product requirements and manufacturing capacities with each other, in order to generate technologically relevant process variants (Teich et al., 2008).

The technologically feasible process variants are generated as a result of feature-based balancing of product requirements (demand vector) with available resources (offer vector). This means, that for each feature of the customer's demand, a machine is assigned automical.

This process information is the basis for the concept used to calculate the manufacturing costs outlined in chapter 2.4.

2.4 Determining manufacturing times and costs

The determination of manufacturing time and costs is divided into two steps. For the accurate time determination, our model relies on the on the REFA time diagram (REFA 1997), which is briefly introduced in figure 1.

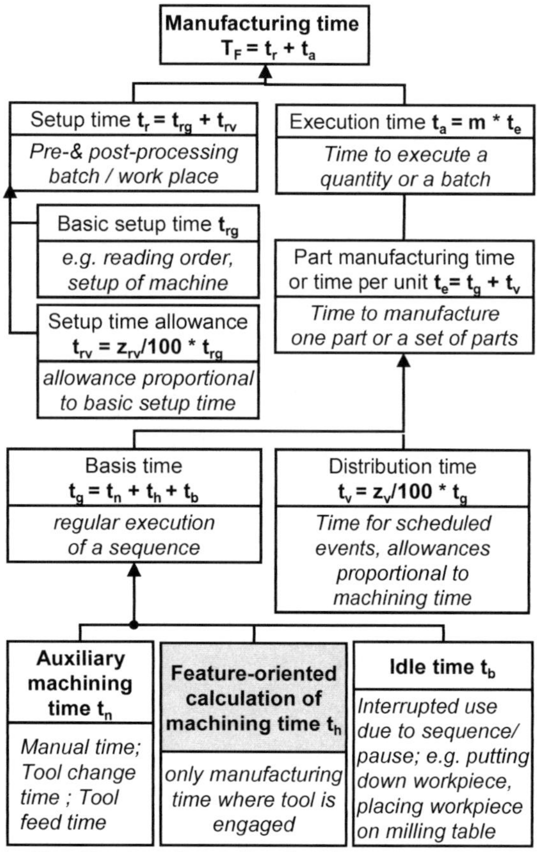

Fig. 1. Classification of manufacturing time.

Machining time t_h is calculated based on features with sufficient accuracy. Time is computed according to the known equations for each manufacturing technology. The determination of manufacturing costs is the second step in the automated offering model introduced in this paper. Costs can be determined using different calculation methods like global estimating of allowances, direct cost calculation or estimating the machine costs per hour.

To achieve sufficient accuracy when calculating the production cost, we chose the paradigm based on machine cost per hour. The function of machine cost per hour includes the addition of overhead costs proportional to the used machine time. Machine costs per hour k_{hM} are calculated using the ratio of the total of machine-related manufacturing overhead cost types per annum $K_{Mach.}$ to possible annual machine time T_N at the planned degree of utilization:

$$k_{hM} = \frac{K_{Mach.}}{T_N}\left(\frac{€}{h}\right) \quad (1)$$

The Machine-dependent manufacturing overhead cost (Mach.-depend. FGK) of a batch corresponds to machine costs per hour k_{hM} multiplied by manufacturing time T_{Fz}.

$$mach.-depend.\ FGK = \sum_{z=1}^{A}\left(T_{F_z} * k_{hM_z}\right) \quad (2)$$

The machine dependent production overhead costs are calculated automatically based on the feature description applied to each, demand and supply. The cost of production can then be determined by the scheme shown in table 1, including direct costs for production material, labor and special costs of production as well as overhead costs for material and production.

Production material	Material costs	Cost of production *(model for feature-oriented estimating)*	Prime costs	Selling price
Material overhead cost *(allowance proportional to production material)*				
Direct labor costs	Manufacturing costs			
Machine-dependent production overhead costs (feature-oriented)				
Residual production overhead costs *(allowance proportional to direct labor costs)*				
Special direct costs of production				
Administration- and sales overhead costs *(allowance proportional to production costs)*				
Special direct selling costs				
Profit margin *(as a percentage of prime costs)*				

Table 1: Basic methodology of cost calculation for estimating

3. CONCLUSION

The project focuses on integrated and automated estimating based on feature modeling and competency-based resource description in virtual production networks. The authors were able to verify that it is possible to draft estimates using a method that is both automated and sufficiently accurate. Generating a rough plan of the technological sequence is entirely sufficient at the estimate stage. Cost calculation is centered on feature-oriented calculation of production time and uses characteristics of machine costs per hour. Implementation of the entire concept on a computer was demonstrated with selected manufacturing examples.

4. REFERENCES

Dürr, H.; Mehnert, J. & Teich, T. (2001) Integration of decentral Developing and Process Planning Competence into the model of Extended Value Chain Management, Proceedings *of DAAAM, International Symposium: Intelligent Manufacturing & Automation*, 24.-27. October, S. 125, Jena, ISBN-Nr.: 3-901509-19-4

Ehrlenspiel, K.; Kiewert, A. & Lindemann, U. (2007) Kostengünstig Entwickeln und Konstruiere, Springer, Heidelberg

Hänel, T.; Shatyan, A. & Dürr, H.(2006) Development and evaluation of competence profiles in non-hierarchical production networks, *Proceedings of FAIM'06, Flexible Automation & Intelligent Manufacturing*, Limerick, June 26th-28th, Ireland, ISBN – Nr.: 1874653933

Lässig, J.; Heinrich, S. & Dürr, H. (2006) Supply Chain Executive Monitor for Controlling and Failure Management in Supply Chains, *DAAAM International Scientific Book* 2006, S. 395-408, Published by DAAAM International, ISBN 3-901509-47-X, ISSN 1726-9687, Vienna, Austria

Mehnert, J. & Dürr, H. (2004) Generation and Evaluation of technological Search Patterns for the target-exact Addressing of potential Suppliers *Proceedings of FAIM'04 Flexible Automation & Intelligent Manufacturing, Toronto*, Canada, July 2004, ISBN – Nr.: 0-662-37218-2

REFA (1997) Wörterbuch der Arbeitswissenschaft, Verband für Arbeitsstudien und Betriebsorganisation (Hrsg.), Carl Hanser Verlag, München

Teich, T.; Zimmermann, M.; Unger, K. & Militzer, J. (2008) Functional Characterization of Resources for Automatic Request Handling, *Proceedings of the Fifteenth International Working Seminar on Production Economics*, Innsbruck (Austria), March 3th-7th, vol. 4, pp. 213-225

Annals of DAAAM for 2008 & Proceedings of the 19th International DAAAM Symposium, ISSN 1726-9679
Editor B. Katalinic, Published by DAAAM International, Vienna, Austria 2008
Make Harmony Between Technology and Nature, and Your Mind will Fly Free as a Bird

THE POPULATION'S MANNER OF SERVICES CONSUMPTION – INDEX OF LIFE QUALITY

DUMITRESCU, C[onstantin] D[an] & DOBREN, F[lavius] A[ndrei]

Abstract: *The paper presents a special vision, about the structure of the quality of life, as an important social life indicator. The material shown as a conclusion, that the social life may be realized as soul of work, or a range of excellence.*
The work is part of a study for companies specializing in recruitment of personnel specialized in various kinds of services. The analysis highlights two important factors affecting the quality of human resource services designed: first, vocational training occupies a special place in human resource training, in choosing a job, and secondly, in terms of training profession grew by five/ 6 times the extension in time.
Keywords: *commercial, quality, consumer, services, goods*

1. INTRODUCTION

The quality of life is determined by diversity, quantity and the quality of services, which are in a continuous increase within the territory of those countries with a modern market economy based on demand and offer. The services in the consumption sphere are addressed either to the completion of the range of goods, or to fulfill those needs that have no correspondent in the offer of material goods; furthermore, they can replace some products, ensuring in this way supplementary advantages, in comparison with the acquisition of the item it had been substituted for (Dickinson, 1989).
One can state that the services in the sphere of consumption are structured this way: independent services and services associated to general goods. Services dedicated to the population's need may be divided in:

- commercial services (paid by population) and
- Free-of charge services (services that are financed by the budget of the state).

Commercial services are fulfillments delivered to population at an equivalent cost through an "act" of selling-purchasing.

2. THE COMPONENTS OF THE QUALITY OF LIVE

After the 1990's, the Romanian society has passed a transitional stage from a centralized economy to a market economy, where took place removals of great depth that had leaded to an incessant increase of these services (Grigorescu &Mihai, 1992).
In this manner, a displacement of the active population towards new professional activities has emerged, from a clumsy inefficient economy to an economy of services, proper for the development countries. The total amount of the systems (commercial, banking, health, educational, touring etc) had to adapt to the new conditions, for Romania be able to adhere the European Union. (Jivan, 1996)
The quality of life derives of the quality of each good/ item, good/ item defined by its ability to satisfy the consumer's needs. Quality usually implies a comparison between the delivered service and that expected by the consumer (Kotler & Dubois, 1992).
One of the items which endorse the improvement of the

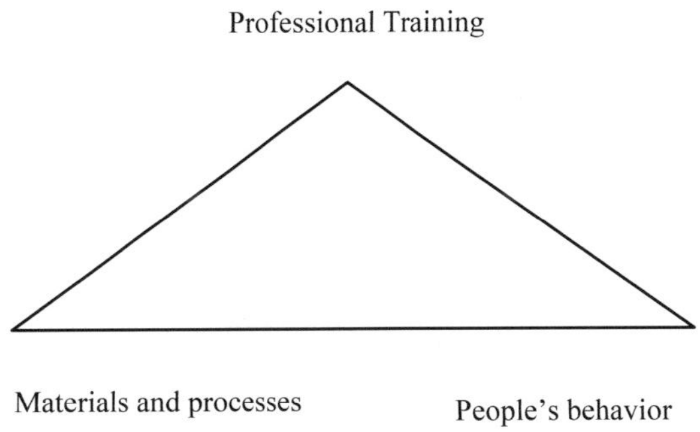

Fig. 1 Elements of obtaining quality

quality in services takes into consideration three groups of elements, marked in the Fig. 1.
This item presented under the shape of a triangle shows that in order to reach a high level of quality; the commercial society has to well- balance the three groups of elements.
An important dimension of the quality of life is the spare time, which is a result of the technical and social development.
Conceived at an extent sense, in contrast with the time dedicated to work, spare time represents that quantum of time meant for a number of activities to which one devotes him/ herself in a free manner, willingly and in favor, either to repose or to entertain him/ herself, or to improve his/ her information, after he had released him /herself from the professional, social and domestic obligations.
The extension of the spare time is determined by the level of the economical and social development, by the policy of promotion in this respect.
Thus, over two centuries, the size of the average length of life was calculated for and included school, transport, but especially time benefit that increases from 5 to 31 years (see Fig.2).

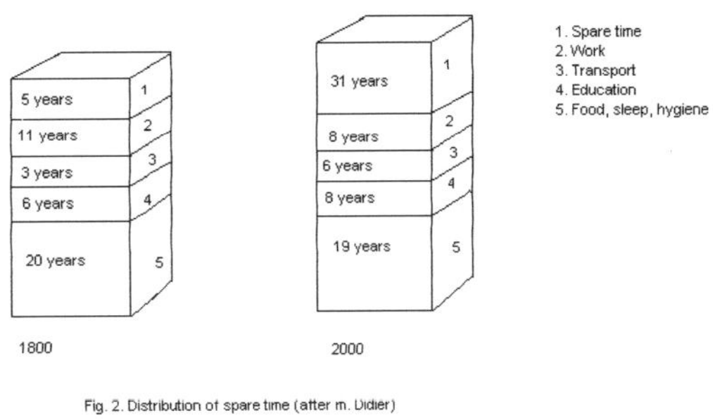

Fig. 2: Distribution of spare time (after N.Didier)

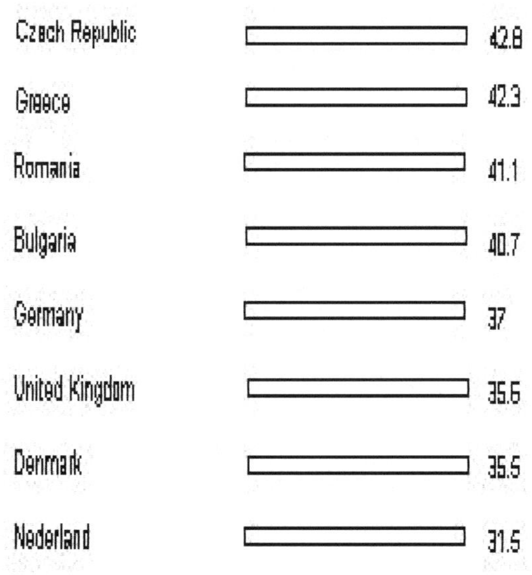

Fig.3 Number of working hours/week in Europe

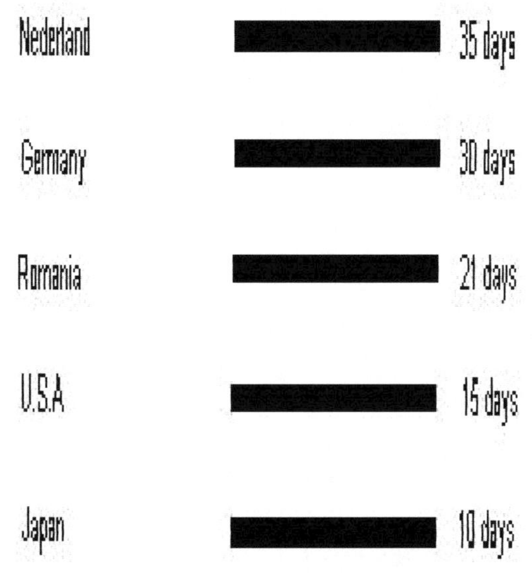

Fig.4 Number of vacation days/ year (without leisure days)

Spare time is daily, weekly (weekend) and yearly (vacation/leave). This problem has to be viewed not as a personal problem, but as a social one.

In this respect, scientific research, the management, information technology, engineering design must be involved in creating all these conditions of spending the spare time

Figure no.3 and figure no4 present the statistics of the present interest regarding the number of working hours and days of vacation in Europe reported to our country.

Concerning the increment in life's quality from the point of view of services, several aspects must be taken into consideration:

- respect for the relation quality- price
- decrease of the time of serving
- diversity in the manner of payment (cash, card, checks leaf, order for payment)
- access, at any moment to any type of service (preferably, where it is possible, for the service "to get" to you)
- wish and pleasure to appeal to a certain service
- state of comfort, good mood after the consumption of the desired service.(Ioncica. & Minciu, 1996).

Among the policies formulated nowadays to ensure quality, the most appreciated of all is the one that places on the first position the satisfaction of the client. Satisfaction is possible through a background of excellent services, continually improving and of efficient use of resources.

According to Kotler's conception, satisfaction is the state of a client, which arises from buying the quality of a service in accordance with his expectations. From such a comparison there should result 4 situations: dissatisfaction, indifference, satisfaction, enthusiasm.

According to Japanese outlook (exposed by S. Yamaguchi) on satisfaction, it appears through the attention offered for quality; we distinguish this way a "hardware" quality (technical characteristics), a software quality (services) and a human quality relation purveyor-client).

3. CONCLUSIONS

1. As a conclusion, the quality of life is linked to the quality of services. Achieving the quality of services implies (according ISO 9004/2):

- to have a social approach
- to be approached the human interrelationships
- to be given the proper importance to the trade mark of the service
- to be personally motivated.

2. The quality of a company can emphasize the quality of services it provides, at a time. It can be seen that in Japan, the number and quality of services compensated with 50% the number of working days a week compared with the United States, and 150 to 150% the number of working days in various states of Europe;

In the end, here are two definitions of quality proper to be taken into consideration:

- Quality is the very soul of work (A. Soljenitin)
- Quality is a range of excellence (Webster Dictionary)

3. The product quality is directly related to the quality of services that accompany them, on the other hand an increase in quality and services have the effect of a final quality social growth, meaning growth in the quality of the relationship producer - beneficiaries.

At the same time, we cannot ignore the inherent increase in product prices, supported by beneficiaries; of the desire to see the more and more diverse needs achieved, beneficiaries \will accept the price difference that occurs. This orientation towards services leads to the reduction in work effort, but develops the training/ instruction activities in the form of diversification and growth in time.

4. REFERENCES

Dickinson, G. *Corporate Risk Management in the Age of Global Networks* in "Strategic Trends in Services" (1989), Harper & Row, New York

Grigorescu, C. & Mihai, St. *The Development and Specialization of Services (1992),* Romanian Academy Publishing, Bucharest

Ioncica, M. & Minciu, R. – *Service's economy(1996)*, Uranus Publishing. Bucharest

Jivan, A. *Modern Services – a challenge for the theory and practice economy (1996),* Mirton Publishing, Timisoara,

Kotler, Ph. & Dubois, B. *Marketing. Management* 7^0Edition (1992), Publi Union Edition, Paris

Annals of DAAAM for 2008 & Proceedings of the 19th International DAAAM Symposium, ISSN 1726-9679
Editor B. Katalinic, Published by DAAAM International, Vienna, Austria 2008
Make Harmony Between Technology and Nature, and Your Mind will Fly Free as a Bird
*Young Researcher and Scientist Paper / * Supervisor, Mentor*

COMPETENCES VERSUS TRUST IN A YOUNG VIRTUAL ORGANIZATION

**DUMITRESCU, D[iana] M[ariana]; POPA, C[icerone] L[aurentiu];
NICA, G[abriela] B[eatrice] & COTET, C[ostel] E[mil]***

Abstract: *Specialized literature in the field of virtual organizations states that the functioning mechanisms of such entities are based around core competences of its members. Therefore, in any given project developed in the virtual organization, partners should always be chosen depending on who has the best suited competences required by the given tasks. Our study wants to show that in the initial phase of the virtual organization's existence, (when members of the network are not all familiar with one another) trust, and not competences, is the building brick for a good cooperation. Nevertheless, trust is build up in time after acknowledging the collaborator's competences, but it is also influenced by personal relations, previous collaboration and constant communication (either formal or informal).*
Key words: *Virtual, Organization, Competence, Trust, Communication*

1. INTRODUCTION

Traditionally, researchers assumed that any organization thrives to be as independent as possible and control its own destiny. But new organizational trends prove the exact opposite. Both profit and non-profit organizations make strategic alliances in order to assure a maximum success rate. Organizational borders start to fade and interdependent networks are more and more numerous (Draft, 1993). One kind of such a network is the virtual organization.

One definition describes the virtual organization as: "a collection of geographically distributed, functionally and/or culturally diverse entities that are linked by electronic forms of communication and rely on lateral, dynamic relationships for coordination." (Desanctis & Monge 1999). Another definition encloses some extra elements: "the virtual organization is a network of entities that act in a unitary fashion during a determined period of time, sharing the member's resources and capabilities based on information technology. Its network-like architecture contains hubs that represent research or innovative teams" (Cotet et al, 2007). By making a summary of numerous definitions, we can conclude that the main characteristics of the virtual organization are: flexible organizational structure, based on knowledge, intensive use of technology, a set of activities concentrated around main competences, free, real time, information flow, multiple coordination points, teams build according to project requirements, minimal control due to high specialization of members. But do people really base their choises on evaluating competences alone? Personal relationships can serve as control mechanisms, because they build up trust and reduce opportunistic behaviour (Kraut et al, 1999). They play an important role in the process of initial trust building. The question that we ask is: do members of virtual organizations choose partners based on evaluating competences, or on trust?

Our study was conducted over a period of 10 months, starting September 2007. The object of study was a young virtual organization, based in Romania, which developed activities in the field of engineering research. The used research methods were: participative observation, qualitative investigation (focus-groups) and a quantitative survey based on questionnaires (Bulai, 2000). The research began with documentation of the existing literature in the field. Then we conducted a paired focus-group with members in the virtual organization, and after analyzing all the gathered information we elaborated a set of 21 questions which represented the questionnaire. This research instrument was applied both in the real and virtual environment, to a number of 28 individuals. This was considered to be the representative group sample for the organization under study. The main research question was: do partners in the virtual organization base their actions on trust or on competences? Additionally, we wanted to find out how trust is built in the virtual working environment.

2. FINDINGS

In order to have a good professional collaboration, a good personal relation with one's partner/collaborator is considered as important by 59,5% of respondents, and as vital by 39.3% of respondents. Therefore, in order to have a successful cooperation in the virtual organization, partners need to establish informal connections with one another. The fact that they see each other vary rarely face to face leads to difficulty in negotiations and building a common understanding over a subject. For any individual it is difficult to work with someone he doesn't know or who he cannot see. Messages sent within the virtual environment lack credibility and power of persuading because they are impersonal. This is why each individual wishes to assign a face, a personality to whom he sends and from whom he receives virtual messages. There is a basic need of establishing human direct contact. Negotiation becomes thus easier; as trust is influenced by the degree one is familiar with its collaborator. Also, the sense of responsibility rises towards someone with whom one shares a personal relation.

In the classification of different aspects that lead to building trust between partners, the first place was attributed to "competences", the second to "previous collaborations", third was "personal relations", fourth "the team in which the person belongs", fifth "the team-building", sixth "the work-shop" and seventh "recommendation by a trusted person". In the process of building trust, a partner can be convinced of his collaborator's competences during a certain collaborating activity (a project with specific tasks), which can be unsuccessful, situation in which trust drops. If the collaboration proves the validity of the partner's competences, then trust rises. Previous collaboration is therefore closely connected with validation of competences, and both contribute to the process of building up trust. Trust derives from previous collaborations in the case when partner's competences are proves and appreciated. The importance of previous collaborations in the process of building up trust must be seen through the opportunity for validating competences. What is interesting is that one form of trust, recommendation from a trusted person was placed last. We can deduce that the form of trust that members in the virtual organization appreciate is of first hand nature that is based on personal experience.

In the analysis of the frequency of communication required for building and keeping trust, 48,8% of respondents agree that they should communicate a few times a week with their partners in order to keep a good future cooperation. This communication should take place even if individuals are not involved at the time in a common project. It is therefore a communication for maintaining a personal relation, which contributes to maintaining trust. 44% of respondents considered the frequency should be "at least once per week". Summarizing the two, 92,8% of participants feel that the effort of constant communication with partners, even if it is an informal one, is vital for future cooperation. As one participant to the focus-group said on the matter "out of sight out of mind". Another form of this effort is represented by the "politeness messages", send at different holidays or special occasions. These are considered as important by 91,7% of respondents. All these types of communication represent the "support activity" within a virtual organization. It is in everyone's interest to maintain cordial relations with as many partners as possible, because this way one can be seen as a prospective partner

Specialized literature claims that the choice of a virtual project partner should be based on the best suited competences. In this situation, a member of a virtual team should always choose the person who has the adequate competences for the given task, no matter if he knows him or not, no matter if he trusts him or not. The key question that members had to answer in order to settle the matter of prioritising between trust and competences was: "In an important/complex project, if you would have to choose between an old partner, with whom you previously collaborated well, and a new partner, who had more resources or a better expertise in the field, who would you choose?". The term of competences was deliberately avoided in order not to influence the respondents.

57,1% of respondents choose the old partner, which proves that despite the possibility of choosing better competences, members consider the risk of un unknown partner too big, especially in an complex project, which by itself has high risks. Made to choose between two compromises, they preferred to base themselves on a safety factor: the familiarity of the partner. The choice was made in the context of a young organization, which hasn't completed the forming of the professional network of relations between its members. Because it still is in the phase of building trust and testing competences, it is understandable why this criteria in momentarily most important.

The relatively large percent of non-responses (14,3%) to this question (which marked the "I don't know/ I don't want to respond" box , therefore not skipped the question entirely), suggests that respondents had problems in making a decision on the matter. If they allocated a short period of time for filling out the questionnaire, they gave quick answers, without interpreting the questions. Faced with this certain question though, as was also noticed in the focus-group, participants were reticent in giving a definite answer, because of the confusion between the manifest and latent levels. Because the majority of previous answers claimed the primary importance of competences, the choice of trust over competences made in this situation created a conflict of principles in participant's minds. From the very beginning of the project of forming a virtual network based on knowledge, members have heard stereotypes such as "the purpose of the project is to build a network of competences". Thus, this belief was suggested to them, which is also in accordance with the theoretical model of the virtual organization.

But man has a sceptical nature. He cannot exclusively rely on theory or recommendation (like recommendation was placed last in the classification of factors which build trust). The individual, thus also the member in the virtual organization, needs his own experience in order to take a certain stand on a matter or on a person. This is why he will not risk, or very rarely will do so, to enter an important project which by itself is risky, together with a unknown person, on whom he cannot be sure he can rely, with whom he doesn't know how well he will cooperate, and about whom he doesn't know the level of competences he claims to have. The association of the term "partner" (in the virtual context) was mostly made with the concept of "collaboration" or "collaborator" (by 41,7% of respondents). Interestingly enough, no association was made between "partner" and "competence" or any relater term. Moreover, 13,1% of respondents associated the word "trust", which again proves the closeness towards trust then towards competences of the "partner" notion in the virtual environment. Other associations were: "communication"- 11,9%, , "project"- 7,1%, "ally", "internet", "e-mail", "network".

The main association suggests a positive notion, referring to common activities. Thus, the partnership is formed in time, by working together. If the partnership in the virtual environment is defined by common actions, than the functioning mechanism of the virtual organization is that of building trust in time, following past collaborations. The association with "ally" is placed in the same analytical understanding. The third most used association (communication) contributes to confirming the idea that partnership is based on trust which is consolidated through formal or informal constant communication. Intensive communication is what makes possible collaboration between two persons geographically dispersed, and thus it's no wonder that in member's minds, to be partner with someone means to be in constant communication with him.

3. CONCLUSION

The choice of trust over competences is valid for the incipient stage in the life of a virtual organization. Nevertheless, trust must be understood as a process in which proving claimed competences is the most important factor, followed by previous collaboration, personal trust and constant communication. In this development stage the manifest level (which claims the importance of competences) and the latent one (which claims the importance on trust) come into conflict in member's mind. If asked theoretically, they will say competences are more important, but in a practical situation, they will instinctively choose trust. Trust is the mechanism which makes possible human socialization; it is the base of formal or informal relations, but a professional collaboration can prove disastrous if trust is not followed and enforced by competence validation. Our further studies will include making another opinion survey on the matter, using the same interrogative items, two years from now. We consider that by than partners will have enough time to get to know each other, and than the process of building mutual trust will have been complete. In this case, partners should rely exclusively on best suited competences when choosing a project collaborator.

4. REFERENCES

Bulai, A. (2007), *The focus-grup method*, Ed. Paideia, ISBN 973-8064-12-7, Bucharest

Cotet C.E., Osiceanu S., Costache B. (2007), *The Management and Marketing of Reserach-Development in architectures of the virtual enterprise*, Ed. Printech, ISBN 978-973-718-893-9, Bucharest

Daft, L.R., Lewin, Y.A. (1993*), Where Are the Theories for the "New" Organizational Forms? An Editorial Essay*, Organization Science, Vol. 4, No. 4, pp. 1-6, ISSN 1047-7039

Desanctis, G., Monge, P. (1999). *Communication Processes for Virtual Organizations*, Journal Information for Organization Science, Vol.10, No. 6, pp. 693-703, ISSN 10477039

Kraut, R., Steinfield, C., Chan, A., Butler, B., Hoag, A., (1999) *Coordination and Virtualization: The Role of Electronic Networks and Personal Relationships*, Organization Science, Vol. 10, No. 6, pp.772-740, ISSN 1526-5455

Annals of DAAAM for 2008 & Proceedings of the 19th International DAAAM Symposium, ISSN 1726-9679
Editor B. Katalinic, Published by DAAAM International, Vienna, Austria 2008
Make Harmony Between Technology and Nature, and Your Mind will Fly Free as a Bird

ECOMATERIALS FOR WOOD PRESERVATION BASED ON ACRYLIC COPOLYMERS AND LIGNIN DERIVATIVES

DUMITRESCU, L[ucia]; MANCIULEA, I[leana]; SICA, M[ihaela] & BARITZ, M[ihaela]

Abstract: *New ecomaterials with biocide activity were synthesized by emulsion copolymerization of acrylic comonomers (ethyl acrylate, butyl acrylate, acrylonitrile, acrylic acid) with poplar, pine and acacia bark extracts, as partial substitutes for different acrylic monomers. Having in view the biocide activity both, for acrylic comonomers and lignin derivatives, the obtained copolymers were biologically tested and proposed as new wood preservation agents.*
Key words: *acrylates, bark, extracts, copolymerization, preservatives.*

1. INTRODUCTION

Wood has been used as an engineering material because it is low in cost, renewable and strong and it requires low processing energy. Wood has many excellent properties that result from its exceptional combination of microstructural, ultrastructural and molecular features. The main drawback is its dimensional instability in the presence of moisture, which can affects also the biological stability.

Wood is formed basically of cellulose, hemicelluloses and lignin, with a minor amount of extraneous materials (tannins, terpenoids, starch, lipids, alkaloids, proteins, etc.).

These three principal constituents make up cell walls and are responsible for most of the physical and chemical properties of wood.

Wood is susceptible to degradation in a number of ways, including rot or decay, insect attack, fire and weathering. Each of these forms of degradation is essentially chemical in nature and can therefore potentially be inhibited by chemical means.

Wood preservation is a complex process involving a wide range of physical, chemical and biological factors, demanding much from the chemicals selected for this purpose (Militz, et al., 1997).

As an alternative to the protection of wood with toxic chemicals that inhibit the activity of deteriogenic organisms, research into chemically modifying the wood substrate as a means of preventing deterioration is a current area of study. The intention is to alter the chemical structure of wood components so that enzymes are unable to degrade the modified holocellulose and lignin, and yet to leave the physical and engineering properties of the material more or less unchanged (Militz, et al., 1997; Hon, 1996).

Based on the variety of functional groups (hydroxyls, carbonyls, carboxyls etc), etherification, esterification, alkylation, hydroxyalkylation, graft copolymerization, crosslinking and oxidation reactions can be conducted to produce ecomaterials with many practical applications (Hon, D.N. S. 1996; Hon, D.N.S.et al., 2001).

2. EXPERIMENTAL

Taking into account these aspects, as well as the main demand regarding the environmental protection, our research was focused on the development of some new wood preservation agents based on acrylic copolymers with extracts obtained from the bark of poplar, pine and acacia trees and copper sulphate.

Characteristic	Poplar extract	Pine extract	Acacia extract
Appearance	brown liquid	brown liquid	Brown liquid
pH- value	4.90	4.72	3.80
Solids, %	45.00	38.50	40.50
Viscosity at 20^0 C, cP	72.50	75.30	69.20
Cellulose, %	29.35	31.85	26.80
Lignin, %	33.55	34.80	36.75
Pentosans	20.20	12.55	18.25

Tab. 1. The chemical characteristics of the poplar, pine and acacia bark extracts

The ideea of the utilization of such lignin derivatives as reaction partners into the emulsion polymerization of acrylic comonomers is based on the observation that the third major wood component, lignin also possess biocide activity and was proved to be the major fixation site for certain wood preservation components, such as copper.

The extracts obtained from the bark of poplar pine and acacia trees have also been used as reactive comonomers in emulsion copolymerization of acrylic comonomers because they possess a reactive chemical potential, caused by the presence in the structure of lignin (as a major wood component) of some functional groups such as: phenolic and alcoholic hydroxyl, carbonyl, carboxyl etc (Hon, 1996; Hon, D.N.S.et al., 2001).

The chemical characteristics of the poplar, pine and acacia bark extracts used in the synthesis, determined following the methodology specific for lignin (Zakis, 1994) are presented in the Table 1.

Both the acrylic copolymers and the copolymers modified by copper sulphate and poplar, pine or acacia bark extracts have been obtained using a semicontinuous technique emulsion copolymerization.

The copolymerization processes have been done in a one litter glass reactor, equipped with a paddle type stirrer, thermometer and cooler, conforming the following standard recipe:

Monomer composition:
- ethyl acrylate, acrylonitrile, butyl acrylate, acrylic acid
- poplar, pine or acacia bark extracts

Emulsifying system:
- ionic and nonionic emulsifier

Initiation system:
-sodium persulphate

Dispersion medium:
- distilled water
- copper sulphate

Copo-lymer	Viscosity at 20^0 C cP	Tensile breaking strenghts MPa	Water absorption of the film (%) after 24h	48h
CP1	36.5	2.5	17.4	37.5
CP2	38.0	3.0	15.5	35.0
CP3	41.8	3.4	14.5	30.6
CP4	52.5	3.8	12.0	25.5

Tab. 2. Chemical characteristics of the acrylic copolymers proposed as wood preservatives.

Percentages of 8% from the acrylic comonomers have been substituted by bark extracts from poplar, pine and acacia trees extracts.

The acrylic comonomers emulsion, with or without bark extracts and copper sulphate, was dozed by a constant ratio, at 80^0C, for 2.5-3 hours into the copolymerization autoclave. At the end of the process, the inner temperature was raised at 90^0 C and kept for an hour, in order to finish the copolymerization.

The chemical characteristics of the acrylic copolymers proposed as wood preservatives are presented in Table 2.

3. RESULTS AND DISCUSSIONS

Considering the biocide activity of acrylic monomers, copper sulphate and the major wood component, lignin, the obtained copolymers were biologically investigated and proposed as wood preservatives.

The wood preservation agents based on the acrylic copolymers (CP1) and the acrylic copolymers with bark extracts from poplar (CP2), pine (CP3) and acacia (CP4) were submitted to the biological testing.

The testing of the biocide activity of the wood preservation proposed agents consisted on treating the sapwood samples by immersion in the wood preservations agents for 30 minutes and exposing the treated samples against the attack of the following microorganisms: *Chaetomium globosum Kunze, Paecilomyces varioti Brainier, Stachibotrys atra Corda, Alternaria tennuis Ness, Trichoderma viridae Person ex. Fries.*

After 28 testing days, the samples were examined using an optic microscope, in order to establish the attack level of the above mentioned microorganisms.

The fungal growth was classified (according to STAS 8022/91) between 0 and 4, as following:

0 = no growth;
1 = trace of growth detected visually;
2 = slight growth or 5-20% coverage of the total test area;
3 = moderate growth or 20-50% coverage;
4 = plenty of growth or above 50% coverage.

The results of the biocide activity of the wood preservation proposed agents is presented in Table 3.

Treatment type	Degree of attack	Note
Treatment with CP1	Samples attacked in extent of 80%	4
Treatment with CP2	Samples attacked in extent of 18%	2
Treatment with CP3	Samples attacked in extent less than 4%	1
Treatment with CP4	Samples attacked in extent less than 2%	1

Tab. 3. The results of the biological testing of the proposed wood preservation agents.

4. CONCLUSIONS

Our research has established the existence of some functional groups (hydroxyl, carbonyl, carboxyl) bonded into the structure of polymers cellulose, hemicelluloses and lignin, main components of the wood structure, which confer to these wood wastes a specific chemical reactivity.

As consequence, bark extracts from poplar, pine and acacia trees, in amount of 8% (based on the total monomer content) were used in the copolymerization of the acrylic comonomers: ethyl acrylate, butyl acrylate, acylonitrile, acrylic acid.

An interesting aspect of this copolymerization process is those regarding the emulsifying capacity of the bark extracts which improve the emulsion stability.

The new preservation agents present better tensile breaking strength and better water absorption of the copolymer film, essential condition for the biocide preservation which depends on the water content and on the easy accessibility of the microorganisms at the wood surface.

These aspects of the emulsion copolymerization of the acrylic comonomers with extracts from poplar, acacia and pine trees, as well as the biocide activity of the copolymers increase the interest in the new research field of emulsion copolymerization of classic monomers with lignin based derivatives and the possibility for obtaining some new wood preservation agents (Dumitrescu et al., 1996).

The biocide activity of the copolymers seems to be influenced by the lignin content of the bark extracts and varies in order: copolymer with acacia extract (CP4)>copolymer with pine extract (CP3)>copolymer with poplar extract (CP2)> acrylic copolymer (CP1).

The new ecomaterials based on acrylic copolymers and bark extracts, will assure a better preservation action than the chemical agents as such, because they also contain lignin derivatives with biocide activity (Dumitrescu et al., 1996).

Our future research will be dedicated to identify and use other wood wastes capable (due their chemical structure) to act as reaction partners in synthesis of bioprotection and finishing agents, which will confer a better access and compatibility with the wood structure.

5. REFERENCES

Dumitrescu L., Popa V. I., Baciu G.& Petrovici V.(1996). Wood preservation agents based on acrylic copolymers and metal complexed lignosulphonates. *Bulletin of the Transilvania University of Braşov*. Series B. 3(38): 63-68. ISSN 1223- 964X.

Hon, D.N.S. (1996). *Chemical Modification of Lignocellulosic Materials.* Mark Dekker, Publishers, New York, NY.

Hon, D.N.S., and N. Shiraishi. (2001). *Wood and Cellulosic Chemistry.* 2nd edition. Marcel Dekker Publishers, ISBN 0824700244, New York, NY.

Militz, H.; Beckers, E.P.J. & Homan, W.J. (1997). Modification of solid wood: research and practical potential.IRG Document WP/97-40098.

Zakis, G. F. (1994). *Functional analysis of lignins and their derivatives.* TAPPI Press, ISBN: 0-89852-258-7,Atlanta, Ga.

Annals of DAAAM for 2008 & Proceedings of the 19th International DAAAM Symposium, ISSN 1726-9679
Editor B. Katalinic, Published by DAAAM International, Vienna, Austria 2008
Make Harmony Between Technology and Nature, and Your Mind will Fly Free as a Bird
*Young Researcher and Scientist Paper / * Supervisor, Mentor*

THE MANAGEMENT –A LEADERSHIP'S SCIENCE USED IN SME IN ROMANIA

DUMITRIU, A[delina]

Abstract: *The management is one of the most important factors of the SME efficient. The economical and social performances are depending to the methodological, decisional, informational and organizational performances. The managers are involved directly in exerting the management processes, adopting and applying of managerial decisions. To the question "HOW DO WE LEAD?" the answer is proper knowledge, the adequate systems, methods and techniques of management. So, a leadership must know the management function, the management processes and relations, the management system, how to innovate, research and discovery, because it was proved so many times that the management is the science of the leadership. The connection: SME-management-leadership must be demonstrate.*
Key words: *management, leadership, organization, coordination, control.*

1. INTRODUCTION

Because it was demonstrated that management is the science of leadership we would like to underline its importance in the Romanian SME environment and the close link between management, the manager, leadership and SME. That is way we tried to give a short definition to management, leading and small enterprises from our point of view.

2. SME IN ROMANIA

Because Romania is in a complex process of economic restructure, some specific trends have appeared as follows: SME are functioning especially in the third part sector because they need a small start up capital, because they can have profit in a short time and because the services area for population is still deficitary. From the statistic data we have obtained results that 68.8% of SME have the activity focused on trade, 14.1% are based on services, 9.4% in industry and 2.7% in transport and 2.4 % in construction. Big part of these enterprises are presented as family association or created by independent individuals. From the total of active SME 97.4% are private, 0.35% are public property and 2.3% are mixed properties (Marin 2002).

Demographically speaking the density of SME has increased reaching in 2006 at 26 SME for 1000 inhabitants but still way under the EU average of more then 50 SME for 1000 inhabitants (Academy of economic studies Bucharest – new issue no 8, 2007).

3. A LEADER MUST DEMONSTRATE

In Romanian SME, as a leader, you need to demonstrate (McCallum 2004):

1. Honesty. Someone told me once that they didn't necessarily agree with me, but they knew I wasn't lying to them. 1 took that as a compliment rather than a complaint;
2. Communication. The only lime I close my u t lice door is when I have a couple of people in my office and we are having a conference call. It's the only time I'm not available to communicate with others outside my office;
3. A positive attitude. You can't lead if you are complaining. You can complain as long as you finish up with a positive suggestion or direction;
4. Dress for success. If you are a service manager, you should dress as a technician manager and not as a banker. If you are an engineer, you should dress as a professional and not a laborer. If you are a job superintendent, you should dress according to the conditions (e.g., building isn't enclosed yet for the heating season) but still professionally because you represent your company on the job site. How you look is an indication on how you do your job;
5. Teach to replace. I think the more knowledgeable the people in my group are, the better our product and the better prepared they will be to lake my job, or fill that same position somewhere else. Plus, I learn from their feedback on what we do it they are thinking like leaders;
6. Mentoring. The better the personnel, the better the product we delivery. Training, sharing your "good and bad" experience, helping others to set their goals all contribute to having the best team for the job. A leader can't do it all himself;
7. Time management. This principal is the cornerstone to getting it done well, on time, and on budget. This skill can carry over to your personal life, too, so that life is more than just work;
8. Delegation. Because you can't do it all yourself, if you achieve the above, giving up portions of the workload and associated responsibilities becomes easy.

4. WHAT IS THE MANAGEMENT?

Over the times it was proved that Management is the science of the leadership and is the most important factor of increasing efficiency and effectiveness of an enterprise. Without a competitive leadership to plan, organize, coordinate and control an activity we can not talk about an economic activity with a profit. Management is a science which studies the processes and relations of management in order to find principles and legitimates to rule and conceiving instruments in order to permit the achievement of the objectives of the firm in conditions of efficiency. By this we have the three big dimensions:

- theoretic dimension – processes and relations of management;
- methodology dimension – discovery of the principles, legitimates and managerial instruments;
- pragmatic dimension – assures the achievement of the objectives in conditions of efficiency.

James K.McCallum considers that management is an art and a science in the same time, a positive and social science. It is an art because it asks a great deal of understanding the phenomenon and of adaptability. It is a science because it was proved it has rules and laws well determined. The four important functions of management are planning, organizing, coordination and control and there are better explained in the scheme below.

Management is:

1. planning (strategic planning on long term period, economical and business politics, operational plans on short term period),
2. organizing (department…, personnel problems, delegation of authority),
3. coordination (professional training, motivation, communication and guidance),
4. control (setting of the standards, measure, correction).

That is why the manager of an enterprise must not be only a person who solves problems but the one identifying the problems and opportunities. Sometimes anticipating problems is better then solving them and can lead to much lower losses.

5. LEADERSHIP BETWEEN SCIENCE AND ART

The antonym for leadership is probably follower. Over the years, I have seen people in responsible charge who better fit the role of a follower rather than a leader because they didn't apply the skills noted above along with several other management skills needed to lead.

For the individual striving to be a leader, or the person wanting to be a better leader, I advise the person to look around at those in responsible charge and write down what they see as good and bad about the other person's management skills. No one has mastered the role of leader yet, so for those wanting to be a leader and a role model, look around and lake notes. There is a saying, "perception is reality," and a good leader takes on the role of being someone others will follow, in part, through perception. Of course, you still have to perform and deliver, but perception has its place in the quest to lead. Being good at what you do frequently takes luck, but no one seems to notice the luck that took place when outlining your business plan, project goals, or job performance ahead of time. People simply perceive you always know what you are doing, and that is why you are the leader.

For example when I named a manager in the company I was expecting him to be perceived as a leader, employees to respect him and follow his instructions. Unfortunately it wasn't like this. He wasn't perceived as a leader because he didn't know how to make himself understood, didn't know to express his ideas even if they were good. The employees kept on asking me what to do and kept on following me. In the end the manager wasn't needed because he didn't fulfill his attributions even if he was well prepared (Russu 2006).

I think the more knowledgeable the people in my group are, the better our product and the better prepared they will be to take my job, or fill that same position somewhere else. Plus, I learn from their feedback on what we do if they are thinking like leaders.

A leader must know :

A. How to fundament and adopt the management decisions?
B. What managerial tools to use?
C. How the quality and efficiency of the adopted decisions is ensured?
D. Do the leader know how to lead and run the organizations?

A leader must not only control very good the managerial notions, strategies, politics, managerial plans but to know how to apply all the theoretic notions and to make himself understood by everyone he is leading. In vain you are a good manager if you do not know how to make yourself understood and to lead the ones you are in charge of.

The management, tends to become the most important factor, of amplifying the efficiency, effectiveness and competition of the organizations. The management can be evidenced and analyzed as management process and as management system (Verboncu 2000).

The most important coordinates of the manager's (leader) work:

- Fundaments and adopts management decisions in the area of prevision, organization, coordination, training and control.
- Use specific decisional mechanisms (acts or processes), depending on the objectives nature, complexity and difficulty, with influences on the decisional problems that follow to be solved.
- Approach varied managerial tools, with varied systems, methods and techniques of management from which there should be missing:

- management systems (profit centers, objectives, budgets, projects and exceptions management):
- general methods of management (diagnosis, delegation, meeting, dashboard)
- specific methods and techniques of management (decisional, of costs management, of creativity stimulation, of analysis and engineering of the organizational and informational system, of making the managers' work efficient
- Use and take advantage of adequate managerial engineering/reengineering methodologies (global or specific methodologies).
- Promote varied management styles.

So "HOW DO A LEADER WORK?" and "What are the roles of the manager?"

To understand better, we will enumerate a few of the manager (leader) work:

- role of disciplining the managers and executants;
- role of responsibility;
- role of stamping of some characteristics of order and rigor;
- role of facilitating the normal functioning of the other managerial components (decisional, informational and organizational);
- role of making efficient the lead field, creating and maintaining some favorable conditions for achieving the objectives;
- role of facilitating the managerial function exercitation;
- role of facilitating the managerial competency manifestation;
- role of management professional.

6. CONCLUSION

SME are economic entities which can not survive without a performing management. Management which, in the end, is a leading science shows us what way must be adopted and especially how to follow the four points: planning, organising, coordinating and controlling (Verboncu & Zalman 2005). As a personal conclusion there is no SME without management, without leading. We wanted to underline that there is a connection between leadership and management in SME, management being an art, a science, a leading process. A leader must know how to explore the possibilities, to coordinate the employees, to control them, search the qualities, must know the management functions, processes and relations, the management system, all the innovation and how to apply them, must have varied managerial knowledge.

7. REFERENCES

Academy of economic studies Bucharest – new issue no 8 (2007). *Review of international comparative management*, ISSN 1582-3458, ASE, Bucharest

Marin, D. (2002). *Romanian economy, SME, what are we integrating with?*, Economic Publishing House, ISBN 973-590-710-0, Bucharest

McCallum J. *American ideas for Romanian managers,* General Management, ISBN 973-95612-4-1, Romania

Russu,C. (2006). *Management of small and medium business,* Economic Publishing House, Bucharest

Verboncu I. & Zalman M., (2005). *Management and performances*, Universitara Publishing House, Bucharest

Verboncu I. (2000). *Managers&management*, Economic Publishing House, ISBN 973-590-425-X, Bucharest.

Annals of DAAAM for 2008 & Proceedings of the 19th International DAAAM Symposium, ISSN 1726-9679
Editor B. Katalinic, Published by DAAAM International, Vienna, Austria 2008
Make Harmony Between Technology and Nature, and Your Mind will Fly Free as a Bird

USING FUZZY REASONING TO GUIDE A MOBILE ROBOT ON A ROUGH TERRAIN

DUMITRIU, A[drian]

Abstract: *The paper presents some author's experiments carried out within the frame of a research project and destined to endow mobile robot modules with small and simple sensors to support navigation. Experiments have proved that acceleration sensors in MEMS technology are very suitable and that Fuzzy logic is an adequate tool for sensor data integration. A Fuzzy controller has been developed and tested on a mobile robot moving on rough terrain.*
Key words: *Robots, Mobile, Navigation, Sensors, Fuzzy.*

1. INTRODUCTION

Navigation problems for autonomous mobile robots are very complex and require a large amount of information from vision and range sensors, to build up a map of environment, and powerful command processors to optimize trajectory. These types of sensors cannot seize robots' inclination angles on rough terrains, in order to make necessary trajectory corrections to avoid turning over. Acceleration sensors seem to be very suitable for this purpose and the main problem this paper tries to solve is to find an efficient tool to integrate data from acceleration sensors with other sensors data, in order to generate an adequate reaction of the robot. Acceleration sensors have been used by (Zhang et al., 2003) for calculating the orientation angles of the modules of a self-reconfigurable mobile PolyBot robot. (Corke et al., 2007) discuss the complementarity of inertial sensors (including acceleration sensors) and vision sensors and describe some fundamental approaches to fusing their outputs and survey the field. Due to its capabilities of interfering and approximate reasoning under uncertainty, Fuzzy Logic seems to be a suitable tool for sensor integration. (Saffiotti, 2001) details the main aspects regarding the use of Fuzzy logic in autonomous navigation for: behaviour design; behaviour coordination; environment modelling; layer integration. (Innocenti et al., 2007) describe a multi-agent architecture with cooperative fuzzy control for a mobile robot, while (Vadakkepat et al., 2007) use a fuzzy behaviour-based architecture to decompose the complicated interactions of multiple mobile robots into modular behaviours at different complexity levels. (Chapman, 1994) presents some very efficient and interesting solutions and C++ routines for implementing fuzzy logic to 8051 8-bit microcontrollers, which have served as a model to do something similar in PBASIC for the 8–bit microcontroller of the BASIC Stamp 2SX board.

The paper presents some author's experiments carried out with a very simple mobile robot, controlled by an on-board 8-bit microcontroller and equipped with an infrared range sensor and a 2-axes acceleration sensor. Original contributions are:

- Use of an acceleration sensor to control inclination angles with respect to two axes;
- Integration of data from infrared range and acceleration sensors and control of robot's movements and behaviour using fuzzy reasoning;
- Design of an adequate fuzzy controller and development of PBASIC software routines to implement fuzzy logic to the 8-bit microcontroller.

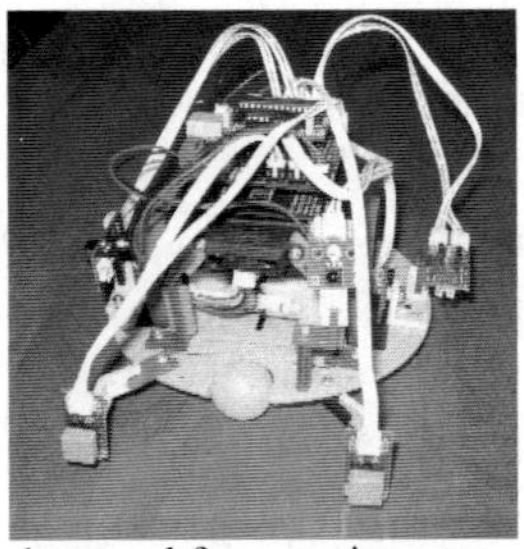

Fig. 1. Wheeled robot used for experiments.

2. ROBOT USED FOR EXPERIMENTS

For experiments with acceleration sensors a wheeled robot, named IZEBOT (Inex Innovative Equipment), has been used. The structure of the body plate of the delivered kit allows building up the robot structure presented in Fig. 1.

IZEBOT uses for each of the two wheels a 6÷9V and 180 mA DC motor and a ratio of 48:1, with an output torque of 0,4 Nm. DC motors are controlled via PWM outputs and LM 293 power drivers. PBASIC PWM commands allow speed control in 255 steps.

Standard configuration includes two touch sensors, three IR reflector sensors and an IR ranger sensor. Each of the two wheels has 8 oval-shaped holes, used as wheel codes, which during rotation generate impulses counted by adequate incremental sensors.

The control module is build around a BASIC Stamp 2SX (8-bit) microcontroller board and includes:

- Two DC motor drivers with status indicators to show motor direction;
- 8 channel analog ports with a 10-bit A/D converter that can accept up to +5 V;
- 7 channel digital I/O ports (P0÷P);
- RS-232 serial port;
- 16 kB memory;

Programs for IZEBOT are developed in PBASIC language and BASIC Stamp environment. Instructions allow access to individual I/O ports, thus, disclosing all the facilities the control module is capable.

This type of robot has been chosen, because integration of the acceleration sensors used, *Memsic 2125* from Parallax Inc., is very rapid and simple, both from hardware and software point of view. This sensor, in MEMS technology, contains, internally, a small heater. This heater warms a "bubble" of air within the device. When gravitational forces act on the bubble, it moves and this movement is detected by very sensitive thermopiles (temperature sensors). On-board electronics convert the bubble position (relative to g-forces) into pulse outputs for the X and Y axes.

Main characteristics:

- Measure 0 to ± 2g on either axis, with less than 1 mg resolution;
- Simple, pulse (PWM) outputs of g-force for X and Y axis, requiring just two I/O pins;

Two sensors, placed in adequate boards, have been mounted on the left and right outsides of the body plate (Fig. 1) using angled joiners and screws and have been connected to 4 available I/O ports, one for each pulse output of g-force for x and y axis of a sensor.

Pulse outputs allow to calculate accelerations with the formula:

$$A = ((T1 / 10) - 0.5) / 12.5\%, \qquad (1)$$

since T2 is calibrated to 10 milliseconds at 25°.

The T1 duration (Memsic output) of each *Pin* is captured, every 10 ms, in *Variable* using a *"PULSIN Pin, 1, Variable"* command for each I/O pin, where the "1" operand indicates that a high-pulse, which begins with a 0 – to - 1 transition, is measured.

Since BASIC Stamp2SX (8-bit) microcontroller has a pulse-time of 0.8 μs, up to12.500 pulses can be counted in 10 ms. 6.250 pulses correspond to a 0g acceleration and 6250+1462/6250-1462 pulses to 1g acceleration. in positive or negative direction of the measured axis.

The GP2D120 infrared ranger module can measure a range from 4 to 30 cm, with an output voltage between 0,4 and 2,4 V, when supplied by +5V. A 10 bit ADC converts sensor output in a proportional number as input to the microcontroller.

A lot of experiments have been carried out (Dumitriu, 2008), based on following assumptions and conclusions:

- Axes of the co-ordinate system attached to the robot are selected in accordance to the wheels' configuration and the moving direction of the robot: X axis is the along the median line of the wheels, pointing in the moving direction, Z axis is perpendicular to the base plate, pointing in the upwards direction and Y axis is along the wheel axis.
- Rotations, the robot can be subject to, are restricted: it must lie only on his wheels and rotations with respect to X and Y axes cannot exceed some limits, to avoid overthrow. Due to these facts the vertical acceleration sensor is redundant. The robot can react correctly using only accelerations measured along the X and Y axes.
- Determination of tilt angles is not of interest, because the number of pulses can be used directly, avoiding unnecessary computations. Thresholds for rotation angles are expressed in pulse numbers: 250 for 10°, 500 for 20°, 730 for 30° etc.
- Robot's behavior was based on following rules: while moving on a flat the robot reacts only to data acquired from other sensors: push-sensors to detect collisions, IR range sensor to avoid obstacles and IR reflector sensors to follow a trajectory. Otherwise the robot reacts to accelerations measured with respect to X and Y axes and, if deviations from the 0g accelerations pulses exceed certain thresholds for one or both of these axes, the robot returns in opposite direction and turns left or right to search for another route.

3. DESIGN OF THE FUZZY CONTROLLER

Since the effect of data evaluation from all types of sensors is the PWM control of the two DC motors in one sense or another, and programs become very complex due to the lot of IF…THEN statements, used to test different input conditions, the author concluded that fuzzy logic could be a very efficient tool. A fuzzy controller has been designed, including:

- 3 fuzzy input variables: distance expressed in ADC units (mean value of 5 measurements); X-axis and Y-axis accelerations (expressed in pulses divided by 8, since experiments have proved that the sensor drift is in the range of ± 8 pulses).

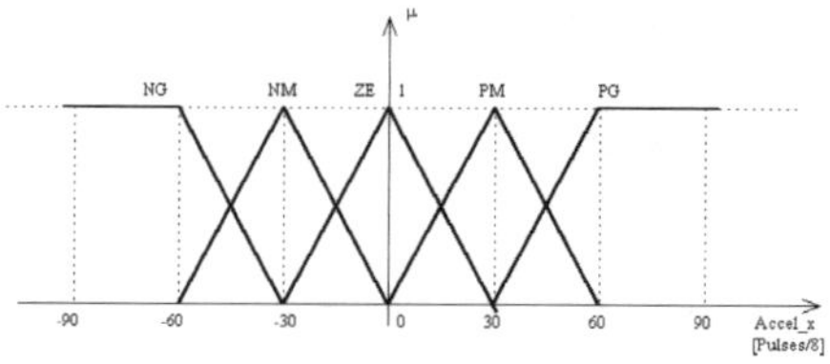

Fig. 2. Membership functions for X-axis acceleration.

Trapezoidal and triangular membership functions (MF) have been selected for all input variables (fig. 2).

- 2 output fuzzy variables: the PWM outputs for the 2 motors, with values in the range of ± 255. Each motor is controlled by 2 processor output pins, one, with positive PWM values for moving in forward direction, the other with negative PWM values for the opposite direction. Singleton MFs have been selected for outputs at 0, ±64, ±128, ±192, ±255 PWM pulses.
- Rules sets, with different numbers and configurations, based on the same principle: as far as the terrain is flat and no obstacles are detected, the robot moves forward with maximum speed. Otherwise it slows down and turns right or left, or moves backward, to search for another more suitable path.

4. CONCLUSION

Studies and experiments presented have proved that fuzzy logic is an efficient tool for a proper and robust reaction of a mobile robot to his environment, seized with different types of sensors. Future work will be dedicated to optimize the fuzzy controller with respect to the membership functions and rules set and to extend the group of sensors used for a proper robot's behaviour.

5. REFERENCES

Chapman, M. (1994). *The Final Word on the 8051*, Chapter 11 "Bumpin' Fuzzies with the 8051", pp.222-255, *Available from*: www.8051projects.net, *Accessed*: 2008-04-15.

Corke, P.; Lobo, J. & Dias, J. (2007). An Introduction to Inertial and Visual Sensing, *The International Journal of Robotics Research,* Vol.26, No.6, June 2007, pp.519-535, SAGE Publications

Dumitriu, A. (2008). Integration of Acceleration Sensors Data in Robot Control Systems, *Proceedings of 2008 IEEE International Conference on Automation, Quality and Testing, Robotics AQTR 2008 (THETA 16*, Poster Session, pp. 83-87, ISBN 978-973-713-248-2, Cluj-Napoca, Romania, May 2008

Innocenti, B., Lopez, B. & Salvi J. (2007). A multi-agent architecture with cooperative fuzzy control for a mobile robot, *Robotics and Autonomous Systems*, 55 (2007), pp. 881-891, Elsevier B.V.

Saffiotti, A. (2001). Fuzzy Logic in Autonomous Navigation. In: *Fuzzy Logic Techniques for Autonomous Vehicle Navigation,* D. Driankov and A. Saffiotii (Ed.), pp. 3-24, Springer-Physica Verlag

Vadakkepat, P., Peng, X., Kiat, Q. B. & Heng, L. T. (2007). Evolution of fuzzy behaviours for multi-robotic systems, *Robotics and Autonomous Systems*, 55 (2007), pp. 146-161, Elsevier B.V.

Zhang, Y et al (2003). Sensor computations in modular self reconfigurable robots, In: *Experimental Robotics VII*, B. Siciliano and P. Dario (Ed.), pp. 276-286, papers from *Eight International Symposium on Experimental Robotics (ISER 2002)*, Springer Verlag

Annals of DAAAM for 2008 & Proceedings of the 19th International DAAAM Symposium, ISSN 1726-9679
Editor B. Katalinic, Published by DAAAM International, Vienna, Austria 2008
Make Harmony Between Technology and Nature, and Your Mind will Fly Free as a Bird
2nd DAAAM-ICAT

THE DEVELOPMENT OF A CUSTOM MAXILLOFACIAL IMPLANT BY MEANS OF RAPID PROTOTYPING

DUMITRIU, D[an]; DRSTVENSEK, I[gor]; IHAN HREN, N[atasha] & BALC, N[icolae]

Abstract: *It is well known for a fact that, in the last decades, the technical field of rapid prototyping has offered aid in medical reconstruction procedures in order to help surgeons reconstruct physical faults and anomalies of their patients.*
Using a combination of Computer Assisted Design (CAD), high medical skills and latest rapid prototyping and manufacturing technologies, it is now possible to help patients with birth defects or who have suffered craniofacial injuries of variable severities, resulting in both aesthetic and functional alterations.
Key words: *rapid prototyping, custom implant, maxillofacial implant, reconstructive surgery*

1. INTRODUCTION

Defects in the craniofacial skeleton are of either congenital (birth defects) or developmental origin (resulting from trauma, infection, tumour, etc.). The purpose of reconstructing abnormalities is mainly cosmetic, in the idea to approximate a normal appearance, but there is also a functional point of view to it. Since they have a strong effect on the facial region, these types of alterations are highly visible, they affect the appearance, and thus the psychological state, social life, and possibility of the patient to found a family, to name a few.

Today there are modern synthetic implants like chin and mandible augmentation implants made of modern plastic materials (acrylates) available, in the shape of contoured two-piece chin implants and angular mandible augmentation implants. A good synthetic material needs to have following properties: biocompatibility, inertness, bone-similar weight or even lighter, capability to generate no artifacts on CT and MRI scans, ease of manufacturing, enough strength to resist functional stress, not expensive and low or no thermal conductivity (Zeilhofer et al.).

Three-dimensional imaging associated with rapid prototyping techniques and availability of alloplastic materials allow for the construction of an angular implant preoperatively (Bran et al.). This paper presents a technique of custom maxillofacial implant manufacturing using a titanium alloy and Selective Laser Melting (SLM), which represents state of the art in prototyping technologies .

2. PROBLEM STATEMENT

A 22 year old male patient was born with Goldenhar Syndrome, a variant of craniofacial microsomia (Goodrich et al.), which has affected the lower left half of his face. His case had been taken care of by an orthognatic surgeon from the Hospital of Ljubljana. The facial bones that have been affected by the malformation are the left zygoma, maxilla and mandible. With the exception of the mandible, the bones have been corrected to a certain extent using bone flaps from a donor site of the patient's body. The mandible has had an angular mandible augmentation implant attached to it in the past, which was made of acrylic plastics material, like the ones presented in the introduction of this paper. Soon after implant surgery the patient developed a bacterial infection at the implant site, requiring ulterior implant removal and treatment. Furthermore, the defect had been enlarged by the necessity of removing additional bone flaps from the infection site.

A solution was required to manufacture a similar implant out of a material that wouldn't allow for bacteria to develop and still successfully provide for a symmetric reconstruction of the person's appearance, while keeping the implant light enough to be functional.

3. IMPLANT MANUFACTURING

The most promising material was titanium, since it is anti-bacterial and strong, while a lot lighter than steel, yet heavier than bone tissue and very expensive. The problems that had to be solved were keeping a low weight and finding a method to manufacture the implant.

3.1 Development from medical data

The beginning was analyzing and processing of the data, in form of CT scans, that had already existed from previous stages of our patient's treatment. The set of CT images have then been converted into a three-dimensional digital model, using Materialise Mimics software, obtaining STL files as output, which can be manipulated and used in most RP technologies to produce real models.

Following was CAD modelling of the implant, performed with several 3D modelling and STL manipulation software packages. The idea was to split the skull in two parts in the middle, mirror the right, healthy, side over the left one and obtain the 3D model of the required implant through Boolean subtraction operations (Figure 1). However, there was a problem: due to the facial bone not being symmetric the defined mirror plane had to be different from the vertical mid-plane. Therefore, it was determined considering the certain well-defined features of the skull, like eye and nose cavities.

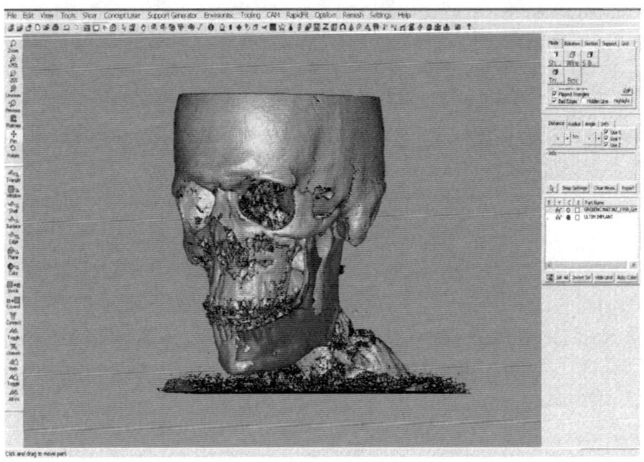

Figure 1. 3D subtraction model (red)

After subtracting, the model was inappropriate for implanting due to medical reasons and further modelling was required using 3D software (Figure 2).

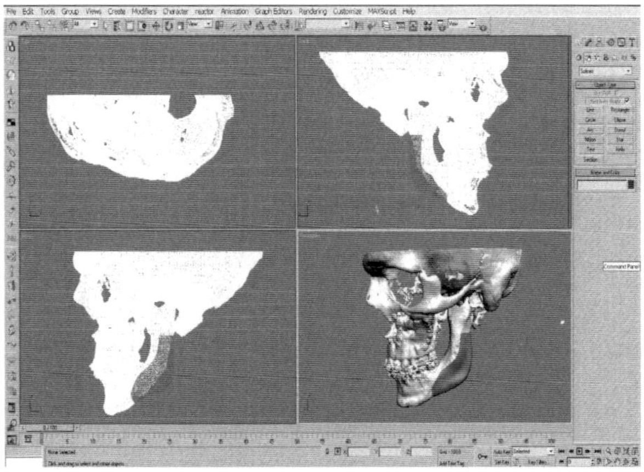

Fig. 2. 3D modeling of the implant

3.2 Production of implant model

After the final inspection of the 3D model (Figure 3), there was need of intermediate real models of the skull and implant, which could be cheaper to manufacture, in order to be tested for dimensional accuracy and analysed by the surgeon, for the phase of operation planning.

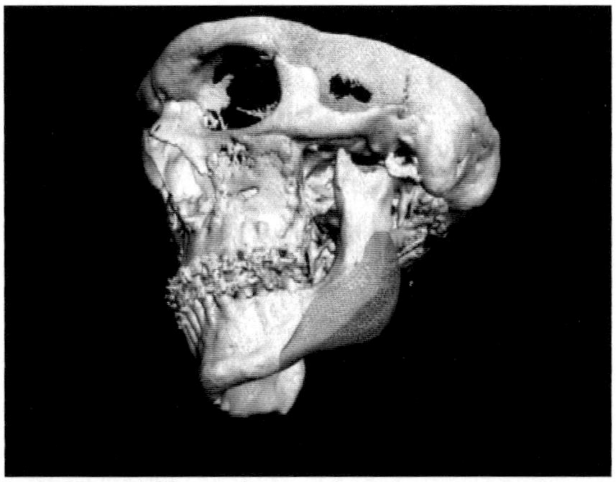

Fig. 3. 3D model inspection

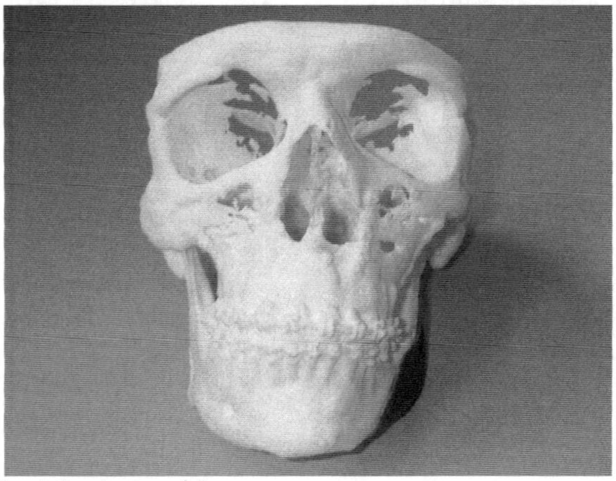

Fig. 4. Implant model

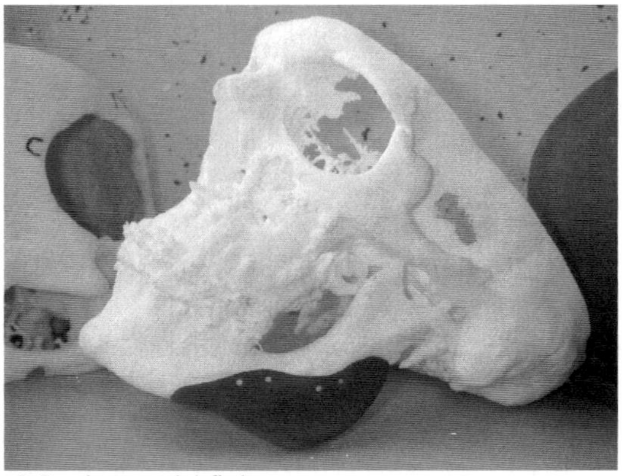

Fig. 5. Implant model fitting

In this sense a model had been developed out of polyamide 2200 using selective laser sintering procedure (Figure 4, 5) (Drestvensek, I.).

3.3 Production of biocompatible implant

The results at the former phase looking promising, the next step would be to produce the final implant out of titanium alloy. The selected method will be state of the art SLM procedure (Wholers, T.). The weight of the implant will measure approx. 6 g, which is quite acceptable, but the compromise was providing it with thin walls (cca 0,7 mm)The powder material will be processed under normal, non-sterile conditions in order to shorten the procedure, and the implant will be sterilized by means of gas sterilization, just prior to the implanting operation.

4. CONCLUSION

The great potential of RP technologies in medical applications is shown once more by the presented case study. Custom made craniofacial implants are expensive and only justify with repairing complex-shaped defects (Binder, W. J., Kaye A.). Further improvements could be directed towards finding solutions to make such implants lighter and cheaper to manufacture.

5. REFERENCES

Binder, W. J., Kaye, A. (1994). Reconstruction of posttraumatic and congenital facial deformities with three-dimensional computer-assisted custom designed implants. Plastic and reconstructive surgery, Vol. 94, 775-785

Bran, S., et al. (2002) *Reconstruction of bone defects with alloplastic biomaterials*. European Cells and Materials

Drstvensek, I., (2004) *Layered Technologies*, ISBN 86-435-0616-8, Faculty of Mechanical Engineering, Maribor, Slovenia

Goodrich et al. (1995) *Craniofacial Anomalies: growth and development from a surgical perspective*, Thieme, New York

Wholers, T. (2006) *Wholers Report*, Wholers Associates, ISBN 0-9754429-2-9, Fort Collins, Colorado, USA

Zeilhofer, H.F., et al.(1997) *Moglichkeiten und Indikationsbereiche der Kohlenstoffaserverstarkten Kunststoffe zur herstellung individueller Implantate fur die Rekonstruktion des Gesichts- und Hirnschadels*. Biomedizinische Technik

Annals of DAAAM for 2008 & Proceedings of the 19th International DAAAM Symposium, ISSN 1726-9679
Editor B. Katalinic, Published by DAAAM International, Vienna, Austria 2008
Make Harmony Between Technology and Nature, and Your Mind will Fly Free as a Bird

INCREASING ECONOMIC COMPETITIVENESS THROUGH INNOVATION

DURAN, I[rina] & DURAN, D[an]

Abstract: *Romania's economy cannot make steady progress unless the economic elements forming it will develop too. The means by which they will develop includes increasing the economic competitiveness of the institutions that generate innovative information on the market. This is the context of the paper development.*
Key words: *institution, competitiveness, innovation, information.*

1. INTRODUCTION

It is well known that productive activities generate an increase in value. Thus, what is at issue here is the fact that these activities should become more efficient, more performing and more competitive. The increase of economic performances is achieved through stimulating creativity and the technological or organizational performances by means of innovation.

The European Union claims and stimulates innovation through particular strategies implemented in the member countries.

2. INNOVATION AND ECONOMIC DEVELOPMENT

2.1. Innovation in institutions

The innovation process is a process from which there ensues a new or improved product that has been launched on the market. All of the companies using **product innovation** fall in this category. Moreover, it does not matter whether the innovation was initially made in the company or in other companies.

We also speak about innovation when a new product or one that has been significantly improved has been introduced in a company. **Process innovation**, as it is called, corresponds to the implementation of a productive process, of a distribution method, of an new auxiliary activity or of a new improved activity. Just as in the case of product innovation, process innovation should be new for the company in question, but it should not be new for the market or for the particular field of activity in which the company is carrying on its activities. (Moraru, 2008)

Innovation can be achieved by resorting to new technologies or to a combination of the technologies that are available – yet using the knowledge achieved in the company is easier and less expensive than changing the technology. The new information that has appeared and has been rightly applied generates innovation. Collected at the right time and rightly managed, the pieces of information will more rapidly lead to innovation in an institution. Technological development by means of new technologies or of a combination of those that are available is more awkward, more time consuming and last but

2.2. The European Union and innovation

In March 2000, the European council met in Lisbon in order to launch a strategic objective for the European Union aiming at strengthening the overall reach of the workforce, the economic reform and social cohesion as part of an economy based on knowledge.

On that occasion, the issue of innovation came up. The discussion about innovation involved two distinct topics, namely: "Creating a European area of research and innovation" and „Creating the proper environment for the inception and development of innovative business related to small and middle-sized institutions (SMI)."

The March 11, 2003 **European Commission Report** addressed to the European Council and Parliament, to the European Economic and Social Committee and to the Committee for Regions, COM (2003) 112, called ***"Innovative politics: updating the approach of the Union in the context of the Lisbon strategy"*** aimed at updating the concept of innovation in the EU and at expressing the multidimensional nature of the innovative phenomenon as well as its implications in economic policies.

The conclusions of the May 13, 2003 **European Council** regarding "strengthening the European innovative politics" called on the Lisbon Strategy once more, on its objective, which was to turn European Economy into the most dynamic economy in the world, based on knowledge and capable of ensuring a lasting economic growth.

The novelty value is the fact that the concept of a steady development is placed in the context of innovation, while innovative activities and technological developments are the key to achieving a steady development. In the same context, innovative politics should be integrated into field policies with a view to improving competitiveness.

3. THE ANALYSIS OF INNOVATIVE INSTITUTIONS IN ROMANIA

The latest data on the level of innovation in Romania is based on the results of a statistic research on innovative activities carried out between 2002 and 2004, coupled with the Community Innovation Survey – CIS 4. The statistic research took place in June 2005, covering a sample of 11,542 companies with more than ten employees.

The following data may be extracted from this study: one out of five companies implemented innovated products and/or services. However, just a few companies chose innovation as a single strategy (1.8% product innovation, 4.6% process innovation). We can observe that 13.3% of the companies resorted to combined innovation.

Types	Total	Product innovation only	Process innovation only	Product and process innovation
Total	5136	472	1203	3461
Small	2834	266	739	1829
Medium	1585	148	343	1094
Large	717	58	121	538

Tab. 1. Innovative enterprises by size class and type of innovation

More than half of the innovative companies are small – due to their adaptability and flexibility. Yet, the large ones have an increased innovative potential because their financial sources are considerably greater and will support the research and development process and later on the innovation process too.

Types	Research of development activity	Acquisition of machinery, equipment and software	Acquisition of other external knowledge
Total	19	62	19
Small	28	67	5
Medium	30	69	1
Large	15	59	26

Tab. 2. Structure of innovation expenditure

Table 2 highlights the fact that out of the total amount of expenses for innovation, the purchase of cars, equipment and software is the most (Miu, 2004)consequential. We would like to underline the fact that a large company does not spend too much money on its own research and development activities since it can afford to acquire outside information; a large company uses up half the amount utilized by small and medium-sized companies on their research and development activities, i.e., only 15%.

Type of information source	Total	Small	Medium	Large
Within the enterprise	38	35	42	45
Suppliers of equipment or software	37	34	42	38
Clients or customers	31	30	33	31
Competitors	19	21	17	16
Consultants or private R&D institutes	5	4	5	10
Universities or other higher education institutions	4	2	5	7
Government or private non-profit research institutes	3	2	5	4
Conference trade fairs, exhibitions	23	20	28	28
Scientific journals and trade/technical publications	24	21	28	28
Professional and industry associations	7	6	7	8

Tab. 3. Weight of enterprises, which indicated the most significant information source for innovation

As one can see in this table, business partners, that is, both the suppliers and the clients, are a very good source of innovation. In order to be competitive, companies should take competition into account too – since competition is an important informational source of innovation. Small companies use this type of information the most. Conferences, fairs, exhibitions and scientific magazines are also instrumental in promoting the innovation process by means of the innovative ideas put forth. The information used for supporting the innovation process has been provided by the company personnel. Learning institutions and research and development institutions provide only up to 5% of the information – and that is certainly a negative aspect and an annoying one as well.

Types	Organisational innovations	Marketing innovations
Total	30	5
Small	31	6
Medium	28	4
Large	28	4

Tab. 4. Weight of enterprises with organisational and marketing innovations

In a comparative analysis, the number of companies with organizational innovations is superior to that of companies with marketing innovations, that is, 30% vs 5%.

Types	Cooperation in total	Romania	European countries	United States of America and other countries
Total	19	15	8	2
Small	16	13	6	1
Medium	19	14	9	4
Large	30	24	15	5

Tab. 5. Weight of innovative enterprises based on cooperation

Out of the overall number of innovative companies, 19% declared that they had achieved innovation through cooperation. The easiest cooperation was with other Romanian companies, next in line was the cooperation with European companies and then came the cooperation with American companies and other companies from other places.

The aforementioned study also highlights the following : 37% of the innovative companies mentioned that the main consequential factor had been the improvement of the quality of the products or of the services, 32% named the growth of the production capacity and only 18% said that the main consequence consisted in the reduction of the negative impact on the environment and on the health system, and in the increase of the work security rate.

There were times when the money factor hindered the innovative process. Out of the innovative companies surveyed, 31% stated that the lack of finances was the main hindering factor, while 30% considered that innovative costs were too high.

4. CONCLUSION

Any performance evaluation, particularly the one accomplished from the point of view of innovation, has a substantial contribution to changing a company's attitude, pointing out to any company the way in which it should increase its competitiveness.

In order to reach the main objective pursued by industrial policy, i.e., increasing competitiveness through innovation, it is highly desirable that each company should pay greater attention to all the factors involved (personal, educational, management, marketing) and to the activities that can generate a steady improvement, productivity increase and added value.

5. REFERENCES

Miu, C. (2004). International Investment Forum, 4th Edition, *Available from:* http://forumworks.ccir.ro *Accessed:* 2008-05-26

Moraru, Gh. (2008). European Union, Innovation and us, *Available from:* htto://www.agir.ro/univers-ingineresc/uniunea_europeana_inovarea_şi_noi_1555.html *Accessed:* 2008-05-26

***http://www.amcsit.ro ("Politehnica" Managerial Agency of Scientific Research, Innovation and Technological Transfer)

***http://www.insse.ro/cms/files/pdf/ro/cap13.pdf (Statistical yearbook 2006)

***http://www.clubafaceri.ro

***htpp://www.mct.ro/ancs_web/index.php (National Authority for Scientific Research)

Annals of DAAAM for 2008 & Proceedings of the 19th International DAAAM Symposium, ISSN 1726-9679
Editor B. Katalinic, Published by DAAAM International, Vienna, Austria 2008
Make Harmony Between Technology and Nature, and Your Mind will Fly Free as a Bird

ECONOMIC EFFICIENCY IN ENERGY RENEWABLE SOURCES MANAGEMENT

DURAN, V[asile]; VARTOLOMEI-M, M[ihaela]; MILOS, T[eodor] & VARTOLOMEI M, M[ihael]

Abstract: *The development and restructuring of energy system has stipulated in the regional development plan and in the framework of energy infrastructure, to discover non-pollutant sources of energy. To develop methods for their settlement represent a real necessity in the current world challenges. Aeolian turbines represent generally a cheaper alternative for energy production than photovoltaic systems, in the region with upper 4 m/s of wind blow. In this paper we shall combine efficiently the demand and the manufacturing of an objective absolutely necessary in an energetically changing world. The financial instruments used for feasibility project study are ANV, IER, and the paper is focused on (ANV) for efficiency evaluation of Aeolian investment. In a multidisciplinary approaching (energetic engineering, automotive, statistics, mathematics, economics) we want to find solutions and spread out the results in order to settle on some clean processes and technologies in environment areas where it is possible.*
Key words: *Aeolian turbines, ERS investment, actualized net value, internal profitability rate, actualization rate*

1. INTRODUCTION

Energy sources are shared in two parts: energy renewable sources (ERS) and energy non-renewable sources (ENRS). ERS represents the resources in the environment, that are regenerating continuously or with certain periodicity and their consumption don't lead to a possible depletion. These kinds of energy sources are constantly renewed or restored and include: sun (solar), internal heat of the earth (geothermal energy), wind (wind power), vegetation (biomass), falling water, tides, and wave motion (hydropower). A clear feature of ERS is their independent existence by any man's consistently activity (Ambros et al., 1999).

The energetic potential of draught is determined with the follow relation (1) (Anton et al., 2007)

$$P = 0{,}6AV^3 \qquad (1)$$

Where: A – represents surface area described by propeller,
V – represents velocity of draught.

For wind velocity between $5-10\,m/s$ the specific power is $75-600\,W/m^2$. Only a part of kinetic energy of wind can be transformed in mechanic energy by Aeolian turbine (Lazarescu, 2003).

The capacity to transform wind energy in mechanic energy is measured by power coefficient C_p. It represents the efficiency of conversion wind energy in mechanic energy (Anton et al., 2007). It is a value less than 0,6. For the highest competitive Aeolian turbines, C_p is equal with 0,4 and the mechanic power obtained is calculated with the expression (2):

$$P_m = 0{,}24\,AV^3 \qquad (2)$$

2. EFFICIENCY EVALUATION

The main economic and financial indicators of ERS are: the specific investment (I_S) – it represents the specific cost of the investment, the cost of ERS on kilowatt-power investment capacity and the price of cost to produce renewable energy. (C_W) -it is calculated for the life time of equipments (Buglea, 2002). Besides, there are exploitation expensive (maintenance, reparation), that are very different for existent technology of producing energy.

About Aeolian installations, the specific investment I_S is determined by the cost of turbo generators and the cost of infrastructure (roads, backgrounds, electric connections). For present, the specific investment in Aeolian equipments is about 1,0 ... 1,2 thousands Euro/kW. The possibility to reduce the specific cost regarding the enhancement of turbo generators performances. But the cost will be reduced in the future years because of technologic development.

The price of energy cost deeply depends on Aeolian resources and the placement of the equipment. The volume of produced energy for the equipment is proportionally with the cube of wind velocity. This means that an error by 10% in wind velocity can make 30% error in energy cost calculation. The price of Aeolian energy cost can be from 4 to 15 Euro/kWh.

This represents a competitive cost with traditional energy sources cost. Regarding other aspects, the expenses on year for maintenance or reparation are about 1,5 - 2,5% from the investment cost. The normal duration for utilization of these equipments is usual between 15 and 20 years. The medium value of duration for utilization the maximum power is between 2.000 and 3.000 h/year.

Taking into account the externalities cost in the price of the cost for energy produced by traditional sources can change the Aeolian equipments competition radically in the favor of the last.

Another aspect is the external cost of the produced energy. Therefore, to produce energy by traditional technologies, electric stations burn fossil fuel, sending out more oxides (SO_2, CO_2, CO, NO_x) and pollutant powders. The nuclear stations harm people's health and entire environment life. Even after the end of functioning, these lands will become "tombs" and will be no more used.

The external effects, leads to external costs that illustrate the real value of the damages or the needed economic effort for avoid these effects producing.

For examples, the external cost of traditional energy sources is very increased. But, also, more powerful solar equipments need more land surface, with an opportunity cost. Aeolian equipments arouse more noises. A simple comparison between total cost of 1 kWh energy (exploitation cost + external cost) for traditional energy sources and renewable energy sources emphasize that the last ones are more competitive from cost aspect too.

The main question that needs a real response is "does really worth to invest in Aeolian investment?" This response can be given only after a technical-economic-financial analysis.

3. PROJECT FEASIBILITY CONDITION

Like any investment project, an energy renewable sources investment project needs to be feasible, that implies a cost-benefit analysis. Therefore, the proposed project worth to be implemented even if the expected effects surpass the achieved endeavors (relation 3 and 4). Thus, when there is a positive net effect the project is feasible.

$$Expected effects \geq Programmed efforts \quad (3)$$

$$Net\ effect \geq 0 \quad (4)$$

$$Net\ effect = Effects - Efforts \quad (5)$$

In the framework of the investment project, the efforts represent the total expenses during life of the project, effects represents total gross income, and net effect represents net income. So, our project is feasible if:

$$ATV > ATE \quad (6)$$

$$ANV = ATV - ATE \quad (7)$$

It results that:

$$ANV > 0 \quad (8)$$

Where: ATV – represents actualized total gross value
ATE – represents actualized total expenses
ANV – represents actualized net value

But efforts and effects evaluation suppose to take into account aspects regarding:

- period of studying (equipments life time T=10 … 20 years);
- actualization process using an actualization rate (*r*);
- opportunity cost;
- the reality of errors;

Cost/benefit analysis suppose a system of indicators as actualized net value, internal profitability rate, actualized total expenses, recuperation time etc.

The optimization of these methods consists in maximizing actualized net value, maximizing internal profitability rate, minimizing actualized total expenses, minimizing yearly expenses, and obtain maximum in safety of consumers' supplying.

The most important aspect is profitability, measured with actualized net value (ANV). Therefore:

$$NI_t = TI_t - TE_t \quad (9)$$

Where NI_t - represents net income estimated for year *t;*
TI_t - represents gross income estimated for year *t*;
TE_t - represents total expenses (costs) estimated for year *t*, including investment costs (I_t) and exploitation cost (C_t), exclusively the amortization sum.

$$TE_t = I_t + C_t \quad (10)$$

For energy source TI_t represents the gains obtained from energy producing process.

$$TI_t = W_t \cdot C_W \quad (11)$$

Where: W_t - represents the volume of produced energy
C_W - represents the opportunity cost for 1 kWh of produced energy.

Furthermore, ANV represents yearly benefit. For the period of time we study, the life time of energy equipments (T) is determined by adding yearly ANV:

$$ANV = \sum_{t=1}^{T} NI_t (1+r)^{\theta - t} \quad (12)$$

Where:T – represents the period of time of the investment;
r – represents the actualization rate (inflation rate or debt rate)
θ - actualization year

Using equation (12) in (9) we obtain equation (13) and (14):

$$ATV = \sum_{t=1}^{T} TI_t (1+r)^{\theta - t} \quad (13)$$

$$ATE = \sum_{t=1}^{T} (I_t + C_t) \cdot (1-r)^{\theta - t} \quad (14)$$

For a ERS project with initial investment cost I_0, the period of time T, ANV is:

$$ANV = \sum_{t=1}^{T} \frac{CF_t}{(1+r)^t} - I_0 \quad (15)$$

Where: *CF* – represents cash-flow, numeral flows foreseen.

$$\text{If } CF_1 = CF_2 = \ldots = CF_T \quad (16)$$

$$\Rightarrow ANV = CF_t \frac{1-(1+r)^{-T}}{r} - I_0 \quad (17)$$

4. CONCLUSIONS

In this paper, author wanted to find if it worth to invest in Aeolian equipments. Like any investment project, our energy renewable sources investment project needs to be feasible. This is why a cost-benefit analysis is very necessary. The ANV instrument is one of the most important efficiency indicators in market economy, where to obtain a more and more profit is a main objective for investors (Popa, 2005). In this context, a solution for optimization the profitability of the ERS project is to maximize ANV criterion. Therefore, the proposed project worth to be implemented if the expected effects surpass the achieved endeavors. Thus, in the project future study we have to find if there is a positive net effect (it represents the general condition of economic feasibility of the investment project). In this situation, our project will be feasible.

5. REFERENCES

Ambros, T.; Arion, V.; Guţu, A.; Sobor, I.; Todos, P.; Ungureanu, D. (1999). *Energy Renewable Sources*, Tehnica-Info Publishing, Chişinău, ISBN 9975-910-79-3

Anton, L.E.; Baya, A.; Milos, T.; Stuparu, A. (2007), *Experimental Hydrodynamics*, "Orizonturi Universitare" Publishing House , ISBN 978-973-638-330-4.

Buglea, A. (2002). *Investments and Their Financing*, West University Publishing, Timişoara, ISBN 973-8433-13-4

Popa, A. (2005). *Investments Efficiency*, Sitech Craiova, ISBN 973-746-059-6

Vartolomei, M. (2008), ***Development of Some Unconventional and Non Pollutant Energy Sources in Economic Efficiency and Sustainable Development Conditions***, COFRET, 11-13 juin, 2008, Nantes, France, ISBN 2.6905267.61.5

Annals of DAAAM for 2008 & Proceedings of the 19th International DAAAM Symposium, ISSN 1726-9679
Editor B. Katalinic, Published by DAAAM International, Vienna, Austria 2008
Make Harmony Between Technology and Nature, and Your Mind will Fly Free as a Bird

THE MATHEMATICAL MODEL OF THE MECHANICAL ANALYZER FOR NON-STATIONARY SIGNAL DECOMPOSITION BASED ON AN ARRAY OF RESONATORS

DUSEK, D[aniel]

Abstract: *This paper deal with possibility of non-stationary signal decomposition by mechanical analyzer based on an array of resonators. The goal of this work is design the mechanical analyzer which could be made by MEMS technology. Simple mathematical model of the array of resonators with mutual coupling is described in this paper. Spectrograms of known and also of random non-stationary input signals calculated by the array of resonators are presented in this paper too. These spectrograms showed quite good ability of non-stationary signal decomposition by the mathematical model.*

Key words: *Cochlea, signal, decomposition, resonators, modelling.*

1. INTRODUCTION

Most of mammalians and therefore also most of humans use mechanical analyzer for decomposition of non-stationary signal for communications every day, because mammalian ear functions just like mechanical analyzer. Mammalian ear transfers input mechanical stimulation into electrical signals which are process subsequently by brain. It is very important that also decomposition of input non-stationary signal into simple frequency components is done simultaneously (Robles & Ruggero 2001, Tianying 2007).

In term of signal decomposition the most important part of the ear is cochlea where basilar membrane with a sense cells is located. The basilar membrane starts to vibrate if it is actuated by external mechanical forces (acoustic pressure). It is important that high frequency tones excited the basilar membrane near its basal end and low frequency tones excite the basilar membrane near its apical end. This interesting character of the basilar membrane is based on its special mechanical behaviour concretely it is based on different width and thickness of basilar membrane along its length. This principle of function of mammalian cochlea was firstly described by Georg von Békésy and it was also verified by experimental measurements (Békésy, 1960).

Today Technologies of miniaturization like MEMS (Micro-Electro-Mechanical-System) and NEMS (Nano-Electro-Mechanical-System) are increasing and thanks to it possibility of design of artificial cochlea in proportion of few millimetres is more and more realistic. A few scientists in the world try to design the MEMS artificial cochlea (White & Grosh 2002, Chen et al. 2006). Most of them design the cochlea-mechanical analyzer just like real basilar membrane it is like membrane with different cross-section along its length. Another possibility of design of the mechanical analyzer can be an array of resonators.

The artificial cochlea based on MEMS technology can have great possibilities of utilization. It can replace today using electronically filters in cochlear implants' speech processors (Rubinstein 2004). The mechanical analyzer may be also used for audio and video data compression instead Short Time Fourier Transformation or instead Wavelet Transformation which need high power processors for real time decomposition.

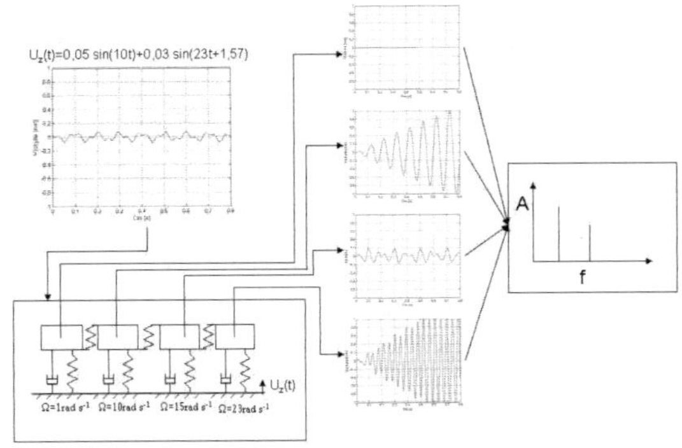

Fig. 1. Diagram of non-stationary signal decomposition based on an array of resonators

This paper deals with verification of ability of non-stationary signal decomposition by mathematical model of the array of resonators.

2. MATHEMATICAL MODEL

The basilar membrane is possible to simulate also by an array of resonators with mutual coupling. Model of every resonator is simple mass with spring and viscous damper. Neighbour resonators are additional connected by another spring.

Diagram of non-stationary signal decomposition based on the array of resonators is shown on figure 1. Principle of function of the array of resonators is very simple. If input signal excites the array of resonators so only that resonators start vibrate whose eigenfrequencies are equal to frequencies included in input signal.

This mechanical system can be described by following system of differential equations:

$$m_1q_1'' + b_1(q_1-q_z') + k_1(q_1-q_z) + k_s(q_1-q_2) = 0$$

...

$$m_iq_i'' + b_i(q_i'-q_z') + k_i(q_i-q_z) - k_s(q_{i-1}-q_i) + k_s(q_i-q_{i+1}) = 0 \quad (1)$$

...

$$m_nq_n'' + b_n(q_n'-q_z') + k_n(q_n-q_z) - k_s(q_{n-1}-q_n) = 0$$

where m_1-m_n are masses of resonators, k_1- k_n are springs of resonators, q_1-q_n, q_1'-q_n', q_1''-q_n'' are displacements, velocities and accelerations of resonators, k_s are springs between neighbour resonators and q_z is a displacement of kinematical excitation. Solution of the equation 1 for every resonator which vibrates by its eigenfrequency is following:

$$Aq_i = (Aq_{zi}\, m_i\Omega_i) / b_i \quad (2)$$

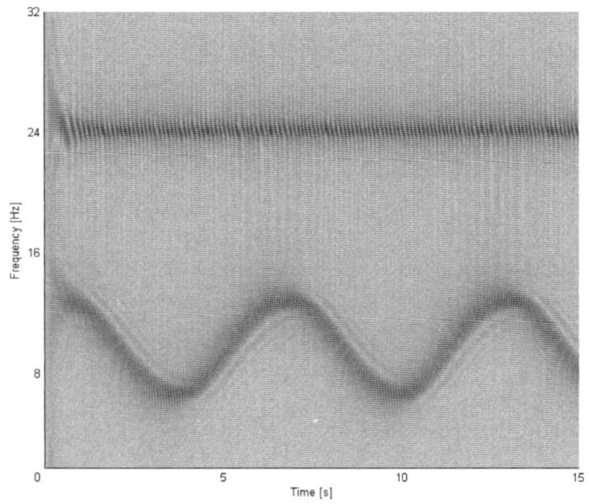

Fig. 2. Spectrogram of known test function calculated by array of resonators

where Aq_i is an amplitude of displacement of i-th resonator and Aq_{zi} is an amplitude of resonant frequency component which is contained in input signal, $\Omega_i = (k_i/m_i)^{0.5}$ is eigenfrequency of i-th resonator.

3. RESULTS

The mathematical model of the mechanical analyzer was tested by known test function first. Test function was compound from stationary and non-stationary part and it can be described by this equation:

$$q_z = A_{z1} \cos ((\omega_1 t)+(a \sin (t))) + A_{z2} \cos (\omega_2 t) \quad (3)$$

where first part of this term is non-stationary part with sinusoidal varying frequency and second part of this equation is stationary part with constant frequency through whole time range.

On the figure 2 is shown spectrogram of input known test signal calculated by the array of resonators. The constants in eq.3 was following: $A_{z1} = 1$ [m], $A_{z2} = 1$ [m], $\omega_1 = 60$ [rad/s], $\omega_2 = 150$ [rad/s], t = 0..15 [s], a = 20 [rad/s].

Parameters of mechanical system were following: Stiffness was same for all resonators' springs k = 1 [N/s]. Frequency range was from $f_1 = 10$ [rad/s] to $f_{190} = 190$ [rad/s] and then mass of resonators must have been also various value from $m_1 = 0.01$ [Kg] to $m_{190} = 2.77\ 10^{-5}$ [Kg]. Stiffness of connection springs between neighbour resonators was same for all with value ks=0.01 [N/s]. Viscous damping of resonators was also same for all b = 5 [Ns/m].

On the figure 2 it is possible to see quite clearly both parts of the input signal- stationary part with frequency 23.88 [Hz] and also non-stationary sinusoidal signal with varying frequency from 6.37 [Hz] to 12.3 [Hz]. The results also showed that for precise decomposition it is need to use quite high value of viscous damping and quite low value of stiffness of connection springs.

After verification of the mathematical model by known test input signal also calculation of spectrogram of random non-stationary signal was made. Default *mtlb* signal from Matlab system was used. Length of the signal was N=4001 [-], and sampling frequency was fs = 7418 [Hz].

Calculation of spectrogram was made in frequency range from 20 [rad/s] to 23000 [rad/s] (it is in range from 3.18 [Hz] to 3662 [Hz]) with frequency step 20 [rad/s]. Parameters of mechanical system were following: Stiffness was same for all resonators' spring k = 1 [N/s]. Mass of resonators had various

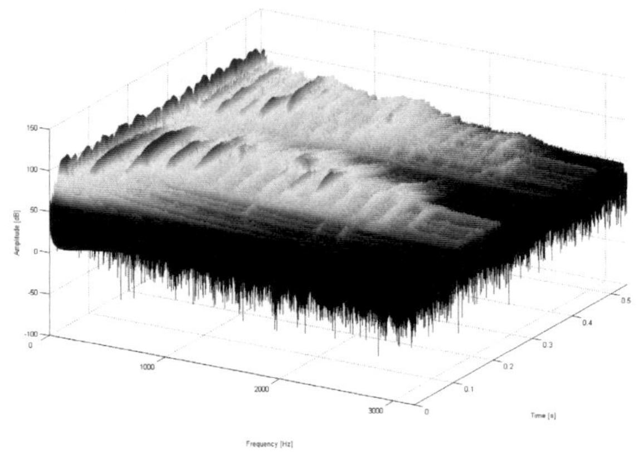

Fig. 3. Spectrogram of random non-stationary signal *mtlb* calculated by array of resonators

value from $m_1 = 0.0025$ [Kg] to $m_{1150} = 1.89\ 10^{-9}$ [Kg]. Stiffness of connection springs between neighbour resonators was same for all with value ks=0.01[N/s]. Viscous damping of resonators was also same for all b=50[Ns/m]. The spectrogram of random non-stationary signal *mtlb* calculated by array of resonators is shown on figure 3.

4. CONCLUSIONS AND FUTURE WORKS

The mathematical model of mechanical analyzer based on the array of resonators was created and also verified by known test function. Results showed that mathematical model is able to decomposed non-stationary signals quite well. But it is need to use quite high value of viscous damping and quite low stiffness of connection springs between resonators. Using of resonators with low value of viscous damping or with high value of connection springs' stiffness causes creation of higher and lower harmonic components and smudge of the results.

The goal of future works is considered of possibility to design the mechanical analyzer which could be produced by MEMS technology.

5. ACKNOWLEDGEMENT

This paper was written with the support by the Grant Agency of Academy of Science of Czech Republic - Project No.: KJB201730802.

6. REFERENCES

Békésy, G. von (1960). *Experiments in Hearing*, McGraw-Hill, New York.

Chen, F.;Cohen, I. H.; Bifano, T. G.; Castle, J.; Fortin, J.; Kapusta, CH.; Mountain, D. C.; Zosuls, A. & Hubbard, A. E. (2006) A hydromechanical biometic cochlea: Experiments and models, *J. Acoust. Soc. Am.*, Vol. 119, No.1, January 2006.

Robles, L. & Ruggero, M. A. (2001). Mechanics of Mammalian Cochlea, *Physiological reviews*, Vol.81, No.3, July 2001.

Rubinstein, J. T. (2004) How cochlear implants encode speech, *Current Opinion in Otolaryngology &Head and Neck Surgery 2004*, Vol. 12.

Tianying, R. (2007). Longitudinal pattern of basilar membrane vibration in the sensitive cochlea, *Proceeding of the National Academy of Science of the United States of America*, January 2007

White, R. D. & Grosh, K. (2002). Design and characterization of a MEMS piezorezistive cochlear-like acoustic senzor, *Proceedings of International Mechanical Engineering Congress and Exposition*, New Orleans.

Annals of DAAAM for 2008 & Proceedings of the 19th International DAAAM Symposium, ISSN 1726-9679
Editor B. Katalinic, Published by DAAAM International, Vienna, Austria 2008
Make Harmony Between Technology and Nature, and Your Mind will Fly Free as a Bird
*Young Researcher and Scientist Paper / * Supervisor, Mentor*

VACUUM CAM TECHNOLOGY TO INCREASE MACHINING ACCURACY

EFTIMIE, A[lina] L[uciana]; EFTIMIE, R[oxana] & EFTIMIE, A[drian] C[atalin]

Abstract: *The paper presents the CAM technology used to save expensive raw material and to achieve maximum dimensional and shape precision. The cutting forces developed during the machining process are distributed on the setting surface. Using this technology, the flatness of the surface is conserved even if the machined layer is stressed by the primary technology. The parallelism of the inner opposite sealing faces of the ring is excellent due to the same machining line which conserves the axes backlash for both pair of faces. The bending forces caused by vacuum distribution are cancelled by opposite support on the setting device. The CAD modelling of the parts and the setting device are presented as application for filled plastic parts and also for lead bronze parts.*
Key words: *modelling, design, CAD-CAM technology*

1. INTRODUCTION

Reciprocating compressor packing case is designed to seal by sliding friction against the rod. As a general design, the packing case is an assembly based on several connected cups containing groups of three cut rings arranged to seal one against the other and all against the rod. Using modern design solution, and engineered plastic field materials, the friction wear decreases, sealing capabilities and reliability are continuously optimized. During the functioning time, the set of rings are wearing and they must be changed by a new set of rings. The general arrangement is shown in fig.1.

The sealing elements – three segment cut rings as fig.2 – are usually made by filled plastic material, as PTFE or PEEK, and they are machined one by one, losing a lot of expensive material.

2. APLICATION FOR TANGENT CUT SEALS

Analyzing the shape of each segment, we can observe – fig.3 – that it is possible a special arrangement of the three segments on a raw material disk to reach maximum useful material and to make possible to machine a wide range of inner and outer diameter from the same primary disk dimension. In this specific case we are not limited to a specific dimension of the raw material diameter, and the raw material stock is minimized (Chu et al., 2006).

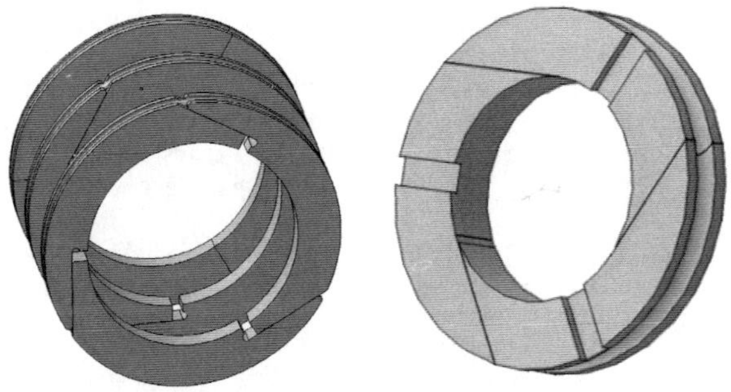

Fig. 2. General arrangement of the sealing rings.

The disks are obtained using lathe cutting technology which provides not enough smooth and flat faces.

The main problem is to fix the primary raw material disk to be machined, because they are not convenient fixing points around them. The problem was solved using a vacuum device to force material against the profiled faces, by atmospheric overpressure. The first face will be sealed and setted on the vacuum device, by O-rings, in deeper grooves than usual, to prevent rubber elastic forces to unseal. The second face is settled directly to the metallic face. It was demonstrated by practice that the pressure is enough strong to fix the ring, and both surfaces will be machined enough smooth and flat.

Different type of rings and materials can be machined using CAM technology and vacuum to set them.

Depending on the sealing target, the segmented rings can be rectangular cross-section or profiled faces (see fig.5).

As a result of different materials tested for the same shape application as PTFE, PEEK or bronze, we have obtained a better clearance of the sealing face position related to the opposite face on the other ring segment. The bronze is more difficult to be fixed and machined because of machining forces

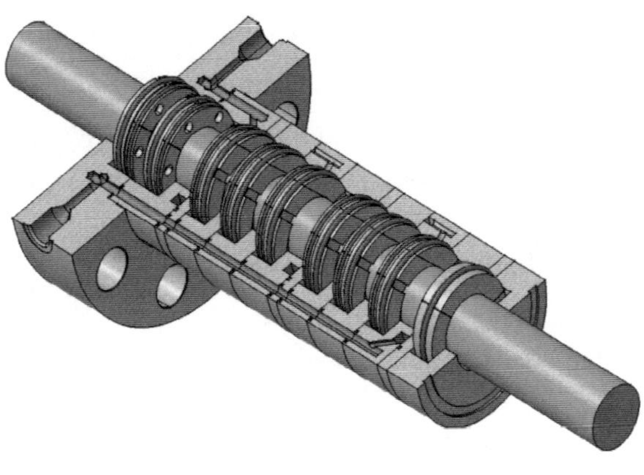

Fig. 1. General arrangement of a gas packing case

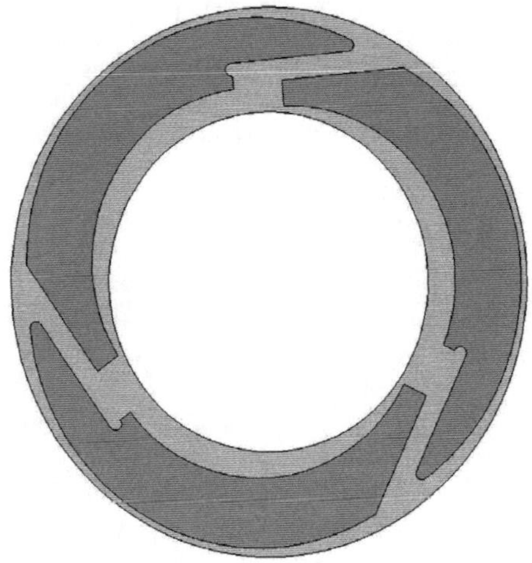

Fig. 3. Special arrangement of the ring segments

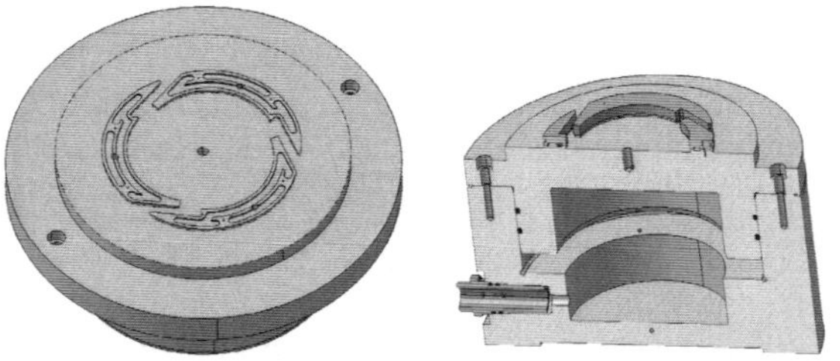

Fig. 4. Vacuum device to fix the part when machining

and less flexibility necessary for the first machined face.

The tangent cut of the ring is necessary to be normal to the side surface to recover the flatness of the three-segment ring when is assembled to work. In this case, both faces of each cut ring will slide against the paired cut ring in respect to the minimum backlash to allow thermal expansion and casual liquid film layer provided by the impure gas compression process.

If the tangent cut is not normal to the face, the composed ring will be space arranged unusable to seal on the faces (Clifford, 2001).

In practice, the liquid and solid impurities can block the internal slide of the segment and external slide of the ring into the pair. The sliding oscillation cleans the surfaces pushing out the impurities. Abrasion is usually damaging the brass sealing rings face and part of solid particles can be included in filled plastic body (Clifford, 2002).

The tangent cut perpendicularity depends on CNC precision and also on machining strategy to generate the cut. If the cut is made extracting material layer by layer, the required smoothness and flatness of the machined faces was not obtained because filled materials are enough plastic to flow, the chips following the cutting edge. The best solution is to finish the faces machining at once all the wall thickness or face area.

The faces are machined as a flat helicoidally speed, machining alternate both faces, two times each. This is the method to eliminate the stressed face of the raw material disk. It was used a 35 mm diameter milling cutter body for 5 mm radius carbide inserts. Machining parameters were established by practice to 2000 rpm and 2000 mm/min linear speed.

It was also better to spread out the plastic by compressed air flow passing through the mill inner channel in the same time with outer air jet. The machining parameters to profile the tangent cut were established to 4500 rpm and 1500 mm/min linear speed for 4 mm diameter carbide end mill. As a different case, the curve face as well as the flat brass rings – radial cut – the machining parameters were established to 4500 rpm and 2200 mm/min linear speed for 6 mm diameter carbide end mill or ball end mill. The test made using steel tool mill provides not enough flatness and smooth faces, required less than 0.01 μm.

We have tested non-filled plastic materials and it was proved that if the material's plasticity increases, the quality of the machined faces is decreasing. In this case the finished face shows as vibrate and rough, inadequate to seal against the pair.

In the same time, the machining forces increase enough to unstuck the disk from the vacuum base plate.

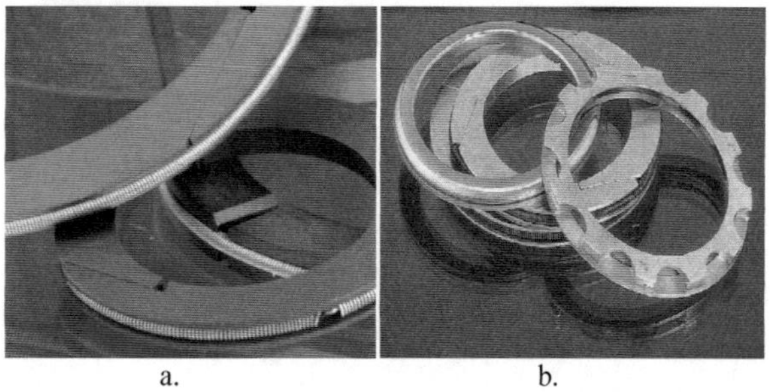

a. b.

Fig. 5. Cut sealing rings – CAM and vacuum technology.

Fig. 6. CNC machining

Only soft working parameters can use to diminish machining forces - 50 % speed (fig.6).

All the machining tests were performed on vertical machining center FADAL 4524. Using a CAM technology and the CNC machine, were provided interchangeable segmented sealing rings, in respect to required clearances and recommended backlash (Landry, 2002). The smoothness of the surfaces was around 10 μm for graphite and glass fiber filled PEEK and PTFE and more than 18 μm for lead filled bronze.

It is not possible to use wet machining because of machining force increase. The disk was unstuck during the machining process. The absolute vacuum under-pressure usually was 50000 Pa (0,5 bar abs) and it was not possible to increase it because of the vacuum pump limit. The active area for vacuum was 705 mm2 for each one of the ring's segment. Considering the friction coefficient for filled plastic, as usual $\mu=(0,2\div0,16)$, the force to fix the full ring was 21,15 N and drops abruptly to 7,05 N, which is one segment fixing force.

During all this machining tests, vacuum was used to reach the friction force between the disk and the vacuum device base plate. After tangential cutting of the ring, the pressure and the friction forces were redistributed on each segment which is set independent.

3. CONCLUSION

This method allows fixing plastic filled parts as well as filled bronze in order to be high speed CNC machined. As a result of machining, the parts are providing accuracy. The process is repetitive, obtaining full interchangeability of the sealing rings. The parts conserve the flatness of the machined face if the vacuum is used as technology to fix the part. The vacuum technology accepts to follow the arrangement of the friction forces during the process, reaching the optimum setting. The smoothness of the faces is fine and fulfils the sliding requirements to seal. This method is recommended as a good technology to machine sensitive materials to conserve flatness during the machining process.

4. REFERENCES

American Petroleum Institute Standard 618, "Reciprocating Compressors for Petroleum, Chemical, and Gas industry Services" American Petroleum Institute, Washington, D.C., fourth Edition, June 1995.

Chu, C.H.; Song, M.C. & Luo, V.C.S. (2006) Computer Aided Parametric Design for 3D Tire Mold Production, *Computers in Industry*, Volume 57, Issue 1, p. 11-25

Clifford, Matthews. *Engineers' Guide to Rotating Equipment – The Pocket Reference*, Professional Engineering Publishing, Londra, (2002).

Clifford, Matthews. *Engineers' Guide to Pressure Equipment – The Pocket Reference*, Professional Engineering Publishing, Londra, (2001).

Landry, J. (2002) Backlash vs. lost motion, Design News for Mechanical and Design Engineers, *Available from*: http://www.designnews.com, *Accessed*: 2008-06-16

Annals of DAAAM for 2008 & Proceedings of the 19th International DAAAM Symposium, ISSN 1726-9679
Editor B. Katalinic, Published by DAAAM International, Vienna, Austria 2008
Make Harmony Between Technology and Nature, and Your Mind will Fly Free as a Bird
*Young Researcher and Scientist Paper / * Supervisor, Mentor*

WEAR TEST CONCEPT TO OPTIMISE THE RIDER RINGS RELIABILITY

EFTIMIE, A[lina] L[uciana]; EFTIMIE, R[oxana] & EFTIMIE, A[drian] C[atalin]

Abstract: *The paper presents the concept to test the wear rate of the test sample from the filled plastic material used for chemical plants air compressor rider rings. The device designed to test the samples, copies the same type of harmonically oscillatory movement as the piston relative to the cylinder liner. The wear of the sample is governed by the sliding friction, contact pressure in the sliding surface and maximum sliding speed. All this parameters can be modified to follow various real conditions during compression function.*
Key words: *methods, optimisation, technology*

1. INTRODUCTION

The reciprocating compressors function is based on the sliding movement of the piston into the cylinder (fig.1). The axis of this dynamic assembly can be vertical or horizontal. The vertical position of the axis leads to minimum wear due to piston assembly weight, which is acting on vertical direction, but the maintenance is more difficult to do because of the height of the assembly which requires crane utilities. The horizontal position of the axis involves minimal maintenance tools to be used, because of minimal height position of the cylinders. Dynamic parts general arrangement of a piston is shown in fig. 2, where we can see marked as crosshatch, the rider rings, made of plastic filled material.

Using engineered plastic filled materials as graphite filled PTFE or PEEK, as well as graphite bronze, we can improve the reliability, decreasing the friction wear. The wear depends on contact pressure between the rider ring lower half face and the cylinder. As an old solution, it was used the oil film to decrease the friction, but the oil usually contaminates the compressed gas, decreasing the final product purity.

It is necessary to establish which material is better to be used for a specific application. This analysis can be done using samples tested in similar conditions as the rider rings when functioning. To simulate the necessary condition for the test, as contact pressure, sliding speed and gas action into the sliding wear process, it is necessary to use a special test device (fig.3).

2. WEAR TEST DEVICE CONFIGURATION

The slide of the piston into the cylinder is governed by

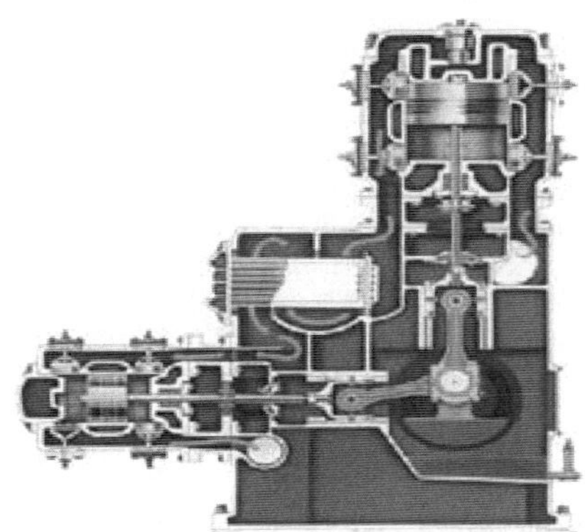

Fig. 1. General arrangement of a reciprocating compressor

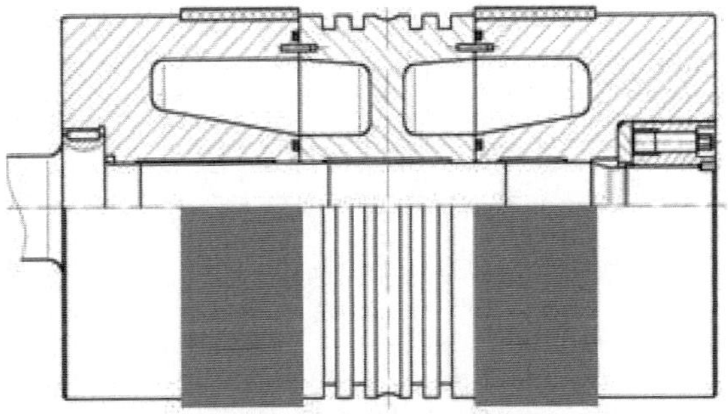

Fig. 2. General arrangement of piston's rider rings

simple harmonic oscillation laws, because the movement is generated by the revolution of a crankshaft which is acting using a connecting rod to a linear translation of the piston. The laws of piston movement are as follows described by equations:

$$y = A\sin(\omega t + \theta_0)$$
$$v = \omega A\cos(\omega t + \theta_0)$$
$$a = -\omega^2 A\sin(\omega t + \theta_0)$$

,where A=half stroke of the piston or crankshaft eccentricity;
ω=angular speed or if T=movement period
$\omega = 2\pi/T$
t=time from the zero movements start

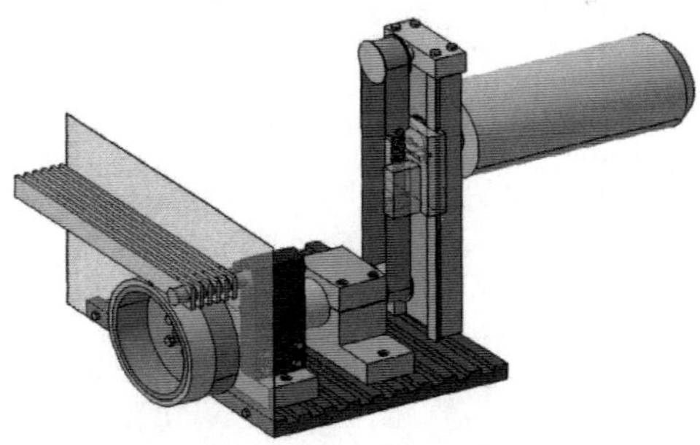

Fig. 3. Sliding wear test device

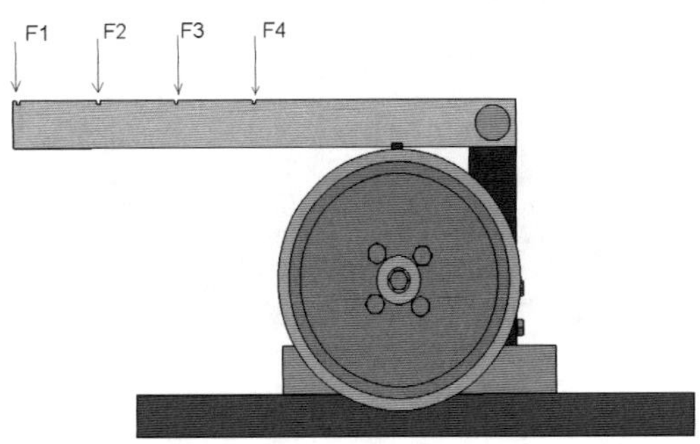

Fig. 4. Forces arrangement

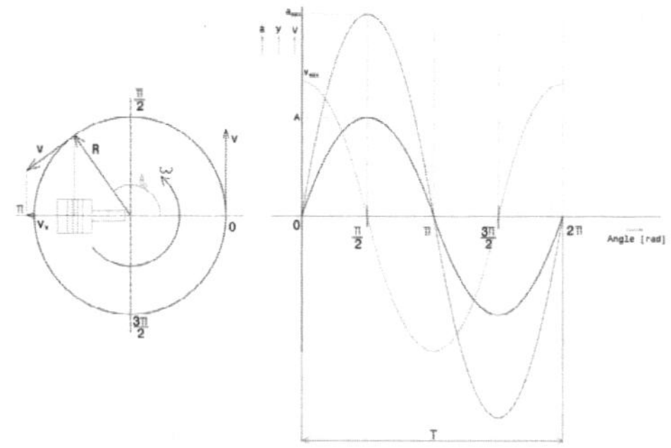

Fig. 5. Simple harmonic oscillation

θ_0=primary angular run out from zero of the start
a =instant acceleration of the piston

This movement is obtained on the test device by the slider, actuated by an adjustable eccentric pin on the motor shaft. The rotating speed of the electric motor can be adjusted to 500rpm reaching the real rotating speed of the compressor crankshaft. The motion is transferred to the cylinder using a flexible tape, stretched by a resort, losing any backlash. The cylinder outer face is made by two different microstructure cast iron sleeves to make possible the test at once two samples on each sleeve. The lever presses the filled plastic samples by forces generated by the lever weight and if necessary, adding specific weight on each lever. The force which acts on the test sample is calculated using simple mechanical laws of the lever.

The system conserves the angular motion of the harmonic movement of the small wheel, but increases the real sliding motion of the test cylinder on the sample, reaching the same stroke for sliding if wheels ratio is well done.

All the parameters of the harmonic oscillation of the piston are transferred identically on the harmonic sliding motion between tests sample and the device cylinder.

The test wheel rotates alternative clockwise and counter clockwise against the test sample reaching the same contact length as the compressor stroke. During one period T of the motion, the sample will act two times on the sleeve surface and the roughness will keep small particles cut from the test sample face. In this case, periodically the sliding characteristics of the contact surface will change continuously, identically to the real case, in respect to the same motion laws.

We can test in the same time four different samples acting on two sleeves with different gray cast iron microstructures. For the sleeves, the quality surface must be identical to cylinder's sliding surface, it means roughness less then R_a=0.4μm.

Fig. 6. a) New rider ring b) Rider ring wear

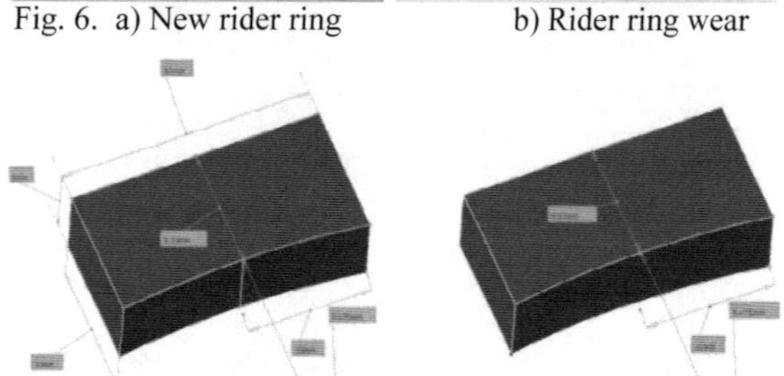

Fig. 7. a) New sample test m_0=2.380g b) Sample test wear m=2.260g 5,04 % mass lost

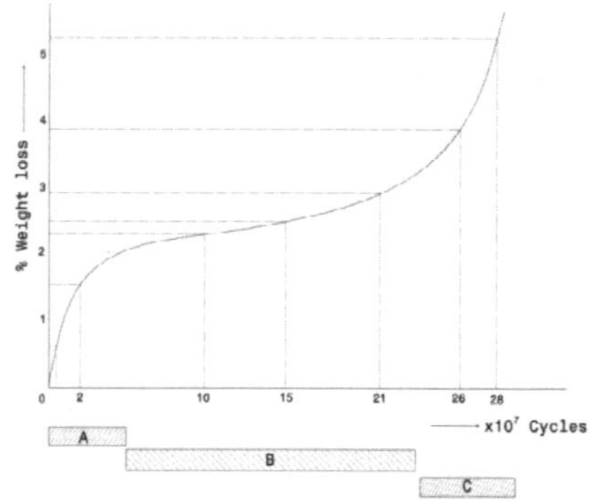

Fig. 8. Experimentally wear diagram for PTFE+25%Gr 500 cycles/min; contact sliding pressure $p_c=4x10^5$Pa

As you can see comparing the wall thickness of a new rider ring (fig. 6a) to a wear rider ring, we can observe that the wear appears to the bottom side and can be appreciated measuring the slot depth. The slots keep the mechanical impurities inside, cleaning the sliding surface and they cannot be removed by the rider rings wear. The primary tests sample dimension and form are shown in fig.7a and the wear is accepted as usually up to 0.5mm, as shown in fig.7b. As usually, the recommended wear of the rider rings is up to 0.5mm, but in practice, it can reach even 2mm. We have tested 25% graphite filled PTFE samples and the percent mass wear rate is show as an experimentally diagram in fig.8. We appreciate that the A area of cycles marked on the fig.8 diagram, shows the sample material transfer on the sliding cylinder surface where it is generating a low friction film layer. On the same diagram (fig.8), the B area shows that the wear is minimized due to the A low friction film. The C area shows that this film is being damaged at a high rate, due to abrasive action of impurities which destroy the low friction film.

3. CONCLUSION

The wear test method allows to establish the sample wear under identically condition of contact pressure, gas parameters and sliding motion parameters as the real case of a piston sliding without lubrication in the cylinder. It can be tested a various couple of materials sliding one against the other. The device is adaptable to be used for various conditions of tests. If there are generated wear diagrams for different couples of sliding material and various conditions, we can extend the tests results to appreciate the wear of rider rings when processing.

4. REFERENCES

American Petroleum Institute Standard 618, "Reciprocating Compressors for Petroleum, Chemical, and Gas industry Services" American Petroleum Institute, Washington, D.C., fourth Edition, June 1995.

Clifford, Matthews. *Engineers' Guide to Rotating Equipment – The Pocket Reference*, Professional Engineering Publishing, Londra, (2002).

Clifford, Matthews. *Engineers' Guide to Pressure Equipment – The Pocket Reference*, Professional Engineering Publishing, Londra, (2001).

www.precisioncompressorparts.com, Accesed 2008.06.09

Lawrence Berkeley National Laboratory Washington, D.C. (2003), Improving Compressed Air System Performance - a sourcebook for industry, Available from: http://www1.eere.energy.gov/industry/bestpractices/pdfs/compressed_air_sourcebook.pdf Accessed: 2008.06.12

Annals of DAAAM for 2008 & Proceedings of the 19th International DAAAM Symposium, ISSN 1726-9679
Editor B. Katalinic, Published by DAAAM International, Vienna, Austria 2008
Make Harmony Between Technology and Nature, and Your Mind will Fly Free as a Bird

CAD MODELLING OF COUPLINGS TO EXTEND RELIABILITY

EFTIMIE, L[ucian]; TIEREAN, M[ircea] & BALTES, L[iana]

Abstract: *The paper will present the method to design a gear tooth and the extended application to the inside gear coupling teeth connection, using CAD design. There are presented the basic elements of the CATIA software application into this case and relative connection to the specific technology. It is obtained a typical modified shape which allows the hub angular oscillation keeping the minimum clearance between the connected teeth.*

Key words: *modelling, CAD-CAM technology.*

1. INTRODUCTION

Gear couplings are as usual, industrial assemblies used to transmit torque of rotating movement and to compensate angular and radial misalignments of the connected end shafts (figure 1). There is a wide range of application for high torque or high speed rotation, used in steel processing, rubber, paper, plastic industries, water and wind power stations, transport and conveyor equipment, or mining applications.

2. APLICATION FOR GEAR COUPLING HUB

It must be used a compact design and reliable solution to avoid the single-point basic design. It is better to use a line contact, but it is very difficult to configure the tooth shape to reach this requirement.

Basic design means involute type teeth, connected as shown in Fig. 1, as a general design assembly made using CATIA V5 software. The connected teeth allow a relative axial movement made by friction of the side connected surfaces of the teeth. The involute is used because the precise machining technology. The clearance between the connected teeth surface manage the angular flexion of this assembly (Alaci et al., 2008).

Using the CAD design – CATIA V5 software, the teeth shape can be generated and modified to accept the angular flexion of the connection.

The contact angle can be extended up to ±5° rotation of the axis, and the contact line is realised on more than 70% length. It is necessary to use nitride steel and a quenched tempered body of the tooth, to realise enough strength (McGinnity et al., 2005).

Firstly, it is necessary to generate the tooth involute shape using specific relations and parameters as shown in figure 2.

Fig.1. General CAD design of the gear coupling assembly.

Fig. 2. Generating the involute tooth.

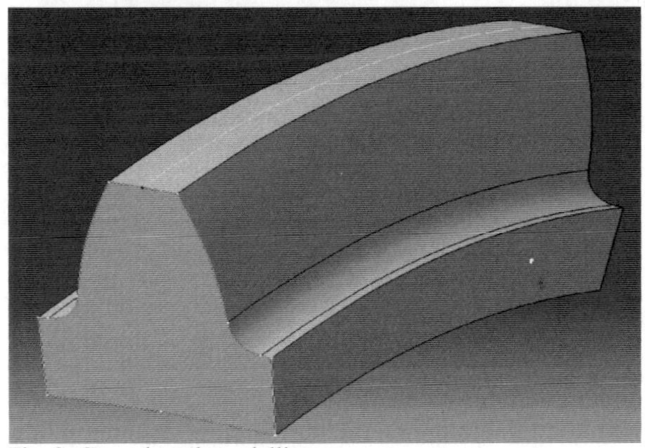

Fig. 3. Curved tooth modelling.

Using the extrusion command of the CATIA software, we generate a curve tooth – figure 3 - to allow the rotation of the hub.

Now there are starting the technical problems, because this fact can be possible for minimum clearance, only if we modify the cross-section of the tooth, as shown in figure 4.

Fig. 4. Complex space curve surface of the tooth.

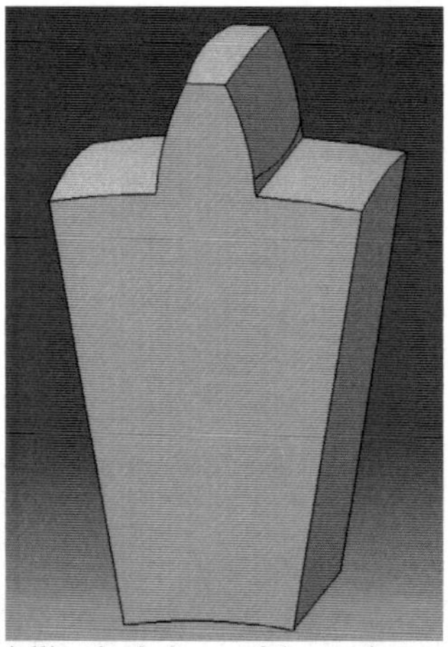

Fig. 5. Modelling both faces of the tooth.

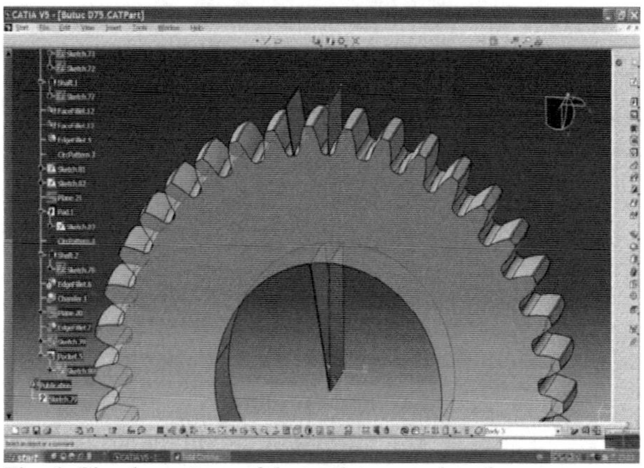

Fig. 6. Circular pattern of the teeth to complete the gear.

Starting from the top of the tooth, we sketch – see figure 4 - the tilting curve as a circle to allow the oscillation on the horizontal contact teeth an we make a pocket to remove material along the involute. The result is a space surface from the same family of involute.

Using specific parameters, we choose the dimension of the top and bottom of the tooth, generating a slice of the future gear – figure 5 and figure 6.

The slice will be repeated as a circular pattern, completing the gear.

General design will be made to obtain the gear hub

Fig. 7. General aspect of the hub.

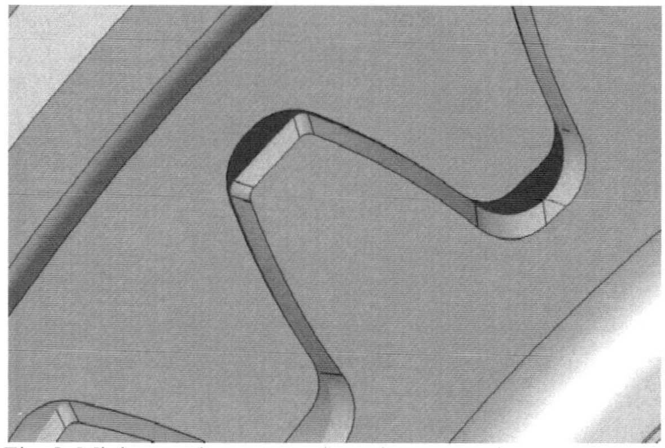

Fig. 8. Minimum clearance and maximum angular misalignment.

assembly in respect to the inner and outer dimensions to realize a good connection to the end shaft.

The connection of the relative teeth will allow the minimum clearance and the necessary oscillation of the assembly in the same time with linear contact – figure 8 (Landry, 2002).

It can be observed that the cross section of the teeth is identical to the medium plane cross section of the involute (Patras-Ciceu et al., 2008).

That means that the contact curve line can be repeated step by step on other two cross sections, symmetrical aligned to the medium plane. That is the situation which allows maximum angular misalignment. Machining of the surface can be made by automatic four axis CNC vertical centre machine Fadal 4020. The surfaces must be smooth and the clearance can be diminished up to 0.02 mm.

3. CONCLUSION

Using CAD design, we provide engineered solutions to achieve better reliability during the gear coupling function.

The backlash between the connected teeth is diminished and optimised to reach the required thickness of the filled oil film. The torque is transmitted mostly by involute surface rolling, minimising the faces sliding friction. The reliability of the coupling is increasing caused by this change of surface contact, similar to ball contact inside a bearing. To obtain a reliable contact of this connected faces it is necessary to increase the hardness of the surface concomitant with body strengths. It means to use nitride alloyed steels hardened and tempered before the surface thermo-chemical treatment. The surface space model will be precisely machined using CAM technology.

4. REFERENCES

Alaci, S.; Amarandei, D.; Ciornei, F.C. & Patras-Ciceu, S. (2008) Simulation of gear rack generation of involute spur gears, *Annals of the Oradea University. Fascicle of Management and Technological Engineering,* volume VII (XVII), p. 1172-1180

Landry, J. (2002) Backlash vs. lost motion, *Design News for Mechanical and Design Engineers*, http://www.designnews.com

McGinnity, M. & Mancuso, J. (2005) New pump coupling reduces effects of torque, misalignment and unbalance, *World pumps,* May, p. 34-37

Patras-Ciceu, S.; Ciornei, F.C. & Alaci, S. (2008) The surface temperature in sliding contacts of gear working flanks, part 1 – Analytical solutions, *Annals of the Oradea University. Fascicle of Management and Technological Engineering,* volume VII (XVII), p. 502-505

CATIA documentation V5R17 2007

Annals of DAAAM for 2008 & Proceedings of the 19th International DAAAM Symposium, ISSN 1726-9679
Editor B. Katalinic, Published by DAAAM International, Vienna, Austria 2008
Make Harmony Between Technology and Nature, and Your Mind will Fly Free as a Bird

MOLD PATTERN TECHNOLOGICAL CAD DESIGN OF THE COMPRESSOR CROSS-HEAD

EFTIMIE, L[ucian]; TIEREAN, M[ircea] & BALTES, L[iana]

Abstract: *The paper explains the CAD technology developed for compressor cross head machining as CAM application to reach the right body. There are applied additional layers necessary to machine the part surface after casting. Also are presented the basic elements of the technology and the layer cross section structure of the part. The technology is explained as an extension of a common sample. The CAD modelling of the cross head application was realised as CATIA V5 CAD design and CAM machining on vertical machine centre Fadal.*
Key words: *modelling, design, CAD-CAM technology.*

1. INTRODUCTION

Reciprocating compressors (fig. 1) usually compress gas inside of cylinder, which inner volume is diminished due to piston movement. The compressed gas goes out through the discharge valves. The new cycle is starting with the admission of the gas, through the suction valve, as a result of the depression caused by the expansion of the inner volume of the cylinder. All the variable volumes of the inner cylinder are obtained as a result of a linear movement of the piston. If the piston is not moving following a straight line, the different supporting elements will wear.

The target of this paper is to show how is generated the additional shallows to reach the necessary shape of a pattern, especially for the compressor's cross head application used in hydrogen processing.

2. APLICATION FOR CROSS HEAD STRUCTURE

The cross-head body is a complex part which allows compensating all efforts of the connecting rod transferring only the axial movement relative to the piston rod.

During the function the cross-head slides, on the cylindrical side faces, covered by low friction material. It is necessary to use a strong, but light design, to force on minimum the oil sliding film. Such a design is presented in the Fig. 2 using CATIA V5 software (Balc et al., 2004).

The part will be made by steel, using green sand casting technology, which requires specific steps to be followed as we show on a simple sample.

Let's suppose that the part is as simple as a cylinder, and applying all the following technological steps.

We analyze the body and choose the splitting plane as symmetry into the horizontal plane (see fig. 3). Having a symmetry system, we can work only on a half body, and we choose to work on the upper side generating the final assembly by symmetry. We add all around the side surface the necessary material to compensate the thermal contraction (figure 3).

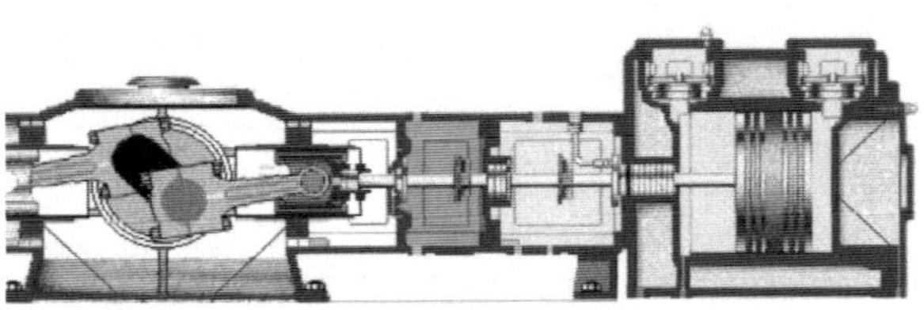
Fig.1. Reciprocating compressor cross-section stage.

Fig. 2. 3D model of the cross-head.

Fig.3. Splitting plane and layers.

Machining will be necessary to obtain a smooth surface. We apply a new layer of material to allow the safe machining of the surface (see fig. 3) (Chu et al., 2006).

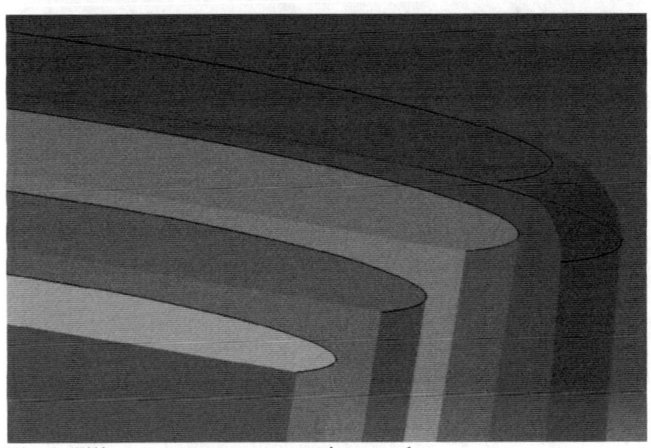
Fig.4. Fillets to prevent contraction cracks.

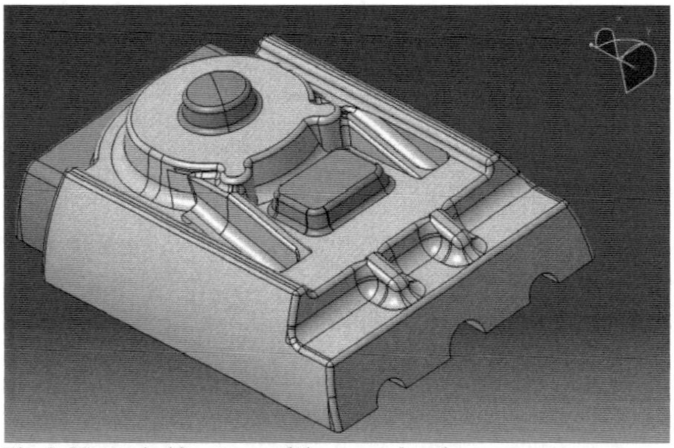
Fig.5. Upper half pattern of the cross-head

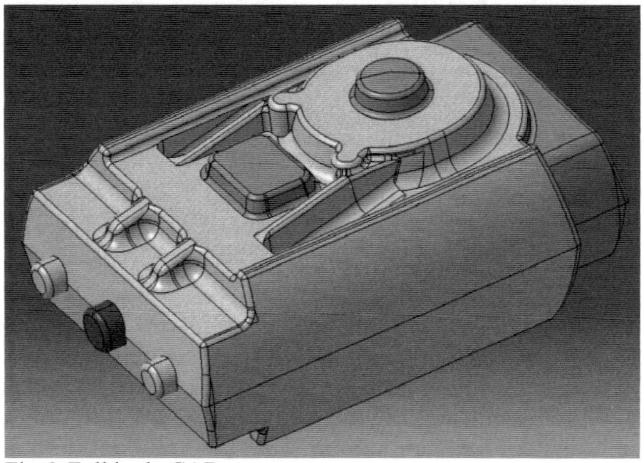
Fig.6. Full body CAD part.

The moulding technology uses cline or taper walls to the extraction direction minimizing the friction to the walls mould cavity (fig. 3) (Madan et al., 2007).

To prevent contraction cracks on the edges of the casting steel when cooling, we apply required fillets on the inner or outer edges (fig. 4). Applying symmetry, we can reconstruct the full body of the sample.

The final analysis will be made on the cross-section of the full body of the sample.

We have applied all the necessary technological steps on the finished part design shown in figure 1, and we have obtained the final shape of the cross-head body, as shown in figure 6. Into this moulding technology, caused by the complex empty inner part, we have applied also some spatial support - red colour represented – for the mould core. This red marked supports will be included both in pattern and core.

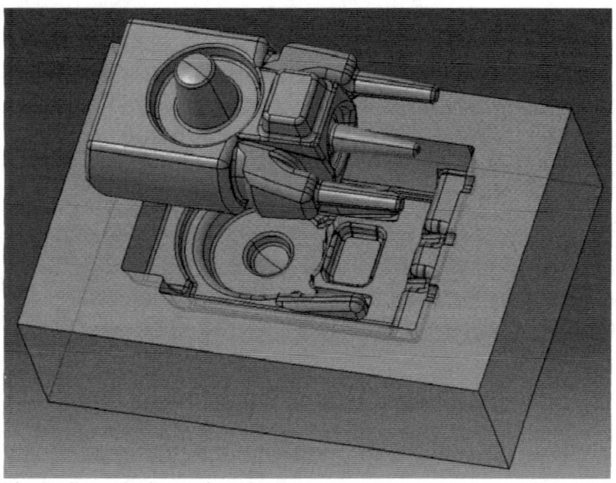
Fig.7. Core design using CAD.

Fig.8. Machining pattern.

By symmetry, we can extend the CAD part to the full body (fig. 7) (Cooper, 2001).

The pattern was produced by wood, because it is cheaper for only one part, using CAM technology on a vertical machining centre Fadal 4525, as shown in figure 8.

It can be observed that all the round surfaces connected to the splitting plane, are not modified as a cline surface because they are enough curved and they are not modifying the cavity when extracting (Homburg et al., 2006).

The full body was oriented one against the other – half by half – using two pins crossing the splitting plane.

The core box was realised using the same technology. It means that we have applied all the necessary layers to allow the thermal contraction, the machining layers, radius to prevent cracks and setting extensions to fix the core inside the mould.

3. CONCLUSION

Using CAD design and CAM technology can be obtained complex and precise patterns.

This is a simple method to produce other more complex parts. The method is applicable to the green sand moulding technology. It is necessary to respect all conventional marking colours for outer surfaces, for inner core marks and splitting plane.

The mould must have correct sized risers to feed rapidly the melted steel into the casting case during solidification. The casting designer will take in consideration the steel flows into the mould, to avoid turbulent flow as well as rapid fill of the mould. Gates and risers are added during the moulding process.

Foundry manufacturing methods specific to green sand moulds, applied for this moulding technology, improves quality of the final products.

4. REFERENCES

Balc, N. & Campbell, R.I. (2004) From CAD and RP to innovative manufacturing, *Computing and solutions in manufacturing engineering*, COSME '04

Chu, C.H.; Song, M.C. & Luo, V.C.S. (2006) Computer Aided Parametric Design for 3D Tire Mold Production, *Computers in Industry*, Volume 57, Issue 1, p. 11-25

Cooper, K. (2001). *Rapid prototyping technology: selection and application*, Marcel Dekker, New York

Homburg, N. & Wellbrock, E. (2006) Knowledge-based manufacturing strategy and methods for foundries, *RTejournal*, Ausgabe 3, p. 1-10

Madan, J.; Rao, P. V. M. & Kundra, T. K. (2007) Computer Aided Manufacturability Analysis of Die-cast Parts, *Computer-Aided Design & Applications*, Vol. 4, No. 1-4, pag. 147-158

Annals of DAAAM for 2008 & Proceedings of the 19th International DAAAM Symposium, ISSN 1726-9679
Editor B. Katalinic, Published by DAAAM International, Vienna, Austria 2008
Make Harmony Between Technology and Nature, and Your Mind will Fly Free as a Bird

CONTRIBUTIONS ON FINAL ELABORATION MODEL OF A FEED KINEMATIC CHAIN (L.C.A.)

ENCIU, G[eorge]; NICOLESCU, A[drian]; STANCIU, M[ihai] & IVAN, I[oana]

Abstract: *The article presents researches realised on an experimental testing stand in order to determine the behaviour and the transfer function for a LCA (feed kinematic chain). The activity is framed in a complex study programme of the dnamic behaviour of the machining centres. There have been determined the stability diagrams both for the mechanisms and for the complete kinematic chain, and then using the identification method has been determined the transfer function which will be used for creating the dynmic model of the machining centre.*
Key words: *LCA, feed drives, identification method, LabView, machine-tool.*

1. INTRODUCTION

Within the framework of a contract of the Machines and Production Systems (MSP) Department, from the Engineering and Management of the Technological Systems (IMST) Faculty, we had to determine the model of a feed kinematic chain for a machining centre for milling and reaming operations (Breaz et al., 2007; Ispas et al., 1999; Neugebauer et al., 2007). The experimental stand is presented in Figure 1 and Figure 2 and has the same mechanisms as the machining centre (Bausic & Diaconu, 2000; Zaeh & Baudisch, 2003).

2. IDENTIFICATION OF TRANSFER FUNCTION OF THE MECHANICAL STRUCTURE

The experimental researches were at different stages and complexity levels and, of course are impossible to be here presented. We present only some aspects about the identification of the transfer function.

Fig.1. The experimental stand

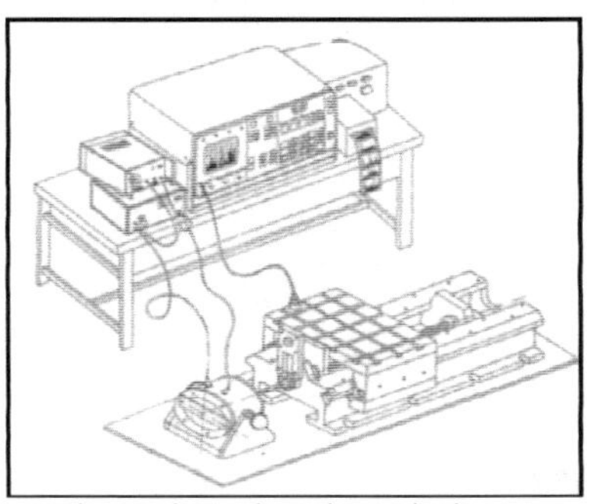

Fig.2. The experimental stand – schematical

One of the reasons of this research was to collect the information necessary to apply an identification method.

For determination the transfer function of the frequency characteristics it was used the method of the smallest ponderous (balanced) squares

Assuming that the process it was described by its frequency characteristics:

$$(j_{wk}) = Re(_{wk}) + jIm(_{wk})\,;\;\; k = 0, \dots p \qquad (1)$$

it was trying the approximation of frequency spectrum that it was determinated with the model help

$$M^{(s)} = \frac{\sum_{k=0}^{a} A_K \cdot S^k}{1 + \sum_{i=1}^{b} B_i \cdot S^i} = \frac{A(S)}{B(S)};\;\; S = Jw \qquad (2)$$

To minimize the error occurred between the real frequency characteristics and the model characteristics the matrix equation should be solved

$$V \cdot X = W \qquad (3)$$

where

$$\begin{aligned} X &= [B_2, \dots B_2 \cdot r1 | B_1, \dots, B_2 \cdot r2 \\ &\quad + 1 | A_0, \dots, A_2 \cdot r3 | A_1, \dots, A_2 \cdot r4 + 1]^T \\ &= \left[U_2, \dots U_2 \cdot r1 \right| \; {}^{0}_{n=n+1} | S_0, \dots, S_2 \cdot r3 | T_1, \dots, T_2 \cdot r4 + 1\right]^T \\ X &= [B_2, \dots B_2 \cdot r1 | B_1, \dots, B_2 \cdot r2 \\ &\quad + 1 | A_0, \dots, A_2 \cdot r3 | A_1, \dots, A_2 \cdot r4 + 1]^T \\ &= \left[U_2, \dots U_2 \cdot r1 \right| \; {}^{0}_{n=n+1} | S_0, \dots, S_2 \cdot r3 | T_1, \dots, T_2 \cdot r4 + 1\right]^T \end{aligned} \qquad (4)$$

In the relations above have been used the notations:

$$S_x = \sum_{k=0}^{p} w_k^x \qquad (5)$$

$$T_x = \sum_{k=0}^{p} [Re(_{wk})] \cdot w_k^x \qquad (6)$$

$$U_x = \sum_{k=0}^{p} [Re^2_{(wk)} + Im^2_{(wk)}] \cdot w_k^x \qquad (7)$$

and the index value $r_{1,\dots}r_2$ has been determined with the relations;

$$r1 = \begin{cases} \frac{b}{2} & b = even \\ \frac{b-1}{2} & b = odd \end{cases} \quad r3 = \begin{cases} \frac{a}{2} & a = even \\ \frac{a-1}{2} & a = odd \end{cases}$$
$$r2 = \begin{cases} \frac{b}{2} - 1 & b = even \\ \frac{b-1}{2} & b = odd \end{cases} \quad r4 = \begin{cases} \frac{a}{2} - 1 & a = even \\ \frac{a-1}{2} & a = odd \end{cases} \qquad (8)$$

For application of the method of smallest balanced squares we need to know the grades *a* and *b* of the polynoms of the numerator and respectively the denominator of the transfer function. Because this is a rare situation, the respective method is applied for more combinations of grades a and b and we have to choose that combination that approximates the best the frequency spectrum determined experimentally.

	a=2,b=3	a=2,b=4	a=2,b=5	a=2,b=6
B2	1.39e-07	2.3e-07	2.3e-07	3.34e-07
B4		8.55e-15	8.55e-15	2.99e-07
B6				7.56e-22
B1	6.93e-05	4.04e-05	2.72e-05	8.06e-06
B3	9.46e-12	5.51e-12	2.49e-12	7.38e-13
B5			-1.21e-19	-3.57e-20
A0	-3.16e-06	-3.02e-06	-2.35e-06	-8.37e-06
A2	-1.02e-12	-1.32e-13	-1.00e-13	-4.56e-13
A1	1.79e-05	1.04e-05	1.05e-05	3.10^e-06

Tab 1.

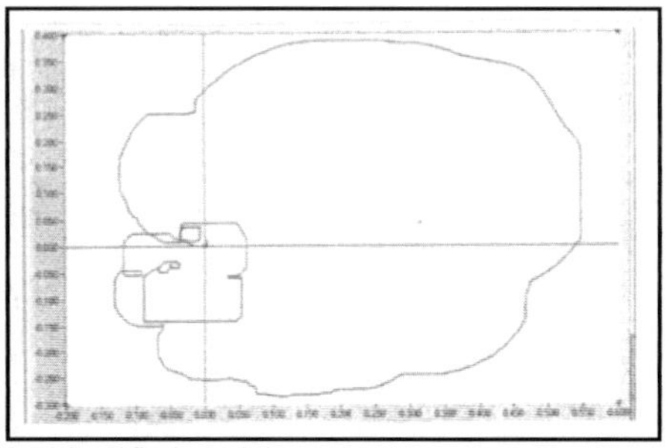

Fig.3. The identified model

For the aplications of smallest methods squares has been realised with a virtual instrument (.vi) in graphic programming Lab VIEW environment. The program required the existence data file contains the value R(ω_k) and Im(ω_k).

In the file of entrance each of two groups of date must be occupy one column, and the dates from same line must be sepparated with special character TAB, have been the ASCII code.

The previously mentioned programme was rolled successively for more combinations of grades a and b of the transfer function polynoms, the resulted values of A and B coefficients for the combinations used are presented in Table 2.1.

The comparation with the identified model (Fig.3.), through the Nyquist diagram has not led to the selection of any variant and that to be proceeded to extend the grade and we have obtained a series of values for the coefficients.

$$H_{mt}(s) = \frac{U_t(s)}{U_m(s)} = \frac{c_t \cdot c_m \cdot (s \cdot \xi_{12} + k_1)}{|\Delta|/s} \qquad (9)$$

Where, $|\Delta|$ is the matrix determinative from (9) relation so the computation of $|\Delta|$ leeds at:

$$\begin{aligned}|\Delta| &= (R+L)\cdot\{(s^2\cdot J_1 + s\cdot\xi_1)\cdot[s^2\cdot J_2 + s\cdot(\xi_1+\xi_{12}) + k_1] + (s^2 + J_2 + s\cdot\xi_2)\cdot(s\cdot\xi_{12}+k_1)\} + c_m\cdot s\cdot c_m\cdot[s^2\cdot J_2 + s\cdot(\xi_1+\xi_{12}) + k_1] = \\ &= (R+L)\cdot[s^2 + J_1\cdot J_2 + s^3\cdot J_1\cdot(\xi_1+\xi_{12}) + s^2\cdot J_1\cdot k_1 + s^3\cdot\xi_1\cdot\xi_2 + s^2\cdot\xi_1\cdot(\xi_2+\xi_{12}) + s\cdot\xi_1\cdot k_1 + s^3\cdot\xi_{12}\cdot J_2 + s^2\cdot J_2\cdot k_1 + s^2\cdot\xi_1\cdot\xi_{12} + s\cdot k_1\cdot\xi_2] + s^3\cdot c_m^2\cdot J_2 + s^2\cdot c_m^2\cdot(\xi_1+\xi_{12}) + s\cdot c_m^2\cdot k_1\end{aligned} \qquad (10)$$

If for easier calculations we note R+L=V we obtain:

$$|\Delta| = V\cdot J_1\cdot J_2\cdot s^4 + [J_1\cdot(\xi_1+\xi_{12})\cdot V + \xi_1\cdot J_2\cdot V + \xi_{12}\cdot J_2\cdot V + c_m^2\cdot J_2]\cdot s^3 + [V\cdot J_1\cdot k_1 + V\cdot\xi_1\cdot(\xi_2+\xi_{12}) + V\cdot J_2\cdot k_1 + V\cdot\xi_2\cdot\xi_{12} + c_m^2\cdot(\xi_2+\xi_{12})]\cdot s^2 + (V\cdot\xi_1\cdot k_1 + V\cdot\xi_2\cdot k_1 + c_m^2\cdot k_1)\cdot s \qquad (11)$$

Replace:

$$\begin{gathered} J_1 = J_m = 0{,}975[kg\cdot m^2] \\ J_2 = J_t = 1{,}357\times 10^{-4}[kg\cdot m^2] \\ J_c = 2{,}43\times 10^{-4}[kg\cdot m^2] \\ d = 3{,}2\times 10^{2}[m] \\ R = 3{,}6\times 10^{-1}[\Omega]\ ;\ L = 0{,}9\times 10^{-4}[H] \\ c_m = 2{,}34\times 10^{-1}[Vs/rad]\ ;\ c_t = 4{,}71\times 10^{-1}[Vs/rad] \\ \xi_m = \xi_1 = 7{,}53\times 7{,}53\times 10^{-4}[Nms/rad] \\ \xi_t = \xi_2 = 3{,}45\times 10^{-4}[Nms/rad] \\ \xi_{12} = 6{,}5\times 10^{-3}[Nms/rad] \\ k_1 = 1{,}02\times 10^{-3}[Nm]\ ;\ k_2 = 6{,}5\times 10^{-3}[Nm] \end{gathered} \qquad (12)$$

It's obtained:

$$H_{mt}(s) = \frac{1{,}116\times 10^{-4}\cdot s + 14{,}235}{8{,}28\times 10^{-8}\cdot s^3 + 1{,}437\times 10^{-5}\cdot s^2 + 1{,}128\cdot s + 61{,}2} \qquad (13)$$

According of structure 1 .c.a. moulded from c.c. engine mechanical – stand structure, the total function will be:

$$H_t(s) = \frac{H_{mt}(s)\cdot H_{ids}(s)}{H_{mt}(s)\cdot H_{ids}(s) + 1} \qquad (14)$$

Made the replacement and adequate calculation have been obtained the final transfer function:

$$H(s) = \frac{a_4\cdot s^4 + a_3\cdot s^3 + a_2\cdot s^2 + a_1\cdot s + a_0}{b_9\cdot s^9 + b_8\cdot s^8 + b_7\cdot s^7 + b_6\cdot s^6 + b_5\cdot s^5 + b_4\cdot s^4 + b_3\cdot s^3 + b_2\cdot s^2 + b_1\cdot s + b_0} \qquad (15)$$

3. CONCLUSIONS

The conclusions of the experimental and research activity integrate in the procedures which have become well-known for the elaboration of the model and for the transfer function of a positioning LCA for machine-tools.

There have been used dynamic testing methods to determine the stability diagrams and through identification the elaboration of a real transfer function. It is to be mentioned that the transfer function obtained through identification has a total different degree and shape than the standard ones used in the theory of the automated systems theoretically allocated to LCA.

We also mention that the stability diagrams obtained from the final transfer function obtained through identification have a behaviour degree in comparison with the stability diagrams obtained in the intial phase, which are between 5-7%.

The developed researches will allow for our department the determination of the transfer functions for positioning LCA through identification method.

4. REFERENCES

Bausic, F. & Diaconu, C. (2000) *Dinamica masinilor (Machines dynamics),* Conspres Publishing, ISBN: 973-99571-3-7, Bucharest

Breaz, R.E., Bologa, O.C., Oleksik, V.S. & Racz, G.S. (2007). Computer Simulation for the Study of CNC Feed Drives Dynamic Behavior and Accuracy, *EUROCON, 2007. The International Conference on "Computer as a Tool"*, pp. 2229-2233, ISBN: 978-1-4244-0813-9, Warsaw, September 2007.

Ispas, C., Mohora, C. & Caramhai, S., (1999) *Simularea sistemelor integrate de fabricate (Simulation of integrated fabrication systems),* Bren Publishing, ISBN: 973-9493-15-7, Bucharest

Neugebauer, R., Denkena, B. & Wegener, K. (2007). Mechatronic Systems for Machine Tools, *CIRP-Annals – Manufcturing Technology*, Vol. 56, Issue 2, pp. 657-686, ISSN: 0007-8506, Imprint ELSEVIER.

Zaeh, M F & Baudisch, T., (2003). Simulation environment for designing the dynamic motion behaviour of the mechatronic system machine tool, *Proceedings of the Institution of Mechanical Engineers, Part B: Journal of Engineering Manufacture*, Vol. 217, No. 7, (2003), pp. 1031-1035, ISSN 0954-4054, Professional Engineerng Publising.

Annals of DAAAM for 2008 & Proceedings of the 19th International DAAAM Symposium, ISSN 1726-9679
Editor B. Katalinic, Published by DAAAM International, Vienna, Austria 2008
Make Harmony Between Technology and Nature, and Your Mind will Fly Free as a Bird

NON–HERTZIAN NORMAL CONTACT OF ELASTIC BODY MODEL BY FINITE ELEMENTS

ENESCU, I[oan]; VLASE, S[orin]; LEPADATESCU, B[adea]; PURCAREA, R[amona] & DUMITRASCU, A[dela]

Abstract: *In this paper is presented an elastic foundation modeling by finite element. The difficulties of elastic contact stress theory arise because the displacement at any point in the contact surface depends upon the distribution of pressure throughout the whole contact. To find the pressure at any point in the contact of solids of given profile, therefore, requires the solution of an integral pressure. The difficulty is avoided in the solids; it can be modeled by a simple Winkler elastic foundation or "mattress" rather than an elastic half-space.*
Key words: *elastic, mattress, modeling, finite element.*

1. ELASTIC FOUNDATION MODEL

The profile, therefore, requires the solution of an integral equation for the pressure. The difficulty is avoided if the solids can be modeled by a simple Winkler elastic foundation or 'mattress" rather than an elastic half-space .The model is illustrated in fig.1. The elastic foundation, of depth h, rests on a rigid base and is compressed by a rigid indenter. The profile of the indenter, z(x, y), is taken as the sum of the profiles of the two bodies being modeled:

$$z(x,y)=z_1(x,y)+z_2(x,y). \tag{1}$$

There the difficulty of elastic contact stress theory arises because the displacement at any point in the contact surface depends upon the distribution of pressure throughout the whole contact. To find the pressure at any point in the contact of solids of given is no interaction between the springs of the model, shear between adjacent elements of the foundation is ignored. If the penetration at the origin is denoted by δ, then the normal elastic displacements of the foundation are given by:

$$\overline{u}_z(x,y)=\delta-z(x,y), \delta>z, \tag{2}$$

$$\overline{u}_z(x,y)=0, \delta\le z. \tag{3}$$

The contact pressure at any point depends only on the displacement at that point, thus

$$p(x,y)=(K/h)\cdot\overline{u}_z(x,y), \tag{4}$$

where K is the elastic modulus of the foundation.

For two bodies of curved profile having relative radii of curvature R' and R'', z(x ,y) we can write

$$\overline{u}_z=\delta-(x^2/2R')-(y^2/2R''), \tag{5}$$

inside the contact area. Since $\overline{u}_z=0$ outside the contact, the boundary is an ellipse of semi-axes a = $(2\delta R')^{1/2}$ and b=$(2\delta R'')^{1/2}$ (Johnson 1985).

The contact pressure by (3), is:

$$P(x,y)=(K\cdot\delta/h)\{1-(x^2/a^2)-(y^2/b^2)\}. \tag{6}$$

which is paraboloid rather ellipsoidal as given by Hertz theory. By integration the total load is:

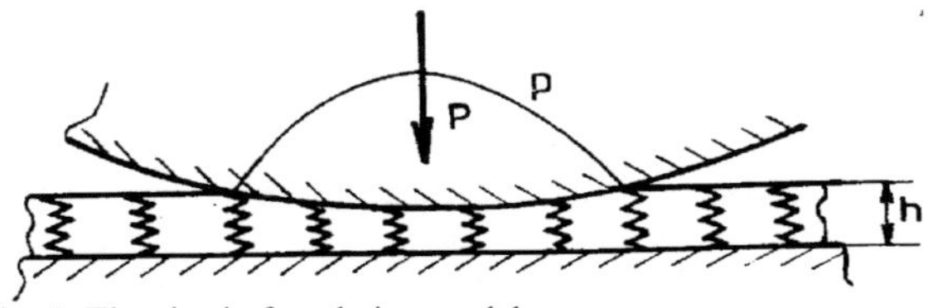

Fig. 1. The elastic foundation model

$$P=K\cdot\pi ab\cdot\delta/2h. \tag{7}$$

In the axes-symmetric case a=b=$(2\delta R)^{1/2}$ and

$$P=\frac{\pi}{4}(\frac{Ka}{h})\frac{a^3}{R}. \tag{8}$$

For the two-dimensional contact of long cylinders:

$$\overline{u}_z=\delta-x^2/2R=(a^2-x^2)/2R, \tag{9}$$

so that

$$p(x)=(K/2Rh)(a^2-x^2), \tag{10}$$

and the load

$$P=\frac{2}{3}\left(\frac{Ka}{h}\right)\frac{a^2}{R}. \tag{11}$$

In the bi-dimensional case (cylinder), K/h=1.8E*/a, and in the axes-symmetric case K/h=1.7E*/a where E* is:

$$\frac{1}{E^*}=\frac{1-\nu_1^2}{E_1}+\frac{1-\nu^2}{E_2}. \tag{12}$$

Equations (8) and (11) express the relationship between the load and the contact width. Comparing them with the corresponding Hertz equations, agreement can be obtained, if in the axes-symmetric case we chose K/h=1.70E*/a and in the two-dimensional case we choose K/h=1.18E*/a. For K to be material constant it is necessary to maintain geometrical similarity by increasing the depth of foundation h in proportion to the contact width a. Alternatively, thinking of h as fixed requires K to be reduced in inverse proportion to a. It is consequence of the approximate nature of the model that the value of K, required to match the Hertz equation are different for the two configurations. However, if we take K/h=1.35E*/a, the value of a under a given load will nod be in error by more than 7% for either line or point contact.

The compliance of a point contact is not so well modeled. Due to the neglect of surface displacements outside the contact, the foundation model gives $\delta=a^2/2R$ which is half of that given by Hertz. If it were more important in a particular application to model the compliance accurately we should take K/h=0.60E*/a; the contact size *a* would then be too large by a factor of $\sqrt{2}$.

2. PNEUMATIC TYRES. TRANSVERSE TANGENTIAL FORCES FROM SIDESLIP AND SPIN

The lateral deformation of the tyre is characterized by the lateral displacement u of its equatorial line, which is divided into the displacement of the carcass u_e and that of the tread u_t. Qwing to the internal pressure the carcass is assumed to carry a uniform tension T. This tension resists lateral deflection in the manner of a stretched string. Lateral deflection is also restrained by the walls, which act as a spring foundation of stiffness K per unit length.

The tyre is deflected by a transverse surface traction q(x) exerted in contact region a≤x≤a .The equilibrium equation is:

$$K_c u_c - T\partial^2/\partial x^2 = q(x) - K_1 u_1, \qquad (13)$$

where K_t is the tread stiffness. The ground is considered rigid (u_2=0) and the motion one dimensional, so that we can drop the suffixes. Equation (13) can then be solved directly throughout in contact region for any assumed pressure distribution. The carcass deflection are clearly not negligible however and it is more realistic to follow von Schilippe (1941) and Temple (1952) who neglected the tread deflection compared with the carcass deflection (u_t=0, u=u_c) as show in fig. 3. Equation (13) then becomes:

$$u - \lambda^2 d^2/dx^2 = q(x)/K_c, \qquad (14)$$

where the relaxation length $\lambda=(T/K_c)^{1/2}$.Tafing the case of side slip first, the displacement within the contact region is given by

$$u = u_1 - \xi \cdot x, \qquad (15)$$

where u_t is the displacement at the leading edge (x=-a).

Outside the contact region q(x)=0 so that the complementary solution to (15) gives:

$$u = u_1 \cdot e^{\{(a+x)/\lambda\}}, \qquad (16)$$

a head of the contact and

$$u = u_2 \cdot e^{\{(a+x)/\lambda\}}, \qquad (17)$$

at the back of the contact.

The foundation model is easily adapted for tangential loading also to viscous-elastic solids (Guangming, 2005).

A one-dimensional model of the resistance of a type to lateral displacement is shown in fig.2.

3. ELASTIC FOUNDATION MODEL BY FINITE ELEMENT

The process is iterative and every date when a node by the possible zone of contact is make in contact, the matrix of stiffness it is modified corresponding (Johnson, 1985).

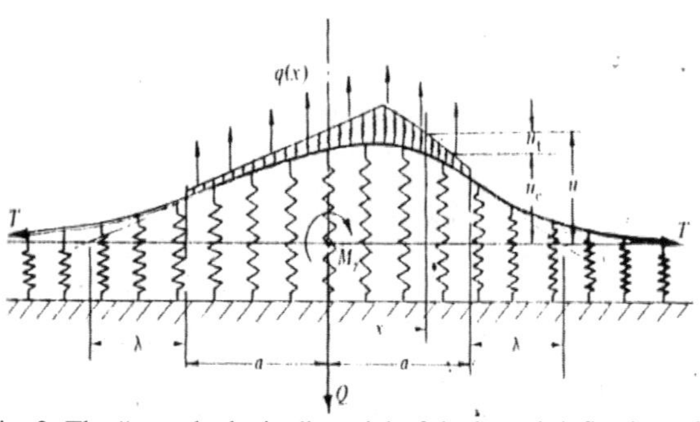

Fig. 2. The “stretched sting” model of the lateral deflection of a tyre.

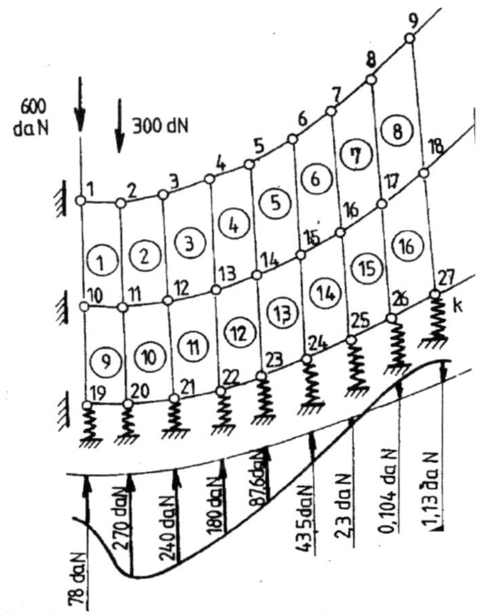

Fig. 3. The elastic foundation model by finite element.

The model is presented in fig.3, the finite plane rectangular elements. In fig.3 is presented the variation of contact pressure between the roll and the rule.

For the 19-27 nodes it was introduced the stiffness (springs) of one constant size for beginning about of Ox,Oy, directions, determinate by the measure of pressure of the 19-27 nodes.

4. CONCLUSIONS

The normal elastic contact could be greatly simplified by modeling the elastic bodies by a simple Winkler elastic foundation rather than by elastic half space. The finite element method are one of the best methods to determinations the pressure of contact.

If the pressure is changed the direction and it is negative and in the anterior node, it is positive, than the limit of the contact zone it's in those case two nodes witch interacted.

If the process is repeated from the intermediate nodes, we find the place where the pressure is changing the sign P>0.
In this way the x coordinate of the respective node represents the semi- breath of contact zone. If every nodes where is in contact, the stiffness matrix is differenced and the maximum stiffness of the elements by who we works carrying on (Rush & Rajkumar, 2000).

The dates are: R = 150 mm, D = 300 mm, b = 40 mm, ν=0.3, E = $2.12\cdot10^5$ Mpa, K = $3\cdot10^8$ Mpa – the maxim stiffness in this model case and from this case of loads the semi-breath is a = 63 mm, (Enescu, 2000).

5. REFERENCES

Enescu, I. (2000). *Aspecte ale mecanicii contactului la rulmenti (Aspects of mechanics contact of bearings)*, Lux Libris Publishing House, Brasov

Guangming, Z. (2005). *Engineering Analysis and Finite Element Methods.* College House Enterprises, LLC, ISBN 0-9762413-1-5, *Available from:* http://www.collegehousebooks.com

Johnson, K.L. (1985). *Contact Mechanics,* Cambridge University Press, Cambridge

Rush, C. & Rajkumar, R. (2000). Analysis of Cost Estimating Processes Used Within a Concurrent Engineering Environment throughout a Product Life Cycle. *7th ISPE International Conference on Concurent Engineering: Researches and Applications*, pp. 58-67, Technomic Inc., Pennsylvania USA, Lyon, France.

Annals of DAAAM for 2008 & Proceedings of the 19th International DAAAM Symposium, ISSN 1726-9679
Editor B. Katalinic, Published by DAAAM International, Vienna, Austria 2008
Make Harmony Between Technology and Nature, and Your Mind will Fly Free as a Bird
*Young Researcher and Scientist Paper / * Supervisor, Mentor*

DEPOSITION SYSTEM FOR THIN FILMS OF SOLAR ENERGY CONVERSION

ENESCU, M[onica]*

Abstract: *The paper presents researches in the field of increasing the efficiency of the solar energy conversion systems. The aim of the research is to design and optimize a system for deposition of thin films for solar energy conversion systems, which used the spray pyrolysis deposition (SPD). For reaching this aim it is used the method of spray distribution modeling. The motive for which it was used this method was for optimization of preparative conditions videlicet substrate temperature, spray rate, concentration of solution etc. For the plane trajectory it was simulated a model of the spray distribution by using the Matlab program. In the conclusion the most efficient deposit film was obtained at the 90 degrees angle of the spray. The enhancement in deposition efficiency and improvement in quality of the thin films can be achieved with automatized systems using to deposit.*

Key words: *deposition, SPD, solar energy, thin film*

1. INTRODUCTION

A significant limit in the development of solar energy conversion systems and in the market implementation and with success in the market implementation for alternative energy systems is made by the higher cost of manufacturing. This problem can be solved through the development of new producing materials and technologies of solar energy conversion systems with low prices. The new development technology of solar energy conversion systems which uses component materials under the form of thin films nanostructured is proved to be an alternative solution for reducing the producing costs.

Over a century many materials were made under the shape of thin films due to their energetic potential and due to their scientific contribution brought because theirs properties. Many studies have been done on spray pyrolysis deposition and preparation of thin films, since the pioneering work by Chamberlin and Skarman (Pramod 1999) in 1966 on cadmium sulphide (CdS) films for solar cells. Later, the simplicity and the reproductivity of the technique offered a new method to form the thin films from noble metals, metallic oxides, spinel oxides, chalcogenides and superconductor compounds.

The deposition technique is used as for obtaining dusts, as for obtaining thin films with structure, morphology and optical properties and are electric controlled using organometallic compounds or salt of the metals as precursor.

Spray pyrolysis (SP) has a number of advantages (Pramod 1999) and is an efficient method: (1) It offers an extremely easy way to dope films with virtually any element in any proportion by merely adding it in some form to the spray solution; (2) Unlike closed vapor deposition methods, SP does not require high quality targets and/or substrates nor does it require vacuum at any stage, which is a great advantage if the technique is to be scaled up for industrial applications; (3) The deposition rate and the thickness of the films can be easily controlled over a wide range by changing the spray parameters, thus eliminating the major drawbacks of chemical methods such as sol-gel which produces films of limited thickness; (4) Operating at moderate temperatures (100-500°C) (fig. 1.), SP can produce films on less robust materials; (5) Unlike high-power methods such as radio frequency magnetron sputtering (RFMS), it does not cause local overheating that can be detrimental for materials to be deposited. There are virtually no restrictions on substrate material, dimension or its surface profile; (6) By changing composition of the spray solution during the spray process, it can be used to make layered films and films having composition gradients throughout the thickness; (7) It is believed that reliable fundamental kinetic data are more likely to be obtained on particularly well characterized film surfaces, provided the films are quiet compact, uniform and that no side effects from the substrate occur.

Thin films of such metal oxide and metallic spinel oxide materials prepared by SPD have matching properties for wide variety of potential applications (Pramod 1999). Chalcogenide semiconductor compounds of group II-VI and V-VI have applications in precise temperature control of laser diodes, optical recording system, electrochemical devices, strain gauges and thermoelectric devices. These thin films have great technological importance owing to their potential applications in photo electrochemical cells, solar selective and decorative coatings, optoelectronic devices and thermoelectric coolers (Pramod 1999). The rest of the paper is structured as follows: in the next section some aspects from related work are presented. In Section 3 we propose and analyze a spray model. Some concluding remarks are provided in the final section.

2. RELATED WORK

In this part of the paper the spray deposit distribution was modeling. The SPD deposition is used in simulation because is a cheap simple and reproducible technique, capable of deposing large or/and figured surfaces in a short time (Okuya et. al. 2004). It is adaptable to any geometry of the surface, with control on the thickness of the films and on the surfaces morphology. SPD is capable of obtaining semiconductor films with good structural, optical and electric properties. The systems for deposition of thin films has a nozzle throught which the precursors solution is sprayed on a substrate heated at 100-500°C. The schematic system for deposition is presented in Fig. 1. The nozzle has to have two translations: one on the X axe and one on Y axe. The third translation and the spraying angle will be incremental.

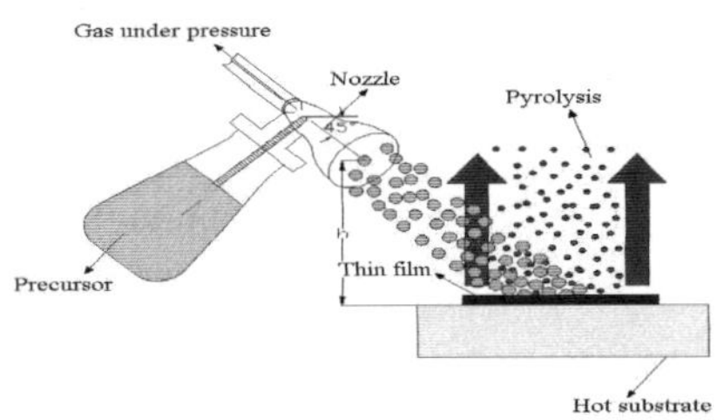

Fig. 1. A scheme of principle of system for SPD

Spray distribution models can be represented by a function z describing the thickness of the liquid film in the two-dimensional space:

$$z = \begin{cases} h(y,v) & y,v \in (h) \\ 0 & y,v \notin (h) \end{cases} \qquad (1)$$

where (y, v) orthogonal coordinate system fixed onto the nozzle, h domain equals the covering area of the spray cone of the nozzle (Ramon et. al. 1996).
Ramon and Clijmans are showed that for a particular cone nozzle the spray deposit distribution could be represented by the Gaussian distribution with principal axes ${}^1y_m$ and ${}^1v_m$ is the perimeter equation of the area treated by nozzle during the spray pyrolysis deposition.

$$g({}^1y,{}^1v) = I_y I_v e^{-\frac{1}{2}\left\{\left(\frac{{}^1y}{\sigma_y}\right)^2 + \left(\frac{{}^1v}{\sigma_v}\right)^2\right\}} \qquad (2)$$

$$I_y = \frac{0.095}{\sqrt{2\pi\sigma_y^2}}, \quad I_v = \frac{0.095}{\sqrt{2\pi\sigma_v^2}} \qquad (3)$$

in which

$$\sigma_y = \frac{{}^1y_m}{2.856}, \quad {}^1y_m = h_n \tan\frac{\alpha_{a,y}}{2} \qquad (4)$$

$$\sigma_v = \frac{{}^1v_m}{2.856}, \quad {}^1v_m = h_n \tan\frac{\alpha_{a,v}}{2} \qquad (5)$$

where h_n vertical distance between nozzle and hot plate at time that the nozzle reaches position, $\alpha_{a,y}$ top angle of the spray cone in the ${}^1y$ direction, $\alpha_{a,v}$ top angle of the spray cone in the ${}^1v$ direction.

3. SIMULATION AND RESULTS

It is necessary an automatization of the deposition system of SPD because for obtaining a higher efficiency of the solar cell is needed a material with good morphology. The morphology of the material is obtained by controlling the deposition parameters: the coating temperature, the spraying rate, the velocity of the spraying. Those parameters are constant parameters during deposition process (Trifa et. al., 2005; Conde-Gallardo et. al., 2005).

Through the modeling of the deposition process it was used a database in which are shown the parameters values proper to the SPD technique. Those parameters are controlled on whole deposition process. In the simulation algorithm we have been used a nozzle with the following characteristics:

$$\alpha_t = 0^o, \ \alpha_{a,y} = \frac{\pi}{10} rad, \ \alpha_{a,v} = \frac{\pi}{10} rad, \ h_n = 300mm$$

A three-dimensional spray model of the nozzle at a liquid pressure of 1 MPa is used in the algorithm to generate the spray deposition of a moving nozzle. To perform the mathematical calculations, the algorithm has been programmed in Matlab. Figure 2 displays the three-dimensional spray deposition patterns as a result of the moving spraying nozzle (rolling and horizontal motions). The trajectories were been calculated in the longitudinal and vertical directions as well and it is showed in fig. 2. In the upper part, the spatial absolute spray liquid amount of the extreme nozzle is represented, while the lowest picture shows the form of the spray drop obtained with constant nozzle height and constant forward speed.

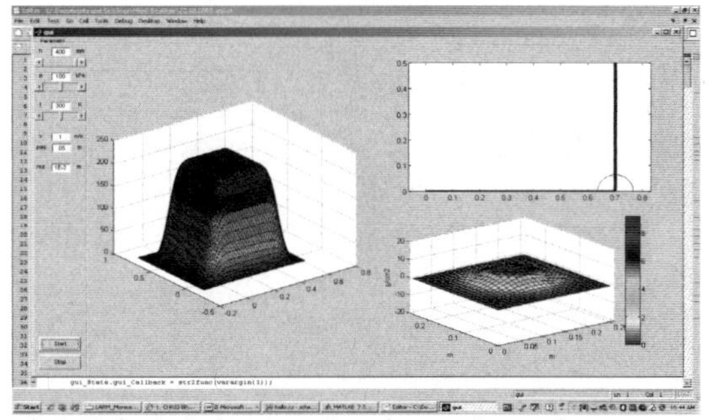

Fig. 2. 3-D deposit shape produced on a plane substrate

4. CONCLUSION

The automatic installation for deposition of nanolayers for renewable energy sources applications will accomplish the spraying process a lot easier and faster, providing the repeatability, a needed feature of nanolayers.

By a 3-D modeling of the shape of the deposit film on a 700x500 mm^2 surface has been analyzed the efficiency of deposition according to the spraying angle for rising the quality of the deposition.

Conclusively the best morphology and the higher efficiency of the deposit film are obtained at 90° of spray angle for a deposition on planar surface.

The aim of the following researches is to model and analyze the SPD deposition for figured surfaces having like aim the rising of the deposition technique efficiency by modifying the deposition parameters.

5. REFERENCES

Clijmans, L.; Ramon H.; Sas P. & Swevers J. (2000) Sprayer Boom Motion, Part 2: Validation of the Model and Effect of Boom Vibration on Spray Liquid Deposition *Journal of Agricultural Engineering Research*, Vol. 76, Issue 2, June 2000, pp. 121-128

Conde-Gallardo, A.; Guerrero, M.; Castillo, N.; Soto, A.B.; Fragoso R. & Cabañas-Moreno J.G. (2005) TiO_2 anatase thin films deposited by spray pyrolysis of an aerosol of titanium disopropoxide, *Thin Solid Films*, Vol. 473, Issue 1, 1 February 2005, pp. 68-73

Okuya, M. ; Nakade, K.; Osa, D.; Nakano T.; Asoka Kumara G.R. Kaneko S. (2004) Fabrication of dye-sensitized solar cells by spray pyrolysis deposition (SPD) technique, *Journal of Photochemistry and Photobiology A: Chemistry* Vol. 164, Issues 1-3, 1 June 2004, pp. 167-172

Pramod, S.P. (1999), Versatility of chemical spray pyrolysis technique, *Materials Chemistry and Physics*, Vol. 59, Issue 3, 15 June 1999, pp. 185-198

Ramon, H., De Baerdemaeker, J. (1996), Spray boom motions and spray distribution: Part 1, Derivation of a mathematical relation, *Journal of Agricultural Engineering Research*, Vol. 66, Issue 1, January 1997, pp 23-29

Ramon, H., Missotten, B., De Baerdemaeker, J. (1996), Spray boom motions and spray distribution: Part 2, Experimental Validation of the Mathematical Relation and Simulation Results, *Journal of Agricultural Engineering Research*, Vol. 66, Issue 1, January 1997, pp. 31-39

Trifa, F.I.; Montavon G. & Coddet C. (2005) On the relationships between the geometric processing parameters of APS and the Al_2O_3–TiO_2 deposit shapes *Surface and Coatings Technology*, Volume 195, Issue 1, 23 May 2005, pp. 54-69

Annals of DAAAM for 2008 & Proceedings of the 19th International DAAAM Symposium, ISSN 1726-9679
Editor B. Katalinic, Published by DAAAM International, Vienna, Austria 2008
Make Harmony Between Technology and Nature, and Your Mind will Fly Free as a Bird
*Young Researcher and Scientist Paper / * Supervisor, Mentor*

THE PROJECT WEB PORTAL QUALITY FUNCTION DEPLOYEMENT

ESSE, M[artin] & MAJERIK, M[arian]

Abstract: *This paper presents the concrete application of Quality Function Deployment (QFD) to the Leonardo da Vinci Improhealth project web portal. The web portal is related to Virtual Healthcare Centre with e-learning, e-implementation and e-improvement services. Using QFD leads to transformation of properties and functions of initial web portal with many improvements. The study attempts to explore voice of consumer's emotional responses and professional responses of members of www.improhealth.org web portal development team and demonstrates all necessary changes to make the web portal be content management based and blind friendly. The paper is based on outcomes gained in the research project KEGA 3-4121-06 "Web Portal of Tools, Methods and Case Studies of Quality Management" and SK/06/B/F/PP-177443 Vocational Education and Training for Quality of Life through eHealthcare & Well-Being.*

Key words: *Quality Function Deployment, web portal, users and technical requirements, House of Quality*

1. INTRODUCTION

The web portals are today a frequent source to obtaining information about something what the people are interested in. So whenever the people need to find and know any information in the area of various fields, they are turning to the internet in lieu of any other source. Web portals influence sundry aspects of the value proposition, are more adaptable and better in marking users (Daetz, Barnard & Norman, 1995).

The Virtual Healthcare Quality Center (VHCQC) allows managers, doctors and healthcare professionals to attain the fundamental knowledge and skills of certified healthcare organization managers via the availability of detailed managerial healthcare glossary, e-learning, e-implementation and e-improvement services, aiming to give a strong incentive and special abilities to markedly improve the quality, efficiency and effectiveness of patient/client oriented healthcare organizational services (Zgodavová, 2006).

In this paper we conducted whose objectives are the following:

- to evaluate the present of the Improhealth VHCQC web portal,
- to recognize and prioritize viewer demands to further improve the effectiveness of web portal,
- to identify and prioritize the technical requirements that can gratify the user's requirements.

2. QUALITY FUNCTION DEPLOYMENT (QFD)

QFD was developed in Japan in the late 1960s by Professors Shigeru Mizuno and Yoji Akao. QFD is tool for designing producing (service) processes in response to customer's demands by transplanting what the customers want into what the organization produces. It is systematic planning process that was created to help a project team bring together and manage all the elements needed to define, design, and produce a product (or delivers a service) that would meet or exceed customer needs.

The major tool through which QFD is implemented is diagram House of Quality (HoQ). The diagram is known as HoQ because its architecture is similar to a house. Like a normal house (which has foundation, walls and roof), HoQ has a number of components (Islam, Ahmed & Alias, 2007), which are:

- customer requirements
- technical requirements
- relationship between customer and technical requirements
- relationship among technical requirements
- prioritized customer requirements
- prioritized technical requirements

Generally, full application of the QFD methodology includes more steps; these steps are incorporated into a popular 4-phase model, the 4 phase are:

I. Product planning: Customer requirements to technical requirements
II. Design planning: Technical requirements to part characteristics
III. Process planning: Part characteristics to process characteristics
IV. Production planning: Process characteristics to production requirements

Clearly, Phase I, Product planning, is the phase with the most leverage (Zgodavová & Lengyel, 2008). It's here that one has the opportunity to "define the right product".

3. RESEARCH METHODOLOGY

We used two approaches for data collection – online survey and focus groups and focus groups included three kinds of respondents, namely:

- Customers (web portal users)
- Developers
- IT experts

The list of technical properties contained 75 technical requirements related to the Improhealth VHCQC web portal (Tab.1). These were divided into six categories: Network, Server, Backup, Security, Application and Programming Language, and Web Design and User Interface. Having obtained the prioritized requirements, these were identified based upon the interview with developers and customers.

The matrix on Fig. 1 shows the relationship between web portal user's and technical requirements and it was used to represent graphically the degree of influence between each

Technical properties (quantity: 75)	
Network *(1-3)* (quantity: 3)	***Security*** *(15-20)* (quantity: 6)
Server *(4-11)* (quantity: 8)	***Application and Programming Language*** *(21-46)* (quantity: 26)
Back up *(12-14)* (quantity: 3)	***Web design and User Interface*** *(47-75)* (quantity: 29)

Table 1. The list of technical requirements (with the quantify date of all technical properties)

Symbol	Meaning	Weight
●	Strong relationship	9
○	Medium relationship	3
Δ	Weak relationship	1

Table 2. The relationship in HoQ diagram

technical requirement. Usually, three symbols are used to represent this relationship (Table 2).

If any technical requirement is not related with any of the users requirements then the corresponding cell in the matrix is kept blank. Once we have identified all possible relationship between every pair of user and technical requirements, we can determine the weight of all the technical requirements. IT experts selected after profound review 50 properties (Tab.3) appointed to processing through QFD method.

After identification of users and technical requirements, we constructed the HoQ. With the help of few IT experts we established the relationship matrix for both types of requirements. HoQ is shown on Fig. 1.

We have obtained the customer´s requests from the questionnaire which was meant for the viewers of the Improhealth web portal. This questionnaire has obtained questions of which we have got a summary about concrete customer´s requests for a more understandable and more simply work with the Improhealth VHCQC web portal.

Following the requirements for realization which we have evaluated from the questionnaire as the most important: Usability, Information Quality, Interactivity, Page Transparence, Sensibleness of Menu, Search Tool, Print Possibility, Colour Scheme, Web portal components position, Blind friendly. We obtained properties from the article "The Study of Kansei Engineering Applied to the VHCQC Web Portal".

4. CONCLUSION

Characteristics of usefulness, usability and desirability play a central role in the good design of all digital products and digital environments such as web portal. The VHCQC web portal that embeds affective appeal is seen to offer user positive experience visiting the web portal services.

Technical requirements
Network
1 Giga byte Ethernet network
Server
4 Media streaming server (audio/video)
5 Database server
9 E-mail server
Back up
10 Continuous back up (mirror)
Security
11 Multi level access
12 Use of id and user password for authentication and authorization
14 Firewall
Application and Programming Language
15 Use HTML (Hyper Text Mark up Language)
22 Minimize flash file of up to 25 Kb
23 Double checking the information before uploading
Web design and User Interface
27 Navigation link of up to 3 words per link
28 Use color to distinguish visited and unvisited links
33 Only one dominant image on one page
34 Use meaningful graphics that relate to the contents
40 Archive old contents
48 Provide a site map
50 Blind friendly

Table 3. The list of selected technical requirements (in the table are shown some properties of total quantify (50 properties) only)

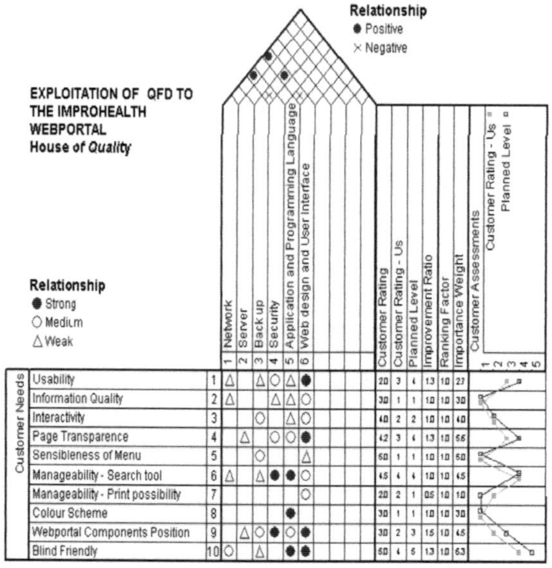

Fig. 1 House of Quality (QFD 2000 software, 1999)

Web portal is an important channel of any organisation through which it disseminates information about the organization. Every web portal has its intended users – customers and they have their own requirements. To maximize the customer satisfaction, organizations are expected to pay adequate attention to fulfill their expectations and requirements. As it has been illustrate in the paper.

5. ACKNOWLEDGEMENT

Our acknowledgement goes to Prof. Kristina Zgodavová, PhD. our PhD. study adviser which markedly contributed in development of this paper and to the team of portal designers: Kristina Lengyelova, Lubomir Lengyel and Slavomir Beznak.

6. REFERENCES

Daetz, D., Barnard & B.Norman, R. (1995): *Customer integration: The Quality Function Deployment (QFD) Leader´s Guide for Decision Making.* Published by John Wiley & Sons, Inc., USA 1995, ISBN-10: 0-471-13277-2

Islam, R., Ahmed, M. & Alias, M.H. (2007): Application of Quality Function Deployment in redesigning website: A Case Study on TV3. *Int. J. Business Information Systems,* Vol.2, No.2, Copyright © 2007 Inderscience Enterprises Ltd., pages number 195-216

QFD 2000 software, For customer LED design, User Manual for Quality Function Deployment Version 2.0, Total Quality Software 1999

Zgodavová, K. (2006). Web Portal of Tools, Methods and Case Studies of Quality Management, *project of Slovak Ministry of Education KEGA 3-4121-06,* 2006

Zgodavová, K. (2006). Improhealth Collaborative SK/06/B/F/PP-177443 Vocatioonal Education and Training for Qaulity of Lifr through eHealthcare & Well-Being, Leonardo da Vinci pilot project, 2006

Zgodavová, K. & Lengyel, Ľ. (2008). Improving Quality, Performance and Cost-Effectiveness of Healthcare Organizations through the "IMPROHEALTH" web portal, In: *Proceedings of the 6th International Conference on Emerging e-Learning Technologies and Applications,*2008, Stara Lesna (Slovakia)

Annals of DAAAM for 2008 & Proceedings of the 19th International DAAAM Symposium, ISSN 1726-9679
Editor B. Katalinic, Published by DAAAM International, Vienna, Austria 2008
Make Harmony Between Technology and Nature, and Your Mind will Fly Free as a Bird
*Young Researcher and Scientist Paper / * Supervisor, Mentor*

PROSPECTS FOR IMPROVING THE ACCURACY OF NATURAL GAS ACCOUNTING AND FOR REDUCING GAS UNBALANCES

FEDORYSHYN, R[oman]; MATIKO, F[edir] & PISTUN, Y[evhen]*

Abstract: *This work deals with the ways to improve the accuracy of natural gas accounting and to reduce gas unbalances in gas transportation systems. The main causes of gas unbalances in gas transportation systems during transportation and supply of natural gas to the consumers were determined and analyzed. Proposals are given to improve the accuracy of natural gas accounting and to reduce gas unbalances in gas transportation systems*
Key words: *Natural gas, accounting, accuracy, unbalances.*

1. INTRODUCTION

In the light of price growth for energy carriers, including natural gas, application of energy saving technologies and saving of energy are top priorities. The problems of energy efficiency can only be solved provided there is an accurate accounting of energy carriers.

Because of the increased attention of gas-transporting and gas-distributing organizations to the accuracy of natural gas accounting and to the gas unbalances the authors analyzed the causes of gas unbalances in gas transportation systems during transportation and supply of natural gas to the consumers and determined the main causes of gas unbalances.

2. ACCURACY OF GAS ACCOUNTING

The insufficient accuracy of gas accounting during gas flow rate and volume measurement is one of the causes of gas unbalances. This accuracy depends on

- the technical base of accounting;
- the normative base of accounting;
- the accounting metrology;
- the accounting personnel.

Pressure differential flow meters and various gas meters are usually used for natural gas accounting. At present we can say that the technical base of accounting meets the accuracy requirements. However, these possibilities of technical means of accounting can be realized and hence a sufficient accuracy of natural gas accounting can be assured provided these technical means are applied properly when constructing an accounting system. Unfortunately, these possibilities of technical base of accounting, as a rule, are rarely used.

New possibilities to improve the accuracy appear in case of application of the new normative base, i.e. implementation of ISO 5167-1,2,3,4:2003 and GOST 8.586.1,2,3,4,5-2005 Interstate Standard developed on the basis of the ISO 5167:2003 as well as "RASKHOD-RU" automated system for calculation and design of pressure differential flow meters (Pistun et al., 2007) and improvement of GOST 30319.0,1,2,3-96 Standard. "RASKHOD-RU" automated system for calculation and design of pressure differential flow meters simplifies the implementation of the above-mentioned standards significantly.

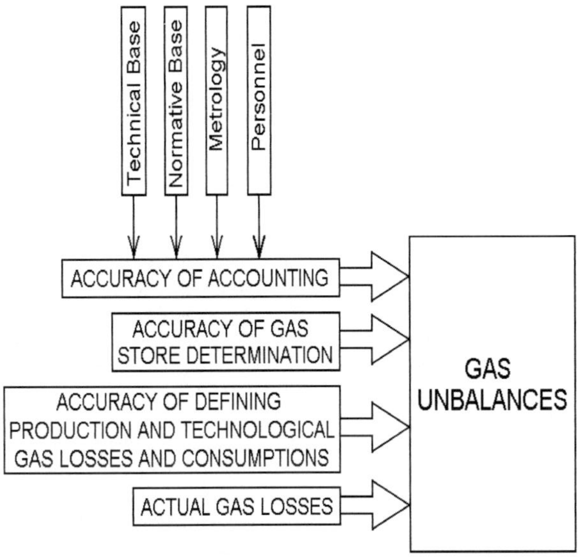

Fig. 1. Causes of gas unbalances.

The point is that the design of the existing gas accounting systems is not based on the condition of minimal uncertainty of natural gas flow rate or volume measurement, and does not take into account additional errors of measurement. Orifice plates optimal as to the accuracy of flow rate measurement are not applied in the pressure differential flow meters.

It should be mentioned that in order to organize gas accounting system on the basis of the pressure differential method correctly and to design such a system from the point of the maximum accuracy of measurement the above-mentioned "RASKHOD-RU" ASCD (Pistun et al., 2007) can be applied.

It is also very important to eliminate or reduce significantly systematic component errors of gas flow rate and volume measurement. These errors define gas unbalances directly. For example for the pressure differential flow meters the systematic component errors due to temperature measurement caused by the placement of the thermometer and by the influence of the ambient temperature can be as high as 0.3 %. For gas meters these component errors can reach a couple of percents. Proper design of gas accounting systems can reduce these errors significantly or even eliminate them.

Adoption of the new normative base of gas accounting is especially important. In particular, application of the new GOST 8.586.1,2,3,4,5-2005 Interstate Standard for the pressure differential method will enable the results of gas flow rate and volume measurement to be corrected by up to 0.5 % in comparison to the old GOST 8.563.1,2,3-97 Standard used before 2008, which is related to the improvement of calculation technique for discharge coefficient of pressure differential device and for gas expansibility factor. Such a correction will automatically lead to reduction of gas unbalances. As far as the normative base for gas meters is concerned it has to be elaborated urgently because a considerable number of additional errors of gas volume measurement are not taken into account at present, which influences gas unbalances directly.

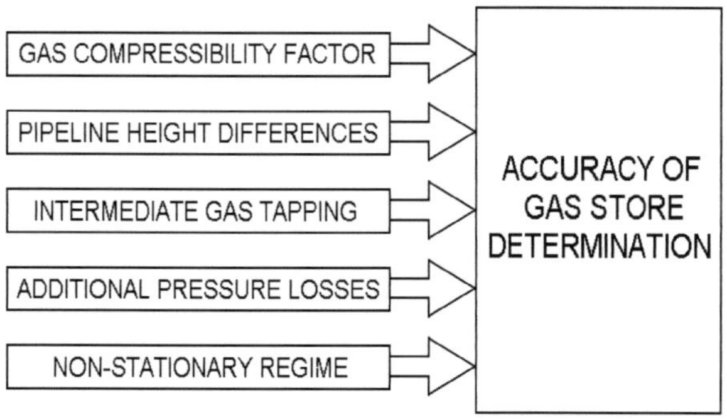

Fig. 2. Factors influencing the accuracy of gas store determination.

3. ACCURACY OF GAS STORE DETERMINATION

Another cause of gas unbalances during transportation in gas transportation systems is inaccurate determination of gas store in main pipelines and other gas pipelines. The existing technique does not take into account a number of important factors that influence directly the accuracy of gas store determination. These factors are variation of gas compressibility factor along the pipeline, height differences between locations of pipeline sections, intermediate gas tapping from the pipeline, additional pressure losses at pipeline fittings and non-stationary regime of gas flow in the pipeline. The first four factors can be taken into account by improving the mathematical models of gas flow process in pipeline and solving the new derived set of equations by a numerical method using a computer. In order to take into account the non-stationary regime of gas flow in pipeline the formalization of the existing non-stationary regimes in operating pipelines has to be done previously. On the basis of the standardized non-stationary regimes the set of non-linear differential equations is solved for given conditions.

A draft improved technique to determine gas store in main pipelines has been developed in Lviv Polytechnic National University (www.lp.edu.ua) in collaboration with the Institute of Energy Audit and Energy Carrier Accounting (www.ieoe.com.ua). According to this technique the gas store in pipelines is determined taking into account the factors mentioned above.

4. ACCURACY OF DEFINING PRODUCTION AND TECHNOLOGICAL GAS LOSSES AND CONSUMPTIONS

Gas unbalances can also be caused by insufficient accuracy of defining production and technological gas losses and consumptions in gas transportation systems during exploitation of the systems. The point is that all these gas consumptions and especially gas losses are not measured directly but calculated according to the existing techniques. Unfortunately, these techniques have a number of significant shortcomings and do not provide sufficient accuracy of calculation of actual volume of gas losses and consumptions. Draft improved techniques for defining the above-mentioned production and technological gas losses and consumptions have been developed in the Institute of Energy Audit and Energy Carrier Accounting. These techniques provide higher accuracy in comparison to the existing ones.

5. ACTUAL GAS LOSSES

Actual gas losses that are not taken into account can also cause gas unbalances during transportation in gas transportation systems. These are gas losses due to leakage in pipelines, fittings and joints. In order to reveal such gas losses direct methods of measurement are applied, which, however, not always can be accomplished. These gas losses could be revealed by an indirect method, i.e. by calculating gas balance in a given restricted pipeline system. However, such calculation can only be accomplished when all the above-mentioned causes of gas unbalances are eliminated or reduced significantly. It is apparent that such gas losses, i.e. unpredictable gas leakage into atmosphere, can only be eliminated after they have been revealed.

6. CONCLUSION

The main causes of gas unbalances in gas transportation systems during transportation and supply of natural gas to the consumers were determined and analyzed. These causes are

- low accuracy of gas accounting;
- inaccurate determination of gas store in main pipelines and other gas pipelines;
- insufficient accuracy of defining production and technological gas losses and consumptions in gas transportation systems during exploitation of the systems;
- actual gas losses (leakage in pipelines, fittings and joints).

In order to eliminate or reduce significantly the causes of gas unbalances in gas transportation systems during transportation and supply of natural gas to the consumers the following measures are proposed

- implementation of the new GOST 8.586.1,2,3,4,5-2005 Interstate Standard;
- application of "RASKHOD-RU" automated system for calculation and design of pressure differential flow meters;
- improvement of GOST 30319.0,1,2,3-96 Standard;
- application of improved technique to determine gas store in main pipelines;
- application of improved techniques for defining the production and technological gas losses and consumptions;
- elimination of actual gas losses due to leakage in pipelines, fittings and joints.

The above-mentioned proposals can actually improve the accuracy of natural gas accounting and reduce gas unbalances in gas transportation systems.

7. REFERENCES

GOST 30319.0,1,2,3-96, *Natural Gas. Methods for Physical Properties Calculation*

GOST 8.563.1,2,3-97, *GSI. Liquid and Gas Flow Rate and Volume Measurement by Means of the Pressure Differential Method*

GOST 8.586.1,2,3,4,5-2005 (ISO 5167-1,2,3,4:2003) *Liquid and Gas Flow Rate and Volume Measurement by Means of Standard Pressure Differential Devices*

ISO 5167-1,2,3,4:2003, *Measurement of fluid flow by means of pressure differential devices inserted in circular cross-section conduits running full*

Pistun, Y.; Lesovoy, L.; Matiko, F.; Markovskyi, D & Lesovoy, R. (2007) *"RASKHOD-RU" ASCD computer programme. Manual.* Publishing House the Institute of Energy Audit and Energy Carrier Accounting, Lviv, *Available from:* http://www.ieoe.com.ua *Accessed:* 2007-05-01

Annals of DAAAM for 2008 & Proceedings of the 19th International DAAAM Symposium, ISSN 1726-9679
Editor B. Katalinic, Published by DAAAM International, Vienna, Austria 2008
Make Harmony Between Technology and Nature, and Your Mind will Fly Free as a Bird

MODEL OF ENVIRONMENTAL LOGISTICS OF THE ENTERPRISE

FIDLEROVA, H[elena] & FIDLER, M[artin]

***Abstract**: The paper deals with the research considering environmentally oriented logistics in Slovak enterprises. We find out that companies of different size view the contribution of an environmental approach to logistics in different ways, but they all recognize that it is important to apply environmental approach in logistics processes. Based on results of research there has been developed a model of environmental logistics of the enterprise, including inputs (materials, personal, energy, machinery and facilities), processes, and outputs (products, pollution, emissions).*
***Key words**: logistics processes, environmental approach, model, sustainability, enterprise*

1. INTRODUCTION

There have been many changes in Slovak companies that are represented by increased international competition, a call for new innovations in products and processes, and the individualization of customer requirements (Virčíková 2001). Customers now insist not only on products being delivered to the right place at the right time, but they also have special requirements concerning quality and environmentally-friendly products, and therefore logistics has to adapt. (Sakál et al 2003; Škara 2005). There are now new concepts in the field of logistics referred to as 'green logistics' or 'reverse logistics' (Červeňan et al 2003; Rodrique et al 2001).

2. ENVIRONMENTAL LOGISTICS

Environmental logistics means taking into account the environmental impact of all of a company's logistical processes. Logistical processes influence the environment in many ways and, in order to recognize them, we consider supply logistics, production logistics, distribution logistics and reverse logistics.

3. RESEARCH ISSUE (PROJECT)

The objective of the research project at the Department of Industrial Engineering and Management, Faculty of Materials Science and Technology, Slovak University of Technology Trnava, for the doctorate thesis *Proposal of Process Model of environmental Logistics of the industry Enterprise (Fidlerová H. 2006)* has been to determine if and how Slovak enterprises apply an environmental approach to their logistical processes, if they have a logistical processes model that takes sustainability into consideration, what contributions could be achieved by green logistics.
The research was carried out by a questionnaire, which was sent to 105 Slovak enterprises, out of which 43.81% responded. The size distribution of respondents was, by percentages, 22% small enterprises (less than 50 employees), 43% middle-sized enterprises (50-150 employees) and 35% large enterprises (more than 150). Half of the participating enterprises had an EMS ISO 14 001 certification.
As is obvious from the research results (Fig. 1), different companies view the contribution of an environmental approach to logistics in different ways. For large enterprises, a good public image and relationship with suppliers was the most important. Small and middle-sized companies rated higher the specific requirements of customers.

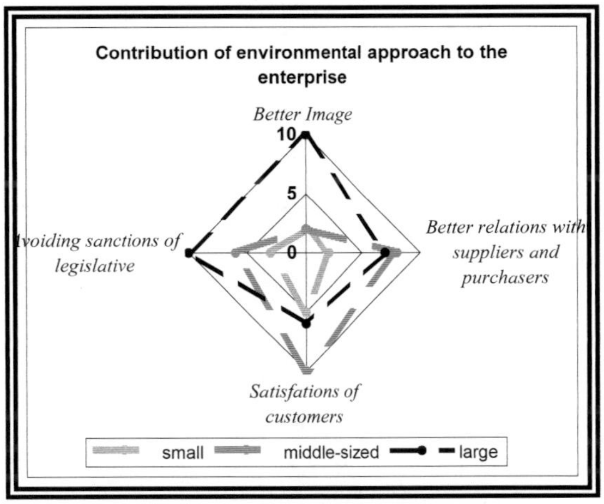

Fig. 1. Contribution of environmental approach to the enterprise

We can say that Slovak firms realized that it is important to apply an environmental approach to logistics and are aware of a potential contribution to the company, but their approach in only partial and the level of implementation in logistical processes is low because their approach to environmental requirements is more defensive or offensive than proactive. Enterprises accept the requirements of legislative pressure (e.g. requirements of Slovak waste laws No. 223/2001, Act No. 245/2003 about Integrated Pollution Prevention and Control (IPPC), etc.) and the competition.

Enterprises with a proactive approach and with an EMS certification reported more contribution, because there some time is needed in order to see a difference.

4 MODEL OF ENVIRONMENTAL (friendly) LOGISTICS OF ENTERPRISE

The objective of a model of environmental enterprise logistics (fig.2) has the following characteristics:
It involves all main logistical processes of an industry enterprise and its environment, along with their hierarchic relations, and it respects the flow of materials and information. It describes the relation between a company's logistical processes and its partners, such as suppliers and customers, and in particular it considers possible influences on the environment.
The main reasons to decide for environmental logistics are the current national and European legislation concerning waste, environmental risks, customer's requirements and public opinion. Without customers and their needs and requirements, firm has no reason to produce goods. Changes in legislative regulations regarding the environment can be divided into two

main groups: laws that place sanctions on a firm if it breaks those laws, e.g. SEVESO II, and legislation that facilitates voluntary decisions to establish an environmental management system from the technical, organizational and administrative points of view, e.g. EMAS.

When establishing a model of environmental logistics, important considerations include inputs (materials, personal, energy, machinery and facilities), processes, and outputs (products, pollution, emissions). An input/output analysis concerning environmental impact should be realised, so that a firm can discover possibilities for improvement in processes and products.

The inputs of environmental logistics include resources (materials, semi-products, human resources, etc.), energy, machines and accessories.

The outputs of environmental logistics include the products requested by customers (in best view, a product with an eco-label in Slovakia EVV), and negative influences on the environment (pollution, emissions, waste).

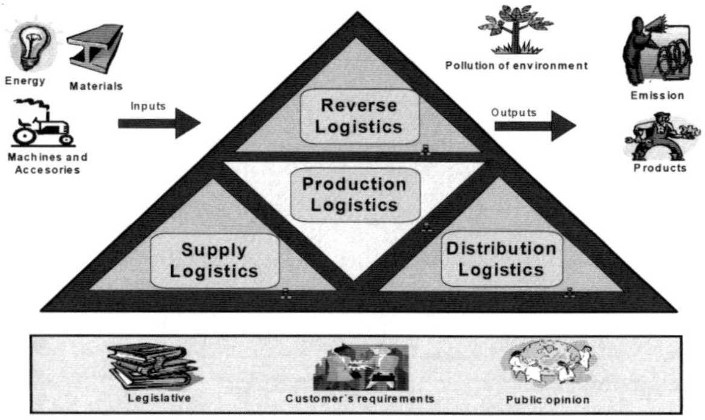

Fig 2: Model of environmental logistics.

Product development and design are not part of the logistics process but their consequences have a considerable influence on logistics. In the course of product development, it is necessary to take into consideration the firm's environmental goals and the possibilities of recycling the product or its components.

Supply logistics (green procurement) ensures the acquisition of resources required for the production process according to environmental and quality standards, reduced energy use, the conservation of resources, and the elimination of hazardous substances in the procurement of raw materials, parts, and semi-products. It is important to determine the most environmentally-friendly form of transport and to prefer suppliers whose products carry an ecolabel, support greener purchasing, take their part of the responsibility for the quality and environmental influences of the materials that make up the final product. When making decisions about output materials and components, one of criterion should be how many waste this particular output will produce and if it could cause toxicity.

Production logistics, the goal of a firm is to produce a product with adequate levels of quality and environmental-friendliness, and get this product to the customer at right time and in the right quantity. It is important to minimalize energy and material expenditure, use techniques that eliminate environmental risk (BAT), avoid unnecessary storage and transport, provide regular training about handling dangerous materials and about the firm's environmental policies, eliminate the use of harmful chemical substances in the manufacturing process and, for substances difficult to eliminate completely, ensure that they do not enter the atmosphere or public waterways.

Packaging provides protection and identification for products according to customer's requirements and current legislation. Less packaging should be used per product and firms should consider ways of recovering it and thus reducing waste. Storage in a firm can be considered as the storage of resources and the storage of products.

Distribution logistics serves to conquer the temporal and space differences between a products production and its use by the customer. In the area of distribution it is important to consider these steps regarding the environment. If it is economic acceptable, choose a kind of transport and transport tours that eliminate negative influences on the environment, maximize the utilization of each transport (truck), and reduce the amount of emission produced by each type of vehicle.

Reverse logistics considers the disposal of waste materials that are the by-products of the production process, packaging and the firm's distribution processes. Reverse logistics deal with the storage and treatment of waste and recycling. It is important to consider the change of used products to other new products with other uses and eliminate the negative influence on the environment within all processes of reverse logistics.

The verification of this model has been realized through personal consultation with specialists from chosen Slovak enterprises.

5. CONCLUSION

There are many ways in which environmental logistics improve competitiveness. The most important include savings in costs (costs of energy and raw materials, costs of transport and handling, costs of waste material treatment), avoiding risk and damages (eliminating ecological accidents and avoiding possible damages to the environment), the enhancement of competitiveness through accepting the customer's requirements, the opinion of the public, who are also possible future customers, and firm culture (the improvement of the firm's image, reliability and confidence in the firm improving personal qualifications).

In the future, the focus of research will be on two main paths, the first will be focused on the definition of reverse logistics processes. This part of logistics is identified as the most under-appreciated. The second will focus on the usage of a prepared process model in information technology. This means that we will enhance the model by new parts, attributes and objects, which will allow the implementation of the model in software applications.

6 REFERENCES

Červeňan, Š. – Vidová, H. – Hološová, H. (2003) *Logistika v praxi manažéra (Logistics in praxis of managers)*, Tripsoft, ISBN 80-968734-1-5, Trnava

Rodrique J-P., Slack B., Comtois C. (2001) *Green Logistics (The paradox of) The handbook of Logistics and Supply – Chain Management*, Pergam, ISBN: 0-08-043593-9, London

Sakál, P. et al. (2005) Project VEGA No. 1/0900/02 Environmentálne orientovaný manažment, marketing a logistika strategických podnikateľských jednotiek. (Environmental Management, marketing and logistics of strategic business units) In *FÓRUM MANAŽÉRA* 1/2005, p. 64-65, ISSN 1336-7773, Trnava

Škapa R. (2005) *Reverzní logistika (Reverse logistics)*, ISBN 80-210-3848-9, Brno

Virčíková, E. - Pálfy, P. (2001) *Environmentálne manažérstvo-teória a metodika* (Environmental management – theory and methodic), Štroffek, 267p., ISBN 80-88896-15-0, Košice

Annals of DAAAM for 2008 & Proceedings of the 19th International DAAAM Symposium, ISSN 1726-9679
Editor B. Katalinic, Published by DAAAM International, Vienna, Austria 2008
Make Harmony Between Technology and Nature, and Your Mind will Fly Free as a Bird

SYNTHESIS OF THE AUTOMATIC CORRECTION SYSTEM OF THE DESIRED SIGNAL PARAMETERS

FILARETOV, V[ladimir] & YUKHIMETS, D[mitry]

Abstract: *In the work the synthesis method of the automatic correction system of the desired harmonic signal frequency for the control system of dynamic object with the variable parameters is proposed. For the realization of this system the addition loop of the regulation of desired signal parameters with the feedback in the amplitude of dynamic error is introduced. As a result, it was possible to ensure the maintenance of a maximally possible frequency of desired signal with the retention of the assigned level of dynamic error amplitude.*
Key words: *control, accuracy, correction, frequency.*

1. INTRODUCTION

Already created a large quantity of synthesis methods of the high-precision control systems, which make it possible to ensure the preset parameters of the control process quality under the conditions of the variability of the control object (CO) parameters (Slotine, 1991; Feng, 1999; Sarangapani, 2006). In this case the use of the systems indicated does not guarantee highest efficiency of technological equipment, since the performance rate of technological operations will be determined by the characteristics of the desired signal of system. Therefore the task so forming desired signal appears, in order to ensure a maximally high speed of CO motion.

Basic problem with the formation of the desired signal is that with a increase in the velocity of CO motion increases also the dynamic error of control, which depends on the parameters of this CO.

In this work we will examine only harmonic signals of form $\alpha^* = A^* \sin(\omega^* t)$. In this case the speed of CO motion will be determined by the frequency of this harmonic signal (the higher the frequency, the higher the speed). Therefore it is necessary to increase this frequency, but in this case to ensure the assigned accuracy of the work of system (amplitude of dynamic error). Since previously this maximum frequency can be unknown, i.e. it is necessary to correct in the process of functioning. This work is dedicated to the development of the correction system of the desired signal frequency.

2. TASK SETTING

Thus, in this work is set and is solve the problem of the synthesis of the automatic correction system of harmonic desired signal for the system of control of dynamic object with the unknown and variable parameters, which makes it possible to ensure a maximally possible frequency of this signal with the restriction of the tolerance level of the dynamic error amplitude

3. DESCRIPTION OF THE CONTROL OBJECT

As CO in the work is used the direct current electric drive, which is described by the differential equation:

$$a_1 J\dddot{\alpha} + a_2 J\ddot{\alpha} + a_3\dot{\alpha} = a_4\, u\ , \qquad (1)$$

where $\dot{\alpha}_i$, $\ddot{\alpha}_i$ is respectively, speed and the rotary acceleration of the electric motor shaft; J is the moment of inertia, in this case $J_{\min} \le J \le J_{\max}$, a_1, a_2, a_3, a_4 is the constant positive coefficients.

For guaranteeing the required margin of stability and satisfactory quality indicators of work with the nominal parameters of drive the linear regulator is used:

$$W_{ki}(S) = (T_1 S + 1)/(T_2 S + 1)\ , \qquad (2)$$

where T_1 and T_2 are the some sufficiently small time constants, determined by the concrete properties of electric drive.

As is known, the greater the amplitude of harmonic signal, the greater the amplitude of the dynamic error $\varepsilon = \alpha^* - \alpha$, where α^* is desired signal, and the greater the moment of the inertia of drive, the greater this amplitude (Dorf, 2001). Therefore regulator (2) ensures the assigned accuracy only with the specific values of the amplitude of the desired signal and moment of the inertia of electric drive (1). In the case of changing these parameters the amplitude of error can change, which requires the correction of the desired signal frequency .

From expressions (1), (2) it is possible to note that even for the simple linear system to obtain the analytical expressions, which describe dependence $\omega^* = f(\varepsilon, A^*, J)$ is sufficiently complicated. Moreover for their use it is necessary to accurately measure the CO parameters, what is the complex and in many instances unfeasible problem. Therefore it is necessary to use another approach, which is proposed in this work.

4. CORRECTION SYSTEM OF THE DESIRED SIGNAL FREQUENCY

In order to produce the correction of the frequency of desired signal in this work it is proposed to introduce additional loop, which will achieve this correction on the basis of the difference between permitted and flowing of the amplitudes of dynamic error (Emelyanov, 1997). The functional diagram of proposed control system is shown to fig.1.

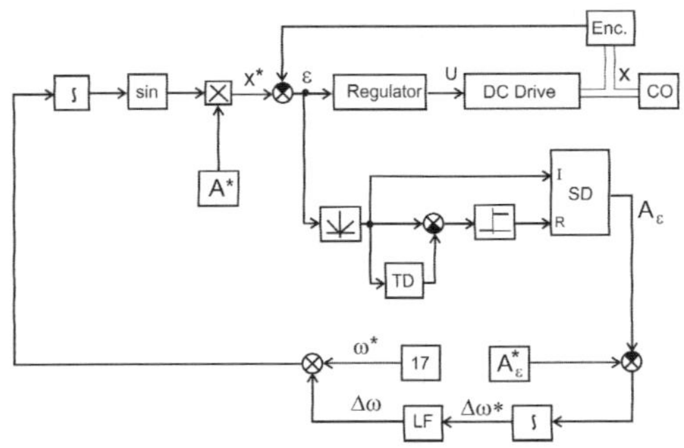

Fig.1. Functional diagram of proposed system

In the figure the following designations are introduced: Regulator is the regulator, described by transfer function (2),

DC drive is the electric drive of direct current, the described by equations (1), Enc is the position sensor of the drive shaft, TD is the device of time delay, SD is the device of storage, LF is the filter of low frequencies.

Basic task of the in the implementation indicated system this creation of the algorithm of the measurement of the current amplitude of dynamic error. Since the signal of dynamic error is harmonic, then this signal will reach maximum at that moment of the time, when the speed of change in its module will change from the growth to the decrease. Therefore it is necessary to determine the moment, when the speed of change ε changes its sign and to preserve instantaneous value ε.

For determining the sign of speed of change $|\varepsilon|$ we will use the expression: $v_\varepsilon = |\varepsilon(t)| - |\varepsilon(t-\tau)|$, where t is the current time, τ is the time delay. It is obvious that when $sign(v_\varepsilon) > 0$ the signal of dynamic error grows on the module, and when $sign(v_\varepsilon) < 0$, then it diminishes. Change in this signal $sign(v_\varepsilon)$ with 1 on - 1 gives the command to the storage device to memorize the instantaneous value of signal $|\varepsilon|$, i.e., the amplitude of dynamic error A_ε. One should say that the value indicated renews two times during the period of a change in the error signal, and between these value it remains constant.

The correction of the frequency of desired signal occurs on the basis of the difference between the permissible and current amplitude of error. If $A_\varepsilon^* - A_\varepsilon < 0$, i.e., the instantaneous value of amplitude ε exceeds the permitted value, then the frequency of desired signal will decrease until the real value of amplitude ε is lowered to the allowed value (until is satisfied condition $A_\varepsilon^* - A_\varepsilon = 0$). And vice versa, if $A_\varepsilon^* - A_\varepsilon > 0$, then the frequency of desired signal will increase.

The additional value of the frequency of desired signal is determined from the following formula:

$$\Delta\dot{\omega}^* = k_i(A_\varepsilon^* - A_\varepsilon), \qquad (4)$$

where ki is a constant positive coefficient.

In order to ensure a smooth change in the frequency of desired signal, signal $\Delta\omega^*$ preliminarily is passed through the filter of the low frequencies of the second order. This approach also makes it possible to ensure the absence of transient processes with a change in the frequency.

5. RESULTS OF THE SIMULATION

For checking the effectiveness of the work of the system of the correction of the frequency of desired signal proposed was carried out mathematical simulation. A study of the work of the developed system was conducted with a change in the amplitude of desired signal from *1.5 rad* to *2.5 rad* and a change in the moment of inertia from *0,0005 kg·m²* to *0,001 kg·m²*, the amplitude of the permissible error was assigned A_ε^* *=0,001 rad.* The parameters of electric drive (1) they had the following values: $a_1 = 0.2$, $a_2 = 20$, $a_3 = 0.02$, $a_4 = 800$. The time constants of regulator (2) had values: T_1*=0,03 s*, T_2*=0,0015 s*. The basic frequency of desired signal was equal $\omega^* = 0.5 s^{-1}$.

In fig. of 2a is shown the process of changing the error amplitude for the systems without the correction (curve 1), also, with the correction (curve 2) of the desired harmonic signal frequency. In this case simultaneously changed the CO inertia moment (J = (0.00075 + 0.00025·sin(0.05t)) kg·m²) and the amplitude of desired signal (fig. 2b).

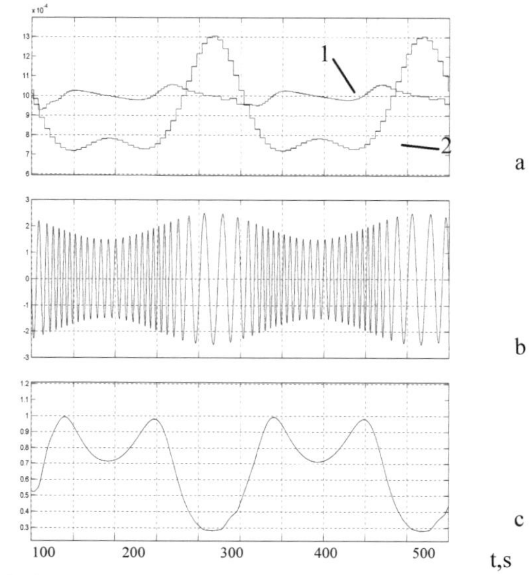

Fig.2. Results of the simulation

As can be seen from these figures with a change in the control object parameters or amplitude of desired signal in the system without the correction of the frequency of desired signal the amplitude of dynamic error can change almost 2 times. In this case it can be both above and it is lower than the permissible value. With the use of additional loop of tuning the frequency of desired signal the amplitude of dynamic error remains in effect constant and equal to the allowed value. In this case the frequency of desired signal can increase in comparison with the base frequency in 2 times (fig. 2c).

6. CONCLUSIONS

In the work is proposed the approach to the formation of the desired harmonic signal frequency for control system of dynamic CO with the variable parameters, which makes it possible to ensure the maximum frequency of this signal with the retention of the assigned dynamic accuracy. The advantage of this approach is simplicity of realization and absence of the need of measuring the CO parameters. The limitation of the synthesized system is the fact that it is intended for the work only with the harmonic desired signals; however, subsequently the developed method can be modified for the work with any forms of demand signals.

7. ACKNOWLEDGEMENT

This work is supported by Russian Fund of Basic Research (Grants № 05-08-33627, 08-08-90018).

8. REFERENCES

Dorf R., Bishop R.(2001) *Modern control system*, Prentice Hall, ISBN-10: 0130306606, USA, 2001.

Feng G., *Lozano R.(1999) Adaptive control*, Reed Educational and Professional Publishing, ISBN 07506 3996 2, 1999.

Emelyanov S.V., Korovin S.K.(1997) *New types of feedback*, Phisical and Mathematical Literature Publishing, ISBN 5-02-015149-1, Moskow, 1997.

Sarangapani J.(2006) *Neural Network Control of nonlinear discrete-time systems*, CRC Press, ISBN 0-8247-2677-4, 2006.

Slotine J.-J.(1991) *Applied nonlinear control*, Prentice-Hall Inc., ISBN 0-13-040890-5, New Jersey, 1991.

Annals of DAAAM for 2008 & Proceedings of the 19th International DAAAM Symposium, ISSN 1726-9679
Editor B. Katalinic, Published by DAAAM International, Vienna, Austria 2008
Make Harmony Between Technology and Nature, and Your Mind will Fly Free as a Bird

NEW APPROACH TO ROBUST OBSERVER DESIGN

FILARETOV, V[ladimir]; ZHIRABOK, A[lexey] & KUCHER, D[imitry]

Abstract: *This paper presents a problem of robust fault detection and isolation in systems described by nonlinear dynamic models with no differential nonlinearities. To solve this problem, so-called logic-dynamic approach is suggested. This approach consists of tree main steps: replacing the initial nonlinear system by certain linear system, obtaining the linear observes, and transforming this observer into the nonlinear one.*

Key words: *fault diagnosis, observer, robustness, logic-dynamic approach, singular values, generalized eigenvalues*

1. INTRODUCTION

There are several approaches to achieve robust fault detection and isolation (FDI) (Frank, 1990; Low, 1986; Patton, 1994; Shumsky, 2007). They are used for nonlinear systems described by models with smooth nonlinearities. So-called logic-dynamic approach to fault diagnosis in systems described by nonlinear dynamic models with no differentiable nonlinearities was proposed in (Zhirabok & Usoltsev, 2002). The main purpose of this paper is to use this approach to obtain the robust observer for such class of systems.

2. LOGIC-DYNAMIC APPROACH

In this paper, the logic-dynamic approach is presented for a class of nonlinear systems described by the equations

$$\dot{x}(t) = Fx(t) + Gu(t) + C\cdot\varphi(Ax(t), u(t)) + Kd(t) + L\rho(t),$$
$$y(t) = Hx(t), \qquad (1)$$

Here x, y, and u are the vectors of state, output, and control; F, G, H, and L are known matrices of appropriate dimensions; A is a row matrix; C is a column matrix: $C[i] \neq 0$ if the i-th component of the state vector contains the scalar function $\varphi(Ax(t), u(t))$, $C[i] = 0$ otherwise. The term $L\rho(t)$ models unknown parameters and unknown inputs to the actuator and to the dynamic process; the evaluation of the vector function $\rho(t)$ must generally be considered unknown. The term $Kd(t)$ models the faults: if there are no faults, then $d(t) = 0$; if a fault occurs, $d(t)$ becomes an unknown function.

The logic-dynamic approach to solve the FDI problem for the system (1) includes the following three steps.
1. Replacing the initial nonlinear system (1) by a linear system.
2. Solving the FDI problem for this linear system and obtaining the linear observer.
3. Transforming the linear observer into the nonlinear one.

On the first step, replace the initial nonlinear system by linear system described by the following equations:

$$\dot{x}(t) = Fx(t) + Gu(t) + Kd(t) + L\rho(t), \quad y(t) = Hx(t).$$

On the second step, the linear observer has to be obtained:

$$\dot{x}_*(t) = F_*x_*(t) + G_*u(t) + J_*y(t), \quad y_*(t) = H_*x_*(t). \qquad (2)$$

Here $x_*(t)$ is the state vector of the observer, $y_*(t)$ is the output vector, F_*, G_*, J_*, and H_* are constant matrices; it is assumed that the matrix F_* is stable. The observer generates the residual $r(t) = Ry(t) - y_*(t)$ for certain matrix R which has to be determined. If there are no faults and $\rho(t) = 0$, then $r(t) = 0$, if a fault occurs, $r(t) \neq 0$.

It is well-known from the linear FDI theory (Frank, 1990; Zhirabok & Usoltsev, 2002) that for the observer design, the matrix Φ such that $\Phi x(t) = x_*(t)$ in the unfaulty case after the response to unlike conditions have died out plays the main role. In the absence of faults, the following well-known set of equations holds (Frank, 1990; Zhirabok & Usoltsev, 2002):

$$\Phi F = F_*\Phi + J_*H, \quad H_*\Phi = RH, \quad G_* = \Phi G. \qquad (3)$$

To ensure the reliable fault detection, the residual $r(t)$ has to be sensitive to the faults and invariant with respect to the unknown inputs $\rho(t)$ that is

$$\Phi K \neq 0, \qquad \Phi L = 0. \qquad (4)$$

According to logic-dynamic approach the equality

$$A = A_*\begin{bmatrix}\Phi \\ H\end{bmatrix} \qquad (5)$$

for some matrix A_* has to be held; it is an additional restriction on the matrix Φ.

To simplify the design procedure, the matrices F_* and H_* are represented in a canonical form as follows:

$$F_* = \begin{bmatrix} 0 & 1 & 0 & \dots & 0 \\ 0 & 0 & 1 & \dots & 0 \\ & \cdot & \cdot & & \cdot \\ 0 & 0 & 0 & \dots & 0 \end{bmatrix}, \quad H_* = [1 \quad 0 \quad 0 \quad \dots \quad 0].$$

On the third step of the suggested approach, it is necessary to transform the obtained linear observer into the nonlinear one. According to (Zhirabok & Usoltsev, 2002) the nonlinear term $\Phi C\varphi(A_*\begin{bmatrix}x_*(t) \\ y(t)\end{bmatrix}, u(t))$ where the matrix A_* is the solution of equation (5) must be added to the right-hand side of the first equation (2) that results in

$$\dot{x}_*(t) = F_* x_*(t) + G_* u(t) + J_* y(t) + \Phi C\varphi(A_* \begin{bmatrix} x_*(t) \\ y(t) \end{bmatrix}, u(t)).$$

3. ROBUSTNESS

The equality $\Phi L = 0$ is a severe restriction and does not fulfill in practice as a rule therefore it is impossible to obtain an observer invariant with respect to the unknown inputs $\rho(t)$. In this case it is necessary to use the robust methods. There are two main approaches to design the robust observer.

1. Singular value decomposition. The matrix L is represented as follows: $L = U_L \cdot \Sigma_L \cdot V_L$ where U_L and V_L are orthogonal matrices, $\Sigma_L = [diag(\sigma_1, \sigma_2, ..., \sigma_n)\ 0]$, $0 \le \sigma_1 \le \sigma_2 \le ... \le \sigma_n$ are singular values of the matrix L. Take the first s columns of the matrix U_L as rows of the matrix Φ. It is known that the Frobenius norm $\|\Phi L\|_F^2$ is minimal for given s. Besides, the matrix Φ must satisfy equations (3)-(5).

2. Generalized eigenvalue-eigenvector problem. The norm $\|\Phi L\|_F^2$ estimates a contribution of the disturbances into the residual. By analogy, the norm $\|\Phi K\|_F^2$ estimates a contribution of the fault. The best choice for the matrix Φ to take into account these norms is the following:

$$J = \frac{\|\Phi L\|_F^2}{\|\Phi K\|_F^2} \rightarrow min.$$

This leads to the equation $w(LL^T - \lambda KK^T) = 0$. This equation gives generalized eigenvalues $\lambda_1, \lambda_2, ..., \lambda_n$ such that $0 \le \lambda_1 \le \lambda_2 \le ... \le \lambda_n$. The first p values gives the optimal solution and corresponding eigenvectors w_1, $w_2, ...,$ w_p are used as rows of the matrix Φ.

4. ROBUST OBSERVER DESIGN

Canonical form of an observer is a limiting factor with respect to robustness. The way to obtain the robust nonlinear observer without this factor based on the logic-dynamic approach is suggested. Consider matrix equations (3) and (5). Find conditions to solve them. Rewrite the first equation as follows:

$$\Phi F = \begin{bmatrix} F_* & J \end{bmatrix} \cdot \begin{bmatrix} \Phi \\ H \end{bmatrix}. \tag{6}$$

It follows from this equation that rows of the matrix ΦF are linear combinations of the matrix $\begin{bmatrix} \Phi \\ H \end{bmatrix}$ rows; this is equivalent to the rank equality

$$rank \begin{bmatrix} \Phi \\ H \end{bmatrix} = rank \begin{bmatrix} \Phi^{\mathrm{T}} & H^{\mathrm{T}} & (\Phi F)^{\mathrm{T}} \end{bmatrix}. \tag{7}$$

The second equation in (3) can be rewritten as follows:

$$\begin{bmatrix} R & -H_* \end{bmatrix} \cdot \begin{bmatrix} H \\ \Phi \end{bmatrix} = 0.$$

This equation has a solution if the rows of the matrices Φ and H are linearly dependent; this is equivalent to the inequality

$$rank \begin{bmatrix} \Phi \\ H \end{bmatrix} < rank(\Phi) + rank(H). \tag{8}$$

Equation (5) is similar to (6) therefore it is equivalent to the equality $rank \begin{bmatrix} \Phi \\ H \end{bmatrix} = rank \begin{bmatrix} \Phi^{\mathrm{T}} & H^{\mathrm{T}} & A^{\mathrm{T}} \end{bmatrix}$ which can be combined with (7) that gives

$$rank \begin{bmatrix} \Phi \\ H \end{bmatrix} = rank \begin{bmatrix} \Phi^{\mathrm{T}} & H^{\mathrm{T}} & (\Phi F)^{\mathrm{T}} & A^{\mathrm{T}} \end{bmatrix}. \tag{9}$$

The main idea to obtain the robust observer is as follows. Obtain the singular value decomposition of the matrix L, in particular, the matrix U_L. Take the first column of this matrix as a row of the matrix Φ and test conditions (8), (9), and $\Phi K \ne 0$ for this matrix. If they fulfill, the task is solved and the matrix Φ is used to build the observer. Otherwise the second column of the matrix U_L is added as a second row in Φ and conditions (8), (9), and $\Phi K \ne 0$ are tested again. This procedure continues until these conditions fulfill. The obtained solution gives the minimal value of the norm $\|\Phi L\|_F^2$.

Generalized eigenvectors can be used instead of the matrix U_L rows by analogy. The matrices F_*, G_*, J_*, H_*, R and A_* can be found from equation (3). A stability of the observer can be obtained by the known methods.

5. CONCLUSIONS

The important problem of robust fault diagnosis in nonlinear dynamic systems with no differentiable nonlinearities has been studied. To solve this problem, so-called logic-dynamic approach has been developed. The new way to obtain the robust odserver minimizing a contribution of the disturbances into the residual has been suggested.

Acknowledgments
This work was supported by Russian Foundation of Basic Researches, grants 07-08-00102 and 07-08-92101.

6. REFERENCES

Frank, P. (1990). Fault diagnosis in dynamic systems using analytical and knowledge-based redundancy – A survey and some new results. *Automatica,* Vol. 26, pp. 459-474, ISSN 0005-1098.

Lou X.; Willsky A. & Verghese G. (1986). Optimally robust redundancy relations for failure detection in uncertain systems. *Automatica,* Vol. 22, pp. 333-344, ISSN 0005-1098.

Patton, R. (1994). Robust model-based fault diagnosis: the state of the art. *Proc. of the IFAC Symposium Safeprocess'94*, Finland, 1994, pp. 1-24.

Shumsky, A. (2007). Redundancy relations for fault diagnosis in nonlinear uncertain systems // *Int. J. AMCS.* Vol. 17. pp.477-489, ISSN 1641-876X.

Zhirabok, A. & Usoltsev, S. (2002). Fault diagnosis for nonlinear dynamic systems via linear methods. *CD ROM Proc. of the 15th World Congress IFAC*, Spain, 2002.

Annals of DAAAM for 2008 & Proceedings of the 19th International DAAAM Symposium, ISSN 1726-9679
Editor B. Katalinic, Published by DAAAM International, Vienna, Austria 2008
Make Harmony Between Technology and Nature, and Your Mind will Fly Free as a Bird

CONSIDERATIONS REGARDING AN UNUSUAL CAUSE OF DATABASE CONCURRENCY EXCEPTION

FILIP, I[oan]; ROBU, R[aul]; SZEIDERT, I[osif] & ROBU, A[ndreea]

Abstract: *This paper presents an unusually cause behind database concurrency exception generated by a disconnected data access using ADO.NET technology. Also a method for overcoming this problem is proposed. The data concurrency exception is common caused by a concurrent access of multiple users to the same records of a data source (usually a database table). However, this exception can occur in some special situation, even in the case of a single user access to database. Such situation is generated by a trigger action, updating directly some rows of the table, mean while DataSet memory-cache information isn't updated. The proposed solution assumes the modification of the Visual Studio auto-generated code (change the UPDATE command automatically generated by the programming tool).*

Key words: *ADO.NET, database, concurrency exception, trigger.*

1. INTRODUCTION

ADO.NET is a technology providing two strategies for data access. The main strategy intended primarily for disconnected data access, and the second – for connected data access (Esposito, 2002). In traditional connected data access strategy, a connection to the database server is done and then interacts with it through SQL queries using the connection. The application stays connected to the database server even when it is not using database services. This strategy assumes operations directly against the database, using a data command object. In this case the connection with database must be explicitly open, execute the command to perform the operation, and the connection is closed. The connected data access strategy is used by many other technologies (Petzold, 2002). However, the main benefit of ADO.NET is provided by the disconnected strategy, using an intrinsic *DataSet* object. The disconnected data access involves the store of data in a *DataSet* object, which is an in memory-cache set of records that can be accessed disconnected from the data source (usually a database). A *DataAdapter* object is used to fill the *DataSet* with records from data source. Also the same *DataAdapter* updates the information to data source.

Therefore, the user modifies data from *DataSet* (disconnected from data source) and then save this changes to data source (commit). The *DataAdapter* object open or close the connection with data source automatically (without an explicitly open or close command). The application automatically connects to the database server when it needs to pass some query and then disconnects immediately after getting the result back and storing it in *DataSet*. The figure 1 presents the two data access strategies (connected access – based on *DataReader* object, and disconnected access – based on *DataAdapter* and *DataSet* object).

Despite of many *DataSet* benefits, this strategy generates a few unexpected problems, related to database concurrency exceptions. The cause of these problems is due by non data-synchronizations between *DataSet* and data source.

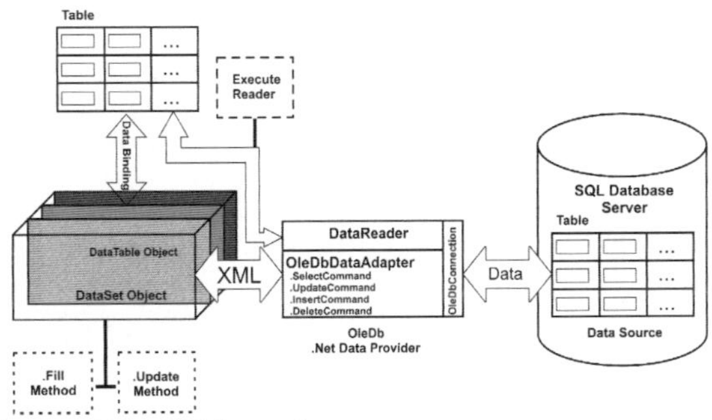

Fig. 1. ADO.NET Strategies.

The cause of these problems is due by non data-synchronizations between *DataSet* and data source (usually a table stored on database server). Usually, data concurrency exceptions occur when multiple users have access to the same data and any single user can update the data without the other users' knowledge. However, this problem can occur even in the case of a single user access, when the application uses triggers fired on the SQL database server side. In the technical literature there are mentioned many solutions that solve the issue in the common case of multiple user access. Those solutions are based on the principles "if a user updated a database table record, another user must refresh the memory-cache information stored in this record according to the new changes and only afterwards it can also update them" or "overwrite all the previously changes" (Esposito, 2002), (Microsoft, 2003), (Papa, 2004).

The main problem is to solve the unusual concurrency exception that occurs in the case of a single disconnected data access (for update/delete one or more rows a of database's table). In this paper will be proved that the Visual Studio tool auto-generated code prohibits the usage of the database trigger based programming. Therefore, a solution must be finding to permit the SQL server side programming based on fired triggers, which represent a useful and powerful programming technique.

2. DATA CONCURRENCY EXCEPTION - STUDY CASE

In order to present a solution to solve the problem of a data concurrency exception occurred for a single data access which fired a database trigger, a study case is presented. There is considered a database table (*TEST*) with two columns: a first column *SSN* (Social Security Number) as a primary key (PK), and a second column *Birth_Day*. The values from second column will be automated updated by a trigger, based on the data from *SSN* (which include de birth day). When a new row is inserted (needing only the value of *SSN* column), a trigger is fired, extracting the information from *SSN* and updating the *Birth_Day* column. The following UPDATE auto-generated command selects a row (using a WHERE clause) uniquely identified by the values of all columns:

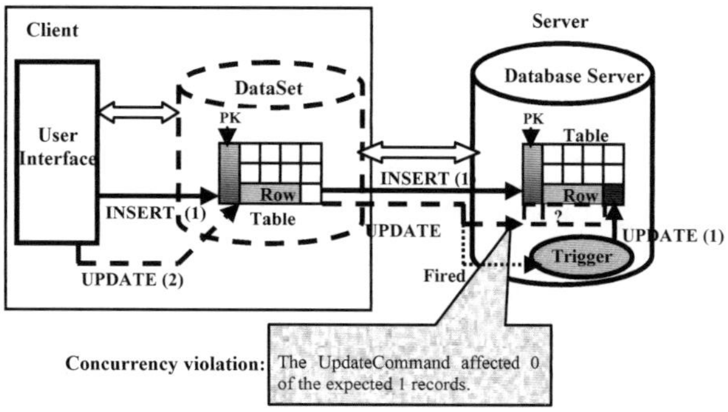

Fig. 2. Database Concurrency Exception.

this._adapter.UpdateCommand.CommandText = "UPDATE STUDENT.TEST SET SSN= ?, BIRTH_DAY = ? WHERE ((SSN = ?) AND ((? = 1 AND BIRTH_DAY IS NULL) OR (BIRTH_DAY = ?)))";

When the client (through a user interface), insert a new row into the table (both the *DataSet* table and data source table – see INSERT operation (1) from figure 2), on the database server side a trigger is fired, and subsequently, for the inserted row, the value of *Birth_Day* column is updated (UPDATE operation (1)). But this update is done only to the data source table (stored on server side). The *DataSet* table row isn't update in concordance with the real value from data source. Then, if the client tries to change this row (updating or deleting- operation (2) – see the figure 2), the self-generated SQL command take into consideration (to select the row) all the columns values of the selected row. These columns values are read from *DataSet* table, but they are different from the real values of table stored in data source. So, the command can't identify a row that can be modified (update or delete) and the fallowing error message occurs: "*Concurrency violation: the UpdateCommand affected 0 of the expected 1 records*". Why? Because no row (record) can be found!

However why concurrency exception? Because the trigger "plays the role of a second virtual user" accessing in write mode the same table row, before the real user, so incoming data changes weren't yet detected by it.

The each row can be also uniquely identified by the value of primary key column. As already stated, in order to allow auto-generate UPDATE and DELETE commands assigned to a *DataReader* object, the data source table must have a column as primary key. Therefore, a table's row can be uniquely identified (selected) by using only this value. For the considered example, the UPDATE command can be modified based on this consideration, as is presented below:

this._adapter.UpdateCommand.CommandText = "UPDATE STUDENT.TEST SET SSN= ?, BIRTH_DAY = ? WHERE (SSN = ?)";

Therefore, it doesn't matter if another column value was changed (into data source on SQL server side), because this value doesn't appear in filtering condition (WHERE clause).

Even if the comments included by Microsoft into the auto-generated file source warn the programmer about a possible incorrect behaviour if the code is changed, the solution works perfectly. Off course, the number of command's parameters must be reduced (therefore, some auto generated code line must be removed). Related to disconnected data access, another small problem can be mentioned: the disconnected data access (using a *DataSet* object) excludes the usage of the default values for the SQL CREATE TABLE command. Suppose that *BIRTH_DAY* column have as a default value the current date. Therefore, on client side (through a user interface – see figure 3), if a *TextBox* object or a cell of a *DataGridView* object (both bound to the *BIRTH_DAY* column of table) are left empty, this means that a NULL value will be inserted in this table cell.

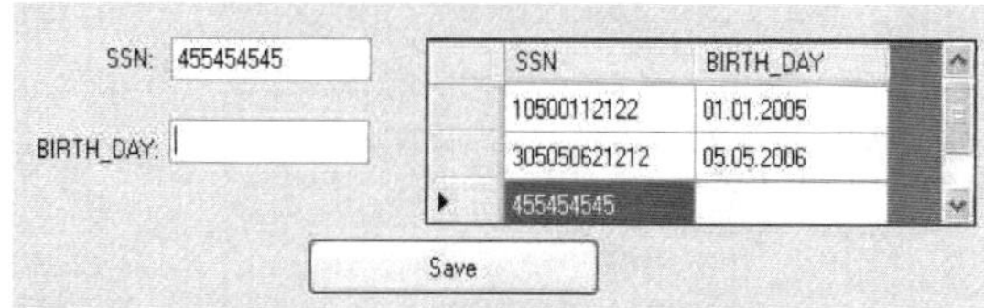

Fig. 3. Empty *TextBox* or *DataGridView* cell.

Therefore, the default values will be ignored! However, to assign a default value to a table column, some code must be written on the client side (for example, assign a value to the properties *Text* of the *TextBox* object).

3. CONCLUSION

As part of .NET Microsoft technology, ADO.NET is a set of components representing a new approach of dynamically retrieved data from a data source, offering several advantages over previous versions of ADO (Lendvai & Shi 2007)), especially due to disconnected data access (Zhang & Hu, 2004). However, the disconnected data access strategy has some disadvantages, generating sometimes "inexplicable" errors. Such error, database concurrency exception, and also a possible solution to solve the problem, are exemplified in this paper.

The proposed solution, in order to avoid this exception generated by a trigger's action updating data on a data source (but not updating data into *DataSet* memory-cache), consist in the change of the auto-generated code. The originality of this approach is the fact that the condition from the "where" clause, of SQL command, which selects the rows updated by the trigger, is more simplified, taking into consideration only the value of the primary key (which uniquely identifies a table row). Moreover, this solution can be used even in some cases of a multiple concurrency data access (when the previously updated row is overwritten by a following update operation). Also, the proposed method, involving the change of auto-generated code, is always valid for all SQL Server (chosen as data source), which allows trigger programming technique. Regarding the problem of a default value specified for a column when a new table is created, the conclusion is that the ADO.NET shifts the programming pole from server side towards client side.

4. REFERENCES

Esposito, D. (2002). *Building Web Solution with ASP.NET and ADO.NET*. Microsoft Press, ISBN: 9780735615786, Redmond, Washington, USA

Lendvai, A. J., Shi, H. (2007). ADO and ADO.NET Object Model Comparisons: A Relational Perspective, *IJCSNS International Journal of Computer Science and Network Security*, Vol.7 No.1, January 2007, pp.330-337, ISSN: 1738-7906

Microsoft. (2003). Tackle Data Concurrency Exceptions Using the DataSet Object. *MSDN Magazine*. April 2003, *Available from:* http://msdn.microsoft.com/en-us/magazine/cc188748.asp *Accessed:* 2008-05-30

Papa, J. (2004). Handling Data Concurrency Using ADO.NET. *MSDN Magazine*. September 2004, *Available from:* http://msdn.microsoft.com/en-us/magazine/cc163924.aspx *Accessed:* 2008-05-30

Petzold, C. (2002). *Programming Microsoft Windows with C#*, Microsoft Press, ISBN: 9780735613706, Redmond, Washington, USA

Zhang, D., Hu, G.H. (2004). Data sharing over the Internet and Web under ADO.NET, *Proceedings of 19th International Conference on Computers and Their Applications*, pp. 286-289, ISBN: 1-880843-50-1, Seattle, WA, Mar. 19-20, 2004

Annals of DAAAM for 2008 & Proceedings of the 19th International DAAAM Symposium, ISSN 1726-9679
Editor B. Katalinic, Published by DAAAM International, Vienna, Austria 2008
Make Harmony Between Technology and Nature, and Your Mind will Fly Free as a Bird

GATES AND BARRIERS DRIVING MECHANISM

FILIPOIU, I[oan] D[an]; SEICIU, P[etre] L[ucian]; MORARIU, Z[aharie] & BARBU, V[alentin]

Abstract: *The driving mechanism is used for gateways, doors and barriers with vertical or horizontal rotation, as an integral part of the open/close system of various precincts: commercial centers, public institutions, garages, parking places etc. We propose a new designed driving gear mechanism with more and better characteristics, smaller overall sizes and the capability of functioning in extreme situations (mechanical or electrical power supply failure).*
Key words: *driving mechanism, gateways, barriers.*

1. INTRODUCTION

The driving mechanism is the main component element of the closing system of some precincts with gates, barriers, doors (commercial centers, public institutions, companies head quarters, garages and parking places).

Figure 1 – 3 present three different mechanisms (1 is the driving mechanism and 2 is the gate, Hamrock et al., 2005).

The actual systems use several types of driving mechanisms: gear, chain drive, belt drive mechanisms, screw-nut systems etc. Usually these systems are expensive and less reliable due to many expensive parts they have (Decker, 2007). There are several companies that produce such systems, but they are CC powered and they have large overall sizes.

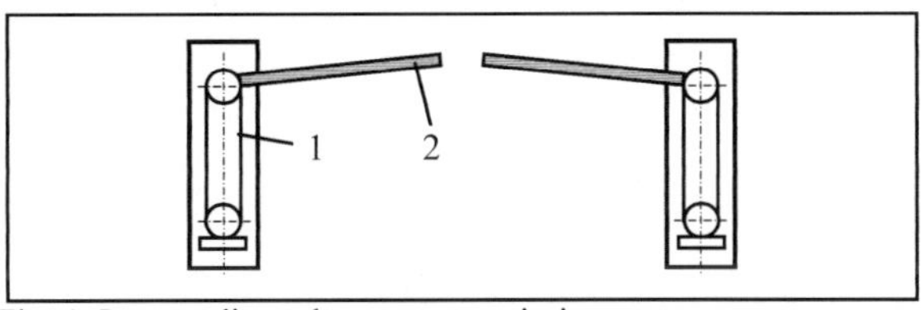

Fig. 1. Intermediary element transmission.

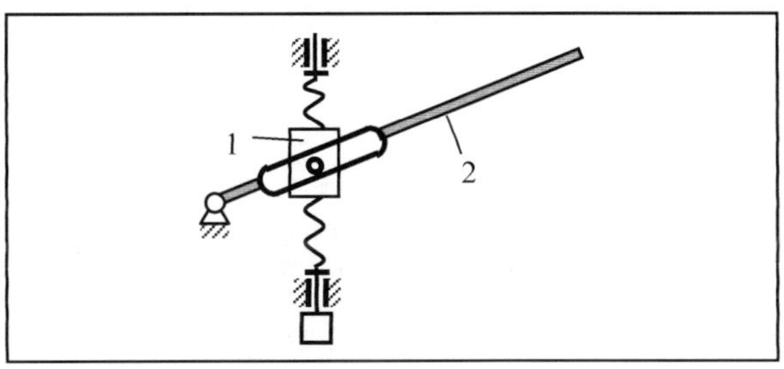

Fig. 2. Screw-nut mechanism.

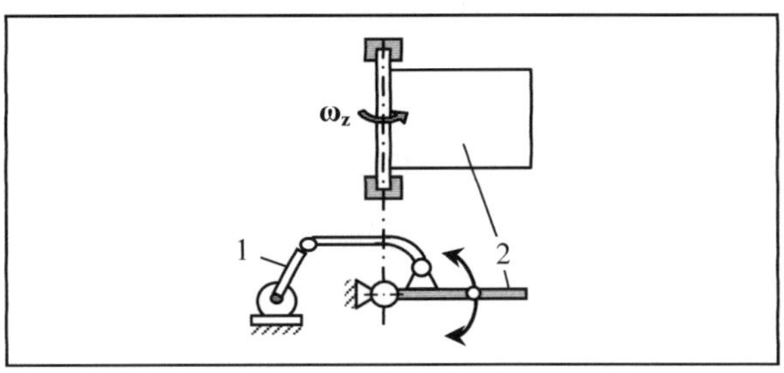

Fig. 3. Hinge joint bars mechanism

2. THE MECHANISM FUNCTIONS

The driving mechanism has the following functions:

- horizontal rotation of gateways and doors and/or vertical rotation of barriers;
- maximum angle between normal closed and normal open working positions: 90°;
- working speed range: 6…12 m/min;
- functioning reliability: 100%;
- switch or remote commands to be performed from maximum 10 m;
- manual operating when electrical power is shut-down;
- IP54 protection level;
- total security of the open/close system.

3. DRIVING MECHANISM CONSTRUCTION

The mechanism (Filipoiu et al., 2006) consists in a gear reducer A and an articulated driving mechanism for the barrier or a quadrilateral mechanism articulated with levers for gates and doors (figure 4).

The decelerator A consists in a worm gear and two helical gears.

The open/close system with gear and supplementary lever transmission, can work in closed or open environment, heated or not, at temperatures between - 20...+ 70° C. It can be placed in or out of buildings and has easy and safe access to the open/close system in case of malfunction.

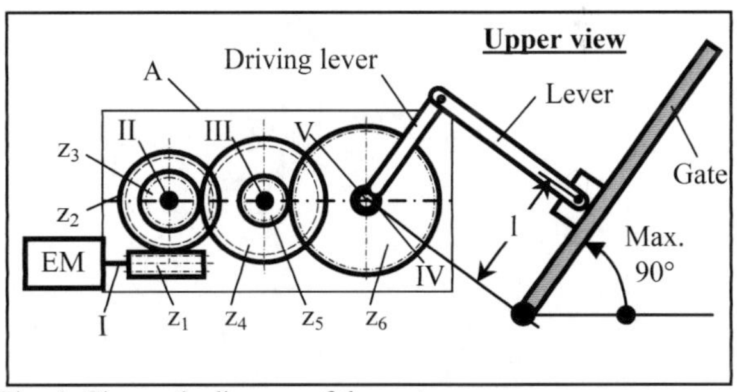

Fig. 4. Cinematic diagram of the system.

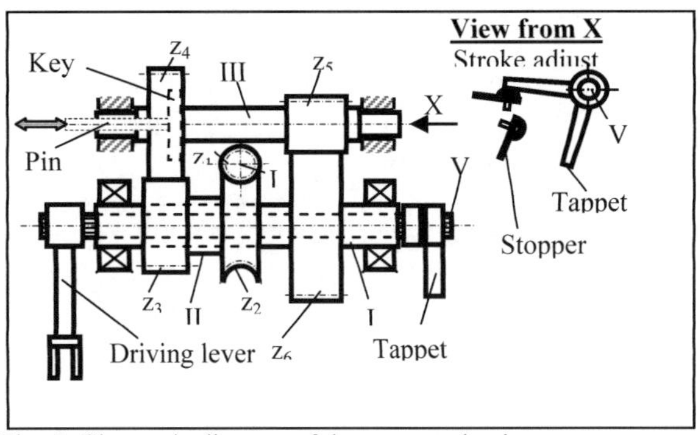

Fig. 5. Cinematic diagram of the gear mechanism.

In case of power shut-down, the system can be manually driven by cutting-off the cinematic chain.

The open/close system has an IP54 protection level against dust and water according with SREN 60529-95.

The open/close system consists of the following parts presented in figures 4 and 5.

The electrical Motor (2MA 63 – 90 W and 860 rpm) has a special construction, different from the standard version. The gear worm is cantilevered on motor's shaft, for minimum size.

The driving mechanism (three speed reduction ratio) has one cylindrical worm gear and two coaxial helical gears, is case mounted with a vertical separation section (Filipoiu & Tudor 2004).

The two bronze bearings of shaft III are mounted in the driving mechanism semi-cases. Due to overall size reasons, shaft I and tubular shaft IV are mounted on ball bearings.

The tubular shaft IV is mounted inside the free shaft II. The exit shaft V has two free ends and it is mounted inside the tubular shaft IV, forming with this a channeled assembly. The shaft's free end in the driving lever area is channeled while the other end is ridged in order to be manually driven.

In extreme situations of malfunction or electrical power failure, the gate has to be manually actuated. In these situations, the cinematic chain is interrupted by moving the key that releases gear z_4 from shaft III. In these conditions it is possible to free rotate the intermediary shaft III. This way, shaft's V striated end can be manually rotated, together with shaft IV, gear z_5-z_6 and shaft III. The helical gear z_3 is fixed.

4. CINEMATIC ASPECTS

Figure 6 presents the synthesis of the quadrilateral articulated mechanism (with the gate shown in two positions).

This diagram is used for computing the lengths, the speeds and the accelerations of all the elements of the system. Figure 7 presents the speed poligon (Seiciu & Stanciu 2002).

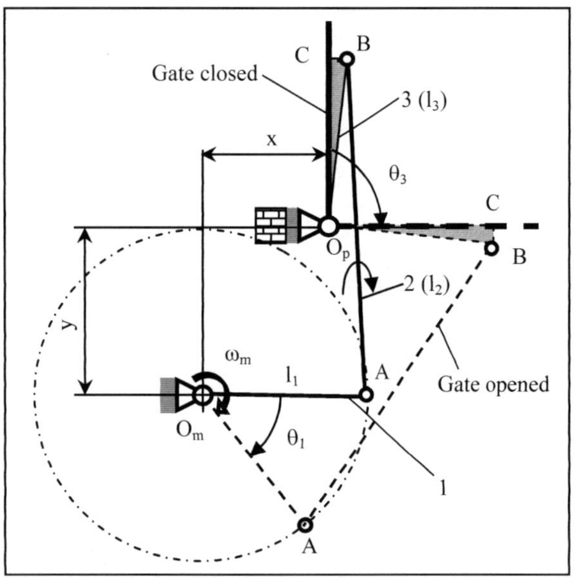

Fig. 6. Mechanism synthesis.

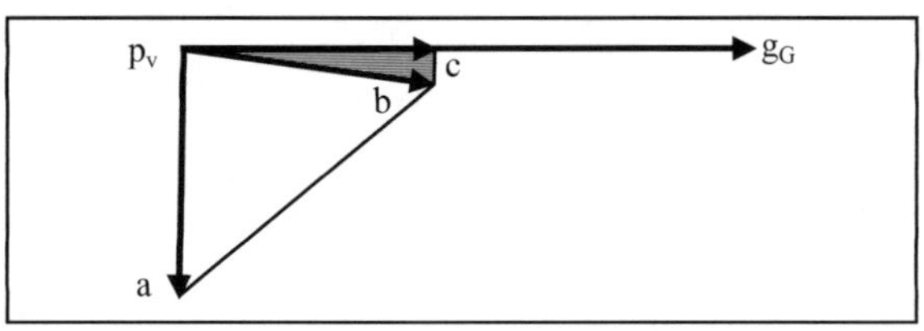

Fig. 7. The speed polygon.

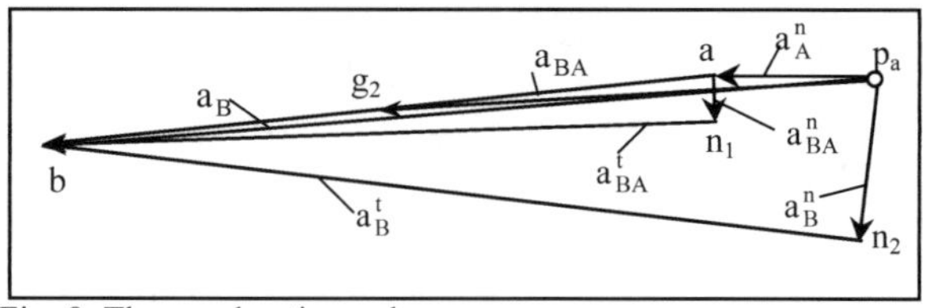

Fig. 8. The acceleration polygon.

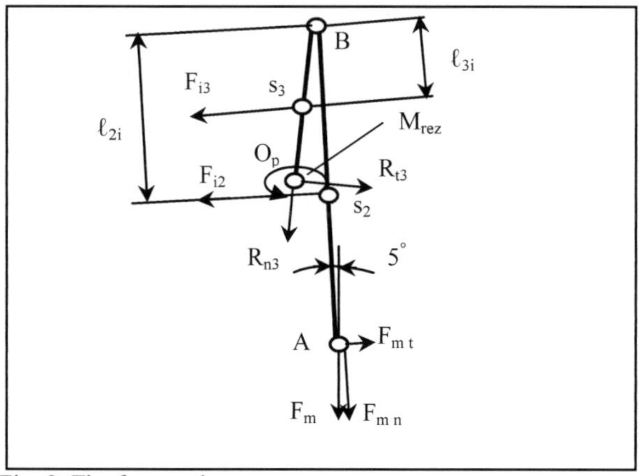

Fig. 9. The force polygon.

Figure 8 presents the acceleration polygon.

Figure 9 presents the force poligon. All the values in figures 6-9 are computed considering the respective scales. The figures 6-9 are drawn for the initial possition: gate is closed and starts to open (Uicker et al., 2003).

The force polygon is used as input data for the resistance calculus (design and verification) of the mechanical parts.

5. CONCLUSIONS

The system is an original product and was developed as a prototype. The innovative features are:

- the system has multiple use;
- small overall size, high efficiency;
- low mass weight;
- switch or remote opening and closing command;
- easy and safe handling and use in emergency cases (i.e. power shut-down).

The manufacturing price is competitive face to other similar systems. The system was produced and tested as part of an innovation project.

6. REFERENCES

Decker, K.-H. (2007). *Machine Elements (in German),* Carl Hanser Verlag, ISBN: 978-3-446-41054-0

Filipoiu, I. D. & Tudor, A. (2004). *Design of the Mechanical Transmissions (in Romanian),* BREN Ed., ISBN: 973-8143-26-8, Bucuresti

Filipoiu, I. D.; Morariu, Z.; Barbu, V. & Chesnoiu, A. (2006). Driving Mechanism for Door Open/Close Systems (in Romanian). *Inovative Tehnology –CM Revue,* Vol. 58, No. 1, January 2006, pp. 13-18, ISSN: 0573-7419

Hamrock, B. J.; Schmid, S. R. & Jakobson, B. (2005). *Fundamentals of Machine Elements,* McGraw Hill, ISBN: 0-07-246532-8

Seiciu, P. L. & Stanciu, S. (2002). *Planar Mechanisms (in Romanian),* BREN Ed., ISBN 973-648-074-7, Bucuresti

Uicker, J. J.; Shigley & Pennock, G. R. (2003). *Theory of Machines and Mechanisms,* Oxford Univ. Pr., ISBN: 978-01-95-15598-3, New York

Annals of DAAAM for 2008 & Proceedings of the 19th International DAAAM Symposium, ISSN 1726-9679
Editor B. Katalinic, Published by DAAAM International, Vienna, Austria 2008
Make Harmony Between Technology and Nature, and Your Mind will Fly Free as a Bird

MACHINING OF RESILIENT WHEELSETS ON WHEEL-TURNING LATHES WITH FRICTION DRIVE - PART 1.: MODELING OF DEFORMATIONS WITH FEM METHOD

FILIPOWICZ, K[rzysztof]; BIEDUNKIEWICZ, W[itold]; KROLIKOWSKI, M[arcin]; & GRZESIAK, D[ariusz]

Abstract: *Article presents FEM modelling and computations of machining process of turning resilient wheel on wheel-turning lathe with friction drive. Presented research is a part of research performed in Institute of Mechanical Technology in Szczecin University of Technology in order to enable regeneration of resilient wheelsets using wheel-turning lathes with friction drive, commonly used for regeneration of heavy railway wheelsets.*
Key words: *Resilient wheels, FEM, turning, technology*

1. INTRODUCTION

Resilient wheels are made to reduce noise, and improve comfort of travelling with lightweight railways. That is why they are commonly used in city rail wagons. Reduction of vibrations is achieved by inserting elastic layer between hard steel wheel tyre, and wheel centre. This solution is very efficient for vibrations damping while wagon runs, but may be a problem during turning of such wheelset on axle-turning lathes. Especially that shape of wheel is complicated and it has to be done very accurately. The problem is negligible while turning wheels without removing them from the car with undertruck wheel-turning lathes. Construction of this lathes simulates conditions of normal work of wheel, what helps to obtain required shape. Problem in this case is that to renovate wheels all wagon must be delivered to machine tool working space and the same it is out of order for all the renovation cycle.

Appearance of friction drive axial lathes became an opportunity to check if they are suitable for regeneration of worn resilient wheel tyres without necessarity of disassembling of wheelsets and without need to exclude all car of use. Friction drive axle-turning lathes were made for heavy railway wheels to avoid a risk of destruction of wheels that were turned with classical lathes with self-locking dog. Using a self-locking dogs while machining causes formation of notches on wheel tyres, what was direct cause of few major accidents of high speed trains.

One of polish companies that builds friction drive wheel-turning lathes asked Szczecin University of Technology to perform research to work out a technology suitable to use this kind of machines for regeneration of resilient wheelsets without disassembling wheels. In order to achieve some preliminary results it was decided to use FEM model to check if turning of resilient wheelsets on friction drive lathe is possible.

2. RESEARCH PREPARATION

First problem was to find out the details of construction of resilient wheels. There are few kinds of construction, but we decided to create model of most flexible wheels in purpose to get the most adverse conditions for machining. Chosen wheels has dumping inserts made as small pieces of elastomer arranged between hub and tyre of wheel with empty spaces between them (as shown on Fig. 1.) what can cause appearing of vibrations during machining.

Fig. 1. Pictorial drawing of chosen resilient wheel construction.

Material properties of steel used for wheel tyres and hubs was taken from European Standards, but material used for elastic inserts is not normalized. Therefore it was necessary to perform tests of material properties of elastic inserts used for resilient wheels to use them for FEM model. This tests was performed in material strength laboratory possessed by Szczecin University of Technology. New and used inserts was examined for check if the properties of elastomer changed during the usage of resilient wheel.

Independently there were performed tests of machining of steel used for wheel tyres. This test was necessary to determine cutting forces appearing during turning of wheelset. Normally machining of wheels is performed with extremely high cutting parameters what requires high power of lathe, and using special tools. Szczecin Technical University does not have enough powerful and big lathes so it was necessary to perform cutting forces tests on small work pieces with low parameters and calculate forces for higher parameters of machining. Another problem is that tools used on wheel-turning lathes are specially designed to be very stiff and durable, and it is impossible to use this tools with standard lathe. Therefore it was necessary to use standard tools with similar cutting edge geometry as tools for wheel-turning lathes for tests. Results of those tests was used to create FEM model of wheel, and check the deformation of resilient wheel in conditions similar with machining process.

3. MODELING AND COMPUTATIONS

Research was performed using computer FEM program COSMOSM 2.95. To obtain possibly accurate results applied model was made of small solid elements that enabled to input forces in distances made by cutting blade during rotation of wheel by angle of 1 degree. This enabled application of load similar to real load during machining process. Model concern machining of one wheel fastened with revolving centre and driven by 3 rollers pressed to wheel with high forces and machined with blade sloped with angle 30 degrees from vertical direction what is feature of particular lathe what facilitates removing chips from working space of the machine.

Force of rollers pressing and their geometrical localisation was taken from technical information of wheel-turning lathe

TUU-1250 which will be used to performing a verification of achieved results.

Preliminary computation was performed to confirm suspicions that deformation of wheel hub is low and can be omitted in further considerations, what allowed to reduce model to elements of tyre and elastic inserts only. Removed hub was replaced by displacements simulating influence of hub on inserts.

First computations was made in coplanar force system perpendicular to wheel axis. Driving rollers, and cutting tool was simulated by forces and fastening was simulated by displacement in the inner side of elastic inserts (as shown on Fig. 2.).

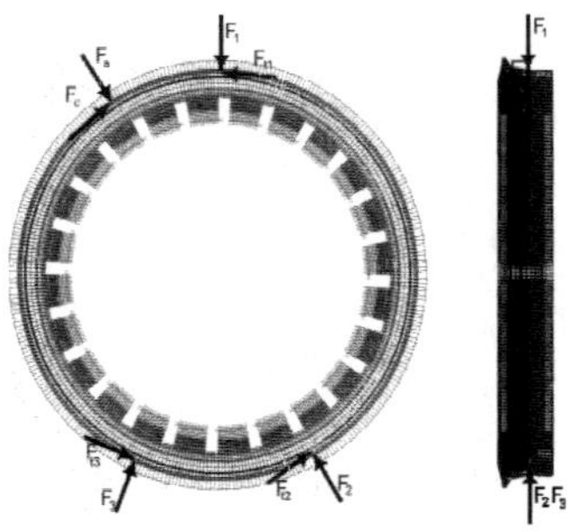

Fig. 2. Placement of forces on first FEM wheel model

Forces F_1, F_2 and F_3 are forces of rollers pressing and their values are:

F_1= 60000N

F_2=F_3=25000N

Forces F_a and F_c are calculated cutting forces and their values are:

F_a=F_c=20000N

Forces F_{t1}, F_{t2} and F_{t3} are applied to counterbalance cutting force F_c

Presented forces was moved along the circumference of wheel to check the influence of empty spaces between elastic inserts to the radial deformation of tyre measured in point of cutting force load. Every time position of each force against other forces was the same. To check one empty space, and one insert there was needed fifteen different analyses every with forces rotated by 1 degree more then previous. Achieved deformations are shown on Fig. 3.

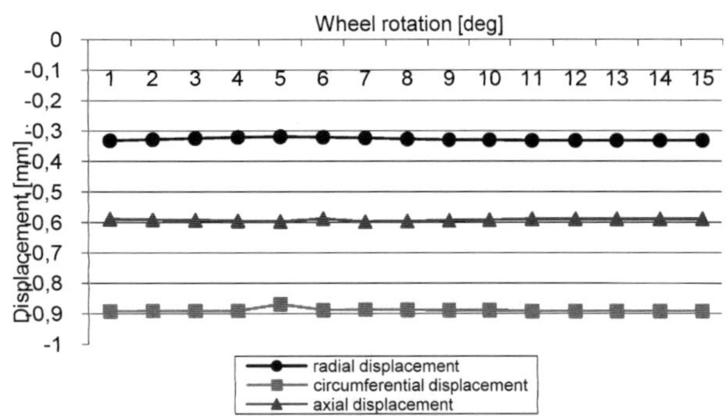

Fig. 3. Displacement of node with applied cutting forces in dependence on angular position of machined wheel.

As one can see influence of empty spaces between inserts on deformation is very small. What is significant on the diagram is that there is a large displacement in axial direction (marked with triangles) although there is no force acting in that direction. Circumferential displacement appeared because main cutting force (F_c) was not fully counterbalanced by rest of the forces, but it has no influence on results of research.

Next step is model with applied axial forces (Fig. 4.).

Values of forces has not been changed only one force has been added, it's axial component of cutting force – F_f=10000N

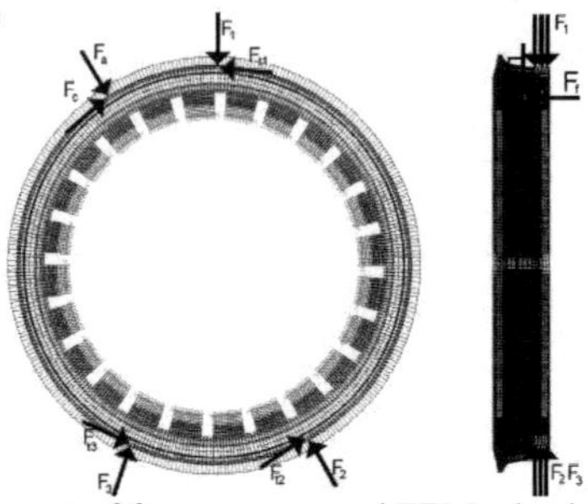

Fig. 4. Placement of forces on second FEM wheel model

This model includes also removing of allowance from machined surface. That is why forces of rollers pressing has been divided into three smaller forces in a row. Simulation of allowance removing has been realized by removing rows of elements along feed of tool machining contour of wheel tyre. This case needed 19 consecutive computations, every one for every removed row of material (elements). Results show dependence of displacements in point of machining on tyre profile (Fig. 5.).

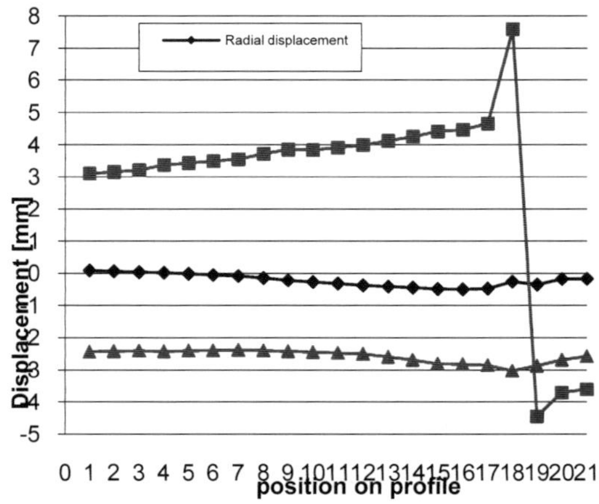

Fig. 5. Displacements of nodes with applied cutting forces in dependence on allowance removing.

4. CONCLUSIONS

The biggest problem appears to be axial deformation of wheel during the machining process not as we firstly expected radial displacement in dependence on discontinuity of elastic inserts. As machining is performed partly in opposite direction axial deformation has direct influence on machined profile.

Next step of the research must be verification of obtained results in real machining.

5. REFERENCES

Collective work(1990) *Poradnik inżyniera – Obróbka skrawaniem, (Engineer's handbook -Machining),vol. II.* WNT, Warszawa

Górski, E. (1999) *Poradnik frezera*. WNT, Warszawa.

Grzesik, W. (1998) *Podstawy skrawania materiałów metalowych*. WNT, Warszawa.

Olszak, W. (2001) *Obróbka skrawaniem i narzędzia, Rozdział 5 – Proces skrawania.* Wydawnictwo Politechniki Szczecińskiej, Szczecin.

Zienkiewicz, O.C.; Taylor, R.L & ZHU. J.Z. (2005) *The Finite Element Method: its Basis and Fundamentals.* Elsevier Buterrworth-Heinemann, ISBN 0-7506-6320-0 ,Oxford

Annals of DAAAM for 2008 & Proceedings of the 19th International DAAAM Symposium, ISSN 1726-9679
Editor B. Katalinic, Published by DAAAM International, Vienna, Austria 2008
Make Harmony Between Technology and Nature, and Your Mind will Fly Free as a Bird

RATIONALIZATION OF THE OPERATION OF THE POWER-SUPPLY SYSTEM OF INTELLIGENT CLOTHING

FIRST ROGALE, S[njezana]; ROGALE, D[ubravko]; DRAGCEVIC, Z[vonko]; NIKOLIC, G[ojko] & RUNKAS, M[artin]

Abstract: *The paper describes an intelligent article of clothing with active thermal protection which changes the value of thermal insulation depending on environmental temperature outside the garment and microclimate inside the garment. On the basis of the results of the performed investigations the first prototype was constructed which has integrated thermoinsulating chambers, necessary sensors, microcontroller and actuators to implement the decisions within the microcontroller assembly. All architecture elements are connected into a harmoniously functioning structure. This paper deals with the investigation results of the rationalization of the operation of the power-supply of the intelligent article of clothing with active thermal protection.*

Key words: *intelligent clothing, micropneumatics, PWM control.*

1. INTRODUCTION

An intelligent article of clothing with active thermal protection has thermal insulation chambers, Fig. 1, with small tubes through which compressed air is conducted. The tube (1) inflates the shoulder insulation chamber, the tube (2) left and the tube (3) right chest insulation chamber. The tube (4) inflates the back and the tube (5) the waistband insulation air chamber. Each chamber has its own sensor for compressed air, the shoulder chamber the sensor (6), the chest one the sensor (7) and the waistband one the sensor (8). The level of air pressure defines the thickness of the thermal insulation chamber and its thermal insulation properties. The thermodynamic parameters within the article of clothing are measured by the sensor (9), while the parameters of the environment are measured by the sensor (10). Depending upon the ratio of the parameters within and outside of the article of clothing, a decision is brought on the thickness of the thermal insulation chamber in question, i.e. on the thermal insulation properties of the article of clothing, achieved by varying the air pressure in the chambers. Additional thermal insulation properties can also be achieved though various combinations of activated and deactivated thermal insulation chambers (Rogale et al., 2007).

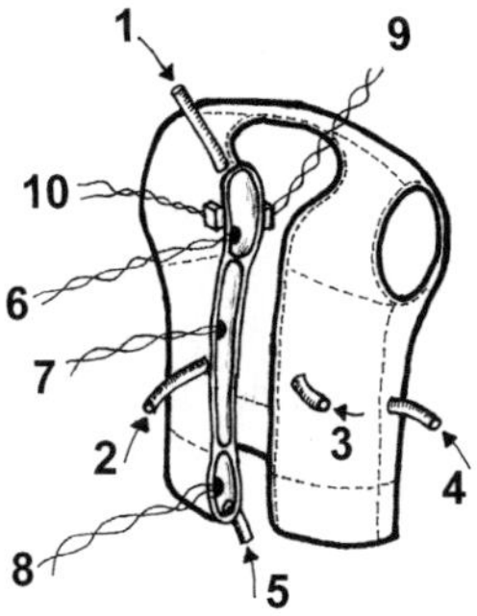

Fig. 1. Basic elements of the insert of an intelligent article of clothing with active thermal protection

Active thermal protection of the first prototype of intelligent article of clothing with active thermal protection is accomplished with versatile combinations of activating the shoulder, breast or waist sealing chamber which supports or disables the so-called chimney effect within the garment.

From the point of view of rational compressed air consumption the activation or deactivation of the chambers means a high consumption of compressed air because some chambers often deflate completely and the others inflate. At the next stage of protection the inflated chambers should be deflated, and the others activated and the like, meaning a higher consumption of electric power for the operation of the microcompressor.

2. TECHNICAL SOLUTION OF THE MICROPNEUMATIC SYSTEM

The operation of the article of clothing with active thermal protection is based on the activation of the thermoinsulating chambers which are supplied with compressed air with the highest pressure up to 50 mbar. There are three groups of thermoinsulating chambers which are activated in various combinations, accomplishing different effects of thermal protection of the garment. Thus, as the actuator system in the intelligent article of clothing with thermal protection a system based on the elements of micropneumatics was selected. Fig. 2 shows the system which has shown the best possible results in the operation of the intelligent article of clothing with thermal protection (Firšt Rogale et al., 2006).

Compressed air is guided via pneumatic tubing manufactured by FESTO, GmbH, Germany, designated as PUN-4x0,75-BL and via L and T plug-in connections designated as QSLM-1/8-4-100 and QSMT-4 to electromagnetic valves 3, 4 and 5 which fill it into thermoinsulating chambers 6, 7 and 8. In such a way the electromagnetic valve 3 after being activated by AS actuator bus allows the flow of compressed air into the shoulder thermoinsulating chamber 6. In a similar way the electrovalve 4 allows inflating the chest thermoinsulating chamber 7, and electrovalve 5 inflates the waistband thermoinsulating chamber 8. Electrovalves 3, 4 and 4 are made by FESTO designated as MZH-3-M3-L-LED.

The RS 235-5784 pressure sensors made by RS Computers, USA, in the thermoinsulating chambers are connected to the thermoinsulating chambers. Signals coming from the pressure sensors are guided to the sensor bus SS, then to the amplifier of measuring signals and to the microcontroller system. A particular thermoinsulating chamber is filled with compressed air after a decision made on the basis of the algorithm of intelligent behavior. The microcontroller activates the microcompressor 1 and at the same time one of three air fill electrovalves 3, 4 or 5 of the associated thermoinsulating chamber 6, 7 or 8. During the operation of the microcompressor and the associated electrovalve the pressure sensor constantly measures pressure value in the associated thermoinsulating chamber. The microcontroller constantly monitors the value of this pressure, and when a value of 50 mbar is reached, the microcompressor also deactivates the air fill electrovalve.

If it is necessary to inflate a particular thermoinsulating chamber, one of the air release electrovalves 12, 13 or 14 is activated. They are made by Matrix and designated as Matrix 821 2/2 NC. Electrovalve 12 is used to deflate the shoulder thermoinsulating chamber 6, electrovalve 13 is used to deflate the chest thermoinsulating chamber, and electrovalve 14 is used to deflate the waistband thermoinsulating chamber 8.

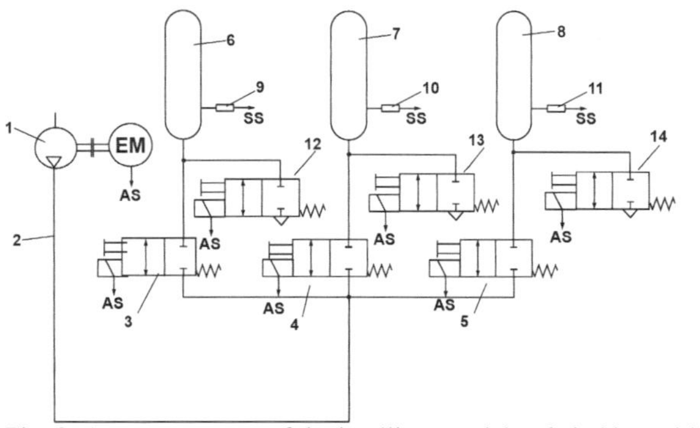

Fig. 2. Actuator system of the intelligent article of clothing with active thermal protection based on the elements of micropneumatics.

If it is necessary to activate several thermoinsulating chambers, program support provides a sequential activation mode so that at the same time only one chamber is filled with air, and only one air fill electrovalve and microprocessor are activated. Only after the first thermoinsulating chamber is filled, the sequential mode of inflating the other thermoinsulating chambers starts. During deflating the thermoinsulating chambers it is possible to activate several air release electrovalves at the same time, since the microcompressor as energy user of heavier currents has not been activated.

3. EXPERIMENTAL PART

For the purposes of rational usage of the electric capacity of the battery assembly, experiments were conducted with combination of a short activation of electrovalves with full voltages and currents lasting for several tens of ms, and afterwards a series of PWM pulses was introduced. This control method provided very interesting and original results obtained and shown by measurements made by a digital memory oscilloscope (Firšt Rogale et al., 2007).

Fig. 3 shows that after the central microcontroller gives a command, the operation of the auxiliary microcontroller is switched on. At its exit it generates a pulse lasting about 60 ms (the first rectangular long-duration pulse); it is used to activate the electrovalves in full force, and afterwards a series of PWM pulses (a series of short wide-modulated rectangular pulses) follows to save energy.

The active part of a PWM pulse with a total duration of 5 ms is only 1 ms and only 20% respectively. The outlet is driven at 24 V, but due to an inductive character of the coil, the voltage from top to top of the pulse is about 27.2 V and it is limited by the protection diodes. Since the active part of a PWM pulse lasts for 1 ms, it may be thought that the electrovalve coil is activated and deactivated by a frequency of 1 kHz.

The presented oscillograms show that it is sufficient a time interval of only 60 ms to activate the electrovalves in full force, and that the electrovalves can be kept in an activated state with PWM pulses in ratios 1/5 in relation to the total duration.

Namely, the electrovalve at a voltage of 24 V consumes the current of 21.4 mA or it needs a power of 0.514 W for an activated state.

It was measured that when the activated electrovalve is kept by PWM pulses the stabilization current is only 1.8 mA so that it needs a power of only 0.043 W to be kept in an activated state.

Thus, when using PWM modulated pulses, 12 times lower power is necessary than the power of the directly activated electrovalve, amounting to 8.4%. In other words, the resources of the drive battery foreseen for the control of electrovalves can be used 12 times longer than the electric control without PWM pulses.

It is expected that the reduction will be about 5 times because the active part of PWM pulses was 20% of the total duration (1 ms in relation to a duration of 5 ms), but it is obvious that the electrovalve coil is not only ohm resistance for direct current, but it represents a complex resistance of inductive impedance which is proportionally high in relation to the frequency of PWM pulses (200 Hz), and still higher from the aspect of the duration of the active part of PWM pulses (1 kHz). This is the reason why the supply current of the electrovalve is lower than expected.

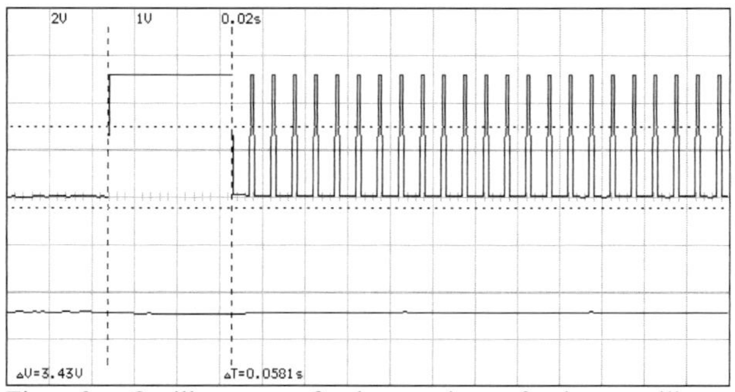

Fig. 3. Oscillogram of the outlet of the auxiliary microcontroller for the activation of electrovalves

Likewise, it has been observed that the electrovalve directly supplied with a voltage of 24 V is rather heated, while the electrovalve supplied with PWM pulses does not show any signs of heating up.

4. CONCLUSION

As a special feature for the needs of this investigation a new approach to the efficient and rational consumption of electric power of the batteries integrated into the intelligent article of clothing with active thermal protection. For these purposes familiar experiences of the so-called PWM control were used. As another novelty the actuator approach of the complex PWM control was used; this consists of two series of PWM pulses. The first in a series of pulses lasts for about 60 ms (first rectangle pulse of long duration) which activates the electrovalves, followed by a series of PWM pulses (a series of short widely modulated rectangle pulses) to save energy.

5. REFERENCES

Firšt Rogale, S. (2007). Intelligent clothing with active thermal protection, Doctoral thesis, 1-224

Firšt Rogale S., Rogale D., Dragčević Z., Nikolić G., Bartoš M. (2006). Technical Systems in Intelligent Clothing with Active Thermal Protection, *Book of Proceedings of the 3rd ITC&DC*, Dragčević, Z. (ur.), Zagreb. Faculty of Textile Technology University of Zagreb, Zagreb, 413-419

Rogale, D. ; Firšt Rogale, S.; Dragčevic, Z.; Nikolić, G. (2007). Intelligent Article of Clothing With an Active Thermal Protection, State Intellectual Property Office of Croatia, P20030727, Zagreb, Croatia

Annals of DAAAM for 2008 & Proceedings of the 19th International DAAAM Symposium, ISSN 1726-9679
Editor B. Katalinic, Published by DAAAM International, Vienna, Austria 2008
Make Harmony Between Technology and Nature, and Your Mind will Fly Free as a Bird

THE COMPUTATION OF THE CALENDER ROLLS DEFLECTION USING THE DIFFERENTIAL CALCULATIONS

FLORESCU, D[aniela]; STAN, G[heorghe] & FLORESCU, I[ulian]

Abstract: *The tendency of incorporating calenders in paper machines is the answer to the needs of the consumers concerning the improvement of quality and uniformity of the paper. Although has been done research concerning the decrease of the level of dependecies between the parameters of the calendering processes and the properties of the final product, an important field remains the one concerning the assurance of uniform loads distributed along the generatrix of the cylinders, and also to improve the structure of the surface of the materials taking into consideration the mechano-rheological conditions. In the area of lamination processes the high quality of the equipments used results in a high quality of final products in respect with its uniformity and thickness.*
Keywords: *calender, deflection, conditions, model, generatrix*

1. INTRODUCTION

There have been conceived a computation model which are using plane elements for the stress transmission, obtained by sectioning the rolls with longitudinal planes. The following hypotheses have been considered:

- the load is constant, uniformly distributed along the calender rolls surface for paper pressing (p_2 = 62 N/mm);
- the model undergoes a plane strain, due to the fact that the shearing strain of the plane element is actually impeded because of the rolls portions which have been removed and which have high stiffness;
- the roll is supported at its ends; the length of the bearings and the bending moments due to the reactions with respect to the planes of the rolls ends are negligible;
- the friction forces are negligible taking into account both their reduced values and their influence on the other parameters. This hypothesis is confirmed by the experimental results.
- the rolls dimensions are: external cylindrical surface: ∅ = 100 × 500 (mm × mm); internal cylindrical surface: ∅ = 40 × 50 (mm × mm); overall length: L = 1100 mm.

2. THE DETERMINATION OF THE SHAPE AND DIMENSIONS OF THE COMPUTATION MODELS

The hypotheses used in conceiving new computation models of a calender roll elements stress lead to structures whose plane dimensions depend on the rolls geometry and on the paper tape width. For the following model we consider the possibility to obtain opposite rolls deflections by introducing a no aggressive work medium, under pressure, within the rolls. Further on we have in view to determine the deflections which appear in the calender rolls, when a uniformly distributed load, occurring during the rolling of the paper tape, acts upon them. A paper bowl with an ellipsoidal internal section is considered. Figure 1 shows the work diagram in the case.
The equation of the ellipsoid surface is:

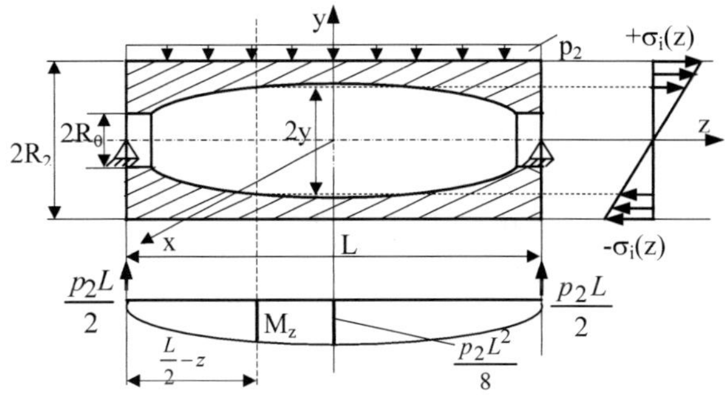

Fig. 1. Diagram of the ellipsoidal internal shape roll.

$$\frac{x^2}{a^2}+\frac{y^2}{b^2}+\frac{z^2}{c^2}=1 \qquad (1)$$

where a = b = R_1 stand for the ellipsoid small semi-diameters. Because the calender rolls are thought to be thick-walled cylindrical bodies, they are characterized by the physical constant β between the external and internal radius $\beta = \frac{R_2}{R_1} \geq 1{,}2 \Rightarrow R_1 \leq \frac{R_2}{\beta}$, in which: R_2 = 50 mm; R_0 = radius necessary for the roll to operate, R_0 = 20 mm; c_1 = basic size, in mm. $c_1 = \frac{L}{2} - h = 470$ mm. $R_1 = \frac{R_2}{\beta} = 41$ mm. The ellipsoid equation in the roll longitudinal plane is: $\frac{y^2}{b^2}+\frac{z^2}{c^2}=1 \Rightarrow y = \pm a\cdot\sqrt{1-\frac{z^2}{c^2}}$ The limiting conditions are: $y = R_0 \Rightarrow z = c_1$. Replacing, we obtain: $c = \pm\frac{R_1 c_1}{\sqrt{R_1^2 - R_0^2}}$ = 538 mm.

3. THE DETERMINATION OF THE DEFLECTION

The equations of the medium fibre deformation are:

$$\frac{d^2f}{dx^2} = -\frac{M_i}{EI_z(x)} \qquad (2)$$

where: M_i – moment of deflection; E – Young's modulus of the roll material; I_z – moment of inertia of the cylinder casing. The moment of deflection in current sections is:

$$M_i = \frac{p_2 x}{2}(L-x) \qquad x = \frac{L}{2} - x \qquad (3)$$

dx = dz; $M_I = \frac{p_{2(L^2-4z^2)}}{8}$ in which: $z \in [o, c_1]$.

$$f^{//} = \frac{p_2L^2 - 4p_2z^2}{2\pi E\cdot\left[(R_2^2-R_1^2)+2\cdot\frac{R_1^4}{c^2}\cdot z^2-\frac{R_1^4}{c^4}\cdot z^4\right]} \qquad (4)$$

In the differential equation (4) we introduce the constants A, B and C: $A = -\frac{R_1^4}{c^4}$; $B = 2\frac{R_1^4}{c^2}$; $C = R_2^4 - R_1^4$.

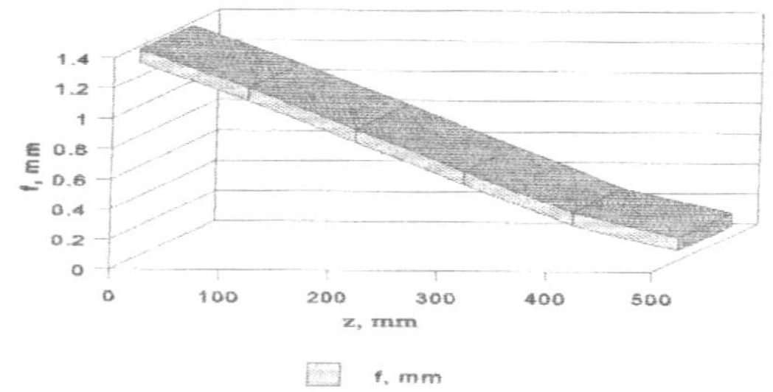

Fig. 2. Calender Roll's Deflection

$$f' = \frac{2A}{\sqrt{\delta}}\left\{\frac{1}{2A\sqrt{B+\sqrt{\delta}}}\cdot\left[arctg\frac{\sqrt{B+\sqrt{\delta}}}{2A}\cdot z - \ln\left|\frac{2A+\sqrt{B+\sqrt{\delta}}\cdot z}{2A-\sqrt{B-\sqrt{\delta}}\cdot z}\right|\right]\right\}$$

where: $\delta = B^2 - 4AC$, with condition $\delta > 0$.

The first boundary condition is: $z = 0 \Rightarrow f' = 0 \Rightarrow C_1 = 0 \Rightarrow$

$$f = C_2 + D\left[zarctg\frac{z}{V} - \frac{V}{2}ln(V^2 + x^2)\right] + \\ + \frac{FM}{K}[(z+K)ln(z+K) - (z-M)ln(z-M)] \quad (5)$$

Second boundary equation is: $z = l \Rightarrow f = 0 \Rightarrow C_2$, where C_1, C_2 are the integrating constants.

$$C_2 = -D\left[larctg\frac{l}{V} - \frac{V}{2}ln(V^2 + l^2)\right] - \\ - \frac{FM}{K}[(l+k)ln(l+K) - (l-M)ln(l-M)]$$

$$f = -D\left[larctg\frac{l}{V} - \frac{V}{2}ln(V^2 + l^2)\right] - \\ - \frac{FM}{K}[(l+K)ln(l+K) - (l-M)ln(l-M)] + \\ + \frac{FM}{K}[(z+K)ln(z+K) - (z-M)ln(z-M)] \quad (6)$$

If $z = 0 \Rightarrow f = f_{max} \Rightarrow f_{max} = -D\left[larctg\frac{l}{V} - \frac{V}{2}ln(V^2 + l^2)\right] - \frac{FM}{K}\left[\begin{matrix}(l+K)ln(l+K) - (l-M)ln(l-M) + KlnK \\ -MlnM\end{matrix}\right]$ (7)

Using MathCAD software we obtain for p_2 = 62 N/mm (pressure of rolling) the values for all the constants are: $A = -3{,}056 \cdot 10^{-5}$; $B = 17{,}689$; $C = 3{,}69 \cdot 10^6$; $K = -9{,}07 \cdot 10^{-6}$; $\delta = 763{,}925$; $\eta = 5{,}073 \cdot 10^{-5}$; $\xi = -5{,}77 \cdot 10^{-4}$; $G = 6{,}733$; $H = 3{,}154$; $D = 0{,}247$; $C_2 = 15397$; $F = -2{,}43 \cdot 10^3$; $V = -1{,}102 \cdot 10^5$. With these values we obtain:

f_{max} = 1.335 mm.

Finally, we prove the results of this method with the Stone – Gaewski formulae and the finite elements method.

Stone's formula (Gaewski, 1955) for calculating the maximum deflection of a roll is eq. (8) where: p_2 – the external rolling pressure, N/mm; E – Young's modulus of the roll material; I_z – moment of inertia of the cylinder casing; D_2 – external diameter of the cylinder casing, in mm:

$$f_{max} = \frac{5p_2L^4}{384EI_z}\cdot\left[1 + \frac{24}{5}\cdot\frac{h}{L} + 2\cdot\left(\frac{D_2}{L}\right)^2\right] \quad (8)$$

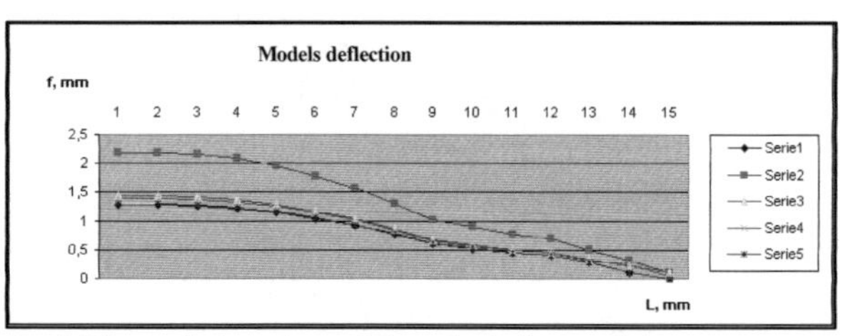

Fig. 3 The variation chart of the calculating models:

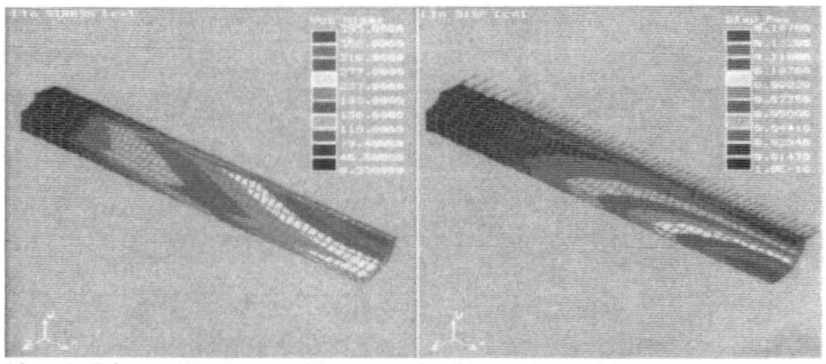
Fig. 4. Linear stress and linear displacement for model

Serie1 = cylinder with constant internal cylindrical shape; Serie2 = bulk cylinder (without internal shape); Serie3 = cylinder with ellipsoidal internal cylindrical shape; Serie4 = cylinder with double cylindrical-truncated cone internal cylindrical shape; Serie5 = cylinder with double truncated cone internal cylindrical shape.

The first term in the square brackets introduces the bending produced by the uniformly distributed load, the second is due to the center bearing deflection, and the third stands for the curvature produced by the shearing forces. The magnitude of this last term is generally 5%, but it can increase up to 10%. Gaevski's formula for calculating the rolls total deflection is:

$$f_{max} = \frac{5p_2L^4}{384EI_z}\cdot\left\{\left[1 + \frac{24}{5}\cdot\frac{h}{L} + 2\cdot\left(\frac{D_2}{L}\right)^2\right] - 8\left[\frac{3}{5} + \frac{12h}{5L} + \left(\frac{2R_2}{L}\right)\right]\right\}$$

After doing the substitutions in the formulae (8), the deflections variation for the rolls under consideration is shown in Figure 3. Figure 4 presents the displacement and stress condition for ellipsoidal internal section roll in COSMOS software.

4. CONCLUSIONS

It is possible to determine the calender roll's deflection using the differential calculations, and prove the results with other methods like Stone-Gaewski formulae or finite element method. The differential method presented in this paper is very useful for the study of the roll's deflection and may also be used in optimization problems. The great volume of work done to develop this method is compensated by the important facilities given to the users, in order to provide an up-to-date for making easier the analytical calculation of the paper bowl.

5. REFERENCES

Florescu, D, Florescu, I. (2005), *The computation of the calender roll deflection using the Stone-Gaevski formulae*, Conference UgalMat, Galati, Romania, October, ISSN 1453-083X, Nr. 2.

Florescu, D, Florescu, I. (2003), *The computation of the calendar roll deflection using the differential calculation*, MOCM 9, Methods and Technique for deformation Measure, University of Bacau, Romania, November, p. 59-63.

Florescu, D. (2004), *The determination of the calender deflections using the video computering analyse*, Conference Equipments, Instalationes and Process engineering, University Politehnica of Bucharest, Romania, p. 157-160, ISBN 973-718-061-5 (in Romanian).

Florescu, D., Iatan, R. (2002), *Theoretical study about the determinations of the calendar rolls deflection,* Machines building, (54), nr.1-2, p. 38-40.

Florescu, I, Florescu, D. (2000), *Research regarding temperature influence over calendar roll cylinders*, Abstract Bulletin of Paper and Technology, april, SUA, vol. 70, no. 10, 10361 (AS).

Annals of DAAAM for 2008 & Proceedings of the 19th International DAAAM Symposium, ISSN 1726-9679
Editor B. Katalinic, Published by DAAAM International, Vienna, Austria 2008
Make Harmony Between Technology and Nature, and Your Mind will Fly Free as a Bird

THE OPTIMIZATIONS OF THE FUNCTIONES OF THE DOUBLE DIAMETER PISTON SOLENOID VALVE

FLORESCU, I[ulian]; FLORESCU, D[aniela] & STAN, G[heorghe]

Abstract: *The theoretical and experimental researches concerning the double diameter piston solenoid valves follow the analytical method and dynamical design which is permitted to obtain an efficient and fast study. It is realized a mathematical dynamical simulation of a solenoid valve with double diameter piston for studies the important physic-mechanical or geometrically parameters influences over the operating mode in different works regime.*
Keywords: *double diameter piston solenoid valves, analytical methods, mathematical dynamical simulation, geometrically parameters*

1. INTRODUCTION

The usage of the control and regulation equipments which can be distance managed, with high performances in different working conditions, such as: high or low temperatures, sliminess, different pressures, or different actioning periods is an evident necessity as a result of the usage, on a large scale,of the automatization for different industrial processes or of a centralised conduction. These equipments should be projected in such a way that they can assure the required parametres for the instalation to which they were given. A well knowledge of their behaviour in different working conditions can be done through experimental trials, in laboratories, for a certain period of time and by using expensive models. For this purpose, the mathematical forming operation of the functioning of those programs and the making of a calculating program, which can be good for studing the influnce of a higher number of geometrical or physico-mechanical parameter, presents considerable advantages for design and research.

2. MATHEMATICAL FORMING OPERATION

I divided and studied the opening and the shutting of the electrovent in phases, starting with the action of the electromagnet on the magnetic casting mould core, its motion in a liquid together with its interior cane which sets in motion the ram and its motion under the pressure influence from its two faces. For shutting the vent the order is reversed.

2.1. The analythical expressions of the forces

The forces which act upon the elements of the electrovent are of different origins: electromagnetical, of pressure, of friction and other onesand are changeable and depend on the number a great number of parametres. **The force of the electromagnet,** changeable according to the position of the magnetic casting mould core in the coil, was experimentally determined using a device which can establish the position of the casting mould core and the right value of the force. The resulted variation curve, figure 1, was estimated with the graph of a rational function. **The pressure force** caused by the pressure variation of the magnetic casting mould core when it moves with high speed in liquid, $F_p = \frac{\pi}{4}\left(d_e^2 - d_i^2\right)\cdot \Delta p(\dot{x})$, where the transverse section of the device in motion can be expressed taking into

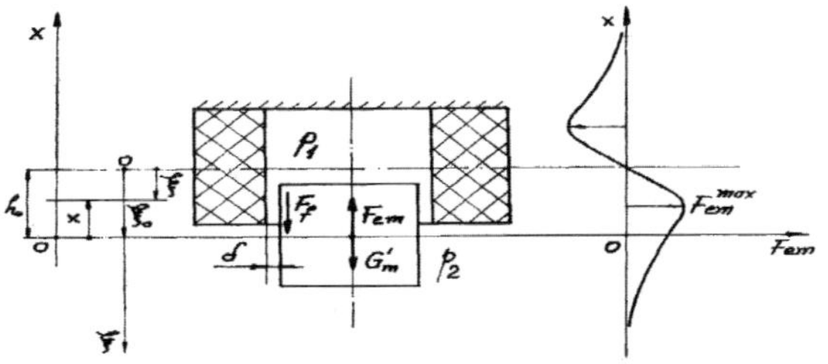

Fig. 1. The variation of the force of the electromagnet

account even the possible longitudinal gunners of the irregular sections which can be tantamount to the gunners with circular sections by equalizing their hydraulic ranges $\Delta p(\dot{x})$ can be determined by equalizing the effluent volume from the circular interstitions of δ_i, δ_e thickeness, figure 1, with the volume replaced by the magnetic casting mould core in motion.

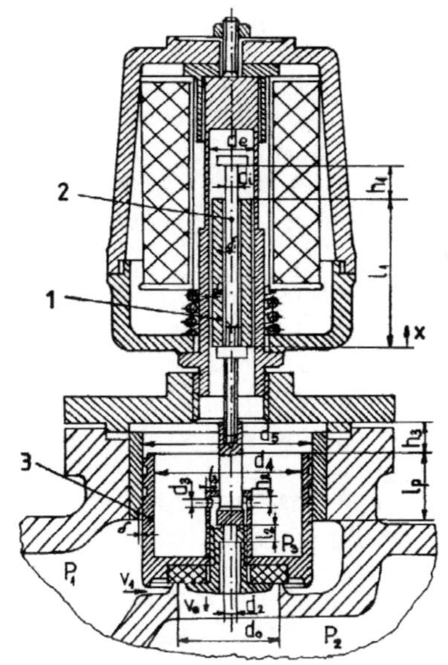

Fig. 2. Electomagnetic Vent

The force of viscid friction, caused by the flowing of the liquid through the interstitions of the magnetic mould core in the opposite way to the movement of the element in motion, can be determined according to the tangent, unitary efforts which may appear, figure 3.

$$F_f = \pi\left[l\eta\dot{x}\left(\frac{d_e}{\delta_e} + \frac{d_i}{\delta_i}\right) + \frac{\Delta p}{2}\left(d_e\delta_e + d_i\delta_i\right)\right] \qquad (1)$$

The pressure force from the interior face of the ram F_{pi} expressed according to the position of the ram from the center, x, considering the distribution of pressures according to the geometrical configuration of the flowing:

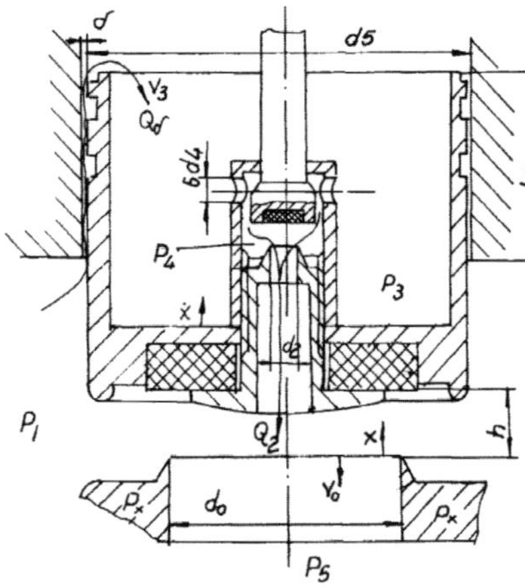

Fig. 3. The pressure force from the superior surface

$$F_{p'} = 2\pi\left(\int_{0}^{r_o} p_2 + \int_{r_0}^{r_5} p(r)\right) r \cdot \mathrm{d}r \qquad (2)$$

where *p(r)* is a linear dependence according to the charging pressure and to the x position of the ram, between the limit pressures.

The pressure force from the superior surface, figure 3, of the ram expressed according to the flowing through the space δ and through the opening with the diameter d_2:

$$F_{p3} = \frac{\pi}{4} d_5^2 \cdot p_3(\dot{x}) \qquad (3)$$

Where p_3 is the pressure from the interior cavity of the ram, figure 5, determined from the sistem of equations established by the motion and conservation laws of the liquid.

The ostensible weight of the pieces in motion, which can be determined taking into account the density of the working liquid and of the material of the device in motion:

$$G_m^{'} = G_m\left(1 - \frac{\gamma}{\gamma_m}\right) \qquad (4)$$

2.2. The determination of the motion equations

Taking into account Newtons fundamental equation $m\ddot{x} = \Sigma F$ for each phase opening and shutting and having in view all the forces which act in that phase, $F = F(t, x, \dot{x})$, projected on the motion direction,we get the the motion equation specific for that phase:

$$\ddot{x} + K_1\dot{x} + K_2 + K_3 = 0 \qquad (5)$$

differential equation of the second type, with changeable coefficients,where $\ddot{x} = \ddot{x}(t, x, \dot{x})$. The integration of these equations which clearly express the accelerations of the volume sets according to time, motion x_n and speed $\dot{x}_n$, was made by using Blaesѕ numerical integration method. The gravel is made step by step, expressing the motion and the speed by using series of Mac-Laurin powersfor short periods of time $\Delta t_n = t_n - t_{n-1}$. When using certain series, we can select only the first two terms, because at low values of the time period of 10^{-3} - 10^{-2} the powers higher than two do not influence the reached values:

$$x_n \cong x_{n-1} + \Delta t_n \cdot \dot{x}_{n-1} + \frac{\Delta t_{n-1}^2}{2} \cdot \ddot{x}_{n-1}(t_{n-1}, x_{n-1}, \dot{x}_{n-1})$$

$$\dot{x}_n \cong \dot{x}_{n-1} + \Delta t_n \cdot \ddot{x}_{n-1}(t_{n-1}, x_{n-1}, \dot{x}_{n-1}) \qquad (6)$$

Replacing the values of the resulted motions and speeds and taking into account the first conditions established for x_o and $\dot{x}_0$, we calculate the motion x_1 and the speed $\dot{x}_1$, with which we calculate the motion equations $\ddot{x}_1$. In the same way, we ingeminate the gravels, determining x_2, $\dot{x}_2$ and then $\ddot{x}_2$ and so on till the value of the motion reaches the limited established value. Starting from the mathematical model made according to the functioning analyses, we made a calculating program which allows the quick monitorization of the influence of some constructive and physicomecanic parameters such as the charging pressure and the slimines. Seeing the way of functioning of the electromagnetic steered vents, one can mention the important influence of a great number of working parameters and we will consider all these parameters in the motion equations of the making elements. The method of integration for all the resulted equations requires a calculating sistem which can follow by phases the behaviour of the electrovent. By well placing the electric contacts of an electric stop which can determine the action time, one can verify the obtained resultsthrough gravels and then the veracity of the mathematic model. For finding out the influence of a certain parameter only the initial value is modified, according to the data of the program. For the studied electrovent we took into account the influence of the sliminess of the position between the ram and the cylinderwhen it moves, of the charging pressure and of the diameter of the vent chair. The calculating program can display a large variety of information: speeds, pressures, accelerations in any moment of the action, which requires the knowledge of the dinamic behaviour of the electrovent, good for design and research.

3. CONCLUSIONS

The paper presents a modern method of design for the control equipments of the type of the steered electromagnetic vents and a method for the analyses of their behaviour in different working conditions or in the situation of the usage of other liquids than those which were established. We made the mathematical model for a certain type of a valve. In adition to the optimum design of those vents one can easy try some simply constructive methods, considering the influence of a great numbers of working parameters. Studing the given results for the influence of the four selected parameters, we can notice that a great influence is given by: the distance between the ram and the cylinder, the sliminess of the working liquid and even the temperature of the environment.

4. REFERENCES

Florescu, I (2006), *Hydraulic*, Publisher Tehnica-Info, Chisinau, Republic of Moldavia, ISBN 978-9975-63-282-9.

Florescu, I, Florescu, D., Olaru, I (2003), *The Mechanic of fluids and hydraulic machines*, Publisher Tehnica-Info, Chisinau, Republic of Moldavia, ISBN 9975-63-222-X.

Florescu, I., Stupu, D. (1996), *The researches concerning the optimum design for the process equipments*, International Conference for scientifical paper, TMCM'96, Iaşi, p.133-136.

Florescu, I., (1993), *The theoretical study about the physic-mechanical or geometrically parameters influences over the piston solenoid valve, The Bulletin of the Oil & Gas Institute*, vol. XLV, Ploieşti, p. 165-168.

Cazacu, M.D., Florescu, I. (1985), *The researches concerning the double diameter piston solenoid valve, The hydraulic and hydro dynamically Conference,* Timişoara, october, p. 37-42 (in Romanian).

Annals of DAAAM for 2008 & Proceedings of the 19th International DAAAM Symposium, ISSN 1726-9679
Editor B. Katalinic, Published by DAAAM International, Vienna, Austria 2008
Make Harmony Between Technology and Nature, and Your Mind will Fly Free as a Bird

TEST STAND AND EXPRIMENTAL RESULTS USING WEAK MIXTURES GASOLINE-HYDROGEN AT SI ENGINE

FODOR, D[inu]; MITRAN, T[udor]; PATER, S[orin] & POLOJINTEF CORBU, N[icolae]

Abstract: *In this paper are presented the testing devices, the block-diagrams as well as the accommodations performed to determine, by comparison with the same S.I. engine, the energetically and economical performances and the investigation of the combustion process in lean mixtures with hydrogen addition. The envisaged research program has been carried out by recording all the parameter required by the investigation.*
Key words: *testing, diagram, hydrogen, mixture, engine*

1. INTRODUCTION

The device on fig. 1 is working as a block diagram.The measuring device (see fig. 2), allowes some researches to be achieved in order to verify the possibility of raising the energetically and economical performances as well as the degree of reducing the polluting exhaust gases at a s.i. engine, by combusting weak mixtures with an addition of hydrogen produced on board of the car. (Furama 1979). The experimental researches also permitted to obtain data concerning the parameters of the duty cycle and those of the combustion process in the test engine, by verifying the possibility to improve the combustion kinetics of the lean mixtures with the proposed solution in the spark-ignition engines. (Matete, & Wolf 1972)

With that it has been studied the engine characteristic comparatively under standard condition and with lean mixtures and hydrogen addition under different operating conditions of the engine, respectively. (Bradford et al. 1984). The envisaged research program has been carried out by recording the following parameter required by the investigation: (xxx, 1979).
- parameters concerning the specific effective charcteristics of the engine;
- parameters concerning the specific indicated charcteristics of the engine;
- coefficient concerning the parameters of the combustion process;
- characteristics concerning the level of the emissions in the exhaust gases;

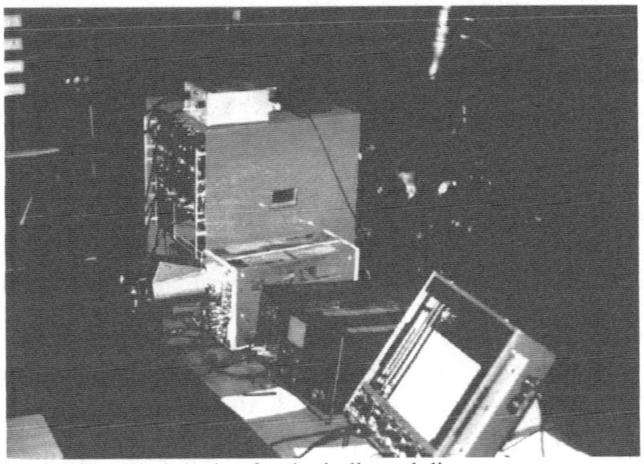

Fig. 1 Measuring device for the indicated diagram

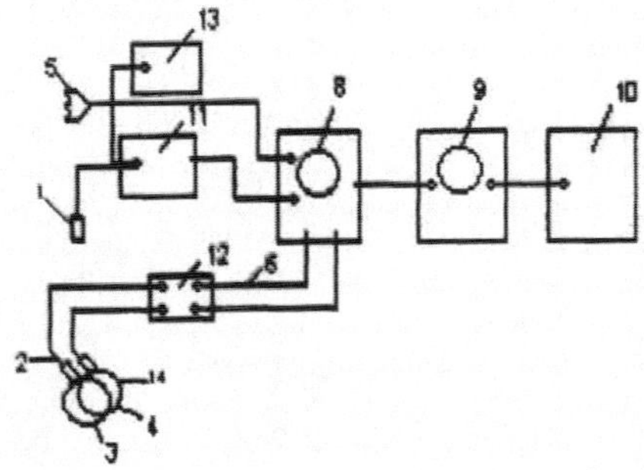

Fig. 2 Measuring device low chart for raising the indicated diagram

2. TESTING BENCH

The tests have been performed on the engine in several stages. The first stage envisages the determination (9, 10, 11) of the engine effective and indicated parameters on the testing bench by means of a braking device endowed with suitable metering equipments as well as with an electronic recorder of the indicated diagram in the engine cylinder, in (p-α) coordinates, concomitantly with the cyclical dispersion both by direct plotting and by photography on the oscilloscope.

The data obtained on theese devices enabled us to process the indicated diagram in p-V coordinates and to determine the parameters of the combustion process. (Lewis & ELBE 1961)

The correlation of the measured pressure depending on the rotation angle of the crankshaft as well as the opening of the inductive transducers type A.V.L. (Fig. 3) firmly mounted on the cylinder block, those transducers taking over the impulses from two coaxial disks mounted on the crankshaft.

The denticulate disk obtained through milling-work marking the angle has been made with a 20° division, and near the upper dead centre has been made symmetrically every 5° in order to accurately pursuit the detachment points. The inductive transducer (5) to indicate the point of ignition (Fig.2) was mounted on the cylinder high-tension lead (1) where the pressure transducer was mounted, too.

Fig.3 Inductive transducers type A.V.L.

Fig. 4 Piezoelectric transducer

The electrical signals obtained from the piezoelectric transducer (1) (Fig.4) are introduced in the load amplifier AVL-3059 (11) which is coupled to the calibration unit AVL-3054 (13). The two inductive transducers AVL (2) are coupled to the time base control unit AVL-4004 (12). The inductive transducer (5) to indicate the point of ignition, together with the two units (11-12) are connected to the Tektronix oscilloscope type 5113(8) that has several calibration positions dependent on the graduation of scale. This oscilloscope is provided with an obscure tube which allows for an EXATTA VAREX camera to be mounted on it, which has allowed the images to be recorded on the scale, dependent on the pressure and the angle of rotation of the crankshaft together with the point of releasing the electric spark, as well as for the cycle of dispersion to be photographed under ill the operating conditions (Fig.5).

In order to obtain more accurate measurements of the pressure change in the cylinder and to determine the parameters of the combustion process, the device has been completed with a NICOLET digital oscilloscope (9) which allows all the elements to be accurately plotted through graphic transposition by means of the X-Y Riken recorder (10). This also permits on the same scale for the pressure curve to be recorded without any combustion under the entire load and speed operating conditions.

The loggings of the indicated diagrams developed with several loads and speeds comparatively for the standard fuel-fed engine and the hydrogen addition fed engine have been made on the A4 sizes, and the pressure unique scale has been chosen 10 bars = 37,5 mm.

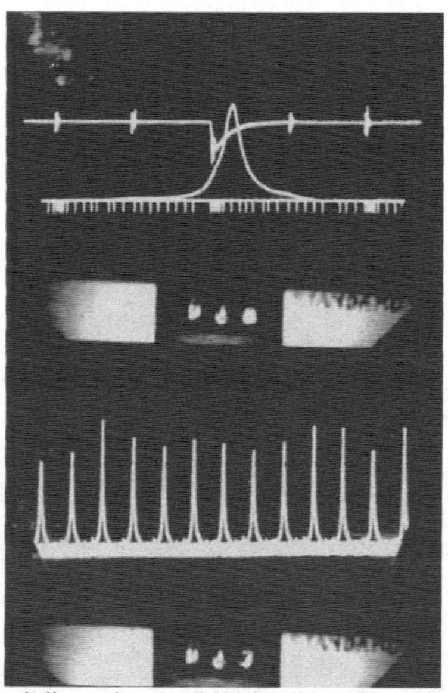

Fig.5 Cyclical dispersion at 100% loading without hydrogen addition

From these loggings one can notice that the pressure change has been maintained lineal dependent on the height of the logging.

In order to obtain the zero pressure line, the pressure of the outset of the compression has been approximated, the error being not bigger than 1%.

The accidental deviations could be rectified by estimating the mean pressure of the mechanical losses that presents a constant value when the engine speed and load are constant.

In this way for the three load operating conditions with the division of 5oo rot/min there have been determined the most probable mean values of the medium pressure of the mechanical loss to which there have been compared all the other pressure diagrams under the respective operating condition of the engine.

Another stage has been the testing of the engine on the car at stabilized speeds on the track as well as in highway traffic. Soat the average speed of 72 km/h, on a 250 km distance, in street traffic it was obtained a fuel consumption of 5,3 l/100 km, at a loading of 50% from the maximum permitted load. The dynamic parameters have also been improved. Thus from 0-100 km/h we got to a 18,3" timing compared to the 22" obtained without our device.

The last testing was made on the roll stand in accordance with the norms 15 CEE as well as the polluting emissions standards.

3. CONCLUSION

By using gasoline-hydrogen mixtures on SI engines with carburetor allows (Fig.6) a substantial reduction of the fuel consumption with 10-15% and At speeds of 40-50 km/h allowed by the urban traffic norms one can see a reduction of the CO between 0,8-1,2%,

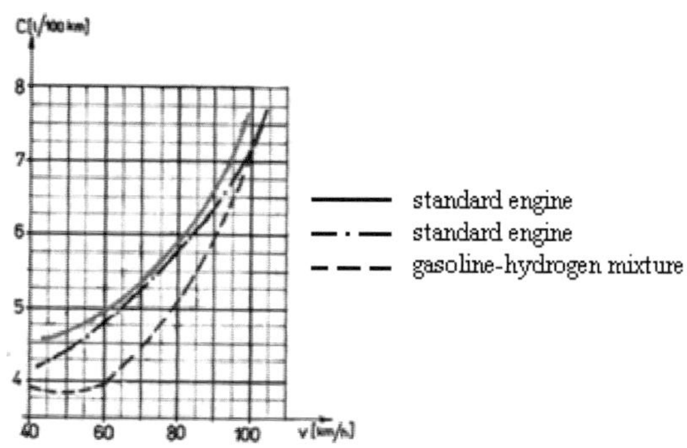

Fig. 6 Comparative consumption diagrams on the roll stand

As a result of the tests performed there arises the necessity for modifying the advance characteristics.

Better results can be obtained on engines with fuel injection systems.

Further reaserches for using fuel-hydrogen mixtures can be extended on diesel engines.

4. REFERENCES

Bradford, W.B.; FINCH, J.G. & PRIOR, Mc.A. (1984), *The Ignition of Some Explosive Mixtures by Modified Oil Discharges.* Journal of the Chemical Society

Furama, S. (1979). *Two-Stroke Hydrogen Injection Engine,* H.E.4, 6

Lewis, B. & ELBE, G.(1961) *Combustion, Flames and Explosions of Gases.* Academic Press

Matete, & Wolf (1972). *Hydrogen and the Big Energy Options.* H.E.4, 2

xxx , (1979). *Hydrogen Update.* In: Hydrogen Progress, nr.9,

Annals of DAAAM for 2008 & Proceedings of the 19th International DAAAM Symposium, ISSN 1726-9679
Editor B. Katalinic, Published by DAAAM International, Vienna, Austria 2008
Make Harmony Between Technology and Nature, and Your Mind will Fly Free as a Bird

STEGANOGRAPHIC TECHNIQUES FOR NETWORK COMMUNICATION

FOELDI, A[ndrei]; MOLDOVEANU, F[lorica] & SOCEANU, A[lexandru]

Abstract: *The steganographic techniques receive nowadays special attention because of their use in solving the problem of unauthorized copying for digital materials. Employing such techniques for the transmission of digital information raise important threats with regard to the security of information systems. The current paper will present known methods for camouflaging data inside normal data channels as well as methods to detect and destroy such data. An analysis of the performance and stealth of two novel ways of creating hidden channels inside normal data channels will conclude this paper.*
Key words: *Steganography, Hidden data channels.*

1. INTRODUCTION

Trying to solve the problem of unauthorized creation and distribution of digital copies for copyrighted material, steganography received an increased interest as a way of marking digital materials helping with the enforcement of the copyright laws. Another use of steganography was signalled in a newspaper reporting on the capture of the members belonging to a terrorist cell who seemed to use apparently inoffensive pictures as a mean of transport for secret information. (Millen, 1999)

The current paper deals with means of data concealment inside normal communication channels. The different ways of information insertion inside TCP/IP channels will be discussed as well as ways to detect and / or destroy such hidden information.

2. TAXONOMY

A data covert channel may exist in the following two situations (Wang & Lee 2005):

- The sending party is able to make changes in the visible space of the receiving party.
- The sending party is able to change when an object is modified relative to the observation made by the receiving party.

¶

The data channels created according to modifications in the data channel as stipulated in the previous paragraph are called:

- spatial channels,
- temporal channels.

¶

A **spatial channel** uses the structure of the transmitted data. The goal is the identification of unused or redundant parts in the normal information being transmitted, data that can be replaced with the secret bits in order to create the hidden channel. It is important to alter the original stream without distorting the normal data being transmitted.

A **temporal channel** modulates the secret information to be transmitted through a usage pattern of the network resources available. The receiving party must know the codification for each pattern used in order to extract the secret message being sent. Another type of temporal channel called a numbering channel uses the counting of network events in order to transmit information.

It is possible to use a combination of the two types of channels described. Such a channel is called **hybrid channel** and employs the use of both spatial and temporal changes on the data being transmitted.

3. CURRENT MECHANISMS

Current methods modify the unused bits along the OSI stack in order to create the covert data channels (Murdoch & Lewis, 2005). The most used bits are those from the network and transport layers of the OSI model. It is common to use a combination of different covert data channels for a secret transmission, the detection and destruction of only one channel allows the continuation of the secret transmission through the rest of the covert channels.

3.1 Parameters of a covert channel

For defining a covert channel and for enabling the comparison of two different covert channels the following parameters are used:

- Invisibility. This parameter can be evaluated using expert knowledge in the field. It measures how difficult is for a network expert to detect the anomalies that indicate a secret communication taking place. An invisible channel is better; no correlation with the actual data being sent can be detected.
- Indistinguishability. It measures the impossibility of separating the secret data from the data it is piggybacking. This parameter is important because it also means that an attempt at detecting and removing the secret channel would have as a consequence the alteration of the original data.
- Bit Rate. It measures the utility of a secret data channel, by expressing the number of bits it can transport. The indication can be in bits per second as it is common for the bit rate parameters. It is sometimes useful to measure this parameter in bits per transport unit used or as secret bits per number of original bits used for piggybacking.

3.2 Methods for creating secret communication channels

Secret communication can take place at each of the OSI layers. This presentation of the possible methods does not intend to be exhaustive. It is an indication of how different redundancies of the protocols used nowadays can be exploited to embed hidden information. (Fisk et al., 2003)

At the **physical layer** the redundancy of control and synchronisation signals permit the insertion of supplementary bits that are not part of the normal communication but can be correctly decoded and interpreted at the receiving station.

For serial lines such a transmission could exploit the use if the CTS – Clear-To-Send and RTS – Request-To-Send signals. These signals are not typically used. An implementation by the author, using a microcontroller application, achieved similar rates for the secret transmission as with a normal transmission taking place over the RX and TX signals. (Földi, 1999)

One other way of creating secret communication at this layer is the manipulation of the electrical signals. A logical 1 will be correctly interpreted if tensions of 5V or 6V are used. But for the secret communication channel a logical 1 with 5V may mean a secret 0 bit and a logical 1 with 6V may mean a secret bit 1 being transmitted.

A secret communication created at this layer has the inconvenience that it can take place only between directly connected stations. Any signal repeater or device that deals with amplifying and forwarding the electric signals further in the network will destroy the secret data embedded. A channel created at this layer typically implies some sort of hardware manipulation.

At the **data layer** the manipulation of the retransmission delay with a minimal value and a maximal value allow the encoding of a secret logical 0 or 1 bit. This channel that employs the modification of the collision detection system can be disturbed by others users that send packets in the network. A solution is that one of the parties involved in the secret communication to jam the communication of the other users. This solution has the inconvenience of minimizing the invisibility parameter for this channel.

The manipulation of the frame length can also allow the inclusion of secret data. The receiving party decodes the frames and forwards the useful data up the network stack, simultaneously striping the hidden bytes from frames and reconstructing the secret communication. A channel created at this layer can be achieved only by using software manipulations, firmware manipulation of the network devices involved in the communication may also be necessary, depending on the channel properties.

In TCP/IP networks the unused bits from the IP and TCP header may be used. In this way a secret communication channel can be created at the **network and transport layer**. In the IP layer there are three unused bits.

A communication channel using the three bits has bad performance from the point of view of the three parameters defined in chapter 3.1. Because the majority of the implementations available set these bits to 0, their manipulation for the secret transmission is easily observable. Using a simple traffic analyser a detailed observation of the secret transmission is possible. With regard to the indistinguishability, the typical behaviour of the network equipment is to reset these bits to default values destroying the secret channel.

Without traffic between the sending and the receiving station, the channel has poor performance from the point of view of its bit rate. Artificially creating traffic between the communication parties may raise suspicions for a network administrator, thus minimizing the invisibility parameter.

Until the implementation of the RFC 3168 dealing with congestion, the TCP header had 6 unused bits. These bits could be used together with the ones from the network layer in order to create a bigger capacity secret channel.

Other implementations available use the manipulation of the fragmentation mechanism or the manipulation of the MTU – Maximum Transfer Unit - to transmit secret information.

At the **session layer** a simple implementation for a covert channel which uses the access to a file system can be imagined. Coding a logical 1 may be represented by the access to a file “a” and a logical 0 as the access to a file “b”. Theoretically a bit rate of up to 110 KB/s is possible for typical nowadays networks supposing that the session is established and a typical read operation is done with a 100 bytes TCP packet. The channel has a good performance also from the point of view of the other two parameters since the file accesses cannot be distinguished from normal accesses.

	ideal	real lab	real 60%
10Mbit	5 KB/s	5 KB/s	2,2 KB/s
100Mbit	50 KB/s	48 KB/s	20,2 KB/s
DSL6000	3 KB/s	*	1,1 KB/s
DSL16000	8 KB/s	*	3,5 KB/s

Tab. 1. Bit rate measured for the checksum covert channel.

The creation of secret communication channels at the **presentation and application layer** implies the use of mechanisms for hiding information in the data exchanged by the applications. This overlaps with the area of (multi) media marking which is not a subject of this paper.

In the next chapter, two new mechanisms are presented. These enable the creation of covert channels that cannot be detected with the common tools of a network administrator.

4. PROPOSED SOLUTIONS

One way of keeping a communication secret is to not raise suspicions. A way to achieve this is by the manipulation of control structures from the network protocols leaving the normal communication intact.

4.1 Covert channel using the Time To Live field

This field is used by the TCP/IP stack to indicate how many hops a packet should survive in a network. By carefully absolving a secret communication setup procedure which involves the creation of a TCP connection between two communicating parties a secret communication can be implemented by embedding one secret byte in the TTL field.

Implementations by the author showed that in typical stable networks the TTL value remains constant after the setup process. A congested network would need frequent resynchronisation steps which would dramatically reduce the bit rate of this channel.

4.2 Covert channel using the Checksum field

The TCP checksum field is used to ensure the correct transmission of the data. By reverse engineering the checksum computation it is possible to pass such data to the TCP/IP stack so that the checksum contains two desired bytes.

The values for the performance of this channel are presented in Table 1. The table presents ideal values, real values measured in the lab and real values measured over a public ADSL – Asymmetric Digital Subscriber Line. Similar values as those presented in Table 1 are observed for the TTL Channel with respect to the fact that only one byte per packet is used for the secret communication.

5. CONCLUSION

The creativity of the implementers is the only limit in discovering new ways for embedding secret information inside a normal network communication. The security of a targeted system depends only of the expertise level of the network administrators, of their capacity to detect anomalies in the network traffic and the possibility to enforce policies that can disturb or disable such secret transmissions.

6. REFERENCES

Fisk, G.; Fisk, M.; Papadopoulos, C.; Neil, J. (2003). Eliminating steganography in internet traffic with active wardens, *Information Hiding*, pp. 18-35, Springer Verlag, ISBN 978-3-540-00421-9, Berlin

Földi, A. (1999). Design and implementation of a remotely controlled alarm system, *Scientific Workshop of the UPB*, Bucharest, May 1999.

Millen, J. (1999). 20 Years of Covert Channel Modeling and Analysis, *Proceeding of the IEEE Symposium on Security and Privacy*, ISBN: 0- 7695- 0176- 1, pp. 113 - 114, Oakland, May 1999, IEEE, California

Murdoch S.; Lewis S. (2005). Embedding Covert Channels into TCP/IP, *Proceedings of the 7th Information Hiding Workshop*, Barcelona, June 2005, Springer V.

Wang, Z.; Lee, R. (2005). New Constructive Approach to Covert Channel Capacity Estimation, *Proceedings of the 8th Information Security Conference (ISC'05)* pp.498-505, Heidelberg, September 2005, Springer Verlag, Berlin.

Annals of DAAAM for 2008 & Proceedings of the 19th International DAAAM Symposium, ISSN 1726-9679
Editor B. Katalinic, Published by DAAAM International, Vienna, Austria 2008
Make Harmony Between Technology and Nature, and Your Mind will Fly Free as a Bird

SIMULATION MODEL FOR THE FINANCING STRATEGIES OF A LEASING COMPANY

FOLCUT, O[vidiu]; CIOCIRLAN, D[oinita] & MUSTEA SERBAN, R[azvan]

Abstract: *The paper presents a model of simulation of the financing strategies directed to the procurement of some technical equipment by a leasing company, based on a minimum cost of the capital. The steps of the algorithm are: filling the introductory fields with basic information, checking the eligibility of the sources based on predefined criteria ,verifying the possibility of financing the project, calculating the cost of each source, creating the hierarchy of sources, choosing the minimum cost source / combination of sources, calculating the final cost. The program allows the decision maker in the leasing company to choose the most effective combination of the financing sources .*
Key words: *Simulation ,model, leasing, strategy*

1. GROUNDING THE FINANCING DECISION

It is often the case that payment sources or tools can be combined in a financing strategy that would lead to a minimum financing cost. Such an approach is recommended when the most advantageous financing tool doesn't fully cover the needs for financing. We will go on by outlining a simulation model of financing strategies for equipment purchases by a leasing company.

1.1 Formulating the problem

Elaborate the financing strategy based on a weighted average capital cost minimum for a purchase transaction of one or several products by a leasing company under the following conditions:

-Products are homogeneous and are purchased simultaneously;
-Products are non-homogeneous and are purchased simultaneously;
-Products are purchased on different dates.

The financing strategy entails awareness of the following information:

-Possible financing sources;
-The access conditions, namely:

- The access time-frame (the period of time since the moment the source was called on till the time the financing fund is assigned);
- The source threshold (the minimum and maximum amounts outlining the limits of source access);
- The accessing costs for a certain source;
- The value of products to be financed;
- The financing period (deadline).

Moreover, access to the respective financing source is conditioned by the business and credit worthiness of the economic actor, namely:

-It should not have outstanding debts to the budget;
-It should undertake a profitable activity;
-It should not have debt arrears or outstanding loans or interests rates, losses or be bankrupt;
-It should be able to refund the credit at deadline and to pay interest;
-It should submit collaterals and insurance.

In order to access a bank credit, it should also submit one of the following collaterals:

-Collateral/mortgage for owned or purchased goods;
-Assignment of debts on cash-ins from other sales contracts (letters of credit);
-Assignment of debts on cash-ins from the insurance policy on the suretyship bond;
-Assignment of debts on cash-ins from the insurance policy of the respective good;

Or one of the following secondary bank guarantees:

-Bank deposit;
-Letter of bank guarantee;
-Bills of exchange and promissory notes;
-Collateral for real estates.

1.2 Problem-solving

Input above-mentioned entry data based on which the proposed simulation algorithm will identify financing sources, which can be called upon according to the access conditions. Thus the algorithm will compare access time-frames for every individual source with the financing deadline imposed by the client. Such comparisons rule out sources with a too high access time-frame. If, for instance, the deadline for which the leasing company wants to obtain financing is of ten days, sources with significantly higher access time-frames cannot be considered.

In the case of accessible sources, the thresholds for every source shall be compared against the value to be financed (the required investment). Should a source or several sources separately comply with the required investment, the unitary cost for every source shall be then computed, and the minimum-cost source will be selected, with the financing to be wholly covered by that source. If no source evinces a threshold that exceeds the required investment, combinations of sources will be taken into account in order to compute the weighted average cost of the sources (Boobyer, 2003).

From a mathematical point of view, this algorithm can be assimilated in a generic form by the following system of equations (Andreica et al., 2003) :

$$\begin{cases} \min(x_1c_1 + x_2c_2 + ... + x_nc_n) \\ x_1 + x_2 + ... + x_n = NI \\ q_1 <= x_1 <= Q_1 \\ q_2 <= x_2 <= Q_2 \\ ... \\ q_n <= x_n <= Q_n \end{cases} \quad (1)$$

Where: x1…xn – represent the amounts to enter financing from sources 1…n;
c1…cn – unit costs of sources 1…n;
q1…qn – minimum values for which sources can be accessed;
Q1...Qn – maximum sources thresholds
NI - required investment.

2. ALGORITHM DESCRIPTION

The algorithm was integrated in the IT product LEASYM made in the Visual Basic.Net programming environment, representing a program product for assisting leasing company managers in adopting financing strategies for investment products requested by clients. The program simulates the selection of attracted financing sources and of self-financing, its ultimate goal being to minimize the financing cost (if sources can wholly support the transaction financing) or the weighted average cost (in the case when source participate up to different extents in covering the required investment).

The product analyzes the main financing sources, among the ones most often used in the real-life environment (Marks et al., 2005) namely:

- Domestic credits;
- Foreign credits;
- Bond issues;
- Joint ventures;
- Self-financing.

Self-financing entails capitalizing on the profits and depreciation funds of the respective economic actor.

Purchased products can be homogeneous or non-homogeneous (in this case, we need to know how many distinctive products are required for delivery and, for every individual type, the number ad value of each product). This information is the foundation for computing the initial required financing that the importer must have.

2.1 Stages in the algorithm for financing cost minimization

Stage 1: Introducing initial data

- **client (beneficiary) data**: *number of products, value per product* (based on which the initial required financing shall be determined), *the down payment* to the importer and the desired *deadline* for product delivery. The amount to be provided by the client as down payment will diminish the importer's required financing
- **specific data** for leasing companies: the thresholds they can reach through various attracted financing sources, the euro-lei exchange rate (needed in the case of foreign credit financing), the *time frames* within which the company gains access to the respective financing sources, as well as the *self-financing threshold* the company has and that can be used to finance the transaction.
- **specific data for every financing source under analysis:** fore foreign or domestic credit financing – *credit period, interest rate, commissions* incurred by the bank, as well as the *type of return* used, with constant annuities or equal return rates; in the case of bond issues, the following entry data are necessary – *nominal bond value, coupon value, deadline, inflation rate and the commission to be paid to the intermediary*.

Stage 2. Checking source eligibility (Checking access time frames for sources)

Access time frames for every individual source are compared against the delivery deadline set by the client. Those sources with access time frames that are below the delivery deadline can be deemed eligible sources, and thus the first filtering is performed (Limoncelli, 2005). The algorithm will henceforth operate on eligible sources.

Stage 3. Checking the possibility of supporting the transaction

If access time frames to the importer's financing sources are within the delivery deadline required by the client and if the amounts obtained from the respective sources allow for the transaction to be financed, this can be fully performed, through one, two or even several attracted sources (cost minimization being the goal) or through combinations of sources, by computing the weight for each of them in the required financing (the weighted average cost). If these conditions are not fulfilled the transaction shall be rejected.

Stage 4. Calculating source costs

During this stage unit costs are calculated for every eligible source that can partly or wholly finance the transaction under work: bond issue, domestic credit financing, foreign credit financing (in this case, the exchange rate should also be introduced, in order to make it possible to compare the cost of this financing with other costs), self-financing (the costs are calculated as an opportunity cost, for instance how much the importer would stand to gain if they deposited the money in the bank instead of covering the respective transaction), joint venture (the cost is again calculated as an opportunity cost, for instance how much the economic actor, associated to the importer, would stand to gain if it deposited the money in the bank instead of investing in the ongoing transaction) (Fabozzi, 2002).

Stage 5. Source hierarchy. Selecting the source / the combination of low-cost sources. Calculating the final minimum cost.

If the self-financing threshold allows for the full financing of the purchase, then this source shall be compared against the attracted accessible source, based on the minimum cost criterion. The self-financing cost is calculated as an opportunity cost – the earning turned in by the company by self-financing the client's investment project. This should amount to at least the one obtained by shareholders from other investments with a comparable risk liability.

The same calculation method is to be applied in the case of joint venture, the difference residing in the standpoint for the respective cost. To put it more clearly, for the associated economic actor, the cost of the invested amount will again be an opportunity cost – the earning obtained by the latter must be at least as high enough as the one obtained in case the associate invests in the respective project instead of other projects with a comparable risk liability.

In case the money obtained from all sources we have access to do not allow for a full financing of the transaction, for every individual case, check if the purchase can also be supported by self-financing resources. If it can, calculate the weight of every source in the required financing and, further on, the weighted average cost of the financing. If no, try supporting the purchase by associating with another economic actor interested in the respective transaction. If even this means cannot cover the purchase, then the transaction shall be rejected, either because of insufficient funds, or because of too high entailed costs.

In case at least one of the sources that the importer has access to can fully finance the transaction, with the other sources supplying lesser amounts than the required financing, the transaction will be undertaken through a combination of financing resources if the unit costs of partial sources are lesser than the cost of the integral source. These combinations are based on the ascending ranking of unitary costs for all possible financing sources.

If the source that can provide full financing for the transaction has the smallest unit cost, then it will make up the optimum financing means. If not, it can have a certain weight in making up the weighted average financing cost.

3. CONCLUSION

The simulation model presented above and the Leasym software based on it are some very important tools for the managers of the leasing companies, helping them to solve a very important issue in their activity: the financing strategy, which becomes more and more difficult to figure in the actual dynamic and unstable global financial market. We undertake to pursue further, extended research along this line in the future.

4. REFERENCES

Andreica, M.;Andreica, C.; Mustea-Serban, I.; Mustea-Serban, R. (2003): *The Financing Decision in Leasing*, Cibernetica MC, ISBN 973-96916-0-9, Bucharest

Boobyer, C. (2003): *Leasing and Asset Finance – Fourth Edition*, Euromoney Books, ISBN 978-1855649859, London

Fabozzi, F.J. (2002): *The Handbook of Financial Instruments,*Wiley Finance, ISBN 978-0471220923, Hoboken, New Jersey

Limoncelli, T. (2005): *Time Management for System Administrators,* O'Reilly, ISBN 978-0596007836, Cambridge

Marks, K.H.; Robbins, L.E.; Fernandez, G.; Funkhouser, J.P. (2005): *The Handbook of Financing Growth: Strategies and Capital Structure,* Wiley Finance, ISBN 978-0471429579, Hoboken, New Jersey

Annals of DAAAM for 2008 & Proceedings of the 19th International DAAAM Symposium, ISSN 1726-9679
Editor B. Katalinic, Published by DAAAM International, Vienna, Austria 2008
Make Harmony Between Technology and Nature, and Your Mind will Fly Free as a Bird

MONITORING THE POLLUTANTS IN A ROMANIAN COMPANY PRODUCING CHEMICAL FERTILIZERS

FORGACIU, F[lavia]; OROIAN, M[aria] & OPELSZ, H[elen]

Abstract: *The paper attempts to define the significant pollutants resulted from the activity of a Romanian company, considered the greatest polluting factory in the area. Its production consists of chemical fertilizers for agriculture. As an everyday result, a series of contaminants deteriorates the air, the earth and the river of the nearby city. To restore the damaged and deteriorated systems, the company takes some measures presented below. The paper also presents specific ways of accomplishing these objectives, including the surveillance of aspects concerning pollutants or wastewater and the cost of modernizing technologies involved.*

Key words: *monitoring pollutants, chemical pollutants, air pollution, wastewater*

1. INTRODUCTION

Nowadays, every act of our industrial society leads to environmental degradation, regardless of intention. Discharges from human activities must be released into the air, the water or the soil. Each of these potential reservoirs can accept a limited amount of physical, chemical or biological substances without significant deterioration. Beyond this point of assimilation, the environment can be deteriorated to the point that sustainable biological growth cannot occur (Câmpeanu, 2004). The quality of the environment is of a great importance for all countries, especially for the developing countries, where the losses of economic production (part of the GNP) and expenditures for the environment preservation are more strongly felt (Caracotă & Caracotă, 2004).

Pollution is defined as the deterioration of the existing state. The physical effect of pollution can be biological (e.g. species change, ill-health), chemical (e.g. the effect of acid rain on lands and buildings surfaces) or auditory (noise). The pollution is a concrete phenomenon, sustainable and largely measurable. From this perspective, literature reveals a complex of pollutant factors, of causes and effects, in the fields of life and the spheres of economic activity. Moreover, pollution arising from one country can cause damage in another country. The effects of environmental pollutants are related to time as well as concentration or intensity. Time factors become increasingly important when low concentrations of pollutants do not produce bad effects until after months or years of exposure (Alley, 2006).

Romania still has an exceptional natural capital as compared to other EU countries, but the pressures to which it is exposed are enormous. Chaotic economic development and irrational exploitation of natural resources lead to a rapid increase of pressure on ecosystems. National policy of environmental protection against pollutants mainly consists in the following: introduction of appropriate technologies and techniques for the pollution detection at source, air resource management, reducing pollutant emissions by achieving the lowest levels and not exceeding the improvement of national integrated monitoring system of air, earth and water quality. Communities must consider these needs not only on short term, but on long term as well (Rojanschi et al., 2002).

After studying various publications related to the recent policies and strategies of sustainable development in Romania, we could find that in recent years, companies have given a great importance to the monitoring of environmental pollutants. The quantities and the degree of hazard pollutant emissions depend on industry level, on the state of used technologies and their performance (Grădinaru, 2001).

The purpose of this paper is to review the main water and air chemical pollutants of X Company specialized in chemical fertilizers production, in order to recognize the opportunities for eliminating, minimizing, reusing or treating these pollutants so that their negative effect on the environment will be minimized. The analytical methods applied are suitable and they offer reliable data. The study highlights the possibilities of reducing pollutants by the implementation of new technological processes and by new ways of collecting, treating and storing the wastewater which cannot be avoided or valorized.

2. MONITORING AIR AND WATER CHEMICAL POLLUTANTS IN A ROMANIAN X COMPANY PRODUCING CHEMICAL FERTILIZERS

X Company from Mures County, Romania, is a joint stock company founded in 1962. At present, it is the only chemical factory specialized in chemical fertilizers production in Romania, and the only producer of photo sensible materials in the country. The company has its headquarters and chemical platform located in the western extremity of the industrial area of Targu-Mures city at a distance of 5 km from downtown. The company totally owns and uses an area of land equal to 147 h. The company has main technological facilities and auxiliary installations, which provide the utilities needed, such as: the reception, storage, packaging and shipping of raw materials and finished products.

X Company is responsible for the emission of various substances, with different noxious effects on humans and the environment. It discharges into the atmosphere compounds based on nitrogen, ammonia, chlorine and sulphur. **The polluting substances in the air** are gaseous or solid, being dangerous for humans, animals and plants. The air disperses the pollutants. This dispersion further dilutes the pollutants to a downwind location. Because of this dispersion proclivity, air pollution can be international in scope. Generally, X Company is in accordance with the regulations imposed by NTPA-001/2002 regarding emissions into the air of most agents, but exceeds the limits permitted for ammonium, nitrogen dioxide and sulphur oxides. The last two oxides are responsible for the so called "acid rains" which affect land-based flora and fauna, reducing local biodiversity. Buildings suffer corrosion from acidic deposits, causing extra maintenance to be carried out. SO_x and NO_x are also implicated in the acceleration of certain diseases, especially respiratory diseases.

Figure 1 shows the main pollutant emissions in the atmosphere and their concentrations over the last 3 years, when a severe and more careful monitoring of the environmental agencies started.

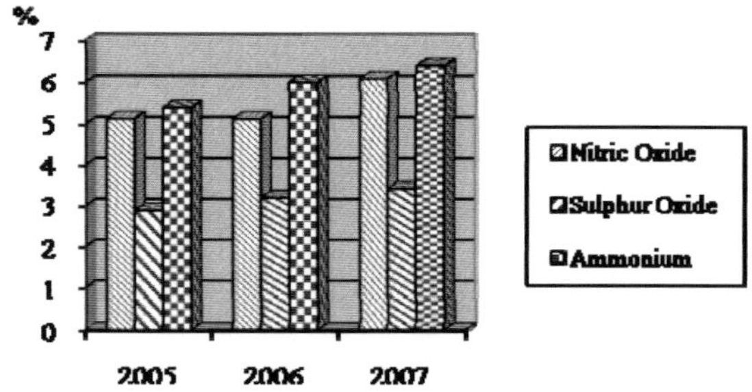

Fig. 1. The concentration of pollutant agents of the X Company in the air

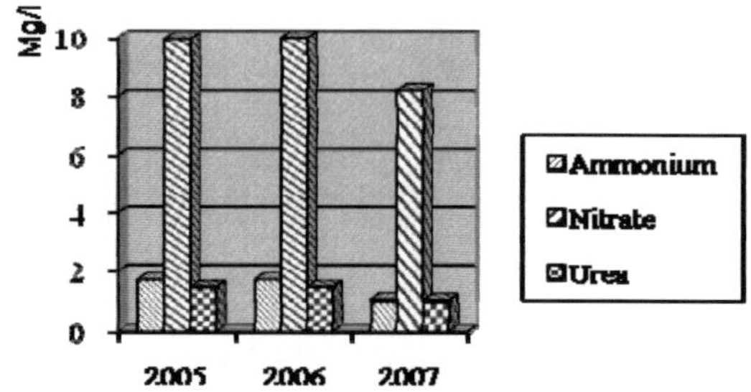

Fig. 2. The concentration of pollutants of the X Company in the local river

As you can see, in 2007, the percentage of substances overcoming the amount allowed for daily average concentration (normally of 1%) was: for ammonium 6.42%, for the nitrogen oxides 6,09% and for oxides of sulphur 3.42%, thus obtaining the highest values as compared to the previous years. To remedy this problem, the company plans a series of investment and modernization works that will be performed over the years to come.

Chemical pollution of water is represented by water penetration of various chemicals, from organics, easily degradable to those with long and persistent toxicity, having as consequence the destruction of the aquatic flora and fauna. Chemical pollution of water may accidentally occur, but most often it occurs due to uncontrolled discharges of various liquid or solid waste residues. Water also acts as a disperser of pollutants, but it is easier to model and predict than air since it flows in a defined channel with a predictable velocity. Water pollution can be of regional interest because of this dispersion. Normally, the municipality will restrict industrial water pollutants.

Industrial production of X Company requires the use and disposal of water. The contact with a process or product can add pollutants to the water. The resulting water is then classified as wastewater. It will either be discharged directly into a receiving artificial lake or into the sewerage system of the municipality, or it will be reused or recycled. In case of X Company, the year with the highest degree of water pollution was 2006, when the factory was close to the limit between the areas with severe risk and medium risk, a situation generated by the real conditions of that year. The main indicators of impurification which produce significant water pollution of the local river are: ammonia (NH_3, NH_4^+), nitrates and nitrites. These pollutants get into the local river by a direct discharge of wastewater from the factory, or from water collected in an artificial lake. Ammonia, at certain concentrations (over 1,0 mg/l) is toxic for fish and wildlife. In 2006, the concentration of ammonium registred a peak of 1.7 mg/l. The highest value was registred in the case of nitrate ions, with a concentration of over 9,99 mg/l, when the normal was of only 2,5 mg/l. The X Company of chemical fertilizers caused an increase in the concentration of pollutants in the local river by 2% to 4%.

In figure 2 the main pollutants discharged into the local river and their concentration are presented.

To reduce the quantities of pollutants, especially ammonium and nitrate ions in wastewater, in 2004, the building of an installation for treatment of these waters with changing ions began. In addition, to reduce the level of the artificial evacuation lake of 30 ha, a new installation for the water evaporation in 2006 was built and put into operation. In order to reduce the amount of suspension of wastewater discharged into the local river, a new investment for filtering wastewater has been started.

The required transitional period was obtained, so all programs of each stage, alongside the necessary investments to comply with environmental rigours, have as deadline the year 2015. Investments made by the company had as source of funding both their own resources as well as loans from the local budgets. In the year 2005, investments have amounted to 16,647,000 EUR and in the year 2006 they reached the value of 19,675,587 EUR. Total estimated overall investment, including operating costs and maintenance for the period 2006-2015 are of 82,555,000 EUR. Investments made by X Company during 2005-2006 had a beneficial effect on pollution, but nevertheless the company could not meet the standards stipulated by the Environmental Protection Agency regarding the impact of pollutant agents resulting from the production process of the company.

3. CONCLUSIONS

Environment is now considered a political and economic priority, both internationally and nationally. In the spirit of the principles of sustainable development, current Romanian laws compel operators to review their attitude, breaking the link between economic growth and environmental deterioration.

In case of the X Company, profiled on the production of chemical fertilizers, exceeds with significant environmental impact are recorded at two chemical pollutants, measured in a systematic way, namely ammonium and nitrates ions. These two are responsible for various effects, affecting both human and local flora and fauna. X Company understands the necessity to ensure life quality for future generations and wishes to comply with the laws in force and environmental standards. To achieve this desideratum, X Company makes huge financial efforts, ensuring the reduction of waste flows resulting from production, by improving its processes.

4. REFERENCES

Alley, R. E. (2006). *Water Quality Control Handbook*, Second Edition, McGraw-Hill Professional Publishing, ISBN: 0071467602 / 9780071467605, New York

Caracotă, D. & Caracotă, C-tin R. (2004). *Contemporary Dimensions of Sustainable and Competitive Development*, Economics Press, ISBN: 973-594-487-1, Bucharest

Câmpeanu V. (2004). *Dimension of European and World Sustainable Development,* Expert Press, ISBN: 973-618-038-7, Bucharest

Grădinaru, I. (2000). *Protecting Environment,* Economics Press, ISBN: 973-590-427-6, Bucharest

Rojanschi, V.; Bran F. & Diaconu, Gh. (2002). *Engineering and Environmental Protection,* Economics Press, ISBN: 973-590-455-1, Bucharest

Annals of DAAAM for 2008 & Proceedings of the 19th International DAAAM Symposium, ISSN 1726-9679
Editor B. Katalinic, Published by DAAAM International, Vienna, Austria 2008
Make Harmony Between Technology and Nature, and Your Mind will Fly Free as a Bird

THE SYNTHESIS OF MANUFACTURING TASK IN FLEXIBLE MANUFACTURING SYSTEMS

FOTA, A[driana]

Abstract: *The fundamental element, which has to be the basis of a flexible manufacturing system design, is the current manufacturing task. On the base of the generalized analytical and global synthesis model of the manufacturing task for designing any flexible system for the round shafts processing, generalized item models, hypothetical and representative items for the family or variant of particular real items have been drawn up. There will be created a data base for the synthesis of the geometric representation from the manufacturing task references, and by the applicative research there will be performed the simulation on the computer for real manufacturing items. The simulating program realized has as objective the application of flexible manufacturing systems for processing round shafts.*
Key words: *system, flexibility, manufacturing, modelling, simulation*

1. INTRODUCTION

The first stage in the design of any flexible manufacturing system is the manufacturing task analysis (Askin & Standrige, 1993). The importance of the manufacturing task analysis and its complexity derive from the fact that flexible manufacturing systems are destined to small series production, characterised by a typological diversity and instability in time. It is estimated, that about 70 % of the actual production is realised by methods, which are specific to small / individual series production. For now, however, in the entire speciality literature, especially in the documentation oriented towards technology, a fabrication task is considered given, and known, exclusively by an old linguistic classification, based on a "morphologic analysis method" (Opitz, 1970). It is considered that the references family, the version, or the reference as technological entity, which is manufactured by mechanical processes have a configuration – geometrical and technological properties given, and known from execution drawings. Alternatively, while designing any flexible fabrication system, the designer has sometimes a partial typology, undefined significantly, insufficiently relevant, and taken from execution drawings, which are available. Under these conditions, the analysis of a manufacturing task opens a large research area, and for now, nobody has elaborated any analytical model of the manufacturing task for the synthesis of flexible manufacturing systems.

Lack of consecrated mathematical models in the design of flexible fabrication systems makes the creation of such systems difficult, with consequences upon their performance. Because it the new property of production systems – "flexibility", is imposed, there can appear some conception errors, and designers cannot anticipate precisely the optimal flexibility degree. As far as the approach method of the large topic of flexible manufacturing systems is concerned, it must be remarked the fact that there have been and there still are different methods and interpretations, but no model practically confirmed has been imposed. The modelling of flexible fabrication systems functioning represents currently the most dynamic and controversial research area in this field, (Tempelmeier & Khun, 1993).

In this paper, the simulation model for real manufacturing items drawn up, and the "Shafts" program having on its grounds, the generalized analytical model of the manufacturing task conceived in paper (Boncoi & Fota, 2000). The "Shafts" program was realised in Visual C^{++} programming language. Essential in simulation modelling, the logical element is set up by data structures that are convenient for the performance of an event.

2. THE MOST LIMITED DATABASE

The geometrical – constructive configuration of the generalized item is obtained from the logical characteristic function by using the database. Before graphical transposition, the logical function is submitted to constraints (conditions and restrictions). Conditions are imposed by construction, functionality, and the assembling process of the real item overall. The main condition is precedence, which requires writing the components of the characteristic logical function strictly ordered. The precedence condition requires only ordering in a monotonously increasing/ decreasing sequence of the diameters conveniently chosen by the designer; lengths of sections result from the constructive – functional conditions following a randomly, unpredictable distribution.

The conditions' question is one of heuristically engineering and expert system. Restrictions are imposed by the same construction or functionality but also by new rules, imposed by the complex synthesis process of the generalized item. There are some important categories of restrictions. The main category of restrictions to precedence restrictions is related. A second category of restrictions is that imposed by the synthesis process.

Lying on the grounds of flexible manufacturing systems design for round shafts processing, the generalized manufacturing task before, has been fixed according to the typological nucleus, which includes the whole range of possible items belonging to an item class, family or variant, limited by restrictions to a required area, in the paper (Abrudan, 1999). There have been followed the stages below. The synthesis method of the current manufacturing task has been drawn up by the analysis and mathematical structure of the items' features. On the base of the generalized analytical and global synthesis model of the manufacturing task for designing any flexible system for the round shafts processing, generalized item models, hypothetical and representative items for the family or variant of particular real items have been set up. The generalized item includes all the constructive-geometrical elements belonging to the multitude of real, factual items in the family or variant described.

These constructive-geometrical components have been ordered in a logical and natural sequence. Thus, six types of generalized item models have been set up, according to the structure and size of the flexible manufacturing system for processing round shafts. These models are the following:

1. The model of a generalized item of compact, typical, asymmetrical, externally configured round shafts family
2. The model of a generalized item of round gap shafts family;
3. The model of a generalized idem of polygonal/ conic round shafts;
4. The model of a generalized item of axles and spindles family
5. The model of a generalized item of threaded shafts family;
6. The model of a generalized item of spherical shafts family

Each item's family or variant represents an ordered literary-numeric (alphanumeric) sequence. All the restrictions associated with the families of the corresponding class have been written, which also will be introduced in the database, next to the corresponding symbol.

3. RESULTS

For the generalized manufacturing task computer simulation, an "assembling" program is set up, which has to associate different database figures, expressing the symbols of the ordered sequences. The program converts the alphanumeric ordered sequence into a succession, joining geometric figures that comply with this ordered sequence.

Running this program on the computer, more item families of the same class, in different variants will be obtained. These variants may be real or theoretical. From an item family the most complex variant is chosen, generally theoretical, hypothetical, fictive, including all the accordingly family features, all the geometrical elements of all variants belonging to the accordingly item family.

3.1 How to Use the Simulation Program

The computer program has been realised in the Visual C^{++} programming language. A database – DB has been conceived, containing the sizes and features of all the elements belonging to the previously fixed round shafts families. The main stages of the program usage are as follows. Through the decision block, called <*option*> the user may automatically select, at any time, from the database, from the six generalized items, the generalized shafts family he or she wants to use. The two windows are visualised on the screen (figure 1):

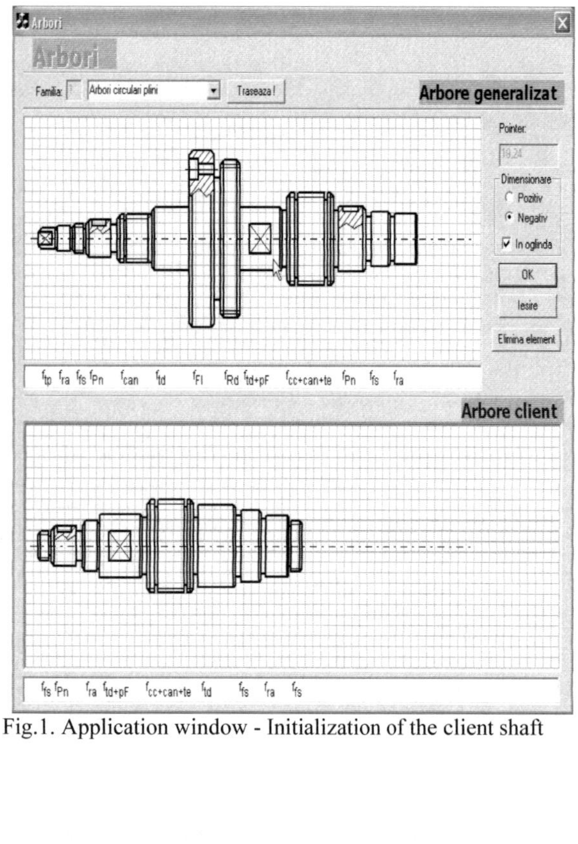

Fig.1. Application window - Initialization of the client shaft

-the generalized shaft window, called "*generalized shaft*" including generalized items for the six round axles' family types;

-the real shaft window, called "*client axle*" which may be automatically generated from the generalized item.

Pressing the key "*outlining,*" the program will automatically generate in the "*generalized shaft*" window:

1.The family selected (example, family round gap shafts;

2.The code bare corresponding to the assigned functions to the elementary geometric components of the database items made and keeping the same symbols (for example: radial-axial bearing section, polygonal section, narrow thread, key transmission, belt canal, grooved section, spacing section, teeth, flange, etc.);

3.From the generalized item any real item type, represented by a model graph, as the ordered sequence determined by the assigned functions to some elementary geometric figures from the database may be obtained.

Generation of real shaft, also called client shaft is performed in the window "client shaft." Initialization of the client shaft elements is performed by selecting the current element of the "generalized shaft" window. From the figure below (figure 1) one application window has been extracted, for exemplifying the construction of any real item type resulted from the generalized items of round shafts families included in the database of the program's that has been used.

4. CONCLUSION

The first essential aspect in using computer simulation of real manufacturing items refers to confronting the flexible manufacturing system designer with a huge volume of information, sometimes unpredictable, uncertain, depending on time, incomplete, which under uncertainty conditions may be appreciated as irrelevant and, consequently, eliminated from the configuring process. Therefore, the need of a strong program is imperious, a program that can be used for a computer network, replacing the classical design method of machine parts and of corresponding technologies. Another advantage of this simulation program use consists in the fact that through the available, easily accessible graphic database, the processing time is considerable reduced, thus reducing the number of designers too. The simulation program contains a database concerning standardized entities and conditions of the flexible manufacturing systems for processing round shafts, providing easy attachment or removal of structures that perform applications. The simulating program realized in this paper has as objective the application of flexible manufacturing systems for processing round shafts, area in which the speciality literature does not offers information.

5. REFERENCES

Abrudan, I. (1996*) - Flexible Manufacturing Systems. Design concept and management*, Dacia Publishing House, ISBN 973-5-0568-4, Cluj-Napoca

Askin, R. G. & Standrige C. R. (1993) - *Modeling and analysis of Manufacturing Systems*, Publisher Wiley Inter-science, ISBN 978-0471514183, New York

Boncoi, Gh. & Fota, A. (2000) - *Family, Variant and Individual Feature Analysis of the Reference Component Parts of the Manufacturing Task for FMS,* International Conference ICMas 2000, University Polytechnic of Bucharest, pp. 413 –424, ISBN 973 – 31 – 1492 – 8, Bucharest.

Opitz, H. (1970) - *A Classification System to Describe Work Pieces,* Pergamon Press Ltd., Oxford

Tempelmaier, H. & Kuhn, H. (1993) - *Flexible Manufacturing Systems*, John Wiley &Sons, Inc., ISBN 78-04713307211, New York.

Annals of DAAAM for 2008 & Proceedings of the 19th International DAAAM Symposium, ISSN 1726-9679
Editor B. Katalinic, Published by DAAAM International, Vienna, Austria 2008
Make Harmony Between Technology and Nature, and Your Mind will Fly Free as a Bird

THE USE OF MODELLING AND SIMULATION IN THE DESIGN OF FLEXIBLE MANUFACTURING SYSTEMS

FOTA, A[driana]

Abstract: *This paper presents the simulation model of the flexible manufacturing system for round shafts processing, using the Delphi software product. It can facilitate the creation of a specific object oriented database, which is useful for the manufacturing system. There have been used object, classes, and lists-type data structures. The program that has been created up allows planning / programming and managing in real time the whole processing system. Simulation has been realized under dynamic working conditions. The evolution as a dynamic system of the flexible manufacturing system condition in time, for round shafts manufacture is described by algorithms or complex procedures, expressing as well as possible, the operation in real environment and time.*
Keywords: *system, flexibility, modelling, simulation, manufacturing*

1.INTRODUCTION

The use of simulation models in the area of large-sized systems is nowadays very widespread. As far as the design of flexible manufacturing systems (FMS) is concerned, it is a frequently used modern procedure of modelling and simulation. In present, the simulation of flexible manufacturing systems is the most dynamic and controversial field of research in this area, (Abrudan, 1996). The absence of consecrated mathematical models in the design of flexible manufacturing systems makes the creation of such systems difficult, with consequences on their performances. In a previously paper (Fota, 2004) one can found elaborate mathematical models used in the dimensioning and the configuration of flexible manufacturing system for circular shafts. The algorithm and simulation program using the Delphi software product have been created through original approaches. The inner structure of flexible manufacturing systems of round shafts processing has been detailed, so that operating systems of all their sub-systems may be calculated and computerized systematically, respecting precision and taking into consideration interactions with external environment.

The applicative research within the scientific research project will be performed through simulations on mathematic models. The applicative character of the project research theme is demonstrated simulating the flexible manufacturing system's performance for circular shafts processing. It is a dynamic simulation, performed in real environment and time, based on a reliable theoretical concept. The simulation models will be validated on the computer and on real flexible manufacturing system for real physical applications. After simulating the performance of the flexible fabrication system, the validity of the models will be confirmed, as well as their utility in the design and the management of the processes in real time and environment. Furthermore, it is considered the use of specific software, and programming languages for design and simulation (Delphi, Visual C^{++}, and Catia) and the development of new software applicable in the design of any flexible manufacturing system, and the development of new, specialized software, which can be applied in the flexible manufacturing systems design. As a dynamic system, the input function concerning the condition of flexible systems for round shafts processing, may be described as algorithms or complex procedures, which can describe as well as possible, the real function.

Using the software Delphi 7, on the base of dimensioning and modelling structures, a simulation program of flexible manufacturing systems for round shafts processing was drawn up. This simulation language contributes to the quality of the graphical animation, visualizing on the screens the simulating behaviour, except the encoded transcription of the simulation model. The programming language used hereby provides the dynamic simulation of discrete technical systems, where also flexible manufacturing systems for round shafts processing are attached.

2. ASSEMBLING PLAN OF A SIMULATION SYSTEM

On the base of the program conceived, the performance simulation in the case of flexible manufacturing systems for round shafts processing is done successively, in order to process the most complex real item (solid round shafts family), an intermediate item (lifting spindles family) and to process a simple item (axles and spindles family), (Boncoi & Fota 2001). Besides, a simulation of simultaneous processing of the three items above is performed – situation that is frequently met in real environments. Within the system, the three items that have been processed have their own technological flow, whose steps take specific times. In figure 1, a computer display capture during the simulation of the three items processing is shown. The whole system's function is described specifying different interactions between processes.

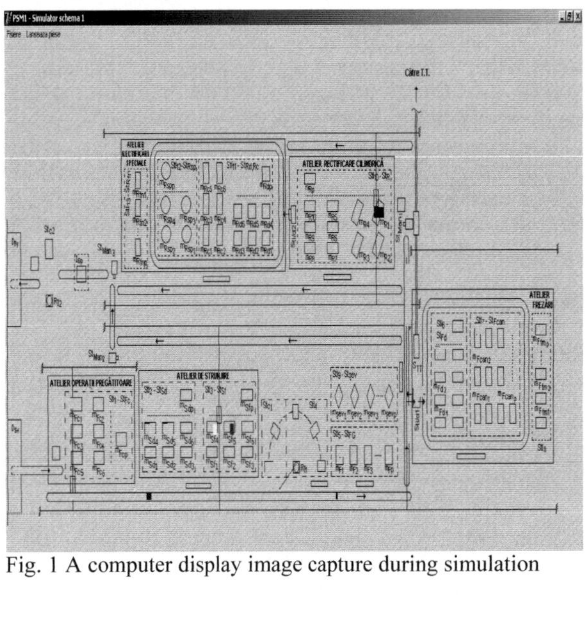

Fig. 1 A computer display image capture during simulation

For flexible manufacturing systems, items in general interact with the other processes, what explains their movement within the system. Simulation is performed describing the movement of the parts, following different processes, according to the tool-machines, to the transporting and handling means, etc., (Tempelmaier, 2006).

The Delphi programming environment is able to supply automatically entities in the system, according to a predefined delivery sequence. This is possible by setting up lists of destination stations and an optional assignment of some properties or variables for each previously described station. In addition, processing times for each working station have been assigned. The simulation program has the following constraints: the choice of a minimum critical path-type trajectory, the selection of those machines of the station that are available, the choice of the most expensive machines of the station and high degrees of loading. Another limit concerns industrial logistic sub-systems belonging to stations that cannot be used. Other constraints are the full use of all handling stations and the achievement of a minimum path algorithm for robots' displacement within the system. Simulation provides definition of some aspects related to manufacturing management, definition of algorithms for transport methods and detection of "tight points" delaying manufacturing flow, or study of breakdown influence upon the process.

3.DELPHI SIMULATION PROGRAM FOR FLEXIBLE MANUFACTURING SYSTEM

Some specialists consider Delphi a version of Rapid Application Development - RAD of Borland Pascal programming environment (Ispas & Mohora 2001). One of the first things to be understood when working in a RAD environment refers to the fact that a Rapid Application Development environment is technology, offering solutions to several IT related questions, based on a series of concepts, components, and protocols concerning their use (Law & Kelto 1991).

Essential in simulation modelling, the logical element is set up by data structures that are convenient for the event's performance. The object oriented modelling outlook first introduces fundamental model categories by counting key model types of each category and, secondly, it specifies the way of getting programmed objects, starting from the real ones.

The object oriented functional networks methodology has been developed in order to incorporate all the aspects concerning a system: structure, functionality, and behaviour. The model has also been developed to provide the user a neat control at any level of a complex system, without losing any aimed aspect. It is also the case of the flexible manufacturing system that has been conceived. This modelling strategy provides graphical representations by visualizing the model. The structure of such a model is based on the concept according to which any real system can be described as an informational part or an acting part.

Subsequently the main data structures that are used will be described, in order to finish simulation. For a better explanation of an object's features, we hereby provide a short example in Delphi (figure 2).

The concept on which this program is based is described below. Each part follows a technological trajectory. In each working station, a processing time is set (each working station has timing devices). When time is out, a warning sound is heard. The program includes a warning manager.

The moment of the warning, the processes find all the useful information about the part that triggered the warning (the part, the active location, the next working station that has to arrive). After having discovered all information, a robot is involved in the process. It takes over the information needed in order to perform the task, from the warning manager.

```
Unit Unit1;
Interface
Uses Windows, Messages, SysUtils, Variants, Classes,
Graphics, Controls, Forms, Dialogs;
Type
        TForm1=class (TForm)
        Private
        {Private declarations}
        Public
        {Public declarations}
        End;
Variable
        Form1: TForm1;
Implementation
{$R *.dfm}
End
```

Fig. 2 Sequence from Delphi simulation program for FMS

4.CONCLUSIONS

The simulation program includes a database concerning parts and conditions of the flexible manufacturing systems for processing round shafts, under a standard structure, providing easy attachment or removal of structures that accomplish the applications. We consider that the research, which will be developed, will produce useful results with impact upon the industrial reality, thus it will restrain the gap between theory and practice, and it will open new directions to research, having as cumulate effect a less degree of uncertainty in the area of flexible manufacturing.

The researches performed within this paper will be directed to the study of flexible manufacturing systems, in order to know their behaviour and their performances very well, and if it is possible, before their physical manufacturing and in order to establish on scientific bases dimensioning models, representation and simulation of flexible manufacturing systems, with direct application in the domain of round shafts. In order to perform the research there will be used software and modern equipment. Moreover, by the project are identified the possibilities of consolidating research of applying it in industry on the purpose of issuing possible further contracts. The researches will be also extended in the field of FMS management and control in real and virtual environment. There will be developed aspects concerning programming with the aid of flexible manufacturing system virtual reality.

The implementing of a flexible manufacturing system in an integrate fabrication system, represents an important result, which leads to the optimization of the entire material and informational flux of the enterprise.

5.REFERENCES

Abrudan, I. (1996) - *Flexible Manufacturing Systems. Design concept and management*, Dacia Publishing House, ISBN 973-5-0568-4, Cluj-Napoca

Boncoi, Gh. & Fota, A. (2001) - *Production Systems*, "Transilvania" University Publishing House, ISBN-973-9474-896, Brasov

Fota, A. (2004) - *Machines Systems Design. Simulation and Modelling*, "Transylvania" University Publishing House, ISBN 973-635-331-1, Brasov

Law, A. M. & Kelto, W. D. (1991) - *Simulation, modelling and analaysis*, McGraw-Hill Book Company, Singapore

Ispas, C. & Mohora, C. (2001) - *Simulation of Product Systems*, Romanian Academy Publishing House, ISBN 973-27-0868-9, Bucharest

Tempelmaier, H. (2006) *Material Logistik. Grundlagen der Bedarfs und Losgrossenplanung in PPS – Systems*, Springer-Verlag, ISBN 978—540-70906-0, Berlin

Annals of DAAAM for 2008 & Proceedings of the 19th International DAAAM Symposium, ISSN 1726-9679
Editor B. Katalinic, Published by DAAAM International, Vienna, Austria 2008
Make Harmony Between Technology and Nature, and Your Mind will Fly Free as a Bird
*Young Researcher and Scientist Paper / * Supervisor, Mentor*

IMPROVING CHANNEL FLOW WITH T-JUNCTION DEFLECTORS SHAPE-OPTIMIZED USING GENETIC ALGORITHM

FUCAK, S[anjin]; CARIJA Z[oran] & MRSA, Z[oran]*

Abstract: *This paper summarizes the optimization process for deflectors designed to lower the water level and ceiling spillage in the S-shaped outflow channel of HPP Vinodol. Parameterized designs of ten deflectors were optimized using a genetic algorithm that evaluated each setup by running 2D CFD simulations. Optimal designs were chosen by considering total pressure loss and flow uniformity. The resulting best case design set showed a 16.7% reduction in energy losses providing a solid basis for future simulations using a 3D fluid flow numerical model of the channel.*
Key words: *genetic, algorithm, deflector, flow, optimization*

1. INTRODUCTION

The HPP Vinodol uses six Pelton turbines with a combined water flow of 18.7 [m^3/s]. The uniquely S-shaped drainage channel that collects the water from the turbines was designed for a much lower throughput and with a less-than-optimal shape for unhindered water flow.

The rising water level under the turbines, caused by flow losses within the channel, is controlled by pumping air into the turbine housing. As the plant often operates with maximum power output, the additional water spillage through channel ceiling vents was deemed inappropriate and another solution was sought.

Ten deflectors were positioned in the channel bends and at T-junctions were designed in order to make the flow more uniform along perpendicular sections of the channel. Numerical modelling and optimization were chosen because the six water inlets, the widening channel profile and the S-bends make the flow in the channel very difficult to estimate analytically.

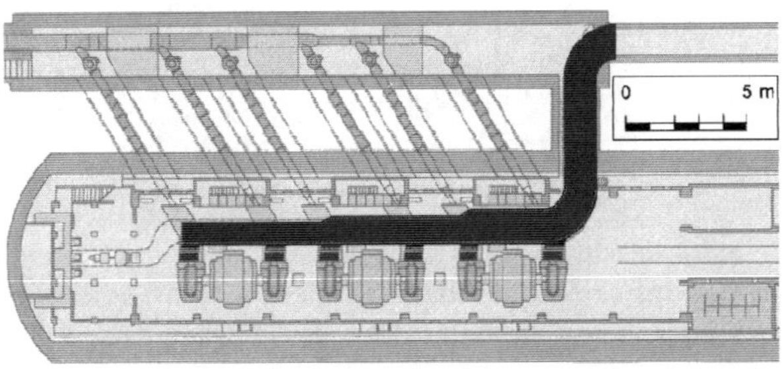

Fig. 1. The 3D model geometry for the computational domain of the S-channel, shown on the engine room schematic.

2. PHYSICAL MODEL DESIGN PARAMETERS

The channel geometry was created according to available original designs. Additional corrections were introduced after a visit to the S-channel in which control measurements and photographs were taken. The final 3D model was imported into the mesher as a geometrical base for creating the computational grid in 3D and 2D fluid flow simulations performed in Fluent 6.2.16. Design restrictions prohibited removal of wall sections in the existing channel geometry due to the armature and structural integrity considerations for the upper levels.

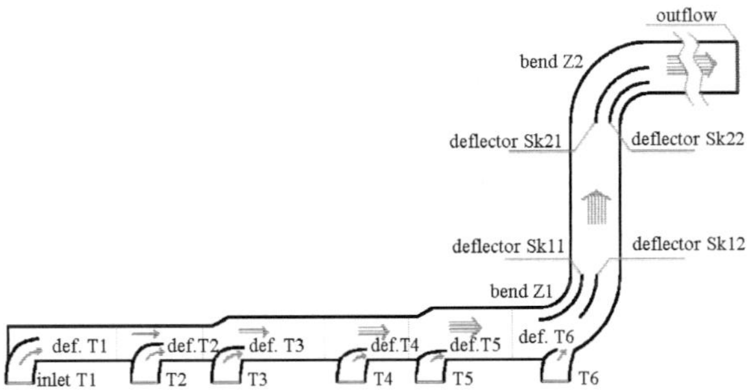

Fig. 2. Nomenclature of computational mesh segments.

A 3D water flow simulation of the actual channel state was performed to define the problem zones by examining result data, streamlines and particle tracks. A 2D intersection in the model Z-plane was chosen as the optimization domain, level with the middle of the trapezoidal velocity inlets T1 through T6 (Fig. 2). The transition from a 3D domain to 2D allowed an efficient comparison of a greater number of deflector setups.

Taking into account the results obtained in the study (*Čarija & Mrša, 2005*), the Sk11 through Sk22 deflectors were positioned according to guidelines set in (*Idel'chik, 1966*). The initial design of the T1 through T6 deflectors was set to follow the streamlines as well as the smoothed lines of the channel widening from 2m to 2.5m and 2.5m to 3m.

A 2D unstructured mesh was used with wall element sizes set to 10mm and maximum interior element size set to 140mm with a 60% growth bias from wall elements. The k-ε turbulent viscosity model was selected for 2D simulations (*Ferziger & Perić, 2002*) and the residual convergence criterion was fixed at 10E-4, as automated adaptions through the Fluent journal script tended to crash the simulation unpredictably.

Individual optimization of each T-junction deflector had shown that the general shape had a lesser impact on energy losses than the aperture of the deflector measured from the channel wall. The deflector tip coordinates were thus chosen as design parameters for deflectors T1 through T6 and the circular arc Sk11 and Sk12 deflectors were defined by their distance from the convex left-hand side of the Z1 bend. Deflectors Sk21 and Sk22 remained in their initially defined positions due to their lesser impact on energy losses.

3. GENETIC ALGORITHM AND OPTIMIZATION

The genetic algorithm (GA) used was a multi-variable single fitness function guided executable of the GALOPPS GA libraries using roulette wheel selection, crossover and mutation (*Goodman D. E. 1996*); (*Goldberg D. E. 1989*). The GA encoded each parameter as a binary gene forming chromosomes that represented each deflector setup. Their fitness function was set as the difference of total pressure in the whole channel (between all six inlets and the outlet) and was calculated in a custom C++ application that monitored the CFD simulation and gathered statistics before returning the fitness value to the GA executable.

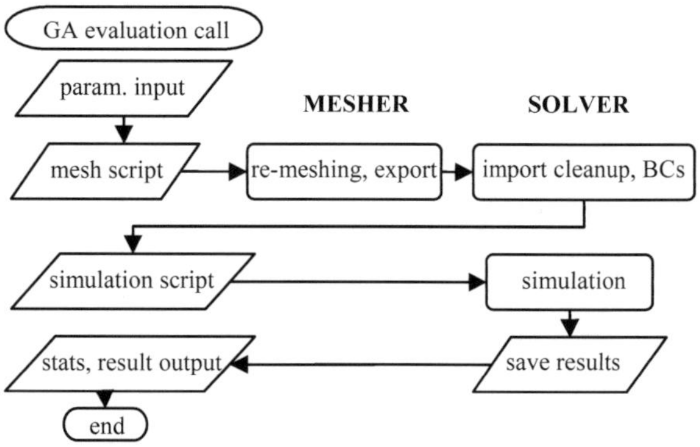

Fig. 3. Block diagram for the fitness function application showing external calls (mesher, solver).

Just the x-coordinates representing distance from the wall were varied for deflectors T1 through T5. Both x and y-coordinates were initially varied for T6, but the x-coordinate shift was omitted in later phases as less influential. Shortening the arc length of deflector Sk11 was tested at the upstream edge and dismissed because the whole 90° arc length.

All parameters were varied in 10mm increments within ranges defined for each GA run which was a population of 32 members over 32 generations.

The fitness function application outlined in (Fig. 3) used a pre-generated mesh for each optimization phase considerably speeding up the meshing part of each chromosome evaluation. The meshing script recalculated the deflector shape from coordinates selected by the GA, altered the pre-generated case mesh accordingly and finally exported it for Fluent. The application then generated the script to run the Fluent simulation and subsequently extract data from result transcripts and output graphical representations for post-processing.

A file with all the models, parameters and results was maintained in the run directory and GA fitness evaluation calls were checked versus existing previously evaluated chromosome results to prevent duplicate simulations. This also enabled GA runs to continue after computer crashes or manual accumulation of results from GA runs on several different machines.

The total pressure loss sometimes differed very little with varying deflector positions thus making another control value necessary. Since the mass flow was constant in all case variations, the surface of the domain area showing velocity magnitudes greater than 4.5 [ms^{-1}] was taken as a measure for flow uniformity.

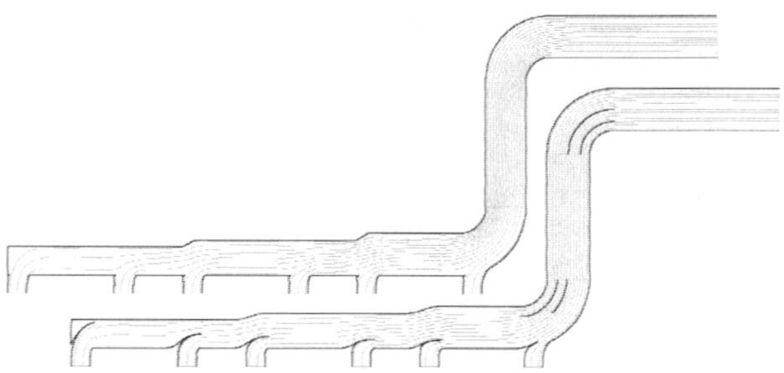

Fig. 4. Final 2D model with best case deflectors (below) and without them (above), showing flow streamlines

4. OPTIMIZATION PHASES AND RESULTS

The optimization was performed in several phases. In the first phase, zero-thickness deflectors were optimized individually for each T-segment, always taking the upstream portion of the channel into account. Due to the upstream influence of each deflector, their results differed from results obtained after attaching downstream channel segments.

The second phase consisted of optimizing the part of the channel upstream from the bend Z1, including segments T1 through T5.

The whole channel was modelled in the third phase using best-case positions from the previous phase and varying the positions of T6, Sk11 and Sk12.

The fourth phase optimized the whole channel again, but varied all the parameters within smaller value ranges as the beneficial ranges were obtained from phases 2 and 3.

The final phase added thickness to the deflectors and extended them so the parameter value ranges had to be expanded again and subsequently narrowed for the final run.

The top results were then checked by varying all the parameters within a small range to verify if the GA runs had indeed found the actual function minimum.

The 2D simulation of the original state shows a total pressure loss of 2615.2 [Pa]. The energy loss is 30250 [W], obtained by multiplying the total pressure loss with volume flow rate. The best case optimization result shows a 16.7% reduction in losses with a d(P_{TOT}) of 2178.5 [Pa], and E_{LOSS} of 25199 [W] in the whole channel computational domain.

d(x)					d(y)	d(R)	
T1	T2	T3	T4	T5	T6	Sk11	Sk12
1800	1000	830	630	590	1250	700	1380

Tab. 1. Best case design parameters in [mm]

5. CONCLUSION

The optimization results show better velocity field uniformity, indicative of the 16.7% reduction in energy losses, and provide the basis for 3D deflector design. The next research step would be the 3D fluid flow simulation of the channel with deflectors extruded to their full height. It would show the gravitational influences and the swirl in the flow enabling the structural design of deflectors under the dynamic body forces during load cycling and maximum turbine output. Validation of 3D CFD results on a scale model of the channel and deflector prototypes would lead to the final design and implementation.

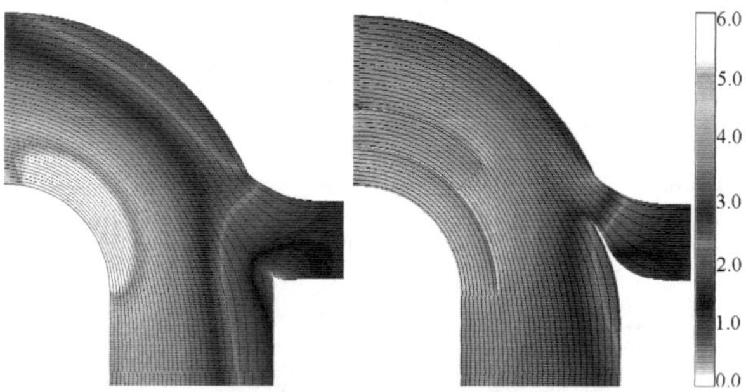

Fig. 5. Velocity magnitudes of the Z1 bend (shaded areas, legend [ms^{-1}]) with streamlines superimposed (black); original state (left), optimized deflectors (right)

6. REFERENCES

Ferziger F. H.; Perić M. (2002) *Computational Methods for Fluid Dynamics*, Springer Verlag, Berlin

Goldberg, D. E. (1989). *Genetic Algorithms in Search, Optimization and Machine Learning*, Addison-Wesley Publishing Company Inc, Boston

Goodman, D. E. (1996) *An introduction to GALOPPS*, Michigan State University, East Lansing

Idel'chik, I. E. (1966). *Handbook of Hydraulic Resistance,* Israel Program for Scientific Translation, Jerusalem

Mrša, Z.; Čarija, Z. (2005). *Hydraulic loss reduction capability analysis for the HPP Vinodol drainage channel*, Tehnički fakultet u Rijeci, Rijeka

Annals of DAAAM for 2008 & Proceedings of the 19th International DAAAM Symposium, ISSN 1726-9679
Editor B. Katalinic, Published by DAAAM International, Vienna, Austria 2008
Make Harmony Between Technology and Nature, and Your Mind will Fly Free as a Bird

MASS CUSTOMIZATION: TRENDS AND RESEARCH

FUERSTNER, I[gor]; ANISIC, Z[oran] & COSIC, I[lija]

Abstract: *This paper presents an overview of the current research results in the field of mass customization worldwide. In the first chapter, general information about main problems concerning mass customization is presented. The following chapters deal with various approaches in the research area. The last chapter deals with the proposed future research possibilities in the area of mass customization in the context of presented overview.*

Key words: *Mass Customization, Product Configuration, Customer, Customizable Services*

1. INTRODUCTION

This paper presents an overview of some of the current research results in mass customization.

Global competition is forcing producing companies to change their activities from a seller point of view towards a buyer point of view, what results in a drastic increase in the number of product variants and costs.

To maintain there competitiveness, companies are modularizing their products and introducing platform concepts, and this transfer from no customizable products to modular products that involve individual customer variants is one of the most important industrial strategies nowadays.

The recent development of *IT* technology enabled the software based product configuration systems that support the process of customized product development. They compose customer specific solutions using the modules based on the customer's requirements.

2. OPEN FRAMEWORK FOR PRODUCT CONFIGURATION SYSTEMS

Many companies develop their own configuration systems, so the required rules for combining product components or modules, as well as the design of the entire product are usually statically implemented in the configuration system. As a result of this approach, any change concerning the product or the system itself, requires comprehensive changes in the configuration system's source code. This direct conversion of the rules into source code also leads to more dependency on software engineers, because other users cannot edit or change the configuration rules. This can be a large psychological barrier for many companies due to the necessity to share know-how with external persons, knowledge engineers and computer scientists. The other problem is the unclear structure of the rules, which leads to hard serviceability. These problems lead towards a development of an open framework that would allow the creation and maintenance of different individual product configuration systems by the engineers themselves without high dependency on computer scientists.

The product configuration solution should include virtual 3D model of the product, standardized services for the support of the system introduction to a company, accurate assessment of costs and profit, unitized software system, which can be tailored to customers and open communication structure between the involved systems (Abramovici at al. 2007).

3. THE ROLE OF CUSTOMER IN MASS CUSTOMIZATION

As the practice and research of mass customization develops, there is increasing understanding of how it can be implemented in terms of manufacturing capability and expertise, data transfer and management, the implementation of business systems, and the development of product architectures.

The so called 'solution space' is defined, as a conceptual 'container' for the matrix of product possibilities that are made available for any given mass customized product to a customer co-designer, established through the assessment of product architecture, range, overall company strategy and manufacturing capability.

Beside this, the added importance of connection with customers is likely to have an equal, if not greater impact on future working methods and technologies. Mass customization by its very nature consists not only of the customizable product offering but of the co-design experience. This experience differs from purchasing a mass produced product as it requires engagement and participation in the creation process, it is this participation that changes the role of the customer from consumer of a product to a partner in a process of adding value. There is a need to develop a conceptual model for mass customized product offerings that encompasses not only the current understanding of 'solution space', but also the wider aspects relating to the co-design experience steaming from the customer co-designers emotional connection with the product and purchasing process (Jordan, 2000, Herd et al., 2007).

4. CONFIGURABLE PRODUCT PLATFORM

The main driving force for platform based development and manufacturing is the possibility to combine customization with economy of scale or the effect off production process in which an increase in the scale of the produced component causes a decrease in the average cost of each unit. The means to achieve this is reuse of common resources (common parts) in multiple customized product variants. By doing this it is possible to create new product variants without having to develop all of its contained parts, just the ones that are variant specific. The rest can be used from already existing products or from a common core of parts in a product family or in different involved families of different brands – the product platform.A product platform is a set of parameters, features, and/or components that remain constant from product to product within a given product family. Configuration of product variants is thus achieved by combining the parts in the platform with variant specific parts.

Although much have been gained with this strategy it has its limitations, and it needs to be further developed in order to prevent the amount of part numbers to be managed in a developing and manufacturing company to grow out of hand. What is needed is an integrating tool that handles all knowledge related to the whole platform system, the relations between contained items and the rules governing the use of the contained knowledge for different purposes during the platform lifecycle. (Johannesson & Gedell, 2007).

5. OPTIMUM DEGREE OF CUTOMIZATION

Usually, the literature regards the degree of customization as how early or how far the customers are integrated in the product's production cycle, or more generally in it's life cycle. The point at which the customers' involvement or input starts is referred to as the "Order Decoupling Point". The earlier the customers' involvement in the product's life cycle is, the higher is the degree of customization. On the other hand, the closer the customer involvement is to the final product and distribution stage, the lower is the degree of customization. The level of customization can be also viewed from a supply chain perspective. A higher degree of customization would entail direct customer involvement starting backwards at the first tier suppliers, while a lower degree of customization occurs when customer involvement is close to the retailers or end users. The stage at which the customers' actual input integrates into the system is referred to as the "Stockholding Decoupling Point".

However, there is a need to determine a certain degree of customization concerning the possible increase of the market share and premium price, while lowering costs, risk level, time to market, and other.

To achieve this, determining the degree of customization from a product structural design perspective is necessary. That is, the degree of customization is determined by breaking down a product into a number of modules or components, and then examining the various options for each module or the extent to which some features within each module can be changed if required (Spahi et al., 2007).

6. CUSTOMIZATION OF SERVICES; TOURISM OFFERINGS

The approach of mass customization offers gain to both consumers and manufacturers of products, but it also has huge potential in the field of services. Tourism, as one of the most propulsive industrial branch, has recently become a very fertile soil to implement the technologies and approaches of mass customization.

The reduction of interest for solid arrangements with fixed and predefined content is noticeable, and the public wants more freedom in the choice of arrangements, which allows users to adapt and correlate their needs and wants with the offer and the possibilities of the destination to which they are traveling to. Large and successful agencies have always been ready to please the special needs and wants of their clients, but that usually went with a much higher price than that of a standard service (Taylor Made Customization), which a normal consumer cannot afford. Global increase of investments in tourism, and with that, the increase of the competition between the tourist service providers, leads towards finding new, hidden market places, which can secure significant profit with the help of widely used information technologies.

The main aspects considering the implementation of mass customization in tourism are therefore interesting for deeper research (Anišić et al., 2008).

7. CONCLUSION

The implementation of modular, standard product configuration systems is much easier and more acceptable than the existing ones.The advantages of the approach are the high quality of the configured productse, the improved serviceability of the knowledge base without the need of involving external engineers and computer scientists. Additionally the flexibility of the software system makes it possible to adopt changes to the knowledge base very fast and easily.

The product envelope model that defines the role and the position of the customer in the mass customization system in its current form is useful in helping to frame the reflection of an experience, but further work is currently being undertaken to both develop the model and structured methods for its use.

The main driving force for platform based development and manufacturing is the possibility to combine customization with economy of scale. This is in practice achieved by reuse of common parts, i.e. the platform, in different customized variants. Configuration of product variants is achieved by combining the parts in the platform with variant specific parts. This strategy has its limitations as it drives an increase of part numbers to be managed in a developing and manufacturing company to grow out of hand. In order to address these problems a new more system oriented and abstract knowledge based approach is needed. The platform model and modeling approach have the potential to enable more efficient product customization without driving growth of part numbers to be managed and to provide more efficient means for reuse of product knowledge for platform development.

In a mass customization system, personalized products can have different choices and customizable features with varying extents. It is important to be able to determine what scope of choices would work best for the organization. Not only that, but it is important to know which set of modules, components and/or feature choices, specifically, need to be expanded or narrowed down based on a set of given criteria and restrictions.

The use of the customization scale has been introduced as an overall indicator for the degree of customization for a particular product. The methodology is not expected to provide an exact solution for the optimum level of customization. However, it should put investors or management on the right track as far as the extent of customization, regarding each component/feature of the product, is concerned. The proposed methodology is a seed to a model that can be further expanded to provide more accurate and practical results. The formulations associated with each goal still needs to be further developed.

The approach of mass customization has also large potential in the field of services, especially in tourism. The reduction of interest for solid arrangements with fixed and predefined content is noticeable, and the public wants more freedom in the choice of arrangements, which allows users to adapt and correlate their needs and wants with the offer and the possibilities of the destination to which they are traveling to. This change of tourism business towards more specialized, flexible network configurations and further consumer integration demonstrates that this industry poses hard challenges, which offer interesting research opportunities.

8. REFERENCES

Abramovici, M.; Ghoffrani, M.; Neubach, M. & Bertram, S. (2007). *KOVIP – Product configuration Software and Services for Virtual Products*, GITO Verlag, Berlin.

Anišić, Z.; Duvnjak, S. & Mitrović, T. (2008). Concept of Customized Tourist Offer of Vojvodina Region, *Proceedings of 3rd International Conference on Mass Customization and Personalization in Central Europa*, pp. 1-5, Palić, Serbia.

Herd, K.; Bardill, A. & Karamanoglu, M. (2007). Designing for co-design: using the product envelope model as a framework for reflection, *World Conference on Mass Customization & Personalization*, Cambrige, USA.

Johannesson, H. & Gedell, S. (2007). Knowledge – Based Configurable Product Platform Models, *World Conference on Mass Customization & Personalization*, Cambrige, USA.

Jordan, P. (2000). *Designing Pleasurable Products. An Introduction to the New Human Factors*, Taylor & Francis, London.

Spahi, S. & Hosni, Y. (2007). Optimum level of ustomization for Mass Customization Systems, *World Conference on Mass Customization & Personalization*, Cambrige, USA.

Annals of DAAAM for 2008 & Proceedings of the 19th International DAAAM Symposium, ISSN 1726-9679
Editor B. Katalinic, Published by DAAAM International, Vienna, Austria 2008
Make Harmony Between Technology and Nature, and Your Mind will Fly Free as a Bird

USING SIMULATION IN RISK MANAGEMENT ANALYSIS

FUNAR, S[tefan] P[etru]; OSICEANU, S[anda] & COSTACHE, B[ogdan] M[ihai]

Abstract: *The internationalization process is mutually favourable, both for the localization country, and for investors. For a better manufacturing risk management of an international manufacturing project in order to provide the means that make program and project success a reality and helping to create horizontal and vertical project communities in a virtual workspace we need a validation tool for project planning. We consider a virtual enterprise meant to organize and get a global prospective of international manufacturing project. Validation of the project risk management gives the organization better flexibility and control on who gets in, who sees what and what tools they have the right to use. We are using Witness software, adapted by us in order to validate the project planner risk analysis for a virtual enterprise project.*

Key words: *risk analysis, project planning, simulation, Witness*

1. INTRODUCTION

We consider here that one of the greatest challenges of a virtual enterprise is to organize and get a global prospective of projects in progress. The international business unfolding confronts the managers of the multinational companies with a series of problems regarding their business opportunities, their position on different markets and the selection of the best markets for a future successful activity (Berkowitz & Rudelius, 1999). The analysis of the manufacturing risk management is very important for the management of the company because it helps the managers to decide on the risk strategy that they could adopt and it warns the managers in time about the deterioration of the business climate and has an important role in fundament some strategies that could be adopted in critical situations (Thomson & Strikland, 1997).

Manufacturing risk management, as a component of the international management, represents the total amount of the measures taken by the companies which become international in view of diminishing this one from their operations.

In the elaboration of a multinational manufacturing risk strategy the companies should take into account the fact that:

the income they obtain doesn't belong exclusively to the foreign company; it must be a profit for the employees and for the economy of the host country as well;

the risk of the investment reduces substantially by running some initial programmes of qualification for the managers in view of a good knowledge of the foreign environment;

they could penetrate the market easier if they would become involved in a series of economical or cultural projects of public interest;

it is less risky to manage the foreign investment with the help of a mixed managerial team with local managers as well, than to let it be managed exclusively by the mother country.

they are "guests" in a completely foreign environment and they must act accordingly;

The international investment of a company can be seen as a portfolio of financial and real assets that it must administrate efficiently (Carrison, 1988).

The first element of the total risk of the international portfolio represents the systematic risk of the international portfolio, a risk that depends on the international market evolution and that cannot be eliminated. The international market has got, in this case, more descriptions: the world market, the international market or a regional market. The second risk element has got a dual character:

On the one hand, from the point of view of the international portfolio, it is about an unsystematic risk which is determined by elements that are specific to every national portfolio. On the other hand, from the point of view of the national portfolio, it is about a systematic risk which is determined by the evolution of every national market. The last element of the total risk has got an unsystematic character which is influenced by elements specific to every national portfolio, and respectively by the total risk of each title.

2. SIMULATION ROLE IN RISK MANAGEMENT

For a better management of the project in order to provide the means that make program and project success a reality and helping to create horizontal and vertical project communities in a virtual workspace we need a validation tool for project planning. Simulation is a highly effective analytical tool for assessing the quality of design of a production system relative to its ability to meet production goals of quantity and quality within constraints of operational complexity and cost. In view of the complexity of typical manufacturing systems and the high level of stochastic variability among their operations, analysis of manufacturing systems is among the most venerable and frequent of simulation application areas. This simulation project was undertaken with the goals of identifying ways of improving the risk management. We had applied our risk improvement algorithm on a case study for a pump cover manufactured at SC Mazak Group SA on a concentrate flexible cell (Cotet & al., 2007). The lab provides project planning and controlling using Primavera (a Primavera Systems software product) and also a discrete material flow simulation for the parts that are machined in the flexible cell system. In order to improve the process, one can perform a project simulation using WITNESS, which provided a realistic model of the manufacturing process and allowed to quickly and inexpensively evaluate a wide range of alternatives.

The PRIMAVERA software assures for the Project the activities flow management in the manufacturing system that we want to manage and optimize the risk management (figure 1). PRIMAVERA uses the results of Witness simulation and modifies the real flow according to this. Use of a simulation model is a surrogate for experimentation with the actual system (existing or proposed), which is usually disruptive, not cost-effective, or simply impossible. Thus, if the model is not a "close" approximation to the actual system, any conclusions derived from the model are likely to be erroneous and may result in costly decisions being made. Validation should and can be done for all models, regardless of whether the corresponding system exists in some form or whether it will be built in the future.

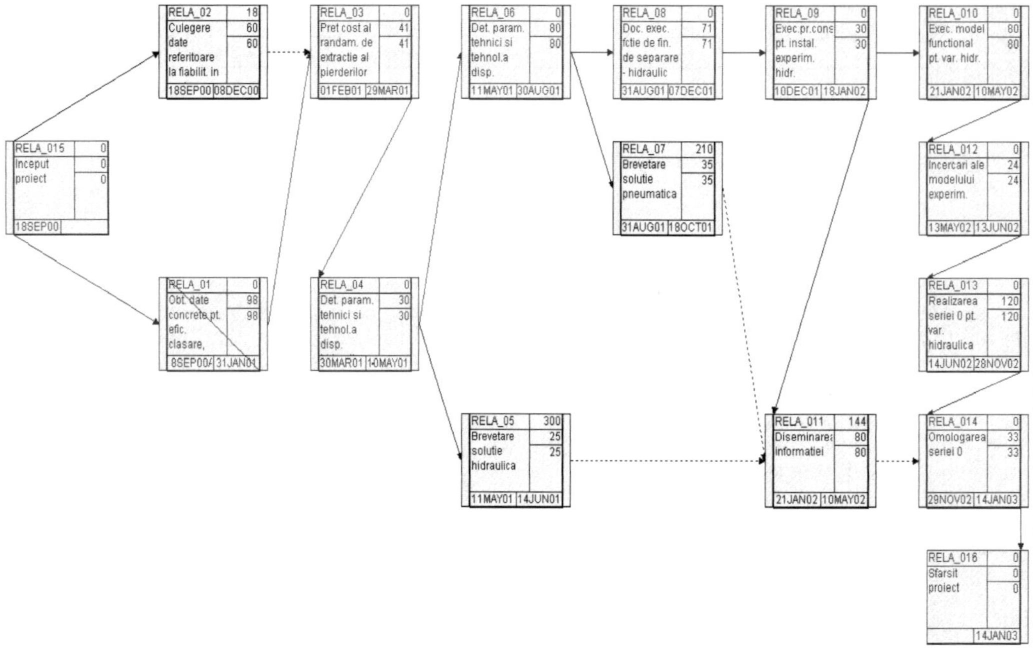

Fig. 1 The activities flow management in the manufacturing system using the PERT view in PRIMAVERA

Working with the simulation, engineers were able to substantially improve routing logic and test mixtures and make other changes as well that they are able to increase productivity by about 50% without any additional expenditures. We propose here project based on a comparative analysis illustrating the risk improvement using material flow simulation for an industrial case study for a specific manufacturing system. First we consider the initial model of the manufacturing project, and then the material flow simulation results and we focus our conclusions on the risk management based on those simulations. After identifying the risk we will focus on our manufacturing system modification in order to improve this critical point flow transit without generate another concentrator in the flow. After that we will need to perform a new simulation of a manufacturing cycle on our modified system in order to validate this solution and we will quantify the risk improvement. For using our project planner related with the material flow simulator in a Virtual Enterprise Architecture we apply the following steps:

Step 0. In the economic lab module (M1) the project of a part manufacturing is designed and transferred using Intranet as data entry for PRIMAVERA in the simulation lab module (M2).

Step 1. In M2 an activity flow for the part is designed using PRIMAVERA.

Step 2. In M2 the PRIMAVERA project is transferred via Intranet as data entry for Witness.

Step 3. In M2 the Witness simulation of the activity flow is performed.

Step 4. In M2 the Witness activity flow is transferred as data entry for Witness material flow simulation.

The results of this simulation give the possibility to identify the risk generated for example by the human resources assigning in the discrete material flow. It is possible now to recognize the elements from M1 as they are in the model used for simulation in M2.

Step 5. In M1 the results of the Witness simulation are transferred to Primavera who is updating task activities and reviewing resources assigning to activities. The result is a new optimized project.

Step 6. The new project is transferred using Intranet as data entry for Witness activity flow in M2 in order to be validated.

3. CONCLUSION

Companies work in collaborative processes within virtual organization to a large extent which means that information concerning product definition needs to be available, communicated and transferred in a not homogeneous information system environment (Rosu & al., 2007).

In this paper we describe an approach between discrete risk management and project planning using an open platform for training the students and engineers to work in the new VE environment imposed by globalization in competition and manufacturing. We consider necessary that we find the training tools for the human resources that will be involved in this new industrial paradigm. The applications like the one presented here may be such a tool. We can validate the results of project planning on a web based system for use within any project driven, document intensive and quality assured organization.

4. REFERENCES

Berkowitz, K., Rudelius R., (1999): Marketing, Richard D. Irwin Ind., 1999.

Carrison, R., (1988): Managerial Accounting (Concepts for Planning Control Decision Making), Homewood, Illinois, 1988.

Cotet, C. E., Dragoi, G. & Carutasu, G. (2007). Material Flow & Process Synchronous Simulation In Concentrate Manufacturing Systems, *Annals of DAAAM for 2007 & Proceedings of The 18th International DAAAM SYMPOSIUM, „Intelligent Manufacturing & Automation: Focus on Creativity, Responsibility and Ethics of Engineers"*, Katalinic, B. (Ed.), pp. 180-181, ISSN 1726-9679, ISBN 3-901509-58-5, Zadar, Croaţia, October 2007, Publisher DAAAM International Vienna 2007.

Rosu, S. M.; Draghici, A. & Guran, M. (2007). *Knowledge Transfer in the Enterprise Business Intelligence*, Annals of DAAAM for 2007 & Proceedings of the 18th International DAAAM Symposium, October 24th - 27th, B. Katalinic (Ed.), published by DAAAM International, Vienna, Austria, pp. 647-648, ISBN 3-901509-58-5.

Thomson, A., Strickland, A., (1997) Strategic Management (Concepts and Cases), Homewood, Illinois, 1997.

Annals of DAAAM for 2008 & Proceedings of the 19th International DAAAM Symposium, ISSN 1726-9679
Editor B. Katalinic, Published by DAAAM International, Vienna, Austria 2008
Make Harmony Between Technology and Nature, and Your Mind will Fly Free as a Bird
2nd DAAAM-ICAT

VISUALISATION OF FDM PROTOTYPES

GAJDOS, I[van]; SLOTA, J[an] & SPISAK, E[mil]

Abstract: *This paper is oriented to visualization methods of Rapid Prototyping prototype produced by Fused Deposition Modeling technology. Practical part is oriented to process solutions of visualization FDM prototype model and application of these methods for tests of mechanical properties. For comparison test of mechanical properties by 3-point bending was executed. Main goal of this work is creation of a virtual model of a prototype without need of creation of a physical model.*
Key words: FDM, Rapid Prototyping, visualization

1. INTRODUCTION

Using rapid prototyping methods we are able to obtain real concept about new product. Rapid prototyping (RP) meets the current needs in the industry to shorten design cycles and improve the design quality. Fused deposition modeling (FDM) is one of the key technologies of RP (Anitha et. al, 2001).With prototype there is possibility to make mechanical and assembly tests and get almost real representation of new developed product. One of the most suitable methods for preparing plastic prototypes is FDM method, that is commercially available as Dimension printers from Stratasys, making prototypes from ABS polymer (Pahole et. al, 2005). One limit of using FDM for preparing prototypes of plastic parts is that these parts are typically thin walled and with the width of extruded fibre (0,511 mm) it is almost impossible to create walls thinner than 1mm or these walls are very brittle. (e.g. snap fit). Another issue with FDM prototypes is surface quality which is dependent on layer thickness (typically 0,254 mm) and build angle (fig.1).

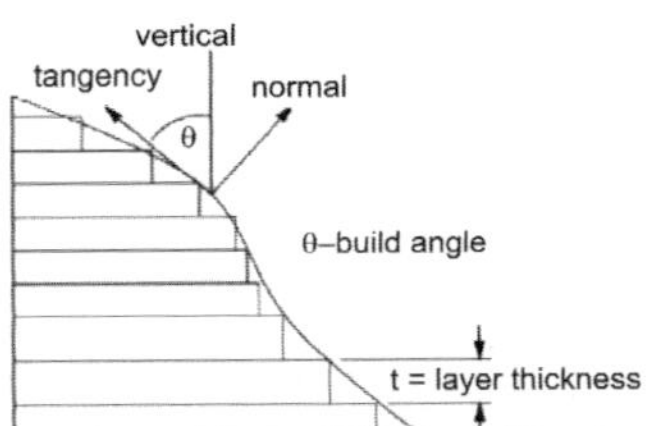

Fig. 1 Building angle and layer thickness (Pandey et al., 2003)

Aim of the paper is to show possibility preparing virtual CAD model of FDM prototype, which can be subsequently used for identify possible build problems, find optimal build angle or for FEM analysis.

2. CREATING VIRTUAL FDM PROTOTYPE

To create virtual CAD model of FDM prototype it is necessary to know paths of printer extrusion head and cross section of extruded fibre. Based on these entry data it is possible to create CAD model with "extrusion feature" in modelling software.

2.1 Analysis of fibre cross-section

Fibre extruded trough printer head has circular cross-section (whit diameter 0, 38 mm), but with whit contact previous layer, the cross-section section is deformed) as shown in figure 2 (Pandey et al., 2003). The visual analysis of specimen broken in 3-point bending test has shown that shape of contact surface between two fibres lying in two layers over each other, has square shape with dimension 0,33x0,33 mm (fig.3).

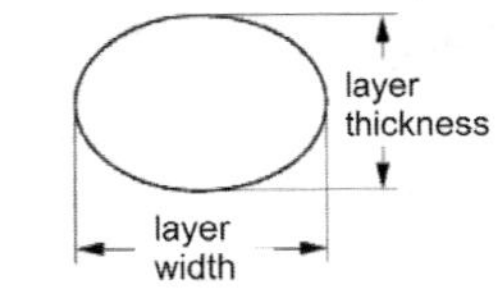

Fig.2 Cross-section of fibre (Pandey et al., 2003)

Based on dimensions obtained from figure 3a, we can determine simplified shape of fibre cross-section in FDM prototype (fig.3b). The shape of fibre cross-section in not exact as in real FDM prototype but sufficient to prepare CAD model of this prototype.

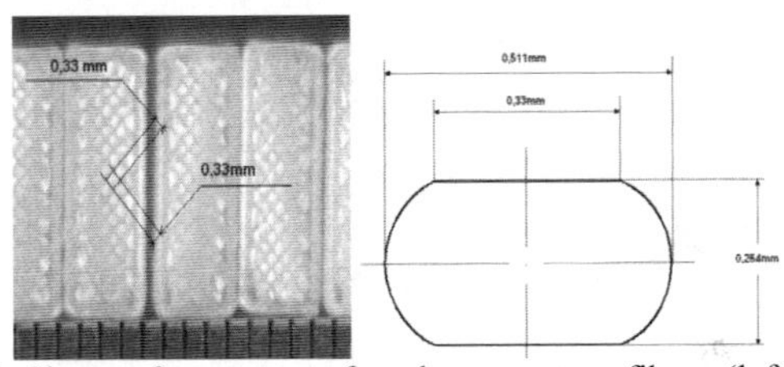

Fig.3 Shape of contact surface between two fibres (left) and simplified shape of fibre cross-section (right) (Boháček, 2008).

2.2 Paths analysis of printer extrusion head

Rapid prototyping processes produce models layer-by-layer. Hence, the geometric model must be first sliced into layers before the physical rebuilding process. Therefore, the slicing algorithm plays a very important role in RP systems. (Haipeng, P. & Tianrui, Z., 2007).

The layer in FDM prototype is created by fibre extrusion from printer head. The path for printer head is for Dimension SST printer defined in "*.CMB" file, generated with help of Catalyst 4.3 software from "*.STL" file.

After loading STL part model into Catalyst 4.3., are generated layers with modelling and support material and printer extrusion head tool paths. Screenshots are made from each layer and saved as pictures in jpeg format figure 4, with decreased colour depth to 1bit (black and white).

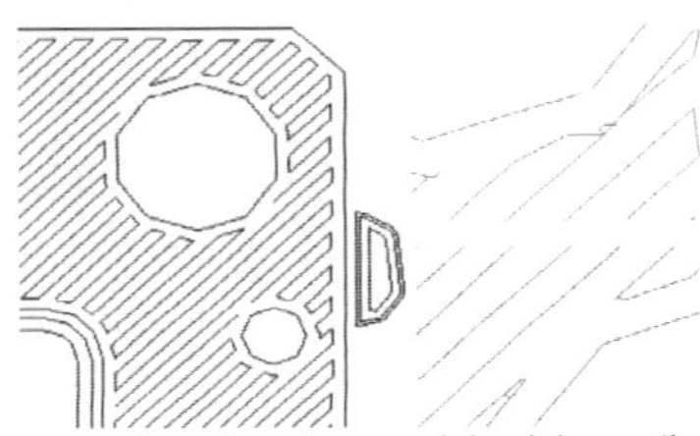

Fig.4 Fragment of tool path saved in *.jpeg (left) and some possible bugs in tool path (right).

From these source images with all layers of the model (converted to 1bit depth), are prepared data in "*.DXF" format, using IMG2CAD 7.0. Possible bugs in tool paths (fig.4 right),are manually fixed in AutoCAD 2007. Fixed tool path in DXF file is consequently imported as sketch into ProE Wildfire 3.0.This sketch then serve as path for "SWEEP – PROTRUSION" function (fig.5) with the cross-section from figure 3 (right).

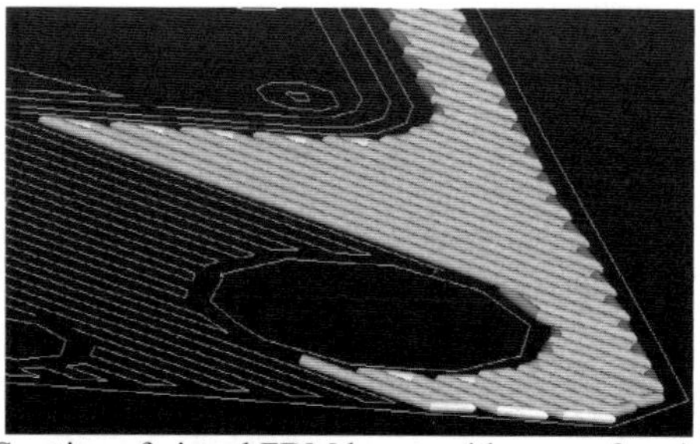

Fig. 5 Creation of virtual FDM layers with sweep function.

Using all imported data (from DXF file) for sweeping cross-section from figure 3, we obtain one layer from FDM prototype as a part from ProE Wildfire 3.0. After preparing all the layers is made assembly of virtual FDM prototype (fig.6)

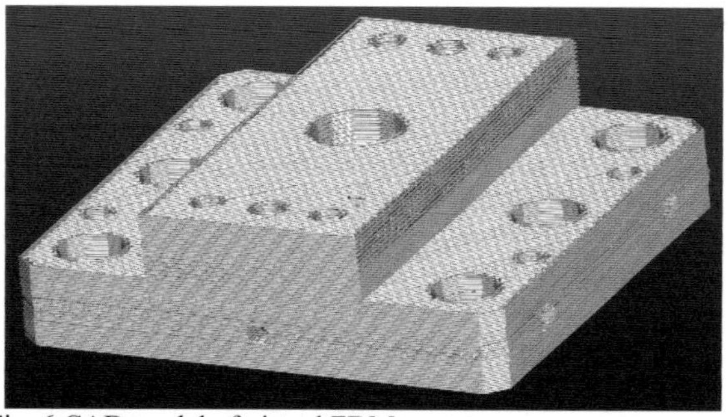

Fig. 6 CAD model of virtual FDM prototype.

3. EXPERIMENT

Visualized FDM prototype model as a CAD model can be used as base for FEM analysis and results of analysis compared with tested specimen. . For this purpose was used specimen for 3-point bending test according to ISO EN 178:2001 (figure 7).

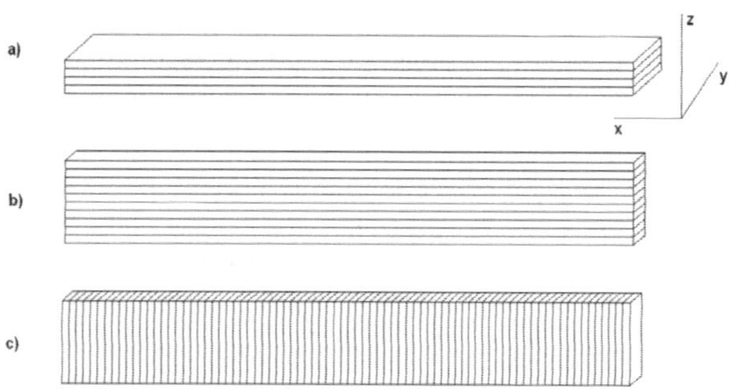

Fig. 7 Layers orientation in specimens for 3-point bending test

The results for 3-point bending test are shown in figure 8, where strong anisotropy of bending strength depending on layer orientation can be observed.

For the A-specimen (fig.7), was also prepared FEM analysis of 3-point bending test, to compare virtual FDM model whit experimental results.

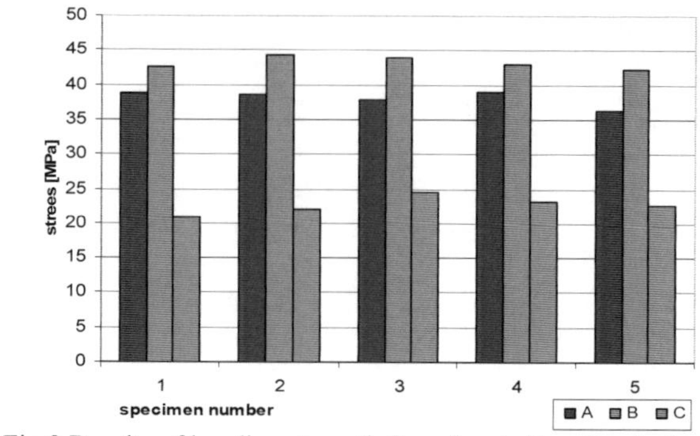

Fig.8 Results of bending strength from 3-point bending test

FEM analysis was carried out in CosmosExpress and the visual comparison of result is displayed in figure 9.

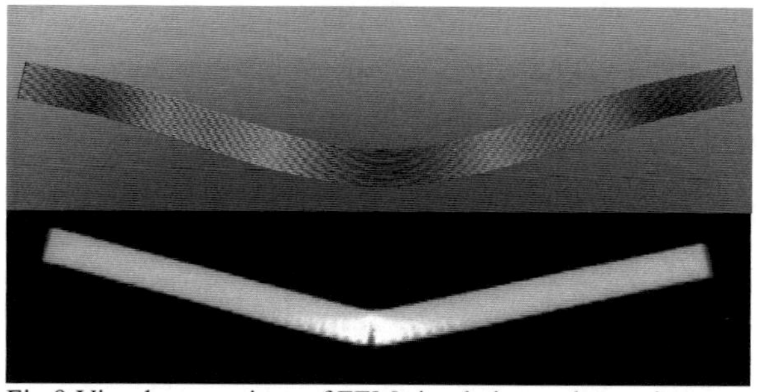

Fig.9 Visual comparison of FEM simulation and experiment

4. CONCLUSION

In this paper was presented one approach for visualization of FDM prototypes. Ability to prepare virtual CAD model of FDM prototype, can help identify possible build problems, find optimal build angle or for FEM analysis.

FEM analysis can be used to find out optimal part orientation especially in large expensive prototypes when customer has demand on strength specific area of the part.

Disadvantage of this method are high time demands to prepare virtual FDM model and high demands on computer performance for FEM simulation. The next step in reengineering proposed procedure is to prepare software which could extract tool paths directly from "*.CMB" file and save them in DXF format.

5. REFERENCES

Anitha, R.; Arunachalam, S. & Radhakrishanan, P.(2001): *Critical parameters influencing the quality of prototypes in fused deposition modeling* In: Journal of Materials Processing Technology, No. 118 , p. 385-388.

Boháček, M. (2008): *Vizuálizácia modelov prototypov vyrobených metódou FDM*, Diploma work - Technical university in Košice, Košice.

Haipeng,P.& Tianrui, Z. (2007): *Generation and optimization of slice profile data in rapid prototyping and manufacturing.* Journal of Materials Processing Technology, Vol. 187-188, p. 623-626.

Pahole, I.; Drstvensek, I.; Ficko, M. & Balic, J. (2005): *Rapid prototyping processes give new possibilities to numerical copying techniques.* Journal of Materials Processing Technology 164–165, p.1416–1422.

Pandey P.M.; Venkata Reddy N. & Dhande S.G.(2003): *Improvement of surface finish by staricase machining in fused deposition modeling* In: Journal of Materials Processing Technology, No. 132 (2003), p. 323-331.

Annals of DAAAM for 2008 & Proceedings of the 19th International DAAAM Symposium, ISSN 1726-9679
Editor B. Katalinic, Published by DAAAM International, Vienna, Austria 2008
Make Harmony Between Technology and Nature, and Your Mind will Fly Free as a Bird
*Young Researcher and Scientist Paper / * Supervisor, Mentor*

INCREASE OF COMPRESSIVE STRENGTH OF MASONRY WALLS BY USING GLASS FIBER STRAPS AND FABRIC

GALIC, J[osip]; KISICEK, T[omislav] & SORIC, Z[orislav]*

Abstract: *This paper describes results of testing unreinforced, reinforced and strengthened masonry walls subjected to vertical compression load. All masonry wall specimens were constructed of solid masonry units (bricks) and cement-lime-sand mortar. They all had the same shape, with dimensions h/l/t=55/51/12 cm. Testing was conducted on different types of wall specimens: unreinforced, reinforced with steel wires, reinforced with GFRP wires, strengthened with horizontal and diagonal glass fiber straps and those strengthened with glass fiber fabrics. Test methods were identical for all wall specimens. Carrying capacity was tested and behavior of each specimen during the testing was recorded. Comparisons of test results were described.*
Key words: *Masonry, Walls, Strengthening, Polymers, Straps*

1. INTRODUCTION

Tests conducted at the Faculty of Civil Engineering, University of Zagreb, were performed on unreinforced masonry (URM), reinforced and strengthened masonry walls. Masonry elements in this research were made of solid masonry units (bricks). Wall specimens were subjected to increasing vertical compression load. Reinforced masonry walls have had one steel or Glass Fiber Reinforced Polymer (GFRP) wire in every second horizontal mortar joints. Unreinforced masonry walls were strengthened with glass fiber straps or fabric.

The aim of these tests was to investigate efficiency of reinforced walls and walls strengthened with glass fiber materials regarding enhancement of compressive resistance when compared to URM walls. Ratio of length and thickness of walls were 4.25. The results have shown that reinforced and strengthened walls have increased compressive resistance, and what is even more important, that they showed more ductile behavior before failure. Strengthening of walls, with glass fiber straps, was more efficient then reinforcing with steel and GFRP wires.

Strengthened masonry columns using FRP products were tested by other researchers too (Bieker at al., 2002, Krevaikas & Triantafillou 2005, Corradi at al., 2007, Tumialan at al., 2001). Research of Bieker et al., 2002, and Krevaikas & Triantafillou, 2005, showed that strengthening of masonry columns with glass fiber fabric increased carrying capacity up to 300%. In these researchs specimens were with aspect ratio of cross setion 1.0 to 2.0 and only strengthened with straps, not reinforced with wires.

2. EXPERIMENTAL INVESTIGATION

Testing of compressive strength was done on six different types of masonry wall specimens with two or three specimens per each type. Dimensions of all wall specimens were *l/t/h* = 51/12/55 cm. One specimen type was unreinforced masonry URM wall and other five specimens were reinforced and strengthened walls. Drawings of specimens are given in Figure 1, and the list of wall specimen's types is: a) Unreinforced (unstrengthened) masonry (URM) walls (1/1, 1/2, 1/3);

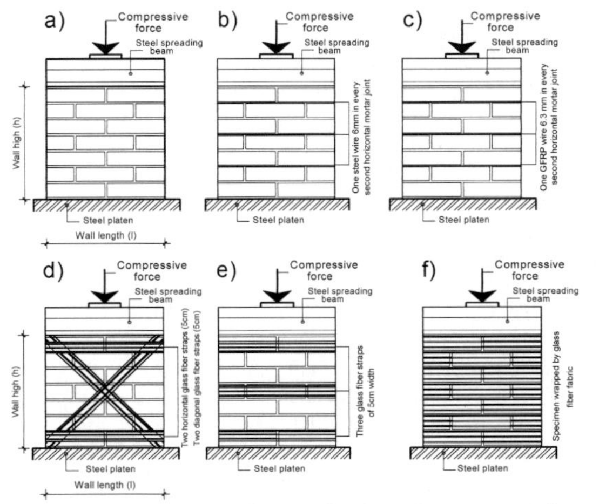

Fig. 1. Shape and types of strengthening of masonry wall.

b) Walls reinforced with steel wires (2/1, 2/2, 2/3); d) Walls strengthened with horizontal and diagonal glass fiber straps (4/1, 4/2, 4/3); e) Walls strengthened with horizontal glass fiber straps at the wall surface (5/1, 5/2, 5/3) and f) Wall strengthened with wrapped glass fiber fabric glued at the wall surfaces (6/1). All fifteen specimens were built from old bricks (Group 1) of 12/25/6.5 cm size. Mean normalized compression strength of masonry units was f_b = 18.58 N/mm^2. Mortar was of cement-lime type (c/l/s = 1:2:8) with mean compression strength f_m = 6.27 N/mm^2. Steel reinforcing bars were diameter Ø6 mm of type S-400 and were placed in every second horizontal joint. GFRP bars were diameter Ø6.3 mm with tension strength f_u = 827 N/mm^2 (according to Castro, P. & Carino, N.J. (1998)). For strengthening of URM walls were used GFRP straps *Mapewrap G UNI-AX 900/60* of manufacturer MAPEI (Italy). Lap length of straps and fabric was 40 cm. The both surfaces of the wall were prepared, i.e. uniformly aligned with mortar for leveling. Straps were glued with epoxy paste *MapeWrap 31* on both sides of the wall.

Test arrangement and data acquisition system is shown in Figure 2. All specimens were tested in the same way with central compressive load according to European norms, EN 1052-1 (1998). Displacements were measured with 7 LVDT's.

3. DISCUSSION OF THE RESULTS

Max. stresses, i.e. stresses at failure, are shown as groups of vertical bars in Figure 3. The significant increase in carrying capacity of walls that were strengthened with horizontal glass fiber straps (5/1, 5/2, 5/3) compared to unreinforced walls (1/1, 1/2, 1/3) is apparent (6.21 MPa compared to 3.83 MPa). The increase in carrying capacity by wrapping specimen with glass fiber fabric (6/1) was somewhat smaller (5.91 MPa). A problem during placing of fabric was the fact that the wrapped fabric of specimen (6/1) could not be tightened as much as the narrow straps of specimens (4/1-4/3) and (5/1-5/3). This resulted in 5%

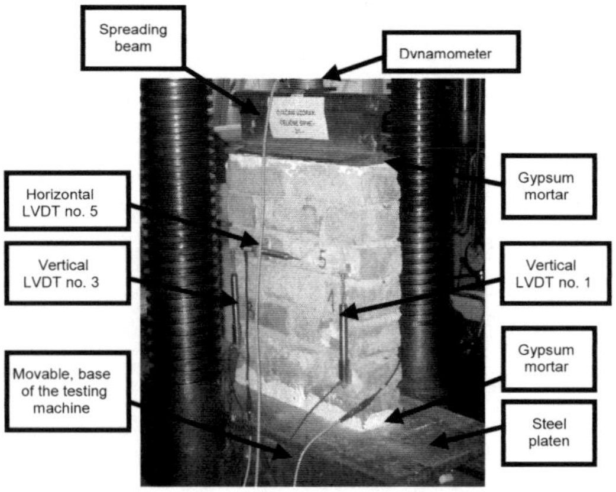

Fig. 2. Test arrangement and data acquisition system

smaller increase in strength comparing to walls strengthened by narrow straps (5.91 MPa compared to 6.21 MPa).

Appearance of first vertical cracks due to compression, in URM walls and walls with horizontal steel or with GFRP reinforcement, happened almost at the similar compressive stresses between 2.4 and 2.7 MPa. After appearance of the first vertical cracks, URM walls due to vertical load, experienced large horizontal deformation. Walls with horizontal steel or GFRP reinforcement in bed joints have shown much smaller (up to 3 times) horizontal deformation. URM walls failed at compressive stress of 3.83 MPa while RM with steel wires failed at 5.23 MPa. Walls reinforced with GFRP wires failed at 4.28 MPa. It can be concluded that horizontal reinforcement restrained horizontal stretching of a wall, and increased failure stresses as well. The failure of reinforced walls occurred because of the stress concentration at the contact of steel wires and bricks. Appearance of the first vertical cracks due to vertical compression, in walls strengthened with glass fiber straps, happened at the stresses between 2.3 and 2.5 MPa. Horizontal and diagonal straps restrained horizontal stretching of a wall after appearance of vertical cracks, so there was not significant increase of horizontal strain. Walls with diagonal straps failed at compressive stress of 5.92 MPa while walls with horizontal straps failed at 6.21 MPa. Considering stresses at failure, glass fiber straps increased carrying capacity of walls by 55 to 65%, comparing with URM walls.

Behavior diagrams of some walls are depicted in Figures 4. Except for the significant increase in carrying capacity of glass strengthened walls, compared to URM walls, the significant increase in ductility is apparent too. The wall that was wrapped with glass fiber fabric has had the highest ductility. Specimen strengthened by wrapping did not fail after reaching the maximum force; instead it resisted the force even after having noticeable vertical cracks.

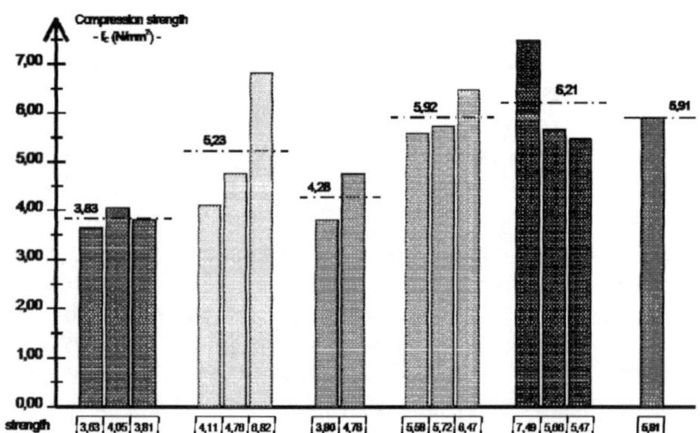

Fig. 3. Compressive strength of masonry wall specimens

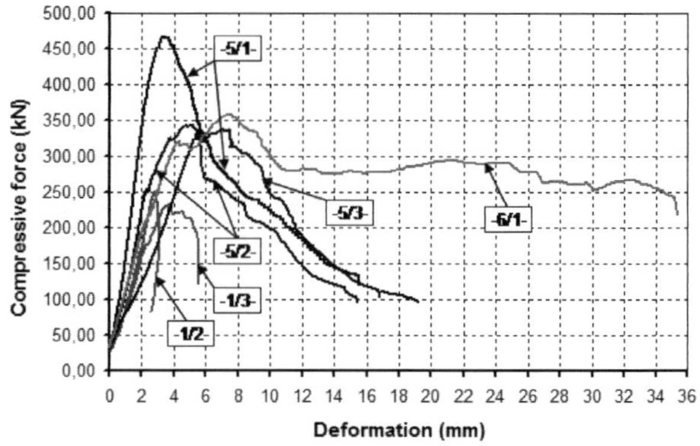

Fig. 4. Parallel diagrams "force – deformation" for URM specimens, specimens strengthened with horizontal glass fiber straps and specimen strengthened with glass fiber fabric.

4. CONCLUSION

The test results showed that reinforced and strengthened walls increased carrying capacity i.e. compressive strength by about 12 to 65 %. The first visible cracks, sideways, at all wall specimens appeared at about 2,3-2.7 MPa. After appearance of the first vertical cracks, URM walls experienced large horizontal deformation. The walls with horizontal reinforcement, have shown much smaller (up to 3 times) horizontal deformation. Horizontal reinforcement restrained horizontal stretching of wall, and increased carrying capacity. Increasing carrying capacity was about 12 % for wall specimens reinforced with GFRP wires and about 35% for wall specimens reinforced with steel wires.At strengthened walls, glass fiber straps restrained horizontal stretching of wall and increased carrying capacity. The test results showed that strengthening of the wall by glass fiber straps or fabric increased carrying capacity by about 55 to 65 %. Besides that, wall specimens that were strengthened with straps have had higher ductility, than unstrengthened ones. Especially high ductility was expressed by the specimen strengthened with wrapped glass fiber fabric (increasing was by eight times).

Increase of compressive strength of walls strengthened with glass fiber straps, was significant but smaller than strengthened columns in previous researchs. Strengthening of walls, with glass fiber straps, was more efficient then reinforcing with steel and GFRP wires. Further research is numerical modeling of these experiments on microplane model for masonry with relaxed kinematic constraint.

5. REFERENCES

Bieker, C.; Seim, W. & Sturz, J. (2002). Post-strengthening of masonry columns by use of fiber-reinforced polymers, *Third International Conference of Composites in Infrastructure*, San Francisco, 2002. Available from: http://www.quakewrap.com Accessed:2004-05-24

Corradi, M., Grazini, A. & Borri, A. Confinement of brick masonry columns with CFRP materials, *Composites Science and Technology*, Available from: http://www.sciencedirect.com Accessed:2007-01-15.

Castro, P. & Carino, N.J. (1998). Tensile and Nondestructive Testing of FRP Wires, *Journal of Composites for Construction*, ASCE 1998, Vol. 2, No. 1, 17-27.

Krevaikas, D. T.; Triantafillou, C. T. (2005). Masonry Confinement with Fiber-Reinforced Polymers, *Journal of Composites for Construction*, ASCE 2005, 9(2), 128-135.

Tumialan, G.; Micelli, F.; Nanni, A. (2001). Strengthening of Masonry Structures with FRP Composites", *Structures 2001*, Washington DC, May 21-23, 2001.

Annals of DAAAM for 2008 & Proceedings of the 19th International DAAAM Symposium, ISSN 1726-9679
Editor B. Katalinic, Published by DAAAM International, Vienna, Austria 2008
Make Harmony Between Technology and Nature, and Your Mind will Fly Free as a Bird

STRATEGY OF ROUNDNESS MEASUREMENT WITH CMM

GAPINSKI, B[artosz]; OCENASOVA, L[enka] & RUCKI, M[iroslaw]

Abstract: *In the paper, the roundness has been chosen as an object of CMM measurement strategy analysis because of wide application of bore-shaft connection. Obviously, the form deviations of those elements (roundness and cylindricity) will affect substantially the way of their work. The measurement of roundness may be performed using both universal and specialized measuring devices, dependent on the assumed uncertainty, price and time of the measurement. The measurement with CMM could give reasonable accuracy of the results in the short time and smaller price, but some parameters like out-of-roundness type, number of probing points and their distribution should be taken into consideration.*

Key words: *CMM, roundness, form deviation, measurement strategy, simulations*

1. INTRODUCTION

The increase of the quality requirements forces the development of the measurement techniques. At present, it is not enough to check merely dimensional tolerances. In many cases also form deviations (like out-of-roundness) should be checked, too. The roundness measurement is performed mainly by means of specialized devices. Additionally, coordinate measurement technique enables such a measurement. In any case, the measuring device should be applied effectively and correctly. It should not generate additional errors coming from inappropriate measurement strategy, and the time of measurement should not be unnecessarily prolonged.

The investigation have been focused on such characteristics of the measurement strategy as number of probing points dependent on the type of out-of-roundness (n-lobing). Additionally, the way of the coordinate system definition underwent analysis.

2. MEASUREMENT STRATEGY

The measurement of roundness became an important issue because of authomatical assembling processes and because common use of various types of connections of two round details. The problems of measurement with specialized devices and universal ones undergo analysis and investigations. Coordinate Measuring Machines (CMMs) are taken into account as a fast and accurate devices able to perform a wide range of tasks (Humienny 2001). Additionally, the achieved results could be complex set of parameters including dimensional and form deviations.

2.1 Measurement uncertainty

Dependent on the measuring task, the measuring device should be chosen properly. The criterion of choice should be the uncertainty of measurement. There are various methods for uncertainty estimation proposed by GUM (ISO/IEC Guide 1995). The specific way of measurement with CMM causes the need to calculate the uncertainty separately for each measurement task.

When the round elements are measured, the uncertainty budget should be estimated separately for different parameters: coordinates of the circle center, diameter of the circle and out-of-roundness (Gapinski & Rucki 2008). Additionally, the rule of $0.1T$ should be taken into consideration which states that the measurement uncertainty should not exceed 10% of assumed tolerances (in exceptional cases, when the tolerance is very small, it could be 20% of tolerance T).

2.2 Coordinate system

The basic procedure of the coordinate measurement is the definition of the coordinate system (Grzelka & Gapinski & Rucki 2006). This should be performed in order to receive all the freedom degrees and to define its position in the measurement space of the CMM. There are many ways to orient the measured detail, and therefore there are many factors that could affect the results of measurement. The correctness of the coordinate system definition and all further measurement, depends in a large degree on the machine operator.

In order to evaluate the position of the circle, the coordinate system related to the detail should be used. However, for the measurement of circle itself, including out-of-roundness measurement, the additional coordinate system should be defined. Its axis should be the axis of the measured circle. When the coordinate axis is declined to the axis of the circle, the measured circle appears as an ellipse. In that case, the increase of some harmonics and the smoothen of others could appear. As a result, the type and the value of out-of-roundness will be deformed, as well as the results of the measurement of diameter (Fig.1; Fig.2).

The results of measurement always depend on the fitting element (ISO/TS 12181). When the diameter is measured, the minimal circumscribed reference circle is the one most affected by the declination of the axis α. On the other hand, the maximal inscribed reference circle is not dependent on the angle α.

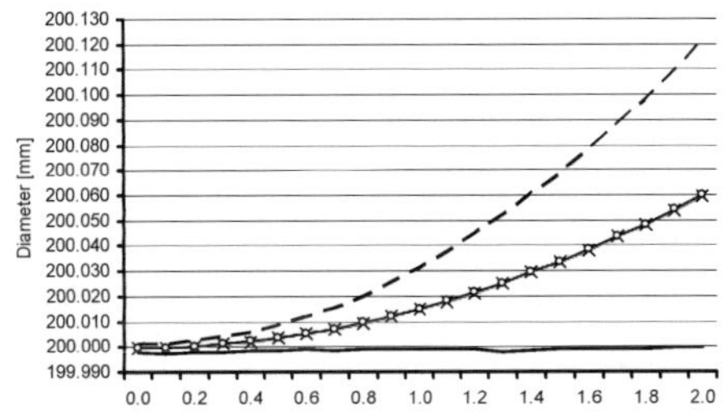

Fig. 1. Influence of the α angle on the measured diameter.

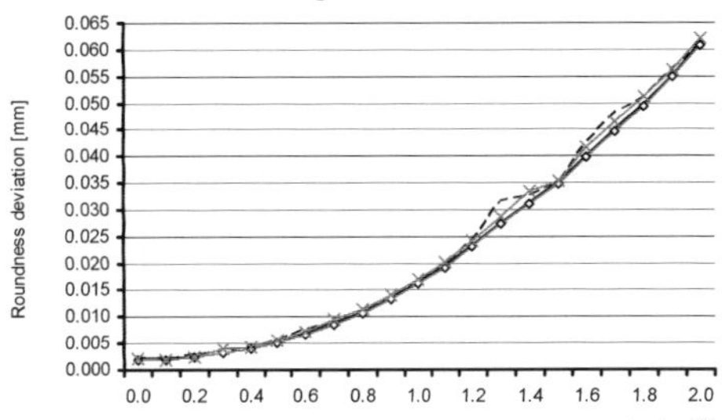

Fig. 2. Influence of the α angle on the results of roundness measurement.

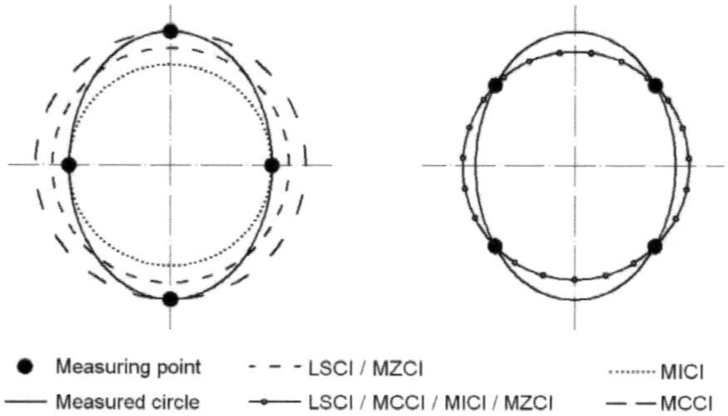

Fig. 3.The results of roundness measurement influented by the distribution of the measuring points and different fitting methods.

2.3 Distribution of the probing points

Number and the distribution of the measuring points could affect the results of the measurement with CMM. In the Fig. 3, there are two examples of the measurement of the detail with ovality type of out-of-roundness. Number of the probing points is 4, which is the minimal number for the circle measurement with CMM. The results are different for different positions of probing points related to the extremal dimensions of oval. Additionally, the best fitting algorithms introduce its own variations.

When the number of probing points is increased, the dispersion of the results is smaller, and after some number the results become stable (Fig. 4). However, it should be noted that the unnecessary increase of the points number will increase the time of measurement with no effect on accuracy, because the dispersion of the calculated results is smaller than the measurement uncertainty.

Figure 4 presents the influence of the number of probing points and the fitting element on the results of the diameter measurement. In the Fig. 4a, there are presented values of the diameter gained from the measurement with different numbers of probing points, while Fig 4b presents the dispersion of those results. It is seen that the large numbers of probing points do not improve the accuracy any more. The same could be observed for the obtained coordinates of the circle center, as well as for the measured out-of-roundness.

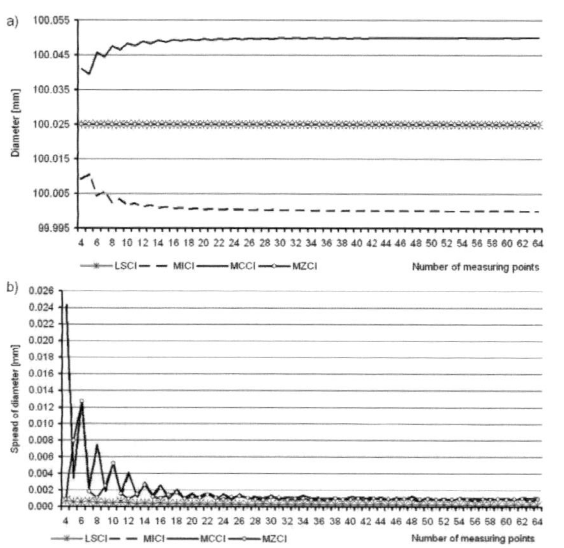

Fig. 4. Example of the symulation results of the ring with diameter Ø100 mm and form deviation 0.025 mm: a) influence of the probing points number on the achieved diameter; b) range of the values for various angle position of the ring related to the coordinate system.

When the operator is to choose the number of probing points, he should take into consideration the expected type of out-of-roundness. It is important to use the number of points different from the number of extremes of measured circle for dominating harmonic. E.g., for 4-lobing it should not be 4, 8 or 12 probing points, according to the following formulas:

$$LP \neq \pm (x \cdot n + 2) \text{ dla } n \geq 5 \quad (1)$$

$$LP \neq \pm (x \cdot n) \text{ dla } n < 5 \quad (2)$$

where:
LP –number of probing points,
n – number of harmonic,
x – coefficient 1, 2, 3, ...

3. CONCLUSION

Coordinate Measuring Machines with appropriate measurement software and measurement strategy are able to replace the specialized measuring devices in many industrial tasks. The strategy of measurement is highly dependent on the knowledge and experience of the operator. Good designed measurement algorithm enable to exploit the CMM in the most effective way with the highest possible accuracy.

Taking into consideration the characteristic factors affecting the results of roundness measurement, it is possible to apply the minimal number of the probing points for the particular task. This number could ensure the measurement with needed accuracy (the uncertainty no higher than 10% of measured detail tolerance *T*) in the shortest possible time. When the tolerances are too small, it seems necessary to use rather specialized devices for roundness measurement, at least for reference checkings.

In the measurement strategy, the choice of reference circle algorithm (fitting element) is also very important. It should be chosen according to the destination of the measured detail (Weckenmann & Knauer 1998). In assembling applications, the base for the holes should be approximated with MICI (maximal inscribed reference circle), while the shaft rather with MCCI (maximal circumscribed circle). However, when the detail is destined for the further machining, the hole should be approximated with MCCI and the shaft with MICI. In each case the evaluation of roundness deviation could be performed with MZCI (minimum zone circles).

4. REFERENCES

Gapinski B., Rucki M. (2008). The Roundness Deviation Measurement with CMM. *2008 IEEE Workshop on Advanced Methods for Uncertainty Estimation Measurement (AMUEM) Proceedings*. 21-22 July 2008 Sardagna, Trento, Italy pp. 108-111 (printed on CD)

Grzelka M., Gapinski B., Rucki M. (2006). The Roundness Deviation Measurement with Coordinate Measuring Machines; *ENGINEERING REVIEW,* Vol. 26 (2006); ISSN 1330-9587; Rijeka, Croatia; pp. 1-6

Humienny Z., ed., (2001). *Geometrical Product Specifications*, Warsaw University of Technology Press, Warsaw, Poland.

Standard ISO/IEC Guide 98:1995 *Guide to the expression of uncertainty in measurement (GUM)*

Standard ISO/TS 12181-1:2003: *Geometrical Product Specifications (GPS) – Roundness – Part 1: Vocabulary and parameters of roundness*; International Organization for Standardization, 2003.

Weckenmann A., Knauer M. (1998). Comparability of coordinate measurements. *Zeszyty Naukowe. Akademia Techniczno – Humanistyczna w Bielsku – Białej. Budowa i Eksploatacja Maszyn. 44/1998* Zeszyt 28 pp. 245-255

Annals of DAAAM for 2008 & Proceedings of the 19th International DAAAM Symposium, ISSN 1726-9679
Editor B. Katalinic, Published by DAAAM International, Vienna, Austria 2008
Make Harmony Between Technology and Nature, and Your Mind will Fly Free as a Bird

CREEP FORCE ON THE WHEEL-RAIL CONTACT AFTER THE START OF ELECTROTRACTION VEHICLE

GAVRILOVIC, B[ranislav] & STANKOVIC, A[na]

Abstract: *Measured creep force curves often decrease at high creepages. This phenomenon cannot be explained with common theories of rolling contact that are based on laws with a constant coefficient of friction. Due to sliding friction, one gets high contact temperatures. If the coefficient of friction is assumed to be decreasing at high temperatures, the calculated creep force curves agree well with measurements. In this paper, computer algorithm for determining temperature, coefficient of friction and creep force for wheel-rail contact evinced at the heating of the electrotraction vehicle of „Serbian Railway". The proposed computer algorithm will be able to be important for the calculating of the creep force after the start of traction vehicle because the high values of traction current and teperature at contact area.*

Key words: *traction current, creep force.*

1. INTRODUCTION

The calculation of wheel-rail contact forces has always been important because of analysis of dinamic behaviour of rail vehicles, their running qualities and forces and stresses acting on the track. In a computer simulation, the computation of wheel- rail forces is repeated many times. Therefore a short calculation time is very important. The exact theory by Kalker (computer programme CONTACT (Kalker, 1990) is not possible to use in the simulations because of its very long calculation time. The simplified theory used in Kalker's programme FASTSIM (Kalker, 1990) is much faster than the exact theory, but the calculation time is also often too long for use in complicated multibody systems.

The analytical solution for the simplified model of line contact was found by Carter already in 1926. Based on his work, Vermeulen and Johnson approximated the solution for point contacts with an elliptical area of contact. Shen, Hedrick and Elkins used the linear theory of Kalker to improve the accuracy of the solution of Vermeulen and Johnson. Their creep force law is fast and efficient while the results agree well with CONTACT and FASTIM computations (Kalker, 1990). Measured creep force curves often decrease at high creepages and the initial slope is much lower than in calculations (cf. Fig. 1a). For creep force calculations, the coefficient of friction is usually assumed to be constant and independent on operating conditions. With this assumption, the decrease of creep force curves cannot be explained. It is well-known that one gets high contact temperatures due to sliding friction. In this work, a temperature-dependent coefficient of friction at the electro-heating will be introduced into usual creep force models. With this extension, the calculated creep force curves agree well with measurements at high creepages.

In this paper we were insisted on the impact of traction current at temperature and creep force on wheel-rail contact after the start of trains because high amounts of traction current. However, the impact of traction current at temperature and creep force on wheel-rail contact is resident. Because we suggest the model and computer programme which can determinate impact of traction current at temperature and creep force on wheel-rail contact in each running regimes of the electrotraction vehicles.

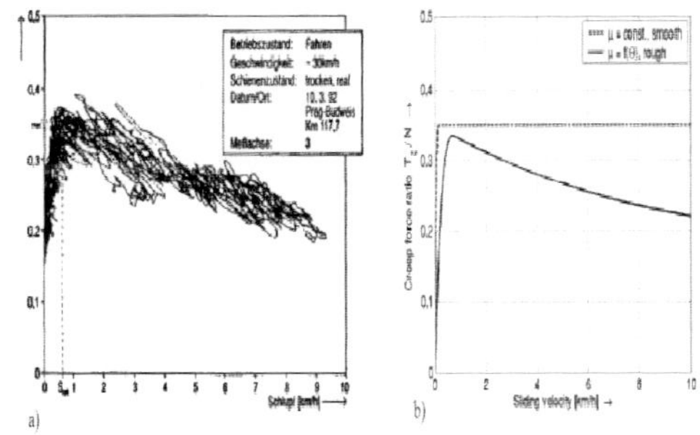

Fig. 1. Measured creep force curve (a), calculation (b)

2. ELECTRIC HEATING

Each electric heating originates in the Joule's law. By this law, the generated thermal power in whichever segment of the monoblock wheel is equiavalent to the committed electric power (Vorsic & Bratina, 2000).

Effectual value of current through observed segment of the monoblock wheel (I) is:

$$I = \int_{U_{Hertz}} J \cdot dS \qquad (1)$$

Where: $\boldsymbol{U_{Hertz}}$ – Hertz's contact area (Atmadzhova, 2004); J- current density through the observed segment of the wheel.

If effectual value of current through observed segment is constant (Vorsic & Bratina, 2000):

$$\theta = \theta_0 + \frac{\left(\frac{I}{2 \cdot k_{ski} \cdot b \cdot \int_{-b}^{b} \sqrt{1 - \frac{x^2}{a^2}} \cdot dx} \right)^2}{c \cdot \rho_m \cdot} \cdot \eta_{th} \cdot \rho \cdot t \qquad (2)$$

Where: θ -utter temperature heating of the monoblock wheel, θ_0- inchoate temperature heating of the monoblock, k_{ski}- updated factor because decrement Hertz's contact area by skin effect, a and b - the longitudinal and transversely semiaxis of Hertz's contact area, η_{th}- coefficient of thermal operation, t - heating period, c - specific warmth of the monoblock wheel, ρ_m - specific density of the monoblock wheel.

The average contact temperature from equvation (2) determinates the impact of traction current at temperature on the abutment of the wheel and the railway track in each running regimes of electrotraction vehicles. The highest values of traction current and average contact temperature has got after the start of the electrotraction vehicles and the trains (Vorsic & Bratina, 2000).

In made analysis we assume that the abutment of the wheel and the railway track is at ambient temperature before entering the area of contact. But after heating period (***t***) at constant operating conditions, the wheel temperature increases due to the continuous frictional heating.

3. TEMPERATURE - DEPENDENT COEFFICIENT OF FRICTION

The coefficient of friction is the most important parameter for the calculation of creep forces. We expect that the contact temperature should have a significant influence on the coefficient of friction According to Bowden and Tabor, friction is a process of continuous destruction and regeneration of microscopic contacts on the roughness peaks (asperities). This process depends mainly on the strength of the material. We assume that the temperature-dependence of the coefficient of friction is similar to that of the yield strength of steel (cf. Fig. 2b). This assumption has been used first by Rick for the investigation of wheel/rail contact (Rick, 1998).

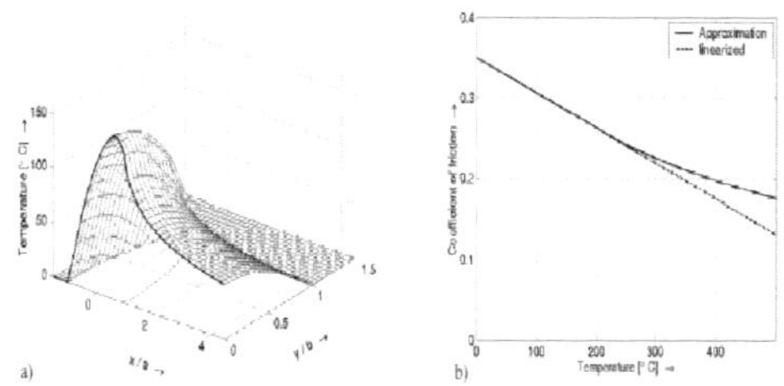

Fig. 2. Surface temperature for elliptical area of contact (a), temperature-dependent coefficient of friction (b).

4. CREEP FORCE LAW FOR FOR WHEEL - RAIL CONTACT

Alredy we accentuated ourselves that the average contact temperature from equvation (2) determinates the impact of traction current at temperature on the abutment of the wheel and the railway track in each running regimes of electrotraction vehicles. Besides, the highest values of average contact temperature has got after the start of the electrotraction vehicles because the highest values of the traction current. Based on research in »Serbien Railway«, the values of average contact temperature amount the few hundred ^{0}C after the start of the electrotraction vehicles.

Based on the algorithm ADAMS/Rail (N.N.,1998), we calculated local temperature and local coefficient of friction for values of traction curren of about 50 A and lesser time t=10 ms. This solution is shown at Fig. 3.

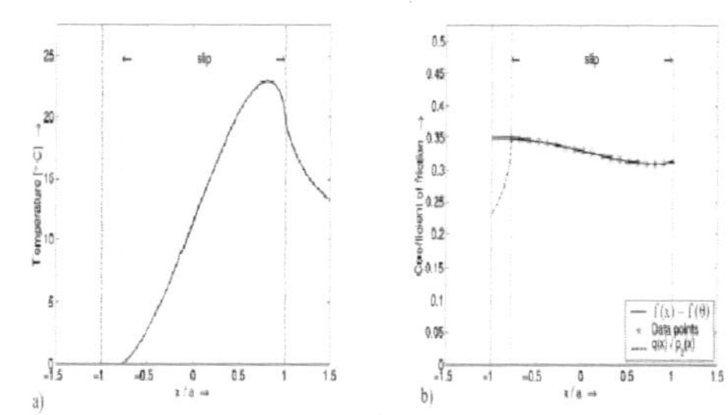

Fig. 3. Local temperature (a) and local coefficient of friction (b) (Traction current is about 50 A).

This solution is compared with adequancy solution of FASTIM (Kalker, 1990). and shown at Fig. 4. Based on the fig. V we shall be able to conclude that the results of presented model for tangential forces caused by longitudinal and lateral creepages for constant temperature-dependent coefficient of friction on wheel/rail contact f are nearly equal results of Kalker's programme FASTSIM (Kalker, 1990).

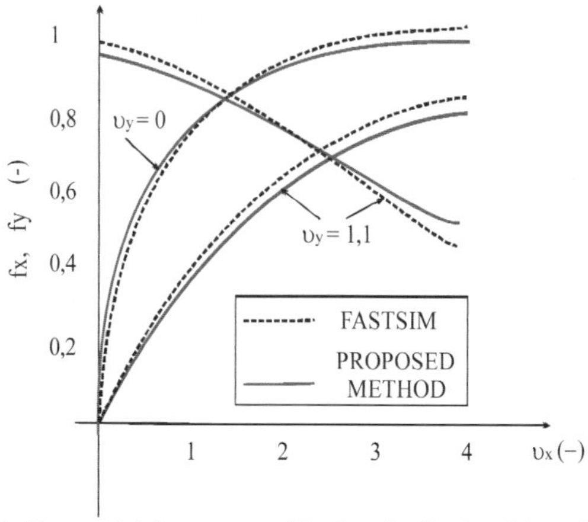

Fig. 4. Tangential forces caused by longitudinal and lateral creepages for f= 0,2 and $v = 5m/s$

5. CONCLUSION

The article is shown modeling influence of traction current on coefficient of friction and creep force. According to Joul's law we were described influence of the traction current on the temperature of contact area and according to Rick we were described influence of the temperature on the coefficient of of friction. Based on this approach and by the algorithm ADAMS/Rail we may calculate very quickly and precisely the influence of traction current on the creep force in each running regimes of the electrotraction vehicles.

The proposed model will be able to be important primarily for the calculating of the creep force after the start of traction vehicle because the high values of traction current and teperature at contact area.

Future research in this area should concentrate on the investigation of the collective influence of traction current and creepages on the coefficient of friction and the creep force under realistic railway operating conditions, including high contact temperature.

6. REFERENCES

Atmadzhova D. (2004): Theoretical and experimental method of determining angle of attack between wheel and rail with rolling stock running in curves, *Proceedings of the XI Railcon '04 - Scientific-Expert Conference on Railway*s, Nis, pp. 9- 14, ISBN 86-80587-60-5, 2006, Nis, Serbia.

Kalker, J.J. (1990). *Three-Dimensional Elastic Bodies in Rolling Contact.* Kluwer Academic Publishers Dordrecht, Netherlands.

N.N. (1998). *Introducting ADAMS/Rail 9.0*, ADAMS/Rail News & Views, No 1, MDI Marburg, Germany.

Rick, F. (1998). *Zur Erfassung der Geschwindigkeitsabhängigkeit des Kraftschlußbeiwertes eines hochbelasteten Rad-Schiene-Kontaktes.* Ph.D. Thesis, TU Clausthal,

Vorsic, J.; Bratina, J. (2000): *»Thermoelectric«,* Faculty for Electrical Engineering and Computer Science, ISBN 86-435-0414-9, Maribor, Slovenia.

Annals of DAAAM for 2008 & Proceedings of the 19th International DAAAM Symposium, ISSN 1726-9679
Editor B. Katalinic, Published by DAAAM International, Vienna, Austria 2008
Make Harmony Between Technology and Nature, and Your Mind will Fly Free as a Bird

CHANGES WHICH APPEAR IN THE IDEAL GAS LAW UNDER THE TEMPERATURES AND PRESSURES USED IN HOT ISOSTATIC PROCESSING

GEAMAN, V[irgil]; MILOSAN, I[oan]& STOICANESCU, M[aria]

Abstract: *Hot isostatic compression (HIP) involves the simultaneous application of pressure and elevated temperature to materials. Under these conditions of heat and pressure, internal pores or defects within a solid body or a powder compact collapse and weld up. Therefore HIP is today used for a lot of applications, like: upgrading castings (removing shrinkage pores in interdendritic space), densifying pre-sintered components, consolidation of powders and interfacial bonding.*
In this paper is given a worked example used in the field of HIPing applications with changing's related to the ideal gas law specific for these domain.
It is also given a mathematical application for concrete experimental conditions in the related domain, very useful for students and engineers who are working with HIPping technologies.
Key words: *isostatic, compression, design, properties, density*

1. INTRODUCTION

An ideal gas is defined as one in which all collisions between atoms or molecules are perfectly elastic and in which there are no intermolecular attractive forces. One can visualize it as a collection of perfectly hard spheres which collide but which otherwise do not interact with each other. In such a gas, all the internal energy is in the form of kinetic energy and any change in internal energy is accompanied by a change in temperature.

An ideal gas can be characterized by three state variables: absolute pressure (P), volume (V), and absolute temperature (T). The relationship between them may be deduced from kinetic theory and is called the:

$$PV = nRT = NkT; \quad (1)$$

where:

n = number of moles;
R = universal gas constant = 8.3145 J/mol K;
N = number of molecules;
k = Boltzmann const. = 1.38066×10^{-23} J/K;
$k = R/N_A$;
N_A = Avogadro's number = 6.0221×10^{23} /mol ;

The ideal gas law can be viewed as arising from the kinetic pressure of gas molecules colliding with the walls of a container in accordance with Newton's laws. But there is also a statistical element in the determination of the average kinetic energy of those molecules. The temperature is taken to be proportional to this average kinetic energy; this invokes the idea of kinetic temperature.

One mole of an ideal gas at STP (STANDARD TEMPERATURE AND PRESSURE - i.e. 298 K and 0.1 MPa) is used widely as a standard reference point for expression of the properties and processes of ideal gases.

The standard temperature is the freezing point of water and the standard pressure is one standard atmosphere.

These can be quantified as follows:
- standard temperature: 0°C = 273.15 K;
- standard pressure: 1 atmosphere = 760 mmHg = 101.3 kPa;
- standard volume of 1 mole of an ideal gas at STP: 22.4 liters.

2. EXPERIMETAL CONDITIONS

The principle feature of HIP is described below: a component is put into a vessel which can be pressurized by a gas (usually argon, but in special cases also helium or nitrogen).

The whole vessel is heated up to approximately 0.7 - 0.8 of the absolute melting point of the material. The very hot gas is then under very high pressure such that it acts like a hot forge isostatically.

Typical gas velocities are around 1000 m/s and one can calculate from the kinetic theory of gases, assuming ideal gas laws to hold, that 10^{30} collisions per square meter and second occur on the component walls.

The density of the gas used usually increases considerably (such that even ideal gas laws may not further be applicable). For instance argon, which has a density at standard conditions of 1.8 kg/m^3 will have a density of 370 kg/m^3 under HIPing conditions at 1500 K and 100 MPa.

For the ideal gas laws to apply, the following requirements must be met:

- the attraction between gas atoms or molecules must be negligible;
- the volume of the atoms or molecules must be negligible in comparison with the volume occupied by the gas;
- the atoms or molecules must behave like perfectly elastic spheres;
- the duration of their collisions must be negligible in comparison with the time between collisions.

The intention here is not to prove that the ideal gas law break down during HIPing but rather that the conditions developed are such that to the above conditions are unlikely to apply. Calculate the mean atom spacing in argon gas under STP conditions and compare this with the mean spacing under typical HIPing conditions of 100 MPa and 1500 K. The density of argon under these conditions given under STP conditions is 1.8 kgm^{-3} (James, 1992; Loh & Sia, 1992).

3. MATHEMATICAL APPLICATION

Usually, in practical applications, we need a lot of experimental data for using them in the easiest way for the mathematical modeling of the processes.

So, under **STP** conditions (i.e. 298 K and 0.1 MPa) the atom spacing d_1 is given by:

$$d_1 = \left(\frac{\text{volume occupied by 1 mol at STP}}{\text{number of atoms in 1 mol}}\right)^{\frac{1}{3}} \quad (2)$$

$$d_1 = (\frac{22,4*10^{-3}}{6*10^{23}})^{\frac{1}{3}}$$

$$d_1 = 3,4*10^{-9}m \qquad (3)$$

$$d_1 = 34\mathring{A}$$

This compares with an atom spacing of about $2\mathring{A}$ in solids. Assuming that the ideal gas laws apply (an assumption which is not in fact justified), then the volume V_2 of 1 mol of gas under HIPing conditions can be obtained from:

$$\frac{p_1V_1}{T_1} = \frac{p_2V_2}{T_2} \qquad (4)$$

where: p_1 = 0.1 MPa (1 atm)
$V_1 = 22.4 \times 10^{-3}\ m^3$
T_1 = 298 K
p_2 = 100 MPa (1000 atm)
T_2= 1500 K.

$$V_2 = \frac{p_1V_1T_2}{p_2T_1}$$

$$V_2 = \frac{1*22,4*10^{-3}*1500}{1000*298} \qquad (5)$$

$$V_2 = 1,1*10^{-4}m^3$$

The atom spacing d_2 under HIPing conditions is then given by:

$$d_2 = (\frac{\text{volume occupied by 1 mole under HIPping conditions}}{\text{number of atoms in 1 mol}})^{\frac{1}{3}}$$

$$d_2 = (\frac{1,1*10^{-4}}{6*10^{23}})^{\frac{1}{3}}$$

$$d_2 = 6\mathring{A} \qquad (6)$$

Despite the fact that this calculation requires an assumption that the ideal gas laws do in fact apply, when the result shows that strictly they cannot, it does illustrate that the atomic spacing in the HIPing gas is much closer to that of a solid or liquid than to that of a gas at STP (Atkinson & Rickinson, 1991; Vogel & Ratke, 1991).

Given the density at STP of 1.8 kgm^{-3} for argon, and the fact that, from the above calculation, the volume is reduced by a factor of:

$(1.1 * 10^{-4})/(22.4 * 10^{-3})$.

Between STP and HIPing conditions the density ρ_2 under HIPing conditions will be:

$$\rho_2 = \frac{\text{mass}}{\text{volume}}$$

$$\rho_2 = \frac{1,8}{1,1*10^{-4}/22,4*10^{-3}} \qquad (7)$$

$$\rho_2 = 370kgm^{-3}$$

This should be compared with the density of water which is 1000 kg m^{-3}. Since increasing the pressure increases the density while increasing the temperature reduces it, there is in fact a temperature at which the argon density is a maximum. This occurs at around 733 K for a pressure of 100 MPa and the density then is approximately equivalent to that of water (Geamăn, 1995, 1996, Matthews, 1980; Wilkinson & Ashby, 1975).

Where the ideal gas law breaks down at high pressure, the Van der Waals equation is used:

$$(p + \frac{a}{V^2})(V - b) = RT; \qquad (8)$$

where:
a and b are constants and
$R = 8.314\ Jmol^{-1}\ K^{-1}$.

4. CONCLUDING REMARKS

There are a lot of phenomena that occur to make a good practice for industrial application in the field of HIPing consolidation for metallic or ceramic materials. But these phenomena are very difficult to be controlled. In the last years a lot of scientists are trying to give more details in this field to help the designers to have more complete data bases for practice use.

Assuming that the ideal gas laws apply, we can calculate:

- the mean velocity of argon gas atoms at 1500 K (typical of HIPping);
- the frequency with which the atoms collide with unit area of the surface of a component under a pressure of 100 MPa and at a temperature of 1500 K.

The mean velocity v can be obtained from the equation relating to the kinetic energy of gas atoms to the thermal energy:

$$\frac{1}{2}mv^2 = \frac{3}{4}kT \qquad (9)$$

where: m- is the mass of each gas atom;
$k = 1.38066 \times 10^{-23}$ J/K (the Boltzmann constant);
T - is the absolute temperature (in this case 1500 K).

The collision frequency f may be obtained from the equation 10 (Atkinson & Rickinson, 1991).

$$f = nv/6 \qquad (10)$$

where: n - is the number of atoms per unit volume and
v - their mean velocity.

This paper can be also a good example for practical application and a guide for engineers who are working in the field of HIPping.

For the future, the intention of the authors is to prepare a mathematical modeling of the processes and to find the easiest way for applying theoretical analysis in design practice.

5. REFERENCES

Atkinson, H.V. & Rickinson, B.A. (1991) - *Hot isostatic processing*, ISBN 0-7503-0073-6, Great Britain, 1991

Froes F.H. (1991) - *Fourth International Conference on Isostatic Pressing*, Journal of Materials, U.S.A. **11**/1991, 52

Geamăn, V. (1995) - *Theoretical and experimental researches regarding to isostatic processing of poured blanks made from non-ferrous alloys* - Ph.D. Thesis - "Transilvania" University of Braşov, 1995

Geamăn, V. (1996) - *Proc. Int. Conf. Hot Isostatic Pressing*, Andover –Massachusetts, U.S.A. 1996, 221-223

James P.J. (1992) - *Isostatic Pressing Technology*, Metals and Materials, **10**/1992, 541 - 546

Loh N.L. & Sia K.Y. (1992) - *An overview of hot isostatic pressing*, Journal of Materials Processing Technology **30**/1992, 45-65

Matthews J.R. (1980) - *Indentation hardness and hot pressing*, Acta metallurgica, **28**/1980, 311 – 318.

Vogel H.J. & Ratke L. - *Instability of grain boundary grooves due to equilibrium grain boundary diffusion*, Acta metallurgica, Vol. 39, 4/1991, 641 - 649

Wilkinson D.S. & Ashby M.F. - *Pressure sintering by Power Law Creep*, Acta metallurgica, Vol.ll, 11/1975 1277-1285.

Annals of DAAAM for 2008 & Proceedings of the 19th International DAAAM Symposium, ISSN 1726-9679
Editor B. Katalinic, Published by DAAAM International, Vienna, Austria 2008
Make Harmony Between Technology and Nature, and Your Mind will Fly Free as a Bird

THE CORRELATION STUDY BETWEEN THE EDUCATIONAL MARKET AND THE ENGINEERING WORK MARKET

GEORGESCU, L[uminita] & DRAGOMIR, M[adalina]

Abstract: *After ten years of decline of the labor force in the engineering field, period of time in which the temptations of this profession has decreased, we can observe that in the present the employers make special efforts to find engineers and even managers. The study analyses some of the factors which have determined the diminution of the interest young people show towards the engineering faculties. We have identified important solutions that can lead to a growing attraction for the technical profiles from which we can mention: the diversification of the educational offer, the permanent updating of the didactic activity and the development of recruiting programs.*
Key words: *Education, Engineering, dynamics, work market*

1. INTRODUCTION

Investing in individuals represents an advantage for them as well as for the society (Suciu, M.C. 2000). This type of investment should take into consideration the reality of the educational market (the specific condition of the work market; the correlation between the educational market and the work market; the demand for a certain socio-professional field).

Young people easily identify the most advantageous professions, as far as the earnings, the work opportunities and the socio- professional statuses are concerned (Georgescu, L. 2007).

2. SURVEY OF THE ENGINEERING EDUCATIONAL MARKET

This information analysis was done by The Romanian Statistics Institute (http://www.insse.ro - 2006), regarding the number of students who want to enter the faculty, on specialization groups, from 1990 until 2005.

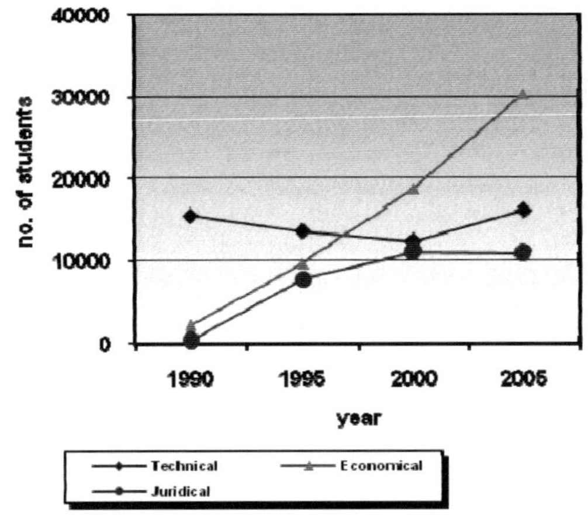

Fig. 1.The dynamics of students opting for university studies, 1990- 2005

From 1990 until 2000 we can see a decrease in the number of students choosing to continue the technical studies, in favor of some other specializations (economical, juridical). After this, a remedy period followed.

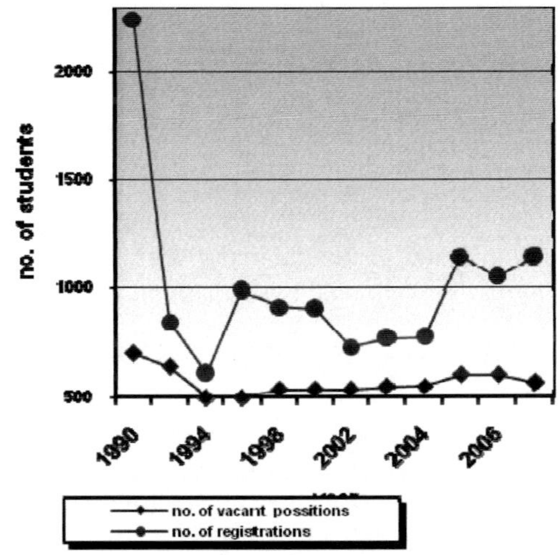

Fig. 2.The dynamics of students opting for the IMST (The Engineering and Management of Technological Systems) faculty, 1990- 2007

With reference to the data received from the IMST faculty's secretariat, regarding the number of candidates and the number of vacant positions registered from 1990 until 2007, we have presented the dynamics of registrations to this faculty.

We can notice that, from 1991 until 2003, youngsters had shown a lowering attraction towards the IMST faculty's specializations. In 2003, the continuous decrease of the candidates' number has lead to the decision of changing the admission configuration from exam to file contest. Candidates seem to prefer this type of admission, which is now applied by particular universities too, as the intellectual effort and the stress caused by an exam, very close in time to the bachelor exam, are diminishing.

Since 2005, the attraction has continuously been growing and growing and we can notice that its level is now, equal to that of the engineering work market.

3. SURVEY OF THE ENGINEERING WORK MARKET

Analyzing the engineering field and the enthusiasm shown by the students regarding the technical faculties, we are assisting alternating periods. Once in five years we can witness periods of a surplus of engineers and periods of deficit.

The decreased attraction towards engineering caused, for a short time, a lower number of students opting for technical faculties. In the last years, the educational request for such professions has increased. We are facing now an improvement of the perspectives offered by these faculties.

Period	Aspects regarding the engineering market	Engineers
1989	• the opportunity to get advantageous jobs • bigger salaries than other professional categories • a relatively high socio-professional status (until 1989, the majority of the managers had been engineers)	Big attraction towards the engineering jobs
1990-2000	• cutting salaries • reduced opportunities for engineers • a relatively increased number of graduates	Reduced attraction towards the engineering jobs
2001-2005	• better salaries and new employment opportunities	Low number of graduated students;
	• a relatively decreased number of graduates	An increasing attraction towards the profession

Tab. 1.The dynamics of the engineering work market, 1989-2005

4. DISCUSSION AND INTERPRETATION OF THE RESULTS

Until 1989, there was a great attraction towards the engineering professions, due to the big perspectives offered by the engineering market. After the 90's, we have witnessed a reduced attraction towards the technical specializations. So, there is a certain gap between the educational market and the work market (students who had started the university studies in 1989, finished the specialization in1994).

There has been a reducing demand for engineers on the work market, in the period of transition, thing which determined an increasing number of students wanting to become engineers. The surplus of offers represents a feature of the educational market's complexity, as well as the result of some transitional situations.

We have the key. Some students may choose to study a technical faculty and in the same time they may simultaneously decide on other much more "attractive" faculties. In other cases, having an engineering diploma, students can opt for another faculty.

They have the key. Those in charge with the educational process, noticing the decreasing work opportunities, should vary the educational offer, through creating new perspectives in training the students for other related, catching jobs and through creating partnerships with the successful existing companies. The board of the IMST faculty saw this perspective and set up new specializations in order to give options which could answer the needs for jobs like: economist engineer, quality engineer and robotics or mechanics-electronics engineer. Since the setting up, almost 60 % of the candidates have one of these specializations as their first option.

The didactic activity should always keep up with the development of science. This can be possible as long as the teacher is an active researcher, an example for all of his students. The IMST faculty has replaced some scientifically obsolete courses by new, up to date ones. These courses are much more attractive for the students and they offer a variety of modern presentation options. They also develop the students' skills and broaden their further options regarding a future career, for example: 3D Modeling and simulation; Computer aided design; Advanced parametric design; Techniques and methods for auditing control; Quality engineering for industrial robots.

We find the key they have. Many Romanian employers make efforts to find the best engineers. (http://www.cariereonline.ro). There are companies which sponsor the technical faculties with money or equipment. They also have recruiting programs for graduates or youngsters in the final years of study. In 2007, The Polytechnic University of Bucharest- IMST faculty in partnership with Dr. Krocher Group set up the Center of Studies, Development and Marketing of Products and Technologies. This centre is specialized on CNC machine tools and it has proved itself to be a great opportunity for students who need to practice their laboratory activities on high generation equipment.

Another example is represented by Renault, a company which recruits and offers scholarships to final year students and to master students.

5. CONCLUSIONS

The rapid changes of technology, the discovery of new materials, the economic situation and the much more flexible international work force emphasize the fact that we should adapt education to the different trends that may appear.

It is estimated that before 2010 (http://www.see-educoop.net), almost half of the new jobs in the UE will be destined to persons with high qualifications. So, the employees will have to improve their ability to adapt to the new requirements the market and the current economy may have. On the other hand, the employers should consider the possibility of creating a much more flexible schedule or that of organizing trainings in order to get the employees' loyalty (http://www.Manpower.ro). Furthermore the employers, as well as the authorities, should encourage the strategic migration, the development of specializations and they should, most of all, invest in formation.

It is compulsory for Romania to elaborate a strategy of educational development, based on permanent formation and in tight connection with the trends existing in the socio-economical development of the country. The strengths and skills one can get due to the professional formation, allow the individuals to be well informed and to properly face changes.

The survey of the work market and adapting the education to its needs may foresee or give solutions to a crisis situation.

The rapid evolution of the main fields of activity requires permanent information, formation and improvement, processes that can be accomplished only by having an extraordinary flexibility of the mind and of the attitudes. If there is no flexibility, it can be educated, only that it may take the time and the sacrifice of many other generations.

6. REFERENCES

Georgescu, L. (2007) *Cercetări privind optimizarea formării RU în sistemele industriale*, Referat III (*Research regarding the improvement of human resources in the context of industrial systems*)

Suciu, M.C (2000) *Investment in Education*, Editura Economică, pp105-107

***http://www.cariereonline.ro/index.php/articles/Unde_ne_sunt_inginerii (Where are our engineers), *Accesed on:2008-08-13*

***http://www.manpower.ro/Study, Regarding Talent Deficit, *Accesed on:2008-08-13*

***http://www.insse.ro - The National Statistics Institute, Statistics Year Book (2006), *Accesed on:2008-08-13*

***http www.see-educoop.net/education_in/pdf/commun-from-commision-oth-rmn-t02.htm, *Accesed on:2008-08-13*

Annals of DAAAM for 2008 & Proceedings of the 19th International DAAAM Symposium, ISSN 1726-9679
Editor B. Katalinic, Published by DAAAM International, Vienna, Austria 2008
Make Harmony Between Technology and Nature, and Your Mind will Fly Free as a Bird

MEASUREMENT OF THE GASES VOLUME DIFFUSED FROM SOLIDIFIED MATERIALS

GHENGHEA, D[an]; NEDELCU, D[umitru] & DODUN, O[ana]

Abstract: *The work presents the researches done to measure the volume of the gases diffused out side from solidified materials from welded joints using for this objective the eudiometer. Test pieces were made with two very used welding processes: shielded metal arc welding (SMAW in AWS codification and 111 in EN codification) and gas tungsten arc welding (GTAW in AWS codification and 136 in EN standards) in argon inert atmosphere. The research was part of activities of a grant type A offered in national contest by Romanian National Authority for Scientifically Research in Universities in 2006.*

Key words: *welding, joint, diffusion, gases, volume*

1. INTRODUCTION

Mechanical characteristics of welded beads are imposed to be equal or higher as those of base materials, both the two elements and the heat affected zone, forming welded joints. At no matter what welded joint, quality acceptance requirements criteria do not permit existence of cracks and the pores are accepted only in specific volumetric dispersion. These two specific welding defects formation are generally assumed to be caused by diffusion of gases that are introduced, during welding process, in melted metals bath or to gases being in base materials from other manufacturing processes.

An objective of all welding technologies is to avoid the contact between welded joint and the atmosphere. The work presents the researches done to measure the volume of the gases diffused out side from welded joints using for this objective the eudiometer. Test pieces were made with two very used welding processes: shielded metal arc welding and gas tungsten arc welding in argon inert atmosphere.

2. LITERATURE CRITICAL REVIEW

In liquid state welding processes the melted metals bath has direct contact with dissociated gases in electric arc (Miclosi et al., 1982). Atomic hydrogen, oxygen, nitrogen and carbon oxide, diffuse or make chemical combinations and participate to solidification process. The theory of welding process assumes that atomic hydrogen has strong influence for welding joints quality and this is why many researchers investigate this topic. Presence of hydrogen embittermen like a common problem for API 5 L steel welded by electric arc was marked (Plascensia,G. et al. 2008) and ultrasonic measurements estimate its penetration rate. Cracks in welded zones produced by residual stress and H^+ diffusion from sulphured hydrogen molecule (Rogante, M. et al. 2006) were investigated. Susceptibility to hydrogen-induced cracking of weld beads on high-strength structural steel with weld beads subjected to hydrogenation under the conditions of cathodic polarization was studded (Malina, J. et al. 2005). The studied specimens fail in the heat-affected zone on the boundary between the main metal and the weld bead in the zone of application of the latter and, especially, near its end. This was explained by a higher hardness of the metal in this zone caused by the specific features of the temperature field formed in the process of welding. The gas tungsten arc welding (GTAW) process is used extensively for welding various grades of stainless steel. The addition of hydrogen to argon shielding gas improves weld pool wet ability, bead shape control, surface cleanliness, and heat input without causing hydrogen-induced cracking (Louthan Jr., M.R. Cannell, G.R., 2005) and it was observed that the use of hydrogen-containing shielding gases for the closure welding of austenitic stainless steel containers for nuclear materials packages does not lead to hydrogen-induced cracking (HIC) of the weld or weld heat-affected zone (HAZ).

Authors screening in technical reports did not find researches about volume of gases diffusion from welding joints and initiate a research about this subject taking account about utilisation of metallic joint in water environments. Results could be used in further researches to identify gases composition and each gas quantity.

3. EXPERIMENTAL PREPARATION

In the experimental part the shielded metal arc welding (SMAW, 111) and gas tungsten arc welding with argon protective inert atmosphere (GTAW, 136) were applied for welding bead deposition using same specific energy. The base materials for welding experiments were made in carbon steel, type OL 44.2, STAS 500/2-80, included in the first welding material group with a very good weld ability. Chemical composition and mechanical features for this type of material are presented in Table 1.

Filler material in welded joint manufacturing was a wire type S10 STAS 1126-90 which chemical composition and mechanical features are presented in Table 2. In shielded metal arc welding process, outside of the wire, a shell of minerals and chemical products like titanium oxide are deposed. In experiments was used the type SUPERTIT with corresponding international codifications: STAS 1125/2 E 43.3.RR.2.2, DIN 1913 E 43.32.RR.6, AWS A 5.1 E 6013 and ISO 2560 E 43.3.RR.22 made at Fro Ductil Buzău, Romania. This shielded metal wire is used for welding of metallic structures, boilers, pipe vessels, railway cars and has Manganese in composition. It gives a stable electric arc with very good starts and restarts.

Chemical element and mechanical feature	Concentration and feature values
C [%]	Maxim 0.20
Mn [%]	1.05 – 1.55
Si [%]	Maxim 0.04
P [%]	Maxim 0.04
Al [%]	Minim 0.025
R_p0,2 [MPa]	255
R_m [MPa]	420 – 500
A5 [%]	Minim 22
KV [J]	28

Tab. 1. Chemical composition and mechanical features of OL 44.2, carbon steel.

The melting process has fine drops, no sputtering and the slag is fluid, with good covering of the welded bed, easy to detach after the process of manufacturing. Real efficiency deposition rate is around 96 %. Chemical composition and mechanical features of welding deposed material are presented in Table 3. Electrode can be used in all welding positions excepting down vertical with inverted polarity (cc$^+$, the plus of welding source at the shielded electrode) or in AC with a no charge minimum voltage of 50 V. SUPERTIT electrode is authorized to be used by following international authorities: Romanian Shipping Authority, Lloyd's Register of Shipping, Russian Shipping Register, American Bureau of Shipping, Veritias Bureau.

Argon utilized like protective atmosphere was in B class with purity 99.990 % and other gasses like nitrogen, hydrogen under 0.0075 %, oxygen under 0.0030 %.

The both welding processes were done with inverter's electrical source type TECHNOLOGY 200 made by TELWIN Company from Italy. It gives a CC current up to 150 A, for direct polarity (cc$^-$ minus source at electrode) or in inverted polarity (cc$^+$ plus source to the electrode).

Experimental pieces for research has following dimensions 100 x 100 x 12 mm, were polished to metallic surface to clean all oxides or superficial layers and using a 2.5 mm diameter wire electrode, welding bead were deposed with small deviation around 0.5 of diameter. From its have been cutting test pieces with dimensions 50 x 5 x 12 mm notated: 1 – test pieces from heat affected zone (zit 1 respectively 2), 2 – test piece made with shielded metal arc welding (SMAW), 3 – test piece made by gas tungsten arc welding process in argon inert protective atmosphere (GTAW). Polishing opposed welded bead edges the same weight for all test pieces was made.

Test pieces have been introduced in the eudiometer for 72 de hours, reading at 1, 10, 24 and 72 hours the volume of diffused gasses. At the end of analyzed period of time once again the test pieces have been weighted but no significant difference have been observed.

4. RESULTS AND DISCUTION

Results given by the experimental activity made in Theory of Welding Processes laboratory at Faculty of Machine Manufacturing, "Gh. Asachi" Technical University of Iasi, are presented in table 4.

Chemical element and mechanical feature	Concentration and feature values
C [%]	Maxim 0.10
Mn [%]	0.4 – 0.6
Si [%]	Maxim 0.3
Cr [%]	Maxim 0.2
Ni [%]	Maxim 0.3
P [%]	Maxim 0.03
S [%]	Maxim 0.03
R_m [MPa]	1180 (ϕ 2–2.5)or980 (ϕ 3.15–5)

Tab. 2. Chemical composition and mechanical features of filler welding wire S 10.

Chemical element or mechanical feature	Concentration or values
C [%]	0.06 – 0.1
Mn [%]	0.4 – 1.7
Si [%]	0.2 – 0.6
S [%]	Maxim 0.04
P [%]	Maxim 0.04
R_p0,2 [MPa]	430 – 490
R_m [MPa]	490 – 550
A5 [%]	Minim 24

Table 3. Chemical composition and mechanical features of welded deposed material.

Test	1 HAZ, ml/100grames		2 SMAW ml/100g	3 GTAW ml/100g
Welding	111	136	111	136
1 hour	0.2	0.1	0.1	0.1
10 hours	2.1	1.9	1.5	0.6
24 hours	4.0	4.1	2.7	2.1
72 hours	5.5	5.3	5.5	3.6

Tab. 4. Experimental results of diffused gases from welded bead and heat affected zones.

Welded joints diffused gasses could arrive from atmospheric humidity, filler materials asked by fabrication or from impurities existing on welding edges.

High temperatures in welding zone create very good conditions for dissociation process of diatomic gasses (H_2, O_2, N_2) respecting following reactions:

$$2\,H \leftrightarrow H_2 + 432.7 \text{ kJ/mol} \quad (1)$$

$$2\,O \leftrightarrow O_2 + 494.6 \text{ kJ/mol} \quad (2)$$

$$2\,N \leftrightarrow N_2 + 712.3 \text{ kJ/mol} \quad (3)$$

In atomic state these gasses has a strong chemical activity and react with melted metal drops from electric arc column and with metallic melted bath formed by base and filler materials. In steels nitrogen combined with alloying elements conduct to nitrides that do not remain in the gas state; oxygen gives oxides with all alloying elements but with carbon gives the monoxide that could be diffused from welding joints; hydrogen do not have chemical combinations with Fe, Ni, Cr, Mn, Si, from metallic materials used in welding joints but could have combinations with sulphur or carbon resulting gasses witch will diffuse from welded joints. These arguments offer the possibility to appreciate that diffused gasses from studied welded joints could be carbon monoxide CO, methane CH_4, sulphurated hydrogen H_2S and H_2.

Experimental data analyse from table 4 shows that shielded metal arc welding process produced a volume of diffused gasses with a value near the industrial electrode producers feature related to diffused hydrogen. For the gas tungsten arc welding in protective inert argon atmosphere, the volume of diffused gasses is smaller and this could be caused by low quantities of hydrogen from inert gasses compared with the hydrogen from shielded electrode used in surrounding atmosphere. The values for diffused gasses from test pieces originate from heat- affected zone, presented are closed to those manufactured with shielded metal arc welding. This shows the presence of gasses in base material and could be one explanation for the cold cracking process in heat-affected zone.

5. REFERENCES

Louthan Jr., M.R., Cannell G.R, (2005), Impact of H_2 in shielding gas for welding austenitic stainless steels, *Welding Journal*, v 84, n 4, p 38-40, ISSN: 0043 - 2296.

Malina, J., Samardzic, I., Gliha, V., (2005) *Materials Science*, v 41, n 2, ISSN:1068-820X.

Miclosi, V. et al. (1982), *Fundamentals of welding processes*, E.D.P, Bucharest, Romania.

Plascensia, G. et al (2008), Estimation rate of hydrogen penetration in a weld API 5L steel pipe, *Defect and Diffusion Forum*, 273-276, pp. 500-505.

Rogante, M. Batistella, P. Cesari, F. (2006). *International Journal of Hydrogen Energy*, v 31, n 5, pp.597-601, ISSN 0360-3199.

Annals of DAAAM for 2008 & Proceedings of the 19th International DAAAM Symposium, ISSN 1726-9679
Editor B. Katalinic, Published by DAAAM International, Vienna, Austria 2008
Make Harmony Between Technology and Nature, and Your Mind will Fly Free as a Bird

MECHATRONICS AND MICRO-MECHATRONICS CONCEPTS AND TECHNIQUES FOR INTELLIGENT INTEGRATING MICRO ENGINEERING

GHEORGHE, I[on] G[heorghe]

Abstract: *The paper presents new mechatronics concepts and solutions for intelligent equipments and advanced robotics with applications of measurement and control in industrial processes from HIGH-TECH environments, for assuring quality levels of the industrial products. Are realised mechatronics systems / micro-systems as parts of obtained assemblies by experimental and behavioural simulation that are based on signals measured and parameters of mathematical and physical models.*

Key words: *micro-mechatronic concepts, intelligent systems, advanced robotics, control and measurement process.*

1. INTRODUCTION

The general concept of intelligent equipments and advanced robotics, brings into the structure, systems and sub-systems MEMS and MECHATRONIC/ MICROMECHATRONIC identified by experimental and behavioral structure simulation, functioning and maintenance and by results of measured signals and mathematical and physical models, but also through destination of applications which will take into account the process or product type from geometrical and precision characteristics, of nature and properties of the materials used, of technical and technological parameters and of process accessibility methods (Dumitriu, 2006).

The general and inovative concept of intelligent equipments and advanced robotics, is realized considering elements structure, components and functional compatibility, that enter in automatization and informatization chart of the ensemble, and depending on micro-movements obtaining, micro-positionings and micro-measurements in coordination, command, control and decision in expert matrix (Gheorghe, 2006).

The concept of intelligent equipments and advanced robotics, integrates the architecture of sensors and micro-sensors in value chain of signal generation and information transfer, actuators and micro-actuators architecture in mobility chain of ensemble (Suzumori, 1991), architecture of electronic and micro-electronic units in signals processing chain, controllers architecture in command chain and coordination process automated and informatized, architecture of process computers in driving process and architecture of software packages in movement chain, measurement, integrated and decisional control, commands, coordination, driving, visualization results, static status monitoring and dynamic status (Brugger, 1993).

2. INTEGRATION OF INTELLIGENT EQUIPMENTS AND ADVANCED ROBOTICS

The intelligent equipments and advanced robotics, have been realized and implemented by author and also in some high-tech level companies.

An example of realized advanced robotics, is "**the concept of mecatronic micro-robot of measurement and dimensional integrated control with five degree liberty with micro-detector with Laser fascicle**", used in metrological and industrial measurement processes, from integrating industry of precision mechanics and mechatronics and showed in the figure 1:

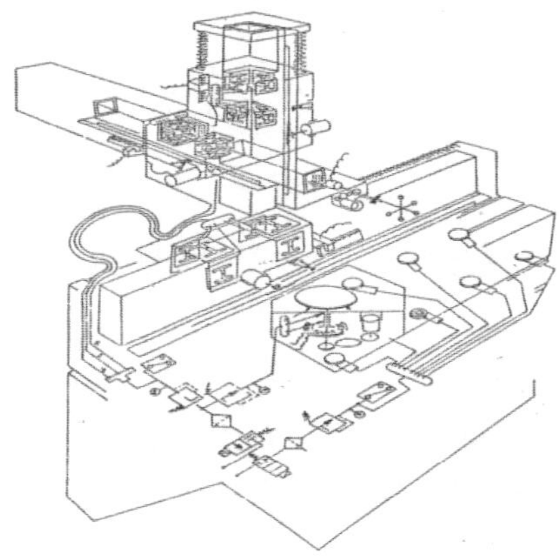

Fig.1 Equipment mechatronic micro-robot of measurement and dimensional control

The constructive mechatronic solutions of the equipment and means of sensors / transducers architecture integration are presented in the figure 2 (Fatikow, 1997):

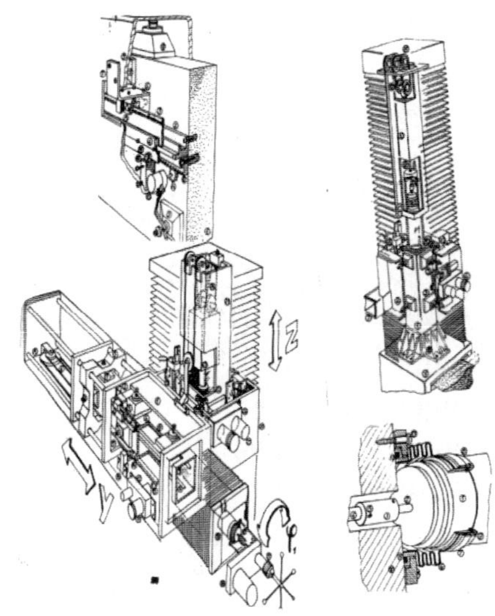

Fig.2. Constructive solutions and architecture integration for sensors / transducers and actuators – Mecatronic micro-robot equipment

Functional chart of mecatronic micro-robot of measurement and dimensional integrated control is showed in the figure 3:

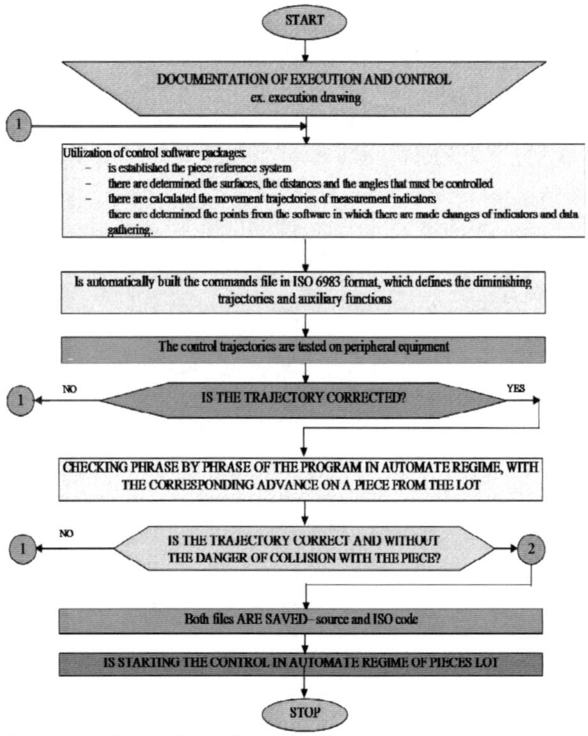

Fig.3. Functional chart of mecatronic micro-robot equipment

3. EXPERIMENTAL DATA AND RESULT REGARDING INTELLIGENT EQUIPMENTS AND ADVANCED ROBOTICS

3.1 Applications and comparison areas of actuators / micro-actuators used at intelligent equipments and advanced robotics

From the laboratory experimentations performed can be synthesize diagrams that realize comparisons between different actuators as: electrical micro-engines and step by step micro-engines, microelectromagnets, piezoelectric micro-actuators, pneumatic and hydraulic micro-cylinders etc.

So, in the figure 4, there is synthesized the maximum speed of the micro-actuators and comparison speed-force:

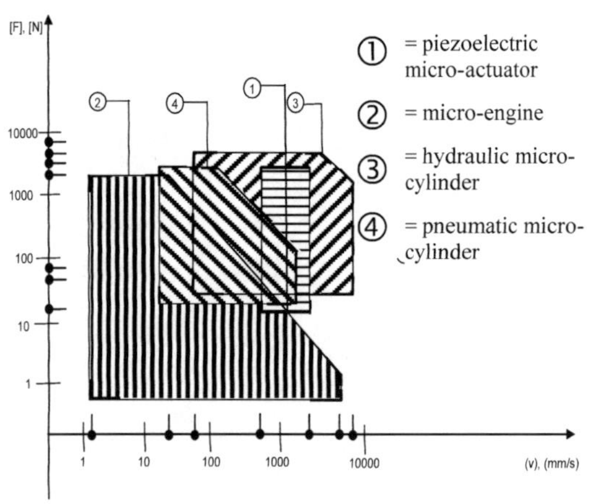

Fig. 4. Comparison speed-force

In the figure 5 is synthesized positioning time for controlled response operating.

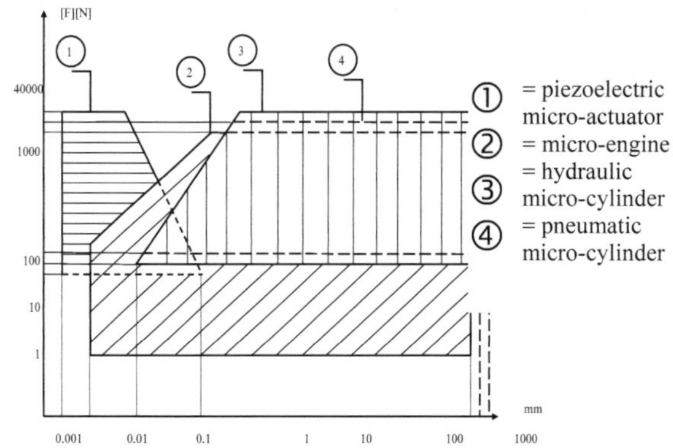

Fig. 5. Positioning time for controlled response operating

3.2 Measurement of deviation from the roundness and / or from rugosity

In the figure 6, there are presented the results of realised measurement in laboratory, of roundness deviation and rugosity, using an intelligent equipment with Laser detector:

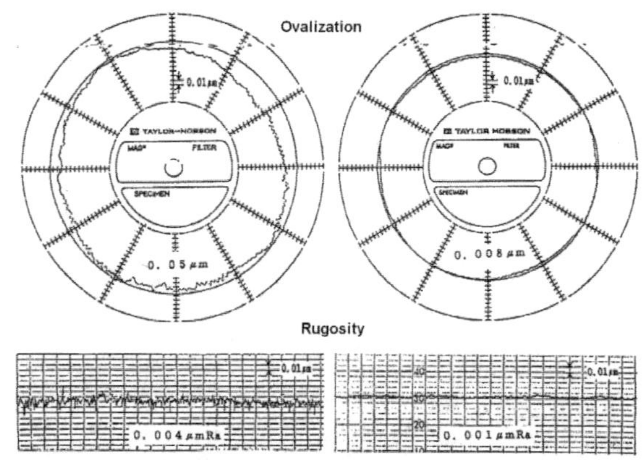

Fig.6. Measurement of ovalisation and rugosity

4. CONCLUSION

In perspective, intelligent equipments technique and advanced robotics, will develop new **methods and mechatronics and nano-mechatronics methodologies which will control at total quality level** of industrial products and of manufacturing processes, contributing in this way to increasing competitiveness on national, European and international level.

5. REFFERENCES

Brugger, J (1993). *Nanopositioniersystem, Microelectronik,* (5) Collection Microsystem technique

Dumitriu, A. (2006). *Mechatronics*; Transilvania University of Brasov Publishing House, ISBN 973-635-429-6, Brasov, Romania.

Fatikow, S. (1997). *The Microsystems and robotics tehnology*, Springer-Verlag Berlin Heidelberg, ISBN 973-31-1387-5, in Germany

Gheorghe, G. (2006). *Handbook of precision mechanic, mecatronic and integronic engineer*; Cefin Publishing House, ISBN 973-87042-6-8, Bucharest, Romania.

Suzumori, K. (1991). *Applying a flexible Microactuator for Miniature Robots*; Proc. of Int. Conf. on Robotics and Automation, Sacramento, CA., pp 1622-1627

Annals of DAAAM for 2008 & Proceedings of the 19th International DAAAM Symposium, ISSN 1726-9679
Editor B. Katalinic, Published by DAAAM International, Vienna, Austria 2008
Make Harmony Between Technology and Nature, and Your Mind will Fly Free as a Bird

FEM ANALYZE FOR AN ARTERIAL IMPLANT DEVICE

GHEORGHIU, H[oria]; CHIRCOR, L[idia] & NICODIM, M[ariana]

Abstract: *This paper presents numerical results predicted by a simulation methodologie for in-vivo performance of a particular type of stent design using finite element analysys(FEM). Typically a stent is mounted and crimped onto a balloon catheter, which is followed by a PTCA (Percutaneous Transluminal Coronary Angioplasty) procedure. The stent is then positioned in the dilated artery by passing it through vasculature mounted onto a PTCA balloon catheter. When the balloon is correctly positioned the stent is deployed by inflating the balloon. With the stent fully deployed the balloon catheter is deflated and removed from the vasculature.*

Key words: *stents, FEM, PTCA, numerical results*

1. INTRODUCTION

Stents are small scaffold-like structures, placed inside occluded blood vessels to restore the vascular lumen and flow conditions. The mechanical performance of stents is dependent on the geometrical design, material employed and fabrication technology used. Typical stent designs can be categorized as follows: metallic mesh balloon-expandable stents (expanded by inflation of an angioplasty balloon), elastic and shape memory self-expandable stents (expanded by temperature change), silicone, and biodegradable devices (ANSYS, 2007). Common fabrication methods used to manufacture stents include laser cutting, braiding, and stereo lithography (Suresh, 1998).

This paper presents numerical results predicted by a simulation methodology for the crimping, deployment and in-vivo performance of a particular type of stent design. Typically a stent is mounted and crimped onto a balloon catheter, which is followed by a PTCA (Percutaneous Transluminal Coronary Angioplasty) procedure.

The stent is then positioned in the dilated artery by passing it through vasculature mounted onto a PTCA balloon catheter. When the balloon is correctly positioned the stent is deployed by inflating the balloon.

With the stent fully deployed the balloon catheter is deflated and removed from the vasculature.

The stent considered in this paper is assumed to be manufactured by laser-cutting the mesh pattern into an annealed 316L stainless steel tube (Bunea & Nocivin, 1999).

The resulting mesh is then electro-polished to give a smooth surface finish (ANSYS, 2007). The stent will remain implanted in the artery for the lifetime of the patient, during which the stent will undergo cyclic loading as a result of the variation of blood pressure due to the beating of the patient's heart.

For the purposes of the fatigue evaluation the pressure variation in the coronary arteries resulting from the beating heart is assumed to be 100-mmHg. It is also assumed that the entire pressure load is transferred from the artery to the stent. Both of these assumptions are very conservative for diseased coronary arteries.

The finite element method is used to analyze the crimping, deployment and fatigue loading in the stent geometry by two-dimensional (2D) finite elements.

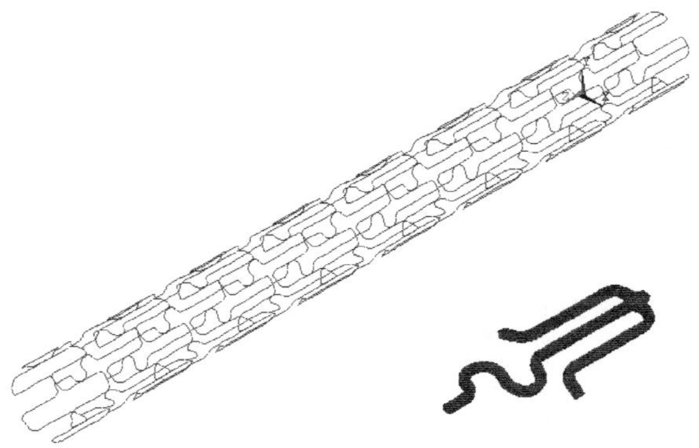

Fig. 1.a. Fig. 1.b.

Fig 1. Geometry GMIT Stent

2. DEVELOPMENT OF 2D FINITE ELEMENTS MODEL

2.1. Material Properties

The level of stress and strain defines the severity of loading of the material in a component. Metals normally have a region in which the strain is directly proportional to stress. In this region the material is said to be linear elastic. The limit of the linear elastic region is called the proportional limit. The strains accumulated at stresses higher than the proportional limit are not elastic (these are referred to as inelastic or plastic strains) because they remains when the material is unloaded. This is the material mechanism by which the stent is able to keep an artery open. Plastic strains accumulate as the stent is expanded by a balloon and remain when the balloon is removed, thereby keeping the stent deployed.

It is assumed that the stent under consideration in this paper is manufactured from 316L stainless steel. Young's modulus was taken as 207 GPa and Poisson's ratio was assumed to be 0.3 (Dieter, 1987).

2.2 Identification of Stent Geometry and Analysis Stages

The selected stent geometry analyzed in this paper is shown in Figure 1a. This design was developed by researchers at GMIT and as with commercially available stents (Suresh, 1998) it processes both axial and circumferential periodic symmetry. The presence of this symmetry implies that the analyses may be restricted to the repeating cell unit as shown in Figure 1b.

The stages of the stent's load history may be broken down into the following load steps:

Load Step 1. Crimping of the stent to a crimped diameter of 1.6 mm.

Load Step 2. Remove crimping tool and allow stent to spring-back to its equilibrium diameter.

Load Step 3. Expansion of the stent to its deployment diameter by inflation of internal balloon to a diameter of 4.25 mm.

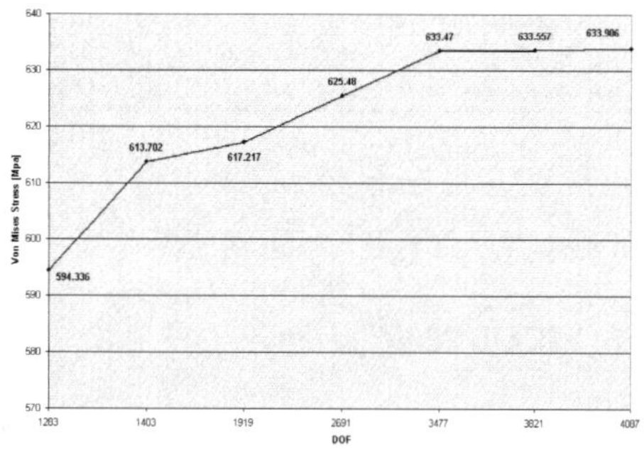

Fig. 2. 2D Mesh Convergence Study.

Load Step 4. Remove balloon and allow spring-back of the stent to its new equilibrium diameter.

Load Step5. Apply pulsatile loading on the outer surface of the stent with a cyclic pressure equivalent to 100 mmHg.

The residual stresses remaining after the first four steps as well as the pulsatile loading in Load Step 5 were used to calculate mean and alternating stresses for the cyclic loading of the stent. These were subsequently plotted on a Goodman diagram to predict the fatigue factor of safety of the stent.

3. FATIGUE LIFE CALCULATIONS

A stent is required to function under high cycle fatigue conditions. Regulatory authorities insist on numerical and experimental evidence that the proposed stent is capable of withstanding at least 400 million cycles of stress. In the results presented in this paper, a 'stress-life' approach was taken for the fatigue life prediction (Fung, 1993). This approach is further supported by the fact that the analysis predicts that no plastic straining takes place during the cyclic loading. Factors of safety were calculated for each model by employing the following formula:

$$n = \frac{S_a}{\sigma_a} \quad (1)$$

where: n = fatigue factor of safety,

$$S_a = \frac{S_e}{\frac{\sigma_a}{\sigma_m} + \frac{S_e}{S_{UT}}}, \sigma_a = \frac{\sigma_{max} - \sigma_{min}}{2},$$

$$\sigma_m = \frac{\sigma_{max} + \sigma_{min}}{2} \quad \text{and} \quad (2)$$

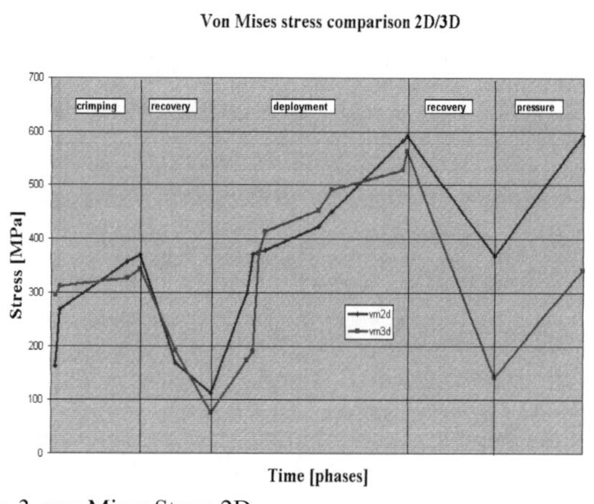

Fig. 3. von Mises Stress 2D

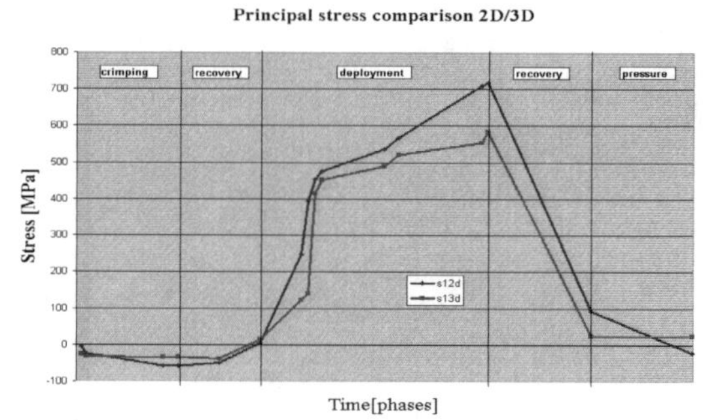

Fig. 4. 1st Principal Stress 2D

σ_{min} is the residual stresses after the deployment is completed, σ_{max} is the sum of the residual stresses after the deployment and the oscillating stresses resulting from the pressure loading, S_e is the fatigue endurance limit of 316L, S_{UT} is the Ultimate Tensile strength of 316L.

4. CONCLUSIONS & RESULTS

The 2D model presented in this paper is a two dimensional representation of a three dimensional structure. Therefore out of plane effects such as fish-scaling are assumed to be insignificant for the two dimensional model. In developing the 2D model the following procedure was used: Consider the stent structure as having a radius r, the repeating unit subtends an angleφ, equal to 360/nrep, where nrep represents the number of the repeating units in the circumferential direction. This implies that the arc length of the curved repeating unit is $r\varphi$. Now consider that the repeating unit is deployed by an amount δr, the angle φ must remain the same which implies that the arc length of the repeating unit will increase to $(r+\delta r)\varphi$. These could be represented in a 2D space as two straight sections having these calculated development lengths.

A mesh convergence study was completed for the 2D models, which helped to identify the optimum mesh density distribution for this type of analysis, which yielded accurate results in acceptable CPU times.

A study was conducted in order to find the optimum mesh density to be used in the 2D analysis. A total of seven mesh densities were investigated. Figure 4 present the variation in the maximum of von Mises stress in the U strut as the total number of degrees of freedom (which is related to the mesh density) increases.

The variation of the stress levels from when the stent is crimped to when the pulsatile loads are applied is presented below. Stress-history plots are presented which indicate the von Mises stress, 1st Principal stress (tension) and 3rd Principal stress (compression)

5. REFERENCES

ANSYS Help System Online Documentation, 2007, ANSYS Inc., 275 Technology Drive, Canonsburg, PA 15317, USA

Bunea, D., Nocivin, A., (1999). *Implant Materials*, Printech Ed., Bucharest, ISBN 973-9475-85-X

Dieter, G.E., (1987). *Mechanical Metallurgy*, SI Metric Ed., McGraw-Hill Book Co., Singapore, , ISBN 0-07-100406-8

Fung, Y.C., (1993). *Biomechanics – Mechanical Properties of Living Tissues*, Second Ed., Springer, ISBN 0-387-97947-6 (New York). – ISBN 3-540-97947-6 (Berlin).

Suresh, S., (1998). *Fatigue of Materials*, 2nd Ed. Cambridge University Press, ISBN 0521578477

Annals of DAAAM for 2008 & Proceedings of the 19th International DAAAM Symposium, ISSN 1726-9679
Editor B. Katalinic, Published by DAAAM International, Vienna, Austria 2008
Make Harmony Between Technology and Nature, and Your Mind will Fly Free as a Bird

SOME ASPECTS OF CAVITATION DAMAGES IN AUSTENITIC STAINLESS STEELS

GHIBAN, B[randusa]; BORDEASU, I[lare]; GHIBAN, N[icolae]; BADARAU, R[odica]; HADAR, A[nton] & SERBAN, N[icolae]

Abstract: *In this paper the authors analyze the manner of cavitation damages in three types of austenitic stainless steels. The SEM and optic microstructures highlight important differences in its behavior. These aspects are discussed in correlation with its chemical composition and the manner of the metallic material response (deformation, crystalline slide or crack). The cavitation attack was carried out using a magnetostrictive vibrating apparatus/device existing in the Laboratory "CAVITATION TESTING", Polytechnic University of Timisoara – Romania.*

Key words: *behavior of cavitation-erosion for austenitic stainless steels, austenitic microstructure influence on cavitation, stereoscopy and scanning electron analysis, chromium and nickel equivalents*

1. INTRODUCTION

The life time increasing of some specific machine elements (like pumps, navy and hydraulic turbine screw) is deeply influenced by the cavitation attack phenomena. So, a high resistance material must be used in very aggressive sea water media. Present paper presents a structural investigation of some austenitic stainless steels exposed to cavitation attack. A correlation between chemical constitution - elaborating technology and mechanical behavior is made, in comparison with other researchers (Frank & Michel, 1995), (Szkodo & Giren, 2004).

2. RESEARCHED MATERIALS, DEVICES AND METHODS

In this paper three types of austenitic stainless steels are analyzed – manufactured at Siderurgic Plant, Resita. These types of stainless steel alloys are classified by Cr and Ni equivalent. Chemical composition, determinated at CEMS laboratory – Bucharest, including Cr & Ni equivalents for structure determination, after Schaffler diagram are: Steel 20/26: 0.07% C; 20.51% Cr; 26.81% Ni; 1.30% Mn; 3.37% Mo; 1.53 % Si, rest Fe - $(Cr)_e$ = 26,175 % $(Ni)_e$ = 30,16 %; Steel 18/13: 0.1% C; 18.32% Cr; 13.22% Ni; 1.96 % Mn; 2.84% Mo; 1.56 % Si, rest Fe - $(Cr)_e$ = 23,5 % $(Ni)_e$ = 17,22 %; Steel 22/27: 0.07% C; 22.07% Cr; 27.28% Ni; 1.08% Mn; 3.02 % Mo; 1.27% Si, rest Fe - $(Cr)_e$ = 26,995 % $(Ni)_e$ = 29,92 %.

Cavitation destruction and surface microscopically study were performed in magnetostrictive vibrating apparatus at Cavitation Laboratory (Polytechnic University of Timisoara). 165 minutes cavitation exposed samples were analyzed by scanning electron microscopy at a Philips Microscope at Polytechnic University of Bucharest. Figures 1-3 show images of eroded microstructure of the experimental steels at different magnifications of the cross-section. Figure 4 indicates the depths of the erosion cavity in the investigated samples. The highest value of depth is present in case of sample 2, with about 400 μm, in comparison with the other two, where the depth is about 80 μm. Figure 5 shows the typical curve of the cavitation erosion (erosion rate vs cavitation attack time).

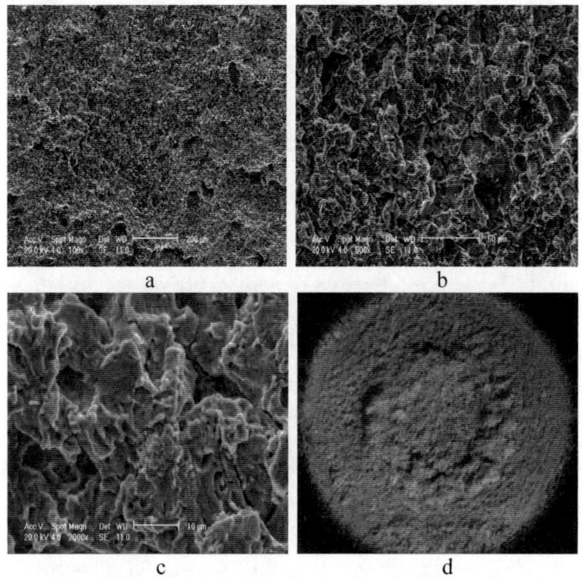

Fig. 1. Imagines of erosion attack produced at steel 20/26 after 165 minutes: a - SEM (x100); b (x500); c (x2000); d - macrostructure x6,4

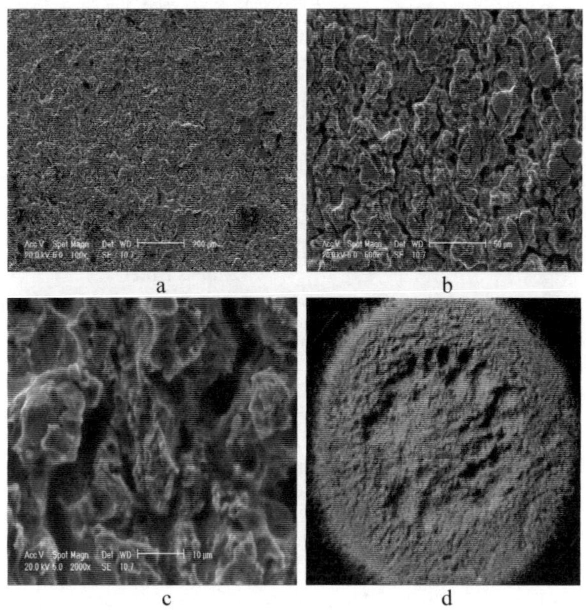

Fig. 2. Imagines of erosion attack produced at steel 22/27 after 165 minutes: a - SEM (x100); b (x500); c (x2000); d - macrostructure x6,4

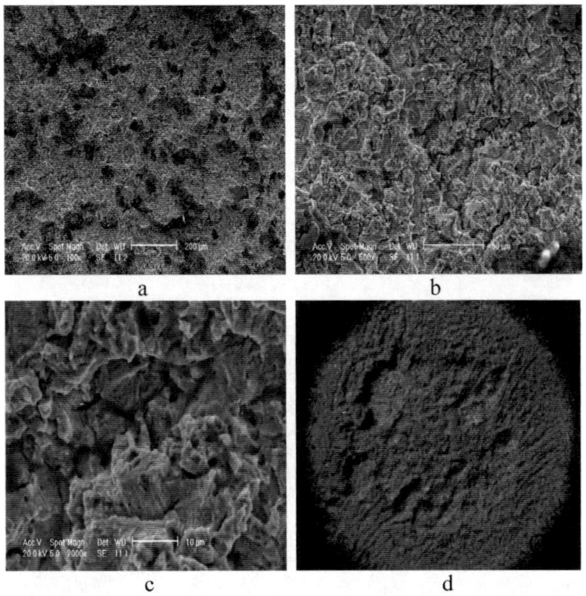

Fig. 3. Imagines of erosion attack produced at steel 18/13 after 165 minutes: a - SEM (x100); b (x500); c (x2000); d - macrostructure x6,4

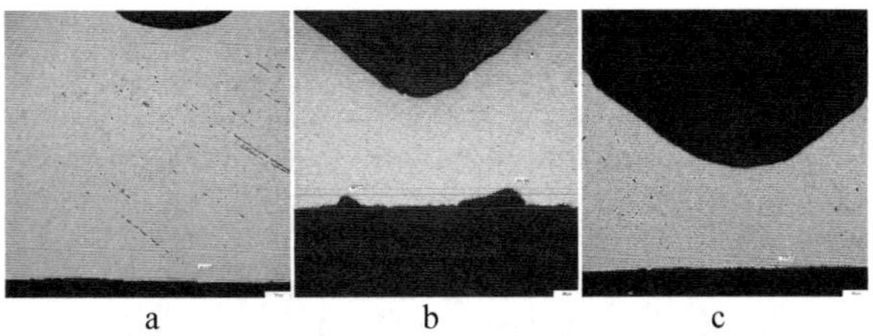

Fig. 4. Transversal cross sections of erosion cavitation samples: a - steel 20/26 (with 68.15 μm depth), b - steel 22/27 (with 283 – 400 μm depth), c - steel 18/13 (with 78 μm depth)

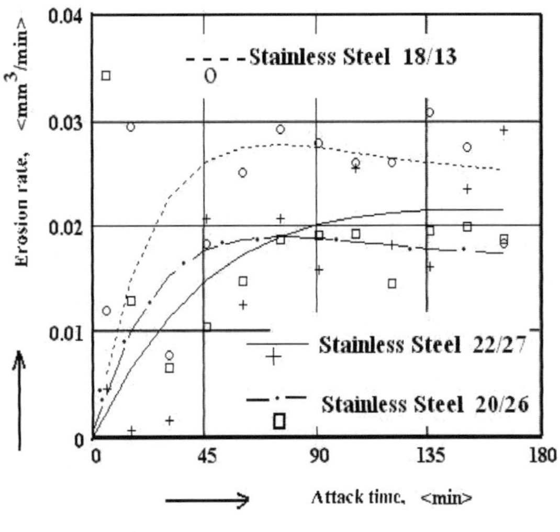

Fig. 5. Erosion rate versus cavitation attack

3. DISCUSIONS

Even if the stainless steel alloys have the same structure, the figures 1 – 3, a – d show different evolutions of structural damages, from the point of view of *caverns* - on one hand, and deformations - on the other hand, previous grain sputtering (or fragments of grains). We consider, as proposed by other researchers (Bojin et al., 2005), (Bregliozzi et al., 2005), (Lebrun & Poirier, 2002), that these phenomena are due not only to Cr – Ni couple, but to other base elements, especially Mn and Si, which produce loosening and growth of grain dimensions. The size of sputtered grains during the cavitation attack is very well reflected in the distribution of the experimental points, face to approximation curves. SEM analysis evidenced the following aspects of the steel erosion: 20/26 alloy (figure 1); the eroded surface presents a mixed aspect of smooth cavitation and relatively big cavitation (30-80 μm diameters); propagation of breaking front by intergranular cracks, the breaking propagation by intergranular cracks sliding paths; 18/13 alloy (figure 2); equal proportions of fine and big cavitations (30-140 μm diameters); the surface with deep secondary intergranular cracks; fragile character breaking with intergranular propagation and sliding paths; 22/27 alloy (figure 3); fragile breaking aspect; mixed propagation aspect of the front by intergranular cracks and cleavage planes; corrosion aspects with intergranular propagation. From the curves evolution (figure 5) results a similar behavior of 22/27 and 20/26 steels, which quantities of $(Ni)_e$ and $(Cr)_e$ are very appropriate. This aspect leads to the idea that the austenitic structure, according to Schaffler diagram, is the same, even that there are major differences between chemical elements concentrations: Mn, Si & Mo.

4. CONCLUSIONS

The behavior of cavitation erosion depends on the chemical constitution of the studied austenitic stainless steel in the same mechanical testing conditions. Scanning electron analysis may offer spectacular observations of cavitation erosion attack for austenitic stainless steels. Present paper gives an explanation of cavitation attack by means of fine structure and fracture front propagation trough grains and grains bounderies. For the same structure, austenite, cavitation attack is deeply influenced by chemical composition, reflected in chromium and nickel equivalents. At 20Cr/26Ni propagation of breaking front is given by intergranular cracks. At 18Cr/13Ni breaking character is due to intergranular cracks and sliding path and at 22Cr/27Ni only brittle behavior may appear, with cracks both intergranular and cleavage surfaces.

ACKNOWLEDGMENTS

The present work has been supported from the National University Research Council Grant (CNCSIS) PNII, ID 34/77/2007 (Models Development for the Evaluation of Materials Behavior to Cavitation).

5. REFERENCES

Bojin, D.; Miculescu, I. & Miculescu, M. (2005). *Microscopie electronica de baleiaj si aplicatii (Scanning electron microscopy and applications)*, Ed. AGIR, ISBN 973-720-019-5;

Bregliozzi, G.; Di Schino, A.; Ahmed, S.I.U.; Kenny, J.M. & Haefke, H. (2005). Cavitation wear behaviour of austenitic stainless steels with different grain sizes. *Wear*, vol. 258, No. 1-4 (704 p.), 2005, pp. 503-510, ISSN 0043-1648;

Frank, J. P. & Michel, J. M. (1995). *La cavitation. Mécanismes physiques et aspects industriels (Cavitation. Physical mechanisms and industrial aspects)*, Presse Universitaires de Grenoble;

Lebrun, J.P. & Poirier, L. (2002). Solutions to improve the surface hardness of stainless steels without loss of corrosion resistance, *ATTS Congress,* Paris, June 2002;

Szkodo, M. & Giren, B.G. (2004). Cavitation erosion of steels processed by CO_2 laser beams of various parameters. *Journal of Materials Processing Technology,* Vol. 157-158, december 2004, pp. 446-450.

Annals of DAAAM for 2008 & Proceedings of the 19th International DAAAM Symposium, ISSN 1726-9679
Editor B. Katalinic, Published by DAAAM International, Vienna, Austria 2008
Make Harmony Between Technology and Nature, and Your Mind will Fly Free as a Bird

THE CALCULUS OF THE THERMODYNAMIC ACTIVITY OF CARBON IN SILICON ALLOYED AUSTENITIC STAINLESS STEELS

GHIBAN, B[randusa]; GHIBAN, N[icolae]; SERBAN, N[icolae]; HADAR, A[nton]; BORDEASU, I[lare] & JIGA, G[abriel]

Abstract: *In this paper, the authors analyse the corrosion processes from thermodynamic point of view for eight types of austenitic stainless steels. It has been determinated the maximum temperature at which the intergranular corrosion appears in the silicon alloyed stainless steels only from thermodynamic considerations and also the influence of the silicon content in the presence of carbon in stainless steels.*
Key words: *intergranular corrosion of stainless steel, thermodynamic activity, silicon alloying influence, chromium and nickel alloyed austenite*

1. INTRODUCTION

The rapid development of austenitic stainless steels (A.S.S.) was stopped in the 70's due to the susceptibility of these materials towards intergranular corrosion (I.C.). The acting way of this corrosion form leads to selective destructions of the grain without the matrix being attacked. The amazing phenomena of I.C. which have been found in the exploitation process were rapid destroyed with the help of metallurgical means. One of the metallurgical methods of diminishing the intergranular corrosion of stainless steels in aggressive media is silicon alloying. The present paper reveals by thermodynamic considerations the influence of silicon content in presence of carbon in the austenitic stainless steels.

2. MATERIALS AND METHODS OF INVESTIGATION

There were elaborated eight stainless steels with the chemical composition given in table 1.

Steel	Chemical compound, %							N ppm	E_{Cr}	E_{Ni}
	C	Mn	S	P	Cr	Ni	Si			
A	0.020	0.87	0.007	0.017	17.9	15.6	0.94	60	20.1	16.4
B	0.015	0.73	0.008	0.015	18.7	14.9	3.39	60	25.5	15.6
C	0.012	0.85	0.009	0.014	17.2	15.6	3.83	268	23.7	16.5
D	0.031	1.48	0.008	0.011	17.1	15.6	4.11	300	24.1	17.2
E	0.030	1.23	0.006	0.011	17.3	14.8	5.34	280	26.1	16.3
F	0.012	1.10	0.006	0.011	18.3	14.8	5.62	108	26.7	15.6
G	0.039	0.94	0.011	0.017	17.3	15.2	4.60	128	25.0	16.6
H	0.050	0.91	0.011	0.015	18.00	15.2	5.03	100	26.3	16.8

Tab. 1. The chemical composition of the experimental stainless steels

From the corrosion behavior of the experimental austenitic stainless steels, not only in nitric acid agents, but also in chlorine agents it has been proven mainly the positive role of the silicon annihilated by the presence of the carbon content. A possible explanation of the operation mode on the behavior of the studied stainless steels is given by the influence of the different alloying elements on the thermodynamic activity of the carbon in the alloyed austenite.

Considering as constant the nickel and chromium contents for all the experimental austenitic stainless steels (respectively 18%Cr and 15%Ni) and the carbon contents for the three experimental stainless steels categories (respectively of 0.02%, 0.03% and 0.05%) it has been determinate the thermodynamic activities of the carbon in the unalloyed austenite (a_C^{γ}), in the nickel alloyed austenite ($a_{C,Ni}^{\gamma}$), in the chromium alloyed austenite ($a_{C,Cr}^{\gamma}$) and also in the complex alloyed austenite ($a_{C,Complex}^{\gamma}$). The formulas used for calculating the carbon's thermodynamic activities, in the unalloyed austenite and also in the alloyed austenite, are the following:

$$\lg a_C^{\gamma} = \lg\left[\frac{N_C}{1-5N_C} + 0.9N_C^2\right], \quad (1)$$

where: N_c – the carbon concentration, in atomic %;
T – the temperature, in Kelvin degrees (K).

$$N_C = \frac{\frac{\%gr.Ei}{A_{Ei}}}{\sum\left(\frac{\%gr.Ei}{Ai}\right)} \quad (2)$$

where: % gr.Ei – the weight concentration of the i element;
Ai – the atomic mass of the i element.

$$\ln a_{C,Ni}^{\gamma} = \ln(0.048\times\%C) + \left(0.525 - \frac{300}{T}\right) - 1.845 + \frac{5100}{T} - \left(0.021 - \frac{72.4}{T}\right)\%Ni \quad (3)$$

$$\ln a_{C,Ni}^{\gamma} = \ln(0.048\times\%C) + \left(0.525 - \frac{300}{T}\right) 1.845 + \frac{5100}{T} + \left(0.248 - \frac{404}{T}\right)\%Cr - \left(0.010 - \frac{9.422}{T}\right)(\%Cr)^2 \quad (4)$$

$$\lg a_{C,Cr}^{\gamma} = \frac{10500}{4.575\times T} - \frac{3.95 - 0.69\times\%C - 0.40\times\%Si}{4.575} + \lg N_C \quad (5)$$

3. RESULTS AND INTERPRETATIONS

The results regarding the calculated values of the carbon's thermodynamic activities corresponding to the experimental stainless steels are given in figure 1 (a and b). Similar behavior was found also by other researchers (Vilenskaya et al., 2004).

From the analysis of the data shown in figure 1 considering thermodynamic activities of the complex alloyed austenite in the experimental stainless steels dependent on the temperature and on the silicon and carbon contents, it results that the silicon is even able to double, at the same carbon content, its activity in the austenite alloyed with chromium and nickel. This fact has major consequences regarding the behavior of the experimental steels under different stress conditions. If in the austenitic stainless steels with a relatively high carbon content (of 0.03%, respectively 0.05%) the silicon presence is positive, from the

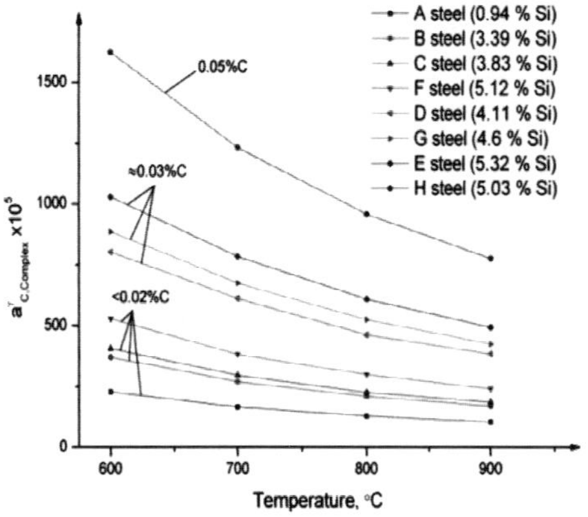

a.

b.

Figure 1 - The thermodynamic activity of the carbon in the

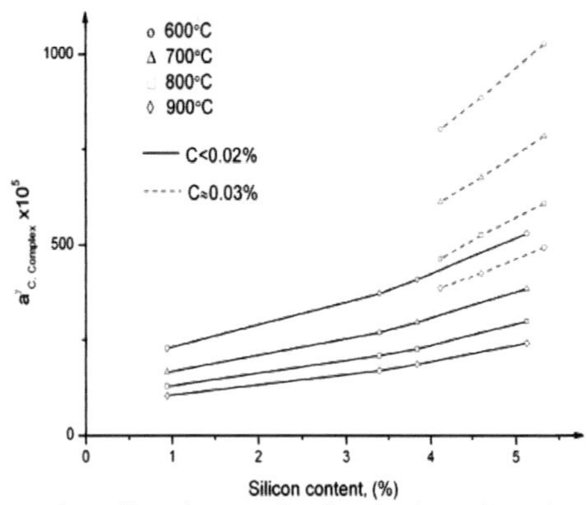

complex alloyed austenite for the investigated steels: a - versus temperature; b - versus silicon content

corrosion resistance under stress point of view, in case of loading the same steels in nitric acid environments the influence of the silicon is neutralized by the carbon contents. The effect of the silicon on the intergranular corrosion resistance appears only if the carbon content is very low (≤ 0.02%).

Considering the dissolution temperature (the precipitation of the first carbides from the stainless steels) given by the formula:

$$\frac{1}{T_S} = \frac{4.575}{\Delta G^0_{M_xC_y}}\left(\ln a^n_M + \ln a_C\right) = \frac{4.575}{\Delta H^0}\left(\frac{\Delta S^0}{4.575} + \ln a^n_M + \ln a_C\right) \quad (6)$$

the only carbides that precipitate in the area of 550 - 850°C are the $Cr_{23}C_6$ type, and in the austenitic stainless steels with ≈ 20%Cr and 9 – 40%Ni we have found the formula:

$$\frac{1}{T_S} = \frac{1}{5500}\left(2.92 - 0.010\%Ni - \ln\%C\right) \quad (7)$$

in which %Ni and %C are the weight percentage of the mentioned elements (nickel, respectively carbon).

Supposing that ΔH^0 and ΔS^0 are not depending on the temperature, from the calculation of the thermodynamic activity of the carbon in the alloyed austenite, we can calculate T_S and therefore we can estimate T_M. (Cihal, 1970) and (Gustavsson et al., 2005) have showed in their work the same behavior.

According to our own data and the information from the literature it has been found, for the determination of the T_M temperature the following formula, considering as constant the nickel content:

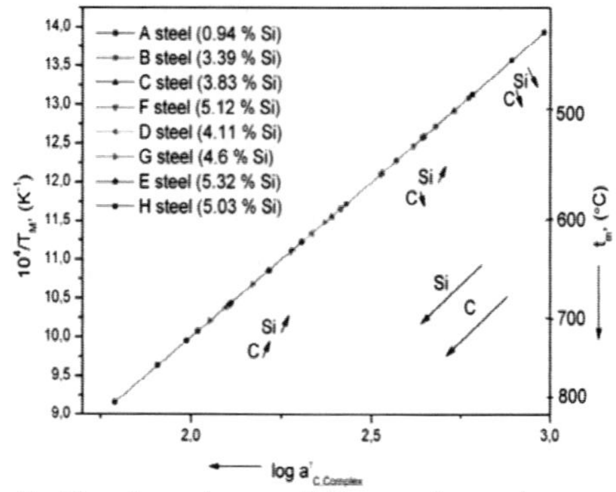

Figure 2. The dependency of the maximum temperature, at which the intergranular corrosion for the experimental austenitic stainless steels appears, on the carbon's thermodynamic activity

$$\frac{10^4}{T_M} = 2 - 4\lg a^{\gamma}_{C,Complex} \quad (8)$$

From the figure 2, in which it's shown the dependency of the maximum temperature at which the intergranular corrosion for the experimental austenitic stainless steels appears, we can clearly see that the silicon moves the T_M temperature to higher values for the low carbon content steels. The temperature is between 500 - 550 °C. This is perfectly according to the experimental results. Therefore the silicon influence on the intergranular corrosion behavior is positive only in the presence of a low carbon content (<0.02%). Similar behavior was illustrated also by (Josh & Sern, 1972), (Million et al., 1995).

4. CONCLUSIONS

This present paper presents the corrosion processes from the thermodynamic point of view. It has been shown that the silicon doubles the carbon's thermodynamic activity in the complex alloyed austenite with chromium and nickel. Based on the existent correlation between the chromium carbides separation domain and the domain of susceptibility at intergranular corrosion, it was determinated the maximum temperature at which the intergranular corrosion appears (T_M) in the silicon stainless steels only from thermodynamic considerations. So, the silicon, by raising the thermodynamic activity of the carbon, it also moves the T_M temperature to higher values for low carbon content steels.

5. REFERENCES

Cihal, V. (1970). Sur la corrosion intergranulaire des aciers inoxydables Cr-Ni (About the intergranular corrosion of the Cr-Ni stainless steels). *Corrosion Trait. Protecion. Finisation*, vol. 18, no. 7, p. 441;

Gustavsson, J.; Andersson, A.M.T. & Joensson, P.G. (2005). A thermodynamic study of silicon containing gas around a blast furnace Raceway. *ISIJ International,* Vol. 45 (2005), 5, pp. 662-668, ISSN 0915-1559;

Josh, A. & Sern, D.F. (1972). Chemistry of Grain Boundaries and It's Relation to Intergranular Corrosion of Austenitic Stainless Steels. *Corrosion*, vol. 28, nr. 8;

Million, B.; Kucera, J. & Michalicka, P. (1995). The influence of silicon on carbon redistribution in steel weldments. *Materials Science & Engineering* (1995), ISSN 0921-5093;

Vilenskaya, T.N.; Lifshits, A.E. & Mikhailov, L.A. (2004). Calculating the interaction of controlled atmospheres with alloyed steels. *Chemistry and Materials Science,* Springer, december 2004, ISSN 0026-0673.

Annals of DAAAM for 2008 & Proceedings of the 19th International DAAAM Symposium, ISSN 1726-9679
Editor B. Katalinic, Published by DAAAM International, Vienna, Austria 2008
Make Harmony Between Technology and Nature, and Your Mind will Fly Free as a Bird

EXTRUSION PROCESS MODELLING OF THE NON-ROUNDED PRODUCTS: A 2-D APPROACH

GHIBAN, N[icolae]; CHELU, G[heorghe]; SERBAN, N[icolae] & GHIBAN, B[randusa]

Abstract: *The non-rounded products plastically formed by direct extrusion often have specific nonuniformity of the mechanical properties either across transversal section, or on the length of the semifinished product. These nonuniformities (observed in laboratory tests) were put in evidence by industrial tests. Modelling using finite elements was used in order to show the difference between the optimum and the non-optimum position of the die cavity during direct extrusion process. It was used a COSMOS M 1.65 programme for modelling. The model allows the study of processes in plastic field only up to the moment in which the first plastic formed zone appears. The stress and elongation distribution has been studied for the quasistationary extrusion process. The results showed the nonuniformity of the total and partial deformation and also of the equivalent stresses for both conditions (the optimum and the non-optimum position of the die cavity).*

Key words: *extrusion process modelling, nonuniformity behaviour, Von Misses criterion of plasticity, die cavity position, finite element analysis*

1. INTRODUCTION

Products with non-rounded complex geometrical sections can be obtained only by extrusion, a plastical forming process, which generally leads to nonuniformities in final product. The main problem in extrusion is the equilibration of the yielding rates and the place of final cavities.

The aim of the present paper is to put in evidence by a modern modelling method the difference between two positions of the die cavity, optimum and respectively non-optimum position, in comparison with other researchers (Altan et al., 1994), (Ulysse, 2002).

2. MATERIALS AND METHODS OF INVESTIGATION

The modelling of the direct extrusion with the finite element method is based on a COSMOS M 1.65 program (COSMOS, documentation, 1997). A structural analytical program was used, which nonlinear subprograms allow the study of the plastic forming process up to the first plastically formed zones and non on the effective plastic forming process in the considered regime. The modelling was established for the direct extrusion process of the "yalle piece", made on brass CuZn39Pb3. Two situations were considered: one situation in which the displacement of the die cavity from the longitudinal axis is null (the so called optimum position), which leads to the uniformity of the mechanical properties of the product and another situation which corresponds to a displacement about 20 mm from the longitudinal axis of the die cavity (the so called non-optimum position).

These situations were considered taking into account other laboratory tests, made in (Ghiban, 1997). The charged curves and the yielding curve of the material were buildt considering: applied pressure of the semifinished product is 20 MPa; Young modulus $E = 7.3 \cdot 10^4$ MPa; plasticity modulus E_{pl}= -0.087 N/m^2; Poisson coefficient $\nu = 0.4$. The whole process was divided into 20 steps, during one second, in order to determinate the dynamics of the process.

The surface model used SHELL 4T elements, which simulated the nonlinear behavior of the medium plastic material. The surface model allows the determination of the distribution of total equivalent stresses and deformations (total, elastic and plastic) for both conditions (optimum and non-optimum position of the die cavity). COSMOS program may offer good results for comparing with other FEM based programs, like COMSOL or QFORM, (Zienkiewicz & Taylor, 2000), (Hartley et al., 1992).

3. RESULTS AND INTERPRETATION

The surface model for the "Yalle piece" is shown in figure 1 (*a* for optimum and *b* for the non-optimum position of the die cavity). The surface model has 218 nodes and 180 elements for the optimum position and respectively, 207 nodes and 170 elements for the non-optimum position of the die cavity. The charging is made linearly at forces in nodes from the end of the semifinished product in the opposite position of the die cavity.

The results regarding the variation of the main parameters of the direct extrusion process by modelling are presented in figures 2, 3, 4 and 5 and summarized in table 1. It can be seen that the surface model describes very well the situation on the rounded ends of the die, because in these places the maximum yield stress is obtained (and so, it is satisfied the Von Misses criterion). This model shows the differences in the distribution of all values for the equivalent stress, total and plastic deformations (both on x and on y direction) and on the angle-deformations. So, one may observe the symmetry of the distribution of all values obtained at the optimum position in comparison with inhomogeneity of all values obtained at the non-optimum position of the die cavity.

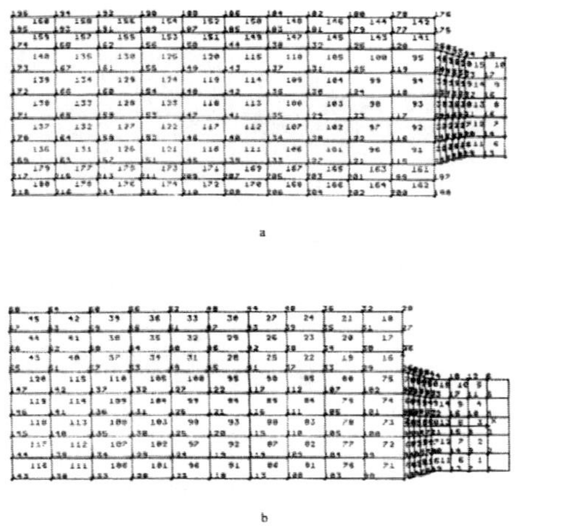

Fig. 1. The map of the elements and nodes of the surface model for optimum (a) and non-optimum (b) position of the die cavity

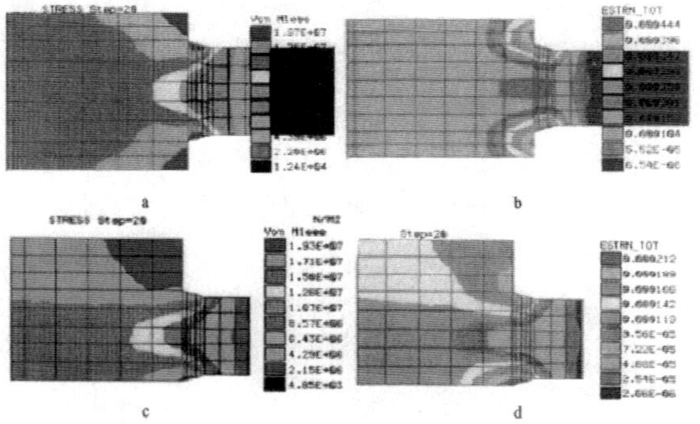

Fig. 2. Equivalent stresses (a, b) and total deformations (b, d) at step 20, optimum position (a, b); non-optimum position (c, d)

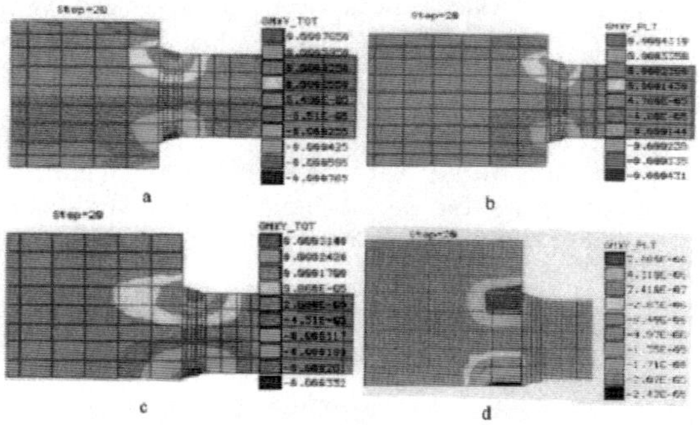

Fig. 3. Total angle deformations (a, c) and plastic angle deformations (b, d); step 20; optimum position (a, b); non-optimum position (c, d)

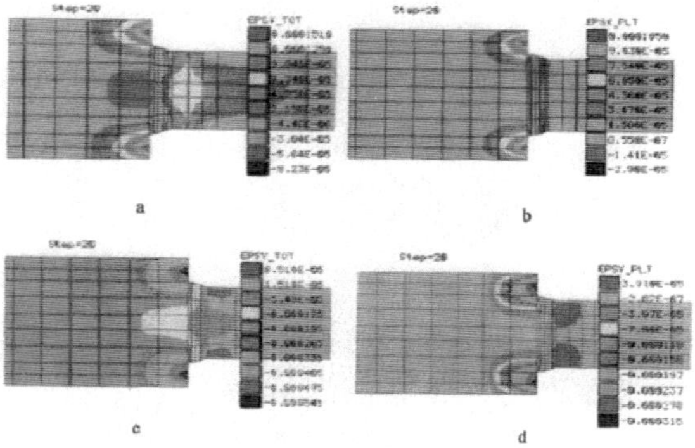

Fig. 4. The total deformations [ε_x (a), ε_y (c)] and plastic deformations [ε_x (b), ε_y (d)] for optimum condition, step 20

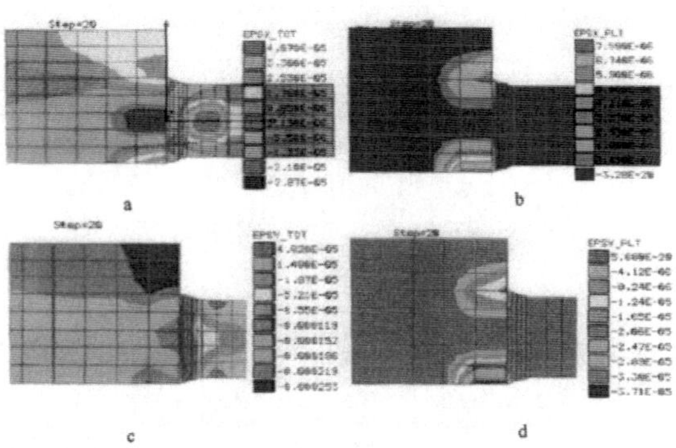

Fig. 5. The total deformations [ε_x (a), ε_y (c)] and plastic deformations [ε_x (b), ε_y (d)] for non-optimum condition at step 20

Value	*Step 1*	*Step 5*	*Step 10*	*Step 16*	*Step 20*
a) optimum position of the die cavity					
Equivalent stress (MPa)	*1.02*	*5.12*	*10.7*	*16.4*	*19.7*
Total deformation (10^{-5})	*1*	*5*	*10*	*16.1*	*44.5*
ε_x *total* (10^{-6})	*0.25*	*1.24*	*2.56*	*4.85*	*15.3*
ε_x *plastic* (10^{-6})	*0.137*	*0.94*	*1.86*	*2.75*	*8.2*
ε_y *total* (10^{-6})	*3.2*	*6.7*	*15.01*	*27.3*	*85*
ε_y *plastic* (10^{-6})	*1.2*	*2.8*	*7.2*	*13.8*	*54.3*
γ_{xy} *(MPa)*	*±0.101*	*±0.61*	*±0.82*	*±2.3*	*±7.65*
b) non-optimum position of the die cavity					
Equivalent stress (MPa)	*0.947*	*4.74*	*9.47*	*14.2*	*19.3*
Total deformation (10^{-5})	*0.942*	*4.7*	*9.42*	*14.2*	*21.2*
ε_x *total* (10^{-6})	*0.14*	*0.94*	*1.72*	*3.78*	*16.7*
ε_x *plastic* (10^{-6})	*0.092*	*0.37*	*0.97*	*2.65*	*7.59*
ε_y *total* (10^{-6})	*4.9*	*8.3*	*15.1*	*34.5*	*48.2*
ε_y *plastic* (10^{-6})	*1.75*	*4.7*	*10.8*	*25.4*	*37.1*
γ_{xy} *(MPa)*	*±0.092*	*±0.34*	*±0.52*	*±1.96*	*±8.15*

Table 1. The process parameters of the direct extrusion by surface modelling

4. CONCLUSIONS

The surface model based on COSMOS M 1.65 program may be successfully used in simulating and modelling the direct extrusion process. Using SHELL 4T elements, the authors put in evidence the difference between process parameters (equivalent stresses, total and plastic deformations, angle-deformation) obtained at direct extrusion of a brass "yalle piece", in two different situations: the optimum and the non-optimum position of the die cavity. These results are used in correct design and manufacture of the non-rounded direct extruded products. Future researches, based on the same program, may be done also for a tridimensional modelling approach.

5. REFERENCES

Altan, T.; Oh, S. & Gegel, H. (1994). *Metal Forming - Fundamentals and Applications*, ASM, 1994;

COSMOS M 1.65 (1997). *Documentation*;

Ghiban, N. (1997). *Studies and experimental researches about the non-rounded extruded profiles*, Doctoral Thesis, University Politehnica Bucharest;

Hartley, P.; Pillinger, I. & Sturgess, C.E.N. (1992). *Numerical Modelling of Material Deformation Processes*, Springer – Verlag;

Ulysse, P. (2002). Extrusion die design for flow balance using FE and optimization methods. *International Journal of Mechanical Sciences*, Vol. 44 (2002), pp. 319-341;

Zienkiewicz, O.C. & Taylor, R.L. (2000). *Finite Element Method*, Elsevier Butterworth-Heinemann, ISBN 0-7506-5049-4.

Annals of DAAAM for 2008 & Proceedings of the 19th International DAAAM Symposium, ISSN 1726-9679
Editor B. Katalinic, Published by DAAAM International, Vienna, Austria 2008
Make Harmony Between Technology and Nature, and Your Mind will Fly Free as a Bird

FEM SIMULATION OF ELECTRODISCHARGE MACHINING FINISHING AIDED BY ULTRASONICS

GHICULESCU, D[aniel] & MARINESCU, N[iculae]

Abstract: *The paper deals with modelling of complex material removal finishing mechanism of electrodischarge machining aided by ultrasonics (EDM+US). Finite Elements Method (FEM) emphasised the role of ultrasonically induced cavitational phenomena in spectacular improvement of machining rate and surface roughness in comparison with classic EDM finishing process that has great instability and consequently low technological performances. Our solutions are based on optimisation of some working key parameters and machined material selection. Improving poor flexibility of EDM+US is the objective of our further researches.*
Key words: *EDM, ultrasonics, simulation, finishing.*

1. INTRODUCTION

The ultrasonic aiding is very functional in EDM finishing due to huge pressure developed by cavitational phenomena ultrasonically induced within very narrow working gap. The cumulative microjets resulted at final of each ultrasonic tool oscillation solve the particular instability of EDM finishing and spectacularly improve its performances.

Finite elements method (FEM) was used to clarify more the intricate removal mechanism of EDM+US focused on two main aspects: (1) machined surface macrogeometry by studying the influence of concave and convex shapes through the volume adjacent to EDM spot; (2) machined surface microgeometry analysing the influence of previous types of machining.

2. MACROGEOMETRY INFLUENCE

The influence of machined surface macrogeometry on temperature distribution and implicitly on crater form produced by discharge and surface quality was studied by FEM.

The types of boundary conditions were: (1) constant temperature surface; (2) constant thermal output flux $\Phi = 0.1$ W/mm^2. Regarding condition (1), the simplifying hypothesis of constant temperature on cathode and anode spot was introduced. It was taken account of Utsumi's measurements concerning spot temperature and Conn's hypothesis of plasma channel narrowing in cathode zone (Van Dijck & Dutré, 1974), experimentally confirmed with enough accuracy (Ghiculescu, 2004). Thus, for commanded pulses and positive polarity, the temperature on cathode spot, t_{spot} was 2550 °C and its radius R_{cs} = 2.5 μm. For relaxation pulses and negative polarity, anode temperature was t_{spot} = 2500 °C and its radius R_{as} = 10 μm.

The thermo-physical characteristics of main components of steels, Feα and Fe_3C most frequently EDMed, whose dimensions are comparable with crater dimensions produced by discharges, were entry data for FEM. The meshing, finer in interest zone, was designed using ANSYS SOLID 70 and 87 elements, which take into account aggregate state changing, due to EDM thermal phenomena.

The dimensions of workpiece were considered of mm order, much greater than EDM spot size. Consequently, the zones with ambient dielectric temperature of 40°C are much greater than ones thermally influenced by discharge - fig. 1, 2.

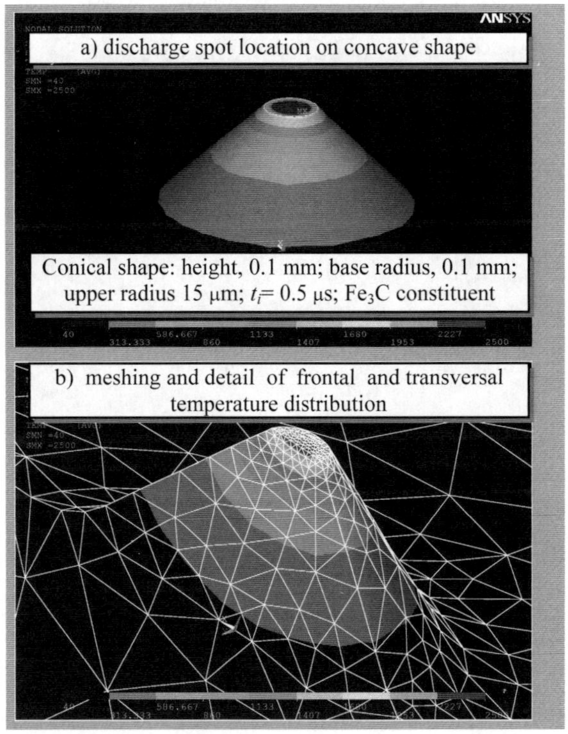

Fig. 1. Temperature distribution on a conical shape after a single relaxation pulse

Crater volume at pure EDM, delimited by boiling isothermal, is 4-5 times lower than that produced by EDM+US, which is bounded by melting isothermal (fig. 1). At classic EDM, gas bubble lifetime surrounding plasma channel is much greater than pulse time. Thus, when hydraulic forces arrive to EDM spot, almost all molten material is already solidified (Van Dijck & Dutré, 1974). At Fe_3C machining, boiling isothermal has 2500°C. Comparatively, at EDM+US, melting isothermal has 1227°C and the volume bordered by this isothermal is roughly entirely removed by cumulative microjets (Ghiculescu, 2004). Those results provided by FEM are in agreement with experimental data informing about machining rate increase of 400-500% through ultrasonic aiding of EDM (Kremer et al., 1991), (Sundaram et al., 2006), (Ghiculescu et al., 2008).

Low roughness could be obtained when EDM homogenous steel with Feα major constituent; FEM results highlighted crater volumes lower with 15-20% than those obtained on Fe_3C.

Crater dimensions when machining cavity shapes are sensible reduced than in case of opposite shapes because discharge energy is dispersed on much greater adjacent volume. Consequently, when machining cavity shapes through EDM+US, the power for actuating the acoustic chain can be increased with positive consequences on particles evacuation from the gap and without damaging fine details of cavity shapes.

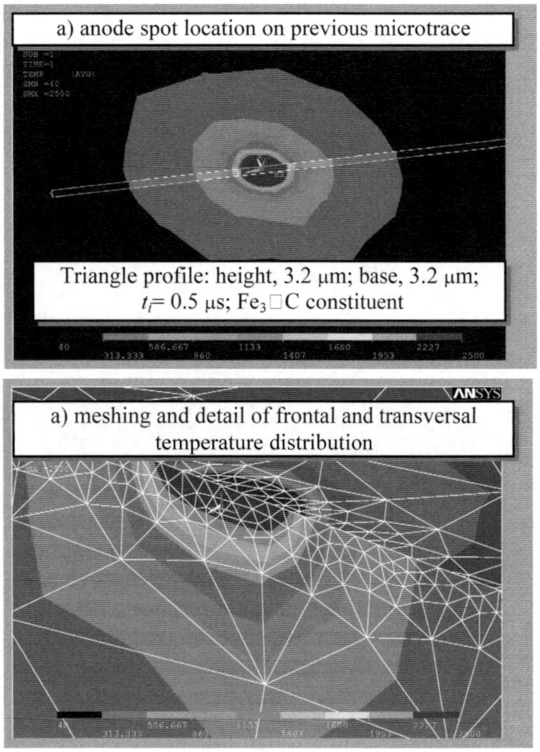

Fig. 2. Temperature distribution on triangle microprofile after a single relaxation pulse

3. MICROGEOMETRY INFLUENCE

The same boundary conditions and hypotheses as in previous section were used for assessing microgeometry influence on EDM+US mechanism. Several cases assimilated with finishing by EDM+US after previous machining as grinding, milling and roughing EDM were studied.

A single discharge simulation on microtrace produced by grinding or fine milling (R_a=0.8μm) is presented in fig. 2 in case of a single relaxation pulse. By comparison with commanded pulse, the relaxation one produces higher surface smoothening due to specific flatter shape of crater. In this case, the margins of this crater have greater sensitivity to cumulative microjets action and need a decrease of consumed power for supplying the ultrasonic chain (P_{cUS}). In experimental conditions, we established a decrease of 30% for P_{cUS} power, relative to commanded pulses case, in order to improve R_a roughness with up to 50%, (Ghiculescu, 2004).

In case of initial EDMed surface, the placement of boiling isothermal pointed out that at classic EDM finishing, the discharge located on a micropeak could not usually remove it entirely if thermal energy is given by a commanded pulse. A single relaxation pulse is able to do that, emphasising again this type of pulse efficiency at EDM finishing. Position of melting isothermal indicates that the action of smoothening produced through cavitational effect of ultrasonic is obvious, taking place by removing a micropeak with single discharge, even in case of commanded pulses and Feα constituent, which has lower thermal absorption capacity.

4. FURTHER RESEARCHES

Working of acoustic chain under the resonance condition is essential for high performances of EDM+US, practically requiring a great number of horn lengths adjustments. This is the main disadvantage of EDM+US technology - lack of flexibility at manufacturing preparation of new complex surfaces – which future researches must be focused on.

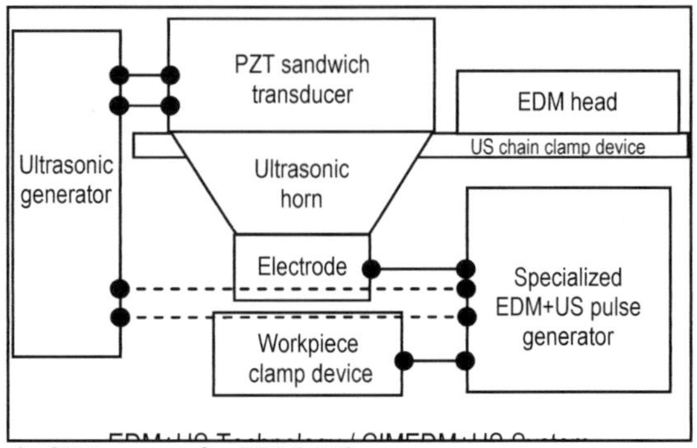

Fig. 3. CIM system for EDM+US

This is the reason for, a computer integrated manufacturing (CIM) is the next stage of our researches (fig. 3). The objective of the CIM system is to reduce the fabrication cycle with at least 50%. Starting from draft of workpiece to be machined, mechanical parts will be achieved using CAD/CAE/CAM system. Simulation of acoustic chain working aims to avoid its iterative lengths adjustments and to obtain the condition of resonance. The selfadjustment frequency of ultrasonics generator is able to compensate the electrode wear and possible changes of acoustic chain characteristics as growing of working temperature. Synchronisation between generators of pulses and ultrasonics will contribute to machining rate increase and volumetric relative wear decrease by producing pulses within stretching ultrasonic semiperiod (Ghiculescu et al., 2008).

5. CONCLUSION

Some aspects of intricate material removal mechanism of EDM+US finishing are elucidated, allowing spectacular increase of machining rate with up to 500% and surface roughness decrease with up to 50%. The adopted technological solutions stand on adjustment of power supply of acoustic head depending on machined macro and microgeometry, machining materials with low thermal absorption capacity and using relaxation pulses. The major inconvenience of EDM+US, poor flexibility, could be eliminated through a CIM system, leading to a large scale application of this technology.

6. REFERENCES

Ghiculescu, D. (2004). *Nonconventional Machinings*, Printech, ISBN 973-652-9758-4, Bucharest, Romania.

Ghiculescu, D., et al. (2008). Precision Increasing at Ultrasonics Aided Electrodischarge Machining, *Proceedings of 6th International DAAAM Baltic Conference "INDUSTRIAL ENGINEERING"*, Kyttner, R. (Ed.), pp. 233-238, ISBN 978-9985-59-783-5, Tallin, Estonia, April 2008, DAAAM Baltic, Tallinn University of Technology.

Kremer, D.; Lhlaubet, C. & Moisan, A. (1991). A study of the effect of synchronizing ultrasonic vibrations with pulses in EDM,
Available from: http://http://www.cirp.net/ *Accessed:* 2008-05-25.

Sundaram, M.; Pavalarajan, G. & Rajurkar, K. (2006). A Study on Process Parameters of Ultrasonic Assisted Micro EDM Based on Taguchi Method,
Available from: http://www.springerlink.com/content/ *Accessed:* 2008-05-25.

Van Dijck, F. & Dutré, W. (1974). Heat conduction model for the calculation of the volume of molten metal in electric discharges,
Available from: http://http://www.iop.org/EJ/ *Accessed:* 2008-05-25.

Annals of DAAAM for 2008 & Proceedings of the 19th International DAAAM Symposium, ISSN 1726-9679
Editor B. Katalinic, Published by DAAAM International, Vienna, Austria 2008
Make Harmony Between Technology and Nature, and Your Mind will Fly Free as a Bird

VON MISES STRESS EVALUATION FOR A MECHANICAL PART USING THE CATIA FINITE ELEMENT METHOD

GHIONEA, I[onut]; MUNTEANU, G[abriel] & BEZNILA, H[ortensia]

Abstract: *The Finite Element Method (FEM) is one of the best existing approaches that accomplish a vast range of engineering calculus and simulation. The method and FEM-based programs becomes a main constituent of modern engineering computer-aided design. A FEM-based analysis is now compulsory for high performance engineering tasks. The current paper aims to analyze a mechanical part's structure with some given constraints. The obtained results will be used further for optimize the part's structure.*
Key words: *finite element method, mesh, Von Mises, stress calculation.*

1. INTRODUCTION

For many equipments and particularly for the machinery manufacturers, the resistence structure design is the most important component that has to be analyzed with FEM instruments; the structure comprises the whole mechanical system having a strictly loads absorbtion, ensure a specific meaning (e.g.: the relative movement between the subassemblies), together with the static and dynamic stability or to guarantee a nominal stiffness established by the designer in his specifications (Constantinescu et al., 2006).

Increasing the strength, the stability and the durability are the main required characteristics of an assembly comprising as well subassemblies, joints and pieces. So, for a specific product, the engineers should always consider constraints like: the number and intensity of the static loads as well as dynamic ones, the maximum strains values, different safety factors (for bucking, breaking or fatigue), the susceptibility at execution, mounting or errors appeared in operation, frequency characteristics, steady afterflow speed, product cycle of life, weight, material and moment of inertia, different loads stiffness, static and dynamic stability or simultaneous loads response (Ispas & Ghionea, 2001).

The FEM analysis of a structure is in fact a numeric computation review. Hence, for a specific geometrical model, particular loads and seat constraints well known will result the required values of the deformation, stress, bearing reaction and natural frequency (Ghionea, 2007).

2. THE FINITE ELEMENT METHOD ANALYSIS

2.1 Initial data

The current analysis comprises the FEM applied for an arm type piece shown in figure 1.

Fig. 1. The shape of the mechanical part.

To accomplish the functional meaning, the structure has an assembly and a connection surface where the load is applied (Barlier & Poulet, 1999).

The solid model has been built in CATIA Part Design module, having a steel material with the characteristics shown in figure 2: Young's modulus of 2×10^{11} N/m^2, Poisson's ratio of 0.266, density of 7,860 kg/m^3, thermal extension ratio of 1.17×10^{-5} K, yield streght of 2.5×10^{8}N/m^2.

By applying the material characteristics, the specification tree is completed with the *Material=Steel* element. To start the FEM analysis, the user can access the CATIA Generative Structural Analysis module to set up the type of statical analysis as Static Case.

PartBody\Material	Steel
`Steel\Steel.1.1\Young Modulus`	2e+011N_m2
`Steel\Steel.1.1\Poisson Ratio`	0.266
`Steel\Steel.1.1\Density`	7860kg_m3
`Steel\Steel.1.1\Thermal Expansion`	1.17e-005_Kdeg
`Steel\Steel.1.1\Yield Strength`	2.5e+008N_m2

Fig. 2. Physical characteristics of the structure steel material.

2.2 The methodology

Although CATIA software defines a default net of nodes in the process named digitization, it is higly recommended for experienced users to edit this net and establish the dimension of the finite element (Size), the maximum deviation of the virtual digitized model from the real model (Absolute sag), the type of the element (Element type) with a double click on OCTREE Tetrahedron Mesh element from the specification tree.

A new dimension of the finite element of 2 mm has been established, a deviation of 1 mm and a linear type of element are the new coordinates. On the every surface of the bearings from the bottom of the structure, the user applies the Clamp restrains as shown in figure 3.

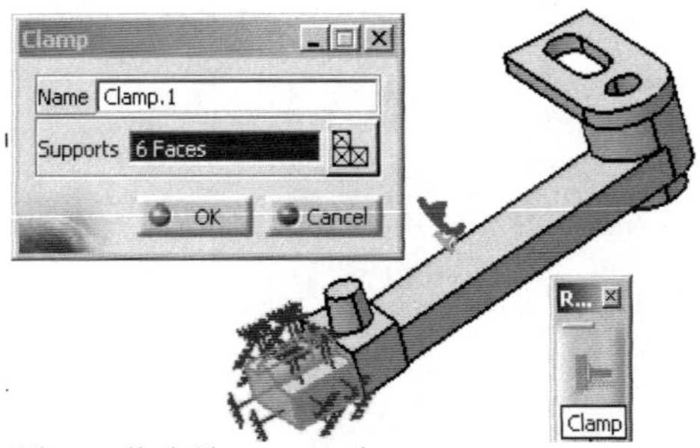

Fig. 3. The applied Clamp restrains.

On the surface of the bore from the upper part of the structure, is applied a distributed force of 150 N from interior to exterior in the opposite direction of the X axis. In the specification tree, the Distributed Force.1 element becomes available.

The force is represented by a four red arrows symbol, and characterized by a value and direction. All the parameters can be introduced in the appropriate fields from the dialog box as figure 4 shows.

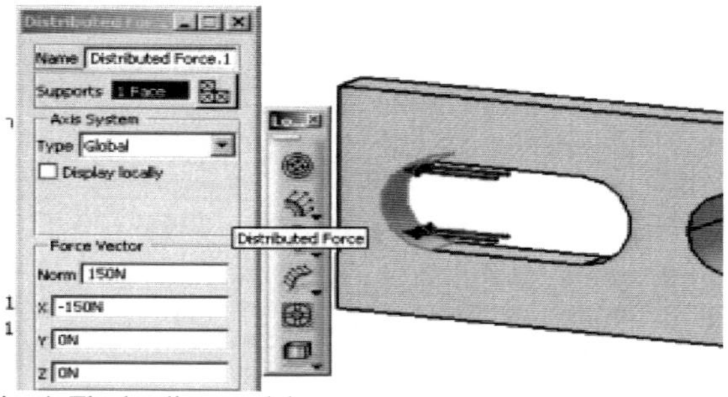

Fig. 4. The loading model.

2.3 The outputs

As the restrains and loading are established, the next step is to calculate the model behaviour running the Compute routine. At the end of the calculus, the user may decide to view the results from the Image menu. The specification tree is being completed based on the inserted images.

The figure 5 shows an output as Von Mises Stress type of image. A variety of parameters might be studied, but the present study analyses the structure behaviour considering the deformation, principal stress and precision.

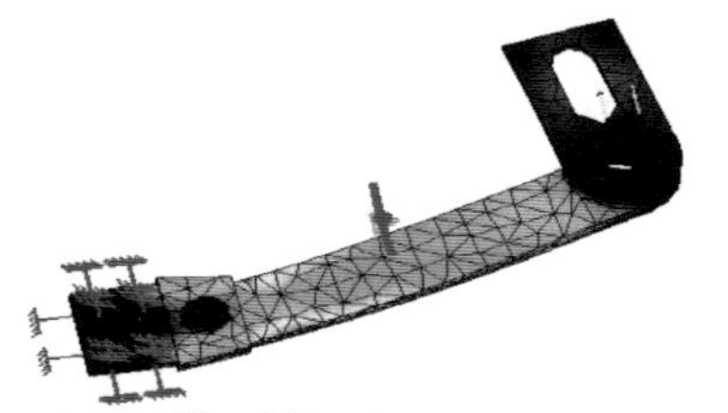
Fig. 5. The outputs – Von Mises image.

In order to identify the maximum and minimum stress values it is used the Information instrument to show the results.

In the Figure 6 is presented the window with stress value as well as the colour scale of Von Mises image. The lowest stress values correspond to the bottom of the scale, close to blue; the higher are the red ones at the scale top; between these are quite safe yellow and green shades.

To draw conclusions, we have picked from the dialog window the choose the exact values: minimum of 0 to a maximum of 2.2×10^8 N/m^2. Knowing the yield strenght of the material of 2.5×10^8 N/m^2, it can be considered that the structure will resist to a distributed load force of 150 N, but the operational safety is nearby the maximum limit and an increase of more than 10% is dangerous and not recommended.

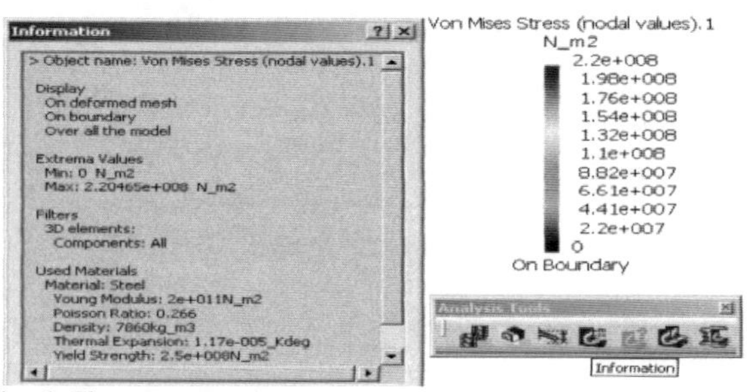

Fig. 6. The structure stress status.

2.4 Optimizing the model

To verify the accuracy of the results we performe a specific test with Precision software instrument as well as locate the zones containing the biggest errors, based on the Image Extrema software instrument (see fig. 7). Hence, the global error is calculated at 41.9%; such error seems to be high but it indicates in fact all the differences between the proposed virtual model and the real structure.

Fig. 7. Estimated errors – extreme values identification.

Therefore, it is imperative to optimize the model running the New Adaptivity software instrument and filling in a suitable value – for the start: 25%. To reiterate the computation in 5 steps is imperative for the analysis to refine model and to reach the established goal. The net is also refined by increasing it's accuracy from 2 to 1 mm with Minimum Size instrument. Definitely, all these improvements of the virtual model increase the computing time for analysis (Ghionea, 2007).

Analysis of the optimized model results shows an enhanced error, but +4.58% more than the initial target of 25%. Overall, an improvement of +12.36% has been achieved, but regarding the Von Mises outputs, the new model analysis shows an increased maximum stress value, from 2.20×10^8 N/m^2 to 2.34×10^8 N/m^2 (figure 8). For an increased precision and higher improved model, the user may continue an iterative analysis and to refine the model. The goal is to decrease the target of the global error, but the outputs of the maximum stress must not exceed the admissible steel yield strength.

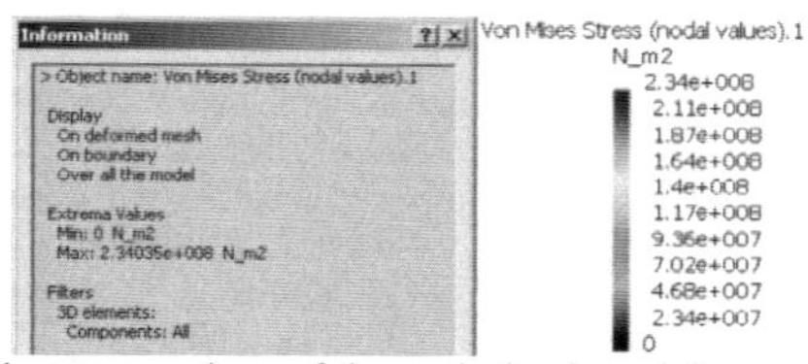

Fig. 8. The stress values of the optimized model.

3. CONCLUSION

On the sequences of CAD – FEM – CAM an iterative process of design – computing – execution is being realised, comprising both analysis and synthesis FEM routines applied to the virtual model or prototype. Experienced and explained above, each reiteration brings a new improvement for the model or prototype until the goal is attained.

4. REFERENCES

Barlier, C. & Poulet, B. (1999). *Mémotech. Génie mécanique, productique mécanique*. Deuxième édition (*Mémotech. Mechanical engineering, mechanical manufacturing)*. Editions Casteilla, ISBN2-7135-2063-0, Paris.

Constantinescu, N. I., Sorohan, Şt. & Pastramă, Şt. (2006). *The practice of finite element modeling and analysis*. Printech Publishing House, ISBN 978-973-718-511-2, Bucharest

Ghionea, I. (2007). A practical approach in the finite element method study of a mechanical part. *Scientific Bulletin, Serie C, Volume XXI, Fascicle: Mechanics, Tribology, Machine Manufacturing Technology*, North University of Baia Mare pp. 251-258, ISSN-1224-3264, , 2007, Baia Mare.

Ghionea, I. (2007). *Proiectare asistată în CATIA v5. Elemente teoretice si aplicatii.Computer aided design in CATIA v5. Theoretical elements and applications.* BREN Publishing House, ISBN 978-973-648-654-8, Bucharest.

Ispas, C., Ghionea, I., (2001). Stude of computer design and management in the conception and development phases of a product. *Optimum Technologies, Technologic* Systems and Materials in the Machines Building Field, University of Bacau, ISSN 1224-7499, Bacau.

Annals of DAAAM for 2008 & Proceedings of the 19th International DAAAM Symposium, ISSN 1726-9679
Editor B. Katalinic, Published by DAAAM International, Vienna, Austria 2008
Make Harmony Between Technology and Nature, and Your Mind will Fly Free as a Bird

ABOUT THE USE OF SHAPE MEMORY ALLOYS IN VIBRATION DAMPING

GILLICH, G[ilbert] R[ainer]; AMARIEI, D[aniel]; GILLICH, N[icoleta]; CHIONCEL, C[ristian] P[aul]; BIZAU, V[iorel] & BIRDEANU, E[lena] D[aniela]

Abstract: *Vibration damping for structures is an important issue, because it mitigates hazards, increases the comfort of people who use the structures, and enhances the reliability and performance of structures. The paper presents the concern of the authors to realize a vibration-isolated workstation for own use. In this mean the authors have both modelled and experimented small scale devices, usable in vibration damping and control. The theoretical models use the spring(s)-damper(s)-mass model. The results of simulations were compared whit experiments made using Ni-Ti alloys with different shapes. We concluded that the used theoretical model present sufficient accuracy for practical issues. Regarding the device's shapes, the best results were obtained for semi-cylindrical elements. The mass of the stiff plate and equipment placed on it has also an important role in the damping system.*
Key words: *Shape Memory Alloy, vibration, damping.*

1. INTRODUCTION

Constructions are normally not sensitive to vibrations. However, they can be affected; light to strong damages can appear, depending on the vibration parameters. More than that, optimal functioning of sensitive equipment depends on low level of vibrations. Regulations are different in various counties (Bratu & Gillich, 2006).

Shape Memory Alloys (SMAs) have the ability to change their shape, stiffness, natural frequency, damping coefficient and other mechanical characteristics in response to a change in temperature and/or stress. SMAs exhibit two phenomena, known as: shape memory effect and superelasticity or pseudoelasticity. These characteristics are a result of a phase transformation between different crystallographic structures of the materials known as the martensite and austenite phases. A SMA is easily deformed in its low-temperature martensitic condition and is returned to its initial shape by heating above the austenite start temperature (As). At temperatures above the austenite finish temperature (Af) pseudoelastic behaviour is observed (Graesser & Cozzarelli, 1991).

This makes SMAs proper to be used as passive damping devices due to their inherent energy dissipation, resulted from the hysteretic phase transformation between austenite and martensite. A study on the use of SMAs for passive structural damping was performed, where three different quasi-static models of hysteresis were reviewed and compared with an experimental investigation of a cantilevered beam constrained by two SMA wires (Thomson et al., 1995). Thermal effects in the SMA constitutive behaviour has to be considered (Fosdick & Ketema, 1998), for instance by including an "averaged" thermal rate dependency. It is possible to realize a physically based SMA model identified from an SMA helical spring response (Yiu & Regelbrugge, 1995). The dynamical behaviour of a SMA bar in a single degree of freedom spring mass damper system was also studied (Feng & Li 1996), where a modified plasticity model (Graesser & Cozzarelli 1991) was used to model the pseudoelastic response of an SMA bar.

2. PSEUDOELASTIC EFFECT

The pseudoelastic behaviour is defined as inducing detwinned martensite from austenite by thermo-mechanical loading. A simplified model that requires specific transition points (points 1 - 4) representing martensitic and austenitic start and finish temperatures, where begins and ends the transformations is presented in figure 1.

In addition to the change in material properties and large recoverable strain during pseudoelastic transformation, hysteresis is also an indicator for energy dissipation during the transformations. This energy dissipation is proportional to the degree of transformation completed during a loading cycle for both complete and incomplete, or partial transformations. These partial transformations are referred to as minor loop hysteresis cycles (figure 2) and complete or full transformations are referred to as major loop hysteresis cycles (Figure 3).

The hysteresis behaviour (represented by points 1 to 4 in figures 2 and 3) along with the stiffness change (represented by k_F and k_R in figure 2 and k_{A-M} , k_M , k_{M-A} and k_A, characteristic for each state phase, in figure 3) of the material during the pseudoelastic phase transformations confirm the availability for our SMA spring-mass system to be used as damping and vibration isolation device.

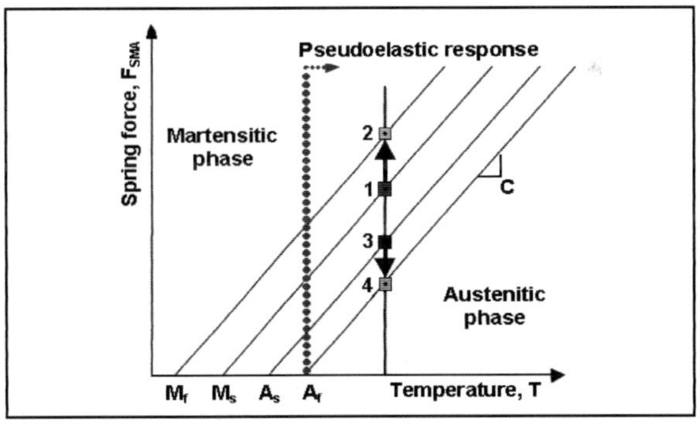

Fig. 1. SMA spring element force-temperature diagram with pseudoelastic loading path.

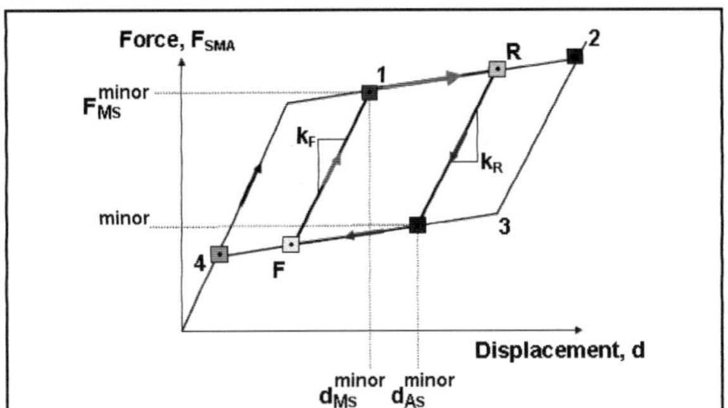

Fig. 2. Force - displacement path for minor loop loading.

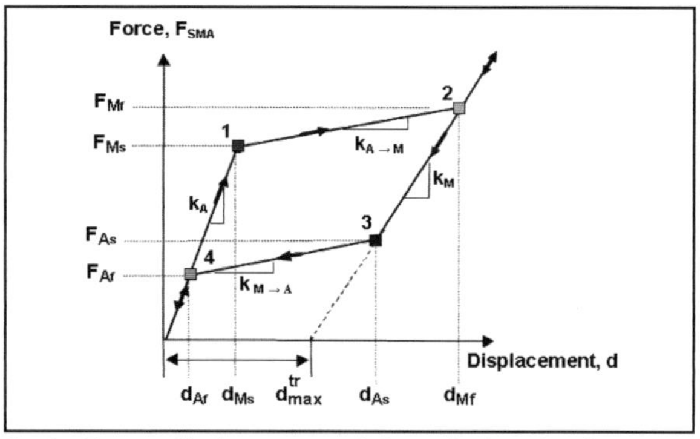

Fig. 3. Force - displacement path for major loop loading.

3. ISOLATED WORKSTATION

Sensitive equipment, for proper use and avoid of defects, needs a good isolation against vibrations. Classical solutions are: the use of tables with heavy plates or implementation of dumpers in the table's structure. These solutions are inadequate because of the high weight or expensive.

The authors have developed for the use in the own university a workstation, schematic illustrated in figure 4, based on the pseudoelastic behaviour of SMAs. The structure is a rigid one, on which are placed two mobile plates assigning a pre-compression of the SMA elements placed between the two mobile plates and the structure. We have tested elements with three different forms: spring, cylinder and semi-cylinder, the last providing the best isolation. Finally we placed them parallel to the long side of the plates, with the open side orientated to the stiff mounting plate.

To determinate the natural frequencies a simplified mechanical model was used (figure 5), where linear springs, frictional and slip elements are used to describe the behaviour in austenitic or martensitic domains or during the austenitic to martensitic respectively martensitic to austenitic transformations. In the figure 5, $d \geq 0$ represents the load cycle, $d<0$ the unload cycle. The four stiffness values k_A, k_M, k_{A-M} and k_{M-A} can be determined experimental, the values for k_I, k_{II} and k_{III} for load and unload cycles by calculus.

Experiments regarding the influence of pre-compression have been made to find a good isolation. Tri-axial transducers placed on the structure and the upper mobile plate have determined the transmissibility of vibrations. We obtained good results in isolation in vertical field, but worst in horizontal field.

Next approaches will be focused to find out shapes which permit isolation in all directions, or if the results are unsatisfying, constructive solutions for integration of SMA elements designed to isolate in all directions. Meantime, theoretical background will be improved, in order to permit accurate design of this kind of isolating workplaces.

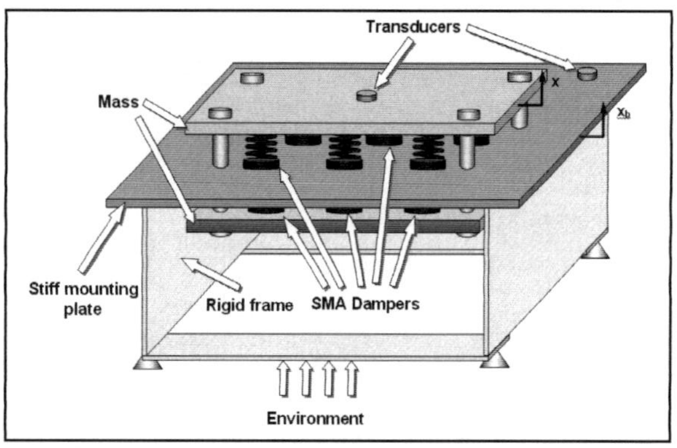

Fig. 4. Isolated workstation for sensitive equipment.

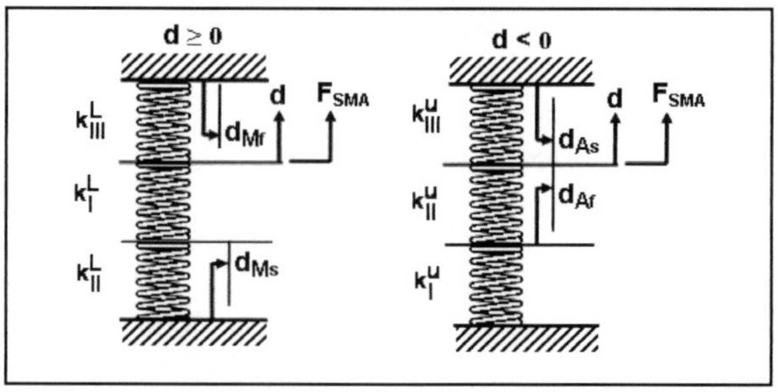

Fig. 5. Simplified mechanical model of SMA behaviour.

4. CONCLUSION

Elements from SMA are proper devices for vibration damping/isolation. Different models describing them, respecting shape, dynamic behaviour and/or temperature changes are available in the literature. Based on that, the authors have developed a work station for sensitive equipment for own use (electron microscopy). The best results for the spring form were obtained for semi-cylindrical elements, having the open side orientated to the stiff plate. Precompression of the spring can affect, positively or negatively the damping performances of the system. The excitation level defines the type of the loop; a higher level determines a major loop, being the favourable situation, while a low level determines a minor loop, where the behaviour of the SMA element is close to them of a linear spring. Because of the fact that the mass intended to be isolated has an important role in the effectiveness of the system, it can be concluded that the performance of the damping system is given by a combination of four parameters: form of the SMA, excitation level, mass and pre-compression. As future research the authors intend to develop specific form-dependent models for the SMA elements.

5. REFERENCES

Amariei, D.; Gillich, G.R.; Frunzăverde, D. & Micloşină, C. (2008). Study on the behavior of Ni-Ti Shape Memory Alloys in order to highlight the Memory Effect, *IEEE International Conference on Automation, Quality Testing and Robotics AQTR 2008,* pp. 517-524, Cluj, May 2008

Bratu, P. & Gillich, G.R. (2006). Reference Values in Evaluating Effects of Vibrations on Buildings and Persons. *Annals of "Eftimie Murgu" University Resita. Engineering Fascicule,* Vol XIII, Nr. 1, October 2006, pp. 111-116, ISSN 1453-7394

Feng Z.C. & Li, D.Z. (1996). Dynamics of a mechanical system with a shape memory alloy bar. *Journal of Intelligent Material Systems and Structures*, 7(4), July 1996, pp. 399–410, ISSN 1530-8138.

Fosdick, R & Ketema, Y. (1998). Shape memory alloys for passive vibration damping. *Journal of Intelligent Material Systems and Structures*, 9(10), October 1998, pp. 854–870, ISSN 1530-8138.

Graesser, E.J. & Cozzarelli, F.A. (1991). Shape-Memory Alloys as New Materials for Aseismic Isolation, *Journal of Engineering Mechanics*, 117(11), November 1991, pp. 2590-2608, ISSN 0733-9399

Thomson, P.; Balas, G.J. & Leo. P.H. (1995). The use of shape memory alloys for passive structural damping, *Smart Materials and Structures*, 4(1), March 1995, pp. 36–41, ISSN 0964-1726

Yiu, Y.C. & Regelbrugge, M.E. (1995) Shape-memory alloy isolators for vibration suppression in space applications. *Proceedings of the 36th AIAA/ASME/ASCE/AHS/ASC Conference: Structures, Structural Dynamics, and Materials*, pp. 3390–3398, New Orleans, April 1995.

Annals of DAAAM for 2008 & Proceedings of the 19th International DAAAM Symposium, ISSN 1726-9679
Editor B. Katalinic, Published by DAAAM International, Vienna, Austria 2008
Make Harmony Between Technology and Nature, and Your Mind will Fly Free as a Bird

APPLICATION OF NUMERICAL MODELING IN ASSEMBLY PROCESS

GODAL, D[aniel]; RUZAROVSKY R[oman] & TARABA, B[ohumil]

Abstract: *In the article is presented numerical modeling of an assembly process of a pneumatic cylinder cover with a cylinder body. The three-jaw gripper with a rotary actuator was used for manipulating with assembled parts. Loads necessary for accomplishing the assembly were obtained by proposing and realizing experiments. Analyzed process was divided into a six phases. The assembly process was modeled as a contact-nonlinear task. The results of numerical analyses are stress-strain fields and the deformed zone in the contact area. The results of numerical simulation show possibility of a particular assembly phase realization. The task was solved by a finite element method using an engineering-scientific program code ANSYS.*
Key words: *computer modeling, assembly, gripper, contact task, stress-strain fields*

1. INTRODUCTION

The article is focused on application of computer modeling of an assembly process in a flexible assembly cell. The article refers to a diploma work: Modeling and simulation of an assembly process of assigned work piece in a flexible cell (Debrecký 2007), which was solved at the Department of Production Devices and Systems. The numerical simulation is used for analysing assembly of a pneumatic cylinder cover. Particular assembly phase realisation and stress-strain states in the contact of gripper jaws with cylinder cover were observed. The cylinder cover is assembled by a three-jaw gripper and a rotary actuator. Three states were analysed: 1) minimal loads state, 2) maximal load state, 3) finding jaw force that produce only elastic deformation state. Assembly operation was divided to a 7 load steps. Parameters of assembly (force, moment, friction coefficient) were obtained experimentally. Building of simulation model, load-steps and stress-strain states are presented. Interpretive tool of numerical simulation was engineering and scientific program Ansys.

2. TEORETICAL BACKGROUND

Transmission of normal forces and moment from gripper jaws to the cylinder cover were modeled as a standard contact task. The contact governinig equations are published in (Zhong 1993). The friction force is given by the Coulomb friction model of normal component of gripping force (Bhavikatti & Rajashekarappa 1994)). There was elastic-plastic material model with ideal plasticity used for the cylinder cover and ideally elastic material model for the material of the gripper jaw. Total strain is given by summary of elastic and plastic relative deformations. Generation of plastic strains was evaluated by hypothesis Huber-Mises-Hencky (Trebuňa et al., 2002).

3. EXPERIMENT

Loading parameters for simulation model were obtained experimentally. Experiments were proposed and realised by authors of article. Force measurement (Fig. 1b, 2. Press). Force necessary for pressing the cover into the cylinder body were measured by pressing the cover into the cylinder against a digital scale.

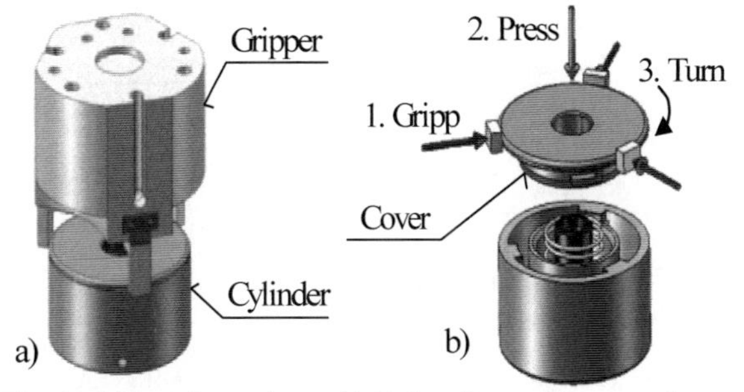

Fig. 1. a) Three jaw gripper, b) Order of assembly operations.

Turning moment measurement (Fig. 1b, 3. Turn). The cover was tightly clamped with clamp chucks. There was an arm fastened on the cylinder body. There was a 0.15 kg weight hung on the arm. Changing the distance between the weight and the cylinder was used for setting the torque necessary for the cover locking. Static friction coefficient measurement. Classical approach was used for measurement of static friction coefficient. Pneumatic cylinder was placed upside- own on a steel plate that could swivel. In a state when concurrent force system disequilibrium was impending (sliding occurred) the angle of repose was measured. Each experiment was composed of six tests. Measured values were evaluated statistically. Results are (assembly loads): pressing force 28 ± 1.15 N, friction coefficient 0.243 ± 0.008, turning moment 0.593 ± 0.006 Nm.

4. SIMULATION MODEL

Geometry of the model has been made in a CAD software Catia. Model was thereafter imported in a *.model format to the Ansys. Only one third of the cover and one gripper jaw were used for simulation. Beam elements (BEAM188, MPC184) were added to the model (ANSYS 10.0 2005). These elements enabled to apply a turning moment.

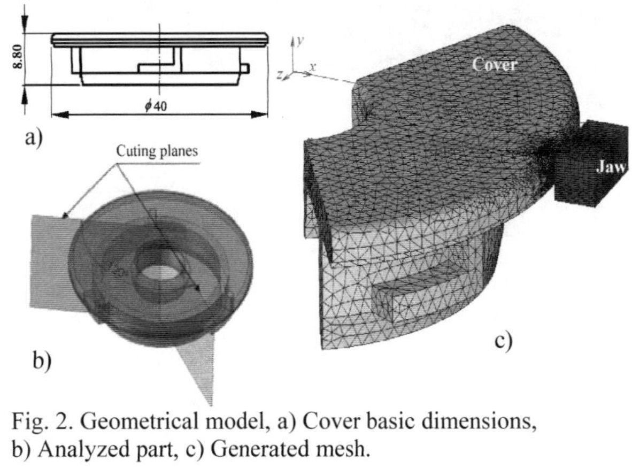

Fig. 2. Geometrical model, a) Cover basic dimensions, b) Analyzed part, c) Generated mesh.

Finite element mesh was intentionally refined in the contact area. Structural elements SOLID185, SOLID187 and contact elements CONTA174, TARGE187 were used. Algoritm of contact calculating was Augmented Lagrange Method. Considered material properties in analyze were: cylinder cover, material copolymer butadien-styren, elastic modulus $E = 2.2$ GPa, Poison's ratio $\nu = 0.35$, yield stres $R_e = 25$ MPa (Ehrenstein 2001); gripper jaw material carbon steel, elastic modulus $E = 210$ GPa, Poison's ratio $\nu = 0.3$.

There were three states analyzed: 1) assembly loads from experiments, 2) maximal loads (producer gripper data HGD-23-A Festo), 3) prediction of loads where elastic material behaviour of cover was accepted only, Tab. 1. Every analyzed state was divided into six time referring load-steps (LS).

Load pro jaw	Jaw slide	Initiation of the contact	Jaw force	Pressing force	Turning moment	Release
Load step	LS 1	LS 2	LS 3	LS 4	LS 5	LS 6
Assembly load	1 mm	1.003 mm	38.4 N	9.3N	0.198 Nm	0
Maximal load	1 mm	1.003 mm	120 N	9.3N	0.300 Nm	0

Tab. 1. Applied loads

The first LS 1 represents sliding the gripper jaw towards the cover to the distance 1mm. In the following step was the jaw approached to the cover so the contact commenced. In the third load step was the force applied on the jaw. Following step represented pushing the cylinder cover into the cylinder body. This step was simulated by aplying a experimenaly obtained force on the lower surfaces of the cover. In the LS 5 was applied turning moment. Two folowing steps were release of loads and pulling the jaw away from the cover surface.

5. OBTAINED RESULSTS

The results of modeled states are presented in Tab. 2. Equivalent Mises stresses higher as 25 MPa were solved in the jaw material (steel).

	Maximal equivalent Mises stress [MPa]					
Load step	LS 1	LS 2	LS 3	LS 4	LS 5	LS 6
Assembly loads	0	3.396	28.31	30.326	25.402	24.316
Maximal loads	0	3.396	31.45	39.535	37.135	24.975

Tab. 2. Maximal solved equivalent Mises stress [MPa].

Stress fields by assembly loads are presented. Fig. 3 shows the equivalent Mises stress field in the contact area for LS 4. In cover material was echieved Mises stress at the level yeild stress 25 MPa. The highest stress 30.326 MPa was solved for jaw body material.

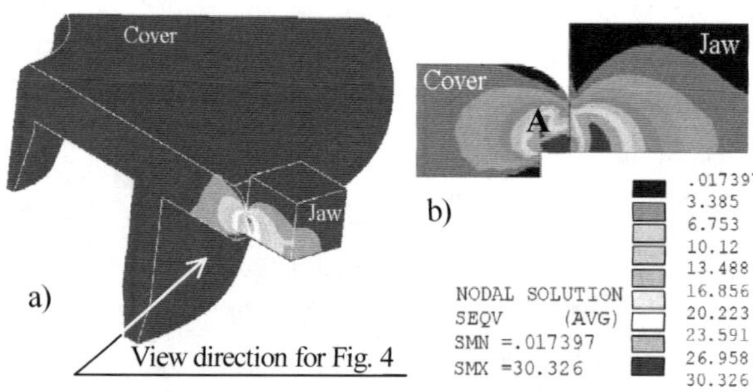

Fig. 3. Equivalent Mises stress field [MPa] for LS 4,
a) Crosssection view, b) Section plane.

The equivalent Mises stress field within the deformed zone of cover is in Fig. 4 presented. Deformed zone is ploted with scale factor 4. The sumar displacement in the section plane (Fig. 3) had maximal value 0.031 mm.

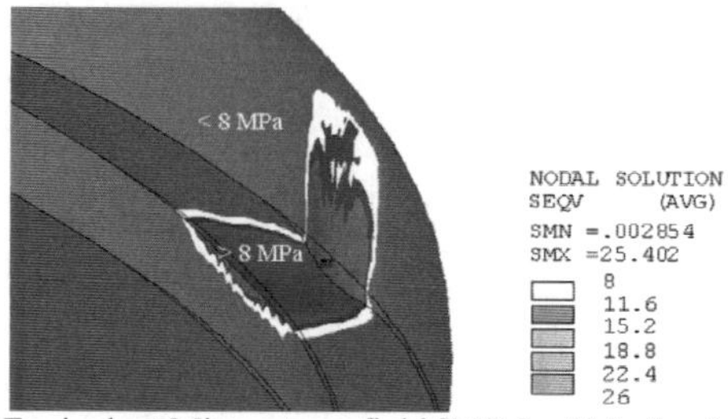

Fig. 4. Equivalent Mises stress field [MPa] within the deformed zone.

The stress-displacement dependance at point A is shown in Fig. 5. The point A is placed in the cover body and shown in Fig. 3. Effects of single load steps on the Mises stress generation at Fig. 5 is shown.

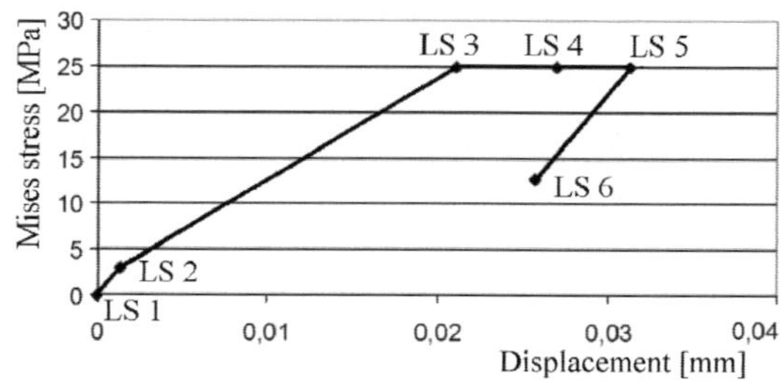

Fig. 5. Stress-deformation dependence at point A.

6. CONCLUSION

The assembly operation is not possible without yield stress exceeding of the cover material, shown the computer simulation. The cover shape after assembly will be in the end deformed and the contact zone will be visible. The assemble loads obtained of the experiment are for gripper HGD-23-A the working parametres. By applying of the maximal loads is the assembly proces always realizable bad the deformed contact area will be larger. The elastic behaviour of plastic cover material will be determined by applying of jaw force 13 N. The jaw force 13 N is to low for assembling of pneumatic cylinder cover with a cylinder body.

The furher research will be oriented at the implantation of computer modeling and numerical simulation into assembly processes in a flexible assembly cell. The main aim of the research is the prediction of load forces solution and their surface impacts at assembled parts.

The research has been supported by VEGA MS and SAV of the Slovak Republic within the project No. 1/3193/06.

7. REFERENCES

Ansys Theoretical Manual, Release 10.0. (2005). *Available from*: http://www.tsne.co.kr/intra/data_center/ansys/theory.pdf *Accessed* 2008-06-23

Bhavikatti, S. S. & Rajashekarappa, K. G. (1994). *Engineering Mechanics*. John Wiley & Sons, ISBN 81-224-0617-3, New Delhi

Debrecký, Ľ. (2007). *Model and simulation of the chosen product assembly process in the flexible cell.* Diploma work. MtF STU Bratislava

Ehrenstein, W., G. (2001). *Polymeric Materials Structure-Properties-Aplycation,* Hanser Publishers, ISBN 3-446-21461-5, Munich

Trebuňa, F., Šimčák, F., & Jurica, V. (2002). *Elasticity and strenght II*, Vienala, ISBN 80-7165-364-0, Prešov

Zhong, Z. H. (1993). *Finite Element Procedures for Contact-Impact Problems*. Oxford University Press Inc., ISBN 0-19-856383-3, USA

Annals of DAAAM for 2008 & Proceedings of the 19th International DAAAM Symposium, ISSN 1726-9679
Editor B. Katalinic, Published by DAAAM International, Vienna, Austria 2008
Make Harmony Between Technology and Nature, and Your Mind will Fly Free as a Bird
2nd DAAAM-ICAT

INJECTION MOULDING OF THERMOPLASTICS IN HYBRID TOOLS

GODEC, D[amir]; SERCER, M[laden] & PILIPOVIC, A[na]

Abstract: *One of the basic disadvantages of rapid prototyping (RP) processes is still the limited number of materials for the prototype production. Moreover, the physics of prototype production is different than the one for e.g. injection moulding. Such state causes application of RP processes for production of elements or complete moulds - Rapid Tooling processes (RT). Different properties of materials for RT mould inserts than classic mould materials could be disadvantage. Therefore, it is necessary to analyze the differences in moulding properties caused by application of RT mould inserts.*
Key words: *injection moulding, rapid tooling, hybrid mould*

1. INTRODUCTION

Application field of rapid prototyping processes for manufacturing of mould elements is newer dated. Therefore the influence of materials of prototype (hybrid) moulds on moulding properties is still insufficiently explored. Many RT processes include use of non-metal or non-ferrous materials opposed to classic mould manufacturing processes. Mould material properties substantially determine final properties of manufactured moulding (Karpatis et al., 1998). Therefore it is necessary to understand the influence of metal hybrid moulds on moulding properties (Gebhardt, 2003). In this study RT mould inserts are produced by Indirect Metal Laser Sintering process. The material used for RT inserts was the *LaserForm A6* steel, and for classic inserts 1.2312 steel. The injection moulding was performed with both inserts, and the properties of moulding were compared.

2. IMLS RAPID TOOLING PROCESS

The IMLS procedure uses powder which consists of metal component as the building material and the polymer component as the binder. In the first phase the insert was made by the usual SLS process which is the so-called *green* phase. The mechanical properties of such pre-form are not sufficient for the application in injection moulding. Therefore, the *green* insert needs solidification in the furnace. The process in the furnace can be divided into three steps: debinding, final sintering (*brown phase*), and bronze infiltration (Dalgarno & Stewart, 2001; King & Tansey, 2003). The IMLS procedure results in very good mechanical properties of the mould inserts (Fig. 1).

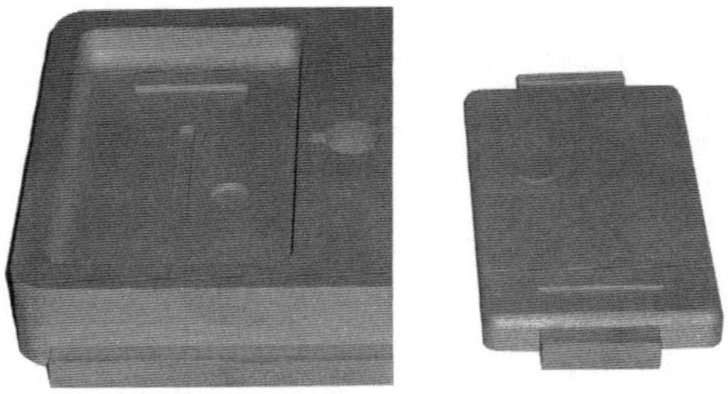

Fig. 1. Experimental *Rapid Tool* mould inserts (Godec, 2005)

The final mould inserts have low porosity, and they can be used for the production of over 100 000 moulded parts. Compared to direct metal laser sintering (DMLS) procedure, the disadvantage is the subsequent treatment in the furnace, and also greater deviations in dimensions (Dalgarno & Stewart, 2001; King & Tansey; 2003).

3. CASE STUDY

Based on the analysis from the literature (Pötsch & Michaeli, 1995; Gao et al., 2001), and performed screening design, it was concluded that there are four significant processing parameters of thin-walled injection moulding: injection time, packing pressure, packing pressure time and melt temperature. The packing pressure time, in case of the cold runner system, is reduced to the gate cooling time. An experiment was done to determine the maximum packing pressure time (in this study the value was 5,0 s). For further analysis, the remaining three significant parameters were used, maintaining maximum value of the packing pressure time.
The evaluation of the influence of prototype (hybrid) and classic mould, was based on the monitoring of several properties of the moulded parts. These included the moulded part surface properties (completeness of the moulded part and flush), mass, dimensional stability of the moulded part, deformation of the moulded part, and tensile impact strength.

3.1 Experimental results

For the analysis, the central composite design with three factors was used. The values of arithmetic means of the observed properties are given in Table 1. The resulting difference in the values of the mass of the moulded parts should be assigned o to the higher compressibility of the materials of the prototype mould inserts. Moreover, the flush on the moulded part produced in the hybrid mould occur at significantly lower packing pressures (Fig. 2.) This occurs also, to a lesser extent, in the results of the dimensional stability of the moulded part (*Dimension_1* and *Dimension_2*).

Moulded part property / Injection moulding parameter	**Arithmetic mean**	
	Hybrid mould	*Classic mould*
Mass (g)	8,26	7,82
Dimension_1 (mm)	100,22	100,03
Dimension_2 (mm)	58,12	57,97
Deformation_1 (mm)	1,59	1,47
Deformation_2 (mm)	0,98	0,93
Tensile impact strength (kJ/m^2)	96,6	123,48
Contact mould temperature (°C)	36,55	32,66
Temperature gradient (°C)	5,76	2,53

Tab. 1. Values of the analysed properties (Godec, 2005)

Higher values of deformations of the moulded parts can be explained by the fact that in the injection moulding, as a rule, higher cavity wall temperature was achieved in the hybrid mould (Fig. 3). The moulded parts in case of hybrid mould left the mould cavity at higher temperatures which resulted in greater shrinkage of the moulded parts.

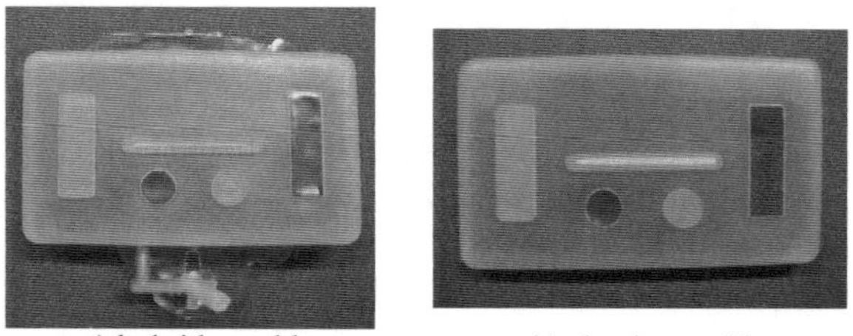

a) hybrid mould b) classic mould

Fig. 2. Occurrence of flush: a) p_p= 600 bar, ϑ_M = 240 °C, b) p_p= 1000 bar, ϑ_M = 240 °C (Godec, 2005)

The reason for such difference lies in the lower heat penetration coefficient of the materials of prototype mould inserts. Lower heat penetration coefficient prevents fast conduction of heat into the interior of the mould, so that higher contact temperature of the cavity wall is achieved. Also, the heating of such wall during a constant injection moulding cycle is faster, so that maximal mould cavity temperature is achieved faster than in the case of classic mould. On the other hand, in Fig. 3 one may observe that the temperature gradient in case of hybrid mould is somewhat greater compared to the classic mould. This can be explained by a higher value of thermal conductivity of the material of hybrid mould inserts, which results in faster change of temperature on the mould cavity wall.

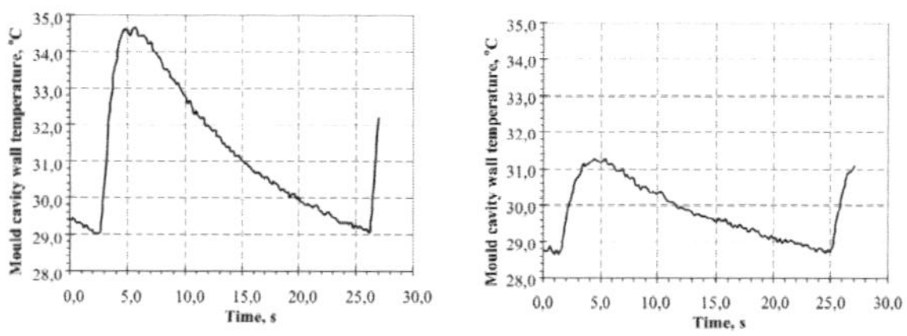

a) hybrid mould b) classic mould

Fig. 3. Mould cavity wall temperature field (Godec et al., 2008)

Impact tensile strength (toughness) of the moulded parts produced in the classic mould is much higher than the toughness of moulded parts made in the prototype mould. The higher toughness of moulded parts made in the classic mould can be explained by the lower temperature gradient of the cavity. In general, the heating and cooling of these mould inserts is slower which is suitable to acquiring higher degree of cristallinity and higher values of toughness.

3.2 Optimizing the processing parameters

Within the research, the optimization (complex and partial) of the parameters with the classic and the hybrid mould was done with the aim of optimizing the observed properties of the moulded parts. As the result of such optimization, the obtaining of the corrected adjustable processing parameters in case of hybrid mould was expected. Based on the results of the optimization, it is possible to provide basic guidelines for the adjustment of the processing parameters in the hybrid mould (Table 2), in order to obtain comparable properties of the moulded parts to those produced in the classic mould.

Parameter	Complex optimization
Packing pressure	+ 10 %
Melt temperature	- 5 %
Injection time	-60 %

Tab. 2. Guidelines for adjusting parameters in hybrid mould (Godec, 2005)

4. CONCLUSION

By comparing the prototype hybrid and classic mould for the production of thin-wall PP moulded parts, it can be concluded that the biggest differences in the properties of moulded parts result from higher compressibility of the prototype inserts (higher mass of moulded parts, larger dimensions of moulded parts, occurrence of flush at lower values of packing pressure), and the difference in thermal properties of mould inserts materials. The material of prototype mould inserts has lower heat penetration coefficient, which results in achieving higher maximum cavity wall temperature, but also higher temperature gradient than in the mould with classic inserts, due to higher thermal conductivity. Such condition causes higher values of shrinkage of the moulded parts and more difficult maintenance of the dimensional stability of the moulded parts. Also, in the prototype mould, lower values of tensile impact strength are achieved. In order to achieve comparable properties of moulded parts made in both moulds, it is possible to define adequate corrections of the most significant parameters of the injection moulding in using the hybrid mould. The work analyses pure influence of the materials of both sets of mould inserts on the properties of the moulds. The possibility of "conformal cooling" in the case of hybrid mould has not been used. Future research will analyze the influence of different configurations of cooling channels on the mould properties. (Godec et al., 2008)

5. REFERENCES

Dalgarno, K. & Stewart, T. (2001). "Production tooling for polymer moulding using RapidSteel process", *Rapid Prototyping Journal*, Vol. 7, No. 3, pp. 173-179, ISSN 1355-2546

Gao, F., Koresawa, H., Narahara, H., Suzuki, H. (2001). Evaluation of thermal environment in plastic injection mold, *Proceedings of ANTEC 2001*, pp. 969-973, ISBN 1-58716-098-6, Dallas, May 2001, Society of Plastics Engineers, Brookfield

Gebhardt, A. (2003). *Rapid Prototyping*, Carl Hanser Verlag, ISBN 3-446-21259-0, München

Godec, D. (2005). The influence of hybrid mould on thermoplastic moulded part properties, *D. Sc. Thesis*, Faculty of Mechanical Engineering and Naval Architecture, Zagreb

Godec, D., Šercer, M., Rujnić-Sokele M. (2008). Influence of hybrid mould on moulded parts properties, *Rapid Prototyping Journal*, Vol. 14, No. 2, pp. 95-101, ISSN 1355-2546

Karapatis, N.P., van Griethuysen, J.P.S., Glardon, R. (1998). Direct rapid tooling: a review of current research, *Rapid Prototyping Journal*, Vol. 4, No. 2, pp. 77-89, ISSN 1355-2546

King, D. & Tansey, T. (2003). Rapid tooling: selective laser sintering injection tooling, *Journal of Materials Processing Technology*, Vol. 132, pp. 42-48, ISSN 0924-0136

Pötsch, G. & Michaeli, W. (1995), *Injection Molding*, Carl Hanser Verlag, ISBN 3-446-17196-7, MünchenEquations are centered and numbered consecutively, from 1 upwards.

Acknowledgement

This work is part of the research included in the project **Increasing Efficiency in Polymeric Product and Processing Development** *supported by the Ministry of Science Education and Sports of the Republic of Croatia. The authors would like to thank the Ministry for the financing of this project.*

Annals of DAAAM for 2008 & Proceedings of the 19th International DAAAM Symposium, ISSN 1726-9679
Editor B. Katalinic, Published by DAAAM International, Vienna, Austria 2008
Make Harmony Between Technology and Nature, and Your Mind will Fly Free as a Bird

THE ALTERNATIVE TO STOCK EXCHANGE FINANCING

GRECU, E[ugenia]

Abstract: *The paper analyses companies' possibility of financing through the capital market, and the Bucharest Stock Exchange (BSE), in particular. So, we look at how companies may enter the primary market with an IPO – Initial Public Offering – and then on the secondary Stock Exchange market. The paper underlines both the advantages and the obstacles of being listed on the BSE for companies and investors, as well as the coordinates for a successful IPO.*
Key words: *financing, shares, stock exchange, advantages.*

1. INTRODUCTION

Developing a business often implies getting help from external resources; theoretically, from three directions. The first one is the so-called own resources, representing the profits reinvestment. The second source of financing is the bank loans. And its alternative is represented by the capital market, which is cheaper and easier to get. That is why companies choose to join the primary market with an Initial Public Offering and then the secondary Stock Exchange market. While in the USA, the highly developed capital market ensures about 70% of the companies' financing and the bank loans only 30%; in Europe, the capital market ensures 40% of the financing (Anghelache, 2004). In Romania, this type of financing is only at the beginning with just 1%.

2. THE ADVANTAGES OF BEING LISTED ON THE STOCK EXCHANGE

The Romanian companies listed on the Bucharest Stock Exchange (BSE) have the following advantages: The most important goal of a public offering made by a company is to attract money from the capital market, through the issue of new securities, shares. But, being accepted for the BSE quotation also facilitates any subsequent capital attracting operation, the issuer thus finding the necessary resources to continue the development of his/her activity. More and more listed companies get the needed financing through the Stock Exchange mechanism, increasing the social capital and asking for issuing premiums, paid by share-holders, who have a discount from the market price. An important advantage of capital markets is represented by the low costs of stock exchange financing. The statistics show that the expenses for the Initial Public Offering (IPO) on the European markets are 4-5% of the issue, while in the United Kingdom the percentage is 7%. In Romania, the IPO's costs varied from 5 to 10%. Anyway, they are small as compared to bank loans, where, in May, the interest for more then five years long loans varied from 12 to 20%. Another advantage of the Stock Exchange is that companies are not requested warranties, so the company may settle its own financing criteria, as compared to banks which generally offer standard products. Also another advantage for the listed companies is the free publicity in the country and abroad. The summary of the transaction from every transaction session, which also contains the name of the companies whose securities are admitted for the BSE Quotation, is sent to domestic and international press agencies. The companies with Stock Exchange quotations enjoy increasing credit. This leads to an increasing market share for them, in the disadvantage of competing companies. Getting more credit may also bring about commercial facilities, better credit and payment terms form suppliers and customers. The transparency of issuers, who have securities admitted for the BSE Quotation, when it comes to the information made public, make it easier to follow the transactions and quotations on the stock exchange market and attracts investors with easy and fair access on this capital market, thus increasing the securities' liquidity. For the quoted companies, the existence of a recognized market value is another advantage in fusion, taking over or any other "joint venture" type operation. Potential creditors of the issuer have the possibility to compare the shares' book value with their market value. The securities accepted for the stock exchange may be easier accepted as warranties for bank loans. Getting involved the employees, by allotting them shares, is another advantage for the stock exchange. The employees may be involved in the following ways: distributing to employees shares of the company's net profit and then paying them by allotting hoarding shares or shares of a new supplementary issue; giving employees preferential rights and/or further rights to buy issuer's shares in a supplementary shares issue. The latter strategy has been successfully applied by one of the blue-chips companies in the stock exchange: Transilvania Bank. Paying the creditors with shares represents another advantage of being listed in the stock exchange. The issuers registered in the BSE Quotation have the possibility to pay their debts to various creditors (suppliers, banks, etc.) with the equivalent of the amount in shares, if the creditors agree. This type of payment is possible as the respective securities have a market value and they are used for this purpose. Increasing the manager's quotation on the market is another advantage of the stock exchange. When the issuer's evolution is good, then the share holders and potential investors consider it the exclusive contribution of the manager.

3. OBSTACLES IN GETING LISTED ON THE STOCK EXCHANGE

An explanation for the small number of IPO to BSE consists in the fact that people are rather aware of the disadvantages of the stock exchange than its benefices. The large companies, which are truly interested in investors, are afraid to get public, mainly because they risk loosing control over and confidentiality of the business (Prunea, 2005). The companies seeking financing through the capital market may get discouraged by the long period of time needed to be listed to BSE with an IPO and by the low liquidity on the secondary market. The small number of IPO on the Romanian market is rather caused by the delayed answer, as it takes at least 6-7 months to draw an offering prospect, to get it accepted and then to make the IPO and get the money. For getting listed on the stock exchange quotation, companies have to pay a lot of commissions to BSE (for the administration of the file, registration and acceptance for transactions, a commission for the transaction of the sale initial public offering and a fee for continuing the transactions). Besides, they have to pay the transaction's intermediary (Anghelache et al., 2005).

4. IMPLICATIONS FOR INVESTORS

So far, most of IPO managed to attract investors, even though they did not get substantial profits in all cases after the listing of the respective companies on the Stock Exchange (Badea, 2005). As regards the level of subscription at IPO and the value of Stock Exchange transactions, it may be noticed that the trends are similar. The trend was expected to last, and, as the market develops, the number of IPO to remain small while the level of subscription in offering to increase even more; As compared to Warsaw Stock Exchange, where the number of IPO is bigger(only this year they are expected to be more than 50), the level of over-subscription is much higher and implicitly the allotment rates are lower(on average about 1%), in Romania, the most solicited offerings had a much higher allotment index (AlumilRom 6.07%, Transgaz in the case of large investors, more than 500000 RON invested: 2.56%).In the conditions of an expanding capital market and the lack of interesting companies on the BSE quotation, the few IPO carried out between 2004 and 2007 through the Stock Exchange system proved highly profitable for the investors that put high amounts on them. The players that subscribed for the IPO during the last years obtained impressive profits in the first day of shares' transaction at BSE. Until now, the most profitable public offering was SSIF Broker (that took place in 2004) as the subscribing investors got 159% profit if they sold in the first day of shares' transaction. The offering of AlumilRom Industry enjoyed the investors, as the profit in the first day was almost 100%.But, the poor beginning of 2008 on the Stock Exchange brought a change in this strategy, because the investors that subscribed for the offering of this year and sold within the first week could recover not even the invested amounts. In the period 2004-2007, the smallest profits that investors subscribing for the public offering could earn were about 10%, while the investors subscribing for IPO in 2008 did not reach such profits. For the Teraplast offering, the price of a right at the end of the fist day of shares' transaction on BSE was 1% smaller than the offering price; while in the case of « Casa de Bucovina », the fist day quotation dropped with 0.5%. For these titles, the second day of shares' transaction was even worse, as the price reduced with 13% as compared to the previous day and with 13.5% as compared to the offering price. The only exception to this rule was Transgaz, which in the first day of rights' transaction brought a profit of 53% for the IPO's subscribers, and in less than a week, the quotation raised to a maximum of 74.55% more than the IPO price. Along with the low interest in the shares of newly listed issuers, it is observed a drastic reduction of the subscription degree for the offerings place in 2008. Contor Group and Teraplast had subscriptions slightly over 100%, while for the STK the percentage was nearly 43%.An exception is « Casa de Bucovina» IPO, where the subscription rate was 142% (www.bvb.ro).

5. IPO'S SUCCESSFUL CONDITIONS

Although Romanian market can not praise with an impressive number of IPO, investors in BSE had both striking successes and losses. So, it is very important to precisely estimate the steps to be taken for the public offering and the listing (Stoica & Gruia, 2006). First of all, a company should choose the moment it goes public, then it should settle the "correct" price in order to be accepted by investors. As far as the moment is concerned, an IPO should be delivered for transaction when the market is increasing, as the quotations tend to increase and investors need new potential securities. It is more difficult to settle the selling price for the new shares, and this process could last 3-4 weeks. Moreover, if the offering price is not well established, on the stock exchange the shares shall be set right in the very first day. But, if the new shares are sold for a "correct" price, the offering shall be over-subscribed and the stock exchange quotation shall increase. The success of a company on BSE after the initial offering depends on three coordinates: the company and its management (interesting field of activity, good financial results, serious managers with long term strategic vision), the offering price (evaluation indicators smaller than the market average or the activity sector of the new issuer) and the Stock Exchange conditions. The over-subscription is the proof that all companies have been attractive because of the activity and the potential of the activity branch, the evolution of the price on BSE reveals the weak points: bad conditions of the market and high offering price. Because of the market conditions, the newly-come companies on BSE in 2008 suffered because of the lack of investors, as in this moment players orient themselves mainly to blue-chips. Also, the listing followed a long period of strong corrections of BSE actions, which reduced both the values of PER indicators and the price/book value. In the case of an IPO, the first objective is considered to be the capital attracted on the market, but the second objective, the one of liquidity in the secondary market of securities is just as important. Unfortunately, as compared to other stock exchanges in the region, the liquidity of these securities is relatively low. Still regarding the liquidity, on the shares' section BSE registered last year a cash flow more than 35 times smaller than the one of the actions carried out in 2007 on the main market of Warsaw Stock Exchange, where a lot of small IPO occurred, as a result of private investments, while the main vector for the large amount of public offerings are the retirement funds. The success of an IPO is also influenced by the company's position on the market (interesting field of activity, good financial results), the quality of the management (serious managers and having long term strategic vision).A company which is the leader of its market share has very good chances to close a successful IPO. Also important for success are the good, increasing financial results in the last year and a trustworthy managerial team.

6. CONCLUSIONS

The main goal of being listed on the stock exchange is to get financing from the free capital market; most of Romanian companies prefer getting financing from bank loans or leasing, rather than the Stock Exchange. But, being listed on the stock exchange brings about irrefutable advantages; the company and its management, the offering price and the time and conditions of the stock exchange are the success elements of an IPO. The small number of IPO on BSE reveals that this is a financing method which barely starts to be taken into consideration by business people in Romania.

7. REFERENCES

Anghelache, G.(2004). *Capital Market Characteristics. Evolutions. Transactions,* Economica Publishing House, ISBN 973-709-065-9, Bucharest

Anghelache, G.; Nan, S.&Badea, D.&Ioaca, A.(2005) *Agents of the Capital Market),* Economica Publishing House, ISBN 973-709-202-3, Bucharest

Badea,D.G.(2005).*Capital Market& Economic Reorganization,* Economica Publishing House, ISBN: 973-590-738-0, Bucharest

Prunea, P.(2005). *Capital Market. Reports of the Risk Challenge,* Economica Publishing House, ISBN: 973-590-544-2, Bucharest

Stoica,V.; Gruia, A.(2006). *Capital Markets and Stock Exchange Products,* University Publishing House, ISBN: 973-749-102-4,Bucharest

Capital Markets, Available from http://www.bvb.ro *Accessed:* 2008-05-22

Annals of DAAAM for 2008 & Proceedings of the 19th International DAAAM Symposium, ISSN 1726-9679
Editor B. Katalinic, Published by DAAAM International, Vienna, Austria 2008
Make Harmony Between Technology and Nature, and Your Mind will Fly Free as a Bird

A MATHEMATICAL APPROACH OF THE TEMPERATURE FIELD IN CASE OF ELECTRON BEAM WELDING

GRECU, L[uminita]; DEMIAN, G[abriela] & DEMIAN, M[ihai]

Abstract: *The aim of this paper is to establish a mathematical model regarding the thermal processes during electron beam welding (EBW). This model is based on the thermal transfer equations during the heating process. We analyze a piece temperature variation in terms of the process parameters. The numerical temperature distribution is obtained by means of a MathCAD application based on the mathematical approach. One can thus notice the way any of these parameters influences the thermal field during EBW.*
Key words: *electron beam, heat transfer, welding parameters*

1. INTRODUCTION

The Electron Beam Welding (EBW) is a welding process which uses the heat coming from a concentrated beam composed of high-velocity electrons impinging upon the surfaces to be joined.

This paper presents a mathematical and numerical model for the 2D temperature distribution which emerges during the stationary case of EBW, based on the equation of heat conduction. The convection and radiations losses are not taken into account in this paper because of the short heating time, and the small area we have considered (consequence of the fixed position of the EB). The numerical scheme which solves the mathematical model – derived by the finite difference method – is transposed into a MathCAD application that offers us a numerical solution for the thermal field. The temperature distribution and its peak value – closely connected to the depth to which the welding is performed – allow us to state that the developed mathematical model is an effective tool in predicting the temperature distribution and the depth to which the welding is performed, as well as in selecting process parameters in such a way as to enable the welding. Temperature distributions and shapes of melted zones at EBW are described using other methods too: using the line and point source theory (Eagar, 2002), and statistical analysis (Koleva & Mladenov, 2000).

2. ELECTRON BEAM WELDING

Among the most important applications of EB technology one could mention: melting, cutting, welding, microscopic scanning, etc. (http://www.twi.co.uk). A wide range of physical processes appear when matter is submitted to an electron bombardment, the main effect being heat generation. The EBW procedure belongs to the category of fusion welding which is characterized by a high intensity, and the phenomena that arise are totally different from those typical to fusion welding conventional processes (Radaj, 1992). Electrons always penetrate matter. At the impact of the beam with the material there is a penetration of the beam to a certain depth and the kinetic energy of the electrons is transferred to the material that is penetrated and transformed into heat upon impact.

On the one hand, the penetration is influenced by the characteristics of the beam itself. On the other hand, it is influenced by the characteristics of the material, practically by the way electrons interfere with the material microstructure.

The maximum penetration depth (the electron range) depends upon a lot of factors: electron beam current, accelerating voltage, electron beam diameter, electron beam power, material density, etc.

Absorptivity is a variable which should be carefully analyzed as it greatly influences the thermal field. It directly influences the material transferred energy amount.

Power absorption per unit of volume can be expressed as:

$$g(y,t)=\eta_A \frac{U\cdot j(t)}{r} f(y,r)=\eta_A \frac{P(t)}{S\cdot r} f(y,r) \tag{1}$$

where: $j(t)$ is the electron current density, U the acceleration voltage, $P(t)$ the power of EB, S the irradiation area of the EB, η_A beam power absorption ratio (normal value for steels ranges: 0.7 – 0.8.). In the above equation, f stands for the distribution function of the absorbed output density.

As electrons interact with material, x-rays, secondary electrons, and backscattered electrons are generated, and these phenomena bring alterations (fractionally cuts off the heat). Due to this loss, the absorption ratio of the beam output is chosen to be 0.8. From experimental studies different empirical formulas for the penetration depth (r) of an EB into matter have been obtained. For $10keV \le eU \le 100keV$, it can be expressed, with the formula:

$$r \approx 2.1\times 10^{-5}\frac{U^2}{\rho}, \tag{2}$$

Experimental data analysis allows us to find out that the distribution function of the absorbed output density is influenced by various factors, like: the distance between the current point and the bombarded area (y), the maximum penetration depth, the atomic number and consistency, etc.

For the distribution function of the output density this paper operates with the dependence described by the following ratio:

$$f(y,r)=1-\frac{9}{4}\left(\frac{y}{r}-\frac{1}{3}\right)^2 \tag{3}$$

The absorbed power distribution per volume unit as being dependant on the electron track inside the material describes some specific curves. One can notice that its maximum value is reached at about 1/3 of the beam track (Ying et al., 2003).

3. MATHEMATICAL APPROACH

We suppose to be welding two identical parallelepiped plates made of the same material. So we consider the workpiece to be a parallelepiped plate. The EB is considered to be a cylinder, of radius R, whose axis is parallel to the united faces, and which penetrates the material. Experimentally it was observed that heat spreading within the material is more intense in the direction of the beam action.

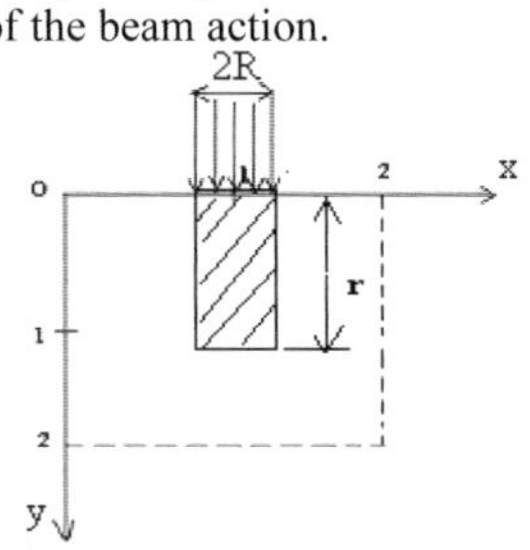

Figure 1. Plane section of the workpiece (2D domain)

Taking this fact into account we, will consider in this approach that the area in which the conduction heat transfer is studied is a two-dimensional one (the section of the workpiece with a plane parallel to the welding direction-Oy axis- passing through the projection of the beam centre on the superior face. The area under consideration is represented in Fig. 1

Considering so the 2D stationary heat transfer equation and an isotropic homogenous body with a constant thermal conductivity, the mathematical model is characterized by:

$$\left(\frac{\partial^2 T}{\partial x^2}+\frac{\partial^2 T}{\partial y^2}\right)=-\frac{1}{\lambda}q(x,y)\ ,(x,y)\in\Omega=[0,2]\times[0,2] \qquad (4)$$

where: λ is the thermal conductivity, q is a source-type term. The boundary condition is: $T(x,y)/_{\partial\Omega}=T_0, T_0$ -temperature of the environment. $q(x,y)$ depends on the energy generated by the EB and on the chosen material. For a uniform power distribution of EB, we model $q(x,y)$ by the following relation:

$$q(x,y)=\begin{cases}\eta_A\frac{U\cdot j}{r}f(y,t), & |x-1|\le R\ and\ y\le r\\ 0\end{cases} \qquad (5)$$

In order to solve (4), we apply the finite difference method (Antia, 2002). To discretize the area, we use 33 equidistant nodes along both axes: $x_i=ih$, $y_j=jh$, $0\le i,j\le 32$. By using the central finite difference formula we get the following linear system of equations:

$$T_{i+1,j}+T_{i-1,j}-2T_{I,J}+T_{i,j+1}+T_{i,j-1}-2T_{i,j}=-\frac{h^2}{\lambda}q_{i,j} \qquad 1\le i,j\le 32$$

where $T_{i,j}=T(x_i,y_j)$.

4. NUMERICAL RESULTS AND CONCLUSIONS

We have implemented the method described in this paper into a MathCAD application in order to obtain numerical solutions for the temperature field and to study the influence of the process parameters on the thermal field.

The numerical results obtained for a 13CrMo4 steel and process parameters: $U=60\,kV$, $I=80mA$ are performed in the following diagrams:

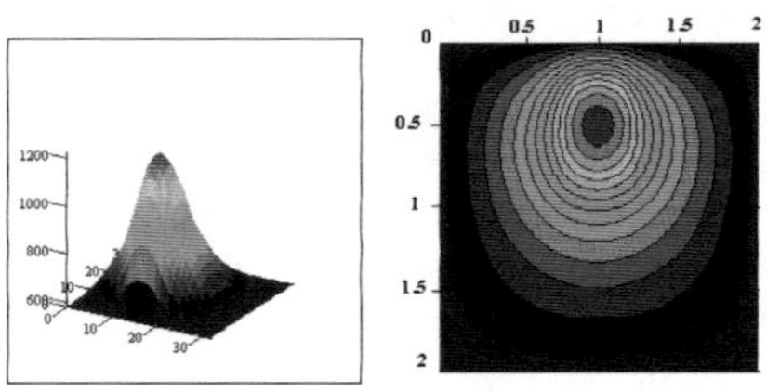

Figure 2. Temperature distribution for beam radius R = 0.2cm.

The first graph shows the nodal values of the temperature for R = 0.2cm, while the second graph shows the condition of the isothermal lines which appear inside the thermal field.

We notice that the temperature does not reach its top value at the surface the beam was incident to, but inside, i.e. at the point corresponding to the 1/3of the electron penetration depth - in agreement with the maximum value of the absorbed power distribution per volume unit (Ying et al., 2003).

The influence of the electron beam radius is studied considering different values of this parameter. For R=1.7, respectively 0.1 the numerical results are performed in Fig. 3, 4.

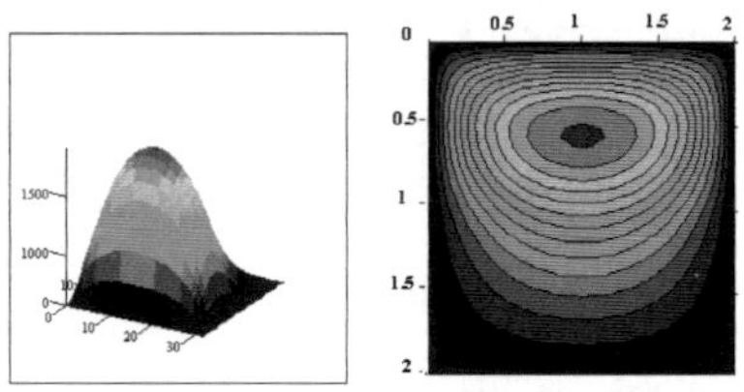

Figure 3. Temperature distribution for beam radius R= 1.7 cm

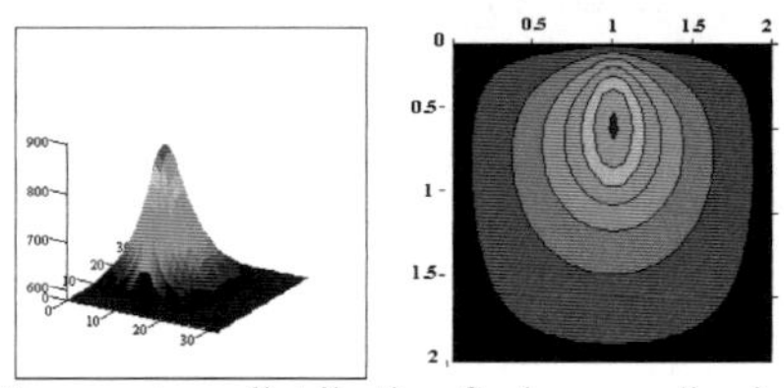

Figure 4. Temperature distribution for beam radius is R=0.1cm

One can notice that the point temperature reaches its peak value has not changed too much (as we have not changed the parameters influencing electron tracks within the material under consideration). A significant modification is connected to isothermal lines and temperature values within the area under discussion.

The results indicate the fact that in any parameter situation, maximum temperature level is under the surface which is directly exposed to the EB action. Temperature distribution and isothermal line form change and depend on the EB radius. These results are in concordance with others in literature, which are obtained by others techniques or experimentally.

The MathCAD application can also be used in the future to study the diagram of the process predicting melting bath formation, and it also offers the opportunity to study the effect of the material thermal properties may have upon the thermal field., and for making comparison studies. In a further work the non-stationary case will be considered and also the 3D case.

5. REFERENCES:

Antia H. M. (2002). *Numerical Methods for Scientists and Engineers*, Birkhauser, ISBN-10: 3764367156.

Eagar, T.W., *Heat flow and laser/electron beam welding*, MIT courses, October 2002.

Koleva, E. & Mladenov, G. (2000). Analysis of the thermal Processes and Shapes of Melted Zones at Electron Beam Welding and Melting, Proceedings of the 4th General Conference of the Balkan Physical Union, 22-25 August 2000, Veliko, Turnovo, Bulgaria, pp.83-96, ISSN 1310-0157.

Radaj, D. (1992). *Heat effects of welding – Temperature Field Residual Stress Distorsion*, Springer – Verlag Berlin, ISBN-10: 0387548203

Ying Qin & Chuang Dong & Xiaogang Wang & Shengzhi, Hao & Aimin Wu & Jianxin Zou, & yue Liu. (2003). Temperature profile and crater formation induced in high current pulsed electron beam, processing , *Journal of Vacuum Science and Technology* 2003, vol. 21, Issue 6, pp 1934-1938, ISSN: 0734-2101.

Annals of DAAAM for 2008 & Proceedings of the 19th International DAAAM Symposium, ISSN 1726-9679
Editor B. Katalinic, Published by DAAAM International, Vienna, Austria 2008
Make Harmony Between Technology and Nature, and Your Mind will Fly Free as a Bird

INFLUENCE OF INTERNET ON CONSUMER BEHAVIOUR IN TOURISM

GRGONA, J[adranko] ; ANDRLIC, B[erislav] & DEL VECHIO, M[anuel]

Abstract: *Consumers in tourism are very heterogeneous social group. For the purpose of implementing marketing plan in this situation, there is indispensability of identifying their behaviour in buyer decision process. Buyer behaviour in tourism is affected by many different factors, yet understanding it is essential task of marketing management. Undoubtedly, one of the most important factors is new technology development on the bidding side, which leads to new consumer behaviour modality. During development of new technologies on the bidding side, it is necessary to determine their affect on consumer behaviour in tourism This paper attempts to identify a level of usage of Internet as an information source and product-purchasing tool by students in eastern Croatia.*
Key words: *Internet, buying decision, tourism, students*

1. INTRODUCTION

Tourism is at the same time one of the most significant yet misunderstood phenomenon nowdays. The globalisation process is characterised by stronger competition and ever-stricter marketing conditions. Therefore, marketing management in tourism and consumer behaviour become a necessity in the struggle for a market and buyers resulting in continuous work in this field.

There is a general lack of information on how these consumers use the Internet for information, booking and purchase of travel products and services. The aim of this paper is to determine influence of Internet on consumer behaviour in tourism and to identify the levels of usage of Internet in student population. By using scientific methods of interview, analysis and synthesis, purpose and terms of certain parts of research are described.

2. CONSUMER BUYING BEHAVIOUR IN TOURISM

The consumers buying behavior in tourism has been always a interesting marketing theme, extensively studied and debated over the last decades. Consumers buying behaviour is behaviour of final customers who buy goods or services for personal consumption. The central point og marketing planning in tourism is to identify how tourists respond to the different marketing stimuli, becacuse they are very heterogeneous social group (Senecic & Grgona, 2006).

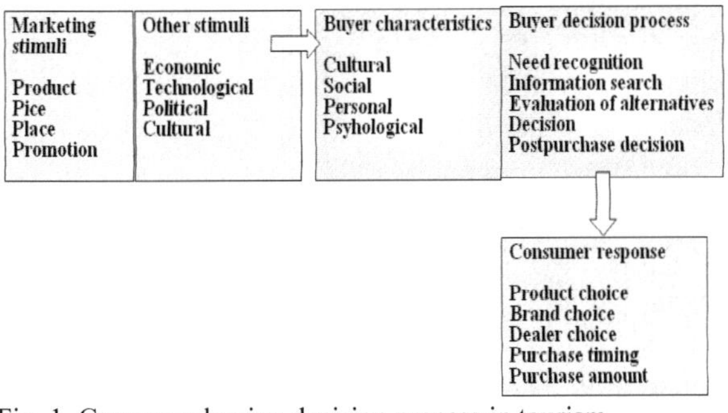

Fig. 1. Consumer buying decision process in tourism

The predominant approach, explaining the fundamentals of consumer behavior, describes the consumer buying process as a learning, information-processing and decision-making activity divided in several consequent steps (Engel, et al., 1995):

1. Problem identification
2. Information search
3. Alternatives evaluation
4. Purchasing decision
5. Post-purchase behavior

The starting point for studying consumer buying behavior is the stimulus response model shown in Figure 1. As this model shows, both marketing and environmental stimuli enter the tourist's consciousness. In turn, the tourist's characteristics and decision process lead to certain purchase decisions. The marketer's task is to research what happens in the tourist's consciousness between the arrival of outside stimuli and the tourists's purchase decisions. As this model shows, a consumer's buying behavior is influenced by internal and external factors. Starting with problem recognition, the tourist passes through the stages of information search, evaluation of alternatives, purchase decision, and postpurchase behavior. As this approach demonstrates, the buying process in tourism starts long before the actual purchase and hasconsequences long afterward. (Kotler, 2001). Although the approach implies that tourists pass sequentially through all five stages in buying a product, tourists sometimes may skip or reverse some stages. However, this model captures the full range of considerations that arise when a consumer faces a highly involving new purchase.

3. THE IMPACT OF THE INTERNET ON BUYING BEHAVIOUR IN TOURISM

Information technology has changed fundamentally the very nature of marketing in tourism. Traditional marketing activities are being translated into their electronic equivalents which gives rise to entirely new and particular form of e-marketing in tourism. The development of the Internet and the World Wide Web in the 1990s as a means for the global sharing of information has opened up a whole range of new possibilities in marketing practice. The Internet serves as a new communication and distribution channel for e-travellers and suppliers of travel services and products (Law et al., 2004). The changes in information technology are at the same time threatening and stimulating for the potentials of business entities to develop enduring relationships with their customers, which sustains the theory of marketing relationship. Internet enables the customer to be in direct contact with business entity any time of day or night, from any location, providing him with all the information necessary for buying. The information can be tailored to the needs of a specific customer. Internet is also an easy tool that enables the customers to find available information on competitive products and services. he Internet provides them with the means to gain immediate access to relevant information of greater variety and depth than has been available previously, about destinations throughout the world;

and to book quickly and easily. For tourism destinations and businesses, it offers the potential to make information and booking facilities available to large numbers of consumers at relatively low cost; it enables them to make large-scale savings on the production and distribution of print and on other traditional activities (e.g. call centres and information centres); and it provide a tool for communication and relationship development with tourism suppliers and market intermediaries, as well as end-consumers. It is also an easy way for the business entities to find new consumers while at the same time rendering their services to the existing ones. For that purpose they use databases containing personal profiles of customers.
Tourism products and services are perfectly suited to online selling because they possess the necessary characteristics that can function in the electronic environment. three main characteristics of products and services are likely to influence Internet usage for both information searching and making online purchases. These are: cost and frequency of purchase; value proposition; and degree of differentiation. (Peterson et al., 1997). Tourism products engage a higher level of involvement, intangibility and higher level of differentiation than other tangible consumer goods and, therefore, are more easily sold through the Web. The development in electronic communication is significant for tourism marketers who sell a perishable, intangible, heterogeneous and high-risk product. The ease of description and commodity-like nature of many travel products (i.e. airline seats or hotel rooms) also favour the development of electronic commerce (Lewis & Semeijn, 1998).

4. RESEARCH RESULTS

The consumer's decision in the tourism sector is very complex because he demands services in several areas of the economy. E-marketing is growing at a dramatic pace and is significantly impacting customer and business market behaviour in tourism. Because of these reasons, main focus of further research is influence of Internet on buying behaviour in tourism. There is a general lack of information of how students use Internet for information, booking and purchase of travel products and services. This research is exploratory, in that it tests ideas about what factors are important to decisions about Internet purchases in tourism. The purpose of research is to examine the relationships between Internet purchasing and tourist factors. Specifically, how do tourist personal factors affect individual intentions to make Internet purchases and actual purchasing behavior? To answer this question, a empirical research was conducted. Data collection took place in April 2008. A sample of 120 students were recruited from all studies of Polytechnic of Pozega to complete a survey that contained questions about influence of Internet on buying behaviour in tourism.

1. Do you you use internet for information and buying in tourism?		
Yes	49	40, 83%
No	68	56, 66%
Without answer	3	2, 51%

Tab. 1. Level of Internet usage

2. Which are the reasons nonuse of Internet for information and buying in tourism?		
I don't have Internet access	16	23,52%
Unsufficient knowledge about Internet possibilities	14	20,58%
Paying risk and privacy	31	45,58%
Other ways of buying	6	10,32%

Tab. 2. Reasons for nonuse of Internet

3. Which other sources you use for in buyer decision process in tourism?		
Radio	3	4,41%
TV	4	5,88%
Magazines	9	13,23%
Friends	24	35,29%
Family	18	26,47%
Company's staff	6	8,82%
Own experience	4	5,9%

Tab. 3. Other sources in buyer decision process

4. Where do you live?		
Pozesko slavonska county	12	17, 64 %
Osjecko baranjska county	20	29, 41 %
Brodsko posavska county	25	36, 76 %
Vukovarsko srijemska county	11	16, 19 %

Tab. 4. Place of living

Male	42	61,76 %
Female	26	38, 24 %

Tab. 5. Tourists segmentation

It is difficult to imagine that today exists big number of students who do not use Internet for information and buying in tourism. Presently there are no relevant information on the degree of e-marketing usage by student users in eastern Croatia. On the basis ofthis research, we can conclude that Croatian students, even though limited by a considerably small number of Internet users, still have to change their consumer habits and to follow the trends of more developed countries.

5. CONCLUSION

The rapid development of modern telecommunication and digital technology, and the resulting Internet technology, provides a number of new possibilities in marketing in tourism. Nevertheless, it seems that the vast majority of students prefer personal contact to make a buying decision.In the tourism sector, however, the use of Internet for this purpose is below average in eastern Croatia.
It is possible to conclude that e-marketing will grove at a dramatic pace and significantly impact customer and business market behaviour in tourism, what will be the subject of future research.

6. REFERENCES

Engel F. J., Blackwell D. R. & Miniard W. P. (1995). *Consumer Behavior*, The Dryden Press, ISBN 0-03-076751-2, Fort Worth

Kotler, P. (2001). *Marketing Management*, Prentice-Hall, Inc., ISBN 0–536–63099-2, New Jersey

Law, R., Leung, K. & Wong, J. (2004). The impact of the Internet on travel agencies, *International Journal of Contemporary Hospitality Management*, Vol. 16, No.2, p.100-107. ISSN 09596119

Lewis, I. & Semeijn, J. (1998). The impact of information technology on travel agents, *Transportation Journal*, Vol. 37 No.4, pp.20-26.

Peterson, R.A., Balasubramanian, S. & Bronnenberg, B.J. (1997). Exploring the implications of the Internet for consumer marketing, *Journal of the Academy of Marketing Science*, Vol. 24, pp.329-346.

Senecic, J. & Grgona J. (2006). *Marketing management in tourism*, Informator, ISBN 953-6286-74-2, Zagreb

Annals of DAAAM for 2008 & Proceedings of the 19th International DAAAM Symposium, ISSN 1726-9679
Editor B. Katalinic, Published by DAAAM International, Vienna, Austria 2008
Make Harmony Between Technology and Nature, and Your Mind will Fly Free as a Bird

THEORETICAL RESEARCH AND EXPERIMENTAL TESTS FOR A NEW VERSION OF TUBING AND CASING THREAD CONNECTION

GRIGORAS, S[tefan]; HADAR, A[nton]; MARIN, C[ornel]; STIRBU, C[ristel] & STOICA, G[ina]

Abstract: *One of the modern constructive solutions for tubing and casing joint is the integral two steps threaded joint. The buttress thread assures the joint functioning. The sealing metal/metal zones are placed, one on the outer of the joint and the second on the inner zone of the box. A plastic material gasket may be used, as anticorrosive barrier, in the inner sealing zone. A central shoulder assures the final contact between the pin and the box and limits the make-up torque. The experimental tests, which we have performed, confirm the theoretical functional parameters and the technological characteristics of the joint elements.*
Key words: *tubing, joint, extractive industry, design, contact.*

1. INTRODUCTION

The extractive industry uses many types of joints for tubing and casing. The integral joint is obtained of the pipes bodies.

Figure 1 presents the aspect of an integral two steps cylindrical thread joint (D_{OP} - the outer pipe diameter; D_{IP} - the inner pipe diameter; CS - central shoulder; OS - outer seal; IS - inner seal; plastic gasket). The thread is placed on two cylindrical steps. The maximum outer diameter D_{OBOX} - and the minimum diameter D_{IPIN} are limited by the bore and the drill diameter (the well technical conditions).

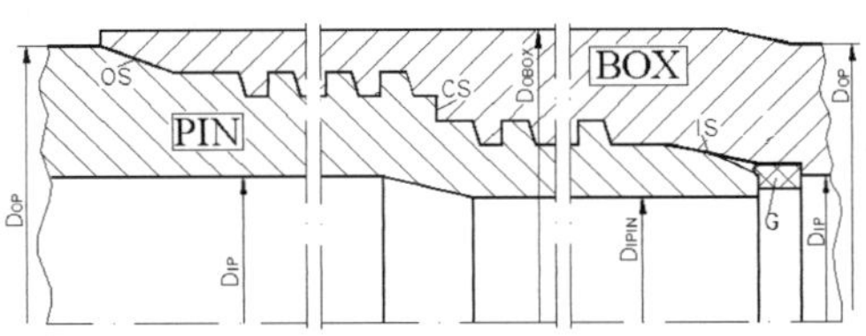

Fig. 1. The aspect of an integral joint

2. THEORETICAL APPROACH

The joint functioning is assured by the thread, by the sealing zones, by the minimum area of the cross section, the wall thickness, by the bending and combined resistance of the critical area and by the thread resistance and screwing capacity. The main functional parameter of any joint is the make-up torque. The cylindrical thread is placed on two steps. The two steps have pitch continuity, for manufacturing process assurance. The thread assures the strength end the montage. The metal/metal contacts represent the sealing zones, presented by figure 2. In the inner contact (fig.2a), the theoretical spherical end of the pin (radius R) is replaced by two tapered zones, AC_I and C_IB. The stiffness of the metal/metal linear contact, of the end of the pin, of the gasket and the elastic deformations of the contact, of the gasket and of the pin generates the axial load F_{aI}. This load must be produced by the thread. The outer seal (fig.2b) is a metal/metal linear contact too. The stiffness of the contact, of the end of the box and of the pin is the main parameter of the axial load of this zone, F_{aO}. This load is produced by the screwing. The maximum

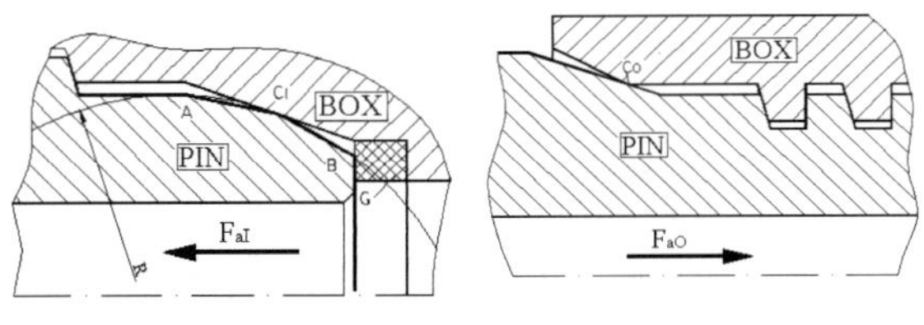

a) – inner contact b) outer contact
Fig. 2. The sealing metal/metal contacts

value of F_{aO} is limited by the box resistance and the minimum value is shown by the sealing functioning.

The contact between the pin and the box, on the central shoulder (figure 3) is the contact which limits the value of the makeup torque. The contact pressure brings about to the axial load F_{aCS} determination, by limiting the pressure values to the material yield limit. The make-up torque is the functional parameter which assures the load on the contact.

The friction forces in all the described contacts, like the friction in the thread, produce the friction global torque, which is the make-up torque:

$$T = T_I + T_O + T_{TH} + T_{CS} \tag{1}$$

where: TI, TO, TTH and TCS are the friction torques in the inner (index I), outer (index O) contacts; in the thread (index TH) and on the contact shoulder, respectively. Another parameter of the joint functioning is his efficiency, defined by the ratio between the critical area of the joint, AcrJ (for traction) and the area of pipe section, AP, (American Petroleum Institute, 1993):

$$\varepsilon = \frac{A_{crJ}}{A_P} \cdot 100 \ [\%] \tag{2}$$

efficiency -defined by (2) - is around 60...70%, for this joint type. Other loads on the joint are the outer pressure, the inner pressure, the traction (column weight), the bending of the column and the thermal lengthen of the column (compression). The thread loads are the nonlinear distribution of traction load along the joint, the wear produced by the screwing and the bending and compression effects. An element of the wall of pipe or joint is represented in figure 4.

The components of equivalent stress in the wall are:
-the normal axial stress: $\sigma_{ax} = f(\sigma_t, \sigma_b)$;

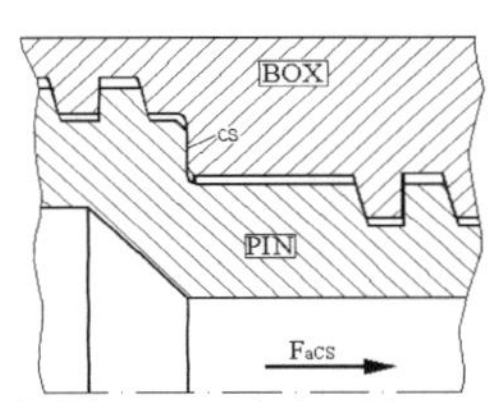

Fig.3. The central shoulder.

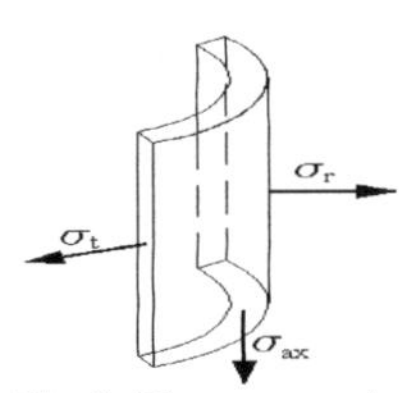
Fig.4. The stresses in the wall of pipe and joint

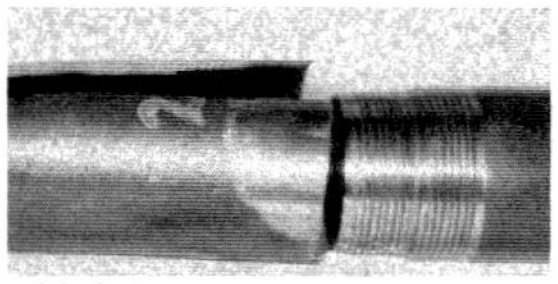

Fig. 5. A destroyed joint

-the normal radial stress: $\sigma_r = f(p_i, p_o, p_c, p_s)$;

-the normal tangential stress: $\sigma_t = f(p_i, p_o, p_c, p_s)$,

where: σ_t is the traction stress, σ_b - the bending stress, p_i - the inner pressure, p_o- the outer pressure, p_c- the equivalent compression pressure produced by the thread and p_s- the pressure produced by the screwing, (Gafitanu et al., 1996). The equivalent stress, calculated using these components limits the load of the pipe and joint (Gafitanu et al., 2001, Stirbu, 2002).

3. EXPERIMENTAL TESTS

Experimental tests have been developed on different lots of joint, for many sizes and for different materials too. The make-up torque was calculated by our theoretical method (American Petroleum Institute, 1993). The aim was to determine the resistance of the joint and the sealing efficiency evaluation. We have developed tests under inner pure pressure. For the initial tests, water and oil were the active fluids. The pressure can be considered as the normal radial stresses in the pipe and joint wall. They stop at the yield limit of the pipes steels (maximum values were 1000 bar). Figure 5 presents the aspect of a joint, destroyed under inner pressure. The sealing functioning was perfect, if the tests keep the values of the makeup torque and of the manufacturing precision. The combined tests represent the experimental evaluation of the functional and strength parameters of the joints, simultaneously for minimum two loads: the inner pressure and the traction force. Table 1 presents the results of the tests, for 3 specimens of the joint. The sealing efficiency is remarkable (Ianus et al., 2007).

Specimen	Inner pressure [MPa]	Traction force [N]	Sealing efficiency
1	70…71	$8 \cdot 10^5$	Losses=0 /3.5 hours
2	71…73	$8.5 \cdot 10^5$	Losses=0 /72 hours
3*	69	$7.8 \cdot 10^5$	Losses=0 /one hour
* - test on joints without gasket.			

Tab. 1. Tests parameter for combined load.

We have created the condition for the experimental tests developed under inner pressure, traction forces and bending of the column, taken over by the joints. The testing fluid was the liquid nitrogen. The angle of the column bending was maximum 10^0/20m column length, (American Petroleum Institute, 1993). In this case, the axial stresses in the joint wall increase with 120 %, and the metal/metal seals lessen the normal loads on the contact surfaces.

The minimum value of the make up torque assures the sealing efficiency, in perfect conditions and the resistance is

Fig. 6. Test under inner pressure, traction force and bended column

perfect, too. Figure 5 shows a test under bending, inner pressure and traction on a column obtained of three joint pipes. The maximum inner pressure was 40 MPa.

4. RESULTS

Figure 6 presents the theoretical axial load repartition on the thread pitches, along a joint between two pipes. The 1 curve represents the load produced by screwing and the curve number 2 is the repartition of the column weight. The superposition of the two component leads to a uniform load distribution. The figure number 7 presents the evolution of make-up torque values for any used steel, according to (American Petroleum Institute, 1993). The experimental tests show that the joint assures the sealing and resistance parameters for well column in the oil or gas extractive industry.

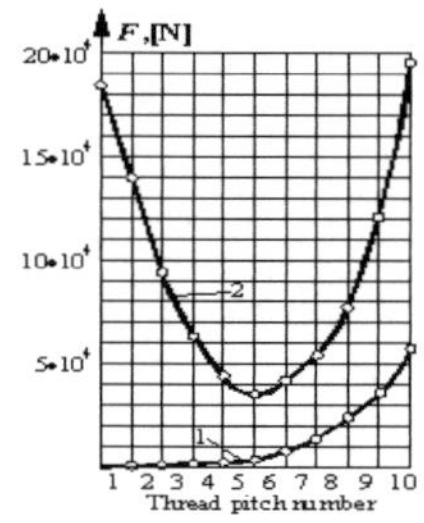

Fig. 7. Axial load distribution

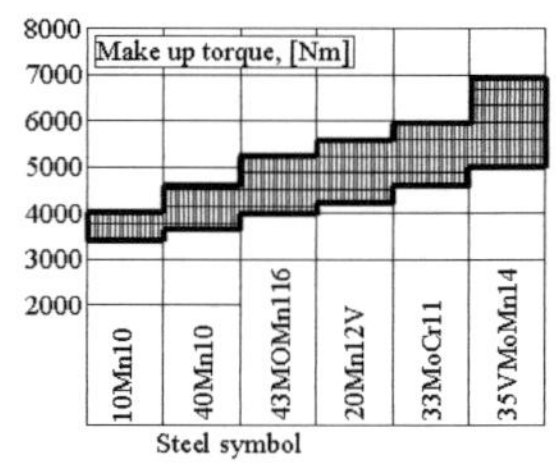

Fig.8. The make-up torque

5. CONCLUSIONS

The presented joint is an original construction, because of its thread, sealing zones and manufacturing process. The dimension of the joint are placed in the same domain like Hydril, Extrem Line, Vam and other companies. Joint performances are greater comparing with all the analyzed joints.The experimental tests on pipes and joint confirm the choice of a type of joint (the technical solution) and the original algorithm of joints design. The tests developed under pure inner pressure show the maximum efficiency of meta/metal seals, the role of anticorrosive barrier of the gasket and the resistance of the pipe and joint.The combined tests (inner pressure and traction forces) determine the diminishing of contact stresses in the metal/metal seals. If the central shoulder is rationally placed and all the length dimensions depend on its position, the optimum functioning of the joint is assured. The manufacturing accuracy has a very important role in the joint evolution. The main advantage of the new joint is the diminishing of thread stresses and the time of the screwing (reduced with 50%).

6. REFERENCES

American Petroleum Institute (1993), Bulletin on Formulas and Calculations for Casing, *TUBING, Drill Pipe, and Line Pipe Properties*, ISO 10400/1993.

Gafitanu, M.D, Grigoras, St. & Stirbu, C. (1996), Original Joint for Casing and Tubing. Theoretical Analysis, *Proceedings of the 26 – The Israel Conference on Mechanical Engineering*, Haifa.

Gafitanu, M.D., Grigoras, S. & Stirbu C. (2001), *Functional Aspects of a Thread Joint for Tubing and Casing*, Second World Trybology Congress, Vienna.

Ianus, G., Grigoras, St., Hanganu, C.L. & Borzan, M. (2007), The Effects of the Damaged Structure of Grease's Soaps on Ball Bearings Vibrations, *Mat. Plast.*, 44 (1), p. 47-50.

Stirbu, Cr. (2002), *Tubing and Casing Joints. Resistance and Functional Aspects,* Meridian Engineering, no.2, Edit. TEHNICA INFO, Chisinau, Moldova.

Annals of DAAAM for 2008 & Proceedings of the 19th International DAAAM Symposium, ISSN 1726-9679
Editor B. Katalinic, Published by DAAAM International, Vienna, Austria 2008
Make Harmony Between Technology and Nature, and Your Mind will Fly Free as a Bird

STUDIES CONCERNING THE MAINTENANCE OPTIMIZATION OF POWER PLANT STEAM CONDENSER

GRIGORE, R[oxana]; BADEA, A[drian] & HAZI, G[heorghe]

Abstract: *Maintenance of the steam condenser is of the great importance for reliable and efficient power generation. The paper presents a mathematical model for optimization the maintenance of steam condenser, by minimizing the total cost of condenser and considering the effect of fouling on the heat transfer surface. The optimization problem is a nonlinear one dimensional problem with restrictions. This problem is solved using gold section method. The proposed method is tested for a steam condenser of a DSL 50-1 turbine.*

Key words: *maintenance, optimization, steam condenser, fouling, total cost.*

1. INTRODUCTION

The steam condenser is a major component of a power plant. The condenser plays two important roles (Badea et al., 2003): it converts the used steam back into water for return to the steam generator as feedwater and it increases the cycle's efficiency.

The performance of steam condenser decreases in time and can be restored to its initial state by performing cleaning operations. If the time period between two cleanings is small, the probability of the appearance of faults due to tube fouling is very small. In the same time, a big number of maintenance actions mean supplementary costs and less delivered energy in National Power System. Under the circumstances, it needs the assurance of a correct maintenance policy and an optimization of time period between two condenser cleanings (Putman, 2004).

2. THE PROPOSED MATHEMATICAL MODEL

The objective function of the optimization method is an economic function and it is represented by total condenser costs due to fouling (Grigore & Hazi, 2008):

- Preventive maintenance cost;
- Corrective maintenance cost;
- Cost due to less energy produced pursuant to the appearance of tube fouling;
- Cost due to production losses during planned and unplanned shutdowns due to fouling.

The variable is represented by the time between two actions of scheduling maintenance.

2.1. Preventive maintenance cost

The relation for one preventive maintenance activity is:

$$C_m = t_{mp} \times r_m \times n + C_{mat} \text{ [m.u./interv.]}, \qquad (1)$$

Where: t_{mp} – the duration of preventive maintenance activity in hours, r_m- the schedule average tariff for an worker in m.u./h, C_{mat} – materials costs which appear for cleaning the condenser in m.u./intervention. The variable of objective function is T_i in hours.

The number of action of scheduling maintenance for steam condenser in one year will be:

$$N_m(T_i)=T/T_i \text{ [interv./year]}, \qquad (2)$$

Where: T – number of operation hours in one year. For a power plant, $T=8760h$. The total cost of preventive maintenance actions became:

$$C_{MP}(T_i) = \frac{T}{T_i}(t_{mp} \times r_m \times n + C_{mat}) \text{ [m.u./year]} \qquad (3)$$

2.2. Corrective maintenance cost

The number of corrective maintenance actions is determinate by the average number of faults $\mathbf{N_d}$ $\mathbf{(T_i)}$. The corrective maintenance cost will be:

$$C_{MC}(T_i) = N_d(T_i) \times C_d \text{ [m.u./year]}, \qquad (4)$$

$$C_{MC}(T_i) = N_d(T_i) \times (t_{mc} \times r_m \times n + C_{matc} + C_{sch}), \qquad (5)$$

Where: C_{sch} – the average cost for a replacement of an damaged tube or equipment in m.u./intervention, C_{matc}- material costs in m.u./intervention, t_{mc} – average time for one corrective maintenance action in hours.

2.3. Cost due to less energy produced pursuant to the appearance of tube fouling

This cost can be quantified thus:

$$C_P(T_i) = p_E \times \delta P \times \frac{T}{T_i} \times \int_0^{T_i - t_{mp}} \Delta P_b(t)dt, \qquad (6)$$

Where: p_i – delivered price of electric energy in system in m.u./kWh, δP – the percentage of power produced delivered in system, T/T_i – number of maintenance actions in one year in maintenance actions/year, $\int_0^{T_i - t_{mp}} \Delta P_b(t)dt$ - loss energy pursuant to the appearance of tube fouling, in MWh/maintenance action.

2.4. Cost due to production losses during planned and unplanned shutdowns due to fouling

The planned and unplanned shutdowns are the periods of preventive maintenance and corrective maintenance. The cost due to production losses during planned and unplanned shutdowns due to fouling is the cost due to energy not produced during these periods. The computation relation is:

$$\begin{aligned} C_E = {} & N_m(T_i) \times P_b \times t_{mp} \times p_E \times \delta P + \\ & + N_d(T_i) \times P_b \times t_{mc} \times p_E \times \delta P \end{aligned} \text{ [m.u./year]}, \qquad (7)$$

Where: P_b – output power in MW.

2.5. The optimization problem

Optimization problem is presented in the next relation:

$$
\begin{aligned}
Z(T_i) &= N_m(T_i)\times(t_{mp}\times r_m\times n + C_{mat}) + \\
&+N_d(T_i)\times(t_{mc}\times r_m\times n + C_{matc} + C_{sch}) + \\
&+p_E\times\delta P\times\int_0^{T_i-t_{mp}}\Delta P_b(t)dt + \\
&+N_m(T_i)\times P_b\times t_{mp}\times p_E\times\delta P + \\
&+N_d(T_i)\times P_b\times t_{mc}\times p_E\times\delta P = \min!
\end{aligned} \qquad (8)
$$

The problem is a non-linear programming problem with inequality restrictions, $4000 \le T_i \le 40000$ [h].

2.6. The consideration of maintenance activity to failure rate of steam condenser

For considering the influence of time between two maintenance activities, it is propose next relation:

$$\lambda_i(T_i) = m + n\times T_i\,[\mathrm{h}^{-1}], \qquad (9)$$

Where: $\lambda_i(T_i)$ – failure rate. For $T=8760$ h, it is consider $\lambda(T)=\lambda$ and for $T=8760/2$ h, $\lambda=\lambda/2$, where λ – failure rate from reliability standard or calculated from exploitation data. In these conditions, it can determine the coefficients *m* and *n*from the next equations system:

$$
\begin{aligned}
\lambda &= m + n\times 8760 \\
\frac{\lambda}{2} &= m + n\times\frac{8760}{2}
\end{aligned} \qquad (10)
$$

2.7.The resolution of optimization problem

The resolution is realized using golden section method. The optimization problem, which is a one-dimensional minimum problem, has form:

$$
\begin{aligned}
&Z(T_i) = \min! \\
&a < T_i < b
\end{aligned}, \qquad (11)
$$

Where: [a,b] is the restriction interval. It is a combination of several step of accelerated search to optimum point and bordering optimum point with Fibonacci method (Hazi, 2004).

It is obtained T_{iopt} – the value for optimum period between two scheduling cleanings in the interval [a_n, b_n], where n – number of iterations and $|a_n - b_n| < \varepsilon$. The golden section method requires no information about the derivative of the function. If such information is available it can be used to predict where best to choose the new point T_i in the above algorithm, leading to faster convergence.

2.8. A case study: optimizing the cleaning frequency of a steam condenser from a DSL 50-1 turbine.

The proposed mathematical model is applied to a concrete case, considering the exploitation data. In the analyzed case, the time between two cleanings for the steam condenser is, conform actual scheduling maintenance, *8760 h.* The costs due to fouling at $T=8760$ h/year are presented in table 1:

Cost	Monetary unit	Value
C_{MP}	\$/year	3144
C_{MC}	\$/year	1876
C_P	\$/year	349365
C_E	\$/year	680032
Z	\$/year	1034517

Tab. 1. Costs due to fouling for steam condenser analyzed, $T=8760$ *h.*

After the optimization method is applied, the optimized time between two action of scheduling maintenance became $T_{iopt}= 9843$ *h* and the costs associated to fouling are presented in table 2.

Cost	Monetary unit	Value
C_{MP}	\$/year	2798
C_{MC}	\$/year	2219
C_P	\$/year	359208
C_E	\$/year	666216
Z_{opt}	\$/year	1030441

Tab. 2. Costs due to fouling for steam condenser analyzed, $T_{iopt}= 9843$ *h.*

The variation of these costs in time are presented in figure 1:

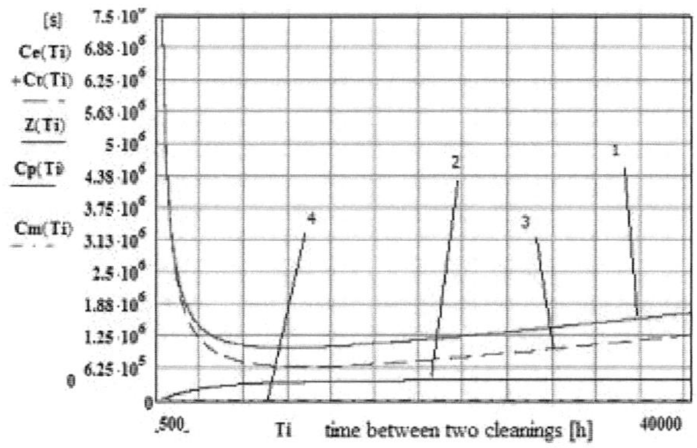

Fig 1. The variation in time: total cost due o fouling (1), cost due to production losses during planned and unplanned shutdowns due to fouling (2), cost due to production losses during planned and unplanned shutdowns due to fouling (3), predictive and corrective maintenance (4)

3. CONCLUSIONS

The optimization algorithm determines:

- The period between two scheduling cleanings;
- The number of cleaning actions required along with their corresponding timings.

The proposed method is a general method, available for all heat exchangers. For application it is necessary to have a base knowledge about exploitation data. It is important to generate a mathematical fouling model for calculate power losses due to appearance of fouling. Comparable with existing strategy to mitigate fouling by managing cleaning, the proposed approach leads to higher energy savings and lower total costs at the level of the entire power plant.

4. REFERENCES

Badea, A.; Necula, H.; Stan, M.; Ionescu, ,L.; Blaga, P.; Darie, G.(2003). *Echipamente şi instalaţii termice,* Editura Tehnică, ISBN 973-31-2183-5, Bucureşti,

Grigore, R.;Badea, A. (2007). Aspects regarding fouling of steam condenser – a case study, *UPB, Scientific Bulletin, Series C: Electrical Engineering*, vol. 69, nro4 pg. 181-189, ISSN 1454-234x

Grigore,R.;Hazi, G. (2008). Contributions regarding the maintenance optimization of power plant steam condenser, *Proceedings of World Energy System Conference*, ISSN 1198-0729, Iasi , june 2008

Hazi, G. (2004) *Tehnici de optimizare în energetică: Teorie, programe, aplicaţii*, Ed. Tehnică – Info, ISBN 9975-63-242-4, Chişinău

Putman R.,(2004). Return on Investment Anaysis: The Economics of Regular Condenser Maintenance, CONCO SYSTEMS,INC. *Available from*: www.concosystems.com

Annals of DAAAM for 2008 & Proceedings of the 19th International DAAAM Symposium, ISSN 1726-9679
Editor B. Katalinic, Published by DAAAM International, Vienna, Austria 2008
Make Harmony Between Technology and Nature, and Your Mind will Fly Free as a Bird

DEDICATED ROBOT-ROBOT COOPERATION

GRIGORESCU, S[anda]; VATAU, S[teliana] & DOBRA, A[ndreea]

Abstract: *The paper presents applications on collaborative robot for massive work-pieces manipulation. The objective is position control of the slave robot motion, synchronized to the master robot motion. The two heterogeneous industrial robots cooperation, working together on parallel trajectories (linear or circular) in the same time, implies kinematic control, Cartesian coordinate's transformation and communication between robot controllers. Dedicated software in Visual Basic environment, achieves commands on robot-robot cooperation.*
Key words: *Cooperation, synchronization, robots, motion.*

1. INTRODUCTION

Cooperation has been the key to success of most human endeavour; the similar incorporation of cooperation in robotic systems is critical to realize the next generation of systems and applications. Interest in cooperating systems arises when the tasks may be too complex for a single system to accomplish, or when building and using several simple systems can be more flexible, fault-tolerant or cheaper than using a single large system. In recent years, cooperative robots have continued to receive a great deal of attention from both the robotics research community and the robotics industry. The cooperation of multiple manipulators can extend the fields where robots can be used effectively. Synchronization, coordination, and cooperation are intimately linked subjects, and sometimes it used as synonymous to describe the same kind of behaviour, mainly in mechanical systems. Typically robot coordination and cooperation of manipulators (Liu et al., 1997) form important illustrations of the same goal. It is desired that two or more mechanical systems, either identical or different, are asked to work in synchrony. In robot coordination the basic problem is to ascertain synchronous motion of two (or more) robotic systems. This is obviously a control problem where, at least for one of the robots, a suitable feedback controller has to be designed such that this robot (slave) follows the other robot (master). This problem is further complicated by the fact that frequently only position measurements of both master and slave robots are available. This partial access to the state of the system has been the reason for developing model-based observers which are integrated in the feedback control loop. In practice, robot manipulators are equipped with high precision position sensors, such as encoders. Meanwhile new technologies have been design for measuring velocities, e.g. brushless AC motors with digital servo-drivers. Nevertheless such technologies are not yet common in applications. Therefore, velocity measurements are often obtained by means of tachometers which are contaminated by noise. Moreover, velocity sensor equipment is frequently omitted due to the savings in cost, volume and weight that can be obtained. For these reasons, a number of model-based robot control methods have been proposed (Nicosia & Tomei, 1990). In these methods a velocity observer is integrated in the control loop, although exact knowledge of the non-linear robot dynamics is assumed, which in practice is generally not available. To overcome this drawback, robust tracking controllers only based on position measurements have been proposed (Rodriguez-Angeles & Nijmeijer 2001). However, all the mentioned before papers deal with the tracking control problem and not with the robot coordination problem. The problem of coordinating (synchronizing) physical systems can be seen as tracking between two physical systems. Although it seems to be a straightforward extension of classical tracking controllers, this problem implies challenges that are not considered in the design of tracking controllers. Most of the tracking controllers are based on full knowledge of the desired reference to be tracked, and no one predicts what would happen in the case of partial knowledge of the reference signal, or how to deal with it. Cooperative manipulation is an important capability for extending the domain of robotic applications. It allows multiple robots to work together in such way that results a significant increase in their overall effective workload and workspace. For many tasks, the use of heterogeneous robots is indicated because of the difficulties of constructing a single robot that has the needed size, strength, dexterity, etc. One such application domain is assembly of large-scale structures, such as terrestrial buildings, planetary habitats, or space solar power structures. Such domains need both heavy lifting capabilities, as well as precise, dexterous manipulation to connect parts together. Another application is in industrial field, where is necessary to handling different kind of object (see fig. 1).

Fig. 1 Cooperation of two angular robots in the flat glass handling (Novotny & Horak, 2008)

2. SYSTEM AND COOPERATION TASK DESCRIPTION

The robot cooperation was studied on the two "Eshed Robotec" robots; these are components of a small scale Flexible Manufacturing System (FMS). The structure of the system consists of three stations: the ASRS, the Milling and the Assembly Workstations. The systems layout is presented in figure 2.

Notations used in figure 2:1-The Storage Carousel; 2-ER VII Robot; 3-ER VII Controller; 4-Conveyor Pallet; 5-Pallet Stop Stations; 6-Conveyor; 7-Conveyor PLC; 8-Milling Machine; 9-Milling Machine CNC; 10-ER V+ Robot; 11-ER V+ Controller; 12-Station Buffers; 13-PC for Station 2; 14-Central PC; 15- Central Controller; 16-PC for Station 3; 17-Scora ER 14 Robot; 18-Assembly Table; 19-XY Motions Table.

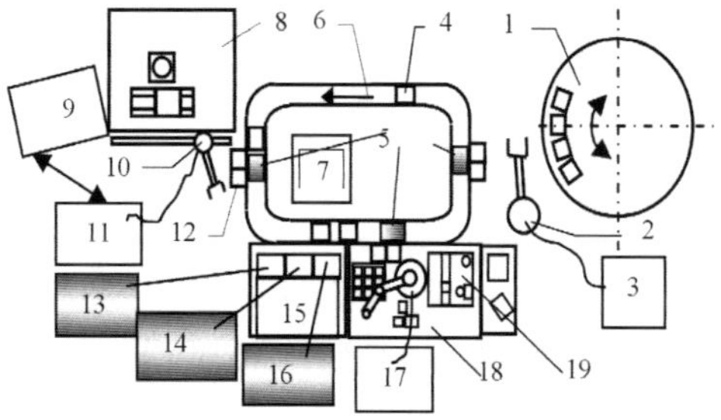

Fig. 2. The Eshed Robotec Systems Layout

The Central and Station PC's carry out different functions, among them is Human-Machine Interface for on-line robotic system visualization, programming and set-up. The control architecture of FMS is hierarchical, the Central Controller achieves the sequences of whole production (sends commands and receives rapports to/from Robot Controller), via RS 232 ports.

The two robots have RRRRR joints, but different mechanical structure and different workspace dimensions.

The collaboration possibilities between the two robots ER VII and ER V+ include motion synchronisation (the same motion in the same time), cooperation (different motions in the same time for a common goal) and coordination master-slave (master robot's motion followed up by slave robot's motion). Due to the relative robot installation, the cooperation refers to the large dimensions pieces manipulation.

The paper's proposed tasks are synchronous motions of ER VII and ER V+ robots, on linear and full circle trajectories. The ER VII is considered the master robot, with programmed poses, which are transformed in slave robot ERV+ position coordinates. The motion speed on both robots is constant and the programmed values are determined experimentally.

3. APPLICATION ALGORITHMS AND RESULTS

The application problems were:

- Measure the base reference system attached to a piece in the workspace of both robots: P1 for ER VII ($^{P1}T_{R1}$) and P2 for ER V+ ($^{R2}T_{P2}$).
- Finding out the transformation matrix $^{P2}T_{P1}$.
- Finding out the transformation matrix:

$$\underline{^{R2}r} = \underline{^{R2}T_{P2}} \cdot \underline{^{P2}T_{P1}} \cdot \underline{^{P1}T_{R1}} \cdot \underline{^{R1}r} \qquad (1)$$

- Programming in ACL Language the master robot for linear and circular movement;
- Programming in ACL Language the slave robot for the same movements, but with the values of points coordinates calculated with the equation (1).
- Testing the algorithm for different robot master positions and motions.
- Programming in ACL Language the master robot for automatic downloading the Cartesian coordinates of poses to the PC.
- Conceiving a program (Robot-Robot Cooperation) in Visual Basic environment (Schneider, 1998) for receiving the data from master controller, computing the transformation coordinates and downloading the data to the slave controller. The input data window of the program is presented in figure 3.
- Programming in ACL Language the slave robot for automatic receiving the Cartesian coordinates of poses from the PC.
- Testing the programs.

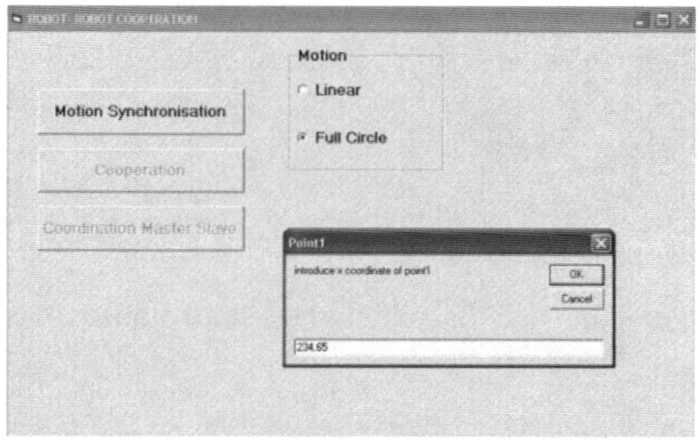

Fig. 3. The input data window of Robot-Robot Cooperation program

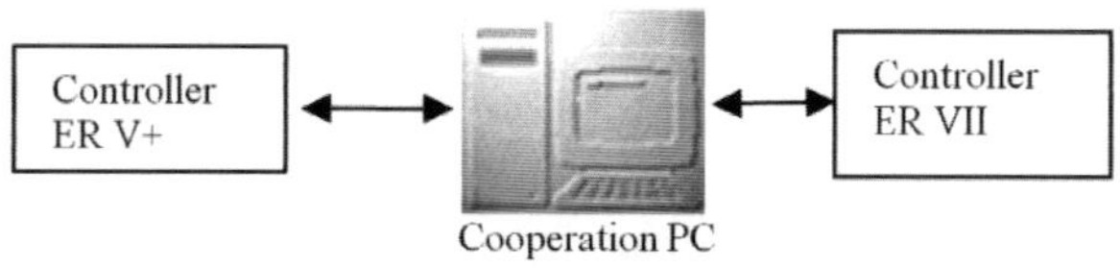

Fig. 4. The control system of robot cooperation

The scheme of the control system is presented in figure 4.

For the full circle motion, there were three poses teach-in, with respect to the master robot world coordinates. The fourth point was calculated in following steps:

- Three points determine a plane. A reference system was attached to this plane.
- The three point coordinates, with respect to the master robot, were transformed with respect to the new reference system. The three points determine a circle.
- The fourth point coordinates were calculated to be on the circle and then transformed with respect to the master robot reference system.

4. CONCLUSION

The motions were programmed only for position, the tool orientation was neglected and it can be a further development task. The ER V+ is more versatile. The robot ER VII was selected to be the master robot, because the motion programming was done on-line. So it can be avoided (minimized) the mechanical blockage of the slave robot arm.

The further development on collaborative robots may include the force and speed control in both master and slave robot controller, motion coordination by a targeted work piece or by a human operator.

5. REFERENCES

Liu, Y.-H., Arimoto, S., Parra-Vega, V., & Kitagaki, K. (1997). Decentralized adaptive control of multiple manipulators in cooperations. *International Journal of Control, Volume 67, Number 5, 20 July 1997, pp. 649-674(26)*

Nicosia, S. & Tomei, P. (1990). Robot control by using only joint position measurements. *Automatic Control, IEEE Transactions on Volume 35, Issue 9, Sep 1990, pp. 1058 - 1061*

Novotny, F. & Horak, M. (2008). Parallel Cooperation of Robots during Handling with Jumbo Glass Sheets. *Available from* http://www.scientific.net/0-87849-387-5/465/ *Accessed:* 2008-04-15

Rodriguez-Angeles, A., & Nijmeijer, H. (2001). Coordination of two robot manipulators based on position measurements only. *International Journal of Control, Volume 74, Number 13, pp. 1311 — 1323*

Schneider, D (1998). *An introduction to programming using Visual Basic 6.0,* Prentice Hall, ISBN 3-13-936428-5, USA

Annals of DAAAM for 2008 & Proceedings of the 19th International DAAAM Symposium, ISSN 1726-9679
Editor B. Katalinic, Published by DAAAM International, Vienna, Austria 2008
Make Harmony Between Technology and Nature, and Your Mind will Fly Free as a Bird

ENERGY EFFICIENCY EVALUATION METHOD

GRIGORIU, M[ircea]; GHEORGHIU, H[oria]; CRAI, A[lina]; DINU, D[oina] & VISAN, D[ana]

Abstract: *The paper presents an original energy efficiency evaluation method, appropriate for the pumping station equipped with variable speed driving motors. The operation specific situations are compared with the most inefficient procedure of operation, using the vanes.*
Key words: *pumping, station, vanes, installation, motors*

1. INTRODUCTION

A general applied solution is using variable speed driving pumps. Pumping equipments and installations energy efficiency improvement represents a key solution, taking into consideration the percentage of about 20% from the total energy consumption.

If for the maximum efficiency of the pumping equipments at the nominal working regime is hardly to obtain significant improvements, taking into consideration the actual performances of the devices, there is an important potential to be evaluated in pumping systems operation optimal adjustment.

There are different ways to operate adjustments of pumps working point, depending to the period of regime changing, the pumps power and dimensions, pumps and installation type, adjustment sharpness, etc.(Dragoi et al, 2007)

For definitive operation point changing, it is preferable to action on the pump itself, by using different types of impellers in the same casing, the same impeller in different casings, or providing a permanent changing to a standard pump type.

For operating point changing for a long period of time, it could be used different types of diaphragms at the pump outlet or in appropriate points of the network. (Grigoriu, 2006)

For short term changes, it is preferable to use special adjustable devices of the installation (vanes, by passes), of the driving engines (variable speed motors), or pump itself (adjustable impellers for axial flow pumps, or same diagonal pumps). (Lupea, 2007)

Generally, the challenge is between adjustable vanes and variable speed motors for short term changes in the majority types of installations.

2. THE PROPOSED ENERGY EFFIENCY METHOD

2.1 Operation point representation

Operation point could be obtain graphically as the intersection between the pump head-flow characteristic curve $H_p(Q)$, equation (1), with the pipe head-flow resulting characteristic curve $H_r(Q)$, equation (2), as it is presented in fig. 1, or analytically solving the equations system representing the polynomial approximation of the mentioned characteristics curves, as follows [2]:

$$H_p=a_1Q^2+a_2Q+a_3 \quad (1)$$

$$H_r(Q) = b_1Q^2+b_2 \quad (2)$$

It is to mention that the pump characteristic represent the resulting head-flow characteristic of one pump or of a whole pumping station, and the pipe characteristic represent the

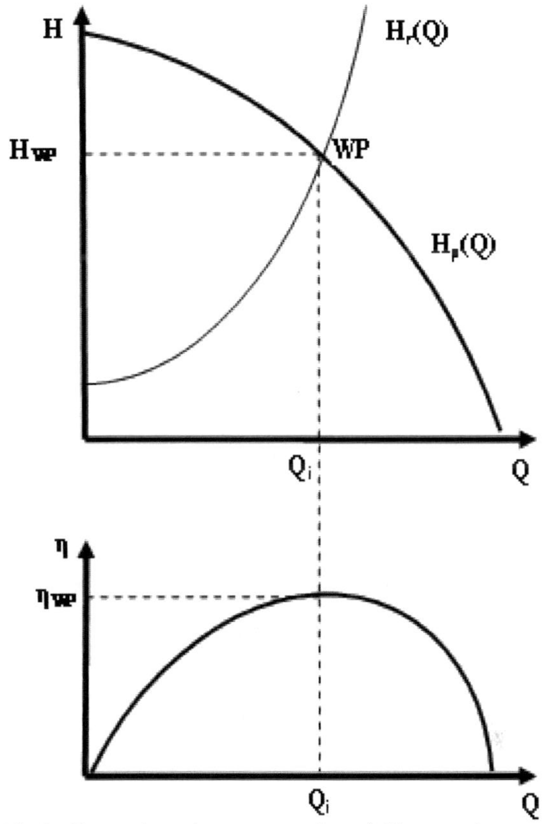

Fig.1 Comparing the vane and variable speed operation adjustment efficiency

resulting head-flow characteristic of one pipe or a whole network. For variable speed driving motor, the pump polynomial form of the head-flow characteristic is (3), and for the pipe polynomial form of the head-flow characteristic is (4)

$$H_p(Q,n)=a_1Q^2+a_2Qn+a_3n^2+a_4Q+a_5n+a_6 \quad (3)$$

$$H_n(Q,x) =(b_0+ b_1)Q2+b_2 \quad (4)$$

where:

Q – flow at the operating point ;

n – rotation speed at the operating point,

$a_1, a_2, ..., a_6$ – polynomial coefficients for the pump head-flow characteristic,

b_0, b_1, b_2 – polynomial coefficients for the network head-flow characteristic.

In fig 2 is represented the operation adjustment of pumping system using the vane and a variable speed driving engine [3] from the initial flow (Q_1), to the requested flow (Q_r), where:

$\eta(Q,n)$ – efficiency characteristic of the pumps, depending the flow and the rotation speed;

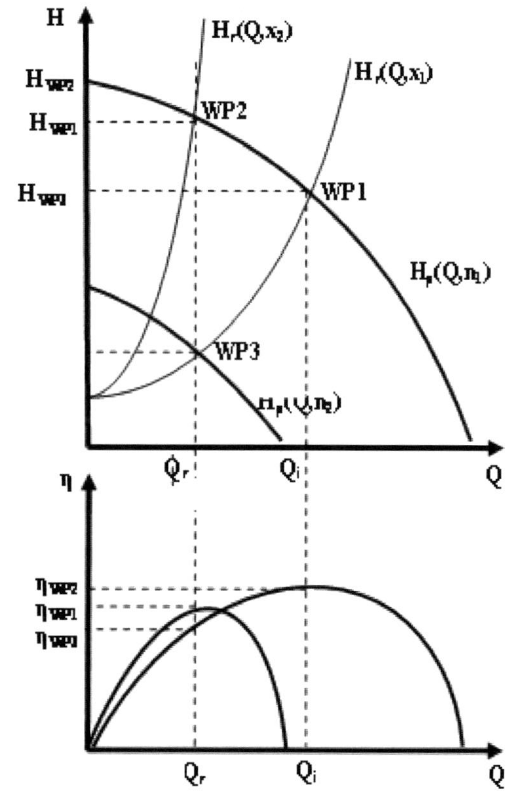

Fig.2 Operation adjustment of pumping system

WP_1 – operation point of the system at the initial flow;
WP_2 – operation (working) point of the system at the requested flow, using vane adjustment;
WP_3 – operation (working) point of the system at the requested flow, using a variable speed engine adjustment;

Analytical representations of the pump efficiency at different rotational speeds are the equation (5) (Grigoriu, 1995).

$$\eta(Q,n)=c_1Q^2+c_2Qn+c_3n^2+c_4Q+c_5n+c_6 \quad (5)$$

where:
$c_1,\ldots,c_6$ represent polynomial coefficients for pump the efficiency characteristic.

For the initial and requested flows, the necessary head and efficiencies are:

$$H_{wp1}=H_p(Q_i n_1), H_{WP2}=H_p(Q_r n_1),$$

$$H_{wp3}=H_p(Q_r n_2),\ \eta_{WP1}=\eta(Q_i n_1),$$

$$\eta_{WP2}=\eta(Q_r n_1),\ \ \eta_{WP3}=\eta(Q_r n_2)$$

To see a procedure or equipment efficiency, it is compared the utile and consumed powers, or energy for a determined period of time (Grigoriu, et al., 2006).

2.2 Vane adjustment efficiency

The utile power is computed for the point WP3 and the consumed power, for WP2.

$$P_{WP3}=\rho g Q_r H_{WP3},\ \ P_{WP3}=\rho g Q_r H_{WP3}/\eta(Q,n_1) \quad (6)$$

2.3 Variable speed engine adjustment efficiency

Both the utile and the power are computed for the point WP3and, defining head efficiency, as follows:

$$\eta_h = H_{WP3}/H_{WP2} \quad (7)$$

The vane adjustment efficiency is:

$$\eta_v=P_{WP3}/P_{WP2} = \eta_h\eta \quad (8)$$

2.4 Variable speed engine adjustment efficiency compared to the vane adjustment

Both the utile and consumed power is computed for the point WP_3 [4]:

$$P_u = P_{WP3} = \rho g Q_r H_{WP3} \quad (9)$$

$$P_c=P_{WP3}/\eta(Q_r,n_2)=\rho g Q_r H_{WP3}/\eta(Q_r,n_2) \quad (10)$$

then, the variable speed adjustment efficiency is equal with the pump itself at Q_r :

$$\eta_s=\eta(Q,n_2) \quad (11)$$

2.5 Variable speed engine adjustment efficiency compared to the vane adjustment

They are compared the consumed power adjusting the flow with a vane (P_{cv}), with the consumed power adjusting the flow with a variable speed device (P_{cs}), by defining a specific efficiency (η_{vs}) or consumed power difference

$$\Delta P_c = P_{cv} - P_{cs}$$

$$P_{cv} = \rho g Q_r H_{WP2}/\eta(Q_r\, n_1),\quad P_{cs}= \rho g Q_r H_{WP3}/\eta(Q_r\, n_2), \quad (12)$$

$$\eta_{vs} = P_{cs}/P_{cv} = \eta_h\, \eta(Q\,,\, n_1)/\ \eta(Q\,,\, n_2) \quad (13)$$

$$\Delta P_c = P_{cv} - P_{cs} = \rho g Q_r\, (H_{WP2}/\eta(Q_r\,,\, n_1) - H_{WP3}/\eta(Q_r\,,\, n_2)) \quad (14)$$

2.6 Energy savings using variable speed engine adjustment

Considering the period of operation time during one year or other period of time, it's easy to calculate the energy savings as follow:

$$E= t\,\Delta P_c =t\,(\,P_{cv} - P_{cs})\ t\,\rho g Q_r\, (H_{WP2}/\eta(Q_r\,,\, n_1) - H_{WP3}/\eta(Q_r\,,\, n_2)) \quad (15)$$

3. CONCLUSIONS

Energy efficiency of the pumping systems is an important factor of municipal infrastructure investments evaluation and the proposed method is really effective, considering that the variation of system operation with vanes represents, in many situations the real solution that is modernized by using variable speed pumps.

5. REFERENCES

Dragoi, G., Funar, St., Solea, M., Cotet, C.E. (2007), Simulation and optimization of a waste processing flux, *Mat.Plast.,* Vol.44, No.1, p.77 - 81

Grigoriu, M; (2006), Pumps and Pumping Systems. Ed. Printech, Bucharest, Romania.

Grigoriu, M. (1995), *Pumps, Fans, Compressors*. Ed. Printech, Bucharest, Romania

Grigoriu, M.; Guzun, B.; Barglazan, M. (2006), Optimal High Powerful Pumped Storage System. *Rev Energetica*, p.487-491, No.11

Lupea, I., Cormier (2007), J., Size and Shape Optimization of a Polymeric Impact Energy Absorber by Simulation, *Mat.Plast.,* Vol.44, No.4, p.339 - 344

Annals of DAAAM for 2008 & Proceedings of the 19th International DAAAM Symposium, ISSN 1726-9679
Editor B. Katalinic, Published by DAAAM International, Vienna, Austria 2008
Make Harmony Between Technology and Nature, and Your Mind will Fly Free as a Bird

PUMPED-STORAGE SYSTEM OPERATION COMPUTER MODELLING

GRIGORIU, M[ircea]; GHEORGHIU, L[iviu]; HADAR, A[nton]; JIGA, G[abriel] & CIOBANU, G[heorghe]

Abstract: *The paper presents the computer modelling of a powerful pumped storage station operating in a complex system, having 2.10 MW, 6/110 kV, 2.3 m³/s, 1000 rot/min. The authors' main contributions are as follows: it is clearly stated that, with some minor modifications on the remote generating side, one hydro power unit could remain closed to the fixed frequency power grid as before, while the other one is to operate at variable speed imposed by the maximum efficiency hydraulic constraints of pumping operation following these modified hydraulic conditions (head or discharged flow). Also, the authors present the operation model of the station.*

Key words: *pump, storage, hydro power, design, efficiency.*

1. INTRODUCTION

The operation of a hydraulic pumping system may be described in stationary regimes based on plotting of head – capacity curves. The rated head and rate of flow is obtained by the intersection of the pump curve by the maximum efficiency with those from the hydraulic network. This *optimal designed situation* reflects the *best mass and energetic equilibrium*.

Knowing the rated regime's parameters in the hydraulic network namely the head H_0 = 247 m and the discharge in the pipes $2Q_0$ = 6 m³/s, it will be calculated the constant *C* of the network with the formula:

$$H_r = h_0 + C(2Q_0)^2 = H_0 \quad (1)$$

It results: $C = 1/6\ s^2/m^5$.

Taking into account the turbo-machinery similitude formulas:

$$Q = k_Q n \quad (2)$$

$$H = k_H n^2 \quad (3)$$

the calculus of the parameters of the extreme regimes means solving the systems of equations (4) (5) and (6), respectively (7) (8) and (9):

$$H_m = (H_0 / (2Q_0)^2)\, Q_m^2 \quad (4)$$

$$H_m = h_m + C\, Q_m^2 \quad (5)$$

$$n_m = (n_0 / Q_0)\, Q_m \quad (6)$$

The results are: discharge Q_m = 5.438 m³/s, head H_m = 202.929 m and speed n_m = 906.4 rot / min.

$$H_M = (H_0 / (2Q_0)^2)\, Q_M^2 \quad (7)$$

$$H_M = h_M + C\, Q_M^2 \quad (8)$$

$$n_M = (n_0 / Q_0)\, Q_M \quad (9)$$

The results are: discharge Q_M = 6.148 m³/s, head H_M = 259.299 m and speed n_M = 1024.6 rot/min.

Analyzing the universal characteristics of the pumps, figure 1, it may be interpolated and estimated the corresponding efficiencies and powers. (Grigoriu, 2006).

The maximum efficiency gains are about $\Delta\eta_{AC}$ = 1.63 % and $\Delta\eta_{BD}$ = 7.23 %. These values give a measure of the effectiveness of the speed control in comparison with the operation at constant speed or the throttling control of the pumps.(Hadar et al., 2007)

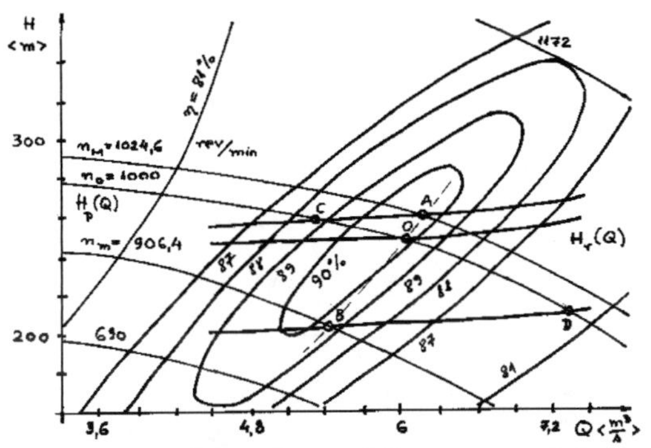

Fig.1. A sequence of the characteristics curves of the two centrifugal pumps in parallel

2. THE POWER ELECTRIC DRIVING SYSTEM

The power electric driving system, shown in the Figure 5, is at variable controlled speed for pumps, operates at specific hydro-constraints, and offers some remarkable advantages. However, the most important among these advantages, remains the higher efficiencies over constant speed or throttling; also we have to add the more reliable working with lower speeds, plus better computer controlled speed changes for preset operating conditions (head, discharge flow) (Grigoriu, 2005).

The method of speed control over the hydraulic turbo-machinery may be used in any hydraulic system, especially for higher power ratings of MW, as it could be seen in this herewith focused practical example (Grigoriu, et al., 2005).

The power electric energy – at variable frequency is to be power plant HPP, working in this particular case with one single generator rated - 24 MVA, 10/110 kV; it is injecting its MVA in islanding conditions of operation, through its second separate power electric overhead line, LEA 110 kV.

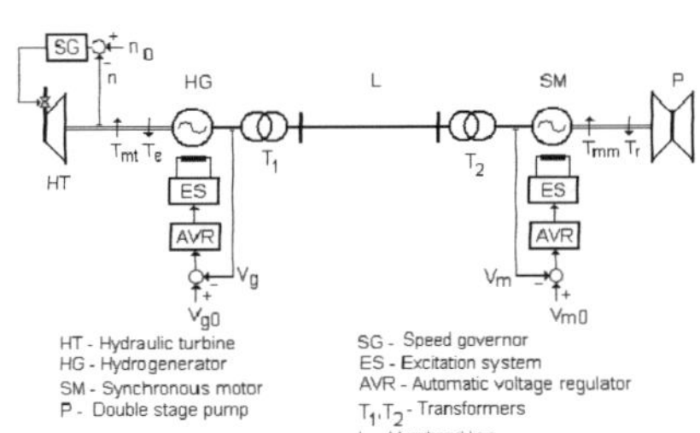

Fig. 2. The power electric simulated system

The other one hydro generator HG from the HPP could remain connected to the system at fixed frequency; the two units are to be separated by one coupling cell deliberately newly-introduced in this scheme, in order to gain elasticity for the new operational regimes. The powerful pumping station PPS - rated 2→10 MW, 110/6 kV, 2→3m^3/s, 1000 rot/min is located at 30 km distance approximately, up in the mountains; it is loaded by the specific hydraulic conditions given by the two lakes' levels, of small and bulky capacity l, L.

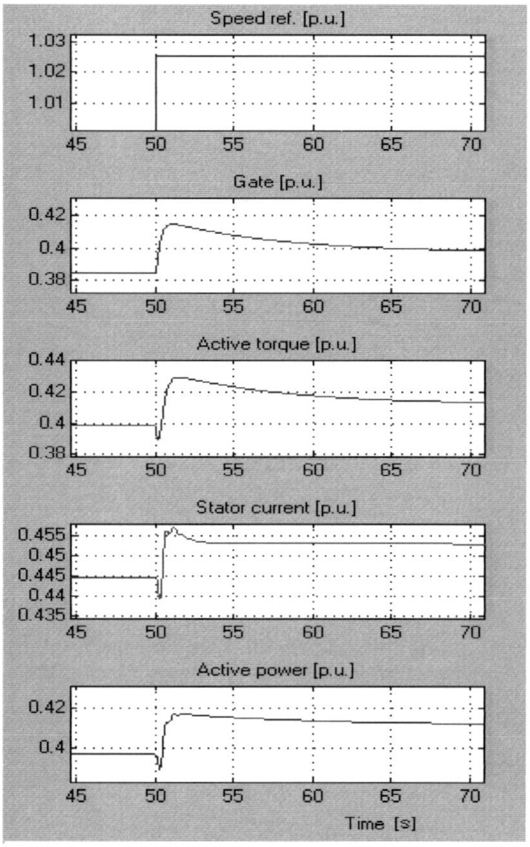

Fig. 3. The main transients on the hydro-generator side

The two levels are to be communicated via GPS to the basic generating HPP, and the whole driving system power electric energy – is working at the requested variable frequency, in order to gain the overall maximum efficiency for the PPS in question (Lupea, Cormier, 2007).

As far as the local conditions for this - quite unique application, we can suggest how favourable these are, represented by the existing infrastructure, so: the neighbouring generating HPP where only one single unit is able to perfect match the requested power for both motor units at the PPS.

3. CONCLUSIONS

The results (fig. 2) released from the simulations are based on the standard synchronous machine's equations. So, one small step-up, average 10 rev·min-1 or equivalent 3% increased during seconds the hydro generator's HG speed, by opening its turbine's gate: the remote motor slower responds by reaching its upper speed limit of 1025 rot/min over some delay, but no more than 10 -15 s.

On the HG side (fig. 2), as a result of increased active torque, promptly increases the stator current and the active power injected at the terminals (small reverse swings probably mark the AVR's small delayed intervention).

On the motor's side at the PPS (fig. 3), as a result in promptly command of increased frequency coming from HG via 110 kV line, clearly we can see the same prompt responses with increased stator current, absorbed active power and dropping in stator voltage (small reverse oscillations probably mark also here the AVR's small delayed intervention).

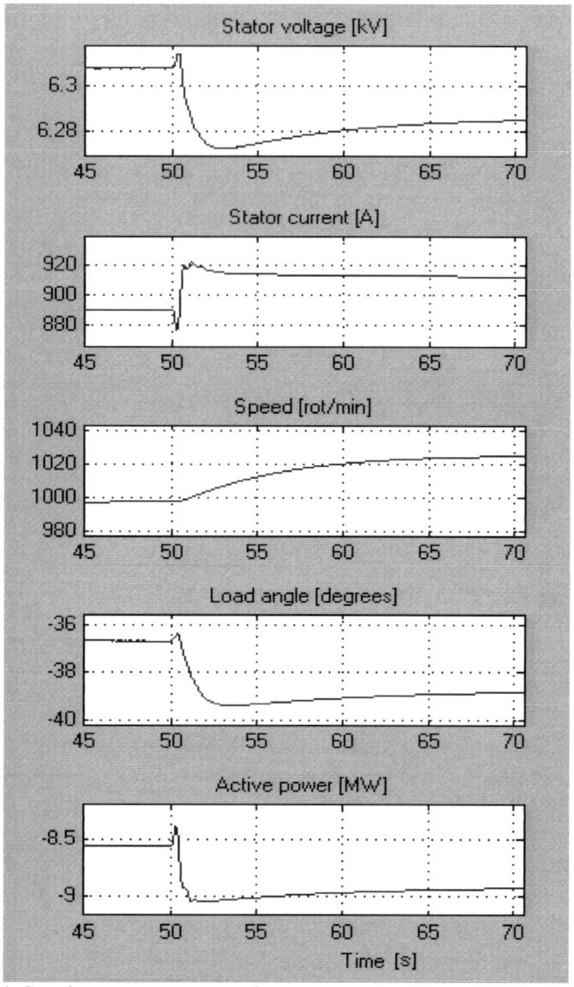

Fig.4. Synchronous motor transients

However, as stated above the motor's response in speed-up is much slower - as a mechanical parameter, while the internal load angle displacement marks one intermediate time evolution.

In conclusion, the discussion over the simulated behaviour for the first time in this particular back-to-back powerful scheme running at variable speed and exploiting the local favourable conditions, as we can hope - this exercise proves to be a successful one.

4. REFERENCES

Grigoriu, M. (2006), *Pumps and Pumping Systems*. Ed. Printech, Bucharest, Romania.

Grigoriu, M. (2005), *Pumps, Fans, Compressors*. Ed. Printech, Bucharest, Romania

Grigoriu, M.; Guzun, B.; Barglazan, M. (2005), *Optimal High Powerful Pumped Storage System*, Rev. Energetica, 487-491 pg., No.11.

Hadăr, A., Constantinescu, I.N., Jiga, G., Ionescu, D.S. (2007), *Some Local Problems in Laminated Composite Structures*, Mat.Plast., Vol.44, No.1, p.354-360

Lupea, I., Cormier, J., (2007), *Size and Shape Optimization of a Polymeric Impact Energy Absorber by Simulation*, Mat.Plast., Vol.44, No.4, p.339 - 344

Annals of DAAAM for 2008 & Proceedings of the 19th International DAAAM Symposium, ISSN 1726-9679
Editor B. Katalinic, Published by DAAAM International, Vienna, Austria 2008
Make Harmony Between Technology and Nature, and Your Mind will Fly Free as a Bird

ALTERNATIVE LOCALIZATION OF DEFECTS USING WAVELETS IN THE DEFECTOSCOPY BY EDDY-CURRENTS

GRMAN, J[an] & SYROVA, L[ivia]

Abstract: *Eddy-current testing is one of the methods of non-destructive material testing. Paper presents alternative method for detection of indications within tested tube. Signal from measuring probe was processed using wavelet transformation. The significant changes of signal correspond to local maxima of calculated matrix of wavelet correlation coefficients on the surface. Paper presents that real defects and construction elements are detected with high reliability using combination of profiles from different measurement frequencies.*
Keywords: *defectoscopy, localization, eddy-currents, wavelets*

1. INTRODUCTION

One of the methods that are useful in non-destructive defectoscopy of heat-exchanger tubing is method based on eddy currents (Ravas, 2001) (Riley at all, 2003). Output signal from measurement probe (in our case differential probe) has two orthogonal components and measurement is usually done using four basic frequencies. One of the possibilities how to analyze signal data is application of wavelet transformation (Addison at all, 1997) (Hamelin at all, 1996). Equation (1) calculates wavelet matrix for one component of measured signal.

$$C_{x(t)}(s,p) = \int_{-\infty}^{\infty} x(t).\psi'(s,p,t)dt \qquad (1)$$

Using equation (2) are results for the both channels connected into one wavelet matrix.

$$M(s,p) = \sqrt{C_x^2(s,p) + C_y^2(s,p)} \qquad (2)$$

Result matrix can be interpreted as image, where local maxima correspond to positions of significant signal changes. This result was expected, because shape of wavelet is very similar to the shape of typical defect indication.

We can easy interpret wavelet coefficients calculated by above mentioned formulas:

- position + scale = indication centre and effective range
- correlation M(s,p) = analogy to the amplitude of indication
- angle of correlation components = analogy to the phase angle of indication

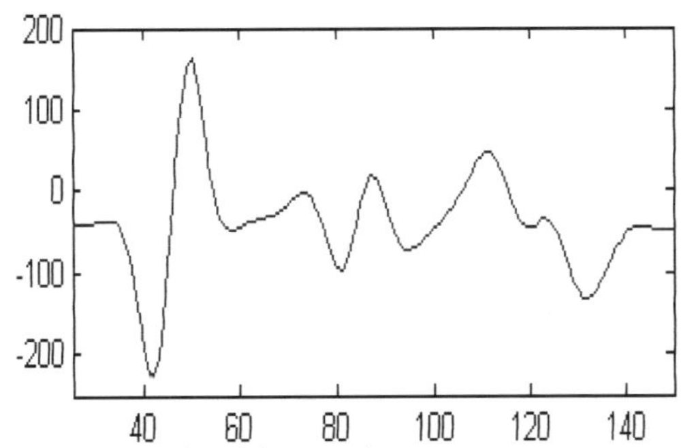

Fig. 1. Original signal from probe – one component

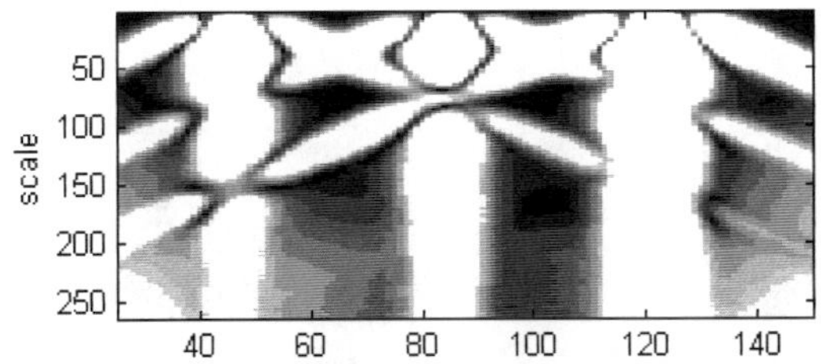

Fig. 2. Wavelet surface – white = higher correlation

In our previous work we decided to use morphological operators to calculate positions of all significant local maxima from matrix or sharply image of wavelet coefficients. Figure 3 shows typical result of mentioned method.

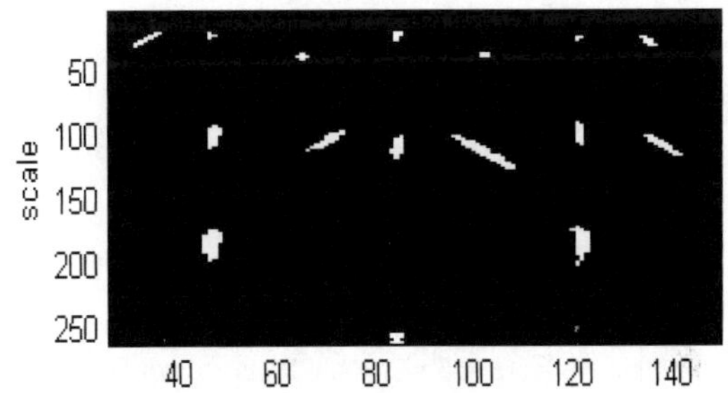

Fig. 3. Positions of local maxima on coefficients surface calculated using morphological operators

2. IMAGE THRESHOLDING

Alternate method how to find local maxima is image adaptive threshold. Local maximum is a global maximum on some local region. Problem is to define suitable regions, not very large and not very small. Figure 4 shows result using only one global threshold. This method is very sensitive to threshold selection.

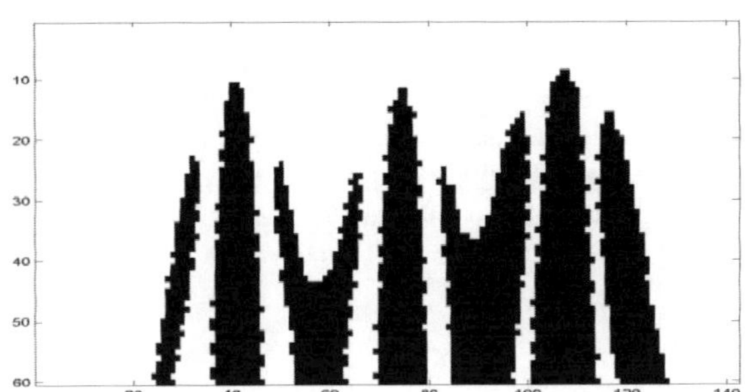

Fig. 4. Global threshold

For our next experiments we have used set of threshold intervals uniformly calculated from min/max projection of wavelet coefficients as shown on figure 5.

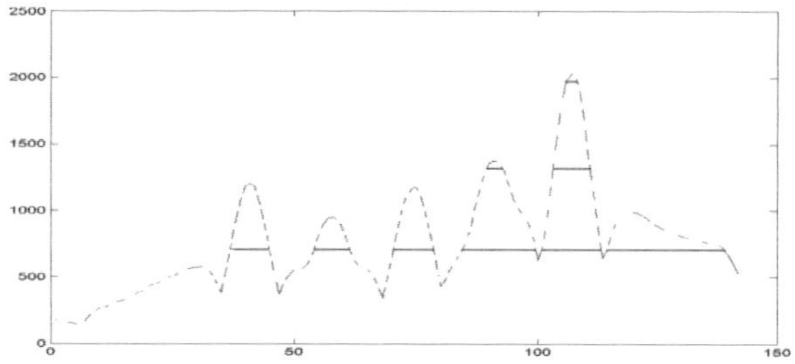

Fig. 5. Set of threshold intervals

Result set of local maxima for image can be calculated as a union of global maxima for all regions and all threshold intervals.

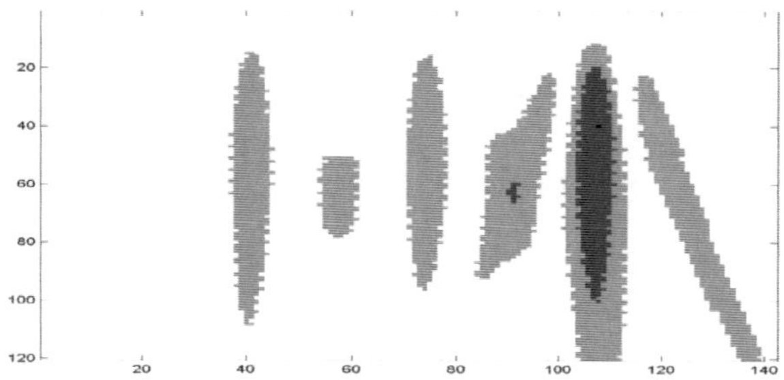

Fig. 6. Regions corresponding to threshold intervals

3. EXPERIMENTS

Local maxima of image brightness calculated using above mentioned threshold reflect all significant changes of signal (called indications). Result list of indications also contains original position within signal, scale coefficient of correlated wavelet, correlation of wavelet and signal (both components), correlation of wavelet and signal components and relative angle of correlation components corresponding to the phase angle of indication.

Every type of indication has its typical values for correlation, wavelet size and phase angle. For example typical parameters for indication of 100% defect in 100 kHz signal are: correlation = 1320, scale = 20 a phase angle 40 degrees. The same profile can be defined for this type of indication at frequencies 25, 200 and 700 kHz and also for all other types of indications.

Result lists of positions of indication from all measurement frequencies are then filtered using know profiles of typical indications. Filtered lists are then connected. Some position is finally marked as indication only if it exists in 3 or more lists.

type\frequency	25kHz	100kHz	200kHz	700kHz
defect 100%	40	40	40	40
defect 72%	51	60	81	120
defect 48%	57	69	117	230
support plate	106	131	40	---

Tab. 1. Typical values for phase angle parameter for selected types of indications

4. RESULTS

Above mentioned method can detect potential indications of defects and structure elements with high reliability. It is sensitive to signal changes not recognized manually by operator. One of the known problems is that not every marked location really corresponds to some defect or construction element. Indication edges typically cause significant change of signal usually marked by above algorithm as local maximum, but this effect was expected. These "false" indications can be eliminated using profiles (interval tests for correlation, size and angle values) and using information from different frequency channels. Low frequencies better indicate structure elements and high frequencies better describe defects.

Classification of mixed indications is based on the data obtained from all measurement frequencies. Mixed indication is indicated as dual (example: 25kHz – support plate with high probability, 200kHz – defect with high probability).

25 kHz	100 kHz	200 kHz	Profile validation
1	1	1	FALSE
15	16	6	---
		18	---
28	28	28	TRUE 100% defect
47	47	37	---
67	66	66	TRUE 72% defect
85	85	85	FALSE
104	104	104	TRUE 48% defect
121	117	116	---
146			---
265	267	266	FALSE
280	279	279	TRUE 20% out defect
296	292	301	---
306		306	---

Tab. 2. Connection of lists of found indications and validation of calculated parameters against profile

5. CONCLUSIONS

Experiments show that real defects and construction elements are detected with high reliability. Above described algorithm automates and formally simulates real testing procedure where operator finds an indication and examines shape of the same indication in different frequencies. Algorithm is very sensitive and result list of indications covers all relevant signal changes. Operator must evaluate indications, which couldn't match any of defined profiles manually. Number of mentioned unclassified indications is small, so detailed analysis of signal can be done more effectively.

6. REFERENCES

Addison, P.S., Sibbald, A. & Watson, J.N. (1997). Wavelet Analysis: a Mathematical Microscope with Civil Engineering Applications. *INSIGHT*, Vol 39, No 7, July 1997.

Hamelin, J.I.Alvarez, D'Attellis, C.E., Mendonca, H., Ruch, M. (1996). A new method for quasi on-line eddy current signal analysis: wavelets transform. *INSIGHT*, vol. 38 No.10, pp.715-717, October 1996.

Ravas, R. (2001). Automation of Evaluation of Measurement in Defectoscopy of Power Steam Generator. In: *Measurement Science Review*, Vol. 1, No. 1, pp. 103-106.

Riley, L. A., Nassersharif, B., Garcia, G. & Schaub, J. (2003). An Automated Testing and Classification System For Identifying Defects in Nuclear Steam Generator Tubes Using a Learning Vector Quantization Neural Architecture. In: *Proceedings of the 2003 Advanced Simulation Technologies Conference*, Society for Computer Simulation International, Orlando, Florida, April 2003.

ACKNOWLEDGMENT

We gratefully acknowledge that the results presented here were solved as partial goals of the research task VEGA No 1/3099/06.

Annals of DAAAM for 2008 & Proceedings of the 19th International DAAAM Symposium, ISSN 1726-9679
Editor B. Katalinic, Published by DAAAM International, Vienna, Austria 2008
Make Harmony Between Technology and Nature, and Your Mind will Fly Free as a Bird

APPLICATION OF SLM FOR OPTIMISATION OF WEIGHT OF SUPPORTING LEVER IN RELATION TO EXPECTED STRENGTH RESISTANCE

GRZESIAK, D[ariusz] & KROLIKOWSKI, M[arcin]

Abstract: *Selective laser melting of metal powders is a method which applies incremental material addition for part creation. This method enables creation of part geometry, impossible for other classical methods because of complexity of form, thin walls, application of hard machinable materials etc. Paper presents application of this technology, belonging to Rapid Manufacturing techniques. SLM is applied to manufacture supporting lever, optimised by the weight, in relation to its expected strength resistance.*
Key words: *SLM, FEM, Rapid Manufacturing*

1. INTRODUCTION

Selective laser melting of metal powders is a relatively new technology of Rapid Manufacturing (Kruth, J.-P... 2005, 2007). It enables significantly more applications in comparison to classical manufacturing methods of metals and alloys. This should be considered however still as „prototyping" rather than real manufacturing. The „Rapid Manufacturing", in this case means, that this incremental part creation technique is offering the parts, which are applicable just after the process, and their material and physical characteristics are fully comparable with conventional production metal parts (Ghany, A... 2006), as the density of SLM produced parts is approx up to 98%. Besides the obvious benefits of incremental metal part creation technology, offering any complexity of geometry and almost no residual stresses in the part, this technology enables to melt together different metals and non-metal powders.

The benefits of SLM method were applied to research work carried out in Centre of New Materials and Technologies at the Szczecin University of Technology to design heavy loaded lever element (fig. 1), which weight should have been minimised without the loss of its strength resistance in predicted directions. The optimisation process was performed in following steps:

1. Digitalisation of existing supporting lever,
2. Modification of construction,
3. FEM analysis and incremental optimisation by repeating 2&3,
4. Manufacturing of element with application of SLM

2. 3D MODEL OF SUPPORTING LEVER

For future optimisation of existing part of the supporting lever, the reverse engineering technique of reconstruction of lever body had to be carried out. Than CAD parametric body was applicable for modification & optimisation.

2.1 Reverse engineering of lever body

First stage of reconstruction of 3D geometry of lever was to transform the shape of the body of the lever to CATIA CAD/CAM system. Lever was scanned to cloud of points at the 3D optical scanner. Next cloud of points (fig. 1) was imported to CATA system. After cleaning and preparation of cloud of points, the shape was reconstructed by specialised shape modules of CATIA.

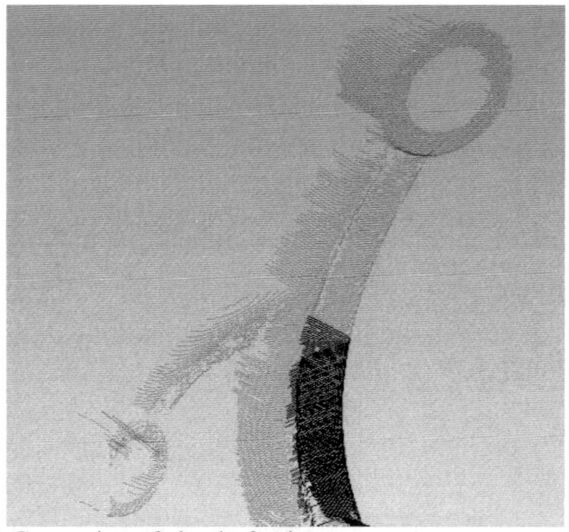

Fig. 1 Generation of cloud of points

Fig. 2 Reconstructed supporting lever

Holes were measured separately at 3D measurement machine and implemented by solid modelling in CATIA modeller. The reconstructed 3D part is presented at fig. 2.

2.2 Modification of supporting lever

In order to minimise the weight of supporting lever a series of square holes was foreseen. The square holes were parameterised to control the wall thickness between holes and both dimensions of holes (fig. 3). This element was a subject of FEM analysis. The loads and boundary conditions were taken from existing part loads in mechanism. This analysis provided information of behaviour and deformations of the lever and local strength resistance in important directions. This analysis also enabled to calculate the influence of hatch parameter on strength resistance parameters and deformations at mounting holes. After series of optimisation steps (fig. 4) where all hatch parameters were adapted to current calculations, the result was achieved, where weight of supporting lever was lowered by 9% from original, without loss of its strength resistance (fig.5).

Obviously, the permissible stresses for the material were not overrun, taking into accounts proper safety factors. One must consider that this optimisation was taken because no further modification of existing solid was possible without loss of stiffness and of course only having in mind SLM possibilities.

Fig. 3 Cut-off of the lever arm with square holes

Fig. 4 FEM analysis - deformation

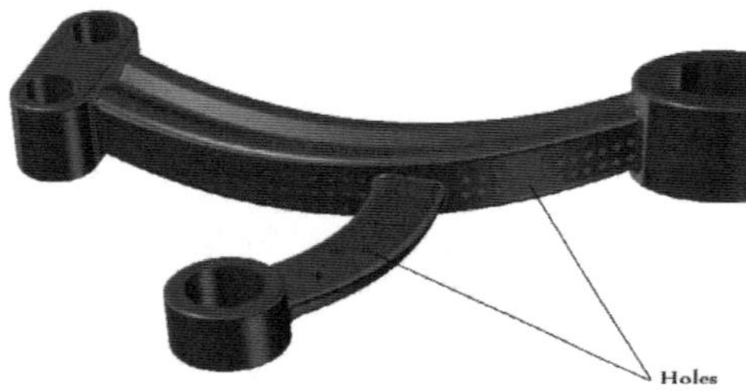

Fig. 5 Part after modification

3. MANUFACTURING OF PROTOTYPE

After optimisation, there was a need to manufacture the final prototype made of stainless steel. First stage was to generate a file with part geometry in compatible format, for future elaboration by special software system provided together with REALIZER II by MCP HEK SLM system. This special software system analyses structure of files with geometry, cleans improper geometry and prepares model for SLM technology. This preparation includes two main parts which are necessary for SLM – slicing into flat surfaces and building so called support bridges which carry the geometry, which is projected besides of the base of model (hanging).

The slice of the model is presented at fig 7. Model with generated support walls is presented at fig. 6. After preparation model of sliced solid and support were transferred into machine computer of REALIZER II, and the manufacturing process was started.

After SLM there was a need to machine holes for proper joint structure with corresponding parts, as the SLM's weak point is rough surface and some inaccuracy caused by the process itself. One can compare it to rough parts after precise casting.

4. CONCLUSIONS

Selective laser melting is a method enabling manufacturing of parts, which geometry is impossible to obtain with application of typical methods like casting.

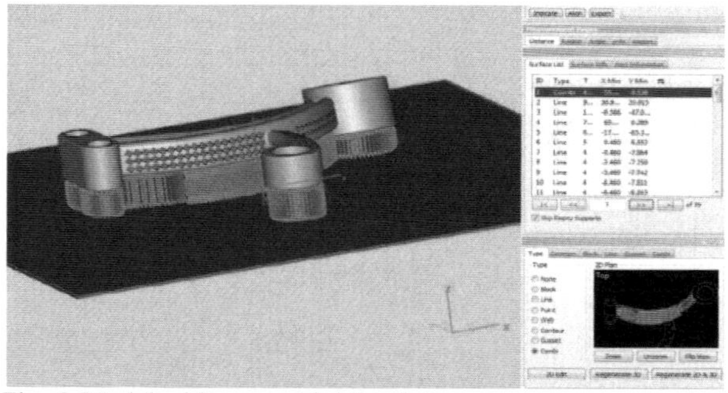
Fig. 6. Model with support bridges below

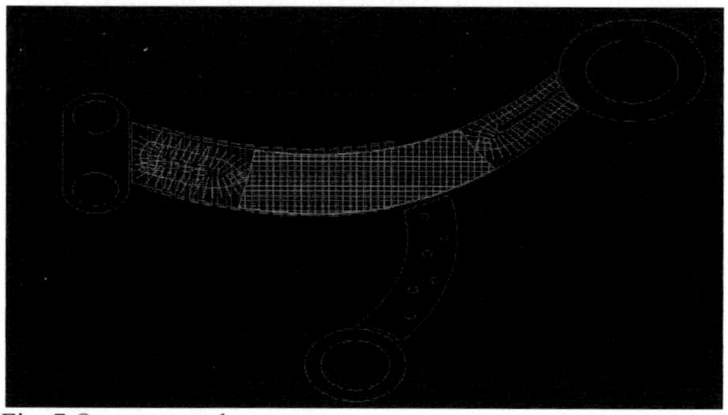
Fig. 7 One support layer

Especially interesting is a powerful possibility to build lightweight constructions with openwork structure which can be only comparable to metal foam casting (Over, C...2001, 2003). There is also possibility to build hybrid material constructions or even composites.

In the paper, the optimisation of weight of loaded part by partial lightweight construction application was presented.

As the result of optimisation, the required strength resistance parameters were kept in foreseen tolerances but the weight of the element was lowered significantly. The applied material was the same as original supporting lever.

Lowering a weight of already quite optimised structure of complex loaded lever by 9% seems very good, but one should remember that SLM is not yet a production technique but can be used for instance to test the prototype of casting of lightweight constructions (foams) without a need to build a metal moulds. Further investigations with application of hybrid materials and different alloys are carried out at our university.

5. REFERENCES

Kruth, J.P. & oth. (2007). Consolidation Phenomena in laser and powder-bed based layered manufacturing. *Annals of the CIRP Vol 56/2/2007,* Elsevier, 730-759

Kruth, J.P. , Vandenbroucke, B., Van Vaerenbergh, J., Mercelis, P., (2005). Benchmarking of different SLS/SLM processesas rapid manufacturing techniques. *Proc. PMI, Gent, CDROM*

Ghany, A., K., Moustafa, S.F., (2006). Comparison between the products of four RPM systems for metals, *Rapid Prototyping J.*, 12/2, 86-94

Over, C., Meiners, W., Wissenbach, K., Hutfless, J., Lindeman, M., 2003,
Rapid Manufacturing of metal parts and tools using laser melting, Proc. WLT
Conference on Lasers in Manufacturing, Munich: 265-269

Over, C., Meiners, W., Wissenbach, K., Lindeman, M., Hammann, G., 2001,
Selective laser melting: a new approach gor the direct manufacturing of
metal parts and tools, Proc. LANE. Erlangen: 391-398

Annals of DAAAM for 2008 & Proceedings of the 19th International DAAAM Symposium, ISSN 1726-9679
Editor B. Katalinic, Published by DAAAM International, Vienna, Austria 2008
Make Harmony Between Technology and Nature, and Your Mind will Fly Free as a Bird

ACHIEVEMENTS OF MANUFACTURING ENGINEERING DEPARTMENT FROM TU CLUJ-NAPOCA IN THE FIELD OF COMPETITIVE AND ECOLOGICAL PRODUCTS

GYENGE, C[saba]; ROS, O[limpia] & FRATILA, D[omnita]

Abstract: *Political and economical changes from the few past years had significant influence over almost all countries economical life. High-developed countries react after these changes in marketing by using high-performance working methods and by raising investments in researching fields. The world competitive manufactures attention in the last years is on the competitive and environment-friendly products development. For that there has been adopted the concept of integrated manufacturing as a strategy for achieving shorter time to market, reduced development costs, and high quality products. Manufacturing Engineering Department from T.U. of Cluj-Napoca started in 1994 to introduce this concept in university and in the Romanian industry. This paper presents our steps and achievements in this field of competitive product development.*
Key words: *assembly, environment, product development, competitiveness*

1. COMPETITIVE PRODUCT DEVELOPMENT STRATEGIES AND METHODOLOGIES IMPLEMENTATION

In order to introduce the modern competitive product development in Romania our Manufacturing Engineering Department from TU Cluj-Napoca elaborated a general strategy (Fig. 1) composed by seven basic steps.

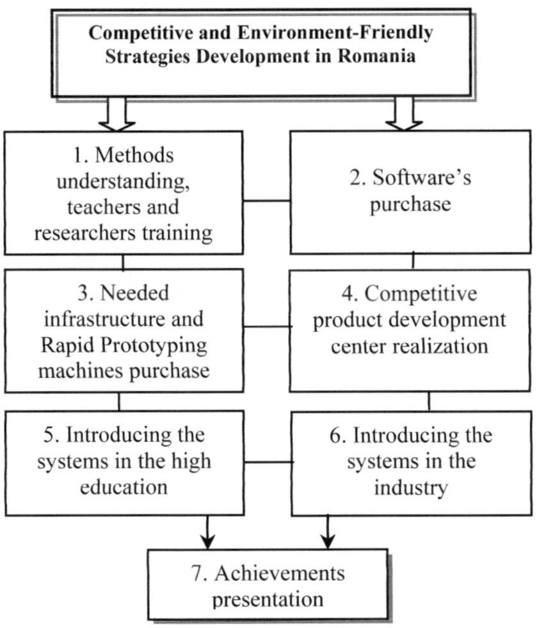

Fig. 1. Competitive and environment-friendly methods and strategies development in Romania.

The necessary software's, DFMA (Design for Manufacturing and Assembly), DFE (Design for Environment), ProEngineer, SolidWorks, Catia, and MIMICS were purchased with the support of Tempus project and SNSF (Swiss National Science Foundation). With this funds and using some facilities offered by the seller company, we succeeded to buy already in 1994 the main DFA and DFMA modules.

For a good knowledge of the competitive product developments new methods, by our students and Ph.D. students, we developed in this way a teaching work plan. According to this, beginning with the 1994-95 academic years we introduced in our curricula special courses in the field of competitive engineering. They were very useful to introduce the DFMA methods in the diploma and Ph.D. students preparing works.

2. COMPETITIVE PRODUCT DEVELOPMENT METHODS PRACTICAL RESULTS

To introduce this new concept in the industry we presented our results to the companies from whom the analyzed products originate, and the resulted discussions about this were very useful to prepare future collaborations, and teaching them in this field.

Taking into consideration the Romanian industry characteristics and our presented results we developed a strategy to choose target products, which were developed in the new competitive manner, considering the production structure, annual production, used materials, the field of utilization, technical equipments, etc.

Considering this strategy we chose from more than six industrial companies different representative products and we sorted them by product types (Fig. 2).

Introduction and implementation of the DFMA methodology in Romania was made as showed in the figure 1. As you can see the successful implementation consists of 7 modules. Using the mentioned programs modules and in concordance of the databases with the Romanian realities we choose several products (sewing machine, water supplier, flat iron, drip coffee, etc.) and analyzed their assembly process (Gyenge et al, 1995).

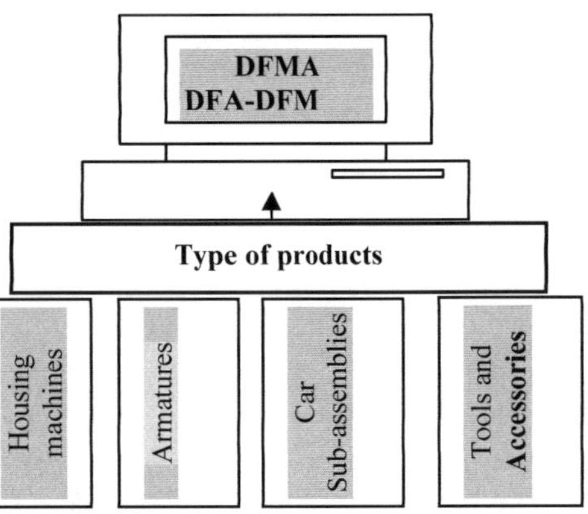

Fig. 2. Products type.

To this presentations took part specialists in part design, department chiefs and directors. The presented materials and analysis had a great interest and the specialists showed their interest in finding the possibilities of implementing this methodology.

From 1994 we are regularly presenting our results as well as at home as abroad, by scientific publications (Ros et al, 2001). The significant results obtained in the field of competitive product development are shown in Table.1

As we can see in the table, after the redesign of the products the number of parts decreased with (10-53)% and the assembly efficiency indicator increased with (5-80)%. In the design the new products we used softwares as ProEngineer and SolidWorks and in testing, the rapid prototyping technologies which where very useful.

	Analyzed products	Number of parts		Assembly efficiency (%)	
		Last	*New*	*Last*	*New*
1.	Ice-cream machine	49	44	32	42
2.	Garden sprinkler	17	8	12	36
3.	External car mirror	22	19	39	41
4.	Hydraulic Distributor	61	46	27	36
5.	Back-flow Preventer	25	19	22	30
6.	Straight-way Cock	47	34	12	16
7.	Garden shears	13	9	16	26
8.	House holding balance	39	30	14	18
9.	Hydraulic automatic valve	50	18	5	29

Tab. 1. Significant results.

3. CHARASTERISTIC CASE STUDY

In the following we will present one example realized at our department, containing the analysis and redesign using DFA (Boothroyd& Dewhurst, 1991) and DFE methodology (Ros et al, 1999) of a garden water supplier, manufactured in mass by Armatura Company from Cluj-Napoca (Fig. 3).

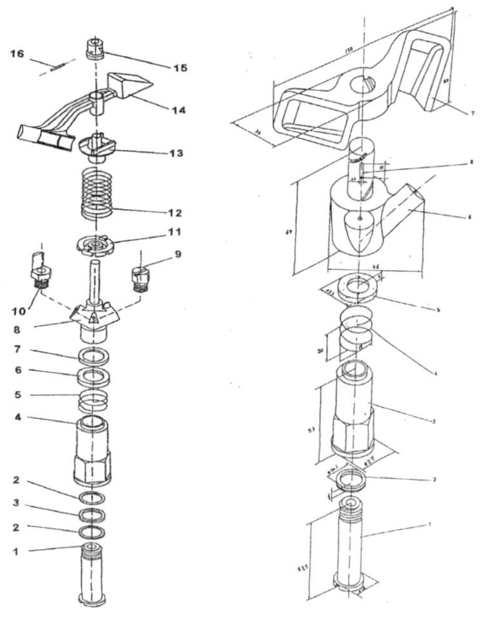

Fig. 3. Exploded scheme of the water-supplier (original and redesigned construction).

Original type of water supplier consists of 16 components. Analyse was made in the following steps (Balc& Gyenge, 1994):

- Detailed description and definition of components by the point of view of shape, size, material, symmetry, functionality, possibility of positioning and fixing, etc.
- Determination of the elements of manual-assembly process
- Determination of time of manual assembly process (catching, orientating, positioning and fixing to the other components).

Analysis of the ecological impact of the garden water supplier in the disassembly phase has been done using the DFE program.. Results are summarized in table 1.

So, the optimal variant from both assembly and ecological point of view is the redesigned one that war previously optimized using the DFA program.

Name	Old variant	New Variant
Assembly efficiency (%)	12	36
Total assembly time (s)	118	40
Number of different parts or sub-assemblies	16	8
Total number of operations (inc. repeats)	23	8
Number of components	17	8
Efficiency index (%)	14%	38%
Assembly time (s)	152.8	67.2
Necessary member of workers	3/ turn	2/ turn
Ecological negative cumulative effect (MET points)	1817.4	1639.2
Disassembly time (s)	170	55

Tab. 2. Comparing data of the analyses.

4. CONCLUSION

Researches presented briefly in this paper facilitate assembly process optimization of products and allows the decreasing of corresponding times and costs. Redesign of some parts was made in order to reduce their complexity, so the previous manufacturing processes of the parts were not influenced negatively.By the results obtained it can be established, that introducing the integrated DFA and design DFE methodology product become environment friendly and competitive on the market and the designing work became more easier and even more productive and efficient. Investments costs will be replayed in short time and parallel with this, qualification of designers and technological engineers will raise a lot.

5. REFRENCES

Bâlc, N. & Gyenge, Cs. (1994). Assembly automation of the sprinklers using DFA, *Proceedings of 5th International DAAAM Symposium*, pp. 31-32, ISBN 3-901509-00-3, Slovenia, October 1994, Katalinic B. (Ed.), Maribor.

Boothroyd G., Dewhurst P. (1999) *Product Design for Assembly*, Wakefield.

Gyenge, Cs., Marcu, V., Gligor, E. & Lörincz, S. (1995). Results of integrated computer aided product development methodology introducing in Romania, *Gépgyártástechnológia,* Vol.IX, pp. 311-8, ISSN 0016-8580.

Ros, O., Gyenge, Cs., Petho, L., Fratila, D. (1999). Analysis of the ecological impact of the products in the disassembly phase, using the DFE program, *Proceedings of the 2nd International Conference on Industrial Tools*, Vol.2, pp.474-477, ISBN 961-90401-4-7, Slovenia, April 1999, Maribor.

Ros O., Fratila D., Varga, A., Goia, C., Development of the integrated parts in the automobiles industry. *Proceedings of MTeM 2001.* p. 203-204, 4-6October 2001. ISBN 973-85354-1-7, Cluj-Napoca.

Annals of DAAAM for 2008 & Proceedings of the 19th International DAAAM Symposium, ISSN 1726-9679
Editor B. Katalinic, Published by DAAAM International, Vienna, Austria 2008
Make Harmony Between Technology and Nature, and Your Mind will Fly Free as a Bird

A MODEL OF HUMAN PERFORMANCE IN INFORMATION SEARCHING TASKS

HA, J[un] S[u] & SEONG, P[oon] H[yun]

Abstract: *In this work, a model of human performance in information searching tasks is proposed. Attention is affected by four factors such as salience, expectancy, value, and effort. Expectancy and value of information are determined based on the operator's mental model, whereas salience and effort are the matter of human-machine interface (HMI) design. Hence, poor design and/or mental model lead to poor performance in information searching tasks. Effects of poor HMI design and poor mental model on operator's performance (perception success) are demonstrated from an experimental study.*

Key words: *human performance, monitoring, detection, selective attention, HMI design, mental model*

1. INTRODUCTION

The main role of operators in control rooms (CRs) of nuclear power plants (NPPs) is generally to supervise and operate the system. Operators' tasks are performed through a series of cognitive activities (Barriere et al., 2000). Among the activities, correct understanding of the situation is frequently considered a crucial key to improving performance and reducing error (Adams et al., 1995). For the accurate and timely understanding of the situation, monitoring and detection should be effectively accomplished. In other words, effective information searching is crucial for enhancing the safety of NPPs.

In this work, a model of human performance in information searching tasks is proposed in order to provide a framework that can be effectively used for evaluating HMI designs and/or developing a training program. Poor HMI design and poor mental model are modelled to most affect operator's performance. Hence effects of poor HMI design and poor mental model on operator's performance (perception success) are demonstrated through an experimental study. In Section 2, the proposed model is described. An experimental study regarding the model is presented in Section 3. Concluding remarks are maded in Section 4.

2. A MODEL OF INFORMATION SEARCHING

Performance in information searching tasks such as monitoring and detection in NPPs is affected by four factors such as salience, expectancy, value, and effort (Wickens & Hollands, 2000). Operators in NPPs may have expectancy and/or values of various information sources more clearly, as their mental models are getting well developed through experiences and training. This is the reason why experts show better performance than novice in information searching tasks (Wickens & Hollands, 2000; Ha & Seong, 2005). Salience and effort are matters to be considered during designing an HMI. Important information should be designed with appropriate level of salience and effort to access. If an information source cannot be distinguished clearly from adjacent information sources, sometimes the information source may be misunderstood by operators. In addition, if it is too difficult to find out an information source important to understand a situation, an operator eventually gives up finding out the information source. Hence, poor performance in monitoring and detection in NPPs is thought to be mainly caused by a poor mental model and/or a poor design of HMI.

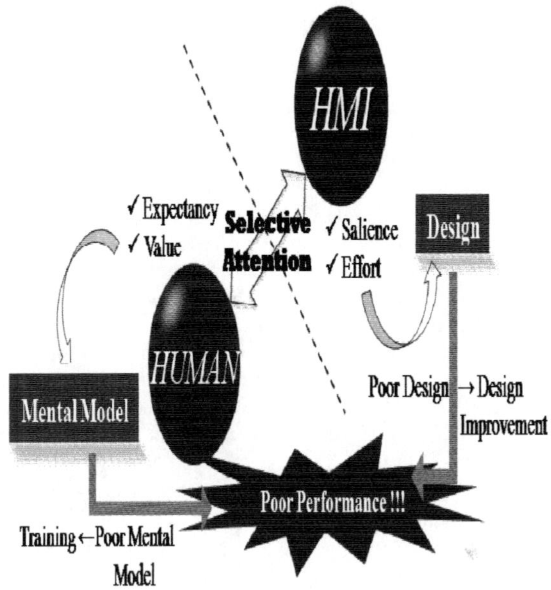

Fig. 1. Amodel of information searching

In complex systems such as NPPs, there are numerous information sources that should be monitored but operators have only limited capacity of attention and memory. Operators selectively attend to important information sources based on their mental models to overcome the limitation. If poor performance in the selective attention is observed, there can be problems regarding the mental model and/or the HMI design. Poor performance in information searching tasks is usually coupled with difficulties such as poor situation awareness, frustration, excessively physical or/and mental load, and so on. If the difficulties are once reported, they are thought to be eliminated by training relevant operators (enhancing operator's mental model) and/or improving the HMI design.

The model can be used as a framework in applications regarding human performance (e.g., especially in evaluating HMI designs or developing a training program). For example, if poor performance results from the mixture of a poor mental model and a poor HMI design, it cannot be found out whether poor performance is originated from poor knowledge on the relevant system or deficiencies in the relevant HMI. Hence deficiencies in HMI can be assessed by evaluating difficulties coupled with poor performance given that the operator's mental model is well-constructed. Developing a training program can be considered in a similar way as well.

3. EXPERIMENTAL STUDY

3.1 Preparaton of the Experiments

The purpose of this experimental study is to investigate the effect of a poor design and/or poor metal model on operator's performance. Subjects conduct complex diagnostic tasks as operators with the FISA-2/PC real time micro-simulator, which simulates a PWR type NPP (Seok & No, 1994). A normal (or acceptable) and a faulty HMI designs are used for this study. The digit number indicators for levels of both loop-A steam generator (S/G (A)) and S/G (B) are intentionally removed from a normal GUI (only bar graph indicators are available), which was identified as a HMI deficiency from the previous study (Ha & Seong, 2007). Also subjects' mental models are controlled by training. The mental model before training is considered poor. The mental model after training is considered well-constructed. Subjects' mental models were scored with four questionnaires about the dynamics of the FISA-2 simulator before conducting experiments. The questionnaires have 100 point scales. The mental model score is then evaluated by averaging the scores of the four questionnaires.

Six tasks including loop-A steam generator tube rupture (SGTR (A)) and loop-B steam line break (SLB (B)) out of fourteen diagnostic tasks are randomly given to subjects. The SGTR (A) and the SLB (B) cases are conducted with the normal GUI and the faulty GUI, respectively. Subjects were provided with the purpose and the procedure of the experiments and the explanation of fourteen diagnostic tasks before experiments. Subjects conduct diagnostic tasks before and after training. The training includes education in the dynamics of the FISA-2 simulator in the cases of the fourteen diagnostic tasks and exercises with the FISA-2 simulations. After an experiment, it is checked as operator's performance whether changes in the S/G (A) and (B) levels are perceived or not during the experiment.

Fifteen graduate students (fourteen males and one female) with nuclear engineering backgrounds of 5.2 years on average participated as operators in the experiments. They had normal or corrected-to-normal vision.

3.2 Results of the Experiment

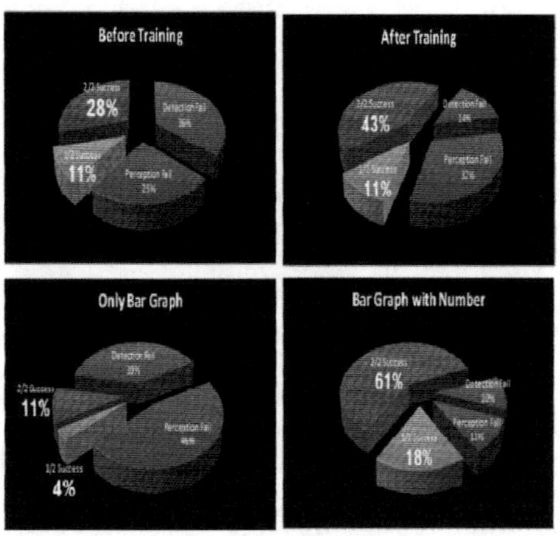

Fig. 2. Effects of poor design and/or poor mental model on operator's performance

Two S/Gs are equipped in the FISA simulator. Hence, there can be 2/2 success and 1/2 success of the S/G level perception, perception fail, and detection fail, as shown in Fig. 2. 2/2 success means that a subject have perceived the trends of a change in both S/G (A) and (B) levels. The detection fail represents that a subject do not figure out the existence of the change.

After training, 2/2 success were increased 15 % (43-28=15) and 1/2 success had no change. As the GUIs is changed from the faulty ane to the normal one, 2/2 success and 2/1 success were increased 50 % (61-11=50) and 14 % (18-4=14), respectively. Hence, the effect of poor HMI design is concluded to be stronger than that of poor mental model in this study.

4. CONCLUSIONS

In this work, a model of human performance in information searching tasks is proposed. Attention is affected by four factors such as salience, expectancy, value, and effort. Expectancy and value of information are determined based on the operator's mental model, whereas salience and effort are the matter of HMI design. Hence human performance in information searching tasks is modelled to be most affected by poor HMI design and poor mental model and an experimental study regarding this model shows poor HMI design affect more than poor mental model on operator's performance (perception success) this study. Hence much effort should be given to HMI design as well as rigorous training programs in order to improve human performance (eventually system safety such as in nuclear power plants). The proposed model is also expected to be helpfully used as a framework in evaluating HMI designs or developing a training program.

5. REFERENCES

Adams, M.; Tenney, Y. & Pew, R. (1995). Situation awareness and cognitive management of complex system. *Human Factors*, Vol. 37, No. 1, pp. 85-104.

Barriere, M.; Bley, D.; Cooper, S.; Forester, J.; Kolaczkowski, A.; Luckas, W.; Parry, G.; Ramey-smith, A.; Thompson, C.; Whitehead, D, & Wreathall, J. (2000). *Technical Basis and Implementation Guidelines for a Technique for Human Event Analysis (ATHEANA),* Rev.01, NUREG-1624, US NRC.

Ha, J. & Seong, P. (2005). An experimental study: EEG analysis with eye fixation data during complex diagnostic tasks in nuclear power plants. In: *Proc. of International Symposium On the Future I&C for NPPs (ISOFIC)*, Chungmu, Republic of Korea.

Ha, J. & Seong, P. (2007). Attentional Resources Effectiveness Measures during Complex Diagnostic Tasks in NPPs. In: *Proc. of International Symposium on Symbiotic Nuclear Power Systems for 21st Century (ISSNP)*, Fukui, Japan.

Seok, H. & No, H. (1994). Development of a Software for the Micro Simulator for Kori Nuclear Power Plant Unit 2. *Nuclear Technology*, Vol. 106, No. 3, pp. 384-396.

Wickens, C. & Hollands, J. (2000). *Engineering Psychology and Human Performance*, 3rd Edition, New Jersey, Prentice Hall.

Annals of DAAAM for 2008 & Proceedings of the 19th International DAAAM Symposium, ISSN 1726-9679
Editor B. Katalinic, Published by DAAAM International, Vienna, Austria 2008
Make Harmony Between Technology and Nature, and Your Mind will Fly Free as a Bird

CONSIDERATION ABOUT ONE WAY HYDROCYCLON WITH RING-SHAPED APERTURE FOR SEPARATION

HADAR, A[nton]; MOTOMANCEA, A[drian]; SZABO, A[dam]; GHEORGHIU, H[oria]; MARIN, D[umitru] & OCNARESCU, C[onstantin]

Abstract: *The hydrocyclon is a very efficient and simple static device, without parts in motion. In this paper, we propose a new type of hydrocyclon having a ring-shaped aperture that permits it to retrieve a part of loosing particles in classic hydrocyclons. That is why this alternative increases the output of this equipment with about 15-17%. We present a model for calculus of the main important aspects of design parameters of this type of hydrocyclon.*

Key words: *hydrocyclon, ring-shaped aperture, hydromass, radial velocity.*

1. INTRODUCTION

This type of equipment named hydrocyclon was invented by E. Britney more than one century ago. It was a real technical revolution because it substituted a lot of very heavy and large equipments with low productivities and outputs. The large scale utilities to separate the solid factions from any kind of hydromass are made to be used in foodstuff industry, pharmaceutical, chemistry, mining, metallurgy and others.

Although the work principle of hydrocyclons remained unchanged, in the last 2-3 decades same improvements were introduced like as a better record between the supply pipe and hydrocyclon body, the decreasing of inside roughness, the improvement of abrasive protection.

The great drawback of actual hydrocyclons consists in just their work principle namely in the existence of two distorted meaning streams inside them, that "disputed" the fine faction. The main descendant stream recuperates from hydromass the most important part of solid factions, while the central hold stream embezzles, respectively draws in over trickling an important part of fine solid factions. This evidence is accepted by all the most important companies through indication of quality parameter "D50" proper for any type of hydrocyclon and that shows the granulometric class with 50% chances to be found again in bold. For example for a hydrocyclon of 10", D50 = 22 - 44 μm.

The hydrocyclon's flows varying between 40l/min-40.000 l/min with supply pressures between 1÷4 bar. The sorting efficiency depends on some qualitative and technical parameters like internal roughness of the lining that must be 0.7÷1.6 μm, the ratio between the supply pipe diameter and hydrocyclon diameter but even the ratio between the supply diameter and trickling diameter (Magyari, 1989).

In connection with the flow inside the hydrocyclon we'll present some novelty considerations unseen by the authors in accessed literature.

The entrance hydromass velocity in hydrocyclon is usually bigger than 10 m/s and the born centrifugal force caused of the rotation of all the fluid volume inside the hydrocyclon conduce to a very fast evolution of all the solid particle to adhere on or in the vicinity of the inside wall of the hydrocyclon. The total moving time of a particle in hydrocyclon is 0.2÷1s. The circular moving from the entrance is changed than in a descendant helicoidally one (because of the continuous hydromass contribution), so in the vicinity of the top of the cone a part of

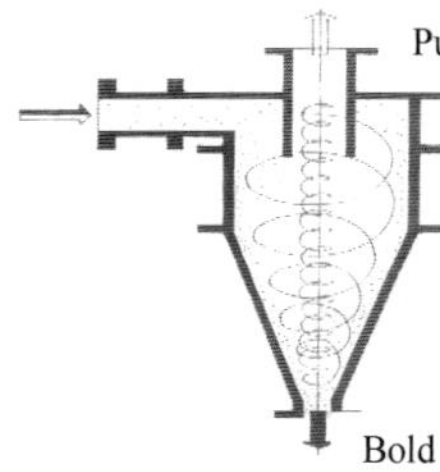

Fig.1. The classic hydrocyclon

the fluid is changed the moving sense in an ascending one like in fig. 1.

It is obvious that a safe boundary between the two streams of different senses does not exist, but certainly they are mutual influence - surely negative: the ascendant helicoidally stream will drive a main part of the thin faction in over trickling. In the same time in a glass hydrocyclon it can observe that because of the rotation movement in the centre of hydrocyclon is born a depression and as a result of this, appears on air column that is also responsible of the ascending stream.

2. THE CONSTRUCTIVE AND FUNCTIONAL DESCRIPTION OF THE NEW HYDROCYCLON

The existence of these perturbations in the classic hydrocyclon is the main cause of this new solution that we propose here. We want to specify that one of the important destinations of this equipment is to produce very pure fluids (water, even air).

In fig.2, we present the scheme of the one-way hydrocyclon with ring-shaped aperture (Jinescu & Sporea, 2007).

It has a cylindrical shape in its active part with descendent one-way inclusively for purified phase and having a ring-shaped aperture for collecting the dense part of the hydromass. The reduction of the evacuation section surface for water is mode in order to maintain the entrance pressure and it doesn't alternate the separation efficiency. In this case the main important problem is to maintain the solid particles in the vicinity of the cylindrical wall until the ring-shaped aperture.

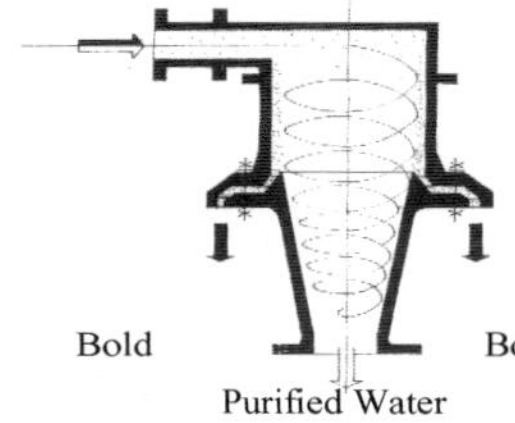

Fig.2. One way hydrocyclon with ring-shaped aperture for separation

That is why on the cylindrical part all the particles will move already on a distance of 0÷30 mm from the inside wall and until the ring-shaped aperture, the centrifugal force will restrict this distance to about 0÷5 mm, so the hole solid part to be separated by ring-shaped aperture. This is thoroughly possible because the dominant force is radial, the ascending stream being present only in conic way where its effect doesn't affect separation, because the solid particles were already eliminated.

3. SOME ELEMENTS OF CONSTRUCTIVE CALCULUS

In order to study the dynamic behavior of a solid particle in fluid is necessary to accept some simplified hypothesis. One of them is that the particle has a spherical shape. In this case we present in fig. 3a the forces on a spherical particle having a descendent movement in a static fluid and in fig. 3b the forces on the same particle in a fluid having a rotation movement.

$$F_w + F_a - G = 0 \quad (4)$$
$$F_{w'} - F_c = 0 \quad (5)$$
$$F_w + F_a - G = 0 \quad (6)$$

where (Voinea et al., 1989):
$F_w = 6\pi\, r\, \eta\, V_e$ the fluid broke force (The Stokes' law)
$F_a = \frac{4}{3}\pi r^3 \rho g$ (Archimedes force); $G = \frac{4}{3}\pi r^3 \rho g$ (The particle's weight); $F_c = m\frac{V_r^2}{R} = \frac{4}{3}\pi r^3(\rho - \rho')\frac{V_r^2}{R}$ (The centrifugal force)

For a hydrocyclon feeding on 1.5 bar pressure, neglecting the damages of hydrocyclon resistance it results an entrance velocity (Hadar et al., 2002):

$$V = \sqrt{2gH} = 17.15 ms^{-1} \quad (7)$$

This tangential velocity will diminish in fluid descendent way and will produce first a rotation movement having:

$$V = \omega R = \frac{\pi n}{30} R = 17.15 ms^{-1} \quad (8)$$

The rotation:

$$n = \frac{30V}{\pi R} = 1639 rot/\min \quad (9)$$

It is possible to obtain the stationary time of a particle in hydrocyclon using the continuity equation

$$V_1 = \frac{\pi d^2}{4} = V_2 \frac{\pi D^2}{4} \quad (10)$$

$$V_2 = V_1 \frac{d^2}{D^2} = 1.1 m/s \quad (11)$$

where: V_1 - the velocity in supply pipe
V_2 - the velocity in cylindrical part of hydrocyclon
d = 50 mm the supply pipe diameter
D = 200 mm the hydrocyclon diameter.

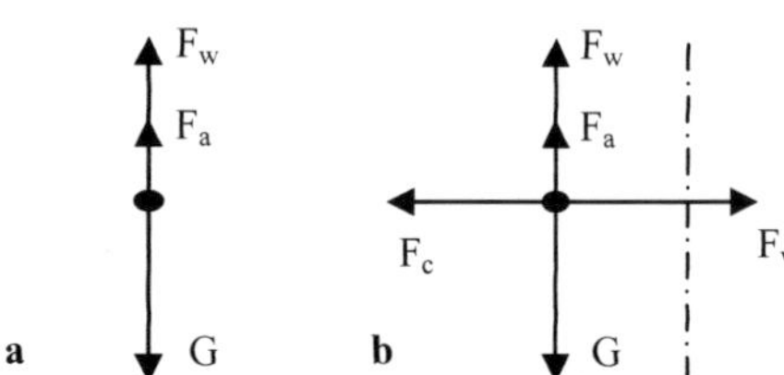

Fig.3. The schemes of the particle forces

For a distance between supply entrance and ring-shaped aperture of 260 mm it is obtained a stationary time of a particle in hydrocyclon

$$t = \frac{0.26}{1.1} = 0.24 ms \quad (12)$$

The route time for a particle of the radial distance (30 mm) until the hydrocyclon wall in order to be picked up by the ring-shaped aperture is calculate using (4) formula for 4a and (5) formula for 4b. Also it is possible to calculate radial velocity.

4. EXPERIMENTAL MEASUREMENTS

The crude sample analysis for a hydromass having $2.64 \cdot 10^3$ Kg/m^3, reveal the next granulometric analysis shown in table 1. (Magyari, 1983).

Granulometric Category [mm]	Weight Extraction v_i	Cumulate Refusals Σv_i[%]	Cumulate Passing 100-Σv_i[%]
1-0.63	1.00	1.00	99.00
0.63-0.32	7.76	8.76	92.24
0.32-0.15	52.36	61.12	38.88
0.15-0.071	36.76	97.88	2.12
0.071-0.032	1.50	99.38	0.62
0.032-0	0.62	100.00	0.00
Total	100.00	-	-

Tab. 1. The granulometric analysis

It is easy to observe that the greatest weight from hydromass is in the interval of 0.071-0.32 mm with 89.12 from whole mass.

4. CONCLUSIONS

Next, we want to present some advantages of this type of hydrocyclon:
- The major advantage of one-way hydrocyclon is that of eliminate the hold stream and controlling of separation process.
- Because of the more simplified circulation inside the hydrocyclon, its flow increases about 20-25% for the same diameter given the used hydrocyclon.
-The accuracy of separation decrease under 10 μm
-The total retriever output increase to 95%.
-The one-way hydrocyclons could separate on one passing mineral with different density because the heaviest would be retrieve in first aperture and the thinnest by the followings.

5. REFERENCES

Hadar, A., Motomancea, A.; Szabo, A.& Gheorghiu, H. (2002), One way hydrocyclon with ring-shaped aperture for separation, *Proceedings of International Conference of Machine Systems, T.C.M.M. No. 40*, Editura Academiei, p. 95-100, Bucharest, Romania

Jinescu, V., Sporea, N. (2007), The Flow Rate of Corotating Twin Screw Extruder. (II), *Mat. Plast.*, Vol. 44, nr. 1, p. 14-17.

Magyari, A. (1983), *Utilaje de preparare a substantelor minerale utile (Equipment for preparing usable mineral substances)*, Editura Tehnica, Bucuresti

Magyari, A. (1989), *Manualul inginerului de mine (Handbook of mining engineers)*, Editura Tehnica, Bucuresti

Voinea, R.; Voiculescu, D. & Simion, F. (1989), *Mecanica Solidului cu Aplicatii in Inginerie (Solid mechanics with engineering applications)*, Editura Academiei Romane, Bucuresti, Romania.

Annals of DAAAM for 2008 & Proceedings of the 19th International DAAAM Symposium, ISSN 1726-9679
Editor B. Katalinic, Published by DAAAM International, Vienna, Austria 2008
Make Harmony Between Technology and Nature, and Your Mind will Fly Free as a Bird

ANALYSIS OF THE QUALITY CONTROL PROBLEMS USING THE WINQSB SOFTWARE PRODUCT

HAMAT, C[odruta] O[ana]; AMARIEI, O[lga] I[oana] & DUMITRESCU, C[onstantin] D[an]

Abstract: *This paper presents, based on a case study, the possibility to use the Quality Control module, one of the 19 modules of the WinQSB product, which builds quality control graphs and performs analyses in connection to it. Thus we analysed a manufacturing process, having the diameter as quality characteristic, using representative analyses and graphs.*
Key words: *quality, graph, analysis, control, simulation.*

1. INTRODUCTION

The industrial development has generated the necessity that, in the manufacturing systems, the information and data collected about quality should be accessible to the decision makers, with the possibility of co-ordinating the actions of different compartments of the company, so that the information could contribute to the identification and elimination of the causes provoking defects.

These conditions imposed, as appropriate methodology for obtaining, processing and analysing quality information from the industrial processes, the grounding of the quality control on methods of mathematical statistics.

A cause of the more difficult affirmation of the statistic control in the industrial practice was represented by the fact that the abundance of the statistic methods presented in the literature accessible to the industrial management was not accompanied by examples of applications (Hamat, 2003).

Consequently, the purpose of this paper is to present the possibility of using a software product, i.e. WinQSB, which allows the solving of the quality control problems simulation.

2. PRESENTING THE PROBLEM

A factory produces pistons. The Quality Control Service intends to introduce the statistic control, and in this respect one removed from the manufacturing process, every 4 hours, samples made of n = 5 pieces. The quality characteristic is the diameter of the pistons, corresponding to the value of 76.12 ± 0.05 mm. The data collected are introduced in a table (figure 1).

For each problem of quality control, the module predefines a set of 14 rules, which can be modified using the options of the Edit menu.

The Quality Control Chart module of the WinQSB software provides 21 types of graphs for variable data, 15 types of graphs for attributive data, histograms, Pareto analysis, curves of the operational characteristics etc (Mihalca et al., 2003).

For this type of problem we may apply the statistic control, using the arithmetic mean and amplitude method. The arithmetic mean measures the stability in time of the process adjustment, while the amplitude measures the stability precision in time of the manufacturing process (Olaru, 1999).

We first perform a statistic analysis (average values, median, dispersion, maximum and minimum values within the group) of the recorded data, by sub-groups and on the whole. The results obtained are shown in figure 2.

Number	Date	Time	Subgroup	Diametru
1	01.07.2008	12.00	1	76.12
2	01.07.2008	12.00	1	76.12
3	01.07.2008	12.00	1	76.11
4	01.07.2008	12.00	1	76.12
5	01.07.2008	12.00	1	76.11
6	01.07.2008	16.00	2	76.10
7	01.07.2008	16.00	2	76.10
8	01.07.2008	16.00	2	76.11
9	01.07.2008	16.00	2	76.11
10	01.07.2008	16.00	2	76.12
11	02.07.2008	12.00	3	76.11
12	02.07.2008	12.00	3	76.12
13	02.07.2008	12.00	3	76.11
14	02.07.2008	12.00	3	76.12
15	02.07.2008	12.00	3	76.12
16	03.07.2008	16.00	4	76.10
17	03.07.2008	16.00	4	76.10
18	03.07.2008	16.00	4	76.10
19	03.07.2008	16.00	4	76.10
20	03.07.2008	16.00	4	76.11
21	04.07.2008	12.00	5	76.11
22	04.07.2008	12.00	5	76.11
23	04.07.2008	12.00	5	76.11
24	04.07.2008	12.00	5	76.11
25	04.07.2008	12.00	5	76.12
26	07.07.2008	16.00	6	76.10
27	07.07.2008	16.00	6	76.11
28	07.07.2008	16.00	6	76.12
29	07.07.2008	16.00	6	76.10
30	07.07.2008	16.00	6	76.10
31	08.07.2008	12.00	7	76.11
32	08.07.2008	12.00	7	76.11
33	08.07.2008	12.00	7	76.11
34	08.07.2008	12.00	7	76.12
35	08.07.2008	12.00	7	76.12
36	09.07.2008	16.00	8	76.12
37	09.07.2008	16.00	8	76.10
38	09.07.2008	16.00	8	76.11
39	09.07.2008	16.00	8	76.12
40	09.07.2008	16.00	8	76.12
41	10.07.2008	12.00	9	76.10
42	10.07.2008	12.00	9	76.11
43	10.07.2008	12.00	9	76.11
44	10.07.2008	12.00	9	76.11
45	10.07.2008	12.00	9	76.12
46	11.07.2008	16.00	10	76.11
47	11.07.2008	16.00	10	76.11
48	11.07.2008	16.00	10	76.11
49	11.07.2008	16.00	10	76.10
50	11.07.2008	16.00	10	76.11
51	14.07.2008	12.00	11	76.12
52	14.07.2008	12.00	11	76.12
53	14.07.2008	12.00	11	76.11
54	14.07.2008	12.00	11	76.12
55	14.07.2008	12.00	11	76.10
56	15.07.2008	16.00	12	76.12
57	15.07.2008	16.00	12	76.12
58	15.07.2008	16.00	12	76.12
59	15.07.2008	16.00	12	76.11
60	15.07.2008	16.00	12	76.11

Fig. 1. The input data of the problem

	20:24:34		Monday	July	28	2008			
Sample	Sample Size	Mean	Median	Midrange	Variance	S.D.	Range	Maximum	Minimum
1	5	76,1160	76,1200	76,1150	0,0000	0,0055	0,0100	76,1200	76,1100
2	5	76,1080	76,1100	76,1100	0,0001	0,0084	0,0200	76,1200	76,1000
3	5	76,1160	76,1200	76,1150	0,0000	0,0055	0,0100	76,1200	76,1100
4	5	76,1020	76,1000	76,1050	0,0000	0,0045	0,0100	76,1100	76,1000
5	5	76,1120	76,1100	76,1150	0,0000	0,0045	0,0100	76,1200	76,1100
6	5	76,1060	76,1000	76,1100	0,0001	0,0089	0,0200	76,1200	76,1000
7	5	76,1140	76,1100	76,1150	0,0000	0,0055	0,0100	76,1200	76,1100
8	5	76,1140	76,1200	76,1100	0,0001	0,0089	0,0200	76,1200	76,1000
9	5	76,1100	76,1100	76,1100	0,0001	0,0071	0,0200	76,1200	76,1000
10	5	76,1080	76,1100	76,1050	0,0000	0,0045	0,0100	76,1100	76,1000
11	5	76,1140	76,1200	76,1100	0,0001	0,0089	0,0200	76,1200	76,1000
12	5	76,1160	76,1200	76,1150	0,0000	0,0055	0,0100	76,1200	76,1100
Overall	5	76,1113	76,1125	76,1112	0,0000	0,0065	0,0142	76,1200	76,1000

Fig. 2. Analysis->Sample Summary

We continue the analysis with the elaboration of certain representative graphs, available in the Gallery menu, first the control charts for the data obtained "through measurement" which can by of several types: control charts for means (X charts – figure 3), control charts for amplitude (R charts– figure 5) and control charts for the sample standard (Axinte, 2007).

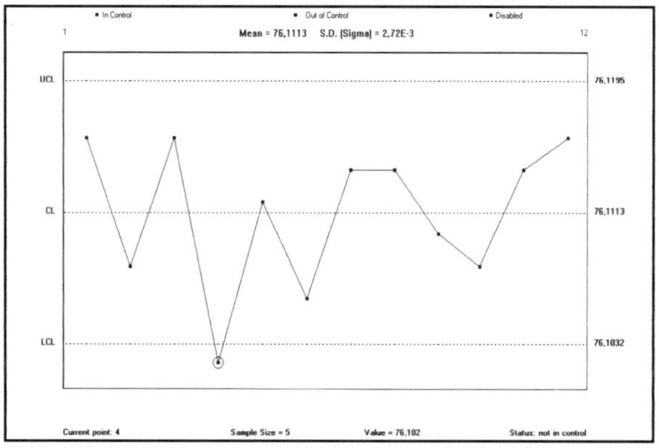

Fig. 3. X-bar (Mean) Chart

We remark in figure 3 that sample 4 is below the inferior limit of 76.1032. We can obtain more information about this sample by making appeal to the Analysis option – Individual Point, in a window of the type shown figure 4. We notice that the average / mean of this subgroup is not under control, because it violates the rules number 2 and 14 in the rules table offered by the software.

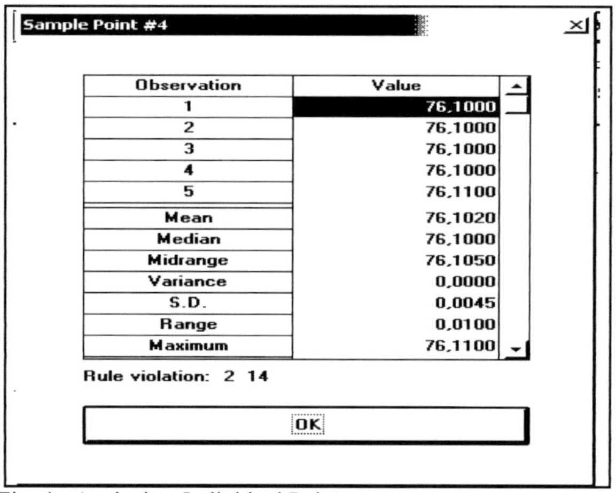

Sample Point #4

Observation	Value
1	76,1000
2	76,1000
3	76,1000
4	76,1000
5	76,1100
Mean	76,1020
Median	76,1000
Midrange	76,1050
Variance	0,0000
S.D.	0,0045
Range	0,0100
Maximum	76,1100

Rule violation: 2 14

OK

Fig. 4. Analysis – Individual Point

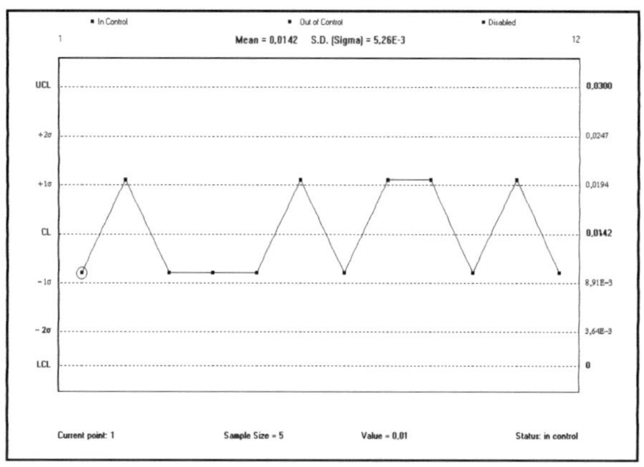

Fig. 5. The control chart for amplitude

The CUSUM graph for average / mean values – CUSUM Chart for Mean, presenting the dispersion of the average dispersion from the central average / mean and we verify if from this viewpoint the values fall within the normal limits. The graph is presented in figure 6, where we remark that all the points are represented in green, and consequently there are no uncontrolled values.

This type of graph is one of the several different modes provided by the product for the verification of the compliance of values with the normal limits (Amariei et al., 2008).

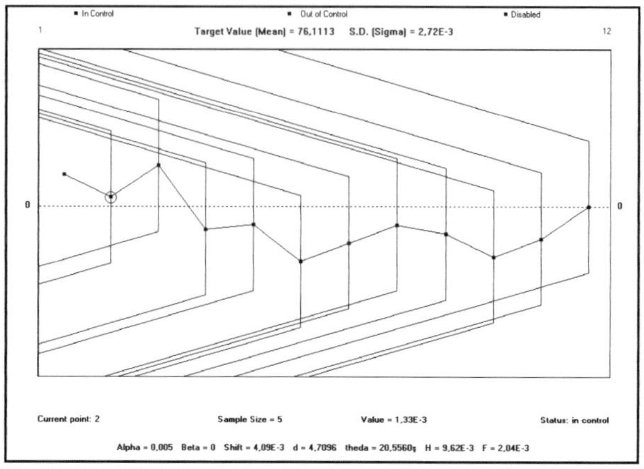

Fig. 6. The CUSUM graph for average / mean values

3. CONCLUSION

From the analysis of the graph corresponding to the mean variation it results that the sample number 4 recorded on 03.07.2008, 16:00 hours, is not under control, and thus we consider that the manufacturing process has not been adjusted appropriately. Moreover, all the other points are situated between the superior and the inferior limit, but we remark a rather large dispersion of the process.

Analysing the graph corresponding to the amplitude variation results that all the points are situated below the superior control limit, indicating the fact that the process provides the necessary precision.

In the present situation, the manufacturing process is considered inappropriate. We must analyse the causes having triggered deviations of the adjustment, using the cause-effect diagram. In order to bring the manufacturing process to the normal operation condition, corrective actions are initiated and implemented, actions meant to eliminate the causes generating defects, after which the statistic control is continued.

4. REFERENCES

Amariei, O.I., Dumitrescu, C. D., Gillich N., Hamat, C., Maloş, C.R.(2008). *Optimizing a manufacturing system after analysis of Extend™ and WinQSB simulation environment,* 2008 IEEE International Conference on Automation, Quality and Testing, Robotics AQTR 2008 - THETA 16th edition - May 22-25, ISBN 1-4244-2577-8, 2008 Cluj-Napoca, Romania,

Axinte, E. (2007). *Elements of quality insurance in industrial engineering*, Demiurg Editorial House, Iasi

Hamat, C.O. (2003). *Quality Management*, Orizonturi Universitare Publishing House, Timisoara, ISBN 973-638-047-5

Mihalca, R; Fabian, C (2003). *The use of the software products - Word, Excel, PMT, WinQSB, Systat*, ASE Publishing House, Bucharest ISBN 973-594-320-4

Olaru, M. (1999). *Quality Management*, ASE Publishing House, Bucharest, ISBN 973-946-202-2

Annals of DAAAM for 2008 & Proceedings of the 19th International DAAAM Symposium, ISSN 1726-9679
Editor B. Katalinic, Published by DAAAM International, Vienna, Austria 2008
Make Harmony Between Technology and Nature, and Your Mind will Fly Free as a Bird

COOLING INSTALLATION WITH LIQUID NITROGEN FOR THE CRYOGENIC TURNING OF AUSTENITIC STAINLESS STEELS

HAMAT, C[odruta] O[ana]; POPOVICI, G[heorghe]; COMAN, L[iviu] & MARTA, C[onstantin]

¶

Abstract: *The paper presents a cooling installation with liquid nitrogen for the cryogenic turning processing put together at the "Eftimie Murgu" University of Resita, Romania.*
The requirement is that the liquid nitrogen be synchronously "drawn" rotation-wise towards the contact area.
By construction, the installation may be used both at turning as well as at drilling and milling when processing austenitic stainless steels.
Key words*: cryogenic, installation, nitrogen, steel, turning.*

1. INTRODUCTION

The cooling installation with liquid nitrogen for the cryogenic turning processing (Resita, Romania variant) is an installation in modular built. The most important issue in building the installation was to bring the liquid nitrogen to the cutting area under liquid form and not as vapours, whose cooling rate is one dimension-order smaller than the liquid nitrogen.

2. AREA OF APPLICATION

The present installation is used to the cryogenic turning of austenitic stainless steels (18-8). By construction, the liquid nitrogen cooling installation can be used for the other basic procedures in metal cutting, i.e. drilling and planning.

3. CONSTRUCTIVE DESCRIPTION OF THE INSTALLATION

From the constructive viewpoint, the designed liquid nitrogen cooling installation (IRC) is made of 4 modules:
MS – the module of liquid nitrogen stocking, with the role to create a super-pressure in the Dewar-type cooling vessel and the transport of the liquid through the installation;
MA – the feeding module, assuring the transfer of the liquid nitrogen form the stocking vessel to the cutting area;
MP – the pressurisation module, with the role to locate and concentrate the liquid nitrogen on the processed part and / or cutting tool;
MR – the module for the adjustment of the liquid nitrogen flow, in accordance with the temperature imposed in the cutting area (Decker, 2004).

A modular view of the liquid nitrogen cooling installation is shown in Figure 1.

4. FUNCTIONAL DESCRIPTION OF THE INSTALLATION

The rotating adapter 19 is conceived so that the liquid nitrogen under pressure flow from the central feeding pipe into a distribution chamber, from where – through radial orifices – it must arrive at the pressurisation nozzle 20.
The orientation of the liquid jet under a certain angle compared to the tangent to the trajectory of the cutting movement (Gornic, 2007) is made with the rotating adapter 19,

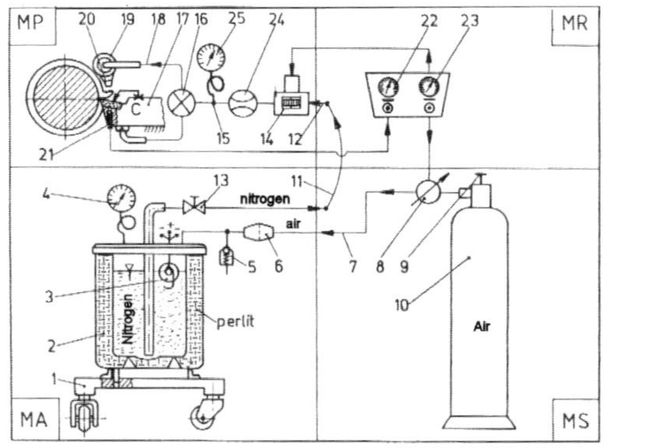

Fig.1. Installation scheme – the Reşiţa, Romania variant

and the concentration of the liquid is assured through the pressurising nozzle 20.

5. RESEARCH

The efficiency of the liquid nitrogen cooling is shown by the manner it is guided through the cooling area to the cutting area. The form of the nozzle is given according to the type of procedure. When turning, circular nozzles are recommended, with the spraying holes located on the diameter.

The angle of the nozzle against the cutting trajectory and the width of the contact area are more important than the distance between the nozzle and cutting edge. (Tapalaga, 1998).

The placement angle must be between 15° and 20° towards the processed part, compared to the tangent to the cutting trajectory. The basic requirement is that the nozzle follows the processed part, and the spraying is in the sense of the part or tool rotation. Moreover, the difference between the main cutting speed (v_{ac}) and the liquid jet speed (v_m) must be small (Badut, 2008):

$$v_c - v_m \rightarrow 0 \quad (1)$$

Only then the liquid is "synchronously drawn" in the rotation sense to the contact area and the cooling is correct (Figure 2).

Fig. 2. Synchronous cooling with liquid nitrogen

Cutting speed v_c [m/min]	<75	75…100	100…125	>125
Flowing rate v_m [m/s]	1.5 -1.75	1.75-2.0	2.0-2.25	2.25-2.5

Tab. 1. Recommended values of the flowing rate at the nitrogen jet depending of the main cutting speed

Besides, one aimed at establishing a regime of efficient cooling by determining the working pressure and the minimum nitrogen flow for a certain cooling degree.

Experimentally one found that the liquid nitrogen starts to flow to the cutting area at the pressure p_a=0.05 MPa (0.5 bar), but the flow is continuous only from the pressure of p_{min}= 0.125 MPa (1.25 bar) and up to the pressure of p_{max}=0.35 (3.5 bar), when the turbulence effect occurs.

The recommended values of the flowing rate of the nitrogen jet, depending on the main cutting speed are given in Table 1 (Muhs, 2007).

The working pressure, depending on the flowing rate imposed for the nitrogen jet is obtained with the relation:

$$p_m = 0.07 \cdot \rho_a \cdot v_m^2 \quad [MPa] \tag{2}$$

Where:

ρ_a [kg / dm³] – the density of the liquid nitrogen;
v_m [m/s] – the flowing rate.

The "pressure - flowing rate" diagram is shown in Figure 3.
The liquid nitrogen flow (debit) is given by the relation:

$$Q_v = 0.05 \cdot v_m \cdot d_0^2 \quad [1/\min] \tag{3}$$

Where:

v_m [m/s] – the flowing rate;
d_0 [mm] – the nozzle diameter.

The "flow - nozzle diameter" diagram is shown in Figure 4.

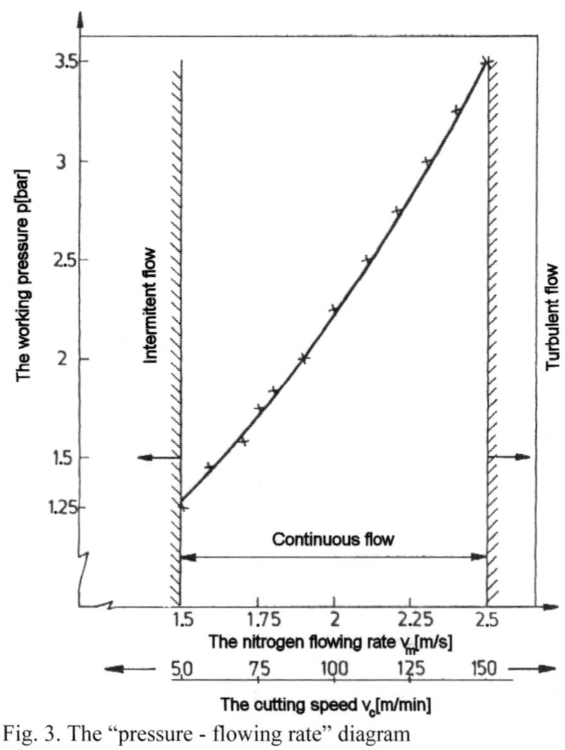

Fig. 3. The "pressure - flowing rate" diagram

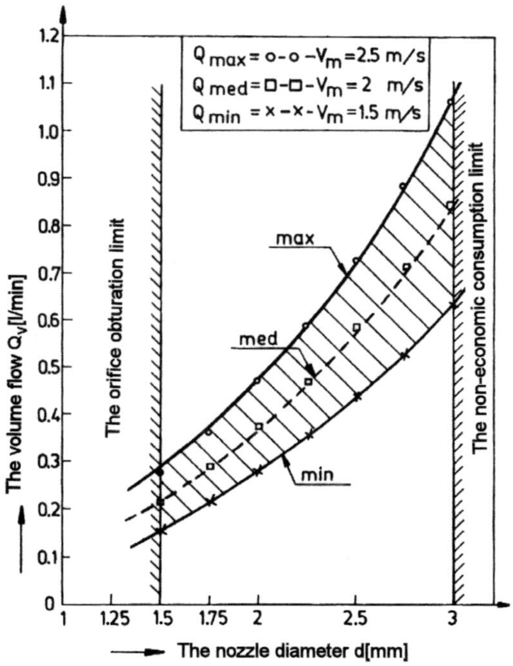

Fig 4. The "flow - nozzle diameter" diagram

6. CONCLUSION

The present installation, compared to other installations known in the field, has the following advantages:

- constructive simplicity and easy operation, as the compression of the liquid nitrogen can be done either from a cylinder, or directly from the supply network, but if the latter is the case, the air must be filtered and dried;
- the nitrogen flow (debit) can be modified within limited ranges, by the fine adjustment performed by the electro-check valve;
- the cryogenic cooling is possible in all three variants (cooling of the processed part, of the tool or of the two simultaneously), by means of the three-way distributor (handle position: 0°, 45° and 90°);
- nevertheless, by construction, the installation has also a drawback, as it does not provide a perfect thermal insulation for the transfer of liquid nitrogen through copper pipes and flexible tubes for cryogenic liquids.

7. REFERENCES

Badut, M. (2008). CAD-CAM-CAE for products. *Techniques and Technology,* issue 3 (35), pp. 72-73, ISSN 1453 8423, Bucharest

Decker, K.-H. (2004). *Maschineelemente–Aufgaben, (Machine Design),* Carl Hansen, Verlag, ISBN 3-446-21525-5, Munchen

Gornic, C. (2007). What means to know how to design? *Techniques and Technology,* issue 5 (35), pp. 38-39, ISSN 1453 8423, Bucharest

Muhs, D. (2007). Roloff/Matek. *Maschinenelemente (Machines Design),* Auflage. Viewegs Fachbucher der Technik Design, ISBN 3-528-54482-1, Augsburg

Tapalaga, I. (1998). *Criogenia în construcţia de maşini. (Cryogenics in Machines Design)*, Dacia Publishing House, No. 2954-332, Cluj-Napoca

Annals of DAAAM for 2008 & Proceedings of the 19th International DAAAM Symposium, ISSN 1726-9679
Editor B. Katalinic, Published by DAAAM International, Vienna, Austria 2008
Make Harmony Between Technology and Nature, and Your Mind will Fly Free as a Bird

NEW SOLUTIONS IN THE CREDITING PROCESS OF AGRICULTURE

HARANGUS, D[aniela]

***Abstract**: In the introduction of this paper are showed the principal elements for analysis taken into consideration in the crediting process of agriculture. An excessive rigour in the analysis of these elements leads to a limitation of access to banking credits for agricultural farms. The method of this research is based to quantification of flexibility degree of banks granting credits to agricultural customers and finding new solutions in the crediting process of agriculture. It results that the totality of risk factors specific to any activity, including the agricultural farm, can be quantified. Starting with the analysis of risk factors and symptoms, this paper proposes new solutions for diminishing the degree of banks risks.*
***Key words**: credit, risk, agriculture, quantification, flexibility.*

1. INTRODUCTION

For a prosperous Europe, the European Union has as priorities sustainable rural development as well as agricultural development. These priorities are motivated by the fact that over 80% of the territory of the European Union is covered by rural areas populated by approximately 1/3 of population. In order to adapt the agriculture to the requirements for the development of rural areas, new financial resources for agriculture have to be found (Rizov, 2005). Besides non-reimbursable financial sources, bank loan should have an important position in the agricultural financing process (Osborne, 2006).

In current banking theory and practice the elements for analysis taken into consideration in the crediting process of agriculture are the following (Harangus, 2008):

- financial performance of the agricultural farms;
- level, structure and quality of the guarantees presented by the agricultural customer upon soliciting credit. East European banks prefer real estate guarantees (mortgage), which is difficult for the bank to asses and less agreed by the agricultural customer;
- amount of expenses necessary for production. The level of credit granted by the bank is established according to the level of such expenses;
- analysis and credit risk assessment for each customer applying the "principle of 5 C's" used in American banking system. The 5 C's represent character, capacity, capital, conditions, and collateral;
- ensuring good conditions for preservation, storing and valuing of agricultural production credited by the bank..

An excessive rigour in the analysis of these elements and lack of adequate banking consultancy and counseling of agricultural customers leads to a limitation of access to banking credits for agricultural farms. For this reason East European banks are considered to have a rigid system of agricultural crediting, based on a static analysis of agricultural activities. A dynamic analysis of the activity of agricultural farms is necessary as well as taking crediting decision by considering a context of internal and external factors which influence agricultural activity.

In the present context of European development the main target is the increase of banking credit for supporting agriculture and sustainable rural development as well as a rapid access to banking credits for agricultural customers (Kostov & Lingard, 2004).

2. RESEARCH AND METHOD

This study is based on information regarding present circumstances of crediting system for agriculture and regarding the sum of agricultural credits in the total volume of credits granted by banks.

The need for credits in agriculture is greater than in other sectors of economy because of the following reasons:

- biological laws governing production process in agriculture generate a much slower regeneration of funds and capital, unlike other areas of production, the need for credits being thus much higher;
- immobilization of funds in agriculture is higher than in other sectors and the capital of the farmers is not sufficient;
- land is expensive but necessary and it represents the main means for production. In the Eastern Europe, buying land represents an increased necessity for farm owners as family farms are much viable and favourable for competitive agriculture;
- agriculture ensures population's alimentary security and it conditions the development of rural space;
- agriculture uses natural agricultural potential of a country.

Despite of the fact that agriculture needs a greater amount of credits than other sectors, the analysis of information regarding evolution and amount of credits for agriculture shows that the amount of credits for agriculture is very small considering the total amount of credits granted by the European banks (between 2.4% and 6.8%), according to data presented in table 1.

In these conditions we consider that the crediting decision for agriculture should be much more flexible. The increased risk factors are specific for agriculture in comparison with other sectors of economy. These risk factors are emphasized by credit risks the banks have to cope with (Loubergé & Schlesinger, 2005).

Countries	Bank credits offered to agriculture	
	€ mill.	% of total credits
France	42,100	2.4
Germany	32,500	2.6
Hungary	1,167	6.8
Romania	677	2.5

Tab. 1. The level of bank credits for agriculture in some European states in 2006.
Source: Data from reports of National Banks from these countries: www.banque-france.fr; www.bundesbank.de; www.mnb.hu; www.bnr.ro, Statistical Section.

Taking into consideration increased risk in crediting agriculture, the decision of a bank to grant a credit has to consider risk factors and risk symptoms manifested in the activity of the agricultural customer. Risk factors specific for any activity refer to:

- overestimating the volume of business in comparison with existing possibilities (uncontrolled expansion);
- non-corresponding capital structure, i.e. large capital of non-current assets and small capital of current assets;
- non-corresponding capitalisation (very small reinvested capital);
- investment projects in execution, too large or too many.

When carrying out a financial and economic analysis for assessing risk factors, banks take into consideration the following risk symptoms that can appear in the activity of the customer soliciting credit:

- financial signals regarding delay of payments, increase of supplies, etc.;
- "creative" accountancy (presenting synthetic data, not justified by the analytical evidence, "arranging" some indices, etc.);
- non-financial signals regarding rejection because of non-corresponding quality, non-observance of contractual deadlines, etc.;
- other signals referring to: resignation of personnel, especially management resignation; court cases;
- unfavorable rumors about the customer, customer's involvement in various financial or corruption scandals;

Considering the above mentioned reasons, we consider that quantification of flexibility degree of banks granting credits to agricultural customers and finding new solutions in the crediting process of agriculture are very important for this research.

The totality of risk factors specific to any activity, including the agricultural farm, can be quantified as follows:

$$\sum_{i=1}^{4} Ri = Rb + Rs + Rc + Rp \tag{1}$$

where *Rb* represents the risk of overestimating the volume of business, *Rs* represents non-corresponding capital structure, i.e. large capital of non-current assets and small capital of current assets, *Rc* is non-corresponding capitalisation (very small reinvested capital), and *Rp* represents investment projects in execution, too large or too many.

If we grant a score from 1 to 15 for each risk factor (*Rb*, *Rs*, *Rc*, or *Rp*), it results that:

$$\sum_{i=1}^{4} Ri \leq 60 \text{ points} \tag{2}$$

It results that the totality of possible risks (T) equals to 60 points, namely:

$$T = 60 \text{ points} \tag{3}$$

In banking practice the appearance of a single risk factor (*Rb*, *Rs*, *Rc*, or *Rp*) is enough to create the danger threshold (D) for the bank. Thus, the danger threshold for the bank is:

$$D = 15 \text{ points} \tag{4}$$

The difference between the totality of possible risks and danger threshold for the bank in granting credit represents the flexibility degree of the bank for credit granting, namely:

$$F = T - D \tag{5}$$

where F represents the flexibility degree in the crediting process, T represents the totality of possible risks, and D is the danger threshold regarding risk for the bank. It results that the flexibility degree of the crediting process (F) can have the following values:

$$15 \leq F \leq 60 \tag{6}$$

Considering the small amount of credits for agriculture in all the credits granted by banks, it results that the flexibility degree in granting credits for this sector is very low (close to 15 points). Under these circumstances, the access to credits for agriculture is much lower than in other economic sectors.

3. CONCLUSION

This theoretic quantification of the flexibility degree in granting credits that I propose, allows for banks to resize and estimate the degree of exigency in the crediting process of agriculture as well as a pertinent assessment of risks the banks face. It also allows for a more efficient management of credit risks.

In banking practice, regardless of credit level granted and risk degree, the bank requires ensuring guarantees to protect from risks. Guarantees requested by the bank (collateral), usually cover the maximum level of debt to the bank, namely the credit plus the interest until the reimbursement of first installment. Under these circumstances, it becomes evident that the lack of flexibility in the crediting process of agriculture is not completely justified. It can be motivated by the attitude of the banks that consider agriculture as a non-attractive sector implying great risks.

Moreover, in the crediting process of agriculture, the banks can find solutions to retrieve granted credits by making use of surrendering compensation rights obtained by the agricultural customer from the specialized insurance companies. These compensations can be obtained from the insurance companies if the agricultural production credited by the bank has previously been insured by the agricultural customer. The banks possess many instruments to protect themselves from risks generated by the agricultural credit. The small volume of credits for agriculture from the total volume of credits granted by European banks implies a revision of the crediting process of agriculture, as well as its greater flexibility.

The research described in this paper proposes to improve the methodology for the analysis of the credit risks by banks. The research will be continued by other studies with the same theme and by a summary contained in a book.

The topics will be placed in the working plan of the Center for Research of the Faculty of Economics, "Tibiscus" University of Timisoara, from the academic year 2008-2009. The results of the research are addressed in particular to the commercial banks that grant credits for agriculture sector.

4. REFERENCES

Harangus, D. (2008). The Dimension of Bank Risks in the Crediting Process of Agriculture, *Proceedings of International Scientific Symposium* (Ed. x (3), pp. 207-212, ISSN 1453-1410, Timisoara, May 2008, Agroprint Publishing House, Timisoara

Kostov, Ph.& Lingard, J. (2004). Risk management: a general framework for rural development, *The Journal of Rural Studies,* No. 19 (2004), pp. 463-476, ISSN 0743-0167

Loubergé, H. & Schlesinger, H. (2005). Coping with Credit Risk, *The Journal of Risk Finance*, Vol. 6, No. 2 (2005), pp. 118-134, ISSN 1526-5943

Osborne, T. (2006). Credit and Risk in Rural Developing Economies, *Journal of Economic Dynamics & Control*, No. 30, (2006), pp. 541-58, ISSN 0165-1889

Rizov, M. (2005). Rural Development under the European CAP: The Role of Diversity. *The Social Science Journal*, No. 42, (2005), pp. 621-628, ISSN 0362-3319

Annals of DAAAM for 2008 & Proceedings of the 19th International DAAAM Symposium, ISSN 1726-9679
Editor B. Katalinic, Published by DAAAM International, Vienna, Austria 2008
Make Harmony Between Technology and Nature, and Your Mind will Fly Free as a Bird

ASPECTS REGARDING THE ROMANIA'S POSITION IN THE GLOBAL COMPETITION OF KNOWLEDGE-BASED ECONOMY

HARANGUS, D[aniela] & DUDA-DAIANU, D[ana] C[odruta]

Abstract: *The objectives of proposed research are fits into the context of reflecting the level of development reached by our country, developments in information and knowledge-based society. We consider relevant for our research the approach and the diagnosis of our country's position in the global competition of the knowledge-based economy, through representative models of strategic analysis. Taken into account the fundamental importance of information and communication technology (ICT) for supporting the development of our country towards the knowledge society and its integration into EU research, the research aims at achieving a "radiography" of ICT sector.*

Key words: *competitive advantage, ICT, competitiveness, knowledge, development*

1. INTRODUCTION

The development of knowledge-based society seems to be the only viable explanation for, and a solution to what Romanian economists called "the paradox of the Romanian economy". The paradox is determined by developments of the country for the past 150 years (except make the period after 1989), through the development stages during the centenary growth's cycle - considering the global cycle and/or European - and the contradiction of development during descendent phases which was small. Economic gap has expanded constantly (under the influence of performance criteria), as the gap of productivity and standard of living gap.

A possible explanation of this paradox is that the Romanian economy has always focused on the least innovative products. This has provided only a temporary competitive advantage, in the short term. This vicious circle can be broken only through investment in implementing the knowledge-based economy, which depends on new factors pf production like inputs, such as knowledge and the transition to a real competitive advantage, as this concept was defined by Michael Porter (1998).

Modern theories clearly describe the relationship between knowledge management and competitive advantage of organizations (Draghici G., 2006). Knowledge management is a complex activity that has two categories of influence in organizations: making the connection between intellectual capacity of organizations - in the form of explicit and tacit knowledge - and directly influence its economic growth (Barclay & Murray, 2000); its fall entirely on strategies, policies and practices at all levels of the organization because of the explicit approach of the activities' knowledge component (Van Der Bly, 2005).

The cognitive substantiation of the activities is finding again at macroeconomic level. The new online economic world is supported by an exponential development of knowledge (Harangus D, 2008).

2. METHODOLOGY

The SWOT analysis can be done in a qualitative or in a quantitative manner, which quantize the position in the space of two axes. In the following analyse is presented the relationship between internal medium and - quantitative external SWOT medium in case of Romanian ICT branch. Quantitative SWOT approach involves the following steps:

1. Listing the main internal factors;
2. The granting of share of these factors, so that the weights should be 1;
3. The granting of scores on a scale which has "0" as a median. Scores "-" means that the internal factor considered represents a weakness of the organization and the "+" have the contrary meanings;
4. The calculation of the total score as sum of the multiplications between considered factors' scores and their weight. The result represents the co ordinate's resultant on SW axis.

$$p_j = \frac{\sum_{i=1}^{20} K_{ij}}{\sum_{j=1}^{20}\left(\sum_{i=1}^{20} K_{ij}\right)} \qquad (1)$$

$$x = \sum_{j=1}^{20} p_j \times n_j \qquad (2)$$

The quation (1) represents the calculation's formula for weight of each factor and the equation (2) represents calculation's methodology for coordinate on Ox axis .

5. Listing the main external factors;
6. The granting of share of these factors, so that the weights should be 1;
7. The granting of scores on a scale which has "0" as a median. Scores "-" means that the internal factor considered represents a threat and the "+" have the contrary meanings;
8. The calculation of the total score in a manner similar to the point 4. The result represents the coordinates' resultant on OT axis.

$$p_j = \frac{\sum_{i=1}^{9} K_{ij}}{\sum_{j=1}^{9}\left(\sum_{i=1}^{9} K_{ij}\right)} \qquad (3)$$

$$y = \sum_{j=1}^{9} p_j \times n_j \qquad (4)$$

The quation (3) represents the calculation's formula for weight of each factor and the equation (4) represents calculation's methodology for coordinate on Oy axis .

The two numbers resulted to point 4 and 8 permit to determine the position of the industry through their signs, in one of the four dial, suggesting a certain type of strategy. The size of that right segment, joining the origin point and the founded point, and its inclination give us an indication regarding the particular strategy from dial's typology.

3. RESULTS OF THE RESEARCH

The results of applying the SWOT analysis model for ICT sector are the following.

3.1. Strengths

- Highly qualified human resources, internationally recognized - more than 115 university with over 60 faculties with this specialty (computers, telecommunications, electronics);
- The convergence of communications and information technology in the new ICT sector: strong, coherent, integrated;
- The ICT's community began to get more "personality" by increasing the involvement of ICT companies, experts and professional associations in their efforts for the development of ICT sector in Romania;
- The capacity of companies to adapt quickly at modern technologies;
- Possibility of rapid development of the sector due to the development of ICT' solutions "from scratch";
- Quick recovery of investment in the field; introduction of ICT solutions increase the value of organizations;

3.2. Weaknesses

- There were strategies which have had a strategic control (causes: inappropriate control systems, conventional thinking, fear of change, inability to allocate resources, hide results unfavorable);
- ICT teams at central and local organizations are still poorly prepared because of weak access to new technologies;
- High level of piracy in the use of IT (approximately 70%)
- Over 62% of organizations with activities in retail sales, telecommunications, finance and public sector have not adopted any measures against fraud on the Internet;
- ICT is not used sufficiently in the public domain (for example, the concept of "outsourcing" for IT services);
- The average salaries of ICT's professionals is small (an average of $ 503 per month) compared with developed countries (between $ 4000 per month and $ 10,000 per month);

3.3. Opportunities

- The trend shown worldwide for building Information Society and the knowledge-based (reducing the technological gap - the digital divide, etc.);
- There is a huge demand for specialists in this field in developed countries (access to leading technology, some of the leave becomes true "ambassadors" of Romania and are living testimony to the high level of preparedness of ICT professionals Romanian);
- EU imposes some conditions that require the development of ICT sector;
- There are numerous international funds dedicated to the implementation of Information Society in developing countries;
- Financial aid for SME's projects on e-commerce, support for companies that develop e-commerce applications.
- Technological progress makes possible the convergence of communications channels.

3.4. Threats

- There is a huge demand for specialists of this field in the developed countries (approximately 30,000 specialists). The departing of Romanian specialists abroad, mainly due to the difference in income and living standards;
- Use of alternative infrastructures require additional investment for operators;
- Competition exists in this area on other markets;
- Increasing crime through electronic means decreased consumer confidence in the use of such means.

4. CONCLUSIONS AND PROPOSES

Based on the methodology previously presented, was presented a graphic version of strategic ahead for the ICT sector so as to support Romania's progress and development of knowledge-based society (figure 1.).

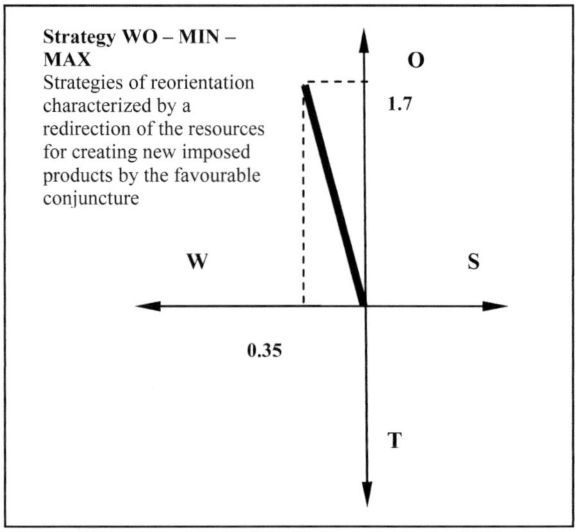

Fig. 1. Determining the future strategic direction based on the SWOT model of analysis (SWOT profile)

Strategic direction to be followed is possible to be implemented by receiving a high level score on Oy axis, which proves the existence of a "volume" and a substantial opportunity for ICT domain in Romania, which is able to support the progress in the knowledge-based economy.

Based on the methodology previously presented, was presented a graphic version of strategic ahead for the ICT sector so as to support Romania's progress and development of knowledge-based society (figure 1.).

By the strategic adopted direction must be overcome the weaknesses in the field of ICT by the fructifying of existing opportunities. The threats are those items which could be reduced through better capitalization of the "strengths", presented above. Such a strategy has become apparent that should be adopted by the resort ministry, as a result of the application of quantitative SWOT analysis. In fact, the strategy is a shift one and is characterized by a redirection of resources in creating new products imposed by the favourable situation of EU accession.

In future researches we will look for new solutions for reducing or even eliminating the weaknesses of the Romanian ICT sector that were outlined in the results of current analyse.

5. REFERENCES

Barclay, R.O.; Murray, P.C. (2000). What is Knowledge Management, *Available from:* http://www.mediaaccess.com /whatis.html, *Accessed:* 2008-02-23

Draghici, G. (2006). New Business Requirements in the Knowledge Based Society, capitol in *Adaptive Technologies and Business Integration: Social, Managerial and Organizational Dimensions*, Editors Cunha, M., M., Cortes, B., Putnik, G., D., Idea Group Publishing, Information Science Publishing, IRM Press, CyberTech Publishing and Idea Group Reference.

Harangus, D. (2008). The New Dimension of Banking Knowledge in the Global Economy, *Proceedings of International Scientific Conference,* (4th Ed. online), pp.374-377, ISSN 1582-5450, Oradea – Romania, May 2008, The University Publishing House, Oradea

Porter, M.E. (1998). *The Competitive Advantage of Nations*, Free Press Edition, ISBN 0-648-84147-9, New York

Van Der Bly, M. (2005). Globalization: A Triumph of Ambiguity, *Current Sociology Journal,* Vol. 53, No. 6, (November 2005), pp. 875-893, ISSN 0011-3921

Annals of DAAAM for 2008 & Proceedings of the 19th International DAAAM Symposium, ISSN 1726-9679
Editor B. Katalinic, Published by DAAAM International, Vienna, Austria 2008
Make Harmony Between Technology and Nature, and Your Mind will Fly Free as a Bird
2nd DAAAM-ICAT

A COMPREHENSIVE BUSINESS MODEL FOR RAPID MANUFACTURED PRODUCTS

HASAN, S[aad] & RENNIE, A[llan]

Abstract: *The use of Rapid Manufacturing (RM) in a mass scale depends on a sustainable business model. The context of RM functionality can be segmented into two scenarios: Scenario 1 being the potential of RM to manufacture products with complex design. Scenario 2 presents that, once identified, if a product is viable to be produced using RM technologies, then how would the supply and demand paradigm work? This paper tries to answer these questions through the means of an e-business platform for RM products. The principle function of the platform is to electronically settle the supply and demand over a virtual trading system.*

Key Words: *Rapid Manufacturing, Supply Chain, Batch Size, E-Business, Reverse Auction*

1. INTRODUCTION

Rapid Manufacturing (RM) is still thought by some people as a mere extension of Rapid Prototyping (RP). In practice, the major difference between the two is in the supply chain where RM products are intended for end use, not just as prototypes. However, a number of companies are undoubtedly performing RM with RP methods (Hopkinson *et al.*, 2006). The use of these technologies for the manufacture of end use products in terms of speed, cost and quality, which is acceptable by the general consumer, is still not in existence today (Hopkinson *et al.*, 2006) and thus, it can be argued that a fully functional supply chain for the RM industry is yet to be realised. This paper classifies the issue in two scenarios; in Scenario 1, it investigates the possible application of RM in the spare parts industry; and in Scenario 2, it analyses the possibility of an e-business model through which the supply and demand paradigm for RM products could function.

1.1 Related Research

There has been previous work published on RM implementation strategies, with Table 1 listing the factors involved in such an RM implementation strategy (Reeves, 2007). Additionally, there has been significant work on cost estimation models for RM technologies, that can provide guidelines regarding cost estimates of RM implementation (Ruffo *et al.*, 2006).

Identifying business benefit	Material selection
Process selection	Process constraints
Process cost	Cost benefit

Tab. 1. RM implementation factors

1.2 Research Methodologies

In order to investigate Scenario 1, the authors have undertaken a series of interviews with appropriate businesses that includes: motorcycle spare parts manufacturers, aerospace supply chain manufacturing businesses, designers and RM consultants. The second part of this research, Scenario 2, involves a feasibility study on the prospect of an e-business model which can settle the supply and demand of RM products electronically.

2. SCENARIO 1

The use of RM could possibly solve some long standing supply chain problems. According to the Pareto rule, 20% of items make up 80% of sales, whereas 80% of items are not in high demand. These are parts where inventory holding logistical costs are high in relation to production costs (Walter *et al.*, 2004). The expense of repairable spares for aircraft engines and avionics contributes greatly to the financial outlay of airline operators. These parts, though low in demand, are critical to operation and their unavailability could lead to excessive down time costs. In the course of this research, it has been identified that it is more expensive to produce spare parts with complex geometry. These products which are high value and low volume are ideal candidates for RM production, provided there is appropriate certification (e.g. in the case of aerospace components). Another potential user of RM technologies is the automotive industry including suppliers of classic motorcycle spares (Hasan & Rennie, 2008). There is definite potential that RM could be used to manufacture some of these spare parts due to the geometrical complexity and exclusive nature of the demand. A situation where parts could be manufactured on demand is more ideal; that is implementing the concepts of lean and just-in-time manufacturing.

2.1 Batch Size

RM could further provide a solution to the batch size problem. It is understood from this research that spare parts manufacturers supplying to the aerospace industry, are often approached with orders for repairable spares comprising very small batch sizes; this could even be down to a batch size of one. Currently, the manufacturers are unable to comply with such orders due to the higher break-even batch size of conventional technologies. However, RM can be used to manufacture such products since it is cost effective for products with small batch sizes compared to producing them using, for example, high speed machining techniques.

3. SCENARIO 2

Once identified, if a product is viable to be manufactured using RM technologies, then the subsequent question that arises is "*How would the supply and demand paradigm work*?" This paper argues that an e-business platform could potentially provide an alternative to the RM industry, in terms of supply chain functionality. E-business or electronic commerce, can be defined as any electronic communication that facilitates the exchange of goods, services or other assets between suppliers and buyers (Stein & Hawking, 2002).

3.1 Reverse auction and e-catalogue

This paper proposes the concept of reverse auction as a method for settling the demand of RM products online. At the heart of any auction model is the concept of personal price elasticity. That is, customers will determine the price depending upon the price/value trade-off (Friedman & Barbara, 1999). A reverse auction can be defined as a process where buyers set up an auction to receive bids from suppliers. Suppliers bid down the price for fulfilling that order (Wyld, 2000). The e-catalogue is another proposed method to buy and sell RM products through the e-business platform. Online catalogues, as a means of selling products, are firmly established in the market (Sweets, AEC info, Barbour Index, etc.) (Amor & Kloep, 2003).

3.2 Potential services:
There are potentially five services that the e-business platform can offer and these are described in Table 2.

Services	Description
Sourcing or discovery	Giving buyers an easy access to a pool of suppliers, a process that can save time and offer buyers better prices due to the increase in competition amongst suppliers.
Demand identification	Suppliers can identify customers and their demands in a free market economy.
Content	Through the e-catalogue, that could contain a vast volume of products, the buyer can acquire the required product.
Transaction	The actual exchange of procurement information, such as purchase orders, between the buyer and supplier.
Promotion	Advertisement for suppliers through the e-catalogue.

Tab. 2. Potential services to be provided by the e-business platform

4. CONCLUSION

The discussions that have been put forward in this paper are concepts resulting from current research on formulating supply chain and implementation strategies for RM technologies. It is expected that the potential services of the e-business model, as described in this paper, could provide an alternative way for the supply and demand of RM products to function. However, the biggest impact it could have is perhaps to increase the existing market size for RM products; that is impacting Scenario 1, increase the potential use of RM.

5. FURTHER RESEARCH

It is understood that certification could be an issue in the widespread uptake of RM usage in the aerospace industry. A sample survey amongst aerospace and other spare parts manufacturers is currently underway to clarify this critical issue and generalise the arguments put forward in this paper.

6. ACKNOWLEDGEMENTS

This research has been part-funded through the European Regional Development Fund *Rapid Manufacturing North West Project* and the Higher Education Innovation Fund Phase 3.

7. REFERENCES

Amor, R. & Kloep, W. (2003). E-Product Catalogues, *Proceedings of the EIA9 Conference on E-Activities and Intelligent Support in Design and the Built Environment*, Istanbul, Turkey, pp.75-82, October 2003, *Available from:* http://www.cs.auckland.ac.nz/~trebor/papers/AMOR03A.pdf *Accessed*: 2008-05-22

Friedman, H.H. & Barbara, J.L. (1999). Dynamic Pricing Strategies for Maximizing Customer Satisfaction. *Available from*: http://www.allbusiness.com/sales/customer-service/161659-1.html *Accessed:* 2008-06-01

Hasan, S. & Rennie, A.E.W. (2008). Virtual Trading System For Rapid Manufactured Products, *Proceedings of the Ninth National Conference on Rapid Design, Prototyping and Manufacturing*, Rennie, A.E.W., Bocking, C.E. & Jacobson, D.M., (Eds.), Lancaster University, Lancaster, UK, June 2008.

Hopkinson, N, Hague, R.J.M. & Dickens, P.M. (2006). *Rapid Manufacturing: An Industrial Revolution for the Digital Age*, John Wiley & Sons Ltd, ISBN 0-470-01613-2, West Sussex, U.K.

Reeves, P. (2007). A Methodology for the Business Implementation of Rapid Manufacturing, *Proceedings of the Eight National Conference on Rapid Design, Prototyping and Manufacturing*, Jacobson, D.M., Bocking, C.E. & Rennie, A.E.W. (Eds.) pp.51-59, High Wycombe, June 2007, MJA Print, Beaconsfield, Bucks, U.K (Publ.).

Ruffo, M., Tuck, C. & Hague, R. (2006). Cost estimation for rapid manufacturing - laser sintering production for low volume to medium volumes, *Proceedings of the Institute of Mechanical Engineers, Part B: Journal of Engineering Manufacture,* Volume 220, Number 9/200, pp.1417-1427, ISSN 0954-4054, Professional Engineering Publishing.

Stein, A. & Hawking, P. (2002). Reverse Auction e-Procurement: A Suppliers Viewpoint, *The Eighth Australian World Wide Web Conference*, Twin Waters Resort, Sunshine Coast Queensland, Australia, July 2002, *Available from: http://ausweb.scu.edu.au/aw02/papers/* refereed/stein/index.html *Accessed:* 2008-02-23

Walter, M., Holmstrom, J. & Yrjola, H. (2004). Rapid manufacturing and its impact on supply chain management, *Logistic Research Network Annual Conference*, Dublin, Ireland, September 2004. *Available from:* http://www.tuta.hut.fi/logistics/publications/LRN2004_rapid_manufacturing.pdf *Accessed:* 2008-04-14

Wyld, D. (2000). The Auction Model, The PriceWaterhouseCoopers Endowment for the Business of Government, *Available from:* http://www.businessofgovernment.org/pdfs/WyldReport.pdf Accessed: 2008-06-04

Annals of DAAAM for 2008 & Proceedings of the 19th International DAAAM Symposium, ISSN 1726-9679
Editor B. Katalinic, Published by DAAAM International, Vienna, Austria 2008
Make Harmony Between Technology and Nature, and Your Mind will Fly Free as a Bird

FINISHING QUARTZ GLASS SURFACES WITH LASER RADIATION: ANALYSES OF THE PARAMETERS FOR PROCESS OPTIMISATION/AUTOMATION

HECHT, K[erstin]; BLIEDTNER, J[ens]; MUELLER, H[artmut] & SCHMIDT, T[homas]

Abstract: *Compared to traditional polishing methods the laser beam is small, flexible and fast. With laser radiation it is possible to finish many outlines or geometries on quartz glass surface in the shortest possible time. Start with 600nm for example just one machining step is necessary to achieve a surface quality below10nm. It's a fact that the temperature developing while polishing determines the reachable surface smoothing. To find out which parameters are important for the laser polishing process and the surface roughness respectively, extensive experiments took place. During these experiments starting and machining parameters were changed and temperatures were measured contact-free.*
The results should enable us to optimise and automatise the polishing process and reduce time and tensions.
Key words: *laser, polishing, finishing, quartz glass, temperature measurement*

1. INTRODUCTION

From state of the art methods of polishing surfaces with laser radiation are already known. So it's possible e.g. to reduce processing time of metallic injection moulding moulds from 30min/cm² to a few seconds per cm² (ILT, 2008). Also in polishing cast materials good surface qualitiy is reached (ILT, 2003). The laser material processing of glass has made good progress whereas the high-precision finish (for optical parts) causes still problems. It's barely possible to reach good surface quality without creating thermal tensions (LZH, 2005).
To finish the quartz glass parts in this present report a 1.5kW CO_2-laser is used. The polishing is carried out by a regional short melting of a thin surface layer with a defocused beam. Using adjusted laser power stock removal is prevented and the smoothing happens just because the surface tension of the melting layer. This tension is responsible that the profile peaks were levelled and the valleys were filled. The analysis should also show, how the starting roughness and the changeable laser parameters influence the final surface quality. It's interesting too how the parameters have to be changed to reach requested roughness values.
The Optimisation of the process parameters should be carried out by measure temperature and surface quality. This quality is detected by the help of a stylus instrument.

2. EXPERIMENTS

Fig. 1 shows the experimental setup. It's a portal system where the laser beam is coupled in and leaded over different mirrors to a scanner system. In this scanner mirrors move the beam in lines with a maximum speed of v_s=3m/s. Because of this high velocity a polishing line is generated on the glass part (part size: 20x20x~3mm³). The feed rate v_f is realized with the axis of the portal system and is diversified in different (polishing) sections (s_1, s_2, s_3). The pyrometer measuring spot for the temperature recording is carried on behind the laser line. The measuring system detects the surface temperature (T) and the developing T-curve is used to optimise the process parameters. In Fig. 4 two T-curves (continuous lines) are visible.

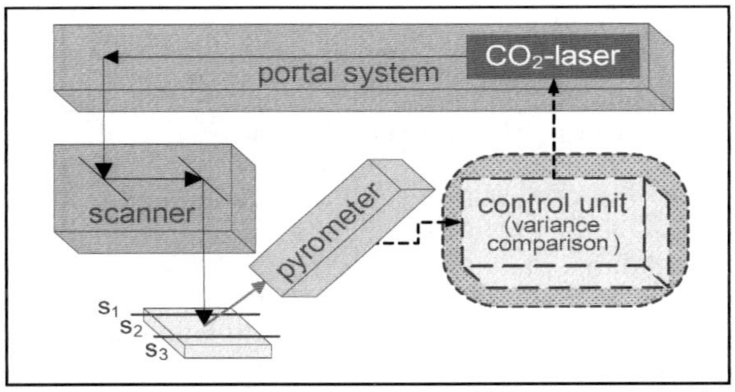

Fig. 1. Scheme of experimental setup, additional: control unit

The task is it first to adjust the parameters feed rate and length of the polishing sections in a way, that T raises steep (s_1), stays constant during the finishing process (s_2) and has just a small peak at the end of the part (s_3). The steep rise guarantees that enough energy is inserted to have a good finishing result from the beginning. To reach uniform polishing results, constant T along the whole surface is deciding. To prevent unnecessary additive tensions in the material and rounding the rear edge of the part a heat accumulation has to be avoided. As you can see regarding the curve V034_C28 (Fig. 4) it's possible to improve the unfavourable temperature distribution from trial V002_C03 by optimise v_f in the 3 sections.
Beside temperature the surface quality is a criterion for process optimisation. To check influence of starting roughness (Ra_1) (before finishing) the parts were divided in three experimental groups (after defined roughness groups).

experimental group	group A	group B	group C
maximum Ra [µm]	1,6	0,8	0,4
roughness group	N7	N6	N5

The optimisation of the process parameters is carried out in every group and during the experiments the following parameters were changed:

- feed rate v_f in 3 sections (already discussed)
- laser output power P, 400...700W
- beam velocity v_s, 400...1000 mm/s

3. RESULTS

After more than 120 tests the influences of the machining parameters and Ra_1 could be studied in detail. Altogether it turned out that the reachable surface quality depends on Ra_1. This connexion gets very obvious regarding group A. It's actually observed that it isn't possible to polish the parts of this group without stock removal. Therefore these parts of group A were no longer considered in this paper. The influence of Ra_1 changes also with variation of the other parameters. Regarding the dependency of laser output power (P) on Ra (Fig. 2) there is no connexion between Ra_1 and final roughness (Ra_2) visible at all. Observable is that with increasing P the surface is smoothed more. In the areas of hight P there is already stock removal.

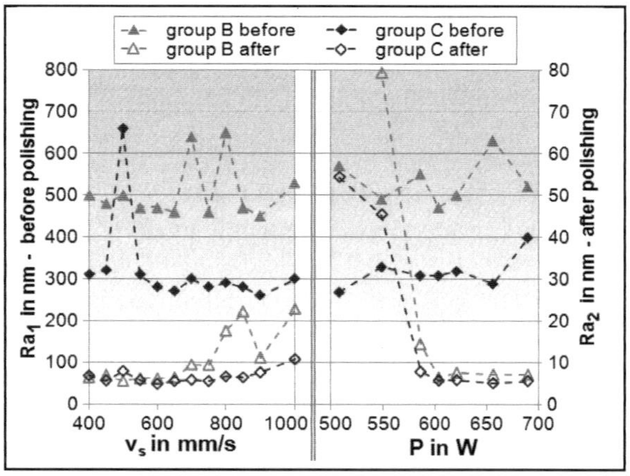

Fig. 2 Surface quality depending on v_s and P

The laser beam velocity (v_s) has also an effect on reachable surface roughness. The influence of Ra_1 stays partly important while diversifying v_s. Considering Fig. 2 it's obvious that Ra-distribution changes only frictional. With higher v_s less energy is inserted into the glass part and so the surface smoothing abates. However it seems that (in the chosen limits) v_s has no such significant influence on the Ra_2 as P for example.
Over all experiments T-curves of every polished part were recorded. They have to be optimised in each group by changing v_f in the three sections. Afterwards their appearance doesn't change much. Just the relation of P and v_s and the maximum temperature (T_{max}) seems still interesting. Due to the pyrometric measurement the measured T-values don't reflect real T in the interaction zone on the glass surface. It's not possible to measure these temperatures contact-free so far. It's assumed that T in the interaction zone is about 2230°C – the vaporisation point of quartz. That's most likely because there is always a light steam of sublimate while polishing. The T-measure-point is selected in a way that it is located always short behind the polishing line. The thickness of the parts is not constant for each group that's why the pyrometer has to be adjusted time and time again. From this it fallows that the distance between measuring spot and interaction zone changes slightly. But even the smallest change of this distance leads to variations in the measured T-values. That's why the position of the curves is not significant but their appearance however is. Fig. 3 shows the curves of T_{max} in relation to P (left) and v_s (right). It's obvious that with higher P the temperature rises. Against that higher v_s causes lower T_{max}. The degree of surface smoothing (Ra_1/Ra_2) follows the T-profile so far. It turned out that the measured T-value can be used for process automation. For this automation the aim is to observe T continuously (online) and adjust parameters accordingly to the T-profile.

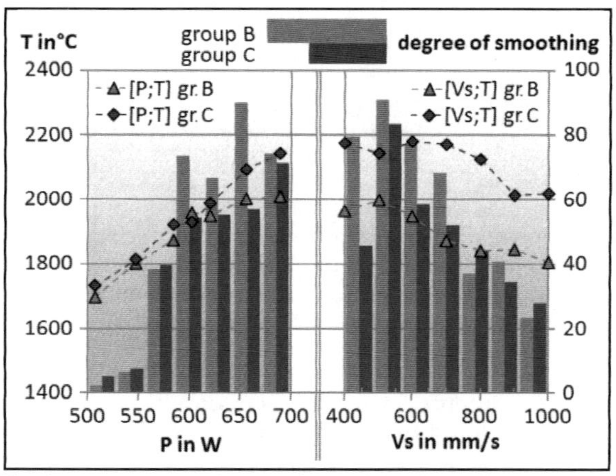

Fig. 3 Temperature, degree of smoothing depending on P, v_s

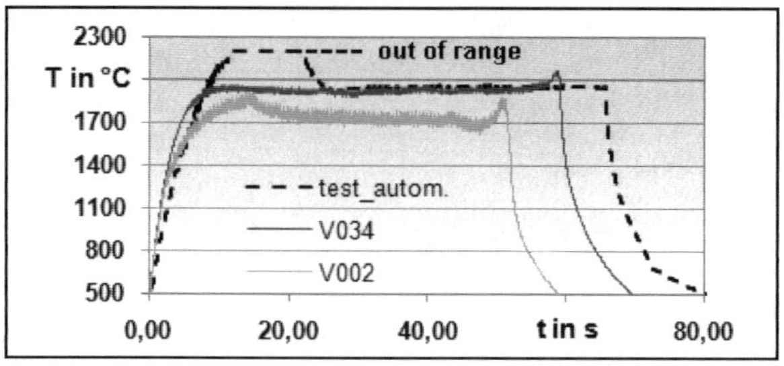

Fig. 4 Optimisation of temperature curve; roughness

4. PROCESS AUTOMATION AND CONCLUSION

Regarding results from experiments laser power is interesting for automation. It has the most significant influence on surface-T and quality. So a laser control system was installed to control P. In the scheme in Fig. 1 you can see aditionally the connection of the devices with the control unit (CU). It's not necessary to describe all technical details of this CU because it isn't part of the investigation. It was constructed for different control tasks and this is just one example of using it.
Return to the present study it's important to know, that the CU works on basis of temperature measurement data were T symbolise corresponding P-values. The pyrometer is connected with the CU and this has a connection to the laser system. You can determine a temperature (/power) control routine. In this case you teach times and temperatures to the CU and P would be adapted accordingly. So there is a permanent comparison between measured T and selected. (Kamann, 2008)
To apply the CU there are some problems to solve. Remember the measuring point of the pyrometer adjusted short behind the polishing line. This "line" is created because the laser needs 50ms to pass the surface. The pyrometer detects the surface temperature with a sampling rate of 250ms. So the laser spot is sometimes nearer and sometimes farther the measuring spot which leads to different measured values. The CU would adapt P permanently and you'll get no constant surface temperature. It's necessary to average the measured values according to the FIFO-method with the CU too. The system works with an average value out of 10 measured values. Regarding Fig. 4 (discontinuous line) it's visible that it's possible to realise the whished T-distribution without different parameters or polishing sections by using the CU. This picture shows another problem still to solve. In the beginning it last too long until the CU and/or the pyrometer recognise the developing surface-T. It even rises out of range and there is a lot stock removal. On the other hand the CU is able to keep T within a limit of 25K.
To solve the problem of overheat maybe different measuring spot positions could be tested. Lay the spot into the polishing line to reduce the influence of heat conduction is one possibility. But altogether further investigations have to fallow. In addition it would be also necessary to develop and apply a construction to hold and adjust the pyrometer high precisely.

5. REFERENCES

Fraunhofer Institut für Lasertechnik-ILT; (2003). Fraunhofer ILT Annual Report 2003, *Available from:* http://publica.fraunhofer.de/eprints/N-35932.pdf, *Accessed:* 2007-08-06

Fraunhofer-ILT (2008). *Laser-Beam Polishing of Injection Molded Tools*, *Available from:* http://www.ilt.fraunhofer.de/eng/100607.html, *Accessed:* 2008-05-06

Kamann, J. *Entwicklung und Aufbau eines Laserleistungs-regelsystems für temperaturgeführte Materialbearbeitungs-prozesse*, 2008, diploma thesis

LZH e.V. (2005) *Innovative manufacturing processes for polishing glass surfaces*, Final Report, *Available from:* http://www.laser-zentrum-hannover.de/en/projects/index.php *Accessed:* 2007-08-06

Annals of DAAAM for 2008 & Proceedings of the 19th International DAAAM Symposium, ISSN 1726-9679
Editor B. Katalinic, Published by DAAAM International, Vienna, Austria 2008
Make Harmony Between Technology and Nature, and Your Mind will Fly Free as a Bird
*Young Researcher and Scientist Paper / * Supervisor, Mentor*

ALGORITHM FOR CHOOSING FACTORY PLANNING METHODS AND TOOLS

HENNERSDORF, S[ibylle] & DOMBROWSKI, U[we]*

Abstract: *Factory planning is an integrative part of corporate planning. It is basis for the realisation of the place where adding value activities are located. The integrative view on factory and corporate planning ensures coordinated planning processes as well as the specific selection of suitable factory planning methods and tools in a systematic and enterprise specific way. Therefore, this paper describes a systematic way to the selection of factory planning methods and tools.*
Key words: *Production Planning, Industrial Engineering, Factory planning, Factory Reorganisation*

1. INTRODUCTION

Nowadays, the management of manufacturing enterprises is required to complete complex tasks as a daily challenge. They are forced to guarantee a successful leadership (management) as well as an effective foresight (operational planning). Against this background the planning of production factors like personnel, material, qualification, organisational knowledge, capital and resources inside a factory gets an important meaning.

For approximately four decades the subtask factory planning is a scientific discipline in German-speaking countries. The term factory is referred to the place where creation of value takes place by production using factors of production. In practice factory planning departments face special tasks. Long-lasting buildings and inflexible machines and transportation systems lead to inflexible structures and complex decisions. Large scale enterprises face this development in order to engage huge planning teams which enacts about experimental know-how. Current results of research in this field contain several approaches. These approaches range from structural adaptive factory (Sudhoff et al., 2006), factory life cycle management (Westkämper et al., 2006) to extended cost effective analysis (Kolakowski et al., 2005). Furthermore, possibilities are analysed concerning the increase of the planning speed, maintenance of the general planning ability as well as for handling increasing planning tasks. Another field of research deals with the requirements for participative planning processes (Dombrowski et al., 2001).

Nevertheless, especially for small and medium sized enterprises (SME) a practice-oriented proceeding to tackle the abovementioned challenges by continuous factory planning is still missing.

Therefore, the Institute for Production Management and Operations Research (IFU) of the Technical University Carolo-Wilhelmina of Braunschweig started the INNOFAP research project. Research objective is the development of a special factory planning approach for continuous planning activities focussed on SME. One of the main subproject deals with the design of an algorithm for choosing factory planning methods and tools. Below, the development of the algorithm is presented.

2. STATISTIC FUNDAMENTALS

However, in research as well as in practice a uniform basic understanding of factory planning is still missing. Especially knowledge about appropriate methods, tools and the systematic usage throughout the factory planning process comprehensively does not exist.

In order to correct these deficits, the IFU developed a structured questionnaire to learn about the actual status of factory planning activities especially in SME. The questionnaire was sent to about 300 German companies in manufacturing industries. The response rate is nearly 4.6%.

The results of the survey show different definitions of the term factory planning. For most SME's factory planning deals with machine and workstation arrangement, production process planning and inventory planning. Other tasks like architectural planning, structural planning, technical equipment planning, environmental influences planning, material flow planning, communication planning, office planning and ergonomics have not got that point (see figure 1).

Another part of the survey deals with the motivation for factory planning activities. The main reason for factory planning activities in SME's is the need for capacity or to much available capacity (e.g. space, technical capacity). In addition to that, the management often forces factory planning activities. The development of key data, economic difficulties as well as other reasons is less important.

What factory planning methods and tools do you use? This question leads to answers in order to analyse the current state of using factory planning methods and tools (see figure 2). Microsoft Office tools (e.g. Microsoft Visio, Microsoft Excel and Microsoft Word) as well as 2-D/3-D-CAD tools are wide spread used in German SME's. Furthermore, paper mock-ups are frequently used.

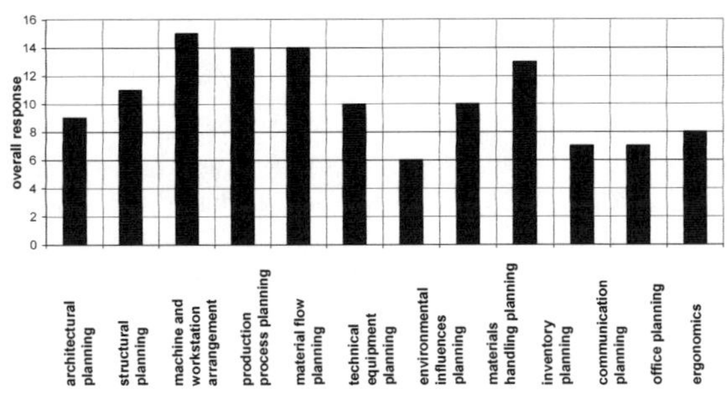

Fig. 1. Definition of factory planning in SME's

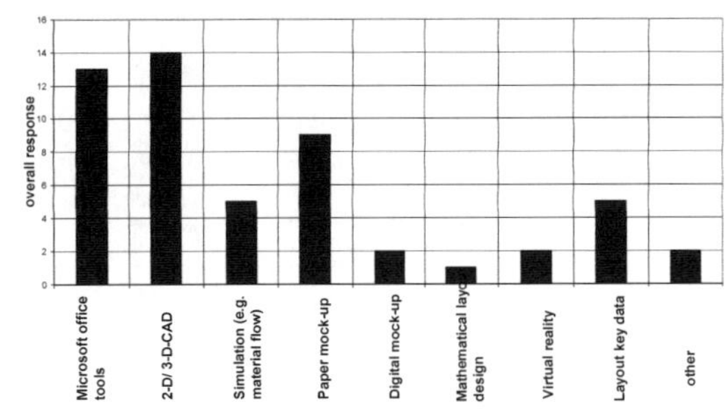

Fig. 2. Use of factory planning tools and methods in practice

Based on these theoretical foundations the IFU developed the factory planning methods and tool matching as a holistic factory planning algorithm. It is based on three components: (1) control-loop for factory planning (Hennersdorf et al. 2007), (2) factory planning tools and methods catalogue and (3) system screen. Below, the factory planning methods and tools catalogue and the system screen are presented.

3. DESIGN OF THE FACTORY PLANNING ALGORITHM

In order to describe "traditional" and current factory planning methods and tools in a systematic and application-oriented way, the IFU developed a comprehensive catalogue of methods and tools. In the following factory planning methods are defined as a structural way, implying an orderly logical arrangement of steps. These steps help to transform a given initial state into an aspired final state. Considering this definition factory planning methods illustrate the way a task is handled (REFA, 1993). In contrast a factory planning tool is a device that typically provides a way in accomplishing the acquisition or processing of information. Therefore, it supports the use of a method. To compile the catalogue the state of the scientific and technical knowledge in the fields of production planning, automation, logistics, enterprise organisation, architecture, civil engineering, facility management, building physics and environmental protection, the official editions, ergonomics, the Supply chain management and the marketing and management was used. Furthermore, the actual requirements of a certain number of SME's were considered by the above-mentioned survey. Hence, an application-oriented development of the catalogue is ensured.

The catalogue of methods and tools for factory planning developed at the IFU contains, beside from the name of the method/tool, a short description, the data-input, the data-output, requirements, qualitative and quantitative results, pro and cons, synonymous methods/tools, a short description of the procedure, further reading as well as references to other tools. Especially the detailed description of the methods in a user-oriented way implies assistance for the employees of the factory planning department. Moreover, the structure of the catalogue can easily be amplified with other methods/tools and supports an enterprise-wide knowledge-management (Bodendorf, 2006).

The factory planning algorithm bases on matching the different tasks of factory planning with the described methods and tools. For the first time this approach was used for matching methods of knowledge-management with the experience-level of a user. Validating this matching-method pragmatically applicableness as well as great problem focus could be stated (Horatzek, 2006).

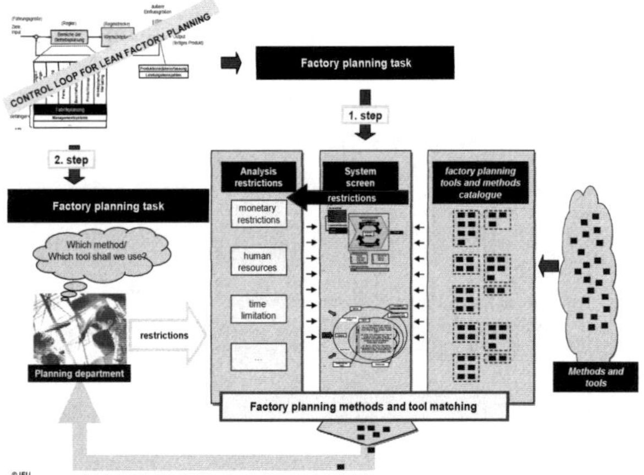

Fig. 3. Factory planning algorithm

As a result of interdependencies between specific tasks of factory planning and applicable restrictions (mostly resource-based restrictions like finance, human resources or time) the former method has to be adopted. Regarding this, the method at hand now contains a two-step approach (see figure 3).

The first step ensures the classification of the existing factory planning task and the consideration of the relevant restrictions. The second step supports the user to answer questions about the identified restrictions using an easy-to-handle user interface. Following, the provided system screen allows the user to get an extract of suitable methods and tools for solving the factory planning task accordingly.

For the first time, the upon introduced algorithm supports especially SME's by recommendations of methods and tools in their factory planning activities.

4. CONCLUSION

At the moment the factory planning algorithm is tested in different industrial projects. The results of the first use cases show potentials for the enlargement and further development of the factory planning algorithm.

Against this background the conceptual design at hand will be improved and enterprise specific adjustments will be integrated in a modular way. At the same time software applications will be developed in order to realise a practical user interface.

In addition, the following research steps of the INNOFAP project includes the integration of a self-adjusting control loop for factory planning and the algorithm for choosing methods and tools. The main goal is the development of a holistic system concept for continuous factory planning activities in SME's.

5. REFERRENCES

Bodendorf, F. (2006). *Daten- und Wissensmanagement (Information and knowledge management),* Springer, 978-3540287438, Berlin

Dombrowski U.; Tiedemann H. & Bothe T. (2001). Visionen für die Digitale Fabrik (Visions for the digital factory). *ZWF Zeitschrift für wirtschaftlichen Fabrikbetrieb. Vol.*96, No. 3 (2001), 96-100, 0947-0085

Hennersdorf, S.; Palluck, M. & Dombrowski, U. (2007). System Approach for Plant Design - Development of a Control Loop for Lean Factory Planning- (2007). 325-326, *Annals of DAAAM for 2007 & Proceedings of the 18th International DAAAM Symposium*, B. Katalinic (Ed.), pp 164, ISBN 3-901509-58-5, ISSN 1726-9679, Zadar, October 2007, DAAAM International, Vienna

Horatzek, S. (2006). *Lebenszyklusorientiertes dezentrales Wissensmanagement in der Nachserienversorgung (Lifecycle-oriented knowledge management),* Shaker, 978-3-8322-5056-0, Aachen

Kolakowski, M.; Reh, D. & Sallaba G. (2005). Erweiterte Wirtschaftlichkeitsrechnung (EWR) (Extended cost effective analysis). *wt, Werkstattstechnik online, No. 4 (2005)*, 210-215, 1436-4980

REFA (1993). *Lexikon der Betriebsorganisation (encyclopaedia of management)*, Carl Hanser Verlag, 978-3446175235, München

Sudhoff, W.; Rimpua, C.; Berlak, J.; Müssig, B. & Redelstab, P. (2006). Strukturadaptive und mobile Fabrikkonzepte (Structure-adaptive and mobile factory concepts). *wt, Werkstattstechnik online,* No. 4 (2006), 162-166., 1436-4980

Westkämper, E.; Constantinescu, C. & Hummel, V. (2006). Fabrik Life Cycle Management. (Factory life cycle management). *wt, Werkstattstechnik online,* No. 4 (2006), 178-182, 1436-4980

Annals of DAAAM for 2008 & Proceedings of the 19th International DAAAM Symposium, ISSN 1726-9679
Editor B. Katalinic, Published by DAAAM International, Vienna, Austria 2008
Make Harmony Between Technology and Nature, and Your Mind will Fly Free as a Bird

APPLICATION OF GRAPHICAL SOFTWARE IN TEACHING OF MICROECONOMICS – LECTURE MONOPOLY

HERCEG, T[omislav]; JAKOVIC, B[ozidar] & GALETIC, F[ran]

Abstract: *This paper investigates how graphical software can be used in the process of teaching microeconomics on example of the lecture Monopoly. Teacher should use simple software or just its simple components in order to make it close to all students. It is especially useful for business schools where students are not expected to know calculus on higher levels of comprehension. Using these programmes teacher can interactively draw graphs of monopoly (in this case), gradually add curves and then shade areas of the interest (using Graph 4.3 software) and finally calculate areas that should be interpreted (Using Mathematica 5.2 or Graph 4.3). In this manner lecture is transformed into an interesting teaching experience for both the teacher and the students.*

Key words: *Monopoly, Teaching techniques, E-learning, Lerner index, Profit rectangle*

1. INTRODUCTION

Students and teachers of microeconomics often find themselves jammed between two confronted positions: one of a teacher, who tries to simplify and illustrate complex microeconomic phenomena using long mathematical deductions, lots of graphical representations and some exercises to prove it, and one of the students, who can hardly grasp all the materials that are to be taught and examined afterwards. Hence it is crucial to use time and concentration in the best possible way. In this paper we will try to simplify some parts of the lecture explaining monopoly.

We made an investigation in the period from 2005-2008 on nine groups of students randomly taken from the overall students of the first year of business economics. In four of these groups e-learning techniques were used and these groups yielded significantly better results (non-elarning groups: average grade bellow 2.9, 1 being the worst and 5 the best, while e-learning groups all had the average grade above 3.3. Average of the average grades in non-learning groups was 2.7, and in e-learning groups 3.7)

In this paper we present techniques that were used in the e-learning groups. This approach would not be applicable for students with good calculus background, but for the students with poor knowledge of both calculus and economics. However, this approach would be applicable for business school and short degree (programme that last for only 2 years or even less) students who do not intend to deepen their theoretic background, but try to achieve practical knowledge as soon as possible.

2. LITERATURE OVERVIEW

Sadler-Smith in his holistic research described learning methods. He said that learning methods can be grouped into at least three categories: (1) autonomous: guided reading; programmed instruction (distance learning, computer based learning), (2) collaborative: games/simulations; projects; case studies; experiential; role playing and (3) dependent: lecture; seminar (Sadler-Smith 1996). We are interested in the latter one for now; how to use new technologies in order to achieve the main education goal, comprehension.

Bose described one of the e-learning pilot projects that resulted in flexible time used for teaching and student oriented techniques that preserve the quality of education while increasing its productivity (Bose 2003).

Galetić, Herceg and Jaković have investigated how to use other teaching software in order to simplify teaching of microeconomics in several fields, with examples given for the complex deductions and how they should be taught. They have as well shown how to use graphical applications in order to explain a concept of concave and quasi-concave functions and its layer curves. (Galetić et al. 2008)

Littig in his paper focuses on eLearning projects supported by the European Leonardo da Vinci programme and gives recommendations for the future development. The innovation itself is not a technological innovation, but only application of the already developed technologies for pedagogical purposes; hence it is called pedagogical innovation (Littig 2006).

Cardinali says that "the interface that allows interaction between end-users and the computer plays a vital role in the effectiveness by which certain system functions. He also states that it is crucial to enable students to use and understand software which make learning process easier more effective since it saves time and increases productivity (Cardinali 1994).

El-Deghaidy and Nouby in their paper on the development of implementation of e-learning techniques in the Egyptian education system made a research using test group and control group of students. Their results show that technology can help in teaching, but it is also important to use the right techniques and to try to eliminate those which are ineffective and to much time consuming. They were among the first to show the relevance of picking up the right technique and that it is not enough to have technology necessary to even begin with the teaching progress (El-Deghaidy and Nouby 2008).

Graham in his paper investigates the history of e-learning and technologies used at the time. He concludes e-learning does not have to increase to quality of the learning process, but the increase in the quality of the e-learning techniques and receptiveness of the learner do have a strong impact on the learning process and make it faster and more amusing (Graham 2004).

Harvey and Beards made a thorough survey of the Scottish higher education system and their use of e-learning. Authors concluded that in order to have an e-learning system that functions, universities have to, after investing in technology, invest in education of the teaching staff how to use it as well as to develop new pedagogical approach (Harvey and Beards, 2003).

3. MONOPOLY LECTURE

Lecturer begins the lecture with informal explanation of the basic tools needed to use the software (Graph 4.3. and Mathematica 5.2.). After explaining the main concepts, the lecturer begins to draw certain monopoly allowing students to choose demand functions and cost functions. It makes the process interactive and thus more interesting.

If students choose demand P = 10 – Q (like students of the authors' of this paper did), teacher starts to draw it, and using some of the software mentioned, deduct marginal revenue. Then students suggest some total cost (let it be TC = 5 + 0.2Q^3). Again we find marginal and average costs using Mathematica (if it were more complex than in this case) and put it on the same graph (Fig. 1.)

After this is done the lecturer starts to deduct the equilibrium state and explain it intuitively. Firstly it has to be shown what would be the equilibrium in the perfectly competitive case and then discuss what would happen if demand flattens. It is a good way to begin to explain Lerner's index, but it is not the part of this paper.

Teacher now shades the areas of interest that they together identified as interesting (Fig. 2.). After that the shaded area is calculated (students will very often forget to deduct fixed costs – now it would be useful to show why this should not be omitted). Students would have to be able to know this number represents the profit of monopolist. Critical points can be added and calculated by the software as well.

Following the lecture, teacher wants to show it is not the only way of finding the profit area. Instead of finding areas under curves, one can find the profit rectangle that corresponds to the previously obtained profits (Fig. 3.). Finally, it will be shown these areas are equal and it holds for any demand and cost functions when they change coefficients in equations.

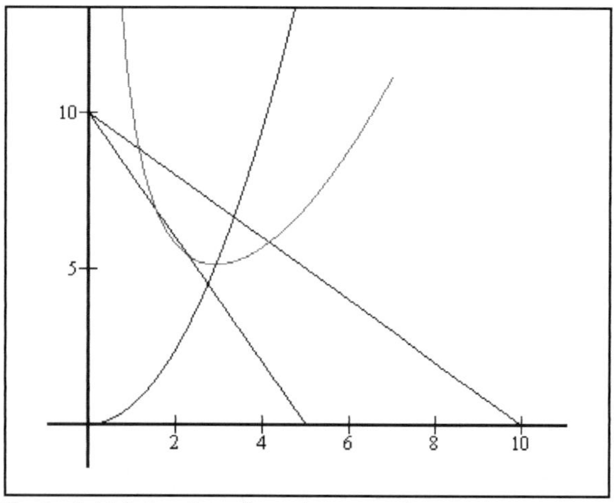

Fig. 1. Initial situation of certain monopoly

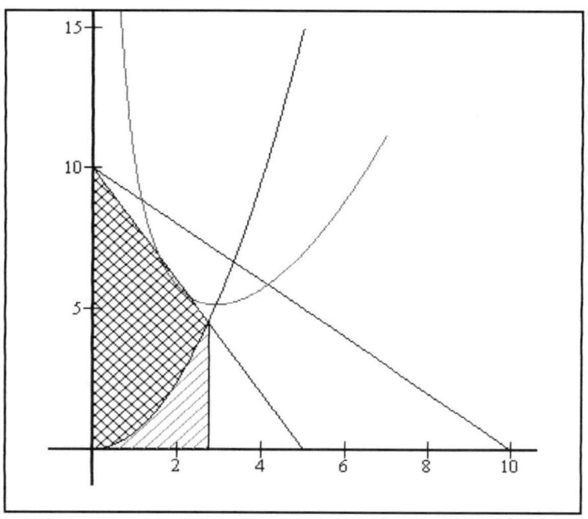

Fig. 2. Profit areas between marginal functions

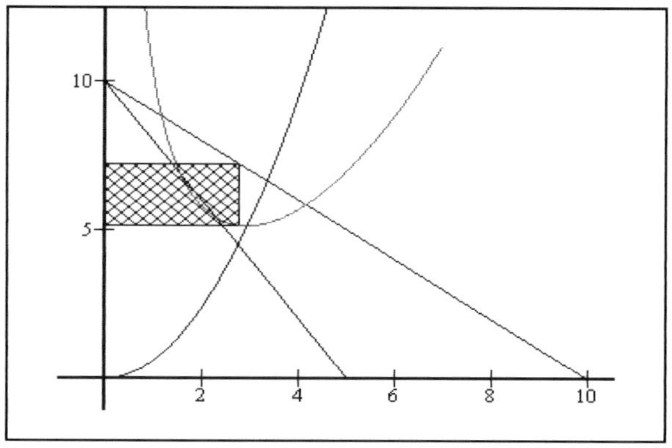

Fig. 3. Profit rectangle

4. CONCLUSION

The aim of this paper was to show the application of e-learning techniques in teaching of microeconomics. Making a survey on 9 different students groups and applying techniques here presented on 4 of them, we found out that on the average e-learning groups had the average grade by a whole 1 grade above the non-e-learning groups. We present a part of one of the lectures used in this process, crucial for understanding of the market structures.

The lecture proceeds as follows: introduction of the main concepts and basics of the software. It is followed by simple example after which students give their own examples. Collaborative work of the lecturer and the student raises the students' level of attention and their comprehension of the subject. Students are also encouraged to use software at home in order to prolong the learning process making it more thorough.

This approach, however, is much more applicable for students with weaker calculus background, like those who attend business schools and short business courses. Advanced students find this approach too straightforward, and thus less interesting. For them other teaching methods should be used.

5. REFERENCES

Bose, K. (2003), An e-learning experience – a written analysis based on my experience in an e-learning pilot project. *Campus-Wide Information Systems,* Volume: 20 Issue: 5; Research paper.

Cardinali, R. (1994). Productivity Improvements through the Use of Graphic User Interfaces. Industrial Management & Data Systems; Volume: 94 Issue: 4. pp. 3

El-Deghaidy, H., Nouby, A. (2008), Effectiveness of a blended e-learning cooperative approach in an Egyptian teacher education programme, Computers & Education 51 (2008) 988–1006

Galetić, F., Herceg, T., Jaković, B. (April 2008): E-learning and Its Application to Microeconomics, *WSEAS Transactions on Information Science & Applications*, Issue 4, Volume 5

Graham, G. (2004): E-learning: a philosophical enquiry, *Education + Training*, Volume:46, Number:6/7,pp.308-314

Harvey and Beards (2003): E-learning in Scottish further and higher education, *Education + Training*, Volume: 46, Number: 6/7, pp.353-360

Littig, P. (2006), New media-supported learning today and tomorrow: recommendations for the next generation of education and training concepts supported by new learning media. *Industrial and Commercial Training*, Volume: 38 Issue: 2.

Sadler-Smith, E. (1996), Learning styles: a holistic approach. *Journal of European Industrial Training,* Volume: 20 Issue: 7; Research paper.

Annals of DAAAM for 2008 & Proceedings of the 19th International DAAAM Symposium, ISSN 1726-9679
Editor B. Katalinic, Published by DAAAM International, Vienna, Austria 2008
Make Harmony Between Technology and Nature, and Your Mind will Fly Free as a Bird

EVALUATION OF THE PITTING CORROSION IN AUTOMOTIVE EXHAUST SYSTEMS

HOFFMANN, C[ristina]; GÜMPEL, P[aul]; ZAMFIRA, S[orin] & BRAUN, B[arbu]

Abstract: The rear section of automotive exhaust systems is mainly subjected to wet corrosion. Various stainless steel grades are used for the end parts, ferritic stainless steels with different contents of chromium and molybdenum as well as austenitic types. For selection of an appropriate material enabling cost-effective constructions the corrosion resistance of different stainless steel grades has to be evaluated. Therefore technological tests are carried out which take into account the specific feature of the corrosion loading in the rear part of the automotive exhaust systems and afford the same corrosion forms as in the reality. In the paper the technological tests which are carried out, the results of the mass loss and the measurement method of the depth of the pits are described. The mass loss and the depth of the pits provide different ranking of the materials. The different corrosion forms which occur demand a combination of methods for the evaluation of the corrosion attack: mass loss, visual investigation, and the measure of the depth of the pits.

Key words: *automotive corrosion, pitting corrosion, corrosion evaluation, length gauge, etching pit.*

1. INTRODUCTION

Automotive exhaust systems are complex constructions with different sections placing different demands on the materials (Cunat P-J. 2000; Yoshiharu I. & Masao K. 2003).

To wet corrosion, the rear section of the automotive exhaust system is mainly exposed. Stainless steel grades are used for the end parts, ferrites with different contents of chromium and molybdenum as well as austenitic types. For selection of an appropriate material enabling cost-effective constructions the corrosion resistance of different stainless steel grades has to be evaluated. Various stainless and heat-resistant steel grades are used to achieve an optimal combination of properties in each section (Seeger B. 2007).

The nature of the corrosion attack is mixed: general attack, crevice corrosion and pitting corrosion till penetrate are occurring. Some of the shallow pits are developing and growing under the corrosion products of the general attack where crevice conditions are created.

We investigated the corrosion conditions in the rear part of the automotive exhaust systems and we evaluated the attack (Gümpel P. 2001). On the outside of the rear muffler the corrosion it's caused by the chemical influence of the road salt projection, which is used in winter on the streets. Inside the system, condensation of combustion gases produces sulphurous acid, sulphuric acid and low levels of hydrochloric acid, creating critical conditions with acidic pH-values (Gümpel P. et al. 1997; Weltens H. et al. 1998). These condensates, combined with an accumulation of chloride ions and deposits of electrochemically active soot particles, result in a substantial wet corrosive impact on the inner surfaces of the components. To compare the suitability of various stainless steels concerning their application in the wet sections of automotive exhaust systems, technological tests are carried out which take into account the specific features of this corrosion impact. The following factors must be considered: a) the wet/dry alternation, b) the impact of a chloride ion-containing acidic medium c) the presence of electrochemically active carbon (occurring as soot particles in the systems) d) simple test assembly e) the same corrosion forms as in the practice.

After the simulation of the corrosion conditions in the end parts of the automotive exhaust systems (Gümpel P. et al. 2004), for a major importance is the evaluation of the corrosion (Heitz E. et al. 1990).

2. EXPERIMENTAL

2.1 Technological Tests

Deep drawn specimens from strip of selected materials (Table 1) were used for the tests. The specimens were pickled and weighted before each test. A specimen underwent several cycles of being filled with corrosive medium and drying. The exposure conditions and the composition of the corrosive medium were varied to optimize the process.

Stainless steel grades	Alloying additions [wt %] Cr	Ni	Mo	Mn
Cr-Steel 1.4512	11.54	-	-	-
Cr-Steel 1.4509	17.59	-	-	-
CrMo-Steel 1.4526	16.86	-	0.983	-
Cr NiMn-Steel 1.4376	18.93	3.33	-	7.7
CrNi-Steel 1.4301	17.88	9.08	-	-
CrNiMo-Steel 1.4404	16.36	11.88	1.84	-

Table 1. Selected materials for test.

2.1.1 Test with Electrolyte of pH 4

The specimens were filled with 10ml of the corrosive medium: acetate buffer solution with sodium chloride and active carbon. They were placed in a climate chamber for 12 hours at 85°C and 50% relative humidity, and then for another 12 hours at 23°C and 50% relative humidity. After 24 hours the electrolyte was replenished. After 12 cycles, the specimens were cleaned, examined for corrosive attack and weighed. The test lasted 48 cycles.

2.1.2 Test with Electrolyte of pH 3

With both Diesel and Otto engine the condensates reach low acidic pH-values under certain conditions. In the test with electrolyte of pH 3, this corrosion factor, i.e. the acidic medium, simulated the most aggressive conditions in auto exhaust system.

The specimens were filled with 20ml of the corrosive medium: citric acid – phosphate buffer solution with sodium chloride and active carbon. They were placed in a climate chamber for 12 hours at 85°C and 75% relative humidity, and then for another 12 hours at 23°C and 75% relative humidity. After 24 hours the electrolyte was replenished. After 12 cycles, the specimens were cleaned, examined for corrosive attack and weighed. The test lasted 12 cycles.

2.2 Pit depth measurements

In order to measure the pit depth we used the MT 2751 length gauge, from HEIDENHAIN. It has ± 0.2 µm accuracy and 25 mm measuring range. Measuring standard is a DIADUR phase grating on Zerodur® glass ceramic with grating period 4 µm. The output signals are TTL levels. The MT 2751 length gauge was interfaced to PIC18F452 microcontroller through the LS736, a 32-bit CMOS counter, with direct interface for quadrature clocks from incremental encoders. We programmed the PIC18F452 microcontroller to acquire, the encoded TTL signals, to process, to display on a LCD and to store them on a PC (Zamfira S. et. al. 2007).

3. RESULTS

3.1.1 Technological test with Electrolyte of pH 4

The nature of the corrosion attack is determined by various effects: both classic pitting and uniform attack are observed. The results of these tests are summarized in Fig. 1.

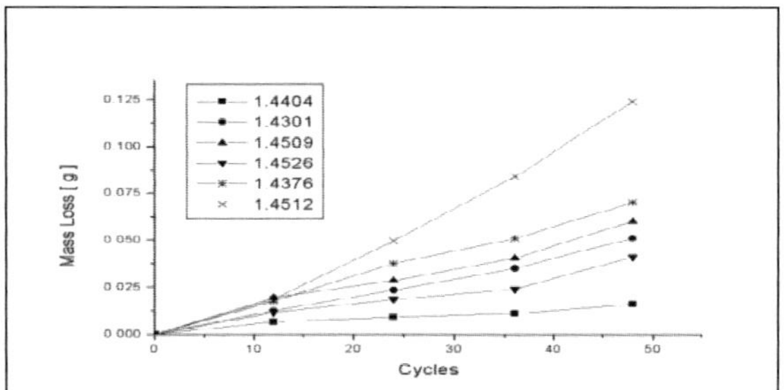

Fig.1 Run of the mass loss curve, the test with electrolyte pH 4

The mass loss and the differentiation among the grades increase with the number of cycles. A relatively good agreement is observed between the mass loss rates and the alloy content of the materials. But based on the pitting attack mechanism the results of the mass loss measurements should always be treated with caution.

3.1.2 Technological Test with Electrolyte of pH 3

For this test the mass loss as well the medium depth of the pits was measured. The mass loss of the specimens in this test is summarized in Fig, 2.

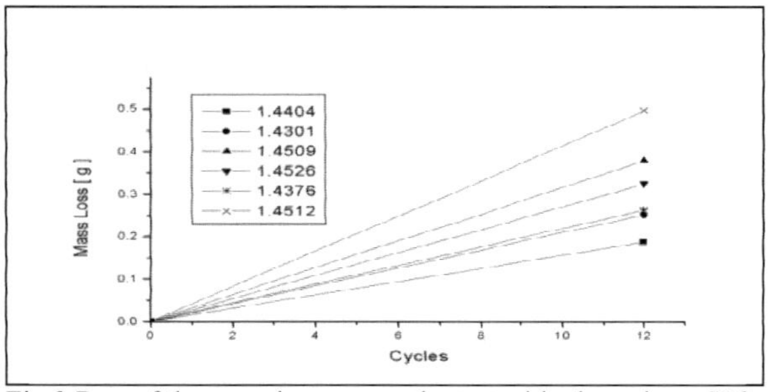

Fig.2 Run of the mass loss curve, the test with electrolyte pH 3

3.2 Etching depth results – Test with Electrolyte of pH 3

In this stage we measured etching depth using our increased precision length measurement system based on the MT 2751 length gauge. We measured negative values for the depth pit, because zero point was put at the border of the pit. In every measurement zero point was fixed. Medium measured depth of the pits was calculated for each samples. In table 2 we present measured medium depth over 30 points for same stainless steel samples, covering a 20 cm^2 surface.

Stainless steel code	Medium depth
CrNiMo-Steel 1.4404	-186.824
CrNi-Steel 1.4301	-262.446
Cr-Steel 1.4509	-365.184
CrMo-Steel 1.4526	-320.981
Cr NiMn-Steel 1.4376	-113.454
Cr-Steel 1.4512	-174.321

Table 2. Medium etching depth for investigated samples

4. CONCLUSION

The technological tests are showing that the mass loss method is suitable to only a limited extent to evaluate a mixed corrosion attack. There in beside the mass loss of the samples a measurement method of the depth of the pits has to be used.

For the stainless steels grades 1.4404, 1.4301, 1.4509 and 1.4526 the mass loss of the samples and the medium etching depth of the samples provide the same material ranking. For the stainless steel grades 1.4376 and 1.4512 the two methods provide different results. The visual investigation shows that at these samples (1.4376 and 1.4512) the uniform attacks prevail.

The different corrosion forms which occur demand a combination of methods for the evaluation of the corrosion attack: mass loss, visual investigation, and the measure of the depth of the pits.

For the next future we shall measure the corrosion surface and together with depth measurements we shall correlate it, with mass loss of the samples, in order to evaluate the pitting corrosion in automotive exhaust systems.

5. REFERENCES

Cunat P-J. (2000). *Stainless Steel Properties for Structural Automotive Applications,* Paper presented on: Metal Bulletin International Automotive Materials Conference, Cologne, 21st to 23st June 2000

Gümpel P. & Arnold N. & Heitz T. & Pscheidel P. (1997). *Chlorinduzierte Korrosion (Chloride Induced Corrosion)* von Edelstahl Rostfrei, Focus Rostfrei 22/1997

Gümpel P. (2001). *Rostfreie Stähle (Stainless) Steels, 3.* Auflage, Expert Verlag, 2001

Gümpel P. & Schiller D. & Arlt N. & Bouchholz D. (2004). *Simulation des Korrosionsverhaltens von nichtrostenden Stählen in Pkw-Abgasanlage (Simulation of the corrosion behaviors of stainless steels in automotive exhaust system)* Automobil Technische Zeitschrift, 4/2004

Heitz E. & Henkhaus R. & Rahmel A. (1990). *Korrosionskunde im Experiment (Corrosion Science in Experimen)t*, 2. Auflage, VCH Verlaggesellschaft mbH, 1990

Seeger B. (2007). *Material for decoupling elements in complex operating conditions*, Vortrag gehalten auf International CTI Forum – Stuttgart 22.11.2007, Material in Exhaust Gas Technology

Yoshiharu I. & Masao K. (2003). *Present and Future Trends of Stainless Steel for Automotive Exhaust System*, Nippon Steel Technical Report No. 88, July 2003

Weltens H. & Garcia P. & Walther H-D. (1998). *Innen- und Außenkorrosion von Abgasanlagen für Pkw – Prüfverfahren (Internal and external corrosion of exhaust systems for passenger vehicles – Test procedures, laboratory and fiels results)*, Labor- und Feldergebnisse, Vortrag, gehalten auf dem Dvm–Tag 1998

Zamfira S. & Braun B. & Turcu C. & Stefanescu O. (2007) *Comparative methods for the dimensional control, using virtual instrumentation* REV2007 4th International Conference on "Remote Engineering and Virtual Instrumentation"June 25-27, 2007, Porto, Portugal.

Annals of DAAAM for 2008 & Proceedings of the 19th International DAAAM Symposium, ISSN 1726-9679
Editor B. Katalinic, Published by DAAAM International, Vienna, Austria 2008
Make Harmony Between Technology and Nature, and Your Mind will Fly Free as a Bird
*Young Researcher and Scientist Paper / * Supervisor, Mentor*

A REFERENCE MODEL BASED APPROACH FOR THE EVALUATION OF INDUSTRIAL PLANT SERVICE AND ASSET MANAGEMENT TOOLS

HOLM, T[imo]; LUSCHMANN, C[hristian]; DELGADO, A[ntonio] & AMBERG, M[ichael]*

Abstract: *Industrial plant services like asset management, plant revision management, and knowledge based maintenance pose challenges, which require the use of appropriate service information systems (SIS). Explicit knowledge of these challenges allows the evaluation of SIS, which can help suppliers to identify future requirements and objectify potential for innovation. This article describes the prototypical challenge based evaluation of a large integrated plant asset and field device management system. During evaluation the problem was disclosed, that the classification methodology is ambiguous. Therefore a solution, a reference model for SIS, is proposed, which maps generic components of SIS to challenges, thereby offering an abstract structure used to evaluate tools in a more objective and easy-to-use way.*
Key words: *Plant Asset Management Tool Evaluation*

1. INTRODUCTION

The most important goal for process industries is the return on investment (ROI). One way to raise this value is to reduce total cost of ownership (TCO) by efficient plant services including integrated maintenance management and execution, asset management and process optimization. This applies to operators of plants as well as to service providers. To achieve efficient plant service, supporting software systems are needed to handle the large number of special tasks involved, for instance outage optimization or knowledge based maintenance.

In contrast to plant operators and service providers the suppliers of SIS run a product business. They have to sell a product, which meets the domain specific challenges and needs of their customers. So for sustainable business success it is most important to know the requirements of customers, for example the special needs of the maintenance engineers who plan plant outages.

To be able to innovate SIS in a planned and systematic way in a first step the challenges posed by the industrial service business have to be identified and structured for application. In (Amberg et al.2008) a concept for the characterization of SIS has been introduced. In this concept challenges and structures as well as a knowledge base of best practices suitable to address any single challenge are presented. The methodology was developed at the Department for Systems Engineering, Siemens AG (CT SE 5). It is based on research work conducted and project experience gained within the context of the industrial solution, service and plant engineering business at Siemens AG. Supportive research work was done by the Department of Information Systems 3 (WI3) at Friedrich-Alexander-University Erlangen-Nuremberg (FAU). An evaluation of a large integrated plant asset management (PAM) and field device management system (FDM) was prototypically executed by the Department of Chemical- and Bioengineering - Chair of Fluid Mechanics (LSTM), Bioprocess Automation Research Group - at FAU to verify the practical applicability.

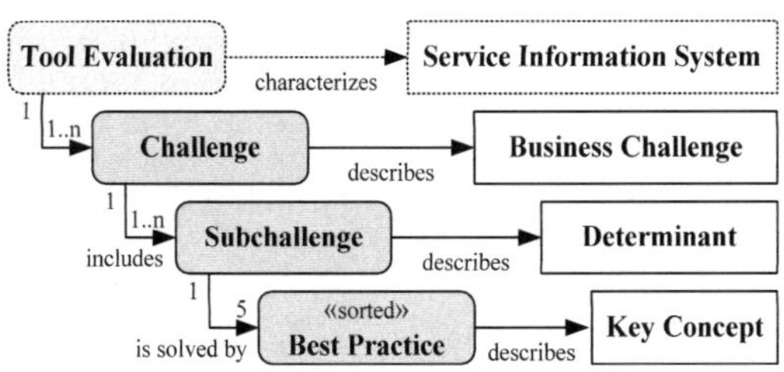

Fig. 1. Meta-model of Classification Methodology.

2. CHALLENGES IN INDUSTRIAL SERVICE BUSINESS

2.1 Three Types of Challenges

Along the life cycle phases of a plant three different types of challenges can be identified: Engineering Challenges (based on Löwen et al., 2005), Project Challenges (Svensson et al., 1999) and Service Challenges (Amberg et al., 2008).

In engineering many types of risks and technical dependencies have to be managed in several disciplines. Part of service engineering for instance is the parameterization of SIS for PAM as well as process monitoring and optimization systems.

Project challenges address issues based on concurrent and cross project engineering, for example change management, multiple language support and user management.

Service Challenges are derived from service execution, for example maintenance, which includes inspection, monitoring, compliance test, function checkout, routine maintenance, overhaul, rebuilding, repair, temporary repair, fault diagnosis, fault localization, improvement, and modification (EN 13306, 2001). In (Ehrenspiel et al., 2007) maintenance for industrial plants is defined as "all technical measures to obtain or restore the functional state" of an industrial plant. Challenges of service execution result mainly from the existence of a plant, for instance the need to respond to urgent malfunctions and the plant itself as data source of SIS (for example supported by asset management systems in practice).

2.2 Structuring Challenges

Every challenge of industrial service business is described in detail by a number of so called subchallenges, which describe a single determinant of the efficiency of supporting SIS regarding the industrial service business. For every determinant described by one subchallenge five different best practices, ordered from Class 0 to 4, have been gathered (example and meaning of classes see Tab. 1), which reflect the corresponding determinant within the SIS (meta-model see Fig. 1). For every challenge, the arithmetic mean of the performed classifications is calculated and finally visualized in a kiviat graph, forming a characteristic evaluation footprint of the SIS (Fig. 3).

2.3 Service Challenges

Service challenges have been indentified analyzing multiple SIS, a number of expert workshops (see Amberg, 2008) as well as know-how gathered in several projects by CT SE 5.

During evaluation, the mapping of concepts implemented in SIS to challenges is a complicated step which is up till now done intuitively producing subjective results. Therefore a generic architecture of SIS is introduced as guideline for the evaluator, contributing to the formalization of the evaluation process by providing a mapping-oriented reference model (Fig. 2).

The reference structure of a SIS and the mapping to the challenges are derived from the plant service process. The model shows all relevant layers of an SIS from processing of field data from the plant up to possibilities to connect to ERP systems. For an efficient service execution there is a need to store, process and change large amounts of datasets coming from the plant. These challenges are grouped via the service challenge *data handling*. The challenge *data processing* deals with this data and characterizes e.g. the level of automation that is provided by the SIS during further processing and data consistency checks that might be executed.

This data has to be pre-processed appropriately before it can be presented to the user. Subchallenges addressing the visualization of service information, are subsumed by the service challenge *view concept*. The service challenge *service know-how reuse* deals with the necessity to formalize and reuse service know-how (Fig. 2 – right).

	General Meaning of Class	Customization - Possibilities for Views on Service Data
Class 4	Generic Support	Definition of user-specific views with integration of data
Class 3	Explicit Support (High Level)	Definition of user-specific views (more than spreadsheet) without integration of data
Class 2	Explicit Support (Low Level)	Definition of user-specific spreadsheet views (e.g. queries)
Class 1	Implicit Support	Limitation of directly supported views for specific users
Class 0	No Concept	No support

Tab. 1. Structure of Subchallenge *Customization.*

2.4 Problem of Mapping Challenges to SIS Components

Especially if large integrated systems are evaluated, the problem occurs, that evaluators can't decide, which component of the SIS has to be assessed, solely based on the description of the challenges. This can be illustrated by the following example, taken from the evaluation of an integrated PAM and FDM system which was executed in cooperation by WI3 and LSTM. In this particular example different results would be reached by two possible component candidates.

The evaluated SIS, a PAM system, is embedded in a process control system and therefore has direct access to all field-devices. The PAM system uses, amongst other things, the process control system's integrated view component. The FDM system in this case reads the diagnostic data directly from the field-devices via a fieldbus or HART and acts as a gateway; simply writing data into a placeholder box surrounded by pictures defined in the SIS view component.

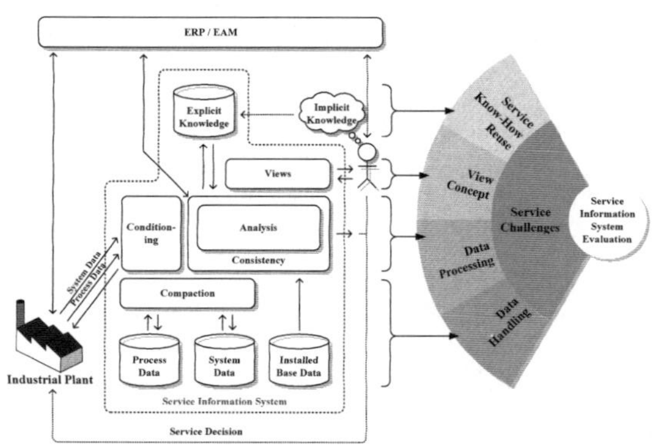

Fig. 2. Generic SIS Reference Model

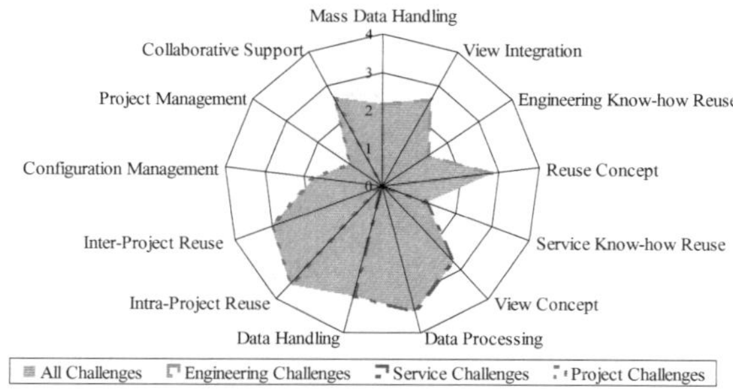

Fig. 3. Kiviat Graph of the Asset Management System.

In case of the subchallenge *customization*, which is assigned to the challenge *view concept*, either Class 0 or 4 can be matched (best practices see Tab. 1). Class 0 is applicable since the FDM system itself has no user definable views. The diagnostic values from a field device are always read completely by the gateway and can't be filtered. On the other hand Class 4 can be matched, because the powerful view component integrated into the PAM tool allows free engineering of views and also the integration of every type of data.

The systematic solution of this problem is the reference model (Fig. 2.), which maps generic components of SIS to challenges. In our example the FDM gateway belongs to the component *Conditioning*, which is evaluated by the challenge data processing. The considered subchallenge *customization* belongs to the challenge *view concept* matched to the component *View* of the reference model. Therefore the *View* component of the PAM system should be evaluated, eventually leading to a Class 4 rating for this subchallenge.

As shown in this example, the challenges combined with the reference model offer an abstract structure used to evaluate tools in a more objective and easy-to-use way.

3. SUMMARY AND OUTLOOK

The results of a SIS evaluation (Fig. 3) can help tool suppliers to specifically identify and evaluate requirements of future versions to match the needs of service engineers and personal to get the capabilities to reduce total costs of plant ownership by service optimization. The evaluation method based on a reference model approach leads to more objectified results by formalizing the evaluation process making design decisions more secure and riskless for SIS developers. It is an easy-to-use practical tool for product managers and service engineers, which closes the gap between plant service business and the product driven business of the service system suppliers.

The formulation and application of the developed methodology is currently being extended to other life cycle phases.

4. REFERENCES

Amberg, M.; Holm, T.; Maier, R. & Maurmaier, M. (2008). Evaluation von industriellen Service-Werkzeugen *Proceedings of 10th Symposium on Entwurf komplexer Automatisierungssysteme (EKA 2008)*

Ehrenspiel, K.; Kiewert, A. & Lindemann, U. (2007). *Cost efficient design,* Springer Verlag, Berlin

EN 13306 (2001) *European Standard: Maintenance Terminology*

Löwen, U.; Bertsch, R.; Böhm, B.; Prummer, S. & Tetzner, T. (2005) Systematization of industrial plant engineering, *atp – Automatisierungstechnische Praxis,* Vol 50, I. 4, pp 54-61

Svensson, D.; Malmström, J.; Pikosz, P. & Malmqvist, J. (1999) A Framework for Medelling and Analysis of Engineering Information Management Systems, *In: Proceedings of ASME Design Engineering Technical Conference,* Las Vegas

Annals of DAAAM for 2008 & Proceedings of the 19th International DAAAM Symposium, ISSN 1726-9679
Editor B. Katalinic, Published by DAAAM International, Vienna, Austria 2008
Make Harmony Between Technology and Nature, and Your Mind will Fly Free as a Bird
2nd DAAAM-ICAT

FABRICATION OF CERAMIC CROSS FLOW FILTERS WITH SOLID FREEFORM FABRICATION

HOMA, J[ohannes] & STAMPFL, J[uergen]

Abstract: *In this paper a method is presented to fabricate monolithic cross flow filters by a solid freeform fabrication (SFF) process. The monolithic cross flow filter with strong undercuts can only be produced with a fugitive wax mold and gelcasting. The production techniques for wax molds (rapid prototyping and injection molding) have different advantages and limitations concerning size, wall thickness, design alterability and capability for mass production.*
Key words: *Ceramic prototyping, Gelcasting, DPF, RSiC*

1. INTRODUCTION

Ceramic wall flow diesel particulate filters are used to reduce the particle emission of diesel vehicles by filtering the exhaust gas through a porous wall. Due to the potential health risk of diesel particulate matter many countries have restrict these emissions. While current legislation does not necessarily require the use of a diesel particlulate filter (DPF), future limitations will only be met by the implementation of DPFs (Adler, 2005).

2. CROSS FLOW FILTERS

Nowadays these filters are mainly made of cordierite and recrystallized silicon carbide (RSiC) in a honeycomb shape by extrusion. The exhaust gas is forced through the porous walls due to the alternate plugging of the parallel channels. Particulate matter is deposited on the walls of the inlet channels and is regenerated by thermal means. The organic matter is burnt off, but the inorganic ashes cannot be removed during regeneration. This leads to the accumulation of the inorganic ashes and finally to a clogging of the filter. Fig.1 demonstrates the scheme of a wall flow filter and the clogging with inorganic ashes.

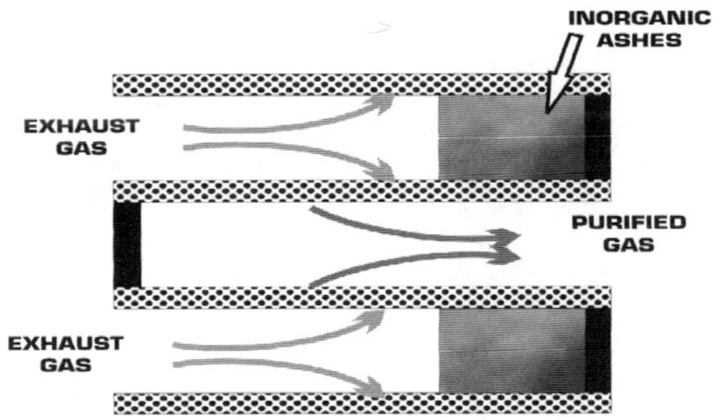

Fig. 1. Scheme of a wall flow filter.

To overcome this problem other designs were developed, where the inorganic ashes can be removed. A cross flow filter (CCF) can manage these ashes by collecting them in an ash chamber. The exhaust gas is also purified through the porous walls and the inorganic ashes can be easily removed by cleaning out the ash chamber, which is attached to the filter. The basic design of this cross flow filter shown in Fig. 2 is derived from the patent WO 2005/033477 A1 (Hoffschmidt et al., 2005).

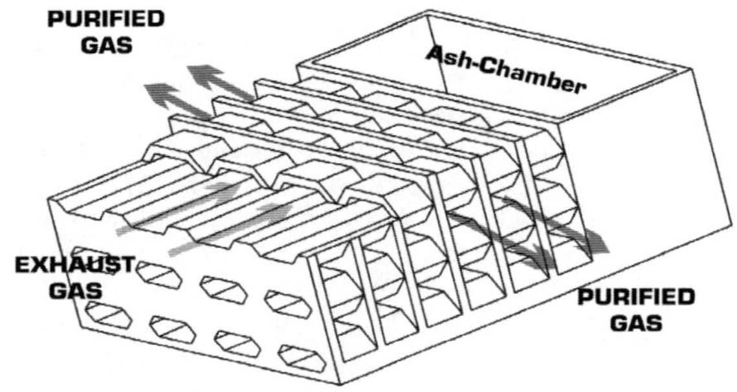

Fig. 2. Scheme of a cross flow filter.

Since the cross flow filter has internal undercuts, this design cannot be fabricated by classical ceramic forming techniques as a monolith. Only SFF processes are capable of producing this design in ceramics. 3D-printing technology was first used to produce ceramic cross flow filters (Parish & Jeffery 1995). The disadvantage of 3D-Printing is the time consuming post processing of removing the unbound powder from the channels. The smaller the channels the harder it is to remove the powder. Furthermore 3D-printing is from the present point of view not able to produce parts on a larger scale for a mass production.

3. INDIRECT FABRICATION METHOD FOR CFF

The technique used for the production of this CCF should fulfill two requirements. One the one hand, design alterations should be cheap and fast and on the other hand the chosen process should be capable of a production on a larger scale. Since RP-processes are at the moment not dedicated for mass production, other ways have to be found. To satisfy both requirements the process should consists of two steps. At first a mold could be the fabricated, which is then filled with a slurry. For the first design alterations a RP-process can be used and once the design is fixed conventional processes can be used for the fabrication of the mold.

Conventional ceramic casting processes are very limited in the mold material and its design. A more generic process is gelcasting, which is a fairly new slurry based forming technique for ceramics. A low viscosity slurry, which consists mainly of particles, solvent and monomers, is poured into a mold and the solidification of the slurry is done by polymerizing the monomers. The polymerization is triggered by a free radical initiator and a catalyst and thereby forming a firm polymer-solvent gel matrix which immobilizes the particles in the desired shape of the mold. After drying the parts may be machined then they are debinded and sintered. One main advantage of gelcasting is the possibility to use nearly any material for the mold. (Janney et at., 1998). Cross flow structures have already been made by gelcasting, but the casting molds were made of steel. (Vaubert et al., 2001) Since the design shown in Fig. 2 has internal undercuts, hard tools can not be used. The cross flow filter needs to be made by the lost mold method, because the constant wall thickness of the hexagonal inlet channels causes undercuts in the outlet channels. Hence, the gelcasting technology has to be combined with a wax mold, which can be melted off to produce the required undercuts.

3.1 Fabrication of the mold via Rapid Prototyping (RP)

Solidscape offers a wax printer, which allows the production of fugitive molds with internal structure. As every additive forming process the wax printer deposits the material layer by layer. Additionally this wax printing technology uses a solvable support material, which allows the production of any desired shape. The build material is a low melting wax, which is inevitable for the gelcasting process. The combination of RP and gelcasting is an ideal fabrication process for very complex ceramic structures. After casting the slurry into the mold and after solidification of the slurry, the part can be demolded by simply melting off the wax. A concept model of a DPF cast in a mold, made by the Solidscape wax printer (BT 612) is shown in Fig. 3. The part is made of recrystallized silicon carbide, which is a very common material for DPFs. The model has a total volume of 25 x 36 x 65 mm³ and a constant wall thickness of 0,5 mm. The hexagonal inlet channels have a cross section of 3,8 mm² and a cell desity of 172 cells per in² (cpsi).

Fig. 3. Concept model of a SiC DPF cast in a RP mold.

CCF with wall thicknesses down to 0,3 mm have been produced by this process. For larger molds the build time becomes very long and furthermore larger parts (larger than 5 x 5 cm²) with fine structures tend to break and curl due to thermal stresses.

3.2 Fabrication of the mold via stacking

Since additive technologies are not suitable for a production on a lager scale and large molds can not be produced with the wax printing technology, other mold fabrication techniques have to be found (Homa, 2007). Gelcasting can not only be combined with the wax printing technology, but also with common wax fabrication techniques (Stampfl, 2002). A very effective way of producing these wax layers in terms of output is injection molding. Since the outlet- and inlet channel do not intersect, the mold can be made by stacking two different sorts of wax layers together, which are both produced by injection molding. After stacking the layers together and sealing the surface, the slurry is poured into the mold and solidified by thermal means. Fig. 4 shows a filter which was cast in a mold, where the layers have been injection molded and stacked together. The filter has a constant wall thickness of 0,45 mm, a filtration area of 0,36 m²/l and the cell density of 28 cpsi. CCF with a volume of up to 1,5 l should be produced in the future by that fabrication technique. Further improvements have to be made to decrease the wall thickness and increase the cell density of the filter.

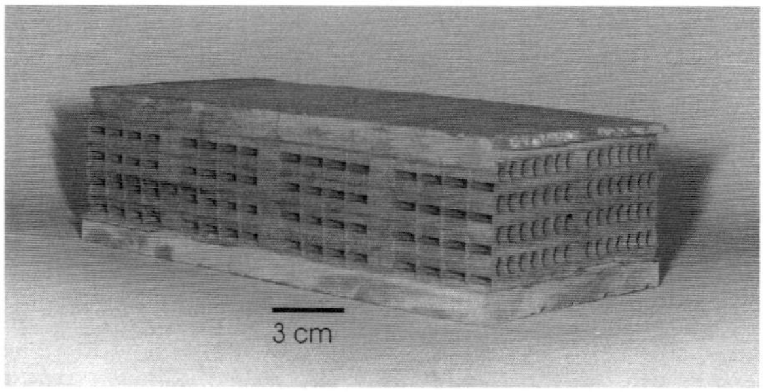

Fig. 4. Concept model of a SiC DPF cast in stacked injection molded wax layers

4. CONCLUSION

Gelcasting in combination with a fugitive wax mold is an appropriate fabrication technique for CCF. Small models can be made with a RP wax printing technology, which makes design alterations very easy, but for a production on a larger scale and for bigger parts this technologies is not suitable. Injection molding can be used to produce two types of wax layers, which are than stacked together to form the mold. Thereby the build speed can be increased by a factor of 1000, but design alteration for injection molding tools are very expensive and time consuming. It takes 4 to 6 weeks to produce a tool. Gelcasting in combination with a fugitive wax mold is a promising fabrication process for complex ceramic parts. Further effort has to be undertaken to achieve fine structures for the injection molded layers, to decrease the wall thickness and the increase the cell density of the resulting DPFs.

5. REFERENCES

Adler, J. (2005). Ceramic diesel particulate filters, *International Journal of Applied Ceramic Technology*, Vol. 2, No. 6, pp. 429-439, ISSN 1546-542X

Hoffschmied, B.; Fend, T. & Miebach, R. (2005). Particulate Filter for an Internal Combustion Engine, PCT/EP2004/010177, WO 2005/033477 A1

Homa, J.; Zellhofer, S. & Stampfl J. (2007). Novel Fabrication Technique for Ceramic Filters, Proceedings of the 10th International Conference of the European Ceramic Society, ISBN: 3-87264-022-4, Berlin, June 2007, Göller Verlag GmbH, Baden-Baden

Janney, M. A.; Omatete, O. O.; Walls, C. A.; Nunn, S. D.; Ogle, R. J. & Westmoreland, G. (1998). Development of low-toxicity gelcasting systems, *Journal of the American Ceramic Society*, Vol. 81, No. 3, pp 581-591, ISSN: 0002-7820

Parish, M. & Jeffery A. B. (1995). Ceramic filter elements with tailored macro- and microstructures, *Filtration and Separation*, Vol. 32, No. 1, pp 31-35, ISSN 0015-1882

Stampfl, J.; Liu, H.; Nam, S.; Sakamoto, K.; Tsuru, H.; Kang, S.; Cooper, A. G.; Nickel, A. & Prinz, F. B. (2002). Rapid Prototyping and Manufacturing by gelcasting of metallic and ceramic slurries, *Materials Science and Engineering A*, Vol. 334, No. 1-2, pp. 187-192, ISSN 0921-5093

Vaubert, V.; Armstrong, B.; Janney, M.; Walls, C.; Stinton, D.; Barra, C.; Nagesvaran, R. & Limaye S. (2001). Design and Development of a Ceramic hot-gas filter for Fossil Energy, Available from: http://www.ornl.gov/~webworks/cppr/y2001/rpt/ 110061.pdf Accessed: 2007-08-02

Annals of DAAAM for 2008 & Proceedings of the 19th International DAAAM Symposium, ISSN 1726-9679
Editor B. Katalinic, Published by DAAAM International, Vienna, Austria 2008
Make Harmony Between Technology and Nature, and Your Mind will Fly Free as a Bird

PATTERN RECOGNITION BY USE OF FUNCTIONS OF INVARIANTS

HORAK, K[arel] & KALOVA, I[lona]

Abstract: *Visual system is the basic tool for non-contact inspection in industries. Such system usually has to detect defects and reject faulty products automatically. This article deals with autonomous visual system for glass bottle inspection, which has to detect non-homogenous regions in ROI in images. Additional system is able to distinguish between regions belongs to defect from regions belongs to manmade patterns, so that products can be made in high quality and cost-effective way. Snag is that synthetic patterns (like a word-sign) and defects (dirt, crack etc.) have similar image characteristics, hence it is difficult to recognize between them. In order to solve mentioned task we introduce novel method based on moment invariants. Each detected non-homogenous region is wilfully modified by so called "circle degradation" algorithm in several steps. Moment invariants corresponding to modified regions are then computed and subsequently functions of these invariants are created. On the basis of these functions, both synthetic patterns and defects can be recognized in a usual classification way.*
Key words: *recognition, pattern, moments, invariants.*

1. INTRODUCTION

Visual system for glass bottles inspection was originally designed as system looking for non-homogenous regions on images (strictly speaking only in region of interest). Each detected region is considered as defect automatically (Horák & Kalová, 2007). For example defect on the bottom of the bottle is shown at the left side of the first figure.

Until quite recently appear bottles with synthetic patterns. These intentional patterns created by manufacturers look like a word-sign, numbers, letters or various geometric entities. Example of manmade word-sign on the image is shown at the right side on the figure 1.

As denoted above, previous visual system locates and evaluates these intentional synthetic patterns in the same way as defects. Classification of such patterns as defects is obvious failure. On this account was designed new system which is able to recognize between synthetic patterns and defects. Unfortunately it turned out that patterns occur on the images in various transformations. That is why the new recognition system has to fulfil the condition of independency on brightness, rotation, translation and scale factor. Problem described above was solved by moment invariants, strictly speaking by functions of moment invariants (see chapter three).

Structure of the paper is following. In the next chapter are described methods for extraction of regions from original image and consequential preparations for computation of feature vector. Chapter three occupies oneself with proposed method for pattern recognition by means of functions of moment invariants. Last chapter is dedicated to brief description of classification method.

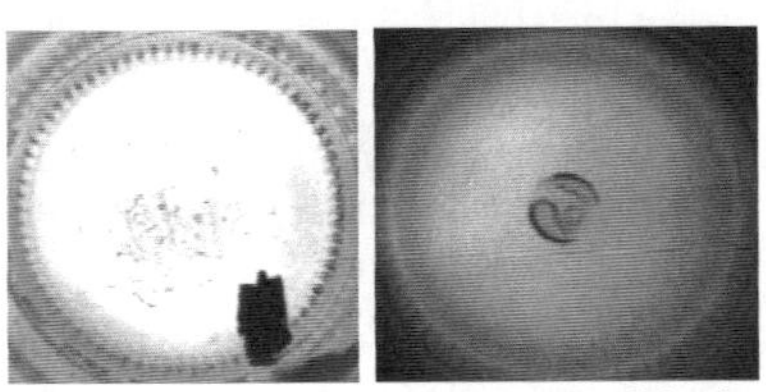

Fig.1. Defect and synthetic pattern on bottom of the bottle.

2. MOMENT INVARIANTS

In order to distinguish synthetic patterns from true defects is necessary to segment all non-homogenous regions from ROI (chapter 2.1) and then compute feature vector (chapter 2.2) suitable for classification process (chapter four).

2.1 Regions extraction

As you can see on the picture below, standard image processing methods are used for detection and extraction of non-homogenous areas in image.

On the original image data (a) is executed edge detection process (b) followed by noise reduction and small objects elimination (c). In the next two steps are done morphological operations closing (d) and filling holes (e). In order to obtain non-homogenous regions from image as rectangle of image data, minimal and maximal coordinates of connected segments are founded and appropriate image region is separated (f) from original image.

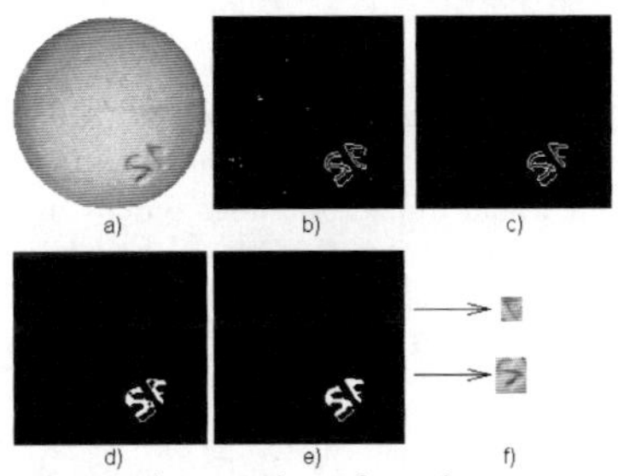

Fig.2. Detection and extraction of non-homogenous regions.

2.2 Moments and moment invariants

Each synthetic pattern on the bottle generally looks like a simple geometric entity, a letter of alphabet or like a word-sign. For distinguish between such simple shapes (but unknown and in very complicated image background) are suitable geometric or complex moment invariants derived from geometric or complex moments respectively (Muharrem et al., 2005).

Geometrical moments are classical tools for pattern recognition during the last 40 years (Hu, 1962), but complex moments were introduced by professor Flusser only recently. It was introduced in total 11 moment invariants from ψ_1 up to ψ_{11}. These invariants are created from complex moments of the second, third and fourth order (Flusser, 2000). In our work are only used moment invariants ψ_1 and ψ_2 defined as follows:

$$\begin{aligned} \psi_1 &= c_{11}. \\ \psi_2 &= \mathrm{Re}\{c_{20} \cdot c_{12}^2\} \end{aligned} \qquad (1)$$

Complex moments c_{pq} required in previous equation are given by formula:

$$c_{pq} = \sum_{Y} \sum_{X} (x + iy)^p \cdot (x - iy)^q \cdot s(x, y). \tag{2}$$

Sums over X and Y in equation above are in the sense of region data $s(x,y)$ and sum $p+q$ determines order of moment.

Only first two invariants ψ_1 and ψ_2 was capable to distinguish between different patterns, because invariants of higher order are strongly dependent on noise in our task (Flusser & Suk, 2003). For that reason their values oscillate and are not stable for the same class patterns.

2.3 Region preparation

Non-homogenous areas like the one depicted on figure 2(f) have several negative features unfortunately. The most significant features are noise, complex background and reflections (see on the figure 2(a) that pattern has light boundary on the one side and dark on the opposite one).

Each and every feature has negative impact on invariants ψ_1 and ψ_2, strictly speaking on their stability and capability of discrimination. For that reason all non-homogenous regions are processed as is shown on the following figure.

Negative properties are removed and relevant features are simultaneously maintained in this way. Outgrowth of depicted operations is higher stability of invariants of the same class patterns. As mentioned above in chapter 2.2, in this moment feature vector has two stable elements (invariants ψ_1 and ψ_2).

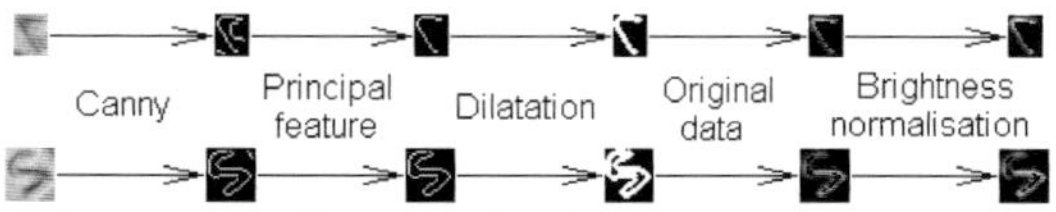

Fig.3. Preparation of extracted regions.

3. FUNCTION OF COMPLEX INVARIANTS

It's obvious, that feature vector with two elements creates two-dimensional feature space, but discriminative capability of such space is relatively low. As shown on next figure, specific defect (e.g. encircled triangle number four) can easily collide with some cluster of known patterns (black stars). By force of experiments was unfortunately discovered, that such collisions happened very often in real scenes, so we had to find a solution.

Elimination of such collisions as just described is possible in following way. Every one extracted non-homogenous region is modified by so called *„circle degradation"* in the N degrees as shown on the next figure (here $N=13$).

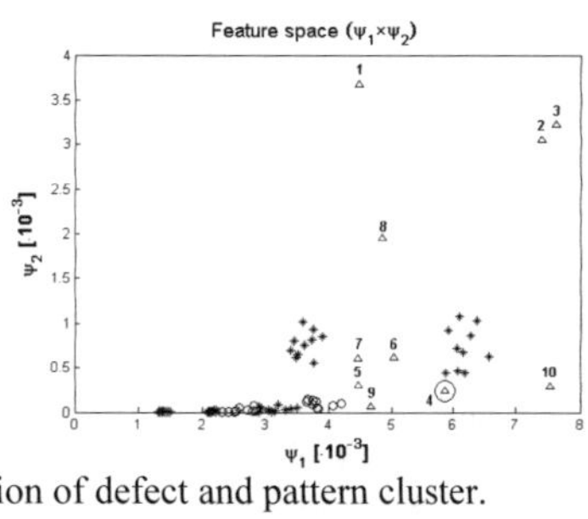

Fig.4. Collision of defect and pattern cluster.

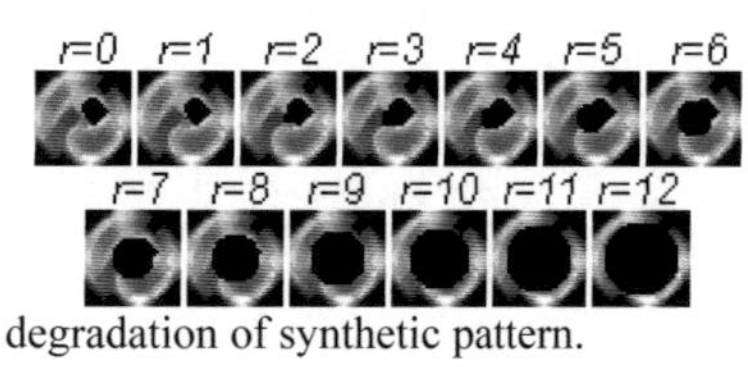

Fig.5. Circle degradation of synthetic pattern.

For every one modified region as depicted above is then computed its own pair of ψ_1 and ψ_2, that means instead of two values ψ_1 and ψ_2 we have two vectors $\psi_1(r)$ and $\psi_2(r)$ with parameter r (degree of degradation). Both vectors $\psi_1(r)$ and $\psi_2(r)$ are so called *„function of moment invariants"* here. Next figure shows vector $\psi_1(r)$ corresponding to previous pattern.

Because of circle degradation produces different function $\psi_1(r)$ (and analogically $\psi_2(r)$) for patterns, which values ψ_1 are similar, this method is able to distinguish between like this patterns. Dimension of feature space is $dim(\psi_1,\psi_2)*N = 2*N$ and has much higher discriminative capability now.

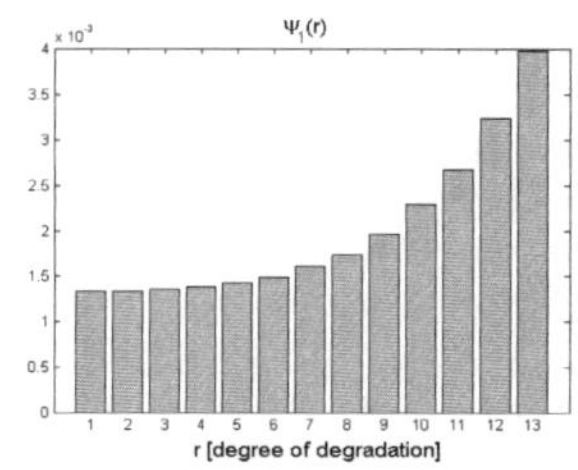

Fig.6. Function $\psi_1(r)$ of moment invariant ψ_1.

4. CLASSIFICATION METHOD

Classification of unknown pattern is based on measuring of absolute distances between its feature vector and feature vectors of the known synthetic patterns. If distance between closest cluster and feature vector of unknown region is lower than specified limit, unknown region is classified as member of nearest cluster. On the contrary, if mutual distance is higher than limit, unknown region is considered as a true defect.

5. CONCLUSION

In this work was briefly introduced visual system designed for pattern recognition on images of glass bottles. Novel method for increasing discriminative capability of feature space was proposed, tried and verified. In the future work, we'll pay attention to more accurate image processing methods described in chapter two, because it appears that accuracy of region extraction and preparation is crucial for correct classification.

6. ACKNOWLEDGEMENT

This work has been supported by the project No. 1M0567 and project No. MSM0021630529.

7. REFERENCES

Flusser, J. (2000). On the independence of rotation moment invariants. *Pattern Recognition*, vol. 33, 2000, 1405-1410, ISSN 0031-3203.

Flusser, J. & Suk, T. (2003). Construction of Complete and Independent System of Rotation Moment Invariants, *Proceedings of 10th International Conference on Computer Analysis of Images and Patterns*, Petkov, N. & M.A.Westenberg, pp. 41-48, Groningen, August 2003, Springer-Verlag, Heidelberg.

Horák, K. & Kalová, I. (2007). Applied Methods For Transparent Materials Inspection. *Computational Modelling of Objects Represented in Images*, Tavares J.M. & Natal J., 221-225, Taylor & Francis, ISBN 0-4154-3349-5, London.

Hu, M.K. (1962). Visual pattern recognition by moment invariants. *IEEE Transactions on Information Theory*, Vol. 8. February 1962, 179-187, ISSN 0018-9448.

Muharrem, M., Kayhan, G. & Tarik, V.M. (2005). Real object recognition using moment invariants. *Academy Proceedings in Engineering Sciences*, Vol 30. December 2005, 765-775, Sahana.

Annals of DAAAM for 2008 & Proceedings of the 19th International DAAAM Symposium, ISSN 1726-9679
Editor B. Katalinic, Published by DAAAM International, Vienna, Austria 2008
Make Harmony Between Technology and Nature, and Your Mind will Fly Free as a Bird

STRATEGIES OF COMPANIES REORGANIZATION

HORJA, I[oana] M[onica]; VANCEA, S[maranda]; IGNAT, A[ndreea] & AVRAM, E[leonora] L[aura]

Abstract: *Economic reorganization and development, processes viewed at macroeconomics level and also at microeconomics level, are definitory for market economy and inseparable for the evolution of the economic reality. At the company level, reorganization and development spreads out in different ways and with different intensity. For reaching his final objective, obtaining profit, the company has to take some decisions about future development, being necessary to look for a growing strategy based on reorganization. The paper tries to present reorganization and development strategies used by the Romanian companies.*

Key words*: merger,acquisition,division, economic growth*

1. INTRODUCTION

Economical reorganization and social development are two inseparable steps of economic reality. At enterprise level, reorganization and development can have different ways.The companies, as legal persons can be formed, can be consolidated or developed, or reduce their activity or can be dissolved or liquidated function of the economic conjuncture favorable or not.

2. BODY OF PAPER¶

Merger, division and the chain companies are ways of reorganization through development which goes to continuing the activity. The restructuring of a company is determined by the influential of many facts, such as: economic concentration, economic development in many countries, market globalization and access to international market.(Matis, 2003) At company level, reorganization is made for economic growth, especially for reaching the main objective of a company, obtaining bigger benefits. The classical ways of reorganizations can be: companies merger or absorption, acquisition, companies division. Merger and division are the main strategies of companieis reorganization. For obtaining benefits companies are put in some decisional situations about their future. For reaching the main object, the profit, an enterprise has to take many decisions about the future. An enterprise tries to increase her market rate in connection with competition or at least to maintain it. This requires suitable concentration strategies or business reduction or even breaking the entity in many parts. The law no. 31/1990 republished in the O.M. no. 33/20.01.1998 defines in the article no.233 the merger: the merger is made by the absorbtion of a company by another or by the merger of two ore more companies for making a new company. In the groups can be a 3rd way of merger, the merger way of internal restructuring, operation who has as an effect the reduction of the companies number in a group, the liquidation of the unprofitable companies an of the companies without aim of activity. (Toma, 2003) So, the merger is an operation whereby one or more entities decide to transfer their patrimony to another entity and to develop their activity together. The merger is one of the capital concentration forms. Merger is a way of capital concentration (economic, technique, human) for surviving or development of two or more entities in a concurrencies market. Division is part of the same category of companies strategies of reorganization, like merger, being the opposite of merger, so lies in dividing the assets of a company by one or more companies beneficiary. Companies beneficiary may exist (division by acquisition) or may be newly set up (division is held by the formation of new companies).

In occidental countries, mergers were the main form capital concentration in the development step of the `70 years. Usually was merger through absorption because of different size of participants and because of many fiscal advantages. In France, mergers were profitable for national economy, but in our days are not considerate any more universal solutions because of the permanent changes in economy and because of little companies which are more dynamic than bigger companies.In Romania merger was used for state capital company's reorganization. Before 1990, Romanian economy had a high rate of concentration by administrative bases. For passing to a transitory economy were necessary two steps: economy transformation by divisions of big companies and economical, commercial and financial grouping of companies through merger.

A strategically develop of an enterprise imposes frequently the revise of the organization, either internal measures or profound measures which affects the company structure. These development ways are part of the strategy which was chose by the company management for increase the competitive advantage for obtaining bigger benefits. One of these ways of development is merger. Merger is based on a few reasons through the most important reason is *earning efficiency*: when two ore more company merged, an enterprise who makes a product who is complementary to the existing product can complete the products going to the growth of total demand. The economies which results by this way are obtained through synergy, the resulted company through fusion is more than summing the parts (scale economies). The economies are obtained through synergy; the new company obtained through merger is more advantageous than the first companies.Synergy is the magic force that allows for enhanced cost efficiencies of the new business. Synergy takes the form of revenue enhancement and cost savings. By merging, the companies hope to benefit from the following: *staff reductions* (as every employee knows, mergers tend to mean job losses, it considers all the money saved from reducing the number of staff members from accounting, marketing and other departments), economies of scale (whether it's purchasing stationery or a new corporate IT system, a bigger company placing the orders can save more on costs, mergers also translate into improved purchasing power to buy equipment or office supplies), *acquiring new technology* for staying competitive(by buying a smaller company with unique technologies, a large company can maintain or develop a competitive edge), *improved market reach and industry visibility* (companies buy companies to reach new markets and grow revenues and earnings, for expanding two companies' marketing and distribution, for giving them new sales opportunities, for improving a company's standing in the investment community: bigger firms often have an easier time raising capital than smaller ones) That said, achieving synergy is easier said than done, it is not automatically realized once two companies merge. Sure, there ought to be economies of scale when two businesses are combined, but sometimes a merger does just the opposite. In many cases, one and one add up to less than two. So, the synergies are supplementary benefits which can be obtained through business combination. Synergies have many forms: *sales synergies* (the new company has all the distributions channels, promotions, common deposits); *operational synergies* (a better using of facilities and human capital, a decrease of unit acquisition costs for raw materials); *investment synegies* (using in common equipments, some activity, common stocks of raw materials); *managerial synergies* (a good management of the absorbent can increase the absorbed management); *financial synergies* (obtaining economy in financial expenditures credits with little interests).When the necessary founds for development aren't enough through merger this thing can be solved. The companies with more liquidity but with less possibility of investments can choose merger. In economic, mergers in fact expressed a willingness regrupation and concentration, they mean above all a union of means and skills and express divisions, on the contrary, the division in other distinct entities, according to different criteria.

When a company decides it wants to aquire or merge with another, it tipically does so by using the services of outside professionals such as attorneys, accountants and valuation experts.(Gaughan, 2002)For knowing if a merger or division is good for the company a detailed analyzes is necessary, a feasibility study undertaken to determine the merger's or division's viability. These results of this study are used to make a decision whether to proceed with the project, or table it. If it indeed leads to a project being approved, it will be used to ascertain the likelihood of the project's success. It is an analysis of possible alternative solutions to a problem and a recommendation on the best alternative. Merger and divisions are made for purpose of improving activity, they should be profitable. But can be situations in which merger or division fails. For avoiding these cases, there is a prealable time of merger and division, time when the companies must know each other. (Dumbravă, 2004) In this time there must be done a lot of studies and analyses by consulting companies such as commercial studies, assets evaluation, company evaluation, financial analyses, and audit. These studies are the base of decision of continuing merger or division or not. Analyze must show the evolution of the company in the future. Before the decision of merger or division there are necessary: financial study (any reorganization or development of a company supposes investments for a long period of time.,these investments must be restructured on factors of production such as from there combination the economic effect should be maximum), technical study (considers only the actual price behavior of the market or instrument, based on the premise that price reflects all relevant factors before an investor becomes aware of them through other channels).This documentation is realized in a few steps: opportunity study (for identification of reorganization possibilities, advantages or disadvantages, economic brands of companies, geographical area, the chances of the new company on the market),prefeasebility study (for identification the market capacity, the human capital, financial prognosis, factors of production) and feasibility study (shows the necessity of reorganization and the economic and financial results which can be obtained in the future).Market analysis plays a major part in a firm's planning activities, it guides decisions on inventory, purchase, work force expansion/contraction, facility expansion, purchases of capital equipment, promotional activities, and many other aspects of a company. Forecasts in these areas must be accurate and decision makers must understand how they were derived. Only after these analyses the decision of merger and division can be taken. The next stage consists of drawing up the merger or division project by the managers of the companies' part at the merger or division. In a merger success can appear problems because of synergies supraevaluation. The development of companies in economy is based on tangible assets. But the intangible assets are also a very important part: know how, innovation, competitive advantage. The intangible assets can create value for the company. In order that the merger or division to be successfully the companies must try to find the value of the intangible assets, the company has to try to stimulate the employers not to control them. Every employee has his own vision and capacity which can bring value for the company.

3. CONCLUSIONS

The Chinese expression for crisis – wei ji – combines the character "risk" with the character "opportunity". Reorganization transactions,such as merger or division,are opportunities that bear some considerable risk.(Bruner, 2004)

In the last years in Romania the company's reorganization followed a growing trend and reached high values in terms of number and total value of publicly available private transactions. Romania's path towards a stable economic environment and its desired EU membership contributed to fueling foreign investors' interests in the local market. Among the most active industry sectors from the investor's standpoint were manufacturing, services, financial services and food beverages.The accession of Romania to the European Union has produced changes in the economic behavior of Romanian entrepreneurs: some of them have chosen to improve the competitiveness of their companies through business partnerships with either financial institutions or strategic investors, some have considered exit options as the most appropriate alternative and others have started to consider acquisitions in neighboring countries.The data points towards a certain conclusion: as the economy is improving, an increasing number of investors is turning and tuning their focus on Romania.In the last few years many mergers and acquisition were in production, construction and services sectors and the main investors were from Austria, Great Britain, USA etc. The company managers can try to save a company in difficulty or can try to develop a company. The paper tries to get a view about the reorganization strategies of companies, to the question whz do so many reorganization strategies fail or what drives the success of those deals that are consummated.If a company is in difficulty, the managers can use a strategy to save the company through a merger or division or even by an acquisition by a group of companies.The reorganization is more and more probable in Romanian economy because of multiple changes.The evolution of an economic system supposes reorganization and development from many points of view such as technological, social, and organizational. The essential factor is information, because information has elements about benefits in general, benefits of some activities, market and costs of the future founds.Mergers and acquisitions can be understood as a response by managers to forces of turbulence in their environment.

6. REFERRING

Bruner.R.,Perella.J.(2004), *Applied Mergers and Acquisitions,* John Wiley and Sons, ISBN 0471395056, San Francisco

Dumbravă, P.(2004),*Patrimonial Modifications*, Universitary Press of Cluj, ISBN 973-610-045-6, Cluj Napoca

Gaughan.P.(2002), *Mergers, Acquisitions & Corporate Restructurings,* John Wiley and Sons, ISBN 0471237299, San Francisco

Matiş,D.(2003),*Special Operations Accounting*, Intelcredo Publishing House, ISBN 973-8197-16-5, Deva

Toma,M.(2003),*Companies reorganization by merger and division.*CECCAR Publishing House, ISBN 973847138, Bucharest

Annals of DAAAM for 2008 & Proceedings of the 19th International DAAAM Symposium, ISSN 1726-9679
Editor B. Katalinic, Published by DAAAM International, Vienna, Austria 2008
Make Harmony Between Technology and Nature, and Your Mind will Fly Free as a Bird

APPLICATION OF THE FINITE VOLUME METHOD TO PROCESSES IN WOOD TECHNOLOGY

HORMAN, I[zet] & MARTINOVIC, D[unja]

Abstract: *In this paper a numerical method (the finite volume method) for solid body stress analysis is presented. In order to demonstrate the methods capabilities, an analysis of stresses and deformation in a wood samples during a drying process and a wood steaming is performed. The development of computer technology and numerical methods have made the research much easier and enabled obtaining information of what is happening inside a loaded product.*
Key words*: wood, drying, steaming, finite volume method*

1. INTRODUCTION

The finite volume method for stress analysis is equally applicable to linear, isotropic, anisotropic, porous and non-linear materials. In what follows an outline of the method is given and some results illustrating the method's abilities are presented. More details about the method can be found in (Demirdžić, Martinović, Horman, at al.,1993,1995,2000,2001,2003).

2. MATHEMATICAL MODEL AND NUMERICAL SOLUTION PROCEDURE

The behaviour of an arbitrary part of a solid, porous body of volume V bounded by the surface S at any instant of time t can be described by equations of momentum (1), thermal energy (2) and mass balance (3) (Demirdžić, at al.,2000)

$$\frac{\partial}{\partial t}\int_V \rho \frac{\partial u_i}{\partial t} dV = \int_S \sigma_{ij} n_j dS + \int_V f_i dV \qquad (1)$$

$$\frac{\partial}{\partial t}\int_V \rho c_q T dV = -\int_S q_j n_j dS + \int_V s_q dV \qquad (2)$$

$$\frac{\partial}{\partial t}\int_V \rho c_m M dV = -\int_S \dot{m}_j n_j dS + \int_V s_m dV \qquad (3)$$

In order to close the system of Eqs. (1) to (3) or (1) and (2), the constitutive relations are used:

– for an elastic, porous, orthotropic material
- for Eq. (1) is

$$\sigma_{ij} = C_{ijkl}\varepsilon_{kl} - \alpha_{ij}\Delta T - \langle \beta_{ij}\Delta M \rangle \qquad (4)$$

where C_{ijkl} are the elastic constant tensor components (the nine non-zero orthotropic elastic constants) (Demirdžić, at al.,2000, Martinović, at al.,2001). The terms in $\langle\,\rangle$ brackets are "active" only for $M \langle M_h$ (M_h is the moisture potential at the fibre saturation point).

- for Eqs. (2) and (3) are heat and mass flux vector

$$q_j = -k^q_{jl}\frac{\partial T}{\partial x_l} + \varepsilon r \dot{m}_j = -\left(k^q_{jl} + \varepsilon r \delta k^m_{jl}\right) - \varepsilon r k^m_{jl}\frac{\partial M}{\partial x_l} \qquad (5)$$

$$\dot{m}_j = -k^m_{jl}\frac{\partial M}{\partial x_l} - \delta k^m_{jl}\frac{\partial T}{\partial x_l} \qquad (6)$$

– for a thermo-elasto-plastic isotropic material (Demirdžić, at al.,1993)
- for Eqs. (1) and (2) are

$$\delta q_{ij} = 2G\delta\varepsilon_{ij} + \lambda\delta_{ij}\delta\varepsilon_{kk} - (3\lambda + 2G)\alpha\delta_{ij}\delta T - \langle \frac{3G\sigma^d_{ij}\sigma^d_{kl}\delta\varepsilon_{kl}}{\bar{\sigma}^2\left(\frac{H'}{3G}+1\right)} \rangle \qquad (7)$$

and constitutive relation (5) for $\dot{m}_j = 0$

In the case of elastic conditions, the expression within the brackets $\langle\,\rangle$ vanishes, and the constitutive relation (7) reduces to the Duhamel-Neumann form of Hooke's law.

By introducing corresponding constitutive relations into governing equations a closed system of 2 or 3 (generally non-linear and coupled) equations with two or three unknown functions of spatial coordinates and time $(u_i, T, M \text{ or } u_i, T)$ is obtained. To complete the mathematical model, initial and boundary conditions have to be specified.

The solution domain is discretized by a finite number of contiguous hexahedral control volumes (CV) or cells of the volume V which are bounded by six cell faces of the area S_j with calculation points P in the CV's centres (Figure1).

The time domain is subdivided into a number of time intervals δt .

Equations (1), (2) and (3) are integrated over time interval δt and over each control volume resulting in a system of $5N$ (generally non-linear and coupled) algebraic equations of the form

$$a_{P_0}\varphi_{iP_0} - \sum_{j=1}^{n} a_{P_j}\varphi_{iP_j} = b_{\varphi_i} \qquad (8)$$

where φ stands for displacement components u_i $(i = 1,2,3)$ or temperature T or moisture potential M , n is the number of cell faces of a control volume (Fig.1).

Systems of algebraic equations (8) are solved by an iterative procedure (Demirdžić&Muzaferija, 1995).

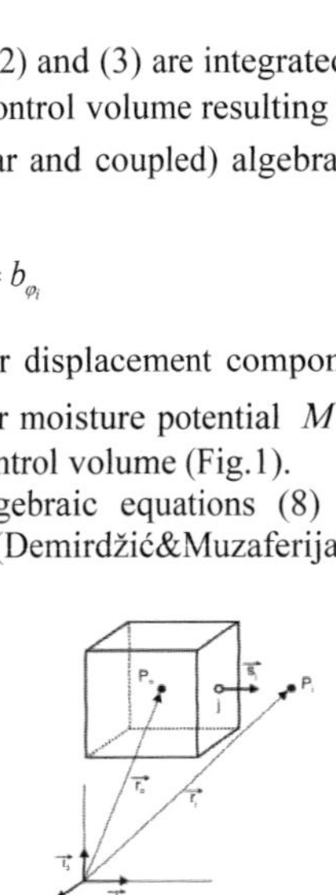

Fig 1. A typical control volume

3. APPLICATION OF THE METHOD

The method described in the previous sections has been applied to a number of both isotropic and orthotropic body deformation problems, few of which will be presented in this paper.

3.1. Numerical analysis of a wood drying process

The beech-wood beams are exposed to the (uniform, unsteady) flow of the hot air in a laboratory dryer with an automatic control of the ambient air parameters.

The temperature and moisture dependent physical properties of the wood are used (Martinović, at al.,2001).

Equations (1) to (3) and the constitutive relations (4) to (6) are used. Fig.2 shows the fields of moisture, stress σ_{xx}, displacements at $t = 159$ h and cross section shape of deformed wood sample, at the end of the drying process (t = 246 h).

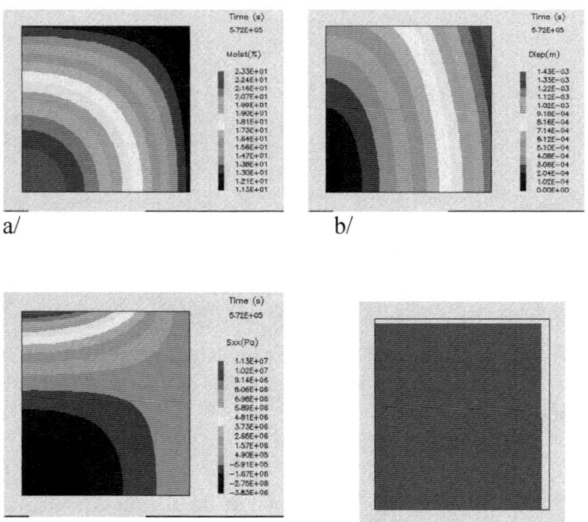

a/ b/ c/ d/

Fig. 2. Fields of moisture, displacement, stress at $t = 159$ h, and cross section shape of deformed wood sample (one quarter of the cross-section) at t = 246 h

The maximal normal stresses are in regions near the sample's surface, and the contraction of the wood sample is 6,5 % and 3,3 % (axis x and y) at the end of the drying process.

3.2. Numerical analysis process of wood heat treatment

The log is exposed to the (unsteady) flow of the steam during his thermal preparation in the veneers production (Horman, at al.,2003).

Equations (1) and (2) and the constitutive relations (7) and (5) (for $m_j = 0$) are used. The problem is considered to be a $2D$ plane strain problem. Fig.3 and Fig.4 show temperature and circular stress histories at three reference sections.

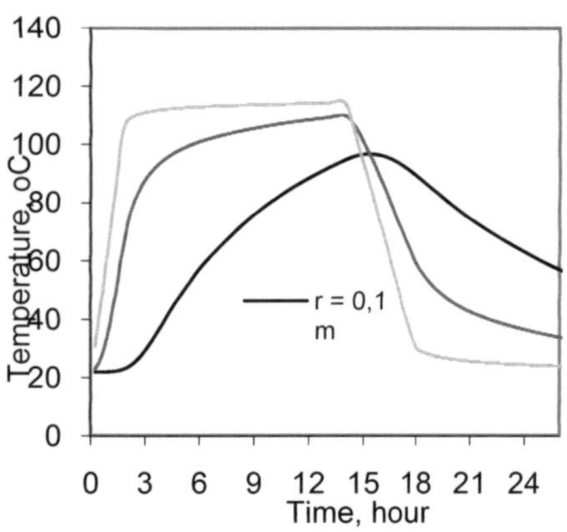

Fig. 3. Temperature histories at three reference sections

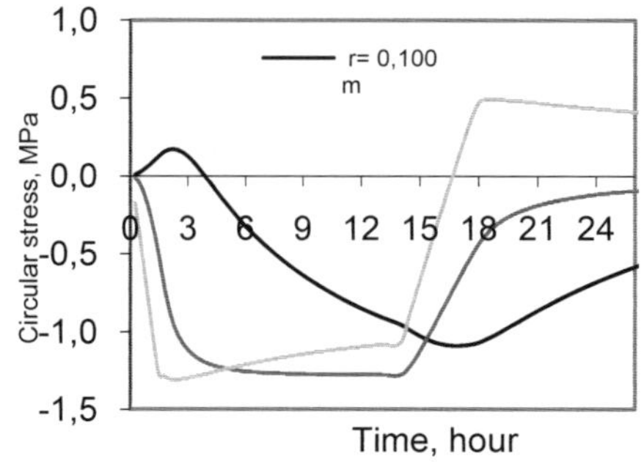

Fig. 4. Circular stress histories at three reference sections

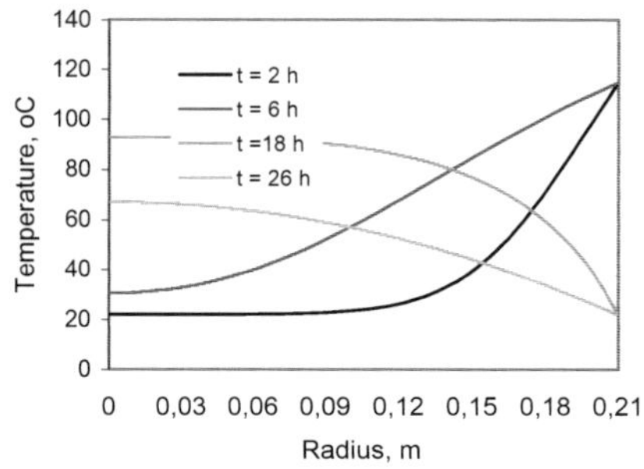

Fig. 5. Temperature profile at $\varphi = const.$

Fig.5 shows the temperature distribution at $\varphi = const.$, in four time instants. The temperature gradients are the largest in the region near the log's surface and this is the region of the largest stress, where the residual stress occurs (Horman, at al.,2003).

4. CONCLUSIONS

The numerical method for stress analysis has been outlined and its applicability to the solution of transient problems involving various porous, orthotropic and elasto-plastic materials has been demonstrated. The mathematical model and the numerical calculation employing the finite volume method presented enable the prediction of the distribution of deformation and stresses in wood during a drying process and wood steaming. The development of computer technology and numerical methods have made the research much easier and enabled obtaining information of what is happening inside a loaded product.

5. REFERENCES

Demirdžić, I.; Martinović, D. (1993). Finite volume method for thermo-elasto-plastic stress analysis, *Comput. Methods Appl. Mech. Engrg.*, 109, 331-349.

Demirdžić, I.; Muzaferija, S. (1995). Numerical method for coupled fluid flow, heat transfer and stress analysis using unstructured moving meshes with cells of arbitrary topology, *Comput. Methods Appl. Mech. Engrg.*, 125, 235-255.

Demirdžić, I.; Horman, I. & Martinović, D. (2000). Finite volume analysis of stress and deformation in hygro-thermo-elastic orthotropic body, *Comput. Methods Appl. Mech. Engrg.*, 190, 1221-1232.

Horman, I.; Martinović, D. & Bijelonja, I. (2003). Numerical Analysis of Process of Wood Heat Treatment, 4th Int. Symposium RIM, Bihać, 443-450.

Martinović, D.; Horman, I. & Demirdžić, I. (2001). Numerical and Experimental Analysis of a Wood Drying Process, *Wood Science and Technology*, 35, 143-156.

Annals of DAAAM for 2008 & Proceedings of the 19th International DAAAM Symposium, ISSN 1726-9679
Editor B. Katalinic, Published by DAAAM International, Vienna, Austria 2008
Make Harmony Between Technology and Nature, and Your Mind will Fly Free as a Bird

CLASSIFICATION OF THE MODELS AND THE MATHEMATICAL PRODUCTION PROGRAMME

HRUBINA, K[amil]; WESSELY, E[mil]; MACUROVA, A[nna] & BALCAK, S[tanislav]

Abstract: *Problem of modelling, classification, simulation mathematical models as well as models of the production program is investigated in the paper. The paper deals with the possibility of mathematical model creation of the manufacturing program. In the process of creation we are going to consider means of production as well as technological equipment of the company.*

Key words: *mathematical model, simulation, production program*

1. INTRODUCTION

The development of science, technology and manufacture requires new control methods and means that in the current stage of knowledge are expressed as "Integrated system of manufacture and enterprise control".

From the viewpoint of the system science, it is essential to investigate systems properties through the behavior of their mathematical models by means of simulations and various modeling techniques.

Mathematical models are created especially for the purpose of their utilization in the control and information systems design whose present development is enabled by informatics methods of a high quality.

It is known that cybernetics deals with the problems of description of real processes behavior and control that show the evidence of considerable complexity. Special attention is given to some difficulties that appear with their formal mathematical description. Classical formal procedures that are based on the knowledge of natural laws and numerical mathematics as well as statistics utilization have a rather limited utilization when applied to the mathematical models in the control.

From the above it follows that the application of classical procedures often leads to the simplifications that are the reason of insufficient adequacy of mathematical description (Sarnovsky et al.,1992).

Mathematical model usually failures if there is a lack in information about the controlled process. In such a case the procedures of the new scientific branch could be applied, i.e. those of the artificial intelligence whose methods are: solution of the problems with constraints, artificial neural networks and fuzzy systems (Zadeh– Polak,1969). By means of these new procedures, the problems of processes control can be solved, in particular non-linear processes. The utilization of these methods to design and implement algorithms is a part of intelligent control (Neuschl et al.,1992).

Machining centres for the selected parts machining with adjacent times optimisation will be taken into consideration.

2. MODELING AND CLASSIFICATION OF MODELS

The models present generally a wide and miscellaneous area of means and ideas that differ from the point of view of mission and according to method of realization as well. The model is a formal or material formation that pictures some points of view of original (example system, modeled system) and that is created for it that is used at study and at intentional influence of the modeled system.

Then the modeling is a process where other system called model is attached to the searched system i.e. the object, the original according to the certain criterions.

The model is an isomorphic or homomorphism picturing objects into the choice aim quantity that transfers a competent picturing and properties of the object (Obona, 1990).

Classification of models

The problem of modeling is very wide and therefore we limit on some points of view of classification:

a) Distribution of models according to degree of abstraction

a1) Nature models (degree of abstraction zero)
a2) Physical models (the first degree of abstraction)
a3) Physical analog (the second degree of abstraction)
a4) Structure model (the third degree of abstraction)
a5) Functional models (higher degree of abstraction)

b) Distribution of models by their use in process of control

b1) Functional models
b2) Directive models
b3) Described models
b4) Structural models

c) Distribution of models by time relations

c1) Prognostic models (extra polar models)
c2) Topical models
c3) Retrospective models

Mathematical models of complicated systems

We can divide mathematical models of complicated systems, complexes with regard to the introduced classification on two classes:

a) deterministic models
b) stochastic models.

Physical models of complicated systems

The most known physical models of complexes dynamics are:

- modes passed
- diffusion models, (Hrubina, 1999; Jadlovska, 2003).

3. MATHEMATICAL MODEL OF MANUFACTURING PROGRAM

3.1 Manufacturing Program selection

Automated manufacturing systems enable to solve optimal conditions of machining and adjacent time minimisation. The task can be formulated as follows: a part with a group of slots is being manufactured. Distances between any neighbouring slots are given. The shortest closed path which will pass through each slot only once is to be selected. Mathematical model of the task resembles different cases occurring in technological processes. Intensification of cutting conditions without corresponding adjacent movements of the tool does not lead to more significant increase of machining productivity.

Process of searching for the sequence how to „detour" the slots system is characterized by its variability. It is necessary to take into consideration design peculiarities of the machine tool (range of table movements, way of changing the tool, etc) as well as design peculiarities of the part (position and dimensions of the slots). Group of slots can have the same diameter and meet the same requirements, thus they can be considered as a group. The sequence can be understood as follows: the solution is being searched gradually and the selection depends on the momentary state of the system, i.e. the previous systems. The characteristic feature of the solved task is that the slot is being determined at each step that follows after the previous one has been machined. Thus, the task is to determine the length of the optimal path from the point i to the point I, passing through, at the moment, each point $j_1, j_2, ... j_k$ once in random order and is not passing through any other slots axise (Etner, 1991; Douglas et al, 2005).

3. 2 Mathematical Model Design

In the process of mathematical model creation, the Euclid norm of the distance of the two points has been used. In order to find optimal variant using dynamic programming, algorithms of discrete programming have been used. With this method, optimisation process is considered to be a multistep process (Hrubina et al., 2002). Characteristic feature of this method is that the next solution always follows from the previous state (passed path to the given point):

$$T(i,j) = c_{i,j} + c_{j,1} \quad (1)$$

From the above it follows, that for the each step (n steps) we are going to select the path with which the sum of step distances plus function of optimal previous values will be minimal.

Let us use denotation $f_n(v_i)$ for the distance between points that corresponds to the minimisation strategy for the path from i if n-steps remain to the final slot.

Then

$$f_n(v_i) = \min\left(c_{ij} + f_{n-1}(p_i)\right) \quad (2)$$

Cyclic productivity of machining in machining centres will be determined from the relationship:

$$Q = \frac{1}{\tau_c} = \frac{1}{\sum \tau_z + \sum \tau_{x.x}} \quad (3)$$

Where $\tau_{z\,z}$ is the basic time needed to machine all the surfaces, $\tau_{x,x}$ is the time of passive running needed to position and swing out the table, change and supply the tool.

Time of the machining cycle for all the surfaces is expressed by the relationship

$$\tau_c = \sum \tau_z + \sum \tau_{x,x} \quad (4)$$

where τ_z is the basic time needed to machine one surface, $\tau_{x,x}$ is the time of the passive running needed to machine the elementary element.

Body parts are machined from the four sides. One group of the machined slots can have the same diameter and accuracy requirements. Several groups of the same slots can be placed on each surface of the work piece. These slots can be machined using three methods.

Analysis of the three mentioned variants enables us to conclude that selection of the sequence of machining can be realised by the two basic methods: parallel and sequential. With the sequential method, each slot is being machined by the tools. Then, after position is being changed, the next slot is going to be machined. With the parallel method, each tool passes through all the slots diameter of which corresponds to the tool. After that, the tool is being changed and cycle repeats (Macurova, 2007; Macura, 2003; Tirpakova, 2000).

4. CONCLUSION

The main contribution of the paper is to generally investigate the problem of modelling and models classification.

Creation of mathematical models of the manufacturing program is considered to be the introductive stage of manufacturing program optimisation.

To realise parametric and structural optimisation of manufacturing program, the database of cutting conditions, cutting tools, technical characteristics of manufacturing device is needed. Database also contains typical technological procedures of elementary surfaces machining as well as their files. Part of the manufacturing program optimisation that is being evaluated is realised with the aim to minimise the selected machining time of the part surface (4). Adequate mathematical model can be realised on PC using simplex algorithm.

5. REFERENCES

Douglas, L. et al (2005). Logistics, Brno, CP Books, 589 p.

Etner, F. (1991). Microéconomie. Prosses Universitaires de France

Hrubina K. et al. (1999). Application of Informatics in the System of Manufacturing Control and Training. *Scientific Project No. 1/4331/97*, Kosice: Informatech Ltd., 297 pages. ISBN 80-88941-09-1

Hrubina K. – Jadlovska A. (2002). Optimal Control and Approximation of Variational Inequalities Cybernetics. *The International Journal of Systems and Cybernetics*. MCB University Press of England. Vol. 31, No 9/10, pp. 1401 –1408

Hrubina K.- Jadlovska A.-Hrehova S. (2002). *Methods and Tasks of the Operation Analysis Solution by Computer.* Kosice: Informatech, Ltd., 324 p., ISBN 80–88941–19–9

Jadlovska A. – Hrubina K. (2002). Algorithm of Estimation Parameters of Process by Laplace Transformation. In: *Proceedings the 5th International Scientific – Technical Conference PROCESS CONTROL 2002, RIP 2002*, Kouty on Desnou, University of Pardubice, p. R103b – 1–4, CD ROM. ISBN 80–7194–452–1

Macurova A. (2007). Aproximation of Functions in Experimental Methods. *Forum Statistikum Slovakum.* Number 2/2007. Year III. pp. 164-167, ISSN 1336-7420

Macura, D. (2003). Ordinary Differential Equations. 1. issue. Presov: FHPV PU, 79 p., ISBN 80-8068-175-9

Macura, D. (2005). *Function of Multivariables* 1. issue. Presov: FHPV PU, 56 pages, ISBN 80 -8068-321-2

Neuschl S. et al. (1992). *Modeling and simulating.* Bratislava: Alfa.

Obona, J. (1990). Systems and System Analysis in Practice Bratislava.

Sarnovsky J. et al. (1992). *Complex Systems Control*, Bratislava: ALFA, 382 pages.

Tirpakova, A. (2000). Using Computers in the Process of Applied Statistics Teaching. *PRASTAN 2000*, SF STU Bratislava, 257 pages.

Zadeh L. A. – Polak E. (1969). *System Theory*, New York: McGraw – Hill.

Annals of DAAAM for 2008 & Proceedings of the 19th International DAAAM Symposium, ISSN 1726-9679
Editor B. Katalinic, Published by DAAAM International, Vienna, Austria 2008
Make Harmony Between Technology and Nature, and Your Mind will Fly Free as a Bird

COAXIAL IMPACT OF ELASTIC BODIES EXPERIMENTAL ANALYSIS OF THE REFLECTED WAVES

HULE, V[oichita]; TARCA, I[oan] & BLAGA, F[lorin]

Abstract: *The paper presents an experimental analysis of phenomena that appear during coaxial impact of two elastic bodies, based on the longitudinal wave propagation theory. The acquired experimental data refers to the verification of the propagation and reflection of mechanical waves in metallic rods theory during their impact at low speed. Data conditioning methods in time, frequency and in scale-frequency range were used to relieve the analyzed phenomena.*

Key words: *impact, reflected wave, wavelet analysis.*

1. INTRODUCTION

The paper presents an experimental study of the coaxial impact of two cylindrical rods having the longitudinal dimensions much greater than those transversals based on the longitudinal wave propagation theory.

An experimental setup for finite length rods was conceived in order to reveal the phenomena that appear during the impact period which are the longitudinal wave's propagation inside rods, analyzing the incident wave and also the reflected wave movements through rods and also the contact time analysis between rods (Brindeu et al., 2003).

For the phenomena analysis which constituted the subject of the experimental studies, software data acquisition in Visual C++ were made. MATLAB software was used for data processing because it has a series of predefined functions useful for the data analysis (FFT and wavelet) (Davis & Nosratinia 1998, Strang 1993, Tsai et al. 2005, Josso et al. 2001).

2. EXPERIMENTAL SETUP DESCRIPTION

A variant with suspended rods was used for the experimental setup. Elastic wires through metallic rings fixed on holders were used to suspend the rods. This solution offers a series of advantages comparing to that one in which the rods are guided. The construction is simple and is suitable for the testing of a wide variety of rods' lengths and sections.

The experimental setup is composed on two suspended cylindrical rods of C45 (OLC45 in Romanian STAS), having 40 mm, respective 35 mm in diameter and 2 m, respectively 6 m in length. Strain gauges were mounted on the rods equally disposed at L = 1 m. They transmit the signals through a Wheatstone bridge to a data acquisition board. The study was made in the low velocity impact zone, with v_0= 2m/s. At the impact moment compress strain occurs, together with mechanic waves which propagate inside the rod material. Compression strains cause the strain gauges to deform proportionally with their magnitude, thus generating electric signals toward Wheatstone bridge.

3. EXPERIMENTAL ANALYSIS OF THE WAVES' PROPAGATION GENERATED AT THE RODS' IMPACT

For the test rig described earlier some experiments were conducted connecting the strain gauges in full bridge. The voltage generated during rods' impact on each strain gauge was measured. Time correlation of the measurements has been realized with a trigger.

Because of the great length of the stricken rod, is difficult to relieve the reflected wave from it's free end. The main cause of this difficulty is that the wave is significantly absorbed along the rod. Beside its attenuation, a series of other effects occurs (side reflections, bending vibrations of the rods) which significantly diminishes the possibility of detecting the reflected wave.

Two sets of 5 measurements were conducted to detect and analyze the reflected wave. In the first set the signal acquired from the first strain gauge (T1) was measured for an impact realized in the conditions mentioned above. The second measurement set was conducted in the same conditions excepting the fact that at the free end of the stricken rod a vaseline film was applied. This layer of viscous material partly absorbs the reflected wave energy thus the reflected signal being diminished. The average of the 5 unabsorbed signals (without vaseline layer – ms_{fa}) and also the average of the 5 absorbed signals (with vaseline layer – ms_a) were computed. Both average diagrams are presented in Fig. 1.

The difference between these signals should be noticed at the moment of time t = 2,313 x 10^{-3} sec., on which (based on computed values and also on conducted measurements) both the separation of the rods and the reflected wave return in the section which corresponds to T1 strain gauge position occurs. Because of the reduced damping caused by the vaseline layer, this difference couldn't be easily noticed on the diagram in Fig. 1. Because of this a frequency analyze program created in MATLAB which computes and displays the frequency spectrum of the signal was realized. The software offers the possibility of signal filtration eliminating the insignificant frequency ranges (from the analyzed signal point of view) after which an inverse transformation from the frequency range to the time range is applied. This way noise can be eliminated (for example the 50 Hz frequency component induced by the a.c. net. Fig. 2 shows the frequency spectrum of the ms_{fa} (black) and ms_a (grey) signals. Three domains appear on the diagram in Fig. 2 in which significant differences between the two signals type occur. In the A domain a phase delay can be observed while in B and C zones a significant amplitude difference can be noticed.

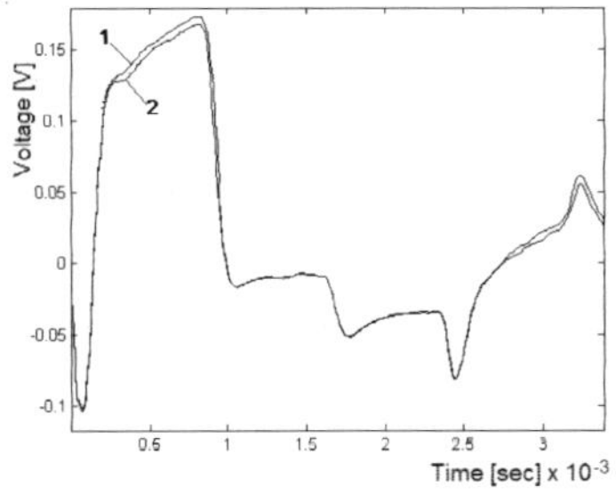

Fig.1. Unabsorbed signals average (1) and absorbed signals average (2) diagram

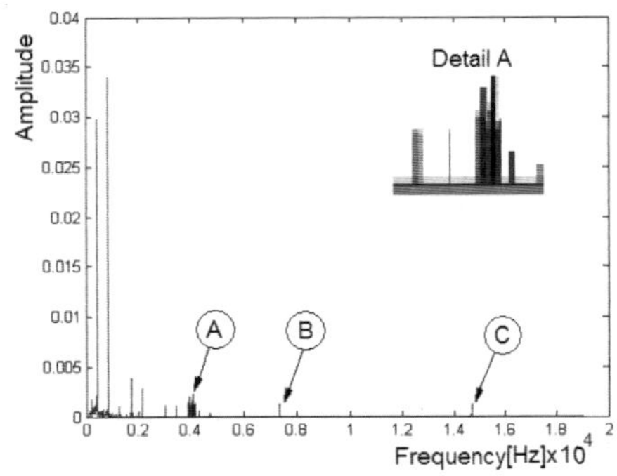

Fig.2. Spectral diagram of the damped signals (grey) and undamped signals (black) for the T1 strain gauge

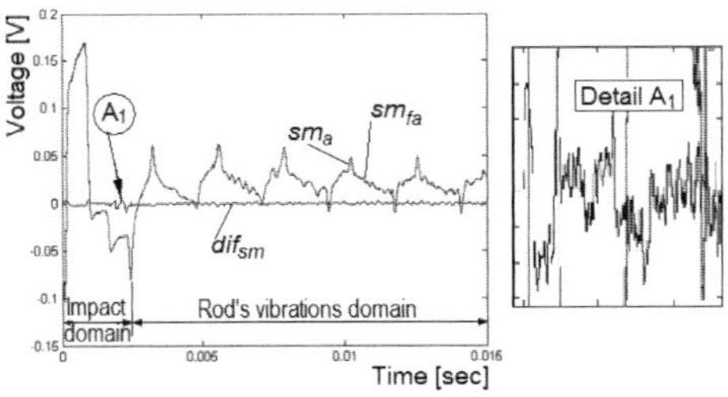

Fig.3. Signal diagram after filtering in frequency range

Keeping these frequencies domains and eliminating the others in which differences were not been noticed, and then applying an Inverse Fourier transforming the diagram presented in Fig. 3 is achieved. (The damped signal is sm_a, the undamped signal is sm_{fa}, the difference between them after an Inverse Fourier transformation is dif_{sm}). In this diagram a significant variation of the difference between the signals (point A1) can be noticed, but such variations can be noticed in other time ranges in which the apparition of the reflected wave is not estimated, such as the range (0 … $1x10^{-3}$) seconds.

The limitation of the Fourier transformation can be noticed in this case; this means that it is useful for signal filtration but presents the major disadvantage of loosing information regarding time localization of the phenomena. To localize the reflected wave in time the dif_{sm} signal was analyzed using the MATLAB software. The diagram shown in Fig. 4 was created with this software. The program uses the „rbio 3.1" wavelet (Reverse Biorthogonal wavelets version 3.1). From the time-level diagram it can be noticed that in the range of interest (around 2,4 x 10^{-3} seconds) a prominent amplitude of the signal exists on the D1 and D2 detail levels. The reflected signal synthesis was realized through the selection of the "rbio3.1" wavelet transformation coefficients corresponding to D1 and D2 details.

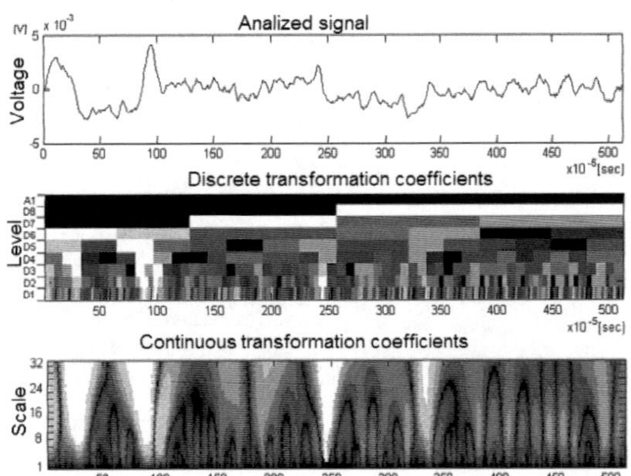

Fig.4. dif_{sm} signal diagram

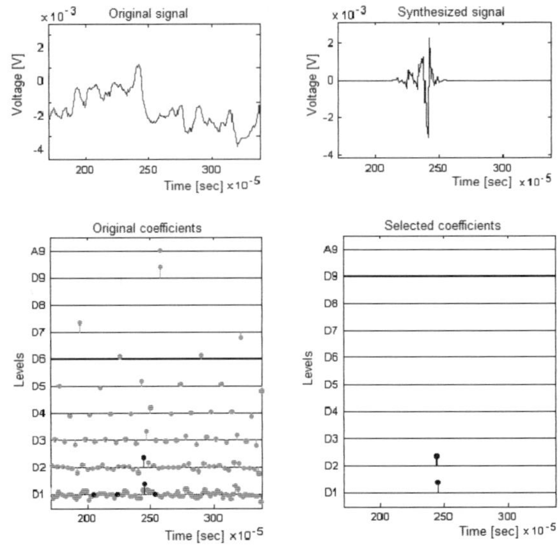

Fig.5. Reflected wave signal syntheses

Fig. 5 shows the steps of the transformation, coefficient selection and the resulted time range diagram. Through the synthesis of the reflected wave signal achieved from the dif_{sm} signal using wavelet transformation, its share in the base signal can be estimated and also the shape of the signal together with the frequency range in which the signal appears.

4. CONCLUSIONS

The parameters used for data acquisition were correctly been defined, the analyzed phenomena being properly relieved.

Because the reflected wave at the free end of the stricken rod measured on the T1 strain gauge is difficult to be detected, a viscous layer was applied on the free end thus observing the differences between the damped and normal reflected signals. This technique didn't offer remarkable results either in time analysis or in frequency analysis. This is why the wavelet analysis (in time-scale range) was used thus resulting the signal filtration and the reconstitution of the component due to the reflected wave.

For the phenomena analysis which constituted the subject of the experimental studies, software that can be used for further researches for data acquisition and signal conditioning was made.

5. REFERENCES

Brindeu, L., Hule, V. & Petcovici, O. (2003). Dynamic model of impact, considering the propagation of the stress waves in the deformable body, *Scientific Bulletin of Polytechnical University Timisoara*, Mechanics Series, Vol. 48, No. 1, 2003, ISSN 1224-6077

Davis, G. & Nosratinia, A. (1998). Wavelet-Based Image Coding: An Overview. *Applied and Computational Control, Signals, and Circuits*, Vol. 1, No. 1, Spring 1998.

Josso, B., Burton, D. R. & Lalor, M. J. (2001). Wavelet Strategy for Surface Roughness Analysis and Characterisation. *Computer Methods in Applied Mechanics & Engineering,* Liverpool John Moores University, Vol. 191, No. 8-10, pp. 829-842

Strang, G. (1993). Wavelet transforms versus Fourier transforms, *Bulletin of the American Mathematical Society,* Vol. 28, No. 2, April 1993, pp. 288-305

Tsai, D. M., Wu, S. K., Chen, M. C. (2005). *Optimal Gabor filter design for texture segmentation using stochastic optimization, Image and Vision Computing*, Elsevier Science B.V.

Annals of DAAAM for 2008 & Proceedings of the 19th International DAAAM Symposium, ISSN 1726-9679
Editor B. Katalinic, Published by DAAAM International, Vienna, Austria 2008
Make Harmony Between Technology and Nature, and Your Mind will Fly Free as a Bird

EXPERIMENTAL RESEARCH REGARDING THE JOINING RADIUS OBTAINED IN THE CASE OF MINIMAL ENERGY AT THE ELECTROHYDRAULIC PROCESS

IACOB, D[umitru]; MIRONEASA, C[ostel] & MIRONEASA, S[ilvia]

Abstract: *The paper presents the results obtained during electrohydraulic experiments on sheets of aluminum and on OLC 10 steel. In this paper we wanted to verify, through experiments, and find analytical expressions to determine the radius between polygon sides by energy, sheet thickens and the angle between the polygonal sides. We can notice that the maximum of joining radius obtained experimentally, moves towers smaller values of the angle between sides.*
Key words*: electrohydraulic, radius, polygon, concentrator*

1. INTRODUCTION

In the case of punching and classic perforation, with punching die on which both tools are rigid, and in the case of the use of the electrohydraulic punching method, the correct determination of the joining radius between two contiguous sides of a polygonal piece, is an important issue in order to obtain the right quality of the processed pieces (Iacob, 1993), (Ciacin, 1987) and (Marcenko, 1979). In the paper (Iacob, 1993), a theoretical study has been carried out, regarding the joining radius which can be obtained in the electrohydraulic perforation and cutting of the pieces, taking into consideration that this radius should be put in the triangle formed by the two sides of the polygon and the segment which joints them.

The electrohydarulic punching research has also been undertaken in Romania (Andrei, 1992) for the sheet deformation and minimal radii piercing (Rusu, 1994).

2. THE EXPERIMENTAL PROGRAM

In order to verify experimentally the proposed relations, in (Iacob, 1993), a discharge chamber has been designed and given shape. The chamber can be used with and without energy concentrator. The concentrator was designed after a modular system; it was made from the combination of four rings with the same conic shape or with different conic shapes, thus obtaining the necessary height of the concentrator. Through the combination of the rings, concentrators of different heights can be obtained, having different α angles, and different areas of the exit sections of the concentrators. Also, concentrators can be obtained, with variable angles on the axial section of the concentrator, for example α-120° in the entering section, α-90° in the median axial section and α-60° in the exit section of the concentrator.

The experimental device was used on the electrohydraulic installation IDEH - 1260, which has the following characteristics: maximum working voltage U – 6000V; inductivity of the discharge circuit – 2,7.

The experimental determinations were made on aluminum sheets with the thickness of 0,5 mm, 1,0 mm and 1,5 mm and on OLC 10 steel sheets with the thicknesses of 0,3 mm, 0,5 mm and 0,7 mm. Regular polygonal holes were made, having the shape of triangles, square, pentagon, hexagon and circle, the perimeter was constantly 60 mm.

In order to find the analytical expressions of the joining radius, a large number of experimental determinations have been made, using the mathematical planning technique of the experiments, finally reaching the following analytical expressions:

- for the aluminum pieces perforated with the concentrator, equation (1):

$$r = 8896{,}15 \cdot E^{-1{,}36} \cdot g^{0{,}804} \cdot \alpha^{-1{,}128}\,[m] \qquad (1)$$

- for the aluminum pieces perforated without the concentrator, equation (2):

$$r = 13{,}289\times10^{6} \cdot E^{-1{,}44} \cdot g^{0{,}891} \cdot \alpha^{-1{,}592}\,[m] \qquad (2)$$

- for the steel pieces perforated with the concentrator, equation (3):

$$r = 681{,}845 \cdot E^{-0{,}909} \cdot g^{0{,}353} \cdot \alpha^{-0{,}96}\,[m] \qquad (3)$$

- for the steel pieces without the concentrator, equation (4):

$$r = 3{,}493\times10^{3} \cdot E^{-0{,}981} \cdot g^{0{,}34} \cdot \alpha^{-1{,}324}\,[m] \qquad (4)$$

where: *r* is the joining radius between the two contiguous sides of the perforated hole, in mm; *E* is the energy accumulated in the condenser battery, in Joules; *g* is the thickness of the sheet, in mm; α is the angle between the polygonal sides, in degrees.

In the case of perforation with the concentrator, the blank is at a distance of H_1 = 410 mm from the explosive wire, and in the case of perforation without the concentrator, the blank is at a distance of H_1+H_2 =140 mm. The angle of the concentrator cone was constant and equal to 60°.

3. EXPERIMENTAL RESULTS

In order to shape the variation graphs of the joining radius of the pieces, the minimal energetic values of discharge, E_{min}, have been determined experimentally, for the perforated aluminum and steel pieces. And for smaller values of the angle between the sides 0° – 60° there have been used minimal energy values corrected with correction coefficients proposed by the author, K and K_A, which are given by the following equation (5),

$$K = \frac{1}{2}\cdot\sin\frac{\alpha}{2}; \qquad K_A = \frac{1}{\sin\alpha} \qquad (5)$$

where: K is the correction coefficient which appears because of the efforts in the material, generated by the transversal plastic waves; K_A is the correction coefficient of the minimal energy value owed to the reduction of the area of the triangle, having an angle $\alpha < 60°$ and the perimeter constant.

The values of the K_A coefficient are available for the general case: when the area of the triangle is reported to the area of a rectangular triangle, when the area does not impose the area of the triangle, when the triangle is part of the polygon. If a certain perimeter is imposed, then the area of the chosen triangle must be linked to the area of the equilateral triangle, which has the maximum area at the same perimeter.

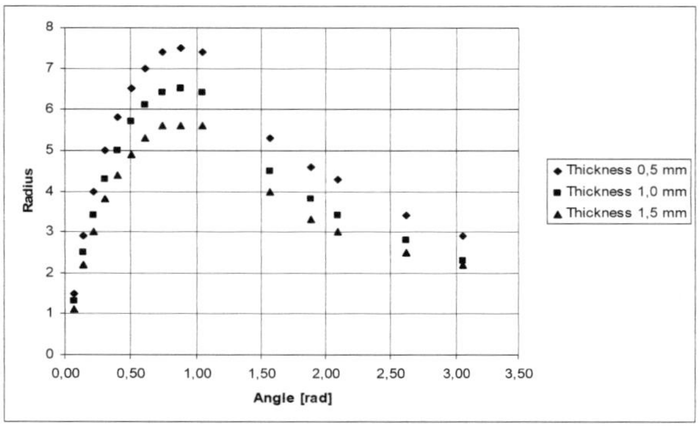

Fig. 1. The value of radius for aluminum - perforation with concentrator

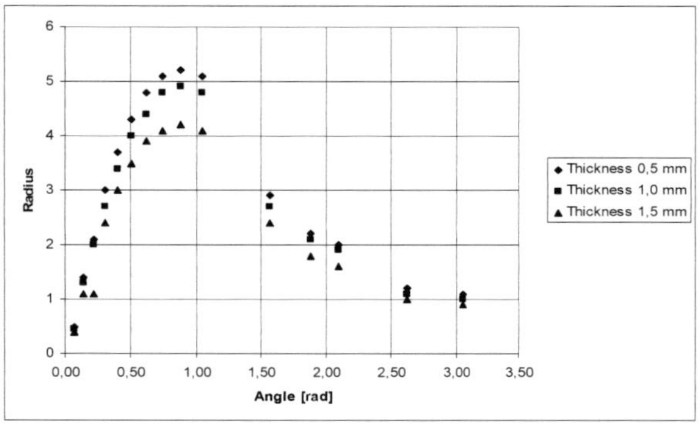

Fig. 2. The value of radius for aluminum - perforation without concentrator

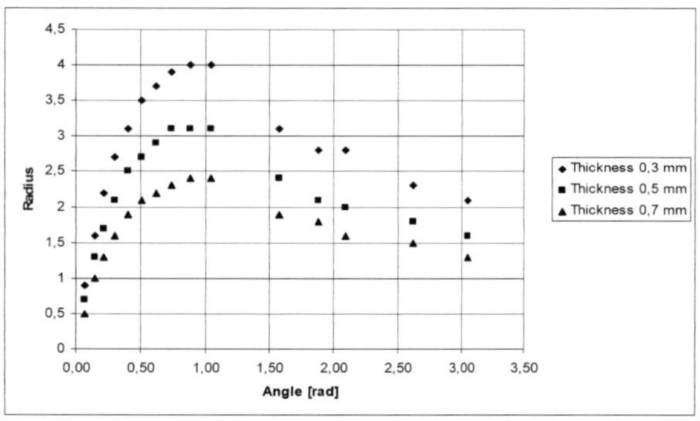

Fig. 3. The value of radius for OLC 10 steel - perforation with concentrator

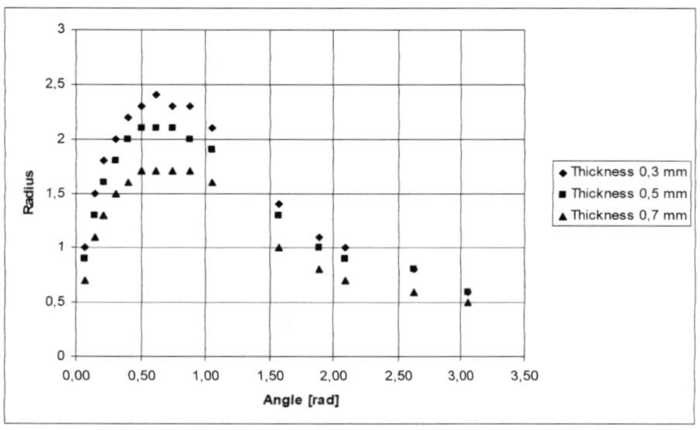

Fig. 4. The value of radius for OLC 10 steel - perforation without concentrator

The complex curves of the joining radius between the sides, obtained for the minimal value of the energy are given in Fig. 1 and Fig. 2 for perforated aluminum pieces and in Fig. 3 and Fig. 4 for perforated steel pieces OLC 10.

Comparing the graph of Fig. 1 and Fig. 2 those of Fig. 3 and Fig. 4 we can notice a very good concordance of variation laws of joining radius obtained experimentally with the joining radius r_T, theoretically determined for angles between 0° and 150°. This confirms the fact that the joining radius r_T theoretically determined for the minimum energy required by punching of bore holes, being a little higher than those obtained experimentally. From the graphs presented we can notice that at values of over 150° of angles between sides, the decrease of joining radius is insignificant.

To obtain lower values of the joining radius for minimum energy in the case of some bigger thickness of material is due to, in this case, the increase of waste rigidity, thus causing a significant increase of minimum energy required by punching. Also, in the case of higher thickness, due to higher rigidity in cross-plane, the cutting conditions are improved, the material separation being made almost simultaneously on the whole outline of piece, and not at first on the middle of sides and later gradually towards peaks, as it happens in the case of small material thickness.

If in the case of radius r_T, the maximum is between 60° and 90° ($76^o23'$), in the case of experimental values, the maximum is around 50°, and in the case of steel pieces punched without concentrator it even reaches smaller values of the angle, of 35°-40°. But even so, the values determined theoretically r_T, showing that the values calculated r_T meet the conditions of electrohydraulic punching. This fact is more valid as it is known that the energy used for punching is higher than the value E_{min}, observing the relation: $E_{min} \leq E \leq 2E_{min}$. So the radii obtained experimentally will be in thus case smaller.

4. CONCLUSIONS

The joining radius between two consecutive sides of a regular polygon, are strongly influenced by the angle between the sides, thus by the rigidity of the blank material.

The minimal energy for perforation and punching increases significantly as the angle between the sides decreases.

For the values of the angle between the sides, under $\alpha = 60^o$, in order to obtain the minimal radius, the discharge energy is increased significantly, because the rigidity of the material increases significantly, and the cutting conditions of the material are not favorable.

The joining radii are greater in the case of perforation with an energy concentrator; this is due to the fact that the material separates faster than the case of perforation without an energy concentrator, which determines a more fragile rupture of the material, without putting it under the deformation process.

It must be said that experimental research of this kind cannot be found in the existent literature.

5. REFERENCES

Andrei, N., (1992), *Research Regarding the Electrohydraulic Deformation Processing of the Piece Form Sheet Metal*. Ph. D. Thesis, I.P. București

Ciacin, V.N., (1987), *Electrohydraulic Shocks at Material Punching Perforation in Machine Fabrication*, Nauka I Tehnika, Minsk

Iacob, D., (1993), *Research Regarding the Fillet Corner Radius at Electrohydraulic Cutting and Punching Perforation*, The Annals of University "Ștefan cel Mare", XII, 2004, No. 6 – 12, Mechanic section, Suceava

Marcenko I.I., (1979), *Hydraulic Shocks at Sheet Punching Perforation. Impulsion Method at Material Machining*, Nauka I Tehnika, Minsk

Rusu, B., Braha, V. & Iacob, D., (1994), *Research of the Minimal Radii in the Electrohydarulic Piercing*, 5th International DAAAM Symposium, pp. 55-56, University of Maribor, 27 – 29 Oct., Slovenia

Annals of DAAAM for 2008 & Proceedings of the 19th International DAAAM Symposium, ISSN 1726-9679
Editor B. Katalinic, Published by DAAAM International, Vienna, Austria 2008
Make Harmony Between Technology and Nature, and Your Mind will Fly Free as a Bird
*Young Researcher and Scientist Paper / * Supervisor, Mentor*

SENSITIVITY ANALYSIS FOR PARALLEL ROBOTS

IBRAIM, C[ristina]; CHELARU, E[lena] S[ilvia] & PREDINCEA, N[icolae]

Abstract. *The objective of the program is to give emphasis about the method for accuracy analysis which is numerical stable and computationally efficient but its rate of convergence can be considered slow. It is also possible that this method can be further extended for any parallel robot structure. The Monte Carlo Simulation method is a combined numerical and statistical method for problem solving. Monte Carlo Simulation is a good choice and the choice of the right distributions depends on the given task. With the Monte Carlo simulation it is also possible to evaluate the tolerance intervals.*
Keywords: *Monte Carlo simulation, joint, variable, program*

1. INTRODUCTION

Numerical methods that are known as Monte Carlo methods can be loosely described as statistical simulation methods, where statistical simulation is defined to be any method that utilizes sequences of random numbers to perform the simulation.

Monte Carlo is now used routinely in many diverse fields, from the simulation of complex physical phenomena such as radiation transport in the earth's atmosphere and the simulation of the esoteric sub nuclear processes in high energy physics experiments.

Statistical simulation methods may be contrasted to conventional numerical discretization methods (Germer et al., 2003), which typically are applied to ordinary or partial differential equations that describe some underlying physical or mathematical system. In many applications of Monte Carlo, the physical process is simulated directly, and there is no need to even write down the differential equations that describe the behaviour of the system. The essential characteristic of Monte Carlo is the use of random sampling techniques to arrive at a solution of the physical problem.

In contrast, a conventional numerical solution approach would start with the mathematical model of the physical system, discretizing the differential equations and then solving a set of algebraic equations for the unknown state of the system.

1.1. Gauss Distribution

The normal (or Gaussian) distribution is one which appears in an incredible variety of statistical applications.

The Monte Carlo simulation consists of many repetitions of the random experiment with a changing random number input list.

The basic idea is to perform the direct kinematical algorithm for a huge numbers of parameters variations. The choice of the right distribution for parameters variation depends on the given task; in this case the variation will be for each parameter a Gauss Distributions.

2. PROGRAM DESCRIPTION

2.1 Data representation

Information about all parameters is stored in KGraph matrix as:

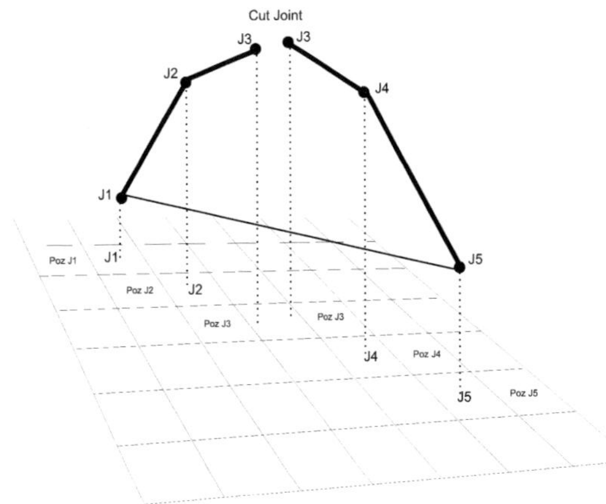

Fig. 1. Data representation

First we will store the reference frame in data structure as following:

KGraph(1,2).u KGraph(7,6).u
KGraph(1,2).v > Reference frame < KGraph(7,6).v
KGraph(1,2).w KGraph(7,6).w

Information about joint i

KGraph(i,i).type - type of Joint >**R**evolute **T**raslation **U**niversal **S**pherical

KGraph(i,i).driving - kind of Joint > 0–driven Joint; 1– driving Joint

KGraph(i,i).teta - initial value for angle theta

KGraph(i,i).a - KGraph(i,i).b - KGraph(i,i).c

KGraph(i,i).alfa - KGraph(i,i).beta - KGraph(i,i).gama

For the first direction

Information about position of Joint i: KGraph(i,i-1).poz this is a vector with 3 elements [x y z]

Information about Inlet orientation of joint i:
KGraph(i,i+1).u KGraph(i,i+1).v KGraph(i,i+1).w

Information about outlet orientation of joint i:
KGraph(i,i-1).x; KGraph(i,i-1).y; KGraph(i,i-1).z

For the second direction

Information about position of Joint i: KGraph(i,i+1) poz this is a vector with 3 elements [x y z]

Information about inlet orientation of joint i:
KGraph(i,i-1).u; KGraph(i,i-1).v; KGraph(i,i-1).w

Information about outlet orientation of joint i:
KGraph(i,i+1).x; KGraph(i,i+1).y; KGraph(i,i+1).z

Information about the Cut Joint

There are two different positions and two different orientations for the Cut Joint NRCJ – Number of cut joint:

- First position - KGraph(NRCJ+1,NRCJ).poz
- Second position - KGraph(NRCJ+1,NRCJ+1).poz

With this method, the local coordinate systems are automatically arranged in the correct order, and in this way, the efficient recursive coordinate transformations can be performed.

2.2 Functions description

The following functions are used in the program in the purpose of reducing the errors (Szatmari, 1999).

Read_Input_Data.m – this function is used to input the parameters

We need to enter the next variables:

- Number of joints
- Number of Cut Joint

For each joint we need to enter the next parameters:

- Type of Joint (**R** for rotation and **T** for translation)
- Kind of Joint (**1** for driving Joint and **0** for driven Joint)

If we have to describe a Revolute Joint, we must enter the next values: parameter for joint transformation Theta, parameters for link transformation: A (Alfa), B (Betha), C (Gama)

If we have to describe a prismatic/translation Joint, we must enter only the distance.

Build_Matrix_Structure.m – this function create the complete structure of information based on the input parameters and transformation rules.

The next transformations will be computed:

First Direction	*Second Direction*
1. Link Transformation between Reference Frame and First Joint	1. Link Transformation between Reference Frame and Last Joint
2. Joint Transformation and Link Transformation until to the Cut Joint	2. Joint transformation and Link Transformation until to the Cut Joint

Calc_errors.m – this function compute the error between the two ccordinate systems of the cut Joint. The error is the distance between the two positions: Pxyz=KGraph(i,i-1).Poz;

Puvw=KGraph(i,i+1).Poz;

DP=sqrt((Pxyz(1)-Puvw(1))^2+(Pxyz(2)-Puvw(2))^2).

Minimisation.m – this function adjust the joint variables one by one until the objective function can not be further reduced.

Gauss_Distribution.m – this function generate random values for each parameter. We have to define here 3 variables: maximum variation for distance, maximum variation for angles - number of variation (the number of points).

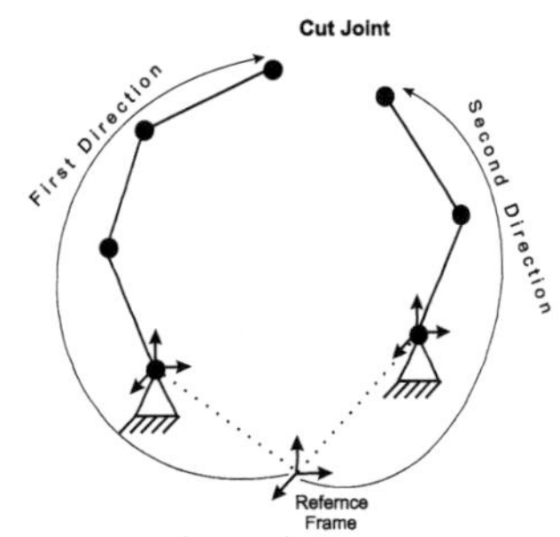

Fig. 2. Coordinate Transformation

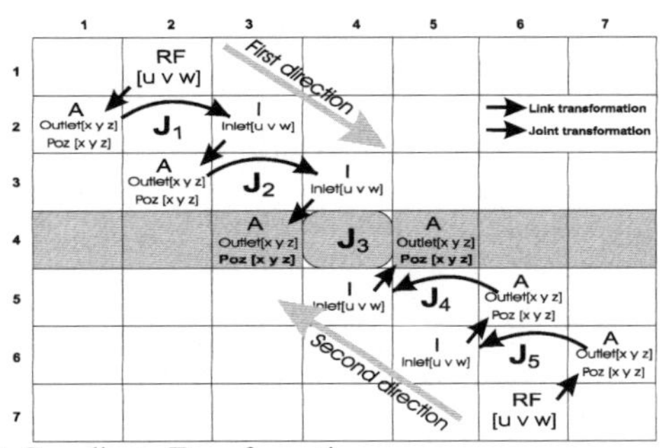

Fig. 3. Coordinate Transformation

	1	2	3	4	5	6	7	8	9	10
A1	-150.0433	-150.1666	-149.9875	-149.9712	-150.1146	-149.8809	-149.8811	-150.0038	-149.9673	-149.9825
	8	2	10	7	4	3	6	9	5	1
B1	-0.0187	0.0726	-0.0588	0.2183	-0.0136	0.0114	0.1067	0.0059	-0.0096	-0..832
	5	6	9	1	4	2	10	8	3	7
C1	0.0294	-0.1336	0.0714	0.1624	0.0692	0.0858	0.1254	-0.1594	-0.1441	0.0571
	4	1	5	7	10	6	9	2	8	3
Alfa1	-0.0400	0.0690	0.0816	0.0712	0.1290	0.0669	0.1191	-0.1202	-0.0020	-0.0157
	1	9	6	3	5	8	10	2	7	4
Beta1	-0.1604	0.0257	-0.1056	0.1415	-0.0805	0.0529	0.0219	-0.0922	-0.2171	-0.0059
	2	10	9	4	8	6	3	7	5	1
Gama1	-0.1011	0.0614	0.0508	0.1692	0.0591	-0.0644	0.0380	-0.1009	-0.0020	-0.0048
	5	1	3	7	4	10	2	6	9	8

Fig. 4. Coordinate Transformation

After that we have to create the order for combinations, Fig. 4 (method to create Monte Carlo Simulation) shows how this algorithm works.

Monte_Carlo_Simulation.m – based on the Gauss Distribution this function store the parameters in data structure. We have after Gauss distribution the vector *A1* with normal distributed values and vector *permutA1* with a random permutation.

For example:

```
for i:1:1000
KGraph (2,2).a=A1(permutA1(i));
....
end
```

CCD_for_Mesh_Analysis.m – this is similar with function CCD_complete.m this time we need to store in matrix information about each point from workspace.

Inverse_Kinematik.m – compute the inverse Kinematical solutions (Li-Chun & Cheng Chen 1993) q1 and q2 starting with the value P(x,y) from mesh surface.

MeshAnalysis.m - this function is described in the chapter 'Mesh analysis over the entire workspace'

3. CONCLUSIONS

With this program the users have a powerful tool to reduce the errors and to identify which parameter have a big influence on the absolute accuracy and witch can be neglected in calibration process (Conrad, 2000). To verify this method presented in this paper, it is applied to 'FÜNFGELENK' robot. The simulation results have proved that for finding a realistic prediction of absolute accuracy.

This article is intended to determine the influence of geometric parameter variation on the absolute accuracy of parallel robots.

4. REFERENCES

Conrad, K. L.; Shiakolas, P.S.& Yih, T. C (2000). "*Robotic calibration issues: Accuracy, Repeatability and Calibration"*, Mediterranean Conference on Control & Automation, Patras, Greece

Costa, M. & Smaby, N. (1997). "*Calibration and precision manufacturing"*, ME 319 Robotics and Vision Lab

Germer, C.; Hansen, U.; Franke, H.-J. & Buttgenbach, S. (2003). *"Development of a 3D-CAD add-in for tolerance analysis and synthesis in micro systems"* Symposion on Design, Test, Integration and Packing of MEMS/MOMS, ISBN: 0-7803-7066-X

Li-Chun, W. & Cheng Chen, C. (1993). "*On the Numerical Kinematic Analysis of general Parallel Robotic Manipulators* Robotics and Automation, IEEE Transactions on, ISSN: 1042-296X

Szatmari, S. (1999). "*Geometrical errors of parallel robots"* Periodica Polytechnica Ser. Mech. Eng. Vol. 43, No. 2

Annals of DAAAM for 2008 & Proceedings of the 19th International DAAAM Symposium, ISSN 1726-9679
Editor B. Katalinic, Published by DAAAM International, Vienna, Austria 2008
Make Harmony Between Technology and Nature, and Your Mind will Fly Free as a Bird

APPEARANCES CONCERNING THE USING OF THE BILL OF EXCHANGE AND THE PROMISSORY NOTE IN ROMANIA

IGNAT, A[ndreea]; AVRAM, L[aura]; HORJA, M[onica] & VANCEA, S[maranda]

Abstract: *Throught define the bill of exchange and the promissory note are titles of commercial credits. It is important to mention the fact that thay appeared in practice as instruments of commercial credits.*

Moreover, due the law number 83/1994, about the bill of exchange and the promissory note, these paper of value are besides titles of commercial credits, modality to guarantee the pay of internal and external price too, the guarantee being insured by the quality of these titles of credits to be the titles rightful executories. So due to 53, 54 and 57 article of law nr. 58/1934 about the bill of exchange and the promissory note reviset and completed by the law nr. 83/1994 these titles have the value of rightful executory titles for capital and accessories.
Key words: *bill of exchange, promissory note, tickets to* order,

1. INTRODUCTION

The choosing of a payment instrument in the scrolling of economic affairs, involves from the managers and of the teams of specialists, depth knowledge in the field of commercial law, banking, financial, especially as each of these instruments have their own particularities of form and background.

This choice also depends on other factors too such as: national legislation, which must be aligned to the international one, by the experience accumulated in time and by the tradition in using such payment instrument, the company's position on the domestic and international markets, the relationship between the trader and his bank or working on family banks, the economic and financial power of the company etc.

Due to the fact that between the dispatch of the goods and that of the amount of cash is a considerable difference of time and space, each contractor seeks, during negotiations, to win as many guarantees: the seller that will pick up in time the value of the goods sold and the buyer that will receive in time the requested quantity of goods at the price and quality parameters established in the contract.

The separation in time of both operations involving the act of selling, respectively the time of delivery of goods by that of payment of price (immediate delivery-postpayment) gave birth to a complex process of creating effects of trade, which constitutes the guarantee of obligation of payment enforcement resulting from sale of goods.

2. BODY OF PAPER

The bill of exchange and the promisorry note are used as instruments of comercial credit for delaying payment, at maturity, to obtain funds in cash, when they are discouted at a bank, or to cover a payment obligation, when the beneficiary endorses them to a third creditor. In order to have the function of the commercial credit instruments, the promissory note and the bill of exchange should always be payable at term, as any other redeemable debenture. Owing to this function, the customer obtains a loan binding himself only with the bill. The sellers who thus on the bill only the signature of the issuing customer, holds a debenture over the customer.

The issue of a promissory note by the importer or the acceptance of a Bill of exchange by the exporter are means of accomplishing personal guarantee.

In the Romanian Commercial Code since 1929, the two instruments were seen as a unique instrument, the promissory note being considered a form of bill of exchange. Here's how it was formulated in Article 270 of the Romanian Commercial Code the definition of bill of exchange: "the bill of exchange including the bond of paiment or the bond the pay to be done at maturity the estabilised amount to the person which appearance…. The bill of exchange including the obligation to pay at maturity a fixed amount of money is called a promissory note too. The bill of exchange including the obligation to make the pay to be done at maturity a fixed amount of money is also caleed policy or treat" (Drăgoescu, 2000).

But the bill of exchange and the promissory note can not be regarded as two forms of the same instrument, for several reasons, namely: the obligation of the issuer in the case of the promissory note differs from the obligation of the issuer in the case of the bill of exchange, more specifically said, the issuer's obligation in the case of the promissory note lies in making itself the payment, and the drawer's obligation, the issuer in the case of bill of exchange, is to ensure the payment, the obligation of payment occurs only in the event of failure of payment or denied payment by the paying shirred.

Thus, if the payer signature as the main cambial obliged appears on the promissory note from the beginning of issue, the payer signature does not appear on the bill of exchange in the moment of issuing the title, it being able not to appear at all materialized on the title specially in that cambial process. „The payment debtor of the promissory note is the principal cambial obliged, since the issuance of the title, while the payment debtor of the bill of exchange becomes the principal cambial obliged only when he accepts the title".

Another reason why you can not put equivalence between the bill of exchange and the promissory note is that, a promissory note can not be delivered to the issuer order, as it is possible in the case of bill of exchange because the promissory would not be "in order" but it will be "at sight".

In other specialized papers they believe that the commercial credit instrument generically named bill of exchange includes two forms: treat, the first form of the bill of exchange in which are engaged in three persons: the drawer, the shirred, and the beneficiary, and "the promissory note, a second form of the bill of exchange is an easier instrument than the treat, because from the three persons participating in a treat remain only two…disappearing the shirred who in his quality as a creditor gave order to pay to the debtor, so the formula "you will pay" used to treat is replaced with "I'll pay "" (Hoanţă, 2001).

In some specialized work is deemed that the treat and the bill of exchange are two different instruments, the bill of exchange is in this sense "a negotiable credit title, and while a payment instrument,...and the treat has the same content as the bill of exchange, but there are three participants: the drawer, the drawee and the beneficiary" (Manolescu, 2006).

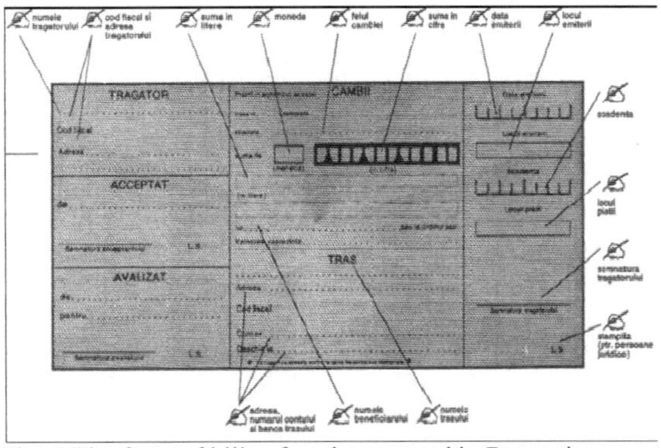

Fig. 1 The form of bills of exchange used in Romania

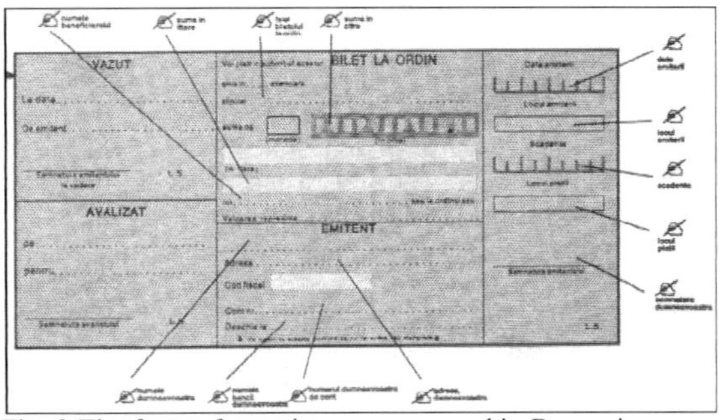

Fig. 2 The form of promissory notes used in Romania

„A promissory note says in effect „I promise to pay you....” and then describes how and when you`re to make payments” (Steingold, 2008).

Another distinction between the bill of exchange and the promissory note refers to the quality of the debtor's signature. The debtor of the promissory note signs the title as the issuer and the shirred debtor in the bill of exchange, signs to accept. „The borrower must sign the promissory note for it to be valid” (Leonard & Warner, 2007).

I am on the same opinion of other authors who believe that the terms bill of exchange and the promissory note are never equal because the bill of exchange is an unconditionated order of payment, while the promissory note is not an order, it is a commitment of payment.

According to the law nr. 83/1994, concerning the promissory note and the bill of exchange, these valuable papers are “guarantee modalities of intern and extern price payment, beside the negotiable credit titles, the guarantee being assured by the quality of these credit titles, being executorial right titles”. This means, that it isn’t necessary to be obtained through justice, the debtor’s execution right, but after the protest advance by the creditor, without complicated formalities, it passes to obliged execution of the debtor exactly on the basis of the title signed by him. Although these instruments have the legal quality of being executorial right titles, many companies in Romania, are using them to deceive their business partners. These companies are economically operators who issue payment instruments without coverage and, implicitly, enter into CIP records with major payments incidents.

Payment Incident Bureau, abbreviated CIP, “is a centre for intermediation and management on behalf of the National Bank of Romania of the specific information of payment incidents both from the bank point of view (drawer in overdrafts, namely drawings from accounts over the available limit) and the social point of view (thievery of payment instruments). The payment incidents are generated by instruments unconform to the regulations of the National Bank of Romania in force”.

The payment incident is that information conveyed towards the Payment Incident Bureau relating to the non-fulfilment exactly or just in time the obligations of the participants before or during the process of settlement of the instrument, notified to the Centre Incident Payments by the reporting persons.

In Romania is not used at all the bill of exchange, the settlements forward being made especially by issuing promissory notes and cheques. But there are many companies in Romania that do not use the promissory notes in the qualities for which they were created, namely as a tool of commercial credit instrument and the guarantee instrument, but as instruments of fraud, becoming a real "public hazard" for business. One of the causes that determines the inclusion of serious business and good payer firms among those issuing payment instruments without coverage "on the rolling tape is modifying the shareholders.

Thus the team that publishes the magazines “FINANCIARUL” and “IAŞI INVEST” have made a study and found out that the same persons who manage companies' with “notable performances" in the field of payment incidents are the shareholders or the associates to other companies too being in banking prohibition or in CIP's sights with major incidents in paying the promissory notes. Every time when a company pays with the promissory note or with the bill of exchange should be made an inspection at Central Incident Payments, especially if the amount is important and if the first deal concluded with the company because it could find information about fraud committed or not, by the business partner.

3. CONCLUSION

The lack of discipline in the Romanian business environment and the desire of some to enrich fraudulently by the issuance of the bills of exchange without coverage and that than have generated major incidents of payment, led to loss the confidence of serious and honest economic agents in using the bills of exchange. Although the two instruments have appeared in practice since the Middle Ages as a commercial credit instruments, as tools of payment and guarantee, they have lately become instruments of fraud and deceit, some became true experts in the issue of instruments of payment without coverage. The fact that in the CIP records appear more than 500 companies from Romania which have generated major incidents of payment regarding to the bills of exchange, and for some companies, the CIP's records include many pages, with dozen or even hundreds of incidents, plus the fact that only a company refused to pay in the last six months promissory notes in total value worth one million euro represents a "proof that there is no confidence in using these tools. It is very important that when a company is put in a position to accept in exchange for the amount of goods delivered a promissory note, to verify at the CIP the financial position of the issuing company of the instrument, because if a company appears in the CIP's records it means that it does not honoured its payment obligations towards its previous creditors and there is a high probability it will not honour any further debts.

4. REFERENCES

Drăgoescu E., (2000), *Currency, financial and credit international relations*, D. Cantemir Publisher, ISBN 973-8024-22-4. Târgu Mureş

Hoanţă N., (2001), *Money and banks*, Economic Publishing House, ISBN 973-590-500-5, Bucharest

Leonard R., Warner R., (2007), *101 Law Forms for Personal Use*, Published by Nolo, ISBN 1413307124, Berkeley CA

Manolescu G., (2006), *Currency and credit,* Published by Foundation, Romania for Tomorrow, ISBN 978-973-725-567-9, Bucharest

Steingold F., (2008), *Legal Forms for Starting & Running a Small Business*, Published by Nolo, ISBN 141330754X, Berkeley CA

Annals of DAAAM for 2008 & Proceedings of the 19th International DAAAM Symposium, ISSN 1726-9679
Editor B. Katalinic, Published by DAAAM International, Vienna, Austria 2008
Make Harmony Between Technology and Nature, and Your Mind will Fly Free as a Bird
2nd DAAAM-ICAT

HEMIFACIAL MICROSOMIA – A CASE REPORT

IHAN HREN, N[atasa] & DRSTVENSEK, I[gor]

Abstract: *Hemifacial microsomia is asymmetry of facial structures (bone and soft tissues). An adult patient with hemifacial microsomia was treated by classical ortognatic surgical procedures (Le Fort I osteotomy and autogenous bone grafting) and by modern surgical technology as distraction osteogenesis of mandible. After these bone procedures the remaining defect of bone and soft tissues was partially compensated by on-lay graft of bone substitute (Medpore mandible on-lay graft), but it was removed after more than one year because of inflammation. After that a custom made titanium angular implant (graft by titanium) was prepared on the basis of computer tomography (CT) scans, Computer Aided Design (CAD) and Rapid Prototyping and Manufacturing technologies.*
Key words: *hemifacial microsomia, bone grafts, xenografts, customized implant manufacturing*

1. INTRODUCTION

The human face holds many functions (eating, speaking, nonverbal communication etc) which combined with its appearance define and affect human psychosocial activities. Severe facial deformities, either congenital, traumatic or consequences of tumour removal are treated by reconstructive surgery procedures. The bone defects in maxillofacial region can be replaced by patients' own bone by different surgical principles as bone grafts or by engineering bone by distraction osteogenesis (Bell, 1992). These different autogenous bone grafts are "golden standard" for reconstruction procedures because they provide osteogenic cells, but they are of limited quantity and connected with risk of complications on donor site (Ellis III, 2003). The allogenic grafts or homografts are taken from another individual of the same species and need a lot of preparations before being used not to induce immunologic response; similar is with bone from unhuman species. So the need for bone substitutes – xenogenic grafts was the generator of great progress in developing inorganic bone substitutes, as biotechnology of bone cells cultivating. All these materials have advantages and disadvantages, but there is no ideal material yet. A good synthetic material needs to have following properties: biocompatibility, inertness, bone-similar weight or even lighter, capability to generate no artefacts on CT and MRI scans, ease of manufacturing, enough strength to resist functional stress, not being expensive and low or no thermal conductivity.

2. OBJECTIVES

The article presents the treatment of a severe facial deformity, where surgical procedures with autogenous bone graft, distraction osteogenesis, prefabricated and custom made xenogenic bone grafts were performed. Twenty four year old mentally healthy man was born with hemifacial microsomia. This is a severe asymmetry of facial bone and soft tissues in vertical, sagital and transverse plane combined with hearing impairment on the affected side (Proffit, 2003). He wasn't treated before his adulthood; all surgical procedures were done in Clinical department of maxillofacial and oral surgery, University clinical centre Ljubljana. He was treated by classical ortognathic surgical procedures (Le Fort I osteotomy and autogenous bone grafting) and by a modern surgical technology as distraction osteogenesis of mandible. After these bone surgical procedures the remaining defect of bone and soft tissues was partially compensated with on-lay xenogenic graft, later replaced with custom made titanium angular implant. His images before and after surgical procedures are presented on fig.1.

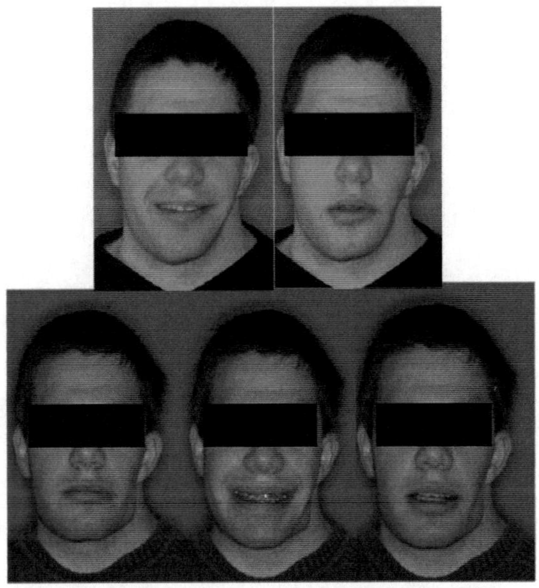

Fig. 1. Patient with hemifacial microsomia before and after surgical treatment (bone surgery only).

3. METHOD

The treatment of an adult patient with hemifacial microsomia has the goal to achieve bone symmetry as good as possible, more difficult is to compensate the soft tissue deficiency. In presented patient the first surgery procedure was producing vertical part of his left lower jaw by distraction osteogenesis (fig.2).

Than his upper jaw was elongated and rotated by LeFort I osteotomy and his autogenous bone grafting (fig. 3). Because of the transverse discrepancy the on-lay xenogenic graft (Medpore mandible on-lay graft) was performed, but it was removed after more than one year because of the inflammation. Than we decided for custom made titanium angular implant (fig. 4), which was prepared on the basis of computer tomography (CT) scans, Computer Aided Design (CAD) and Rapid Manufacturing technologies (Laboratory for Intelligent Manufacturing Systems, of the University of Maribor, Faculty of Mechanical Engineering).

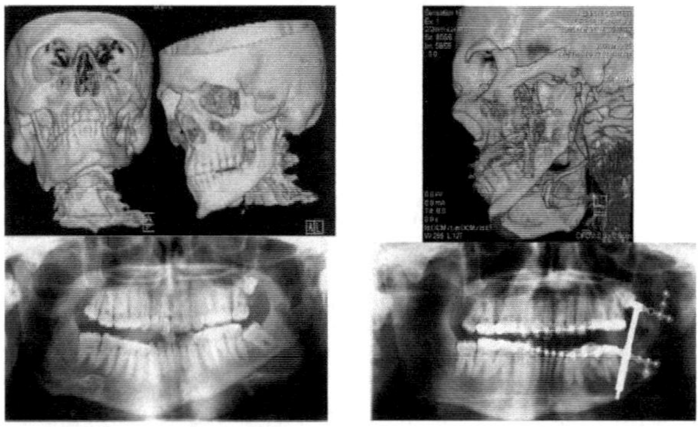

Fig. 2. CT images and roentgenograms before and after distraction osteogenesis (mechanical device for distraction is still in the lower jaw).

4. IMPLANT PRODUCTION

The production of bone implants starts by capturing a three-dimensional data set of the problematic area (scull, face, mandibular area...). Usual and the most common way is transformation of sets of CT or MRI two-dimensional pictures into a three-dimensional, digital model. The model is then used as a base on which the modeling of defective – missing area takes place. If the defect is positioned in an area that has its "mirror image" on another side of the body than the form of the implant can relatively easily be produced by means of Boolean operators. In a case of "mirror less" features some more sophisticated methods and dedicated software have to be used to finish the implant. Presented case study deals with mirroring and some special finishing methods performed by dedicated animation software. Mirrored model of the jaw was used as a reference for modeling of the implant using standard CAD approaches. In a case of a jaw and other bones with attached muscle tissues simple Boolean subtraction can not be used to model a final shape of the implant. Positions of muscle attachments need to be taken in consideration, therefore the mirrored model is only used as guidance for shaping cross section polylines. The polylines are later used to guide a modeling of implant surfaces that shape the final form of the implant.

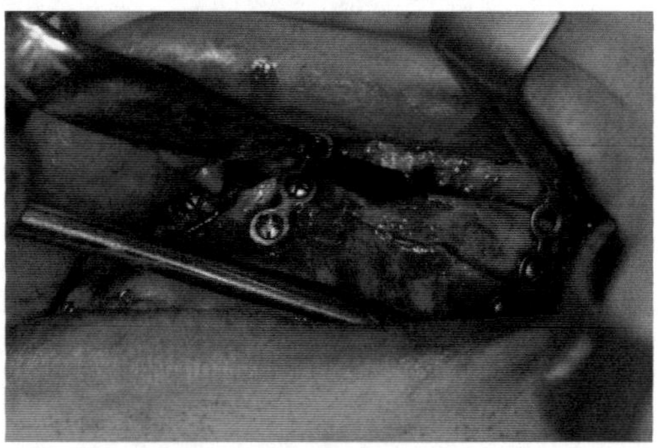

Fig. 3. Elongating of the upper jaw with autogenous bone graft.

Finished digital model is then manufactured by one of the Rapid Manufacturing (RM) technologies. RM products are usually made of titanium or cobalt – chrome alloys since these are at the moment the only biocompatible materials available for RM technologies that can be directly used as implants. Research is going on in the field of so called Bio-plotters that are able to manufacture parts from bio-compatible and even biodegradable materials but there has not been a commercial brake- through yet what makes these technologies useless for surgical praxis.

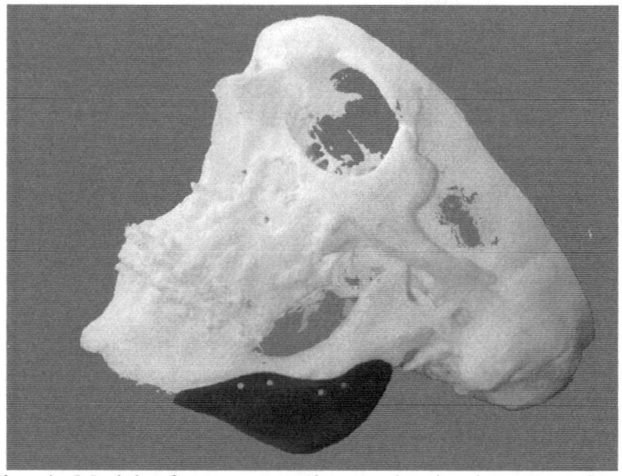

Fig. 4. Model of custom made angular implant on printed model of patient's head skeleton.

5. DISSCUSION

The presented case report is an example of different surgical procedures used to correct a facial deformity with different surgical procedures. With standard orthognatic procedures (osteotomies of the bone, their correction in the position and volume) we can achieve better facial function and harmony, but not perfect (Cheng, 1998). The distraction osteogenesis is very innovative approach for forming patient's own bone (Samchucov, 2000) but at least in one dimension the quantity is not ideal (Bell & Guerrero, 2007). So there is still the need for xenografts as bone substitutes – in the majority of the facial deformity cases for esthetic correction of bone or soft tissue defects.

6. CONCLUSIONS

The medical needs and contemporary technological development are the fields that will be in close relation in the future in many fields of the medicine. For facial deformities, in spite of different surgical approaches, there is still a need for development of materials for xenogenic bone grafts and the technologic facilities can nowadays prepare custom made bone implants to achieve better esthetical results.

7. REFERENCES
8.

Bell, WH. (1992). *Modern practice in orthognatic and reconstructive surgery.* W. B. Saunders company, Philadelphia.

Bell, WH. & Guerrero, CA. (2007). *Distraction osteogenesis of the facial skeleton.* BC Decker Inc, Hamilton

Cheng, LHH.; Roles, D. & Telfer, MR. (1998). Orthognatic surgery: the patients' perspective. *Brit J Oral Maxillofac Surg* 36., 261-263,

Ellis III, E. (2003). Surgical reconstruction of defects of the jaw. In: Peterson,; Ellis,; Hupp, & Tucker. *Maxillofacial surgery*, 4rd edition. Mosby, New York, 646-659,

Proffit, WR. & Turvey, TA. Dentofacial asymetry. In: Proffit, WR.; White, RP. & Sarver, DM. (2003). *Contemporary Treatment of Dentofacial Deformity*. Mosby, St. Louis, 574-621,

Samchucov, ML.; Cope, JB. & Cherkashin, AM. (2000). *Craniofacial distraction osteogenesis*. Mosby, St Louis.

Annals of DAAAM for 2008 & Proceedings of the 19th International DAAAM Symposium, ISSN 1726-9679
Editor B. Katalinic, Published by DAAAM International, Vienna, Austria 2008
Make Harmony Between Technology and Nature, and Your Mind will Fly Free as a Bird
2nd DAAAM-ICAT

THREE-DIMENSIONAL ANALYSIS OF THE FACIAL SOFT TISSUES

IHAN HREN, N[atasa] & POGACAR, V[ojko]

Abstract: *For the diagnostic procedures and treatment planning of facial deformities we are using standard imaging diagnostic procedures focused on facial bones, but the appearance is determined also by their soft tissues. So the analysis of the face as integrity is problem partially solved by standard photograph, but their analysis is very difficult and subjective, so the modern technology is developing the computer system for face soft tissue analysis on the basis of digital facial scans of the face surface. These images of the face are done by laser scans or by, using optical white-light-scanner Atos II.. The method of three-dimensional scanning and digital analysis of the face are developing together maxillofacial surgeons and engineers in design lab.*
Key words: *face, three-dimensional analysis*

1. INTRODUCTION

Maxillofacial surgeon, as esthetic plastic surgeon and orthodontist for the diagnostic procedures and treatment planning of facial deformities use standard diagnostic procedures as x-ray images and computer tomography scans, plaster models of their teeth and standard photography of the face. Plaster models of teeth are good tool for planning esthetic and functional bite after treatment, but they are not sufficient for whole facial area. Standard roentgenographs as lateral and posterior projection of the head with standard analysis (lateral cephalogram) had been developed to good golden standard for diagnostic, but the difficult cases need also computer tomography scans with significant radiation risk (Khambay et al., 2002). Also these x-ray pictures are mainly for bone analysis and they offer only little information about the soft tissues of the face (for example there is the good information of the nose, upper lip and chin, but only in lateral view). So the analysis of the face as integrity is problem partially solved by standard photography (Moss et al., 2003). The analysis of the face photos is very difficult and subjective, routinely evaluate transverse discrepancy clinically and on 2D photographs. However, 2D photograph analysis of a 3D structure as the face is, provides incomplete data and does not account for differences in facial depth and shape (Moss et al., 1994, Miller er al., 2007). So the modern technology is developing the computer system for face soft tissue analysis on the basis of digital facial scans of the face surface. With this method we can reproduce the facial form in 3D to a greater precision. The whole face or different parts of the face can be evaluated for asymmetry and other facial characteristics (Da Silveira et al., 2003). These facial surface images are done by laser or optical scanners and by three-dimensional digitizing method.

2. METHOD

The innovative method of three-dimensional scanning and digital analysis of the face is developing together by Faculty department of maxillofacial and oral surgery, Medical faculty in Ljubljana with Engineering Design Lab of the University of Maribor, Faculty of Mechanical Engineering.

The three-dimensional scanning is done by optical white-light-scanner Atos II. The precision of that technologic scanner is very high for facial surface structure. Taking scans is very fast, only one scan is necessary for analysis and all that is advantage for patients. There is also very few unwanted data. Only the heavy light is a little disturbing but for very short period. The posture of the head is natural (Chiu & Clark 1991). The two examples of these face images are presented on fig. 1.

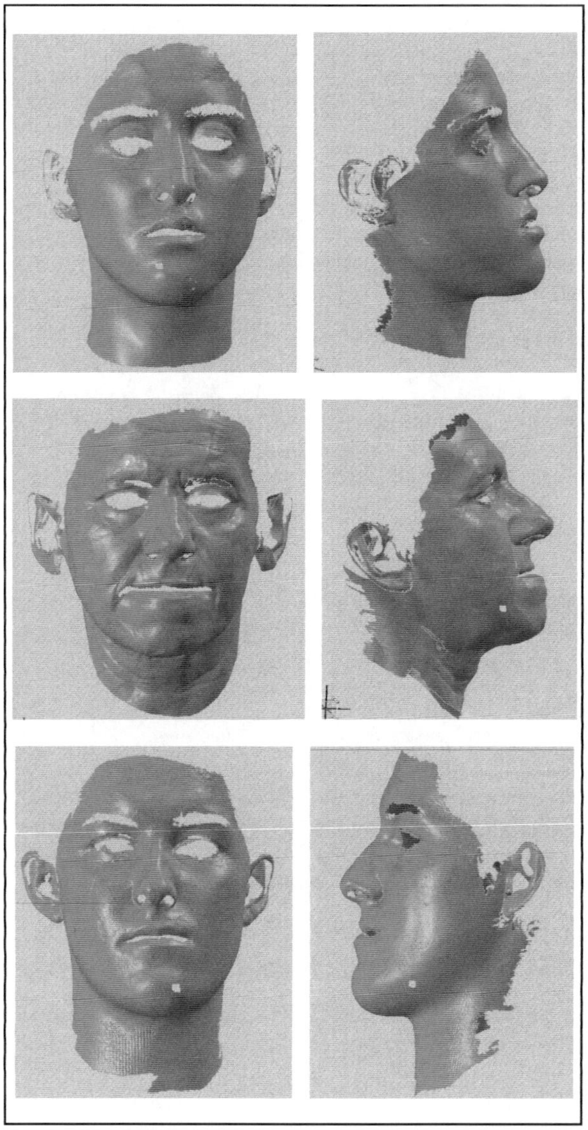

Fig. 1. The facial scans of normal man face and man face with mandibular prognathism.

The 3D data was imported to a reverse modeling software Atos II. and Polyworks, but later we planed to develop more focused user fiendly software for maxillofacial analysis. Each scan of the face (left and right images) was processed in order to remove unwanted data, registered and merged to produce a complete facial image. The facial shell was aligned to two planes: the mid-sagittal plane (Y-Z) and the inner cantus of the eyes (X-Z). The facial shell was divided into three parts; the upper part was defined as the part of the face above the inner cantus plane, the middle part ranged from the inner cantus plane to the plane through the outer commissures of the lips and the lower part was below this plane. To check for left/right symmetry the face was mirrored on the mid-sagittal plane.

3. DISSCUSION

Similar methods as our developing one are using in some medical centres in the world (Nute & Moss 2000, Ji et al., 2002). The Konica/Minolta Vivid 910 (20) is the example of laser scanner (Da Silveira et al., 2003) and these methods are disadvantaged because of possible laser damages. Also photo scanner as 3DMD is commercially available.

The 3D data also can be analysed by reverse modeling software package, Rapidform ™ 2006 (@ INUS Technology Inc, Seoul, Korea). But in spite off that we decided to develop our own method comparable by them. The main reason is to use the advantages of cooperation of people of two very different faculty departments and that relationship can produce in the future not only the system for facial scanning, but also user-friendly software for a lot of research and clinical work; to determine the normal standards for women' and man' faces in Slovenia. This could be the base for assessment of facial deformities, better treatment planning and assessment of the surgical methods and clinical results.

4. CONCLUSIONS

The medical needs and contemporary technological development are the fields that will be in close relation in the future. The analysis of facial soft tissues in safe way for patients will be developing. Our innovative method is one example of these solutions. It respects the basal postulations and enables to develop individual needs as research method and treatment modality.

5. REFERENCES

Chiu, CS. & Clark, RK. (1991). Reproducibility of natural head position. *J Dent* 19., pp. 130,

Da Silveira, AC.; Daw Jr, JL.; Kusnoto, B.; Evans, C. & Cohen, M. (2003). Craniofacial application of three-dimensional laser surface scanning. *Journal of Craniofacial Surgery* 14., pp. 449-456.

Ji, Y.; Zhang, F.; Schwartz, J.; Stile, F. & Lineaweaver, WC. (2002). Assessment of facial tissue expansion with three-dimensional digitizer scanning. *Journal of Craniofacial Surgery* 13., pp. 687-92,

Khambay, B.; Nebel, JC.; Bowman, J.; Walker, F.; Hadley, DM. & Ayoub, A. (2002). 3D stereophotogrammetric image superimposition onto 3D CT scan images: the future of orthognathic surgery. A pilot study. *International Journal of Adult Orthodontic and Orthognatic Surgery* 17., pp. 331-341,

Marmulla, R.; Hassfeld, S.; Luth, T. & Muhling, J. (2003). Laser-scan-based navigation in cranio-maxillofacial surgery. *Journal of Craniomaxillofacial Surgery* 31., pp. 267-277,

Miller, L.; Morris, DO. & Berry, E. (2007). Visualizing three dimensional facial soft tissue changes following orthognathic surgery. *Eurpean Journal of Orthodontics* 29., pp. 14-20,

Moss, JP.; Linney, AD.; McCance, AM. & James, DR. (1994). The use of three-dimensional optical scanning techniques in dental and related fields. *Dental Update* 21., pp. 386-390,

Moss, JP.; Ismail, SF. & Hennessy, RJ. (2003). Three-dimensional assessment of treatment outcomes on the face. *Orthodontics and Craniofacial Research* 1., pp. 126-131,

Nute, SJ. & Moss, JP. (2000). Three-dimensional facial growth studied by optical surface scanning. *Journal of Orthodontics* 27., pp. 31-38,

Annals of DAAAM for 2008 & Proceedings of the 19th International DAAAM Symposium, ISSN 1726-9679
Editor B. Katalinic, Published by DAAAM International, Vienna, Austria 2008
Make Harmony Between Technology and Nature, and Your Mind will Fly Free as a Bird
*Young Researcher and Scientist Paper / * Supervisor, Mentor*

CONCEPTUAL DESIGN AND SYSTEM OF GYRO CABLEWAY CONTROL

ILCIK, S[tefan]

Abstract: *The paper deals with the theory, construction, technology and automation of the control of a mechanical recuperative cableway with a mechanical flywheel. This is the first world's conceptual technical design in the sphere of forest cableways which is oriented towards the recovery of energy as the source of traction for forests cableways, by the use of gravity force during gravity skidding of a carriage and load.*
Key words: *gyro-cableway, recuperation, accumulation, energy*

1. INTRODUCTION

Currently used cableways pose a considerable technical problem, because they function on the principle of the combustion process, which causes a high loss of energy due to low efficiency of the combustion engine. For the above reason, current forest cableways are fully energy dependent on driving fuels (Ferenčík et al, 2008). Having in mind this problem and increasing requirements for the utilization of driving fuels in cableway skidding in forest management, Technical University in Zvolen is conducting fundamental research into energy-saving recuperative cableways using a hybrid propulsion system. The conceptual technical designs are protected by patent applications PP 0053-2006, PP 0080-2007, PP 0103-2007, PP 0108-2007, PP 0109-2007, PP 0037-2008 owned by the Technical University in Zvolen. (Štollmann & Ilčík, 2007).

2. THEORY OF THE GYRO-CABLEWAY

The new conceptual technical design according to the patent application PP 0108-2007 is represented by a gyro-cableway which uses a mechanical flywheel as a basic construction element functioning as a mechanical accumulator. From a physical point of view, it is a homogeneous solid of revolution rotating on a symmetry axis in a vacuum casing with magnetic bearings, which reduces aerostatic and mechanical friction and greatly increases the efficiency of this device.

The mechanical accumulator consists of a system of mass points and its total energy can be calculated as the sum of kinetic energies of all its mass points using the following formula. (1).

$$E_k = \sum_{i=1}^{n} \frac{1}{2} \cdot J \cdot \omega^2 \tag{1}$$

The theory and operation mode of the gyro-cableway is in more detail illustrated using an actual example situation, where during the phase of gravity skidding of a carriage and load the mechanical flywheel gains charging mechanical energy E_{vn} = 4433659.983 J, and during the phase of hauling up it consumes discharging mechanical energy E_v = 819137.54 J needed for hauling the empty carriage back up the slope. If, for example, a mechanical flywheel weighing 100 kg has its inertia moment J = 1,28 kg.m^2, the mechanical flywheel angular velocity ω_{vn} in the phase of gravitational skidding can be counted according to formula (2).

$$\omega_{vn} = \sqrt{\frac{2 \cdot E_{vn}}{J}} \tag{2}$$

$$\omega_{vn} = \sqrt{\frac{2 \cdot 4433659.983}{1.28}} = 2632.032\ rad \cdot s^{-1}$$

The calculated angular velocity is used for the determination of the number of revolutions of the mechanical flywheel per one minute during gravity skidding according to formula (3).

$$n_{vn} = \frac{\omega}{2\pi} \cdot 60 \tag{3}$$

$$n_{vn} = \frac{2632.032}{2 \cdot 3.14} \cdot 60 = 25146.802\ r \cdot \min^{-1}$$

With the charging energy E_{vn} = *4433659.983 J* gained during gravity skidding of the carriage and load down the slope, the mechanical flywheel will rotate at the speed of 25146.802 revolutions per minute. The kinetic energy which the mechanical flywheel gains during the phase of carriage and load gravity skidding is then used for hauling the empty carriage back up the slope.

The total kinetic energy of the mechanical flywheel E_k, which remains accumulated in the mechanical flywheel after the empty carriage has been hauled up the slope, can be determined on the basis of the difference between the kinetic energy in the phase of gravitational skidding E_{kvn} and the kinetic energy in the phase of hauling the empty carriage up the slope E_{kv} using formula (4). On the basis of the kinetic energy of the mechanical flywheel we can determine its angular velocity ω_k after hauling the carriage back up the slope using formula (5).

$$E_k = E_{kvn} - E_{kv} \tag{4}$$

$$\frac{1}{2} \cdot J \cdot \omega_k^{\ 2} = \frac{1}{2} \cdot J \cdot \omega_{vn}^{\ 2} - E_{kv}$$

$$\omega_k^{\ 2} = \omega_{vn}^{\ 2} - \frac{2 \cdot E_{kv}}{J}$$

$$\omega_k = \sqrt{\omega_{vn}^{\ 2} - \frac{2 \cdot E_{kv}}{J}} = \sqrt{\omega_{vn}^{\ 2} - \left(\frac{2}{J} \cdot E_{kv}\right)} \tag{5}$$

$$\omega_k = \sqrt{2632.032^2 - \left(\frac{2}{1.28} \cdot 819137.54\right)}$$

$$\omega_k = 2376.487\ rad \cdot s^{-1}$$

The number of revolutions of the mechanical flywheel per one minute after the empty carriage has been hauled back up the slope can be calculated using formula (6).

$$n_k = \frac{\omega_k}{2 \cdot \pi} \cdot 60 \qquad (6)$$

$$n_k = \frac{2376.487}{2 \cdot 3.14} \cdot 60 = 22705.289\, r \cdot \min^{-1}$$

In the case of the discharge energy E_v = 819137.54 J, which was spent for hauling the empty carriage back up the slope, the revolutions of the mechanical flywheel decrease to the value of 22705.289 rpm. The revolutions of the mechanical flywheel which are utilized for hauling the empty carriage back up the slope can be calculated using the following formula (7).

$$\Delta n_v = n_{vn} - n_k \qquad (7)$$

$$\Delta n_v = 25146.802 - 22705.289$$

$$\Delta n_v = 2441.513\, r \cdot \min^{-1}$$

During hauling the empty carriage back up the slope, the revolutions of the mechanical flywheel decrease by 2441.513 revolutions per minute. After the empty carriage has been hauled back up the slope, the mechanical flywheel still accumulates a great part of kinetic energy E_k, which corresponds to 22705.289 revolutions per minute. The calculated examples show, that after hauling the empty carriage back up the slope, the mechanical flywheel still stores a great number of revolutions which the mechanical recuperative cable device can utilize for further operations.

3. CONSTRUCTION AND TECHNOLOGY

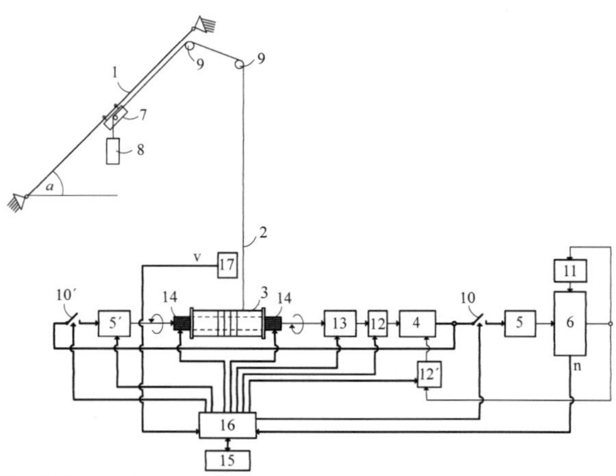

Fig.1 Gyro-cableway.

The Gyro-cableway is a mechanical recuperative cableway device with a flywheel (Fig.1), which utilizes the gravitational energy of a carriage 7 and load 8 accumulated during the phase of gravity skidding for the actuation of an electric voltage generator 4. The generator of electric voltage 4 drives an electric motor 5 which is used to drive a mechanical flywheel 6. The mechanical flywheel 6 stores the gravitational energy of the carriage 7 and load 8, which is here transformed into kinetic energy. During the hauling phase, the mechanical flywheel 6 drives the electric voltage generator 4 which is used to drive the electric motor 5′. The electric motor 5′ drives a winding-up device 3 which hauls the empty carriage back up the slope (Štollmann & Ilčík, 2007).

4. AUTOMATION OF CONTROL

Automatic control of carriage speed is ensured by means of the control circuit 16, which is the principal control unit of the mechanical recuperative cable installation with a flywheel (Božek, 2007).

The automatic control of carriage speed by means of the control circuit 16 operates as follows. During the phase of gravity skidding of the load down the slope, the control circuit 16 engages the clutch 12 and switches on the electric switch 10. Subsequently, the control circuit 16 releases the brakes of the braking system 14 in the winding up device 3 and changes appropriate gears in the transmission mechanism 13 in dependence on the information received from the mechanical flywheel 6 about the number of its revolutions per one minute.

The speed of gravity skidding of the carriage 7 and load 8 down the slope is maintained at a constant appropriate level in metres per second by means of the control circuit 16, which acquires the information about actual speed of the carriage from the carriage speed sensor 17, and on the basis of this information the speed of the carriage 7 and load 8 is slowed down by means of the transmission mechanism 13 and braked to stop by means of the breaking system 14. If the desired speed is exceeded, the control circuit 16 activates the braking system 14 in the winding up device 3, and vice versa, if the speed of gravity skidding of the carriage 7 and load 8 is lower than the desired speed, it releases the braking system 14 in the winding up device 3. In this way the gravity skidding of the carriage 7 and load 8 down the slope is controlled and maintained at a constant desired speed. The speed of hauling the empty carriage back up the slope is controlled by means of the control circuit 16 through the electric motor 5′. (Štollmann & Ilčík, 2007).

5. CONCLUSION

At present the Technical University in Zvolen is conducting fundamental research into energy-saving recuperative cableway installations. This research represents basic preparation for applied research and design of recuperative cable installations. In the next step, the University plans to produce the first prototype of recuperative cable installation.

The knowledge in this paper comes from solving grant project VEGA No. 1/3523/06 entitled Research into New Technical and Technological Principles of Machines for Skidding, and grant project KEGA No. 3/6448/08 entitled Transformation of Development Shops and Laboratories Supported by the Ministry of Education of the Slovak Republic.

6. REFERENCES

Božek, P. (2007). *Virtual Control Systems,* In: *Strategic management of the manager′s practice*, pp. 612-618, Tripsoft, ISBN 978-80-89291-04-5, Trnava

Ferenčík, M.; Messingerová, V. & Stanovský, M. (2008). *The Mounty 4000 Forest Cableway in the Calamity.* Proceedings of International Scientific Conference Logging Technologies in Salvage Felling, pp. 27-32, ISBN 978-80-213-1791-8, Praha 18-20. 6. 2008, CZU, Praha

Štollmann, V. & Ilčík, Š. (2007). *Mechanical Recuperative Cable Installations with a Flywheel – Gyro-cableway*, patent application 0108-2007, Int. Cl.: B66B 7/00, Industrial Property Office of the Slovak Republic

Štollmann, V. & Ilčík, Š. (2007). *Gyro-cableways – Theory of Flywheels, Final Report of a Partial Research Assignment Nr.* 1/3523/06, p. 15, TU Zvolen, SIG. SLDK VSL 275, Zvolen

Štollmann, V. (2007). *Cableway Skidding – Innovations of Higher Orders*, p. 95, TU Zvolen, ISBN 978-80-228-1794-3, Zvolen

Annals of DAAAM for 2008 & Proceedings of the 19th International DAAAM Symposium, ISSN 1726-9679
Editor B. Katalinic, Published by DAAAM International, Vienna, Austria 2008
Make Harmony Between Technology and Nature, and Your Mind will Fly Free as a Bird
2nd DAAAM-ICAT

RAPID PROTOTYPING USE IN MANUFACTURING COMPONENTS OF A MEDICAL LASER DEVICE

ILIESCU, M[ihaiela]; COMANESCU, B[rindus] & NUTU, E[mil]

Abstract: *Knowing that Rapid Manufacturing (RM) reefers to group of technologies developed in order to improve the manufacturing cycle of a product and that, Rapid Prototyping (RP) represents a very important element of it, the paper deals with some real, practical aspects of 3D printing rapid prototyping technology. In medical applications (urology, abdomen surgery, etc.), laser treatment, has known a continuous extend so, the parts of such a laser device need a special attention in manufacturing. This paper presents the steps carried out and the problem solving, performed by ink jet printing rapid prototyping technique, on a ZPrinter 310 Plus. A finite element study on the temperature filed, within the most important parts of the considered device, is also presented.*

Key words: *rapid prototyping, physical model, 3D printing, temperature field simulation*

1. INTRODUCTION

The nowadays competition, both in industry, medicine, education etc. resulted into the development of several technologies, known as Rapid Manufacturing, whose main objective is (http://www.efunda.com, 2008) to shorten the design and production cycle. Many times, a prototype is often required for testing the product's manufacturing specifications or for an evaluation of its characteristics. So, Rapid Prototyping is worth to be used.

Rapid Prototyping represents (http://en.wikipedia, 2008) the automatic construction of physical object using solid freeform fabrication. It "takes" virtual design from computer aided design or animation modeling software, transforms it into thin, virtual, horizontal cross-sections and then, creates each cross-section in physical space, one after the next, until the model is finished. The virtual model and the physical one correspond almost identically.

In medicine, laser treatment has known a high extent and, thus, specific laser devices had to be designed and manufactured. This paper presents the application of ink jet printing rapid prototyping technology in designing and manufacturing components of a laser device used in urology - for splitting kidney stones and in surgery - for fixing damaged bones. As it is *an innovative laser device*, rapid prototyping is of great help, specially because of the fact that the real components' materials are very expensive ones and the least designing and/or manufacturing mistake costs a lot – money, effort, material, time.

2. EXPERIMENTAL RESEARCH

2.1 Experimental System

The drawing of the laser device studied is presented in figure 1 and the parts to be prototyped are indicated by an arrow. The laser involved is solid state holmium one, with a THC:YAG rod emitting at 2100 nanometer, with 20 watts power and 250÷350 microseconds pulse duration.

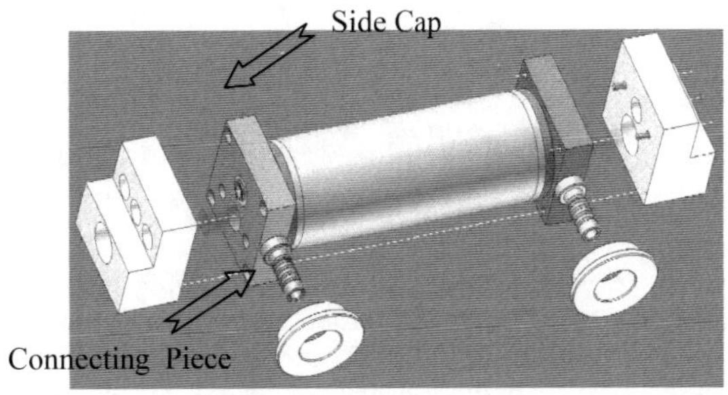

Fig. 1. Drawing of the laser device.

One of the rapid prototyping technologies is the 3D printing, meaning (http://www.csa.com, 2008) layers of fine powder are selectively bonded by printing an adhesive from the inkjet print-head in the shape of each cross-section, as determined by a CAD file.

The *technological system* used consists in: ZPrinter 310 Plus (Z Corporation) – printing machine; a compressed air cleaning enclosure; an electric oven and *the materials* used for rapid prototyping are: zp®131 powder (high performance composites for tough parts and very good resolution); zb60 – binder solution and z-max – high strength epoxy.

2.2 Experiments

The experiments – meaning obtaining the prototypes – were carried out according to the above mentioned aspects. The ZPrint software was set for 0.01 mm layer thickness and the time for prototype printing was about 21 minutes.

Some images, taken while ink-jet prototyping are shown in figure 2 (a, b) – as computer print screen and in figure 3 (a, b) - as real process, on ZPrinter.

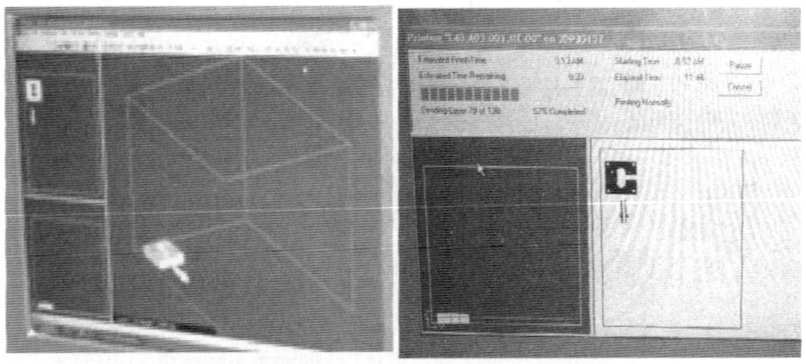

Fig. 2. Ink-jet prototyping – computer screen images

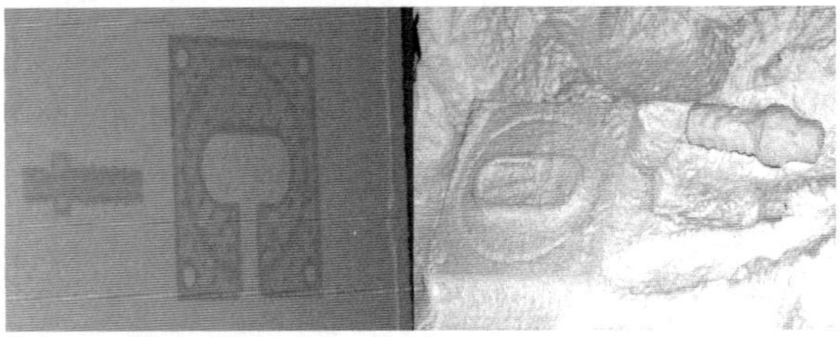

Fig. 3. Ink-jet prototyping – Zprinter images

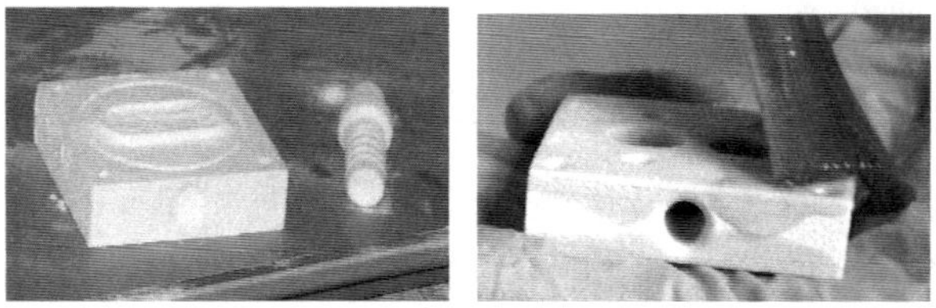

Fig. 4. Intermediate stages– after 3D printing

Fig. 5. The obtained prototypes

When the printing process was over (as mentioned by the software), the parts were removed from the printing machine. They were cleaned off the remaining powder and, after a little dry into the oven, were impregnated with epoxy strength then, finally dried into the oven. These above mentioned facts are illustrated by figure 4 (a, b).

An image of the final prototypes is shown in figure 5, where it can be noticed that finally, there were obtained two prototypes of each part – needed for the laser device (as seen in figure 1). It is worth mentioning the fact that the parts are joined together by thread assembly type but, the thread could not be obtained in rapid prototyping (it was almost impossible to extract the models out of the powder without damaging the threaded part). So, the threading operations, to join the parts together, were carried out on a drilling machine.

2.3 Experimental Results

Similarly, by ink-jet printing rapid prototyping, there were obtained all the components of the laser device. When assembling them together, there could be noticed some problems, most of them dealing with parts' design.

So, through each of side cap's main central holes there is special glass tube that isolates the laser flash and, respectively, active medium (Holmium) from the other components. Through the connecting piece there is the entrance/exit of a cooling fluid – unionized water, that maintains the appropriate temperature of laser body.

The major problem, while joining the prototyped parts, was that the cooling fluid could not "get" into both of the glass tubes, as they were too long and their end restricted the fluid access. Another problem was that the shape of the parts "flushed" by the cooling water did not fit tight so, there was fluid spread away. Thus, there was pointed out the need of using special insulating elements, such as O-rings.

3. TEMPERATURE SIMULATION

Considering the fact that all the rapid prototyping process was done in order to discover eventual mistakes in device's designing, it has been considered useful to simulate temperature within the side cap, knowing that the cooling fluid's temperature, into the exit connecting piece must not exceed a specific value, for example, 37°C.

Thus, it was carried out a simulation of the temperature field (Sushil et al., 2006), with ANSYS software. The model digitization was a 10 nodal tetrahedron and the considered heat transfer types were convection (from cooling fluid to metal) and conduction (from metal to metal).

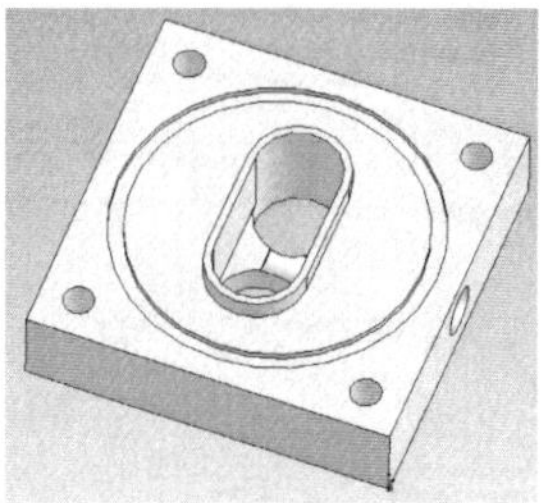

Fig. 6. Temperature simulation's initial data

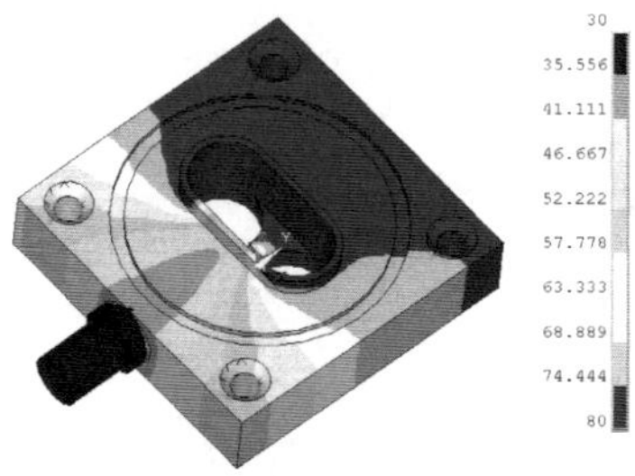

Fig. 7. Temperature simulation's results

The temperature into the main holes of the side cap (around the special glass tubes – flash and active medium) were considered to be of 75°C and 80°C, while the cooling water entrance temperature was considered to be 27°C - as shown in figure 6. The ANSYS simulation's results are presented in figure 7 – noticing that exit fluid's temperature was below 37°C

4. CONCLUSION

Using 3D printing technology of Rapid Prototyping it was possible to discover some errors in components' designing of a laser device. So, it was possible to save time (about one month) and money but, more important, not to waste very expensive special materials.

The temperature simulation, with ANSYS software, pointed out the need to reduce entrance cooling fluid's temperature below 27°C so that, the laser system be able to work well.

Further research and use of other (than ink-jet printing) Rapid Prototyping technique, could be done if several changes in component parts would be required. Thus, Rapid prototyping should be used for reducing product development cycle.

5. REFERENCES

Wang, C.; Park H. S. Rapid Manufacturing, *Available from:* http://www.csa.com/discoveryguides/rapidman/overview *Accessed:* 2008-03-08.

Rapid Prototyping, *Available from:* http://en.wikipedia.org/wiki/Rapid_prototyping *Accessed:* 2008-03-07.

3D Printing, *Available from:* http://en.wikipedia.org/wiki/3D_prototyping *Accessed:* 2008-03-07.

Rapid Prototyping, *Available from:* http://www.efunda.com/processes/rapid_prototyping/intro.cfm *Accessed:* 2008-03-10.

Z CORPORATION, *Available from:* http://www.zcorp.com/Products/3D-Printers *Accessed:* 2008-03-13.

Sushil Sharma, Bijaya Adak, Brian Rusthoven (2006), *Long-Term High-Velocity Erosion of Glidcop in DI Water,* APS, Argonne National Laboratory, USA.

Annals of DAAAM for 2008 & Proceedings of the 19th International DAAAM Symposium, ISSN 1726-9679
Editor B. Katalinic, Published by DAAAM International, Vienna, Austria 2008
Make Harmony Between Technology and Nature, and Your Mind will Fly Free as a Bird

COMPARISON BETWEEN PLASMA ARC CUTTING AND LASER CUTTING

ILII, S[anda] & MUNTEANU, A[driana]

Abstract: *At present, the industry has the possibilities to process almost any type of material, with maximum of technical and economical efficiency, by using plasma cutting and laser cutting methods. In this paper, plasma arc cutting and laser cutting processes are analyzed. The discussed topics include material applications, productivity, cost and influence of cutting on the material quality in the edge zone.*

Key words: *Plasma, laser, cutting, comparison.*

1. INTRODUCTION

The machining technologies based on the thermal effect of plasma and those based on the laser beam have an important place in the field of unconventional technologies. At present, plasma arc machining (PAM) and laser beam machining (LBM) represents the most modern machining technologies used in the industry domains such as: machines manufacturing, electronics, aeronautics etc., due to their advantages, to the fact that they allow machining of the high-alloy refractory and stainless steels till exotic metals with maximum of productivity, to the automation capacity, to the low expenses towards traditional techniques, and also due to the quality obtained of the surface material.

In the machine manufacturing industry, *the plasma*, as a "*tool*", is used especially in cutting operations, coating, welding, melting, and assistance of the mechanical processing operations such as turning, threading, drilling, grooving etc., in order to improve the machinability of various materials (Marinescu et alt., 2000); (McGeough, 2002).

Laser beam machining process is used in cutting operations, drilling, engraving, welding, heat treating, cladding, vapor deposition, tracing etc.

2. OPERATING PRINCIPLE

2.1 Plasma Arc Cutting

The plasma/ion beam machining is based on the thermal or chemical effects generated in the contact zones between ions or plasma and the accessible surfaces of the workpiece (Slătineanu et alt., 2004). The electric energy is used to form the jet of plasma in the presence of a plasmogen gas. The plasmogen gas (primary gas) must accomplish the following conditions: it ensures protection to the incandescent electrode against the oxidation process; to be neutral toward the material of the workpiece. The mono-atomic inert gases, witch are today the most used to produce thermal plasma (Argon, Helium, Azoth, air etc.), accomplish these conditions. Since in the process the thermal energy is delivered, the material in solid state is heated, melted and then boiled. The interatomic material bond-breaking is realized by thermal way (Nemchinsky & Severance, 2006).

In accordance with the Welding Handbook, „the plasma arc cutting process severs metal by using a constricted arc to melt a localized area of a workpiece, removing the molten material with a high-velocity jet of ionized gas issuing from the constricting nozzle".

In today's industry it is a widely used process to cut different metals and, combined with computer numerically controlled (CNC) machines, to cut alloy plate into desired shapes and prepare weld angles.

Normally, a plasma arc cutting system has a power supply, an arc starting circuit, and a torch. The power source and arc starter circuit are connected to the cutting torch through leads and cables that supply proper gas flow, electrical current flow, and high frequency to the torch to start and maintain the process. The arc and the plasma stream are focused by a very narrow nozzle orifice (Nemchinsky & Severance, 2006).

2.2 Laser cutting

In the case of laser cutting, a gas jet propagating in vertical direction is used to blow out the molten material that is produced due to heating of the workpiece by the laser beam at the momentary end of the curve. This process is usually called 'laser fusion cutting' or 'gas assisted laser cutting'. If the assisting gas jet has a considerable content of oxygen, the gas jet does not only have a mechanical effect as blowing out liquid material, but leads also to an additional heating of the workpiece due to the exothermic reaction of oxygen, for instance with steel. In this case, with the same laser power, an improved thickness can be cut or a higher processing speed can be obtained.

3. COMPARISON BETWEEN PAC AND LBM

The first and most important issues to consider are those of cost and material application, influence of cutting on the material quality in the edge zone and productivity.

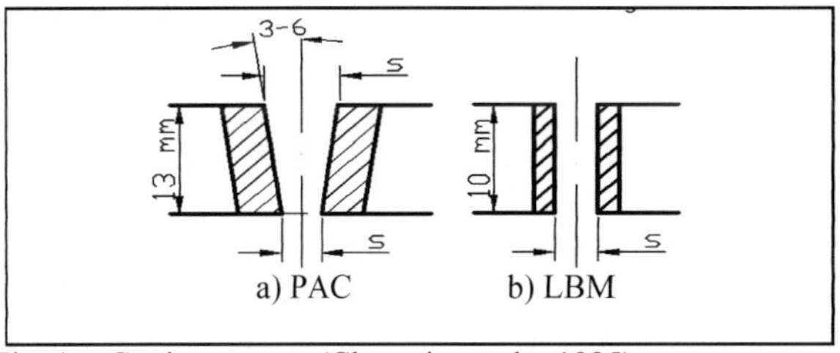

Fig. 1. Cutting aspects (Chaussier et alt., 1995).

Regarding the accuracy obtained, previous work (Marinescu et alt., 2000) shows that laser cutting machines are more accurate than plasma cutting machines. Also, towards LBM, in the case of PAC machining, its applications to many parts shapes are severally limited.

As we can observe in table 1, grater productivity can be obtained using the plasma for cutting instead of using lasers.
Plasma cutting requires electrically conductive material. Lasers, in contrast, because they use highly focused light, can cut both electrical conductive and non-conductive materials.

Machining type	Productivity (maximum) Q [mm^3]	Specific productivity [mm^3/min]	Precision (max. value) [mm^3/minKW]	Roughness (min. value) [μm]
PAC	10^5	10^3	0,5	25
LBM	10	10	0,01	1,6

Tab. 1. PAC and LBM process performances (Mihăilă, 1999)

Criteria	Plasma Cutting	CO_2 Laser cutting
Minimum size of the cutting slit (kerf width)	0.006",depending on cutting speed	0.002"
Cut surface appearance	Cut surface will show a striated structure	
Degree of cut edges to completely parallel	Good; occasionally will demonstrate conical edges	Fair, will demonstrate non-parallel cut edges with some frequency
Processing tolerance	Approximately 0.002"	Approximately 0.02"
Degree of burring on the cut	Only partial burring occurs	
Thermal stress of material	Deformation, tempering and structural changes may occur in the material	
Forces acting on material in direction of gas or water jet during processing	Gas pressure poses problems with thin workpieces, distance cannot be maintained	

Table 2. Precision of process (Marinescu, 2000)

Thus, a very large range of materials could be machined using the laser beam: mild and carbon steel, stainless steel, alloy steel, tool steel, aluminum alloys, cooper alloys, titanium, plastics, rubber, ceramics, composites, wood, glass etc.; however care must be taken in choosing the correct type of laser. Laser cutters are often used on exotic metals and minerals that would quickly dull or damage conventional blade or wire cutters. These included such items as hardened ceramics and precious gemstones. Laser cutting is an excellent choice for cutting Acrylic/Perspex. The cut edge results appear smooth and at 90° angle to surface.

PAC process was imposed especially due to the fact that it allows the cutting of high-alloy refractory and stainless steels with maximum of productivity, through the automation capacity, through the low expenses towards traditional techniques, and also due to the quality of the cut and low thickness of the heat affected zone (HAZ), within 1.50 mm.

Previous work (Sommer, 2000) on PAC and LBM processes shows minimal differences between PAC and laser cutting as regards to *heat distortion and heat affected zone.*

Also, regarding *metal removal rates,* only PAC process can compare with conventional machining, having the greater removal rate (about 300 in^3/hr.) while LBM is so low in material removal rate (about 0.0004 in^3/hr.) that it should be considered only for special applications where no better process can produce the part.

Analyzing the *possible surface damage*, both PAM and LBM processes presents damage on the machined surface caused by the intense heat generated by the processes. Such damage generally must be removed by subsequent operations, machining, grinding, or honing for example (Weller, 1984).

Regarding the *dimensional control* issue, LBM process can produce parts with tolerance quality comparable to conventional milling (0.03 mm.), while PAM process cannot be controlled to close dimensions and its use is restructed to simple roughing cuts (1,27 mm.) (Weller, 1984).

Before buying a PAC and Lasers machine, purchase & running costs, along with *long term* maintenance must carefully be considered. Referring to workload & Cost, in many instances, a laser has to run for 20+ hours a day to justify its existence and the amount of workload has to be carefully considered. A PAC machine can often be an ideal alternative tool with flexibility and low costs requiring a much lower utilization.

Material		PAC	LBM	Material		PAC	LBM
Metals and Alloys	Aluminum	A	B	Non-metals	Ceramic	D	A
	Steel	A	B				
	Super Alloys	A	B		Plastic	C	B
	Titanium	B	B		Glass	D	B
	Refractory	C	C				

A – Good application; C – Poor; B – Fair; D – Inapplicable.

Table 3. Materials applications (Weller, 1984).

Another factor that must be considered is the space issue. Lasers need a very large floor area due to the additional equipment which is essential for its operation.

The following *disadvantages* must be taking into considerations before choosing one of the PAM or LBM processes. *PAC machining case*: edges rounding, as a result of a low power of plasma generator; high coefficient of roughness on one of the cutting surfaces, as a result of turbulence effects; appearance of burs and drops as a result of a too high cutting speed; appearance of thermal affected zone in witch high stresses can be developed and cracks can appear; radiations emissions and noise.

There are a few *disadvantages on laser cutting*, also. The material being cut gets very hot, so in narrow areas, thermal expansion may be a problem. Distortion can be caused by oxygen, which is sometimes used as an assist gas, because it puts stress into the cut edge of some materials; this is typically a problem in dense patterns of holes. Lasers also require high energy, making them costly to run. Laser cutting produces a recast layer in the kerf that may be undesirable in some applications. Lasers are not very effective on metals such as aluminum and copper alloys due to their ability to reflect light as well as absorb and conduct heat. Neither are lasers appropriate to use on crystal, glass and other transparent materials.

4. CONCLUSION

This paper summarizes the PAC and LBM processes performances. The theoretical comparison study was made taking into consideration the following issues: material applications, productivity, cost and influence of cutting on the material quality in the edge zone. After analyzing the specific technical literature that we had access to, the authors came to the conclusion that laser beam machines offer more accuracy and precision. On the other hand, PAC machines have a significant advantage over laser machines in terms of acquisition cost, maintenance and productivity. However, both laser and plasma arc cutting machines offers good results depending on the type of material being cut and of the results needed.

5. REFERENCES

Chaussier, J.F.; Gredey, D.; Bermer, D.; Vannes, A.B.; Rigolet, M. (1995). *La découpe par procédés «haute énergie» Laser - Jet d'eau – Plasma – Elecro - érosion*, Mechanical Industries Technical Centre (CETIM), ISBN 2-85400-356-X

Marinescu, N.I.; Nanu, D.; Lăcătuş, E. et.alt. (2000). *Manufacturing processes*, Optoelectronics National Institute, Bucharest, Romania

McGeough, J.A. (2002) *Micromachining of Engineering Materials*, New York – Basel: Marcel Dekker, Inc.

Mihăilă, I.V. (1999). *Non-conventional Technologies,* West Printery, ISBN 973-9329-52-7,Oradea, Romania

Nemchinsky, V.A.; Severance, W.S. (2006). What we know and what we do not know about plasma arc cutting. *Journal of Physics D: Applied Physics*, Vol. 39, No. 22, November 2006, R423–R438

Slătineanu, L.; Nagîţ, Gh.; Dodun, O. et.alt. (2004). *Non-traditional manufacturing processes*, Tehnica-Info Publishing, ISBN 9975-63-164-9, Kishinev

Sommer, C. (2000). *Non-Traditional Machining HandBook, Unit 1: Fundamentals of Non-Traditional Machining,* Advance Publishing. Ing., ISBN 1-57537-325-4, Houston

Weller, E.J. (1984). *Nontraditional Machining Processes*, Society of Manufacturing Engineers, ISBN 0-87263-133-8 Dearborn (Michigan, USA)

Annals of DAAAM for 2008 & Proceedings of the 19th International DAAAM Symposium, ISSN 1726-9679
, Editor B. Katalinic, Published by DAAAM International, Vienna, Austria 2008
Make Harmony Between Technology and Nature, and Your Mind will Fly Free as a Bird

ESTIMATION OF SPEED AND DISTANCE OF AN APPROACHING VEHICLE

INAGUMA, T[akanao]; NISHIYAMA, D[aisuke] & NAKATANI, H[iromasa]

Abstract: *We propose a driver support system that estimates the relative speed and distance of an approaching vehicle, and judges whether the driver can change lanes safely or not. We set two cameras at the rear end of a vehicle, and apply the principle of stereoscopy to three-dimensional measurements. We detect white lines on the roadway to identify the lanes, and set a vehicle search area in the image region between two adjacent lines to detect the feature points of a vehicle. When a vehicle is detected, the three-dimensional distance is calculated using the correspondence between the right and left images. If the distance is short and the relative speed is high, the system displays an alert to avoid potential collisions.*
Key words: *Driving, Vehicle, Stereo, Image*

1. INTRODUCTION

It is important for drivers to obtain the surrounding information of their vehicles. However, human eyes cannot grasp the whole view at a time. Thus, computer vision systems for traffic analysis and monitoring have been proposed (Atkouciunas et al., 2005), (Coifman et al., 1998). They set stationary cameras above the road for obtaining road traffic information. In our study, we plan to set multiple cameras around a vehicle and analyze the road traffic surrounding the vehicle. At present in this work, we set two cameras at the rear end of a vehicle to check the blind spot areas in the rear view. Our system detects vehicles that are running from behind in the adjacent lane, and can judge whether the driver can change lanes safely.

2. SYSTEM OUTLINE

We set two video cameras in the rear of a vehicle, and their height is set to 1.1m from the ground and the length of the baseline between the cameras is 0.6 m. Figure 1 shows input images that are captured by each camera with the rate of 30 frames/sec in 360 x 240 pixels. At present, we assume that the images are captured in daylight and the adjacent lane is showing its bounding two lines. Under those conditions we detect vehicle regions in the adjacent lane and find image feature points of the vehicle. We use two images to find the corresponding points between two images and calculate the three-dimensional coordinates.

Left image Right image

Fig. 1. Input from stereo camera

2.1 Lane Detection

In this paper, we focus on detecting a vehicle that is running in the adjacent lane. First, we need to identify the adjacent lane. Since lanes are separated on a road with white lines, we detect white lines from the image. We detect white lines using the Sobel edge detector (Gonzalez & Woods, 2002) and the Hough transformation (Duda & Hart, 1972).

Using the Sobel operator, we obtain edge strength and edge direction. We eliminate small edge regions from the edge image, and feed the resulting regions into the Hough transformation to detect straight lines. There are many edges that do not belong to the white lines on the road. Therefore, we limit the voting range by the direction of the edge. Since the angles of two white lines that we want to detect can be limited in a certain range in this work, we vote the pixel in Hough space only if the gradient of an edge is nearly equal to that of the two lines. Thus we only vote such edges that have the gradient θ in the range of $10^\circ < \theta < 80^\circ$ or $190^\circ < \theta < 260^\circ$. After voting in the Hough space, we search for the local maximum points to determine the white lines. Two white lines are detected and the adjacent lane region is obtained in between the two lines (Fig. 2).

Fig. 2. Horizontal edges inside a lane region.

2.2 Vehicle Detection

Vehicle regions are detected by finding long horizontal edges and feature points such as headlights, a license plate, and tires. Since the intensity of such shadow that exists under a vehicle is lower than its neighborhood, there should be a strong horizontal edge. Therefore, we detect those edges in the adjacent lane to find the boundary between the vehicle and the road.

We use the horizontal component of the Sobel operator to detect horizontal edges. After detecting horizontal edges, we need to examine whether they are contained in a vehicle region because they may be contained in a road sign or in the shadow of buildings or other vehicles. Thus, we set a vehicle search region based on horizontal edges. In Fig. 3, the white rectangle shows the vehicle candidate.

In that region, we find feature points by the Harris operator (Harris & Stephens, 1988). The number and the variance of the feature points are examined to determine whether the region is the vehicle region. If the number of the feature points is large and those points are scattered throughout the region, the region is recognized as a vehicle region.

Fig. 3. Vehicle detection.

2.3 Calculation of the Distance

When a vehicle region is determined in one image, we find in the other image such points that correspond to a vehicle region by template matching. Then we can calculate the three-dimensional distance of the approaching vehicle using the feature points, the corresponding points, and the camera parameters (Horn, 1986)

The calculated three-dimensional distance is used to judge the risk of changing lanes. Either if the distance between vehicles is very short or if the distance is short and the relative speed is high, the system warns the driver that the risk becomes high.

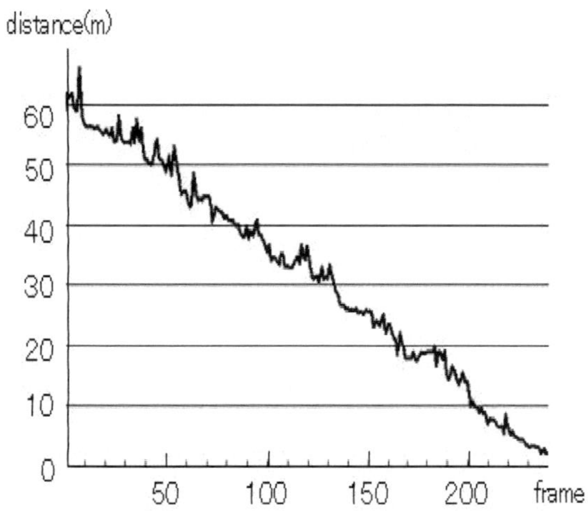

Fig. 4. Relative distance to the approaching vehicle.

3. EXPERIMENTAL RESULTS

The system was tested on a highway on a fine day. At the first frame of the image sequence, the distance to an approaching vehicle on the right lane was 60m. Then, the vehicle was approaching and passed us. Figure 4 shows the distance between the vehicles. The system also can show the top view of the surroundings (Fig. 5). The black rectangle indicates our vehicle and the white rectangle indicates the approaching vehicle. This view helps the driver to recognize the vehicle in the adjacent lane. At the 73rd frame, the system successfully warned the driver not to leave the lane.

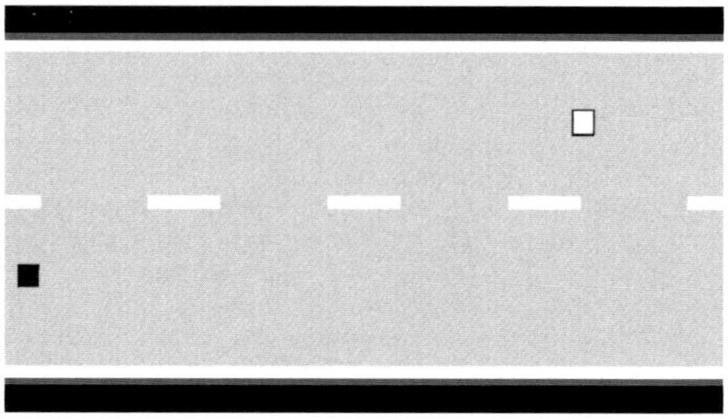

Fig. 5. Locations of the two vehicles.

4. CONCLUSION

In this paper, we proposed a system that can detect a vehicle that is running in the adjacent lane and estimate the relative distance of the approaching vehicle. Using the distance, the system judges whether a driver can change lanes safely or not. Since we use white lines to search for the adjacent lanes, the system cannot find the lane if the road bends on a curve. Moreover, the system may miss vehicle regions if there is no shadow under a vehicle. Those problems are left for our future work.

5. REFERENCES

Atkouciunas, E., Blake, R., Juozpavicius, A., Kazimianec, M. (2005) Image Processing in Road Traffic Analysis, *Nonlinear Analysis: Modelling and Control*, Vol. 10, No. 4, pp. 315-332.

Coifman, B., Beymer. D., McLauchlan, P., Malik, J., (1998) A Real-Time Computer Vision System for Vehicle Tracking and Traffic Surveillance, *Transportation Research Part C 6*, Vol. 6, No. 4, pp. 271-288.

Duda R. O. & Hart. P. E. (1972) Use of the Hough transformation to detect lines and curves in pictures, *Communications of the ACM*, vol. 15, pp. 11–15.

Gonzalez, R. C. & Woods, R. E., (2002) *Digital Image Processing, 2nd Ed*, Addison-Wesley, Reading, Mass.

Horn, B. K. P., (1986) *Robot Vision*, MIT Press, Cambridge, Mass..

Harris, C. & Stephens, M., (1988), A Combined Corner and Edge Detector, *Proceedings. of 4th Alvey Vision Conference*, pp. 147–151.

Annals of DAAAM for 2008 & Proceedings of the 19th International DAAAM Symposium, ISSN 1726-9679
Editor B. Katalinic, Published by DAAAM International, Vienna, Austria 2008
Make Harmony Between Technology and Nature, and Your Mind will Fly Free as a Bird
*Young Researcher and Scientist Paper / * Supervisor, Mentor*

ASPECTS REGARDING THE TRIBOLOGICAL BEHAVIOUR OF SOME MAGNETRON SPUTTERED TI-SI-C COATINGS

IONESCU, C[ristian]; MUNTEANU, D[aniel] & MUNTEANU, A[lexandru]*

Abstract: *The research of the Ti-Si-C thin films deposited on different types of substrates has attracted an increasing interest in the science world, because of the extraordinary properties showed by these new materials, which combine in a favorable way the characteristics of the metallic and ceramics materials. The present paper presents an overview on Ti-Si-C system and some experimental tribological results obtained with the Ti-Si-C coatings, deposited on high-speed steel substrate, at low temperatures. There are presented the tribological characteristics (wear behaviour, friction coefficients) of the deposited films, data obtained with a "ball on disk" CSM type Tribometer.*
Key words: *coatings, deposition, Ti-Si-C, wear, friction*

1. INTRODUCTION

The subject of the deposition of Ti-Si-C type thin films presents a wide interest at world level, fact revealed by the numerous articles and monographs published in the specialty literature (Barosum, 2000; Lopez et al., 2007; Emmerlich et al., 2004; Barosum, 2001; Eklund et al., 2005; Hogberg et al., 2005; Veprek, 1997). Despite of the fact that the available data regarding Ti-Si-C system (MAX phase, generally) have an increasing trend, the most of the researchers encourage the further investigation on this system.

The research regarding the Ti-Si-C system has drawn toward a new family of materials, i.e. MAX phases, Ti_3SiC_2 being the most investigated, (Barosum, 2000). The successful synthesis of MAX phases in the form of thin films has attracted the attention over these unique materials. The ternary compound known as the MAX phase ($M_{n+1}AX_n$, where M = transition metal; A = A group element; X = C and/or N; n=1-3) and, particularly, Ti_3SiC_2 phase, has attracted a considerable attention lately (Lopez et al., 2007; Emmerlich et al., 2004; Barosum, 2001) because of the extraordinary properties. In these materials, (Lopez et al., 2007; Barosum, 2001), the metallic properties, such as electrical and thermal conductivity and thermal shock resistance, are combined with ceramics properties, such as good oxidation resistance, a refractivity and high decomposition temperature (1800°C). MAX phase materials are interesting from the technological point of view, because of ductility and workability, (Emmerlich et al., 2004). This material presents, also, a high rupture resistance and a very low friction coefficient. According to (Emmerlich et al., 2004), thin films containing MAX phases, particularly Ti_3SiC_2, can become essentials in applications such: electrical contacts and coatings with wear protection role. The big disadvantage is the fact that the thin films containing MAX phases can be obtained at deposition temperatures higher than 700°C, (Lopez et al., 2007; Emmerlich et al., 2004; Eklund et al., 2005). Because of this high temperature, the choice of the substrates becomes very important, because these do not have to change the composition and especially the structure at the deposition temperature. In this way, (Lopez et al., 2007), the research programmes should be developed in order to solve this big disadvantage. On the other hand, lower deposition temperatures lead to the formation of some nanocomposite materials, considered promising for protective coatings from the electrical contacts, because of the high wear, corrosion resistance and high conductibility, (Eklund et al., 2005; Hogberg et al., 2005).

The technique used for Ti-Si-C films deposition on high-speed steel substrates is the one of sputtering. The sputtering technique is (Munteanu et al., 2007) a deposition method of thin films in vacuum, where the deposition particles in the form of neutral atoms or neutral atoms groups, which have energy between 10 eV and 40 eV, are obtained by vacuum sputtering of the solid-state deposition material.

In the last ten years, a relevant number of papers regarding the deposition of nanostructured films by sputtering were published (Lopez et al., 2007; Emmerlich et al., 2004; Veprek, 1997). The goal of these researches is the obtaining of hard films, but also tenacious, thermal stable and witch should present low friction coefficients. In the most of the cases (Lopez et al., 2007), the motivation of the researches was directly correlated with the influence that the crystalline grains have over the properties, so over the performance of these multifunctional coatings, knowing the fact that the final effects will be as relevant as the grain size is smaller. For example, from the mechanical behaviour point of view, (Lopez et al., 2007), a decrease of the grain size leads to the improvement of the mechanical resistance and toughness, by the process of blocking and stopping the dislocation pinning. Nanostructured polycrystalline materials were obtained, consisting of atoms with different chemical reactivity, witch then form different types of phases that do not exist naturally. During the deposition operation, phase segregation take place, creating multiphase materials, in witch, for example, crystalline phases are encircled by other types of phases, in the crystalline grain lattice. A nanocomposite material is created this way, such the case of the well-known Me-Si-N system (where Me = Ti, W, V), in which Me-N nanocristallites are incorporated in a Si-N amorphous matrix, (Veprek, 1997).

2. EXPERIMENTAL DATA

The Ti-Si-C thin films studied in the present paper were deposited on AISI M2 high-speed steel substrates, using the magnetron sputtering technique. The principal parameters of the sputtering operation were: substrate temperature in the deposition chamber was maintained at 300°C, substrate bias voltage –50V, argon flow 100 sccm, target – substrate distance 65 mm, Ti target current varied between 0,25 – 0,5A.

Sam No.	Composition, at.%				Atomic formula
	Ti	Si	C	O	
A2	51.14	2.62	43.09	3.25	$TiSi_{0.05}C_{0.84}$
A4	27.19	5.29	65.32	2.2	$TiSi_{0.19}C_{2.40}$
A6	48.79	5.01	44.24	1.96	$TiSi_{0.10}C_{0.90}$
A15	78.63	8.63	2.65	10.09	$TiSi_{0.10}C_{0.03}$

Tab. 1. The composition of the coatings

Table 1 presents the composition of the deposited films. The highest Ti content and the lowest C content were registered in the case of sample A15, while the highest C content corresponds to the sample A 4.

The wear behaviour was appreciated using a CSM Instruments Tribometer. The tribometer also gives information regarding the friction coefficients. The principal parameters of the wear test were: method – rotative "ball on disk", wear radius 5 mm, rotation speed of the disk 4,8 cm/s, normal force 5 N. The working principle of the tribometer is the following: the static partner (the ball) is loaded over the sample with a constant force. While the disk is rotating with the specified speed, the friction forces that appear at the ball – sample contact are measured using a sensor. The wear for ball and sample is calculated on the base of the volume of lost material during test. This simple method facilitates the study of the friction and of the wear behaviour of almost any type of solid-state material, by changing time, load, speed, temperature, humidity.

3. RESULTS AND DISCUSSIONS

Table 2 presents the results (friction coefficients, wear) obtained after the wear test. The A2 and A6 samples presented an insignificant wear, which could be considered equal with zero. This insignificant wear is a result of the high hardness registered in the case of the two samples, from figure 1 resulting clearly the strong correlation between the wear behaviour and the hardness of the analyzed coatings.

Figure 2 presents the correlation between the wear behaviour and the residual stress state. It can be observed that a very good wear behaviour is associated with low values of the residual stress.

Regarding the friction coefficients, figure 3 shows that the lowest value of the friction coefficient was registered in the case of sample A4, sample for witch it was registered the highest value of mass loss during wear.

4. CONCLUSIONS

The paper presented some general aspects regarding the deposition of Ti-Si-C films on high-speed steel substrates by magnetron sputtering. Taking into account the tribological and the mechanical characteristics of the Ti-Si-C coatings synthesized in this research, it is possible to conclude the fact that there is a strong correlation between the tribological properties (wear, friction) and hardness, residual stress state. The further research should follow the obtaining of coatings with different chemical composition, in order to see how it changes the tribological behaviour of the coatings, if compare with the coatings from this paper.

Sample no.	Thickness μm	C/Si	Medium Friction coeff.	Wear rate mm3/N/m
A2	0.9	16.44	0.45797	≅0
A4	0.3	12.34	0.23547	1.38·10-5
A6	0.6	8.83	0.54169	≅0
A15	2	0.31	0.56149	1.02·10-5

Tab. 2. The thickness and tribological characteristics of the coatings

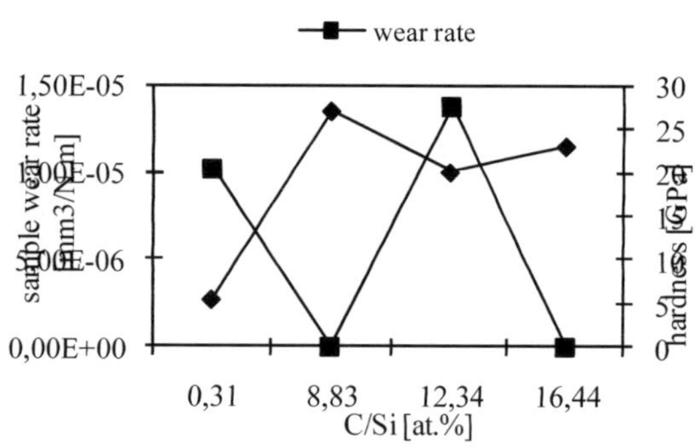

Fig. 1. Sample wear rate and hardness as a function of the C/Si at. ratio

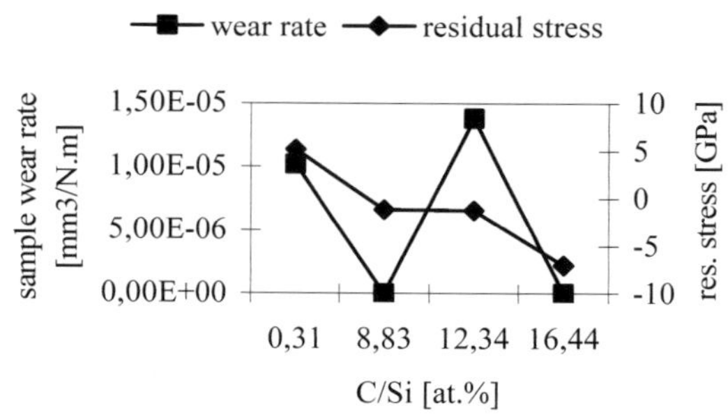

Fig. 2. Sample wear rate and residual stress as a function of the C/Si at. ratio

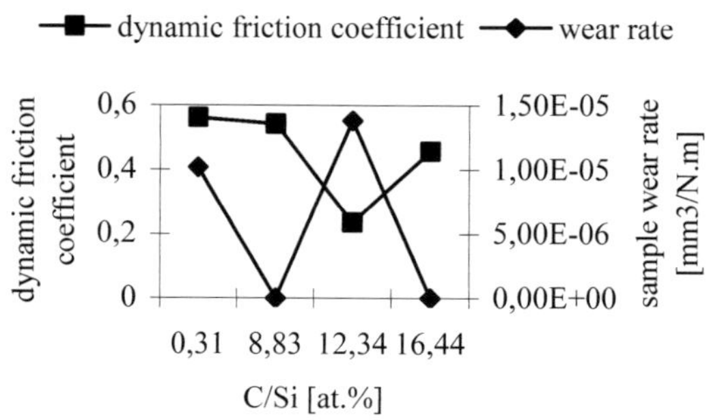

Fig. 3. Sample wear rate and dynamic friction coefficients as a function of the C/Si at. ratio

5. REFERENCES

Barsoum, M. (2000): The $M_{n+1}AX_n$ phases: A new class of solids: Thermodynamically stable nanolaminates, *Progress in Solid State Chem.*, vol. 28, no. 1, pp. 201-281

Barsoum, M. (2001): The MAX Phases: Unique New Carbide and Nitride Materials, *American Scientist*, vol. 89, no. 4, (Jul.-Aug. 2001) pp. 334-343

Eklund, P. et al. (2005): Structural, electrical and mechanical properties of nc-TiC/a-SiC Nanocomposite thin films, *Journal of Vac. Science and Technol.*, B.23, (Nov. 2005) pp. 2486-2495

Emmerlich, J. et al. (2004): Growth of Ti_3SiC_2 thin films by elemental target magnetron sputtering, *Journal of Applied Physics*, vol. 96, no. 9, pp. 4817-4826

Hogberg, H. et al. (2005): Growth and characterization of MAX phase thin films, *Surface Coat. Technol.* vol. 193, issues 1-3, pp. 6-10

Lopez, C. et al. (2007): Magnetron sputtered Ti-Si-C thin films prepared at low temperatures, *Surface and Coatings Technology*, doi: 10.1017/j.surfcoat.2007.01.025

Munteanu, D. et al. (2007): *Ti-Si-C and Ti-O-C type coatings obtained by reactive magnetron sputtering*, Transilvania University of Brasov Publishing House, ISBN 978-973-635-931-6, Brasov, Romania

Veprek, S. (1997): Conventional and new approaches towards the design of novel superhard materials, *Surface Coat. Technol.* vol. 97, issues 1-3, pp. 15-22

Annals of DAAAM for 2008 & Proceedings of the 19th International DAAAM Symposium, ISSN 1726-9679
Editor B. Katalinic, Published by DAAAM International, Vienna, Austria 2008
Make Harmony Between Technology and Nature, and Your Mind will Fly Free as a Bird

TRAINING SYSTEM FOR WELDING OPERATORS

IONESCU, G[abriela] & IONESCU, O[ctavian]

Abstract: *The training system developed is used for training simultaneously welders using different types of welding equipment such as MIG, TIG, Stick or Oxi-Gas by using effectively only one trainer, saving energy, filler and base materials, avoiding accidents or damages to the welding equipment. It consists of one computer network, inertial measurement units (IMU) which could be installed on different types of welding guns and specialized software. Trainees are operating the welding guns equipped with IMUs connected to their computers and must set the welding parameters according to the needs of application received from the trainer. The trainer is supervising through the network the activity of the trainees. The next step of development for the system is to establish a knowledge database and an inference engine and thus an expert system capable to provide the feedback directly to the trainee.*

Key words: *training, virtual, welding, inertial, network*

1. INTRODUCTION

The process of training welding operators is difficult and requires a complex system of equipments and expensive resources such us filler material, base material and electric power. The trainer has to supervise every trainee, to understand the mistakes made by each of them and to provide feedback for each individual. Due to the large number of companies producing welding equipment and the various welding processes developed, the training system should be capable to be used for a large variety of equipments and to cover a wide area of processes.

2. STATUS OF INTERNATIONAL RESEARCH AND DEVELOPMENTS

There are sustained efforts to solve the problem and to develop training systems for welders. The rising price of electric power and welding materials, the increased need for training welding operators for mastering various welding processes in a short period of time are powerful factors which are pressing for the development of virtual training systems (www.vrsim.net, www. virtualworldlets.net). Companies like MATRA Systems, Lincoln Electric and Miller have developed variants of training stations for welders; however, they are focused more on training welders to use their equipment. There are also several attempts to create virtual training systems for a specific method of welding such us Stick, Oxi-Gas, GMAW, Laser or Plasma (Fast et al., 2004, Collen & De Jong 2008, Heston, 2008). Despite those efforts, there are not developed systems capable of training simultaneously operators for different welding processes. To improve the process of training and to make it more effective it is necessary to develop an adaptable training system which combines the computer capabilities to be connected in a local network, collecting a large quantity of data from different welding stations and to use interactive the expertise of the trainer. On the screen, the trainer would receive data from one of the training posts, assess the work of trainee in real time, evaluate the progress comparing records of previous work, sent the required corrections to the student, and switch to another trainee records to analyze and supervise his progresses.

3. TECHNICAL SOLUTION ADOPTED

To solve the problem and to develop a system capable to be interfaced with various welding equipment the team from Petroleum and Gas University of Ploiesti conducted research and focused on the most important problems that should be monitored during the training of welders. The parameters chosen to be supervised during the training are: trajectory, velocities, accelerations and tilt of the welding gun on three axis, welding current and volume of shielding gases.

To provide a better interaction between the trainer and the trainees a local area network (LAN) was established within the welding laboratory and the trainer had been provided with full right of access to the workstations used by trainees.

Each training station is provided with a computer and a display, several welding guns from which the trainee has to chose the one adequate to the application received, a welding mask and a pad. The position of pad is well established and correlated with the initial position of the welding gun equipped with the IMU. When starting, the trainees are pressing a button and veriffing the correlation between the working pad and the welding gun and once a week the technician responsible for laboratory is calibrating each system.

The welding gun is equipped with a laser diode which is activated when the traine starts the welding process. The light spot generated by the laser diode is reflected from the pad and allow the trainees to follow the movement of welding gun. The welding mask is provided with a protective filter which protects the eyes of trainees.

Each trainee receives a different problem to solve and randomly the trainer supervises the records of the work done by trainees. The schematic of the system developed is presented in Fig. 1. The movement of the welding gun is monitored by using a complete Inertial Measurement Unit with 6 degree of freedom which consists of single triple axis accelerometer MMA 7260Q

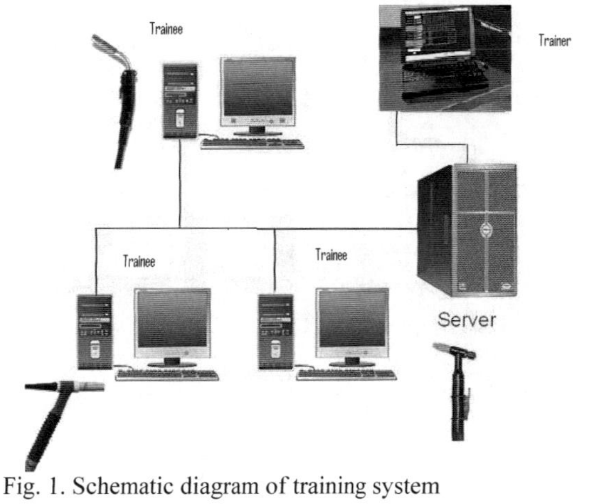

Fig. 1. Schematic diagram of training system

Fig. 2. IMU with 6 degree of freedom

and three MEMs gyroscopes for roll pitch and yaw presented in Fig. 2. All measurements are transmitted over a simple TTL level serial connection via a 4 pin header onboard.
The connection to the computer is realized using a MAX 232 circuit. On the display the trainer would have information concerning the position of the welding gun and its tilt as is presented in Fig. 3.
Also, variations of velocity and accelerations are recorded and compared with a threshold value established after several sets of trials. Whenever the students are manipulating in a wrong manner the welding gun the supervisor receive a warning and in real time is able to contact the trainee and to check which the cause of the mistake was.
In addition to the data concerning the pitch, roll, yaw and trajectory the trainer would receive information concerning the main parameters set by the trainee on the welding machine such as current amplitude and volume of shielding gas.
The software developed for this application provides the opportunity for the trainer to receive objective information and track the progresses made by the students, to recognize the systematic mistakes and to correct them.

4. RESULTS OBTAINED

Usually using classic training methods one trainer was able to supervise up to 8 trainees and there were necessary up to 3 weeks of training to achieve good results. By using the virtual training system one supervisor was able to supervise up to 20 trainees and in one week it was possible to detect and correct typical mistakes made by each student.
The virtual welding system simulates well the real welding processes and thus in a sort time, students were able to maneuver correctly the welding gun, to correlate the movement of gun with the trajectory provided by the supervisor and to set the right welding parameter. After one week of training, students were able to operate correctly the real welding equipment saving in this way filler and base material, shielding gas, electrical power and not the last avoiding accidents and damages produced to welding equipment.

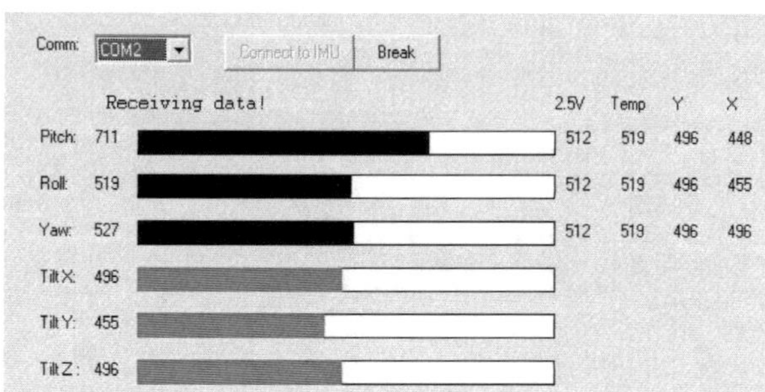

Fig. 3. Trajectory parameters collected from each trainee

5. FURTHER DEVELOPMENTS

Although the results are impressive there is more work to be done to improve the welding training system. One of the most important directions is to develop an expert system capable to analyze and assess the activity of the students and to signal to the trainer and to the trainee when mistakes are made. In this way the work of the trainer would be easier and the training time could be reduced significantly. The team from Petroleum and Gas University is working in this moment to develop a knowledge database and an inference engine and thus to provide the basic elements needed for the expert system. Developing the knowledge database it is a difficult process and would require several months of sustained work and furthermore would be an open system since welding processes are continuously developed.
Another important direction is to diversify the types of welding equipments available in the laboratory of the University and thus to provide a wide area of training for students. We are working with specialized companies for acquiring various welding guns and torches and thus solving this issue.
Another path for improvement is to develop a system which would simulate the specific sound/noise which occurs for the selected welding process and to deliver it to the students through headphones. Thus, the students would be better trained to correlate the sound specific for every welding process and able to recognize troubles associate with sounds, when occurre.

6. CONCLUSIONS

The virtual training system provides a real asset for the University or for any other training center for professional welders. It is versatile and could be used for any welding process. Also the developed system is important for a better and safer training of the students, for lowering considerably the costs and period of training. Trainees using the computerized system are more confident, they are doing fewer mistakes due to the fact that they are not afraid of damaging the equipment or wasting materials. The trainer is more relaxed since the probability of work accidents is very low and the most important has the time of analyzing the records of trials executed by trainees. Although the costs of implementing such a system are relatively high, for an institution of education which usually prepare hundreds of students every year it is an whortier investment. By simulating welding no gas is used, no metal is consumed, and no waste is created, students get more training time, and instructors get better data.

7. REFERENCES

Collen, A, De Jong. (2008). *Virtually Assembled, Available from* http://www.autofieldguide.com/articles/019804.html *Accessed*: 2008-05-07

Fast, K.; Gifford, T. & Yancey, R. (2004). *Virtual training for welding, Mixed and Augmented Reality*, pp: 298 – 299, ISMAR 2004, Third IEEE and ACM International Symposium on Volume, Issue, 2-5 Nov. 2004

Heston, T. (2008). *Virtually welding*, Training in a virtual environment gives welding students a leg up, The FABRICATOR®, *Available from* http: //www.thefabricator.com/ArcWelding/ Arc Welding _Article.cfm/ID=1878 *Accessed*: 2008-04-07

*** (2008). *Train better welders faster,* VRSim webpage, *Available from* http://www.vrsim.net/products/ simweldermain.html *Accessed*: 2008-04-14

*** (2008). *Virtual Reality Welding Trainer, Resource News,* VR Interfaces, *Available from* http://www. virtualworldlets. net, *Accessed*: 2008-05-02

Annals of DAAAM for 2008 & Proceedings of the 19th International DAAAM Symposium, ISSN 1726-9679
Editor B. Katalinic, Published by DAAAM International, Vienna, Austria 2008
Make Harmony Between Technology and Nature, and Your Mind will Fly Free as a Bird

INCREASING THE EFFICIENCY OF WATER SUPPLY SYSTEMS BY OPTIMIZING THE ELECTRICAL ENERGY CONSUMPTION

IONESCU, G[heorghe] C[onstantin] & IONESCU, D[aniela] S[maranda]

Abstract: *The present paper shows a few solutions of increasing the efficiency of water supply systems, by reducing the electrical energy consumption, which should also be correlated with the implications of the investment and exploitation costs, the extent of used surfaces, the volume and the quality of the built surfaces, taking into account that both the equipments and the surfaces built need a lot of energy for their construction.*
Key words: *water, supply, energy, electrical, optimisation*

1. INTRODUCTION

Each m^3 of water insured for the consumption includes a huge quantity of electrical energy necessary for the transportation, treatment, distribution, different internal technological processes of the water supply system. (Ianculescu & Ionescu 2002).

The practice in water supply systems shows that for 1.000 m^3 of treated water and delivered for human consumption are used 60 up to 600 kW. The very large limits of the interval mentioned (1 to 10) show the huge availabilities existent in the projection of water supply systems, regarding the insurance of a minimum of electrical energy consumption.

Finally, the electrical energy expenditure depends on the specific character of the area, the specific character of the water supply source, the treatment technology used, the means by which the water is transported to the consumer.

Next we will show some solutions which should be taken into consideration in establishing the projecting the water supply system stages in order to obtain a minimum of energy consumption.

The solutions will be displayed in the following stages:

- The elaboration of the water supply system scheme, according to the character of the area and the source;
- The implementation of some treatment technologies with low energy expenditure;
- Bad water management to the consumer.

2. THE ELABORATION OF WATER SUPPLY SYSTEM SCHEME

It is known the fact that huge quantities of electrical energy, used in water supply systems, are spent during the pumping phases.

In this way, when the consumption – per system – is around 500 - 600 kW/1.000 m^3 of delivered water, approx. 200 - 300 kW/1 000 m^3 are used for the treatment station, the rest are used for the pumping. If we take into consideration that in a treatment station, the consumption is due to pumping (reagents pumping, water for washing, etc.), we realize that in a water supply system the energy consumption used for pumping is substantial. On the other hand, there are certain water supply systems, having as sources storage basins, for which the energy consumption has reached 50-60 kW/1.000 m^3 for the water delivered to human consumption. Hence the huge energetic advantage the complex water arrangement presents, with all the storage basins and their use as water supply source. Besides other pumping phases, there are also other advantages:

- Increasing the degree of insuring the discharge;
- The possibility of employing a more sophisticated treatment scheme;
- A simpler exploitation.

The storage basins – used as a water supply source – have a well established character from this point of view.

3. THE IMPLEMENTATION OF SOME TREATMENT TECHNOLOGIES WITH LOW ENERGY EXPENDITURE

The water treatment problem is directly connected with the quality of surface or underground waters, from which sampling is taken. The better the quality of water (I class category, according to STAS 4706-74), the easier the treatment, thus the electrical energy consumption is lower.

For lower quality waters category, the treatment is more difficult (in many stages) and, therefore the energy consumption is higher. Maintaining a correspondent water quality is a desideratum wished by water management authorities, through the measures of intensifying the rhythm of achieving and operating the waste water treatment plants, water coming form industrial sources and urban zones.

Without going too much into detail we will try to illustrate a few ways, which introduced into today's projecting technology, could lead to direct and indirect methods of saving the energy.

Such a method would consist in using some materials or treatment reagents, which have similar proprieties or which allow appreciable reductions of the classical materials and reagents.

These would be:

- Coarse sand – secondary product from the exploitation of some cyanide deposits, instead of classic quartz sand;
- The use of some adjuvant polyacrylamides type (med sol produced by ICPAO – Mediaş or "Petru Poni" – Iaşi), which used in very small quantities – 0,1÷ 0,3 mg/l – allow reductions of coagulation reagents doses (aluminum sulfate) of the order of ten mg/l;
- The use of some substances with superior treating proprieties as the ozone, powder or granulated activated charcoal, which allow being used as sources of water supply and natural waters with a high degree of pollution. Although the substances mentioned above entail electric energy consumption, sometimes – even substantial, the total energetic balance can be in their favor,

especially in the areas where there is no other source of water.

Another method of reducing the electrical energy consumption in treatment plants consists in adopting some treatment technologies or methods of exploiting appropriate for the final purpose, for example:

- The use of suspension decanters by the recirculation of the suspension layer with the help of some hydro-ejectors with reduced load loss (ICPGA or ICB type); this installation type concentrates the mix, reaction and clearing in a sole installation, eliminating the rapid or slow propeller stirrer necessary in usual situations;
- Providing a decanter by-pass pipe and treating the water directly through filters; it can be applied almost all year round, for many water streams in our country; in this way the coagulation reagents consumption are reduced, thus indirect energetic savings are achieved;
- The use of technologies which involve stages of biological treatment, able to replace the stages of high energetic consumption (ozonization, ultra-violet treatment etc.)

4. JUDICIOUS MANAGEMENT OF WATER FOR CONSUMPTION

This is considered to be the area in which there's a need to insist on a change of mentality regarding the way water is perceived. Water must be considered a raw resource that includes a large quantity of energy and that must be managed with the same care the other energy-rich raw materials are managed with: iron, wood, cement etc. (Ionescu 1996)

The following are needed to this end:

- Improvements to the way reinforcements, sanitary installations and water distribution networks are built, in order to reduce water loss; presently, in some cities or farms, the loss percentage is as high as 20 ÷ 25%, and even higher in some cases;
- On industrial plants and agricultural and zoo-technical centers a large scale implementation of not only recirculation, but methods of successive use of water, starting with the technological processes that are more demanding with regard to water quality and on to the technological processes that have less rigorous water quality limits.

Maybe, in a not so distant future, modifying the industrial technological processes in order to reduce water consumption on one hand and the pollution level in waste water on the other – in other words, developing "Dry technologies" for different technological processes – will also become a subject of discussion.

In this way, the effect will be more complex, resulting in direct energy savings as well as indirect savings, by protecting the quality of the water sources that will be assumed by the consumers downriver.

5. MEANS OF REDUCING ELECTRICAL ENERGY CONSUMPTION

In order to use electrical energy more efficiently, there are methods that can be applied in the design phase of a water supply system as well as in the case of already operating targets (Ionescu 2004, Ionescu 2003).

The main means for reducing electrical energy consumption are:

- Choosing the unit power and number in such way as to avoid a close ratio between the unit powers and the total electrical power absorbed by the water supply system; otherwise an over-sizing of the electrical energy supply system occurs in order to provide enough energy for starting the electrical engines, an over-sizing that leads to additional electrical energy loss; also, the units that function simultaneously musts me established in a precise manner;
- Choosing the adequate electrical tension (0,4 or 6 kV) for the pumping mechanisms engines (the 160-315KW range permits options), in relation to the total power absorbed by the target system, taking into account that higher tensions lead to less energy losses in the energy transport system, at the same level of power consumption;
- Using the variable rotation speed of the pumping mechanisms, in the water supply systems that don't have buffer-accumulations;
- Placing electrical energy delivery and distribution facilities (including substations) in the load centers of the energy consumers; the number and size of the distribution points is also correlated to these load centers;
- Using local reactive energy compensation systems (improving the power factor;
- Establishing technological measurements in the water supply system in such way as to reduce – or even eliminate – electrical energy consumption, at peak load times in the power system, when the quality parameters of the supplied electrical energy are low, due to strain;
- Partial or total automation of the technological processes involved in water supply;
- Adequate training for the staff handling the exploitation of the water supply system;
- Eliminating water losses within the water supply system, including consumer premises.

The means for reducing electrical energy consumption must be correlated to the effects on investment and exploitation costs, extent of used surfaces, the volume and quality of the constructions, taking into account that the equipments as well as the constructed buildings involve consumption of energy.

6. REFERENCES

Ianculescu, O., Ionescu, G., (2002), *Water Supply Systems,* Ed. MatrixRom Publishing House, Bucharest, Romania.

Ionescu, G., C., (2003), Methods of Adjusting the Supply Discharge of Water Storage Reservoirs, *The XXXVIII edition of The Installations National Conference,* Sinaia, Romania.

Ionescu, G., C., (1996), Water Supply Installations, *University of Oradea Publishing House*, Oradea, Romania.

Ionescu, G., C., (1996), The Optimization of Electrical Energy Consumption in Water Supply Systems, *The XXX edition of The Installations National Conference*, Sinaia, Romania.

Ionescu, G., C., (2004), *The Optimization of the Hydraulic Installations within the Water Supply Systems,* Ed. MatrixRom Publishing House, Bucharest, Romania.

Ionescu, G., C., (2003), *Contributions to the Study and Optimization of the Hydraulic Installations Reliability within the Water Supply Systems in Urban Areas,* PhD Thesis.

Ionescu, G., C., (2004) *Water Supply Installations*, Ed. Matrix Rom Publishing House, Bucharest, Romania.

Pislarasu, I., Rotaru, N., Teodorescu, M., (1981) – *Water Supply Systems,* Ed. Technique Publishing House, Bucharest, Romania.

The Plumber magazine collection.

Annals of DAAAM for 2008 & Proceedings of the 19th International DAAAM Symposium, ISSN 1726-9679
Editor B. Katalinic, Published by DAAAM International, Vienna, Austria 2008
Make Harmony Between Technology and Nature, and Your Mind will Fly Free as a Bird

THE IMPACT OF INTELLIGENT SYSTEMS OF MANUFACTURING ON CORPORATE SOCIAL RESPONSIBILITY: A CONCEPTUAL MODEL

IONESCU, S[orin] ; DUMITRESCU, C[orina] & MILITARU, GH[eorghe]

Abstract: *This study provides a conceptual support based on the using of the intelligent system of manufacturing for improving the social responsibility of the manufacturing companies. Our findings illustrate that a company can take a competitive advantage from using the intelligent systems of manufacturing and improvement the quality of life for citizens. This study offers insights to managers interested in enhancing the efficiency of manufacturing companies because the driver of good manufacturing is efficiency. Our research suggests that it is possible to find mechanisms by which social responsibility might enhance profitability by examining the impact of intelligence systems of manufacturing on social responsibility. In further research we will develop an empirical research to test our hypotheses and validate the model.*
Key words: *e-manufacturing, corporate social responsibility*

1. INTRODUCTION

This paper focuses on the social responsibility of the manufacturing companies particularly those which are using the intelligent system of manufacturing (**ISM**). The purpose of this study is to provide a conceptual model which highlights the impact of the intelligence systems of manufacturing on the corporate social responsibility (**CSR**).

The deductive method used in this research was based on literature review and companies reports. This method allowed the identification of the determinants of CSR and construction a conceptual model. In addition, the theoretical model is the result of a research project focused on improving the corporate social initiatives that use core competences to enhance social welfare.

This paper is divided into five sections. Following this introductory section, the second section explores the key concepts and literature review. The third section presents the determinants of social responsibility and hypothesis. The next section provides a theoretical model for the social responsibility of the ISM. The final section provides some limitations and suggestions for future research and concluding remarks.

2. KEY CONCEPTS AND LITERATURE REVIEW

Corporate social responsibility is a key differentiation factor that can be sustained in the longer term. The key factor is to make CSR uptake a voluntary process. A working definition of CSR may be used from the World Business Council for Sustainable Development. "CSR is the continuing commitment by business to behave ethically and contribute to economic development while improving the quality of life of the workforce and their families as well as the local community and society at large" (Collier & Esteban, 2007).

The basic social responsibility of a manufacturing company is to stay in business. Leading companies are looking to replace the traditional ways of manufacturing by improving manufacturing quality, quick response to the market, and lower unit costs. Using computers to integrate manufacture, planning and linking together a variety of operating and management systems has significantly simplified the collecting, storage and processing of data about key processes in manufacturing. At the core of this transaction is the recognition that corporate values have to be taken into consideration more seriously and be incorporated throughout the company.

CSR has the potential to shape the strategic context for companies and can exploit longer-term opportunities. The real advantages will only be realised once CSR is tied to business strategy. Furthermore, improving abilities in creative business strategy offers new opportunities for companies to realize efficiency through collaboration.

Social dimensions of competitive context are factors in the external environment that significantly affect the underlying drivers of competitiveness in those places where the company operates. For example, in the auto industry Volvo has chosen to make safety a central element of its competitive positioning, while Toyota has built a competitive advantage from the environmental benefits of its hybrid technology. Toyota is to concern about automobile emissions by the hybrid electric/gasoline vehicle that has produced competitive advantage and environmental benefits (Porter & Kramer, 2006).

Intelligent manufacturing systems improve the efficiency with which renewable and non-renewable resources are used. Technology can be used for solving many problems of the society. In the following paragraphs, we will present some of the main intelligent systems of manufacturing.

Computer-integrating manufacturing (**CIM**) is a method of manufacturing in which the entire production process is controlled by computer. This system provides a better interaction among various components and functions of manufacturing (Stephen et al., 2007).

Computer-aided manufacturing (**CAM**) is a programming tool that allows a company to manufacture physical models using computer-aided design (**CAD**) programs.

Computer-aided quality assurance (**CAQ**) is the engineering application of computers and computer controlled machines for the definition and inspection of the quality of products.

Enterprise resource planning (**ERP**) systems integrate several data sources and processes of an organization into a unified system. The two key components of an ERP system are a common database and modular software design (O'Brain & Marakas, 2007).

3. DETERMINANTS OF CSR AND HYPOTHESIS

Our literature review about the intelligent system of manufacturing has identified four determinants of social responsibility. They have been drawn deductively from the literature about intelligent systems of manufacturing. These determinants are: clean manufacturing processes, energy efficient processes, more flexibility and autonomy in processing modules that compose manufacturing systems, and technology innovation in manufacturing processes.

Clean manufacturing processes (F1). Manufacturing processes designed to minimize environmental impact by using the minimum amount of energy and raw materials possible, increased reuse, recycling and recovery, producing of more sustainable products, and producing limited waste or emissions (green technology). Increasing the life of products through better design is pollution prevention.

Energy efficiency of manufacturing processes (F2). For some industries, gas and electricity represent a major portion of total operating costs. Reducing energy consumption can significantly decrease these expenses and global warming impact. Using less energy reduces carbon emissions and other air pollutants that result from the burning of fossil fuels.

Flexibility and autonomy (F3). Competition in the manufacturing industry over the next decade will be focused on the ability of being flexible and rapid response to unanticipated changes while keeping costs at the mass-production level. The goal of this system is to produce and deliver customized products rapidly while keeping costs at the mass-production level. The key for adjusting the manufacturing capability is to reconfigure the system, developing and integrating new functions when necessary (Heizer & Render, 2006).

Technology innovation in manufacturing processes (F4). Technology today can deploy new business processes that support environmentally sound practices and reduce costs by conserving energy and other natural resources. Other technologies like enterprise resource planning (ERP), logistics applications, supply chain management applications, product lifecycle management (PLM) also enable businesses to be more social responsible (Senguta et al., 2006).

Determinants of social responsibility can have both positive and negative effects on CSR. Nevertheless, since drivers have been positively related to CSR and long-term performance of companies, we propose that:

***H1**: Clean manufacturing and energy efficient of manufacturing processes have a positive impact on CSR*
***H2**: Flexibility and technology innovation in manufacturing processes will have a significant effect on the CSR and long-term success of companies*

4. THE RESULTS OF RESEARCH

The main result of this research based on both the hypothesis mentioned above and the determinants of CSR is the theoretical model of the impact of ISM on CSR. This model is depicted in Fig. 1.

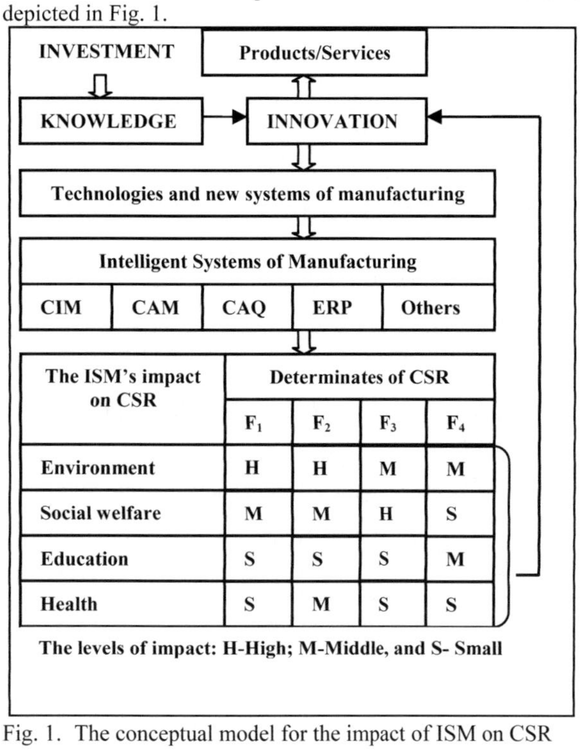

Fig. 1. The conceptual model for the impact of ISM on CSR

This model indicates that there is a positive relationship between investment in knowledge, innovation and development of new intelligent systems of manufacturing, and the social responsibility of the manufacturing companies.

The major contribution of this theoretical model to managers and professionals is that it offers a background to understand how an intelligent system of manufacturing can increase the efficiency and social responsibility of the manufacturing companies. Although it is difficult to measure the impact of each ISM on the CSR, the findings from this study show that all have a considerable influence on the CSR.

The results of the study shown in Fig 1 emphasize the impact of the determinants on CSR. Clean manufacturing processes (F1) and energy efficiency of manufacturing processes (F2) have high impact on environment while flexibility and autonomy (F3) has high impact on social welfare. The impact of technology innovation in manufacturing processes (F4) is lower because is difficult to quantify.

The driving force behind the growth of advanced manufacturing is the development and implementation of advanced technologies and processes. The future of advanced manufacturing depends on both continuous investment in research (knowledge) and well trained labour. These systems enable industry to contribute to global wealth creation. This finding has theoretical and practical implications in that, the manufacturing companies wish the improvement of their social responsibility through higher investments in ISM.

5. CONCLUSIONS

This study helps managers to understand the way in which ISM influence CSR. For this reason four types of determinants were identified, namely, the clean manufacturing processes, energy efficient of manufacturing processes, flexibility and autonomy, and the technology innovation in manufacturing processes. These determinants exert different impacts on CSR.

In this research the authors have noticed that CSR as one important part of manufacturing systems is not appreciated enough. Therefore, the conceptual model was necessary to integrate the influence of all determinants on CSR. From this model it is observed that investments in knowledge increase the innovation potential with direct influence on technologies and new systems of manufacturing especially in intelligent systems of manufacturing.

In further research we will develop an empirical research to test our hypotheses and validate the model. The development of the model will allow the quantification of impact levels.

6. REFERENCES

Collier, J. & Esteban, R. (2007). Corporate social responsibility and employee commitment. *Business Ethics: A European Review,* Vol. 16, No. 1, (January 2007), pp. 19-33, ISSN 0962-8770

Heizer, J. & Render, B. (2006). *Operations Management,* Pearson Prentice Hall, New Jersey, ISBN 0-13-185755-X

O'Brain, J. & Marakas G. (2007). *Enterprise Information Systems,* McGraw-Hill/Irwin, New York, ISBN-10: 0-07-110710-X

Porter, M. & Kramer, M. (2006). Strategy and Society. The link between Competitive Advantage and Corporate Social Responsibility. *Harvard Business Review,* December 2006, pp. 78-92, ISSN 0017-8012

Senguta, K. Heiser D. & Cook L. (2006). Manufacturing and Service Supply Chain Performance: A Comparative Analysis. *The Journal of Supply Chain Management* Vol. 9, pp. 4-15, ISSN: 1523-2409

Stephen H. Cumming M. & Philips A. (2007). *Management Information Systems. For the information age,* McGraw-Hill/Irwin, New York, ISBN 007305223X

Annals of DAAAM for 2008 & Proceedings of the 19th International DAAAM Symposium, ISSN 1726-9679
Editor B. Katalinic, Published by DAAAM International, Vienna, Austria 2008
Make Harmony Between Technology and Nature, and Your Mind will Fly Free as a Bird
*Young Researcher and Scientist Paper / * Supervisor, Mentor*

USING PATTERN RECOGNITION TECHNIQUES TO DETERMINE PROFITABILITY ON FINANCIAL MARKETS

IONESCU, S[tefan] A[lexandru]; GHEORGHE, C[amelia] M[onica] & ZAHEU, I[oana]

Abstract: *The paper aims at characterising the Romanian companies listed on the Bucharest Stock Exchange by taking into account a number of financial ratios and market variables. These are grouped in five categories (liquidity, solvency, efficiency, profitability, and market value), and the analysis follows their changes in time, trying to evaluate their development possibilities in time. The research method empoyed is cluster analysis, while the period under analysis is 2002 – 2006, with annual data used. The cluster analysis progressively groups the companies, according to two main criteria: objects' similarity within the groups and groups' differentiation. We use, as clustering procedures, the single linkage method and the Ward's method and we also construct a number of clusters using the k-means method. The results show that Romanian companies do not display any industrial specificities that are reflected in their financial performance. Still, further research is needed in order to provide categorical arguments in favor of this conclusion.*
Key words: *pattern recognition, data analysis, cluster analysis, stock exchange, portfolio analysis.*

1. INTRODUCTION

Analysis of the financial performance of companies, both from the accounting and business development from the perspective on the capital market, is for financial investors, but also strategic, a permanent fixture, depending on the accuracy of this analysis, ultimately, the structure of portfolios they hold and the yields obtained.

One of the methods used to formation of diversified portfolios is considering inclusion in the same portfolio of companies coming from various sectors of activity. The logic of this method departs from the recognition that the prices of financial securities on the capital market and yields generated by them are influenced by a number of risk factors, which affect not the same as all companies.

We propose in this work, to make a first step towards the construction of standards sector by seeking to achieve a grouping of companies listed on the Bucharest Stock Exchange, depending on their financial performance. Subsequently, we intend to notice the evolution over time of the companies analysed by reference to such groups made up. Dozen of indicators are used to make predictions, but very little work consistently in time. Investing in companies that have done well in a certain period of time can make a profit higher than the use of indicators classics as S&P500. (Damodaran, 2003)

2. PREVIOUS RESEARCH

Sharpe is the one that opened the way to theoretical developments in the centre who share these risk factors specific to the firm and variable following the formation of portfolios, on the one hand, and systematic, affecting all companies on the same capital market (Sharpe, 1963).The latter are considered to be unvariable nationally, but variable at international level.

King was the first which showed a common movement of stock prices for companies in the same industry, which is stronger than the market effect (King, 1966). Using cluster analysis for grouping companies according to their performance can be found from the work produced by Jensen (1971), Meyers (1973) and Livingston (1977), who tries to identify the natural groupings of U.S. companies along sectors of activity from which they originate. All the three authors mentioned using a relatively small set of companies and industries, and their results are satisfactory (Livingston, 1977)

3. METHODOLOGY OF CALCULATION

Applying this method to the Romanian capital market seems at first glance, a fairly easy task. Nevertheless, reduced the number of companies traded on the Bucharest Stock Exchange (65 currently, those two classes of stock exchange) and relatively low financial information regarding the task they are much more difficult. Too, believe that the distribution companies in terms of trade and often the presence of a very small number of companies in a particular sector raises additional application of such methods.

A method of grouping or classification of similar entities according to a series of features that are considered each entity is represented by cluster analysis. In this type of analysis, which belongs to the group multivariety analysis methods, the problem is that the classification entities in a number of groups, called clusters, where the entities in the clusters are more homogeneous than entities in different clusters. This sort of analysis in the discovery of natural cluster according to a specific criterion internally, without knowing in advance party entities considered to these clusters. Criterion internally that is used in cluster analysis is based on a measure of similarity or approach between the entities concerned.

Applying the cluster analysis raises, in general, two issues: the first is that the methods of measuring the distance between the entities concerned. We note entities by i and j, given the coordinates (X_{1i}, X_{2i}), respectively (X_{1j}, X_{2j}), depending on the X_1 and X_2 considered. A survey i is considered to be closest (more similar) of j remark then remark k if D (i, j) <D (i, k). The variables are standardized before applying cluster analysis. The second problem concerns the methods of formation of clusters, they may be divided into two broad categories, depending on assumptions which are based on the nature and results:

- hierarchical clustering methods, including clustering methods through aggregation and clustering methods by division for each of the procedures being used to prepare clusters, among which are the most popular single linkage method, complete linkage method, or Ward's method.
- iterative methods of clustering also include a series of algorithms, of which the most popular and use the algorithm are the K-means, k-medoids algoithm, CLARA algorithm, or Fuzzy c-means clustering. As iterative method we use non-hierarchical grouping, known mostly as k-means clustering is a priori principle fixing the number of groups to be obtained and

grouping objects (cases) so as to steal the biggest differences between media groups (statistically tested on the basis of an ANOVA type procedure).

After finding a corresponding way of separating Ω set elements on the prediction classes found, the main task for the discriminante analysis is to decide regarding membership of the n classes of new objects from the multitude Ω or to make predictions concerning the affiliation of these objects.

4. DATA USED 2002-2006

Analysis of this work is carried out for the period 2002-2006, for a number of 45 companies listed on the Bucharest Stock Exchange (BSE). Although the number of companies listed on the BSE varied during the aforesaid from 65 in 2002 to 58 in 2006, we selected only those who were listed permanently, without being withdrawn from trading in that period. In each case we used eight variables, defined as: liquidity ratios (current ratio); solvability ratios (debt-to-equity ratio, total debt ratio); efficiency ratios (total assets turnover); profitability ratios (return on assets, return on equity, net profit margin); indicators of market value earnings per share)

5. RESULTS OBTAINED

Applying first time the simple aggregation method, we see that the number of clusters can be identified for all those five years review is low, ranging between two and four. For all the years, it worth to underline the grouping of the "Financial Investment Companies" (FIC) in a separate cluster, which highlights the characteristics of this type of investment on the market. In this division cluster of companies in those four years, the identification of sectorial groups is not possible, in the same cluster can be found companies from different sectors: machine tools, chemical industry, textiles, etc.

Applying the Ward's amalgamation method leads to the differentiation of the same five-FIC, again they form a cluster for each of the years taken into account. Since the method uses standard deviation as an instrument for measuring the distance between the entities, the new clusters are more numerous and include a fewer number of companies. Nevertheless, can not speak about the formation of cluster at sectorial level, and even less than their stability over time, companies "moving" from a cluster to another over the period studied. This suggests that the Romanian capital market performance and financial are fluctuant, and, of course, we can not talk about the existence of sectorial performance features companies. Figure 1 (processed in Statistica 7.0) illustrates the dendrogram for the year 2006, the other may not be submitted to the lack of space available.

To confirm the results obtained by applying hierarchical amalgamation methods, we also used the k-means algorithm, while each of the years considered a number of three clusters, which included companies with different performance characteristics. The three clusters formed for those four years shows different attributes, which leads us back to conclusion that the Romanian market can not talk about stability in time of

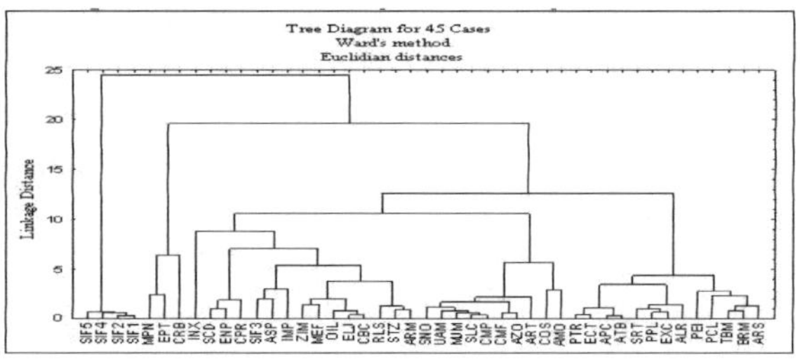

Fig. 1. Formation of clusters Ward's method (2006)

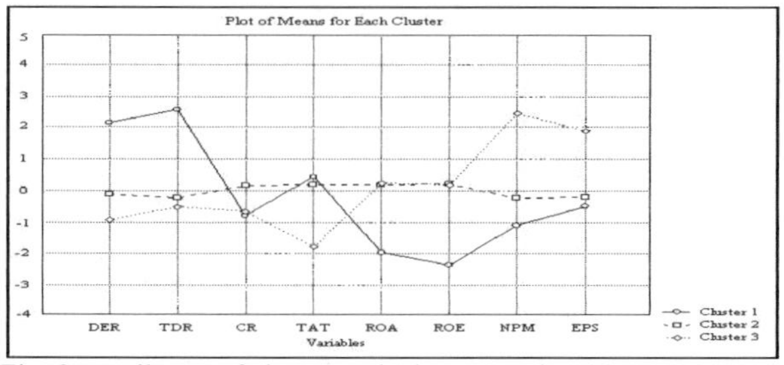

Fig. 2. Attributes of clusters - the k-means algorithm (2006)

financial performance of companies Thus, for the year 2006 (see Figure 2 – processed in Statistica 7.0) we see the formation of a cluster characterized by low solvency, low profitability and low yields on the BSE, another one characterized by the average performance for all aspects of activity and another one with high profitability, accompanied by low efficiency. In the last cluster we find again only FICs.

6. CONCLUSIONS

Pseudo-industry groups formed from cluster analysis do not match the performance of industry classifications on an out-of-sample basis (Chan et. al, 2007). The K-means algorithm has provided us an interesting and correct classification of the 45 firms, without being able to establish ranks according to the industry. The only cluster that has gathered companies in the same sector was one that contained FIC. This was somewhat predictable because these companies are investing similar, in the same areas or even in the same company. Because BSE is very young compared with markets in America and Western Europe, these differences appear more overtoned in Romania. With discriminant analysis we calculated the classification of functions that may help in future predictions.

7. WHAT TO DO

The research presented will be continued by including other amalgamation methods, on the one hand, and using discriminant analysis in order to better distinguish between such groups formed. At the same time, identifying industry clusters of companies in the Romanian capital market represents a first step towards building a system of performance rating.

Extending the base of indicators and determining the cluster for each year in part would be useful for an analysis of evolution in the dynamics, an observation of the clusters modification from one year to another and for explanations of the migration of objects from one cluster to another.

8. REFERENCES

Chan, L.K.C.; Lakonishok, J. & Swaminathan B. (2007) „Industry classifications and the comovement of stock returns" *Financial Analysts Journal*, vol.63, no.6, November/December, 2007, pp56-70, ISSN : 0015-198X

Damodaran, Aswath. (2003) *„Investment Philosophies: Successful Strategies and the Investors Who Made Them Work"*, John Wiley and Sons, ISBN : 0-471-34503-2, NY

King, B.F. (1966) „Market and Industry Factors in Stock Price Behavior", *Journal of Business*, vol. 39, no.1, January, 1966, pp139-190, ISSN: 0021-9398

Livingston, M. (1977): „Industry Movements of Common Stocks", *Journal of Finance*, vol.32, no.3, June 1977, pp861-874, , ISSN: 0022-1082

Sharpe, W. F. (1963) „A Simplified Model for Portfolio Analysis" *Management Science*, vol 9, no.1, January, 1963, pp277-293, ISSN : 0025-1909

Annals of DAAAM for 2008 & Proceedings of the 19th International DAAAM Symposium, ISSN 1726-9679
Editor B. Katalinic, Published by DAAAM International, Vienna, Austria 2008
Make Harmony Between Technology and Nature, and Your Mind will Fly Free as a Bird

SEVERAL PROPERTIES OF A FRICTION STIR PROCESSED LAYER OF ALUMINIUM

IORDACHESCU, M[ihaela]; IORDACHESCU, D[anut] & SCUTELNICU, E[lena]

Abstract: *The paper presents several properties of cold rolled AA6061-T4 alloy after Friction Stir Processing, FSP, a derivative from Friction Stir Welding, FSW. The "cold working" conditions were used and the investigation aimed the material characterisation through microhardness and residual stresses analysis. It was found that by FSP of AA6061-T4 produces the material effective surface modification characterized by material softening and compressive residual stresses, improving the fatigue resistance.*
Key words*: friction, processing, aluminium, microhardness, stresses*

1. INTRODUCTION

Recently, a derivative from Friction Stir Welding - FSW, the so called Friction Stir Processing – FSP was proved as being useful for inducing directed, localized, and controlled materials properties in any arbitrary location of components. A rotating cylindrical tool with a pin to heat the material by friction is used. The tool pin stirs the plasticized material, and a processed material layer results when it longitudinally moves. The process is carried out in solid state at low temperature, typically below 0.3 of the aluminium alloys fusion temperature (T_f), (Staron et al., 2004). The thickness of the modified layer is directly related to the tool pressure, whereas it is inversely affected by the rotation speed of friction tool and the traverse speed, (Colligan, 1999). The microstructure and properties of the modified layer depend on the amount of frictional heat generated by processing (Mishra & Ma 2005). The processed layer microstructure is complex and highly dependent on the position within the processed zone. This is due to the large local variations in the plastic flow and to the thermal history. In case of FSP of as cast AA6061 aluminium alloy the processed zone is deformed and under the thermal effect recovers or recrystallizes in fine grain microstructure (Iordachescu et al., 2007).

Residual stresses presence after FSW of different aluminium alloys was previously reported. These can influence the structure service performance with respect to fatigue and corrosion properties. However, the literature contains contradictory information about weld-induced residual stresses and the analysis of stress distribution, (Staron et al., 2004), and (Mishra & Ma 2005). Therefore, residual stress analysis after FSW/FSP of aluminium alloys still needs in deep investigations.

The present investigation aimed the FSP material characterisation through microhardness and residual stresses analysis. The paper presents several FSP induced properties into cold rolled AA6061-T4 aluminium alloy, demonstrating that the changes are beneficial for improving the fatigue resistance. The main process parameters were adjusted to reduce the frictional heat; ensuring the "cold processing". Thus, a larger basin shape processed layer results, and lower residual stresses are expected to occur, than in the hot working case.

The "*cold working", and the "hot working" conditions*, respectively, were previously defined through the ratio ω/v, where ω is the rotation speed, [rpm] and v_a – the travel speed, [mm/min], of the processing tool (Iordachescu et al., 2007).

Shoulder diameter [mm]	15
Pin diameter [mm]	5
Pin length [mm]	4
Rotation speed ω, [rpm]	1120
Travel speed v_a, [mm/min]	320
Friction pressure (up-setting force), [kN]	25
Pin angle, [^{0}]	2.5

Tab. 1. FSP tool characteristics and process parameters.

2. EXPERIMENTS

FSP tool characteristics and the process main operating parameters as the friction pressure (tool upsetting force), tool rotation and travel speed are presented in Table 1. The investigation was conducted on cold rolled AA6061-T4 plates, 10x200x300 mm, using the "cold working" conditions ($\omega/v_a = 3.5$).

Vickers microhardness measurements were performed on transverse sections of the processed layer by using a micro-hardness tester at 100 g load and 15 s dwell time. The X-ray, peak shift $\sin^2\psi$ technique with CrKα radiation was used to determine the residual stresses by measuring the surface strains.

2.1. Hardness Field in the Processed Material

Fig. 1 shows the microhardness distribution in the transverse section of the single layer processed of AA6061-T4 alloy, corresponding to different depths. The microhardness of the stirred zone, SZ, is considerably lower than that of the parent metal (which is situated in a range of 95-105 HV), varying from the top to the bottom of the processed layer. Generally, the hardness depends on the precipitate distribution rather than grain size. It is likely that low hardness in the SZ can be attributed to dissolution of the precipitate during FSP. The width of this soft zone depends on peak temperature and thermal cycle duration.

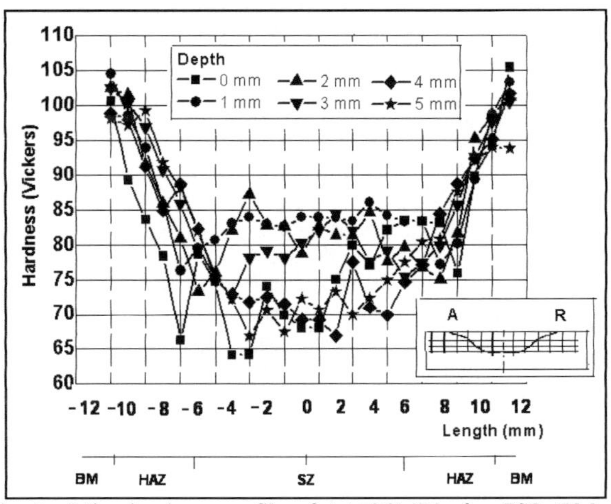

Fig. 1. Microhardness profile of AA6061-T4 after FSP: BM-Base Metal; HAZ - Heat Affected Zone; SZ - Stirred Zone; A - Advancing side; R - Retreating side.

The retention of the processed region at a certain temperature for prolonged period generates considerable softening. According to previous research (Dalle Donne et al., 2001), this is due to the temperature range going by about 450°C within the SZ of AA 6XXX aluminium alloys. In case of AA 6061-T4, this temperature is sufficient to dissolve all precipitates; when the processing ends, the cooling rate is sufficiently rapid to retain alloying elements in saturated solid solution. An abrupt change in hardness between Base Metal (BM) and Heat Affected Zone (HAZ) is evident at a distance of ±7 mm from the FSP layer centre. At a distance of about ±7 mm from the centre of the processed layer, the hardness dips to a value of about 65 HV, and 70 HV for advancing and retreating side, respectively (Fig. 1). This indicates that the naturally age hardened parent metal has been overaged as a result of the processing induced frictional heat. HAZ softening is explained by the coarsening of the metastable phases due to temperature increase during processing. The loss of hardness in HAZ can be due to the dissolution and/or coarsening of Mg_2Si phase, the main strengthening phase in Al-Mg-Si alloys.

2.2. Residual stresses distribution

Three principal sources of residual stresses are the shrinkage of the processed material, its quenching, and possible phase transformation, respectively. Generally, shrinkage causes the appearance of tensile stresses, whereas quenching and phase transformation cause compressive stresses.

Residual stresses values at the top and bottom side of the friction stir processed layer, respectively, were made by X-ray diffraction, on a Ω – diffractometer, using the $\sin^2\psi$ – method. The resolution in the in-plane direction was approximately 2 mm. The residual stress values are obtained considering that for a given variation *(Δd)* of the metal lattice interplanar distance (*d*), due to residual stresses, the *(θ)* angle described by the Bragg law will vary; differencing this relationship it results:

$$\frac{\Delta d}{d} = -\cot\theta \times \Delta\theta \tag{1}$$

Therefore, if the interplanar distance and the Bragg angle of the material without residual stresses are d_0 and θ_0, respectively, and the diffraction angle of the X-rays is φ, the strain measured on a direction (ψ) parallel to the specimen surface is:

$$\varepsilon_{\varphi\psi} = \left(\frac{\Delta d}{d}\right)_{\varphi\psi} = -\cot\theta_0 \times \left(\theta_{\varphi\psi} - \theta_0\right) \tag{2}$$

Writing this as a function of the principal stresses (σ_1 and σ_2) and the stress along the φ direction, and differencing with respect to $\sin^2\psi$, it results

$$\sigma_x = \frac{E}{1+\nu}\left(\frac{\partial\varepsilon_{\varphi\psi}}{\partial\sin^2\psi}\right), \tag{3}$$

the residual stresses, of tensile or compressive state; E is the Young modulus and ν the Poisson ratio.

Since the theoretical variation of the residual stresses on the thickness of a plate heated on one side is well known, it was satisfactory to measure the stress on the opposite sides of the plates ("top" and "bottom", Fig. 2). The residual stress distribution found on the transverse direction of the processed layer was inhomogeneous. For the whole examined specimen, "M" like stress distribution was obtained on the plate top surface, and only compressive residual stresses were found. The maximum values of these are located at the HAZ-SZ interface. The residual stresses in other zones are smaller than these maximum values, and the BM adjacent to the HAZ, as well as the processed layer contains compressive residual stresses of low values.

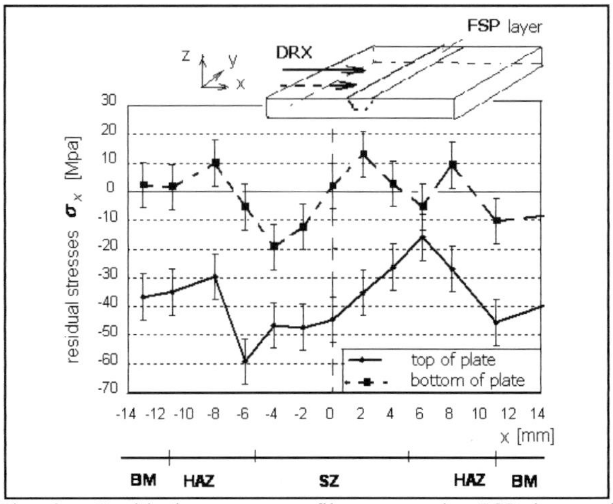

Fig. 2. Residual stresses profile across the FSP layer of AA6061-T4.

Compressive residual stresses on the surface of the processed layer indicate that they are caused mainly by phase transformation. The "M" like stress distribution is still found at the bottom part of the processed plate, but the residual stresses values are close to those corresponding to the initial stress state of the sheet material, and are both tensile and compressive.

The magnitude of the compressive residual stresses reached in HAZ, measured on the plate top surface, is caused by the process heat input in the "cold working" conditions. The resulted relatively wide HAZ and the lower material cooling rate have produced also the relatively low magnitude of the compressive residual stresses in the processed zone. In general, the compressive stresses can be beneficial by reducing the crack propagation speed, and by increasing the loading capability of the structure (where applicable).

3. CONCLUSIONS

- FSP on AA6061-T4 makes an effective surface modification characterized by material softening and compressive residual stresses.
- The hardness loss reflects the material softening and increases the material behaviour at dynamic loading.
- The compressive residual stresses obtained by FSP are beneficial, improving the material fatigue resistance; further investigation in this area is envisaged.

4. REFERENCES

Colligan, K. (1999). Material Flow behaviour during Friction Stir Welding of Aluminium, *Welding Journal*, Vol. 75, No. 7, Jul. 1999, pp. 229s-237s, ISSN 0043-2296.

Dalle Donne, C.; Lima, E.; Wegener, J.; Pyzalla, A. & Buslaps, T. (2001). Investigations on Residual Stresses in Friction Stir Welds, *Proceedings of 3rd International Symposium on Friction Stir Welding,* Kobe, Japan, 27 - 28 September 2001, TWI (UK), pdf/CDrom.

Iordachescu, M.; Iordachescu, D.; Scutelnicu, E.; Vilaca, P. & Ocana, J. L. (2007). Aluminium friction stir processing – roughness vs. macro/microscopically results, *Welding in the World,* Vol. 51, Sp. Iss., Jul. 2007, pp. 441-448, ISSN 0043-2288.

Mishra, R.S. & Ma, Z. Y. (2005). Friction Stir Welding and Processing, *Materials Science and Engineering*; Reports, Vol. 50, No. 1-2, Aug. 2005, pp. 1-78, ISSN 0927-796X.

Staron, P.; Kocak, M.; Williams, S. & Wescott, A. (2004). Residual stress in friction stir-welded Al sheets, *Physica B: Condensed Matter,* Vol. 350, No. 1-3, S. 1, Jul. 2004, pp. e491-e493, ISSN 0921-4526.

Annals of DAAAM for 2008 & Proceedings of the 19th International DAAAM Symposium, ISSN 1726-9679
Editor B. Katalinic, Published by DAAAM International, Vienna, Austria 2008
Make Harmony Between Technology and Nature, and Your Mind will Fly Free as a Bird

RESEARCH CONCERNING MACHINE-TOOL ACCURACY BASED ON DYNAMICS BEHAVIOUR

ISPAS, C[onstantin]; ANANIA, F[lorea] D[orel] & ZAPCIU, M[iron]

Abstract: *The paper presents a possibility to reduce the errors of high speed Gantry machine tools by changing CAD and CAM models according with dynamical behaviour of the machine tools. The study results of the machine dynamics were used into homogeneous transformation matrix. Software was developed to generate compensation surfaces, curves or points of the pieces in order to generate a better NC programme*
Key words: *positiong accuracy software, machine errors, dynamic behaviour, CAD-CAM compensation.*

1. INTRODUCTION

High speed machining (HSM) is a key which enables technology to be used in an increasing number of industries. In the aerospace industry, structural components are increasingly being machined as monolithic structures from a single billet. (Ispas C-tin et al. 2007) The results are drastically reduced part counts, assembly costs, and even maintenance costs.

In the present paper, an approach is developed to optimize NC programs by implementing the machine tools errors on each axis into algorithm. Therefore, the CAD (Computer Aided Design) systems can generate optimized surfaces based on real and ideal pieces surfaces for CAM (Computer Aided Manufacturing) advanced systems.

2. MACHINE TOOLS ACCURACY

Numerous error origins affect tool tip position. Among the key factors that affect the accuracy of this relative position are the geometric errors of the machine tool and thermal effects on the machine tool axes (Marinescu et al. 2002). Other error origins are the resolution and accuracy of the linear measuring system, elastic deformation of drive components, inertia forces when braking/accelerating, friction and stick slip motion, the servo control system and cutting force and vibration (Anania at al., 2007).

For a multi-axis machine, the calibration should include each axis and its roll, pitch, yaw, squareness and positioning error in the workspace (Ispas et al., 2006). The static working load and the mass of the workpiece being machined produce distortions that result in positioning errors in the machine tools.

The following Fig. 1 shows the error origins of multi-axis machine tools and their high level relationships. Broadly, machine tools errors can be divided into two categories: systematic errors and random errors. Systematic errors can be described and are predictable based on some mathematical models. Random errors are difficult to model and to compensate.

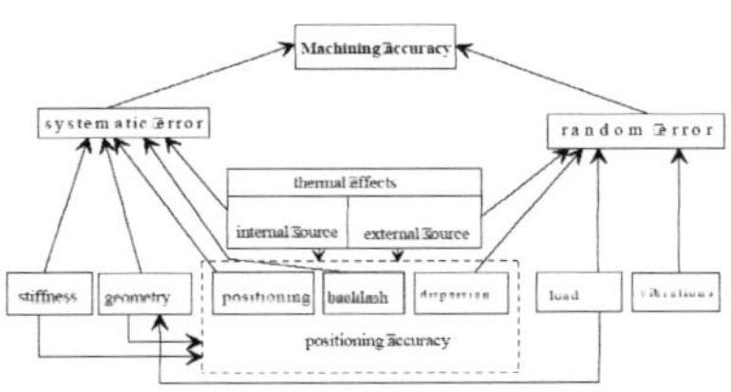

Fig. 1. Total error sources of machine tools.

The real position of tool tip in space will be translated and rotated after each axis of Cartesian systems. The tool tip position will be translated on X with x_r (cumulated machine tool linear errors on X axis), on Y with y_r (cumulated machine tool linear errors on Y axis), on Z with z_r (cumulated machine tool linear errors on Z axis) and rotated with Ψ (cumulated machine tool angular errors on X axis), with φ (cumulated machine tool angular errors on Y axis) and with θ (cumulated machine tool angular errors on Z axis). In reality this position can not be measured by machine tools command system so it can not be corrected.

For transformation from Cartesian system *OXYZ* into Cartesian system *O'X'Y'Z'* the homogeneous transformation matrixes were used.

The transformation vector from point O to point O' I results as:

$$T_{OO'} = R_x \cdot R_y \cdot R_z \cdot T_r . \tag{1}$$

$$T_{OO'} = \begin{bmatrix} (\cos\psi \cdot \cos\varphi - \sin\psi \cdot \cos\theta \cdot \sin\varphi) \cdot r_x + \\ + (\cos\varphi \cdot \sin\varphi + \sin\psi \cdot \cos\theta \cdot \cos\varphi) \cdot r_y + \\ + \sin\psi \cdot \sin\theta \cdot r_z \\ (\sin\psi \cdot \cos\varphi + \cos\psi \cdot \cos\theta \cdot \sin\varphi) \cdot r_x - \\ - (\sin\psi \cdot \sin\varphi - \cos\psi \cdot \cos\theta \cdot \cos\varphi) \cdot r_y - \\ - \cos\psi \cdot \sin\theta \cdot r_z \\ \sin\theta \cdot \sin\varphi \cdot r_x + \sin\theta \cdot \cos\varphi \cdot r_y + \cos\theta \cdot r_z \\ \end{bmatrix} \tag{2}$$

The C++ software takes in account the geometrical errors (linear and angular) measured.

The errors generated by other factors can also be easily introduced into calculus. The generated elements are used to change CAD and CAM model.

Based on this compensate surface an NC program should be generated for high speed surface machining. The compensate errors plus machine tool errors should be nearly 0 (ideal machine tool).

3. STUDY CASE OF A GANTRY MACHINE TOOL

In this paper, it is presented a study case of a GANTRY milling machine tool. There are established some methods to study the dynamical behaviour the machine tool into working environment. There were measured some systematic errors and random vibration using a Vibroport41 devices.

Under continuous machining conditions, two types of vibration occur as a result of movement between a workpiece and tool: external vibration (result of interference force transmitted through the foundation into the machine); self-excited vibration, (the machine system oscillates, basically, at one or more natural frequencies, when no external forces are acting). (Valdes et al. 2006)

For this study case were used the „Frequency spectrum" and „Transfer" functions. The acquisitions were made with a vibroport 41 device and two piezoelectric accelerometer Schenck type AS-020 S/N: 0022FE7H (fig.1). This equipment is located to the National Research Centre for Performances of Technological Systems - Optimum into University POLITEHNICA of Bucharest.

The measurements points are presented in figure 3. The numbers represent the position of the accelerometer and the arrows represent the direction of measurement. The letter "A" represents the point of the impact for transfer functions measurement.

An example of data obtain a for transfer functions measurements on the machine structure (accelerometer in position 2, direction on Y machine axis and impact in A after Y direction) is presented in fig 4. After analysing all measured data it was identified the next frequency proper 12.5Hz; 25Hz; 50Hz, 62.5Hz, 75Hz.

The dynamical study was made for the machine tool in the next working condition: machine stopped; the main spindle speed:1500 rot/min, 3000 rot/min, 5000 rot/min, 7500 rot/min (fig.5), 10000 rot/min, 20000 rot/min of and for linear movements with speed:: 1000 mm/min, 5000 mm/min and 15 0000 mm/min..

After analyzing the data it was identified a number of frequencies which are repeating almost into all measurements. So this frequency could influence directly the performances of the machine tool. The frequencies, their amplitude in µm and direction are presented in table 1. These frequencies were measured on the machine structure in different points on different direction, but their effects are on the tool.

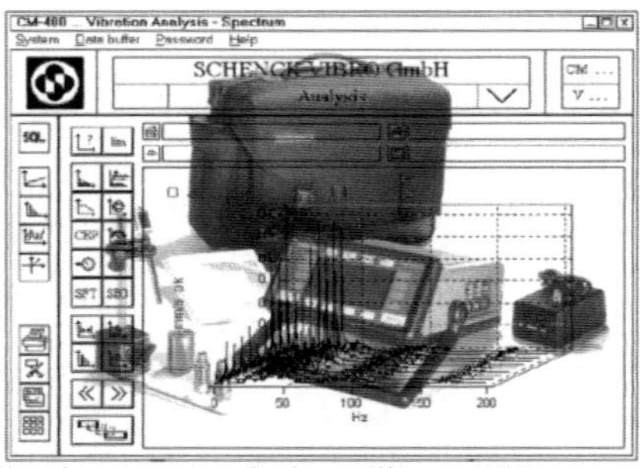

Fig.2. Vibration measure device -Vibroport 41

Fig.3 Measurement points on the Gantry machine tool.

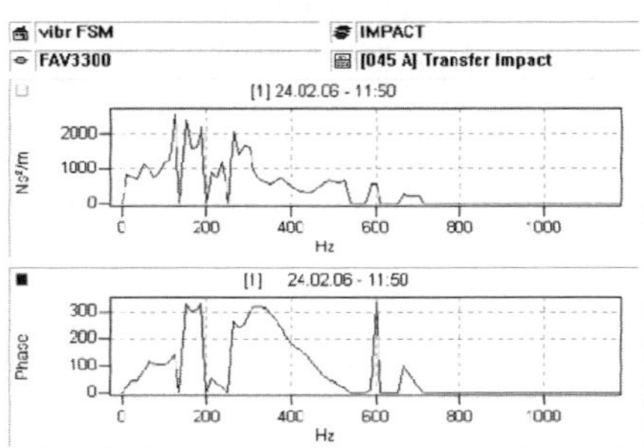

Fig.4.Transfer function: data graphical representation using Vibroport software

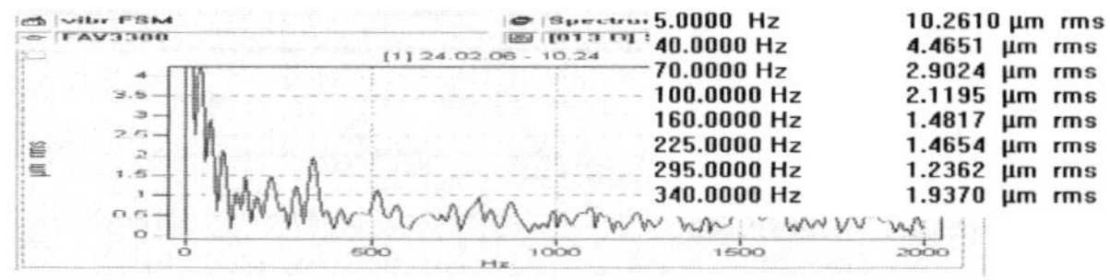

Fig.5. Frequnciees spectrum for 7500 rot/min with accelerometer

Frequencies	Displacement	Accelerometer position and direction
20,00 Hz	2,1237 μm	2z
33,75 Hz	0,2152 μm	1x
35 Hz	0,0683 μm;	1x
41,25 Hz	1,3785 μm	2z
55 Hz;	1,5505 μm	1x
55,75 Hz	1,5732 μm	1y
57,5 Hz;	1,0719 μm	2z

Table 1. Important frequencies

The amplitude value of the measured frequencies can be used as imput data into C++ correction software.

5. CONCLUSIONS

The solution that we proposed in this paper has the advantage of making a CAD correction of the pieces surfaces, function of the machine tool cumulated errors.

In the dynamical study presented in this paper it were established the direct influences of the external frequencies in combination with the frequencies proper to the machining accuracy. Combining experimental measures with a theoretical study, the dynamical errors of the machine tool can be accuracy defined.

The errors correction by modifying the CAD surfaces is not depending of machine tool CNC system or the machine tool systems. It can be easy implemented on all type of machine tool, even for the high speed. The errors values used for calculus were measured with high precision instrument, independently from the machine tools.

The results of this study were used as input data for tool correction software of the machined surface of the pieces.

6. REFERENCES

Anania F.D., Ispas,C.; Mohora, C.(2007) *An Algorithm for CAD correction of the work pieces based on machine tool errors*, Proceedings Annals of DAAAM for 2007, 24-27 octomber, Zadar, Croatia, issn1726-9679, ISBN 3-901509-58-5,

Ispas C-tin. Anania F.D., Mohora C.(2007) *Experimental research concerning machining accuracy for a gantry machine tool,* The 5th international conference of advanced manufacturing technologies ICAMaT , 12-14 iulie 2007, sibiu, romania, ISSN 1843-3162 ISBN

Ispas, C-tin., ANANIA, F.D., Dogariu, C-tin., TILINA, D. C-tin., (2006). *Contributions about Gantry machine tool geometric accuracy improvement by laser alignment* - The International Conference of the Carpathian Euro-Region Specialists in Industrial Systems - BAIA MARE

Marinescu I., Ispas C., Boboc D.(2002) – Handbook of Machine Tool Analysis, United States of America, ISBN 0-8247-0704-4, 002, 2002

Valdes E., Greffioz A, Dequidt A.(2006) - *Differents modeles de comportament dynamique d'une machine-outil durant sa phase de conception*, 4-emes Assies Machines et Usinage Grand Vitesse, 8,9 juin 2006, ENSAM, Aix-en-Provence, www.lsis.org/AssisesMUGV, *Accessed on*: 2008-08-09

Annals of DAAAM for 2008 & Proceedings of the 19th International DAAAM Symposium, ISSN 1726-9679
Editor B. Katalinic, Published by DAAAM International, Vienna, Austria 2008
Make Harmony Between Technology and Nature, and Your Mind will Fly Free as a Bird

RESEARCHES AND SOLUTIONS FOR OPTIMISING TECHNICAL PROBLEM WITH TRIZ THEORY

ISPAS, C[onstantin]; MOHORA, C[ristina]; TILINA, D[ana] & PARASCHIV, M[arius]

Abstract: *The car manufacturing field of application is a priority not only on the national, but also on the international plan with major implications in the quality of everyday people life, and also on the environmental protection. Taking into consideration the last decade concurrency and the will of the cars manufacturers to monopolies the market, they make big efforts to find innovative high quality technical solutions. The paper presents the research made by the authors with TRIZ theory to improve the innovative dampers performances.*

Key words: *innovation, TRIZ, dampers*

1. INTRODUCTION

The concurential character in the car manufacturing field of application, concentrated on the globalisation and internationalisation, requires the application of some efficient methods and instruments in design and achievement of the new products. So, the introduction of some inventive and new methods requires a more complete satisfaction of the customers, regarding the technical performances, economical conditions, the price and the market launch at the right time and place.

The degree of innovation and the quality of the solutions depend almost 90% of what Pahl & Beitz (Pahl & Beitz; 1996) called "The development of solving principles" which belonged to the design process. That's why this stage is also named "the key stage" in the development of a product.

Taking into account the importance of this one, along the time, the researchers had developed more than 100 methods, instruments and techniques of creativity.

The research objective is to find a technical solution for the development of a new type of damper by applying some instruments of integrated methodology for solving problems called TRIZ.

2. GENERALS INFORMATIONS

The primary purpose of the shock absorber is to control spring and suspension movement. For the absorbers presented in the paper, this is accomplished by turning the kinetic energy of suspension movement into thermal energy, or heat energy, dissipated through the hydraulic fluid.

Shock absorbers are basically oil pumps. A piston is attached to the end of the piston rod and works against hydraulic fluid in the pressure tube. As the suspension moves up and down, the hydraulic fluid is forced through tiny holes, called orifices, inside the piston. However, these orifices let only a small amount of fluid through the piston. This slows down the piston, which slows down also the suspension movement (Figure 1).

The resistance of the shock absorber depends on the speed suspension and the number and size of the orifices in the piston. All modern shock absorbers are velocity sensitive hydraulic damping devices - meaning the faster the suspension moves, the more resistance the shock absorber provides.

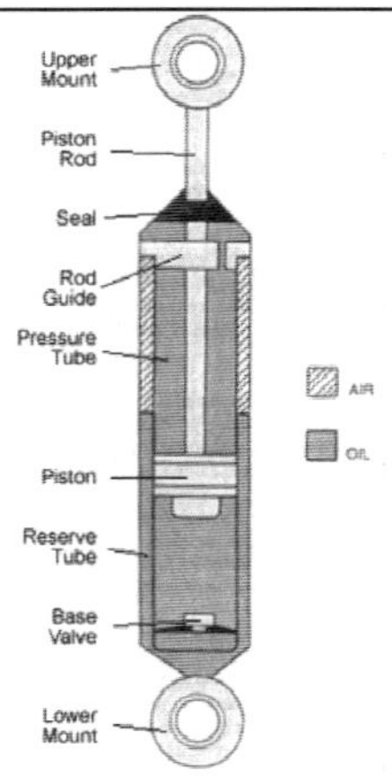

Fig. 1. The Monroe damper.

Because of this feature, shock absorbers are made for the various road conditions.

As a result, shock absorbers reduce the rate of:

- Bounce,
- Roll or sway,
- Brake dive and Acceleration squat.

Shock absorbers work on the principle of fluid displacement on both, the compression and extension cycle. A typical car or light truck will have more resistance during its extension cycle then its compression cycle. The compression cycle controls the motion of a vehicle's unsprung weight, while extension controls the heavier sprung weight.

3. THE IDENTIFICATION AND THE SOLUTION OF THE PROBLEM

3.1 The problem

The damper influent directly the breaking efficiency, the direction and also the cars behaviour along the road.

Because the dampers wear is gradual and influence the wear of another's car components, the experts consider that these must be verified at each 20 000km.

Taking into account that a damper makes about 500 movements/km and after 20.000km about 10 millions movements, it is easy to imagine these wear degree.

A study achieved in the auto-services in Romania showed that because the roads bad quality, the wear on different components is very high at a smaller number of kilometres than the guaranteed one (about 80 000km).

Taking into account all these elements it was studied the possibility of the dampers improvement with the help of TRIZ (Ispas et al.; 2005).

3.2 The solution

In the first stage the problem that must be solved was clear defined. Generally speaking, some problems are not always clearly defined, and all necessary information are not enough obvious as know. That's why the problem will be very difficult to solve (Domb; 1997).

In order to clarify the problem, first, it was used the instrument called "Innovative Situation Questionnaire". As a result of ISQ application it was established that irrespective of the damper type the following conditions must be accomplished:

- To assure a corresponding absorbing of the coachwork and of the car wheels;
- To have a large (big) period of function;
- The weight and dimensions must be as small as possible;
- A simple construction;
- An easy assembling in the car suspension;
- To assure a correspond absorbing in different conditions of use.

The conclusion after using ISQ was that the best way to solve the problem is to modify some damper components.

Trying to solve the identification problem, some methods of TRIZ theory were utilized.

The first instrument chose was "Little smart people", a creative method very simple to apply: we have to imagine that a conflict zone from the interior of a system (a car, a device, a mechanism) represents, in fact, a lot of people in miniature. Utilizing this method appears that:

- Many small peoples are microscopic.
- Each one can be different as age, colour, sex, etc.
- This people are able to fulfil the received orders (moving, actions, etc.).

Then, a more complex approach of the problem was made by applying the following TRIZ instruments: The Contradiction Matrix, The Separation Principles and the method that associate the Su-Field Modelling and the 76 Standard Solutions (Cordova; 2002).

TRIZ considers that generally a problem supposes a conflict between two parameters. For example, x and y are two parameters that are in conflict and require a compromise. As x is improved, y is affected in a wrong way.

The instrument known as the Contradiction Matrix reunites 39 parameters and 40 inventive principles. The parameters are disposed horizontally (parameters that must be improved) as well as vertically (parameters that are affected). In the intersection points of these are mentioned the 40 principles of witch application should solve the conflict between these parameters. In this case, the parameters are:

- 27. Reliability
- 34. Reparability.
- 35. Adaptability.
- 36. Complexity.

The physical contradictions are the system requires which are reciprocally excluded. This situation is caused by a function, process or component (an object which must be heavy and light in the same time). The principles of separation are used to eliminate these contradictions. TRIZ introduced this instrument with the aim of facilitating the solving of technical problems, when there is a physical contradiction. The most important principles of dividing are:

- In time.
- In space.
- Of an object in its components.
- According to the conditions.

The last used method is Su-Field Modelling and the Solving Standards. TRIZ mentioned the fact that generally, the transition of an initial system A to an improved system B can be described as a transformation of a Su-Field structure.

When we built a Su-Field Model we must respect the following steps:

- The identification of the elements (S1, S2 and F) and the links between them.

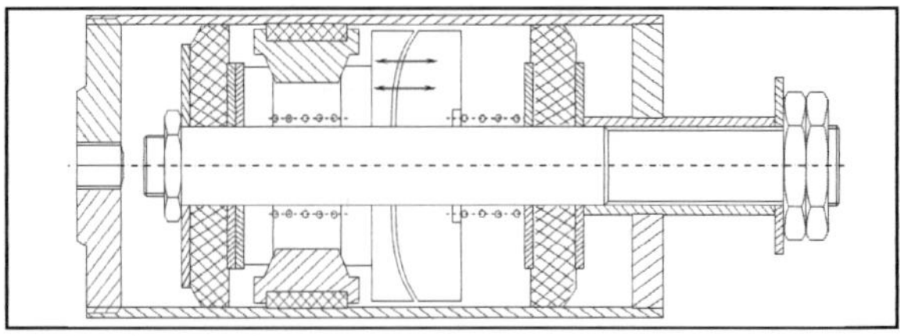

Fig. 2. The new damper.

- The construction of the initial Su-Field Model.
- The identification of the solution taking into account the 76 Standards.
- The development of the concept based on the found solution.

The solving standards are structured in five classes as follows:

1.The composition and the decomposition of a Su-Field Model.
2.The evolution of a Su-Field Model.
3.The transition to the super-system and micro-level.
4.Measuring and detecting standards.
5.Auxiliary Standards.

After we applied all of these TRIZ methods the result was a new type of damper which can be adapted for a larger variety of cars (Figure 2) (Tilina; 2008).

4. CONCLUSION

The cars manufacturing need a continuous products improvement and therefore a rise of the creative potential of the people that are in charge with the conception in order to be competitive on the market. The "traditional" instruments that support the creativity encourage the analysis and implicit the solving of the problems from all the points of view and possibly lead to a wide spectre of solutions. Moreover, these methods are based on human competence. Because of that, the creativity has remained an aleatory process, fact that doesn't enable an immediate found of solutions.

Even the producers guaranteed the function at normal parameters of the dampers for about 80 000 km, the research in the auto-services showed that because the roads bad quality, their life cycle of is smaller, irrespective of the damper type or the producer.

Taking in account all that, we applied TRIZ method and we had simplified the construction and improved the dynamic dampers characteristics. The result was a new ecological type of damper with really new characteristics like: a high outturn (200% compared with the classic one), easy to manipulate; low weight and the possibility to be set in different positions.

5. REFERENCES

Cordova, L. (2002). *Contribution to a methodological approach of the innovation process: Application of TRIZ theory to the systems product - method – process,* Doctor's degree thesis, University Luis Pasteur, Strasbourg

Domb E. (1997). *How to Help TRIZ Beginners Succed* , Invention Machine User Group Conference, New Orleans, LA USA.

Ispas C., Mohora C., Cioroianu D., Tilină D.(2005). *Considerations regarding the simulation of creativity in forming specialists*, 3rd Balkan region conference on engineering education, Sibiu, România

Pahl, G., Beitz, P. (1996). *Engineering Design. A Systematic Approach"*, Springer, Berlin

Tilina, D. (2008), *Contributions concerning with application of methodical integration in products conception*, Doctor's degree thesis, UPB.

Annals of DAAAM for 2008 & Proceedings of the 19th International DAAAM Symposium, ISSN 1726-9679
Editor B. Katalinic, Published by DAAAM International, Vienna, Austria 2008
Make Harmony Between Technology and Nature, and Your Mind will Fly Free as a Bird

THE INFLUENCE OF THE TECHNOLOGICAL PARAMETERS ON THE MACHINING RATE AT THE EDM WITH THICK ELECTRODE

IURAS, E[milia]; TANASA, R[aluca]; CARAUSU, C[onstantin] & PRUTEANU, O[ctavian]

Abstract: *The purpose of this work is to present some experimental results regarding the influence exerted by the technological parameters on the machining rate of the electrical discharge machining with a thick electrode, in case of the small dimensions surfaces machining. On the other hand, the optimal conditions for which it could be obtained a maximum value for technological parameter, productivity Q_P have been evaluated.*
Key words: *EDM machining, productivity, thick electrode.*

1. INTRODUCTION

Regarded by means of technical-economics indicators, the electrical discharge machining (EDM) is one of the processes that achieved a high degree of performance, which assures a high machining rate, with minimal shape and dimension errors. Because of developing and appearance of new industrial fields, but also, of increasing of the requests regarding the productivity, and the economical efficiency, the necessity to optimize the EDM machining process has appeared (Kunieda et.al., 2003) (Momalis et.al., 2004).

Increasing of the productivity could be achieved by means of increasing of the discharging energy W_i; the experimental researches show that the increasing of this energy could be achieved only by the increasing of the discharging current I and the pulse time t_i; and also by means of the increasing of the discharging frequency f.

2. EXPERIMENTAL CONDITIONS

2.1 Materials, tools and working conditions

The test pieces required for experiments has been made of steel with the following chemical composition: 6.16 % wolfram, 4.58 % molybdenum, 4.34 % chrome, 1.65 % vanadium, 0.94 % carbon, 0.63 % siliceous, 0.343 % manganese, 0.0029 % sulphur, 0.027 % phosphor, 0.57 % cobalt, 0.309 % nickel.

The pieces used for experiments had a prismatic shape, with the length L=18 mm, the width l =13 mm and the thickness g=7 mm.

Thee tool–electrode is made of cooper, having also a prismatic shape, with the active zone having the following dimensions: length L= 28 mm, width l = 3.5 mm.

2.2 Working Equipment

For performing the experiments, an EDM machine tool endowed with a thick electrode, with numerical control, type FORM 20ZNC (manufactured by Charmilles Company) was used (User' s Manual FORM 20ZNC Charmilles, 2001).

2.3 Planifying of the experiment

In order to perform the experimental plan, the Taguchi method for planify the experiment was used. In order to develop the experiments, a complete factorial plan type 2^3 has been chosen (Ionescu et.al., 2004), (Pillet, 1992).

The main technological factors that could influence the objective function - the productivity of EDM machining Q_P - are: the intensity of the electric current I, the pulse time t_a, the non-pulse time t_b, the working depth h, the polarity of tool-electrode and tool - piece P User' s Manual FORM 20ZNC Charmilles, 2001), (Kunieda et.al., 2003).

Thus, for the achieved experimental plan, the working depth h=2 [mm], and the polarity of tool - electrode P (-) and tool–piece P (+) were retained as constant, while the values of the parameters current I, duration of a pulse t_a and duration between pulses t_b were considered at two levels (Pillet, 1992), (User' s Manual FORM 20ZNC Charmilles, 2001).

The values established for I are 8A and 12A, the pulses t_a have been chosen at the following two levels: 50μs and 300μs and for the duration between pulses t_b 100μs and 500μs.

3. EXPERIMENTAL RESULTS

The experimental plan and also the results of tests are shown in table 1.

The mathematical model (described by means of the Taguchi's method) of the machining rate $Q_{p,}$ that describes the influence of the independent factors and the interactions on the objective function could be described by the expression (1) (Ionescu et.al., 2004), (Pillet, 1992):

$$Q_P = 3.48015 + [-1.62116 \;\; 1.62116]\cdot I + [-1.29700 \;\; 1.29700]\cdot t_a + [0.74204 \;\; -0.74204]\cdot t_b + {}^tI\begin{bmatrix} 0.96492 & -0.96492 \\ -0.96492 & 0.96492 \end{bmatrix}\cdot t_a + {}^tI\begin{bmatrix} -0.43762 & 0.43762 \\ 0.43762 & -0.43762 \end{bmatrix}\cdot t_b + {}^tt_a\begin{bmatrix} -0.43762 & 0.43762 \\ 0.43762 & -0.43762 \end{bmatrix}\cdot t_b \quad (1)$$

After applying the Taguchi's method, it could be noticed that the objective function – the machining rate Q_P is maximum for the displacement of the factors at the following levels: I at level 2; t_a at level 2; t_b at level 1.

By the studying the interactions between factors, it could be noticed that the objective function is maximum for the combinations: I_2t_{a2}, I_2t_{b1}, $t_{a2}t_{b1}$.

No	Essay factors						Q_P - exp.
	I	t_a	t_b	$I\cdot t_a$	$I\cdot t_b$	$t_a\cdot t_b$	[mm³/min]
1.	1	1	1	1	1	1	2.15441
2.	1	1	2	1	2	2	0.89939
3.	1	2	1	2	1	2	2.17240
4.	1	2	2	2	2	1	2.20973
5.	2	1	1	2	2	1	3.57463
6.	2	1	2	2	1	2	2.10414
7.	2	2	1	1	2	2	8.98734
8.	2	2	2	1	1	1	5.73916

Tab1. Experimental plane and the values of parameters Q_P

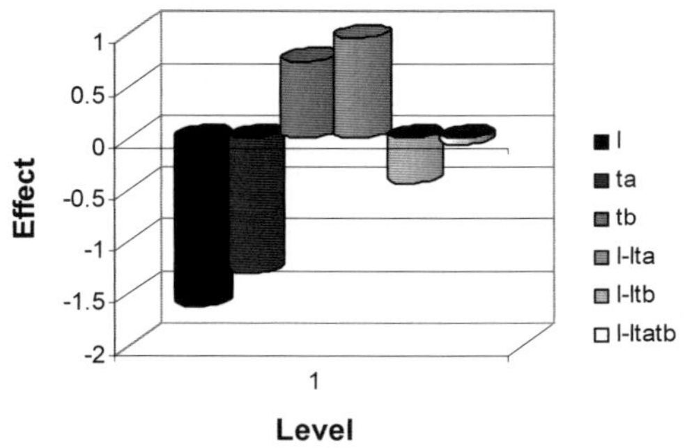

Fig. 1. Effects of the factors and interactions at level 1 on the parameter Q_P

In order to obtain a suggestive image (fig. 1) concerning the effects of the factors taken into account, and the interactions of the factors on the objective function, the level 1 (I=8A, t_a=50μs, t_b=100μs) was examinated.

From the figure 1, one can notice that the greater influence exerted by the factor *I*, followed by the effects and interactions $t_a, It_a, t_b, It_b, t_at_b$.

The theoretical values obtained by means of the matrix model $Q_P\sim$ for those eight experimental points, and also for the experimental values, for Q_P are presented in table 2.

Graphical representation of the values for Q_P and $Q_P\sim$ on the same diagram can be seen in the diagram from figure 2.

By the study of the graphic included in figure 2, one can notice a small difference between the values measured and those obtained by means of the model; this means that there is a good approaching between the theoretical model and the experimental values.

On the other hand, one can notice the smaller influence of the non-controlled factors (for example, the disturbing factors) and a very good accuracy of measures performed for the parameters taken into account.

No.	Q_{Pexp}	Q_{Pteor}	Residuum
1.	2.15441	1.77065	0.38375
2.	0.89939	1.28314	-0.38375
3.	2.17240	2.55615	-0.38375
4.	2.20973	1.82597	0.38375
5.	3.57463	3.95839	-0.38375
6.	2.10414	1.72038	0.38375
7.	8.98734	8.60358	0.38375
8.	5.73916	6.12291	-0.38375

Tab. 2. The values of the theoretical and experimental answers and residuums

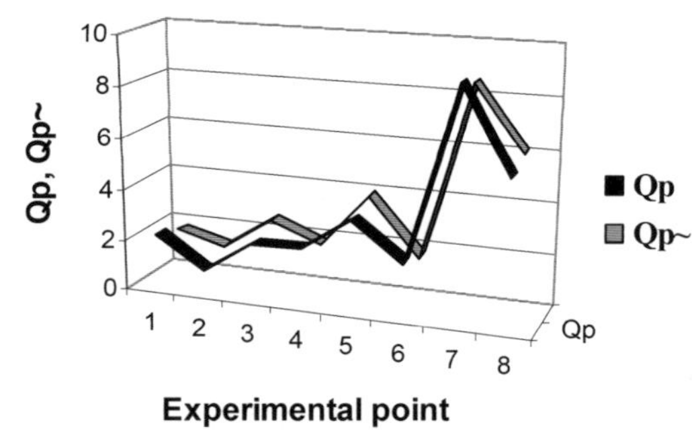

Fig. 2. Values of the machining rate $Q_P\sim$ and Q_P

Source	Variability	Variant Disp.	F	F_α α=70 [%]	Significance $F > F_\alpha$
I	21.0255	21.0255	17.84	3.852	Sig.
t_a	13.4578	13.4578	11.42	3.852	Sig.
t_b	4.40504	4.40504	3.73	3.852	Sig.
I t_a	7.44863	7.44863	6.32	3.852	Sig.
I t_b	1.53210	1.53210	1.30	3.852	Insig.
t_a t_b	0.02944	0.02944	0.02	3.852	Insig.
Res.	1.1781	1.1781			
Total	47.8985				

Tab. 3. Analyzing of the variance of the parameter

One must find out the value "*F*" according to Fisher law for each factor and interaction. In order to estimate which of the factors has significance, the calculated values are compared with the values included in standard tables (Snedecor Test) for a level of trust 70%. The obtained results are shown in table 3 (Ionescu et.al., 2004), (Pillet, 1992).

In accordance with the data included in table 2, the values of the coefficient Fisher are greater than those displayed in tables, for a certain level of trust of 70 %, for the parameters *I*, t_a, t_b, and also in case of interactions It_a.

All those parameters become significant, and the interactions t_at_b, It_b, for that $F < F_\alpha$ could be considered as not significant and for this reason they could be neglected (Ionescu et.al., 2004), (Pillet, 1992).

The matrix model could be written in a simplified shape, after the elimination of the insignificant interactions as follows (equation 2):

$$\begin{aligned} Q_P &= 3.48015 + [-1.62116 \;\; 1.62116]\cdot I + \\ & [-1.29700 \;\; 1.29700]\cdot t_a + [0.74204 \;\; -0.74204]\cdot t_b + \\ & + {}^t I \begin{bmatrix} 0.96492 & -0.96492 \\ -0.96492 & 0.96492 \end{bmatrix} \cdot t_a \end{aligned} \qquad (2)$$

4. CONCLUSION

Following the experiments, and analyzing the results with Taguchi's method, the problem of the optimization of the EDM process with thick electrode is achieved in this work. Thus, could be formulating the following conclusions.

Analyzing the graphics of influence of the factors and interactions, can be estimated that the maximum machining rate could be achieved if we adjust the factors taken into account at the following levels: *I* at level (12A), t_a at level 2 (300 μs) and t_b at level 1 (100 μs). Accepting a trust level of 70%, the significant factors are: *I*, t_a, t_b and interaction It_a.

A greater discharging energy, required for EDM process, implies a greater intensity of current and also a greater pulse time. Increasing the discharging energy has as effect the local.

5. REFERENCES

Ionescu , R. et.al. (2004). *Planification of the experiments efficiency and quality*, Publishing House Agir, ISBN 973-8466-43-1, Bucharest, Romania

Kunieda, M., et.al. (2003). *High speed 3D Milling by dry EDM.* Annals of the CIRP, PP. 147-150, vol. 52, no.1, Swissland

Momalis, A.G. et al. (2004). *Two–stage electro-discharge machining fabricating superhard cutting tools*. Journal of Materials Processing Technology, pp. 318-225, no. 146

Pillet, M. (1992). *Introduction aux plans d'experiences,* Clamency Publisher, ISBN 2-7081-1442-5, France

User' s Manual FORM 20ZNC Charmilles, (2001). Edipresss Imprimeriers Reunies s.a./Renes, CTSA, Geneva

Annals of DAAAM for 2008 & Proceedings of the 19th International DAAAM Symposium, ISSN 1726-9679
Editor B. Katalinic, Published by DAAAM International, Vienna, Austria 2008
Make Harmony Between Technology and Nature, and Your Mind will Fly Free as a Bird
*Young Researcher and Scientist Paper / * Supervisor, Mentor*

SOME QUALITY ASPECTS REGARDING ADVANCED CERAMICS MECHANICAL PARTS FOR PRECISION MACHINERY

IVAN, I[oana] C[armen]; ENCIU, G[eorge] & GHINEA, M[ihai]*

Abstract: *If until 1980 advanced ceramics were indicated just for "static" applications (where dynamical phenomenon does not appear), due to their fragility, now, new generations of advanced ceramics are used even inside of car box gear. A lot of industries use advanced materials in order to improve their achievements or to respond these demands regarding ecology, row materials poverty etc. Advanced ceramics represent a wide known and used category of these kinds of new materials. Here we try to present a part of a large research regarding the behavior of advanced ceramics that can be used in various mechanical applications. There is a comparative analysis (ceramic / steel) with FEM in order to emphasis the better behavior of advanced ceramics.*

Key words: *ceramics, FEM, precision*

1. INTRODUCTION

Advanced ceramics represent a wide range of very useful materials if they are very well processed with sophisticated technologies (HIP-Hot Isostatic Pressing) and if there are used high quality ceramic powders.

Before using these expensive technologies, it is important that it all researches and all comparative analysis (ceramics-traditional materials) to be done (Ghinea, 1999). Sometimes is better and cheaper to use Alumina ceramics (cheaper) instead Zirconia ceramics (expensive), or to use a ceramic materials obtained by extrusion or CIP (Cold Isostatic Pressing) instead HIP.

Of course, today there are a lot of specialists in mechanics that do not agree to use advanced ceramics for these parts that work under dynamic conditions, frictions, or if there is a contaminated working area. For them, classical materials (metals, steel, and composites) can still fix the problem. In certain cases a numerical method can make obviously the difference between advanced ceramics and other material. FEM analysis offers the possibility to obtain a complex and precise simulation especially for complicate geometry and shapes (Ispas et al., 1990). In this paper there were brought out few aspects from a research based on FEM analysis, which present the advanced ceramics advantages near to steel behavior (Dogariu, 1994; Ghinea & Vieru, 1997).

It was used a simple shape, a ring, that can be worked in a lot of mechanical applications. In table 1 there are the most important mechanical properties used in FEM analysis (for ceramic material and steel, also) and in figure 1 it can be observed the geometry of the ring and the finite element modeling. The chosen advanced ceramic material was Alumina (Al_2O_3) and ordinary steel.

Property	Steel	Alumina
Elasticity modulus E, N/m^2	$2,1\cdot10^{11}$	$4\cdot10^{11}$
Density, ρ kg/m^3	7850	3940
Poisson ratio, ν	0,3	0,23
Coefficient of thermal expansion, α, 10^{-6}K^{-1}	6,2	7,4

Tab. 1. Mechanical properties for Alumina and steel

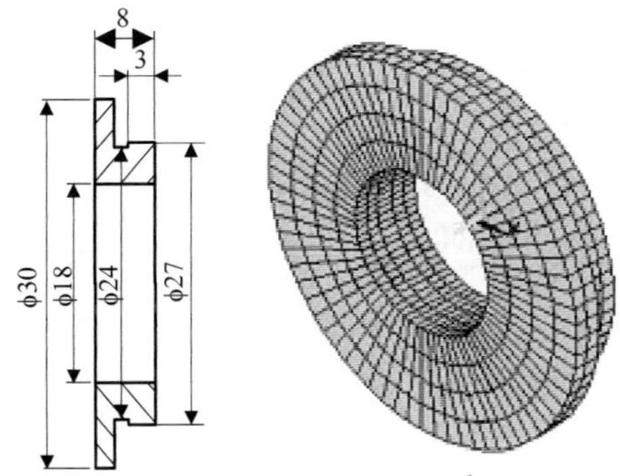

a *b*

Fig. 1. The ring geometry (*a*) and his FEM discretization (*b*)

2. MODAL ANALYSIS

In FEM analysis equilibrium general equation is:

$$[M]\{\ddot{u}\}+[C]\{\dot{u}\}+[K]\{u\} = \{F(t)\} \quad (1)$$

where [M] is masses matrix, [C] - damping matrix, [K] - rigidity matrix, $\{\ddot{u}\}$ - acceleration vector, $\{\dot{u}\}$- speed vector, {u} - displacement vector, F(t) - external loads vector (Dogariu, 1994). Of course, for static phenomenon

$$F(t) = F = \text{const.} \Rightarrow \{\ddot{u}\} \text{ and } \{\dot{u}\}=\{0\} \quad (2)$$

so,

$$[K]\{u\} = \{F\} \quad (3)$$

Obviously, [M] is known from ring geometry and materials properties and also, if there are concentrated masses, [M] can be determinate. [C] matrix can be obtained from experimental determinations and simulations (with Matlab), [K] matrix results also from materials properties (E and G) and ring geometry. The vectors $\{\ddot{u}\}$, $\{\dot{u}\}$ and {u} remain unknown. If [C] = 0 means that the system will work without dumping.

In the case of dynamical analysis by FEM a modal analysis it will considerate ({F}=0) and also, when a dynamical solicitation appears, ({F}≠0), afterward it will be established the displacements {u(t)}, speeds $\{\dot{u}(t)\}$ and accelerations $\{\ddot{u}(t)\}$, in nodes and stresses (α(t)) which help to define the ring/system fatigue behavior.

Modal analysis supposes the vibration proper modes and frequencies, and these can be established based on solving of the general system that described the part behavior. In this system all values of the time dependent loads vector are null.
In modal analysis proper modes are obtained (which, if they are the same with the work frequencies the resonance phenomenon can appear), the infinite displacements for undumped structures and huge displacements for dumped structures can also appear (Kunkel et al., 1990). A good result means that proper frequencies do not be similar with the work frequencies, and that they can be obtained by any parameters variation. So, if

rigidity K increases, frequency f increases, or if mass m increases frequency f decreases (soft materials).

3. FINAL RESULTS AND CONCLUSIONS

In order to obtain a lot of numerical data linked by ring vibration modes, there were calculated lots of proper frequencies, but for our research only the first three vibration modes were chosen (fig.2, 3, 4, *a* results for metal ring, and *b* results for Alumina ring). The reason was that due to a large rigidity, these first three proper frequencies are already high (small part with a good fixation on X,Y,Z directions). The most important conclusion is the first vibration proper frequency for

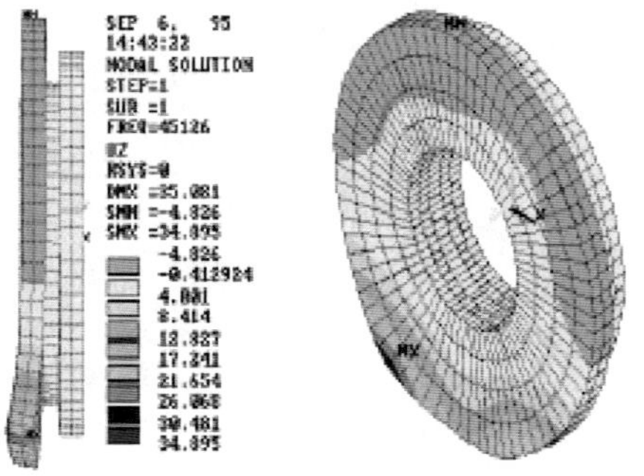

a

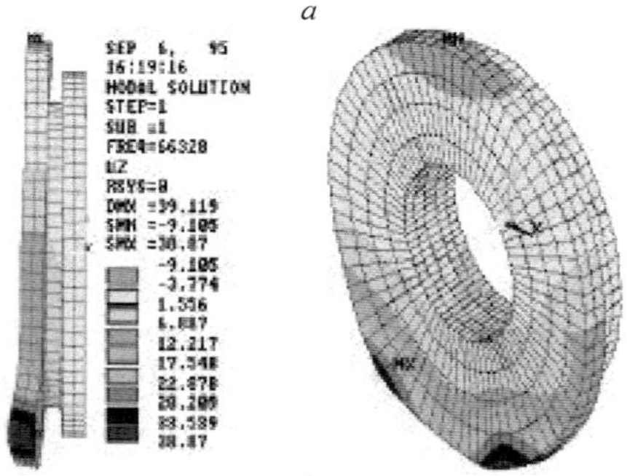

b

Fig. 2 The first vibration mode (*a*- metal, *b* – Alumina)

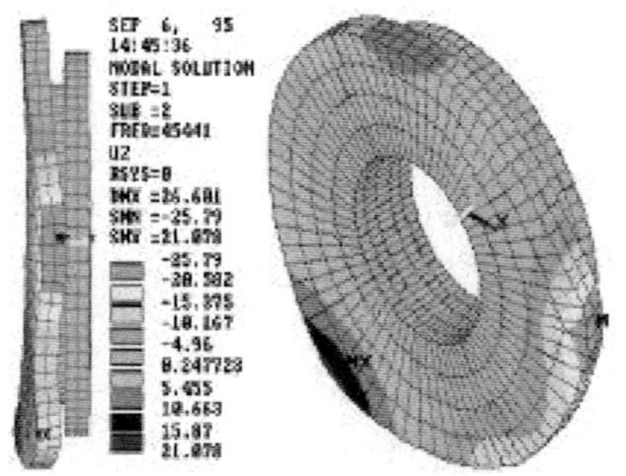

a

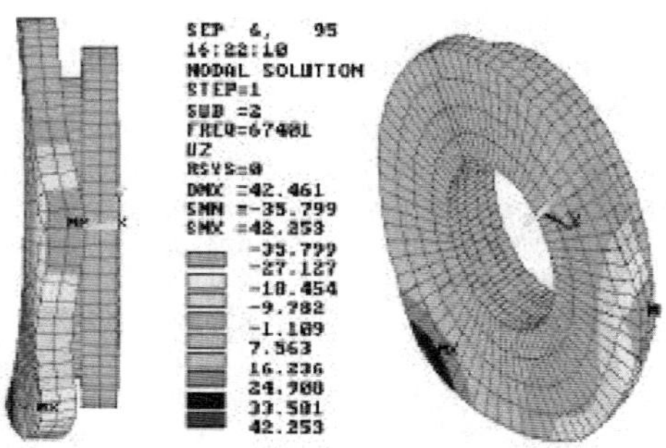

b

Fig. 3 The first vibration mode (*a*- metal, *b* – Alumina)

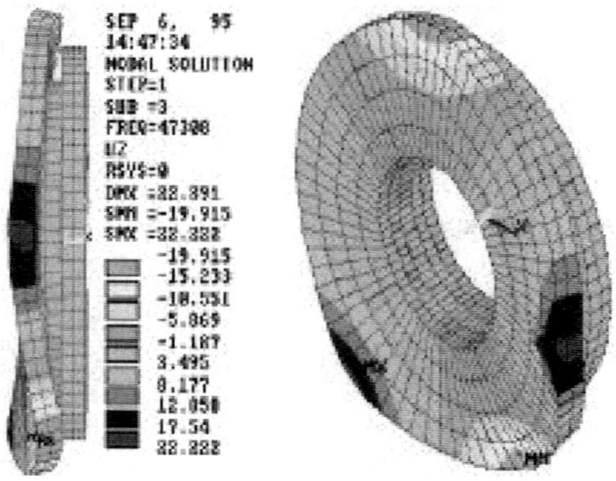

a

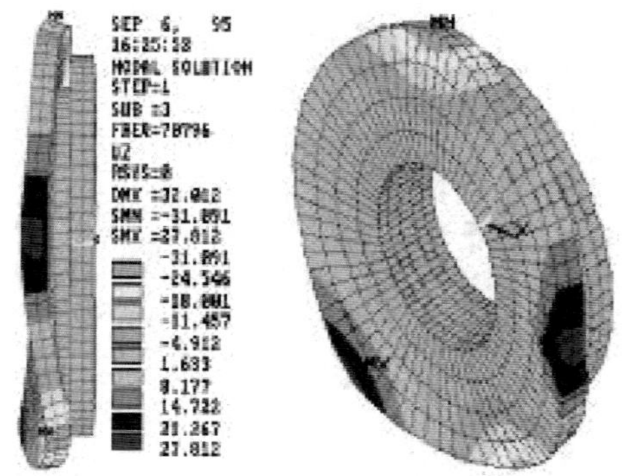

b

Fig. 4 The third vibration mode (*a*- metal, *b* – Alumina)

Alumina is with 35% higher then steel, and in this case a resonance phenomenon cannot appear.

Finally, it is obviously that advanced ceramics can successfully replace any classical material, in any mechanical application. Especially, top industrial activities like mechatronics, robotics, and IT branches, can fully use ceramic materials even in these applications where dynamic phenomenons often appear. The entire research, that contains also this FEM analysis, presents a lot of studies regarding advantages of ceramics properties, new obtaining technologies of ceramics part used in several mechanical assemblies. After that it will follow the ceramics parts utilization inside several precision machinery which are present in our research centers or laboratory.

4. REFERENCES

Dogariu, C. (1994). *Theoretical and Experimental Researches Regarding the Concrete Utilization in Machine Tools Construction*, PhD. thesis, University POLITEHNICA of Bucharest, Romania.

Ghinea, M, & Vieru, A. (1997) *The Behavior of Ceramics Used in Mechanics,* In: Interceram (International Ceramic Review), 1/97, Verlag Schmid GmbH, pp.16-22.ISSN 0020-5214.

Ghinea, M. (1999) *Theoretical and Experimental Researches Regarding the Advanced Ceramics Utilization in the Machine Tools Construction*, Ph.D. thesis, University POLITEHNICA of Bucharest, Romania.

Ispas, C.; Ghinea, M.; Dogariu, C. & Vieru, A.(1994) *Technical Ceramics Used in Machine Tools Building,* Proceedings of Materials Engineering conference, Haifa, Israel, pp.64-71,

Kunkel, H.A.; Locke, S.& Pikeroen, B. (1990). *Finite-Element Analysis of Vibrational Modes in Piezoelectric Ceramic Disks,* Ultrasonics, Ferroelectrics and Frequency Control, IEEE Transactions on Volume 37, Issue 4, pp.316 – 328, DOI 10.1109/58.56

Annals of DAAAM for 2008 & Proceedings of the 19th International DAAAM Symposium, ISSN 1726-9679
Editor B. Katalinic, Published by DAAAM International, Vienna, Austria 2008
Make Harmony Between Technology and Nature, and Your Mind will Fly Free as a Bird
*Young Researcher and Scientist Paper / * Supervisor, Mentor*

SIMI RFID IDENTIFICATION SYSTEM IN FMS LOGISTICS

IVAN, I[oana] C[armen]; GHINEA, M[ihai] & ENCIU, G[eorge]*

Abstract: *This work is part of a project on the potential contribution of electronics and information technologies to the logistics in flexible systems. The presentation and the practical realisation of the Romanian concept SIMI is included in an applications' structure implemented in a Romanian enterprise. The applied technology is the active technology RFID. The results obtained are reflected in the hardware network, the software application of identification and management and also in the User interface.*
Key words: *RFID, SIMI, Reader, Tag, TagART*

1. INTRODUCTION

The omnipresent barcode labels that triggered a revolution in identification systems some considerable time ago, are being found to be inadequate in an increasing number of cases including the Flexible Machining Systems (FMS) logistics. The technically optimal solution is *RFID systems* (Radio Frequency Identification) (Günther et al., 2008; Seslija & Tešić, 2006).

The potential applications for this technology are numerous for FMS logistics including in modern terms AGVs, robots, conveyors, etc. (Enciu, 2002; ***ASAP Hitchcock Publishing). For example the RFID system increases flexibility and improves throughput in factories, significantly reducing the capital tied up in materials in the manufacturing processes. This flexibility catered for very limited installation efforts with low educational costs for the production staff (Lau et al., 2000).

2. SIMI – ROMANIAN RFID SYSTEM

SIMI is an integrated software and hardware system, designed and realized by dr. George Enciu, which is able to identify, locate and manage industrial parts, devices, equipments and personnel (inside classic machining systems or inside FMS) a defined area using RFID technology; is able to connect access control, alarm and locking systems. The system offers simple practical, easy to use interface that allows you to control allocation, movement and security of any asset. The system employs the latest Radio Frequency (RF) technology applying it in a simple form to pinpoint an event. Using miniature transmitters physically secured on a part, device, tool it can recognise the current whereabouts, who owns or uses it and where in the building plant it is allowed to go and at what times. This allows the user to fully customise the systems entire operation. A unique transmitter 'tag' is installed onto or into a high value item and its details are programmed into a database together with all the details pertaining to the make, and other comparable model number, serial number, parts allocation and office allocation etc of the tagged asset(s). Even pictures of the parts manufactured or owner can be included.

3. SIMI HARDWARE STRUCTURE

Through the network of readers-ReaderNet, which are all allocated positions on a scale floor plan of the factory, any of the following transmissions states are relayed to the database.

The types of data transmissions typically available in the system are:

- the part management transmission
- the security transmission
- the maintenance transmission

The sophisticated software modules TagARTIdent, TagARTMon analyses the received data and will, depending on the system set up & configuration, log the event, raise any appropriate alarm, or track the asset and display its zoned location on a pre defined map, among many other possible operations. This is done by active tags that each transmits a unique ID at a factory pre-set time interval. Multifunction readers, appropriately situated in the area where position monitoring has to be conducted, interpret signals transmitted by the tags. Through the receiver network it is possible to develop an audit / management trail of the item as it moves within the monitored area. Various comprehensive reports can be tailored and produced from this audit trail.

4. SIMI STRUCTURE

In the Fig.1 is presented a real ReaderNet designed and installed by us in a Romanian plant in two phases, the hardware being under the general control of SIMI.

The Reader (Fig.3) has the following functionality and properties:

- RF Module (RF Receiver and Demodulator),
- Micro-controller,
- LED indicators on the output connectors.

The unit receives RF data on a frequency of 433.92MHz. Such data is demodulated, interpreted and validated by the microprocessor. If the data is valid, relevant data is output to in a serial format at a baud of 57 600 bps. Although the voltage levels are 5V TTL levels, it is compatible with all RS232 devices.

More than one reader can be connected on a ReaderNet. The device is aligned to receive RF at a frequency of 433.92MHz. The bandwidth is 20MHz. Size 84mm x 40mm x 19mm. Input/Output Connections 2 x RJ45 Sockets.

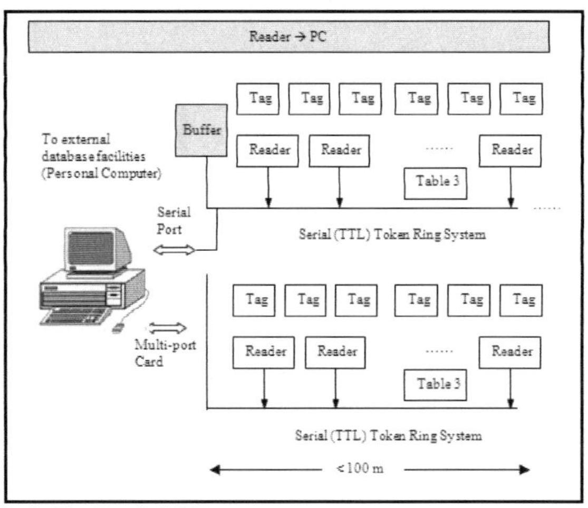

Fig.1. The ReaderNET concept

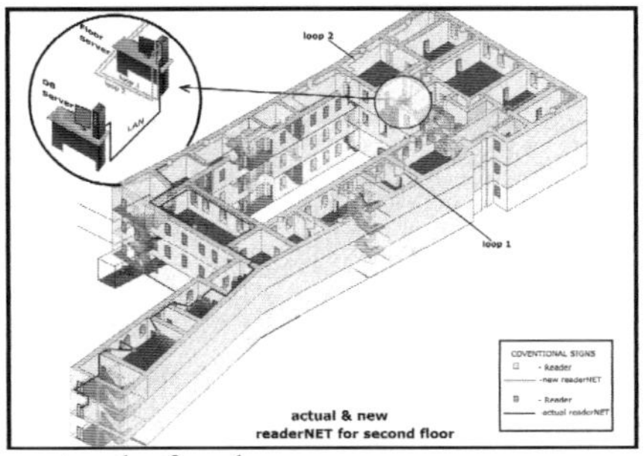

Fig.2.The network of readers

Fig.3. Multifunctional Reader

5. GENERAL INFORMATION ABOUT SOFTWARE

The application suite is comprised of 3 software modules:

5.1. ReaderNet Server (Administrator Application)

This application is the live link to the ReaderNet. There will be a physical connection between the ReaderNet and the Server PC, either via a COM Port or USB port. This application needs to be started with a Parameter indicating the COM Port to connect to, whether it is a Physical COM Port or a Logical one.

A sample start-up for the application is "TagArtident_Readernet_Server.exe <COM=1>" (indicating that the PC will be connect to the ReaderNet using COM1).

Other Start-up parameters include: <AUTO=Y> Sets the application to Auto-Start monitoring the ReaderNet <DELAY=n>, Sets the application to start monitoring the ReaderNet n seconds after it opens (this must be used in conjunction with <AUTO=Y> above).

The ReaderNet Server is an application that needs to be installed on a SECURE Server PC.

5.2. Readernet Monitor (Administrator Application)

This application is the Rules Monitor application. It monitors ALL activity detected by the ReaderNet Server and performs any operations specified by the user according to the rules specified.

The rules are set using the User Application, under the Group Rules Screen. Here the user can specify what activities can be monitored, (e.g. a tag entering a zone), and what actions to take.

This application also needs to be installed on a SECURE Server PC. It can be installed on the SAME PC as the ReaderNet Server, but this is NOT a requirement.

5.3. User Application

The user application is the user's interface to the SIMI. This application is used to do the following: add users and other tagable parts, tools etc, monitor violations in the system (i.e. broken rules), view tags in zones, report on Activity, set Rules and actions relating to users and assets.

SIMI can be customised to meet user's specific requirements. SIMI uses an open MS-SQL database and has the ability to call other programs allowing users to perform their own system integration; can be a completely wireless system, hard wired or a combination of the two RFID tags are assigned to people and parts, tools, devices, so that they can be located in real time.

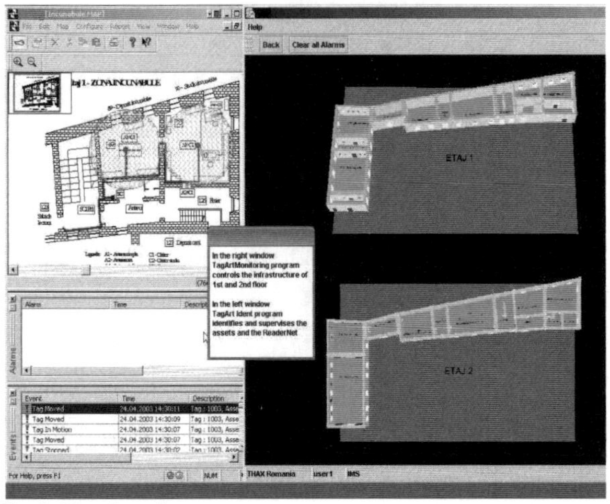

Fig.4. User interface

6. CONCLUSIONS

For the first time we introduce in Romanian mechanical field and particularly in flexible assembling systems the RFID concepts. We consider that this will be one of the most important ways to improve the FMS performances. Like in PC's, there is a trend towards application programmable and intelligent units implemented with database and modern protocols like TCP/IP (e.g. for Internet connection).

Features of SIMI for improving logistics in manufacturing:

- it uses high and medium distance communication (50-100m),
- it is very difficult to fake,
- it allows the identification and management of a great number of devices, parts, etc without a human operator,
- it allows localization and identification of assets in large areas,
- online localization of items,
- history of movement and inventory of parts for any desired time interval.

The research, the realisation and the implementation presented in this paper allow for the future the application in different fabrication structures of the concept and technology of the RFID.

7. REFERENCES

Enciu, G. (2002). Transport and assembly systems and RIFD. Basics and applications, *International Conference on Manufacturing Systems ICMaS 2002,* pp. 385-389, ISBN 973-27-0932-4, Bucharest, October 2002, The Romanian Academy Publishing House, Bucharest

Günther, O.; Kletti, W. & Kubach, U. (2008). *RFID in Manufacturing*, Publisher Springer Berlin Heidelberg, ISBN 978-3-540-76453-3, Berlin

Lau, H.C.W.; Ning, A.; Chan, F.T.S. & Ip, R.W.L (2000), Design and development of a flexible workflow supply chain system. *Proceedings of the 2000 IEEE International Conference on Management of Innovation and Technology ICMIT 2000.* Vol. 2, pp.792-796, ISBN: 0-7803-6652-2,

Seslija, D. & Tešić, Z. (2006). RFID Middleware as a connection between manufacturing processes and enterprise level information system, *FACTA UNIVERSTATIS, Series Mechanical Engineering*, Vol. No.1, pp.63-74, UDC 681.518:65.011.56, Publisher SCIndeks, Novi Sad

November 2000, Singapore

***RFID machine vision drive gains at Sony (1995), *ASAP Hitchcock Publishing Company Assembly*, pg.68, ISSN: 1050-8171, Vol. 38, No.4, April 1995

Annals of DAAAM for 2008 & Proceedings of the 19th International DAAAM Symposium, ISSN 1726-9679
Editor B. Katalinic, Published by DAAAM International, Vienna, Austria 2008
Make Harmony Between Technology and Nature, and Your Mind will Fly Free as a Bird

IMPROVEMENTS IN THE DESIGN OF THE MACHINES-TOOLS AND THE IMPACT ON THE MANUFACTURING SYSTEMS

IVAN, I[oana] C[armen]; ISPAS, C[onstantin]; PARASCHIV, M[arius] D[aniel] & TILINA, D[ana] I[uliana]

Abstract: *In the mechanical engineering industry field, the most used are the open chain kinematic structures. The new directions for researchers are pointed towards the study and the introduction of the parallel kinematics structures into different application and the study of high speed machining and the implementation of both into production systems. Authors will analyse the main both advantages and disadvantages of the implementation of the two new study directions and their impact to the production system.*
Key words: *machine-tool, parallel structures, machining manufacturing.*

1. INTRODUCTION

New research directions in the mechanical field are the study of the parallel kinematic structures and the implementation study of the high speed machining in the industrial applications. This fact is due to the fact that there are necessary to be done some improvements in the technological systems.

As industrial companies are more and more interested to offer high quality products, competitive on the market, the research regarding design and technology focuses on the finished product and on its utility resented by the customers; the utility is measured by the accomplishment of their clients' needs and demands. Never the less, companies are also interested in the new manufacturing methods in order to lower the production costs and in the same time to increase their profit.

Classical machine-tools are often behind the times in the industrial competition for offering the best product on the market with the lowest production costs and with a high profitability rate. As this classical structure of machine-tools overcome sometime with difficulty the industrial market demands, improvements need to be brought in the machine-tools domain (Neugebauer et al. 2000).

2. IMPROVEMENTS IN THE DESIGN AND TECHNOLOGY

Improvements need to be achieved in every field of the machine-tools, like accuracy, precision, speed, flexibility in operations.

A competitive industrial product means a product with a shorter time for the development phase, a high quality product which fulfils the client's demands with success and last but not at all least, especially for the manufacturing companies, a product with low production costs, as low as possible in a strong correlation with the two points mentioned above.

Trying to achieve some of the objectives mentioned previously, the researchers have obtained important results like the appearance of new structures in the conception of machine-tools and new machining methods.

The parallel kinematic machine-tools structure is one of the new structures used in the recent design of the machine-tools.

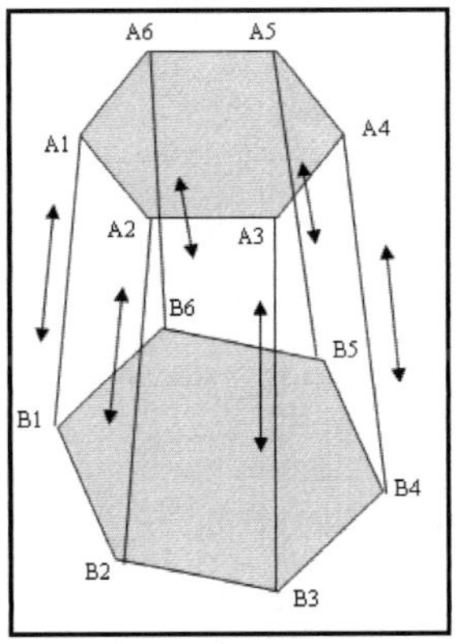

Fig.1. The basic parallel kinematic structure

This structure, best known as the Stewart platform, has six degrees of freedom. It consists of a rigid base (A1,...A6) connected through six identically, extensible jointed arms to a mobile spindle platform, where it is placed the end-effector (B1,...B6); the structure is pointed in Figure 1.

Several prototypes of machine-tools with the structure based on a Stewart platform, commonly called hexapods, have been produced and tested by machine-tools manufactures from different state, like Germany, US, France, Sweden.

The parallel structures are naturally more resistant then the serial ones, as the load is distributed on all arms and because for certain architectures the arms are loaded only with axial loads (Merlet, 2006).

The high speed machining is a method of a high productivity for the rough machining and the final machining of the small dimension pieces and also for the final machining and super final machining of all type dimensions pieces (e.g. obtaining cavities in aluminium structures for aero-spatial industry) (Ispas, 2006).

High speed machining does not represent only a spindle rotation, but an assembly that involves every component of the machine-tool. The structure of the machine-tool must be as rigid as possible to be able to support, in certain cases, the machining efforts, and especially the strong accelerations. The mobile parts should be as light as possible and in the same time they have to be strongly ribbed to obtain a good balance between rigidity and inertia.

3. STRONG AND WEAK POINTS OF EACH TYPE OF THE IMPROVEMENTS

The newly structures with parallel kinematics have significant advantages relative to conventional structures based on serial kinematics, advantages like:

* large field of activity domains; * higher speed; * higher accuracy, because they are more rigid and because the arm errors are calculated as an average, in comparison with the ones from the classical machine-tools, errors which cumulate; * high

structural rigidity; * low friction in joints, offering a longer life time (Wavering, 1998).

The parallel structures, in spite of their advantages regarding the accuracy, the rigidity and the large area of applications, they have also disadvantages that need to be considered, some weak points like:

- singularities and low stiffness outside the working volume,
- large machine size for a given working volume,
- programming and calibration difficulties of the kinematics linkages for achieving the desired tool path accuracy.

The implementation of high speed machining in the production process brings along some important modifications regarding different components of the manufacturing system:

- tools: are the first step that allowed the progress toward the high speed machining; the tool has to prove a very high chemical stability for increasing the number of the machined pieces but it also increases the prices of the tool,
- spindles: are essentials for the high speed machining and they have to function for a long period of time with very high speed and to provide a high power not only for high speed, but also for a lower speed,
- machining centres: the machine-tool has to take and to absorb important accelerations and decelerations compared to the classical machining, without endangering its base structure and preserving a long life time function and a high precision also,
- numerical control: its quickness is imperative. If the machine stops for waiting dates from the computer, unacceptable vibrations appear for the quality of the surface,
- lubrication: can be realised through the centre of the tool at a high pressure, as the classical system is inefficient because of the air ring which is created around the tool, caused by the high speed,
- human training: appears the necessity for training the human operators and readapting their working habits for a high speed machining centre.
- noise and price: noises are higher then the ones in the classical machining and the price of a high speed machining centre is higher, more then twice, but the price for the production time is at least twice less.

There are two important characteristics of the high speed machining which influence the manufacturing process: the first characteristic is the increase of dynamic efforts; the second characteristic is the quality of the machined pieces.

Regarding the application domains, the high speed machining has revolutionised the production in some industrial fields like the automobiles production, the aero spatial field (light alloy, composite materials).

4. IMPACT ON THE MANUFACTURING SYSTEMS

The improvements brought to the manufacturing systems change the approach of the companies towards the market and the approach to the engineers training.

Companies are interested into producing competitive products, which is translated by the diminution of the production development cycle, by a higher product quality and last but not least by a lowering of the manufacturing costs.

Those three important objectives make important changes in the design and manufacturing methodologies and work tools of products and systems.

These three important parameters, time quality and costs, that need to be taken into consideration for a successful company, bring along important modifications and improvements to the manufacturing methodologies and the fabrication systems. Each one of these parameters can and will be translated into technical terms for a clear comprehension, as it follows.

A shorter time for obtaining a product is gained by improving and increasing the speed of the machining, of the production system (Chanal et al. 2006).

A higher quality of a future product, compared to the ones which can be produced with existing equipments of a manufacturer, is provided by improving and increasing the accuracy of the operations from the technological chains.

Lower costs for production, in the same time not diminishing the quality, are obtained by improving and increasing the life cycle of the machining tools.

5. CONCLUSIONS

Following the actual design methodology and the level of research and development in engineering on high speed, high quality and low costs, training of future engineers has to speed up in quality, reactivity, team work and flexibility, in order to satisfy the new needs that appear and to find new better solutions for old and present problems. The new achievements and demands in the technical and technological area bring changes and improvements on the engineers training, which is an important part of the manufacturing systems.

It is needed that human operators are properly trained to use, to maintain the actual technical and technological level and to bring new ideas, to create and to innovate.

The advantages of the parallel kinematic structures must allow in the future the achievement of the objectives established for the high speed machining, including high speeds and accelerations.

Although it can be noticed a fast increase of the parallel structures number, only a very small number of them are functioning into the production systems from the industry, while the majority of these structures is remaning in the university or researching centres laboratories which have realised the study and researches. This discreapancy has not yet been explaind so far and a serie of questions appears, like: which are the reasons of the industrial manufacturers' actions or better inactions? or, what is holding the industrials to use parallel structures more often in the production systems.

This study will be the subject and the study for a future paper in which will be closely studied the connexion, if it exists one between the industrial applications and the universitary researches.

6. REFERENCES

Chanal, H. E., Duc, E,. & Ray, P. (2006). A study of the impact of machine tool structure on machining processes, *International Journal of Machine Tools and Manufacture*, Vol.46, Issue 2, (Febr. 2006), Pp 98-106, ISSN 0890-6955

Ispas, C. (2006). *Machines outils pour l'usinage à grande vitesse* (Machine-tools for high speed machining), 4eme Assies Machines et Usinage a Grande Vitesse, 2006, ENSAM, Aix-en-Provence, France.

Merlet, J.P., (2006). *Parallel Robots*, ISBN-10 1-4020-4132-2, Springer Publisher.

Neugebauer, R., Wieland, F. & Ihlenfeldt, S., (2000). Comparison of parallel structure concepts for five-axis machining, *Journal of Manufacturing Prrocesses,* Society of Manufacturing Engineers 2000, ProQuest Information and Learning Company.

Wavering, A. J.(1998). Parallel Kinematic Machine Research at NIST: Past, Present, and Future, *First European-American Forum on Parallel Kinematic Machines Theoretical Aspects and Industrial Requirements,* August 1998, Milan

Annals of DAAAM for 2008 & Proceedings of the 19th International DAAAM Symposium, ISSN 1726-9679
Editor B. Katalinic, Published by DAAAM International, Vienna, Austria 2008
Make Harmony Between Technology and Nature, and Your Mind will Fly Free as a Bird

EFFECTS OF THE TRANSNATIONAL CORPORATIONS ACTIVITY FROM THE SUSTAINABLE DEVELOPMENT PERSPECTIVE

IVAN, M[ihail] V[incentiu] & IACOVOIU, V[iorela]

Abstract: *Transnational corporations, through the foreign direct investments flows, are the main instrument of globalization, affecting, through direct and indirect effects, the economic, social and natural environment in the host country. Starting from this idea and analyzing the Human Development Index in relation to the inward FDI stock per capita for some of the CEE countries, we emphasize the contributions of the foreign direct investment inflows reflected upon the whole given by Human-Society-Natural Environment, proving that the transnational corporations may sustain or hinder the making of an economy based on ethics and ecology.*

Key words: *Corporation, effects, sustainable, development, and responsibilities*

1. INTRODUCTION

Most of the authors consider that the sustainable (human) development is possible only in case there is a stable balance between the economic, social and natural environment, the coverage areas thereof and the related contextures. Another aspect that most of the experts agreed is that transnational corporations (TNC), by means of the foreign direct investments (FDI), are the forwarders of the globalization, representing the main source of the technical progress and, deriving there from, of the development of highly competitive global economies characterized by "the importance of technology' and "the speed of changes" (Roşca, 2006).

What seems to be less obvious and consequently highly controversial is that in their fever for profit and the urge to maintain their competitive advantages that flaw rapidly, the major transnational corporations leave aside the ethical, moral and social issues or place them in the background. In this respect, J. Stiglitz (2003) points out that the impact of globalization on the emerging countries describes very controversial aspects. On one hand, the impact materializes in benefits that consist in higher rhythms of economic growth, the significant reduction of the transport and communication costs, the elimination of the artificial barriers against the circulation of goods, capitals, services, labor force, knowledge and technologies. On the other hand, the globalization's acts of defiance bear several negative effects, such as: deepening the disparities between the rich and the poor; life that became more insecure; domestic companies that turned vulnerable following the foreign companies competition. This context raises the question *whether and up to what extent the activity of the transnational corporations affects this balance?*

2. ENTERPRISES RESPONSABILITIES

Considering that the act of promoting the sustainable development involves a new vision over the functions and responsibilities of the enterprises, European Union introduced the concept of "corporate social responsibility" according to which the enterprises are called to *voluntarily assume new and important liabilities and social responsibilities that do not derive from the demands of obtaining profits* (EC, 2001).

"Internal" Aspect	"External" Aspect
Investments in the human capital: training the employees to permanent education; providing good personnel training; promoting more complex relations between work, family and leisure time.	**Integrating the company within the local community**: granting job places, incomes, cooperation with other enterprises.
Non-discriminatory practices against the ethnic minorities	**Respecting the human rights**: promoting fair moral, legal and political relations at regional level and with the media outside the community
Work safety and protecting the employees' health at the working place	**Voluntary involvement** in respecting the requests of a clean global environment.
The management of the impact on the environment and the natural resources: cutting the natural resources consumption, mitigating pollution, cutting the water waste, practicing the win-win opportunities relation (good for business and environment as well)	

Tab. 1.New enterprise responsibilities in the context of the sustainable development

These new responsibilities target both the "internal" aspect, corresponding to the enterprise-related activities and the "external aspect", in other words, that related to the interaction between the companies and the actors beyond the enterprise doors (table 1).

As a consequence, following the growth of the TNC economic power based on an increased globalization, the European Union, by the practiced policies, aims at developing the social and moral behavior of the enterprises in view of creating a new corporation type to play an increasingly more active role in building a "trans-industrial "society. In this new society, the TNC role changes radically, the corporations turning into a "global servant" with responsibilities in respect of the "living" embodied by the Human-Society-Environment relation (Roşca, 2006).

As from this approach on the corporation role, we propose to emphasize the effects of the TNC over the Central and Eastern European countries (CEE) from the sustainable development point of view. In this respect, we shall analyze the *Human Development Index* in relation to the FDI attracted by the CEE countries (*inward FDI stock per capita*).

3. DATA ANALYSIS

3.1 Human Development Index

Nowadays, in order to emphasize the economic progress by taking into account the social and environmental aspects as well is used the Human Development Index (HDI) that quantifies the three main dimensions: *a long and healthy* life (the life expectancy index); *knowledge* (the education index); *a standard of decent life* (the GDP index).

Countries	1990	1995	2000	2005
Slovenia	0.851	0.857	0.891	0.917
Czech Republic	0.845	0.854	0.866	0.891
Hungary	0.813	0.817	0.845	0.874
Poland	0.806	0.822	0.852	0.870
Slovakia	-	-	-	0.863
Estonia	0.813	0.792	0.829	0.860
Bulgaria	0.794	0.785	0.800	0.824
Romania	0.777	0.772	0.780	0.813

Tab. 2. HDI trend for the CEE countries, 1990-2005

According to the HDI level, countries are grouped into three main categories: high human development (1.000-0.800), medium human development (0.799-0.500) and low human development (0.499-0.000). Pursuant to the data included into the latest Human Development Report (2007/2008), the Central and Eastern European countries recorded an ascending HDI evolution, so that in 2005 they enroll within the highly human developed countries (table 2).

Several main issues are to be observed concerning the comparative analysis of the HDI level corresponding to the CEE countries. Firstly, during the beginning of the period (1990) all countries, except for Romania (0.777) and Bulgaria (0.794), were listed within the highly human developed countries, recording HDI values that exceeded 0,800. Secondly, the place occupied by every country at the end of the period (2005) resembles to that occupied during 1990, except for Estonia, that falls and Poland that marks one position.

The HDI survey must take into consideration the fact that, as per the experts, this composite index does not outline the aspects concerning the deterioration of the natural environment. Consequently, the Human Development Index may experience a long term improvement, under the deterioration of the natural environment. In this respect , it is to be remarked the evolution of several developed countries, namely Japan, Austria, Italy and Ireland that recorded, during 1990-2003, an ascending HDI trend simultaneously with the dramatic aggravation of the environment conditions (Carbon Dioxide emission per capita increased with 30% in Ireland, 25% in Austria, 22% in Japan and 16% in Italy).

The lack of HDI relevance in measuring the sustainable progress, meaning that it doesn't comprise the aspects related to the natural environment perceived as integrant part of the Living embodied by the Human-Society-Environment relation, is also reflected by the evolutions registered by the CEE countries submitted to analyses. For instance, Slovenia, the only Central and Eastern European country where the degree of air pollution has not reduced during 2003 as opposed to 1990, records an HDI increased ascending trend, ranking, during 2005, the first in the hierarchy of the analyzed CEE countries, registering an index level that exceeds 0,900. Meanwhile, countries inside which the pollution level decreased significantly (around 50%), namely Bulgaria, Estonia and Romania, record significantly reduced HDI values.

3.2 The correlation between the HDI and the FDI

The correlation analyses between the HDI and the inward FDI stock per capita, in 2005 (table 3), allow the emphasis of some aspects we consider relevant. Firstly, Slovenia, ranking the 5th in what concerns the foreign capital inputs, ranks the first place in relation to the sustainable development. As compared to the former, Estonia experiences the opposite trend, ranking the 1st as per the attracted FDI level (7,059USD/capita) and only the 6th as per the sustainable development (0.860). Secondly, except for the two above-mentioned situations the rankings held by the countries submitted to analyses as concerns the foreign capital inputs equal or stand close to those held in relation to the HDI level.

Countries	HDI		Inward FDI stock/capita	
	Value	Rank	USD/capita	Rank
Slovenia	0.917	1	2,469	5
Czech Republic	0.891	2	5,504	3
Hungary	0.874	3	6,015	2
Poland	0.870	4	1,590	6
Slovakia	0.863	5	2,676	4
Estonia	0.860	6	7,059	1
Bulgaria	0.824	7	982	7
Romania	0.813	8	806	8

Tab. 3. The correlation between the HDI and the inward FDI stock per capita (2005)

We have to underline the case of Romania and Bulgaria, countries ranking the bottom (position number 8, respectively 7) in the hierarchy of the countries submitted to analyses, both under the aspect of the foreign capital inputs and the HDI level. We believe that, considering the previously presented aspects, this evidence suggest that lacking massive foreign direct investment inputs, the economic and social results are significantly lower than those registered in other Central and Eastern European countries, like Hungary and the Czech Republic.

Also, it has to be noticed the case of Estonia too, country where the transnational corporations, by means of the achieved FDI, sustained the economic progress (as for 2005 the FDI stock represented 90% of GDP) without bearing though, a significant positive impact on the social component.

4. CONCLUSION

The surveys performed in case of the CEE countries confirm the aspects that Stiglitz outlined, with the only mention that the intensity of the TNC activities impact on the economic, social and natural environment from the host country differentiates according to the level and the structure of the foreign capital inflows, as well as the realistic conditions existing within the implanting economy.

In our opinion, through direct or indirect impact on the economic, social and natural environment of these countries, the transnational corporations played an active role in the sustainable development of the CEE countries. On one hand, they have sustained the economic growth, thus generating positive effects on the social environment. On the other hand, negative effects have been registered at both the social level and the economic environment too.

Consequently, by means of the impact over these three components of the sustainable development and the relations in-between, the foreign direct investment inflows generated by the TNC may sustain or hinder the making of an economy based on ethics and ecology.

5. REFERENCES

Roşca, I.Gh. (2006). *The knowledge society*, Economic Publishing House, pp.205;494-496, Bucharest

Stiglitz, J.E. (2003). *Globalization and its discontents*. Economic Publishing House, pp.30-47;378, Bucharest

*** *Green Paper, Promoting a European framework for corporative responsibility* (2001), European Commission, pp.8, Luxembourg

*** *World Investment Report 2006: FDI from Developing and Transition Economies: Implications for Development*, UNCTAD, New York and Geneva

*** *Human Development Report 2007/2008: Fighting climate change: Human solidarity in a divided world*, UNDP

Annals of DAAAM for 2008 & Proceedings of the 19th International DAAAM Symposium, ISSN 1726-9679
Editor B. Katalinic, Published by DAAAM International, Vienna, Austria 2008
Make Harmony Between Technology and Nature, and Your Mind will Fly Free as a Bird

STRATEGIES RELATED TO THE POSITIVE CONTRIBUTIONS OF THE FOREIGN DIRECT INVESTMENTS

IVAN, M[ihail] V[incentiu] & IACOVOIU, V[iorela]

Abstract: *Starting from the theories that point out that the potential effects associated to foreign direct investments depend mainly on the host country existing conditions, in this paper, by analyzing the correlation between the inward FDI stock per capita and the Global Competitiveness Index, we emphasize the fact that in the CEE countries the massive foreign capital input oriented towards activities incorporating a higher content of local resources and most of all, technology and knowledge, could encourage the improvement of the existing production factors quality and creation of some specialized production factors. Therefore, in order to efficiently instrument the accomplishment of the aimed objectives by making use of FDI, the CEE countries must apply adequate strategies.*

Key words: *strategy, investment, capabilities, development, and quality*

1. INTRODUCTION

Within the European integration, the new member states are facing a major decision, namely choosing the right path to follow. The choice involves not only setting the objective, but also identifying the directions in view of reaching the target and drafting a unitary and coherent strategy. This must proceed from a realistic evaluation and a proper, directing of the existing potential, capitalizing in the same time the opportunities given by the EU integration. Taking into consideration the insufficiency of the internal resources, the own efforts targeting a competitive production can be intensified and completed by attracting foreign direct investment (FDI) inflows, especially towards the intensive activities within the capital and technology areas, generating a bigger value added. Although obvious, this scenario is not easy to accomplish. Moreover, practice pointed out that although "desired", the FDI positive contributions are not always effectively "achieved" due to the lack of a unitary, coherent and realist strategy, based on the objective analysis of the host country definite conditions and the characteristics of foreign capital inflows. Consequently, a major issue is raised: *What are the characteristic elements of some strategies that encourage the manifestation of the FDI positive contributions?*

2. THEORETICAL ASPECTS

The literature distinguishes three sets of foreign direct investments, as per the reasons the economic agents (especially the transnational companies) achieve investments abroad: market-seeking investment, resource-seeking investment, and efficiency-seeking investment (Dunning, 1994). In our opinion, this distinction is based on identifying certain primary motivations, namely transnational corporations (TNC) strategic imperatives, according to which the foreign investor targets certain essential attributes of the host country (table 1).

The international investment motivations registered alterations following the changes that affected the world economy, respectively trade liberalization, competitiveness growth, telecommunication and information technology development, increasing globalization.

FDI types	Main investment motivations	Main characteristics of host country
Market – seeking	- to establish a strong position in the market of the host country; - to achieve access to a new regional market.	- the national market potential; - economic integration (internationalization).
Resource – seeking (Strategic asset-seeking)	- to achieve access to natural and human resources - to achieve access to national research and technological expertise/capabilities	- availability of natural resources ; - the skill quality of production labor. - availability of scientific knowledge; - the level of development of innovation and R&D activities.
Efficiency - seeking	-access to low-cost input factors (in order to improve the group's competitiveness)	- availability of low-cost input factors (labor, energy, raw materials)

Tab. 1.Foreign direct investments typology – synthesis

The TNC motivations migrated away from the access to local resources and the reaction regarding the local markets protection towards competitiveness related-issues (costs and efficiency), the access to strategic actives (research-development capabilities) and to the liberalized markets. Recent research reveal that, although the market growth (market-seeking) and the cost of the production factors (the efficiency-seeking) are the main motivations of the multinational companies to invest in Central and Eastern Europe (Manea & Pearce 2004), an increased importance is granted to the access to information and the infrastructure (transport, telecommunications) degree of development (Pournarakis & Varsakelis 2004).

John Dunning and Michael Porter, developing the theories concerning the investment development path and the competitive advantages stages, underlined the importance of proper strategies in accomplishing the migration from the competitive advantage based on the production factors endowment to the creation of specialized production factors. In this respect, we consider suggestive the empirical data concerning the development of the CEE countries competitive advantages by means of attracted FDI contributions.

3. EMPIRICAL DATA

The foreign direct investments have been thought as "a new Marshall plan" for the CEE countries. The extent to which that plan seemed realistic, materialized in positive effects once applied, is given by the success of the CEE countries in strengthening their competitive advantages, namely improving the quality of the existing production factors and the developing several competitive advantages based on specialized factors.

Countries	Inward FDI stock/capita (2006)		Global Competitiveness Index (2006)		
	USD/ capita	Rank	Score	Rank (global)	Rank (CEE)
Estonia	9,451	1	5.12	25	1
Hungary	8,095	2	4.52	41	5
Czech Republic	7,557	3	4.74	29	2
Slovakia	5,627	4	4.55	37	4
Slovenia	3,726	5	4.64	33	3
Poland	2,715	6	4.30	48	6
Bulgaria	2,682	7	3.96	72	8
Romania	1,897	8	4.02	68	7

Tab. 2.The inward FDI stock/capita and the GCI (2006)

In this respect, empirical data regarding the Global Competitiveness Index (GCI) emphasizes the fact that only few states, new members of European Union, registered notable successes in what concerns their economic competitiveness (table 2).

In the hierarchy of the 125 countries analyzed in the Global Competitiveness Report (2006), Estonia, which, at the level of year 2005 had already accumulated a significant FDI stock (over 90% of GDP) ranks 25 (the level of GCI is 5.12) before some traditionally developed countries, old EU members (Spain, Portugal and Italy). A quite promising position is also occupied by the Czech Republic (rank 29) that overtaking countries such as Portugal, Italy, Greece and South Africa, this country also receiving considerable FDI inflows as well as Hungary and Slovakia which occupy honorable places, 41st and 37th , out of the 125 analyzed states. In the same time, Slovenia that attracted FDI inflows comparatively lower occupies place 33rd (the level of GCI is 4.64) overtaking the majority of CEE countries which demonstrates that, the quality of the foreign direct investments is essential in order to maximize the positive contributions of foreign capital inflows.

The presented empirical data emphasize the existence of a strong relationship between the FDI inflows registered by the CEE countries and their level of competitiveness. Thus, the states that have accumulated a significant stock of direct foreign investments (Estonia, Hungary and the Czech Republic) have also registered the highest levels of competitiveness, alongside with Slovenia, while the countries disfavored in what concerns the infusions of capital (Romania and Bulgaria) rank the last regarding their economic competitiveness as well.

We have to underline the cases of Slovenia and Hungary that register big discrepancies between the two variables. Thus, Slovenia occupies place 5th in what concerns the level of FDI inflows, registering a much better position with respect at its economic competitiveness (rank 3). Also, Hungary placed on the second position in what concerns the FDI inflows occupies a lower rank (5) in respect with the level of GCI.

Except for the aspects related to the index calculation (for example, Hungary fell 6 positions during 2006 as opposed to 2005, due to the deterioration of the macroeconomic climate) we consider that the disparities recorded in Hungary and Slovenia, as well as the other analyzed countries are mainly due to the applied strategies that influenced both the quantity and the quality (structure) of the attracted FDI, determining their impact on the economic and social environment.

Considering the presented theoretical aspects, we appreciate that along the process of turning the Central and East European Economies into functional markets, the quality of the FDI inflows came as a condition for advancing towards the stages in which the investments and the innovative capacity represent the development engine. Consequently, the countries inside which the attracted FDI completed, at a certain point, their quantity, gained leading positions concerning their economic competitiveness. Within these countries, Slovenia marks a particular case as from the very beginning has followed the qualitative aspects of the FDI inflows (namely attracting Greenfield investments), targeting their impact on the local competencies. Therefore, although if reported to received foreign capital places behind most of the analyzed countries, as concerns competitiveness, Slovenia surpassed most of them (ranking after Estonia and the Czech Republic) due to the increased quality FDI impact achieved within the economy. So, the states inside which the act of attracting FDI has been assumed as a priority, namely Estonia, Hungary, the Czech Republic, Slovakia and Slovenia, materialized in promoting realistic and coherent policies based either on attracting massive FDI inflows (Estonia, Hungary, the Czech Republic and Slovakia), or on the qualitative aspects (Slovenia), strengthened the competitive advantages, successfully advancing towards the competitive advantages stage, based on innovation, typical of the developed economies.

4. CONCLUSION

The theories point out that the FDI potential positive and negative contributions to the receiving economy vary according to the FDI type, maturity, country of origin, also depending of the characteristics of the host country.

Taking into attention the positive evolutions registered in the last years, in European integration context, we consider that, in Romania, as well as the other CEE countries (e.g. Estonia, Hungary, the Czech Republic, Slovakia and Slovenia), the massive penetration of the foreign capital mostly oriented towards the activities that incorporate a higher content of local resources and most of all, technology and knowledge, could encourage the improvement of the existing production factors quality and the creation of specialized production factors.

In this respect, we appreciate that the long-term development strategy has to be oriented toward the improvement of the human and technological capabilities, through application of some measures, such us: the increase of the investments targeting educational and research activities; the stimulation of local initiative; the stimulation of local companies to invest in activities engendered higher value added; the stimulation of clusters development.

Consequently, considering the here above presented aspects, we appreciate that the maximization of the FDI positive contributions implies the promotion of adequate policies, oriented towards an intelligent use of the foreign capital inflows as a development strategy instrument.

5. REFERENCES

Dunning, J.H.(1994). Re-evaluating the benefits of foreign direct investments. *Transnational Corporations*, Vol.3, No.1, February 1994, pp.18-20,

Manea, J.& Pearce, R.(2004). Industrial restructuring in economies in transition and TNCs' investment motivations. *Transnational Corporations*, Vol.13, No.2, August 2004, pp.12-16

Pournarakis, M.& Varsakelis, N.(2004). Institutions, internalization and FDI: the case of economies in transition. *Transnational Corporations*, Vol.13, No.2, August 2004, pp.90-92

***Global Competitive Index - Report 2006. *World Economic Forum*, New York

***World Investment Report 2003: FDI Policies for Development: National and International Perspectives. *UNCTAD*, New York and Geneva

***World Investment Report 2006: FDI from Developing and Transition Economies: Implications for Development. *UNCTAD*, New York and Geneva

Annals of DAAAM for 2008 & Proceedings of the 19th International DAAAM Symposium, ISSN 1726-9679
Editor B. Katalinic, Published by DAAAM International, Vienna, Austria 2008
Make Harmony Between Technology and Nature, and Your Mind will Fly Free as a Bird

INTELLIGENT FUNCTIONS FOR CAPP PROCESSORS

IVAN, N[icolae] V[alentin]; IVAN, M[aria] C[ornelia]; GAVRUS, C[ristina] & OANCEA, G[heorghe]

Abstract: *The present paper is focused on CAPP processors and is aiming at the CAD/CAM integration. The result of these researches is the intelligent function FuICuPre destined to the determination of the technological routes for cutting processes of the surfaces that belong to cast iron parts. The function is based on the modelling of certain industrial experience aspects and of original mathematical models regarding the working parameters. All these models are integrated in an ensemble of programs written in AutoCAD/ AutoLISP creating the function mentioned above. The paper presents the application about the determination of the technological capability for the processes of a technological route. In the future, the research will be developed, aiming at new functions for CAPP processors.*
Key words: *CAPP, CAD/CAM, Expert system, Manufacturing.*

1. INTRODUCTION

Specialized literature confirms that CAPP problem is a difficult one viewed through CAD/CAM integration (Chang et al., 1998). There are two process planning methods: variant and generative systems (Kumar, 2005); (Kalpakjian & Schmid, 2006); (Amirouche, 2004). They have certain common problems. Thus, the structure of a CAPP processor contains three important segments (Ivan et al., 2004): the case identification processor, the technological processor and the post processor (Figure 1). The technological processor must solve the problems: process-capability analysis; determination of the dimensions of the part being in different processing stages; determination of the cutting parameters; data systematization; processor function and common data basis management in accordance with Figure 1; computer-user dialog. In this way, specialized literature is very poor regarding software functions.

2. INTELLIGENT FUNCTIONS FOR CAPP

2.1 Industrial Experience Modelling

For the machining methods of the technological route the technological capability parameters must be identified (Figure 2). This is a problem based on the industrial experience (Chang et al., 2006); (Kalpakjian & Schmid, 2006); (Ivan et al., 2004).

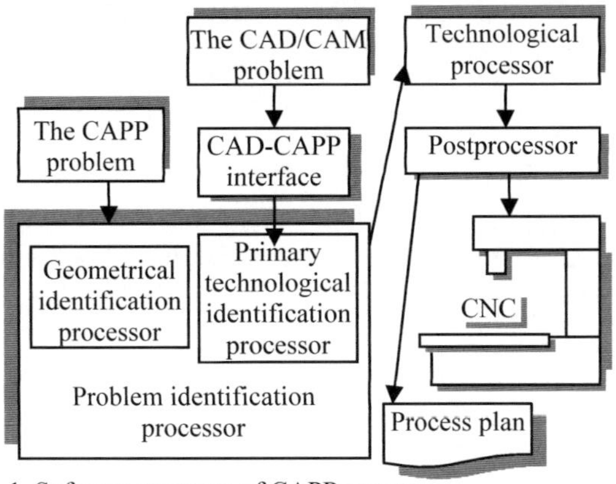

Fig. 1. Software structure of CAPP processor.

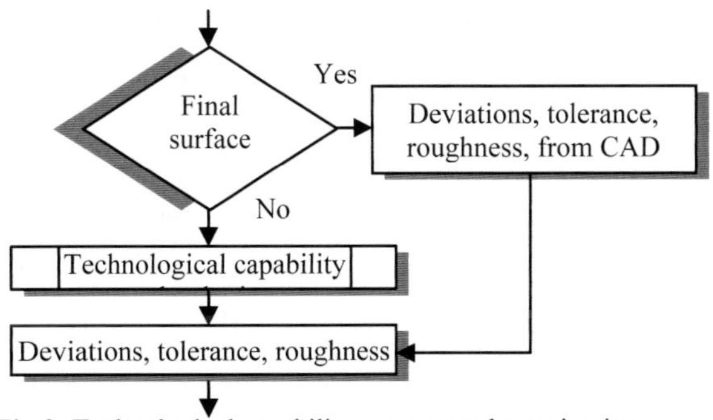

Fig.2. Technological capability parameter determination.

The industrial experience can be modelled in accordance with the matrix from Table 1. The current machining method is those being analyzed at a given moment and the previous machining method is the process that precedes the current one. Every current machining method can be preceded by m processes (m variable). For every machining method there are q capable accuracy classes (q variable) and s capable roughness values (s variable). The elements from Table 1 have the following meanings: MaMe1…MaMen refer to the name of the n machining methods available in a certain company (for example Drilling, Core Drilling); Xc1…Xcn refer to the numerical codes of the machining methods (for example 42, 44); Xp11…Xp1m, Xpn1…Xpnm represent the numerical codes of the possible previous machining methods (for example 42 is the numerical code of the previous machining method for the current machining process Core Drill/ 44); CP11…CP1q, CPn1… CPnq represent the capable dimensional accuracy classes related to the current machining methods (for example the capable dimensional accuracy classes for the current machining method Drilling/ 42 are 10, 11, 12, and for Core Drill the classes are 8, 9, 10); R11…R1s, Rn1…Rns represent the capable roughness related to the current processes (for example the capable roughness values for the current process Drilling/ 42 are 6,3 and 12,5).

2.2 *FuICuPre* Function

Machining processes contain linked pairs of current-previous machining methods. The present function, named ***I***ntelligent ***Fu***nction for ***Cu***rrent-***Pre***vious Machining Methods (***FuICuPre***), is conceived for solving CAPP problems for such machining method pairs.

Current process (alpha code)	Current process (numeric al code)	Numerical code of the possible previous process (variants)			Interval of the classes of accuracy-capability for current process			Interval of the surface roughness-capability for current process		
MaMe1	Xc1	Xp11	.	Xp1m	CP11	.	CP1q	R11	.	R1s
MaMe2	Xc2	Xp21	.	Xp2m	CP21	.	CP2q	R21	.	R2s
……….	…...	……	.	……..	……	.	……	…...	.	…...
MaMen	Xcn	Xpn1	.	Xpnm	CPn1	.	CPnq	Rn1	.	Rns

Tab. 1. Industrial experience matrix-EI.

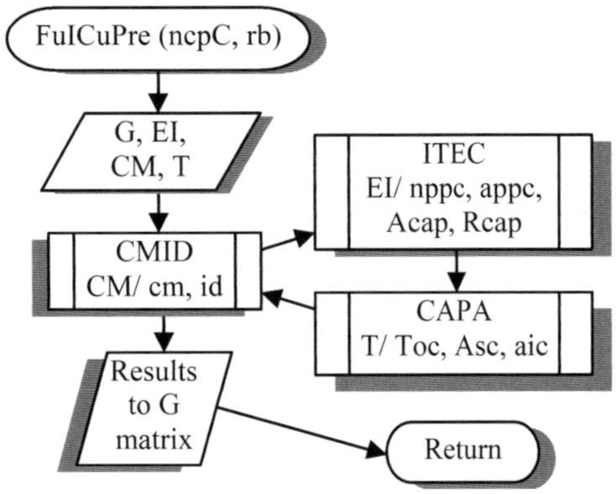

Fig. 3. ***FulCuPre*** function algorithm.

The necessary initial data: the current process numerical code; the rough billet type; the geometry that must be obtained as a result of the current machining method (diameter, roughness and accuracy class). The results obtained by activating ***FulCuPre*** function are: the cutting material for the current process; alpha and numerical codes of the previous machining method; the geometry and accuracy to be obtained as a result of the previous process.

FulCuPre function (Figure 3) appeals three other functions (subprograms) specialized on the determination of the cutting material and intermediary technological dimensions and of the technological capability parameters (capable tolerances, deviations, roughness). The meanings of the variables from Figure 3 are: ncpC- current process numerical code; rb- rough billet type; G-central matrix (result matrix); EI-industrial experience matrix (database); CM-mathematical model matrix for calculating the cutting material (database); T-fundamental tolerance matrix (database); CMID-subprogram for determining the cutting material and intermediary technological dimensions; cm-cutting material; id-intermediary technological dimension; ITEC-subprogram for identifying the technological capability for previous processes; nppC-previous process numerical code; appC-previous process alpha code; Acap-capable accuracy class for the previous process; Rcap-capable roughness for the previous process; CAPA-subprogram for determining the technological capability parameters; Toc, Asc, aic-tolerance, upper and lower deviations at the dimension obtained by the previous process. Other necessary information can be obtained from the G matrix that memorizes initial data and also the results obtained by activating all the CAPP subprograms. Table 2 presents a part of G matrix. The technological route (machine routing) must be known, in accordance with the group technology concept or the user option (Ivan et al., 2004).

3. APPLICATION

Processing the hole $\emptyset 15^{+0{,}018}$, roughness 1,6 μm and length 18 mm, requires: drilling (code 42), core drilling (code 44) and reaming (code 45). After activating ***FulCuPre*** function, the results from Table 3 have been obtained.

Machine routing (alpha cod and numeric. cod)	Surfaces geometry (d, l)	Accuracy of processed surfaces: tolerances, lower and upper deviation)	Surface roughness	Cod of machined surfaces (internal, external, flat)	Cutting paramet. (v, s)

Tab. 2. Central/ general matrix -G.

Machin. routing (alpha and numeric. code)	Surfaces geometry (d; l)	Accuracy of processed surfaces: tolerances,lower and upper deviation)	Surface roughness	Code of machined surfaces (internal, external, flat)
42	14; 18	+0,11; 0	12,5	-1
44	14,8; 18	+0,043; 0	6,3	-1
45	15, 18	+0,018; 0	1,6	-1

Tab. 3. Application result matrix.

4. CONCLUSION

CAPP stage is the difficult bridge for passing from CAD to CAM when talking about the integration of the designing with manufacturing. The result of the research is the software function ***FulCuPre***. This function facilitates the CAD/CAM integration, being focused on the use of concepts such as: concurrent engineering, expert systems and also constructive-technological entity (Ivan & al., 2007). The function ensures the easy taking over of the data from CAD stage and their use in CAPP and CAM stages. The function has generalization, allowing the establishment of the technological capability parameters (tolerances, dimensional deviations, roughness) for the cutting processes of any technological route. The function also allows the establishment of the dimensions of the part being in different intermediary processing stages. The space available allowed the presentation of the application regarding the determination of the technological capabilities parameters. AutoLISP language has been used because it is very much alike LISP, which, as it is known, is a programming language specific for artificial intelligence.

5. ACKNOWLEDGEMENTS

This research has been determined and supported by PN-II Program 71-133/ 2007, having the title: Expert Systems for Technological Process Optimization-ESOP. The authors express their gratitude to all partners for the collaboration.

6. REFERENCES

Amirouche, F. (2004). *Principles of Computer-Aided Design and Manufacturing,* Prentice Hall, ISBN 0-13-064631-8, New Jersey, USA

Chang, T. C.; Wysk, R. A. & Wang, H. P. (1998). *Computer-Aided Manufacturing, 2nd ed.,* Prentice Hall, ISBN 0-13-754524-X, New Jersey, USA

Chang, T. C.; Wysk, R. A. & Wang, H. P. (2006). *Computer-Aided Manufacturing, 3rd ed.,* Prentice Hall, ISBN 0-13-142919-1, New Jersey, USA

Ivan, N. V.; Berce, P.; Dragoi, M. V.; Oancea, G.; Ivan, M. C.; Balc, N.; Lancea, C.; Udroiu, R.; Vasiloni, M.; Mihali, M. & Ivan, C. (2004). *Sisteme CAD/CAPP/CAM, teorie si practica (CAD/CAPP/CAM Systems Theory and Practice),* Editura Tehnica, ISBN 973-31-1530-4, Bucuresti, Romania

Ivan, M. C.; Chicos, L. A. & Gavrus, C. (2007). *Conceptul de entitate constructive-tehnologica (Constructive-Technological Entity in Product Development)*, Editura Univ. Transilvania, ISBN 978-973-635-829-6, Brasov, Romania

Kalpakjian, S. & Schmid, S. R. (2006). *Manufacturing Engineering and Technology, 5th ed.,* Prentice Hall, ISBN 0-13-148965-8, New Jersey

Kumar, M. & Rajotia, S. (2005). Development of generative CAPP system for axisymmetric components for a job shop environment. *The International Journal of Advanced Manufacturing Technology*, Vol. 27, No. 1-2 (November 2005), pp. 136-144, ISSN 1433-3015

Annals of DAAAM for 2008 & Proceedings of the 19th International DAAAM Symposium, ISSN 1726-9679
Editor B. Katalinic, Published by DAAAM International, Vienna, Austria 2008
Make Harmony Between Technology and Nature, and Your Mind will Fly Free as a Bird

EMULATING A PLC ON A MICROCONTROLLER

IVANESCU, N[ick]; BORANGIU, T[heodor]; BROTAC, S[orin] & CIUPITU, L[iviu]

Abstract *— The paper describes an efficient method for process control using microcontrollers, based on emulating the functioning of a* ***Programmable Logical Controller*** *(PLC) and interpreting of logic control diagrams (commonly named* ***Sequential Function Charts*** *– SFC) . The chosen software solution allows even quasi-simultaneous interpretation of several SFC's together with* ***Instruction List*** *(IL) type programs. The results were tested for programming of a CNC machine using an ATMEGA microcontroller based circuit*
Key words*: process control, programmable controller, logical diagrams, microcontrollers.*

1. INTRODUCTION

It is well-known that the programmable logical controllers (PLC), used for controlling industrial processes, use standard programming languages like Sequential Function Chart (SFC), Ladder Diagram (LD), Instruction List (IL) and so on. But in practice, in some situations, using a PLC is inefficient or even impossible; instead a microcontroller can handle the control process and at the same time reduces the cost of the controlling system.

Our goal was to develop a basic source code (in C language) for ATMEL microcontrollers and some algorithms in order to emulate a PLC cycle and interpret SFC diagrams and IL's. We couldn't find any public or commercial project to manage this issue so we considered it can be very useful for automation engineers.

2. PLC CYCLE

In order that a microcontroller to act similar to a PLC, first of all it must emulate a PLC execution cycle. Any PLC has the following execution cycle (Arzen, 1994):

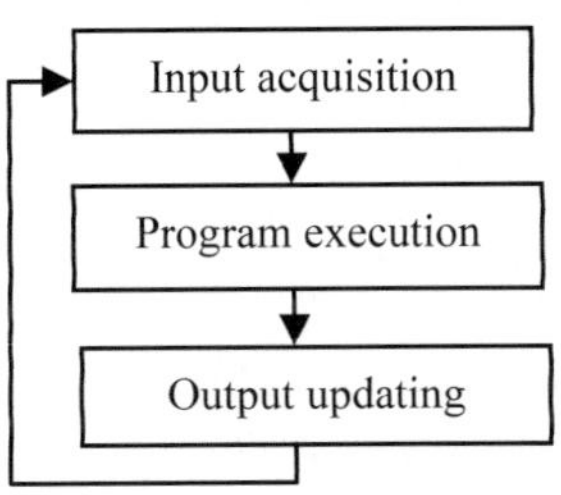

Fig.1. PLC Cycle

In order to emulate the stages belonging to a PLC cycle the microcontroller has to infinitely execute a high-level function similar to the one above, that simply call another 4 functions that implements the 3 stages presented previously (Ivanescu & all, 2002):

```
Void execution_cycle (void){
  input_aquisition();
  SFC_execution();
  IL_execution();
  output_update(); }
```

2.1 Timer functions

The first element that must be implemented when trying to emulate a PLC is to create a system **timer**. As 32 bit timer is not available, timer can be "*extended*" by incrementing a 16 bit variable at every **timer overflow interrupt** obtaining similar results as a hardware 32 bit timer.

The main functions regarding time created inside the project are:

- void **system_timer** (void): increments the extension variable every time the hardware 16 bit timer overflows.
- void **reset_timer** (void): reset the 16 bit timer registers and also the extension variable.
- void **set_timer** (word *time_value*): give the extension variable + timer registers the value of parameter *time_value*.
- word **get_timer** (void): returns the current combined value of extension variable + timer registers as a 32 bit value.

Normally these functions can be used to determine time related events appearing in the controlled process. One big advantage of a microcontroller versus a PLC is the time resolution; very small periods of time can be measured (like µseconds). This is efficient in applications where input signals vary very fast or when important events must be identified in very short time.

2.2 Processing of I/O signal

In microcontrollers digital inputs are acquired automatically at every clock pulse and stored in special I/O registers so inside the *input_aquisition()* function the value of these input registers are transferred in dedicated variables from RAM memory.

Regarding the output update, although digital outputs can be accessed directly to change their values, this approach is not. In fact several memory locations (variables) are dedicated to temporarily store the output values, during the SFC execution stage.

2.3 SFC interpretation and execution

The most complex part of the PLC cycle emulation is the SFC interpretation. In general more than one SFC chart can be interpreted and executed in this project (Ivanescu & Brotac, 2001). A logical diagram consists of steps, transitions, actions, divergences and convergences.

The basic ideas used in this project to accomplish SFC diagrams execution are described below. To determine which steps are active at every cycle an array of **function pointers** are declared and used in the following manner:

- The number of pointers in the array is equal to the number of separate SFC diagrams and every pointer is used to store the address of the function associated to the current active step of the associated SFC diagram.
- The general *execution_cycle* function will test the pointers values and will execute only the functions pointed by non-NILL pointers.

Every "**step - associated actions"** combination is implemented as a C function. Inside such a function the user must insert the correct code to realize the following objectives:

1) execution of "pulse" type associated actions only in the first execution cycle in which the function is called. In order to achieve this, every "**step - associated actions**" function must be associated with a logical variable (named *first_cycle_step_i*) that must be true only at the first cycle when function is executed. Otherwise it should be false.
2) Execution of "non-pulse" type associated actions in every cycle in which the function is executed.
3) Test of all the associated transition conditions. If any of it is found true, the pointer associated with the respective SFC is updated with the address of the function associated with the next step in the SFC diagram.
The analogy step – C function is shown in Fig. 2.

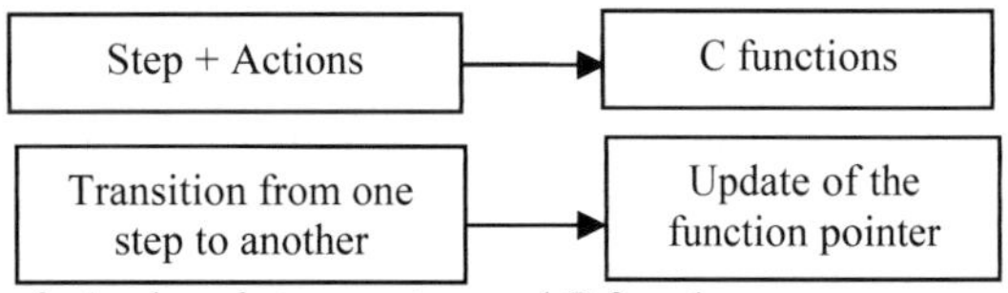

Fig. 2. Analogy between steps and C functions

To conclude this part of the presentation, the simplified structure of a function associated with a step is presented below (in pseudo code):

```
void step_i (void) {
   if first_cycle_step_i = true then{
      execute_pulse_actions;
     step_i_status = TRUE;
   }
  execute_non_pulse_actions;
  for j:=1 to total_number_of_transition_conditions do
       if condition_j = true then
          { update_correspondent_function_pointer;
           if parallelism_element_follows then
              update_function_pointers_for_virtual_SFC
          step_i_status = FALSE;
          jump_out_of_the_funcion;
          }
}
```

2.4 Developing IL type diagrams

IL is another standard programming language for PLC's. Typically, an IL program looks like a sequence of IF .. THEN instructions (Ivanescu & Rosu, 2008). In order to obtain such a program from a logical diagram, the following algorithm can be used (Web & Reis, 2002):

- Every step of the logical diagram is associated with a Boolean variable (practically with a bit). This variable is true when the corresponding step is active and false otherwise.
- For every step 2 types of code lines will be written:
  - 1st type: **if** (associated bit = 1) **then** (executes associated actions)
  - 2nd type: **if** (associated it = 1) and (transition condition = true) **then** (reset associated bit) and (set bit associated with next step)

All these lines are executed by the microcontroller at every cycle; in this way code will practically act like a SFC interpreter, being a variant to the C functions solution presented previously.

3. PROGRAMING A CNC MACHINE

In order to experiment the algorithms and the methods conceived, a full C program was developed for an ATMEGA 8515 Atmel processor, part of an electrical circuit that was sending specific programs to a Computer Numerically Controlled (CNC) machine. We determined here 4 automatons that should have to run simultaneously:
1) Eeprom manager
2) CNC programming manager
3) LCD manager
4) Keyboard manager
Their SFC diagrams were developed; e.g., for managing information display on LCD the following SFC diagram was drawn out (simplified):

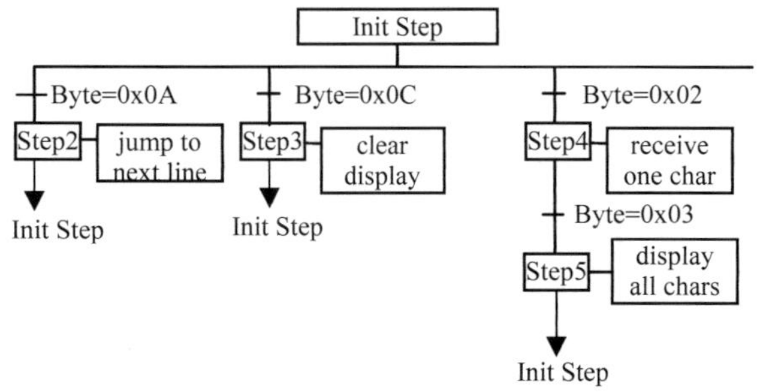

Fig. 3. SFC for controlling the LCD display

For all of these steps 5 functions were developed in the C program, in accordance with the method presented in the previous chapter. Similar to LCD automaton, the other automatons are designed and implemented. The results of the whole process were very good. All 4 SFC diagrams worked perfectly together. It was extremely difficult to imagine a non-logical approach to this process.

4. CONCLUSIONS AND FUTURE WORK

The methods and the basic programs developed are new and facilitate the usage of a microcontroller for controlling different devices or even industrial systems. Anyway there are some limitations that must taking into account when starting to develop a program of this type:
- reduced number of inputs/outputs
- limited memory dimension
- possibility of efficient design of additional circuits,

But some advantages must be also mentioned, like very short execution cycle, very low cost of the processor.
The future work will be concentrated in two directions:
1) Development of actual project by adding different features like: exit actions, step timeout, step minimum time and others.
2) Development of a user interface that allows creation of logical diagrams and generates code for a microcontroller.

5. REFERENCES

Arzen K., (1994). "Grafcet for Intelligent Supervisory Control Applications, *Automatica IFAC*, vol. 30, no. 10

Ivanescu N. & Brotac, S. (2001). "A distributed and configurable architecture for controlling industrial processes," *Preprints of IMS 2001*, Poznan, Poland

Ivanescu N., Borangiu Th. & Brotac, S. (2002). "Programmable controlers, theory and solved problems", *Printech*, Bucharest, Romania

Web J. W. & Reis R. A., (2002). "Programmable Logical Controllers, Principles and Applications", *Pearson Education*

Ivanescu, N. & Rosu A. (2008), *Modernizing a Fodder Plant*, Proceedings of DHMS conference, Athens, Greece

Annals of DAAAM for 2008 & Proceedings of the 19th International DAAAM Symposium, ISSN 1726-9679
Editor B. Katalinic, Published by DAAAM International, Vienna, Austria 2008
Make Harmony Between Technology and Nature, and Your Mind will Fly Free as a Bird
*Young Researcher and Scientist Paper / * Supervisor, Mentor*

NURBS PROCEDURES IN REVERSE ENGINEERING

IVANOV, M[arian] & PUPAZA, C[ristina]*

Abstract: *Reverse Engineering (RE) software provides methods to generate patch structures quickly, but manual organization of the model is always required to obtain the optimal patch layout. The paper investigates the possibility of using advanced CAD functionalities based on NURBS procedures for model recovering and optimal path generation in an early stage of a RE attempt. The polygon model of a human face was processed. Mesh patches were obtained, checked and edited, surfaces were generated and a solid model was obtained. This attempt proved that actual modeling technologies help the designer to close the gap between traditional RE packages and CAD systems. The procedure was tested on free form shapes, but it is also useful for mechanical components with complex geometries.*

Key words: *NURBS, procedure, reverse engineering*

1. INTRODUCTION

Reverse Engineering (RE) methodologies are fairly mature. However, one of the most important issues is still how to adequately define and manipulate surface details, in order to accurate represent and easily modify the original shape.

NonUniform Rational Spline (NURBS) are the ultimate RE output for CAD-CAM-CAE applications. These surfaces can be constructed based on the CAD entities from the curve phase or by using polygon meshes for surface fitting. The main objective of the NURBS surface definition phase in RE is to prepare a patch structure of quadrangular shapes for surface construction and for further utilization of the model. The patches can be drawn on the polygon model manually, semi-automatically, or automatically based on a target patch count and the curvature of the model. Although most RE software packages provide an automatic method to generate patch structures quickly, manual organization of the patch structure is always required to obtain the optimal layout, because, in many cases features such as edges and primitives cannot be recognized and patches do not adequately define the surface curvature (Raja & Fernandes, 2008). Examples of multilevel approximation have been reported (Möller, 2004), focusing on virtual reality simulations. Recent papers present an algorithm of NURBS surface fitting for reverse engineering (Dan & Lancheng, 2006), but the algorithm is computational intensive and the model has to be divided. Each patch has to be digitized separately, which is not the best choice. Many RE challenges are in fact modeling capabilities needed to process the 3D captured data. At present these functionalities are already implemented in advanced CAD systems (Lombard, 2008) and the designer has to use them in a rapid and efficient manner. The present paper deals with new procedures implemented in CAD systems used in processing RE models. A human face model was recovered and details were improved using NURBS technology in SolidWorks 2008.

2. SHAPE REPREZENTATION USING NURBS

NURBS are mathematical representations of 2D or 3D objects, which can be standard shapes or free-form shapes. Advantages of using NURBS representation are: • flexible design to a wide range of shapes; • compact expressions that can be evaluated and displayed quickly; • useful especially in 3D modelling, allowing the designer to easily manipulate control vertices and control curvature and the smoothness of contours; • defined from little data; • used to represent exactly the conic curves, with the added advantage of local control; • easy and stable evaluators, breaking and joining, degree elevation, and affine invariance.

NURBS enables local modification. Changing a control point affects the shape of the curve only in the neighbourhood of that point. The actual range of the change depends on the degree used. NURBS lets the CAD system to interpolate a large number of points while maintaining a low-degree basis function. This helps eliminate twists and spikes that often result from using high-degree polynomials (Piegl & Tiller, 1997). NURBS are generally characterized by four features: control points, weights, degree, and knot vector (Fig. 1).

NURBS curve C(u), which is a vector-valued piecewise rational polynomial function, is defined as (Altmann, 2005)

$$C(u) = \frac{\sum_{i=0}^{n} w_i\, P_i\, N_{i,k}(u)}{\sum_{i=0}^{n} w_i\, N_{i,k}(u)} \qquad (1)$$

where w_i are weights, P_i - control points (vector), $N_{i,k}$ - normalized B-spline basis functions of degree k. These B-splines are defined recursively as

$$N_{i,k}(u) = \frac{u - t_i}{t_{i+k} - t_i} N_{i,k-1}(u) + \frac{t_{i+k+1} - u}{t_{i+k+1} - t_{i+1}} N_{i+1,k-1}(u) \qquad (2)$$

and

$$N_{i,k}\begin{Bmatrix} 1 & \text{if } t_i \le u \le t_{i+1} \\ 0 & \text{otherwise} \end{Bmatrix}, \qquad (3)$$

where t_i are the nodes forming a knot vector $U\{t_0, t_1, ..., t_m\}$. The definition of the NURBS curve (1) can be rewritten using rational basis functions $R_{i,k}(u)$ (4).

$$C(u) = \sum_{i=0}^{n} P_i\, R_{i,k}(u) \qquad (4)$$

The NURBS surface is defined in a similar way. NURBS are actually regarded as a standard way to create and represent complex objects and free form shapes.

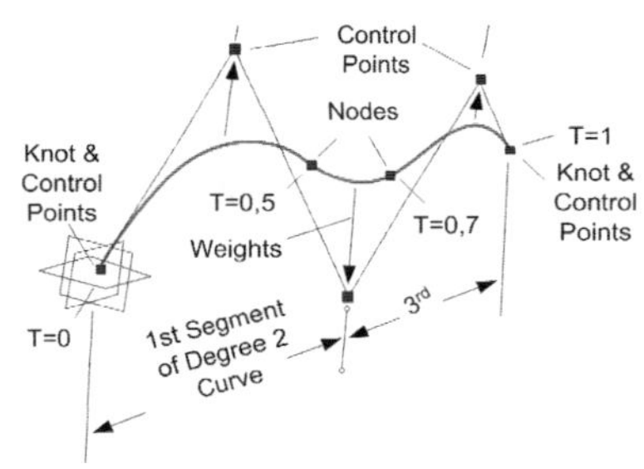

Fig. 1. The definition of a NURBS Curve.

3. MODEL RECOVERING AND INPROVEMENT

The point cloud model of a scanned human face was transformed into a 3D image file created in Virtual Reality Modeling Language and saved in a WRL - ASCII text format. This file contains only the coordinates of the measured points and viewpoint coordinates for the initial view of the model. The most important operations for optimizing the file were noise reduction and gap cleaning, abnormal surface cleaning, polygon and mesh refinement and polygon mesh decimation.

3.1 Mesh recovering and cleanup

A visual inspection of the model was proceeded before mesh preparation and cleanup. Figure 2 shows the WRL format, which was transformed into a polygon model.

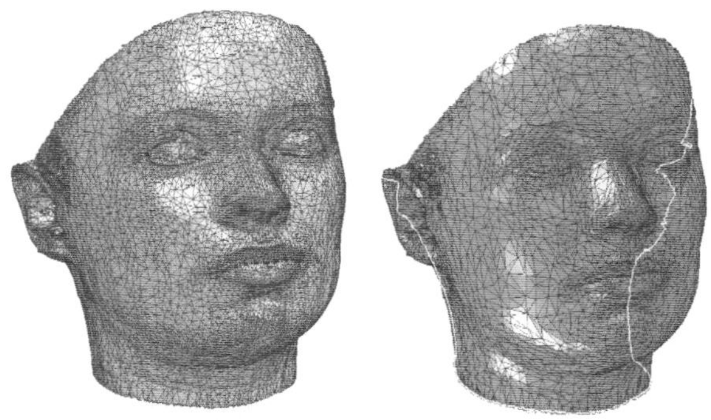

Fig. 2. Imported WRL format and polygon model.

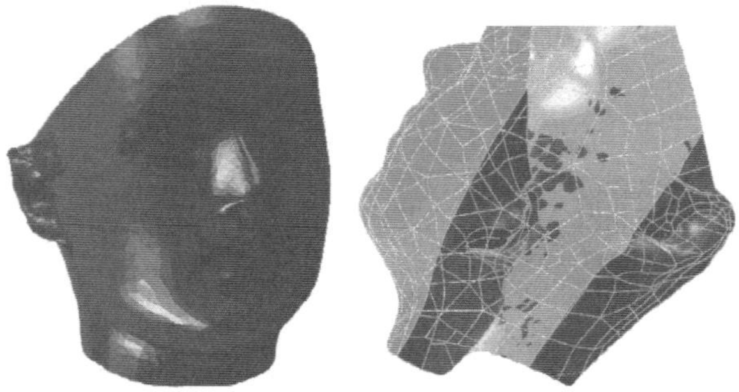

Fig. 3. Individual submeshes identification.

All holes were automatically detected and filled. In order to reduce the dimensions of the model and to emphasize the modelling procedure some simplifications were done. This helped speeding up surface generation of the model, which contains no symmetrical features. Individual submeshes were identified by colors (Fig. 3). The sensitivity was adjusted to obtain a desired submesh separation, which was improved in a later stage.

3.2 Surface extraction and solid model generation

After geometry recovering and mesh refinement the NURBS patches were obtained. When generating patches guided creation of the faces was chosen to better control the shapes. Editing of the patches was needed because the nose of the model proved to be distorted.

The geometry and topology of the model were checked with 'zebra stripes' option, in order to find the face imperfections, difficult to detect otherwise. This time no irregularities or breaks in the model were found. Figure 4 shows stripe checking options and editing of the patches. New curves and control points were added on the NURBS surfaces in order to allow improving local details. The geometry of the model was updated by means of new parameter definition and interactively changing control points. When changing the local shape of the model tangency constraints were imposed (Fig. 5).

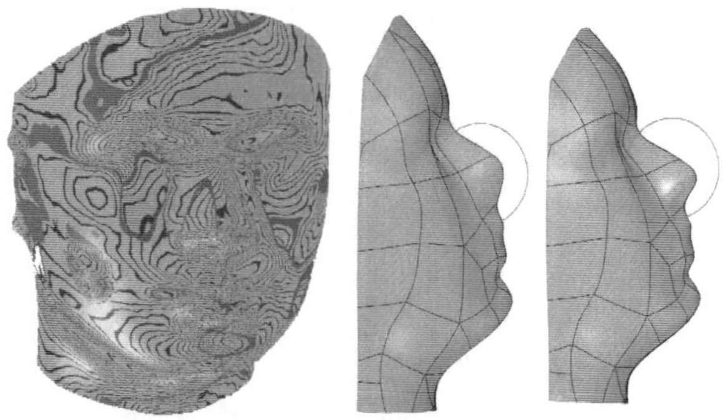

Fig. 4. Checking options and editing of the patches.

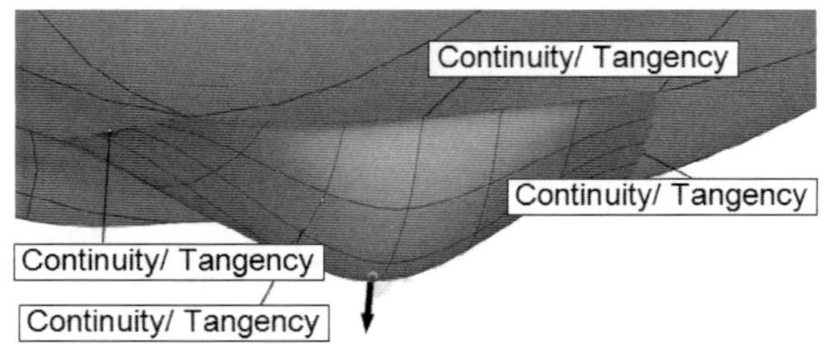

Fig. 5. Local shape changes and tangent constraints.

Surface extraction converted all the regions into faces. When the base model was completed and the surfaces were turned into a solid, the model was saved in an STEP format.

4. CONCLUSION

Most RE applications need at least two software packages to complete the data processing chain: one for RE data manipulation and the other for the geometric modeling work, which is based on CAD entities or NURBS surfaces. This attempt proved that advanced modeling technologies help the designer to reduce the processing time and close the gap between traditional RE packages and CAD systems. The procedure was tested on free form shapes, but it is useful for mechanical components with complex geometries as well. The designer has to carefully check that the geometry of the model remained unchanged, controlling the number of control points of the grids when constructing NURBS patches, although automatic model recovering and cleanup functionalities are user friendly. Further work is in progress for retrieving and editing the parameter definition of the surfaces in order to integrate the procedure in an optimization loop.

5. REFERENCES

Altmann, M. (2005). About Nonuniform Rational B-Splines - NURBS. *Available from:* http://www.cs.wpi.edu/~matt/courses/cs563/talks/nurbs.html. *Accessed:* 2005.05.05

Dan, J.; Lancheng, W. (2006). An Algorithm of NURBS Surface Fitting for Reverse Engineering, *Int. Journal of Advanced Manufacturing Tech.*, Springer-Verlag, DOI 10.1007/ s00170-005-0161-3, No. 31, pp. 92-97, London

Lombard, M. (2008). *SolidWorks. Surfacing and Complex Shape Modelling.* Wiley Publishing Inc., ISBN 978-0-470-25823-1, Indianapolis, Indiana

Möller, D.P.F. (2004). Computational Modeling an Simulation for Industry. *Proceedings of the 16th European Simulation Symposium*, György Lipovszki, István Molnár (Ed.), SCS Press, ISBN 1- 56555-286-5.

Piegl, R.; Tiller, W. (1997). *The NURBS Book*, Springer-Verlag, ISBN 3-540-61545-8, Berlin Heidelberg

Raja, V.; Fernandes K.J. (2008). *Reverse Engineering. An Industrial Approach,* Springer-Verlag, ISBN 978-1-84628-855-5, London

Annals of DAAAM for 2008 & Proceedings of the 19th International DAAAM Symposium, ISSN 1726-9679
Editor B. Katalinic, Published by DAAAM International, Vienna, Austria 2008
Make Harmony Between Technology and Nature, and Your Mind will Fly Free as a Bird

MENTOR PROFILE FOR HUMAN RESOURCES IN SMEs

IZVERCIANU, M[onica]; DRAGHICI, A[nca] & GUTA, C[armen]

Abstract: *The detection of requirements mentoring, of human resources in the small and medium enterprises (SMEs) from the West part of Romania before and after the Romania pre-accession to the European Union, had as specific objective detecting the mentor's profile. Often the mentor is a greater person with more experience, who has the role of model, compatriot, defiant, guide or animator.*
Key words: *mentor, profile, skills, mentoring, HR.*

1. INTRODUCTION

The paper presents the research results gained during our participation in the Leonardo da Vinci European programs:

- FORCREST project (Izvercianu, 2004), and
- TACITUS project (Izvercianu, 2005).

The University "Politehnica" of Timisoara team, from Romania, through the present paper authors, has detected the mentoring requirements in the case of human resources' (employees) of the small and medium enterprises (SMEs), from the West part of Romania, before and after the Romania accession to the European Union. In this context, we depict the research scenario that was followed for defining the mentor's profile and skills. Some important issues of the research results and conclusions will be presented in the end of the paper.

2. MENTORING

The small and medium enterprises development, in the economic point of view, is the research and innovation engine and it triggers the changes in the entire society. In this way the enterprise becomes strongly linked to concepts such as: technical innovation, systemic integration and increasing competitiveness. The new and continuous changing environment reports forward the classical interpretations and traditional concepts about enterprise that begin to remain without senses (Izvercianu, 2002).

During lifetime, humans train and develop, and this training and development leads to a skilful and effective adaptation to environment and manipulation of the environment. Society fosters and facilitates these activities of its members, but also channels and controls them through socialization and education so to produce outcomes which contribute and are acceptable for the society. People continue training throughout life, whether encouraged or not, whether the outcomes are valued or not, at work and at home (Beardwell & Holden, 2001).

The mentoring process facilitates the learning – to –learn of the employees and contributes to the process of meaning – making in the organization and hence to its responsiveness to environment, while meeting the development needs of employees. The mentees are not the only beneficiaries of mentoring: mentors also gain very much from being challenged to understand their jobs and the organization, and to find ways of helping their mentees share this understanding and work efficiently.

By mentoring they may also find that they need mentoring, too. Mentors draw upon their own networks to give experience and support to their mentees, and encourage them to develop networks of their own. In this way, the practice and benefits cascade through their organization (Torrington, & al., 2002).

3. THE RESEARCH SCENARIO AND RESULTS

The research regarding the identification of mentor profile, for the human resources involved in the small and medium enterprises environment, had as objective: defining the mentor's profile and identifying the skills requirements for a mentor.

That objective has risen from a vast analysis developed upon 32 small and medium enterprises from the West part of Romania. The analysis was developed at:

- Medium level, in the context of the work collectively (or work team);
- Micro level, for each person / employee (Foster-Turner, 2006).

Defining the mentor's profile - The first question to arise is: who are the mentors?

We consider that mentors may be friends, parents, colleagues, teachers as well as historic and contemporaneous personalities.

Age has no great importance in mentoring; although it is sure that the person who participates in a mentoring program is, in general, younger than mentors, that is to say innovations advance so fast and personal experience learn so many things that often young people have the opportunity to make mentors greater workers.

The mentoring relation works well if both parties spend time together and if they want it. A short mentoring relation means a long way for the mentee: a relative small investment of the mentor can suppose a significantly positive difference in the career and in the life of the mentee (Foster-Turner, 2006).

Study sample - Knowing who can be the mentors, for mentoring study were randomly chosen from the 32 small and medium enterprises a group of 8 enterprises. These enterprises have the next identified features (Table 1).

No.	Enterprise*	No of subjects in the group	The group structure by gender		Professional experience [years]		
			Female	Male	0-5	5-10	>10
E1	SP	14	8	6	4	3	7
E2	SP	15	5	10	7	2	6
E3	TP	15	15	-	2	1	12
E4	TP	17	17	-	1	3	13
E5	TE	16	1	15	-	3	13
E6	AM	18	2	16	-	3	13
E7	AM	20	1	19	1	6	13
E8	AM	21	3	18	-	7	14
TOTAL	8	136	52	84	15	28	93

Tab. 1. The structure of groups - *) Shoe production (SP); Textile production (TP); Thermo- electrics (TE); Apparatus & machines (AM).

The research methodology - For the first research objective we have chosen a qualitative method, namely the group's phenomenological analysis and we have used "the non-direct centred group interview technique". The topic of the group interview was based on the following questions:

- "Which are, in your opinion, the skills that a mentor must have?"
- "Which is the mentor's profile?"

Regarding the skills that a mentor must have, we have encountered many opinions, but considering the group's final opinion we kept 19th of them.

Each opinion was scored on the following scale from 1 to 10, according to the level of importance considered by each member of the group from Insignificant – Grade 1 to Extremely important – 10, with intermediate values. From the mathematical point of view, the decreasing ordering of the mentor's skills importance in students' opinion was made depending on the average value of opinions K_i. This was calculated with the following formula:

$$K_i = \frac{\sum_{j=1}^{n} a_{ij}}{n} \quad (1)$$

where P_i = skill i; i = 1…m; S_j = interviewed subject j ; j = 1…n. The skill value P_i in subject's opinion S_j: P_i , $S_j \rightarrow a_{ij}$ according with the information presented in Table 2. In our actual case were retained: P_i skills where i= 1……19 (m = 19) and S_j subjects where j = 1…136 (n = 136).

The research results regarding mentors' skills and profile – Followed the mathematical model presented before for all the opinions regarding the mentor's necessary skills, there have been resulted the decreasing ordering as results of the arithmetic means of grades given by the 136 subjects answers that are shown in Table 3. For each skill there have been deteminated the correspondent average formula. The result was the matrix of grades given by the subjects S_i answer (opinion) for each abilities P_i.

It can be observed that from the initial list made of 19 skills, were retained only the first ten skills, because under the 6 grade the skill importance becomes very less essential.

Proceeding in the same way and using the research method and the mathematical calculus the have been determinated the mentor's profile as it is described in Table 4. In the final part of the research, seven features of the mentor's profile were considered as essentials from the praxiology point of view.

Skills	Subjects S			Sum	Average
	S_1	…	S_n		
P_1	a_{11}	…	a_{1n}	Σa_{1j}	$K_1 = \frac{\sum_{j=1}^{n} a_{1j}}{n}$
P_2	a_{21}	…	a_{2n}	Σa_{2j}	$K_2 = \frac{\sum_{j=1}^{n} a_{ij}}{n}$
…	…	…	…	…	…
P_m	a_{m1}	…	a_{mn}	Σa_{mj}	$K_m = \frac{\sum_{j=1}^{n} a_{mj}}{n}$

Tab. 2. The matrix of grades given by the subjects S_i to the skills P_i

Skill P_i	Mentor's skills	$\sum_{j=1}^{n} a_{ij}$	K_i
1	They know what they are talking about, usually they are good in the activity they develop	1202	8.838
2	They give honest answers	1195	8.786
3	They are natural, they do not hurt with their answers or behaviors	1124	8.264
4	They do not intimidate, they are easily accessible	1110	8.161
5	They formulate questions actively	1041	7.654
6	They make possible, they take care, they act, and they facilitate	1002	7.367
7	They give a constructive and positive feed-back	975	7.169
8	They are disposed to discuss, to argue, to talk openly	968	7.117
9	They are interested in the mentee in a personal way, in a genuine worry	835	6.139
10	They guide subtly but checking that the mentee is able to take decisions	817	6.00

Tab. 3. The research results regarding mentors' skills

Features of mentor's profile	$\sum_{j=1}^{n} a_{ij}$	K_i
1. Relevant experience not to provide answers but know how to listen, to formulate questions and to give feedback.	1237	9.09
2. Sharing the leadership practice	1200	8.82
3. Disposition to reduce the mentee's learning curve.	1192	8.76
4. Disposition that the trained persons overcome them.	1150	8.45
5. Values that enrich the personalized relation	1100	8.08
6. Vocation and consideration of the mentoring like a contribution or legacy for the mentorized.	1095	8.05
7. High level of respect in all situations or proposals.	1051	7.72

Tab. 4. The research results regarding mentor's profile

4. CONCLUSIONS

The mentoring relations are not spontaneous and their occurrence is not automatic. It was outlined a kit of skills, which enabled us to design the mentor's profile. During the future phases we intent to disseminate the research results. The research for the mentor's profile determination was grounded on a group of 136 subjects. The paper was finalized with the mentoring concept analysis and development in economical environment.

5. REFERRING

Beardwell, I.; Holden. L. (2001). *Human Resources Management,* Ed. Pearson Education, ISBN 0273643169, Harlow, England.

Foster-Turner, J. (2006). *Coaching and mentoring in health and social care: The essentials of practice for professionals and organisations.* Oxford, UK.

Izvercianu, M.(2002). *Marketingul resurselor umane,* Ed. Solness, ISBN 973-8472-17-2, Timisoara, Romania.

Izvercianu, M.(2004). Leonardo da Vinci Programme, Forcrest Project

Torrington, D.;Hall, L. &Taylor S. (2002). *Human Resources Management,* Ed. Pearson Education, ISBN 0273646397, Harlow, England.

Annals of DAAAM for 2008 & Proceedings of the 19th International DAAAM Symposium, ISSN 1726-9679
Editor B. Katalinic, Published by DAAAM International, Vienna, Austria 2008
Make Harmony Between Technology and Nature, and Your Mind will Fly Free as a Bird

ASPECTS REGARDING THE EVOLUTION OF FRANCHISE SYSTEM ON ROMANIAN MARKET

IZVERCIANU, M[onica]; MIHARTESCU, A[na] A[ndreea]; MAZILESCU, C[risanta] A[lina]; DUPTA, P[avel] E[rnest] & STAICU, F[lorentiu]

Abstract: *This paper objective is to identify the tendencies that develop on the Romanian franchise market, correlated with international evolution and to inventory the needs and preferences of potential clients, in order to diversify and improve the franchise system. The paper presents the results of a quantitative research made on an illustrative population sample (50 subjects) at The Franchise Market, in February 2008 at Bucharest. The research was made in a national number 91-069, PN II/2007. The research made for objectives accomplishment was mainly a* ***descriptive transversal*** *research.*
Key words: *franchise, quantitative research, descriptive research, market study.*

1. INTRODUCTION

Romania has some features which anticipate a "boom" of franchise. After 50 years of etatism, Romania is a country where the entrepreneurial culture new, and to the entrepreneurs suits "to a fit" the possibility to receive all know-how in a package and to be trained by the franchisors in all managerial areas implied by a new business.

The admission in European Union, in 2007, enforced commercial and ecological standards for which 80% of Romanian SMEs are not ready, yet. So, the purchasing of a franchise which works respecting international rules becomes tempting. The growing competition makes many entrepreneurs to choose to cooperate with a company nationally or internationally recognized, respecting a recipe that was verified and validated on other markets.

In the first part of paper is presented the research methodology, and in the second part are emphasized the research results. In the paper are emphasized the advantages and opportunities of a franchise system, correlated with the perceptions of subjects investigated.

2. THE RESEARCH METHODOLOGY

The Research Methodology used was an interdisciplinary one, centered mainly on the research methods specific to marketing and mathematical statistics, with transmission and transformation of gathered information in statistical data. The research made to accomplish the objectives was mainly a descriptive research (Bacali, 2002), transversal research, the information gathered being **quantitative**. For acquisition of data needed for accomplish the objectives of the research contract were used both **secondary** and **primary** sources.

As method of **primary data** (Kotler, 2006) acquisition we used the classical survey based on two types of questionnaires applied to the representative subjects' samples, franchisors and possible franchisee. The elaboration of questionnaires, depending of the two populations analyzed – franchisors and franchisees – was made by a team made of the paper authors and also of other specialists from the economical community in Timisoara. The data from **secondary sources** were obtained from National Institute of Statistics, presented in Romanian Statistic Yearbook.

For to choose and apply a sampling method were considered the next factors: the financial resources allotted, the time allotted to the research, the human resources available, the size of target population, the sampling error, etc.

The method of information acquisition used was the INVESTIGATION, and the research tool was the QUESTIONNAIRE.

The interest populations considered were: franchisors who participated as exhibitioners at The Franchise Market, in February 2008 at Bucharest and potential franchisees who participated as buyers.

The sample dimension was determinate based on the statistical formula:

$$n = \frac{t^2 \times p(1-p)}{\Delta\Omega^2 \pm \frac{t^2 \times p(1-p\ )}{N}} = 50 \qquad (1)$$

Where the meanings are:

- t – Coefficient which correspond to the probability with which the results are guaranteed. (in statistical tables);
- p – the proportion of sample components which have the studied feature (usually "p" value is not known, it can be considered as equal with 0,5 to make the dispersion maximal);
- $\Delta\Omega$ – the accepted limit of error (± 3 %);
- N – The size of total aimed population, named also universe population (1000 participants la at The Franchise Market, in February 2008 at Bucharest).

The sampling method chosen was the **random stratified sampling**, which consist of: dividing the collectivity in groups (1000 subjects) based on preset criteria – incomes (250-2000E) and age (18-60 years), and also the activity area (manufacturing, services) of companies that offer the franchise– from every group (50 subjects) were randomly extracted a sample which was interviewed.

The questionnaire elaboration is one of the main elements which influence the research success. The questionnaire used is made of 20 questions opened and closed: 3 introductive questions, with cu general character, 7 questions regarding the present behavior of subjects, 5 questions regarding motivations, needs, attitude, 5 identification questions, with socio-demographic character.

The Synthesis of research methodology is presented in figure 1.

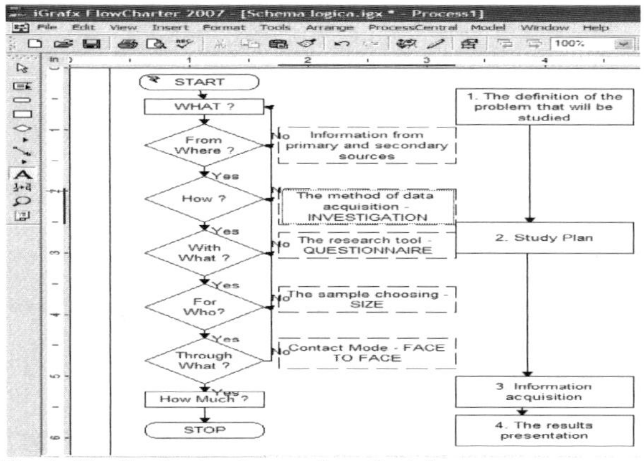

Fig. 1. The Synthesis of research methodology.

3. THE RESULTS OF MARKET STUDY

After this market study was made we obtained the following results:

- For many of investigated subjects, the franchise is a thing completely new, but this does not impair them to be interested by a franchise business. Only 14% of investigated subjects have a franchise sistem.
- 30 % of investigated subjects are hesitant regarding the acquisition of a franchise system, and 24% do not intent to buy one in the next year.
- 90% of investigated subjects are interested in a franchise buying in services area especially of the retail ones (28%), because these request minimal investments and the profit is rapid, and 10% want to invest in production area. The percent is small because the costs are higher, and the profit will be obtained during medium or long term. In this percent, 61% are interested to buy a franchise in food industry, because in Romania 55-60% form monthly incoming of a family is spent for food.
- 54% of investigated subjects are willing to pay an admission tax of maximum 10000 E.
- 55% of investigated subjects want to finance the activity with combined incomes (loans, own sources financing), so they are not willing to take the risk.
- 42% of investigated subjects interested in a franchise acquisition have ages between 24 and 35 years (age flexibility) and they graduated university studies.

After the market study accomplishment we made a compared analysis of opportunities offered by a franchise system which will be acquisitioned and of threats which hang over it. The results of this analysis are shown in table 1.

Opportunities offered by the franchise system	**The threats which hang on the franchise system**
Known Brand, requested on market, with notoriety **(30%)**	
Immediate Profit, Accessible price, Profitableness **(25%)**	Financial conditions (high costs, low profit, high royalty fee) **(25 %)**
Low costs, Permissive admission taxes **(10%)**	Excessive admission taxes **(25%)**
Flexible franchise conditions (installment admission tax, assets renting) **(7%)**	Conservative conditions **(3%)**
Low competition (exclusivity in a certain area) **(7%)**	Future excessive competition through franchise extending **(20%)**
Minimal investment for a maximal profit **(5%)**	Small investment / low profit **(3%)**
An achieving organizational structure **(5%)**	Lack of informational support and information **(5%)**
Technical assistance from the franchisor (HR training, technical consultancy) **(5%)**	Lack of: training and transparency **(6%)**
Minimal expenses for assets **(3%)**	High costs for assets **(3%)**
Stability, business continuity **(3%)**	Negative evolutions on market due to the international economic recession **(10%).**

Tab. 1. Results of compared analysis

Observation: for compared analysis opened questions were used and therefore, there were many different answers which lead to subjectiveness of codification and to interpretable answers.

After the compared analysis of opportunities and threats which "rule" the potential Romanian franchisees and franchisors results the next conclusions:

- The subjects were divided in two segments: the ones that have access to funds and want an investment with immediate profitability, and the ones with limited financial resources do not assume the investment risks, which lead to the Romanian society polarization in reach and poor people.
- 10% of investigated subjects consider admission taxes as being too permissive, due to their own financial power.
- 20% of investigated subjects manifest prudence regarding a franchise buying, due to the world economic recession and to excessive competition.

4. CONCLUSION

The franchise market in Romania is in continuous formation and transformation, because after many years when the economy was "sustained" by lohn system companies and which, due to globalization and Romania adhesion at European Union, the companies were forced to relocate their production units in other areas (Asia). For this reason, many investors were forced to find solutions to carry on their activity. The main problems a franchisor / franchisee confront with a Romanian market are:

- Lack of training and information of Human Resource in the franchise area.
- Romanian people has not the financial resources that can be immobilized on medium or long term
- After the `90s Romanian people give a special attention to brand (30% consider brand as an opportunity).

The main opportunities brought by the Romania adhesion at European Union, perceived by the investigated subjects are: legislation which will be according the European one (24%) and free circulation of Human Resource, information and materials (23%).

Why a franchise system is attractive to the investigated subjects? Because it is a solution tried and which brings immediate profit. More, a franchise does not ask for innovation. This means that there are not research-development costs.

In the future we propose to continue the research in order to identify the evolution of franchise system on the Romanian market, in conditions of international economic recession.

5. REFERENCES

Bacali, L (2002) – *Marketing*, Dacia Publishing House, ISBN 973-35-1387-3, Cluj-Napoca

Kotler, P.& Keller L. (2006) – *Marketing Management*, Prentice Hall Publishing House, ISBN 0131457578, New Jersey

The Romanian Statistics Yearbook (2007), I.N.S. Publishing House , ISBN 1220 -3246, Bucuresti

The National Research Contract - *Partenership for Exelency Research in Development of Entrepreneurship Skills and of Competitive Human Capital in Economy and Society Based on Knowledge and Innovation,* number 91-069, PN II/2007.

http://www.fbb.ro/home/franciza-arhiva/Referat-franciza.htm *Accesed:* 2008-06-10

Annals of DAAAM for 2008 & Proceedings of the 19th International DAAAM Symposium, ISSN 1726-9679
Editor B. Katalinic, Published by DAAAM International, Vienna, Austria 2008
Make Harmony Between Technology and Nature, and Your Mind will Fly Free as a Bird

THE FEATURES OF A COMPETITIVE PROFESSIONAL TRAINING

IZVERCIANU, M[onica]; POPA, H[oria] & MOCAN, M[arian]

Abstract: *The detection of required features for the competitive professional training of human resources in the small and medium enterprises (SMEs) from the West part of Romania before and after the Romania pre-accession to the European Union, had as specific objective identifying the training features for small and medium enterprises' employees.*
Key words: *human resources, features, training, mentor, competitiveness.*

1. INTRODUCTION

In the context of the Leonardo da Vinci European program, through the projects FORCREST and TACITUS, the team from the University "Politehnica" of Timisoara, Romania, through the present paper authors, has detected the *professional training features* for human resources' from the small and medium enterprises (SMEs), from the West part of Romania, before and after the Romania's accession to the European Union.

2. THE PROFESSIONAL TRAINING FEATURES

Romania is in a period of transformations and transition to a competitive economy that demands another set of features for human resources professional training (Izvercianu, 2006). The great complexity of training contributes to development. Training is not synonymous with it, but development cannot take place without learning (Whittaker, 1992). Investments in the time required for continuing professional development should be regarded as being as important as investments in other activities. (Pickard, 1999), (Quinn, 2003).

3. PROFESSIONAL TRAINING IDENTIFICATION FEATURES FOR EMPLOYEES OF SME

These objectives have risen from a vast analysis developed upon 32 small and medium enterprises and at next levels:

High level - when the enterprises' strategic objectives and their socio-organizational evolution were analyzed;

Medium level - in the context of the work collectively (or work team);

Micro level - for each person/ employee.

Study sample - The participant subjects in this study are from 32 small and medium enterprises from different activity domains: Shoe production (5), Textile production (6), Thermo-electrics (2), Apparatus and machines (3), Transports (4), Service (12).

The research methodology - For this research objective we have chosen a qualitative method, namely the group's phenomenological analysis and we have used "the non-direct centred group interview technique", combined with the "questionnaire technique". The thematic of the group interview refers to:

Their position, opinion about an efficient vocational training course/program;

What they have to suggest to the trainer in order to attend the desired results after a vocational training course/program.

No.	Professional training features	Σ	K_i
F1	The vocational training need	1480	8.31
F2	The courses/programs have to be practice oriented, with case studies and examples of good practice	1425	8.00
F3	The trainer has to present the theoretical part of the courses/programs in a motivating and attractive manner so that the students will be attentive during the whole course/program	1379	7.74
F4	The training program has to be done only when a specific practical situation requires new knowledge for the employees	1322	7.42
F5	All available facilities for the training courses/programs have to be at the students' disposal	1115	6.26
F6	The trainer has to use efficiently the time allocated for the training courses/programs	1093	6.14
F7	The trainer has to be able to answer with high competencies and in due time, to all students' questions for satisfying their needs	1078	6.05

Table 1. The ranking for training features

At the end the groups retained 12 training features, but after the 178 subjects assigned grades on a scale from 1 (very, very less important) to 10 (extremely important), to determine the importance level of every training features, it was observed that for grades under 6, the importance level is negligible, that is the reason for there were kept only 7 training features (table 1).

From the mathematical point of view, the decreasing ordering of the professional training features importance in students' opinion was made depending on the average value of opinions K_i. This was calculated with the following formula:

$$K_i = \frac{\sum_{j=1}^{n} a_{ij}}{n} \qquad (1)$$

Where F_i = feature i; i= 1...m (m=12); S_j = interviewed subject j; j = 1...n (n=178). The feature value F_i in subject's opinion S_j: F_i, $S_j \rightarrow a_{ij}$.

F1-The research sample consists of adult persons who are aware of their professional training basic need as a long life learning process and who have different arguments to motivate them. It must be remarked that:

1. Young people/employees, with a professional experience from 0 to 5 years, who represent 11.8% from the total subjects of the sample; (21 persons from 178), are motivated to **accumulate knowledge** because this will help them to find a new, and better job which will satisfy them better. The

described situation implies a non-fidelity attitude of the young employees regarding the enterprise and, of course, a wrong behaviour of the enterprise regarding their young employees

2. The subjects/employees with a professional experience from 5 to 10 years that represents 25.8% from the total subjects of the sample; (46 persons from 178) were more careful with the answers regarding the personal motivation linked to the vocational training:

- 58.7% (27 responders from 46 participant groups) have motivated the **technological progress** stands at the base of the professional training. We can remark that the 27 subjects/employees belong to those enterprises which produce apparatus and machines and to those enterprises with a thermo-electrics profile. Those justify the subjects/employees answers.
- 36.9% (17 responders from 46 participant groups) have motivated their need for vocational training as a necessity for **work-place stability**, respective for a better paid job. We can remark that the majority of them are employees in textiles industry and services where, is well known, that the level of remuneration is low. Another remark is that women, who dominate the jobs' occupation in this industry and who have comments about the level of their salaries, have declared and sustained the priority on having a stabile work-place.
- 4.4% (2 responders from 46 participant groups) with a professional experience from 5 to 10 years have motivated their need for vocational training according to their own **professional code**.

3. The employees with more than 10 years of professional experience (not necessarily in the same enterprise) represent 63.48% of the groups' participants (111 responders from the total of 178 subjects) and they have the following motivations: 25.22% (28 responders from 111 subjects) were motivated by their **job attestation or by immediate, concrete promotion**; 19.8% (22 responders from 111 subjects) of the responders had priorities in **developing their carrier**, the majority were devoted to their enterprise and well motivated; 54.98% (61 responders from 111 subjects) were motivated by the **work-place stability** and the technological progress which they have to face in order to gain stability. It is worthy to note that from the 178 subjects, 65 were women, who fear that from a specific age it is difficult to find a job corresponding with their occupation.

F2 - The courses have to be practical, concrete with punctual solving. The vocational training is addressed to adults which have a formed personality, who are aware of their vocational training, but who, in majority, want a punctual, practical education, in their professional area (or in adjacent area). Approximately 75 % (133 persons) of the responders wish to **acquire knowledge for solving the problems they confront with or knowledge necessary for their job attestation or promotion**. It has been noticed that young persons (who search another job, adequate to their tendencies) have also underlined the need of theoretical context understanding on a bigger range of information. This group attitude is correlated with the next category of features.

F3 - The theoretical part of the courses must be presented by the formative staff in a stimulating manner for students in order to keep the attention up during the course. Once the people get older, **new barriers** regarding the theoretical accumulations appear, concerning new competencies acquirement. This fact was remarked into the studied groups. Some of them with more than 10 years of experience in their professional field have imposed their point of view in the frame of the discussion concerning the theoretical informational content reported to the practical informational content. This point of view was favourable to the practical informational content. The subjects' need to inform and train themselves was restricted, beaked by their **limited availability for the theoretical accumulations**, in favour of the practical accumulations.

F4 - To this idea, of **practical specific problems training**, 70.2% responders have totally adhered (125 subjects from 178 subjects), the others, 29.8 % (53 subjects) agreed the idea, but they exhibit the need of a continuous training, from different reasons, namely: some want a professional career and have the disposition of continuous information, others want to accumulate new competencies for finding a well paid job for which experience and good professional knowledge is required. It is worthy of note that **women**, justifying the lack of time and also the number of domestic responsibilities, have adhered in proportion of approximately 90 %, to this "feature " stating that the "training has to be made just when concrete situations require that" (Zadel, 2006).

F5 - The subjects participating to this study have identified some **material, financial and informational facilities** for a good learning development, the presence of a competent trainer who has to answer "in a concrete and precise way" to their questions and the theoretical support needed for every student. Regarding the financial resources only 14.0% (25 from 178 subjects) are disposed to pay the course fee themselves, the others want the enterprise to invest in their vocational training.

F6 - The future students want to get useful professional information, in an **optimal period of time** for them, because their time budget for training is limited: professional duties, personal problems, etc. which means that the trainer must use the time efficiently.

F7 - Into the rational component of learning, subjects keep in view the learning development by competently trainers who fulfil their implicit expectations from the informational, professional point of view, at the appropriate moment.

4. RESEARCH CONCLUSIONS

The interviewed subjects consider that professional training is a key factor that guarantees their career development. Each employee who participated at the study has different motivations for satisfying her/his vocational training needs (a better job, promotion, new competencies owed to technology's progresses etc.). Women and most of the men with a professional experience more than 10 years need vocational training for work-place stability. The subjects/employees expressed their need for "just-in-time" training and with subjects of most interest for the moment. The subjects/employees have great exigencies for the trainers and for their professional competencies. 14% of the employees have financial disposal for the courses/programs payment. The cognitive behaviour has influenced the learning rhythm and the type of learning method (theoretical or practical competencies learning). The paper was finalized with the training features inventory for small and medium enterprises human resources.

5. REFERRING

Izvercianu, M.(2006). *Human Resources Marketing (Marketingul resurselor umane)*, Ed. Solness, ISBN 973-8472-17-2, Timisoara, Romania.

Pickard, J. (1999). Emote possibilities-sense and sensitivity, *People Management*, Vol. 5, 48-56, ISSN 0143-5124

Quinn, B., A. (2003). Building a profession: A sociological analysis of the internship development program, *Journal of Architectural Education*, vol. 56 (4), 41-9, ISSN-0097-8507

Whittaker, J.(1992). Making a policy of keeping up to date, *Personnel Management*, March, 28-34, ISSN-0036-6439

Zadel A. (2006). Impact of Personality and Emotional Intelligence on Successful Training in Competences, *Managing Global Transitions. International Research Journal,* vol. 4, no. 4, ISSN 1581-6311, 363-376.

Annals of DAAAM for 2008 & Proceedings of the 19th International DAAAM Symposium, ISSN 1726-9679
Editor B. Katalinic, Published by DAAAM International, Vienna, Austria 2008
Make Harmony Between Technology and Nature, and Your Mind will Fly Free as a Bird

MODEL FOR DETERMINING THE ENTREPRENEURIAL PROFILE

IZVERCIANU, M[onica]; PUGNA, A[drian]; NEGRU-STRAUTI, G[abriela] & TION, M[onica]

Abstract: *The paper presents a research regarding the entrepreneur profile from 5 West Region in Romania having in view that entrepreneur's role from countries which currently are in transition period to free economy and free market is becoming increasingly important. The Entrepreneurs may be considered as "trail blazers" in establishing new activity branches, which, by themselves determine major structural changes in economy. Thus, the entrepreneurial activity is in the center of economical innovation spirit, productivity improvement, economical competitiveness and new jobs generating.*

Key words: *entrepreneurship, questionnaire, sample, survey, model*

1. INTRODUCTION

The study presented below is part of a larger analysis of the entrepreneurial phenomenon in Romania, a study that is to be carried on by contract no. 91-069/14.10.2007, Excellence Research Partnership for Development of Entrepreneurial Aptitudes and of Competitive Human Capital inside the Economy and the Society Based on Knowledge and Innovation. Currently, the entrepreneurship has a bigger impact then ever in the past 100 years. Recent studies shows that entrepreneurship became a significant "way of life", a career for many, more than 4% of adults trying to start a business. Thus, Entrepreneur's role, considering the Entrepreneur as a person characterized by a high degree involvement in introducing novelty or innovating the present, is increasingly becoming important in countries which currently are in transition period to free economy and free market (Chell, 2001),(Vaduva, 2004).

2. RESEARCH METHODOLOGY

The study has in view determining some specific characteristics to Romanian Entrepreneurs from 5 West Region. Thus, in order to draw up the Entrepreneur's profile, first were identified the factors which influences the decision to be, or to became, his own employee. The differences in earning regarding being your own employee or being some other's employee, have a central position (Izvercian, 2002). Also, it was utilized a large set of variables in order to describe the factors which influences the efficiency of these two situations, their relative risk, against the individual's preferences and abilities (Douglas, 2002).

The survey has been done based on a questionnaire submitted to enterprisers-managers. The survey – from designing point of view – represented a hybrid between a stratified random sampling with unequal probabilities and a controlled semi-random sampling – aspect which was imposed by organizational motives taking in account restrictions and objective limitations of this kind of research. The sample can be considered representative, not necessarily by size, but rather by afterwards procedures of stratifying amelioration and by successive characteristics recalibrations (by main activity – CANE code – and by number of employee).

The research used a regional database and had as dependent variables, transition to self-employment and sometimes longevity in business. The typical descriptive variables category includes: age, gender, education, earning, owned capital, previous professional experience, marital status and cultural system from which Entrepreneurs emerged.The study sample has been chosen out of 1321 subjects that have been mailed, out of which only 122 (9.235%) from all of them answered, the sample being as a randomly chosen sample.

3. SAMPLE CHARACTERISTICS

Taking in account the firm's dimensions, one can see that micro-enterprises prevail, representing 52.28% from the total investigated firms, followed by small-enterprises representing 32.74% and medium-enterprises representing 14.98%.

From juridical form point o view, the most often set up firms are the ones with Limited status, representing 94.5% from total SMEs established, following the tendency recorded on national level, where 92.15% are of Limited type.Due to the array of characteristics analyzed through the survey, the sample size was established on first instance based on linear characteristic, in unrepeatable variant (Isaic-Maniu, 2004):

$$n \cong \frac{N \cdot z_{1-p}^2 \cdot \sigma_b^2}{N \cdot \delta^2 + z_{1-p}^2 \cdot \sigma_b^2} \qquad (1)$$

where z_{1-p} are Laplace's variable values, for a probability $\alpha = 1 - p$, where α is the significance level, in our case α = 0.001 (p = 99.9%), and therefore $z_{0.001} = 3.09$; σ_b^2 - is the binary characteristic's variance, that is $\sigma_b^2 = \alpha \cdot (1-\alpha)$. In order to reduce statistical uncertainty, maximum variance has been utilized, therefore the sample was designed at the worst case scenario, that is in our case $\sigma_b^2 = 0.5 \cdot (1 - 0.5) = 0.25$; δ^2 is the probable error, established in our case at a level of 4%. After calculation, a sample size of n = 120 units (questionnaires) was obtained. To reduce further the uncertainty of the research, there has been a supplementary increase of sample size by 8 %, that is approximately with 9 units (questionnaires).Finally 7 questionnaires were scrapped due to some uncorrectable errors, the survey being performed on 122 questionnaires. The questionnaires distribution per Counties from 5 West Region in Romania is presented in table 1.Regarding activity branch, the firms structure from sample is presented in table 2. Thus, the most SMEs activate in Wholesale and retail CANE section (42.62%), Real estate transactions CANE section (18.05%) and Manufacturing CANE section (14.75%).

No.	County	Number of answers	Percentage from total [%]
1	Arad (AR)	34	27.86
2	Caras-Severin (CS)	16	13.11
3	Hunedoara (HD)	21	17.21
4	Timis (TM)	51	41.82
	Total	122	100

Tab. 1 Questionnaires distribution per Counties from 5 West Region in Romania (Romanian Statistical Yearbook, 2006)

No.	CANE section	No. of answers	Percentage from total [%]
1	Manufacturing	18	14.75
2	Construction	9	7.37
3	Wholesale and retail	52	42.62
4	Hotels and restaurants	7	5.73
5	Transport, storage and communications	8	6.55
6	Real estate transactions	22	18.05
7	Other activities	6	4.93
	Total	122	100

Tab. 2 SMEs structure from sample as per CANE sections from 5 West Region in Romania

4. STUDY RESULTS

The determinant factor in entrepreneurial activity is the Entrepreneur. Hence the necessity and importance of studying the Entrepreneurs from different point of views. Hereinafter are presented the elements which define the profile of Romanian Entrepreneurs from 5 West Region, presented in table 3
From the total of questioned individuals (122 Entrepreneurs), 40.25% have age between 31-40 years, these ones having already knowledge and necessary experience in business. 24.74% of younger individuals, between 26-30 years, are involved in entrepreneurial activities due to their enthusiasm and willing to take risks. Entrepreneurs with age between 41-50 years are in proportion of 23.13%. The Entrepreneur's categories with ages over 50 years and under 25 years, are rather small (only 5.47% and only 6.41% respectively). The average age of the questioned Entrepreneurs is 41 years and the average age on when they've start up the business is 33 years.
Another criterion, gender, shows that the most Entrepreneurs are males (67.29%), while women are in a proportion of 32.71%.
From type of studies point of view, the vast majority of the respondents have graduated from an University (69.32%), which can be hypothesized as being the premise for their success in entrepreneurship activity. From these, 69.32% have graduated from a technical university type and 34.53% have graduated from an economical university type. The ones specialized in IT domain represents 6.50% and 13.19% have other specializations.From the questioned Entrepreneurs, only 35.24% have graduated lately a managerial course, which is not surprising at all having in view that they are not only owners but also managers, buyers, sellers, car drivers etc, invoking this way the lack of time.
Regarding the motives for which the Enterprisers decided to develop their own business, these are in order of importance: necessity to develop a self-safety feeling for 58.23% of them, willingness to obtain higher earnings for 52.76% of them and opportunity arising for 48.18% of them.
By correlating managerial readiness with how the firm's results are evaluated, it was concluded that 14.35% of the individuals which were able to take courses after graduating from university evaluate the firm's results as being "very good" and only 9.23% from the category which wasn't able to follow this type of courses evaluate that the firm's results are "very good".
On the one hand we can conclude that as the firm's dimension grows, the problems which the enterprisers-managers are confronting with, also grow and therefore they must use a more "scientific" management and they to find time to prepare themselves in managerial area. On the other hand this correlation can draw the conclusion that after taking courses in managerial area, the enterprisers became more performing, thing reflected in the dimension of the firm they own/manage (Nicolescu, 2006).

No.	Parameter taken in account	Elements of characterization
1.	Entrepreneur's gender	Male enterprisers are the majority (67,29%).
2.	Entrepreneur's age	Category 31-40 years (40.25%) is dominant, that is mature individuals. Average age at the business start up is 33 years.
3.	Educational level	Over two-thirds (69.32 %) had graduated from an University, which shows a high level of education among Entrepreneurs
4.	Professional readiness	Entrepreneurs with technical readiness are dominant (45.78%).
5.	Managerial readiness	Only 35.24 % of individuals from the sample had graduated a managerial course after completing their studies, which shows their relatively low managerial knowledge and abilities.
6.	Business start up motives	58,23% of the questioned individuals placed the necessity of self-safety and family-safety feelings on first place. Hence there is a multidimensional motivation

Tab. 3 Romanian 5 West Region Entrepreneur's characteristics

5. CONCLUSIONS

Thus, it can be drawn up a Romanian Entrepreneur from 5 West Region "portrait" as being a mature male, 41 years old on average, which starts up the business at an age of approximately 33 years, with good professional readiness (mainly technical or economical) and with a desire of a better earning or in order to develop self-safety. Determination in starting up and developing a new business – recognizing the opportunities offered in the environment, combining the resources in order to fructify the opportunities, developing a strategy to utilize the resources in order to exploit the opportunity and structuring an organization capable of applying that strategy – all of them having roots in enterpriser's behaviors and decisions. Future research will include mastering an action plan to lower entrepreneur's age and expanding this survey on national level.

6. REFERENCES

Chell, E. (2001). *Entrepreneurship: globalization, innovation and development*, Thomas Learning, London

Douglas, E.J. & Shepherd, D.A. (2002). *Self-employment as a career choice: Attitudes, entrepreneurial intentions, and utility maximization,* Entrepreneurship Theory and Practice, 26(3) p. 81-90

Isaic-Maniu Al. (2004). *Statistics (in Romanian)*, Editura Universitară, Bucureşti

Izvercian, M. (2002). *The Human Resources Marketing (in Romanian)*, Solness, Timisoara

Nicolescu, O. et al. (2006). *The White Charter of SMEs from Romania (in Romanian)*, Olimp, Bucureşti

Vaduva, S. (2004). *The Entrepreneurship (in Romanian)*, Editura Economică, Bucureşti

Romanian Statistical Yearbook (2006), *Available from*: http://portal.feaa.uaic.ro/C7/Biblioteca/Document%20Library/Anuarul%20Statistic%202006/ASR_2006_integral.pdf Enterprise Activity, pp.600-605 *Accessed:* 2008-06-12

Annals of DAAAM for 2008 & Proceedings of the 19th International DAAAM Symposium, ISSN 1726-9679
Editor B. Katalinic, Published by DAAAM International, Vienna, Austria 2008
Make Harmony Between Technology and Nature, and Your Mind will Fly Free as a Bird

ENTERPRISE LEARNING MANAGEMENT SYSTEM DEVELOPMENT GUIDELINES

JANCIC, Z[oran]

Abstract: *Current learning management systems on the market have significant limitations when used in corporate environment. This paper is an analysis of the most common problems with LMS systems in an entreprise environment based on many elearning projects done by Algebra d.o.o. in the past six years. LMS system architecture and user interface imperfections proved to be very time-consuming, which caused great losses in financial construction of large elearning projects. Possible solutions are presented in a form of guidelines for system architects and developers. Main goal of this article is to help LMS system architects in the development of efficient LMS systems. It is a result of many elearning projects that demonstrated common problems within the elearning system.*
Keywords: *elearning, education, online, corporate, enterprise*

1. INTRODUCTION

Learning management system is a software platform for online education, commonly known as elearning. Besides being a platform used for running elearning courses, it has a great number of other purposes such as users and groups management, course repository management, scheduling educations, monitoring user's progress in enrolled courses, reporting, collaboration, integration with other systems, etc. In corporate environment there are extensive reporting requirements, high security requirements, integration with other systems, a need for handling large groups of users, etc. Based on data gathered during many common elearning projects done by Algebra d.o.o., almost none of these requirements are covered well enough in all current LMS systems on the market. Some of the companies included in the research are Algebra POU, Zagrebačka banka (member of UniCredit group), Visoka škola za sigurnost, Croatian civil engineering Institute, Croatian Postal Bank Inc., Hrvatska elektroprivreda (national electric company) and many more.

2. LEARNING MANAGEMENT SYSTEMS TODAY

2.1 Corporate and educational environments

The first thing we should consider is the environment for learning management system utilization. Schools, colleges and other educational institutions often have requirements for online education that differ from those of big corporations. Businesses are transformed into knowledge-based businesses (Maier, 2007). Electronic transmission of information is easy and inexpensive, but simple availability of information does not necessarily mean that business performance needs have been solved (Allen, 2002).

2.2 User interface

It is important that users do not spend a lot of time on learning how to use educational technologies (Bates, Pool, 2003). Most complicated tasks require most sophisticated solutions. Let's say, for example, that there are a hundred thousand users of one system and a manager has to make a course progress report for ten thousand of chosen ones. Choosing a combination of appropriate search criteria and adequate user controls is of greatest importance. Most systems allow filtering user searches by a very large number of parameters but application controls can be very complicated. Instead, there should be a reasonable number of filtering options while application controls remain simple to use and respond fast to user interactions.

The most common mistake is simplifying user interface to the extreme. That could lead to a leak of useful and even necessary features. For example, extremely easy and fast reporting interface is of no use if it doesn't enable you to make the required report.

2.3 Reporting

Reporting tools are most extensively used by education managers. They provide ways to track user progress during the educational processes, evaluate quality of educational resources and learning system. Furthermore, complicated statistical and other analyses can be done using sophisticated reporting modules. Most learning systems possess only the essential reporting abilities. Tables with user results for a particular course are not applicable in other situations. For example, a manager could wish to see the number of successfully completed courses for a group of students over the past few months. Many of learning system providers create a lot of similar reports just to obtain interesting numbers for marketing brochures. Therefore, the mere number of reports is not always the right indicator of reporting module abilities. Some companies that develop learning systems went a step further by developing special modules for configuring and creating custom reports. For reporting generator module to be efficient, the backend database has to be well organized. Getting the right data from poorly organized data sources can be very time consuming and, in some situations, even impossible. The other problem is the complexity of such reports creating tools. In most learning systems, a user needs knowledge in databases, SQL scripting and, most often, programming in some high level computer language.

2.4 Managing elearning users in corporate environment

Managing users in learning managing system could be simple and straightforward in a situation when the learning manager handles a small number of employees. Unfortunately, that is rarely the case. In most situations, especially in very large corporations, handling a large number of education users could be very difficult if the learning system does not have proper capabilities for such tasks. Most learning management systems have a way of grouping users, but in most cases, one cannot use those groups in every aspect of user management. For example, an educational manager can enroll a user group in a course by selecting the group and clicking an "enroll" button. However, when it comes to reporting there is no way to use those groups to create reports. Another example would be using groups to separate users in collaboration events such as chat rooms and forums, but still not being able to utilize those groups for segmenting users to apply LMS system permissions.

2.5 Integration with other systems

Integrating LMS system with other systems in a corporation is usually the most expensive part of elearning projects. Such integration can include integrating existing user database, security policies, HR or CRM software, portals and other systems with LMS. Incompatibility among systems is very common but, fortunately, there are several solutions that can be used to resolve this problem. World leading LMS systems already use some of these solutions in one way or another. Time and expenses needed to integrate LMS with other systems can be cut down 10% or more if the appropriate technology is used.

3. GUIDELINES FOR LEARNING MANAGEMENT SYSTEM DEVELOPMENT

3.1 Improvements of user interface

Instead of balancing between "easy to use" and interface with a lot of features, special modules for configuring reporting and other user interface tools should be developed in a learning system. That can be done in several ways. For example, user could configure his interface for his particular tasks. In other case, system administrator or higher manager can configure a system for entire departments or even entire companies. Module's configuration settings can be kept in a database or in other data storages like configuration XML files. Such special features require extra development but provide seamless adopting a system to corporate particulate needs. One of the best approaches is to develop "what you see is what you get" type of configuration tools where a user can use "drag-and-drop" method to place any interface parts onto some particular web page of the system. Such system is used in some portal technologies like Websphere portal and Sharepoint portal.

3.2 System integration technologies

A common way for integrating LMS system with other systems in a corporation is by exposing LMS's application programming interfaces (API) to other applications. These API's provide LMS's programming objects to developers so they can use them to integrate specific functions with other systems. For example, let's say we have to list all users involved in elearning education in HR software and LMS system has API function for getting a list of users. HR application developer can call this existing function to get the list into HR application. HR application programmer has to do very little programming to achieve that. Same principle can be used for many LMS functions. Another example would be if a manager wanted to enroll a user in an elearning course from HR software he's using most of his working time. Developer could call LMS APIs enroll function if such a function exists.

A problem with incompatibility between programming languages can emerge, though. For example LMS can be written in Java and HR software in C#. The solution is to use technologies like web services that can communicate in both languages. Web service is a way to call some system's API function regardless of programming language used for function implementation. Web service can be written in one language and program that calls web service in another. Both languages should communicate through SOAP (simple object access protocol) using XML messages as a common format for transferring simple or complex data between systems. Modern systems are built in three-tierd architecture. Data layer is used for storing and managing data. Typically, data layer would be a database system. Presentation layer consists of user interfaces that present data to user and provide controls for user interaction with the application. Business logic layer is where we find web services in most systems. This layer connects the other two layers to form an application.

Some of modern LMS systems provide web services API's that can be used for most of LMS system functions. It is very important to build web services according to widely adopted SOAP standards. Otherwise, web services and applications that call them, can be incompatible regardless of the common protocol used to transfer data. These standards are still evolving. Most recent version at the moment is SOAP 1.2.

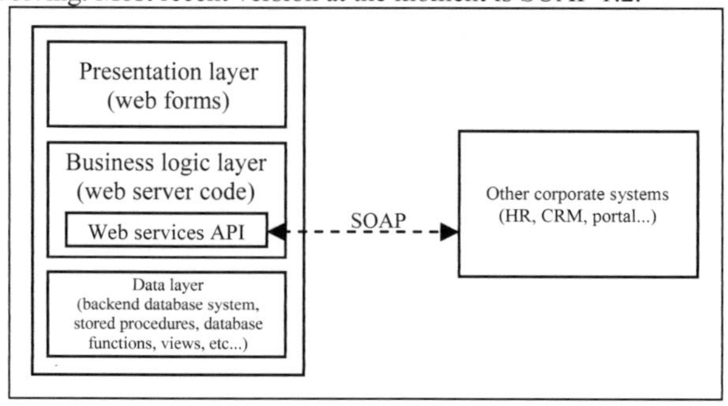

Fig. 1. Example of integrating LMS with other corporate systems

3.3 Reporting system

From a model of "transformative assessment" point of view, assessment is the purposeful gathering and use of data to ensure that the application of findings and the dissemination of results would substantially change and enrich the learning experience (McGee, Carmean, Jafari, 2005.).

To speed up the process of development, there are some useful reporting solutions on the market. Such solutions involve powerful tools for making complex database queries. LMS reporting module should be configurable and yet simple. There are a few possible approaches. One is to make a module for creating custom reports. This should be done in such a way that a user doesn't need to have much technical knowledge, i.e. like writing database queries. Second approach is to make a set of complex reports that can be configurated in many ways. User is presented with reports that contain all of the data about some object but he can filter them to get only the data he needs. This is similar to first approach because the user is creating custom report in the end. However, in second approach, creation of these reports can be much simpler. Reports should be presented in a form of charts and graphs if that's appropriate. While charts and graphs are not as precise for showing numeric data as a table, they are excellent for displaying trends and relationships (Kapp, 2003).

4. CONCLUSION

LMS systems built upon these guidelines can significantly improve usability of LMS system. That means less time used for managing educational process, cheaper and faster ways for integration with other systems in a company, more education processes maintained using LMS user interface only. In big environments where great number of people are involved in an elearning process, results will be more appreciable than in small and middle-sized organizations.

5. REFERENCES

Allen, M. (2007). *Michael Allen's Guide to E-learning,* Wiley, ISBN 978-0471203025, New Jersey, USA

Bates, A.W.; Gary P. (2003). *Effective Teaching with Technology in Higher Education.* Jossey-Bass, ISBN 0-7879-6034-9, San Francisco, USA

Kapp, K. M. (2003). *Winning E-Learning Proposals.* J.Ross Publishing, ISBN 1-932159-04-5, Boca Raton, Florida, USA

Maier, R. (2007). *Knowledge Management Systems: Information and Communication Technologies for Knowledge Management.* Springer Berlin Haidelberg, ISBN 978-3-540-71407-1, New York, USA

McGee, C.; Jafari, (2005). *Course Management Systems for Learning: Beyond Accidental Pedagogy,* Information Science Publishing, ISBN 1-59140-512-2, Harshey, USA

Annals of DAAAM for 2008 & Proceedings of the 19th International DAAAM Symposium, ISSN 1726-9679
Editor B. Katalinic, Published by DAAAM International, Vienna, Austria 2008
Make Harmony Between Technology and Nature, and Your Mind will Fly Free as a Bird

RELIABILITY ASSESSMENT SUPPORT IN UML

JEDLICKA, M[artin]; MORAVCIK, O[liver]; SCHREIBER, P[eter] & TANUSKA, P[avol]

Abstract: *Reliability is an important quality attribute of software systems. It is essential to assess the reliability in early stages of software development – analysis and design. The main trend for modelling of software systems modelling is Unified Modelling Language (UML). Hence, the UML should be natural modelling technique also for reliability assessment purposes. This paper discusses the work done in this area, evaluates common approaches of reliability assessment using UML modelling techniques and outlines our future work in this area.*

Key words: *Reliability, Reliability Assessment, UML, UML models.*

1. INTRODUCTION

The evaluation and testing of system in early stages of software development lifecycle is becoming important field of software engineering. The research in testing of functional requirements is well developed but the testing of non-functional requirements is still in the beginning. One of the most important non-functional requirements, which cannot be missed out when designing software system, is reliability. There exist several traditional approaches for reliability assessment of software systems but they have these main disadvantages:

1. Assessment of the reliability at the very end of software development lifecycle
2. Reliability evaluation of system as whole
3. Various reliability modelling techniques (reliability block diagrams, fault trees, Markov chains, etc.)

Our research focuses on developing the new reliability assessment approach which will evaluate the reliability of the system in early development stage using UML notations for modelling the assessed software system. First step of the research was to explore the work done in this field and to confirm the hypothesis that UML notations are appropriate for reliability assessment.

2. RELIABILITY ASSESSMENT USING UML MODELS

2.1 Approach Cortellessa et al.

This approach (Cortellessa et al., 2002) can assess reliability from early stage of software development lifecycle. It uses annotated UML diagrams:

1. Use Case diagram
2. Sequence diagram
3. Deployment diagram.

Use Case diagram is annotated as follows:

- q_1 and q_2 represent the probabilities of requesting services from system for users u_1 and u_2
- P_{11} and P_{12} represent that user u_1 requests the functionality f_1 represented by use case uc_1 or f_2 represented by use case uc_2

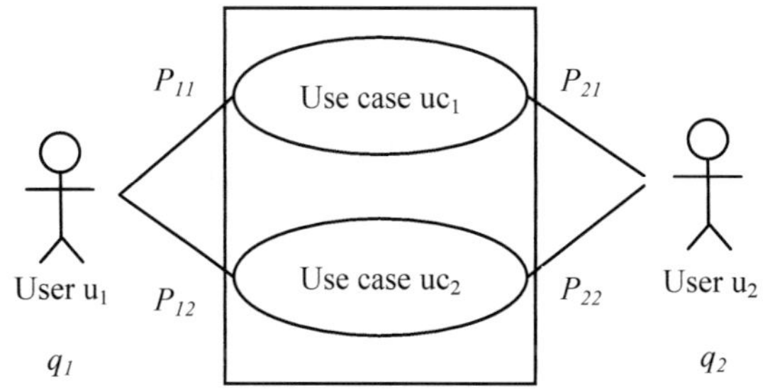

Fig. 1. Annotated Use Case Diagram

Then the probability of executing general use case x is:´

$$P(x) = \sum_{i=1}^{m} q_i \cdot P_{ix} \qquad (1)$$

where m is the number of user types.

Then they assume that for each use case there are specified all relevant Sequence Diagrams representing main scenarios each use case. These sequence diagrams have the non-uniform probability distribution according to following equation:

$$P(k_j) = p(j) \cdot f_j(k) \qquad (2)$$

where $f_j(k)$ is the frequency of the k-th overall the sequence diagrams referring to the j-th use case and parameter $P(k_j)$ represents the probability of a scenario execution.

In sequence diagram, the approach focuses on the interval where the component is busy. For example, in Figure 2 the interval, when component C_3 is busy, is between the moment when interaction l_1 enters the component and interaction l_2 leaves the component. The total number of busy periods of component C_3 is 2.

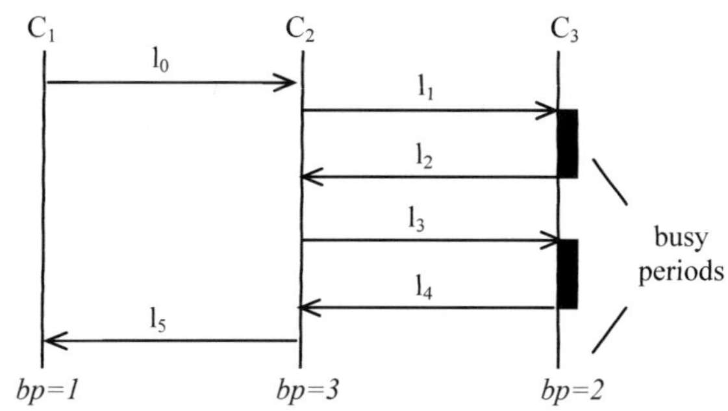

Fig. 2. Annotated Sequence Diagram

Then the estimate of the probability of failure θ_{ij} of the component C_i in the scenario j represents the next equation:

$$\theta_{ij} = 1 - (1 - \theta_{ij})^{bp_{ij}} \qquad (3)$$

where bp_{ij} the number of busy periods that component C_i has in the scenario j.

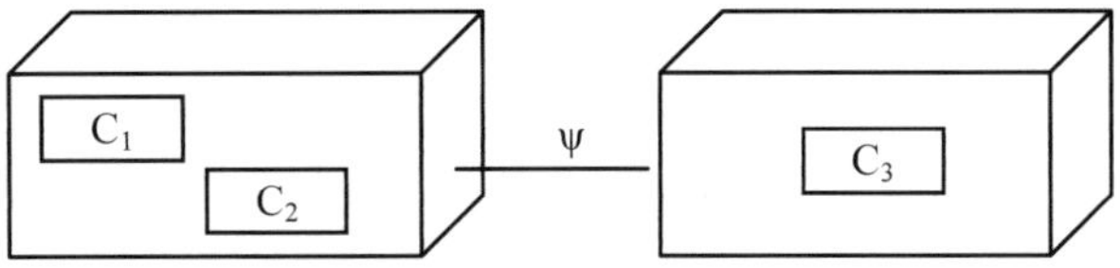

Fig. 3. Annotated Deployment Diagram

The deployment diagram in Figure 3 includes the same components from the sequence diagram. The subject of failure probability ψ_i is every communication between components (l, m) through the connector i. and $|Interact(l, m, j)|$ represents the number of interactions between components l and m from the scenario (sequence diagram) j. e.g. there exist 2 interactions exchanged between C_2 and C_3 but only 1 interaction between C_1 and C_2.

Then the contribute ψ_{lmj} to the reliability of communication between all these components over the connector i in the scenario j is:

$$\psi_{lmj} = (1-\psi_i)^{|Interact(l,m,j)|} \quad (4)$$

Finally the reliability of the whole system is combination is combination of equation (3) and (4)

$$\theta_S = 1 - \sum_{j=1}^{K} p_j \left(\prod_{i=1}^{N} (1-\theta_{ij})^{bp_{ij}} \cdot \prod_{(l,i)} (1-\psi_{lij})^{|Interact(l,i,j)|} \right) \quad (5)$$

2.2 Approach of Rodrigues et al.

This approach can also be used in early stages of software development lifecycle for component-based systems. As basic modelling techniques, the scenarios and a Message Sequence Charts (MSC) are used (Rodrigues et al., 2005). The framework of the approach shows the Figure 5.

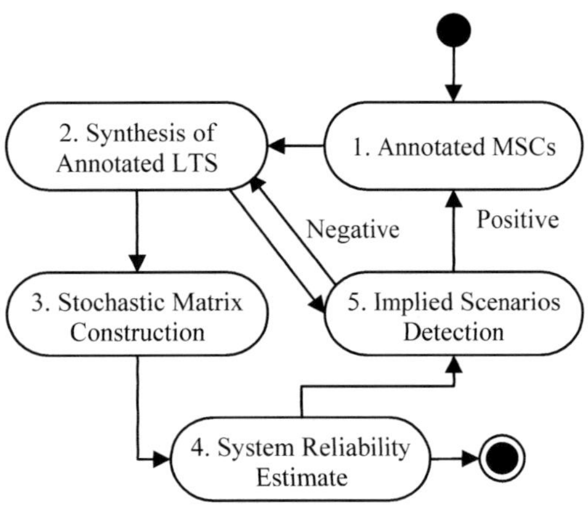

Fig. 5. Framework

First step is annotation the MSCs with two kind probabilities:

1. Probability of transitions between scenarios PTS_{ij}
2. Reliability of the components R_C.

PTS is simply the probability that scenario S_j will be executed after executing scenario S_i which value comes from an operational profile for the system (Musa, 1993). It is annotated on High-level Message Sequence Charts (HMSC).

The component reliability R_C is the probability that component C will successfully perform its service which is invocated by message from another component. The execution time of the service is not significant for component reliability assessment. Component reliability is annotated on Basic Message Sequence Charts (BMSC).

Second step is to made synthesis of Labelled Transition System (LTS) from annotated scenarios. This step consists of several sub-steps to construct the architecture model of the system.

In the third step the architecture model is interpreted as Markov model. First the mapping of the probability weights from architecture model into a square transition matrix is constructed.

In the fourth step the Cheung approach (Cheung, 1993) is used to determine the reliability prediction of the whole system.

In last step, the implied scenarios are searching. These scenarios can be found as the result of the fact that in the system there can exist such set of components that don't communicate exclusively through the interfaces described and that can exhibit via unspecified traces when running in parallel (Uchitel, 2004). Founding of these scenarios has impact on step 1 and 2 of the approach's framework.

4. CONCLUSION

The present research made in this area shows that UML modelling techniques are applicable for reliability assessment of software.

We made also some thesis and assumptions:

- Sequence diagrams will be the core modelling technique for system architecture and behaviour
- We assume that developing of new UML extensions will be necessary for reliability assessment
- Our new approach will combine traditional reliability techniques with UML modelling
- The whole process of our reliability assessment approach will be automatic as much as possible
- The new tool which will cover our approach will be developed
- The approach will be developed with focus on control systems software

In this paper we discuss the area of software reliability and its support in UML. We also introduce the today's main approaches and concepts in this area, make the short evaluation of them and outlook the basis of our future work in this area.

5. REFERENCES

Cheung, R. C. (1980) *A User-Oriented Software Reliability Model*. IEEE Transactions on Software Engineering, vol. 6(2), pp. 118–125

Cortellessa, V., Singh H., Cukic, B. (2002). *Early reliability assessment of UML based software models*. Proceedings of the 3rd international workshop on Software and performance, Rome, Italy

Musa, J. D. (1993). *Operational profiles in software-reliability engineering*. IEEE Software, vol. 10, no. 2, pp. 14-32

Rodrigues, G. N., Rosenblum, D. S., Uchitel, S. (2005). *Reliability prediction in model-driven development*. Proceedings of the 8th international conference MoDELS pp. 339-354, Montego Bay, Jamaica

Uchitel, S., Kramer, J., Magee, J. (2004). *Incremental Elaboration of Scenario-Based Specifications and Behavior Models Using Implied Scenarios*. ACM Transactions on Software Engineering and Methodologies, vol. 13, pp. 37-85

Annals of DAAAM for 2008 & Proceedings of the 19th International DAAAM Symposium, ISSN 1726-9679
Editor B. Katalinic, Published by DAAAM International, Vienna, Austria 2008
Make Harmony Between Technology and Nature, and Your Mind will Fly Free as a Bird

OPEN-LOOP CONTROL AND DATA ACQUISITION OF A BIOMECHANICAL RESEARCH DEVICE USING LabVIEW DEVELOPMENT ENVIRONMENT

JELECANIN, J[elena]; PALM, S[tjepan] & MRCELA, T[omislav]

Abstract: *The solution of supervisory control of a device for specific purpose in biomechanical research of soft tissues is presented in this paper. An outline of an electropneumatic device is laid out in order to explain the specific requirements for the data acquisition system according to working conditions, mobility, accuracy and speed. The concept of implementing sensors and computer control and measurement with electropneumatic device is given from the viewpoint of an automatic control theory. The characteristics of actuator and sensors are given and important relations are explained. Data acquisition system consisting of a portable hardware and a custom-made LabVIEW software support is presented. Finally, the advantages of the described system are given and the possible applications for different types of systems are explored.*

Key words: *Control, LabVIEW, biomechanics, pneumatics.*

1. INTRODUCTION

Automation of an electropneumatic device especially designed for a specific purpose in biomechanical research of human tendons is the subject of this paper. The device can be represented by a pneumatic scheme (Fig. 1.). In terms of automatic control, the device can be described as an electropneumatic system for force control. Since force control systems are widely used in materials testing equipment (Krivts & Krejnin, 2006), open-loop scheme does not differ from those described in literature.

However, the device was designed for a specific purpose and there are specific requirements on it performance and design. In order to answer to these demands, supervisory control system, consisting of portable data acquisition (DAQ) card and LabVIEW software support, is conceived. In general, the purpose of a data acquisition system is to capture and analyze some sort of physical phenomenon from the real world and it may also produce electrical signals simultaneously. Programming language is used to build an application from the driver software which is then ready for an end user to easily control and acquire data from the hardware – a custom instrument built specifically for the user's needs. (Curtis, 2002)

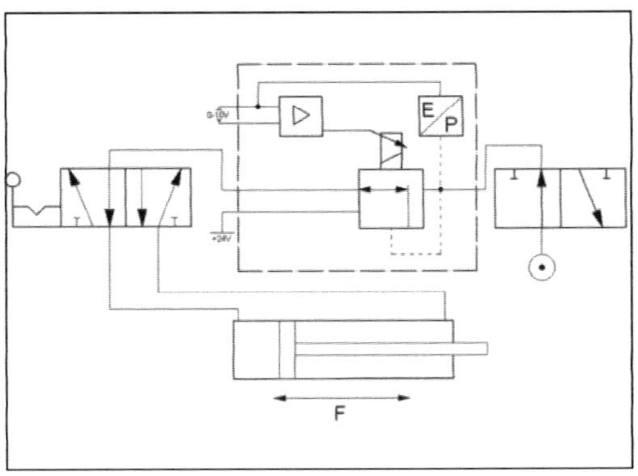

Fig. 1. Pneumatic scheme of the device

2. ELECTROPNEUMATIC DEVICE

Pneumatic control is a solution for process control by use of pneumatic executive and control elements. Correlation between executive and control elements towards completing the given task is shown by pneumatic scheme (Nikolic, 2002). When observing the electropneumatic device shown in Fig. 1. as an object of automation, the following characteristics are of importance; control variable is pressure, and the force and the displacement are variables that need to be measured. Fig. 2. shows functional scheme of a pneumatic system for force control. Actuator controls the force and the displacement on the executive element.

Executive element is a double-acting pneumatic cylinder that can perform useful act in both directions, forward and backwards. Relation (Nikolic, 2002) for backwards force in N is:

$$F_2 = A_2 \cdot p_{do} - A_1 \cdot p_1 - F_T \qquad (1)$$

where p_{do} is supply pressure in Pa, p_1 stagnated air pressure in an irradiated chamber in Pa, F_T force of friction in N, A_1 and A_2 back and front area of a piston in m^2.

Executive element is driven by mechanic direction control valve with proportional pressure actuator that ensures precise pneumatic control. The pressure supplied by actuator is directly proportional to the controlled variable. Relation for force F applied to the executive element can be derived by simplifying relation (1):

$$F = k_F \cdot p_{do} \qquad (2)$$

where k_F is the force constant dependent on the useful area and the position of cylinder, and p_{do} is directly proportional to electrical (voltage) signal applied on the actuator:

$$p_{do} = k_p \cdot U_p \qquad (3)$$

where U_p is supply voltage in V and k_p the pressure constant that equals to 1. Relation (3) is valid for force control on the executive element.

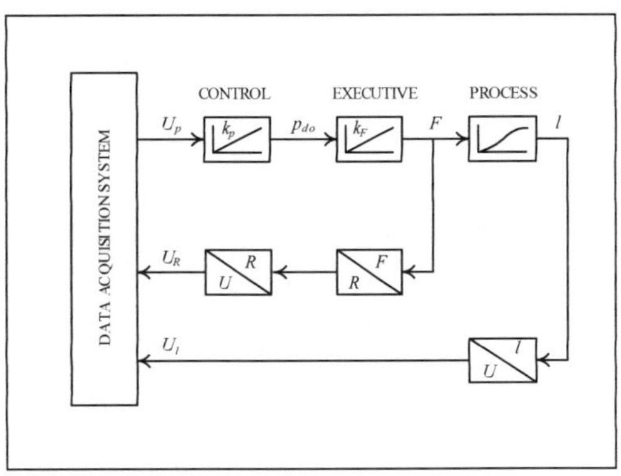

Fig. 2. Functional scheme of a system for force control

3. FORCE AND DISPLACEMENT SENSORS

Force applied by the cylinder is measured by linear measuring pressure sensor and is proportional to the resistance R that can be indirectly measured by measuring current through secondary resistor with known resistance R_s:

$$F = k_R \cdot R = k_R \cdot \frac{R_S}{U_S} \cdot U_R \quad (4)$$

where k_R is the resistance constant.

The displacement of cylinder is measured using a resistive displacement sensor. This basically simple mechanical device is crucial to the validity of measurement because the output from the sensor gives actually the position of the cylinder and not the sample itself (Bozicevic, 1980). The voltage output U_l in V is proportional to the displacement l in m:

$$l = k_l \cdot U_l \quad (5)$$

where k_l is the displacement constant.

4. SOLUTION OF SUPERVISORY CONTROL

The purpose of supervisory control system is to control the force applied by actuator to the executive element and acquire data measured by sensors. Data acquisition system consists of portable DAQ card attached on to the device (Fig. 3.) and a custom made LabVIEW software support.

DAQ card is connected to the portable computer by USB cable. Analog output and input channels are used for communication with actuator and sensors. Force control is conducted by voltage control from 0-5 V equivalent to pressure from 0-5 bar.

In order to establish communication between the DAQ card and the computer, it is necessary to have adequate software support. Compute has to know how often to measure and in what time intervals (Valter, 2008). Software support in LabVIEW is developed using two interfaces simultaneously.

Block diagram shown in Fig. 4. is the code of the program in which the commands are represented by blocks. Once connected together, these blocks enable the flow of information acquired using hardware to user interface called Front Panel. Program consists of drivers for communication with hardware, the main loop and blocks for displaying and writing results to measurement file. The main loop consists of control and measurement loops that execute simultaneously.

Front Panel (Fig. 5.) contains controls for setting up control parameters such as data about the specimen, desired force and measurement time. Ones the program has been stopped, the acquired data is displayed using charts and also automatically saved in Excel document for further processing.

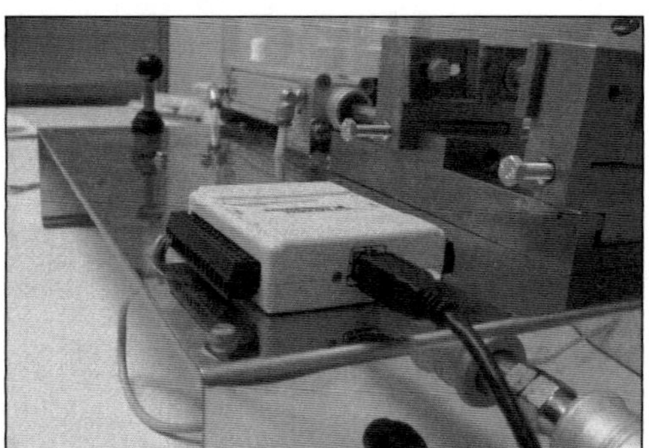

Fig. 3. Portable data acquisition hardware attached to device construction

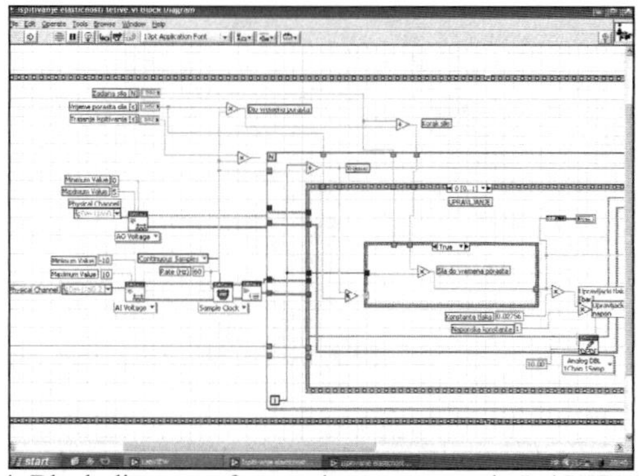

Fig. 4. Block diagram of supervisory program in LabVIEW

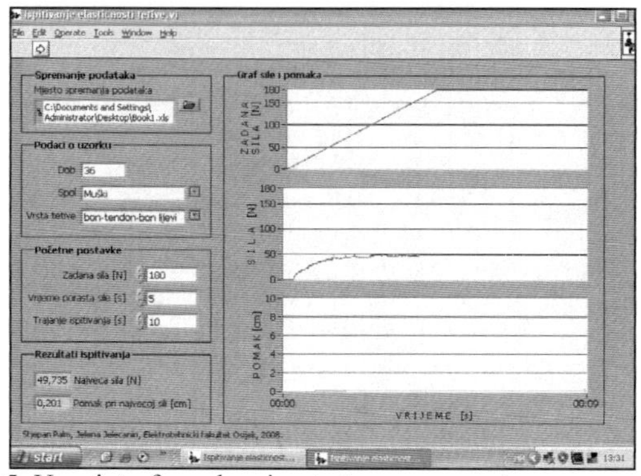

Fig. 5. User interface showing test measurement

5. RESULTS

Experimental system was tested in biomechanical laboratory on specimens of human tendons. During test measurement, satisfactory results are indicated as shown on Fig. 5. Desired force ramp and measurement time are set with high accuracy and response time. Measurement of quantities is determined to be reliable.

6. CONCLUSION

A custom made supervisory control system is necessary for automation of a device designed for a specific purpose. The solution of supervisory control described in this paper consists of a DAQ card connected to a computer and a LabVIEW software. This solution is simple are relatively cheap but in the same time required functionality is achieved. Modularity is achieved by choosing a portable DAQ card that can be easily replaced or connected to another computer. Presented concept can be used in automation of almost any type of system.

7. REFERENCES

Bozicevic, J. (1994). *Temelji automatike 2 (Fundamentals of Automation Part 2)*, Skolska knjiga, Zagreb

Curtis, J. (2002), Introduction to Data Acquisition, In: *The Mechatronics Handbook*, Bishop, R. H., (Ed.), Chapter 44, CRC Press, ISBN 0-8493-0066-5, Boca Raton, USA

Krivts, I. L & Krejnin, G. V. (2006). *Pneumatic Actuating Systems for Automatic Equipment: Structure and Design: Structure and Design*, CRC Press, ISBN 0-8493-2964-7, Boca Raton, USA

Nikolic, G. (2002). Pneumatika (*Pneumatics*), Skolske novine, ISBN 953-160-174-7, Zagreb

Valter, Z. (2008). *Procesna mjerenja (Process Measurements)*, Elektrotehnicki fakultet Sveucilista J. J. Strossmayera u Osijeku, ISBN 978-953-6032-59-4, Osijek

Annals of DAAAM for 2008 & Proceedings of the 19th International DAAAM Symposium, ISSN 1726-9679
Editor B. Katalinic, Published by DAAAM International, Vienna, Austria 2008
Make Harmony Between Technology and Nature, and Your Mind will Fly Free as a Bird

PROBLEMS OF PRECISE MEASUREMENT WITH TWO-POINT AIR GAUGES

JERMAK, C[zeslaw] J[anusz] & RUCKI, M[iroslaw]

Abstract: *In the paper, the factors important for exact calculation of the metrological properties of air gauge have been specified. Some results of the investigations on their influences have been presented. The attention has been paid to the two-point air gauges and their characteristics. The comparative analysis with one-point air gauges have been performed, too.*

Key words: *air gauges, metrology, measurement.*

1. INTRODUCTION

Technological development continually forces metrological sciences to improve measurement devices. The air gauges, though they have been used for a dimensional measurement for decades, are also a subject of investigations in order to be improved. They are widely applied in many branches of Mechanical Engineering, especially in quick inspections, measuring automatons and in-process control (Zelczak 2002). The main merits of the air gauges are: high resolution, possibility of non-contact measurement, self-cleaning of the measured surface, high accuracy, high resistance to the external conditions influence and adaptability for the wide range of different measuring applications (Farago 2004; *Pneumatische Längenmessung* 2000). Those merits cause that air gauges are able to work in hard industrial conditions.

2. FACTORS INFLUENCING THE METROLOGICAL PROPERTIES OF THE AIR GAUGE

The static characteristics of the air gauge depend directly on the feeding pressure p_z, diameters of inlet and outlet nozzles, and on the outflow coefficients. It could be described by the following formula (Glukhov 1971):

$$p_k = \frac{p_z}{1+\left(\dfrac{\alpha_p A_p}{\alpha_w A_w}\right)^2} \qquad (1)$$

where:

p_k , p_z - pressure in the measuring chamber, and feeding pressure,

A_p , A_w - the outflow surface for flapper-nozzle area and for inlet nozzle; A_p depends on the diameter of the measuring nozzle d_p and the slot width s:

$$A_p = \pi d_p s \qquad (2)$$

$$A_w = \frac{1}{4}\pi d_w^2 \qquad (3)$$

d_w - the inlet nozzle diameter,

α_p, α_w - the flow-through coefficients for flapper-nozzle area and for inlet nozzle.

The Figure 1 presents metrological properties z_p (measuring range) and s_o (middle point of the measuring range) of the air gauges with various inlet nozzle d_w for different values of feeding pressure.

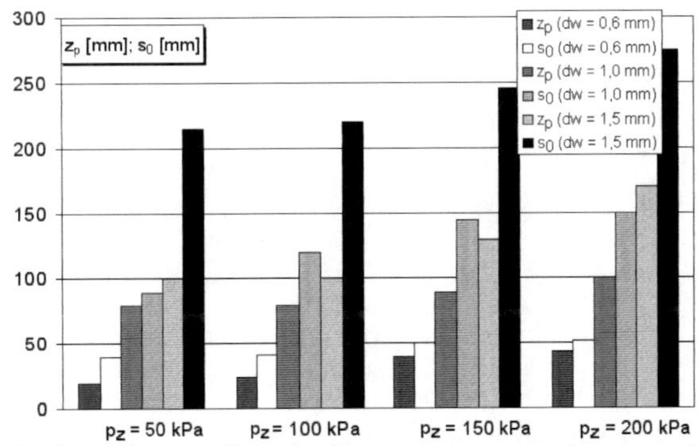

Fig. 1. Influence of the feeding pressure p_z on the measuring range z_p and middle point of the measuring range s_o of the air gauges with various inlet nozzles d_w.

It should be noted also, that the declination of the measuring nozzle axis from the perpendicularity to the measured surface causes the changes of the value of outflow surface A_p and, hence, the metrological properties of the air gauge. Additionally, there are many parameters usually not taken into consideration that could change the outflow coefficients. Those are among others:

- phases or roundness of the inlet and outlet nozzles orifices,
- geometrical correction of the measuring nozzle flat head,
- outer diameter of the measuring nozzle (related to the diameter of its orifice).

3. INVESTIGATIONS ON THE TWO-NOZZLE SYSTEMS

In many cases, the air gauges are built into plugs with two measuring nozzles as a sensing elements, as it is shown in the Fig. 2 (Farago 2004). Here, the pressured air is outgoing from two nozzles d_{p1} and d_{p2}, and the back-pressure depends on the summed up measuring slot $s=s_1+s_2$.

In the investigational set for two-jet plugs, two measuring nozzles of the same dimensions were connected to one measuring chamber. The width of slot s may be then changed simultaneously for both measuring nozzles ($s_1 = s_2$) simulating the symmetrical position of the air plug in the measured workpiece. To investigate the influence of asymmetrical plug position on the measurement results, the asymmetrical changes of both slots are possible. Fig. 3 presents an example of the behaviour of two-jet air plug.

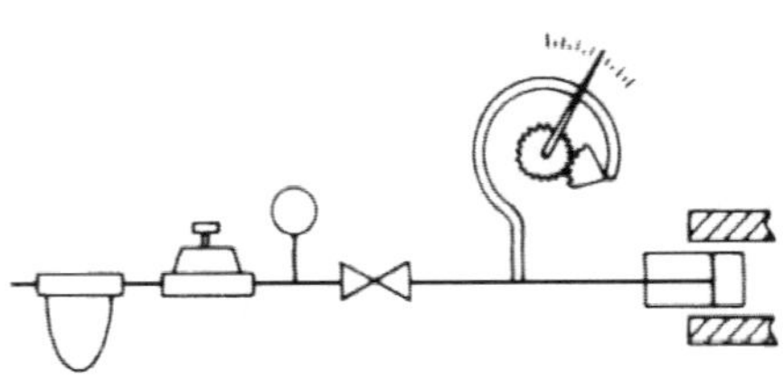

Fig. 2. Directly indicating Bourdon tube type pressure gauge.

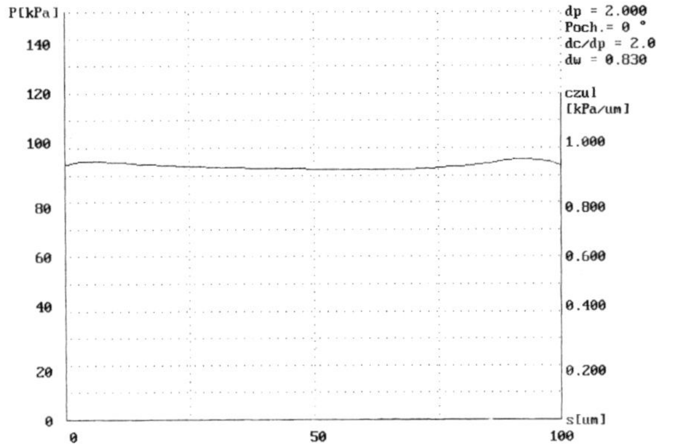

Fig. 3. Graph of the pressure p_k in two-jet plug versus the slot $s=s_1+s_2$.

In the investigations, several sets of the nozzles were manufactured ensuring the sensitivity of the air gage in the range from 1.3 down to 0.4 kPa/μm. Accurate values of sensitivity K were calculated for the maximal linearity errors of δ_{lin}=0.75% and 1.0% related to the Mean Square Line.

The orifice diameter of measuring nozzle was d_p = 2.014 mm. In the double-jet systems, the second measuring nozzle was a little different d_{p1} = 2.010 mm because of technological limitations. However, such a small difference did not affect the results of measurement.

Additionally, the investigations were performed for the nozzles with wide and with narrow head surfaces. A nozzle head surface coefficient was introduced as the relation of its outer diameter to the inner diameter (Fig. 4):

$$D_c = d_c/d_p \tag{4}$$

where:

d_p – nozzle orifice diameter,

d_c – outer diameter of the measuring nozzle.

In the investigations, the measuring nozzles with diameters d_{p1} = 2.010 mm, d_{p2} = 2.014 mm were used. The surface head coefficient was D_c = 2 and D_c = 3. Parameter D_c could not be directly inserted into the formula (1), but its influence on the metrological properties of the air gauge is obvious. It was examined, and the results if investigations were published (Jermak 2004). Similar influence could be observed in two-point air plugs.

4. COMPARATIVE ANALYSIS WITH THE SINGLE-NOZZLE AIR GAUGES

The main question to be answered was: in what extend the static metrological parameters of the double nozzle (d) air gauge differ from ones of the single nozzle (s). In order to answer this question, the first hypothesis was assumed that the sensitivity of one-nozzle gauge is equal to the sensitivity of two-nozzle gauge: $K_2 = K_1$. It could be presented in other way: $K_2 - K_1 = 0$. The alternative hypothesis was reversed: $K_2 \neq K_1$. The results of the analysis are shown graphically in the Fig. 5 (for linearity errors δ_{lin} =0.75% and 1.0%). They proved that the

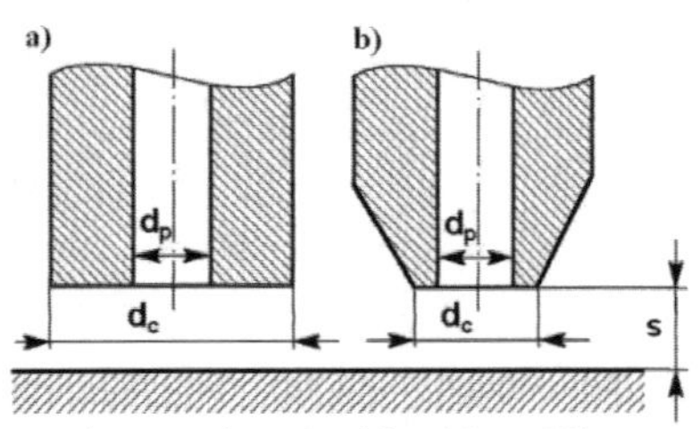

Fig. 4. Measuring nozzles: a) with wide and b) narrow heads.

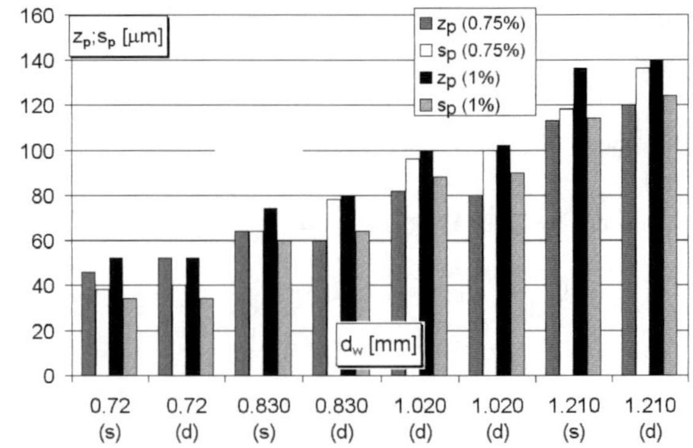

Fig. 5. Measuring range z_p and their initial values s_p for single and double nozzle air gauges (approximated with linearity δ_{lin} of 0.75% and 1.0%).

first hypothesis was correct: the sensitivity of both systems could be considered as equal. The differences between the characteristics of single-nozzle and double-nozzle air gauge never extended 10%. It should be noted, however, that for the smaller assumed linearity error, the sensitivity of the graph is a little higher, but the measuring range shorter respectively.

The analysis of calculated parameters of metrological characteristics (sensitivity K and measuring range z_p) for the single and double nozzle air gauges shows that the double systems tend to have a little higher sensitivity in a little wider measuring range. However, this trend is not too significant and has some exceptions (e.g. for inlet nozzles 0.830 and 1.020 mm, with linearity error δ_{lin}=0.75%).

In general it could be stated that no substantial difference between the characteristics of single-nozzle and double-nozzle air gauge system were found. They may be applied in the same work conditions with no need of new investigations. It is possible, hence, to perform investigations on the simple single-nozzle air gauge and then just add the second nozzle in the working industrial application.

5. CONCLUSION

The air gauges are widely applied in many branches of Mechanical Engineering, especially in quick inspections and in-process control in hard industrial conditions. Sometimes open jet air plugs contain two or more nozzles joined together in a measuring chamber. The comparative analysis of single-nozzle and double-nozzle air gauges proved that no substantial difference between the single-nozzle and double-nozzle air gauge system could be found. The differences did not exceed 10% of the maximal value. Hence, they may be applied in the same work conditions, and the metrological properties determined for single-nozzle air gauge could be relevant for the double-nozzle with the same geometry.

6. REFERENCES

Farago, F.T.; Curtis, M.A. (2004). *Handbook of Dimensional Measurement,* Industrial Press Inc., New York

Glukhov, V.I. (1971), Calculation of the High Pressure Air Gauges Characteristics, Vol. 6, 1971, *Journal "Measuring Techniques"*, p. 107 (in Russian)

Jermak, Cz.J.; Rucki, M. (2004), Influence of the Geometry of the Flapper – Nozzle Area in the Air Gauge on its Metrological Properties, No. 1860, 2004, *"VDI-Berichte"*, p. 385-393

Pneumatische Längenmessung (2000), German Standard on air gauging DIN-2271, 1977, 2000. (in German)

Zelczak, A (2002). *Pneumatic Dimensional Measurements*, WKL Ltd., Warsaw (in Polish)

Annals of DAAAM for 2008 & Proceedings of the 19th International DAAAM Symposium, ISSN 1726-9679
Editor B. Katalinic, Published by DAAAM International, Vienna, Austria 2008
Make Harmony Between Technology and Nature, and Your Mind will Fly Free as a Bird

CONSIDERATION ON CLIMATE ENERGY EFFICIENCY IMPACT

JIGA, G[abriel]; GRIGORIU, M[ircea]; CIUCA, I[on] & VLASCEANU, D[aniel]

Abstract: *The Union commitment to reduce the Greenhouse Gases (GHGs) with 20 %, to provide 20 % from the total energy production from renewable sources (ERS) and to reduce the energy consumption with 20%, have a major effect on the natural resources protection of the Planet and on the GHGs mitigation, taking into consideration the estimation that about 80 % of the emissions are provided by energy production. Energy efficiency is one of the favorite solutions to reduce both the energy consumption and e to mitigate GHGs emissions. The paper presents the energy end climate effect of a new heating systems operating control solution.*

Key words: *consumption, resources, efficiency, emission.*

1. INTRODUCTION

The proposed controlled operation system, which is compared with the actual operation systems with three ways vane proposed system has the main advantage the adjustment in the same time both circulation flow and pump speed system.

Its simple schema is presented in fig. 1. It coordinates the operation of circulator pump and control valve. Depending on the control signal issued by the higher-level heating controller, the two control valves adjust the resulting volume flow rate pumped through the consumer installations (Dimonie et al 2007), (Dragoi et al., 2007).

The supply temperature is increased by the higher-level controller. This results in hydraulic savings regarding both volume flow rate ΔQ and discharge head ΔH, whose product is proportional to the corresponding electrical power savings for the circulator pump, considering electrical power equation

$$P_{el} = K + Q(t) \cdot H(t) \cdot dt \qquad (1)$$

where:

P_{el} - electrical necessary power,

K - constant which describes the efficiency of circulator pump and frequency inverter as well as water density and acceleration due to gravity,

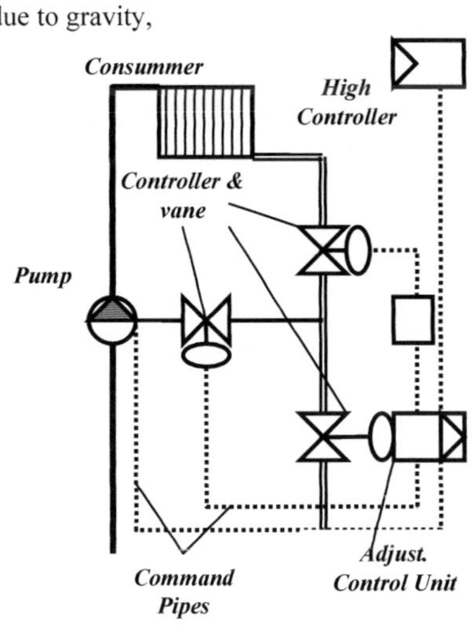

Fig.1. Heating system operation command
$Q(t)$ - flow-time characteristic,
$H(t)$ - Heat-time characteristic.

Thermal output of the heating circuit is calculated for the design point as the product of volume flow rate Q and temperature ΔT .

$$P_{th} = 1.163 \cdot Q \cdot \Delta T \qquad (2)$$

where:

P_{th} - thermal power output.

The innovative control concept consists in regulating thermal output by variation of the temperature differential and the volume flow rate, considerably reducing the water volume to be pumped through the heating circuit.

The radiator diagram illustrates that volume flow rate can be reduced while thermal output at the consumer installations stays the same, if the supply temperature is increased accordingly.

$$Q_{total} = Q_{2m} + Q_{1s} \qquad (3)$$

where:

Q_{total} - volume flow rate of the radiator,

Q_{2m} - volume flow rate from mixing pipe,

Q_{1s} - volume flow rate from water supplier.

The controller's heating curve is shifted parallel by a fixed amount towards higher supply temperatures. The controller adds the fixed amount ΔT to the supply temperature set point (parallel shift of the heating curve). Due to the increased temperature differential, the nominal volume flow rate is reduced by 25% in all cases, resulting in a new design point for the circulator pump (Grigoriu, 2005).

System valve authority is constant over the entire operating range. This correlation is reflected in the system control curve and explains why the system control curve runs parallel to the system curve.

Depending on the external temperature, the operating point of the system controlled circulator pump moves along the system control curve. As external temperature rises, the operating point moves towards lower volume flow rates and lower discharge heads (Grigoriu, 2006).

Under the influence of external heat load, the operating point also moves along the characteristic curve of the differential pressure controlled pump. Depending on the pump curve set, the discharge head either remains constant (Δp = constant) or decreases with decreasing volume flow rate (Δp = variable). All the familiar features of the variable-speed pump and the thermostatic valves are retained.

The differential pressure at nominal load:

$$\Delta p_n = c \cdot Q_n^2 \qquad (4)$$

Under part-load conditions, the differential pressure at the balancing valve is reduced as follows (Differential pressure at partial load):

$$\Delta p_t = c \cdot (FQ_n)^2 \quad (5)$$

where:
F - factor describing the partial load level (0-100)%.

It results the differential pressures at nominal and partial load for branch circuits:

$$\Delta p_t = F^2 \cdot \Delta p_n \quad (6)$$

Therefore, the ratios of the differential pressures are:

$$\frac{\Delta p_{ti}}{\Delta p_{tj}} = \frac{\Delta p_{ni}}{\Delta p_{nj}} \quad (7)$$

If parallel shift of the heating curve is planned, new system will always reduce the volume flow rate for the design point by 25 %, based on the volume flow rate the system designer has calculated for a conventional mixing or injection-type system.

The system has to be configured prior to commissioning. This is done by entering the value of the volume flow rate determined for a conventional mixing or injection-type system into the commissioning software, together with the nominal system diameter of the system (parameterization). The system automatically determines the circulator's discharge head required for the design point. To do so, the circulator pump is started up with its minimum discharge head. A corresponding volume flow rate is produced in the heating circuit, which is measured by the system controlled valve at the main feed manifold.

2. ESTIMATED ECONOMIC EFFECTS FOR A SPECIFIC EXAMPLE

The following analysis shows that for a specific case of heating system, it is selected a circulation centrifugal pump, which is controlled on one hand by the proposed control system and on the other hand by a conventional system. The pipe curve of the heating circuit is calculated as the quotient of the pump head (H) and the square of the volume flow rate (Q) (BOA-Systronic, 2008):

$$k_{heating} = \frac{H}{Q^2} \quad (8)$$

$k_{heating}$ - pipe curve of the heating circuit constant.

The volume flow rate for the design point of the conventionally equipped main feed circuit is

$$Q_{nc} = 13\frac{m^3}{h}. \quad (9)$$

At this volume flow rate, the pump can provide a maximum head of approximate H=8m. For this case, the following system constant is calculated for the heating circuit:

$$k_{heating} = \frac{8}{13^2} = 0.0473 \quad (10)$$

The example is based on a temperature differential of ΔT = 20 K, and the heating curve of the higher-level controller was shifted by 3.5K (parallel shift), resulting a reduction of 25 % of the volume flow rate in the design point.

$$Q_n = 0.75 \cdot Q_{nc} = 0.75 \cdot 13 = 9.75\frac{m^3}{h} \quad (11)$$

$$H_n = k_{heating} \cdot Q_n^2 = 0.0473 \cdot 9.75 = 4.5m \quad (12)$$

This new design point Q/H → 9.75m^3/s / 4.5m can be handled by a smaller selected pump. For the present example, investment costs for the circulator pump are reduced.

The external temperatures measured were plotted for the period mentioned. The diagram shows an average external temperature of approx. 10°C in the period monitored. Depending on the location, the heating circuits are selected for an external temperature of approx. –12°C to –15°C. Regulations stipulate that the building must be heated until the external temperature reaches approx. 16°C. The temperature difference between the design point and the switch-off temperature of the heating system is [(-12°C) – (+16°C)] = 28°C. The temperature difference between the design point and the average temperature is [(-12°C) – (+10°C)] = 22°C. The ratio between both values is 22/28 = 0.79. This means that in the monitored period both heating circuits on average only require about 21% of the thermal output they were selected for.

3. SOME ELEMENTS OF CONSTRUCTIVE CALCULUS

The mean values of the volume flow rates measured in the two heating circuits are calculated:

Controlling system	Flow rate (m3/h)	Ratio(Qsn/Qn)
3-way configuration	Q n = 11.2	0.29
New System	Q sn = 3.2	

With the new system, the pump is operated on average at only about 66% of the nominal discharge head at design point:

Controlling system	Differential pressure (H)	Ratio
3-way configuration	H n = 4.4	0.66
New System	H sn = 2.9	

The pump input power, and thus its power consumption, is proportional to the product of the discharge head and volume flow rate.

The new system operates the pump at reduced volume flow rates and discharge heads. The pump draws much less power from the electricity grid. The new system was shown to save approx. 70% in electrical energy:

Controlling system	Input power (W)	Ratio
3-way configuration	P n = 561	0.33
New System	P sn = 184	

4. CONCLUSIONS

The most important ecologic effect of the energy efficiency application is the greenhouse gasses mitigation, reducing the electrical energy consumption and, by consequence production. Burning fossil fuels produces CO_2 emissions of roughly 0.53 kg per 1 kWh of electrical energy produced. Thanks to the drastic reduction in power consumption, the new system therefore makes a positive contribution to environmental protection.

5. REFERENCES

Dimonie, D., Radovici, C., Serban, S., Taranu, A., Vasilievici, V. (2007), Biodegradable compounds from regenerable resources for packings, *Mat.Plast.,* Vol.44, No.2, p.148-155

Dragoi, G., Funar, St., Solea, M., Cotet, C.E. (2007), Simulation and optimization of a waste processing flux, *Mat. Plast.,* Vol.44, No.1, p.77 - 81

Grigoriu, M. (2005), *Pumps and Pumping Installations*, Ed. Printech, Bucharest

Grigoriu, M. (2006), Remote Pumped-Storage, Back to Back Variable Speed Operated, WEAC, Session A9- 340-344, Polito, Italy

*** BOA-Systronic (2008). *Systematic Savings on Energy and Costs*, KSB Know-How series, Germany

Annals of DAAAM for 2008 & Proceedings of the 19th International DAAAM Symposium, ISSN 1726-9679
Editor B. Katalinic, Published by DAAAM International, Vienna, Austria 2008
Make Harmony Between Technology and Nature, and Your Mind will Fly Free as a Bird

DESIGN AND EXPERIMENTAL ANALYSIS FOR A NEW CONFIGURATION OF A SAFETY BARRIER DESTINED TO ABSORB THE IMPACT ENERGY

JIGA, G[abriel]; VLASCEANU, D[aniel]; NUTU, E[mil]; GHIBAN, B[randusa] & VISAN, D[ana]

Abstract: *The safety of road traffic is a complex problem, which by a great number of specialists in the field is studied. The common purpose of this research is to improve as much as possible the actual concept of street traffic safety on one hand, to develop, and to integrate new systems and high-performance equipment, which could lead to the improvement of indices characteristics for the road traffic. The European Union proposed to diminish till 2010 a half of deathly accidents in comparison to those produced in 2003, by two different ways: on one hand a tightening of standard foreknowledge linked to traffic rules and a development of new technologies in order to increase the traffic safety on the other hand.*
Key words: *barrier, impact, analysis, traffic, safety*

1. INTRODUCTION

Because important progress should be done in order to decrease the traffic accidents (especially those with serious injuries), an important role comes to researchers from the automotive industry. (Bayton, et al., 2007)

The impact energy produced during the collision with a safety barrier has to be absorbed by a deformable material able to dissipate this energy by deformation. Starting from these considerations one concludes that the application of rubber cushions on some barrier regions with a high injuriousness degree should contribute to impact attenuation as effect of the collision between the car and the safety barrier.

In the last years, parallel to a continuous development of numerical calculus systems, specialized software programs were developed for the simulation and analysis of the impact behavior of these structures. These codes are used by all automotive manufacturers for the simulation of the collision tests as well as for other tests linked to the passive safety.

The structure modeling at impact type loading represents one of the most complex and difficult demands for the structural analyst, involving the modeling of dynamic action, material properties, eventual car – barrier interaction and sensitivity influence of material high strain speeds.

2. EXPERIMENTAL ANALYSIS

2.1 Preparation and testing of specimens for the determination of strain-stress characteristics

The testing of the specimens has been achieved in conformity with the SR ISO37 Standard, which is identical to ISO 37/1994 International Standard – hard rubber or thermoplastic – determination of stress-strain characteristics at tensile tests (ASTM, 2001).

For the determination of tensile strength, dumb bells specimens were used, according to the standard presented above. The geometry of the specimens is presented in fig.1.

The standardized thickness of the middle part of the specimen must be 2 ± 0,2 mm, and the length of the specimen L was considered to be 25 ± 0,5 mm. The samples have been protected against all external influences (light, heat, humidity).

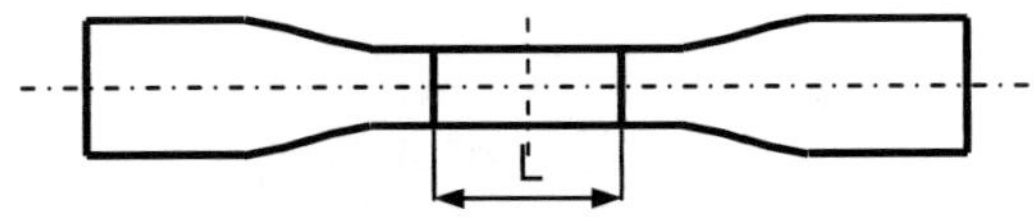

Fig.1 Shape of the specimen

Three different variants were proposed to be tested. From a qualitative point of view, all the variants had as basic elastomer the styrene-butadiene rubber (SBR), reinforced with different materials able to induce different properties, a special accent on the elastic behavior it is being put on (Ren & Vesenjak, 2005).

The compositions for the three variants of rubber are presented below.

- First variant: blend of rubber with 30 PHR powder, 200 PHR regenerated rubber and 20 PHR filler;
- Second variant: blend of rubber with 150 PHR powder, 100 PHR regenerated rubber without filler;
- Third variant: blend of rubber with 100 PHR regenerated rubber and 100 PHR filler;

The rest of materials specific for the rubber composition are protective, cure and plastification agents.

The fundamental characteristics for the three variants are listed in Table 1.

Variant	Hardness	Fracture strength	Ultimate tensile strain	Tensile Young modulus
	[°ShA]	[MPa]	[%]	[MPa]
1	70 - 71	900	420	200
2	78 - 79	380	150	220
3	68 - 70	550	220	240

Tab. 1 Elastic and mechanical characteristics

2. 2 The experimental stand for impact simulation

Following the analysis of a series of safety barriers existent on our national roads, one could observe that their performances have not been improved during the last decades, being considered too stiff and consequently leading to inacceptable decelerations during impact.

It is necessary to introduce new technical solutions in order to obtain the new safety barriers with greater possibilities to absorb energy during impact. One of these solutions may be to attach some modular elements manufactured from rubber, which could diminish the impact effects during collision.

For designing certain safety barriers in accordance with Union European Standards, one has to take into account some aspects regarding damping characteristics of different types of rubber, which should be applied on the safety barriers surface.

Results of researches regarding static and dynamic mechanical behavior of safety barriers components are presented (Jiga, et al., 2007).

In figures, 2 and 3 are presented the actual and the improved solution of a safety barrier where a rubber strip has been applied, in the middle of a span, in order to increase the absorption of energy.

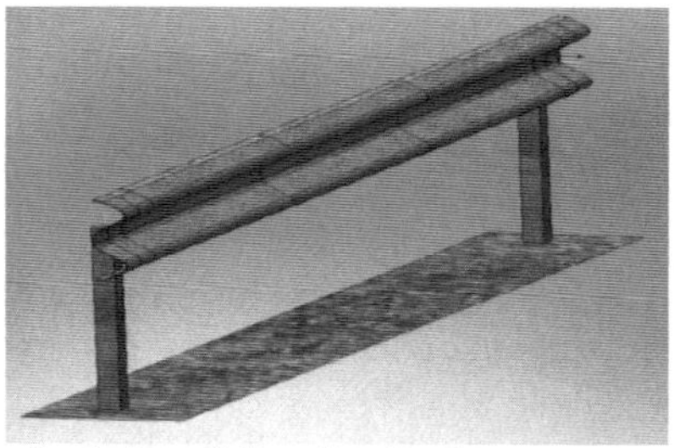

Fig. 2 Actual solution of the safety barrier

Fig. 3 Safety barrier with rubber applied on the traffic side

As an evaluation criterion for the experimental researches, the static and dynamic strains were considered. The tests were made in order to build up a safety barrier prototype for its future use in our country (Bratu, et al., 2007).

The modular components of composite rubber, having different thicknesses were made by S.C. Arteca S.A. This material was chosen because of its low cost fabrication.

The experimental tests were made on a stand that allowed fixation of the safety barrier at its ends, control of falling height of the impactor and measurement of displacement in the impact point. The displacements were determined using the strain gauge technique. A Spider 8800 bridge having eight channels and a data acquisition card was used for displacements measurement during the impact tests, which were collected from a displacement cell.

Three different configurations for safety barriers were tested:

a) safety barrier plated with 10 mm thick rubber;
b) safety barrier plated with 20 mm thick rubber;
c) non-plated safety barrier.

The safety barrier was fixed in both ends and the impact took place in its center (Figure 4).

Fig. 4 The safety barrier fixed for tests

Fig. 5 The data acquisition system

3. RESULTS

Case	H [m]	v [m/s]	Δt [s]	δ din [mm]	F [N]	F*Δt [Ns]
a	1	14.91	0.014	12.06	25039.45	351.83
	2	21.52	0.016	16.73	31078.7	507.83
	3	26.35	0.015	22.32	42422.04	621.96
b	1	15.22	0.016	10.54	21953.04	359.09
	2	21.52	0.017	15.89	29160.3	507.83
	3	26.35	0.016	20.48	39412.46	621.96
c	1	15.37	0.018	13.80	19812.93	362.66
	2	21.52	0.012	15.31	42158.46	507.83
	3	26.35	0.012	25.33	53496.13	621.96

Tab. 2 – The results for tested samples

The data presented in table 2 showed that the impact force corresponding to the 3 m falling height in case of non-plated safety barrier is 26% and 36 % greater than the ones corresponding to the safety barrier plated with 10 mm and 20 mm thick rubber respectively.

Between the two plated barriers, an increase of impact damping with 8% can be obtained in case of safety barrier plated with 20 mm thick rubber.

The displacement at the time of impact in the case of 20 mm thick plated safety barrier is 9% smaller than the 10 mm thick plated one.

4. CONCLUSIONS

A method for evaluation of the material deterioration consists in determination of the amount of energy dissipated in the material during impact. The experimental results showed that a considerable amount of impact energy is absorbed by the composite rubber layer.

5. AKNOWLEDGEMENTS

This research was supported by AMCSIT Politehnica through the research project CEEX 211/2006 ELMOSTPRO. The financial support is gratefully acknowledged.

6. REFERENCES

American Society for Testing and Materials (ASTM), (2001), *Standard specification for sheet steel, zinc coated or zinc–iron alloy-coated by the hot dip process*. Pennsylvania: ASTM International;

Bayton, D.A.F., Jones, T.B., Fourlaris, G. (2007), *Analysis of a safety barrier connection joint post-testing*, Materials and Design, April, p.1-7.

Bratu, P., Jiga, G., Vlasceanu, D., (2007), *"Methods for the analysis of dynamic performance of composite modular systems attached to safety barriers"*, A.S.T.R.- The Academic Days, November 28-30.

Jiga, G., Vlasceanu, D., Baciu, F., (2007) Analysis of a roll over protective structure using the finite element method, *The 18th International DAAAM Symposium*, ISSN 1726-9679, Croatia.

Ren, Z., Vesenjak, M., (2005), *Computational and experimental crash analysis of the road safety barrier*. Eng Failure Analalysis;12(6), p.963–73.

Annals of DAAAM for 2008 & Proceedings of the 19th International DAAAM Symposium, ISSN 1726-9679
Editor B. Katalinic, Published by DAAAM International, Vienna, Austria 2008
Make Harmony Between Technology and Nature, and Your Mind will Fly Free as a Bird

HYBRID ARCHITECTURE FOR A SINGLE-PRECISION ARITHMETIC PROCESSOR

JURCA, L[ucian]; GONTEAN, A[urel]; ALEXA, F[lorin] & VASAR, C[ristian]

Abstract: *This article presents a way to build a very fast single-precision arithmetic processor which can perform addition and subtraction in floating-point and all the other operations by using the binary logarithms. Format converters are used to ensure floating-point inputs/output for the logarithmic subunit. Pipelined structures of 6 stages for the logarithmic core and 3 stages for the floating-point one will be used to optimize the clock and to facilitate the implementing of DSP algorithms.*
Key words: *Floating Point, Logarithm Number System.*

1. INTRODUCTION

The increase of integration density has permitted the development of the logarithmic-number-system (LNS) processors out of which we mention (Coleman et al., 2000) and (Arnold, 2001), but in these the main difficulty was to implement the addition and subtraction operations. Avoiding this disadvantage and at the same time keeping the qualities of both floating-point (FP) and LNS can be achieved through the design of a hybrid unit which combines the attributes of the FP processor with logarithmic arithmetic. A solution in this direction had already been proposed in (Lai, 1993), where addition and subtraction were performed in FP and multiplication, division, square root and all the other operations in LNS. But the format conversions FP-LNS and LNS-FP were slow because in the linear-interpolation algorithm only non-redundant adders were used.

To improve the format conversions, a new architecture was proposed in (Jurca et al., 2007) where the method of redundant summation of partial products with other inputs was applied because a multiplication and a series of additions occur in sequence in the conversion algorithm. Moreover, corrections of one or two LSB were applied in some memory locations of the log and antilog look-up tables content. In this way, the conversions became two times more accurate than in (Lai, 1993), e.g. the conversion error was kept under 1.5×10^{-7}, while the FP single-precision error is still 1.2×10^{-7}.

The aim of this paper is to offer an alternative of classical FP units because, for some particular applications and with new improvements, the logarithmic and hybrid units can run faster than the floating-point ones.

Thus, in section 2 we present the pipelined architecture of the logarithmic subunit and the conversion algorithms FP-LNS and LNS-FP.

For implementing the floating-point addition/subtraction, a classical 3-stage pipelined subunit synchronized with the logarithmic one (4ns per stage) was designed. This means that in our case we don't need a high level of parallelism for the data paths and thus we can save area. These aspects will be discussed in section 3. To facilitate the comparison with related works, in our design the propagation delays through different gates were adopted corresponding to 0.5-µm CMOS technology and all the *PSpice* models in our digital simulations were settled according to this.

In section 4 we will give an example of how we can implement a simple DSP algorithm using a hybrid processor, we will also show the main directions of our future work and we will conclude the paper.

2. LOGARITHMIC SUBUNIT

The stages of the logarithmic subunit and the conversion algorithms will be briefly presented in this section because a more detailed description can be found in (Jurca et al., 2007).

With the same hardware, the 6-stage pipelined structure presented in Fig.1 can perform either multiplication or division for two FP operands, A and B, depending on the bit-line S_{OP} that acts on the third stage, as we can see in equations (1):

$$A \times B = \text{antilog}(\log A + \log B)$$
$$A / B = \text{antilog}(\log A - \log B) \qquad (1)$$

The linear interpolation algorithm for the direct conversion is based on equation (2):

$$\log(1+y) \cong y + E_y \pm \Delta E_y \times y_2 \qquad (2)$$

where y is the fractional part of the 23-b significand, out of which the least significant 12 bits represent y_2; the values E_y of the function log(1+y)-y for $2^{23-Ny2} = 2^{Ny1} = 2^{11}$ points were memorized in internal ROM, as well as the values ΔE_y, for its derivative function too, memorized in internal ROM'. For the conversion LNS-FP the algorithm is similar and only the ROM content changes.

The unit includes two format converters FP-LNS for the two operands and one LNS-FP converter which produce the operation result in floating-point format. In fact the direct conversions are fused with ALU operation in the very purpose to reduce the non-redundant additions to the one performed in ALU. The Wallace trees are built with 4:2 compressors.

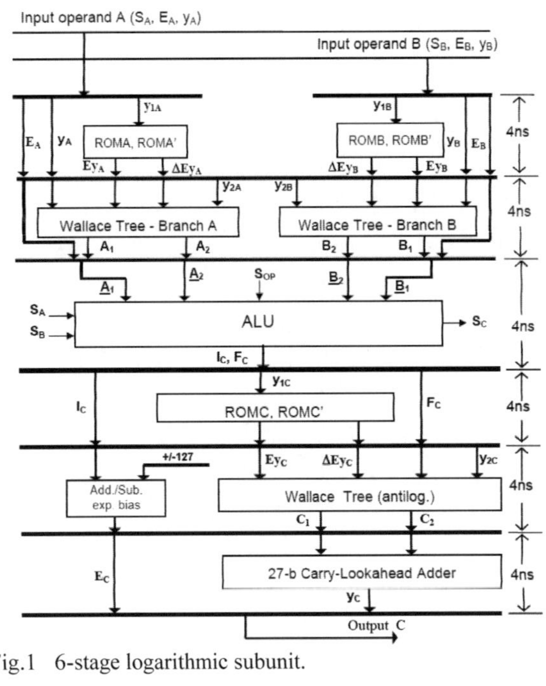

Fig.1 6-stage logarithmic subunit.

Of course, at the output of the second stage the latched data are in carry-save form (A_1, A_2, B_1, B_2) and the binary logarithms will never be produced explicitly among the pipelined structure. The exponents E_A and E_B are concatenated as integer part of the data ($\underline{A}_1$, $\underline{B}_1$) which is to be operated in fixed-point in ALU. The digital simulations proved that the propagation delays across all stages were 4ns.

3. FLOATING-POINT SUBUNIT

The first stage of the floating-point subunit (the alignment of mantissas) is shown in Fig.2. The adders "ExpA-ExpB" and "ExpB-ExpA" work simultaneously and the MSB of "ExpA-ExpB" (whose fan-out was multiplicated with the block Mlt) will select the bigger exponent and the associated significand (mantissa) at the output of the stage. It will select the positive exponent difference also to command the barrel-shifter BS which shifts to the right the other significand.

The second stage is an adder/subtracter circuit (Fig.3). Adder1 and Adder2 are 25-b adders because the sign bit and the implicit 1 are attached to the 23-b significand. In our design we used very fast 2-level hybrid adders (Jurca & Maranescu, 2004), with five (1+2×2) 8-b carry look-ahead adders (CLA, with input carry 0 and 1 respectively) plus in the most significant position two 9-b CLA, these ones on the 1st level and a carry select mechanism on the 2nd level. The same type of adder was used in the third stage of the logarithmic subunit.

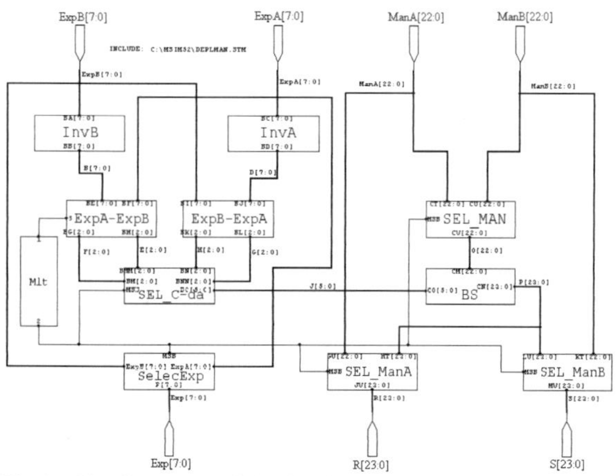

Fig.2 The first stage of the floating-point subunit.

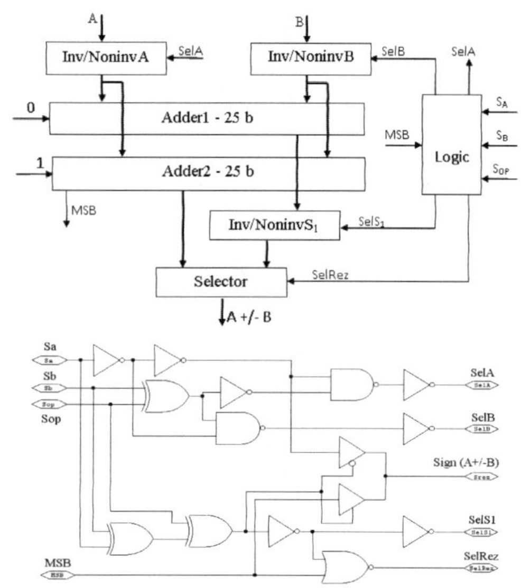

Fig.3 The second stage of the floating-point subunit.

The normalization of the final result and the adjusting of the exponent are done in the third stage (Fig.4). In the case of subtraction, the number of leading zeros, counted with LZC, is subtracted from the transmitted exponent and the output of the barrel-shifter BS will be selected. In the case of addition, the data will be either shifted one bit to the right, or unshifted.

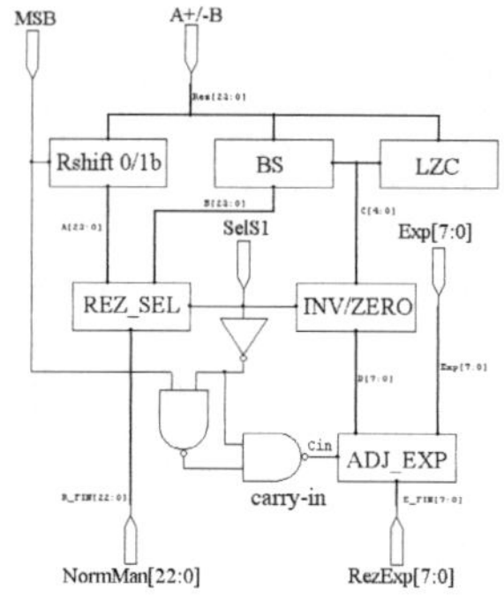

Fig.4 The 3rd stage of the FPSU.

4. APPLICATIONS. CONCLUSIONS

A DSP algorithm can be very easily implemented in the hybrid unit as we can see in Fig.5 where the 20th and 29th steps of a 20-term Livermore-loop computing are presented. The total time is only 140ns (35 cycles), but our aim is to improve further the hybrid unit, by reducing its latency and by implementing 3-operand multiplication and shift capability in logarithmic ALU for square root or any power capabilities. A remarkable feature of this unit is that it performs a division faster than the most recent FPUs (Nikmehr & al., 2007).

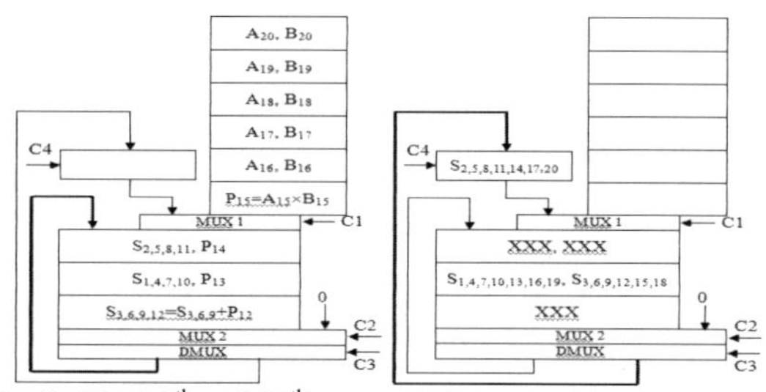

Fig.5 The 20th and 29th steps of the sum of 20 products.

5. REFERENCES

Arnold, M. (2001). A Pipelined LNS ALU, *Workshop on VLSI. Proceedings,* pp. 155-161, ISBN: 0-7695-1056-6, Orlando, FL., USA, April 19-20, 2001, IEEE Computer Society.

Coleman, J.N.; Chester, J.E.; Softley, C. & Kadlec, J. (2000). Arithmetic on the European Logarithmic Microprocessor, *IEEE Transactions on Computers,* Special Edition on Computer Arithmetic, Vol. 49, No. 7, pp.702-715, July 2000, ISSN: 0018-9340.

Jurca, L. & Maranescu, V. (2004). A New Way to Build a Very Fast Binary Adder, *Scientific Bulletin of the Politehnica University of Timisoara. Transactions on Electronics and Communications*, Tom 49 (63), Fascicola 1, 2004, pp. 193-198, ISSN: 1583-3380.

Jurca, L.; Gontean, A.; Alexa, F. & Curiac, D.I. (2007). Proposal to Improve Data Format Conversions for a Hybrid Number System Processor, *Proceedings of the 11th WSEAS International Conference on COMPUTERS*, Agios Nikolaos, Crete Island, Greece, July 26-28, 2007, ISSN: 1790-5117, ISBN: 978-960-8457-95-9.

Lai, F. (1993). The Efficient Implementation and Analysis of a Hybrid Number System Processor, *IEEE Transactions on Circuits and Systems*, Part II, Vol. 40, No. 6, June 1993, pp. 382-392, ISSN 1057-7130.

Nikmehr, H.; Phillips, B. & Lim, C.C. (2007). A Fast Radix-4 Floating-Point Divider with Quotient Digit Selection by Comparison Multiples, *The Computer Journal,* Vol. 50 Issue 1, pp.81-92, Jan. 2007, Oxford University Press. ISSN: 0010-4620.

Annals of DAAAM for 2008 & Proceedings of the 19th International DAAAM Symposium, ISSN 1726-9679
Editor B. Katalinic, Published by DAAAM International, Vienna, Austria 2008
Make Harmony Between Technology and Nature, and Your Mind will Fly Free as a Bird

IMPORTANT FACTORS OF CORRUGATED CARDBOARD'S QUALITY OF PUNCHING RESISTENCE

JURECIC, D[enis]; BABIC, D[arko] & LAJIC, B[ranka]

Abstract: *The construction and production of transport boxes involves a lot of parameters that need to be fulfilled for a box to meet its requirements. One of the most important parameters is punching resistance. It is known that the corrugated cardboard punching resistance depends on, for instance a number of paper layers and paper quality used for corrugated cardboard production. Almost certainly it also depends on cardboard humidity. Four types of two-waved five-layered corrugated cardboard shall be used for examination purposes, and each type of cardboard will be exposed to different humidity (97%, 85%, 75%, 55% and 35%) in order to test corrugated cardboard punching resistance. After the test the punching spot shall be microscopically observed.*

Key words: *corrugated cardboard, punching resistance*

1. INTRODUCTION

Corrugated cardboard is defined as *cardboard made of one or more layers of paper used for a wave and glued between two or more single- or multi-layered flat areas* (Stričević, 1983). Corrugated cardboard used for packaging in the food industry must be glued with gluten, not with siliceous glue. Packaging used for overseas transport must be waterproof and corrugated cardboard packaging for frozen products must be resistant to low temperatures (those below 0°C) and humidity and must not be slippery (Rodin, 1991). In other words, the choice of the type of paper used in corrugated cardboard manufacturing will largely depend on the packaging type (transport or commercial). High-grammage paper types are used for making transport packaging, while low-grammage ones fit commercial packaging better.

2. EKSPERIMENTAL

Dynamic punch test is used to determine mechanical resistance of corrugated cardboard to direct mechanical hits: the test tells us how strong those hits can be before a specific corrugated cardboard type breaks (Hanlon, 1989). Four types of five-layered corrugated cardboard were tested: 2BT, BŠT, 2TŠ and T2Š. The material tested is specified in Table 1.

A *Franck* dynamic punching device was used to test the dynamic punching resistance of corrugated cardboard and its relation to humidity (www.techlabsystems.com, 2007). The samples were cut into 175x175mm pieces, as required by the standard. The cut samples were then exposed to different levels of humidity: 97%, 85%, 75%, 55% and 35%. Five large bags containing solution of varying humidity levels (35%, 55%, 75%, 85% and 97%) were used in the process. A container of liquid was placed in a bag and so were 20 samples of each corrugated cardboard type. The bag was then sealed and the samples were exposed to a particular humidity level for 5 days.

After the samples had been exposed to humidity for 5 days, 10 punches were made on the obverse and 10 on the reverse for each humidity level and each corrugated cardboard type. After punching all the samples, the average was calculated. The total reached 400 punches (200 on the obverse and 200 on the reverse).

	Corrugated cardboard type			
	2BT	**BŠT**	**2TŠ**	**T2Š**
Flat layer type – obverse	white testliner	white testliner	testliner	testliner
grammage [g/m^2]	130	130	132	132
Flat layer type – reverse	white testliner	testliner	testliner	šrenc
grammage [g/m^2]	130	115	115	100
Mid-layer type	testliner	šrenc	šrenc	šrenc
grammage [g/m^2]	115	100	100	100
Wavy layer type	fluting	fluting	fluting	fluting
grammage [g/m^2]	105	105	105	105

Tab. 1.Specification of the material tested

2.1 Determining humidity in laboratories

Humidity can be extracted from corrugated cardboard by putting it in kilns for a longer period of time at 105 – 120°C. Corrugated cardboard dries until it reaches a constant weight. The cardboard is then taken out of the kiln and weighted. The difference in weight before and after the drying process tells us how much water it contained. A round corrugated cardboard sample of 10 cm in diameter is placed under a 2.5 cm column of distilled water of temperature around 20°C for 30 minutes. The difference in weight between the wet and the dry corrugated cardboard must not exceed 1.25 g for the outer layer of paper or 2 g for the inner one. Absorption of corrugated cardboard is tested in the following way: a sample of 15x15 cm is immersed completely in a water container for 10 minutes. The quantity of water absorbed can be determined by weighing the sample before and after the immersion. The test results are called *immersion numbers* and they express grams of absorbed water after the immersion in standard room conditions.

2.2. Testing the dynamic punching resistance of corrugated cardboard - Puncture Test

The test is used to determine the behaviour of corrugated cardboard when exposed to direct mechanical hits. Toughness of corrugated cardboard and its resistance to tearing is determined on the basis of that test: the test tells us how strong those direct mechanical hits can be before a specific corrugated cardboard type breaks, which is very important if a corrugated cardboard packaging contains fragile objects (Talbi, N., Batti, A., Ayad, R., Guo, Y. O., 2008),. Three different samples of corrugated cardboard were tested: 2K, 2K/3 and 2K/e. Each cardboard sample was tested 10 times on the obverse and 10 times on the reverse.

3. RESULTS AND DISCUSSION

3.1. Results

The results show that the maximum work required for a punch-through on the obverse is 7.58 J when the humidity level is 75%. The minimum work required is 7.33 J when the humidity level is 35%. The maximum work required for a punch-through on the reverse is 7.82 J when the humidity level is 55%. The minimum work required is 7.58 J when the humidity level is 97%. It is, therefore, to be concluded that the optimum humidity levels for a punch-through on the obverse are between 97% and 75%, while the optimum humidity levels for a punch-through on the reverse are between 85% and 55%.

The results show that the maximum work required for a punch-through on the obverse is 7.25 J when the humidity level is 85%. The minimum work required is 7.11 J when the humidity level is 35%. The maximum work required for a punch-through on the reverse is 7.44 J when the humidity level is 55%. The minimum work required is 7.11 J when the humidity level is 97%. It is, therefore, to be concluded that the optimum humidity levels for a punch-through on the obverse, as well as on the reverse, are between 85% and 55%. When the humidity level is 97%, a larger amount of work is required for a punch-through on the obverse.

The results show that the maximum work required for a punch-through on the obverse is 7.18 J when the humidity level is 55%. The minimum work required is 6.82 J when the humidity level is 97%. The maximum work required for a punch-through on the reverse is 7.25 J when the humidity level is 85%. The minimum work required is 7.13 J when the humidity level is 75%. It is, therefore, to be concluded that the optimum humidity levels for a punch-through on the obverse are between 55% and 35%, while those for a punch-through on the reverse are between 85% and 35%.

The results show that the maximum work required for a punch-through on the obverse is 7.04 J when the humidity level is 55%. The minimum work required is 6.69 J when the humidity level is 35%. The maximum work required for a punch-through on the reverse is 7.38 J when the humidity level is 55%. The minimum work required is 7.03 J when the humidity level is 35%. It is, therefore, to be concluded that the optimum humidity levels for a punch-through on the obverse, as well as on the reverse, are between 97% and 55%.

3.2. Discussion

The results for the BŠT type are good when compared to those for 2BT, i.e. the average values of work required at all humidity levels on the obverse and the reverse are higher than 7.1 J. The two other types of material tested (2TŠ and T2Š) gave almost the same results for a punch-through on the obverse (less than 7 J at all humidity levels), while their results for a punch-through on the reverse are slightly better, ranging between 7 J and 7.4 J. However, when compared to 2TŠ, T2Š yields better results when the humidity levels are higher than 55%. The amount of work required for a punch-through on the 2BT type is the largest of all, on the obverse and on the reverse alike. BŠT comes second and 2TŠ and T2Š could both hold the third place, since the results are very similar. At all humidity levels, 2BT and T2Š are more resistant to punching on the reverse, while BŠT is more resistant to punching on the obverse only when the humidity level is 97%. 2TŠ is more resistant to punching on the obverse only when the humidity level is 55%. The punching resistance of the 2BT, BŠT and T2Š types on the obverse is proportionate with the humidity level, but for the 2TŠ type it varies inversely with humidity. The punching resistance on the reverse varies inversely with humidity for 2BT, BŠT and 2TŠ, while it is proportionate with the humidity level for T2Š. The test shows that the results obtained for the obverse of the 2BT type are the best when the humidity level is between 97% and 75%. The best values for the work done on the reverse are obtained when the humidity levels are between 85% and 55%. The greatest difference between the values obtained for a punch-through on the obverse and that on the reverse of 2BT is 3.7% at the humidity level of 55%, and the smallest difference is 1.3% with the humidity level of 75%. The results obtained for the obverse and the reverse of the BŠT type are the best when the humidity level is between 85% and 55%. The values obtained for a punch-through on the reverse exceed those on the obverse only when the humidity level is between 85% and 35%. When the humidity level is 97%, a larger amount of work is required for a punch-through on the obverse. The greatest difference between the values obtained for a punch-through on the obverse and that on the reverse of BTŠ is 3.76% at the humidity level of 55%, and the smallest difference is 0.14% with the humidity level of 35%. The test shows that the results obtained for the obverse of the 2TŠ type are the best when the humidity level is between 55% and 35%. The best values for the work done on the reverse are obtained when the humidity levels are between 85% and 35%. The values required for a punch-through on the reverse exceed those on the obverse at all humidity levels except that of 55%, in which case a slightly larger amount of work is required for a punch-through on the obverse. The greatest difference between the values obtained for a punch-through on the obverse and that on the reverse of 2TŠ is 4.74% with the humidity level of 97%, and the smallest difference is 0.13% with the humidity level of 55%. The results obtained for the obverse and the reverse of the T2Š type are the best when the humidity level is between 97% and 55%. The values obtained for a punch-through on the reverse exceed those on the obverse at all humidity levels. The greatest difference between the values obtained for a punch-through on the obverse and that on the reverse of T2Š is 5.97% with the humidity level of 85%, and the smallest difference is 4.6% with the humidity level of 55%. The 2BT type yielded the best results in the test and could, therefore, be said to stand out from other tested cardboard types. The second best type would be BŠT, while 2TŠ and T2Š yielded very similar results and are considered to be of the same quality.

4. CONCLUSION

However, the research done on the material and the analysis of the results obtained did not result in any linear or square function that could illustrate the trend of increasing or decreasing amount of work required for dynamic punching. The results remain unpredictable. Nevertheless, it could be concluded that the best results for all the types tested were obtained at the humidity levels between 85% and 55%, as the practice has clearly confirmed.

5. REFERENCES

Hanlon, J. F., (1989). *Handbook of Package Engineering*, McGraw Hill Book Company, pp. 1-6, New York

http://www.techlabsystems.com/en/datasheets/corrugated/puncture-testers-web.pdf, Accessed: 2007-03-25.

Rodin A., (1991). *Packaging of cardboard,* pp. 265-266, Progres, Zagreb

Stričević N. (1983). *Modern Packaging 2,* pp. 31-32, Školska knjiga, Zagreb

Talbi, N., Batti, A., Ayad, R., Guo, Y. O. (2008), An analytical homogenization model for finite element modelling of corrugated cardboard*Composite Structures, Available online 11 April 2008*

Annals of DAAAM for 2008 & Proceedings of the 19th International DAAAM Symposium, ISSN 1726-9679
Editor B. Katalinic, Published by DAAAM International, Vienna, Austria 2008
Make Harmony Between Technology and Nature, and Your Mind will Fly Free as a Bird

INVESTIGATIONS OF PERSONAL HUMAN ERGONOMIC DESIGN IN AUTOMOTIVE INTERACTION

JURUM KIPKE, J[asna]; KOVACEVIC, D[razen] & SUMPOR, D[avor]

Abstract: *The working activity of body interaction between humans and the ergonomic structure of automobiles is related to respective working loads, including strain. The loads people are subjected to when approaching and leaving road motor vehicles, when they occupy different working postures form a wide and still insufficiently studied bio-mechanical and ergonomic area. Carrying, lifting and handling of cargo often result in relatively high loads. Loads that can occur during human work in most cases are related to the non-ergonomic working position. The purpose of the carried out ergonomic research is to better design the working ambient system, correct and sustainable body posture with bio-mechanically more accurate performance of the movements of people in interaction with the automobile.*

Key words: *Digital anthropometry, ergonomics, biomechanics of work*

1. INTRODUCTION

Physiological anthropology is a branch of science which studies the functions and processes of living organisms and their parts and organs. From this aspect, the main task of the physiological anthropology is to study the differences among people, and how and in which way the single body parts work together with the aim of achieving some of its working functions (Muftić et al., 2001).

In human organism no property is constant for a longer period of time. Human body consists of heterogeneous materials and its properties are different for different parts. The gender is also a very significant factor in designing a computer 3D model (Mijović et al., 2001).

The human body is very complex and it is practically impossible to describe all the movements that a body can produce. The complexity of movements results from the fact of a large number of degrees of freedom of the movements and therefore many assumptions and simplifications are introduced into the analyses of the human movements in space.

The problem of adjusting the activities to humans is studied through the analysis of the static and dynamic workloads, researching the groups of the most favourable postures and extremities during work.

In order to solve the problems related to the body mobility a correlation between the biomechanical and anthropometric research is necessary. All body movements consist of different basic movements of its single parts and in single joints. Thus the success of complex movement depends on the type and scope of these individual movements. The body as a living organism is a much more complex system than any today known technical construction, with a huge number of interacting elements. Observing the body parts of different masses and dimensions as parts of the kinematic chain, and analyzing different body postures in resting and in motion from the aspect of biomechanics, it is possible to find the optimal types of movements as part of a technological process, thus preventing the unnecessary fatigue or illness.

2. BIOLOGICAL ANTHROPOMETRIC ANALYSIS

Anthropometric measures that are of relevant interest for ergonomic considerations will depend on the type of working activity and on the requirements of the working process, most in the so-called condition of active work. Therefore, the considerations include all those dimensions of the human body that form the sum of the relations subject to performing of certain work task, as well as the environment in which the human stays and works. Thus, every ergonomic anthropometric research requires the application of some specific measures as well as definition of some new factors that are necessary to determine the morphological specifics of human body, in relation to a certain purpose (Maver et al., 1976).

The list of ergonomic anthropometry is continuously supplemented by measures that do not only indicate the possibility of adequate action of the tested system, but can also perform a certain task given to the subject, in the most adequate and optimal working posture. It is necessary to determine also the diverse possibilities of taking the most favourable work position, and movement in the given space having at the same time in mind also the considerations of the space dimensions, sizes and shape of the work environment. Such research and the resulting solutions have to completely satisfy the basic bio-mechanical, biomedical and ergonomic requirements (Muftić & Milčić 2001).

3. BIOMECHANICAL ANALYSIS

Acquiring precise knowledge about the dimensions and parameters of the human body and the creation of the biomechanical model that will best simulate the actual conditions is the basic task during computer biomechanical movement analysis. The basic values necessary to describe the movement are anthropologic measures, i.e. the lengths of the human body segments, segmental mass and their distribution according to the selected coordinate system.

Since the human skeleton contains: 95 joints with one degree of freedom of movement, 80 joints with two degrees of freedom of movement, and 75 joints with three degrees of freedom of movement, which results in a total of 250 degrees of freedom of movement, the entire complexity of the kinematic and dynamic study of the human skeleton is understandable. This is significant because the mobility of the computer-designed 3D model depends on the degrees of freedom of movement of its basic components (Muftić et al., 2001).

Precise analysis of the human body movement within a three-dimensional space is a very complex and demanding task. In numerous examples of research, the factors of movement are usually simplified to such an extent that there are obvious deviations of the computation results from the measured ones. Therefore, there is also the need to find more adequate digitized methods that, apart from certain methodological benefits would insure also more reliable results (Baksa et al., 2007).

4. RESEARCH RESULTS

For the research of personal body interaction and ergonomic harmony while performing the work of opening the back door, putting the travelling bag into the trunk and closing the car doors, a male subject was selected, 190cm tall, with a mass of 83kg, and the selected test car was compact, B segment, Hyundai, Getz model. By using the new and advanced contactless 3D body scanner "*Body*SABA 0.7.", the digitally generated three-dimensional model of the subject was used to read all the relevant anthropometric measures, and Table 1 presents characteristic anthropo-measures of the standing work posture.

Symbol and name of anthropometric measures		Values of anthropo-measures (cm)
A	Standing height	190,0
B	Height of eyes (standing)	179,0
C	Height of shoulders (standing)	156,5
D	Height of elbow above floor	118,0
E	Height of knee of a standing person	54,8
F	Arms span	201,0
H	Length of forearm with hand	51,8
I	Width of shoulders	49,8
K	Bust measurement (chest)	24,3
L	Thigh width	34,5
V	Foot length	29,5
X	Foot width	10,6
Y	Hand length	21,5

Tab. 1. Characteristics of anthropometric measures of the subject for standing work posture.

In order to make a computation of individual biomechanical characteristics of work and ergonomic analysis of the work performance of the tested measuring subject "**SABA***Lab*" system was used, of three-dimensional measurements and analysis of the work of humans, the so-called MoCap system, "*Vato***SABA** 2.1.", as can be seen from the characteristic kinematograms in Figure 1.

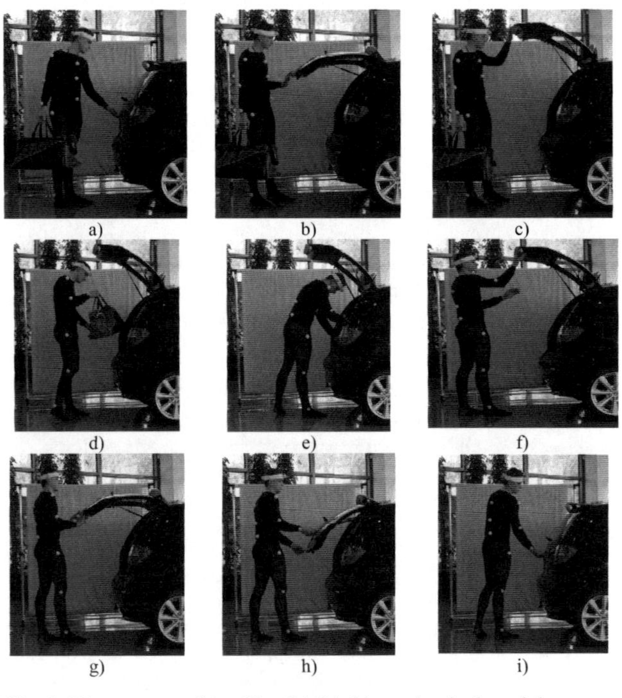

Fig. 1. Kinematographics VatoSABA biomechanical model of subject in characteristic working postures

Sequential kinematogram	Measuring value of angle (⁰)	
	α	β
a)	142,15 (lh)	178,94 (lh)
b)	96,42 (lh)	124,88 (lh)
c)	110,34 (lh)	155,31 (lh)
d)	134,39 (lh)	168,73 (lh)
e)	141,29 (lh)	172,43 (lh)
f)	132,42 (rh)	140,84 (rh)
g)	118,22 (rh)	121,11 (rh)
h)	110,58 (rh)	181,49 (rh)
i)	140,65 (rh)	149,94 (rh)

Tab. 2. Measuring values of angles of the working extremity of the subject.

Table 2 gives a presentation of measurement values of angles α between the forearm and upper arm and angles β between the hand and the forearm of the working extremity. Since the subject handled a travelling bag during the analysis, and it was a person usually using the right hand, the working procedure of opening the door with left hand (lh) was understandable and the closing with the right hand (rh).

The results that range within the limits 96.42^0 to 142.15^0 for angle α and 121.11^0 to 181.49^0 for angle β, show the ergonomic compatibility of the tested subject and the automobile.

5. CONCLUSION

By knowing the anthropometric measures and with the application of computer equipment, computer 3D graphic software and instruments, e.g. the system of "*Body*SABA" and "*Vato*SABA" systems, it is possible to perform very efficiently and very fast the ergonomic modelling and the analysis of dimensions and forms of elements of the working environment system in order to make judgement on the spatial performance of the ergonomic design of the working automotive environment with regard to the diversity of the motorist and their understanding of the comfortable working posture.

The carried out bio-mechanical analysis, based on anthropometric research and the use of computer bio-mechanical models, allows digitization of the actual activity and results in the researched indicators that contribute to the knowledge on more acceptable and better design of the working ambient system of automobiles in interaction with the customers.

The guidelines of future research are directed to the possibility, based on the software-processed video-recording, of carrying out adequate ergonomic analysis of any form of physical effort in humans, of classified values to adequate statistical distribution of the defined domicile as well as world population.

6. REFERENCES

Baksa, S.; Muftić, O. & Sumpor, D. (2007). Computer Aided Ergonomics Analysis of Exhibition-transport element, *Proceedings of 3rd International Ergonomics Conference, Ergonomics 2007,* Mijović, B. (Ed.), pp. 249-257, ISBN 987-953-98741-4-6, Stubičke Toplice, Jun, 2007, Zagreb.

Maver, H.; Rudan, P. & Tarbuk, D. (1976). *Practical work in anthropology,* Ergo. metode, Ant. Bib., Zagreb

Mijović, B.; Ujević, D. & Baksa, S. (2001). Visualization of Anthropometric Measures of Workers in Computer 3D Modeling of Work Place. *Collegium Antropologicum,* 25., 56., (2001), 639-650.

Muftić, O.; Veljović, F.; Jurčević-Lulić T. & Milčić D. (2001). *Fundamentals of ergonomy,* Mašinski fakultet, Sarajevo.

Muftić, O. & Milčić, D. (2001). *Ergonomy and safety,* Visoka škola za sigurnost na radu, Zagreb, Iproz, Zagreb

Annals of DAAAM for 2008 & Proceedings of the 19th International DAAAM Symposium, ISSN 1726-9679
Editor B. Katalinic, Published by DAAAM International, Vienna, Austria 2008
Make Harmony Between Technology and Nature, and Your Mind will Fly Free as a Bird

MOBILE AUTOMATED EQUIPMENT FOR WIND BLADE PRODUCTION

KACHAKOV, Z[harko]; SAMAKOSKI, B[lagoja] & SOKOLOSKI, Z[latko]

Abstract: *New concept of integration of various processes in mobile automated equipment for production of modern wind blades is discussed in this paper. These processes are Automated Fiber Placement, Filament Winding and CNC cutting and milling. With this concept the composite wind blades should be manufactured at the place of installation of the wind turbines. The comparative simulations of the influence factors between today's classical manual methods of production of composite wind blades, and the new concept developed at MIKROSAM Company, shows a decrease in the overall price of the wind blades by up to 20% using the latter method.*

Key words: *composite, wind blade, wind turbine, fiber placement, mobile.*

1. INTRODUCTION

Wind energy is becoming an ever more competitive source of energy. The continuous optimization of the production of wind turbines is essential. The rotor blade plays a critical role in the optimization of the modern wind turbine. The power output from a wind turbine is (Burton at al., 2001):

$$P = \frac{1}{2} C_p \rho A U^3 \quad (1)$$

where C_P is the power coefficient, ρ is the density of the air, A is the rotor swept area, and U is the wind speed.

The effective usage of the wind turbine, which is directly correlated with the final energy price, varies with the cube of the average wind speed (wind class), and the rotor swept area, which is the square of the length of the rotor blade. Over the last 10 years there has been a continuous increase in the rotor diameter of commercially available wind turbines from around 30 m, to more than 120 m. A doubling of the rotor diameter (rotor blade length) leads to an increase in the power output by a factor of 4.

The direct participation of production price of the rotor blades in the total production price of the large wind turbines is 10% (E-composites Inc., 2005). If we add the cost of the special transportation of the large wind blades (> 30 m in length) to that, then the optimization in the production, and the finding of solutions for reduction of the transportation costs, can lead to an increase in the price-performance ratio.

Since today's modern wind blades are entirely produced of composite materials (fiber reinforced plastics), the research at the MIKROSAM Company has provided a solution of automated - mobile - modular equipment for different processes in the production chain for on-site fabrication of large wind blades.

2. DESIGN VS. MANUFACTURING

The design and the manufacturing processes for wind blades are so interrelated, that one cannot be discussed without the other.

The main design factors are: the profile (airfoil); resin type (thermoset or thermoplastic) and the content; fiber curvature; length and weight of the blade; speed of the wind and the method of control of the rotor speed (National Research Council, 1991).

Manufacturing processes that are in use today include the labor intensive manual processes as Dry and Wet Hand Lay-up, and automated processes, such as Filament Winding and Tape and Fiber Placement.

The aerodynamic efficiency is directly related to the chosen airfoil. The influence of the airfoil is increasing with the increase of the wind speed and the length of the blade. So, if the blade is designed without concave segments and in one piece, for the small wind turbines (<10 m length), the automation can be achieved by using the Filament Winding Machine. The large wind blades are designed for ultimate wind speeds, and the influence of the more efficient airfoil (with concave segments), the length and the weight of the blade are significant. These blades are usually manufactured from two shells, which are bonded on an internal spar or shear web reinforcement.

The usage of the automation processes for the large wind blades consists of fabrication of the central spar by Filament Winding process, and manufacturing of the shells on Tape or Fiber Placement Machine.

The use of the Automatic Tape and Fiber Placement Machine is more efficient due to higher use of thermoplastic resins as matrices (CORONET, 2004), replacing the thermoset resins, and skipping the consolidation process in an autoclave.

It is possible to consolidate the thermoplastic resins during the automatic layup of the thermoplastic pre-impregnated tape, reducing the time of production and cost of the heating in the autoclave.

The introduction of the carbon fiber, as a replacement of the glass fiber in some elements in the structure of the large rotor blades, has led to improvements in the control of the volume and the lay-up directions during the manufacturing (Harrison at al., 2000).

The price of the carbon fiber is much higher than that of the glass fiber, but due to its lower specific weight and higher mechanical properties, the large rotor blade has a better price to performance ratio.

3. MOBILE AUTOMATED EQUIPMENT

Previously, we have been discussing the need of automation of the manufacturing processes for the large wind blades.

The wind turbines are installed on sites which provide conditions for continuous working, i.e. which are able to provide optimum wind speed during the whole year.

Usually, those sites are at hard-to-reach places, on some hills for on-shore wind farms, where the transportation of the large structures is very difficult and expensive.

The new concept (Fig.1.), developed at MIKROSAM Company, introduces a mobile, hybrid and modular automated equipment, ready for easy transportation and installation near to the wind farm sites.

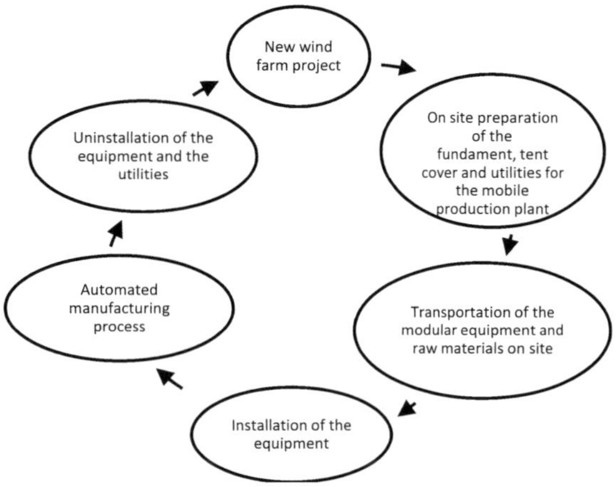

Fig. 1. Flow chart of the new concept for on-site production.

Fig.2. shows one model of such a hybrid, mobile equipment for wind blade manufacturing, which can be used for different processes on two parts or molds of the wind blade.

The multi-bridge construction allows working on different sides of the blades, and can be fully automated and synchronized from one place.

Fig. 2. One model of the hybrid, mobile equipment for wind blade manufacturing.

Depending on the needs, this equipment includes several integrated processes, which may be separated in following categories:

a) Automated Fiber, Tape or Fabric Placement process for the production of the complex shapes of the shells on the two-part molds for the large wind blades. Depending on whether a thermoset or a thermoplastic resin system is used, the process includes consolidation in the place, avoiding the relatively long process of consolidation in separate autoclaves.

b) Filament winding process for the production of one-part blades, when the airfoil shape permits the use of this process. This process is also used for the production of the central reinforcement spars mounted in the blades.

c) CNC cutting and milling process for final machining due to the final design and functionality on the finished blade. The cutting and milling process can also be used for production of the molds for the two-part wind blades.

d) Automated measuring and control of the materials. This includes the measuring and control of the molds and raw materials, but also non-destructive inspection (NDI) of the final parts.

These different processes can be included in the same equipment with hybrid construction, and can be fully automated and controlled from one place. The control of the different processes includes coordination of many parameters where it is integrated as part of one centralized process of the production of the whole wind blade.

Once the blade is designed, the design is approved by complete analysis of the materials and FEA analysis is performed on the loads and strengths, and then the automated manufacturing, introduced in this paper, can result in production of the structure of the composite wind blade with the same or close enough shape and structure as the original design.

The automation introduces more freedom in the design of the composite structure, using the benefits of the anisotropic characteristics of the composites entirely. The Automatic Fiber Placement process gives the certain quantity and direction of the reinforcement and matrices, without unnecessary layers, and with quality bonding, i.e. better utilizing of the materials in correlation to the loads. There are even more benefits of this technology when more expensive materials, such as the carbon fiber are used in the production of the wind blades.

Using the integrated measuring, control and testing of the final product, the designed shape and structure can be easily approved.

This new concept discussed in this paper is correlated with the further research in MIKROSAM Company for development of an advanced process and motion control used for these or similar complex automated systems.

4. CONCLUSION

For further development of the wind energy, there is a need for new processes that can lead to the decrease of the final price of the energy produced from the wind.

Today's wind blades are projected for a life time of 20 years, with maintenance of the moving parts, and one of the most exposed parts during the working of the wind turbine is the blade. Therefore, the production of highly reliable wind blades is one of the priorities in the wind turbine production.

This automation will reduce the material waste, increase the uniformity of the structure of the wind blade and it will increase the repeatability. Minimizing the need for skilled manual labor, and manual processes will also maximize the production.

The introduction of the mobile hybrid automated equipment in the production of the complex parts from composite materials, such as wind blades, will reduce the overall price of the production, transportation and maintenance of the wind turbine, and will increase the working life up to 30 years, which is today's goal in the industry.

The first cost simulations lead us to a decrease in the price of the wind blade up to 20% by using this concept, compared to the classical methods.

MIKROSAM Company is in negotiations for production of the prototype of such equipment, which should confirm the results of the simulations.

5. REFERENCES

Burton, T.; Sharpe, D.; Jenkins, N. & Bossanyi, E. (2001). Wind Energy - Handbook, John and Wiley & Sons Ltd, ISBN 0471489972, Great Britain

CORONET (2004). Thermoplastic Composites in Europe to 2025. *Available from*: www.coronet.eu.com. *Accessed:* 2008-05-13

E-composites Inc. (2005). *Growth Opportunity in Wind Energy Market 2004-2010: Materials, market and Technologies*, E-composites Inc., USA

Harrison, R.; Hau, E. & Snel, H. (2000). *Large Wind Turbines – Design and Economics,* John and Wiley & Sons Ltd, ISBN 0471494569, Great Britain

National Research Council (1991). *Assessment of Research Needs for Wind Turbine Rotor Materials Technology*, National Academy Press, ISBN 0309583187, Washington D.C., USA

Annals of DAAAM for 2008 & Proceedings of the 19th International DAAAM Symposium, ISSN 1726-9679
Editor B. Katalinic, Published by DAAAM International, Vienna, Austria 2008
Make Harmony Between Technology and Nature, and Your Mind will Fly Free as a Bird

TRANSPARENT OBJECTS DETECTION AND COUNTING

KALOVA, I[lona] & HORAK, K[arel]

Abstract: *The aim of this project is an automatic determination of the number of glass cut stones in production batch. One possibility is using of precision weights but the accuracy is deficient today. The other way is try to use a visual system for this purpose. The stones are sprinkled on measuring area and this is watched by camera. The main problem is detection of transparent objects (clear stones) and overlaying of stones. These problems were solved by suitable scene preparation, choosing of light source and lighting methods and due to exact way of image processing and evaluation.*

Key words: *camera, lighting, transparency, stone, algorithm.*

1. INTRODUCTION

Glass stones have dimensions from 1 to 4 mm and they will be randomly placed on area 300x300 mm (contact and overlaying of stones are considered too). Stones are from different materials and colours - clear glass, coloured transparent materials, opaque stones. Objects can have reflective surface. One batch include from 600 to 3000 stones (according to the size of stones) with the same colour and size (size difference max 0.2 mm). Stones are in three different phases of cutting (see below).

In this phase of problem solution the experimental workplace was put together. Several lights (implementation, wavelength) and cameras (colour with resolution 3Mpix, BW with resolution 400kpix) were tested. First algorithms were designed. Accomplished experiments and algorithms are described in next chapters.

2. ACCOMPLISHED EXPERIMENTS

2.1 Experiments with light source

Provided samples of stones (different size, colour, materials and phase of processing, (Hauβecker, H. & Geiβler, P., 1999)) were used for examination of optical attributes mainly. The biggest problem is how to illuminate clear stones. These experiments led to the design of the most suitable light source. Several obtained images with different sources of light and different methods of lighting (bulb, fluorescent lamp LED backlight, LED light ring, etc.) are presented on Fig.1.

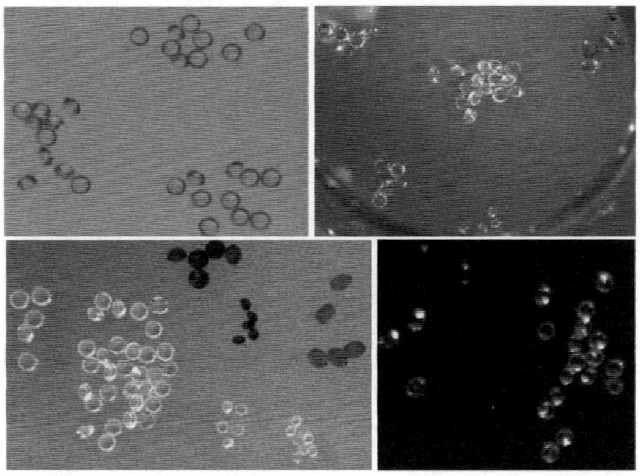

Fig. 1. Experiments with light sources.

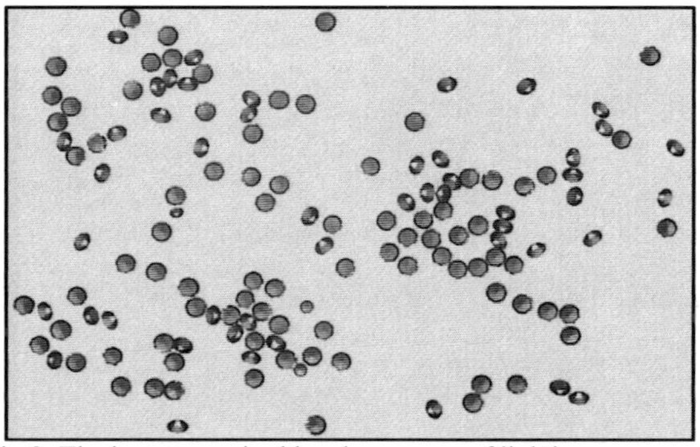

Fig.2. The image acquired by chosen way of lighting.

The best results were obtained by using infrared LED diodes blocks and transmission method of lighting. Stones are placed on the dull groundglass, diodes blocs are under the ground and whole scene is monitored by camera from above. The infrared light is partial absorbed in glass stones and this is the way that transparent even clear stones are darker on the lighter background.

2.2 Image analysis for three different processing phases

The cut glass stones are used for jewellery production. It is necessary to do sums in three different phases of stones treatment. The stones differ in shape and optical attributes (see Fig.3). It was necessary scarify the chosen lighting method from the point of view of variance of phases and also colour and transparency of stones. It is important to obtain similar images as far as possible. As you can see on the images below, designed way of lighting fulfils these conditions.

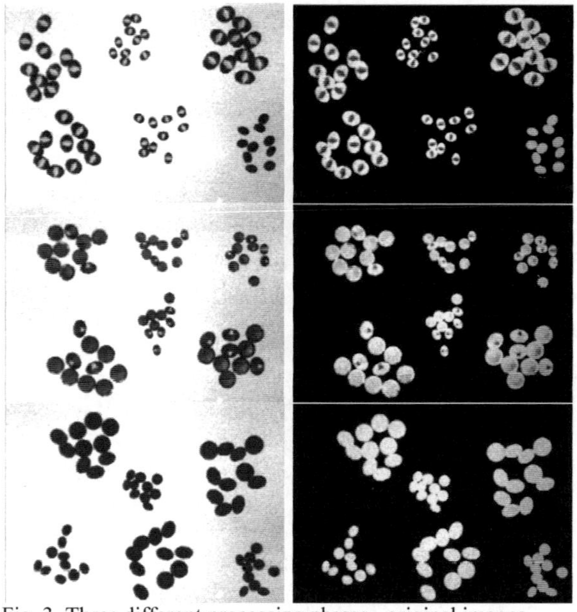

Fig. 3. Three different processing phases - original images (left), partial preprocessing (right)

2.3 Analysis of stones placement

The successful adding depends on proper stones spacing on measured area. If the stones are cumulated on small area too much, it is impossible to distinguish separate stones by algorithm (even by human, see Fig.4 left). Therefore it is required to provide stones severance. It can be done by vibration of groundglass. The same scene after succussion is on Fig.4 right).

Fig.4. Improper layout (left), preferable layout (right).

3. ALGORITHMS, IMAGE PROCESSING

The biggest problems for counting bring clear stones from first phase of processing (Fig.3 top). There are composed from two octahedral pyramids, the floor projection is almost ellipse. Due to their shape, the transmigrant light reflects and stones look like two sickles on the images (Fig.5). The wrong image preprocessing and segmentation can cause the separation of these two parts and inclusion two stones instead one.

Beyond stones create clusters still. Although the groundglass was shaken, several stones can be laying close together even overlap still. These clusters have to be analysed more thoroughly.

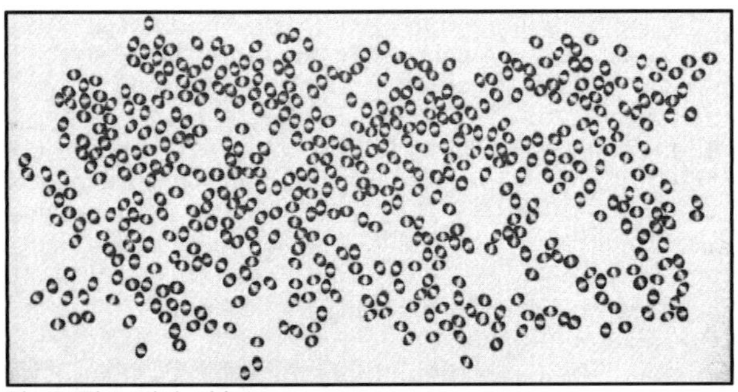

Fig.5. The original image

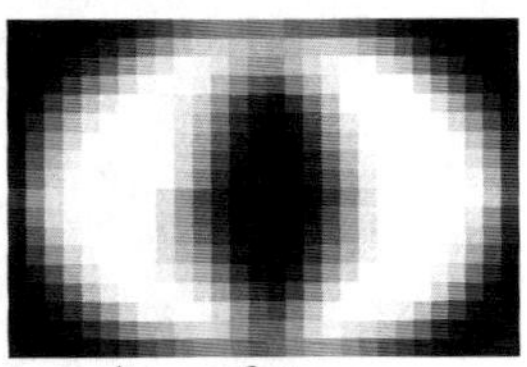

Fig.6. The average specimen of stone.

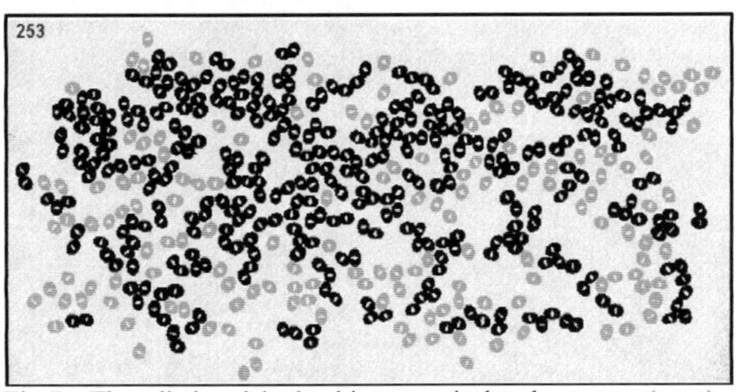

Fig.7. The distinguished objects – isolated stones (gray), regions consist of more stones (black).

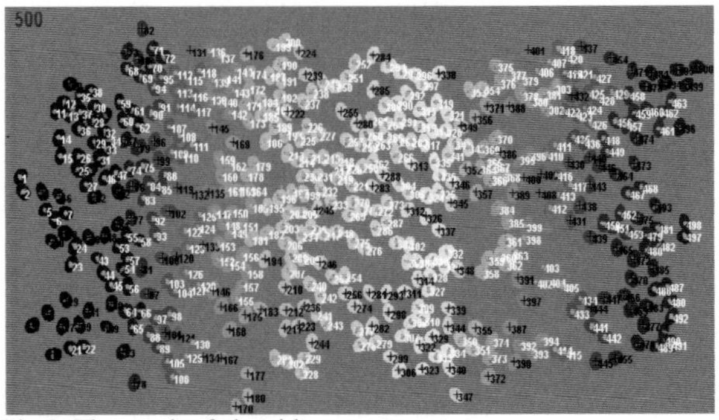

Fig.8. The result of algorithm.

After original image loading (Fig.5), there are executed an operations leading to the image adjustment and preprocessing - contrast increasing, image filtering, thresholding, negation etc.; (Russ, J.C., 1994; Vernon, D., 1991). The individual connected objects are labelled and important properties of regions are measured (mainly area, position and orientation). The proportion of one stone is determinate according to the most frequent regions area. Then the average specimen is found (Fig.6).

The knowledge of average specimen area enables to count individual stones (regions with approximately the same area, gray on Fig.7). The near regions with roughly half area are put together and other small objects are excluded. The normalized two-dimensional cross-correlation (Lewis, 1998; MathWorks, 2003) with calculated specimen is applied to more complicated regions (bigger area, black on Fig.7).

The distinguished separate stones and total number of stones are results of described algorithm (see Fig.8).

4. CONCLUSION

From realized experiments (described above) results that the problem of automatic stones counting by visual system is solvable. Even stones from clear glass are sufficiently contrastive on taken images. Chosen lighting method is suitable for all stones variation. The described algorithm (chapter 3) is naturally a little bit complicated in reality but the main idea was mentioned. Error rate is lower than 0.5 % depends on stones spacing.

5. ACKNOWLEDGEMENTS

This work was supported by the Ministry of Education of the Czech Republic under projects 1M0567 and MSM0021630529. This support is very gratefully acknowledged.

6. REFERENCES

Hauβecker, H. & Geiβler, P. (1999). *Handbook of Computer Vision and Applications*, Academic press, ISBN 0-12-379771-3, San Diego

Lewis, J.P. (1998). Fast Normalized Cross-Correlation, *Available from:* http://www.idiom.com/~zilla/Papers/nvisionInterface/nip.html *Accessed:* 1998-02-19

MathWorks (2003). Registering an Image Using Normalized Cross-Correlation. *Available from:* http://www.mathworks.com/products/demos/image/cross_correlation/imreg.html *Accessed:* 2003-03-03

Russ, J.C. (1994). *The image processing handbook*, CRC Press, ISBN 0-8493-2516-1, Boca Raton

Vernon, D. (1991). *Machine Vision: Automated Visual Inspection and Robot Vision*, Prentice Hall International (UK) Ltd, ISBN 0-13-543398-3, New York

Annals of DAAAM for 2008 & Proceedings of the 19th International DAAAM Symposium, ISSN 1726-9679
Editor B. Katalinic, Published by DAAAM International, Vienna, Austria 2008
Make Harmony Between Technology and Nature, and Your Mind will Fly Free as a Bird

FEATURES OF DESIGN OF A PLUNGER TRANSMISSION

KARAKULOV, M[aksim]

Abstract: *The construction of a new kind of mechanisms for transformation of movement - plunger harmonic transmissions and principle of its operation is considered. The analysis of areas of such mechanisms usage is carried out on the base of which the task are identified, for effective solving of them the plunger reducers can be applied.*
Key words*: planetary reducer, speed mechanism, plunger transmission.*

1. INTRODUCTION

Cam mechanisms have a wide circulation in technical equipment. With their help practically any law of movement of a pusher that allows solving the majority of problems of mechanization and automation of the control by drives in many areas of the industry can be received. But in cam mechanisms one pair of interacting surfaces is used. While for reproduction of uninterrupted one-way movement is necessary to have sequentially interacting pairs of the conjugate surfaces. Such overlapping of functions allows receiving a new kind of mechanisms for transformation of movement – the plunger transmission which is a mechanism combining attributes and structure of two types of cam mechanisms: the central cam mechanism and gearing. Such combination allows expanding considerable area of possible application of the received mechanisms.

All mechanisms have distinctions of a design; therefore have various areas of effective application. Therefore revealing of such areas is the basic problem is solved by the researcher of a new kind of mechanisms.

The area of application of a mechanism is determined by its qualitative parameters. Therefore this problem has the difficult decision. However, qualitative parameters of a mechanism determine on geometry of gearing. Therefore the solution of a problem is obliged to be made with carrying out of the analysis of gearing.

2. ANALOGUES

In Krainev's classification (Krainev, 1987) the closest analogue to the considered mechanisms can be the mechanism of cam - rack transmission which is named also harmonic rack transmission. In the work of this mechanism the principle of harmonic transformation of movement is also used, and elements of gearing represent separate parts.

The basic difference of plunger transmission from the considered analogue is a replacement rack for a wheel of internal gearing that causes changes in regarding kinematical interaction and geometry of gearing, allowing receiving the transmission. The transmission can be attributed to the tooth-cam mechanism the special case of which is considered to be plunger transmission.

Scientists also pay attention to revealing of areas of effective application of tooth-cam mechanisms.

The development this direction began with the research of frictional planetary mechanisms with intermediate solid of revolution. Because of increase of loadings which are necessary for transferring with the help of these mechanisms, such devices were supplemented with gearing. As a result of the development of this direction some authors suggest to use a circular profile of elements of such gearing, replacing a pusher for a solid of revolution: a ball or a roller that allows to receive, depending on the form of the connected surfaces of internal gearing, (hypo-) epicycloidal (Dumoulin, 1989) or (hypo-) epitrochoidal (Riffel, 1981) engagement of harmonic tooth-cam transmission. In such a transmission cam, moving with a speed V_1, forces the pushers to execute as balls, the parts united in one group with a separator, to make motion. The balls, entering in engagement with a conducted part of the mechanism - a wheel of internal gearing, force it to make movement with a speed V_2. Thus formed, the geometry allows receiving a low dimension of clearance between elements since as a result of such representation practically exact engagement possessing low losses of capacity due to application of solid of revolution is formed. But it has also a number of lacks connected with a low level of manufacturability of production and high level of contact stress in gearing. The essential lack of such transmissions especially appearing in transmissions with solid of revolution and eccentrics of big sizes is an unbalance of weights and connected with it noise, palpation and the increased deterioration. It is necessary to note that the full disposal of sliding, i.e. its replacement for the revolution in transmissions of such a type, because of a difference of ways, passing by a solid of revolution *g* on a surface of the generator of waves and a surface of a conducted link, is a challenge the decision of which leads to a low level of manufacturability of a profile of elements of internal gearing of an output link.

The similar direction of researches is a research of harmonic transmissions with segment gears. Their main difference is the division of a deforming gear (DG) into the segments containing from one up to three teeth (Schmitt, 1994). With such change of a design, most likely, authors are going to increase a resource of transmission. But it has an essential lack in a low level of manufacturability of production of segments of a gear.

3. DESIGN OF A PLUNGER TRANSMISSION

Theoretical and experimental researches of plunger transmissions (Jastrebov, 1962) were started in 1960 and are still being researched our days.

The reducer equipped with a plunger transmission (fig. 1), contains gear 1, rigidly fixed on a ring-separator 2.

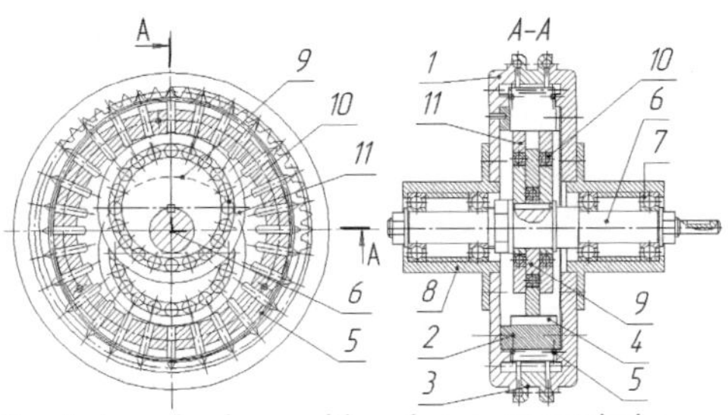

Fig. 1. The design of reducer with a plunger transmission

The mobile gear 3 is interfaced with plungers 4 and gear 1. Plungers 4 are located in regular intervals on a circle and are fixed by a ring-separator 2, fixed on gear 1 from angular moving from each other, and from axial moving by the covers of the case. The role of which is carried out with the gear 1. In the grooves of plungers 4 inextensible flexible elements (IFE) 5 with identical cross-section are placed. In the central hole of the gear 1 the shaft 6 is placed. It is fixed from radial and axial moving by bearings 7, which are placed in the cups 8 rigidly fixed on gear 1. On the eccentric disks 9 which are placed and fixed from axial moving on the shaft 6, bearings 10 are placed with the disks 11 fixed on them.

The plunger reducer works as follows. The rotation of the shaft 6 together with the eccentric disks 9 fixed on it which influence by bearings 10 on the disks 11 rotating around their own axes and simultaneously around axis of the shaft 6, cause deformation of the inextensible flexible element 5, and the plunger 4 forces to make motion. The plunger, being in conjunction with disks 11, comes into contact with a mobile gear 3 and gear 1, creating two opposite zones of gearing located in one plane, and a mobile gear 3 starts to make rotary movement around the axis of the shaft 6. Plungers 4, which are located in a free zone of engagement, are deduced from contact with a tooth of a gear by force of deformation of inextensible flexible elements 5.

4. THE COMPARISON OF PLUNGER TRANSMISSION WITH ANALOGUES

The basic criteria of compare of transmissions are the efficiency, mass and inertial characteristics. Except for the specified criteria, allowing to lead a quantitative estimation of a degree of quality of transmission, the area of its effective application can be determined and with the help of comparison of the quality indicators determined by features of its design.

One of features of a design of plunger transmissions is absence of a plenty of the details making movement in the viscous environment of an oil case of a reducer.

It is especially important at work of transmission in conditions of the low temperatures leading to substantial growth of viscosity of oil. The low level of power loss for overcoming of viscosity of oil allows using such transmissions to a wide temperature range without application of special expensive oils.

Application in plunger transmission of a harmonic principle of transformation of movement allows receiving a new kind of harmonic mechanisms. Due to features of kinematic interaction in such transmission there are no the parts making planetary movement in its classical understanding.

It enables to apply in their design not only mechanical wavegenerators, but also hydraulic and gaseous generators (Karakulov et. al., 2006). The mechanisms received thus are engines with a harmonic principle of deformation IFE, instead of transmissions (Ivanov, 1981). It considerably expands area of an effective using of mechanisms of such type.

In fig. 2,*a* diagrams are plotted, its allowing to lead comparison of plunger transmissions with analogues.

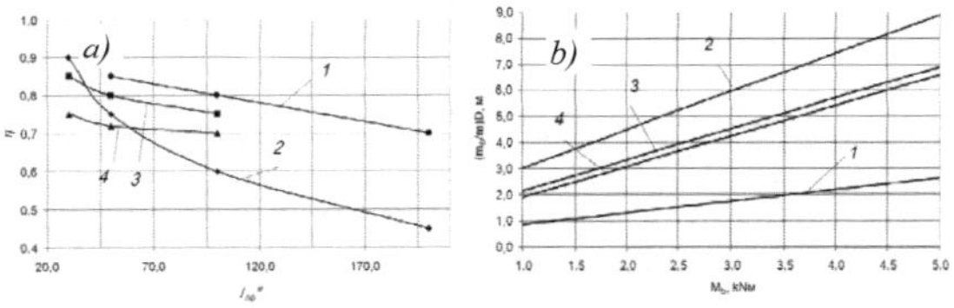

Fig. 2. The compare of transmissions:
1 – harmonic with DG, *2* – classical planetary, *3* – plunger transmission with UDE, *4* – plunger transmission with IFE

For comparison is used value of efficiency η to its dependences on the transmission ratio of reducer (from the generator of waves *h* to a gear *b* at a motionless separator *e*). In fig. 2,*b* diagrams are plotted, which allowing to compare transmissions with help by inertial characteristics in their dependence on the torque on output shaft M_b. As a measure inertia expression $(m_B/m)D$, where m_B is the total weight of rotating parts of transmission, *m* is a weight of all details of transmission, *D* is the maximal diametrical dimension of transmission in this case was applied. For comparison the results are published in works (Krainev, 1987; Ivanov, 1981) were used.

The simultaneous analysis fig. 2,*a* and fig. 2,*b*, would allow to draw a conclusion that efficiency of plunger transmissions with undeformable element (UDE) is a higher, than at the similar transmissions with IFE, but transmission with UDE have higher inertia. Because (at equal transmission ratio) the design of transmissions with UDE will consist of two lines of plungers that makes heavier a rotor and makes its work of more inertial (in comparison with transmission with IFE, which allowing to use one line of elements of gearing). It is necessary to mark, that efficiency and inertial characteristics of harmonic transmissions with DG are out of competition with other considered transmissions, but it have low durability of DG as basic detail of such transmissions. It considerably narrows area of their application.

The fig. 2,*a* and fig. 2,*b*, can serve only for a tentative estimation of transmissions. And the final choice of the scheme should be accompanied by the careful analysis industrial and the working costs connected to application of compared transmissions.

5. CONCLUSION

From the resulted data follows, those due to small inertia and rather low efficiency plunger transmissions are expedient for applying in drives with rather easy and frequently reverse an operating mode, with low frequencies of rotation of shaft and the transmission ratio belonging to an interval from 10 up to 60 rpm, and in drives working at low temperatures. At long inclusion or a permanent work under loading (especially in case of use mechanical wavegenerator) in high-power drives there can be a necessity for installation of the devices providing heat removal.

REFERENCES

Dumoulin, J. (1989). *Method for making an epicyclic speed reducer*, United States Patent №4838741, 06.1989

Ivanov, M. (1981). *Harmonic gearing*, Vyshaja schkola, Moscow

Jastrebov, V. (1962). Theoretical research of plunger transmission, *Izvestia vuzov. Mashinostroienie*, №8'1962, 27-36

Karakulov, M.; Popkov, E. et. al. (2006). *The plunger gaseous drive*, Russian Federation Patent №2278979, 11.2004

Krainev, A. (1987). *The directory about mechanisms*, Mashinostroienie, Moscow

Riffel, H.; Black, T. (1981). *Pancake planetary drive*, United States Patent №4282777, 08.1981

Schmitt, H. (1994). *Epicyclic roller gear*, Patent WO 94/12812 (WIPO), 06.1994

Annals of DAAAM for 2008 & Proceedings of the 19th International DAAAM Symposium, ISSN 1726-9679
Editor B. Katalinic, Published by DAAAM International, Vienna, Austria 2008
Make Harmony Between Technology and Nature, and Your Mind will Fly Free as a Bird

AUGMENTATION AND INTEGRATION OF STANDARDIZED MANAGEMENT SYSTEMS

KARAPETROVIC, S[tanislav]

Abstract: *Two aspects of the development and application of management systems (MSs) based on international standards are discussed, namely their augmentation and integration. Augmentation refers to the use of specific MS standards (e.g., ISO 10002) to strengthen and improve a single component of a minimal-requirement MS (ISO 9001 quality MS). Integration denotes the often-applied practice of the unification of minimal-requirement or augmented MSs in an organization. A brief theoretical and empirical discussion on the standardization of MSs is provided first. An illustration of several augmentative standards follows, particularly focusing on their ability to foster augmentative and traditional integration. Some relevant empirical results and a framework featured in the ISO integration handbook are given.*

Key words: *Quality Management, Integrated Management Systems, ISO 9000, ISO 10000, ISO 14000*

1. INTRODUCTION

Although the year leading to this DAAAM symposium has been full of various new developments in the world of standardized management systems (MSs), the notions of their augmentation and integration seem to have finally come to the fore. On the augmentation side, a number of new standards that can be applied to create stand-alone MSs or to enhance a particular subsystem of the overarching MS (see, e.g., Karapetrovic, 2008A, for additional features of these standards), were either published or are being developed. For instance, in November 2007, the International Organization for Standardization (ISO) has completed the "complaint system for customer satisfaction" (CSCS) series of quality management system (QMS) standards, with the publication of ISO 10001 on codes and ISO 10003 on dispute resolution to complement the existing ISO 10002 on complaint handling (see, e.g., Dee et al., 2004, for more detail on the CSCS series). The systems based on these and other "Augmentative Standards" (AUGSs), e.g., ISO 19011 on MS auditing and ISO 10012 on measurement MSs, can be effectively integrated, both amongst themselves ("augmentative integration") and with other standardized MSs ("traditional integration"). This integration feature of AUGSs holds a particularly promising future in the research and practice of standardized MSs (Karapetrovic, 2008A).

Further increasing the potential scope for the integration of standardized MSs in organizations, ISO also announced the development of two brand-new management system standards (MSSs) this year, namely the ones for road safety and energy management. However, this "avalanche" of MSSs has been "channelled" with the July 2008 publication of the ISO book on their integrated use (ISO, 2008). This book was very much needed, especially since the past research seemed to have largely focused on the models of integrated management systems (IMSs), rather than on the integration methodologies, which were particularly required in practice (Karapetrovic & Jonker, 2003).

This paper is a much-streamlined and updated version of Karapetrovic, 2008B. It continues with a discussion on the development and application of MSSs, in other words the standardization of MSs. An overview of four AUGSs, emphasizing their augmenting features and integrative context, follows. Subsequently, an illustration of some empirical results and the new ISO handbook on integration forms the analysis of integration methodology issues. The conclusion further addresses the importance of augmentation and integration in the research and practice of standardized MSs.

2. STANDARDIZATION

Apart from the AUGSs, two other groups of MSSs are being written by ISO and other standardization bodies, and subsequently implemented in various organizations (Karapetrovic et al., 2006). "Assimilating Standards" (ASSIMSSs) contain the minimal-requirement frameworks that organizations apply in order to establish general function-specific or stakeholder-focused MSs. Examples of ASSIMSs include ISO 14001 and CSA Z1000 for environmental and safety management, respectively, as well as ISO 27001 and ISO 28000 for information and supply chain security, respectively. In addition, new "Ascending Standards" (ASCSs), such as ISO/TS 16949 for the automotive industry and ISO 22000 for food safety, have joined the generic ISO 9004 and ISO 14004 and numerous Business Excellence Models (BEMs), to form the third group by providing additional criteria for all MS components in order to improve the overall performance level of an organization.

It is also interesting to illustrate some related empirical results, which point out the need for augmentation and integration in order to capitalize on the application of these MSSs and the respective MSs. The results come from a survey of 529 ISO 9001-registered companies in the Spanish region of Catalonia, with 353 of those having the ISO 9001 certificate only and exhibiting a 20% response rate, and the remaining 176 organizations having registered to at least ISO 9001 and ISO 14001 and responding with a 33% rate (see Karapetrovic et al., 2006, for full details on the survey). When asked to rank their preference for the future application of either AUGSs, ASSIMSs, BEMs or an alternative not to add any new standards or models, most companies from the ISO 9001-only group placed the first priority on AUGSs (31%), followed by ASSIMSs (27%) and BEMs (24%), with only 18% of the companies indicating the first preference for the "add nothing" option. Not surprisingly, companies with multiple registrations preferred ASSIMSs at 43%, BEMs at 22%, then AUGSs at 20%, and finally 15% ranked the "add nothing" option first. Looking at such preferences for the standards overall, as well as for AUGSs and ASSIMSs specifically, it is clear that augmentation and integration alike will be in demand.

3. AUGMENTATION

Since the standards from the CSCS series and ISO 19011 provide a representative sample of both the established and upcoming AUGSs, these four documents will be briefly illustrated here. The AUGSs that have been in existence for a number of years now seem to have been used successfully worldwide. For instance, Clark, 2007, reports that the Financial Ombudsman Service in the United Kingdom applies ISO 10002 as "*the framework for handling over 100,000 complaints per year*", while Karapetrovic et al., 2006, found that about a third of ISO 9001-registered companies in Catalonia conduct their internal audits according to ISO 19011. Karapetrovic et al., 2006, survey results indicate that a similar trend may continue for the recently-developed AUGSs, e.g., ISO 10001 / 10003.

The shortest standard in the CSCS family, ISO 10001 covers the processes for creating and using product-related promises made to an organization's customers (Dee et al., 2004). Unlike ISO 10001, which facilitates the implementation of a simple framework for code establishment, ISO 10002, ISO 10003 and ISO 19011 provide guidance for the setup of a complete complaint-handling, dispute-resolution, and auditing MS, respectively, including all the necessary MS elements, such as MS performance evaluation and review. The CSCS standards are characterized by identical structures (Dee et al., 2004), which are fully compatible with ISO 19011.

Several of the AUGSs' characteristics make the related augmentative systems very flexible for different integration possibilities. Unlike the "traditional" integration, which is by definition cross-functional and must combine at least two overall MSs, integration of augmentative systems can be accomplished in both the horizontal and the vertical direction (Karapetrovic, 2008A). For example, an augmentative system established in accordance with any of the CSCS standards can be vertically integrated within an overall CSCS (Dee et al., 2004), then at the next level up within, for instance, an ISO 9001-based or with an ISO 28000-compliant MS, and also within the overarching IMS. This integration can be done regardless of which underlying model is used to provide the framework for the higher-level function-specific MS or IMS (Karapetrovic and Jonker, 2003). Using the structure of ISO 10001 as an example, Karapetrovic, 2008B, illustrates how the framework underlying the CSCS standards can be integrated into the "process model" of ISO 9001 and similar MSSs, the "PDCA approach" of ISO 28000 and analogously-structured MSSs, as well as any standardized IMS model.

However, the real advantage of AUGSs rests with the horizontal integration, where systems are used to augment each other and are combined more naturally (Karapetrovic, 2008A). For example, an ISO 10001-based system can be established by building on the MS components from other AUGSs, e.g., ISO 10002 and its sections on "*commitment*" (5.1), "*policy*" (5.2), "*responsibility and authority*" (5.3), and "*management review*" (8.6), as well as ISO 19011 to setup internal auditing. The same can be said for complaint-handling and dispute-resolution systems, which can be integrated themselves and use ISO 19011 for the auditing component of this integrated complaint-resolution system. In turn, the auditing system in an organization can use ISO 10001 for setting up codes related to its performance, as well as ISO 10002 and ISO 10003 for dealing with any complaints or feedback on auditing.

4. INTEGRATION

Due to the long existence of the various ASSIMSs, such as ISO 9001 (21 years) and ISO 14001 (12 years), the integration of the corresponding minimal-requirement systems has been widely practiced. For instance, the available research (Douglas & Glen, 2000; Karapetrovic et al., 2006) shows that a significant majority of companies with two or more ASSIMSs certificates integrate their standardized MSs. The survey illustrated in Karapetrovic et al., 2006, specifically focused on the integration methodology issues in those organizations, such as the resources and techniques used in the integration, as well as the auditing of their IMSs. The results include the prevalence of the process mapping for MSs and the analysis of the "common elements" of MSSs as the integration methods applied by 93% and 92% of the respondents, respectively (Karapetrovic et al., 2006).

Also addressing a number of integration methodology issues is the brand-new ISO handbook on the "Integrated Use of MSSs" (ISO, 2008). By providing a flexible eleven-step framework for the integration of the requirements of both the generic ASSIMSs and the industry sector-focused ASCSs into the organizational MS, the handbook clearly differentiates the integration of standardized MSs from the integration of MSSs (see, e.g., Karapetrovic and Jonker, 2003, for a further discussion on this issue). Guiding an organization through the integrative implementation of one or more MSSs, the handbook suggests the *understanding* and tailored *structuring* of both the *MS* (in Chapter 1 and Subsection 3.4.1, respectively) and the *MSS(s)* being implemented (in Chapter 2 and Subsection 3.4.2, respectively), followed by the *mapping* (3.4.3) and *incorporation* (Section 3.5) *of the MSS(s) requirements into the MS*. The subsequent steps of *maintenance and improvement* (3.6) and *applying the lessons learned* (3.7) can loop back to any one of the first three steps, namely *leading* (3.1), *determining the scope* (3.2) *and planning* (3.3) *the integration.*

5. CONCLUSION

At least six standards originally or currently belonging to the ISO 10000 series represent examples of a new "species" of augmentative standards and have an outstanding potential for application and integration within organizational MSs (e.g., see Karapetrovic, 2008A). Although the benefits of the development of new MSSs are sometimes questioned, the integration of standardized MSs seems to be set for growth as an area of both practice and research. One important related development is the recent publication of the ISO book "The Integrated Use of Management System Standards" (ISO, 2008).

6. REFERENCES

Clark, E. (2007). Britain's Ombudsman Service Follows ISO 10002 for Complaints Handling, *ISO Mgmt. Sys.*, Jul.-Aug.

Dee, B., Karapetrovic, S., Webb, K. (2004). As Easy As 10001, 2, 3, *Quality Progress*, Vol. 36, No. 6, pp. 41-48

Douglas, A. & Glen, D. (2000). Integrated Management Systems in Small and Medium Enterprises, *Total Quality Management*, Vol. 11, No. 4/5/6, pp. S686-S690

ISO (2008). *The Integrated Use of Management System Standards*, ISO, ISBN 978-92-67-10473-7, Geneva

Karapetrovic, S. (2008A). Integrative Augmentation of Standardized Management Systems, *International Journal of Quality Science*, Vol. 2, No. 1, pp. 15-22

Karapetrovic, S. (2008B). IMS: Focus on ISO 10000 Augmentative Standards, *Total Quality Management and Excellence*, Vol. 36, No. 1-2, pp. 1-8

Karapetrovic, S.; Casadesus, M. & Heras, I. (2006). *Dynamics and Integration of Standardized Management Systems: An Empirical Study*, Documenta Universitaria, ISBN 84-935231-2-7, Girona

Karapetrovic, S.; Jonker, J. (2003). Integration of Standardized Management Systems: Searching for a Recipe and Ingredients, *Total Qual. Mgmt.*, Vol. 14, No. 4, pp. 451-459

Annals of DAAAM for 2008 & Proceedings of the 19th International DAAAM Symposium, ISSN 1726-9679
Editor B. Katalinic, Published by DAAAM International, Vienna, Austria 2008
Make Harmony Between Technology and Nature, and Your Mind will Fly Free as a Bird

PROCESS ANALYSIS AND RELIABILITY EVALUATION

KARAULOVA, T[atyana]; PREIS, I[rina] & PRIBYTKOVA, M[arina]

Abstract: *Present paper introduces the reliability analysis of prototype production line process. The modelling method was used for process analysis considering complex dependences among various parts of the system. On the base of created IDEF0 model the reliability evaluation of the process was completed by contemporary reliability analysis (RA) techniques such as FTA, FMEA, and RBD. As the reliability analysis has been performed, it was possible to determine major system problems and to assist in implementation of design changes and process improvements.*
Key words: *Production line analysis, FMEA, FTA, RBD.*

1. INTRODUCTION

Reliability is the probability that a system will operate successfully for a specified period of time, under specified conditions, when used for the manner and purpose for which it was intended (MIL-HDBK-338B, 1998).

Reliability is a crucial element to business success; therefore the reliability analysis and its methods are developed as a main tool to facilitate companies to achieve their goals.

For Estonia the sphere of reliability engineering is rather new, however it is successfully developed for decades in US and Europe. For instance, in 1996-2000, the National Technology Agency of Finland (Tekes) financed a Competitive Reliability technology programme with the aim of increasing the international competitiveness of Finnish industry by developing techniques for more convincing control of production systems' reliability. (Kenneth Holmberg, 2001). Present study considers the process of real retractors' production line existing at AS Norma Estonia. A retractor is the main part of the car seatbelt mechanism. The whole process of retractors' assembling is long and consists of many consecutive operations. The layout of the line is illustrated in the Figure 1. All operations of components assembling are manual and all the testing operations are automated. The line is divided into 13 working stations responsible for certain type of process operation. All the stations are jointed into a conveyor.

The aim was to create model of the production process with follow-on analysis of the production line reliability applying various RA methods.

2. PROCESS ANALYSIS

System analysis starts with development of process model using IDEF0 methodology.

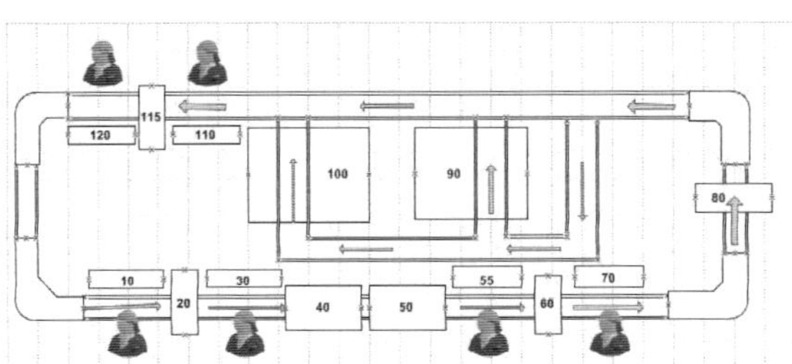

Fig.1 Layout of the retractors' production line

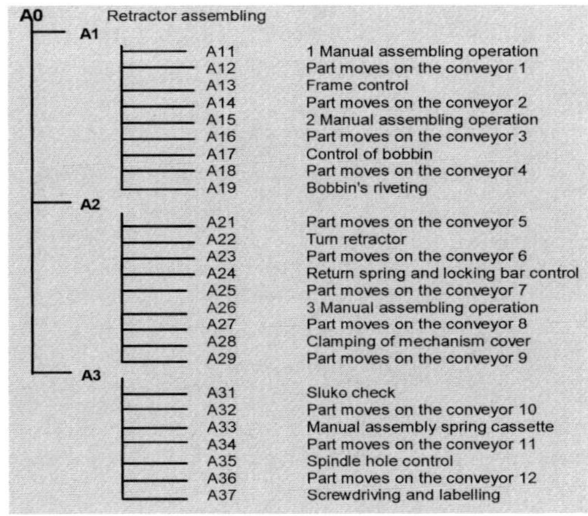

Fig.2 Process description – Node Tree Diagram

This model of every activity then decomposed into more detailed diagrams with further representation in form of Node Tree Diagram (NTD).

Figure 2 shows the conceptual IDEF0 model structure. As the process analysis done structure has been analysed and simulation completed, the reliability analysis is ready to be carried out.

3. RELIABILIRY ANALYSIS

Realistically, it is impossible to avoid all feasible failures of a system or a product on the design stage, so one of the goals of reliability engineering is to recognize the most expected failures and then to identify appropriate actions to mitigate the effects of those failures (Lendvau, 2004). As shown in Figure 3, for the system safety analysis the qualitative and quantitative methods were used. All they are interrelated and help to understand the logical structure of failure modes of a system.

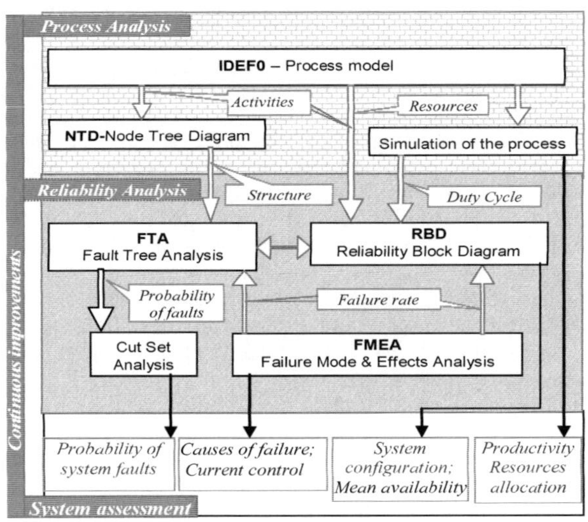

Fig.3 Layout of production line reliability analysis

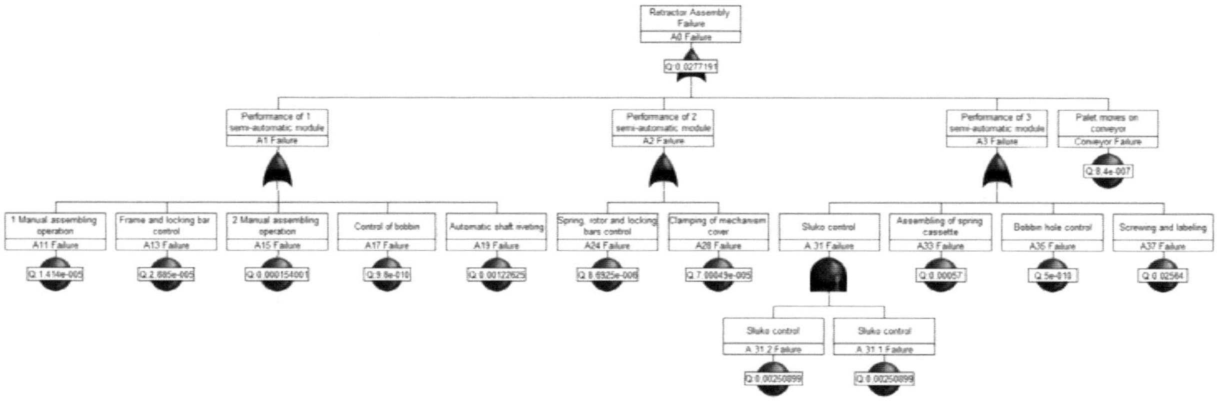

Fig. 4 Fault Tree Analysis of the retractors production line

The reliability, *R(t),* of the component is the probability of a component surviving to a time *t* and is expressed as

$$R(t) = \frac{number\ surviving\ at\ instant\ t}{number\ at\ time\ t = 0} \qquad (1)$$

Failure rate - expected rate of occurrence of failure or the number of failures in a specified time period. This parameter is wide used in reliability analysis.

The probability of system failure, or system unreliability, corresponds to the probability of successful system maintenance, or system maintainability (Salvendy, 2001).

3.1 Failure Mode and Effect Analysis (FMEA)

A typical failure modes and effects analysis incorporates some method to evaluate the risk associated with the potential problems identified through the analysis. The most common used method, Risk Priority Numbers, is described next. The risk priority number provides a qualitative numerical estimate of the design risk. This number is then used to rank order the various concerns and failure modes associated with a given design as previously identified in the FMEA. RPN is defined as the product of three independently assessed factors: Severity (S), Occurrence (O) and Detection (D).

$$RPN = (S) \times (O) \times (D) \qquad (2)$$

3.2 Fault Tree Analysis (FTA)

The FTA technique is a method for block diagramming constituent lower level elements. It determines, in a logical way, which failure modes at one level produce critical failures at a higher level in the system (MIL-HDBK-338B, 1998). The technique is useful in safety analysis where the discipline of block diagramming helps prevent an oversight in the basic FMEA.

FMEA and FTA have three main differences: boundaries of the analysis, direction of analysis, and presentation of the analysis process and results. FMEA deals with single point failures, is built bottom-up, and is presented as a rule in the form of tables. FTA analyzes combinations of failures, is built top-down, and is visually presented as a logic diagram. By taking into account combinations of failures, FTA avoids the obvious shortcomings of FMEA. (Bluvband et al., 2005).

In figure 4 is shown FTA of the production line by structure introduced in figure 3.

One of the most powerful tools of the results analysis in FTA is a Cut Set. A Cut Set is any group of fault tree initiators which, if all occur, will cause the TOP event to occur. In addition a minimal cut set definition exists. A minimal cut set is a least group of fault tree initiators which, if all occur, will cause the TOP event to occur (Clemens, 2002).

3.3 Reliability Block Diagram (RBD)

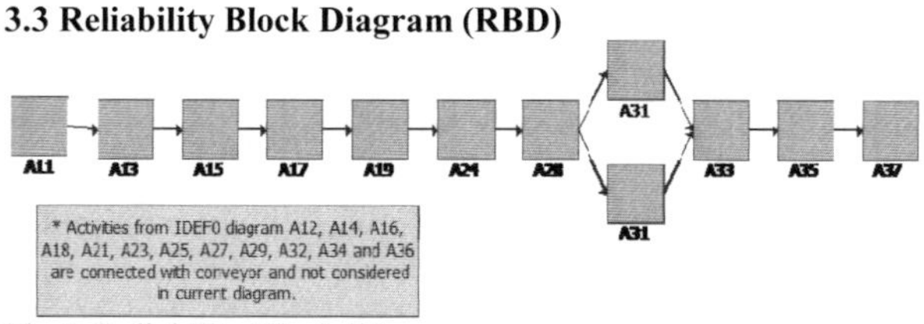

Fig.5. Reliability Block Diagram

RBD performs the system reliability analyses using block diagrams to show network relationships. The structure of the reliability block diagram defines the logical interaction of failures within a system that are required to sustain system operation. Each module from IDEF0 model (A11, A12,...A37) is defined by a separate block Figure 5. RBD analysis allows take into consideration duty cycle and failure rate for every block of the system.

Fault trees and reliability block diagrams are both symbolic analytical logic techniques that can be applied to analyze system reliability and related characteristics. Although the symbols and structures of the two diagram types differ, most of the logical constructs in a fault tree diagram (FTD) can also be modelled with a reliability block diagram (RBD).

4. CONCLUSIONS

An integrated modelling method based on system modelling and complemented with reliability evaluation mechanism has the capability to analyse and design manufacturing systems. The tool developed to analyse the production process enables the companies to analyse the processes as a whole and its parts and get efficient prognosis for production process reorganization.

5. REFERENCES

Bluvband, Z.; Polak, R. & Grabov,P. (2005) Bouncing Failure Analysis (BFA): The Unified FTA-FMEA Methodology

Lendvay, M (2004) Dependability Assurance of Industrial Production Processes, *Proceedings: Science in Engineering, Economics and Education,* Budapest

MIL-HDBK-338B (1998) Military Handbook, *Electronic Reliability Design Handbook*

Holmberg, Kenneth, (2001) – Competitive Reliability *1996-2000, Final report, National Technology Agency*

Clemens, P. (2002) -, *System Safety Scrapbook* , 9th Edition Sverdrup Technology, Inc.

Salvendy G., (2001) – *Handbook of Industrial Engineering: Technology and Operation Management*, A Wiley-Interscience. ISBN 0-471-33057-4

Annals of DAAAM for 2008 & Proceedings of the 19th International DAAAM Symposium, ISSN 1726-9679
Editor B. Katalinic, Published by DAAAM International, Vienna, Austria 2008
Make Harmony Between Technology and Nature, and Your Mind will Fly Free as a Bird
*Young Researcher and Scientist Paper / * Supervisor, Mentor*

USER INTERFACE MODEL OF INTERACTIVE EDUCATIONAL SOFTWARE

KARUOVIC, D[ijana] & RADOSAV, D[ragica]*

Abstract: *At the University of Novi Sad a research has been conducted in the field of preschool children needs when using computers and designing the software just for them. It has been noticed that there is no unique technology for making software aimed users of a specific age. Therefore, a lot of software for preschool children has been developed as a part of the subject Projecting the Educational Software and this dissertation is showing an interactive software model for learning colours and shapes. Creating the software that depends on users is the main aim. The mathematical model of the interactive software according to the Fitts' Law has been presented, too.*
Key words: *interface, education, software, interactivity, child*

1. INTRODUCTION

When we want to talk about the interactive software, the first thing to do is to define the meaning of interactivity in education. It is, also, impossible to imagine the educational software without any interactivity. Interactive-communicative aspect could be taken as one of the teaching aspects among other two: social-generational and individual aspect of personality development (Bratanić, 1990). The process of teaching is a creative act, whose purpose has to be fulfilled by developing the user interface of the educational software.
The software in the educational field represents the intellectual technology and it is called the educational computer software (ORS - abbreviation in Serbian for *educational computer software*, OS for *educational software*), which needs program languages and tools and a specific organization in teaching and learning, which are based on logic and pedagogy, (Nadrljanski, 2000).
This problem is not just pedagogically founded, but also cybernetic, psychological, socially, (Radosav, 2005.). Theories and projects in the field of man-computer-interaction are giant and various. Software that are available at the very moment depend on hardware characteristics. Here is the aim to create software that depends on users. Further, a universal model of user interface is being researched – no matter what a research field, age of a user or his/her computer knowledge are.
If we want to be specific about these researches, we need to explore advantages of user interface in the following: improving educational effects of OS; developing cognitive, affective and sense and movement skills.

2. HCI RESEARCH

Computers give children lots of possibilities for learning. *Alliance for Childhood* thinks that it is of a great importance for children to talk to their parents, friends, teachers and other members of society to find their own place in it. Preschool age is the time when children are told the stories, when their imagination is in a constant growth, while the educational software is just a circle of programmed, the same actions on the same place that tell the children what should they do, considered by the members of Alliance for Childhood. This problem is easy to solve if the main goal of educational software aimed at preschool children is to evolve the creativity and imagination at users, and ideal would it be to design such software where the child would be allowed to express itself in a way that is only possible with the software.
The significant research projects in the world connected to the given issue are: ClearBoard, MICOLE (Multimodal collaboration environment for inclusion of visually impaired children) Projekat Pinball, Projekat StarCraft, Projekat Gender Related Game Interface Study, Project LISTEN's Reading Tutor, SearchKids, Animal Blocks, PETS (Personal Electronic Teller of Stories), Classroom of the Future, KidPad.

3. METHODOLOGICAL FRAMEWORK OF RESEARCHING

Our research, which is done at University of Novi Sad (Serbia), should give the answers about the role and effects of user interface in interactive educational software aimed at a child during the preschool period. To conduct the research there has been used software, which is made as the graduate work of students at the Technical faculty "Mihajlo Pupin" in Zrenjanin, University of Novi Sad. In this dissertation the model of user interface, the mathematical model of universal knowledge testing and interactive software model for learning colours and shapes were created. The program of education work for children between 3 and 7 has been organized in Serbian, Hungarian, Slovakian, Romanian, Russian, Croatian and Gipsy at Vojvodina region. The sample for this research was made of 349 interviewees aged 6-7 in Zrenjanin, Novi Sad and Subotica.

4. CREATING THE MATHEMATICAL MODEL OF INTERACTIVE OS

Children are specific users of computers and they can be divide into 3 common groups of typical computer users: the children who do not have computers at home and have not used them before, the children who do have computers at home but they use them a little and the children who do have computers at home and use them often, mostly for playing games. Therefore the performance index has been introduced that will give the picture of user abilities to approach the objects and the time needed to accomplish the determined task successfully.
It has been concluded that the following components influence to knowledge gain: ability of a user to use the software; complexity of the task; time set to solve the task, exactitude while solving the task. All these components can be unified and their correlation can be concluded based on Fitts' Law. The researchers of this project have taken out next mathematical rules of the user interface model in educational software based on (J.P.Hourcade, 2003). Mathematically, the Fitts' Law can be applicable to the testing of the learning effects with the following interpretation:

$$IP = IS * a/t, \qquad (1)$$

where
IP – performance index, ability of a user to approach the object, a – variable that marks the number of approaches to the object,t – time of approaching to the object (or time of solving the task), and IS – complexity index of the task that can be counted behalf of the formula:

$$IS = V/S, \quad (2)$$

where

S – width of an object, V – object movement speed.

When the previous mathematical presentation is applied to test the knowledge in interactive software, the next equality could be done:

$$EU = IP \cdot T \quad (3)$$

where

EU – achieved learning effect, T – percentage of successfulness in solving tasks

When the formulas (1), (2) and (3) are combined, it appears the following:

$$EU = \frac{V}{S} * \frac{a}{t} * T, \quad (4)$$

The effect of learning presented like this is universal and it can be done in software that only has static objects and in software that has dynamic objects. The interpretation of the presented model in the case of software with still objects is: The effect of learning presents the possibility of a user to solve the number of questions given in the test (S) in given time (t) with the number of attempts (a) and with a certain punctuality percentage (T). The interpretation of the presented model in the case of software with moving objects is: The effect of learning presents the possibility of a user to approach the given object with the width (S) and speed (V) a times in given time (t) and with a certain punctuality percentage (T).The resulted effect of learning in software that contains the dynamic component presents the number of correct answers in the test, regurarly ordered elements on the screen and similar tasks in the unit of time.In divers modul realisations for examination of knowledge the index of performance can give the incorrect results which are not the right parameters for user performances because the limits of users reading abilities ocurr, further comprehension abilities, abilities to click a mouse to specific objetcs on the screen. These limits are easy to avoid if: the made software fits the user age – without using the text on the screen when knowing that users cannot read; the software is created to match the cultural signs of users; different groups of index performance results are formed for users with different abilities when using the computer; the test are always made in the same conditions for all the interviewees – the same hardware and software platforms, the same number of interviewees, ...

5. PROJECTING THE USER INTERFACE MODEL OF INTERACTIVE EDUCATIONAL SOFTWARE

In the process of creating the interactive user interface of educational software there have been isolated 5 connected modules (Karuović, 2007).

The module of animated intro - In the introduction part of the software the user performance index is measured and the users get the feedback of their own performance index. This is the first news according to traditional educational software.

The module of the presentation part of educational software content - This module has been made in two forms, which is the next new thing to traditional software: content presentation by using moving objects and content presentation by using non moving objects.

The module of knowledge testing - Knowledge testing in this module is realised in a very new and unique way. It is created to be variable, so that it can be adjustable to the users' needs. Following the performance index that is gained in the introduction part of the software, 15 tasks have been set, which are the same in design and activities for all the users but which are different in moving speed of objects and here we have 3 groups of users (the ones with IP number less than 50, the ones with the IP number between 50 and 100 and the ones with the IP number bigger than 100). The time for testing is limited to 3 minutes. The users are placed in 3 groups, where the speed of moving objects in the first group is 3 times lower than the speed of moving objects in the third group. This kind of knowledge testing is needed when realising the educational computer software, because not all the users are of the same abilities with the computer usage and that can drastically influence their learning effects. The intellectual skills are here not of importance, so we must resort this evaluation of learning effects. This is the most important news while designing the user interface of educational software.

The module of a helping hand - The helping hand part is available at any moment during the work. Before every module starts, the users are given information what is expected from them and how to use the software. The help is taught the users to be led to the cognition and to be guided when using the software.

The module of ending animation -This module contains the information of authors and technical characteristics of software.

6. THE RESULTS OF THE RESEARCH

It has been already mentioned that the specific performance index was based on the users successfulness to catch with the mouse the object that is 70 p wide and 50p/s fast in a period of 30s. After the performance index of interviewees had been examined, the experimental research started. The children used the control software first, which was not realised on recommended user interface model, and its topic was the same as in the experimental software, which is shapes and colours. The results of their successfulness are: Zrenjanin - 91.29, Novi Sad - 92.22, Subotica - 93.18. According to the results we can assume that the percentage of correctness while testing the received knowledge with the control software group was a little bit above 90%. After using the software made in the first phase, the children used the software, which was created after the user interface model. We assume that the percentage of correctness while testing the knowledge in software that has the test, which is applied according to the performances of users, is significantly higher than using the control software group and leads to 98%.The results based on the given performance index and correctness percentage, after testing the knowledge through the module for knowledge examination with the user interface model, told us to approach the examination of learning effects.

6. CONCLUSION

Studies, like this dissertation, are of big help to develop qualitative and more usable software for all age groups so that the computer can be taken as a basic educational tool. What we all need in our further work is to expand the researches to even bigger circle of users with the material and technical support of competent ministry and other institutions in the social community.

7. REFERENCES

Bratanić M. (1990). *Micro pedagogy: Interactive-Communicative Aspect of Education*, Školska knjiga, Zagreb

J.P.Hourcade (2003). *User Interface Technologies and Guidelines to Support Children's Creativity, Collaboration, and Learning, Dissertation submitted to the Faculty of the Graduate School of the University of Maryland*

Karuović D.,Radosav D. (2007) *User interface model and guidelines to support children's learning by the interactive educational software*, 30th International Convention MIPRO 2007, May 21 –25, 2007 - Opatija, Croatia

Nadrljanski Đ. (2000). *Educational software - hypermedia systems,* Technical faculty "Mihajlo Pupin", Zrenjanin

Radosav D. (2005). *Educational software and authorial systems,* Technical faculty "Mihajlo Pupin", Zrenjanin

Annals of DAAAM for 2008 & Proceedings of the 19th International DAAAM Symposium, ISSN 1726-9679
Editor B. Katalinic, Published by DAAAM International, Vienna, Austria 2008
Make Harmony Between Technology and Nature, and Your Mind will Fly Free as a Bird

INTEGRATING TAGUCHI PRINCIPLE WITH NEURAL NETWORK TO MINIMIZE INGATE VELOCITY FOR ALUMINIUM ALLOYS

KASALA, J[ozef] & MASIAR, H[arold]

Abstract: *Used numerical simulators based on FDM method provide powerful means of analyzing various phenomena occurring during the casting process to predict better final design, but still not the optimum design. The filling process has a large number of parameters that may affect the quality of the castings. Some of these parameters are controllable while others are noise factors. The purpose of this paper is to demonstrate how the application of computational fluid dynamics (CFD) modelling can be used to develop an effective optimization process for runner system design. The present paper illustrates a novel concept of integrating the Taguchi principle with neural network system for prediction of ingate velocity during the mould filling operations to minimize ingate velocity in vortex runner system.*

Key words: *aluminium castings, Taguchi method, neural networks, runner system design, vortex gate*

1. INTRODUCTION

Computers are beginning to have some capability of optimizing a filling system design. In most cases, it helps in decision making, where industrial expertise is not sufficient. However, simulation software is often inefficient, especially in case where a large number of parameters needs to be examined. The resulting large number of simulation runs coupled with lengthy execution times per run (on the order of hours or even days) may render such investigations totally impractical.

The runner system design, named a vortex-gate, has been explored for aluminium gravity casting to control flow of the liquid metal below the critical velocity. In a study of a vortex gate concept to find important dimension of the vortex runner system and minimize the outlet velocity, the integrating Taguchi principle with neural networks system has been developed.

2. BACKGROUND AND CURRENT STATUS

Recent work by Campbell (2003, 2004) has demonstrated the effect of liquid metal flow and the surface turbulence on the reliability of aluminium castings. It has been found to be important to minimize the surface turbulence during filling a mould to attain reliable mechanical properties of the castings. Extremely high velocities at which the liquid metal enters the mould is damaging to metal. The theoretical background of critical velocity concept by Campbell has been confirmed experimentally for liquid aluminium by Runyoro at al. For nearly all liquid metals this critical velocity is close to $0.5ms^{-1}$, involving aluminium alloys (Campbell, 2003).

The idea to use a novel runner system, named a "vortex gate", for uphill gravity pouring was introduced by Campbell (2003). The vortex gate was investigated by Hsu at al. A computational fluid dynamics and real-time X radiography were used for these studies. The main role of the real-time radiography was to verify the results of computational modelling. In this work, simulations show flow to dampen the circular motion quickly, implying that the internal losses in the flow as a result of turbulence are an overestimation in contrast with the video radiography. The results presented that the vortex gate has the potential to reduce the velocities to get quiet filling of melt into moulds. There are no rules to calculate and optimize the vortex gate dimensions in references.

Recently, research on runner and gating systems has included a growing number of papers on optimization algorithms, the focus being to generate routines to assist the designer in the work of mould and part design. Prasad at al. developed the artificial neural network system to generate the process parameters for the pressure die casting process. With this network, the selection of the process parameters can be carried out by any inexperienced user without prior knowledge of the die casting process.

In the work by Karunakar & Datta, an attempt was made to predict major casting defects like cracks, misruns, scabs, blowholes and air-locks by using back-propagation neural networks from the data collected from a foundry. The neural network was trained with parameters like green compression strength, green shear strength, permeability, moisture percent and melting conditions as inputs and the presence/absence of defects as outputs. After the training was over, the set of inputs of the casting that is going to be made was fed to the network and the network could predict whether the casting would be sound or defective. Thus the neural network makes a forecast about the nature of the casting just before the pouring stage.

The abductive neural network analysis method is used by Lee & Lin for simulation and optimization of runner system parameters for multi-cavity moulds. It has been shown that prediction accuracy in an abductive network is much higher than that in a traditional network. Abductive neural analysis based on the abductive modelling technique is able to represent complex and uncertain relationships between injection analysis results and runner and gating systems design.

Sulaiman & Gethin used a network for metal flow analysis in the pressure die casting process to predict the metal flow characteristics in the filling system by simplifying the complex Navier–Stokes equation.

All abovementioned works have indicated that the methodology of artificial neural networks can be used to replace the trial-and-error technique.

3. DESIGN OF EXPERIMENTS FOR GRAVITY CASTING SIMULATION RUNS

Design of experiments (DoE) procedure was used to systematically organize experiment runs to improve processes.

Level/Factor	Ingate diameter [mm]	Outlet diameter [mm]	Outlet length [mm]	Inlet velocity [ms^{-1}]
1	90	20	90	1.5
2	105	27.5	110	2
3	120	35	125	3

Tab. 1. Gating system parameters and their levels

It involves a fraction of the possible parameter combinations for a given experiment, which results in conducting a minimum number of experiments without losing significant information. This combination fraction is chosen according to rules and statistic matrices called Taguchi's orthogonal arrays. The DoE procedure is divided into three stages: experiment design, experiment running and statistical analysis.

DoE methodology was applied to the vortex gate optimization. Experiment design involves the choice of design parameters and parameter levels, as well as the choice of the appropriate orthogonal array according to desired resolution. The parameter ranges of the design variables are given in Table 1. The chosen orthogonal array defines the number of casting simulation runs to conduct and the parameter level values, which are carried out in the second stage of DoE. Simulations for the present work were conducted using Flow3D software. For the experiment, L_9 orthogonal array with four columns and nine rows was used. The experimental layout for four gating system factors using L_9 orthogonal array is shown in Table 2. During the third stage, analysis of variance (ANOVA) determines which parameters are statistically important along with their influence in the process. Preliminary simulation runs proved that the most significant factor that affecting the value of velocity in the outlet of the vortex runner are:

- outlet diameter
- inlet velocity

4. PREDICTIVE NEURAL NETWORK MODEL

An artificial neural network's (ANN) main characteristics are architecture, which defines the way that neurons are connected to each other, and training algorithm, which determines the way that weight factors are corrected.

In the proposed ANN architecture, neurons are arranged in layers. The first layer is the input layer and its neurons are the same in number as the input parameters (ingate diameter, outlet diameter, outlet length, inlet velocity), while the last layer has as many neurons as the output parameters (outlet velocity). In the ANN model the training algorithm of back-propagation was used. Table 2 illustrates the runner and gating system parameters used for flow simulations. The outlet velocities from the simulations were used as the output parameters to train proposed back-propagation neural network in Figure 2.

Pre-processing of input signals prior to input to the neural network is carried out as follows to improve convergence. All input and output data are scaled so that they are confined to a subinterval of (0.1 ... 0.9). Each input or output parameter X is normalized as X_n before being applied to the neural network, according to the following equation

$$X_n = [0.8\,((X - X_{min})/(X_{max} - X_{min}))] + 0.1 \qquad (1)$$

where X_{max} and X_{min} are the maximum and minimum values of the data parameter X.

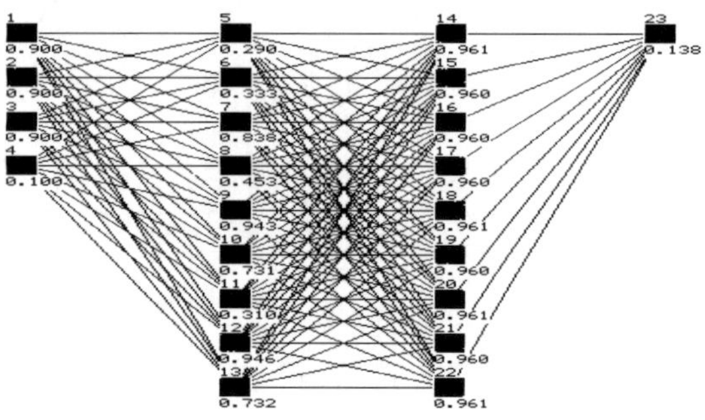

Fig. 2. Proposed back-propagation neural network

Experiment number/ Parameter level	Ingate diameter	Outlet diameter	Outlet length	Inlet velocity
1	1	1	1	1
2	1	2	2	2
3	1	3	3	3
4	2	1	2	3
5	2	2	3	1
6	2	3	1	2
7	3	1	3	2
8	3	2	1	3
9	3	3	2	1

Tab. 2. Experiment plan using L_9 orthogonal array

Other network architectures were also tried with different training algorithms and different numbers of hidden layers and hidden neurons.

5. CONCLUSIONS

From realized experimental tests on the proposed vortex runner system and back-propagation neural network the following conclusions can by drawn:

- based on the modelling of the neural network, the relationships between the vortex gate parameters and outlet velocity of the melt can be obtained
- neural network models can replace casting simulation software to design the vortex gate dimensions
- the predictions by the network within the input range agree closely with the values obtained from the simulations
- the accuracy of the tested networks can be different but within the acceptable limit.

Acknowledgements

The authors wish to acknowledge the support of the grant agency of the Ministry of Education of the Slovak Republic under the contract DAAD 0300/2008

6. REFERENCES

Campbell, J. (2004). *Casting Practice, The 10 Rules of Castings*, Elsevier Butterworth Heinemann, Oxford

Campbell, J. (2003) *Castings,* Butterworth Heinemann, Oxford

Hsu, F.Y.; Jolly, M. R. & Campbell, J. (2006). Vortex-gate design for gravity casting. *International Journal of Cast Metals Research*, Vol 19 No 1, p. 38-44

Karunakar, D.B. & Datta, G.L. (2007). Prevention of defects in castings using back propagation neural network. *Int J AdvManuf.Technik* http://www.springerlink.com/content/q725746273820461/ *Accessed* 2007-11-13

Lee, K.S. & Lin, J.C. (2006). Design of the runner and gating system parameters for a multi-cavity injection mould using FEM and neural network. *Int J Adv Manuf Technik*, Vol. 27 p. (1089–1096)

Prasad, K.D.V.; Yarlagadda, J.& Chiang, E. Ch. W. (1999). A neural network system for the prediction of process parameters in pressure die casting. *Journal of Materials Processing Technology,* 89–90, p. 583–590

Runyoro, J.; Boutorabi, S.M. & Campbell, J. (1992). *Trans AFS,* 100. p. 225-234

Sulaiman, S.B & Gethin, D.T (1992). A network technique for metal flow analysis in the filling system of pressure die casting and its experimental verification on a cold chamber machine. *J. Eng. Manuf.* Vol 206, No 4, p.261–275

Annals of DAAAM for 2008 & Proceedings of the 19th International DAAAM Symposium, ISSN 1726-9679
Editor B. Katalinic, Published by DAAAM International, Vienna, Austria 2008
Make Harmony Between Technology and Nature, and Your Mind will Fly Free as a Bird

AUDIT AND BUSINESS MANAGEMENT SYSTEM

KNEGO, N[ikola]; KOVACEVIC, Z[oran]& LAZIBAT, T[onci]

Abstract: *Setting up and maintaining a business management system requires audits. Undertaking an audit is not for free. Subsequently, the main issue is the cost effectiveness of such activities, and the influence of audits on a business management system. This paper presents an analysis of a survey conducted on forty-one Croatian companies in which at least one management system was present. The survey contained questions relating to the internal audit process, as well as the opinion on the influence of audits on business. In order to create positive effects of an internal audit, it is necessary to carefully design the process and implement it with the aim of obtaining real results which would be the basis for system improvement.*

Key words: *audit, internal, management, system, process, business*

1. INTRODUCTION

The paper concentrates on audits of business management systems. The existing management systems in an organization include systems such as corporate rules, management of quality, the environment, health, informational safety and occupational safety. In Croatian professional literature, the word audit is most commonly translated as surveillance, evaluation, estimate, review, or inspection. According to the series of ISO 9000 standards, audit is being defined as a systematic, independent and documented process for the purpose of gathering evidence and its objective evaluation in order to determine the degree to which audit criteria have been fulfilled (Tricker, 2005).

Internal auditing is an independent, objective assurance and consulting activity designed to add value and improve an organization's operations. It helps an organization accomplish its objectives by bringing a systematic, disciplined approach to evaluate and improve the effectiveness of risk management, control, and governance processes (Spencer Pickett, 2005).

Audits are divided to compliance audits conducted in order to obtain a guarantee that the activities have been implemented according to the rules (for example, registration audits, tax audits, audits to establish compliance with legal standards); and performance audits directed towards three fields (compliance with the rules, effectiveness of the rules, and the appropriateness of the rules in order to achieve goals). According to scope, audits are divided into product, process and system audits. The scope of the audit describes the degree and the limit of the audit, such as physical locations, organization units, activities and processes which need to be audited, as well as the period covered by the audit. By combining the mentioned categories, six types of audit can be distinguished. In reality, audit type depends on the goals and the relations between the auditor and the audited (Spencer Pickett, 2006).

2. STUDY OF AN AUDIT'S INFLUENCE ON BUSINESS

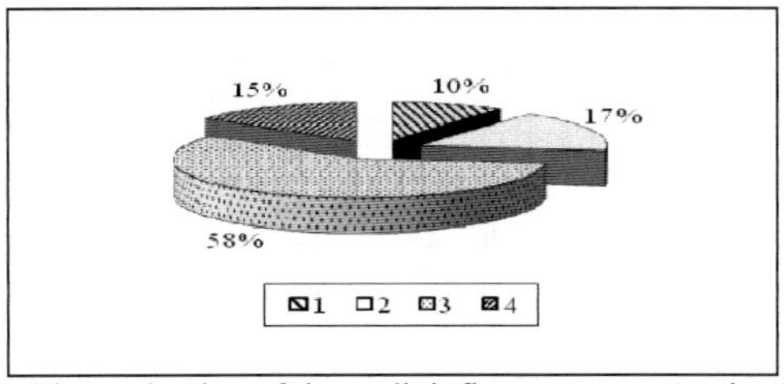

Fig. 1. The evaluation of the audit influence on cost reduction

Observation of the influence of audits on business was conducted on the basis of a survey analysis. The survey was conducted in July 2007, with the participation of forty-one Croatian enterprises which possess at least one management system certificate (ISO 9001, ISO 14001, OHSAS 18001). The survey included the inquiry into several sets of data concerning the respondents: general information on the company, intensity of the audit, support to the process, investment in competencies and response to the audit results, individual information and principles related to the internal audit process, evaluation of indicators of business improvement on the basis of the audit, and the implementation of a supplier's audit.

Regarding the influence of the audit on cost reduction, the survey offered marks from 1 to 4, with 1 signifying that the reduction did not exist, 2 meaning that it most likely existed, 3 that it definitely existed but was hard to prove, 4 that it existed and could be proved.

Based on the analysis of respondents' answers presented in Figure 1, it can be concluded that the surveyed companies share a prevalent opinion about the existence of the audit influence on cost reduction, but that it is hard to prove.

Similar results about the positive influence on risk reduction favour the hypothesis about the existence of an audit influence on cost reduction.

Since only in 15% of the cases the existence of methodology to prove the cost reduction on the basis of an audit was confirmed, and in additional 58% of the cases it was certain that audits influenced cost reduction, it is obvious that there is room for improvement and development of the mentioned methodology.

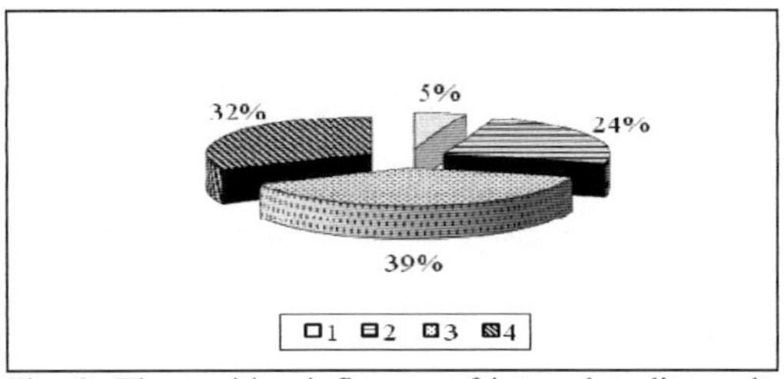

Fig. 2. The positive influence of internal audit results on risk reduction

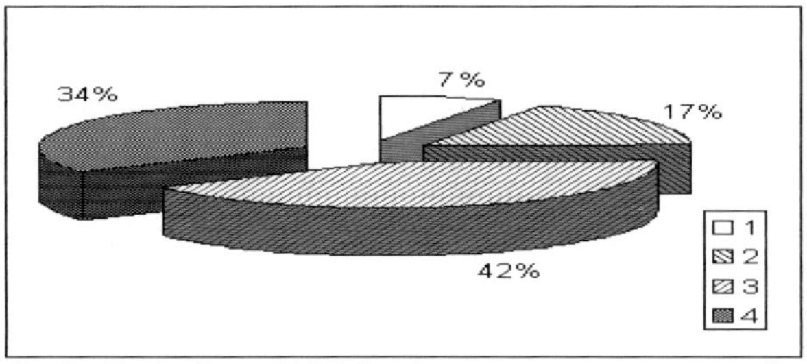

Fig. 3. Evaluation of high-level management support to the internal audit process

The survey confirmed that various factors contribute to the positive influence of audits on business:
1) High-level management support to the internal audit process;
2) Investments into auditors' competences;
3) Existence of improvements in the internal audit processes;
4) Response to the implementation of activities based on audit results.

The Figure 3 shows the support of high-level management to the internal audit process. The results show that the lowest mark was given in 7% of the surveyed companies, and in almost 80% of them high marks were awarded (3 and 4), which indicates the existence of high-level management support to the process of internal audit.

Marks presented in Figure 4 were awarded by taking into account the implementation of repeated education of internal auditors and the existence of leading auditors: 1 "retraining" was not done, 2 "retraining" was done every two years or less, 3 "retraining" was done every year, plus if there were leading auditors present another point was added, so the total lowest mark was 1, and the highest 4.

Figure 4 shows that the lowest marks (1 and 2) were present in almost 50% of the cases.

It is also necessary to point out that the analysis of the survey results showed less dependence of this factor on the positive influence of the audit on business. It is to be assumed that higher levels of competence can be achieved by the auditors' own initiative and self-education, which of course needs management support, as well as the more frequent implementation of audits. Figure 5 shows the marks regarding response to the implementation of activities based on audit results.

These marks include analysis, as well as planning and implementation of necessary activities of correction, prevention and improvement. It turns out that higher marks were awarded to the implementation of subsequent activities in almost 80% of the cases, which points to the fact that in most cases the audited organisations give satisfactory importance to audit results.

3. INTERNAL AUDIT PROCESS AND POSSIBILITIES OF ITS IMPROVEMENT IN THE FUTURE

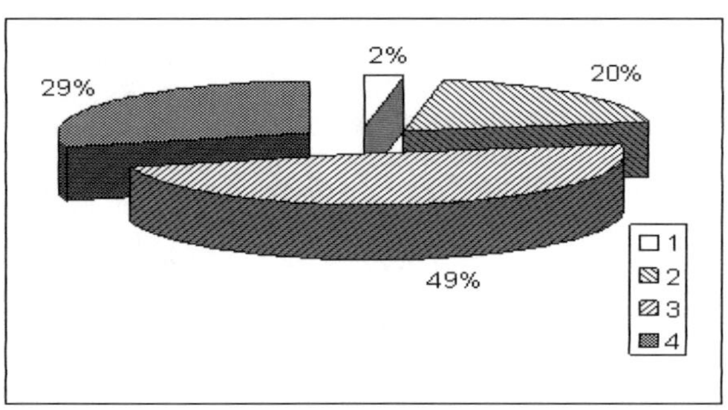

Fig. 4. Evaluation of investments in auditor's competences

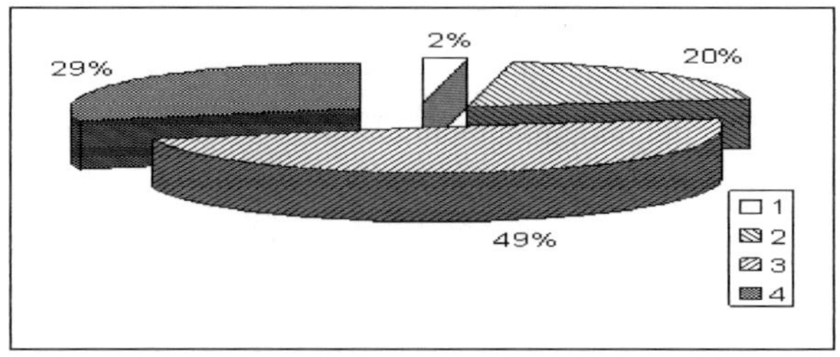

Fig. 5. Response to the implementation of activities based on audit results

The internal audit process can be divided into several phases: planning of the internal audit cycle, planning and preparation of an individual internal audit, implementation and reporting after an individual internal audit, checking any incongruence and collective reporting about the internal audit cycle. Internal audits require support which, above all, includes competent internal auditors. Planning, preparation, implementation and reporting an internal audit requires time and funds for the operative implementation of activities (Georgiades, 2006).

In order to ensure effective progress and management of the internal audit process, as well as all necessary resources and methods of supervision, measurement and analysis of the process, high-level management support is necessary, as well as the existence of an adequately documented procedure to manage the internal audit process. Of course, internal audits cannot be realised without the support of the management of the audited company, especially during the planning, preparation, implementation and reporting stages (Cangemi & Singleton, 2003). Also, IT support can significantly improve the effectiveness of the management and implementation process, by supporting the process activities and enabling prompt availability of internal audit notes (Gallegos et al., 2004).

4. CONCLUSION

The conducted survey confirmed that audits ensure cost reductions. However, since in only 15% of the cases some sort of methodology existed which could prove cost reduction as a result of the audit, and in extra 58% of the cases it was certain that audits influenced cost reduction, there is room for improvement – the mentioned methodology needs to be developed.

The following factors have an impact on the positive influence of audits on business: investment in auditor competencies, improvement of the internal audit processes, responses in the form of activities based on audit results, high-level management support to the internal audit process.

5. REFERENCES

Cangemi, M. P. & Singleton, T. W. (2003). *Managing the Audit Function: A Corporate Audit Department Procedures Guide*, Wiley, ISBN 0 471 28119 0, Chichester

Gallegos, F., Manson, D. P., Senft, S. & Gonzales, C. (2004). *Information Technology Control and Audit*, Auerbach, ISBN 0 849 32032 1, Jersey

Georgiades, G. (2006). *Audit Procedures, 2007*, CCH, Inc., ISBN 0 808 09044 5, Chicago

Spencer Pickett, K. H. (2005). *The Essential Handbook of Internal Auditing*, Wiley, ISBN 0 470 01316 8, Chichester

Spencer Pickett, K. H. (2006). *Audit Planning: A Risk-based Approach*, Wiley, ISBN 0 471 69062 X, Chichester

Tricker, R. (2005). *ISO 9001:2000 Audit Procedures*, Butterworth-Heinemann, ISBN 0 750 66615 3, Boston

Annals of DAAAM for 2008 & Proceedings of the 19th International DAAAM Symposium, ISSN 1726-9679
Editor B. Katalinic, Published by DAAAM International, Vienna, Austria 2008
Make Harmony Between Technology and Nature, and Your Mind will Fly Free as a Bird

MOBILE ROBOT-FOOTBALL PLAYER CONTROL VIA THE ARTIFICIAL LIFE METHOD

KOBRIN, A[lexandr] & SINYAVSKIY, O[leg]

Abstract: *The technique of the training of the neural nets for controlling the football-player robot has been created. In this technique we used a framework of theory of the artificial life and self-learning neural networks.*

Key words: *Artificial life, neural networks, robot controlling*

1. INTRODUCTION

The game «Virtual football» is the model of a real football-game (fig. 1). Two teams of football player robot models play on the special virtual football pitch. This work considers the model of the game is used at virtual football competitions, which are carried out at the Moscow State University (Okhotsimsky et al, 2000). Each team consists of five virtual robots, which are capable to kick the ball with their bodies. The game is discrete in time. The behavior algorithm of the team should give out two controlling signals for each step for every robot: change of its angle of turn and speed. Maximum speed, acceleration and the change of an angle of turn are limited in order to make the model more realistic. A robot is not capable to move backwards. Full information of the current game can be obtained from the server at any moment: coordinates, speeds, angles of turn for all players and for ball, as well as the geometry of the field.

Virtual football sets the problems of the collective controlling of the group of agents which do not compete among themselves. However, the effect of the collective management could be achieved as a first approximation through forcing each robot to carry out some local task on the field, for example the task of pursuing the ball. A virtual player should kick the ball with its body with a highest possible force in such a way that at the moment of impact robot should be at the correct side of the ball to force the ball to move in the necessary direction (for example, towards the opponent's gate). It is important to make the impact at minimal time in order to outstrip the opponent's players.

2. INFORMATION

2.1 Possible approaches

There is a variety of algorithms for solving this control problem (Johnson J, 1998). Among them we can mention the group of the methods based on classical laws of kinematics and on the theory of automatic control (Kim et al, 1998). These algorithms allow calculating the point and the time of impact as a first approximation with high computing speed. However, an increase in the accuracy of the calculations considerably decreases the speed, because of the great nonlinearity of the problem. Besides, the given methods do not take into account time discontinuity of the model results in various errors of management.

In addition, we have tested control method based on simple artificial neural networks, which had been learned with the help of the backpropagation of error method (Sinyavsky & Kobrin, 2006). These nets with two or three layers of neurons with a sigmoid activation function were not capable to satisfactory solve this control problem. This was due to the fact that net needed a huge amount of input samples for approximating the control signals on a wide range of input's amplitudes for different robot's and ball's position.

2.2 Artificial life method

At the same time with construction of neural nets, trained with the help of a teacher, we carried out our research about an opportunity to control the football-player robot with the help of self-learning neural nets within the framework of the theory of artificial life (Mark A. Bedau , 2003). The program (fig. 2), that emulates a certain virtual environment, has been created. It looks like a football pitch in which football-player robots live (further they are referred to as agents).

The agents possess many qualities of live organisms. They are able to be born, to die and to reproduce. In addition, they are capable to react to the changes of their environment and to interact with it. Each agent has a limited stock of the energy which can be spent on its moving, reproduction, or it can be changed as a result of influence of negative external factors. If the energy of the agent drops to zero, this agent dies and disappears from the virtual environment. The ball in the given realization of the artificial life serves as an equivalent of food, i.e. the agent needs to touch the ball in order to receive the energy. Accordingly to it, only that agent lives for a long time who can kick the ball frequently enough. The models of robots are similar to the «Virtual football» project models. Each living robot is controlled with the help its own neural net that has a structure of two-layer artificial net. The parameters of condition of the ball and the player are given to the inputs of the network; the output of the net is the controlling signal of the angle of turn change of the robot. Thus, behavior of the player and its ability to get food and to survive are determined by structure of matrices of weights of its neural net.

At the initial stage of functioning of the artificial life, the neural net weights of agents are initialized by some random values (the behavior of the players is chaotic and senseless; their impacts with the ball are unpredictable). Agents quickly spend their initial stock of the energy and die. The program supports a certain minimal count of the players living in the environment and constantly initializes new agents with the random matrices of weights. If the player will suddenly obtain some structure of weights that allows it to kick the ball frequently enough, he can reproduce himself if he has accumulated enough energy. At the reproduction the agent generates a descendant to whom he gives a half of his energy and copies into it his neural network. During copying the matrices of the weights a mutation can occur, that is a random change of one or several values of the weights in the neural network of the descendant. Thus we implement genetic algorithm in our framework (Kjellström, G. , 1996). Soon, there will be a steady population of agents in the enviroment, capable to survive for a long time. The number of the population varies near some balance position which is determined by the amount of the food produced in the virtual environment at a time unit. Members of the population are similar in their neural net structures. However, the presence of the mutations results to a competition inside the population. An extinction of less adapted agents improves a quality of the population as a whole.

To speed up the search of the first steady population, it is required to considerably simplify the problem of the survival for the agents. For this purpose the ball can be fixed motionlessly in the center of the field and its radius can be increased. To make impacts with the ball more appropriate for the football player, it is necessary to make a ball "inedible" from the one side. For example, if there is a training of a left-hand team, the player will receive a portion of energy only if it kicks the ball from the left side, i.e. towards to the opponent's gates. Impact with a ball from the right side, on the contrary, will harm the player - it will take away a portion of its energy. If we will compare the player with a predator, the ball with a victim, and if we will constantly improve abilities of the victim, forcing it to evade from the predator, it will be possible to achieve the qualitative solving of the problem of kicking the ball. After each period of an approximately hundred generations in the steady population of agents, it will be necessary to make gradually the problem of their survival more complicated: we should assign to the ball a greater initial speed and reduce its size. The agent's neural nets soon will be able to successfully solve the high-grade problem of the movement control (fig. 3), in many cases predicting a trajectory of the ball, taking into account reflection of the objects from the football field boards, etc.

3. FIGURES

Fig. 1. «Virtual football».

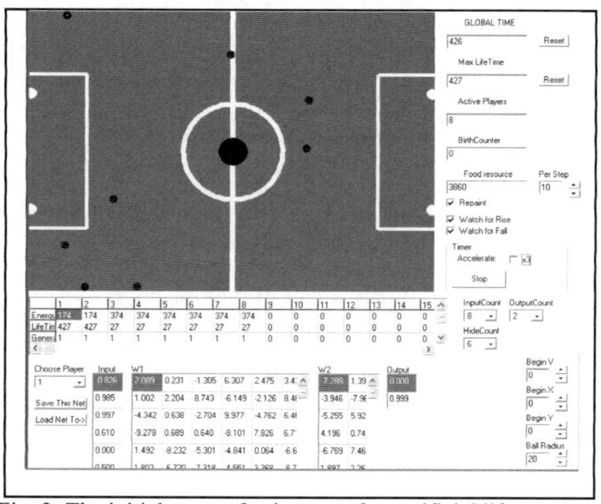

Fig. 2: The initial stage of existence of an artificial life

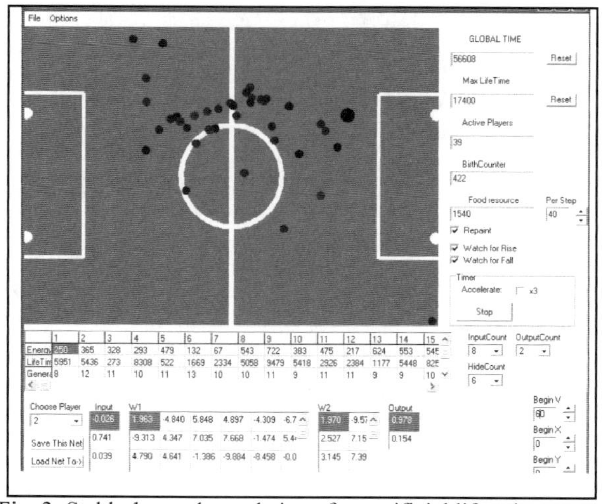

Fig. 3: Stable-learned population of an artificial life robots

4. CONCLUSION

The results of the works in the field of self-learning neural networks allow us to state, that algorithms of functioning of the artificial life can be applied to the construction of control for a wide variety of objects, which optimum controlling functions are impossible or too labor-consuming to elaborate with the help of other approaches (for example, it is difficult for controlling multi-agent systems, anthropomorphous robots, etc.) (Genci & Yasuo, 2004). With satisfactory quality of controlling, the speed of the calculation of the controlling signal with the help of the artificial life approach increased more than 50 times in comparison with the methods based on classical laws of kinematics and on the theory of automatic control. The experiments have helped to find out optimum parameters of the virtual environment. In order to achieve a greater biological similarity of the artificial agents, we plan to move from classical multilayered networks to the neural networks with arbitrary, spontaneous structure.

5. REFERENCES

Genci Capi, Yasuo Nasu, "Optimal Control of Humanoid Robot in Long Distance," aina,pp.559, 18th *International Conference on Advanced Information Networking and Applications* (AINA'04) Volume 2, 2004

Johnson J, 1998 . Robotic football, artificial life, and complexity, *Artificial Life and Robotic,* Springer Japan

Kim J.-H., Kim K.-C., Kim D.-C.,Kim Y.-J, and Vadakkepat P., "Path planning and role selection mechanismfor soccer robots," presented at IEEE Int. conference on *Robotics and Automation* - , Leuven, Belgium, 1998.

Kjellström, G. , 1996. "Evolution as a statistical optimization algorithm". *Evolutionary Theory* (11): 105–117.

Mark A. Bedau , 2003. Artificial life: organization, adaptation and complexity from the bottom up *TRENDS in Cognitive Sciences*. Retrieved on 2007-01-19

Okhotsimsky D.E., Pavlovsky V.E., Plahov A.G, Tuganov A.N. (2000), Simulation system for robotic soccer game modeling, *Mobile robots and mechatronic systems*, Lomonosov MSU, Moscow, 2000, pp.192-203.

Sinyavsky O.Y., Kobrin A.I., Research opportunities of management models of the mobile robot – football player with the help of neural net algorithms, in *FIRA RoboWorld Congress 2006, Robots in education, entertainment and service, Proceedings*. University of Dortmund, Germany 2006, 225p.

Annals of DAAAM for 2008 & Proceedings of the 19th International DAAAM Symposium, ISSN 1726-9679
Editor B. Katalinic, Published by DAAAM International, Vienna, Austria 2008
Make Harmony Between Technology and Nature, and Your Mind will Fly Free as a Bird

MODELING OF PROTEIN ENZYMATIC EXTRACTION FROM AMARANTH FLOUR AND SEED

KODRIKOVA, K[lara], KOLOMAZNIK, K[arel] & ADAMEK, M[ilan]

Abstract: *The topic of the article is oriented at modeling of amaranth protein enzymatic hydrolysis. The purpose of the hydrolysis is to enrich the solid phase with very valuable amaranth protein which could be used not only in food industry to produce new products rich in high-quality protein, but also in pharmaceutics, medicine and other branches. Regarding to diverse character of chosen materials (amaranth flour and seed) various hydrolysis course is expected, therefore a different model is built for each of the stuff. The designed mathematical models are verified by experimental measurement.*

Key words: *Modeling, amaranth, enzymatic extraction*

1. INTRODUCTION

Amaranth, a plant falling into the group of pseudo-cereals, starts to be a goal of the research for its unique specific properties presently. Study of this crop leads to ascertainment of remarkable content of quality protein with high portion of essential amino-acids and other considerable substances, like squalene and flavonoids. (Paredez 1994) Most of the so far published publications deal with amaranth as a food supplement for healthy population or as a possible compound in special diets of diabetic patients or people with protein allergy; new amaranth products are conducive against civilization diseases. On the other hand, amaranth starts to be used also in different areas -medicine, pharmaceutics and cosmetics.

The success of amaranth products application for normal utilization direct depends on possibility of particular parts separation. The aim is to obtain products of high purity, abreast with process conditions which do not raise the price of. However, papers dealing with separation are published only rarely as well as optimalization of single technology process using advantage of process engineering is seldom solved.

It is known, that the separation of particular compounds is possible by hydrolysis. The purpose of our present experiments is to enrich the solid phase with the vegetable protein and opposite – enrich the solid phase with valuable starch. The kinetics of these reactions and modeling of concentration fields are presented in our paper.

2. MATHEMATICAL MODELS OF PROTEIN ENZYMATIC EXTRACTION

2.1 The kinetics of protein enzymatic extraction from the amaranth flour

As mentioned above, the purpose of the hydrolysis is to enrich the solid phase with required compound – protein or starch. With regard to very small size of amaranth flour particles, is eventual that hydrolysis reaction is not broken by diffusion. On this condition we can apply the first order mechanism for the description of hydrolysis kinetics. On the assumption that the conversion speed (y) direct depends on irreducible starch (protein) fraction, it is possible to write following equation:

$$\frac{dy}{d\tau} = k(1-y) \quad (1)$$

By integration (1) we obtain:

$$y = 1 - B^{-k\tau} \quad (2)$$

B is equal 1 on starting condition y(0)=0. By setting out the natural logarithm of non-reacted portion *–ln(1-y)* to the time *(τ)* we acquire a straight line whose outline equals the speed constant of hydrolyzed starch (protein).

2.2 The kinetics of protein enzymatic extraction from the amaranth grain

During the enzymatic hydrolysis of amaranth flour the final heterogeneous mixture contains an aqueous solution of hydrolysed protein and a starch mud containing fixed protein. The separation of the heterogeneous reaction compound is possible only by using a centrifuge, because simple filtration proceeds very slowly.

By this reason we were engaged in hydrolysis experiments of the amaranth grain on condition of its permanent preservation. This technique enables a relative easy separation of the heterogeneous mixture by the pure filtration after finishing the reaction. For the description of the diffusion protein extraction from the amaranth seed we used following model (Crank 1975):

$$\frac{1}{D}\frac{\partial c(r,\tau)}{\partial \tau} = \frac{\partial^2 c(r,\tau)}{\partial r^2} + \frac{2}{r}\frac{\partial c(r,\tau)}{\partial r} \qquad 0 < r < a, \quad \tau < 0 \quad (3)$$

$$\frac{\partial c}{\partial r}(0,\tau) = 0 \quad (4)$$

$$-SD\frac{\partial c}{\partial r}(a,\tau) = V_0 \frac{\partial c_0}{\partial \tau}(\tau) \quad (5)$$

$$c(r,0) = c_p \quad (6) \qquad c(a,\tau) = \varepsilon c_0(\tau) \quad (7) \qquad c_0(0) = 0 \quad (8)$$

The equation (3) describes the concentration field inside the amaranth grain during the extraction process, (4) presents the axis symmetry of the concentration field, (5) responds the solid-state balance equality of the surface diffusion flow with the accumulation speed of the hydrolysed protein in the aqueous solution. (6) is a complete mixing condition of the aqueous phase and (7,8) are starting conditions.

Through the application of dimensionless quantities we get:

$$C = \frac{c}{c_p}, \quad R = \frac{r}{a}, \quad Fo = \frac{D\tau}{a^2}, \quad C = \frac{c_0}{c_p} \quad (9)$$

Finally, we obtain dimensionless model of the diffusion extraction:

$$\frac{\partial C(R,Fo)}{\partial Fo} = \frac{\partial^2 C(R,Fo)}{\partial R^2} + \frac{2\partial C(R,Fo)}{R\partial R} \qquad 0 < \mathrm{R} < 1, \ \mathrm{Fo} > 0 \quad (10)$$

$$\frac{\partial C(0,Fo)}{\partial R} = 0 \quad (11) \qquad -\frac{\partial C(1,Fo)}{\partial R} = \frac{Na}{3\varepsilon}\frac{\partial C_0(Fo)}{\partial Fo} \quad (12)$$

$$C(R,0) = 1 \quad (13) \qquad C(1,Fo) = C_0(Fo) \quad (14) \qquad C_0(0) = 0 \quad (15)$$

The analytical solution of the equations (9) to (11) was published for the temperature field of spherical body (Carlslaw 1959). By the modification of the published solution for concentration field in the amaranth grain we acquire:

$$C=\frac{c}{c_p}=\frac{\varepsilon}{Na+\varepsilon}-\frac{2Na}{3\varepsilon R}\sum_{n=1}^{\infty}\frac{\left[\frac{Na^2}{\varepsilon^2}q_n^4+3\left(2\frac{Na}{\varepsilon}+3\right)q_n^2+9\right]\sin(Rq_n)\sin(q_n)}{\frac{Na^2}{\varepsilon^2}q_n^4+9\left(\frac{Na}{\varepsilon}+1\right)q_n^2}e^{-Foq_n^2} \tag{16}$$

$$C_0=\frac{c_0}{c_p}=\frac{1}{Na+\varepsilon}-\frac{6Na}{\varepsilon}\sum_{n=1}^{\infty}\frac{e^{-Foq_n^2}}{\frac{Na^2}{\varepsilon^2}q_n^2+9\left(\frac{Na}{\varepsilon}+1\right)}$$

$$Fo=\frac{D\tau}{a^2} \tag{17}$$

The protein concentration fields in the seed of spherical shape are shown in the following pictures Fig.1 to Fig. 2. Axis x represents the radius of a seed (R), axis y shows the concentration (c) of the enzyme solvent and finally, axis z presents non-dimensional time Fo. The differences between varied dimensionless consumption of the enzyme solution are visualized in interactive modeling environment Matlab.

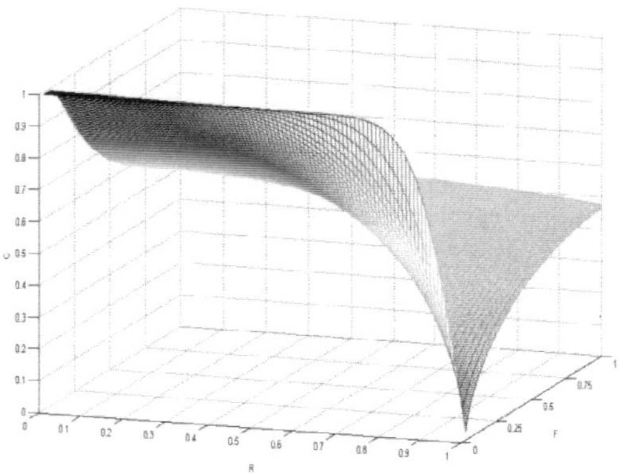

Fig. 1. Concentration field for the dimensionless solvent consumption Na = 1

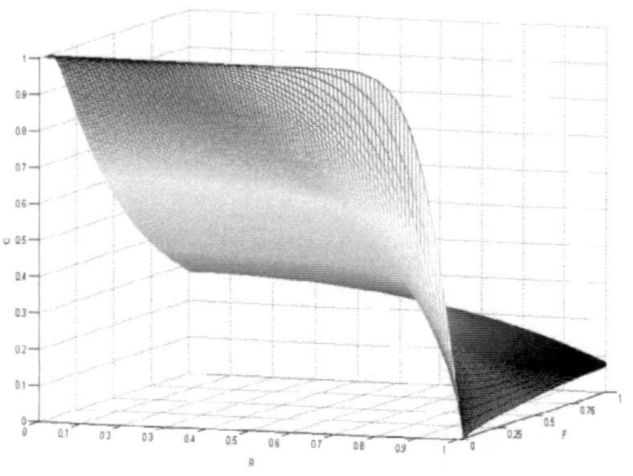

Fig. 2. Concentration field for the dimensionless solvent consumption Na = 10

In the first picture, you can see the concentration field inside the amaranth grain by the elected dimensionless solvent consumption 1. In the second case in figure 2 is well seen, how the concentration field changes with high-voted consumption of enzyme – Na equals to 10.

3. EXPERIMENTAL SET-UP

3.1 Protein enzymatic hydrolysis of the amaranth technical flour

The experiments were focused on the hydrolyze effectivity of amaranth flour protein with usage of several readily available enzymes in the market. Protein liquefaction of the technical flour runs under slight reaction conditions (temperature, neutral pH) which were proposed:

- rate flour/water = 1/20
- speed of reaction mixture warming 2 °C. min^{-1}

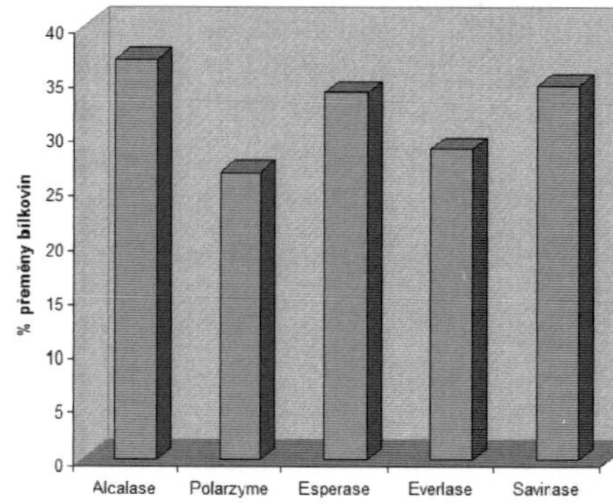

Fig. 3. Percentage of protein conversion using several enzymes by the 5 h hydrolysis and 50°C temperature

The experiments proceed in variable hydrolyze times (minimum 1 hour, maximum 5 hours) and variable temperatures (30°C to 50°C). Into the boiling flask was 5 g of the flour weighted out, after that 100ml of distilled water was added and the mixture was stirred in the water bath and simultaneously warmed up. After achieving of the specified temperature 500μl of enzyme solution was tacked with a pipette. Enzyme hydrolysis was running for required time and by rated temperatures. After it the reaction alloy was centrifuged with 6000 rpm. Liquid and solid phase were analyzed for the total residue content and finally screened using TKN method. Examples of the experiments results are presented in the table II. and showed in the graphs above.

Out of the experiments results it is evident, that the highest efficiency by the enzyme liquefaction of the amaranth technical flour enzyme Alcalase proved. By the five hours hydrolysis and 50°C got almost 37% of protein liquid.

4. CONCLUSION

In this paper, the modeling of amaranth raw material (flour and seed) is solved using Matlab computer environment. Amaranth flour particles have very small size so that the enzyme hydrolysis is not broken by diffusion; hence the mathematical model is simpler. On the other hand, amaranth seed has a diffusion bar presenting by its skin. The grains mathematical model was built in spherical coordinates. The models were verified by experimental measurements, which acknowledged the advantages of the whole grain hydrolysis. Fulfilling the condition of seed permanent preservation after protein extraction we considerably simplify the process of mixture separation which leads to an economical profit when industry used.

Acknowledgements

The work has been supported by the grant VZ MSM 7088352102 the Ministry of Education of the Czech Republic. This support is very gratefully.

5. REFERENCES

Carlslaw, H.S. and Jaeger, J.C., (1959). *Conduction of heat in solids,* Oxford at the Clarendon Press,.

Crank, J, (1975). *The mathematics of diffusion,* Clarendon Press, Oxford 2nd Edition.

Paredes-Lopez, O., (1994). *Amaranth – biology, chemistry and technology,* Library of Congress Cataloging in Publication Data.

Annals of DAAAM for 2008 & Proceedings of the 19th International DAAAM Symposium, ISSN 1726-9679
Editor B. Katalinic, Published by DAAAM International, Vienna, Austria 2008
Make Harmony Between Technology and Nature, and Your Mind will Fly Free as a Bird

A FORMAL VERIFICATION METHOD OF FUNCTION BLOCK DIAGRAMS WITH TOOL SUPPORTING: PRACTICAL EXPERIENCES

KOH, K[wang] Y[oung]; JEE, E[un] K[young]; JEON, S[eungjae]; SEONG, P[oong] H[yun] & CHA, S[ungdeok]

Abstract: *Function Block Diagram (FBD) is a standard application programming language for the Programmable Logic Controller (PLC) and currently being used in the development of a fully-digitalized Reactor Protection System (RPS) under the Korea Nuclear Instrumentation & Control System (KNICS) project. Therefore, rigorous verification of the FBD programs is indispensible. In this paper, we propose a formal verification technique of FBD programs used in KNICS RPS and its supporting tool, FBDVerifier, which performs model checking and gives us counterexamples displayed in the form of timing graph to enhance their readability. Several important errors of FBD programs were found and reflected to the next version. Large scale case study showed that the proposed method is effective and practical.*
Key words: *Function Block Diagram (FBD), Programmable Logic Controller (PLC), Formal Verification, Model Checking*

1. INTRODUCTION

A fully-digitalized Reactor Protection System (RPS), which is called IDiPS (Integrated Digital Protection System), is being developed under Korea Nuclear Instrument and Control System (KNICS) project in order to be used in newly constructed nuclear power plant and also in the upgrade of existing analog based RPSs. Programmable Logic Controller (PLC) and its representative programming language Function Block Diagram (FBD) are used to control many kinds of functional processes of IDiPS. Because the software embedded in a RPS is very crucial to the safety of a nuclear power plant, rigorous safety demonstration of FBD programs is indispensible.

Several different approaches to formalize PLC programs have been presented in (Bani Younis & Frey, 2003), but a few works have been performed for FBD programs. In (Baresi et al., 2000) a toolset called PLCTOOLS has been introduced. The FBD programs are modelled and described as High Level Timed Petri Nets (HLTPN) which is used for validating the design and generating the code. PLCTOOLS supports the design validation and code generation processes, but it does not support formal verification such as model checking, a major function of *FBDVerifier*. In (Vyatkin & Hanisch, 2000) the controller code in FBD format and the overall system are translated to Singal-Net-Systems (SNS) (Starke, 2000). On the combined model of plant and controller modelled by SNS, model checking is performed using Signal/Event System Analyzer (SESA) which is a model checker for Signal-Net models (Starke & Rosh, 2000). This approach, however, follows IEC 61449 while our approach follows IEC 61131, and it is therefore inapplicable to our FBD programs.

In this paper, we propose a formal verification technique of FBD programs; FBD programs are defined formally in compliance with IEC 61131-3, and then the programs are automatically translated into Verilog model by the supporting tool, *FBDVerifier*, which is developed to support the FBD verification framework, and finally the model is verified using Cadence SMV, and counterexamples are displayed in the form of timing graph to enhance their readability. This approach is described in Fig. 1.

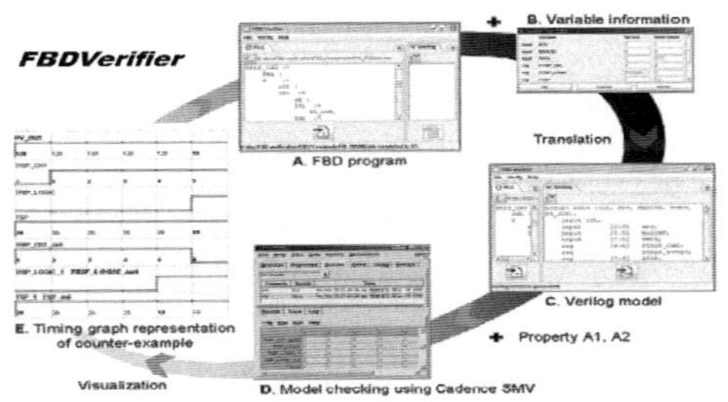

Fig. 1. FBD verification framework supported by *FBDVerifier*

2. FORMAL VERIFICATION OF FBDS

Verilog (IEEE, 2003) is one of the most popular Hardware Description Languages (HDL) used by integrated circuit designers. In order to verify FBD programs, we chose Verilog as a verification language because the semantics of FBD is similar to those of Verilog so that FBD can be translated into Verilog efficiently. We used model checking technique to verify FBD programs and chose Cadence SMV (McMillan, 2001) as a model checker to verify Verilog models generated from FBD programs.

2.1 Translation Rules

Rule 1. Module declaration: A *module* is the principal design entry in Verilog. The first line of a module declaration specifies the name of the module and list of input/output ports.

Rule 2. Variable type decision: All variables are declared with their type, bit size, and name in Rule 2. Each variable in the FBD is mapped to one of the Verilog variable types; *input, reg, wire and output*. A variable v_i □ V_I has *input* type if its value is transmitted from external input. A variable v_i has *reg* type if its value needs to be stored internally. The *reg* variables hold their values which will be used at the next scan cycle operation. Variables whose values do not need to be stored internally are declared as *wire* variables. A variable v_o □ V_O has output type if it is designated as an external output of the module.

Rule 3. Initialization of reg variables: The *reg* variables are initialized in Rule 3. Verilog operator '<=' means the assignment to l-value. Usually initial values of *reg* variables are specified in the FBD program. If not, they are determined by a user or have default values.

Rule 4. Output assignment for each wire and output variable: In Rule 4, the target FBD represented by a set of connected function blocks is translated into assignment statements from top to bottom in accordance with the execution order of the FBD. While a function which does not have internal state is mapped into a Verilog operator, a function block which stores internal state is mapped into a Verilog module.

Rule 5. Stored value assignment for reg variables: The stored values are assigned to the reg variables in Rule 5. @ **(posedge** clk) means the positive edge of the clock signal, i.e., the beginnings of each cycle. As the updated value of a *reg* variable becomes visible at the next time unit, new value is read at next cycle.

Rule 6. Insertion of properties: The template **always begin - end** is generated automatically and properties are embedded by a user after automatic generation of a Verilog model.

2.2 *FBDVerifier*

FBDVerifier automates the FBD verification framework described in Fig. 1. It takes LDA files, which is FBD storing format of a tool pSET (POSCON R&D Center, 2007) as inputs, and then converts the FBD programs into Verilog model. User can adjust bit sizes and initial values of variables during the translation. After Verilog translation is completed, a user inserts properties to be verified into the Verilog model and executes Cadence SMV with one click in *FBDVerifier*. When a target FBD has a lot of variables, a generated counterexample is often lengthy and complex. It is very hard to analyze a counterexample only with the table-style trace view provided by Cadence SMV. In many cases of our case study, we had to examine several hundreds rows of variables and many columns of steps in order to analyze a counterexample. To enhance readability of counterexample, timing graph form is displayed. Variables are highlighted in different color and shape for better visualization. With this function, counterexample analysis time was significantly reduced.

3. CASE STUDY

We applied the proposed method to Bistable Processor (BP), Coincidence Processor (CP) and Automatic Test and Interface Processor (ATIP) which are safety-critical subsystems of IDiPS and their information is described in Table 1. The software design specification document for the RPS has approximately 700 pages and the FBD programs for the RPS are composed of approximately twenty thousands function blocks and nine thousands variables. The Verilog model generated from the FBD for the RPS consists of more than fourteen thousands of lines totally.

We divided found errors into 5 categories according to their causes. *Incorrect logic* errors are the most serious errors caused by incorrect logic. *Omission* means errors caused by omission of specification. *Ambiguous logic* errors are possible errors which can be serious or not according to the environment constraints. *Incorrect FBD* category represents the cases that FBD specification is wrong while SDS is correct. *Incorrect SDS* errors are the opposite cases. The verification results are summarized in Table 2.

Subsystems	Page # of natural lang. spec.	# of function blocks	# of variables	Line # of Verilog model
BP	190	1,335	1,038	7,862
CP	163	1,623	820	3,085
ATIP	365	18,359	7,024	3,401

Tab. 1. RPS subsystem information

		BP	CP	ATIP
# of properties		216	83	126
Found errors	Incorrect logic	14	6	23
	Omission	0	2	0
	Ambiguous logic	4	0	3
	Incorrect FBD	13	5	9
	Incorrect SDS	16	0	11
Total # of errors		47	13	46

Tab. 2. Summary of verification results

4. CONCLUSION AND FUTHER STUDY

We proposed a formal verification technique for FBD which is a commonly used PLC programming language. We suggested the translation rules from an FBD program into a Verilog model and developed the *FBDVerifier* tool to automate the FBD verification framework. We performed model checking for the Verilog models generated from FBD programs successfully. It is almost impossible to verify FBD through manual translation due to huge amount of work and time. Automatic translation function of *FBDVerifier* made it possible for us to verify huge size of real case FBD programs, and its counterexample visualization function reduced the time and effort of analyzing counterexamples significantly. By using *FBDVerifier*, we could accomplish FBD verification and counterexample analysis for the whole RPS within 3 man-months which is relatively short time.

But, we used manual abstraction techniques to make the verification feasible against state explosion problems and current *FBDVerifier* receives only LDA file format used in pSET. It is therefore necessary to construct systematic abstraction method instead of manual abstraction and to extend *FBDVerifier* to be able to deal with various FBD storing formats.

5. REFERENCES

Bani Younis M & Frey G. (2003). Formalization of Existing PLC Programs: A Survey, *Proceedings of Computing Engineering in Systems Applications (CESA)*, July 9-11, 2003, Lille

Baresi L; Mauri M; Monti A & Pezze M. (2000). PLCTools: Design, Formal Validation, and Code Generation for Programmable Controllers, *Proceedings of the IEEE Int Conf on System, Man, and Cybernetics,* pp. 2437-2442, October 8-11, 2000, Nashville

IEEE Standard Hardware Description Language Based on the Verilog hardware Description Language. IEEE, 2003

KNICS (Korea Nuclear Instrumentation and Control System Research and Development Center). *Available from:* http://www.knics.re.kr/english/eindex.html

McMillan K.L. (2001). Getting started with SMV, Cadence Berkeley Labs

POSCON R&D Center. (2007). pSET (POSCON Software Engineering Tool), *Available from:* http://rnd.poscon.co.kr

Starke P & Roch S. (2000). Analysing Signal-Net Systems. Report, Humboldt University Berlin, Institut fur Informatik, August 2000

Starke P. (2000). Symmetries of Signal-Net Systems, *Proceedings of the Workshop on Concurrency, Specification and Programming,* pp. 285-297, October 9-11, 2000, Berlin

Vyatkin V & Hanisch H.-M. (2000). Modeling of IEC 61499 function blocks - a clue to their verification, *Proceedings of the XI Workshop on Supervising and Diagnostics of Machining Systems,* pp. 59-68, March 12-17, 2000, Karpacz

Annals of DAAAM for 2008 & Proceedings of the 19th International DAAAM Symposium, ISSN 1726-9679
Editor B. Katalinic, Published by DAAAM International, Vienna, Austria 2008
Make Harmony Between Technology and Nature, and Your Mind will Fly Free as a Bird

VIRTUAL INDUSTRIAL CLUSTERS - A NEXT STEP MODEL FOR CLUSTERIZATION OF CROATIAN SMEs

KOLAKOVIC, M[arko] & SISEK, B[oris]

Abstract: *In complex and turbulent conditions of today's global business environment small and medium enterprises (SMEs) more intensely experience the competition of the large company – technologically integrated and more marketing savvy. For the survival of SMEs in all branches of the economy today exists different solutions. One of the most prominent is strategic linking in clusters. In general, by networking and clustering SMEs can compensate for the disadvantages of size, achieve larger levels of productivity and innovation and so contribute not only to increased individual competitiveness, but also further the development of competitiveness of regions in which they exist. We propose to expand this model to accommodate intermittent solutions with temporary partners through the intervention of web brokerage systems to create virtual organizations equivalent to competing large enterprises.*
Key words: *cluster, virtual organization, strategic networking, small and medium enterprises, competitiveness.*

1. INTRODUCTION

Clusters and virtual organizations exist alongside other modern organizational forms (virtual teams, networks and outsourcing models, strategic alliances, partnerships etc.) and tend to replace the more traditional organizational structures. A major driver for this development lies in the uncertain and turbulent business climate.

Under such business ambivalent conditions of unpredictability of competitors' actions combined with unified business processes and methods dynamically interconnecting SMEs fare a better chance of success.

Dynamic strategic linking and networking of enterprises thus enables modern SMEs to optimize costs, increase efficiency of utilization of capacity, and maximize flexibility when responding to changes in market conditions.

In addition to the above microeconomic perspective, dynamic strategic linking and networking has implications for political economy, in particular for economic development and industrial innovation policy. This indeed supports and further develops the ambitions of the Croatian regulators and policymakers at the regional and national level.

In particular, it allows our clustered SMEs to comprehensively integrate with existing industrial complexes, educational and scientific institutions, along with the private sector consulting, financial, and investment activities. Such a process transforms the originally geographically defined clusterization by region and country into boundaryless and global *virtual industrial clusters.*

2. DEFINITION OF CLUSTERS AND VIRTUAL ORGANIZATIONS

One of the most frequently used definitions of clusters is that of Michael E. Porter who defines clusters as „geographically concentrated, horizontally and vertically interconnected specialized enterprises - suppliers, service providers and connected institutions - which in certain areas connect their core competencies for joint product creation" (Porter, 1998). More simply said, clusters are groups of interconnected business entities with superior performance and committed to achieving common goals.

On the other hand, virtual organizations are specific forms of network organizations whose existence is enabled by modern information and communication technologies (the Internet / intranets / extranets) that facilitate business cooperation between entrepreneurs and enterprises through space, time, and organizational boundaries.

Virtual organizations can be defined as groups of enterprises, parts of enterprises, or sole proprietors who mutually network for the purpose of specific temporary or permanent business activities.

Inversely to partnerships in the technical sense, individual members of virtual organizations keep their legal independence with respect to each other, but appear as cohesive business entities to the market.

Virtual organizations are apparently nonhierarchical organizational forms of independent enterprises that individually decide about network entry where they exchange raw materials, equipment, information, knowledge, technology, research and development, products and services.

Thus both the cluster concept and the construct of the virtual organization are new business models of strategic linking and networking of unique activities.

The difference is that the clusters are predominantly (but not necessarily) composed of small and medium enterprises and mostly have regional character. Even in the context of European integration, though, the notion of region remains a relative concept.

Similarly the definition of SME in Europe differs (e.g. many Croatian large enterprises may be categorized as only SMEs in the forthcoming EU classification).

Other important characteristics of clusters are:

- superior level of success of enterprise members,
- implementation of modern technology and innovativeness,
- clear expectations and commitment to a common goal as guiding principle for participation,
- construction and development of mutual trust,
- willingness to share knowledge actively,
- the active participation in cluster management activities (e.g. the management of the network and the cluster, coordination of activities of the cluster and management of external linkages),
- voluntary and independent integration and participation of the members, etc.

These cluster characteristics are equally valid for the concept of the virtual organization!

Furthermore, through the virtual organization it is possible to achieve the increase of loyalty of buyers, market share, sales, profit, competitiveness, growth rate and individual product value. These are also primary objectives of clusters.

Very common areas of cooperation of subjects included in the cluster concept are (Horvat & Kovacevic, 2004):

- promotion and marketing,
- procurement of raw and production materials,
- cooperation in the area of distribution (especially during of new market development) and its maintenance,
- separation of similar activities between different enterprises and engagement of additional partners for such activities (e.g. IT),
- training and development of expert human potential,
- cooperation in the area of research,
- information exchange concerning competition, etc.

Indeed, the concept of the virtual organization, through its construct of specialization of core competencies and outsourcing covers the same. These similarities and overlaps demonstrate a very close relationship between the two conceptions.

In fact today, with the growth of clusters and the involvement of new, ever more distant members any geographic definition disappears in favor of virtualized business activities coordinated by virtual clusters.

Important characteristics of this forthcoming model of virtual clusters can be identified as follows:

- flexibility, adaptability and modularity of member activities,
- narrow focus of members on very specialized core competencies,
- significant employment of information and communication technology,
- mutual confidence and partnership attitude of participants in virtual networks,
- nonexistence of clearly defined time and place of business activities,
- sharing of resources, capacities, infrastructure and costs,
- relentless pursuit of excellent performance and perfected responsiveness to buyers,
- knowledge sharing and aspirations towards the constant increase of knowledge – the *learning organization.*

3. THE IMPORTANCE OF CLUSTERS FOR DEVELOPING COUNTRIES AND BASIC CHALLENEGS AND RESPONSIBILITIES OF GOVERNMENTS IN CLUSTER DEVELOPMENT

Clusters and virtual organizations enable business entities in numerous ways otherwise unavailable to them individually. They are the direct result of close cooperation of business entities with complementary services and public institutions which ensure research, training and other specialist opportunities (Kolakovic, 2006).

Further development of clusters would enable countries and regions to better respond to the resource and timing requirements of investment. Additionally, clusters support the pursuit of social goals and provide promotional capacity for regions or countries.

Superimposing the virtual organization onto the cluster concept provides the additional major advantage of elegantly addressing the challenges of an increasingly uncertain market environment.

VOs thus enable quick market access to new markets, provide activities in much shorter time with considerably lower costs and at lesser risk. They increase the share of sales and profit, they improve competitiveness and consumer loyalty. They also open new opportunities for better geographic and professional complementarity among partners, and reduce need for employment of new people (Lazibat & Kolakovic, 2004).

Basic tasks of governments in the creation of strategy for modern virtual clusters are therefore:

- creation of the legal and administrative conditions for the promotion of competitiveness;
- planning and the execution of promotion concepts that support the development of clusters;
- encouragement of national competitiveness for purposes of higher quality and increased innovation.

Concrete effects of such clusters for regions and the countries could be summarized as follows:

- accelerated growth of industrial production,
- improved economic restructuring and transformation of traditional industrial complexes
- stimulation of development of entrepreneurship and the strengthening of the sector comprised of small and medium sized business entities,
- growth of competitiveness of the entire national economy including exports,
- revitalization of traditionally underdeveloped and less densely populated regions,
- linking with clusters of neighboring countries on the basis of common interest – internationalization and the virtualization of clusters.

4. CONCLUSION

The above cluster concept represents a solution of high quality and acceptance for the promotion of SMEs' competitiveness in developed countries already. Regrettably, in Croatia the process of clusterization is still rather unpopular. The ensuing resistance is mostly due to ignorance, bad experiences with and distrust in partnering.

Recent results, however, show some positive movement of Croatian SMEs towards the acceptance of undeniable benefits of dynamic strategic linking and networking. The Croation government (especially through its Ministry of the Economy) has recognized progress in this area and has thus initiated various programs with the purpose of directly encouraging such advances.

In this context the concept of the virtual organization represents one logical next step towards regional SME networks expanding internationally with compatible partners not limited to SMEs but comprised of any type of enterpise, including large global entities, parts thereof, as well as numerous associated financial and research institutions. Members of such virtual clusters jointly perform certain temporary or permanent business activities, make decisions jointly, keep their legal independence and outperform traditional entities in similar business activities.

5. REFERENCES

Croatian Ministry of Economy, Labor and Entrepreneurship & German Society for Technical Cooperation (2007). *Cluster Management, Part A,* Handbook

Horvat, D. & Kovacevic, V. (2004). Clusters – a way to competitiveness, ISBN CERA PROM p.l.c., Zagreb

Croatian export offensive, Strategy for period 2007.-2010., Croatian Ministry of Economy, Labor and Entrepreneurship

Kolakovic, M. (2006). *Entrepreneurship in knowledge based economy*, ISBN 953-6895-32-3, Sinergija, Zagreb

Lazibat, T. & Kolakovic, M. (2004). *International business in global condition*, ISBN 953-6895-19-6, Sinergija, Zagreb

National Competitiveness Council (2003). *Suggestions for improvement of competitiveness*, Expert Commission for Regional Development and Clusters, Zagreb

Porter, M. (1998). Clusters and the new economics of competition, *Harvard Business Review*, 76(6), p. 77-90.

Annals of DAAAM for 2008 & Proceedings of the 19th International DAAAM Symposium, ISSN 1726-9679
Editor B. Katalinic, Published by DAAAM International, Vienna, Austria 2008
Make Harmony Between Technology and Nature, and Your Mind will Fly Free as a Bird

OPTIMIZATION OF THE MANUFACTURING PROCESS ROUTE OF MACHINE COMPONENTS AIDED WITH ACTIVITY BASED COSTING

KOLOSOWSKI, M[ariusz]

Abstract: *The research problem presented in the paper concerns the choice of the optimal manufacturing process route that is characterized by the lowest cost level for required product quality. To this end the appropriate procedure of the choice and the model of product self cost and quality cost calculation based on the Activity Based Costing method have been worked out. The procedure of cost calculation has been shown on the example of the production of a gear wheel. Optimization of the manufacturing process route of machine components in conditions of piece and small batch production based on performed calculations has been carried out – optimal solutions have been pointed out. Worked out method may be used to analyse other processes within production system in enterprise, not only quality assurance processes.*
Key words: *cost, calculation, quality, ABC, optimization.*

1. INTRODUCTION

Contemporary enterprises have to make competitive advantage in a number of fields of activity. In a customer's viewpoint, quality, time of delivery and product's price are crucial factors of competitiveness. That is why a problem of economics of quality is widespread in literature and gains more importance (Omachonu et al., 2004; Tkaczyk & Jagła, 2001; Tsai, 1998).

Efficient management of quality assurance processes depends on accurate cost data of these processes. Piece and small batch production is characterized by high share of indirect cost within production total cost. Moreover, that share shows permanently an increasing trend. That is why Activity Based Costing (ABC) is the best tool that ensures high precision of cost calculation, determining cost structure and identifying causes of cost (Kaplan R. & Atkinson A.A., 1998). The essence of Activity Based Costing and examples of its practical application are generally and theoretically presented in literature (Gunasekaran et al., 1999; Kaplan R. & Atkinson A.A., 1998; Nowak et al., 2004; Tsai, 1998). However, it is impossible to find algorithms of calculation applied to peculiarity of the industrial branch and conditions of enterprise's economic activity. It especially concerns small and medium enterprises manufacturing products in small batches or according to the order.

The characteristic of piece and small batch production are several alternative variants of the production process route. Design and production engineers, who do not get information coming from Activity Based Costing, are not able to analyse the possibility of designing a variant characterized with the lowest cost simultaneously fulfilling quality requirements and both cost and time constraints. In this point of view problem of optimization of the manufacturing process has not been solved up to now.

2. THE COST CALCULATION MODEL

Production self cost determination aims at calculating the unit cost and showing its structure. Incorrect division of indirect cost causes taking wrong decisions. The most common mistakes are (Matuszek & Kołosowski, 2004):

- lowering the cost of product in piece and small batch production,
- reassessment of the cost of product in conditions of large batch production.

That is why sometimes the product generates loss, although the calculation points at its profitability. Implementing Activity Based Costing enables avoiding mistakes mentioned above. The costs calculation model, enabling determining quality cost, based on the ABC method is presented in the figure 1.

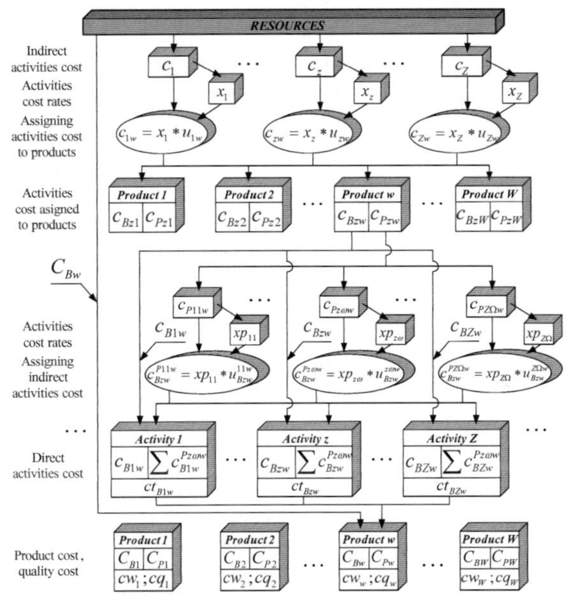

c_z – activity "z" cost,
x_z – activity "z" cost rate,
u_{zw} – number of units of activity "z" output measure assigned to product "w",
c_{zw} – activity "z" cost assigned to product "w",
c_{Bzw} – direct activity "z" cost assigned to product "w",
c_{Pzw} – indirect activity "z" cost assigned to product "w",
$xp_{z\omega}$ – indirect activity "zω" cost rate,
$u_{Bzw}^{z\omega w}$ – number of units of indirect activity "zω" output measure assigned to direct activity "z",
$c_{Bzw}^{Pz\omega w}$ – indirect activity "zω" cost assigned to direct activity "z",
ct_{Bzw} – total direct activity "z" cost assigned to product "w",
C_{Bw} – generic direct cost assigned to product "w",
C_{Pw} – total indirect cost (total direct activities cost assigned to product "w"),
cw_w – total product "w" cost,
cq_w – product "w" quality cost,
$z=1,\ldots,Z;\ \omega=1,\ldots,\Omega;\ w=1,\ldots,W.$

Fig. 1. The cost calculation model based on the ABC method

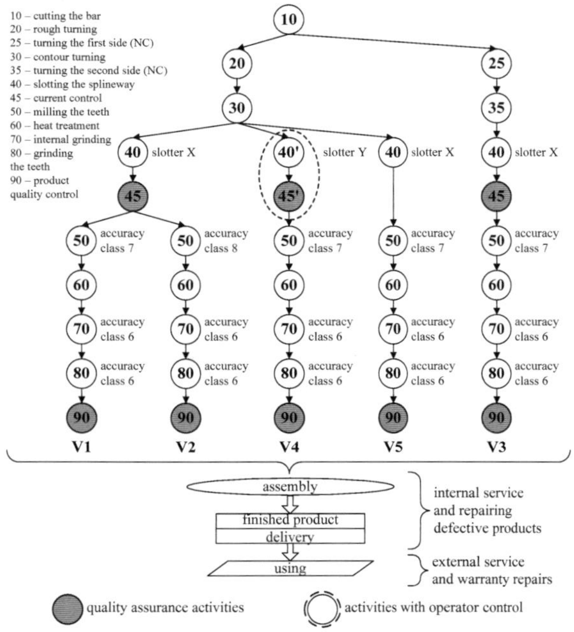

Fig. 2. Exemplary variants of a gear wheel manufacturing process route including quality assurance activities

3. CHOICE OF THE OPTIMAL VARIANT OF THE MANUFACTURING PROCESS ROUTE

The suggested method applied to choice of the optimal variant of the manufacturing process route consists of following stages:

a) determining alternative variants of the process route,
b) selecting activities within analysed variants,
c) establishing activities outputs measures,
d) calculating product cost and quality cost for all variants,
e) estimating failure cost for all variants,
f) determining production process period for all variants,
g) choice of the optimal variant (Pareto optimum).

4. EXAMPLE

Example of choice of the optimal variant of the manufacturing process route concerns production of a gear wheel (z = 50, m_n = 2.0). Rolled bar is an input material. Volume of production is 55 units. Variants of manufacturing process route differ from each other in following properties: structure of activities, machining stations and accuracy of machining (fig. 2).

The result of the calculation for variants V1-V5 is presented in the table 1. Pareto optimum analysis on the basis of the cost and period of the production order is shown in the figure 3. Variants V4 and V5 form Pareto optimum set.

	Cost [cost units]	V1	V2	V3	V4	V5
1	Material cost	5920,15	5920,15	5920,15	5920,15	5920,15
2	Labour cost	3715,13	3829,39	3601,83	3559,47	3513,77
	inc. quality cost	458,78	458,78	458,78	309,70	257,42
3	Internal failure cost	1890,90	1556,50	1596,10	1729,20	3461,15
	inc. internal failure cost linked to product unit	34,38	28,30	29,02	31,44	62,93
4	Indirect cost	11303,41	11250,07	14548,45	11302,38	11210,02
	inc. indirect quality cost	1783,20	1783,39	1773,82	1764,21	1683,55
5	Total product self cost (55 units)	22829,59	22556,11	25666,53	22511,20	24105,09
	inc. quality total cost	4132,88	3798,67	3828,70	3803,11	5402,12
6	Product unit cost	415,08	410,11	466,66	409,29	438,27
	inc. quality unit cost	75,14	69,07	69,61	69,15	98,22

Tab. 1. Cost calculation for analysed variants

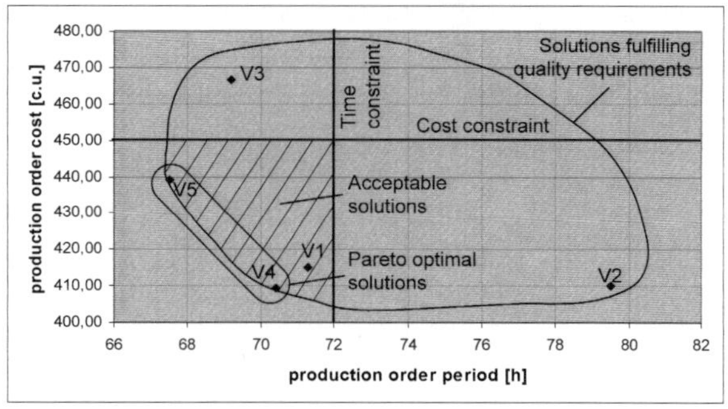

Fig. 3. Pareto optimum analysis

5. CONCLUSION

Taking into consideration cost and time constraints variants V4 and V5 are Pareto optimal solutions. It is important to indicate that the main assessment criterion is the product self cost, not the quality cost. It results from the fact, that variants of the process route may differ from each other in quality assurance activities structure and accuracy of machining – it influences forming failure products and its cost. That relationship has been proved in the presented example.

Presented method enables calculating quality cost according to individual product, not only general business. The result of the calculation is the information concerning the effectiveness of taken activities, incured cost and processes that require streamlining. The information aids making decisions connected with improving quality processes in enterprise.

Received results point at the need to continue the research on it. Complex analysis of other processes within production system in enterprise, not only quality assurance processes, is possible using the presented method. It is justified to work out the Activity Based Management system to manage processes and activities to increase the profits. This system should include analysis of factors which would generate cost, activities analysis and efficiency measurement.

6. REFERENCES

Gunasekaran A.; Marri H.B. & Grieve R.J. (1999). Activity based costing in small and medium enterprises. *Computers & Industrial Engineering*, 37, 1999, pp. 407-411, ISSN: 0360-8352.

Kaplan R. & Atkinson A.A. (1998). *Advanced Management Accounting, Prentice Hall*, ISBN: 0-13-262288-2, Englewood Cliffs, New Jersey.

Matuszek J. & Kołosowski M. (2004). Applying the Activity Based Costing method to improve quality management processes in enterprise, *Proceedings of International Conference Náradie - Tools 2004*, pp. 222-225, ISBN: 80-227-2043-7, Kočovce, Slovak Republic, April 2004.

Nowak E.; Piechota R. & Wierzbiński M. (2004). *Rachunek kosztów w zarządzaniu przedsiębiorstwem (Cost account in business management)*, PWE, ISBN: 83-208-1492-8, Warsaw.

Omachonu V.K.; Suthummanon S. & Einspruch N.G. (2004). The relationship between quality and quality cost for a manufacturing company. *International Journal of Quality & Reliability Management*, Vol. 21, No. 3, pp. 277-290, ISSN: 0265-671X.

Tkaczyk S. & Jagła J. (2001). The economic aspects of the implementation of a quality system process in Polish enterprises. *Journal of Materials Processing Technology*, No. 109, pp. 196-205, ISSN: 0924-0136.

Tsai W.-H. (1998). Quality cost measurement under activity-based costing. *International Journal of Quality & Reliability Management*, Vol. 15, No. 7, pp. 719-752, ISSN: 0265-671X.

Annals of DAAAM for 2008 & Proceedings of the 19th International DAAAM Symposium, ISSN 1726-9679
Editor B. Katalinic, Published by DAAAM International, Vienna, Austria 2008
Make Harmony Between Technology and Nature, and Your Mind will Fly Free as a Bird

HIGH SPEED CUTTING OF GREY CAST IRON

KOPAC, J[anez] & SOKOVIC, M[irko]

Abstract: *In our paper is presented the procedure of grey cast iron high speed cutting and the comparison tests of different cutting materials. We will present the machines and the cutting tools, which have been used for the testing. The procedure is fine and rough turning and the grey cast iron mark is SL 25 (Slovenian designation). The entire test is presented in diagrams and photographs.*

Key words: *Grey Cast Iron, HSC, Cutting tool, Cubic Boric Nitride.*

1. INTRODUCTION

Machines are the first important part, which is necessary for modern manufacturing. Today's economy or market demands are completely clear; the lowest possible costs are demanded and the delivery times are short. The machine, who would achieve such demands, is a CNC machine-tool. With the high speed cutting is presenting the executive phase of development.

An appropriate cutting tool selection is the basic foundation for all mentioned, because is the machining time and with this the machining costs dependent from it. The problem in tool choosing is at the high speed cutting of grey cast iron, because the iron quality and composition is changing along the ingot.

2. TESTING PRESENTATION

2.1 The material

Grey cast iron is among all castings the most frequent. The casting is made from grey crude iron with another refound in the special furnaces called cupola. It has a grey break, because the carbon is separated as graphite. The chemical structure is usually: 2,7 - 4,2 % C, 0,5 - 3,5 % Si, to 1,3 % Mn, to 1 % P and 0,04 - 0,15 % S. The alloyed grey cast iron can also contain: Cr, Ni, Mo, V, Ti, and Cu, which improve its technical characteristics (Maleric 2007).

2.1 HSC machine

The machining measurements have been achieved in the LABOD laboratory at the Faculty of mechanical engineering in Ljubljana. The machine used for testing was a MORI SEIKI SL-153; CNC turning centre with rotary tools, presented on the lower figure.

Fig. 1. HSC machine – MORI SEIKI SL-153.

2.3 Cutting tools

For the testing we used some standard CNMA 120412 turning cutting tools made from carbide – group K20, which is the intermediate value between the hardness and toughness. The cutting tools were: Walter WAK 20, Sumitomo AC 700G and 300G.

2.4 The cooling lubricant

For the cooling and the lubrication during the test execution we used the cooling lubricant fluid Vasco 5000. The fluid is a capacity cooling and lubricating medium without chlorine on the base of vegetable oil mixed with water. It is suitable for hardly cutting machining of alloyed steels and aluminium.

During the test we also used the INDUTEC® device for minimal cooling-lubrication (Kopac 1991, 2000).

3. TESTING PROCEDURE AND RESULTS

3.1 Rough machining – bark peeling

The rough machining was performed as a cut along the total workpiece length. The cutting parameters were: cutting depth 3 mm, feeding 0,25 mm/vrt, cutting speed 450 and 300 m/min, starting diameter 60 mm, cutting angle 95°.

The measurements were taken at different cutting speeds, but at the same feeding and cutting depth. The criterion was: VB = 0,3 mm or Ra = 3,2 μm.

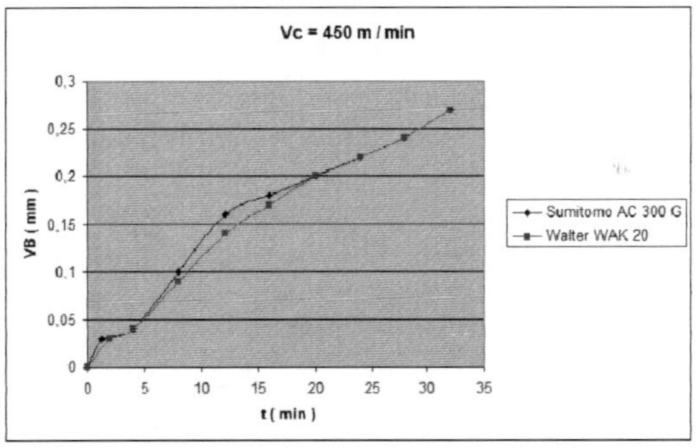

Diag. 1. The comparison of two cutting tools at the same cutting speed.

The diagram is presenting us the comparison between two cutting tools, AC 300G and WAK 20, at the same machining speed. We noticed that the results are quite equal. Such results can be a consequence of the workpiece crust structure or is the result dependent on the unsuitable cutting tool choice.

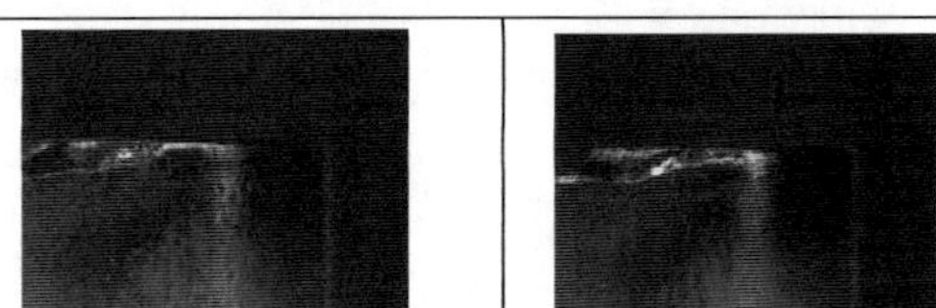

1 mm from the edge t = 32 min VB = 0,27 mm VBmax = 0,33 mm Vc = 45 0m/min WAK 20	1 mm from the edge, t = 32 min VB = 0,27 mm VBmax = 0,3 mm Vc = 450 m/min AC 300 G

Fig. 2. The demonstration of the cutting tools consumption at rough machining.

3.2 Fine machining

The fine machining was executed with the next machining parameters: cutting depth 0,4 mm, feeding 0,2 mm/vrt, cutting speed 600 and 800 m/min, starting diameter 54 mm, cutting angle 95°. The cooling was executed with the cooling lubricant liquids and oil fog, and a small part without cooling – dry machining.

The measurements were taken at different cutting speeds, but at the same feeding and cutting depth. Like in the precedent testing we set the criterion: VB = 0,3 mm or Ra = 3,2 µm.

From the test results we can notice that the best cutting conditions turn up at the turning without cooling lubricants or at dry machining. In production these means an additional saving and smaller influence on the environs with different chemicals, like cooling lubrication fluids.

If we look at the tests with the cutting tool Sumitomo AC 700 G, we can notice that is the case completely reverse as at the Walter WAK 20 tests. Measurements show, that the cutting is better, if we cool the tool with the cooling lubricant fluids at lower cutting speeds and perform the dry machining at higher cutting speeds.

Also interesting is the conservation comparison between the two cutting tools. For the comparison we chose the test results at the cutting speed of 600 m/min, with the use of cooling lubrication fluids and without it.

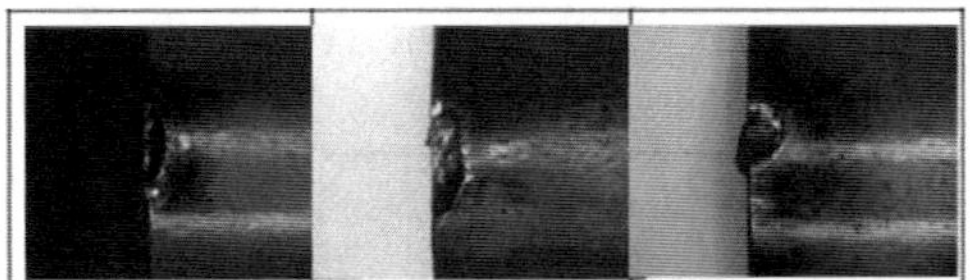

t = 63,63 min, VB = 0,3 mm Vc = 600 m/min WAK 20 With the use of cooling lubricant	t = 69,2 min, VB = 0,36 mm Vc = 600 m/min WAK 20 Without cooling lubricant	t = 42,75 min, VB = 0,32 mm Vc = 600 m/min WAK 20 Cooling with the oil fog

Fig. 3. The WAK 20 tool free plane wear at turning with the same parameters but different types of cooling.

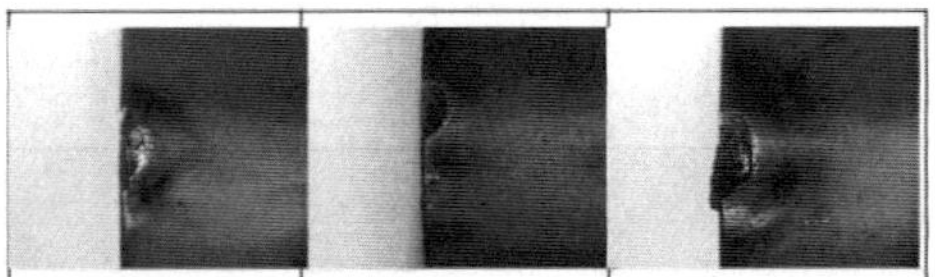

t = 39,19 min, VB = 0,31 mm Vc = 600 m/min AC 700 G With the use of cooling lubricant	t = 26,5 min, VB = 0,3 mm Vc = 600 m/min AC 700 G Without cooling lubricant	t = 35,85 min, VB = 0,31 mm Vc = 600 m/min AC 700 G Cooling with the oil fog

Fig. 4. The AC 700 G tool free plane wear at turning with the same parameters but different types of cooling.

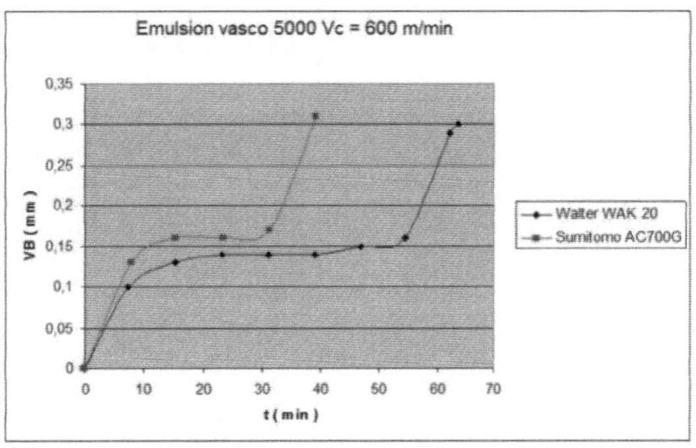

Diag. 2. The conservation comparison between the two cutting tools; with the use of cutting lubricant.

The next two diagrams (Diag. 2 and 3) are showing the conservation times comparison for the two cutting tools, considering the cutting edge wear VB whit the use of cooling lubrication fluid Vasco 5000 and without it.

Trough the entire measurement was the cutting tool Walter WAK 20 showed as more quality, this is clearly visible on the diagram. The tool has better conservation for approximately 24 minutes.

The cutting tool Walter WAK 20 again showed as better. The findings can be connected with the preliminary tool tests, where we already noticed some advantages of the Walter tool. Results are better for approximately 40 minutes.

From the acquired results we can deduce, that the use of the cutting tool Walter WAK 20 is more efficient, above all because of his conservation at dry machining (Maleric 2007).

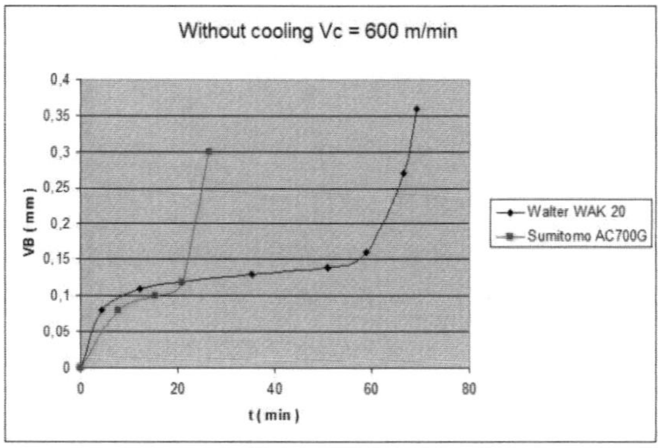

Diag. 3. The conservation comparison between the two cutting tools; without cooling.

4. CONCLUSION

Today's industry depends upon the competitiveness; this can be achieved with fast response time as also with short machining times. Many factors influence the machining time, one of these is the machined material along with the use of different cutting materials. We must pay attention at the use or choice of cutting tools, because certain materials can react between themselves (Marek 1963).

Trough the entire test and measurements the cutting tool Walter WAK 20 was better than the Sunitomo AC 700G. The cutting tool is the most suitable for the grey cast iron machining at the dry machining, for such treatment is the tool proved as optimal.

5. REFERENCES

Kopac, J. (1991). *Odrezavanje (Cutting),* Fakulteta za strojništvo (Faculty of Mechanical Engineering), Univerza v Ljubljani, ISBN 86-7217-089-X, Ljubljana

Kopac, J. (2000). *Odrezovalni stroji (Cutting machines),* Fakulteta za strojništvo, Univerza v Ljubljani (, ISBN 961-6238-32-9, Ljubljana

Marek, E. (1963). *Mehanska tehnologija II (Mechanical technology II),* Višja tehniška šola v Mariboru, Maribor

Maleric, S. (2007). *Primerjava obdelovalnosti sive litine SL 25 z različnimi rezalnimi orodji (The comparison of SL 25 grey cast iron machining with different cutting tools),* Diplomska naloga visokošolskega strokovnega študija, Fakulteta za strojništvo, Univerza v Ljubljani, Ljubljana

***www.sts.si/arhiv/tehno/struzenje/st1.htm, *Accessed on: 2008-09-13*

Annals of DAAAM for 2008 & Proceedings of the 19th International DAAAM Symposium, ISSN 1726-9679
Editor B. Katalinic, Published by DAAAM International, Vienna, Austria 2008
Make Harmony Between Technology and Nature, and Your Mind will Fly Free as a Bird

CLAMPING FIXTURES IN CELL MANUFACTURING

KOSTAL, P[eter]; MATUSOVA, M[iriam] & CHARBULOVA, M[arcela]

Abstract: *Philosophy of flexible manufacturing cell dedicated for small batch production present production system that manipulates with workpieces and produce from semi products individual parts for concrete final products. Base of the complex automatic production is automation of technological process control, automatic transport, manipulation, feeding, workpiece and tool exchange. Industrial production develops, the new generation production systems – intelligent production systems. Development of new generation production machines and systems demand that they are equipped by adequate fixture devices. These fixture devices are developed with machine together, to don't decrease their production facilities and their time and performance utilizations. Efficiency of fixture device in automated production systems underlies using efficiency of entire production systems.*

Key words: *manufacturing, clamping, fixtures, cell,*

1. INTRODUCTION

Cell manufacture is characterized by wide economical effect, which is nearly the same as economical effect of mass or high series production. Cell production also ensures flexible change of manufacturing assortment, which can be changed following to the market requirements.

Modern production cells has in many aspect inbuilt intelligent control system on different level. This intelligent control system ensures better using of such devices. It also ensures manufacturing costs minimization, as well as manufacturing quality increasing. Such intelligence in these days includes various control algorithms. Such algorithms are able to optimise manufacturing cell activity following to the various actual external factors. These factors are influencing to the manufacturing process inside of the cell and in connection with manufacturing process, can these factors influence to the whole factory and its parts. Very important parts of intelligent manufacturing cells are its peripheral devices, such as for example clamping devices. These devices have to design, that they are not able to limit abilities of whole manufacturing cell. Better say, whole design process of all peripheral devices have to simple integration of new intelligent functions in to the manufacturing cell and also to the manufacturing process. Such peripheries have to be equipped by proper actuators and sensors to be able effective cooperate with manufacturing cell control system. We are building such manufacturing cell on our department. This cell will ensure manufacturing and assembly operations and we are also developing some peripheral devices such as clamping device with some intelligent functions.

2. MODULAR CLAMPING SYSTEM

The modular clamping system contains a group of normalized parts. From these parts is possible making the compact unit for needed part clamping. The clamped part is usually more complex.

The standard structure of modular clamping system is defined by part list. Number and type of used parts is changeable according a concrete disposition, workpieces set, manufacturing conditions and manufacturing organization.

Fig. 1: Using of modular clamping system in industry

The modular clamping system parts construction is optimized to part number minimizing and maximizing of usability. According of this optimization we achieve the short upsetting time of clamping fixture and higher rigidity of clamping too. The final effect of these achievements is higher accuracy of machining. (Matúšová & Hrušková 2007)

The using of modular clamping systems is one from more important rationalization process in flexible manufacturing. These kind of clamping fixtures are successfully used in industry and helps achieve more effective manufacturing (Fig. 1). Their advantages are shown mainly in small and medium batch production. Possibility of their usage is given by its flexibility. (Danišová et al 2006)

3. INTELLIGENT CLAMPING FIXTURE DESIGN

A robotic work place is realized in laboratories of institute. This work place is equipped by SMC industrial robot with storage system. Both this systems are connected with feeding rotary device. The system is equipped by automated gripper changing system and it is also equipped by pneumatic clamping system, which will contain some basic functions of intelligent clamping system.

Pneumatic driven clamping device will be placed in the working area of industrial robot. This device will be designed by these requirements:

- Clamped dimension to the 70mm – there will be no manipulation with parts, which will overreach this dimension. Robotic arm bearing capacity is 100N.
- Synchronized clamps movement – this requirement is necessary because of rotary parts clamping
- Open clamps signalisation – to prevent manipulation with clamped part, or to prevent putting the part in the fixture, when there is one part already
- Part clamped in fixture device signalisation – prevent to the collision of part which are placed in the fixture and which are gripped in robotic arm
- Clamped shapes variability – clamping of various shapes – cylinder, square
- Using of changeable clamps – clamps will be changed with help of the gripper
- Proper clamps on the fixture signalisation – to prevent clamping of oval shaped part with clamps not used for oval shaped parts

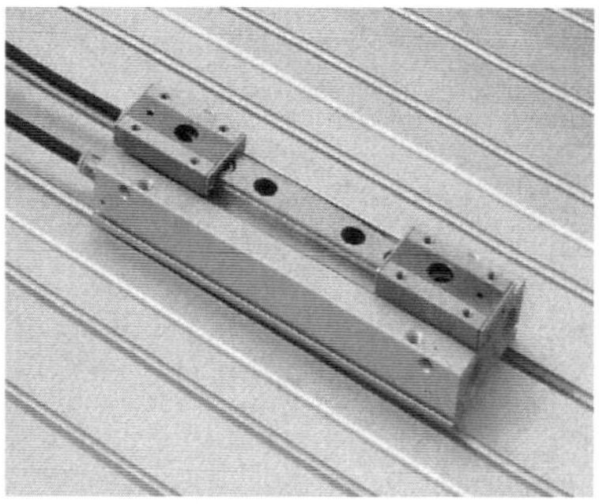

Fig. 2: Clamping device MHF2

- Clamping force regulation ability – it is necessary to use various clamping force for various types of pars.

According to these requirements was chosen the pneumatic base for proper clamping device. The basic building element of the fixture will be double clamp fixture with marking MHF2. Firm SMC produced the fixture. The design of the fixture is known as design with synchronized, parallel driven clamps movement. Chosen fixture has maximal clamping dimension 80 mm. More precisely we chose clamping fixture MHF2 20D2 (Fig. 2.)

Fixture device is pneumatic driven. Two double acting cylinders realize fixture acting. One pressed air input will open the clamps and other one input will close them. Clamps synchronization movement is ensured by geared transfer mechanism. One or two sensors can monitor clamps position. Sensors can be mounted to the fixture with help of grooves. Chosen pneumatic actuator advantages are for example simple design, safeties, reliability, vide clamping dimension, undemanding maintenance, high lifetime small sizes and low costs. Working pressure can be set up from 0,1 to the 0,7 MPa, what means that clamping forces will rise from 30N to the 220N on each clamp. (Danišová & Zvolenský 2006)

3.1 Clamps presence scanning

Also design of monitoring system of claps presence on the fixture body, was created following to some conditions. Most important conditions influenced to the design are sensor sizes, sensor simplicity, its reliability, lifetime, rapid and exact scanning.

For actual needs (it is only important to scant clamps presence, not the type of clamps) is the most proper solution, using of micro switchers. Micro switchers will be placed at clamps holder. Biggest advantage of such scanning is price of micro switchers. Sensors will be placed on the pneumatic fixture body. But we cannot do any holes in to the actuator body, that why switchers will be glued. In the future, after design finishing phase, after short fixture running, switchers will be replaced by other more complicated and effective system, which will ensure the clamps presence scanning. (Košťál & Velíšek 2007)

3.2 Workpiece presence scanning

There are different ways how to scan presence of work piece in the fixture clamping area. Various types of sensors such as optical sensor, inductive sensor, camera scanning system and so on can scan this position. Most effective solution is using of optic sensor, which will be placed in the middle of clamping dimension. Big advantage of optic sensor against to the inductive sensor or against to the capacity sensor is its longer scanning range and there is also a possibility of various material kinds scanning. Camera scanning systems are for this solution vain complicated. For this concrete solution there will be used optical reflex sensor with axial optic, with advantage of optical beam regulation.

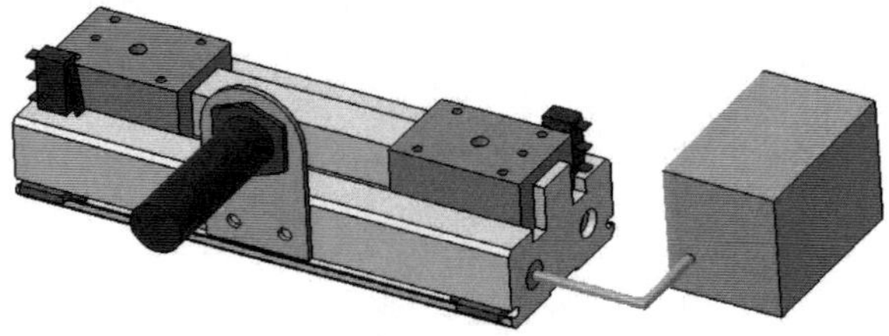

Fig. 3: Model of clamping fixture equipped by sensors

Optical reflex sensors are working at principle of optical performance scanning. They are comparing set optical performance value with actual value coming from the environment. Such sensors are able to scan object of various sizes and complexity. Sensor will be placed on the fixture device with help of assembly clip, which will be glued on the side of the fixture device body. (Koš ál et al 2007)

4. CONCLUSION

Intelligent clamping systems can help to increase reliability of some operations. Automated manufacture can be done more flexible. Such clamping systems are more expensive and complex as classical fixtures, because they have integrated some automated control and measuring units. Intelligent clamping systems are able to prevent to some failures, which are created during the production in automated manufacture and assembly. Designed intelligent sensors will equip pneumatic clamping system. Sensors will scan clamps presence, working product presence; clamps placement and also clamping force regulation.

The best way to workpiece detection is by our meanings to use of an optical sensor. The capacitive or inductive sensors have small range of detection, and detection by camera is very complicated and expensive solution for this case.

Acknowledgment

The research has been supported by VEGA MS and SAV of the Slovak Republic within the project: **VEGA 1/3164/06 - Using of intelligent fixtures in manufacturing and assembly processes.**

5. REFERENCES

Danišová, N; Pastierovič, M. & Fidler, B. (2006) Nové generácie upínaèov vo výrobnom procese. (New generation of clamping fixtures in production) *Vedecké práce MtF STU v Bratislave so sídlom v Trnave.* pp. 21-26 ISSN 1336-1589. - È. 21 (2006),

Danišová, N.; Zvolenský, R. (2006) Automated flexible manufacturing systems. *Proceedings of KOD 2006,* pp. 253-256, ISBN 86-85211-92-1, Palic, 30-31. maj 2006. - Novi Sad : Fakultet tehnièkih nauka

Koš ál, P.; Matúšová, M. & Velíšek, K (2007) Modeling of clamping fixtures. *Academic Journal of Manufacturing Engineering.* pp.65 – 68, ISSN 1583-7904. - Vol. 5, No. 2

Košťál, P.; Velíšek, K. (2007) Intelligent clamping fixtures for box shaped workpieces. *Proceedings of 7th. International Multidisciplinary Conference.* ISSN 1224-3264. - Vol. XXI Baia Mare, Romania, May 17-18, 2007 (2007). - Baia Mare : North University of Baia Mare, 355-358, I. diel

Matúšová, M.; Hrušková, E (2007) Database of fixture mechanisms. *Proceedings of RaDMI 2007* pp. 613-617 ISBN 86-83803-22-4 , 16-20 September 2007. - Trstenik : High Technical Mechanical School of Trstenik,

Annals of DAAAM for 2008 & Proceedings of the 19th International DAAAM Symposium, ISSN 1726-9679
Editor B. Katalinic, Published by DAAAM International, Vienna, Austria 2008
Make Harmony Between Technology and Nature, and Your Mind will Fly Free as a Bird

PRODUCT DEVELOPMENT THROUGH MULTI-CRITERIA ANALYSIS

KOSTANJEVEC, T[omaz]; POLAJNAR, A[ndrej] & VUJICA HERZOG, N[atasa]

Abstract: *The trend regarding the development of mass production product is forecast by the use of a developed model of multi-criteria analysis. In this analysis a multi-dimensional space is used, which enables the establishment of a time argument in comparison to product parameters. The establishment of product parameters is achieved by means of internal company information and market needs. In this way it is possible to establish an innovative environment for the development of new products.*
Key words: *product, development, analysis, forecasting*

1. INTRODUCTION

Modern companies require constant investment into development. The development of the equipment for product development is also very wide, while the use of such equipment is still not as widespread as it should be (Crawford & DiBenedetto 2008). As a line of business, product development has played an important role in production engineering by researches within the global industry experience, design and analysis (Cooper 2001), product design (Clark & Wheelwright 1993) and creativity used in product development (Goldenberg & Mazursky 2004).
On the other hand, product development also includes research within the market with regard to customer needs, product positioning and segmentation, product forecasting and test marketing. The production and engineering precision, combined with the marketing approach, both of which were focused on customer needs and production capability, proved very successful (Griffin & Hauser 1996). In parallel with the development of established equipment, researchers discovered the correlation of new product success by establishing the communication between marketing and production engineering as the most important elements necessary for success. New challenges and opportunities are reflected in global markets, global competitiveness, the global spread of engineering knowledge and with communication technologies. The use of multi-criteria analysis with product development represents a new challenge and an opportunity in design research and new product forecasting (Glavač & Ren 2007).

2. THEORETICAL BACKGROUND

In the last decade companies were focused on new product development on the basis of satisfying customer needs. Researchers in the field of marketing were convinced that understanding customer needs and improving the transfer of these needs to product manufacturers was the key to success.
Under unchanged conditions a product will prove profitable if it supplies customers with greater satisfaction, is among the leading products in the market, and has low production and development costs. The concept of joint or common dealing of product parameters is included in the conjoint product analysis. For example, a company should have a strategy for dealing with technology and with implementing methods for understanding opportunities that are to be offered to customers by means of offered products, as well as with recognising

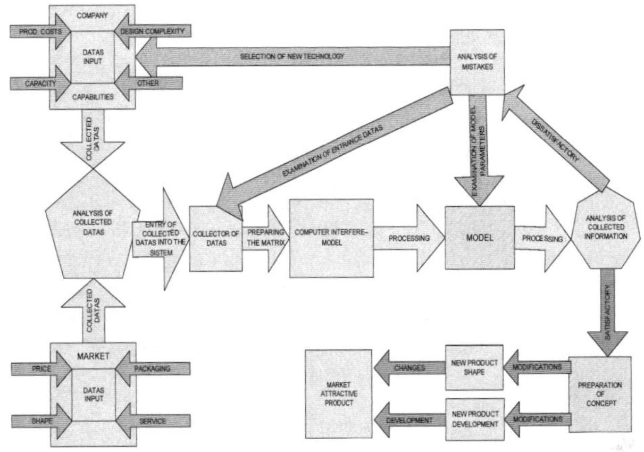

Fig. 1: Conceptual scheme of product development

where demands and expectations are not being fulfilled. The process of multi-criteria product analysis also includes product platforms. It has been recognised in numerous industry branches and companies that it is far more effective to develop products in platforms. From the point of view of customers these platforms supply companies with the possibility of adapting to customer demands and market needs (Ulrich & Eppinger 2000).

3. MODEL OF MULTI-CRITERIA ANALYSIS FOR PRODUCT DEVELOPMENT ASSISTANCE

The conceptual model with data capture and information flow is shown in Figure 1. The data is collected from the market with a data collecting form and the data flows into the data collector together internal - production data. The combined data is processed in the model, and in this way the model anticipates trends on the basis of individual parameters as well as the common trend. The possibility of development determination also depends on the body's centre of gravity which is time-dependent. Possible feedback connections are also shown on the conceptual Figure 1 and serve as corrections or modifications to the model.
The idea about multi-dimensional analysis of product acceptability in the market was born from watching two-dimensional graphs showing the dependence of the dependent variable from the independent one. The independent variable represents time, the dependent one is derived from the observed and most representative parameters – e.g. prices, sale quantities, input into development on individual product, costs of manufacturing. Because of the incapability of demonstrating individual parameters on one graph the concept or model is presented, which could eliminate that weakness. Due to longer non-changeability of products and technology in the inspected branch most of the products observed remained the same in that time or changed minimally. The calculated trend on the basis of the gravity centre is represented graphically and mathematically and applies for future years.

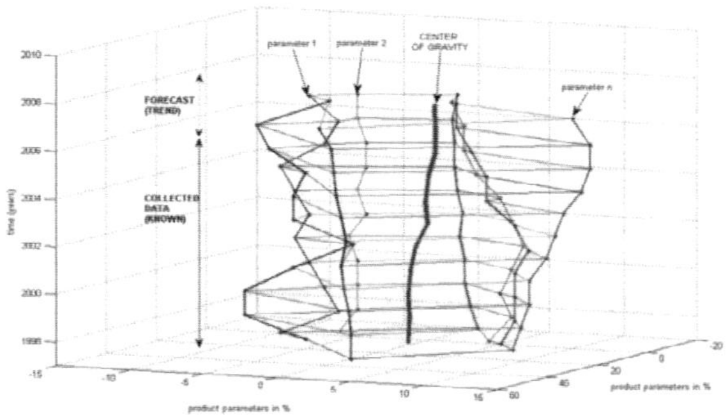

Fig. 2: Outline scheme of the model

Presentation of results with a polar way of data demonstration is improved since it shows the trend and not just information. The centre of gravity of the surface shape enclosed by polar coordinates of observed parameters is calculated. The shape contains a set of n-triangles and the sum of the *n*-centers of the mass of those triangles is the mass centre of one observed time parameter. An algorithm draws and calculates polar coordinates for individual periods between the year *m-1* and the adjacent variable, and the year *m+1* and the adjacent variable. The year is selected as a time argument typical for the branch since the product and technology changes are relatively slow. The representation of the centre of mass in a polar way (Eq.1), (Eq.2) and (Eq.3) provides complete information related to an individual time period.

$$T_{pol_m} = \rho_m e^{j\gamma_m} \qquad (1)$$

$$\rho_m = \sqrt{x_m^{\;2} + y_m^{\;2}} \qquad (2)$$

$$\gamma_m = tg^{-1}\left({y_m}/{x_m} \right) \qquad (3)$$

Time as the only independent variable is represented by the central axis in the graph, which is the result of model visualization. The displacement of dependent variables on time shows their importance and development upon time parameter.
There are an arbitrary number of dependent variables in the model. They are configured in the circle around time in the form of uniform distribution. In the observed changing of the trend it has been ascertained that linear accommodation is inaccurate.
These are higher order curves that are genomically imposed and on the bases of polynomial curves and Fourier rows (harmonic analysis in Eq.4) they form a trend for each parameter separately (curve accommodation is higher than 95%).

$$s_n(x) = \frac{a_0}{2} + a_1 \cos\frac{2\pi x}{T} + a_2 \cos 2\frac{2\pi x}{T} + \ldots + \frac{a_n}{2}\cos n\frac{2\pi x}{T} + $$
$$+ b_1 \sin\frac{2\pi x}{T} + b_2 \sin 2\frac{2\pi x}{T} + \ldots + b_{n-1}\sin(n-1)\frac{2\pi x}{T} \qquad (4)$$

4. DISCUSION

The observed model shows the aggregate variable to be the centre of gravity. Analysis of data in the future allows linear review or quadratic review of the trend, including an arbitrary form of establishing a trend based on lines of respective independent variables (Figure 3). The common trend of spatial lines allows for analysis of a common product development trend. It depends on the respective field whether data from the market or from the company is more important.
When the model is ideal, all dependent variables may be presented arbitrarily at optional angles, and further broken down into their dependent variables. When reviewing independent variables this may be done in a classical circle diagram. The system is limited by the values of the dependent parameters of the product, which represent the limit value.

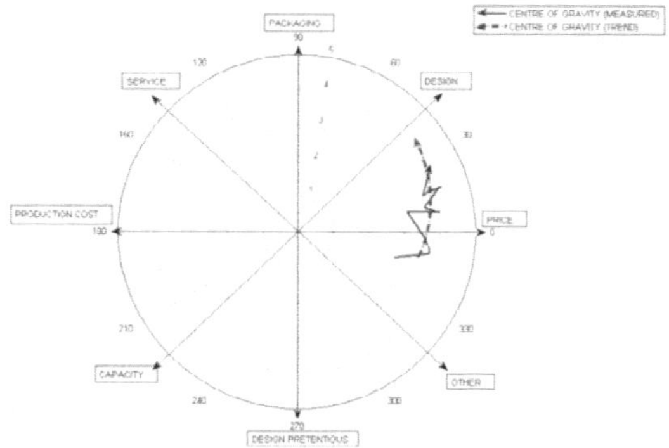

Fig. 3: Polar graph of the movement of the centre in the observed years of specific product (real case study)

This limit value is the limit which the company is not allowed to surpass in a respective product development in considering the model, as doing so would impair its competitiveness, which must be monitored with regard to all parameters of the product. In this part the model looks to the finite elements analysis, where outer limits are recommended for development and establishment of individual product parameters - summarised according to (Glavač & Ren 2007). Figure 2 shows the »skeleton« multidimensional graph of product development. Limits to the system are set uniformly with boundary lines of observed parameters. On the Figure 3 (which was made from real case study) we see, that importance of price is moving toward design. Attention of the specific firm should be focused on design in next few years.

5. CONCLUSION

Life cycles of technologies, products and processes are becoming ever shorter, so technological foresight is a very important aspect of their planning. In a time when foretelling the development of products is difficult and the price of error as steep as it is, the article offers a solution for the development of products through multidimensional analysis. The developed model of product design through multi-criterial analysis can be used in the development of virtually any mass produced product. Model has a lot of potential and a lot of space for further investigations and real case study modifications.

6. REFERENCES

Clark, K. B., Wheelwright, S. C. (1993) *Managing new product and process development: text and cases.* New York [etc.]: The free press: Simon & Schuster. XV, 896 p. ISBN 0-02-905517-2

Cooper , R. G. (2001) *Winning at new products: accelerating the process from idea to launch.* 3rd ed. Reading (Mass.): Perseus,. XIII, 425 p. ISBN-13 978-0-7382-0463-5

Crawford, C. M., Di Benedetto, C. A. (2008) *New products management.* 9th ed. Boston [etc.]: McGraw-Hill/Irwin. XVIII, 558 p. ISBN-13 978-0-07-352988-2

Glavač, M., Ren, Z. (2007) *Multicriterial optimization of a car structure using a finite element method.* Journal of Mechanical engineering, vol. 53, No. 10, p. 657–666.

Goldenberg, J., Mazursky, D. (2004) *Creativity in product innovation.* 3rd printing. Cambridge (U. K.) [etc.]: Cambridge university press. XIII, 224 p. ISBN 0-521-80089-7

Griffin, A. J., Hauser J. R. (1996) *Integrating R&D and marketing: a review and analysis of the literature.* Journal of product innovation management, vol. 13, No. 3, p. 191–215.

Ulrich, K. T., Eppinger, S. D. (2000) *Product design and development.* 2nd ed. Boston: Irwin/McGraw-Hill. XXVI, 358 p. ISBN 0-07-229647-X

Annals of DAAAM for 2008 & Proceedings of the 19th International DAAAM Symposium, ISSN 1726-9679
Editor B. Katalinic, Published by DAAAM International, Vienna, Austria 2008
Make Harmony Between Technology and Nature, and Your Mind will Fly Free as a Bird

QUADRATIC APPROXIMATION OF CHARACTERISTIC POLYNOMIAL OF SYMMETRIC POSITIVE DEFINITE TOEPLITZ MATRIX

KOSTIC, A[leksandra] & COHODAR, M[aida]

Abstract: *In this note we present a modified Newton's method for computing the smallest eigenvalue* λ_1 *of a symmetric and positive definite Toeplitz matrix. This method based of the Taylor series characteristic polynomial of a symmetric and positive, definite Toeplitz matrix. The characteristic polynomial has been approximated with the polynomial of the second order from the Taylor series because it is easy to develop* $p_n(\lambda)''$ *from specific structure Toeplitz matrix from Gohberg-Simencul formulae.*

Keywords: *eigenvalue problem, Toeplitz matrix, modified Newton's method, characteristic polynomial.*

1. INTRODUCTION

The problem of finding the smallest eigenvalue λ_1 of a real symmetric, positive definite Toeplitz matrix T_n is of considerable interest in signal processing. Given the covariance sequence of the observed data, Pisarenko (Pisarenko,1973) suggested a method which determines the sinusoidal frequencies from the eigenvector of the covariance matrix associated with its minimum eigenvalue. The computation of the minimum eigenvalue of T_n was studied in, e. g. (Cybenko&Van Loan,1986; Kostic,2004; Mackens&Voss,1997,2000; Melman,2006).

In this paper we propose new modification of Newton's method for the computation of the roots of the characteristic polynomial of a real symmetric positive definite Toeplitz matrix. To compute the Newton step in (Mackens&Voss,2000; Mastronardi&Boley,1999), these methods use a recursion for the evaluation of the characteristic polynomial and its derivative which requires $O(n^2)$ flops to solving the Yule-Walker equations. Melman in (Melman,2006) show that this recursion can be replaced by a shorter computation, involving the computation of the trace of an appropriate matrix, and, after solving the Yule-Walker equations, requiring only $O(n)$ flops.

As it is pointed out, the paper presents new modified Newton method which is based on the Taylor expansion of the characteristic polynomial. Specified structure of the Toeplitz matrix provides relatively simple and cheaper calculating the second derivatives of function in sense of flop's numbers. In this way, we replace the characteristic polynomial with the polynomial of the second order from the Taylor series, which faster convergences than classical Newton method in enough small neighborhood of the smallest eigenvalue.

In Section 2 we present the basic properties of Toeplitz matrices and the notation we will use. In Section 3 we compute the new Newton's method and in Section 4 we compare the methods. In Section 5 we present further research.

2. PRELIMINARIES

The $(i,j)^{\text{th}}$ element of an n x n symmetric Toeplitz matrix T_n is given by $t_{|i-j|}$ for some vector $\left(1,t_1,\ldots,t_{n-1}\right)^T \in R^n$. The matrix T_n satisfies $JT_nJ=T_n$ and is therefore centrosymmetric. We use I for the identity matrix and J for the exchange, or "flip" matrix with ones on its southwest-northeast diagonal and zeros everywhere else. We note that for any $\lambda\in R$, the matrix $(T_n-\lambda I)$ is symmetric and centrossymmetric, whenever T_n is. In what follows, an important role is played by the so-called Yule-Walker equations. For an n x n symmetric Toeplitz matrix T_n, defined by $\left(1,t_1,\ldots,t_{n-1}\right)$, this system of linear equations is given by $T_n y^{(n)}=-t$ where $t=(t_1,\ldots,t_n)^T$. There exist several methods to solve these equations. Durbin's algorithm solves them by recursively computing the recursively computing the solutions to lower-dimensional systems, provided all principal submatrices are non-singular. This algorithm requires $2n^2+O(n)$ flops.

3. THE (MODIFIED) NEWTON'S METHOD FOR THE CHARACTERISTIC POLYNOMIAL

The characteristic polynomial for the symmetric matrix T_n is given by $p_n(\lambda)$:=det(T_n -λI). If Newtton's method were used to compute the roots of this polynomial, then the following lemma gives a convenient expression for the Newton step.

Lema1. The Newton step for the characteristic polynomial $p_n(\lambda)$ of an n x n matrix T_n at $\lambda=\bar{\lambda}$, where $\bar{\lambda}$ is not one of the eigenvalues of T_n, is given by:

$$N(\bar{\lambda})=-\frac{p_n(\bar{\lambda})}{p'_n(\bar{\lambda})}=\frac{1}{trace\left(T_n-\bar{\lambda}I\right)^{-1}}. \tag{1}$$

As we apply Gohberg-Semencul formulae (Gohberg&Semencul,1972)for the inverse of a Toeplitz matrix we use the next definition. With the first row of T_n given by $(1,t^T)$, we first define

Definition 1. Let $q(\lambda)$:= $-1+\lambda+t^T\,(T_{n-1}-\lambda I)^{-1}t$ the secular function. With $w=-JT_{n-1}t$, the Gohenberg-Semencul formula for the inverse of T_n is then given by:

$$T_n^{-1}=-\frac{1}{q(0)}\left(M_1M_1^T-M_o^TM_0\right), \tag{2}$$

where

$$M_0=\begin{pmatrix} 0 & 0 & \ldots & 0 \\ w_1 & 0 & \ddots & 0 \\ w_2 & w_1 & \ddots & \ddots & \vdots \\ \vdots & \vdots & \ddots & 0 & 0 \\ w_{n-1} & w_{n-2} & \ldots & w_1 & 0 \end{pmatrix}, M_1=\begin{pmatrix} 1 & w_{n-1} & w_{n-2} & \ldots & w_1 \\ 0 & 1 & w_{n-1} & \ldots & w_2 \\ \vdots & \ddots & \ddots & \ddots & \vdots \\ 0 & & \ddots & 1 & w_{n-1} \\ 0 & 0 & \ldots & 0 & 1 \end{pmatrix}.$$

Melman in (Melman,2006) shows the following theorem:

Theorem1. The Newton step for the characteristic polynomial $p_n(\lambda)$ of an n x n symmetric positive definite Toeplitz matrix T_n at $\lambda=\bar{\lambda}<\lambda_1$ is given by:

$$N\left(\bar{\lambda}\right)=\frac{q(\bar{\lambda})}{\sum_{j=1}^{n}(2j-n)w_j^2}, \tag{3}$$

where q is as in Definition 1. and the first row of T_n is given by $(1,t^T)$ and $w=(w_1,\ldots,w_{n-1})=-J(T_{n-1}-\bar{\lambda}I)^{-1}t$. For compactness of writing, we have set w_n=1.

On this way, Melman (Melman,2006) avoided the use of recursion for development of the first derivatives of function and minimized development of the first derivation of characteristic polynomial to $O(n)$ flops. In this manner, there are assumptions for efficiency applying of the Newton method, as well as the method with double Newton step. We have idea to moreover apply Gohberg-Semencul formulae for development the second derivative of the characteristic polynomial. Namely, we need to approximate our characteristic polynomial with the polynomial of the second order based on the Taylor series. Let

$$p(\lambda):=p_n(\bar{\lambda})+p_n'(\bar{\lambda})(\lambda-\bar{\lambda})+\frac{p_n''(\bar{\lambda})}{2!}(\lambda-\bar{\lambda})^2 \tag{4}$$

is approximation of characteristic polynomial p_n whereby $\bar{\lambda}$ is not eigenvale of the matrix T_n. So, we can find approximated value of the smallest eigenvalue using solution of the equation:

$$1+\frac{p_n'(\bar{\lambda})}{p_n(\bar{\lambda})}(\lambda-\bar{\lambda})+\frac{p_n''(\bar{\lambda})}{2!\,p_n(\bar{\lambda})}(\lambda-\bar{\lambda})^2=0. \tag{5}$$

Coefficient $a:=\frac{p_n''(\bar{\lambda})}{2!\,p_n(\bar{\lambda})}$ can be easy calculated, because it holds $2\cdot a=\left(trace\left(T_n^{-1}\right)\right)^2-trace\left(T_n^{-2}\right)$. It is obviously from $trace\left(T_n^{-1}\right)=\sum_{i=1}^{n}\frac{1}{\lambda_i}$ and $trace\left(T_n^{-2}\right)=\sum_{i=1}^{n}\frac{1}{\lambda_i^2}$, where λ_i, i=1,2,…,n are all eigenvalues of the matrix T_n. To calculate $trace\left(T_n^{-2}\right)=\sum_{i=1}^{n}\frac{1}{\lambda_i^2}$ we need to find some elements of matrix T_n^{-1}. Without loss of generality we will consider case vhen n is even. As matrix T_n has specific structure, each element $c_{i,j}$ located out of the main and southwest-northeast diagonal of matrix $T_n^{-1}=\frac{1}{q(0)}\left(c_{i,j}\right)_{i,j=1,2,\ldots,n}$ are repeated four times in the matrix and two times on these diagonals. On this way, during calculation these elements, it is needed only quarto of them, that results to cheaper algorithm. Required elements can be calculated by using following recursions:

$$\begin{aligned} c_{i,i-k}&=c_{i+1,i-k+1}+w_i\cdot w_{i-k}-w_{n-i}\cdot w_{n-i+k},\\ c_{i,i-k-1}&=c_{i+1,i-k}+w_i\cdot w_{i-k-1}-w_{n-i}\cdot w_{n-i+k+1},\end{aligned} \tag{6}$$

for k=0,2,…,n-4, i=n-1,…,((n+k)/2)+1.

For above proposed calculation we spent (n-2)n flops. As it holds $trace(T_n^{-2})=\frac{1}{q^2(0)}\sum_{i=1}^{n}\sum_{j=1}^{n}c_{i,j}^2$, that additionally gives $n^2/2$ flops. That means, for new step it is necessary $7n^2/2+O(n)$ flops. Although method is expensive, it gives good results as we are enough closely to eigenvalue of matrix. If it would be combined with Newton method with double step it gives improving that numerical results show.

4. NUMERICAL RESULTS

To test the method we considered the following class of Toeplitz matrices:

$$T_n=m\sum_{k=1}^{n}\eta_k T_{2\pi\theta_k}, \tag{7}$$

where m is chosen such that the diagonal of T_n is normalized to t_0=1. $T_\theta=(t_{i,j})=(cos(\theta(i-j)))$ and η_k and θ_k are uniformly distributed random in the interval [0,1] (cf. Cybenko and Van Loan). Table 1 contains the average number of Durbin steps needed to determine in 100 test problem each of the dimension n=32,64,128,256,512 and 1024.

dim	New method	Method dnm
32	5.47	5.74
64	5.89	6.3
128	5.59	6
256	5.99	6.59
512	6.5725	7.03
1024	6.8425	7.57

Table 1. The average number of Durbin steps

Method with double Newton step is denoted with dnm in Table 1.

5. CONCLUSION AND FUTURE RESEARCH

In this paper we present modified Newton method for characteristic polynomial of symmetric positive definite Toeplitz matrix which approximates characteristic polynomial with second order polynomial from Taylor series. On this way, we improved previously developed algorithm what we also have expected, because we used second derivative of characteristic polynomial. In other words, approximation of characteristic polynomial with second order polynomial, as well as development of second derivative of characteristic polynomial are the new results and the main contribution of the paper.

It would be very interested to approximate the even and odd polynomial of the second order on above pointed out manner but with much more using property of symmetry of the Toeplitz matrices. Also, results would be applied on methods based on secular function and its Padé approximation.

6. REFERENCES

Cybenko, G. & Van Loan, C.F. (1986). Computing the minimum eigenvalue of a symmetric positive definite Toeplitz matrix, *SIAM J. Sci. Stat. Comput,* Vol. 7, No. 1, pp. 123-131. MRO819462(87b:65042)

Gohberg, I.C. & Semencul, A. A. (1972). The inversion of finite Toeplitz matrices and their continual analogues. (Russian), *Mat. Issued.* Vol. 7., No. 2(24), pp. 201-223. MR0353038 (50:5524)

Kostić A. (2004) *Verfahren zur Bestimmung einiger extremaler Eienwerte einer symmetrischen Toeplitz Matrix,* Shaker Verlag, ISBN 3-8322-3235-4, Aachen

Mackens , W. & Voss, H. (2000). Computing the minimum eigenvalue of a symmetric positive definite Toeplitz matrix by Newton-type methods. *SIAM J. Sci. Comput.,* Vol. 21. No. 4, pp. 1650-1656 MR1756049 (2001g:65043)

Mackens , W. & Voss, H. (1997). The minimum eigenvalue of a symmetric positive definite Toeplitz matrix and rational Hermitian interpolation. *SIAM J. Matr. Anal. Appl,* Vol. 18. pp. 523-534

Mastronardi, N & Boley, D.(1999). Computing the smallest eigenpair of a symmetric positive definite Toeplitz matrix. *SIAM J. Sci. Comput.* Vol. 20, No. 5, pp. 1921-1927. MR1694690

Melman, A. (2006). Computation of the Newton step for the even and odd characteristic polynomials of a symmetric positive definite Toeplitz matrix. *Mathematics of computation.* Vol. 75, No 254, pp. 817-832, S 025-5718(05)01796-5

Pisarenko, V. F. (1973) The retrieval of harmonics from a covariance function. *Geophys. J. R.. astr. Soc.* Vol. 33., pp. 347-366

Annals of DAAAM for 2008 & Proceedings of the 19th International DAAAM Symposium, ISSN 1726-9679
Editor B. Katalinic, Published by DAAAM International, Vienna, Austria 2008
Make Harmony Between Technology and Nature, and Your Mind will Fly Free as a Bird

RESEARCH OF SYSTEMS' UNCERTAINTY

KREMLJAK, Z[vonko]

Abstract: *Uncertainty is a characteristic or a property of a system. The structure of a system is defined as a selection of elements, connections between them and boundary of the system. Elements within the boundary of the system are more densely interconnected than the elements outside the system. In the paper we deal with the parametric (mostly used in decision making process) and structural uncertainty. Usually the managers are faced with the cognitive barriers to understand the structural uncertainty, because appropriate knowledge about uncertainty and its consequences is missing.*
Keywords: *organisational systems, parametric uncertainty, structural uncertainty, decision making.*

1. INTRODUCTION

State uncertainty is defined as the inability of an individual to understand and foresee possible changes of elements and their relationships in outside environment. State uncertainty depends on the complexity of the system. The more it is complex the higher, due to bonded rationality and basic inability to predict future, is the understood uncertainty.

Effect uncertainty is defined as the inability of an individual to predict the effect of change in outside environment on the organisational system. Decision making individual knows the state of the environment but does not know the cause-effect relationship between certain environmental changes and influences these changes will have on the system in question.

Uncertainty is the essential characteristics in which every system works. The importance of uncertainty for operations of organisational systems is proven by empirical research which examined the importance of uncertainty in business systems. Schulz (2001) showed that unreliability influences the transfer of knowledge in organisational systems. Unreliability lessens the synergies between marketing, development and production. High unreliability negatively influences the process of new product development.

2. DEFINITION OF SYSTEMS

Response uncertainty is defined as the inability to foresee the consequences of the response. This is a case when the decision making individual knows the response must be made but cannot know what kind of consequences his decisions will bring. In Fig. 1 the division of systems with the help of Venn diagram is shown.

Formal systems are abstract systems which are examined especially from theoretical point of view. Their elements are abstract notions like numbers, symbols, letters, and points. The relationship between these elements is defined by a system of rules which are called axioms. **Natural systems** are systems which were created independently of human will or wishes; their behaviour, development and growth abide certain natural laws. Here belong biological and **ecological systems**. **Social system** is comprised of groups of people which associate with the aim to reach certain goals. Social systems are defined by patterns of behaviour which change dynamically.

The intersection between formal systems and natural systems represent **technical systems**. These are systems which were designed by people which themselves are not a part of the system. Characteristic of technical systems is that they are based on the same laws as the natural systems, or that natural laws are used for achieving functionality and quality. Technical systems are defined by their functionality, quality of operations and adaptability. Clear boundaries of the system, well defined elements and relationships between them enable traditionalistic treatment of technical systems. It is possible to explain the workings of a technical system by taking it apart to its constituent components, where every element has stable incoming and outgoing characteristics (Ljung 1987).

According to the definition criteria technical systems are deterministic, which means that to inputs in technical system belong precisely specified outputs. This means that operations of technical system must be definite; there must be no doubt about the realisation of foreseen function at a specified quality. Technical systems are not completely deterministic but it is clear that uncertainty in operations of technical system clearly limits its functionality and quality. Uncertainty of technical system is usually defined with probability levels, where we presume that the probability levels are objectively definable.

Organisational systems represent an intersection between technical and social systems. Their characteristics are that technical systems are strategically important resources in the system, but the operations of the entire system are not subject to natural laws. Operations of organisational system can be described with certain patterns and principles; these can be changed, because people on the basis of interpretation of patterns and principles form their own decisions and behaviour, which leads to establishment of new patterns and principles. Organisational systems comprise of business systems, production systems, defence systems, logistics systems. Their characteristics are openness, complexity and stochasticity. Activities between elements create patterns of behaviour of the whole system which cannot be understood or foreseen on the basis of knowledge about characteristics of individual elements. Relationships between elements are difficult to define, for elements can change their behaviour according to the information about operations of other elements. This means that it is impossible to control all the aspects of operations of organisational system and estimate the consequences of all available options. Organisational system is not entirely stochastic.

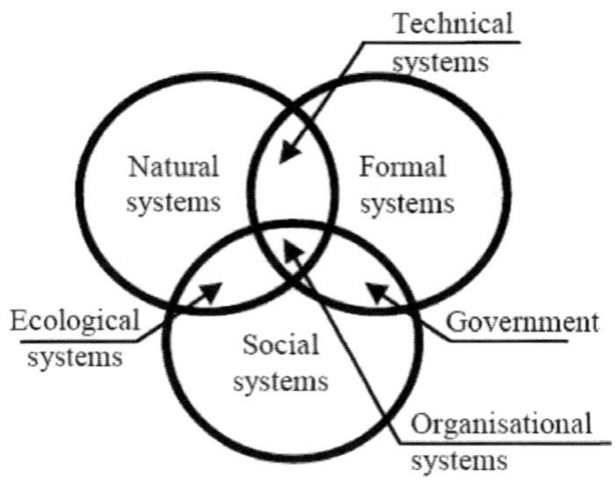

Fig. 1. Venn diagram of division of systems.

3. UNCERTAINTY IN ORGANISATIONAL SYSTEMS

Organisational systems are result of human planning and management. Planning and management in organisational systems can be illustrated with decision making process. This process is subject to uncertainty as the basic characteristics of organisational system.

Dealing with uncertainty in organisational systems can be started by traditional examination of differences between certainty, risk and uncertainty. Certainty in organisational system can be mentioned, when a competent individual reaches decisions connected with operations of organisational system, where he has all the knowledge about possible states in the future and these states are completely independent of activities this kind of system is performing. This kind of organisational system is adaptable, for it is possible to prepare for all eventual states.

The concept of risk in organisational system means that it is possible to objectively define the levels of probability of a state or event. It means that a competent individual knows all possible states in the future and probability that these states will be realised.

Reality in organisational system is usually difficult to describe with the concepts of certainty and risk. The concept of uncertainty is much more suitable for describing the state in organisational systems (Laviolette & Seaman 1994).

Parametric uncertainty represents a type of uncertainty which can be mathematically mastered. Let $U(a; c; \pi)$ be a function, which encompasses activities (a) and expected results (c). Let $\{a\}$ be a group of possible activities, which can be performed, $a = (1, ..., A)$. Let $\{s\}$ be a group of possible states $s = (1, ..., S)$ and $\{c_{as}\}$ a group of results which come from reciprocal operations between activities and states. Let $\{\pi\}$ be a group of subjective probabilities connected to the results and states. Parametric uncertainty can now be clearly defined. It means that the person who makes decision has a full spectrum of activities that can be performed, knows all possible states, and knows all the possible results of all activities. Therefore, he has complete knowledge regarding the structure of the decision making problem $\{a\}$, $\{s\}$, $\{c_{as}\}$. Uncertainty concerns only subjective parameters of possibility $\{\pi\}$.

Structural uncertainty means that the decision maker has incomplete knowledge on the structure of the problem. Or, put another way, he has incomplete knowledge about the three, $\{a\}$, $\{s\}$, $\{c_{as}\}$. For structural uncertainty is characteristic that it is impossible to possess knowledge about all possible consequences. It is important to point out that structural uncertainty means that states in the future are not independent of activities. This means that an individual, who is making decisions, can change his opinion on possible future states according to the information regarding consequences of the activities. Future states are therefore a construct with dynamic characteristics.

Structural uncertainty is not the only name in scientific literature. Kouvelis (1999) talks about radical uncertainty when pointing out that in organisational systems it is impossible to know in advance which knowledge will be developed and which combinations of diffused knowledge will be important in certain circumstances. Unforeseeable and radical uncertainty draws attention to important factors of uncertainty in organisational environment, for example complexity, cognitive limitations of an individual and basic human inability to predict future. The concept of bounded rationality is based on a premise that the complexity of organisational systems causes uncertainty which cannot be managed by individuals; they can only adapt to it. The reality in organisational systems is so complex that a person is limited in processing information. Bounded rationality thus means that a person can never have so much knowledge to transfer structural uncertainty objectively into parametric uncertainty. If it were possible to reduce reality to parametric uncertainty it would be so complex that it would be impossible to examine in its entirety. Bounded rationality highlights cognitive limitations of an individual to process large numbers of activities, alternatives and consequences.

A factor which most prominently presents uncertainty of organisational systems is a human inability to predict future. Future can be treated only as a part of the present. This means that when dealing with organisational systems one cannot have objective knowledge about the future. Structural uncertainty is the result of complexity and openness of organisational system and the inability of people to understand future.

Additional point of view regarding dealing with structural uncertainty represents the fact that a part of uncertainty comes from the environment of organisational system. Scientific literature established a term environmental uncertainty. Organisational systems are open and their survival depends on events in environment (Jauch & Kraft 1986). Milliken (1987) warns about specific inconsistencies in understanding the term environmental uncertainty. It must be distinguished between 'objective' and understood environmental uncertainty. The first describes the state of the environment. The second describes the understanding and experiences of an individual, who must make decisions and is trying to understand the state of environment. The concept of comprehended uncertainty means that environmental uncertainty is not an objective phenomenon but a result of human understanding of the environment. Milliken divides environmental uncertainty into state uncertainty, effect uncertainty and response uncertainty.

4. CONCLUSION

Extensive conceptual discussions and empirical research prove that understanding the concept of uncertainty is the key to understanding the operations of organisational system. The operations of every organisational system are subject to circumstances of uncertainty, therefore, the task of decision making individuals is to design such organisational systems, which will be able to successfully adapt to uncertain conditions. In spite of the knowledge about the importance of organisational system management in the circumstances of uncertainty, there are surprisingly few approaches and techniques which support the decision making process in uncertain circumstances. The main contribution of our work is in the development of new heuristic tools which support decision making process. The challenge for the future is in upgrade of mathematical procedures – informatisation.

5. REFERENCES

Jauch, L. R. & Kraft, K. L. (1986). Strategic management of uncertainty. *Academy of Management Review,* Vol. 11, No. 4, 777-790

Kouvelis, P. (1999). Global Sourcing Strategies under Exchange Rate Uncertainty. In: *Quantitative Models for Supply Chain Management*, Tayur, S.; Ganeshan, R. & Magazine, M., (Ed.), 625-667, Kluwer Academic Publishers, Boston

Laviolette, M. & Seaman, J. W. (1994). The efficacy of fuzzy representations of uncertainty. *IEEE Transactions on Fuzzy Systems*, Vol. 2, No. 1, 4-15

Ljung, L. (1987). *System identification*, Prentice-Hall, New Jersey

Milliken, F. J. (1987). Three types of perceived uncertainty about the environment: state effect, and response uncertainty. *Academy of Management Review*, Vol. 12, No. 1, 133-144

Schulz, M. (2001). The uncertain relevance of newness: organizational learning and knowledge flows. *Academy of Management Journal,* Vol. 44, No. 4, 661-681

Annals of DAAAM for 2008 & Proceedings of the 19th International DAAAM Symposium, ISSN 1726-9679
Editor B. Katalinic, Published by DAAAM International, Vienna, Austria 2008
Make Harmony Between Technology and Nature, and Your Mind will Fly Free as a Bird

CODING A LIBRARY FOR THE CREATION OF AUDIO AND VIDEO STREAM FILES FOR THE INTERNET USING OPEN SOURCE PROJECTS

KRISTALY, D[ominic] M[ircea]; SISAK, F[rancisc]; BUJDEI, C[atalin]; SINTEA, S[orin] & UNGUREANU, D[elia]

Abstract: *This paper is set to describe what tools are necessary and how it can be developed a library for the creation of audio and video stream files for the internet. The solution contained here-in was developed for a sharing application included in the e-learning platform of VET TREND, Leonardo da Vinci project. The paper includes the architecture of this library and explains how the FLV files are created.*
Key words: *audio, video, stream, library, .NET.*

1. INTRODUCTION

The e-learning platforms destined for the internet developed nowadays must include modern tools for communication between the teacher/tutor and students.

The usual communication channels on such a platform are internal messaging, e-mail, whiteboard, chat, blog and forum.

The communication can be:
1) one-way (only one actor can speak, the other/s will just listen) – this type of communication is used when presenting a course in the old fashioned manner; useful for transmiting new information in a very short time (e-courses, whiteboard, blogs);
2) two-ways (an usual dialog, both actors can send messages) – used in chatrooms, forums, e-mails; this type of communication can be used succesfully to clear up aspects that are more difficult to understand and/or when the actors cannot meet in person.
The latter category can be split into:
1) asynchronous communications: the period between the sending of the message and the receiving of the answer is greather than in an usual, face-to-face, conversation. The sender will post a message now and will have to check later for the response. This type of communication is profitable when the schedules of the actors cannot be synchronized or when the response needs a thorough documentation. Some of the tools that implement this type of communication are: e-mail, forums and blogs;
2) synchronous communications: the conversion progresses in a ping-pong manner; the sender writes a message and the receiver answers immediately. This type of communication is, in most cases, more productive than the asynchronous one, but both actors must be available in the same time. The tool that molds on this type of communication is the chat.

With the development of Web 2.0 technologies, the users' expectations grew: new tools are required to offer to both tutors and students exciting, new means of communication, in order to improve the teaching/learning process. Just to mention a few such tools:

- wikis – a collection of webpages and a mechanism that allows the user to contribute;
- whiteboard – the tutor/teacher can explain by drawing on a virtual blackboard;
- virtual classroom – a tool for audio-video conference with an integrated chatroom (Kristaly et al, 2008.)

In teaching information technology related courses, for example, the teacher must instruct the students how to use different software applications, which is not very easy to do whitout the visualization of the teacher's screen. For this, an application sharing tool is priceless. This type of application usually runs on a one-way communication channel, from the teacher to one or more students.

The sharing application tool that will be considered in this paper allows the teacher to share with the students the image of a software application running on his machine, by means of video streaming. The application was developed for the e-learning platform of a European Leonardo da Vinci project – VET TREND. The teacher starts a server program and selects the process that he wants to broadcast. The clients link to the server from an activex component that is embedded into a webpage of the e-learning platform.

The problem with this particular application is that it doesn't offer the possibility to record the activity of the teacher for later viewing. A new module must be developed that allows the recording of the activities in a multimedia format, playable on the internet.

2. ASSESING THE CHOICES

The first decision to be taken it is the file format in which the image and sound will be stored. The chosen format must allow streaming over the internet and must be easy to implement into the new module.

There are four formats that come to mind when streaming is involved: Microsoft's Windows Media Video (WMV), Real Media Video (RM, RAM), Apple's Quicktime (MOV), Adobe's Flash Video (FLV).

One very important aspect it's to keep the total cost of the module as low as possible. This can be achieved by using open source project and to minimize the changes in the already existing code of the sharing application.

The considered e-learning platform includes a virtual classroom developed using Adobe Flash technologies, able to play external Flash files, so it comes natural to choose the FLV format for the recording.

The video encoding format chosen is Screen video, which uses the zlib compression for the frames. For the audio data, the MP3 format was chosed, for its great compression ratios.

The FLV files are transferred over RTMP connections with the Adobe Flash Media Server™. RTMP, or *Real Time Messaging Protocol*, is a proprietary protocol developed by Adobe Systems for streaming audio, video and data over the internet between a Flash player and a server. A flash video player published for Flash Player 7 or above can also play FLV files directly with MIME type *video/x-flv* (libflv, 2008).

The FLV format it's a good choice also for the fact that is platform independent: all the data is stored in big-endian byte order. Also, there are many free and open-source players for Flash movies. For example, *JW FLV media player* for embedding a FLV stream into a webpage, or *FLV Player* for a stand-alone player. A FLV file encodes synchronized audio and video streams and consists of one single audio-video stream.

The second decision it's related to the programming language that will be used to create this module. The sharing application being written in Visual C++ 6.0, as a COM component, the FLV library was developed in Visual C# 2008 as a DLL library.

3. LIBRARY'S ARCHITECTURE

The figure 1 illustrates the architecture of the library, as well as the flow of data.

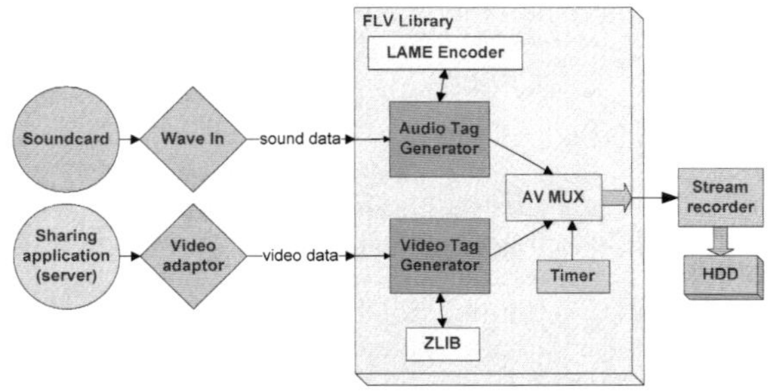

Fig. 1. The architecture of the FLV library.

The sound is captured using the *WaveIn API* and fed into the *Audio Tag Generator*, where is converted into MP3 format by the *LAME Encoder*. The output of the generator consists of FLV audio tags.

The frames captured by the sharing application are prepared by the video adapter for the *Video Tag Generator* which produces FLV video tags. To compress the blocks in the *Screen Video Packets*, the ZLib is used (Lame, 2008.)

The *AV MUX* creates the FLV stream and synchronizes the sound with the image. A *Stream recorder* writes the generated FLV stream to a file on a persistent storage device.

4. IMPLEMENTATION DETAILS

4.1 The FLV file format

The FLV file format it's very simple, as shown in the following figure (Adobe Systems, 2008.)

FLV Header
FLV Body
Previous Tag Size #0 – UI32 – Always 0
FLV Tag #1 (Audio, Video or ScriptDataObject)
Previous Tag Size #1 – UI32 – Size of Tag #1
....
FLV Tag #N (Audio, Video or ScriptDataObject)
Previous Tag Size #N – UI32 – Size of Tag #N

Fig. 2. The FLV file format.

The FLV tags are sections of the file that contain the audio, video or script information. These tags also contain the timestamp data that controls the way the file will be played.

The video data it is segmented and encapsulated in FLV Video tags. The audio data is also segmented and encapsulated in FLV Audio tags. After each tag, a 32 bits unsigned int stores its size (Adobe Systems, 2008.)

One video tag contains information about one frame:

- Frame type: specifies if the frame is a keyframe (the frame is seekable and it is stored in its whole), inter frame (not all the frame data it's store, only the changes from the previous frame), disposable inter frame, generated keyframe or video info/command frame.
- Codec used: Sorenson H.263, ScreenVideo, On2VP6, AVC
- Video frame payload, different for each codec type.

For the developed library, the Screen video codec is used. This is quite easy to implement and it is fast (Munoz, 2008.)

An audio tag defines what type of audio data is stored in the attached sound data: sound format (PCM, MP3, Nellymoser, G.711, AAC or device-specific sound), sound rate (5.5 kHz, 11 kHz, 22 kHz or 44 kHz), sound size (8 bit or 16 bit), sound type (mono or stereo), sound data (the payload in the format specified by the sound format). For the FLV library, the audio data will be compressed using MP3; in this way, the size of the file will be significantly smaller (Cardoso, 2008.)

4.2 Compressing the video data

The video data is encapsulated into a *Screen Video Packet* that contains the width and height of the frame in pixels and the width and height of a *block* (Adobe Systems, 2008.)

To compress the video data, the frame is split into blocks. Blocks have width and height that range from 16 to 256 in multiples of 16. The block size must not change except at a keyframe. Each block is compressed with the open source ZLIB

If the frame is a inter frame, not all the blocks need to be written, just the ones that changed from the previous frame.

4.3 Compressing the audio data

The audio data is captured from the primary sound device and it is compressed in MP3 format using the LAME encoder (Lame, 2008.)

The main issue is how to synchronize the sound with the video; the idea is to interlace the video and audio tags so none of them will load faster than the corresponding one.

The MP3 data it's organized in frames that contain a fixed number of samples (576 or 1152, depending on the sampling rate). The frames must be written in the FLV audio tag entirely, whitout fragmenting them, so a block of MP3 sound data always contains a number of samples that is a multiple of 576 or 1152 (mp3-tech, 2008.)

The ideal number of samples can be determined by dividing the sampling rate by the video frame rate. This number must be rounded to a multiple of 576 or 1152, so full MP3 frames will be written to the file.

To keep the MP3 streaming in sync with the video playback, the MP3 frames must be distributed as uniform as possible among video frames and to provide appropriate SeekSamples values in the audio data header.

5. CONCLUSION

The developed library can be use in any software project, not only in the context presented in this paper.

The use of the FLV format is the best solution at present time for streaming audio and video content over the internet: it can be very easily embedded in webpages or Flash applications and offers very good performances.

6. REFERENCES

Cardoso, I. *C# MP3 Compressor*, http://www.codeproject.com (May, 1st 2008)

Kristaly, D.M.; Sisak, F.; Truican I.; Moraru, S.A. & Sandu, F. (2008). *Web 2.0 technologies in web application development*, Proceedings of the 1st PETRA Conference – Workshop PTLIE, Athens, Greece

Munoz, I. *A full-duplex audio player in C# using the waveIn/waveOut APIs*, http://www.codeproject.com (May, 1st 2008)

SWF Format Specification, Adobe Systems, April 2008, electronic version

Video File Format Specification – Version 9, Adobe Systems, April 2008, electronic version

***http://klaus.geekserver.net/libflv, *Accessed on*: 2008-05-03

***http://lame.sourceforge.net, *Accessed on*: 2008-05-01

***http://www.mp3-tech.org , *Accessed on*: 2008-04-23

Annals of DAAAM for 2008 & Proceedings of the 19th International DAAAM Symposium, ISSN 1726-9679
Editor B. Katalinic, Published by DAAAM International, Vienna, Austria 2008
Make Harmony Between Technology and Nature, and Your Mind will Fly Free as a Bird

CONSIDERATIONS AND USE OF THE GENETIC ALGORITHM IN OPTIMIZATION OF STANDARD FIXTURE COMPONENTS

KRSULJA, M[arko]; BARISIC, B[ranimir] & CAR, Z[latan]

Abstract: *The accurate clamping and locating is essential in factory automation process. Fast localization of strategic placement of structure elements for 3-2-1 fixture based on genetic algorithm and milling tool path is presented. Travelling salesmen method is used for a 2D tool path search. A final design of an optimised placement is also shown with improved overall workpiece force and deflection influence.*
Key words: *GA, fixture, milling, locating, tool*

1. INTRODUCTION

Mechanical standard components for factory automation are commonly used in today modern factories. They consist of components for workpiece holding, supporting, positioning, locating and components for tool guiding. Fig. 1 shows obtained design for fixture. The main problem is the supplementary time, difficulties in handling of the workpieces, obtaining optimal solution of automation.

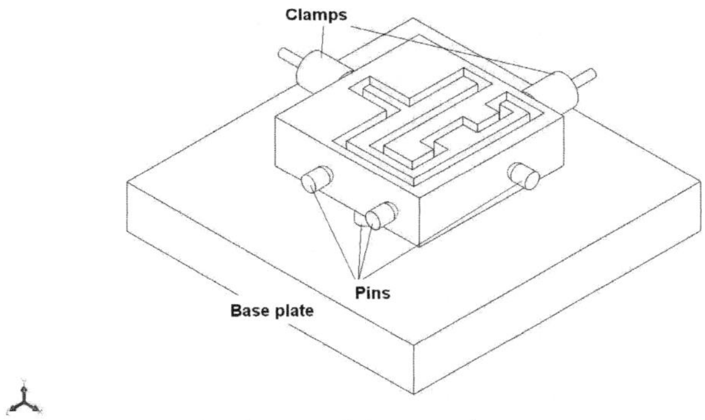

Fig. 1. Fixture design for one of the examples.

Nowadays stability in milling process should be performed in the planning phase where it is much easier to implement changes. It is easier to control material and process specifications by the use of optimization processes. Artificial intelligence can be used to perform backward computing to redefine the milling process. Product on market is under severe pressure concerning costs and it is required to be precise with narrowed tolerances. Locating system can be a bottleneck in small and medium size production if not planned carefully. By adequate positioning of locators and clamps the mechanical properties of material can be affected. And systems of a higher generation of automation can be implemented. Reducing the error caused by elastic deformation of loaded fixture-work piece. Wrong positioning of the workpiece may cause dimensional problems and deformation in machining.

The clamping fixture construction is usually adapted to specific situation. In order to create a good accuracy of the milled part the interaction of tool work piece and locating system must be taken in account. The process stability and sensitivity analysis should be conducted before starting of the production. Positioning and clamping are the basic functions of fixtures, and supplementary time includes the time needed for the manipulation of workpiece and tools. Fixture clamping sequence also influences the part location errors (Anand & Shreyes 2003). The optimum layout of devices and machines in the manufacturing systems is one of the basic requirements for the design of flexible manufacturing systems (FMS), since good solutions in designing such a system are prerequisite for its efficient functioning and low costs of operation (Ficko et. Al. 1998). It has been estimated that 20% to 50% of the manufacturing costs are due to handling of workpieces: proper planning of devices can reduce the above costs for 10% to 30% (Tompkins et. Al. 1996). The FMS include devices and machines which usually do not have firmly determined dimensions and distances (Heragu & Kusiak 1988). The layout can be optimized with the respect of the relations with the neighboring devices, cost function, mutual dependencies (pin positioning, workpiece orientation, immediate vicinity of clamps, tool path, workpiece geometry etc.).

2. FIXTURES AND MILLING

In milling operations fixtures are used for locating and holding of the workpiece.

2.1 Fixtures

Fixtures are components used in a milling operation for control of the workpiece movement. Wrong clamping or pin placement can influence the final workpiece accuracy. For a relatively rigid workpice the localized elastic deformations cause it to undergo rigid body translations and rotations which alter its location with respect to the cutting tool. (Bo Li & Shyres N. Mekolte 1998). It is important to reduce the displacement that rises under the influence of bigger cutting forces. In this paper a model is presented that a process engineer or fixture designer can use to understand the impact of fixture sequence on workpiece location error. In Fig. 2 components used in fixtures are shown.

2.2 Milling

Milling is a process of metal removal; principle is that the multiple cutting edges rotate in a spindle. Teeth of a milling cutter alternately engage and disengage from the workpiece. Milling machines use machine tools that can rigidly hold and rotate a cutter while feeding a workpiece into the cutter. The workpiece is fixed with standard components on the milling table. During process a consideration has to be taken for influence of milling tool on the fixture (force, feeds, chatter, vibrations etc.). Rigidity of the fixture setup influences the feed rate of milling process.

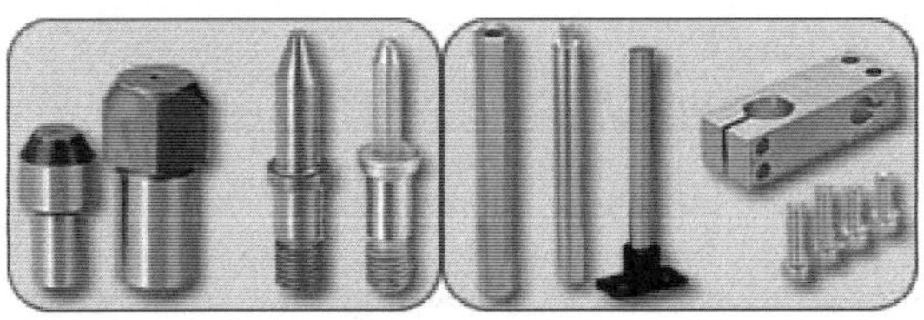

Fig. 2. Typical pin and clamp components.

3. GENETIC ALGORITHM

We used the traveling salesmen method for optimizing the tool path. Parameters ware inserted in the algorithm, TSP method was selected, and city positions were defined. Number of used cities was in example 1 (34 cities), example 2 (25 cities), example 3 (34 cities). The program went trough simulation and listed results. The obtained path like in Fig. 3 represented the best tool path. In the Fig. 4 it is shown that with the complex shape (number of cities) a bigger number of generations are needed to find an optimal tool path. Today computers can solve up to 100 cities in a very short time, so it is not a problem. Also with a bigger tool head the number of cities can be reduced. Together the cities represented a mesh for obtaining the desired shape. The best tour represents the best tool path; a tour is a closed path which traverses each city only once. The problem is to find a tour of minimum length, for different city configurations (shape). For genetic algorithm we used: population size of 10000, mutation of 3%, maximum generation 10000000 (ending condition), nearby cities 5, group size 5, nearby city odds 90%, and random seed 0. One of the problems in the generation of final shape was the selection of nearby city odds (80%, 85%, 90%) and group size (3, 4, 5) that resulted in different shapes. Different results for same problem, but a simple solution is to take fixed parameters in order to get uniform results in experiments.

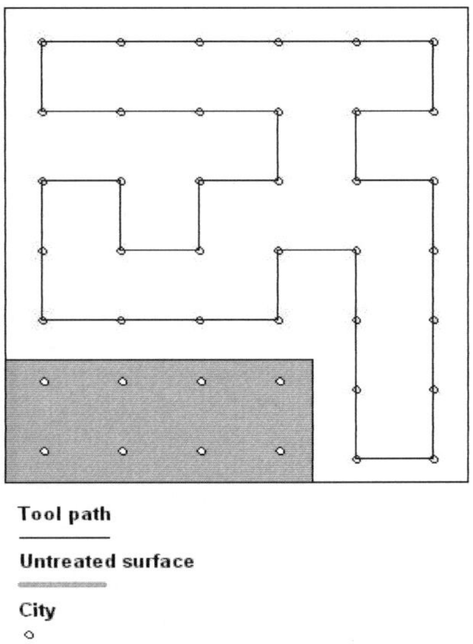

Fig. 3. Generated tool tour of example 1.

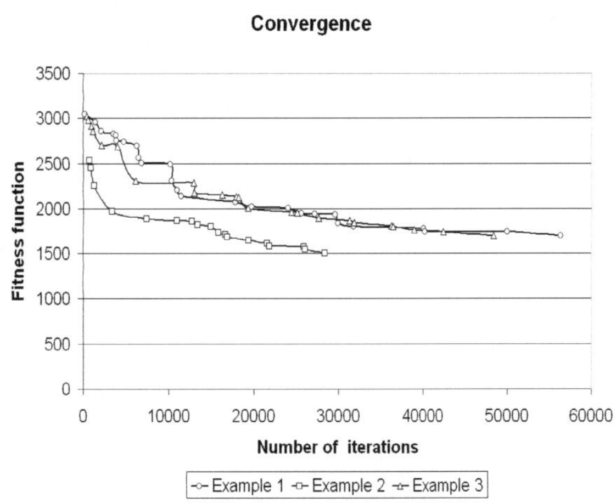

Fig. 4. Convergence of shape complexity.

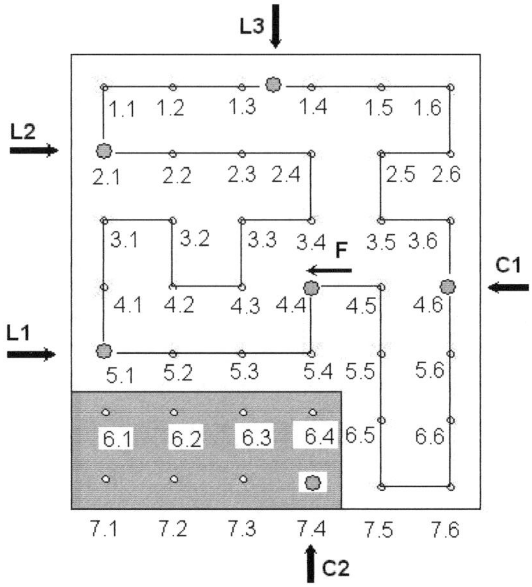

Fig. 5. Structure of optimal fixture placement.

4. RESULTS AND DISCUSION

Obtained optimal clamping mesh with placed fixture components is shown in Fig. 5. Under the workpiece are located three supporting pins. The force enters and continues from point 4.4 to 5.4 and continues to 5.1 the pin L1, L2 and L3 have been accordingly placed, in order to minimise initial influence of tool contact forces. Other directions have been balanced with clamps C1 and C2.

5. CONCLUSION

Process planning contains careful setup planning with consideration of manufacturing features. Mechanical standard components for factory automation affect the overall cost and quality of part machining. Tool movement influences the final quality; an approach has been presented for constraining the workpiece which can help the optimal layout. The result is a better clamping setup. Different criterions could be used to find the optimal solution but fixed conditions and different examples of similar geometry ware used.

6. REFERENCES

Ficko, M.; Brezocnik, M. & Balic, J. (2005). On designing of layout of flexible manufacturing systems, *Proceedings of the 16th International DAAAM Symposium 2005,* Katalinic, B. (Ed.), pp. 115-116, ISBN 3-901509-46-1, Opatia, Croatia, October 2005, DAAAM international, Vienna

Heragu, S. Kusiak, A. (1988). *Machine layout problem in flexible manufacturing systems,* Operations research, Vol. 36, pp. 258-268, ISSN: 0030-364X

Tompkins, J. A.; White, J. A.; Bozer, Y. A.; Frazelle, E. H.; Tanchoco, J.M. & Trevino, J. (1997). *Facilities Planning,* John Wiley & Sons, ISBN 0-471-41389-5., New York

Bo, L. Shryes N. Melkote (1998). *Improved workpiece location accuracy trough fixture layout optimization,* International Jurnal of Machine Tools, Manufacture Vol. 39, (1999) 871-883

Anad R., Shreyes N. M. (2003). *Analysis of the effect of fixture clamping sequence on part location errors,* International Jurnal of Machine Tools, Manufacture Vol. 44, (2004) 373-382

Annals of DAAAM for 2008 & Proceedings of the 19th International DAAAM Symposium, ISSN 1726-9679
Editor B. Katalinic, Published by DAAAM International, Vienna, Austria 2008
Make Harmony Between Technology and Nature, and Your Mind will Fly Free as a Bird

AN EMBEDDED STEREO VISION SENSOR FOROBSTACLE AND LANE MARKER DETECTION

KUBINGER, W[ilfried]

Abstract: *This paper describes design and realization of a real-time embedded stereo vision sensor. The sensor delivers information on obstacles and lane markers in front of an autonomous vehicle. The sensor was successfully tested at the DARPA Urban Challenge 2007 in Victorville, CA, USA.*
Key words: *Embedded Stereo Vision, Autonomous Vehicle*

1. INTRODUCTION

The DARPA Urban Challenge 2007 was the third in a series of competitions the US Defense Advanced Research Projects Agency (DARPA) had held to foster the development of autonomous robotic ground vehicle technology for various – mostly military – applications (Gourley 2007).

The focus of the 2007 event was shifted from desert to an urban environment. The urban area selected was the (abounded) housing area at the former George Air Force Base in Victorville, CA, USA. The challenge for the participating teams was to complete a complex 60 mile urban course with live traffic and had to be completed in less than six hours. During the course, the vehicles had to cope with traffic intersections, live traffic, road following, 4-way-stops, lane changes, three-point turns, U-turns, blocked roads, and pulling in and out of a parking lot (DARPA 2006).

This paper presents the design of a real-time embedded stereo vision sensor, which was used for obstacle and lane marker detection. This sensor was part of the sensor suit of RED RASCAL and contributed to the autonomous capabilities of this vehicle.

The paper is organized as follows. Section 2 describes the system concept of the autonomous vehicle RED RASCAL. The embedded stereo vision sensor is presented in section 3. Section 4 summarizes the achieved results and section 5 concludes the paper.

2. THE VEHICLE RED RASCAL

The team SciAutonics / Auburn Engineering successfully participated in the Grand Challenges 2004 and 2005 with their ATV-based vehicle RASCAL (Travis et al., 2006).

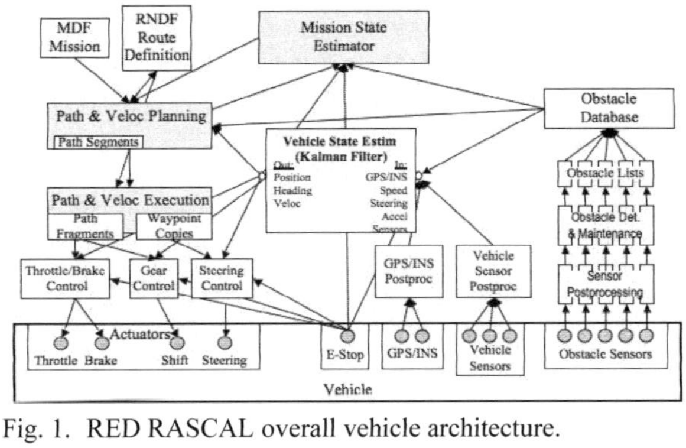

Fig. 1. RED RASCAL overall vehicle architecture.

For 2007, the team built a new vehicle based on the lessons learned from the former Challenges. The new vehicle was a modified Ford Explorer and was adapted for "drive-by-wire". The vehicle remained "street legal" and readily drivable on public highways. It was called RED RASCAL to reflect its close relation to its predecessor.

2.1 Vehicle Architecture

The high-level architecture of RED RASCAL is shown in Fig. 1. The vehicle is at the bottom. On the lower right, sensors send data up to a series of processing stages. On the lower left, actuators on the vehicle receive commands from the control logic. In the centre is the Vehicle State Estimator, which continually makes optimal estimates of the vehicle's position, velocity, heading, roll, pitch, and yaw, based on inputs from the GPS, IMU, vehicle sensors, and obstacle sensors. The vehicle state estimates are used by all parts of the control system, including the Path Planner.

At the top of the diagram is the Mission State Estimator. It determines what phase of a mission the vehicle is in and identifies when unique behaviours such as passing or parking are required. When the vehicle is driving in an urban environment, the key challenge that must be addressed is situational awareness. Not only must the vehicle follow a path and avoid moving obstacles, but it must also demonstrate correct behaviours depending on the situation. The Estimator treats each type of situation as a state and uses state diagrams to cope with behaviours and transitions (Bevly et al., 2007).

2.2 Sensor Suite

Sensors were needed for both localization (GPS, IMU, etc.) and perception (obstacle and road detection). A strategy of redundancy was employed to provide measurements from some sensors when others were not available or in the event of the failure of a particular sensor. The main sensor for vehicle localization was a single antenna Navcom SF-2050 DGPS receiver with Starfire satellite-based corrections provided by Navcom. A Honeywell HG1700 tactical grade 6 degree of freedom IMU was used to measure translational accelerations and angular rates. A NovAtel Beeline RT20 dual antenna GPS system was used for obtaining the initial vehicle orientation, as well as longitudinal and lateral velocities (Bevly et al., 2007).

The vehicle used several types of environmental sensors for obstacle and vehicle avoidance: LIDAR sensors and the embedded stereo vision sensor. The capabilities of these sensors overlapped to provide the redundancy desired to protect against sensor failure and improve reliability of measurements. As a near- and mid-range (4m to 20m) obstacle sensing system, the embedded stereo vision system was used.

3. EMBEDDED STEREO VISION SENSOR

The embedded stereo vision sensor detected objects up to a distance of 20m and provided information about lane markers and lane borders too. The stereo vision sensor consisted of a pair of Basler A601f monochrome cameras with a resolution of 656 (H) x 491 (V) and a quantization of 8 bits/pixel.

Fig. 2. The Embedded Stereo Vision Sensor integrated into RED RASCAL (left – stereo head, right – DSP box).

The frame rate of the sensor was 10 frames per second to cope with the real-time requirements of the vehicle. Both cameras were connected by a 400MBit-FireWire network to an embedded vision system.

The embedded system was based on a Texas Instruments TMS320C6414 DSP running at 1GHz. The operating system was DSP/BIOS II from Texas Instruments. The embedded vision system was responsible for the synchronous acquisition of both images, for execution of the computer vision algorithms, and the communication with the vehicle central brain via an Ethernet interface using UDP sockets (Hemetsberger et al., 2006).

The stereo head of the sensor was mounted on the dashboard while the DSP box was located in the trunk of the autonomous vehicle.

The main task of the stereo vision sensor was the detection of obstacles, lane markers, and lane borders in front of the vehicle. For obstacle detection, a fast stereo matching method was used. Furthermore, the bouncing of the vehicle was predicted and compensated in the stereo images to improve detection of obstacles and lane markers. For extraction of lane markers and lane borders, only the right camera image was used. The image was searched on different levels of resolution for significant markings and borders on the road. Identified lane markings and borders were classified, labeled, and reported back to the vehicle brain (Travis et al., 2006).

Additionally, the stereo vision sensor contained of a debugging interface for real-time logging of the right sensor input image. Extracted obstacles, lane markers, lane borders, and some internal states were logged for field-testing and evaluation, too.

4. RESULTS

During the qualification runs for the urban challenge event at Victorville, the embedded stereo vision sensor was used for obstacle and lane marker detection. An example of some detected obstacles (during sensor setup tests) is shown in Fig. 3. Figure 4 presents an exemplary output of detected lane markers on a road with live traffic.

Due to an accident in the second test run, the vehicle RED RACSAL was not able to qualify for the final event on November 3, 2008, the Urban Grand Challenge. However, our embedded stereo vision sensor did an excellent job during all pre-tests and test runs. Our sensor had proven to be an excellent and valuable assistance system for autonomous vehicles.

5. CONCLUSIONS

This paper presented design and realization of a real-time embedded stereo vision sensor. The sensor was designed to deliver information on obstacles, lane markers, and lane borders in front of an autonomous vehicle. The sensor was successfully tested at the DARPA Urban Challenge 2007 in Victorville, CA, USA as part of the vehicle RED RASCAL from the team SciAutonics / Auburn Engineering.

Just recently we have started two new projects for advanced driver assistance systems in next generation intelligent cars. In these projects we plan to disseminate and extend the used technology for next generation intelligent sensor devices.

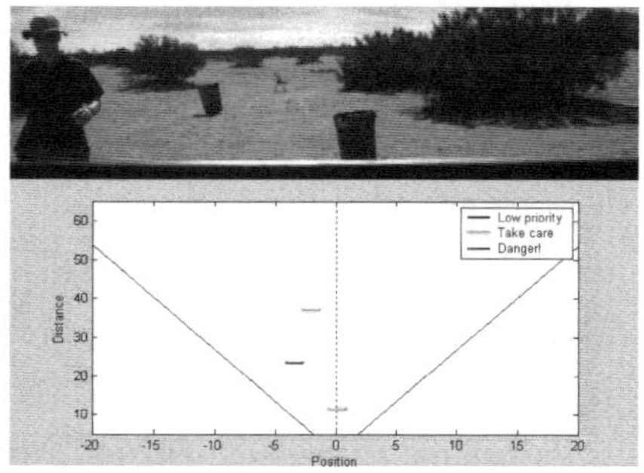

Fig. 3. Results of Obstacle Detection.

Fig. 4. Results of Lane Marker Detection.

6. ACKNOWLEDGEMENTS

The author would like to thank Mr. Daniel Baumgartner, Mr. Christoph Loibl, and Mr. Christoph Schönegger from the University of Applied Sciences Technikum Wien, Austria for their excellent support during that project. Furthermore, we would like to thank all people of the SciAutonics / Auburn Engineering team, Thousand Oaks, CA, USA for their valuable assistance during this project.

The research leading to these results has received funding from the European Community's Seventh Framework Programme (FP7/2007-2013) under grant agreement n° ICT-216049 (ADOSE).

7. REFERENCES

Bevly, D.M., Porter, J. (2007), Development of the SciAutonics / Auburn Engineering Autonomous Car for the Urban Challenge, *SciAutonics, LLC and Auburn University College of Engineering Technical Report*, June 1, 2007.

DARPA (2006), Defense Advanced Research Projects Agency (DARPA) Announces Third Grand Challenge - Urban Challenge Moves to the City, *DARPA News Release*, May 1, 2006.

Gourley, S.R. (2007), DARPA 'Urban Challenge' results show the way forward, *BATTLESPACE C4ISTAR TECHNOLOGIES*, pp. 13-20, ISSN 14651157

Hemetsberger, H., Kogler, J., Travis, W., Behringer, R., Kubinger, W. (2006), Reliable Architecture of an Embedded Stereo Vision System for a Low Cost Autonomous Vehicle, *Proceedings of the IEEE Intelligent Transportation Systems Conference (ITSC)*, pp. 945-950, Toronto, Canada, September 17-20, 2006

Travis, W., Daily, D., Bevly, D.M., Knoedler, K., Behringer, R., Hemetsberger, H., Kogler, J., Kubinger, W., Alefs, B., (2006), SciAutonics-Auburn Engineering's Low-Cost High-Speed ATV for the 2005 DARPA Grand Challenge, *Journal of Field Robotics*, Vol. 23, No. 8, pp. 579–597.

Annals of DAAAM for 2008 & Proceedings of the 19th International DAAAM Symposium, ISSN 1726-9679
Editor B. Katalinic, Published by DAAAM International, Vienna, Austria 2008
Make Harmony Between Technology and Nature, and Your Mind will Fly Free as a Bird

INTEGRATED APPROACH FOR AUTOMATIC ASSEMBLY, PACKAGING AND DISASSEMBLY PLANNING

KUNICA, Z[oran]; VRANJES, B[ozo] & ZORC, D[avor]

Abstract: *The paper tries to establish an integrated approach related to planning of assembly, packaging and disassembly processes, in the frame of entire product life cycle. The approach covers considerations of shared and specific features of mentioned technologies, on the process design and process execution levels, with special emphasis on applicable software, commercial or still underdeveloped.*
Key words: *assembly, packaging, disassembly, planning.*

1. INTRODUCTION

Modern, fair-minded approach considers complete product life cycle, "from needs to absorption, from raw materials to recycling, from cradle to grave" (Prasad, 1996), despite present deviations which often happen in profit-only driven product realisations. So mentioned modern trend requires product's integrated considerations, not only from the viewpoint of design, manufacturing and usage (Shen & Su 2008), but also including what happens to a product after it had been used.

The phenomenon of a product does not exist anymore in light of technical or economic duration only, but in sense of its life cycle, which is motivated by efforts for rational use of natural resources and environment protection. By terminology (cradle-grave), it is obviously established analogy between life-less material artifacts and living systems (which in turn shows perhaps further level of our adoration of material and fetishism, neglecting humanity).

A typical part of product life cycle contains the following processes: 1. assembly, 2. packaging, 3. consumption, 4. collecting, 5. disassembly.

The most of the products consist of more than one part, so they must be assembled. After assembly, the products are packed and distributed to consumers. Environmental and other reasons require collecting and disassembly of used products.

Present global market and consumerism establish controversy: we are called to keep the nature and planet, at the same time in which we buy, use and throw away newer and newer products (according to a manufacturer's instructions on disposing of a recent, brand new video-camera, a user should simply brake the camera housing).

It is obvious that solving of such situation will have a strong impact both on society (industrial, trade, legislative) and private (individual) levels, implying increased collaboration, responsibility, obligations and discipline.

It can be assumed, for example, that the collecting system would be much more comprehensive, so the shopping centres will have large areas for receiving used products, including their packages (responsibilities of retailers and customers).

Manufacturing plants would have output and input areas for finished and used products. The manufacturing plant would be responsible for its own product disintegration. It is natural, because the manufacturer may have the largest competences (knowledge, facilities) regarding particular product. Despite this, the other closed-loop organisational solutions are also possible (recycling centres – particularly specialised, product specific ones), but it is largely a matter of transportation cost.

2. APPROACH

Let us imagine a single manufacturing plant – a spot, where all activities of producing of finished goods, as well as disassembling, occur. Each activity may have (for example, because of clean-room, or technological requirements), or may have not (reasons: putting down the equipment investment costs, too low capacity), its own machinery.

At engineering and technology level, of course, the goal is to achieve appropriate effectiveness through automation, on the levels of: product (and its package) design, process/system planning and process execution (Fig. 1.).

The technical systems aimed at automatic assembly, packaging and disassembly are, despite significant differences (different roles in a production process/product life-cycle), closely related technologies: they have handling as the major activity, so hardware is mostly the same.

Regarding the levels of automation, it is needed to stress a discrepancy between assembly and packaging, at one side, and disassembly, at the other side. The discrepancy – backwardness of disassembly, exactly shows trend and need for changing of the relationship between industry (engineering work) and natural environment, meaning not only the changing of planning content, but also planning approach (Kunica & Zorc 2006).

The planning content related to assembly and disassembly (Sony Corp. 1996) is given by Fig. 2 (Kunica & Vranjes 1999). In a plan generation, as given, the sequence of planning steps is not mandatory in its entirety. Besides, assembly and disassembly are not always inversive.

Commercial software still does not cover complete planning content, nor is satisfactory automated and structured (a lot of interactive work; design procedures are sometimes redundant, for example in designs of assembly and mechanism). Sometimes, there are even issues regarding software documentation incompleteness.

It can be stated that the use and development of the required planning software support may be viewed from at least three points:

1) General CAD(/CAE/CAM) software (such as Catia and Pro/ENGINEER);
2) Engineering design/planning problem-specific (CAE) software (such was Delmia, now within Catia, and Tecnomatix, now with UGS; COSIMIR; CapePack);
3) Device programming/control software (developed by equipment manufacturers).

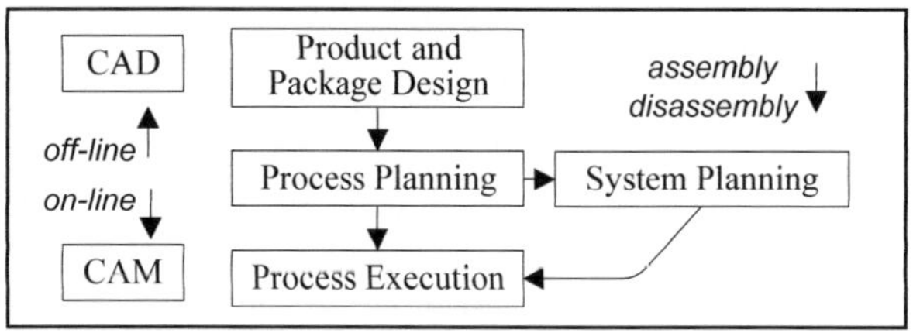

Fig. 1. Main stages of product realisation.

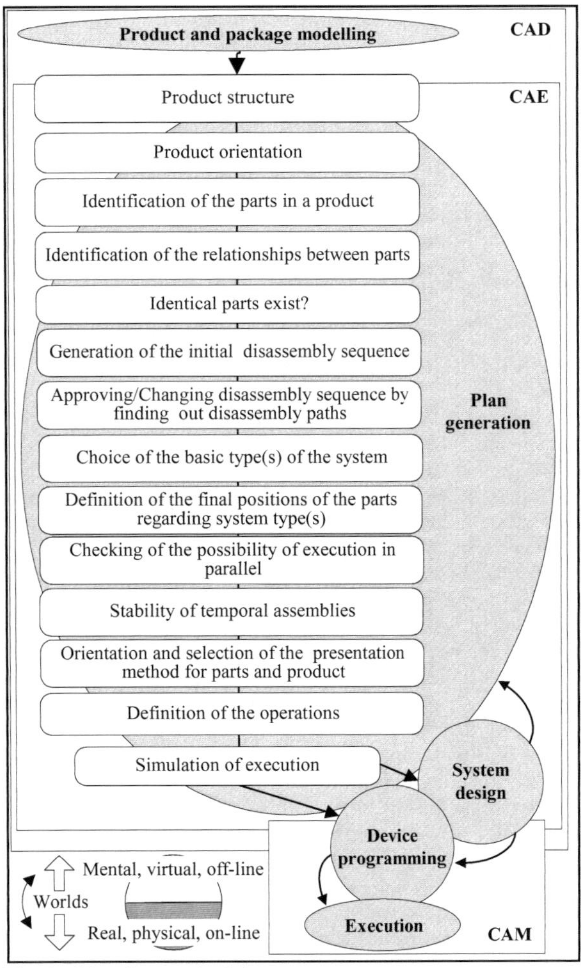

Fig. 2. Planning methodology.

The use of software includes its ability of large-scale visualisation, computing and virtualisation, which is adequate for engineering mental work of various complexity (from trivial to very creative; the treatment of large assemblies where standard interfaces become insufficient). However, in last ten years the development was pretty slow, from technological euphoria and optimism in the second half of 90's, through failures, to today's situation of moderate optimism. For example, Adept's Production Pilot software suite, several years ago, although as advertised, very advanced even in today's terms, was not successful. One of the main reasons may be too little market for such product. Why would a manager at assembly plant buy such complex software if his robots need only minor code modifications in their programmes per year? In contrary, the need for off-line programming tool is of essence for planning environments (system integrators).

The trade interest of manufactures helps designers and planners to import equipment in various CAD file formats. Although formats significantly vary and are differently supported from software to software, even greater issue is a lack of postprocessors, suitable to quicker, straightforward planning accomplishment. Very often, software is developed as an ad-on to a main production programme of a particular hardware manufacturer.

3. EXPERIMENT

Some parts of the mentioned approach are considered (Topalović, 2007) using the following experimental apparatus: the appropriate product/assembly, software Catia/Delmia and robot AdeptSix 300.

Two variants of automatic disassembly system (based on robot Adept) are developed using Catia/Delmia. Prior to system design, an analysis (disassembly plan) resulted with six possible product's (natural) orientations (which have an effect on: stability of temporal assemblies, disassembly sequence and kind and number of disassembly operations). For process detailing and system design, the orientation which offered two of maximum four disassembly sequences, is chosen. In last "manual" disassembly planning step, the most convenient disassembly sequence is chosen between two (using heuristic criterion, however).

Catia/Delmia session included: modelling of parts, assemblies and mechanism (of product, robot, gripper and other components of environment); disassembly process design and simulation; postprocessing (to obtain a programme code uploadable to the controller of the real robot).

Both variants of the disassembly system are based on ordered environment (product to be disassembled was always taken from the same place/position). In the first variant the intention was to achieve closed assembly-disassembly process cycle, while in the second variant, disassembly system was considered independently regarding assembly process.

4. CONCLUSION AND FURTHER WORK

The next generation of closed-loop discrete production/consumption/recycling systems brings further requirements for higher levels of automation, modularity and integration, and the same for their planning. In terms of that arises the importance of development and use of adequate planning approaches and software tools that will enable straightforward accomplishment of engineering task in the fields of automatic assembly, packaging and disassembly planning. In further work of special interest will be considering of integration of general CAD software, CapePack and laboratory automatic equipment, as well as inclusion of vision ability in disassembly process (enabling different disassembly scenarios and mechanical behaviours).

5. REFERENCES

Kunica, Z. & Vranjes, B. (1999). Towards automatic generation of plans for automatic assembly, *International Journal of Production Research*, Vol. 37, No. 8, pp. 1817-1836, ISSN 0020-7543

Kunica, Z. & Zorc, D. (2006). Designing experimental framework for multi-agent autonomous mechatronic processes, *Annals of DAAAM for 2006 & Proceedings of the 17th International DAAAM Symposium*, Katalinic, B. (Ed.), pp. 213-214, ISBN 3-901509-57-7, Vienna, November 2006., DAAAM International Vienna

Prasad, B. (1996). *Concurrent Engineering Fundamentals - Integrated Product Development, Vol. II*, Prentice-Hall, Inc., ISBN 0-13-396946-0, Upper Saddle River

Shen, W. & Su, D. (2008). Enabling technologies and frameworks for collaborative intelligent manufacturing, *International Journal of Production Research*, Vol. 46, No. 9, pp. 2329- 2331, ISSN 0020–7543

Sony Corporation (1996). *DAC Disassembly Evaluation/Design Manual*, Sony Corporation

Topalović, D. (2007). *Integrated automatic disassembly system*, Graduation work (in Croatian), Faculty of Mech. Eng. and Naval Architecture, University of Zagreb, Zagreb

ACKNOWLEDGEMENT

The paper is an outcome of the scientific project "Modelling of mechanical behaviour for assembly, packaging and disassembly", supported by the Ministry of Science, Education and Sports of the Republic of Croatia.

Annals of DAAAM for 2008 & Proceedings of the 19th International DAAAM Symposium, ISSN 1726-9679
Editor B. Katalinic, Published by DAAAM International, Vienna, Austria 2008
Make Harmony Between Technology and Nature, and Your Mind will Fly Free as a Bird

SPECIFIC FEATURES IN PRESSURE MICROSENSORS DESIGN

LACHE, S[imona] & LUCULESCU, M[arius]

Abstract: *As MEMS components, the pressure microsensors have specific features regarding design and manufacturing, taking into account the MST principles. Due to the relatively high cost of the prototype, it is strongly recommended to use analytic models for simulation. This paper presents some aspects regarding design approaches and results of the finite element modelling of a pressure microsensor. The analysis determines the stress and strain distribution along the membrane due to different pressure loads. These results would allow the precise positioning of piezoresistors on the pressure sensor.*
Key words: *MEMS, microsensor, modelling, analysis, design.*

1. INTRODUCTION

Microsystems domain, known in the literature also as MEMS (Micro-Electro-Mechanical Systems) has been internationally acknowledged only two decades ago, even if it is based on preoccupation regarding sensors and actuators miniaturization that has been started more then forty years ago. As a result of the rapid progress microelectronics has lately achieved, the domain has been developed, nowadays it finds several applications in many fields: medicine, biology, chemistry, automotive industry. This work focuses on pressure microsensors, as component part of MEMS. In their design two main features are taken into consideration: the maximum non-destructive pressure and the sensitivity (Bistue et al., 1997). Works have been carried on calculation and prediction of sensitivity, which is now possible by a set of simple equations (Shih-Chin & Chengkuo, 2001).

Since they have specific features (dimensions in the millimetre range and below, forces other then gravity dominate due to the scaling effects, new materials involved - with different mechanical characteristics) the whole design and manufacturing process differs from the one we are acquainted with from the macro-world. There are specific design techniques and micro-fabrication methods developed in order to realize microsystems able to respond to complex functions (mechanical, electronic and control functions).

2. SPECIFIC FEATURES AND DESIGN LEVELS

Microsystems (MEMS) design requires some different description and detail levels: in the first place, the documentation regarding the microsystem specific features and needs has to be achieved, together with the assessment of different micro-fabrication possibilities (Fatikow & Rembold, 1999). If the designed device will become a commercial product, the calculation of the fabrication cost is needed, as well. Secondly, the following steps have to be performed for each proposed version of the designed device: the system division in components, materials choice, operations sequence setting-up for each component, assembly methods set-up and product compact packaging.

It is emphasize that, due to the prototypes high cost, the analytical model development – for simulations – is strongly recommended. Thus the research and development cost is significantly reduced. Microsystems modelling and analysis is a very complex issue. Modelling is present at different levels and uses a large variety of formulations, depending on the level.

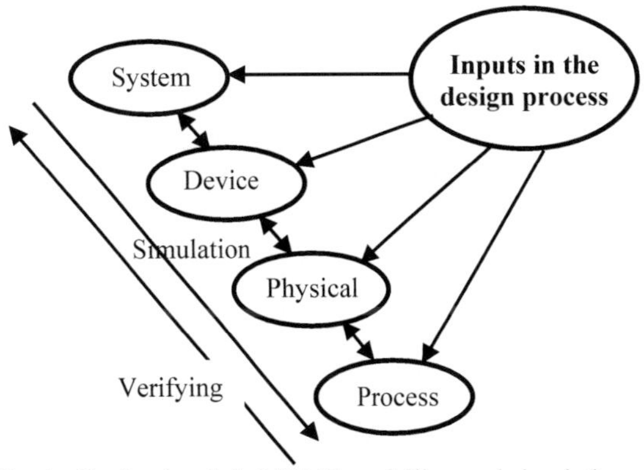

Fig. 1. Design levels in MEMS modelling and simulation.

Four design levels are identified (Poppinger, 1991): system level, device or component level, physical level and process level. Among these levels there is a bilateral information exchange.

3. CONSIDERATIONS REGARDING THE PRESSURE SENSOR DESIGN

The operating principle of a pressure sensor is simple and well known. It consists of a silicon membrane developed on a substrate (Fig. 2) by bulk micromachining and using the selective chemical etching techniques (Gardner et al., 2001).

When a pressure is applied, the membrane curves, the curvature measured being proportional to the pressure value. Fig. 2 also presents the scheme of a resistive pressure sensor, having the piezo-resistors integrated into the membrane; they change their resistance proportional to the applied pressure. The resistance change, measured with a Wheatstone bridge, shows how much the membrane has been curvetted.

The design procedure, based on the literature [3], is described in Fig. 3. The desired sensor properties are recorded inside the design specifications. The mechanical structure is then set up, considering the limitations of the manufacturing process. The geometrical model is developed which represents,

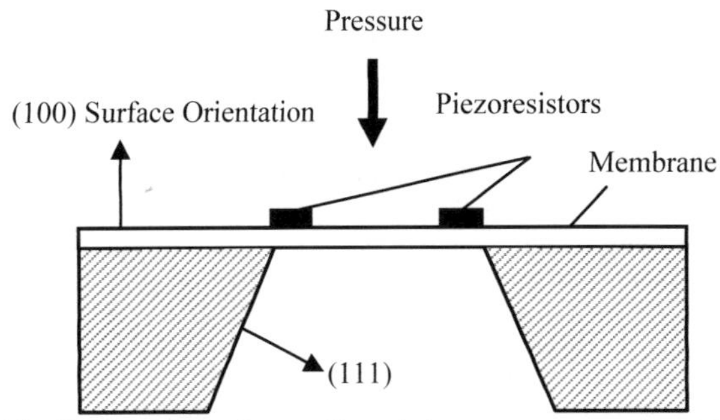

Fig. 2. Operating scheme of a resistive pressure sensor obtained by bulk micromachining of silicon (100).

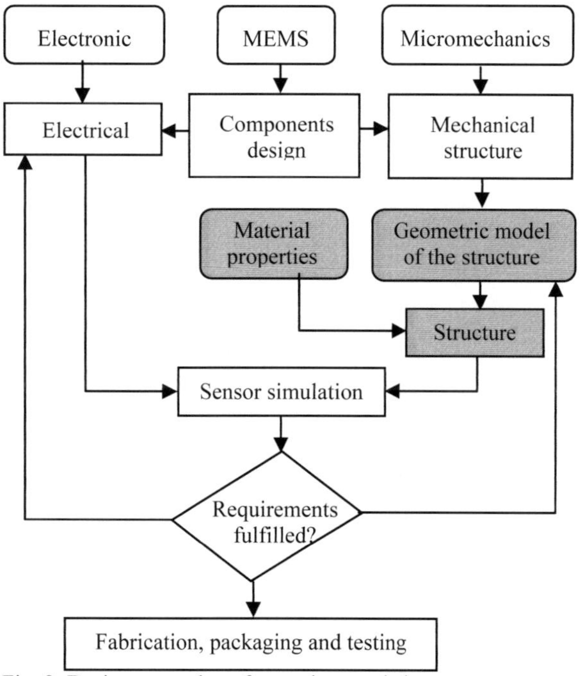

Fig. 3. Design procedure for a piezo-resistive pressure sensor (Poppinger, 1991).

together with the material properties, the solid model for finite element analysis. The analysis results in stress distribution along the membrane due to different pressure loads. These results would allow the precise positioning of piezo-resistors on the pressure sensor. Thus it yields the optimal geometry for the maximum sensor sensitivity. Further, the finite element analysis results are used for manufacturing process optimisation and for determining the electronic components best placement on the substrate (considering integrated MEMS is developed). The analysis is performed both in static and dynamic regime, the later being useful for resonance frequencies calculation and components behaviour investigation at different frequencies.

Whatever the design analysis is performed, one has to bear in mind that the dimensional reduction determines a considerable change in the requirements regarding the material properties, together with an increase of the surface forces with respect to the volume forces. This leads to increased friction and wear. Unlike the macro-systems, where the inertial effects are prevalent, for structures with dimensions of millimetres or below the surface phenomena change their dynamic behaviour. In this sense, many problems may occur due to the Van der Waals forces, the structure mechanical strength, the surface state and the micro-friction elements.

4. ANALYSIS AND RESULTS

The finite element model consists of the free part of the membrane, subjected to the pressure. The main feature taken into consideration is the maximum non-destructive pressure, which is related to the following design variables: dimensions of the membrane and mechanical properties of the selected material. The microstructure has the following dimensions: 7.5 mm (length) × 5 mm (width) × 0.1 mm (height); the active part of the membrane is along the length of 2.5 mm. The material properties for silicon, necessary for modelling, are: Young modulus, Poisson ratio and density. The structure is modelled with volumes and discretized in solid finite elements. The constraints are considered along the two lateral areas that simulate the parts of the membrane fixed on the substrate. The analysis is performed for a pressure range from 2kPa to 40Mpa.

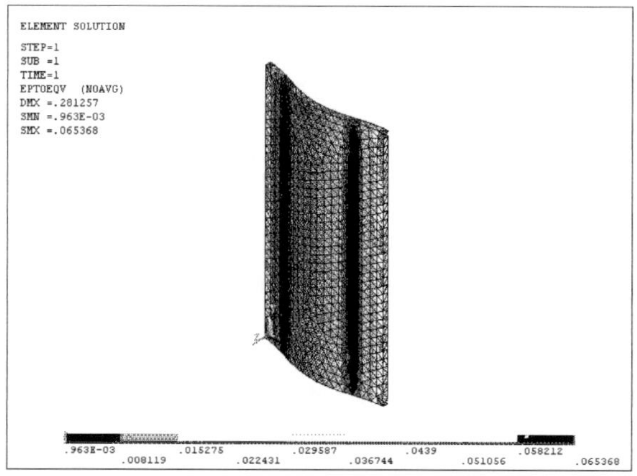

Fig. 4. Structural strain distribution for the sensor membrane.

Von Misses stress and strain distributions have been calculated (Fig. 4).

5. CONCLUSION

Pressure microsensors play an important role in microsystem technology, related to several advantages such as being cheep, having good resolution, precision, linearity and stability. In order to obtain reliable products, able to be integrated with electronic and signal processing components on a single chip, it is important to pay attention to the design of micromechanical part. The paper describes the overall design procedure and discusses the results obtained by performing finite element analysis and simulation of the sensor membrane. These results would allow the precise positioning of piezo-resistors on the pressure sensor. Thus it yields the optimal geometry for the maximum sensor sensitivity. Further, the finite element analysis results are used for manufacturing process optimisation and for determining the electronic components best placement on the substrate.

6. ACKNOWLEDGMENT

This work was supported by the Romanian National Council for Scientific Research from Higher Education, within the framework of a research project (financial support acknowledgment goes here).

7. REFERENCES

Bistue G. et al. (1997). A design tool for pressure microsensors based on FEM simulations, *Sensors and actuators*. vol. 62, no. 1-3, pp. 591-594 Elsevier Science, Lausanne ISSN 0924-4247.

Fatikow, S.; Rembold, U. (1999). *Tehnologia microsistemelor şi robotică* (trad lb. engleză), Ed. Tehnică Bucureşti; (*Microsystem Technology and Robotics*, Technical Press, Bucharest.)

Gardner J.W.; Varadan V.K.; Awadelkarim O.O. (2001). *Microsensors MEMS and Smart Devices*, John Wiley&Sons ISBN 047186109X.

Poppinger, M. (1991). Entwurf piezoresistiver Drucksensoren, *Mikroelektronik, No.5,* Fachbeilage Mikroperipherik; (Design of piesoresistive pressure sensors, *Microelectronics, No.5,* Fachbeilage Mikroperipherik.)

Shih-Chin G.; Chengkuo L. (2001). Analytical solutions of sensitivity for pressure microsensors, *Sensors Journal IEEE*, vol. 1, no. 4, Dec 2001, pp. 340-344, ISSN 1530-437X.

Annals of DAAAM for 2008 & Proceedings of the 19th International DAAAM Symposium, ISSN 1726-9679
Editor B. Katalinic, Published by DAAAM International, Vienna, Austria 2008
Make Harmony Between Technology and Nature, and Your Mind will Fly Free as a Bird

A GENETIC ALGORITHM FOR SINGLE SOURCE CAPACITATED FACILITY LOCATION PROBLEM

LAI, K[im] T[eng]; LUONG, L[ee H.S.] & MARIAN, R[omeo M.]

Abstract: *Single source capacitated facility location problems are basic location-allocation models used in supply chain, computer network, transmission power designs and machine scheduling, amongst others. It is a constrained combinatorial optimization problem which is NP-hard. While genetic algorithm (GA) has been successfully applied to many combinatorial optimization problems, it has limited success when applied to capacitated facility location problems. This paper proposes a GA solution methodology to single source capacitated facility location problems.*

Keywords: *Genetic Algorithm, Single Source Capacitated Facility Location Problem*

1. INTRODUCTION

Fixed charge facility location problems (FCFLP) form the basis of most location-allocation supply chain models (Daskin, 1995), and has industrial applications in computer network design and machine scheduling, amongst others (see Sridharan, 1995; Daskin, 1995). The problem involves selecting a number of plants each with its own fixed costs and capacity constraints from a set of potential locations to serve a predetermined set of customers at the lowest costs. The problem is NP-hard (Daskin, 1995). There are two variants of FCFLP: a) uncapacitated facility location problem (UFLP), and b) capacitated facility location problem (CFLP). Under CFLP, two variants can be distinguished: demand is sourced from a single source (SSCFLP) and from multi-sources (CFLP).

CFLP is solved by decomposing the problem into an allocation problem (which can be solved by polynomial time transportation algorithm) and a location problem (where the capacity constraints is initially relaxed, transforming it to a UFLP, which has many ready solutions). For SSCFLP, the transportation algorithm is not applicable as it leads to fractional denominations (i.e. more than one facility may be allocated). The allocation problem of SSCFLP requires a knapsack algorithm which is itself a pseudo-polynomial time algorithm or the NP-hard Generalized Assignment Problem

In this paper we use genetic algorithm (GA) to solve SSCFLP. GA is a metaheuristics which derives its inspiration from natural evolutionary process of the survival of the fittest and has been successfully applied to many large scale combinatorial optimization problems (see Gen and Cheng, 2000).

The next section provides a brief review of the literature on SSCFLP followed by the problem statement of this paper. Section 4 describes the proposed solution methodology and section 5, a case-study. In section 6 we present our conclusions.

2. LITERATURE REVIEW

Several solution methodologies have been proposed for SSCFLP (see Daskin, 1995). (Sridharan,1995) presents a survey of research conducted in capacitated plants up to 1995. (Klose & Drexl, 2005) document some of the major solution methodologies that have been employed for SSCFLP which include: Lagrangian relaxation, metaheuristics comprising very large scale neighborhood, GRASP, tabu search, evolutionary algorithms, simulated annealing and local search procedures.

In papers using GA for SSCFLP, (Cortinhal & Captivo, 2004) analyze four different approaches to using GA and conclude that GA is not suitable for SSCFLP. (Jaramillo et al., 2002) apply GA to various location problems which include UFLP and CFLP, however, found that GA is too excessively time consuming for CFLP and abandon the computational test, although GA is found to be effective for UFLP. In a more recent study in (Marian et al., 2008) GA is successfully applied to distribution network problems which involve CFLP.

The successful application of GA to location-allocation problems can be found in p-median problems, which is similar to UFLP, except that they do not involve fixed charges and the number of facilities is predetermined. p-median problems are NP-hard; however, have less variables and constraints then UFLP, CFLP or SSCFLP. Initial application of GA to p-median problem was not successful, however recent research have adopted different combinations of chromosomes coding, population and parent selections, crossover and mutation, and hybrid GA to successfully apply GA to the problem. (Bozkaya et al., 2002) designs each gene to be an index of a p-median vertex and show that GA gives better result than vertex substitution. (Lim and Xu, 2003) use fixed-length subset GA. (Correa et al., 2004) assigns vertices to the nearest median that is not already full.

The above literature review shows that GA is generally not efficient in solving CFLP and SSCFLP. However, (Marian et al., 2008) show that it can be effectively applied to CFLP. It has also been successfully applied to p-median problems, despite the initial failure. This indicates that the coding of chromosomes, selection of genetic operators, parent and population is significant for the application of the methodology.

3. PROBLEM STATEMENT

The problem statement in this paper is to select a number of plants (candidates) from a set of candidate locations to meet all the demand of demand nodes (customers) at the lowest fixed and variable transportation costs (1), subject to single sourcing (2), supply at open facility must be equal or more than demand (3), capacity constraints (4), and integral constraints (5,6). The mathematical formulation is as follows:

Inputs and Decsion Variables

f_j = fixed cost of locating at candidate site j ; h_i = demand at node i
d_{ij} = distance from demand node i to candidate location $_j$;
a = cost per unit distance per unit demand
X_j = 1if we locate at candidate site j, 0 if not;
Y_{ij} = demand at node i that is served by a facility at node j
k_j = capacity of a facility at candidate site j if a facility is located

Minimize: $$\sum f_j x_j + a \sum \sum h_i d_{ij} y_{ij} \quad (1)$$

Subject to: $$\sum y_{ij} = 1 \ \forall_I \quad (2)$$

	C1	C2	C3	C4	C5	C6	C7	C8
P1	10	20	15	30	20	10	25	30
P2	30	25	10	30	25	30	25	15
P3	25	30	20	20	15	40	30	25
P4	50	40	30	40	40	50	30	50
P5	60	50	40	50	30	40	40	20

Transportation Costs between customers and plants

	C1			C2			C3			C4		
	w	%	∑%	w	%	∑%	w	%	∑%	w	%	∑%
P1	91	27	27	61	25	25	56	23	23	71	21	21
P2	71	21	48	56	23	48	61	25	48	71	21	42
P3	76	23	71	51	21	69	51	21	69	81	24	66
P4	51	16	87	41	17	86	41	17	86	61	18	84
P5	41	14	100	31	14	100	31	14	100	51	16	100
Total	330	100		240	100		240	100		335	100	

	C5			C6			C7			C8		
	w	%	∑%	w	%	∑%	w	%	∑%	w	%	∑%
P1	81	21	21	81	28	28	66	22	22	61	19	19
P2	76	20	41	61	22	50	66	22	42	76	24	43
P3	86	23	64	51	18	68	61	20	62	66	20	63
P4	61	17	81	41	14	82	61	20	82	41	13	76
P5	71	19	100	51	18	100	51	18	100	71	24	100
Total	375	100		285	100		305	100		315	100	

w = adjusted transportation costs magnitude

Cumulative percentage for roulette wheel selection

Roulette wheel number		
C1	22	P2
C2	12	P1
C3	67	P3
C4	82	P4
C5	56	P3
C6	35	P2
C7	15	P1
C8	82	P5

Assignment of open facilities

Fig. 1. llustration of roulette wheel allocation

$$Y_{ij} \leq X_j \ \forall_{i,j;} \quad (3)$$

$$\sum h_i Y_{ij} \leq k_j X_j \ \forall_j \quad (4)$$

$$X_j = 0, 1 \ \forall j \quad (5)$$

4. SOLUTION METHODOLOGY

Chromosome representation is very significant in GA to obtain optimal or good solutions (Gen and Cheng, 2000). (Ruiz et al., 2006) show that having genes in the initial population close to the optimal solution assist in the convergence of chromosomes. We adopt an integer-based chromosome representation since only one plant can serve one customer.

Our chromosome encoding approach differs from others in the area of allocation of plants to customers. In the literature, the allocation is randomly generated. The random approach leads to too diverse a search space requiring high computational time for convergence. We use the concept of roulette wheel selection to asign plants to customers based on the reverse magnitude of the transportation costs (plants to customers). This approach skews the selection to plants with lower transportation costs but allow those with higher transportion costs (which may have low fixed costs) to be selected. This assists in obtaining good solutions in the initial population. Fig.1 shows a plant-to-customer allocation using the roulette wheel selection. The encoded chromosome is checked for capacity constraints and new chromosomes are coded to replace those which fail the capacity test. The resulting chromosomes are all feasible chromosomes. A randomly generated two-point crossover approach with single randomly generated gene mutation is adopted. The population selection is based on the 'incremental population' approach.

5. A CASE STUDY

A case study is conducted to apply the GA to a problem involving 150 customers and 50 plants. The result of the case study shows the convergence of the chromosome to near optimal solution within a relatively short time (185 seconds).

The parameters for the case study are randomly generated as follows: Locations (x,y axis) using a grid of size 50 by 50. Transportation costs are derived from the Euclidean distance using a one unit of cost per unit distance, operating costs are randomly selected from (1, 5), fixed costs (10,000, 30,000), capacity of plants randomly distributed from 2 * (total demand of customers), and demand of customers from (10,100).

Fig. 2 presents the graph of the maximum, minimum and average costs over 200 generations. It may be relevant to note that the slope of the graph for the minimum costs curve is not as steep as expected as the integer-based roulette wheel selection chromosomes comprised of 'good' solutions in the initial population which also assist in earlier convergence.

6. CONCLUSION

This paper described the development of a GA which demonstrates that GA can provide an effective and efficient solution methodology to solve large scale SSCFLP within acceptable computational time, contrary to conclusions arrived at from earlier researches. This is significant as SSCFLP forms the basis of a number of optimization models for real world problems, such as location-allocation supply chain models, which involve large search space in which mathematical techniques and other metaheuristics are currently still found to be inefficient for large scale problems.

$$Y_{ij} \geq 0, 1 \ \forall_{i,j} \quad (6)$$

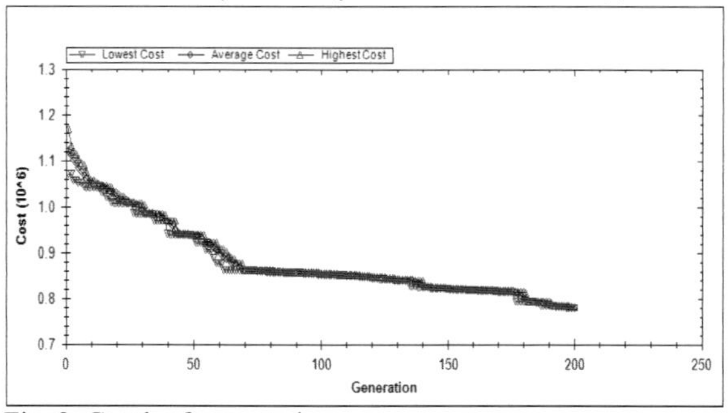

Fig. 2. Graph of case study

7. REFERENCES

Bozkaya, B.; Zhang, J. & Erkut, E. (2002). An efficient genetic algorithm for the *p*-median problem, In *Facility location: Applications and theory*, Drezner, Z.& Hamacher, H. (Ed.), pp.179–205,Springer Verlag Berlin Heidelbeg, ISBN 3-540-21345-7, New York

Correa,E.S.;Steiner,M.T.A.;Freitas, A.A. & Carnieri, C. (2004). A genetic algorithm for solving a capacitated *p*-median problem. *Numerical Algorithms,* Vol. 35, (2004) pp 373–388

Cortinhal, M.J. & Captivo, M.E. (2004). Genetic algorithms for the single source capacitated location problem. In *Metaheuristics: Computer Decision-Making*, Resende, M.G. & de Sousa, J.P. (Ed) (2004) pp 187-216 Kluwer Academic Publisher BV, ISBN 978-1-402-07653-3, The Netherlands

Daskin, M.S. (1995). *Network and Discrete Location: Models, Algorithms, and Applications*, John Wiley & Sons, Inc, ISBN 0-471-01897-X, New York

Gen, M. & Cheng, R. (2000). *Genetic Algorithms & Engineering Optimization,* John Wiley & Sons, Inc, ISBN 978-0-471-31531-5, New York

Jaramilo, J.H.; Bhadury, J. & Batta, R. (2002). On the use of genetic algorithms to solve location problems. *Computers and Research,* Vol.29, (2002), pp 761-779, ISSN 0305-0548

Klose, A. & Drexl, A. (2005). Facility location models for distribution system design. *European Journal of Operational Research,* Vol. 162, (2005), pp 4–29, ISSN 0377-2217

Lim, A. & Xu, Z. (2003). A fixed-length subset genetic algorithm for the *p*-median problem, *Lecture Notes in Computer Science,* Vol. 2724, (2003), pp 1596–1597

Marian, R.M.; Luong, L.H.S. & Akararungruangkul, R. (2008). Optimisation of distribution networks using Genetic Algorithms.Part1–problems modeling and automatic generations of solutions, *Int.J.Manufacturing Technology and Management,* Vol.15(1),(2008),pp64-83,ISSN 1368-2148

Ruiz, R.; Maroto, C. & Alcaraz, J. (2006). Two new robust genetic algorithms for the flowshop scheduling problem. *Omega*, Vol. 34, (2006), pp 461–476, ISSN 0305-0483

Sridharan, R. (1995). The capacitated plant location problem. *European Journal of Operational Research,* Vol. 87 (2), (1995), pp 203–213, ISSN 0377-2217

Annals of DAAAM for 2008 & Proceedings of the 19th International DAAAM Symposium, ISSN 1726-9679
Editor B. Katalinic, Published by DAAAM International, Vienna, Austria 2008
Make Harmony Between Technology and Nature, and Your Mind will Fly Free as a Bird

OPTIMISATION OF CNC MILLING FILES SINCE CAD PHASES

LANCEA, C[amil]; IVAN, N[icolae] V[alentin]; CHICOS, L[ucia] A[ntoneta] & OANCEA, G[heorghe]

Abstract: *Presently it is well known that CNC milling is a very used method for manufacturing of the 3D complex parts. The paper presents an original software for generating the CNC files, from the roughing phase to finishing, for 3 axis milling machines with ball-nose mills. CNC files syntax is compatible with FANUC controllers.*
Key words: *CAD, CAM, NC milling, Tool path generation.*

1. INTRODUCTION

The aim of a modern manufacturing is to obtain high precision parts as quickly as possible and with lower costs too. A significant element of this problem is the CNC program generation. Within the CNC program, the tool path has a major importance because the entire machining time depends on it. Therefore along the time (Dragomatz & Mann, 1997) a lot of researches have been aimed to obtain the best tool path for CNC machining (Song & Aoyama 2008) (Stenerson & Curran, 2007).

This paper presents a software module that generates the CNC program for machining complex 3D surfaces (fig. 1) on 3 axes milling machines. The geometrical information is taken within the CAD phase (Ivan et al., 2004) and generates technological results by using its own mathematical models. Beside the facilities presented above, the software is also able:

- to generate the tool path for roughing and finishing;
- to determine the stepover (fig. 1);
- to indicate the minimal value of the tool diameter.

2. TOOL PATH DETERMINATION

The tool path will be determined in a very short time and depends on several factors: the current phase, the stepover or the tool radius.

2.1 The current phase

The current phase can be a roughing phase or a finishing phase. After indicating the cutting depth for the current phase the program is able to generate a new surface, above the initial surface in accordance with the depth value.

2.2 The stepover

The stepover can be defined as the value of the tool diameter which is engaged in a cutting (fig. 1) or as the distance between two successive passes of the tool path. Usually the stepover should be 70% to 80% of the tool diameter.

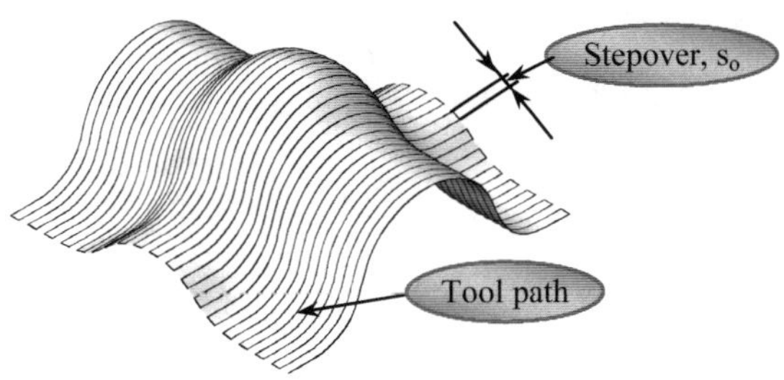

Fig. 1. Tool path and stepover for surface machining.

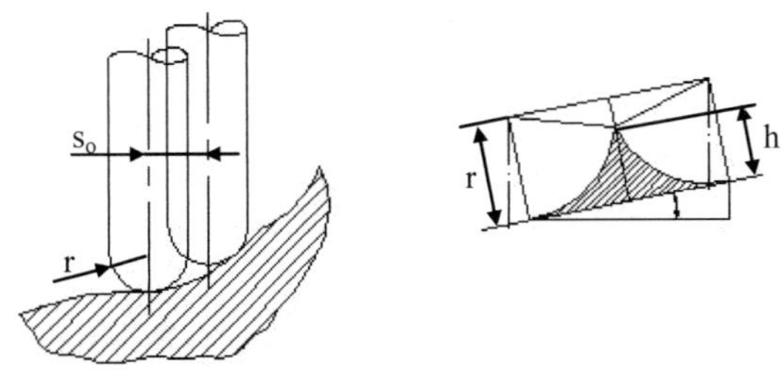

Fig. 2. Stepover determination.

The stepover value is determined with an original expression (1) that depends on the part roughness (cusp high - h) and the tool radius (r) (fig. 2):

$$s_o = 2\sqrt{2hr} \quad (1)$$

2.3 The tool radius

The tool radius will be determined by the software as follows:

- the surface is sectioned with a set of parallel planes generated along of two rectangular directions;
- the result of this intersection will be composed of a parallel curves set;
- each curve will be analyzed and where a concavity is found the software generates a circle through three successive points of the current curve;
- along the concavity curvature a set of circles will be generated, through three consecutive points (fig. 3);
- after analysing a curve the minimal circle radius will be saved;
- the next curve will be analysed as well and so on;
- after analysing all curves the lowest radius is saved;
- the indicated tool diameter has to be the first standard diameter that is less than double of the calculated radius.

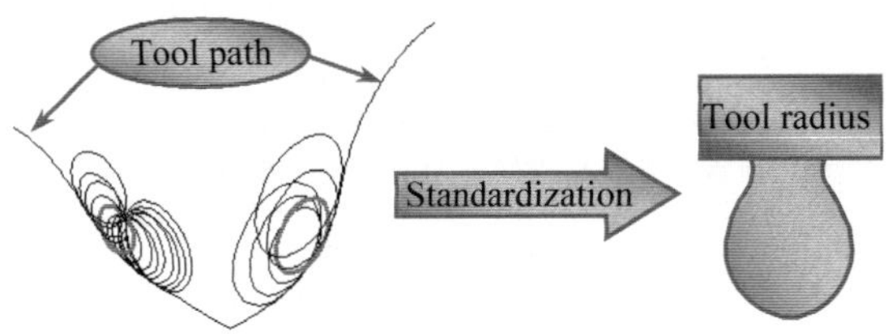

Fig. 3. Minimum tool radius determination.

3. CNC PROGRAM GENERATION

After determining of the tool path, stepover and tool radius the software analyses if the designed surface can be machined on a 3 axes milling machine. If the surface passes the machining test the software calculates the manufacturing parameters (speed, feed etc.). When all of these parameters will be saved (Lancea, 2004) the system starts the procedure for generating the CNC program, when the task "CNC program generation" is selected from the software menu (fig. 4).

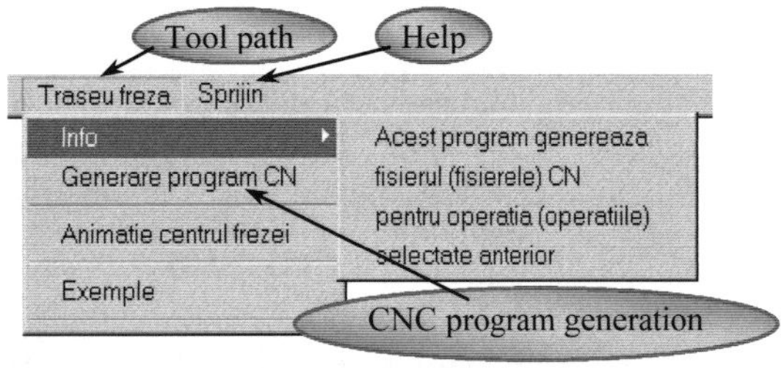

Fig. 4. Menu for generating the CNC program.

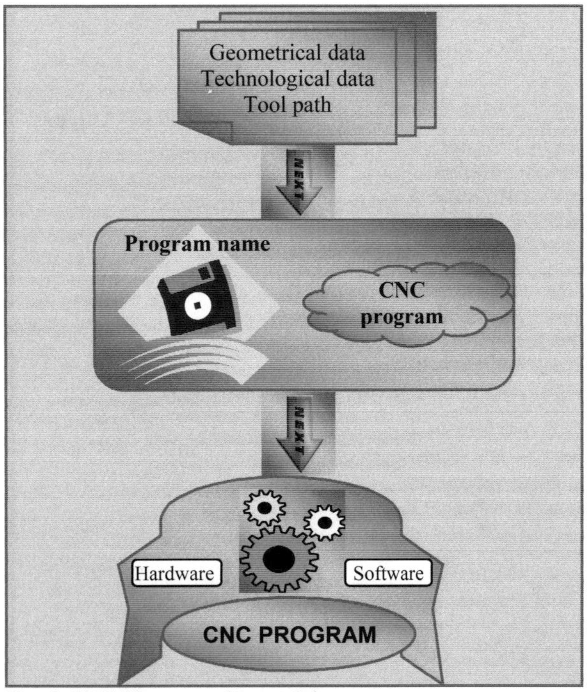

Fig. 5. CNC file generation module.

The module for generating CNC files takes all the information prepared within the CAD phase and starts to write the content of the CNC files. Each CNC file may be completed with other particular information saved as comments (fig. 5).

The CNC module of this software was developed in AutoCAD, as a Visual LISP application, completed with own menus (fig. 4) and dialog boxes (fig. 6 and fig.7).

The software gives the possibility to insert different comments having maximum 123 characters at the beginning of each CNC program.

After writing the comment a new dialog box is lunched for saving the program (Fig. 7). The CNC program files have a *.fnc extension because the software was developed for a CNC machine with a FANUC CNC controller.

The beginning sequence of the generated CNC file is presented in figure 8.

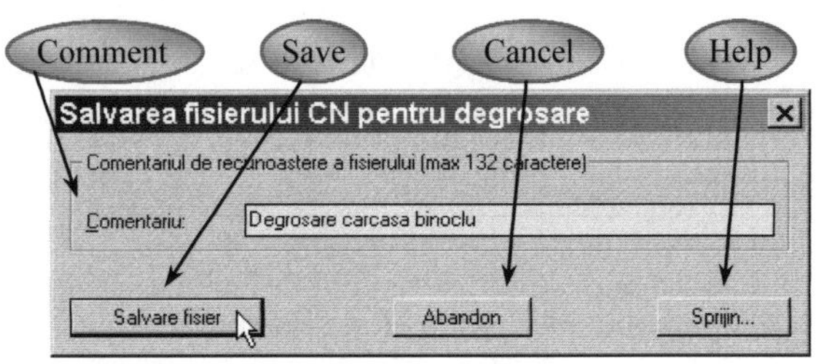

Fig. 6. Dialog box for inserting a comment.

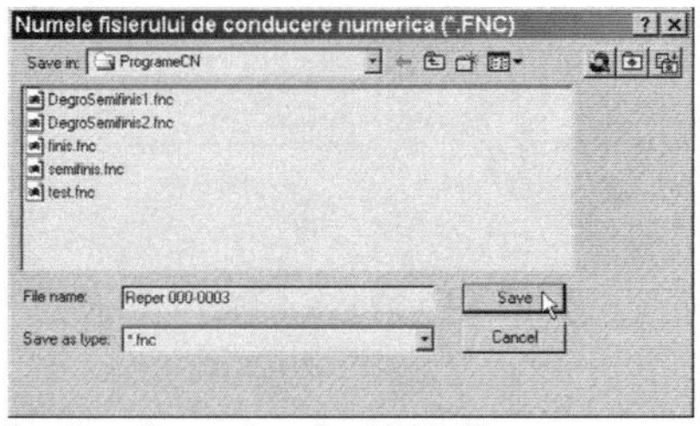

Fig. 7. Dialog box for saving the CNC files.

```
(Degrosare carcasa binoclu)
[BILLET X100 Y100 Z34]
[EDGEMOVE X0 Y0 Z0]
[TOOLDEF T1 D12 T2 D10 T3 D8]
G91 G28 Z20
M06 T01
M03 S2000
G90 G0 X0 Y0
Z10
G1 Y0 Z0 F1000
X0 Y0 Z-3.93
X1.35 Y2.23 Z-4.1
X2.77 Y4.45 Z-4.27
X4.59 Y6.68 Z-4.45
```

Fig. 8. CNC Program sequence.

4. CONCLUSION

The software presented above is used for generating the CNC files for milling, in a friendly dialog manner. An advantage of this software, in contrast with other software systems, is that it gives in very short time information about the milling path (for roughing and finishing), the stepover value and the optimum tool diameter, for ball-nosed mills with inserts.

As further researches we intend to extend this program for CNC machines with 4 axes in order to cover the machining of the very complex surfaces too.

5. ACKNOLEDGEMENTS

This research has been supported by CEEX Project 41/2005. The authors express their gratitude to all partners for the fruitful collaboration.

6. REFERENCES

Dragomatz, D. & Mann, S. (1997). A Classified Bibliography of Literature on NC Tool Path Generation, *Available from:* http://www.cgl.uwaterloo.ca/~smann/Papers/survey2.pdf *Acessed*: 2008-05-29

Ivan, N.V.; Berce, P.; Drăgoi, M.V.; Oancea, Gh.; Ivan, M.C.; Bâlc, N. et al. (2004). *Sisteme CAD/CAPP/CAM, teorie şi practică* (*CAD/CAPP/CAM Systems, Theory and Practice),* Editura Tehnică, ISBN 973-31-1530-4, Bucureşti, Romania

Lancea, C. (2004). Considerations about the Cutting Force Variation when Milling with Ball-nose Tools. *Academic Journal of Manufacturing Engineering,* Vol. 2, No. 2, (May 2004) pp. 55-59, ISSN 1583-7904

Song, C. W. & Aoyama, H. (2008). Development of CAM system to Estimate Tool Wear of ball-end-milling, In: *Manufacturing Systems and Technologies for the New Frontier,* Mitsuishi, M., Ueda, K. & Kimura, F., (Ed.), pp. 409-412, Springer London, ISBN 978-1-84800-266-1, London, UK

Stenerson, J. & Curran, K. (2007). *Computer Numerical Control Operation and Programming - Third Edition,* Pearson Prentice Hall, ISBN 0-13-111547-2, New Jersey, USA

Annals of DAAAM for 2008 & Proceedings of the 19th International DAAAM Symposium, ISSN 1726-9679
Editor B. Katalinic, Published by DAAAM International, Vienna, Austria 2008
Make Harmony Between Technology and Nature, and Your Mind will Fly Free as a Bird

A GENETIC FUZZY REAL-TIME EXPERT SYSTEM IN TOBACCO INDUSTRY BANJALUKA

LATINOVIC, T[iho]; JOKANOVIC, S[imo] & ROGIC, M[iroslav]

Abstract: *This paper presents an approach to modeling genetic fuzzy real-time expert diagnostic system for PLC controlled manufacturing system in Tobacco Industry in Banjaluka. The knowledge is the description of the functional and operational logic embodied in the PLC. A more usable form compared to that held in the mind of manufacturing system designers themselves. These approaches to modeling inspired by biological evolution are called evolutionary computation. It contains the design and engineering knowledge about the manufacturing system to be diagnosed.*
Key words: *Genetic, Fuzzy, Expert System, PLC*

1. INTRODUCTION

In modern production systems we need automation and flexibility. Many complex manufacturing systems are controlled by Programmable Logical Controllers (PLC) (J. Jarvis a & D. Jarvis 1996). This is because that PLC's are adaptable, modular, user-friendly and acquired at low cost. However, because of PLC's inflexible programming system, their capability in fault detection and diagnosis is limited. Therefore, automatic algorithms and intelligent system approaches need to be investigated to improve the efficiency of diagnosing complex PLC controlled manufacturing systems.

In this paper, we employ knowledge-based techniques to implement. The diagnostic system is constructed mainly by associating the machine states (indicated by PLC signals) with the possible faults. The association which is represented as diagnostic knowledge in the diagnostic system, is acquired by an artificial method or using model-based methods.

2. PLC CONTROL OF MANUFACTURING SYSTEMS

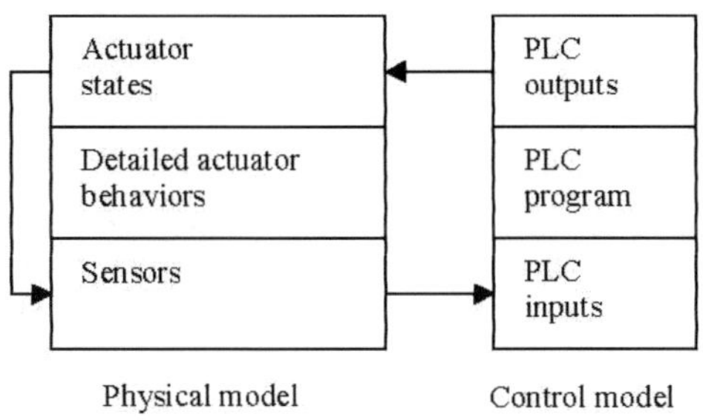


```
For (ever)
{
    Read all PLC inputs;
    Evaluate the PLC program;
    Set all PLC outputs;
}
```

Fig.1. the model for a PLC controlled manufacturing system

In a manufacturing system, PLC is used to control the behaviors of the system. The operating actions of the system and the sequence of these actions were edited beforehand into the control program by the manufacturer.

The control program sets a series of operations of the manufacturing system, which tells the PLC how to control a system.

The control loop of a PLC and the overall model of a PLC controlled manufacturing system can be described as in Fig. 1 (J. Jarvis a & D. Jarvis 1997).

3 THE DIAGNOSTIC SYSTEM DESIGN

From Fig. 1 we can know that, when diagnosing a PLC controlled manufacturing system, we always have to consider both of its two sub-systems, i.e., the control system (PLC) and the physical system (the system being controlled), which are linked by wiring diagrams. These wiring diagrams are maintained separately from the PLC program and the pneumatic and hydraulic circuit diagrams those typically constitute the documentation for the physical system. Usually in order to understand the behavior of the manufacturing system, we need to understand how both of its two sub-systems behave and how they interact with each other (J. Jarvis a & D. Jarvis 1996). The features of the evolutionary computation are that its search or optimization is conducted
(1) Based on multiple searching points or solution candidates (population based search),
(2) Using operations inspired by biological evolution, such as crossover and mutation,
(3) Based on probabilistic search and probabilistic operations, and
(4) Using little information of searching space, such as deferential information mentioned in section 3.3. Typical paradigms which consist of the evolutionary computation include GA (genetic algorithm), ES (evolution strategies), EP (evolutionary programming), and GP (genetic programming).
Fig. 2 is the general structure of a knowledge-based real-time diagnostic system for PLC controlled manufacturing systems. It is the extension of an existing diagnostic system we developed for a Flexible Manufacturing System (FMS) (W. Hu, 1995). The system can continuously acquire data from the PLC, identify possible faults, search for their causes and suggest corrective actions

4. DIAGNOSTIC KNOWLEDGE ACQUISITION

4.1 Artificial knowledge acquisition

In a PLC controlled manufacturing system, there are some alarms for the purpose of self-protection of the system. These alarms protect the manufacturing system from working in an error condition. Knowledge about such a fault alarm can be described as fault <– (signal, state) {, (signal, state)}

Where {} means it may be the combination of multiple signals.

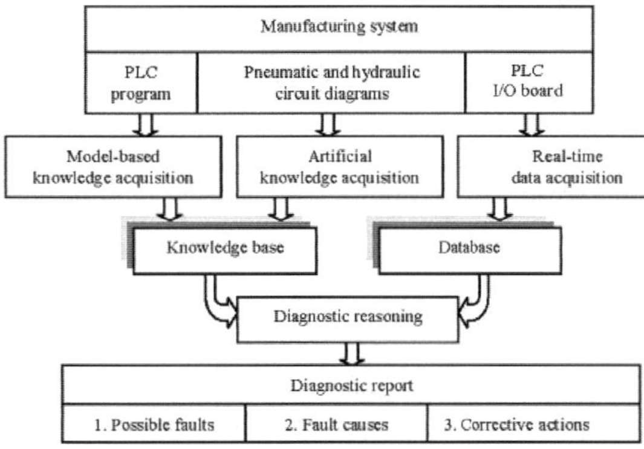

Fig. 2. The general structure of the diagnostics systems

4.2 Model-based knowledge acquisition the behavior of a PLC controlled manufacturing system is defined in its PLC program (Z. S. Matthias 1996).

4.3 Model based on PLC control logic. In this model, all variables associated with PLC control logic are described in a binary form.

4.4 Model based on PLC control sequence. This model consists of a certain number of system states and state changes in the PLC control sequence.

4.5 Model based on logic programming. This model was first developed and used in REVISE (J. J. Alferes & L. M. Pereira, 1996), a non-monotonic reasoning system that revises extended logic programs (C. V. Damasio, Luis Moniz Pereira, & Michael Schroeder, 1997).

5. DIAGNOSTIC REASONING

When the manufacturing system fails to work, diagnostic reasoning is performed and produces a diagnostic report in the end. The reasoning mechanisms are:

5.1 Diagnostic reasoning for logic control faults having had the logical expression at a fault state of the machine using the logical diagnosis model. (T. Latinović, Z. Konjević, D. Obradović, K. Bošnjak, 2004)

5.2 Diagnostic reasoning for sequential control faults
Under normal operating conditions, the PLC controls the manufacturing system according to the sequence of actions. At the same time, each step in the control sequence is monitored by the watch-dog-timer in PLC. (T. Latinovic, K. Bosnjak, M. Ostoja, T. Mladen, 2001)

5.3 Diagnostic reasoning based on logic programming

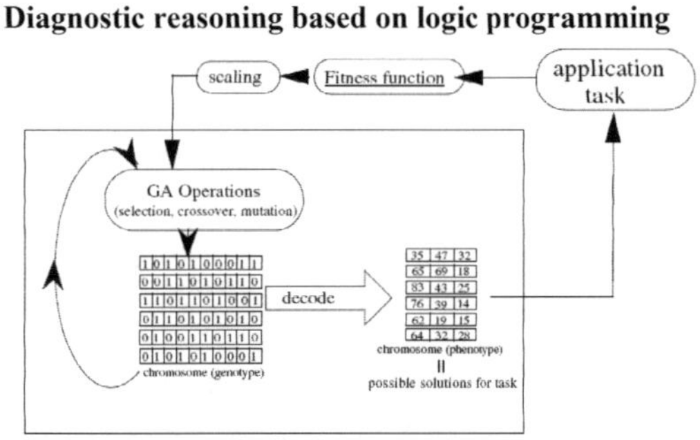

Fig. 3 The flow of Genetic Algorithm data

Fig. 3 shows the flows of Genetic algorithm process and data, respectively. Six possible solutions are expressed in bit code in Fig. 3. These evaluation values are fitness values.

6. CONCLUSIONS

Manufacturing systems present an important domain of diagnostic application. The development of automated diagnostic techniques and systems can help to minimize downtime and maintain an efficient output. The knowledge-based genetic fuzzy expert real-time diagnostic system is developed to meet this need.(Hideyuki TAKAGI,997). It can continuously acquire data from the PLC, identify possible faults, search for their causes and suggest corrective actions.
In Tobacco Factory in Banjaluka this system is implemented in Mollins Mark 5 machine for cigarette making.

From the author's point of view, future work will concentrate to

- Refine the diagnostic knowledge acquisition methods and reasoning algorithms, so as to improve the efficiency of the diagnostic system.
- Investigate models that incorporate PLC control on continuous processes of the manufacturing systems, implementing a systematic integrated methodology for prediction, monitoring and diagnosis.
- Define an embedded diagnosis system approach which will integrate the diagnostic models in the PLC's, so that faults can be diagnosed in real time.

7. REFERENCES

C. V. Damasio, Luis Moniz Pereira, & Michael Schroeder, *"REVISE: Logic Programming and Diagnosis*," Proc. of the Conf. on Logic Programming and Non-monotonic *Reasoning LPNMR97, 1997, LNAI 1265, Springer-**Verlag***

Hideyuki TAKAGI *Introduction to Fuzzy Systems, Neural Networks, and Genetic Algorithms*, Kyushu Institute of Design,1997

J. Jarvis a & D. Jarvis, *"Life Cycle Support for PLC Controlled Manufacturing Systems"*, in A. Storr and D.H. Jarvis (Eds), Software Engineering for Manufacturing Systems: Methods and CASE Tools, Chapman & Hall, 1996.

J. Jarvis and D. Jarvis, *"Simulation of a PLC-Controlled Assembly Line,"* Proc. of 9th European Simulation Symposium, 1997, pp. 342-346.

J. J. Alferes & L. M. Pereira, *Reasoning with Logic Programming*, 1996, LNAI 1111, Springer-Verlag.

T. Latinović, Z. Konjević, D. Obradović, K. Bošnjak *"PQM (Process Quality Model) for the analysis, improvement and control of supply chain systems in tobacco factory* (15 international conference on systems science) , Poland, 2004, p. 4-353 4-259

T. Latinovic, K. Bosnjak, M. Ostoja, T. Mladen: *Expert system like auditor for prediction failure and efect analysis in industrial systems* (DEMI 2001, conference Banja Luka, Bosnia and Herzegovina), p 371- 376

W. Hu, *Research on a Quality-Control-Based Fault Diagnosis System in a Flexible Manufacturing Environment*, PhD Thesis, Huazhong Uni. Of Sci. & Tech., 1995.

Z. S. Matthias, *"Model-based Diagnosis of PLC Controlled Assembly Equipment,"* Proc. of 6th Int. Conf. On Data and Knowledge Systems for Manufacturing and Engineering (DKSME'96), Tampe (Arizona), USA. 1996, pp. 11-22.

Annals of DAAAM for 2008 & Proceedings of the 19th International DAAAM Symposium, ISSN 1726-9679
Editor B. Katalinic, Published by DAAAM International, Vienna, Austria 2008
Make Harmony Between Technology and Nature, and Your Mind will Fly Free as a Bird

GENERIC DATA CONVERSION BASED ON IEC'S GENERIC DATA ACCESS

LENDAK, I[mre]; ERDELJAN, A[leksandar] & VUKMIROVIC, S[rdan]

Abstract: *Different software components in a utility management system can operate on different data models. This paper describes a mechanism which allows easier software integration through data model conversion. It will be shown that if the software components implement the IEC Generic Data Access interface, then it is possible to describe the conversion rules in human readable XML.*
Key words: *data conversion, engine, XML, GDA.*

1. INTRODUCTION

The International Electrotechnical Commission's (IEC) Generic Data Access (GDA) (OMG, 2005) standard describes a simple interface for accessing data in a data schema independent manner and intended for Utility Management Systems (UMS). Although GDA has been available for a few years, it is seldom implemented and seldom written about in scientific papers. Previous publications (Lendak et al., 2007) had shown that GDA is applicable in the field of Utility Management Software (UMS).

One recurring problem in this field is the integration of heterogenous systems. The process of integration usually involves the process of data model conversion, as software vendors usually have different data models. Even when there are standards – as the IEC's Common Information Model (IEC, 2006) for power systems – they are usually used for data exchange and not for internal data representation.

Once implemented, GDA allows access to any data model in a standardised manner (Erdeljan & Čapko 2007). The aim of this paper is to show a data conversion engine which is located between two GDA servers and converts data between them. The engine itself is generic and customizable through an XML (W3C, 2001) configuration file for any two data models published via GDA interfaces.

Further research of this problem could result in conversions in fewer iterations and with better performance.

2. THE PROBLEM

As already stated in the introduction, data conversions and migrations are everyday activities in the field of Utility Management Systems integration. There are mature standards which enable solution providers to publish their data in a generic manner. One such solution is the IEC 61970-403 Generic Data Access standard.

If two software solutions have different internal data representations, but both implement the GDA interfaces, then it is possible to convert data between them in a generic fashion.

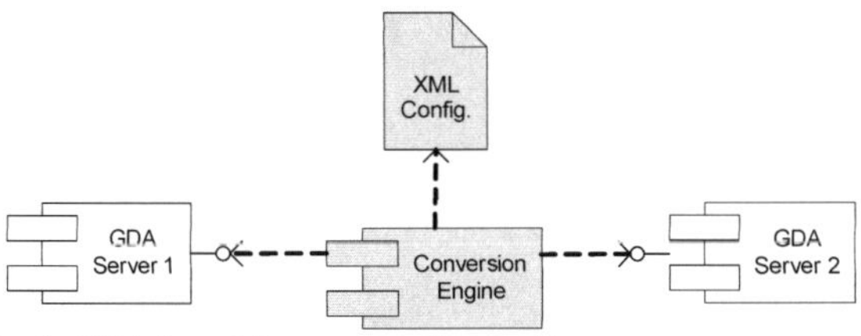

Fig. 1. GDA based data conversion architecture

Figure 1 shows the location of the proposed data convesion engine. This paper will deal with the two shaded components only, namely the conversion engine and its XML configuration. We assume that the GDA servers are readily available.

3. THE SOLUTION

The data conversion engine has to fetch the data from *GDA server 1* (S1) (see Figure 1), convert it and then send to *GDA server 2* (S2). The configuration file tells the engine how to fetch the data, how to convert and where to put it. Before proceeding to discussing the format of the XML configuration and the internal structure of the engine, it is necessary to discuss the various situations which can arise in such data conversions.

- One to one – mapping a single resource from S1 to a single resource in S2 – e.g. mapping a CIM Switch instance to a Switchgear instance in another data model
- One to many – creating many resources from a single
- Many to one – creating one resource from many
- Many to many – creating many resources from many – different from many one to one mappings

The first, simplest scenario will not be discussed in this paper as it can be solved by any of the next three conversion scenarios.

The solution of each scenario should consist of at least a *query* for fetching the data from S1 and a set of *rules* for converting and consequently sending the data to S2.

For simplicity, none of the XML configurations contain attributes and attribute conversion rules.

3.1 One to many

The query in this scenario should fetch one or many resources of a single type (e.g. Breaker). The conversion rule defines how to create one or many resources of different types from a single resource from the set of fetched resources. The destination resources (if more than one) can be of different types. It should be possible to solve this scenario in a single iteration. Listing 1 shows a configuration XML which fetches a single "BREAKER" instance and instructs the conversion engine to generate two new resources (a Breaker and a Disconnector) from it.

```
<query>
  <sources>
    <resource type="BREAKER" id="10001" />
  </sources>
</query>
<operations>
  <operation type="insert">
    <destination type="Breaker" />
  </operation>
  <operation type="insert">
    <destination type="Disconnector" />
  </operation>
</operations>
```

Listing 1. One to many configuration XML

The rule implicitly contains the information that for each source all destination resources must be created.

3.2 Many to one

The query in this scenario should fetch more than one resource which can, in general, be of different types. The conversion rule defines how to create a *single* resource from the set of fetched resources.

```
<query>
  <sources>
    <resource type="TRANSF" id="10001"/>
  </sources>
  <associations>
    <association link="TRANSF.WINDING"/>
  </asscociations>
</query>
<operations>
  <operation type="insert">
    <source type="TRANSF" />
    <destination type="Transformer" />
  </operation>
  <operation type="update">
    <source type=" TRANSF.WINDING" />
    <destination type="Transformer" />
  </operation>
</operations>
```

Listing 2. Many to one configuration XML

Listing 2 shows the configuration used for configuring the engine for many to one conversion scenarios. In this example, a source transformer (TRANSF) is fetched along with its windings (WINDING). The links between the source resources are defined by a list of associations (OMG, 2005). From the set of fetched resources, a single destination transformer is created. Note that there are multiple operations in the destination definition section of Listing 2, but they all operate on the same resource (Transformer).

3.3 Many to many

This is the most difficult scenario and the most general which, if successfully solved, can solve the previous three scenarios. The query in this scenario fetches more than one resource, which can be of different types. The conversion rule defines how to create a set of destination resources, which can also be of different types. The definition of these rules, in general, is a lot more complicated than any of the previous, as it can not be implied based on the configuration which destination resource should be created from which source resource and in which order.

```
<query>
  <sources>
    <resource type="TRANSF" id="*" />
  </sources>
  <associations>
    <association link=" TRANSF.WINDING" />
  </asscociations>
</query>
  <operations>
   <operation type="insert">
    <source type=" TRANSF " />
    <destination type="PowerTransformer" />
   </operation>
   <operation type="insert">
    <source type=" TRANSF.WINDING" />
    <destination type="TransformerWinding" />
   </operation>
</operations>
```

Listing 3. Many to many configuration XML

The query is the simplest part of this scenario, as it can be the same or very similar to the source definition shown in the previous (many to one) scenario. The sample configuration shown in Listing 3 instructs the engine to fetch a transformer and its windings from the source model and to create their equivalents in the destination model. The difference between this configuration and the one shown in Listing 2 is that in this scenario more than one resource is created in the destination model (see the two insert operations in Listing 3).

Another important difference is the use of "*" in the source definition. This character tells the engine to fetch all resources of specified type.

3.4 The engine

The "many to many" scenario is more complicated than the previous two but, once implemented, allows solution providers to solve all data conversions between two different data models published via GDA interfaces. In this section we will discuss the architecture of the programmatic solution and the complexity of the applied algorithm.

The engine fetches data from the source model, converts it and puts into the destination model. Once the data is fetched with DAF "get" queries, each *operation* should be carried out for each subset of the source data.

In the "many to many" scenario, the engine performs an insert operation for each source transformer (TRANSF) and then for each of their windings (TRANSF.WINDING). The number of operations for a power transformer with three windings is in this scenario four (per transformer). Each operation is carried out in order for each source transformer. This practically means that the formula for calculating the number of operations can be defined by (1).

$$O(n) = NS * (SO + \sum_{1}^{n} SL_i * O_i) \qquad (1)$$

In (1) NS is the number of sources (e.g. number of source TRANSF instances), SO is the number of operations performed on the sources, SL_i are the number of sources on the ith level (e.g. in Listing 3 the number of source windings), O_i is the number of operations performed on them and n is the number of associations in the source definition section of the XML.

4. CONCLUSION

The data conversion engine discussed in this paper is standards-based, simple and has adequate performance for off-line data conversions. Further work in this field could involve finding a more time and resource efficient algorithm, which would convert data in less iterations. Another necessary improvement of this algorithm is the implementation of mathematical formulas for attribute value calculus.

5. REFERENCES

Aleksandar Erdeljan, Darko Čapko (2007), *Analysis of DAF and GDA specifications usability for UMS object model access,* 51st International Symposium Interdisciplinary Regional Research, May 2007, Novi Sad, Srbija

IEC Task Group 13 (2006), *Energy Management System Application Program Interface (EMS-API) - Part 301. Common Information Model (CIM) Base - Revision 4*

Lendak I., Erdeljan A. & Zarić S. (2007), *A generic DAF server architecture for water distribution systems*, 9th International Symposium Interdisciplinary Regional Research, May 2007, Novi Sad, Srbija

Object Management Group (2005), *Utility Management System (UMS) Data Access Facility Specification*, v2.0.1, July 2005

W3C (2001) "XML Base W3C Recommend.", *Available from*: http://www.w3.org/TR/xmlbase/, Accessed: 2008-08-01

Annals of DAAAM for 2008 & Proceedings of the 19th International DAAAM Symposium, ISSN 1726-9679
Editor B. Katalinic, Published by DAAAM International, Vienna, Austria 2008
Make Harmony Between Technology and Nature, and Your Mind will Fly Free as a Bird

CONTRIBUTIONS FOR INCREASING THE PRODUCTIVITY OF CNC MACHINES

LEPADATESCU, B[adea]; BUZATU, C[onstantin]; DUICU, S[imona]; DUMITRASCU, A[dela] & ENESCU, I[oan]

Abstract: *Productivity on CNC machines is an important factor in using them for machining parts in a flexible technology. Many factors are involved on the productivity and accuracy of producing parts like: cutting parameters, tool used, stiffness coefficient, holding devices. In this paper is taken into account only the influence of the corner radius r_ε of the tool insert on the part accuracy and productivity. It is known that the machine program for CNC machines is made for a particular corner radius of the tool insert. But, sometimes, it is happen that the machine operator is forced to work with other corner radius of the tool insert that was in the initial machine program of workpiece. In this case if is working with other corner radius of tool insert, the accuracy of machining and part profile will be not as is demanded in technical documentation of part. For this reason in the paper are shown what are the modifications that appear on the workpiece surface and what are the changing that is necessary to do to maintain the part accuracy without stopping the running of the production.*
Key words: *accuracy, productivity, correction, modeling.*

1. INTRODUCTION

The computer numerical control lathe is used to perform several operations on different surfaces of the workpieces with complex shapes. Usually these workpieces require a high surface finish and dimensional accuracy. In the program for numerical control are set up for different surfaces different cutting tools with different corner radius according with the accuracy that is asked by the documentation (Avallone & Baumeister 2003).

But for different reasons it is happen the operator is forced to use in machining a part with a tool with other corner radius that was initial in the machine program. In the paper are shown what will be happen if it uses this cutting tool radius and what will be the influences on the part accuracy and on the concordance between the dimensions of the different zones of the part. Also, there are shown the modifications that the operator has to make direct on the panel of machine tool to obtain a successfully machining in these new conditions.

In these cases are two interventions that have to be done. One is directly on the machine tool by the machine operator without changing the machine program and other is in the machine program itself when a great accuracy of machining is required. In both cases are shown the corrections that have to do direct on the machine panel by the operator, the deviations that appear after these corrections in linear and circular interpolations.

In the calculations was taken into account for the tool path the theoretical point of the cutting tool tip. It was noted that if is working with other corner radius that was in the initial machine program is not obtained the shape identically with that was in the part drawing. If the part tolerances permit these dimensional modifications it is not necessary to modify the initial machine program (Boothroyd & Knight 1989).

After were calculated the equations for corrections and deviations is given an example for a particular type of the tool insert.

2. THE IMPORTANCE OF THE TOOL CORNER RADIUS FOR THE MACHINING ACCURACY

The values of tool corner radius are very important when machining are the work pieces with surfaces in steps without undercutting. The values of the fillet radius is the identically with the corner radius of the cutting tool. In Fig.1a is shown the value of the deviation Δ which appears when is used other corner radius $r_{\varepsilon 2}$ instead that of $r_{\varepsilon 1}$ which is use in the initial machine program. It is noted that if two parts are to be assembly, there can not make it completely, because the point B can't reach point A due to the different corner radius of the cutting tool. In Fig.1b is the same situation but for the machining parts with radius at corners.

In Fig.2 is presented an example of a part that have to be machined on CNC lathe and this has linear interpolation on the path AB and GH, circular interpolation on the path EF, turning to obtain diameters on the path BC, DE, HI and flat surfaces on the path CD and FG. For each of these trajectories have to make modifications to maintain the accuracy that is demanded for the shape of part.

In Fig.3 are shown the influences of the corner radius cutting tool modifications on the machining accuracy of a part without linear and circular interpolations. If is working with the cutting tool with corners radius $r_{\varepsilon 2}$ instead of $r_{\varepsilon 1}$ that was in the initial machine program is necessary to do some corrections by the machine operator on the machine panel in order to obtain the part accuracy that is required. These corrections are A_x and A_z that are given by the following equation, (Dumitrascu et al., 2007):

$$A_x = (r_{\varepsilon 2} - r_{\varepsilon 1})(ctg\frac{\varepsilon_r}{2}\cos\gamma - \sin\gamma - 1) \quad (1)$$

$$A_z = (r_{\varepsilon 2} - r_{\varepsilon 1})(ctg\frac{\varepsilon_r}{2}\sin\gamma + \cos\gamma - 1) \quad (2)$$

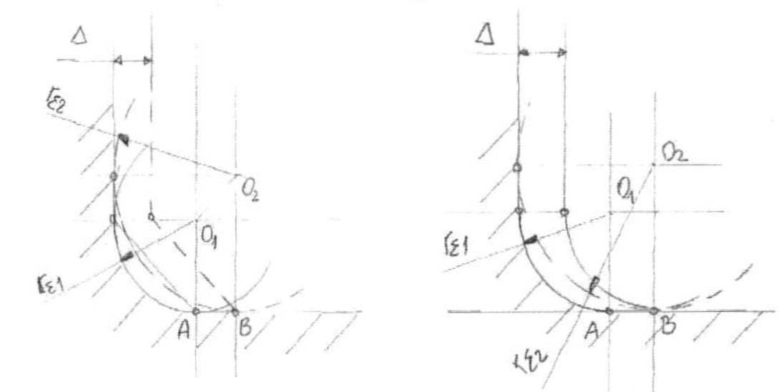

a) b)

Fig.1. Assembly of parts with corner radius and conical surfaces (a) and both with corner radius (b).

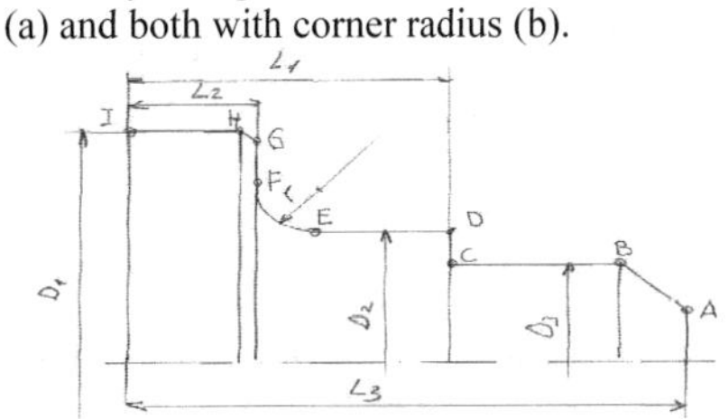

Fig. 2. Example of a part with different type of surfaces.

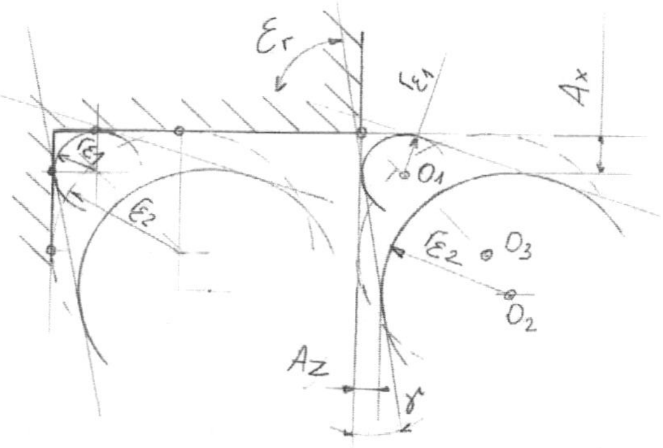

Fig. 3. Example of turning a part without linear and circular interpolations.

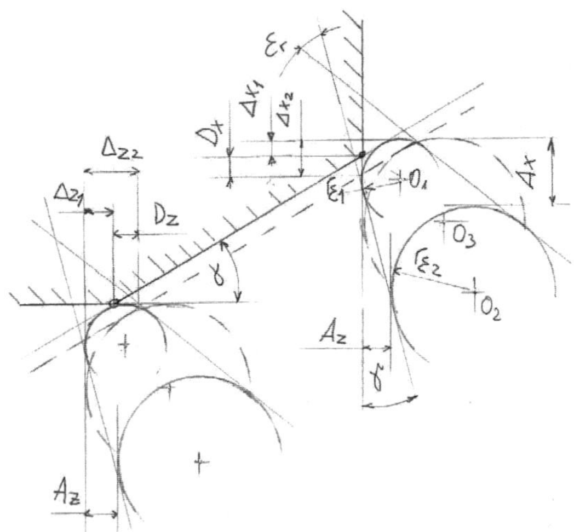

Fig. 4. Deviations D_x and D_z at turning with linear interpolation.

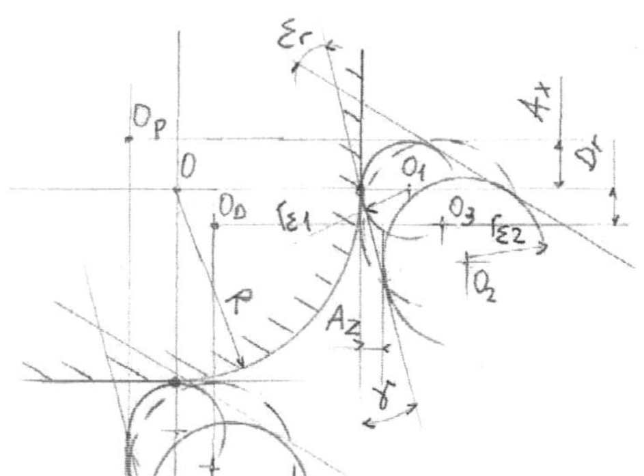

Fig. 5. The calculation of the deviation D_r at circular interpolation.

In equations (1) and (2), the γ is the angle between the part flat surface and the tool insert edge and ε_r is the tool included angle.

With these corrections will be obtained the diameters and flat surfaces as the part drawing is required, but the radius between the flat surface and cylindrical surfaces is now equally with $r_{\varepsilon 2}$ which is bigger that was in the initial program.

In Fig.4 is shown the calculation of the deviations D_x and D_z which appears at linear interpolation after were made the corrections by the operator direct on the machine tool panel if is used an insert with corner radius $r_{\varepsilon 2}$ instead of $r_{\varepsilon 1}$ which was in the initial machine tool program (Reshetov & Portman 1989):

$$D_x = (r_{\varepsilon 2} - r_{\varepsilon 1})(1 - tg\frac{90^0 - \alpha}{2}) \quad (3)$$

$$D_z = (r_{\varepsilon 2} - r_{\varepsilon 1})(1 - tg\frac{\alpha}{2}) \quad (4)$$

In Fig.5 is presented the calculations of the deviation D_r in the case of circular interpolation that have to be done by the operator direct on the machine tool panel when in used an insert with a corner radius $r_{\varepsilon 2}$ instead of $r_{\varepsilon 1}$ that was in the initial machine program ($D_r = r_{\varepsilon 2} - r_{\varepsilon 1}$).

If after the modifications on the machine tool panel the part dimensions are not like in documentation will be necessary to make modifications in the initial machine program.

3. APPLICATION FOR THE CASE OF TURNING WITH TRIANGULAR INSERTS

If is used at turning a cutting tool with a triangular insert with the next values: $\varepsilon_r = 60^\circ$; $r_{\varepsilon 1} = 0.8$; $r_{\varepsilon 2} = 1.2$; $\alpha = 30^\circ$; $\gamma = 0^\circ$ will appear the next situations:

1. For the part surfaces when is not used linear or circular interpolation (Fig.3) the operator will do directly on the machine tool panel the corrections (Reshetov & Portman 1989), $A_x = 0.293$ mm and $A_z = 0$. The part will have the dimensions exactly with the drawing requirements, only the fillet radius will be changed to the value of $r_{\varepsilon 2} = 1.2$ mm.
2. If is used the linear interpolation (Fig.4) after is done the same corrections will appear the next deviations: $D_x = 0.169$ mm and $D_z = 0.293$ mm.
3. At turning with circular interpolations (Fig.5), after were made the corrections A_x and A_z the value of radius R will have a deviation $D_r = 0.4$ mm (Simon & Mares, 2005).

If the deviations are inacceptable then it will be necessary to change the initial machine program.

4. CONCLUSIONS

In practical applications for different reasons are used other corner radius than were established in the initial machine program. In these cases, it is useful to know what they are the corrections which it has to be done by the operator direct on the machine tool panel to obtain a workpiece according with the technical documentations. The paper presents what are these corrections and in this case the productivity of machining will increase because the machine tool will not be necessary to be stopped till will be supplied with the inserts according with the initial machine program.

5. REFERENCES

Avallone, E.A. & Baumeister, TH. (2003). *Mark's Standard handbook for Mecahanical Engineers,* Tenth Edition, McGraw-Hill, New York

Boothroyd, G. & Knight, W.A. (1989). *Fundamentals of Machining and Machine Tools,* Edition New York: Marcel Dekker

Simon, A.E. & Mares, Gh. (2005). *Surface Technology,* "Transilvania" University of, Brasov

Dumitrascu, A.E.; Buzatu, C.; Duicu, S. & Enescu, I. (2007). Theoretical and applicative contributions to flexibility of technology at turning on CNC machine tools. *"Recent Advances in Systems, Communications & Computers" Conference*, pp. 325-329, ISSN 1790-5117, Hangzhou, China, April 6-8. *Available from:* http://www.wseas.org

Reshetov, D.N. & Portman, V.T. (1989). *Accuracy of Machine Tools,* New York, American Society of Mechanical Engineers.

Annals of DAAAM for 2008 & Proceedings of the 19th International DAAAM Symposium, ISSN 1726-9679
Editor B. Katalinic, Published by DAAAM International, Vienna, Austria 2008
Make Harmony Between Technology and Nature, and Your Mind will Fly Free as a Bird

CULTURAL AND CREATIVE TOURISM AS THE FACTOR OF THE RURAL COMMUNITY DEVELOPMENT-THE CASE OF NAKOVANA, PELJESAC PENINSULA

LETUNIC, S[tijepo]; DRAGICEVIC, M[arija] & BRAUTOVIC, H[elena]

Abstract: *New tourist destinations are able to attract tourists by the origin of tourist supply. Active participation in cultural events is extremely attractive. The development of this kind of alternative tourism is just in its infancy in the Republic of Croatia, especially in rural areas. The aim of the paper is to present at one of the potential touristic destination with rich cultural resources that have not yet been valorized. The aim of the paper is also to show the way it could be done. In that sense the paper gives the current situation, possibilities and the future actions necessary for the development of cultural and creative tourism on Peljesac Peninsula, in place of Nakovana.*

Key words: *cultural and creative tourism, possibilities, actions, Peljesac Peninsula, Nakovana*

1. INTRODUCTION

Attractions have been viewed by many as central to the tourism process. They are often the reason for visiting a particular destination (Richards, 2002). Strategies and program of tourism based on cultural-historical heritage is relatively new throughout the World. Cultural tourism is the fastest growing market in Europe, although empirical arguments are rare. Culture is becoming the basis for the "economy of experience", and cultural tourism its major part (Richards, 2002). Every destination is unique in its own way, and this uniqueness is what attracts tourists. Tourism based on the cultural and historical heritage is an excellent way to put history and art in the middle of the economic growth of a destination (Brook, 2000). Active participation of tourists in cultural programs, i.e. a creative component of a cultural tourist offer is an extremely interesting segment of a tourist offer. The programs to include cultural and historical heritage into a tourist offer should help some Croatian areas in developing new destinations or in improving existing environmental characteristics and cultural destinations. Unfortunately, in the Republic of Croatia this does not still gain enough attention. The aim of this paper is to point out to the possibilities of the development of cultural and creative tourism on Peljesac peninsula, in the place called Nakovana, whose extraordinary cultural potential has not yet been evaluated from the tourism angle.

2. CULTURAL AND CREATIVE TOURISM

Cultural tourism can be defined as a type of tourists' special interest in the search for a more extensive participation in new cultural experiences, esthetic, intellectual, emotional and psychological values (Xie & Wall, 2001). Today's tourists are seeking to actively participate in cultural events. This is possible in the area of creative tourism. In this matter, tourist buys dreams, i.e. escape from everyday life is commercialized (Schouten, 2002). It is much cheaper to develop creative than cultural tourism. It is based on conveying local skills and experience to the tourists in a suitable environment, and it does not require a developed infrastructure, which is not the case with cultural tourism. Visitors seek for a unique experience, funny and unusual adventures which provoke curiosity and rouse emotional reactions (Schouten,1995).

Physical remains of the past, whether it is a landscape, site or a building, can be presented to visitors in many ways. The real challenge is to find a way to make remains of the past present in today's everyday life and interesting to the tourists at the same time. Creative tourism is becoming more and more the main factor of the development of the outlying districts. There are very successful examples of the cultural and creative tourism development in the world. Programs of including cultural heritage in tourist supply should help Croatia at the development of new destinations or improve exisiting ones. It could attract new visitors and stimulate local economy. Creative production processes can also attract enterprises, and individuals involved in the cultural sector (Richards & Wilson, 2006). The main purpose of Croatian tourism development should be creation of sustainable development based on unique cultural heritage.

3. THE POSSIBILITIES OF CULTURAL AND CREATIVE TOURISM DEVELOPMENT IN NAKOVANA

Peljesac, peninsula in the South Dalmatia is the region with many different cultures existing throughout history (Peljesac tourist monography, 2003). In the Western part of the peninsula there is a place called Nakovana, consisting of two villages, Upper and Lower Nakovana. Nakovana is situated in the ecological area and it has a rich fond of endemic plants and rare animals. In the close vicinity to the villages there are unique historical sites like Spila, where the 8000 year Neolithic remains have been found, an intact Illyrian shrine from the 4th century B.C., the Illyrian hill-fort Gradina, the Illyrian tombstones, medieval chapels, and a family house of a torpedo inventor Ivan Lupis Vukic. Both villages are unfortunately abandoned and mostly devastated. One can conclude that the whole place is abandoned due to the fact that only one family lives during the entire year. The other owners of the family houses which are in better state, visit the village just occasionally. The climate is typical Mediterranean and convenient for development of tourism.

3.1. Existing cultural attractions and their state

3.1.1. Spila

With its three halls, the cave was a shelter already 8000 years ago. It represents a very rare finding and the richest finding from the Illyrian time in the World. A great number of artifacts were found, witnessing a life of Illyrians of that time. It is mostly about the ceramic dishes, imported mainly from Greece and Greek colonies (Forenbacher & Kaiser, 2003). The foundings are concentrated around stalagmite which dominates in the cave. It has been symbol of fertility for Ilirians in Nakovana. The collected amounts of the artifacts would be sufficient to open a middle-size museum. Unfortunatelly, the site is not prepared for the tourists' visits, therefore there are no organized visits to the site and there is not any financial interest from which the local community would make any profits. However, there was an exhibition of some of the artifacts found in Spila in the Rector's Palace in Dubrovnik, where many prominent researchers and scientists from around the World as well as from Croatia participated, and which aroused huge interest with the public. This only witnesses a great importance of the finding.

3.1.2 Hill Grad

This place had outstanding geo-strategic position from which the whole Adriatic see could have been controlled. The area was populated by the Illyrians during the ancient times, and they strengthened Grad by building terraces and walls, thus turning it into one of the most fortified towns in the Adriatic. Illyrians used this place to start their war and pirate crusades. Today, only the fundaments of the Grad are visible. The Illyrian masters of this exquisite city are buried in tumulus graves, which are in the close vicinity.

3.1.3. Illyrian tumuli

In the footage of Grad, there are about hundred Illyrian tumuli, called "Greek tumulus", which are artificial mounds over a grave. Some of them are of a monumental size, more than twenty meters in diameter and about four meters of height. They had been made from stone in the form of circle ground-plan. There are about 80 tumuli found in the Nakovana area. Limited number of tumuli points to the conclusion that they have not been built only for important persons. Unfortunately, they have all been excavated and destroyed in the search of treasure.

3.1.4. Fort on Spila Gubavica

It can be said that this area is one of the most preserved Illyrian tumuli. This pile is what has remained from an Illyrian fort which was built to monitor the Peljesac channel. Today, it is an extraordinary belvedere.

3.1.5. Upper and Lower Nakovana villages

Upper and Lower Nakovana villages were first mentioned in the 14th century, although there had been some settlements even before. They can be considered as a true jewel in the autochthones architecture. During the Second World War the the Lower Nakovana village was burnt by the Italian army. After Peljesac came under German occupation most population exiled to Sahara desert, in El Shattt, Egypt. After the Second World War most population migrated to other parts of the world, like New Zealand and Australia. Few important persons have their origin in village Nakovana. One of them is Ivan Lupis-Vukovic, the torpedo inventor, and the first Croatian in New Zealand, Pavle Lupis. Although many buildings are devastated, and are need reconstruction, some houses have been preserved, as well as the traditional equipment that has been used in everyday living.

3.1.6. Nativity of the Blessed Virgin Mary Church

This is church was mentioned in the 14th century for the first time although it probably originates from the 12th or the 13th century, which would mean that this is the oldest church where the mass is still celebrated. The church is still in good condition thanks to the inhabitants who take good care for the church. The feast of Nativity of the Blessed Virgin Mary, the 8th September attracts many local people, emigrants to far away countries, domestic and foreign tourists. In the past, up to five hundred inhabitants from neighboring places used to visit Nakovana on the 8th September. After the mass and procession, they have continued the feast with some social activities and the traditional dance. Since 2004 this tradition has been continued.

3.2. Development directions

The beginnings of the tourist valorization of Nakovana could be primarily directed towards creativity in culture. The tourists would be able to experience the past by actively participating in different kinds of workshops and organized excursions in Nakovana area. It demands additional education of human resource. With minor financial investments, some of the existing buildings could be restored and used as an adequate accommodation for a certain number of tourists. Next phase of development should stimulate the project of the cultural and creative tourism development. Strategic plan for the development has to be made and some operative plans as well, in cooperation with local inhabitants and associations. The state could be a significant factor in stimulating investments, participating in financing and supporting the project in any other ways. Tourist board should be involved in stimulating and improving of original values and attractions. Local community, i.e. local inhabitants should be actively included into the provision of services. It is necessary to improve the existing infrastructure and telecommunication. An urgent reconstruction of the village into its original state is necessary too, in order to preserve the authenticity of the place. Building apartment complex and violation of autochthony is out of question. Building of all other side objects of the tourist offer should be managed in accordance with the principles of sustainable development. Beside the main road which is satisfactory, there are many side roads in a very poor condition, and some of the sites are impossible to reach at all. The tourist development of Nakovana village could be aided by the fact that it is located in vicinity of Viganj and Loviste, who have accommodation capacities and other tourist facilities.

4. CONCLUSION

Especially attractive segment of tourist supply is tourist active participation in cultural events. That is a creative component of tourist supply. Development of this alternative form of tourism can improve the attractive features of destination and bring multiplicative financial benefits. Although these programs are rather new in the world, there are many successful examples of the development of cultural and creative tourism in the local communities. Unfortunately, the abundance of the cultural resources has not yet been used to a full extent in the Republic of Croatia. There are still many potentially interesting and attractive tourist destinations that have not valorized their resources yet. One of them is village Nakovana with an extreme cultural wealth that could be categorized as very rare, but not valorized in the tourism sense. The village is attractive because of its location, authentic architecture, numerous historical findings and tradition. It is necessary to induce a tourist valorization of this extremely interesting area.

5. REFERENCES

Brook, R. (2000). Kulturno povijesna bastina hrvatskih mjesta u funkciji turizma: izazovi, mogucnosti, potencijali, *Turizam*, (Cultural and historical heritage of the Croatian places in the function of tourism: challenges, possibilities, potentials, *Tourism*), Vol.48., No., 4., 2000., pp. 407-412.

Forenbacher, S. & Kaiser, T. (2003). *Spila Nakovana (The Cave of Nakovana)*, V.B.Z., Zagreb

Schouten, F. (2002). Razvoj proizvoda za lokalitete naslijeda: studija slucaja, *Turizam*, (Development of the product for the heritage localities: case study, *Tourism)*, Vol. 50., No. 3. ,pp. 285-293

Richards, G. (2002). Od kulturnog do kreativnog turizma: europske perspektive, *Turizam,*(From cultural to creative tourism: European perspectives, *Tourism*), Vol. 50., No. 3., pp. 229-236

Richards, G. & Wilson, J. (2006). Developing creativity in tourist experiences: A solution to the serial reproduction of culture? *Tourism Management*, No. 27., pp. 1209-1223

Richards, G. (2002). Tourism attraction systems exploring Cultural Behaviour, *Annals of TourismResearch,* Vol. 29., No. 4., pp. 1048-1064

Schouten, F. (1995). Improving visitor care in heritage attractions, *Tourism Management*, Vol. 16., No. 4., pp 259-261

Xie, P.E. & Wall, G. (2001). Iskustva kulturnog turizma na Hainanu: prostorni prerazmjestaj etnosela, *Turizam*, (The experience of the cultural tourism on Hainan: space reconfiguration of ethno-village, *Tourism*), Vol. 49., No. 4., pp. 375- 382

Peljesac, tourist monography, Zagreb, 2003.

Annals of DAAAM for 2008 & Proceedings of the 19th International DAAAM Symposium, ISSN 1726-9679
Editor B. Katalinic, Published by DAAAM International, Vienna, Austria 2008
Make Harmony Between Technology and Nature, and Your Mind will Fly Free as a Bird

IT APPLICATIONS FOR MANUFACTURING OPERATIONS AND CONTROL

LEWANDOWSKI, J[akob] & KITTL, B[urkhard]

Abstract: *This paper deals with IT systems used in the area of manufacturing. In face-to-face interviews companies were questioned about their software solutions in eight defined areas. The results focus on the range and level of supported functions. Furthermore, satisfaction of enterprises with implemented software and their opinions concerning integrated systems in the area of manufacturing were a main part oft the study.*
Keywords: *Manufacturing Execution System (MES); Production Planning and Control; IT survey.*

1. INTRODUCTION

Today there are approximately 70 different MES software solutions offered for the German-speaking market (Wiendahl et al., 2007), and many more specialised applications which cover single MES functionalities. This shows that many software vendors try to answer the call for IT in the area of manufacturing. But in contrary to ERP systems there is no general agreement about the design of an MES.

The first step was to popularise the idea of MES. Over the past decade papers concerning the definition of MES and its functionalities have been published (MESA, 1997; VDI 2006). But the awareness of the term MES in the German speaking industry is still quite low (Wiendahl et al., 2007; Lewandowski et al. 2008). Further publications about practical examples, where MES solutions were implemented, were more focused on the advantages in general (cp. Kozian, 2000) rather than on the realisation of MES functions with software. Only the topic of interfaces between applications in the area of MES and across its boundaries is a slightly more developed (ISA, 2000).

For this reason two surveys within a study of the Vienna University of Technology were carried out in order to get a better idea about the state-of-the-art concerning IT in the manufacturing industry. This paper focuses on the implemented systems in layer 3 (manufacturing operations and control) of the functional hierarchy by ISA 95 (ISA, 2000). Furthermore the interpretation of the data basis will be the start for future researches.

2. APPROACH

The main part of the study is based on face-to-face interviews with twenty-two polled enterprises. In addition an online survey was carried out in order to increase the number of questioned companies to fifty-one.

2.1 Objectives

All objectives of the study focus on the area of MES:

- Identification of the used IT systems.
- Finding out the level of automation and integration of IT systems.
- Assessing the demand for IT systems.
- Looking for potentials in the companies' software landscape.
- Identifying the awareness of new IT solutions like MES.

2.2 Target Group

The regional origin of the enterprises was restricted to Austria. Moreover the study was primarily aimed at small to medium sized companies. The number of employees in the face-to-face interviews ranged approximately from 50 to 500 employees, whereas there was no limit in the online survey. All of the polled companies belong to the producing sector.

2.3 Survey Method

During the preparations for the study a theoretical framework was developed. It is based on the tasks and functions of the "Aachener model for production and control" (Schuh, 2006) combined with the structure of the functional hierarchy of ISA 95 (ISA, 2000). Furthermore the model was cross-checked with further American and European literature like MESA (MESA, 1997) or the VDI report (VDI, 2006).

This framework was used in the face-to-face interviews in order to build up a guideline for the interview procedure and to structure and document the information afterwards. The duration of the interviews ranged from 70 up to 180 minutes (with an average of 117 minutes). In contrast to the online survey, there were no predefined answers used in the face-to-face interviews.

3. RESULTS

In this chapter the results of both surveys are included. Percentage figures in this chapter always refer to the online survey and will be annotated by the experiences and analyses of the face-to-face interviews. Differing results in the two surveys will be marked separately.

The use of IT in the various areas of the manufacturing environment is diverse. Table 1 shows the used IT systems in eight manufacturing areas.

	MS Office	ERP system	ERP extension	Other standard SW	individual SW	no SW
Order management	70%	70%	44%	44%	44%	0%
Resource management	57%	43%	35%	17%	39%	13%
Maintenance management	64%	41%	27%	45%	32%	5%
Material management	59%	76%	41%	24%	38%	0%
Deployment management	64%	44%	24%	44%	32%	4%
Quality management	77%	46%	23%	35%	46%	0%
Document management	64%	41%	23%	32%	36%	0%
Performance analysis	76%	40%	44%	48%	44%	0%
Average	66%	51%	33%	36%	39%	3%

Tab. 1: Systems in use.

ERP extension stands for additional individually developed functionalities of the standard. For the purpose of the survey standard software modules like scheduling, shop floor data collection or an integrated MES are subsumed under "other standard software". In contrast, individual software stands for software which is specially developed for a certain functionality of one company.

In general MS Office products are the most used software type, followed by manufacturing-related models of ERP systems. Only in three areas (resource, maintenance and deployment management) companies work without any IT system. This outcome is a confirmation of the face-to-face interviews. There were also different systems used in the area of manufacturing but there were more companies which quoted to use no software for single functionalities like maintenance or in production control. It looks as if the area of manufacturing is well covered by IT systems, but the reality is different. Analyses of the face-to-face interviews, as well as the online survey, demonstrate that the reported systems in use show only half the truth. The range and level of supported functions is very important and differs a lot from company to company.

In the area of **order management** IT support (mainly ERP systems) ends after rough planning. 74% of the companies send their production orders as a printed document to the job-shop. Concerning completion-confirmation-data the majority of the companies' shop floor data is entered manually. Automated data collection is realised by 50% of the companies. In general completion-confirmation-data is seen as very important, but companies have some problems in the aggregation of the data.

In 87% the companies' **resource management** is covered by a software system. Slightly more than half of the enterprises are quite satisfied with the IT support and one third demands more automated data collection. Approximately 9% are not happy with their current software solution.

Although the companies stated that they support **maintenance** with several systems the level of automation is rather low. 73% of the enterprises have a manual interface for this functionality and only 23% demand more automated data collection. Approximately 60% of the companies feel comfortable with the IT support and will not invest in this area. In the face-to-face interviews many companies claimed maintenance to be an important function, but only few of them were using an integrated software tool.

In most cases **material management** functions are covered by production-related ERP modules. Furthermore the coverage of this area is very good. Material status or booking are well supported functionalities as well as the integrated inventory management. The area of tracking and tracing is also IT supported in many companies, but in in the face-to-face interviews users voiced that they were not satisfied with the provided functionalities. The topic RFID seems not to be important as almost no company uses RFID. Additionally the use of B2B, which was questioned only in the face-to-face interviews, turned out to be a marginal area of interest.

The most supported function of **deployment management** is the registration of attendance time. In this area 92% of the companies use an IT system. But only half of the companies use a software tool in order to plan the human resources. The willingness of the companies to invest in this functionality was approximately 30%. In many cases specialised systems are used for the deployment management which are seldom integrated. MS Office seems to be very important in this section, but it is always used in combination with other systems.

Similar to the deployment management there are also a lot of specialised systems used in the area of **quality management**. Moreover, the rate of MS Office products in use is very large, but ERP systems and individual software are also used quite often. The manipulation of the data is supported by 70% of the IT systems. A data transfer to another system is established by half of the enterprises. The readiness to invest is limited to quality inspection and documentation systems.

Document management is supported quite well by IT systems. Nearly 70% of the companies cover a huge range of functions with their software tools.

In the area of **performance analysis** every company records one or more performance indicators. Noticeable is that only one third of the companies quoted to use OEE. Furthermore, the enterprises polled in the face-to-face interviews stated that performance indicators demanded by the management level differ from the ones needed in the manufacturing area and are harder to provide due to missing integration or absence of IT systems.

4. CONCLUSION

In the area of manufacturing companies are using many different software applications. A reason for this situation is that enterprises have implemented IT systems in several manufacturing areas and different periods. Each solution was focused on the special needs of that area and realised with existing or best fitting software systems. As a result specialised vendors emerged, which provide IT applications for single manufacturing functions. But many companies have recognised the potentials of sharing the information of one IT system with others. On the other hand enterprises are quite satisfied with the present applications and in particular smaller companies do not want to replace their systems by an integrated MES software.

5. FUTURE RESEARCH

In order to find a solution for the problems reported in the chapters before, the Institute of Production Engineering of the Vienna University of Technology has built up a test facility, in which different software scenarios can be simulated. The idea is to work with an IT framework where several applications can be integrated with the help of standard interfaces of ISA 95 (ISA 95).

6. REFERENCES

Instrument Society of America (2000). *Enterprise-Control System Integration, Part1: Models and Terminology*, ISA, 1-55617-727-5, North Carolina

Kozian, D. (2000). Anwendernutzen von MES / User benefit of MES, *Available from*: http://www.wonderware.de *Accessed:* 2008-08-01

Lewandowski, J.; Buhl, M. & Kittl, B. (2008) IT Applications in Production Planning and Control, *Proceedings of the ICE-B*, Filipe, J. et al. (Ed.), pp. 285-291, 978-989-8111-58-6, July 2008, INSTICC, Porto

Manufacturing Execution Systems Association (1997). MESA White Paper #06: MES a High Level Vision, *Available from:* http://www.mesa.org *Accessed:* 2008-08-01

Schuh, G. (2006). *Produktionsplanung und -steuerung / Production planning and control*, Springer, 3-540-40306-X, Berlin Heidelberg

Verein Deutscher Ingeneure (2006). VDI 5600 Blatt 1 Fertigungsmanagementsysteme / Manufacturing Execution Systems, *Available from:* http://www.vdi.de *Accessed:* 2008-08-18

Wiendahl, H.-H.; Mussbach-Winter, U & Kipp, R. (2007). Marktspielgel Business Software, MES-Fertigungssteuerung 2007/2008 / Reflection of the business software market, MES-shop-floor control, Fraunhoferinstitut Produktionstechnik und Automatisierung/Trovarit AG, 978-3-938102-08-4, Stuttgart/Aachen

Annals of DAAAM for 2008 & Proceedings of the 19th International DAAAM Symposium, ISSN 1726-9679
Editor B. Katalinic, Published by DAAAM International, Vienna, Austria 2008
Make Harmony Between Technology and Nature, and Your Mind will Fly Free as a Bird

FINANCIAL DECISION-MAKING FOR THE INVESTMENT OF COORDINATE MEASURING MACHINE

LIM, M[ei] & WONG, H[ung] K[un]

Abstract: *A traditional mould making company encountered challenge with advanced new products, such as lead frames. Although the production lines have gradually been enhanced by the use of advanced manufacturing machines, such as NC lathes, CNC machining centre, and robots, conventional tools were still being utilised by the inspection division. The accuracy and the production speed are not meeting customers' demands. A task force was to investigate the feasibility of implementing an automatic measuring system. The major concern is the investment of a Coordinate Measuring Machine (CMM). This machine aims at bringing the mould inspection process from conventional manual operations with a variety of gauges to automatic measurement and data processing. The advantages of introducing this machine and economic benefits were considered. The detailed evaluation and analysed results are presented by simulated data, which finally determines that the manual CMM and a CNC CMM as a suitable investment decision.*

Key words*: Decision, Measuring, Manufacturing, Finance.*

1. INTRODUCTION

This company has been established since ten years ago and is located in Australia. The main activity of the business is to produce a variety of moulds for plastic injection and die-casting. In the early stage, the company designed and manufactured simple moulds for home wares, such as soap cases and bottle lids. With the advent of industrial automation and precision machines during the past decade, the mould technology was significantly upgraded by some emerging industries, such as computer industry. The company has since invested advanced equipment on the design and production lines, such as CAD/CAM, CNC machine tools and automated warehouse. At present, quality assurance of the products is, however, their major concern, which can promote not merely the company reputation, but also generate more business. The company utilizes a variety of gauges and meters for dimensional inspection of the moulds. They started to realise the bottleneck of their inspection technology, as the conventional measuring tool was too labour intensive and time consuming. A large amount of measurement uncertainties caused by human errors and conventional gauges also have negative impacts on the operations. After investigation on the implementation of coordinate measuring machines (CMM) for inspection room and wined confidence of customers, a task was formed to implement the decision making of investing a CMM (Dong et al., 2002).

2. DEFINE COMPANY'S PROBLEMS

The company makes a variety of moulds. Original inspection of dimensional accuracy of the mould relied entirely on the operator's proficiency in the use of gauges and meters. As the making of moulds became more complex as required by emerging industries, measured items and difficulties increased drastically. A large amount of measurement uncertainties caused by human errors and conventional gauges have negative impact of the company such as weaken confidence from inspectors due to the lengthy measuring time. Further, the continuous change of gauge for completing a single mould has also caused concern. The problems that the company encountered are as follow:

- difficulties in evaluating the capital expenditure and its contribution to the operation;
- difficulties in estimating the rate of operation;
- difficulties in quantifying quality improvements and reducing production time (Dong et al., 2002).

Table 1 below shows the different types of workpieces and the time required for measurement by using conventional measuring tools (CMT), a manual CMM (M-CMM), and a CNC-CMM in this manufacturing company.

Workpiece Type	Number of features to be measured	CMT	M-CMM	CNC-CMM
Transmission case	Diameter: 40 Hole post.: 30 Surface dist.: 3	180 min.	40 min.	35 min.
Jig for Boring	Index angle:16 Diameter: 4 Surface dist.: 9	300 min.	15 min.	10 min.
Casing	Co-axiality: 4 Parallelism: 2	180 min.	40 min.	30 min.
Video Tape Cassette	Hole interval:17 Diameter:10 Surface dist.: 8	260 min.	30 min.	3 min.
Joint	Angle: 13 Intersetion:7 Diameter:5	120 min.	20 min.	20 min.

Tab. 1. The time required to make workpieces by using three different types of CMM.

3. FINANCIAL METHODS - NET PRESENT VALUE & INTERNAL RATE OF RETURN

In this analysis, cost measurements were considered for comparison between the three alternatives, which are:

1. The conventional measuring tools (CMT);
2. Manual CMM; and
3. CNC-CMM.

	CMT	Manual CMM	CNC CMM
Equipment item	Graplate, height gauges, height master, micrometer, caliper, etc.	F3905 with MICROPAK 220, touch-trigger probe	FN905 with MICROPAK 2810, touch-trigger probe, etc.
Ratio of measuring time	4.0	1.0	0.7
Total investment	44K	280K	336K
Cost depreciation (8 years linearly)	5.5	27.5	42
Interest	2.6	16.8	20.2

payment(6 % for the first year)			
Running cost	0.9	13	16.8
Inspection cost	491	75.6	52.9
Total cost per year	500	125	87.5

Tab. 2. The details and cost for all three equipments are summarized. (All units in $1000)

In order to determine the operational costs, several important issues were considered:

- Ratio of measuring time: 4.0 (conventional method) to 1.0 (manual CMM) to 0.7 (CNC-CMM);
- Cost depreciation: 8 years by linear method;
- Running cost: cost of equipment maintenance, electric power, and material consumption are estimated such as conventional method technique is about 2 % of equipment value and CMM is 5 % of equipment value;
- Labour cost involved in inspection: average labour cost per head: $ 78/hour;
- Working hours: 1680 hours/year (15% idle time);
- Amount of work: 100% of operating capacity;
- Interest rate: 6%.

To perform the analysis, estimation of *Net Income after Tax* of the company per year was undertaken. The estimation was directly related to the use of measurement equipment by the company. The situation of the company is that a conventional manual inspection operation is still in use, it causes bottom-neck in the inspection technology. For this reason, it is assumed that a relative low sales volume thus a low Cash-In value to start with in order to fit in the real situation.

For **CMT**, Cash In (Net income before tax) = $550 K p.a., (compared with Cash Out calculated as $500 K p.a.)

If CMM technology is introduced in the company, since it has been suggested that CMM can dramatically reduce the time taken in inspection process, it is believed that production efficiency could also be improved. On the other hand, with this improvement in production, sales as well as net income (cash in) could be increased accordingly.

For **Manual CMM**, cash in is expected to grow up by 20%, and Cash In = $660 K p.a., with Cash Out = $125 K p.a.

For **CNC-CMM**, an extra 30% of increase in cash income is expected due to more time can be saved in the inspection process by this method, therefore, Cash In = $825 K p.a., with Cash Out = $87.5 K p.a.

Other useful assumptions for NPV calculations:

- Federal company tax rate = 30%
- MARR or discount rate set by the company = 15%
- Economic life for the all the alternative equipment is 8 years
- Straight line depreciation is employed with no salvage value

An "after-tax income" analysis that takes taxation and depreciation into account is then carried out and the results are summarized below:

	CMT	Manual CMM	CNC-CMM
Initial cost	44K	280K	336K
Annual income (after tax)	36.65K	382.75K	528.85

Tab. 3. The details of the initial cost and annual income (after tax) for three equipments are shown above.

Year	Cash In	Cash Out	Dep.	Tax In	Tax	Net In
0		-336				-336
1	825	87.5	42	695.5	208.65	528.85
2	825	87.5	42	695.5	208.65	528.85
3	825	87.5	42	695.5	208.65	528.85
4	825	87.5	42	695.5	208.65	528.85
5	825	87.5	42	695.5	208.65	528.85
6	825	87.5	42	695.5	208.65	528.85
7	825	87.5	42	695.5	208.65	528.85
8	825	87.5	42	695.5	208.65	528.85

Tab. 4. The final best financial results of CNC CMM approach were provided above. (All units in $1000)

Net Present Value (NPV): This is found by solving for NPV in the following equation (Petty, 2008):

$$NPV = \sum_{t=1}^{n} \frac{ACF_t}{(1+k)^t} - IO \qquad (1)$$

NPV = $120.5 K (CMT)
NPV = $ 1.42 M (manual CMM)
NPV = $ 2.04 M (CNC CMM)

Internal rate of return (IRR): This is found by solving for IRR in the following equation (Petty, 2008):

$$IO = \sum_{t=1}^{n} \frac{ACFt}{(1 + IRR)t} \qquad (2)$$

IRR = 82.6% (CMT)
IRR = 136.7% (manual CMM)
IRR = 157.3% (CNC CMM)

As NPV and IRR are the only attributes in the analysis, it was concluded that the best alternative is the CNC CMM because it has the biggest NPV among the others. When capital rationing exists, the company should select the set of project with the largest NPV (Peacock et al., 2003, Nally & Willig, 2007).

4. CONCLUSION

It was concluded that the CNC CMM technique is the best option compared to the other two and should be considered by the mould making company in relation to cash flow issues. The resulting improvement in product quality will enhance the reputation of the business.

These two financial methods give us the same accept-reject decision on an individual project, but may rank alternative projects differently. In addition, they can incorporate risk into the capital budgeting evaluation, either by adjusting the cash flows or by building risk into the discount rate.

5. REFERENCES

Dong, C., Zhang, C., & Wang, B. Prediction and Compensation of Dynamic Errors for Coordinate Measuring Machines. *Journal of Manufacturing Science and Engineering*, Vol. 124, No. 3 (August 2002), pp. 509-514, 1087-1357.

Nally, J. & Willig, S. (2007). *Good manufacturing practices for pharmaceuticals*, 6th Ed., Informa Healthcare, 9780849339721, New York.

Peacock R., Martin P., Burrow M. (2003), *Financial Management, 3rd Edition,* Pearson Education, 1740097874, Australia.

Petty, W. (2008). *Financial management: principles and applications*, 5th Ed., Pearson Education, 9780733993503, Australia.

Special thanks to the anonymous manufacturing company for providing all relevant data to researchers. Metadata withheld due to confidentiality requested by the company.

Annals of DAAAM for 2008 & Proceedings of the 19th International DAAAM Symposium, ISSN 1726-9679
Editor B. Katalinic, Published by DAAAM International, Vienna, Austria 2008
Make Harmony Between Technology and Nature, and Your Mind will Fly Free as a Bird

METHOD OF FUZZY FAULT TREE ANALYSIS

LISJAK, D[ragutin]; CALA, I[vo] & STEFANIC, N[edeljko]

Abstract: *This paper presents the possibility of extension of the classical fault tree analysis FTA with the fuzzy fault tree analysis FFTA. The integration of the FFTA within the tree fault is presented as well as the example of the simulation of the fuzzy sets basic events with affiliated functions of affiliation. Some linguistic rules are defined on which basis fuzzy prompted surface of the reliability of intermediate events is attained. The expressions for calculation of total reliability of unwanted or top event FTA in combination with calculation of the fuzzy accountability are presented.*
Key words: *fault tree analysis, fuzzy ,reliability.*

1. INTRODUCTION

Modern industries are made from complicated technical systems which require new possibilities in the approach of their maintaining. The features as *availability* and *reliability* are more and more presented as main criteria during the process of the maintenance definition. Modern maintenance strategies are based on the reliability estimates of the elements of complex systems with the purpose of the on-time replacement of same systems before the fault occurs. One possibility for the determination of the reliability of the complex systems is Fault Tree Analysis - FTA which represents deductive *top-down* analysis structured by the terms of the events. FTA methods requires definition of the unwilling event which is positioned on the top of the tree, through which arms potential component and subsystem faults are checked, i.e. procedural faults which could cause the unwilling event. For reliability calculation of the unwilling event by classical FTA method, time independent reliabilities of the all system elements must be known in advance. However, some of the system elements could have eg. timely dependent reliability, what timely influence change of the unwilling event. Due to this fact, this paper presents the proposal for the extension of the classical (FTA) methods with the Fuzzy Fault Tree Analysis (FFTA) method which solves the before mentioned problem.

2. CLASSICAL FTA

The analysis by the fault tree method FTA which is presented in the picture 1., is one of the basic methods for analysis of the reliability and security estimates of technical systems. Determination and analysis of reliability of complex technical systems by the fault tree method represents deductive procedure of identification of all relevant possible causes of reliability and determination of their interactions (Veseley & Dugan, 2002). The basic condition and necessity in determination of the application of the reliability analysis by the fault tree method is definition of the unwilling or top event (TE1) of the observed technical system. *Basic-events* (BE1, BE2, IE, BE3) which cause the top event make the arms of the tree. By the definition of the top event within the fault tree and by passing its arms, the possible subsystem faults and components are searched for, as well as procedural mistakes which could cause the top event. The system entities which are named gates (AND, OR) are found within the fault tree. They enable or disable propagation of the fault logic on the top of the table. It is important to mention that the fault tree does not represent the model of all possible causes of the fault of technical system, but covers only those faults which are given priority and estimation of the biggest possibility for the top event. It should be mentioned as well, that the fault tree is not itself the quantitative model, but also qualitative model which is subject to quantitative evaluation (Goldberg & Dugan, 2002).

3. FUZZY FTA

The fuzzy fault tree analysis (FFTA) represents the extension of the classical fault tree analysis (FTA) method and the base for its application could be found in the next assumptions:

- There is no clear border between the potential fault and correct way of the system work.
- The fault reliability cannot be precisely calculated due to the data absence or noise in existing data.
- The reliabilty estimation of the system is based on the subjectivness and it is expressed with the linguistic terms as: high, middle, law.

Picture 1. represents one of integration possibilities of FFTA model within the classic FTA model fault tree. The simulation assume reliability calculation of the top event partially by FFTA (until level IE1) i partially by FTA model. Subjective reliability estimations on the base of linguistic expressions (eg. high, middle, low) are translated for all events (BE1, BE2 i IE1) into *fuzzy sets* mapping by *membership functions*. The subjective reliability of the fault on the base of linguistic expressions of every of events (BE1, BE2, IE) is defined within intervals 0 and 1.

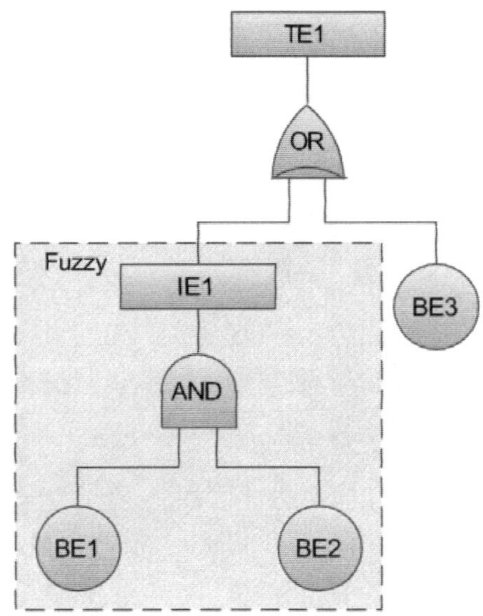

Fig. 1. Integration of *fuzzy* FFTA within the fault tree

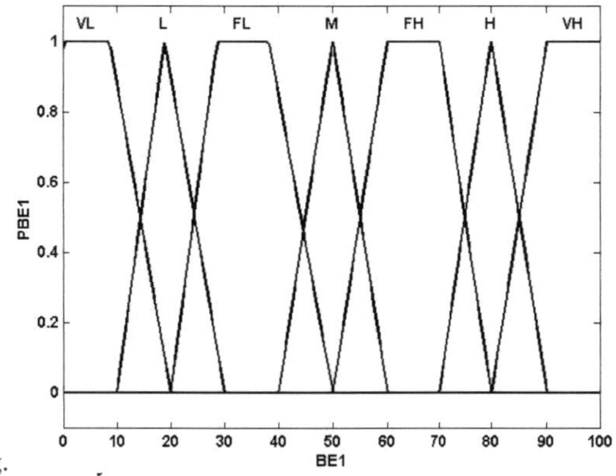

Fig.

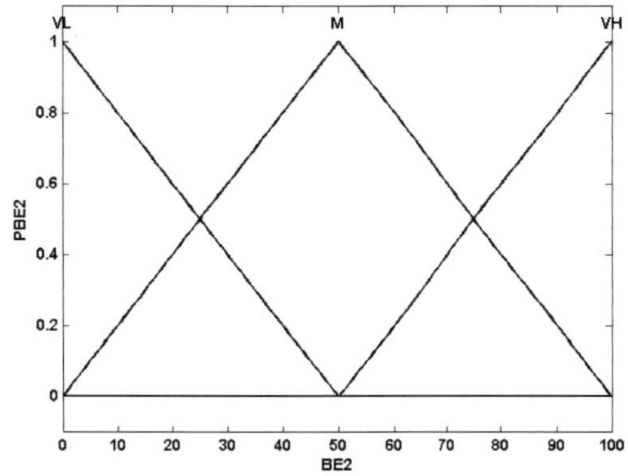

Fig. 3. Fuzzy sets for BE2

Picture 2. and 3. represents simulations of fuzzy sets for BE1 and BE2 events with associatied functions for every linguistic expression. The reliability of IE event is calculated by the application of the fuzzy aplicants of multiplication and substraction (Menzies, 1999; Ross,2004; Suresh at al., 1996) what is generally presented by the expression (1).

IE = *fuzzy*(BE1,BE2) (1)

Simulation of the reliability surface IE1 (picture 4.) is gained on the basis of definition of next *fuzzy* rules where:

1: *If* **BE1** *is* **M** *And* **BE2** *is* **M** *Then* **IE1** *is* **M**
2: *If* **BE1** *is* **H** *And* **BE2** *is* **VL** *Then* **IE1** *is* **M**
3: *If* **BE1** *is* **L** *And* **BE2** *is* **VH** *Then* **IE1** *is* **H**
4: *If* **BE1** *is* **M** *And* **BE2** *is* **VH** *Then* **IE1** *is* **VH**

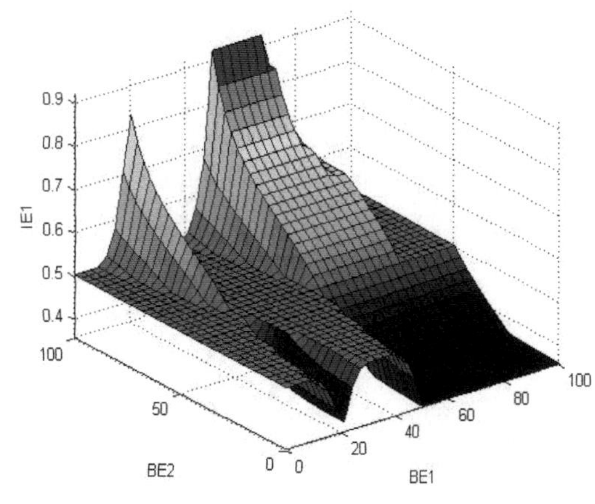

Fig. 4. *Reliablity surface* IE1

- VL -**V**ery **L**ow,
- L -**L**ow,
- FL -**F**airly **L**ow,
- M -**M**edium,
- FH -**F**airly **H**igh,
- H -**H**igh,
- VH -**V**ery **H**igh.

According to the picture 1. the total reliability of the top event u (TE1) is calculated in accordance with the expression (2):

TE1 = 1 – [(1 – IE) * (1 – BE3)] (2)

On the base of the defined rules presented in table 1., calculated *fuzzy* reliabilities of the event IE1 and total reliability of the top event TE1. This paper assumed predefined reliabilities of the event BE3.

Fuzzy				Reliability	
Rule	BE1	BE2	IE1	BE3	TE1
1	60	50	0.50	0.10	0.55
2	70	40	0.43	0.20	0.54
3	40	90	0.50	0.30	0.65
4	65	90	0.74	0.40	0.84

Tab. 1. The example of the calculation of the event TE1

4. CONCLUSION

The reliability calculation of some complex technical system, and detection of potential faults by the fault tree method is complex due to the next facts:

1. Design of the fault tree structure is more complicated as the system is complicated.
2. The absence of the precise quantitative data on reliability of any event within the fault tree structure.

The first problem is possible to solve more or less successfull by the application of different software equipment. As the FTA method requires quantitative data for reliability calculation of the top event, the second problem is possible to solve successfully by the application of the FFTA method of the fault tree which represents the extension of the classical FTA method. As it is presented in simulation examples, by the application of the *fuzzy* logic, it is possible to get quantitative reliabilities for the events within the fault tree which have dinamic caracter, eg. they are changable during the time. The events which does have such caracter are described with linguistic (in practise confirmed) rules, and by the application of fuzzy applicators their quantitative reliability is determined, as well as determination of the top event.

5. REFERENCES

Goldberg,F., & Roberts, N., (1981). *Fault Tree Handbook*, U.S. Nuclear Regulatory Commission, Washington.

Menzies, T., (1999). *Knowledge maintenance: the state of the art*, The Knowledge engineering Review, Vol. 14., Issue 1.

Ross,T.J., (2004). *Fuzzy logic with Engineering Applications*, 2nd ed., John Wiley & Sons.

Suresh, P.V., Babar, A.K., & Raj, V.V., (1996). *Uncertainty in fault tree analysis: A fuzzy approach*, Fuzzy Sets and Systems, Vol. 83., Issue 2.

Veseley, W., & Dugan, J., (2002). *Fault Tree Handbook with Aerospace Application*, NASA, Washington.

Annals of DAAAM for 2008 & Proceedings of the 19th International DAAAM Symposium, ISSN 1726-9679
Editor B. Katalinic, Published by DAAAM International, Vienna, Austria 2008
Make Harmony Between Technology and Nature, and Your Mind will Fly Free as a Bird
*Young Researcher and Scientist Paper / * Supervisor, Mentor*

THE EFFECT OF WIRE RODS QUALITY ON MECHANICAL PROPERTIES OF DRAWN WIRES

LISKOVA, J[ana]; TITTEL, V[iktor] & ZELENAY, M[iroslav]

Abstract: *This article is oriented to the effect of wire rods quality on the mechanical properties of drawn wires. During our experiment, the wire rods from two different producers were used as input materials. The material from producer A was used to make high- tensile strength drawn wire and the material from producer B was planned to its making too. Required parameters of finished wire were diameter and tensile strength limit. These values were detected: values of bends, values of torsions and elongation after drawing process. It was detected and defined mathematically formula of material B for strengthening by processing, which can be used by application on other products. These determinations will be applied by appraisal of applicability of material B for other products.*
Key words: *wire rod, drawn wires, tensile strength, program SPC-PC IV.*

1. INTRODUCTION

Mechanical properties of cold drawn wires are influenced by all mechanical properties of wire rods above all and by technological conditions of cold drawing. Not only the diameter of wire, but also the mechanical properties of wires are changing during the drawing process because there is an increase of strength properties and a decrease of plastic properties. Therefore, every change of input material WR will evoke the need of availability attest of using technology, eventually its change (Tittel, 2006).
Main goal of the experiment was to verify WR availability from the new producer and compare the reached properties after drawing to standard reached values from current used material. These wires were produced from WR quality C76D from producer A and are suitable in the present.

2. ACTUAL PRODUCTION TECHNOLOGY

Technology of drawn wires production is following (Lišková, 2007):

- Pickling of WR in HCl by two sinks. Goal of pickling is to eliminate the oxides, which are created by rolling process, from WR
- Phosphating and boraxing. Goal of this surface treatment is to create an appropriate layer on wire surface which serves as lubricant carrier. This layer also protects wire against corrosion for a short term (few days).
- Drawing from Ø 5.5 mm to Ø 2.8 mm. This (first) drawing process from Ø 5.5 mm to Ø 2.8 mm, consists of six passes. Total reduction by first drawing process is 74.09 %. Drawing speed was 7 m.s^{-1}. Drawing was performed on the accumulation drawing wire machine.
- Patenting with using austenitisation process and cooling in lead bath for acquirement necessary composition applicable to drawing.
- Surfacing - continuous phosphating and boraxing.
- Drawing from Ø 2.8 mm to Ø 0.83 mm. This (second) drawing process consists of 11 passes and total reduction is 91.21%. Drawing speed was 12 m.s^{-1}. Drawing was performed on the straight-drawing wire machine. Values of partial and total reductions are put in table No.1.

Nr. of draw	Diameter of die [mm]	Partial reduction ε_d [%]	Total reduction ε_c [%]	Real tensile strength [MPa]
1	2.48	21.55	21.55	1 401.8
2	2.20	21.31	38.26	1 443.3
3	1.96	20.63	51.00	1 524.1
4	1.75	20.28	60.93	1 612.9
5	1.56	20.54	68.96	1 691.1
6	1.39	20.61	75.36	1 774.5
7	1.25	19.13	80.07	1 885.5
8	1.12	19.72	84.00	1 998.3
9	1.01	18.68	86.99	2 138.9
10	0.91	18.82	89.44	2 274.1
11	0.83	16.81	91.21	2 329.9

Tab.1. Measuring values by second drawing on straight-drawing wire machine from ∅ 2.8 mm to ∅ 0.83 mm

$$\varepsilon_d = \frac{d_0^2 - d_1^2}{d_0^2} \bullet 100 \text{ [\%]} \quad (1)$$

$$\varepsilon_c = \frac{D_0^2 - d_n^2}{D_0^2} \bullet 100 \text{ [\%]} \quad (2)$$

d_0 – input diameter of wire before die, d_1 – output diameter of wire after die, D_0- begginig diameter of wire, d_n- finishing diameter of wire (Bača & Bílik, 2000)

3. DESCRIPTION OF THE EXPERIMENT

We tried to find out the parameters differences of drawn wires which were produced from two sorts of WR in our experiment. We found out that the mechanical properties, chemical composition and also metallography were very similar without much difference. Also the quantity of processing was about the same.
There was retained production technology such as by WR from producer A by experiment of checking another WR. Experiment consisted in exact evaluation of tensile strength after drawing process, patenting and the following - second drawing process and also in more detail evaluation of mechanical properties, which were not required by a customer, such as: values of bends, values of torsion and elongation after drawing so that it is possible to confirm not only the availability of this material for producing but also the comparison of both WR sorts in whole technological process. We also followed the values of worn dies and occurrence of wire breakages which gave a vision about way- and conditions- of drawing. Request on drawn wires is according to EN STN 10204.

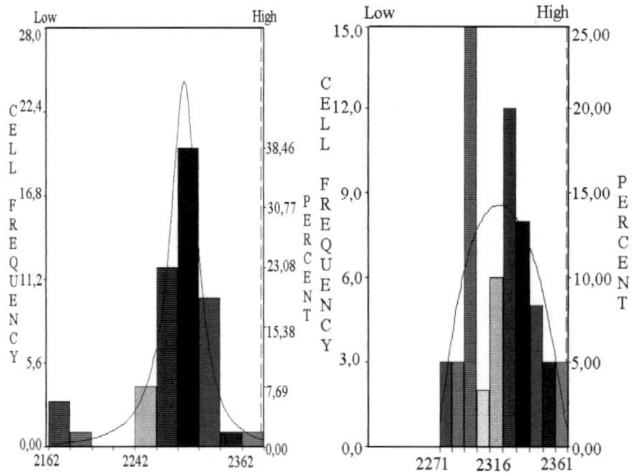

Fig. 1. Tensile strength after final drawing on material A and *B*

No. of test	Material A			Material B		
	Values of bend	Values of torsion	Elongation [%]	Values of bend	Values of torsion	Elongation [%]
1	18	33	4.8	17	43	4,6
2	16	38	4.9	18	33	4,9
3	20	35	5.1	14	38	4,8
4	23	45	4.7	16	38	5,0
5	18	38	5.1	19	32	4,4
6	16	45	4.1	21	35	4,4
7	17	36	5.0	17	35	4,4
8	18	36	4.9	16	42	4,9
9	17	43	4.6	21	41	4,5
10	16	42	4.9	18	40	5,0
Min.	16	33	4.7	16	32	4,1
Max.	23	45	5.1	21	43	5,0
Average	**17.9**	**39.1**	**4.81**	**17.7**	**37.7**	**4.69**

Tab. 2. Measuring values of the number of bends, torsions and elongation on wire ∅ 0.83 mm

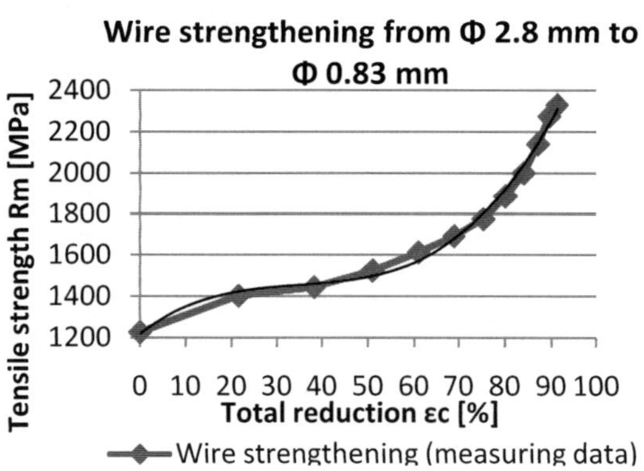

Fig. 2. Diagram of treatment compatibility

Formula of wire strengthening (polynomial of 3. level):

$$R_m = 0{,}0043\varepsilon_c^3 - 0{,}4469\varepsilon_c^2 + 17{,}192\varepsilon_c + 1216{,}4 \quad (3)$$
$$R^2 = 0{,}9936$$

4. MEASURING DATA VALUATION

Tensile strength of wire. By measuring there was detected, that required values R_m = (2 160 - 2 360) MPa were fulfilled in both of materials acc. to specification. Material A had the average value 2 314 MPa and material B 2 279 MPa. The variation of tensile strength was greater by material B. It follows that values of tensile strength are better than by material A, however, both are completely suitable. By reason of obtaining tensile strength is not necessary to change the technology (diameter of half product, number of passes, material change, or patenting change). Fig. 1 was made by help of the program SPC- PC IV, which is a statistical program for appraisal of capability process. Limit value of Cpk is 1 (Zelenay, 2006). Cpk for material A was 0.9 and for B 1.13. It follows that material B is more capable. **Elongation.** Acquired average values of elongation by final wire for material A were 4.81% and 4.69% for material B. It follows, that plasticity ratio by both of materials is almost the same, however, material A shows better plasticity. **Values of torsions.** Average value for material A is 39.1 and 37.7 for material B. Following result is that material A is better. **Values of bends.** The wire was tested by bending with angle 90°. Average value for material A is 17.9 and for material B it is 17.7. Both of materials reached almost the same results. **Treatment of compatibility.** There is a regression curve on fig. 2. It follows that the total reduction and tensile strength are statistically interdependent.

5. CONCLUSION

Results were acquired by processing of both wires rod sorts, while time between material A and material B processing was only 10 days. For comparison, there were 49 tons of wires rod used and consisted of 25 tons from material A and 24 ton from material B, which is sufficient value for attest and arbitration. Based on found results we can say that the results of material A were better than the results of material B in general. On experiment were used two sorts of WR quality C76D but each wire was from another melt.

Following determined values, tensile strength of wire and values, which were made for plasticity of material check, we can state that both materials are suitable to fulfill the requirements which are given for this specification of high-strength drawn product.

In addition, the determined values or the pass of consolidation or values on final wire can serve to appraisal of these materials availability to produce other products- wire spring and ropes.

6. REFERENCES

Bača, J. & Bílik, J. (2000) *Technology of forming*. STU, Trnava

Enghang, P. (2004). *Steel wire technology*. Repro Örebro University, Sweden

Lišková, J. (2007). *Influence of wire rods quality on mechanical properties of drawn wires*. Graduation theses. MTF STU, Trnava

Tittel, V. (1992). *Wire forming from non-standard steels..* Academic dissertation thesis. MTF STU, Trnava

Tittel, V. (2006). *Material for production of steel ropes. In: Steel ropes*. TU, ISBN 80-8073-639-4, Košice

Zelenay, M. (2006). *Tools for drawing of steelcoord wires.* Graduation theses. MTF STU, Trnava

Annals of DAAAM for 2008 & Proceedings of the 19th International DAAAM Symposium, ISSN 1726-9679
Editor B. Katalinic, Published by DAAAM International, Vienna, Austria 2008
Make Harmony Between Technology and Nature, and Your Mind will Fly Free as a Bird

GRAPHICAL USER INTERFACE FOR ARTIFICIAL NEURAL NETWORKS IN MEDICAL IMAGING

LUCULESCU, M[arius] C[ristian] & LACHE, S[imona]

Abstract: *Situated on the borderline of some important domains – engineering – medicine – computer science, involving knowledge regarding biosystems structure and functionality as well as knowledge and skills of technical and IT systems, the paper presents some results of research on Computer-Aided Diagnosis (CADx) used in medical imaging. Being a part of a highly important inter- and transdisciplinary study, combining aspects from biosystems (human visual system), image acquisition and processing (medical imaging), artificial intelligence techniques (neural networks) and information management (databases), the paper focuses on aspects regarding developing of a parametric Graphical User Interface (GUI) for designing, training, testing, optimizing and using Artificial Neural Networks (ANN) for image classification/recognition.*
Keywords: *Artificial, neural, network, imaging, diagnosis.*

1. INTRODUCTION

Computer-Aided Diagnosis (CADx) in medical imaging is a highly important inter- and transdisciplinary topic; it combines aspects from biosystems (human body), image acquisition and processing (medical imaging), artificial intelligence techniques (neural networks) and information management (databases).

By using CADx, an increasing in the initial and evolutional identification precision of disease is obtained, together with short diagnosis time, accuracy, consistence, a high confidence coefficient in results interpreting and an improvement in the prevention and treatment management.

The huge number of advantages offered by this type of diagnosis and the challenge regarding top researches on image acquisition, processing, classification/recognition using artificial intelligence methods, put together in the benefit of the human health care, have lead to such a study.

2. PROBLEM FORMULATION

2.1 Problem statement

Images of some diseases have a great variety of shapes and textures and sometimes they are hard to be identified and recognized even by experienced doctors. In these situations, a CADx system becomes a very useful tool not only for doctors, but also for researchers. That is why we are trying to develop and optimize such a system based on intelligent image classification/recognition algorithms and validating it on diagnosis of human visual system diseases, namely the macular ones.

2.2 The actual knowledge stage in the domain

The analysis of the present stage showed an extremely important interest, the computer-aided diagnosis based on image recognition being one of the major research topics within the framework of the medical imaging (Erickson & Bartholmai, 2002); (Summers, 2003). This interest is motivated by the care shown towards the people's health state, the increasing of the medical services quality and accessibility, the cost reduction of long-term treatment and hospitalization, of extra tests and of the number of surgical treatments by an early prevention or discovery of diseases. A minute analysis of the specialty literature stated that in the field of medical imaging, there have been carried out numberless research studies concerning the computer-aided diagnosis of breast cancer (Hall, 2007), of the pulmonary nodules from breast X-ray pictures (Shiraishi et al., 2003) and CAT SCANs (Lawler et al., 2003), of the cerebral aneurysms (Yokoyama et al., 2007), some areas being better covered, other areas less covered, just like the case of retinal imaging.

The in-depth analysis carried out has permitted the highlighting of certain unexplored areas towards which was oriented the research (e.g., the elaboration of some parametric identification methods in order to cover a wide range of diseases, the use of texture descriptors for images analysis a.s.o.) and of some elements that can be improved (the sensibility and specificity values as parameters defining the system performance in detecting the possibly affected areas and the recognition of the disease present in that particular area).

3. PROBLEM SOLUTION

3.1 System description

The structure of the CADx process used for macular diseases classification/recognition is presented in Fig.1. (Luculescu & Lache, 2008).

The image of the retina is acquired using a retinal investigation camera and transferred to computer. An image processing module is used for different image quality improvements or geometrical transformations.

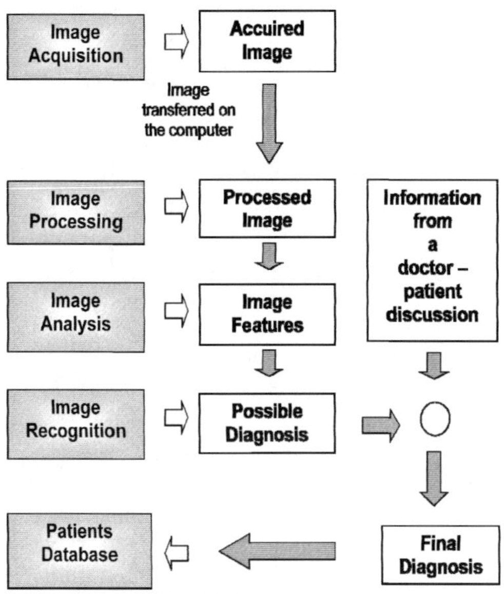

Fig. 1. The structure of the CADx process.

A certain vector of image features (seven 2D moment invariants and six values that describe a region by quantifying its texture content (Gonzales et al., 2004)) is then automatically extracted and the obtained values are applied to the inputs of the artificial neural network (ANN) used for image classification/recognition. The ANN generates a response that represents a set of four possible diagnoses, ordered by the level of certitude. The doctor can establish the final diagnosis after a discussion with the patient, taking into account personal and heredo-collateral antecedents. The results are stored in a database, as a patient medical record, together with personal information and recommended treatment.

In the design phase of the system, the image recognition module using ANN has as main objectives to find out what network structure, training algorithm and training parameters are most suitable for this kind of application. That is why we have developed in Matlab® a parametric GUI for configuring, training, testing and optimizing the ANN.

3.2 ANN configuration

The used ANN is a feedforward total connected multilayer perceptron network type with 3 or 4 layers of neurons (1 or 2 hidden layers). The number of inputs is determined by the number of values representing features and it can be selected from the main window of the GUI. Using RGB images, for example, and seven 2D moment invariants plus six texture descriptors, the number of inputs is 39 (13 for each color component). The structure of the network can be modified by selecting one or two hidden layers and choosing the number of neurons in each layer. The number of diagnostics from the database will determine the number of outputs.

The transfer function can be selected from a list that contains *tansig* - hyperbolic tangent sigmoid transfer function, *logsig* - log sigmoid transfer function and *purelin* - linear transfer function. Because the teacher output vector is composed from 0 and 1 values (1 for selecting the correct diagnosis), *logsig* and *tansig* functions will be preferred considering the interval in which they are taking values.

The training parameters and the stopping conditions can be also modified from the GUI.

3.3 ANN training and testing

Training process started for a database containing 25 images of 15 diagnostics. The process can be stopped manually by user or automatically if a stop condition is fulfilled. At every stop the ANN is tested and a report regarding the number of classified diagnostics, training time and square mean error is displayed. The network can also be tested on a specific database containing images that were not used in the training phase. For example, all 25 images of the initial database were classified in less than 1000 epochs.

Due to the complexity of the shapes, distribution and texture of the images representing macular diseases it was necessary to test different training algorithms.

3.4 ANN optimization

In order to obtain the best results, an optimization is needed, regarding the structure of the network, the training algorithms and parameters.

Optimization criteria take into account maximizing the number of classified images (first priority – main objective), minimizing the training time (second priority) and minimizing the square mean error (third priority).

Restrictions refer to: the ANN type - total connected feedforward multilayer perceptron network; image database (training set) - 25 RGB images, representing 15 diagnostics; ANN architecture - 39–X–15 with 3 layers of neurons, where X is the variable number of neurons in the hidden layer; training process - 15 cycles with 30.000 epochs each, results being stored at each 1000 epochs.

As optimization parameters were considered the training function, the transfer function and the ANN architecture.

Twelve training function were pretested, six of them being finally used for analysis: *Traingd* - Gradient descent backpropagation, *Traingdm* - Gradient descent with momentum backpropagation, *Traingda* - Gradient descent with adaptive learning rate backpropagation, *Traingdx* - Gradient descent with momentum and adaptive learning rate backpropagation, *Trainrp* - Resilient backpropagation and *Trainscg* - Scaled conjugate gradient backpropagation. Three transfer function were tested: *tansig* – bipolar sigmoid, *logsig* – unipolar sigmoid, *purelin* – linear function. Three architectures were used, with 10, 39, 78 neurons in the hidden layer.

3.5 Results

Best results were obtained for the *trainscg* training function, *tansig* transfer function and a 39-78-15 architecture (39 neurons in the input layer, 78 in the hidden layer and 15 in the output layer), even the training time is higher.

Then, more images were appended to the database and the training process continued cumulative until all the diagnostics were classified. Finally, the network was tested with a set of images some of them representing similar diagnostics or normal macula status. The specificity obtained was 90,91%.

Regarding performance, the percentage of images correctly detected as having potential diseases *(sensitivity)* and the percentage of diseases images correctly recognized *(specificity)* were better in our system comparing to others (Luculescu & Lache, 2008).

4. CONCLUSION

The developed GUI is a very useful tool in CADx research that allows designing and optimizing amazing image classification/recognition systems based on artificial intelligence techniques. As future research, new development directions will focus on system improvement by adding new image features that can be used as inputs in the neural network and adapting the system for recognizing images from other domains (industrial for example).

5. REFERENCES

Erickson, B.J. & Bartholmai B. (2002). Computer-Aided Detection and Diagnosis at the Start of the Third Millennium. *Journal of Digital Imaging*, No. 15, pp. 59–68.

Gonzales, R. et al. (2004). *Digital Image Processing Using Matlab*, Pearson Prentice Hall, ISBN 0-13-008519-7.

Hall, F. M. (2007). Breast Imaging and Computer-Aided Detection. *The New England Journal of Medicine*, Vol. 356, pp. 1464-1466.

Lawler, L.P. et al. (2003). Computer Assisted Detection of Pulmonary Nodules: Preliminary Observations Using a Prototype System with Multidetector Row CT Data Sets. *Journal of Digital Imaging,* No. 16, pp. 251–261.

Luculescu, M. & Lache, S. (2008). Computer-Aided Diagnosis System for Retinal Diseases in Medical Imaging, *WSEAS TRANSACTIONS on SYSTEMS Journal*, Issue 3, Vol. 7, March 2008, pp.264-276, ISSN 1109-2777.

Shiraishi, J. et al. (2003). Computer-Aided Diagnosis for Distinction Between Benign and Malignant Solitary Pulmonary Nodules In Chest Radiographs: ROC Analysis of Radiologists' Perform, *Radiology*, No. 227, pp. 469–474.

Summers, R.M. (2003). Road Maps for Advancement of Radiologic Computer-Aided Detection in The 21st Century. *Radiology*, No. 229, pp. 11–13.

Yokoyama, R. et al. (2007). Development of an Automated Method for the Detection of Chronic Lacunar Infarct Regions in Brain MR Images. *IEICE Trans. on Inf.and Systems,* No. E90-D, pp. 943-954.

Annals of DAAAM for 2008 & Proceedings of the 19th International DAAAM Symposium, ISSN 1726-9679
Editor B. Katalinic, Published by DAAAM International, Vienna, Austria 2008
Make Harmony Between Technology and Nature, and Your Mind will Fly Free as a Bird

THE USAGE OF MATH GAMES THEORY IN THE OPTIMIZATION OF WORKING TIME OF A TOP MANAGER

LUNGU, F[lorin]; ABRUDAN, I[oan] & OTEL, C[alin]

Abstract: *The aim of this paper is to shows some results obtained through applying the math games theory in the analysis of the optimal structure of working time of managers reported to the salary's incomes. Based on a linear mathematical model it was obtained an optimal structure of the daily working time of a top manager reported to the salary.*
Keywords: *game, optimization, manager, time, activities*

1. INTRODUCTION

Game theory is a branch of applied mathematics which approaches the issue of optimal behavior in games with 2 or *n* persons. This theory studies the way in which strategic interactions between rational players trigger results according to the preferences (or utilities) of those players. Game theory is used in many scientific fields, from biology to psychology or sociology. Game theory has drawn the attention of experts in the computer field due to its application in the field of artificial and cybernetics intelligence.

In a short summary of game theory evolution, the following events have to be mentioned (Lungu, 2006):
- In 1713, James Waldegrave turns the minimax principle as solution to a cards' game, with 2 persons; this principle has been used since the beginning of the XXth century;
- In December 1926, the mathematician von Neumann demonstrates minimax theorem; he is considered the discoverer of game theory;
- In 1938, the economist Oskar Morgenstern joins von Neumann in his try to offer a theory on game theory; in 1944, both of them publish *The Theory of Games and Economic Behavior*;
- In 1950, John Nash develops a definition of optimal strategy for games with more partners, which will further be known as 'Nash equilibrium'.

As one can see, though the game theory has only recently been put on logic and mathematic bases, one can identify approaches on the subject from many centuries ago. No matter if the approach should be philosophic, military, and economic or from the perspective of the organizations' study, maximizing the human wellbeing represents one of the core elements. And the strategies used in attaining this goal, the logic used in many situations can be analyzed based on the game theory.

As shown above, the basic premise of the game theory is the fact that the players (the agents) act rationally. This implies the following (Lungu, 2006):
- the players can evaluate the eventual results
- the players can plan the means of obtaining the results
- the players can choose those actions which generate the results they prefer, if the other players also act.

An aspect which is worth mentioned is the fact that through rational decision, choice does not necessarily mean the result of a mathematical calculation.

The simple variants of the games assume the existence of some pure strategies - in which each player looks for the optimal variant at each action of the adversary. In reality this type of situations is rarely to be found. More often are to be encountered the situations in which mixed strategies are used – meaning the balance between more pure strategies. Another important criterion refers to the number of the game repetitions. From this point of view the games can be repeated – when the sequence of the game is passed through repeatedly, or singular.

For the first situation the following aspects are important:
- the players base their decisions by taking into consideration the previous evolution of the game;
- the players also take into consideration the possibilities of the further evolution of the game.

2. INFORMATION – SCIENTIFIC RESEARCH

In this paper we used the mathematical mechanism from the zero-sum game's theory, who gives the solutions in the confliction problems. The principles from that part of the games' theory that we will use it are given below (Lungu & Abrudan 2005):
1) The games theory is a theory of conflicts where are shown at least two conflicting interests promoted by two players (partners, rivals). What a player loses the other one gains it.
2) The player that loses is called minimizing because he wants to minimize the lost and the player that gains is called maximizing because he wants to maximize the benefit.
3) The optimum situation accordingly with the games theory between the two players shows a level of the bargain respectively of the loss, acceptable for the both players. This level of bargain/loss is called the game value. From the confronting of the two opposite tendencies, it appears an equilibrium situation (game value).

The aim is concretized through next steps:
- the assimilation of the managerial activity with a mathematical game;
- the solving of the math game and the interpretation of results.

The management activity for a company is based on some activities who are necessary to be achieved and who form the management functions. These functions was postulated by Henry Fayol, who identified five functions of management: forecasting, organization, order, co-ordination and control (Simionescu, 2002).

The frequency apparition of the functions in the activity of managers depends by the management system of the company and by the level of manager. For the top managers, the forecasting and co-ordination are the most important (have the

No.	Structure of work time	Percent from work time [%]
1	Programmed meetings	59
2	Desk work	22
3	Unexpected (non-programmed) meetings	10
4	Phone conversations	6
5	Inspections in company (control)	3

Tab. 1. The structure of working time for a top manager

greatest weight of the working time and the organization and control functions have a little weight).

Opposite, for the low-level managers, the most important are the organization and the control functions.

On the other side, using the Mintzberg researches, the usage of the daily working time for a top manager has the structure like in Table 1 (Tuturea, 1999).

The research has been determined the optimal structure of a daily working time for a top manager, using math games theory and taking in consideration the Mintzberg model.

The problem consists in to divide the resource between activities so that to have an equilibrium between salary and working time.

This paper approaches the problem of the management efficiency from the mathematical perspective, from quantitative point of view.

For to can use the math games theory in the solving of this problem, it was assimilated the activity of managers with a math games with two players, so:
- they are two opposite tendencies: the desire to gain more and the desire to work shortly;
- the two players are: salary (S) and work time defined through the five activities (A). The salary is the maximizant player (tends to increase) and the working time is the minimizant player (tends to decreases);
- the value of the game is an equlibrium situation that limits the working time without the decreasing of the salary under an acceptable level;
- the matrix of the game which contains the looses and the gains of each player is given in Table 2.

The values of the matrix was obtained in (Lobontiu, 2007) through interviewed of 18 of top managers through the North-West Region of Romania. The values represents, through a scale from 1 to 5, the level of the salary reported to the corresponding activity.

The mathematical model is an linear model who minimizes the working time.

The goal function is:

$$\min \sum x_i \text{ , } i=1,\dots,5 \qquad (1)$$

The restrictions are:

$$\begin{cases} 5x_1+4x_2+3x_3+2x_4+1x_5 \ge 1 \\ 4x_1+5x_2+3x_3+2x_4+1x_5 \ge 1 \\ 5x_1+2x_2+3x_3+1x_4+2x_5 \ge 1 \\ 5x_1+3x_2+4x_3+2x_4+1x_5 \ge 1 \\ 4x_1+4x_2+1x_3+5x_4+2x_5 \ge 1 \\ 5x_1+2x_2+4x_3+1x_4+3x_5 \ge 1 \\ x_1 \ge 0 \\ x_2 \ge 0 \\ x_3 \ge 0 \\ x_4 \ge 0 \\ x_5 \ge 0 \end{cases} \qquad (2)$$

where:

x_1 – percent from working time dedicated to programmed meetings;

x_2 – percent from working time dedicated to desk work;

x_3 – percent from working time dedicated to non-programmed meetings;

x_4 – percent from working time dedicated to phone conversations;

x_5 – percent from working time dedicated to inspections in company.

	A1	A2	A3	A4	A5
S1	5	4	3	2	1
S2	4	5	3	2	1
S3	5	2	3	1	2
S4	5	3	4	2	1
S5	4	4	1	5	2
S6	5	2	4	1	3

Tab. 2. The matrix of the game

Solving the model and making the variable changes, we obtained following final results (non-zero variables):

$$\begin{aligned} v &= \frac{1}{z} = \frac{1}{0.245} = 4.08 \text{ - value of the game} \\ X_1 &= 4.08 \cdot 0.175 = 0.72 \qquad (3) \\ X_2 &= 4.08 \cdot 0.052 = 0.22 \\ X_4 &= 4.08 \cdot 0.017 = 0.06 \end{aligned}$$

3. CONCLUSIONS

So, as conclusions, based on the answers given by the 18 top managers interviewed, the optimal structure for a daily working time of a managers is:
- programmed meetings - 71%
- desk work - 21%
- phone - 6%

We can observe that for to assure an optimal ratio between the fifth activities, the inspections and the unexpected meetings must be at very low level or to miss.

The basic aim of management is the increasing of the organization efficiency, that means the ratio between obtained effects and efforts. For to decrease or to eliminate the causes which provoke deficiencies in the work of managers (like inadequate structure of working time or often exceed of the duration of daily work), is necessary a programming and organization on the scientific basis of managers work.

The apply of the conclusions can help to the increasing of the efficiency of managers work, in the conditions of assuring of an optimal structure of the managerial activities and having in view the minimization of the working time.

The obtained results can be the subject of a future research, taking into account the answers of more managers and more restrictions. Anyway, the results of this research can be a starting point for a manager in to organize the daily working activities. Through the respect of the obtained percents it can increases the efficiency of the managers work.

4. REFERENCES

Lobontiu, G., (2007). *Studies and researches on the management of small business in the Maramures County, Romania*, PhD. Thesis, Technical University of Cluj-Napoca, Romania.

Lungu, F., Abrudan, I., (2005). Research on the apply of math games theory in the simulation of Flexible Manufacturing Systems (FMS), *Proceedings of 4-th International Conference on the Management of Technological Changes*, Chania, Greece, ISBN 960-8475-05-8, pag. 59.

Lungu, F., (2006). *Simulation of Flexible Manufacturing Systems with help of mathematical games theory*, Ed. Dacia, Cluj-Napoca, Romania, ISBN 978-873-35-2140-2.

Simionescu A. (2002). *General management*, Ed. Dacia, Cluj-Napoca, ISBN 978-873-35-2138-1.

Tuturea, M., (1999). *Management basis. Course support*, University “Lucian Blaga” Sibiu, Romania.

Annals of DAAAM for 2008 & Proceedings of the 19th International DAAAM Symposium, ISSN 1726-9679
Editor B. Katalinic, Published by DAAAM International, Vienna, Austria 2008
Make Harmony Between Technology and Nature, and Your Mind will Fly Free as a Bird

SOME RESEARCHES UPON THE TOOL-PIECE CONTACT GEOMETRY IN THE ROLLING PROCESS

LUPESCU, O[ctavian]; POPA, R[amona]; POPA, I[onut]; IOSUB, A[drian] & LUPESCU, O[vidiu]

Abstract: *The authors propose in the present paper a mechanical treatment method for pieces with metallic matrix using superficial hardening rolling process. The curved radius of the attack surface of the tool is determined in an analytical way, tacking into account only the elastic deformations. Considering both the elastic and plastic deformations of the piece one had determined the equation of the contact ellipse of the deformation focus.*
Keywords: *rolling, superficial deformation, hardness, contact ellipse*

1. PROBLEM STATEMENT

The superficial cold plastic deformation (SPD) is a mechanical treatment method of the metallic materials that allows us to obtain the mechanic characteristics improvement (Lupescu et al., 2004), the expensive materials changeover with any other cheaper, hardened using SPD process, as well as the thermo and thermo-chemical secondary treatments changeover because of their amount prices.

This technical paper shows also a surface geometry determination of the deformation focus, in case of the rolling hardening with disk rolls (Lupescu et al., 2001).

2. APPLICATION AREA

The researches objective was to determine the tool-piece surface contact projection on the tangent plan, in their initial contact point.

The literature (Lupescu, 1999) shows that process is able to be applied to cylindrical pieces, especially to big diameters (>100 mm), for individual series and also mass production series. The degree hardness resulted on superficial layer (until 1 – 1, 5 mm) exceed in many cases with 10% (Lupescu & Baciu, 2002) the obtained results of the thermo-chemical treatments.

3. METHOD USED

First of all, there were established the curved radius taking into account the elastic deformation which affect the rolls. For the second step, the deformation focus surface results taking into consideration the plastic and elastic deformations aboard the working piece (figure 1).

Using the general relation to establish a surface curved radius elastic deformed (Lupescu et al., 2005):

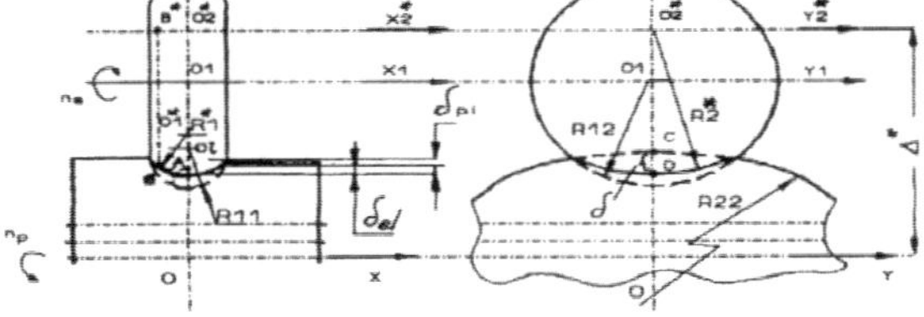

Fig. 1. The general tool-piece contact scheme.

$$\frac{1}{R_{1,2}^*}=\frac{1}{2}\left(\frac{1}{R_{I,1(2)}}-\frac{1}{R_{II,1(2)}}\right) \tag{1}$$

the authors suggest for the curved radius of the roll attack surface in the case of his elastic deformation, the following relations:

$$\frac{1}{R_1^*}=\frac{1}{2}\left(\frac{1}{R_{I,1}}-\frac{1}{R_{II,1}}\right)=\frac{1}{2}\times\frac{1}{R_{I,1}};R_1^*=2R_{II} \tag{2}$$

$$\frac{1}{R_2^*}=\frac{1}{2}\left(\frac{1}{R_{I,2}}-\frac{1}{R_{II,2}}\right)=\frac{1}{2}\times\frac{1}{R_{I,1}};R_2^*=\frac{2R_{12}R_{22}}{R_{22}-R_{12}} \tag{3}$$

Working in a Cartesian system OXYZ because of the elastic roll deformations under the F force action, his initial radius R_{11} and R_{12} belonging to the principal two plans will be modified to R_1^* and R_2^* with their centers displacement from O_1 to O_1^* and from O_2 to O_2^*. Using the next notation, the total contact deformation will be:

$$\overline{CD}=\delta=\delta_{el}+\delta_{pl}\,,\qquad \text{thus:} \tag{4}$$

$$\Delta^*=R_2^*+R_{22}-\delta\,,\qquad \text{so:} \tag{5}$$

$$O_1^*A=\sqrt{R_1^{*2}-\overline{AB}^2}=\sqrt{R_1^{*2}-x^2}\,,\qquad \text{and:} \tag{6}$$

$$O_1^*O_2^*=R_2^*-R_1^*,\qquad \text{and:} \tag{7}$$

$$\overline{BB}^*=\left[\left(R_2^*-R_1^*\right)+\left(\sqrt{R_1^{*2}-x^2}\right)\right] \tag{8}$$

The deformation surface equation, belonging to the $O_2^*X_2^*Y_2^*Z_2^*$ system is:

$$y_2^{*2}+z_2^{*2}-\overline{BB}^{*2}=0 \qquad \text{and so:} \tag{9}$$

$$y_2^2+(z-\Delta^*)^2-\left[\left(R_2^*-R_1^*\right)+\sqrt{R_1^{*2}-x^2}\right]^2=0 \tag{10}$$

In these conditions, the contact surface curve can be determined with the system equations:

$$\begin{cases} y^2+(z-\Delta^*)^2-\left[(R_2^*-R_1^*)+\sqrt{R_1^{*2}-x^2}\right]^2=0 \\ y^2+z^2-R_{22}^2=0 \end{cases} \tag{11}$$

The outline intersection of two bodies (tool-piece) projected on the tangent plan in the initial contact point is an ellipse whose expression was obtained from (11):

$$y^2 + \left[\frac{R_{22}^2 + \Delta^2 - \left[\left(R_2^* - R_1^*\right) + \sqrt{R_1^{*2} - x^2}\right]^2}{2\Delta^*}\right] - R_{22}^2 = 0 \qquad (12)$$

The equation of the real contact surface in the elastic-plastic deformation process was solved using the overlap effects principle.

4. RESULTS

Thus, on the first step, it was admitted only the static contact of the elements. The ellipse from the tangent plan results like projection of the determination focus surface (figure 2) with (a) and (b) semi-axis established by (12); the small semi-axis (b) is the y value from this equation, when x = 0 and the big one (a) is the limit value of x when the above equation ensure the real values for y.

On the second step it was simulated the axis movement of the contact ellipse along the longitudinal speed direction of the deforming roll.

On the left side, the roll came into contact with the working piece, on the existing surface from the last step. On the right side the roll remain in contact with the piece on a reduce surface, because of his plastic deformation.

Thus, on the right side, the intersection outline is an elliptic contour having (a_I) semi-axis, taking into account the elastic contact only. The semi-axis (b) is unchangeable. So:

$$a_I = a^* \sqrt[3]{\frac{3F}{E^{`}\sum\zeta}}, \quad \text{where:} \quad a^* = \left[\frac{2k\xi(e)}{\pi}\right]^{1/3} \qquad (13)$$

On the third step, one simulates the tool rotation either, along side OY axis. The new surface contour is represented in figure 3. In this case also, behind the roll, it remains a reduced contact surface, resulted only due to the plastic piece deformation.

Thus, during the rolling process, the real contact between tool-piece is produced on a contact surface having the following curved sectors:

- AB, remained from the initial unmodified ellipse, whose equation is done after relation 12;
- BC, ellipse arc which has a semi-axis (b) and the other semi-axis (a_I) resulting from relation 13. The analytically expression of these outline is:

$$\left(\frac{x}{a_I}\right)^2 + \left(\frac{y}{b}\right)^2 - 1 = 0 \qquad (14)$$

- CD, ellipse arc having (a_I) and respectively (b_{II}) semi-axis, calculated from the elastic pure contact conditions:

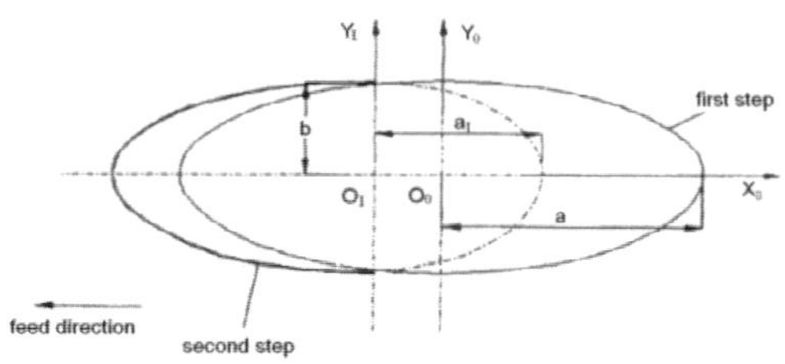

Fig. 2. The tool-piece form contact surface, in static contact, neglecting their rotation motion.

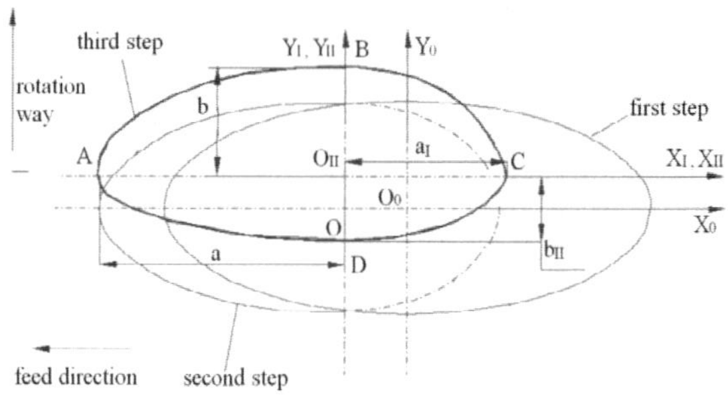

Fig. 3. The real tool-piece contact surface.

$$b_{II} = b\sqrt[3]{\frac{3F}{E^{`}\sum\rho}}, \quad \text{where:} \quad b^* = \left[\frac{2\xi(e)}{\pi k}\right]^{1/3} \qquad (15)$$

The contact expression outline on this sector becomes:

$$\left(\frac{x}{a_I}\right)^2 + \left(\frac{y}{b_{II}}\right)^2 - 1 = 0 \qquad (16)$$

- DA, ellipse arc with (a) semi-axis and the other equal with (b_{II}), when the analytically contour expression becomes:

$$\left(\frac{x}{a_{II}}\right)^2 + \left(\frac{y}{b_{II}}\right)^2 - 1 = 0 \qquad (17)$$

In these conditions, the mathematic expression became:

$$z = \Delta^* - \sqrt{\left[R_2^* - R_1^*\right] + \sqrt{R_1^{*2} - x^2 - y^2}} \qquad (18)$$

with X and Y values being throughout the outline delimited by the ellipse arcs AB, BC, CD and DA, anterior defined.

5. FURTHER RESEARCHES

The determined focus surface afterwards will allow: the contact surface pressure establishment, the slides speeds between the couple elements, the power friction consumed during the process and friction parameters establishment using the lubricant film calculus.

6. REFERENCES

Lupescu, O., (1999), *Surfaces polishing through plastic deformation and steels cold hardening proces (Netezirea suprafetelor prin deformare plastica si durificare la rece a otelurilor),* Junimea, ISBN 973-37-0471-7, Iasi

Lupescu, O., et al., (2001), The device choosing for rolling the exterior cylindrical surface using the utilities theory, *Proceedings of Modern Techniques, Quality, Reorganization*, vol. 1, pg. 229-232, Chisinau

Lupescu, O., Baciu, M., (2002), The characterization of mechanical treatment at SPD at cold through rolling as a tribological process, *Proceedings of Meridian Engineering*, 12 (3), pg. 29-32, Chisinau

Lupescu, O., et al., (2004), Structural modification of the Ol 25 Steel in the Cold SPD Process Through Rolling, *Academic Journal of Manufacturing Engineering*, 2(4), pg.43-47, Timisoara

Lupescu, O., et al., (2005), About the establishment of surface geometry for the SPD process by rolling, *Proceedings of Machine – Building and Technosphere of the XXI Century*, Tom. 4, 12-17 September., 255-260, Sevastopol

Annals of DAAAM for 2008 & Proceedings of the 19th International DAAAM Symposium, ISSN 1726-9679
Editor B. Katalinic, Published by DAAAM International, Vienna, Austria 2008
Make Harmony Between Technology and Nature, and Your Mind will Fly Free as a Bird

THE INFLUENCE OF MODIFIERS ON THE CAMPACTED GRAPHITE IRON CHARACTERISTICS

LUPINCA, C[inca] I[onel]

Abstract: *In this paper we have presented the results of the compacted graphite iron produced in an induction furnace from a metallic charge, made up of synthetic pig iron. This operation was made using complex modifications for the compacted graphite modification from the FeSiMgCe(RE) category, respectively from the FeSiMgCaCe(RE) category, and for the graphite modifications, from the FeSiSr category. We also presented the representative results for the four of the compacted graphite irons charges.*
Keywords*: Compacted graphite irons; Rare Earths.*

1. INTRODUCTION

The producing of the pig irons, used in order to modify and to obtain vermicular graphite, was made in an induction furnace, of a low frequency and with a capacity of 250 kilos and the treatment of the cast irons in order to obtain the compacting of the graphite separations, was made in a casting pot with 1000 kilos capacity, by applying the Sandwich procedure.

The metallic charge of the furnace was composed of synthetic pig iron (Lupinca et al., 2005), obtained in a arc electric furnace, starting with a charge of scrap irons, carefully selected regarding their trace elements contents.

In order to obtain the compacted graphite we used two modification technological variants regarding the type of nodulizing (compacting) modifiers used.

Thus, the following nodulizing (compacting) modifiers were used:

I. FeSiMgCe(RE) (nodulizing modifier used, mainly, in the obtaining of ductile graphite iron (Lupinca, Riposan, 2006); in order to obtain the vermicular graphite, we had to use a smaller quantity of modifiers than in the treatment of the pig iron in order to obtain the nodular graphite;

II. FeSiMgCaCe(RE) (nodulizing - compacting modifier, with a complex chemical composition, recommended by the producer to obtain compacted graphite iron.

For all the cast iron charges studied, in order to obtain a graphite modification, we used the same type of graphitizing modifier (FeSiSr), in sufficient percentage, in order to obtain the silicon percentage necessary. This paper represents just a small part of my personal studies and it aims the production of casted pieces from CGI-s, from the engine construction industry, in the casted pieces plant of U.C.M. Resita (Engine Construction Plant Resita).

2. EXPERIMENTAL RESULTS

Charge no.	C	Si	Mn	P	S
1.	3.59	2.29	0.17	0.032	0.019
2.	3.54	2.93	0.56	0.055	0.020
3.	3.45	2.65	0.24	0.035	0.015
4.	356	3.48	0.64	0.046	0.014

Charge no.	Cu	Ni	Cr	Mo	Mg
1.	0.080	0.050	0.060	0.020	0.033
2.	0.018	0.013	0.0055	0.004	0.020
3.	0.060	0.080	0.080	0.020	0.028
4.	0.020	0.010	0.020	0.003	0.015

Tab. 1. Chemical composition of the compacted graphite irons[%]

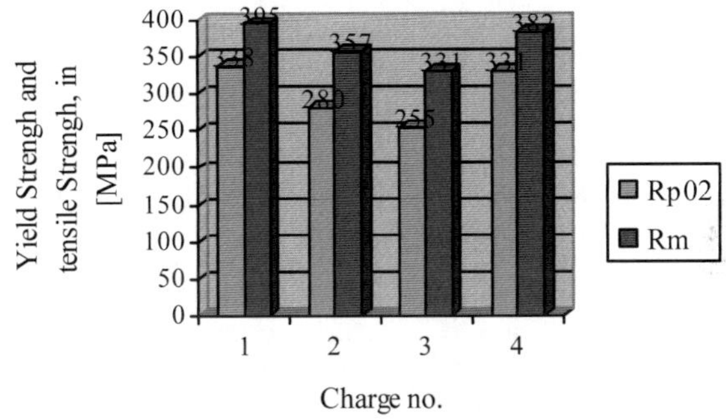

Fig. 1. Yield strength and tensile strength of the compacted gaphit irons.

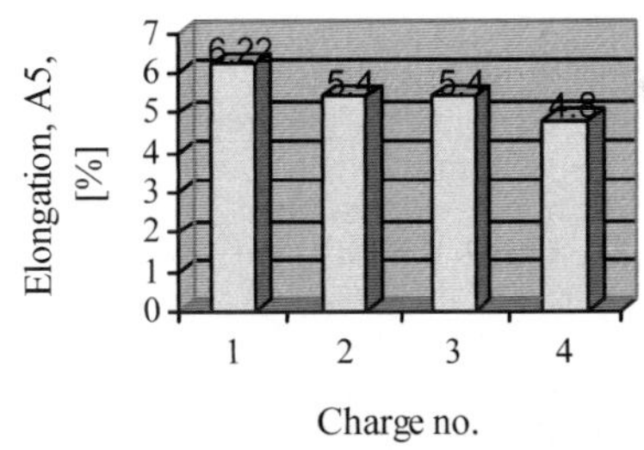

Fig. 2. Elongation of the compacted graphite irons.

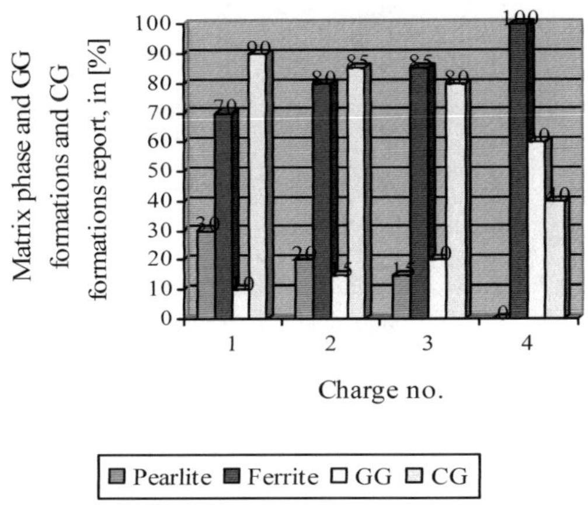

Fig. 3. Graphic representation of the CGI matrix phases and of the CGI graphite separations, in CGI.

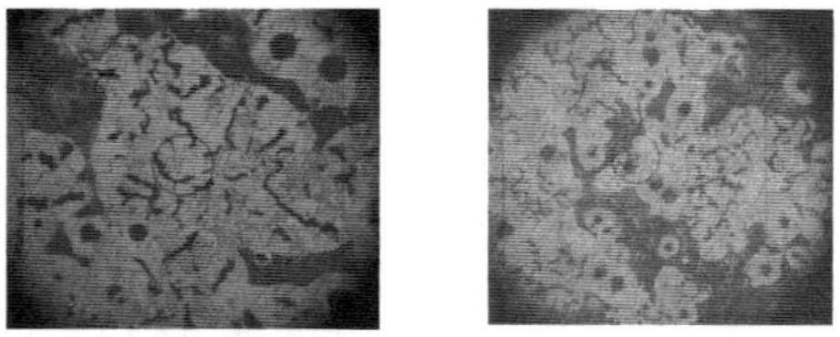

A. B.

Fig. 4. Metallographic structure of CGI-s, modified with FeSiMgCe and FeSiSr: A. Charge 1; B. Charge 2 (Etchant: Nital 4%) (100X).

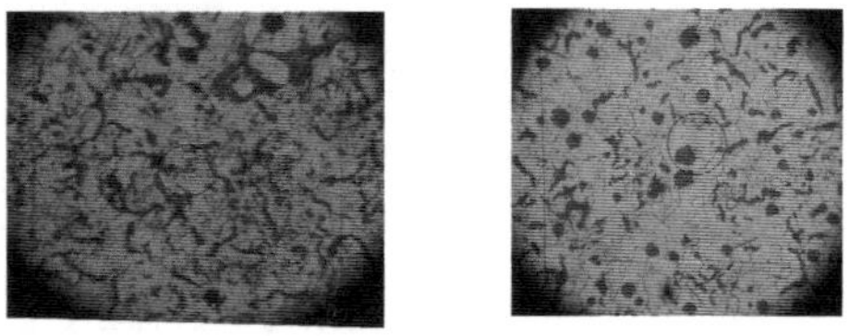

A. B.

Fig. 5. Metallographic structure of CGI-s, modified with FeSiMgCaCe and FeSiSr: A. Charge 3; B. Charge 4 (Etchant: Nital 4%) (100X).

3. CONCLUSIONS

In the following paragraphs we will present a comparative analysis of the mechanical characteristics and structural values of the compacted graphite irons obtained:

∗ Higher values of the yield strength were obtained in the CGI, obtained through modification by the compacting of the graphite separations with a slightly greater supplement of the FeSiMgCe(RE), charge 1, compared to charge 2, where we used (keeping in mind the explanations already presented) a less higher supplement of graphite modifier (FeSiSr); lower yields strength resulted for the CGI, charge 3, obtained through a nodulizing modification with FeSiMgCaCe(RE) and through a graphite modification with a lower supplement of FeSiSr, compared to charge 4.

∗ The slightly different supplements of graphite modifier FeSiSr, in the four CGI charges presented, are determined by the different silicon concentrations in the cast irons, before the modifications which in their turn, may be determined by the charge of the furnace and also by the heating and melting technology conditions of the charge and the producing of the cast irons.

∗ The same considerations may also be mentioned in the case of the tensile strength values analysis of the CGI, keeping the proportion from the yield strength values.

∗ On ensemble, we may conclude that the CGI charges 2 and 3, obtained through modification with FeSiMgCe(RE) + FeSiSr, respectively FeSiMgCaCe(RE) + FeSiSr, present lower yield strength values and strength values, and the CGI-s, charges 1 and 4, obtained in similar modification conditions, slowly higher values and of the yield strength and the tensile strength (the values being very close to one another).

∗ On the other side, the elongation of tensile presents same particularities regarding the values obtained from the presented CGI-s, and these are: slightly higher values of the elongation are obtained in the CGI obtained through modification with FeSiMgCe(RE) + FeSiSr - charge 1, but which are not also higher than the CGI-s values presented in different scientific papers (Dawson, 1999), (Gusser et al., 2001); we also obtained low values of the elongation from the CGI - charge 4, obtained through modification with FeSiMgCaCe(RE) + FeSiSr.

∗ Even if the CGI charges 2 and 3 were obtained by two modifications values, the elongation values are equal.

Finally, we may conclude that these results, regarding the mechanical properties values, are determined by the cast irons structural characteristics, respectively by the report of the matrix phases, by the graphite separations shape, by the report between the nodular graphite separations and the compacted graphite separations.

Thus the modifiers used to compact the graphite separations due to their chemical compositions, present a strong nodulizing action of the graphite separations and that only the used concentrations and the chemical, the termic and the structure parameters of the cast irons introduced in the modification process, in a liquid state, lead to these results.

∗ The CGI-s obtained through modification with FeSiMgCe(RE) + FeSiSr-charges 1 and 2, present, studied only at the surface studied, close concentrations of pearlite (30%, respectively 20%), and rest, ferrite.

∗ The same two charge of CGI present very close concentration of compacted graphite separations (90%, respectively 85%) and of nodular graphite separations (10%, respectively 15%).

∗ It is, also, mentioned the fact that on the CGI charge analyzed 2, compacted graphite separations represent 85% of the graphite quantity observed, but which also present a particular characteristic: the global notions of compacted graphite contains (in percentage), 10% of graphite separations type "coral", 15% graphite separations type "rosette" and the rest, just classical compacted graphite separations.

∗ The charge 3 and 4, of CGI-s, obtained by modification with FeSiMgCaCe(RE) + FeSiSr, present on the surfaces analyzed, different concentrations of the matrix phases (ferrite: 85%, respectively 100%, and pearlite: 15%, respectively 0%) and different reports between spheroidal graphite separations (20%, respectively 60) and compacted graphite separations (80%, respectively 40%).

∗ This type of structural characteristics, obtained from the studied cast irons, lead to the conclusion that only the treated charges 1 and 2, with FeSiMgCe(RE) + FeSiSr and the charge 3 of the cast iron, treated with FeSiMgCaCe(RE) + FeSiSr, may be introduced in the compacted graphite irons category.

∗ The cast iron - charge 4, treated with FeSiMgCaCe(RE) + FeSiSr, because of the same structural characteristics, may be considered closer to the spheroidal graphite iron (GGG), because it presents a concentration of over 50% (60%) of spheroidal graphite separations in the structure, and the apparition of the compacted graphite separations may be caused by the following causes:

- The modifier supplements used and indicated by the producer in the producing of compacted graphite irons (FeSiMgCaCe(RE)), are not the right ones. When producing the charge 4 of CGI, we used a supplement indicated by the producer.

4. REFERENCES

Dawson, S. (1999). *Compacted Graphite Irons: Mechanical and Physical Properties for Engine Design*, VDI Conference on Materials in Powertrain (Werkstoff und Automobilantrieb), Dresden, Germany, www.tupy.com.br.

Guesser, W., Schroeder, T., Dawson, S. (2001). *Production Experience with Compacted Graphite Iron Automotive Components*, AFS Transactions, IL, USA, www.tupy.com.br.

Lupinca, C.I., Riposan, I., Chisamera, M., Barstow, M. (2005). 109th AFS Casting Congress, *Foundry Experience using Synthetic Pig Iron (Produced in EAF) in Ferritic Ductile Iron Production*, AFS Transactions, V 113 Paper 05-108 (05), p. 1 - 13, St. Louis, IL, USA, 2005.

Lupinca, C.I., Riposan, I. (2006). The 4th Symposium with the International Participation - KOD 2006, *The influence of the metallic charge on the structure and the properties of ductile irons*, p. 341-344, ISBN 86-85211-92-1, Palić, Serbia i Crna Gora, May 2006.

Annals of DAAAM for 2008 & Proceedings of the 19th International DAAAM Symposium, ISSN 1726-9679
Editor B. Katalinic, Published by DAAAM International, Vienna, Austria 2008
Make Harmony Between Technology and Nature, and Your Mind will Fly Free as a Bird

MEDIUM OPTIMIZATION OF PYROLYSIS REACTOR

LUPU, C[iprian] & AZZOUZI, M[essaouda]

Abstract: *This paper deals with the supervision of an industrial plant by taking a petrochemical process as a study case; it is a new optimization procedure that was applied on a pyrolysis reactor which is fed in by the petrol as a primary matter and produces the ethylene, one of its several products. The medium optimization principle was tackled in this work, starting from a set of stationary system data that depends on the petrol feed-in source selected in each case under certain probability estimations. So the received petrol quality plays an important role in the produced ethylene concentration.*
Keywords: *optimization, pyrolysis reactor, Simplex algorithm, petrol separation, modelling.*

1. INTRODUCTION

Parametric optimization is one of the most important fields in mathematical research (Popescu et al., 2004). The first optimization technique, which is known as steepest descent, goes back to Gauss. Historically, the first term to be introduced was linear programming, which was invented by George Dantzig in the 1940s. The classical optimization cannot always achieve effective supervision goals especially in the actual great plants, which have many interactions between the process variables and its control parameters. We proposed a new approach of optimization which we called the medium optimization; it can be useful in the complex systems, multi-model process, and especially in systems with stochastic behavior, the open proposed idea of our future research (Koncsag, 1988).

2. PLANT DESCRIPTION

The actual study case analyses a pyrolysis reactor installed in the Piteşti petrochemical plant, one of the greatest Romanian petrol separation and extraction areas; the plant is fed-in by three external petrol sources, and that is causing the diversity of the received petrol quality, its concentration, its quantity and its flow, the fact that influences the chemical characteristics of the reactor products and more than that the system operation conditions (Popescu et al., 2006). So the idea is to adapt the reactor so that it could operate in a medium interval which must have a relative position between the different operation constraints, and because of this, we have proposed the linear programming procedure to resolve the optimization level problem, see figure (1).

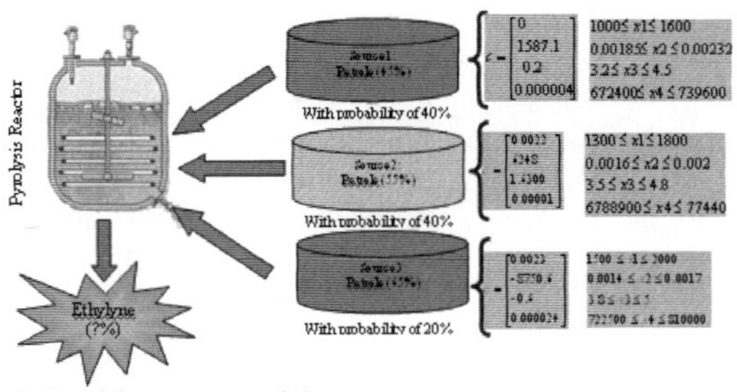

Fig. 1. Feed-in sources of the reactor

3. MEDIUM OPTIMIZATION APPROACH

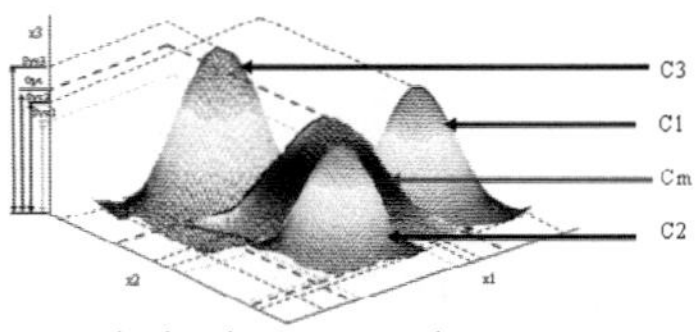

Fig. 2. Medium optimization approach

Linear and non linear problems are situated in the critical area of optimization which becomes "difficult" at the crossroads of combinatorial optimization and continuous linear and non-linear programming. A symbolic scheme in figure (2) shows the principle of the medium optimizations in this application by using linear programming (Glicksman, 2001).

One of the best-known methods for solving linear programming problems in real number is the method of Simplex. In theory, it has a no polynomial complexity and is therefore supposed inefficient; but it describes a systematic procedure that can be applied to solve linear programming problems. However, in practice it is the contrary, it is a good method (Melloul et al., 2004).

4. NUMERICAL CALCULATION AND RESULTS

SISCON is a recent application designed for the great systems optimization and efficient control; it is a software that was developed in C++ (Builder), and was dedicated to the ACPC laboratory of the Faculty of Automatic Control and Computer Science of "Politehnica" University of Bucharest.

Our goal is to optimize the concentration of the produced ethylene considered as the controlled reactor output depending on the received primary matter composition; we propose an initial value of this output as a medium value of the plant output in normal operation, where y_0=30%, is the lowest constraint of the plant output and we try to obtain the highest concentration of ethylene by using the medium optimization approach. We consider a stationary system that defines the reactor operation according to four inputs; the petrol flow, the steam flow, the pressure and the temperature represented by the linearized vector x. The proposed constraints are represented by the vector b, the parameters vectors c_i corresponding to each system considered as a case of selected feed-in, and which were obtained with the help of SISCON software, the calculated medium parameters vector c_m, and the system optimized output which is the ethylene concentration given by $y=c*x$. The mediation operator is $m'=[0.4\ \ 0.4\ \ 0.2]$. The mediation operator helps to deduct the medium value of the vector c depending on the weighting of each feeding source selection; figure (3) shows the constraints intervals of the optimal model, resulted from simplex method approach. A summarized part of the program results is illustrated below; it represents the calculations of the realized MATLAB program, by using the simplex algorithm in linear programming, where the first three systems are the partial ones and the forth system represents the medium one.

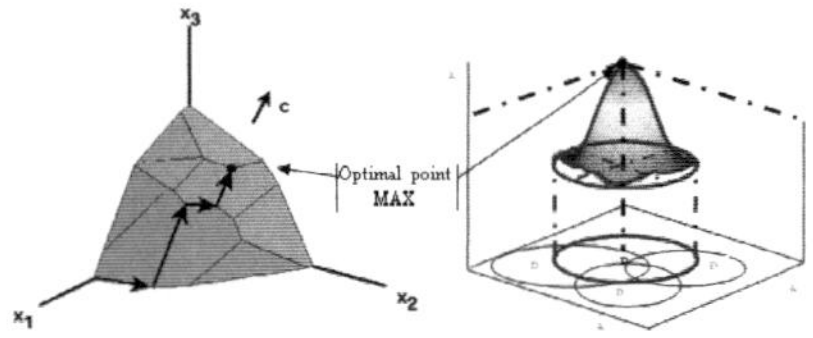

Fig. 3. Simplex method approach

System 1

```
Ai =
   1   0   0   0
   0   1   0   0
   0   0   1   0
   0   0   0   1
bi' =
  1.0e+005 *
   0.0160   0.0000   0.0000   7.3960
ci' =
  1.0e+003 *
   0.0000   1.5871   0.0002  -0.0000
Values of variables:
x(1)= 1600.000000
 x(2)= 0.002320
 x(3)= 4.500000
 x(4)= 739600.000000
The objective value at the optimal point
y= 27.019649
```

System 2

```
Ai =
   1   0   0   0
   0   1   0   0
   0   0   1   0
   0   0   0   1
bi' =
  1.0e+005 *
   0.0180   0.0000   0.0000   7.7440
ci' =
  1.0e+003 *
   0.0000   6.2814   0.0016   0.0000
Values of variables:
x(1)= 1800.000000
 x(2)= 0.002200
 x(3)= 4.800000
 x(4)= 774400.000000
 The objective value at the optimal point
y= 56.733036
```

System 3

```
Ai =
   1   0   0   0
   0   1   0   0
   0   0   1   0
   0   0   0   1
bi' =
  1.0e+005 *
   0.0100   0.0000   0.0001   8.1000
ci' =
  1.0e+003 *
  -0.0000  -8.7506  -0.0006   0.0000
Values of variables:
x(1)= 1000.000000
 x(2)= 0.002300
 x(3)= 5.000000
 x(4)= 810000.000000
 The objective value at the optimal point
y= 17.399551
```

System 4: is the medium system

```
Ai =
   1   0   0   0
   0   1   0   0
   0   0   1   0
   0   0   0   1
bi' =
  1.0e+005 *
   0.0156   0.0000   0.0000   7.6760
ci' =
  1.0e+003 *
   0.0000   1.3973   0.0006   0.0000
Values of variables:
x(1)= 1560.000000
 x(2)= 0.002268
 x(3)= 4.720000
 x(4)= 767600.000000
 The objective value at the optimal point
y= 36.412110
```

From these results we managed to obtain a gain of 6.41% in the concentration of the ethylene produced by the reactor, compared to a medium production obtained in the normal operation, which proves the correctness and effectiveness of the research and of the solution proposed in the paper.

5. CONCLUSION

- The large framework of this paper is automatic control applications in petrochemical industrial plants, where the optimality analyses in stationary system variables was deeply studied;
- A new approach of optimization was proposed, which helps to adapt the system in case of its model multiplicity with the help of the linear programming procedure;
- The actual study can be extended to stochastic operation regime in pyrolysis reactor, as it characterizes the majority of petrochemical processes; more than that it proposes the idea of minimal risk calculation, between each of the partial combinations and the optimal medium one, which requires the non linear programming methodology and exactly Gradient-Rosen algorithm development. Figure (4) shows the dimensional projections between the partial systems and the medium one, an idea which should be analyzed in our near future applications.

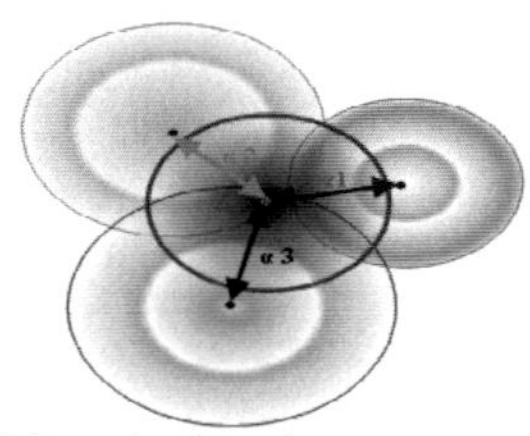

Fig. 4. Minimal risk projection view

6. REFERENCES

Glicksman, A. (2001). *An Introduction to linear Programming and the Theory of Games*, published by COURIER DOVER, ISBN:0486417107, USA

Koncsag, C. (1988). *Chemical Reactors: volum. I*, Edition OVIDIU, ISBN 43-7654-09-321, Bucharest

Melloul, K.; El Kamel, A.; Borne, P. (2004). *Linear Programmation and applications; tutorial and exercises,* Editions TECHNIP, ISBN 2-7108-0844-7, Paris

Popescu, D.; Lupu, C.; Petrescu, C.; Matescu, M. (2004). *Conrol systems of industrial processes*, Edition PRINTECH, ISBN 973-718-016-X, Bucharest

Popescu, D.; Stefanoiu, D.; Lupu, C.; Petrescu C.;Ciubotaru, B & Dimon, C. (2006). *Industrial Automation*, Edition AGIR, ISBN 973-720-093-4, Bucharest

Annals of DAAAM for 2008 & Proceedings of the 19th International DAAAM Symposium, ISSN 1726-9679
Editor B. Katalinic, Published by DAAAM International, Vienna, Austria 2008
Make Harmony Between Technology and Nature, and Your Mind will Fly Free as a Bird

METHODS FOR NONLINEAR PROCESSES CONTROL: CONTROL STRUCTURES BASED ON NONLINEARITY COMPENSATORS

LUPU, C[iprian]; TANASA, V[alentin]; UDREA, A[ndreea] & CRACIUNESCU, P[uiu]

Abstract: *In order to achieve improved real-time control of a class of nonlinear systems, two structures based on the inverse model and with nonlinearity compensator are presented. The use of these structures imposes solving some specific problems like: determination of the process characteristic, construction of inverse model, robust control law design. The applicability of the methods is proved using a real-time structure with a RST control algorithm. Using a software implementation and an experimental platform and analyzing the obtained results leads to the conclusion that the presented methods provide efficient control performances.*
Key words: *nonlinear, control, design, process characteristic.*

1. INTRODUCTION

Various papers and researches target the inverse model control approach; a few of these can be mentioned: (Tao & Kokotovic, 1996), (Pajunen, 1992), (Dumitrache, 2005) etc.

In these researches there have been proposed several types of structures based on inverse model. In accordance with those results, this paper comes up with two very simple and efficient structures presented in Figures 1 and 2. Here, the inverse model is reduced to the process (nonlinear) characteristic.

The first solution (parallel structure) considers the addition of two commands: the first "a feedforward command" generated by the inverse model command generator and the second, generated by a classic, simple algorithm (PID, RST).

The first command, based on the process static characteristic, depends on the set point value and is designed to generate a corresponding value that drives the process's output close to the imposed set point. The second (classic) algorithm generates a command that corrects the difference caused by external disturbances and, accordingly to the set point, by eventual bias error caused by mismatches between calculated inverse process characteristic and the real process.

The second solution (serial structure) has the inverse model command generator between the classic algorithm and the process. The inverse model command generator acts as a nonlinear compensator and depends on the command value. The (classic) algorithm generates a command that, filtered by the nonlinearity compensator, controls the real process.

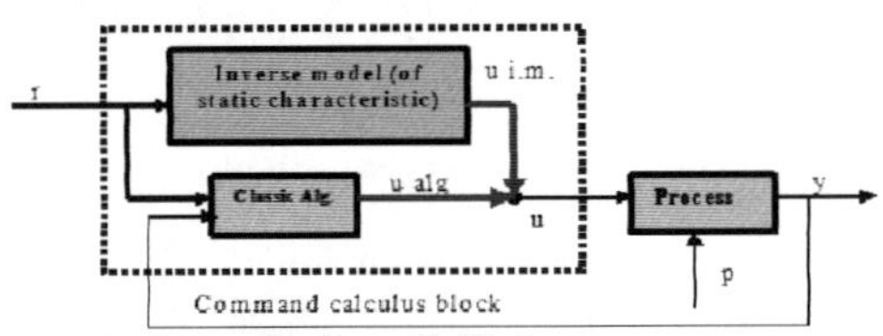

Fig. 1. Proposed scheme for "parallel" structure

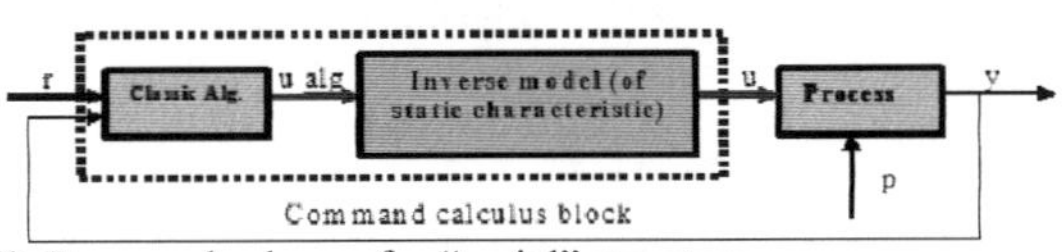

Fig. 2. Proposed scheme for "serial" structure

The presented solutions propose treating the inverse model mismatches that "disturb" the classic command as some algorithm's model mismatches. This approach imposes designing the classic algorithm with a sufficient robustness reserve.

In Figure 1 and 2, the blocks and variables are as follows: Process – physical system to be controlled; Command calculus – unit that computes the process control law; Classic Alg. – control algorithm (PID, RST); y – output of the process; u – output of the Command calculus block; u alg. – output of the classic algorithm; u i.m. – output of the inverse model block; r – system's set point or reference trajectory; p – disturbances.

Related to classical control loops, both solutions need addressing some supplementary specific aspects: determination of static characteristic of the process, construction of inverse model, robust control law design.

In next sections we will focus on the most important aspects met on designing of the presented structure.

2. CONTROL DESIGN PROCEDURE

For the first structure the specific aspects of the control design procedure are: a) determination of process's (static) characteristic, b) construction of command generator, c) robust control law design of classic algorithm. The second structure imposes following these steps: a) determination of process's characteristic, b) construction of nonlinearity compensator, c) designing the classic algorithm based on "composed process" which contains the nonlinearity compensator serialized with real process. These steps are more or less similar for the two structures. For the (a) and (c) steps is evident; for (b) the command generator and nonlinearity compensator have different functions but the same design and functioning procedure. Essential aspects for these steps will be presented.

2.1 Determination of process characteristic

This operation is based on several experiments of discrete step increasing and decreasing of the command $u(k)$ and measuring the corresponding stabilized process output $y(k)$ (figure 3 (a)). The command $u(k)$ cover all (0 to 100%) possibilities. Because the noises are present, the static characteristics are not identically. The final static characteristic is obtained by meaning of all correspondent positions of these experiments. The graphic between two "mean" points is obtained using extrapolation procedure.

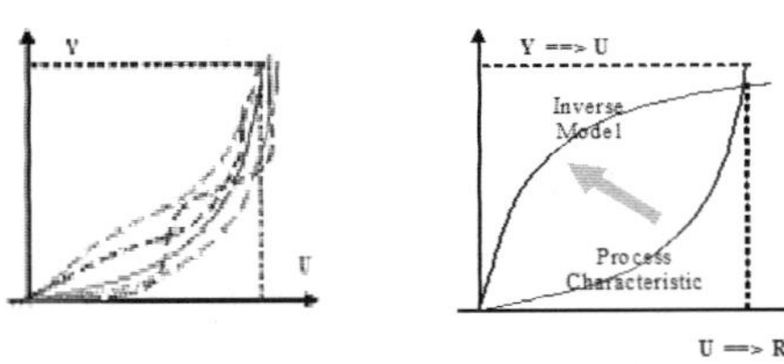

Fig. 3. (a) - left - Determination of process characteristic. Continuous line represents the final characteristic. (b) - right - Construction of nonlinearity compensator

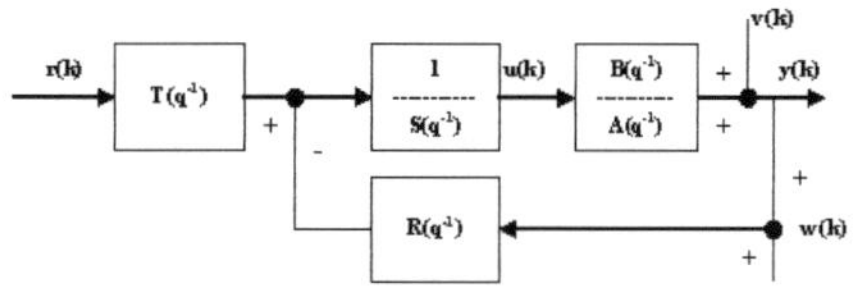

Fig. 4. RST control algorithm structure

According to system identification theory, the dispersion of process trajectory can be found using (Ljung & Soderstroom, 1983). This can express a measure of superposing of noise onto process, process's nonlinearity etc. and is very important for the control algorithm robust design.

2.2 Construction of nonlinearity compensator (generator)

This step deals with the process's static characteristic „transposition" operation. Figure 3 (b) presents this construction. According to this, *u(k)* is dependent to *r(k)*. This characteristic is stored in a table; thus we can conclude that, for the inverse model based controller, selecting a new set point *r(k)* will impose finding in this table the corresponding command *u(k)* that determines a process output *y(k)* close to the reference value.

2.3 Control law design

The control algorithm's duty is to eliminate the disturbances and differences between the inverse model computed command and the real process behavior. A large variety of control algorithms can be used: PID, RST, fuzzy etc., but the goal is to have a very simple one. For this study we use a RST algorithm. This is designed using pole placement procedure (Landau et al., 1997). Figure 4 present a RST algorithm structure.

The identification is made in a specific process operating point and can use recursive least square algorithm (Landau & Karimi, 1997).

This approach allows the users to verify, and if is necessary, to calibrate the algorithm's robustness. Next expression presents "disturbance-output" sensibility function.

$$S_{vy}(e^{j\omega}) \overset{def}{=} H_{vy}(e^{j\omega}) = \frac{A(e^{j\omega})S(e^{j\omega})}{A(e^{j\omega})S(e^{j\omega}) + B(e^{j\omega})R(e^{j\omega})}, \quad \forall \omega \in R \qquad (1)$$

Based on this value (Landau et al., 1997), some robustness measures like gain margin, modulus margin can be calculated.

For the presented structures, for classic algorithms, it is imposed that the gain margin to be greater or equal to the process static characteristic dispersion: $\Delta G \leq \sigma$; a controller that has sufficient robustness was designed.

3. EXPERIMENTAL RESULTS

We have evaluated the achieved performances of the proposed control structures using a laboratory experimental installation, presented on Figure 5 (left). Here, the main goal is to control the pressure process, which has a nonlinear characteristic presented in Figure 5 (right). Three real-time software applications, which can be connected with the experimental installation have been designed and implemented using National Instrument LabWindows/CVI and a data acquisition card.

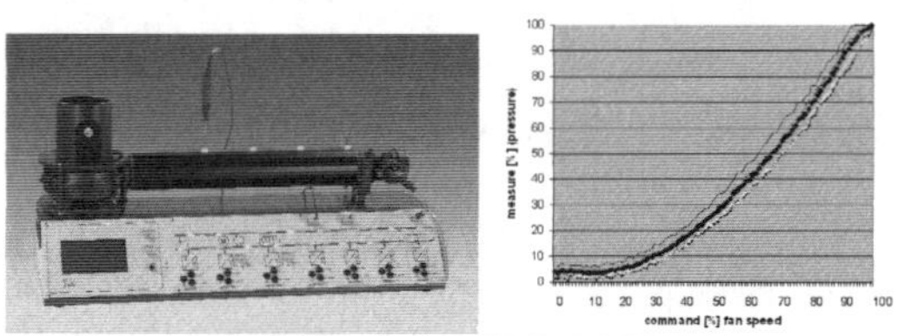

Fig. 5. Experimental installation ELWE LTR 701

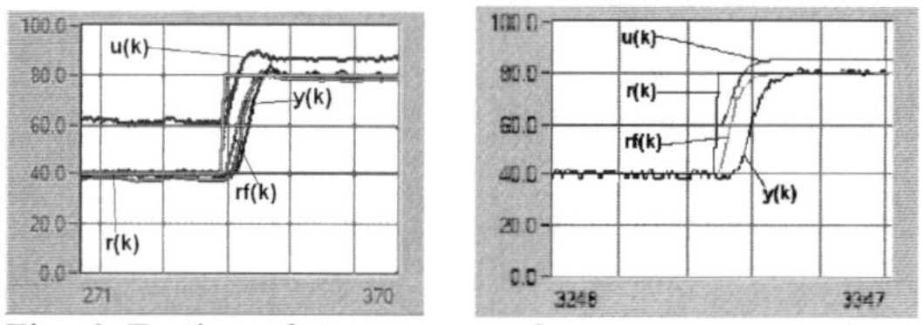

Fig. 6. Testing of structures performance on set point tracking (left – serial, right - parallel)

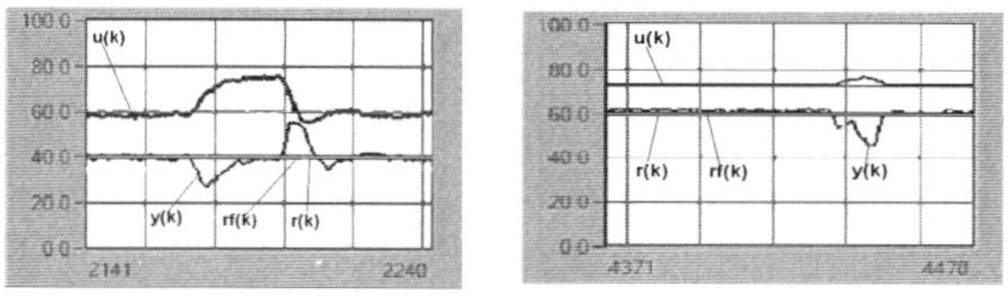

Fig. 7. Testing of structures performance on disturbance rejection (left – serial, right - parallel)

The tests made during real-time functioning prove the structures stability and performances at set point changing. Figures 6 and 7 present few of these tests.

On these figures the evolution curves are represented using the notations: r(k) – set point; rf(k) – filtered set point; y(k) – process output; u(k) – control structure output (total command). In these tests it can be observed that:

- Both structures are globally stabile and track the set point;
- "Serial" structure tracks the set point very precise but can be sensible at important disturbances ;
- "Parallel" structure doesn't track the set point very precise but can be insensible at important disturbances.

4. CONCLUSIONS

This paper proposes two structures as solution for nonlinear processes control. For each component of these structures, there are presented the design methods. These are based on experimental tests and classic identification and close loop pole placement.

The performances of the classic algorithm command component, is evaluated using robustness criterions.

In experimental results section there are presented the evaluated results obtained using a real time software implementations of proposed control structures. The tests prove good performances of the two structures.

In next researches, we will try to improve these solutions which are limited from the nonlinearity class point of view.

5. REFERENCES

Dumitrache, I., (2005). *Numeric control*, Politehnica Press, Bucuresti, ISBN 973-8449-72-3.

Landau, I.D. and A. Karimi, (1997). Recursive algorithm for identification in closed loop: a unified approach and evaluation, *Automatica*, vol. 33, no. 8, pp. 1499-1523.

Landau, I. D., R. Lozano and M. M'Saad, (1997). *Adaptive Control*, Springer Verlag, London, ISBN 3-540-76187-X.

Ljung L., Soderstroom T., *Theory and Practice of Recursive Identification*, MIT Press, Cambridge, Massashusetts, 1983, ISBN 10 - 0-262-12095-X.

Pajunen, G., (1992). Adaptive control of wiener type nonlinear system, *Automatica* vol. 28, no. 4, pp. 781-785.

Tao, G., and Kokotovic P., (1996). *Adaptive control of systems with actuator and sensor nonlinearities*, Wiley, N.Y., ISBN-10 047115654X.

ACKNOWLEDGEMENT

Work supported by IDEI and PARTENERIATE Research Programs of Romanian Research,Development and Integration National Plan II,Grants 1044/2007(Idei), 84/2007 (Parteneriate)

Annals of DAAAM for 2008 & Proceedings of the 19th International DAAAM Symposium, ISSN 1726-9679
Editor B. Katalinic, Published by DAAAM International, Vienna, Austria 2008
Make Harmony Between Technology and Nature, and Your Mind will Fly Free as a Bird

ECONOMETRIC MODEL STRUCTURE AS A SUPPORT TOOL IN REAL PROPERTY MARKET PARAMETERS FEATURING

MACH, L[ukasz]

Abstract: *The article includes a complex process of econometric model construction, analysing dependencies within real estate market. The process allowed selecting variables which determine significantly the interdependencies in real estate sale/purchase and constructing a model of appropriate qualitative parameters. The main task of the model is to support construction investors, or stock market investors in making optimum decisions for the real estate purchase/sale. Econometric model with parameter features may also be applied as a monitoring tool for economic market conditions in banking analyses from the macroeconomic point of view.*
Keywords: *econometrics, optimization, decision, property, market*

1. INTRODUCTION

Determining the real estate market interdependencies is a very important issue as far as the long term stock market investment, construction or economic investments are concerned. In market conditions it is very difficult to set parameters due to a difficulty in obtaining primary data concerning market units dealing with the real property sale/purchase. To estimate a real property value it is necessary to make a research thesis – "What variables and to what degree is the real estate price influenced by?" Trying to answer the research question, the article presents a process of economic model construction, which might prove to be helpful in the real property market analysis. Article structure includes a detailed description of three main econometric model construction stages (variables choice, initial analysis and linear regression) as well as the actual procedure of econometric model construction according to parameters, depending on interdependencies within the real property markets.

2. STAGES OF ECONOMETRIC MODEL CONSTRUCTION

The econometric model construction process involves 3 stages (Błaszczuk, 2006):

- Stage I – choice of variables;
- Stage II – preliminary analysis;
- Stage III – regression analysis.

Each of the said stages is required for a correct econometric procedure. However, due to the kind of issue under analysis, every individual stage should be of dedicated character (Churchill, 2002).

2.1 Stage I – choice of variables

The first stage includes a choice of potential explanatory variable, which might significantly influence the analysed phenomenon, and a determining of the variables type. Kind of variable (quantitative, ordinal, nominal) shall define the procedure of further model construction stages (Kaczmarczyk, 2003). Preliminarily, the following variables have been determined for the variables set: X_1 – flat surface (quantitative var.); X_2 – number of rooms (quantitative var.); X_3 – floor on which a flat is located (quantitative var.); X_4 – location evaluation (ordinal var.); X_5 – finish standard (ordinal var.); X_6 – building construction year (quantitative var.); X_7 – number of storeys in the building a flat is located in (quantitative var.).

These determiners are to estimate statistically the value of a real property object (explained Y value). The next step is to code variables according to the following principles:

- for nominal scale – every feature is coded within one column by means of letter or numerical symbols, i.e. finish standard with values: 1-high, 2-average, 3-basic.
- for ordinal scale – features coded by means of ranges, including the sequence of appearance of a given feature, e.g. at employment of the so called binary coding.
- for quotient scale – coding is based on entering appropriate feature values for each structure in appropriate column (i.e. we do not enter the non coded values, but actual ones).

2.2 Stage II – preliminary analysis

Preliminary statistic analysis was carried out mainly for the coding correctness check up and to find out whether there are any outlier observations as compared with the tested variables (i.e. with value range exceeding (or lower) a few times from the basic data).

The process of checking whether there are outlier observations was executed by means of plot boxes. The plot boxes are formed on the basis of descriptive statistic values; they are therefore applied for ordinal features only. High cohesiveness and transparency are its advantages; however they transfer a large amount of information of average value, variability scope and asymmetry of tested variables distribution. The next step was to check up what is the variables distribution and the central tendency measures were calculated. To check up the variables distribution, amount charts were drawn up for the purpose of histogram construction. A second way to check up the variable distribution was to check assumptions on the normal variables distribution by means of hypothesis testing. As the testing under consideration involves different kinds of variables, various tests were applied (Kołmogorow-Smirnow and Sharipo-Wilk tests). Choice of a test was determined by the number of trials. Determining histograms in economic phenomena it is necessary to take into account a histogram right-hand skewness as in such a case it is necessary to find a logarithm of primary data (Mynarski, 2006).

2.3. Stage III – regression analysis

Due to the fact of choosing 7 explanatory variables the regression analysis for the constructed econometric model was carried out in the following steps (Dittmann, 2004):

- Step 1 – Testing of co linearity of explanatory variables – this step involves determining correlation matrixes.
- Step 2 – Testing the total variables significance. This step involved a defined and research zero hypothesis of the following form:

H_0: $\beta_1 = \beta_2 = \beta_3 = \beta_4 = \beta_5 = \beta_6 = \beta_7 = 0$

and an alternative hypothesis:

H_1: not all β_i (i=1,2,3,4,5,6,7) are equal with 0.

- Step 3 – Model remainder analysis. The aim of the step is to find out whether the model remainder randomness is involved.
- Step 4 – Calculating determination coefficient.
- Step 5 – Verification of the econometric model.

3. DETERMINING THE ECONOMETRIC MODEL FOR REAL PROPERTY MARKETS – CASE STUDY

The analysed example shall involve econometric model construction procedure. The model shall describe interdependency between a flat price and chosen determiners (7 factors). The primary model was defined as follows:

$$Y = \beta_0 + \beta_1 x_1 + \beta_2 x_2 + \beta_3 x_3 + \beta_4 x_4 + \beta_5 x_5 + \beta_6 x_6 + \beta_7 x_7 + \varepsilon \quad (1)$$

with:
y – flat price,
x_1 – flat area,
x_2 – rooms number,
x_3 – storey on which the flat is located,
x_4 – location evaluation (1 – very good, 2 – good, 3 – average),
x_5 – finish standard (1 – very good, 2 – good, 3 – basic),
x_6 – building construction year,
x_7 – the number of storeys in the building where the flat is located,
ε - randomness factor.

First it was checked whether *y* variable includes outlier observations. A plot box diagram was drawn up to determine it on the basis of parameters from chart no. 1.

Points on the diagram above value 325000 (values > 1,5 IQR) were treated as outlier data and have been removed.
Then a histogram for the price variable was determined (fig. 1).

After a graphic analysis was carried out it was found out that the tested histogram has a normal distribution (i.e. it is unimodal, of a bell shaped curve and symmetric). Its symmetric character causes that there is no need to find the logarithm of the data. The next step included a procedure of coding qualitative data i.e. evaluation of location, finish standard. Coding is a step which is necessary for further joint analysis of qualitative and quantitative data.

The smallest observation result within the limit of 1,5 IQR	Q1	Median	Q3	The highest result observed within 1,5 IQR
118500	189500	210000	250000	325000

Tab. 1. Parameters for the plot-box diagram (variable *y*)

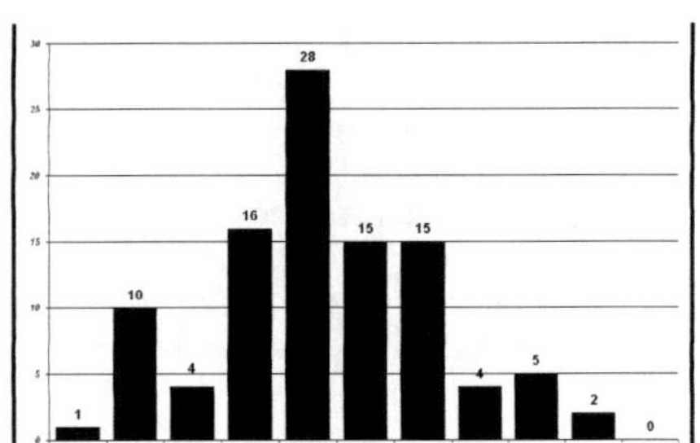

Fig. 1. Histogram of *y* variable.

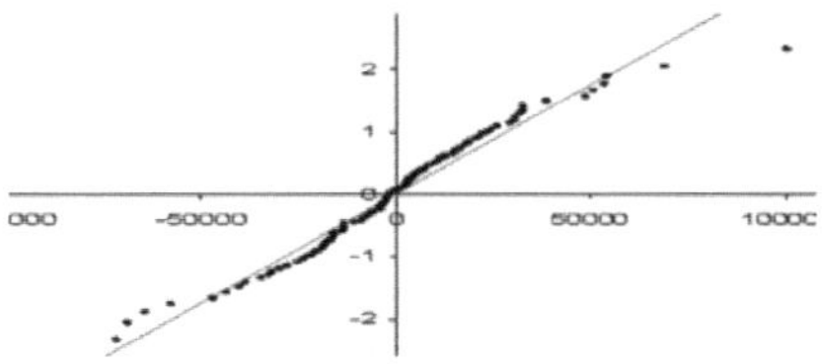
Fig. 2. Normal probability plot

After a preliminary analysis, the next procedure is to determine interdependencies between variables. To obtain it a correlation matrix was determined. The most correlative influence with the variable *price for one square meter of flat* have variables such as: *area flat* (R^2=0,80) and *the number of rooms* (R^2=0,65). The highest correlation between explanatory variables appears in case of variables: *area flat* and *the number of rooms* (R^2=0,77).
The next step is to determine joint linear regression.

The results analysis indicates that variables such as: *area flat, finish standard, construction year,* are variables which statistically, significantly influence *Y* variable. The last stage is to check randomness of the model remaining part (Fig. 2).
The final econometric model is as follows:

$$Y = \beta_0 + \beta_1 x_1 + \beta_5 x_5 + \beta_6 x_6 + \varepsilon \quad (2)$$

Adaptation coefficient (R^2) of a model defined in this way is 0,66, which must be acknowledged as a very good value. Using this kind of econometric model definition in actual conditions is confirmed by the efficiency of its work.

4. CONCLUSION

Parameters of interdependencies in real property market indicate that among 7 variables only 3 influence significantly a flat's price. Trying to define the flat value it is sufficient to take into account the *flat surface, standard,* and, *year of construction.* The other variables such as the number of rooms, storey on which a flat is located, location evaluation and the number of storeys in a building do not influence the flat's price significantly. The main obstacle in forming an econometric model for real estate market is the difficulty to collect primary data which might provide parameters for the analysed phenomenon. The constructed model may be helpful for the financial investors, construction investors and for the banks in estimating real estate loans, being a decision support tool. Further research shall include a monitoring of validity of a constructed model, and if necessary it will be optimized.

5. REFERENCES

Błaszczuk, D. (2006). *Introduction to forecast and simulation*, in Polish: *Wprowadzenie do prognozowania i symulacji*, Wydawnictwo Naukowe PWN, ISBN 83-01-14648-1, Warszawa

Churchill, G.A. (2002). *Marketing Research. Methodological Foundations*, in Polish: *Badania marketingowe. Podstawy metodologiczne*, Wydawnictwo Naukowe PWN, ISBN 83-01-13820-3, Warszawa

Dittmann, P. (2004). *Forecasting in enterprise*, in Polish: *Prognozowanie w przedsiębiorstwie*, Oficyna Wydawnicza, ISBN 83-89355-35-3, Kraków

Kaczmarczyk, S. (2003). *Marketing Research,* in Polish: *Badania marketingowe,* Polskie Wydawnictwo Ekonomiczne, ISBN 83-208-1433-2, Warszawa.

Mynarski, S. (2006). *Analysis of market and marketing data*, in Polish: *Analiza danych rynkowych i marketingowych*, Oficyna Akademii Ekonomicznej w Krakowie, ISBN 83-7252-316-9, Kraków.

Annals of DAAAM for 2008 & Proceedings of the 19th International DAAAM Symposium, ISSN 1726-9679
Editor B. Katalinic, Published by DAAAM International, Vienna, Austria 2008
Make Harmony Between Technology and Nature, and Your Mind will Fly Free as a Bird

RANGE TRACKING IN WIRELESS NETWORKS

MACHEDON PISU, M[ihai]

Abstract: *The principle of tracking by triangulation, successfully applied for the Global Positioning System (GPS), can also be used for the wireless networks deployed inside buildings for determining the position of moving targets. The radio propagation phenomena affect the radio link in a way that is difficult to predict. In this paper, the quality of the radio transmissions is tested in different scenarios and a correlation with the distance is found. The localization algorithms used must be scalable and such algorithms are Maximum Likelihood (ML) and Weighted Centroid Localization (WCL). The proposed solution combines a logarithmic correlation with a ML-based algorithm. The tracking error is determined by using field tests and simulations.*

Key words: *communication technology, wireless systems, simulation engineering, sensors*

1. INTRODUCTION

The recent development of wireless sensor networks (WSN) makes it possible for applications such as industrial monitoring and control, automations in buildings, or target tracking to be implemented with low cost, low power and reduced complexity. GPS cannot comply with these requirements but its principle of triangulation can be used with other wireless technologies for range tracking inside buildings. Position information can play a major role in wireless networks, and many methods for localization within WSNs have been proposed. Similar to triangulation tracking, these methods calculate the position of a mobile node based on the locations of more than three reference nodes and the distance between a mobile node and a reference node can be determined by the received signal strength or quality, which is evaluated with the link quality indicator (LQI). Such estimation depends on the signal propagation phenomena, which affect the radio link quality. A proper correlation should adapt to the unpredictable changes in LQI. Also, the localization errors are greatly reduced by using the best localization algorithm. Combining these strategies with the results from the field tests, a simulator is developed and run.

2. THE CORRELATION BETWEEN DISTANCE AND LQI

Different values of LQI are obtained from packets sent at the same distance. Averaging these values gives unsatisfactory results. Using the Probability Theory, the value of LQI for that distance is better determined in the following way:

$$LQI_{FINAL} = (\sum P(LQI_i) * LQI_i) / \sum P(LQI_i) \quad (1)$$

Where LQI_i are the LQI values, $P(LQI_i)$ the probability of a LQI value to occur for that distance, and it is equal to $1/ k_{MIN}$ if $NP(LQI_i) >= NP_{ALL}/k_{MIN}$, where $NP(LQI_i)$ is the number of packets received with that LQI value and NP_{ALL} is the total number of packets sent (as seen in Fig. 1), where k=2,3,4...(integer values).

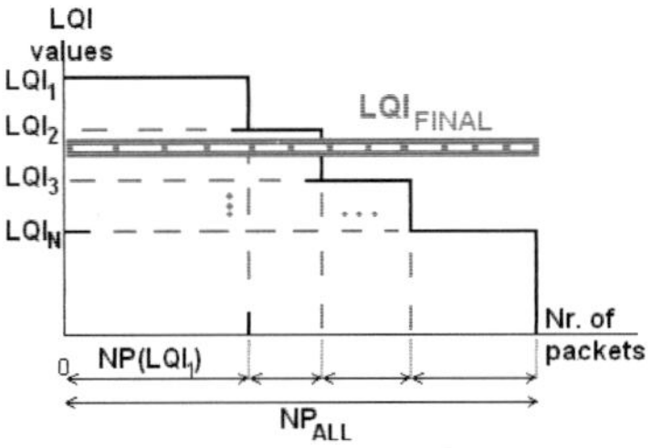

Fig. 1. The final LQI value for the same distance

LQI measurements provide the means to estimate the performance of the link. The changes in LQI do not depend only on the communication distance, but also on factors such as the transmission medium and the surrounding environment. The effects of these factors for radio propagation are related to attenuation, multipath and interference. A correlation between LQI and distance is needed in order to adapt to these effects, and by approximating the LQI value as close as possible to the measured distance, it should be achieved (Fig. 2). The power correlation refers to the Friis' free space transmission equation, where the power at the receiver follows an inverse square law related to the distance value. Due to dynamic and uncertain propagation conditions, this approach does not work, and a more adaptive solution is given by the logarithmic correlation. The link performance is tested in different indoor and outdoor scenarios and the errors obtained with the two correlations are compared.

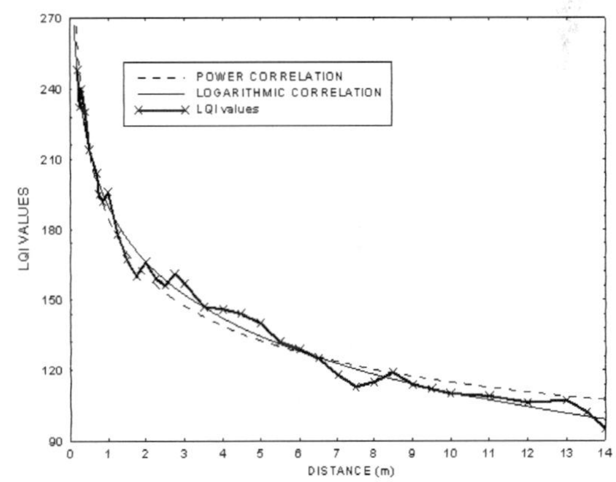

Fig. 2. LQI Power vs. Logarithmic correlation, up to 14 m

For a 10m x 10m grid, the logarithmic correlation gives a mean error of 1.49% and a maximum error of 3.9%. The power correlation mean error is 2.28% and the maximum is 5.47%. The formulas for logarithmic (2) and power correlation (3) are the following (with d as distance):

$$-77.9 * \log(d) + 190 = LQI \quad (2)$$

$$185 * d^{-0.206} = LQI \quad (3)$$

3. IMPLEMENTING LOCALIZATION METHODS IN WIRELESS NETWORKS

Before implementing the localization method, the algorithm used for tracking must be tested in order to establish its precision. ML estimates the position of the target by minimizing the differences between estimated and measured distances, by using the minimum mean square error (MMSE). An algorithm that uses ML can determine the target's position, for 10m distance between the neighbour reference nodes, in the following way (D_i represents measured distances):

$$X_{TARGET} = [(D_4^2) - (D_3^2) + 100] / 20 \quad (4)$$
$$Y_{TARGET} = [(D_1^2) - (D_4^2) + 100] / 20 \quad (5)$$

The WCL algorithm proposes a method in which the distances measured are encapsulated as weighted functions: $w = 1 / D^k$ where k=1,2,3…(integer values). The target's position is estimated (X_i, Y_i are reference node coordinates):

$$X_{TARGET} = (\sum X_i / D_i^k) / \sum 1 / D_i^k \quad (6)$$
$$Y_{TARGET} = (\sum Y_i / D_i^k) / \sum 1 / D_i^k \quad (7)$$

ML and WCL algorithms are tested within a 10m x 10m grid, with 4 reference nodes, one in each corner and 121 points are used for position measurement (Fig. 3).

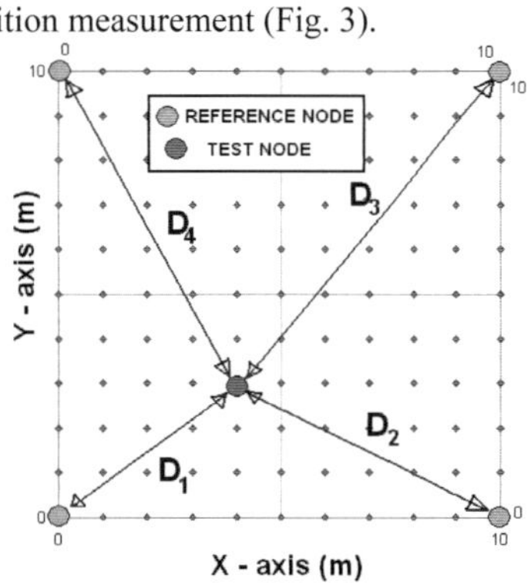

Fig. 3. Principle of triangulation: estimating a node's position when distances D_1, D_2, D_3, D_4 are determined.

Both algorithms are scalable but the tracking error should not exceed 1%. The tracking errors are illustrated in Figures 4 and 5:

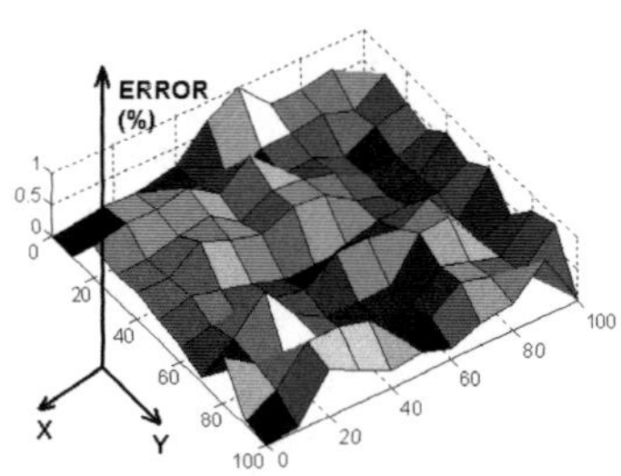

Fig. 4. Tracking error distribution with ML, for a 100m x 100m test grid

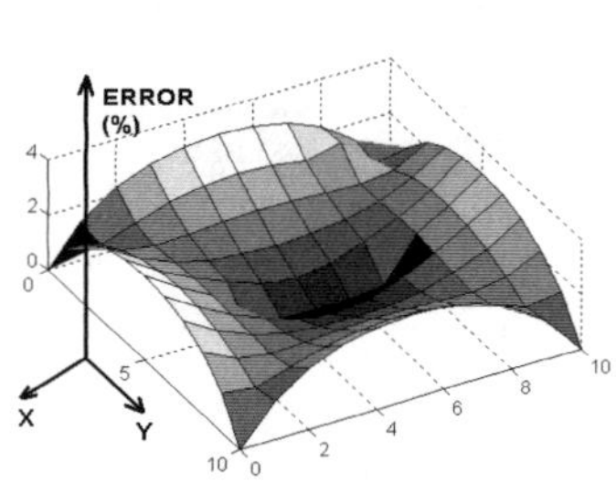

Fig. 5. Tracking error distribution with WCL, for a 10m x 10m test grid, for k = 1

For a 100m x 100m test grid, the maximum tracking error of ML is 0.94 %, while the mean error is 0.39%. The tracking errors for WCL are far greater, the maximum exceeding 3%, while the mean error is around 1.7%, for a 10m x 10m test grid.

4. SIMULATION RESULTS

WSNs are special networks which deploy hundreds of tiny sensors. Due to range limitations, a range-based approach for tracking in WSNs gives poor results. A more adequate approach should be a range-free tracking with a large number of reference nodes. Therefore, tracking simulation uses 121 reference nodes for a 100m x 100m test grid, with a distance of 10 m between neighbour reference nodes. There are 100 test points and their position is determined by tracking with the most near four reference nodes, for which the LQI value is the greatest (Fig. 6).

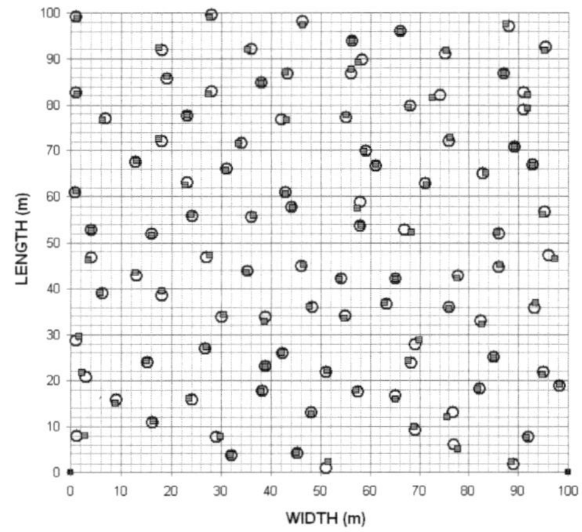

Fig. 6. Tracking simulation in a 100m x 100m test grid (ring – real position range, square point – estimated position)

The errors provided by the field tests have shown that the logarithmic correlation and the ML algorithm can be used for precise tracking in wireless networks. The proposed solution combines the two methods within the simulator, which gives a maximum error of 1.91% and a mean error of 0.97%.

5. CONCLUSION

Range tracking is one of the main applications that can be implemented with wireless sensor networks. The present research has provided the solutions for obtaining precision in position estimation, and as an effect of the simulation, we can see that *GPS for large buildings* is possible by considering the proposed tracking method.

6. REFERENCES

Blumenthal, J., Grossmann, R., Golatowski, F. & Timmermann, D. (2007), Weighted Centroid Localization in ZigBee-based Sensor Networks, *Proceedings of Intelligent Signal Processing*, pp. 14-17, ISBN: 978-1-4244-0830-6

Ferrari, G., Medagliani, P., Di piazza, S.& Martalo, M. (2007), Wireless Sensor Networks: performance analysis in indoor scenarios, *Eurasip Journal on Wireless Communications and Networking*, Vol. 2007, No. 1, pp. 41, ISSN 1687-1472

Machedon-Pisu, M., Szekely, I., Gavrus, R. (2008), Efficient Data Propagation Techniques and Security Concerns in Low Rate Wireless Personal Area Networks in Outdoor and Indoor Scenarios, In: *OPTIM 2008*, Vol.3: Industrial Automation and Control, pp.201-207, ISSN 1842-0133

Annals of DAAAM for 2008 & Proceedings of the 19th International DAAAM Symposium, ISSN 1726-9679
Editor B. Katalinic, Published by DAAAM International, Vienna, Austria 2008
Make Harmony Between Technology and Nature, and Your Mind will Fly Free as a Bird
2nd DAAAM-ICAT

DEVELOPMENT OF RAPID PROTOTYPE MODELS FOR TEMPORAL BONE DISSECTION SIMULATION

MAEKITIE, A[ntti]; PALOHEIMO, K[aija] S[tiina]; PALOHEIMO, M[arkku]; KANERVA, J[ukka]; BJOERKSTRAND, R[oy]; RAMSAY, H[ans] & TUOMI, J[ukka]

Abstract: *Temporal bone dissection using human cadaver tissue is an essential part of Ear-, Nose-, and Throat surgical training. We used various Rapid Prototype (RP) techniques and various materials to develop a 3D-temporal bone RP model. An experienced ear surgeon evaluated their suitability for surgical training. The RP models were found suitable for the teaching of anatomic landmarks, basic principles and preoperative planning of otologic surgery. Resolution of microsurgical details still needs to be further developed. The RP temporal bone model will be further developed as a training tool.*
Key words: *Ear, surgery, training, stereolithography, selective laser sintering, 3D-printing*

1. INTRODUCTION

Practice is an essential part of otologic (ear) surgical training. It enables the acquisition of adequate sensorimotor skills and allows for the fine-tuning of cognitive processes under the guidance of experienced surgeons, to perform successful otologic surgical procedures without causing injury to the patient (Sewell et al., 2005). Mastoidectomy is a fairly common otologic surgical procedure performed to control middle ear and mastoid pathologic conditions. In mastoidectomy the surgeon drills through the mastoid part of the temporal bone (Figure 1) and removes spongiform bone to gain access and to remove inflammatory tissue from the mastoid and middle ear. The role of practice is especially significant in mastoidectomy as the surgeon is working in the immediate vicinity of minute, multifaceted and delicate structures such as auditory ossicles, lamellar bone, blood vessels and nerves.

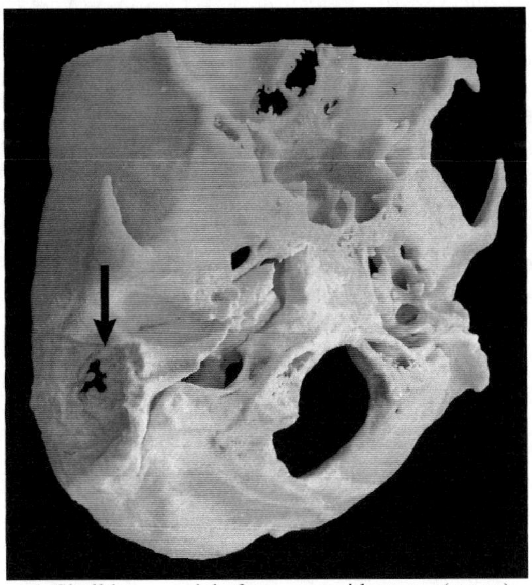

Figure 1. Skull base model after a mastoidectomy (arrow)

Currently surgeons-in-training mostly practice in temporal bone laboratories (Zirkle et al., 2007) using cadaver temporal bones which are obtained through local donations or tissue banks. Apart from the obvious ethical challenges to be overcome for removing parts from dead bodies there is always at least a perceived risk of acquiring a disease from the cadaver. Further, there are no two cadaver dissections alike and repetition of unsuccessful procedures or having many trainees perform the procedure on a temporal bone dissection with the same anatomic structures is not possible. Also, with the limited supply and availability of practice material the opportunities to perfect one's skills do not arise.

Computer Based Education (CBE) methods, including interactive CD-ROMs and internet-based applications have recently been increasingly accompanied by virtual simulators which, utilizing technologies from the computer game industry, provide a relatively life-like 3-dimensional impression from several angles and haptic response for the surgeon (Kuppersmith et al., 1997; Wiet et al., 2002; Wiet et al., 2005). In the simulator the trainee uses both hands, one for drilling and the other to remove the resulting bone dust through rinsing and suction (Sewell et al., 2007). The advantages of the virtual simulators are the ability to adjust operational parameters as well as to simulate various disease states. They also provide ample opportunity for repetition, give immediate constructive and documentable feedback, and provide a non-threatening environment to learn from one's mistakes (Sewell et al., 2007; Mason et al., 1998; Basdogan et al., 2007; Fried et al., 2007). On the down-side, simulators are expensive and to-date technically imperfect. There are also problems with providing an exact match to reality in drilling bone.

Although CBE methods inarguably have their place in surgical training, real hands-on removal of mastoid tissue maintains its role, and real human tissue used in a cadaver laboratory has yet to be replaced by other materials. The possibility of solving some of the above mentioned problems by utilizing bone-like materials and rapid manufacturing technologies has not been adequately investigated (Löppönen et al., 1997). In this study we aimed at developing a RP temporal bone model for otologic surgical training.

2. RESEARCH DESIGN

The skull of a healthy male volunteer was radiographed by Cone Beam Computed Tomography (CBCT, Planmeca Oy, Finland). CBCT results in a tense pile of parallel grey-scale images of the "sliced" radiographed object thus depicting e.g. skull bone structure. Grey-scale values were used to distinguish bone structure from soft tissue. This segmentation was done to approximate exact bone boundaries. Topologically organized segments were interpolated to form a 3D-representation of the object. The 3D-representation was further converted into stl-format. OsiriX open source software was used for these operations. The stl-model was modified to cover temporal bone ("ear") areas. Corresponding RP-models were commissioned from four Finnish RP-service providers. The materials used were polyamide (PA), aluminium-filled polyamide (Al-PA), a liquid photopolymer (Photopolymer) and a gypsum-based composite powder (Composite). The manufacturing techniques used were Stereolithography (SLA), Selective Laser Sintering (SLS) and 3D-printing (3D-P) (Table 1).

Material	Color	Techn.	Trade name	Layer thickness
PA	White	SLS	PA2200	0,1 mm
Al-PA	Grey	SLS	Alumide®	0.1 mm
Photo-polymer	Trans-parent	SLA	Somos® 11120	0.1 mm
Composite	Blue-White	3D-P	ZP®131	0.089 mm

Tab. 1. Materials and techniques used in the study.

An experienced ear surgeon performed cortical mastoidectomy on each of the different models and evaluated their suitability for temporal bone dissection training. In addition, four residents from the Otolaryngology – Head and Neck Surgery training program at the Helsinki University Hospital also performed a mastoidectomy using the RP models and provided a summative evaluation of the models.

3. RESULTS

The size and macroscopic structure of the models were comparable to a real temporal bone. The "feel" of drilling was fairly close to real bone in most of the studied models. The best result regarding this was achieved using the one made of composite powder.

The microstructures of the temporal bone models were not precise and some of the surgically and anatomically significant structures (lamellae, tympanic membrane, ossicles) were missing from the models. These problems and limitations have to be solved before the studied models can replace cadaver temporal bones in surgical training. However, the RP models were found suitable for the teaching of anatomic landmarks, basic principles and preoperative planning of otologic surgery. Resolution of microsurgical details still needs to be further developed. Various materials with different mechanical properties were also found to provide different handling environments.

4. CONCLUSIONS

The RP temporal bone model may be further developed as a training tool. We are currently improving the parameters concerning resolution and material properties of our model. We are also investigating the possibility to combine different manufacturing techniques with RP, to overcome the absence in the present models of some of the delicate but anatomically and surgically significant structures.

5. ACKNOWLEDGMENTS

This study was supported by grants from the University of Helsinki and the Finnish Funding Agency for Technology and Innovation (Tekes). The authors want to thank M.Sc.(tech) Pekka Paavola for the photograph.

6. REFERENCES

Basdogan, C.; Sedef, M.; Harders, M. & Wesarg, S. (2007). VR-based simulators for training in minimally invasive surgery. *IEEE Comput Graph Appl*, Vol. 27, (2007) 54-66

Fried, M. P.; Uribe, J. I. & Sadoughi, B. (2007). The role of virtual reality in surgical training in otorhinolaryngology. *Curr Opin Otolaryngol Head Neck Surg*, Vol. 15, (2007) 163-169

Kuppersmith, R. B.; Johnston, R.; Moreau, D.; Loftin, R.B. & Jenkins, H. (1997). Building a virtual reality temporal bone dissection simulator. *Stud Health Technol Inform,* Vol. 39, (1997) 180-186

Löppönen, H.; Holma, T.; Sorri, M., Jyrkinen, L.; Karhula, V.; Koivula, A.; Ilkko, E.; Laitinen, J.; Koivukangas, J.; Oikarinen, J. & Alamäki, O. (1997). Computed tomography data based rapid prototyping model of the temporal bone before cochlear implant surgery. *Acta Otolaryngol Suppl*, Vol. 529, (1997) 47-49

Mason, T. P.; Applebaum, E. L.; Rasmussen, M,; Millman, A.; Evenhouse, R. & Panko, W. (1998). The virtual temporal bone. *Stud Health Technol Inform*, Vol. 50, (1998) 346-352

Sewell, C.; Morris, D.; Blevins, N. H.; Barbagli, F. & Salisbury, K. (2005). Quantifying risky behaviour in surgical simulation. *Stud Health Technol Inform,* Vol. 111, (2005) 451-457

Sewell, C.; Morris, D.; Blevins, N. H.; Barbagli, F. & Salisbury, K. (2007). Evaluating drilling and suctioning technique in a mastoidectomy simulator. *Stud Health Technol Inform,* Vol. 125, (2007) 427-432

Wiet, G. J.; Stredney, D.; Sessanna, D.; Bryan, J. A.; Welling, B. & Schmalbrock, P. (2002). Virtual temporal bone dissection: an interactive surgical simulator. *Otolaryngol Head Neck Surg* , Vol. 127, (2002) 79-83

Wiet, G. J.; Schmalbrock, P.; Powell, K. & Stredney, D. (2005). Use of ultra-high-resolution data for temporal bone dissection simulation. *Otolaryngol Head Neck Surg*, Vol. 133, (2005) 911-915

Zirkle, M.; Roberson, D. W.; Leuwer, R. & Dubrowski, A. (2007). Using a virtual reality temporal bone simulator to assess otolaryngology trainees. *Laryngoscope*, Vol. 117, (2007) 258-263

Annals of DAAAM for 2008 & Proceedings of the 19th International DAAAM Symposium, ISSN 1726-9679
..., Editor B. Katalinic, Published by DAAAM International, Vienna, Austria 2008
Make Harmony Between Technology and Nature, and Your Mind will Fly Free as a Bird
*Young Researcher and Scientist Paper / * Supervisor, Mentor*

THE STUDY OF KANSEI ENGINEERING APPLIED TO THE VHCQC WEB PORTAL

MAJERIK, M[arian] & ESSE, M[artin]

Abstract: *This paper describes a concrete application of Kansei Engineering to the Virtual Healthcare Quality Centre web portal (www.improhealth.org). The study of website evaluation focuses on set of emotion and attempts to explore user's responses to Virtual Health-Care Centre web portal and investigate its relationships with website designers. This study was recently launched internationally and we involved managers of healthcare organization, healthcare personnel, and general public interested in healthy life style also with visually impaired people as future possible users of web knowledge portal. Finally, the finding of this relationship will enable the research to strategize a guideline to the design of attractive virtual environment for healthcare interested parties. The paper is based on outcomes achieved in the research project KEGA 3-4121-06 "Web Portal of Tools, Methods and Case Studies of Quality Management" and SK/06/B/F/PP-177443 Vocational Education and Training for Quality of Life through eHealthcare & Well-Being.*

Key words: *Kansei Engineering, The Virtual Healthcare Quality Centre web portal, quality evaluation, improvements*

1. INTRODUCTION

Qualities of usefulness, usability and desirability play a central role in the good design of all digital products and digital environments such as web portal. The Virtual Healthcare Quality Centre web portal (VHCQC) embeds affective appeal is seen to offer consumer positive experience visiting the website.

2. WHAT IS KANSEI ENGINEERING

"What looks good? For me, what looks good is what 'feels' good," renowned architect Michael Graves was quoted as saying in a newspaper interview.

Kansei Engineering is a method for translating feelings and impressions into product features, invented in the 1970s by Professor Mitsuo Nagamachi (Dean of Hiroshima International University). Kansei Engineering can "measure" the feelings and shows the relationship to certain product properties. In consequence, products can be designed to bring forward the intended feeling. The Kansei Engineering began with sensory or sensitivity engineering at the starting point. However, current Kansei Engineering is the trans-disciplinary engineering that extends over the humanities, social science and natural science. Filtering, acquiring information, estimating, recognizing, modelling, making relationship, producing, giving information, presenting and etc. are the contents of Kansei (Grimsaeth, 2005).

2.1 An Example

Web sites can be described with adjectives like transparent, simply for using and elegant. But, which parameters influence these impressions? Kansei Engineering can show to what extent product attributes have an effect on these impressions. Target values for the mentioned product properties can be derived.

3. GOAL OF THE STUDY

The study attempts to explore user's emotional responses to VHCQC web portal (www.improhealth.org) (Zgodavová, 2006) and demonstrates all necessary changes to make the new web portal www.fsev.tnuni.sk/improheath be better. This aim is in line with the idea of Kansei Engineering that is to participate consumer's emotion into web portal design. An application of Kansei Engineering to the VHCQC Webportal consists of more steps; the most important are decision strategy, correction of Kansei words, setting of scale of Kansei words, correction of explored web portal samples (properties), and analysis using numerical evaluation, explanation improvements to the designer (Zgodavová, 2006). We want locate the concrete features (from huge quantify of all properties) which are interesting for final portal's users and make them better (from the affective point of view). The methodical background was based on Kansei measuring through the words of user's potential emotional requirements for improvements.

4. THE SELECTION WEB PORTAL PROPERTIES AND KANSEI WORDS

Each product has too many properties and functions which can be evaluated and developed for the better using and viewing (Zgodavová, 2006). We selected the group of concrete properties; we obtained knowledge for composing following four group (Fig.1) of web portal's properties:
Manageability, Attractiveness, Security and Readability.

Altogether fourteen portal's properties were identified in consultation with the IT experts. These are divided into four categories (four groups of portal's properties and). Details are shown in Table 1.

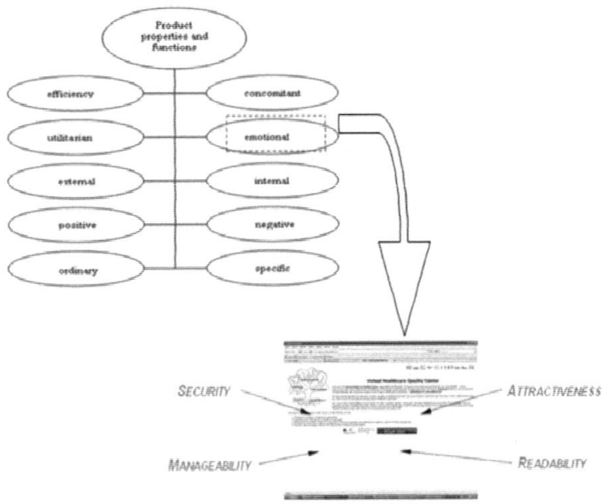

Fig.1. The set of web portal's properties applied in the Kansei redesign ("Product Properties and Functions" by Zgodavová, 2002)

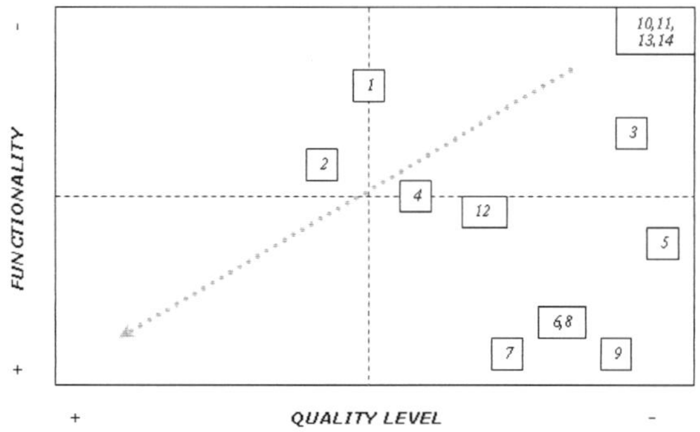

Fig. 2. The diagram of VHCQC web portal's properties required for improvements (The number marks of properties are equipollent with the Table 1)

After identification of web portal's properties we span the semantic space. The most common way of Kansei Engineering measuring is through words (Kansei word pairs), because words are tool for man's feeling expression (Grimsaeth, 2005; Nagamachi, 2007). We made the list which included a big set of multifarious words related to the web portal. In our study we selected following word pairs to better evaluating properties of explored portal:

- complicated / elementary
- confused / synoptic
- difficult / easy
- unpractical / practical
- boring / elegant
- ordinary / modern
- inaccessible / accessible (for blind people)
- professional / unprofessional
- appealing / afflictively
- sophisticated / simple
- dismal / positive
- absent / present
- meaningful / futile

We used this semantic space for evaluating of each property from the previous step. For example, on the uppermost right edge (Fig. 2) are situated properties which were evaluated as the most inopportune properties of web portal (number marks in Fig.2 are taken from Table 1). The VHCQC web portal quality evaluation was realized by IT experts and portal's users. The axis of diagram (Fig. 2) was named "Functionality" and "Quality level". Functionality includes all technical requirements necessary for the precise portal's behaviour. The second parameter "Quality level" embraces all emotional and affective perception of potential portal's users.

5. THE RECEPTION OF THE IMPROVEMENTS OF THE VHCQC WEB PORTAL

Goal of our study was reception of the improvements for the better user's affective perception. We received 14 improvements for each portal property. The particular improvements are shown in Table 1. This study presents preview part of the Quality Function Deployment of the VHCQC web portal. The Quality Function Deployment is a tool for designing product to user needs by translating what the user wants (Zgodavová et all, 2002).

6. CONCLUSION

The study presents initial Kansei structure which was described by principal steps of VHCQC web portal evaluation. Results of the study have identified semantic space of Kansei words representing human Kansei responses to web portal. The relations of Kansei and web portal designs were also visible.

THE VHCQC PORTAL'S PROPERTIES AND FUNCTIONS / THE VHCQC PORTAL'S IMPROVEMENTS	
Manageability	
1.Search	*Search tool* situated on the main page
2. Print	Add the tool for *direct print*
3. Understandability of menu	*Transformation* of menu
4. Difficult discrimination of un-/visited links	*Use colour* to distinguish visited and unvisited links
5. Portal lucidity	Perform the *tree-structure* of portal
Attractiveness	
6. Graphics	Use *meaningful graphics* that relate to the contents
7. Colour scheme	*Better colour* variety
8. Images position	*one dominant image* on one page only
9. Frames	*Avoid using too many frames* on one page
Security	
10. Access	Permit *multi level access*
11. Inopportune state for authentication and authorization	Use of *id and user password*
12. Programming security	Provide *programming security* of the highest degree
Readability	
13.Blind friendly	Add *the sound tracks* of human voice
14. Multilingual readability	*Multilingual copywriter*

Table 1. The list of the VHCQC web portal's improvements obtained through Kansei Engineering

Further studies will be performed, as to find associations between these Kansei Words and detail web portal designs. This is to achieve the research target which is to formulate guideline that translates human Kansei to web portal design. This paper provides principal methodical level of web portal quality evaluation by way of selected Kansei words. Consideration on the universal and localized Kansei features of the web portal environment will be considered in our other research by using concrete multi-variation statistical methods.

7. ACKNOWLEDGEMENT

Our acknowledgement goes to Prof. Kristina Zgodavová, PhD. our PhD. study adviser which markedly contributed in development of this paper and to the team of portal designers: Kristina Lengyelova, Lubomir Lengyel and Slavomir Beznak.

8. REFERENCES

Grimsaeth, K. (2005). Kansei Engineering, Linking Emotions and Product Features, *Available from:* http://www.ivt.ntnu.no/ipd/fag/PD9/2005/artikler/PD9%20 Kansei%20Engineering%20K_Grimsath.pdf *Accessed:* 2008-07-03

Nagamachi, M. (2007). Perspectives and new trend of Kansei/Affective Engineering, in: *Proceedings of the 1st European Conference on Affective Design and Kansei Engineering,* 18-20 June, 2007, Lund University and Linkoping University

Zgodavová, K. et all (2002). *The Quality Professional,* Technical University in Kosice, ISBN 80-7099-845-8, Kosice (Slovakia)

Zgodavová, K. (2006). Web Portal of Tools, Methods and Case Studies of Quality Management, *project of Slovak Ministry of Education KEGA 3-4121-06,* 2006

Zgodavová, K. (2006). Improhealth Collaborative SK/06/B/F/PP-177443 Vocational Education and Training for Quality of Life through eHealthcare & Well-Being, Leonardo da Vinci pilot project, 2006

Annals of DAAAM for 2008 & Proceedings of the 19th International DAAAM Symposium, ISSN 1726-9679
Editor B. Katalinic, Published by DAAAM International, Vienna, Austria 2008
Make Harmony Between Technology and Nature, and Your Mind will Fly Free as a Bird

THE INFLUENCE OF THE TONER STRUCTURE ON THE QUALITY OF BLACK-WHITE DIGITAL PRINTING

MAJNARIC, I[gor]; BOLANCA, S[tanislav] & BOLANCA MIRKOVIC, I[vana]

Abstract: The actual black-white digital printing techniques have been analyzed in this article. It is to assume, that different composition and structure of black ink will establish different thicknesses of ink film on identical printing substrate, namely it enables different quality of monochromatic prints. For the analyses the representative digital printing techniques have been used: Electrophotography, Inkjet, Digital Screen Printing and Digital Waterless Offset. The ability of the black toner adhering to the printing substrate has been analyzed by the measurements of the optic inking density (X-rite) which is supplemented by the image analysis (Personal IAS). The results show that EA toner allows maximum quality establishing the thickest ink film (D=2,14) and the smallest oscillation (1,41 µm) of the defined line.
Key words: *digital printing, density, ink, image analysis*

1. INTRODUCTION

In recent times, the majority of the digital printing machines development has been reduced to the investigation of the new formulation and the finest possible toner structures which result in better printing properties and higher productivity. The aim of this work is the analysis of the quality of the black-white digital printing and recognizing how different toner types adhere to the non absorbent printing substrate. Indirect determination of thickness of the ink film is based on the optical density measurement. Optical density (D) is defined in form of logarithmic ratio: $D=\log 1/\beta=\log I_o/I$. Reflectance factor ($\beta$) is the ratio of light intensity, (I) of light remitted by the ink film in relation to the intensity of light (I_o) remitted by the blank paper.

2. THEORETICAL PART

The digital printing techniques are characterized by the unique digital input data which define the contents of the future printing form, i.e. the elements of the final print. The digital printing has been developed into two directions: Computer to Press and Computer to Print (Goldmann, 2004). The characteristic of the black–white digital printing, is achieving the coloration by different types of coloration carriers. They are polymerised toners and new colorants, pigment dispersion, encapsulated pigment technology and dye. (Williams, 2001).

In the digital waterless offset black pigments are tiny particles of carbon black or pure elementary carbon (20-150µm) whose share is 10-30%. The basis of black–white digital screen printing ink is soy oil (>6%), petroleum solvent (<10%), glycerol (<5%), black pigment (<10%), alkid resins and water. The particles of the carbonate black toner are less than 1µm. The basic principle of EP (electrohotoraphy) is founded on photoelectric effect (Bolanča et al., 2005). In 80% of the cases the toners are powders. Two component toner of 1st generation is a mixture which contains: resins (80–90%), coloured pigments (5-15%), substances for charge control (1-3%) and additives (Thompson, 2004). Toner of the 2nd generation is based on process EA (emulsion aggregation). The process of EA enables exact position control of waxes within the toner particles. The printer fusers melt the toner particles by heating, making it sticky and able for adhering to paper even on the lower fusing temperatures (Suzuki et al, 2003). The pigment particles of the liquid EP toners have star like shapes (1-2µm). For directed movement of ElectroInk, the uniformly dispersed agents (polar molecules) for increasing the electric conductivity are responsible (Majnarić et al., 2007). Ink Jet more recent pigment dispersions are made on the basis of uncapsulated technology where each pigmented particle is comprised with the oil coating adapted for good penetration into the printing substrate. In this way the corresponding surface smooth coating is obtained which increases the optical reflectance of the printed layer (Larson et al., 2002).

3. EXPERIMENTAL PART

In this work, the printing quality of black–white digital printing systems which is applied in office business and editing has been investigated. The digital printing machines have been compared for that: EP (Hp Indigo Turbostream, Xerox 250, HP LaserJet 9050), Piezo Ink Jet-a (Epson 2400 photo), digital screen printing (RISO RZ 970 E) and digital offset printing (Heidelberg Quickmaster 46 DI). They all use especially composed ink. For the needs of investigation, the special printing form in PDF was generated. It contains: the standard achromatic illustration, wedge with 99 patches (in the range from 1 to 100%) positive and negative micro elements, and positive and negative lines of the standard thickness of 650 µm. By means of the densitometer X-rite DTP 41 the optical inking density of the printed ink layer was determined, which directly showed the binding of toner to the printing substrate, i.e. the possibility of printing the broadest possible range of the achromatic inking. With the system for image analysis (Personal IAS) the dimensional exactness of the printed reproduction. The calendering natural paper substrate with the high degree of whiteness, designed for digital printing was used for printing (Splendogel EW 115 g/m2).

4. RESULTS AND DISCUSSION

In figure 2 the curves of the dependence of the inking density (D) on total screen value (1-100% screen value), have been presented. All the curves have the increasing form which is uniform. The adapted printing substrate for the digital printing has the greatest inking density on Xerox (D_{MAX}=2,14). Indigo TS in greatest part (1-90% screen value) has the smallest inking density. The exception is Epson 2400 which gets the maximal inking density at 85% screen value ($D_{85\%RTV}$=0,89) but it does not succeed to keep it in higher tones.

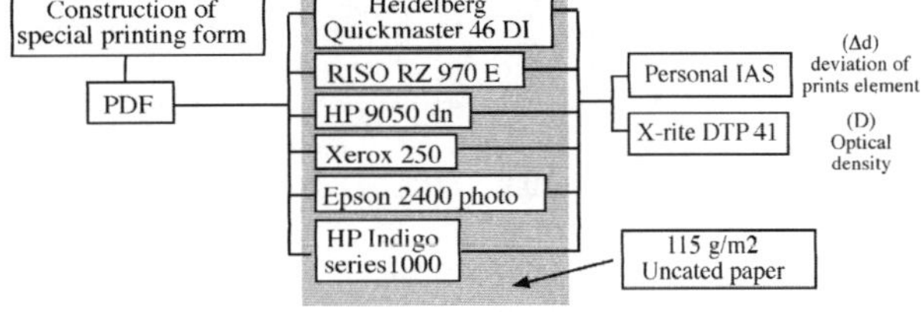

Fig. 1. Schematic presentation of the performed experiment

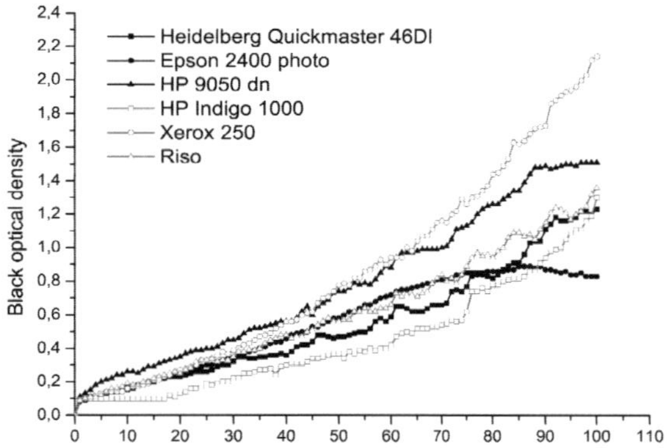

Fig. 2. Dependence of the inking density (D) on screen value for the printing substrate in the range from 1-100% screen value

The liquid pigment inks (Epson 2400) in great layers cause the opposite effect, i.e. too big ink penetration into the substrate which causes the capillary blooding and closing the darkest screen values. The quality of the surface layer is emphasized which is visible at low screen values. The greatets inking density is achieved by HP 9050 ($D_{10\%RTV}$=0,26), after that follow Riso, Epson, Xerox and Heidelberg. In these printers, the growing trend in dependence curves $D/RTV_{1-10\%}$ is visible, which means that there are differences in creation of the lighter tones. On Indigo there is no difference in inking density and the areas 1-15% screen value are unchanged. Such characteristic movement in value can be explained by the usage of the software linear LUT, by which propositions the dot gain is eliminated. In reproduction of high screen values, the impressions of Xerox stand out more considerably. The nearest to it are the impressions of HP 9050 DN. It is characteristic fot the darker tone areas, that the smallest inking density has Epson 2400. On the formed layer in the zone of the medium quality there are the digital machines Riso, Indigo, Heidelberg QM, which have the maximal inking densities in the area between D=1,23 and D=1,36. In relation to the linear curves, only Xerox realizes the increase, i.e. the possiblity of the more qualitative reprodcution of darker tones. By the image analyses Personal IAS positive and negative lines are increased and measured. Results are presented in table 1.

In relation to the desired positive original line, the increase of reproduction was noticed at: Indigo TS (Δd=35,35 µm) and Epson 2400 (Δd=40,81µm). The other printers generate the decrease of the e.g. Xerox (Δd=-1,41µm) Heidelberg QM 46 (Δd=-17,87µm), HP 9050 (Δd=-23,89 µm) and RISO (Δd=-83,44 µm). For the reproduction of the negative lines the decrease of the desired size is applicable. The smallest change appears at Heidelberg QM 46 (Δd=-33,31 µm), while the greatest one at Epson 2400 (Δd=-127,93 µm), Riso (Δd=-136 µm) and Indigo TS (Δd=-173,95 µm).

5. CONCLUSION

The investigations have shown their justification in the scientific sense and in application. The different types of digital black white machines generate different ink layers on the adapted non coated printing substrate for digital printing by the application of the specific toners. The liquid toner based on the uncapsulated pigment, will realize the different deviations in the inking density on different screen values (in the range from 10% screen value). For the same screen areas, the black liquid EP ink realizes minimal deviations within the light and medium tones ($\Delta D_{1-10\%}$=0,01; $\Delta D_{45-55\%}$=0,04), but great deviations in dark areas as well (ΔD_{91-100}=0,36). Paste offset ink realizes identical deviation in inking density in lighter and medium

PRINTERS		Measured width of 650 µm line
EPSON 2400	POZ	690,81 µm
	NEG	523,07µm
HP 9050 dn	POZ	626,11 µm
	NEG	555,56 µm
HEIDELBERG QM46	POZ	632,17 µm
	NEG	616,69 µm
HP INDIGO	POZ	685,35 µm
	NEG	474,05 µm
RISO RZ970 E	POZ	567,56 µm
	NEG	514,00 µm
XEROX DC 250	POZ	648,59 µm
	NEG	519,02 µm

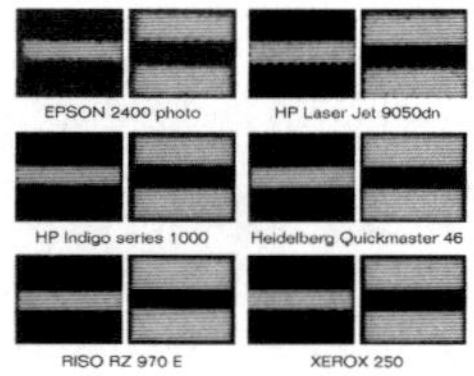

Tab. 1. The thickness of the reproduced positive and negative 650 µm lines

tones ($\Delta D_{1-10\%}$=0,06; $\Delta D_{45-55\%}$=0,06), while in the darker areas it grows for 100% ($\Delta D_{91-100\%}$=0,12). The inking densities of the water-oil black emulsion toner behave according to the whole screen value. With the increase of the screen value, the range of inking densities constantly grows; for the light areas ($\Delta D_{1-10\%}$=0,10), medium areas ($\Delta D_{45-55\%}$=0,13) and for the dark areas (ΔD_{91-100}=0,21). In relation to the tested toners, the powder two component toner will oscillate greatly in its light areas ($\Delta D_{1-10\%}$=0,15). In the middle area, the oscillation is also important ($\Delta D_{45-55\%}$=0,19), while in the darkest area the saturation and constancy appear (ΔD_{91-100}=0,02). The toner obtained by EA enables the highest quality in black white printing. The inking density deviation grows proportionally with the increase of the screen value ($\Delta D_{1-10\%}$=0,07; $\Delta D_{45-55\%}$= 0,24; $\Delta D_{91-100\%}$=0,41). In this way the thickest ink coating is achieved (ΔD_{max}=2,14) while the tone values remain open. The densitometric results are confirmed by the image analysis. Liquid inks do not succeed the preciseness of the defined printing elements, so the small gain of the positive printed elements was noticed, i.e. the greater decrease of negative printing elements. The positive elements printed with powder toner reproduce smaller line, and only the toner obtained by emulsion aggregation corresponds most closely to ideal width. Algorithm of printing elements formation influences that movement. The constructions of simple single coloured units have lower resolution in relation to the multicoloured ones. Riso printed lines oscillate multi dimensionally. The tolerance threshold in quality is very low which enables the printing of black white applications with lower demands.

6. REFERENCES

Bolanča, S.; Majnarić, I. & Pigac, S. (2005) Digital Printing with Increased Ink Layers, *Graphic Arts Technology design comunications:* Z. Bolanča, M. Mikota, (Ed.) 131- 142, University of Zagreb Faculty of Graphic Arts, ISNB 953-96020-3-3, Zagreb.

Goldmann, G. (2004) *The World of Printers*, Oce Printing Systems GmbH, ISBN 3-00-001081-5, Dusseldorf.

Larson, J. R.; Gipson, G. A. & Schmidt, P. S. (2002) *Liquid Toner Materials, HandBook of Imaging Materials,* 2nd edition, A.D. Diamond and D. S. Weiss (Ed.). ISNB 0-8247-8903-2

Majnarić, I.; Bolanča, S. & Golubović, K. (2007). The Influence of ElectriInk Pigmentation on the Quality of Indirect Digital Printing, *Proceedings of 11th International Conference on Printing, Design and Graphic Communications*, Z. Bolanča (Ed.), p.p. 85-89, ISBN 978-953-96020-7-7, Zadar, Croatia.

Suzuki, C.; Takagi, M. & Inoue, S. (2003) Toner Characteristics and Xero Interactive Performance of EA Particles with Specific External Additives, *Proceedings of International Conference on Digital Printing Technologies NIP 19, IS&T*, p.p.134-137, ISBN 0-89208-247, New Orleans, Louisiana, USA.

Thompson, B. (2004). *Printing Materials: Science and Technology*, Pira International, ISBN 1 85802 981 3, Surrey.

Williams, C. (2001). Printing Inks, Digital Demand: *The Journal of Printing and Publishing Technology 6(1)*, (Pira International) 40-47, ISBN 1471-5694

AUTHOR INDEX

A

B

C

M

N

O

P

R

S

T

U

V

W

Y

Z

SUBJECT INDEX

A

B

C

D

E

F

G

H

I

J

K

L

M

N

O

P

Q

R

T

U

V

W

X

Z

Annals of DAAAM for 2008 & Proceedings of the 19th International DAAAM Symposium, ISSN 1726-9679
Editor B. Katalinic, Published by DAAAM International, Vienna, Austria 2008
Make Harmony Between Technology and Nature, and Your Mind will Fly Free as a Bird

MTF STU Trnava expresses thanks to Supporters and Sponsors of 19th DAAAM International Symposium „Intelligent Manufacturing & Automation" DAAAM-2008

Supporters:

Town of Trnava
Mesto Trnava

Self-governmental region of Trnava
Trnavský samosprávny kraj

Sponsors:

Výskumný ústav jadrovej energetiky, a. s. Trnava

BEKAERT, a. s. Hlohovec

Jadrová a vyraďovacia spoločnosť, a. s. Jaslovské Bohunice

SLOVNAFT, a. s. Bratislava

PRVÁ ZVÁRAČSKÁ, a. s. Bratislava

DRILL, s. r. o. Bratislava

Nadácia Ján Korca Trnava

Železničné opravovne a strojárne, a. s. Trnava

INVEX, s. r. o. Trnava

Slovenské elektrárne, a. s., člen skupiny ENEL Bratislava

KESA Trnava

Nadácia ARX Jaslovské Bohunice

SEPS, a.s. Bratislava

IBOK

VUZ PI SR

TECHNODAT, CAE – systémy, spol. s r.o. Trenčín

ZLIEVÁREŇ TRNAVA s.r.o.

Závody všeobecného strojárstva – holding, a. s. Dubnica n./Váhom

RESEARCH IS WHAT I'M DOING WHEN I DON'T KNOW WHAT I'M DOING.

WERNHER VON BRAUN

- Acceptance without proof is the fundamental characteristic of Western religion, Rejection without proof is the fundamental characteristic of Western science. Gary Zukav, "The Dancing Wu Li Masters"
- Any sufficiently advanced technology is indistinguishable from magic. Arthur C. Clarke
- Basic research is like shooting an arrow into the air and, where it lands, painting a target. Homer Adkins
- Every revolutionary idea - in science, politics, art, or whatever - evokes three stages of reaction in a hearer: It is completely impossible - don't waste my time. It is possible, but it is not worth doing. I said it was a good idea all along.
- Every successful person has had failures but repeated failure is no guarantee of eventual success.
- I do not believe that it is possible to create living art out of anything but the direct visual experience of nature, combined with the heritage of a Tradition, unless it be by the practice of magic ritual. Michael Ayrton 1921-1975
- In theory, there is no difference between theory and practice. In practice, however, there is.
- Murphy's Law of Research: Enough research will tend to support your theory.
- Never worry about theory as long as the machinery does what it's supposed to do. R. A. Heinlein
- No committee could ever come up with anything as revolutionary as a camel anything as practical and as perfectly designed to perform effectively under such difficult conditions. Laurence J. Peter
- Not until a program has been in production for at least six months will the most harmful error then be discovered.
- Practical sciences proceed by building up; theoretical sciences by resolving into components.[Saint] Thomas Aquinas ca 1225-1274
- • Putt's Law: Technology is dominated by two types of people: Those who understand what they do not manage. Those who manage what they do not understand.
- • Rule of Creative Research: (1) Never draw what you can copy. (2) Never copy what you can trace. (3) Never trace what you can cut out and paste down.
- Science is the meeting place of two kinds of poetry: the poetry of thought and the poetry of action. George Agostinho da Silva
- Science is what happens when preconception meets verification.
- So, He was a great teacher. But what did he publish? From an Israeli academic
- The goal of Computer Science is to build something that will last at least until we've finished building it.
- The goal of science is to build better mousetraps. The goal of nature is to build better mice.
- The habit of apprehending a technology in its completeness-, this is the essence of technological humanism, and this is .what we should expect education in higher technology to achieve. I believe it could be achieved by making specialist studies the core around which are grouped liberal studies which are relevant to these specialist studies. But they must be relevant; the path to culture should be through a man's specialism, not by-passing it.... A student who can weave his technology into the fabric of society can claim to have a liberal education; a student who cannot weave bis technology into the fabric of society cannot claim even to be a good technologist [Lord] Eric Ashby 1904-
- Theft from a single author is plagiarism. Theft from two is comparative study. Theft from three or more is research.
- To get to know, to discover, to publish-this is the destiny of a scientist. Frangois Arago 1786-1853
- To know / To know what to do / To do it / To make it known. Savoir / Savoir faire / Faire Faire savoir. Department of Physics, Trinity College, Dublin
- Watson's Law: The reliability of machinery is inversely proportional to the number and significance of any persons watching it.
- What is a magician but a practicing theorist? Obi-Wan Kenobi
- Whenever anyone says, "theoretically," they really mean, "not really". Dave Parnas
- There is no national science just as there is no national multiplication tables; what is national is no longer science. Anton Chekhov (1860-1904)

AFTER ALL IS SAID AND DONE, A HELL LOT OF A LOT MORE IS SAID THAN DONE

DAAAM International Vienna
Karlsplatz 13/311, A-1040
Wien, Austria, European Union

ISBN 978-1-5108-4000-3

19th DAAAM International Symposium on Intelligent Manufacturing and Automation (DAAAM 2008)

Focus on Next Generation of Intelligent Systems and Solutions

Trnava, Slovakia
22-25 October 2008

Volume 2 of 2

Editor:

B. Katalinic

ISBN: 978-1-5108-4000-3

Printed from e-media with permission by:

Curran Associates, Inc.
57 Morehouse Lane
Red Hook, NY 12571

Some format issues inherent in the e-media version may also appear in this print version.

Printed by Curran Associates, Inc. (2017)

For permission requests, please contact DAAAM International Vienna
at the address below.

DAAAM International Vienna
Karlsplatz 13/311, A-1040
Wien, Austria, European Union

Phone: 0043-1-58801-31121
Fax: 0043-1-58801-31199

secretary@daaam.info

Additional copies of this publication are available from:

Curran Associates, Inc.
57 Morehouse Lane
Red Hook, NY 12571 USA
Phone: 845-758-0400
Fax: 845-758-2633
Email: curran@proceedings.com
Web: www.proceedings.com

ISSN 1726-9679

DANUBE ADRIA ASSOCIATION FOR AUTOMATION & MANUFACTURING

DAAAM INTERNATIONAL VIENNA

www.daaam.com

ANNALS OF DAAAM FOR 2008 &
PROCEEDINGS

OF THE 19TH INTERNATIONAL DAAAM SYMPOSIUM
"INTELLIGENT MANUFACTURING & AUTOMATION:
FOCUS ON NEXT GENERATION OF INTELLIGENT SYSTEMS AND SOLUTIONS"
22-25TH OCTOBER 2008, TRNAVA, SLOVAKIA
(SECOND EDITION)

ORGANIZED BY:
DAAAM INTERNATIONAL VIENNA
SLOVAK UNIVERSITY OF TECHNOLOGY
FACULTY OF MATERIALS SCIENCE AND TECHNOLOGY IN TRNAVA
VIENNA UNIVERSITY OF TECHNOLOGY, UNIVERSITY OF APPLIED SCIENCES
TECHNIKUM, AND AUSTRIAN SOCIETY OF ENGINEERS AND ARCHITECTS - ÖIAV 1848

UNDER THE AUSPICES OF: THE DANUBE RECTORS' CONFERENCE &
RECTORS' HONOR COMMITTEE OF DAAAM INTERNATIONAL

SUPPORTED BY FESTO

EDITOR: B.[RANKO] KATALINIC

S T U
M T F

☙ DAAAM INTERNATIONAL TABULA GRATULATORIA ❧

Happy Birthday to:

Mr Manuel Del Vechio (25 Year) / Dr Radek Matusu (30 Years) / Mr Daniel Novoselovic (30 Years) / Dr Jun Su Ha (35 Years) / Dr Andreas Kandelbauer (35 Years) / Mr Kan Wongkasem (35 Years) / Professor Tiberiu Dobrescu (40 Years) / Dr Snjezana First Rogale (40 Years) / Dr Gabriela Patrascu (40 Years) / Dipl.-Ing. Jana Stefankova (40 Years) / Professor Miran Brezocnik (45 Years) / Mr Sergey Vasilevich Koukhta (45 Years) / Professor Neculai Eugen Seghedin (45 Years) / Professor Anton Hadar (50 Years) / Dr Miroslav Matysek (50 Years) / Professor George Constantin (50 Years) / Dr Kvetoslava Resetova (50 Years) / Professor Predrag Cosic (55 Years) / Professor Franc Cus (55 Years) / Professor Ivan Budic (60 Years) / Professor Ilija Cosic (60 Years) / Professor Malik Kulenovic (60 Years) / Miran Milfelner (60 Years)

University of Rijeka (35 Years) / Middle East Technical Universiy (45 Years) / Sevastopol National Technical University (45 Years)

Faculty of Mechnical –Engineering – Univesity of Sarajevo (50 Years) / Tallinn University of Technology (90 Years) / Austrian Society of Engineers and Architects (160 Years)

Congratulations to:
Professor Evgenij Valentinovich Pashkov on his election to the post of the Rector of SevNTU, Sevastopol
Professor Ivan Budic on his election to the post of the Dean of Polytechnic of Pozega, Pozega

Welcome to the 4808th DAAAM Author:
Since 1990 **4808** authors and co-authors published their scientific papers in the Annals & Proceedings of DAAAM International Symposium

Special thanks to concern **Festo** for establishing of **"Festo Young Researchers and Scientist Support Scholarship"** for financial support of next generation of Researchers and Scientist at DAAAM international conferences

With love and appreciations
World-wide DAAAM International Community

Who is in the field, and not in the DAAAM he/she does not exist.
(Jyri Papstel 1996)

EDITOR'S NOTE This publication was reproduced by the photo process, using the manuscripts and soft copies supplied by their authors. The layout, the figures and tables of some papers did not conform exactly to the standard requirements. In some cases was the layout of the manuscripts rebuild. All mistakes in manuscripts there could not been changed, nor could the English be checked completely. The readers are therefore asked to excuse any deficiencies in this publication which may have arisen from the above causes. The editor and the DAAAM International Vienna are not responsible either for the statements made or for the opinion expressed in this publication.

DAAAM International is a member of CrossRef. Contributions in DAAAM Scientific Book 2006 and DAAAM International Journal of Simulation Modelling are cross-referenced and available on Internet. DAAAM International publications are cited and referenced in the: Inspec Database, Cambridge Scientific Abstracts, and ISI Scientific Proceedings Thomson Reuters. Papers are available at Thomson Gale.
ANNALS OF DAAAM FOR 2008 & PROCEEDINGS OF THE 19TH INTERNATIONAL DAAAM SYMPOSIUM, VOL 19, NO1,
Abbreviated key title: ***Ann. DAAAM proc. int. DAAAM Symp.***
ISSN 1726-9679, ISBN 978-3-901509-68-1, Editor B.[ranko] Katalinic
Published Annually by DAAAM International Vienna, Vienna 2008

Layout & Design: B.[ranko] Katalinic / Technical Editors: Cesarec P.[aulina], Kettler R.[oman], Alispahic, A.[smir], Brötzenberger R.[ene], Fuchs M.[artin], Richter M.[artin], Rupar M.[iljan], Saliger A.[lex]; Trauzettel, U.[lrich]
DAAAM Home Page Administrators: Malisa V.[iktorio] / Stopper M.[arkus] Graphics used in this Publication: M.C. Escher
Publisher: DAAAM International Vienna, TU Wien, Karlsplatz 13/311, A-1040 Vienna, Austria
E-mails: president@daaam.com katalinic@mail.ift.tuwien.ac.at / DAAAM homepage: http://www.daaam.com http://www.daaam.org

THE 19TH INTERNATIONAL DAAAM SYMPOSIUM

"INTELLIGENT MANUFACTURING & AUTOMATION:

FOCUS ON NEXT GENERATION OF INTELLIGENT SYSTEMS AND SOLUTIONS"

22-25TH OCTOBER 2008, TRNAVA, SLOVAKIA

ORGANISED BY:

Vienna University of Technology	DAAAM International Vienna	Slovak University of Technology MTF Trnava	University of Applied Sciences Technikum Vienna	Austrian Society of Engineers and Architect

UNDER THE AUSPICES OF
DANUBE RECTORS' CONFERENCE &
RECTORS' & PRESIDENTS' HONOR COMMITTEE OF DAAAM INTERNATIONAL FOR 2008

ECATERINA ANDRONESCU	RECTOR, UNIVERSITY POLITEHNICA OF BUCHAREST
JAVIER UCEDA ANTOLÍN	RECTOR, UNIVERSIDAD POLITÉCNICA DE MADRID
VLADIMÍR BÁLEŠ	RECTOR, SLOVAK UNIVERSITY OF TECHNOLOGY BRATISLAVA
GABRIELE BEIBST	RECTOR, JENA UNIVERSITY OF APPLIED SCIENCE
FLOELLA BENJAMIN	CHANCELLOR, THE UNIVERSITY OF EXETER
LAJOS BESENYEI	RECTOR, UNIVERSITY OF MISKOLC
ALEKSA BJELIS	RECTOR, UNIVERSITY OF ZAGREB
VLADIMIR BOBAL	VICE-RECTOR, TOMAS BATA UNIVERSITY UNIVERSITY ZLIN
ION BOSTAN	RECTOR, TECHNICAL UNIVERSITY OF MOLDOVA
ANTONIO CASTORANI	RECTOR, POLYTECHNIC OF BARI
TOMÁŠ ČERMAK	RECTOR, TECHNICAL UNIVERSITY OF OSTRAVA
ANTON CIZMAR	RECTOR, TECHNICAL UNIVERSITY OF KOSICE
FRANCE A. CÓRDOVA	PRESIDENT, PURDUE UNIVERSITY
GIOVANNI DEL TIN	RECTOR, POLITECNICO DI TORINO
JERZY DEMBCZYŃSKI	RECTOR, POZNAN UNIVERSITY OF TECHNOLOGY
MIKHAIL I. DEMCHUK	RECTOR, BYELORUSSIAN STATE POLYTECHNIC ACADEMY
AKOS DETREKŐI	RECTOR, BUDAPEST UNIV. OF TECHNOLOGY AND ECONOMICS
JOZEF GAWLIK	RECTOR, CRACOW UNIVERSITY OF TECHNOLOGY
SERGEY N. GRIGORYEV	RECTOR, MOSCOW STATE UNIVERSITY OF TECHNOLOGY
IAN GOULD	CHANCELLOR, UNIVERSITY OF SOUTH AUSTRALIA
BORIS V. GUSEV	PRESIDENT, RUSSIAN ACADEMY OF ENGINEERING, MOSCOW
VÁCLAV HAVLÍČEK	RECTOR, CZECH TECHNICAL UNIVERSITY IN PRAGUE
SAID IRANDOUST	PRESIDENT, ASIAN INSTITUTE OF TECHNOLOGY
ZEJNEL KELMENDI	RECTOR, UNIVERSITY OF PRISHTINA
MICHAEL L. KHEIFETZ	VICE-RECTOR, POLOTSK STATE UNIVERSITY
DO-YEON KIM	PRESIDENT, UNIVERSITY OF ULSAN
MARKKU KIVIKOSKI	RECTOR, TAMPERE UNIVERSITY OF TECHNOLOGY
GORDANA KRALIK	RECTOR, STROSSMAYER UNIVERSITY OF OSIJEK
VLADO MAJSTOROVIC	RECTOR, UNIVERSITY OF MOSTAR
SALVATORE MARZANO	RECTOR, POLYTECHNIC OF BARI
RADU MUNTEANU	RECTOR, TECHNICAL UNIVERSITY OF CLUJ-NAPOCA
HASAN MURATOVIĆ	RECTOR, UNIVERSITY OF SARAJEVO
CONSTANTIN OPREAN	RECTOR, LUCIAN BLAGA UNIVERSITY OF SIBIU
IVAN PAVIC	RECTOR, UNIVERSITY OF SPLIT
KAREL RAIS	RECTOR, BRNO UNIVERSITY OF TECHNOLOGY
ÁKOS RÉDEY	RECTOR, UNIVERSITY OF PANNONIA
NICOLAE ROBU	RECTOR, THE "POLITEHNICA" UNIVERSITY OF TIMISOARA
PAVEL V. ROSLYAKOV	VICE-RECTOR, MOSCOW POWER ENGINEERING INSTITUTE - TECHNICAL UNIVERSITY
IVAN ROZMAN	RECTOR, UNIVERSITY OF MARIBOR
DANIEL RUKAVINA	RECTOR, UNIVERSITY OF RIJEKA
PETR SAHA	RECTOR, TOMAS BATA UNIVERSITY UNIVERSITY ZLIN
TAKESHI SASAKI	PRESIDENT, UNIVERSITY OF TOKYO
SUPACHAI SAMAPPITO	PRESIDENT, MAHASARAKHAM UNIVERSITY
FRITZ SCHMÖLLEBECK	RECTOR, UNIV. OF APPLIED SCIENCES TECHNIKUM, VIENNA
SERGEY V. SEREBRYANNIKOV	RECTOR, MOSCOW POWER ENGINEERING INSTITUTE - TECHNICAL UNIVERSITY
VIKTOR V.SHALAI	RECTOR, OMSK STATE TECHNICAL UNIVERSITY
PETER SKALICKY	RECTOR, VIENNA UNIVERSITY OF TECHNOLOGY
SERGEY M. STAZHKOV	VICE-RECTOR, BALTIC STATE TECHNICAL UNIVERSITY
WILFRIED STOLL	PRESIDENT, CHAIRMAN OF SUPERVISORY BOARD - FESTO
NAM PJO SUH	PRESIDENT, KOREA ADVANCED INSTITUTE FOR SCIENCE AND TECHNOLOGY
PEEP SÜRJE	RECTOR, TALLINN TECHNICAL UNIVERSITY
WIESLAW TRAMPCZYNSKI	RECTOR, TECHNICAL UNIVERSITY OF KIELCE
DZEMO TUFEKCIC	RECTOR, UNIVERSITY OF TUZLA
ANTE UGLESIC	RECTOR, UNIVERSITY OF ZADAR
ION VISA	RECTOR, TRANSILVANIA UNIVERSITY OF BRASOV
BORIS A. YAKIMOVICH	RECTOR, IZHEVSK STATE TECHNICAL UNIVERSITY, IZHEVSK
WOJCIECH ZIELIŃSKI	RECTOR, SILESIAN UNIVERSITY OF TECHNOLOGY, GLIWICE

Annals of DAAAM for 2008 & Proceedings of the 19th International DAAAM Symposium, ISSN 1726-9679
Editor B. Katalinic, Published by DAAAM International, Vienna, Austria 2008
Make Harmony between Nature and Technology and your Mind will fly free as a Bird

DEANS' AND ACADEMIC DIRECTORS´ HONOR COMMITTEE FOR 2008

REVIEWING COMMITTEE FOR 2008

ORGANIZING COMMITTEE FOR 2008

Annals of DAAAM for 2008 & Proceedings of the 19th International DAAAM Symposium, ISSN 1726-9679
Editor B. Katalinic, Published by DAAAM International, Vienna, Austria 2008
Make Harmony between Nature and Technology and your Mind will fly free as a Bird

Scientific DAAAM International Committee for 2008

Sponsors & Supporters

City of Trnava / FESTO / Self-governmental Region of Trnava

BEKAERT, a.s. Hlohovec
Denona d.d.
DRILL, s.r.o. Bratislava
HSTEC d.d., Zadar
Ing. Mária Čunderlíková - KESA –Trnava
IBOK
INVEX, s.r.o. Trnava
JAVYS, a.s. Jaslovské Bohunice
KHS AG - Filling and Packing - Worldwide
NADÁCIA ARX - Jaslovské Bohunice
NADÁCIA JÁNA KORCA Trnava
PRVÁ ZVÁRAČSKÁ, a.s. Bratislava
SLOVENSKÁ ELEKTRIZAČNÁ PRENOSOVÁ SÚSTAVA, a. s. Bratislava
SLOVENSKÉ ELEKTRÁRNE, a.s., člen skupiny ENEL Bratislava
SLOVNAFT, a.s. Bratislava
TECHNODAT, CAE - systémy, s.r.o. Trenčín
VÚJE, a.s. Trnava
VUZ PI SR
ZLIEVAREN TRNAVA
ŽOS, a.s. Trnava
ZVS HOLDING, a.s. Dubnica nad Váhom

Annals of DAAAM for 2008 & Proceedings of the 19th International DAAAM Symposium, ISSN 1726-9679
Editor B. Katalinic, Published by DAAAM International, Vienna, Austria 2008
Make Harmony between Nature and Technology and your Mind will fly free as a Bird

INTERNATIONAL PROGRAM COMMITTEE FOR 2008

Annals of DAAAM for 2008 & Proceedings of the 19th International DAAAM Symposium, ISSN 1726-9679
Editor B. Katalinic, Published by DAAAM International, Vienna, Austria 2008
Make Harmony between Nature and Technology and your Mind will fly free as a Bird

YOUNG RESEARCHERS´ AND SCIENTISTS´ INTERNATIONAL COMMITTEE FOR 2008

The main aim of this International Committee is international promotion of talented young researchers and scientists. Target group are: young engineers, master students, young master, doctoral students, young doctors, talented graduate students and other tolerant and talented young persons from one of the fields of interests of DAAAM International. The young winners of Best Paper Award of DAAAM International are also the member of this committee. The candidate has to be under age 35 years and has to be author or co-author of paper published in one of DAAAM publications. The candidate has to be nominated by the supervisor/professor/team leader. The nomination (short CV, research: status, field, course, targets, results) has to be send to the President of DAAAM International.

Annals of DAAAM for 2008 & Proceedings of the 19th International DAAAM Symposium, ISSN 1726-9679
Editor B. Katalinic, Published by DAAAM International, Vienna, Austria 2008
Make Harmony between Nature and Technology and your Mind will fly free as a Bird

DAAAM International Industry, Business and Administration Committee for 2008

Agarwal, Beni Prasad
Akhonin, Andrey A.
Andreic, Andrej
Arsin, M.
Bader, Udo
Beer, Herbert
Beer, Michael
Birgersson, M.
Blaschko, Christian
Brajinovic, Marijan
Bruck, Peter
Buljan, Jakov
Burgjelez, Ivana
Carder, M.W.
Carlbom, Terri
Celar, Stipe
Cosic, Pejo
Coskun, Deniz
Crafer, Roger C.
Crnjac, Miljenko Mico
Culjak, Vlado
Dadic, Mihovil Braco
Darko, Csaba
Denona, Marko
Dicak, Josip
Dikic, Masha
Divic, Antun
Foltin, A.
Freiberger, Dietrich
Gloser, Manfred
Goja, Zeljko
Grabenweger, Johann
Haas, B.
Haas, Peter
Hanslik-Kupsa, R.
Haslauer, H.
Hebemägi, A.
Holjevac, Nikola
Ivanisevic, Darko
Jukic, Ivan
Kalinyaprak, Volkan
Katalinic, Ana
Katalinic, Mihovil
Keiner, Wolfgang
Knezevic, Mirjana
Kohlmaier, Martin
Kremnjak, Zvonko
Krezic, Davor
Kunstmann, Ruediger
Lahoda, Thomas
Lengauer, Markus
Lerch, Zoran
Levstek, Anton
Limbeck, Matthias
Ljuhar, Davul
Loktev, Alexander A.
Lörincz, S.
Magyar, Gabor
Marasovic, Tomislav
Marincic, Valentina
Martak, Lothar V.
Mayer, A.
Mazuran, Zeljko
Milovac, Ante
Moisseev, Iouri
Moll, Wolfgang
Morinaga, H.
Mortin, M.
Moskovenko, Igor B.
Muresan, Vasile
Murinde, Victor
Nader, Hans
Nakamoto, Nabuyuki
Nanasi, Jozef
Neuper, Peter
Niwinski, Janusz
Novels, Mike
Pfliegel, Reinhard
Pichugin, Vladimir
Pieslinger, Kurt
Poettinger, Klaus
Rametsteiner, Werner
Raming, Otto
Redtenbacher, Guenter
Reichel, Peter
Richler, Hannes
Santek, Mladen
Sobolewski, Andrzej
Szava, Tibor
Tsydenov, Dorzhi T.
Widtmann, Georg
Worener, K.
Yoshida, H.
Yoshida, Susumu

The Members of Permanent Committee of the University Network of DAAAM International for 2008 are:

Member organisation	***Delegate***
DAAAM International Vienna	*President Univ. Prof. Dipl.-Ing. Dr. techn. Dr.mult. h.c. Branko Katalinic*
Danube Rectors´ Conference	*Professor Dr. sc. Franci Cus*
Budapest University of Technology and Economics, Budapest	*The members of Permanent Committee have to be nominated by Budapest University of Technology and Economics*
Josip Juraj Strossmayer University of Osijek	*Professor Dr. sc. Dragomir Krumes*
Kobe University, Kobe	*Director, Chair and Professor Kanji Ueda*
Moscow State University of Technology "Stankin", Moscow	*Professor Dr. Sci. (tech.) Member of Russian Academy of Science Juri M. Solomentsev / Univ. Prof. Dipl.-Ing. Dr. Sergey Egorov*
Romanian-American University of Bucharest, Bucharest	*Univ.Sen.Lect. Dipl.-Ing. Dr. techn. George Carutasu Univ.Sen.Lect. Dipl.-Math. Dr. techn. Cezar Botezatu*
Slovak University of Technology Bratislava	*Professor Dr. Marian Vesely Professor Ladislav Javorcik*
Tallinn Technical University	*Professor Rein Kyttner*
Technical University of Brno, Brno	*Senator Prof. Ing. Jaroslav Balate, DrSc. Doc. Ing. Vladimir Vasek, CSc.*
Technical University of Cluj-Napoca, Cluj-Napoca	*Dean Prof. Dr.Eng. Petru Berce, Prof.dr.Eng. Csaba Gyenge*
Technical University of Kosice, Kosice,	*Prof. Ing. Karol Florian, DrSc., / Prof. Dr. Emil Wessely*
University Politehnica of Bucharest, Bucharest	*Professor Dr. Eng. Horia Gheorghiu Lecturer Dr. Eng. Costel Emil Cotet*
University of Maribor, Maribor	*Professor Dr. sc. Joze Balic*
University of Miskolc, Miskolc	*Professor Dr Gabor Terstyanszky*
University of Mostar, Mostar, B & H	*Rector Prof. Dr. Vlado Majstorovic, / Prof. Dr. sc. Vojo Visekruna*
University of Prishtina, Prishtina, Kosovo	*Vice-Rector Prof. Dr. Ahmet Geca / Prof.-Dr. Bajrush Bytyci*
University of Rijeka, Rijeka	*Professor Josip Brnic, Ph.D. / Professor Zoran Mrsa, Ph.D.*
University of Split, Split	*Rector Professor Dr. Ivan Pavic*
University Ovidius of Constanta, Constanta	*Professor Dr. Eng. Mihael Chircor*
University of Zagreb, Zagreb	*Rector Professor Dr. Aleksa Bjelis*

Annals of DAAAM for 2008 & Proceedings of the 19th International DAAAM Symposium, ISSN 1726-9679
Editor B. Katalinic, Published by DAAAM International, Vienna, Austria 2008
Make Harmony between Nature and Technology and your Mind will fly free as a Bird

DAAAM INTERNATIONAL NETWORK:
BACKGROUND, HISTORY, AIMS, STRUCTURE, ACTIVITIES & FUTURE

Branko KATALINIC
President of DAAAM International
www.daaam.com / www.daaam.org

0.0 Contents:

1.0 BACKGROUND

The DAAAM International was founded and based on the idea and initiative of Professor B. Katalinic, on November 5th, 1990 at the Vienna Symposium, which was organised in order to mark the 175th anniversary of the Vienna University of Technology.
The purpose of DAAAM International is to provide a world forum for engineers, scientists and industrial managers to present and discuss the current status and impact of advanced manufacturing and automation.
During the last seventeen years, the activities of DAAAM International have been supported by scientists and experts from over fifty countries all over the world. In the course of the last five years, DAAAM International has focused activities on the regional and global long-term co-operation between (technical) universities. The result of this initiative is the founding of the University Network of DAAAM International Vienna and DAAAM International Network for Advanced Technologies which was founded on 2002-10-25 in Vienna.

2.0 AIMS

The main aims of DAAAM are:

- To organise an annual DAAAM International Symposium each year.
- To found and organise the University Network of DAAAM International Vienna as a strategic alliance and permanent international open platform for long-term co-operation and partnership in all fields of technology, research and education.
- To organise long-term projects, initially known as Inter-university Doctoral Studies
- To organise a pilot project, initially known as the Virtual University.
- To enable and promote the exchange of people, scientific, and technical information.
- To enable and promote international co-operation among scientists and engineers.
- To enable and promote R & D, education and industrial application.
- To co-operate with other national and international organisations in matters of common interest.

3.0 ORGANISATION OF DAAAM INTERNATIONAL VIENNA

DAAAM International Vienna

President:
Branko **Katalinic**

Vice presidents for 2008	Joze **Balic** / Andrew **Kusiak** / Oliver **Moravcik** / Toma **Udiljak** / Kanji **Ueda** / Koloman **Ulrich** / Karol **Velisek**
Honour Vice presidents for 2008	Jaroslav **Balate** / Klaus **Meissner** / P. C. **Pandey** / Janez **Peklenik** / Emil **Wessely**
International Senior Advicer for 2008	Franci **Cus** / Vladimir **Filaretov**/ Malik **Kulenovic**
Secretary General for 2008	Paulina **Cesarec**
Team 2008	Branko **Katalinic** / Paulina **Cesarec** / Liane **Höller** / Roman **Kettler** / Markus **Stopper**

Presidents and Vice Presidents* of National DAAAM Organisations for 2008

Jaroslav Balate	*Czech Republic*
Arne Bilberg	*Denmark*
Jens Bliedtner	*Germany*
Iosif Breido	*Kazakhstan*
Glen Bright	*Republic of South Africa*
Costel Emil Cotet*	*Romania*
Fernando M. M. Cruz	*Portugal*
Mahmud Dikic	*Canada*
Jianhua Dong	*P.R. China*
Igor Drstvensek	*Slovenia*
Joost Duflou	*Belgium*
Vladimir I. Dynnikov*	*Brazil*
Sergey Egorov	*Russia*
Jan Goossenaerts	*Holland*
Johann Grabenweger	*Brazil*
Csaba Gyenge	*Romania*
Qing-Yao Han	*P.R. China*
Peter Holmstedt	*Sweden*
Velko Iltschev	*Bulgaria*
Mohammad A. Irfan	*Pakistan*
Ashutosh Khanna*	*India*
Yurij Kondratenko	*Ukraine*
Petri Kuosmanen	*Finland*
Andrew Kusiak	*USA*
Rein Kyttner	*Estonia*
Marcin A.Krolikowski	*Poland*
Luis Norberto Lopez de Lacalle	*Spain*
Vlado Majstorovic	*Bosnia and Herzegovina*
Ali Meghdari	*Iran*
Karim Mustafizul	*Malaysia*
P. C. Pandey	*India*
Hong-Seok Park	*Korea*
Gilberto Hererra Ruiz	*Mexico*
Kemajl Stuja	*Kosovo*
Dmitry Svirsky	*Belarus*
Kok Kiong Tan	*Singapore*
Mario T. Tabucanon	*Thailand*
Victor Timoschenko	*Moldavia*
Kanji Ueda	*Japan*
Toma Udiljak	*Croatia*
Erol Uyarr	*Turkey*
Ernesto Lo Valvo	*Italy*
Karol Velisek*	*Slovakia*
George Christopher Vosniakos	*Greece*
Emil Wessely	*Slovakia*

DAAAM International Coordinators for 2008

Miroslav **Badida**	*Technical Environmentalistics*
Ivan **Baranek**	*Production Technologies*
Joze **Balic**	*Intelligent Manufacturing*
Ivo **Cala**	*Maintenance*
Dragan **Covic**	*Management*
Costel Emil **Cotet**	*Discrete material flow management*
Franci **Cus**	*High Education Quality Management*
Mahmud **Dikic**	*Assembly Automation*
Zvonko **Dragcevic**	*Textile and Garment*
Alexey **Eliseev**	*Internet Lab Sharing and Networking*
Sergey **Egorov**	*Securities & Financial Management*
Vladimir F. **Filaretov**	*Teleoperators*
Johann **Grabenweger**	*Material Sciences*
Manfred **Gronalt**	*Logistics*
Branko **Katalinic**	*Intelligent Manufacturing Systems*
Viktorio **Malisa**	*Mechatronics / Robotics*
Klaus **Meissner**	*Precision Engineering*
Zoran **Mrsa**	*Optimal Shape Design*
Jozef **Nanasi**	*Automotive Industry*
Markus **Stopper**	*Production Information Technology*
*Ante **Uglesic***	*History of Artefacts*
Kanji **Ueda**	*Biological Manufacturing Systems*
Ante **Uglesic**	*History of Artefacts*
Koloman **Ulrich**	*Production Technologies*
Heinz **Weigl**	*Simulation*

NB: The further International Coordinators for different topics/fields are needed and welcome, if you want to be one of them please contact the president of DAAAM International

4.0 DAAAM Symposiums Clubs

Since founding of DAAAM International Symposium in 1990 **4808** authors and co-authors published at least once their scientific papers in the Annals & Proceedings of DAAAM International Symposium. The members of the club are the persons which presented and/or published their papers at all / not les than 15 / not les than 10 / not less than 5 DAAAM International Symposiums:

All DAAAM Symposiums Authors´ Club

Balic, Joze / Cus, Franci / Katalinic, Branko / Polajnar, Andrej / Veza, Ivica / Wessely, Emil

Members of 15+ DAAAM Symposiums Authors´ Club

Balate, Jaroslav / Boyadjiev, Iliya Kirilov / Buchmeister, Borut / Budic, Ivan / Gyenge, Csaba / Novakovic, Branko / Udiljak, Toma / Velisek, Karol

2009 Candidates for 15+ DAAAM Symposiums Authors´ Club

Following DAAAM authors are candidates and have a chance to join the 15+ Club in 2009
Alpek, Ferenc / Kulenovic, Malik / Micieta, Branislav / Vasek, Vladimir

Members of 10+ DAAAM Symposiums Authors´ Club

Adamczak, Stanislaw / Agic, Darko / Badida, Miroslav / Bobal, Vladimir / Brezocnik, Miran / Buble, Marin / Capek, Jan / Cohodar, Maida / Cyklis, Jerzy / Dragcevic, Zvonko / Drstvensek, Igor / Dulcic, Zelimir / Gojo, Miroslav / Grizelj, Branko / Hodolic, Janko / Janac, Alexander / Janacova, Dagmar / Jerbic, Bojan / Kheifetz, Michael / Kljajin, Milan / Knego, Nikola / Kolomaznik, Karel / Kostal, Peter / Koukhta, Sergei Vasilievich / Krolikowski, Marcin Andrzej / Krumes, Dragomir / Kubalcik, Marek / Kundrak, Janos / Kunica, Zoran / Kuric, Ivan / Leber, Marjan / Mahovic Poljacek, Sanja / Majetic, Dubravko / Majstorovic, Vlado / Malakov, Ivo / Math, Miljenko / Matysek, Miroslav / Meissner, Klaus / Milfelner, Matjaz / Mrsa, Zoran / Mursec, Bogomir / Nanasi, Jozef / Neumann, Petr / Nikolic, Gojko / Pahole, Ivo / Papstel, Jyri / Park, Hong Seok / Pavic, Ivan / Rogale, Dubravko / Siemiatkowski, Mieczyslaw / Stopper, Markus / Szalay, Tibor / Visekruna, Vojo / Voloder, Avdo / Vranjes, Bozo / Weiss, Zenobia / Zajac, Jerzy

2009 Candidates for 10+ DAAAM Symposiums Authors´ Club

Following DAAAM authors are candidates and have a chance to join the 10+ Club in 2009
Acko, Bojan / Ancau, Mircea / Anna, Vojtech / Brnic, Josip / Canadija, Marko / Cosic, Predrag / Cotet, Costel Emil / Dukovski, Vladimir / Karaulova, Tatyana / Kittl, Burkhard / Kralikova, Ruzena / Markos, Sandor / Mikac, Tonci / Palcic, Iztok / Pandilov, Zoran / Peterka, Jozef / Popescu, Diana / Poppeova, Viera / Seghedin, Neculai / Sysala, Tomas / Ueda, Kanji / Zorc, Davor / Zuperl, Uros

Members of 5+ DAAAM Symposiums Authors´ Club

Abhary, Kazem / Abram Zver, Marta / Abrudan, Ioan / Adamek, Milan / Andrlik, Vladimir / Anisic, Zoran / Apostol, Bogdan / Aracic, Stjepan / Arz, Gusztav / Aurite, Traian / Babic, Darko / Bachvarov, Angel / Badjgholi, Farsad / Balc, Nicolae / Barisic, Branimir / Baumann, Christian / Bekes, Jan / Belansky, Jan / Belohoubek, Pavel / Belous, Vitalie / Berce, Petru / Bibu, Marius / Bijelonja, Izet / Bleicher, Fritz / Bliedtner, Jens / Bohez, Erik L J / Bolanca, Stanislav / Bolanca, I / Bolanca, Zdenka / Bolcu, Dumitru / Brezak, Danko / Budic, Hrvoje / Bunjaku, Avdyl / Cala, Ivo / Car, Zlatan / Carija, Zoran / Carutasu, Nicoleta Luminita / Carutasu, George / Casalino, Giuseppe / Catrina, Dumitru / Cekic, Ahmet / Celar, Stipe / Chramcov, Bronislav / Cieloszyk, Janusz / Ciglar, Damir / Ciucescu, Doru / Constantin, George / Cosic, Ilija / Costea, Aurel / Covic, Dragan / Cristea, Luciana / Crnekovic, Mladen / Croitoru, Christian / Cselenyi, Jozsef / Cubonova, Nadezda / Dabacan, Mircea / De Filippis, L A C / Deacu, Liviu / Dimitrovski, Krste / Dolga, Lia / Dolga, Valer / Dostal, Petr / Dostalek, Petr / Dudas, Illes / Dumitras, Catalin Gabriel / Durakbasa, M Numan / Dvorak, Jiri / Dvorakova, Jana / Egorov, Sergey / Emert, Rudolf / Erdeljan, Aleksandar / Farcas, Cristian / Feld, Mieczyslaw / Ficko, Mirko / Filaretov, Vladimir F / First Rogale, Snjezana / Fratila, Domnita / Gatalo, Ratko / Gavlas, Josef / Gawlik, Jozef / Gecevska, Valentina / Ghionea, Adrian / Grabenweger, Johannes / Grgona, Jadranko / Grman, Jan / Gronalt, Manfred / Gudac, Goran / Guenther, Ulrich / Guran, Marius / Gyani, Karoly / Hajduk, Mikulas / Halbritter, Erno / Hernandez, Oscar / Hesselbach, Juergen / Hlava, Jaroslav / Hlavenka, Bohumil / Hofreiter, Milan / Honczarenko, Jerzy / Hradocky, Ladislav / Hrman, Miljenko / Hrubina, Kamil / Hruska, Frantisek / Hruskova, Erika / Ivakovic, Cedomir / Ivancic, Ivan / Jablonski, Wojciech / Jaehn, Hendrik / Janecki, Darius / Jardzioch, Andrzej / Jurecic, Denis / Jurum Kipke, Jasna / Kachanak, Anton / Kasac, Josip / Kerekes, Laszlo / Knezevic, Petar / Knosala, Ryczard / Kolakovic, Marko / Kopac, Janez / Krys, Nikolai / Krzyzanowski, Jozef / Kubinger, Wilfried / Kurtovic, Mate / Lalic, Bojan / Lange, Dirk / Lazea, Gheorghe / Lazibat, Tonci / Lee, Gyu Bong / Lendak, Imre / Lerch, Zoran / Lipa, Zdenko / Ljubetic, Juraj / Lopez De Lacalle, Luis Norberto / Ludovico, Antonio Domenico / Luong, Lee H S / Maglic, Leon / Majdandzic, Niko / Majernik, Milan / Malopolski, Waldemar / Mandic, Lidija / Marchelek, Krzysztof / Marusic, Vlatko / Matjasic Fris, Mateja / Matusu, Radek / Mehta, Narinder Kumar / Meszaros, Imre / Mihoci, Kristijan / Monostori, Laszlo / Morar, Liviu / Moraru, Sorin Aurel / Moskovenko, Igor B / Motika, Roman / Mrocheck, J / Muransky, Juraj / Nagy, Zoltan / Navratil, Zdenek / Navratil, Pavel / Nedelcu, Dumitru / Nekrassov, Grigori / Novoselovic, Daniel / Nowicki, Bogdan / Obsieger, Boris / Olaru, Adrian / Olteanu, Ciprian / Olteanu, Felicia / Oluic, Cedomir / Opruta, Dan / Oravec, Ladislav / Osanna, Peter Herbert / Osmani, Hysni / Otto, Tauno / Pajor, Miroslaw / Pandza, Krsto / Parpala, Lidia Florentina / Pasic, Mugdim / Patrascu, Gabriela / Pavletic, Dusko / Pavlic, Mile / Perinic, Mladen / Perutka, Karel / Peshkoun, Sergei Vasilievich / Petkovic, Darko / Petreus, Dorin / Pierzchala, Wieslaw / Pinteric, Andreja / Pisla, Arian / Pitica, Dan / Pivonka, Petr / Pokorny, Pavel / Poljak, Krunoslav / Pomenic, Loreta / Popa, Marcel Sabin / Popa, Cicerone Laurentiu / Popescu, Sorin G / Powalka, Bartosz / Prodan, Dan / Prokop, Roman / Prostrednik, Daniel / Pruteanu, Octavian Vasile / Przybylski, Wlodzimierz / Rachieru, Nicoleta / Rados, Jerko / Radu, Mihaela / Ravas, Rudolf / Repanovici, Angela / Rezic, Snjezana / Rizescu, Sabin / Robotin, Radu / Rogic, Kristian / Ros, Olimpia / Rovan, Mario / Rugescu, Radu Dan / Ryniewicz, Andrzej / Safran, Mario / Saks, Alo / Samardzic, Ivan / Samek, David / Sanchez, J A / Saric, Tomislav / Sebastijanovic, Slavko / Seda, Milos / Seminsky, Jaroslav / Seucek, Ivan / Sihn, Wilfried / Simunic, Juraj / Sinay, Juraj / Sisek, Boris / Skenderi, Zenun / Skoczynski, Waclaw / Slota, Adam / Sluga, Alojzij / Sokansky, Karel / Somodji, Zeljko / Sopta, Luka / Sostar, Adolf / Srdjak, Miroslav / Staicu, Stefan / Stan, Gheorghe / Stefanic, Nedeljko / Stoic, Antun / Stuja, Kemajl / Sugar, Peter / Syrova, Livia / Sysel, M / Szabo, Otto / Szadkowski, Jan / Szelig, Karoly / Tabucanon, Mario T / Teich, Tobias / Tkalic, Mladen / Tomas, Andrija / Tudjarov, Boris / Turkalj, Goran / Ujevic, Darko / Uricek, Juraj / Varga, Gyula / Vavrik, Ivan / Viducic, Ljiljana / Vincze, Markus / Virkus, Uwe / Vitecek, Antonin / Vitez, Ivan / Vrljicak, Zlatko / Wantuch, Edward / Weigl, Karl Heinz / Zamfira, C Sorin / Zeljkovic, Milan / Zhirabok, Alexey / Ziberna Sujica, Milena / Zichil, Valentin / Zivic Djurovic, Marijana / Zmijewski, Wladyslav / Zvirinsky, Vladimir

2009 Candidates for 5+ DAAAM Symposiums Authors´ Club

Following DAAAM authors are candidates and have a chance to join the 5+ Club in 2009
Adamczyk, Zbigniew / Axinte, Eugen / Bajic, Drazen / Baksa, Sarajko / Bana, Viktoria / Banabic, Dorel / Banyai, T / Benes, Petr / Benic, Djuro / Bliznak, Michal / Borza, Alexandru Ioan / Brajlih, Tomaz / Brandes, Arndt / Budin, Rajka / Buerger, Wolfgang / Bulgaru, Marius / Bytyci, Bajrush / Carp Ciocardia, Daniela Craita / Chodzko, Marcin / Comsa, Dan Sorin / Corluka, Venco / Cuesta, Eduardo Gonzales / Cukor, Goran / Curtu, Ioan / Curtu, Djurdjica / Dlapa, Marek / Dobras, Branka / Dolecek, Vlatko / Dolinay, Jan / Dragoi, George / Duerr, Holger / Dumitriu, Adrian / Dvorackova, Marie / Dzaferovic, Ejub / Eftimie, Lucian / Egorova, Tatyana / Ekinovic, Sabahudin / Evin, Emil / Gabrijelcic, Helena / Gallo, Mikulas / Gallova, Stefania / Gersak, Jelka / Ghionea, Ionut / Giannoccaro, Rosanna / Gil, Stanislaw / Gotovac,

Sven / Greguss, Pal / Grizelj, Dejan / Grubisic, Dragana / Grzesiak, Dariusz / Hadar, Anton / Haertig, Guenter / Hait, Alexander L / Hajrizi, Edmond / Hebbal, Shivanand / Hrzic, Dragana / Iltschev, Velko / Jadlovska, Anna / Jain, Pramod K / Janzekovic, Marjan / Jermak, Czeslaw Janusz / Johanyak, Zsolt Csaba / Jovic, Franjo / Jurda, Pavel / Kavran, Zvonko / Keran, Zdenka / Kirchfeld, Maria / Kiss, Akos / Klapka, Jindrich / Kodacsy, Janos / Kolibal, Zdenek / Kosmol, Jan / Kovacevic, Branko / Kovacevic, Drazen / Kovacic, Miha / Kowalski, Marek / Kozak, Drazan / Krajci, Vesna / Kuenanz, Klaus / Kuzle, Igor / Kuzman, Karl / Lache, Simona / Lamikiz, Aitzol / Ljubic, Franjo / Ljuhar, Davul / Macku, Lubomir / Manriquez, Jose / Marek, Bartosz / Marincic, Valentina / Markic, Brano / Marunic, Gordana / Matusova, Miriam / Mera, Mircea / Mihelic Bogdanic, Alka / Mlinaric, Tomislav J / Moeller, Klaus / Mohora, Cristina / Mrcela, Tomislav / Mudronja, Vedran / Mudura, Pavel / Nagla, Kuldeep Singh / Nardin, Blaz / Neage, Irina / Nedeff, Valentin / Nemes, Toderita / Neslusan, Miroslav / Oberlaender, Rudolf / Osmera, Pavel / Palaghita, Niculaie / Paralic, Jan / Parpala, Radu Constantin / Pavlickova, Jirina / Pawlukowicz, Piotr / Pechacek, Frantisek / Perez, Loseba M / Peric, Nedjeljko / Petrescu, Valentin / Piculin, Srecko / Pidany, Jan / Pintilie, Gheorghe / Pisla, Doina / Pizmoht, Petja / Plancak, Miroslav / Ploj, Tone / Popescu, Ion / Radic, Jure / Rajkovic, Borivoje / Rubio, Eva M / Rucki, Miroslaw / Rusu, Lazar / Sabol, Tomas / Salihu, Avdi / Salopek, Ivana / Samanovic, Josip / Sariguil, Erdal / Sebastian, Miguel Angel / Sebastijanovic, Nebojsa / Sebo, Dusan / Senchilo, Igor Arkadievich / Severincu, Miha / Siminiati, Dubravka / Simionescu, Gheorghe / Situm, Zeljko / Skoko, Miroslav / Skolud, Bozena / Skunca, Marko / Sladek, Jerzy / Sliwka, Janusz / Smutny, Lubomir / Sogorovic, Danijel / Sokhan, Serg / Solc, Frantisek / Solenicki, Goran / Starcevic, Gordan / Stefanek, Michal / Stefanuta, Ioan / Stjepanovic, Zoran / Strgar Kurecic, Maja / Sunje, Aziz / Surace, Rossella / Sutlovic, Igor / Svirsky, Dmitry / Tache, Florin / Takakuwa, Soemon / Tarca, Ioan / Tarca, Radu Catalin / Tierean, Mircea / Tomic, Drazana / Topliceanu, Liliana / Toth, Andras / Uhlik, Alexander / Ungur, Petru / Ungureanu, Catalin Sorin / Vaclavek, Pavel / Vaida, Liviu / Valter, Zdravko / Vaupotic, Bostjan / Viteckova, Miluse / Vrdoljak Raguz, Ivona / Vujasinovic, Edita / Vujica Herzog, Natasa / Vukmirovic, Srdan / Vukovic, Ivan / Weiland, Kristian / Weinert, Klaus / Wong, Hung Kun / Zebala, Wojciech / Zizka, Jan / Zjakic, Igor / Zurek, Witold

Authors of DAAAM 2008 - NewComer Club

Abrudan, Adrian / Abrudan, Gheorghe / Adamcikova, Andrea / Adamescu, Dorina / Agrianidis, Anastasios / Aisman, David / Alecu, Aurel / Alexa, Florin / Alexandrescu, Marius / Alexei, Adrian / Alionte, Cristian Gabriel / Almeida, Henrique / Alutei, Adrian / Alvarez, Roberto / Amariei, Olga Ioana / Amberg, Michael / Anania, Florea Dorel / Andersson, Paul / Andrei, Gabriel / Andrejcak, Imrich / Andres, Solomia / Anghel, Alina / Anghel, Cornelia / Anthymidis, Konstantinos / Antonescu, Ion / Ardelean, Felician / Argesanu, Veronica / Arion, Aurel / Aryassov, Gennady / Atanasiu, Costica / Atanasiu, Virgil / Avram, Calin / Avram, Eleonora Laura / Avram, Laura / Ayyaz, Ahmad / Azzouzi, Messaouda / Baba, Marius Nicolae / Badarau, Rodica / Badea, Adrian / Badea, Cristinel / Badescu, Loredana Anne / Baeckstroem, Mikael / Bagu, Constantin / Bajcicak, Martin / Bajramovic, Rasim / Bal, Carmen / Bal, Nicolaie / Balan, Corneliu / Balasoiu, Victor / Balcak, Stanislav / Balcu, Ion / Bancila, Daniel / Banciu, Felicia / Bandrabur, Diana Lilia / Banea, Mariana / Banerjee, Soumya / Banica, Mihai / Bara, Camelia / Bara, Lucian / Bara, Vasile / Barborak, Oto / Barbu, Teodora / Barbu, Valentin / Barglazan, Diana / Barglazan, Mircea / Barta, Jozef / Bartolo, Paulo / Bassoli, Elena / Baya, Alexandru / Beca, Paul / Bejan, Mihai / Beles, Horia / Bena, Irina / Benak, Michal / Bertrand, Philippe / Beznak, Matej / Beznila, Hortensia / Biclesanu, Cornelia / Biedunkiewicz, Witold / Bienert, Jan / Bilic, Bozenko / Bilik, Jozef / Birau, Nicolae / Birdeanu, Elena Daniela / Biriescu, Marius / Biris, Cristina / Birsan, Dan / Birsan, Iulian Gabriel / Bisu, Claudiu / Bizau, Viorel / Bizdoaca, Nicu / Bjoerkstrand, Roy / Blaga, Florin / Blaga, Vasile / Blidisel, Rodica / Bode, Florin / Bogoi, Alina / Bogovic, Tomislav / Boiangiu, Costin Anton / Boiangiu, Mihail / Boicea, Alexandru / Boieriu, Camelia / Bolanca Mirkovic, Ivana / Bolfa, Traian Eugen / Borangiu, Theodor / Borawska, Patrycja / Bordea, Nicolae / Borza, Ioana / Borza, Paul Nicolae / Bota, Josip / Botezatu, Elena / Botezatu, Mihai Alexandru / Boyadjiev, Ilia / Bozek, Pavol / Brabie, Gheorghe / Bratu, Anca / Braun, Barbu / Bretcu, Angela / Bria, Vasile / Brindasu, Paul Dan / Brodnianska, Zuzana / Brotac, Sorin / Bucur, Gabriela / Budimir, Verica / Budzik, Grzegorz / Buidos, Traian / Bulej, Vladimir / Bunjaku, Januz / Buracu, Victor / Buransky, Ivan / Burca, Ioan / Burger, Wolfgang / Butnariu, Silviu / Buzera, Marius / Caganova, Dagmar / Cahuc, Olivier / Calbureanu, Madalina / Calefariu, Gavrila / Camacho, Ana / Cambal, Milos / Cananau, Dan Cristian / Cananau, Sorin / Capko, Darko / Caplovic, Lubomir / Carausu, Constantin / Carbunar, Mihai / Carcea, Ioan / Carp, Iulia / Carutasu, Nicoleta / Catana, Dorina / Catana, Madalin / Catana, Razvan / Catas, Adriana / Cazacu, Maria / Cehajic, Arijana / Cernecky, Jozef / Cervenansky, Jaroslav / Cha, Sungdeok / Charbulova, Marcela / Charvatova, Hana / Chaus, Alexander / Chelariu, Romeu / Chelaru, Elena Silvia / Chelu, Gheorghe / Cherciu, Mirela / Cherlea, Ioan Valeriu / Chicos, Lucia Antoneta / Chioncel, Cristian Paul / Chioreanu, Nicolae / Chira, Flavia / Chircor, Lidia / Chiriacescu, Sergiu / Chirilus, Alina Alexandra / Chirita, Bogdan / Chiru, Anghel / Chita, Monica Anca / Choi, Hung Won / Chwajol, Grzegorz / Cibulka, Viliam / Cican, Grigore / Cioata, Vasile George / Cioban, Horia / Ciobanu, Bogdan / Ciobanu, Camelia / Ciobanu, Gheorghe / Ciobanu, Ioan / Ciobanu, Mircea / Ciocirlan, Doinita / Ciofu, Ciprian / Cionca, Dan / Ciotea, Margareta / Circiumaru, Adrian / Ciuca, Ion / Ciuca, Sorin / Ciupitu, Ion / Cociu, Nicolae / Cocorada, Sorin / Coculescu, Cristian Catalin / Coculescu, Cristina / Cohal, Viorel / Colic, Mirsad / Coman, Daniela / Coman, Liviu / Comanescu, Adriana / Comanescu, Brindus / Comanescu, Dinu / Constantin, Emil / Cooper, Edward / Copilus, Cristian / Cornescu, Viorel / Cosic, Kristina / Cotetiu, Adriana / Cotetiu, Radu / Cotrut, Cosmin Mihai / Covic, Daniela / Cozminca, Mircea / Craciun, Dan / Craciunescu, Puiu / Crainic, Nicolae / Cretu, Dan / Crisan, Aurel / Crisan, Gheorghe / Crisan, Lia / Croitoru, Bogdan / Croitoru, Catalin / Croitoru, Sorin Mihai / Cronskaer, Marie /

Crudu, Liviu / Cruz, Fernando / Cseki, Liviu Bogdan / Cuciumita, Cleopatra / Curak, Ivan / Curkovic, Milan / Cus, Franci / Dado, Miroslav / Dancsecz, Gabriella / Danihelova, Anna / Danisova, Nina / David, Ioan / David, Konstantinos / Davidescu, Arjana / De Agustina, Beatriz / Deaconescu, Andrea / Deaconescu, Tudor / Deaky, Bogdan / Dehelean, Nicolae / Del Vechio, Manuel / Delgado, Antonio / Demian, Gabriela / Demian, Mihai / Denzinger, Joachim / Dermanovic, Ivana / Diaconescu, Dorin / Diaconu, Ilie / Diaconu, Nicolae / Dichiu, Daniel / Dima, Bogdan / Dima, Dumitru / Dini, Gino / Dinita, Ion / Dinu, Doina / Dinu, Gabriela / Dobra, Andreea / Dobre, Daniel / Dobren, Flavius Andrei / Dodun, Oana / Dogaru, Florin / Doicin, Cristian / Domankova, Maria / Domuta, Cristian / Donevski, Davor / Dorian, Marin / Dormal, Thierry / Doroftei, Ioan / Dosoftei, Constantin Catalin / Dragan, Dinu / Draghia, Marius / Draghici, Sorin / Dragomir, George / Dragomir, Madalina / Dresvyannikov, Denis / Dreucean, Mircea / Drosescu, Radu / Druica, Elena / Duicu, Simona / Dujic, Lucijana / Dumitrascu, Adela / Dumitrescu, Constantin Dan / Dumitrescu, Corina / Dumitrescu, Diana Mariana / Dumitrescu, Lucia / Dumitriu, Adelina / Dumitriu, Dan / Dupta, Pavel Ernest / Duran, Dan / Duran, Irina / Duran, Vasile / Durica, Jan / Duris, Rastislav / Dusek, Daniel / Dvornic, Andrei Iulian / Dzimbeg Malcic, Vesna / Dziubek, Tomasz / Eberst, Otto / Eftimie, Adrian Catalin / Eftimie, Alina Luciana / Eftimie, Roxana / Elezi, Dervish / Enciu, George / Enescu, Ioan / Enescu, Monica / Esse, Martin / Ester, Jozef / Evans, Kenneth / Failli, Franco / Farcane, Nicoleta / Faur, Cosmin / Fedoryshyn, Roman / Fejzullahu, Xhemajl / Fetecau, Catalin / Fialka, Miloslav / Fidler, Branislav / Fidler, Martin / Fidlerova, Helena / Filip, Ioan / Filip, Iulian / Filipowicz, Krzysztof Z. / Florea, Claudia / Florescu, Daniela / Florescu, Iulian / Foeldi, Andrei / Folcut, Ovidiu / Forgaciu, Flavia Lelia / Fota, Adriana / Frietsch, Markus / Fucak, Sanjin / Fulconis, Farcane / Gabor, Gianina / Gaese, Thomas / Gajdos, Ivan / Galetic, Fran / Galic, Josip / Gapinski, Bartosz / Gatto, Andrea / Gavrilovic, S. Branislav / Gavrus, Cristina / Geaman, Virgil / Geca, Ahmet / Georgescu, Luminita / Gerard, Alain / Ghenghea, Dan / Gheorghe, Camelia Monica / Gheorghe, Gheorghe / Gheorghe, Ion Gheorghe / Gheorghiu, Liviu / Ghiban, Brandusa / Ghiban, Nicolae / Ghiculescu, Daniel / Ghinea, Mihai / Ghinea, Mihalache / Ghinea, Valentina Mihaela / Ghiorghe, Adrian / Ghita, Constantin / Ghitescu, Teodor / Gitea, Manuel / Godal, Daniel / Godec, Damir / Gogu, Grigore / Goh, Gyoung Tae / Goia, Ioan / Golubovic, Kristijan / Gombos, Mircea / Gombos, Radu / Gontean, Aurel / Gordan, Cornelia / Grebenisan, Dan / Grecu, Eugenia / Grecu, Luminita / Grigoras, Stefan / Grigore, Roxana / Grigorescu, Costin / Grigoriu, Aurelia / Grigoriu, Mircea / Guempel, Paul / Guiman, Maria Violeta / Guta, Carmen / Ha, Jun Su / Hajduchova, Luba / Hamat, Codruta Oana / Han, Changhyo / Harangus, Daniela / Haruta, Ovidiu / Hasan, Saad / Hazi, Gheorghe / Herceg, Tomislav / Hoffmann, Cristina / Hoghova, Katarina / Holm, Timo / Homa, Johannes / Horak, Karel / Horja, Ioana Monica / Horja, Ioana Monica / Horman, Izet / Horvath, Stefan / Hribernik, Ales / Husanu, Valerica / Iacovoiu, Viorela / Iancu, Anica / Iancu, Carmen / Ibraim, Cristina / Ignat, Andreea / Ignaton, Elemer / Ihan Hren, Natasa / Ihan Hren, Natasha / Ilcik, Stefan / Ileana, Ioan / Ilies, Paul / Iliescu, Mihaiela / Iliescu, Nicolae / Ilii, Sanda / Inaguma, Takanao / Ionescu, Alexandru / Ionescu, Cristian / Ionescu, Daniela Smaranda / Ionescu, Gabriela / Ionescu, Gheorghe Constantin / Ionescu, Mircea Dan / Ionescu, Nicolae / Ionescu, Octavian / Ionescu, Sorin / Ionescu, Stefan Alexandru / Iordachescu, Danut / Iordachescu, Mihaela / Iordan, Andrei / Iosub, Adrian / Istrat, Nicolae / Itu, Calin / Itul, Tiberiu / Iuras, Emilia / Ivan, Ioana / Ivan, Ioana Carmen / Ivan, Maria Cornelia / Ivan, Mihail Vincentiu / Ivanescu, Andrei Nick / Ivanescu, Nick / Ivanov, Marian / Ivascu, Costel / Ivetic, Dragan / Jadaneant, Mihai / Jancic, Zoran / Jedlicka, Martin / Jee, Eun Kyoung / Jelenc, Lara / Jeon, Changsu / Jeon, Seungjae / Jirkova, Hana / Jivanescu, Iulia / Jokanovic, Simo / Jones, Jason / Jula, Mihaela / Jurca, Lucian / Jurchela, Alin / Kachakov, Zharko / Kanerva, Jukka / Karabenciov, Adrian / Karakulov, Maksim / Karaoglanidis, George / Karapetrovic, Stanislav / Karuovic, Dijana / Kasala, Jozef / Kii, Konstantin / Kisicek, Tomislav / Knevez, Jean Yves / Kobrin, Alexandr / Koh, Kwang Yong / Kolenic, Frantisek / Kolosowski, Mariusz / Koniar, Jan / Konobritskiy, Dmitry / Koptioug, Andrei / Kostanjevec, Tomaz / Kostic, Aleksandra / Kramar, Vadim / Kramberger, Slavko / Krasniqi, Fehmi / Krizanova, Gabriela / Krupa, Marek / Krupka, Lothar / Kucher, Dimitry / Kwak, Hyun Soo / Laatsit, Toomas / Laget, Bernard / Lai, Kim Teng / Lajic, Branka / Lampart, Marek / Lancea, Camil / Langova, Nada / Latifi, Ahmet / Latinovic, Tiho / Laureyns, Isabelle / Lavrov, Valentin / Lazar, Liviu / Lazarescu, Cezara / Lee, Kyungkeun / Leemet, Tonu / Letinic, Svjetlana / Letunic, Stijepo / Lewandowski, Jakob / Lica, Dumitru / Lim, Mei / Lisjak, Dragutin / Liska, Jan / Liskova, Jana / Loeser, Carsten / Loncar, Damir / Lorenzen, Leif Erik / Lovasz, Erwin Christian / Luca, Gheorghe / Luchin, Milenco / Lungu, Andy / Lungu, Bogdan / Lungu, Ion / Lungu, Mihai / Lungu, Romulus / Luong, Lee / Lupeanu, Mihaela / Lupescu, Ovidiu / Lupinca, Cinca Ionel / Lupu, Ciprian / Lupulescu, Nouras / Luschmann, Christian / Machedon Pisu, Mihai / Maekitie, Antti / Magheti, Ioan / Majdik, Andras / Majerik, Marian / Majnaric, Igor / Maksarov, Viacheslav / Maksuti, Rrahim / Malciu, Raluca / Maleshkov, Stoyan / Malina, Jiri / Malos, Raul / Mamulea, Adina Ana / Mancia, Aurora / Manciulea, Ileana / Mandru, Dan / Manea, Adriana Sida / Maniu, Laura / Marascu Klein, Vladimir / Marc, Gheorghe / Marcu, Cosmin / Marcu, Florin / Margineanu, Dan / Marin, Alexandru / Marin, Cornel / Marin, Dumitru / Marin, Gheorghe / Marin, Marta / Marin Pantelescu, Andreea / Marinac, Antun / Marincel, Petra / Marinescu, Mirel / Marinescu, Niculae / Marinescu, Sorin Alin / Markowska, Olimpia / Markowski, Tadeusz / Maronek, Milan / Marta, Constantin / Martinescu, Ionel / Martinovic, Dunja / Maruf, Hasan / Masar, Alojz / Masarova, Renata / Masek, Bohuslav / Masiar, Harold / Matiko, Fedir / Matsi, Birthe / Matulja, Dunja / Mazilescu, Crisanta Alina / Mazuru, Sergiu / Meciarova, Julia / Meckl, Peter / Menyhardt, Karoly / Merticaru, Vasile Jr. / Meudtner, Andreas / Mezentsev, Mikhail / Micu, Ioan / Miechowicz, Stawomir / Mihai, Valia Maria / Mihail, Laurentiu / Mihaila, Stefan / Mihailidis, Athanassios / Mihartescu, Ana Andreea / Mihet Popa, Lucian / Mihic, Srdan / Mihljevic, Viktor / Mihok, Jozef / Miklosevic, Kresimir / Mikota, Miroslav / Mikulic, Zelimir / Milcic, Diana / Militaru, Gheorghe / Milos, Teodor / Milosan, Ioan / Minica, Mirela / Mircea, Octavian / Mironeasa, Costel / Mironeasa, Silvia / Misik, Ladislav / Misulic, Marko / Mjaku, Malush / Modog, Traian / Modrea, Arina / Molcho, Gila / Moldovan, Cristian / Moldoveanu,

Alin / Moldoveanu, Florica / Morariu, Zaharie / Moroianu, Corneliu / Morovic, Mira / Mrcelic, Zeljka / Mrenes, Mircea / Mrvac, Nikola / Mudrik, Jozef / Muhaxheri, Mustafe / Munguia, Javier / Muntean, Angela / Muntean, Sebastian / Munteanu, Adriana / Munteanu, Daniel / Munteanu, Eliza / Munteanu, Gabriel / Munteanu, Irena / Munteanu, Mihaela Violeta / Munteanu, Sorin Ion / Muriqi, Ali / Musca, Elena / Musca, Gavril / Mustea Serban, Razvan / Mutka, Alan / Nad, Milan / Nagi, Mihai / Nagy, Ramona / Nakatani, Hiromasa / Nanasi, Tibor / Nanu, Neculai / Nascak, Lubomir / Nastac, Silviu / Neagu, Dumitru / Nedelcu, Adrian Valentin / Nedelescu, Mihai / Negoitescu, Arina / Negrea, Virgiliu Dan / Negru, Mihai / Negru Strauti, Gabriela / Neidenbach, Norbert / Nerantzis, Ioannis / Nesvadba, Petr / Ngoc Anh, Tran / Nicodim, Mariana / Nishiyama, Daisuke / Nizetic, Josip / Nizic, Berislav / Nutu, Emil / Ocenasova, Lenka / Ocnarescu, Constantin / Olah, Iosif / Olarescu, Alin / Olt, Jueri / Olteanu, Vlad Gabriel / Opelsz, Helen / Opetuk, Tihomir / Oroian, Maria / Ostoia, Daniel / Otel, Calin / Paclisan, Dana / Pacurar, Razvan / Paecurar, Razvan / Palm, Stjepan / Paloheimo, Kaija Stiina / Paloheimo, Markku / Panduru, Lucian / Pantea, Ioan / Paraschiv, Marius Daniel / Pardanjac, Marjana / Park, Jin Woo / Parv, Luminita / Parvu, Felix / Pastrama, Stefan Dan / Patachia, Silvia / Pavlic, Daniel / Pavlovic, Ivana / Pavlovic, Tomislav / Pelau, Corina / Perminova, Anastasia / Pernevan, Ioan / Pernevan, Maria / Pernis, Rudolf / Petcu, Romulus / Petre, Mihai / Petrescu, Ligia / Petrlic, Dario / Petropoulos, Georgios / Pikna, Lukas / Pilipovic, Ana / Pirtea, Marilen / Pistun, Yevhen / Pleniceanu, Aristica / Podhorsky, Stefan / Poenaru, Violeta Domnica / Polakovic, Milos / Pop, Alin / Pop, Nicolae / Pop, Nicolae Alexandru / Pop Lupu, Anca / Popa, Adina / Popa, Elena / Popa, Ionut / Popa, Ramona / Popa, Viorel / Popentiu Vladicesu, Florin / Popescu, Cristina / Popescu, Daniel / Popescu, Jeni / Popescu, Mihaela / Popescu, Radu / Popescu, Rodica / Popescu Mitroi, Maria Monica / Popovici, Gheorghe / Prajova, Vanesa / Preda, Ion / Preda, Oana / Predoiu, Constantin / Prejmerean, Mihaela / Prem, Florica / Pribytkova, Marina / Prichici, Mariana Adriana / Pricope, Cornel / Prisacaru, Gheorghe / Prostean, Gabriela / Prostean, Octavian / Pruna, Liviu / Purcarea, Ramona / Racic, Ifigenija / Radakovic, Nikola / Radman Pesa, Anita / Radosav, Dragica / Radu, Adrian / Radu, Ciprian / Radu, Mioara / Raducanu, Bogdan / Radulescu, Alexandru / Radulescu, Bruno / Radulescu, Florin / Radulescu, Irina / Radulescu, Mara / Raennar, Lars Erik / Rahau, Loredana / Raicu, Lucian / Ramsay, Hans / Rat, Nadia Ramona / Recio Perero, Pablo / Rejda, Rudolf / Rennie, Allan / Ridzon, Martin / Rindasu, Venera / Risovic, Stjepan / Risteiu, Mircea / Rittidech, Sampan / Robu, Andreea / Robu, Raul / Rohan, Rodica / Roman, Costel / Romanescu, Maria Camelia / Roosimoelder, Lembit / Rosca, Ileana / Rosca, Ioan / Rosu, Dorin / Rotaru, Horatiu / Rubio, Eva M. / Rugescu, Dragos Ronald / Runkas, Martin / Rus, Alexandru / Rusnac, Lucian / Rusu, Lucian / Ruzarovsky, Roman / Ruzicka, Jan / Rybansky, Rudolf / Ryu, Kwangyeol / Saenz De Pipaon, Jose / Salminen, Kai / Samak, Samoil / Samakoski, Blagoja / Sandru, Diana / Sandu, Ion / Santos Martin, Francisco / Sarbu, Flavius / Saulescu, Radu / Savin, Adriana / Savu, Lucian / Scarneciu, Ioan / Schor, Vladimir / Schuermeyer, Jan / Schumann, Christian / Scukanec, Danijel / Scurtu, Dan / Scutaru, Maria Luminita / Scutelnicu, Elena / Seong, Poong Hyun / Serb, Adrian / Serban, Nicolae / Sercer, Mladen / Sergiu, Nicolae / Sewell, Neil Thomas / Shin, Seung Ki / Shirobokov, Konstantin / Sica, Mihaela / Siketova, Katarina / Silivestru, Valentin / Sima, Ion / Simeunovic, Nenad / Simion, Ionel / Sindila, Gheorghe / Sintea, Sorin / Sinyavskiy, Oleg / Skala, Tibor / Skalova, Ludmila / Slota, Jan / Smurov, Igor / Soare, Vasile / Sobolak, Mariusz / Sobota, Robert / Soceanu, Alexandru / Sokoloski, Zlatko / Spataru, Andrei Cristian / Spisak, Emil / Staicu, Florentiu / Stamate, Valentin / Stamatescu, Grigore / Stampfl, Juergen / Stan, Andrei / Stanasel, Iulian / Stanasel, Oana / Stanciu, Anka / Stanciu, Mariana / Stanciu, Mihai / Stanciu, Valerian / Stanescu, Cristina / Stanescu, Marius / Stanescu, Titu / Stanic, Nikolina / Stankovic, Ana / Starko, Kirill / Starshev, Denis / Stefan, Constantin / Stibritz, Guido / Sticlaru, Carmen / Stimac, Igor / Stipancic, Tomislav / Stirbu, Cosmin / Stirbu, Cristel / Stoerzner, Falko / Stoia, Dan Ioan / Stoica, Gina Florica / Stoica Guzun, Anicuta / Stoicanescu, Maria / Stollmann, Vladimir / Stroita, Daniel Catalin / Stuparu, Adrian / Suba, Roland / Suchomel, Jozef / Suciu, Lenuta / Surcel, Ioan / Suriansky, Jozef / Szeidert, Iosif / Szekely, Iuliu / Szentes, Balazs / Tabara, Iulian / Tallis, Andrea / Talpos, Ioan / Tamas, Levente / Tanasa, Raluca / Tanasa, Valentin / Tane, Nicolae / Taratinsky, Ilya / Tarcolea, Mihai / Tarnita, Dan / Tarnita, Daniela / Tatar, Olimpiu / Tatarciuc, Monica / Tataru, Mircea Bogdan / Tatucu, Iancu / Teodorescu, Horatiu / Tera, Melania / Teusdea, Alin / Thierheimer, Diana / Tiba, Zsolt / Tilina, Dana / Tilina, Dragos / Timar, Maria Cristina / Tion, Monica / Tittel, Viktor / Toca, Alexei / Todericiu, Ramona / Tofan, Mihai / Tokar, Adriana / Toma, Victoria / Tomasegovic, Tamara / Tomic, Mladen / Tomoiaga, Tiberius / Tomuletiu, Elena Adriana / Tonoiu, Sergiu / Toth Tascau, Mirela / Tothova, Maria / Trso, Martin / Tulbure, Adrian / Tuomi, Jukka / Turna, Milan / Turnova, Zuzana / Udrea, Andreea / Udroiu, Razvan / Ulrich, Koloman / Ungur, Patricia / Ungureanu, Dan / Ungureanu, Delia / Ungureanu, Miorita / Ungureanu, Nicolae / Ursu, Ioan / Vacarescu, Cella Flavia / Vacarescu, Valeria / Vacek, Vladimir / Valcuha, Stefan / Van Voorthuysen, Erik Jan / Vancea, Smaranda / Vartolomei, Mihaela / Vartolomei M, Mihael / Vasar, Cristian / Vasicek, Vesna / Vasilache, Simona / Vaxevanidis, Nikolaos / Veres, E Mircea / Viglasky, Jozef / Vilag, Valeriu / Villeta, Maria / Vintila, Georgeta / Visan, Dana / Visan, Daniel / Vitalariu, Anca / Vlad, Angelica / Vlad, Dorin / Vlase, Auremia / Vlase, Sorin / Voicu, Andreea Raluca / Voicu, Vasilica / Voicu, Vlad / Voisan, Emil / Volf, Irina / Vologdin, Sergey / Vujic Ziljak, Jana / Vujicic Tomic, Brankica / Vulpe, Anca Diana / Vulpe, Marius Cristian / Walekova, Gabriela / Weiss, Edmund / Wimpenny, David / Winkler, Sebastian / Wolf, Edwin / Wolny, Ryszard / Wongkasem, Kan / Yadroitsev, Igor / Yassin, Adrian / Yiangkamolsing, Chana / Zaharia, Sebastian Marian / Zaheu, Ioana / Zamfira, C. Sorin / Zelenay, Miroslav / Zeqiri, Fitim / Zezelj, Dragan / Zgardea, Emanuel / Zimmermann, Matthias / Zitinski Elias, Paula

5.0 IN MEMORIAM OF NICE DAAAM COLLEAGUES AND DEAR FRIENDS

Professor Dr Jyri Papstel (1940- 2004), Tallinn Technical University, Tallinn Estonia, Vice President of DAAAM International, Founder and President of DAAAM Estonia, Founding member of the pilot projects Interuniversity Doctoral Studies and Virtual University. Member of: International Organizing / Program / Scientific / Review Committees. President of Award Committee. Author. Honored with Medal of DAAAM International Vienna.

M. Sc. Dipl.-Ing. Ivan Ivancic (1952-2004), Ericsson - Tesla, Zagreb, Croatia, International Coordinator for the field Maintenance, Member of International Organizing / Program / Review Committees.

Professor Dr Zdenko Kordic (1949-2004), University of Mostar and University of Split, Founder and First Rector of the University of Mostar, Member of Rectors´ Honor Committee, Founding member of the pilot projects Interuniversity Doctoral Studies and Virtual University. Member of International Program and Scientific Committees. Author. Honored with Medal of DAAAM International Vienna.

6.0 DAAAM INTERNATIONAL MEDAL

On the occasion of 10th DAAAM International Symposium, DAAAM International introduced the Medal of DAAAM International "for excellence in the science, technology, and international scientific co-operation in the framework of DAAAM International Vienna" as a highest recognition for individual persons and institutions. The medal is designed by B. Katalinic, produced by Moscow State Mint Institute and sponsored by Moscow State University of Technology "Stankin" and Cassandra Trust – Ireland. Maximal one person and/or one institution are honoured by medal in one year.

7.0 UNIVERSITY CHARTER OF DUBROVNIK

The text of the University Charter of Dubrovnik: We, the representatives of different Universities and Institutions, after two years preparation at the meeting initiated and organized by The Danube Adria Association for Automation & Manufacturing, DAAAM International under the auspices of The Danube Rectors' Conference on the occasion of the Eighth International DAAAM Symposium "Intelligent Manufacturing & Automation" held at the International Center of Croatian Universities (University of Zagreb) in Dubrovnik, Croatia, decided to found the University Network of DAAAM International Vienna as a strategic alliance and permanent open platform for long-term co-operation and partnership in all fields of technology, research and education.

The co-operation has to be continuously organized and is based on the partnership, friendship, tolerance, appreciation and positive experience of international scientific co-operation within the framework of DAAAM International during the last eight years. The co-operation can use different flexible forms, structures and aspects. The practical co-operation will be project-based organized. The highest priority is given to the organization of two long - term projects, initially known as: Interuniversitary Doctoral Studies and a pilot project Virtual University. Signed by the founding members in Dubrovnik, Croatia, on the twenty-third of October 1997

8.0 DAAAM INTERNATIONAL NETWORK FOR ADVANCED TECHNOLOGIES

The text of the Network founding document: We, the representatives of different Universities, Institutions, and Firms at the meeting initiated and organized by The Danube Adria Association for Automation & Manufacturing, DAAAM International Vienna under the auspices of The Danube Rectors' Conference on the occasion of the Thirteenth International DAAAM Symposium "Intelligent Manufacturing & Automation: Learning from Nature" held at the Vienna University of Technology and Austrian Society of Engineers and Architects in Vienna, Austria, decided to found the DAAAM International Network for Advanced Technologies as a strategic alliance and permanent open platform within European Research Area for long-term co-operation and partnership in all fields and aspects of research, developing, transfer, education and use of advanced technologies.

The co-operation has to be continuously organized and had to be based on the partnership, friendship, tolerance, appreciation and positive experience of international scientific co-operation within the framework of DAAAM International during the last fourteen years. The co-operation can use individual flexible project oriented forms, structures and aspects and support the young researchers' mobility as well. The practical co-operation will be project-based organized. The highest priority is given to the organization of long - term international cooperation projects focused on advanced technologies for balanced development and harmony between men, technology and nature. The network has to be platform for dissemination of relevant information and ideas, supporting international partnership and encouraging of cooperation initiatives. Signed by the founding members in Vienna, Austria, European Union, on the twenty-fifth of October 2002

Activity No. 1: The conference bags and flags for DAAAM 2003 in Sarajevo are sponsored by the Faculty of Textile Technology, University of Zagreb, and organised by Professor Z. Dragcevic as the first activity in the frame of DAAAM International Network for Advanced Technologies.

9.0 DAAAM INTERNATIONAL NETWORK FOR MECHATRONICS AND ROBOTICS

The text of the Network founding document: We, the representatives of different Universities, Institutions, and Firms at the meeting initiated and organized by The Danube Adria Association for Automation & Manufacturing, DAAAM International Vienna under the auspices of The Danube Rectors' Conference on the occasion of the Fifteenth International DAAAM Symposium "Intelligent Manufacturing & Automation: Globalisation - Technology - Men - Nature" held at the Vienna University of Technology and Austrian Society of Engineers and Architects in Vienna, Austria, decided to found the DAAAM International Network for Mechatronics and Robotics as a strategic alliance and permanent open platform within European Research Area for long-term co-operation and partnership in all fields and aspects of research, developing, transfer, education and use of advanced technologies.

The co-operation has to be continuously organized and had to be based on the partnership, friendship, tolerance, appreciation and positive experience of international scientific co-operation within the framework of DAAAM International during the last fifteen years. The co-operation can use individual flexible project oriented forms, structures and aspects and support the young researchers' mobility as well. The practical co-operation will be project-based organized. The highest priority is given to the organization of long - term international cooperation projects focused on advanced technologies for balanced development and harmony between men, technology and nature. The network has to be platform for dissemination of relevant information and ideas, supporting international partnership and encouraging of cooperation initiatives. Signed by the founding members in Vienna, Austria, European Union, on the fifth of November 2004

10.0 DAAAM INTERNATIONAL SYMPOSIUMS (1990-2008)

One important activity is organising the DAAAM International Symposium. The main objective of the Symposium is to provide a world forum, which takes place annually in Central Europe to exchange knowledge, experience, results and information related to various aspects of the advanced manufacturing and modern automation.

The scope of the Symposium covers scientific, technological and practical concepts concerning research, development and realisation of advanced manufacturing and automation concepts, offering a unique opportunity for experts to meet and exchange ideas. The Symposium is concerned with the current status of automation and manufacturing, methods and results of research and development, as well as application problems and publishes proceedings accordingly.

2008	**19th International DAAAM Symposium "Intelligent Manufacturing & Automation: Focus on Next Generation of Intelligent Systems and Solutions"**
Date & Place	22-25th October 2008, Trnava, Slovakia
Organized by	DAAAM International Vienna, Slovak University of Technology Faculty of Materials Science and Technology in Trnava, University of Applied Sciences Technikum Vienna, Vienna University of Technology - Department of Production Engineering / Intelligent Manufacturing Systems (IFT-IMS), ÖIAV 1848 Austrian Society of Engineers and Architects
Under the auspices of	Danube Rectors' Conference & Rectors' Honor Committee of DAAAM International
Organizer	Katalinic, Branko (Chair) / Moravcik, Oliver (Co-Chair) / Cesarec, Paulina / Dragcevic, Zvonko / Höller, Liane / Kettler, Roman / Knezevic, Sime / Resetova, Kvetoslava / Stefankova, Jana / Stopper, Markus / Velisek, Karol
Publication	Annals of DAAAM for 2008 and Proceeding of the 19th International DAAAM Symposium, ISSN 1726-9679, ISBN 978-3-901509-68-1
Referred/indexed in	ISI Scientific Proceedings Thomson Reuters
Editor	B.[ranko] Katalinic
2007	**18th International DAAAM Symposium "Intelligent Manufacturing & Automation: Focus on Creativity, Responsibility, and Ethics of Engineers"**
On the occasion of	Ograrised in Order to Celebrate 70th Birthday of Dr Wilfried Stoll
Date & Place	24-27th October 2007, Zadar, Croatia
Organized by	DAAAM International Vienna, University of Zadar, University of Applied Sciences Technikum Vienna, Vienna University of Technology - Department of Production Engineering / Intelligent Manufacturing Systems (IFT-IMS), ÖIAV 1848 Austrian Society of Engineers and Architects
Under the auspices of	Danube Rectors' Conference & Rectors' Honor Committee of DAAAM International
In cooperation with	Croatian National Tourist Board
Organizer	Katalinic, Branko (Chair) / Uglesic, Ante (Co-Chair) / Cesarec, Paulina / Dragcevic, Zvonko / Höller, Liane / Kettler, Roman / Knezevic, Sime / Stopper, Markus / Stuja, Kemajl / Wenger, Monika
Publication	Annals of DAAAM for 2007 and Proceeding of the 18th International DAAAM Symposium,

	ISSN 1726-9679
Referred/indexed in	ISI Scientific Proceedings Thomson Reuters
Editor	B.[ranko] Katalinic
2006	**17th International DAAAM Symposium "Intelligent Manufacturing & Automation: Focus on Mechtronics and Robotics"**
On the occasion of	150th Birthday of Nikola Tesla (1856-1943) Inventor of Mobile Robot Teleoperation
Date & Place	8-11th November 2006, Vienna, Austria
Organized by	DAAAM International Vienna, Vienna University of Technology - Department of Production Engineering / Intelligent Manufacturing Systems (IFT-IMS), ÖIAV 1848 Austrian Society of Engineers and Architects , University of Applied Sciences - Technikum
Under the auspices of	Danube Rectors' Conference & Rectors' Honor Committee of DAAAM International
In cooperation with	Vienna Convention Bureau
Organizer	Buchmeister, Borut / Dragcevic, Zvonko / Egorov, Sergey / Höller, Liane / Katalinic, Branko (Chair) / Lazinica, Aleksandar / Malisa, Viktorio / Sagmeister, Nicole / Stuja, Kemajl / Viktorik, Hans / Viktorik, Silvia
Publication	Annals of DAAAM for 2006 and Proceeding of the 17th International DAAAM Symposium, ISSN 1726-9679, ISBN 3-901509-57-7
Referred/indexed in	ISI Scientific Proceedings Thomson Reuters
Editor	B.[ranko] Katalinic
2005	**16th International DAAAM Symposium "Intelligent Manufacturing & Automation: Focus on Young Scientists and Researchers"**
On the occasion of	15 Years of DAAAM International Vienna, 1990-2005
Date & Place	19-22nd October 2005, Opatija, Croatia
Organized by	DAAAM International Vienna, Vienna University of Technology - Department of Production Engineering / Intelligent Manufacturing Systems (IFT-IMS), ÖIAV 1848 Austrian Society of Engineers and Architects
Under the auspices of	Danube Rectors' Conference & Rectors' Honor Committee of DAAAM International
In cooperation with	Vienna Convention Bureau
Organizer	Anisic, Bojan / Barisic, B. / Blazevic, Dalibor / Buchmeister, Borut / Car, Zlatan / Dragcevic, Zvonko / Egorov, Sergey / Grabenweger, Johannes / Gronalt, Manfred / Höller, Liane / Huber, Ingrid / Jurkovic, Z. / Katalinic, Branko (Chair) / Lazinica, Aleksandar / Malisa, Viktorio / Merdan, Munir / Mikac, Tonci (Co-Chair) / Mirjanic, Valentina / Nanasi, Jozef / Schaumberger, Werner / Stopper, Markus / Stuja, Kemajl / Udiljak, Toma / Widtmann, Georg
Publication	Annals of DAAAM for 2005 and Proceeding of the 16th International DAAAM Symposium, ISSN 1726-9679, ISBN 3-901509-46-1
Referred/indexed in	ISI Scientific Proceedings Thomson Reuters
Editor	B.[ranko] Katalinic
2004	**15th International DAAAM Symposium "Intelligent Manufacturing & Automation: Globalisation - Technology - Men - Nature"**
On the occasion of	**Organised in Memoriam on Professor Dr Jyri Papstel (1940- 2004)**
Date & Place	3-6th November 2004, Vienna, Austria, European Union
Organized by	DAAAM International Vienna, Vienna University of Technology - Department of Production Engineering / Intelligent Manufacturing Systems (IFT-IMS), ÖIAV 1848 Austrian Society of Engineers and Architects
Under the auspices of	Danube Rectors' Conference & Rectors' Honor Committee of DAAAM International
In cooperation with	Vienna Convention Bureau
Organizer	Bartak, Harald; Buchmeister, Borut; Dragcevic, Zvonko; Egorov, Sergei; Frank, Willy; Grabenweger, Johann; Gronalt, Manfred; Höller, Liane; Hrelja, Igor; Katalinic, Branko (Chair); Kettler, Roman; Kordic, Vedran; Lazinica, Aleksandar; Löschner, Adam; Magyar, Teresa; Malisa, Viktorio; Merdan, Munir; Müller, Simon; Nanasi, Jozef; Piller, Johannes; Reinisch, Roland; Roth, Michael; Saliger, Florian; Schaumberger, Werner; Schmidt, Thomas; Stuja, Kemajl; Trauzettel, Ulrich; Udiljak, Toma; Widtmann, Georg; Wirths, Thomas.
Publication	Annals of DAAAM for 2004 and Proceeding of the 15th International DAAAM Symposium, ISSN 1726-9679, ISBN 3-901509-42-9
Referred/indexed in	ISI Scientific Proceedings Thomson Reuters
Editor	B.[ranko] Katalinic
2003	**14th International DAAAM Symposium "Intelligent Manufacturing & Automation: Focus on Reconstruction and Development"**
On the occasion of	155 Years of Austrian Society of Engineers and Architects ÖIAV 1848 / 45 Years of the Faculty of Mechanical Engineering University of Sarajevo
Date & Place	22-25th October 2003, Sarajevo, Bosnia and Herzegovina

Organized by	DAAAM International Vienna, University of Sarajevo - Faculty of Mechanical Engineering, Vienna University of Technology - Department of Production Engineering / Intelligent Manufacturing Systems (IFT-IMS), ÖIAV 1848 Austrian Society of Engineers and Architects
Under the auspices of	Danube Rectors' Conference & Rectors' Honor Committee of DAAAM International
In cooperation with	Federal Ministry of Education, Sciences and Culture - Vienna, ÖIAV 1848 Austrian Society of Engineers and Architects
Organizer	Katalinic, Branko (Chair), Kulenovic, Malik (Co-chair), Widtmann, G., et al.
Publication	Annals of DAAAM for 2003 and Proceeding of the 14th International DAAAM Symposium, ISSN 1726-9679, ISBN 3-901509-34-8
Referred/indexed in	ISI Scientific Proceedings Thomson Reuters
Editor	B.[ranko] Katalinic
Information	Contact: DAAAM International Vienna (Katalinic B.) president@daaam.com www.daaam.com
2002	**13th International DAAAM Symposium "Intelligent Manufacturing & Automation: Learning from Nature"**
On the occasion of	150 Years of Austrian Society of Engineers and Architects ÖIAV 1848
Date & Place	23-26th October 2002, Vienna, Austria
Organized by	DAAAM International Vienna, Vienna University of Technology - Department of Production Engineering / Intelligent Manufacturing Systems (IFT-IMS), ÖIAV 1848 Austrian Society of Engineers and Architects
Under the auspices of	Danube Rectors' Conference & Rectors' Honor Committee of DAAAM International
In cooperation with	Federal Ministry of Education, Sciences and Culture - Vienna, ÖIAV 1848 Austrian Society of Engineers and Architects
Organizer	Katalinic B. (Chair), Buchmeister B., Doganac A., Dragcevic Z., Gersak J., Grabenweger J., Kordic V., Ljoljic B., Nanasi J., Stopper M., Stuja K., Widtmann G., et al.
Publication	Annals of DAAAM for 2002 and Proceeding of 13th International DAAAM Symposium, ISBN 3-901509-29-1
Referred/indexed in	ISI Scientific Proceedings Thomson Reuters
Editor	B.[ranko] Katalinic
2001	**12th International DAAAM Symposium "Intelligent Manufacturing & Automation: Focus on Precision Engineering"**
On the occasion of	10 Years of Jena University of Applied Science / 150 Years of ÖIAV 1848 Austrian Society of Engineers and Architects
Date & Place	24-27th October 2001, Jena, Germany
Organized by	DAAAM International Vienna, Jena University of Applied Science, Vienna University of Technology - Department of Production Engineering / Intelligent Manufacturing Systems (IFT-IMS), Austrian Society of Engineers and Architects ÖIAV 1848
Under the auspices of	Danube Rectors' Conference & Rectors' Honor Committee of DAAAM International
In cooperation with	Thuringian Ministry of Sciences, Research and Culture - Erfurt,
Organizer	Katalinic B. (Chair), Meissner K. (Co-chair), Beez S., Buchmeister B., Doganac A., Dragcevic Z., Fidi L., Gersak J., Grabenweger J., Jacob M., Kordic V., Ljoljic B., Nanasi J., Pachmann A., Schoele H., Selke K., Stopper M., Stuja K., Walther J., Widtmann G.
Publication	Annals of DAAAM for 2001 and Proceeding of 12th International DAAAM Symposium, ISBN 3-901509-19-4
Editor	B.[ranko] Katalinic
2000	**11th International DAAAM Symposium "Intelligent Manufacturing & Automation: Men – Machine - Nature"**
On the occasion of	150 Years of ÖIAV 1848 Austrian Society of Engineers and Architects / 40 Years of Faculty of Engineering – University of Rijeka / 10 Years of DAAAM International
Date & Place	19-21st October 2000, Opatija, Croatia
Organized by	DAAAM International Vienna, Vienna University of Technology - Department of Production Engineering / Intelligent Manufacturing Systems (IFT-IMS)
Under the auspices of	Danube Rectors' Conference & Rectors' Honor Committee of DAAAM International
In cooperation with	Federal Ministry of Education, Sciences and Culture - Vienna, ÖIAV 1848 Austrian Society of Engineers and Architects, Union of Croatian Universities
Organizer	Katalinic B. et al.
Publication	Annals of DAAAM for 2000 and Proceeding of 10th International DAAAM Symposium, ISBN 3-901509-13-5
Editor	B.[ranko] Katalinic
1999	**10th International DAAAM Symposium "Intelligent Manufacturing & Automation: Past – Present - Future"**
On the occasion of	10th International DAAAM Symposium, Last International DAAAM Symposium in this Millennium

Date & Place	21-23rd October 1999, Vienna University of Technology - Department of Production Engineering / Intelligent Manufacturing Systems (IFT-IMS), Austrian Society of Engineers and Architects ÖIAV 1848, Vienna, Austria
Organized by	DAAAM International Vienna, Vienna University of Technology, Department of Production Engineering / Intelligent Manufacturing Systems (IFT-IMS)
Under the auspices of	Danube Rectors' Conference & Rectors' Honor Committee of DAAAM International
In cooperation with	Federal Ministry of Sciences and Traffic - Vienna, ÖIAV 1848 Austrian Society of Engineers and Architects
Organizer	Katalinic B. (Chair), Adamczak S., Alpek F., Balate J., Balic J., Beer M., Buchmeister B., Celar S., Cus F., Dinnyes C., Dragcevic. Z., Duflou J. R., Egorov S., Fidi L., Grabenweger J., Gyenge C., Hajrizi E., Kerekes L., Kordic V., Kordic Z., Ljoljic B., Llorente J. I., Meissner K., Nanasi J., Papstel J., Park H.-S., Stuja K., Tabucanon M. T., Udiljak T., Ueda K., Vranjes B., Wessely E.
Publication	Annals of DAAAM for 1999 and Proceeding of 10th International DAAAM Symposium, ISBN 3-901509-10-0
Editor	B.[ranko] Katalinic
1998	**9th International DAAAM Symposium "Intelligent Manufacturing, Automation & Networking"**
On the occasion of	50 Years of the Technical University of Cluj - Napoca 150 Years of ÖIAV 1848 Austrian Society of Engineers and Architects Founding of Permanent Committee of University Network of DAAAM International Vienna
Date & Place	22-24th October 1998, Technical University of Cluj - Napoca, Cluj - Napoca, Romania
Organized by	DAAAM International Vienna, Technical University of Cluj - Napoca, Vienna University of Technology - Department of Production Engineering / Intelligent Manufacturing Systems (IFT-IMS)
Under the auspices of	Danube Rectors' Conference & Rectors' Honor Committee of DAAAM International
In cooperation with	Ministry of Research and Technology - Bucuresti, Federal Ministry of Sciences and Traffic - Vienna, Journal of Intelligent Manufacturing, ÖIAV 1848 Austrian Society of Engineers and Architects
Organizer	Arsin M., Beer M., Bejan M., Berce P., Borzan M., Buchmeister B., Calinescu O., Car Z., Celar S., Dusa P., Fidi L., Gheorghe I., Gramescu T., Gyenge C., Hajrizi E., Ivan N.V., Katalinic B. (Chair), Kerekes L., Lazea G., Ljoljic B., Lörincz S., Nanasi J., Nemes S., Popescu S., Pusca M., Szava T., Udiljak T., Vranjes B.
Proceedings	ISBN 3-901509-08-9
Editor	B.[ranko] Katalinic
1997	**8th International DAAAM Symposium "Intelligent Manufacturing & Automation"**
On the occasion of	Founding of the University Network of DAAAM International Vienna
Date & Place	23-25th October 1997, ICCU - International Centre of Croatian Universities (University of Zagreb), Dubrovnik, Croatia
Organized by	DAAAM International Vienna, University of Zagreb, Vienna University of Technology, Department of Production Engineering / Intelligent Manufacturing Systems (IFT-IMS)
Under the auspices of	Danube Rectors' Conference & Rectors' Honor Committee of DAAAM International
In cooperation with	Ministry of Science and Technology - Zagreb, Federal Ministry of Sciences and Traffic - Vienna, Journal of Intelligent Manufacturing, ÖIAV 1848 Austrian Society of Engineers and Architects, Union of Croatian Universities, Croatian Society - KoREMA - Zagreb.
Organizer	Benic D., Brunsko V., Budic I., Burgjelez I., Celar S., Gojo M., Hajrizi E., Ivancic I., Jerbic B., Katalinic B. (Chair), Kovacic B., Majstorovic V., Mencer H.J., Nanasi J., Pavic I., Polic-Bobic M., Udiljak T. (Co-Chair), Veza I., Visekruna V.
Proceedings	ISBN 3-901509-04-6
Editor	B.[ranko] Katalinic
1996	**7th International DAAAM Symposium "Product & Manufacturing: Flexibility, Integration, Intelligence"**
On the occasion of	1,000 Years of Austria
Date & Place	17-19th October 1996, Vienna, Austria
Organized by	DAAAM International Vienna, Vienna University of Technology, Department of Production Engineering (IFT-IMS)
Under the auspices of	Danube Rectors' Conference
Organizer	Katalinic B., Celar S., Fabian E., Nanasi J.
Proceedings	ISBN 3-901509-02-X
Editor	B.[ranko] Katalinic
1995	**6th International DAAAM Symposium "Intelligent Manufacturing Systems"**
On the occasion of	50 Years of Technical University of Krakow
Date & Place	26 - 28th October 1995, Poland

Organized by	DAAAM International Vienna; Technical University of Krakow Technical University of Kielce; Vienna University of Technology - Department of Production Engineering / Intelligent Manufacturing Systems (IFT-IMS)
In cooperation with	IMECO; Ministry of Education, Warsaw; Ministry of Sciencees and Technology, Warsaw; Federal Ministry of Sciences, Technology and Art, Vienna; BIT, Bureau for International Research and Technology Cooperation, Vienna
Organizer	Adamczak S., Duch A., Gawlik J., Katalinic B., Osanna H. P., Sladek J., Wantuch E., Rusiniak K., Zebala W.
Proceedings	ISBN 3-901509-01-1
Editor	B.[ranko] Katalinic
Co-Editors	Osanna H.P., Gawlik J.
1994	**5th International DAAAM Symposium "Automation and Metrology: Challenge and Chance"**
On the occasion of	700 Years of City of Maribor, 35 Years of Faculty of Technical Sciences
Date & Place	27-28th October 1994, Maribor, Slovenia
Organized by	DAAAM International Vienna, University of Maribor, Faculty of Technical Sciences
In cooperation with	IMECO; Ministry of Sciencees and Technologies, LjubljanaFederal Ministry of Sciences, Technology and Art, Vienna
Organizer	Balic J., Cus F., Katalinic B., Novak D., Osanna H. P., Sujica M. Z.
Proceedings	ISBN 3-901509-00-3, ISBN 86-435-0084-4
Editors	Katalinic B., Osanna H.P., Balic J., Cus F.
1993	**4th International DAAAM Symposium "Automation and Metrology: Man, Technology, Environment"**
On the occasion of	750 Years of City of Brno
Date & Place	16-18th September 1993, Brno, Czechia
Organized by	DAAAM International Vienna; Technical University of Brno, Faculty of Mechanical Engineering
In cooperation with	IMECO, IUTAM
Organizer	Balate J., Dvorak J., Kadrnozka J., Katalinic B., Lacko B., Osanna H. P., Skopal M., Vavrin P.
Editors	Katalinic B., Osanna H.P:
1992	**3rd International DAAAM Symposium "Flexible and Intelligent Automation"**
Date & Place	4-6th November 1992, Budapest, Hungary
Organized by	DAAAM International Vienna; Technische Universität Budapest, Department of Production Engineering
In cooperation with	IMECO
Organizer	Alpek F., Arz G., Katalinic B., Osanna H. P., Szelig T., Jerzsabek P.
Editor	B.[ranko] Katalinic
1991	**2nd International DAAAM Symposium "Flexible Automation"**
Date & Place	9-13th December 1991, Strbske Pleso, Slovakia, Czechoslovakia
Organized by	DAAAM International Vienna; House of Technology, Bratislava; Technical University of Kosice, Faculty of Mechanical Engineering, Department of Automated Production Systems
Organizer	Sebo D., Wessely E., Kovacik J., Katalinic B.
Proceedings	ISBN 80-233-0225-6
1990	**1st International DAAAM Symposium „Automatization and Measurement Technique“**
On the occasion of	175 Years of Vienna University of Technology
Date & Place	5th November 1990, Vienna, Austria
Organized by	DAAAM International Vienna, Vienna University of Technology - Department of Production Engineering / Intelligent Manufacturing Systems (IFT-IMS), ÖIAV, Sections: Measurement and Automation, ASESA, Robotics
Organizer	Beer R. J., Osanna H. P., Katalinic B.
Editor	B.[ranko] Katalinic

11.0 BEST PAPER / PRESENTATION DAAAM AWARD

Since 1999 the DAAAM International Vienna recognizes excellence of contributions in a number of ways. There are three categories for papers:

- Best Paper Award of DAAAM International for the best paper presented at Symposium,
- DAAAM International Award for Best Oral Presentation of Paper,
- DAAAM International Award for Best Visual Presentation of Paper.

The Awards are made at the Closing Section and Best Paper Award Ceremony of the Symposium takes part on Friday at 17:00 hours. The Award Committee of DAAAM International makes the evaluation of presented papers during the Symposium. The final decisions are made by the President of the Award Committee of DAAAM International. Former Presidents of the Paper Award Committee of DAAAM International are: 2008 Professor, Master of Sport and Cosmonaut Alexey Eliseev, 2007 Professor Kanji Ueda & Professor Vladimir Filaretov, 2006 Professor Vladimir

Filaretov, 2005 Professor Ueda, 2004 Professor Franc Cus, 2003 Professor Jyri Papstel, 2002 Professor Kanji Ueda, 2001 Professor Toma Udiljak, 2000 Professor Jyri Papstel, and 1999 Professor Kanji Ueda. The President and the final list of the members of the Committee will be announced at Opening Ceremony of the Symposium. All (co)authors of awarded papers are getting the certificate, the paper (main author) additionally getting Best Paper Medal. The layout of the Best Paper Medal you can see at the cover of this publication. The papers which authors are member of the Award committee and the papers which are not presented at Symposium are excluded from competition and can not get the Award.

12.0 DAAAM INTERNATIONAL YOUNG RESEARCHER AND SCIENTIST AWARD

DAAAM International Vienna wants to recognize excellence of contributions of young researchers and scientists, and their supervisors and mentors. Starting with the 1st European DAAAM International Young Researchers´ and Scientists´ Conference, and 18th DAAAM International Symposium "Intelligent Manufacturing and Automation: Focus on Creativity, Responsibility and Ethics of Engineers", which are held 24-27th October 2007 in Zadar, Croatia, the best young researchers and scientists will be honoured by "DAAAM International Young Researchers and Scientist Award" Awards are possible in following categories:

Related to:	Categories:
Paper:	Best Paper / Best Oral Presentation / Best Graphic Presentation / Best Poster Presentation / Best Demonstration
Supervisor / Mentor:	Best Supervisor/Mentor / Best Team Manager / Best Project Manager / Best Leader / Best Female Supervisor
Quality of Project / Research:	Best Industrial Project / Best Research Project / Best Application / Best Innovation / Best Theory
Contest:	Interdisciplinarity / European Relevance / Future Oriented / Ethics / Harmony with Nature
Thesis:	Best Bachelors Thesis / Best Master Thesis / Best Diploma Thesis / Best MBA Thesis / Best PhD/Doctor Thesis
Team:	Best Team / Best University Team / Best Firm / Best International Team / Youngest Team

In the case of extraordinary contributions which can not be expressed by above mentioned recognitions is possible to make a Special Award. The Awards are made at the Closing Section of the Symposium. The final decision is made by President of the Award Committee of DAAAM International for actual year. Former Presidents of the Award Committee are recognised international leading experts and professors. All awarded authors and co-authors of paper are getting attractive **DAAAM International Best Award Certificate.** The Award Certificates are personal, very nice professional designed, framed and ready to hang in your room. Authors' team is getting one medal named: **"DAAAM International Young Researcher and Scientist Award".** This Medal is designed by: Branko Katalinic; Sabine Wulffert, and Igor Drstvensek who made the prototype and first edition of medals in 2007. This medal is sponsored by concern Festo.

Very Important: The papers which authors are member of the Award Committee and the papers which are not presented at Symposium are excluded from competition and can not get the Award.

13.0 FESTO YOUNG RESEARCHERS AND SCIENTISTS SUPPORT SCHOLARSHIP

Since 2007 concern Festo gives, by courtesy of Dr. Wilfried Stoll, support scholarship to excellent young researchers and scientists to present own research results at European DAAAM International Young Researchers´ and Scientists´ Conferences, and DAAAM International Symposiums "Intelligent Manufacturing and Automation". This scholarship covers conference fee, accommodation and (partly) travel expenses. Conditions for wining this scholarship are: candidate must come to conference and must present own research results, candidate must have high quality research and presentation, and candidates have to come from different countries. The winners are getting special "Certificate of Attendance" with special text " ...participated as winner of Festo Young Researcher and Scientist Support Scholarship". The awards are made at the closing award ceremony, and given by high representative of concern Festo. The first representative for 2007 was Ing. Wolfgang Keiner, general manager Festo Austria.

In 2007 thirty four young researchers and scientists from 11 countries won Festo Young Researcher and Scientist Support Scholarship. The 2007-winners are: Babic, Anita / Bagimov, Igor / Bartos, Stanislav / Beatrice, Nica / Belosevic, Ivan / Bodini, Ileana / Dokupil, Radek / Florina, Anghel / Hong, Dongphyo / Hursa, Anica / Jeong, Donghwa / Kim, Insu / Klevakin, Vladimir / Kovács, Zoltán / Lancini, Matteo / Lasunon, OnUma / Lorenc, Vlastimil / Loun, Kaia / Mączka, Janusz Józef / Mikhailov, Maxim / Mogilnikov, Pavel / Moise, Cristian / Noskovicova, Lucia / Ovari, Nora / Parpala, Radu / Pelikan, Radek / Stjepanovic, Goran / Sysala, Tomas / Tache, Florin / Vesely, Libor / Vesely, Milos / Vidović, Nataša / Zbranek, Pavel / Zuev, Alexandr.

14.0 BEST DIGITAL PHOTO DAAAM AWARD

Importance of digital photography for research, publishing and social relations is rapidly increased during last time. For that reason we are going to establish DAAAM International Photo Gallery with photos which are related to the symposium and our activities (place, people, situation, art, special effects). The gallery will be free accessed and searchable. For this project we are looking for amateur photographs among the participants to be official photographs of DAAAM International and member of DAAAM International Photo Team. Digital photos can be given during the symposium to the organizer, or send after symposium by e-mail. All photos which are technically good made will be included in the gallery. On the next symposium the up to three authors of nicest photos will get Best Digital Photo Award.

15.0 BEST TASTE AND SMELL OF DAAAM AWARD

"Best Taste and Smell of DAAAM International": Many of colleagues in DAAAM International network are excellent hobby cooks, liquor maker or winemaker. Each year we are making small tasting of their specialities under the motto "Best taste and smell of DAAAM International". Authors are encouraged to bring some self-made specialities to DAAAM symposium. The best will be honoured with "Best Taste and Smell DAAAM International Award".

16.0 DAAAM INTERNATIONAL SPECIALIZED CONFERENCES

2008	**2nd International Conference on Additive Technologies, DAAAM International Specialized Conference**
Founder and Chair	Drstvensek, I.
Date & Place	September, 17th till 18th, Hotel Primus, Ptuj, Slovenia
Organized by	DAAAM International Vienna/Interesansa (RAPIMAN – Rapid Prototyping and Innovative Manufacturing Network)/ University of Maribor - Faculty of Mechanical Engineering/University of Ljubljana, Institute for Innovation and Development
Organizers	Drstvensek, Igor (Chair) / Katalinic, Branko (Co-Chair) / Dolinsek, Slavko (Co-Chair) / Valentan, Bogdan / Brajlih, Tomaz / Bauer Fuzir, Gabriela / Reinhardt, Andrea / Ficko, Mirko / Pogacar, Vojko /
Proceedings	ISBN 3-901509-72-0, ISSN1992-5085
Editors	Drstvensek I.[gor] & Katalinic B.[ranko] & Dolinsek S.[lavko]
Published by	DAAAM International Vienna
2007	**1st DAAAM International Specialized Conference on Additive Technologies**
Founder and Chair	Katalinic B.
Date & Place	20th April 2007, Celje, Slovenia
Organized by	DAAAM International Vienna / RAPIMAN – Rapid Prototyping and Innovative Manufacturing Network / Vienna University of Technology - Department of Production Engineering / Intelligent Manufacturing Systems (IFT-IMS), / University of Maribor , Faculty of Mechanical Engineering - Department of Production Engineering / ÖIAV 1848 Austrian Society of Engineers and Architects / University of Applied Sciences - Technikum
Organizers	Drstvensek, Igor (Chair) / Katalinic Branko (Co-Chair) / Valentan, Bogdan / Brajlih, Tomaz / Tercek, Tomaz / Reinhardt, Andrea / Ficko, Mirko / Pogacar, Vojko / Dolinsek, Slavko
Proceedings	ISBN 3-901509-61-5, ISSN1992-5085
Editors	Drstvensek I.[gor] & Katalinic B.[ranko]
Published by	DAAAM International Vienna

17.0 DAAAM INTERNATIONAL CONFERENCES "ADVANCED TECHNOLOGIES FOR DEVELOPING COUNTRIES"

2006	**5th DAAAM International Conference on Advanced Technologies for Developing Countries, ATDC-2006**
Founder and Chair	Katalinic B.
Date & Place	28-30 June 2006, Rijeka, Croatia
Organized by	University of Rijeka, Faculty of Enineering, DAAAM International Vienna
Organizers	Mikac T.[onci] (Chair), Car Z.[latan] (Co-chair), Barisic B.[ranimir], Cukor G.[oran], Ikonic M.[ilan], Pavletic D.[usko], Perinic M.[laden], Blazevic D.[alibor], Jurkovic Z.[oran]
Proceedings	ISBN 3-901509-54-2 (Book), ISBN3-901509-55-0 (CD), ISBN 953-6326-45-0
Editors	Mikac T.[onci] & Katalinic B.[ranko]
Published by	University of Rijeka, Faculty of Engineering, DAAAM International Vienna
2005	**4th DAAAM International Conference on Advanced Technologies for Developing**

	Countries, ATDC-2005
Founder and Chair	Katalinic B.
Date & Place	September, 21-24th, 2005, Slavonski Brod, Croatia
Organized by	Faculty of Mechanical Engineering, Josip Juraj Strossmayer University of Osijek; DAAAM International Vienna
On the occasion of	30th Anniversary of Josip Juraj Strossmayer University of Osijek
Organizers	Budic, Ivan (Chair); Kljain Milan (Co-Chair); Kozak, Drazen (Secretary); Katalinic, Branko (Chair of Program Committee), Raos, Pero; Saric, Tomislav; Stojic, Antun; Simunovic, Goran; Novoselovic, Daniel; Karakasic, Mirko
Proceedings	ISBN 3-901509-49-6 ISBN 953-6048-29-9
Editors	Kljain M.[ilan], Katalinic B.[ranko] & Budic I.[van]
Published by	DAAAM International Vienna; Faculty of Mechanical Engineering Slavonski Brod Josip Juraj Strossmayer University of Osijek
2004	**3rd DAAAM International Conference on Advanced Technologies for Developing Countries, ATDC-2004**
Founder and Chair	Katalinic B.
Date & Place	June 23-26, 2004, Split, Croatia
Organized by	Faculty of Electrical Engineering, Mechanical Engineering and Naval Architecture, University of Split; DAAAM International Vienna
Organizers	Ivica Veža (Chairman); Željko Domazet (Co-Chairman); Branko Katalinic (Co-Chairman); Tonći Grubić (Secretary), et al.
Proceedings	ISBN 3-901509-41-0
Editors	Katalinic B.[ranko] & Veza, I.[vica]
Published by	DAAAM International Vienna; Faculty of Electrical Engineering, Mechanical Engineering and Naval Architecture, University of Split
2003	**2nd DAAAM International Conference on Advanced Technologies for Developing Countries, ATDC-2003**
Founder and Chair	Katalinic B.
Date & Place	25-28th June 2003, Tuzla, Bosnia and Herzegovina
Organized by	Faculty of Mechanical Engineering, University of Tuzla; DAAAM International Vienna
On the occasion of	40th Anniversary of Mechanical Engineering Study in Tuzla
Organizers	Tufekcic, Dzemo (Chair); Katalinic, Branko (Co-Chairman); Widtmann, Georg (Co-Chairman); Topcic, Alen (Secretary); Kapetanovic, Izrudin; Alagic, Sandira; Alic, Izet; Buljubasic, Indira; Hadzic, Seniha; Mehmedovic, Muhamed; Mujic, Elvedin; Nezirevic, Salahudin; Saric, Emir; Kapetanovic, Irena
Proceedings	ISBN 3-901509-35-6
Editors	Katalinic B.[ranko] & Tufekcic, Dz.[emo]
Published by	DAAAM International Vienna; University of Tuzla - Faculty of Mechanical Engineering
2002	**1st DAAAM International Conference on Advanced Technologies for Developing Countries, ATDC-2002**
Founder and Chair	Katalinic B.
Date & Place	12-14the September 2002, Slavonski Brod, Croatia
Organized by	Faculty of Mechanical Engineering, Strossmayer University of Osijek; DAAAM International Vienna
On the occasion of	40th Anniversary of Mechanical Engineering Study in Slavonski Brod
Organizers	Majdandzic, Niko (Chair), Kljajin M. (Co-Chair), Kozak D. (Secretary), Cicak M., Galeta T., Hnatko E., Ivandic Z., Lujic R., Maglic L., Maticevic G., Raos P., Stojic A., Saric T., Simunovic K., Simunovic G., Vujcic D.
Proceedings	ISBN 3-901509-32-1; ISBN 953-7611-64-X
Published by	DAAAM International Vienna; Strossmayer University of Osijek - Faculty of Mechanical Engineering; Grafika Osijek
Editors	Katalinic B.[ranko] & Kljajin M.[ilan]

18.0 DAAAM INTERNATIONAL WORKSHOPS "COMPUTER AIDED SYSTEMS AND TECHNOLOGIES"

This workshop is based on the bilateral co-operation between two similar departments of the Faculties of Mechanical Engineering of the University of Zilina, Slovakia and Cracow University of Technology - CUT, Cracow, Poland. Those activities are supported by DAAAM National Organisations of Poland and Slovakia, and DAAAM International. In order to install this workshop as stabile long-term regional platform for the *CA Systems and Technologies* fields it was decided to found:

Permanent Committee of Workshop:

Permanent Members:

I. Kuric *(University of Zilina, Zilina, Slovakia),*
B. Katalinic *(Vienna University of Technology and DAAAM International, Vienna, Austria),*
J. Duda *(CUT, Cracow, Poland).*

Academic Year	Members	Institutions
2008/2009	B. Barisic	*University of Rijeka, Croatia*
	J. Audy	*Edith Cowan University, Australia*
	L. Goldhahn	*FH Mittweida, Germany*
	J. Marcinčin	*TU Kosice, Slovakia*
	J. Matuszek	*ATH - University of Bielsko Biala, Poland*
	S. Legutko	*PUT Poznan, Poland*
	N. Ungureanu	*University of Baia Mare, Romania*
2007/2008	B. Barisic	*University of Rijeka, Croatia*
	R. Cotetiu	*University of Baia Mare, Romania*
	L. Goldhahn	*FH Mittweida, Germany*
	J. Marcinčin	*TU Kosice, Slovakia*
	J. Matuszek	*ATH - University of Bielsko Biala, Poland*
	J. Peterka	*TU Bratislava, Slovakia*
	N. Ungureanu	*University of Baia Mare, Romania*
2006/2007	B. Barisic	*University of Rijeka, Croatia*
	R. Cotetiu	*University of Baia Mare, Romania*
	L. Goldhahn	*FH Mittweida, Germany*
	J. Marcinčin	*TU Kosice, Slovakia*
	H. Sjolander	*University of Kristianstad, Sweden*
	P. Šugár	*TU Zvolen, Slovakia*
	N. Ungureanu	*University of Baia Mare, Romania*
2005/2006	R. Cotetiu	*University of Baia Mare, Romania*
	J. Marcinčin	*TU Kosice, Slovakia*
	J. Matuszek	*University of Bielsko Biala, Poland*
	S. Legutko	*PUT Poznan, Poland*
	P. Šugár	*TU Zvolen, Slovakia*
	N. Ungureanu	*University of Baia Mare, Romania*
2004/2005	L. Goldhahn	*FH Mittweida, Germany*
	J. Madl	*CVUT Prague, Czech Republic*
	J. Marcincin	*TU Kosice, Slovakia*
	E. Pay	*University of Baia Mare, Romania*
	D. Plinta	*University of Bielsko Biala, Poland*
	G. Varga	*University of Miskolc, Hungary*
2003/2004	D. Georgiev	*TU Varna, Bulgaria*
	L. Goldhahn	*FH Mittweida, Germany*
	J. Manlig	*TU Liberec, Czech Republic*
	J. Marcincin	*TU Kosice, Slovakia*
	G. Varga	*University of Miskolc, Hungary*
2002/2003	D. Kumicakova	*University of Zilina, Slovakia*
	S. Legutko	*PUT Poznan, Poland*
	J. Marcincin	*TU Kosice, Slovakia*
	I. Neaga	*University of Loughborough, U.K.*
	A. Pendrey	*TU Birmingham, U.K.*
	J. Peterka	*STU Trnava, Slovakia*
	L. Soos	*STU Bratislava, Slovakia*
2001/2002	N. Cubonova	*University of Zilina, Slovakia*
	S. Legutko	*PUT Poznan, Poland*
	J. Marcincin	*TU Kosice, Slovakia*
	I. Neaga	*TU Iasi, Romania*
	J. Peterka	*STU Trnava, Slovakia*
	L. Soos	*STU Bratislava, Slovakia*
2000/2001	N. Cubonova	*University of Zilina, Slovakia*
	S. Legutko	*PUT Poznan, Poland*

	J. Marcincin	*TU Kosice, Slovakia*
	I. Neaga	*TU Iasi, Romania*
	J. Peterka	*STU Trnava, Slovakia*
	L. Soos	*STU Bratislava, Slovakia*
1999/2000	S. Legutko	*PUT Poznan, Poland*
	J. Marcincin	*TU Kosice, Slovakia*
	I. Neaga	*TU Iasi, Romania*

The main aim of the Permanent Committee is to prepare the DAAAM International workshops oriented on CA systems and technologies and propagate all accompanying activities during academic year (projects, WEB journal, common publications, PhD seminars, exchange of students and lectures, etc.). The list of workshops organised during the past.

2008	**13th DAAAM International Workshop "CA Systems and Technologies"**
Date & Place	6 – 8th October 2008, Cracow, Poland
Organized by	CUT Cracow, University of Zilina, DAAAM International Vienna
Organisers	Duda J., Kuric I., Katalinic B.
Proceedings	ISBN 978-83-7242-481-5
Editors	Duda J., Kuric I., Katalinic B.
2007	**12th DAAAM International Workshop "CA Systems and Technologies"**
Date & Place	17 – 19th October 2007, Žilina, Slovakia
Organized by	University of Zilina, CUT Cracow, DAAAM International Vienna
Organisers	Kuric I., Duda J., Katalinic B.
Proceedings	ISBN 3-901509-56-9
Editors	Kuric I., Duda J., Katalinic B.
2006	**11th DAAAM International Workshop "CA Systems and Technologies"**
Date & Place	2 – 4th October 2006, Cracow, Poland
Organized by	CUT Cracow, University of Zilina, DAAAM International Vienna
Organisers	Duda J., Kuric I., Katalinic B.
Proceedings	ISBN 3-901509-56-9
Editors	Duda J., Kuric I., Katalinic B.
2005	**10th DAAAM International Workshop "CA Systems and Technologies"**
Date & Place	12 – 14th December 2005, Strecno - Zilina, Slovak Republic
Organized by	University of Zilina, CUT Cracow, DAAAM International Vienna
Organisers	Kuric, I., Duda J., Katalinic B.
Proceedings	ISBN 3-901509-51-8
Editors	Kuric I., Katalinic B.
2004	**9th DAAAM International Workshop "CA Systems and Technologies"**
Date & Place	5 - 6th October 2004, Cracow, Poland
Organized by	CUT Cracow, University of Zilina, DAAAM International Vienna
Organisers	Duda J., Kuric I., Katalinic B.
Proceedings	ISBN 83-7242-320-2
Editors	Duda J., Katalinic B.
2003	**8th DAAAM International Workshop "CA Systems and Technologies"**
Date & Place	29th September - 1st October 2003, Strecno - Zilina, Slovak Republic
Organized by	University of Zilina, CUT Cracow, DAAAM International Vienna
Organisers	Duda J., Kuric I., Katalinic B.
Proceedings	ISBN 3-901509-39-9
Editors	Kuric I., Katalinic B.
2002	**7th DAAAM International Workshop "CA Systems and Technologies"**
Date & Place	8-9th October 2002, CUT Cracow, Poland
Organized by	CUT Cracow, University of Zilina, DAAAM International Vienna
Organisers	Duda J., Kuric I., Katalinic B.
Proceedings	ISBN 3-901509-33-X
Editors	Duda J., Katalinic B.
2001	**6th DAAAM International Workshop "CA Systems and Technologies"**
Date & Place	25-26th September 2001, Sulov, Slovak Republic
Organized by	University of Zilina, CUT Cracow, DAAAM International Vienna
Organisers	Duda J., Kuric I., Katalinic B.
Proceedings	ISBN 3-901509-27-5
Editors	Kuric I., Katalinic B.
2000	**5th DAAAM International Workshop "CA Systems and Technologies"**
Date & Place	28-29th September 2000, Cracow, Poland

Organized by	CUT Cracow, University of Zilina, DAAAM International Vienna
Organisers	Duda J., Kuric I., Katalinic B.
Proceedings	ISBN 3-901509-16-X
Editors	Duda J., Kuric I., Katalinic B.
1999	**4th DAAAM International Workshop "CA Systems and Technologies: Advanced Methodology of Process Planning"**
Date & Place	12-13th October 1999, Zilina, Slovak Republic
Organized by	University of Zilina, VUT Vienna, CUT Cracow, DAAAM International Vienna, DAAAM Slovakia, DAAAM Poland
Organisers	Kuric I., Katalinic B., Duda J., Poppeova V., Marcincin J.
Proceedings	ISBN 3-901509-12-7
Editors	Kuric I., Katalinic B.
1998	**3rd DAAAM International Workshop "CA Systems and Technologies"**
Date & Place	29-30th September 1998, Cracow, Poland
Organized by	CUT Cracow, University of Zilina, DAAAM Poland, DAAAM Slovakia, DAAAM International Vienna
Organisers	Kuric I., Katalinic B., Duda J., Poppeova V., Marcincin J.
Editor	Kuric I.
1997	**2nd International DAAAM Workshop "CA Systems and Technologies"**
Date & Place	19th March 1997, Zilina, Slovakia
Organized by	University of Zilina, Faculty of Mechanical Engineering, DAAAM International Vienna; DAAAM Slovakia
Organizer	Cubanova N., Janac A., Kuric I., Micieta B., Wessely E.
Editor	Kuric I.

19.0 DAAAM INTERNATIONAL WORKSHOPS "INTELLIGENT MANUFACTURING SYSTEMS"

Founder and chair of DAAAM International workshop on Intelligent Manufacturing Systems is Professor Emil Wessely, Technical University of Kosice.

2002	**4th DAAAM Workshop "Human Factor and Environmentalist"**
Date & Place	5th December 2002, Kosice, Slovakia
Organized by	Department of Environmental Studies and Operations Management of Mechnaical Engineering Faculty of Technical University of Kosice, DAAAM International Vienna
Organized in cooperation with	Slovak Technical Museum of Kosice Oesterreichische Sued-Ost Institut - Vienna
Organisers	Badida M., Wessely E., Bosak M., Majernik M., Lumnitzer E., Kralikova R., Evin E., Rovnak M., Hricova B., Karaszova I., Liberkova L., Verebova H.
Supported by	Action Austria - Slovakia
Proceedings	ISBN 3-901509-37-2
Editors	Katalinic B., Wessely E.
2001	**3rd DAAAM Workshop "Intelligent Manufacturing Systems"**
Date & Place	29 November 2001, Kosice, Slovakia
Organized by	Technical University of Kosice, DAAAM International Vienna, DAAAM Slovakia
Organized in cooperation with	Action Austria-Slovakia, Faculty of Mechanical Engineering - Technical University of Kosice, Slovak Community for System Integration, Slovak Mechanical Engineering Association
Organisers	Wessely E.(Chair), Badida M., Seminsky J., Evin E., Ljoljic B., Kordic V., Pecha C., Rovnak M., Taticek F.
Supported by	Action Austria - Slovakia
Proceedings	ISBN 3-901509-31-3
Editors	Katalinic B., Wessely E.
2000	**2nd DAAAM Workshop "Intelligent Manufacturing Systems"**
Date & Place	30 November 2000, Kosice, Slovakia
Organized by	Technical University of Kosice, DAAAM International Vienna, DAAAM Slovakia
Organized in cooperation with	Action Austria-Slovakia, Faculty of Mechanical Engineering - Technical University of Kosice, Slovak Committee for Management - ZVTS Bratislava, Slovak Mechanical Engineering Association
Organizer	Wessely E.(Chair), Seminsky J., Badida M., Evin E., Paralic J., Knoflicek R., Hradocky L., Bosak M., Kozak L., Vargova J.
Supported by	Action Austria - Slovakia
Proceedings	ISBN 3-901509-18-6
Editors	Katalinic B., Wessely E.

1997	**1st DAAAM Workshop "Intelligent Manufacturing Systems"**
Date & Place	28th November 1997, Kosice, Slovakia
Organized by	Technical University of Kosice, DAAAM International Vienna, DAAAM Slovakia
Organized in cooperation with	Action Austria-Slovakia, Faculty of Mechanical Engineering - Technical University of Kosice, Slovak Committee for Management - ZVTS Bratislava, Slovak Mechanical Engineering Association
Organisers	Wessely E. (Chair), Seminsky J., Evin E., Kovacik J.
Proceedings	ISBN 3-901509-09-7
Editor	Seminsky J.[aroslav]

20.0 DAAAM INTERNATIONAL WORKSHOPS

As an extension of the Symposiums, International Workshops focussing on key issues are arranged. A number of these are detailed below.

1999	**1st International Regional DAAAM-CEEPUS Workshop on Intelligent Machines and Technologies in the 21st Century**
Date & Place	27-29th May 1999, Miskolc, Hungary
Organized by	Faculty of Mechanical Engineering, University of Miskolc in co-operation with DAAAM International Vienna
Proceedings	ISBN 3-901509-14-3
Organizer	Cselenyi J., Alpek F., Katalinic B., Kalmar L., Terstyanszky G., Kerekes I., Siposs I., Peter J.
1996	**4th International DAAAM Workshop "CAD & Intelligent Manufacturing Systems"**
Date & Place	22nd February 1996, Kosice, Slovakia
Organized by	DAAAM International Vienna; DAAAM Slovakia; Technische Universität Kosice, House of Technology, Informatech
Proceedings	ISBN 80-967223-8-7
Organizer	Katalinic B., Seminsky J., Wessely E.
1995	**3rd International DAAAM Workshop "Intelligent Manufacturing Systems"**
Date & Place	31st May 1995, Kosice, Slovakia
Organized by	DAAAM International Vienna; DAAAM Slovakia; Technische Universität Kosice, KAVS; Slovak-Austrian Society Kosice
Organizer	Katalinic B., Seminsky J., Wessely E.
1995	**2nd International DAAAM Workshop "Integration of Applications"**
Date & Place	15th May 1995, IBM Austria, Vienna, Austria
Organized by	DAAAM International Vienna; IBM Österreich; IBM Central Europe and Russia; AKS
Organizer	Beer M., Hansjakob K., Katalinic B., Kaufmann A., Lerch Z., Steinbichler J., Zucker R.
1994	**1st International DAAAM Workshop "Automatic Identification in Theory and Praxis",**
Date & Place	28th April 1994, Vienna University of Technology, Vienna, Austria
Organized by	DAAAM International Vienna; DATALOGIC; Vienna University of Technology, Department of Production Engineering, IFT-IMS
Organizer	Bruck P., Katalinic B., Lerch Z.
1991	**1st International DAAAM Seminar "Concurrent Engineering"**
Date & Place	17-18th April 1991, Kosice, Czechoslovakia
Organized by	Schlumberger Technologies, CAD/CAM Division – Wien; Department for Automatic Production Systems, Faculty of Mechanical Engineering, Technical University of Kosice
Organizer	Wessely E., Katalinic B., Ljuhar D.

21.0 INTERNATIONAL CONFERENCES ORGANIZED IN COOPERATION WITH DAAAM INTERNATIONAL AND/OR DAAAM NATIONAL

2008	**4th International Textile, Clothing and Design Conference, ITC&DC 2008**
Date & Place	5-8th October 2008, Dubrovnik, Croatia
Organised by	Faculty of Textile Technology University of Zagreb
Under the auspices of	Croatian Ministry of Science, Education and Sports, University of Zagreb, Croatian Chamber of Economy
In cooperation with	DAAAM International Vienna; AUTEX Ghent, Belgium; The Textile Institute, Manchester, Great Britain; The International Textile Academy, New Delhi, India; Croatian Academy of Engineering; Croatian Society of Chemical Engineers; Society of Dyers and Colourist, Bradford, Great Britain; INA Oil Industry, Zagreb, Croatia; CARNet, Zagreb, Croatia
Organisers	Dragcevic Z. (Chair), Vuljanic N. (Co-Chair), Bogovic S., Hursa A., Mauser B., Runkas M., Simoncic K.N., Vujasinovic E.

Proceedings	978-953-7105-26-6
Editors	Dragcevic, Z.[vonko]
Published by	Faculty of Textile Technology University of Zagreb
2008	**1st International Conference "Vallis Aurea":Focus on Regional Development**
Date & Place	19th September 2008, Pozega, Croatia
Organized by	Polytechnic of Pozega, DAAAM International Vienna
Organizers	Budic, I. (Chair) / Katalinic, B. (Co-Chair) / Andrlic, V. / Cesarec, P. / Culo, I. / Del Vechio, M. / Fajdic, J. / Svitlica, B. / Zima, D.
Proceedings	ISBN 978-3-901509-60-5, ISBN 978-953-98762-7-0
Editors	Katalinic, B.[ranko]
Published by	Polytechnic of Pozega, Croatia and DAAAM International Vienna
2008	**6th International Conference of DAAAM Baltic "Industrial Engineering"**
Date & Place	24-26th April 2008, Tallinn Technical University, Estonia
Organized by	Tallinn University of Technology, Faculty of Mechanical Engineering, DAAAM Estonia
Under the auspices of	DAAAM International Vienna
In cooperation with	DAAAM International Vienna , BALTECH Consortium, Estonian Academy of Science, Federation of Estonian Engineering Industries, Association of Estonian Mechanical Engineers
Proceedings	ISBN 978-9985-59-783-5
Referred/indexed in	ISI Proceedings
Editor	Kyttner R.
2006	**3rd International Textile Clothing & Design Conference, ITC&DC 2006**
Date & Place	8-11th October 2006, Dubrovnik, Croatia
Organized by	Faculty of Textile Technology, University of Zagreb
Under the auspices of	Croatian Ministry of Science, Education and Sport; University of Zagreb; Croatian Ministry of Economy, Labour and Enterpreneurship; Croatian Chamber of Economy
In cooperation with	DAAAM International Vienna; Association of Universities for Textile, Ghent, Belgium; The Textile Institute, Manchester, Great Britain; The International Textile Academia New Delhi; Croatian Academy of Engineering; Croatian Societs of Chemical Engineers; AATCC, United States of America; IFKT, Reutlingen, Germany; INA Oil Industry, Zagreb, Croatia; City of Dubrovnik, Croatia
Organizers	Dragcevic Z. (Chair), Vuljanic N. (Co-Chair), Bobovcan M., Bogovic S., Geratovic I., Hursa A., Jandric Dj., Martek M., Matusic K., Mauser B., Petrunic I., Vujasinovic E.
Proceedings	ISBN 953-7105-12-1
Editors	Dragcevic, Z.[vonko]
Published by	Faculty of Textile Technology, University of Zagreb
2006	**11th International FMTÜ Conference for Young Researchers: Focus on IMS and Environmentally friendly technologies**
Date & Place	24-25 March 2006, Cluj -Napoca, Romania
Organized by	Technical Section of Transylvanian Museum Society (EME)
Under the auspices of	DAAAM International Viena
In cooperation with	Hungarian Academy of Sciences
Organizers	Gyenge Csaba - president of Scientific Committee Bitay Eniko - president of Organizing Committee
Proceedings	ISBN 973-82331-50-7
Editors	Gyenge C.[saba] & Katalinic B.[ranko]
Published by	Transylvanian Museum Society (EME) - external section of Hungarian Academy of Science
2006	**5th International Conference of DAAAM Baltic "Industrial Engineering – Adding Innovation Capacity of Labour Froce and Entrepreneur"**
Date & Place	20-22nd April 2006, Tallinn Technical University, Estonia
Organized by	Tallinn University of Technology, Faculty of Mechanical Engineering, DAAAM Estonia
Under Auspices of	DAAAM International Vienna
In cooperation with	DAAAM International Vienna , BALTECH Consortium, Estonian Academy of Science, Federation of Estonian Engineering Industries, Associations of Estanian Mechanical Engineers
Proceedings	ISBN 9985-894-92-8
Referred/indexed in	ISI Proceedings
Editor	Kyttner R.
2004	**3rd International Conference Business Systems Management UPS 2004**
Date & Place	23-24th September 2004, Mostar, Bosna and Herzegovina

Organized by	University of Mostar Faculty of Mechanical Engineering and Computing, DAAAM International Vienna, Austrian Society of Engineers and Architects OEIAV-1848, Scientific and Technical Forum of Mostar
Under Auspices of	Presidency of Bosnia and Herzegovina
Organizers	Majstorovic V. (Chair), Katalinic B. (Chair), Covic D., Visekruna, V.
Proceedings	ISBN 3-901509-24-0
Editors	Majstorovic V., Katalinic B., Covic D., Visekruna, V.
2004	**The 5th International Workshop on Emergent Synthesis**
Date & Place	May 24-25, 2004, Budapest, Hungary
Organized by	Computer and Automation Research Institute, Hungarian Academy of Sciences
Chair	L. Monostori, B. Katalinic, K, Ueda
2004	**4th International DAAAM Conference "Industrial Engineering – Innovations as Competitive Edge for "**
Date & Place	29-30th April 2004, Tallinn Technical University, Estonia
Organized by	Tallinn Technical University - Department of Machinery, DAAAM Estonia, DAAAM International Vienna
In cooperation with	Baltic Rector's Council; Association of Estonian; Mechanical Engineers, Federation of Estonian Engineering Industries
Proceedings	ISBN 9985-894-59-6
Editors	Papstel J., Katalinic, B.
2003	**The 1st European-Pacific Congress Globalization: Interaction of Economies, Cultures, Technologies, and Nature,**
Date & Place	23-26th September, 2003, Vladivostok, Russia
Organised by	Association "International Assembly of Capitals and Cities"; DAAAM International Vienna; Far Eastern State Technical University; Far Eastern National University; Maritime State University; International Engineering Academy Primorsk Chamber of Commerce & Industry; Far Eastern Branch of the Russian Academy of Sciences; Russian PEN Center; Vladivostok City Administration; Vladivostok Program Committee; Vladivostok State University of Economics
Organizers	Yu.M. Kopylov (Chairman); (Co-Chairman); G.S. Vishnyakova (Co-Chairman); I.V. Abrosjkin (Co-Chairman); Branko Katalinic (Co-Chairman); M.M. Sokolov, (Co-Chairman); A.A. Akhonin (Coordinator); R.G. Kazmiruk (Secretary)
Proceedings	in Print
2003	**The 8th Scientific Conference of Young Technical Researches, FMTU-2003**
Date & Place	21-22nd March 2003, Cluj-Napoca
Organised by	Technical Section of The Transilvanian Museum Association
Organizers	President of Scientific Committee: Professor Dr Csaba Gyenge
Proceedings	953-96408-8-1
Editors	Gyenge C.[saba]
2002	**1st International Textile Clothing & Design Conference, ITC&DC**
Date & Place	6-9th October 2002, Dubrovnik, Croatia
Organized by	Faculty of Textile Technology, University of Zagreb
Under the auspices of	DAAAM International Vienna; Association of Universities for Textile, Ghent, Belgium; The Textile Institute, Manchester, Great Britan; International Textile Academia; University of Zagreb; Croatian Academy of Engineering
On the occasion of	40th Anniversary of University Education and Scientific Research in the Field of Textiles and Textile Technology in Croatia
Organizers	Dragcevic Z. (Chair), Vuljanic N. (Co-Chair), Grundler D., Hursa A., Krmpotic N., Rolich T., Rogale D., Vujasinovic E.
Proceedings	953-96408-8-1
Editors	Dragcevic, Z.[vonko]
Published by	Faculty of Textile Technology, University of Zagreb
2002	**8th International Conference on Soft Computing: MENDEL 2002**
Date & Place	5-7th June 2001, Brno, Czech Republic
Organized by	Brno University of Technology, Faculty of Mechanical Engineering, Institute of Automation and Computer Science; DAAAM International Vienna, Austria; Foundation for Advanced Information Technology GENOPS, Ltd
Organizers	Ošmera P., Šeda M., Dvořák J., Roupec J., Matousek R.
2001	**5th International Conference Modern Technologies in Manufacturing MTeM 2001:** Assembly and disassembly, CIM, Cold Metal forming technologies, Concurrent Engineering, Ecological Design, Flexible Manufacturing Systems,Industrial Automation,,

	Nonconventional technologies, Rapid Prototyping, Quality Assurance, Robotics
Date & Place	4-6th October 2001, Cluj-Napoca, Romania
Organized by	Technical University of Cluj Napoca, Faculty of Machine Building, DAAAM International Vienna, S.C. Fortpres Cluj-Napoca, S.C. Plastor Oradea, AGIR- General Association of Engineers from Romania
Organizers	Gyenge C. (Chair) et all.
Proceedings	ISBN 973-85354-1-7
Editors	Gyenge C.
2001	**7th International Conference Maintenance 2001**
Date & Place	25-26th June 2001, Opatija, Croatia
Organized by	Croatian Maintenance Society
In cooperation with	DAAAM International Vienna, Union of Croatian Engineers
Organizers	Ivančić I., Cala I., Frković D., Sacevic T., Podgorčić D., Madžarević P., Kristovic D.,Poljak K., Kudelić, R., Ugarković Ž., Susanj B., Pavlovic A.
2001	**7th International Conference on Soft Computing: MENDEL 2001:** Evolutionary Computation, Genetic Programming, Fuzzy Logic, Rough Sets, Neural Networks, Fractals, Bayesian Methods
Date & Place	6--8th June 2001, Brno, Czech Republic
Organized by	Brno University of Technology, Faculty of Mechanical Engineering, Institute of Automation and Computer Science; DAAAM International Vienna, Austria; Foundation for Advanced Information Technology GENOPS, Ltd
Organizers	Ošmera P., Šeda M., Dvořák J., Roupec J., Matousek R.
2001	**2nd International Conference Business Systems Management UPS 2001**
Date & Place	31st May - 2nd June 2001, Mostar, Bosna and Herzegovina
Organized by	University of Mostar Faculty of Mechanical Engineering, DAAAM International Vienna, Austrian Society of Engineers and Architects OEIAV-1848
Organizers	Majstorovic V. (Chair), Katalinic B. (Chair), Covic D., Visekruna, V.
Proceedings	ISBN 3-901509-26-7
Editors	Majstorovic V., Katalinic B., Covic D., Visekruna, V.
2000	**6th International Conference Maintenance 2000**
Date & Place	19-20th June 2000, Opatija, Croatia
Organized by	Croatian Maintenance Society
In cooperation with	EFNMS - European Federation of National Maintenance Societies, DAAAM International Vienna, Union of Croatian Engineers
Organizers	Ivančić I., Cala I., et al.
2000	**5th International Conference on Soft Computing: MENDEL 2000:** Evolutionary Computation, Genetic Programming, Fuzzy Logic, Rough Sets, Neural Networks, Fractals, Bayesian Methods
Date & Place	7-9th June 2000, Brno, Czech Republic
Organized by	Brno University of Technology, Faculty of Mechanical Engineering, Institute of Automation and Computer Science; DAAAM International, Vienna, Austria; Foundation for Advanced Information Technology GENOPS, Ltd
Organizers	Ošmera P., Šeda M., Dvořák J.
Proceedings	ISBN 80-214-1609-2
2000	**2nd International Conference on Industrial Engineering – Activities Targeted to SME**
Date & Place	27 - 29th April 2000, Tallinn Technical University, Estonia
Organized by	Tallinn Technical University - Department of Machinery, DAAAM Estonia, DAAAM International Vienna
Under the auspices of	DAAAM International Vienna
In cooperation with	Baltic Rector's Council; Association of Estonian; Mechanical Engineers, Federation of Estonian Engineering Industries
Organizers	Kyttner R. (chair) Katalinic B., Kulu P., Papstel J., Reedik V., Riives J.
Proceedings	ISBN 9985-894-21-9
Editors	J. Papstel / Katalinic B.
1999	**District Heating Control ´99 – Control of Extensive Heating Systems**
Date & Place	26-27th October 1999, Zlin, Czech Republic
Organized by	Brno University of Technology, Faculty of Technology in Zlin; Czech Power Plants – Ltd.; Association for District Heating of Czech Republic; EUROHEAT & Power, Unichal – Brussels; DAAAM International, Vienna, Austria; DAAAM - Czech Republic
Organizers	Balate J., Vasek V., Vachutova I., Sysala T.
Proceedings	ISBN 80-214-1461-8

1999	**5th International Conference on Soft Computing: MENDEL ´99:** Evolutionary Computation, Genetic Programming, Fuzzy Logic, Rough Sets, Neural Networks, Fractals
Date & Place	9-12th June 1999, Brno, Czech Republic
Organized by	Technical University Brno, Faculty of Mechanical Engineering, Institute of Automation and Computer Science; DAAAM International, Vienna, Austria; Foundation for Advanced Information Technology GENOPS, Ltd
Organizers	Ošmera P., Šeda M., Dvořák J.
Proceedings	ISBN 80-214-1131-7
1999	**5th International Conference Maintenance 99**
Date & Place	7- 8th June 1999, Opatija, Croatia
Organized by	Croatian Maintenance Society
In cooperation with	DAAAM International Vienna, EFNMS - European Federation of National Maintenance Societies
Organizers	Ivančić I., Poljak K., Ugarković Ž., Frković D., Podgorčić D., Madžarević P., Kuzmanić A. Kudelić, R.
Proceedings	ISBN 953-97615-1-4
Editors	Ivančić I., Čala I.
1999	**XXIVth Seminary ASR 2000: Instruments and Control**
Date & Place	28th April 1999, Ostrava, Czech Republic
Organized by	VŠB - Technical University Ostrava, Faculty of Mechanical Engineering, Institute of Automation Equipment and Control; DAAAM National - Czech Republic, Zlín, Czech Republic; IMECO - Czech National Committee; Committee of Applied Cybernetics and Information Technology Ostrava
Organizers	Farana R., Smutny J., Szlachta D., Fojtik D., Piechota H.
Proceedings	ISBN 80-7078-774-0
Editors	Farana R., Smutny J., Szlachta D., Fojtik D., Piechota H.
1999	**XXIIth Seminary ASR ´99 : Computers on Measurements, Diagnostics and Control**
Date & Place	22nd May 1999, Ostrava, Czech Republic
Organized by	VŠB - Technical University Ostrava, Faculty of Mechanical Engineering, Institute of Automation Equipment and Control; DAAAM National - Czech Republic, Zlín, Czech Republic; IMECO - Czech National Committee
Organizers	Smutný L., Farana R., Vítecek A., Baláte J.
Proceedings	ISBN 80-7078-666-3
1998	**3rd International Conference „ New Trends in Automation of Energetic Processes ´98"**
Date & Place	19 - 20th May 1998, Zlín, Czech Republic
Organized by	Technical University Brno, Faculty of Technology in Zlín; Czech Power Company, Ltd.- CEZ, a.s.; Association for District Heating, Czech Republic; STU Bratislava, Faculty of Electrical Engineering and Information Technology; Slovak Power Plants, Ltd. - SE, a.s., Bratislava; DAAAM International Vienna, Austria; DAAAM National - Czech Republic
Organizer	Balate J., Vasek V., Sysala T. & Co.
Proceedings	ISBN 80-214-1094-9
Editor	Balate J., Sysala T.
1998	**4th International Mendel Conference: MENDEL ´98** on Genetic Algorithms, Optimization Problems, Fuzzy Logic, Neural Networks, Rough Sets
Date & Place	24 - 26th June 1998, Brno, Czech Republic
Organized by	Technical University Brno, Faculty of Mechanical Engineering, Institute of Automation and Information Technology; DAAAM International, Vienna, Austria; Foundation for Advanced Information Technology GENOPS, Ltd.
Organizer	Osmera P., Seda M., Simoník I.
Proceedings	ISBN 80 – 214 - 1199 - 6
1998	**XXIth Seminary ASR ´98 : Computers on Measurements, Diagnostics and Control**
Date & Place	22nd May 1998, Ostrava, Czech Republic
Organized by	VSB - Technical University Ostrava, Faculty of Mechanical Engineering, Institute of Automation Equipment and Control; DAAAM National - Czech Republic; IMECO - Czech National Committee
Organizer	Smutny L., Farana R., Vítecek A., Balate J.
Proceedings	ISBN 80-7078-559-4
1998	**9th International Conference on Information and Intelligent Systems IIS´98**
Date & Place	23-25th September 1998, University of Zagreb, Faculty of Organization and Informatics Varazdin, Croatia
Organized by	University of Zagreb, Faculty of Organization and Informatics Varazdin & DAAAM

	International Vienna
Organizers	Aurer B., Katalinic B., Kero K., et. al.
Proceedings	UDC: 007:681.3:651(082)
Editors	Aurer B., Logozar, R.
1998	**3rd National DAAAM Conference "Science ´98"**
On the occasion of	80 Years of Tallinn Technical University
Date & Place	September 1998, Tallinn Technical University, Estonia
Organized by	DAAAM Estonia, Tallinn Technical University, Department of Machinery
Organizers	Papstel J.
Proceedings	ISBN 9985-894-02-2
Editor	Papstel J.
1997	**4th International Conference Maintenance 97**
Date & Place	16-17th October 1997, Opatija, Croatia
Organized by	Croatian Maintenance Society
In cooperation with	DAAAM International Vienna, Croatian Academy of Science and Art, Zagreb EFNMS - European Federation of National Maintenance Societies
Organizers	Ivančić I., Čala I., Vidić J., Udiljak T., Ugarković Ž., Širola D., Podgorčić D., Keser A.
Proceedings	ISSN-1331-4114
Editors	I. Ivančić
1997	**1st International Conference "Industrial Engineering - Actual Activities"**
Date & Place	25-27th September 1997, Tallinn Technical University, Tallinn, Estonia
Organized by	Tallinn Technical University - Institute of Machinery, DAAAM Estonia Tallinn, DAAAM International Vienna
Under the auspices of	DAAAM International Vienna
In cooperation with	Baltic Rectors´ Council, Association of Baltic Material Societies, Federation of Estonian Engineering Industries, Association of Estonian Mechanical Engineers
Organizer	Kulu P. (vice-chair), Kyttner R. (general chair), Papstel J.(vice-chair), Katalinic B., Kübarsepp J., Reedik V., Riives J.
Proceedings	ISBN 3-901509-07-0
Editors	Papstel J., Katalinic B.
1997	**2nd International Conference "Enterprise in Transition"**
On the occasion of	1700 Years of the City of Split
Date & Place	22-24th May 1997, Split - Brela, Croatia
Organized by	University of Split, Faculty of Economics, Split, Croatia; DAAAM International Vienna
Under the auspices of	The Government of the Republic of Croatia, The Croatian Chamber of Economy, Bank of Zagreb
Organizer	Alfirevic N., Buble M., Derado D., Goic S., Katalinic B., Mijacika D., Pavic I. (Chair), Reic Z. (Secretary)
Proceedings	ISBN 953-6024-14-4
Publisher & Editor	Faculty of Economics, University of Split, Split
Co-Publisher	DAAAM International Vienna
1997	**3rd International Mendel Conference on Genetic Algorithms MENDEL '97**
Date & Place:	25-27th June 1997, Brno, Czech Republic
Organisation:	Technical University of Brno, Faculty of Mechanical Engineering, Institute of Automation and Computer Science; DAAAM International Vienna; Foundation Advanced Information Technology; GENOPS, Ltd.
Organizer:	Osmera P., Seda M., Simonik I.
1996	**The Second National DAAAM Conference SCIENCE '96**
Date & Place:	1996, Tallinn, Estonia
Organisation:	Tallinn Technical University - Estonia, DAAAM International Vienna
Organizer:	Papstel J. & Co
Proceedings:	ISBN 9985-59-041-4
Editor:	Papstel J.
1996	**2nd International Mendel Conference on Genetic Algorithms MENDEL '96**
Date & Place:	26-28th June 1996, Brno, Czech Republic
Organisation:	Technical University of Brno, Faculty of Mechanical Engineering, Institute of Automation and Computer Science; DAAAM Czech Republic & DAAAM International Vienna; Foundation Advanced Information Technology; AutoCont Brno Ltd. Computer Systems
Organizer:	Osmera P. & Co.

Proceedings:	ISBN 80-214-0769-7
1996	**13th Conference BIAM '96**
Date & Place:	18-20th June 1996, Zagreb, Croatia
Organisation:	KoREMA, Croatian Society for Communications, Computing, Electronics, Measurement and Control, DAAAM International Vienna
Organizer:	Surina T., Vranjes B., Katalinic B., Jerbic B., Cosic P., Husic S., Kovacic Z., Kunica Z., Majetic D., Momcilovic D., Schauperl Z., Stambuk-Borsic N., Stefanic N., Udiljak T.
Proceedings:	ISBN 953-6037-12-2
1996	**Nanotechnologie - Nanometrologie**
Date & Place:	26th April 1996, Brno, Czech Republic
Organisation:	Technical University of Brno, Faculty of Mechanical Engineering, Vienna University of Technology; DAAAM Czech Republic
Organizer:	P. H.Osanna, J. Svejcar, J. Vackar, M. Skopal
1996	**Rechenmaschinen in Messungen, Diagnostik und Steuerung**
Date & Place:	18th April 1996, Ostrava, Czech Republic
Organisation:	Technical University of Ostrava TU-VSB, KAKI Ostrava, DAAAM Czech Republic, IMECO Czech Republic
Organizer:	Smutny L., Farana R., Vitecek A.
1995	**„New Trends in the Automation of Energetical Processes"**
Date & Place:	17-18th May 1995, Zlin, Czech Republic
Organisation:	TU-Brno / FT-Zlin, KART, DAAAM Czech Republic & DAAAM International Vienna, and others
Organizer:	Balate J., Janacova J., Kubalcik M., Sysala T., Neumann P., Vasek V.
Proceedings:	ISBN 80-214-0654-2

22.0 DAAAM & CEEPUS JEP

Actual Project: HR-108 CEEPUS Project "Concurrent Product and Technology Development - Teaching, Research and Implementation of Joint Programs Oriented in Production and Industrial Engineering"
Network Co-ordinator: ***Barisic Branimir - University of Rijeka***

Project Code: ***Project Duration:*** ***Project Coordinator:***	**HR-108** ***2006-*** ***Barisic Branimir - University of Rijeka***
2006/2007 Name:	**Concurrent Product and Technology Development**
Project Partner 06/07	Department of Production Engineering, Faculty of Engineering, University of Rijeka, Rijeka, Croatia / Department of Production Engineering, Faculty of Mechanical Engineering and Naval Architecture, University of Zagreb, Zagreb, Croatia / Institute of Production Engineering, Vienna University of Technology, Vienna, Austria / Department of Manufacturing Engineering, Budapest University of Technology and Economics, Budapest, Hungary / Department of Production Engineering, Faculty of Mechanical Engineering, Czech Technical University in Prague, Prague, Czech Republic / Department of Production Engineering, Faculty of Mechanical Engineering, University of Miskolc, Miskolc, Hungary /Institute of Mechanical Technology, Poznan University of Technology, Poznan, Poland / Department for Material Forming Technologies and Surface Engineering, University of Novi Sad, Novi Sad, Serbia / Department of Machining and Automation, Faculty of Mechanical Engineering, University of Žilina, Žilina, Slovakia / Department of Production Technologies and Systems, Faculty of Mechanical Engineering, University of Ljubljana, Ljubljana, Slovenia.
Organizer 06/07	**Barisic B.[ranimir],** Math M.[iljenko], Katalinic B.[ranko], Szalay T.[ibor], Kreibich V.[iktor], Varga G.[yula], Legutko S.[tanislaw], Plancak M.[iroslav], Kuric I.[van], Kuzman K.[arl]
2007/2008 Name:	**Concurrent Product and Technology Development**
Project Partner 07/08	Department of Production Engineering, Faculty of Engineering, University of Rijeka, Rijeka, Croatia / Department of Production Engineering, Faculty of Mechanical Engineering and Naval Architecture, University of Zagreb, Zagreb, Croatia / Institute of Production Engineering, Vienna University of Technology, Vienna, Austria / Department of Manufacturing Engineering, Budapest University of Technology and Economics, Budapest, Hungary / Department of Production Engineering, Faculty of Mechanical Engineering, Czech Technical University in Prague, Prague, Czech Republic / Department of Production Engineering, Faculty of Technology, Tomas Bata University in Zlín, Zlín, Czech Republic / Department of Production Engineering, Faculty of Mechanical Engineering, University of Miskolc, Miskolc, Hungary /Institute of Mechanical Technology, Poznan University of Technology, Poznan, Poland / Department for Material Forming Technologies and Surface Engineering, University of Novi

	Sad, Novi Sad, Serbia / Department of Machining and Automation, Faculty of Mechanical Engineering, University of Žilina, Žilina, Slovakia / Department of Production Technologies and Systems, Faculty of Mechanical Engineering, University of Ljubljana, Ljubljana, Slovenia.
Organizer 07/08	**Barisic B.[ranimir],** Math M.[iljenko], Katalinic B.[ranko], Szalay T.[ibor], Kreibich V.[iktor], Lukovics I.[mrich],Varga G .[yula], Legutko S.[tanislaw], Plancak M.[iroslav], Kuric I.[van], Kuzman K.[arl].
2008/2009 Name:	**Concurrent Product and Technology Development**
Project Partner 08/09	Department of Production Engineering, Faculty of Engineering, University of Rijeka, Rijeka, Croatia / Department of Production Engineering, Faculty of Mechanical Engineering and Naval Architecture, University of Zagreb, Zagreb, Croatia / Institute of Production Engineering, Vienna University of Technology, Vienna, Austria / Department of Manufacturing Engineering, Budapest University of Technology and Economics, Budapest, Hungary / Department of Machining and Assemmbly, Faculty of Mechanical Engineering, Technical University Ostrava, Ostrava, Czech Republic /Department of Production Engineering, Faculty of Mechanical Engineering, Czech Technical University in Prague, Prague, Czech Republic / Department of Production Engineering, Faculty of Technology, Tomas Bata University in Zlín, Zlín, Czech Republic / Department of Production Engineering, Faculty of Mechanical Engineering, University of Miskolc, Miskolc, Hungary / Institute of Mechanical Technology, Poznan University of Technology, Poznan, Poland / Department of Mechanical Technology and Metrology, Faculty of Mechatronics and Machinery Design, Kielce University of Technology, Kielce, Poland / Department for Material Forming Technologies and Surface Engineering, University of Novi Sad, Novi Sad, Serbia / Department of Machining and Automation, Faculty of Mechanical Engineering, University of Žilina, Žilina, Slovakia / Department of Production Technologies and Systems, Faculty of Mechanical Engineering, University of Ljubljana, Ljubljana, Slovenia.
Organizer 08/09	**Barisic B.[ranimir],** Math M.[iljenko], Katalinic B.[ranko], Szalay T.[ibor], Cep R.[obert], Kreibich V.[iktor], Lukovics I.[mrich],Varga G .[yula], Legutko S.[tanislaw], Adamczak S.[tanislaw], Plancak M.[iroslav], Kuric I.[van], Kuzman K.[arl]

PAST DAAAM – CEEPUS Joint European Annual Projects

Project Code:	***A-104***
Project Duration:	***1999-2004***
Project Coordinator:	***Branko Katalinic - Vienna University of Technology***
2003/2004 Name:	**Intelligent Manufacturing and Automation**
Project Partner 03/04	Institute of Production Engineering, Vienna University of Technology, Vienna, Austria / Department of Manufacturing Engineering, Budapest University of Technology and Economics, Budapest, Hungary / Department of Manufacturing Engineering, Technical University of Cluj-Napoca, Cluj-Napoca, Romania / Department of Technology, Faculty of Mechanical Engineering and Naval Architecture, University of Zagreb, Zagreb, Croatia / Department of Technological Devices and Systems, Faculty of Material Science and Technology Trnava, Slovak University of Technology Bratislava, Trnava, Slovakia / Institute of Production Engineering, University of Maribor, Maribor, Slovenia / Department of Environmental Studies and Process Control, Technical University of Kosice, Kosice, Slovakia / Department of Machine Technology, The Silesian University of Technology, Gliwice, Poland / Department of Production Engineering, Faculty of Engineering, University of Rijeka, Rijeka, Croatia / Institute of Mechanical Technology, Poznan University of Technology, Poznan, Poland
Organizer 03/04	**Katalinic B.[ranko],** Alpek F.[erenc], Balic J.[oze], Gyenge C.[saba], Kosmol J.[an], Legutko S.[tanislaw], Mikac T.[onci], Velisek K.[arol], Wessely E.[mil], Udiljak T.[oma]
2002/2003 Name:	**Intelligent Manufacturing and Automation**
Project Partner 02/03	Institute of Production Engineering, Vienna University of Technology, Vienna, Austria / Department of Manufacturing Engineering, Budapest University of Technology and Economics, Budapest, Hungary / Department of Manufacturing Engineering, Technical University of Cluj-Napoca, Cluj-Napoca, Romania / Department of Technology, Faculty of Mechanical Engineering and Naval Architecture, University of Zagreb, Zagreb, Croatia / Department of Technological Devices and Systems, Faculty of Material Science and Technology Trnava, Slovak University of Technology Bratislava, Trnava, Slovakia / Institute of Production Engineering, University of Maribor, Maribor, Slovenia / Department of Environmental Studies and Process Control, Technical University of Kosice, Kosice, Slovakia /

	Department of Machine Technology, The Silesian University of Technology, Gliwice, Poland / Department of Production Engineering, Faculty of Engineering, University of Rijeka, Rijeka, Croatia / Department of Industrial Robotics, University of Applied Sciences - Fachhochschule Technikum - Wien, Vienna, Austria / Institute of Mechanical Technology, Poznan University of Technology, Poznan, Poland
Organizer 02/03	**Katalinic B.[ranko],** Alpek F.[erenc], Balic J.[oze], Gyenge C.[saba], Kosmol J.[an], Legutko S.[tanislaw], Mikac T.[onci], Stopper M.[arkus], Velisek K.[arol], Wessely E.[mil], Udiljak T.[oma]
2001/2002 Name:	**Assembly Automation in Manufacturing Engineering**
Project Partner 01/02	Vienna University of Technology, Budapest University of Technology and Economics, Silesian University of Technology, Slovak Technical University Bratislava / Trnava, Technical University Cluj-Napoca, Technical University of Kosice, University of Maribor, University of Zagreb
Organizer 01/02	**Katalinic B.,** Alpek F., Kosmol J., Velisek K., Gyenge C., Wessely E., Balic, J., Jerbic B.,
2000/2001 Name:	**Assembly Automation in Manufacturing Engineering**
Project Partner 00/01	Vienna University of Technology, Slovak Technical University Bratislava / Trnava, Technical University Cluj-Napoca, Technical University of Budapest, Technical University of Kosice, University of Maribor, University of Zagreb
Organizer 00/01	Katalinic B., Velisek K., Gyenge C., Alpek F., Wessely E., Balic J., Jerbic B.
1999/2000 Name:	**Assembly Automation in Manufacturing Engineering**
Project Partner 99/00	Vienna University of Technology, Slovak Technical University Bratislava / Trnava, Technical University Cluj-Napoca, Technical University of Budapest, University of Maribor, University of Zagreb
Organizer 99/00	**Katalinic B.,** Velisek K., Gyenge C., Alpek F., Trzan F., Jerbic B.

Project Code:	***RO-01***
Project Duration:	***2002/2003***
Project Coordinator:	***Csaba Gyenge, Technical University of Cluj-Napoca***
2002/2003 Name:	**Computer-Aided Systems for Manufacture and Measurement of Machine Elements**
Project Partner	Department of Manufacturing Engineering / Technical University of Kosice, Department of Manufacturing Systems / University of Maribor, Faculty of Mechanical Engineering, Institute for Production Engineering / Poznan University of Technology, Faculty of Mechanical Engineering, Institute of Mechanical Technology / Kecskemet College, Faculty of Technology

Project Code:	***PL-1***
Project Duration:	***1999-2002***
Project Coordinator:	***Stanislaw Adamczak - Technical University of Kielce***
2001/2002 Name:	**Computer-Aided Systems for Manufacture and Measurement of Machine Elements**
Project Partner	Technical University of Kielce, Vienna University of Technology, University of Zilina, Faculty of Mechanical Engineering of The Technical University of Kosice, Faculty of Production Technologies of The Technical University of Kosice in Presov, University of Maribor, Technical University of Ostrava, University of Zagreb, Technical University of Bratislava, Technical University of Prague, Technical University of Brno
Organizer	**Adamczak S.,** Osanna P. H., Obmascik M., Wessely E., Vasilko K., Cus F., Vrba V., Mahovic S., Janac J., Dvorak R., Vackar J.
2000/2001 Name:	**Geometrical Surface Structure of Machine Parts, Constitution, Methods for Measurement and Assessment, Influence on the Operational Capacity of Mating Machine Parts**
Project Partner	Technical University of Kielce, Vienna University of Technology, University of Zilina, Technical University of Kosice, University of Maribor, Technical University of Ostrava, University of Zagreb, Technical University of Bratislava, Technical University of Prague, Technical University of Brno
Organizer	**Adamczak S.,** Osanna P. H., Obmascik M., Wessely E., Cus F., Vrba V., Mahovic S., Janac J., Dvorak R., Vackar J.
1999/2000 Name:	**Geometrical Surface Structure of Machine Parts, Constitution, Methods for Measurement and Assessment, Influence on the Operational Capacity of Mating Machine Parts**
Project Partner	Technical University of Kielce, Vienna University of Technology, University of Zilina, Technical University of Kosice, University of Maribor, Technical University of Ostrava, University of Zagreb, Technical University of Bratislava, Technical University of Prague, Technical University of Brno

Organizer	**Adamczak S.,** Osanna P. H., Obmascik M., Wessely E., Cus F., Vrba V., Mahovic S., Janac J., Dvorak R., Vackar J.

Project Code:	***RO-01***
Project Duration:	***1998/1999***
Project Coordinator:	***Csaba Gyenge, Technical University of Cluj-Napoca***
1998/1999 Name:	**Development and Application of Environment-Friendly Technologies in Manufacturing Engineering**
Project Partner	Technical University of Cluj-Napoca, Technical University of Budapest, University of Maribor, Technical University of Kosice, Technical University of Bucharest
Organizer	**Gyenge C.,** Alpek F., Balic J., Anna V., Iatan F.

Project Code:	***PL-1***
Project Duration:	***1998/1999***
Project Coordinator:	***Stanislaw Adamczak - Technical University of Kielce***
1998/1999 Name:	**Metrology in Quality Assurance Systems**
Project Partner	Technical University of Kielce, Vienna University of Technology, University of Zilina, Technical University of Kosice, University of Maribor, Technical University of Ostrava, University of Zagreb
Organizer	**Adamczak S.,** Osanna P. H., Obmascik M., Wessely E., Cus F., Vrba V., Mahovic S.

Project Code:	***H-78***
Project Duration:	***1998/1999***
Project Coordinator:	***Alpek,F.***
1998/1999 Name:	**Process Integrated Measurement Techniques and Methods for Quality Management**
Project Partner	Technical University of Budapest, Vienna University of Technology, Cracow Institute of Technology, Slovak Technical University Bratislava, University of Maribor
Organizer	**Alpek F.,** Osanna H. P., Sostar A., Gawlik J., Kralik,M.

23.0 DAAAM INTERNATIONAL BOOKS

The DAAAM International initiated publishing of books in the field of advanced manufacturing and automation. There are three series: "Manufacturing technology", "DAAAM International University Book" and "Management Sciences". Till now the following books are published:

DAAAM Publishing series - Manufacturing technology:

1. **Joze Balic:** ***"Contribution to Integrated Manufacturing"***, DAAAM Publishing series - Manufacturing technology, Editor: B Katalinic, ISBN 3-901509-03-8, DAAAM International Vienna, Vienna 1999 (Promotion at DAAAM 1999 - Vienna)
2. **Borut Buchmeister:** ***"English-Slovene Glossary of Terms used in Production Engineering and Production Management"***, DAAAM Publishing series - Manufacturing technology, Editor: B Katalinic, ISBN 3-901509-17-8, DAAAM International Vienna, Vienna 2000 (Promotion at DAAAM 2000 - Opatija)
3. **Joze Balic:** ***"FMS: Development Structure, Operation, Handling, Tooling"***, DAAAM Publishing series - Manufacturing technology, Editor: B Katalinic, ISBN 3-901509-20-8, DAAAM International Vienna, Vienna 1999 (Promotion at DAAAM 2001- Jena)
4. **Joze Balic & Niko Majdandzic:** ***"Digit@l F@ctory"***, DAAAM Publishing series - Manufacturing technology, Editor: B Katalinic, ISBN 3-901509-67-4, DAAAM International Vienna, Vienna 2008 (Promotion at DAAAM 2008-Trnava)

DAAAM Publishing series - University Book:

5. **Vlado Majstorovic:** ***"Upravljanje proizvodnjom i projektima – Production and Project Management"***, (in Croatian) DAAAM International University Book, Editor: B Katalinic, ISBN 3-901509-23-2, DAAAM International Vienna and University of Mostar, Mostar – Vienna, 2001 (Promotion at International Conference in Mostar, May 31st, 2001)

6. **Dragan Covic, Vlado Majstorovic, Vojo Visekruna:** ***"Poslovni sustavi – Business Systems"***, (in Croatian) DAAAM International University Book, Editor: B Katalinic, ISBN 3-901509-22-4, DAAAM International Vienna and University of Mostar, Mostar – Vienna, 2001 (Promotion at International Conference in Mostar, May 31st, 2001)

DAAAM Publishing series – Scientific Monograph:

7. **Iztok Palcic,** Industrial Clusters, DAAAM International, Editor: B. Katalinic, ISBN-10 3-901509-80-1; ISBN-13: 978-3-901509-80-3, DAAAM International Publishing, Vienna, 2007

DAAAM Publishing Series – Management Science:

8. **Dragan Covic:** ***"Inventivnost Tima – Vodic za menedzere"***, (in Croatian) DAAAM International University Book, Editor: B Katalinic, ISBN 3-901509-25-4, DAAAM International Vienna and University of Mostar, Mostar – Vienna, 2003. (Promotion October 2003, Sarajevo – Mostar)

9. **Zvonko Kremljak:** ***"Decision making under risk"***, DAAAM International University Book, Editor: B Katalinic, ISBN 3-901509-44-5, DAAAM International Vienna, Vienna 2004 (Promotion at DAAAM 2004 – Vienna)

10. **Zvonko Kremljak & Borut Buchmeister:** ***"Uncertainty and Development of Capabilities"***, DAAAM Publishing Series – Management Science Editor: Branko KATALINIC, ISBN 3-901509-55-0, DAAAM International Vienna, Vienna, 2004 (Promotion at DAAAM 2006 – Vienna)

DAAAM International Scientific Book

11. ***DAAAM International Scientific Book 2002,*** ISBN 3-901509-30-5, hard cover, 713 pages, Publisher DAAAM International Vienna, Editor: B.[ranko] Katalinic, Vienna, 2002. (Promotion of the book at DAAAM 2002 Vienna, 2002-10-23/26)
12. ***DAAAM International Scientific Book 2003,*** ISSN 1726-9687, ISBN 3-901509-36-4, 654 pages, hard cover, Publisher DAAAM International Vienna, Editor: B.[ranko] Katalinic, Vienna 2003. *Abbreviated key title: DAAAM int. sci. book* (Promotion of the book at DAAAM 2003 Sarajevo 2003-10-22/25)
13. ***DAAAM International Scientific Book 2004,*** ISSN 1726-9687, ISBN 3-901509-38-0, 730 pages, hard cover, Publisher DAAAM International Vienna, Editor: B.[ranko] Katalinic, Vienna 2004. *Abbreviated key title: DAAAM int. sci. book* (Promotion of the book at DAAAM 2004 Vienna 2004-11-03/06)
14. ***DAAAM International Scientific Book 2005,*** ISSN 1726-9687, ISBN 3-901509-43-7, 728 pages, hard cover, Publisher DAAAM International Vienna, Editor: B.[ranko] Katalinic, Vienna 2005. *Abbreviated key title: DAAAM int. sci. book*
15. ***DAAAM International Scientific Book 2006,*** ISSN 1726-9687, ISBN 3-901509-47-X, 692 pages, hard cover, Publisher DAAAM International Vienna, Editor: B.[ranko] Katalinic, Vienna 2006. *Abbreviated key title: DAAAM int. sci. book*
16. ***DAAAM International Scientific Book 2007,*** ISSN 1726-9687, ISBN 3-901509-60-7-X, 694 pages, hard cover, Publisher DAAAM International Vienna, Editor: B.[ranko] Katalinic, Vienna 2007. *Abbreviated key title: DAAAM int. sci. book*

Book published under the auspices of DAAAM International

17. **Balc, N.; Gyenge, Cs. & Berce, P: *"Design for Competitive Manufacturing",*** Published under the auspices of DAAAM International, ISBN 973-7898-31-1, Alma mater Publishing, Cluj-Napoca, 2006.

Several other book projects are in process, if you want to join one of them or propose to write a new book please contact the Editor.

24.0 INTERNATIONAL JOURNAL OF SIMULATION MODELLING

Title	**International Journal of Simulation Modelling**
Abbreviated key title	**IJSIMM Int. j. simul. model.**
ISSN	ISSN 1726-4529
Editors	B.[orut] Buchmeister (Editor-In-Chief); B.[ranko] Katalinic (Editor)
Founded	2002
Published	Quarterly by DAAAM International Vienna
Citation and reference	Article published in Journal are crossreferenced and online available on Internet. Additionali they are cited and referenced in the: Inspec Database and Cambridge Scientific Abstracts
Info and Link	http://www.ijsimm.com, http://www.daaam.com,

25.0 20 YEARS DAAAM INTERNATIONAL 1990-2010 PROJECT

Aims: The main aims for future development of our organisation are to increase meaning, importance, quality, relevance and prestige of DAAAM International Network and members in international academic and scientific community. To achieve those aims it is necessary to improve quality and quantity of existing activities and to organize new creative future oriented activities using enormous potential of knowledge, abilities, prestige, relevance and creativity of our members.

Who: Each person in DAAAM International community is welcome to give his/her own contribution to achieving defined aims and can be responsible for one initiative or activity on the level of DAAAM International.

Focus: Focused on upper aims and the fact that we are going soon to celebrate first 20 years of our organisation we are starting an open platform under the name "20 Years of DAAAM International Project". The project is open platform the main parts and topics of the project are:

- To complete and publish Cumulative Bibliography of DAAAM International including all publications: proceedings of DAAAM Symposiums; proceedings of DAAAM Workshops; proceedings of ATDC conferences; DAAAM International Scientific Books; International Journal on Simulation Modelling; Books.
- Founding and organisation of web DAAAM International Archive
- Founding and organisation of web DAAAM International Photo Gallery
- Completing and publishing exclusive monograph "20 Years of DAAAM International"
- Founding, completing and publishing "Who is who in DAAAM International" (Founding members, members of committees since 1990, authors, honours, awards, personal contributions, ...)
- improve quality and quantity of existing activities of DAAAM
- to motivate old and animate new people to take a part in our activities
- better use of human resources
- including of younger researchers in our activities
- sponsoring and partnerships
- to increase dissemination of information about our activities
- use of synergies of DAAAM Network in science, research, education and international scientific and academic cooperation
- awards and recognition for DAAAM International on the occasion of celebrating of anniversary of your institution
- to increase quality of our publications
- increase citing and indexing of our publications
- publishing of journals, monographs and student books
- international scientific and academic cooperation with joint application for co-operation of institutions and mobility of teachers and students
- cooperation in industrial and research projects
- DAAAM International certifications of Labs
- Marketing

- DAAAM International Experts data bank
- Cyber space and DAAAM
- Spectrum of Activities to celebrate 20 Years of DAAAM International in 2009
- and many of yours ideas

Call for cooperation: Please think about your wishes, possibilities, and field of interest and propose me one to three initiatives you or your team want to start, organize, manage and be responsible in the frame of DAAAM International. I will pick out of your proposal one initiative which is at the moment the most important from the point of view of DAAAM International. I will give you additional information and suggestions to support your initiative and make it relevant and usable for DAAAM International and synchronised with other activities.

Contact: I am looking forward to see soon your ideas, creativity and initiatives. Please be free to contact me at president@daaam.com I will be happy to hear from you.

26.0 DAAAM - INTERNET & CONTACT ADDRESS

Additional information about DAAAM activities your can obtain from DAAAM homepage: http://www.daaam.com and http://www.daaam.org If you are interested to take a part in DAAAM activities please contact:

✉ PROF. DR. BRANKO KATALINIC, DAAAM INTERNATIONAL PRESIDENT,
TU-WIEN, KARLSPLATZ 13/311, A-1040 WIEN, AUSTRIA, EUROPE
Phone: 0043 1 58801 31121, *Fax#* 0043 1 58801 31199,
E-mails: president@daaam.com, katalinic@mail.ift.tuwien.ac.at

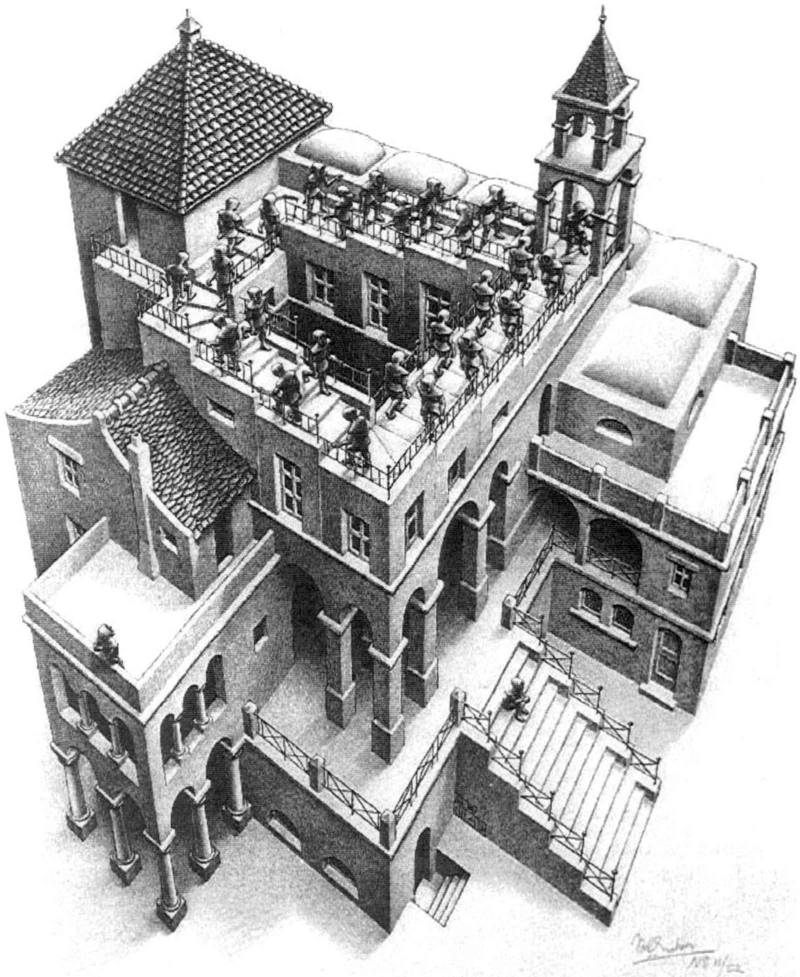

Annals of DAAAM for 2008 & Proceedings of the 19th International DAAAM Symposium, ISSN 1726-9679
Editor B. Katalinic, Published by DAAAM International, Vienna, Austria 2008
Make Harmony Between Technology and Nature, and Your Mind will Fly Free as a Bird

CONTENTS

D

I

N

O

P

R

S

W

Y

Annals of DAAAM for 2008 & Proceedings of the 19th International DAAAM Symposium, ISSN 1726-9679
Editor B. Katalinic, Published by DAAAM International, Vienna, Austria 2008
Make Harmony Between Technology and Nature, and Your Mind will Fly Free as a Bird

INFLUENCE OF LASER POWER OUTPUT ON THE QUALITY OF COLOUR IMAGING

MAJNARIC, I[gor]; MODRIC, D[amir] & GOLUBOVIC, K[ristijan]

Abstract: *This paper investigates commercial 4-beam laser diode array with 830 nm laser wavelength output which is used for creating DAD virtual image on OPC photoconductors, i.e. the future printing elements. It is possible to generate different sizes of screen elements on photoconductor, and consequently in printing process, with the controlled variation of output laser power (from 1 μW to 12 μW). The results have been processed colorimetrically and with the defining of color difference CIE LAB ΔE with characteristic solid tones and 50% screened area. Results obtained are interesting because of the fact that changes in prints are not fully proportional to laser power, differences also exist between different separations (especially in yellow colour), which is caused by the chemical composition of toner.*

Key words: *digital colour offset, CIE LAB ΔE, image analysis*

1. INTRODUCTION

Commonly used imaging method in electrophotographic printing process is DAD (discharged area development) method where the photoconductor surface is exposed to prior defined electromagnetic wavelength (energy) which causes local photoelectric effect in CGL (Charge Generation Layer) layer. The result is a spatially localized electropositive potential on the photoconductor surface. This localization (electrostatic charges) corresponds to the latent image (charged light image) on the photoreceptor drum. Well defined light rays formed printing elements to be colored later in the developing process by electropositive toner which adheres selectively on the discharged areas of the surface, thereby making the latent image visible.(Kipphan, 1994)

Laser diode is common in the laser head construction and by varying the laser output strength it is possible to control the photoconductor potential difference, influencing the formation quality of the thinnest image elements (digital screen dot).(Landa, 1994)

2. THEORY

Productivity of electrophotographic machines is in direct correlation with operation principles of the exposure device (laser head). The construction of the exposure device is decisive for the quality of the color reproduction.

Main goal of our investigation was to analyze the influence of laser power on quality of color prints and behavior of half-tone image according to this modified conditions. We expected that laser power increase would generate the screen element enhancement. To accomplish this analysis we used the machine internal software to increase laser power. (Chatow, 2001)

Laser beam is focused at the center of the pixel (matrix super pixel 6 x 6 is used in formation the smallest screen element). It reduces the complexity of images from hundreds of thousands of pixels to only a few hundred superpixels. Bell shaped curves on figure 1 represent levels of charge on photoreceptor which is proportional to the distribution intensity of laser beam in TEM_{00} mode. Dashed line is the threshold voltage which defines charge on developer drum. The ink is deposited on

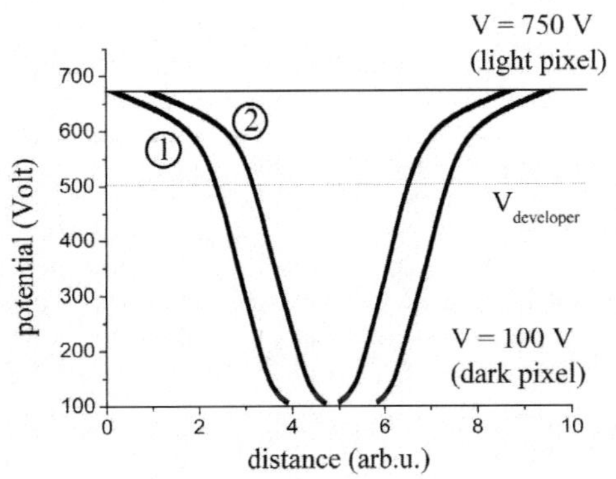

Fig. 1: Figure summarizes the influence of the applied voltage on developer drum and laser power on the image formation. Two horizontal lines on graph represent the level of the applied potentials on photo-imaging plate. Curve 1) corresponds to higher and 2) to lower laser power. Horizontal dashed line represents voltage on the developer drum. (Livne, 2003)

photoreceptor where the area below the developer line intersects with area above the laser curve. It is evident that rising of the laser power enhances the area around the center which results in larger halftone dot.

Laser power primarily affects the size of halftone dots and not their thickness which makes it very effective in controlling of the midtones output.(Goldman, 2004)

3. EXPERIMENT

In this work we analyzed the influence of the laser diode power change on multicolor reproduction. Experimental prints were made with previously calibrated electrophotographic machine HP Indigo TurboStream on the standard fine art paper. Variation of output laser power was applied while other relevant electrophotographic parameters retain constant by means of introducing the bypass control. For colorimetric measurements we applied spectrophotometers X-rite DTP 41 and ColorShop (determination of ΔE CIE Lab) software. Two calibration areas (solid tone and 50% screen value) were thoroughly analyzed for colorimetric purpose. The results are presented in two dimensional forms for the primaries (CMY) and for the secondary colors (RGB). For determination of laser strength influence on print quality it is important to measure the primary and the process colors on printed model with spectrometer. Each primary color is presented with two characteristic patches (solid tone and 50% screen patch). Their deviation from the calibration print, i.e. their color difference (ΔE CIE Lab) which is the result of the laser strength variation, is presented in figure 2.

It was compared with visual estimation which presents colors shift regarded to the calibration print. During our research we have controlled the possible optical system defocusing and the change of focus position of the writing head which could considerably influence the size and the shape of laser spots. The defocusing influences the negative appearance of banding; especially during the illumination of HDI (High Definition Images).(You, 2004)

4. RESULTS AND DISCUSSION

For determination of laser strength influence on print quality it is important to measure the primary and process colors with spectrometer on the printed model. Each primary color is presented with two characteristic patches (solid tone and 50% screen patch). It was compared with visual estimation which presents colors shift regarded to the calibration print: (the upper part of the graph (gray area) presents the darker hue compared to the calibration, while the lower part of the graph (white area) presents lighter hue). Variation of the laser strength generates the visible hue deviation when printed on the fine art paper, (average coloring of all the tones is ΔE_{MAX} - ΔE_{MIN} = 6.90). For all applied laser powers the solid tones, reproduced on fine art paper, reveal tone deviation (ΔE_{MAX} - ΔE_{MIN} = 0.93) which could be observed by measuring with optical spectrometer. This is mostly evident on green and yellow prints ($\Delta E_{100\%}$ = 1,1).
The average deviation of the screened hue colors is mostly prominent on yellow print ($\Delta E_{50\%}$ = 20,2), while magenta print exhibits the value ($\Delta E_{50\%}$ = 7,0) and cyan print ($\Delta E_{50\%}$ = 6,3). This behavior influences the maximum color difference in green prints ($\Delta E_{50\%}$ = 19,9) and red prints ($\Delta E_{50\%}$ = 16,8) which is the result of color mixing of the process colors (CMY).
In respect to the calibration the smallest aberration appears with the strength application of the laser 9 (cyan and magenta) and laser 1 (yellow). Fine quality reproduction of the secondary color prints depends on laser powers.
Our results point that optimum choice should be laser 8 for violet blue, laser 2 for green and laser 1 for red. We designated laser powers values with numbers representing the applied laser powers (e.g. laser 6 means that laser operates in 6 μW regime). The voltage difference between the illuminated and non illuminated areas is approximately 600V. In this way the latent printing form is formed which can accept the optimal quantity of developer material (100% color coating on print corresponds to standard classical lithographic offset printing).
Concentration of the yellow pigment in liquid ElectroInk (special electrophotographic liquid toner for Digital Color Offset) is higher, compared to other ElectroInks (cyan, magenta and black). The increase of pigment particles portion in liquid toner requires the increase of electric conductivity of ElectroInk, which directly depends on the quality of the ink adherence on photoconductor. Excessive fraction of this substances results in thicker ink layers and enhanced dot gain, which is predominantly observable for higher tone values.

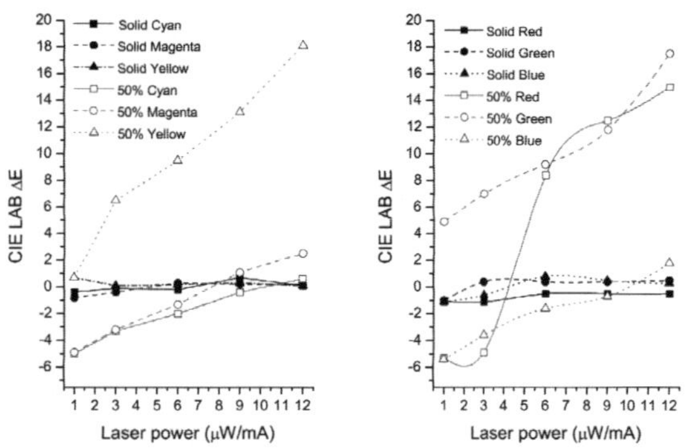

Fig. 2. CIE LAB ΔE color differences for cyan, magenta and yellow (right graph) and for red, green and violet blue (left graph). Part of the graph with ΔE>0 presents the darker hue compared to the calibration, while the lower part of the graph presents the lighter hue

5. CONCLUSION

The printed light and middle tones formed by varying the laser power in the above mentioned process suffer ten times greater tone change compared to the solid patches.

By comparing the calibrated images and the images obtained with laser power regulation (from 1 to 12 μW), it was noticeable that the process inks in solid patch had the minimal change (ΔE_{cyan}=0,8, $\Delta E_{magenta}$= 0,9 ΔE_{yellow}=1,1). It is similar with the secondary inks in solid patch ΔE_{red}=0,9, ΔE_{green}=1,1, ΔE_{blue}=0,8). Nevertheless, it is evident that ΔE is not zero even at nominal laser power. Reason for this behavior lies in the fact that we compare test prints with calibrated print. Our investigation is a part of larger analysis of our printing system during which we needed some reference in the form of the calibrated print.

In the screen reproduction (50% screen value) the expected stronger influence of laser strength on the reproduction is noticeable. These results demonstrate the initial idea that the initiated laser power change affects alternation of screen value. Yet, this method must be applied selectively for every color (process or secondary). Change of the laser power influences the minimal deviation in coloring the solid tones (100% saturated inks), but it influences the obvious change of the printed screen value (increase of laser power generates dimensional enlargement of the screen elements mostly on yellow, red and green).

Additional attention must be paid to the yellow prints which are troublesome in the sense of reproduction due to the reasons mentioned above. For this reason the yellow separation demands monitored control of laser power which complicates the print quality maintenance. This leads to certain screen printing defects due to mechanical damage of photoconductor surface which can be solved with regular replacement of photoreceptor with the new one.

This effect can be compensated with increasing the laser power maintenance of the exact (desired) middle and low hues.

6. REFERENCES

Chatow, Udi, (2001), The Fundamentals of Indigo's Digital Offset Colour Printing Process and How it Rivals Mechanical Offset Printing, *Proc. DPP 2001: Eye on the Future*, p.p. 21-25., ISBN / ISSN: 0-89208-233-X, Antwerpen, Belgium, May 13, 2001., IS&T Antwerpen, Belgium

Goldmann, G. (2004) *The World of Printers*, Oce Printing Systems GmbH, ISBN 3-00-001081-5 Dusseldorf, p.141.

Kipphan, H. (2001) *Handbook of Print Media*, Springer - Verlag Berlin, ISBN 978-3-540-67326-2, Heidelberg, p. 90.

Landa, B. (1994), Digital Offset Color -Today and Tomorrow, *Proc. SPIE*, 2171: 2, ISBN: 9780819414663, San Jose, CA, USA, May 1994

Livne, H. Plotkin, M., High Speed Laser Scanning Unit Using 12-Beam Laser Diode Array and Image Tracking System (ITT) for High Quality Color Printing, *IS&T's NIP19: International Conference on Digital Printing Technologies*, pp. 472-475., ISBN / ISSN: 0-89208-247-X New Orleans, Louisiana, September 2003.

You, J. Kim. H. Han, S., (2004) Banding Reduction in Electrophotographic Printer", *Proc. IS&T NIP 20th International Conference on Digital Printing Technologies*, pp. 470-473. ISBN / ISSN: 0-89208-253-4, Salt Lake City, Utah

Annals of DAAAM for 2008 & Proceedings of the 19th International DAAAM Symposium, ISSN 1726-9679
Editor B. Katalinic, Published by DAAAM International, Vienna, Austria 2008
Make Harmony Between Technology and Nature, and Your Mind will Fly Free as a Bird

OPTIMALIZATION OF TECHNOLOGICAL PARAMETERS OF FLOW FORMING PROCESS

MALINA, J[iri]; JIRKOVA, H[ana] & MASEK, B[ohuslav]

Abstract: *The extreme price escalations of raw materials, energies and human labor lead to a progressive tendency to find new and more effective methods. One of this progressive method is the spin extrusion (*Neugebauer et al. 2002, Neugebauer et al. 2001*). An alternative use of the experimental spin extrusion machine is for flow forming. The aim of this work was in the first step to try to apply the flow forming on the forming of a thin walled semi product and verify the relevancy of this method. The second step was to try to find suitable technological parameters for the hollow shaft.*
Keywords: *flow forming, hollow shaft, mechanical properties, micro-hardness, annealing*

1. INTRODUCTION

Flow forming (Awiszus et al. 2005, Awiszus et al. 2005, Ufer et al. 2006*)* uses three rollers for the reduction to final diameters. These three rollers are not driving and their rotation is attained through the friction between the wrought rotating semi-product and the shaping rollers Fig. 1.

The optional parameters for this forming technique are, apart from material, the size of reduction and speed of feed. During the experiment, the influence of the forming parameters on the quality of the final product was obtained.

2. THE INITIAL SEMI-PRODUCT

A semi product made of hot rolled thin-walled 16MnCrS5 steel tube was used with an initial diameter of 60 mm and wall thickness of 6 mm (Table 1).

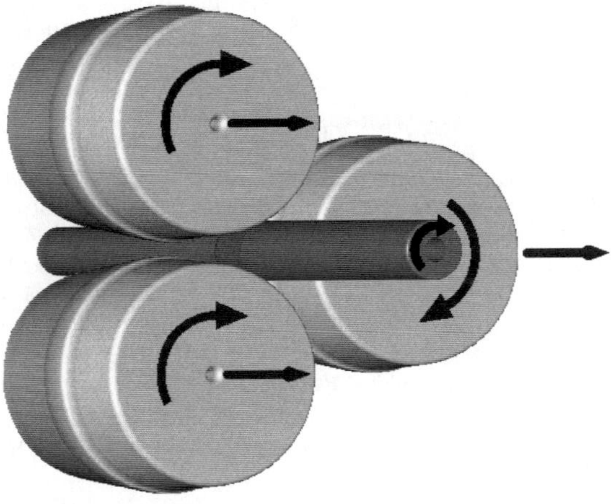

Fig. 1. Schematic illustration of flow forming.

element	C	Si	Mn	Cr	S	P
%	0.16	0.4	1.2	1	0.03	0.03

Table 1. Chemical composition of 16MnCrS5 steel.

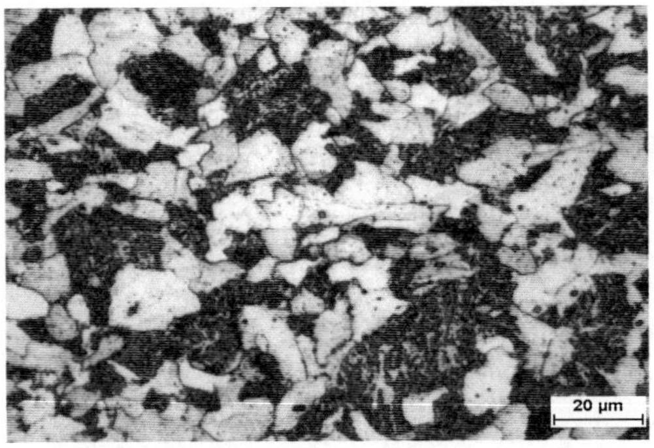

Fig. 2. The ferrite and pearlite structure in the middle of the wall of initial semi-product (cross section).

The 16MnCrS5 steel is low alloyed manganese-chromic steel with good hardening capacity for cementation. It is primarily used for medium stressed motor vehicle components. The initial ferrite pearlite microstructure had an average grain size of about 10±5 µm (Fig. 2)

3. EXPERIMENT

Firstly the initial formability of the material without heat treatment was tested. The material had the following mechanical properties (Table 2).

The material in this state was found to be unsuitable for this forming technology. It was impossible to use the material in this state for production and it was necessary to use soft annealing.

The initial material was soft annealed at 700°C for 30 to 180 min (Fig. 3).

HV 0.2	R_m [MPa]	R_{p02} [MPa]	A_{5mm} [%]
238	687	680	23

Tab. 2. The micro-hardness and mechanical properties of the initial material.

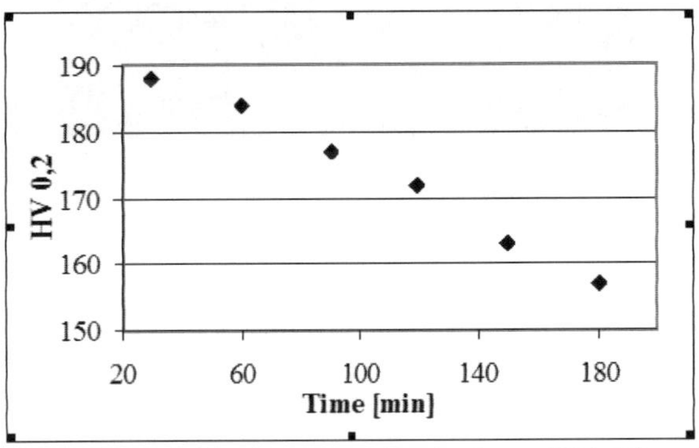

Fig.3. The relation between micro-hardness HV 0,2 and annealing time.

Annealing time [min]	HV 0.2	R_m [MPa]	R_{p02} [MPa]	A_{5mm} [%]
60	184	582	375	41
180	157	476	312	44

Tab. 3. The micro-hardness and mechanical properties of the initial material, after soft annealing.

On the basis of the resulting relationship between micro-hardness and annealing time and the metallographic analysis, the initial material was annealed at 700°C for time 60 and 180 min.

The annealing time of 60 minutes was chosen, because after this time the lamellar pearlite can transform to globular, leading to increased formability of the material.

The annealing time of 180 min provided the possibility of comparing the forming process on the material with inferior mechanical properties (Table 3) with globular pearlite structure. Mechanical properties were measured on the micro tensile specimens.

Annealing for 180 minutes led to the steep reduction of yield and ultimate strength, and increasing ductility A_{5mm} to 44 %.

While forming to the required size, the feed speed was varied and its influence on the surface quality was observed. In this case the surface of the material was not turned before forming. The feed speed was 1, 2 and 3 mm/rev.

On the basis of the surface analysis, a feed speed of 2 mm/rev was chosen as the optimal speed. This feed led to the most balanced surface quality (Fig. 4)

Both variants of the heat treated materials were reduced to the same dimensions (Fig.5). The size of reduction and speed of feed (2 mm/rev) were also the same. The forming process was successful for the first material variant (annealed for 60 min), and for the second variant (annealed for 180 min). The materials and technological properties were found to be suitable for reaching the required goal, which was the production of a hollow shaft (Fig.6).

The final product was cut into sections. The structure, micro hardness and mechanical properties were found for the smallest and middle diameters, which were where the material was reduced. The results of the microstructure analysis did not prove the structure refinement in the first or the second material variant.

The mechanical properties were relatively similar for the smallest diameter Ø 37 mm and the middle diameter Ø 42 mm, but there were big differences between materials (Table 4).

Fig.4. The influence of the feed speed on the surface quality A) Feed speed 1 mm/rev B) 2 mm/rev C) 3 mm/rev.

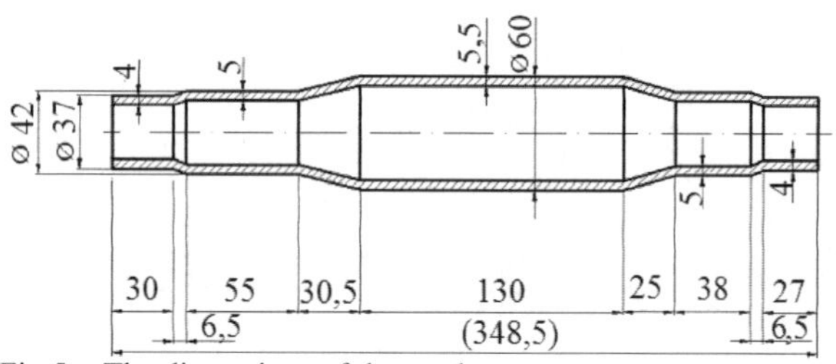

Fig.5. The dimensions of the product.

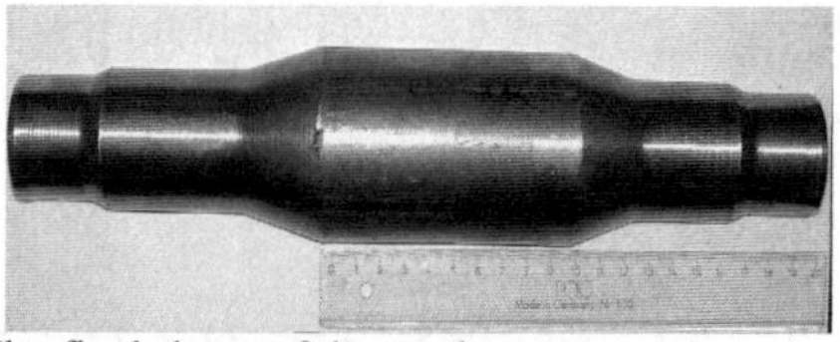

Fig.6. The final shape of the product.

Diameter	Annealing time [min]	HV 0.2	R_m [MPa]	R_{p02} [MPa]	A_{5mm} [%]
Ø 37	60	248	750	736	27
	180	230	647	645	26
Ø 42	60	245	739	716	27
	180	221	624	622	26

Tab. 4. The micro-hardness and the mechanical properties in relation to size of reduction and annealing time.

4. CONCLUSION

The results of the experiment verified that an experimental machine developed primarily for spin extrusion is suitable for the flow forming process, and that it is possible to produce a viable product with this technology. This product has many potential uses in the construction area. It can, because of the minimal amount of scrap created during production, contribute to the reduction of material costs.

On the material annealed at 700°C/60 min an increase of ultimate strength by about 29% and yield strength by about 96% was obtained. For the forming material annealed at 700°C/180 min, a 36% increase of R_m and 106 % of R_{p02} was achieved. All with residual ductility A_{5mm} higher than 25 %.

In the future, the fatigue properties of the whole semi-product will be tested.

5. REFERENCES

Awiszus, B. (2005) *Erweiterung der Formgebungsgrenzen durch inkrementelle zyklische Umformung am Beispiel des Drückwalzens.* (*The extension of the forming limits with use incremental cyclical forming on the example of flow forming).* The conference transcript for the final colloquium of DFG (German Research Council) main theme 1074 "Extension of formings' limits in the forming process.", Aachen

Awiszus, B. & Meyer, F. (2004) *Der Einfluss technologischer Kenn-größen und Parameter auf den quasistationären Zustand beim Drückwalzen -Beschleunigung der FEM durch modifizierte Anfangswerte. (The influence of the technological parameters on the quasistatic state in Flow forming described with FEM with help of modified beginning conditions).* Progress report for DFG (German Research Council) main theme: Modeling of the incremental forming. Technical University Chemnitz.

Neugebauer R.; Glass R.; Kolbe M. & Hoffmann M. (2002) *Optimisation of processing routes for cross rolling and spin extrusion.* Journal of Materials Processing Technology, Volumes 125-126, Pages 856-862

Neugebauer R.; M. Kolbe & R. Glass (2001). *New warm forming processes to produce hollow shaft*s. Journal of Materials Processing Technology, Volume 119, Issues 1-3, Pages 277-282

Ufer R., (2006). *Modellierung und Simulation von Drückwalz-prozessen, (Modelling and Simulation of Flow forming process)* ISBN 3-937524-43-6, Zwickau

ACKNOWLEDGEMENTS

This paper includes results obtained within the project 1M06032 Research Centre of Forming Technology.

Annals of DAAAM for 2008 & Proceedings of the 19th International DAAAM Symposium, ISSN 1726-9679
Editor B. Katalinic, Published by DAAAM International, Vienna, Austria 2008
Make Harmony Between Technology and Nature, and Your Mind will Fly Free as a Bird
*Young Researcher and Scientist Paper / * Supervisor, Mentor*

THE IMPACT OF PUBLIC MANAGEMENT OVER THE POPULATION OF RESITA CITY

MALOS, R[aul]; AMARIEI, O[lga]; ANDRES, S[olomia]; RINDASU, V[enera] & MINICA, M[irela]

Abstract: *We tried in this paper to point out the impact of public management above Reşiţa's city population, starting with a research divided in two components: a quality study (a made pre-research) and a quantity study. The investigated subjects are the population of Reşiţa city. The research questionnaire is made with 10 question witch from three of them are identification question, referring to sex, age and occupation of the questioned subjects. The sample for the quality study, are 60 persons. The validity of the quality study is realized by the quantity study by using a survey in witch the building of the sample is made by benchmarks method. In our case the sample obtained is 158 persons. The conclusion we obtained are shown in the paper*
Key words: *management, sample, survey, study, bureaucracy*

1. INTRODUCTION

It's well known that in Romania, after 1989 has been initiated an administrative reform meant to lead at it's adaptation at economical and social changes, but the situation remained the same, because of the budgetary restriction and the absence of political determination and experience regarding the alternative administration structures (Mustăcilă et al., 2005). The efficiency of managerial processes in public institutions is determined by the quality of public functionaries and their labor conscription mainly depends by knowing and applying public management. This assume the possibility of understanding the characteristics whom with it identifies, the analysis of organizational structures in public area, knowing the relationships between it's components, knowing the role, competence, task and responsibility witch every civil servant has it (Drucker,2001). With the object of knowing the impact of public management over the population, in this case Reşiţa city population, it has been made a research appealing to interrogation of this population with the help of a questionnaire that contains questions connected by the main activities of the most important institutions in public administration. This investigation is very important for the decisional factor in order to understand the population's attitude regarding public area, because the public manager has to take care about population when fundaments institution's objectives. Researches and studies in this area has been made, but they concerns the efficiency of human resources in public administration and they have been connected to professional training of human resources, the quality of administrative matter, the capacity of public administration to resolve the tasks in useful time, admitting that there is a problem regarding the intercourses between civil servants and citizens or public administration human resources and society.

2. THE PRESENTATION OF THE QUESTIONAIRE

Hello, I am doing a investigation over the impact of public management over the population of Resita city and I am asking you to have the kindness to answer t the following questions:

1. Are you familiar with the institutions from public administration?
   a) Yes
   b) No
2. How do you appreciate the quality of services in public administration?
   a) Not so good
   b) Satisfying
   c) Good
   d) Very good
3. How do you appreciate the bureaucratic system in public administration?
   a) Not good
   b) It runs you from one to another
   c) Efficient
4. How do you appreciate the efficiency of human resources in public administration?
   a) Very efficient
   b) Not so good comparing to the responsibilities
   c) Efficient
5. How frequently do you have contact with civil servants?
   a) daily
   b) 1-2 times per week
   c) 1-2 times per month
   d) Occasional
6. Witch institution of public administration do you think that adopts a professional attitude?
   a) The city hall
   b) The area council
   c) The financial administration
7. How do you consider the quality of management in public administration?
   a) good
   b) Satisfying
   c) Not so good
8. Your gender?
   a) Male
   b) Female
9. How old are you?
   a) Under 20 years
   b) Between 20-30 years
   c) Between 30-40 years
   d) Between 40-50 years
   e) Over 50 years
10. Your occupation?
   a) Pupil- student
   b) Over the age of occupation
   c) Worker
   d) Others
   e) Intellectual

Thank you for the amiability of answering to this question and have a nice day.

3. THE QUALITY STUDY (A MADE PRE-RESEARCH)

The quality study consist in collecting and analyzing the psychical and sociological elements that allows explaining of the attitudes, motivations and the behavior of all implied in the study (Olteanu, 2002).

The questioned subjects are population of Resita city compressed in a representative sample. A quality study precedes a quantity study because the information obtained is not enough for elaborate hypothesis.

The results of the quality study depend of the number of questioned persons, sampling, effective display of the study, and presentation and interpretation of the data obtained. For making a correct sampling we have to list the main criterions and for each criterion we list the possible levels. By listing the criterions and making a junction between them we have obtained a sample for 60 persons. The questioning of the 60 persons is considered a pre-research made with the purpose of finding out witch is the percent of persons who believes that the quality of public management is good. Function of this percent, witch represents the appearing frequency of studied phenomenon; we will determine the size of the sample at the city of Resita level, sample that is used in the quantity study.

For a better representatively we questioned the population of every quarter in Resita city. For this the statistical data have been used, more exactly the census made in 2006.

4. THE QUANTITY STUDY

The validation of hypothesis formulated by the quality study is made through the quantity study. For this we can use two categories of investigation, the census and the questioning. As main steps in elaborate an investigation through a questioning we remind: building the sample, determining the size of the sample, making the questionnaire, administrating the questionnaire, analysis of obtained data, making the synthesis rapport.

The determining of the sample is made different function of polling rate:

$$R = n / N \times 100 \qquad (1)$$

R = poll rate
n = studied population
N = mother population

If the sounding rate is less then 14,3% we have the situation of a non exhaustive poll, and if the poll rate is more then 14,3 we have an exhaustive poll.

In this case it has been determined a non exhaustive poll, whom size is given by the relation:

$$n = \frac{t^2 \times p \times (1-p)}{\Delta\omega^2} \qquad (2)$$

n=minimal size of the sample
t=coefficient that correspond to the probabilities whom with it is guaranties the results
p= proportion in the sample that posses the studied characteristic
$\Delta\omega$= accepted limited error

The sample used in our quantity study is:

$$n = \frac{t^2 \times p \times (1-p)}{\Delta\omega^2} = \frac{1.96^2 \times 0.8833 \times (1-0.8833)}{0.05^2} = 158 \qquad (3)$$

From the questionnaire we see that 53 persons have chosen variant “a” at question number 7, meaning a percent of 88.33. Dividing with 100 we obtain 0.8833 (the p from the formula).

5. THE INTERPPRETATION OF THE RESULTS

At the question “Are you familiar with the institutions from public administration?” 100 % of questioned persons have answered affirmatively, this thing maintaining over the criterions of age, gender and occupation.

Asked about the quality of services in public administration, most of them, 42,3 % answered that this is satisfying, and they are under 20 years old and the persons between 30-40 years old.

When asked “How do you appreciate the bureaucratic system in public administration?”, 48,3 % of them said that this is not good, and they are persons under 30 years old, while 13,4 % of them considered the bureaucratic system OK

Referring to the efficiency of civil servants, 53,4 % from the questioned persons said that they are efficient, and they are persons between 40-50 years, while 5 % of them considers the civil servants not well trained comparing to the asked responsibilities

Asked how frequently they are in contact with public servants, 56,6 % of investigated persons have answered that 1-2 per week, and they are persons between 20-30 years and those who are between 30-40 years old

Questioned “Witch institution of public administration do you think that adopts a professional attitude? “, 58,4 % of those investigated have chosen the financial administration, and lees of them , 1,6% have chosen the area council.

Referring to the quality of management in public administration, 84% of questioned people have said that this is good, while only 1,6% of them considers that this is satisfying.

6. CONCLUSION

As we seen from the research the quality of public management is connected to the efficiency of civil servants. The public manager has to be continuously in the know with people attitudes about the public area, in order to improve the relationship between public institutions and people. The positive outcomes are the propagation of efficiency while the negative outcomes means higher costs in training the civil servants and the reticence of public servants in adopting the new. Because of these, in the future, public manager has to focus on people, because investment in people makes the institution efficient.

The manager of public administration has to invest in the development and training of civil servants, create evaluation system of performance, to offer them development opportunities in career.

The new public management assumes decentralization of human resources management and adopting flexible practices regarding paying at national level.

Very important in the new public management is information management (Androniceanu, 2004). Information is that one who brings new elements of knowledge about past, present and future (Dănăiţă et al., 2004). In order to make possible measuring of the performances by the management, are necessary qualitative and quantitative information regarding inputs and outputs.

Hopefully the conclusions of this research will lead to the improvement of the bureaucracy system and the next step will be the growth of efficiency in public area. Those are essential conditions for regional development, based on EU funds, and provide new insights regarding regional development management.

7. REFERENCES

Androniceanu, A.; (2004). *The New Public Management,* Ed. ASE, Bucureşti

Drucker, P.; (2001). *The Decisional Factor Efficiency*, Ed. Destin, Bucureşti

Dănăiaţă, I; Bibu, A; Predişcan, M (2004) *Management.-Theoretical Bases*, Ed. Universităţii de Vest, Timişoara

Mustăcilă, I; Rotaru, I; Thernes, K, (2005) *Public Institution Management,* Ed. Eftimie Murgu, Reşiţa

Olteanu, V. ; (2002) *Management – Marketing,* Ed. Ecomar, Bucureşti

Annals of DAAAM for 2008 & Proceedings of the 19th International DAAAM Symposium, ISSN 1726-9679
Editor B. Katalinic, Published by DAAAM International, Vienna, Austria 2008
Make Harmony Between Technology and Nature, and Your Mind will Fly Free as a Bird

KAPLAN TURBINE BLADE DESIGN WITH OCTAVIAN POPA METHOD

MANEA, A[driana] S[ida]; BARGLAZAN, M[ircea]; STROITA, D[aniel] C[atalin] & BALASOIU, V[ictor]

Abstract: *This paper presents the design of axial turbine runners based on the method of Professor Octavian Popa The method permits to obtain the aerodynamic characteristics of the airfoils obtained trough arbitrary parameters which enter in the conformal mapping chosen function. Through the modification of the parameters we obtain different airfoils which appertain to the same family.*
Keywords: *airfoil, axial turbine, conformal mapping*

1. INTRODUCTION

For the projection of axial hydraulic turbines the method of conformal mapping permits the determination of runner blade airfoil on the base of some computations which permits the theoretical estimation of the airfoil energetical and cavitational behaviour. Although the calculus method suppose the consideration of the potential flow, the results obtained theoretically, permit the analysis of the possible performances of the airfoil in operation in real flow.

The method, of conformal mapping used at the design of the axial turbines runners, have the base on the O. POPA method (Popa, 2004). This method permits to obtain the aerodynamic characteristics of the airfoils obtained through arbitrary parameters which enter in the conformal mapping of the chosen function. Through the modification of the parameters we obtain different airfoils which appertain to the same family.

In the case of a conformal mapping the velocity circulation doesn't change. The transformation must be made as the trailing edge to correspond to a stagnation point on the generative circle.

If we know the dynamics on the external unit disc domain, through a conformal mapping function we can obtain the movement in the physical plane on the airfoil external domain.

The method of conformal representation can be applied for airfoils networks, accomplishing the conformal mapping of the external of a circle network on the external domain of a airfoils network (Stepanov, 1962).

At the design of an axial turbo machine, it intersects the runner with co-axial cylinders. The evolute of a cylindrical section on the tangent plane forms a linear airfoils network. The movement trough this airfoils network is planar.

2. O. POPA METHOD

The method, of Professor Octavian Popa for airfoils networks, suppose to consider the airfoils boundary appertaining to the network represent the image through the conformal mapping function.

Conformal to O. Popa method, the parameter m used in the design is the solution of the transdescendent equation:

$$\frac{l}{t} = \frac{2}{\pi}\left\{(\cos\lambda)Arth\left(\frac{2m}{M}\cos\lambda\right) + (\sin\lambda)arctg\left(\frac{2m}{M}\sin\lambda\right)\right\} \quad (1)$$

λ - install angle of network

$$M = \sqrt{1 + m^2 + 2m^2\cos(2\lambda)} \quad (2)$$

The network is obtained through parametric presentation:

$$X = \frac{1}{2} + X_L(\psi) \quad (3)$$

$$Y = \frac{a_0}{2} + \sum_{n=1}^{N}\left[a_n\cos(n\psi) + b_n\sin(n\psi)\right] \quad (4)$$

The coefficients of the network depend on the coefficients a_n and b_n of the isolated airfoil, on the network λ and on network pitch to chord ratio t/l.

A plane airfoils network is described through geometrical parameters of the airfoil from the network: relative arrow f/l, relative thickness d/l and through parameters which define the emplacement of the airfoil in the network: install angle λ and pitch to chord ratio t/l, where the pitch to chord ratio is obtained by the number of runner blades (Anton, 1979). The emplacement angle of the airfoil in the network is $\beta = \pi/2 - \lambda$, and the smooth of the network is the ratio l/t.

For the construction of the runner blade, we start from a generative airfoil with the known coordinates x, y and we obtain the coefficients of the trigonometrical polynomial approximation using the formulas:

$$a_{n=}\frac{1}{N}\sum_{p=0}^{2N-1}Y_p\cos(p\frac{n\pi}{N})$$
$$b_{n=}\frac{1}{N}\sum_{p=0}^{2N-1}Y_p\sin(p\frac{n\pi}{N}) \quad (5)$$

N=6; n=0,1,2,3,4,5,6

3. THE CHOOSING OF THE GENERATIVE AIRFOIL

In order to construct the runner blade, we chose as a generative airfoil a theoretical airfoil, for which were obtained the coefficients a_n and b_n of the trigonometrical polynomial. The pressure distribution calculated with O. Popa method is presented in fig. 1.

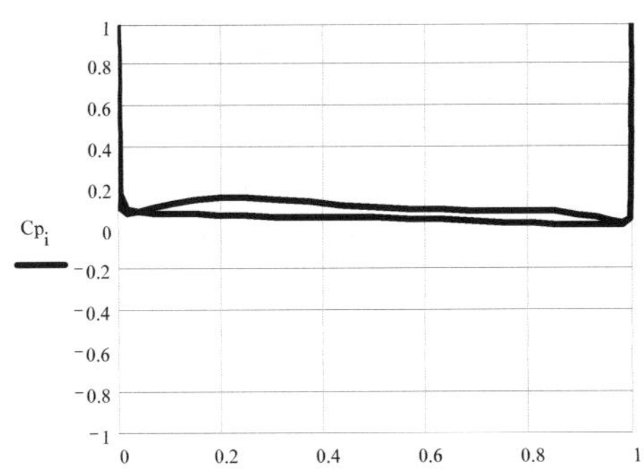

Fig.1. Pressure distribution on a theoretical airfoil for $\alpha=0^0$

4. THE DESIGN OF RUNNER BLADE

For a turbine, with the characteristics presented in table we choose as a generative airfoil a theoretical airfoil and the thickness and the arrow were calculated for the characteristic dimensions.

Calculus net head	H_c=18 m
Maximum flow	Q_c=35 m^3/s
Nominal power	P=5662,5 kW
Runner diameter	D=2500 mm
Speed	n=230,8 rot/min

Tab.1. Turbine caracteristics

The blade was calculated with O. Popa method (Popa, 2007) in 11 sections having the following elements:

Radius [m]	Chord length [m]	f/l	d/l	β_s [°]
0.5	0.53915	0.052	0.13	36.56811
0.575	0.60762	0.0375	0.12	29.41753
0.65	0.67285	0.024	0.11	24.98855
0.725	0.73486	0.0185	0.1	22.0816
0.8	0.79362	0.015	0.09	19.85673
0.875	0.84916	0.0123	0.08	18.03378
0.95	0.90145	0.011	0.07	16.6774
1.025	0.95051	0.0096	0.06	15.43556
1.1	0.99634	0.008	0.05	14.50612
1.175	1.03893	0.007	0.04	13.41805
1.25	1.07829	0.0063	0.03	12.42049

Tab.2. Calculated blade

In fig.2. – 4. are presented the variation diagrams of kinematic and geometric parameters (Barglazan, 1999) for theoretical airfoil used in the design of the turbine runner:

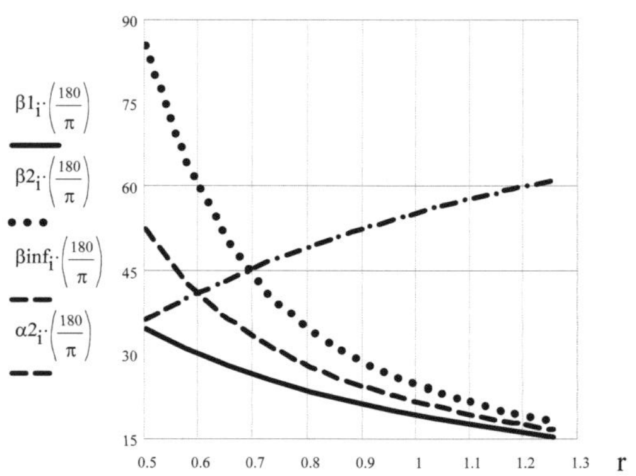

Fig.2. The variation of kinematic parameters along the radius

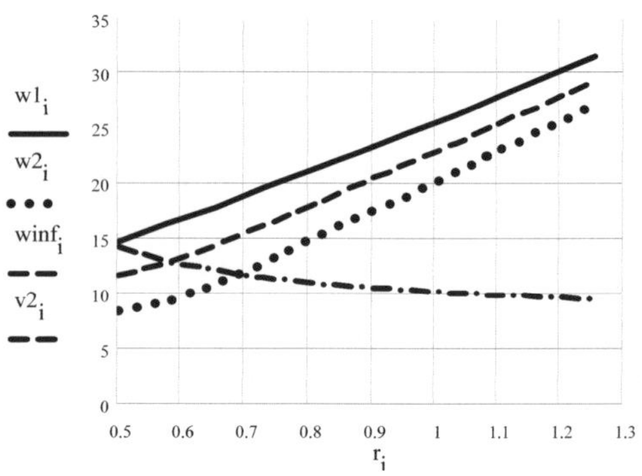

Fig. 3. The variation of the velocities along the radius

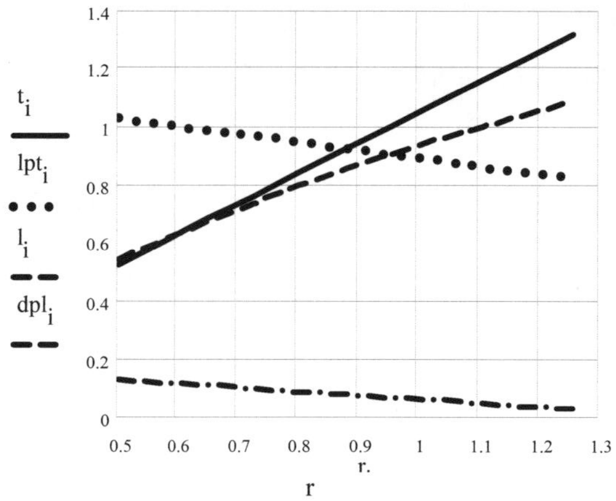

Fig. 4. The variation of parameters t, l/t, l and d/l along the radius

The runner obtained for turbine characteristics, using the O. Popa method is presented in fig.5.

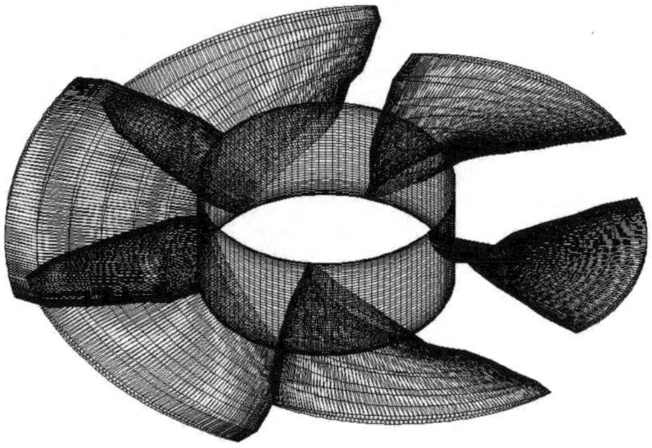

Fig. 5. The runner

5. CONCLUSIONS

The O. Popa blade design method proposed for hydraulic machinery permits the using of the same generative airfoil for different blade sections. The airfoil is approximated analytically and it's characteristics also through the coefficients of the trigonometrical polynomial approximation of the contour.

6. REFERENCES

Anton, Ioan, (1979). *Hydrodynamic Turbine*, Ed. Facla, Timişoara,

Bărglăzan, Mircea, (1999). *Hydraulic Turbine and Hydrodynamics Transmisions*, Ed. Politehnica, Timişoara, ISBN 973-9389-39-2.

Popa, Octavian, (2007). *Fluid Mechanics, vol I, II*, Tempus Timişoara, ISBN 978-973-88147-2-1, Vol I, ISBN 978-973-88147-3-8, Vol II, ISBN 978-973-88147-4-5.

Popa, Octavian, (2004). *Fluid Mechanics, Chosen scientific papars, vol I, II*, Tempus Timişoara, vol.I, ISBN 973-97842-8-3, 973-97842-9-1, vol.II, ISBN 973-97842-8-3, 973-87105-0-2.

Stepanov, G. I., (1962). *Hydrodynamics Airfoil Network,* Gosudarstvenoe, Izdatelstvo fiziko-matematiceskoi literaturî, Moskva.

Annals of DAAAM for 2008 & Proceedings of the 19th International DAAAM Symposium, ISSN 1726-9679
Editor B. Katalinic, Published by DAAAM International, Vienna, Austria 2008
Make Harmony Between Technology and Nature, and Your Mind will Fly Free as a Bird

METHOD FOR ONLINE DETERMINATION OF THE LATHE DYNAMIC CHARACTERISTICS

MAREK, B[artosz]; SKOCZYNSKI, W[aclaw] & ROSZKOWSKI, A[ndrzej]

Abstract: *The aim of the work was devise multipoint method for lathe vibration measure between tool and workpiece. The new idea of vibration method measure for turning workpieces was presented. The method demands to apply three contactless sensors. The main feature of the method is cancelling of roundness deviation influence on vibration measure. It was described two variants of the method: with and without feed. The conception of sensors placement on the test stand was presented.*

Key words: *lathe vibrations, multipoint vibrations measure method*

1. INTRODUCTION

The accuracy of turning and the quality of turned surfaces depends on dynamic properties of lathe structure. In spite of many years research, assessment methods for lathe dynamic properties often gives ambiguous results. In most of the cases designers made a variant evaluation of structures, based on a priori taken indexes or reference parameters. The relative dynamic susceptibility was taken as a such index (Weck 1992).

The unification of measurement methods of dynamic characteristics is the one of the problems. The most often methods of dynamic characteristic measurements are based on one point measures. That has common disadvantage: there are run radial out dominant components in vibration spectrum. It means that all lathe dynamic characteristics based on vibration spectral analysis depends on run radial out components. Harmonic frequencies of them correspond to component of rotation speed of workpiece and its harmonics.

A new, multipoint measurement method of vibrations between tool and workpiece was developed. The method based on three point measures with contactless sensors. It can be velocity or displacement sensors. The sensors (Fig.1) have to be. fixed on plane perpendicular to axis of rotation. It is important that the sensors are positioned with the same angle φ Sensor's axes of symmetry should cross one point close to axis of workpiece rotation.

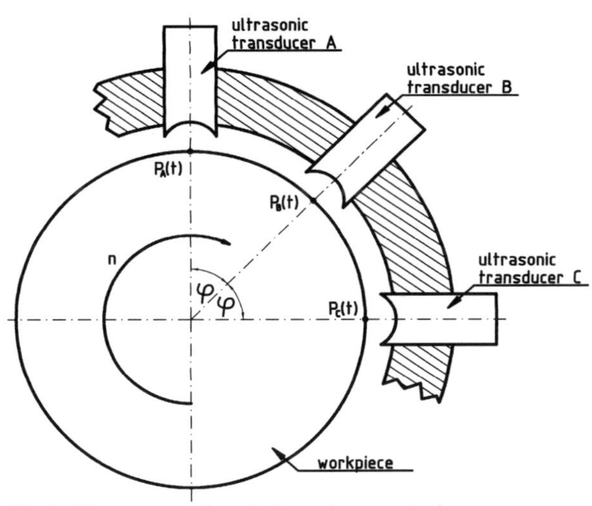

Fig. 1. The sensor and workpiece placement schema

The sample rate selection depends on spindle rotation speed n and angle φ of sensor placement.

2. THE MEASURE METHOD WITHOUT FEED

First variant of the method overrides a lathe's feed (Fig.1.). The spindle rotation speed n is established. The $P_A(t)$ point position should be moved to the $P_B(t)$ point position with whole number of steps. It is demanded the whole number of samples during workpiece φ angle rotation. The sample rate f_s should be a whole number also. The A, B and C transducers are displacement sensors that give deviation from the base distance. Then difference between $P_A(t)$ and $P_B(t+l)$ samples describes a vibration signal components without influence of roundness profile. The A or B direction of these components depends on subtraction order. The B and C probes case is analogous. The A and C direction vibration components can be described as follow:

$$m_A(t+l)=P_A(t)-P_B(t-l) \tag{1}$$
$$m_C(t+l)=P_C(t+l)-P_B(t) \tag{2}$$

The l parameter is the number of samples per φ angle rotation.

3. THE MEASURE METHOD WITH FEED

Calculate of the Z-axis shift between two points of Z-shifted helixes: $P_A(t)$ and $P_B(t+l)$ (Fig.2). The helix is described by three parametric equations:

$$y(t) = r\cos(t-\varphi) \tag{3}$$
$$x(t) = r\sin(t+\varphi) \tag{4}$$
$$z(t) = t\,h / (2\pi) \tag{5}$$

where r is workpiece radius, φ is beginning angle of helix (sensor placement angle) and h is a helix pitch.

The h parameter described the Z-shift of any established helix point during full rotate about Z-axis. The h(φ) distance between $P_A(t)$ i $P_B(t+l)$ points and analogous $P_B(t)$ i $P_C(t+l)$ can be described as:

$$h(\varphi) = h\,\varphi / (2\pi) \tag{6}$$

Calculation of the h(φ) distance enable to estimate its influence on measure accuracy. When it was estimated that h(φ) has a negligible effect on accuracy, the following formulas (1,2) can be applied. In case of h(φ) is important, it has to be calculated roundness profile along helix line with h pitch. In order to calculate of roundness profile we can use the general three point method with Fourier transform (DFT-G3P) described by (Gao & Kiyono, 1997).

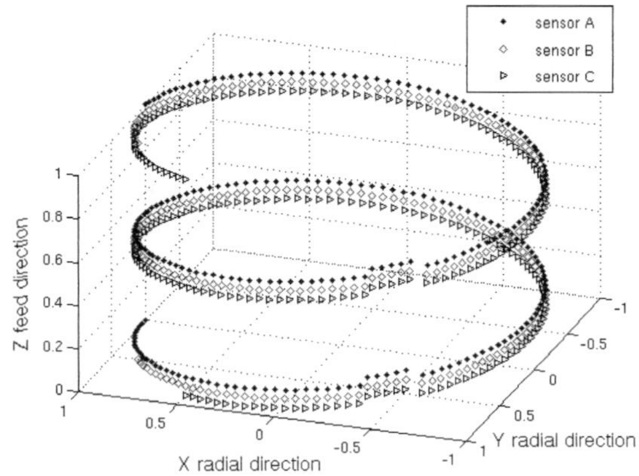

Fig. 2. A measure points placement on the side area of workpiece – variant with feed

4. THE MEASURE METHOD OF ROUNDNESS PROFILE (DFT-G3P)

The DFT-G3P method (Discrete Fourier Transform Generalized Three-Point Method) is based on sequential three point method for profile roundness measure (Gao & Kiyono, 1997). The original method is suited for roundness profile reconstruction and eliminated from measure run radial out components. A sensor cooperates with side area of turning workpiece. The sensor placement should fulfill assumptions mentioned in introduction. Measure signals from A, B and C sensors can be described as follow:

$$m_A(t) = r(t)+e_X(t) \quad (7)$$
$$m_B(t) = r(t-\varphi)+e_Y(t)\sin(\varphi)+e_X(t)\cos(\varphi) \quad (8)$$
$$m_C(t) = r(t-2\varphi)+e_Y(t)\sin(2\varphi)+e_X(t)\cos(2\varphi) \quad (9)$$

where t is sequentially measured sample, and t described the angle of workpiece rotation between two samples. The $e_X(t)$ and $e_Y(t)$ expressions means run radial out components on X and Y axes respectively. A differential signal from A, B and C sensors can be describe as (10) expression and Fourier transform of these expression as follow (Hamming 1997):

$$m_3(t) = m_A(t) - 2\cos(\varphi)m_B(t) + m_C(t) \quad (10)$$
$$M_3(\omega_n)=[\, 1-2\cos(\varphi)-\exp(j\omega_n\varphi)+\exp(j2\omega_n\varphi)\,]R(\omega_n)$$
$$n = 2,3,...,N-1 \quad (11)$$

If $m_3(t)$ is treated as a system response, and rotation of workpiece (roundness profile r(t)) is treated as an excitation, then Fourier system transmitation can be described as:

$$H_3(\omega_n) = M_3(\omega_n) / R(\omega_n), \quad (12)$$

where $R(\omega_n)$ is Fourier transform of r(t) roundness profile. Based on expressions above, it can be described the roundness profile r(t) as an inverse Fourier transform of $M_3(\omega_n)$ and $H3(\omega_n)$ quotient, thus (Gao & Kiyono, 1997):

$$r(t)=F^{-1}\{M3(\omega_n) / [\, 1-2\cos(\varphi)-\exp(j\omega_n\varphi)+\exp(j2\omega_n\varphi)\,]\} \quad (13)$$

After calculation of the roundness profile r(t), the shift between the r(t) and signal from sensor A, B or C can be found by the correlation methods. After determination the shift s between the r(t) and signal from sensor, it can be subtracted $P_A(t)$ and r(t+s). The procedure has to be repeated for signals from all sensors. Differential signals calculated as above should be without influence of roundness profile.

5. TEST STAND – THE CONCEPT

A test stand for multipoint vibration measure between tool and workpiece was designed. The schema presented on Fig.3. shows the way of sensors could be positioned on test stand. The sensors holder makes possible to mount different types of sensors eg. ultrasonic transducers, eddy current sensors etc., with different angles: 30 °, 60°, 90° i 120°.

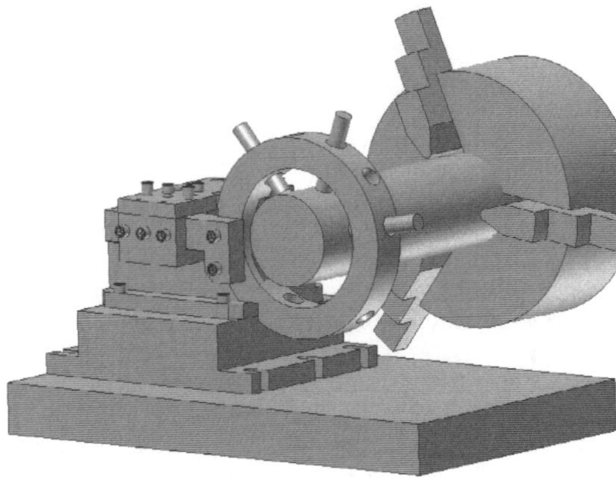

Fig. 3. The contactless sensors holder mounted on lathes toolrest.

The proposed solution is flexible. It makes possible measures of vibration between tool and workpiece with turning. The next step of our researches is carry out turning tests.

6. CONCLUSIONS

The spectrum of lathe vibrations between tool and workpiece includes several components. Usually it can be identified the rotation speed component and its harmonics which are spectrum dominants. One of the sources of those harmonics is roundness profile deviation. In case of the overlapping spectrum bands of those components and useful signal, traditional filtration methods are useless. In this case the developed measure method makes possible reduction or removal of run radial out harmonics from the vibration signal.

7. REFERENCES

Gao, W. & Kiyono S. (1997). On-machine roundness measurement of cylindrical workpieces by the combined three-point method. *Measurement*, Vol. 21, No. 4, (1997) 147-156

Gao, W.; Yokoyama Jun, Hidetoshi Kojima, Satoshi Kiyono (2002*): Precision measurement of cylinder straightness using a scanning multi-probe system,* Journal of the International Societes for Precision Enineering and Nanotechnology, 26 (2002), pp 279-288

Manfred Weck (1992): *Werkzeugmaschinen - Fertigungssysteme BAND 4 Messtechnische Untersuchung und Beuerteilung,* VDI-Z Verlag, Dusseldorf, 1992

Richard Wesley Hamming (1997): *Digital Filters,* Third Edition, Courier Dover Publications 1997, ISBN 048665088X

AKNOWLEDGMENT

A scientific project founded from the budget resources appropriated to science in the years 2005 to 2008 as a research project.

Annals of DAAAM for 2008 & Proceedings of the 19th International DAAAM Symposium, ISSN 1726-9679
Editor B. Katalinic, Published by DAAAM International, Vienna, Austria 2008
Make Harmony Between Technology and Nature, and Your Mind will Fly Free as a Bird

ANALYSIS OF FACTORS AFFECTING THE DECISION TO PURCHASE HOLIDAY PACKAGES BY TOURISTS THROUGH TRAVEL AGENCIES

MARIN-PANTELESCU, A[ndreea] & MANIU, L[aura]

Abstract: *The decision to purchase a product tour by a customer is based on a number of factors, such as the price of the tourist product, the confidence provided by the agency, promotional offers, product quality, facilities and gratuities offered by the agency, the recommendations received from relatives and friends. Given these considerations, the paper aims towards the achievement of direct marketing research at travel agencies, to assess factors that influence the decision to purchase package tours. It aims to identify, measure and the ranking according to their importance. Finally, it will present concrete solutions to increase the number of tour packages purchase by tourists through travel agencies.*

Key words: *travel, agencies, decision, marketing, research*

1. INTRODUCTION

The tour-operator travel agency trades in organizing and selling services packages or their components, directly or through its dealers.

Lately, the changes that have taken place on the tourist market have given birth to new ways of purchasing holiday packages. We can mention here the internet, on line booking services, the systems of global distribution and the electronic systems of distribution.The decision to buy tourist packages through a travel agency depends on a series of factors, such as: the price of the tourist product, if the agency is considered reliable or not, promotional offers, the quality of the product, facilities and free services offered by the agency, recommendations from relatives and friends.

Taking into account all these elements, the purpose of this paper is to investigate the factors that can influence the decision to purchase holiday packages. We would like to do this research because we have the belief that the identification of these factors and the measurement of their importance will help the travel agencies to understand their consumers better. Only through this approach the travel agency could maintain its position as a major distributor of the holiday packages.

The creation of the decision to purchase the tourist services is studied with a great deal of interest along the time. Thus, (Wearne, N. & Baker,K., 2002) sustains that people, according to their personality, choose different tourist products. Furthermore, the pressure coming from their relatives and friends represents another factor which influences the process of purchasing in tourism (Holloway, Ch., 2004).

The research made in the field of motivation and decision to purchase in tourism revealed two factors, " the distance to the holiday destination" and " the price of the destination" (Nicolau, J. & Más, Fr., 2006) as being decisive for consumers in most of the cases. The research on decisions to purchase holiday packages made by parents and their adolescent children (Carr, N., 2006) shows that these choose holidays according to different factors. Thus, while parents want a safe holiday destination, teenagers do not want anything organised in advance or they have no exaggerated expectations. The relations between the purchasing behaviour of the tourists, the quality of the services and satisfaction coming after experimenting the tourist product were studied by (Bigné, J., 2001). He concluded that satisfaction is in strict connection with the offered quality, and the wish to purchase tourist services is related to satisfaction. Analysing the actual stage of knowledge in this field, it can be observed an evolution of the perspectives on consummer's behaviour, the motivation and the factors that influence the decision to purchase a tourist product. But, the factors must be understood and taken into account in the development of the tourist activities, according to (Marin-Pantelescu, A & Maniu, L, 2007).

2. METHODS PROPOSED FOR DECISION MAKING SYSTEM

The research on the factors affecting the decision to purchase the holiday package was made using the computerized program SPSS (Statistical Package for Social Sciences) wich reports the frequence of responses, the valid percentages cumulated to each question in the questionnaire. Defining the decisional problem - The problem we made the moment we started this study was to identify the factors that affect the decision to purchase tourist products, if these differ as importance and what impact could have their identification on the activity of the travel agencies. The purpose of the research - The general purpose to achieve through this research was to identify, measure and put them in a hierarchy according to their importance. The objectives of the research - to identify the decisional factors to purchase a tourist product and to measure and put each of them in a hierarchy according to their importance. Specific hypotesis were the price and the quality of the offered products are the most important factors that affect the decision to purchase tourist products and the recommendations from relatives and friends became more and more important in purchasing a holiday package.

The study was made at the stand of a travel agency from Bucharest participating at that moment to the International Tourist Fair at Romexpo, Bucharest, the 19th Edition, within the period of 27-30 March 2008, through a questionnaire of 27 questions, out of which 4 questions were open and 24 close. The sample of the survey was made appealing to an aleatory method of sampling – simple aleatory sampling method, whose implementation led to a number of 318 questionnaires.

3. CASE STUDY (PROJECT NO.182, CNCSIS, PN-II-RU-TD-2008-3 'DIVERSIFYING AND PERSONALIZING TOURIST'S SERVICES IN THE CONTEXT OF GLOBALIZATION')

Regarding the profile of the responding people, it was observed the various regions the ressponding people come from, however, most of them being from Moldavia (23%) and Bucharest (20%). Regarding their repartition on gender, 49% of them are males and 51% are females. Their repartition on age is that approximately 46% of the responding people are under 35 years old, 59% are married, 43% have high education and only

7% have no school. Finally, regarding their standard of living, 25% have incomes within 751 – 950 Euro.

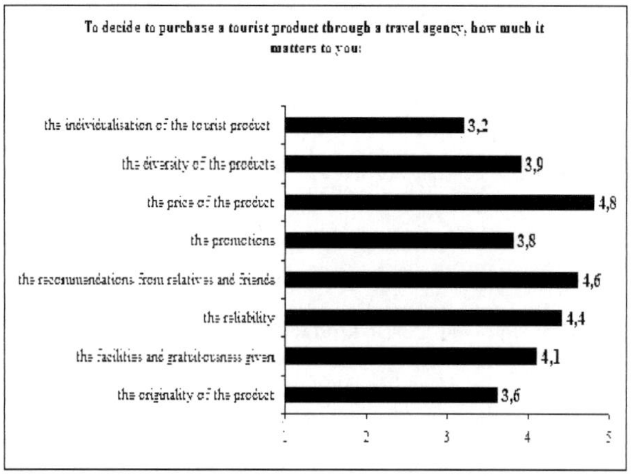

Fig.1. Decision making system

Given the socio-demographic coordination described above, it can be made the analysis of the results of the direct marketing research.

Thus, the decision to purchase tourist products was studied in connection with a series of features, such as: the reliability offered by the travel agency, the promotions offered, the facilities and gratuitousness given, the price of the tourist product, its originality and diversity, and its degree of individuality.

The final result for each factor was calculated using the semantic differential in five stages (1 representing the fact that they appreciate that factor very little, and 5 that they appreciate it very much), and the results are presented in the graph below:

The values coming after this analysis, close to the maximum of 5, prove the importance of these factors in taking the decision to purchase the holiday package through a travel agency. The highest score was obtained by the price of the tourist product (4,8), this being the most important decisional factor.

The Romanian are very affected by the price and for this reason the travel agency has to take into account the contracts negotiated with the providers of tourist services and the amount of the commission applied to these services. On the second place there are the recommendations from relatives and friends (4,6) because the tourist services are immaterial and thus, the important thing is to experience the product and share this with the others.

The most reliable persons in tourism are those who went to a specific destination and share this experience. If these are friends, relatives or colleagues they are even more reliable and their opinion matters.

On the third place there is the reliability offered by the agency (4,4). This thing happens because not in rare occasions the tourists faced problems at the holiday destination or the airport and they could not complain to anybody or could not take the money back, etc.

The facilities and gratuitousness are on the fourth place and obtain a good score of 4,1 owning to the Romanians incomes which are inferior to those in other countries, and this determine them to give a great importance to those elements.

The diversity of the products represents the fifth factor to affect the decision with a score of 3,9. It is normal for a tourist product rich with entertainment and tourist attractions to be more attractive than another one that offers only accommodation.

The promotions, which represent an important technique in sales are on the sixth place, with a score of 3,8. On the seventh place there is the factor related to the originality of the product, with a score of 3,6.

A new tourist product creates, at the beginning, a little reticence from tourists because they do not know what expectations to have related to the holiday destination. After this psychological limit is surmounted , they will experience it. In the end, on the last place there is the factor related to the individualisation of the product, with a score of only 3,2, which means that the tourists are indifferent to this factor.

The consumers do not take into account the individuality of the product when they take the decision to buy it. We believe that for taking into account this factor, consumers must have a certain standard of living , a certain tourist experience which are not determinant in our country at this moment.

4. CONCLUSION

The research on the modality the consummers take the decision to buy the tourist product has evolved considerably in the recent years.

In the marketing study it could have been seen wich are the most important factors that affect the decision to purchase the holiday package through a travel agency.

We created a hierarchy according to the score obtained through the semantic differential and We commented each factor. Thus, as the research showed, the order of the factors that affect the decision to purchase the holiday package through a travel agency is the following: the price of the tourist product, recommendations from relatives and friends, the reliability offered by the agency, facilities and gratuitousness, the diversity of the services included in the holiday package, the promotions, the originality of the product and the level of individualisation.

We hope that this research will represent a starting point for the travel agencies that want to know better their consumers. We hope that the travel agency will maintain its major role as a distributor of holidays, but for this, it needs to offer reliability, high quality services at attractive prices and new and various products.

Even if our research achieved its goal, future marketing studies must be done in order to observe the evolution of the factor referring to the individualisation of the tourist products. Maybe, in the future, the Romanian tourists will appreciate more this factor.

5. REFERENCES

Bigné, J. (2001). Tourism image, evaluation variables and after purchase behaviour: inter-relationship, *Tourism Management*, Elsevier, Vol. 22., No. 6, December 2001, pp. 607- 616, ISSN 0261-5177

Carr, N. (2006). A comparison of adolescents' and parents' holiday motivations and desires, *The Journal of Tourism and Hospitality Research*, , Vol. 6, No. 2, Feb. 2006, pp.129- 143, ISSN 1467-3584

Holloway, Ch. (2004). *The marketing for tourism*, 4th Edition, Prentice Hall, ISBN 0 273 68229 6, London

Marin- Pantelescu, A (2007). Diversifying tourist services and personalizing them with the help of travel agencies, *The 2007 International Conference on Tourism,* pp. 26-28, ISBN: 978-973-594-991-4, Bucharest, November 22- 24, 2007, Romania

Marin- Pantelescu, A.; Maniu, L., (2007). The integration of the Romanian travel services within the European travel market, *The 2007 International Conference on Tourism,* pp. 24-25, ISBN: 978-973-594-991-4, Bucharest, November 22- 24, 2007, Romania

Nicolau, J.; Más, Fr. (2006). The influence of distance and prices on the choice of tourist destinations: The moderating role of motivation, *Tourism Management*, Elsevier, Vol. 27, No. 5, October 2006, p.982-996, ISSN 0261-5177

Wearne, N.; Baker, K. (2002). *Hospitality Marketing in the e-Commerce Age*, 2nd Edition, Hospitality Press, ISBN 1 86250 511 X, Australia

Annals of DAAAM for 2008 & Proceedings of the 19th International DAAAM Symposium, ISSN 1726-9679
Editor B. Katalinic, Published by DAAAM International, Vienna, Austria 2008
Make Harmony Between Technology and Nature, and Your Mind will Fly Free as a Bird

STEP FUNCTION METHOD USED IN CALCULATING CONTINUOUS FOOTING FOUNDATION PLACED OF ELASTIC AND DISCRETE SOIL

MARIN, C[ornel]; HADAR, A[nton]; GRIGORAS, St[efan]; MARIN, A[lexandru] & STANESCU, M[arius]

Abstract: *The actual graphical methods used by engineers when plotting the stress distributions in elastic beams are based on integrating the differential equations of stresses for each beam segment. The resulting integration constants are obtained by imposing boundary conditions. Using MATHCAD, the proposed analytical method uses the step function $\Phi(x-a)$ which introduces a compact form of the stresses and displacements expressions. The constructive optimization in field of elastic footing foundations placed of elastic soil can be done easier.*
Keywords: *continuous foundations, step function method, numerical methods, stress, displacement.*

1. INTRODUCTION

This paper presents an alternative method of expressing the variation of the bending moment $M_{iy}(x)$ and translations $w(x)$ of elastic beam placed on elastic and *discrete* soil using the MathCAD STEP FUNCTION $\Phi(x-a)$, (Marin, 2007). This method is an alternative to WINKLER method for elastic beams placed of elastic and continuous soil (Manoliu, 1986).

The practical application is the design of a continuous footing foundation placed under four columns. The axial compressive force in the columns is P. The layout of the continuous footing loading is represented in fig.1.

The properties of soil under the footings lead to a loading model having as reactions the distributed loads q_0 depending on translation (Manoliu, 1986, Gruia & Haida, 1990).

For this loading model, the condition between the internal soil reactions V_k and the corresponding translations w_k will become (Marin, 2007):

$$V_k = k_s \cdot w_k \qquad (1)$$

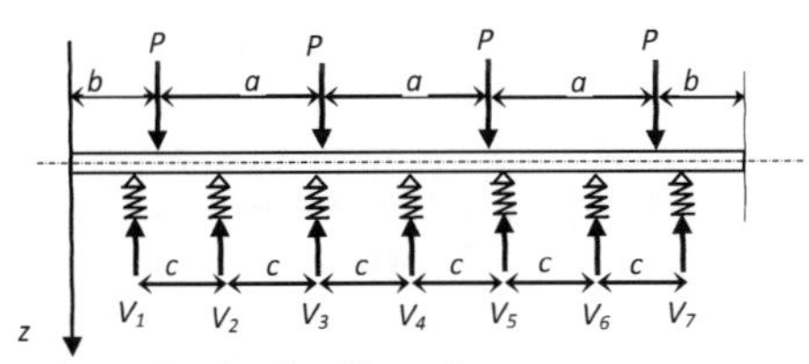

Fig 1.Continous footing loading scheme

2. ANALYTICAL EXPRESSIONS

One considers a beam subjected to bending. The beam is characterized by length L and constant bending stiffness EIy. There are 4 different load types, presented in fig. 2 (Marin & Marin, 2006, Constantinescu et al., 2005):

-Bending moment N, at distance a from the left end of the beam;

-Concentrated force P, at distance b from the left end of the beam;

-Uniform distributed load q_0 which acts on a beam segment delimitated by the distances e and f.

The differential equation of translations $w(x)$ and rotations $\varphi_y(x)$ corresponding to a cross-section is (Marin, 2006):

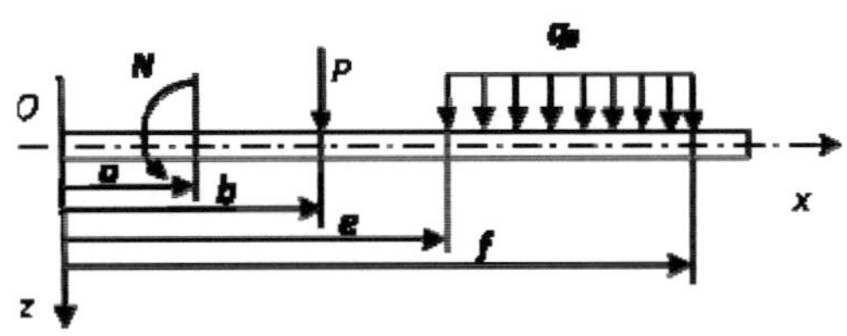

Fig 2. Beam model loaded with different loading types

$$\frac{d^2w}{dx^2} = -\frac{M_{iy}}{EI_y} \qquad (2)$$

Integrating twice the differential equation (2), one obtains after the first step, the rotations function $\varphi_y(x)$ and then the translations function $w(x)$:

$$\varphi_y(x) = \frac{dw}{dx} = \varphi_0 - \int_0^x \frac{M_{iy}(s)}{EI_y} ds$$
$$w(x) = w_0 + \varphi_0 \cdot x - \int_0^x \int_0^t \left(\frac{M_{iy}(s)}{EI_y} ds \right) dt \qquad (3)$$

The analytical expressions for the cross sectional resultants considering the loading types from fig. 2, using the step function Φ are:

For shear force:

$$T_z(x) = -P \cdot \Phi(x-b) - q_0 \cdot (x-e) \cdot \Phi(x-e) + $$
$$+ q_0 \cdot (x-f) \cdot \Phi(x-f) \qquad (4)$$

For bending moment:

$$M_{iy}(x) = -N \cdot \Phi(x-a) - P \cdot (x-b) \cdot \Phi(x-b) -$$
$$- q_0 \cdot \frac{(x-e)^2}{2} \cdot \Phi(x-e) + q_0 \cdot \frac{(x-f)^2}{2} \cdot \Phi(x-f) \qquad (5)$$

For the cross-sectional rotations functions:

$$EI_y \varphi_y(x) = EI_y \varphi_0 + F(x)$$
$$F(x) = N \cdot (x-a) \cdot \Phi(x-a) + P \cdot \frac{(x-b)^2}{2} \cdot \Phi(x-b) +$$
$$+ q_0 \cdot \frac{(x-e)^3}{6} \cdot \Phi(x-e) - q_0 \cdot \frac{(x-f)^3}{6} \cdot \Phi(x-f) \qquad (6)$$

For the cross-sectional displacements functions:

$$EI_y w(x) = EIw_0 + EI_y \varphi_0 + W(x)$$
$$W(x) = N \cdot \frac{(x-a)^2}{2} \cdot \Phi(x-a) + P \cdot \frac{(x-b)^3}{6} \cdot \Phi(x-b) +$$
$$+ q_0 \cdot \frac{(x-e)^4}{24} \cdot \Phi(x-e) - q_0 \cdot \frac{(x-f)^4}{24} \cdot \Phi(x-f) \qquad (7)$$

3. NUMERICAL APPLICATION FOR COMPUTING DIAGRAMS

The application's task is to obtain the expressions of required diagrams for the continuous footing foundation represented by the model in fig. 1 using the step function. The required diagrams will be:

-shear force diagrams $T_z(x)$;

-bending moment diagrams $M_{iy}(x)$;

-cross-sectional rotations distribution $\varphi_y(x)$;

-cross-sectional displacements distribution $w(x)$ (settlements of the soil under the foundation).

The unknown reactions V_1, V_2, ..., V_7 will be obtain writing the following equations of equilibrium and deflections (Marin, 2006):

$$\begin{aligned}
&\sum F_{zs} \downarrow = V_1 + V_2 + V_3 + V_4 + V_5 + V_6 + V_7 \\
&\sum \overleftarrow{M}_{7s} = V_1 \cdot (r_7 - r_1) + V_2 \cdot (r_7 - r_2) + V_3 \cdot (r_7 - r_3) + \\
&\quad + V_4 \cdot (r_7 - r_4) + V_5 \cdot (r_7 - r_5) + V_6 \cdot (r_7 - r_6) \\
&1-2-3: \quad EI\left[w_1(r_3 - r_2) - w_2(r_3 - r_1) + w_3(r_2 - r_1)\right] = \\
&= \left[W(r_1)\cdot(r_3 - r_2) - W(r_2)\cdot(r_3 - r_1) + W(r_3)\cdot(r_2 - r_1)\right] \\
&2-3-4: \quad EI\left[w_2(r_4 - r_3) - w_3(r_4 - r_2) + w_4(r_3 - r_2)\right] = \\
&= \left[W(r_2)\cdot(r_4 - r_3) - W(r_3)\cdot(r_4 - r_2) + W(r_4)\cdot(r_3 - r_2)\right] \\
&3-4-5: \quad EI\left[w_3(r_5 - r_4) - w_4(r_5 - r_3) + w_5(r_4 - r_3)\right] = \\
&= \left[W(r_3)\cdot(r_5 - r_4) - W(r_4)\cdot(r_5 - r_3) + W(r_5)\cdot(r_4 - r_3)\right] \\
&4-5-6: \quad EI\left[w_4(r_6 - r_5) - w_5(r_6 - r_4) + w_6(r_5 - r_4)\right] = \\
&= \left[W(r_4)\cdot(r_6 - r_5) - W(r_5)\cdot(r_6 - r_4) + W(r_6)\cdot(r_5 - r_4)\right] \\
&5-6-7: \quad EI\left[w_5(r_7 - r_6) - w_6(r_7 - r_5) + w_7(r_6 - r_5)\right] = \\
&= \left[W(r_5)\cdot(r_7 - r_6) - W(r_6)\cdot(r_7 - r_5) + W(r_7)\cdot(r_6 - r_5)\right]
\end{aligned} \tag{8}$$

where: r_k (k=1,...7) is distance from the left end of the beam to reaction V_k;

$W(x)$ is the function defined by relation (7) for this particular case.

3.1 Hypothesis 1 –low stiffness soil

Replacing in MathCAD the parameters values: $P=10kN$; $b=1m$; $a=2\ m$; $c=1m$; $k=10^7 N/m$; $EI=10^6\ N{\cdot}m^2$ the following values of reactions and variation diagrams for the foundation (fig. 3) are obtained:

$$\begin{pmatrix} V_1 \\ V_2 \\ V_3 \\ V_4 \\ V_5 \\ V_6 \\ V_7 \end{pmatrix} = X \ ; \ X = \begin{pmatrix} 8.696 \\ 3.682 \\ 5.571 \\ 4.103 \\ 5.571 \\ 3.682 \\ 8.696 \end{pmatrix} \tag{9}$$

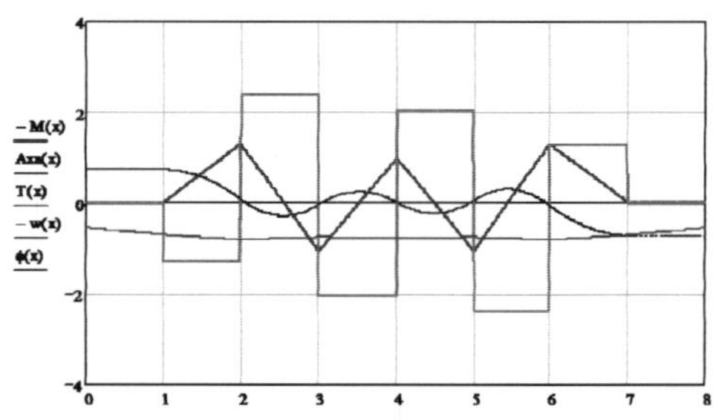

Fig 3. Beam diagrams for hypothesis 1

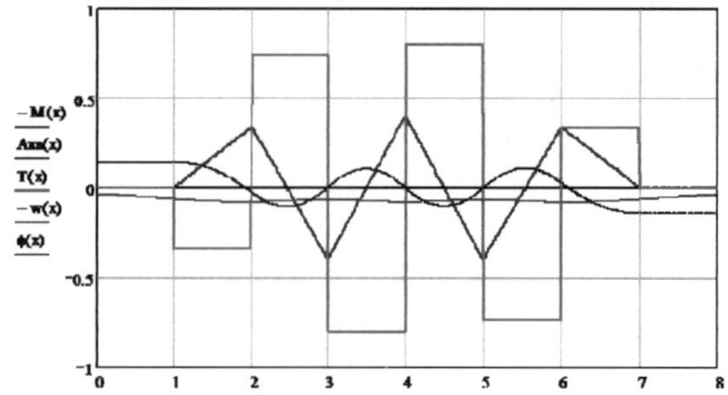

Fig 4. Beam diagrams for hypothesis 2

3.2 Hypothesis 2 –high stiffness soil

Replacing in MathCAD the same parameters values: excepting $k=108N/m$ the following values of reactions and variation diagrams for the foundation (fig. 4) are obtained:

$$\begin{pmatrix} V_1 \\ V_2 \\ V_3 \\ V_4 \\ V_5 \\ V_6 \\ V_7 \end{pmatrix} = X \ ; \ X = \begin{pmatrix} 9.661 \\ 1.078 \\ 8.457 \\ 1.609 \\ 8.457 \\ 1.078 \\ 9.661 \end{pmatrix} \tag{10}$$

4. CONCLUSIONS

This method is well suited for the constructive optimization of the structure, using the simulated values and varying different parameters: in the current application, the stiffness k was changed for optimization purposes.

As noticed by authors, the fact that the step function method uses simple operation expressions, it's obvious that the numerical applications are being solved fast with a minimum number of computational cycles. Because traditional methods use integral expressions which usually are solved by means of numerical methods, they imply a large number of computational cycles, causing slower results.

5. REFERENCES

Constantinescu, I.N., Picu, C., Hadar, A., Gheorghiu, H. (2006) *Rezistenta materialelor (Strength of materials)*, BREN Publishing House, Bucharest (in Romanian).

Gruia, A. & Haida, V. (1990), *Geotehnică şi fundaţii (Geotechnics and foundations)*, "Traian Vuia" Politechnic Institute, Timisoara (in Romanian).

Manoliu, I. (1986), *Geotehnică şi fundaţii (Geotechnics and foundations)*, Technical Publishing House, Bucharest (in Romanian).

Marin, C. (2006), *Rezistenţa materialelor şi elemente de teoria elasticităţii (Strength of materials and elements of theory of elasticity)*, Biblioteca Publishing House, http://fsim.valahia.ro/cursuri.html (in Romanian).

Marin, C. (2007), *Aplicaţii ale teoriei elasticităţii în inginerie (Engineering applications of theory of elasticity)*, Biblioteca Publishing House http://fsim.valahia.ro/cursuri.html (in Romanian).

Marin, C., Marin, A. (2006), Metoda analitica pentru calculul deplasărilor şi rotirilor barelor drepte supuse la încovoiere (Analytical method for calculus of slope and deflection in straigth beams), *SIMEC 2006*, UTCB Bucharest (in Romanian).

Annals of DAAAM for 2008 & Proceedings of the 19th International DAAAM Symposium, ISSN 1726-9679
Editor B. Katalinic, Published by DAAAM International, Vienna, Austria 2008
Make Harmony Between Technology and Nature, and Your Mind will Fly Free as a Bird

THE ROLE OF THE SHAPE IN INDUSTRIAL DESIGN

MARIN, D[umitru]; RAICU, L[ucian] & SIMION, I[onel]

Abstract: *The paper presents the role of the shape generally and the cylindrical shape particulary in industrial products design, as well as turning of a cylindrical shape into a hyperboloidal shape. The theoretical transformation study of a cylindrical shape in a hyperboloid shape and vice versa is simply from mathematical point of view and it leads to multiple practical applications such as: pressing and grasping mechanisms, rolls, cutting and abrasive tools with variable geometry, Linear to Angular Displacement Devices (LADD) etc. with a series of advantages (low cost, high output, low noise, absence of lubricant).*

Keywords: *transformation, shape, cylinder, hyperboloid, mechanism.*

1. INTRODUCTION

The shape concept appears in almost all design definitions, formulated by different schools or personalities. For exemple "The International Council of the Societies of Industrial Design" (I.C.S.I.D) gives following definition: "The design is a creative activity which consists in determination of the shape properties of the industrially made objects. The object shape properties, include not only exterior characteristics, but also all the structural relations which make a coherent unity from an object or from an objects system, regarding the point of view of both the producer and the consumer." (Quarante, 1994)

The cylindrical shape is the most frequently used shape in structure of industrial products (shafts, spindles, cylindrical bars and pipes, cylinder bores etc.). It is either manufactured by machine tools through lathing, drilling, boring or through wrapping from sheet, plastic materials etc., the rotation cylinder is the most technological shape, therefore it is more easily achieved than other shapes (conical, spherical, hyperboloidal shape etc.). Also the cylindrical shape can be turned onto a hyperboloidal shape.

Consequently, we suppose a rotation cylindrical shape, such as Fig.1. The circles C_1 and C_2 with radius R are fastened between themselves through inextensible generatrices G. The circles C_1 and C_2 can be in practice rings or disks and the generatrix G can be rigid bars. When the circle C_1 is turned in its plane all around the cylinder axis with a certain angle and the circle C_2 remains fixed, then the cylinder from Fig.1 turns into a rotation hyperboloid (Fig.2).

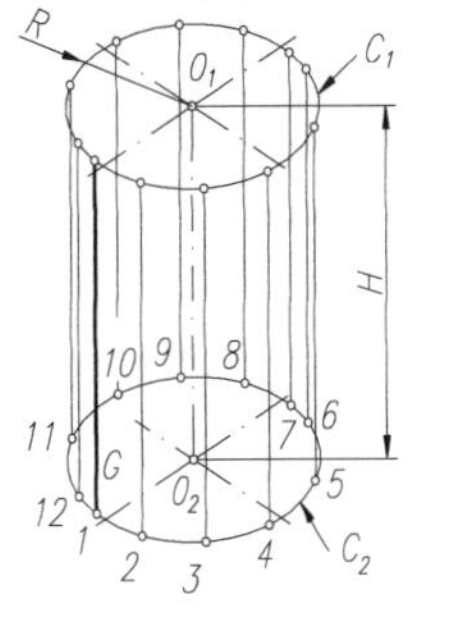

Fig.1. The rotation cylinder

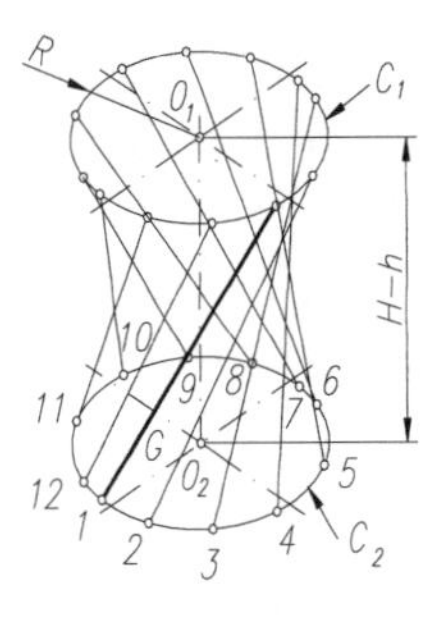

Fig.2. The rotation hyperboloid

The rotation hyperboloid from Fig. 2 is a surface generated by a straight line (the generatrix G), which rotates all around O_1O_2 axis. The generatrix G doesn't intersect the axis O_1O_2 and isn't in a parallel direction with it.

2. CYLINDER TURNING INTO HYPERBOLOID. CALCULATIONS AND REMARKS.

For simplification, we suppose that a new position of the one generatrix G (segment AB) is A_1B_1; then the circle C_1 turns and changes one's place with distance h and the cylinder turns into (changes in) hyperboloid such as in Fig.3.

In succession, we calculate displacement h. From the rectangular triangle $b'e'a_1'$ can write (Marin et al., 2006):

$$b'e'^2 = b'a_1'^2 - e'a_1'^2 \tag{1}$$

With dimensions from Fig. 3, formula (1) becomes:

$$(H-h)^2 = H^2 - 4R^2\sin^2\alpha \tag{2}$$

After calculations, finally we obtain:

$$h = H - \sqrt{H^2 - 4R^2\sin^2\alpha} \tag{3}$$

The value of h displacement depends on invariable parameters H and R of cylinder and the variable angle α .

We distinguish following cases:

a) H>2R.

The radical is positive and theoretically the circle C_1 can be rotated with the angle $\alpha = \pi/2$ and the cylinder becomes a cone that is the generatrix G intersects the O_1O_2 axis.

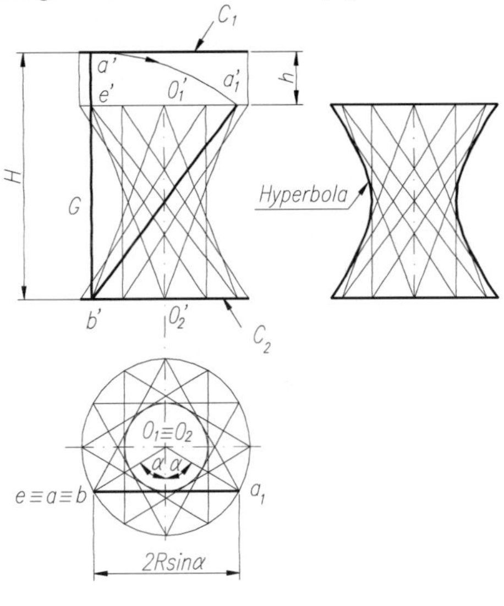

Fig. 3. Rotation hyperboloid; dimensions for calculations.

b) H=2R.

The formula 3 becomes:

$$h = 2R(1-\cos\alpha) \quad (4)$$

For $\alpha = \pi/6$; $2\alpha = \pi/3$ and $h = R(2-\sqrt{3}) \cong 0{,}27R$.

For $\alpha = \frac{\pi}{6}$; $2\alpha = 2\frac{\pi}{6}$ and $h = 2R(1-\frac{1}{2}) = R$.

For $\alpha = \frac{\pi}{2}$; $2\alpha = \pi$ and $h = 2R = H$, the circle C_1 coincides with the circle C_2, it is a theoretical case.

c) H<2R.

Because $H^2 - 4R^2\sin^2\alpha \geq 0$, then $\alpha \leq \arcsin\frac{H}{2R}$ and the rotation angle α is limited.

For example if $H = R \Rightarrow \alpha \leq \frac{\pi}{6}$.

A rotation hyperboloid can be generated, also, by a hyperbola, which rotates all around O_1O_2 axis; the hyperbola and the axis are in the same plane (see Fig.3).

3. APPLICATIONS

If the circle C_1 and C_2 in Figure 1 are special rings made from special fibre fastened between themselves through a very strong elastic fibres, then is achieved a Linear to Angular Displacement Device (LADD). Several elements LADD fastened among them can help a robot to raise the leg and it can step (Fig.4). In Fig. 5 is represented a Concentric Linear to Angular Displacement Device (CLADD), made from two concentric LADD. An electric motor turns an extremity of the interior LADD and the other extremity is fastened to an extremity of an exterior LADD. The advantages of LADD elements are: low weight, low noise and the absence of lubricant (Mennitto & Buehler, 1996).

The reduction of fluid flow, even its closing, when the fluid flows through an elastic pipe. This can be realized by spinning of a textile threads bunch fixed around of the pipe with two cylindrical disks. (see the circles C_1 and C_2 in Fig. 1). The cylindrical fragment changes into hyperboloidal fragment with variable section (Plahteanu & Belous, 2000).

Pressing, grasping mechanisms; when the ring C_1 turns and the ring C_2 remains fixed, can be created axial forces in a direction or the other. The cylinder changes into hyperboloid and the distance between rings decreases or the hyperboloid ctanges into cylinder and the distance between rings increases (Marin, 2007).

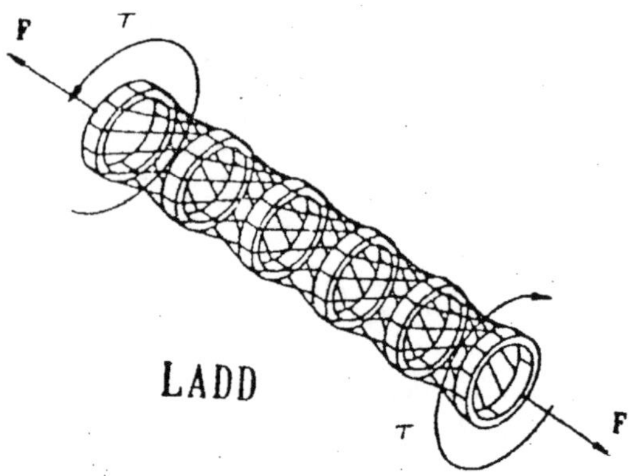

Fig .4. Linear to Angular Displacement Device (LADD)

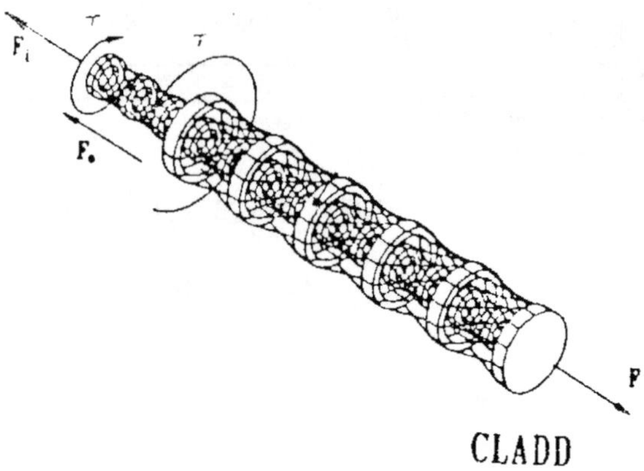

Fig.5. Concentric Linear to Angular Displacement Device

Rolls, rotors, adjustable abrasive or cutting tools based on the hyperboloid's property to change its profile until it becomes cylinder.

Different ornamental items, for example a lamp-shade, can be changed from a cylinder in many hyperboloid shapes (Fiell & Fiell, 2000).

4. CONCLUSIONS

The change of rotation cylinder in rotation hyperboloid and vice versa is simply from the theoretical point of view and beneficially from practical point of view.

The choosing of rotation angle 2α and displacement h depends on: constructive parameters H and R of the mechanism, the force which must be obtained, the kinematics and dynamic characteristics of the mechanism.

In different constructive shapes, the principle described in this paper can be the technical solution for obtaining of linear motion and of technological forces for various mechanisms (Raicu, 2002; Simion, 2005).

For the further research, we can calculate and plot the velocity and acceleration variation of the designed mechanisms, starting from formula (3), through the derivation of the axial displacement h and we can use the mechanism in the best dynamic conditions.

5. REFERENCES

Fiell, Ch.& Fiell, P. (2000). *Industrial design A-Z*, Taschen GmbH, ISBN 3-8228-6310-6, Koln.

Marin, D.; Raicu, L. & Radulescu, C. (2006). *Geometrical shape - a design creative resource*, Scientific Bulletin, University POLITEHNICA of Bucharest, Serie D: Mechanical Engineering; Vol.68, No 3, pp 55-62, I.S.S.N 1454-2358, Bucharest.

Marin, D. (2007). *Design Industrial; Designul Formei (Industrial Design; Shape design)*, Editura BREN, ISBN 978-973-648-706-4, Bucharest.

Mennitto, G. & Buehler, M (1996). *A compliant articulated robot leg for dynamic locomotion*, Robotic and Autonomous System – No18.

Plahteanu, B. & Belous, V. (2000). *Efecte geometrice in creatia tehnica (Geometrical Effects in Technical Creation)*, Editura Performantica, ISBN 973-8075-01.7, Iasi.

Quarante, D. (1994). *Elements de design industriel*, Polytehnica 2e Edition, ISBN 2-84054-018-5, Paris.

Raicu, L. (2002). *Grafic si vizual intre clasic si modern (Graphics and Visual between Classic and Modern)*, Editura Paideia, ISBN 978-596-062-1, Bucuresti.

Simion, I. (2005). *AutoCAD for Engineers*, TEORA USA LLC, ISBN 1-59496-033-X, Wisconsin, USA.

Annals of DAAAM for 2008 & Proceedings of the 19th International DAAAM Symposium, ISSN 1726-9679
Editor B. Katalinic, Published by DAAAM International, Vienna, Austria 2008
Make Harmony Between Technology and Nature, and Your Mind will Fly Free as a Bird

NEW STRATEGIES FOR DESIGNING ASSEMBLIES

MARIN, G[heorghe]; PETRESCU, L[igia] & DOLGA, L[ia]

Abstract: *This paper briefly presents the main techniques used nowadays in the Computer Aided Design (CAD) in general and especially in the computer aided design of the mechanical assemblies. The authors outline the main advantage of the Computer Aided Design over the classical, human design, and, then they try to explain some of the very powerful capabilities offered by the latest CAD environments.*
Ke words: *computer, assemblies, design, modeling.*

1. INTRODUCTION

The complex transformations, which have occurred in the field of designing and manufacturing products, consist in reducing the renewal and supply cycle, increasing requirements on functionality and technical and quality level, reducing dimensions of the production batches, parallel to the increase of the production systems flexibility. These transformations have determined factories to change their obsolete industrial strategies, to seek for permanent innovation and implementation as soon as possible of new technologies from the research-development field to the production one.

In the last decade, throughout the world, the production facilities in the goods field have been brought up-to-date beginning from the activity of designing, production planning and processes checking to the systems that provide measurements of the production process parameters or measurements of the dimensional, functional and shape product parameters.

It is well known that in the economically developed countries (which hold "de facto" the supremacy in efficiency, quality and performances), using computer in research-designing activity, production, process checking and quality certification has no longer been considered as a *natural matter* but as a *necessity*, a necessary prerequisite that is not always an enough condition for market survive.

This writing intends to survey the main strategies, solutions used at present in the field of computer-aided design, generally, and mechanical assemblies design, particularly.

2. NEW COMPUTER-AIDED DESIGNING STRATEGIES FOR MECHANICAL ASSEMBLIES

We can say without being wrong that a today's most widespread and appreciated working tool of engineers from all over the world is the computer. This extraordinary development of the information science is due to the unique qualities of the computer in comparison with human being (Fig. 1).

Indeed, the large successful use of the computer in almost all activity fields including engineering has been resulted from its very large memorizing capacity, impressive computing speed under computing accuracy conditions many times more than sufficient, visualization facilities as well as "patience" and "working capacities". The computer-aided industrial processes lead not only to an automation of same activities carried on by specialized employees but also to the introduction of the entire available technological knowledge for that production line in the research-designing-planning-production-checking process. (Boothroyd, G. 2005).

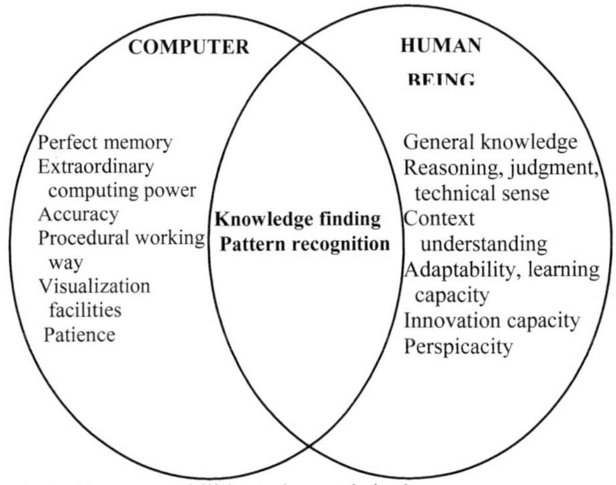

Fig. 1. Computer abilities to human being's.

Nowadays, the CAD (*Computer-Aided Design)* systems market is one of the largest and most dynamic in the world and there are millions of applications

Regardless of the application type or the platform which those applications are created for we can notice the permanent effort of the producing companies for developing new programs that offer the possibility to reduce to a minimum the designing time, the production expenses as well as the designing effort.

On that context (Rao, P. N. 2004), we can notice some important trends for CAD systems development:

a.Applications integration: allows the model achievement, description in detail, its analysis and finally manufacturing having one virtual model, using the same interface; in this way the operator's acquaintance with those applications is facilitated, being avoided the date files translations. The unique product model includes all the necessary information for each process and iterations take place efficiently and quickly. In fact, in this case we are considering a CAE *(Computer-Aided Engineering)* application including modules for geometric modeling, structural analysis, mechanism analysis and simulation, finite element analysis for different requirement types, for dynamical Internet data publishing, for documents management, etc.

b.Visualization facilities: have already reached **a** professional level (Shipulsky, M. 2007), including options like *radiosity, ray tracing*, photorealistic mapping, parametric animation, possibility **to** manipulate cameras, etc.

c.Interconnection and team work: in today's world of different designing systems and of necessity to convey projects, a designing environment should be as open as possible. The operations change is essential among systems that place at disposal and share resources for designing process, either among teams within the same company or among research groups from all over the world. In this way, the designing environments have modules allowing the control on the drawing standards; various symbols are stored in one single library for an easy maintenance and an efficient using (Lamarche, B.; Rivest R. 2007).

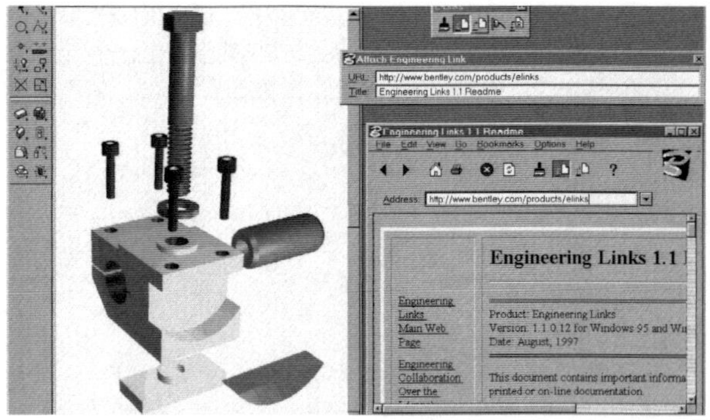

Fig. 2 – Including an Internet address (URL) into virtual model.

d.Digital signature: by applying such a signature on a certain area of the project the professional people can offer the legal access to the create drawings. The digital signatures allow the authentication of the stored drawings and provide information on the designer identity and putting up-to-date made after project publication. By means of the programs, the signature falsification or modification is not possible.

e. On-line data change and Internet connection: allow the connection to different parts catalogues or the Internet seeking of a certain producer whose parts match the in-process project, in this way the designing time and efforts are shortened. When the item is found it can be included in the model (Fig. 2), thus the data change could take place among the designing team members, among corporations and even among distributors (by including URL addresses in the model).

f. Hybrid modeling: represents the capacity to model 3D solids using both *wireframe* entities or solids and surfaces; all these describe the same geometric solid and have the advantage of a mixed solver (parametric and functional); also, the hybrid modeling includes the capacity to edit and modify the model under various techniques not depending on the way the model was achieved but on the operator necessity at one time.

Primitives and model features handling: graphically shows the tree structure of model achievement history and allows modifications of parameters and position of any element in the tree (Fig. 3). In case of modifications causing an impossibility to achieve elements dependent on the modified one, the system warns the operator very fast.

Using subassemblies technique (modular designing): reduces the complexity of the designing process for it allows to focus the effort on a single component at a moment (different assembly parts will be modeled one by one and are to be put together in the final stage) and also for the fact that the same part may be used several times in the same assembly or in different assemblies. (one feature modification will automatically appear in all positions where the feature appears). That working way removes project date redundancy and makes fit different parts of the assembly (*** 2008).

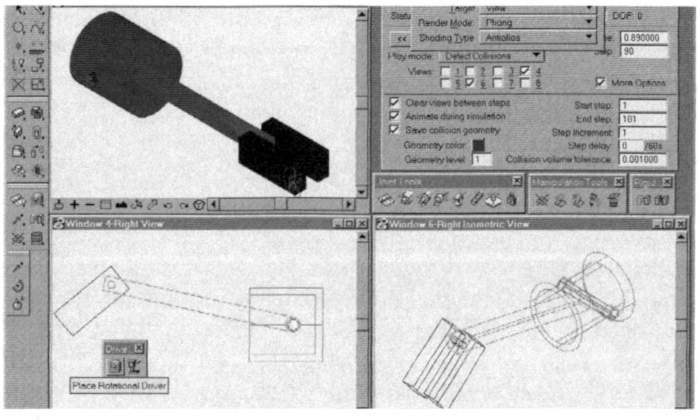

Fig. 3 – Drawing an assembly from previously designed parts (modular designing) by using Part Manager.

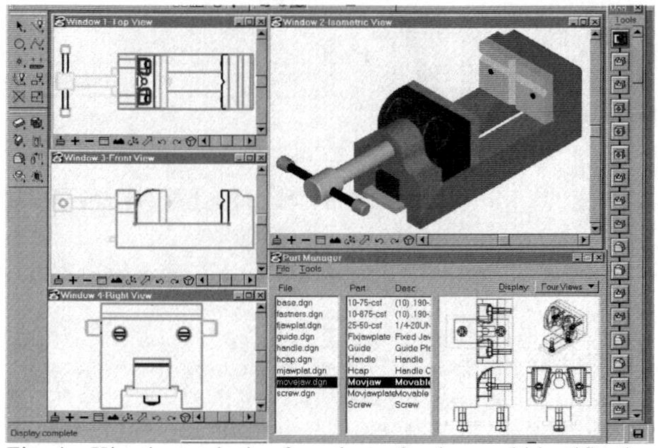

Fig. 4 – Kinetics analysis of crank mechanism and identification of possible collisions among assembly parts.

The assembly *Part Manager* (Fig. 3) allows parts achievement and their modification within the assembly by using one of the techniques "from complex to simple" (the designer makes a general assembly plan and a hasty draft of every parts and hands them over to other people to finish and at the end he is to put together all the parts) or "from simple to complex" (each part is separately designed and then it is attached to the assembly) which best suit to the chosen purpose.

g. Relative movements can be defined for the parts of an assembly and thus, there is the possibility to detect the eventual collisions that may appear among moving parts (Fig. 4). Also, there is possible to attach couplings to another subassembly so that each may have its own coupling system (very useful thing when the movement of some subassembly parts is desired).

3. CONCLUSION

The mechanical field designing is subject to some powerful pressures because of the increasing competitions and market globalization, more and more increased clients' requirements and quickly developing technology. As a result, companies have to be more productive, more responsible and flexible in order to survive and have success. The CAD mechanical packages become more efficient in terms of their cost. They are production-orientated, easy to use and allow data management, issuing materials list, parametric surfaces and solids modeling, various analysis and simulations, numeric control, etc.

There are also achieved interfaces according to the international and national standards, other applications and CAD/CAM/CAE systems, which are our next goals of research.

4. REFERENCES

Boothroyd, G. (2005). *Assembly Automation And Product Design Second Edition,* CRC Press Taylor& Francis Group, ISBN I-57444-643-6, Broken Sound Parkway, New York, USA.

Lamarche, B.; Rivest R. (2007). *Dynamic Product Modelling with Inter-Features Associations: Comparing Customization and Automation. Computer Aided Design and Applications,* Vol. 4, No. 6, (June 2007) p. (877-880), ISSN 1686-4360

Rao, P. N. (2004). *ACAD/ CAM Principles and Applications,* Tata McGraw Hill, ISBN 0070583730, Broken Sound Parkway, New York, USA

Shipulsky, M. (2007). *Successful design for assembly,* Available from: http://www.assemblymag.com/CDA/Articles/Feature_Article/BNP_GUID_9-5-2006_A_10000000000000059386 Accessed: 2008-04-10

*** (2008). *Design for assembly,* Available from: http://www.teamset.com/design-for-assembly-2.html Accessed: 2008-06-11

Annals of DAAAM for 2008 & Proceedings of the 19th International DAAAM Symposium, ISSN 1726-9679
Editor B. Katalinic, Published by DAAAM International, Vienna, Austria 2008
Make Harmony Between Technology and Nature, and Your Mind will Fly Free as a Bird

THE CYLINDER INTERSECTIONS VISUALIZATION IN TEACHING DESCRIPTIVE GEOMETRY

MARIN, G[heorghe]; PETRESCU, L[igia] & DOLGA, L[ia]

Abstract: *Starting from the development of the CAD systems and computer based visualization techniques; the paper presents the authors' opinions about the computer's use in teaching descriptive geometry. The subject was illustrated by representing and imagining different cylinder intersections. The authors conclude only good skills in 3D modeling could make this didactic method to be efficient and, consequently, learning of 3D computer-aided modeling should precede the instructional stage in advanced descriptive geometry.*
Keywords: *cylinder, intersection, teaching, descriptive geometry.*

1. INTRODUCTION

The conception, development and illustration of technical information by means of graphical method represent an ultimate part of the engineering activity. As a result, training and developing the skill to "mentally 3D see", to imagine and create best spatial shapes for certain functions, are essential for training the future mechanical engineer.

The descriptive geometry (Biran, A.; Lopez-Pulido R. 2005) is one of the difficult common disciplines, which the first year student at mechanical faculty has to face with. For the most students the secondary school *knowledge* in this discipline is almost non-existent. S*kills* in the mentally spatial modeling of shapes, in the process of drawing these mentally modeled shapes in plan graphical views, the spatial view, are at the beginning at very various levels, from medium, seldom very developed, to non-existent or almost non-existent levels. As a subject matter in the first year, the descriptive geometry is intended to train and develop these skills.

Computer aided designing and spatial visualization systems can be successfully used in teaching descriptive geometry. However, these techniques are not conceived for didactic purpose because their use in training activity requires significant efforts.

2. ASPECTS OF CAD TECHNIQUES USED IN THE DESCRIPTIVE GEOMETRY STUDY

Using the CAD techniques implies the assimilation by students not only of the descriptive geometry but also of the methods for geometrically modeling and defining CAD features.

Nevertheless, teaching the descriptive geometry should take into consideration its essential specific purposes: improving visually geometric reasoning, training the spatial view and ability to transfer information between 2D and 3D space mutually.

Under the above mention considerations, the study of a number of bibliographical books and papers in the field, the study plans in different worldwide engineering universities and their own attempts, the authors consider as necessary the assimilation of the 3D-CAD modeling techniques before learning the descriptive geometry advanced ideas (solids representations, solids and evolutes intersections).

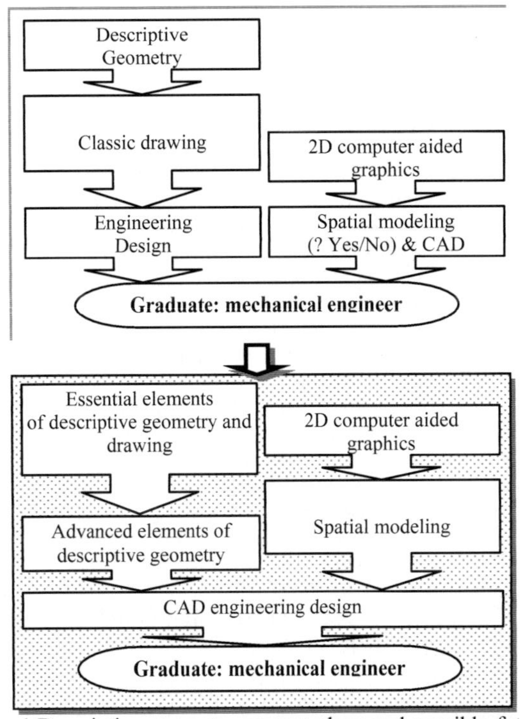

Fig. 1 Descriptive geometry present place and possible future sequence of graphic disciplines in training the mechanical engineer.

Approaching the spatial modeling should naturally be preceded by good knowledge of the 2D-CAD techniques. Under these conditions, the descriptive geometry study should consist in two training stages; the first including essential elements and the second the advanced elements. The second stage comes after assimilation of the basis elements in conventional plan views and in the plan and spatial computer aided views, respectively (fig. 1).

In order to acquire efficiency in teaching the descriptive geometry, the CAD techniques should be implemented according to a strict plan based on a unitary discipline outlook and a structure, with clearly defining the purpose and modeling methods used, and creating some best working algorithms. The student should always be aware of what it is modeled, for what purpose, by which methods, what is the operations sequence, what it is essential to be noticed during the process and what he should finally obtain.

Teacher should define these elements under the specific conditions of the descriptive geometry.

The experts settle two descriptive geometry-learning strategies:

▪Visual learning described by a great trust in- and an especial using of the imagination and mental visualization capacity;

▪"Haptic" learning strategy based on additional tactile and

moving elements and references. There is an opinion that a quarter of persons has a "haptic" learning approach and does not have quickness and intuition in the correct mental visualization of objects.

The persons natively gifted with such capability are considered to use almost exclusively the first strategy. They can easily elaborate reasoning based on mental imagines exclusively. The classic method of training in the descriptive geometry field has very good results in their case. For those individuals who resort to the "haptic" strategy using computer in the descriptive geometry learning process may be very useful.

3. THE ADVANTAGES OF USING COMPUTER FOR THE DESCRIPTIVE GEOMETRY

The total reproduction of a geometric construction requires at least two images followed by their exclusively mental combination. The CAD spatial modeling allows straight visualization in the virtual three-dimension space, model examination from different space points, using colors, selective visualization of some model parts, quickly and simply getting orthogonal views.

The classic working technique of the descriptive geometry allows graphic study of a single case. A new case means a new graphic construction. By means of computer, generalizations of the studied problem are possible by relocating and rotating some geometric elements.

The CAD techniques for solids modeling allow the combination of the simple geometric shapes in order to generate complex shapes, thus being possible to achieve intersection curves. The automat generation of any main and/or auxiliary orthogonal projections, in views and sections, with separating the visible and invisible edges, are particularly valuable facilities offered by computer.

4. GENERAL RECOMMENDATIONS FOR SOLVING PROBLEMS

In the descriptive geometry, the computer may be used in two ways: for problems 2D solving according to the manual methods, the computer replacing only the drawing tools, and for problems 3D modeling, respectively, by emphasizing 3D solutions (Ryan, D. L. 1991). Considering the large industrial application, the authors have proposed the cylinder intersections study: The oblique cylinder can be modeled in four steps, by the extrusion of a circle along a line parallel to the generatrix, followed by its upper part sectioning. Use the same method to draw the second oblique cylinder.

The study of the intersection curves of the two cylinders is based on the interference solid construction (Chao, Y. et al., 1999) and not on the Boolean operations (reunion, intersection, difference). Thus, this method allows keeping the original cylinders.

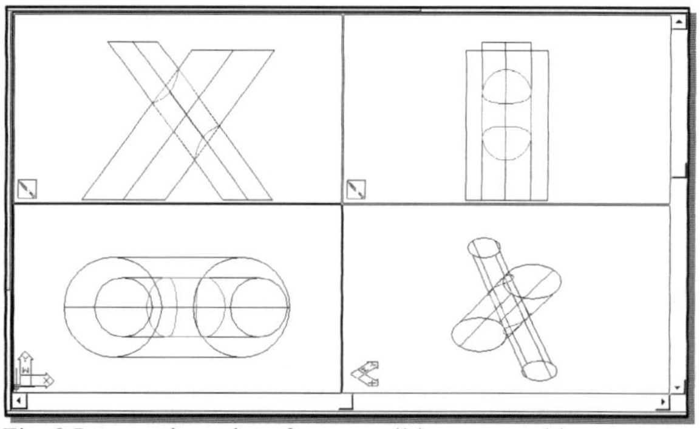

Fig. 2 Intersection – interference solid as a new object.

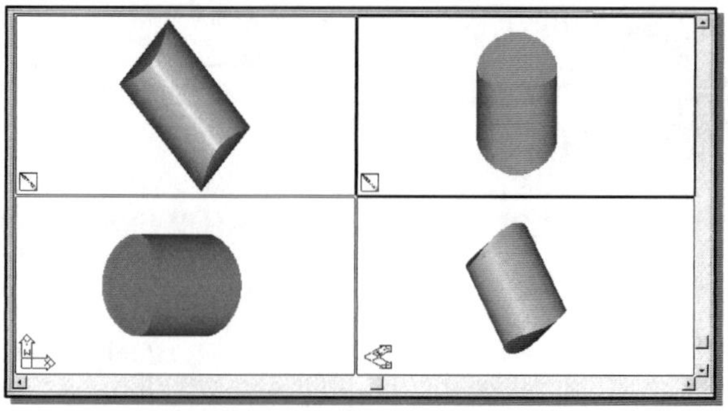

Fig. 3 The interference solid.

The program compares the two solids sets and requests the option for creating the interference solid as a new object.

In order to generate composite solids by the reunion and difference of the two cylinders, apply the desired Boolean operation. The resulted composite solid can be studied by freezing all the layers excepting the study one (fig. 4).

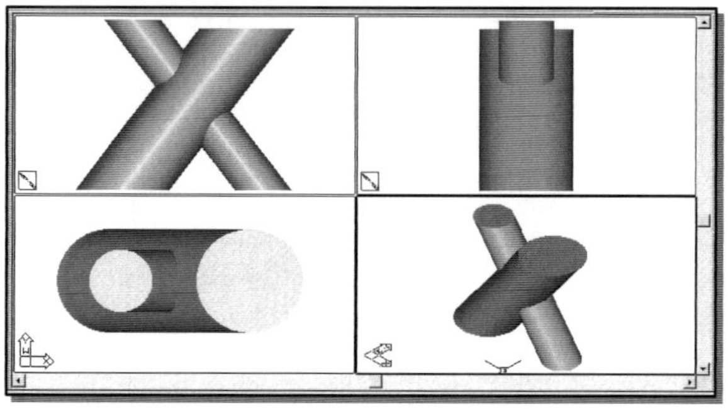

Fig. 4 The reunion composite solid.

3. CONCLUSION

One of the main "disadvantages" of the plan views specific to the descriptive geometry is the difficulty to suggest space intuitively (Standiford, K. 2001), (Standiford, K., et al., 2001).

Using the computer in study of the descriptive geometry can be an efficient method providing a correct application. This method offers large generalization possibilities being attractive and nice at the same time. However, this progressive step requires changes in the sequence of teaching the engineering graphic disciplines, during two year of study, in two stages. The second stage comes after assimilation of the basis elements in conventional plan views and in the plan and spatial computer aided views, respectively.

4. REFERENCES

Biran, A.; Lopez-Pulido R. (2005). *An Analytical introduction to Descriptive Geometry*, Available from: http://meeng.technion.ac.il/Studies/PDF_Files/Descriptive.pdf *Accessed:* 2008-01-15

Chao, Y.; Watanabe, T.& Minemura, K. (1999). Inferring solid objects with cylindrical surfaces from orthographicviews, *Proceedings. Third International Conference on Computational Intelligence and Multimedia Applications 1999 ICCIMA '99, 1999.* pp. 191-195 , ISBN 0-7695-0300-4, New Delhi, India

Standiford, K. (2001). *Descriptive Geometry,* Thompson Delmar Learning, ISBN 0766811239, New York, USA

Standiford, K.; Standiford, D. (2001). *An Integrated Approach Using AutoCAD,* Thompson Delmar Learning, ISBN 0766811239, New York, USA

Ryan, D. L. (1991). *CAD/CAE Descriptive Geometry,* CRC Press, ISBN 0849342732, New York, USA

Annals of DAAAM for 2008 & Proceedings of the 19th International DAAAM Symposium, ISSN 1726-9679
Editor B. Katalinic, Published by DAAAM International, Vienna, Austria 2008
Make Harmony Between Technology and Nature, and Your Mind will Fly Free as a Bird
*Young Researcher and Scientist Paper / * Supervisor, Mentor*

EVALUATION OF FORCES IN INDENTATION PROCESSES CONSIDERING GEOMETRIC PARAMETERS OF THE WORKPIECE

MARIN, M[arta]; DE AGUSTINA, B[eatriz]; CAMACHO, A[na] & SEBASTIAN, M[iguel] A[ngel]

Abstract: *New technological trends in manufacturing processes are focused on improving the integration of the systems, the process accuracy, reducing costs and times and increasing the sustainability of the environment. Metal forming processes are one important group among them indentation processes present some characteristics that make them especially interesting. In this study an indentation process was simulated by the Finite Element Method, assuming plane strain conditions. Friction between the workpiece and the punch was neglected assuming that a good lubrication was realized. Under these conditions a reduction of 5% was applied by the punch for different values of the shape factor. Results show that the greater the height and the width, the greater the required forces, keeping constant the rest of the conditions in the process.*
Keywords: *Indentation Process, Shape factor, Forces, Finite element Model*

1. INTRODUCTION

Among the different compression processes, indentation was considered because of its similarity to other processes such as forging, that is widely used in the industry field. Usually, indentation processes are used as a method for estimating material hardening (Minh-Quy, 2008), (Kucharski & Mróz, 2007). However, in this work, indentation process will be considered as a manufacturing process (Camacho et al., 2008), (Sebastián & Camacho, 2007).

In conventional compression processes the flow of information is mainly restricted to design and manufacturing (Camacho et al., 2006a); (Camacho et al, 2006b). For this reason, it is important to see how such processes are influenced by the size of the workpiece. Based on the use of dies with simple shapes it is intended to analyze what parameters of the indentation processes can be modified in order to achieve more economic processes (Camacho et al., 2007).

In this work, the dimensions of the punch will remain constant while the dimensions of the billet are changed. The approached problem was studied assuming ideal conditions such as plain strain ones. Different cases were analyzed according to different geometrical values of the workpiece by means of the Finite Element Method (FEM).

2. METHODOLOGY

2.1. Parameters

At first, a rectangular section of the punch is considered, where B is the width. This dimension remains constant throughout the whole study.

Similarly, the workpiece is considered a rectangular billet but its dimensions (w, h, width and height respectively) change throughout the analysis (see Figure 1). These geometrical parameters are chosen as follows:

$$w = B + 2\cdot x = B + 2\cdot n\cdot B = B\cdot(1+2\cdot n) \quad (1)$$
$$h = x = n\cdot B \quad (2)$$

The value of the parameter n changes from 1 to 5. From now on the ratio w/h is used in order to quantify the influence of the geometry of the billet and it is called the "shape factor".

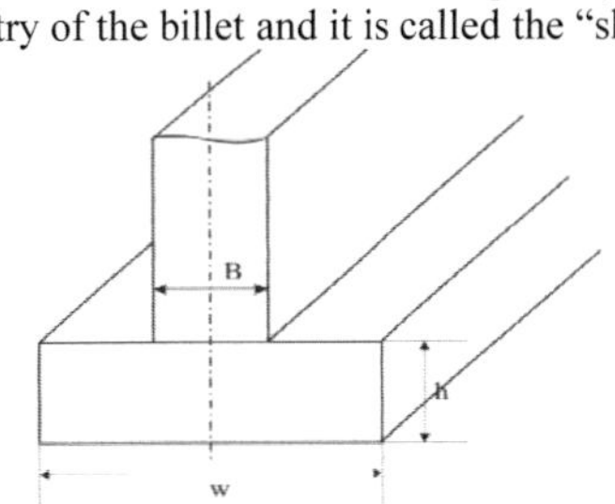

Fig. 1. Geometry of punch and billet

In all the cases, the friction at the punch-workpiece interface is considered null and the reduction in height applied is 5%.

2.2. Cases of Study

In a first approach and according to its previous definition, different shape factors are obtained varying the values of n (Table 1). This will be the first group of cases to be analyzed, and it means a simultaneous increase of both dimensions, the width and the height of the billet.

Afterwards, a second group of geometries of the workpiece were studied. Assuming different values of the height but keeping each one constant, and varying the width of the billet while the parameter n is gradually increased, it is possible to obtain the geometric values of Table 2, where the second group of cases to be analyzed are summed up.

	n				
	1	**2**	**3**	**4**	**5**
w/h	3B/B	5B/2B	7B/3B	9B/4B	11B/5B

Tab. 1. First group of cases: shape factors for different values of n

h = B; 2B; 3B; 4B; 5B
w = 3B; 5B; 7B; 9B; 11B

Tab. 2. Second group of cases

3. FINITE ELEMENT MODEL

This study was completed using the Finite Element Method (FEM). A general purpose software of implicit methodology, ABAQUS/Standard (Hibbit et al., 2007), was used. The puch was designed as a rigid part and the workpiece was modeled as a deformable body. The type of element is CPE4R and consists of a continuous, plain strain, linear interpolation and reduced integration element. The billet was modeled with an aluminum alloy. The type of material was considered as a strain hardened one. A linear behaviour of strain hardening was assumed for the material. Finally, a Coulomb friction model was assumed by the Finite Element Software. As a first approach in this work, a frictionless problem was assumed for all the cases, simulating a well lubricated process.

4. RESULTS

In this work the variable that was analyzed is the required force to develop the indentation processes. As it was explained, this study was divided into two groups of cases. In the first one, forces were obtained for the shape factors given in Table 1, and the results are showed in the figure 2.

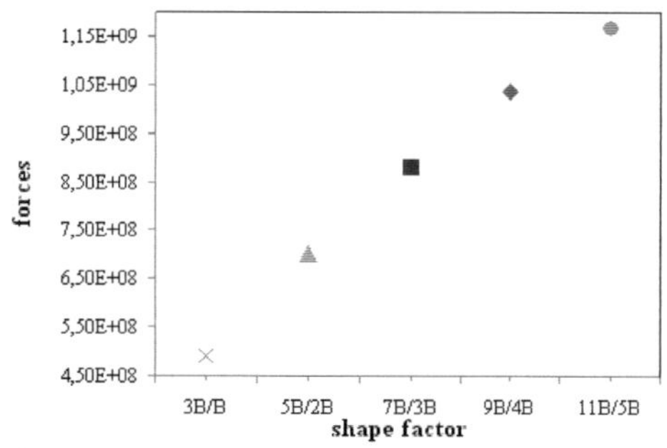

Fig. 2. Forces for the first group of cases

In this figure it can be observed that the greater the shape factor, the greater the forces required to carry out the process, so it means that as the dimensions of the workpiece increase, the necessary energy also increases for the same conditions of the process.

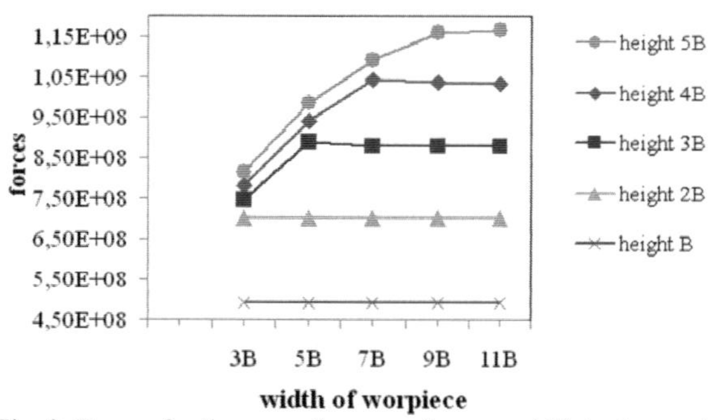

Fig. 3. Forces for the second group of cases: width in the x axis

For the second group of cases, the influence of the width and the height of the billet on the forces were studied (see Table 2). The figure 3 represents the forces that were obtained and the width was represented in the x axis. The analysis was carried out in the same conditions as in the first group of cases.

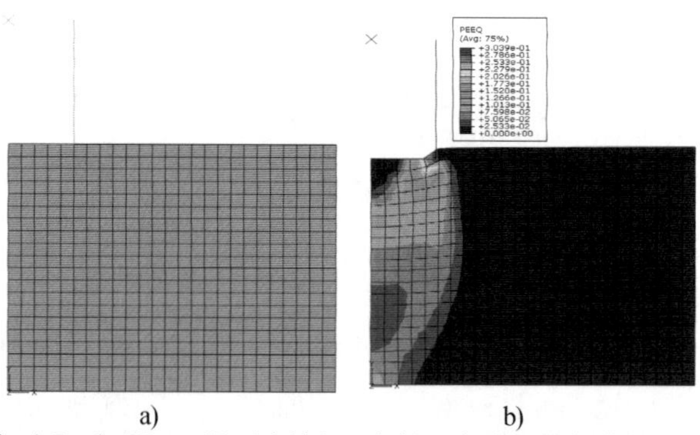

a) b)

Fig. 4. For $h=2B$, $w=5B$: a) initial mesh; b) equivalent strain diagram

This figure shows that the greater both the width and the height, the greater the required force, until a value where forces keep constant and there is not variation with the width. The graph shows that this value is reached when $n = 4$. This behaviour means that if $n \geq 4$ forces do not depend on the width, for the conditions of the problem that were analyzed.

An example of the numerical simulation that was developed, is shown in figure 4. In this illustration the case where $h = 2B$ and $w = 5B$, the mesh of the model and a strain diagram are represented.

5. CONCLUSIONS

This paper studied the influence of geometrical parameters on the forces of an indentation process. In a first analysis, a shape factor was chosen as the geometric parameter, in order to evaluate the influence of a simultaneous variation of both dimensions, height and width. In a second analysis, the dimensions of the workpiece were changed independently. The first conclusion is that the obtained forces increase when the dimensions of the workpiece increase simultaneously. Looking at the influence of each dimension, it is observed that the greater the height, the greater the force, and also the greater the width, the greater the force. But this effect is greater when there is a change in the height. In fact, starting from a value *n,* force is not affected by the width. Therefore, for each geometry of kind of workpiece this, it would be possible to predict the value of the force starting from that width, if the rest of conditions keep constant. Additionally, taking into account these results the punch may be positioned more efficiently. In future works this indentation process will be analyzed under different conditions. Regarding, the influence of the geometrical parameters on variables such as the contact distributions should be studied. It is thought that having a good knowledge with regard to all these factors it will be possible to improve the efficiency of this process.

6. REFERENCES

Camacho, A.M.; Marín, M.; Sevilla, L. & Sebastián, M.A. (2007). Analysis of axisymmetrical compression processes by the finite element method, *Proceedings of the 9th International Conference on Numerical Methods in Industrial Forming Processes,* J.M.A. Cesar de Sá; A.D. Santos (Ed.), pp.1011–1016, ISBN 978-0-7354-0416-8, Oporto, June 2007, Portugal.

Camacho, A.M.; Marín, M.; Domingo, R. & González, C. (2006a). Analysis of open die flat forging processes in plane strain conditions by FEM, *Proceedings of the 17th International DAAAM Symposium "Intelligent Manufacturing & Automation: focus on mechtronics and robotics"*, Katalinic, B. (Ed.), pp. 75-76, ISBN 3-901509-57-7, Vienna, November, 2006, Austria.

Camacho, A.M.; Marín, M.; Rubio, E.M. & Sebastián, M.A. (2006b). Analysis of upsetting processes by the Finite Element Method, *DAAAM International Scientific Book 2006,* Katalonic, B. (Ed.), 93 – 100, Austria, ISBN 3-901509-47-X.

Camacho, A.M.; Vallellano, C.; Sebastián, M.A. & García-Lomas, J. (2008). Análisis mediante el MEF de los efectos de la geometría del extremo del punzón en procesos de forja localizada-incremental. *Anales de Ingeniería Mecánica*, Vol. 2, pp. 279-285.

Hibbitt, D.; Karlsson, B. & Sorensen, P. (2007), *ABAQUS v6.7, User´s Manuals*, Providence (RI).

Kucharski, S. & Mróz, Z. (2007). Identification of yield stress and plastic hardening parameters from a spherical indentation test. *International Journal of Mechanical Sciences*, Vol. 49, pp. 1238-1250.

Minh-Quy, L. (2008). A computational study on the instrumented sharp indentations with dual indenters. *International Journal of Solids and Structures,* Vol. 45, pp. 2818-2835.

Sebastián, M.A. & Camacho, A.M. (2007). Geometrical study and basis for the analysis of localized-incremental forging processes by FEM, *Proceedings of the 2nd* ICNFT: 2nd *International Conference on New Forming Technology*, Vollersten, F.; Yuan, S. (Ed.), pp. 295-304, ISBN 978-3-933762-22-1, Bremen, September 2007, Germany.

Annals of DAAAM for 2008 & Proceedings of the 19th International DAAAM Symposium, ISSN 1726-9679
Editor B. Katalinic, Published by DAAAM International, Vienna, Austria 2008
Make Harmony Between Technology and Nature, and Your Mind will Fly Free as a Bird

SIGNIFICANCE OF SMALL MUNINCIPALITIES FROM THE ASPECT OF MANAGEMENT

MARINAC, A[ntun]

Abstract: *Existing territorial organization of local self-government in the Republic of Croatia is based upon a large number of small local units, especially small municipalities. Along the many problems small municipalities encounter (insufficiency of personnel and finance capacities), they still take a significant place in the territorial constitution of the Republic of Croatia (they are better for implementation of immediate democracy, and make budgetary expenditure easier to control). In those small municipalities the citizens have found a local identity. That reality should be considered and existing local self-government units should be enabled to perform their tasks, aimed at becoming efficient and close to the citizens. .*
Keywords: *local self-government units, small municipalities, scope of self-government, management, territorial constitution*

1. INTRODUCTION

Today there is not a democratic country in the world that lacks local self-government units in its administrative structure. In compliance with the article 133, paragraph 1 of the Constitution of the Republic of Croatia, local self-government units are municipalities and towns. In Croatia almost nobody has conducted surveys on small municipalities.
But regarding the fact that there are around 400 municipalities in the Republic of Croatia at the moment, and more than 350 of them are small municipalities (with up to 5000 citizens), research target in this paper is to provide the answer on their significance.
In order to achieve this, following methods have been used: methods of data acquisition (internal, external), methods of processing data (manual processing of data, machine processing of data), methods of data analysis (analysis, synthesis, induction, deduction, method of comparison, and method of compilation, mathematical and statistical methods).

2. ABOUT LOCAL SELF-GOVERNMENT IN GENERAL

It is a fact that world's state systems use local management systems as their auxiliary structures and through them achieve their local goals rather than they could achieve them using just their own organs. (Brunčić, 2001). By so doing local self-government is not a counterbalance to state government, but a replacement for functioning of state authority. The simplest and generally acceptable definition of local self-government is that it is a level of reign closest to citizens with the role of representing significance and statement of the local (Lauc, 2001).
Introduction chapter of European Charter of Local Self-Government talks about "the right of citizens to participate in the conduct of important public affairs". This right is one of the democratic principles mutual to all member countries of The European Council, and it is possible to be exercised at local level.By so doing local self-government is not a counterbalance to state government, but a replacement for functioning of state authority. (Table 1.)

	COUNTRY				
	The Parliament (legislative power)				
	The government (executive power) Courts (power)				
COUNTRY	Government				
LEVEL	• ministry				
	• central State Offices				
	• state administrative organizations				
			COUNTY		
			County Assembly		
			County chief		
			Hall		
REGIONAL LEVEL	state administrative organizations		Administrative bodies:		
	in counties		• departments		
			• services		
					TOWN/ MUNINCIPALITY
					Council
					Mayor
					Chief of municipal
LOCAL			Administrative bodies		Hall
LEVEL	Branch offices		outside centers		Administrative bodies:
	of state administration				• departments
					• services
					unique administrative department
	STATE ADMINISTRATION		REGIONAL ADMINISTRATION		LOCAL ADMINISTRATION

Tab. 1. Political and administrative structure of the Republic of Croatia

Regarding the fact that local self-government becomes the right of citizens, its constitution depends upon their decisions. Only at local level and thus in small municipalities citizens are regularly in connection with their environment, and whereby the most invited to manage municipalities directly and through elected representatives in the councils of their municipalities (Hrženjak, 2004). From a normative approach Croatian local self-government does not necessary differ from European countries. With no major problems the principles of local self-government have been taken over as the part of the reign i.e. self-government closest to the citizens, participation of citizens in decision-making directly through their elected representatives or indirectly, the right of citizens to participate in the important part of public affairs, autonomy of local units in defining the internal constitution, financial independency of local units etc.

3. SCOPE OF SELF-GOVERNMENT OF SMALL MUNICIPALITIES

Role and significance of small municipalities arise from their scope of self-government. According to the article 134, paragraph 1 of the Constitution of the Republic of Croatia local self-government units and also small municipalities conduct affairs related to: organization of settlements and housing, town and urban planning, utility services, child-care, social welfare, primary health protection, education and primary-school education, culture, physical culture and sport, consumer protection, protection and development of natural environment, and fire-protection and civil defence, all in the aim of meeting citizens' interest and needs. We believe that the activity "participation of citizens" in local affairs is crucial in small municipalities. Citizens should be active in local initiatives through: search for income recourses, conduct of daily problems, suggestion of priorities in the use of scarce public recourses, control of local bodies and functionaries providing them with possible solutions, so municipalities are enabled to make decisions in an open and transparent way which contributes to the improvement of the local self-government function.

4. SIGNIFICANCE OF SMALL MUNICIPALITIES

Small municipalities take significant place in social and political life of the Republic of Croatia because in them, almost with no exception, there has been strong identification of citizens with their municipalities (Antić, 2005). That can be a source of a strong motivation for their participation in local management, and with right citizens from the areas of small municipalities believe, in compliance with the guaranteed right to local self-government, they are better at making decisions for that municipality than someone who lives outside its area. It is undisputable that small municipalities have also some advantages over bigger local self-government units. From the aspects of theory of democracy it could be proved that smaller the municipalities are, more democratic they are, because citizens can participate in local affairs in greater amount. In numerous small municipalities citizens estimate that they are able to control public money expenditure in greater amount, gathered by collecting taxes or by capitalization of the asset of the local self-government, than they could do it in bigger local self-government units. (Marinac, 2006).Small municipalities along with development of utility services should gradually become growth factors in series of social activities (education, health care, culture etc) and economy. To be able to achieve this they ought to: be efficient and economical, satisfy citizens' requirements, behave rationally and economically, earn and not only collect, lead active politics and not react only to the problems, be enterprising and honest, work in compliance with the law and own decisions, impartial and work towards everyone's benefit, with efficient surveillance from the middle level (Koprić, 2001). Even though it seems it should not significantly interfere with the existing territorial organization, due to identification of citizens with their local self-government, the problem remains open in the context of Croatia joining European Union in which decentralization and regional development are top valued. Therefore it is necessary to make assumptions for regionalization of Croatia inside of which the identity of small municipalities would be also recognized. Instead of cutting down some of the small municipalities, we believe that conditions should be created to enable small municipalities to independently estimate opportunities for their existence i.e. their possible voluntary collaborations, joining or merging, but always and exclusively at will of the local population.

5. BASIC PROBLEMS OF SMALL MUNICIPALITIES

Parallel data about the size of municipalities show that there are also a great number of small municipalities in some European countries. In the Republic of Slovenia there are more than half, out of 193 municipalities (per 2 million citizens), with less than 5000 citizens. In Hungary there are 55% of municipalities with less than 1000 citizens, and in the Czech Republic 8%. In Swiss municipalities there are approximately 2.333 citizens, and in Croatian there are 3.220. In France there are more than 36.000 municipalities, and there are even smaller ones with even less than 100 citizens (Hrženjak, 2004).

According to the Law of Areas of Counties, Towns and Municipalities of the Republic of Croatia, territorial constitution of the Republic of Croatia consists of 20 counties, the City of Zagreb with special status, 126 towns and 429 municipalities.(Lukeš-Petrović, et al, 2005) The basic problem of small municipalities is not their size, but their administrative and financial capacity.

In order to minimize the differences among local units we believe it is necessary to find the relation in which a local unit is still small enough to be close to the citizens and able to efficiently solve local priority issues, and yet big enough that its administrative capacities are satisfying, that it is able to ensure that in its area and with its asset is possible to provide with sufficient amount of public money which may be used for providing with public goods whose provision is supervised by local units.

6. CONCLUSION

The main role of small municipalities is to research citizens' needs and use as cheap as possible organization to respond to them creating conditions for economy growth and efficient management of the poor recourses they have in disposition.

But orientation towards solving practical problems of citizens is, at this moment, more important than economy frame of small municipalities itself. In today's conditions small municipalities must learn to depend onto their own resources, in order to be able to compete with others in competition for public goods.

Everything that is conducted in small municipalities would be purposeless if it was not conducted in the interest of citizens, therefore they must be at dispose to citizens in the way that their organization will be adjusted towards meeting citizens' needs, which would finally result in citizens' satisfaction with life quality in their living area.

We believe that at this moment it is not advised to cut down small municipalities, but create conditions for independent fulfilment of obligations which arise from their self-government scope, strengthening their capacities, human ones firstly, but also technical, spatial, information, financial etc.

7. REFERENCES

Antić T. (2005). Reforms of local self-government – challenges and limitations, *Newsletter for local democracy* No 8, *Available from*: http://www.stina.hr/ Accessed: 05-27-2008

Brunčić, D. (2001). *State and Local Governance*, *Local Self-Government: Croatian and Dutch experiences*, Croatian Institute for Local Self-Government Osijek and Interkerklijk Vredesberad Haag, Osijek

Hrženjak, J.(2004). Local and regional self-government of the Republic of Croatia, *Informator*, Zagreb

Koprić, I.(2001). Frameworks, accents and perspectives of current decentralization in Croatia, *Croatian Public Administration*, numbers 3-4, Zagreb, pp. 411-454.

Lauc, Z. (2001). *Key Notions of Local Self-Government*, Local self-government: Croatian and Dutch experiences, Croatian Institute for Local self-government Osijek and Interkerklijk Vredesberad Haag, Osijek

Lukeš-Petrović M., Mesarić, H. & Nikolov J.(2005). *A guide to the Croatian system of local and regional self-government, United Nations Program for Development* (UNDP), Croatian Institute for Local Self-Government (HILS), str. 3.

Marinac, A.(2006). *Place and role of small municipalities with reference to municipality of Kaptol*, Master's paper, Faculty of Law Osijek

Annals of DAAAM for 2008 & Proceedings of the 19th International DAAAM Symposium, ISSN 1726-9679
Editor B. Katalinic, Published by DAAAM International, Vienna, Austria 2008
Make Harmony Between Technology and Nature, and Your Mind will Fly Free as a Bird

MODELLING AND EXPERIMENTAL RESULTS AT LASER BEAM MACHINING FINISHING OF LASER FORM ST-100 COMPOSITE MATERIAL

MARINESCU, N[iculae] & GHICULESCU, D[aniel]

Abstract: *The paper deals with modelling of material removal finishing mechanism through laser beam machining (LBM). The experimental results of LBM surface levelling in case of workpieces made from a particular composite material, LaserFormST-100, by Selective Laser Sintering are also presented. LBM finishing is able to decrease the surface roughness down to R_a=0.5 μm at a single passage of laser spot over machined surface, correlating the material absorption capacity and values of technological parameters.*
Keywords: *laser finishing, modelling, composite material.*

1. INTRODUCTION

Laser Beam Machining (LBM) assures one of the highest energy density ever known and consequently, is able to machine almost any type of material in respect of its thermal absorption capacity.

The LaserFormST-100 is a composite material obtained by Selective Laser Sintering (SLS). The major inconvenience of this rapid forming process consists in poor quality of surfaces microgeometry, which has grainy aspect and small irregularities at boundary zones between deposed layers. This is due to starting state of rough composing materials, which initially were in powder phase.

Thus the workpieces made through SLS could not be used in several applications, e.g. active parts of injection moulds, because of their high surface roughness. In these cases, reduced R_a surface is required, which could be obtained by a very effective finishing process as LBM. Nevertheless the inhomogeneous structure concerning so different thermal physical characteristics of these composite material components generates great difficulties in establishing the optimal values of the technological parameters.

2. MODELLING OF LBM FININSHING

In order to attain the required high level of surface smoothening through LBM, only a superficial melting has to be achieved, respectively only the peaks of surface microgeometry. The melted material fills the valleys of microtopography, thus decreasing the surface roughness.

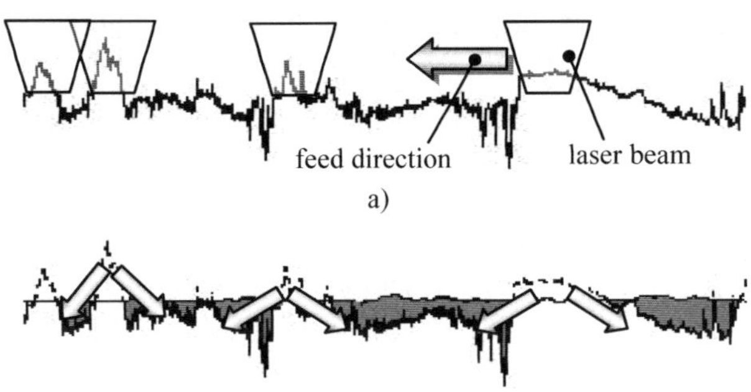

a) initial state; b) final state after laser finishing
Fig. 1. Surface microgeometry profile

The basics of LBM finishing mechanism are illustrated in fig. 1. An example of initial state of profile microgeometry is presented in fig.1.a, characterised by high roughness. After the passage of laser beam over machined surface, a decrease of surface roughness is achieved through flowing of superficial molten layer into microvalleys - fig. 1.b.

The essence of laser finishing technology consists in correlation of thermal energy absorption capacity of machined material - depending on structure and thermo-physical characteristics of material as well as initial roughness of surface - with technological parameters, which basically contribute to the energy density (q_E) brought on the surface to be machined, calculated with relation (Marinescu, 2003):

$$q_E = \frac{P}{d \cdot v_a} \tag{1}$$

where: P is power of laser radiation [J/s]; v_a - feed rate of laser spot [mm/s]; d - laser spot diameter [mm].

By adjusting the standoff distance, correct defocalisation of laser beam can be obtained, i.e. position of focal point is above machined surface (fig. 2), with major importance in reaching optimum value of laser spot diameter (d), and implicitly, energy density along with high machining rate. Uneven Gaussian distribution of energy on laser spot needs the overposition of laser traces as critical technological parameter (Marinescu et al., 2008).

Currently, there are two mechanisms, with corresponding phenomena, and different values of technological parameters, modifying the machined surface during laser processing: *Surface Shallow Melting* (SSM) and *Surface Over Melting* (SOM). The principle of each mechanism is determined by different levels of power (P) and feed rate (v_a) (Burakowski & Wierzchrow, 1999), (Ramos et al., 2001).

In case of SSM, when machining with high feed rate (v_a) and low power (P), then only thin material layer is melted. This molten volume flows toward microvalleys due to gradient of superficial tension and gravity force.

The SSM mechanism has particular deployment in case of metallic sinterised surfaces, which currently, are frequently encountered. This specific structure can be considered an aggregate of spheres (fig. 3). A micropeak corresponds to the upper part of sphere and a microdepression matches the space formed by three tangent spheres. During machining, the laser beam attacks the upper parts of the spheres and these surfaces rapidly arrive to the melting temperature.

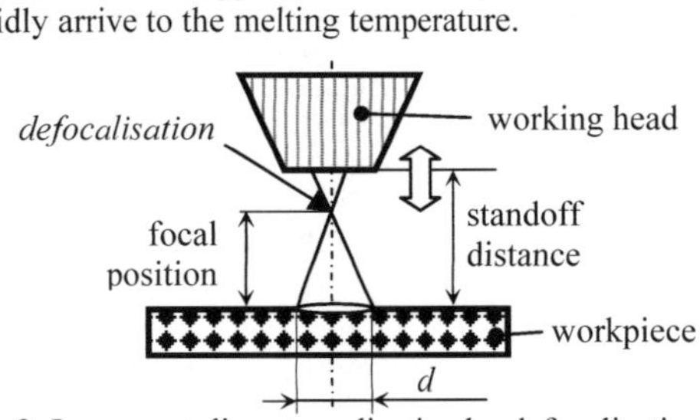

Fig. 2. Laser spot diameter adjusting by defocalisation

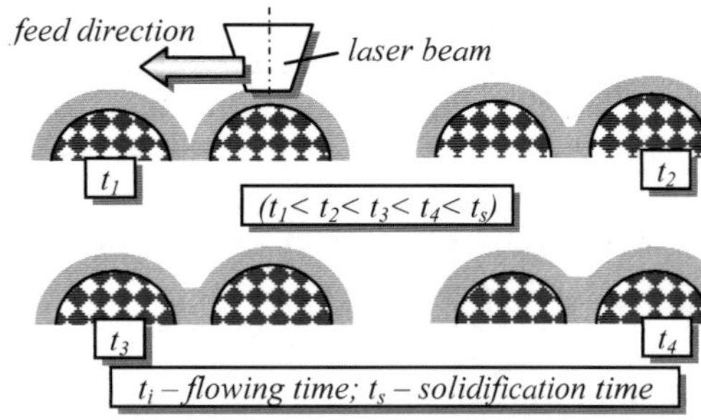

Fig. 3. Stages of LBM finishing at sinterised material

In case of SOM mechanism, the feed rate of laser beam is decreased, leading to an overheating of the surface. Hence deep melting of contact surface with laser beam is produced, resulting increasing of initial roughness. Thus, laser spot movement determines flowing of molten material along the tension gradient direction, till solidification front, creating waves at its proximity (Ramos et al., 2001).

The structure of LaserFormST-100 includes a copper surface layer (10-30 μm), bronze and stainless steel particles in depth with high and respectively, low absorption capacity. Final composition is roughly 40% bronze and 60% stainless steel. LaserFormST-100 has similar mechanical properties to P20 ones (C - 0.35%, Si – 0.4%, Cr – 1.2%, Mo – 0.35%).

The influence of laser radiation penetration depth (h) on surface microgeometry is presented in case of LaserFormST-100 (fig. 4). It can be noticed four stages of melting mechanism produced by LBM: *Stage* 1 - h is lower than microvalleys level (R_z roughness) and SSM occurs, determined by low values of feed rate and power. Thus roughness decreasing is produced; *Stage* 2 - h is greater than R_z roughness and first type of SOM occurs; *Stage* 3 - h is considerably greater, bronze particles being melted; *Stage* 4 - h is greater than 30 μm, steel particles being also melted, mixing with melted bronze particles; *Stage* 5 – h in continuous growing determines maximum level of Fe oxidation from the surface and R_a roughness incessant increase.

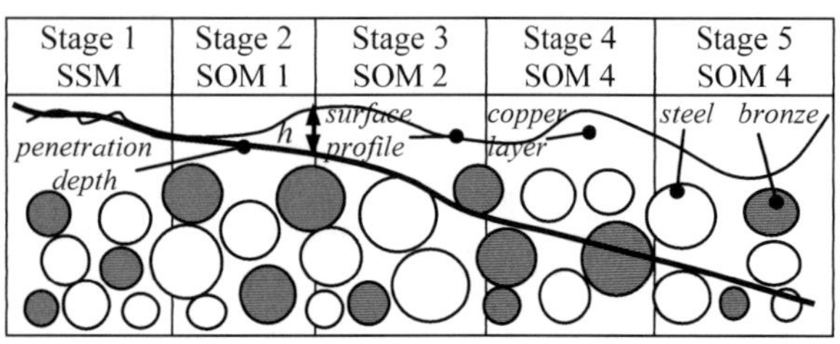

Fig. 4. Stages of melting mechanism for sinterised material

3. EXPERIMENTAL RESULTS

The experimental researches were completed on a CO_2 laser installation with exit maximum power of 2500 W, part of the Technological Park of Zamudio, Spain. The working head featured ZnSe focalisation convex lens, with 200 mm focal distance, and 0.4 mm minimum spot diameter. The roughness obtained on work samples was measured by an electronic device and then computer aided processed in order to obtain 3D microtopography. In order to emphasise, the effect of laser finishing, the workpiece was machined by LBM only on the centre, the sample margins remaining in the initial state (fig. 5).

The main physical properties of LaserFormST-100, influencing the laser thermal mechanism are: density 7700 kg/m^3 (at 23 ^{o}C), thermal conductivity 49 W/m^{o}C (at 100 ^{o}C) and 56 W/m^{o}C (at 200 ^{o}C).

We used 1500 W laser power, 22 mm for standoff-distance, 1000-1400 mm/s feed rate, and 0.2-0.4 mm over traces.

Surface roughness (R_a) was obtained in the range of 0.501-1.191μm, starting from R_a= 7.51 μm, initial SLS roughness.

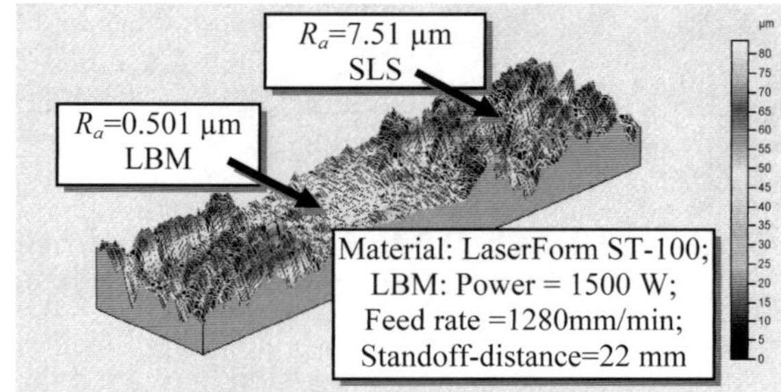

Fig. 5. 3D Microtopography of LaserFormST-100

The energy density (q_E) used at LaserForm ST-100 was relative high because the dominant component, stainless steel, has low absorption capacity. In order to avoid SOM degeneration of the surface smoothening process, due to bronze sensibility to over melting, we adopted the following solutions for technological parameters: low values for standoff distance and medium values for feed rate (v_a) and power (P).

4. FURTHER RESEARCHES

The objective of our researches is obtaining nanometric values for R_a surface roughness for homogenous materials. This could be achieved through an accurate computer control of the process focused on correlation of thermal absorption capacity with technological parameters and, on the other hand, by elaboration of specialised software, necessary to generate the optimum trajectory of laser spot. The software is based on an iterative algorithm to obtain surface roughness maximum decreasing at each passage of laser spot. At each step, the parts of surfaces with high protrusion are identified by sectioning the current 3D microtopography with a horizontal plane, zones covered by laser spot. The software transfers the coordinates of high profile surface parts, i.e. laser spot path to machine code instructions followed by CNC of laser installation.

5. CONCLUSION

Even materials as LaserForm ST-100, with such grainy structure, great initial roughness and difference between thermal properties of composing particles could be efficiently finished by computer aided LMB. The module of laser spot path optimisation of dedicated software assures the iterative attack of microgeometry protuberances. The module of technologic optimisation correlates the materials absorption capacity with working parameters using a large experimental data base. Therefore LBM finishing could lower R_a roughness in nm range, remarkably extending its application field.

6. REFERENCES

Burakowski, T. & Wierzchrow, T, (1999). *Surface Engineering of Metals*, CRC Press, ISBN 0-8499-8255-4, Boston, USA.

Marinescu, N.I. (2003). *Machining Technologies with Beams and Oscillations*, Bren, ISBN 973-648-128-X, Bucharest, Romania.

Marinescu, N.I.; Ghiculescu, D. & Anger, C. (2008). Correlation between laser finishing technological parameters and materials absorption capacity, *Available from:* http://www.springerlink.com/content/ *Accessed:* 2008-05-25.

Ramos, J.A., et al. (2001). Surface Roughness Enhancement of Indirect-SLS Metal Parts by Laser Surface Polishing, *Available from:* http://home.att.net/ *Accessed:* 2008-05-25.

Ramos, J.A., Bourell, D.L., & Beaman, J.J. (2001). Reducing surface roughness of metallic freeform-fabricated parts using non-tactile finishing methods, *Available from:* http://www.inderscience.com/ *Accessed:* 2008-05-25.

Annals of DAAAM for 2008 & Proceedings of the 19th International DAAAM Symposium, ISSN 1726-9679
Editor B. Katalinic, Published by DAAAM International, Vienna, Austria 2008
Make Harmony Between Technology and Nature, and Your Mind will Fly Free as a Bird

INFLUENCES OF THE GROWTH OF THE CARBON DIOXIDE EMISSIONS DEPENDING ON THE PROCESS OF PHOTOSYNTHESIS IN TIMISOARA

MARINESCU, S[orin] A[lin]; RUSNAC, L[ucian] & DOBREN, F[lavius] A[ndrei]

***Abstract**: This article presents Timisoara biotope's inability to adsorb the emissions of carbon dioxide emitted by local pollutants. The maximum quantity of carbon dioxide that can be adsorped by plants through photosynthesis is 35% in summer months and under 6% of the total number of emissions in Timisoara during winter, when the CO_2 emissions increase and only resinous plants are in the vegetation period.*
***Keywords**:dioxide, pollution, photosynthesis, emissions, combustion;*

1. INTRODUCTION

This paper is an integral part of a more extensive study that represents a case study for the city of Timisoara, which highlights the statistics and mathematical modeling of the carbon dioxide emissions and of the absorption of the carbon dioxide through the photosynthesis.

Carbon dioxide recorded a significant quantity increases from one year to another, caused mainly by emissions from burning fossil fuels and reducing photosynthesis process as a result of grubbing-up forests (Skiba et al., 2002).

The most important groups of anthropogenic pollution sources are burning in industrial processes, individual heating systems of housing, traffic, auto, cylinders with spray, etc.(Ionel et al., 2004).

The vaste majority of pollutants are the product of combustion processes in large heating centrals, of power system and of the combustion of fuel in internal combustion engines. This is the reason why sources like the industry and the auto traffic occupy the front places in the research related to the air quality (Ionel et al., 2004).

Inevitable, CO_2 is a product of burning fossil fuels; the limitation of CO_2 emissions can be done only by changing the fuel used, finding alternative energy sources, by increasing efficiency of combustion plants (Wehner et al., 2004).

CO_2 is a greenhouse gas, international laws relating to sustainable development impose limitations at country level regarding the amount emitted into the atmosphere.

Together with the increased importance given to quality of life, the level of pollution has sharply increased in the last two decades (http://www.mmediu.ro/arhiva.htm; 2008)

Different causes may be found in the evolution of technology, transport systems with internal combustion engines, which have become subject of discussions related to environmental protection. In literature it is specified that the activity of transport is responsible for about 33 to 35% of total CO_2 emissions, representing the 5th factor in rank which contributes to creating the effect of global heating of the planet; transport holds such 7% of the polluting sources generating this effect (Ionel et al., 2004).

Road transport is the key reason for the deterioration of the environmental factors. In the centers of large urban areas, road traffic is responsible for about 90-95% of the concentrations of carbon monoxide and lead found in air, for 60-70% of the nitrogen oxides and hydrocarbons, and for a significant percentage of particles under suspension (Ionel et al., 2004).

Nevertheless, paradoxically, while the problems of pollution caused by road transport become critical, application and use of vehicles continues to grow. The standard of living wrongly interpreted and that infrastructure has evolved without control judicios only on economic considerations, reinforces unsustainable, have turned the car into an indispensable factor, which may be interpreted as a necessary evil (Ertl et al., 1999).

Internal motors used in transport, generate an air pollution per unit of energy more than any other consumer of energy. Road transport occupies a major share, being responsible for many harmful effects of cross-biological systems (Ionel et al., 2004).

2. CASE STUDY CO_2 EMISSIONS IN TIMISOARA

This case study presents and highlights in table 1 the status of the carbon dioxide emissions corresponding to the months of January, April, July and October 2006, as well as the CO_2 emissions from road transport, Colterm, and other combustion sources (including waste incineration) in relation to carbon dioxide absorbed through the process of photosynthesis for the city of Timisoara.

Clean vegetation function is to influence the physical and climatic factors in their environment. The influence we talk about depends on the volume and its crown, on the leaf density, the number and thickness of tree branches and age.

Months	CO_2 Colterm[t] emissions	CO_2- road transport emissions	CO_2 of other combustion	CO_2 adsorption photosynthesis [t]	Rest of CO2 [t]
January	82517	23381.17	79397.29	596.0877	184699.4
April	35239	23381.17	33906.72	8933.360	83593.53
July	3311	23381.17	3185.821	10509.189	19368.8
October	30362	23381.17	29214.11	7654.413	75302.86
Total	151429	23381.17	79397.29	27693.05	184699.4

Tab. 1: The CO_2 emissions in Timisoara

Clean vegetation function is to influence the physical and climatic factors in their environment. The influence we talk about depends on the volume and its crown, on the leaf density, the number and thickness of tree branches and age. Their influence can be especially felt in the absorption of $C0_2$, interception of precipitation, absorption of solar energy, powerful sound absorption, emission of oxygen and of the filter for the solid particles in the air.

The absorption of the solar energy in the leaves by assimilation of the chlorophyll can be materialized through access to the photosynthesis.

The process of photosynthesis represents the transformation process of the raw sap into the sap developed at the leaves through chlorophyll when light, with the absorption of carbon dioxide and release of oxygen.

In general, the leaves of plants (Ciupa et al., 2005) assimilate carbon dioxide with an intensity of approximately 20 mg/ dm^2 of the leaves/ 1 hour; $1m^2$ gathered around the leaves. 10g organic substances/ day. The absorption of the carbon dioxide is approximately. 30 mg/dm^2 / hour.

Oxygen, the most important substance for life, is produced mostly through biological processes. The research (1) as a result "in the process of formation of 1 tons of timber, the forest issue 1.3 tons oxygen-free, one hectare of forest productivity average oxygen produces 6-8 tons/ year in the hot days, with the sun, a hectare forest high productivity eliminate 180 to 220 kg per day oxygen" (Ciupa et al., 2005)

The next three charts present the status of the carbon dioxide emissions for the data presented in table1 (time range) and the carbon dioxide absorption by plants for the same time range.

The first chart represents the evolution of CO_2 emissions according to sources (road transport, Coltherm and other burning and incineration installations) for the analyzed periods.

The second chart presents the evolution of the adsorption capacity of the CO_2 through the process of photosynthesis.

The third chart presents the weight of the CO_2 adsorption through the photosynthesis process depending on the amount of emissions of CO_2 .

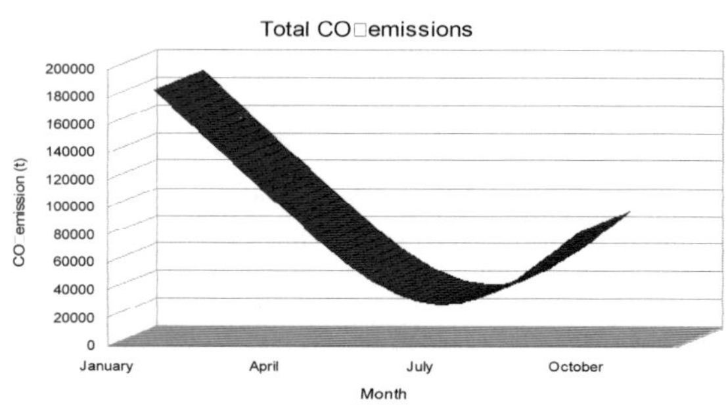

Fig.1 The evolution of CO_2 emissions according to sources

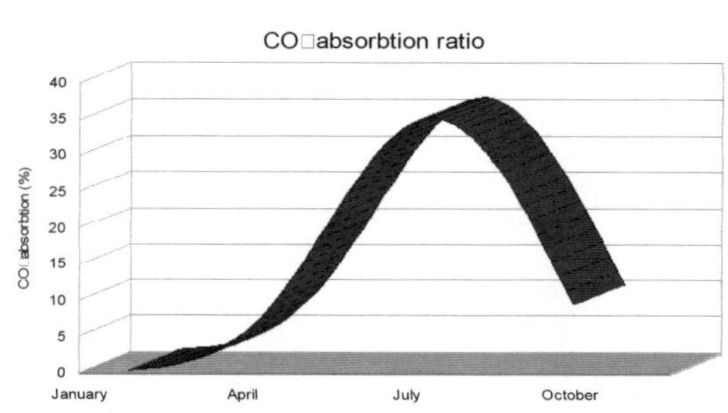

Fig 2. The evolution of the adsorption capacity of the CO_2 through the process of photosynthesis

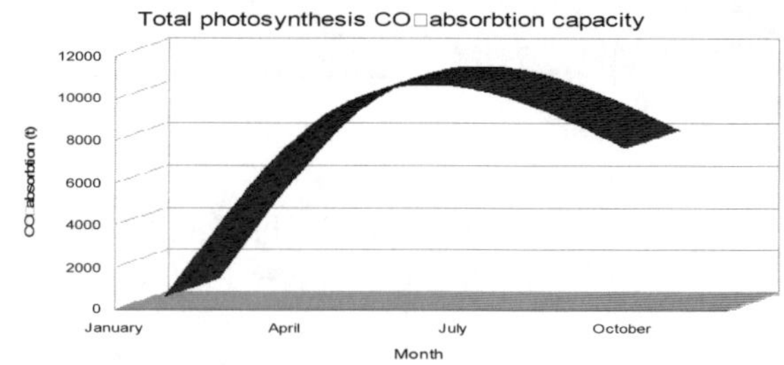

Fig. 3 The weight of the CO_2 adsorption through the photosynthesis process depending on the amount of emissions of CO_2

3. CONCLUSIONS

The issue of green space for parks and vegetation is very acute in Timisoara. Because of the heavily increase of the population, the green area for a person is net inferior to the one reglemented in the international norms. Timisoara, although once called the city of parks (Ciupa et al, 2005), is nowadays behind many other cities, although some areas for fitting out green spaces could be achieved through concrete decommissioning of industrial units and stores closed for over 10 years that do not produce anything but pollution.

In conclusion, analyzing the quantities of carbon dioxide absorbed by plants during photosynthesis, it is maximum 35% in summer months (July) when this process (photosynthesis) is very high because the functions of plant vegetation are maximum in this period and under 6 % In winter months when the period of plant vegetation existed only resinous (the other plants stopped during the winter having vegetative functions).The small percentage of the carbon dioxide adsorped in relation to quantities of carbon dioxidegenerated byvarious sources, sources quoted in the paper (the transport and other combustion installations) is materialized by the increasing pollution in the past 10 years, tripling the population and its needs, by broadening the areas inhabited by sometimes reducing green spaces and doubling the number of vehicles are issues that are raised not only at the city level but throughout the world. In this case, we are moving faster, with larger steps towards an ecological collapse following the pace of increase in pollution in the past 10 years in relation to the decrease of green spaces, which could lead to another phenomenon like "Copsa Mica" Romania, the only other type.

4. REFERENCES

Ciupa V., Radoslav R., Oarcea C. & Oarcea Z. (2005). *Timisoara verde. Sistemul de spatii verzial Timisoarei – Green Timisoara. The green spaces of Timisoara,* Marineasa Printing House, Timisoara

Ertl G., Knözinger H & Weitkamp J. (1999). Environmental Catalysis, Weinheim

Ionel I., Popescu Fr., Oprisa-Stanescu D.P., Bisorca D. & Gruescu Cl. (2004). *Energoecologia combustibililor fosili. Teme experimental- Energoecology of fossil fuels. Experimental subjects,* pg. 130, ISBN 973 625 186 1, Politehnica Printing House, Timisoara

Skiba, Yu.N. & Davydova Belitskaya, V. (2002). Air pollution estimates in Guadalajara City, 153–162., Environ. Model. Assess. 7

Wehner, B., Philippin, S., Wiedensohler, A., Scheer, V. & Vogt, R. (2004). *Variability of non-volatile fractions of atmospheric aerosol particles with traffic influence. Atmospheric Environment* 38 (36), 6081–6090 *Available from*: http://www.mmediu.ro/arhiva.htm ; *Accessed*: 2008-06-02

Annals of DAAAM for 2008 & Proceedings of the 19th International DAAAM Symposium, ISSN 1726-9679
Editor B. Katalinic, Published by DAAAM International, Vienna, Austria 2008
Make Harmony Between Technology and Nature, and Your Mind will Fly Free as a Bird
*Young Researcher and Scientist Paper / * Supervisor, Mentor*

"CROENERGO$_2$" – FUTURE OF RENEWABLE ENERGY SOURCES

MARJANOVIC, T[omislav] & SCUKANEC, D[anijel]

Abstract: *Use of renewable energy sources is in line with the global strategy of sustainable development. Using such sources of energy increasingly contributes to the security of supply and reduce dependence of the energy sector from foreign countries, reducing shares greenhouse gas emissions and supporting sustainable development of rural areas. Reducing the dependence on fossil fuels and import of energy will increase security of energy supply, allowing the realization of new opportunities and enterprise economy. Furthermore, renewable energy sources are helping to increase industrial competitiveness and have positive impact on regional development and employment. The Republic of Croatia as the candidate country for EU should as soon as possible to organise the production of electricity from renewable energy sources. "CroEnergO$_2$" represents thel student model of energy strategy development of the Republic of Croatia to 2030. This project consists of systematic scientific studies and analyses of the Croatian natural resources. This is a potential in the energy systems of renewable energy sources for electricity production.*
Key words: *Energy, renewable, development, sources, biofuel.*

1. INTRODUCTION

Humanity gained energy through thousands of years by burning wood and other biomass for cooking, shaping metals and heating. Since medieval times man adapted rivers for river plants and windmills to grind corn. Energy from renewable sources is not a novelty. The history of the first hundred or more years of production of electricity shows us that apart from development of energy from the water almost all production technologies are based on the burning of coal, oil and gas, and in recent times nuclear fission. These thermal technology productions of electricity have significant advantages, and that is the ability to convert them to gain energy of high concentration which is accordingly of low volume requirements. Established technologies were subject to continuous and intensive development. They are mainly designed to connect to each other on associated network systems, starting from 1920, at a regional, national or international level. The goal was to connect and supply easily available and reliable electricity with the appropriate combination of power plants at the lowest possible costs. Social and environmental impacts of supply electricity until recently had limited importance.

2. THE THREAT TODAY

During the past 15 years more and more people believe in the existence of global warming and climate changing. The influence of the manufacturing processes mentioned before is linked to this phenomenon. However, there is common agreement that the production processes that create harmful gases, mainly carbon dioxide and hydrocarbons, threaten the future stability of our global climate (Kalea, 2007). Global increase in production of all major quantities of gases is now intensively controlled and measured, especially carbon dioxide, as a byproduct of most manufacturing processes, including the production of electricity, energy consumption in households and transportation. According to the International organization for Energy (IEA) up to 2030 global carbon dioxide emissions will be 52% higher compared to those of 2005, unless the world reduces its energy consumption. Even if developed countries begin to implement policies related to climate change, the IEA considers that the global emissions of carbon dioxide will continue to grow to about 30% during this period.

3. PROTOCOLARY REGULATIONS

International Protocol of Kyoto from 1997., (ratified in 2005.), sought to reduce global emissions of CO_2 by 5% to 2012. U.S. is interested in reducing CO_2 emissions, but their non-participation in the achievement of this aim reduces the credibility of the agreement of Kyoto. Since the production of electricity is one of the key sources of CO_2 emissions, it is not surprising that the governments around the world focused on the activities of electricity supply to reduce CO_2 emissions. This focus on the production of electricity is slightly easier than the focus on transportation, manufacturing and households, which are many, diverse and have greater political interaction. One of the most significant contributions to reducing emissions from electricity production is the introduction and development of the production of electricity from renewable sources, which are expected to over time have an increasing share in the production of electricity. Especially in Europe, the anticipation of the production of electricity from renewable sources which are now seen as guidelines, will probably soon become mandatory.

4. RENEWABLE ENERGY SOURCES

Renewable energy sources, not including hydro, give less than 1% of the total energy needed. This share should significantly increase in the future because non renewable energy sources are decreasing, and their adverse impact is becoming greater in the past few decades. The sun delivers to Earth 15 thousand times more energy than humanity in the current phase could manage to spend, but despite that some people on Earth are freezing. This can be seen as a starting poing that renewable sources can and must be better exploited and that we don't need to worry about energy after fossil fuels. The development of renewable energy sources (particularly from wind, water, solar and biomass) is important for several reasons: renewable energy sources have very important role in reducing emissions of carbon dioxide (CO_2) into the atmosphere. Reduction of CO_2 emissions into the atmosphere is a policy of the European Union, so it can be expected that Croatia will have to accept that policy; increasing shares of renewable energy sources increases the sustainability of energy systems. It also helps to improve the safety of delivery of energy in a way that it reduces dependence on energy imports of raw materials and electricity, and it is expected that the renewable sources of energy will become economically competitive conventional sources of energy in the mid to long term. Several technologies, particularly wind energy, small hydro exchange, energy from biomass and solar energy are economically competitive. Other technologies are dependent on market demand to become economically cost-effective in comparison to classical sources of energy. The process of acceptance of new technologies is very slow and always seems a bit abstract. The main problem for installation of new technologies is the price. It raises the cost of energy in the first few years up to a level of complete non cost effectiveness in relation to other commercially available sources of energy. A large share of energy production from renewable sources is the result of environmental awareness of the population, which despite the initial economic non cost effectiveness sets up technology for the production of "clean" energy.

5. THE RESULTS OF SCIENTIFIC STUDIES

The most interesting renewable energy sources (wind energy, solar energy, water energy, hydro energy, bioenergy and other) are the main theme of this scientific study. Scientific studies have showed the development of the energy sector in Croatia in the systems of renewable energy sources and made by projects (models) of renewable energy sources for electricity production and bio-technology as well as the solutions which include the application of "Greenfuel 3D Matix System Technology" for the production of biofuels and "CroSunSystem" – Solar system for more efficient utilization of solar energy in the production process of electricity (Kulušić et al., 1994). Strategy for the Energy Sector Development, as a result of these scientific studies, covers the period up to 2030. The period covers present and future technologies, changing relations and ways of management of energy, the period in which Croatia will be a non memer, as well a member state of the European Union. The period up to 2010 will be different from the period 2010.-2020. and significantly different than the period 2020.-2030., according to the tasks that must be resolved, as well as options for solutions. Many questions have no answers today; and especially unknown is the technological development and the dynamics of the introduction of new technologies for commercial usage. Their breakthrough on the market will depend on a multitude of related factors: visibility and estimates of environmental damage caused by the existing energy technologies, the readiness of individuals to pay more for clean energy, energy crises created by economic reasons, geo-political or purely physical reasons, the degree of development of new technologies the readiness of the Government in the developed countries to support the research, development and demonstration of new technologies, including large industries and financial institutions. These new and expensive technologies may seen inappropriate for a small country such as Croatia, but just such small countries with relatively non well-developed infrastructure may have the most use of these decentralised and clean technologies in that way they will reach the energy system of tomorrow via a shortcut, avoiding the energy intense phases through which the developed countries have gone through. In the strategy of energy development in Republic of Croatia reasonable goals have been set up during the entire period, regardless of insecurity which might have exist for any reason, and relate to the harmonization of sustainable development and organised system of management of energy (Potočnik, 2003). These are: increasing energy efficiency, secure supply, diversification of energy sources, the use of renewable energy sources, the real price of energy, the development of energy markets and environment protection.

6. PLANS FOR THE FUTURE

The strategy is starting from today's technology in the whole process of production, transformation, transmission distribution and consumption of energy, and current capabilities and characteristics of primary and transformed forms of energy. The role of fossil fuels will change, and it is not unrealistic and too visionary to expect that by mid-next century final consumer will predominantly use electricity and hydrogen energy (Udovčić, 2005). The production of electricity is a long-term open question of today and it is not possible to give unambiguous answers. Its expectable that the scenario of priorities and primary forms for the production of electricity will change and be strengthened by environmental conditions and how the new technologies will reach commercial applicability. This the energy development strategy of Croatia recognizes as a strategic support to natural gas in the first ten years, and opens the possibility that in the middle or in the second half of the period includes a nuclear power plant, if the technology is sufficiently safe and acceptable to citizens. Of importance to the strategy, regardless of what will come into consideration in the second part of the period, prominent are the hydrogen energy technologies. They will significantly change the appearance of today's energy system. Their implementation is expected to fuel cars, airplanes and de-centralised production of electricity (Government of Republic of Croatia 2002).

7. CONCLUSION

A sudden increase for energy on the global level and the impact which is done on the environment and society in general, leads to review the needs of today dominant role of fossil fuels. That is why more and more attention devoted to renewable energy - energy sources that are preserved in nature and updated in whole or in the part. These are the energy of water, waves, tidal, wind, biomass, solar radiation, geothermal energy and energy from wood. Nowadays, energy from renewable sources have been collected from energy which is obtained from the so-called conventional energy sources - fossil or nuclear fuel. This energy has many advantages - reducing the influence of the energy sector on the environment, availability, create new jobs and the development of local economy. As a result, the use of renewable energy sources have been representing sustainable energy option, and it is one of the three pillars along with energy efficiency and cleaner production on which rests overall sustainable development of energy industry to humanity today and in the future. So far, this scientific research project has been accomplishing the analysis conducted for utilization of renewable energy sources in Republic of Croatia and its shares in the total production and consumption of energy. Moreower, it has been conducted systematic analysis of natural resources for production of electricity in systems of renewable energy sources in accordance with the scientific codex for preservation of nature and peoples living and working environment.

8. REFERENCES

Kalea, M. (2007). *Električna energija (Electical energy)* Kigen, 953-6970-39-1, Zagreb

Kulišić, P.; Vuletin, J.; Zulim, I. (1994). *Sunčane ćelije (Solar cells)*, Školska knjiga, 953-0-31631-3, Zagreb

Potočnik, V. (2003). *Obnovljivi izvori energije i zaštita okoliša u Hrvatskoj (Renewable energy sources)*. Ministarstvo zaštite okoliša i prostornog uređenja Republike Hrvatske (Ministry of Environmental Protection, Physical Planning and Construction), 953-6793-17-2, Zagreb

Udovčić, B. (2005). *Elektroenergetski sustav (Electroenergetic system)*, Kigen, 953-6970-18-X, Zagreb

Vlada Republike Hrvatske (Government of Republic of Croatia) (2002). Strategija energetskog razvitka Republike Hrvatske (Strategy of energetic development of Republic of Croatia), *Available from:* http://www.nn.hr/ *Accessed:* 2008-03-10

Annals of DAAAM for 2008 & Proceedings of the 19th International DAAAM Symposium, ISSN 1726-9679
Editor B. Katalinic, Published by DAAAM International, Vienna, Austria 2008
Make Harmony Between Technology and Nature, and Your Mind will Fly Free as a Bird

DAMAGE ANALYSIS OF ALUMINIUM - STEEL BIMETALS

MARONEK, M[ilan] & CAPLOVIC, L[ubomir]

Abstract: *Aluminium- steel bimetlas used in metal production are very often made by explosion welding due to strict demands on low transient resistance between aluminium and steel and good mechanical properties of the joint. This paper deals with reasons of aluminium steel bimetal damage. The key factor is the energy coming from explosion welding process represented by welding parameters and its influence on formation of brittle intermetallic compounds of aluminium and iron. There were observed several intermetallic compounds with different thickness and containing cracks in weld joint interface and their presence was evaluated as the principal reason of bimetal damage.*

Keywords: *bimetal, explosion welding, damage, quality*

1. INTRODUCTION

Aluminium steel bimetals are often used in metal production industry due to their good mechanical properties and very low transient resistance between aluminium and steel. They can be made by several technologies among which belongs also an explosion welding. However, the combination of aluminium and steel is difficult to weld due to possible formation of intermetallic compounds between aluminium and iron which are hard and brittle. There is why the some bimetal manufactures uses titanium as an interlayer between aluminium and steel.

An explosion welding process uses high speed impact and despite of extremely short welding time (typically milliseconds) there is generated enough heat in the joint interface to produce melted areas and to form intermetallic compounds if the welded metals are prone to their creation. That is why there is necessary to control the heat input by explosion welding parameters, e. g. detonation velocity, dynamic collision angle, set-up distance etc. (Hudák, 1995), (Kwang-Jin, 2007).

The bimetals subjected to the analysis in this paper were damaged after a very short time of their work. Having known some of the process variables, there could have been calculated other variables allowing to estimate whether welding parameters had not been set too high with a risk of intermetallic coumpound formation (Marônek. 1995). The proof of their presence in the weld joint interface was carried out by examination methods described below.

2. INVESTIGATION PROCEDURE

The first part of analysis was to determine the detonation velocity of an explosive charge. In order to get the straight propagation of detonation wave, the detonator initiates the primary charge with high detonation velocity at first and this charge consequently initiates the main charge having detonation velocity significantly lower. The front of detonation wave forms an angle γ with an edge of primary charge (Fig. 1). Detonation velocity of the main charge is described by formula:

$$v_{d2}=v_{d1}.\sin\gamma \quad (1)$$

where v_{d2} is the detonation velocity of main charge and v_{d1} is the detonation velocity of primary charge. Having known the angle γ and detonation velocity of primary charge v_{d1}, there is possible to estimate the main charge detonation velocity and to compare it with the recommended detonation velocity.

The next step used light microscopy, microhardness measurements in the joint interface and electron probe microanalysis (EPMA).

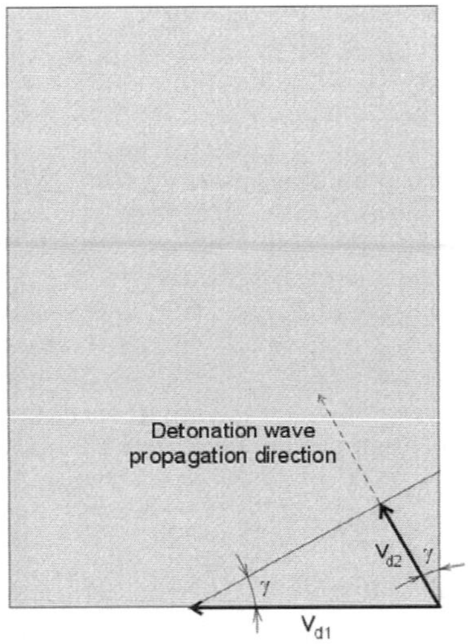

Fig. 1 Estimation of detonation wave propagation direction

3. RESULTS

A surface of damaged bimetals had wavy pattern typical for explosion welding joints (Fig. 2).

Fig. 2 Wavy patterned bimetal surface

The value of γ angle measured on damaged bimetal (Fig. 3) was 24°.

Detonation velocity of plastic explosive used as the primary charge (Tvarex 4A) was 7300 $m.s^{-1}$. Hence the detonation velocity of the main charge (Sypex 14A, loose consistency) was determined according to formula (1) to 2969 $m.s^{-1}$.

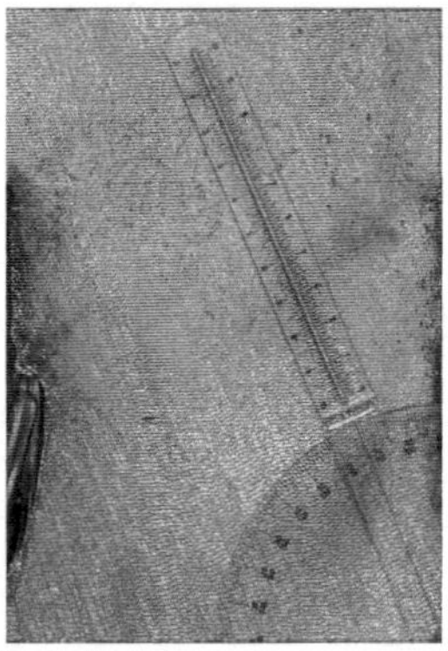

Fig. 3 Determination of angle γ

The recommended detonation welocity according to welding procedure specification was 2200 to 2300 m.s^{-1}.

It was obvious the detonation velocity used was 29 to 35 % higher than recommended and the occurrence of intermetallic compounds could have been predicted.

Predicted presence of intermetallic compound was proved by light and electron microscopy analysis. There was observed continual layer of approximately 50 µm thick brittle phase (Fig. 4) with. This layer included cracks and there had been observed significantly plastically deformed islands of metal matter.

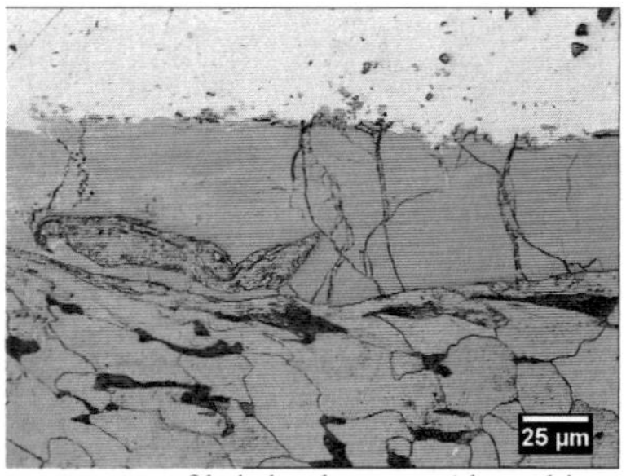

Fig. 4 Microstructure of brittle phase on Al-steel interface

The microhardness measurements showed, that the hardness of brittle phase was 400 to 700 HV 0,1. This microhardness values significantly exceeded the hardness of both basic materials (Fig. 5).

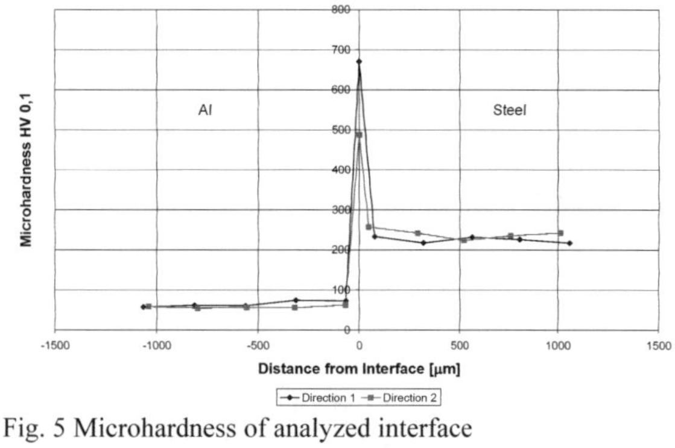

Fig. 5 Microhardness of analyzed interface

A test sample for the impact test had been prepared from the weld joint. Having broken the test sample, the fractographic analysis of revealed fracture surface as well as the the chemical composition of the fracture surface by EDS analysis had been carried out. The morphology of fracture surface (Fig. 6) was transcrystallic with brittle splitting. The chemical analysis detected that the element composition corresponded to intermetallic phase FeAl (Brundle, 1992), (Westbrook, 2000).

Fig.6 Fracture surface with characteristical transcrystalline cleveage of intermatallic compound FeAl

4. DICSUSSION

Detailed analysis of thin intermetallic phase showed, that a fine microcrystalline metal compound was squashed into the crack area. That means the created layer cracked immediately after its formation due to intensive deformation. Consecutively small volumes of welded materials were pressed into the cracks and aterwards in the process of impact load they represented only a small barrier during the crack propagation. That is why the fracture surface had such a heterogenous character. From the chemical composition point of view the layer was unambiguously identified as the intermetallic phase FeAl, which has low mechanical properties.

5. CONCLUSION

The obtained results prooved, that the main cause of continuous highly brittle phase FeAl formation having significant influence on mechanical properties degradation of weld joint fabricated by explosion welding with parameters mentioned and revealed above was the high kinetic energy of used explosive charge. This energy caused the local overheat of weld interface with temerature exceeding the aluminium melting point. Due to this fact there came up appropriate cicrcumstances for brittle phase formation.

The reason of high energy input during explosion welding could have been in using explosive charge with higher detonation velocity, thicker explosive charge or higher charge density. This finally led to higher flyer plate impact velocity and thus higher kinetic energy transforming after collision with parent plate to heat.

This paper was realised with the support of KEGA 3/4157/06 and APVV 0057-07 grants.

6. REFERENCES

Brundle, C.Richard; Evans, Charles A. Jr. (1992). *Encyclopedia of Materials Characterization - Surfaces, Interfaces, Thin Films*, Butterworth, ISBN 0-75069-168-9, Boston

Hudák, J., Marônek, M., Turňa, M. (1995). Quality evaluation of Al- steel weld joitnt fabricated by explosion welding. *Zváranie*, Vol. 44, No. 6, 131 - 133. ISSN 0044-5525

Kwang-Jin L, Shinji K, Takashi A, Tomokatsu A. (2007) Interfacial microstructure and strength of steel/aluminum alloy lap joint fabricated by magnetic pressure seam welding, Materials Science and Engineering A 471, 95–101

Marônek, M.: Presence of melted areas in the interface of weld joint fabricated by explosion welding.(1995). In.: *Zborník vedeckých prác MtF STU v Bratislave*. Vol 3., 45 - 50. ISBN 80-227-0810-0

Westbrook, J.H., Fleischer, R.L.. (2000). *Intermetallic Compounds, Volume 2 - Basic Mechanical Properties and Lattice Defects of Intermetallic Compounds*, John Wiley & Sons, ISBN 978-0-471-61175-2

Annals of DAAAM for 2008 & Proceedings of the 19th International DAAAM Symposium, ISSN 1726-9679
Editor B. Katalinic, Published by DAAAM International, Vienna, Austria 2008
Make Harmony Between Technology and Nature, and Your Mind will Fly Free as a Bird

LASER BEAM WELDING OF NITROOXIDATION TREATED STEEL SHEETS

MARONEK, M[ilan]; BARTA, J[ozef]; KOLENIC, F[rantisek] & DOMANKOVA, M[aria]

Abstract: *The paper deals with laser beam welding of nitrooxidation treated low-carbon steel sheets, which are characteristic by increased mechanical properties and corrosion resistance. The experiments proved the suitability of laser beam welding for welding treated steel sheets. Visual inspection, mictrostructure analysis and microhardness measurements did not reveal any irregularities in weld joints. The mechanical testing of weld joints is currently a subject of further research.*
Keywords: *laser, welding, nitrooxidation treatment, steel.*

1. INTRODUCTION

Lowering fuel consumption goes hand in hand with vehicle weight reduction. One of the possible ways to reach this goal is to use materials with low specific density and another one is to use materials with increased mechanical properties. Nitrooxidation treated low-carbon steel seem to be an effective alternative to common steel sheets due to its mechanical and chemical properties (Lazar, 2007).

Nitrooxidation treated low-carbon steel sheets have, in comparison to untreated material, many advantages. Improvement of the maximum strength of 41% can be observed and surface microhardness is increased by 653 %. Formability of these steel sheets is increased by 9% by Erichsen cupping test. Nitrooxidised steel showes also 10-times increased resistance to atmospheric corrosion compared to basic state steel. Nitrooxidation has a beneficial influence on formability as well as corrosion resistance (Kuníková, 2005), (Lazar, 2005).

In this article nitrooxidation treated steel sheets are used as a possible replacement for some ordinary sheets, used in industry where high corrosion resistance and surface hardness is required. On the other hand the positive corrosion resistance brought by surface oxidic layer causes problem during arc and resistance welding. The laser beam welding appears to be an alternative for welding this kind of material.

2. EXPERIMENTAL

Thin low-carbon steel sheets Cr 01 ISO 17/12N49-69 (DC 01 EN 10130-91) of 1 mm thickness were used for investigation. Table 1 shows the chemical composition of this material. This base material was treated by nitrooxidation in different parameters whereby the difference was in nitridation and oxidation temperature and time (table 2). Fluidised bed was composed of Al_2O_3 with grain size of 120 μm. The waft of the fluidised layer during treatment was provided by gaseous ammonia, during oxidation using a vapour of distilled water, supplied to the furnace chamber (Marônek, 2005).

ISO code	C [max %]	Mn [max %]	P [max %]	S [max %]
DC 01	0,1	0,45	0,03	0,03

Tab. 1. Chemical composition of examined steel

Nitridation time	45 min
Nitridation temperature	540 °C
Oxidation time	5 min
Oxidation temperature	380 °C

Tab. 2. Nitrooxidation parameters

In contrast to zinc coated sheets ((Sejč, 2004), laser welding was chosen as a suitable joining method for this type of treated sheets . The welding parameters are shown in table 3.

Laser type	Ferranti Photonics AF 8 CO_2
Protective gas	Ar 99,996% (18 l/min)
Welding speed v_s[mm/s]	30, 40, 50, 60
Laser power [W]	2000

Tab. 3. Laser welding parameters

3. RESULTS

Welding speed [mm/s]	Cross-section
30	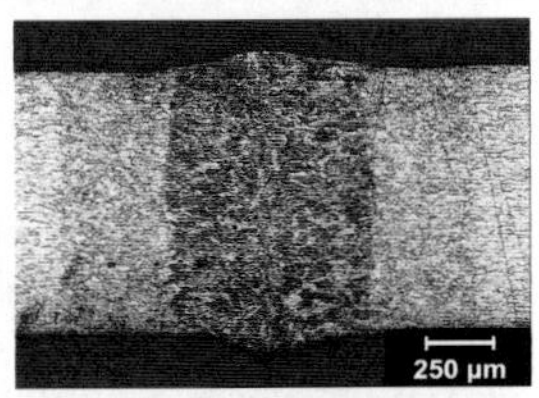
40	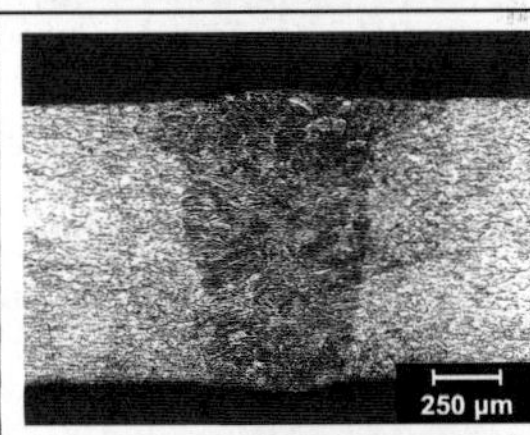
50	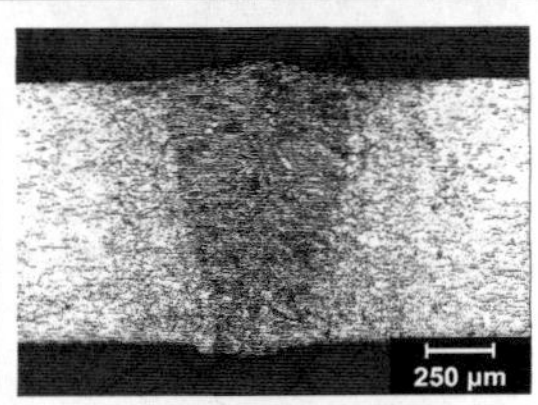
60	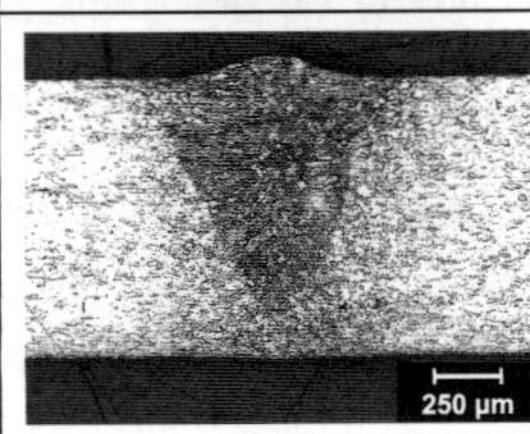

Tab. 4. Weld cross-section appearance

Weld surface	Weld root
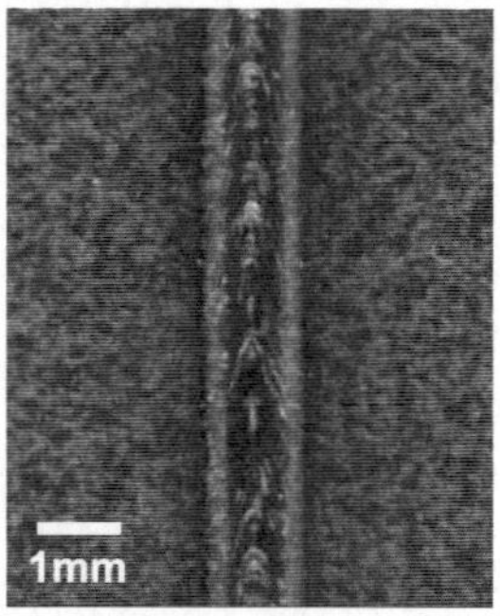	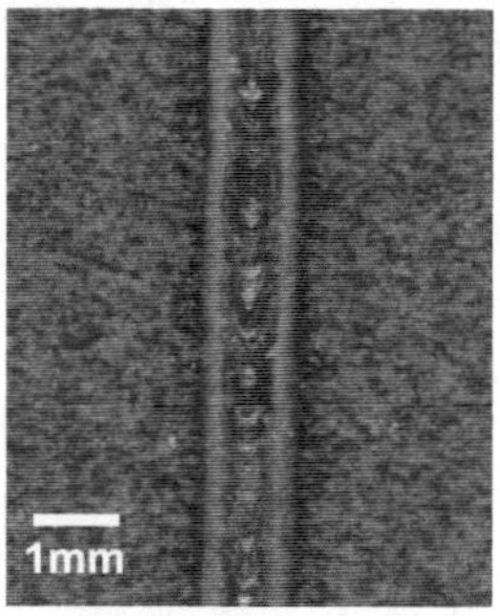

Tab. 5. Weld surface and root appearance

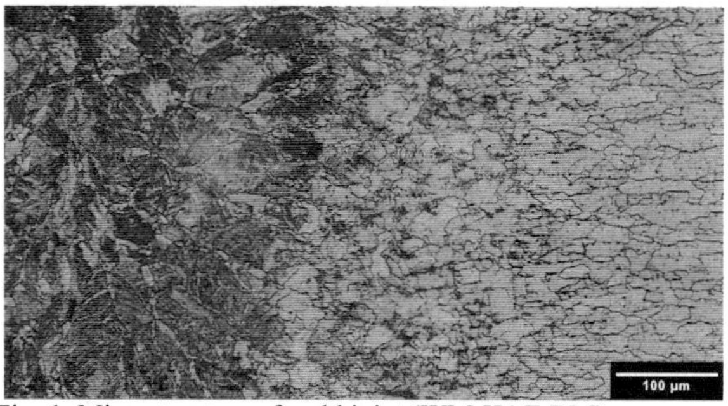

Fig. 1. Microstructure of weld joint (WM-HAZ-BM)

Welding speed was a primary variable parameter in range from 30 to 60 mm/s. Majority of specimens has shown undesirable lack of penetration in 60 mm/s welding speed (Table 4). The specimens welded by welding speed of 30 mm/s had excess weld width as well as the heat affected zone. These welding speeds were evaluated as unsuitable.

The optimal shape of weld in cross-section was achieved at welding speeds 40 and 50 mm/s however in 50 mm/s welding speed, a very small lack of penetration was observed. It can be assumed that the optimal welding speed for this material is between 40 and 50 mm/s and it will be one of the subjects for further research. The appearance of weld surface and weld root for welding speed of 40 mm/s is shown in Table 5.

The nitrooxidation treatment produced a characteristic surface layer about 300 µm in thickness. This was composed of two zones – about 70 µm thick compound layer, with continuous thin Fe_3O_4 a Fe_2O_3 layer of approximately 3 to 5 µm in thickness and ε phase ($Fe_{2\text{-}3}N$), about 10 µm in thickness. The welding process destroys this nitrooxidation layer and affects treated material as well as the corrosion resistance.

The microstructure of weld joint is shown on Fig. 1. where weld metal, heat affected zone and base metal can be seen. The microstructure analysis exposed the acicular ferrite and ferrite precipitated along boundaries of columnar crystals. Deep analysis proved that nitrides coming from nitrooxidation process were dissolved by laser beam welding process up to distance of 1 mm from the weld joint boundary. However, the structure showed no abnormalities in phase composition.

The microhardness measurements (Fig. 2.) showed maximum microhardness of weld metal 323 HV 0.1 in upper section and 355 HV 0.1 in root part of the weld. The base metal microhardness was approximately 150 HV 0.1.

Microhardness didn't show significant differences in top and root section of the weld joint. All in all the specimen of 60 mm/s welding speed had the highest values and differences of microhardness due to the highest cooling rate. On the other hand the specimen with lowest welding speed showed a well-balanced progress of microhardness due to the lowest cooling rate.

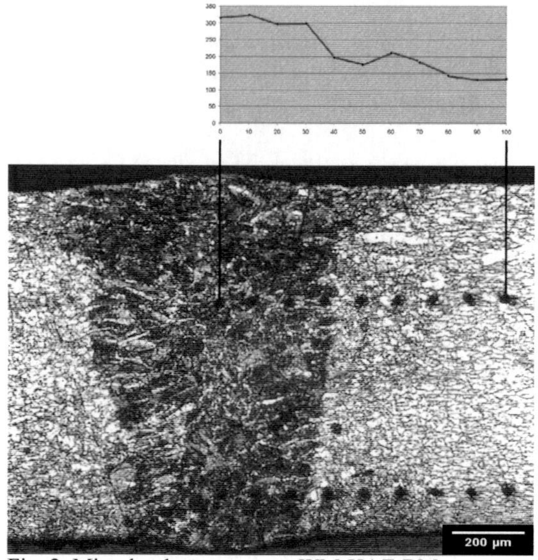

Fig. 2. Microhardness progress WM-HAZ-BM

5. CONCLUSIONS

In contrast to arc and resistance welding (Sejč, 2004) laser beam welding appearesto be a suitable welding method for welding nitrooxidation treated low-carbon steel sheets.

The recommended welding parameters for CO_2 laser and 1 mm material thickness are 2000 W beam power, welding speed from 40 to 50 mm/s, beam defocus 0 mm and protective gas Ar 99,996% with 18 l/min flow rate.

Based on the mirostructure analysis and microhardness measurement which did not reveal any irregularities or abnormalities in weld metal and the heat effected zone it can be assumed that, the mechanical properties of the weld joints, currently subject to further research, will confirm our assumptions concerning weld quality.

6. ACKNOWLEDGEMENT

This paper was realised with the support of KEGA 3/4157/06 and APVV 0057-07 grants.

7. REFERENCES

Kuníková, T; Lazar, R; Dománková, M; Mokošová, Ľ; Kamasová-Mareková, Z. (2005) Evaluation of corosion resistance of nitrided and nitrooxidised low-carbon steel. *Proceedings of CO-MAT-TECH 2005 13th International Scientific Conference, pp.* 671-678, ISBN 80-227-2286-3, *Trnava Slovak Republic, October 2005*, STU, Bratislava

Lazar, R.; Marônek, M.; Dománková, M. (2007) Low carbon steel sheets treated by nitrooxidation process. *Available from:* http://www.strojarstvo.sk/docwww/SK/285/285.pdf *Accessed:* 2008-06-17

Lazar, R; Dománková, M; Kotras, P (2005) Analysis of nitrooxidation layer of deep-drawing sheet metals with using TEM and X-Rays. *Available from:* http://www.mtf.stuba.sk/docs//internetovy_casopis/2005/mimorc/lazar.pdf *Accessed:* 2008-06-17

Marônek, M.; Lazar, R.; Dománková, M.; Kotras, P. (2005) Microstructure analysis of weld joints from nitrooxidation treated low carbon steel sheets. Zvárač, 2, 1, 24-28, ISSN 1336-5045

Sejč, P. (2004) Optimalization of selected MIG/MAG brazing parameters of zinc coated steel sheets. *Zváranie – Svařování* 2004, 53, 3, 57-62, ISSN 0044-5525

Annals of DAAAM for 2008 & Proceedings of the 19th International DAAAM Symposium, ISSN 1726-9679
Editor B. Katalinic, Published by DAAAM International, Vienna, Austria 2008
Make Harmony Between Technology and Nature, and Your Mind will Fly Free as a Bird

SIMULATION OF THE AUSTENITIC MANGANESE STEEL BALLS CASTING AND COMPARISON WITH THE EXPERIMENTAL RESULTS

MARTA, C[onstantin]; DOROFTEI, I[oan]; PRISACARU, G[heorghe]; HAMAT, C[odruta]; SUCIU, L[enuta] & ZGARDEA, E[manuel]

Abstract: *Starting from the fact that the austenitic manganese steel casting (OAM) into ball shapes is difficult, as the pipe cavity is placed at the ball's surface or very close, we focused our research on the positioning of the casting defect into the ball's interior and as close as possible to the core. We performed simulations on the balls casting, and based on the simulation results we created the casting technology for 12-ball shapes, also presenting the practical results of the casting, which are almost identical to those of the simulation.*
Keywords*: austenitic, manganese, steel, casting, simulation.*

1. INTRODUCTION

Hadfield steel with very different chemical compositions but observing the ratios $M_n/C= 10$ and $Cr/C = 0,8... 1,92$ has a very wide utilisation in the casting of parts that have to resist to abrasive wearing (in very hard conditions) and to corrosion (Sporea & Crainic 2005). Some of the work-parts obtained by OAM casting are also the balls for grinding the hard rocks in mills with balls. One analysed the exploitation behaviour of manganese steel balls (Ivan&Mladen 2001) compared to that of the balls made of stainless steel, cast nickel alloy, or chilled iron. In the case of hard or very hard materials grinding one obtained more reduced wear by using the manganese steel balls. The high resistance to wear and tear, the small prices of OAM balls, the world shortage of ferroalloys, all this encouraged the research in the field. The technical aspects, i.e. steel contraction of 2,5-3% provoking large cavity and implicitly the influence of casting defects on the life duration of the ball were improved through simulations having led to the finding of certain solutions of cavity concentration to the core of the ball. (Cristea 2006). The simulation was performed on a 12-ball sand mould, with gate end and central feeding. The results of the research can be applied to the entire range of OAM balls, for diameters between 60 – 130 mm, and also to other parts (Suratnam 1998). The simulation was done with Magmasoft (MAGMA GIESSEREITECHNOLOGIE 2005).

2. MAGMASOFT SIMULATION PROGRAMME

This simulation programme represents for the caster and for the designer an instrument enabling them to rapidly test a great number of options, it also allows the selection of the optimal combination for improving and optimising of the casting process. The use of the MAGMASOFT software imposes the following main compulsory stages : pre-processing of start initial data of the simulation, calculus of the simulation with the selection of the adequate solver (simulator), post-processing with the presentation of the simulation results. Then one defines the components of the casting system according to classes of materials, i.e.:

- Cast Alloy;
- Sand Mould;
- Feeder;
- Gating and Inlet

Figure 1 shows the presentation of the class of materials

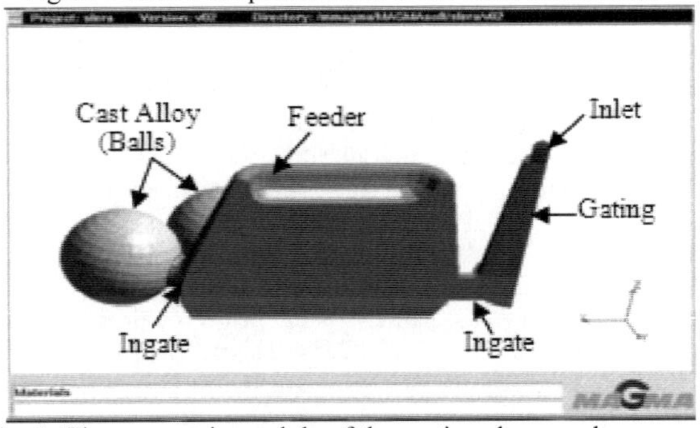

Fig. 1. The geometric module of the casting shape and definition of groups of materials

This denomination allows the identification of components in the casting process. It is not the material that is named, but only the function of each component in the casting process.

The definition of the components is done considering the geometry of the part subjected to simulation, more precisely the 100-mm diameter sphere, cast from the alloy which will be defined as class of materials, i.e. G1X20CrMn130- make steel, in sand mould defined also as class of materials, based on theoretical studies about casting simulation in moulding mixtures for balls for 100-mm spheres. The casting of balls is made in 12-ball sand moulds, with central feeding and end gate for every 6 balls. For the situations in which the casting system is symmetrical, like in our case, in order to reduce the simulation times, the software allows simulation only on a quarter of the cast mould, according to Fig.1. The simulation is thus performed for a quarter of the inlet, half of the feeder and 3 balls. The software allows the calculation of the interior volume of the mould, which is of 62853 cmc. For the simulation of solidification one established the check points of the mould filling and part solidification, which allows a certain closeness to reality. The results of simulation for the casting ensemble are represented through the porosity criterion, and we remark the shrinkage and the maximum porosity respectively, according to the colour code in the inlet and in the feeder Figure 2.

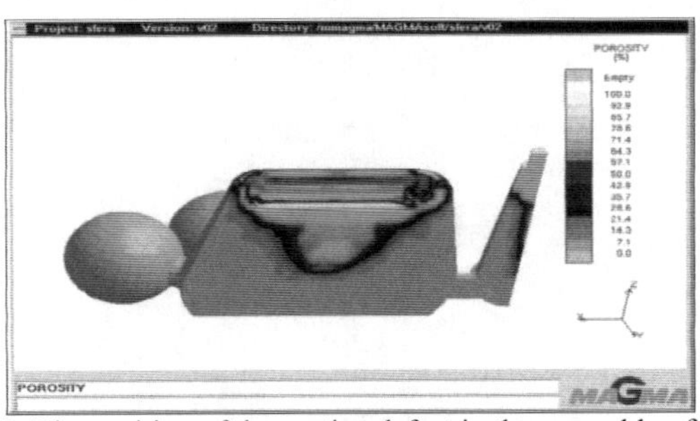

Fig. 2. The position of the casting defect in the ensemble of the cast shape represented through the porosity criterion

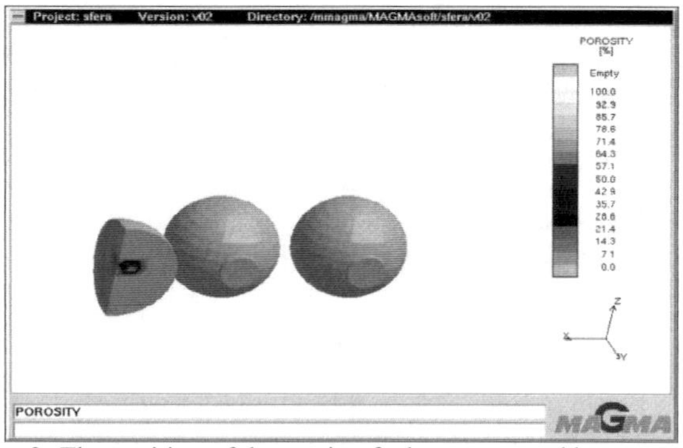

Fig. 3: The position of the casting faults represented by "porosity" criterion

The casting temperature is of 1440^0 C, the filling time being of 10 s. The balls cast in these moulds under the above conditions should exhibit a pipe cavity as small as possible, placed in the central area of the part. In order to visualise the results of the simulation and the location of the casting defect in the balls, one of the balls was sectioned, the porosity being approximately assessed by means of the colour bar in the right side. All the results of casting are 3-D. The casting defect is in the interior of the cast part and is defined through the porosity criterion, ranging between 20-23%. This variant is interesting because the defect is within the part. It can be stated that the casting fault defined by porosity criteria (Figure 3) is place in the centre of the work part.

3. PRACTICAL TESTS FOR BALLS CASTING OF THE SPHERE WITH 100 MM DIAMETER

For highlighting and establishing the importance of the MAGMASOFT simulation system, castings were performed for the 100-mm balls in similar conditions with that of MAGMASOFT programme. After the casting process the ball was sectioned and it had a hole similar (Figure 4) with that obtained by simulation (Figure 5), in the case of the GX120 CrMnl30 steel. The cavity has the following dimensions : along the vertical diameter – 15 mm, and along the horizontal diameter - around 10 mm. The maximum depth of cavity is of about 7 mm. The volume of the cavity determined also through liquid insertion is of 5,5 cm^3. The volume of the cavity / shrinkage is estimated based on the fact that the work is done on sectioned parts. Compared to the volume of the part, the volume of the cavity represents 1.051% of the total part volume. With the help of the technology designed we cast 1500 balls. For the verification of results the weight of balls was also measured. The weight variations ranged between 4.370 kg and 4.470 kg, which confirms the validity of the casting technology. Another imposed issue refers to the sphericity of the cast balls. One performed measurements along 3 diameters and the deviations were between +2 and -2 mm

Fig. 4. The cavity from the cast ball of steel T120CrMnl20 type

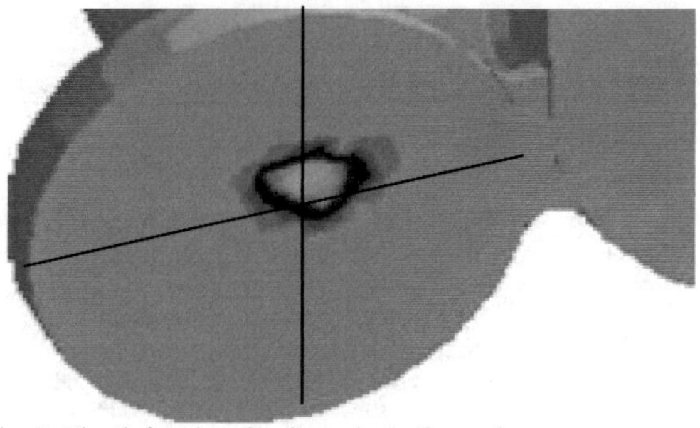

Fig. 5. The defect / cavity from the ball cast from GX120CrMnl30 steel simulated by porosity criterion

4. CONCLUSION

By using the simulation software the traditional casting tests are much reduced. One avoids the refuse cast parts and assures the quality of the finite product since the beginning of manufacture. Within the elaboration of a casting project one can elaborate several versions containing different variants of the casting system, so that the improvement of the casting system be done only through modification of the geometry and parameters of the casting process, the analysis of results, their comparison and the selection of the optimum solution, thus avoiding the doubling of work and saving time and money. The use of the technology realised on the basis of simulation does not trigger the obtaining of the most perfect and the less expensive part. The simulation does not take into account the elaboration technology, the technologies of mould extraction or the modality of application of the appropriate heat treatment, these being parameters of the manufacturing process. Another three very important factors in the case of the OAM casting, i.e. the temperature at the beginning of casting and the casting times, the mould temperature, the materials use are integrated into the pre-processor and have a decisive influence on the results. The simulation method using MAGMASOFT and having as idea the placement of the contraction to the core of the ball is original and along with the obtaining of more reduced holes is a consequence of the appropriate casting technologies used, in order to have a spherical shape, which triggers a uniform wear and tear of the sphere's surface and the increase of the exploitation duration. The results of the simulation are valid only of the austenitic manganese steel with the following chemical composition: C = 1,05 – 1,35, Cr = 1,50 – 2,50, Mn//C ratio = 10.

5. REFERENCES

Cristea, C.; (2006) *Theoretical and experimental studies on the casting of steel balls with high contents of manganese for ore-grinding crushers)*, D. P. Publishing House, ISBN code 973-30-1522-9, Bucharest

Ivan, Şt., & Ghiţă, M. (2001) Symposium, Western University, *On the duration of exploitation of wear-resistant steel of the T110Mnl30 Type*.pag.400-405, ISSN 1453 – 7394, 9-12 November, 2006,Timişoara

Sporea, I.; Crainic, N. & Mladen, M, (2005) On casting of austenitic manganese steel in wear-resistant parts, *Annals of the University of Oradea, Mechanics fascicle,* pag.479-486, ISSN 1011-2855, 12-15 May, 2005, Oradea.

Suratman R.; (1998) *Alloy Design and Casting Practice of Hadfield's Manganese Steel,* Metallurgical Science and Technology, Vol. 1, 1990, pag. 822 – 840, 12-16 July, 1998, Hawaii

*** MAGMA GIESSEREITECHNOLOGIE GMBH, *Copyright 1997, 1998, 1999, 2000, 2001 2002;*

Annals of DAAAM for 2008 & Proceedings of the 19th International DAAAM Symposium, ISSN 1726-9679
Editor B. Katalinic, Published by DAAAM International, Vienna, Austria 2008
Make Harmony Between Technology and Nature, and Your Mind will Fly Free as a Bird

INFLUENCE OF THERMAL FIELD IN THE GMAW PROCESS: MODELLING AND COMPARISON WITH EXPERIMENTAL RESULTS

MARTA, C[onstantin]; DOROFTEI, I[oan]; SUCIU, L[enuta]; TATUCU, I[ancu]; PRISACARU, G[heorghe] & HAMAT, C[odruta]

Abstract: *The goal of this paper is to determine the thermal field in the parts subjected to welding, considering the electric ark as the mobile welding source, with the help of the Ansys modelling and simulation software, and to compare the simulated dimensions with the values resulted from the experimental determinations. Mathematically, this means determining the T(t,r), where t represents a random moment, and r is the value of the position vector of a random point on the part subjected to welding, or in a Cartesian system, of the function:* $T = (t, x, y, z)$.

Key words: *welding, thermal field, simulation, validation.*

1. INTRODUCTION

As in the literature one has already studied the equations of the mathematical model of the welded seam formation, it is only a question of monitoring the evolution of the thermal field in components (basic material), with the purpose of observing the possible modifications of the welding bath, due to the change of the welding regime parameters. For the study one used two methods: an analytical method, consisting in the explicit determination of the thermal field with the help of the formulae from the literature, and a numerical method with finite elements, which will be compared to the results obtained experimentally, for the validation of the utilised software. The safety requirements and the high prices of components are the reason for the use of simulation, and thus manufacture may be optimised if the residual strains occurred as a result of welding influence the deformations appeared at welding through ulterior heat treatments.

2. ANALYTICAL DETERMINATION OF THE THERMAL FIELD

The literature presents equations of the thermal fields, the most frequently encountered in the welding processes, and with their help one may calculate the temperature in a certain point from a welded part, at any moment, both in the welding period, and in the cooling one.

One calculated the thermal fields at different moments in time, with the help of the equation (1), used for permanent mobile source of high power and speed, for the case of plate-shaped bodies:

$$T = \frac{P}{v_s \cdot g \cdot \sqrt{4\pi \cdot c \cdot \rho \cdot t}} \cdot e^{\frac{-r^2}{4at}} \cdot e^{-bp \cdot t} \qquad (1)$$

r[m]	0,06	0,06	0,06	0,06	0,06	0,06	0,06	0,06	0,06	0,06	0,06
t[sec]	0	2	4	6	8	10	12	14	16	18	20
T[⁰C]	20	1434,30	1013,019	827,307	716,554	641,004	583,13	541,74	506,766	477,792	453,814

Tab. 1. The analytical calculus o the thermal field at the middle of the welded seam

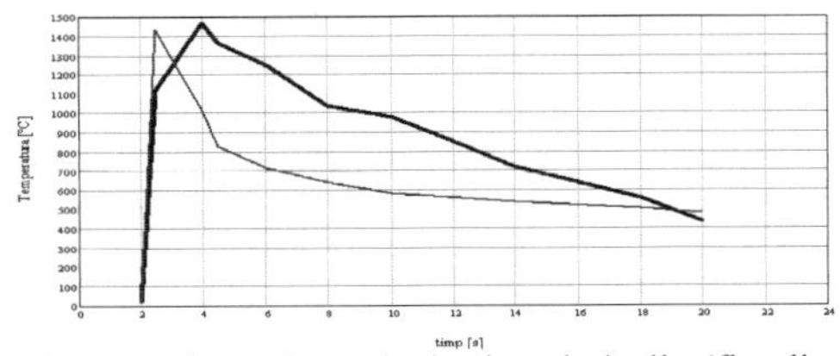

Fig. 1 Comparative values obtained analytically (fine line) and experimentally (big line)

One performed calculation for three distinct positions along the welded seam, with the MATHCAD calculus software and they are presented under a table form, in Table 1, for a medium position of the welded seam, and in Figure 1 these values are compared to the results obtained experimentally.

3. NUMERICAL PROCEDURE

The thermal analysis was performed in non-stationary (transitory) regime. In the case of such an analysis we must establish the duration of the analysis (TIME). Keeping in mind that the length of components is of 120 mm, and the welding speed is of 6 mm/s, from a simple calculus it results that the duration of welding is of 20 seconds. The Ansys software determines the value of the dimensions in knots within certain time intervals, beginning with the initial moment $t_0=0$. Between each two moments consecutively determined there is a time step Δt. In order to obtain a very high precision of the simulation results, it is necessary that this step be as reduced as possible, but this leads to a very long calculus time and consequently to the increase of the file dimensions. Moreover, one imposes a minimum value of the time step Δt, so that during it the thermal waves should cross each finite element. For an optimum precision of results, Δt=0.2 seconds. Knowing that the welding duration is of 20 seconds, one will have 100 time steps. The geometric model of the components (1) crossed by the electric ark (2), produced by electrode (3), during welding, is protected by the protective gas curtain (5), passing through the gas nozzle (4), is presented in Figure 2.

The theoretic study of the welding bath shape was approached by numerous researchers, among whom the Russian researchers Rîkalin and Prohorov, who elaborated equations of the isothermal surface corresponding to the melting temperature, surface.

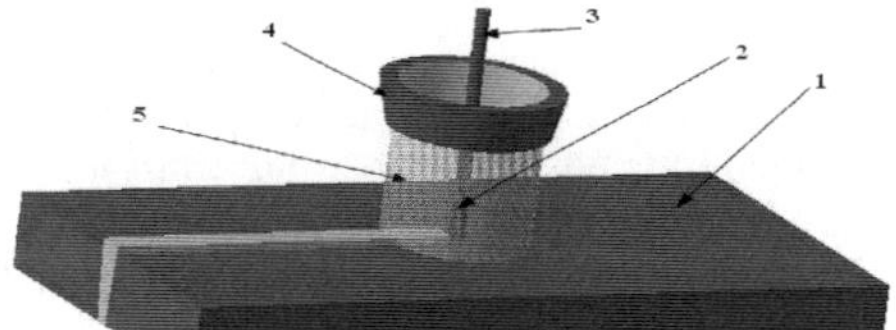

Fig. 2. The geometric model

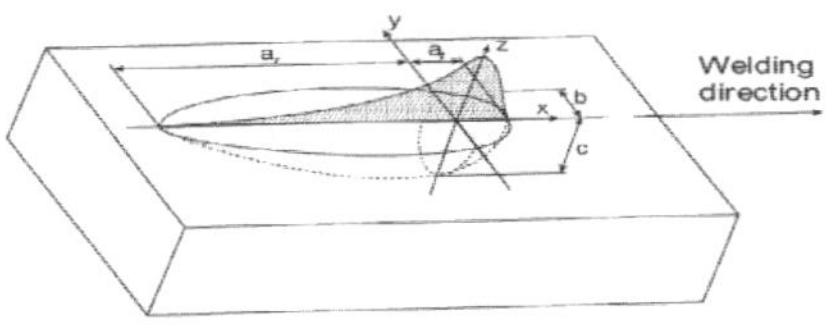

Fig.3 The Goldak model of the double ellipsoid

The issue is very complex, as each welding procedure requires a separate study, the profile in longitudinal and transversal section of the welding bath being determined by a series of factors connected to the characteristics of the procedure, the basic metal and added material, the nature of the electrode, the manner of source motion, etc. For arc welds, accurate results are obtained with a power density distribution in which surfaces of constant power density are ellipsoids and on radial lines the power density obeys a Gaussian distribution (Goldak et al., 2007). Nevertheless many authors use as tridimensional model a double ellipsoid suggested by the English researcher Goldak, whose model is presented in Figure 3. The software allows the visualisation of the temperature distribution, both in 2D co-ordinates and in 3D ones, through an isometric representation, according to the same established time periods.

4. RESULTS AND DISCUSSION

The software allows the introduction of the properties of material according to temperature, which leads to the correct estimation of the evolution in space of the thermal field. (Mughal et al.,2005). After the modelling and simulation of the welding processes in an environment of protective gases, it is absolutely necessary to proceed to the confirmation of the theoretical premises by means of the experimental programme in real, effective welding conditions. Thus one measured the temperature in three areas situated along the welded seam, with the help of thermocouples of the R 87%Pt-13%Rh/Pt type, with the diameter of 0.5 mm which can measure temperature up to 1800^{0}C, in a short period. The situations of the temperature variation in different points on the welded seam, simulated with the help of the Ansys software and calculated analytically, were compared to the results obtained experimentally. One may notice the accuracy with which the curve of the simulated values follows the curve of the measured values, compared to those obtained through analytical calculus.
The application of the finite elements method, as a preliminary method of the analysis for the thermal transfer, the estimation of temperatures in the welded joints and extension of the thermal influence areas in the heterogeneous welded joints constitute a very actual and useful solution for the checking and optimization of the welded technologies, correlating the power produced by the electric arc and welding speed (Wang et al., 2003). If we note that the maximal values of the temperature surpass the ones given by the specialized literature, the diminution of the arc power is imposed, maintaining the same welding speed or the increase of the welding speed is imposed maintaining the same power of the electric arc.
Thus, the finite elements analysis allows a correlation of the electric arc power with the welding speed and the establishment of the optimal welding technology so that the final joint should correspond to the desired characteristics, (Dilthey & Pavlyk 2005).
One presented the values of the temperatures measured experimentally T_M, and those obtained by simulation, noted T_{MEF}, for the same periods of time, according to Table 2.

.

Timp(s)	0	2	4	6	8	10	12	14	16	18	20
T_M [^{0}C]	20	1515	1466	1367	1250	1037	890	721	555	434	311
T_{MEF}[^{0}C]	20	1535	1421	1327	1200	998	851	673	505	400	278

Tab. 2. Comparative temperature values at the middle of the welded seam

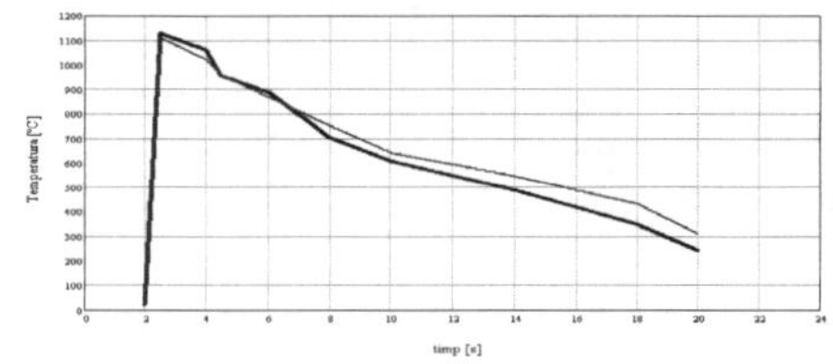

Fig. 4 Comparative values obtained by simulation (fine line) and experimentally (big line)

The experimental checking of the temperatures distribution in the welded joints confirms the results obtained through the finite elements analysis which means that the numerical modeling with finite elements of the thermal transfer and the simulation of the continuous process represent useful instruments in establishing the temperatures values, their variations in different areas of the welded joint and the estimation of the welding bath dimensions and of the thermal influence area.

5. CONCLUSION

The necessity of knowing the thermal field in the welded joints is extremely important in the study of the residual tensions and deformations occurring in the welded structure.
The application of method of the finite elements, as preliminary method of the thermal transfer analysis, estimation of temperatures in the welded joints and extension of thermal influence areas in the welded joints constitutes a very modern and useful solution in the verification and optimization of welding technologies, correlating the power developed by the electric ark and the welding speed.
If one finds that the maximum values of the temperature exceed those furnished by the literature, it is necessary to reduce the power of the ark, maintaining the same welding speed, or to increase the welding speed maintaining the same power of the electric ark.
The study refers to the possibility of quantifying the modifications which take place due to the temperature variation in the material to be welded and in the welded joint during and at the end of the welding process.
The situation of the temperature variation in different points on the welded joint was simulated with the help of the Ansys program and the experimental results juxtaposed with those obtained by calculation/simulation can validate the accuracy of the calculation program.

6. REFERENCES

Dilthey, U., Pavlyk, V., *Integrative MIG/MAG Welding Process Simulation,* (2005), In: Proceedings of the 4th German-Japanese Seminar "Materials, Processes and Components", Juli, 2005, Karlsruhe.

Goldak, J., Chakravarti, A., Bibby, M. (1985), *A double ellipsoid finite element model for welding heat sources,* IIW Doc. 212-603-85.

Goldak, J.; M. Bibby, J. Moore, R. House, & B. Patel, (2007), *Computer modeling of heat flow in welds*, Metallurgical and Materials Transactions B, ISSN: 1073-5615, Springer Boston, pg. 587-600.

Mughal, M. P., R. A. Mufti, H. Fawad, (2005*), Deformation modelling in layered manufacturing of metallic parts using gas metal arc welding: effect of process parameters,* Proceedings of the Institution of Mechanical Engineers, Part B: Journal of Engineering Manufacture, ISSN: 0954-4054, pg. 1499-1509.

Wang, G., Huang, P. G., and Zhang, Y. M., (2003), *Numerical Analysis of Metal Transfer in GMAW*, University of Kentucky, source: Internet.

Annals of DAAAM for 2008 & Proceedings of the 19th International DAAAM Symposium, ISSN 1726-9679
Editor B. Katalinic, Published by DAAAM International, Vienna, Austria 2008
Make Harmony Between Technology and Nature, and Your Mind will Fly Free as a Bird

COMPARISON OF NUMERIC SIMULATED AND EXPERIMENTAL MEASURED RESULTS OF SETTING DEFORMATION

MARTINKOVIC, M[aros]; SOBOTA, R[obert] & KAPUSTOVA, M[aria]

Abstract: *All numerical models of alloy forming based on finite elements methods can lead to excellent results, but it is necessary to verify them by experimental results. The simple way as to verify numerical result is to compare them with experimental result. Bulk forming leads to grain boundaries deformation. To obtain value of strain is very simple by stereology – measurement of degree grain boundaries deformation.*

Key words: *forming, setting, simulation, quantification, stereology.*

1. INTRODUCTION

Macroscopic effects of deformation in the volume forming are not corresponding fully with microscopic structural chances. Different degree of deformation depending on the area of specimen was observed (Martinkovič 2005). Degree of deformation heavily influences the materials properties due to deformation strengthening and materials behaviour during forming depends on the materials properties. Finite elements methods can lead to excellent results of deformation in each place of deformed parts, but it is necessary to verify them (their parameters) by experimental results. There are several microscopic methods (Forejt 2006) to obtain degree of strain in material structure, but measurement of grain boundaries deformation by means of grain boundaries orientation measurement was not used to recent time. The grain boundaries orientation direct depend on the grain deformation.

2. EXPERIMENTS

The verification of numerical model of bulk forming was made on two types of materials: steel S235JRG1 (1.0036) and aluminium alloy AW-6063A (6063A) and two types of forming: cold and warm bulk forming. For numeric simulation MSC.SuperForge simulation software was used (http:// www.mscsoftware.com).

2.1 Numeric simulation

The parameters of numeric simulation for steel specimen was: forging on crank press, tool temperature 20°C, friction coefficient 0,4, cylindrical bar from steel 1.0036 with diameter 14 mm and high 29 mm, temperature 20°C in basic state, high 16,7 mm after deformation. The results of numeric simulation of deformation – effective plastic strain in longitudinal section in the middle of cylindrical bar was observed.

The parameters of numeric simulation for aluminium alloy specimen was: forging on crank press, tool temperature 20°C, friction coefficient 0,4, cylindrical bar from aluminium alloy 6063A with diameter 25 mm and high 49 mm, temperature 240°C in basic state, high 30 mm after deformation. The results of numeric simulation of the process – effective plastic strain re in longitudinal section in the middle of cylindrical bar was observed. In this case the temperature in longitudinal section in the middle of cylindrical bar was observed too, because increasing temperature within process can lead to recrystallization effect.

2.2 Setting

Cold forming of the real part from steel was realized at the same basic parameters as are described hereinbefore (see chapter 2.1). Eccentric press LEXN 100 was used.

Warm forming of the real part from aluminium alloy was realized at the same basic parameters as are described hereinbefore (see chapter 2.1). Eccentric press LEXN 100 was used.

3. MICROSTRUCTURE ANALYSIS

The structure of steel material was observed with about 200x magnification of light microscope on metallographic cut of longitudinal section in the middle of cylindrical bar. The metallographic cut was mechanical grinded and polished, chemical etched in 3% HNO_3 alcohol solution. An example of steel structure in basic state is on Fig. 1, an example of steel structure after deformation is on Fig. 2.

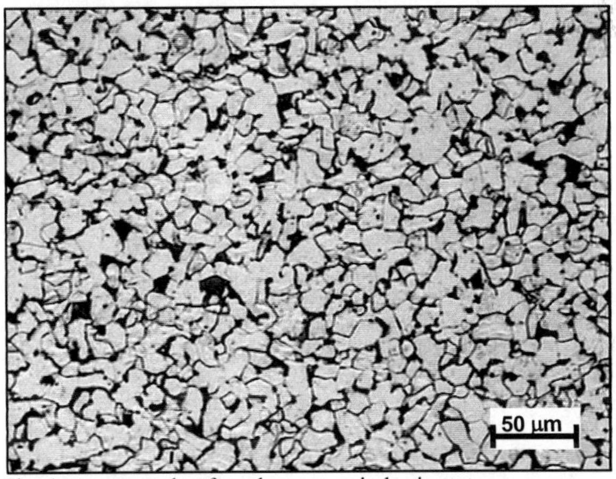

Fig. 1. An example of steel structure in basic state

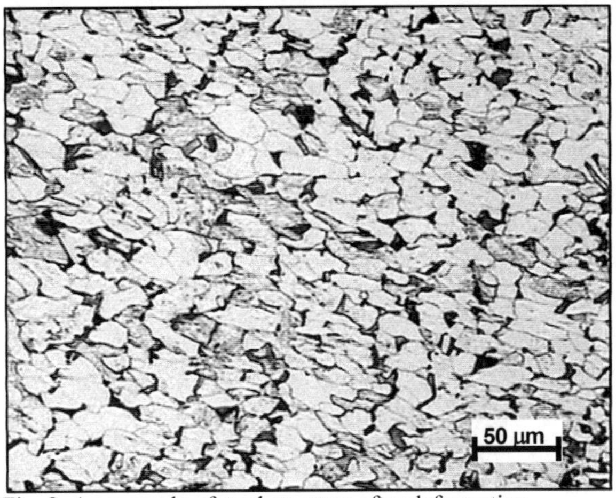

Fig. 2. An example of steel structure after deformation

The structure of aluminium alloy material was observed with about 50x magnification of light microscope on metallographic cut of longitudinal section in the middle of cylindrical bar. The metallographic cut was mechanical grinded and polished, chemical etched in Keller etcher (1,5 ml HF, 3 ml HNO_3, 100 ml H_2O). An example of aluminium alloy structure in basic state is on Fig. 3, an example of aluminium alloy structure after deformation is on Fig. 4.

Fig. 3. An example of aluminium alloy structure in basic state

Fig. 4. An example of aluminium alloy structure after deformation

3.1 Quantitative analysis

The strain of probes on their sections was obtained by stereological measured by measurement of degree grain boundaries deformation – degree of grain boundaries orientation (Saltykov 1970; Underwood 1970). The method of oriented test lines was used. Test lines were placed perpendicular and parallel to the grain boundaries orientation direction effected by straining (Russ & Dehoff 2000). From the relative number (number to unit of length) of parallel test lines intersections with grain boundaries $(P_L)_P$ and perpendicular lines ones $(P_L)_O$ was total relative surface area (area to unit test volume) $(S_V)_{TOT}$ of grains estimated according equation (1) and planar oriented part of relative surface area $(S_V)_{OR}$ of grains estimated according equation (2). The relative measure precision was always smaller then 10% with reliability 90%. Degree of grain boundaries orientation was estimated as $(S_V)_{OR}$ to $(S_V)_{TOT}$ ratio.

$$(S_V)_{TOT} = (P_L)_O + (P_L)_P \qquad (1)$$

$$(S_V)_{OR} = (P_L)_O - (P_L)_P \qquad (2)$$

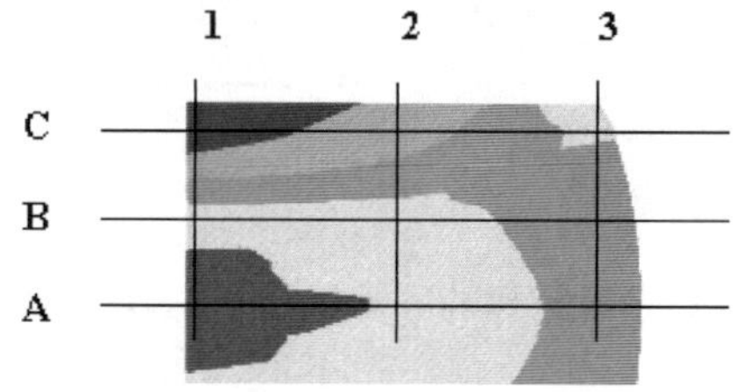

Fig. 5. Positions of deformation analysis

The results of grain boundaries orientation measurement in different places (see Fig. 5.) of deformed bulk of steel specimens are in Table 1, the ones of deformed bulk of aluminium alloys specimens are in Table 2.

Position	Orientation [%]		
	1	2	3
C	30	23	40
B	42	40	35
A	62	44	43

Tab. 1. Measured grain boundaries orientation in different places of steel specimen

Position	Orientation [%]		
	1	2	3
C	19	27	40
B	18	28	33
A	0	20	32

Tab. 2. Measured grain boundaries orientation in different places of aluminium alloy specimen

4. CONCLUSION

The effective strain in each places of steel specimen obtained by numeric simulation was in very good coincidence with the experimental results (Table 1.). The effective strain obtained by numeric simulation of aluminium alloy specimen in the middle of specimen bulk was greater then experimental ones (Table 2.). In these places the temperature of the material during deformation increased over the recrystallization temperature 250°C and recrystallization was passed and the deformation of grain boundaries was decreased. In other places the coincidence was observed. The results of numeric simulation are valid. The utilization of stereology metallography allow very simple and effective experimental estimation of plastic deformation degree in various places of bulk formed parts and to verify numerical model by comparing this results with numeric simulated ones.

This work was supported by Slovak Republic Ministry of Education VEGA Grant No. 1/3192/06.

5. REFERENCES

Forejt, M. & Píška, M. (2006). *Teorie obrábění, tváření a nástroje, Theory of machining, forming and tools,* CERM, 80-214-2374-9, Brno

Martinkovič, M. (2005) Quantitative analysis of plastic deformed material structure, *Proceedings of Forming 2005,* pp. 175-180, 80-248-0888-9, Lednice, 14.-17.9.2005, VŠB-Technická univerzita, Ostrava

Russ, J., C. & Dehoff, R., T. (2000). *Practical stereology,* Plenum Press, 0-306-46476-4, New York

Saltykov, S., A. (1970). *Stereometričeskaja metallografia, Stereometric metallography*, Metallurgia, 3-11-1 88-70, Moskva

Underwood, E., E. (1970) *Quantitative stereology*, Addison-Wesley Pub., Mass

Annals of DAAAM for 2008 & Proceedings of the 19th International DAAAM Symposium, ISSN 1726-9679
Editor B. Katalinic, Published by DAAAM International, Vienna, Austria 2008
Make Harmony Between Technology and Nature, and Your Mind will Fly Free as a Bird

IDENTIFICATION OF MAIN PROBLEMS OF SOFTWARE DEVELOPMENT AUTOMATION

MASAR, A[lojz] & MASAROVA, R[enata]

Abstract: *Process of software development is described very well. Despite this, it is still very risky and gruelling. Some problems have been solved by introducing the object oriented approach, by visualization of ideas and code using of the UML, introducing decomposition via the MDA and by creating new tools, which support these approaches. This position paper names some general problems of today's software development and attempts to set basic evaluation criteria.*
Key words: *MDA, UML, UsiXML, transformation.*

1. INTRODUCTION

It is clear that the development of software requires the collaboration of different professionals. Different groups of professionals use different jargon, methods, techniques, means and tools. They also are interested in the software in distinct way.

Despite these differences, they use common sequencing of activities:

- Gathering of requirements.
- Evaluation and analysis.
- Design of solution.
- Produce output.
- Release output.
- Maintain changes.

These activities are used during every stage and at every level of abstraction event though they usually have various extents or iterations.

We concentrated on applications. The application is software, which:

- Is given to solve some problem of particular domain or to support activities of specific user process.
- Has at least one interface (to the user or to another system).
- Optionally is able to retain some information.

We looked at the process of creating application from a possibility its automation utilizing today's technology and tools point of view.

2. REVIEW OF PRESENT STATE

2.1 Approach

One of the respected concept of software development is the Model Driven Architecture (OMG, 2003). MDA provides the approach for, and enables tools to be provided for:

- Specifying a system independently of the platform that supports it.
- Specifying platforms.
- Choosing a particular platform for the system, and transforming the system specification into one for a particular platform.

MDA is successfully adopted for design and construction phase of the domain and the persistence. Also is usable for transition from the design to the persistence. It is supported by

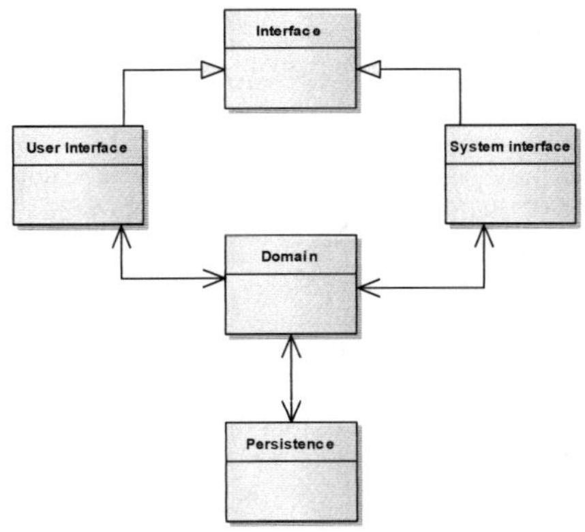

Fig. 1. Application

commercial tools. In spite of this, an adoption to the requirements gathering, evaluation, analysis and transition evaluation, analysis and transition to the design is still poor.

Activities of creating interfaces, notably the activities of creating the user interface, have some specialties (Jespersen & Linvald, 2003) which complicate the adoption of this approach to them. There is not any widely accepted and used standardised approach to develop the user interface.

The system interface is not so complicated. The functionality of majority applications is usually deterministic and the communication respect technological standards which determine communication rules of systems. The specialization of these standards prevent simply adopt MDA approach also to the field of creating the system interface. The starting point of solution would be acceptance of more general standards like the ESB (Chappell, 2004).

2.2 Method of expression

It is very important to properly express the outputs from and the inputs to the activities. This enables easier way of artefact transformation for each part of application and for every activity, generation of output/input from/to activities and increase chance of automation.

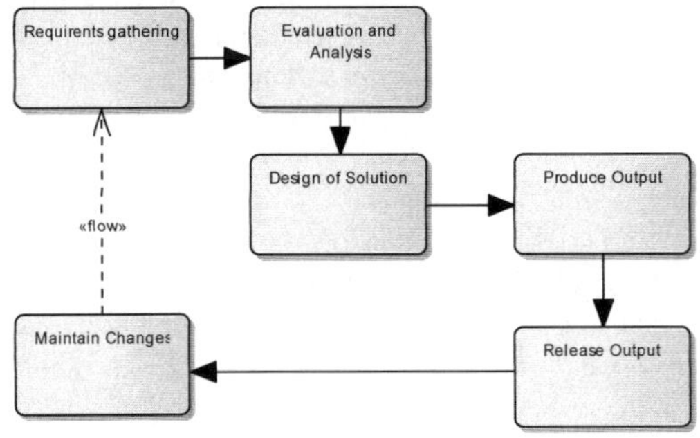

Fig. 2 Activities

The most adopted formalism is the UML (OMG, 2007). The UML is the way of model application structure, behaviour, architecture, business process and data structure. However, this standard doesn't tolerably cover activities oriented to the interfaces. It is not oriented to support early activities of software development too. These are covered by special means of expression which are not integrated to the one concept.

The UsiXML is the low level formalism to describe user interface. It is based on XML and is currently being submitted for standardization. The UsiXML describes the UI for multiple contexts of use such as Character User Interfaces (CUIs), Graphical User Interfaces (GUIs), Auditory User Interfaces, and Multimodal User Interfaces. Application can be described in a way that preserves the design independently from peculiar characteristics of physical computing platform.

2.3 Tools

There are a lot of tools which support the UML and the MDA. Whilst the UML is supported very well, the support of the MDA is restricted to the expressing capability of the UML. Each tool realizes the transformations by its own way. They suffer from round trip problems too.

3. STANDING PROBLEMS

We inspected some other existing methodologies, methods, techniques and tools. We attempted to determine some weighty gaps of software development. We based our point of view on the high level abstraction of application, on the main activities of development process and on the possibility of adopting the MDA approaches.

We paid attention to the following evaluation criteria:

- Correctness.
- Maintainability.
- Comprehensibility.
- Support of automated transformations or code generations.
- Amount of effort to make output against of effort to make things by usual way.
- Round trip ability.
- Support of visualizations.
- Support of existing standards.
- Support of testing outputs and artefacts.

We discovered that serious lack of standardization lay mainly in the area of creating user interface and its binding to the domain. The transformation of requirements to the form convenient for automation was appointed as second area, which would be explored.

3.1 User interface

We found some promising approaches such as UIOAD concept (Doroodchi et al., 2006) or the MDA compliant environment for developing UI (Vanderdonckt, 2005). The UIOAD assumed relatively direct transformation from the UI to the domain. Therefore was quite limited. The environment described in (Vanderdonckt, 2005) was directed only to user interface. None of them was widely adopted or standardised.

3.2 Binding UI and domain

Some work had been done in this area (Belenguer et al., 2003) or (Drori, 2003). However, we did not discover any solid solution. These two areas have different main concerns and solve problems using different approaches. The presented information via user interface had sometimes different structure from the structure which was conformed to the domain. Therefore, direct transformation from one to another was difficult. The transformation required some new piece of information from opposite area to do its job successfully. Due to the new approach would be required.

The whole user interface need not be transformed to the domain and vice versa not whole domain must be prepared to the representation via user interface. The part of user interface, which could be good candidate of transformation is usually named presentation logic.

The research of this area would be the job of the next work.

4. CONCLUSION

We found, that today's way of creating applications mainly suffers from the lack of:

- Suitable formalism to catch requirements
- Possibility of transform vague expression of requirements to the more formal form which would be suitable for MDA style decomposition.
- Standardised way of creating interfaces. Especially this problem is visible for the user interface.
- Standardised tools which would support automation of producing outputs of early activities.
- Insufficient formalism to create transformations from inputs early activities to its outputs.

.

5. REFERENCES

Belenguer, J.; Parra, J.; Torres I. & Molina P. J. (2003). HCI Designers and Engineers: It is possible to work together? *Proceeding of INTERACT 2003 Workshop*, Harning, M. B. & Vanderdonckt, J. (Ed.), pp. 14-19, Zürich, September 2003, Institut d'Administration et de Gestion (IAG), Université catholique de Louvain (UCL), Louvain-la-Neuve

Chappell, D. (2004). *Enterprise Service Bus*, O'Reilly Media, Inc., ISBN 978-0596006754

Doroodchi, M.; Farahani, B. K. & Moravej, M. (2006) User Interface Oriented Application Development (UIOAD). *Proceedings of World Academy of Science, Engeneering and Technology*, Volume 16, November 2006, 216-220, ISSN 1307-6884

Drori, O. (2003). Integration of HCI Needs with SE Methods Using OODPM Methodology, *Proceeding of INTERACT 2003 Workshop*, Harning, M. B. & Vanderdonckt, J. (Ed.), pp. 20-26, Zürich, September 2003, Institut d'Administration et de Gestion (IAG), Université catholique de Louvain (UCL), Louvain-la-Neuve

Jespersen, J. W. & Linvald, J. (2003). Investigating User Interface Engineering in the Model Driven Architecture, *Proceeding of INTERACT 2003 Workshop*, Harning, M. B. & Vanderdonckt, J. (Ed.), pp. 63-66, Zürich, September 2003, Institut d'Administration et de Gestion (IAG), Université catholique de Louvain (UCL), Louvain-la-Neuve

OMG; (2003). MDA Guide Version 1.0.1, *Available from: ftp://ftp.omg.org/pub/docs/omg/03-06-01.pdf Accessed:* 2007-06-25

OMG. (2007). UML 2.1.2 Superstructure and Infrastructure, *Available from: http://www.omg.org/spec/UML/2.1.2/ Accessed:* 2007-06-25

Vanderdonckt, J. (2005). A MDA-Compliant Environment for developing User Interfaces of Information Systems. *Proceedings of the 16th Conference on Advanced Information Systems Engineering*, Pastor, O. & Cunha, J. F. (Ed.), pp. 16-31, ISBN 3-540-26095-1, Porto, June 2005, Springer-Verlag, Lecture Notes in Computer Science, Vol. 3520, Berlin.

Annals of DAAAM for 2008 & Proceedings of the 19th International DAAAM Symposium, ISSN 1726-9679
Editor B. Katalinic, Published by DAAAM International, Vienna, Austria 2008
Make Harmony Between Technology and Nature, and Your Mind will Fly Free as a Bird

PI-LIKE FUZZY LOGIC CONTROLLER WITH LOOK-UP DECISION TABLES AND FUZZY CONTROL GAIN

MASTACAN, L[ucian]

Abstract: *In real time applications for fuzzy control, to avoid the complex algorithm computing of fuzzy rules inference, look-up decision tables are usually implemented. The crisp control values from a look-up table are computed off line. From a look-up table is possible to extract only a finite number of crisp control values. To increase the control performance, coarse and fine look-up decision tables are introduced in many applications. The paper presents a new method to implement a PI-like type fuzzy logic controller to obtain the best control performance.*
Key words: *fuzzy logic, controller, look-up table, temperature control, performance.*

1. INTRODUCTION

Fuzzy logic enables us to incorporate human intelligence into automatic control by means of fuzzy algorithms based on the intuition and experience of the skilled operators (Lee, 1990). The experience of operators working with the process allows finding the fuzzy controller structure and the linguistic rule base. Then the fuzzy controller parameters can be tuned by an on - line optimization to obtain a better behaviour of the plant. Knowing the actual state of the controlled process, an operator will modify the controller parameters to improve the process performance (Driankov et al., 1996). PI-like type fuzzy logic controllers are introduced to obtain the best control performance (Mastacan et al., 1997). To implement a fuzzy logic controller for a real time application, the fuzzy algorithm computation time is an important criterion.

Usually look-up decision tables are used in order to decrease the time computation, especially in the fuzzy inference reasoning and output defuzzification. All computations are carried out beforehand (off-line) and the resulting look-up tables are stocked in memory. In real time control, only quantizations are performed and the corresponding look-up table values are obtained. When output quantization is linear, the values contained in the look-up tables are multiplied by different constants and then applied to the process as control variables. The PI-like type fuzzy control algorithm can be improved by a new method introducing the fuzzy control gain to adjust the value of the control change. The proposed method is implemented to control the process temperature.

2. NEW PI-LIKE TYPE FUZZY LOGIC CONTROLLER

Fuzzy control is based on conditional linguistic rules and approximate reasoning. A fuzzy logic controller can be designed starting from a linguistic model which emulates the behaviour of a skilled operator working with the process.

The operator takes a decision to modify the actuating signal "u" observing the system error "e" and the system error change "Δe". The PI-like type fuzzy logic controller algorithm is based on the "*if–then*" rules in the form (Driankov et al., 1996):

$$\boldsymbol{R_i}\text{: if error } e_i \text{ is } A_i \text{ and error change } \Delta e_i \text{ is } B_i \text{ then control change } \Delta u_i \text{ is } C_i \quad (1)$$

To obtain the control value u_k the control change Δu_k is added to the former control value u_{k-1}:

$$u_k = u_{k-1} + \Delta u_k \quad (2)$$

A new PI-like type fuzzy logic controller is proposed to be implemented as follows:

$$\Delta U_k = (E_k \times \Delta E_k) \circ R_{F1} \quad (3)$$

$$G_k = (E_k \times \Delta U_k) \circ R_{F2} \quad (4)$$

$$U_k = U_{k-1} + G_k \Delta U_k \quad (5)$$

where:
- U_k is the fuzzy controller output;
- ΔU_k is the change in the fuzzy controller output;
- E_k is the fuzzyfied process output error;
- ΔE_k is the fuzzyfied derivative error of the process output;
- G_k is the control gain;
- R_{F1} is the *error - derivative error* fuzzy relation of the controller;
- R_{F2} is the *error - change in control* fuzzy relation of the controller;
- $\times$ is the Cartesian product;
- $\circ$ is the compositional rule of inference based on Mamdani reasoning.

If E_k, ΔE_k, ΔU_k and G_k are obtained as fuzzy singletons, respectively as e_k, Δe_k, Δu_k and g_k, R_{F1} and R_{F2} are in the form of two look-up tables **U** and **G**, the controller fuzzy output u_k is computed as

$$\Delta u_k = \mathbf{U}(e_k, \Delta e_k) \quad (6)$$

$$g_k = \mathbf{G}(e_k, \Delta u_k) \quad (7)$$

$$u_k = u_{k-1} + \alpha g_k \Delta u_k \quad (8)$$

where:
- Δu_k is the element of **U** at the position $(e_k, \Delta e_k)$;
- g_k is the element of **G** at the position $(e_k, \Delta u_k)$;
- α is a constant depending of the number of points in which are divided the change in control universe of discourse and control gain universe of discourse.

The output defuzzification is based on the centre of area method (Driankov et al., 1996). The universes of discourse of the variables e_k, Δe_k, Δu_k and g_k are quantized into 11 points in the quantized set Q ={-5; -4; -3; -2;-1; 0; +1; +2; +3; +4; +5}. The quantized values are associated to a fuzzy partition with six terms: NEGATIVE BIG (NB), NEGATIVE SMALL (NS), ZERO (ZO), POSITIVE SMALL (PS), POSITIVE BIG (PB) and ANY (A). Because e_k, Δe_k, Δu_k and g_k are quantized into a finite number of values, the look-up decision tables **U** and **G** have finite numbers of control values. To improve the control performances the fuzzy logic controller must to manipulate two look-up tables for each control variable, one for a coarse control and the other for a fine control. These look-up tables differ for a control variable by the step of quantization. Four look-up decision tables are introduced, two for a coarse control $\mathbf{U_c}$ and $\mathbf{G_c}$ and the other two for a fine control $\mathbf{U_f}$ and $\mathbf{G_f}$. Using the fuzzy control algorithm (8) is possible to obtain from the finite number of control values of the look-up decision tables a variable number of fuzzy control values and a very small change in the fuzzy logic controller output signal which is necessary to get zero steady state error. Equation (8) is typical for a PI-like fuzzy logic controller. Because the universes of discourse of all the variables are quantized into 11 points, the look-up decision tables have 11 values for the control change Δu_k and 11 values for the control gain g_k. These values are numbers in the quantized set Q = {-5; -4; -3; -2;-1; 0; +1; +2; +3; +4; +5}. In the control algorithm (8) by multiplying $g_k \Delta u_k$ are obtained 121 different values for the control variation.

Starting from the values of Δu_k and g_k, it is possible to obtain any value of the control variables in a given universe of discourse, by using a linear interpolation in the **U** and **G** look-up tables (Mastacan & Olah, 1996), (Mastacan & Olah, 1998).

3. EXPERIMENTAL RESULTS

The configuration of the fuzzy logic controller for temperature control is given in Fig. 1.

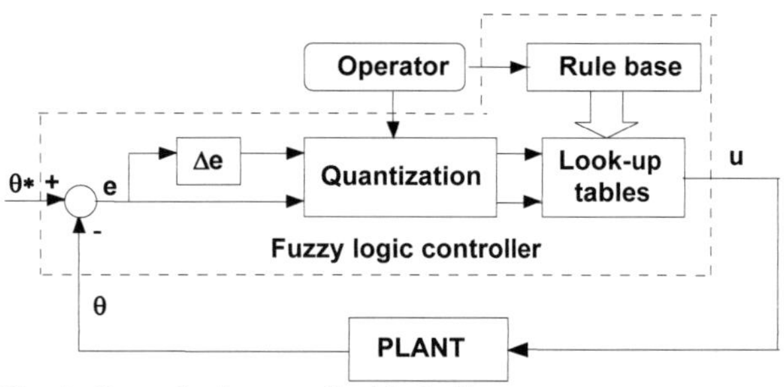

Fig. 1 Fuzzy logic controller block diagram

The PI-like fuzzy logic controller is implemented on a PC in LabVIEW software from National Instruments. It is used to control the temperature of an electric oven with infrared bulbs (Mastacan, 2008). Fig. 2 presents the temperature fuzzy control of the electric oven when a classic PI-like fuzzy control algorithm is implemented. Even two look-up decision tables are used, for coarse and fine control, the fuzzy control cannot be adjusted to the proper values because of the crisp values obtained from the look-up tables.

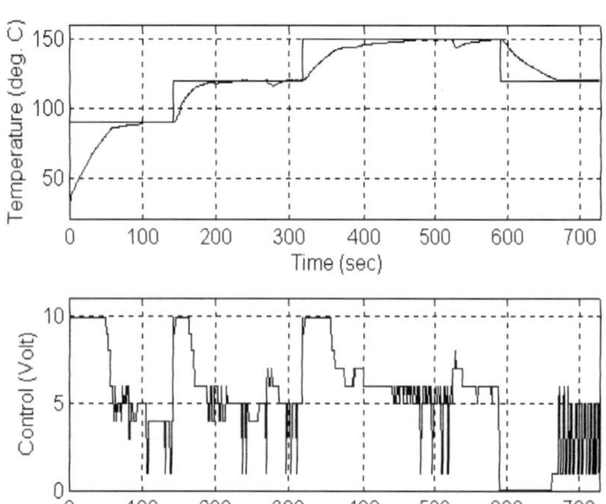

Fig. 2 Temperature control with classic PI-like fuzzy control algorithm.

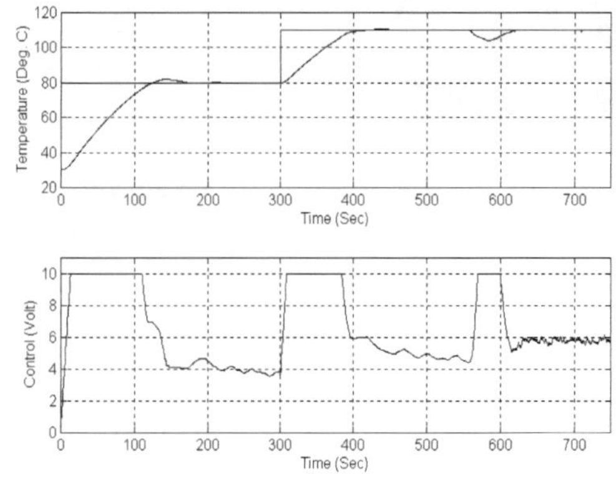

Fig. 3 Temperature control with the new PI-like type fuzzy control algorithm.

To improve the fuzzy logic controller performance, the operator will adjust the parameters from the look-up decision tables by trials and errors. Fig. 3 presents the temperature fuzzy control after several fuzzy logic controller parameters tuning by trials and errors. Finally, the operator actions to tune the fuzzy logic controller allow the fuzzy control system to get the best performance. The temperature fuzzy control performance is the best because the fuzzy logic controller output signal can reach the proper values to get the desired goal. This can be observed from the fuzzy logic controller control signal variation in fig. 3.

4. CONCLUSION

A new method to compute the fuzzy control variables from the look-up decision tables is introduced. With the new PI-like fuzzy control algorithm the best temperature control performance is obtained. The time response is short, the overshoot is small and no steady state error is present. The fuzzy control system performance is the best for the change of the set point and also for the step change in the perturbation.

5. REFERENCES

Driankov, D., Hellendson, H. & Reinfrank, M. (1996). *An Introduction to Fuzzy Control*. Ed. Springer-Vela, ISBN 978-3-540-60691-8, Berlin.

Lee C.C. (1990). Fuzzy Logic in Control Systems: Fuzzy Logic Controller -Part I and II. *IEEE Transaction on Systems, Man, and Cybernetics,* 20(2), 404 -435, ISSN 1083-4419.

Mastacan L. & Olah I. (1996). Fuzzy Rules Interpolation in Look-Up Decision tables, *Proc. of the EUFIT'96, Forth European Congress on Intelligent Techniques and Soft Computing*, vol. 1, pp. 661-664, ISBN 3-896531-87-5, Aachen, Germany, September 1996, ELITE and Verlag Mainz, Aachen.

Mastacan, L., Olah, I., Anita, L.L. & Peiu, T. (1997). Fuzzy Logic Controllers Design: A Case Study. *Proceedings of INES 97, IEEE International Conference on Intelligent Engineering Systems*, pp. 233-238, ISBN 0-7803-3627-5, September 1997, Published by IEEE, New York.

Mastacan L. & Olah I. (1998). *Fuzzy Control by Interpolation in Mamdani Type Look-up Tables, Proc. of EUFIT'98, 6th European Congress on Intelligent Techniques and Soft Computing,* vol. 2, pp. 905-908, ISBN 3-89653-500-5, September 1998, Aachen, Germany, ELITE and Verlag Mainz, Aachen.

Mastacan, L. (2008). Fuzzy Logic Control System for Thermal Process, *Polytechnic Institute Bulletin of Iași – Machines Construction,* Tome LIV (LVIII), Fasc. 2, pp. 483-490, 2008, ISSN 1011-2855.

Annals of DAAAM for 2008 & Proceedings of the 19th International DAAAM Symposium, ISSN 1726-9679
Editor B. Katalinic, Published by DAAAM International, Vienna, Austria 2008
Make Harmony Between Technology and Nature, and Your Mind will Fly Free as a Bird

DRAFTS OF TECHNICAL-ECONOMICS SYSTEMS NECESSARY TO QUALITY ENSURANCE

MATEESCU, L[iviu] M[ihail]

Abstract: *The Technical-Economics System reflects the integrator vision of an element or elements group having technical and economical specific features. The Technical-Economic System (STE) is an entity having the objective to manufacture some product assortment at the amount, structure and quality established in a technical, economical and socio-human medium, generated and kept up, with a view to obtain fluxes of positive hard cash for a long time. Quality at present isn't a preoccupation only for the organizations but also for every person, becoming an education and culture element, and that allows a better understanding of its necessity and the knowledge of what is better. The quality is a behaviour problem that implies activity insistence by resumption, return, perseverance and labour comfort.*

Key words: *The Technical-Economic System, quality draft, objectives, performance, economical structure.*

1. INTRODUCTION: CONTEXT ELEMENTS

The Technical-Economic System is a modern concept modern draft which reflects the integrator vision on am elements or group of elements which have both technical and economical specific features. It is represent by any system allowing the taking over, stocking, processing, turning some resources into bodily or not bodily goods, with economical importance and includes technical and technological elements. The Technical-Economic Systems (TES) are characterized by constructive elements as well as the connections which create interdependences with other systems or singular elements. The connections can be physical or informational.

This definition suits to industrial organizations, too. If we analyse the socio-economical situation in its materially reproducible aspect, we'll find the existence of a productive socio-system spectrum which in a direction of the movement fits in productive Macro systems and in the other one it is structured in Microsystems (Ionescu, 1997). Analyzing in this vision the industrial organizations, we find on owe hand their aggregation in industrial branches and implicitly in entire economy where it exists as subsystems submitted to the system laws and on the other hand, a structure in production subsystems and administrative subsystems.

A technical, economical and social system type black-box, isn't an abstract cybernetic system but a socio-economical one, which differs from other systems of this category by a series of specific features where became conspicuous the direct and aware presence of man on command, control, adjustment positions and derived from this – the dynamic character in continuous improvement of the TES.

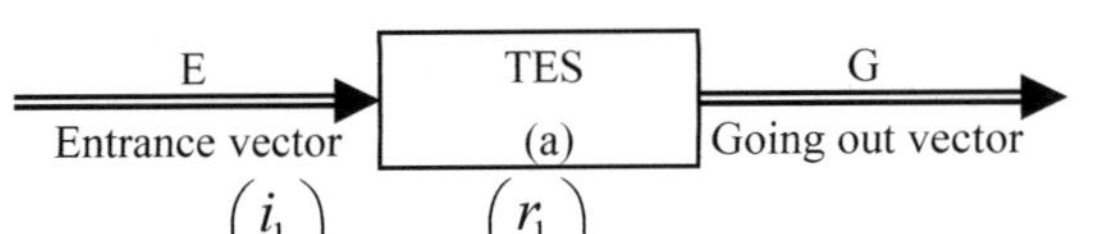

$$I = \begin{pmatrix} i_1 \\ i_2 \\ \vdots \\ i_n \end{pmatrix} \text{ or } R = \begin{pmatrix} r_1 \\ r_2 \\ \vdots \\ r_n \end{pmatrix}$$

Fig. 1. Functional scheme of industrial TES

The elements of entrances space $\mathbf{i}_i$, based on a command / order implementation, are submitted to the system activity $\mathbf{a}_i$, followed by the goings out $\mathbf{r}_i$, that can be type goods and services or information, which in their turn will be entrances for other productive or consumption system (Filip, 2004).

A TES activity goes on as part of some subsystems whose structure general valid can be organized thus:

- The technological subsystem (T) consists of technological means ensemble which take part in the raw material and know ledges / information processing necessary to ensure the subsystem work, corresponding to the established parameters;
- The technical subsystem (TH) consists of the means ensemble (less the technological ones) which takes part directly in the production going on;
- The organizational subsystem (O) includes the labor engaged with the co-ordination in time and space of the production elements having also in view the relation established on this purpose;
- The economical-financial subsystem (EF) administrates patrimonialy all the TES goods, from provisioning the raw material and materials to production sale;
- The creative-innovator subsystem (CI) consists of the labour engaged in these activities and the necessary material endowment;
- The informational management subsystem (IC) is made up from the labour engaged in the informational management process and the necessary material endowment. It is functional and informational opposable to all other TES systems. The activity it carries on consists in putting forward the decisions which settle the other subsystems activity and co-ordinate their relation in order to implement the planned tasks.

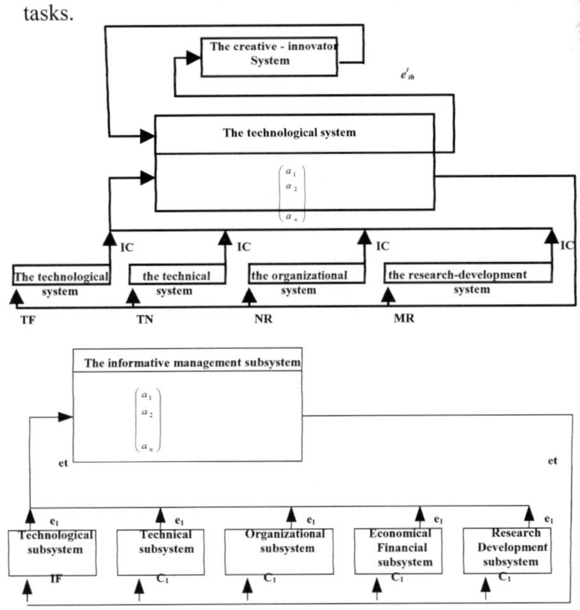

Fig. 2. The relation established between each subsystem and the other subsystem of TES

A technical-economic system can be considered as well an organization group, united by a common interest, an informational network type internet / intranet, a bodily or not bodily real estate objective of an organization (considered solidary) or an activities group of the organization or more activities of the same type but having a particular specificity.

2. THE QUALITY DRAFT

The quality notion has a long history in the humanity consciousness. Witch the progress registered in industry the quality isn't only a philosophical draft but an economical one, to. The quality role developed appreciably, becoming an important saving factor. In the present outlook, the quality has become a general concern of all the organization and it is achieved by management. The quality specific feature, represent a product feature which differentiates it from other and comes from its utilization value, namely from the product feature of answering the social needs. It express precisely the way of achieving the usefulness, the satisfaction level felt by the product user. In the quality management is fallowed their identification and measurement, in order to know the level where the producer offer frames in the users demands. The analyze of the products quality needs first their division on quality categories which internal have common typological specific features. To establish the goods and services specific features it is necessary to know all those involved in the products accomplishment and utilization, which are: the client, the supplier and the society on its whole. In the significance theory, these are the ones who care state the case for the product. For everybody, the quality becomes manifest during the utilization period, but the quality analyze can't be summarized only to this period but it must spread to the others, too: conception and manufacture, commercial accomplishment and utilization period (which materializes in actual fact the product usefulness, characterized by achieving the planned parameters and reliability). It comes out that the producer can't make abstraction of competitiveness and the labour included in the product (when analyzing the quality and the usefulness), that the beneficiary considerates first the product usefulness and the society is interested in the consumers production and in the existence of a real market competition. Starting from this assumption we can define quality as an image of the studied good made up of typological qualitative specific features and other specific features less meaningful. By product quality it mustn't by understood only its technical level but also the aspects connected to the product impact on the environment, its efficiency and competitiveness created suitable to the producer interests (Alistair et al., 1999).

3. THE QUALITY DRAFT EXTENSION

A wider defining of the quality draft in economy could be connected to three coordinates: efficiency, environment, and functionalism. The quality is image characterized by economic, ecological and functional elements. At present there are two tendencies of goods quality assessments. The first one connects the quality only to the good utilization and analyses separately its efficiency and the second tendency takes into consideration the efficiency included in quality.

The first tendency can't be accepted because it hasn't the ability to define the good quality for all its coordinates, as it is considerated in the quality management and in the standard ISO 9000/2000.

Although the "quality" notion as human behaviour fundamental elements comes out since the beginning of its history, the quality draft has included in the management science for only two centuries and the actions for quality materialized to each organization level by objective and organizational structures. The organization management grew richer with a new field – the quality management.

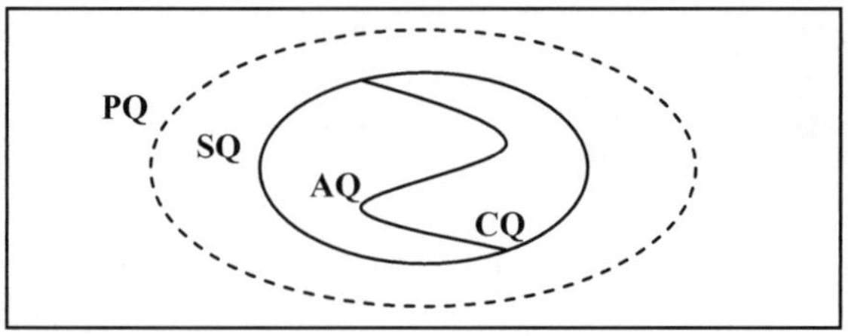

Fig. 3. The definition given by standard to the "quality management"

The management structure consists of the following elements: the quality ensurance AQ and the quality control CQ, which are united in a quality system SQ, elaborated under the basis of a quality policy PQ, pat of the strategical management.

4. CONCLUSIONS

The present orientations for quality defining know the following aspects: ***the orientation to the product*** – in this case quality is defined as representing the product quality specific features ensemble; t***he orientation to the production processes*** – here the quality is considered from the producer view; ***the orientation to cost*** – in this case the product quality is defined through the agency of cost and saling prices by cost / quality criterion; ***the orientation to the user*** – the good quality represents the ability of answering the user demands who estimates the good quality. The quality represents the consumer's demands concerning the functionality price, delivery date, safety, reliability, environment compatibility and services. Two quality hypostasis have a decisive role in meeting the customer's demands: ***conception quality (planning)*** and ***concordance quality***. The product quality is often expressed by technical and commercial quality. Although some hypostasis can be more important at a given moment, it is thought that all of them contribute to ensure the product quality on the whole.

TES must adopt a policy to define the position it wants to own by quality in the market. In keeping with the standard ISO 9000, the policy in the quality field represents a TES directions and general aims regarding quality. The 11th Committee of European Organization for Quality put forward a direction which shows that the elaboration of the policy quality is implemented from upside to below, under the basis of the firm general policy. The policy is interpreted afterwards by the managers from the other levels in order to be transposed in objectives specific to different activity fields. The direction does not exclude the possibility of working out the polity from below to upside. In this case the direction recommends that the firm should take into account the following elements: the orientation to the customer, the continuous improvement and staff instruction in this field. The implementation of a quality system requires the elaboration of the TES quality policy which is made by the summit management. Within the framework of the policy are settled the quality objectives which reflect the position proposed by the unities to own in the market.

5. REFERENCES

Alistair Inglis, Peter Ling, and Vera Joosten, Delivering Digitally: Managing the Transition to the Knowledge Media, Kogan Page, London, 1999.

Filip, N., Morariu, C.O., Popescu, I., The Quality Management and Engineering [Ingineria şi Managementul Calităţii], Editura Universităţii „Transilvania", Braşov, 2004, ISBN 973-635-271-4.

Ionescu Sorin Cristian, Industrial excellence – exercise and theory of quality [Excelenţă industrială - practică şi teoria calităţii], Editura Economică, Bucureşti, 1997.

CEC-Commission of the European Communities, Towards a European Research Area, Brussels, 18 January 2000.

Annals of DAAAM for 2008 & Proceedings of the 19th International DAAAM Symposium, ISSN 1726-9679
Editor B. Katalinic, Published by DAAAM International, Vienna, Austria 2008
Make Harmony Between Technology and Nature, and Your Mind will Fly Free as a Bird

DATA MINING IN PRODUCTION MANAGEMENT AND MANUFACTURING

MATSI, B[irthe]; LOUN, K[aia]; OTTO, Tauno] & ROOSIMOELDER, L[embit]

Abstract: *Importance of data collecting and storing is growing in engineering sectors. Conventional databases have been developed to integrated systems, but in terms of e-Manufacturing due to increasing amount of data feeds from smart dust sensor networks novel analysis methods are needed. Data mining implementation in production management and manufacturing as an alternative method is discussed in this paper. Integrating a data mining framework within the manufacturing information system will provide improved manufacturing decision making process and enhanced productivity.*

Key words: *Data Mining, e-Manufacturing, Databases, Industrial Information Systems*

1. INTRODUCTION

Traditionally analysts have performed the task of extracting useful information from recorded data. But the increasing volume of data in modern business and science calls for computer-based approaches. As data sets have grown in size and complexity, there has been an unavoidable shift away from direct hands-on data analysis toward indirect, automatic data analysis using more complex and sophisticated tools. Nowadays the modern technology has made data collection as an almost effortless task. However, the captured data needs to be converted into information and knowledge to become useful. Data Mining (DM) is the entire process of applying computer-based methodology, including new techniques for knowledge discovery from data (Kantardzic, 2003).

DM is a general term, which encompasses a number of techniques to pick out useful information from large data files and enables to sort it. It is a new powerful technology for analyzing data from different perspectives and summarizing it into useful information. It has been also said that DM is the analyses of (often large) observational data sets to find unsuspected relationships and to summarize the data in novel ways that are both understandable and useful to the data owner. Nowadays it is also very often known as Knowledge Discovery in Databases (KDD) or simply Knowledge Discovery (KD), which enables to identify trends within data that go beyond simple analysis and through the use of sophisticated algorithms; users have the ability to identify key attributes of business processes and target opportunities. DM is predicted to be "one of the most revolutionary developments of the next decade", according to the online technology magazine ZDNET News (February 8, 2001). In fact, the MIT Technology Review has chosen the DM as one of 10 emerging technologies that will change the world.

DM has been used widely by companies with a strong consumer focus like retail, financial, communication, and marketing organizations, where it enables to determine relationships among "internal" factors such as price, product positioning, and "external" factors such as economic indicators, competition, and customer demographics (Hand et al., 2001).

There are many different possibilities for using DM but the most frequent examples encompass mainly these applications:

- Rate customers by their propensity to respond to an offer.

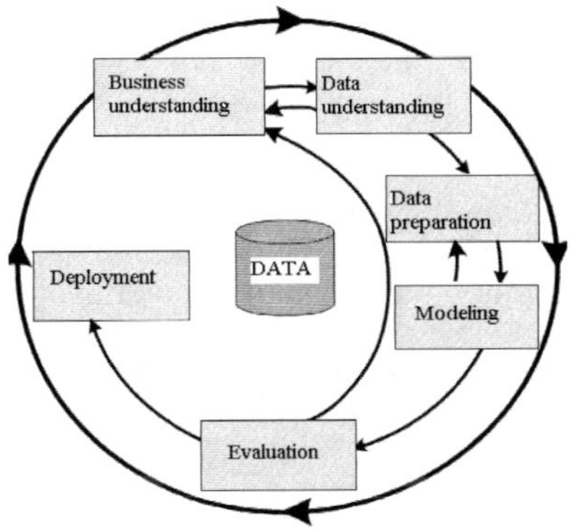

Fig. 1. Typical data mining process.

- Identify cross-sell opportunities.
- Detect fraud and abuse in insurance and finance.
- Estimate probability of an illness re-occurrence or hospital re-admission.
- Isolate root causes of an outcome in clinical studies.
- Determine optimal sets of parameters for a production line operation.
- Predict peak load of a network.

The general Data Mining Process (DMP) includes six phases that address the main issues in DM (Lentzsch, 2007). All these phases fit together in cyclical process and cover the full DM process (see Fig 1).

In fact, the knowledge discovery activity could become the key factor to innovation and business success (Kusiak & Smith, 2007). Therefore DM could have a great potential also in manufacturing.

Data mining implementations in production management and manufacturing is the key question of this paper.

2. DM IMPLEMENTATION IN PRODUCTION MANAGEMENT AND MANUFACTURING

2.1 Data mining in manufacturing

Products and components generate a data trail across lifecycle phases such as market analysis, design engineering, manufacturing, and service. DM algorithms extract knowledge from this large volume of data leading to significant improvements in the next generation of products and services. Integrating a DM framework within the manufacturing information system enables to improve manufacturing decision making process and enhance the productivity. It enables to analyze enterprises opportunities and employee's skills and competences, find relations between enterprises, customers and subcontractors, and make consequences based on different data conjunctions.

In order to implement the DMP to the production management and manufacturing sector, it is important to understand the business problem – what we would like to do with all that data and also to understand what that data is all about. It is obvious that all data need to be in an easily accessible format and available from one central database. It is often the case that relevant data files are stored in several locations and in different formats and need to be pulled together before analyses. The extracted information and knowledge can assist the engineers as their reference and basis for advanced investigation of the root causes of the defects (Larose, 2005). In current case study is used data about the enterprises, their technological capabilities, and employees' competences from three different databases: Metnet, Innomet and Innoclus (see Fig 2). For data analyzing the data understanding and preparation is unavoidable phases in DMP. Also there is need to assess the data for the DM project. The several aspects which should be considered are the followings:

2.2 Relevant factors covered by data

For making a DM project worthwhile, it is important that the data contain all relevant factors/variables and is mutually joinable. Therefore the data was separated into different tables according to logical themes. For example, one table includes all information about the enterprises contacts, other enterprises technological capabilities, etc. It is smart to hold the data in different tables, because it simplify the data understanding and facilitate later the DMP. In order to joint the data between these thematically separated tables, the common identifications (*ID-s*, for example: enterprise_id, sector_id etc) were worked out. These ID-s are needed for connecting one enterprise data from different tables.

In the figure 2 is pointed out in which tables all the data were divided and from which primary source the data is available from.

2.3 Handling noisy data

The term „noisy" in DM refers usually to errors in data or also sometimes to missing data (Hastie et al., 2001). In this study that is the problem we have to handle. It results from the data collection. All data about the enterprises capabilities were gathered without multiple-choices. Initially the question options were not defined and therefore every enterprise answered to the questions differently. In order to understand the answers unambiguously, the data synchronization was unavoidable. This solution for employee's skills and competences and enterprise technological capabilities was simpler, as the multiple-choices were worked out and enterprises answered to the questions by doing the suitable selections.

2.4 Gathering enough data

It is obvious that the more complex patterns and relationships we would like to find with data mining, the more records required to find them. There is self-evident difference, when we are analyzing ten, hundred or all Estonian machinery enterprises. It is important to point out, that in our case study all machinery enterprises have been included and that information has been gathered and will be used in different analyses by DM.

3. DM ANALYSES

When the data will be gathered into one central database and also would be structured and therefore easily under-standable, we could use the DM in many different applications. We could build up the models, which would be able to predict different important indications for better and more effective production management. For example, it could be possible to create the predictive DM model for investigating the competences, which could be needed for most effective product development. In addition it could be possible to classify enterprises for different clusters based on different technological capabilities and etc. Therefore the DM implementation is also effective in manufacturing sector and certainly is necessary for improving enterprises productivity and innovation in product development and manufacturing.

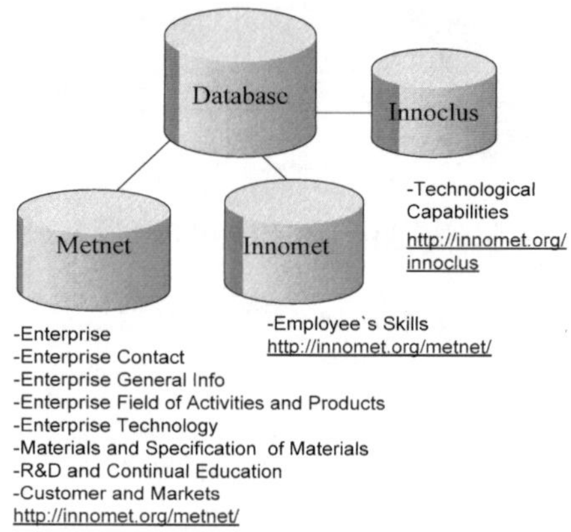

Fig. 2. Database background.

4. CONCLUSIONS

Data mining is a powerful tool, needed when amounts of data increase rapidly. In addition, it could be used also for complex analysis at a country level in sector of machinery, metal and apparatus engineering. The implementation of DM could be useful for analyzing and updating existing databases in a process of development collaborative e-Manufacturing information system. In addition, its implementation could give the powerful effect for machinery enterprises productivity and innovation in product development and manufacturing. Therefore the future research is targeted to increase the proactivity of the system. If we add data feeds from embedded systems reporting technological capability, the DM is one of the most promising methods to handle the information thus increasing productivity of management and innovation in the collaboration network.

After the main database will be created (for example based on three existing databases: Metnet, Innomet and Innoclus) and "noisy" data will be eliminated. The aim is to use all these gathered data about the enterprises and those technological capabilities and employees' skills for making and experimenting some DM models in order to increase these enterprises productivity and innovation in product development. On the other hand DM could be also useful from product improvement and repair process improvement perspectives to be able to determine the most frequent repairs by product, the factors that contribute to a failure type, and the correlations between failures.

5. ACKNOWLEDGEMENTS

The work has been supported by Estonian Science Foundation grant G6795.

6. REFERENCES

Hand, D.; Mannila, H.; Smyth, P. (2001). *Principles of Data Mining*, MIT Press, Cambridge, MA.

Hastie , T.; Tibshirani, R. & Friedman, J. H. (2001). *The elements of statistical learning: Data mining, inference, and prediction*. New York: Springer.

Kantardzic, M. (2003). *Data Mining: Concepts, Models, Methods, and Algorithms*. John Wiley & Sons.

Kusiak, A.; Smith, M. (2007). *Data mining in design of products and production systems, Annual Reviews in Control, 31*. 147–156 A.

Larose Daniel T. (2006). *Data Mining Methods and Models*. John Wiley & Sons Inc., United States of America.

Lentzsch, K. (2007). *Introduction to Clementine and Data Mining*. SPSS Inc.

Annals of DAAAM for 2008 & Proceedings of the 19th International DAAAM Symposium, ISSN 1726-9679
Editor B. Katalinic, Published by DAAAM International, Vienna, Austria 2008
Make Harmony Between Technology and Nature, and Your Mind will Fly Free as a Bird

NEURAL NETWORK PREDICTION OF AN OPTIMUM SHIP SCREW PROPELLER

MATULJA, D[unja] & DEJHALLA, R[oko]

Abstract: *A neural network (NN) has been trained to enable the choice of an optimum ship screw propeller geometry. The idea was to avoid the mistakes that can occur in the use of open water diagrams, as well as to save time required to run various computer programs. The network provides the optimum diameter, pitch ratio and thrust for a given case of delivered power, propeller revolution, advance velocity, blade number and expanded area ratio, within the treated design range. The necessary training data were obtained using the polynomials representing the Wageningen B-screw series. The NN was used as a black box, and the data were approximated to a multiple variable function. After the training, the NN was tested with satisfactory results, especially for diameter and thrust. In addition to that, the processing speed is not affected by the number of cases to compute. This makes the neural network a reliable tool, fit to predict the principal geometric features of an optimum propeller.*
Key words: *ship screw propeller, optimization, neural network*

1. INTRODUCTION

The choice of an optimum ship screw propeller is a persistent problem in naval architecture. The selection in the preliminary design stage is usually based upon some of the standard series data. The Wageningen B-screw series is considered to be the most extensive and the open water diagrams were reduced to a polynomial form, applying regression analysis (Oosterveld &Van Oossanen, 1975). The next step, treated in this paper, is the generation of a neural network (NN) able to predict the geometry parameters of an optimum propeller. Once a consistent database has been created, the network has been trained and tested within the operational range. Although the NN proved to be efficient, it can be improved by extending the training database. The optimization of the network structure will be the subject of subsequent researches.

2. WAGENINGEN B-SCREW SERIES

The choice of the optimum propeller is based upon the open water test data of the B-series propellers, which gives non-dimensional results for thrust and torque (Carlton, 1994). The thrust coefficient K_T and the torque coefficient K_Q are plotted against the advance ratio J.

Applying regression analysis, (Oosterveld &Van Oossanen, 1975), the open water diagrams of the B-series propellers can be reduced to a polynomial form, for a Reynolds number of $2 \cdot 10^6$:

$$K_T = \sum_{n=1}^{39} C_n (J)^{s_n} (P/D)^{t_n} (A_E/A_0)^{u_n} (Z)^{v_n} \quad (1)$$

$$K_Q = \sum_{n=1}^{47} C_n (J)^{s_n} (P/D)^{t_n} (A_E/A_0)^{u_n} (Z)^{v_n} \quad (2)$$

To extend the calculation on Reynolds numbers up to $2 \cdot 10^9$, the following corrections have been included:

$$K_T(R_n) = K_T(R_n = 2 \cdot 10^6) + \Delta K_T \quad (3)$$

$$K_Q(R_n) = K_Q(R_n = 2 \cdot 10^6) + \Delta K_Q \quad (4)$$

Corrections ΔK_T and ΔK_Q were also taken from Oosterveld &Van Oossanen, 1975. Using these polynomials, calculations have been made to obtain data necessary to train the NN. For different propeller revolutions N and delivered power P_D, variations of advanced velocity V_A, expanded area ratio A_E/A_0 and blade number Z have been considered. In this manner the geometry parameters (diameter D and pitch ratio P/D) of the optimum propeller for each case have been calculated, as well as thrust T.

In this case, an optimum propeller implies the one with the highest efficiency, without considering the limitations imposed by the ship's hull form.

3. NEURAL NETWORKS

Artificial neural networks are developed as models of neural processing in the brain. They involve networks of simple processing elements (artificial neurons) which can exhibit complex global behaviour, determined by the connections between the processing elements and element parameters. A neuron is an information-processing unit that is fundamental to the operations of a neural network (Haykin, 2005).

The networks have the possibility of learning, which means that after having them trained, the networks will be able to solve a given task in the optimal sense.

The cost function C is an important concept in learning, as it is a measure of how close are the obtained and the optimal solution. Learning algorithms search through the solution space (back propagation) in order to find a function that has the smallest possible cost. The most commonly used cost is the mean-squared error. The example of a neural network learning scheme is shown in Fig. 1.

The NeuroSolution software (2008 NeuroDimension, Inc.) has been applied. It offers different possibilities of neural network generation. The NeuralExpert mode selects the neural network size and architecture that will likely produce a good solution, considering the problem type (Classification, Prediction, Function approximation or Clustering) and the size of the user's data set, but it allows even some more advanced operations, such as changing the network structure properties, cross validation and genetic optimization. The NeuralBuilder mode centers the design specifications on the specific neural network architecture the user wishes to build.

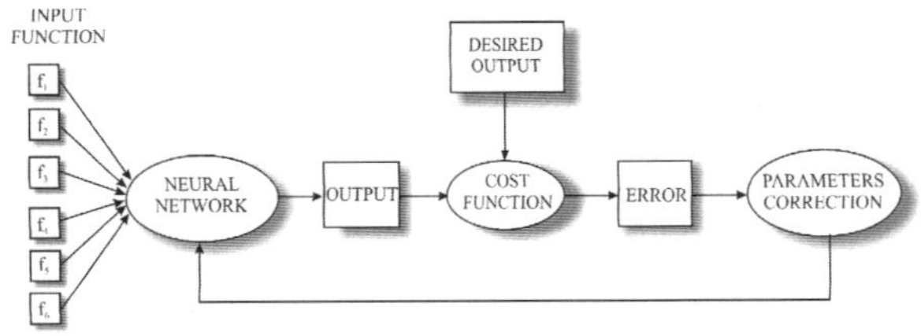

Fig. 1. Example of a neural network learning scheme.

4. NN TRAINING METHODOLOGY

The NeuralExpert mode of the NeuroSolutions software has been applied in this case, and the function approximation network type has been tried, as well as the prediction type.Since some sample data are required to train the network, calculations have been done using the program based on the B-series polynomials, and a database containing optimum propeller parameters was created. The covered data range is reported in Table 1. The values P_D, N, V_A, Z and A_E/A_0 have been selected as input data, while D, P/D and T were chosen as desired output. Once the network is trained, these quantities will be the parameters to be evaluated. Considering the number of the chosen variables, the number of data required for successful learning had to be quite large. The final database contained 555 sets of input and desired output data.

The training was completed for the network types of function approximation and prediction in 1000 epochs. At first the mean squared error curve for cross validation indicated an increasing trend with the passing of epochs, instead of decreasing. An assumption was made that the random setting of input and output data had caused that issue, so the data were roughly sorted ascending considering the delivered power and thrust. The function approximation network type has been chosen as the most appropriate, since the prediction type would have requested much more data for accurate results in this case. With this ultimate solution the training was performed again, with satisfactory results.

Parameter	Design Range
N, min^{-1}	58.5 ~ 199
P_D, kW	3215 ~ 54570
A_E/A_0	0.40 ~ 1.10
Z	3 ~ 7
D, m	3.3 ~ 11.5
P/D	0.5 ~ 1.4

Tab. 1. NN parameter search space for optimum B-series propeller design.

5. RESULTS

The results of the training are best presented by the learning curves in Fig. 2 and Fig. 3. Comparing the Output and Desired Curves for D and P/D, a fair agreement can be noticed. The mean squared error of training and cross validation is acceptably low. This error could be additionally reduced by extending the training database, or by changing some of the network's structure properties (Taylor, 2006).

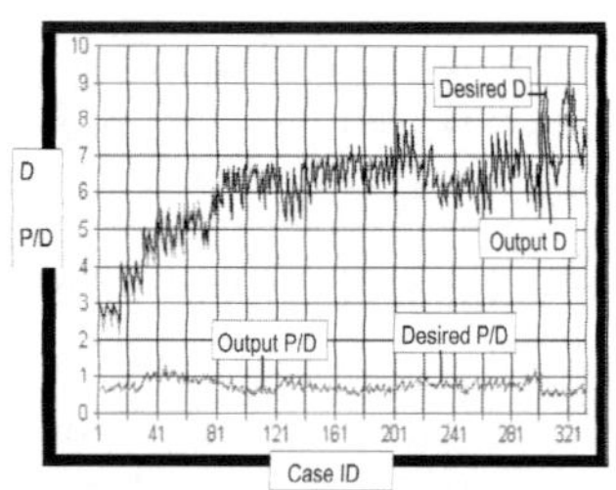

Fig. 2. Output vs. Desired: Diameter and pitch ratio

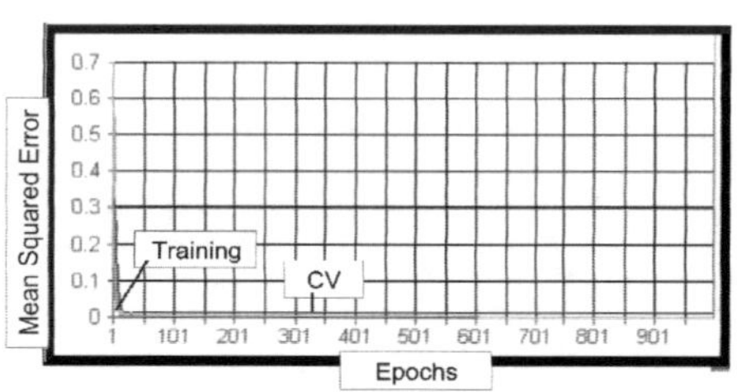

Fig. 3. Learning curve

Inputs					B-series results		
Z	A_E/A_0	V_A, m/s	N, min^{-1}	P_D, kW	D, m	P/D	T, kN
6	0.61	6.50	88.0	11867.0	7.265	0.889	1135.34
4	0.50	5.50	112.0	6530.0	6.269	0.683	717.89
5	0.85	12.95	159.0	19975.0	5.749	1.061	1236.22

Tab. 2. B-series inputs and outputs for NN testing.

Out D	Out PitchR	Out T
7.232412702569	0.848404609159	1166.170313506178
6.350609476750	0.697525870175	719.701651486002
5.824777936648	1.143807301086	1261.010084349461

Fig. 4. NN outputs.

The network can be tested by giving some new input values, in order to obtain the matching outputs. The training inputs can not be used for testing.

The testing can be performed on an arbitrary number of samples. Three sets of data have been randomly chosen for the testing within the parameter search space. They are indicated as inputs in Table 2. For comparison, the optimal propeller characteristics have been calculated using the B-series computer program. These results are reported in Table 2, while the NN outputs are shown in Fig. 4.

Comparing the calculated results to the outputs estimated by the NN, a coincidence between the values can be noticed. Although no acceptable error value has been defined, the variations are within 1.3% for D, 7.8 % for P/D and 3% for T. This means that the diameter difference is within a few centimeters, which is precise enough for a preliminary design stage, considering the treated data range.

6. CONCLUSION

The advantage of using the NeuroSolutions NN to predict the optimum ship screw propeller is the processing speed, once the database requested for training is prepared. So, if the geometry of a single propeller is to be evaluated, the B-series calculation might be a better solution. But if more evaluations have to be performed, the skills of the neural network obviously prevail, as the time needed for the B-series calculations increases with the number of calculations. The achieved level of accuracy provides reliable results, which makes the network suitable for quick processing of a large amount of data, as well as for an approximate estimation of optimum propeller geometry.

Since the generated NN can be additionally improved, as the next step of research in this area the treated data range will be extended to allow the choice of the optimum propeller for any boat, without power or dimension restrictions in the software.

7. REFERENCES

Carlton, J.S. (1994). *Marine Propellers and Propulsion*, Butterworth-Heinemann Ltd, ISBN 0 7506 1143 X, Oxford

Haykin, S. (2005). *Neural Networks, A Comprehensive Foundation*, Second Edition, Ninth Indian Reprint, Pearson Education, Inc., ISBN 81-7808-300-0, Singapore

Oosterveld, M. W. C. & Van Oossanen, P. (1975). *Further Computer-Analyzed Data of the Wageningen B-Screw Series, International Shipbuilding Progress*, Vol. 22, pp. 251-262

Taylor, B. J. (2006). *Methods and Procedures for the Verification and Validation of Artificial Neural Networks*, Springer Science + Business Media, Inc., ISBN-13: 978-0-387-28288-2, ISBN-10: 0-387-28288-2, Fairmont, USA

NeuroSolutions, The Neural Network Software, 2008 NeuroDimension, Inc., http://www.neurosolutions.com *Accessed:* 2008-05-26.

Annals of DAAAM for 2008 & Proceedings of the 19th International DAAAM Symposium, ISSN 1726-9679
Editor B. Katalinic, Published by DAAAM International, Vienna, Austria 2008
Make Harmony Between Technology and Nature, and Your Mind will Fly Free as a Bird

ROBUST STABILIZATION OF INTERVAL PLANTS USING PI CONTROLLERS

MATUSU, R[adek]

Abstract: *The primary requirement in all control applications is the stability of the closed-loop system. However, if the controlled object is supposed to be described as an interval plant family, the situation can be more complicated, because the closed-loop stability must be ensured for all possible combinations of uncertain parameters. In this case, so-called robust stability is the issue of interest. This contribution deals with robust stabilization of interval plants using fixed proportional-integral (PI) controllers while the determination of the stability regions for parameters of these controllers are based on the combination of computing/plotting the stability boundary locus and on the sixteen plant theorem.*

Key words: *Control, Robust Stabilization, Interval Plants.*

1. INTRODUCTION

The mathematical model containing interval parameters is a common tool for description of many industrial processes. This approach helps to compensate the simplifications made during modelling, imprecise knowledge of plant parameters or its variability. However, an interval system itself is only the beginning. The important step consists in designing the controller, which ensures some desired properties of the control loop for the whole interval plant family. And definitely the fundamental requirement of all users is the robust stability of the feedback control system. Despite the existence of many advanced control technologies, the contemporary industrial practice clearly prefers the use of standard PI and PID controllers with fixed parameters and so an easy and effective way of PI/PID tuning is still very topical, especially in case that these algorithms are able to cope with various uncertain conditions.

The main aim of this paper is to present a method of determination of robustly stabilizing PI controllers with fixed parameters for interval systems (Tan & Kaya, 2003) and to demonstrate its capabilities on an example for third order interval plant.

2. COMPUTATION OF STABILIZING PI CONTROLLERS

A possible approach to calculation of stabilizing PI controllers based on plotting the stability boundary locus is proposed in (Tan & Kaya, 2003). The method supposes the classical closed-loop control system with controlled plant:

$$G(s) = \frac{B(s)}{A(s)} \tag{1}$$

and PI controller:

$$C(s) = k_P + \frac{k_I}{s} = \frac{k_P s + k_I}{s} \tag{2}$$

First, one needs to use the substitution $s = j\omega$ in the plant (1) and subsequently to decompose the numerator and denominator of this transfer function into their even and odd parts:

$$G(j\omega) = \frac{B_E(-\omega^2) + j\omega B_O(-\omega^2)}{A_E(-\omega^2) + j\omega A_O(-\omega^2)} \tag{3}$$

Then, the expression of closed-loop characteristic polynomial and setting the real and imaginary parts to zero lead to the equations:

$$\begin{aligned} k_P &= \frac{\omega^2 A_O(-\omega^2) B_O(-\omega^2) - \left(-A_E(-\omega^2)\right) B_E(-\omega^2)}{-\omega^2 B_O^2(-\omega^2) - B_E^2(-\omega^2)} \\ k_I &= \frac{\left(-A_E(-\omega^2)\right)\left(-\omega^2 B_O(-\omega^2)\right) - \omega^2 A_O(-\omega^2) B_E(-\omega^2)}{-\omega^2 B_O^2(-\omega^2) - B_E^2(-\omega^2)} \end{aligned} \tag{4}$$

Simultaneous solving of these relations and plotting the obtained values into the (k_P, k_I) plane result in the stability boundary locus, which splits the (k_P, k_I) plane up to the stable and unstable regions. The determination of the stabilizing one(s) can be done via a test point within each region. Furthermore, this technique can be embellished with the Nyquist plot based approach from (Söylemez *et al.*, 2003) to avoid potential problems with proper frequency gridding. In this refinement, the frequency axis can be divided into several intervals by the real values of ω which fulfill:

$$\mathrm{Im}\left[G(s)\right] = 0 \tag{5}$$

Such intervals are then sufficient for testing.

3. IMPROVEMENT OF THE METHOD FOR INTERVAL SYSTEMS

So far, the area of stabilizing controller coefficients for a given plant with only fixed parameters can be computed. However, the paper (Tan & Kaya, 2003) has improved the stabilization also for interval plants using the simple idea of its combination with the sixteen plant theorem (Barmish *et al.*, 1992), (Barmish, 1994). In compliance with this principle, a first order controller robustly stabilizes an interval plant

$$G(s,b,a) = \frac{B(s,b)}{A(s,a)} = \frac{\sum_{i=0}^{m}\left[b_i^-, b_i^+\right] s^i}{s^n + \sum_{i=0}^{n-1}\left[a_i^-, a_i^+\right] s^i}; \quad m < n \tag{6}$$

where $b_i^-, b_i^+, a_i^-, a_i^+$ are lower and upper bounds for numerator and denominator parameters, respectively, if and only if it stabilizes its 16 Kharitonov plants, which are defined as:

$$G_{i_1,i_2}(s) = \frac{B_{i_1}(s)}{A_{i_2}(s)} \tag{7}$$

where $i_1, i_2 \in \{1,2,3,4\}$; and $B_1(s)$ to $B_4(s)$ and $A_1(s)$ to $A_4(s)$ are the Kharitonov polynomials for the numerator and denominator of the interval system (6), respectively.

Remind that the Kharitonov polynomials e.g. for an interval polynomial:

$$B(s,b) = \sum_{i=0}^{m} \left[b_i^-; b_i^+ \right] s^i \tag{8}$$

can be constructed using the upper and lower bounds of interval parameters according to the rule (Kharitonov, 1978):

$$\begin{aligned} B_1(s) &= b_0^- + b_1^- s + b_2^+ s^2 + b_3^+ s^3 + \cdots \\ B_2(s) &= b_0^+ + b_1^+ s + b_2^- s^2 + b_3^- s^3 + \cdots \\ B_3(s) &= b_0^+ + b_1^- s + b_2^- s^2 + b_3^+ s^3 + \cdots \\ B_4(s) &= b_0^- + b_1^+ s + b_2^+ s^2 + b_3^- s^3 + \cdots \end{aligned} \tag{9}$$

The stabilization of an interval plant is grounded in the stabilization of all 16 fixed Kharitonov plants together, and so the final stability region is given by intersection of all partial regions.

4. AN ILLUSTRATIVE EXAMPLE

Suppose that controlled process is described by interval transfer function:

$$G(s) = \frac{[1; 2]}{s^3 + [3; 4]s^2 + [5; 6]s + [7; 8]} \tag{10}$$

and the objective is to find all possible robustly stabilizing PI controllers.

First, consider e.g.:

$$G_{1,1}(s) = \frac{B_1(s)}{A_1(s)} = \frac{1}{s^3 + 4s^2 + 5s + 7} \tag{11}$$

as the first of 16 Kharitonov plants. The equation (4) here takes the concrete form:

$$\begin{aligned} k_P &= 4\omega^2 - 7 \\ k_I &= -\omega^4 + 5\omega^2 \end{aligned} \tag{12}$$

Using of (5) and consequent stability test for two obtained intervals lead to the range of the frequency $\omega \in (0; 2.236)$, which is necessary for computing/plotting the stability boundary locus. The analogical procedure has been done generally for all 16 Kharitonov plants. However, in this specific case, the locus of only 8 systems is enough to investigate, because the nominator of (10) takes only two extreme values and the construction of Kharitonov polynomials would be redundant here.

The fig. 1 provides the graphical representation of the stability boundary locus for 8 Kharitonov plants, while the fig. 2 brings closer look to the intersection, which constitutes final stability region for the interval plant (10). An arbitrary pair of (k_P, k_I) from inside of this stability region would entail robustly stable control system.

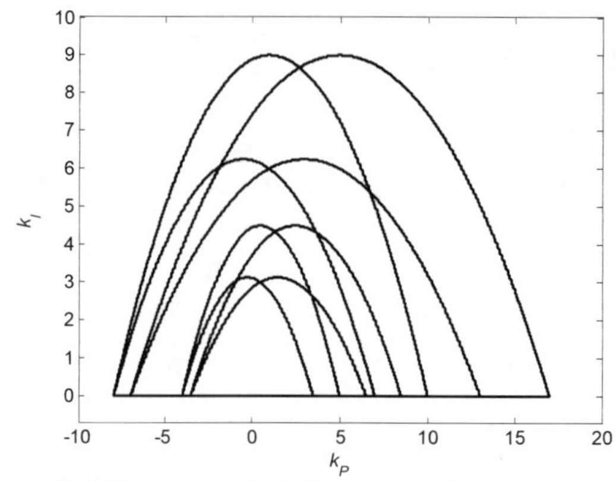

Fig. 1. Stability regions for 8 Kharitonov plants

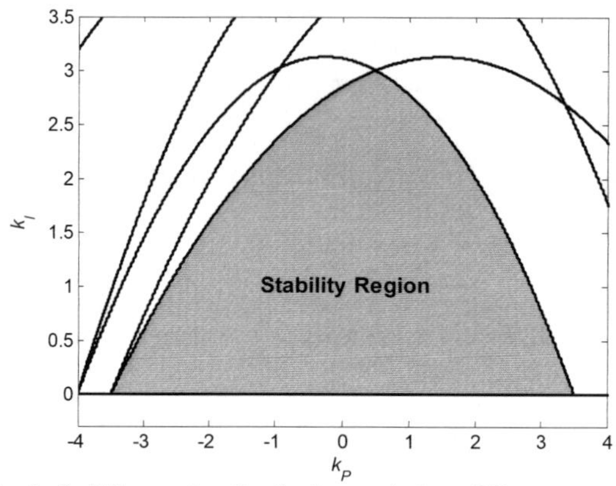

Fig. 2. Stability region for the interval plant (10)

5. CONCLUSION

The contribution has presented a possible robust stabilization technique for interval systems using PI controllers with fixed parameters. The third order interval plant has been successfully stabilized in the illustrative example.

6. ACKNOWLEDGEMENT

The work was supported by the Ministry of Education, Youth and Sports of the Czech Republic under Research Plan No. MSM 7088352102. This aid is gratefully acknowledged.

7. REFERENCES

Barmish, B. R. (1994). *New Tools for Robustness of Linear Systems*, Macmillan, New York, USA.

Barmish, B. R.; Hollot, C. V.; Kraus, F. J. & Tempo, R. (1992). Extreme point results for robust stabilization of interval plants with first order compensators. *IEEE Transactions on Automatic Control*, Vol. AC-37, pp. 707-714.

Kharitonov, V. L. (1978). Asymptotic stability of an equilibrium position of a family of systems of linear differential equations. *Differentsial'nye Uravneniya*, Vol. 14, pp. 2086-2088.

Söylemez, M. T.; Munro, N. & Baki, H. (2003). Fast calculation of stabilizing PID controllers. *Automatica*, Vol. 39, No. 1, pp. 121-126.

Tan, N. & Kaya, I. (2003). Computation of stabilizing PI controllers for interval systems, In: *Proceedings of the 11th Mediterranean Conference on Control and Automation*, Rhodes, Greece.

Annals of DAAAM for 2008 & Proceedings of the 19th International DAAAM Symposium, ISSN 1726-9679
Editor B. Katalinic, Published by DAAAM International, Vienna, Austria 2008
Make Harmony Between Technology and Nature, and Your Mind will Fly Free as a Bird

NEW IMPLEMENTATION OF THE MOBILE ORDERING SYSTEM

MATYSEK, M[iroslav]; ADAMEK, M[ilan]; NEUMANN, P[etr] & MATULIK, T[omas]

Abstract: *The work goal was to create an effective tool for supplying companies sales representatives to make their activities easier and more effective. It represents the communication improvement and acceleration between the branch and the salesman, the limiting of human factor influence on the order processing regarding errors, the acceleration of order processing, live offer related to the stock inventory, and last but not least, the quicker information feed for the customers and for salesmen.*
Key words: *ordering system, PDA, smartphone, GSM*

1. INTRODUCTION

The Razat Style Ltd. products distribution is carried out by menus of the ROE method (Remote Order Entry). Sales representatives visit their customers and generate new orders which are transferred to the central office then. There takes place the order processing in the stock information system and goods forwarding to the customers. Sales representatives do not use any computing device. They exclusively use the "pen and paper" method. There is no unified products catalogue according that representatives would offer goods. The web pages as an alternative offer for pertinent parties concerned are man-made and its updating is depending therefore on the human factor. The proposed solution is not a replacement of the current information system (onward abbreviated as IS) but exclusively its superstructure implementing the new way of entering and processing orders. The proposed system will be mentioned as Mobile Ordering System (MOS) thereinafter. Data are moved ahead to current IS for further processing according to the current procedures via data interface. The proposed system should replace the current "pen and paper" way of accepting orders.

2. BASIC RESOURCES

2.1 Operational System Pocket PC 2002

Comparing the user interface graphic design between older operational systems Windows CE and Pocket PC 2002, we can relate the comparison to the difference between Windows 98 and Windows XP (Trnečka, 2003). The similar comparison would result from encapsulated technologies here the differences are not so evident at a glance but they are all the more significant. The new operational system is much more sophisticated in terms of graphics and in color what will contribute to the better control and the user comfort. The new Microsoft technology namely provides the users with more technological possibilities especially in wireless connection in the range from local networks (802.11a,b,g) and personal networks (Bluetooth) till remote networks (CDPD, CDMA, GSM or by means of duplex solutions using existing operator networks). The Pocket PC 2002 accompanying software includes also the new Server ActiveSync technology which enables the direct server synchronization of mail, calendar and contact between Pocket PC 2002 based equipment and Exchange 2000 servers. The ActiveSync Server is supplied as a Microsoft Mobile Information 2002 Server component and its design support the fixed and wireless interconnections.

2.2 Communication Tool Compaq iPAQ 3870

Supplied Software: Microsoft Mobile Pocket PC 2002, iPAQ Task Manager, IBM via voice command and control, iPresenter, PowerPoint Converter for Pocket PC, Dashboard Encryption, Java VM, eWallet and iPAQ Reference Guide. Communication Possibilities: Integrated Infrared IrDA and Bluetooth technology.

2.3 Communication Tool HP smartphone iPAQ 510

Supplied Software: Windows Mobile 6.0, Windows Media Player 10, Internet Explorer, Outlook/Mobile Office with Word Mobile, Excel Mobile and PowerPoint Mobile. Communication Possibilities: Integrated quad-band GSM/EDGE, WiFi 802.11b/g and Bluetooth technology.

2.4 Development Tools

Microsoft® eMbedded Visual Tools 3.0 is a complete development environment for applications and software components for platforms marked as Windows Powered, for instance for the class Pocket PC devices. The successor of that development environment is Microsoft eMbedded Visual C++ 4.0. The big advantage is represented with eMbedded Visual Tools 3.0 autonomy what means that we do not need any other development environment, for Visual Studio or the newest Visual Studio .NET version. The component part of eMbedded Visual Tools 3.0 installation are Software Development Kit supporting modules aimed at the device we are developing the application for. We keep the modules for Pocket PC and Handheld PC at our disposal. It is possible to extend the installation with the modules for the platform Smartphone 2002 and Pocket PC 2002. These modules are supplied in the form of SDK (Software Development Kit).

We can separate the eMbedded Visual Tools 3.0 development environment in two almost independent blocks:

- eMbedded Visual C++ 3.0,
- eMbedded Visual Basic 3.0.

3. COMMUNICATION SYSTEM

3.1 Application Server

The application server is built up on the Linux platform and all its software accessories are free of charge. We have chosen that policy because of MOS implementation cost reduction. The other reason for such policy was the fact that individual users are not supposed to access application server so that the user-friendliness of the operational system and application program is not so important. The whole user side of the server servicing is solved via web interface. The user can access it from any computer connected to the Internet.

We have selected the MySQL database system for data holding on the application server (Lacko a, 2003). It is a wide-spread data server above all in the area of web applications and it is fully adequate for our needs.

The application itself is written in the PHP 4 programming language which is frequently used for such application development and moreover, it is free of charge even for the commercial use.

3.2 Client Terminals

The reason for the Pocket PC 2002 platform selection was its simple intuitive operating, its effortless accessibility and its consequent extensibility. We also had a respect to a fact that the Razat Style Ltd. Company uses the Microsoft operational systems so that its sales force does not need to get into a new user interface. Last but not least argument for our decision making was the experience and abilities of the realization team and the platform program set-out.

The programming tool eMbedded Visual Tools 3.0 was chosen for the MOS client part programming (Krčmář, 2003). This comprehensive tool offers enough means for the MOS client part realization. As mentioned hereinbefore, it includes two programming languages each of them designated for different area of application. eMbedded Visual Basic 3.0 did serve as a very quick tool for the user interface creation and for establishing accesses to the local database. Nevertheless as revealed itself during the project processing, it has its disadvantage in speed of application when dealing with higher data volumes, for instance during searching the catalogue or its listing on the screen. That problem was solved with the help of individual program code parts optimization and reducing the set of processed data in one moment (Lacko b, 2003).

We also used the language eMbedded Visual C++ 3.0 namely for the creation of quick functions not included in Visual Basic and for the module for data synchronization with the application server. That module is not a part of client terminals basic application. The external application triggered from the main application and running independently on it serves for that purpose.

4. PRACTICAL SOLUTION

4.1 Application Server

It was necessary to create a separate data storage area for MOS because the current IS is a locked program system without any intervention possibility for the development team. There was composed a store mechanism for remote access in the data storage area from mobile terminals for bidirectional data synchronization – recording the up-to-date catalogue and clients address list in terminal and transfer of new orders and new clients from the terminal to the application server. There is a "front-end" component with simple intuitive interface (web application) at the application server. This component facilitates process with stored data. The stored data process means mainly orders browsing, new clients added to the system by the sales representative approval, review, summary report and statistics generation, user accounts administration, catalogue browsing and relevant picture to the catalogue items allocation. Only users with official account on the application server have access to that application. The authorization process asks for user name and password. Each user has his particular role influencing his rights according to the data access. The sales representative creates in this application his personalized catalogue with his own description for individual items. Data access is possible from any computer with web browser installed and with the Internet connected.

4.2 Client Terminals

Each member of sales force is equipped with a pocket computer of class Pocket PC with relevant user software and with a mobile phone supporting the fast data transfer based on the GPRS technology. There is an actual branch stock state stored in the client terminal. The branch is related to its client address book and state is related to the time of terminal record. The sales representative has possibility to update the stock state and client address book via a secure connection to the application server or directly via intranet at the branch site at any time. The orders are recorded directly in the PDA and then locally stored. Only one order at most can be related to the particular client from the address book. That order can be in the semi-finished state for later editing. The finished orders are marked as closed and it is not possible to edit them any more. Any sale representative can send new created orders for processing to the application server from anywhere and at any time providing there is the GSM signal coverage. Only finished orders no semi-finished orders are sent to the server. The successfully transferred orders are deleted from the terminal. The application offers the addition of a new client in the address book. The information about new clients is sent together with new orders related to those clients. Pocket PC serves also as an offer catalogue for presentation at the client site. The application facilitates the items searching according to various criteria and also creation of filtered catalogues. The application offers also the possibility to download the relevant picture to the particular item from the application server. There is an export function for the catalogue in various data format (CSV, XML and HTML) in the software as well and it makes it possible to store the exported catalogue in the client computer (Bradley, 2000). The application start in the PDA is secured with the access password.

4.3 Data Converter

There was necessary to create a component "Data Converter" as an extension of IS which ensures the data exchange between the current IS and MOS. The data transfer from IS to MOS includes the stock actual state export, customer address book export and accepted and processed orders export. The export is limited only on data relevant for orders creation. The data transfer from MOS to IS includes the insertion of newly accepted orders from sales representatives and from new clients.

5. CONCLUSION

There was a remarkable increase in the accuracy and speed of orders processing after the MOS system implementation. The big asset of the new MOS system is an immediate possibility for a sales representative to find out the actual goods stock state and accommodate the count and price for individual article. That MOS system quality represents the financial advantage not only for the Razat Style Company but even for their customers. It has also proved that the PDA article presentation saves often time and money consuming paper catalogue printing and distribution in case of improvements and innovations. With the new MOS system, the innovations can reach the customer much earlier and for the more reasonable price.

6. ACKNOWLEDGEMENTS

This work was supported by the Ministry of Education of the Czech Republic in the range of research projects No. MSM 7088352102.

7. REFERENCES

Bradley, N. (2000). *XML - complete guide,* GRADA, ISBN 80-7169-949-7, Praha.

Krčmář, D. (2003). *Programming .NET applications in Visual studio,* Computer Press, ISBN 80-7226-569-5, Brno.

Lacko, L. a (2003). *SQL - Finished solutions,* Computer Press, ISBN 80-7226-975-5, Brno.

Lacko, L. b (2003). Development applications for mobile arrangements, *Available from:* http://download.microsoft.com/download/8/6/c/86c09926-affc-4e14-bec0-3c45cd989436/Vyvoj_aplikacii_pre_mobilne_zariadenia.pdf *Accessed:* 2008-06-11

Trnečka, I. (2003). *PDA pocket computers for everyone,* Computer press, ISBN 80-86593-15-0, Brno.

Annals of DAAAM for 2008 & Proceedings of the 19th International DAAAM Symposium, ISSN 1726-9679
Editor B. Katalinic, Published by DAAAM International, Vienna, Austria 2008
Make Harmony Between Technology and Nature, and Your Mind will Fly Free as a Bird

APPROACHING THE MODELLING OF A HUMAN RESOURCES SELECTION INFORMATION SYSTEM FROM THE PERSPECTIVE OF ITS USERS

MAZILESCU, C[risanta] A[lina]; MIHARTESCU, A[na] A[ndreea] & BANCIU, F[elicia]

Abstract: *The objective of this study is the modeling of a personnel selection information system from the perspective of its users, by using UML (Unified Modeling Language) as modelling language. The functioning of system involves the successive implementation of several activities: post-analysis, personnel recruitment, sorting applications, evaluation of knowledge and capacity of candidates, selection interview, etc. Study results are out in a theoretical proposal of an information system for selecting personnel. The advantage of using UML is the possibility of cooperation between two specialists: the computer scientist, who will design and implement the system and the specialist in the field of application, who can build the exterior image of the system, from the perspective of its users.*
Key words: *modelling, information system, personnel selection*

1. INTRODUCTION

In the present economic environment, the study and elaboration of a personnel selection program, as well as its application within an organization, is a means of stimulating the performances and the efficiency of organizations, and also a basic tool for their change and development.

An information system (IS) is a work system made of people, materials, technology and information, which is based on information processing (selection, storage, analysis and communication) in order to achieve the organization's objectives. Porter and Weick were the first to discuss the issue of developing information systems correlated with the organization's needs (Porter, 1985). Henry C. Lucas has emphasized the importance of the personal parameters and of the users' attitude when it came to achieving the performances of an information system, and R.W. Zmud has studied the level of the success of implementing an IS in organizations, setting the bases for the studies regarding the impact of an IS on the user human resource (Dumoulin, 1986).

In what regards the I.S. typology, Alter (1999) has identified 8 categories (office, communication, processing transaction-related information, etc.), while R. Reix (2002) suggests another classification of the IS, based on their functionality (trade management, production, accounting and financial management, and human resources). J.M. Peretti (1994), as a result of a study conducted in France, has identified 5-6 information applications in the human resource field. Although the computerization has had a rapid evolution within organizations, the functional applications in human resources management focuses on budget control (Weiss, 2003).

The elaboration of a model implies the simplification of reality, allowing a better understanding of the system that is to be developed, visualizing the system as it is or as it should be, specifying the data and behaviour structures and supplying a guide for building the system. The modelling of an I.S. may be approached from several perspectives: either by appealing to the designated scientific methods and following a strict succession of stages, either by appealing to empirical methods which explore new techniques with the purpose of modelling the system (Reix, 2002).

The objective of this study is the modeling of a human resources selection information system from the perspective of its users, employing UML (Unified Modelling Language) as the modelling language.

2. MODELING AN I.S. WITH THE HELP OF UML

UML is not a method, but an object-oriented modeling language, used mostly in the banking field and for financial services, in the field of telecommunications, transportation, but also for information systems within enterprises. UML-based modelling means to approach the system from several perspectives (4+1 Kruchten views), the use case perspective being the starting point and the direction we shall approach in this study. UML-based modelling, from the perspective of the activities it entails, is based on the cooperation between two specialists: a specialist in the activity field for which the information system is designed (in our case, the human resources management specialist), who must build the exterior image of the system, from the users' point of view, and a computer specialist, who must design and implement the system.

This paper shall approach the modeling of an information system for personnel selection from the perspective of the human resources specialist who translates the models resulted from the user's expressed needs in a language similar to that of the computer scientist's, but maintaining, at the same time, a language understandable by the clients and users of the system.

A use case is a means of representing the various possibilities to use the system and to emphasize one of its functionalities, usable by an actor. A diagram of use cases defines the system, the actors, the use cases and the connections between the actors and the use cases. A model of the use cases is defined by means of use cases diagrams, by textual description of the use scenarios and by describing these scenarios through sequence and collaboration diagrams.

The *actors* are persons or peripherals fulfilling a role or occupying a position in the functioning of the system. In the case of the selection system, the persons benefiting from the functioning of the system, who supply data to the system or who use the information system, are: the personnel and the manager of the human resources department, the enterprise manager, the candidates taking part in the selection, and even the personnel of the enterprise;

Identifying the use cases within the selection system

A use case is an action sequence, when the actor interacts with the system in order to achieve an aim. In what regards the selection system, we have identified the following use cases:

1. **the case of the job analysis**, when the purpose is to establish the aptitudes, skills, capacities and knowledge needed by the job holder. The job analysis aims at collecting, structuring and checking the information connected to a job position in order to identify the job requirements and specifications.
2. **the case of establishing the performance criteria and choosing the prediction tools**, when aiming to identify the professional efficiency criteria and choosing those test based on

which one can predict the future job performance of the candidate. Based on the information obtained after analysing the job position, one chooses those tests based on which predictions regarding the future job performance of the candidate can be made. The predictors are the psychological tests chosen and used for assessing the capacities of the persons investigated. This stage, when the performance criteria are identified and the predictors are chosen, is necessary for the further validation of the selection program.

3. **the case of the recruitment**, when the aim is to sort candidacies based on biographical data. Sorting candidacies aims to achieve an initial selection of the candidates, based on the analysis of their biographical information, of their letters of intent and their letters of recommendation.

4. **the case of the assessment** – aims at assessing the candidates' capacities and knowledge. Assessing field knowledge and measuring the candidates' capacities is a stage that aims at investigating the candidates' personality in order to identify the aptitudes and capacities needed to fulfil their job tasks. In order to increase the objectivity of the assessment results, the personality investigation is conducted by means of several psychological tests, which either supplement one another, either jointly confirm the results obtained.

5. **the case of the interviewing** – the case of the selection interview aims both at completing and verifying the information on the candidate obtained in the previous stages, as well as identifying the candidate's motivation level.

6. **the case of the decision to employ**, when it is decided whether to reject or hire the candidate. The selection of the candidates is the result of a decision made by the organization's manager according to the biographical information, according to the results of the capacity and knowledge assessment of the candidates and according to the information obtained during the selection interview.

7. **the case of the selection procedure assessment**, when the purpose is to validate the selection performed by correlating the results of the selection tests and the work performance after being hired

8. the case of the decision regarding the utility of the selection, when it is established to what extent has the selection reached its aims, in reference to, first of all, the costs implied, and the results of the persons selected.

9. **the case of reassessing the selection program**, which aims at the reassessment of the selection system in order to improve it from a functional point of view.

The description of a use case is carried out through scenarios which define the logical process of the interactions that form the case. We can define simple scenarios or detailed scenarios which also feature variants of the use case and even errors. The description is done in a simple manner, comprehensible to the persons in the field of application, and it mentions what the actor does and what the system does. Below you have the description of one of the use cases settled for the selection system, that is the case of the recruitment.

Use case: Sorting candidacies
Purpose: initial selection of the candidates
Actors: the company recruiter
Initial point: The candidate sends in the biographical data
Final point: The candidate is rejected or accepted for the stage of assessing their capacity and professional knowledge
Description of the process: As a result of the interest or willingness to occupy the job position for which the enterprise has made a recruitment add, the candidate informs himself/herself of the requirements the organization has regarding the submission of one's candidacy. The candidate has 3 possible ways of sending in the biographical data requested by the organization:

1. over the Internet, when they access a web page where they shall fill in a candidacy form with the data requested and shall then send the form directly to the company interested;
2. by mail, after having previously obtained the specific forms to fill in with their biographical data;
3. by personally submitting the candidacy file directly with the organization's head office.

The recruiter gathers all the biographical data of all the candidates in a data basis. In order to do so, they either access their e-mail (electronic mail), or enter the data on electronic support, the candidacies sent by mail, or those submitted directly at the company's head office. In order to carry out the pre-selection of the candidates based on the biographical data, the recruiter must consult the literature that deals with the assessment criteria and their share, identified as significant for the pre-selection stage. If the information wanted cannot be found, a human resources specialist shall conduct a research within the company, in order to identify the respective criteria. The recruiter shall enter into the system the data regarding these criteria and shall then sort the candidacies based on a hierarchy according to the biographical index calculated for each candidate. The biographical index is obtained by adding the values of the modality of the biographical criteria taken into account. After obtaining the candidates' hierarchy, the recruiter shall communicate the results, and shall send the rejection letters to the rejected candidates, and the letters programming the candidates for the next selection stage to those who have been accepted to continue with the selection process.

Based on textual descriptions, the scenarios may be illustrated through sequence diagrams which allow a visualization of the temporal aspect of the interactions and emphasize their meaning. The modeling process continues with the elaboration of the object classes, where the information analyst comes into play.

3. CONCLUSION

Study results are out in a theoretical proposal of an information system for selecting personnel. The findings of the study established the advantage of using UML in modeling of system information, because it was found necessary the collaboration of two specialists from two different fields: the computer scientist, who will design and implement the system, and the specialist in the field of application, who can build the exterior image of the system, from the perspective of its users.

In what regards the human resources selection, modeling an information system aims to increase the efficiency of the personnel-related activities by quickly solving certain administrative tasks related to the structuring of the information on the personnel hired and on the candidates, to increase the efficiency of the personnel-related decisions, to internationalize communication between employers and candidates, to reduce the costs implied by personnel recruitment and selection etc.

The theme discussed has aimed to personalize the research on the information systems in the personnel selection field, the paper focusing on the modelling of a personnel selection information system. This system may be just a module within a human resources information system, if the objective is to extend the system to all the personnel–related functions.

4. REFERENCES

Alter, S. (1999). *Information Systems:Management perspective*, Addison-Wesley, Indianapolis

Dumoulin, C. (1986). *Management des systèmes d'informati*on, Editions d'organisation, Paris

Peretti, J.M. (1994). *Ressources humaines*, Ed. Vuibert, Paris

Porter, M. (1985). *Competitive Advantage: Creating and Sustaining Superior Performance*, Free Press, New York

Reix R., (2002). *Systèmes d'information et management des organisations,* Librairie Vuibert, Paris

Weiss, D. (2003). *Ressources humaines*, Editions Organisation, Paris.

Annals of DAAAM for 2008 & Proceedings of the 19th International DAAAM Symposium, ISSN 1726-9679
Editor B. Katalinic, Published by DAAAM International, Vienna, Austria 2008
Make Harmony Between Technology and Nature, and Your Mind will Fly Free as a Bird

JOB SATISFACTION AND JOB MOTIVATION OF ROMANIAN YOUNG PROFESSORS

MAZILESCU, C[risanta] A[lina]; MIHARTESCU, A[na] A[ndreea] & POPESCU-MITROI, M[aria] M[onica]

Abstract: *This research is based on the analysis of the job motivation and job satisfaction among young professors. The 60 subjects are professors working in the Romanian educational system. Two scales have been used: Job Diagnostic Survey (by Oldham and Hackman), referring to the job characteristics, and Job Descriptive Index by Smith, Kendall and Hulin, referring to job satisfaction. In what regards the job's motivating potential, there are no significant differences between the university and pre-university system, and the values of the job characteristics indicate a low motivating potential of the job as a professor. In what regards the job satisfaction, the wage is the factor that dissatisfies the most, and at the opposite end were the relationships with the colleagues and superior.*
Key words: *job, satisfaction, motivation, job characteristics*

1. INTRODUCTION

Job satisfaction has become one of the most important and researched aspects of organizational psychology. The need to make labor more efficient has contributed to increasing the appreciation of the human factor within organizations, which led to an increase of interest in job satisfaction.

Starting with this perception of social reality, research has focused on the problems related to the job motivation and satisfaction of the professors in the pre-university and university educational system.

The theories and the models regarding work satisfaction have highlighted four of its essential determiners (personality, values, work situation, social influence) and many other specific determinants. The values reflect the convictions of a person regarding the results of his/her work.

The research on Hackman and Oldham's model (1976, 1980) supports the hypothesis of the connection between the job characteristics and the work results. Kim, Price, Mueller and Watson (1996), and Pearson and Chong in 1997 have proven there is a link between job characteristics and work satisfaction, but the causal connection is not totally clarified. It is believed that between the two aspects there is an indirect (mediated) relation.

In what regards the extrinsic conditions, studies show that the wage is a predictor of work commitment (and of the intention to leave (Lum, Kervin, Clark, Reid and Sirola, 1998).

Work satisfaction and the effort made are predictors of occupational stress, of professional performance, and of the tendency to leave the organization (Jamal & Cooper, 1999).

The studies mentioned, on the precedents and consequences of work motivation and on its implication at the individual and organizational level highlight the present nature of the issue within scientific research.

The research objectives:

1. to determine the motivating potential of the job position as professor (pre-university and university assistant);
2. to identify the factors that influence the professional satisfaction of young professors;
3. to identify the connection between the motivation potential of the job position as a professor and the work satisfaction of professors in the university and pre-university educational system.
4. to identify the differences between the professional satisfaction of professors in the university educational system and those in the pre-university educational system.

2. THE METHOD

2.1. Subjects

For the investigations proposed, we have used two groups of subjects. The first one was made of 30 persons (20 male and 10 female), professors (university tutors and assistants) in the polytechnic university educational system of Timisoara (Romania), with seniority between 1-5 years. The average age was 27.5 years, with a minimum age of 24 years and a maximum of 33 years.

The second group was made of 30 subjects, technical subject professors in the pre-university educational system, with an average age of 29 years.

Both groups of subjects have been selected by using simple random sampling, by drawing lots.

2.2. Instruments

In order to study the relation between the motivating potential of the job and work satisfaction (together with its components), we have used the following assessment instruments:

J.D.S *(Job Diagnostic Survey).* This scale has been elaborated by Hackman and Oldham who wanted to measure the workers' perception of the job, reported to each of its basic dimensions. The scale (in its reduced variant) has 15 items and is used to measure the following 5 dimensions: variety of aptitudes, task identity, task meaning, autonomy, feed-back.

JDI (Job Descriptive Index) and JIG (Job in General)
JDI is a questionnaire elaborated by P.C. Smith, Kendall and Hulin (1969) and adapted to the Romanian population by H. Pitariu (1994), which measures work satisfaction by means of 5 of his facades: the work itself, relationship with the boss, wage, promotion, and relationship with work colleagues. In 1985 another JIG scale was added, regarding the job in general.

3. RESULTS

In order to calculate the motivating potential of the job as a professor, we have started from the following formula, indicated by the authors of the JDS scale: MPS equals the average of the first three characteristics (skills variety - S.V, task identity - T.I., task significance - T.S.) multiplied by autonomy - A and feedback - Fb:

$$MPS = [(S.V.+T.I.+T.S.)/3] \times A \times Fb \qquad (1)$$

The average of the JDS answers and the values of the motivating potential of the job for the two groups of subjects (by university and pre-university educational system) are listed in table 1.

Educational systems	skills variety	task identity	task significance	Auto nomie	Feed back	MPS
pre-university	5,32	4,55	4,73	5,24	5,24	131,5
university	6,30	4,31	4,91	5,54	5,55	157,7

Tab. 1. The average values of the characteristics of a professor job position of technical subjects (university and pre-university educational system)

By analyzing the average values of the job characteristics we notice that their hierarchy is similar for both groups taking part in the study, and the values of the same characteristics of the job do not differ significantly in the pre-university environment from the university one. In what regards the hierarchy of the job characteristics, the differences between the most appreciated characteristic (task variety) and the least appreciated one (task identity), are also insignificant ($\chi^2 = 2.37$)

The theoretical values of MPS may vary between 1 and 343. In comparison with the average of the theoretical values of MPS (m=172), we notice that the MPS specific to the professor job position (technical subjects) in the pre-university and university educational environment, is below the theoretical MPS average.

In what regards the job satisfaction, Smith, Kendal and Hulin have taken into consideration five aspects: the works itself, relationship with the superior, wage, promotion possibility and relationship with the colleagues. Starting from the maximum theoretical score for each JDI scale (54) and taking into account the average (27) as being the basic satisfaction, we can say that the scores below 13.5 indicate a strong dissatisfaction, the scores below 27 but over 13.5 are representative for a moderate dissatisfaction, and those below 40.5 represent a moderate satisfaction, and those over 40.5 show a strong satisfaction.

Table 2 shows the values of the five aspects studied both in the university and pre-university educational system.

By analyzing the data we notice that the satisfaction of the job in general has the highest values, the job as a professor being considered very satisfactory by both groups of professors. Except for one difference, (the satisfaction of the possibility of being promoted) the differences between the university and the pre-university environment are insignificant.

In what regards promotion, the university assistants are more satisfied by the perspective of having access to a university hierarchy than the young professors in the pre-university educational system, the differences between the two groups being statistically significant ($\chi^2 = 5.45$; $p < 0.05$).
For the entire group of professors taking part in the survey (N=60), the wage is the factor indicating a strong dissatisfaction, while for the professors in the pre-university environment, the lack of promotion possibility is the second dissatisfactory factor. Otherwise, the factors taken into account are considered moderately or very satisfactory.

By studying the correlations between the motivation potential of the job as a professor and the satisfaction given by work itself and by the job as a professor in general, we elaborated table 3 - significant correlations from the statistical point of view and the thresholds of significance corresponding to these correlations.

Educational systems	Work	Wage	Promotion	Superior	Collegues	Job in general
pre-university	42,03	7,67	19,33	38,80	37,53	41,67
university	39,80	7,86	36,83	37,17	33,93	41,07

Tab. 2. Comparative presentation of the factors involved in assessing professional satisfaction within JIG and JDI scales

MPS	Work	Job in general
pre-university	r = .445 p=0.05	r = .422 p=0.05
university	n.s.	n.s.
MPS	r = .293 p=0.05	n.s.

Tab. 3. Correlations between MPS (university and/or pre-university) and the satisfaction given by job itself and by job in general

4. CONCLUSIONS

This research aims, mainly, at identifying the work characteristics with a high motivating potential and which influence the job satisfaction of young professors in the Romanian pre-university and university educational system.

In conclusion, for young professors, „task variety" is the most valuated characteristic of the job and it refers to the multitude of tasks entailed by a professor's activity: designing didactic activities, transferring knowledge, management of a pupils/students group, assessing competences, research activities, project elaboration etc. For the group studied, the motivating score of the job as a professor is below the theoretical averages.

In what regards the connection between MPS and the job satisfaction felt by the group of professors taking part in this study, we notice a difference between the university and high school professors. While for the young university professors the MPS is correlated with the satisfaction of work itself and the satisfaction of the job as a professor in general, for the high school professors, neither one of these correlations are significant.

In what regards the factors influencing the job satisfaction of the professors, we list, according to their importance: "the work itself", "the relationship with the superior" and "the relationship with the colleagues", and for the university assistants "the possibility to be promoted" as well. The wage is the factor that produces the strongest dissatisfaction, both for high school professors, as well as for the university assistants.

This study may be continued and completed with a study of job's stressors and the results of both may set the foundation of a program regarding the increase of attractiveness of the job as a professor in Romania, thus putting to good use the information regarding professional motivation and satisfaction of young professors. Being part of the poor countries which contribute to the forming of the elite in the rich countries can only draw our attention to the importance we should grant to the needs of young people, in general, and their professional needs in particular.

5. REFERENCES

Hackman, J.R. & Oldham, G.R. (1976). Motivation through the design of work: Test of a theory, *Organisational Behavior and Human Performance*, 16, 250-279

Hackman, J.R. & Oldham, G.R. (1980). *Work Redesign*, *Reading*, MA: Addison Wesley, 8, 121-124

Jamal,M. & Cooper,C.L. (1999). Job Stress, Type-A Behavior, and Well-Being: A Cross-Cultural Examination in *International Journal of Stress Management,* 6, 57-67

Kim,S.W.; Price,J.L.; Mueller,C.W &Watson,T.W. (1996). The determinants of career intent, *Human Relations*, *49*, 947-976

Lum L.; Kervin J.; Clark K.; Reid F. & Sirola, W. (1998). Explaining nurse turnover intent: job satisfaction, pay satisfaction or organisational commitment, *Journal of Organisational Behavior*, 19, 305-320

Pearson, C.A.L, Chong, J. (1997). Contributions of job content and social information on organizational commitment and job satisfaction, *Journal of Occupational and Organizational Psychology* Vol. 70, nr 4, 357-74.

Annals of DAAAM for 2008 & Proceedings of the 19th International DAAAM Symposium, ISSN 1726-9679
Editor B. Katalinic, Published by DAAAM International, Vienna, Austria 2008
Make Harmony Between Technology and Nature, and Your Mind will Fly Free as a Bird

COMPUTER APPLICATION FOR DECISION-MAKING SUPPORT IN MANUFACTURING TECHNOLOGY

MECIAROVA, J[ulia] & DADO, M[iroslav]

Abstract: *The selection of cutting fluid is often observed to be a multi-criteria decision-making problem with conflicting and diverse objectives. This contribution presents a current state of the development of computer application for optimal selection of cutting fluids in process planning stage with regard to minimization of health and environmental risks. The system requirements and principal structure of developed software tool are described. The proposed tool automates the cutting fluid selection and further it will serve for educational purposes.*
Key words: *CAPP, cutting, fluid, optimization, risk*

1. INTRODUCTION

Due to the replacing of classical education methods and putting emphasis on the development of virtual education and distance learning, many researchers nowadays deal with the use of computer applications, multimedia or hypermedia. It is an effective way how to improve the qualitative level of manufacturing technology education.

Computer aided process planning is the systematic determination of manufacturing methods and operation details by which parts can be produced economically and efficiently from raw materials to finished products (Zhao, 2001). Selection of optimal cutting fluid is one of the many process planning tasks. Traditional approaches of cutting fluid selection are based on functional requirements and cost performance (Šugár, 2006). Due to the increasing number of laws and directives governing industrial safety and environmental protection, the use of cutting fluids is putting intense economic pressure on manufacturing companies because cutting fluid is important source of health and environmental risks (Mádl, 2002). Optimal selection of cutting fluid, based on health and environmental risks, is an effective path how to reduce occupational diseases of employees, minimize the environmental pollution and improve cost performance of cutting fluids (Skerlos, 2007).

The purpose of this paper is to present a computer application that automates the selection of cutting fluid while minimizing the risk to machine operator and environment.

2. DEVELOPMENT OF THE APPLICATION

The principal structure of the application called "CF System" is created by three modules: database module, user interface module and process optimization module (see Fig. 1.).

The application is created in Microsoft Visual Basic 2008 Express Edition and is designed to run in Windows 2000/XP using Microsoft .NET Framework 2.0.

2.1 System requirements and installation

The "CF System" is designed to run in Windows 2000/XP, so the minimum hardware requirements are determined by these operating systems. In addition, it has to be installed Microsoft .NET Framework 2.0 and DAO 3.7+. Minimum hardware requirements are: processor Intel Pentium III 500 MHz, 128 MB RAM, 20 GB HDD. The "CF System" can be installed by following the installation steps in *Setup.exe*.

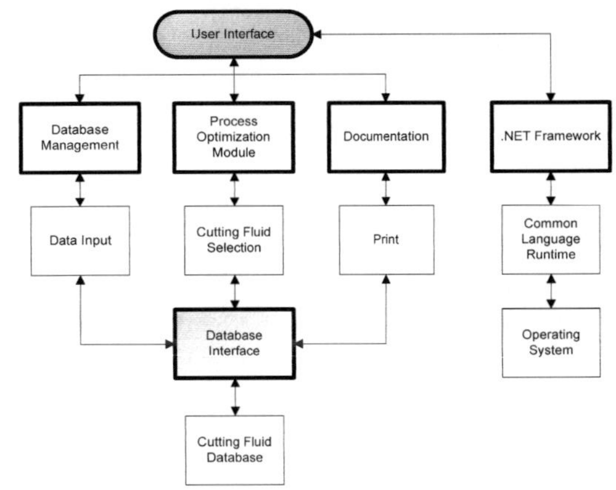

Fig. 1. Structure of the application.

2.2 Database module

Data included in sections of MSDS (Material Safety Data Sheet) will serve as input data for the development of a cutting fluid database created by using of Microsoft Access.

Structure of the database is given by these tables:

- identification of manufacturer,
- identification of product/cutting fluid,
- identification of health and environmental risks – health hazard symbols and R-phrases,
- toxicological information (acute toxicity – oral, dermal, inhalation; subchronic toxicity; chronic toxicity – mutagenicity, carcinogenicity, toxicity for reproduction),
- ecotoxicological information (aquatic toxicity, biodegradability, bioaccumulation).

In the relational database it is important to specify the relationships between individual tables (Fig. 2.):

- one manufacturer may have more products (cutting fluids),
- some product can have more health hazard symbols, R-phrases and toxicological information,
- health hazard symbol can have more R-phrases,
- R-phrase can be used by more health hazard symbols.

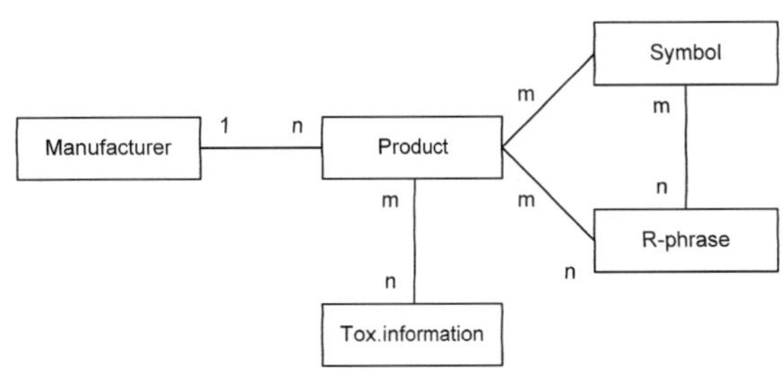

Fig. 2. Scheme of relationships between tables.

2.3 User interface module

This module is designed for controlling the operation of the whole application, for data input, editing and deleting as well as for reporting the results (printing).

The user can log into the application by entering the user name and password. Then by choosing *Search* (Fig. 3.), the user can view the list of cutting fluids, manufacturers, symbols, R-phrases and toxicological information. The data can be added, edited, deleted or printed. The application can be extended in further development and new data may be added by choosing *Add New* in Main Form. After that, Data Input Form appears. The user can add the name of a cutting fluid, manufacturer, the health/environmental hazard symbols, R-phrases, toxicological and ecotoxicological information of a cutting fluid. Saving of the added data is automatic.

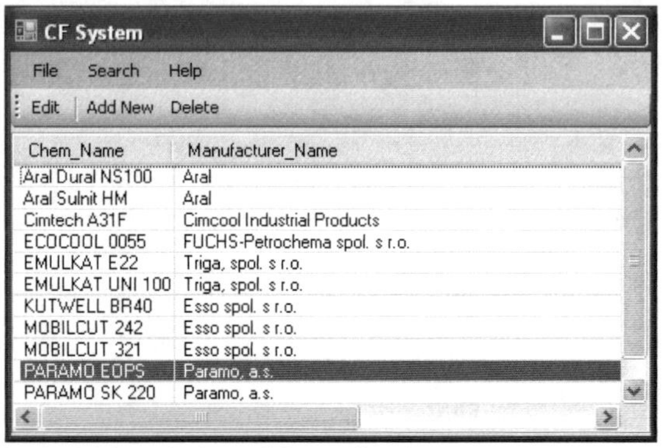

Fig. 3. Main Window of the application.

2.4 Optimization module

The optimization module is the main part of the application. The user can launch this module by choosing *File* and then *Cutting Fluids Assessment*. The model of cutting fluid assessment is based on the similarity comparison.

The principle is to compare the health/environmental properties of a given cutting fluid, which are represented by R-phrases with the health/ environmental properties of a reference cutting fluid (with the maximum number of R-phrases).

Similarity is defined in terms of a distance metric, most often Euclid distance (d_E) or relatives of the Euclid distance. The similarity index between the fluids is defined as follows:

$$d_{E(x,y)} = \sqrt{\sum_{i=1}^{n} g_i (x_i - y_i)^2} \tag{1}$$

where: g_i is the weighting factor of the *i*-th cutting fluid property, x_i, y_i are values of the *i*-th cutting fluids property.

It indicates how much the given cutting fluid is similar to the reference cutting fluid. The bigger the index is, the more different the cutting fluids are and the less harmful a cutting fluid to the health of workforce and environment is. If d_E equals zero, it means that the cutting fluids are identical. Some weighting factors for individual properties are shown in Table 1. The basis for the weights allocation of each R-phrase or their combination takes into account the seriousness of the resultant health/environmental effects.

Some example of health and environmental risk assessment of a given cutting fluid is shown in Fig. 4.

R26, R27, R28, R50	6
R26/27, R26/28, R27/28, R50/53	6,1
R26/27/28	6,2
R34, R41, R42, R43, R65	7
R35, R42/43	7,1
R48	8

Tab. 1. Examples of weighting factors for health/environmental risks assessment.

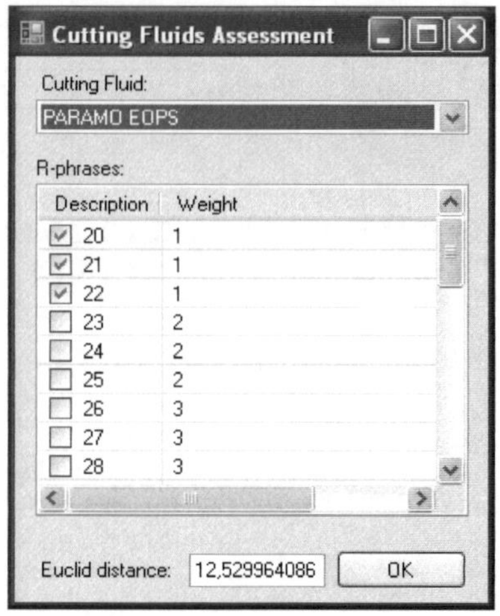

Fig. 4. Cutting Fluids Assessment Window.

The user can compare two cutting fluids by the number of Euclid distance. Higher number means that the cutting fluid is less harmful for the health of workers and the environment than the other one.

3. CONCLUSION

It is obvious that the selection of the appropriate cutting fluid has to be governed not only by functional requirements and cost performance but as well health/environmental hazards (Sokovic, 2001). This paper has introduced software tool for cutting fluid evaluation based on health and environmental risk criteria. The tool enables the calculation of overall score to measure health/environmental performance of a given cutting fluid. Possible utilisation of developed tool includes predictive assessment of health/environmental impacts of cutting fluid in production process planning and valuable instrument in education process. Further research will focus on implementation of developed software application as a separate module into some existing CAPP systems.

Acknowledgement

The authors wish to thank the financial support of the Ministry of Education of the Slovak Republic, in framework research project KEGA No. 3/5209/07: "*Manufacturing technology education by virtual excursions*."

4. REFERENCES

Mádl, J. & Koutný, V. (2002). How to Select Suitable Fluid. *Manufacturing Technology,* Vol. 2, No.2, pp. 45 – 53, ISSN 1211-4162.

Sokovic, M. & Mijanovic, K. (2001). Ecological aspect of the cutting fluids and its influence on quantifiable parameters of the cutting processes. *Journal of Materials Processing Technology,* Vol. 109, No. 2, pp. 181-189, ISSN 0924-0136.

Skerlos, S.J. (2007). Prevention of Metalworking Fluid Pollution, In: *Environmentally Conscious Manufacturing,* Kutz, M. (ed.), pp. 95-122 , John Willey & Sons, ISBN 0-471-72637-0, Delmar, New York.

Šugár, P. (2006). Optimization of the cutting fluid selection from the human risks point of view. *Acta Mechanica Slovaca,* Vol. 2, No. B, pp. 439 – 442, ISSN 1335-2393.

Zhao, Z. (2001). Multi-Level Decision Making for Process Planning in CIM Systems, In: *Computer Aided Design, Engineering And Manufacturing Systems,* pp. 107-140, CRC Press, ISBN 0-8493-0994-8, Boca Raton.

Annals of DAAAM for 2008 & Proceedings of the 19th International DAAAM Symposium, ISSN 1726-9679
Editor B. Katalinic, Published by DAAAM International, Vienna, Austria 2008
Make Harmony Between Technology and Nature, and Your Mind will Fly Free as a Bird

KINEMATICAL ANALYSIS OF AN UPPER LIMB PROSTHESIS

MENYHARDT, K[aroly]; NAGY, R[amona] & LUCA, G[heorghe]

Abstract: *A kinematical model was realized based on the physical model of an upper limb prosthesis. The results of the analysis were compared with the data gathered with an APAS measurement system on the physical prosthesis. We focused our attention on the displacement, velocity and acceleration for the prosthesis' center of gravity. The results validate the prosthesis' behavior under normal load conditions.*
Key words: *kinematics, prosthesis, modeling, analysis.*

1. INTRODUCTION

The main purpose of this study is to compare the kinematical parameters from the mechanical model (figure 1a), the 3D simulation (SolidWorks/CosmosMotion) and the physical data gathered, with a professional APAS measurement system, from a prosthesis designed (figure 1b) and developed at CMPICSU Research Center from "Politehnica" University of Timisoara by the first author.

The kinematic study was initially done in order to compare the motion of the human upper limb and the prosthesis. Further studies of the mechanical model presented in the current paper helped us understand the motion of the prosthesis.

This study was developed from the necessity of quantifying the basic movements of the prosthesis in order to make it run smoother and more natural, like the human upper limb (Pravin, 2006).

Throughout the literature there are few papers that present the mechanical model and compare it with a physical prosthesis, realized by the authors (Duraisamy et al., 2006), (Jadhav & Krovi, 2003). In most cases, the kinematical analysis is made on an idealized anatomical model where the muscles are replaced by given forces. In the current study the analysis is done on a real mechanical model that is analogue for the developed prosthesis.

A concrete and in detail analysis will permit to specify the conditions for the working regime (Lloyd et al., 2000). In addition, graphics that evaluate the phenomena are traced putting in evidence the validity of the model.

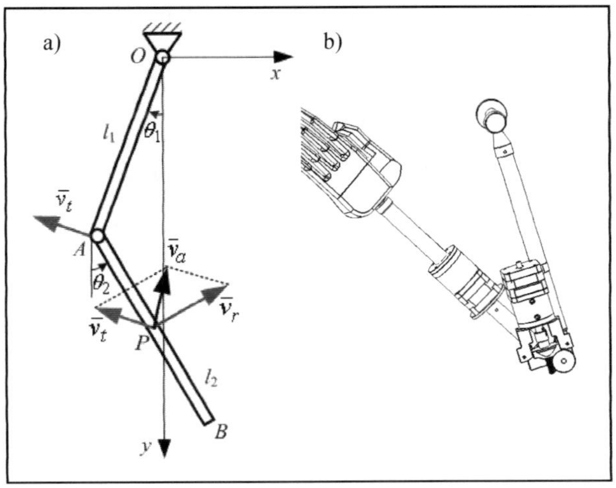

Fig.1. Kinematical model of the arm prosthesis.

2. MECHANICAL APPROACH

The stability and motion of the prosthesis is influenced by lots of factors, sometimes even the least significant factor could affect the outcome of the task (Raikova, 1992), (Veeraraghavan et al., 2004).

Because of the difficulties in calculus, the prosthesis was modeled as a two bar mechanism (figure 1a) where OA (the arm) executes a rotation around joint O, due to change in the prosthesis' center of gravity. The AB segment (forearm with hand) executes a planar motion, driven by the connected arm and elbow motor. The angular velocity $\dot{\theta}_2$ was considered constant throughout its range of motion (the prosthesis is able to do a 150 degrees flexion in the elbow).

Considering the position for the center of gravity being:

$$x_c = \frac{l_1 \cdot \left(-\frac{l_1}{2} \cdot \sin\theta_1\right) + l_2 \cdot \left(\frac{l_2}{2} \cdot \sin\theta_2 - l_1 \cdot \sin\theta_1\right)}{l_1 + l_2}$$
$$y_c = \frac{l_1 \cdot \frac{l_1}{2} \cdot \cos\theta_1 + l_2 \cdot \left(\frac{l_2}{2} \cdot \cos\theta_2 + l_1 \cdot \cos\theta_1\right)}{l_1 + l_2} \quad (1)$$

θ_1 was determined as a function of θ_2:

$$\sin\theta_1 = \frac{l_2^2}{l_1 \cdot (l_1 + 2 \cdot l_2)} \cdot \sin\theta_2 \quad (2)$$

Trough various calculus and restraint conditions, e.g. the center of gravity remains always on the Oy axis, results that its velocity is that from equation (3) and its curve Fig.2.

$$v_c = \frac{-\left(\frac{l_2^4 \cdot \dot{\theta}_2 \cdot \sin(2 \cdot \theta_2)}{4 \cdot \sqrt{[l_1 \cdot (l_1 + 2 \cdot l_2)]^2 - (l_2^2 \cdot \sin\theta_2)^2}}\right)}{l_1 + l_2} - \frac{\dot{\theta}_2 \cdot \frac{l_2^2}{2} \cdot \sin\theta_2}{l_1 + l_2} \quad (3)$$

The acceleration for the center of gravity for the prosthesis' mechanical model is shown in equation (4) and figure Fig.3.

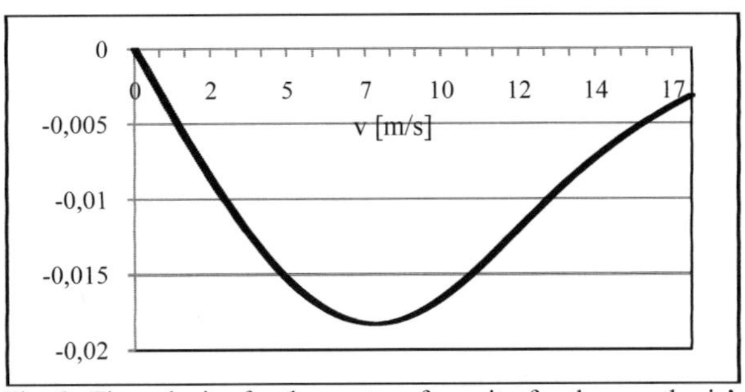

Fig. 2. The velocity for the center of gravity for the prosthesis' mechanical model

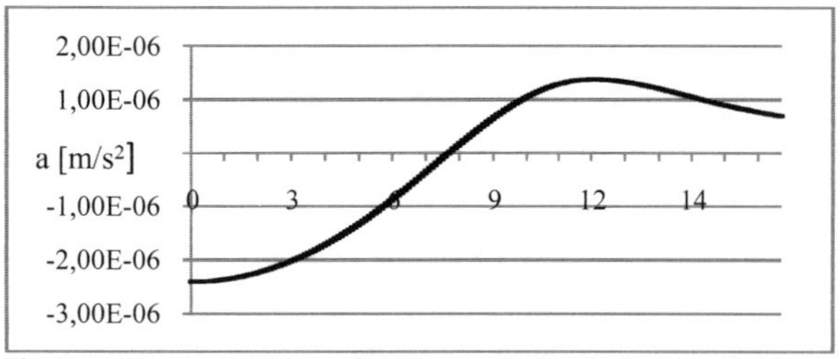

Fig.3. The acceleration for the center of gravity for the prosthesis' mechanical model.

$$a_c = \frac{-\ddot{\theta}_1 l_2^2 \cdot \sin\theta_2 - \dfrac{l_2^4 \cdot \dot{\theta}_2^2 \cdot (\cos\theta_2)^2}{\sqrt{[l_1 \cdot (l_1 + 2 \cdot l_2)]^2 - (l_2^2 \cdot \sin\theta_2)^2}}}{2(l_1 + l_2)} - \frac{\dot{\theta}_2^2 l_2^2 \cos\theta_2}{2(l_1 + l_2)} \quad (4)$$

All the other linear and angular kinematical parameters were determined for each point of the mechanism. These parameters helped in establishing the trajectory and the workspace for the prosthesis in order to assess the possible movements. The kinematical analysis of the movement refers at the quantitative description of body segments without taking into account the forces that generate the movement.

3. RESULTS OF THE ANALYSES

In this paragraph are presented the results of the kinematical analysis from different points of view/interest. In all the studied cases the time needed to perform the prosthesis movement was slightly longer, the artificial "muscles" (DC steppers) being slower than the human muscles. The trajectories and exercises were the same in all cases, generated by computer and by the upper limb prosthesis.

For the time interval [10, 27] seconds the curve from the theoretical model almost overlaps with the curve from the physical model. The differences in the amplitude are due to the simplification of the model (uneven distribution of the masses in the physical model) and the measurement uncertainty. The curves are presented in figures Fig.4 and Fig.5.

The resulted data from the analysis for a movement cycle of the upper limb/prosthesis or for a succession of cycles can be used to analyze the movement pattern (Karamanidis et al., 2003). By evaluating this movement a joint, motor or coordination deficit can be outlined or quantified and efficient strategies of compensation can be made based on these values (Murray & Johnston, 1998). By determining the kinematic parameters (linear and angular acceleration), is possible to know the inertial forces for further dynamical studies.

However, long-term cyclic testing is necessary to ensure reliability and a valid assessment of the efficacy of the prosthesis can only occur by clinical trial.

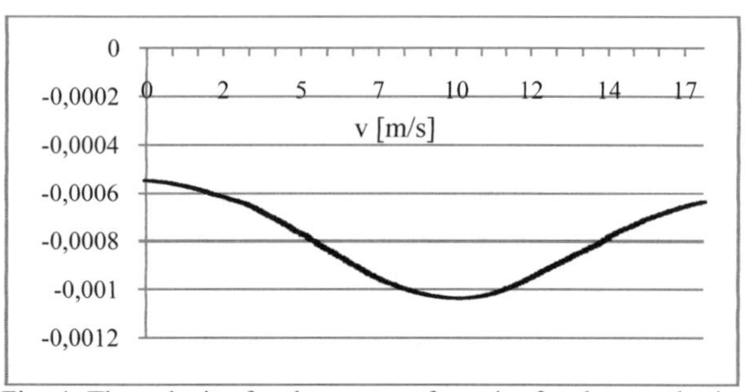

Fig. 4. The velocity for the center of gravity for the prosthesis measured with the APAS system.

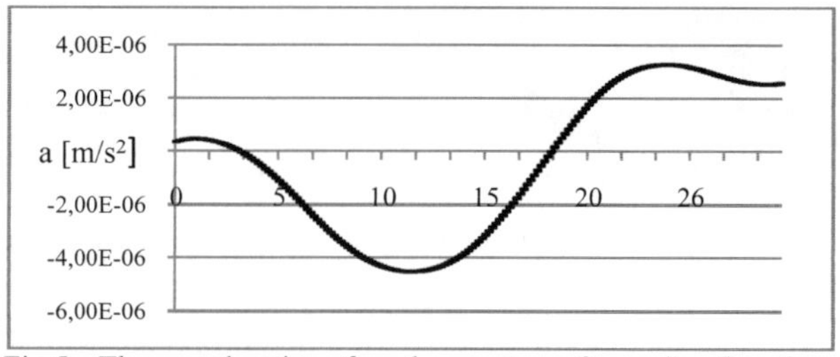

Fig.5. The acceleration for the center of gravity for the prosthesis measured with the APAS system.

4. CONCLUSION

This performed mobility study is useful in outlining the differences between the movement of a human arm and a prosthesis. The gathered results have been filtered out and processed in order to correct and enhance the prosthesis's unnatural movement. Every movement scheme has different solutions based on the forces that appear and the prosthesis' orientation and position. Determining the orientation and position in real time without the use of external equipment brings forward new obstacles that need to be surmounted.

The results validate the physical prosthesis and its behavior in normal load conditions.

Comparing the results from the mathematical approach, mechanical simulation and the measured data from prosthesis and human upper limb, could help us improve the control of the prosthesis' motor by feeding adequate input parameters to the microcontrollers.

The next steps in the research will include sensorial feedback and adaptation of the microcontrollers to these inputs.

5. REFERENCES

Duraisamy, K.; Isebor, O.; Perez, A.; Schoen, M.P.; Naidu, D.S. (2006). Kinematic Synthesis for Smart Hand Prosthesis, *Biomedical Robotics and Biomechatronics, BioRob 2006. The First IEEE/RAS-EMBS International Conference on*, Volume , Issue , 0-0 0 Page(s):1135 – 1140

Jadhav, C., and Krovi V. (2003). In-Vivo Estimation of Unknown Upper-Limb Kinematic Parameters, *11th National Conference on Machines and Mechanisms, (NaCoMM-2003)*, Delhi, India, December 18-19, 2003.

Karamanidis K.; Arampatzis A, Bruggemann G.P. (2003). Symmetry and reproducibility of kinematic parameters during various running techniques, *Medicine and science in sports and exercise* vol.35, 6, pp. 1009-1016 Lippincott Williams & Wilkins, Hagerstown, **ISSN** 0195-9131

Lloyd D.G., Alderson J., Elliott B.C. (2000). An upper limb kinematic model for the examination of cricket bowling: A case study of Mutiah Muralitharan, *Journal of Sports Sciences*, Volume 18, Number 12, 1 , pp. 975-982(8)

Murray I.A., Johnson G.R. (1998). Upper limb kinematics and dynamics: the development and validation of a measurement technique, *Proceedings of Fifth International Symposium on the 3D Analysis of Human Movement*, Chattanooga, Tennessee. 2 –5 July.

Pravin N. (2004). Development of Quantitative Measures for Characterization of Upper Limb Dysfunction, M.S. Thesis, Department of Mechanical & Aerospace Engineering, SUNY Buffalo.

Raikova R. (1992). "A General Approach for Modelling and Mathematical Investigation of the Human Upper Limb", Journal of Biomechanics, 25, 857–867.

Veeraraghavan A., Roy-Chowdhury A.K., Chellappa R. (2004). "Role of shape and kinematics in human movement analysis," in *Proceedings of the IEEE CVPR '04*, Washington, DC, USA, vol. 1, pp. 730–737.

Annals of DAAAM for 2008 & Proceedings of the 19th International DAAAM Symposium, ISSN 1726-9679
Editor B. Katalinic, Published by DAAAM International, Vienna, Austria 2008
Make Harmony Between Technology and Nature, and Your Mind will Fly Free as a Bird

COLLABORATIVE ENGINEERING AND DESIGN AUTOMATION FOR ADVANCED MATERIALS IMPLEMENTATION IN RAILWAY SUPPLY INDUSTRY

MERTICARU, V[asile jr.]; RECIO PERERO, P[ablo] & CARAUSU, C[onstantin]

Abstract: *The paper presents an application of collaborative engineering, respectively of collaborative new product development, using CAD instruments for design automation into a PDM environment, in order to replace some classical steel structures with new material solutions, with better characteristics, in the Romanian railway supply industry. A particular case study is described in the paper, referring to the conception and CAD activities developed in collaborative manner for an assembly product for railway industry, made of aluminum alloys and designed to be assembled by welding operations.*
Key words: *design, automation, materials, welding, transportation.*

1. INTRODUCTION

In the nowadays context of the European Union economy, there can be predicted that the Romanian railway industry will experience an important renovation and the products manufactured in the railway supply industry will have to align to the European requirements. One of the main issues in this sense is represented by the need to replace some classical steel structures with new material solutions, such as aluminium alloys, having better characteristics and a lot of advantages in use. In these conditions, the specialists from railway manufacturing industry and from the research centers from universities can assume time consuming tasks such as new product solutions' conceiving, execution of drawings and technical documentation, conception of the products' aesthetics and ergonomics, simulation of the products behavior through FEM (Finite Element Method), generation and testing of the manufacturing technologies.

2. ENGINEERING REFERENCE

As engineering reference, the Spanish railway industry was chosen to be subjected to study, because at the present time, the railway sector is experiencing also in Spain an important renovation that comes bound up with the strong increase of the demand of transportation due to a couple of clearly identifiable phenomena: the phenomenon of the people from the big cities, changing their houses toward their metropolitan areas, and the necessity of daily traveling; the search for a viable alternative to the air transportation, for not very long distances (Guillén-Carames et al., 2007). Therefore, there have been clearly identified two different directions of growth in the Spanish railway sector: the proximities transportation and the high-speed transportation. There has been noticed that administration actors which have the responsibility of offering public transportation services mostly chose to invest in a type of infrastructures that despite of being expensive and difficultly profitable, bring as a result a better interconnection among urban cores moderately distanced and between the center and the periphery of the big cities and, consequently, an improvement in the flows of people and goods that finish in determining the growth and enabling of the economic activities, mainly due to the great capacity of transportation, such as for passengers' volumes, very superior to that of the highways, to the competitive times of travel that can be obtained, and to their environmental smallest impact. Considering the new works and investments mainly in the high-speed railway transportation and in the urban and proximity transportation of the main metropolitan areas of Spain, the forecasts show that the demand volume of railway material and products will be very important in the next period, bringing serious difficulties of adaptation for the big productive companies in the field (Recio-Perero, 2008).

On the other hand, at the present time, about 80% of the orders that the administrations of the European railway sectors are launching, stipulate in the corresponding sheets of specifications the requirement that the structure must be manufactured in aluminum alloy. Lately, the aluminum has consolidated as a clear alternative to the steel in the production of boxes of railway vehicles. In the high-speed railway transportation segment, the aluminum is used in 100% of the cases. In tram and metropolitan transportation segments, the aluminum is located around 90%, while in regional and suburban train it descends around 60%. Anyway, also in the suburban trains, mainly when speaking of electric traction, the aluminum overcomes a lot over the steel in the orders of railway material during the last years (Recio-Perero, 2008).

In relation to the new visions upon Concurrent Engineering (Kusiak 1992) and upon Integrated Engineering (Usher et al., 2005), respectively upon concepts and systems as CIM (Computer Integrated manufacturing) (Singh 1995), CIE (Computer Integrated Enterprise) or CII (Computer Integrated Industry), the main advantages brought by a system of product information management, respectively by a PDM or PLM solution, in the new product development are: the time for realizing a new product (Time to market) is reduced; the productivity of the design/calculation process is improved; the design and manufacturing accuracy is improved; an innovative content of the products is determined; data security is provided; a better control upon changes is provided. In the environment of PDM/PLM, cPD (Collaborative Product Development) begins the first phase of the new product's lifecycle (Ausura & Deck 2003) and, as initial stage, it is an approach which intend capturing, organizing, coordinating and controlling all the information related to the new product development, including all the functional requirements, geometry, specifications, characteristics and manufacturing processes, for providing a shared common view upon the development of the product to accomplish the requirements and for creating a unique vault of product information accessible throughout the product entire lifecycle.

3. OBJECTIVES, METHODS & INSTRUMENTS

The reason that directed the decision towards the design of a product assembled by welding of big and medium aluminum structures has been represented by the increased impact and use that this material is having in diverse sectors of the industry, but with special emphasis in the European railway sectors.

The group of products that has been chosen as subject of the case study for the collaborative product development is represented by the back headboard of a modern suburban train. The headboard is the assembly that represents the mechanical interface between the railway carriage's structure and the coupling elements. The coupling elements are the mechanical elements whose function is to provide the joint between railway carriages.

The studied back headboard, as well as the rest of the components that constitute the structure of a railway vehicle, is usually manufactured starting from plates of medium or big thickness, cut at stipulated dimensions and then joined by welding.

In the case of aluminum structures, one of the big advantages is the possibility of using profiled bars extruded in cold conditions. The aluminum profiles give the possibility to build a structure more optimized in relation to the mechanical efforts the structure is subjected to.

Considering the experience in the CAD area of the authors and of their collaborators from the universities and from Romanian industrial partners, AutoCAD-2007 and SolidWorks-2007 have been chosen as CAD instruments and PDM-Works have been chosen as product data management instrument in the collaborative engineering activities.

4. RESULTS OF COLLABORATIVE CAD

The above mentioned headboard, designed in collaborative manner, is formed of 26 parts and an example of CAD result for this product is shown in Fig. 1. For the above presented reasons, all the parts in the designed assembly were stipulated to be manufactured of Anticorodal-110, which is a medium strength aluminum alloy with Mg and Si, having good machine-ability, excellent corrosion resistance in inland atmosphere and excellent TIG/MAG weld-ability.

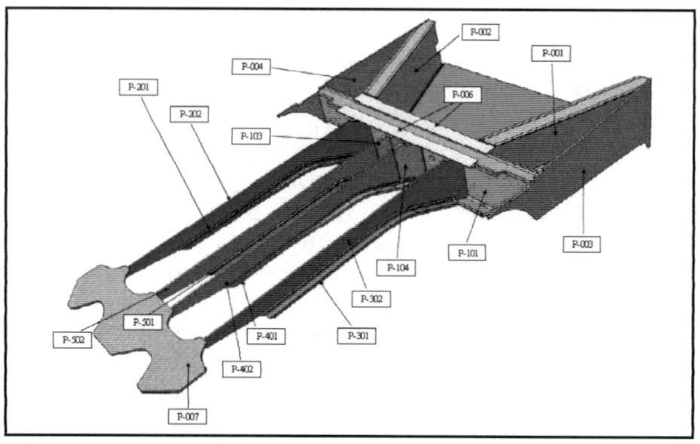

Fig. 1. Example of CAD result, for the designed headboard.

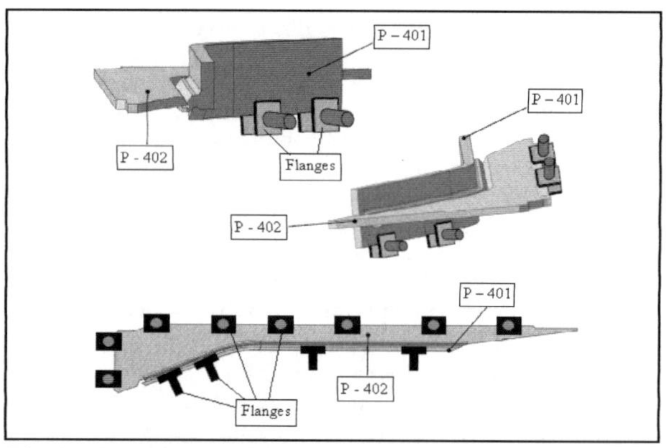

Fig. 2. Example of positioning technological scheme for welding operation.

For the assembly product subjected to the case study, also the manufacturing technical documentation was elaborated, the above mentioned CAD instruments being used in this stage for elaborating the technological schemes for parts positioning and the details working schemes for each welding phase and operation in the assembling process route.

In Fig. 2, there is exemplified such a technological scheme for parts positioning for welding operations.

In Fig. 3, there is exemplified one technological detailed scheme for welding operations.

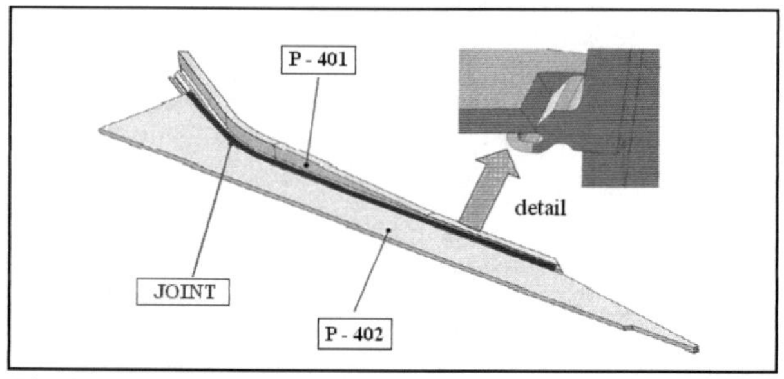

Fig. 3. Technological detailed scheme for welding operation.

5. CONCLUSION

An application of collaborative engineering, respectively of product development, using CAD instruments into a PDM environment is presented in the paper. So, the paper proposes a collaborative methodology for developing some type projects, specific to the machine manufacturing industry. A case study is described in the paper, referring to the CAD activities developed for an assembly product for railway industry, made of aluminum alloys and designed to be assembled by welding operations. The proposed development manner of such type projects consists in the realization, by the collaborative partners and using CAD software, of parts from the product, in making information exchange and in realizing assemblies by working on the network, in realization of the technical changes imposed by the product design (shapes, colors, presentation materials), in simulation of the loading conditions using FEM, in development of the manufacturing technologies for the complex parts in the project and is intended to be extended in the research centers from universities and at industrial partners.

6. REFERENCES

Ausura, B. & Deck, M. (2003). The "new" Product Lifecycle Management systems: What are these PLM systems? And how can they help your company do NPD better? *Available from:* http://www.pdma.org/visions/jan03/plm.html, *Accessed:* 2007-12-12.

Guillén-Carames, J. et al. (2007). *El régimen jurídico del sector ferroviario en España (The legal regime of the railway sector in Spain)*, Thomson-Aranzadi, ISBN 978-84-8355-197-4, Madrid.

Kusiak, A. (1992). *Concurrent Engineering: Automation, Tools, and Techniques*, John Wiley & Sons Inc., ISBN: 978-0-471-55492-9, New York.

Recio-Perero, P. (2008). *Train Carriage System Joint Factory - Final project* (coordinator V. Merticaru jr.), Universidad de Valladolid-Spain & Technical University "Gh. Asachi" of Iasi-Romania.

Singh, N. (1995), *Systems Approach to Computer-Integrated Design and Manufacturing*, John Wiley & Sons Inc., ISBN: 978-0-471-58517-6, New York.

Usher, J. M.; Roy, U. & Parsaei, H. (2005), *Integrated Product and Process Development: Methods, Tools, and Technologies*, John Wiley & Sons Inc., ISBN: 978-0-471-15597-3, New York.

Annals of DAAAM for 2008 & Proceedings of the 19th International DAAAM Symposium, ISSN 1726-9679
Editor B. Katalinic, Published by DAAAM International, Vienna, Austria 2008
Make Harmony Between Technology and Nature, and Your Mind will Fly Free as a Bird
*Young Researcher and Scientist Paper / * Supervisor, Mentor*

ELECTRONIC-BEAM WELDING LARGE THIKNESS

MEZENTSEV, M[ikhail]

Abstract: *There is a problem connected with production of defect-free metal welded joints featuring large thickness through use of electronic-beam welding, and these defects may be associated with the specifics of joints formation. For solving this problem is used method of N. A. Olshansky, and welding take in magnetic field. Using this technological trick, it is possible to achieve prevention probability defect formation.*
Key words: *Beam, electron, welding, defect.*

1. INTRODUCTION

EBW of horizontal joints and, probably, vertical ones in bottom-up direction must be more perspective form viewpoint of joint formation than EBW in lower position. Horizontal position of the welding bath allows improving conditions of welded joint formation and intensifying the process of gas and vapors of metal removing from the fusion channel. (Kaidalov, 2004)

In MPEI under guidance of Professor N. A. Olshansky, there has been developed a method of EBW using a vertical joint on a vertical wall under acceleration voltage up to 60 kV. This method allows matching the direction of the liquid metal transfer and direction of the gravity force so that the fusion depth won't be restricted by the combination of forces influencing the liquid metal in the welding bath as it occurs in case of welding in lower position. There has been developed a technology of plate steels and titanium alloy welding, including application of formation devices for the apex of the joint and the back plate used for the bottom run (Leskov, 1980).

To implement measures targeting to improve formation stability, we need to know forces influencing the liquid metal in the welding bath, as well as conditions, under which the metal runs out of them. (Martynov 1988)

Let's consider the forces using a simplified model (Fig. 1) featuring the following assumptions: the fusion channel is a horizontal cylinder, from the front and rear of the product made the welding bath has equal curvature radii, with the latter curvature radius being infinitely large. To analyze the joint formation, let's see equilibrium of the liquid metal in the welding bath on the border of the fusion channel. In this case, the metal in the welding bath is affected by the following:

1. Metal static pressure:

$$p_g = \rho g h \tag{1}$$

where ρ is density of the liquid metal, kg/m^3; g = 9.8 m/s^2 is free fall acceleration; h is a depth of the welding bath;

2. Pressure resulted from the surface tension forces:

Fig. 1 Method of calculation of the forces influencing the welding bath when using EBW with horizontal ascending beam: 1 – The welding bath; 2 – The welded joint; 3 – The gas vapor channel; 4 – The beam axis; 5 – The butt.

$$p_\sigma = \sigma \cdot (1/R_1 + 1/R_2) \tag{2}$$

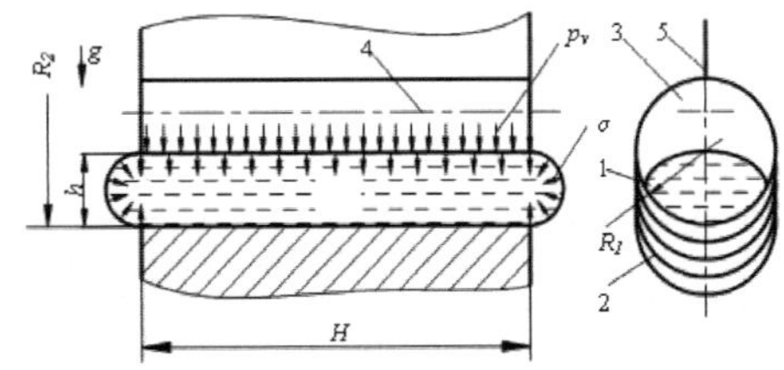

Fig. 1 Method of calculation of the forces influencing the welding bath when using EBW with horizontal ascending beam: 1 – The welding bath; 2 – The welded joint; 3 – The gas vapor channel; 4 – The beam axis; 5 – The butt.

where σ is a surface tension coefficient, N/m, R_1, R_2 are curvature radii of the liquid film, m;

3. Pressures resulted from vapor recoil reaction p_V.

The metal static pressure and that from vapor recoil reaction try to push out the melt from the channel, and the pressure resulted from the surface tension forces holds the welding bath in place, so, upon reaching a certain value, the metal film ruptures, and the melt runs out from the fusion channel, which results in defects such as vacuum cavities. Thus, the total pressure acting on the liquid metal may be written as follows:

$$p = p_\sigma - p_g - p_V \tag{3}$$

and this, taking into account (1) and (2) leads to:

$$p = \sigma \cdot (1/R_1 + 1/R_2) - \rho g h - p_V \tag{4}$$

where, if $p \geq 0$, the melt is held in place by forces without defect formation.

In case of melting X12CrNiTi182 grade steel with thickness H = 0.08 m, let's take the metal density ρ = 7900 kg/m^3; the depth of the welding bath h = 0.002 m; the surface tension coefficient for iron at temperature of 1700 °C σ = 1.76 N/m. For satisfactory formation of the bottom run the recommended radius of the channel shall be 0.0015...0.002 m, let's assume $R_1 = 0.0018$ m and the channel curvature radius $R_2 = \infty$. According to scientific researches, the pressure in channel in case of deep fusion changes, by various estimates, within $10^2 \ldots 10^4$ Pa (Sukach et al., 1989), and let's take $p = 10^3$ Pa. Under such assumptions, the liquid element of the bath is affected by the pressure (4):

$$p = 1.76 \cdot \left(\frac{1}{0.0018} + \frac{1}{\infty}\right) - 7900 \cdot 9.807 \cdot 0.002 - 1000 = -177.166 \text{ Pa}$$

preempting the melt form the fusion channel. (Lankin, 1978)

To prevent defect formation, it is necessary to increase pressure of the surface tension forces p_{σ}, or decrease pressure of vapor recoil reaction p_V and metal static pressure.

The equation of pressure from the surface tension forces (2) includes the coefficient of surface tension σ, which is a constant specific to a particular material. Decrease of channel radius R_1 may lead to channel collapse and such defects as cavities, as well as to increase of variation in fusion depth and defects in bottom run. Thus, the most appropriate way is to change channel curvature radius R_2. This is realized through induction of the local lateral magnetic field in the activity field of electronic beam over the whole thickness of the butt between welded parts.

Using this technological trick, it is possible to achieve such shape of the welding bath that the liquid metal influenced by the gravity forces will rush to its center (Fig. 2). Thus, the metal static pressure influencing melt elements located near the rear and front surfaces of the channel respectively may be read as follows:

$$p_g = \rho g \Delta h \tag{5}$$

as Δh is low in value, then it is may be ignored. Thus, the equation (4) may be transformed to the form;

$$p = \sigma \cdot (1/R_1 + 1/R_2) - p_V \tag{6}$$

the magnetic field was generated by a stray field of a solenoid equipped with a ferromagnetic core with clearance. However, this method may be applied only for paramagnetic materials because in case of ferromagnetic material magnetic flux line will be «pulled-in» into the material and in a vapor dynamic channel the field will disappear.

The value of magnetic flux from the stray field of the solenoid is not possible to get by the calculation method. Its distribution law was found in experiments.

The measurements and calculation estimates were obtained only for the component of the magnetic flux of the field generated by the coil that was directed perpendicular of the welding velocity vector and responsible for deflection of the electron beam along the butt.

As a result, there were obtained the values of the magnetic flux vs. distance to the center of the clearance of the ferromagnetic core (Fig. 3).

Fig. 2 Method of calculation of the forces influencing the

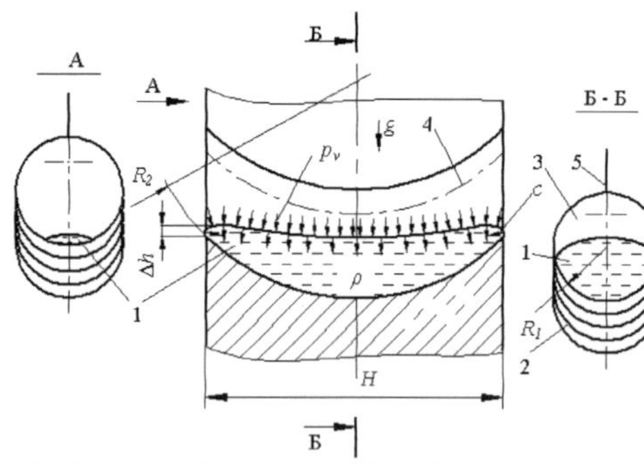

welding bath when using EBW with horizontal ascending beam under condition of the lateral magnetic field: 1 – The welding bath; 2 – The welded joint; 3 – The gas vapor channel; 4 – The beam axis; 5 – The butt.

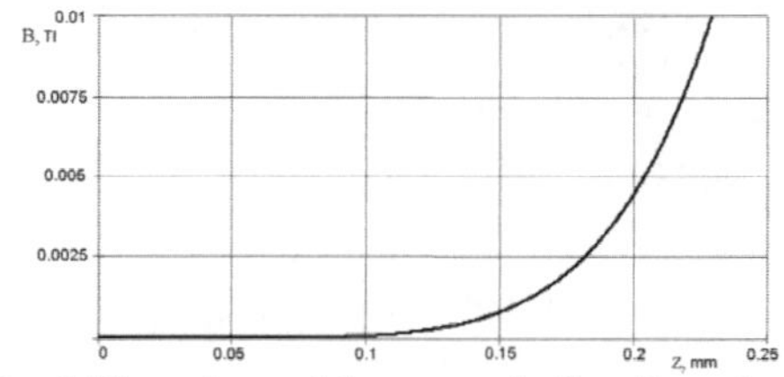

Fig. 3 The values of the magnetic flux B vs. the distance from the center of the clearance of the ferromagnetic core.

Change of the magnetic flux depending on the distance from the center of the solenoid was approximated with a power function looking like:

$$B(z) = A_0 \cdot z^{A_1} \tag{7}$$

where $A_0 = 60$, $A_1 = 5.9$.

When reviewing the action of the magnetic field (7) on the axial electron of the beam, there were obtained expressions describing behavior of the beam in the fusion channel:

$$x(z) = x_0 + tg(\gamma_2(z)) \cdot z \tag{8}$$

where x_0 - location of the electron at the entrance in the area of magnetic field activity, angle γ_2 – angle, with which the electron beam exits to the area of magnetic field activity, counted from $0x$ axis:

In this case, assuming that $R_2 = 0.07$ m, and upon other conditions unchanged, we find that (6):

$$p = 1.76 \cdot \left(1/0.0018 + 1/0.07\right) - 1000 = 2.921 \text{ Pa} \geq 0$$

which prevents preemption of melt out of the fusion channel and also enables to obtain high quality welded joint.

2. CONCLUSION

EBW of horizontal joints large thickness in magnetic field prevent probability defect formation as generated curvilinear beam forms curvilinear channel fusion penetration, thereby preventing metal run-out.

3. REFERENCES

Kaidalov, A.A. (2004). *Electronic-beam welding and adjacency technology*, 2-nd edition, « Environmentally friendly technology », ISBN 966-8409-03-5, Kiev, 260 pp.

Leskov, G.I. & Zhivaga, L.I. Auto welding. *The joint formation of steel welded joint featuring large thickness through use of electronic-beam welding in different attitude position,* №10, (1980) page numbers (1-5)

Martynov, V.N. (1988). *Engineering technology of electronic-beam welding horizontal beam heavy carcass details from steel and titanium alloy,* Moscow, number page (16)

Sukach, K.A. & Zhadkevich, M.L., etc. Welding fabrication. *Specificity formation of bottom defects titanium thickness through use of electronic-beam welding*, №4, (1989) page number (13-14)

Lankin, U.N. Auto welding. *Definition temperature and steam pressure in the fusion channel using of electronic-beam welding*, №2, (1978) page number (16-19)

Annals of DAAAM for 2008 & Proceedings of the 19th International DAAAM Symposium, ISSN 1726-9679
Editor B. Katalinic, Published by DAAAM International, Vienna, Austria 2008
Make Harmony Between Technology and Nature, and Your Mind will Fly Free as a Bird

NEW TRENDS IN THE IMPROVEMENT OF PRODUCTION PROCESSES

MICIETA, B[ranislav] & STOLLMANN, V[ladimir]

Abstact: *The article deals with new trends in the improvement of production and logistics processes. In the first part, the article deals with the importance of improvement of processes. Reduction of costs represents a very important factor in increasing the competitiveness of companies in a sector with strong competition. Planning and controlling of processes enables responding to condition in the market in a short period of time. In the second part of the article, an approach used in improvement of production processes will be described. In the third part, the authors refer to new and perspective methods used in their department as well as to contributions in this field. In the last part, new ways in the improvement of production processes will be described.*

Key words: *production processes, digital factory, operation costs, industrial engineering*

1. INTRODUCTION

In recent years, a great progress has already been made in the field of production processes. This article brings a view on its various methods with focus on a new research in the Department of Industrial Engineering in Zilina.

2. PROBLEM DESCRIPTION

Enterprise in strong world competition has to bring high-quality products at permanently low prices into market.

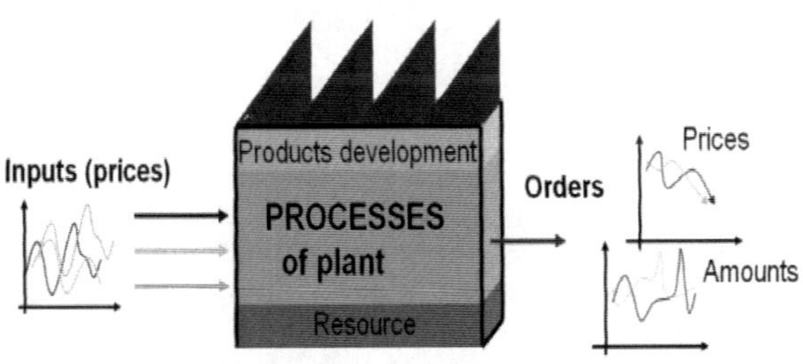

Fig. 1. Diagram of enterprise reaction to the environment conditions

Enterprise can react to oscillation of raw materials input prices and output production quantity with the target of reaching minimum prices by using three different methods:

1. Adjusting the construction of products.
2. Adjusting production technologies to the production quantity.
3. Using apposite proposal of design and optimizing production processes.

From the economical point of view, flexible reactions to perpetual changes are possible only with suitable tools of production processes optimization.

3. SOME METHODS OF PRODUCTION PROCESSES OPTIMIZATION

Many supporting tools falling into the field of Industrial engineering have been developed for the field of production systems design and optimization of production processes.

One of the best-known philosophies for the field of production processes is JIT (Just in Time).

2.1 What is JIT

Just-In-Time (JIT) production systems, invented by Taiichi Ohno, is a management philosophy that calls for reducing work-in-process inventory to improve the manufacturing cycle, increase the speed of information exchange and upgrade productivity. It is aimed at "a close alignment of production and market requirements". Small lot production, frequent delivery ofparts and components, levelling of production volume,and reduction of set-up times are the essential featuresof the system. 4Some others are the use of grouptechnology, total preventive maintenance, statisticalprocess control, cross-trained employees and uniform work loads (Mittal 2007).

2.2 TQM

Total Quality Management is a management approach that originated in the 1950's and has steadily become more popular since the early 1980's. Total Quality is a description of the culture, attitude and organization of a company that strives to provide customers with products and services that satisfy their needs. The culture requires quality in all aspects of the company's operations, with processes being done right the first time and defects and waste eradicated from operations (Nayantara ,2008).

2.3 Information systems

ERP philosophy has evolved from MRPII philosophy. MRPII philosophy evolved from MRP philosophy. It is important to understand the difference between each term.

Effective ERP requires that integrated management processes extend horizontally across the company, including product development, sales, marketing, manufacturing, and finance. It must extend vertically throughout the company's supply chain to include the acquisition of raw materials, suppliers, customers, and consumers (Bubeník, 2004).

2.4 Digital Factory

Data from workroom have to be handled online so that we can immediately respond to any product aberrations. We can create various scenarios by changing various internal and external parameters and observe how the system will behave and how it will reflect in customer requirements (Paur ,1996). This system enables us to make decisions before actual need of production. This way we can solve questions of suitable technology and profitability of production. The digital factory finds its application in all sectors with high level of competition and complex production processes and restriction of cost and rise of productivity – which are one important aspect of organization survival (Gregor, 1998).

Possibilities of recent **financial cost reduction** and CIM data Reports from March 2003 are demonstrated in Fig 1:

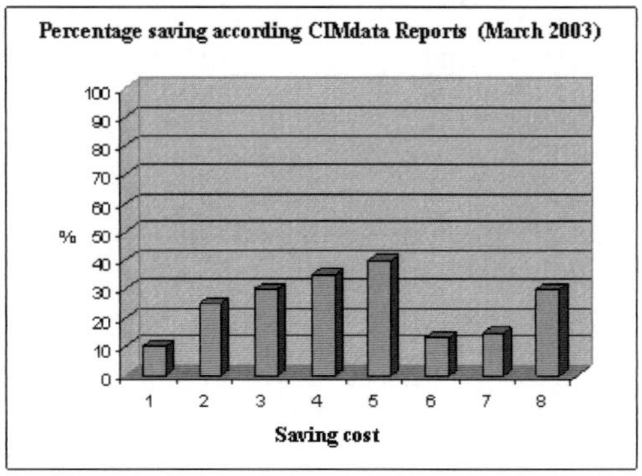

Fig. 3. Saving after implementation of the digital factory

Description of picture:

1. Cost savings due to reduction in inventory – 10%.
2. Facility saving from optimized plant layouts – 25%.
3. Cost saving due to improved labour utilization in product manufacturing – 30%.
4. Saving cost due to optimized material flow– 35%.
5. Reduced number of Workstations count, Tools, Machine – 40%.
6. Reduction in overall production – 13%.
7. Increased production throughput - 15%.
8. Time to Market Reduction – 30%.

4. RESEARCH IN OUR DEPARTMENT

Our department deals with a research of approach leading to optimization of production processes in the field of digital enterprise. Its target is to develop a progression enabling to achieve a process of production approaching the required optimum. Basic principle of optimization methodology is demonstrated in figure 4. It is based on a relation between products parts, processes and resources. This approach enables the best adjustment of production processes considering products construction as well as available production mechanisms. In methodology, the aforementioned principles of digital factory and industrial engineering are being used. Likewise, our department focuses on integrating these classical industrial engineering methods into software environment of a digital factory. Whereas the important point is the creation of integrity among examined elements.

Fig. 4. Process of production systems planning

Significance of this activity lies in considerable production effectiveness increase due to better reaction to changes of inputs and outputs. This leads to an increase in companies competitiveness.

5. WHAT IS NEW

Tools of digital factory are very promising in the modern enterprise. At present, their implementation is being put into effect in big companies, mostly in automotive and aeronautics industry. They are also being counted on in the implementation in middle companies in the future. Department of Industrial Engineering activities chart the results in the following fields of research:

1. New approaches of processes control.
2. Designing methodology in the environment of a digital factory.
3. Automation of assembly processes.

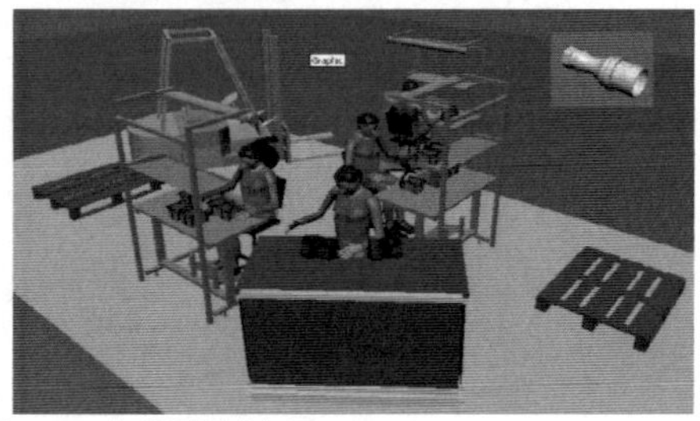

Fig. 5. Layout in DELMA

6. FURTHER RESEARCH

Future research is focused mainly on the following fields:

- integration of software base with tools of industrial engineering,
- development of specific areas of digital factory,
- ergonomic aspects of production and assembly processes.

7. CONCLUSION

This article enables to create a new view on the field of production processes improvement. This view is significant mainly for companies with high complexity of production processes. The focus of this research lies in methodology of processes improvement in the environment of a digital factory. The Department of Industrial Engineering research is intended to continue in the future.

8. REFERENCES

Bubeník, P.; Bubeníkova, E. & Korbel, P. & Nagy, P. (2004). Informačné technológie pre podnikovú prax, Information Technologies for Company Practise, Žilinská univerzita, ISBN 80-8070-288-8, Žilina, Slovakia

Gregor,M.; Halušková,M. & Hromada,J. & Košturiak,J. & Matuszek,J. (1998) . Simulation of Manufacturing System, Politechnika Lodzska, ISBN 83-87087-11-4 , Poland

Nayantara Padhi (2008), The Eight Elements Of TQM, Available from: http://www.isixsigma.com/library/content/c021230a.asp Accessed:10 September 2008

Mittal,S.;Wang ,H.P. (2007). Simulation of JIT Production to Determine Number of Kanbans, Springer London, ISSN 0268-3768,

Paur, J.; Gregor, M. & Laššák,V. & Košturiak,J. (1996) . Industrial Engineering and Manufacturing Enterprise, Univerzita Žilina, ISBN 80-966996-5-2, Žilina, Slovakia

Annals of DAAAM for 2008 & Proceedings of the 19th International DAAAM Symposium, ISSN 1726-9679
Editor B. Katalinic, Published by DAAAM International, Vienna, Austria 2008
Make Harmony Between Technology and Nature, and Your Mind will Fly Free as a Bird

CURRENT TENDENCIES IN THE ROMANIAN PUBLIC UNIVERSITY QUALITY MANAGEMENT

MIHAI, M[aria] V[alia] & DAVID, O[ana]

Abstract: *In the 19th and 20th centuries the University represented a locus of excellence. It seemed absurd to validate the quality of higher education, since the professors were examples of scientific and human quality. The new academic context, under pressure as a result of rapid extension and lack of funds, induces questions about the quality of tertiary education. In order to meet the minimum quality standards, to attain its goals of being competitive, every higher educational institution needs to develop its quality management system.*
The present article proposes to describe the evolution of the Romanian universities from the moment the socialist system was destroyed and the democratic system enabled the implementation and development of quality management programmes. .
Key words: *quality management system, quality, accreditation, public university*

1. INTRODUCTION

The concept of quality guarantee appeared in the USA during World War II, and was applied in the 1950s in the arms, spatial and nuclear industry. Later on, with the success scored by the ISO 9000 standards, the concept extended to the most diverse sectors of activity. A new client-supplier relationship developed, based on the customer's confidence in his supplier which led to the method of quality system certification.

In Romania, the issue of quality in education was out of the question before 1990. The educational system was centralized and strictly hierarchical, all the measures taken being compulsory at the inferior levels, their declared goal being to improve the learning system.

Important changes occurred after 1990 in the Romanian educational system, particularly the academic one. The most powerful one was that of the private universities that triggered competition among universities, leading to an increase in the number of students, but also to a crisis of budgetary resources for the public higher education institutions. This determined relaxation of state control and higher university autonomy.

The financial resources crisis accelerated the competition among universities, not only within the private sector, but also within the public sector, especially with a view to obtaining extra-budgetary funding resources.

2. CONTAIN PAPER

Central and Eastern European universities applied the accreditation system, following the American model, in order to ensure the minimum quality standards. However, at present, each and every educational institution has to continue its growth, to pursue its way towards competition and excellence and these are not possible without a system of quality management. Therefore, the accreditation and quality management system are complementary and synergic (Cristea, 2001).

The accreditation system establishes whether there are acceptable conditions for the initiation of a study programme, or whether the programme is to be ended as a consequence of deterioration. The moment the programme is authorized, it can benefit from quality management and from a continuous progress.

Ever since 1993, accreditation has been legalized through normative acts that have amended the legislative framework and have made possible the development of an institutional culture of quality in education, as well as of the protection of culture beneficiaries.

The National Council of Academic Evaluation and Accreditation was established, whose aims are to accredit the education suppliers and the study programmes. There are also evaluation commissions which are specialized in different domains. Every five years they assess the academic competence and performance.

A step ahead in orienting the state universities towards the client and the labour market was represented by the implementation of a new budgetary fund repartition basis in 1999.

This new system of fund distribution meant a radical, unprecedented transformation since the financing quota results from two categories of indices: quantitative and qualitative.

If during the first three years of the new type of financing mechanism, i.e. 1999-2001, the funds allocated for higher education institutions were only quantitatively based by the annual budgetary law, i.e. according to the number of students and the equivalent student cost, starting with 2002, the funds are conditioned by the meeting of qualitative indices:

- Percentage of occupied job positions;
- Quota represented by the professor and reader positions from the total teaching jobs;
- Percentage of teaching staff under 35 years of age;
- Ratio of teaching staff who are PhDs.

These are not the only qualitative indices that have been introduced for the distribution of basic finance allocations. Indeed, they measure the *premises* of quality, not the quality per se. More or less, their effects represent adjusting the procedure of money allocation with costs volume.

The quality of the didactic process is influenced by the quality of other processes or activities that take place in a university: scientific research, publishing, academic management, social services or other university activities.

These aspects have been taken into consideration since 2003, when another 11 indices are added to the initial four.

In 2005 the quality indices are readjusted and revised; they were grouped as follows:

- Teaching staff indices (5 items);
- Indices of the impact of scientific research on the didactic process (3);
- Material basis indices (4);
- University management indices (4).

The quality index notion was introduced in order to include in the university funding methodology an incentive and as a corrective component; thus, if the unitary equivalent student number parameter expresses the *volume of training services* delivered by the university to its students, by means of the quality indices the premise of the *delivered service quality* (from a financial perspective) is also taken into account.

The practice of conditioning the allocation of budgetary funding to the achievement of qualitative criteria confirms the present Romanian state policy towards an educational system adjusted to the standards of the EU concerning the academic benchmarks. In this respect, we can assert that, if in 1999-2002 the funding was 100% quantitative-based (equivalent student number), in 2003-2004 the percentage diminished to 87.3%, the difference of 12.7% representing allocation adjustment to qualitative indices; this level reached 15% in 2005 and 20% in 2006 and will continue to raise in the next years, as the tendency is clearly quality oriented, as far as budgetary allocation dimension (Cristea, 2001).

The quality orientation adopted by universities is necessary not only in the matter of accessing budgetary funds; it has an important impact on the increasing number of students able and willing to pay higher fees, since they are aware of the benefits derived from these schooling expenses. (The number of applicants for the entrance examinations and their further evolution represents a direct consequence of the credibility of a university, of its cultural potential, of its interaction with the economic and social context, in a word, it is a positive sign that the academic activity follows the right path).

An essential part in defining the quality of academic institutions has been played by the research of the Dutch academics, members of the "*Quality Assurance in HCO*" study group, coordinated by Prof. Petra van Dijk from the Polytechnic University of Rotterdam; they were preoccupied with elaborating an international quality assessment guide, and they identified six dimensions of higher education quality:

1. The legal dimension – the extent to which legal regulations and specific higher education procedures are applied;
2. The professional dimension – congruence between teaching staff activity and professional standards specified and accepted for each upper education domain/specialty;
3. The economic dimension – reaching the goals with a certain amount of expenses, such as costs, time, equipment, personnel.
4. The clients' demands – the extent to which services meet the demands, expectancies and needs of potential clients.
5. The labour market – the extent to which higher education institutions are capable of rapid adjustment to the labour market demands;
6. The organizational development – academic institution ability to define and implement the organizational strategies in keeping with the educational system requirements and objectives.

A highly contested notion can be derived from the list of higher education quality dimensions, a notion was mentioned before in this paper, namely, that of *client*, in other words the *beneficiary* of the academic services.

According to quality international standards SR ISO 8402-95, *the client* is defined as „the recipient of a product (material or unsubstantial) delivered by the supplier". Some more recent works on higher education quality provide a more comprehensive meaning for *client*, i.e. „interested party in achieving a product or service".

The implementation of a quality-oriented strategy is feasible through quality management, notion which in conformity with SR ISO 8402-95 represents "all the activities of the general management function, which determine the quality policy, its objectives and liabilities and which are implemented within the quality system by means such as quality planning, quality control, quality insurance and quality improvement".

Hence, quality management involves the implementation of a *quality system* through: quality planning, quality control, quality insurance and quality improvement.

Considering the necessity of quality education insurance through a legislative framework which would permit the development of an institutional culture of quality education and the protection of the education-beneficiary, taking into account the need of a change in the present situation when Romania was among the few european countries that do not have an established mechanism of educational quality insurance, the Law concerning the insurance of quality in education was promulgated; the law creates the legislative basis for the implementation of the management of quality in education.

The system of quality insurance of the higher education finds itself in full process of clarification and finality, the legislative framework once being institutionalized at the national level, the mechanisms of quality insurance will determine every higher education institution to become responsible for quality insurance in all activities, at all levels, in conformity with the standards in this domain.

The introduction of quality management represents a major change in the culture of every institution that is why the staff and the established structures are generally reluctant to implementing it. Resistance to change is characteristic not only of higher education institutions, but of all organizations. What is specific to public universities is how ‚successful' they prove to be in matters of conservatism and resistance to change. Their traditional structure and systems, developed along centuries, are the main resistance generators (Dinca, 2002).

Alongside the above-mentioned historical conservatism, change resistance in higher education institutions can be connected to other, more „contemporary" causes, such as the way they are organized and managed, for example, aims and objectives which are not clearly conceived by the university management, which leads to obscure results or misunderstandings, lack of perspective and direction so that actions can reach their goals, absence of a coherent action map that can offer alternative solutions, etc.

The introduction of quality management systems requires the development of negotiation skills within the educational system, the improvement of the means capable of reaching consensus of opinion, as far as the suggestions and propositions of the higher education community is concerned. Last, but not least, we need to plan and pursue a system of benchmarks that may assist us in the permanent evaluation of university performance.

3. CONCLUSION

The quality of the higher education is and must be a landmark of the didactic and research activity; it proved to be the central issue in the creation of the European Space of Higher Education.

For Romania, as signatory of Bologna agreement and as member of the European Union, it is obvious that quality can be implemented only by conforming to the standards required by the Higher Education European Space.

4. REFERENCES

Cristea L, *University Quality Management*, 10th International Symposium „Alternative Economic Strategies", Era, 2001

Cristea L, *The quality of education act,* Alternative Economic Strategies, Era, 2001

Dincă G, *Financial Management and institutional Relationships with Civil Society,* UNESCO-CEPEX, Bucuresti, 2002

*** Law 88/2003, concerning the accreditation of higher education institutions and diploma acknowledgement

*** Emergency Decree 75/2005, concerning educational quality insurance

Annals of DAAAM for 2008 & Proceedings of the 19th International DAAAM Symposium, ISSN 1726-9679
Editor B. Katalinic, Published by DAAAM International, Vienna, Austria 2008
Make Harmony Between Technology and Nature, and Your Mind will Fly Free as a Bird

A NEW CUTTING TOOL DESIGN FOR RAISING THE EFFICIENCY OF TURNING PROCESS

MIHAIL, L[aurentiu]

Abstract: *This article presents a robust and reliable design solution for the automation of the cutting insert exchange that can raise the efficiency of the turning manufacturing systems. The productivity provided by this technical solution help the manufacturing companies to raise their return of investments in a competitive and profitable manner.*
Key words: *efficiency, automatic exchange, turning tool.*

1. INTRODUCTION

According with the newest approaches on the machining sector, the costs of the cutting tool in the overall cutting cost is approximately 3% of the manufacturing costs (Sandvik 2005). Although this relatively small percentage, the influence of this cost factor on the production process economics and on the overall product's cost is considerable. The metal cutting production overall cost is, generally, composed of the sum of costs with: cutting tools, holding, fixturing, measuring equipment, machine tools,workpiece materials, labour costs and overheads. The economic performance of the cutting tool is, generally, made up of: tool life, cutting data, chip control, reliability, tool changing and inventory. Improving one of the variables in the metal cutting process has a very noticeable effect on the productivity, and thus the return on the total investment. The return on investment can be improved by several means – one of them is the use of modular cutting tools and the quick change systems. In this manner, the time that the machine is stoped will be smaller.

Setting up the machine for a new production run can be improved through the use of modular tools. The use of modular tools, where only a part of the tool is changed through a quick change, with an accurate coupling, changes the setting up time. Thanks to the newest inventions a new cutting edge is available in the right position to take over machining in seconds or, at maximum, in minutes. This fact may affect the planning of cutting data for the metal cutting. When tools needs a long time for changing, a more careful balance had to be kept between tool life and tool changing time. Tools were run at lower cutting data to make them last longer for less frequent tool changes. This aspect means obviously a poor utilization of the metal cutting time. Modular tooling breaks the bad circle which affects both setting up and tool changing times. Stoppages are reduced dramatically to the extent that cutting data is optimized, with no time needrd to change the tools. The universality of the proposed cutting system can reduce the inventory of the cutting tools, too.

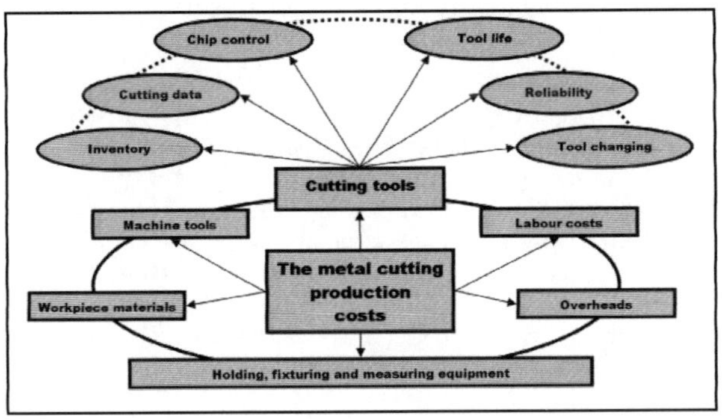

Fig. 1. The metal cutting production costs components with the aspects that regard the economic performance of the cutting tool, after (Rabezzana & Ruggero 2001).

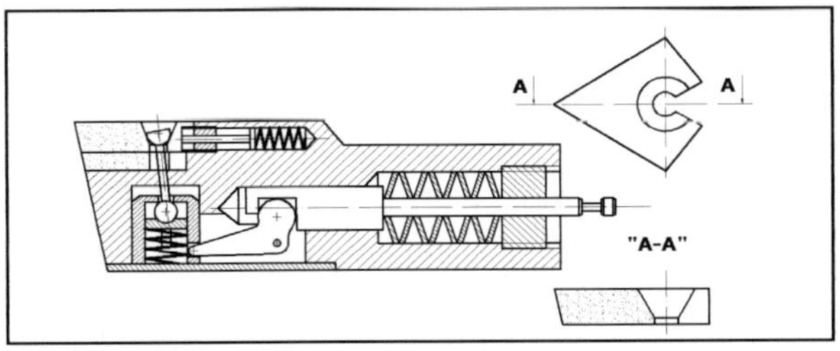

Fig. 2. A solution for a quick-change system (Кудряшов & Кукушкин 1988).

2. THE PRESENTATION OF EXISTENT SOLUTIONS

Within this article there will be taken in consideration only the first aspect: the cutting tool with his tool holding (see figure 1). There was already presented the fact that the modularity of the cutting tool system and the quick change systems have a very big importance as regards the productivity and, in consequence, the economics of cutting process. In consequence, starting from this premises, there were studied the actual commercial quick change systems and some trials from several producers and there was designed a totally new solution, that may be patented and applied as quick change system, for the turning process.

For the solution presented in the figure 2, it is needed a special type of cutting insert. In consequence, even if the constructed solution allow a certain "automation" of the cutting insert's change, there is not feasible – this form of cutting insert may be produced, but this is not a commercially design. Supplementary, the proposed mechanism to change automatically the cutting insert is not very reliable and safe.

The solution for the automatic exchange of the cutting insert presented in the figure 3 as a schetch was tried by Sandvik Coromant several years ago. It seems that this solution was not 100% reliable. In consequence, for the actual technical

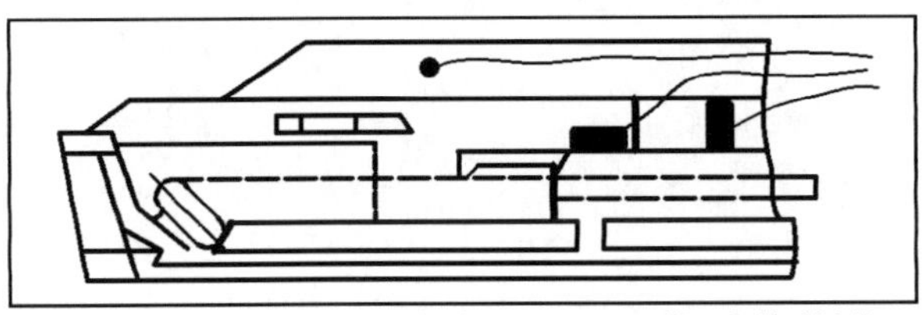

Fig. 3. A solution for a quick-change system (Sandvik 1994).

needs, there was not the best solution, so, it was not extended at commercial level. The same producer (Sandvik 2005) has another cutting tool design solution that is widely used under the name - Clamping system CoroTurn RC (Rigid Clamp), and the entire unit is denominated - Coromant Capto Cutting Unit. The power of this solution is the fact that the worn cutting edge can be exchanged very rapid, manually, with another cutting unit. After that the insert will be indexed or exchanged "off - line" – during that time the machine can run, starting to generating a new surface. The entire clamping unit is inserted in the machine tool's turret and there will be exchanged only the utile part of that cutting system.

3. THE PROPOSED SOLUTION

The proposed solution is a designed solution that allows the exchange of the cutting insert in approximately 2 seconds, time in witch the worn insert is tacked away, and another cutting insert is fixed in the ready to work place.

If it is made a comparison with the other designs, already presented, about the proposed solution it can be affirmed the next aspects: is reliable, is flexible, is cheaper, it raise the efficiency of the operation, is rigid enough, has a certain universality, can be automated and included in a flexible machining system.

Unfortunately an indexing option for the cutting tool, to transform this designed prototype in a 100% automated solution is not reliable because it does not assure the rigidity needed to work in safe conditions. A supplementary moving (rotating) element, in the working area of the cutting tool means nothing else but a rigidity decrease that transforms all the system in an useless one. Because of the lack of rigidity the accuracy of the machined surface and the safety of work will be strongly affected. The rigid structure is conferred by the centering and fixing solutions of the system's components.

Even if it is affected by this aspect, the proposed solution may be superior to the actual Coromant Capto solution because there it is not a cutting unit that must be exchanged manually, so, in consequence, the changing time remains very short. The cutting inserts magazine can be fueled with inserts at the start of each shift. The worker must take the worn cutting insert from the chips tank, after each change. In this manner, those inserts can be used to fuel the cutting insert's magazine, stored in an appropriate way to allow the alimentation of the cutting tool with an appropriate, unworn, cutting edge. The need to recuperate the cutting insert from the chips tank would not be a new task for the operator because of the relatively new requirements of the environment protection that requires the appropriate utilisation of recyclable materials. So, the worn cutting inserts must be recuperated from the swarf.

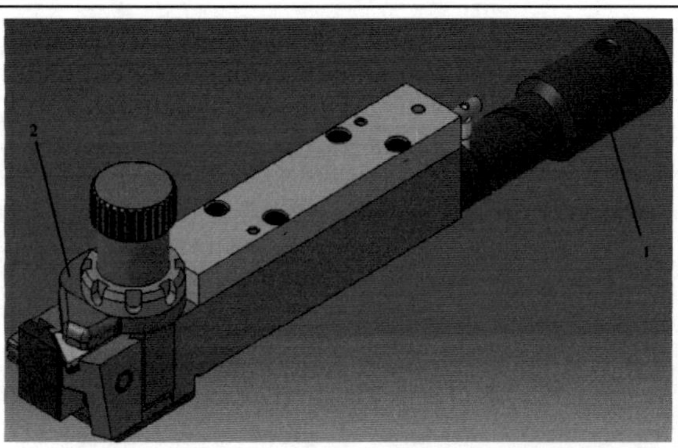

Fig. 4. The original solution (named "Gun Turning Tool") for the automation of the cutting insert's exchange at the turning process, designed in CATIA V5.

Neither the figure 4 (Dassault 2003), which presents the innovative solution, neither the text references that present his functioning mode are not intended to be exhaustive for protection of the idea reasons.

Briefly, the designed solution function in the next manner: by rotating the element (1) by 900 degrees (two times and a half) in the clockwise direction the clamping system (2) will be elevated, together with the inserts magazine, situated on the clamping system. A stick that is commanded by the spatial helical slot will help to exchange the worn insert with a new one, from the magazine. By rotating the element 1 in the anticlockwise direction, there will be fixed the new insert by the clamping unit (2) and the stick will sustain the cutting insert from his backside, to close the radial cutting force's component (that cannot be sustained in totality by the fixing clamping force and by the friction coefficient) by the rigid clamping system, practically, the movements will be annulated; in this manner the cutting insert's wear will not be affected by the supplementary vibrations. After that description it is easy to see the significance of the system's denomination "Gun Turning Tool".

The presented solution is a manual one, so, for exchanging of the cutting insert, it is needed a longer interruption of the cutting process (all the movements of the machine's tool mobile elements must be stopped, the protection screen must be opened, the rotation must be done manually, and, after that the machining cycle can be restarted). Practically the represented solution was conceived for traditional turning tools. For the computer controlled cutting machines, there can be used a linear pneumatic motor, that action a gear. In this manner there will be needed only some seconds to exchange the cutting insert (5 or 6 seconds) and another period of time to stop and to start the cutting cycle (10 -15 supplementary seconds), or the cutting insert's tool life can be programmed at the end of machining of one part, and, in this manner, there will not be needed any supplementary time interval.

The proposed solution is conceived to be utilized at the finishing turning operations, for generation of the external surfaces. It can be estimated that the solution is rigid enough to support roughing operations even for strong materials.

The chip's control can be helped with an appropriate cutting fluid flow, capable to flush the swarf in a convenient direction. The presented solution uses cutting inserts with a clearance angle of zero degrees. An uncontrolled long chip can damage the cutting tool system and the generated surface.

4. CONCLUSIONS

There can be affirmed that the market concurrence pushes the engineers to bring new ideas to raise the efficiency of the manufacturing process. The present article presents an approach that follows this tendency. By using the proposed design idea, there will be reached an improved return on investment that will allow a bigger economic performance.

5. REFERENCES

Dassault Systèmes (2003). *CATIA Version 5 Release 12 Print CD,* Dassault Systèmes, France

Кудряшов, Е.А.; Кукушкин, А.Н. (1988). Stanki i Instrument, Резец с автоматичесой сменой режущей пластины (en. *Cutting tool with automatical exchange of the insert),* No. 9, 0ct. 1988, pages 34-38, УДК 621.9.025.

Rabezzana, F.; Ruggero, C. (2001). *Manuale degli utensili (en. The Cutting Tools Handbook),* Techniche Nuove, ISBN: 88 – 481 – 1217 X, Italy

Sandvik Coromant (1994). *Modern Metal Cutting – a practical handbook,* AB Sandvik Coromant, ISBN 91-972299-0-3, Sweden.

Sandvik Coromant (2005). *Metalcutting Technical Guide,* AB Sandvik Coromant 2005.10, C-2900:3 ENG/01, Sweden.

Annals of DAAAM for 2008 & Proceedings of the 19th International DAAAM Symposium, ISSN 1726-9679
Editor B. Katalinic, Published by DAAAM International, Vienna, Austria 2008
Make Harmony Between Technology and Nature, and Your Mind will Fly Free as a Bird

MODELING AND SIMULATION OF A SPHERICAL BEARING MOUNT

MIHAILIDIS, A[thanassios]; PUPAZA, C[ristina]; NERANTZIS, I[oannis] & KARAOGLANIDIS, G[eorge]

Abstract: *The paper deals with FEM modeling and simulation of a spherical bearing mount used in Formula SAE race cars, where performance, reliability and safety are important requirements. Calculations are presented to evaluate the load carrying capacity of these critical components, and to support the engineer to find a light weight but also durable design. Several load cases were performed and advanced contact technology procedures were used presuming severe form of boundary nonlinearities. The attempt proved the possibility of the reduction of the car total weight. Remarks regarding links with optimization procedures integrated in solvers are also included.*

Key words: *FEM, contact, bearing, car, race.*

1. INTRODUCTION

Contact technology is a powerful functionality implemented in advanced solvers which assists the user to obtain realistic simulation results for complex assemblies and geometries. Formula SAE race cars components such as spherical bearing mounts are special machine elements requiring performance, reliability and safety. These assemblies are used more than 70 times in a race car design. Modeling and simulation are necessary in order to verify the suitability to the applied load and the possibility of using smaller size mounts for future releases of the race cars. Although recent structural reports regarding the mechanical systems of race cars are available (Hill et al., 2001) (Johnson, 2003), no information about spherical bearing mounts and bolted highly-prestressed joints are included. Finite element analysis for bolted joints (Kim et al., 2007), taking into account the pretension and friction with different types of contact elements has been reported, but simplified models were included in the assembly in order to reduce the dimension of the model. The paper deals with modern contact technology procedures applied on a spherical bearing mount, emphasizing the efficiency of this functionality. Remarks regarding possible links with optimization procedures integrated in solvers are also included.

2. SPHERICAL BEARING MOUNT

The A-arms, the pushrods, the shock absorbers and various levers of modern formula SAE race cars are connected to the chassis, bell cranks or wheel uprights by ball joints, also known as spherical bearings. The spherical bearing mount considered in this study is supplied by Askubal which provides the requiring technical data (Askubal, 2008).
Figure 1 shows the bearing mount. The outer ring is press-fitted to a carrying ring which is welded to the carrying part (i.e. the A-arm or the pushrod etc.). It is secured axially by deforming the lips, which have been manufactured at both faces of the carrying ring. This design has proved to be compact, light, reliable and cost effective. No further investigation is needed. The inner ring is typically attached in a U-holder or between two thin metal sheets with a thickness of 1,5 – 3 mm, by using two identical spacers and a bolt. The cylindrical portions of the spacers have a thickness of just 1 mm and are press-fitted, 0,015 mm oversize, in the inner ring.

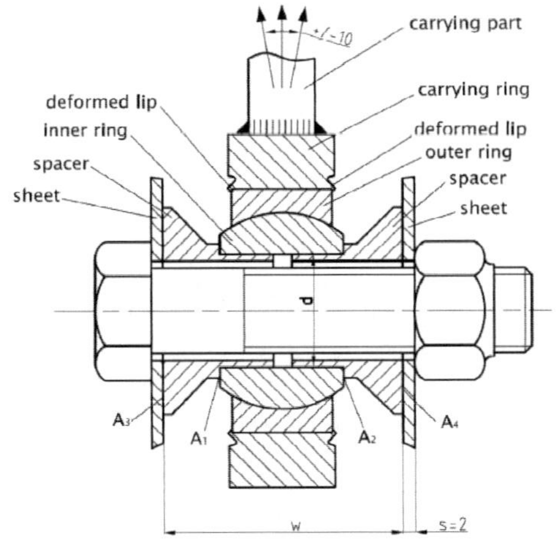

Fig. 1. Spherical bearing mount design.

Due to their small thickness, their effect on the clearance of the spherical bearing can be neglected. The load is transmitted from the spherical bearing to the sheets entirely by the friction on the faces A1, A2 and A3, A4. The required normal force is applied by the bolt, which is mounted through the sheets and the spacers. The racing versions of spherical bearing mounts are made of hardened steel and they operate without clearance. Their attachment has to fulfill the following design requirements: the connection must be completely clearance free, the A-arms, the push-rods and the shock absorbers should be easy to disassemble and reassemble.

3. MODELING AND SIMULATION

3.1 CAD and FEM models

The CAD model was completed using Autodesk's Mechanical Desktop and Inventor, modifications were done in SolidWorks 2007 and solved with ANSYS. Friction coefficients with a value of 0,12 were included in the model for the faces A1, A2 and A3, A4 and between the outer and inner ring a 0,04 value was considered. Corresponding materials were properly assigned, as follows: the spacer - 30CrNiMo6 with $R_{p0,2}$=1050N/mm^2, and the inner ring - 100Cr6 hardened steel.

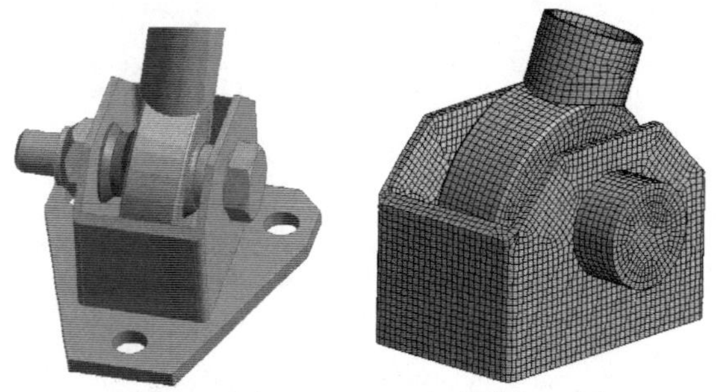

a. CAD model b. FEM model

Fig. 2. CAD and FEM models.

Several load cases were performed. Finally, the bolt pretension was 20 KN, and the force acting on the carrying part was 2kN, when the carrying part is 10° right positioned. Figure 2 shows the FEM model which contains 144490 nodes and 36868 elements. Contact sizing, refinement and hexahedral dominant mesh options were chosen in order to obtain a sufficient refined mesh. The final error estimation energy, which gives information for optimizing the mesh density was also processed and controlled. No penetration between parts was detected.

3.2 Contact nonlinearities and FEM procedure

Contact conditions in assemblies are severe form of nonlinearities. The algorithm used for solving boundary or contact nonlinearities is the Newton-Raphson procedure. The finite element discretization yields a set of simultaneous equations:

$$[K]\{u\}=\{F^{a}\} \tag{1}$$

where [K] – coefficient matrix; {u} – vector of unknown DOF values, $\{F^{a}\}$ - vector of applied loads. Equation (1) is nonlinear because [K] is a function of unknown degrees of freedom values and can be written (Bathe, 1996):

$$[K_i^T]\{\Delta u_i\}=\{F^{a}\}-\{F_i^{nr}\} \tag{2}$$

$$\{u_{i+1}\}=\{u_i\}+\{\Delta u_i\} \tag{3}$$

where $[K_i^T]$ - Jacobian matrix (tangent matrix); i – subscript representing the current equilibrium iteration; $\{F_i^{nr}\}$ - vector of restoring loads corresponding to the element internal loads. The solver options were: iterative, contact stiffness updating after each iteration, weak springs included to facilitate solution, preventing numerical instability, force convergence value, contact interface treatment: add offset value, no ramping.

3.3. Results

Figure 3 contains a plot of the equivalent von-Mises stress, on the pre-tensioned bolted joint. When processing the results, the sheet bracket was hidden in order to show the contact stress distribution. The maximum von Misses stress was 473 MPa and the maximum normal stress 137 MPa. The highest value of the stress acts on spacers, parts manufactured from materials with excellent strength characteristics: tensile yield strength σ_{yield}=789 MPa, and tensile ultimate strength σ_u=1050 MPa.

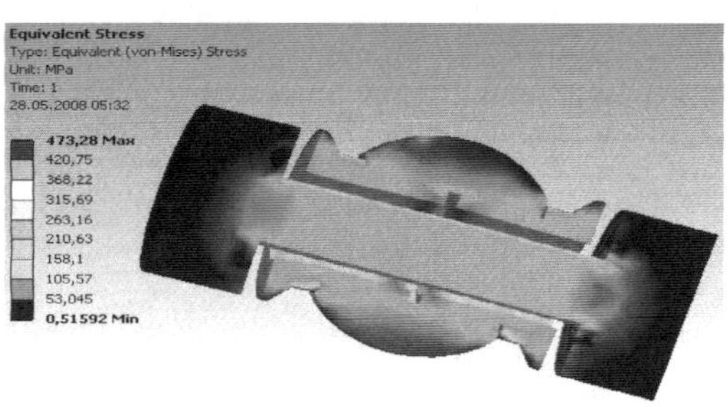

Fig. 3. Equivalent stress in the cross-section.

3.4. Verification

Because the big number of spherical mounts included in the SAE race cars and the high levels of the stress obtained multiple checks were performed. The theoretical value for the displacement along Z axis was 0,0246 mm and the value obtained through simulation was 0,0241 mm (Fig. 4.). The error can be considered even lower taking into account that the pre-tensioned bolt assembly was modeled as a bar under tension.

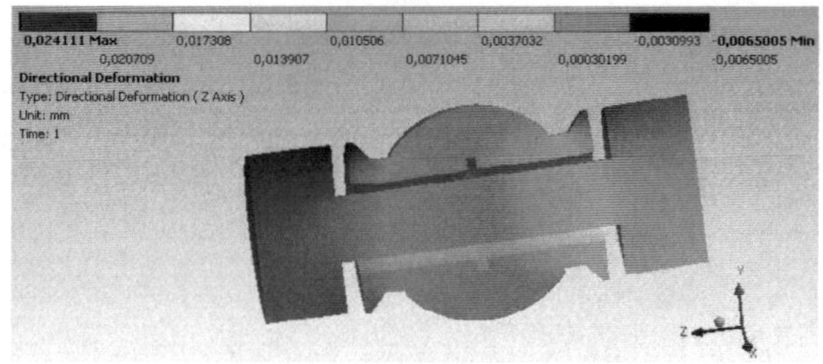

Fig. 4. Displacements along Z in the spherical bearing mount.

This allows the reduction of the spherical bearing mount size without an increased potential for failure, but additional checks regarding the dynamical behavior of the assembly are necessary, especially when the load is applied with shock.

4. COUPLING CONTACT ANALYSIS WITH OPTIMIZATION PROCEDURES

Recent contact features implemented in solvers allow the user to rapidly obtain stress evaluation in complicated assemblies, but handling and maintaining the parameters definition is not easy. Usually, when transferring data between CAD-CAE systems the definition of the parameters in a text format is not available anymore. The DS option allows the definition of geometrical parameters in a late stage of the design without returning to the CAD system. This is an important advantage because parametrical optimization procedures can be accessed.

5. CONCLUSION

The research revealed the possibility of using smaller size bearing mounts for future releases of the race cars, which means the reduction of the total weight. Contact technology supports the designer in rapidly evaluating solutions, and allows the reduction of the total weight of a product in cases were reliability, safety and integrity are required. The procedure has the following advantages: it is simple and fast; no assumptions have to be done regarding the load input and no empirical factors have to be used. Mesh quality in the contact region is controlled, model parameterization is possible and the user can edit the solver input file. Future development of the procedure in optimization loops has also been investigated. The possibility of applying the same procedure to internal spur gears was already done and promising results are in progress.

6. REFERENCES

Askubal (2008). Rod ends and spherical bearings. *Available from:* http://www.askubal.de. *Accessed:* 2008-04-12

Bathe, J.K. (1996). *Finite Elements Procedures*, Prentice-Hall Inc., ISBN 0-13-301458-4, New Jersey, USA

Hill, J.; Binderup, A.; McBride, H.; Nimmergut, B. (2001). American Solar Challenge, PrISUm Odyssey Structural Report, *Available from:* http://xnet.rrc.mb.ca/solacar/downloads/structuralReport.pdf. *Accessed:* 2008-03-06

Johnson, D. (2003). American Solar Challenge 2003 Structural Report. Team Lxu, Yale University 10 Marston Hall, Ames, Iowa 50011. *Available from:* http://www.eng.yale.edu/TeamLux/TLProject_Web/docs/4-JL_Structural_Report.pdf. *Accessed:* 2008-03-10

Kim, J.; Yoon, J. C. & Kang, B-S. (2007). Finite element analysis and modeling of structure with bolted joints. *Applied Mathematical Modeling*, Vol. 31, Issue 5, May 2007, p. 895-911. Elsevier Inc., *Available from:* http://www.sciencedirect.com/. *Accessed:* 2008-04-1

Annals of DAAAM for 2008 & Proceedings of the 19th International DAAAM Symposium, ISSN 1726-9679
Editor B. Katalinic, Published by DAAAM International, Vienna, Austria 2008
Make Harmony Between Technology and Nature, and Your Mind will Fly Free as a Bird

THE IDENTIFICATION OF CUSTOMER'S REQUIREMENTS IN CHOOSING OF A PRODUCT'S MODEL

MIHARTESCU, A[na] A[ndreea] & MAZILESCU, C[risanta] A[lina]

Abstract: *The objective of this paper is the development of a mathematical model for decision making, when a product is chosen. The paper presents the results of a multidisciplinary research developed on a representative sample of population (30 subjects) and accomplished through an informatics' model. The research method used was the investigation, and the research tool used was the questionnaire, which was applied on a HTML interface. The research was made in the framework of a national research contract, no. 3/1.10.2007, PN IIRU. Based on results of this experimental research, we propose, for the future, the extension of the sample at national level and the model generalization, so that it can be applied for a variety of products.*

Key words: *requirements, customer, multidisciplinary research, product.*

1. INTRODUCTION

Today, the conception methods are preponderant and they have an important place in product's development. Any enterprise can afford to spend money and to loose time. The enterprises must design, produce and sell EFICIENTLY. This is the reason for a company must answer to its customers' requirements as exact as possible.

In the first part of paper is presented the research methodology and a short description of informatic model for business data processing of information provided by customer for conception process.

In the second part of paper is shown a mathematical model for decision making, when a product is chosen.

2. THE RESEARCH METHODOLOGY

For the development of informatic model for business data processing of information provided by customers, a multidisciplinary research was made. The research consisted in optimization of different methods and techniques, which are specific to the following domains: conception, marketing, psychology - sociology, ergonomics, etc. The main processes to develop the model business data (Broadbent, 2003) processing of information provided by customer are: Acquisition, Processing, Using-Reusing and Dissemination.

For customer information acquisition I will use a questionnaire, for which I used as informatics' support an interface HTML.

For stocking the information provided by the customer I propose the usage of a data base, for example MySQL.

The information processing will be made with StatGrafics, and for data integration in the conception process is used QFD (Quality Function Deployment).

The methods presented are not competitional but complementary. With StatGraphics are processed the information provided by customer, and with QFD the customer's needs-requirements are transformed in achievable technical specifications, and so these are integrated in conception process.

The Research Methodology used was a multidisciplinary one, centred mainly on the research methods specific to marketing and mathematical statistics, with transmission and transformation of gathered information in statistical data.

The method of information acquisition (Kotler, 2006) used was the INVESTIGATION, and the research tool was the QUESTIONNAIRE. A HTML interface was used for the questionnaire application on an (experimental) sample made of 30 subjects.

The sampling method chosen was the **random sampling**. For its implementation it were took count of the next preset criteria – age (18-40 years), occupation (different activity domains) and gender (50% women and 50% men); they are customers of a company from Timisoara, which produces furniture. The questionnaire used is made of 20 close questions and 17 questions regarding the product features.

3. THE RESULTS OF RESEARCH

After the acquisition and processing of data provided by customers, were obtained information which was transposed in specific requirements of customers (Goldstein, 2005) which must be satisfied by a product, and which are shown in table 1.

- **The economic requirements** are those which impose the quality of the product at the minimum cost.
- **The functional requirements** refer to what is expected from the product, which are the global function and the service functions that have to be carried out by the product.
- **The ergonomical requirements** refer to the characteristics of the product which bring about additional comfort in usage.
- **The ecological requirements** refer to the product manufacture, use and integration within the natural environment, as well as its recycling when is no longer used.
- **The aesthetical (Roy 2007) requirements** refer to the impact of the product over the customers senses (subjectively expressed), the most remarkable being the visual effect correlated with the product market image.

Based on the results obtained, the company decided to produce 5 product variants (computer bureau), which to be offered to the company customers. The customers must decide to buy a variant or another from the 5 product variants that company offered, depending on the most important requirements, in each customer opinion.

No.	Requirements	Symbol	Results %
1	Economical	R_1	35%
2	Functional	R_2	25%
3	Ergonomical	R_3	15%
4	Ecological	R_4	15%
5	Aesthetical	R_5	10%

Tab. 1. Specific requirements of customers.

R_j / V_i	R_1 u.m.	R_2	R_3	R_4	R_5
V_1	150	9	VG = 1	No = 0	VS = 0
V_2	186	15	S= 0,33	No = 0	S =0,33
V_3	300	20	S= 0,33	No = 0	S =0,33
V_4	225	20	S= 0,33	Yes = 1	G =0,66
V_5	200	29	G= 0,66	Yes = 1	S = 0,33
Kj	3	2	1	2	1

Tab. 2. Correlation between the product's variant (V_i) and the customers' requirements (R_j)
Note:VG = Very Good=1;G= Good=0,66; S = Silly=0,33; VS = Very Silly=0.

In table 2 is shown the correlation between the customers' requirements and the product's variants offered by the company.

For making the decision to choose a product variant the next phases must be covered:
1) Establishment of importance coefficient (K_j) for every requirements; after establishment of importance coefficient must be given grades there is a qualitative appreciation;
2) Determination of coefficient γ and of concordance matrix Γ;
3) Determination of coefficient δ and of discordance matrix Δ;

Concordance coefficient γ is determined as it follows:

$$\gamma = \frac{\sum_{j=1}^{n} k_j^C}{\sum_{j=1}^{n} k_j}; \ V_k \text{ preferred } V_l \qquad (1)$$

Discordance coefficient δ is determined using the formula:

$$\delta = \begin{cases} 0 \\ \max a_{lj} - a_{rj} \end{cases} \quad V_r \text{ preferred} V_l; \ V_l \text{ preferred } V_r \qquad (2)$$

a_{lj}, a_{rj} – the grades given to the variants V_l and V_r for j criterion.

For optimal variant choosing (minimum or maximum) the following formula is applied:

$$a_{ij} = \frac{x_{ij} - \min x_{ij}}{\max x_{ij} - \min x_{ij}} \text{ or } a_{ij} = \frac{\max x_{ij} - x_{ij}}{\max x_{ij} - \min x_{ij}} \qquad (3)$$

$$a_{11} = \frac{300-150}{300-150} = 1; \ a_{21} = \frac{300-186}{300-150} = 0{,}76 \ldots a_{ij} = \ldots$$

After the calculus iteration and of grades given, the table componence becomes:

R_j / V_i	R_1 u.m.	R_2	R_3	R_4	R_5
V_1	1	0	1	0	0
V_2	0.76	0.3	0,33	0	0,33
V_3	0	0.55	0,33	0	0,33
V_4	0.5	0.55	0,33	1	0,66
V_5	0.66	1	0,66	1	0,33
Kj	3	2	1	2	1

Table 3. Correlation between the product's variant (V_i) and the customers' requirements (R_j).

In the variants set is introduced a surpassing relation. We consider that variant V_r surpasses variant V_l if:

$$\gamma (V_r; V_l) \geq p \qquad (4)$$
$$\delta (V_r; V_l) \leq q$$

where p and q are limiting values imposed by the decision maker.
Based on concordance and discordance matrix and on relation (4) we considered:

1) p=1; q=0; p+q=1
$\gamma_{ij} \geq p$ and $\delta_{ij} \leq q \Rightarrow V_i$ preferred V_j
$\gamma_{43}=1 \geq p$ and $\delta_{43}=0 \leq q \Rightarrow V_4$ preferred V_3
$\gamma_{53}=1 \geq p$ and $\delta_{53}=0 \leq q \Rightarrow V_5$ preferred V_3
2) p=0.88; q=0.12; p+q=1
$\gamma_{54}=0.88 \geq p$ and $\delta_{54}=0 .33 \geq q$, in this case condition (4) is not accomplished
3) p=0.66; q=0.34; p+q=1
$\gamma_{54}=0.88 \geq p$ and $\delta_{54}=0 .33 \leq q \Rightarrow V_5$ preferred V_4
$\gamma_{42}=0.66 \geq p$ and $\delta_{54}=0 .26 \leq q \Rightarrow V_4$ preferred V_2
$\gamma_{52}=0.66 \geq p$ and $\delta_{52}=0 .10 \leq q \Rightarrow V_5$ preferred V_2

Concluding, the V_5 variant surpasses all other variants. The product model in V_5 variant has a medium price, with a maximum points rating regarding its functionality and with a medium points rating regarding ergonomics.

4. CONCLUSION

The model shown helps customers to choose from many products assortments existing on the market, and also helps producers to offer these products which corresponds with customers requirements and expectations expressed in market studies made before the product development.Under the present conditions of the world economy where mass production has been replaced with ordered production, personalized products are actively present on the international market. On account of this reason, the customer's involvement in the products design stage is needed.

For the future we propose to develop the informatics model and to generalize an informatics support which helps the decision making in the shortest time and which is applicable in a wider range of activity domains.

5. REFERENCES

Broadbent, M. & Weill, P. (2003). *The Implications of Information Technology Infrastructure for Business Process Redesign, MIS Quarterly*, Vol. 23, No. 2, Jun., 2003, pp. 159-182, Published by: Management Information Systems Research Center, University of Minnesota

Goldstein,D. & Lee,Y. (2005). *The Rise of Right-Time Marketing,* Journal of Database Marketing & Customer Strategy Management 12, no. 3/2005.

Kotler, P.& Keller L. (2006). *Marketing Management*, Prentice Hall Publishing House, ISBN 0131457578, New Jersey

Roy, R.; Baguley, P. & Reeve, L. (2007). Understanding the Link between Aesthetics and Engineering in Product Design, *Proceedings CIRP Design Conference,* pp.155-164, ISBN: 3540698191 ISBN13: 9783540698197, Berlin, march 2007, Springer-Verlag Berlin

http://www.hindawi.com/getarticle.aspx?doi=10.1155/jamds/2006/10936, *Accessed: 2008.07.20*

Annals of DAAAM for 2008 & Proceedings of the 19th International DAAAM Symposium, ISSN 1726-9679
Editor B. Katalinic, Published by DAAAM International, Vienna, Austria 2008
Make Harmony Between Technology and Nature, and Your Mind will Fly Free as a Bird

DATA STRUCTURES FOR ROAD CONDITION AVI FILE VIDEO AUGMENTATION

MIHIC, S[rdan] & IVETIC, D[ragan]

Abstract: *The paper addresses issues which road engineers experience while using the ROad Measurement and Data Acquisition System (ROMDAS). The ROMDAS system collects and analyses the road-condition state through video and the discrete data acquired by measurement devices. Videos are stored separately from the corresponding data. Thus, road engineers have to search the videos manually in order to find details of interest provided by data analysis. We have designed Augmented Video stream Framework (AVF) that enables creation, search and playback of augmented AVI files for effective road surveying. This paper focuses on AVF data structures used for video data augmentation. The AVF uses a type system similar to C++ and Java programming languages and offers encapsulation of arbitrary data.*

Key words: *Data, retrieving, augmentation, AVI video, ROMDAS.*

1. INTRODUCTION

The public institution Road Center of Vojvodina (Centar za puteve Vojvodine, CPV) is using ROad Measurement and Data Acquisition System (ROMDAS) (Bennett et al., 2007) for road inspection and maintenance. The ROMDAS system consists of several measuring devices (gyroscope, GPS receivers etc.), a video camera mounted on a vehicle, and software to process the collected discrete data. The video camera captures video into AVI format. The measuring devices capture the discrete data about the physical characteristics of the road-condition state such as: road roughness, transverse profile and rut depths, traffic density etc. After the completed survey run, the measured data are processed and analyzed.

ROMDAS captures the visual road-condition state with the video camera mounted on the vehicle. This video represents a valuable source of information for road managers and road engineers since it provides visual feedback of the collected discrete data. Unfortunately, the video is stored separately from the discrete data acquired by the measuring devices. Therefore, road engineers have to search the video manually in order to find details of interest provided by data analysis. This is a tedious and error prone task. Hence our approach has been to integrate both data and video in order to support more effective and comfortable information retrieval. The integration is carried out by encapsulation of both the video and the data into one file in a way that facilitates data storage and communication. The data used for video augmentation are called augmented data.

Augmented Video stream Framework (AVF) was designed for ROMDAS road-condition state video augmentation with discrete data collected by ROMDAS measurement devices. The AVF provides a full search of augmented data according to the data properties. The implementation of the AVF is based on Microsoft DirectShow for synchronized playback of the basic video and augmented data (Mihić, 2007).

The proposed AVF framework was created to solve one specific problem of road engineering. It was designed, however, to be extensible and applicable on a wide range of information systems.

This paper focuses on data structures that encapsulate the augmented data.

2. VIDEO AUGMENTATION

The problem which we were to solve can be classified as part of the so-called field of video augmentation. Studies have shown that video stream augmented with data provides a deeper understanding of captured reality and promotes active watching (Correia & Chambel, 1999). Additionally, the augmented data provide consumers with the ability to acquire new knowledge and advocate content-oriented video access and retrieval.

One of frequently used approaches for video augmentation is annotated video. An annotated video is a video augmented with annotations. By annotations Schroeter et al. (Schroeter et al., 2007) mean: descriptions, notes, subjective comments and various observations that can be attached to the video document without actual document modification. The variety of annotated information is limited by the annotator's knowledge, and it is subjective (Schroeter et al., 2007).

Another popular approach for video augmentation is augmented video. An augmented video is the result of augmenting a certain video with non-perceivable data captured at the time of the recording. Usually, 3D computer generated objects are rendered on top of the video and merged into the video stream.

The ROMDAS system can be classified as an annotated video system with characteristics of augmented video. The measured discrete ROMDAS data are closely related to the video content and therefore inseparable from the video.

Agosti and Ferro abstracted the definition of annotation to include all forms of video data augmentation. Annotations are divided by their correlation to the video content into: content enrichment and stand-alone documents. The former regards annotations as closely related to the video content and therefore inseparable from the video. The latter regards annotations as real documents and autonomous entities that maintain some sort of connection with the video content (Agosti & Ferro, 2007). By their definition, the ROMDAS system can be classified as a stand-alone annotation system, whereas the nature of the augmented data poses that the system should be a content enrichment annotation system.

3. AUGMENTED DATA STRUCTURES

Microsoft's Audio Video Interleave (AVI) was used as implementation multimedia container format (MCF). AVI was chosen because ROMDAS video camera captures video into AVI format. Additionally, our comprehensive study of commercial MCFs (Mihić, 2007) showed that other MCFs offer very similar features, and almost all of them are suitable for augmented data encapsulation and storage. In the early stages of development there were considerations to create a new specifically designed MCF, like in ANNODEX system (Pfeiffer et al., 2003). Since our study showed that existing MCFs are suitable, that approach was abandoned.

For security reasons and to achieve effective data storage, it has been required to restrict access to augmented data and enable augmented data compression while maintaining compatibility.

The next two approaches for augmenting data into MFC are common: interleaving or non-interleaving. The former approach is suitable for applications where video streaming is needed, and the latter is suitable for applications where high compression ratio and effective data encryption is needed. AVI MCF does not support interleaving of augmented data and therefore the latter option was chosen.

AVI file is organized into small pieces called chunks. Chunks are identified by their name (FOURCC) and they can be nested. Specification provides a chunk named 'INFO' for additional data description. The AVI specification requires that AVI file parsers should ignore any unknown 'INFO' sub chunk (CORPORATE Microsoft Corp. 1991). According to this constraint augmented data encapsulated in 'INFO' chunk should maintain compatibility and therefore would enable unauthorized consumers to watch the basic video. We created an 'INFO' sub chunk named 'AUGD', since several sub chunks are already defined by the specification. All the augmented data are encapsulated into this chunk. The augmented data are compressed and encrypted before encapsulation. Arbitrary compression and encryption algorithms can be used.

The augmented data are structured in an object-oriented way. The data are described using a type system similar to those in C++ and Java programming languages (Fig. 1). AVF defines: bool, int, double, string, image, audio stream and video stream as atomic types, as same as in (Romero & Correia, 2003). All the data are encapsulated into classes. AVF type system supports class inheritance. The concrete value of the measured discrete ROMDAS data in certain time interval is represented by the Object class. We have chosen time intervals instead of frame/frame intervals because of the nature of the discrete ROMDAS data. Namely, real-time video capturing poses constraints on achievable frame rate, and the time resolution of measurement devices goes as low as 10^{-3}s. If we would have used frames/frame intervals, valuable measured data could not have been described using our system.

The implemented time resolution of AVF framework was set to 10^{-8}s – the lowest value usable in commercial multimedia presentation frameworks.

Although the AVF type system was designed to describe discrete ROMDAS data, it can also describe arbitrary data structured in an object-oriented manner.

4. CONCLUSION

The ROad Measurement and Data Acquisition System (ROMDAS) collects and analyses the road-condition state through videos and the discrete data acquired by specific measurement devices. Separation of video and data storage forces road engineers to search the video manually in order to find details of interest.

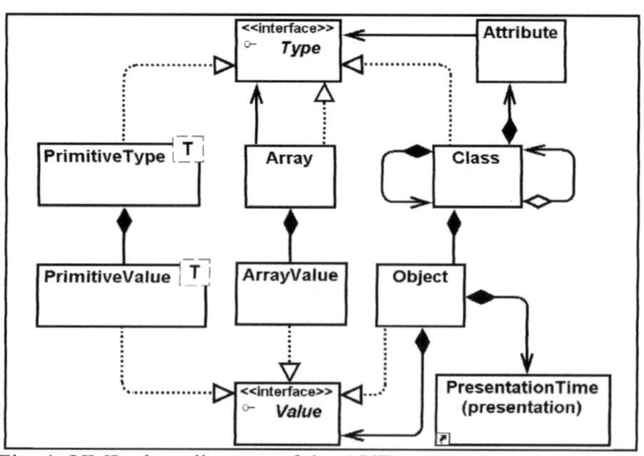

Fig. 1. UML class diagram of the AVF type system.

We concluded that the augmented video should be a self-contained entity allowing the full data search according to data properties. A hybrid video augmentation system was designed: Augmented Video stream Framework (AVF). The AVF enables creation, search and playback of self-contained augmented AVI files for effective road surveying. The AVF approach is valuable far beyond the application area of road maintenance.

This paper introduced AVF data structures used for video data augmentation. The AVF uses type system similar to C++ and Java programming languages and offers encapsulation of arbitrary data in an object-oriented manner. Supported AVF atomic types are: bool, int, double, string, image, audio stream and video stream. Time intervals were used as synchronization units between the video and the augmented data.

The ROMDAS video augmentation was carried out using only discrete atomic types because ROMDAS measurement devices only capture the discrete data. In the future we plan to enhance ROMDAS videos with continual media (e.g. supplemental video of road roughness state).

Further research will be conducted towards creation of augmented videos based on ontologies of different applications.

Acknowledgements This research was supported by IT Project No. 13013, financed by the government of Republic of Serbia.

5. REFERENCES

Agosti, M. & Ferro, N. (2007). A formal model of annotations of digital content. *ACM Transactions on Information Systems (TOIS)*, Vol. 26, No. 1, (November 2007), Article No. 3, 57 pages, ISSN: 1046-8188

Bennett, C.; R., Chamorro, A.; Chen, C.; Solminihac, De.H. & Flintisch, G.W. (2007). *Data Collection Technologies for Road Management,* The World Bank, East Asia Pacific Transport Unit, Washington, D.C.

CORPORATE Microsoft Corp. (1991). *Microsoft Windows multimedia programmer's reference,* Microsoft Press, ISBN: 1-55615-389-9, Redmond, WA, USA

Correia, N. & Chambel T. (1999). Active video watching using annotation. *Proceedings of Seventh ACM international Conference on Multimedia (Part 2),* pp. 151-154, ISBN: 1-58113-239-5, Orlando, Florida, USA, October 1999, ACM, New York, NY, USA

Mihić, S. (2007). *Augmented Video stream Framework.* M.Sc. thesis (in Serbian), Faculty of Technical Sciences, Novi Sad, Serbia

Pfeiffer, S.; Parker, C. & Schremmer, C. (2003). Annodex: A Simple Architecture to Enable Hyperlinking. *Proceedings of the 5th ACM SIGMM international Workshop on Multimedia information Retrieval.* pp. 87-93, ISBN: 1-58113-778-8, Berkeley, California, November 2003, ACM, New York, NY, USA

Romero, L. & Correia, N. (2003). Mixed reality hypermedia: HyperReal: a hypermedia model for mixed reality. *Proceedings of the Fourteenth ACM Conference on Hypertext and Hypermedia*, pp. 2-9, ISBN: 1-58113-704-4, Nottingham, UK, August 2003, ACM, New York, NY, USA

Schroeter, R.; Hunter, J.; Newman, A. (2007). Annotating Relationships between Multiple Mixed-Media Digital Objects by Extending Annotea. *Lecture Notes in Computer Science*, Vol. 4519/2007, (June 2007) pp. 533-548, ISSN: 0302-9743

Annals of DAAAM for 2008 & Proceedings of the 19th International DAAAM Symposium, ISSN 1726-9679
Editor B. Katalinic, Published by DAAAM International, Vienna, Austria 2008
Make Harmony Between Technology and Nature, and Your Mind will Fly Free as a Bird
*Young Researcher and Scientist Paper / * Supervisor, Mentor*

AUTOMATED CONTROL SYSTEM OF FESTO MODULAR PRODUCTION STATIONS

MIKHAILOV, M[axim] & LAVROV, V[alentin]

Abstract: *The project purpose is design of the automated control system by modular industrial complexes of firm FESTO - MPS (Modular Production Station) which use for training in the field of industrial automation. The control system consists from: the hardware in which quality the module of discrete input-output ADVANTECH PCI-1756 and a program part in the form of the developed software for the control of manufacturing processes - MPS-CT - MPS Control Tool.*
Key words: *control system, training, mechatronics, FESTO.*

1. INTRODUCTION

Use of computer and information technologies for control of technological processes at the modern industrial enterprise allows reducing quantity of the working personnel and defines new, increased requirements to training of technical specialists. The specialist is obliged to lead service and control of the complex automated systems for what knowledge of their mechanical arrangement and principles of functioning of their basic components are necessary for him. Hence, process of training of experts in the field of manufacture automation should have complex character and cover not only an information component of a control system, but also its electronic and mechanical part.

For realization of training in the field of industrial automation are apply complexes on the basis of modular industrial stations FESTO – MPS (Modular Production Station). In particular, these stations use in the International educational project "Synergy" (Mogilnikov et al., 2007). MPS represent prototypes of industrial technological complexes, realize real processing procedures and consist of real industrial components: sensors, engines, pneumatic and electromechanical linear drives, solenoids, etc. There are 12 types of modular stations which carry out a various procedure, and at interconnection in certain sequences can form fabrication lines of a demanded configuration.

2. INFORMATION

2.1 Project substantiation

Control system of the stations, existing at present, is based on the industrial controller S7-300 of firm Siemens programmed by means of software package Siemens STEP7. This control system demands enough high qualification and as it was found out in the course of its operation in educational process, it has a number of shortcomings which do its inconvenient for training at its initial stages:

- Complexity and "closeness" of a control system of station for trained, i.e. impossibility at the initial stage in the evident shape to study features of work of industrial elements that complicates understanding of main principles of their control.
- For modeling even the simplest processes long studying of specific programming languages is necessary.
- There is a risk of breakages of the equipment as a result of erroneous programming.
- Dearness of a control system.

But, together with it, the existing control system has also a number of essentially important advantages:

- Possibility of programming of any process sequence thanks to presence of the flexible and developed programming languages of high level.
- Possibility of an off-line control modular station.
- Presence of means of simulation of work of the controller and diagnostic aids.
- The maximum nearness to real industrial systems.

Therefore there was a necessity for creation of a control system by stations which would make programming and training process more visual and convenient. Being based on experience of application of existing system, and also on the analysis of merits and demerits following basic criteria to which there should correspond the developed system which application is supposed in educational process are defined:

- Simplicity and visualization of control of modular stations.
- Visualization and "openness" of construction of an electric and interface part of operating system.
- Replacement of the difficult environment of programming by the simple and evident form for programming of sequence of operations.
- Presence of possibility of realization of any process sequence.
- Use of the minimum set of the means necessary for educational process.
- Lower cost of system in comparison with the existing.

At use in educational process the offered system is not intended to replace existing, and should add it in respect of studying of principles of control and the organization of similar systems at the training initial stage.

2.2 System definition

The offered control system consists of the hardware on the basis of the module of discrete input-output and a program part in the form of the developed environment of programming for the control of technological processes (Lavrov, 2003). In aggregate with operated stations it is mechatronic system (it's shown in fig.1).

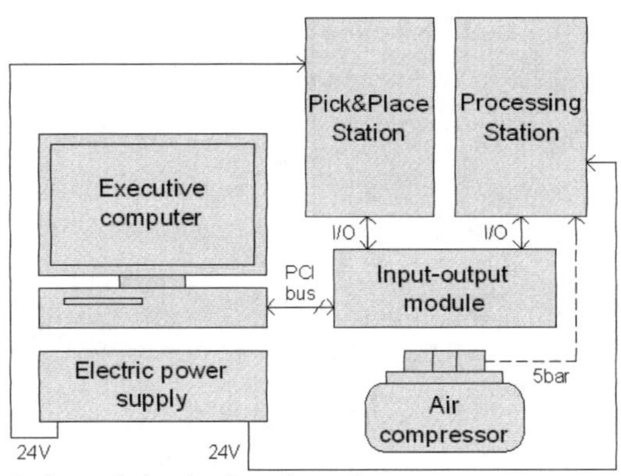

Fig. 1. General circuit of mechatronic system

The basic parts of the presented mechatronic system are:

- Processing Station (Ebel & Pany, 2006, a) and Pick&Place Station (Ebel & Pany, 2006, b) which are banked in a uniform industrial line and represent a mechanical component of mechatronic system.
- Module of discrete input-output (the control system hardware).
- Operating computer with the installed software (the program part of a control system).
- Additional equipment: a power unit on 24V DC and the air compressor on 5 bar.

2.3 Hardware

The control system hardware is based on the module of discrete input-output PCI-1756 of firm ADVANTECH. The interface for electric connection of elements of station to this module has been realized by means of connecting terminal-block module ADVANTECH ADAM-3951. The hardware of an offered control system serves for the decision of following problems: reception of discrete signals from sensors of operated stations, transformation of the received signals for processing by the operating computer and an output of sequence of the control signals generated by a program part of system to executive elements of stations. Executive elements and sensors of the presented modular industrial stations are supervised and controlled by discrete electric signals; each station has 8 discrete inputs and 8 discrete outputs. Signals from sensors of stations enter in the input-output module, whence, after transformation, are transferring to the operating computer. By means of the developed control software the computer generate control signals. These signals enter in the input-output module, transform and output to executive elements of stations.

2.4 Program part

The program part of system represents developed software MPS-CT - MPS Control Tool which serves for the assignment of algorithm of work of stations, control of a technological process and its monitoring. The primary problem solved by the application in the course of functioning is processing of the signals received by the module of input-output from sensors of stations and generation of their certain sequence for an output by means of the module.

Before programming the technological process breaks into steps (Auslander et al., 2002), each of which represents the finished technological procedure of station (for example, detail fixing, separator turn, turn on of the conveyor engine etc.). In application MPS-CT programming is carried out in a special window (it's shown in fig.2) and consists in forming of the block diagrams containing the information on carried out actions on each step.

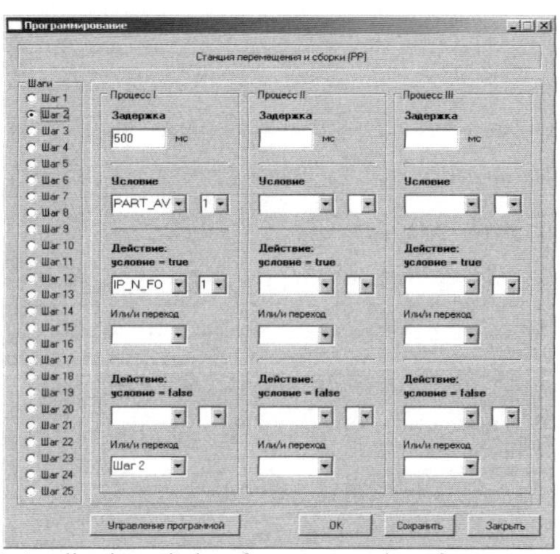

Fig. 2. Application window for programming of steps

The window form gives possibility of programming to 25 steps, each of which can include to three parallel processes which are carried out simultaneously. The sequence which can consist of following actions is meant process:

- The assignment of the time interval defining a time delay of the beginning of performance of a current step.
- Check of a condition of coincidence of a state of the chosen sensor with a set value.
- At condition carrying out: is made adjustment of a state of an executive element in a set value or transition to the set step of algorithm. At condition not carrying out is performance similar action, but with other set parameters.

Control of the made programs is carried out in an application window, simulating the operator panel of stations where are provided operations of activation, stops, reboot and step-by-step execution of the program. The user, having arranged in an application working area the windows for each station, has an opportunity to operate at once all made programs simultaneously.

MPS-CT is the simple and flexible tool for the assignment of algorithm of a technological process for any modular industrial station, allowing programming any combination of technological actions, including the actions which are carried out simultaneously, and also gives the chance to organize branching of algorithm and the elementary cycles.

2.5 Practical application

Practical application of the developed control system means its use in educational process for studying of principles of control by modular stations and the organization of similar systems. The program part of system can use as the sample for studying of principles of control on the basis of discrete input-output and as the sample for design of similar systems by the trained. At expansion of possibilities of its hardware and program part, this system, if necessary, can be used in the field of control of real industrial plants.

3. CONCLUSION

In the project has been developed and realized the automated control system of FESTO modular industrial stations, consisting of hardware and program parts and satisfying to the set criteria. As the hardware is used the module of the discrete input-output ADVANTECH which implement are input, transformation and output of control signals. The program part of system is realized in the form of the developed software MPS-CT which serves for the assignment of algorithm of work of stations, control of a technological process and its monitoring. The developed control system it will be used in educational process for studying of principles of control by modular stations and the organization of similar systems.

4. REFERENCES

Mogilnikov, P.; Bagimov, I.; Klevakin, V. & Mikhailov, M. (2007). International Educational Project "Synergy", *Proceedings of the 18th International DAAAM Symposium*, Katalinic, B. (Ed.), pp. 473-474, ISBN 3-901509-58-5, University of Zadar, Zadar, Croatia, October 2007, DAAAM International Vienna, Vienna

a. Ebel, F. & Pany, M. (2006). *Processing Station,* Festo Didactic GmbH, ISBN 678864, Denkendorf, Germany

b. Ebel, F. & Pany, M. (2006). *Pick&Place Station,* Festo Didactic GmbH, ISBN 678813, Denkendorf, Germany

Lavrov, V. (2003). *Computer Control of Mechatronic Systems,* BSTU "Voenmeh", St.-Petersburg, Russia

Auslander, D. M.; Ridgely, J. R. & Ringgenberg, J. D. (2002). *Control Software for Mechanical Systems,* Prentice Hall PTR, ISBN 0-13-786302-0, New Jersey, USA

Annals of DAAAM for 2008 & Proceedings of the 19th International DAAAM Symposium, ISSN 1726-9679
Editor B. Katalinic, Published by DAAAM International, Vienna, Austria 2008
Make Harmony Between Technology and Nature, and Your Mind will Fly Free as a Bird
*Young Researcher and Scientist Paper / * Supervisor, Mentor*

MODELING AND SIMULATION OF INDUCTION MOTOR FOR DYNAMIC BEHAVIOUR TESTING USING SPECIFIED LOAD

MIKLOSEVIC, K[resimir]; SPOLJARIC, Z[eljko] & VALTER, Z[dravko]

Abstract: *In this paper model for testing induction motor dynamic behaviour using direct current (DC) generator as discontinuous load is presented. Basic mathematical formulation which describes induction motor and DC generator is given. Existing laboratory model is described. Simulation model of induction motor, DC generator and model of alternating current source is made in Matlab-Simulink. Essential parameters for starting simulation of used induction motor and DC generator are given. Dynamical characteristics of induction motor as results of simulation are shown.*
Key words: *motor, dynamic, load, model, simulation.*

1. INTRODUCTION

The main problem of this paper is building simulation model which is based on real laboratory model. Laboratory model (Fig. 1) consists of induction motor (AM), DC generator, power source for motor and generator. Simulation model will be used for testing induction motor dynamic behaviour during motor load. Before model building the experiment is done on three phase induction motor. DC loading generator is in solid mechanical connection with induction motor. While changing resistance on the resistor for loading the generator, behaviour of induction motor is observed. This is necessary for understanding dependence of motor main parameters such as: motor current, torque, slip, power, power factor, efficiency and motor losses. The analysis is done with assumption of constant sinus voltage and constant frequency of induction motor power source. Simulation model is made by the mathematical model.
Modeling and simulation are performed in Matlab Simulink in program part called Simpower System Blockset which is used for simulation of electrical drives. Similar simulation of dynamic behaviour of induction (asynchronous) motor is done by (Ertugrul, 2002) in LabVIEW program. Basic equations which describe induction motor and DC generator are given in (Bose, 2002), (Leonhard, 2001) and (Jadric, Francic, 1995). The dynamic behaviour of electric motors is a difficult topic to study without a computer-aided tool (Ertugrul, 2002). With analysis of mathematical model expensive and unwanted testing of original can be avoided (Jovic et al., 2006).

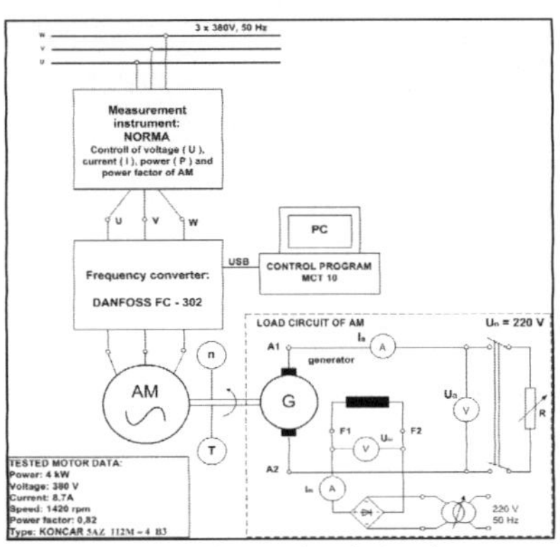

Fig. 1. Laboratory model for testing induction motor.

2. MATHEMATICAL MODEL

Mathematical model consists of differential, algebraic and logic equations. They present the starting point before simulation on computer. That provides experimenting on mathematical model instead original. The mathematical model to be used is tailored to the needs of controlled drives (Leonhard, 2001).

2.1 Induction motor

One of the most fundamental principles of induction machines is the creation of a rotating and sinusoidally distributed magnetic field in the air gap (Bose, 2002). The speed difference between rotating magnetic field of stator and rotor speed is needed for creation of electromagnetic torque. The mathematical model derives from equivalent scheme of induction motor (Fig. 2a) which consists of resistances, self inductances and mutual inductances. The induction machine may be viewed as a system of electric and magnetic circuits which are coupled magnetically or electrically (Boldea & Nasar, 2002). The mathematical model consists of several groups of equations in matrix and differential form as given in (Jadric, Francic, 1995):

- Voltage equations for stator and rotor:

$$\begin{aligned} \boldsymbol{u}_{abcs} &= \boldsymbol{R}_s \boldsymbol{i}_{abcs} + \frac{d\boldsymbol{\psi}_{abcs}}{dt} \\ \boldsymbol{u}_{abcr} &= \boldsymbol{R}_r \boldsymbol{i}_{abcr} + \frac{d\boldsymbol{\psi}_{abcr}}{dt} \end{aligned} \tag{1}$$

where $\boldsymbol{u}_{abcs}$ and $\boldsymbol{u}_{abcr}$ are voltage vector matrixes of three phase system for stator and rotor, $\boldsymbol{R}_s$ and $\boldsymbol{R}_r$ are resistance matrixes of stator and rotor, $\boldsymbol{\psi}_{abcs}$ and $\boldsymbol{\psi}_{abcr}$ are flux vector matrixes of stator and rotor, $\boldsymbol{i}_{abcs}$ and $\boldsymbol{i}_{abcs}$ are current vector matrixes of three phase system for stator and rotor.

- Flux equations for stator and rotor:

$$\begin{aligned} \boldsymbol{\psi}_{abcs} &= \boldsymbol{L}_{abcs}\boldsymbol{i}_{abcs} + \boldsymbol{L}_{abcsr}\boldsymbol{i}_{abcr} \\ \boldsymbol{\psi}_{abcr} &= \left(\boldsymbol{L}_{abcsr}\boldsymbol{i}_{abcs}\right)^T + \boldsymbol{L}_{abcr}\boldsymbol{i}_{abcr} \end{aligned} \tag{2}$$

where $\boldsymbol{L}_{abcs}$ and $\boldsymbol{L}_{abcr}$ are induction sub matrixes of stator and rotor, $\boldsymbol{L}_{abcsr}$ is mutual inductivity sub matrix.

- Mechanical (motion) equation:

$$J\frac{d\omega_m}{dt} = M_m - M_l \tag{3}$$

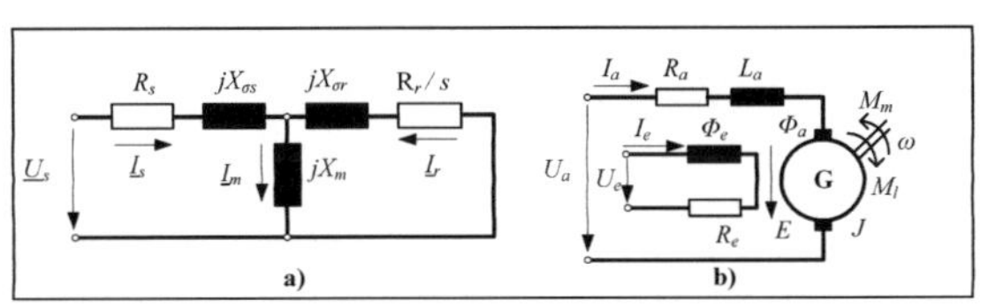

Fig. 2. Equivalent scheme of AM a) and DC generator b).

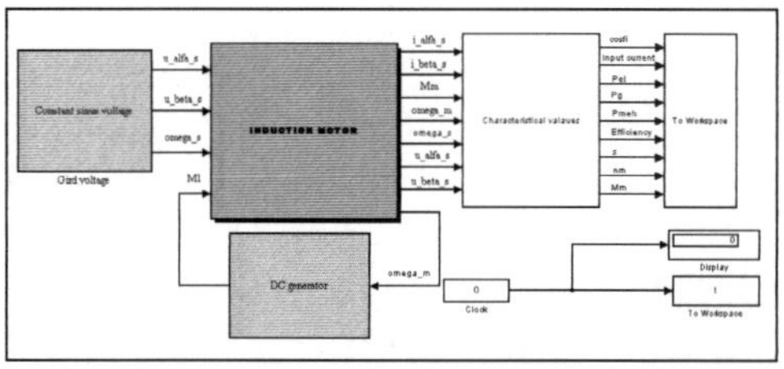

Fig. 3. Simulation model of induction motor load circuit.

- Electromagnetic torque equation:

$$M_m = \frac{1}{2}(\boldsymbol{i}_{abcs})^T \frac{d\boldsymbol{L}_{abcsr}}{d\vartheta_m}\boldsymbol{i}_{abcr} + \frac{1}{2}(\boldsymbol{i}_{abcr})^T \frac{d(\boldsymbol{L}_{abcsr})^T}{d\vartheta_m}\boldsymbol{i}_{abcs} \quad (4)$$

where J is moment of inertia, ω_m is angular motor speed, M_m is electromagnetic torque, M_l is load torque, and ϑ_m is difference angle between rotor and stator position.

2.2 DC generator

Functioning of DC generator can be explained by using two electrical circuits (Fig. 2b). Exciting (field) circuit creates magnetic flux and in armature circuit voltage is induced. DC generator can be described by using following equations for armature and field circuit as given in (Leonhard, 2001):

$$U_a = R_a I_a + L_a \frac{dI_a}{dt} + E$$
$$U_e = R_e I_e + N_e \frac{d\Phi_e}{dt} \quad (5)$$

where U_a is armature voltage, R_a is armature resistance, I_a is armature current, L_a is armature inductance, E is induced voltage, U_e is field circuit voltage, R_e is field circuit resistance, I_e is current of field circuit, N_e is number of field windings, Φ_e is field flux. Motion equation of DC generator is similar to equation (4) of induction motor.

3. SIMULATION MODEL AND PARAMETERS

Simulation model of induction motor load circuit consists of model of power source, induction motor model, model of DC generator and model of discontinues resistance connected with DC generator which represents load for induction motor (Fig. 3). The simulation model of separately excited DC generator is determined by explicit equations and in this way directly implemented in the model. Usage of Park transformation, on the equations mentioned in mathematical model of induction motor, provides creation of simulation model (Fig. 4). This new explicit equations are directly implemented in the simulation model. In Fig. 3 the block which presents characteristic values provides information of all main motor values (current, torque, speed, etc.) needed to display wanted characteristics. Main

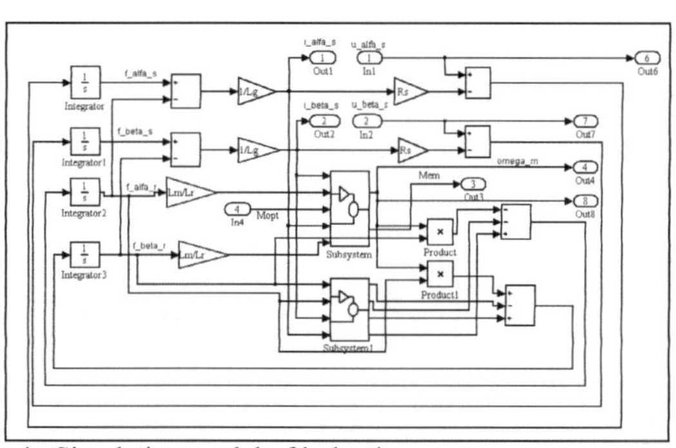

Fig. 4. Simulation model of induction motor.

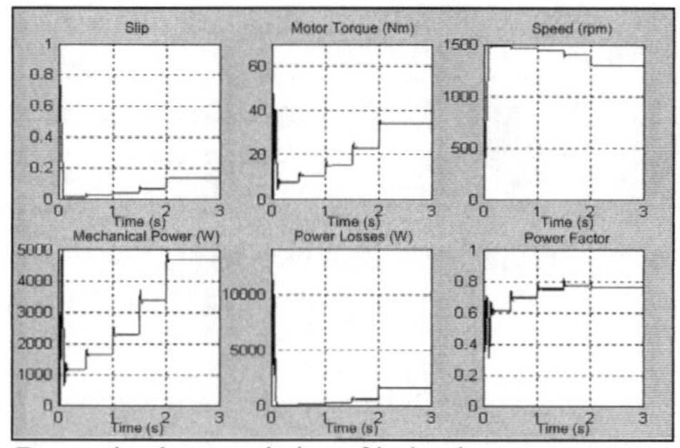

Fig. 5. Dynamic characteristics of induction motor.

parameters of induction motor and generator are determined by measuring and using manufacturer catalogues. Before simulation start, a wide offer of numeric integration algorithms are available. Integration methods are determined by the model of electromotor drive. User evaluates which method will suit him the best. After creation the Simulink models, and before simulation starting it is necessary to determine characteristic simulation parameters (tolerance, integration step, etc.).

4. RESULTS

Some results of simulation are shown on Fig.5. They present all important variables necessary for understanding motor dynamics during motor starting and loading in four steps (each step presents load of 1,5 kW). Instant values of characteristics variables (slip, motor torque, speed, etc.) are possible to monitor during the simulation performance. From given results motor static characteristics can be derived.

5. CONCLUSION

Usage of different kinds of programs, such as Matlab, provides complete dynamic state analysis and testing of induction motor by using simulation model instead original. Programming in Matlab Simulink represents a relatively simple task which consists of combining "in" and "out" ports of different subsystems which make the system. This model can be used for testing different motor types. The used model shows considerable accuracy in the area of nominal point. The largest deviation occurs in determination of motor slip. Further research will include complete evaluation of real and simulation model, testing of model for different types of load and motor and determination of dynamical parameters. For analysis of some tasks M-files will be used. It should be mentioned that usage of simulation model without good understanding of actual physical model gives wrong results.

6. REFERENCES

Boldea, I. & Nasar, S.A. (2002). *The Induction Machine Handbook* , CRC Press LLC, ISBN 0-8493-0004-5, Boca Raton, Florida, USA

Bose, B.K. (2002). *Modern Power Electronics and AC Drivers,* Prentice Hall PTR, ISBN 0-13-016743-6, Upper Saddle River, New Jersey, USA

Ertugrul, N. (2002). *LabVIEW for Electric Circuits, Machines, Drives and Laboratories,* Prentice Hall PTR, ISBN 0-13-061886-1, Upper Saddle River, New Jersey, USA

Jadric, M. & Francic, B. (1995). *Dynamic of Electric Machines (Dinamika električnih strojeva),* Graphis, ISBN 953-96399-2-1, Zagreb, Croatia

Jovic, F.; Flegar, I. & Slavek, N. (2006). *Tehnical Process Modeling (Modeliranje tehničkih procesa),* Grafika, ISBN 978-953-6032-56-2, Osijek, Croatia

Leonhard, W. (2001). *Control of Electrical Drivers*, Springer, ISBN 3-540-41820-2, Berlin, Heidelberg, New York

Annals of DAAAM for 2008 & Proceedings of the 19th International DAAAM Symposium, ISSN 1726-9679
Editor B. Katalinic, Published by DAAAM International, Vienna, Austria 2008
Make Harmony Between Technology and Nature, and Your Mind will Fly Free as a Bird

INFLUENCE OF THE PRINTING TECHNIQUE ON THE QUALITY OF THE DIGITALLY SHOT COLOUR PORTRAIT

MIKOTA, M[iroslav]; PAVLOVIC, I[vana] & MRVAC, N[ikola]

Abstract: *The paper deals with the influence of choosing the printing technique on the quality of the perception of digitally shot portrait photograph. In the experimental part standard Munsell Colour Checker Table and portrait photograph were shot and edited in the same way and than printed in piezo ink jet, bubble jet, sublimation and digital laser photographic technique. L, a, b values and* ΔE_{00} *for each system phase are determined and visual estimation of printed portrait photographs based on skin colour reproduction and general impression is done. L, a, b values and* ΔE_{00} *calculations for skin colours show that shooting and basic editing don't make excessive changes in skin colours, but the highest changes are visible after printing. Tonal CMY printing (sublimation, digital photographic printer) shows smaller differences in skin colours than halftone printing (piezoelectric ink jet and bubble jet printer). However, all printed portraits are visually marked as portraits that please professional standards.*
Key words: *photography, portrait, printing, skin colours*

1. INTRODUCTION

Digital photography system is the most applicable photographic system current in all areas of amateur and professional photography as well as in the area of portrait photography. Digital photography system has three main phases: shooting, editing and realization of photographs. Realization is mostly achieved through different digital photography printing techniques. Shooting and editing of the digital format of the photograph primary belong in the area of the author's creative approach to the photography while the realization itself is primary technical phase of the system in which photographer can choose technique and the materials for gaining a photograph of desired characteristics. Portrait is one of the oldest, the most used and the most demanding photographic motives (London et al., 2007). In the case of colour portraits, if the right white balance with the right light source is chosen, the basic and the most measurable characteristic is the reproduction of the skin colour. As the system is creative, visual mark of the general impression of the printed photographs needs to be considered, and the aberration of the printed photographs (its skin colours) has to be determined in the relation to original values and in the relation to creative phases of the system (shot and edited photographs). Digital format of the photograph gained like that actually becomes creatively defined original. The aim of this paper is to analyze the influence of choosing the printing technique of portrait photography on the quality of the perception of digitally shot photograph with the right white balance.

2. THEORETICAL PART

2.1 Digital photography portrait system

The system of digital portrait photography has three basic phases - shooting with digital photographic camera, editing of the digital format of the shot photograph and its realization (printing). During the shooting photographer defines basic parameters of the picture both in technical and creative way. One of the basic characteristics of the digital photographic system is the possibility of defining the white balance in dependence to the light source temperature. With corresponding way of defining exposure elements in this case, it enables better reproduction of the skin tones (while shooting) from the classical photography systems especially with less standard light sources. Basic editing of the shot portrait photographs considers redefining of the colour space from the wider Adobe RGB colour space to the output friendly narrower sRGB colour space by rendering with the perceptual intent (Nelson, 2007).

2.2 Digital photography printing

Shot and edited photograph can be realized with different printing techniques among which dominant are: piezoelectric ink jet printing, bubble jet printing, sublimation printing and printing with digital laser photographic printer (Mikota et al., 2006). Ink jet printing techniques have halftone printout and presume printing with four colours (cyan, magenta, yellow and black) or more (mostly two cyan and two magentas). Printing with sublimation and digital laser photographic printer gives real tone picture with the subtractive synthesis (cyan, magenta, yellow). To make whole impression of the photograph made by halftone techniques adequate to the one made by tonal printing, resolution of the halftone printing has to be significantly higher (around 1200 dpi) in the relation to resolution of the halftone printing (around 300 dpi).

3. EXPERIMENTAL PART

Standard Munsell Colour Checker colour table was shot with the leica format digital photographic camera Canon EOS 5D and with the lighting from the photographic camera with Canon Speedlite 430 flashlight for the measure estimation. White balance was determined by photographic camera and manually set to 6000 K. ISO speed was 100/21 ISO and the exposition elements were determined with TTL camera system, by segment metering on the skin colours and over all metering of other colours. The photograph was shot with aperture 5.6 and exposure time of 1/60 s in the finest JPEG format with resolution of 12.7 MP. In the same way the face portrait for the visual estimation was shot. Editing of the digital format of photographs included changing the colour space from Adobe RGB to sRGB with rendering with perceptual intent. All photographs were printed with piezoelectric ink jet printer Epson Stylus Pro 1290, bubble jet printer Canon Pixma iP 4000, sublimation printer Olympus Passport P-400 and digital laser photographic printer Agfa d-lab.3 on corresponding materials (table 1).

Substrate	Grammage (gm^{-2})	Lumination (%)	Effect	Printer
Epson Premium GPP	255	95	Glossy	Epson 1290
Canon PR-101	245	96	Glossy	Canon iP 4000
Olympus P-4PASE	250	96	Glossy	Olympus P-400
Agfa Sensatis	250	93	Glossy	Agfa d-lab.3

Table 1. Used materials.

Light and dark skin colours on the printed photographs of Color Checker Table were measured by spectrophotometer X-rite Color Digital SwatchBook. Colour difference ΔE_{00} was determined in relation to colours of the original and colours of edited photograph (Luo et al., 2001; Knešaurek & Brozović, 2005). L, a and b values of the shot and edited photographs of colour checker table for the colours of light and dark skin colour were determined by Adobe PhotoShop CS3 program. Portraits were visually marked in conditions defined by ISO 3664 standard. Photographs were estimated by 20 observers with marks from 1 to 5 by criterion of the skin colour reproduction and general impression of the photograph.

4. RESULTS

Table 2 shows L, a and b values for light and dark skin colour for Color Checker Chart, shot, edited and printed photographs and ΔE_{00} colour differences. Figure 1 shows CIE Lab chromatic diagram for light and dark skin colour through single phases of digital photography system. Table 3 shows results of visual estimation of printed portrait photographs.

5. DISCUSSION

Table 2 and chromatic diagram (Fig. 1) show that shooting with the standard light source, corresponding white balance, light metering and the basic editing don't make excessive changes in skin colour and also, that these changes are in the tolerance limits even with the direct colour comparison ($\Delta E_{00}<6$). However, generally, in these phases of photographic portrait system changes in light skin colour are more visible than changes in dark skin colour. The highest changes in skin colours are visible after printing. Piezoelectric ink jet printing shows the highest colour changes in skin colours in the relation to colour achieved with shooting and editing as well as in the relation to shot standard. Changes made with other printing techniques are quite smaller and in tolerance limits (Table 2). Tonal CMY printing shows smaller colour differences for skin colour than halftone printing. However, chromatic diagram shows differences in skin colour reproduction depending on the chosen printing technique what points out that the major influence on the effect of portrait photography has chosen printing technique (Fig. 1). Table 3 shows high compatibility of the skin colour reproduction marks and the general impression marks. Nevertheless, differences in marks of skin colour reproduction and general impression are consequence of the fact that skin colour is not the only factor that influences on the general impression of the photographs. Visual marks in both criteria also show slightly better impression of the portrait photographs printed with tonal printing but also that all printed portraits are marked as portraits which please professional standards. Good visual marks by skin colour criteria can also be explained with the small aberrations in lightness of skin colour of the printed photographs in relation to standard values (Table 2) because at colours close to achromatic area lightness markedly influences on the general impression of the colour.

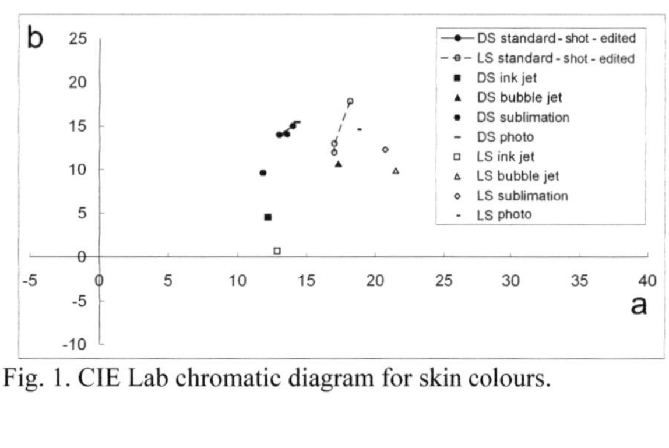

Fig. 1. CIE Lab chromatic diagram for skin colours.

Colour	Phase	L	a	b	ΔE_{00} standard	ΔE_{00} edited
dark skin (DS)	standard	37,99	13,56	14,06	0	0,5897
	shot	38,00	13,00	14,00	0,4476	0,8105
	edited	38,00	14,00	15,00	0,5897	0
	piezo ink jet	43,62	12,19	4,52	8,4575	8,9148
	bubble jet	39,50	17,31	10,65	4,6099	4,8643
	sublimation	36,66	11,84	9,60	3,1845	3,7439
	photo	40,39	14,31	15,43	2,2638	2,0984
light skin (LS)	standard	66,71	18,13	17,81	0	4,1311
	shot	66,00	17,00	13,00	3,0081	1,7529
	edited	68,00	17,00	12,00	4,1311	0
	piezo ink jet	66,95	12,89	0,68	11,7854	8,1916
	bubble jet	66,86	21,95	12,72	5,1451	3,8044
	sublimation	65,90	20,69	12,33	4,7344	2,8871
	photo	68,45	18,74	14,57	3,2577	1,7251

Tab. 2. L, a, b values of skin colours through single digital photography system phases and ΔE_{00} colour differences in relation to shot standard and edited photograph.

Technique	Mark (skin colour)	Mark (general impression)
Piezo ink jet	4	4
Bubble jet	4,3	4,2
Sublimation	4,8	4,7
Digital Photo printer	4,8	4,7

Tab. 3. Visual marks of portraits.

6. CONCLUSIONS

Skin colours change through all phases of the digital portrait photography system. However, changes made in the phase of shooting and editing are relatively small and since these phases are creative, colours (portraits) defined in that way can be observed as the originals. In the phase of printing author-photographer expects as small as possible changes, especially in the skin colour, in relation to colours defined through creative phases of the system. However this investigation showed that printing significantly influences on the quality and the impression of the portrait. Among investigated printing techniques as optimal printing techniques for printing colour portrait photographs shot with corresponding white balance CMY printings (printing with digital photographic and sublimation printing) are pointed out. Although CMY printing techniques are pointed out as better ones, they have limitations in printing formats, so gallery portrait photographs are often made with ink jet printing techniques. Because of that, further investigation will be concentrated on gallery black and white and colour portrait printing with different ink jet techniques, materials and software solutions.

7. REFERENCES

Knešaurek, N. & Brozović, M. (2005). The influence of substrate on tonal artwork reproduction, In: *Graphic Arts: Tecnology, Design, Communications – Scientific Book 2005*, Bolanča, Z. & Mikota, M., (Ed.), pp. 51-58, Faculty of Graphic Arts Zagreb, Ogranak Matice hrvatske Senj, Pulp and Paper Institute Ljubljana, ISBN 953-96020-3-3, Zagreb

London, B.; Stone, J. & Upton, J. (2007). *Photography (9th edition)*, Prentice Hall, ISBN-10: 0131752014, New Jersey

Luo, M. R.; Cui, G. & Rigg, B. (2001). The Development of the CIE 2000 Color-Difference Formula: CIEDE2000. *Color Reasearch & Application*, 26(2001), (August 2001) 340-350, ISSN 0361-2317

Mikota, M.; Mitrović Kukoč, B. & Pavlović, I. (2006). Comparison of Ink Jet Gallery Art Photography Prints, In: *Graphic Arts: Tecnology, Design, Communications Scientific Book 2006*, Bolanča, Z. & Mikota, M., (Ed.), pp. 51-62, Faculty of Graphic Arts Zagreb, Ogranak Matice hrvatske Senj, Pulp and Paper Institute Ljubljana, ISBN 953-96020-5-X, Zagreb

Nelson, P. (2007). *Photographer's Guide to Color Management: Professional Techniques for Consistent Results*, Amherst Media, ISBN-10: 1584282045, New York

Annals of DAAAM for 2008 & Proceedings of the 19th International DAAAM Symposium, ISSN 1726-9679
Editor B. Katalinic, Published by DAAAM International, Vienna, Austria 2008
Make Harmony Between Technology and Nature, and Your Mind will Fly Free as a Bird

OPPORTUNITIES AND LIMITATIONS OF THE INTELLIGENT SYSTEMS OF MANUFACTURING ADOPTION: ROMANIAN CASE

MILITARU, GH[eorghe] & DUMITRESCU, C[orina]

Abstract: *The authors present in this paper the results of an explorative study whose purpose is to determine the opportunities and limitations of the intelligent systems of manufacturing adoption. The paper focuses on: (1) the identification of both the factors which impede or limit the investments in the ISM and the risks of the companies when they purchase such assets; (2) the impact of these technologies on the traditional production companies. The research method is a comparative analysis of interviews conducted in several Romanian manufacturing companies. The authors found out that despite of all money, energy, and time spent by companies trying to automate their factory, ISM is still an unfulfilled promise for many.*
Key words: *Intelligent Systems of Manufacturing, flexibility*

1. INTRODUCTION

Global competition and rapid change of customer requirements are forcing major changes in the production styles and configuration of manufacturing organizations. Increasingly, traditional centralized and sequential manufacturing planning, scheduling, and control mechanisms are being found insufficiently flexible to respond to the changing of production styles and highly dynamic variations in product requirements (Hernandez-Matias et al., 2007).

The Romanian manufacturing companies, as well as those from the former communist states in South-East Europe, need to assimilate new technologies, after they have been privatised and reorganized. These companies have to make great investments in the Intelligent Systems of Manufacturing (**ISM**) in order to get through the pressures created by the global competition and the technological gap between developed countries and those with emergent economies.

Investments in modern manufacturing technologies can provide competitive advantages for the companies on the long run. Nevertheless, the Romanian manufacturing companies have purposes on the short run especially, such as: the maximization of the profits, capitalization or the accelerated recovery of the funds advanced in the privatization process by the owners. This behaviour is punished by the market. For instance, the exports of manufactured goods produced by the Romanian companies decreased by 40 %, in the first trimester of 2008 comparing to the same period in 2007.

The authors try to identify the factors which impede or limit the investments of the Romanian companies in the *Intelligent Systems of Manufacturing*. Therefore, the main purposes of this paper are: (1) the determination of the extent in which the Romanian companies adopt the new manufacturing technologies, (2) the identification of the risks when they assimilate these technologies, and (3) the impact of these technologies on the traditional manufacturing companies.

The paper is organised as it follows. After a brief introduction there is a presentation of the way in which the investments in the intelligent system of manufacturing might generate competitive advantages for the manufacturing companies. Then, a study based on a sample of specialists from few manufacturing companies is presented in the third section of the paper. The fourth section is dedicated to the presentation of both the results and the limits of the study. Finally, the last section presents the main conclusions and the directions of the future research.

2. THE INVESTMENTS IN ISM CAN PROVIDE COMPETITIVE ADVANTAGES FOR COMPANIES

In this section, we will explore in more details the literature on the intelligent systems of manufacturing adoption and its impact on business performance.

In Fig.1 is illustrated the impact of the investment in intelligent systems of manufacturing on the traditional manufacturing companies.

The traditional approaches limit the reconfiguration capabilities of the manufacturing systems. Reducing costs often means to use expensive equipment. At the same time, to increase sales means being able to respond quickly to market requirements even if the equipments and people might be less used. Low inventory levels mean less cash tied up on the shelves but customers want product options and more products means more inventory. Short cycles contribute to cost reduction and revenue increases (Kuzgunkaya & ElMaraghy, 2008).

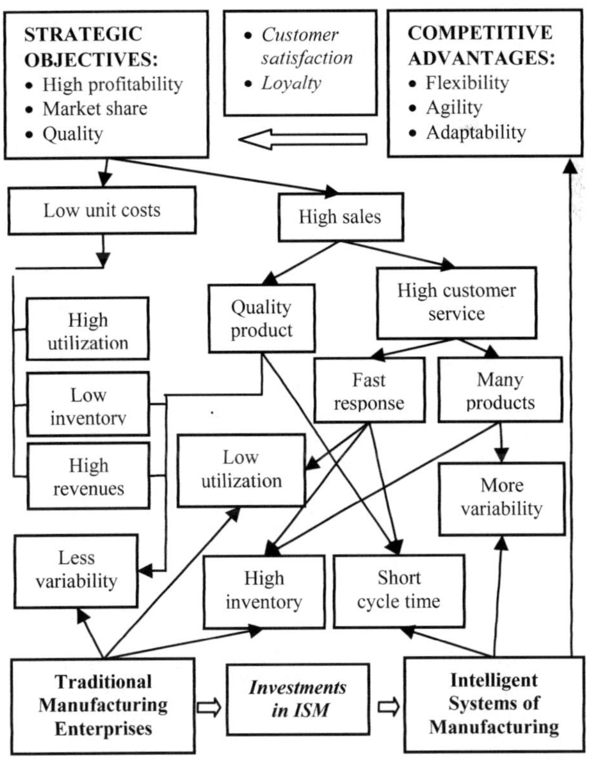

Fig. 1. Investments in the ISM can provide competitive advantages for the companies

Computer Integrated Manufacturing (**CIM**) is a new production approach in which the functions of design and manufacturing are rationalized and coordinated using computer, communication, and information technologies. CIM has the capability to largely or entirely automate flexible manufacturing by coordinating work cells, robots, automatic storage and retrieval facilities and material handling systems (Hsich & Jiang, 2007).

The competitive advantage of CIM in manufacturing companies comes from its ability to develop a large quantity of new products quickly, to produce small production runs of custom-made items efficiently and maximize the flexibility of the manufacturer in responding quickly to changes in the environment (Heizer & Render, 2006).

Computer Aided Design (**CAD**), Computer Aided Manufacturing (**CAM**), Flexible Manufacturing System (**FMS**), Robotics together with Knowledge and Information Based Systems and Communication Networks are expected to respond effectively to the manufacturing requirements that are becoming highly integrated and complex (Halevi & Wang, 2007).

3. RESEARCH DESIGN AND METHODOLOGY

The research method is a comparative analysis of interviews conducted in several Romanian manufacturing companies. We conducted in-depth interviews with production and project managers. Before the interviews, we developed a questionnaire focused around our research question.

4. ANALYSIS AND RESULTS

The analysis involved iterating between theory and data. We first conducted an in-depth examination of the initial interview data to uncover key themes, then a second-level data analysis, which included follow-up interviews. During this process, we reviewed theory on the identification of both the factors which impede or limit the investments in the ISM and the risks of the companies when they purchase such assets. Results of the exploratory analysis are reported in Table 1.

Our exploratory analysis revealed two key findings. First, we recognized that despite all the money, energy, and time spent by companies trying to automate their factory, ISM is still an unfulfilled promise for many. We have investigated the problem and identify the primary obstacles to move rapid adoption of ISM technologies. Some of ***the barriers to ISM adoption*** are presented below:

1. *Top management commitment*. ISM installation must start from the top with a commitment to provide the necessary time, money, and other resources needed to make the changes that ISM requires;
2. *Management perception and attitude*. ISM requires a new perspective on the part of management - maybe even a new philosophy. Top management, industrial and manufacturing engineers must change their way of thinking and develop new skills;
3. *Integration challenge*. To take full advantage of ISM benefits, the entire manufacturing process from product design to procurement, production scheduling, management, production and delivery must by integrated;
4. *Lack of planning*. Lack of understanding of the technology and suitable infrastructures to support the new technology, inappropriate matching of technology to organizational strengths and weaknesses will all contribute to top management failure to appreciate the promise of ISM;
5. *ISM requires flexible organizational structure*. The tendency of the most managers from manufacturing companies to use less flexible organizational structures and rigid corporate rules are significant barriers to ISM adoption.

Antecedents of ISM adoption	*%*	*Results*
Management commitment	60	High
Organizational structure	30	Flexible
Customer satisfaction	76	Important
Loyalty	20	Important
Organizational learning ability	30	High
Efficiency	70	Very important
Flexibility^{++}	90	High
Technical skill	65	Very important
Information technology	78	Advanced
Agility^{+}	95	Very important
Adaptability^{+}	93	Very important

Tab. 1. Results of the exploratory analysis (selection)

Secondly, the data revealed problems (***risks***) that may lead to failure in ISM adoption: partially obsolete facilities, user hostility, dynamic volume and mix, shortage of technical skill, incompatibility between systems, varieties of process options, and facilities with mixed processing.

5. CONCLUSIONS

Two strategic approaches about the adoption of the intelligent systems of manufacturing (ISM) might be outlined on the basis of both the results of the study and the literature reviewed by the authors. The first approach is *proactive* and it is applied by the companies which want to be the first that exploit the potential of the most advanced manufacturing technologies. These companies act in industrial sectors with a development rhythm on the long run comparable with capital cost. They have digital integrated systems, massive capital investments and a considerable market share.

The second strategic approach is *reactive* and it is adopted in most of the manufacturing companies from the emergent economies, among which Romania as well. These companies have low wages costs, but the lower this competitive advantage is the more interested in assimilating manufacturing technologies. In Romania the manufacturing companies impede or limit the adoption of these technologies because of the very high technological gap, the low investments in the information technologies, and the lack of knowledge about the ISM or the high opposition to changes.

Future research in this field ought to be focused on both the industrial sectors in which these companies act and the identification of such mechanisms which accelerate the adoption of the advanced manufacturing technologies available in both developed countries and emergent economies.

6. REFERENCES

Halevi, G. & Wang, K. (2007). Knowledge based manufacturing system (KBMS), *Journal of Intelligent Manufacturing,* vol. 18, no. 4, pp. 467-474, ISSN: 1572-8145

Heizer, J. & Render, B. (2006). *Operations Management,* Pearson Prentice Hall, ISBN 0-13-185755-X, New Jersey

Hernandez-Matias J.; Vizan J. ; Hidalgo, A. & Rios, J.(2006). Evaluation of techniques for manufacturing process analysis, *Journal of Intelligent Manufacturing*, vol. 17, no. 5, pp. 571-583, ISSN: 1572-8145

Hsich, S. & Jiang, Z. (2007). Evaluating investment-return-based design for AMHS in semiconductor manufacturing systems, *Journal of Intelligent Manufacturing*, vol. 18, no. 6, pp.617-639, ISSN: 1572-8145

Kuzgunkaya, O. & ElMaraghy, H. (2008). Economic and strategic perspectives on investing in RMS and FMS, *International Journal of Flexible Manufacturing Systems*, vol. 19, no. 3, pp. 217-246, ISSN: 0920-6299

Annals of DAAAM for 2008 & Proceedings of the 19th International DAAAM Symposium, ISSN 1726-9679
Editor B. Katalinic, Published by DAAAM International, Vienna, Austria 2008
Make Harmony Between Technology and Nature, and Your Mind will Fly Free as a Bird

THE CORPORATE SOCIAL RESPONSIBILITY AND THE SUPPLY CHAIN: AN EMPIRICAL INVESTIGATION

MILITARU, GH[eorghe] & IONESCU, S[orin]

Abstract: *The authors investigated the availability of the managers from Romania in order to integrate the practices of social responsibility in their actions. The research has been carried out on a sample of companies which make a supply chain and it has been developed on three directions: behaviour ethics, social investments and environmental protection. Using linear regression method and the data collected from the questioned managers it has been observed that the behaviour ethics of the partners that form the supply chain, the social investments and the environmental protection are positively correlated to the long run economic success of the whole chain. Nevertheless, it has been found out that many managers prefer to act in order to earn immediate profits.*
Key words: *CSR, ethical behaviour, supply chain*

1. INTRODUCTION

The purpose of this paper is to provide an overview of the major issues related to integrate the corporate social responsibility (**CSR**) with supply chain. It is the result of a research project focused on improving the socially and environmental responsible performance across manufacturing and supply chain.

This study investigates the relationships between the ethical behaviour of all players in the supply chain, the social and environmental initiatives and the long-term success of a company such as customer loyalty and brand reputation. It suggests that CSR may lead to significant gains in social welfare.

Structuring a responsible supply chain has the power to yield significant benefits. Lowering the risk profile of the supply chain can potentially decrease the costs of the company, help it wines new contracts and remain responsive to growing customer expectations (Paulraj & Chen, 2007). In fact, improved corporate responsibility enhances the brand reputation and customer loyalty. Most important, the company can begin to wring even greater strategic value out of its supply chain, providing it with a long-term competitive advantage (Tan & Tracey 2007).

In the subsequent sections, we develop the conceptual background and advance hypotheses that specify the impact of corporate social responsibility across the supply chain on long-term success of a company such as customer loyalty, and brand reputation. Using data from a survey of Romanian firms that have initiated CSR programs, we estimate a regression model to test the hypotheses. We conclude with discussion and implications of our findings.

2. THEORY DEVELOPMENT AND HYPOTHESES

In this section, we will explore in more details the literature on corporate social responsibility and supply chain. It is important to know that, to be successful, a company needs to look for competitive advantages beyond its own operations into the value chain of suppliers, manufacturing organizations, distributors, financial institutions, and customers.

Integrating corporate responsibility with supply chain management can be challenging. Supply-chain management is the integration of the activities that procure materials and services, transform them into intermediate goods and final products, and deliver them to customers. Competition is no longer between companies; it is between supply chains (Senguta et al., 2006).

Ethical decisions are critical to the long-term success of any company. However, the supply chain is particularly susceptible to lapse, as the opportunities for unethical behaviour are enormous. With sales personnel anxious to sell, and purchasing agents spending huge sums, the temptation for unethical behaviour is substantial. Many salespeople become friends with customers, do favours for them, take them to lunch, or present small (or large) gifts. It may be a challenge to determine when gifts become bribe. Many companies have strict rules and codes of conduct that limit what is acceptable. Some companies have developed principles and standards to be used as guidelines for ethical behaviour (Heizer & Render, 2006). Therefore, we formally hypothesize that:

***H1**: Ethical behaviour of partners from a supply chain is positively correlated to the long-term success of the chain*

CSR has the potential to shape the strategic context for company and can exploit longer-term opportunities. The real advantages will only be realised once CSR is tied to business strategy. The key to generating customer loyalty is to deliver high customer value. For example, Volvo's core positioning has been "safety", but the buyer is promised more than just a safe car; other benefits include a long-lasting car, good service, and a long warranty period (Paulraj & Chen, 2007).

A lot of attention is being paid to the topic of integrating social and environmental concerns into core business practice. For some industries, gas and electricity represent a major portion of total operating costs. Reducing energy consumption can significantly decrease these expenses. Using less energy reduces carbon emissions and other air pollutants that result from the burning of fossil fuels. In addition, the companies also contribute to society through their social investment.

Some companies have implemented "green" initiatives to make their corporate operations more environmental friendly. More and more companies may examine the global environmental impact of their supply chains as information becomes public.

The finding that there is weak relationship between CSR and profitability is consistent with a market equilibrium in which firms invest in socially responsible projects until the marginal returns decline to the overall market rate of return. However, companies that are profitable are more likely to engage in more CSR activities.

In fact, corporate social responsibility enhances the brand reputation and customer loyalty. Most important, a company can begin to wring even greater strategic value out of its supply chain, providing it with a long-term competitive advantage (Porter & Kramer 2006). Hence, we formally hypothesize that:

***H2**: The improvement of the corporate social responsibility across the supply chain by social investment and environmental protection will have a significant effect on the long-term success of a company (customer loyalty and brand reputation)*

3. RESEARCH DESIGN AND METHODOLOGY

The research context for this study is the retail companies and their investments in corporate social responsibility. To test these hypotheses, we collected data from a sample of 26 respondents (managers) using self-report survey. Additionally, based on qualitative interviews, and published case studies. The main sources were in-depth individual interviews and annual reports of companies. There interviews helped to confirm the variables of importance, and provide a practical perspective on CSR initiatives across the supply chain.

Using the results of previous section we can specify the theoretical regression model. In this case, we consider the following model:

$$Y_i = \beta_0 + \beta_1 X_{1i} + \beta_2 X_{2i} + \beta_3 X_{3i} + \varepsilon_i \quad (1)$$

Where: Y_i is the long-term success of a company such as customer loyalty and brand reputation (dependent variable), X_{1i} is the ethical behaviour of all players in the supply chain (independent variable), X_{2i} is the social investment across all players in the supply chain (independent variable), X_{3i} is the extent in which a player in the supply chain is committed to reduce environmental impact (independent variable) and ε_i represents the error term.

We used regression analysis (*Ordinary Least Squares*) to obtain numerical values for the coefficients of the regression equation. Then using the regression results we can test the hypotheses. The first step to use the t-test consists of set up null (H_0) and alternative (H_A) hypotheses. From regression equation the one-sided hypotheses are set up as: (1) H_0: $\beta_1 \leq 0$; H_A: $\beta_1 > 0$ (2) H_0: $\beta_2 \leq 0$; H_A: $\beta_2 > 0$ and H_0: $\beta_3 \leq 0$; H_A: $\beta_3 > 0$. Then we can calculate t-values for each of the estimated coefficient in the equation. The larger in absolute value this t-value is, the greater the likelihood that the estimated regression coefficient is significantly different from zero.

Further on we choose a level of significance and therefore a critical t-value about 5 percent as the level of significance with which we want to test. There are 26 observations in the data set that is going to be used to test these hypotheses, and so there are 26-3-1 = 22 degrees of freedom. At a 5 percent level of significance, the critical t-value is t_c =1.717. We consider that the level of significance is the same for all the coefficients in the same regression equation.

4. ANALYSIS AND RESULTS

Now we must run the regression by using a computer package (SPPS) and obtain an estimated t-value. Results of the regression model are reported in Table 1.

The results show support for our H_1 (β_1 = 26.42 and t = 3.7), because 3.7 > 1.717 and the sign of the multivariate regression coefficient β_1 is positive then we reject the null hypothesis. Therefore ethical behaviour of partners from a supply chain is positively correlated with the long-term success of the whole chain.

In the case of the social investment across all players in the supply chain, we reject the null hypothesis, because 2.4 > 1.717 and the sign of the coefficient is positive. Then the social investment will have a significant effect on the long-term success of a company. We suggest that a company by improvement the social responsibility enhances the brand reputation and customer loyalty. Therefore a lot of attention is being paid today to the topic of integrating social concerns into core business practice.

Independent variable	*Estimated coefficients*	*Estimated standard errors*	*t-value*
The ethical behaviour of all players in the supply chain	26.42	7.13	3.7
The social investment across all players in the supply chain	9.85	4.1	2.4
The degree to which a player in the supply chain is committed to reducing environmental impact	12.8	4.3	2.977

Tab. 1. Regression results

For the reduction of environmental impact, as expected, we find that t-value calculated (t = 2.977) is greater than the critical t-value (t_c = 1.717) and the calculated t-value has the sign implied by β_3. Thus, we can reject the null hypothesis.

5. DISCUSSION AND CONCLUSIONS

In this research the authors have noticed that most firms consider their socially responsible actions in the same way that they perceive the most of traditional business activities. Instead of altruistically sacrificing profits, they engage in a more limited - but more profitable - set of socially beneficial activities that contributes to their financial goals.

The findings confirm that the companies are realizing the added value of being socially responsible such as lower energy consumption and environmental protection. Firms that are the most profitable are the most able to sacrifice profits in the public interest. However, the opportunity cost of sacrificing profits may also be the greatest for these firms, assuming that they could otherwise invest the resources in their businesses and earn similarly high returns.

From these results, we can develop some implications. First of all, firms have additional moral or social responsibilities to commit resources to environmental protection. It is therefore important to influence the company's suppliers in the reduction of their environmental impacts and improve the environmental value of its brand. Secondly, many companies contribute to society through their social investment. The companies are sacrificing profits in the social interest. Thirdly, ethical decisions are critical to the long-term success of any company. Finally, we recognize that our empirical study only deals with data collected from a small sample. Therefore, future research may be directed toward testing our hypotheses in companies from different industries.

6. REFERENCES

Heizer, J. & Render, B. (2006). *Operations Management,* Pearson Prentice Hall, ISBN 0-13-185755-X, New Jersey

Paulraj, A. & Chen, I. (2007). Environmental Uncertainty and Strategic Supply Management: A Resource Dependence Perspective and Performance Implications. *The Journal of Supply Chain Management*, (Summer 2007), pp. 29-42, ISSN: 1523-2409

Porter, M. & Kramer, M. (2006). Strategy and Society. The link between Competitive Advantage and Corporate Social Responsibility. *Harvard Business Review*, (December 2006), pp. 78-92, ISSN 0017-8012

Senguta, K. ; Heiser D. & Cook L. (2006). Manufacturing and Service Supply Chain Performance: A Comparative Analysis. *The Journal of Supply Chain Management*, pp.4-15, ISSN: 1523-2409

Tan, C. & Tracery, M. (2007). Collaborative New Product Development Environments: Implications for Supply Chain Management. *The Journal of Supply Chain Management*, (August 2007), pp. 2-15, ISSN: 1523-2409

Annals of DAAAM for 2008 & Proceedings of the 19th International DAAAM Symposium, ISSN 1726-9679
Editor B. Katalinic, Published by DAAAM International, Vienna, Austria 2008
Make Harmony Between Technology and Nature, and Your Mind will Fly Free as a Bird

THE ADJUSTMENT OF THE OPTIMUM OPERATING POINT OF A PUMP DEPENDING ON THE REQUIREMENTS OF THE WATER SUPPLY

MILOS, T[eodor]

Abstract: *For many pumping systems appears the problem of the automatic adjustment of the operating point correlated with the requirements of the water supply. In the conditions of our days equipments the adjustment of the operating point can be done so that the pump efficiency to be maximum and the energy consumption to be minimum. The results presented in the paper were obtained from a test facility that contains a pump with variable rotational speed. The professional software commands the adjustment of the rotational speed so that the operating point is optimum with respect to the energetic consumption and the pressure value to the consumers to be maintained in the initial recommended limits.*
Key words: *pump, optimum, efficiency, energy, water supply.*

1. INTRODUCTION

Economical and social development of the cities of Romania leads to capital changes in the daily and the annual water consumption. The water supply is assured by one or more pumping stations. The adaptation of the operating parameters of the pumping stations to the dynamic of the consumption and the most efficient exploitation is a serious challenge for the decision factors of water delivery (Sanks, 1998).

2. CHARACTERISTICS OF THE TESTED PUMP

For this study in a first phase there were reproduced the energetic parameters Q, H, P_{sh} with the reference data given by the pump producer. The curves from the catalogue were made at 3 different rotational speeds. The rotational speeds were: n_1 = 2900 rot / min as a maximum reference speed (100 %), n_2 = 2320 rot / min (80 %) and n_3 = 1740 rot / min (60 %). Reading the values of head and power from the catalogue for flow rate between 0 and 11 [l/s], the analytical curves were interpolated using polynomial functions (third degree for head and second degree for power). In figure 1 and 2 is shown pumping head vs. flow rate and pump efficiency vs. flow rate.

With these two categories of curves we can build the pump operation characteristic which gives us information about the optimum operating point of the pump (Gandhi et all, 2001). The identification of the equal efficiency curves was made analytical, using a developed program and the graphic postprocessor program used was developed in AutoCAD under AutoLisp. The results obtained for the pump operation characteristic are shown in figure 3.

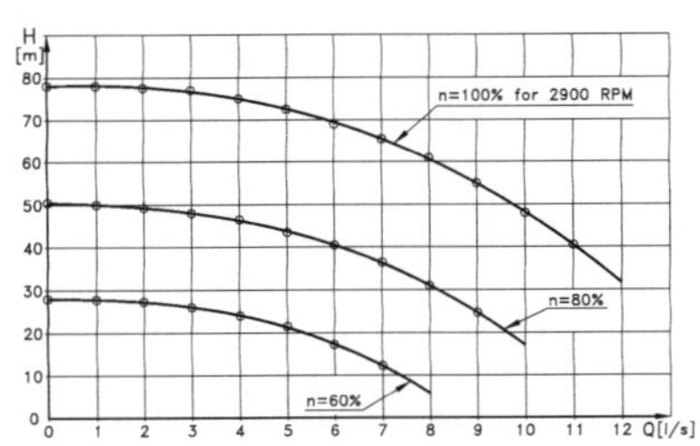

Fig.1. Pumping heads versus flow rate interpolated.

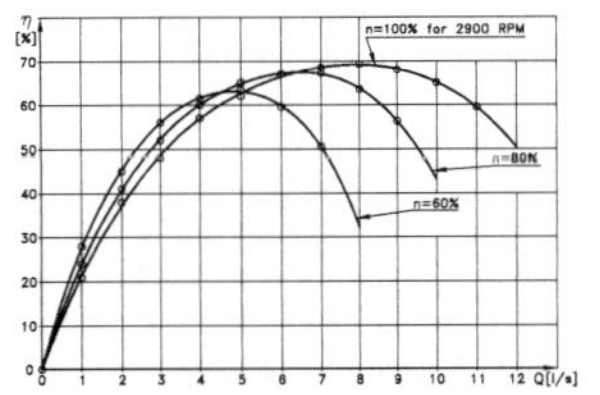

Fig.2. Efficiencies versus flow rate for the three rotations speed.

The pipe network operation characteristic, H_{pn}, represents the amount of the losses from the hydraulic loop exterior to the pump and it is calculated with the formula, (Anton et all, 2002):

$$H_{pn} = H_g + C_{hl} \cdot Q^2 \quad (1)$$

The hydraulic losses constant, C_{hl}, is calculated for an operating point measured from the pump characteristic knowing that $H_{pn}=H$ and from equation (1) results, (Alexandrescu, 2003):

$$C_{hl} = \frac{H_{pn} - H_g}{Q^2} \quad (2)$$

For the installation being a closed hydraulic circuit, $H_g = 0$. The other points of the pipe network operation characteristic are calculated giving values to flow rate from zero to the maximum value. Having an installation with variable rotational speed, the pipe network operation characteristic can be obtained adjusting the flow rate through the modification of the rotational speed.

3. A CASE STUDY ON ENERGY SAVING

The pumping station behavior results is obtained by overlapping the pump operation characteristic at variable rotating speed on the pipe network operation characteristic. A minimum flow rate Q_{min}, a medium flow rate Q_{med} and a maximum flow rate Q_{max} can be estimated by taking into account the requirements of the consumers from a network supplied by the pumping station. The distribution of these flow rates in the exploitation is expressed in percents through a coefficient, C_p, which results from a statistical study. Also, for the closed hydraulic loop from this case we assume that it has to ensure a constant pressure in conformation with a constant pumping head, H_{rq} = 50 m, for the following domains of flow rate, (Kudo, 1994):

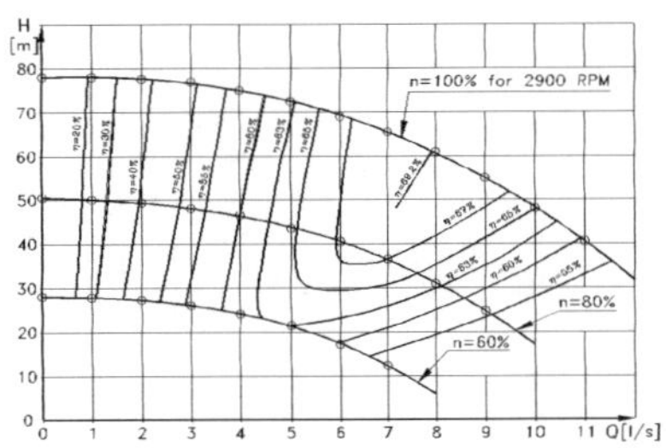

Fig.3. Pump characteristic with equal efficiency curves

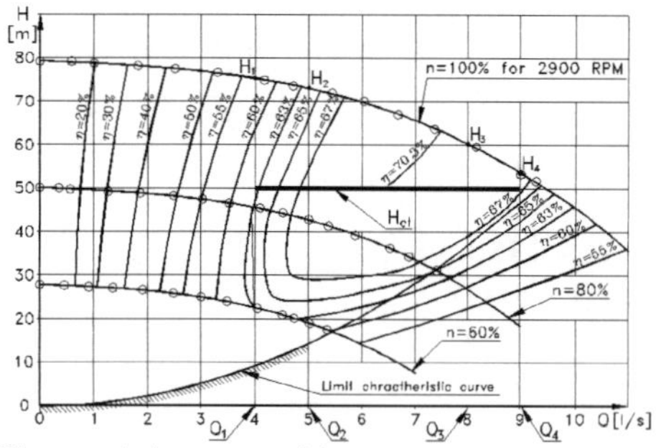

Fig.4. Characteristic curves of the pump

$\Delta Q_{min} = 4 \div 5$ l / s with a coefficient of $C_{pmin} = 20$ %
$\Delta Q_{med} = 5 \div 8$ l / s with a coefficient of $C_{pmed} = 65$ %
$\Delta Q_{max} = 8 \div 9$ l / s with a coefficient of $C_{pmax} = 15$ %

The corresponding flow rates for the three domains are: $Q_1 = 4$ l/s; $Q_2 = 5$ l/s; $Q_3 = 8$ l/s; $Q_4=9$ l/s. To these flow rates are corresponding the pumping heads H_1, H_2, H_3, and H_4 according to the pumping head characteristic for the maximum rotational speed. If the pumping station would not have the possibility to adjust the rotational speed then the flow rate adjustment on the consumers would be realized only through the outlet vane and on the given domain will be provided a covering pumping head between 75 m and 53 m, although in the pipe network would be sufficient a 50 m pumping head. Even if the pump efficiency is relatively good on this domain it can be proved that a big part of the consumed energy is unnecessarily wasted. If it is to take into consideration the possibility of adjusting the rotational speed then it becomes possible the adjusting of the installation so that the pumping head to be constant, $H_{rq} = 50$ m, and provides to any consumer from the network optimum operating conditions. On the industrial installation the adjusting through outlet vane is realized on the consumer and the adjustment through rotational speed is performed in the pumping station. The operating of the pump on the three domains of flow rate it observes that the efficiencies and the pumping heads are sensible different from an operating point to another. That is way in the following section it is presented an averaging method of the power on the three exploitation domains. The shaft-power of pump result from the formula:

$$P_{sh} = \frac{P_u}{\eta_p} = \frac{\rho g Q H}{\eta_p} \qquad (3)$$

Important is the electrical power, P_{em}, consumed in the power station and which is affected by the efficiency of the electrical motor $\eta_{em} = 85$ %. Two efficiency curves are used for the calculus. The first one corresponds to the operating pump at maximum rotational speed $\eta_{100\%} = f(Q)$ and the second one $\eta_{50m} = f(Q)$ results from the characteristic curve for $H = H_{rq} = 50$ m. For a good averaging of the powers on the considered domains it is computed the electrical consumed power in 25 points for each interval of 1 l/s. The medium power will be the average of the powers on each considered domain. The equation deduced is:

$$\Delta P_{em} = \frac{\rho g Q}{\eta_{em}} \cdot \left(\frac{H_{100\%}}{\eta_{100\%}} - \frac{H_{50m}}{\eta_{50m}} \right) \qquad (4)$$

The averaging of the power on the three domains it is realized with the equation:

Flow rate domains	C_p [%]	$P_{sh100\%}$ [kW]	P_{sh50m} [kW]	$(\Delta P_{em})_{med}$ [kW]	ΔE [kWh]
$Q_1 \div Q_2$	20	6,34	3,98	2,36	4142
$Q_2 \div Q_3$	65	7,31	5,37	1,94	11052
$Q_3 \div Q_4$	15	8,08	7,18	0,89	1175

Tab. 1. The results obtained for the three domains

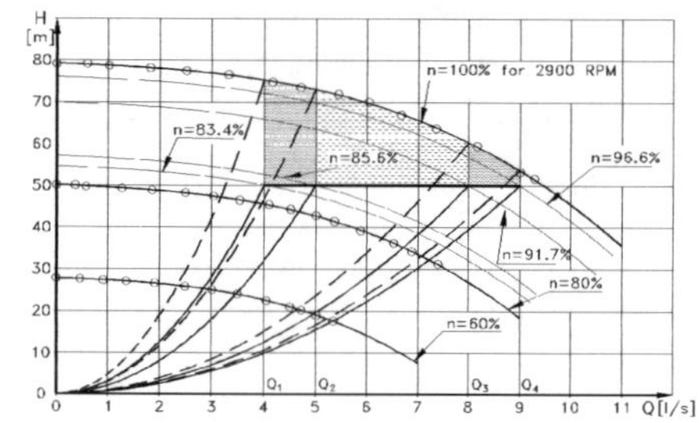

Fig.5. Exploitation domains and the limit of characteristics

$$(\Delta P_{em})_{med} = \frac{\sum_{k=0}^{N} \Delta P_{em_k}}{N+1} \qquad (5)$$

The saving of consumed energy depends on the working time according to the equation:

$$\Delta E = (\Delta P_{em})_{med} \cdot N_{dy} \cdot N_{hd} \cdot C_p \;\; [\text{kWh}] \qquad (6)$$

where N_{dy}=number of the days from a year (365), N_{hd}=number of hours from a day (24). The results obtained for the three domains are presented in table 1.

The researches will be continued for establish Research will be continued for determining the proportion of optimal adjustment with control gate versus the speed.

4. CONCLUSIONS

This study show that flow rate adjustment through the variation of the rotational speed leads to a significant energy saving.

The modality how the automatic adjustment of a pump operating in a pumping station so that the pumping head remains constant indifferent of the operating flow rate determine first the economic advantage through the energy saving realized in time.

The hydraulic network is not overcharged anymore at extreme pressures, which could lead to damages in case of uncontrolled maneuvers.

The averaging method of the economic efficiency on the three operating domains allows a much precise appreciation than the estimation at the medium flow rate of each domain.

5. REFERENCES

Alexandrescu, A., (2003), Concerning optimization's working of the pumping station for water feedings, *18th International Conference on Hydraulics and Pneumatics*, Prague, Czech Rep., ISBN 80-02-01567-3, pp. 263-268.

Anton L., Baya A., Milos T.& Resiga R., (2002), *Experimental Fluid Mechanics,* Vol. 1, Publishing house „University Horizons", Timisoara, Romania, ISBN 973-8391-72-5.

Gandhi B.K., Singh S.N.& Seshadri V., (2001), Performance Characteristic of Centrifugal Slurry Pumps, *Journal of Fluid Engineering,* June 2001, Volume 123, pp. 271-280

Kudo K., (1994), Japanese experience with a converter-fed variable speed pumped-storage system, *The International Journal on Hydro-power & Dam,* March 1994, pp 27-32

Sanks R.L., (1998), *Pumping Station Design*, Second Edition Publishing house „ Butterworth-Heinemann", Woburn, MA-USA, ISBN 1-7506-9483-I.

6. ACKNOWLEDGEMENTS

The present work has been supported by National Centre of Management Programs through Project No. 1365/21-041/2007 and Grant type "IDEI", CNCSIS Department, Contract No. 35-68/2007.

Annals of DAAAM for 2008 & Proceedings of the 19th International DAAAM Symposium, ISSN 1726-9679
Editor B. Katalinic, Published by DAAAM International, Vienna, Austria 2008
Make Harmony Between Technology and Nature, and Your Mind will Fly Free as a Bird

THE MANUFACTURING OF A CRYOGENIC S. G. CAST IRON ALLOYED USED IN AUTOMOTIVE INDUSTRY

MILOSAN, I[oan]; GEAMAN, V[irgil] & STOICANESCU, M[aria]

Abstract: *The paper presents the manufacturing of a cryogenic special S.G. cast iron used in automotive industry. The material used was a Mo-Ni-Cu S.G. cast iron, which it was elaborated in an induction furnace. Nodular changes were obtained with the "In mold" method, with the help of prealloy FeSiCuMg with 10-16% Mg, added into the reaction chamber in a proportion of 1.1% of the treated cast iron. After elaboration, it was made a heat treatment and then a maintaining on a low temperature. It is pointed out the influence of some factors (the temperature and the maintained time at the isothermal level and at low temperature) over the structure and properties of the studied materials used in automotive industry.*

Key words: *cast iron, A.D.I., cryogenic, heat treatment.*

1. INTRODUCTION

Earlier papers (Simon, 1996) have shown what a major importance represents the studies about the bainitic S.G. cast irons obtained by heat treatment, especially the isothermal hardening (Pratt, 1997).

Austempered Ductile Iron (A.D.I.) obtained by heat treatment and isothermal hardening is the material which combines a lot of superior attributes of the classical Austempered Ductile Irons or forged iron (Cox, 1986), being in a serious competition with the iron used by the moment in the automotive industry (Milosan, 1998).

This materials describes a range of nodular ductile iron which have been subjected to an isothermal heat treatment to produce an essentially bainitic structure in this material (Schissler et al., 1993).

Austempered ductile iron (ADI) has received considerable attention from researchers and manufactures during the past years for its unique properties (which combines high strength and good wear resistance) and low cost. It has been used for many applications and there have been a number of papers published which refer to the structure and properties of ADI (Liu, et. al., 1995).

ADI's have the highest strength in the family of cast irons, the actual properties depending on the metal composition and the heat treatment conditions.

The ADI's behavior of low temperature is one of the most important study when we want to determinate the phase transformation in solid state of this material and this paper present some aspects of this (Aranzabal et. al., 1994).

The paper presents the manufacturing of a cryogenic special S.G. cast iron used in automotive industry and belongs to the dates about the influence of some technological factors over the phase transformation in solid state(in the case of the samples made of A.D.I. type S.G. cast iron (Darwish, & Elliott, 1993).

2. MATERIALS

The studied cast iron has the following chemical composition (% in weight): 3.63% C; 2.68% Si; 0.45% Mn; 0.012% P; 0.0036%S; 0.056% Mg; 0.45 Mo; 0.24% Ni; 0.19% Cu.

This cast iron was made in an induction furnace.

Nodular changes were obtained with the "In mold" method, with the help of prealloy FeSiCuMg with 10-16% Mg, added into the reaction chamber in a proportion of 1.1% of the treated cast iron.

3. HEAT TREATMENT

The heat treatment was done by using a resistor furnace (Schissler et al., 1993).

The parameters of the heat treatment done were the following: the austenizing temperature, t_A=900 [°C]; the holding time at austenizing temperature, τ_A=30 [min]; the temperature at isothermal level, t_{iz} = 300 and 400 [°C]; the holding time at the isothermal level, τ_{iz} = 30, 40, 50, 60 and 50 [min].

All these 2 experimental lots A (t_{iz} = 300° C) and B (t_{iz} = 400° C) were performed at isothermal holding in salt-bath (55% KNO_3 + 45% $NaNO_3$), using a resistor furnace, being the cooling after the isothermal holding was done in air.

4. EXPERIMENTAL RESULTS

After the heat treating there was used 20 samples (V-notched specimens):

- for determining the resistance of the material to the crack propagation (KCV) at room temperature (+20°C);
- for determining the resistance of the material to the crack propagation (KCV) at low temperature (-70°C);
- for determining the harness of material (HB) at room temperature (+ 20 °C).

The influence of heat treating over the mechanical properties of A.D.I (results of the tests) is presented in figures 1 to 3.

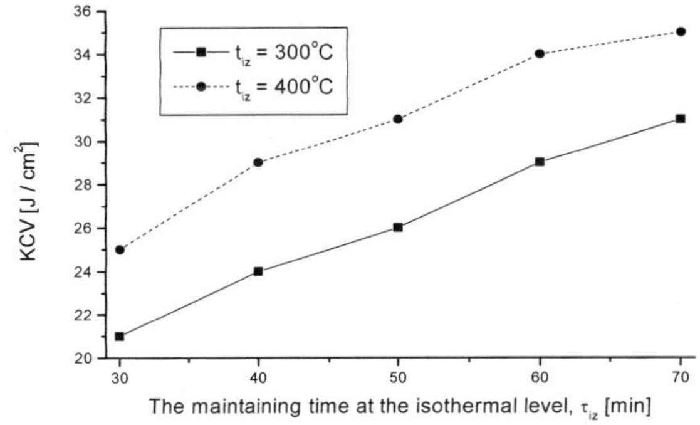

Fig. 1. The influence of the heat treatment over the impact strength properties of A.D.I. at the room temperature (+ 20°C).

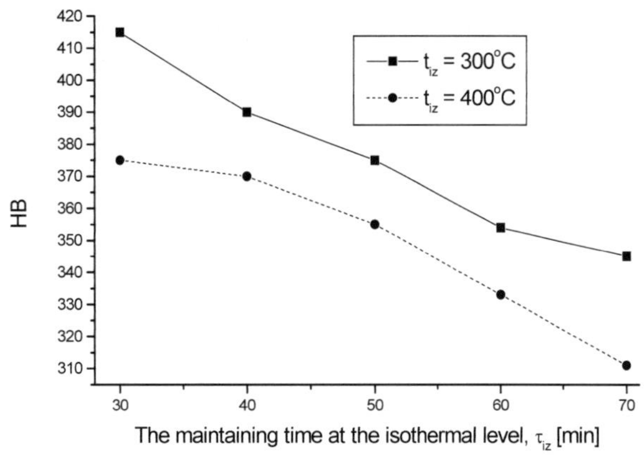

Fig. 2. The influence of the heat treatment over the hardness (HB) properties of A.D.I. at the room temperature (+ 20°C).

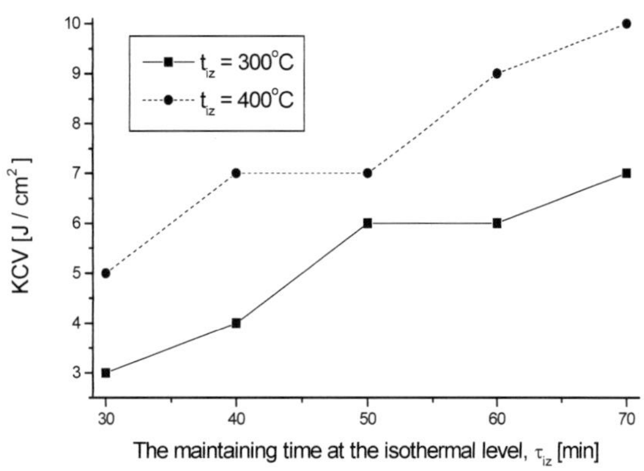

Fig. 3. The influence of the heat treatment over the impact strength properties of A.D.I. at low temperature (- 70°C).

It can be certainly observed a normal evolution of the values for mechanical characteristics (Liu et.al., 1995).

When the heat treating parameters are changing and the temperature is growing for isothermal maintenance than the hardness (HB) are decreasing and the impact strength (KCV) are increasing (Milosan, 2008).

When maintaining time at the same temperature of the isothermal level is increasing then the hardness (HB) is decreasing and the impact strength (KCV) are increasing.

This evolution of the mechanical properties is determined by the structural changes reported to the parameters of the heat treating.

In the case of lot A structure can be constituted of inferior bainite, residual austenite and martensite. These constituents are determining high values for hardness (HB), and less high for the impact strength (KCV). Together with increasing the level of the isothermal maintenance temperature inside the structure will appear the superior bainite and the martensite will disappear (Aranzabal et al., 1994).

In the same time there can be observed a general characteristic about the studied lots: less maintaining time for the isothermal variation provides higher values of the hardness (HB) but lower for the impact strength (KCV). This can be explained by the time of the isothermal level maintenance, followed by air cooling at the room temperature, is increasing the proportion of martensite, a constituent which is determining higher values for HB and lower KCV in the structure of the lots

In the case of the samples for determining the impact strength characteristics at low temperature (fig. 3) and the variation of the austenite proportion, it can be ascertained that the impact strength (KCV) are decreasing in lot B compared to lot A (Darwish et al., 1993).

This can be explained by the significant decreasing of the austenite stability, its transformation into martensite resulting in the variation of the values for the impact strength characteristics.

5. CONCLUSIONS

The main results are summarized as follows:

(a) The classical isothermal bainitic heat treatment combines a lot of superior attributes used in the automotive industry. The values of HB and KCV, assure a good comportment to fatigue strength.

(b) Very important are the variations of the heat treatments' parameters over the values of the mechanical properties. It is possible to obtain an acicular structure with high values for mechanical properties. That is a good reason for the replacement of the iron used by the moment in the automotive industry.

(c) The increasing of the level temperature for the isothermal maintenance from 300°C to 400°C is leading to an increase of the values for the impact strength (KCV) and a decrease of values of the hardness (HB).

(d) The increasing of the maintaining time at the same temperature of the isothermal level is producing a decrease of values for the hardness (HB) and an increase of values for the impact strength (KCV).

(e) The transformation of some austenite percentage into martensite is the resulting of the significant decreasing of the austenite stability at low temperature

(f) In the future, the research plan contains to develop of some specific automotive components which include parts made from S.G. cast iron type A.D.I.

6. REFERENCES

Aranzabal, J., Gutierrez, I. & Urcola, J.J (1994). Influence of heat treatments on microstructure of austempered ductile iron. *Materials Science and Technology,* vol. 10, (Aug. 1994), pp. 728-737

Cox, G.J. (1986). The effect of austempering time on the properties of high-strength S.G. iron, *The British Foundryman,* (June 1986), pp. 215-219

Darwish, N. & Elliott, R. (1993). Austempering of low manganese ductile iron. Part 2 Influence of austemnitising temperature. *Materials Science& Technology*, v9, (July 1993), pp.586-602.Simon, D. (1996). ADI- a new material for the automotive engineer, *Foundry Trade J.,* 2, (February 1996) pp. 66-67

Liu, Y.C., Schissler, J.P., Chabout, J.P. & Vetters, H. (1995). Study of The Structural Evolution of Austempered Ductile Iron (ADI) during Tempering at 360°C, *Metallurgical Science & Tech.* 13, (January, 1995) pp. 12-20

Milosan, I. (1998). The influence of the heat treatment parameters over the microstructures and properties of a S.G. cast iron with special properties, *Proceeding of the Romanian Aviation Academy,* 2, pp. 116-119, ISSN 1453-0139, September 1998, Academy Publisher, Brasov

Milosan, I. (2008). Some Aspects About Mo-Ni-Cu Low-Alloy S.G. Cast Iron Type A.D.I, *Metalurgia International*, 5, pp. 54-55, (May 2008). ISSN 1582-2214

Pratt, A. (1997). Autempered and reborn, *Materials World*, v. 5, n. 3, (March 1997), pp.142-143

Schissler, J.M. Choaubout, J.P., Guridi, J.R. Procede d'obtention de pieces d'usure en fonte G.S. bainitiques a hautes caracteristiques mecaniques par traitement thermiques en four a lits fluidizes, *Proceeding of 60th World Foundry Congress*, pp. 20.1-20.11, 1993, Haga

Annals of DAAAM for 2008 & Proceedings of the 19th International DAAAM Symposium, ISSN 1726-9679
Editor B. Katalinic, Published by DAAAM International, Vienna, Austria 2008
Make Harmony Between Technology and Nature, and Your Mind will Fly Free as a Bird

FORCED VIBRATIONS IN FEEDING KINEMATIC CHAIN AND ROUGHNESS OF SURFACES MANUFACTURED BY MILLING

MINCIU, C[onstantin]; CROITORU, S[orin] M[ihai] & CONSTANTIN, G[eorge]

Abstract: *In the first part of this paper, results of the experimental research related to the measurement of vibrations maintained by the variable cutting forces along longitudinal and transversal feed directions of a universal milling machine are presented. In the second part of the paper the connection between the vibrations maintained by the variable cutting forces along longitudinal feed direction and the roughness of the surfaces manufactured by milling is presented.*
Key words: *Cutting, Forced vibrations, Roughness, Milling.*

1. INRODUCTION

In the technical literature, there are few researches of studying the technological kinematic chains (main and feed kinematic chains) from the dynamic point of view considering a unitary theory. The feed kinematic chain was treated unilaterally regarding only the self vibrations due to friction or the vibrations induced by gears, neglecting other causes that determine the inconstancy of the feed motion (Minciu et al., 2000).

This paper proposes methodologies for experimental research (Minciu et al., 1995) related to the measurement of vibrations maintained by the variable cutting forces, and also the connection between the vibrations maintained by the variable cutting forces along longitudinal feed direction and roughness of the surfaces machined by milling (Stănescu, 2000).

2. RESEARCH ON FORCED VIBRATIONS

Experimental research was done on a stand consisting of a milling machine with vertical spindle FV-1, a table with stress gauges for cutting forces measurement and a seismic captor for measuring the forced vibrations in case of milling with longitudinal feed, using face milling cutters. The experiments were done using two face milling cutters with the diameters and number of teeth d_1 = 80 mm, d_2 = 40 mm, z_1 = 12, z_2 = 3. Workpieces were made of OLC 45 (Romanian steel).

During the experiments, the parameters of cutting regime (cutting speed, feed and cutting depth), milling cutter diameter and number of teeth varied (Minciu, 1971).

The milling cutters were ROMASCON type (Romanian patent) having teeth enforced with metal carbide K20 sort plates brazed on the conical body of the tooth-cutter.

For the milling cutter having z_1 = 12 teeth and d_1 = 80 mm the experimental conditions are presented in tests 1.1*a* to 1.9*b* and for the milling cutter having z_2 = 3 teeth and d_2 = 40 mm the experimental conditions are presented in tests 3.1*a* to 3.9*b* from Table 1. The following notations were considered: *w* – feed speed, *rpm* – spindle number of revolutions per minute, *t* – cutting depth. Consequently, in order to describe the longitudinal vibrations when milling with these two face milling cutters, 36 tests were done. There are some observations:

- all diagrams correspond to maintained periodical oscillatory movements (forced vibrations);
- the diagrams are repeated by the frequencies corresponding to the spindle rpm, $v_{aş}$, and milling cutter's tooth action, v_z;

Test No.	*w* mm/min	rpm	*Test no.*	*w* mm/min	rpm
1.1a/b*		154	3.1*a*/*b**		315
1.2a/b*	23.5	315	3.2*a*/*b**	23.5	622
1.3a/b*		622	3.3*a*/*b**		1230
1.4a/b*		154	3.4*a*/*b**		315
1.5a/b*	47.5	315	3.5*a*/*b**	47.5	622
1.6a/b*		622	3.6*a*/*b**		1230
1.7a/b*		154	3.7*a*/*b**		315
1.8a/b*	95	315	3.8*a*/*b**	95	622
1.9a/b*		622	3.9*a*/*b**		1230
* a: *t* = 0.5 mm; b: *t* = 1 mm					

Tab. 1. Cutting regime parameters.

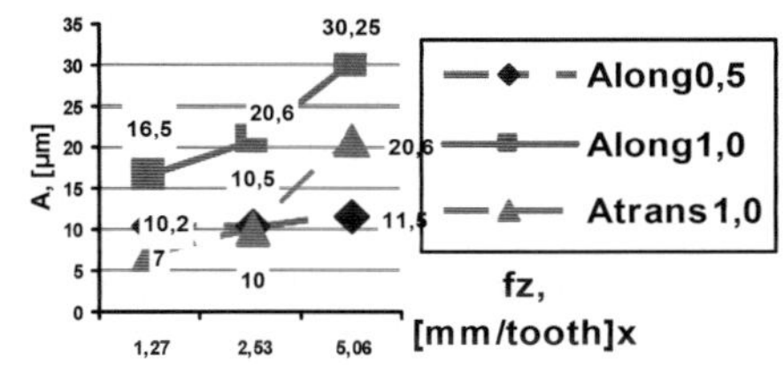

Fig. 1. Tool 1: z_1 = 12 teeth; $v_{aş}$ = 38.7 m/min; v = 2.56 Hz.

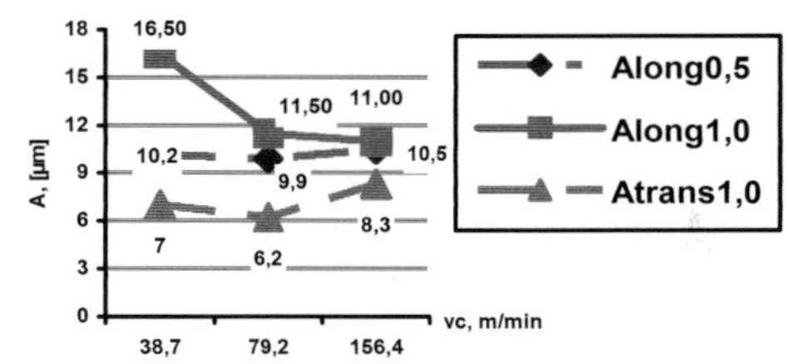

Fig. 2. Tool 1: z_1 = 12 teeth; f_z = 0.127 mm/tooth.

- in order to underline the periodical character of the studied dynamic phenomenon the periods corresponding to one rotation of the milling cutter were evidentiated and also the teeth of the milling cutter were numbered;
- amplitude of the forced vibration due to one tooth is much smaller than the amplitude due to one revolution of the tool.

Figure 1 (all figures in this paper are just samples) presents the influence of feed upon the amplitude of the forced vibrations due to the cutting forces, case of face milling cutter in longitudinal milling. Similar graphics showing the influence of cutting speed are presented in Fig. 2.

2.1 Conclusions

Some conclusions result from this research:

- amplitude of the longitudinal forced vibrations increase with feed per tooth, no matter the cutting speed, from 1.5 to 3 times for an increase of the feed per tooth of 4 times; increase of transversal forced vibrations amplitude is approx. 1.2 times;
- increase of cutting depth has a great influence upon the amplitude of longitudinal forced vibrations: 2-3 times for cutting depth increase of 2 times;

- influence of the cutting speed upon the vibrations amplitude is very interesting: decrease of the vibrations amplitude with the increase of the cutting speed; this can be explained by increase of the cutting temperature, which leads to the decrease of the cutting forces, which are the excitation of the elastic system;
-amplitude of transversal forced vibrations is less than the longitudinal ones because of the smaller cutting forces and higher rigidity along this direction.

3. FORCED VIBRATIONS VS. SURFACE ROUGHNESS

Experimental research was done in the same conditions and using the same stand as presented before. For roughness measurement was used a Romanian electronic roughness meter.

Using the results concerning the amplitude of forced vibrations, their influence upon the roughness of the surfaces obtained by milling is determined. The experimental research considered the following variables: cutting speed, feed and cutting depth, milling cutter diameter and number of teeth (Minciu et al., 2002).

Measurement of the roughness was made by means of an inductive transducer, by direct palpation of an instrument having a diamond point. The transducer is in contact with the work surface. During the measurements, the diamond point moves automatically along the reference direction, explores the roughness and the electronic device determines the arithmetical average R_a. The diamond point is surrounded and protected by a metallic sheet being also the reference surface.

In case of the end milling cutter with the number of teeth $z_1 = 12$ some of the diagrams are shown in Fig. 3.

In case of the end milling cutter having $z_1 = 12$ teeth the following conclusions can be underlined:
- no matter the cutting speed, the roughness R_a rises together with the growth of the feed per tooth *fz* from 1.5 to 3 times for a growth of the feed of 4 times;
- the same variation of the roughness R_a and the amplitude of induced longitudinal vibrations A_{long} is observed, the percentage of the growths of the two variables are slightly the same.

For the end milling cutter $z_2 = 3$ teeth, some of the obtained diagrams are shown in Fig. 4.

In this case the conclusions are almost the same as previous: roughness R_a increases with the growth of the feed per tooth *fz* from 2.5 to 4 times. These values are substantially greater than in the case of the end milling cutter having $z_1 = 12$ teeth. This means that the decrease of the tooth number is an inconvenient factor for the workpiece roughness. The same influence was observed for the amplitude of the induced vibrations;

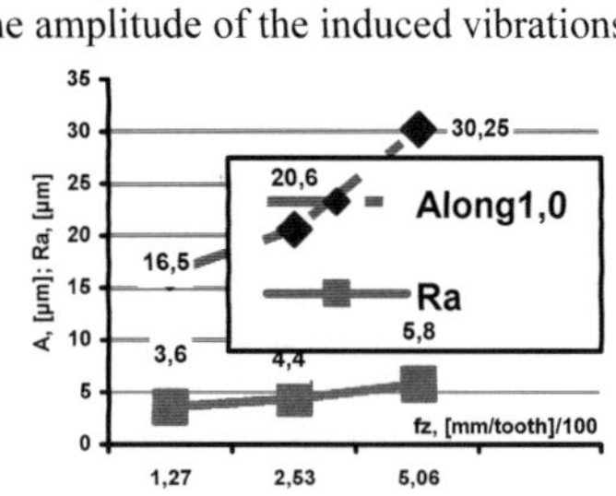

Fig. 3. Tool 1: $z_1 = 12$ teeth; $v_c = 38.7$ m/min; $\nu = 2.56$ Hz.

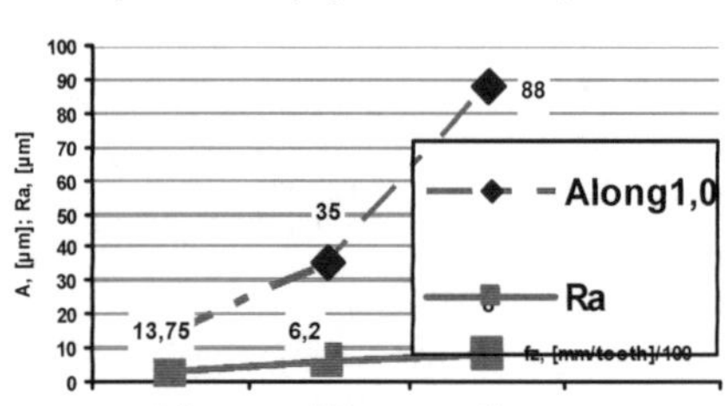

Fig. 4. Tool 2: $z_2 = 3$ teeth; $v_c = 39.58$ m/min; $\nu = 5.25$ Hz.

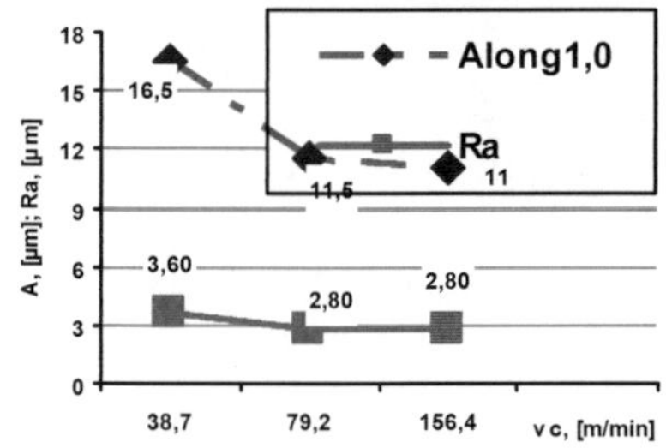

Fig. 5. Tool 1: $z_1 = 12$ teeth; $fz = 0.127$ mm/tooth.

- again is observed the same variation with percentage growth slightly the same of the roughness R_a and the amplitude of the induced longitudinal vibrations A_{long};
- in case $v_c = 79.2$ mm/min ($\nu = 10.36$ Hz) it was observed a great growth of the roughness, due to the dynamic instability (resonance) of the dynamic system for this frequency, compared to the other two cutting speeds. For the same feed, the roughness is 2 times greater.

The influence of the cutting speed is presented in Fig. 5.

3.1 Conclusions

Concerning the analysis of the results of the experimental research the following conclusions and observations must be underlined:
- the same growth of the roughness R_a and the amplitude of the induced longitudinal vibrations, A_{long}, was observed;
- roughness R_a is deeply influenced by the feed per tooth *fz*: for a rise of the feed of 4 times, the roughness growth was of 1.5 to 4 times for both the end milling cutters;
- the cutting speed, v_c, has a minor but convenient influence upon the roughness: the rise of the cutting speed leads to a diminish of the roughness, even if this diminish is not very important;
- as expected, the number of teeth of the end milling cutter (determining the teeth cutting simultaneously, z_{sim}) influences the variation range of the roughness R_a (see Figs. 4 and 5);
- having the same causes, it was expected and it was proved experimentally the "jump" of the roughness of about 2 times, the same as amplitude, in case of resonance (milling with the tool $z_2 = 3$ teeth at $n_c = 622$ rev/min).

5. REFERENCES

Minciu, C. (1971). Măsurarea preciziei dinamice a lanţului cinematic de rulare cu ajutorul captorilor seismici pentru vibraţii torsionale (Measurement of dynamic precision of the rolling kinematic chain by means of seismic captors for torsional vibrations), *Construcţia de Maşini*, No. 2/1971, Bucharest, Romania.

Minciu C.; Velicu Şt. & Croitoru S. (1995). *Bazele aşchierii şi generării suprafeţelor (Fundamentals of cutting and surface generation)*, University "Politehnica" of Bucharest.

Minciu, C.; Dumitrescu, E. & Stănescu, I.I. (2000). Vibrations in the feeding kynematic chain, *TCMM*, No. 41, ISBN 973-31-1492-8, 973-31-1494-4, Edit. Tehnică, Bucharest, Romania.

Minciu, C.; Croitoru, S.M. & Bălan, E. (2002). Aspecte privind influenţa vibraţiilor forţate asupra rugozităţii suprafeţelor prelucrate prin frezare (Aspects regarding the influence of forced vibrations upon roughness of surfaces processed by milling), *Construcţia de Maşini*, No. 6, Bucharest.

Stănescu, I. I. (2000). *Cercetări teoretice şi experimentale privind precizia şi dinamica lanţului cinematic de avans (Theoretical and experimental researces regarding the precision and dynamics of the feed kinematic chain)*, PhD Thesis, University "Politehnica" of Bucharest.

Annals of DAAAM for 2008 & Proceedings of the 19th International DAAAM Symposium, ISSN 1726-9679
Editor B. Katalinic, Published by DAAAM International, Vienna, Austria 2008
Make Harmony Between Technology and Nature, and Your Mind will Fly Free as a Bird

OPTIMIZED ARCHITECTURE MANUFACTURING SYSTEMS USING WITNESS MODELS AND SIMULATION

MINCIU, C[onstantin]; GANDILA, S[anda] & ANGHEL, F[lorina]

Abstract: *We propose here an algorithm used in our researches for optimizing manufacturing architectures. This algorithm is based on Witness software, an interactive graphic simulation mode with artificial intelligence that allows a non – specialist in the simulation field to build models for complex operators. The language allows the representation of an operation by creating imagines for the parts and vehicles flow and for activities from different working zones.*

Key words: *model, simulation, flow concentrator.*

1. INTRODUCTION

Witness contains a series of elements for discrete manufacturing (assembly marks) but in the mean time one can do a continuous systems modelling.

One can specify variables and attributes and the order of operations can be based on times, priorities or other criteria defined by the user. Simulated actions, from beginning and ending of simulated events, can be used for programming structures type like: For-Next, While-End, Go to-Label and use mathematics and logical operators.

It also has an animation module build when defining a model. A characteristic for Witness is the ability of stocking an unlimited number of variables and attributes.

Witness Optimizer module was projected in order to significantly reduce the period of time necessary for automatic search of the optimum solution.

The number of the modeled parameters that will be optimized is big and it includes: manufacturing times, operator's number, the vehicles speed, etc.

Witness Optimizer is not only looking for the optimum solution but it also makes a rapid analyze. By utilizing intelligence algorithms that measure and react to every assay, the best solution is usually found after trying at least 10% of all possible combinations. Witness Optimizer users can make a rigorous analyses for different production sceneries in a flexible system production. Below we give some of the most important advantages of the program:

- An intelligent help in production optimization;
- Fast confirmation of good solutions;
- Bigger amount of time for making more creative solutions;
- Confirmation that the solution found is solid, statistically speaking.

Witness Optimizer gives graphic or table results.

POWERPAQ is another module of simulation language Witness. It has functions that allow an automatic simulation by defining a simulation category of parameters which will be subjected to supplementary tests determining the optimum set of characteristics. Using more optimization techniques one can identify blockages and sensible point in the system, test options and develop more efficient alternatives.

2. MODELLING THE MATERIAL FLOW

When modeling a Witness discrete value material flow managing system we studied this system's behavior for a period of 240 hours functioning, with 3 shifts of 8 hours daily, one operator for each shift with monitoring role.

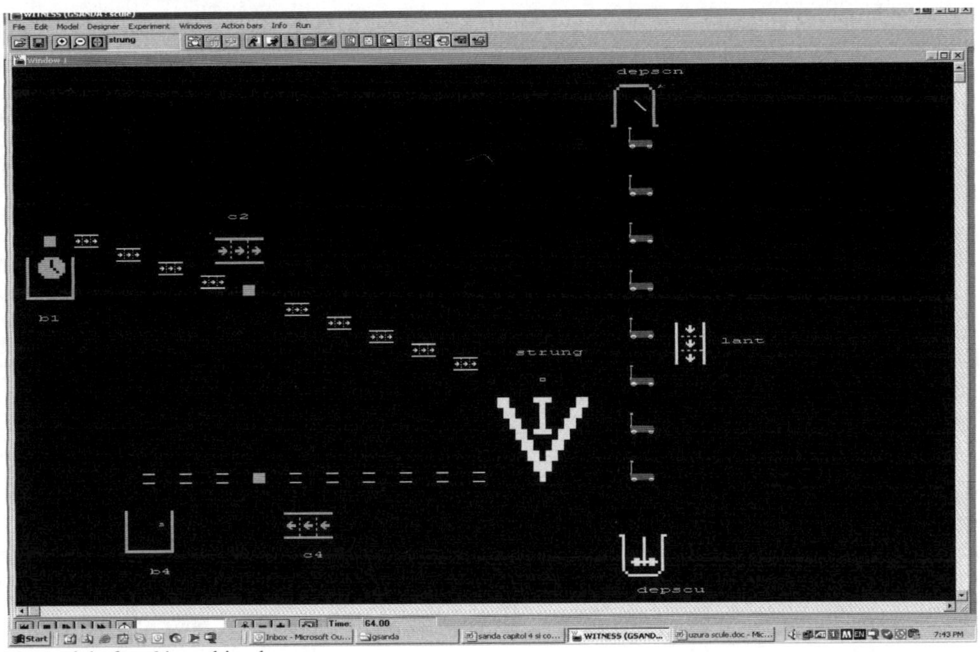

Fig. 1. System's model after 64 working hours.

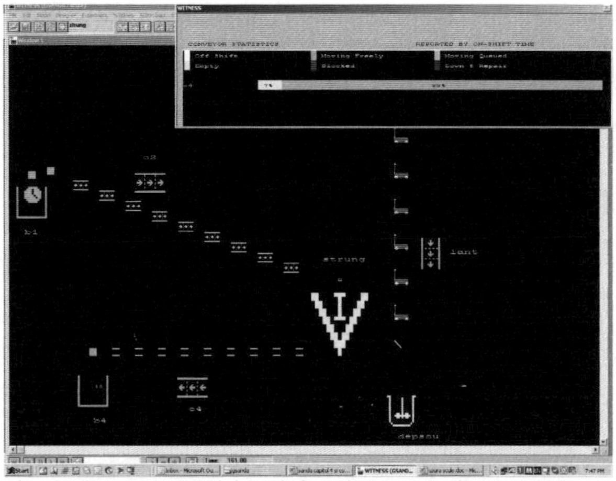

Fig. 2. System's model and function report for c4 conveyor after 161 working hours.

The development for a non-linear model process of a production system containing a tool store or not was made by giving information of medium value behavior based on standard operation curve.

Received commands are particularized by receiving moment and quantity. Both are modeled as random values with a given rate and a standard deviation. System's working performance is determined by planned capacity as an external entrance and MTBF (mean time between failure) as well as MTR (mean time to repair) taken from practical knowledge and observation of the production process Exit values are resolved command and unfinished production for the working system. After doing a calibration with a set of parameters the model behavior on a long period of time is similar to practical working systems behavior described thru operating curve. These parameters were determined thru simulation analyses (Bley, 1999). In practice, especially in lot production, real curve object oriented are different from operating curves based on medium values by 5-7 %. The model is composed from four partial models: the reference model is made of two partial models, information model and objects model. The information describing models is taken by method and activities models; these last two partial models form the reference model frame. The activities from the activities model are attributed to the individual phases. Here we can find data given by the simulations from CATIA V5R16 (Lee, 1999) and Deform (Patrascu, 2004).

Discrete material flow simulation from the studied system can be done in real time in Witness, in accelerated time (if we want to visualize in time behavior) or slowed down time (if we want to see a detailed actions that causes a critical moment during function). As earlier mentioned this is an accelerated simulation, lasts 240 hours, which aims for a medium time optimization of these flow's functioning. In figure 1 we can see the system's model after 64 hours of functioning. In figure 2 is given an example of a function report for c4 conveyor after working 161 hours. The meaning of the references from the functioning reports is as follows:

blocked = the system is blocked
broken down = = the system is interrupted
busy = the system is in function
busy/fill/empty = the system is in function / full / empty
empty = the system is empty
idle = the system is inactive
moving freely = the system is moving without constrains
setup = the system is being setup
wait repair labour = the system is waiting to be repaired
wait setup labour = the system is waiting to be setup
waiting parts = the system is waiting for parts

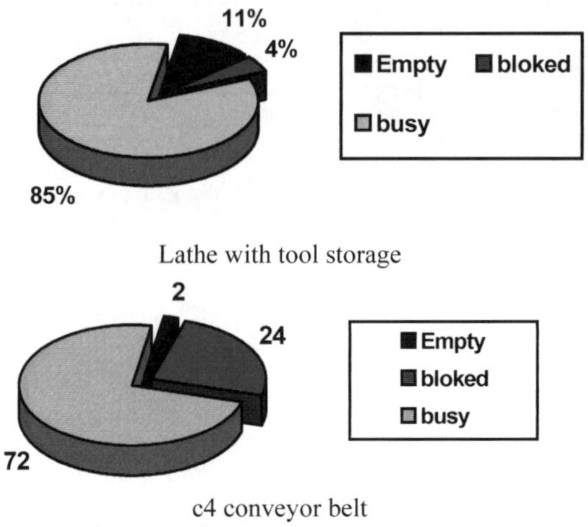

Fig. 3. Usage degree of the system.

Based on the Witness reports and on calculus algorithms we determined that c4 conveyor is a material flow concentrator.

3. CONCLUSION

In this paper we illustrate a manufacturing architecture optimizing algorithm by a case study used in our research grants. The case study is basically a concentrated manufacturing system (Cotet & al., 2007) where we increase the productivity by eliminating the flow concentrator. As one can see in figure 3 we can do that if we double c4 conveyor capacity and we obtain a 20% increasing of the manufactured parts number for the studied period of time. In conclusion based on the data given by the process simulation through flow simulation we can increase the productivity with 20% for the studied system.

4. REFERENCES

Bley H. & Wuttke C.C.(1999), „Multiple Use of Simulation Models for Production Systems", Institute of Production Engineering / CAM, University of Saarland – Germany, 1999.

Cotet, C. E., Dragoi, G. & Carutasu, G. (2007). Material Flow & Process Synchronous Simulation In Concentrate Manufacturing Systems, *Annals of DAAAM for 2007 & Proceedings of The 18th International DAAAM SYMPOSIUM, „Intelligent Manufacturing & Automation: Focus on Creativity, Responsibility and Ethics of Engineers"*, Katalinic, B. (Ed.), pp. 180-181, ISSN 1726-9679, ISBN 3-901509-58-5, Zadar, Croaţia, October 2007, Publisher DAAAM International Vienna 2007.

Cotet, C. E., Dragoi, G. & Abaza, B., F. (2007). Multipolar synchronous material flow & process simulation in diffused manufacturing systems, *Annals of DAAAM for 2007 & Proceedings of The 18th International DAAAM SYMPOSIUM, „Intelligent Manufacturing & Automation: Focus on Creativity, Responsibility and Ethics of Engineers"*, Katalinic, B. (Ed.), pp. 179-180, ISSN 1726-9679, ISBN 3-901509-58-5, Zadar, Croaţia, October 2007, Publisher DAAAM International Vienna 2007.

Lee, K. (1999). Principles of CAD/CAM/CAE Systems, *Addison Wesley Longman, Inc.*, ISBN 0-201-38036-6, USA.

Patrascu, G. (2004). 3D Simulation of Turning Process using FEM Software, Proceedings of the International Conference on Manufacturing Systems ICMaS 2004, Constantin, I., Ghionea, A., Constantin, G. (Ed.), pp. 297-300, ISBN 973-27-1102-7, Bucharest, 2004 October, Editura Academiei Romane, Bucharest

Annals of DAAAM for 2008 & Proceedings of the 19th International DAAAM Symposium, ISSN 1726-9679
Editor B. Katalinic, Published by DAAAM International, Vienna, Austria 2008
Make Harmony Between Technology and Nature, and Your Mind will Fly Free as a Bird

A COMPARATIVE STUDY OF TEMPERATURE FIELDS IN X52 STEEL BUTT WELDS

MIRCEA, O[ctavian]; VISAN, D[aniel]; BANEA, M[ariana] & CONSTANTIN, E[mil]

Abstract: *This paper presents a comparative study of theoretical and experimental temperature fields on butt-welding. A theoretical method using finite element analysis has been used for the temperature prediction in the X52 welded joints. A non-linear transient thermal analysis was used to obtain the global temperature history generated during the welding process. The temperature fields were plotted function of the distance from the weld centre, at different time steps. Several measurements and visualization of the temperatures distribution have been made using infrared thermography. Temperature values for both used methods, in the cross section of the welded joints, on the lower and upper side of welded joint in the selected points are presented. The errors were calculated for all cases and error curves were plotted.*
Key words: *temperature, field, analysis, stress, welding.*

1. INTRODUCTION

While welding joins the components of a structure together, the complex thermal cycles from welding result in formation of residual stresses in the joint region, and deformation/distortion of the welded structure. Both weld residual stress and distortion can significantly impair the performance and reliability of the welded structures. They must be properly dealt with during design, fabrication, and in-service use of the welded structures. At the basis of deformations and stresses analysis lays the thermal field during welding, measured or determined numerically, as well as the microstructural transformations induced by the thermal field. Finite element simulations in order to predict temperature fields, residual stresses and deformation due to welding were performed by many investigators [Goldak et al. 1984; Lindgren, 2001; Komanduri & Hou, 2000; Radaj, D. 1992].
The data necessary for modelling were established after manufacturing a submerged arc welded joint in a single stage, using as base material the X52 steel [Mircea et al., 2003]. As the thermophysical properties of X52 steel (specific heat, thermal expansion, thermal conductivity and diffusiveness, emissivity etc.) vary with the temperature, this condition was taken into account for modelling.

2. THEORETICAL THERMAL FIELDS

In finite element modelling of the thermal field, the following simplifying hypotheses were employed:

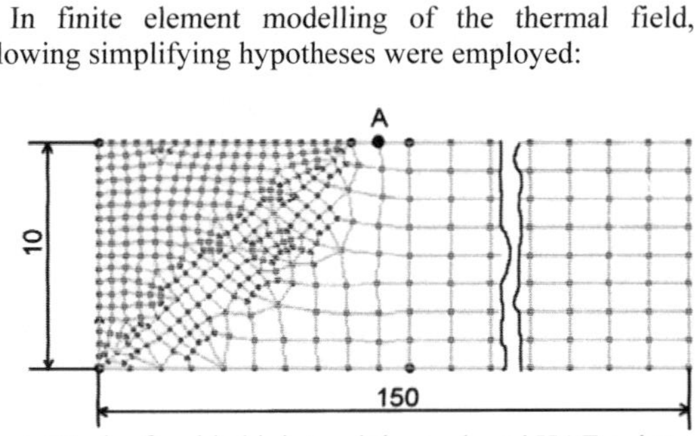

Fig. 1. Mesh of welded joint and the analyzed HAZ point.

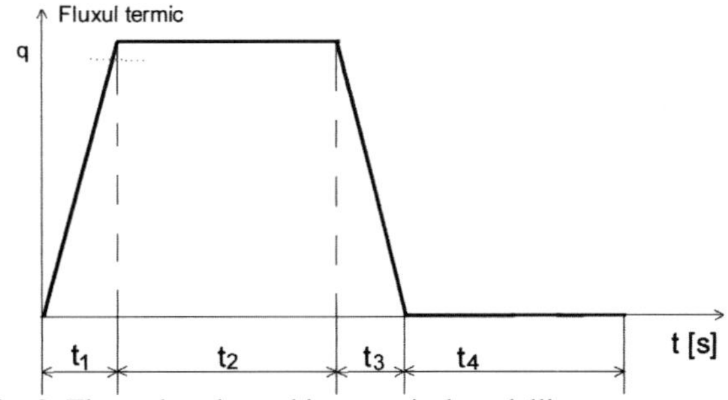

Fig. 2. Thermal cycle used in numerical modelling.

- the base material is isotropic;
- thermophysical characteristic depend on the temperature;
- the thermal flux presents a Gaussian distribution;
- heat losses into the environment, due to convection and radiation, are considered;
- the movement of the fluid in the pool is ignored;
- the influence of the latent heat at melting and solidification is ignored.

During the analysis, the variation in time of temperatures and the expansion of the heat affected zone (HAZ) were monitored for a butt welded joint. In this case, for the analysis of the temperatures distribution, the source is considered to be ellipsoidal.
The extension of the thermal source influence on the two sides of the sheet, its dimensions and concentration were taken into account. The value of the thermal flux was calculated, according to the power of arc and the distribution factors, and it was uniformly distributed in the points located in the deposited metal area.
The finite element analysis (FEA) was carried out in two steps. A non-linear transient thermal analysis was conducted first to obtain the global temperature history generated during the welding process.
A stress analysis was then developed with the temperatures obtained from the thermal analysis used as loading to the stress model. The welding was assumed to be symmetric so that only one half of the model was analyzed.
The mesh used in the thermal and structural analysis is illustrated in figure 1.
For the thermal cycle analysis, a full heating and cooling cycle was considered at different periods (t_1 = 0.1s, t_2 = 5s, t_3 = 0.1s and t_4 = 1000s), as presented in figure 2.

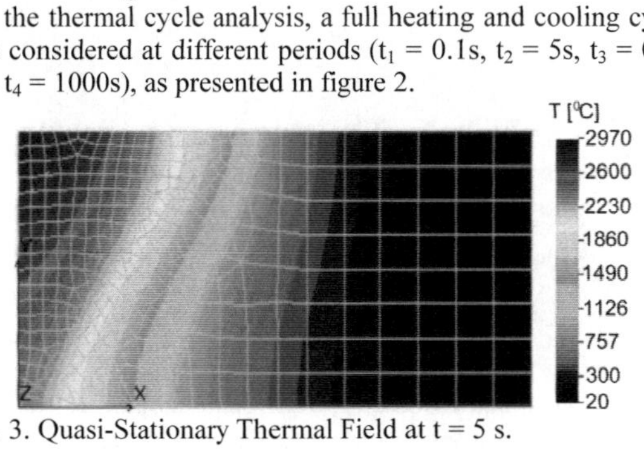

Fig. 3. Quasi-Stationary Thermal Field at t = 5 s.

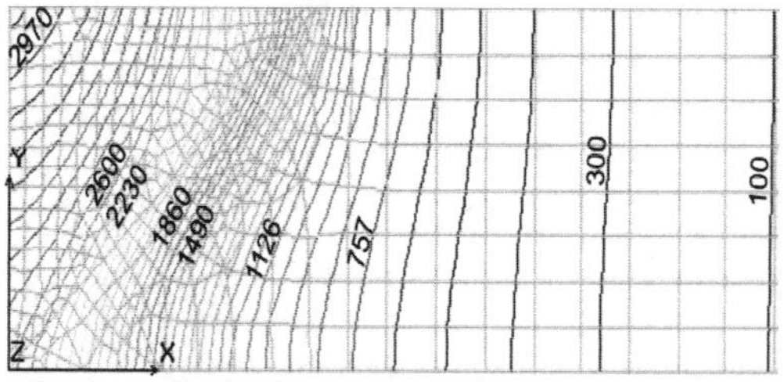

Fig. 4. Isothermals distribution at t = 5 s.

For numerical modelling, the group element PLANE 2D was used, which is a bi-dimensional isoparametrical element with four nodes. In the thermal transfer analysis, each node has a single degree of freedoms and a certain temperature.
The thermal field analysis was performed on two different cases, namely:
A – Material characteristics depending on temperature and zero convection;
B – Material characteristics depending on temperature and by convection on the cooling curve.
The distribution of temperatures in different moments of the welding process was monitored. In figures 3 and 4 are presented the thermal field and the isothermals distributions at t = 5 s for different moments of the welding process simulation for case A.
At the beginning of the process, during the transitory stage, a reduced thermal source influenced area is noticed, the thermal field being dependent on time. For process stabilization, during the quasi-stationary stage, the maximum temperature of 2970°C is reached at step t = 5s.
The welding pool undertakes changes as the thermal field stabilizes. If the images taken at various moments of simulation are compared, it was noticed that, as compared to the initial stage (transitory stage, t_s = 2.5s), the pool expands during the stabilized, quasi-stationary stage (t = 5s).

3. THERMOGRAPHED THERMAL FIELDS

By using infrared thermography it was obtained the thermal cycle for submerged arc welded joint of two sheets made of 10 mm thick X52 steel.

This is presented as a thermogram including the temperatures spectra around the thermal source. The image thus obtained presents the thermal cycle at a specific moment, the thermography equipment could not take measurements according to specific time. Determinations were done at room temperature, emissiveness being of 0.80 and the thermography distance of 2 m. The thermography had been done on opposite side of the weld.

Figure 5 presents the lines, according to which the thermal cycles inside the thermal source were drawn, the forefront of the thermal source and its rear side.

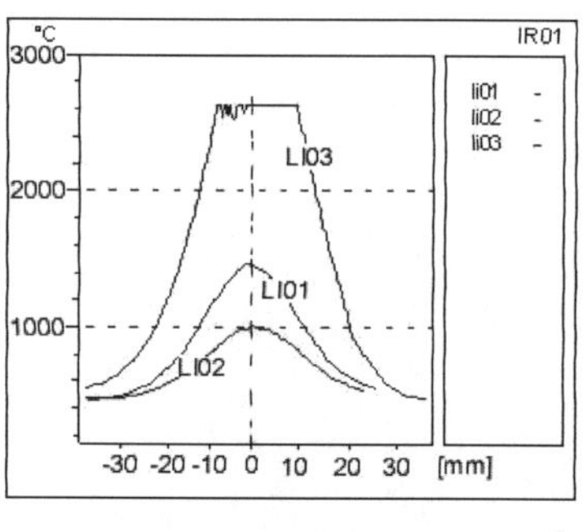

Label	Value
IR : max	2993°C
IR : min	31°C
LI03 : max	2632°C
LI03 : min	475°C
LI01 : max	1486°C
LI01 : min	389°C
LI02 : max	1012°C
LI02 : min	468°C

Fig. 5. Temperature transversal variation in various areas of the welded joint: LI01 in front of the thermal source; LI02 behind the Thermal Source; LI03 in the Middle of the Thermal Source.

Distance to the welding axis [mm]	Temperature determined with FEM [°C]	Temperature determined by thermography [°C]	Error ε [%]
0	2970	2993	0,76
4	2588	2900	10,75
8	1723	1800	4,27
12	505	600	15,83
20	56	60	6,66

Tab. 1. Temperatures value and the error for thermography and FEM.

The ellipsoidal shaped thermal source is to be remarked, the more concentrated isothermals are to be found in the front of the thermal source and more extended when found behind it, the maximum temperature of 2993°C, and the temperature of 31°C at which determinations were made. Infrared thermography allows determination of temperatures in one spot or in various spots of the welded joint, as well as on a surface. From the thermogram, the temperature of a certain isothermal can be determined, and compared to the temperature obtained by finite elements numerical modelling, which is subsequently reproduced on a simulator of thermo mechanical cycles. The temperatures values for the two methods employed, according to the welding axis in the points considered, are presented in table 1.

4. CONCLUSION

Numerical modelling of the thermal field using the finite elements method is a commonly used method which, by its results, gives distributions of the thermal field and values of temperatures close to those experimentally measured. The maximum temperature reached during process stabilization, for a stationary regime, is of 2970°C being located within the interval 1500...3000°C mentioned in the literature. Thermal cycles have been drawn, according to the distance to the centre of the thermal source, on both sides of the sheet and for different time frames, the obtained distributions and values showing a lower temperature on the opposite part to the thermal source, due to the heat dissipation in time and to the time necessary for conveying heat by conduction. Temperature differences determined on the two sides decreases with increasing the distance to the heating source axis. From the experimental determination of the thermal field by infrared thermography, it can be noticed that the employed method gave real results, the values and distributions being close to those mentioned in the literature. Higher temperatures were obtained experimentally, difference that may be explained by the simplifying hypotheses used in numerical modelling.

5. REFERENCES

Goldak, J., Chakravarti, A. & Bibby, M., (1984). A new finite element model for welding heat sources. *Metallurgical Transactions B* 15B, pp. 299-305.

Lindgren L. E., (2001). Finite element modelling and simulation of welding, Part2: Improved material modelling. *Journal of Thermal Stresses,* Volume 24, pp. 195-231.

Mircea, O., Banea M. D., Visan D. & D. Birsan, (2003). Technological Aspects for the One-Pass Flux Submerged Welding. *The Annals of "Dunarea de Jos" University of Galati, Fascicle XII Welding Equipment and Technology*, ISSN 1221-4639, pp 37-40.

Komanduri R & Hou Z B (2000). Thermal analysis of the arc welding process. Part I: general solutions *Metall. Mater.Trans.* B 31 pp. 1353–1370

Radaj, D. (1992). *Heat Effects of Welding: Temperature field, Residual Stress, Distorsion,* Springer-Verlag Berlin Heidelberg, Germany.

Annals of DAAAM for 2008 & Proceedings of the 19th International DAAAM Symposium, ISSN 1726-9679
Editor B. Katalinic, Published by DAAAM International, Vienna, Austria 2008
Make Harmony Between Technology and Nature, and Your Mind will Fly Free as a Bird

STUDY ON THE DEPENDENCE BETWEEN THE ERICHSEN COEFFICIENT BY ANISOTROPY AND HARDENING COEFFICIENTS

MIRONEASA, C[ostel] & MIRONEASA, S[ilvia]

Abstract: *The characterization of the thin sheet materials used in the drawing process is very important. There are different methods to establish the material properties. The paper presents the results of experimental researches during Erichsen and tensile tests. The goal of the research was to establish a relation to determine the Erichsen indicator using the anisotropy and the hardening coefficients. The mathematical models of dependence was built on data from 141 samples tested, using three statistic equation types. After statistic tests, it resulted that the polynomial regression equation expresses more faithfully the nature of the dependence.*
Key words: *Erichsen coefficient, anisotropy, hardening, tensile test*

1. INTRODUCTION

The drawing test using Erichsen method is a technological test that allows characterization of the capacity of deformation for metallic materials. The sample is stretched on two directions in conditions of lesser thinning. The Erichsen indicator (coefficient) *IE*, has for quantitative expression the material behavior, when a mechanical stretch is applied in order to obtain a format for the piece (Mironeasa et al. 2001).

Lankford, in 1950, established that the capacity of the material is in correlation with the anisotropy coefficient, *r* and the hardening coefficient *n*. Teodorescu (Teodorescu, 1994) presents a history of the developments made by Pomey G, in 1976, by Schneider G, in 1967 and by Lloyd D.H in 1962.

Due to the rolling process by which they are obtained, the metal sheet presents anisotropy of the mechanic properties, according to the rolling direction. The material anisotropy has a great influence on the deformation process, especially at deep drawing and, for this reason, has ben studied by many researchers as presented by (Barlat et al., 1991), (Hosford, 1972), (Karafillis & Boyce, 1993), (Gutt, 2000). The anisotropy coefficient has a significant influence in deformability appreciation of the sheet at deep drawing and at the drawing of complex and asymmetrical pieces. High value of the coefficient indicates a good lateral contraction and a high strength at undulation, parameters that are favorable at very deep drawing. As the anisotropy coefficient has high values, the height of the piece at drawing increases.

The structural alteration in the material, due to plastic deformation, causes important alterations of the mechanical and technological properties. The resistance appears in the material structure when the action of the exterior force will be large, with the formation of new sliding planes. As the deformation increases, the resistance to traction, the yield limit and hardness increase too, whereas the plasticity properties decrease. Therefore the material becomes harder, but less plastic, a phenomenon called hardening.

The phenomenon of sheet hardening, due to plastic deformation, is a complex one, based on the dislocation theory. In accordance with this theory, hardening would be due to move braking and at the same time to the movement of a great number of dislocations by barriers and obstacle formats in material even in the deformation process.

A high value of the hardening coefficient indicates that the material supports a maximum uniform deformation, and the maximum local deformations are lower due to the deformation process that appears during the Erichsen test.

As the value of hardening coefficient increases, the sheet would have a better behavior during the drawing process, especially for the pieces obtained in the automotive industry, where the predominant strain is biaxial stretching. The hardening coefficient is used to appreciate the thin sheet due to complex drawing for asymmetric pieces.

2. EXPERIMENTAL RESEARCH

The test was made on an Erichsen hydraulic machine, type LUH1, with force capacity of 100 kN and 10kN chucking force. The tensile tests were made on equipment Tinius-Olsen with the force domain between 3 kN and 150kN. A strain gauge type Epsilon 3542-050 M-020-ST, B1 class was used. The elongation speed was constant.

The samples were extracted from the sheets with different thicknesses. The sheets studied are thin table of cold rolled steel for drawing, in accordance with the standard SR EN 10130 + A1: 2000, with thickness variance from 0,75 mm to 2,13 mm.

The axis of the samples was at 90° from rolling direction. The samples were taken from different places of a roll of sheet.

The chemical composition of the materials was not taken into account, but the sheets are from steel destined to drawing processes. The limits of the parameters obtained after tensile tests and the Erichsen tests are presented in Table 1.

Application of the statistic method follows in order to obtain a mathematical model which would express more faithfully the nature of the dependence.

After the error elimination, three equation types were taken into account: linear (1), exponential (2) and polynomial (3). The subscripts of the Erichsen coefficient reflect the equation type.

$$IE_L = 13{,}695 - 1{,}219 \cdot r + 4{,}216 \cdot n \tag{1}$$

$$IE_E = 15{,}779 \cdot r - 0{,}185 \cdot n\ 0{,}076 \tag{2}$$

$$IE_P = 30{,}72 - 2{,}638 \cdot r - 145{,}714 \cdot n - 0{,}44 \cdot r^2 + 16{,}184 \cdot r \cdot n + 278{,}26 \cdot n^2 \tag{3}$$

The accuracy of the two types of regression equations is expressed by the indicators presented in Table 2.

The dispersion *s* explains that the linear equation displays the lowest average value of the quadric residue and the coefficient of variation C_v % indicates that there was a small variation between the values obtained experimentally and those calculated by equations with a linear regression, respectively exponentially.

The intensity relation between *IE* and independent parameters is given by the coefficient *R*. The share of the parameters that influences concomitantly, from each equation type is given by R^2.

Limits	Parameters					
	g [mm]	R_e [N/mm²]	R_m [N/mm²]	r	n	IE [mm]
Minimal	0,75	149,1	286	1,228	0,181	10,6
Maximal	2,13	239	352	2,99	0,225	13,3

Tab. 1. Parameter limits

Indicator	Values		
	IE_L	IE_E	IE_P
s	0,413	0,438	0,416
C_v %	3,447	3,655	3,467

Tab. 2. The indicators expressing the precision of regression equations

Indicator	Values		
	IE_L	IE_E	IE_P
R^2	0,626	0,451	0,659
R	0,791	0,671	0,812

Tab. 3. The values of the determination and multiple correlation coefficients

Variable	r	n	IE
R	1	-0,111	-0,788
N	-0,111	1	0,162
IE	-0,788	0,162	1

Tab. 4. The values of correlation coefficients

The coefficient R^2 gives the statistical weight observation for the independent parameters. The values of these indicators for each type of equation are shown in Table 3. The value R = 0,812 obtained for the polynomial equation indicates that the regression equation of this type offers the best mathematical representation for the determination of *IE* coefficient depending on the hardening and anisotropy coefficients.

The value obtained for R^2 shows that the coefficients of hardening and anisotropy have a combined influence on the Erichsen coefficient with a weight of 65,95%.

The significance of the correlation coefficient for the polynomial equation type was checked by the Fisher test, which has the highest value of this factor. The result $F_c = 134{,}633 > F_{critic} = 3{,}07$ indicates that the correlation coefficient is relevant for the level of signification $\alpha = 0{,}05$ and for the freedom level $v_1 = 2$ și $v_2 = 139$.

Furthermore, the significance of the regression equation coefficients of polynomial type is tested by the Student test. The results of the test for $t_{0,05;\,140} = 1{,}976$ indicate that:

$$|a| = 30{,}72 > |\Delta a| = 0{,}052;\ |b_1| = 2{,}638 > |\Delta b_1| = 0{,}013;$$
$$|b_2| = 145{,}714 > |\Delta b_2| = 0{,}473;\ |b_3| = 0{,}44 > |\Delta b_3| = 0{,}013;$$
$$|b_4| = 16{,}184 > |\Delta b_4| = 0{,}064;\ |b_5| = 278{,}26 > |\Delta b_5| = 0{,}473$$

It is noted that the absolute value of the linear equation coefficients, a, and b_i (i = 1…5) are superior to the confidence interval of the calculated coefficients and it is considered that they are significant.

After applying the Fisher test for checking of the verisimilitude of the polynomial model, $F_c = 0{,}618 < F_{critic} = 0{,}757$. The value that was obtained indicates that the mathematical model given by the exponential regression is reliable.

The graph representation of the dependence between the Erichsen coefficient IE – hardening coefficient and anisotropy coefficient is presented in Fig. 1. The processing of statistical results aimed at determining the values of the correlation coefficients between the variables studied, whose values are presented in Table 4.

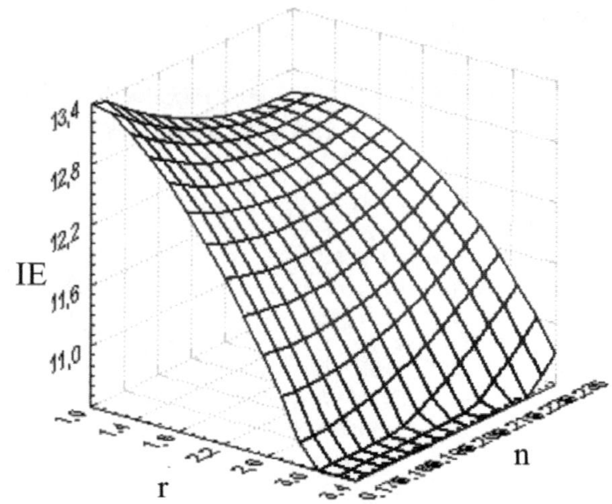

Fig. 1. The dependence IE = f (n, r)

From an analysis of the correlation of the Erichsen coefficients values, it can be noted that there is a strong reverse relation between *r* and *IE* and a weak link between *n* and *IE*. The Erichsen coefficient is influenced by the anisotropy in proportion of 78,80% and less and only 16,20% by the hardening coefficient.

3. CONCLUSIONS

Deformability appreciation of thin steel sheet for drawing process cannot be achieved on the basis of a single parameter or a single type test. Each parameter in part provides specific information about a material behavior, but does not characterize it entirely.

In the case study of dependence of the *IE* according to *r* and *n*, the mathematical model of polynomial type offers the best mathematical representation.

Using only the tensile test, it is easy to calculate the drawing coefficient IE with the polynomial equation determined in this paper. Thus, it is not necessary to run two different tests. For the next research to create a mathematical model we will take into account: the thickness of the sheet, the yield limit R_e and the breaking strength R_m.

4. REFERENCES

Barlat, F., Lege, D.J., Berm, J.C., (1991), *A Six Component Field Function for Anisotropic Materials*, Int. J. Plasticity 7, pp. 693.

Teodoresu, M. Al., (1994), *Contributions Regarding the Study of Deep Drawing Fabrication,* Plastic deformation, vol.1, no. 2. The Study and Research Center for Plastic Deformation, University Lucian Blaga, Sibiu

Gutt, G., (2000), *Modern Methods for Material Testing and Characterization*, vol. II, Editura Universității Suceava

Hosford, W.F., (1972), *A Generalized Isotropic Yield Criterion.* J. App. Mech. 39, pp. 607

Karafillis, A.P., Boyce, M.C., (1993) *A General Anisotropic Yield Criterion Using Bounds and Transformation Veighting Tensor*, J. Mech. Phys. Solids, 41/1859

Mironeasa, S., Gutt, G., Mironeasa C., (2001), *Experimental Erichsen Coefficient Determination Using the Metal Chemical Composition of Steel Sheet*, Acta Universitatis Cibiniesis, Vol. XLIV, Technical series, pp. 96÷100, Sibiu

Annals of DAAAM for 2008 & Proceedings of the 19th International DAAAM Symposium, ISSN 1726-9679
Editor B. Katalinic, Published by DAAAM International, Vienna, Austria 2008
Make Harmony Between Technology and Nature, and Your Mind will Fly Free as a Bird

STUDY OF THE RELATIONSHIP BETWEEN THE DRAWING COEFFICIENT AND HARDENING RATIO

MIRONEASA, S[ilvia] & MIRONEASA, C[ostel]

Abstract: *In the design of the drawing processs, different equations are used that allow for modeling of the deformation of the piece. One of these coefficients is the drawing coefficient. The determination of the drawing coefficient involves tests or use of equations. The paper presents a modality to determine the drawing coefficient using the results of the tensile test. The equation was determined using experimental data for the hardening coefficient and an equation for the determination of the drawing coefficient. The mathematical model was tested usig statistics. Averkiev's equation is difficult to use. The purpose of the paper is to establish a simple equation to determine the drawing coefficient using the data form tensile tests.*
Key words: *drawing coefficient, hardening coefficient, deformation, tensile test*

1. INTRODUCTION

Knowledge of the conditions under which the material passes during processing from the status of elastic state to plastic presents importance in the analysis of the deformation processes.

In the case of traction – uniaxial tensile, the passage from the elastic state to the plastic state occurs when strain exceeds the limits of the material's flow.

The appreciation of the ability of deformation for thin steel plates can be carried out on the basis of their mechanical properties: yield limit R_e, breaking strength R_m, breaking elongation, necking ratio, properties determinted by tensile tests.

For the plastic deformation process a specific tensile strain is necessary in the structure of the material. During the tensile test, this is possible if the strains have values between R_e and breaking strength R_m.

Generally, as the Romanian literature mentions, (Gâdea & Petrescu, 1983), (Teodorescu et al, 1987), (Teodorescu, 1992), it can be appreciated that a material, which has a large breaking elongation, a small yield point and great tensile strength, has also the mechanical properties that are favorable for a good behavior deformation. A small value of the ratio between yield limit R_e and breaking strength R_m indicates also a good behavior for the thin sheet during the deformation process.

The hardening process increases the yield point to a greater extent than the tensile strength (Cao et al., 2000). The difference between these two factors (R_m - R_e) determines the decreasing of the material deformation capability.

The breaking strength is a representative factor that characterizes the mechanic strength under normal conditions and also for the deformation in the elastic-plastic field where information about the admittance of maximum value of the force during deep drawing process is given. This strength characterizes the maximum tension that can be supported by material without breaking. The increase of the value over the maximum limit (R_e) produces necking of the samples.

The deformation strength of the material can be appreciated using the value of breaking strength R_m, equation 1, where: F_m is the maximum value supported by the sample before breaking; S_0 is the initial area of the sample in the calibrated zone. The yield limit is another factor of the deformability that is determined by the ratio between force F_c at which there occurs the flowing of material, and the area S_0 of the samples in the calibrated zone, equation 2.

$$R_m = \frac{F_m}{S_0} [N/mm^2] \tag{1}$$

$$R_e = \frac{F_c}{S_o} [N/mm^2] \tag{2}$$

The tensile test of the thin metal sheet consists in stretching the samples along the longitudinal axis using an increased force. This test allows for the determination of the mechanical characteristic of the thin metal sheet.

The hardening ratio R_e/R_m can characterize the hardening of the material under strain conditions during the plastic deformation process (Mironeasa & Mironeasa, 2004). For the majority of materials, the lesser the hardening flow ratio, the greater the value of relative necking is, and the plasticity of the material is superior (Iliescu, 1977).

Low values of the hardening ratio assure a good stability for the piece format after deep drawing. The bigger the hardening ratio, the higher the value of the hardening coefficient is.

The hardening ratio can be determined by the ratio between the maximum force before sample breaking and the value of the force when the material is flowing. In these conditions, the dimensions and the measure errors of the sample occur.

The forces of the values are determined during the same tensile test so the influence of the sample dimensions cannot influence the hardening ratio.

The drawing coefficient characterizes the degree of the total deformation of material in the drawing process, provided that the pieces should be transformed from plate to cylindrical forms in a single operation, without deliberate thinning of the material. There are different approaches to determine the drawing coefficient. In his paper (Averkiev, 1989) proposes an equation that allows calculation of the coefficient using the results form tensile tests, equation 3, where: r is the anisotropy coefficient; n is the hardening coefficient.

$$m = \frac{d}{D} = 0{,}8169 \cdot \left(1{,}2247 \sqrt{\frac{1+r}{2+r}} \cdot \frac{R_e}{n+1} \cdot \left(\frac{2{,}72}{n} \right)^n \cdot n^{(n+1)} \right)^{-0{,}0994} \tag{3}$$

2. EXPERIMENTAL RESEARCH

The experimental researches were made on 655 samples from special steel use in drawing processes. The samples were extracted from the sheets with different thicknesses (0,42 – 3,00 mm).

The axis of the samples was at 90° from the rolling direction.

Limits	Parameters				
	R_e [N/mm²]	R_m [N/mm²]	r	n	m
Minimum	136,45	281	0,63	0,172	0,533
Maximum	328	387,5	3,215	0,236	0,575

Tab. 1. Limits value of the parameters determined

Indicator	s	C_v %	$s_{\bar{y}x}$	$s_{\bar{y}x}$ %	F_c	F_{critic}
Value	0,005	0,901	0,0002	0,035	0,724	0,879

Tab. 2. Statistic precision indicators

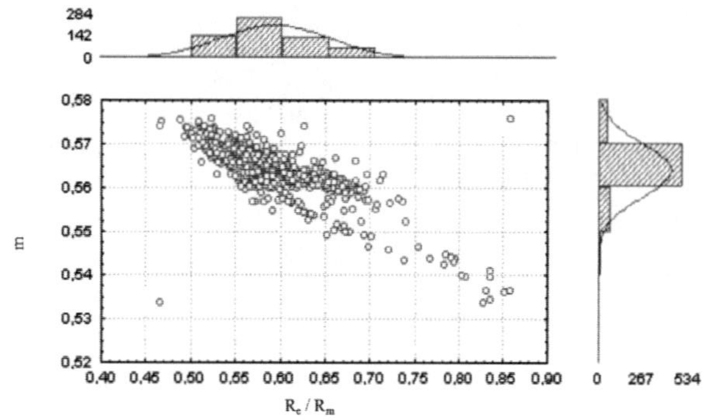

Fig.1. Dependence between m and R_e/R_m

The tensile tests were made on Tinius-Olsen equipment with the force area between 3 kN and 150kN. A strain gauge type Epsilon 3542-050 M-020-ST, B1 class was used.

The elongation speed was constant. The nature of the material composition was not taken into account.

The limits of the parameters obtained after tensile tests and for the hardening ratio are presented in Table 1. To determine the dependence between the hardening ratio and the drawing coefficient, the data were analyzed using statistic analysis. The dependence between the studied parameters is presented in Fig. 1.

Using the regression method, in order to determine the dependence, a linear equation was established, which allows for easy determination of the hardening coefficient, equation 4.

$$m = 0{,}614 - 0{,}085 \cdot (R_e/R_m) \quad (4)$$

The equation coefficients' significance of the linear equation was tested with the help of the Student test. The results of the test $t_{0,05;\,653} = 1{,}965$ indicate that:

$|a| = 0{,}614 > |\Delta a| = 0{,}0002$; $|b| = 0{,}085 > |\Delta b| = 0{,}0001$

The values of equation coefficients a, and b are superior to the trust interval, so these coefficients are significant.

The values of the indicators that predict the precision of the regression equation are presented in Table 2, where: $s_{\bar{y}x}$ is the standard error of the regression curve; $s_{\bar{y}x}$ % is the imprecision of the regression curve.

The intensity between m and R_e/R_m is given by the correlation coefficient. The value r = -0,85 indicates a strong relationship. The minus sign indicates that the relation for m and R_e/R_m is reversed.

The significance of the correlation coefficient was verified by the Student test for a significance for $\alpha = 0{,}05$ and n-2 liberty degree. After the test, $t_c = 41{,}381 > t_{0,05;\,653} = 1{,}962$, the result indicates that the correlation coefficient is significant. Besides, the relationship between m and R_e/R_m was tested using R^2. The coefficient indicates that the drawing coefficient is influenced by R_e/R_m in a proportion of 72,38%. The value of C_v%, variance coefficient indicates that the variance between the values obtained using experimental models and the values obtained using calculated value is negligible. If the Fisher test is applied in order to establish the reliability of the linear equation, the result $F_c = 0{,}724 < F_{critic} = 0{,}879$, indicates that the mathematic model is reliable.

3. CONCLUSIONS

The determination of the drawing coefficient needs other specific tests. The 3rd equation (Averkiev, 1989), used to determine the drawing coefficient is a possible solution, but difficult to use because it requires the calculation of anisotropy and the hardening coefficient. To determine these coefficients the dimension of the sample at the beginning and at the end must be considered. This procedure introduces errors because the measurements of the samples at the end are influenced by necking. A solution is to eliminate the samples' dimension measurements. If the hardening ratio is used, the dimensions of the samples are eliminated from calculus and from errors, too.

The drawing coefficient is a factor frequently used in practice because of its simplicity and convenience. The numeric value of the drawing coefficient allows for the determination of other indicators that characterize the material deformation.

Using only the tensile test, it is easy to calculate the drawing coefficient with the linear equation determined in this paper. Between the drawing coefficient and the hardening ratio, an inverse proportion relation is established; height values for hardening coefficient determine small values of the drawing coefficient. If the drawing coefficient is decreased, then the material allows for great deformations.

If it is necessary to determine the drawing coefficient of the materials we need a different test, for which punches dies and pressing machines are necesary. Using the proposed equation (4), the materials can be characterized form the point of view of drawing capability, using only the tensile test.

Because the drawing coefficient is influenced by the hardening ratio in proportion of 72,38%, it is necessary to establish, and take into account, other factors that contribute to the deformation.

In our future research undertakings, the thickness of the sheet will be looked into.

4. REFERENCES

Averkiev, A.I., (1989), *The Technology of Deep Drawing*, Maşinostroenie, Moskva

Cao, J., Yao, H., Karafillis, A. and Boyce, M.C., (2000), *Prediction of Localized Thinning in Sheet Metal Using a General Anisotropic Yield Criterion*, International Journal of Plasticity, Vol. 16/9, pp.1105-1129, August.

Gâdea, S., Petrescu, M., (1983), *Physical Metallurgy and the Metals Study*, Editura Didactică şi Pedagogică, Bucureşti

Iliescu, C., (1977), *Punching and Cold Forging Technology*, Editura Didactică şi Pedagogică, Bucureşti

Mironeasa C. and Mironeasa S., (2004), *Research Regarding The Dependence Between the Drawing Coefficient and Different Parameter Used at the Thin Sheet Deformations*, The Annals of University "Ştefan cel Mare", XII, 2004, No. 1 – 6, Mechanic section, pp. 25-28, Suceava

Teodoresu, M.Al., et al., (1987), *Processing by Plastic Cold Deformation*, vol. I, Editura Tehnică, Bucureşti

Teodorescu M. Al., (1992), *Metal deformation. The Methodology to Establish the Propriety of Deformation of the Materials*, Sophia – Antipoli

Annals of DAAAM for 2008 & Proceedings of the 19th International DAAAM Symposium, ISSN 1726-9679
Editor B. Katalinic, Published by DAAAM International, Vienna, Austria 2008
Make Harmony Between Technology and Nature, and Your Mind will Fly Free as a Bird

USING EMTP-ATP AND ANSYS PROGRAMS FOR DETERMINATION OF DIELECTRIC STRESS OF MEDIUM VOLTAGE CABLES

MISAK, S[tanislav] & CHMELIK, K[arel]

Abstract: *Cables lines are important components of the reliability system at medium voltage levels. The insulating system of the cable line is stressed by some degrading factors during its operating mode. A temperature, a kind and amplitude of voltage, humidity and mechanical stress are main degrading factors. The dielectric stress of medium voltage cable line 6AYKCY 3 x 70/16 is determined by EMTP-ATP and ANSYS program for running state and 1phase arc-earth fault. The electric intensity distribution is simulated by ANSYS program in dielectric of cable in the concrete. Results are used as enters dates for ANSYS program from simulations in EMTP-ATP. Some results from mathematical modelling are introduced within this paper.*

Key words: *Diagnostic, Dielectric, Cable, EMTP-ATP, ANSYS.*

1. INTRODUCTION

The main part of medium voltage electrical industry systems is operated with cables lines. All plastic cables on the voltage level 1, 6, 10 and 22 kV are used for the transmission of the electric energy mainly.

The damage of cables can be caused by the mechanical stress and then by the humidity penetration. The insulating system can be damaged by the electrical and the temperature stress at normal and fault states. Last mentioned fault states are with the short-period, but the activity of these states can accelerates degrading process of insulating system and cause of disruptive discharge in the durability life. This fact is caused by the inception of partial discharges (Slaninka, 1989). Cables lines with plastic materials are sensitive on the inception of partial discharges. It is desirable in order that electro-insulating material must have high specific dielectric strength and the maximum of dielectric stress has to be at the allowed limit.

The maximum dielectric stress is influenced by form of the electric field at specific voltage. Where, the electric field is defined by the cable construction and by parameters of individual parts of dielectrics. The intensity of electric field is not constant in the heterogeneous filed of cable line, but the intensity of electric field is dependent on the position data at the cable dielectric.

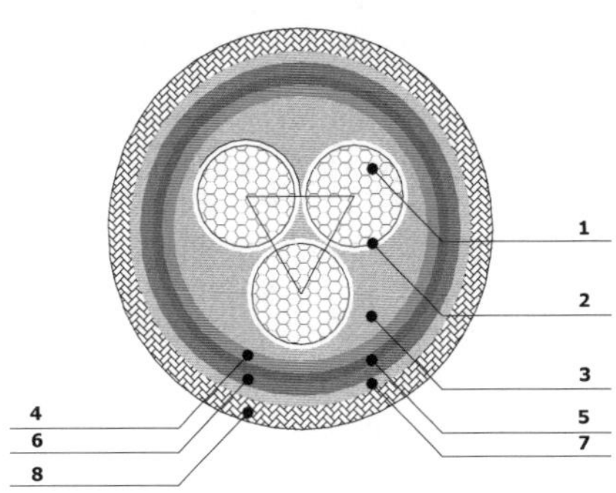

Fig. 1. The section of the cable 6AYKCY 3 x 70/16.

The main target of the electric field solving is location of the position with the maximum value of the electric field intensity for variety of running and fault states of the cable line.

Results from this analysis are applicable for the optimizing of the cable line design and the medium voltage system too.

2. MATHEMATICAL MODEL OF MEDIUM VOLTAGE CABLE IN EMTP-ATP AND ANSYS

Results from mathematical modelling of the electric field of the medium voltage 3phases enclosing pipe 6AYKCY 3x70/16 are introduced within this chapter. The electric field of the mentioned cable line is simulated with using program ANSYS for the running state. The place with maximum dielectric stress is located with the program ANSYS too. The section of the cable 6AYKCY 3 x 70/16 is shown in the Fig. 1, where "1" is aluminum core, "2" PVC isolation, "3" cushioning cover, "4" inner sheath, "5" tape1, "6" cooper concentric conductor, "7" tape2 and "8" is PVC outer sheath.

The mathematical model of the cable line 6AYKCY 3 x 70/16 was created in the program ANSYS as a next. The core of ANSYS program uses principle of Finite Element Method. The cable is modelled as a geometrical dissociation of individual layers with the specification of basic parameters. Individual layers are defined by material density ($kg \cdot m^{-3}$), specific thermal conductivity ($W \cdot m^{-1} \cdot K^{-1}$), specific resistance ($W \cdot m$), specific heat capacity ($J \cdot kg^{-1} \cdot K^{-1}$) and relative permeability (-). The solution of the electric field distribution was realized for the no-load state of the cable at supply by source 6kV, 50 Hz. The result of the simulation is shown in the Fig. 2. From Fig. 2 is evident, that the maximum dielectric stress is on the core surface in the cores intersection. It is state, when the applied voltage is line-to-line voltage. The maximum dielectric stress value was defined by the simulation about electric field intensity 1,93 $kV \cdot mm^{-1}$. The maximum value of dielectric stress, which is guaranteed by producer for the PVC insulation, is in the interval (2,8-3,3) $kV \cdot mm^{-1}$. It is evident, that the maximum value of dielectric stress was not exceeded. On the other side, it is necessary to respect, that this analysis was realized for the running state.

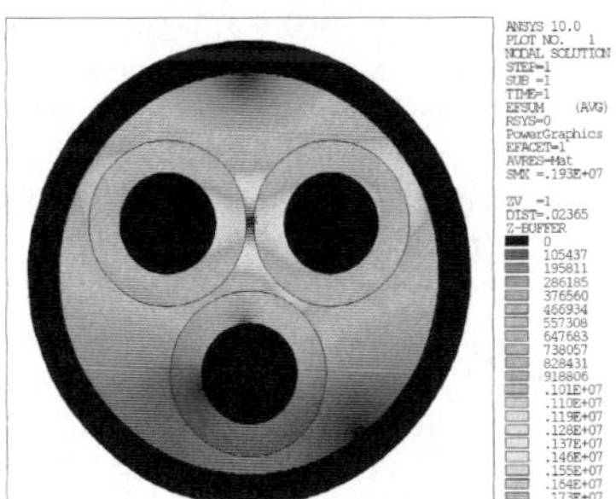

Fig. 2. The solution of the electric field distribution at cable 6-AYKCY 3x70/16, no-load state, running state.

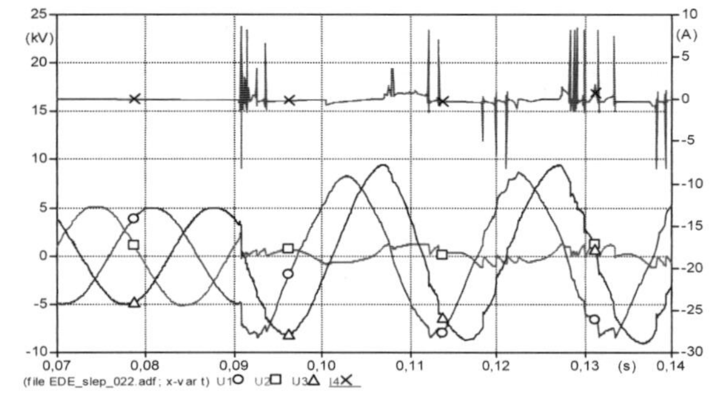

Fig. 3. Courses of systems voltages in the fault point (U_1, U_2, U_3 (kV)) and current by electric arc-earth fault (A) generated by EMTP-ATP.

In the case inception of the 1polar earth-fault in the medium voltage system is insulation of the cable line extremely stressed by increasing of the amplitude voltages at phases without fault. The voltage in the phase with the fault is limited by resistance of the electric arc (See Fig. 3). The mathematical model of the tested medium voltage system with the cable lines was created in the program EMTP-ATP. The mathematical pre-processor ATPDraw was used in the concrete. This model was used for the definition of earth-fault influence on the cable line insulationThe cable line has size 2000 meters with the total capacitive current c. 2A. 1polar earth-fault was simulated as a 1phase arc earth-fault; the mathematical model of electric arc respects Mayr theory in the EMTP-ATP (Mišák, 2007). Final courses of instantaneous values of voltages at arc earth-fault are shown in the Fig. 3. The course of current by fault is shown in the Fig. 3 too. All courses were obtained with the using mathematical modeling in the EMTP-ATP and verified by real experiments in the medium voltage system Power Station Dětmarovice. The increasing of voltage without fault to line-to-line voltage 6kV and drop of voltage with fault to value limited by the resistance of the electric arc is evident from the Fig. 3.

The new analysis distribution of the electric field intensity was realized repeatedly for the case inception of the arc earth-fault in the medium voltage analyzed system. Analysis was executed by the program ANSYS too.

The solution of the electric field distribution at cable 6-AYKCY 3x70/16 for the fault state is introduced in the Fig. 4. There is representation of the extreme stress with the amplitude 3,34 kV·mm^{-1}. This value is over the limited allowed dielectric stress.

Furthermore, the dielectric stress of the electric insulation is increased in the case, when initial conditions are changed by time - dependence resistance of the electric arc (Van der Sluis, 2001).

3. MATHEMATICAL MODEL OF CABLE IN ANSYS

For the examination of electric stress of an insulation system, there is necessary to find time and space distribution of the electric field intensity. It is familiar that electric field with intensity vector has irrotational character, also term conservative is used. We can obtain two forms of Maxwell equation – differential or integral – related to an intensity of electric field vector, electric induction and charge, with the mutual conversion by the help of Stokes' integral theorem. Direct mathematic solution with vector of the intensity is complicated, so it can be done with the help of scalar electric potential (Haňka, 1975; Vanderline, 2005). It is a single-valued function of location, continuous at a boundary, what is favourable to numerical modelling.

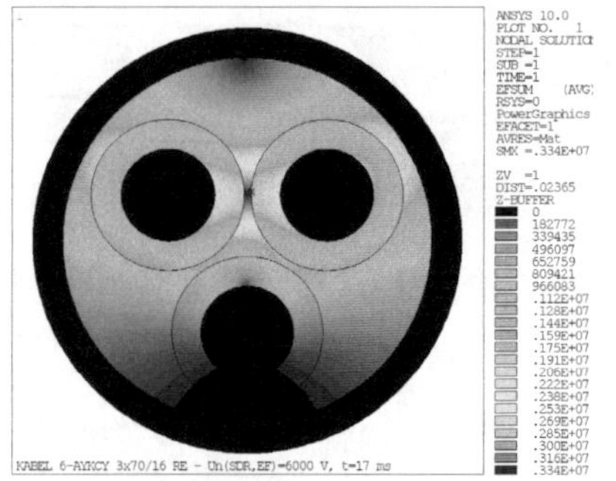

Fig. 4. The solution of the electric field distribution at cable 6-AYKCY 3x70/16, fault state, arc earth-fault.

Physical model was simplified by omitting the tape areas 5 and 7 in Fig. 1. The cable is assumed to be infinity long and solved problem is two-dimensional. There were used two-dimensional, eight-node elements PLANE121 with one degree of freedom (voltage) at each node. This element is based on the electric scalar potential formulation, and it is applicable to a 2-D electrostatic and time-harmonic quasi-static electric field analyses. Model was realized with a number of 56 637 nodes and 20 278 elements. Next simplification is a constant permittivity.

Defined materials are considered as a linear – independent to the field quantities and isotropic – the same in all of directions.

Problem is solved as a range of static analyses for defined instants of time and particular voltage amplitudes. It is assumed three-phase cable and it means three time-dependent instantaneous voltages or potentials at each phase conductor.

Self and mutual capacitances of a general multi-electrode system with a certain configuration can be solved by the Simpson's mirroring method and by calculation of capacitance coefficients, also with the aid of FEM or directly with Ansys internal macro CMATRIX.

4. CONCLUSION

The possibility of using mathematical programs for the technical diagnostic of insulation system was verified within this paper. Methods of mathematical modelling were used for the optimizing of cable line design and medium voltage system for running and fault states too. Results of simulation were verified within real experiments at medium voltage system in the Czech Republic.

5. REFERENCES

Mišák, S. (2007). Mathematical Model of Electric Arc at High Voltage Systems. *Proceedings of the 18th International DAAAM Symposium, Katalinic, B. (Ed.)*, vol. 18, October 2007 page numbers (14-17), ISSN 1726-9679.

Haňka, L. (1975). Theory of Electromagnetic Field, SNTL, Praha.

Vanderline, J. (2005). Classical Electromagnetic Theory. Kluwer *Academic Publishers*, 1-4020-2699-4, USA.

Slaninka, P. (1989). Cable Technic 1, SVŠT, ISBN 80-227-0026-6, Bratislava.

Van der Sluis, L. (2001). Transients in power systems, John Wiley & Sons Ltd., ISBN 0-471-48639-6, England.

Annals of DAAAM for 2008 & Proceedings of the 19th International DAAAM Symposium, ISSN 1726-9679
Editor B. Katalinic, Published by DAAAM International, Vienna, Austria 2008
Make Harmony Between Technology and Nature, and Your Mind will Fly Free as a Bird

ABSORPTION PROPERTIES OF THE NOISE BARRIERS MADE OF SCRAP TIRES

MISIK, L[adislav] & RADVANSKA, A[gata]

Abstract: *The aim of the paper is to show the possibilities of scrap tires reuse to manufacture the noise barriers. Recycled tire crumbs and other additives can be incorporated to the new material to form a noise-absorptive barrier panel. Using recycled rubber tire as noise absorption material has been proved to be satisfactory. Intact tire can be filled into the noise barrier to provide air gap for the powder tire and give sufficient reduction of sound between 250 Hz and 500 Hz which are the dominant frequencies of traffic noise. The panel with tire components as absorption material can effectively reduce noise reflection. Compared to fragile conventional absorption material that is easy damaged, absorptive material made of waste tire is economical, highly durable and effective without any adverse effects. The study demonstrates that rubber tire has great potential to be further developed into a practical absorption material for reducing transportation noise.*

Key words: *rubber, noise, barrier, traffic, absorption*

1. INTRODUCTION

Vehicles in motion produce the friction between the vehicle's body and the air touching the vehicle. Such friction renders an aerodynamics effect that noise will be generated because of the gradient in the air pressure field induced by the friction. This pressure field will propagate to generate noise that can be heard at significant distances. Additionally, the contact of grooved tires on pavement surfaces occurring at high speeds creates a substantial sound pressure field as well as engine operations and exhausts systems. The acoustic spectrum of traffic noise, originated by moving vehicles, is of multiple frequencies. The majority of the spectrum falls within the frequency range of 250 Hertz and 4000 Hertz (Lapcik, 1998). The noise within this frequency range can be easily heard by the human ear, and can cause great discomfort. To control the propagation of this traffic noise, common practice is to build noise barriers along highways so that noise will be contained and absorbed within barriers, and will not propagate to any significant distance.

2. MATERIAL OF NOISE BARRIERS

The most of highway noise barriers are built with pre-cast concrete or concrete blocks or slabs. The study shows that these barriers are of very high acoustic reflectivity (95% and above) (Bulletin of the Acoustical and Insulation Materials Association, 1974) and of low sound absorption for the frequency band of highway noise between 250 Hertz and 4000 Hertz. Thus the effectiveness of concrete noise barriers in controlling vehicle noise is not satisfactory. With the drastic increase in highway traffic in the last two decades, the effort to develop new and better noise-reduction barriers for highways as well as airport and other applications has been intensified. It is predictable that such intensification will continue because noise poses an increasingly environmental threat. In recent years, some notable progress has been made in this respect. The polycarbonate noise reduction panels were developed in the U.S.A. The polycarbonate plastic has been tested to shelter a jet engine in New York airport (The Wall Journal, 1997). Another development is the noise barrier system made from lightweight hollow panels (Carsonite) made of tongue-and-groove planks of reinforced composite material filled with crumbed tire rubber. Traditional noise barrier walls have a flat surface. Now new designs are experimented with non-flat surface textures (Zhu, Carlson, 1999). These newly developed noise barriers exhibit a much better performance than concrete with respect to the capability of sound absorption and transmission loss, but the noise reduction is not the only criterion. In fact, there are other crucial criteria in constructing noise barriers. These criteria include cost effectiveness, technology maturity, durability, low cost and convenience in installation, and in maintenance and repair, and last but not least its aesthetics. The conventional concrete noise barriers meet those criteria very favorably. Polycarbonate plastic or composite noise barriers are very costly, and much less competitive in those criteria in comparison to concrete ones. This is why so far the progress made in replacing concrete noise barriers with aforementioned new noise reduction materials is very limited.

2. PARAMETERS OF NOISE BARRIERS

The most critical parameter in characterizing the capability of a material of how well it can absorb sound or noise is called the acoustical absorption coefficient (AAC). A sound wave carries certain amount of the energy called sound energy.

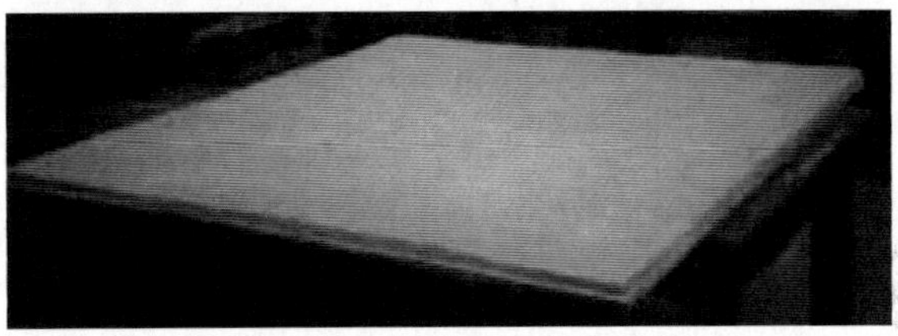

Fig. 1. Crumb rubber based noise barrier panel

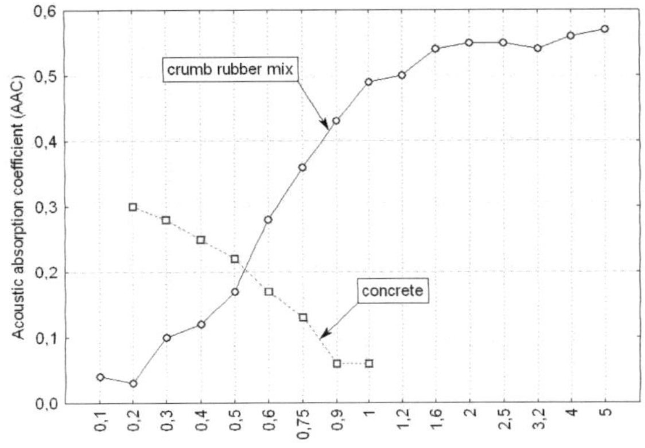

Fig. 2. Acoustic absorption coefficient versus frequency

When a sound wave hit a material, portion of the sound energy will be reflected or "bounced" back. From the noise reduction point of view, the lesser sound energy being bounced back, the better the effect of noise reduction. A value AAC=0 means sound energy being reflected completely, and a value AAC=1 means that all the sound energy is absorbed by the material, which is the best in noise reduction. A crumb rubber based noise barrier panel, manufactured as a mixture of 80% crumb rubber and 20% of bonding agent with a mixed size of rubber particles using the second and third spray devices (Fig. 1) has the acoustical absorption coefficient versus frequency was obtained and is plotted in Fig. 2, compared to the acoustic absorption coefficient for concrete noise barriers. It can be seen that the crumb rubber based specimen shows superiority in acoustical absorption (Zhu & Carlson, 1999, a).

3. PROPERTIES OF NOISE BARRIER

Rubber is notorious for its high *flammability* and the dense smoke which is produced when it burns. The noise barrier made from this material could ignite as a result of such incidents as grass fires, accidents, or vandalism. To reduce rubber's susceptibility to these concerns, flame and smoke retardants are available that can be added to the mixture during the manufacturing process. Recycled rubber tire material has been found to be *nontoxic* under leachate testing. However, additives, such as binders, retardants, coatings, and coloring, included in the mix to form and enhance the material, can create potential toxicity problems. These additives are, in some cases, proprietary with the specific formulations kept in confidence by the manufacturer. Rubber material, on its own, does not have sufficient *rigidity* to be considered as a structural component of a noise barrier panel. Therefore, bonding agents must provide adequate stiffness to enable the panels to be considered strong enough to withstand wind loading, or the rubber material must be firmly attached to a suitable stiffener, such as channel backings, cores, or casings. The structural strength of the panel must be verified through load testing on a production panel. Rubber and some binders tend to *oxidize* over time when exposed to the elements. They may also be susceptible to certain chemical or petroleum products. This increases the potential of premature disintegration of the panels. If concrete is used as a binder, concrete modifiers and special treatment of the crumb rubber are required before they will bond properly to each other. This is particularly important when these panels are exposed to salt, cold weather, and flexing for a long period of time. To optimize the bond between the rubber crumb particles, it is necessary to ensure that the rubber crumb is new or has been protected from the elements. The binders used should be stable under prolonged exposure to ultraviolet light. The manufacturing process should ensure that each rubber particle is completely encapsulated by the binder. If cement is used, the rubber surface should be treated or impregnated with a bonding agent compatible with both the rubber and the concrete. Or, the concrete should contain modifiers that will allow it to firmly bond to the rubber and be able to stand the test of time. Some coatings have a tendency to oxidize prematurely, particularly when used in conjunction with certain pigments. If the surfaces of the noise barrier panels are being manufactured to be sound absorptive, the coatings may clog the surface openings thereby reducing the Noise Reduction Coefficient (NRC). The recyclability of the final product may have been reduced drastically by the type of additives needed to alter the physical properties of the panel so that it can meet the various fundamental requirements for an effective, safe, and durable noise barrier product.

4. TREATMENT OF NOISE BARRIER

Barriers constructed of recycled rubber materials are limited to shapes obtainable through molding of their components. The surface texture of such panels is also influenced to some degree by the density and porosity of the rubber (Fig. 3). (Zhu & Carlson, 1999, b)

Fig. 3. Rubber noise barrier surface texture

5. CONCLUSION

The rubber tire has great potential to be further developed into a practical absorption material for reducing transportation noise. Most importantly, it shows the possibility to transforming the problem of waste pollution from traffic vehicles into the advantages of solving the related noise pollution from the traffic vehicles themselves. Traditional sound barriers technology is often expensive to install, and made from concrete or wood materials, limiting traditional noise barriers as a solution to noise pollution. The above mentioned technologies use a recycled rubber compound to form a panel mixed with additives or as the center in a composite channel design. Installing such noise barriers is inexpensive and straightforward. The technologies are durable and outperform traditional noise barrier materials. Using scrap tires in the sound barrier is more energy efficient than using the tires for combustion fuel or placing them in landfills. This sound barrier technology effectively reduces both noise and waste pollution in an innovative way – it removes scrap tires from the waste stream entirely.

6. REFERENCES

Albany Airport to Test Noise Abatement Product. (1997). *The Wall Journal*, Vol. 31, September-October/1997, pp. 5-6.

Bulletin of the Acoustical and Insulation Materials Association, Park Ridge, Illinois, 1974.

Lapcik, L. (1998). *Recycled Polymer Based Noise Absorbing Composite Materials*, Proceedings of 5th International Conference on Composites Engineering, July 1998, pp.515, Las Vegas, USA, ISBN 0-7333-0504-0

Noise Barrier Materials and Surface Treatments, Federal Highway Administration, U.S. Department of Transportation, *Available from*: http://www.fhwa.dot.gov/environment/noise/5.htm *Accessed:* 2007-04-12

Zhu, H. & Carlson, D.D. (1999). A Spray Based Crumb Rubber Technology in Highway Noise Reduction Application, Rubber Pavements Association, 1999, *Available from*: http://www.rubberpavements.org/library/spray_based/images/fig1.jpg *Accessed:* 2007-01-11

Zhu, H., & Carlson, D.D. (1999). A Spray Based Crumb Rubber Technology in Highway Noise Reduction Application, Rubber Pavements Association, 1999, *Available from*: http://www.rubberpavements.org/library/spray_based/noisebar.html *Accessed*: 2007-02-28

ACKNOWLEDGEMENTS

The author would like to acknowledge the support of Scientific Grant Agency of the Ministry of Education of Slovak Republic, Commission of mechanical engineering, metallurgy and material engineering, for their contribution to project 1/3174/06.

Annals of DAAAM for 2008 & Proceedings of the 19th International DAAAM Symposium, ISSN 1726-9679
Editor B. Katalinic, Published by DAAAM International, Vienna, Austria 2008
Make Harmony Between Technology and Nature, and Your Mind will Fly Free as a Bird

PROPERTIES AND APPLICATION OF SCRAP TIRE PYROLYSIS PRODUCTS

MISIK, L[adislav] & RADVANSKA, A[gata]

Abstract: *The paper deals with the waste tire pyrolysis processes and the conversion of waste tires into value-added products such as olefins, chemicals and surface-activated carbon. There are discussed primary barriers for its successful application, both economic and technical. The capital cost is high, and the products from pyrolysis do not have sufficient value and must compete with commodity materials. However, certain technological innovations should be made to this barrier in the near future. Developments of less costly techniques or processes for higher value added products would enable pyrolysis to become a profitable alternative for waste tire recycling. Pyrolysis is known for low emissions to the environment.*

Key words: *pyrolysis, activation, charcoal, activated carbon, oil*

1. INTRODUCTION

Waste tire pyrolysis involves the thermal degradation in the absence of oxygen. The benefit of this application is the conversion of waste tires into value-added products such as olefins, chemicals and surface-activated carbon.

The basic process of tire pyrolysis consists of following steps:

- Chipped tires are heated to 600 – 800°C in the absence of oxygen; Primary products are pyrolytic gas, oils and charcoal;
- The oils and charcoal go through additional processes to manufacture secondary, value-added products (Tab. 1).
  - Charcoal upgrading is implemented in a closed-loop activation step that yields an activated carbon and eliminates undesirable by-products and emissions.
  - Upgrading the charcoal produces high-surface-area activated carbon in several grades.
  - Ash-free oil is turned into high-quality carbon black by using the furnace process.
  - As an alternative, oils can be separated into valuable chemical feedstock by distillation.

The amount and composition of those three fractions depends on process parameters (temperature, heating velocity, pressure, time, material size, etc.), in volatile fraction, the controlling parameter is condensation temperature (Wojtowicz, 1996).

2. PRODUCTS OF TIRE PYROLYSIS CONVERSION

The primary products (Tab. 1) are essentially low molecular weight olefins and char. The problem of their application is their low price in the market. On contrary, other chemicals are valuable, but the yield is low. High quality carbon black and surface activated carbon are also valuable but there is no particular price advantage for the same quality products from traditional processes. The valuable chemicals from pyro-gas or oil are generally high molecular weight substances. The purification of high molecular weight substances is expensive, so the greatest drawback is the price of the final product (Tukachinsky et al., 1996).

Primary Products	Weight %	Content	Secondary Products
Pyro-gas	10 - 30	Hydrogen, CO_2, CO, Methane, Ethane, Propane, Propene, Butane, other hydrocarbons, app. 1% of Sulfur	-
Oil	38 - 55	High aromatic (Mw 300 – 400, low sulfur content app. 0.3 - 1.0%) Aromatics, Alkanes, Alkenes, Ketones, Aldehydes	Carbon Black
Charcoal	33 - 38	>15 % of Ash (ZnO) 3 -5 % of Sulfur	Activated carbon

Tab. 1. Pyrolysis products

3. TECHNOLOGIES

The approach to obtain higher value products of pyrolysis can be made through the development of new technologies. Current economic barriers of tire pyrolysis are low quantities of pyro-products and the low molecular weight of obtained olefins, to be curable and moldable for their further application (Mw has to be greater then 15,000). Olefins are typically produced in small quantities because the process temperature is high. At high temperature, vulcanized rubbers are quickly decomposed to low molecular weight olefins (Mw 300 - 400). High molecular weight compounds can be generated by low temperature pyrolysis. However, lower temperature will require longer process times. There are new technologies being developed, that can help the commercialization of the pyrolysis.

Microwaves can heat objects more uniformly than conventional heating methods. Microwave heating requires shorter heating times. Microwave pyrolysis will result in relatively high molecular weight olefins and a high proportion of valuable products such as ethylene, propylene, butene, aromatics, etc. The short process time also contributes to a reduction in the process cost. Moreover, for microwave heating, the shape of the tire chip is less important compared to the requirements of conventional heating. Whole tires or larger chips can be processed using microwave pyrolysis, which greatly reduces pre-processing cost. A. I. Isayev in USA has patented a method which minimizes heating and uses sonic energy to break down sulfur-carbon chemical bonds in tires. Chipped tires are heated to about 200°C, and then subjected to 20,000 Hz of ultrasonic energy (just above the highest frequency the human ear can discern). The rubber is transformed from a solid to a highly viscous fluid within milliseconds. With additional curative agents the viscous material can be moulded into new products (Sangdo, Earnest, 1997) Supercritical water can be used to controllably depolymerize the rubber compounds. This approach requires lower temperatures (approx. 400°C) and shorter processing times. Tire compounds are decomposed to high molecular weight olefins (Mw 1,000 - 10,000), or oils (max. 90%) (Nobeyuki, 2007). Because of the expensive supercritical water equipment, this application would require a relatively large initial cost (Wojtowicz, 1996). Use of catalysts can reduce processing temperature or time. As shown in the above applications, reduced temperature and time can result in either higher molecular weight olefins or an increasing proportion of valuable substances. The advantage of catalysts is that no new equipment or knowledge is required.

One approach to reduce processing cost is to operate at a high process temperature with the use of a special catalyst. Approximately 3.2% of zinc-oxide is added to tire compounds, and the zinc-oxide remains in the char. To produce surface active carbon, the remaining zinc must be removed from the surface, and high temperature processing is able to facilitate this (Chen et al., 1995). The long polymer chains of the rubber decompose at high temperatures to smaller hydrocarbon molecules. When the pyrolysis is performed under vacuum, the spectrum and quality of products obtained is distinct from the other (usually atmospheric pressure) pyrolysis process (Roy et al., 1995). The advantage of a reduced pressure is that secondary decomposition reactions of the gaseous hydrocarbons are limited. Preliminary studies of the tire vacuum pyrolysis process were performed with a bench scale reactor and with cross-ply tires as feedstock. The decomposition of the elastomer in the tire is complete at a pyrolysis temperature of 420°C. A further increase of the pyrolysis temperature does not change the yields of oil, pyrolytic carbon black and gas. Tire particles are fed semi-continuously into the reactor. The pyrolytic carbon black produced is removed from the reactor by an Archimede screw which simultaneously acts as a vacuum seal. The heavy and light oils are condensed in two successive scrubbers. Typical yields are as follows: 55% oil, 35% carbon black and inorganics and 10% gas. The right function of flash pyrolysis reactor is controlled by extremely fast heat supply (approx. 2 sec.) by plasma burner into the pyrolysed material, maintenance of required temperature by electrical heating elements, short residence time of vapor in the reaction zone and fast cooling of the product. The advantage of flash pyrolysis is, that the waste, or waste tires can be transformed into the products of higher energetic value – oil, respectively pyro-gas, depending on the process temperature. The main drawback is that waste material has to be shredded to required size to obtain fast reaction process and simple separation of solid fraction. The gas samples are led via heated pipes to the analysis unit consisting of a FTIR and a gas chromatograph.

4. PROPERTIES OF PYROLYSIS PRODUCTS

Distillation of the pyrolytic oil yields approximately 20% light naphtha, 6.8% heavy naphtha, 30.7% middle distillate, and 42.5% bottom distillation residue. Benzene, toluene, xylene and other benzene-derivatives were identified in the naphtha fraction, as well as a valuable chemical, dl-limonene, which was found to be present with a concentration of 15% by weight. The pyrolytic light naphtha has a relatively high concentration of sulphur, mercaptans and nitrogenous compounds due to the thermal decomposition of the additives originally present in the tires as vulcanization agents. The relatively high levels of sulphur, nitrogenous, olefinic and diolefinic compounds in the pyrolytic light naphtha make it an unsuitable blend for gasoline. Reforming processing is required to convert it to a high value gasoline component. Comparison of the pyrolytic naphtha and commercial petroleum naphtha indicates that the pyrolysis light naphtha is a more complex mixture than the petroleum naphtha. Fossil fuel is basically composed of homologous series of compounds such as n-alkanes, iso-alkanes and anti-iso-alkanes. On the contrary, pyrolysis light naphtha is a heterogenous mixture of various compounds with higher isomerization which were produced during the tire thermal decomposition. Another potential application for the pyrolytic oil is the fabrication of coke. It was confirmed earlier that coal tar recovered by thermal decomposition of coal can easily be used in electrode coke manufacturing. The composition and character of the pyrolytic oil are basic to the quality of the coke and hence its potential usage. Sulphur content and metallic constituents in the feedstock have an important effect on the quality of the coke. The metallic constituents in coke, in particular vanadium, are almost as important as sulphur in determining the coke quality. The presence of nitrogen in the coke is the result of the thermal decomposition of additives originally used in tires, such as organic accelerators, antidegradants and antiozonants, for example sulfenamide and nitrile compounds. The asphaltenes content of the oil is sufficiently high and the viscosity is suitable for the transportation of the oil. The toluene insolubles content is too low to affect the quality of the coke. Pyrolytic oil has almost the same carbon content as the usual petroleum feedstock. However, high carbon content results in a higher yield and a better quality of coke (Roy et al., 1995).

5. CONCLUSION

Pyrolysis can transform scrap tires, usually considered as waste material, into a variety of useful products – oil, pyrolytic carbon black, activated carbon or syn-gas. The oil can be separated into different fractions: naphtha, light and heavy oil and a distillation residue. All these products have a commercial value. It was found that whatever the properties considered, the pyrolytic oil gave similar effects to those of the commercial aromatic oil and a mere substitution of the commercial aromatic oil by the pyrolytic oil could be considered in the compounds studied without significant differences, either in the processing behavior (flow and curing) or in the properties of cured items. A proper choice of the pyrolysis or of post pyrolysis treatment yields a pyrolytic carbon black which is close in its properties to commercial rubber-grade carbon black. An additional potential market for carbon black is filler for road asphalt. The commercial value of the products can make the tire pyrolysis process both ecological friendly and economical attractive.

6. REFERENCES

Chen, D. T.; Perman, C. A.; Riechert, M. E. & Hoven, J. (1995). Depolymerization of tire and natural rubber using supercritical fluids. *Journal of Hazardous Materials.* No. 44, pp. 53-60, ISSN 0304-3894

Isayev, A. I.; Chen, J. & Tukachinsky, A. (1995). Novel Ultrasound Technology for Devulcanization of Waste Rubbers. *Rubber Chem Tech.* No. 68, pp. 267-280 May-Jun 1995, ISSN 0035-9475

Nobeyuki Itoh. Waste Tire Recycling Plant Producing High-Performance Activated Carbon. *Available from*: http://www.p2pays.org/ref/11/10504/html/biblio/htmls/pyh1.htm Accessed: 2007-04-12

Roy, C.; Rastegar, A.; Kaliaguine, S.; Darmstadt, H. & Tochev, V. (1995). Physicochemical properties of carbon blacks from vacuum pyrolysis of used tires, *Plastics, Rubber and Composites Processing and Applications*, No. 23/1995, pp. 21-30, ISSN 09598111

Sangdo, P. & Earnest, F. G. (1997). Statistical study of the liquefaction of used rubber tyre in supercritical water. *Fuel.* Vol.76, No. 11/1997, pp. 999-1003, ISSN 0016-2361

Tukachinsky, A.; Schworm, D. & Isayev, A. I. (1996). Devulcanization of waste tire rubber by powerful ultrasound. *Rubber Chem Tech.* Vol. 1. No. 69/1996 pp. 104-114, ISSN 0035-9475

Wojtowicz, M. (1996) *The Manufacture of Carbon Black from Oils Derived from Scrap Tires.* EPA 68D98117

ACKNOWLEDGEMENT

The author would like to acknowledge the support of Scientific Grant Agency of the Ministry of Education of Slovak Republic, Commission of mechanical engineering, metallurgy and material engineering, for their contribution to project 1/3174/06.

Annals of DAAAM for 2008 & Proceedings of the 19th International DAAAM Symposium, ISSN 1726-9679
Editor B. Katalinic, Published by DAAAM International, Vienna, Austria 2008
Make Harmony Between Technology and Nature, and Your Mind will Fly Free as a Bird

THE HEAT TRANSFER DURING THE COMPRESSION PROCESS IN AN INTERNAL COMBUSTION ENGINE

MITRAN, T[udor]; PATER, S[orin]; FODOR, D[inu] & POLOJINTEF CORBU, N[icolae]

Abstract: *At the beginning of the compression process the fresh charge is at a lower temperature than the one of the inner walls of the combustion chamber. The heat transfer takes place from the inner walls to the working fluid. The temperature of the fresh charge rises during the compression process and at one moment equals the one of the walls. Further gas temperature continues to rise and heat transfer takes place from the working fluid to the walls. So compression is a complex process with variable polytropic exponent. This paper proposes a model to calculate heat transfer during the compression process.*
Key words*: heat transfer, compression, combustion engine*

1. INTRODUCTION

The scheme of the heat transfer process is presented in figure 1.

where: - t_f [^{0}C] – the temperature of the working gas
- t_{iw} [^{0}C] – the temperature of the cylinder inner walls
- t_{iw} [^{0}C] – the temperature of the cylinder inner walls
- t_{mw1} [^{0}C] – the temperature of the cylinder exterior walls
- t_{mw2} [^{0}C] – the temperature of the cylinder block inner walls
- t_c [^{0}C] – the temperature of the cooling agent

The heat transfer between the working fluid and the inner walls of the combustion chamber is a convective one, than through the walls of the cylinder and the cylinder block heat transfer is conductive with an inner thermal resistance (Mollenhauer, 1997). From the walls of the cooling system to the cooling agent heat transfer is also convective (Pishinger et al., 1989).

The general equation of heat transfer is:

$$\dot{Q} = \frac{\Delta t}{R} \cdot S \tag{1}$$

where: - $\dot{Q}$ [W] – the thermal flux
- Δt [^{0}C] – the thermal gradient
- R [$m^{2 \cdot 0}C/W$] – the thermal resistance
- S [m^2] – the heat exchange surface

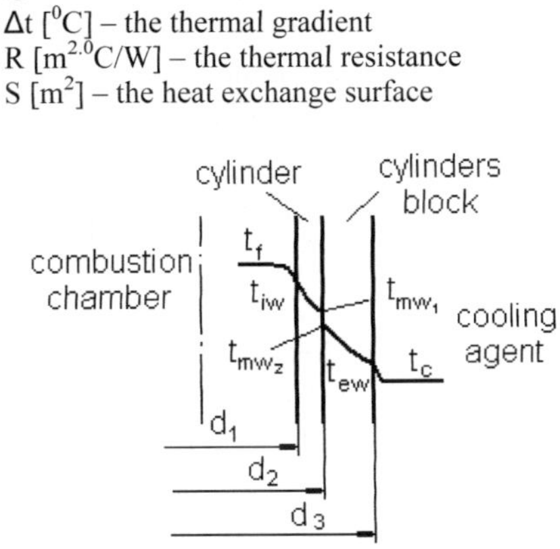

Fig. 1. The scheme of the heat exchange process

The problem is to determine thermal resistances in the convective heat exchange processes between the fluids (working gas and cooling agent) and the engine walls (van Basshuysen & Schafer, 2004).

The empirical formulae used to calculate the convective coefficient (Cirkov, Sitkei, Woschni) (Grunwald, 1980) are determined in the conditions of a flow inside a manifold and can not be applied with sufficient precision for the combustion chamber which is closed during the compression process.

A more precise method to determine heat exchange in an i.c.e. is useful in order to predict fluid maximum temperature inside the combustion chamber. This is important because NO_x emissions depend on this maximum temperature inside the combustion chamber. Formation of NO_x increases at temperatures over 1700K.

2. THE CALCULUS OF THE THERMAL RESISTANCES

The thermal flux in heat exchange between the working gas and the cooling agent is (see fig.1) (Stefanescu et al., 1983):

$$\dot{Q} = \frac{(t_f - t_c) \cdot S_i}{R_1 + R_2 + R_3 + R_4 + R_5} \tag{2}$$

where: - S_i [m^2] – instantaneous heat transfer surface

$$R_1 = \frac{1}{\pi \cdot d_1 \cdot \alpha_1} \tag{3}$$

d_1=D – the inner diameter of the cylinder
α_1 – the convective coefficient in the heat transfer between working gas and cylinder walls

$$\alpha_1 = \frac{Nu_f \cdot \lambda_f}{l_f} \tag{4}$$

λ_f – the thermal conductivity of the working gas
l_f – the characteristic length

$$Nu_f = 0.023 \cdot Re^{0,8} \cdot Pr^n \tag{5}$$

Nu, Re, Pr – Nusselt, Reynolds, Prandtl non-dimensional criteria applied to the working gas

$$R_2 = \frac{1}{2 \cdot \pi \cdot \lambda_{Ft}} \cdot ln \frac{d_2}{d_1} \tag{6}$$

λ_{Ft} – the thermal conductivity of cast iron

$$R_3 = \frac{\delta}{\left(0{,}08 \cdot \frac{\lambda_{Ft} \cdot \lambda_{Al}}{\lambda_{Ft} + \lambda_{Al}} + 0{,}96 \cdot \lambda_a\right)} \qquad (7)$$

$\delta=\delta_1+\delta_2$
δ_1, δ_2 – the medium roughness of the two surfaces
λ_{Al} – the thermal conductivity of aluminium alloys
λ_a – the thermal conductivity of air

$$R_4 = \frac{1}{2 \cdot \pi \cdot \lambda_{Al}} \cdot \ln \frac{d_3}{d_2} \qquad (8)$$

$$R_5 = \frac{1}{\pi \cdot d_3 \cdot \alpha_2} \qquad (9)$$

$$\alpha_2 = \frac{Nu_c \cdot \lambda_c}{l_c} \qquad (10)$$

λ_f – the thermal conductivity of the cooling agent
l_f – the characteristic length

$$Nu_c = \left(0{,}35 + 0{,}56 \cdot Re^{0{,}52}\right) \cdot Pr^{0{,}3} \qquad (11)$$

Nu, Re, Pr – Nusselt, Reynolds, Prandtl non-dimensional criteria applied to the cooling agent

It is possible, by applying rel. (1), to determine all the temperatures presented in fig. 1.

3. THE COMPARISON BETWEEN CALCULATED AND EXPERIMENTAL DATA

To compare the calculated with experimental data it is necessary to calculate the instantaneous pressure inside the combustion chamber:

$$p_i = \frac{m \cdot R \cdot T_f}{V_i} \qquad (12)$$

m [kg] – the mass of the working gas (is constant)
R [J/kg·K] – the ideal gas constant
T_f [K] – the working gas temperature
V_i [m^3] – the instantaneous volume of the combustion chamber

The measurements were taken on one cylinder water cooled diesel engine.

The comparison between measured and calculated pressure inside the combustion chamber during the compression process is presented in figure 2.

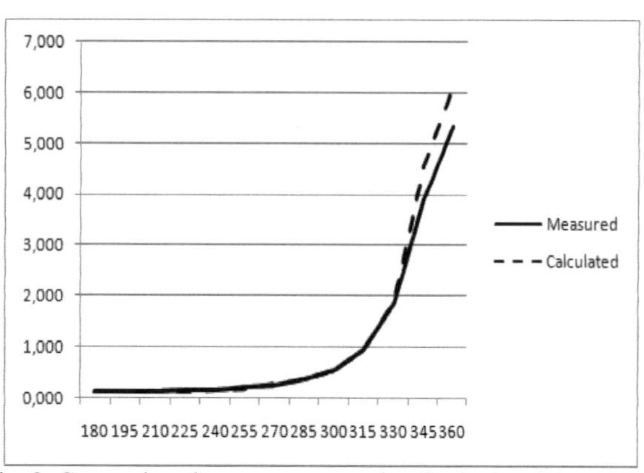

Fig. 2. Comparison between measured and calculated data

The differences between the two sets of data are less than 20%, that's a sufficient precision in the terms of heat exchange processes.

The general problem of heat exchange in i.c.e. is difficult to solve also because of the variable volume (surface) of the combustion chamber. One of the assumptions made in calculus was that the surface of heat exchange don't include piston head surface.

In order to calculate thermal flux, temperature of cooling agent and the rate of flow were measured at the entrance and the exit from the engine. This means that the heat exchanged with the piston is not taken in consideration.

The instantaneously heat exchange surface is:

$$S_i = \pi \cdot D \cdot \{l_c + r \cdot [1 - \cos\alpha + (\Lambda/4)(1 - \cos 2\alpha)]\} \qquad (13)$$

where: - D [m] – the cylinder bore
- l_c [m] – the dead volume length
- r [m] – the crankshaft radius
- Λ=r/b – the conrod ratio
- b [m] – conrod length (the distance between the centres of the small-end and the big end eyes)
- α – the rotational angle of the crankshaft

In the proximity of TDC this assumption leads to relative great differences between measured and calculated data.

4. CONCLUSION

The model presented in this paper is good enough to describe heat transfer in the compression process for this particular engine.

It would be interesting to determine a general relation for the convective coefficient of the heat transfer between the working gas and the inner walls of the combustion chamber. This general relation should be applied for a particular engine by using a correction coefficient.

The correction coefficient can be determined for each specific engine only on test benches.

For the heat transfer during burning, detention, exhaust and intake processes other general relations must be taken in consideration.

5. REFERENCES

van Basshuysen, R. & Schafer, F. (2004). *Internal combustion engine,* SAE International, ISBN 0-7680-1139-6, Warrendale PA

Grünwald, B. (1980). *Teoria, calculul si constructia motoarelor pentru autovehicule rutiere,The Theory, the calculus and the construction of the engines for automotives,* Ed. Didactica si pedagogica, Bucuresti

Mollenhauer, K. (1997). *Handbuch Dieselmotoren, Diesel Engines Handbook,* Springer, Berlin

Pishinger, R.; Krassnig, G.; Taucar, G. & Sams, Th. (1989). *Thermodynamic der Verbrennungskraftmaschine. Die Verbrennungskraftmaschine.The Thermodynamic of the Internal Combustion Engine. The Internal Combustion Engine, Reeissue Volume5,* Springer, Vienna

Stefanescu, D.; Leca, A,; Luca, L.; Badea, A. & Marinescu, M. (1983). *Transfer de caldura si masa, Heat and Mass Transfer,* Ed. Didactica si pedagogica, Bucuresti

Annals of DAAAM for 2008 & Proceedings of the 19th International DAAAM Symposium, ISSN 1726-9679
Editor B. Katalinic, Published by DAAAM International, Vienna, Austria 2008
Make Harmony Between Technology and Nature, and Your Mind will Fly Free as a Bird

MATHEMATICAL MODEL AND STATISTICAL ANALYSIS OF THE ELONGATION OF THE STEEL J55 API 5CT BEFORE AND AFTER THE DEVELOPMENT OF THE PIPES

MJAKU, M[alush]; KRASNIQI, F[ehmi]; MAKSUTI, R[rahim] & ELEZI, D[ervish]

Abstract: *Object of this study is the tin of quality from the steel quality J55 API 5CT and the process of pipe forming ∅139.7x7.72 mm, and ∅219.1x7.72 mm with rectilinear seam. Aim of this paper is to study the impact of deformation level in the cold and mechanical properties of the steel coils before and after the formation of the pipes. For the realization of the project we have used the planning method of the experiment. We have built the mathematical model for the experiment with one index (elongation A_2) and with one factor (level of deformation in the cold) and with few levels and two blocks (before and after the formation of the pipes). Applying this work, the results obtained in an experimental method are shown in the table and are processed in an analytical way, implementing the one factored experiments.*
Key words: *One-factor, steel coils, pipe, elongation (A_2)*

1. INTRODUCTION

During technological process of pipe production with rectilinear seam entrance, a factor with significant impact is plastic deformation in the cold which is realized based on the deformation forces in inflexion throughout formation process of pipe calibration. It is more likely that the impact will be bigger as long as diameter of the pipe is smaller. To invent and assess this impact in mechanical attributes, extension in pulling, we have planned the experiment in three conditions of the material: preliminary steel coil, pipe Ø139.7x7.72 *mm* and pipe Ø219.1x7.72 *mm* [1]. These three conditions, express three levels (1, 2 and 3) of quality factor"deformation level". For each level there have been conducted 5 experiments in inflexion [3]. Specimens have been taken in direction of pipe's axis and experiments have been conducted based on application of fortuity's criteria. Calculating indicator is percentage of elongation (A_2), marked with *y*.

Table 1. Results

Reiterations / Levels	1	2	3
1.	29	28	27
2.	30	27	27
3.	35	27	28
4.	31	27	28
5.	36	27	30
Sum y_{i+}	161	136	140 $y_{++} = 437$
Average values $\overline{y}_{i+}$	32.20 $\overline{y}_{1+}$	27.20 $\overline{y}_{2+}$	28 $\overline{y}_{3+}$

2. MATHEMATICAL MODEL AND STATISTICAL ANALYSIS

2.1. Mathematical Model

Mathematical model which is predicted to reflect such a study is composed from a system by *n* equations forms (Pantelic, 1976) :

$$y_{ij} = \overline{m} + a_i + \varepsilon_{ij} \tag{1}$$

The formulas for calculation of round constant in which are based all observing results of index/indicator y ($\overline{m}$) and effects ($\overline{a}_i$) are:

$$\overline{m} = \frac{1}{n} \cdot y_{++} \qquad \overline{a}_i = \frac{1}{p} y_{i+} - \overline{m} \tag{2}$$

Based on values from table 1 and formulas (2) we will have:

$$\overline{m} = \frac{1}{15} 437 = 29.13$$

$$\overline{a}_1 = 32.20 - 29.13 = 3.07$$
$$\overline{a}_2 = 27.20 + (-29.13) = -1.93$$
$$\overline{a}_3 = 28 + (-29.13) = -1.13$$

With replacement of effects values in equations (1) mathematical model will have this form:

$$\begin{aligned} y_{1j} &= 29.13 + 3.07 + \varepsilon_{1j} \\ y_{2j} &= 29.13 - 1.93 + \varepsilon_{2j} \\ y_{3j} &= 29.13 - 1.13 + \varepsilon_{3j} \end{aligned} \tag{3}$$

2.2. Statistical Analysis

2.2.1. Variance Analysis

Total sum of the squares of differences (deviations) of the measured values from the average is composed by two components (Kedhi, 1984):

$$S = S_g + S_p \tag{4}$$

Value of summary of error squares S_g is:

$$S_g = \sum_{i=1}^{\mu} \sum_{j=1}^{p} y_{ij}^2 - \frac{1}{p} \sum_{i=1}^{p} y_{i+}^2$$
$$= \sum_{i=1}^{3} \sum_{j=1}^{5} y_{3,5}^2 - \frac{1}{5} \sum_{i=1}^{3} y_3^2 = 12849 - \frac{1}{5} 64017 = 45.60$$

In similar method we will have also the value of deviation of experimental mistake.

$$S_p = \frac{1}{p} \sum_{i=1}^{\mu} y_{i+}^2 - \frac{1}{\mu \cdot p} y_{++}^2 = \frac{1}{5} \sum_{i=1}^{3} y_{i+}^2 - \frac{1}{3 \cdot 5} y_{++}^2 = \frac{1}{5} 64017 - \frac{1}{15} 190969 = 72.14$$

2.3. Control of Hypothesis, upon equality of the effects

For this is required control of hypothesis based the equality of the effects a_i. According to the equation (2), Hypothesis of equation of the effects H_o, will take the form [Douglas & Mongomery, 2000):

$$\sum_{i=1}^{\mu}\overline{a}_i = 0 \qquad H_0 : a_1 = a_2 = ... = a_\mu = 0 \qquad (5)$$

Alternative hypothesis is:

$$H_1:\ a_i \neq 0 \qquad (6)$$

Reason of change	Sum of squares	No. of DOF	Average square of deviations
Processing	$S_p = 72.14$	$\mu - 1 = 2$	$s_p^2 = 36.07$
Reasons of the case	$S_g = 45.60$	$n - \mu = 12$	$s_g^2 = 3.80$
Sum of deviations	$S = 117.74$	$n - 1 = 14$	

Table 2. Summary table of variance analysis

Value of calculated Fisher's criteria is :

$$F_c = \frac{s_p^2}{s_g^2} \qquad (7)$$

$$F_c = \frac{36.07}{3.80} = 9.49$$

For level of importance $\alpha = 0.05$ limit value of Fisher's criteria:

$$F_{t(\alpha);2;12} = F_{t(0.05);2;12} = 3.89\ ;\ F_c = 9.49 > F_t = 3.89$$

Then, with level of importance $\alpha = 0.05$ hypothesis H_0 is rejected and effects $a_i (i = 1,2,3)$ are accepted.

2.4. Comparison of the effects

2.4.1. Comparison of the effects according to minimal valid difference

To emphasize which levels are with important changes, first is required to calculate minimal valid difference $\Delta_{ik}(\alpha)$ for level of importance α = 0.05.

$$\Delta_{ik}(\alpha) = \sqrt{s_g^2\left(\frac{1}{p_i} + \frac{1}{p_k}\right)(\mu - 1)F_{(\alpha;\mu-1;n-\mu)}} = \sqrt{3.80\left(\frac{1}{5} + \frac{1}{3}\right)\cdot 2\cdot 3.89}$$
$$= 4.95$$

Based on the criteria (8) levels of effects "*i*" and *"k"* factor, so it compares a_i and a_k:

$$|\overline{a}_i - \overline{a}_k| > \Delta_{ik}(\alpha) \qquad |3.07 - (-1.93)| = 5 > 4.95$$
$$|\overline{y}_{i+} - \overline{y}_{k+}| > \Delta_{ik}(\alpha) \qquad |32.20 - 27.20| = 5 > 4.95 \qquad (8)$$

from application of this criteria result that:

$|\overline{y}_{1+} - \overline{y}_{2+}| = |32.20 - 27.20| = 5 > 4.95$, between levels 1 and 2 it has important impact

$|\overline{y}_{1+} - \overline{y}_{3+}| = |32.20 - 28| = 4.20 < 4.95$, between levels 1 and 3 it has not important impact $|\overline{y}_{3+} - \overline{y}_{2+}| = |28 - 27.20| = 0.80$ <4.95, between levels 3 and 2 it has not important impact

2.4.2. Comparison of the effects according to collective criteria of deviations

In this way "first type of mistake" to revoke a true hypothesis would be: 1- 0.857=0.142 (and no more 0.05). To avoid this increment of mistake we should use other criteria, Duncan's collective criteria of deviations, which will be described bellow. For case when number of proves/experiments *p* in every level is same, standard mistake is calculated (Kedhi, 1984):

$$s_{\overline{y}_{i+}} = \sqrt{\frac{1}{p}s_g^2} = \sqrt{\frac{1}{5}\cdot 3.80} = 0.87 \qquad (9)$$

By statistical tables, for $\alpha = 0.05$ and number of degrees of freedom f = n-μ=15-3=12, are with row for q=2, 3 valid deviation: $r_{0.05(2;12)} = 3.08$ and $r_{0.05(3;12)} = 3.23$

With valid deviations r_α and standard mistakes of levels, calculation of minimal valid deviations according to the formula:

$$R_q = r_\alpha(q, f)\cdot S_{\overline{y}_{i+},q} = 2,3,...,\mu \qquad (10)$$

$$R_2 = 3.08 \cdot 0.87 = 2.67 \quad \text{and} \quad R_3 = 3.23 \cdot 0.87 = 2.81$$

Minimal valid deviation will be:

$$\overline{y}_i - \overline{y}_k \geq R_q \qquad (11)$$

Now the comparison between levels of averages which are systematized in groups can be done:

$\overline{y}_{1+} - \overline{y}_{2+} = 32.20 - 27.20 = 5 > 2.81 = R_3$, q=3-1+1=3

$\overline{y}_{1+} - \overline{y}_3 = 32.20 - 28 = 4.20 > 2.67 = R_2$, q=3-2+1=2

$\overline{y}_{3+} - \overline{y}_{2+} = 28 - 27.20 = 0.80 < 2.67 = R_2$, q=2-1+1=2

3. DISCUTION/ CONCLUSIONS

Due to the plastic deformation, in cold, which is exercised upon the laminated tin, in warm, during the pipe formation and calibration it came to the strain hardening of steel's quality J55 API 5CT as a consequence of dislocations' formation and blockage.

Hypothesis H_0 of effects equation: $a_1 = a_2 = a_3 = ... = a_1 \cdot \mu = 0$ doesn't exist, while alternative hypothesis H_1 exist at least for one effect $a_i \neq 0$.

As the experimental calculated values of elongation in percentage *(A_2)* of $F_c > F_t$, with importance level α = 0.05 is accepted, effects ($a_i = 1, 2, 3$) are not zero.

Since the effects' difference for elongation in percentage *(A_2)* of levels "*i*" and "*k*" of factor's level $\overline{a}_i$ and $\overline{a}_k$ is more larger than minimal valid difference $\Delta_{ik}(\alpha)$ for importance level α = 0.05, we have: $|\overline{a}_i - \overline{a}_k| \geq \Delta_{ik}(\alpha)$, therefore it is accepted that levels *"i"* and *"k"* have important differences based on their impact in the experimental results.

While effects' difference of two pairs (1, 2) and (1, 3), with exception of pair (3, 2), of averages of arithmetical values watched in *p* levels probations *"i" and "k"* are larger than minimal valid deviations R_q, so: $\overline{y}_i - \overline{y}_k > R_q$. Therefore, from this analysis we can conclude how important are the differences of level's of the two pairs during the research of elongation in percentage *(A_2)* Results are done in "Laboratori mekaniko-metalografik IMK", Ferizaj- Kosovo.

4. REFERENCES

Standard, API Specification 5CT, Washington 2000.

V. Kedhi, *Metoda të planifikimit dhe të analizës së eksperimenteve, (Methods of planning and analysis of experiments)* Politeknik Faculty, Tiranë 1984.

Standard, ASTM-A370, Washington 2000.

Douglas C. Mongomery, *Controllo statistico di qualità, Parte III:(Statistical controll of quality, Part III),* McGraw-Hill, 2000.

I. Pantelic, Uvod u teoriju inzinjerskog eksperimenta (Basic theory in engineetring experiments), Radnicki Universitet , Novi Sad 1976.

Annals of DAAAM for 2008 & Proceedings of the 19th International DAAAM Symposium, ISSN 1726-9679
Editor B. Katalinic, Published by DAAAM International, Vienna, Austria 2008
Make Harmony Between Technology and Nature, and Your Mind will Fly Free as a Bird

ASPECTS REGARDING THE DEVELOPMENT OF THE ENTREPRENEURSHIP IN ROMANIA

MOCAN, M[arian]; POPA, H[oria]; IZVERCIANU, M[onica]; PUGNA, A[drian] & HAMAT, C[odruta]

Abstract: *This paper will present an evaluation of the Romanian economy and the Romanian companies development, of the problems with which entrepreneurs and employees fight. A review will be made regarding the evolution of the work force market, of the companies management and their adapting to the present realities. Also, there possible solutions for solving problems and for implementing the change process starting from the economical environment realities and from the general world economy evolution will be identified.*

Key words: *Economy, evolution, entrepreneurship, change*

1. INTRODUCTION

The economic development of Romania after 1990 and especially from the last decade has been directly connected to the perspective of joining the European Union . From the moment Romania effectively joined the EU on the 1st of January 2007, the development has been more rapid not only because of the existent opportunities from the Romanian market but also because of the development gap between this one and the Western countries. Although Romania is part of the world economic circuit, some evolutions have been in contradiction with the global evolution.

Economic evolution between 2000-2007 was positive (see table 1). According to the data provided by the National Bank of Romania and by the National Statistics Institute (NSI 2007), a continuous improvement of the analyzed base indicators can be observed (economic growth, inflation rate and unemployment rate).

This evolution has lead to a substantial economic revitalization and to creating some better living conditions for the country's population. It is estimated that, from 2010, Romania will enter the Schengen area and from 2014 it will adopt the Euro as its national currency. As regards the level of the Gross Domestic Product (GDP), in 2007 it has reached the sum of 121 billions € (from 97 billion in 2006), meaning 5.720 €/person. It is a very dynamic growth if we take into account the fact that, in the year 2004 the GDP was of 2.800 € /person and in the year 2000 it was only of 1.800 € /person.

The sectors that have had a substantial growth in the last period, were the ones from the fields of constructions, of Communication and Information Technology, of automotive and of the real estates.

Year	Economic growth [%]	Inflation rate [%]	Unemployment rate [%]
2000	2.1	40.4	10.5
2001	5.7	30.3	8.8
2002	5.1	17.8	8.4
2003	5.2	14.1	7.4
2004	8.4	9.3	6.3
2005	4.1	8.6	5.9
2006	7.9	4.67	5.2
2007	6.0	6.57	3.9

Tab. 1. Evolution of the macro-economical indicators

The arrival of 2 important companies (Nokia in Cluj Napoca and of Ford in Craiova) has lead to an unprecedented activation of the Romanian economy.

2. THE ENTREPRENEURSHIP IN ROMANIA

The strong economic development from the last period has brought to the entrepreneurs new problems with which they haven't been used until now. The most important problems that they must face regarding the market are the following:

a – a work force crisis in almost all of the sectors of the national economy and in all the professions and jobs. The most affected sectors are constructions, commerce, hotel industry and processing industry. The work force crisis is more accentuated in big cities such as Bucharest, Timis, Sibiu, Cluj, and Constanta. According to some studies realized by some recruiting companies, approximately half of the Romanian companies have aim at increasing their number of employees in the following 3 months; consequently, this pressure regarding the lack of work force will grow even further. The reasons that have led to this crisis are the following:

- Massive migration of the work force towards the developed countries of the European Union even before but also especially after the year 2007. It is estimated that approximately 2 billion persons are hired in the developed countries of the European Union, but especially in Spain and Italy, in the fields of constructions, social services, industry and tourism. The main reason of the massive departures is given by the difference between the salaries from these countries and from Romania. In Table 2, the level of the net salary from Romania in the last 3 years and the predictions for the next 3 years are presented according to theNSI (2007).
- Rapid growth of the Romanian investments, but especially of the foreign investments. In the "Study Regarding the Investment Attractiveness of the South-Eastern Europe 2008" made by Ernst & Young, Romania, is on the first position with an attractiveness level of 52% (Badarau I, 2008). From the 216 interviewed managers, most of them see Romania as the most favorable country for establishing businesses. The comparison has been made between Romania, Turkey, Bulgaria, Greece, Croatia, Serbia, Bosnia Herzegovina and Cyprus.

b – Bureaucracy and the taxation are some other reasons of worry for the entrepreneurs. The high number of forms that must be completed, the number of accounts towards which they have to pay a lot of contributions, the deficit activity of the state institutions that work with entrepreneurs lead to waste of time and to the increase of the entrepreneurs' feeling of uneasiness. It is a real problem already acknowledged by the government that is trying to make a clear evidence of all these taxes and duties and to diminish their number.

Year	2005	2006	2007	2008	2009	2010
Salary [€]	204	243	313	338	378	417

Tab. 2. Level of net salary

Although, globally, from a fiscal point of view, Romania is an attractive country, the high number of taxes and duties, sometimes divided in ridiculous sums lead to an exaggerated waste of time and resources for the companies.

c - The legislation from the work code gives an extremely high power to the employers in comparison with that of th employees. The modifications of the Work Code from 2004 and 2005 have lead to some extremely restrictive stipulations for entrepreneurs regarding sacking, the work period, the vacations and also other rights of the employees. By changing the work code, an attempt was made to implement some European Union specific norms, but they would only be valid for a country with a developed economy, and not for one as Romania has. Given the reduction of the unemployment rate and the increase of the work force demand, the possibilities for negotiation have significantly diminished.

d – Joining the European Union and the necessity of accomplishing the specific norms that are in place, has lead to a lot of challenges and important changes for the entrepreneurs. Eliminating the customs taxes for the Romanian products is of good omen. The European market, a market of 462 millions people, will be directly accessible to the Romanian products with no restrictions. This is also valid for the European products that want to reach the Romanian market.

The European legislation regarding the protection of the environment is extremely restrictive and many of the Romanian companies would not be able to obesrve it. Already in 2007, 14 104 companies went bankrupt in Romania (Barzoi V, 2008). Most of the bankruptcies from 2007 – 5 801 companies, meaning 41% of all – have been registered in the commercial field (retail, lump commerce and distribution), and this is due to the great chains of retail that entered the market. For the year 2008, the international and the native retailers have announced the opening of 56 supermarkets, 29 hypermarkets, 2 stores type cash and carry and 77 discount shops. On the following places for bankruptcy, are the companies from the agricultural field (1093 companies), from the field of constructions (1066 companies) and of the food and drinks industry (1064 companies).

the optimism of the Romanian entrepreneurs, is quite high.. In a research made by Eurochambres Economic Survey (Vaschi M, 2008), a percentage of 53.8% from these think that the business environment will evolve positively in 2008, only 18.9% think that the business environment will become worse, while 27.3% think that there will be no major changes in comparison with 2007. More optimistic than the Romanians, were only the entrepreneurs from the Czech Republic – 54.5% and from Bulgaria 66.4%, while the weight of the skeptics is higher in Hungary (23.8%) and Slovakia (25.5%).

3. SOLUTION PROPOSED

Adapting a firm's activity to the demands of an ever-changing market must be done by using top notch changing management. It is necessary that a firm develops SMART objectives (Specific, Measurable, Adequate, Realistic, and Tangible) together with new ideas, enthusiasm, and a strong leadership (Mocan M. 2007). The stages that a transformation process must have and the conditions it must uphold are:

a – Defining your purpose and objectives. This is done by starting from the idea that every firm is different and every transformation process is unique. The essence of transformation must be based on the enhancement of the firm's performances and on reaching a superior level of development. This must lead to in-depth transformations based on radical changes.

b – Designing a graph of change development. This is achieved by clearly defining the necessary activities for this process, starting from the objectives defined at the previous step and deciding which the priorities are.

c – Defining milestones. We refer to the 'breaking' of the changing process into several pieces so that the changing process will be accepted more easily by the organization members. an evaluation and monitoring of the changes performed so far will be made at these milestones; changes can be made according to the problems that occurred during the process. Having short term successes can boost the employees' moral and make them get more easily involved in the future changing process.

d – Using efficient communication and efficient leadership. The changing process can be easily accepted if its necessity is understood. A direct, efficient communication can lead to a better understanding of the benefits, of the necessity of this process and of the important impact it will bring to the firm and its members. It is best if the emotional side of things is accentuated-like, for instance, the need for adapting to the new conditions.

e – Maintaining the rhythm and stimulating employee enthusiasm. The two aspects (rhythm and enthusiasm) are interdependent. A process that takes too long leads to loss in patience and therefore to loss of enthusiasm on behalf of the employees. They need to be periodically stimulated by encouraging them to participate with ideas to the changing process. It is not necessary for them to give big solutions, just solutions to small scale processes. It can be said that 4 kinds of ideas are important, those who answer the following questions:

- Why is change needed? Ideas that are born from the motivation of change are essential both on a large as on a small scale.
- What needs to be changed? In this case both the major and the minor processes that will be changed must be defined.
- Who needs to be changed? In this case we are talking about the people, their functions and their place in the firm.
- How does it need to be changed? Practical solutions that are utilized during the whole process are the most valuable out of the whole changing process.

d – Personalizing changes and increasing employee learning capabilities. In order to increase conviction and trust in the employees a firm must find and promote in its turn new models that promote the new changes and values towards which the firm is heading. About the learning abilities of adults: they depend on individual personalities as well as on the internal values that the firm promotes in the organization. A very high success rate offers the employees the possibility of learning new things and then attributing specific tasks to this new knowledge.

4. CONCLUSION

Romania is in a very favorable point in its development now. Joining the EU was an opportunity for the economical environment in this county. During this period entrepreneurs get confronted with new situations that about which they have no experience. The way in which they will deal with the problems depends on their personality, on the knowledge they have or on the knowledge they have and their courage to change the things that no longer work.

5. REFERENCES

Badarau I (2008), Romania is nr. 1 target for the foreign investors, *Ziarul Financiar, 18.04.2008, pag. 3.*

Barzoi V (2008*),* Over 14.000 out of the market, *Capital, nr. 5, febuary 2008, pag. 4*

Mocan M et al (2007). *Management*, Eurobit Publisher House, ISBN 978-973-620-307-7, Timisoara Romania

National Statistic Institute (2007) *Statistics Annuary of Romania*, Bucharest

Vaschi M (2008), The young's rise the level of salaries on the market, *Capital, nr. 14, April 2008, pag. 1*

Annals of DAAAM for 2008 & Proceedings of the 19th International DAAAM Symposium, ISSN 1726-9679
Editor B. Katalinic, Published by DAAAM International, Vienna, Austria 2008
Make Harmony Between Technology and Nature, and Your Mind will Fly Free as a Bird

NETWORKING SYSTEM FOR REMANUFACTURING TECHNOLOGY

MOK, H[aksoo]; LEE, K[yungkeun]; RYU, K[wangyeol]; JEON, C[hangsu]; HAN, C[hanghyo] & KWAK, H[yun] S[oo]

Abstract: *To save a manufacturing energy and to reuse some materials of products for our earth we tried to research for remanufacturing processes of some automobile parts. In many countries, specially Germany and USA, are researched remanufacturing technologies for long time. In this paper we suggest the international cooperation methods to improve on this technology and its networking system.*
Key words: *Remanufacturing, Networking, Process, Product Design, DB*

1. INTRODUCTION

In the world there are many remanufacturing companies. To give lots benefits to customers, who want to use some remanufactured products for saving the maintaining cost, the cost of remanufactured products and the information of remanufacturing companies should be supplied to customers directly.

For this supply many remanufacturing companies have to be connected in the networking system. To establishment the networking system successfully we have to find out the remanufactured products, prices of remanufactured products, remanufacturing companies, supply duration of remanufactured products etc.

2. NEED OF GLOBAL NETWORKING FOR REMANUFACTURING

To save the manufacturing cost and energy we have to use the used parts and materials. After a new product is used, either it will be reused some parts or it will be disposed as a waste.

In some countries the used products are reused very well as a remanufactured part, but in many countries they can not yet be reused (Seliger, 2004). To reuse the used products in the world we have to get a good networking system among many countries.

To make a global networking, all information about the remanufacturing companies and remanufactured products have to be grasped (Nasr & Hilton 2004). With this system the customer can access this system to check and buy the remanufactured products (Kaebernick, 2008).

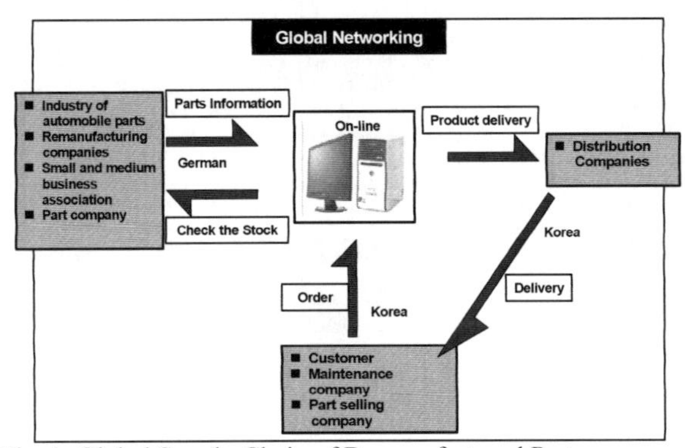

Fig. 1. Global Supply Chain of Remanufactured Parts

3. KEY OF GLOBAL NETWORKING SYSTEM

3.1 Analysis of Remanufacturing System

We have to analyze many networking systems, which were developed in many remanufacturing companies. By the analysis of the flow of the remanufacturing system in a company the problems can be found according remanufacturing process time and cost. And then we can select an optimal process and develop a more advanced remanufacturing system.

Each remanufacturing system can be compared by its characteristics (Steinhilper, 2005). From this comparison we can extract the cooperation between countries.

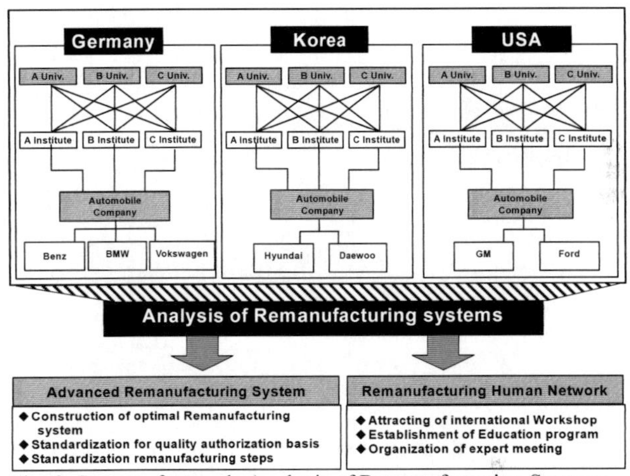

Fig. 2. Remanufactured Analysis of Remanufacturing Systems

3.2 Technology for Enlargement of Remanufacturing

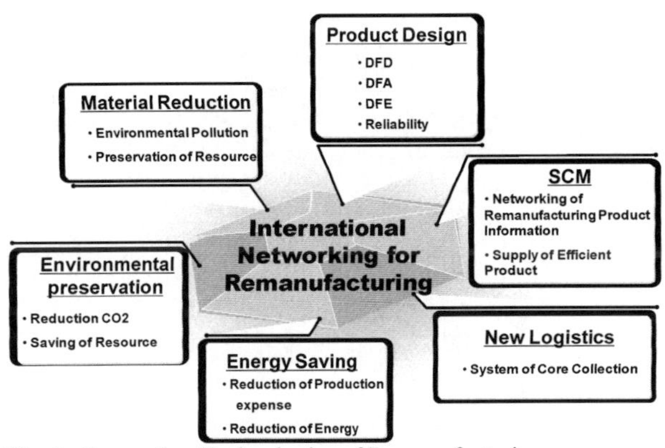

Fig. 3. Connection Technologies of Remanufacturing

To get a good international networking for remanufacturing we need some technologies: product design, material science, supply chain management, environment protection and so on.

In these technologies the product design has the important roll for remanufacturing. During the design of a product we have to consider disassemblability, assemblability, recycling and energy consumption for the reusing of the used products.

Fig. 4. Functions of DfSP

4. CONSTRUCTION OF INFORMATION DATABASE

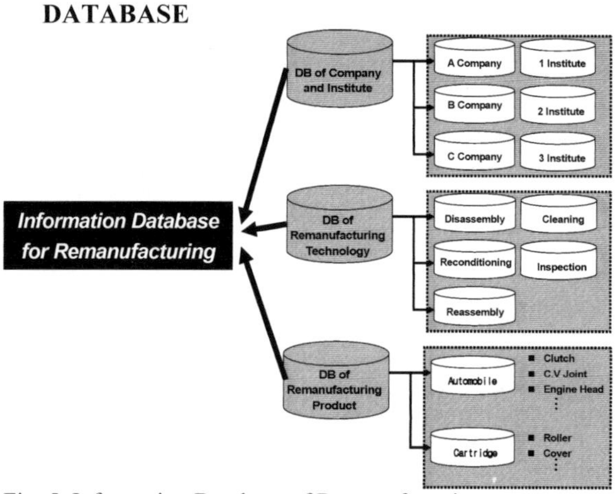

Fig. 5. Information Database of Remanufacturing

To be a global networking system we have to make a database by using of the remanufacturing companies, the current remanufacturing technologies and the kinds of remanufactured products (Fig 5).

By the integration of these database systems the customers, who want to buy remanufactured product, can use this system more usefully.

We planed the co-operation research for remanufacturing between 3 countries: Berlin University(TU) in Germany, University of Michigan in USA and Pusan National University in Korea (Fig 6).

In this co-research program, the main objects are the establishment of design rules of remanufactured products and the grasping of information for remanufacturing companies in three countries (Mok, 2007).

From this research we will get a way to improve the technology of remanufacturing by the exchanging of technology and educational programs.

5. CONCLUSION

To rescue our earth we have to reuse the used products and parts like their original functions. For it the remanufacturing is very important. Each country has different remanufacturing systems.

To save more remanufacturing cost a global networking system has to be constructed. In this networking system remanufacturing technology can be exchanged. The customers can have more merits for repairing used products.

ACKNOWLEDGMENTS

This paper was supported by ILIC at Pusan National University and Cleaner Production Technology Development of Ministry of Knowledge Economy (2007-N006-00), Korea.

5. REFERENCES

Kaebernick, H. (2008). Reuse, Recycling and Remanufacturing Impediments for Industrial Implementation, *15th International Conference on Life Cycle Engineering.*

Mok, H. S. (2007). International Networking for Remanufacturing, *Inter. Workshop on Sustainability in Manufacturing.*

Nasr, N.; Hilton, B. (2004). Design for Remanufacturing, *Global Conference on Sustainable Product Development and Life Cycle Engineering.*

Seliger, G. (2004). Global Sustainability-A future Scenario, *Global Conference on Sustainable Product Development and Life Cycle Engineering.*

Steinhilper, R. (2005). Remanufacturing, *KNCPC, KOREA.*

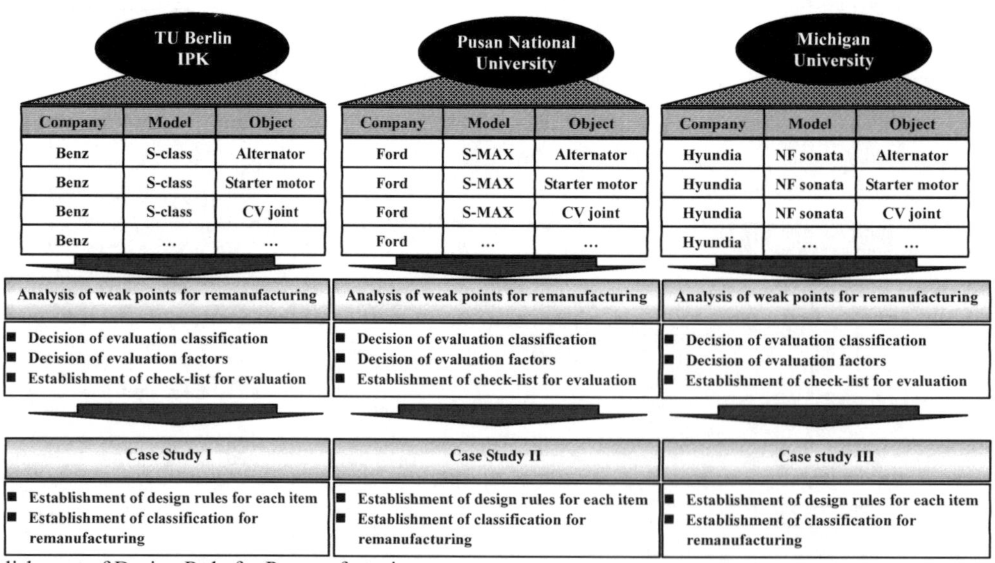

Fig. 6. Establishment of Design Rule for Remanufacturing

Annals of DAAAM for 2008 & Proceedings of the 19th International DAAAM Symposium, ISSN 1726-9679
Editor B. Katalinic, Published by DAAAM International, Vienna, Austria 2008
Make Harmony Between Technology and Nature, and Your Mind will Fly Free as a Bird

PARALLEL ROBOT WITH VARIABLE LINK LENGTH MECHANISMS

MOLDOVAN, C[ristian] & DOLGA, V[aler]

Abstract: *A parallel robot is a device for performing manipulations, where the end effector is connected to the base via multiple kinematic chains. Instead of using bar mechanisms, variable link length mechanisms and a spherical joint are used in the structure of the robot. In order to achieve this, some modifications were made in the classic structure of the parallel robot, the parallelograms are replaced with variable link length mechanisms.*
Key words: *mechanism, robot, link, joint, delta.*

1. INTRODUCTION

In this paper the authors are offering an alternative solution to the construction of a parallel robot. The presented solution consists in replacing the bar mechanism used in the parallel robots construction with a variable link length mechanism.

Also the main kinematic schema of the robot is presented in the paper, including the resulting workspace of the robot.

Theoretical works related to parallel mechanisms, and particularly hexapods, date back to centuries ago, when English and French geometricians were obsessed with polyhedra.

In 1928 James E Gwinett developed a parallel robot for the entertainment industry. This was a spherical parallel robot.

A decade later and only seventeen years after the term "robot" was coined, a new parallel robot was invented for automated spray painting by Willard L.V. Pollard.

In 1954, Dr. Eric Gough builds the first octahedral hexapodat Dunlop Rubber Co., Birmingham, England.

In 1965, Stewart's famous paper appeared in the proceedings of the (British) IMechE. In that paper, Mr. Stewart describes a 6-DOF motion platform for use as a flight simulator.

In 1971, a patent is granted to Klaus Cappel regarding a motion simulator. The patent was filed on December 7, 1964, at which time Mr. Cappel was unaware of Gough's invention (or of Stewart's paper which was not yet published).

2. THE DELTA PARALLEL ROBOT

The authors were inspired by the most successful parallel robot, the DELTA parallel robot designed by Reymond Clavel professor at EPFL - École Polytechnique Fédérale de Lausanne. He comes up with the brilliant idea of using parallelograms to build a parallel robot with three translational and one rotational degree of freedom. In 1999, Dr. Clavel is presented with the Golden Robot Award, sponsored by ABB Flexible Automation, for his innovative work on the DELTA parallel robot.

The basic idea behind the Delta parallel robot design is the use of parallelograms. A parallelogram allows an output link to remain at a fixed orientation with respect to an input link. The use of three such parallelograms restrains completely the orientation of the mobile platform which remains only with three purely translational degrees of freedom. The input links of the three parallelograms are mounted on rotating levers via revolute joints. The revolute joints of the rotating levers are actuated in two different ways: with rotational (DC or AC

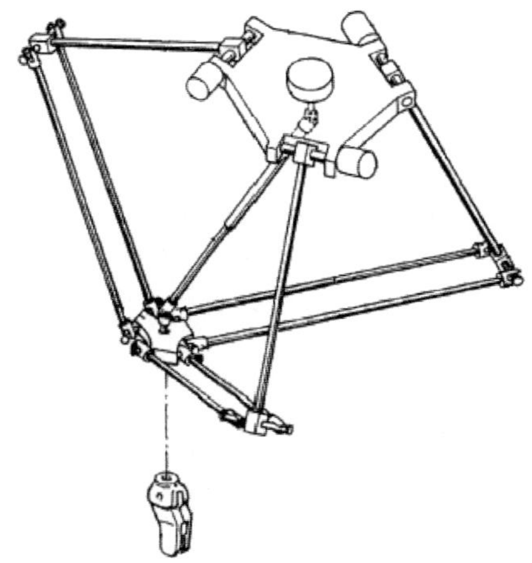

Fig 1. Schematic of the DELTA robot

Servo) motors or with linear actuators. Finally, a fourth leg is used to transmit rotary motion from the base to an end-effector mounted on the mobile platform.

The use of base-mounted actuators and low-mass links allows the mobile platform to achieve accelerations of up to 50 G in experimental environments and 12 G in industrial applications. This makes the Delta robot a perfect candidate for pick and place operations of light objects (from 10 gr to 1 kg). Ideally, its workspace is the intersection of three right circular tori. The Delta robots available on the market operate typically in a cylindrical workspace which is 1 m in diameter and 0.2 m high. In Fig.2. are the parallelograms wich are the basis elements of the DELTA robot recognizable. This figure demonstrates how orientation can be achieved using parallelogram mechanisms. (Clavel, 1991)

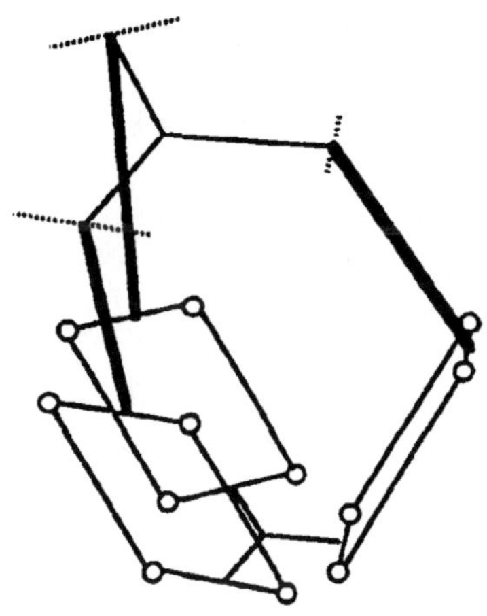

Fig 2. Kinematic schema of the DELTA robot

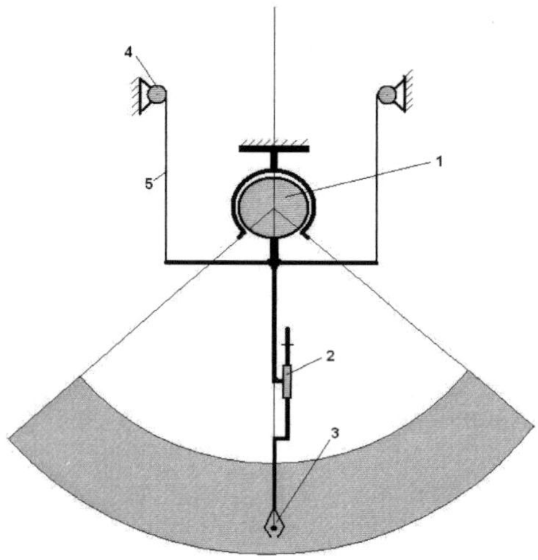

Fig. 3. Schematic of the robot with variable link length mechanism

3.PARALLEL ROBOT WITH VARIABLE LINK LENGTH MECHANISMS

The robot has as central piece a spherical joint (1). This introduces 3 degrees of freedom to the end effector. The spherical joint is attached to the case and considered as a fixed element. To this spherical joint is attached a translation joint (2) trough a bar element. Of the translation joint (2) is the end effector (3) attached trough a bar element. Also in the case are the three motors (4) fixed which serve to positioning the end effector. The motors are linked to the end effector trough flexible and inextensible wires (5)(Perju, 2001). The roller from the motor and the flexible wire build a mechanism with variable link length (Perju, 1971). The 3 motors are situated at an angle of 120 degrees from each other. The positioning of the end effector being made by the simultaneous actuation of the motors.

The kinematic schema is presented in Fig.3.

With the three motors is the positioning achieved on a spherical surface with fixed radius of the end effector and with the translation joint is the depth positioning achieved, resulting a spherical workspace (Lovasz, 2000).

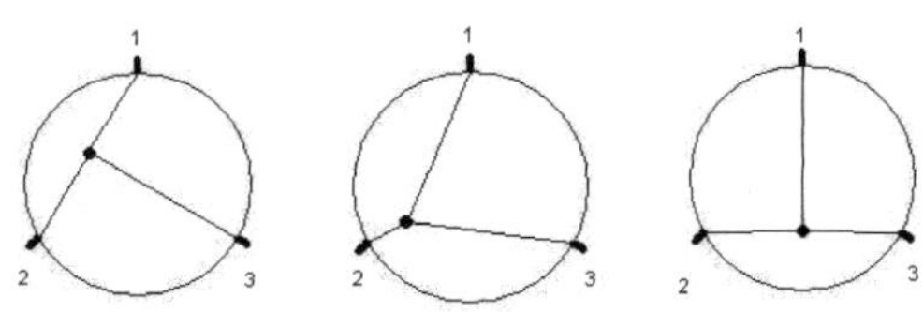

Fig. 4. End effector positions with motors 1-2 actuated, motor 2 actuated, motor 2-3 actuated.

An other type of robot with mechanisms with variable link length is presented in figure 5.

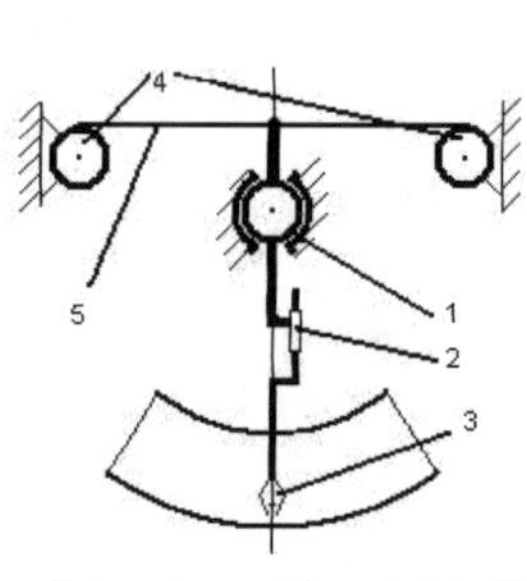

Fig. 5. Schematic of the robot with variable link length mechanism.

The main elements remain the same:
(1) Spherical joint
(2) Translation joint
(3) End Effector
(4) Motor
(5) Flexible Wire

The difference between the first version and the second version is that only the position for the actuation of the mechanism has been changed, the main schema of the mechanism remaining the same.

4. FUTURE WORK

The authors consider that the kinematic schema stage could be in a real robot transformed and practically realised.

Another area where such robots could be used is the stepper robots. In this case of advantage is the spherical joint which takes over the weight of the body, actuation not being needed at the moment when the walking robot uses the robotic foot for support. The loads that are supported by such a structure are bigger than on the classic parallel structure.

Another ideea is to create a small compact cell, with own microcontroller, which can communicate with other similar cells, all controlled by the main computer, in order to achieve movement of the walking robot. The software in the main computer should be adaptable for walking on 2, 3, 4 or 6 legs, based on how many cells the walking robot has.

5. CONCLUSIONS

Parallel robots are usually faster than traditional articulated robots, since the motors can be mounted on the base, thus saving weight. They are also stronger than serial robots because the end effector is connected to more links.

Another benefit is that the error of the end effector is less than the errors of serial robots since the errors are averaged.However, parallel robots are usually more limited in the workspace; for instance, they generally cannot reach around obstacles (Bruyninckx, 2005).

This paper presents an alternative kinematic schema to the parallel robot, based on mechanisms with variable link length, tough the structure needs future optimization because research is in the beginning stages. The introduction of the variable link length mechanisms also replaces the use of classical linear motors.

6. REFERENCES

Bruyninckx, H. (2005). Parallel robots. *Available from:* http://www.roble.info/robotics/parallel/pdf/ParallelRobots-1.pdf, *Accessed*: 2008-03-15

Clavel, R. (1991). Conception d'un robot parallele rapide a 4 degres de liberte. *PhD thesis, Ecole Politechnique Federale de Lausanne*

Lovasz, E-C.; Perju, D. & Mesaros-Anghel, V (2000). On the mechanisms synthesis of centroidal type. *Proceedings of IFToMM International Symposium, Liberec*

Perju, D. (1971). Planar mechanisms for leading a point on a given curve (ro: Sinteza mecanismelor plane pentru conducerea unui punct pe o curba data). *PhD thesis, Politechnic Institute Bucharest*

Perju, D.; Modler, K-H; Lovasz, E-C & Mesaros-Anghel, V (2001) . A new type of function generating mechanism with variable link length. *Proceedings of IFToMM International Symposium, Bucharest.*

Annals of DAAAM for 2008 & Proceedings of the 19th International DAAAM Symposium, ISSN 1726-9679
Editor B. Katalinic, Published by DAAAM International, Vienna, Austria 2008
Make Harmony Between Technology and Nature, and Your Mind will Fly Free as a Bird

HIGHLY-SCALABLE SERVER FOR MASSIVE MULTI-PLAYER 3D VIRTUAL SPACES BASED ON MULTI-PROCESSOR GRAPHICS CARDS

MOLDOVEANU, A[lin]

Abstract: *3D virtual spaces will probably be the next Internet user interface presentation and interaction paradigm. Their growth, despite huge interest from users, is still hindered by servers' scalability issues and costs. Traditional architectures, based on clusters of single processor machines, tend to be complicated in terms of software development and very costly as deployment and operating.*
We present a brand new architectural solution for a massive multi-player server which uses the new graphics cards (GPUs) available on market. Unlike traditional approaches, the presented solution is highly multiprocessor-based.
Prototype-based experiments indicated that the solution is feasible and may represent an important turnaround in the development of simpler and much cheaper massive multi-player server architectures for various types of virtual spaces.
Key words: *virtual spaces, massive multi-player, GPU multiprocessing, CUDA.*

1. INTRODUCTION

Massive multi-player online games and other types of virtual spaces, like exhibitions, museums, virtual stores, etc., are already attracting a huge number of users. The technology and its applications are rapidly evolving and expanding. It is highly probable that 3D virtual spaces will be the next generation of Internet interfaces, replacing partially or totally the web browsers or instant messengers.

The architecture for such applications is client-server oriented, with the server managing the whole content of the virtual world, providing the connectivity and user interface for the clients (Bogojevic, 2003).

An important problem that developers face is the amount of workload such a server must handle. The work is characterized both by the number of tasks, which depends directly on the number of online users, and the complexity of the tasks, which depends on the particularities of the virtual world.

2. CURRENT SOLLUTIONS

Due to the inherent real time and fun oriented aspects of virtual worlds, which leads to complex clients and minimized communication, scaling virtual worlds is quite different from scaling other types of applications (Waldo, 2008).

Current solutions are traditionally designed based on single processor machines – with very serious limitations to the number of online users a computer can handle. To overcome this, clusters or servers, each handling (Nguyen, 2003) distinct parts of the virtual world, are created (Morillo, 2006). This solution is scalable to some degree but is very costly as hardware also involves complicated software architecture (Ferretti, 2005). This is the main reason that current high-end massive multi-player applications (like MMORPGs) are extremely costly to develop and also involve some serious operating costs.

All most recently developed RAD toolkits or frameworks for creating virtual worlds (www.multiverse.net; www.activeworlds.com; www.openskies.net) still follow this traditional architectural approach.

Another solution, more like a workaround, is to impose limitations to the content and interactions in the virtual world, therefore reducing as much as possible the complexity of the tasks the server must handle. Of course, in the end, this affects the quality and possibilities of that virtual space.

3. OUR SOLUTION

In this paper we present a brand new architectural solution for a massive multi-player server which uses the new graphics cards (GPU) available on market. Unlike previous solutions, the one presented in this paper is highly multiprocessor-based.

All the NVIDIA GPUs chipsets from the G8X series onwards, including GeForce, Quadro and the Tesla lines, are targeted specifically not only for the graphics market but for the computing market as well (especially Tesla). Similar solutions are offered by competitor companies, such as ATI. Each such GPU includes a great number of streaming processors (in the order of hundreds) organized in a SIMD architecture that can be programmed for general purpose computing. Therefore, such a GPU can simultaneously run many threads, each of them accessing different areas in the memory of the GPU and following its independent execution. Many applications can get a huge performance and scalability boost from this architecture if their design can accommodate it and we decided to give it a try to solve the limitations of the current massive-multi-player servers.

The architecture of a massive multi-player server is quite complex and is beyond the scope of this article. We focus here on our solution of using GPU multiprocessing in such a server.

Besides the original concept of using the new GPUs multiprocessing power for the massive multi-player server, the challenge is to find an efficient way to distribute the workload to the streaming processors included in the GPU. To obtain high scalability and speed, several aspects must be considered:

- Distribution/coordination, which is mainly directed by the CPU, must be extremely simple and fast.
- The granularity/complexity of individual tasks must not be very high in order to obtain very good client-server latency.
- The distribution of the workload over the streaming processors must adapt in real time to the number of online users or the density of users in a region of the virtual world.
- The tasks and their coordination, benefit from, but also are in some ways limited by details of the GPU architecture; for example:
    - fast shared memory;
    - the bus bandwidth and latency between the CPU and the GPU may be a bottleneck;
    - recursive functions are not supported and must be converted to loops.

We decided to move the highly computational expensive tasks on the GPU streaming processors. Basically, these are the world logic computational operations such as collision detection, physics calculations, etc.

Tasks division:

- The work division rules rely on the fact that a user will mostly interact with the environment, entities and other users in his proximity or in a certain region.
- The world will be divided in regions:
  - regions results from a logical, server-side division of the virtual space;
  - a region contains the essential information about the environment, objects and users within;
  - a region receives a group of streaming processors from the GPU; more or less streaming processors are dynamically assigned to a region, based on its computational needs depending mostly of the number of users within;
  - the division of the world in regions is mostly statically, since any re-division may take some time with memory transfers, therefore must be avoided for real-time performance;
  - when a user or object moves from one region to another, its essential data is moved between regions;
  - for runtime smoothness and to have the effect of a seamless world, regions should be partially overlapped at their common boundary, therefore the users that are moving between such regions will do this in a transparent manner.
- The server side work-flow is divided into frames. A frame is a very short time amount, which, as a design choice can be fixed or variable. During a frame:
  - user input is taken from the input queues and passed to the central GPU control code, which will pass it to the current user's region;
  - the logic for a region will be responsible for creating the computational tasks for that region, which will be delivered to the processors assigned to that region; for example, for collision detection, a tasks will be created for each pair of moving objects;
  - the streaming processors will execute the individual tasks;
  - once all the tasks finish (in the case of variable length frames) or the frame duration is elapsed (for fixed frame design), results are placed into output queues.

The solution is adaptive, as the number of processors assigned to a region can be changed dynamically, as needed. If more users move into a region, that region will receive more processors, while the regions from where users left will lose some processors.

We have tested the proposed solution by creating a server prototype, during a workshop in the Computer Graphics laboratory from University POLITEHNICA of Bucharest. The architecture was not difficult to implement using the CUDA toolkit (http://www.nvidia.com/cuda; http://en.wikipedia.org/wiki/CUDA).

The test application involved complex physical interactions in a 3D environment. The tests showed that the system performance doesn't actually degrade much with the increase in the number of users and objects, at least to the extent we were able to test in the laboratory. The GPUs streaming processors were actually able to handle a huge amount of computing-intensive calculations.

4. RESULTS AND CONSEQUENCES

Since all the computational-extensive tasks that can be parallelized can be potentially moved in this way to the multiprocessors GPUs, a single computer with such devices can handle a much bigger number of users, resulting in huge reduction of costs.

The software architecture is much simpler than in the case of the classical, multi-computer server-farm approach.

As scalability is concerned, we have to consider that:

- Current GPUs can contain up to 256 streaming processors (NVIDIA) and future versions will certainly increase this numbers a lot.
- Current PC mainboards can support up to 4 GPUs; this number might also be increased in the future.
- Hence, current PCs can include up to 1024 programmable GPU streaming processors and incoming generations might probably include from thousands to even millions.

Limitations can occur even with this architecture.

For some applications there might be computing intensive tasks that don't map well on the SIMD GPUs' processing.

Even outside this case, one single PC might not be enough for a server with a huge number of users. The full virtual world coordination and other logical processing might get expensive enough to not be handled by a single PC. The architecture can be extended with the classical multi-computer approach, but still, the needed number of computers will be significantly lower, probably even with a degree, because most of the computational expensive tasks are delivered to the GPUs thousands of streaming processors.

If the proposed architectural solution will be successful, we can see, in the near future that the tens of hundreds of computers server farms for MMORPGs being replaced with a significantly lower number of groups of computers with multiprocessors GPUs.

5. CONCLUSION AND FUTURE WORK

We must still investigate the maximum number of users that a single computer with multiple multiprocessor GPUs can handle, for various types of games and other 3D virtual spaces.

Slightly different solutions regarding the workload distribution may be needed depending on the type of MMO game/application.

If the results will be good enough, a RAD framework or toolkit for MMO virtual 3D spaces will be developed based on this basic described architecture, aiming to allow developers to create, deploy and operate custom 3D virtual spaces at very low costs.

6. REFERENCES

Bogojevic, S.; Kazemzadeh, M (2003). The Architecture of Massive Multiplayer Online Games, *Available from:* http://graphics.cs.lth.se/theses/projects/mmogarch/ *Accessed 2008-07-01*

Ferretti, S. (2005). Interactivity Maintenance for Event Synchronization in Massive Multiplayer Online Games. *Technical Report UBLCS-2005-05*, March 2005

Morillo, P.; Orduña, J.; Fernández, M. (2006). Workload Characterization in Multiplayer Online Games. *ICCSA (1) 2006*, pp 490-499

Ta Nguyen; Binh Duong; Suiping Zhou. (2003). A dynamic load sharing algorithm for massively multiplayer online games. *The 11th IEEE International Conference on Computational Science and Engineering*, pp 131-136

Waldo, J. (2008). Scaling in Games and Virtual Worlds. *Communications of the ACM,* Vol 51,No 08/2008, pp 38-44

Multiverse Platform Architecture. *Available from:* http://update.multiverse.net/wiki/index.php, *Accessed:* 2008-08-21

ActiveWorlds, *Available from:* http://www.activeworlds.com/, *Accessed:* 2008-08-21

Openskies MMOG SDK, *Available from:* www.openskies.net, *Accessed:* 2008-08-21

Annals of DAAAM for 2008 & Proceedings of the 19th International DAAAM Symposium, ISSN 1726-9679
Editor B. Katalinic, Published by DAAAM International, Vienna, Austria 2008
Make Harmony Between Technology and Nature, and Your Mind will Fly Free as a Bird

MULTI VARIANT PROCESS PLANS DESIGN IN RELATION TO THE EUROPEAN MARKET

MONKA, P[eter] & MONKOVA, K[atarina]

Abstract: *The article deals with the use of computer aided process planning in relation to the characteristics of the European market. It introduces the method of "multi-variable process plans" dedicated to engineering activities taking place prior to the production process.*
The aforementioned method was founded by the authors at the Technical University in Košice and it enables a favourable provision of technological and production documentation by built-in instruments for adopting decisions in shorter time. This form of production preparation allows a greater flexibility of manufacturing units when searching for the means of effective production processes with the required economic results.
Key words: *Computer Aided Production, Process Plan Desing, Process Planning, Production Information System.*

1. INTRODUCTION

Production firms can use for selection of production software all variations among two extremes:

1. Complex systems
2. Independent solution for every enterprise activity

First one is for many small enterprises inaccessible by reason of expensive price. Second of them generally dispose only by flat possibility of interconnection to other information systems.

Basics problems deals by authors for constitution of manufacturing information system (MIS) are :

- arrangement for obtain of advantages of both extreme
- integration MIS in environs with specialized systems (CAD/CAM ...)
- possibility of data sharing by co-operators

Existing solution for improve this state authors cannot found in academic sources and also in sphere of production software producers.

One of first steps was detection of typical European enterpriser and selecting of key requests of these unites for manufacturing information system.

2. GENERAL STRUCTURE OF BUSINESS UNITS IN THE EUROPEAN UNION

The studies show that European micro companies constitute a substantial part of the European market as they comprise 92 per cent (17.82 million business units) of the overall number of companies and employ 39 per cent of the employees. Small and medium size companies together comprise 7.5 per cent of the overall number and employ 30.3 per cent of the employees. The rest ((0.2 per cent production unites and 30.2 per cent employees)) is covered by large companies. (Ackerman, 2007)

Other results of this same study show that micro companies has the disposal of a free potential of 20 per cent of the productivity and 15 per cent profitability.

These are very important characteristics which describe a distinct ability of the dynamic growth production and the possibility of effective evaluation of micro company instruments basically 'over a night'.

On the basis of the analysis of potential system users securing the computer support of the computer aid process planning, it can be said that it is the micro companies that constitute the significant part of the enterprise subjects.

The specifications of this type of enterprise units imply diametrically different demands on information systems from the normal setting of IS appropriate for large and medium size companies:

- Simple implementation
- Possibility of modular concept for covering all necessary areas
- Reliable and secure data formats and structures
- Possibility of flexible data sharing with IS of the purchasor and supplier
- Possibility of a trouble-free extension to the needed modules
- Securing the possibility of a relatively fast transfer to IS higher levels when necessary
- Reasonable price (Last but not least)

From the point of view of the information system constitution, one of the basic demands is the security of the possibility of creating the information relation between the enterprise unit subjects on the European market in the way the exchange of the required information in a dataway is secured. (Kuric et al., 1999)

From the point of view of the demand variety of the individual companies, the conditions were assessed when proposing IS:

- The system must be able to work with the possibility of the user view on the production process from several angles,
- Enterprise subject should be limited when launching new products to the production process as little as possible
- It should be applicable for a wide range of business
- It should be modular

Thus, the system was built for the product definitions and their parts from the point of view of three technological approaches: Individual, Type and Group.

3. THE BASIS FOR DRAFTING THE COMPUTER AIDED PROCESS PLANNING

Computer Aided Process Planning (CAPP) represents activities leading to a creation of production documentation and the details of material equipment for the production process.

The cost structure analyses for small and medium series production indicate a significant ratio of CAPP in the production costs composition. From the point of view of these analyses it is very important to pay a considerable attention to the CAPP area which can, as a result, influence the output costs of a product and its quality in a great measure. (Suresh, 1998)

When designing the new product the aim is to secure or increase its technical value not only by systematization of the production process but also by increasing the level of the supporting tools for the rational processing of the production documentation and data needed for planning. (Crow, 1992)

The production system design can be logically divided into parts (Arn, 1975):

1) Product design
   a) Design procedures
   b) Design methods
   c) Value analysis
2) Manufacturing design
   a) Technological structure planning
   b) Process planning
   c) NC programming
3) Studies of labour and production costs
   a) Work analyses
   b) Work measurement
   c) Wage schemes

4. MULTI – VARIANT PROCESS PLANNING

The theory of multil-variant process planning deals with the production process (during its project phase, also during the production) as a homogenous whole, including technological and labour processes organised via various possible parallel phases in the way the final product could be processed in the optimized way for the set conditions whilst fulfilling all the demands required by a consumer. On the basis of this theory it is possible to create combination possibilities of various techniques used in individual process plannings based on the strategy aimed at achieving the specific goal of the production unit. The main objective of this theory is:

- Creation of the unified definition environment for all the factors immediately influencing the result of production process
- Flexible interface which enables bidirectional exchange of the required information with all surrounding systems

Via the unified definition environment the philosophical and conceptual unity is secured within the whole issue falling into the formation area of multi-variable process planning, a distinct classification product constituent and the laws of production sequence for operation projection which allow the use of several possibilities designed by information system based on this theory. (Békés, 2002)

Flexible interface of the system must enable an effective work in the production environment in the way all the individual relevant systems creating heterogeneous information system (CAD/CAM application, wage records, accounting, material management,...) have the inter-connection secured via the suitable interfaces in order to prevent the errors caused by data redundance, human factor, but also to reduce the response time to a minimum. Really tested Multi-Variant Process Plannig system was originated by interconnections for wide variety of CAD/CAM systems (models, CL data & NC programs etc.) and various methods of technological approaches for multi-variant process plan design correspondent to requests of European plants.

5. CONCLUSION

On the basis of the aforementioned theory characteristics the information system was created and applied into real production conditions in the computer aided process planning consisting of approximately 6,000 components.

The given product was a result of the co-operation between a German company, providing investments and co-operation of the activities, and Slovak companies providing a technical process planning and the production of a final product.

From the very beginning of the project the established IS served for a suitable analyzing of individual real database objects (components, substructures, structures, finished product), i.e. new analytical tools were created when required. Established solution serves the purpose of easier and faster assigning of the process parameters, shortening of the computer aided process planning documentation time in real production conditions, and it also supports the effective utilization of the production plant based on the model matematization of object variation of the computer aided process planning, fulfilling the combination of the required characteristics within the given production conditions. Output system data can be used for processing of the details for the warehouse, economic and wage records as for their control and optimization.

The main contributions of assigning IS, elaborated on the basis of the multi-variable process planning in the real manufacturing conditions, can be summerized as follows:

- Reduction of the variability of warehouse stock (at the first application by nearly 30 per cent)
- Immediate information about the product elaboration
- Fast acquisition of the details via interfaces for the wage records and accounting
- Elastic analytical tools enabling the adoption of better decisions
- Acquisition of the statistical values of parameters applicable to plan production in the future

The software tool is created in the way to be easily implemented to an already existing information company structure via flexibly adjustable interfaces. It is also user-friendly, developed with the characteristics of GUI, typical for OS MS Windows, so that the basic grasp of its functioning does not require expensive trainings. Of course, if the maintanance of this system is to be productive, it must be familiarized with the given philosophy and possibilities of tactic and strategy planning, through which the production can be optimized.

Presented manufacturing information system is unique in possibility cooperation by CAD/CAM system (practically with any from known) and connectivity to other systems (accounting, stock, wages etc.). This concept brings advantages manly for micro companies:

- Modular conception
- Flexible interconnections to partners
- Possibility of cooperation with wide variety of external softwares
- Convenient price level

The ground tasks for near future in focus of authors are:

- Research for general format of process plan data
- Investigation of production environs in other European countries
- Study of graphical features use for process planning

6. REFERENCES

Ackerman, J. (2007): presentation of contribution *Integrative Planning of Logistics Structures and Production Plants in Competence Cell-Based Networks*, conference Advances in Production Engineering, Warsaw, 2007-06-14

Arn, E.A.(1975). *Group technology*, Springer Verlag, Berlin, ISBN 3-540-07505-4

Békés, J. (2002) Similarity, variant and generative process planning, *Available from:* http://fstroj.utc.sk/journal/engl/papers/016_2002.pdf *Accessed:* 2008-05-02

Crow, K (1992) Computer-Aided Process Planning, DRM Ass, *Available from:* http://www.npd-solutions.com/capp.html, *Accessed:* 2008-05-12

Kuric, I. –Matuszek, J. – Debnár, R. (1999) *Comuter Process Planning in Machinery industry*, Technical University Lodz, Bielsko-Biala, Poland, ISBN 83-87087-00-9

Suresh N.C., Kay J.M. (1998) *Group technology and cellular manufacturing: State-of-the-Art Synthesis of Research and Practice*, Kluwer Academic Publishers, Massachusetts, USA, 568 p. ISBN: 978-0-7923-8080-1

Annals of DAAAM for 2008 & Proceedings of the 19th International DAAAM Symposium, ISSN 1726-9679
Editor B. Katalinic, Published by DAAAM International, Vienna, Austria 2008
Make Harmony Between Technology and Nature, and Your Mind will Fly Free as a Bird

MODELLING OF COMPLEX-SHAPED COMPONENTS IN THE SCOPE OF REVERSE ENGINEERING

MONKOVA, K[atarina] & MONKA, P[eter]

Abstract: *The contribution deals with the modelling of the part with difficult shapes within the Reverse Engineering. Many of the hand-made prototypes are created without drawing documentation and so the manufacturing of such real parts is very complicated. The data digitizing can considerably decreases the economic cost and greatly affects the efficiency of the part manufacturing.*

Key words: *3D model, scanner, digitizing, complex-shaped*

1. INTRODUCTION

A considerable number of companies in Slovakia produce components on the basis of component drawings which can be drawn manually but nowadays most often by a computer technology. The software using 3D technique notably facilitates the work of constructors and designers. At the design stage of a new product, a 3D model is created at first then the analyses and simulations of the production process are conducted. After the fault elimination, drawing and technological documentation is completed and only on its basis is the component being produced. (Hatala et al., 2008)

The problems occur when the actual component exists, e.g. in the form of a hand-made model, is complex-shaped and its geometric characteristics are unknown. The production of other (exactly the same) components is in this case problematic, especially where a mass production is concerned. The dimension characteristics are essential in order to create a 3D model, drawing or to produce the component itself. Acquiring further data about the component, processing of this data, e.g. by a computer becomes a necessity, and only on the basis of acquired data is the drawing preparation and direct generation of CL data for the component production on NC machine possible.

Nowadays the research regarding the data acquisition on the shapes and dimensions of the real (three-dimensional) objects is in the stage of rapid development. Precise, fast and non-contact methods are significant in many industrial applications including the quality control, surface control or visual systems on assembly lines. (Valíček & Hloch 2006) They are no less important in recognizing 3D objects, in securing the area or in navigation. With respect to the required precision and quality of the subsequent production in mechanical engineering, the most precise and advantageous method of acquiring information on the topography of the real product appears to be a 3D digitization of the examined object, using modern technical devices within the scope of the so-called technological 3D scanning. This method is known as Reverse Engineering.

2. MODELLING OF COMPLEX-SHAPED PART

"Reverse Engineering is the process of the existing component, composition or product duplication without having the drawings, documentation or a computer model at the disposal." Reverse Engineering includes all the activities which enable the determination of the function principle of the product, idea or technology originally used at the product's development. It is possible to use it in order to master the process of design or to use it as a foundation for the process of redesign in the way to make the following feasible (Babjak, 2006):

- monitoring and evaluation of the mechanisms which enable the product's functionality,
- analysis and research into internal processes of the mechanical product,
- a comparison of the existing solutions with own ideas allowing improvement proposals.

Within the scope of Reverse Engineering several techniques of acquiring data on the solid geometry are being used, for instance triangulation (both active and passive triangulation, measuring systems with theodolite, focusing techniques, techniques of "shapes from shading"), optical interferometry, time of flight measurement of modulated light and others. (Hrabovský, 2002)

The digitization of real objects is possible due to scanning equipment which enables the conversion of the real three dimensional objects into a digital form. The principle of the majority of these equipments is based on the scanning the object's surface in its discrete points and it follows that the digitized object is presented on a computer as a large number of points in space, i.e. the so-called point cloud. Scanners differ from one another especially in the way the scanning of the object's surface points is implemented. The scanning equipment can be divided according to whether the scanning technology is contact or a non-contact one. The former concerns 3D scanners and stationary co-ordinate measurement systems CMM (Control Measuring Machine). This category offers the digitization equipment ranging from 3D desktop equipment to the systems used for measurement of large objects of several meters in size. The latter, non-contact systems of measurement, i.e. scanners, generally operate on laser or optical principle. For the most part, a choice of the scanner type depends on the requirements posed on the accuracy of homogeneity between a real and a digitized model. Further, another important factor when selecting a scanner is the scanning time. The fastest scanners are the laser ones. Also, a significant factor is the size of the scanned component or, possibly the mobility of the scanning equipment. The majority of scanners are limited by the scanning space in which it is possible to scan. 3D scanners are usually constructed to be able to scan objects as large as 50 cm. For more sizeable objects the larger scanning equipments are produced.

Apart from hardware devices a substantial role at digitization of 3D objects is played by software equipment. Individual scanning equipments use own software for processing of the scanned data, however these need to be transformed several times and eventually transferred into a neutral format (IGES, STEP,...) which CAD/CAM systems can operate with. (Dúbravčík, 2005)

An example for the processing of a complex shaped component in Reverse Engineering can be a template for winding of the stator of electromotor (See fig. 1).

Fig.1 A template for winding of the stator of electromotor

It is a real component, which used to be produced abroad in a way that its finite shape underwent a hand grinding into an anti-template but the drawing documentation of the resultant topography was not available. The average delivery time was longer than 3 month.

A 3D scanner Roland Picza LPX 250, in which all digitizing operations are controlled by a program Dr. PICZA, was selected for scanning. This software enables the scanning of data to be optimized, edited and transferred to NURBS surfaces. Then the data can be exported to STEP, STL or IGES formats for further processing in 3D programs. Considering the surface of the original component was too reflexive (as it was polished), it was necessary to decrease its gloss values, e.g. by spray-painting it with a gray undercoat colour. At the same time, it was essential to evenly apply the sprayed layer as this factor may also affect the approximation rate of a created model toward its original and a finite accuracy of the component created on the basis of a virtual 3D model.

However, after the scanning the virtual surfaces were not even and the transition curves not smooth. Also, the technological elements such as, for instance, holes were not scanned in a sufficient accuracy but only as the surfaces indicating the position of these elements.

For further processing it was necessary to export the acquired data from the software Dr. Picza3 and subsequently import them into Pixform software. This software allowed

- to translate one complex shaped surface through a cloud (a grid) of scanned points, whilst the accuracy of coverage depended on the number of selected checkpoints,
- to modify the polygonal meshes by means of editing control points, polygon edges and surfaces (removing, moving or adding new surfaces),
- a reduction of a polygonal meshes, i.e. a reduction of the number of polygons in the meshes, however, at the expense of the quality and display fidelity,
- to fill the cracks which arose at the scanning in polygonal meshes on the basis of the NURBS surface definition, and repair a partially insufficient representation of the scanned data,
- to partially polish the obtained model, however, not with a sufficient accuracy

Geometric data describing the established surface were neither applicable for various types of analyses, nor for a CL data generation. As a result, it was necessary to export them again in IGES, STEP or STL format and import them to a selected CAD/CAM system, in order to further process them. Pro/Engineer system was selected as a CAD/CAM system based on the experience and in connection with the software availability. It is advantageous to utilise the surface operations in this system for work with complex shaped design. For this purpose, sections were created on an imported model and the interpolating or approximating curves, defining the profile of "top surface" in the section plane, were translated through a point set via a mathematical apparatus. Approximating curves of Bezier and Spline types were used most often at work as they best represented the imported template shape in the parallel planes. The curves were covered with a coat surface which was created as Pro/Engineer system's own element so the geometric data describing this surface were readable also for CAM system area. As with the curves, it was also possible to control, analyse and modify the curvature and "smoothness" of the selected surface.

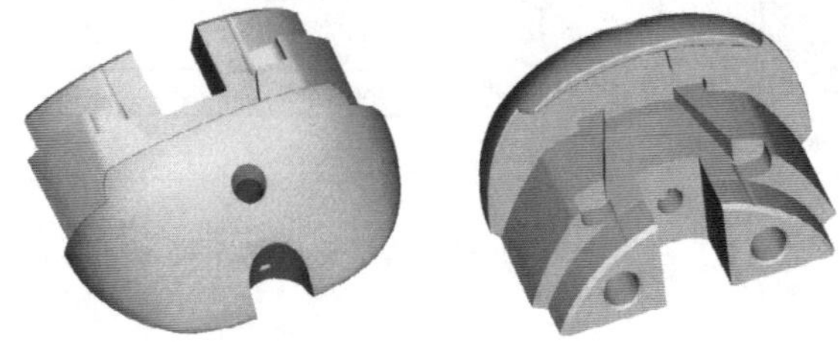

Fig.2 The final shape of the model

In the process of finalising of a 3D model version, various techniques and tools were used, which a user is offered by a selected CAD/CAM system Pro/Engineer. The final version of a 3D model created without geometric and drawing definition is on the Fig. 2.

3. CONCLUSION

Created 3D model was compared by geometry with the scanned shape in Pro/Engineer. The spaces between measured points were 1 mm and the tolerance was 0.1 mm. It can be said that 98 % of the surfaces were inside tolerance. On the basis of created 3D model it was made the new physical part by Rapid Prototyping method. The cast plastic part was compared with the original real part by means of 3D measuring equipment and then it was used in real situation. It was possible to allege that the model corresponds to the real steel part in required accuracy and so CL data were generated. The obtained data were transformed and applied as NC program for controller of concrete CNC machine in company. After the creating of 3D model and after the generating of NC program, the terms of delivery were shortened about 98 % (from 180 days on 2-5 days), the number of stored templates decrease about 50 % and the price of the parts made in Slovakia derogated about 60 % compared with original foreign supplier.

The creation of a 3D model also allows conflicting situations predicting not only at the machining but also at their putting together to assemblies resulting in the reduction of the preparation time, expenses and the quality enhancement of production. One modelled component can thus become the basis for its simple modification and subsequent production of other, type similar components, e.g. within a group technology. The company uses several types of these templates and so it will be created other of them by similar manner in the future.

4. REFERENCES

Babjak, S. (2006). The planning of reverse engineering in quick product development system I., *Transfer of innovation*, 9/2006, p. 59-61, ISSN 1337-7094.

Dúbravčík, M. (2005). The equipment of digitizing, *Transfer of innovation*, 8/2005, p. 52-54, ISSN 1337-7094.

Hatala et al. (2008). Application suitability of power-jet cutting technologies for constructional steel, *Scientific Bulletin: Machine Manufacturing Technology*, vol. 22, serie c, 5/2008, p. 211-214, ISSN 1224-3264.

Hrabovský, M. (2002). Optical methods in experimental mechanics, *Proceedings of the 40th International conference experimental stress analysis*, CTU in Prague, ISBN 80-01-02547-0, Czech Republic, 6/2002, Prague

Valíček, J. & Hloch, S. (2006). The optical identification of surface topography machined by abrasive waterjet, *Fine mechanics and optics*, vol. 51, no. 11-12 /2006, p. 320-322, ISSN 0447-6441

Annals of DAAAM for 2008 & Proceedings of the 19th International DAAAM Symposium, ISSN 1726-9679
Editor B. Katalinic, Published by DAAAM International, Vienna, Austria 2008
Make Harmony Between Technology and Nature, and Your Mind will Fly Free as a Bird

SOLUTIONS FOR DEVELOPING AND INTEGRATING SOFTWARE TOOLS FOR E-LEARNING SYSTEMS

MORARU, S[orin] A[urel]; UNGUREANU, D[elia]; SANDU, F[lorin]; GRIGORESCU, C[ostin] & KRUPKA, L[othar]

***Abstract**: Within the Leonardo da Vinci project, VetTrend, a software platform was developed for e-Learning purpose. The system of applications is built around a central portal (uPortal) and permits the access of a large type of users. Inside the platform the teachers could develop and integrate courses. The students are able to view these courses and communicate each other or with the teachers, using the integrated software tools: chat, discussion forum, whiteboard. Also, we present a newsletter engine application, with which newsletters can be created and sent to different subscribers. The application permits to users to create or delete newsletter articles, to select a visual template of the newsletter's layout and to send the newsletters to its subscribers.*

***Key words**: e-learning, communication, software, system*

1. INTRODUCTION

The main idea is to create a single application that integrates a variety of other tools (Moraru et al, 2006). The system was developed for the Leonardo da Vinci project: VetTrend (Valorisation of an Experiment-based Training System through a Transnational Educational Network Development). The advantage is that all the tools could be executed from the same place. Also, the system have the role to keep the information together, to present it to the users and allow the users to communicate between them, using different tehniques. The communication is a key-point into the process of learning; inside the system diferent types of communication tehniques have been developed and implemented by the authors: chat, discussion forum, whiteboard tool, newsletter engine etc.

2. GENERAL ARCHITECTUTE

The Learning Management System (LMS) manages the delivery of self-paced, e-learning courses. The LMS lets you publish courses and place them in an online catalog. Learners log into the LMS using a browser, select courses from the catalog and launch them. The LMS tracks the learners' activities with the courses (Mustica & Moraru, 2003). Also, the LMS provides online reports for each course and learner. The LMS structure is presented in the figure 1.

3. RESULTS: INTEGRATED SOFTWARE TOOLS

The e-learning platform has some facilities created as software tools integrated inside an *uPortal* application (Moraru et al, 2004). Bellow are described the *Chat* and the *Discussion forum*.

The *Chat* application has been developed using the Java technology (JSP and Servlet). JSP pages have been used for creating the application interface. A Servlet application was developed for processing the information. The application could be used independently of any other application or it could be used inside a system of applications. In our case, we have decided to use the application inside the uPortal system.

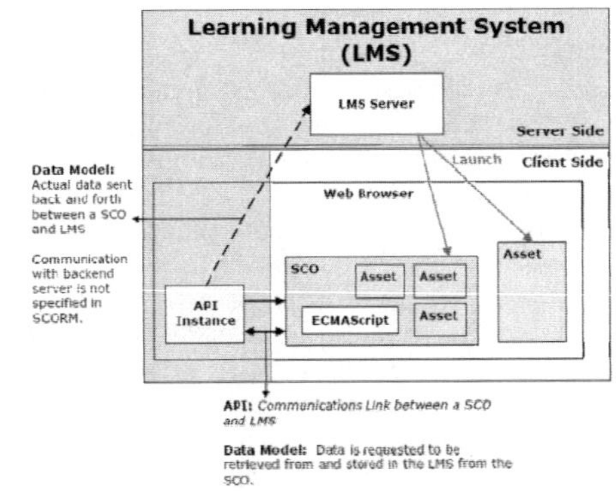

Fig. 1. LMS architecture

The application configuration is done by adding different parameters to the URL path that is used to access it. In this way could be set the name of the user that enters the chat application. Also, the groups at which the user has access could be configured.

Inside the application, all groups appear into a bar on top of the interface. At every group corresponds a button control. The user has the possibility to change from one group to another. Every group is like a different chat room; it exist a different conversation for every group.

The user has the possibility to do a private conversation with any online user from a group.

The *Discussion forum* application has been also integrated inside the uPortal system. In this way we could use the users from the uPortal system inside the forum module. The authentication into the forum module is done automatically, with the current username, when the current user accesses the link to the forum module. The administrator users from the uPortal system have the role of administrator inside the forum module.

In the figure 2 is presented a screenshot of the discussion forum module.

Fig. 2. Discussion forum application screenshot

The *Newsletter engine* application has the goal to offer a solution for a common problem, the communication between a sender (person or institution) and its subscribers (individual members or groups of members interested in a common subject) through the use of personalized e-mail messages that can be sent using the application. The outcome is a browser-based application, for sending out customized emails to its subscribers, emails containing more than plain text. The emails are customized for each client if the data of the client is available.

The structure of each web-page from the application is based on the following template (begining from the top):

- includes the „header.php" file that contains the information about the application's page, metadata such as: author, application name, copyright, linked style sheets, javascript files and the settings for connecting to the database (imported from the „connect.php" file);
- includes the „meniu.php" file that contains the structure of the menu;
- the content of each page such as: welcome message (for the first page), user form (for the page for adding a new user), the new article form (in the page for creating a new article) etc;
- includes the „footer.php" file that contains the bottom part of each page.

The user that doesn't have access to the toolbar functionalities will view an empty toolbar. The user is able to see what is displayed into the drawing area but he couldn't interact with it (do modifications). Having attached to the *Whiteboard* the *Chat* tool is very useful because the person who does the drawings can explain in the same time what he is doing.

In order to be able to access the facilities of the application, the users must log in. They must have a valid username and password. They must enter them in the login form and click on the "Submit" button. Adding a new user can be done by an administrator, and requires the first and last name of the user, a valid email address, a username and password (requested twice, to make sure it is not written wrong). The user's level can be chosen between 1-Admin and 2-User, the difference being that only the first type has full rights on the application.

The email field has a validation to make sure the user entered a valid email address. A JavaScript function checks to see if the email has the right format. The format must be something like this „user@server.[subdomain].domain". The second part of the email can also be the IP address of the host server.

Creating a new article needs to introduce a title, a main body, and an optinal signature. The text area for the body of the article is a WYSIWYG (WhatYouSeeIsWhatYouGet) JavaScript text editor, which allows rich content to be typed in. For saving the current order of the articles in the database, AJAX was chosen because it allows more dynamic user to machine interaction, making the sorting a pleasant experience.

Fig. 3. Personalizing newsletter page

The hover effect of the article is obtained using JavaScript. Scriptaculous is a dynamic user interface that allows the user to interact with the content of the page by dragging and dropping, hovering, and also allowing the content to have a transparency like appearance (Kristaly & Moraru, 2006).

Adding subscribers into the database can be done in two ways: (i) the user can select to enter the contacts one by one, by typing in their title (Mr., Ms., etc), their name and email address; (ii) importing the subscribers from a TAB delimited file.

After the order in which the articles will appear has been selected, the newsletter is ready to be sent. The mailout is done in three simple steps, each one with its own importance: selecting the template, personalizing the newsletter (as shown in figure 3), and sending the newsletter. After the emailing is done, the application displays a report with the number of emails sent.

4. CONCLUSION

The architecture of a system platform developed for e-Learning within the Leonardo da Vinci project - VetTrend - was presented. A central portal (uPortal) includes the developed applications, allowing the access of a large type of users.

The created facilities – chat, discussion forum, whiteboard, etc - were developed and integrated inside the platform by the authors. The students can view the courses and communicate among them or with the teachers, using the integrated software tools, the main purpose of this system being to gather the information together and to facilitate the users' communication.

The Newsletter application can be used for keeping contact with several lists of people at the same time. Inside an university, it could be used to contact each group (or year) of students, or all at the same time, or it could be used to communicate separately with the students and teachers. It would also be useful for contacting the students that are studying abroad (usually with scholarships), or the ones that are preparing to study for a period of time in a foreign country, or in another university.

5. REFERENCES

Kristaly D.M., Moraru, S.A. (2006). Java technologies for model-view-controller architecture, *Proceedings of the 10th International Conference on Optimization of Electrical and Electronic Equipment "OPTIM'06"*, Vol. III, pp. 175-178, ISBN 973 635 705-8, 978 973 635 705-3, Moeciu-Brasov, May 2006, Ed. *Transilvania* University, Brasov.

Moraru S.A., Diaconu, I., C. Bujdei, C. (2006). ICT Use for Remote Experiments in Electric and Electronic Domain, *Proceedings of the 15th International Scientific And Applied Science Conference "ELECTRONICS ET'2006"*, book 3, pp. 201-2008, ISBN 954-438-566-5, Sozopol, Bulgaria, Sept. 2006, Technical University Sofia & Technical University Delft, Sofia.

Moraru, S.A., Pelcz, A., Perniu, L. (2004). Java and e-Learning - a view of a tool and a context for ICT products, *Proceedings of the 9th International Conference on Optimization of Electrical and Electronic Equipments "OPTIM2004"*, Vol. IV, pp. 241-244, ISBN 973-635-285-4, Braşov, May 2004, Ed. *Transilvania* University, Brasov.

Mustica M., Moraru S.A. (2003). ECampus – An E-Learning Management System as a Web Based Application, *Proceeding of the 12th International Scientific and Applied Science Conference "ELECTRONICS ET'2003"*, Book 4, pp. 190-193, ISBN 954-438-374-3, Sozopol, Bulgaria, Sept. 2003, Technical University Sofia & Technical University Delft, Sofia.

http://java.sun.com/product/ *and* .../javaee/javaserverfaces , *Accessed:* (2008-05-15)

Annals of DAAAM for 2008 & Proceedings of the 19th International DAAAM Symposium, ISSN 1726-9679
Editor B. Katalinic, Published by DAAAM International, Vienna, Austria 2008
Make Harmony Between Technology and Nature, and Your Mind will Fly Free as a Bird

EXPERIMENTAL COMPARISON OF CONVENTIONAL AND HIGHLY ABRASIVE ABRADING AGENTS

MORAVCIKOVA, J[ana] & LIPA, Z[denko]

Abstract: *The grinding belongs to the most widespread finishing methods of metal machining. The contribution deals by experimental comparison of conventional and highly abrasive abrading agent at plunge-cut grinding. The goal is to find the optimally setting of the abrasive and dressing parameters, by which they will be achieved qualitative and economically the best parameters of machined surface of work piece. The criterion is the surface roughness and economy of production. Grinding wheel made by the company Tyrolit is used as the conventional abrasive agent. As a highly – abrasive abrading agents of the grinding wheels made by companies Atlantic and Hermes were used. The results of the experiments show that there is a significant difference between the conventional and the highly – abrasive abrading agents.*
Key words: *grinding, abrading agents, dressing, roughness.*

1. INTRODUCTION

Important role in increase of technical level and precision of production is industry, where's a lot of grinding components. To the precision grinding is requirement not only quality grinding machine, dresser wheel but abrading agent, too. Is very important to find the optimally setting of the abrasive and dressing parameters, by which they will be achieved qualitative and economically the best parameters of machined surface work piece. By trend of the most accurate dressing of grinding wheels in small-series production are diamonds CNC forming wheels (Janáč et al., 2004). The dressing of abrasive wheels after defined number of period and by using different abrading agents' type has a big influence on quality of machining surface. The criterion is the surface roughness parameter Rz. For realization of experiment is proposed plunge-cut grinding and grinding wheels with conventional and high abrading agent. The dressing period of grinding wheels depend to abrading agent.

2. CONVENTIONAL AND HIGHLY ABRASIVE ABRADING AGENTS

Between conventional abrading agents belong grinding wheels on basis of alumina (Al_2O_3). Standard alumina and semi-high-grade alumina is made of bauxite. High-grade alumina is made of pure alumina oxide in electric furnace at temperature about 2000 °C (fig. 1). Between conventional abrading agents belong grinding wheels on basis of silicon carbide (SiC), too (Humár, 2005).

Between highly abrasive abrading agents we can to include micro-crystal sintered alumina. Fine crystalline grain structure allows on increasing grain wear only chipping his small particles, and grinding-grain edges stay always sharp (Kolárik, 2005).

Development in field of grinding wheels on basis Al_2O_3 arrived to new type of grinding wheels - SG (sol-gel). SG abrasive grain has higher hardness and better toughness. SG abrasive grain consisted from submicron particles. These particles can to drop out from abrasive grains during grinding. This micro-crystal fracturing of SG abrasive grains leads to different wear process of SG grains in comparison with conventional grain. SG abrasive grain is suitable for grinding of hardened steels with hardness above 60 HRC. (Neslušan, 2003)

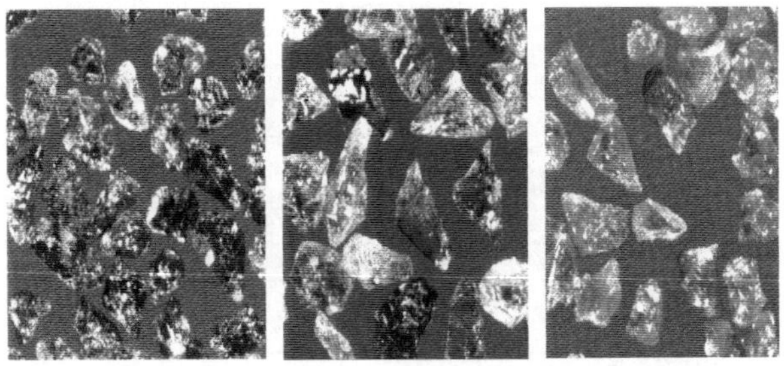

a) b) c)

Fig. 1. Types of alumina: a) standard alumina, b) semi-high-grade alumina, c) high-grade alumina.

3. EXPERIMENTAL RESULTS

For experiment was used CNC machine, by whose we can to regulate abrasive and dressing parameters too. Work pieces used in the experiment were produced from material 100Cr6, hardened and tempered to 58±4 HRC (Hazlinger & Moravčík, 2007).

As a conventional abrading agent, with which were compared the results of grinding by highly abrasive abrading agents, is grinding wheel of company TYROLIT 508x22x254 with specification 97A100L6V112.

As a highly abrasive abrading agent, is grinding wheel of company ATLANTIC 508x20x254 with specification EB3120-K7VB and grinding wheel of company Hermes with specification 3SK3 100/1 M-N5 V09G (fig. 2).

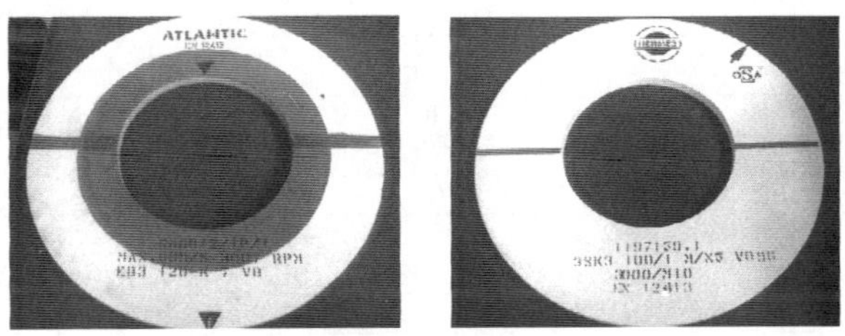

a) b)

Fig. 2. Highly abrasive abrading agent: a) Atlantic, b) Hermes.

3.1 Surface roughness by using conventional abrading agent

Surface roughness specified by drawing documentation $R_Z = 4$ μm. The period of dressing is 25 work pieces. The measuring was realised periodically in 10 consecutively following period of abrasive wheel dressing. Measuring data are in the Table 1.

Average roughness after dressing of grinding wheel: $R_Z = 2{,}689$ μm. Average roughness before dressing of grinding wheel: $R_Z = 2{,}279$ μm (fig. 3). By increasing of dressing period on 30 parts – work pieces, incurred the burning on the work piece surface, which were inaccessible for the further processing.

Work piece period of dressing	Tyrolit 97 A 100 L 6V 112 R_Z [μm]
1	2,61
25	2,21
1	2,49
25	2,23
1	2,64
25	2,33
1	2,44
25	2,17
1	2,88
25	2,45
1	3,05
25	2,45
1	2,88
25	2,22
1	2,65
25	2,28
1	2,72
25	2,29
1	2,53
25	2,16

Tab. 1. Surface roughness by using conventional abrading agent.

3.2 Surface roughness by using highly abrasive abrading agent Atlantic and Hermes

The period of dressing is 70 parts. Measuring data are in the Table 2.

Work piece period of dressing	Atlantic	Hermes
	R_Z [μm]	
1	2,78	2,93
70	2,12	2,04
1	2,69	2,71
70	2,08	2,01
1	3,33	2,83
70	2,23	2,10
1	3,49	2,55
70	2,29	1,99
1	2,56	2,58
70	2,01	2,08
1	2,42	2,44
70	1,98	1,73
1	3,04	2,49
70	2,11	1,88
1	2,65	2,37
70	2,03	1,84
1	3,10	2,67
70	1,86	1,96
1	3,17	2,58
70	1,72	2,03

Tab. 2. Surface roughness by using highly abrasive abrading agents Atlantic and Hermes.

Atlantic - average roughness after dressing of grinding wheel: R_Z = 2,923 μm. Average roughness before dressing of grinding wheel: R_Z = 2,043 μm. By increasing of true period on 80 parts appeared the problems with stability of prescribe geometrical dimensions.

Helmes - average roughness after dressing of grinding wheel: R_Z = 2,617 μm. Average roughness before dressing of grinding wheel: R_Z = 1,937 μm. The stability of prescribe geometrical dimensions was good by observance of dressing period 70 parts. After longer time of grinding were the burnings on the surface of work piece.

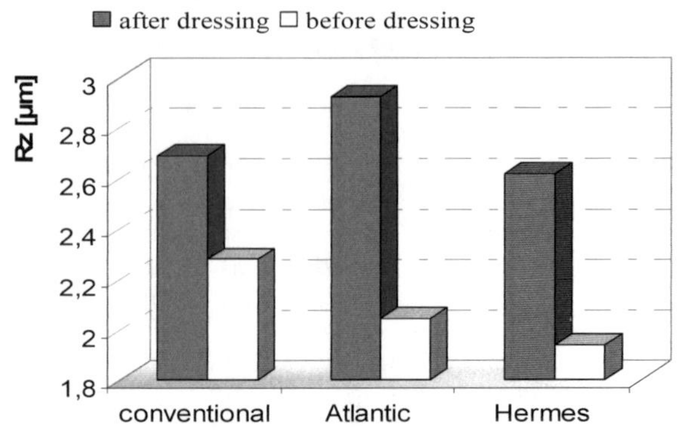

Fig. 3. Roughness comparison after and before dressing.

4. CONCLUSION

The contribution deals by experimental comparison of conventional and highly abrasive abrading agent at plunge-cut grinding. During the experiment was found, that the Atlantic grinding wheel is the best for our grinding conditions and a type of machining work piece. This grinding wheel doesn't achieve the best average roughness, but its geometric dimension stability was comparable with Hermes grinding wheel during the whole time period of wheel dressing. The better quality of the surface by reason of burning, which may to arise during grinding, decides for the selection of Atlantic grinding wheel. The major criterion of conventional and highly abrasive abrading agent comparison was the surface roughness and economy of production. The number of work pieces, after which is needs to dress the grinding wheel, was given on 25 by using of conventional abrading agent. The depth of cut at dressing was 35 μm. The number of work pieces, after which is again needs to dress the grinding wheel, was given on 70 by using of highly abrasive abrading agent. The depth of cut at dressing was only 15 μm. We achieved the significant saving of abrading agent and smaller wear of pulley dresser.

ACKNOWLEDGEMENT: This paper was elaborated in the frame of project: VEGA 1/4111/07

5. REFERENCES

Janáč, A.; Bátora, B.; Baránek, I., LIPA, Z. (2004). *Technológia obrábania. Machining technology.* STU, ISBN 80-227-2031-3, Bratislava.

Humár, A., (2005). *Technológie 1, Technológie obrábění 3 část. Technology 1, Machining technology 3.* Brno. Available from:http.//www.sjf.tuke.sk/ktam. *Accessed:* 2007-05-05.

Kolárik, M. (2005). *Nové trendy presného brúsenia a orovnávania. New precision grinding and dressing trend.* In Strojárstvo – Strojírenství, No. 11, pp. 42 – 43.

Neslušan, M. (2003). *Technicko - ekonomické aspekty aplikácie SG brúsnych kotúčov pri brúsení kalených ocelí. Technical and economic aspects of the application SG grinding wheels by grinding of hardened steels.* In Strojárstvo – Strojírenství, , Vol. 6, No. 7, pp. 54.

Hazlinger, M.; Moravčík, R. (2007). *Degradačné procesy a predikcia životnosti. Degradation processes and lifetime prediction. Elektronické skriptum.* Trnava: AlumniPress, ISBN 978-80-8096-031-5.

Annals of DAAAM for 2008 & Proceedings of the 19th International DAAAM Symposium, ISSN 1726-9679
Editor B. Katalinic, Published by DAAAM International, Vienna, Austria 2008
Make Harmony Between Technology and Nature, and Your Mind will Fly Free as a Bird

POWER LOSS REDUCTION OF THE MAGNETO HYDRO DYNAMIC NAVAL TRUSTER

MOROIANU, C[orneliu]; GROZEANU, S[ilvestru]; CIOBANU, C[amelia]; BEJAN, M[ihai] & SAMOILESCU, G[heorghe]

Abstract: *The magneto hydro dynamic naval thruster has a pipe, parallel with ship axis and isolated walls. A pair of electrods creates an electric current through sea water. The electric current is perpendicular on pipe axis. A magnetic field is applied, perpendicular in the pipe, near electrods. A Laplace force will evacuate the sea water through a noozle when the magnetic field will interact with the current. The ship is propulsed by the reactive force. The electrods have a boundary effect. Some of electric current is dispersed outside the electrods. This dispersed current in not under magnetic field action anymore. The dispersed current increases the power loss through Joule effect.*
The intensity of dispersed currents can be reduced using an isolated restrictive blade, parallel with pipe axis.
Key words: *electric loss seawater sunderwater*

1. INTRODUCTION

An M.H.D. naval thruster is a homo polar electrical engine using a pipe full with sea water as a moving drive. The pipe is parallel with ship longitudinal axis. The bow has an inlet nozzle and the stern an outlet nozzle. Perpendicular on pipe axis two electrodes produces an electric current through water. A magnetic field is also applied. The result is a Laplace force. It moves the sea water through this pipe – fig.1.

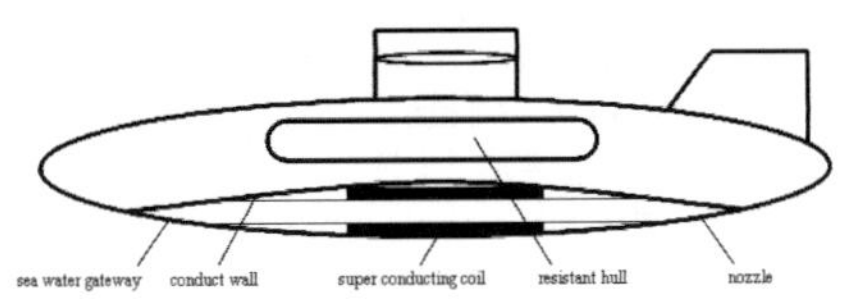

Fig.1. Magneto hydro dynamic naval thruster.

The length of the electrodes and the magnetic field application area being shorter than the pipe, a certain part of the current will be dispersed in the the exterior of this area producing a boundary effect.

It consists in a current circulation outside of electrodes. This dispersed current escapes from magnetic field influence. In this way it increases the lost power. To limit this effect, some restrictive blades from insulating materials are placed parallel with electrodes. This specific zone is extra electrode space (Shatrov, 2006).

These blades reduce dispersed currents, increase system electrical resistance and decrease power loses with several percents – Fig.2.

The R_i is the resistence between electrodes when the dispersed current is blocked.This space is the intra electrode gap. The gap between electrodes is 2a, the gap length is 2b, η=a/b, the power developed in the intra electrode gap is P_i and restrictive blade length is l. Solving the Laplace equation results that the intra electrodes gap potential is (Gavrila, 1999):

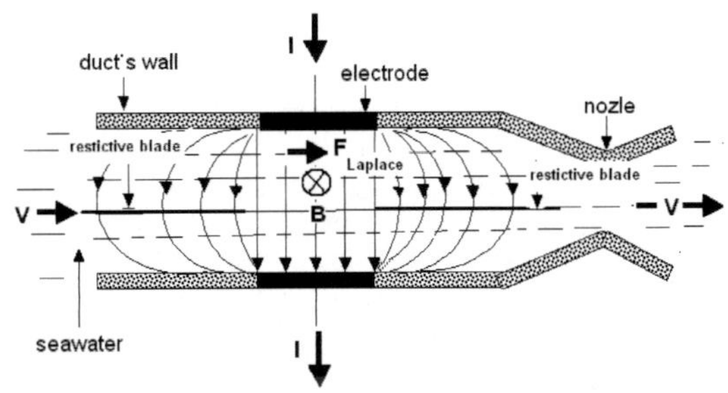

Fig.2. Restrictive blades.

$$V(x,y) = \sum_{n=1}^{\infty} A_n e^{-\frac{n\pi x}{a}} sin\frac{n\pi x}{a} - \frac{U}{2a} \quad (1)$$

and extra electrode gap potential is:

$$V'^{(x,y)} = \sum_{m=1,3,5..}^{\infty} B_m e^{-\frac{m\pi y}{2a}} sin\frac{m\pi x}{2a} \quad (2)$$

Boundary conditions determine A_n and B_m coefficients. Total electrodes – seawater system resistance with restrictive blades is:

$$R = \frac{U}{I} = -\frac{U}{2\sigma h\left(\int_0^b \frac{\partial V(x,y)}{\partial x}dx + \int_l^{\infty} \frac{\partial V'(x,y)}{\partial x}dx\right)} \quad (3)$$

where σ is the electrical conductivity of seawater and h is the electrode height.

Intra electrode gap resistance with circulation of outside current limited by blades is:

$$R_l = \frac{R_i}{1+2\eta\frac{\left(1-e^{-\frac{\pi}{\eta}}\right)}{\pi\left[\frac{32}{3\pi}\left(1-e^{-\frac{\pi}{\eta}}\right)e^{\frac{\pi l}{2a}}-e^{-\frac{\pi}{\eta}}\right]}} \quad (4)$$

Power loses through dispersed currents using limitation blades

$$\Delta P_l = 2P_i\frac{\eta\left(1-e^{-\frac{\pi}{\eta}}\right)}{\pi\left[\frac{32}{3\pi}\left(1-e^{-\frac{\pi}{\eta}}\right)e^{\frac{\pi l}{2a}}-e^{-\frac{\pi}{\eta}}\right]} \quad (5)$$

Previous relation reveals without doubts that resistance increases and the power loss decreases when limitation blades length is increased. Increasing limitation blades length also increases frictions and hydro dynamic loses. The optimum limitation blades length should be find and used (Lin, 1991).

This study presents several experiments in order to confirm the mathematical model of increasing resistance of electrodes – sea water system with restrictive blades increase. The aim is blades optimal length.

2. EXPERIMENTAL DEVICES

We performed several experiments which relate resistance and blades length. Blades were fixed in vertical position, parallel with pipe axis outside intra electrode gap. In Fig.3.a and 3.b, there are an upper view and a frontal section.

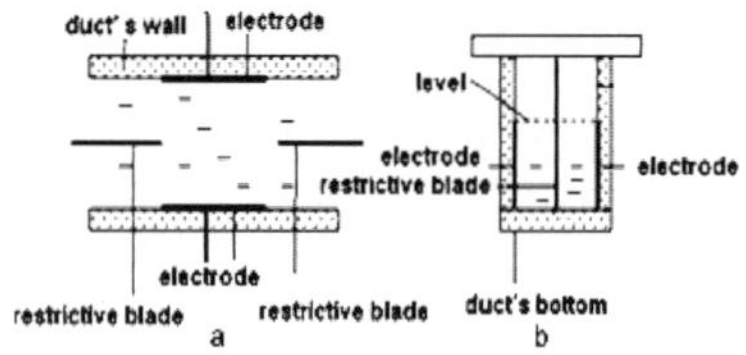

Fig.3. Device with restrictiv blades for resistance determination.

Replacing separation blades "l" length was realized using longer ones. Resistance measurements were performed in A.C. with a 500Hz frequency so that the results not are being perturbed by electrochemical processes from DC. In order not to affect sea water conductivity by temperature gradient during experiments, the experimantal system was built having a thermostat. The resistance between the pair of electrodes placed in seawater is compared with a reference resistance.

A system for small current variation measurement (using a compensate miliammeter with the aid of D.C source) was used in order to measure the finest variation resistance produced by parameter modifications. (Antoniu, 1998).

A diode was introduced into miliammeter and this compensation being in DC, in such way that in this circuit branch (were the miliammeter is fitted) only continuous current will exist – Fig.4.

3. EXPERIMENTAL RESULTS

The distance 2a between electrodes was a constant 80 mm value, the depth 25 mm, the salinity 25 g/l and the electrodes length between 0.5 – 8 cm. Experimental resistence values for different blades and electrodes lengths are presented in Fig.5.a. Powers loses are presented in Fig 5.b.

Diagrams reveal that increasing limitation blades length reduces the power loss and also increases sea water- electrodes system resistance.

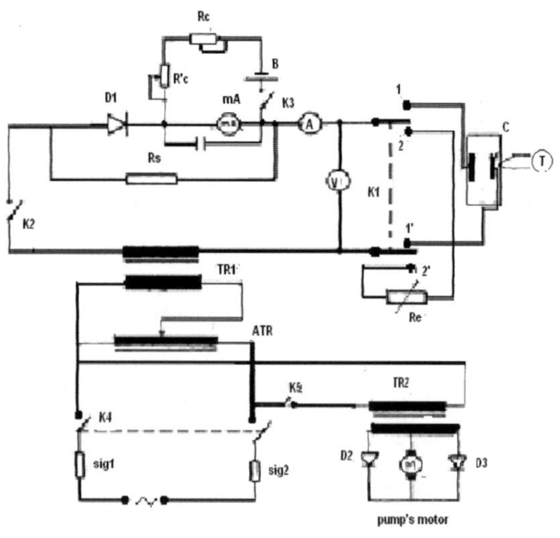

Fig.4.Mesuring electric circuit.

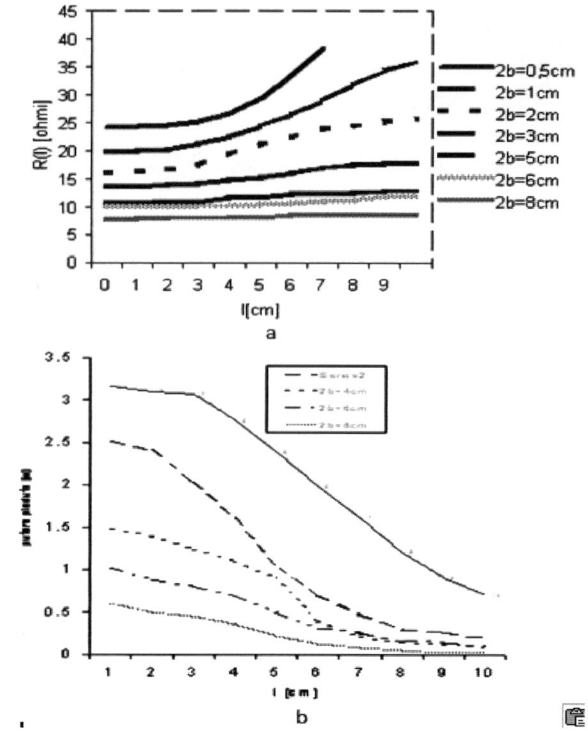

Fig.5. Resistance and lost power related to restrictive blade length.

4. CONCLUSIONS

A bigger $\eta = a/b$ value produces a significant resistance increase. For small lengths of limitation blades (aproximative equal with electrodes length 2b) the increasing of the resistence is small. An signifiant resistence increase was obtained making limitation blade length double to electrode length.

If electrodes length is bigger than gap between them, the resistance is bigger and lost power is smaller.

Limitation blade length must be no more than twice electrodes length. Overcomming this value increase the hydro dynamic loses (Mocanu, 1981).The utilization of restrictive blades increase the system efficiency.Future researches must establish the optimal length of restrictive blades so that the electrical power loss to be minimal without increasing too much the hydro dynamic loss.

5. REFERENCES

Antoniu M. (1998). *Masurari Electronice (Electronical Measurements) Vol.I*, Editura "Gh.Asachi", p.256-259, ISBN 973-9178-22-7, Iasi

Gavrila, H., Centea, O. (1999). *Teoria moderna a campului electromagnetic si aplicatii*, (*Modern theory of electromagnetic field and applications*) Editura All, p.221-224, ISBN, 973-571-257-1, Bucuresti

Lin T.F.; Gilbert J.B.; Roy G.D.(1991). Analyses of magnetohydrodynamic propulsion with seawater for underwater vehicles. *Journal of Propulsion and Power*, Vol.7 Nov.-Dec., p. 1081-1083, ISSN 0748-4698,

Mocanu C.I., (1981). *Teoria cimpului electromagnetic,(Theory of electromagnetic field)* Editura Didactica si Pedagogica, p. 821-840, Bucuresti

Shatrov V.; Gerbert G. (2006). *On magnetohydrodynamic drag reduction and its efficiency*, *Magnetohydrodynamics*, Vol. 42, No. 2/3, p. 181-186, ISSN 0024-998X

Annals of DAAAM for 2008 & Proceedings of the 19th International DAAAM Symposium, ISSN 1726-9679
Editor B. Katalinic, Published by DAAAM International, Vienna, Austria 2008
Make Harmony Between Technology and Nature, and Your Mind will Fly Free as a Bird

THE SOOT AND NITROGEN OXIDE REDUCTION USING WATER/HEAVY FUEL OIL EMULSION

MOROIANU, C[orneliu]; MUNTEAN, A[ngela]; SAMOILESCU, G[heorghe]; BANDRABUR, D[iana]; BEJAN, M[ihai] & CIOBANU, C[amelia]

Abstract: *The soot and oxides of nitrogen (NO_x) emissions from marine engines are significant on a global level. Soot and oxides of nitrogen emissions participate in the formation of photochemical smog and acid rain. The challenge is to control soot and NO_x emissions without increasing fuel consumption and smoke. Ther is much interest in the potential of utilizing oil/water emulsion in liquid-fueled combustors for pollutant reduction and enhanced fuel economy. In this work it is preferred to emulsify by using an ultrasonic method in a hydrodynamic generator, because the fuel and water produce a very finely dispersed emulsion. The graphic transposition of the processes of fuels and oil/water combustion is defined by the combustion graphology, developed in a simulator. The graphic representation of the combustion processes development for a droplet of liquid fuel used in the industrial combustion may be made by means of the so-called "combustion oscillogram".*
Key words: *water/heavy fuel soot nitrogen oxide reduction*

1. INTRODUCTION

Every year approximately 10 million tones of nitrogen oxide (NO_x) are produced by ships. The world fleet include 55% slow speed diesel, 40% medium speed diesel and 5% other engines. Slow speed diesel engines produce higher NO_x emissions. The photochemical smog and acid rain are also produced by NO_x emissions. Introducing water into the combustion chamber the temperature of burning is reduced due to the vaporation process. There are two ways to introduce water. The first way is through air intake, using humidification and the latter way is by water/fuel emulsion. The latter way mentioned above reduces smoke, while humidification increases it. Water/fuel emulsion and direct water injection have the maximum effect on NO_x reduction. This paper presents the resultates of burning the emulsified and non emulsified fuel droplet on the laboratory condition(simulator). The emulsification of the water with oil was performed by means hydrodynamic liquid generator.

2. THE WATER/FUEL EMULSIONS GENERATION USING ULTRASOUNDS

The high intensity ultrasound generation in liquid is realized using a hydrodynamic whistle. The hydrodynamic whistle for liquids, Fig.1, consists on a tapered nozzle (1), with a segment (2) placed in front of it at 0.3 – 1 mm. The segment is fixed in one or two nodal points (Jinescu, G. 1983).

Passing through the nozzle, the liquid flow hits the segment fixed on one end on the bracket. At about 12-15 bar pressure it resonates at frequency (Popa & Iscrulescu, 1983):

$$f=\frac{22.4\cdot d}{4\cdot\sqrt{3}}\cdot\frac{1}{l^2}\cdot\sqrt{\frac{E}{\rho}}\ \ [Hz] \qquad (1)$$

where
l – is the segment length [m]; d – segment thickness [m];
E – Young's modulus [N/m^2]; ρ - density.

Fig.1.The hydrodynamic generator for liquids; 1– snout, 2– flexible segment.

If the resonating segment is made of, the resonance frequency is

$$f=5.4\cdot 10^5\cdot\frac{d}{l^2}\ \ [Hz]. \qquad (2)$$

The ultrasound frequency depends on the flow pressure. The hydrodynamic generator irradiates ultrasounds in the working environment as the following relation reveals:

$$f=\frac{v}{h}\cdot 0.5\ \ [Hz] \qquad (3)$$

where
v – is the flow rate in the nozzle [m/s]; h – the distance between the nozzle and resonating flexible segment [m].

The nozzle and the vibrator segment are situated in a resonant chamber with an acoustic form (Jianu C. 1996).

3. THE EXPERIMENTAL APPROACH

The experimental approach has two stages: in the former stage is the combustion of non-emulsified fuel and the latter one is the combustion process of oil/water emulsion (for the lock of space I didn't write the combustion oscillogram of non-emulsified fuel).The emulsions have different water mass fractions in fuel. The analyses have been performed in special laboratories. The fuel drops bathing at pre-established diameters was realized using a special device. The same temperature and air conditions were used for emulsified and non emulsified fuel drop combustion. The process was realized in free convection (Law, 1997). The remarks were directly made with the naked eye, and the processes were registered using a special camera.

4. THE WATER/FUEL EMULSION FUNDAMENTAL CHARACTERISTICS

In order to state the quality and the steadiness of an ultrasound emulsion its properties will be compared to those of a non emulsified fuel shown in Table 1. The water/fuel emulsion has been stored for 45 days. The determinations revealed a good stability in time. There is no separation of liquid components danger. We also performed tests in order to determine the water/fuel emulsion behavior in heat-proof vessels. Emulsions were stored for 60 and 150 days at 50^0, 60^0 and 80^0C temperatures. The water drop dimensions and the percentage analyses were also performed.

5. THE WATER/HEAVY FUEL EMULSIONS COMBUSTION OSCILOGRAMS

Fuel combustion graphology is a new technical and scientific field (Ghia 1991). It transposes the fuel combustion processes in a simulator using graphics. It is an easy way to establish the ignition and combustion characteristics and relations between combustion conditions and fuel specifications. The combustion oscillogram is a graphic of combustion process development for a liquid droplet. This diagram shows the time variation t of radiation intensity I of volatile substances caused as a result of combustion changed into an electric signal by means of an optical-electronical system equipped with a photoelectrical cell. On a standard condition, the oscillogram establishes self-ignition delay τ_i, the combustion time of volatile matters τ_v, the cenosphere combustion time τ_c, the maximum radiation intensity for cenosphere combustion I_c, the maximum radiation intensity for volatile matter combustion I_v, the energy radiation resulting from cenosphere burning, transformed by photocell into electric energy E_c, etc The standard condition of the combustion tests are mainly specified by the combustion chamber geometry, the temperature, the pressure inside chamber, the emulsion temperature, the fuel admission system, the air flow conditions around droplet and the initial and average diameters of the droplet.

The propulsion systems for ships and the steam generators fitted up on the oil tankers run with residual heavy fuel RME 25, RMF 25, RMG 35 range according to ISO 8217/1995. Fig.2 shows the combustion oscillogram for a water emulsified RMG 35. Fuel drop having the following characteristics:

– water content 3.4%; lower thermal power 39950 kJ/kg.

The results for a drop with a D_0=1.9 mm diameter ignited at T_{fo}=1023 K temperature, T_a=293 K ambient temperature, T_c=300 K fuel temperature and Re=135 are:

– self-ignition delay τ_I = 260 ms; volatile matters burning time τ_v=1070 ms; cenosphere burning time τ_c=1150 ms; volatile matters radiated power E_v=700620 u.c.*; cenosphere radiated power E_c=33870 u.c.* (*u.c- conventional units)

Fig. 3 prezents the combustion oscillogram for a water emulsified droplet RMG 35having the following characteristics:

– water content 7.5 %; lower thermal power 39050 kJ/kg.

The results for a drop with a D_0=1.9 mm diameter ignited at T_{fo}=1023 K temperature, T_a=293 K ambient temperature, T_c=300 K fuel temperature and Re=135 are:

- self-ignition delay τ_I = 288 ms; - volatile matters burning time τ_v=720 ms; - cenosphere burning time τ_c=732 ms;- volatile matters radiated power E_v=549000 u.c.*; - cenosphere radiated power E_c=2530 u.c.*.

Fuel characteristics	Non mulsified fuel	Water/fuel emulsion
Lower thermal power (kJ/kg).	38650 ± 845	36920 ± 845
Water content (%)	0 – 1	7.5 – 10.5
Sulphur content (%)	2.5 – 3.5	2.2 – 3.3
Conradson coke content (%)	12 - 20	11.5 - 19
Density (kg/m^3).	960 - 980	960 - 985
Viscosity at 80^{0}C (^{0}E)	14 - 26	14 – 25.5
Average diameter of water drops (μm)	-	3 – 4.4
Water drops with smaller than 4.8μm diameter content	-	80 – 90.6
Viscosity at injection nozles (^{0}E)	2.5 - 3	3.2 – 5.8

Tab. 1. Fuel characteristics.

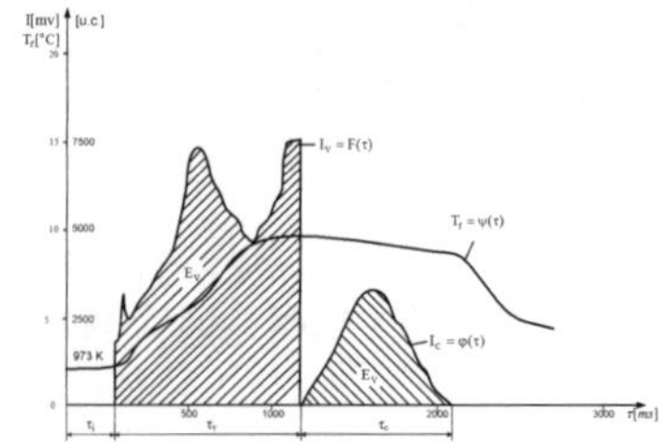

Fig.2. The combustion oscillogram for a RMG 35 fuel drop/water emulsion w = 3.4(%).

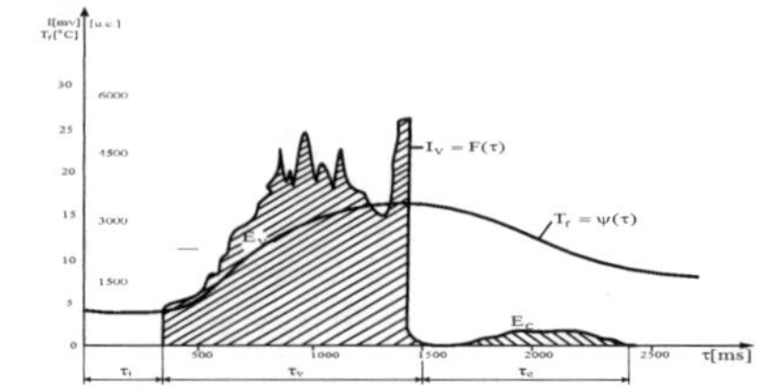

Fig.3. The combustion oscillogram for a RMG 35 fuel drop/water emulsion w = 7.5%.

In both cases the curve $I_v = f(\tau)$ has a gap in the middle part. It is due to water vaporization wich leads to the decrease of combustion temperature. (a potential factor of NO_x reduction). Testing the 3.4 % emulsified fuel the $I_v = f(\tau)$ volatile matters radiated power is bigger than E_c cenosphere radiating power.

By testing the 7.5 % emulsified fuel the radiated power $I_v = f(\tau)$ of volatile matters is a little smaller and the radiating power E_c of cenosphere is much reduced. The reason is the secondary vaporization. The explanation relies on the fast heating of water dispersed inside individual fuel droplets. The internal water droplets undergo a spontaneous vaporization of water, causing a violent change of water droplets into steam. On the other side, the vaporization causes a rapid expansion of surrounding oil droplets, bathing the oil into a large number of small fuel droplets. This process is secondary atomization.

Experiments show that a 7.5% water/residual RMG 35 fuel emulsion has better combustion properties than a 3.4% water/residual RMG 35 fuel emulsion. The explanation relies on secondary atomization effect.

6. CONCLUSIONS

The initial droplet strain subjected to water vapors action breaks into smaller droplets. The I_c low values of E_c and τ_c lead to reduced unburn carbon losses and to decreasing of carbon black (soot) quantity. The increasing of cenosphere burning performance as result of secondary atomization. The introduction of water into the combustion chamber reduces the combustion temperature due to the absortion of energy for vaporization. Humidification can reduce the NO_x emissions.

7. REFERENCES

Ghia V. (1991). *Combustion Graphology of Fuel Oil, Sci. Tech. Electrotehnica Et Energ.,Tome 36*, p 379-396, Bucuresti.

Jinescu, G. (1983). *The hydrodynamic process and special equipments,* Editura Didactică şi Pedagogică, Bucureşti

Law, C. K. (1997). *Combustion Science and Technology*, Vol. 17, p.29-38.

Jianu C. (1996). *The combustion of fuels in sound field,* Ed. U.P., Bucuresti.

Popa, B. & Iscrulescu, V. (1983). *The combustion processes in sound field*, Editura Academiei, Bucureşti.

Annals of DAAAM for 2008 & Proceedings of the 19th International DAAAM Symposium, ISSN 1726-9679
Editor B. Katalinic, Published by DAAAM International, Vienna, Austria 2008
Make Harmony Between Technology and Nature, and Your Mind will Fly Free as a Bird

FILLERS' CONTENT INFLUENCE ON THE MECHANICAL PROPERTIES OF THE GLASS MAT REINFORCED POLYMERIC COMPOSITE

MOTOC LUCA, D[ana] & TEODORESCU, H[oratiu]

Abstract: *The paper focuses on the influence of the different volume fraction of various types of additives on the mechanical performance of polymeric/glass fibre composites. The values are compared with a "bulk" composite structure in order to size the mechanical properties improvement or degradation. The results indicate that for a certain quantity of additives embedded the mechanical properties of the structure do not diminishes and the structures can be used in low-cost engineering applications. Results also indicated that future research towards significant improvements in tensile, compressive or other types of testings should focus on the optimization of fibre strength or conceive of hybride structures, such as the combination among different types of fiber reinforcement architectures.*

Key words: *fiber, glass, composite, mechanical, properties.*

1. INTRODUCTION

The world of thermoplastic composite materials, despite its trend toward nano scale, seems to leave some spaces to what was called 10 years ago "high-performance" advanced composite structures. Nowadays, these types of materials have shifted to "cost-performance" engineering composites, with a widespread use especially due to their advantages like: stampable products due to the polymeric material, excellent price-performance ratio, use of commercially available products, simple configuration of the manufacturing technology, etc.

In the present, ecological concern has resulted in a renewed interest in natural materials and issues such as recyclability and environmental safety are becoming more and more important for introduction of new materials and products. These lead to changes in the research interest, and with respect to the paper's subject to the replace of the non-renewable glass fibres with vegetable fibres (Garkhail, Heijenrath & Peijs, 2000). The data analysis methods, such as statistical investigation, have shifted as well, but still remaining one of the most used tool for experimental vs. theoretical predictions comparisons, even in case of the use of neural network implementation (Esfandiari, 2008).

Fiber composites are very sensitive to the compression loads (Schneider & Lauke, 2007). The behaviour of these structures under compression is much more critical than tensile or bending no matter their application area. This is the reason for which such testings are conducted carefully, the determination of the compressive modulus and strength being a little more difficult to predict in case the measuring devices are not equipped with proper sensitive transducers.

The present work focuses on the mechanical properties of the random E-glass fibres composites, reinforced with different volume fraction of various types of fillers. More specifically, the paper focuses on the experimental values determined in uniaxial tensile and compressive tests as a part of a more extensive work focused on these classes of composite materials (Luca Motoc & Soica, 2008). These experimental results provide useful information for design purpose regarding to the improvements that can be done with respect to their mechanical characteristics, especially when the structure is being employed in aggressive media environments.

The paper contains only a simple comparison of the experimental data retrieved with the theoretical values provided by the most comprehensive model from the literature, and this is due to restrictions implied (Jannerfeldt, et. al., 2001). The introduction of a new reinforcement efficiency factor depending on the fillers content will be left as a further study.

Due to their specificity, the structures approached are referred usually as SMC (e.g. Sheet Moulding Components) even the term may have another meaning, such as Soft Magnetic Materials. It's beyond to the subject of the present paper to debate on the terms' meaning, or to present the individual characteristics of each structure. Nonetheless, it is worth to mention that the Fe particles embedded along with the E-glass fibres mat into the polymeric matrix material exhibit relatively low magnetic properties. The previous mentioned are being on course to be developed, as a natural consequence of these experimental trials along with structures' composition modification.

2. EXPERIMENTAL ANALYSIS

2.1 Materials

In this study were used random, long E-glass fibre mats in a combination with a polymeric matrix from DSM Composite Resins (Switzerland), the latter being choose due to its availability and very good fibre wetting and impregnation properties. The fibres - MultiStrat^{TM} Mat ES 33-0-25 supplied by Johns Manville, USA, were made up from multidirectional continuous E glass fibres, leading to a 60% volume fraction in the composites' panels. The fillers considered were $CaCO_3$ and Fe particles embedded into different volume fraction into the composite's matrix material (5 %, 10 % respectively).

2.2 Testing methods

The samples' mechanical properties were evaluated at room temperature, samples being shaped using a standardized form and subjected to tensile and compression loading conditions, all of them with a speed of 1 mm/min using a LS 100 Plus device from Lloyd Instruments Ltd. The results obtained applying a statistical analysis was considered as corresponding to the mean values.

3. RESULTS AND DISCUSSION

Both testing types lead to something that one may call at a first sight an "awkward" result. A deep looking and recalling the different types of materials involved as additives may solve the dilemma. The comparisons are being done with respect to the simple glass fibres mat and polymeric matrix composite materials, without fillers.

With respect to the tensile tests, seems that the increase of the mineral filler content lead to an improvement of the Young modulus. The same holds in case of the Fe particles embedded into the structure. But unfortunately, these types of composite materials are rather used on compression than traction.

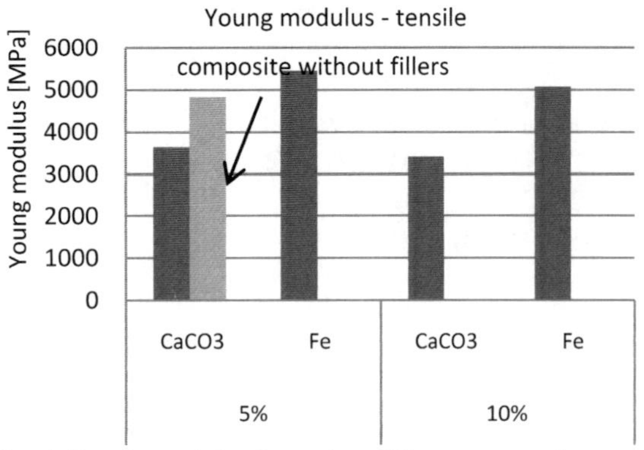

Fig. 1 Tensile modulus for various filler content and type

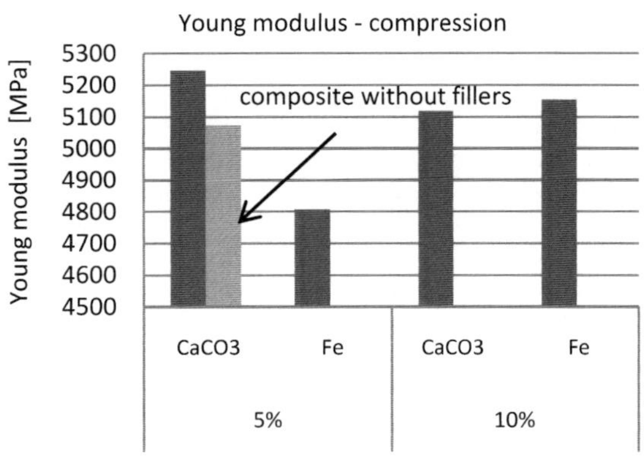

Fig. 2 Compressive modulus for various filler content and type

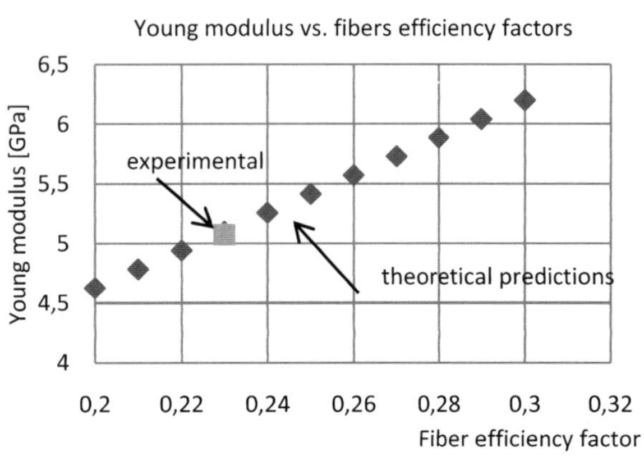

Fig. 3 Theoretical vs. experimental comparison of the Young modulus (tensile) for various fiber efficiency factors

The tensile-strain curves (not presented herein) exhibit some nonlinearity of the response. The nonlinear effect is the result of progressive failure due to various micro-failure modes, such as matrix cracking, fibre debonding and breakage. The composites failed in a brittle manner with a suddenly load drop.

The compressive moduli measured are higher than their tensile values, meaning the fact that no experimental errors were induced into the results. They differ from fillers to fillers, revealing an improvement in case of Fe particles along with the content increase, whereas a decrease for the mineral fillers, the same, along with the increase of their volume fraction content.

In figure 1 is being presented the experimental values retrieved for the Young modulus during the tensile tests conducted, whereas in figure 2 are being presented the results for the compressive tests. Figure 3 represent the variation of the most comprehensive theoretical model from literature used to approximate the stiffness of the composite having almost randomly oriented reinforcements. The model is known as Cox-Krenchel model it's written as, under the assumption of elastic fibers embedded into an elastic matrix material:

$$E_c = (1 - V_f) \cdot E_m + \eta_f \cdot \eta_{fod} \cdot V_f \cdot E_f \qquad (1)$$

where E_c is the modulus of the composite material, V_f the fibre volume fraction, E_m the modulus of the matrix material and E_f the modulus of the fibers. The parameters η_f and η_{fod} are reinforcement efficiency factors related to the fibre aspect ration (length to diameter ratio) and the fibre orientation distributions, respectively. In the general case, both parameters range from 0 to 1. For random reinforced fibers the η_{fod} parameter is set to 0.38 and for a fiber faction volume of 0.6 (this case) seems that the experimental retrieved value for the composite structure without fillers, in tensile tests, lead to an efficiency factor related to the fiber aspect ratio of 0.23 presumming that the length and diameter of the fibers are unknown values. Unfortunately, this theoretical model does not encompass the influence of some supplementary aided filler in the random dispersed fibres polymeric composite structure.

4. CONCLUSIONS

The tensile and compressive mechanical properties modification due to the presence of various types of fillers (e.g. metals, non-metals) embedded into different volume fraction into a random long glass fibres mat polymeric composite materials were studied with the aim of sizing their influence on the investigated parameter.

As expected, there is a significant effect of the fillers' volume fraction on the mechanical properties and no certain anticipated behaviour in the mechanical performance with the increase of this content was found.

The values will be used to tailor other types of composite structures, and special attention will paid due to the fact that a price-competitive character, usually aimed in industry for such materials, do not lead to the improvements of the mechanical performance.

The subject of the paper may seem to a be something that was already done, but it's opening a niche with respect to the needs of modifying the theoretical models in order to encompass the influence of various content of fillers embedded into the classical composite structures to improve their mechanical characteristics.

5. REFERENCES

Esfandiari, A. (2008). The statistical investigation of mechanical properties of PP/natural fibres composites. *Fibers and Polymers*, Vol. 9, No. 1, 48-54.

Garkhail, S. K., Heijenrath, R. W. & Peijs, T. (2000). Mechanical properties of natural-fibre-mat-reinforced thermoplastics based on flax fibres and polypropylene. *Applied Composite Materials*, Vol. 7, 351-372.

Jannerfeldt, G., et. al. (2001). Matrix modification for improved reinforcement effectiveness in polypropylene/glass fibre composites. *Applied Composite Materials*, Vol. 8, 327-332.

Motoc Luca, D. & Soica, A. (2008). Mechanical behaviour of 3-phase polymeric composites subjected to static loading conditions. *Available from:*http://innomet.ttu.ee/daaam.

Schneider K. & Lauke B. (2007). Determination of the compressive properties of fibre-reinforced polymers in the in-plane direction according to ISO 14126. Part 2: A critical investigation of the failure behaviour. *Applied Composite Materials*, Vol. 14, 177-191.

Acknowledgement
The research was supported from grant ID_135, 108/1/10/2007, CNCSIS, Romania.

Annals of DAAAM for 2008 & Proceedings of the 19th International DAAAM Symposium, ISSN 1726-9679
Editor B. Katalinic, Published by DAAAM International, Vienna, Austria 2008
Make Harmony Between Technology and Nature, and Your Mind will Fly Free as a Bird

SIMULATION OF GAS FLOW IN A PRIVATE BURNER MIXING PIPE

MRENES, M[ircea]

***Abstract**: The study goal is to build a numerical model, which allows future optimization of shape and dimension for the mixing pipe of a self-aspirating burner. This optimization is requested by the local industry of burners. The KE-EP turbulence model contained in the Phoenics CFD code is used. The measures for increasing convergence of the calculus process are presented together with practical tips to minimize errors. A body fitted coordinates model is developed, which allows obtaining velocity and pressure distribution, kinetic energy of turbulence variation and streamlines. An elliptic flow zone is identified within the pipe and suggested methods to minimize this area are presented.*

Key words: *simulation, burner, mixing, pipe.*

1. INTRODUCTION

Most common low power burners use a gas-air mixture that is obtained by using the potential energy of gas transformed into kinetic energy in order to aspirate the air from the atmosphere. The mixing process takes place in a pipe, well known in the literature. (Otto, 1989) Numerical modeling is used to obtain the velocity field, the streamlines, the pressure distribution and other characteristic parameters. One studies the shape of gas jet and its modifications produced by different mass flow. The analytic solution can only provide information about the mean velocity of outgoing mixture. The numeric approach gives the velocity distribution in studied sections. Similar studies were performed by other authors on different burner shapes and by using other simulation software. (Bode et al., 2006; Hodor et al., 2006)

2. NUMERICAL METHOD

In order to solve the problem the general fluid flow simulation code called Phoenics was used. The BFC (Body Fitted Coordinates) system allows defining a circular symmetric section with 10° center angle.

The upstream part of the discretization grid is presented in figure 1. The downstream part has 12 equal regions distributed on 100 mm. There are 20 cells on radial axis X and 22 cells on Z-axis that is parallel with the main flow direction. The case is treated in two dimensions so the number of cells in Y direction is one. On radial axis X the dimension of the cells is distributed in accordance with a power law with 1.5 as exponent, in order to allow a higher accuracy in central area where the gas jet is mixing with the air. The same distribution also affects the Z-axis. The fine discretization grid in central zone must be used because of the high velocity gradient, which appears at the interface between the gas jet and the ambient air. Without this measure, the calculus process will have a strong divergent behavior, which cannot be managed even with the most severe relaxation.

The flow is elliptic, steady state and the KE-EP turbulence model is used. (Ferziger & Milovan, 1999; Patankar, 1980) The first order differentiation scheme, linear, hybrid is used. The boundary conditions are of three kinds: inlet, opening and wall. The first entrance is for combustible gas that has the parameters computed with another numerical model. The air inlet velocity is calculated for a gas-air mixture fraction which correspond to the given excess air coefficient. The boundary conditions can be specified in numerical simulation only for individual cells or groups of cells that forms regions. (Miyauchi et al., 1996) In this case, one calculates that the nozzle occupies the first cell and the air-aspirating hole stands from cell 13 to cell 19. The friction between gas and internal walls is activated using a logarithmic function. The number of sweeps is limited to 600, with one iteration on each frame. A severe rate of relaxation is used because of the presence of a high velocity gradient in radial direction. This is a typical case when the divergence usually occurs at the beginning of calculus process.

The variable initialization is a general good security measure for convergence acceleration and computer time decrease. In current case, if this recommendation is not respected, a strong divergence occurs which produces calculus abortion due to the impossibility of evaluation of the differential equations. The prescribed initial values are: P1=10 N/m^2; U1=0,1 m/s; W1=10 m/s; EP=3 J/s; KE=1 J.

In the beginning of the calculus process, a strong divergence occurs, that is why the relaxation factor for x velocity must be intensified to obtain artificial stabilization. After 300 sweeps, when the pressure field is stabilized and the instantaneous velocity curves decrease, the relaxation for x velocity may be released. When the relative error curves decrease and the calculated values for velocity components show a good stability, one can try to increase the relaxation factor in order to increase the convergence. Even if all these observations are applied, an acceptable solution is not obtained at the first run. In this case, the problem must be restarted using as initial values the values obtained from the previous run. Applying this procedure to the present model one obtains relative errors less than 0.1%. The computing time is

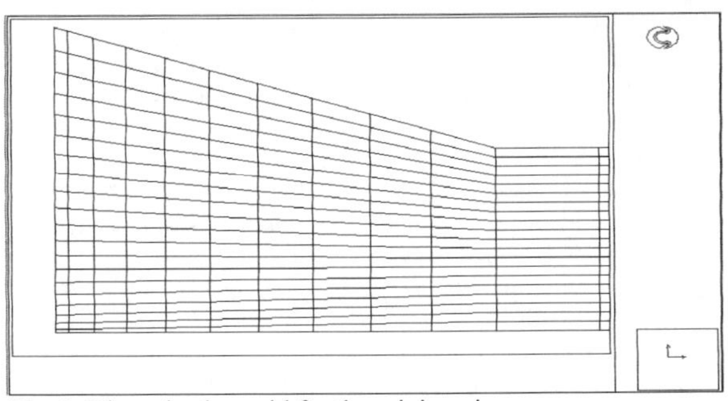

Fig. 1. Discretization grid for the mixing pipe.

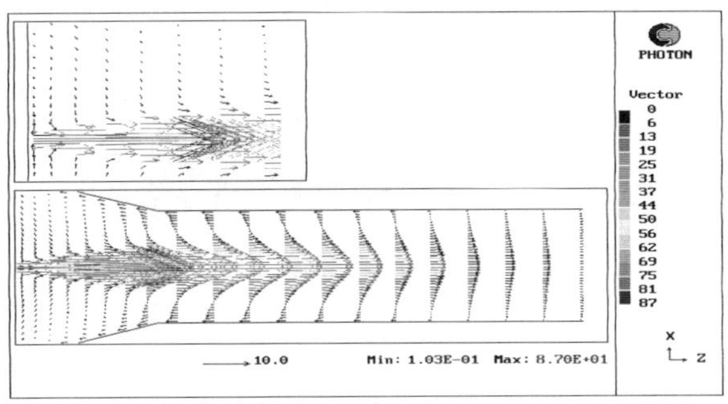

Fig. 2. Velocity field for the mixing pipe

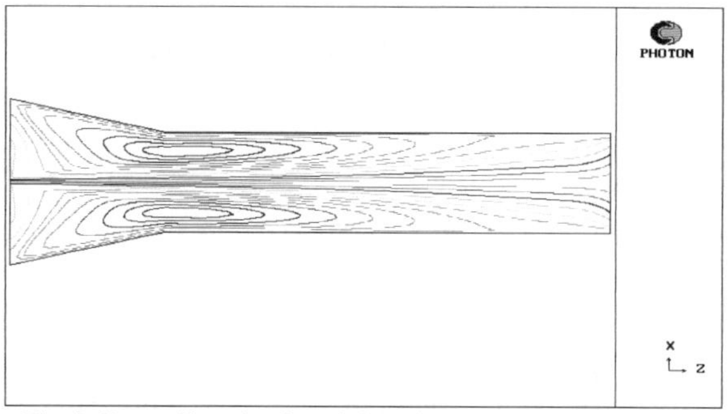

Fig. 3. Streamlines for the mixing pipe.

around 2 minutes on a Pentium 1.60 GHz machine and the convergence is obtained after 800-1000 sweeps.

3. RESULTS AND DISCUSSIONS

The velocity vectors field is presented in figure 2. The inlet methane velocity is 87 m/s but it decreases substantially in the conic zone of the mixing pipe, then it becomes almost uniform as it goes towards the cylindrical zone. Near the walls, there is a gas movement in the opposite way of the main flow direction observed mainly in the upstream cylindrical zone. This phenomenon is obvious when analyzing the streamlines presented in figure 3 where the elliptic flow zone appears clearly. This zone can be reduced by modifying the pipe geometry in order to obtain a lower pressure for air aspiration and to ensure better mixing conditions. The presence of the elliptic flow zone can be observed experimentally by voluntary gas ignition in the mixing pipe. In some conditions, a noisy unstable toroidal flame appears in the elliptic flow zone. The position of this flame depends on the gas mass flow.

The streamlines allow finding out the minimum pipe length for which the main jet touches the internal walls and the velocity becomes almost uniform in the perpendicular section. The figure 2 also shows that the incoming air begins the mixing process as soon as the gas comes out from the nozzle where the air streamlines are almost perpendicular on the pipe axis. It becomes obvious that using a solid disk, which controls the amount of air that enters, can better control the air admission. In this case, the initial direction of air is radial and the energy consumption for ejection decreases.

Figure 4 shows that the pressure has negative values near the air inlet of about 60 Pa. These values have a good correspondence with those experimentally measured. The lowest pressure is present near the central gas jet where the gas speed is higher. Near the downstream zone, a higher pressure occurs which is necessary to allow the mixture motion through the burner body.

Figure 5 shows the kinetic energy of turbulence contours. The higher values are obtained in a cylindrical zone around the central gas jet where a high gas-air relative velocity occurs. As soon as the velocity decreases and the air and gas are mixing, the kinetic energy of turbulence value goes down. The rate of dissipation of

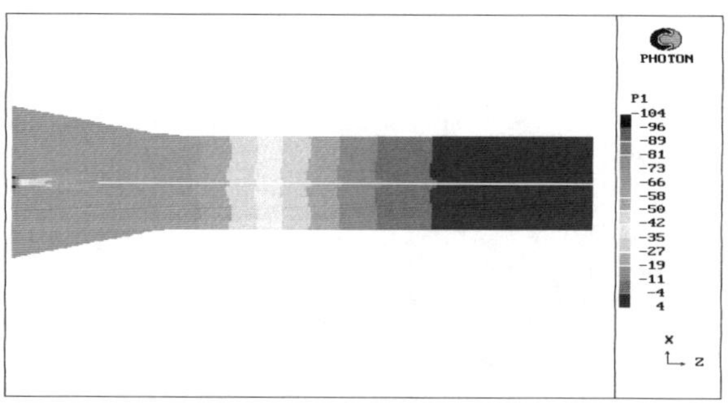

Fig. 4. Pressure field for the mixing pipe.

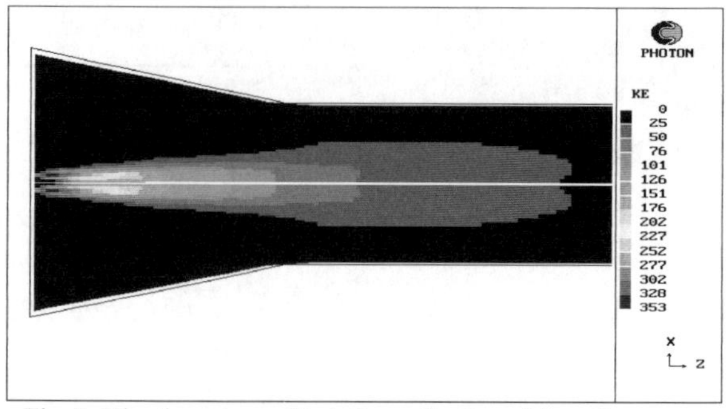

Fig. 5. Kinetic energy of turbulence for the mixing pipe.

the kinetic energy has the same distribution and it has higher values near the central gas jet.

4. CONCLUSIONS

The presented model of gas and airflow in a mixing pipe allows the calculation of velocity vectors, pressures, streamlines and kinetic energy of turbulence in order to determine the characteristics of this assembly. The existence of a toroidal elliptic flow zone that was numerically determined has an important influence on mixture formation. The difficulties involved in solutions convergence and the measures to prevent divergence were presented.

Based on this model the future research will study the self-aspiration process for different pipe dimensions. If some minor modifications are made, the model can also be used for radial air access. Further research will be made for different shapes and different configurations of the mixing pipe. The grid will be refined for better precision. The nozzle for the combustible gas will be modified to achieve different jet shapes. The future goal is to design an improved burner with better mixing conditions at low gas pressure. Numerical study is the best approach because in a short time many configurations can be analyzed. In the next step of the research, the shapes and dimensions of the mixing pipe suggested by designers will be numerically modeled and studied and the optimum solution for each case will be used to build a new prototype.

5. REFERENCES

Bode, F.; Hodor, V.; Madarasan, T. & Unguresan, P. (2006). Investigation on turbulence related to a swirling burner within its back pressure combustion chamber, *Proceedings on 2nd workshop on vortex dominated flows achievements and open problems,* Bernad, S.; Muntean, S.; Susan-Resiga, R. (Ed.), pp. 141-147, Tom 51 (65) Fascicle 3, Special Issue, ISSN 1224-6077, June 30 - July 1, 2006, Bucharest, Romania.

Ferziger, J.H.; Milovan, P. (1999). *Computational methods for fluid dynamics*, Springer, ISBN 3-540-65373-2.

Hodor, V.; Bode, F.; Madarasan, T. & Unguresan, P. (2006). CFD prediction of a burner gasodynamic geometry within its back pressure combustion chamber, *Proceedings on 2nd workshop on vortex dominated flows achievements and open problems,* Bernad, S.; Muntean, S.; Susan-Resiga, R. (Ed.), pp. 135-141, Tom 51 (65) Fascicle 3, Special Issue, ISSN 1224-6077, June 30 - July 1, 2006, Bucharest, Romania.

Miyauchi, T.; Tanhashi, M.; Suzuki, M. (1996). Inflow and Outflow Boundary conditions for Direct Numerical Simulations, *JSME International Journal,* series B, vol. 39 No. 2, pp 305-314, May 1996.

Otto, N. M. (1989). *Infrared Burner*, U.S. Patent No 4878837, Int. Cl. F23D 14/12.

Patankar, S. V. (1980). *Numerical Heat Transfer and Fluid Flow*, Hemisphere Publishing Corporation, ISBN 0-07-048740-5, Washington D.C.

Annals of DAAAM for 2008 & Proceedings of the 19th International DAAAM Symposium, ISSN 1726-9679
Editor B. Katalinic, Published by DAAAM International, Vienna, Austria 2008
Make Harmony Between Technology and Nature, and Your Mind will Fly Free as a Bird

PRINCIPLES OF MECHATRONICAL MODELLING OF MACHINE AGGREGATES

MUDRIK, J[ozef] & NANASI, T[ibor]

Abstract: *Modelling of mechatronic systems requires systematic approach. There is no general agreement about the definition, structure and classification of subsystems of mechatronic equipment. In the present paper general principles of modelling of mechatronic systems representing machine aggregates are outlined. Machine aggregate is treated as complete integration of electronic control subsystem, electric driving subsystem and mechanical working subsystem.*

Key words: *mechatronics, modelling, machine aggregate, time constant, dynamics.*

1. MECHATRONICS IN TECHNOLOGY EQUIPMENTS AND SYSTEMS

The machine aggregate in Fig. 1 represents dynamic system to drive a plant and to control the technological process and displays schematics of a modern controlled system. There are three subsystems:

- the (electric) drive, i.e. the (electro)motor and the gear
- the driven mechanical equipment that represents equipment for electromechanical energy conversion, the actual technological process and the product of the process
- the control system performing an optimal control of the machine aggregate

By the quality of the mechatronic system, i.e. by the static or dynamic characteristics of the subsystems are assigned the productivity of the process, the quality (operational stability, accuracy, speed) of the process and the product as well as the static/dynamic loading of the aggregate mechanical, electrotechnical and electronical units. Also the level and type of mechanical load, accuracy of gears, characteristics of drive and control, overall quality etc. obviously influence the motor and control performance conditions (Mudrik, 2000; Mudrik & Naď 2007).

2. THE MACHINE AGGREGATE AS A MECHATRONIC SYSTEM

A machine aggregate and its intended control functions respecting the mutual energetic interaction of subsystems is a mechatronic system consisting of

- supply and power converter of some kind
- electric AC or DC drive with proper kind of electromotor
- a plant subsystem
- control electronics (analog or digital, if digital than programmable microcontroller system).

All but the last subsystems are power subsystems. The control subsystem is an information subsystem. Hence, from another point of view, a mechatronic system is an integration of following subsystems:

a power electromechanic system of machine aggregate and plant generating the torques and forces needed by the process, under prescribed speed, position etc.

b energy supplying power electronic system modifying the electrical energy constant parameters of the primary sources to values of the converter postulated by the process

c information control electronics.

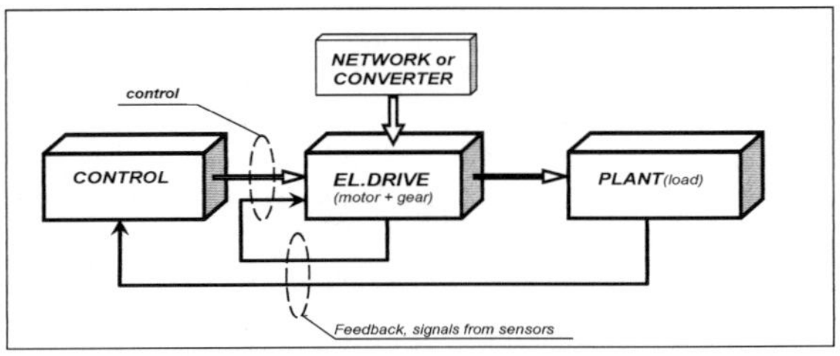

Fig. 1. Block diagram of the machine aggregate.

The power electromechanic system together with the energy supplying power electronic system performs an electromechanical energy conversion. The goal is an optimal control with respect to the technological process or the dynamics of the aggregate as a whole.

Machine aggregates with controlled drives often need a multi-level hierarchical control. In the basic level of, say a speed system, the angular speed of the motor/drive is controlled by a speed controller, perhaps with the aid of a subsidiary current controller (Kratochvíl & Březina 2005).

The control of both current and speed loop controllers can be designed starting with the current controller at the most internal loop in a number of ways. To design the position control systems, the speed control system with speed feedback loop designed above becomes subsidiary to a position control loop. Cascade, parallel and feedback groupings of controllers are available, just to refer to some of the design procedures. In the higher control levels of technology control, operational quantities/parameters are managed, with the goal to keep the conditions of the process optimal. The system approach finds our mechatronic aggregate to be merely a subsystem, which is internally controlled part of the whole system, see Fig. 2.

The mechatronic system MS is an integrated system consisting of three subsystems: electronic control subsystem ECS, electric driving subsystem EDS and mechanical working subsystem MWS

$$MS = ECS + EDS + MWS.$$

The integration covers design, construction, operation and maintenance with respect to optimal static as well as dynamic

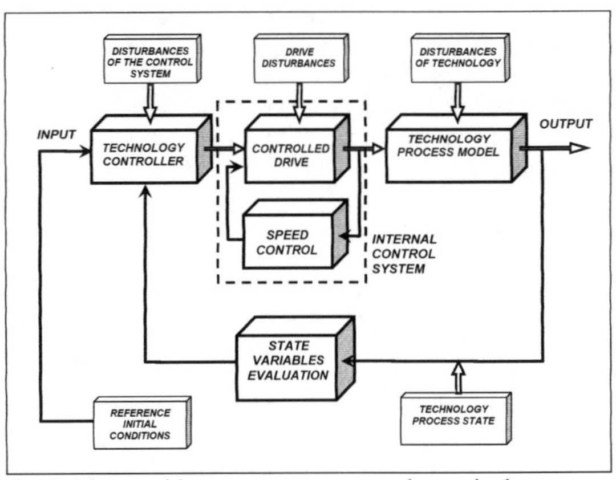

Fig. 2. The machine aggregate as a mechatronical system.

attributes of the MS:

- ECS performs the optimal control of the MS heading the MS objectives: product or/and process
- EDS is an energy electric-to-mechanic converter
- MWS executes the final production or process, following the MS objectives.

Any mechatronic system is not just a marriage of electrical and mechanical systems and is more than just a control system, it is a complete integration of all of them.

3. MODELLING THE MACHINE AGGREGATE

Transition from a real machine aggregate to its model is only possible if ideas on the goal of the system, on the states of the system, on the foregoing analysis, on specification of characteristic features for the designed functions of the system are taken into the consideration. There are lot of cases, when the model and simulation are the only way to get some knowledge. Two main aspects of the approach to the task are:

- Creating a proper model (Modelling)
- Working properly with the model (Simulation).

Writing the mathematical model in a mathematical form is the first step and is deduced from the proper theory of the investigated subject, namely from known physical laws from all the relevant branches.

The next step is writing the model code as a package of main program and supporting subprograms, using the results of the first step. After the verification of the main program and the whole program package, simulation experiments follow as the step three. The fourth step is verification of the results of the previous steps, probably using some kind of experimentation with physical models, or experimentation with the real object or process modelled.

Synthesis of computational and experimental methods is the basis for modern experimental work and is the most effective method for analysis and synthesis of mechatronical systems.

The whole sequence to create a machine aggregate mathematical model as described above has three parts:

1. Modelling of energetic interactions among the subsystems
   - Formulate in natural language and describe mathematically all the individual construction parts of the machine aggregate and interactions among them, Fig. 1.
   - Create the main program based on the previous point and the subprograms for the main program.
   - Perform identification measurements on the model, i.e. perform the simulation experiments and postprocess them. Make statement, how truly the model substitutes the real object/process. Understand the numbers, tables, graphs etc. generated, describe the results in the natural language.
   - Make an expert opinion on the model. The goal is to get the best possible working and simple model of the reality. The quality of the model depends on the quality definition, as an example the quality might be a compromise between the best possible stability of the model, its minimal computational time and accuracy of results.
2. Reducing the model to formulate the control laws
   - Create a linearized model of control, say, for the machine aggregate with working point shifted within a small displacement zone.
   - Create a linearized model in the working part of the torque-speed characteristics.
   - Create a nonlinear model by the constants assigning method for the nonlinear model using the regression principle.
3. Creating the control law. Verification within the validity area of the reduced model is based on simulation experiments.

Numerical or analytical simulation is optional, depends on the CAMS program (program system, package) in use. The MATLAB can work analytically but perhaps its main domain is the discrete simulation. The MATHEMATICA works analytically. As for the dynamic modelling and simulation, the key view is whether the integration of differential equations describing the mechatronical system is performed by symbolic or numerical integration.

4. MECHATRONIC TASKS IN THE MACHINE AGGREGATE DYNAMICS

The recent works in the machine dynamics deal mainly with discrete parameters, lumped mechanical system models and also with FEM models. Let us have a look into the recent decades to display what is to be done in next ones. The recent works of experts in drive systems treated in-depth the phenomenon of their own, while the mechanics of the plant was treated with a reasonable overlap of both branches. The same is valid for the experts in mechanics. As for experts in control, this branch is rich in experts and literature, both in analogue and digital types of control. Even in well done works from electrical (controlled) drives the mechanical subsystem has been assessed and modelled with reduction to single or two rotating bodies, using an ordinary differential equation of the 1st or 2nd order (Kratochvíl et al., 2005).

Some very specific problems emerge due to the above conventional approach to predominantly mechatronical problems. One of them is the time constant: The shortest time constants of a common control subsystem may be in units of 10^{-3} sec. The shortest time constant of a electrical part of the drives is within orders 10^{-2} - 10^{-1} sec, while for usual mechanical subsystem may be within orders 10^{0} -10^{1} sec. This difference is the source of compatibility problems even for mechatronical system with the most trivial control subsystems. Future research and models have to develop means to cover the mismatch at time constants of physically different subsystems.

5. CONCLUSION

The structure of the machine aggregate interpreted as mechatronic system was analyzed, respecting the interdisciplinary nature of individual steps leading to final global model of mechatronic system as well as respecting the interactions between mechanical and control parts of the drive and the parameters of the plant.

The authors greately acknowledge the financial support for this work by the research projects KEGA-2/4154/06 and AV-4/0102/06.

6. REFERENCES

Kratochvíl, C. & Březina, T. (2005). Complex Driving Systems, In: *Simulation Modelling of Mechatronic Systems*, Březina T. (Ed.), pp. 111-119, Brno UT, ISBN 80-214-3144-X, Brno.

Kratochvíl, C. & Březina, T. (2005). Modelling and Analysis of Dynamic Properties of Small Electromechanical Drive System I, *Proceedings of the 9thInternational Seminar of Applied Mechanics,* Zeszyty naukove Politechniky Slaskej, No. 28/5, pp. 89-98,Wisla, Poland.

Kratochvíl, C.; Březina, T. & Procházka, F. (2005). Modelling and Analysis of Dynamic Properties of Small Electromechanical Drive System II, *Engineering Mechanics*, Vol. 12, No. A1, pp. 61-69.

Mudrik, J. (2000). Resonant Phenomena in Electromechanical System with Gearings, *Engineering Mechanics*, Vol. 7, No. 4, pp. 299-308, ISSN 1802-1484.

Mudrik, J. & Naď, M. (2007). Mechatronical Approach to Machine Dynamics, In: *International Symposium – Mechatronika 2007*, pp. 99-102, Trenčianske Teplice, ISBN 80-214-0879-9.

Annals of DAAAM for 2008 & Proceedings of the 19th International DAAAM Symposium, ISSN 1726-9679
Editor B. Katalinic, Published by DAAAM International, Vienna, Austria 2008
Make Harmony Between Technology and Nature, and Your Mind will Fly Free as a Bird

LOGISTICS OF MATERIAL FLOW IN FLEXIBLE MANUFACTURING AND ASSEMBLY CELL

MUDRIKOVA, A[ndrea]; HRUSKOVA, E[rika] & Velisek, K[arol]

Abstract: *Complex of logistic forms a logistic system. These activities are majority and they have to be done that workpiece will be on the suitable position with respect to optimum conditions. Each of logistic actions as transportation, packing, material manipulation, storage, energic and information flow make provision for ergonomic requests and influence manufacturing process like complex. Field of Institute of Manufacturing System and Applied Mechanics interest is building of flexible manufacturing – assembly cell with defined areas conformable with logistics principles and actions.*

Key words: *logistics, material flow, assembly, manufacturing cell.*

1. INTRODUCTION

Automatisation has important vocation in the development of machine production. It brought new quality to production, new level of productivity, distinctive increasing of flexibility in production whereupon final increasing of quality and efficiency. Yet 50th years ago was evident that biggest potential of manufacturing was in promt reaction possibility of market requirements. Idea of constantly production costs decreasing and of making easier the workpiece production to different, led to development of flexible manufacturing system.

Last years the cell production is one of the most important studies of production organisation. By cell production is created relations between machines, there are economize production time and space. Manufacturing cells connect and study many advantages from other kinds of production as f. e. from mass production it is production frequency and from part production it is mainly production flexibility.

2. FLEXIBLE MANUFACTURING SYSTEMS

High elaborateness of technological process is characteristic for production and their secrecy. Conception of production automatisation is different by production characteristic. Automatised manufacturing systems are manufacturing devices with variant automation level. It means level with manufacturing or not manufacturing activities and with variant level of particular subsystems integration. Flexible manufacturing system (FMS) is exerted name for automatised manufacturing systems in repetitive production where is necessary high flexibility rate of manufacturing programme variance. FMS - Flexible Manufacturing System must provides for firstly follow three activities:

Manufacturing – concrete technological proces,

Directing – synchronizing of mainly and accessory activities, functions of all devices that are providing for workpiece production inside FMS. One part of this area is controlling of:

- Production accuracy – particular trend of workpiece accuracy is predicted on the base of periodical enumerations
- Monitoring and diagnostic of transport and storage – resulting of storage places, of transport ways for workpieces, tools, measuring tools and accessories that are necessary for manufacturing process. (Zvolenský et al., 2008).

By organisation structure the flexible manufacturing systems are divided into:

- Flexible manufacturing cell,
- Flexible automatised line,
- Flexible automatised operation,
- Flexible automatised plant.

Flexible manufacturing cells content technological device with program directing with equipments of technological process automation that works autonomous. There are realised various cycles with connecting possibility to higher system. Workpiece containers with technological palletes, fixture devices, devices for changing tools, for waste removal and diagnostics are the automation accessories. Flexible manufacturing cell is associated with industry robot while creates robotised technological complex. Present cells are constructed as manufacturing or as assembly. They are designed with using external manipulation devices for manipulation with semifinished products or workpieces. (Zvolenský & Javorová, 2006).

3. FLEXIBLE MANUFACTURING CELL IN INSTITUTE OF MANUFACTURING SYSTEM AND APPLIED MECHANICS

Prototype of flexible manufacturing cell Fig. 1that is in Institute of Manufacturing System and Applied Mechanics presented multifunctional manufacturing – assembly device introduces new philosophy in creating new complex production. Different from present flexible manufacturing and assembly system is in the operation area. In this operation space the final product (workpiece) is not only manufactured but assemblied too. Final workpiece is assemblied from workpieces that are manufactured in system and from workpieces that are transported to system from backround. Important innovation symbol of this concipated manufacturing – assembly system is that for manipulation and assembly is not use external industrial robot. Changeable grippers for manipulation and assembly are positioned at structure last unit of device – in this case on vertical axis parallel with operatined spindle. Realisation and utilisation of this manufacturing – assembly system conception designed as flexible manufacturing – assembly cell brings smaller built surface, lower economics and higher coefficient of manufacturing device utilisation. (Košťál et al., 2005)

It contents of these base parts:

- shelf storage device,
- rotary feeding device,
- multifunctional industry robot,
- rotary fixture device.

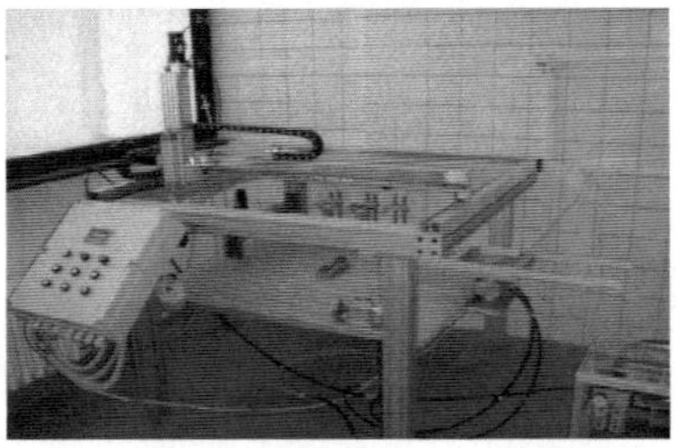

Fig. 1 Flexible manufacturing – assembly cell

Fig. 2 Shelf storage device

Shelf storage device (Fig. **2**) is used for storage semifinished workpieces, particular workpieces and final product. By shelf storage device particular palletes are chosen and positioned to rotary feeding device. This device provides for palettes feeding into and from operational space of multifunctional industry robot. By robot are realised manufacturing and assembly operations. Manufacturing system is built on the table construction with serious kinematics that contains three electric powers. Frame is built of aluminium profiles. Three linear systems create carthesian operational place. The device is directed by industry robot. (Horváth & Danišová, 2008)

For manipulation and assembly are using changeable grippers positioned at the last element. System using of efector change increases productivity and universality of complete system.

Manufacturing process of workpiece consists five periods:

- Storage of semifinished products and its transport to manufacturing device,
- Semifinished product machining to final workpiece,
- Manipulation with particular elements,
- Assembly of particular elements to final product,
- Manipulation with final product and its storage before expedition. (Košťál, 2006)

4. MATERIAL FLOW

Generaly for the resultfull running of manufacturing process is important to provide for continuous flow the material, raw material, semifinished products, and etc. By term *material flow* we mean organised material moving from production input throught all periods of storage, transport and manufacturing processes to output of final products. Analysis of material flow is one of the main points of the final analysis of the manipulation with materials. Next, main proposition of complete material flow analysis is for the process of material moving too. Kind, number, capacity, weight, shape and dimensions influence the kind of manipulation and defined the requirements for manipulation, transport and storage of manipulated material and product.

Principles of racional manipulation with material:

- To create striaght, as short as possible transport ways,
- To design manufacturing process and solution of disposition with reference to optimal material flow,
- If it is possible, to use gravitation for material moving,
- To make optimal quantity, sizes and weights of transported units,
- To design all manipulation, transport and storage operations as effectively as possible.
- To resolve manipulation completely it means from the view of final factory and in relation with out – plant material flows,
- To use possibility of type manipulation device,
- To speculate about optimal using of transport and manipulation devices (minimum 60%),
- To respect ergonomic and syfety requirements.

For the description of material and information flow in the production of this flexible manufacturing cell must be defined it means to choose type range of the workpiece.

5. CONCLUSION

Failure – free, effectively and synoptical operations of material flow are provided for information flow. Information flow is depended on rationalization of material flow. It is in central position in logistics. Before all kinds of material movement could be and follows information. Every this steps are necessary to information flow attend and brings that information on that is possible to adopt a decisions. Mojority of problem resolution is in the optimalisation of material and information flow and their application into highest level of automation and mechanisation into manipulation and transport processes. In the material flow projecting, here is necessary to know that aim of the project is not transporting and storaging material because the price of these operations is high and the material rate is not higher by them.

Paper was created thanks to national project VEGA 1/3193/06 Multifunctional manufacturing – assembly cell.

6. REFERENCES

Horváth, Š.; Danišová, N. (2008). Power elements of flexible manufacturing and assembling systems, *Proceedings of International Doctoral Seminar 2008*, pp. 130 – 136, Smolenice, Slovakia, 18. – 20. May 2008, ISBN 978 – 80 – 8096 – 058 – 2, AlumniPress Trnava

Košťál, P.; Javorová, A. & Zvolenský, R. (2005). Flexible assembly system with automated tool changing, *Proceedings of 6th International conference Systems reducing hazards in operational processes of Machines and Equipment.* KOMTECH 2005, pp. 155-162, ISBN 83 – 922 681 – 4 - 8, Poland, 15. – 17. November 2005, Komdruk – Komag Sp. z.o. o., Zakopane

Košťál, P. (2006). Clamping device for manufacturing cell. - VEGA 1/3164/06. *In: Materials Science and Technology* [online]. - ISSN 1335-9053. - Roč.6, č. 3

Zvolenský, R.; Javorová, A. (2006). Flexible manufacturing and assembly cell with automated tool changing system. *In: RaDMI 2006: Proceedings on CD-ROM. International Conference.* Budva, Montenegro, 13-17.September 2006. - Trstenik: High Technical Mechanical School of Trstenik, 2006. - ISBN 86-83803-21-X. - pp. 1-6

Zvolenský, R.; Ružarovský, R.; Košťál, P.(2008). Design of automated disassembly devices. In: *MicroCAD 2008: International Scientific Conference. [Miskolc]*, 22-23 March 2008. Section K: Machine and Construction Design. - Miskolc : University of Miskolc, 2008. - ISBN 978-963-661-812-4 Ö. - ISBN 978-963-661-822-3. -pp. 51-56

Annals of DAAAM for 2008 & Proceedings of the 19th International DAAAM Symposium, ISSN 1726-9679
Editor B. Katalinic, Published by DAAAM International, Vienna, Austria 2008
Make Harmony Between Technology and Nature, and Your Mind will Fly Free as a Bird

EXAMINATION OF BOUNDARY LUBRICATION FRICTIONAL COEFFICIENT

MUMINOVIC, A[dil] & COLIC, M[irsad]

***Abstract**: Contact bodies such as gears and bearings have contact, if not being completely separated by the lubricant's oil film. Disadvantage of lubricants or undesirable lubrication conditions due to large asperity is incomplete separation of two solids. Wear of the gears contact line points at impossibility of permanent separation of two solid bodies with lubricant's oil film. Tribology process that leads to contact surface wear, is especially expressed at worm gears in consequence of higher sliding share. Energy loss and surface wear are exponentiated with sliding share, which causes essential thermal loads in oil film of lubricated contact. One of most important expectations for power transmission lubricants is viscosity stability of lubricant, which depends on temperature. Beside other working conditions parameters, lubricant used has very important influence on boundary lubrication frictional coefficient. For examination conducted in this research we used, as per origination: mineral and synthetic substance oils. Our goal was to determinate functionality between frictional coefficient and working conditions (lubricants, asperity, loads).*
***Key words:** contact, lubricant, friction, wear*

1. INTRODUCTION

Contributed paper gives experimental apptroach to defining the boundary lubrication frictional coefficient, ie. a component of solid bodies mechanical friction coefficient at conditions of boundary lubrication. Contact of two solid bodies, during gears contact line sliding represents a component of lubricated solids friction across surface roughness peaks ie. boundary friction lubrication. This friction is taken into account while calculating frictional coefficient in area of mixed thermo-elasto-hydrodinamic (TEHD) lubrication theory (Hu et al 2001, Wang et al 2002). Frictional force determination, as outcome of two solids contact requires knowledge of contact surfaces frictional coefficient μ_{KT} (Gelinck & Schipper 2000). Value of boundary lubrication frictional coefficient, due to this kind of friction mechanical complexity, is acquirable in experiment only and in especial contact working conditions. Measuring of boundary lubrication frictional coefficient is done at velocities as low as 2-5 min-1. According (Bouche 1991) hydrodynamic pressure cannot be developed in oil film, and is disregarded at such velocities. According to measuring results (Bouche 1991) presented earlier, temperature has no significant influence on boundary lubrication frictional coefficient for lubricants and combination of materials used in this research, and large asperity enlarges the value of frictional coefficient. According to some authors (Bouche 1991, Opalic 1984) boundary lubrication frictional coefficient is decreasing in the course of sliding velocity increase. Sliding velocity influence on boundary lubrication frictional coefficient was not analyzed for experiments conducted in this research, since rotation velocity enhance implies also sliding velocity enhance, as well as enhancing of summary velocity in contact, which constructs substantial lubrication pressure. Accordingly, measurement results of boundary lubrication frictional coefficient presented here do not take influence of sliding velocity into account. This disregard admits small mistakes, since boundary lubrication friction share during increase of velocity becomes insignificant.

2. EXPERIMENT EXAMINATION

As mentioned earlier, calculating frictional coefficient in area of mixed lubrication requires knowledge of contact surfaces boundary frictional coefficient. Boundary lubrication frictional coefficient values are acquired in experiments conducted on model device developed by (Opalic 1984) according to Timken in Laboratory for machine elements at Faculty of mechanical engineering in Zagreb, Croatia. Figure 1 shows measuring device diagram. Rotation of disc 1 manufactured of quenched temper steel C.4320 and abraded to surface roughness given in table 1, effectuates relative sliding velocity to motionless test-tube specimen 2, manufactured from brass CuSn12. Normal force at the contact place of disc 1 and specimen 2 is substantiated through unequal poles arms system 3 and bob 4. Disc 1 is driven by electro-motor 5 where the rotation velocity is calibrated by the transistor converter 6 for the frequency regulation "Mitsubishi" (type A500); number of rotations is measured by nonconcact digital tachymeter "Sanwa" (type SE-100). Lubrication supply to the place of contact is accomplished through delivery injector 9 by use of the pump 8, while the regulation of lubricant's temperature was conducted with two thermostats (type TR-10) in heating and cooling subsystem of lubricant 10. Temperature control at lubricant intake out of delivery injector was conducted with a contact thermocouple 12 "Technoterm" with accuracy of ±0.5%, which was connected to the measurement amplifier 13 and the computer 14. Friction coefficient was determined on the basis of measured value of torsion moment on measurement axle tree 11 (friction force) and known value of normal force on test disc 1 in the place of contact.

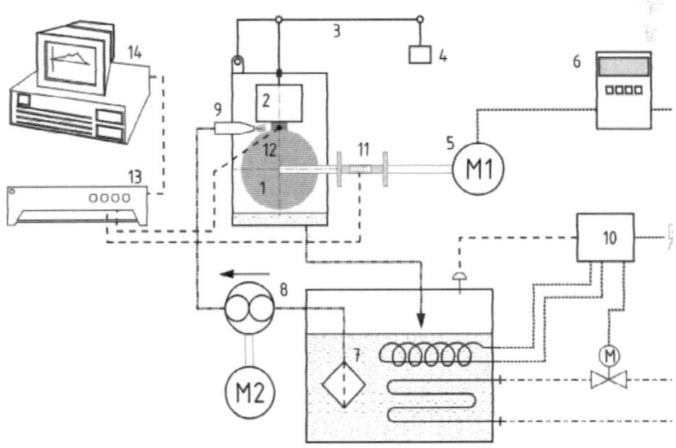

Fig.1. Diagram of measuring device developed by Opalić according to Timken, (1- test disc, 2- tube specimen with its hold, 3- arm system, 4- bob, 5- electro-motor of the measurement axle tree, 6- transistor converter, 7- oil cleanser, 8- oil pump, 9- injector, 10- lubricant heating and cooling subsystem, 11- measurement axle tree with strain gauges, 12- thermocouple, 13- multichannel amplifier, 14- personal computer)

Torsion moment measurement was conducted by tension stress measurement method, with use of strain gauges 11 connected to measurement amplifier 13 and computer 14. Accuracy of torsion moment was ±0.8%.

3. MEASUREMENT RESULTS ANALYSIS

Table 1 expresses measurement plan and conditions of experiment conducted. Measurement was effectuated for per origination two lubricant kinds – mineral and synthetic, as well as two surface roughness qualities for the loads applied.

Disc's radius R* (mm)	Disc's rotation velocity n_d (min-1)	Lubricant 's temperature υ_u (°C)	Surface roughness (mμ)			Maximum Hertzian contact pressure P_0 (MPa)	Measurement of solid bodies friction coefficient μ_{KT}	
			Ra_1	Rp_1	Rp_1+Rp_2			
25	5	50	0,3	0,71	2,2	65 210 300	Mineral matter lubricant	Synthetic matter lubricant
			0,6	1,38	3,6			

Table 1. Conducted experiment conditions.

Figure 2 shows experimental funcionality of boundary lubrication frictional coefficients for two solid bodies μ_{KT} and two lubricant kinds (mineral and synthetic) in terms of maximum Hertzian contact pressure p_0, for disc velocity n_d=5 min^{-1} and surface roughness quality with parameters shown on figure and table 1 (CLA surface roughness R_a, where the sum of disc 1 and motionless test-tube specimen 2 maximum surface roughnesses is $R_{p1} + R_{p2}$). Funcionality appointed between solid bodies frictional coefficients and maximum pressure in terms of working conditions conducted during experiment can be transferred into TEHD model`s system of equations for mixed lubrication and solve The Energy Equation.

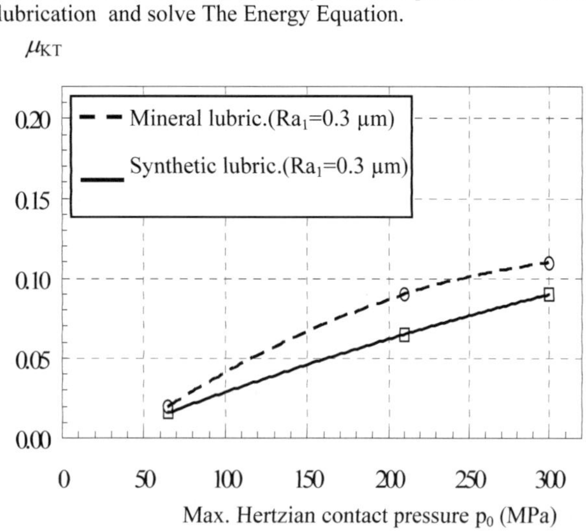

Fig. 2. Friction coefficient of solid bodies (Ra_1=0.3 μm; material CuSn12 / C.4320, nd = 5 min-1, temperature of lubricant ϑu = 50 °C, Disc radius R* =20 mm)

Figure 3 shows measured boundary lubrication frictional coefficients for two solid bodies μ_{KT} and two kinds of lubricants (mineral and synthetic) in terms of working conditions given on figure 1 and surface roughness quality with following parameters: R_a= 0.6 μm, where the sum of disc 1 and motionless test-tube specimen 2 maximum surface roughnesses is $R_{p1}+ R_{p2}$ = 3.6 μm. Functionality between solid bodies frictional coefficients and maximum pressure established in terms of working conditions expresses influence between lubricant used and value of boundary lubrication frictional coefficient.

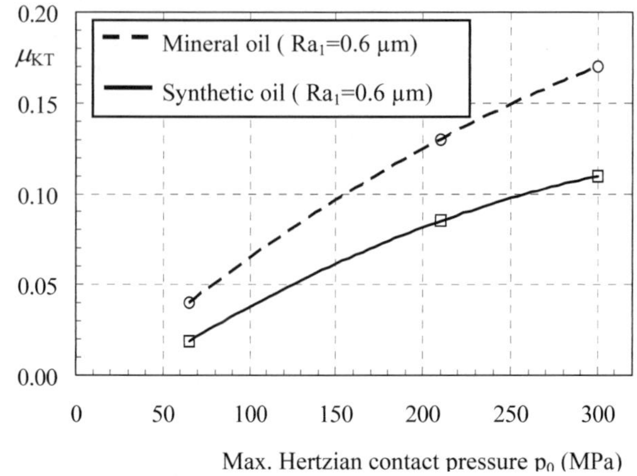

Fig. 3. Friction coefficient of solid bodies (Ra_1=0.6 μm; CuSn12 / C.4320, rotation velocity n_d = 5 min^{-1}; oil temperature ϑ_u = 50 °C, disc radius R^* =20 mm)

4. CONCLUSION

Measurement of solid bodies boundary lubrication frictional coefficient for both kinds of oils used shows following well known lawfullnesses:

- Load increase ascends value of boundary lubrication frictional coefficients for solid bodies.
- Synthetic oil produces lower values of boundary lubrication frictional coefficients in relation to mineral oil. This is especially expressed while parameters defining surface roughness quality ascends (Figure 2).
- Lower surface asperity of the disc, clearly leads to lower frictional coefficient of solid bodies.
- Friction coefficient of solid bodies significantly depends on load applied. This is explained by the fact that even during low disc´s rotation velocities for example nd=5 min^{-1} oil film can be formed if loads applied on solid bodies are low, which clearly contributes to minimizing boundary lubrication frictional coefficient.

5. REFERENCES

Bouche, B. (1991). Reibungszahlen von chneckengetrie-beverzahnungen im Mischreib-ungsgebiet (Worm gears boundary coefficient during mixed lubrication), Dissertation, Bochum,

Gelinck, E. R. M. & Schipper, D. J. (2000). Calculation of Stribeck curves for line contacts, *Tribology International*, 33, 175-181,

Hu, Y. ; Wang, H.; Wang, W. & Dong, Z. (2001). A computer model of mixed lubrication in point contacts, *Tribology International*, 34, 65-73

Opalic, M. (1984). Prilog istrazivanju opteretivosti bokova puznih kola puznih prenosnika (Addition to examination of worm gears teeth flanks load capacity), Dissertation, Zagreb, 1984.

Wang,Y.; Zhang, C. ; Wang, Q. J. & Lin, C. (2002). A mixed TEHD analysis and experiment of journal bearings under severe operating conditions, *Tribology International*, 35, 395-407

Annals of DAAAM for 2008 & Proceedings of the 19th International DAAAM Symposium, ISSN 1726-9679
Editor B. Katalinic, Published by DAAAM International, Vienna, Austria 2008
Make Harmony Between Technology and Nature, and Your Mind will Fly Free as a Bird
2nd DAAAM-ICAT

A SOFT COMPUTING-BASED MODEL FOR RM SELECTION AND COST ESTIMATION

MUNGUIA, J[avier]

Abstract: *Soft-computing (SC) denotes the class of Artificial Intelligence technologies such as: fuzzy logic (FL), neural networks (NN), relational databases and rule-based decision making, which are used to deal with uncertainty and fuzziness on a wide variety of problems like decision making for process selection. This paper depicts a SC-based system for the selection of Rapid Manufacturing (RM) processes according to 2 main dimensions: general feasibility (FL-based) and cost estimation (NN-based). The aim is to asses RM feasibility according to a set of user-defined conditions.*
Keywords: *Artificial intelligence, Rapid Manufacturing selection*

1. INTRODUCTION

Rapid Manufacturing refers to those technologies capable of producing a final part directly from a CAD source, via layer-wise deposition (Bibb, et al. 1999). The selection of the most suitable RM process is dependent on factors such as build envelope, accuracy, material, build speed and other machine-related parameters. With around 22 manufacturers marketing between 40 and 60 models of RP machines (Wohlers 2006), the development of computer-based selection systems has been a recurrent topic in RP related literature (Masood & Soo 2002).

2. PREVIOUS ATTEMPTS FOR RP SELECTION

Systems based on relational databases were introduced by Muller (1996), Campbell and Bernie (1996). The former was rather focused on finding the best materials- process combination expressed in percentages by applying the so called "Benefit Value Analysis". Later on Phillipson (1997) from Arizona State University introduced the RP Advisor also based on relational databases applying MS Access.

Lan et.al (2005) introduced the Fuzzy synthetic evaluation method for the treatment of quantitative and quantitative data for comparison and weighting purposes by splitting criteria factors into: Technology, Geometry, Performance, Economy and Productivity, considering a couple of factors for each process in order to compare them. Nevertheless the final result provides a RP ranking process similar to previous systems which lack of further information on costs, materials or build time.

The Spanish Rapid Manufacturing Association ASERM(2006) launched a web-based RM selector which distinguishes between two options: the use of additive processes for prototype construction or the use of rapid tooling methods for final parts and pre-series. However the system still lacks of costing, materials information and a more comprehensive input since it's based basically in generic data regarding the part's build volume, overall tolerance and type of material.

When it comes to selecting a RM process for a specific application, conventional heuristically rule-based systems fall short in providing an effective yet feasible solution for a given need. Furthermore, none of the previous systems provides a cost estimation or economic assessment to aid during the selection of the most appropriate process. The model presented in this paper aims at filling this gap for a more efficient selection

3. THE PROPOSED MODEL

The Rapid Manufacturing advice system (RMADS) is intended to support designers during the earlier design stages to asses the possibility of using RM as an alternative for production. The model is spitted into 2 modules:

•Module 1. General feasibility evaluation
•Module 2. RM cost estimation

3.1 Module 1

Similarly to Lan (2005) the evaluation follows a Database-Decision making- Final ranking approach. The module makes use of the following elements:

System interface: It supports numerical values or linguistic terms which are split into: Quantitative (Q), Normalized qualitative (NQ) and Individual Qualitative (IQ) (Figure 1). As a result of the modularity each activated option will lead to the development of a vector with values normalized to the range 0-1 depending on the parameter selected.

Expert system and its Rule-base: The expert's knowledge is represented as a series of IF-THEN –ELSE statements. A total of 500 rules are executed when all the options of the system are activated, however less options may be selected.

Fuzzy inference system: This stage consists of representing all the goals and constraints as fuzzy sets (Zimmerman 1991) obtained by mapping the membership values based on the selection of linguistic variable by the user (Mahesh 2005). Depending on the variable chosen (e.g. Surface Roughness, Tolerances, Mechanical properties, etc) the user will be presented a number of suitable options (e.g. high/average/low, etc.). For the RMAD system a range of three grades is used for each parameter.

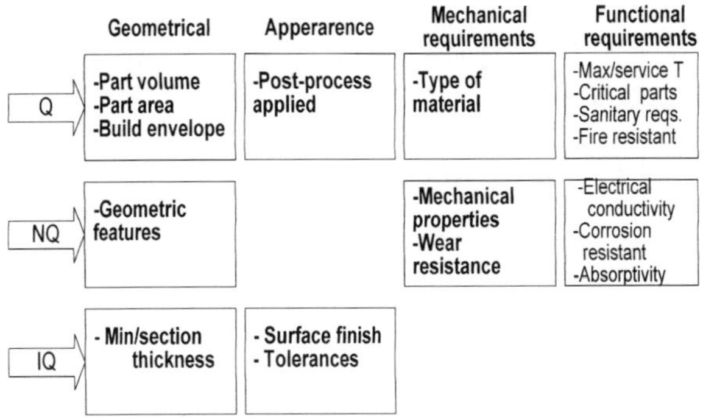

Fig. 1. Data input accepted by the model

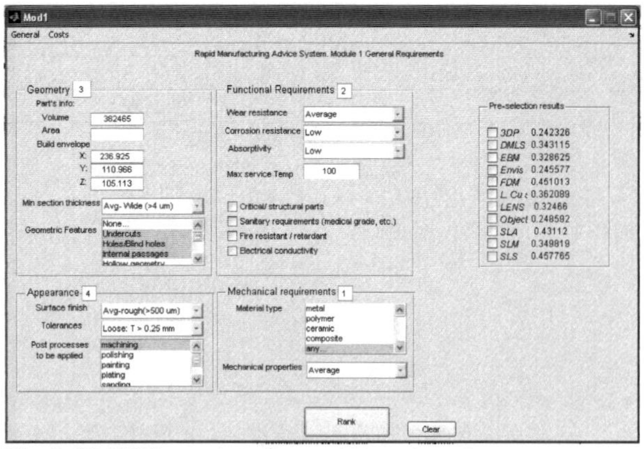

Fig. 2. RMADS user interface and its main options

Database: There are two different databases: 1) RM Process parameters and individual machine information. 2) Materials database. Data has been gathered from available vendor's info, related literature as well as from other online catalogs. Data is stored in an Ms Access database and called from the system by the Matlab database toolbox.

Aggregation and ranking module: This module performs the final decision based on the parameters and options selected. Since qualitative information is normalized to 1 and qualitative information is treated with fuzzy numbers, this module's task is to perform a fuzzy-decision. Although there exist a number of different approaches for ranking fuzzy sets, each one with different advantages and drawbacks, for this paper the transformation method(Lan 2005) is applied.

3.2 Module 2

Cost estimation is performed by means of a NN approach for each different RM process. The proper architecture is selected after testing a number of NN configurations with different number of hidden layers, input parameters and different learning and training algorithms.

Not all NN architectures are applicable to all RM processes. For instance, Selective Laser Sintering costs have been adequately modeled with a 3-3-1 architecture, trained with Levenberg-Marquardt algorithm and a case base of 100 parts. However Stereolithography may need a different number of input parameters of inter-unit connection mechanisms (Cavalieri, 2004).

The suitability of each neural network has been tested by two different performance indicators: The generalization factor:

$$Gf = {}^{k}\!/_{n} * 100 \tag{1}$$

where n= validation set, and k= estimation error under 5%, and the mean absolute performance error:

$$\frac{1}{n}\sum_{i=1}^{n}\left(\frac{|\text{Estimated cost}_i - \text{Actual cost}_i|}{\text{Estimated cost}_i}100\right) \tag{2}$$

4. THE MODEL RESULTS

After preliminary test of the model with sample mechanical parts, it has been observed an ability to provide a final RM processes ranking that is coherent with the input parameters selected and their level. Fig. 2 shows an example of the user interface for a sample part, and Fig. 3 shows the NN-based cost calculation for the same component.

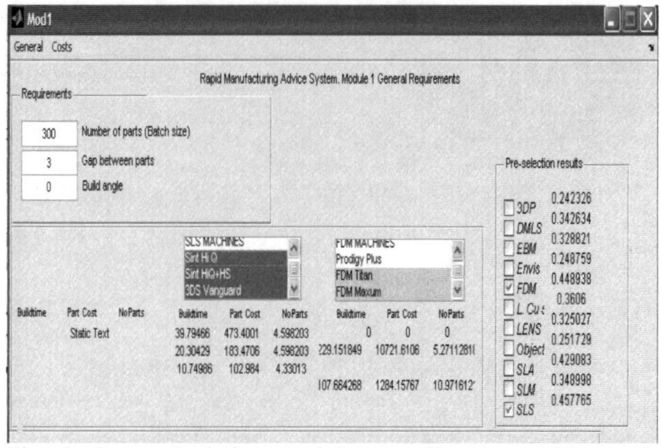

Fig. 3. Cost calculation module according to machine and process selected

5. CONCLUSIONS

Although the proposed model for RM selection is intended for use within machinery elements, its use can be extrapolated to other applications and conventional manufacturing processes.

Fuzzy logic has been of great value to model information regarding design parameters which are imprecise but have a severe influence on the final decision.

Regarding neural networks, special attention must be paid to training and verification data. Since the NN approach still remains an intensive trial and error approach it is crucial to perform correlation analysis to input data in order to evaluate its real effect on cost, besides checking the quality of the data used to train and validate the networks.

6. REFERENCES

ASERM, (2008) www.aserm.net/flexman. Accessed May-08

Bibb, R. et al (1999) Development of a rapid prototyping design advice system. J.Int. Manufacturing, 10, 331-339.

Campbell, R. & Bernie, M (1996) Creating a database of rapid prototyping system capabilities. Journal of Materials Processing Technology, 61, 163-167.

CavalieriI, S et al.(2004) Parametric vs neural networkmodels for estimation of production costs: A case study in the automotive industry. Int. J. Produc. Economics, 91, 165-177.

Hornberger, L. (1993) Rapid Prototyping Program. Santa Clara, California, Santa Clara University.

Lan, H. (2005) Decision support system for rapid prototyping process selection through integration of fuzzy synthetic evaluation and an expert system. Int. J. Production Research, 43, 169-194.

Mahesh, M. (2004) Benchmarking for comparative evaluation of RP systems and processes. Rapid Prototyping Journal, 10, 123-135.

Masood, S. H. & Soo, A. (2002) A rule based expert system for rapid prototyping system selection. Robotics and Computer-Integrated Manufacturing, 18, 267.

Muller, D.(2003) A Knowledge-based Approach to Rapid and Reliable Tool Design. BIBA Rapid Prototyping Group.

Phillipson, D. (1997) Rapid prototyping machine selection programme. Proc.6th European Conference on Rapid Prototyping and Tooling. Nottingham, UK.

Wohlers, T. (2006) Wohlers report 2006: rapid prototyping & manufacturing state of the industry annual worldwide progress report / Terry T. Wohlers. Colorado: Wohlers.

Zimmerman, H. (1991) Fuzzy Set Theory and its Applications, Boston,M.A USA, Kluwer.

Annals of DAAAM for 2008 & Proceedings of the 19th International DAAAM Symposium, ISSN 1726-9679
Editor B. Katalinic, Published by DAAAM International, Vienna, Austria 2008
Make Harmony Between Technology and Nature, and Your Mind will Fly Free as a Bird

ABOUT EKMAN CURRENTS IN SHALLOW WATERS

MUNTEAN, A[ngela]; MOROIANU, C[orneliu]; BEJAN, M[ihai]; BANDRABUR, D[iana] L[ilia] & SAMOILESCU, G[heorghe]

Abstract: *A very important component of Black Sea water circulation on the Romanian seashore is the Ekman currents. In present Ekman currents calculus is made for high depth basins. This paper presents results about Ekman current in shallow marine waters with constant depth. Navier-Stokes equations are used as mathematical model. The boundary conditions are particular for Black Sea waters. These are first results regarding Ekman currents in Romanian Black Sea shallow waters using established relations.*
Key words: *Ekman current in shallow waters*

1. INTRODUCTION

An accurate understanding of wind-driven circulation of Black Sea water on the Romanian seashore is required. Ekman curent is an important elementof the wind-driven circulation. Romanian Black Sea shore is characterized by a large region with shallow depth. We focus on the study of Ekman currents for shallow basins, using Navier-Stokes equations with friction. Boundary conditions and assumptions are made for Black Sea waters. One aim are to compute Ekman currents characteristics as velocity and direction. Flux and direction of the Ekman transport calculus is another aim of this study.

2. THE EQUATIONS OF MOTION FOR THE WIND-DRIVEN MARINE WATER CIRCULATION

The equation of motion used in oceanography is the Navier-Stokes equation. Among many forces which can be take in consideration in this equation, the most important are: gravity, Coriolis force and, for wind-driven circulation, the friction force. It can be used a coordinate system with the two horizontal axes *Ox* and *Oy* on plane tangent to the Earth with *Ox* to the east and *Oy* to the north. The third axis *Oz* is vertical up. In generaly, the friction stress is given by

$$\tau = \mu\frac{\partial u}{\partial z} = \rho\vartheta\frac{\partial u}{\partial z}.$$

This unit force acts on th surface between two neighboor layers which are moving at different speeds. The friction tends to slow down the faster and to speed up the slower (Pickard & Emery 1990).

The friction force component in *Ox* direction is given by

$$F_x = A_x\frac{\partial^2 u}{\partial x^2} + A_y\frac{\partial^2 u}{\partial y^2} + A_z\frac{\partial^2 u}{\partial z^2}.$$

where eddy viscosity for different directions A_x, A_y and A_z are used. In the last scalar equation in Navier-Stokes system all the terms are very smaller than the pressure one and *g* (Stewart 2002).

That's why all these terms can be ignored except the pressure term and *g*. The vertical equation is, in fact, the hydrostatic equation. When friction is included, the first two equations in the Navier-Stokes system are

$$\begin{cases} \dfrac{du}{dt} = fv - \alpha\dfrac{\partial p}{\partial x} + F_x \\ \dfrac{dv}{dt} = -fu - \alpha\dfrac{\partial p}{\partial y} + F_y \end{cases} \quad (1)$$

where F_x, F_y are the components of friction force per unuit mass in the fluid.

The forces in equilibrum are: the Coriolis force, pressure force and the friction one (Knauss 2005).

Taking in consideration that it exists few informations on the manner in which ρA_z does vary with depth ρA_z is assumed to be constant and

the friction force per unit mass = $A_z\frac{\partial^2 u}{\partial z^2}$.

Boussinesq approximation is used in oceanography and then ρ can be take as a constant.

The horizontal equations in Navier-Stokes system can be written

$$\begin{aligned} fv + \alpha\frac{\partial \tau_x}{\partial z} &= fv + A_z\frac{\partial^2 u}{\partial z^2} = \alpha\frac{\partial p}{\partial x} \\ -fu + \alpha\frac{\partial \tau_y}{\partial z} &= -fu + A_z\frac{\partial^2 v}{\partial z^2} = \alpha\frac{\partial p}{\partial y} \end{aligned} \quad (2)$$

The vertical equation reduced to the hydrostatic equation.

In this situation, the motion is due of two forces:

- the wind friction
- the pressure force given by the density variation.

The velocity can be written as a sum of two terms:

$$u = u_E + u_g,\ v = v_E + v_g \quad (3)$$

where u_E, v_E are the components of the velocity associated with the vertical shear friction (Ekman velocity) and u_g, v_g the components of the geostrophic velocity (Pond & Pickard 2003). Ekman was the first who studied the effect of the frictional stress on the sea water circulation due to the wind blowing over the sea. The equations of motion can be written

$$\begin{aligned} f(v_g + v_E) + A_z\frac{\partial^2(u_E + u_g)}{\partial z^2} &= \alpha\frac{\partial p}{\partial x} \\ -f(u_g + u_E) + A_z\frac{\partial^2(v_E + v_g)}{\partial z^2} &= \alpha\frac{\partial p}{\partial y} \end{aligned} \quad (4)$$

But $fv_g = \alpha\frac{\partial p}{\partial x}$. Then, the first equation in (4) system becomes

$$fv_E = -A_z\frac{\partial^2 u_E}{\partial z^2},$$

because the term $A_z\frac{\partial^2 u_g}{\partial z^2}$ is very small.

For the system of differential equation obtained up, Ekman used some assumptions (Lacombe 1965):

- no boundaries;
- infinitely deep water;

- A_z constant;
- a steady wind blowing for a long time;
- homogeneous water at the sea surface level since density depends only on pressure;
- f constant.

We don't use the second assumption. We considered the basin with a finite depth, equal with H. The equations of motion are

$$f v_E + A_z \frac{\partial^2 u_E}{\partial z^2} = 0, -f u_E + A_z \frac{\partial^2 v_E}{\partial z^2} = 0 \quad (5)$$

In order to achieve the system solutions we need the boundary conditions. We assume the wind blowing in direction given by an angle α_1 with the north direction (the direction of Oy axis) and the wind stress magnitude is τ. The boundary conditions can be written

$$\rho_w A_z \left(\frac{du_E}{dz}\right)_{z=0} = \tau\, sin\alpha_1$$
$$\rho_w A_z \left(\frac{dv_E}{dz}\right)_{z=0} = \tau\, cos\alpha_1 \quad (6)$$

where ρ_w is the density of marine water.
Those are the conditions for the surface basin. At the bottom

$$u_E = v_E = 0, \quad (7)$$

for $z = -H$ [2]. We use the Gongenheim and Saint-Guilly method (Lacombe 1965) with $w = u_E + i\, v_E$.

We use the assumption that u_E, v_E depend only on the level (z). It was multiplied the second equation of (4) by i and the two equations are added. It obtained a differential equation with a single unknown, w:

$$\frac{d^2 w}{dz^2} - i \frac{f}{A_z} w = 0.$$

With $a^2 = \frac{f}{2 A_z}$, the equation (7) becomes

$$\frac{d^2 w}{dz^2} - 2i\, a^2 w = 0. \quad (8)$$

We solved the differentiaa equation (8), with the boundary conditions (6) and (7) [4]. It was obtained

$$w = \frac{\tau e^{i\left(\frac{\pi}{4} - \alpha_1\right)}}{2\sqrt{2} a \rho_w A_z (ch\, 2aH + cos2aH)} \left[\left(e^{2aH} + e^{2aHi}\right) e^{(1+i)az} - \left(e^{-2aH} + e^{-2aHi}\right) e^{-(1+i)az}\right] \quad (9)$$

We did a study regarding the variation of wind-driven current with the sea level and, for the same level, with the depth.

The magnitude of velocity is given by the formulas

$$V = \frac{\tau}{\sqrt{2} a \rho_w A_z} \sqrt{\frac{ch2a(z+H) - cos2a(z+H)}{ch2aH + cos2aH}} \quad (10)$$

The magnitude of Ekman velocity for the same stress wind, at every level z depends on the basin depth. The angle θ between the velocity of surface Ekman currents direction and north direction is given by the formulas

$$tan\theta = \frac{sh2aH - sin2aH\, tan\left(\frac{\pi}{4} - \alpha_1\right)}{sh2aH\, tan\left(\frac{\pi}{4} - \alpha_1\right) + sin2aH} \quad (11)$$

We study the variation of the velocity and the angle with the basin depth [3]. For the same z, we conclude that the velocity magnitude increases for values less then about H = 1180 m. After this, the magnitude decreases a little. The angle, too, increases to 47^0, for about H = 1200 m. After ths, the angle decreases to 45^0, value which corresponds to the infinite ocean basin depth [4]. It is known that, for deep basin, the surface current flows at 45^0 to the right of the wind direction in northen hemisphere.

With the wind stress magnitude given by $\tau = \rho_a C_D W^2$ [1], where ρ_a is the of air, C_D the drag coefficient and W is the wind speed, the Ekman current speed is

$$V = \frac{\rho_a C_D W^2}{\sqrt{f A_z}\, \rho_w} \sqrt{\frac{ch2a(z+H) - cos2a(z+H)}{ch2aH + cos2aH}} \quad (12)$$

Using the values $\rho_a \cong 1.3\, kg/m^3$, $C_D = 1.4 \times 10^{-3}$, $\rho_w = 1000 kg/m^3$, $A_z \cong 10^{-1}\, m^2/s$ [1], and it is considered the region of Romanian Black Sea shore, one can takes $\varphi = 45^0$, and $f = 2\,\Omega\, sin\varphi = 10^{-4}\, s^{-1}$, the (11) formulas becomes

$$V = 5{,}75 \times 10^{-4} W^2 \sqrt{\frac{ch2a(z+H) - cos2a(z+H)}{ch2aH + cos2aH}} \quad (13)$$

If V_0 is the magnitude of surface Ekman current , the surface urrent and the wind speed are related as

$$\frac{V_0}{W} = 5{,}75 \times 10^{-4} W^2 \sqrt{\frac{ch2aH - cos2aH}{ch2aH + cos2aH}} \quad (14)$$

The net transport of Ekman current is given by the formulas

$$\Phi = \frac{\tau}{2\mu a^2} \frac{chaH - cosaH}{\sqrt{cos^2 aH + sh^2 aH}} \quad (15)$$

With the same values used in (12) formulas, the net flux becomes

$$\Phi = 0.0182\, W^2 \frac{chaH - cosaH}{\sqrt{cos^2 aH + sh^2 aH}} \quad (16)$$

In the shallow basin, the angle between the wind direction and the Ekman transport i

$$tan\theta_1 = \frac{ch2aH + cos2aH - 2\, cosaH\, chaH}{2\, sinaH\, shaH} \quad (17)$$

3. CONCLUSIONS

The depth of marine basin is very important for all Ekman current characteristics as theoretical results reveal. Marine activities require Ekman current characteristics in Black Sea shallow waters. The core of this study was the determination of relations for this particular zone. Further studies can be orientated to validate the theoretical model through physical determinations.

4. REFERENCES

Knauss, J. (2005). *Physical Oceanography*. Waveland Press, inc., ISBN 1-57766-429-9, Long Grove, Illinois

Lacombe, H. (1965). *Physical Oceanography*, Gautier–Villars, Paris

Pickard, G. & Emery, W. (1990). *Physical Oceanography*. Buttenworth-Heinemann, ISBN 075062759X, Oxford

Pond, S. & Pickard, G. (2003). *Introductory Dynamical Oceanography,* Butterworth Heinemann, ISBN 0 750 62496 5. Butterworth Heinemann, Vancover

Stewart, R. (2002). Introduction to Physical Oceanography, *Available from:* http://www.ig.utexas.edu/library/ *Accesed:* 2008-01-15

Annals of DAAAM for 2008 & Proceedings of the 19th International DAAAM Symposium, ISSN 1726-9679
Editor B. Katalinic, Published by DAAAM International, Vienna, Austria 2008
Make Harmony Between Technology and Nature, and Your Mind will Fly Free as a Bird

EMPIRICAL MODELLING OF THE ELECTRON BEAM HARDENING

MUNTEANU, A[driana]; COTEATA, M[argareta] & NEAGU, D[umitru]

Abstract: *The success of one experiment depends firstly of the input parameters, how we establish the main work condition. Thus, starting with the input parameters we can simulate the output parameters and to optimized the model. In the present paper we try to present different methods of modelling used in order to obtain, for the same values of the input parameters, different mathematical expression of the one output parameter. We intend to cumulate the theoretical knowledge for an overview of the different mathematical methods to solve and to obtain the dependence of the output parameter on the input parameters. The methods used are the Taguchi methods, the least-squares method and for the same problem we use specialized software.*

Key words: *electron beam, mathematical model, hardening.*

1. INTRODUCTION

Due to present industrial technological requirements, the nonconventional technologies were applied more and more often. The non-conventional technology manufacturing today face greater challenges than ever and offers new and intelligent solution in case of some manufacturing process. One of the nonconventional techniques is the electron beam process. Electron beam machining is a nonconventional method of machining which has been used since 1879 when W. Croockes melt his own platinum anode with electron beam.

It is well known that the thermal process using the electron beam technology is based on the changing the kinetic energy of an electron beam with high speed into heat. The electron beam can be considered as thermal tool of a concentrated energy that melts the workpiece material, but in case of few methods (electronolithography, irradiation etc) didn't works like a thermal tool (Rykalin et al., 1988).

Electron beam techniques have developed in many areas. Trends on activities carried out by researchers depend on the interest of the researchers and the availability of the technology. Among the processing methods using the electron beam, we could mention: drilling, welding, cutting, surface cleaning and degasification, the superficial heat-treating, surface micro-alloying and coating (or plating) by melting, doping, electron beam metalizing, electronolithography etc.

2. EMPIRICAL MODELLING

In this paper we are focused on electron beam hardening, that use extremely high-energy input to austenitize a very thin surface layer. The bulk of the substrate remains cool and provides an adequate heat sink for "self-quenching".

Their advantages includes: a) minimal workpiece distortion, b) the ability to selectively harden zones of a surface and better process control, c) ability to harden areas inaccessible to conventional induction techniques, d) repeatability and e) high speed (Neagu, 2001).

The question is what method we can apply, using the same values for the input parameter, to obtain empirical models to express an output parameter.

If we try to realize some experiments of the electron beam hardening, in classical manner, a complete set will suppose a number of **3125** experimental data (5^5).

Similar research was carrying on by the Dr. Eng. D. Neagu. The model of the process function, in case of the electron beam hardening, obtain by him is build upon this classical experimental plan. For example, when the hardening band-width L_{HV} is considered as output parameter, the input parameters are the following: the working distance (L_l), transversal deflecting angle (β), the electron beam current (I_{FE}), accelerating voltage (U_a) and linear speed of the workpiece (V_m) (Neagu, 2001).

The solution obtained in this condition was:

$$\begin{gathered} L_{HV} = 303638{,}546 - 4121{,}22L_1 - 10099{,}492I_{FE} - \\ 8576{,}248U_a + 34036{,}882V_m - 261{,}387\beta - \\ -0{,}864L_1I_{FE} + 89{,}805L_1U_a + 50{,}014L_1V_m + \\ +0{,}938L_1\beta + 257{,}142I_{FE}U_a + 256{,}132I_{FE}V_m - \\ -9{,}658I_{FE}\beta - 946{,}969U_aV_m + \\ +5{,}188U_a\beta + 18{,}939V_m\beta \end{gathered} \tag{1}$$

So that, in order to obtain similar results we can use the Taguchi method as thus, to reduce the number of experimental data.

The Taguchi method supposes that if we modify a limited number of parameters the chances to optimize the system results are increased when the varying parameters are carefully selected, with the condition that this factor is relevant reported to optimization criterion (Goupy, 1990). The high energy machining processes as the electron beam machining process are interrelated directly with a system built on a plurality of input parameters. Taking into consideration such parameters as the material chemical composition, the electron beam energy and the electron hardening phenomena and their action on the working system, they are transformed in relevant outputs, respectively in the products characteristics (Dagnelie, 2000).

The matrix model of the system includes the effects of the main factors: working distance, transversal deflecting angle, the electron beam current, accelerating voltage, and linear speed of the workpiece, on the output parameter the hardening band width - L_{HV}.

A general model can be written as in equation 2:

$$\begin{aligned} L_{HV}{}^{t} = M &+ [E_{Ll1}\ E_{Ll2}\ E_{Ll3}\ E_{Ll4}\ E_{Ll5}]\cdot[L_l] \\ &+ [E_{IFE1}\ E_{IFE2}\ E_{IFE3}\ E_{IFE4}\ E_{IFE5}]\cdot[I_{FE}] \\ &+ [E_{b1}\ E_{b2}\ E_{b3}\ E_{b4}\ E_{b5}]\cdot[\beta] \\ &+ [E_{Ua1}\ E_{Ua2}\ E_{Ua3}\ E_{Ua4}\ E_{Ua5}]\cdot[Ua] \\ &+ [E_{Vm1}\ E_{Vm2}\ E_{Vm3}\ E_{Vm4}\ E_{Vm5}]\cdot[Vm] \end{aligned} \tag{2}$$

Were: $L_{HV}{}^{t}$ is the theoretical answer of the system; M is the general average; $E_{Ll1}\ E_{Ll2}\ E_{Ll3}\ E_{Ll4}\ E_{Ll5}]\cdot[L_l]$ is the effect of the working distance; $[E_{IFE1}\ E_{IFE2}\ E_{IFE3}\ E_{IFE4}\ E_{IFE5}]\cdot[I_{FE}]$ is the effect of the electron beam current; $[E_{b1}\ E_{b2}\ E_{b3}\ E_{b4}\ E_{b5}]\cdot[\beta]$ is the effect of the transversal deflecting angle; $[E_{Ua1}\ E_{Ua2}\ E_{Ua3}\ E_{Ua4}\ E_{Ua5}]\cdot[Ua]$ is the effect of the accelerating voltage; $[E_{Vm1}$

E_{Vm2} E_{Vm3} E_{Vm4} E_{Vm5}]·[Vm] is the effect of the workpiece linear speed;

For our specific output parameter - the hardening band width - we have the following Taguchi model (in this case we neglect the interferences between input parameters):

$$\begin{aligned} L_{HV} &= 2047{,}66 + [120{,}14\ -3{,}75\ -69{,}17\ -74{,}33\ 27{,}11]\, L_l + \\ &+ [61{,}78\ -60{,}53\ -93{,}07\ -61{,}49\ 153{,}32]\, I_{FE} + \\ &+ [40{,}39\ 41{,}10 - 0{,}16\ 24{,}33\ -81{,}32]\beta + \\ &+ [-15{,}68\ 57{,}38\ 19{,}06\ 55{,}73\ -116{,}48] U_a + \\ &+ [-38{,}11\ 119{,}93\ 99{,}37\ -21{,}85\ -127{,}74] V_m \end{aligned} \quad (3)$$

The other case is the least-squares method. We must to consider the fact that sometimes the values of the output parameters of a process can be affected by the aleatory errors, this would make senseless the selection of one function y=y(x) that describe all the n values obtained in the experimental program (Munteanu & Ilii, 2008).

The complete set of the experimental data can be described like a plurality of different type of function y(x). The least-squares method indicates the best function y(x) for our set of data. This method permits to determine the most likely values of the coefficients for the prior function based on the theoretical consideration regarding the phenomena of the process for which we obtain the experimental data.

The software program used in present paper permit to determine a relation of conection between the independent output parameter y and some liniar input parameter $x_1, x_2, \ldots x_n$, a relation writen as:

$$y = f(x_1, x_2, \ldots, x_n) \quad (4)$$

The software gives a best function to express the process influeces, but also offers the possibility to choose between more forms of mathematical expression that can model the answer.

For our output parameter L_{HV} (hardening band width) we shall have five distinct functions. The control criterion is the Gauss sum. So that, the best function is polynomial type (the 2nd degree polynomial function) because the Gauss sum is the smallest (Gauss sum: S = 146973,1).

$$\begin{aligned} L_{HV} &= 99733{,}02 + 1359{,}74 L_l - 6959{,}73 I_{FE} + 382{,}32 I_{FE}^2 - \\ &- 5039{,}55 U_a + +59{,}279 U_a^2 - 6974{,}093 V_m + \\ &+ 1818{,}07 V_m^2 - 250{,}89\beta + 1{,}377\beta^2 \end{aligned} \quad (5)$$

But to obtain a better image of the influence of the each input parameter and an easier ordering of them, we prefer to choose the power function:

$$L_{HV} = 4{,}978 \cdot 10^{-13} L_l^{\,4{,}064} I_{FE}^{\,-8{,}708 \cdot 10^{-2}} U_a^{\,5{,}806} V_m^{-0{,}271} \beta^{-0{,}270} \quad (6)$$

In this case, the GWBASIC language program takes in consideration only five distinct functions like: the exponential function, 1nd and 2nd degree polynomial function, power function and hyperbolic function. Related to the experimental condition we can obtain the regression functions in the conditions to accomplish the request imposed by the law of the minimum Gauss sum (Cretu, 1992).

Using software Datafit we obtain another dependence relation, considered by the software as best expression, an exponential function:

$$L_{HV} = e^{-0{,}004 L_l + 0{,}017 I_{FE} + 0{,}015 U_a - 0{,}104 V_m - 0{,}002\beta + 7{,}551} \quad (7)$$

If we modified the input parameter of the electron beam hardening and using appropriate software of experimental data processing (Datafit) we can see the variation of the output parameter depending on the considered input parameters. A graphical representation related to the mathematical model (7), but without the variation of the all input parameters.

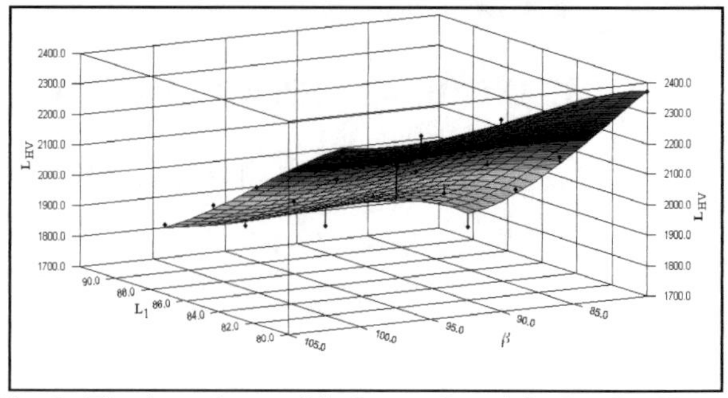

Fig. 1. The dependence of the L_{HV} on L_l and β when I_{FE}, U_a and V_m remain constants

When we analyse the different models we can see that the easiest way to represent the influence of different parameters regarding the one or another output parameter is the mathematical modelling. In this way we can emphasize the influence of each input parameter on the hardening band-width that we considered for our study.

3. CONCLUSION

The optimization of the system answer supposes to find the maximum of all possible answers, since we are concerned to obtain a bigger hardness of the superficial layer, a good sized of the deep penetration and a grater hardening band-width.

The other mathematical modelling methods cannot consider the interaction between the main variables, while the Taguchi matrix model permits besides the interaction between the input parameters, also some other aspects. In the future we will considered to extend the theoretical and experimental research in order to improve the mathematical model so that to obtain the best solution.

The poupose of this paper was to find the best relation to expres the hardening band width using different methods and, in the same time, to compare these dependence relations.

4. REFERENCES

Cretu, Gh. (1992). Base of the experimental data (in Romanian). Publishing House Rotaprint, Iassy

Dagnelie, P. (2000). The planification of the experiments: the chose of the experimental divice and method (in Franch). *Journal of the Statistics French Societe*, No. 141 (1-2), (5 - 29)

Goupy, J. L. (1990). Comparative analysis of the different experimental plans (in Franch). Journal of the applied statistics, Vol. 38, No. 4, (5 - 44), *Available from:* http://books.google.com/books?hl=ro&lr=&id=yiHjQlbN8A8C&oi=fnd&pg=PA9&ots=Q-I28Dz8Xx&sig=S6sb2Srs9xV23kuiqC8S7wRlQew#PPA10,M1 *Accessed:* 2007.05.10

Munteanu, A. & Ilii, S.M. (2008). Taguchi model versus mathematic model in the electron beam process. *Scientific Bulletin of Politechnica Iasi.* Tomul LIV (LVIII), Fasc.1-3 (May), (83-89)

Neagu, D. (2001). Electron beam hardening (in Romanian). Publishing House Printech, ISBN 973-652-377-2, Bucharest

Rykalin, N.; Uglov, A.; Zuev, I. & Kokora, A. (1988). *Laser and electron beam material processing, handbook,* Publishing House Mir Publishers, ISBN 5-03-0000-23-2, Moscow

Annals of DAAAM for 2008 & Proceedings of the 19th International DAAAM Symposium, ISSN 1726-9679
Editor B. Katalinic, Published by DAAAM International, Vienna, Austria 2008
Make Harmony Between Technology and Nature, and Your Mind will Fly Free as a Bird

DESIGN OF ROBUST ACTIVE CONTROL FOR FLUTTER ALLEVIATION

MUNTEANU, E[liza] & URSU, I[oan]

Abstract: *Flutter is an in-flight, self-excited oscillation of flexible structures caused by energy in the airstream absorbed by the lifting surface. Being a very harmful phenomenon for the flight safety, the present paper addresses the problem of flutter prevention in an aeroservoelastic context. An active flutter control is performed using a trailing edge control surface Taking into account the lack of robustness of the LQG control, in a second step of synthesis the control is improved by applying a Linear Quadratic Gaussian/Loop Transfer Recovery (LQG/LTR) procedure. Numerical simulations confirm a significant beneficial increasing of the critical flutter velocity of the controlled system in comparison with the passive system, both in the case of nominal system and various perturbed, uncertain type systems.*
Key words: *active control, aeroelasticity, flutter, wing.*

1. INTRODUCTION

Certification regulations require that any certified aircraft is free of static *divergence* and *flutter* for the range of permitted flight speeds. Flutter is just dynamic, rapidly developed, oscillatory instability of a non-rigid wing that occurs, at a limiting velocity – the flutter velocity –, and caused by the energy in the airstream. To meet the regulatory requirements, one can use either passive or active techniques. The first ones increase the structure weight and in certain situations are not feasible. On the other hand, active techniques suppress flutter without redesign and adding mass.

Two are the main targets in the present work. Firstly, the active control of the flutter is thought as an *additional* functionality of the flight control surfaces (ailerons, rudder, and tailplane) and of their actuation system; the same additional functionality can be thought in the case of other control surfaces, such as a flap. Similar approaches are already well-known (Block & Strganac, 1998), but they differ mainly by the *mathematical models* used. Thus, the essential features for airfoil's aerodynamics, in a 2-D framework, are described by a *quasi-steady* model for the lift, pitching moment and hinge moment; an aggregate aeroservoelastic model having three degrees of freedom (the plunge *h*, the pitch α and the flap angle β) is so obtained. A first order equation of the electrohydraulic servo states the control variable u – current to servovalve – providing flap angle β. We use basic flutter farmework given by Theodorsen, in which the harmonic *unsteady* motion of the airfoil is assumed; the celebrated Theodorsen's function takes into account the lag effects of the unsteady aerodynamics at different values of a reduced frequency.

Indeed, in the work, our target concerns, secondly, the interest for the designing a robust active control law. Various approaches are encountered in the literature of the field. Olds (Olds, 1997) proposed LQR control and Bail (Bail, 1997), considering the same model of Olds, used LQG control and H_∞ control. A recent work (Silva & Júnior, 2006) again appeals the Olds model and exploits the resources of LMI control to ensure certain system robustness. In the present work, we are interested in proving the advantages of the LQG/LTR method, as a robust control approach applied in view of increasing flutter speed or flutter alleviation.

2. THE MATHEMATICAL MODEL

The three degrees of freedom, the plunge *h*, the pitch α, and the flap angle β, the linear spring with constant stiffness K_h, the torsional spring with constant stiffness K_a, and the flap spring with constant stiffness K_b, the semichord b and the nondimensional distance c from elastic axis. The control flap is located at the trailing edge; the control variable u will be considered as a flap hinge torque, added to aerodynamic flap torque T. The airfoil is so subjected to the three unsteady aerodynamic loads: L, M and T and the airfoil's equations of motion are firstly written as the second order system

$$M\ddot{Y}(t)+KY(t)=\begin{bmatrix}-L\\-M\\-(T+u)\end{bmatrix},\quad Y:=\begin{bmatrix}h/b\\ \alpha\\ \beta\end{bmatrix} \tag{1}$$

Following closely (Olds, 1997), the equation (1) can be rewritten, via successive using of Wagner and Jones approximations for Theodorsen's function, Fourier transform and Duhamel integrals, in the form of first order, state space model. The problem considered herein is a LQG optimal one, defined by the system

$$\begin{aligned}\dot{x}(t)&=Ax(t)+B_1w(t)+B_2(t)u(t)\\ z(t)&=C_1x(t)+D_{11}w(t)+D_{12}(t)u(t)\\ y(t)&=C_2x(t)+D_{21}w(t)+D_{22}(t)u(t)\end{aligned} \tag{2}$$

where $x(t)$ is the state, $z(t)$ is the controlled output, $y(t)$ is the measured output, $w(t)$ is the disturbance input (external and internal), and $u(t)$ is the control input. The state vector is given by

$$x(t)=\left(\dot{h},\,\dot{\alpha},\,\dot{\beta},\,h,\,\alpha,\,\beta,\,b_1,\,b_2,\,a_1,\,a_2\right)^T \tag{3}$$

where the last four variables are the so-called aerodynamic-lag states used to describe the "state" of the fluid as a result of aerodynamic load action on the airfoil. The external and internal components of the perturbation vector $w(t)$ are the aerodynamic disturbances and sensor noise vector.

The controlled output $z(t)$, in the following, will concern the whole system state as well as the control variable $u(t)$. As the measured output will be taken the first six components of the state vector.

3. LQG/LTR CONTROL SYNTHESIS

The goal is to find a control such that the system is stabilized and the control minimizes a cost function, which means implicitly a minimizion of the effect of disturbances on the controlled output and, in fact, an active alleviation of the flutter. Using a state-estimator and a control law the system (2) becomes

$$\begin{aligned} \dot{x}(t) &= Ax(t) + B_1 w(t) + B_2(-K_c \hat{x}(t)) \\ z(t) &= C_1 x(t) + D_{11} w(t) + D_{12}(-K_c \hat{x}(t)) \\ \dot{\hat{x}}(t) &= K_f C_2 x(t) + K_f D_{21} w(t) + (A_0 - D_{22} K_c)\hat{x}(t) \end{aligned} \tag{4}$$

It is well known that the LQR control is strongly robust. The weak point of the LQG control is its lack of robustness. A strategy of counteract this drawback (Munteanu & Ursu, 2008) is to choose the filter gain K_f so that the closed-loop LQG/LTR system (having open loop L_{LQG} transfer matrix, containing Kalman filter matrix) recovers internal stability and some of the robustness properties (gain and phase margins) of the LQR design (with open loop L_{LQR} transfer matrix).

$$\begin{aligned} L_{LQG}(s) &= \left[-\left(sI - A - K_f C_2 - B_2 K_c\right)^{-1} K_f C_2\right]\left(sI - A\right)^{-1} B_2 \\ L_{LQR}(s) &= -K_c\left(sI - A\right)^{-1} B_2 \end{aligned} \tag{5}$$

4. NUMERICAL RESULTS AND CONCLUSIONS

The aforedescribed control law was brought to the proof in numerical simulations of the system (2) in the presence of LQG/LTR control law. The nominal values of the system parameters are given in Table 1. By a trail and error process, the following value of the LQG/LTR parameter was established: $q_{LTR} = 8.5 \times 10^6$. The noise matrices was chosen as

$$\Theta = 0.001 I_{6x6}\,,\ \Sigma = 1000 I_{9x9} \tag{6}$$

The weighting matrices Q_j and q_R are fited as

$$Q_j = diag\begin{pmatrix}1 & 1 & 1 & 4000 & 10000 & 1 & 1 & 1 & 1 & 1 & 0\end{pmatrix},\ q_R = 0.005 \tag{7}$$

To determine the flutter occurence, some open-loop simulations, with various configurations listed in Table 1, was performed.

First, we chosen a nominal configuration and fixed the control gain and the filter gain at their nominal values. In the second step, some of parameter configurations are randomly change with maximum ± 20 % and the nominal controller was tested.

For all tested configurations, the two controllers, LQR and LQG/LTR give similar results (with differences smaller than 0.01 %); the exception that improves the rule is configuration 4, in which case the LQG/LTR controlled system seems to work worse than LQR controlled system. These numerical simulations validate this LQG/LTR procedure as a methodology of preserving the robustness of LQR controlled systems.

The obtained enhancing of system robustness is better than that shown by using Linear Matrix Inequalities (LMI) techniques in another work (Silva & Júnior, 2006) where only a parameter variation of 10%, involving exclusively system stiffness, was considered.

As we can see in Table 2, the flutter velocity has some significant increases in the case of controlled systems in comparision with the passive systems in the presence of some models uncertains.

Parameter [SI]	Nominal configuration	Configuration 1		Configuration 2	
		Value	%	Value	%
m	128.72	141.59	+10	109.41	-15
ρ	1.23	1.11	-10	1.05	-15
S_α	23.5	25.85	+10	19.98	-15
S_β	1.47	1.69	+15	1.25	-15
I_α	26.84	24.16	-10	30.87	+15
I_β	0.67	0.57	-15	3.70	+15
$K_h \times 10^{-5}$	3.22	3.70	+15	2.74	-15
$K_\alpha \times 10^{-5}$	2.69	2.28	-15	3.09	+15
$K_\beta \times 10^{-5}$	1.67	1.84	+10	1.92	+15

Parameter [SI]	Configuration 3		Configuration 4		Configuration 5	
	Value	%	Value	%	Value	%
m	102.98	-20	154.46	+20	154.46	+20
ρ	1.48	+20	1.48	+20	0.98	-20
S_α	18.80	-20	28.20	+20	28.20	+20
S_β	1.76	+20	1.18	-20	1.18	-20
I_α	21.47	-20	32.21	+20	21.47	-20
I_β	3.86	+20	2.58	-20	3.86	+20
$K_h \times 10^{-5}$	2.58	-20	3.86	+20	3.86	+20
$K_\alpha \times 10^{-5}$	3.23	+20	2.15	-20	2.15	-20
$K_\beta \times 10^{-5}$	1.34	-20	2.00	+20	2.00	+20

Tab 1. List of parameters configurations

Configuration	Pasiv $V_{flutter}$	LQG/LTR $V_{flutter}$
Nominal configuration	297.36	309.81
Configuration 1	260.34	264.47
Configuration 2	371.27	389.14
Configuration 3	305.02	335.44
Configuration 4	235.37	238.01
Configuration 5	312.01	314.88

Tab 2. LQG/LTR increasing of flutter velocity

Thus, our main target to obtain a realistic robust active control law was fulfilling. Moreover, by using a state-estimator, the number of state variables needs to be effective measured fall to a half in comparision with the LQR controlled system.

A future work will focus on adaptive control synthesis based on online model identification.

Sacred with pain and tears to Letu.

5. REFERENCES

Bail, T. R. (1997). A disturbance-rejection problem for a 2-D airfoil exhibiting flutter, *Available from:* http://scholar.lib.vt.edu/theses/available/etd321019204972 1391/unrestricted/etd.pdf. *Accessed: 2008-07-24*

Block, J. J. & Strganac, T.W. (1998). Applied active control for a nonlinear aeroelastic structure, *J. of Guidance, Control, and Dynamics*, Vol. 21, No. 6, pp. 838-845, Nov-Dec 1998

Munteanu, E. & Ursu, I. (2008) Piezo smart composite wing with LQG/LTR control, *IEEE International Symposium on Industrial Electronics*, CD publisher, ISBN 978-1-4244-1666-0, UK, June – July 2008, Cambridge

Olds, S. (1997). Modeling and LQR control of a two-dimensional airfoil, *Available from:* http://scholar.lib.vt.edu/theses/public/etd-412813204972 1391/etd.pdf. *Accessed: 2008-07-24*

Silva, S. & Júnior, V. L. (2006). Active flutter suppression in a 2-D airfoil using Linear Matrix Inequalities techniques, *J. of the Brazilian Society of Mechanical Sciences and Engineering*, vol. 28, no. 1, pp. 84-93, Jan/Mar 2006, ISSN 1678-5878

Annals of DAAAM for 2008 & Proceedings of the 19th International DAAAM Symposium, ISSN 1726-9679
Editor B. Katalinic, Published by DAAAM International, Vienna, Austria 2008
Make Harmony Between Technology and Nature, and Your Mind will Fly Free as a Bird

MACRO SOLIDIFICATION SIMULATION DURING CONTINUOUS CASTING OF ALUMINUM ALLOYS SLABS

MUNTEANU, S[orin] I[on]; CIOBANU, I[oan] & CRISAN, A[urel]

Abstract: *The paper presents the researches about macro solidification simulation in the specific conditions of continuous casting. Researches are focused on aluminum alloys slabs cast by direct casting and continuous electromagnetic casting. A finite difference mathematical model is used to create a software to simulate aluminum alloys macro solidification process. Overall objective of the software is to predict the temperature distribution in the ingot section, in function of alloy properties and caster working parameters. With this program it can be established the optimum working parameters for caster in order to obtain good ingots without expensive training in industrial conditions.*

Key words: *modeling, simulation, solidification, casting, aluminum.*

1. INTRODUCTION

Continuous casting and especially electromagnetic casting is a very complex process The design and operation of continuous casting machines needs sophisticated models and software to analyze and control caster parameters. Many theoretical and experimental researches were developed model and simulate the complex phenomena in continuous casting. (Abdel, I. N. K. & Khadraoui, Y. ,2004) (Bermudez, A.,et al.2003) On the other hand small and medium foundries for that use DC or EMC need simple software to operate their casters.

Establishing the shape and position of solidification front, at a continuous cast slab (DC or EMC) is a dynamic process that goes after some time on a stationary state

The 2D mathematical model presented in this paper, based on previos models of the authors (Munteanu, S. I., et al.,2005) allows reproducing the dynamics of this process and calculating the stationary state but also analyzing the transitory phase.

2. SOLIDIFICATION MATHEMATICAL MODEL ¶

The model describes the cooling and solidification of a static poured slab that solidifies in a range of temperature. The slab considered infinite on y direction is divided in rectangular elements with section x_i z_i , like in Fig. 1. The time is also divided in small periods $\Delta\tau$. As the elements get smaller and the time period, shorter the simulation precision gets better.

The mathematical model is based on the following hypothesis:

- The heat transmission inside the ingot is made only by conductivity;
- The heat transfer between the ingot and the environment is made by convection;
- The convection currents inside the liquid and at the solid - liquid interface are ignored;
- The volume variation at solidification is ignored, only the density variation between liquid and solid is taken into consideration;
- The alloy solidification happens in a range of temperature T_L - T_S, (hypothesis valid for the alloys with no eutectic);
- The pouring and feeding processes are ignored.

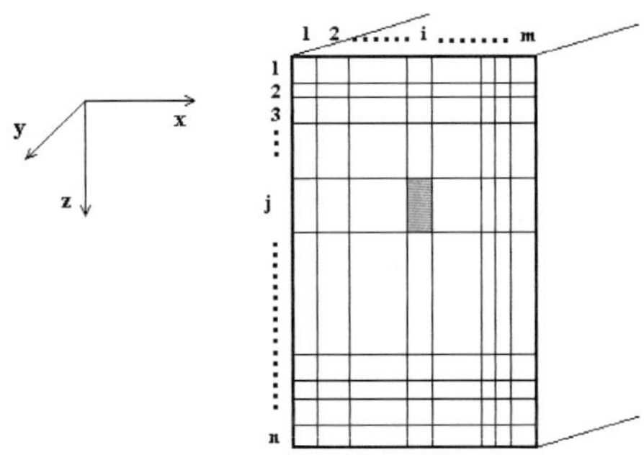

Fig. 1. The mode of deviding the slab section in small elements.

2.1 Static model equations

The static model is based on the heat transfer between one element "i,j" and its neighbours depending on their temperatures and thermal characteristics at the end of the previous time period. Material characteristics as density, specific heat, thermal conductivity are temperature function with a linear variation in the domain between liquidus temperature T_l and solidus temperature T_s:

$$\rho_{ij}^k = \begin{cases} \rho_S & T_{ij}^k \le T_S \\ \rho_S + \dfrac{T_{ij}^k - T_S}{T_L - T_S} \cdot (\rho_L - \rho_S) & T_S < T_{ij}^k < T_L \\ \rho_L & T_{ij}^k \ge T_L \end{cases} \quad (1)$$

To calculate the temperature of an element " i,j " at the time moment τ_{k+1} I used the thermal balance. The heat released by element " i,j " to the neighbour elements on each unit of thickness, in a time interval $\Delta\tau$ at the moment τ_k :

$$\begin{aligned} \Delta Q_{ij}^k = z_j \cdot \left[\alpha_{i-1j,ij}^k \cdot \left(T_{ij}^k - T_{i-1j}^k\right) + \alpha_{ij,i+1j}^k \left(T_{ij}^k - T_{i+1j}^k\right)\right] \cdot \Delta\tau + \\ + x_i \cdot \left[\alpha_{i-1j,ij}^k \cdot \left(T_{ij}^k - T_{i-1j}^k\right) + \alpha_{ij,i+1j}^k \left(T_{ij}^k - T_{i+1j}^k\right)\right] \cdot \Delta\tau \end{aligned} \quad (2)$$

where $\alpha_{i-1j,ij}$ - heat transfer index between " $i-1,j$ " element and " $i\ j$ " element;

Writing the thermal balance for the situations which can appear in temperature evolution of the element " i,j " between the moment τ_k and τ_{k+1} we can obtain 9 equations to calculate the final temperature of the element " i,j " at the moment τ_{k+1}. After a time interval $\Delta\tau$ a temperature map can be calculated, that is used as initial temperature for the next step. Based on this model a computer program was created, which can simulate the cooling and solidification of an ingot knowing the cooling conditions at the surface.

2.2 The model for continous cast ingots (DC casting)

To use this model for continuous casting it must be adapted to consider the variable length of ingot and continuous inflow of liquid metal. The result is a mesh with a high number of elements and depending on time. On the other hand the most interesting part is the head of the ingot near the inductor because in this region the temperature gradient is higher and thermal flux are more intensive. Also in this region it will established the solidification front. For this reason the mesh is limited at a convenient distance under inductor (Fig.2a). In this case the thermal balance is changing, a part of heat being eliminated from studied area by mass transport.

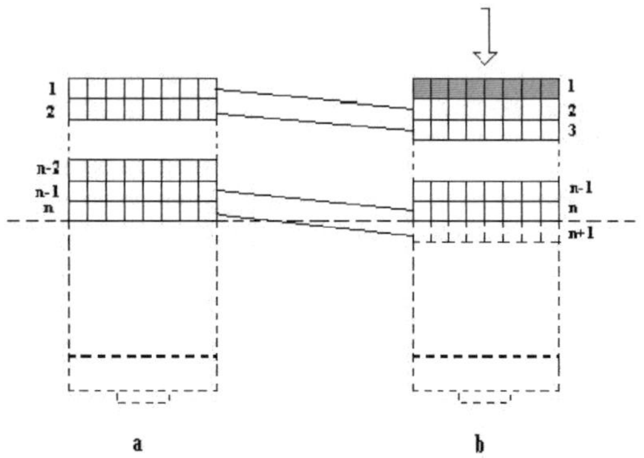

Fig. 2 The model for ingot displacement

2.3 Model improved for electromagnetic casting

The current density induced in a point inside the ingot depends on local coordinates and on geometry and relative position of inductor and screen. Because of skin effect the electromagnetic field and induced currents are strongly attenuated, becoming null at a distance of some skin depth of surface. After establishing the steady state the current distribution inside the ingot can be calculated and represented as an array $|J_{ij}|$. So we can calculate the heat generated by Joule effect in each element of the mesh.

This heat will be considered for each element in the thermal balance.

Magnetically driven flow in the molten pool modifies the heat transfer and the solidification conditions resulting in a homogenisation of temperature in liquid metal pool. In a stationary state the magnetically driven flow configuration is stable and can be calculated like an array $\left|\vec{v}_{i,j}\right|$.For simply introduce the flow effect in thermodynamic model we calculated an equivalent conductivity which produce the same thermal effect like liquid metal movement.

It must be mentioned that the conductivity changes are calculated only for liquid pool, thinking that in the mushy zone the liquid movement is more slowly.

3. RESULTS OBTAINED BY SIMULATION

The model presented allows the simulation of temperature distribution in a longitudinal section of a continous cast ingot. Initial data consist of thermal properties of purred alloy and cooling medium, geometry of system, and working parameters. Because of symmetry it can be studied only a half of section to minimise the computing time. Fig. 3 presents temperature distribution in the longitudinal section of an electromagnetic cast slab of aluminium alloy Al 6063 in the steady state of electromagnetic casting computed using the model presented above.

Analysing the temperature distribution and the shape of mushy zone it comes out:

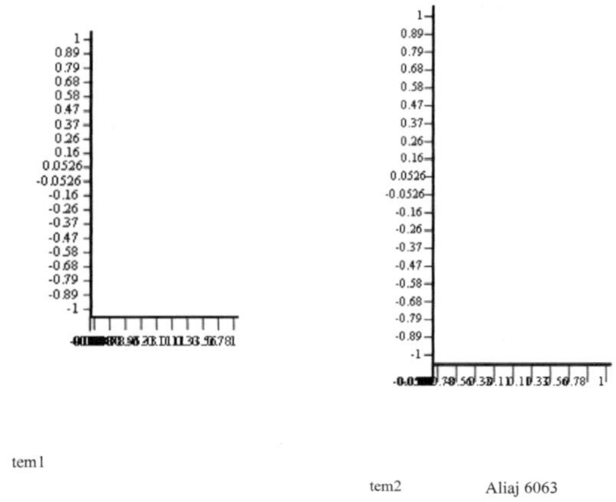

Fig.3Temperature diagram in a longitudinal section of an DC (left) and EMC(right) slab.

- The temperature distribution is similar with reported measurements in industrial conditions or pilot caster (Prasso et al, 1995), (Vives & Ricou, 1995).
- The depth of liquid pool is proportional to extract velocity.
- The induced current and electromagnetic stirring contribute to flatten the solidification front, homogenise the depth of mushy zone and protect the border of meniscus to freeze.

4. CONCLUSION

This research represent a part of an extended research about casting simulation developed in a consortium of 4 Romanian Universities.

The model described in this paper represents a simple and efficient method to compute cooling and solidification of continuous cast slabs. It can simply evidences the influence of cast alloy characteristics and cooling conditions on solidification speed and so to predict the microstructure of cast ingot.

Future researches follow to improve the model and to validate the software so to became a useful tool in controlling continuous casting process.

5. REFERENCES

Abdel, I. N. K. & Khadraoui, Y. (2004). A numerical simulation of the DC continuous casting using average heat capacity, *Scandinavian journal of metallurgy*, vol. 33, n°6, 2004, p. 347-354 ISSN 0371-0459

Bermudez, A.,; Muniz, M. C & Saldago, P. (2003). Asymptotic approximation and numerical simulation of electromagnetic casting, *METALL. MAŢER. TRANS B.,* vol.34, Feb. 2003, p.83-91, ISSN 1073-5615

Munteanu, S. I.; Ciobanu, I. & Crisan, A. . (2005). The macro-solidification simulation for a bar with U section cast in eutectic cast iron - *Metalurgia International*, nr.6, 2005, p.38-50, ISSN 6461-9579

Prasso, D. C.; Evans, J. W.& Wilson, I. J. (1995). Heat Transport and Solidification in the Electromagnetic Casting of Aluminum Alloys. I. Experimental Measurements on a Pilot-Scale Caster, *METALL. MATER. TRANS. B,* 26B,1995, p.1243-1251, ISSN 1073-5615/83

Vives, C.& Ricou. R. (1995) - Experimental Study of Continous Electromagnetic Casting of Aluminium Alloys - *METALL. MATER. TRANS. B* 16B, 1995 p.377-384, ISSN 1073-5615/83.

Annals of DAAAM for 2008 & Proceedings of the 19th International DAAAM Symposium, ISSN 1726-9679
Editor B. Katalinic, Published by DAAAM International, Vienna, Austria 2008
Make Harmony Between Technology and Nature, and Your Mind will Fly Free as a Bird

APPROACH ON IMPLEMENTING A PLM SOLUTION INTO A COLLABORATIVE ENGINEERING NETWORK FOR PRODUCT DESIGN AUTOMATION

MUSCA, G[avril]; MUSCA, E[lena] & MERTICARU, V[asile jr]

Abstract: *The authors propose and exemplify in the paper an approach for developing new products in a collaborative manner, by using CAD (Computer Aided Design) instruments into a PLM (Product Lifecycle Management) environment. In this sense, the realization of a collaborative network composed by technical universities and industrial productive enterprises is nominated. The purpose of the mentioned network is the generation of type projects and new products using the concepts of CAD/CAM/CAE and data management at the network's level by using a PLM solution.*

Key words: *PLM, design, automation, stamping, collaborative.*

1. INTRODUCTION

There are well known the necessities of the industrial economic agents for conceiving, designing and producing qualitative, economical and aesthetical products. A key factor in satisfying these needs is collaborative engineering in product development and a solution for this is the implementation of a PLM system. Usually, in the approach of implementing a PLM solution, a multitude of problems occur. There is intended to nominate in the paper some efforts existing in the Technical University "Gh. Asachi" of Iasi from Romania (Musca et al., 2008) for developing a collaborative network for the implementation of a PLM solution in order to create conditions for a better collaboration between the research teams of specialists from universities and industrial economic agents, for solving the problems identified in the computer aided conception, starting with the activities of CAD (Merticaru et al. 2008)., design and ergonomics of the product, cinematic simulation of the product in the design phase, simulation of the functioning behavior using FEA, conception of the manufacturing technology CAM.

2. PLM SYSTEMS REFERENCE

The companies that develop product design activities are capable to generate the technical documentation of the products to a great extent, but often these companies do not register some properties of the components as they can be easily found and especially reused. That is why the project engineers have problems in accessing the needed information (Usher et al., 2005). Difficulties occur in the ability of the project engineers to efficiently manage the product information.

The information systems for the management of the technical information have the capacity to administrate both the properties associated with each product or with the components of a product and the technical documentation of the projects by relating these types of information (Grieves 2006). The way the PDM (Product Data Management) systems cover these requirements is based on stocking the information just once, providing data integrity, monitoring and controlling of the design data changes or updating. Any information from the system can be distributed toward any department based on the priorities of the importance degree established for each user. When a change is made upon any data from the system, this will be automatically registered by the system, keeping the date and author of the change. Since the information generated by a company becomes more and more voluminous, the necessity of establishing a classifying and managing solution is imperative. This capacity for classifying and managing information is the fundamental requirement for a PDM/PLM system.

The main characteristics of a system of information management regarding the life-time product are (Ausura & Deck, 2003): reduces the time for realizing a new product (Time to market); improves the productivity of the design/calculation process; improves the design and manufacturing accuracy; determines an innovative content of the products; provide data security; provide a better control upon changes.

3. THE DEVELOPED COLLABORATIVE NETWORK AND PLM IMPLEMENTATION

For implementing, in the economic environment specific to the Romanian industry, a PLM solution into a network formed by researchers belonging to some universities and specialists from manufacturing enterprises, certain stages are necessary, within them the following outstanding: the knowledge acquirement and the development of skills for using CAD systems; to know the concept of data management and to establish some procedures for implementation at the level of functional areas into an industrial enterprise; establishment of the soft and hard structure of the system used for data management; the training of the staff involved in the activity of products' conceive, design and manufacturing.

The research team the authors belong to, have passed through the above mentioned stages, both for their own team and for their industrial partners. After the training of several groups of participants in the PLM program, groups belonging to the partners, there have been observed that a vital direction for the success of the approach is to form new specialists in the areas of interest, respectively firstly CAD and CAM and secondly in Computer-Aided Engineering (CAE) or data management for the Product Lifecycle. The best results have been obtained with teams formed by researchers from universities and students.

At the level of a PLM network, functional within an environment heterogeneous as education and preoccupations, one of the main requirements is to form a motivated and efficient virtual team. The premises of forming a team, efficient in work in the virtual environment, base on: adequate level of knowledge for accomplishing the tasks for the main stages of the product lifecycle, CAD, CAM, relationships with the suppliers and customers; motivation of the team working by co-interesting and counsellorship in the domain of PDM/PLM. There have been found that partial accomplishment of some premises is not sufficient; so the financial co-interesting of the team members can be inefficient if the motivation of the team working does not exist. The accomplishment of this premise is difficult for the teams of the partners from industry, even if the presumed results are known and noteworthy. Even in the research environments or in the universities, some malfunctions can be found, caused, accordingly to the authors' opinion, by: insufficient number of persons with abilities in the CAD/CAM/CAE area; inefficient education of the specialists in this domains and of the students, as the work with a CAD system requires to go through a training stage that must be individually realized; differences between the team members regarding the knowledge level, which lead to difficulties in collaborative activity.

Considering the experience gathered in the CAD area by the authors, the implementation of the PLM solution started with the design of some elements necessary for the industrial partners. For the design stage, *SolidEdge-v20* has been used. Some of the projects realized within this approach by a team formed of professors and students, are combined stamping tools used for processing sheet metal. The use of this type of constructive projects as support for the implementation of the PLM solution is justified by: the necessity of managing the information for this type project; this example is applied for managing the design data for a set of stamping tools, able to realize several types of sheet metal parts, using normalized elements such as plates, columns, bushes, clamping and fixing elements; the precise management of the design activity, for realizing some modular stamping tools, by modifying the active elements and keeping the supporting constructive elements (plates, columns, bushes, clamping and fixing elements); the extending of the data management process in the area of stamping tools manufacturing, providing the reducing of the necessary of work by reusing some components normalized at the level of the manufacturing department. Each working team, activating in the CAD stage of the product lifecycle, has a manager, an administrator of the computer network, designers, verifiers and drafters.

In Fig.1 is exemplified a CAD result, for a combined stamping tool, for cut-out and draw, used for processing sheet metal parts.

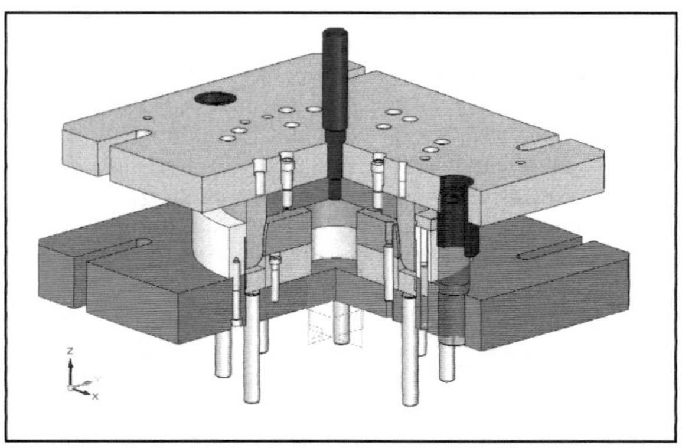

Fig. 1. Combined stamping tool, for cut-out and draw.

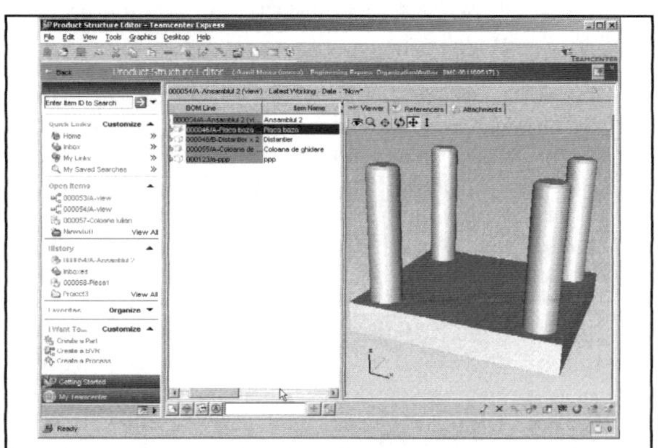

Fig. 2. The access to Product Structure Editor module in TcX.

TeamCenter Express was adopted as environment for the implementation of the PLM solution. The development of the design procedures was done with a team that realized the design of the components on the established network, with the resources managed by the *TcX* system. The database assigned to the project is automatically realized, when saving each component of the project. These components represent parts, drafts, sub-assemblies and the final assembly. The access to the components of the project is possible with the aid of a component of the *TcX* system, *Product Structure Editor*, as it is shown in Fig.2.

For this stage, there are realized the procedures for the management of the design data, procedures which can be appealed by the project manager. These procedures display the project's status and the activity of each member of the team.

There are in course to be elaborated the work-flow procedures for computer-aided management of the assignment of the work tasks (design, verifying, drafting, sign-up). For the product structure, taken over from the bill of materials of the final assembly, there are designed procedures for data management, for the management of the design activities and for the work stages for product development.

4. CONCLUSION

The paper proposes a collaborative engineering methodology for developing some type projects, specific to the machine manufacturing industry. The realization of projects for some stamping devices, dies and injection moulds, using some computer aided design software, uses the experience and the specialty knowledge of the industrial partners and the abilities of the teams from universities in the field of CAD software at international level.

The development of design knowledge at the level of the collaborative network is assured by integrating the CAD, CAM and CAE modules into an integrated solution, into an open environment, leading to a greater efficiency of integrating information in the designed products and processes.

These approaches are stipulated in the phase of product conception but also in that of presenting and realizing some visualization elements, product promotion and presentation catalogues, using specialized software and compatible to the software used in computer aided design.

5. REFERENCES

Ausura, B. & Deck, M. (2003). The "new" Product Lifecycle Management systems: What are these PLM systems? And how can they help your company do NPD better? *Available from:* http://www.pdma.org/visions/jan03/plm.html, *Accessed:* 2007-12-12.

Grieves, M., (2006), *Product Lifecycle Management. Driving the Next Generation of Lean Thinking*. McGraw-Hill, ISBN 0-07-145230-3, New York.

Usher, J. M.; Roy, U. & Parsaei, H. (2005), *Integrated Product and Process Development: Methods, Tools, and Technologies*, John Wiley & Sons Inc., ISBN: 978-0-471-15597-3, New York.

Merticaru, V.jr; Musca, G & Recio-Perero, P. (2008). Using CAD instruments for Collaborative Product Development in PLM Environment, *Proceedings of Annual Session of Scientific Papers "IMT Oradea" – May 2008 & Annals of the Oradea University, Fascicle of Management and Technological Engineering*, Vol.VII(XVII), ISSN 1583-0691, Oradea, Romania.

Musca, G; Musca, E. & Merticaru, V.jr (2008). Implementing a PLM Strategy in Collaborative Network for Products Development. *Academic Journal of manufacturing Engineering,* vol.6, issue 3/2008, pp.99-106, ISSN 1583-7904, Timisoara, Romania.

Annals of DAAAM for 2008 & Proceedings of the 19th International DAAAM Symposium, ISSN 1726-9679
Editor B. Katalinic, Published by DAAAM International, Vienna, Austria 2008
Make Harmony Between Technology and Nature, and Your Mind will Fly Free as a Bird

THE UNIVERSITY'S FINANCIAL MANAGEMENT – THE NEED OF FINANCIAL RESOURCES DIVERSIFICATION IN HIGHER EDUCATION

MUSCALU, E[manoil]; TODERICIU, R[amona] & TITU, M[ihail]

Abstract: *We have tried to approach in this paper the problem of creating income just from the perspective of getting new founds. The specialists say that, "a performing university implements a politic focused on saving the founds, starting from the idea that in any organization, it doesn't matter the size, develop in time certain activities which can not be justified but are very expensive." It is compulsory that all the procedures from the university to be periodically revised and the one that do not correspond for a successful strategy of organization of the activities which generate income"*

Key words: *management, financial management, quality, higher education.*

1. INTRODUCTION

In order to obtain a new character, focused on change, a university needs a high level of financial resources and especially founds of which the university has to use them freely. To increase the financial support becomes a major priority because all over the world the Government tries to reduce the budget for the higher education institutions. Having in view these conditions a permanent preoccupation of the universities is the development of new strategies of diversification the extra budgetary financial resources, the analyses proves that the number of student will decrease and the students which pay the tuition fees will not be able to assure a sustainable financial resource. The research founds, the partnerships with other Universities from Romania and from abroad could be considered to be essential for the new strategy of diversification of the financial resources. Now, for all the Romanian universities regardless of size of profile the main and the vital problem in developing the activities is to attract supplementary founds which together with the founds from the budget to lead to a continuous development of the institution and consolidate its position on the market. Building a diversified financial base in an university supposes to create a portfolio of clients for meeting the expenses which grow continuously. An extended financial resource portfolio gives to the university the possibility to overcome the restrictions enforced by a unique financial resource. This stage it is characterized "as well through the University's special preoccupation of identifying the new financial sources – alternative sources- for compensating the government contribution, through the efficient use of the available resources but also through the ministry's attempts of improving the system of assigning the budget for the universities, of refining the formulas of financing ". As long as "the university's main mission, which characterize the past few centuries, was to create a capable administration in order to "manage" the government, the government accepted to be the only financial support having in view the conditions in which the universities had as main purpose to satisfy the need of qualified personnel for its services". The universities' purpose has been change in the past few decades, becoming institutions focused on satisfying the needs of the productive sectors and of the society in general. In the same time with this change "the government stops being the main and the only beneficiary of the public services from the higher education and therefore the main and the unique financial resource". (Todericiu, 2007)

2. THEORETICAL CONSIDERATION. WAYS OF ATTRACTING SUPPLEMENTARY FOUNDS

Attracting supplementary founds for the Romanian higher education has at this stage the following essential characteristics: the means of attracting founds are specific for each university, and inside the universities, the faculties have different preoccupations and potential having in view the profile and the forms of education (economics, law, sociology and psychology, informatics, foreign languages) for developing tuition fee education; t*o attract founds it is a conjugated action of the entire managerial team* from the university and faculty's level, the main role being held by the Rector; t*he university's prestige and the higher education quality* are very important in attracting founds. Starting from the need of attracting supplementary financial resources in the situation of the diminished financial support from the government, David Warner says that, each university has to diversify the financial resources and he emphasizes the development of some activities which "generate income". Of course we wonder which are these activities which will "generate income" for the universities? It is considered that, are some aspects of the "activities which generate income" all the generated incomes beside the financial support provided by the Ministry: *research incomes, external grants, incomes from the lucrative activities (business) and the tuition fees and the administrative taxes.*

From David Warner's point of view it is very important that before starting any activity in order to attract extra budgetary resources the top management of the *university has to assure itself that all the basically activities are managed accordingly and function as efficient as possible.* It is essential for the financial management that the universities to be aware of the fact that the easiest – but not the only one – way of obtaining the resources it is that of *maximizing the incomes from the base activities.* The academic management specialists say that the most used ways of attracting supplementary founds are:

a. Private institutions which benefit of the higher education services and products and the next few arguments come to sustain this statement: when new technologies are implemented the general academic training reduces the training or recycling period; the trained persons have a bigger productivity, productivity which it is transferred to the other employees and generate positive effects over the entire organization; the society in general (what justify the financial support from public founds) and the enterprises absorb the largest part of the external benefits generated by the graduates of the higher education system; the first and the most important beneficiary of the scientific and technique progress which it is made, are the universities and the enterprises. In a state or private organization "there are individual benefits of the one who graduated a higher education program: *the high value of the academic diploma on the labor market,* in the OCDE's countries, the participation rate on the labor market of the academic diplomats it is with 15% higher then the average. The unemployment rate it is 4,7% for academic diplomats compared with 8,5% the average, the academic diplomats' salaries are higher then the salaries of the one with a lower qualification".

b. Attracting founds through the scientific research development: the income obtained by the Romanian universities out of scientific research in the past few years it is growing (it is said that the higher education institutions without research it is not a true university) being essential for supporting, strengthening the teaching activity, especially for the senior year students and the graduates being a resource of new and important incomes through exploiting the created intellectual property.

c. Financial support through sponsorships: at the Romanian university level the sums obtained through sponsorships are insignificant because of the fact that the legislation does not encourage the firms to sponsor the cultural actions which could be sponsored but the sponsor has to be found and attracted in this process, a factor which depends by the university's capacity of persuasion and through diverse motivations to obtain financial support.

d. Financial support through European projects: since our country has been accepted between the European community's countries the Romanian universities have new financial opportunities. Now in the university's budget of expenses, the European projects' contribution it is a low percentage out of the total amount of the incomes. There are many ***financial resources*** and include organizations like: *The European found of Regional development "The European social found" (having in view the economic performances of a region, the two of them could be combined as creating the Found of Investment in the Academic Objective).* (Todericiu, 2007)

e. Tuition fee training: the tuition fees for the higher education could have positive consequences not only for the universities but for students too. But in the future, the students who pay will be more concerned to analyze the study period costs compared to the future incomes, without neglecting the possibility of non-monetary benefits after graduation. For now the Romanian universities collect low taxes and do not take into consideration the market's factors when they establish those tuition fees.

The development of the tuition fee education has to be a real preoccupation at the level of each higher education institutions; this creates a financial resource and an important component of the institution's image abroad regarding the receptivity, sensibility and adaptability of it to the social and economic environment for training and professional reconversion of the higher education graduates.

f. International students: the enrolment of the international students it represents one of the priorities around the world being a supplementary income resource. Having in view the increasing number of international students it is compulsory that in our country the tuition fees to be established by each university, the amount should be settled having in view the university's strategy and not to be imposed by the Ministry. For the success of these actions the universities have to ***be aware of*** what exactly *the international students are interested in, what make them choose one university or another.*

The Great Britain embassy's specialists, after a study made said that, choosing one university or another depends by the next aspects: *the training quality; the employment perspectives; the accessibility at social facilities; personal security; good accommodation conditions; climate conditions near the one from the origin country; qualification recognized internationally; the teaching system similar to the origin country eventually taught in a language of world wide spread; short term classes.* (Oprean & Ţîţu, 2008)

From this perspective Romania has to solve many of its problems like: *to restructure the curricula at European level, study programs in foreign languages, to improve the accommodation conditions in order to attract a high number of international students.* It is very important to identify the international student's market. In the IDP report it is mentioned that the students which would like to study abroad take their decision having in view the next factors: *the training quality, the employment perspectives after graduation, the training to be accessible, the life style and personal security.* The British Consulate has a different view. They consider that the next few factors are necessary for recruiting the international students: *"the qualification (diploma)after graduation has to have international recognition and appreciated by the student's origin country; the duration of the training has to be as short as possible; the classes to be in English (French could be an alternative); the course quality should be good; the curricula should be relevant for the students aspiration regarding the carrier (for example, first of all business, finance, computers, engineering, medical sciences, and design); good teachers and facilities; life support and transportation expenses"* The British consultant do not take into consideration the life style problems as being very important saying that, the foreign students come to study not to have fun! The Romanian universities whose strategy foresees the international students recruitment as a key way of increasing the incomes, and they should try to clarify the next few problems***:*** *What means of recruiting do they have use? What form of organization brings the best results knowing that in these situations the speed it is essential? What level of investment it is necessary and which are the necessary time spaces for having success? What social and academic problems the international students confront with?* From our point of view clarifying those problems it is essential in order to implement the new coherent strategies of attracting international students. (Ţîţu & Oprean, 2007)

g. Study loans – modern financial source for the higher education;

h. The programs entirely paid;

i. The lucrative activities;

j. Services selling and renting the facilities.

3. CONCLUSIONS

It is very important that the universities to think about the next few aspects: Which it is the corresponding percentage from the university's incomes which come from each one from the above identified areas? Is there an optimum state and if yes which is that? Which of these areas could produce the largest profit, from example the excess of the income compared to the expenses? How easy is for a institution to organize all kinds of activities of which costs are equal or bigger than the income? Which are the areas which could produce the largest income without any specific usage which could be used by the university? Which of the profit areas could be extended with minimum effort? For now doesn't exist a clear answer to these questions, but, they could be taken into consideration by the universities. (Muscalu, 2004), (Todericiu, 2007)

4. REFERENCES

Muscalu, E. (2001). *Universitary Strategies*, Eficient Publishing House, Bucharest

Oprean, C. & Ţîţu, M. (2008). *Quality Management in the Economy and Organisation Based on Knowledge,* AGIR Publishing House, Bucharest

Todericiu, R. (2007). *Universitary Management from the European and Globalising Perspective*, ULBS Publishing House, Sibiu

Tîtu, M. & Oprean, C. (2007). *Quality Management*, University of Piteşti Publishing House, ISBN 978-973-690-646-6, Piteşti

Tîtu, M. & Oprean, C. (2007). *Strategic Management,* University of Piteşti Publishing House, Piteşti

Annals of DAAAM for 2008 & Proceedings of the 19th International DAAAM Symposium, ISSN 1726-9679
Editor B. Katalinic, Published by DAAAM International, Vienna, Austria 2008
Make Harmony Between Technology and Nature, and Your Mind will Fly Free as a Bird

INTRODUCING THE PRODUCT LIFE CYCLE AND V – MODELL IN THE LEASING COMPANIES

MUSTEA - SERBAN, R[azvan]; FOLCUT, O[vidiu] & CIOCIRLAN, D[oinita]

Abstract: *In order to increase the competitiveness of the leasing companies, the paper proposes the implementation of system engineering in the procurement process, applied both to their own business and to the technical aspects of the financed assets. The paper is structured on three "pillars": the product life cycle management, the V Modell methodology and the process orientation. The implementation of the project management methodologies adjusted to the procurement activities (integrated V- Modell) allows the leasing companies to follow the product life cycle for the financed assets, in order to be aware of risks derived from the client's financial requirements such as technological structure of the product, the proper buy- back period and the needs for service.*

Key words: *leasing, V – Modell, system, process,*

1. INTRODUCTION

The legislation in force and the prospects of its evolution in the near future entail both at national and European levels, the aggregation of leasing companies into corporations able to cater both to the demands of market rigors and to the customers' complex requirements. Beyond consumption leasing there is a customer segment with a tremendous financial potential, which primarily requires solutions and not a mere purchase variant.

The accrued experience in the field of car leasing cannot be transferred as such to the new fields. The essential difference lies in the operational risk management strategy. Thus, the following are known in the case of vehicles: the functions of the respective good, means of use, necessary qualification. Taking this risk, coupled with the existence of a secondary market or of collateral security, makes total exposure acceptable (or even minor – in most cases).

Several critical points need to be outlined for industrial machinery and plant (Walker & Hampson, 2003):

- They are part of a manufacturing **system**;
- Technical, functional specifications and performance levels – are not necessary part of the toolkit of leasing company analysts;
- The secondary market is difficult to tap into directly – companies specialize din remarketing are used as a rule, when the contract is not secured through buy-back.

Consequently, the leasing company has either the option of specializing in certain fields – for instance metalwork machinery leasing – or of defining and following a procurement model through which it can tap into a wider range within the new fields.

Specialization in itself leads to market atomization and, consequently, goes against the consolidation trend evinced by strong companies and otherwise supported by most Central Banks.

The remaining eligible variant is defining complex procurement models based on project management, since any client calling for a solution (not necessarily a good) shall be the user of a production system that is in fact property of leasing companies.

In other words, limiting and controlling operational risk from the standpoint of leasing companies can only be undertaken with their involvement in technical issues – as assumed owners.

Thus one of the major differences from credit financing becomes apparent, namely: if in the case of a manufacturing system, the client submits a business plan (the project) that the bank checks from a financial and technical point of view, in the case of leasing, the company will interact with the client in drawing up the business plan, through its own procurement model.

These methodologies are not world novelties, since they have been used by Defense Ministries and Public Administration in European countries in order to have an effective and efficient control over public money engaged in defense or public infrastructure procurement.

2. INTRODUCING THE PRODUCT LIFE CYCLE AND V – MODELL IN THE LEASING COMPANIES

The leasing company is – objectively – under obligation to understand the life cycle of the product it finances. If in the case of cars this visibility is still necessary, especially in the case of used cars, taking into account real estate development and re-technologization, it turns into a critical desideratum in approaching a new leasing product.

Having emerged in the field of IT product development, the V-model is the German standard for federal administration and defense projects.

Its applicability for leasing companies stems primarily from the status of ownership. Thus, when setting up a new financial product, the frequency of "sale" for this product in time shall be considered. This frequency is determined by the user's (functional and design) requirements to be found in the left branch of the V-model.

Product Life Cycle Management

The market evolution stages can be identified simultaneously with the life cycle (Westland, 2007). Thus:

i. Market formation – the latent demand for a certain product category is outlined by introducing the new product.
ii. Market expansion – other companies enter the market, and the number of informed customers for the product category increases.
iii. Market fragmentation – the entire field is subdivided into countless competing groups due to crowding competition.
iv. Market consolidation – companies start leaving this field, due to fierce competition, dropping prices and decreasing profits.
v. Market termination – clients do not support the same level of demand and companies abandon the product.

In our approach, the product life cycle management is two-fold applicable:

- *Internally, by drawing up the financial product strategy to be launched on the market.*
- *Externally, by validating the product life cycle advocated by the client.*

For instance: if a customer requires a technological line for the specialized manufacturing of metallic structures, their plan will be adjusted / interactively remodeled, by taking into account the product life cycle (which can be long-lived yet also exposed to operational and market risks that the leasing company decides to accept or not).

The V-Modell Methodology

To summarize, this methodology is a graphical rendering of the life cycle and comprises the main steps to be taken according to the project evolution, so that the desired outcome is secured (Droschel & Wiemers, 1999). In practice, V-Modell describes the activities and outcome it needs to put forth while developing the system.

Even though initially the "system" meant software, modern approaches in numerous other fields involve the design of a system at the same time with the product/service to be supplied on the market. Thus the company redefines its internal processes, aiming at setting up a unitary, coherent and functional whole that will provide the market – under desired circumstances – precisely with the desired product, and not solely with the product itself (Boehm, 1981). This integrated approach ensures visibility of costs and outcomes as early as the initial stages and, moreover, the ability to control the economic phenomenon throughout the life cycle and to mitigate exposure due to strategy and decision-making errors.

The left branch of the V-Modell represents the flow of specifications, where system requirements are defined. The right branch is the test flow as per specifications. The peak – where the two branches meet – corresponds to the development itself (the actual emergence of the product). The V representation allows for monitoring the mapping of demand specifications onto their testing. The differentiation between validation and verification resides in the fact that verification is performed by comparing the project execution stage with the company's internal specifications that are specific for the product under analysis, while validation entails an external comparison of the product against the requirements of the customer/market and decision-makers within the company.

Thus, the V-Modell defines a unitary product development procedure, and at the same time describes "who" has to do "what" and "when" according to every stage. Product installation occurs when the latter is checked for all predefined design requirements. The operational qualification matches the validation of the product's operational requirements as defined by the customer and by the leasing company's decision-makers, as well as by the technological, legal or commercial limitations at the given time. Performance qualification concludes the V model branch and attests the validity of the requirements of the specified user / market at the onset of the development process.

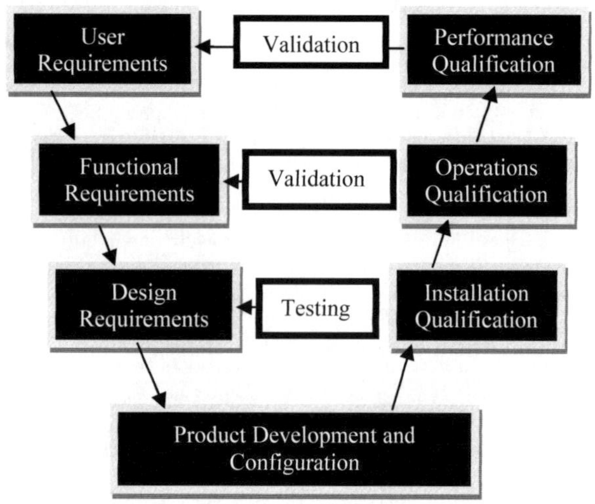

Figure 1. Representation of the V-Modell methodology

Adjusting the V-modell to the actual situation involves extending or restricting the outlined stages for the two branches function of current needs. Figure nr. 1 has provided you with a summary of this methodology.

Process-Orientation

Although loosely used, the concept of process proves its usefulness only once it has been properly understood. Applying models to company strategies needs to draw on a good definition of processes.

The process (in the business sense) – is a sum of activities conducive of value by turning inputs into higher-value outputs. Both inputs and outputs can be physical objects or information, and transformations can be carried out by people, machines or complex systems comprising both people and adequate tools.

Processes are classified as follows (Nelis, 2008):

1. Management processes – that govern the operation. Typical examples are "strategic management" and "company operation".
2. Operational processes – are those giving rise to value flows and that are part of the operation core. Examples: processes whereby materials are prepared to enter the manufacturing flow or, in the case of leasing companies, the process of analyzing request files.
3. Support processes – they back main processes. Examples: Accounting, Staff recruitment, etc.

Any process can be broken down into several sub-processes, each with their own attributes but that contribute to fulfilling the objective of the process they stem from. Process analysis is performed by breaking them down into activities that do not include and decision-making and, therefore, don't need to be further broken down.

In order to undertake a business process company resources are used and the process is at the same time limited by constraints – of various natures - , so that target achievement should comply with all conditions imposed by the business environment.

One of the tools used in process analysis is the ICOR (inputs-constraints-outputs-resources) diagram. Its usefulness is obvious when the super-process is traced back starting at the bottom (the bottom-up approach – i.e. from sub-processes upwards), and it is necessary to coordinate the input and output along the string of processes.

3. CONCLUSION

The applicability of project management in the leasing businesses stems from systems engineering, and it has become a necessity once the complexity of a leasing operation is recognized and the need for planning in the project management sense – a technique equally well defined in SE (systems/software engineering).

As a natural outcome of this BUSINESS-SYSTEMS connection, another major standard can be applied in the leasing industry, namely the CMMI – Capability and Maturity Model Integrated. It outlines the path to follow in order to achieve higher performance levels in developing complex systems. We undertake to pursue further, extended research along this line in the future.

4. REFERENCES

Boehm, B. (1981): *Software Engineering Economics*, Prentice-Hall Inc., ISBN 978-013822122, London

Droschel,W.;Wiemers,M. (1999): *Das V-Modell 97*, Oldenbourg, ISBN 978-3486250862, Berlin

Nelis, J. (2008): *Business Process Management – Second Edition,* Butterworth-Heinemann, ISBN 978-0750686563, Oxford

Walker, D.;Hampson, K. (2003): *Procurement Strategies: A Relashionship - based Approach,* Wiley-Blackwell, ISBN 978-0632058860, Oxford

Westland, J. (2007): *The Project Management Life Cycle,* Kogan Page, ISBN 978-0749449377, London

Annals of DAAAM for 2008 & Proceedings of the 19th International DAAAM Symposium, ISSN 1726-9679
Editor B. Katalinic, Published by DAAAM International, Vienna, Austria 2008
Make Harmony Between Technology and Nature, and Your Mind will Fly Free as a Bird

MOBILE 3D OBJECT MEASUREMENT SYSTEM BASED ON ACTIVE TRIANGULATION METHOD

MUTKA, A[lan]; NIZETIC, J[osip]; CURKOVIC, M[ilan]; STAROVESKI, T[omislav] & UDILJAK, T[oma]

Abstract: *This paper describes a mobile 3D object measurement system based on active triangulation method. By using an off-the-shelf video projector and professional digital camera, good 3D reconstruction is obtained which can be used in many practical applications. A new iterative closest point calibration method (ICPC) is implemented which provides automatic system calibration and improves robustness. The system is particularly suitable for medical applications like body capturing or human face reconstruction.*
Key words: *Face reconstruction, 3D triangulation, Vision systems, Image processing*

1. INTRODUCTION

3D shape reconstruction has been an important field of research for many years. There are a lot of reconstruction methods based on triangulation like basic stereo vision, stereo vision based on simple laser light source (Lubeley et al., 2005) or coded light (Wu et al., 2006). All these methods are divided in two basic groups – passive and active. Passive reconstruction method generates results from two or more images without influencing the scene with additional energy. Algorithms for passive reconstruction are very complicated and have problems with robustness. Active methods are using and additional energy source like a laser stripe or texture generated by video projector. Additional energy improves reconstruction robustness and provides much better results. Our method is based on active stereo vision using a video projector as additional energy source and Canon EOS professional digital camera. The Video projector generates red stripes which are moved over the measured object and images are taken by the camera.

The most difficult task in 3D shape reconstruction using projected stripes is stripe peak detection. In (Forest et al., 2004) laser stripe peak detection method using a FIR filter is presented. The peak of projected stripes must be detected in subpixel precision to obtain satisfactory results. This article describes a new method for peak detection based on kernel shapes. 3D shape reconstruction points calculated using triangulation equations (Labuz, 1988) contain a lot of falsely detected points, which must be recognized and removed using advanced filters like Beziere or LSP (Bourke, 2008). Surface mesh triangulation is generated using additional algorithms like *Delaunay* the triangulation method.

In section II. the system configuration for 3D shape reconstruction is introduced. Section III. describes kernel shapes for line detection and methods for locating peaks in subpixel precision. The filtering method for obtained 3D points is presented in section IV. CogniLine digitalization software, automatic calibration and results are presented in section V.

2. SYSTEM CONFIGURATION

System configuration for 3D shape reconstruction is presented in Fig.1. Camera and projector have aligned optical axes in order to simplify the geometry of getting 3D points. *DistanceD*, *DistanceO*, *DistanceY* and *DistanceX* are system parameters which are obtained by calibration. System calibration is extremely important to get correct results and new algorithms for additional tuning are developed to simplify and improve the calibration procedure (*Iterative Closest Point Calibration - ICPC)*. The video projector generates vertical lines which are shot with digital camera and processed by line extraction image processing algorithms.

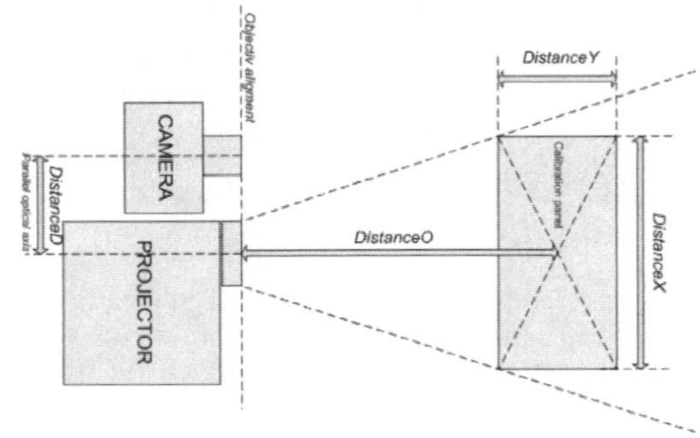

Fig. 1. System configuration for 3D shape reconstruction

3. SUBPIXEL PEAK DETECTION

Detecting the projected stripe can be improved by using controlled light conditions or manipulating camera parameters that control light adjustment (Fig. 2.).

Red stripe detection is implemented in HSV color space. Experimental results show that the most accurate detection is obtained using just the V component of HSV space. Image is also converted to black and white using additional threshold to decrease computational time.

Fig. 3. shows kernel shapes that can be used for vertical line detection. Experimental results show that shape and orientation can have a significant effect on final detection results. The kernel most commonly used in image processing (Fig. 3e.) generates only 65% of maximum detected line points produced by our kernel (Fig. 3b.).

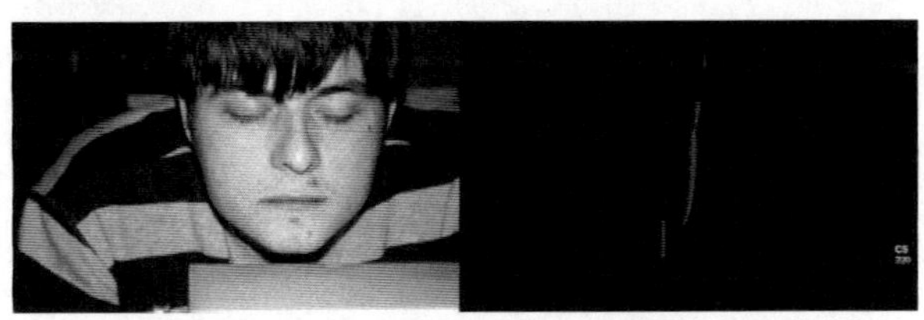

Fig. 2. Projected red stripe on face in controlled light conditions

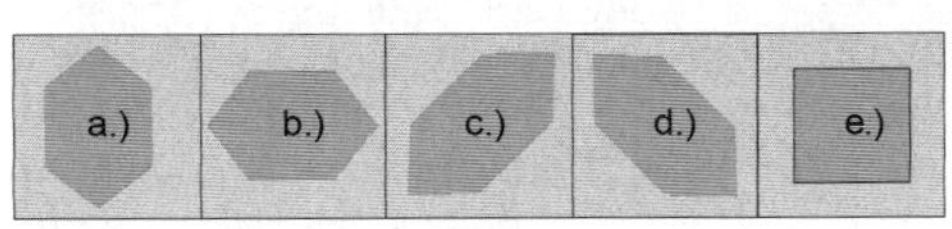

Fig. 3. Kernel shapes for line detection

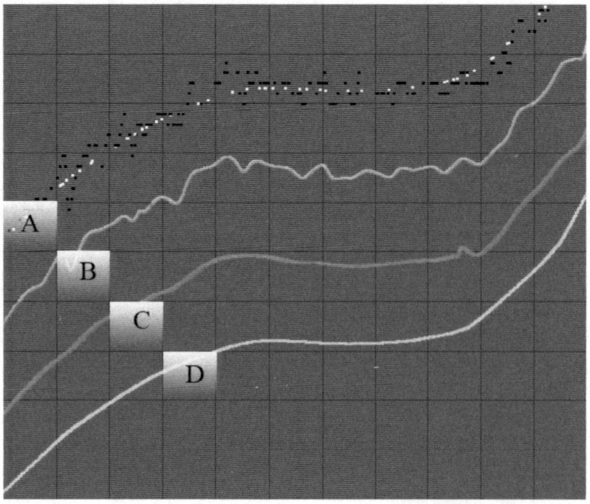

Fig. 4.a) The initial 3D points, b-d) Bezier R=10, 50, 100

The vertical line must be detected in subpixel precision to obtain acceptable results. First, the optimal central pixel position is obtained using the weighted average algorithm (Fisher & Nadiu 1996). In the second step Blais and Rioux linear peak detector (Blais & Rioux 1986) is used since it provides the best results for subpixel position estimation (Fisher & Nadiu 1996). Due to its linear nature, the algorithm is fast and accurate what makes it suitable for multiple image processing.

4. 3D POINTS FILTERING

From detected 2D subpixel line points and known system parameters (Fig.1.), the triangulation method generates 3D points in space. Generated 3D line points usually diverge an ideal continuous line (Fig. 4a.). Applying algorithms based on Bezier curves (Bourke, 1996) and Least-square projection (LSP) methods, the obtained points are converted to a smooth line or surface (Fig. 4b.-4d.).

Since the standard Bezier method approximates every point with other points from the point cloud, this can in some cases lead to unwanted consequences. One of them is great computational complexity when a large point cloud is used. To preserve the authenticity of the surface, every point is approximated using points from given radius R. Complexity difference between the standard Bezier method and Bezier limited with radius R is presented in equation (1).

$\Theta new(n)$ ≅ ***Fehler! Textmarke nicht definiert.*****Fehler! Textmarke nicht definiert.Fehler! Textmarke nicht definiert.** $\left(\frac{R}{n}\right)^d \cdot \Theta std(n)$, d-dimension $\in \{1,2\}$ (1)

Fehler! Textmarke nicht definiert.**Fehler! Textmarke nicht definiert.Fehler! Textmarke nicht definiert.**$\Theta std(n)$ - number of FLOPS in standard Bezier

$\Theta new(n)$ - number of FLOPS in Bezier with radius R

5. COGNILINE

CogniLine is a software solution intended for 3D surface reconstruction, especially body parts digitalization. The great advantage of this system over other solutions on the market is the accessibility, mobility, quality and robustness. Surface reconstruction can be obtained using any digital camera and video projector. A small deviation in projector angle, or a drift from the ideal system geometry, significantly affected the results. Therefore system configuration and calibration is simplified and results are improved using special algorithms for parameter tuning.

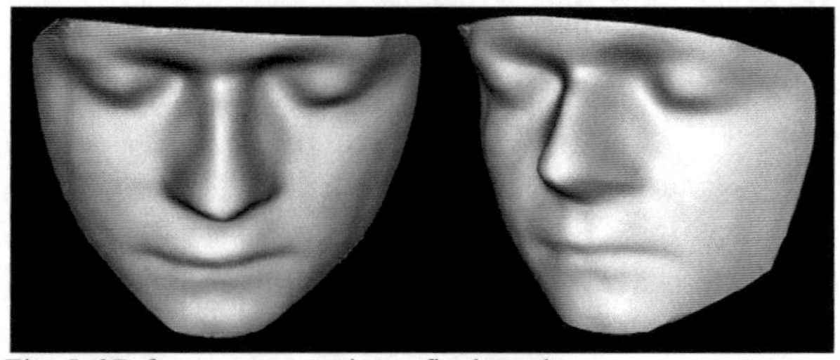
Fig. 5. 3D face reconstruction – final result

The ICPC method is based on matching two clouds of points using *Iterative Closest Point(ICP)* algorithm. ICPC matches a predefined virtual 3D point model with the real 3D point model reconstructed from measurements and the system is automatically calibrated to obtain a perfect match.

For the purposes of displaying the obtained results, it is necessary to convert 3D points into mesh. VTK (Visualization Toolkit) (Schroeder et al., 2007) is used to triangulate and filter points using special correction filters to get satisfying visualization results. Final 3D face reconstruction result is presented in Fig. 5.

6. CONCLUSION

This paper describes a simple and robust 3D object measurement system based on active triangulation method. By using off-the-shelf video projector, professional digital camera and advanced CogniLine software, satisfying results for practical applications are obtained. Information from the real world must be collected accurately to get as much information as possible, and then additional processing is carried out to get desired result. CogniLine is tested on face reconstruction problem but it can be used for any medical and other practical application where a simple and robust method for 3D reconstruction is needed.

7. REFERENCES

Blais, F & Rioux, M. (1986). *Real-Time Numerical Peak Detector,* Signal Processing 11*, 145-155, 1986.*

Fisher, R.B.; Nadiu, D.K. (1996). *A Comparison of Algorithms for Subpixel Peak Detection,* Dept.of Artificial Intelligence, University of Edinbourgh, UK, http://citeseer.ist.psu.edu/782699.html

Forest, Josep; Salvi, Joaquim; Cabruja, Enric & Pous, Carles (2004). Laser stripe peak detector for 3D scanners. A FIR filter approach, *Proccedings of ICPR'04*, Cambridge, UK.

Labuz, J. *Triangulation of surface points with cameras and projectors,* Dept. of Electr. & Comput. End., South Carolina Univ.,Columbia, SC, The proceedings of the 20. Southeastern Symposium on System Theory, 1988.

Lubeley, D.; Jostshulte, K.; Kays, R.; Biskup, K.; & Clasbrummel, B. (2005). Laser pattern based 3D object measurement system for medical applications, *Proccedings of EMBEC'05*, Prague.

Paul Bourke: *Bézier curves,1996. Available from:http://astronomy.swin.edu.au/~pbourke/curves/bezier Accessed:* 2008-07-24

Schroeder, William; Martin, Ken; Lorensen, Bill (2007) *The Visualization Toolkit An Object-Oriented Approach to 3D Graphics, 4th Edition,* Kitware, Inc. , ISBN 1-930934-19-X

Wu, H.B.; Chen, Y.; Guan, C.R. & Yu, X.Y. (2006). *Measurement Technology by Structured Light Using Stripe-Edge-Based Gray Code,* Journal of Physics: Conference Series 48 (2006) 537-541

Annals of DAAAM for 2008 & Proceedings of the 19th International DAAAM Symposium, ISSN 1726-9679
Editor B. Katalinic, Published by DAAAM International, Vienna, Austria 2008
Make Harmony Between Technology and Nature, and Your Mind will Fly Free as a Bird

NUMERICAL SIMULATION OF TURNING TOOL VIBRATION DURING CUTTING PROCESS

NAD, M[ilan]; DURIS, R[astislav] & ESTER, J[ozef]

Abstract: *The dynamical analysis of straight turning tool during machining process is performed. Numerical simulations of turning tool vibration for different cutting state are realised by finite element method (FEM). The cutting forces following from turning process are used as loading forces to excite the computational model. The effects of all cutting force components (tangential, axial and radial) are considered. In the numerical simulation of cutting process of straight turning, the harmonic time-dependency of all forces components is considered. The amplitude-frequency characteristics of tool tip deflection are analysed for different parameters of cutting process.*
Key words: *turning, cutting, force, vibration, FEM.*

1. INTRODUCTION

Machining is a term that covers a large collection of manufacturing processes designed to remove unwanted material, usually in the form of chips, from a workpiece. Machining is used to convert the basic geometrical shapes or shapes manufactured using different technologies (castings, forgings a.o.) into desired shapes, with size and finish specified to fulfil design requirements. Generally, it can be said that almost every manufactured product has components that require machining. Therefore, this collection of processes is one of the most important among the basic manufacturing processes because of the value added to the final product. Machining processes can be performed by using a wide variety of machine tools. One of the most-known machining processes is turning. Turning is defined as a machining process generating external surfaces of revolution by the action of a cutting tool on a rotating workpiece, usually in a lathe. These include facing, straight turning, taper turning, grooving and cut-off, threading, tracer turning, boring, a.o. These machining processes are realized by tools which have specific shapes. The requirements on design and production of the machines and tools with higher performance, materially and economically effective, are increasing in the recent years. The increasing of performance is mostly achieved by increasing of operating (cutting) speeds and cycles. The requirements on material effectiveness tend to the machining equipments and facilities with decreasing stiffness. All these changes affect on their dynamical properties and within the context of this state, the machining equipment and facilities are often loaded by external dynamical effects producing undesirable dynamic difficulties. The cutting velocities, cutting forces, chip creation process, stiffness of the system MTW (Machine-Tool-Workpiece) are the most significant effects influencing the dynamics of system of MTW and machining process (roughness of machined surface, tool wear, damage of tool or workpiece, noise level a.o.).

Generally, the tools are the most critical members of the MTW system. The most serious problems occur when the frequency of periodical change of some significant effect of machining process is near to some of the natural frequencies of machining tool - resonant state (Nánási, T., 2006).

Numerical analysis of the effect of cutting process on dynamical behaviour of straight tool during turning is presented in this paper.

2. FORMULATION OF THE PROBLEM

2.1 Cutting forces during turning

Analysis and mathematical formulation of cutting forces is a very complicated problem. Many different factors influence the magnitude and time-dependency of cutting forces. The dominant effects for determination of cutting forces are cutting resistance, cutting speed, longitudinal feed speed, shapes of cutting edges, chip creation, a.o. Cutting force is a dynamical phenomenon, i.e. it is time-dependent quantity. In the first period, the cutting force is growing to the maximum value. Maximum value is characteristic for separating of chip element from basic materials. In next period, the cutting force is decreased, but no decreased to non-zero value. Before release of sliced chip element the tool starts to deform the next chip element.

The typical dynamic character of cutting force acting on tool is evident (see Fig. 1) from experimental measurements (Beňo & Maňková, 2004).

During turning, the cutting force consists of three components:

- F_c -tangential component - primary motion,
- F_f -axial component - feed motion,
- F_p -radial component - cross feed motion.

Then, the resulting cutting force is defined by equation

$$F = \sqrt{F_c^2 + F_f^2 + F_p^2}\ . \tag{1}$$

On a basis of experimental results, the individual cutting forces can be calculated (λ_s = 0°, κ_r = 45°, γ_0 = 15°) using following expressions (Janáč et al., 2006)

$$\begin{aligned} F_c &= k_c S = k_c f a_p, \\ F_f &= (0{,}3 \div 0{,}4) F_c, \\ F_p &= (0{,}4 \div 0{,}5) F_c, \end{aligned} \tag{2}$$

where k_c - specific cutting resistance, f - magnitude of feed motion, a_p - dept of cut, λ_s - tool cutting inclination, γ_0 - tool orthogonal rake angle, κ_r - tool cutting edge angle.

The cutting forces used for numerical simulation of turning tool vibration are considered in the form of time harmonic function. This function approximately corresponds to curve of cutting force obtained experimentally (Fig.1). Amplitude of any cutting force depends on cutting conditions and we can suppose

$$F_{a,i} = (0{,}15 \div 0{,}25) F_{i,mean} \tag{3}$$

and each of the cutting forces can be expressed

$$F_i = F_{i,mean} + F_{a,i} e^{i\omega t}\ , \tag{4}$$

where $i = c,\ f,\ p$ and ω is angular frequency of forced forces.

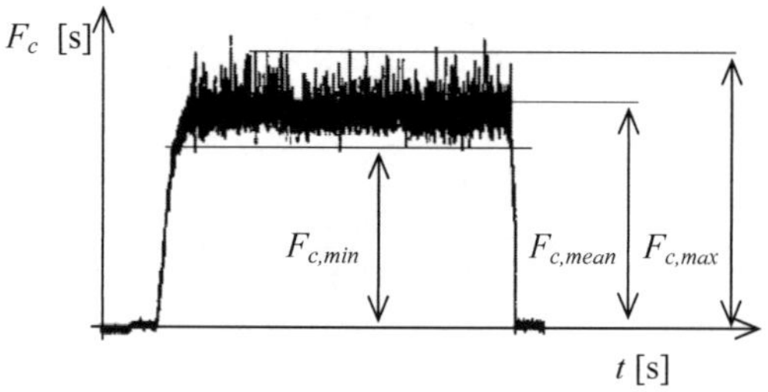

Fig. 1. Time-dependence of cutting force

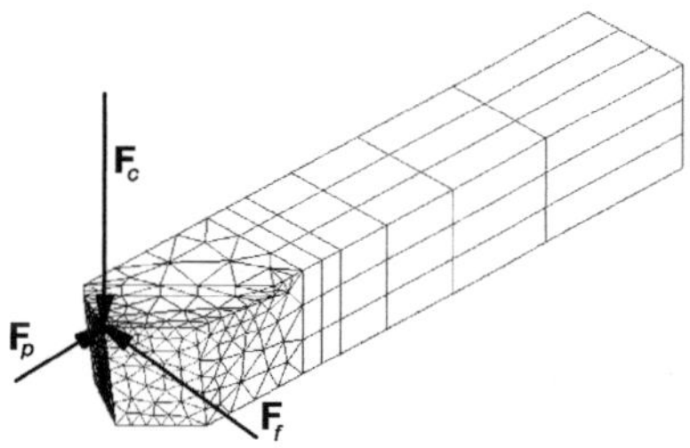

Fig. 2. FEM model of turning tool.

2.2 FEM model of turning tool

Generally, the turning tool can be considered as cantilever beam structure. One side of this tool is clamped in a tool post and other end of tool is free. The cutting forces are acting on the free side of turning tool. Numerical simulations of turning tool vibration during cutting process are realized using the three dimensional computational model of tool (Fig. 2).

The equation of motion describing the forced vibration of turning tool in matrix notation is expressed by

$$\mathbf{M}\ddot{\mathbf{q}} + \mathbf{B}\dot{\mathbf{q}} + \mathbf{K}\mathbf{q} = \mathbf{F}, \tag{5}$$

where **M**, **B**, **K** are mass, damping, resp. stiffness matrices, **F** is vector of exciting forces and $\mathbf{q}, \dot{\mathbf{q}}, \ddot{\mathbf{q}}$ are vectors of displacements, velocities, resp. accelerations.

The response of turning tool to the acting cutting forces (for which harmonic form is considered) is realized using harmonic analysis. The vector of exciting forces on the right-side of equation (5) must be considered in the form $\mathbf{F}_a e^{i\omega t}$, where $\mathbf{F}_a$ is a vector of amplitudes of exciting forces.

3. RESULT OF NUMERICAL SIMULATIONS

The analysis was performed (Ester, 2008) for turning tool (25×25×140 mm, STN 22 35 14) with angles $\lambda_s = 0°$, $\kappa_r = 45°$, $\gamma_0 = 15°$. Material parameters: $\rho = 8300$ kgm^{-3}, $E = 210$ GPa, $\mu = 0{,}3$. The tool is fixed in the tool post - free part of tool has the length $l_f = 50$ mm.

The cutting forces are calculated by using the equations (2) and their values for different cutting parameters are shown in the Table. 2. The amplitudes of the cutting forces are calculated using the equation (3).

No. of mode shape	Natural frequency [Hz]	Type of mode shape
1	7162	1st bending mode - *y*
2	8327	1st bending mode - *x*
3	16553	1st torsion mode - *z*
4	24838	1st longitud. mode - *y*
5	27284	2nd bending mode - *y*

Tab 1. The natural frequencies and mode shapes.

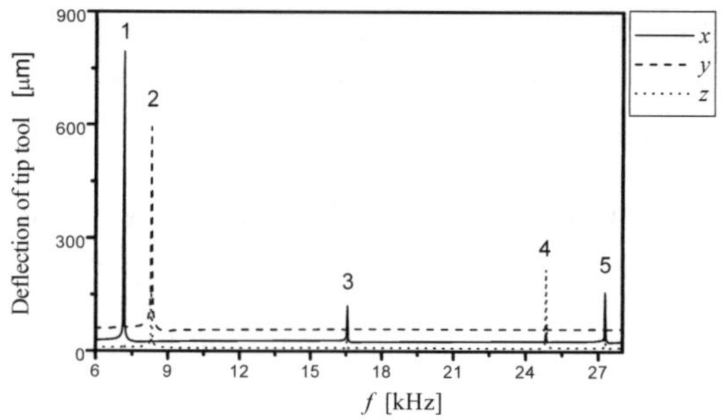

Fig. 3. A-F characteristics of tip tool deflection (15% of amplitude of cutting forces; $a_p = 3{,}0$ mm; $f = 0{,}1$ mm; $l_f = 50$ mm).

Parameters of cutting		Cutting forces F [N]		
depth	a_p [mm]	3	3	3
feed	f [mm]	0,1	0,25	0,35
F_c [N]		1030	1420	1840
F_p [N]		515	710	920
F_f [N]		412	568	736

Tab 2. The cutting forces for different parameters.

The harmonic analysis is performed numerically using FEM (Žmindák & Grajciar, 2004).

The selected results obtained by using harmonic analysis are presented on amplitude-frequency (A-F) characteristics (Fig. 3) of tip tool deflection. The similar results are obtained for other parameters, not presented in this paper.

4. CONCLUSION

The dynamic analysis of turning tool vibration during machining process is presented in this paper. The A-F characteristics confirm the existence of resonant states for some conditions of cutting process. At the resonant state the extreme amplitudes of tip tool are occurred (Fig. 3). From these reasons, the cutting process is inappropriate and this state leads to inaccuracy of geometrical shape of machined parts, increasing of surface roughness, increasing of tool wear, abnormal vibration in the system MTW. In many cases this state leads to destruction of elements of MTW system.

The numerical simulations and calculation procedures presented here provide relatively quick, cheap and effective virtual tool to determine machining tool parameters and cutting processes. Results obtained are used to optimize the cutting tool in the design stage and to establish optimal parameters of machining process.

The author is grateful to thank for the financial support for this work under projects AV-4/0102/06, KEGA-3/4154/06.

5. REFERENCES

Beňo, J.; Maňková, I. (2004). *Technological and material machining factors,* Vianela, ISBN 80-7099-701-X, Košice (in Slovak)

Ester, J. (2008). *The effect of turning process on dynamical behaviour of cutting tool,* Diploma Thesis, STU, Trnava (in Slovak)

Janáč, A.; Lipa, Z. & Peterka, J. (2006). *Theory of machining,* STU, ISBN 80-227-2347-9, Bratislava (in Slovak)

Nánási, T. (2006). *Exotic natural frequencies of coupled beams,* Proc. of 11th Int. Acoust. Conf. - Noise and Vibration in Practice , Žiaran, S. (Ed.), pp. 71-74, ISBN80-227-2425-4, Kočovce, June 2006, STU, Bratislava

Žmindák, M.; Grajciar, I. (2004). *Modelling and computation in FEM-Part I – Modelling in ANSYS.* VTS Žilinská univerzita, ISBN 80-968823-5-X, Žilina (in Slovak)

Annals of DAAAM for 2008 & Proceedings of the 19th International DAAAM Symposium, ISSN 1726-9679
Editor B. Katalinic, Published by DAAAM International, Vienna, Austria 2008
Make Harmony Between Technology and Nature, and Your Mind will Fly Free as a Bird

THE DETERMINATION OF THE OPTIMAL CIRCULATION FLUIDS SPEEDS THROUGH THE HEAT EXCHANGERS

NAGI, M[ihai]; ILIES, P[aul] & NEGOITESCU, A[rina]

Abstract: *In the paper is presented the mode of establishing the optimal circulation speeds of the two fluids through compact heat exchangers made by aluminium. These heat exchangers equip various internal combustion engines. Based on experimental researches are mapped out the energetic efficiency variation versus the fluid circulation speeds and the energetic efficiency variation versus the volume efficiency. It can be observed that to a single speed of a fluid corresponds a single speed of the other fluid for which the energetic efficiency is maximal. The optimal values of the two fluids circulation speeds are obtained.*

Key words*: fluid, heat, volume, efficiency, rib.*

1. INTRODUCTION

In order to choose a heat exchanger from various possible solutions it would be preferred that device (Badea&Necula, 2000) which transmits the constrained heat flow $\dot{Q}$ with a minimum consumption of energy (power P) for the two fluids circulation through the device, and which occupies the lower volume.

The decrease of the heat exchanger volume can be obtained by increasing the circulation speeds of the two fluids (for an existing device) or modifying the heat transfer surfaces by choosing surfaces with turbulence generators (Kays&London, 1984). Both methods lead to the increase of the consumed power for fluids circulation.

The thermal flow transmitted between the two fluids into a heat exchanger depends almost entirely on the two fluids speeds whether their entrance temperatures are constant. The increase of the fluids circulation speeds also increases the required power for fluids circulation.

Under these circumstances the problem is to establish the optimal circulation speeds of the fluids in order to obtain a solution as economic as possible for the selected device (Badea&Necula, 2000; Negoitescu, 2003).

2. THE COMPARISON OF THE HEAT EXCHANGERS

In order to compare various designs of heat exchangers are used the following criterions (Nagi & Negoitescu, 1998; Badea&Necula, 2000):

- *The total thermal transfer coefficient.*

It is a basic criterion according to which the thermal flow is related to the heat transfer surface A and the medium temperature difference Δt_m between the two fluids.

$$k=\frac{\dot{Q}}{A\cdot\Delta t_m}\quad\left[\frac{W}{m^2K}\right] \tag{1}$$

-The volume efficiency "ε_v"

If in relation (1) the surface of heat transfer is replaced by the volume V [m^3] it will be obtained:

$$\varepsilon_v=\frac{\dot{Q}}{V\cdot\Delta t_m}=\frac{k\cdot A}{V}\quad\left[\frac{W}{m^2K}\right] \tag{2}$$

Both criterions of comparison give information about the thermal performances by the total thermal transfer coefficient but give no information about the consumed energy in order to obtain this coefficient value.

-The energetic efficiency "ε_e"

Under these circumstances the thermal flow transmitted for a temperature difference between the two fluids equal to one degree is related to the requisite power for fluids circulation through the device:

$$\varepsilon_e=\frac{\dot{Q}}{\Delta t_m\cdot P}=\frac{k\cdot A}{\dot{V}_1\cdot\Delta p_1+\dot{V}_2\cdot\Delta p_2}\quad\left[\frac{W}{KW}\right] \tag{3}$$

Where:

$\dot{V}_i$ [m^3/s]-the fluids volume flows;

Δp_i [Pa]-the fluids pressure losses (i=1,2).

The last criterion can be rigorous applied only when the thermal power ($\dot{Q}/\Delta t_m$) of the heat exchangers for the constrained volume flows is the same.

From this point forward the optimal circulation speed of the two fluids represents the pair of values (W_1, W_2) that leads to the maximum values of the energetic or volume efficiency.

As the pressure losses in compact heat exchangers on the oil or water side are relative large the requisite power for fluids circulation is large too.

3. EXPERIMENTAL RESEARCHES

By experimental researches were established he optimal circulation speeds of water and oil through an oil cooler with corrugated ribs on the oil side (21 oil tubes).

In Figure 1 is plotted the energetic efficiency variation versus the oil circulation through the cooler tubes. It can be observed that to a single speed of the oil corresponds a single speed of the water for which the energetic efficiency is maximal. Thus, the optimal values (w_u, w_w) of oil and water are obtained.

The less the oil speed is, the more important is to choose correctly the value of the water speed. For an oil speed between 0.6…0.8 m/s, the economic efficiency can decrease below the half optimal value, if the value of the water speed is too large. The water and oil circulation speeds can not be independently selected. Each oil speed corresponds to a single economic value of the water speed.

The optimal circulation speeds of water and cooling fluid for a car radiator with corrugated ribs on the air side were also established by experimental researches (Nagi & Lelea, 1997). In Figure 2 is plotted the energetic efficiency variation versus the air circulation speed through the radiator.

Fig. 1. The energetic efficiency variation versus the speeds of oil and water, ε_e=f(w_w, w_a)

Fig. 2. The energetic efficiency variation versus the air and water speeds, ε_e=f(w_w, w_a)

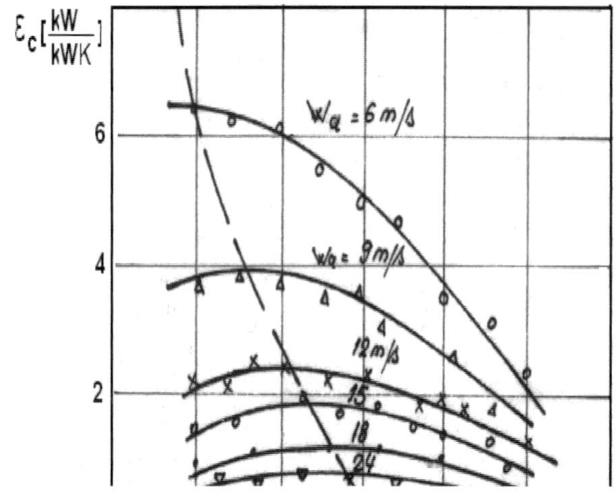

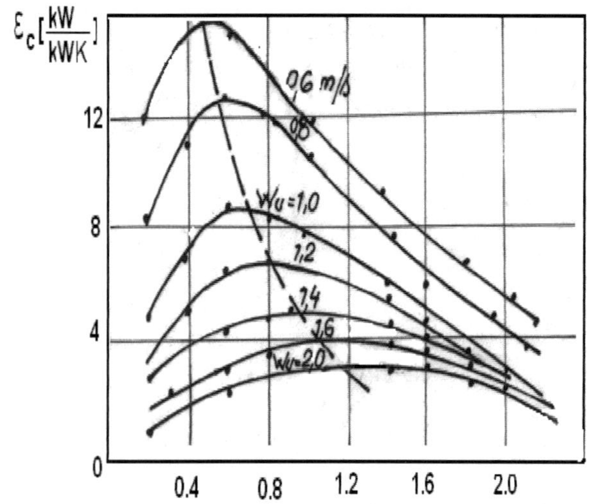

From Figure 2 it can be observed that a single oil speed has a corresponding single economic water speed. The less the air speed is, the more important is to choose correctly a value for the water speed.

In Figure 3 is plotted the energetic efficiency variation versus the volume efficiency for a car radiator. The optimal circulation speeds of water and air result from the diagram which is presented in Figure 3.

4. CONCLUSION

For a heat exchanger selection, from various possible solutions, it will be preferred that device which transmits the constrained thermal flow for a price as low as possible, works under safety conditions and occupies the smaller volume.

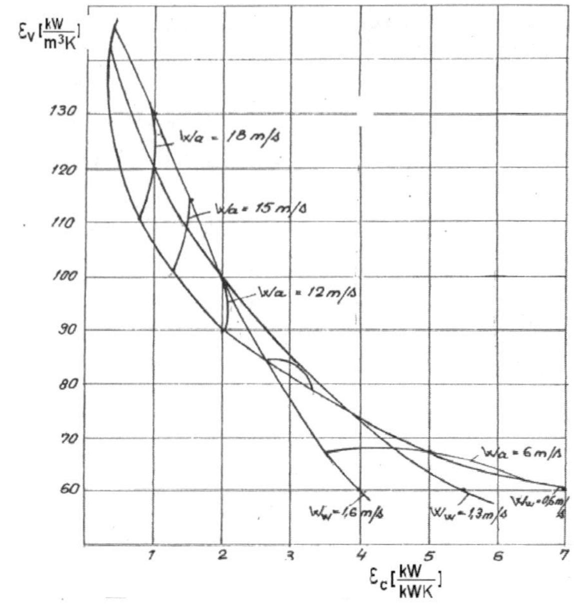

Fig. 3. The energetic efficiency variation versus the volume eficiency for a car radiator

Due to the complex phenomenon that appear in the case of compact heat exchangers, only the experimental researches can answer the questions regarding the thermal and fluidodynamic performances of these types of devices. The optimal circulation speeds through the device can be established by experimental researches. The circulation speeds of fluids can not be independent chosen. By breaching this optimal correlation can lead to important decrease of the energetic efficiency value in comparison with the maximum value.

The experimental results that were obtained and presented above can be used by researchers only for devices and similar working conditions.

5. REFERENCES

Badea, A., Necula, H. (2000), *Schimbătoare de căldură. (Heat Exchangers)*, Editura A.G.I.R., Iași

Kays, M., London, A. (1984), *Compact Heat Exchangers*,Third Edition, McGraw-Hill.

Nagi, M., Lelea, D. (1997), Experimental investigation on heat exchangers for automobiles. *MVM Proceedings*, Vol. 23, pp. 45-47, Kragujevac

Nagi, M., Negoiţescu, A.S. (1998), *Calculul şi construcţia instalaţiilor termice (Design and Development of Thermal Equipments)*, Editura "Eftimie Murgu". ISBN 973-97754-9-7, Reşiţa

Negoiţescu, A.S. (2003), Researches about the Heat Transfer in the Heat Exchanger Extend Surfaces, *6th International Conference on Accomplishments of Electrical and Mechanical Industries DEMI*, ISBN 99938-623-8-X, Banjaluka

Annals of DAAAM for 2008 & Proceedings of the 19th International DAAAM Symposium, ISSN 1726-9679
Editor B. Katalinic, Published by DAAAM International, Vienna, Austria 2008
Make Harmony Between Technology and Nature, and Your Mind will Fly Free as a Bird

A MATHEMATICAL MODEL FOR ROUGHNESS IN CASE OF VIBROROLLING PROCESS

NAGIT, G[heorghe]; FETECAU, C[atalin]; NEGOESCU, F[lorin]; IOSUB, A[drian] & AXINTE, E[ugen]

Abstract: *In this paper, a mathematical model, obtained from geometrical considerations, in case of vibrorolling cold plastically deformation processes, for roughness parameter R_a, was developed. This model takes into account all input factors which characterize the vibrorolling process. It can be used for process estimation when balls are being employed, but also when spherical crests from diamond are used. The model is validated by experimental researches on different materials.*
Key words: *vibrorolling, plastically deformation, roughness.*

1. INTRODUCTION

In case of industrial processing, it is very important to know roughness parameter R_a or R_z, which are currently determined by measurements.

The roughness parameter R_a can be considered the main parameter of surface condition. Calculation of average deviation for R_a can be done using the following relation (Nagit, 1997):

$$R_a = \frac{1}{L_1}\int_0^{L_1} |y|dx \qquad (1)$$

where: L_1 is the length of surface sector where the roughness is determined and y corresponds to surface profile, determined in XOY coordinate system. OX axis coincides with profile medium line.

2. DETERMINATION OF THE MATHEMATICAL MODEL

Starting from the geometrical similarity of turning process, where cutting tools with radius in the tool centre are used and from the mathematical model (2) of roughness (Braha et al., 2003) one will analyze figure no. 1.

$$R_a = m - \sqrt{r^2 - \frac{f^2}{4}} - \frac{2r^2}{f}\left[2\cdot\arcsin\frac{2f}{r} - \sin\left(2\arcsin\frac{2f}{r}\right)\right] + \\ + \frac{r^2}{2}\left[2acrscin\frac{\sqrt{r^2-m^2}}{r} - \sin\left(2\arcsin\frac{\sqrt{r^2-m^2}}{r}\right)\right] \qquad (2)$$

where: r is the center tool radius, f is the work feed and m is a parameter defined as the distance from center point of the tool radius to medium line and has the following value (Nagit et al., 1998):

$$m = \frac{\sqrt{4r^2 - f^2}}{4} + \frac{4r^2}{f}\arcsin\frac{f}{2r} \qquad (3)$$

In case of vibrorolling process there appears the sphericity radius ED having the value $d_b/2$ and not radius of the tool point as in the case of the turning process.

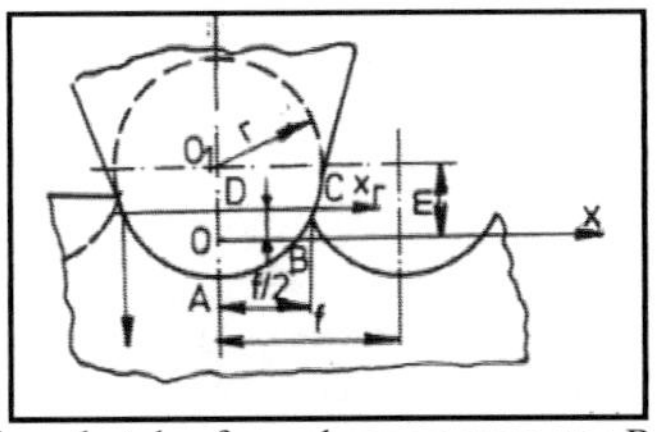

Fig. 1. Calculation sketch of roughness parameter R_a.

In case of spherical diamond tool, radius will be r_y. The feed f will be replaced with L_α, which is the minimal distance projection between elementary cell points L on feed direction.

$$L_\alpha = L\cdot\cos\alpha \qquad (4)$$

or

$$L_\alpha = L\cdot\sqrt{\frac{1}{1+tg^2\alpha}} \qquad (5)$$

where α is the network declination angle, determined by its tangent (Kvackaj, 2006):

$$tg\,\alpha = \frac{2\cdot e\cdot i}{\pi\cdot d_s} \qquad (6)$$

where: $2e$ represents the oscillation amplitude of deformation element;

$i = \frac{n_{cd}}{n}$ is synchronism rapport between double courses number in one minute of deformation element (n_{cd}) and workpiece rotation speed;

d_s is the workpiece diameter.

It is well known that in case of vibrorolling a micro-relief with a channel-network appears or a completely new micro-relief can be obtained, depending on the characteristics and properties of the work-piece that have to be improved (functional characteristics or surface quality (Nagit et al., 2003).¶

In case of completely new micro-relief, hexagonal or tetragonal elementary cells may be formed. Under these conditions, L represents the biggest distance between opposite corners of elementary resulted cells and takes the following values (Braha et al., 2003):

$$L = \sqrt{2}\cdot f_\gamma \qquad (7)$$

for micro-relief with tetragonal cells and

$$L = \frac{2}{\sqrt{3}} f_\gamma \qquad (8)$$

for hexagonal cells, f_γ is the tool progress distance in the direction of network cell inclination angle γ.

$$f_\gamma = \sqrt{\left(\frac{\pi \cdot d_s \cdot \{i\}}{i}\right)^2 + f^2} \qquad (9)$$

where $\{i\}$ represents the fractional part of synchronism rapport p, and f is the work feed.

If these parameters are taken into account, L_α can be expressed as follows:

$$L_\alpha = \frac{2 \cdot \pi \cdot d_s}{i} \cdot \sqrt{\frac{1}{3} \frac{\pi^2 d_s^2 \{i\}^2 + f^2 i^2}{\pi^2 \cdot d_s^2 + (2ei)^2}} \qquad (10)$$

for hexagonal cells and

$$L_\alpha = \frac{\cdot \pi \cdot d_s}{i} \cdot \sqrt{2 \cdot \frac{\pi^2 d_s^2 \{i\}^2 + f^2 i^2}{\pi^2 \cdot d_s^2 + (2ei)^2}} \qquad (11)$$

for tetragonal cells, respectively.

Considering vibrorolling case, feed value f is replaced in relation number (1) by distance value L_α and radius of the tool point with half of ball diameter $d_b/2$; the relation of R_a parameter thus becomes:

$$R_a = \left(m - \sqrt{\left(\frac{d_b}{2}\right)^2 - \frac{L_\alpha^2}{4}} - \frac{d_b^2}{2L_{\alpha a}}\left[2\arcsin\frac{L_\alpha}{d_b} - \sin\left(2\arcsin\frac{L_\alpha}{d_b}\right)\right] + \frac{d_b^2}{8}\left[2\arcsin\frac{2\sqrt{\frac{d_b^2}{4} - m^2}}{d_b} - \sin\left(2\arcsin\frac{2\sqrt{\frac{d_b^2}{4} - m^2}}{d_b}\right)\right] \qquad (12)$$

Relation (12) represents the mathematical model for roughness parameter R_a if taking into consideration a few parameters of vibrorolling behaviour (the pressing force is not taken into account; this parameter has influence on stamped channel depth and on remanent deformation of vibrorolled parts).

3. EXPERIMENTAL VALIDATION

For the experimental validation, a vibrorolling device attached on universal lathe *SN 400x100* was used.

The test pieces were realized by five different steels with a chemical composition as follows:

OLC35-0.34%C, 0.6%Mn, 0.2%Si, 0.3%Cr, 0.3%Ni, 0.3%Cu.
OLC45-0.44%C, 0.6%Mn, 0.2%Si, 0.3%Cr, 0.3%Ni, 0.3%Cu.
OLC15-0.16%C, 0.6%Mn, 0.2%Si, 0.3%Cr, 0.3%Ni, 0.3%Cu.
OL60-0.4%C, 0.6%Mn, 0.37%Si.
18MnCr10-0.2%C, 0.2%Si, 1%Mn, 1%Cr.

The test pieces were machined in advance by turning in the same working conditions at diameter: d_s = 58 mm. The roughness value after this operation was: R_a = 1.5-1.7 μm.

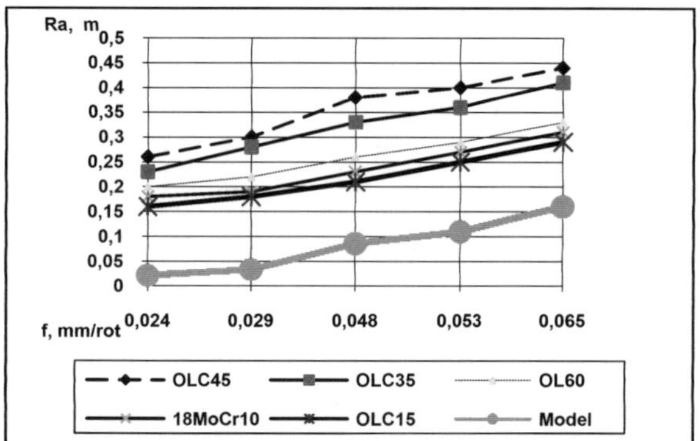

Fig. 2. Feed influence on surface roughness.

The input parameters of vibrorolling process were: n_{cd} = 470 cd/min, n = 100 rot/min, 2e = 1.6 mm, d_b = 15.85 mm, F = 400 N. The feed f was modified so as to obtain a completely new micro-relief.

¶ f(mm/rot)	0,024	0,029	0,048	0,053	0,065
OLC45	0,26	0,3	0,38	0,4	0,44
OLC35	0,23	0,28	0,33	0,36	0,41
OL60	0,2	0,22	0,26	0,29	0,33
18MnCr10	0,18	0,19	0,23	0,27	0,31
OLC15	0,16	0,18	0,21	0,25	0,29
Model(12)	0,021	0,033	0,086	0,11	0,16

Tab. 1. Experimental and theoretical values of roughness for external vibrorolled cylindrical surfaces.

The results obtained are presented in table number 1 and a graphical representation may be analyzed in figure number 2. Constants: n=100 rot/min; n_{cd}=470 cd/min; 2e=1,6mm; d_b=15.85 mm; F=400N; d_s=58mm.

4. CONCLUSION

Experimental results and mathematical model respect the same tendency of roughness variation. For small values of feed, the model fits better for steel with low hardness, while for big values of feed, hard steels correspond better to the mathematical model (fig 2.).¶

These unconformities apear because pressing force is not taken into account. The mathematical model must be revised with a coefficient which describes the dependencies between material hardness and pressing force.

5. ACKNOWLEDGEMENT

The authors gratefully acknowledge the financial support offered to researchers of this paper by Romanian Ministry of Education, Research and Youth, through CEEX Grant no. 62/2006 – BIOMEC.

6. REFERENCES

Braha, V., Nagit, Gh. & Negoescu, F. (2003). *Cold plastic deformation technology (in Romanian),* Technical and scientific publishing house CERMI, ISBN 973-667-025-2, Iasi-Romania

Kvackaj, I. (2006). Development of bake hardening effect by plastic deformation and annealing conditions. Availble from:http://public.carnet.hr/metalurg/Metalurgija/2006_vol_45/No_1/MET_45_1_051_055_Kvackaj.pdf *Accesed:* 2008-05-16

Nagit, Gh., (1997). *Theoretical and experimental contributions on vibrorolling process (in Romanian)*, PhD thesis, Technical University Gheorghe Asachi of Iasi

Nagit, Gh., Braha, V., Slatineanu, L. & Musca, G. (1998). Experimental researches concerning pressure force influence and lubrication on roughness at vibrorolling. *Technical Institut's Bulletin of Iasi*, Vol. XLIV (2003), pp. 391-396, ISBN 1011-2855

Nagit, Gh., Dodun, O., Negoescu, F. & Ignatescu, E.(2003). Cold superficial plastic deformation-a solution to replace heat treatment in small laboratory, *Proceedings of Baia Mare-North University,* Vol. XVII, pp. 227-231, ISSN 1224-3264, Baia Mare-Romania

Annals of DAAAM for 2008 & Proceedings of the 19th International DAAAM Symposium, ISSN 1726-9679
Editor B. Katalinic, Published by DAAAM International, Vienna, Austria 2008
Make Harmony Between Technology and Nature, and Your Mind will Fly Free as a Bird

RELIABILITY OF SIMPLE PNEUMATIC SYSTEMS

NANASI, T[ibor]

Abstract: *An alternative to conventional analysis of the reliability of technical systems are methods, which allow for multi-state nature of elements and subsystems, resulting in less conservative and more realistic reliability estimates. In this paper elements of simple pneumatic productive system are modelled as Markov process to allow for multi-state nature of the possible failure mechanism. Numerical results demonstrate the potential for improvement over the conventional analysis.*
Key words: *Reliability, Markov Process, Failure Modes, Valves, Pneumatic Systems.*

1. INTRODUCTION

Pneumatic systems have reputation of extremely high reliability when compared to hydraulic or mechanical equipment. Producers estimate the life cycle of various types of valves, which are key elements of pneumatic systems, up to as high numbers as fifty million of cycles (Smith, 2001). Production lines are usually designed without any redundancy except for situations, where extremely high requirements are laid on the safety of the system (Birolini, 1999). Basic reliability model of such systems has usually the form of series system when expressed by conventional reliability block diagram. High number of elements in serial layout has destroying effect on the overall reliability even in case of excellent reliability of individual element. Conventional reliability model gives conservative estimate of the global reliability of the complete system. Moreover, individual components are in most cases modelled as two state elements which are either fully functional or in failure. However, real subsystems of technical systems may admit many intermediate partially functional states between the two extremes (Papazoglou, 1998). This is manifested by the existence of several different failure modes and failure mechanisms that contribute to the loss of functionality. Another obvious fact is, that complex systems are sufficiently robust in withstanding the

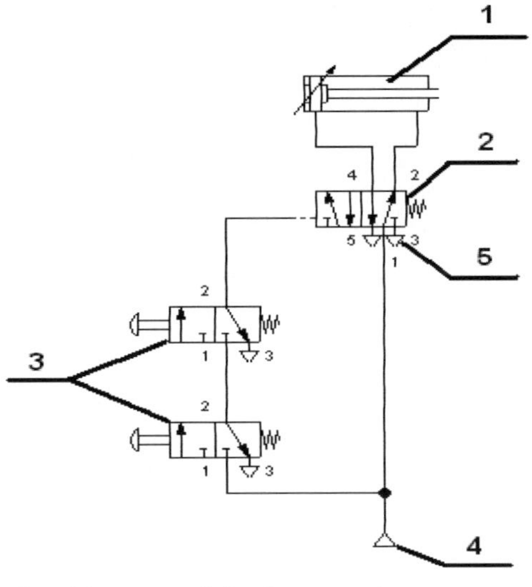

Fig. 1 Subsystems of simple pneumatic system.

Nr.	Functionality of the system	Operator`s action	Button`s state	Valve opened
1	1	1	1	1
2	1	0	0	0
3	0	0	0	1
4	0	1	1	0
5	0	1	0	0
6	0 (1)	1	0	1
7	0 (1)	0	1	0
8	0	0	1	1

Tab 1. Possible states of 3/2 valve.

failure of those subsystems, that are not absolutely vital for their functionality (Kececioglu, 2002a; Dekýš et al., 2004).

More realistic alternative to conventional analysis are methods, which allow for multi-state nature of elements and subsystems, resulting in less conservative and more realistic reliability estimates. In this paper elements of simple pneumatic productive system are modelled via Markov process to allow for multi-state nature of the possible failure mechanism. Numerical results demonstrate slight improvement over the conventional analysis.

2. MULTISTATE FAILURE MODEL

Let us consider the textbook example of simple pneumatic production system, see Fig. 1. To simplify the analysis, the pneumatic press is decomposed to two subsystems. Connecting links and the air supply are neglected in our simplified analysis. First subsystem is the pair of identical 3/2 valves with push-buttons (marked as 3 in Fig. 1), the second one is the linear actuator together with the controlling 5/2 valve buttons (marked as 1 and 2, respectively, in Fig. 1). The valve of the first subsystem is required to be open while the push-button is pressed by the operator and to be closed when push-button released. Internal state of the push-button should be changed due to the action of the operator and the valve should react appropriately. Possible states of the first subsystem under above simplifications are summarised in Table 1 together with the evaluation of the functionality of the system.

Bracketed values at the second column and rows 6 and 7 are acceptable, when spurious opening or spurious closure of the valve is acceptable. Typical case would be a single mission system. If many cycles of steady faultless operation are required, then the values in bracket should be refused. States of the first subsystem can be defined as follows:

STATE 1: Nr.1 &2 (fully functional system)
STATE 2: Nr. 6 & 7 (failed valve and push-button)
STATE 3: Nr. 5 & 8 (failed push-button)
STATE 4: Nr. 3 & 4 (failed valve).

To set up the corresponding Markov process, the transition probabilities between states are to be introduced. For system without repairs the appropriate failure rates are for assumed failure modes as follows:

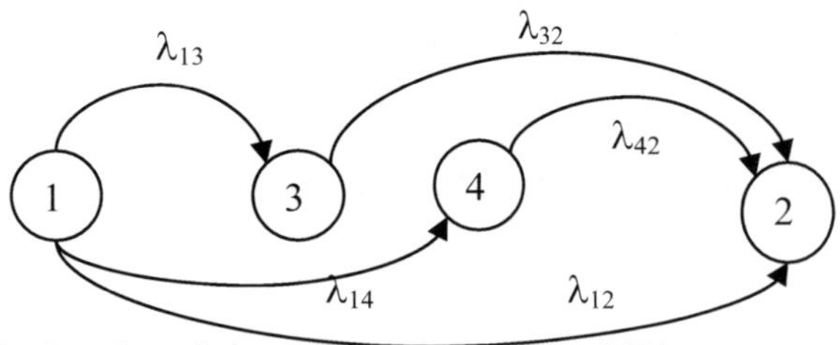

Fig. 2. State diagram with transitional probabilities.

λ_{13} – failure rate of the push-button
λ_{14} – failure rate of the valve
λ_{42} – failure rate of the push-button after the valve failed
λ_{32} – failure rate of the valve after push-button failed
λ_{12} – failure rate because of unspecified cause.

We remind, that failure rates λ_{ij} express the conditional probabilities ($\lambda_{ij}\,\Delta t$) of the occurence of a failure between time t and $t+\Delta t$ provided that failure did not happen until time t.

3. NUMERICAL RESULTS

Markovian model corresponding to state diagram in Fig. 2 results in Chapman-Kolmogorov differential equations (Birolini, 1999; Kececioglu, 2002b), which are a set of four coupled first order linear differential equations for probabilities $P_i(t)$ of states $i=1,2,3,4$.

$$\begin{bmatrix} P_1(t) \\ P_2(t) \\ P_3(t) \\ P_4(t) \end{bmatrix}' = \begin{bmatrix} -(\lambda_{12}+\lambda_{13}+\lambda_{14}) & 0 & 0 & 0 \\ \lambda_{12}\Delta t & 0 & \lambda_{32} & \lambda_{42} \\ \lambda_{13}\Delta t & 0 & -\lambda_{32} & 0 \\ \lambda_{14}\Delta t & 0 & 0 & -\lambda_{42} \end{bmatrix} \cdot \begin{bmatrix} P_1(t) \\ P_2(t) \\ P_3(t) \\ P_4(t) \end{bmatrix}$$

The Chapman-Kolmogorov system was solved by direct numerical integration for initial conditions

$$P_1(0)=1,\ P_2(0)=0,\ P_3(0)=0,\ P_4(0)=0,$$

and for transient probabilities $\lambda_{13} = \lambda_{42} = 0.1\ 10^{-6}$, $\lambda_{14} = \lambda_{32} = 0.9\ 10^{-6}$ and $\lambda_{12} = 0.09\ 10^{-6}$ estimated ad hoc according to data given in (Smith, 2001). Fig. 3 shows probabilities $P_1(t)$, $P_1(t)+P_2(t)$, $P_1(t)+P_2(t)+P_3(t)$ and $P_1(t)+P_2(t)+P_3(t)\ )+P_4(t)=1$ as they are changing with the time. For convenience the time scale is running in thousands of hours and the transient probabilities λ_{ij} are given in failures per million hours.

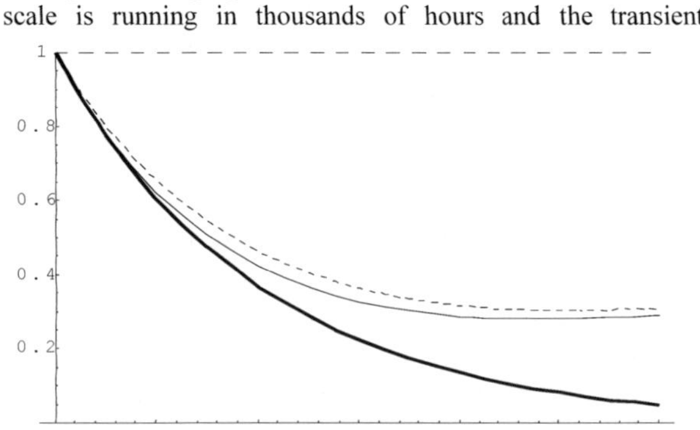

Fig. 3 Combined probabilities for multi-states.

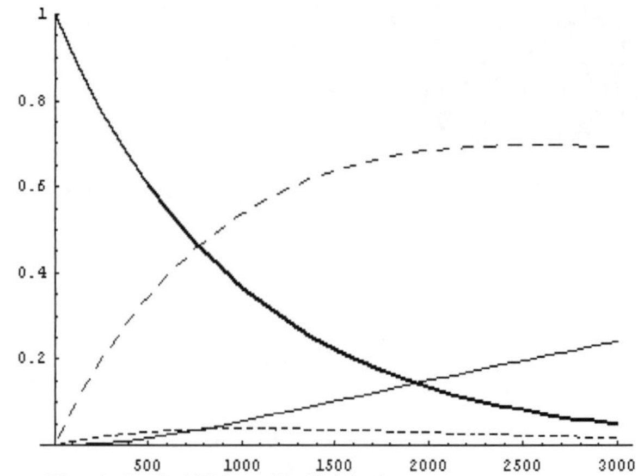

Fig. 4 Probabilities of individual states.

The thick curve in Fig. 4 corresponds to the traditional reliability function $P_1(t)$, thin line gives $P_2(t)$, lower dashed curve is the $P_3(t)$ and the upper long dashed curve corresponds to the time dependence of the probability $P_4(t)$ of state 4. These illustrative results demonstrate that allowing for multi-states has the potential of increasing the conservative estimate of reliability of systems. The formulation the Markov process for the second subsystem is not treated here, as it is a complete analogy of the first subsystem except for different numerical values of transitional probabilities.

4. CONCLUSION

The present paper demonstrates the usefullness of treating the components of pneumatic systems as multi-state elements. For properly defined states and corresponding transition probabilities it is possible to set up Markov process leading to Chapman-Kolmogorov equations. Key element of such analysis is the realistic definition of individual states, which is always highly dependent on detailed knowledge of actual failure modes of the system under consideration. Due to lack of detailed experimental data future research might be focused on development of consistent methods to estimate the transitional probabilities.

The author greatly acknowledges the financial support for this work by the research projects KEGA-2/4154/06 and AV-4/0102/06

5. REFERENCES

Birolini, A. (1999). *Reliability Engineering. Theory and Practice*. 3rd Edition, Springer-Verlag, ISBN 3-540-66385-1, Berlin

Dekýš, V.; Sága, M. & Žmindák, M. (2004). *Dynamika a spoľahlivosť mechanických sústav*, VTS ŽU, ISBN 80-969165-2-1, Žilina

Kececioglu, D. (2002a). *Reliability Engineering Handbook Vol. 1*, DEStech Publications, ISBN 1-932078-00-2, Lancaster, USA

Kececioglu, D. (2002b). *Reliability Engineering Handbook Vol. 2*, DEStech Publications, ISBN 1-932078-01-2, Lancaster, USA

Papazoglou, I. A. (1998). Functional Block Diagrams and Automated Construction of Event Trees. *Reliability Engineering and System Safety*, Vol. 61, pp. 185-214

Smith, D. J. (2001). *Reliability, Maintainability and Risk. Practical Methods for Engineers*, 6th Edition, Butterworth – Heinemann, ISBN 0-7506-5168-7, Oxford

Annals of DAAAM for 2008 & Proceedings of the 19th International DAAAM Symposium, ISSN 1726-9679
Editor B. Katalinic, Published by DAAAM International, Vienna, Austria 2008
Make Harmony Between Technology and Nature, and Your Mind will Fly Free as a Bird

ANALYSIS OF WORKING PERFORMANCES AT DAMAGED VIBRATION ISOLATION DEVICES

NASTAC, S[ilviu]

Abstract: *Based on the structural behaviour influences of dynamic actions, and according to the structural vibration with their parameters modifications under dynamic loads, it could be estimate which, where and how will be the effect quantification of the damages generated by the external loads on a structure. In this paper it was described a variant of response equalisation method, combined with experimental tests, according to, it will be evaluate the demotion degree of a certain system which acquire damages due to the external loads such as vibrations. Basics of this study consists of the stiffness and dampings modifications at one time with structural fatigue, and these modifications leads to a sensible changes of the vibration measured on a structure.*
Key words: *Vibration, Isolation, Damage, Performance.*

1. INTRODUCTION

The main purpose of this research, regarding the isolation performances of the passive systems, consist by the estimation of the deep connection between the structural and behavioural demotion, for antivibrational devices, and, on the other side, the diminishing of the isolation degree, for the same working conditions. During the measurements, both in the laboratory tests, and "*in situ*" behaviour evaluations, it was observed that the damages level growing up with the exploitation time period, and the damages especially appears at the elastic elements. Taking into account the major influence of these elements on the global isolation degree of the insulated ensemble, it was appear the idea of damage level evaluation based on the isolation degree (or transmissibility) of the dynamic system. The main advantages of this method are: non-destructive method, increased safety, reduced costs. Another major advantage of this method consist of the possibility of damages evaluation as a continuous process, and detection the begining of an important failure until these had a critical value and disturbed the dynamic and the integrity of the isolation system. It have to be said that this procedure could be framed into the *Structural Health Monitoring Concept*, that enables *Conditions - Based Maintenance* at structures through diagnosis of the status current health during exploitation (Johnson et al., 2002).

2. RESEARCH SUPPOSITIONS

In the Figure 1 is presented the schematic diagram of the basic model, a rigid bodies system with three DOF underpined to the ground and between them by the visco-elastic elements (Bratu, 1990; Bratu, 2000; Harris et al., 2002). This is a full basic model, with all the linkages, constants and loading forces. In this shape, the model could simulate a large area of technical systems, from the point of view of dynamical behaviour.

The dynamic equations of this model was developed with the approaching of the next hypothesis (Nastac, 2004)

- ✓the entire system are supposed as a rigid bodies ensemble (m_i), with vertical translations $x_i(t)$ for each mass;
- ✓the linkages from the masses to the ground are maded with the visco-elastic elements, with linear characteristics;
- ✓the elastic linkages have the characteristic rigidities k_{ij}, and internal dampings b_{ij};
- ✓the dynamic external loads (excitations) have innertial type and is generated by the forces F_i;
- ✓in the previous notations the indexes i, j denotes the masses (1, 2, 3) and the ground (0).

Taking into account the previous hypothesis, the dynamic equations of the model are

$$M\ddot{X} + B\dot{X} + KX = F \quad (1)$$

where $\ddot{X}$ denote accelerations vector, $\dot{X}$ denote velocities vector, X denote displacements vector, M denote matrix of masses, B denote matrix of dampings, K denote matrix of stiffness, F denote external forces vector.

The transfer function matrix of the three DOF model, presented in Figure 1, is

$$H(s) = \begin{bmatrix} m_1 s^2 + b_1 s + k_1 & -b_{12}s - k_{12} & -b_{13}s - k_{13} \\ -b_{12}s - k_{12} & m_2 s^2 + b_2 s + k_2 & -b_{23}s - k_{23} \\ -b_{13}s - k_{13} & -b_{23}s - k_{23} & m_3 s^2 + b_3 s + k_3 \end{bmatrix}^{-1} \quad (2)$$

where the new terms have next expressions

$$\begin{aligned} b_1 &= b_{10} + b_{12} + b_{13}; b_2 = b_{20} + b_{12} + b_{23}; b_3 = b_{30} + b_{13} + b_{23}; \\ k_1 &= k_{10} + k_{12} + k_{13}; k_2 = k_{20} + k_{12} + k_{23}; k_3 = k_{30} + k_{13} + k_{23}; \end{aligned} \quad (3)$$

The analysis of the damage influences about the spectral composition of the amplification functions for each degree of freedom - vertical translations of the m_i masses - will be maded supposing only the direct linkages between the nearest rigid bodies. This kind of model answer to majority types of technical systems.

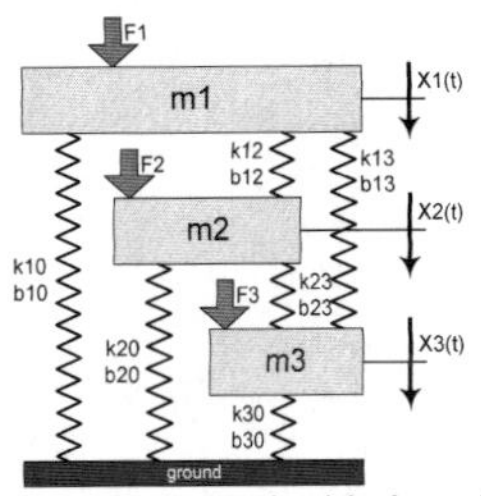

Fig 1. The basic model for the embedded equipments antivibratory isolation

3. TEST APPLICATION

For the first approach of numerical simulation, it was computed the FRFs of the three DOFs system. Even if the transmissibility functions are most sensitives to changes in mass, damping, and stiffness like a frequency response functions, the last are prefered because it's more facilely to evaluate. In the Figure 2 it is depicted the frequency response functions for the three degree of freedom of the system presented in Figure 1. In the Table 1 it is presented the basic values of the model constants for the numerical application.

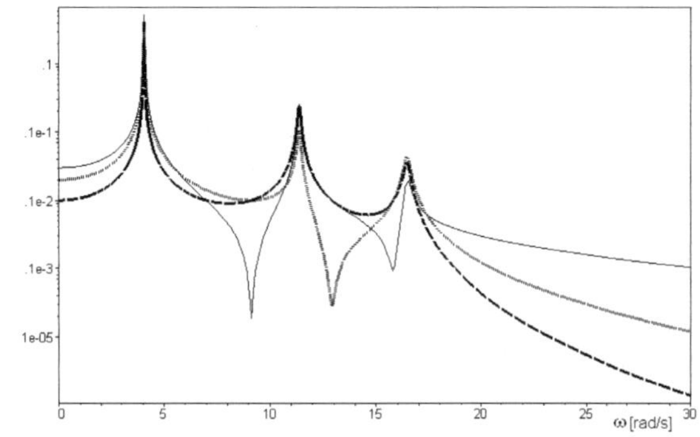

Fig 2. The FRFs magnitudes of the three DOFs system

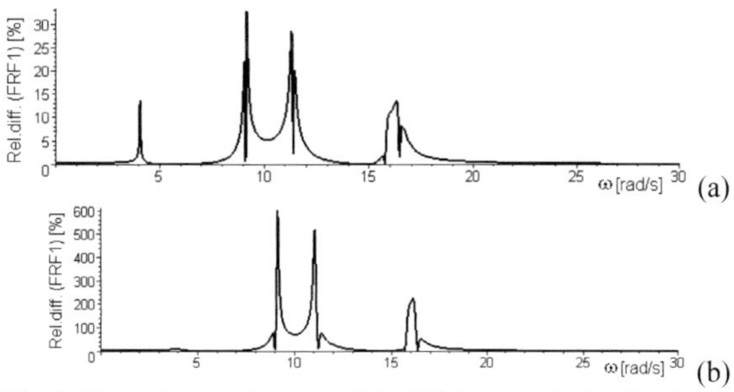

Fig 3. The relative changes of the FRFs magnitudes for the first DOF. (a) for 1% loss in k_{12} (b) for 10% loss in k_{12}

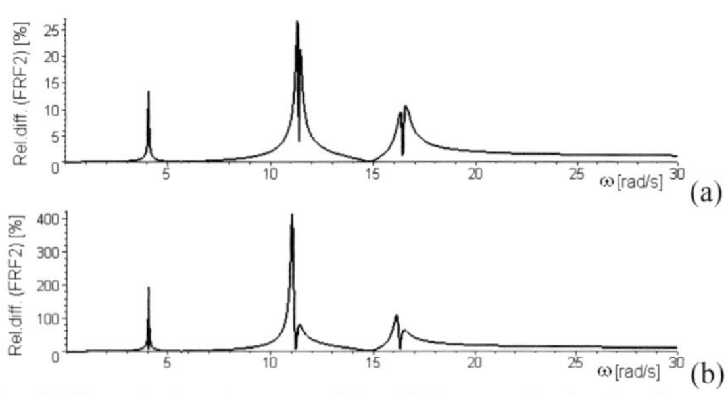

Fig 4. The relative changes of the FRFs magnitudes for the second DOF. (a) for 1% loss in k_{12} (b) for 10% loss in k_{12}

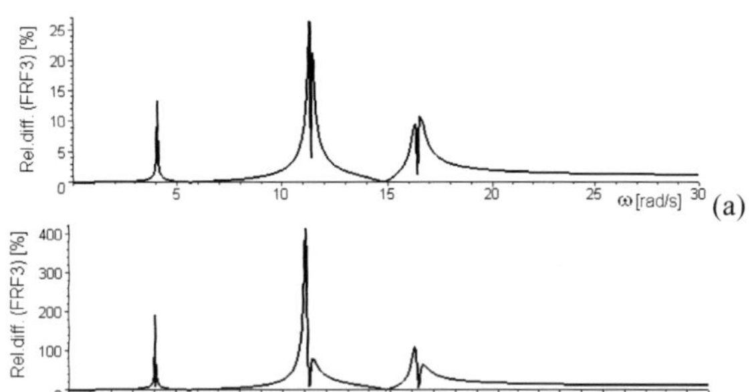

Fig 5. The relative changes of the FRFs magnitudes for the third DOF. (a) for 1% loss in k_{12} (b) for 10% loss in k_{12}

Parameters	Values	Units
m_1 ; m_2 ; m_3	12 ; 12 ; 12 ;	[kg]
k_{10} ; k_{12} ; k_{13} ; k_{20} ; k_{23} ; k_{30}	0;1000;0;0;1000;1000;	[N/m]
b_{10} ; b_{12} ; b_{13} ; b_{20} ; b_{23} ; b_{30}	0 ; 1 ; 0 ; 0 ; 1 ; 1 ;	[N s/m]
F_{10} ; F_{20} ; F_{30}	1 ; 0 ; 0 ;	[N]

Tab 1. The basic values for the numerical application.

The analysis was developed for damages influences identification on the system dynamic characteristics (i.e. FRFs). As a damaged element it was proposed the visco-elastic linkage between m_1 and m_2. This element has k_{12} stiffness and b_{12} damping. In this paper it was presented only the analysis cases for constant dampings of all the linkages. Thus that, for the braked element, it was varied only the stiffness values. It was assume three main cases, which are 0.1%, 1%, and respectively 10%, losses in k_{12} stiffness.

In Figures 3...5 it is depicted only the relative changes of the FRFs, separately for each of the three DOFs system. Every figure contain two set of diagrams, with the next significations: (a) the FRF changes for 1% loss in k_{12}, and (b) the FRF changes for 10% loss in k_{12}. The case of 0.1% loss in k_{12} was not depicted because the real changes of the FRFs were not significant. In the Table 2 it was presented the values of the FRFs relative changes, evaluated for the three cases, at the reference frequency - this value was adopted after a qualitative and quantitative evaluation of the FRFs magnitudes of the original system. This value derived from the excitation pulsation value of 30 rad/sec.

Analysis	*Damages level*	*Ref. Frequency* [Hz]	*Differences* [%]
case I	0,1 % loss in k_{12}		0.0128346204
case II	1 % loss in k_{12}	4.774789113	0.12806582
case III	10 % loss in k_{12}		1.253263361

Tab 2. The relative changes of the FRFs at the reference frequency $(\omega = 30\ rad/s)$.

4. CONCLUSION

Comparative analyses of the diagrams from Figures 3...5, denote a certain sensitivity of the system DOFs frequency response functions. This sensitivity could be observed both at the natural frequency shift, and at the maximum values on these frequencies. If the shift of the natural frequencies is very small, the relative increasing or decreasing of the FRF magnitude at these frequencies acquire high values even for the low values of stiffness losses. The values from the Table 2, and the diagrams from Figures 3...5, show that the maximum sensitivity is on the resonance frequencies area. From the Table 2 result that outside the resonance area, the relative changes acquires low and very low values - under a size order comparative with the stiffness loss value, but supposing a reference value of frequency, it could be evaluate the rigidity demotion degree, by means of FRFs measuring. After this step of the study, it will be analysed the dampings losses influences, separately and together with the stiffness losses. The entire data sets will be re-evaluate, tunning and validate on the instrumental laboratory tests, on the structures with damage level control.

5. REFERENCES

Bratu, P. (1990). *Insulation Elastic Systems for Machines and Equipments,* Editura Tehnica, ISBN 973-31-0234-2, Bucharest, Romania

Bratu, P. (2000). *Elastic Systems Vibrations*, Editura Tehnica, ISBN 973-31-1418-9, Bucharest, Romania

Harris, C.M. & Piersol, A.G. (2002). *Shock and Vibration Handbook*, 5th Edition, McGraw Hill

Johnson, T.; Adams, D. & Schiefer, M. (2002). An Analitical and Experimental Study to Assess Structural Damage and Integrity Using Dynamic Transmissibility, *The Proceedings of the 20th International Modal Analysis Conference*, pp. 472-476

Nastac, S. (2004). *Contributions for Dynamic Behaviour of the Antivibrational and Antiseismical Passive Isolation Elastic Systems*, A Dissertation submitted to the University "Dunarea de Jos" of Galati, Romania, for the Master of Science in Mechanical Engineering Degree

Annals of DAAAM for 2008 & Proceedings of the 19th International DAAAM Symposium, ISSN 1726-9679
Editor B. Katalinic, Published by DAAAM International, Vienna, Austria 2008
Make Harmony Between Technology and Nature, and Your Mind will Fly Free as a Bird

SOME INFLUENCES ABOUT THE REGENERATIVE ENERGY EFFECT IN THE OPTIMAL FUNCTIONING OF THE LINEAR ACTUATORS

NASUI, V[asile]; LOBONTIU, M[ircea]; BANICA, M[ihai] & COTETIU, R[adu]

Abstract: *The paper approaches a major problem in the functioning of the electro mechanic linear actuator on which this capacity in the transitory regime depends, that is the effect of the regenerative energy. It is essential taking into account the technical and technological importance of the actuators in the intelligent flexible industrial systems of refinement.*
Key words: *linear, actuator, systems, regenerative, mechatronic.*

1. INTRODUCTION

The technical and technological evolution towards the integration in mechatronic included the stages of development of the module of linear acting of the actuator type marked by the integration of the microprocessors. The development of the mechatronic technology within this context contributed to the development of the actuators as elements of execution capable with this which have besides the cinematic chain, energetic and informational material. In the dynamic study of the actuators we have to consider the whole actuators, in spite of the effects the transitory functional regime (Catrina & Constantin, 2002).

Thus results the necessity of the analysis of the influence of the energy regenerated in the system as a consequence of the periods of deceleration in the functioning on all the mechanic, electrical, electronically components (Dumitru et al., 2004). Suppose there is a form of kinetic energy recovery system which is considerably less efficient at gathering energy and using it again, particularly at high peak power and energy flow levels. The optimization problem for the actuator is formulated with material, packaging, and performance constraints. Since loads are continuously being accelerated and decelerated, actuators absorb energy as frequently as they output energy, but power is required from the supply regardless of the direction of power flow in the actuator. The absorbed power is simply dissipated in the actuator or power supply (Boiangiu, 2003).

2. THE ECONOMIC WORKING REGIME

The correct definition of the economic functioning regime of an electro mechanic linear actuator must take into account, on one hand, the clues that characterize the economic regime. The main component parts of an actuator (fig. 1), including the electric engine of acting, the transmission itself, the filling and the control devices. The parameters which characterize the economic regime are the training electric power and the efficiency of all the structural components. (Nasui, 1993).

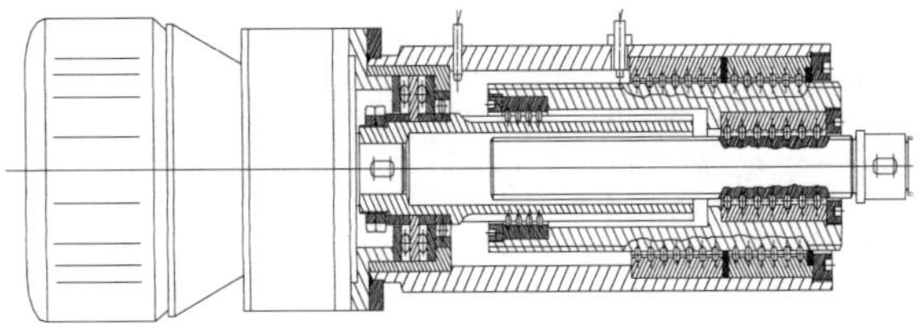

Fig. 1. The telescopic electro mechanic linear actuators.

Saving the electric energy can be done by important measures in the whole chain of the actuator system starting from the filling point, passing to the control one or the regulating one and going on with the acting engine and the cinematic transmission.

The basic problem in order to obtain a profitable regime is to correctly establish the nominal power according to the efficiency, the transmission report, the increase or decrease in the speed and the cinematic moment. In the situation of the actuators, to establish acting power is very important owing to the conditions of transitory regime of work.

The opportunity of the research results from the fact that the authors are in general for real functioning situations with under load power. When calculating the overall efficiency of the actuator's transmission we approximate taking into consideration the only losses that occur while the friction within the mechanism, neglecting these depending among others on the degree of covering as well (Nasui, 2006).

To take into consideration all the losses we write the relation of the overall efficiency according to the total / overall moment M_t and to that clue to the constant ΔM_c and variable ΔM_a mechanical losses as it follows:

$$\eta_t = \frac{M_t}{M_t + \Delta M_c + \Delta M_a} \tag{1}$$

If we define: the loading coefficient, $K_i = M_t / M_n$ and the loss coefficient, $K_p = \Delta M_t / M_t$, using the relation (1) results:

$$\eta_t = \frac{\eta_n}{1 - \eta_n K_p \left(1 - \frac{1}{K_i}\right)} \tag{2}$$

For this, we make real assessment of the value of the efficiency of the actuator's transmission considering the overall of the power losses within the actuator and for partial loading. For the transitory regimes of functioning, the efficiency can be calculated deterring the two components of the moment of losses within the transmission according to the relation (2) including the dynamic moment (Dumitru et al., 2008).

The experimental determination of power losses specific to the mechanical transmission of an actuator has shown its dependence according to the degree of loading and functioning regime. The minimal value of these losses represents 2 - 6 % out of the transmission power, load nearly to the nominal one but not exceeding it.

Establishing the loading coefficient is necessary in view of the choice of the acting engine and of the calculation of the overall efficiency of the mechanical transmission of the actuator.

Thus results the great importance of the correct establishment of the functioning in this functioning regime by mathematical modeling of these influences (Maties et al., 2000).

3. THE REGENERATED ENERGY IN THE ACTUATOR AND ITS EFFECTS

This regenerated energy in the system results from the weight by amplifications from the energy supplier. At the actuators, in order to reduce the stopping time of the acting system, it is frequently applied the dynamic braking or recuperative braking in continuous power regime. The electric machine in this case functions as a generator transforming the rotation kinetic energy or the potential energy of the working machine into electric energy debit ate dons the braking resistance. The whole of the kinetic energy of the weights in movement is transformed in electric power which in its turn is dissipated on the circuit, being controlled in this way.

The catching up braking is similar to the dynamic braking with the difference that the electric engine generates in the continuous power flux. The potential energy is transformed by the engine in electric energy recuperating by itself and being used by other consumers.

The pattern of dynamic control is shown in figure 2. At the actuators with asynchrony engines in recuperating energy regime, the braking is obtained for speeds of the superior's rotor, the speed of the rotator field, generating within the net of active power, but continuing to obtain reactive power.

Because the actuators have a significant control, the recuperative energy has a major part which can influence the optimum functioning of the whole system. Regenerative energy is transferred from the motor load through the amplifier to the power supply during deceleration. If this energy is not managed it will boost the voltage which may damage the driver.

Since the electrical energy stored within the motor is small, it usually can be neglected. However, the driver must be able to handle the motor and load mechanical energy $E_R = 1/2 \cdot J \cdot \omega^2$, less the energy dissipated by the cable and motor, $E_R = I^2 \cdot R \cdot t$.

$$E = \frac{1}{2} \cdot J \cdot \omega^2 (4{,}2 \times 10^5) - I^2 \cdot R \cdot t \qquad (3)$$

$$t = \frac{J \cdot \omega}{M_T \cdot I} \left[\frac{2 \cdot \pi}{60} \right] \qquad (4)$$

where:

E – amplifier regenerative energy capacity;
J – inertia of motor plus load;
ω - velocity load;
I – current limit setting of driver;
R – resistance of cable and motor;
t – deceleration time, [sec.];
M_T – motor torque constant.

The energy handling capacity of the driver power supply must beat least the value calculated. (Aerotech, 1990) suggests a 50% margin on this value. Also, give consideration to simultaneous deceleration in multi-axis systems. The actual systems focus on the energy recovery by dynamic or regenerative braking and less on the direct recovery of kinetic energy and on its rendering at the appropriate moment.

The first systems produce heat which is then internally or externally dissipated and they are not energetically efficient. Regenerative recovery depends on the use of appropriate batteries, non ecological.

Also these systems of electronic regulation require regulation equipment, compatible with the type of acting electric engines. All these make the kinetic systems performances of recovery successful as their weight is smaller and smaller.

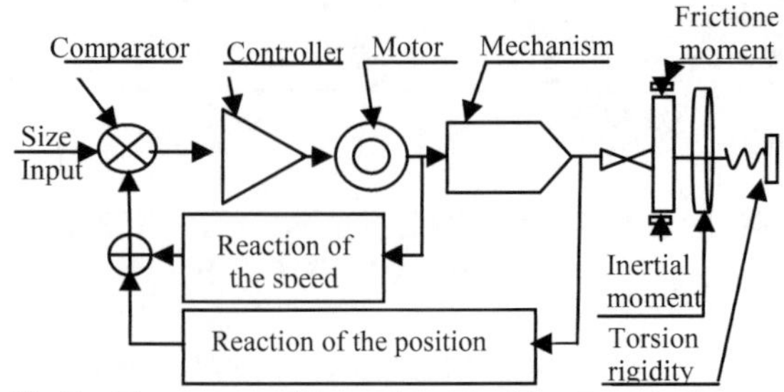

Fig.2. The pattern of dynamic control for the actuators.

4. SUMMARY AND CONCLUSION

The research focuses on the improvement of the energetic performances, such as superior efficiency and the achievement of structural changes especially regarding the solutions for transmission mechanisms of movement, and its control. We propose original innovative solutions from this point of view with new mechanism actuator of movement (Nasui, 1993).

The development perspectives aim optimum solutions equipped with mechanisms of transmission and transformation of the movement with high efficiency and fiability. The problem consists in solving an incompatibility among the functional and constructive of the components of the system actuator.

The conception and its manufacture assisted on the computer has as application field the assembly of the process of developing new products, covering the conception aspects, manufacture and the link between them. The control of the movement of the system actuator is the key factor on which depend the capacity and their availability. Owing to the dynamic regime particular to the functioning of the actuators, it is necessary to know the influence of the regenerated energy in the system in order to take the technical measures most appropriate as far as the performances of the control equipment of the actuators are concerned.

When designing a linear motion system, it is necessary to consider the effect of the variables operation will have on performance. In the future the researches can continue for the development of new applications on other types of mechanical transmission using this method and different modular control laboratory.

5. REFERENCES

Boiangiu, T. (2003). *Advanced Robot Motion Control,* Editura AGIR, ISBN 973-8130-98-0, București.

Catrina, D., & Constantin, G. (2002). Evolution of Some Kinematics Parameters of the Feed Kinematics Chain at High Slide Speeds in CNC Machine Tools, *Proceedings of the International Conference on Manufacturing Systems,* pp. 97-100, ISSN 0035-4074, October 2002, Editura Academiei Romane, Bucuresti.

Dumitru, D. & Strajescu, E. (2004). Performant Optimization of Feed Driving at CNC Machine Tools, *Proceedings of the International Conference Optimization on Manufacturing Systems,* pp. 175-178, ISBN 973-27-1102-7, October 2004, Editura Academiei Romane, Bucuresti.

Maties, V., Mandru, D., Tatar, O. & Maties, M. (2000). *Actuatori in mecatronica, Actuators for Mechatronics*, Editura Mediamira, ISBN 973-9358-16-0, Cluj-Napoca.

Nasui, V. (2006). *Actuatori liniari electromecanici, Linear Electromechanic Actuators*, Editura Risoprint, ISBN 973-656-813-X, Cluj Napoca.

Nasui, V. (1993). Patent RO 106.284. B1.

*** AEROTECH (1990). *Motion Control Product Guide*, Aerotech, Inc., USA

Annals of DAAAM for 2008 & Proceedings of the 19th International DAAAM Symposium, ISSN 1726-9679
Editor B. Katalinic, Published by DAAAM International, Vienna, Austria 2008
Make Harmony Between Technology and Nature, and Your Mind will Fly Free as a Bird

A CONCEPTUAL DESIGN APPLICATION BASED ON A GENERALIZED ALGORITHM PART I. GENERATION OF SOLVING STRUCTURAL VARIANTS

NEAGOE, M[ircea]; DIACONESCU, D[orin]; JALIU, C[odruţa] & SAULESCU, R[adu]

Abstract: *Based on a new algorithm variant of the conceptual design modeling, proposed by the authors in a recent paper (Diaconescu et al., 2008), an application of this algorithm is analyzed on a product of motor-reducer type. According to the new variant, this first part of the paper exemplifies the generation of the reducer solving structural variants and the next part will exemplify the reducer concept establishment.*
Key words: *conceptual design, product concept, reducer.*

1. INTRODUCTION

Based on the analysis of the literature, regarding product conceptual design, a generalized modeling algorithm is proposed by authors in a recent paper (Diaconescu et al., 2008). This paper exemplifies (in two parts) an application of this algorithm on a product of motor-reducer type, which consists of a *planetary speed reducer* and an *electric motor.* This motor-reducer is used in a *rotational platform* on which a parabolic antenna is installed; the motor-reducer must drive the platform in different *angular positions*, under imposed conditions of *accuracy* and *stability*. The first steps of the algorithm up to the generation of solving structural variants inclusive are approached in this paper.

2. ON THE REQUIREMENTS LIST

The following *requirements* (main objectives) regarding the motor-reducer come out from the requirements list (of the global product formed by the parabolic antenna assembly:
(1) The stable maintenance of the platform angular position: the transmission must be blocked when the motor is disconnected.
(2) The energy source: *electrical.*
(3) The reducer must be a *simple planetary gear* with a *kinematical transmission ratio*: $|i| = 100 \pm 1.5\%$.
(4) The reducer admissible *minimum efficiency*: $\eta_{min} = 0.4$ (the motor torque must be amplified at least $\eta_{min} \cdot |i| = 40$ times).
Four *optimization objectives* are associated regarding the minimization of: **A.** *production costs*, **B.** *friction losses*, **C.** *radial size* and **D.** *axial size*. These (secondary) objectives are used as *evaluation criteria* and have the following relative weights: $A \approx 4B \approx 6C \approx 8D$.

3. MOTOR-REDUCER GLOBAL FUNCTION

The motor-reducer global function is a sub-function of the platform *global function*; its VDI symbolical structure (VDI, 1997), illustrated in Fig. 1, contains the following sub-functions:
M1: Connecting the material (platform) to the mechanical energy; M2: The material angular displacing; M3: Registration of the material angular position; E1: Voltage connecting / disconnecting; **E2**: *Transformation* of the electrical energy into mechanical energy + transmission blocking while the motor is disconnected + the speed *primary reducing*; E3: Transmission of mechanical energy with a new speed *reducing*; E4: *Ramification* of the mechanical energy into: *useful energy* and *lost energy*; I1: *Ramification* of the input information into: *emission of the starting signal* and *transmission of the information* regarding the desired angular position; I2: Commanding the execution of the starting and stopping signals; I3: Reception of the information regarding the angular positions: *desired* and *actual values* and their comparison;
I4: Emission of the stopping signal when the actual and desired positions become equal.
The **E2** sub-function (Fig.1), which designates the motor-reducer global function, is illustrated separately in Fig. 2,a.

4. MOTOR-REDUCER FUNCTION DETAILING

The structure of the motor-reducer function, illustrated in a VDI symbolic variant in Fig. 2,b, contains the sub-functions:
E21: Transformation of the electrical energy into mechanical energy; E22: Movement blocking while the motor is disconnected; E23: Modification of the mechanical energy parameters: speed reducing and torque amplification; E24: Transmission of the mechanical energy (unchanged) between the satellite-gear and central shaft.
The motor-reducer's solving structures will be generated based on the structure from Fig. 2,b; this stage contains: 1) generation of the solving structural variants and 2) establishment of the solving structures, by the kinematical configuration (synthesis) of the obtained variants and by the elimination of the variants whose technical characteristics don't fulfill quantitatively the requirements list.

5. SOLVING STRUCTURAL VARIANTS GENERATION

From the research on the sources of existent solutions, as catalogues of physical effects, catalogues of usual functions solutions, prospects, patents etc. (Cross, 1994; Dieter, 2000; Pahl & Beitz, 1995; Ulrich & Epinger, 1995), were obtained following conclusions regarding the sub-functions E21,…,E24: 1) All these sub-functions are of KSP (Known Solving Principles) type; 2) The potential principle solutions of sub-functions are entirely known in the effects plan and are partially known in the plan of the effects carriers' configuration (a part of the found configurations can be reconfigured).

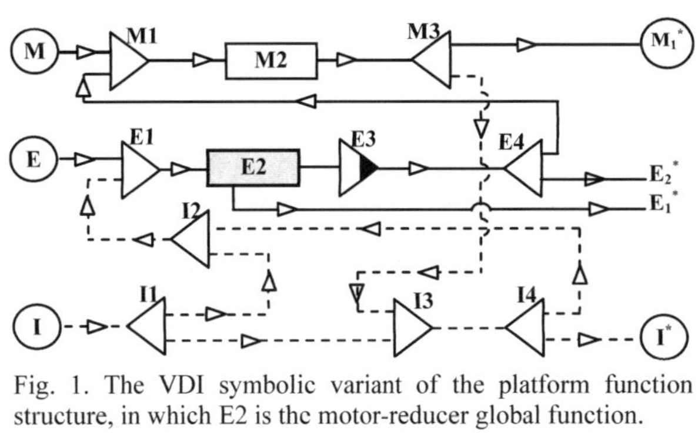

Fig. 1. The VDI symbolic variant of the platform function structure, in which E2 is the motor-reducer global function.

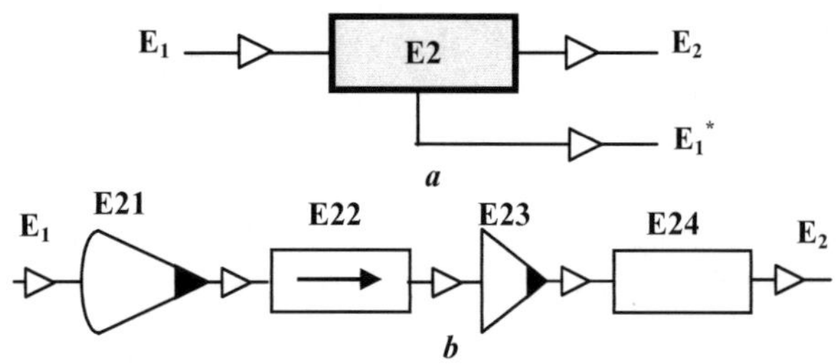

Sub-func.	*Physical effects*	*Potential principle solutions*		
E21	Electro-magnetic force	1.1 *D.C. Motor*	1.2 *Asynchronous AC motor*	1.3 *Synchronous AC motor*
E22	Columbian friction	2.1 *Motor with brake*	2.2 *Irreversible coupling*	2.3 *Irreversible gear pair*
E23	Lever effect	3.1 *Planetary involute gear*	3.2 *Planetary cycloid gear with pins*	3.3 *Planetary transmission with chain*
E24	Lever effect	4.1 *Schmidt linkage semicoupling*	4.2 *Schmidt pin semicoupling*	4.3 *Green (Oldham) coupling*

c

Sub-function / *'olving 'ariants*	*E21*	*E22*	*E23*	*E24*
SR1	*1.2*	*2.1*	*3.1*	*4.1*
SR2	1.2	2.1	3.1	4.3
SR3	1.2	2.1	3.2	4.2
SR4	1.2	2.1	3.2	4.3
SR5	1.2	2.1	3.3	4.1
SR6	1.2	2.1	3.3	4.2
Example: variant SR1 = 1.2+2.1+3.1+4.1				

d

ig. 2. *a)* The motor-reducer *global function*; *b)* The global ınction *structure* in a VDI symbolic variant; *c)* The ıorphological matrix for *solving* structural *variants* generation;) Examples of *solving* structural *variants*.

For exemplification, the results that were found are systematized in the simplified morphological matrix from Fig. 2,c. The product solving structural variants can be generated as compatible combinations between the partial solutions from Fig. 2,c; for example, 6 distinct structural variants were illustrated in Fig. 2,d: SR1, …, SR6.

6. CONCLUSIONS

The previous obtained variants (Fig. 2,d) fulfill the requirements list qualitatively certainly and quantitatively uncertainly; thus, from these variants, there will be considered as solving structures (of the motor-reducer function: E2) only those whose technical characteristics fulfill the requirements list both qualitatively and quantitatively.

The qualitative schemes of the six variants from Fig. 2,d are represented in a simplified way (without representing the motor) in Fig.3. For these schemes, the second part of the paper will establish the main technical performances: the numbers of teeth (from $|i|$ =100± 1.5 %) and then, the efficiency and the torque amplification ratio.

The solving structures of the function E2 will be nominated by the variants that block the transmission when the motor is disconnected and accomplish the requirements: $|i| = 100 \pm 1.5$ % and $\eta \geq 0.4$; then, by technical and economical evaluation will be established the product concept.

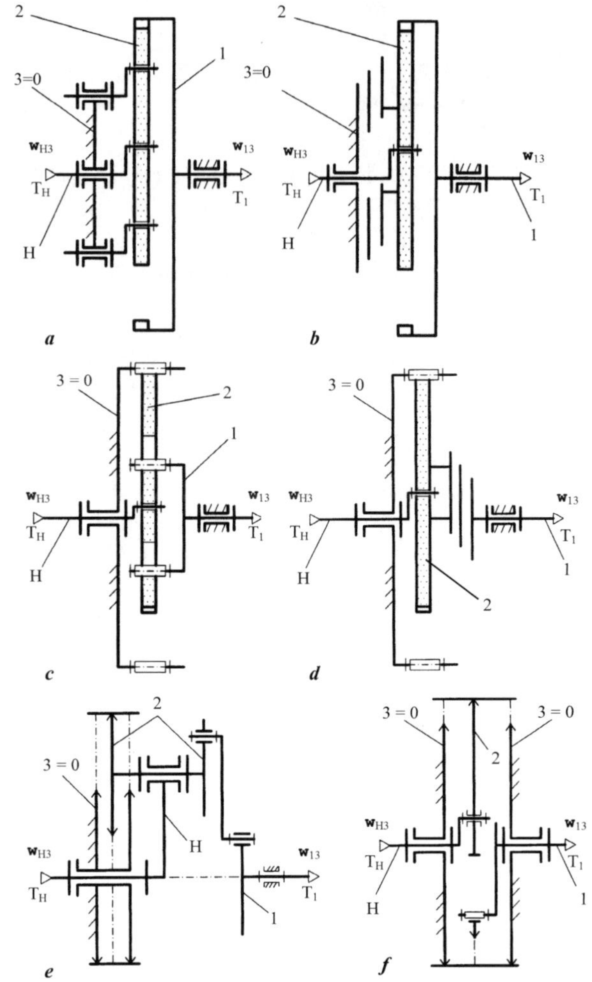

Fig. 3. Simplified schemes (without motor) of the solving structural *variants*: a) SR1, …, f) SR6 .

7. REFERENCES

Cross, N. (1994). *Engineering Design Methods*, J. Wiley & Sons, ISBN 0-4719-4228-6, New York

Diaconescu, D. et al. (2008). *On a Generalized Algorithm of the Technical Product Conceptual Design*, The 19th International DAAAM SYMPOSIUM "Intelligent Manufacturing & Automation: Focus on Next Generation of Intelligent Systems and Solutions", 22-25th October 2008 (*accepted paper*)

Dieter, G. (2000). *Engineering Design*, McGraw Hill, ISBN 0-07-116204-6, Boston

Pahl, G. & Beitz,W. (1995). *Engineering Design*, Springer, ISBN 3540504427, London

Ulrich, K. & Epinger, S. (1995). *Product Design and Development*, McGraw-Hill Inc. ISBN 0-07-113742-4, New York

Verein Deutscher Ingenieure V.D.I (1997). *Richtlinien 2221 and 2222*

Annals of DAAAM for 2008 & Proceedings of the 19th International DAAAM Symposium, ISSN 1726-9679
Editor B. Katalinic, Published by DAAAM International, Vienna, Austria 2008
Make Harmony Between Technology and Nature, and Your Mind will Fly Free as a Bird

A CONCEPTUAL DESIGN APPLICATION BASED ON A GENERALIZED ALGORITHM PART II. SOLVING STRUCTURES SELECTION AND EVALUATION

NEAGOE, M[ircea]; DIACONESCU, D[orin]; JALIU, C[odruţa] & SAULESCU, R[adu]

Abstract: *Based on a new algorithm variant, proposed by the authors for the conceptual design modeling, an application of this algorithm is exemplified on a product of motor-reducer type. According to the new variant, this second part of the paper continues the application from the first part, establishing the product solving structures and selecting the product principle solution (product concept) by fine evaluation.*
Key words: *conceptual design, product concept, reducer.*

1. INTRODUCTION

The qualitative schemes of the six variants, generated in the first part of the paper (Neagoe et al., 2008), are illustrated (Fig. 1) in a simplified way. Among these schemes of variants, the solving structure will be selected by the kinematical configuration (i.e. synthesis of the teeth numbers and establishment of the efficiency and the torque amplification ratio). Obviously, the solving structures of the motor-reducer function will be nominated by the variants that block the transmission when the motor is disconnected and achieve: $|i| = 100$ and $\eta \geq 0.4$; one of them will be selected by evaluation as product concept (Pahl & Beitz,1995; Ulrich & Epinger, 1995).

2. SOLVING STRUCTURES ESTABLISHMENT

In order to establish the solving structures (from the illustrated variants, Fig. 1), first the synthesis of the number of teeth is made from the condition: $|i| = 100 \pm 1.5$ %. Then, on the basis of the known efficiencies of the gear pairs with fixed axes, the efficiencies of the proposed reducers are calculated in the two possible actuation cases (direct and inverse) and the amplification ratio of the input torque for the direct actuation is established. If the efficiency for the inverse actuation is null or negative, then the analyzed reducer transmits the power irreversibly and, therefore, the motor's brake becomes superfluous.

The case of the SR1 and SR2 variants. For each of the planetary reducers from Fig. 1,a and b (consisting of an involute internal gear pair 1-2 and of a synchronic coupling 2-3), the condition of obtaining the transmission ratio can be written as follows (Diaconescu & Duditza, 1994,a and b):

$$i = i^3_{H,1} = \omega_{H,3}/\omega_{1,3} = 1/(1-i_0) = +100; \quad (1)$$

$$i_0 = i^H_{1,3} = \omega_{1,H}/\omega_{3,H} = i^H_{1,2} \cdot i^H_{2,3} = (+z_2/z_1)\cdot(+1) = +z_2/z_1 \quad (2)$$

where i_0 is the kinematical internal ratio of the planetary gears from Fig. 1,a,b (i.e. the ratio of the gearbox with fixed axes derived from planetary gear – set by motion inversion).

On the limit (for the internal involute gear pairs), admitting that $z_1 = z_2 + 4$, the following values are obtained from relations (1) and (2): $z_1 = 400$, $z_2 = 396$ and $i_0 = +0.99$.

Considering that the planetary gears from Fig. 1,a and b have the interior efficiency $\eta_0 = \eta^H_{1,3} = \eta^H_{1,2} \cdot \eta^H_{2,3} = 0.99$, the following efficiencies are obtained (Diaconescu & Duditza, 1994,a and b):

$$\eta = \eta^3_{H,1} = (-\omega_{1,3}\cdot T_1) / (\omega_{H,3}\cdot T_H) = (1-i_0)/[1-i_0\cdot(\eta_0)^w] = 0.5025;$$

$$\eta_{inv} = \eta^3_{1,H} = (-\omega_{H,3}\cdot T_H)/(\omega_{1,3}\cdot T_1) = [1-i_0\cdot(\eta_0)^x]/(1-i_0) = 0\,;$$

$$w = \mathrm{sgn}(\omega_{1,H}\cdot T_1) = \mathrm{sgn}[(\omega_{1,H}\cdot T_1)/(-\omega_{1,3}\cdot T_1)] = +1;\ x = -w = -1.$$

Because $\eta_{inv} = 0$, the planetary reducers from Fig. 1,a and b don't need brake when the motor is disconnected.

As results, the variants SR1 and SR2 (Fig.1,a and b) have the following properties: a) the input angular speed is reduced 100 times: $\omega_1 = +\omega_H/100$; b) the input torque is amplified 50.25 times: $T_1 = -i\cdot\eta\cdot T_H = -50.25\cdot T_1$; c) if the motor doesn't work, transmission is blocked without brake; d) the axial overall size is reduced and the radial overall size is relatively big, e) the manufacturing technology is relatively simple, but needs high accuracy. This means that each of the variants SR1 and SR2 is a solving structure of the motor-reducer function.

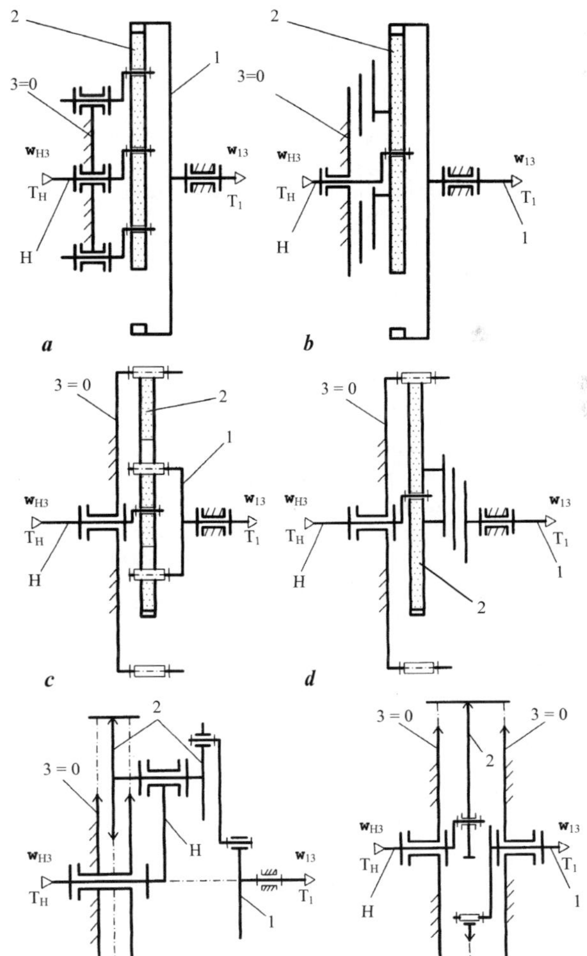

Fig. 1. Simplified schemes (without motor) of the solving structural *variants*: a) SR1, … ,f) SR6 (Neagoe et al., 2008).

Solving structure	SR1	SR2	SR3	SR4	SR5	SR6
Fig. 1	*a*	*b*	*c*	*d*	*e*	*f*
TECHNICAL CHARACTERISTICS						
1. *The numbers of the gears' teeth*	$z_1 = 400$ $z_2 = 396$	$z_1 = 400$ $z_2 = 396$	$z_2 = 100$ $z_3 = 101$	$z_2 = 100$ $z_3 = 101$	$z_2 = 100$ $z_3 = 101$	$z_2 = 100$ $z_3 = 101$
2. *The reducing ratio for the input speed*	**100**	**100**	**–100**	**–100**	**–100**	**–100**
3. *The efficiency for the direct actuation* η	$\eta_{H1} = 0.5025$	$\eta_{H1} = 0.5025$	$\eta_{H1} = 0.584$	$\eta_{H1} = 0.584$	$\eta_{H1} = 0.449$	$\eta_{H1} = 0.449$
4. *The efficiency for the inverse actuation* η_{inv}	$\eta_{1H} = 0$	$\eta_{1H} = 0$	$\eta_{1H} = 0.293$	$\eta_{1H} = 0.293$	$\eta_{1H} = -0.212$	$\eta_{1H} = -0.212$
5. *The amplification ratio for the input torque*	**50.25**	**50.25**	**58.4**	**58.4**	**44.9**	**44.9**

Fig. 2. The technical characteristics of the solving structures.

		SR1		SR2		SR3		SR4		SR6	
Criterion	w_k	N_k	$w_k \cdot N_k$	N_k	$w_k \cdot N_k$	N_k	$w_k \cdot N_k$	N_k	$w_k \cdot N_k$	N_k	$w_k \cdot N_k$
A	**0.649**	8	5.192	8	5.192	8	5.192	8	5.192	10	6.490
B	**0.162**	8	1.296	8	1.296	9	1.458	9	1.458	7	1.134
C	**0.108**	7	0.756	7	0.756	8	0.864	8	0.864	9	0.972
D	**0.081**	8	0.648	7	0.567	8	0.648	7	0.567	8	0.648
***Place*:**	Sum:	**4**	7.892	**5**	7.811	**2**	8.162	**3**	8.081	**1**	**9.244**

Fig. 3 Concept selection (SR6) by fine evaluation of the solving structures SR1, SR2, SR3, SR4 and SR6.

The case of SR3 and SR4 variants. For each of the planetary reducers from Fig. 1,c and d (consisting of the synchronic coupling 1-2 and the cycloid gear pair with rollers 2-3), the condition of obtaining the transmission ratio can be written:

$$i = i^3_{H,1} = \omega_{H,3}/\omega_{1,3} = 1/(1-i_0) = -100; \quad (3)$$

$$i_0 = i^H_{1,3} = \omega_{1,H}/\omega_{3,H} = i^H_{1,2} \cdot i^H_{2,3} = (+1)\cdot(+z_2/z_1) = +z_2/z_1. \quad (4)$$

On the limit (for the internal cycloid gear pairs with rolls), considering that $z_3 = z_2 + 1$, the following values are obtained from relations (3) and (4): $z_2 = 100$, $z_3 = 101$ and $i_0 = +1.01$.
Starting from the premise that each of the planetary gears from Fig. 1,c and d has the interior efficiency $\eta_0 = \eta^H_{1,3} = \eta^H_{1,2} \cdot \eta^H_{2,3} = 0.995 \cdot 0.998 = 0.993$, the following efficiencies are obtained:

$$\eta = \eta^3_{H,1} = 0.584\ ; \quad \eta_{inv} = \eta^3_{1,H} = 0.293.$$

Because $\eta_{inv} > 0$, the planetary reducers from Fig. 1,c and d need brake when the motor is disconnected. As results, the variants SR1 and SR2 are solving structure too.
The case of the SR5 and SR6 variants. The planetary chain reducers from Fig. 1,e and f (consisting of a synchronic coupling 1-2 and of a chain transmission 2-3) have the same kinematics as the previous reducers; the efficiency relations and numerical values remain also unchanged, except for the internal efficiency η_0 which becomes $\eta_0 = 0.988$. Therefore, the chain planetary reducers have the following efficiencies' values :

$$\eta = \eta^3_{H,1} = 0.449\ , \quad \eta_{inv} = \eta^3_{1,H} = -0.212.$$

Because $\eta_{inv} < 0$, the planetary reducers from Fig. 1,e and f don't need brakes when the motor is disconnected.
As results, the variants SR5 and SR6 have the following properties:
a) the input angular speed is reduced 100 times: $\omega_1 = -\omega_H/100$; b) The input torque is amplified 44.9 times: $T_1 = -i \cdot \eta \cdot T_H = +44.9 \cdot T_1$; c) if the motor doesn't run, the transmission can be blocked without brake; d) the complexity is relatively reduced e) the axial overall size is relative reduced and the radial overall size is relatively big in the case of scheme e, f) the manufacturing technology is relatively simple.

This means that both variants SR5 and SR6 are solving structures of the motor-reducer function.

3. CONCLUSIONS

The principle solution or the motor-reducer concept must be identified among the previous 6 solving structures.
Therefore, the generated structures are further ordered through a technical and economical evaluation (fine evaluation, Fig. 3).
The main features of the solving structures that were previously considered are systematized in Fig. 2; because the structure SR5 has the biggest radial overall size and the minimum efficiency, this variant was eliminated.
The remained solving structures are ordered in Fig. 3, considering that the 4 criteria are following different weights: A $\approx$ 4B $\approx$ 6C $\approx$ 8D (fine evaluation).
With the marks from Fig. 3, it results that the principle solution is designated by the structure SR6 (Fig. 1,f).
This principle solution contains three feasible modules (a motor without brake, a chain reducer and a Schimdt semi-coupling with rolls) and represents the input entity in the embodiment design phase.

4. REFERENCES

Diaconescu, D.V.& Duditza, Fl. (1994,a). *Wirkungsgradberechnung von zwangläufigen Planetengetrieben. Teil I: Entwiklung einer neuen Methode*. Antriebstechnik 33 (1994) 10, S. 70-74

Diaconescu, D.V.& Duditza, Fl. (1994,b). *Wirkungsgradberechnung von zwangläufigen Planetengetrieben. Teil II: Weitere Beispielrechnungen und Vorteile*. Antriebstechnik 33 (1994) 11, S. 61-63

Neagoe, M. et al. (2008). *A Conceptual Design Application Based on a Generalized Algorithm. Part I. Generation of the Solving Structural Variants*, The 19th International DAAAM SYMPOSIUM "Intelligent Manufacturing & Automation: Focus on Next Generation of Intelligent Systems and Solutions", 22-25th October 2008 (*accepted paper*)

Pahl, G. & Beitz,W. (1995). *Engineering Design*, Springer, ISBN 3540504427, London

Ulrich, K. & Epinger, S. (1995). *Product Design and Development*, McGraw-Hill Inc. ISBN 0-07-113742-4, New York

Annals of DAAAM for 2008 & Proceedings of the 19th International DAAAM Symposium, ISSN 1726-9679
Editor B. Katalinic, Published by DAAAM International, Vienna, Austria 2008
Make Harmony Between Technology and Nature, and Your Mind will Fly Free as a Bird

MOBILE DATA ACQUISITION AND TELE-TRANSMISSION BY PDA

NEDELCU, A[drian] V[alentin]; SANDU, F[lorin]; BORZA, P[aul] [Nicolae] & MORARU, S[orin] A[urel]

Abstract: *The paper is addressing the tele-monitoring with a cost-effective mobile solution that concentrates data acquisition (DAQ), logging, processing, interpretation, packing and transmission in a general purpose software implementation based on National Instruments (NI) LabVIEW, using tiny specialized NI DAQ boards on small portable Personal Digital Assistants (PDA). Concrete implementation details are given for the case of bio-medical mobile remote monitoring.*

Key words: *PDA, DAQ, Virtual Instruments, Mobile Telecom*

1. INTRODUCTION

The problem addressed by the authors is the integration of a system for data acquisition, logging and/or tele-transmission that can be personal and portable - e.g. by a patient (Perednia, 1995). Most of the pre-existing comparable solutions are highly specialized and proprietary, being not always very affordable. The accomplished configuration is using common PDA-s that have a plugged-in micro DAQ card – general purpose and with common interface (CF – “Compact Flash”), a high-capacity common memory card (SD – “Secure Digital”) and run an universal instrumentation software – NI LabVIEW (LV). LV can perform not only data acquisition and logging but also processing (e.g. digital filtering, identification-classification, compression etc) and communication.

If the PDA has also “SmartPhone” capabilities, it can also transfer the data via mobile communications (e.g. towards a central server). If not, an inexpensive alternative (that was integrated by the authors) is the connection to the owner’s mobile phone that can be controlled, for instance, via AT commands (for dial-up, GPRS and/or SMS transmissions). Local data-logging can be used for symptoms’ collection and detailed diagnosis, while mobile tele-transmission can be used mainly for updates and alarms (e.g. in case of patient’s status deterioration).

2. STATE OF THE ART

As demonstrated (e.g. by Philips “Motiva” Interactive Healthcare Platform – 2007), home tele-monitoring is the basis of remote patient management technology and can be enhanced with two-way communications that allow better personalization and support. Personal means of communication, like mobile phones and SmartPhones or PDA-based “communicators” are more and more involved in different ways of monitoring, logging (Boxwala et al, 2004), tracing but also messaging (including alarms when threshold values are reached) and notifications of clinical decisions (De Clercq, 2006) – all these in a way that should be tailored to the patient’s capability and needs.

For remote health-care, following the famous monitor invented by Norman Holter, many portable devices were developed for acquisition, logging, processing and transmission of bio-potentials (for ECG, EEG) as well as pulse, oximetry etc. Digital signal processing (DSP) enable morphologic analysis (e.g. – for the ECG – analysis of the P wave, of the QRS complex, of the T and U waves) and allow diagnosys of heart rate variability, bundle branch blocks etc.

3. RESEARCH COURSE

The authors developed different tele-monitoring systems in the last 8 years. A “twin-microcontroller” solution had one microcontroller allocated to the control of data acquisition and processing and another one to the management of communications (interfacing and control of a GSM modem). Such dual-processor solutions are common even to SmartPhones (only recently DSP and general purpose processing were brought to a common core).

3.1 System’s architecture

The most recent system implemented by the authors, with the architecture depicted in Fig. 1, consists of a 4 channels, 200kS/s DAQ card, namely NI‘s CF 6004, connected in the Compact Flash slot of a HP iPAQ 2210 PDA. The authors have used in their test configuration a portable ECG probe consisting of electrodes and signal conditioning circuitry. Because most Compact Flash PDAs currently available on the market do not embed mobile communications technologies such as GSM/UMTS or WiFi, which would allow the system to transfer the acquired data directly to a remote server, the authors decided to use a common mobile phone (e.g. the one of the patient) as a GSM/GPRS modem. The PDA is connected to the mobile phone through Bluetooth and is using standard AT commands to transfer a message (that incorporates the acquired data and/or specifically parameters computed out of them) to the mobile phone that should transmit these data to a remote server using GPRS (and/or at least SMS). In the author’s vision this server could be a Hospital Server which stores data about patients with cardiac problems in a database (MySQL and Java were used for the pilot programming).

3.2 Technical Solution and Results

The software for both data acquisition and transmission was implemented using LV’s PDA Module, a subset of NI LV functions specifically designed for PDAs. The data acquisition is performed using NI DAQmx Base, a set of drivers and functions for the DAQ boards made by NI. A DAQ task has to be defined, specifying attributes such as sampling rate, used channels, number of samples etc. This is accomplished using the DAQ Configuration Utility included in LV.

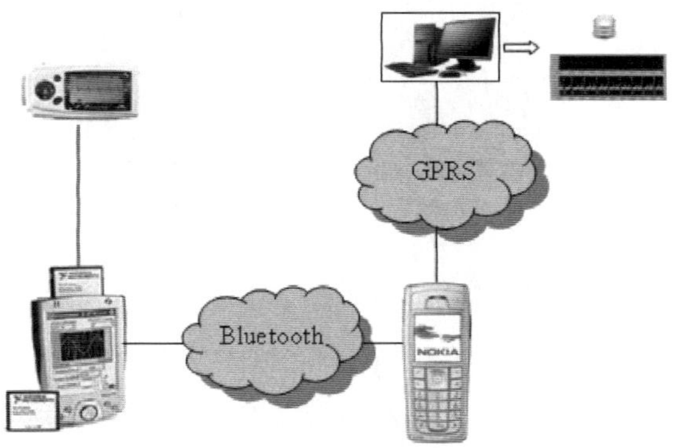

Fig. 1. Test configuration for mobile ECG DAQ

A measurement cycle (Fig. 2) consists of starting a specific DAQ task (using DAQmxBase Start Task.VI), reading data from the DAQ card (using DAQmxBase Read.VI) in a loop until the user stops the system or until an error occurs and then stopping the DAQ task (using DAQmxBase Stop Task.VI).
The acquired data can be viewed both online, on the PDA screen using a Waveform Chart (see the VI Panel on the screen, in Fig.3), or offline by logging the data into spreadsheet files (XLS format). The maximum frequency in the spectrum of an ECG signal is 100Hz (Rijnbeek, 2001); a sampling rate of 250 S/s has been chosen to meet Nyquist's criterion. A relevant measurement cycle implies acquiring 1000 samples which can be stored in 10 KB spreadsheet files. The log file obtained after 24 hours of continuous data acquisition (required for monitoring cardiac activity over a longer period) would be 216 MB large. Generated files which are as big as 10 KB can be directly transferred over GPRS to a remote server (which in the authors' vision could be a Hospital Server). If the acquired signal needs to be monitored over a longer period of time, thus resulting in larger files which are not fit for transfer over GPRS, these files can be stored on the SD card of the PDA and then "downloaded" to a PC in order to be analysed by a cardiologist.
The transfer of the acquired data over Bluetooth to the patient's mobile phone (which acts as a GPRS modem) is handled by a subVI (Fig. 4) which opens a Bluetooth connection to that phone (using the LV function Bluetooth Open Connection.VI). The Bluetooth Open Connection function receives two main parameters, namely the MAC address of the mobile phone to which one wants to connect and the channel number (which in the case of the Bluetooth dial-up service is 0). The subVI then sends a sequence of standard AT commands for initiating and performing a GPRS transfer.

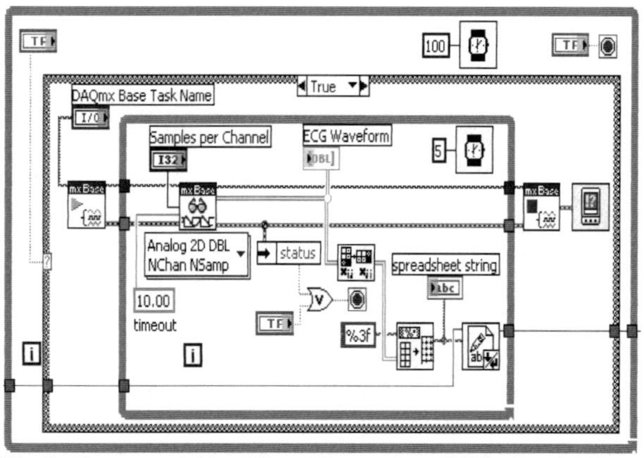

Fig. 2. Data acquisition Virtual Instrument (ECG_Graph.VI)

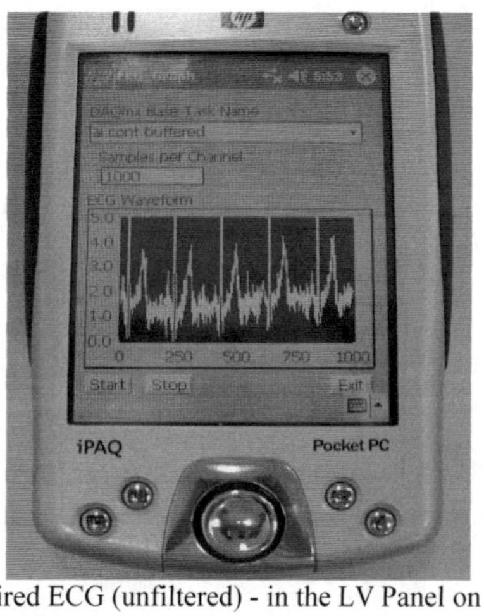

Fig. 3. Acquired ECG (unfiltered) - in the LV Panel on PDA

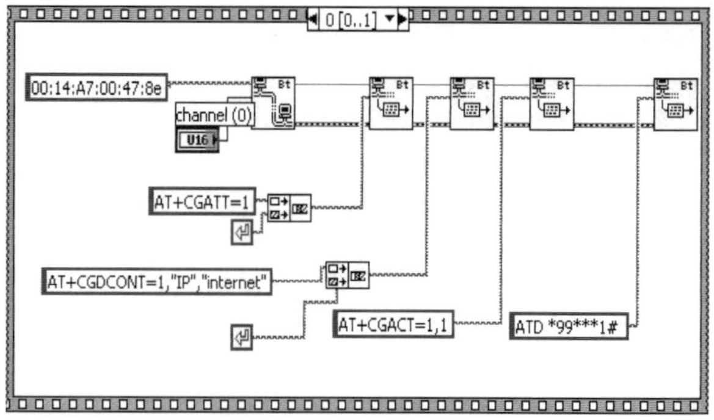

Fig. 4. SubVI for Bluetooth transmission using AT Commands

These commands are sent over Bluetooth by calling LV's Bluetooth Write.VI function. They are as follows:

AT+CGATT=1 - attaches the mobile unit to the GPRS network
AT+CGDCONT=1,"IP","internet" – defines a mobile operator specific PDP context
AT+CGACT=1,1 – activates that context
ATD *99***1# - connects to the GPRS network

4. CONCLUSIONS AND FURTHER RESEARCH

The present implementation proved the possibility of using powerful instrumentation hardware and software (NI CF6004 and NI LV) embedded in a very miniaturized format (for PDA) for mobile tele-monitoring. The research was financed by the Romanian Minstry of Education and Research, through its National Center for Programs' Management, in the frame of the 4th Program - "Partenerships in Prioritary Domains". The project is named „BIOMED-TEL" („Acquisition of Bio-medical Signals and their Tele-transmission via Mobile Computing Equipment") - Contract PNII – P4, nr. 11-057/2007 - Code PO-04-Ed1-R0-F5. It is co-ordinated by „Transilvania" University of Brasov-Romania. Further development is centered on AVR 32 bit microcontrollers produced by Atmel (3 times more computational powerful compared to the standard ARMv5 – due to its pipeline architecture, with 50% more efficient in memory usage, with 3 times less power consumption). Linux OS would ease programming (and NI LV for Linux could also be used, to overcome specific DSP complexity and to offer some protocol stacks) and would allow the development of an open source solution, which could be further extended by other researchers.

5. REFERENCES

Boxwala, A.A., et al. (2004), GLIF3. *A Representation Format for Sharable Computer-Interpretable Clinical Practice Guidelines*, Journal of Biomedical Informatics, Vol. 37, Issue 3, June 2004, pp 147-161 - ISSN: 1532-0464

De Clercq, P.A., et al (2006). *Approaches for Creating Computer Interpretable Guidelines that Facilitate Decision Support*, Journal of Artificial Intelligence in Medicine, Vol. 31, Issue 1, pp 1 – 27 - ISSN 0933-3657

Perednia, D.A.; Allen, A. (1995). *Telemedicine Technology and Clinical Applications*, Journal of the American Medical Association, Vol. 273, Issue 6, February 8, 1995 - pp 483-488, ISSN 0098-7484

Philips Medical Systems (2007) „Motiva" Interactive Healthcare Platform, Available on: http://www.medical.philips.com/main/products/telemonitorite/products/motiva/; *Accessed on: 2008-04-12*

Rijnbeek, P.R., et al (2001). *Minimum Bandwidth Requirements for Recording of Pediatric Electrocardiograms*, Circulation, Journal of the American Heart Association, Vol. 104, N°25, pp 3087-3090 - ISSN 0009-7322

Annals of DAAAM for 2008 & Proceedings of the 19th International DAAAM Symposium, ISSN 1726-9679
Editor B. Katalinic, Published by DAAAM International, Vienna, Austria 2008
Make Harmony Between Technology and Nature, and Your Mind will Fly Free as a Bird

SOME ASPECTS OF TECHNOLOGIES FOR OBTAINING THE SAMPLES FROM COMPOSITE MATERIAL WITH LAYERS AND PARTICLES REINFORCEMENT

NEDELCU, D[umitru]; CARCEA, I[oan] & GHENGHEA, L[aurentiu]

Abstract: *The aim of the paper is to establish the main steps of technologies to obtain the sample of composite material with layers reinforcement (figure 1) and particles reinforcement (figure 2). In general the methodology (for layers reinforcement) consist in: the premould of a rectangular metallic support on which we shall align the uniformly tensioned wave; the premould will be introduced in the steel mould and the whole assembly will be heated in an inert atmosphere of Argon; the inner surfaces will be protected with refractory paint based on aluminum oxide; the contact surfaces between the mould-piston will be covered with a refractory paint; after removing the slag, the liquid matrix is poured over the premould from the mould and the piston goes down; the solidification; the electrolytic polishing and finishing of surfaces; half of the batch of the test bars broken by traction will be boiled in caustic soda; accurately measurements on the electron-scan microscope; calculations the thickness of the aluminum carbide layer.*
The main steps of technology with particle reinforcement (figure 2) consist of: introduction the Si-C particles into the aluminum liquid matrix, mixed, heating the all mixture, cooling of all mixture into the crystallization, solidification, the extraction of composite material reinforcement.
Key words: *mould, particles, layers, reinforcement, aluminum.*

1. INTRODUCTION

The main problem of this paper is to establish the technology steps for obtaining the sample of composite material with layers reinforcement and particles Si-C reinforcement.
At worldwide level there are a certain number of universities and research institutes that have a "tradition" in the activity specific to this domain. We have to mention here that the materials required for these research studies are at accessible prices and have got suitable physical and mechanical properties (Hamann, R., Fougeres, R., 1991).
The engineers can design composites quite rigid and strength, in order to answer the most strictly specified demands, providing the number of layers, kind and thickness (matrix and reinforcement), the packing sequence and the fibers orientation.
Until now we develop some research concerning the characteristics of composite material using the pressure technology (references, Cobzaru. P., 2004 & Gramescu, T., 2002).
Now we want to use the liquid forging technology for obtaining the composite materials with layers reinforcement (fig.2, references Ciocardia C., 1991 & Cazimirovici, E., 1990) and to establish the technology steps for obtaining the samples of composite materials with particles reinforcement (figure 1).
The new idea consists in determining the optimum quantity of aluminum form matrix at the interface level, without diminishing the reinforcing capacity of the composite materials.
For further research we plan the following two main research parts:
a). Experimental research regarding the:
-tensile strength of the MMC for various technological parameters of liquid forging;
-fractographical research studies regarding the tensile strength of the MMC for various technological parameters of liquid forging;
-metallographic research studies of optical and electronic microscopy regarding the damaging of the composite materials;
b). Experimental research concerning the:
-development of mathematical model for damaged composite analysis under the dynamic loading conditions;
-the analysis of damaged layered composite during the dynamic conditions loading;
-the dynamic study of damaged composite materials with particles reinforcement;
-the analysis of interior and interface cracks;
-the analysis of different damages upon the internal tensions;
-the analysis of damaged upon the characteristics of composite materials using destructive methods (optical microscopy) and nondestructive methods (ultrasonic testing, eddy current testing).

2. TECHNOLOGY USE

The methodology and techniques that will be used for obtaining the composite materials with layers reinforcement consist in running the following principal phases (figure 1):
-the premould consists of a rectangular metallic support on which we shall align the uniformly tensioned waves;
-the premould will be introduced in the steel mould made T20MSNC260 and the whole assembly will be heated in an inert atmosphere of Argon at temperatures of 25°C, 600°C, 900^{0}C;
-the inner surfaces, obtained by splintering, that make contact with the liquid matrix, will be protected with a layer of refractory paint based on aluminum oxide applied at 150°C;
-the contact surfaces between the mould-piston and the mould-bottom lid will be covered with a refractory paint based on graphite in industrial alcohol in order to provide a perfect impermeability;
-aside, in an oven, there will be prepared the liquid metallic matrix with the following chemical compositions: a-Al100%, b-Al75Zn25;
-casting temperatures 700°C, 750°C, 800°C, 850°C, 900°C will be selected such as the fluidity determined with the Kerry spiral should be maximal;
-after removing the slag, the liquid matrix is poured over the premould from the mould and the piston goes down until it is reached a pressure of 25MPa;
-at the end of solidification, the newly formed composite will be removed;
-according to the standards in force there are taken several test bars and there are selected longitudinal and transversal surfaces to be analyzed;
-the test bars will be grouped in batches for microscopic metallographic research studies, inclusively for the electron-scan microscope and for their breaking by traction;

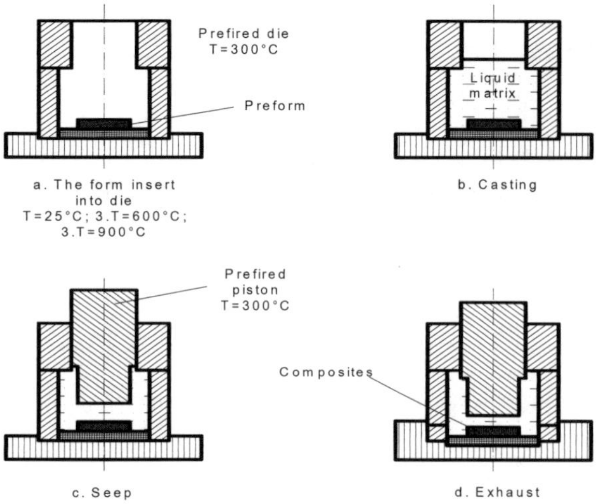

Fig. 1. Liquid forging – layers composite material obtaining.

-after the electrolytic polishing and finishing, the surfaces prepared for analysis are chemically attacked with Lacombe and Finkeldey reactive;
-half of the batch of the test bars broken by traction will be boiled in caustic soda in order to destroy the metallic matrix infiltrated among the fibers. Afterwards, the layer of aluminum carbide will be removed from the fiber surface by cleaning in ultrasound field;
-in the next phase we shall make accurately measurements on the electron-scan microscope in order to quantify the effort bearing diameter of the carbon fibers and the length of the reinforcement extractions from the matrix;
-by crystallographic calculations it is got the thickness of the aluminum carbide layer.
This is a feed-back working procedure that ensures the possibility of an objective measurement of the outputs – an essential condition for getting high-quality results.
The methodology and techniques that will be used for obtaining the composite materials with particles reinforcement consist in running the following principal phases (figure 2):
-introduction the Si-C particles into the aluminum liquid matrix;
-mixed all the mixture using one palette with driving belt;
-heating the all mixture;
-cooling of all mixture into the crystallization;
-solidification;
-the extraction of composite material reinforcement.

The future research consists of obtaining the samples and making the following:
-determination of the tensile strength curve for certain well-determined technological conditions and reinforcement volumes;
-the fractal analysis of the breaking surfaces;
-IR spectroscopy (chemical structure, links and so on).

3. CONCLUSIONS

The technical impact could be evaluated through the intern impact that the new composites will have over the substantial growth of productivity and production's quality, through the possibility of using the technology and other categories of composites, with the possibility of subsequent research studies in connection with composites with organic matrix, composites with reinforcing particles and hybrid composites.

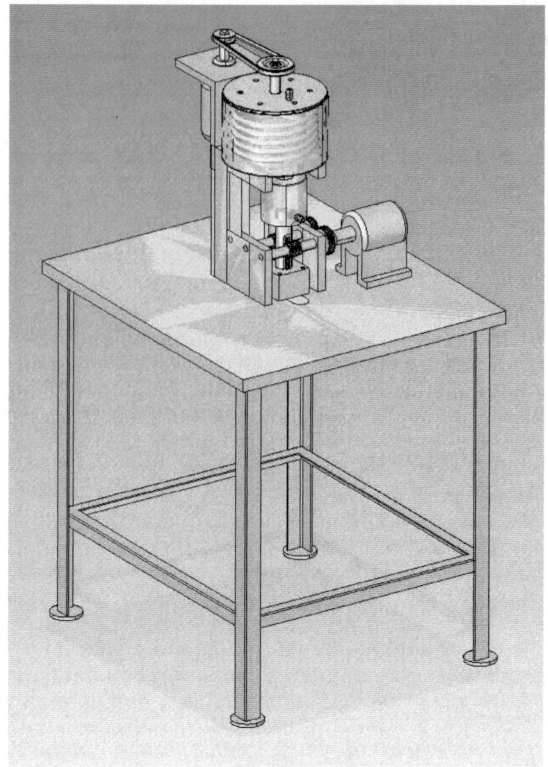

Fig. 2. The general view of equipment for obtaining composite materials with particles reinforcement.

The economic impact could be evaluated through an increased productivity and a much better quality of products, having consequences over the company's profit.

4. REFERENCES

Cazimirovici, E., et al (1990). *Teoria si tehnologia deformarii prin tragere, The theory of cold plastic deformation technology,* Technical Publishing, pp. 160, ISBN 973-31-0266-0, Bucharest

Ciocardia, C. et al (1991). *Tehnologia presarii la rece, The cold plastic deformation technology,* EDP Bucharest, pp. 437, ISBN 973-30-2314-0, Bucharest

Cobzaru, P. (2004). *Materiale compozite, Composite materials,* EDP Bucharest, pp. 34, 38, ISBN 973-30-1445-1, Bucharest

Hamann, R., & Fougeres, R., (1991). *Proceedings of the 12th Riso International Symposium on Materials Science, Roskilde*, Denmark, pp. 373, Roskilde

Gramescu, T., (2002). *Tehnologii de mecanica fina, The fine mechanics technology,* Tehnica-Info Publishing, pp. 261-262, ISBN 9975-63-165-7, Chisinau

Nedelcu, D., Pruteanu, O.V., (1998). *Aspecte ale formarii canelurilor exterioare prin deformare plastica la rece utilizand metoda Taguchi, Some aspects concerning the cold plastic deformation of exterior grooves using Taguchi method,* Tehnica-Info Chisinau, pp. 34-38, ISBN 9975-910-96-3, Chisinau

Pruteanu, O.V., Nedelcu, D.,(1996). *The stress analysis on deformation roll,* Proceedings of 7th International DAAAM Symposium, Katalinik, B. (Ed.), pp. 361-362, ISBN 3-901509-02-X, Vienna, Austria, October 1996, Vienna

Annals of DAAAM for 2008 & Proceedings of the 19th International DAAAM Symposium, ISSN 1726-9679
Editor B. Katalinic, Published by DAAAM International, Vienna, Austria 2008
Make Harmony Between Technology and Nature, and Your Mind will Fly Free as a Bird

SOME ASPECTS CONCERNING THE MEASURING OF FORCES FOR DEFORMATION DIFERENT PROFILES OF COMPOSITE MATERIALS

NEDELCU, D[umitru]; TABACARU, L[ucian]; CIOBANU, R[omeo]-M[ihai] & CRETU, G[heorghe]

Abstract: *This technical paper presents an instrument designed to measure the deformation forces occurring at the deformation of different profiles from composite materials. Was take into account the cold plastic deformation of exterior grooves where the inside part could be from composite materials aluminum-carbon fibers because the high resistance to deformation. The inside part could be obtained using the liquid forging which is our team research target into one national research project. It is known that the classical meter of the values of the strain gauges is an electronic tens meter, an apparatus with limited channels, a carrier frequency that cannot be changed and designated directly. On dynamic systems, the electronic tens meters cannot read enough number of values, being not able to draw the right conclusions upon the studied phenomenon. This is why this virtual meter of deformation forces was created. The virtual tens meter may operate on static systems and on dynamic systems.*
Key words: *measuring, forces, profiles, composite material*

1. INTRODUCTION

In order to make the gauging, we used an equipment presented in figure 1. In the boring of the bearing (3) of the machine-tool MURF 32 was introduced a shaft (8) similar to that of the machine on which the deforming rollers are fixed, shaft that comes in contact with a prism system (4) fixed on the board (7) mounted with the screws (10) on the arm (5). The bearing (3) is laid on a rigid board (2) that is on three load cells DT 210.00 located at 120° on the body (9) of the universal machine used for tests of static strength of materials MU200. The pillars (6) are oriented diagonally and serve to guide the arm (5) (that moves via helicoidally transmissions) the function of which is to apply the charge (the static load).

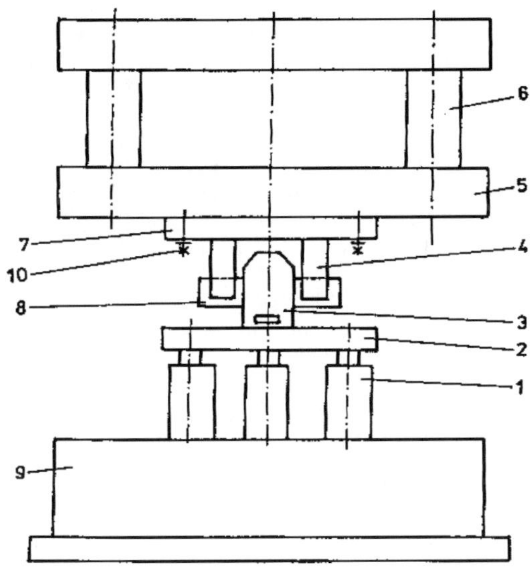

Fig.1. The gauging equipment.

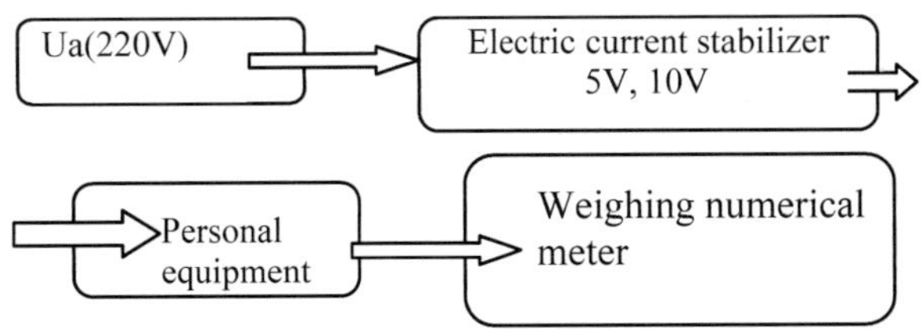

Fig. 2a. Line to be measured.

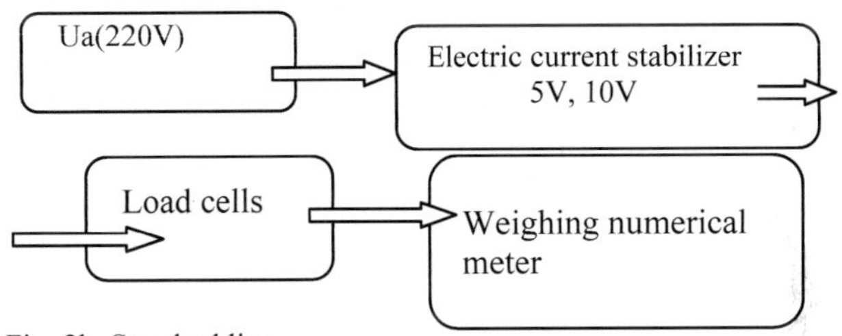

Fig. 2b. Standard line.

The strain gauges EA-06-125BT-120 are located on the opposite sides (serially linked to these sides) of the bearing (3) forming thus two half decks. The measurement of the mechanic sizes is not without errors owed to disturbing factors that cannot be evaluated by calculation, but only appreciatively (Ciocardia, C et al, 1991).

For gauging, we used two lines of action presented in figure 2, where the figure 2a shows the line to be measured and figure 2b the standard driving line (Cazimirovici, E., et al, 1990).

The supplied voltage was U_a= 10,0036[V].

We previously gauged the measuring chain so that at the minimal load (F_{min}=0[daN]) the meter should indicate a value close to zero, and at the nominal load (F_{nom}=20.000[daN]) the meter indicator should show the maximal value.

Further to the attempts made, the results were passed with values for three test cycles, upon loading and unloading, for variation stages of the load applied of 1000 [daN]. The signal was measured in [10^{-6} V]. The error of repetitiveness is good being lower than 5 units both when loading and when unloading.

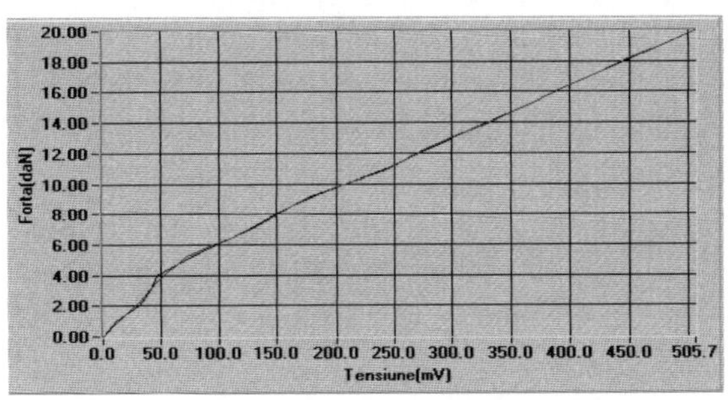

Fig. 3. The gauging and interpolation curve.

Thus we obtained the gauging chart of figure 3.

2. TECHNOLOGY USE

The acquisition plate will make the acquisition on one channel only, that is the channel 0 (figure 4).
To eliminate the noise, metal-braided cables are used, and the signal will be acquired on integration intervals on which an average of the signal values will be made. The acquired signal will be amplified with a value that will bring it to the output parameters that we need.
The classic meter (Pruteanu, O.V., Nedelcu, D., 1996) of the values of a strain gauge is the electronic tens meter. The electronic tens meter is a device with a limited number of channels, a carrier frequency that cannot be changed and designated directly. The device contains identical measurement channels and a common unit of supply and designation, being able to be connected to any measurement channel. The measurement is made by transforming via a resistive strain gauge, the value to be measured into the variance of an electric parameter.
The disadvantages of the meter appear when we want to make a measurement in a dynamic system. In a dynamic system, we do not have the possibility to read a sufficient number of values on the device scale, as the variance of the value to be measured may be too quick and this prevents the drawing of solid conclusions on the respective phenomenon. What is less convenient is the fact that the read values cannot be memorized. In the dynamic system, there appear problems when the value measurement value varies into values that cannot be comprised in a very precise measurement scale (Cobzaru, P., 2004). *Because of these advantages, the use of a virtual meter became a necessity.*
When making the virtual tens meter, we used for the interface of the software the programming environment Lab Windows compiled under Borland C++, where the sources for the acquisition system AX5210 were written.
The virtual tens meter has two ways of operating: acquisition in a static system, acquisition in a dynamic system.
Acquisition in a static system: a number of values are read from the strain gauge during the integration intervals, values that will be memorized in a vector. The integration intervals are the intervals on which the read values will be mediated to eliminate any possible noise.
Acquisition in a dynamic system: a number of values are read from the strain gauge during the integration intervals, values that will be memorized in a vector and displayed in a chart. Every integration interval will be displayed and conclusions will be drawn on the signal.
The user, after turning on the program, will see a panel with options for choosing the operating system (static, dynamic), the hard setting of the acquisition plate and the standard curve. In case there is a direct entrance into the operating mode, the settings taken will be the implicit ones.

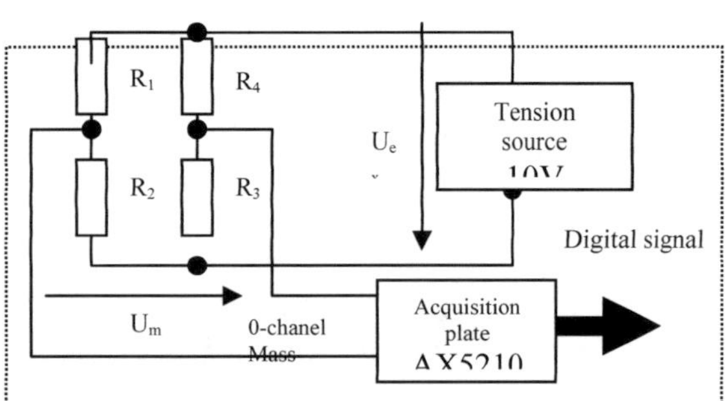

Fig. 4. Measurement circuit.

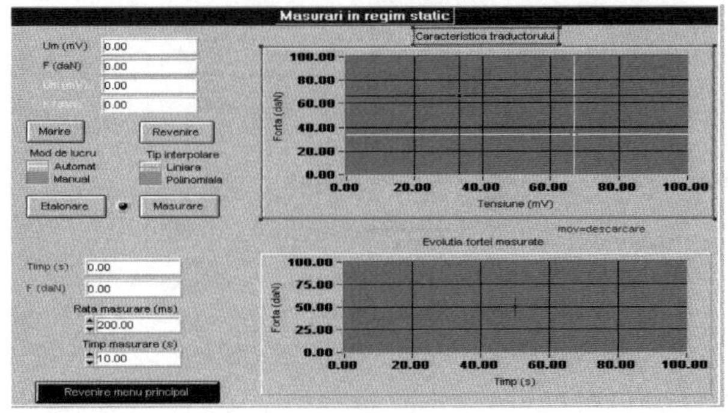

Fig. 5. Panel in the static system.

The acquisition panel (figure 5) in a statically system contains: indicators of the position of on charts; the control for the selection of the interpolation type; control for the selection of the operating type; control for the bridge balancing; control for starting the measurement; control for the selection of the acquisition time; control for the selection of the measurement rate; control for coming back to the main panel; charts for viewing the standard and the force evolution;
As concerns the *interpolation time* there are two options: linear and polynomial. For the *polynomial* option, the interpolation is done according to a 14 degree polynomial. This interpolation is achieved by the "Poly Fit" function. The function looks for coefficients that best represent the curve to follow (standard loading curve).
The future research consists of using this equipment for measurement of different profiles made from others material like steel, iron and so on. Also the improvement of this equipment will be the future research.

3. CONCLUSIONS

This equipment has two ways of operating: acquisition in a static system, acquisition in a dynamic system.
This equipment has a very low measurement errors, may be used, after adequate changes.
This equipment can be use to measure forces in entire cold plastic strain area and also in the metal cutting area.

4. REFERENCES

Cazimirovici, E., et al (1990). *Teoria si tehnologia deformarii prin tragere, The technology and theory of cold plastic deformation,* Technical Publishing, pp. 160, ISBN 973-31-0266-0, Bucharest

Ciocardia, C. et al (1991). *Tehnologia presarii la rece, The technology of cold plastic deformation,* EDP Bucharest, pp. 437, ISBN 973-30-2314-0, Bucharest

Cobzaru, P. (2004). *Materiale compozite, Composite material,* EDP Bucharest, pp. 34, 38, ISBN 973-30-1445-1, Bucharest

Nedelcu, D., Pruteanu, O.V., (1998). *Aspecte ale formarii canelurilor exterioare prin deformare plastica la rece utilizand metoda Taguchi, Some aspects concerning the cold plastic deformation of exterior grooves using Taguchi method,* Tehnica-Info Chisinau, pp. 34-38, ISBN 9975-910-96-3, Chisinau

Pruteanu, O.V., Nedelcu, D.,(1996). *The stress analysis on deformation roll,* Proceedings of 7th International DAAAM Symposium, Katalinik, B. (Ed.), pp. 361-362, ISBN 3-901509-02-X, Vienna, Austria, October 1996, Vienna

Annals of DAAAM for 2008 & Proceedings of the 19th International DAAAM Symposium, ISSN 1726-9679
Editor B. Katalinic, Published by DAAAM International, Vienna, Austria 2008
Make Harmony Between Technology and Nature, and Your Mind will Fly Free as a Bird

OPTIMAL DESIGN OF NONFERROUS CORRUGATED DIAPHRAGMS

NEGOESCU, F[lorin]; IOSUB , A[drian]; AXINTE, E[ugen] & SANTOS MARTIN, F[rancisco]

Abstract: *In this paper the authors purpose a research about the influence of the geometrical parameters of the corrugated diaphragms about their behavior by loading with a uniform pressure. The most important problems are: the establishing the optimal manufacturing technology, and to study how geometrical parameter influences the diaphragm behavior. The results are presented as a matrix model and for validation is used the Snedecor test.*
Key words: *corrugated; diaphragm; deformation; elastic; transducers*

1. INTRODUCTION

Regarding the construction of the measure appliances, we affirm, that because the property of corrugated diaphragms to measure the value of inflection at relative small efforts, as elastic elements, those are used in top technological areas, such as aero-spatial industry, military industry, control and automation for industrials process. Applicable areas for that type of diaphragm are large, so that the modest enumeration for contrivances in which construction are used, may contain some tenthly appliances and are include example part of the most variety technical fields, beginning with the regulators of temperature used at hatcheries and end with measure indicators of high and speed of airplanes. It is important to remark the fact that, the diaphragms may also be used as separating tool between two medias and for elastic tightens, allowed the transmission of movement from medias with pressure or vacuum.

2. EXPERIMENTAL SETUP AND RESEARCHES

2.1 Targets and the problem statement

The main objectives and targets of this theoretic and experimental research are:

1. Establishing the influence of geometrical parameters on the value of the arrow of deformation's diaphragms under the action of pressure
2. Determination of the influence of diaphragm geometrical parameters on the linear characteristics and establishing the effects of the facts and the interactions of those
3. The experimental validation of theoretic data .

2.2 The actually stage of researches

The advantage of those diaphragms, are (Negoescu et al., 2002): characteristics closer to linear, sow a shorter variation for effective surface; give possibilities for designer to adopt easy the characteristics of some given dates; bigger sensitivity give possibilities to keep constant effective surface of the diaphragm when centre of the diaphragm are moving along to plan in which are the diaphragm situated; biggest arrows without appearance of permanent deformations, the characteristic by simply modifying the height H of the undulations.

(in Fig.1 is shown an example of a corrugated diaphragm)

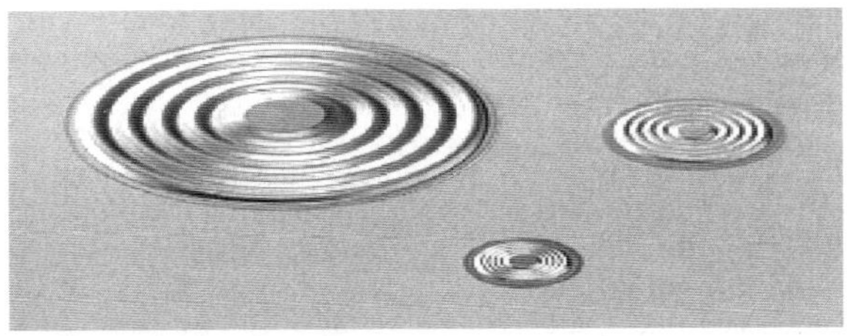

Fig 1. Corrugated diaphragm

In the design of corrugated diaphragms, designer it is interested not only by the size of deformation, but also by the linearity of relation between deformations – pressure. Although that linearity of the relation is useful on all area, that thing it is right only for certain portion of that. Pursuant to SAMA Standard (Scientific Apparatus Makers Association), who is affiliated at Instrument Society of America, usually it is measured at an non-linearity and experimented as linearity; represent the maximum deviation between the obtained curve and a straight line. Principal problem clung of studying the corrugated diaphragm it is the determination of the principal characteristics, witch mean to determine a relation between the arrow or deformation „W" and the force or pressure „P" witch product her, in the case of given form of corrugate.

$$W = f(P) \tag{1}$$

Is one of the most important base parameters of the analytical and elastic diaphragms the characteristic of elastic diaphragms can be express, according the action load, by a relation as (1), and the graphic representation of characteristic can be linear, call constant or non-linear. Some of those elastic diaphragms presenter a linear characteristic only on a certain part linear areas, specific to the corrugate metallic diaphragms depending on geometrical parameters whence characterized it is necessary the following resources.

In choosing the materials destination to execute the elastic diaphragms have to be taken in consider all the facts which characterize operating system (request of the material, work medium, the behavior of the material and eventually change in time of some of those characteristics), the possibilities of supply, the character of production.

2.3 About the design of our experiment

For this paper, the planning and drive the experiments he accomplished taking count of a series of appearances considerate accepted on world plans. To the matrix construction used the experiment with notation Yates (-1, accordingly the inferior level of a respective factor + 1, of superior level of a factor) her originality consist in the fact as the permits the procurement directly the general formula ale of average effects, carry is calculated totalized the all answers of the experiences, preciously of the signs of the factors from each experience and divide the total in number attempts effectuate (Pillet, 1992; Alexis,1999&Vigier,1993).

For the qualitative of a determination sizes on experimental path he accomplished a number finite of experimentations with a sufficient accuracy, maintaining same conditions of menstruation. The gathering of the date and the statistical their remaking he achieved taking count of the general valid methods and using specific programs of computer (Negoescu et all,2002)
To accomplish the porpoise objectives was betake to realize the tests pursuant to the requests of factorial orthogonal complete experiment at two levels, such 2^n ($2^4 = 16$ tests).
The experiments had place for each combination of those two level, for those four independent variables take in consideration (the active diameter of diaphragm D, the thickness of material h, the amplitude of corrugated H and the number of those n), as the table 1.

For the quantification of the admeasurements on the experimental path, are shall accomplished a number finite of measurements with a sufficient accuracy maintaining same conditions of measuring (Kenett, & Shelemayu ,1998). Established most probable value of moderate size, carry estimates the true value is shall utilized the law of natural repartition, and the accuracy of measuring shall be evaluate of the average quadratic his error the standard error.
When the linearity is big (or non-linearity is low) the relation signal out – signal in can be consider a straight line, and so the sensibility can be consider constant. On the other hand, when the linearity is low, (or the non-linearity is big) the sensibility depend by the in – signal. Prosecution of the experiments take place pursuant to the matrix of experiments which was presented in table 2.

The symbol	The factor of test	Level 1	Level 2	Unit of gauge
		-1	+1	
D	The diameter's diaphragm	60	43	mm
h	The thickness of material	0,06	0,12	mm
H	The amplitude's corrugated	0,52	0,87	mm
n	The number's corrugated	6	4	-

Table 1. The factors and levels of test

	The Factors				Pressure [bar]	Elastic Deformation [mm]
	D	H	h	n		
1	-1	-1	-1	-1	0,12	1,98
2	-1	-1	-1	+1	0,12	0,98
3	-1	-1	+1	-1	0,12	1,84
4	-1	-1	+1	+1	0,12	0,78
5	-1	+1	-1	-1	0,12	1,84
6	-1	+1	-1	+1	0,12	0,87
7	-1	+1	+1	-1	0,12	1,68
8	-1	+1	+1	+1	0,12	0,6
9	+1	-1	-1	-1	0,12	0,83
10	+1	-1	-1	+1	0,12	0,32
11	+1	-1	+1	-1	0,12	0,41
12	+1	-1	+1	+1	0,12	0,16
13	+1	+1	-1	-1	0,12	0,88
14	+1	+1	-1	+1	0,12	0,33
15	+1	+1	+1	-1	0,12	0,45
16	+1	+1	+1	+1	0,12	0,17

Table 2. The experimental matrix

2.4 The new mathematic model

The results obtained are expressed in figure 2

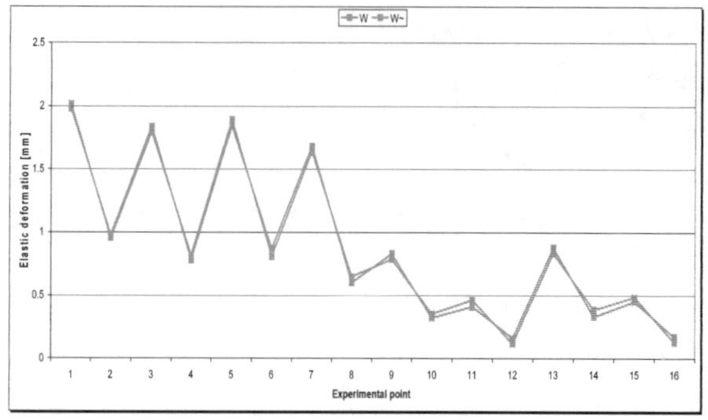

Fig. 2 The graphic representation experimental values and theoretical conformable arrow of deformation

By analyzing the graph from figure 2 is certain a very good preachment between the values determinate experimental (measured) and the theoretical ones so, in consequently, the values are low, fact which evidence the low influence of interactions on answer.

3. RESULTS ,DISCUSSIONS AND NEXT STEPS

The target for the experiment is to find a optimal configuration for a set of facts taking analyze, configuration which are lead to maximize the elastic deformation it is recommends for independent facts that type of matrix D = [1 0], H=[1 0], h=[1 0]. It can be told that the rigidity of the diaphragm grow if the diameter and the thickness of the materials have small values, and the amplitude of corrugates grow, in time to sensibility of diaphragm for same value at the mentioned parameters go low. So is confirmed the fact that sensibility became low in the same time with growing the rigidity. .
The optimization of metallic corrugated diaphragms, is adverted to touch the maximum technical performances, only that satisfy the concomitance the all criteria enforced is difficult, sometimes just impossible, because one factors can be antagonistic his subjective. With all these an optimum design is admissible and she assure to obtain the maximum performances, in certain conditions.

4. ACKNOWLEDGEMENTS to the Romanian Ministry of Education, Research and Youth & CNCSIS .This paper is financed by Research Grant type AT code 28.

5. REFERENCES

Alexis, J. (1999) *Metoda Taguchi în practica industrială, (Taguchi Method in Industrial Practice)* Editura Tehnică, ISBN 9975-63-176-x Bucureşti;Romania

Kenett,R & Shelemayu,Z (1998) *Modern Industrial Statistics* ,Brooks & Cole, ISBN 0-534-35370-3,Pacific Grove CA-USA,

Negoescu, F., Braha V.& Nagîţ, Gh. (2002), *Membrane metalice ondulate - Elemente de proiectare şi tehnologii de fabricare.(Metallic Corugated Diaphragms-Design and manufacturing)* Editura Tehnica-Info, ISBN 9975-63-176-2, Chişinău;

Pillet, M. (1992), *Introduction aux plans d'experiences par la methode Taguchi.* Les Editions d'Organizations, ISBN 9975-63-176-2, Paris,;

Vigier, M. (1993), *Pratique des plans d'experiences. Methodologie Taguchi.* Les Editions D'Organizations, ISBN 9975-63-176-2 Paris.

Annals of DAAAM for 2008 & Proceedings of the 19th International DAAAM Symposium, ISSN 1726-9679
Editor B. Katalinic, Published by DAAAM International, Vienna, Austria 2008
Make Harmony Between Technology and Nature, and Your Mind will Fly Free as a Bird

RESEARCHES CONCERNING THE INFLUENCE OF THE TECHNOLOGICAL PARAMETERS ON 42MOCR11 SURFACE ROUGHNESS

NEGOESCU, F[lorin]; NAGIT, G[heorghe]; IOSUB, A[drian]; SARBU, I[onel] & AXINTE, E[ugen]

Abstract: *In this paper, the influrnce of technological parameters on the roughness surface of 42MoCr11 workpiece in case of vibro-hardening process is studied. The influences of the hardening tool on the workpiece surface and the interactions between process parameters are studied. A mathematical model which describes the process is developed and validated.*

Key words: *vibro-hardening, superficial plastic deformation, roughness*

1. INTRODUCTION

The cold superficial plastic deformation of the metals is a chipless shaping method. One or more tools, without cutting edges, perform a movement and a pressing action on the workpiece surface. Between the workpiece and the tool there is a relative translation or rotation motion (Nagit, 1997). The cold plastic deformation process can be applied to metallic materials and has the following effects: superficial layer hardening, obtaining of new texture, inside of the workpiece there appears a remnant stress, modification of the temperature in the workpiece, transformation of the phase and modification of the material properties (Nagit et al., 1998).

The superficial plastic deformation by vibro-hardening as a superficial treatment contributes to the improvement of the physical and mechanical properties of the workpiece depending on the material nature and structural modifications (Kvackaj, 2006). For the experiments it is used a steel-42MoCr11 with the following chemical composition: 42MoCr11-0.4%C, 0.2%Si, 0.6%Mn, 1%Cr, 0.2%Mo.

2. PRESENTATION OF THE EXPERIMENTAL EQUIPMENT

The experimental equipment is consisted of machine tool, devices used to measure the process forces, roughness parameters and hardness of the workpiece after the vibro-hardening process (fig. 1).

Fig. 1. The equipment used for the experiments.

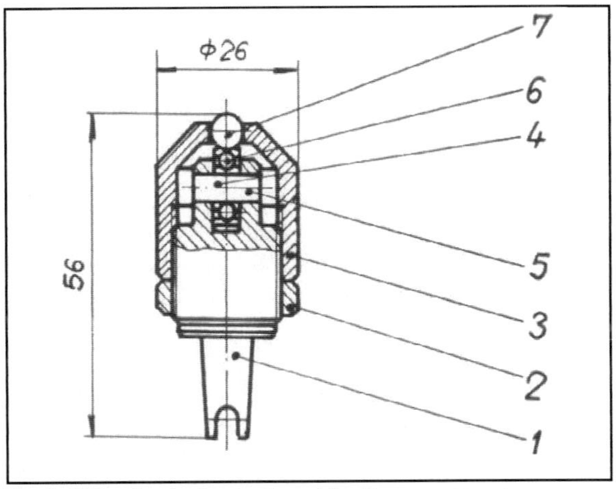

Fig. 2. Vibro-hardening tool used for the experiments.

The objectives armed at this experiment are:

- To establish the technological parameters on the roughness surface.
- To establish the interactions among factors.
- To develop a mathematical model, in a matrix form, able to describe the process.

The tool used in the experiments is presented in flgure 2. The bearing 6 is fixed in the tool body 1 by bolt 5. The ball 7 is maintained and fixed in bush 3 by the bearing 6. The bush can be changed according to ball diameter. A stable position during the process can be ensured by the clutch 2 (Nagit et al., 2003).

The device is designed so as to bear a force between 20 and 1000 N. The working stroke can be adjusted between 0.3 and 0.7 mm. The roughness of the samples used in the experiments was 6 μm. The experiment is a complete orthogonal factorial experiment 2^n (2^4=16 attempts), the Yates algorithm is used in order to establish the research points.

¶The interpretation of the factor effects can be done easier if the model, which characterizes the process, is expressed in a matrix form (Braha et al., 2003). One may express opinions and observations on the influence of the input parameters among the process.

The code of the parameter	Test factor	Level 1	Level 2	Unit of gauge
		-1	+1	
1	The amplitude of the oscillations	0.65	3	mm
n	Rotations per minute	80	315	rot/min
f	Feed	0.024	0.132	mm/rot
P	Deformation force	10	30	daN

Tab 1. The factors and level of tests.

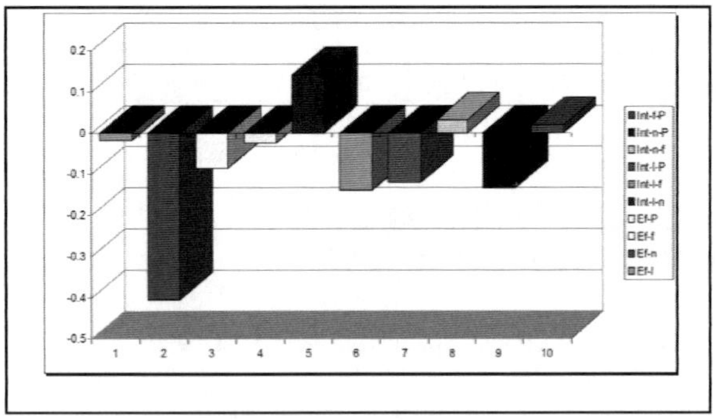

Fig. 3. Effects and interactions between process parameters at level 1 for 42MoCr11 material.

$$Ra_T = M + l + n + f + P + l*n + l*f + l*P + n*f + n*P + f*P \quad (1)$$

$$\begin{aligned} Ra_T = M &+ [E_{l1} \quad E_{l2}] \cdot l + [E_{n1} \quad E_{n2}] \cdot n + [E_{f1} \quad E_{f2}] \cdot f + [E_{P1} \quad E_{P2}] \cdot P + \\ &+ {}^t l \cdot \begin{bmatrix} I_{l1n1} & I_{l1n2} \\ I_{l2n1} & I_{l2n2} \end{bmatrix} \cdot n + {}^t l \cdot \begin{bmatrix} I_{l1f1} & I_{l1f2} \\ I_{l2f1} & I_{l2f2} \end{bmatrix} \cdot f + {}^t l \cdot \begin{bmatrix} I_{l1P1} & I_{l1P2} \\ I_{l2P1} & I_{l2P2} \end{bmatrix} \cdot P + \\ &+ {}^t n \cdot \begin{bmatrix} I_{n1f1} & I_{n1f2} \\ I_{n2f1} & I_{n2f2} \end{bmatrix} \cdot f + {}^t n \cdot \begin{bmatrix} I_{n1P1} & I_{n1P2} \\ I_{n2P1} & I_{n2P2} \end{bmatrix} \cdot P + {}^t f \cdot \begin{bmatrix} I_{f1P1} & I_{f1P2} \\ I_{f2P1} & I_{f2P2} \end{bmatrix} \cdot P \end{aligned} \quad (2)$$

$$\begin{aligned} Ra_T = 1.6206 &+ [-0.0181 \quad 0.0181] \cdot l + [-0.4043 \quad 0.4043] \cdot n + \\ &+ [-0.0856 \quad 0.0856] \cdot f + [-0.0243 \quad 0.0243] \cdot P + \\ &+ {}^t l \cdot \begin{bmatrix} 0.1418 & -0.1418 \\ -0.1418 & 0.1418 \end{bmatrix} \cdot n + {}^t l \cdot \begin{bmatrix} -0.1393 & 0.1393 \\ 0.1393 & -0.1393 \end{bmatrix} \cdot f + \\ &+ {}^t l \cdot \begin{bmatrix} -0.1206 & 0.1206 \\ 0.1206 & -0.1206 \end{bmatrix} \cdot P + {}' n \cdot \begin{bmatrix} 0.0318 & -0.0318 \\ -0.0318 & 0.0318 \end{bmatrix} \cdot f + \\ &+ {}' n \cdot \begin{bmatrix} -0.1343 & 0.1343 \\ 0.1343 & -0.1343 \end{bmatrix} \cdot P + {}' f \cdot \begin{bmatrix} 0.0193 & -0.0193 \\ -0.0193 & 0.0193 \end{bmatrix} \cdot P \end{aligned} \quad (3)$$

Interpretation of the results is facile if it is transferred into a diagram which illustrates the effects and the interactions of the factors mentioned in table 1.

Analyzing figure 3 one may observe that the factors "f" and "P" presente the most important effects.

3. RESULTS AND CONCLUSIONS

Taking into account Fisher criterion, the influence of factors can be illustrated, as can be observed in table 2.

As one may observe from Fisher test, the revolution of the workpiece has a significant importance, with big influences on the surface quality. When it comes to the interactions between the factors, one may notice that **l*n, l*f, l*P** and **n*P** are playing a very important role. Experimental determination of the roughness in case of vibro-hardening process, taking into account process parameters, shows that the roughness decreases significantly after vibro-hardening process [4].

8	Variable	F	α=99% f_α	α=85% f_α
l	S_l=0,0052	0,07	16,26	3,16
n	S_n=2,6163	**37,19**	**16,26**	**3,16**
f	S_f=0,1173	1,66	16,26	3,16
P	S_P=0,0095	0,13	16,26	3,16
l*n	S_{l*n}=0,3220	**4,57**	**16,26**	**3,16**
l*f	S_{l*f}=0,3108	**4,41**	**16,26**	**3,16**
l*P	S_{l*P}=0,2328	**3,30**	**16,26**	**3,16**
n*f	S_{n*f}=0,0162	0,23	16,26	3,16
n*P	S_{n*P}=0,2889	**4,10**	**16,26**	**3,16**
F*P	S_{f*P}=0,0060	0,08	16,26	3,16

Tab 2. The aria of analyses for 42MoCr11.

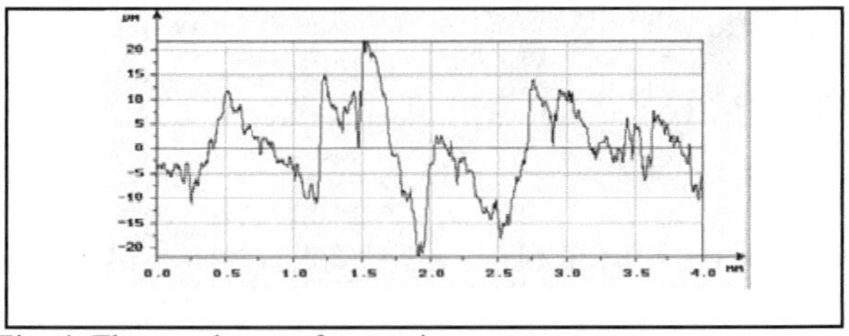

Fig. 4. The roughness after turning process

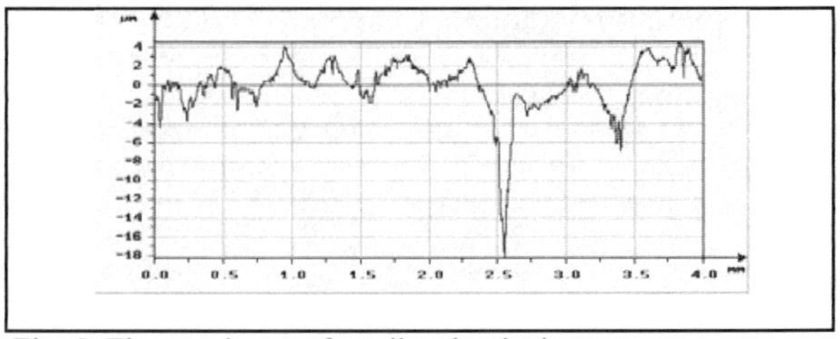

Fig. 5. The roughness after vibro-hardening process

Increase of pressing force has a good influence on the geometric parameters and the quality of the workpiece up to a limit value. After this limit value, the influences are negatives.

Increase of workpiece rotational speed and of feed, usually, has a negative influence on the surface quality. It is necessary to limit the rotational speed in accordance with the frequency of oscillations. The oscillations amplitude is described as having a positive influence regarding quality surface, when its value increases, but only in accordance with the other parameters of the process. If oscillations amplitude increases, many zones of workpiece material may remain unprocessed and, therefore, the surface quality decreases.

As a final conclusion, it is recommended to use small feeds in order to obtain an optimal number of tool crossings on the workpiece surface, decreasing roughness parameter. The feed used has to be optimally chosen in order to ensure at the same time a good productivity.

4. ACKNOWLEDGEMENT

The authors gratefully acknowledge the financial support offered to researchers of this paper by Romanian Ministry of Education, Research and Youth, through CNCSIS Grant no. 28, AT type.

5. REFERENCES

Braha, V., Nagit, Gh. & Negoescu, F. (2003). *Cold plastic deformation technology (in Romanian),* Technical and scientific publishing house CERMI, ISBN 973-667-025-2, Iasi-Romania

Kvackaj, I. (2006). Development of bake hardening effect by plastic deformation and annealing conditions. Availble from:http://public.carnet.hr/metalurg/Metalurgija/2006_vol_45/No_1/MET_45_1_051_055_Kvackaj.pdf *Accesed:* 2008-05-16

Nagit, Gh., (1997). *Theoretical and experimental contributions on vibrorolling process (in Romanian)*, PhD thesis, Technical University Gheorghe Asachi of Iasi

Nagit, Gh., Braha, V., Slatineanu, L. & Musca, G. (1998). Experimental researches concerning pressure force influence and lubrication on roughness at vibrorolling. *Technical Institut's Bulletin of Iasi*, Vol. XLIV (2003), pp. 391-396, ISBN 1011-2855

Nagit, Gh., Dodun, O., Negoescu, F. & Ignatescu, E.(2003). Cold superficial plastic deformation-a solution to replace heat treatment in small laboratory, *Proceedings of Baia Mare-North University,* Vol. XVII, pp. 227-231, ISSN 1224-3264, Baia Mare-Romania

Annals of DAAAM for 2008 & Proceedings of the 19th International DAAAM Symposium, ISSN 1726-9679
Editor B. Katalinic, Published by DAAAM International, Vienna, Austria 2008
Make Harmony Between Technology and Nature, and Your Mind will Fly Free as a Bird

VELOCITY FIELD UNIFORMITY ANALYSIS OF AN ELECTROSTATIC PRECIPITATOR

NEIMARLIJA, N[amir] & DZAFEROVIC, E[jub]

Abstract: *This work presents an attempt to apply the Finite Volume Method to analyze the velocity field quality in an Electrostatic Precipitator (ESP). The uniformity of the velocity field at the entrance and outlet plane of the electrical section is analyzed using some statistics. The available data about the experimental investigation velocity field of the actual precipitator were used to help determining boundary conditions and also for comparing purpose. The performered analysis showed qualitatively good agreement between the calculation values and the field measurements data but also that the ESP needs some improvement in the flow pattern.*

Key words: *velocity field quality, uniformity, numerical calculation, precipitator box.*

1. INTRODUCTION

The flow field characteristics of an electrostatic precipitator are very important for the particles collecting process inside precipitator box. Usually the problem of an insufficient efficiency of precipitator can be related to no uniformity of the velocity field distribution in the precipitation space. For these reasons, proper design of flow control devices within an ESP is a critical design function. The electrostatic precipitation theory considers that gas flow through the precipitation space should be uniform with respect to the standards, usually ICAC (Institute of Clean Air Companies) conditions required to be fulfilled (Dumont & Mudry, 2003). Modeling in some geometry ratio, usually performered under adiabatic and without voltage conditions, is still common method to obtain the necessary information. An alternative to this approach is the use of numerical methods and the mathematical modeling of the process what is our approach too.

Numerous papers on electrostatic precipitation phenomena exist ranging from the experimental and theoretical investigation electrical characteristics, in the past, to the complex numerical models of ESP's developed during last decade. The most of them research only some aspects of ESP's problem. They include also papers which deal with hydrodynamic problems in electrostatic precipitators. Some of them consider ambient air as media in numerical modeling of ESP (Dumont & Mudry, 2003) or simply offer the review of some ESP's problems (Parker, 1997). Several documents refer to the same ESP as the subject of this analysis: a numerical analysis an older version of the ESP (Teskeredzic & Dzaferovic, 1999), part of thesis (Neimarlija, 2006) and the test measurements data of the actual ESP (Internal documents, 1995-2003). The latter one contains data of the air velocity distribution analysis conducted on the considered actual ESP at Thermo Power Plant (TPP) Kakanj.

The actual electrostatic precipitator of unit 5 in Thermo Power Plant Kakanj consists of two parts (figure 1). Because both precipitators are equal, symmetric related to vertical plane which separates them, we chose one, left side. The main elements of precipitator are the precipitator box, inlet and outlet duct with perforated plates and turning vanes, dust hoppers, the electrode pairs, and so on.

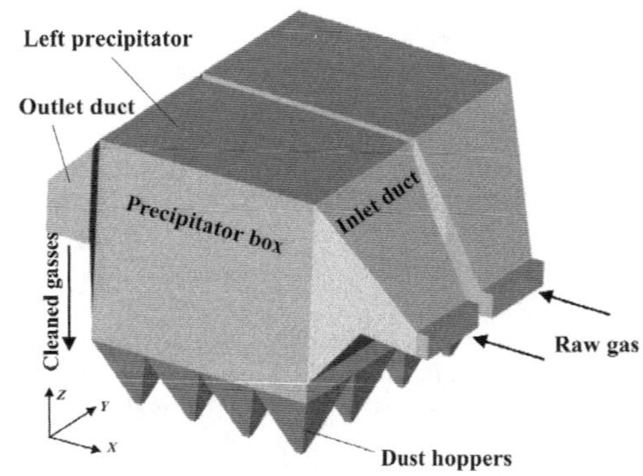

Fig.1.Electrostatic precipitator of a TPP Kakanj, unit 5.

The main data of the ESP are: number of the sections 4; the precipitator box width and height are 10.2 and 11.3 m, respectively; distance between the collecting electrodes is 400 mm; distance between the discharge electrodes is 495 mm; the flow rate of the gas is 543000 Nm^3/h (both sides); the applied voltage is 50 kV.

For obtaining more realistic results, in case of numerical modeling ESP's flow pattern, it is better to present actual geometry as authentic as possible. But, because of real reasons (computer limits), it was necessary to introduce some approximations of the geometry. We neglected the hoppers and the discharge electrodes while the so called "sigma" collecting electrodes being observed as the plain plates (as smooth walls). However, even with these simplifications we can expect valuable and useful results.

2. NUMERICAL MODEL

Mathematical model consists of averaged equations of the mass and the momentum balance. The standard k-ε model is used for turbulence modeling. Closed form of these equations is impossible to find for real configurations such as an ESP's complex geometry. The implemented finite volume method (Demirdzic, 1995) uses vectors and tensors expressed through their Cartesian components. Possibility of the use the unstructured mesh with polyhedral control volumes are very attractive properties of the numerical model easing creation numerical mesh of the complex geometry. Also, all dependent variables are stored at the cell center. Central differencing scheme is used for diffusive terms. For convection upwind difference scheme combined with central difference scheme are used. The segregated approach is used to solve the resulting set of coupled non-linear algebraic equations, for each dependent variable what leads to the set of decoupled linearized algebraic equations for each dependent variable. More information about the numerical method implemented here, especially about discretization schemes, verification and efficiency, can be found in (Demirdzic & Muzaferija, 1997). The very efficient computer code is built to perform the calculation.

3. RESULTS OF CALCULATION

In order to realize numerical calculation as first task it is necessary to perform discretization of the domain space. The discretization procedure resulted by dividing the domain space on 578170 control volumes with refined numerical mesh near location of perforated plates. The main support vertical beams are also defined with respect to their significant dimensions and so influence on the air flow current. The minor construction elements contained in the ESP are neglected.

To complete input data for the calculation purpose we need to know boundary conditions and physical properties of the media. The boundary conditions should be as closer as possible to the test conditions by reason for obtaining more comparable numerical values. The volumetric flow rate in amount of 108.3 m^3/s is calculated on the basis of averaged velocity $\bar{v}$ = 0.94 m/s and known cross-section of the electrostatic precipitator. Hence, one is obtained the average velocity at the inlet of precipitator of 13.48 m/s. The air density (ρ = 1.188 kg/m^3) and the molecular viscosity (μ = 1.824 10^{-5} Pas) of the air are considered as a constant. In the following presentation of results we are focused mainly on the velocity vectors at the inlet precipitator box plane.

In figure 2 are shown contours of air velocity through inlet plane, obtained by calculation and measuring. For the sake of easier comparison, the contour levels of numerical calculation are adjusted to contour levels of measure values. There is some similarity between contours of two velocity fields. We could notice from the contours in the figure 2 that calculation velocities exceed to some extent measured one what is mainly due to the boundary conditions. The influence of vertical central beams on velocity fields is visible on both contours.

For the left precipitator, the experimental analysis showed that the average velocity at inlet plane is 0.94 m/s, and deviation from average velocity is 38.27%, and at outlet plane average velocity is 0.78 m/s and deviation 21.92%. In relation with standard it is clear that these deviations are significant.

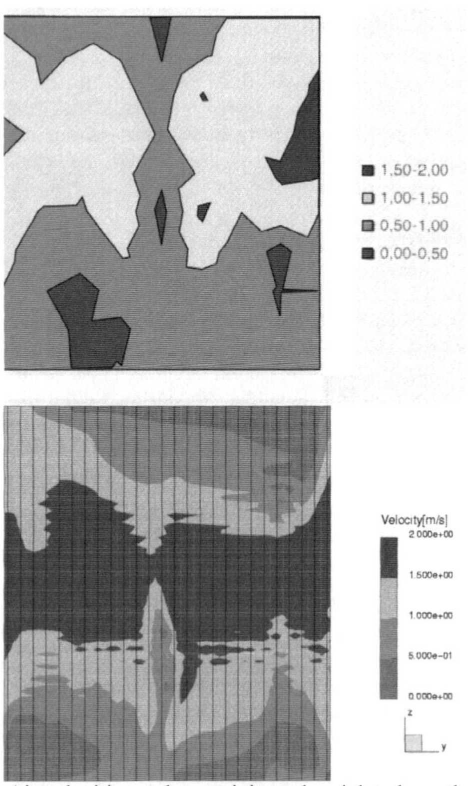

Fig. 2. Air velocities at the precipitator box inlet plane: the measured (above) and calculation (down) values.

	< +15%	+ (15 – 40)%	> +40%
Inlet	62.94	22.21	14.84
Outlet	79.13	19.50	1.36
	$\bar{v}$	σ	S (%)
Inlet	1.26	0.44	34.92
Outlet	0.96	0.16	16.67

Table 1. Statistical parameters for hydrodynamic calculation.

Three top rows from table 1 provide a flow distribution statistics for the ESP for the inlet and outlet plane. Obviously, ICAC 115% conditions are approximately satisfied only for outlet plane, while this deviation at inlet cross-section is above 14%. In second column of table 1 are given the participation percentages of velocities lower of 115% from$\bar{v}$, and in the next column the percentages velocities which value is between 115% and 140% of$\bar{v}$. In third column are the rest of velocities, part which values exceed 140% of$\bar{v}$.

In three bottom rows from table 1 are given data about another parameter S (variation coefficient) defined as the ratio between standard deviation σ and average velocity $\bar{v}$ times 100%. The numerical values of the parameter S are similar to those obtained with actual field measurements. For the inlet cross-section, numerical and measure values of S are 34.92% and S = 38.27%, respectively, while for outlet cross-section in numerical calculation S = 16.67%, and value on the measure basis is S = 21.92%. We see that numerical calculation gives some better values of parameter S. However, these values are also above prescribed value for parameter S of 15%.

The investigation of small velocities was also conducted. The percentage of air velocities higher of 60% and lower then 85% of $\bar{v}$ is about 20% in both planes. The fraction of smaller velocities (under 60% of$\bar{v}$) is relatively high, near 14%, in case of inlet plane.

4. CONCLUSION

The results of numerical analysis confirmed that there is unsatisfactory uniformity of flow field inside considered actual electrostatic precipitator and the certain flow corrections have to be made. Numerical modeling has advantages because offers detailed data in each control volume what is very hard to obtain by test measurements or by physical model.

By other side, in case of electrostatic precipitators, there is a necessity for development new and/or improvement present numerical methods. The future investigation will go to the direction comprising some others phenomena.

5. REFERENCES

Demirdzic, I. (1995). *Numerical Mathematics*, Mechanical engineering faculty of Sarajevo

Demirdzic, I. & Muzaferija, S. (1997). Introduction to Computational Fluid Dynamics, Lecture notes, Mechanical Engineering Faculty, University of Sarajevo

Dumont, B. J. & Mudry R. G. (2003). Computational Fluid Dynamic Modeling of electrostatic precipitators, *Electric Power Conference* 2003, 05 March, 2003

Internal documents of TPP Kakanj (Unit 5), (1995 – 2003)

Neimarlija, N. (2006). Numerical modeling fly ash precipitation in electrostatic precipitator, Doctoral thesis, University in Sarajevo, Sarajevo, may 2006

Parker, K.R. (1997), Technological advances in high-efficiency particulate collection, *Proc.Instn.Mech.Engrs*. Vol.211, Part A

Teskeredzic, A. & Dzaferovic, E. (1999), Final Report for the project: High-performance scientific computing in fluid dynamics, (Ref.-No. GM 2048)

Annals of DAAAM for 2008 & Proceedings of the 19th International DAAAM Symposium, ISSN 1726-9679
Editor B. Katalinic, Published by DAAAM International, Vienna, Austria 2008
Make Harmony Between Technology and Nature, and Your Mind will Fly Free as a Bird

GEOCENTRIC EXTRAVERSION VERSUS ETHNOCENTRIC INTROVERSION: A CLASH IN MULTINATIONAL HUMAN RESOURCES MANAGEMENT

NICA, G[abriela] B[eatrice]; AURITE, T[raian] & DUMITRESCU, D[iana] M[ariana]

Abstract: *In the new context of globalization, in terms of understanding the power of human resources, we consider very important to put into debate the impact of ethnocentric and geocentric approaches in human resources management. In this context our papert tries to put together / match two different worlds: introverted and extroverted, namely organizational members' beliefs (Jungian archetype) as a potentially malleable target of managerial intervention in the background of the different Western and Eastern European Cultures. We consider here the Jungian archetypes as a vehicle for understanding organizational beliefs, while the promotion of values means the practices and procedures used to increase value-related awareness and action*

Key words: *geocentric extraversion, ethnocentric introversion, Jungian archetype, organizational culture.*

1. INTRODUCTION

In our paper, we will put into debate the theorethical models describing two important types of organizational culture (ethnocentric and geocentric) according archetypal Jungian symbols: introversion and extraversion to match the best human resources policy for each organizational culture. We consider that such a diagnosis at all the hierarchical levels will allow us to better integrate the human resources typology in the organizational culture. The globalisation of the economy is creating occasions for the clashing of cultures. From manners and ways of making friends to attitudes toward the environment, different cultures espouse different values. Over the long haul, these values-and their differences-may prove as important as new technologies in determining the future.

2. GEOCENTRIC EXTRAVERSION VERUS ETHNOCENTRIC CULTURE

What is organizational culture? Organizational culture is the collection of norms, values, beliefs, expectations, assumptions, and philosophy of the people within it.

Organizational culture concerns the deeply rooted beliefs, values, and norms shared by the members of an organization. While these beliefs and values generally are not directly visible, they drive and are reflected in the actions taken by an organization. Organizational culture has been linked to economic performance and organization viability/success (Denison & Mishra, 1995).

The cooperative team culture type values flexibility and has an internal focus. This culture type is sometimes referred to as the human relations type because of its emphasis on commitment, cohesion and morale (Quinn, 1992).

Cultural complexes structure emotional experience and operate in the personal and collective psyche in much the same way as individual complexes, although their content might be quite different. (Singer & Kimbles, 2004). Cultural complexes also tend to be bipolar: extraversion and introversion orientation.

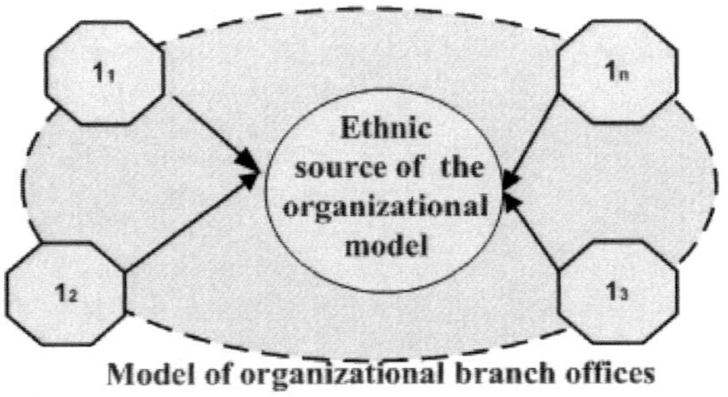

Fig 1 Ethnocentric Corporation

The label ethnocentric derives from the fact that the management is based on strictly one nation's set of criteria, that is, the country of origin of the corporation. Identity is mediated through the values, language and higher management of the same national derivation, found in the mother company and at the higher levels of the different branches of the corporation. It is an asymmetrical regime in which general company policy is planned from the centre, from whence the implementation of the different branches is monitored. The company's decision taking is thus controlled by the mother company and a group of managers of the same nationality (figure 1) Ethnocentric as hierarchical culture reflects bureaucracy and stability. Enforced roles, rules and regulations are emphasized. Ethnocentrism is a tendency to view people unconsciously by using our own customs as the standard for all judgements. The ethnocentric behavior maybe creates the problems in inter-cultural relations since the differences at this level are conceivably very great.

The translational or geocentric companies have a similarly high degree of international coordination on production, but where there is a greater degree of independence observable across branch companies, and greater orientation towards satisfying the demand among local markets. The heads of the branch companies enjoy greater autonomy and decision-taking is more horizontal, tending to be carried out among branch managers. Management by process is emphasised as well as the increased need for communication between the management of the different branches (figure 2).

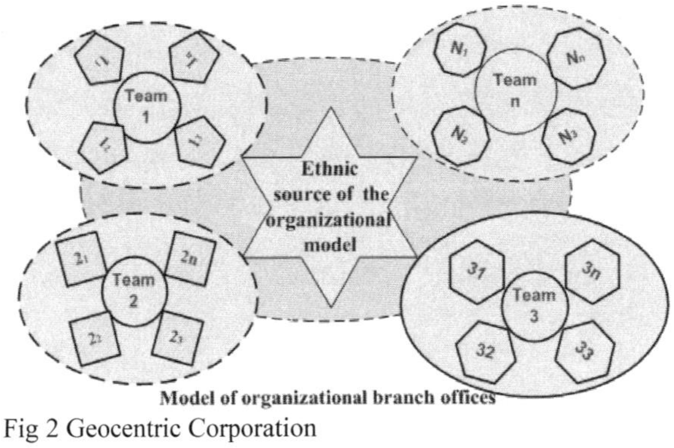

Fig 2 Geocentric Corporation

Geocentric specifically, rational culture reflects an orientation toward efficiency and profit. The primary emphasis is upon planning, productivity and goal clarity.

The two differences between ethnocentric and geocentric corporations in terms of orientation are a home-country orientation on the part of the parent company for ethnocentric model – introversion and a world orientation for geocentric model – extraversion.

Carl Gustav Jung was a Swiss psychiatrist and founder of the school of analytical psychology. Jung believed that symbol creation was a key in understanding human nature. He wanted to investigate the similarity of symbols that are located in different religious, mythological, and magical systems which occur in many cultures and time periods.

The first dimension of behaviour personality type measures are Introversion - Extraversion. The terms are a measure of where we get our energy. Introverts get their energy from within and lose energy quickly when having to deal with a lot of people (figure 3). Extraverts get their energy from the outer world and feel drained when they are by themselves too long (Figure 4).

The Introverts need to make the effort to speak out and take the initiative in social situations (Briggs- Myers, 1984) The Introvert prefers the intimacy of talking in depth with a few people, and then tires rather quickly.

Think of Extraverts like solar cell batteries (Read & all., 1953 - 1979). They need to be in the sun to get charged up. Introverts are more like rechargeable batteries. They need to be by themselves to restore their energy, and then they can go back out into the world of people. Flexibility values refer to spontaneity, change, openness, adaptability and responsiveness.

The Extravert is gregarious and enjoys meeting a lot of different people, and feel energized by the experience The Extraverts need to learn to go within and develop their thoughts in quiet and solitude.

We considered that the geocentric corporation prefer the extraverted attitude, describing it in such favourable terms as outgoing, well-adjusted, while the introverted attitude is dubbed self - centred. On the other hand, in the ethnocentric corporation, at least until recent times, the introverted attitude has been the prevailing one.

Psychological type refers to our ways of adapting to the world around us and to what goes on within us.

We all have to live in the world as well as with ourselves; therefore it is important to develop the attitudes and mental functions that optimally facilitate both adaptations. Adaptation means conscious choice: the possibility of choosing rather than being compelled in one direction or another.

The "inner" world refers to what happens in us: the thoughts, fantasies, dreams, and emotions that spontaneously come into consciousness or that we intentionally focus on. Our primary attitude/orientation is introverted when the "in here" energizes us, draws our attention, and give life meaning.

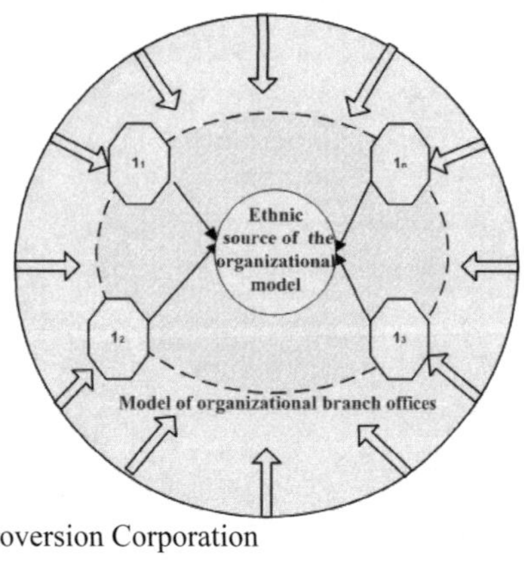

Fig 3: Introversion Corporation

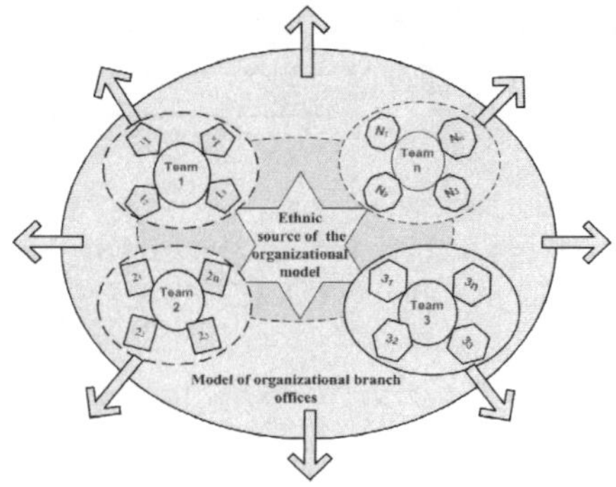

Fig 4 Extraversion Corporation

People whose natural attitude is introverted typically like, need, and seek time to pay attention to, to ponder, and to process what goes on "inside," in the "inner world".

Paying attention to what happens "outside" is more difficult for people who naturally prefer introversion. People who naturally prefer introversion typically know themselves (i.e., their inner experiences) but often are not well adapted to the "outer" world. The "outer" world refers to what happens around us, "out there:" other people, events, things, and the physical world. Our primary attitude/orientation is extraverted when the "out there" energizes us, draws our attention, and gives life meaning. In other words, our point of reference is "outside" ourselves. People whose natural attitude is extraverted typically like, need, and seek a lot of involvement with people and the surrounding environment. Paying attention to what happens "inside" is more difficult for people who naturally prefer extraversion. Persons who naturally prefer extraversion know the world, but often don't know much about them, that is, what goes on inside them.

3. CONCLUSION

Human resources culture is a multidimensional enigma that envelops the organization. Every member of the organization contributes to the culture in some manner. The history, style of leadership, structural stability, level of work-force empowerment and the ability to adapt to a changing environment all contribute to the culture of an organization.

Specifically, developmental culture relies on adaptability and readiness to attain growth, innovation, and creativity. Group culture sees cohesion, teamwork, and morale as means to foster development, empowerment, and commitment of human resources. We analyze here the geocentric corporation based on extraverted attitude, versus the introverted attitude dubbed self – centred prevailing in the ethnocentric corporation.

4. REFERENCES

Briggs Myers Isabel, *Gifts Differing*. Palo Alto, CA: Consulting Psychologists Press (1984).

Denison Daniel R. and Mishra Aneil K. (1995) Toward a Theory of Organizational Culture and Effectiveness *Organization Science*, Vol. 6, No. 2 (Mar. - Apr., 1995), pp. 204-223

Quinn, James Brian (1992), Intelligent Enterprise: A Knowledge and Service Based Paradigm for Industry. N.Y.: The Free Press

H. Read, M. Fordham, G. Adler, and Wm. McGuire, (1953-1979), *The Collected Works of C.G. Jung*. 20 vols. Bollingen Series XX, translated by R.F.C. Hull, edited by. Princeton University Press, Princeton,

Singer, T. & Kimbles, S., (2004) The Cultural Complex, Published by Brunner - Routledge

Annals of DAAAM for 2008 & Proceedings of the 19th International DAAAM Symposium, ISSN 1726-9679
Editor B. Katalinic, Published by DAAAM International, Vienna, Austria 2008
Make Harmony Between Technology and Nature, and Your Mind will Fly Free as a Bird

OPTIMAL LINEARIZATION METHOD FOR DYNAMIC ANALYSIS OF NON-LINEAR ROTORS

NICOARA, D[umitru]

Abstract: *The present paper deals with the dynamical analysis and optimization of non-linear rotor-bearing systems. The method of the optimal linearization or equivalent linearization allows the determination of some elastic stiffness matrices by which a linear model can be built. Its dynamic behavior is equivalent to that of the non-linear system.*
The optimization program written by the authors allows the optimal determination of the design parameters both by the direct study of the non-linear model and by the obtaining of an linear model equivalent with the non-linear model.
Key words: *non-linear rotor, optimization, finite element.*

1. INTRODUCTION

The rotor vibrations are strongly dependent on the rotor geometry, on the bearings type and on the excitation forces type. Usually the performances of the mechanical systems are improved by methods, which lead to the increasing of the dynamic stiffness. In the case of linearized model of the bearings we can build the optimization problems for rotor-bearing systems by coupling the finite element method (FEM) to the methods of non-linear programming with constraints (Nicoara et al, 2000). The goal is the diminishing the vibrations by the maximizing of the dynamic stiffness. To do this we need to find out the design parameters: the position of the bearings and the diameters of the shaft (different diameters for several segments of the shaft). In the case of non-linear rotors the optimization problem is formulated by the "energy" of the response as the objective function.

2. MODEL OF NON-LINEAR ROTOR

The model consists of a rotor treated as a continuous elastic shaft with several rigid disks, supported on the bearings with a non-linear behavior. In the study of the lateral motion of the rotor, the displacement of any point is defined by two translations (v, w) and two rotations (φ_y , φ_z). The model use the beam C1 finite element type based on Timoshenko beam model. The beam finite element has two nodes. In the case of the dynamic analysis four degrees of freedom (DOF) per node are considered: two displacements and two slopes measured in two perpendicular planes containing the beam, (Krämer, 1993).

In the case of linearized model of rotor bearing-systems, when the fluid film bearings are analyzed, the physical model of fluid film bearing may be simplified as a linear element.

It is well known that the behavior of both lubricated journal bearings and rolling element bearings is strongly non-linear and can cause rotors to behave in a non-linear way. In this paper the nonlinearities are involved only in the "elastic" part of the system. The non-linear bearings have a cubic non-linear term (Genta 1993), where the force–displacement relation for a non-linear spring element can be written as a function of the complex displacement z by the law

$$f(z) = k\left(1 + |z|^2\right)z \tag{1}$$

This law is particularly well suited for modeling some rolling element bearings, in particular preloaded angular contact bearings but has a more general application.

The linear part of the equation (1) can be introduced directly into the stiffness matrix.

The global mass matrix and the damping matrix are the same as in the linearized model of the bearings, but the global stiffness matrix contains non-linear terms $k + \hat{k}u^2$ in nodal displacements, due to the stiffness matrix of the bearings. Thus, the global stiffness matrix can be written as the sum of two matrices by isolating the non-linear part

$$[K(\{x\})]^{nl} = [K] + \left[\hat{K}(\{x\})\right]^{nl} \tag{2}$$

where the first matrix is the stiffness matrix of the structure which refers to the shaft and to the constant terms of the bearing stiffness and the second matrix appears due to the non-linearity of the bearings.

The equations of motions of anisotropic rotor-bearing systems which consist of a flexible non-uniform axisymmetric shaft, rigid disk and anisotropic bearings are obtained in second order form, by assembling the element matrices and may be written as

$$[M]\{\ddot{x}\} + [C]\{\dot{x}\} + [K(\{x\})]^{nl}\{x\} = \{F\} \tag{3}$$

Where $\{x\}$ is the global displacement vector, whose upper half contains the nodal displacements in the y-x plane, while the lower half contains those in z-y plane, and where the global stiffness matrix is defined by the equations (2)? The positive definite matrix $[M]$ is mass (inertia) matrix, the skew symmetric matrix $[G]$ is gyroscopic matrix, and the non-symmetric matrices $[C]$ and $[K]^{nl}$ are called the damping and the stiffness matrices, respectively.

3. NUMERICAL SOLUTION

The numerical time response solution for the non-linear system is calculated using the Wilson-θ method in conjunction with an iteration procedure. In the non-linear rotor's case, the stiffness matrix

$$\left[\tilde{K}\right]^{nl}\{x\}_{t+\theta\Delta t} = \left\{\tilde{F}\right\} \tag{4}$$

have non-linear terms, which depend on the values of the elements in vector $\{x\}_{t+\theta\Delta t}$.

Equation (4) is a set of non-linear algebraic equations now. Therefore an iteration procedure is utilized in conjunction with the Wilson-θ method to find $\{x\}_{t+\theta\Delta t}$ and then the displacements $\{x\}_{t+\Delta t}$.

4. OPTIMIZATION MODEL

In this section we propose a passive optimization model for non-linear rotor-bearing systems in dynamic range, i.e. to minimize the amplitudes of unbalanced response (synchronous harmonic). The goal is the diminishing the vibrations by the maximizing of the dynamic stiffness. To do this we need to find out the design parameters in order to minimize certain coast functional. The design parameters are the geometrical building elements as the distances between the bearings, s_i, and the diameters of the shaft, d_i. In the case of non-linear rotors the optimization problem is formulated by the "energy" of the response as the objective function:

$$\min_{s} \frac{1}{T_{max}} \int_{0}^{T_{max}} \{u\}^T \{u\}\, dt \ ,\ \{u\} = \begin{Bmatrix} v \\ w \end{Bmatrix} \qquad (5)$$
$$s^i \le s \le s^s$$

Numerical example. For the rotor model represented in Figure 1, we propose a numerical example with the value $\hat{k} = 10^{14}\, N/m^3$, and the rotor data given in (Nicoara et al., 2000).
The variation of the cost functional with distance between bearings, both for the linear model $(\hat{k} = 0)$ and non-linear model, is represented in Figure 2. The optimization has been done for spin speed of the shaft Ω=8000 rpm and *nstep*=4096.

The numerical solution of the non-linear system (3) is calculated using the method described in section 3.

The optimal value of the distance between bearings obtained from the solving of the optimization problem, with the cost functional the energy, is $s_{opt} = 0.25$ m.

5. EQUIVALENT LINEARIZATION MODEL

The linearization of the non-linear model using the optimization principle means the replacing of the non-linear system (3) by an equivalent linear system.

$$[M]\{\ddot{x}\} + [C]\{\dot{x}\} + [K]_{eq}^{lin}\{x\} = \{F\} \qquad (6)$$

The equivalent linear system is obtained by finding of an equivalent stiffness matrix $[K]_{eq}^{lin}$ which replaces the non-linear r stiffness matrix $[K]^{nl}$. The elements of the equivalent stiffness matrix are obtained by minimizing the cost functional defined as an error between the numerical solution of the non-linear system and the linear system solution i.e., the equivalent linear system is found by solving the following optimization problem

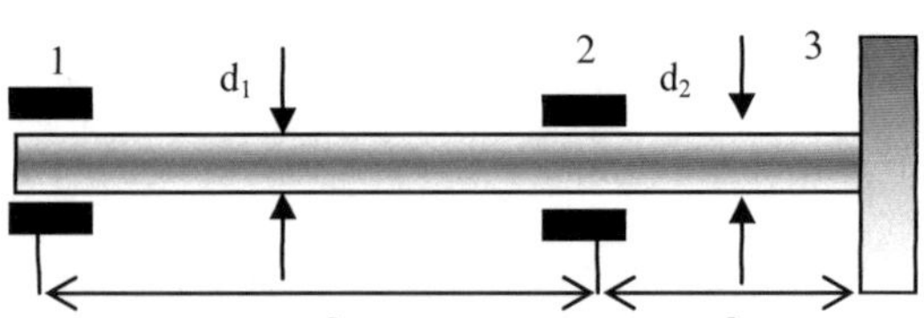

$$\min \sum_{i=1}^{N} \left(\{x(t)\}^{num} - \{x(t)\}^{lin} \right)^2 \qquad (7)$$
$$k_{yy}^{i} \le k_{yy}^{lin} \le k_{yy}^{u} \ ; k_{zz}^{i} \le k_{zz}^{lin} \le k_{zz}^{u}$$

Fig.1. Rotor bearing system configuration

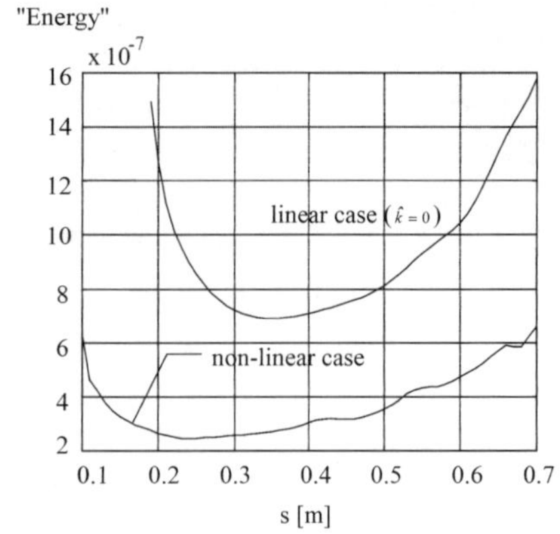

Fig. 2. The variation of the energy

The numerical solution of the non-linear equation (3) was found using the Wilson-θ method in conjunction with an iteration procedure, shown in section 3. The author elaborated several computer codes in MATLAB programming language. The optimization problem is solved by using the method BFGS, Broyden-Fletcher-Goldfarb-Shanno.

Numerical example. We consider again the rotor-bearing system from Fig. 1. We shall determine an equivalent linear system for the optimized non-linear system that means for the system with a determined distance between bearings by resolving the optimization problem (5). Thus, numerical solution $\{x(t)\}^{num}$ is calculated for the optimal configuration of the bearings, s_{opt} = 0,25 m. The equivalent stiffness matrices, for station 1 and 2, obtained by solving of the optimization problem (7) are, respectively

$$[k]_{eq}^{lin} = \begin{bmatrix} 6.0674 \times 10^8 & 0 \\ 0 & 3.9393 \times 10^8 \end{bmatrix} (N/m) \text{ (bearing 1)}$$

$$[k]_{eq}^{lin} = \begin{bmatrix} 8.4330 \times 10^8 & 0 \\ 0 & 6.8278 \times 10^8 \end{bmatrix} (N/m) \text{ (bearing 2)}$$

6. CONCLUSION

The method of the optimal linearization or equivalent linearization allows the determination of some elastic stiffness matrices by which a linear model can be built. Its dynamic behavior is equivalent to that of the non-linear system.

The optimal linearized model is easier to be used at the balancing of the multi-disk-rotor systems supported on bearings with non-linear flexibility.

The optimization program written by the authors in MATLAB language allows the optimal determination of the design parameters both by the direct study of the non-linear model and by the obtaining of a linearized model equivalent with the non-linear model.

7. REFERENCES

Genta, G., (1993). *Vibration of Structures and Machines: Practical Aspects*. Springer-Verlag, New-York, Berlin

Krämer, E., (1993) *Dynamics of Rotors and Foundations*. Springer Verlag, Berlin

Nicoara, D.; Cotoros, D. & Munteanu, M. (2000). External optimization of the rotor-bearing systems, *Proceedings of the 11th International DAAAM Symposium,* Katalinic, B. (Ed.), pp. 331-332, ISBN 3-901509-13-5, Opatjia , Croatia, October 2000, DAAAM International Vienna

Annals of DAAAM for 2008 & Proceedings of the 19th International DAAAM Symposium, ISSN 1726-9679
Editor B. Katalinic, Published by DAAAM International, Vienna, Austria 2008
Make Harmony Between Technology and Nature, and Your Mind will Fly Free as a Bird
*Young Researcher and Scientist Paper / * Supervisor, Mentor*

DESIGN OF A SAFETY MECHANISM FOR MACHINE TOOLS

NICODIM, M[ariana]; VLASCEANU, D[aniel] & BACIU, F[lorin]

Abstract: *This paper presents a safety mechanism used for quick withdrawal, during the automatic cycle, of the vertical feeding slide of surface grinding machines type RPV or RPVR, in case of accidental or forced interruption of electrical energy during processing. At that moment, a collision may appear between the grinding wheel and the working piece due to the clearance of the feeding kinematical chain.*

The safety mechanism is conceived for mounting on a rocker shaft grinding machine but it can be also used on any surface grinding machines type RPV or RPVR.

Key words: *safety, mechanism, grinding, energy, mechanical.*

1. INTRODUCTION

Nowadays, the worldwide MT (Machine-Tool) continues to be a domain in which the fundamental or construction solution news is in a continuous evolution. In our country, the concerns of the MT constructors and of the researchers in the field, especially, are framing in the general context of the worldwide effort aimed to introduce the CN and recently the CA in the structure of the most various technological equipments. (Catrina, et al., 1993).

In order to obtain a better level of usage and understanding, this paper aims to describe the construction, the operating principle, the role and the advantages of a self-conception safety mechanism. This mechanism is intended for the RPX-AC400-CNC rocker shaft grinding machine used for grinding the flat surfaces of the "K" and "D" engine rocker shaft. The working piece has five flat surfaces on one side and another five on the other side and it will be simultaneously grinded with five abrasive wheels. The grinding machine has been conceived using the most advanced technologies from all over the world and it is equipped with numerical control (CNC 802D).

2. SAFETY MECHANISM: CONSTRUCTION, OPERATING PRINCIPLE AND ADVANTAGES

The surface grinding machines with vertical axis and linear table type RPV or with rotary table type RPVR for processing plan-parallel parts in automatic cycle are well known. In case of interruption of electrical energy during processing, these grinding machines cannot assure the operator's safety or the safety of the machine.

The article describes a safety mechanism implemented within the kinematical feeding chain of the vertical slide from the RPX-AC400-CNC rocker shaft grinding machine.

The safety mechanism is used for quick withdrawal, during the automatic cycle, of the vertical feeding slide, in case of accidental or forced interruption of electrical energy during processing. In case of electrical energy failure, rechargeable batteries from the hydraulic diagram will simultaneously feed two single-actuating cylinders causing the withdraw of the whole group – vertical slide, trapezoidal threaded screw - screw nut, worm reducer – avoiding the collision between grinding wheel and the working piece.

The purpose of this invention is to assure quick withdrawal, in automatic mode, of the whole group – vertical slide, trapezoidal threaded screw - screw nut, worm reducer – avoiding the collision between grinding wheel and the working piece, collision possible due to the clearance of the feeding kinematical chain.(Ispas, et al., 1997)

The safety mechanism implemented in the slide feeder of the feeding kinematical chain from surface grinding machines will eliminate the mentioned disadvantages. In case of accidental or forced interruption of electrical energy during processing, in order to eliminate the chance of collision between grinding wheel and the working piece the safety mechanism will use two single-actuating cylinders which will lift up the whole group: vertical slide feeder, trapezoidal threaded screw - screw-nut, worm reducer – on the slide ways existing on the prop of the grinding machine.

When restarting the automatic mode of the grinding machine, the two hydraulic cylinders are released by actuating the hydraulic distributors and the disk springs of the two hydraulic cylinders will enable the clamping for all the group to its original position with respect to prop of the grinding machine.

Figure 1 presents an example of construction for the safety mechanism implemented to the RPX-AC400-CNC rocker shaft grinding machine.

The safety mechanism implemented in the structure of the kinematical chain in case of vertical axis grinding machines type RPV and RPVR consists of: upper plate 1 – provided with

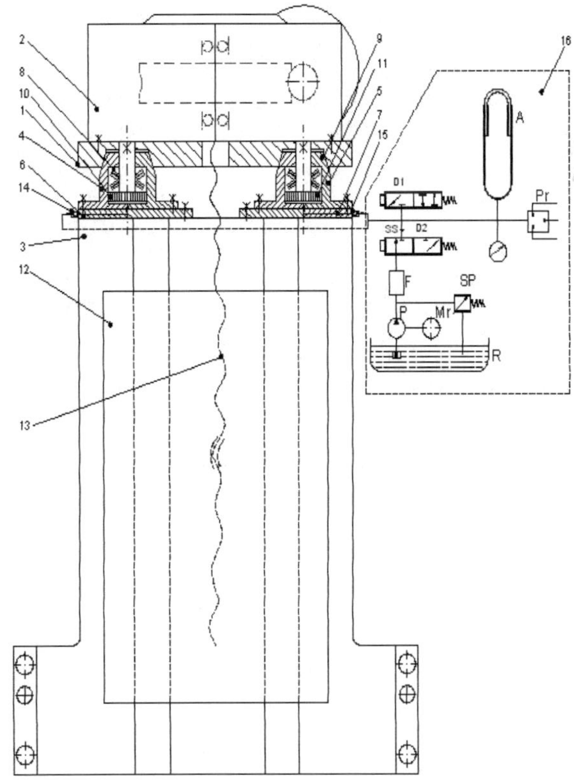

Fig. 1 Safety mechanism

two taper holes symmetrically disposed with respect to vertical axis and the mounting and alignment holes with respect to worm reducer 2. On the upper side of the prop 3, there are two single-actuating hydraulic cylinders 4 and 5 mounted with the fixing plates 6 and 7. Each two single-actuating hydraulic cylinders 4 and 5 has a conical face on which the upper plate 1 is supported using disc springs 8 and 9 that are pressing down the pistons 10 and 11 inside the hydraulic cylinders 4 and 5. In order to avoid the collision between the grinding wheel (driven on by the vertical slide 12) and the working piece, using the trapezoidal thread screw 13 the quick feeding with pressured oil from the rechargeable battery A (part of the hydraulic group 16) is assured.

3. DESCRIPTION OF THE RPX-AC400-CNC

RPX-AC400-CNC rocker shaft grinding machine has three linear axes actuated by the CNC. The feed is being carried out by the alternating current engines, realizing a continuous variation of the number of revolutions by CNC equipment. The engine of the vertical feeding action is an engine with a trammel and on the ball-bolt of the vertical feed further electromagnetic trammel is provided.

The longitudinal feed of the kinematical chain (X axis) - the bench has an eigen feed mechanism actuated by the numeric command of the equipment. The mechanism consists of: servo-engine feeder, toothed drive belt and ball-bolt. The screw nut of the ball-bolt is assembled on the bench. The measurement of the bench displacements are carried out with a transducer at the end of the ball-bolt by means of a clutch.

The transversal feed of the kinematical chain (Z axis) actuated by the numeric command equipment consists of: servo-engine feeder, toothed drive belt and ball-bolt. The nut of the ball-bolt is mounted in the prop.(Bruda, et al., 1980)

The vertical feed of the kinematical chain (Y axis) actuated by the numeric command equipment consists of: servo-engine feeder, toothed drive belt and ball-bolt. The nut of the ball-bolt is mounted in the vertical slide. (Kief, H.B., Hasen C., 1999)

The prop (3) is a cast construction guided along the transversal frame by means of rolling slide bars – roller tappet type. This ensures the stone set positioning ("D" and "K" valve-rocker shafts) in respect with mass of the machine on which the alignment and clamping device and the knurl head are assembled, by means of the transversal feed mechanism (Z axis). The alignment accuracy in case of processing along the axis of the piece is assured.

The vertical slide (12) of the RPX-AC400-CNC rocker shaft grinding machine ensures the stone set alignment adequate to the processed mark ("D" and "K" valve-rocker shafts) with respect to mass of the machine ("Y" axis). The radial alignment accuracy in case of processing along the longitudinal axis of the piece is ensured.

The main shaft is mounted within the vertical slide body. A set of stones adequate to the processing mark are mounted on the bevel of the shaft. The knurl head is aligned with respect to bench of the machine, the axis of the knurl head alignment points being perpendicular on the displacement direction X and parallel with the bench mirror. One of the points is fixed, as a reference axis, and the second one orientated in respect with operator is adjustable on the thread. In the knurling process, the 5-stoned set acquires in the radial section the emission profile that will be processed on the specific valve-rocker shafts type "K" and "D".

The alignment and clamping device ensure the piece alignment with respect to mass of the machine: longitudinal (along the X axis), transversal (along the Z axis) and vertical (along the Y axis) of the 2 sets of 10 pieces each. The hydraulic installation (16) is placed behind the longitudinal frame, in the engine, pump, and command and monitoring device for the correct usage of installation. (Nakazawa H., 1994).

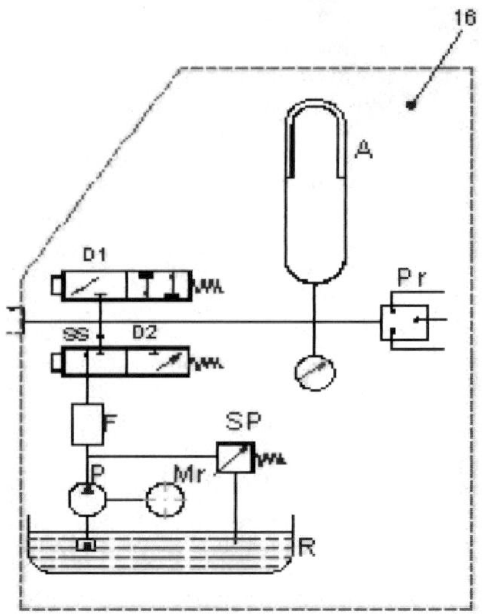

Fig. 2 Hydraulic installation

Legend:

R – receiver	D1 – hydraulic distributor
S – suction basket	D2 – hydraulic distributor
P – pump	SS – direction valve
ME – electric motor	M – manometer
SP – pressure valve	PR – pressure switch
F – filter	A - rechargeable battery

The hydraulic installation ensures operating in the duty cycle in case of piece clamping and loosening machine.

4. CONCLUSIONS

The security mechanism help the functional integration – related to the security – for the process that needs to be protected, into an efficient structure.

Beside the fact that it represent a rigid and compact construction, the implementation of this security mechanism on every plane grinding machine type RPV or RPVR assures the adjustment of the geometrical axes of the trapezoidal threaded screw with respect to slide ways of the vertical feeding slide, by means of the conical surfaces existing in the upper plate and single-actuating hydraulic cylinders. The security mechanism also eliminates the chance of collision between the grinding wheel and the working piece in case of accidental or forced interruption of electrical energy.

The security mechanism contains actions to be taken, equipments and also effects: safety and stability. It is the collocation of this moment, passing through the integrated mechanism phase, and going to the future collocation "security system".

5. REFERENCES

Bruda, D., Dobrescu, T.; Chivulescu, C.; Pana, T.; Guta, R., (1980), *Special control problems of feed drives of d. c. motors with wide speed variation range*, Polytech Inst of Cluj-Napoca, p C. 13-C. 15

Catrina, D., et al., (1993), *Machine tools with NC*, University POLITECHNICA Bucharest.

Ispas C., Predincea N., Ghionea A., Constantin G. (1997) *Machine tools*, Ed. Tehnica, Bucharest.

Kief, H.B., Hasen C., (1999) *"NC/CNC Handbook"*, Verlag, Munchen Wien.

Nakazawa H., (1994), *Principles of Precision Engineering*, Oxford University Press

Annals of DAAAM for 2008 & Proceedings of the 19th International DAAAM Symposium, ISSN 1726-9679
Editor B. Katalinic, Published by DAAAM International, Vienna, Austria 2008
Make Harmony Between Technology and Nature, and Your Mind will Fly Free as a Bird
*Young Researcher and Scientist Paper / * Supervisor, Mentor*

TECHNIQUE FOR CAMERA CALIBRATION

NOSKOVICOVA, L[ucia] & RAVAS, R[udolf]*

Abstract: *In this paper we presented technique for camera calibration. Under the camera calibration we understand determination of intrinsic and extrinsic camera parameters. These parameters are obtained for computed image coordinates from real scene. Technique presented in this paper based on Zhengyou Zhang method is very suitable for calibration.*
Key words: *camera calibration, focal length, intrinsic and extrinsic parameters.*

1. INTRODUCTION

A camera consists of an image plane and a lens which provides a transformation between object space and image space. This transformation cannot be described perfectly by a perspective transformation because of distortions which occur between points on the object and the location of images of those points. However, the model may only be an approximation to the real relationship. Standard camera calibration technique is based on the pinhole camera model where the distortions are not modelled. The pinhole camera performs a perspective transformation. Precise distance measurements for a set of reference points are the necessary burden of the calibration procedure. Various methods form camera calibration can be found in the literature.

We present a camera calibration method that uses calibration object (chessboard) and computes intrinsic and extrinsic camera parameters. This procedure uses Zhengyou Zhang method and computes error variable based on distance between scans points and computes points.

2. CAMERA CALIBRATION

We used a pinhole camera model for this procedure. The pinhole model is an approximation suitable one for many computer vision applications. This model describes the transformation from world coordinate system (X, Y, Z) to ideal camera image coordinate (x, y).

The camera performs a linear transformation from the 3D projective space to the 2D projective space. The projection is carried by an optical ray reflected from a scene point **X** or originating from a light source. The optical ray passes through the optical center **C** and hits the image plane at the projected point.

2.1 Mathematical model

A camera is simulated by the pinhole camera model. The relationship between a 3D point $\tilde{X}$ and its image projection $\tilde{x}$ is given by

$$s\tilde{x} = K[R \quad \mathbf{t}]\tilde{X} \tag{1}$$

where *s* is an arbitrary scale factor, (R; **t**), called the extrinsic parameters, is the rotation and translation which relates the world coordinate system to the camera coordinate system and K, called the camera intrinsic matrix. Without loss of generality, we assume the model plane is on Z = 0 according to the world coordinate system. From the models equation we have

$$s.\begin{bmatrix} u \\ v \\ 1 \end{bmatrix} = K.[r_1 \quad r_2 \quad r_3 \quad \mathbf{t}].\begin{bmatrix} X \\ Y \\ 0 \\ 1 \end{bmatrix} \tag{2}$$

By abuse of notation, we still use **X** to denote a point on the model plane, but $\mathbf{X} = [X\ Y]^T$ since Z is always equal to 0. In turn, $\tilde{X} = [X\ Y\ 1]^T$. Therefore, a model point **X** and its image **x** is related by a homography **H**

$$s.\tilde{x} = \mathbf{H}.\tilde{X} \quad \rightarrow \quad \mathbf{H} = K.\ [r_1\ r_2\ r_3\ \mathbf{t}] \tag{3}$$

As is clear, the 3 x 3 matrix **H** is defined up to a scale factor. (Hartley & Zisserman, 2000)(Heikkilä & Silvén, 1997)

2.2 Constraints on the intrinsic parameters

Given an image of the model plane, a homography can be estimated. Let's denote it by $H = [h_1\ \ h_2\ \ h_3]$, we have

$$\mathbf{H} = [h_1 \quad h_2 \quad h_3] = \lambda.K.[r_1 \quad r_2 \quad \mathbf{t}] \tag{4}$$

where, λ is an arbitrary scalar. Using the knowledge that r_1 and r_2 are orthonormal, we have

$$\begin{aligned} h_1^T K^{-T} K^{-1} h_2 &= 0 \\ h_1^T K^{-T} K^{-1} h_1 &= h_2^T K^{-T} K^{-1} h_2 \end{aligned} \tag{5}$$

These are the two basic constraints on the intrinsic parameters, given one homography. Because a homography has 8 degrees of freedom and there are 6 extrinsic parameters. (Sonka at al, 2006)(Hornberg, 2006)

2.3 Solution camera calibration based on Zhengyou Zhang method

Now the effective solution of the camera calibration problem will be described (Zhang, 1998). Let

$$B = K^{-T}.K^{-1} = \begin{bmatrix} B_{11} & B_{12} & B_{13} \\ B_{12} & B_{22} & B_{23} \\ B_{13} & B_{23} & B_{33} \end{bmatrix} =$$

$$= \begin{bmatrix} \frac{1}{\alpha^2} & -\frac{\gamma}{\alpha^2\beta} & \frac{v_0\gamma - u_0\beta}{\alpha^2\beta} \\ -\frac{\gamma}{\alpha^2\beta} & \frac{\gamma^2}{\alpha^2\beta^2} + \frac{1}{\beta^2} & -\frac{\gamma(v_0\gamma - u_0\beta)}{\alpha^2\beta^2} - \frac{v_0}{\beta^2} \\ \frac{v_0\gamma - u_0\beta}{\alpha^2\beta} & -\frac{\gamma(v_0\gamma - u_0\beta)}{\alpha^2\beta^2} - \frac{v_0}{\beta^2} & \frac{(v_0\gamma - u_0\beta)^2}{\alpha^2\beta^2} + \frac{v_0^2}{\beta^2} + 1 \end{bmatrix} \tag{6}$$

Note that B is symmetric, defined by a 6D vector

$$\boldsymbol{b} = [B_{11}, B_{12}, B_{22}, B_{13}, B_{23}, B_{33}]^T \quad (7)$$

Then, we have

$$h_i = [h_{i1},h_{i2},h_{i3}]^T \qquad h_i^T B.h_j = v_{ij}^T.\boldsymbol{b}$$
$$v_{ij} = [h_{i1}h_{j1},h_{i1}h_{j2}+h_{i2}h_{j1},h_{i2}h_{j2},h_{i3}h_{j1}+h_{i1}h_{j3},h_{i3}h_{j2}+h_{i2}h_{j3},h_{i3}h_{j3}] \quad (8)$$

Therefore the two constraints (5), from a given homography, can be rewritten as 2 homogeneous equations in $\boldsymbol{b}$:

$$\begin{bmatrix} v_{12}^T \\ (v_{11} - v_{22})^T \end{bmatrix}.\boldsymbol{b} = 0 \quad (9)$$

If $\boldsymbol{n}$ images of the model plane are observed, by stacking $\boldsymbol{n}$ such equations as $V.\boldsymbol{b} = 0$ where V is a $2\boldsymbol{n}$ x 6 matrix. If $\boldsymbol{n} \geq 3$, we will have in general a unique solution b defined up to a scale factor. Once b is estimated, the camera intrinsic matrix K can be computed as

$$\begin{aligned} v_0 &= (B_{12}B_{13} - B_{11}B_{23})/(B_{11}B_{22} - B_{12}^2) \\ \lambda &= B_{33} - [B_{13}^2 + v_0(B_{12}B_{13} - B_{11}B_{23})]/B_{11} \\ \alpha &= \sqrt{\lambda/B_{11}} \\ \beta &= \sqrt{\lambda B_{11}/(B_{11}B_{22} - B_{12}^2)} \\ \gamma &= -B_{12}\alpha^2\beta/\lambda \\ u_0 &= \gamma v_0/\beta - B_{13}\alpha^2/\lambda \end{aligned} \quad (10)$$

Once K is known, the extrinsic parameters for each image is readily computed

$$\begin{aligned} r_1 &= \lambda K^{-1}h_1 \\ r_2 &= \lambda K^{-1}h_2 \quad \text{with} \quad \lambda = 1/\|K^{-1}h_1\| = 1/\|K^{-1}h_2\| \\ r_3 &= r_1 \times r_2 \\ \mathbf{t} &= \lambda K^{-1}h_3 \end{aligned} \quad (11)$$

3. EXPERIMENTAL RESULTS

The algorithm was tested on real data. The algorithm involves estimating a singular value decomposition of a small $2\boldsymbol{n}$ *x* 6 matrixes, where $\boldsymbol{n}$ is the number of images. Images were obtained from the calibrated CCD camera. We used colour CCD camera Mitsubishi CCD-400E. The calibration object was a picture composite of 9 x 7 squares with 29mm diameter. From the object calibration we obtained 80 point coordinates. These point coordinates were input data for calibration algorithm. Results of this algorithm are coordinates values (tab. 1) of computed points, intrinsic and extrinsic camera parameters. Five images of the plane under different orientations were taken. From each image was chosen one point.

u_s [pix]	u_c [pix]	v_s [pix]	v_c [pix]
550	549.5	398.5	398.4
293	293.9	148.7	148.6
520	518.9	189.8	189.5
592	592.7	334.3	334.3
421	421.1	347.0	346.7

Tab. 1 Values of scanned points and points computed by the presented algorithm

Δu [pix]	Δv [pix]
0.5	0.1
- 0.9	0.1
1.1	0.3
- 0.7	0
- 0.1	0.3

Tab. 2 Values of error variable

The coordinates values of computed point were testing. The test criterions were based on criterion functions. The criterion is based on distance between coordinates of the scanned point and the computed point.Computed points were compared to scanned points and an error variable was estimated. The error variable (tab. 2) is the distance between coordinates scanned points and coordinates computed points.

$$\Delta u = u_s - u_c \quad a \quad \Delta v = v_s - v_c \quad (12)$$

The camera calibration is good for points where values of the error variable are minimal. This algorithm was very sensitive to the choice of the input points. When we chose inadequate point (not corner point from chessboard), the calibration was incommensurate and the errors variable are over tolerance.

4. CONCLUSION

In this paper, we have described a camera calibration technique based on Zhenyou Zhang method. The algorithm computes intrinsic and extrinsic camera parameters. This procedure uses explicit calibration for mapping 3D coordinates to image coordinates. The experiments in the last section showed that this algorithm is very efficacious. The real data have been used to test the proposed technique, and very good results have been obtained. Computed points corresponded with scanned point. These approaches don't require an expensive calibration apparatus and an elaborate setup. The method is very suitable for many applications.

5. REFERENCES

Hartley, R.; Zisserman, A.(2000): *Multiple View Geometry in Computer Vision.* the University Press, ISBN 0-521-62304-9, United Kingdom, Cambridge.

Heikkilä, J.; Silvén, O. (1997). A four-step camera calibration procedure with implicit Image correction. *Proceedings of Computer Vision and Pattern Recognittion, IEEE Computer Society Conference*, San Juan, Puerto Rico, june 1997

Hornberg, A. (2006). *Handbook of Machine Vision.* WILEY-VCH Verlag GmbH & Co.KGaG, 776. ISBN 3-527-40584-4, Weinheim.

Sonka, M.; Hlavac, V.; Boyle, R. (2006). *Image Processing, Analysis and Machine Vision,* Thomson, ISBN 0-495-24438-4, United States of America

Zhang, Z. (1998). A flexible new technique for camera calibration. *Technical report* MSR-TR-98-71.

ACKNOWLEDGMENT

We gratefully acknowledge that the results presented here were solved as partial goals of the research task VEGA No 1/3099/06.

Annals of DAAAM for 2008 & Proceedings of the 19th International DAAAM Symposium, ISSN 1726-9679
Editor B. Katalinic, Published by DAAAM International, Vienna, Austria 2008
Make Harmony Between Technology and Nature, and Your Mind will Fly Free as a Bird

CROATIAN PERSPECTIVES IN METAL CASTIG PRODUCTION

NOVOSELOVIC, D[aniel]; BUDIC, H[rvoje]; SOLENICKI, G[oran] & BUDIC, I[van]

Abstract: *If Croatian foundries want to be serious producers of castings for western and oversea industrially developed countries it is necessary to solve a great number of different organizational and technological problems. The most important problem is quality of castings. Relating to increase of the success of the foundries on the market, there is a potential possibility of utilization of domestic and international sources of financing. Through the above specified sources of financing it is possible to purchase the equipment, technology and machinery, which contributes to expansion of production activities, product processing and implies the effects on the human factor, which will in such a way accommodate to new business principles and become capable and qualified work force on the market.*
Key words: *efficiency, casting quality, international sources of financing.*

1. INTRODUCTION

Casting is one of technologies for achieving of a specific form of a metal objects by pouring a molten metal into a mould and allowing it to cool. The molten metal takes shape and size of the mould cavity and retains it after solidification (Bonačić-Mandinić & Budić, 2001).

Casting technology is one of the oldest and most efficient ways of product forming. In spite of diverse, highly competitive and new production technologies, enormous demand for castings, due to their simple manufacturing process, will continue High productivity and easiness of production of replicas make it extraordinarily suitable for serial and mass production. Frequently however it is also the only technology available for manufacture of highly complex components with internal cavities (such as motor blocs and similar), or for production of large and heavy machine casings. (Budić, 2006).

2. CHARACTERISTICS OF CASTING IN CROATIA

To make casting technology competitive, high requirements regarding quality and accuracy of castings, their mechanical and other properties has been established, because only up to 5% of defective castings are permitted (i.e. only random error).This means that production of good castings requires great skill, even more because process of formation of casting cannot be observed visually, because the melt fills in the closed mould.

In countries in transition among which Republic Croatia also belongs, casting production has diminished since 1990 even up to 50% (depending on kind of castings). Depiction of the casting production in Republic Croatia from 1990 until 2003 is shown in Fig. 1. (Budić & Bonačić Mandinić, 2004). Namely, with disintegration of the market, it came to significant downfall of production of castings for domestic machine building industry, with some parts of this production ceasing to exist.

Production of castings designed for Western markets remained the same, or shows signs of mild growth. If we want to become notable casting producer for Western and overseas industrial and developed countries, it is necessary to solve a large number of different organizational and technological tasks, casting quality management being the most important of all.

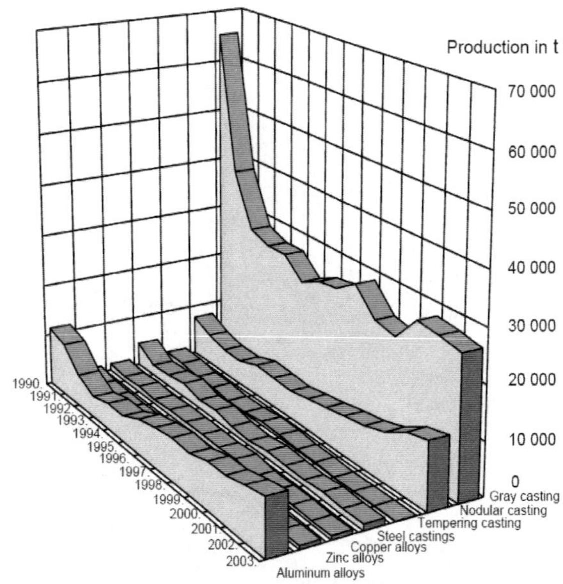

Fig. 1. Depiction of the casting production in Republic of Croatia from 1990 until 2003.

Present situation in huge majority of Croatian foundries is far from good (in spite of huge export potential), which enables them only to survival. There are no possibilities of major development and progress. Namely, western markets, where there is a system of unity of technological and commercial functions, system of fractal production (product development in one country, assembly in another, while products find their customers mostly in third country markets) is presupposition for further development.

To advance foundries to that technological level, not only high quality understanding of global function, but also top technological expertise about produced castings is required. These requirements cannot be introduced into production process without extraordinary effort and engagement of the country's own scientific and human potential.

This opinion is also shared by relevant world experts and financial institutions for which realized profit represents measure of success and criteria for financing of the projects. To achieve this goal, it must be strived to acquire optimum of production parameters. Therefore, the casting does not have to be the best by its quality and the cheapest by the price, but must be competitive, i.e. better than the other products available on the global market.

Hence, the increase of productivity of the foundries could be achieved by research of influential parameters on quality of the castings and development of the matrix of the non-conformances in the casting industry. Quality has been selected as primary criteria, because it represents the most important peculiarity of the market competition.

3. CHARACTERISTICS OF CASTING INDUSTRY IN DEVELOPED COUNTRIES

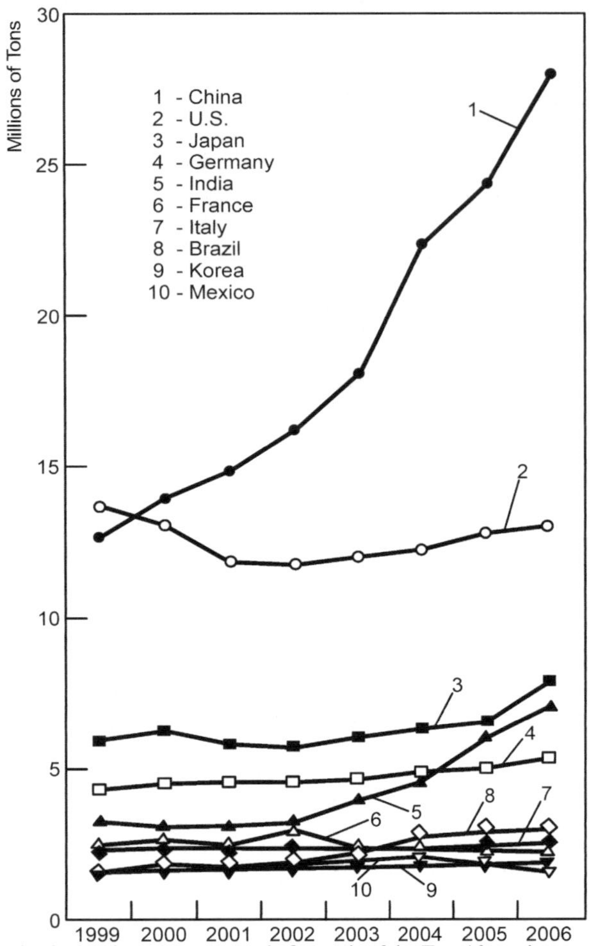

Fig. 2. Casting tonnage trends for each of the Top 10 casting-producing nations

Production of castings in industrially developed countries will not increase (slight downfall is to be expected). A certain re-structuring is to be expected within casting industry itself. (Bonačić Mandinić, 1995). In Fig. 2, we can see production of castings for ten biggest world producers. Staff Report (2007). Significant utilization of gray casting in automobile industry will occur. It will be replaced by aluminum alloy castings, while steel casting will be greatly replaced by nodular casting. Tempering casting will mainly loose market positions. Competitiveness of polymer materials will continue.

Regarding new materials (ceramics, composite materials, polymer concrete and similar), it cannot be expected that they will replace metal, primarily because in most cases their properties are complimentary and not competitive, so that they will be used together with classical materials in new constructions.

4. INCREASE OF THE SUCCESS OF THE FOUNDRIES ON THE MARKET

Relating to increase of the success of the foundries on the market, there is a potential possibility of utilization of domestic and international sources of financing.

One of the more important things is the Law on Stimulation of Investments brought by Croatian Parliament on December 8th , 2006. It regulates incentives for investment of domestic and foreign judicial and physical persons with the goal of economic growth, development and realization of economic policy of the Republic of Croatia, its inclusion into international trends of trade and strengthening of its competitiveness.

Pursuant to this law, custom and tax benefits and allowances for expenditures arising from opening of new working places relating to investments, allowances for training courses for investments, stimulating measures for development and subventions for large investments having significant economic interest have been provided. Important are also international sources of financing provided through cooperation with EU and the World Bank.

Through the above specified sources of financing it is possible to purchase the equipment, technology and machinery, which contributes to expansion of production activities, product processing and implies the effects on the human factor, which will in such a way accommodate to new business principles and become capable and qualified work force on the market.

Use of new and innovative products and technologies, the acquirement of the new business processes leads to increase of quality of the output of the enterprise and reduction of all expenditures. In this way production activities in the segment of the economy are spreading, which naturally leads to better business practice and successful appearance on the market.

Capital and joint venture investments with domestic and international partners are also possible, as well as creation of business clusters and networks.

Such sources of financing would stimulate development of revenue activities and create more enticing business environment, which would lead to increase of total revenue and profit, and result in growth of the enterprise in the economic sense of word.

5. CONCLUSION

The precondition for Croatian foundries to become respectable manufacturer of castings for Western and overseas companies in industrial and developed countries is to solve a large number of different organizational and technological tasks. Problem of quality of castings is of the primary importance. Therefore, it is necessary to increase the efficiency and improve quality of the castings, which in near future (up to forthcoming 5 years) would enable Croatian foundries more significant participation on the Western markets. To increase of the success of the foundries on the market, there is a potential possibility of utilization of domestic and international sources of financing

The presented results come out of the scientific project (Increase of Efficiency and Quality of Castings – Croatian Perspectives), realized with support of the Ministry of Science and Sport of the Republic Croatia.

6. REFERENCES

Bonačić Mandinić, Z.; Budić, I. (2001). *Fundamentals molding technology – Part I* , Strojarski fakultet u Slavonskom Brodu, ISBN: 953-6048-19-1, Slavonski Brod

Bonačić Mandinić, Z. (1995). *Increase of Efficiency and Quality of Castings – Croatian Perspectives*, Znanstveno stručni ljevački skup, Zagreb 1995

Budić, I.; Bonačić Mandinić, Z. (2004). *Fundamentals molding technology – Part II*, Strojarski fakultet u Slavonskom Brodu, ISBN: 953-6048-28-0, Slavonski Brod

Budić, I.; (2006). *Special procedures of casting - Part I* , Strojarski fakultet u Slavonskom Brodu, ISBN: 953-6048-35-3, Slavonski Brod

Staff Report (2007). 41st Census of World Casting Production—2006, *MODERN CASTING (December 2007). 22-25. Availabe from: http://www.moderncasting.com/*

Annals of DAAAM for 2008 & Proceedings of the 19th International DAAAM Symposium, ISSN 1726-9679
Editor B. Katalinic, Published by DAAAM International, Vienna, Austria 2008
Make Harmony Between Technology and Nature, and Your Mind will Fly Free as a Bird

MODELLING AND SIMULATION IN THE CONTEXT OF DISTRIBUTED MANUFACTURING SYSTEMS

NYLUND, H[asse]; SALMINEN, K[ai] & ANDERSSON, P[aul]

Abstract: *This paper discusses modelling and simulation in the context of distributed manufacturing systems, which consist of manufacturing entities with different roles and manufacturing domains connecting the entities. The role of modelling and simulation as a part of digitally presented distributed manufacturing systems is presented. It is intended to lead to a more efficient understanding of design and development activities, offering a wider view of the information and knowledge of the system.*
Key words: *Manufacturing, Simulation, Modelling, Digital, Distributed*

1. INTRODUCTION

This paper presents a model of distributed manufacturing systems. It aims to improve manufacturing activities in a dynamic environment in which the requirements and possibilities are constantly changing. The model describes a manufacturing system with entities having different roles and the domains connecting the entities. The use of modelling and simulation, from ideas and visions to customer satisfaction, is discussed. Modelling and simulation, as part of a digitally presented manufacturing system, enables more efficient solutions to be achieved, in which the information and knowledge needed is available in an accurate form to all related parties, regardless of their time and location.

2. MODEL OF MANUFACTURING SYSTEMS

The distributed manufacturing system consists of manufacturing entities of products, resources, and orders. The division is based on the Holonic Manufacturing Systems (HMS) reference architecture Product-Resource-Order-Staff Architecture, (PROSA) (Van Brussel et al., 1998). The entities are connected with domains of process, production, and business; see Fig. 1. Descriptions of the roles and domains are:

- *Products* define the capability requirements, i.e. what the system must be able to do.
- *Resources* have the capabilities that exist, i.e. what the system is able to do.
- *Orders* define the number of different capabilities needed and how much and when they are needed.
- *Business domain* should react to customer demands, providing the right products and services.
- *Process domain* connects products and resources and it is responsible for the capabilities needed.
- *Production domain* is responsible for manufacturing the right orders at the right time.

Despite having different roles, the entities are similar in structure, consisting of digital, virtual, and real parts, as well as a communication part (Nylund et al., 2008). The digital part holds the information and knowledge that can be presented precisely. The real part represents what exists physically, while the virtual part is a computer model of the real part. The communication part enables the entities to co-operate.

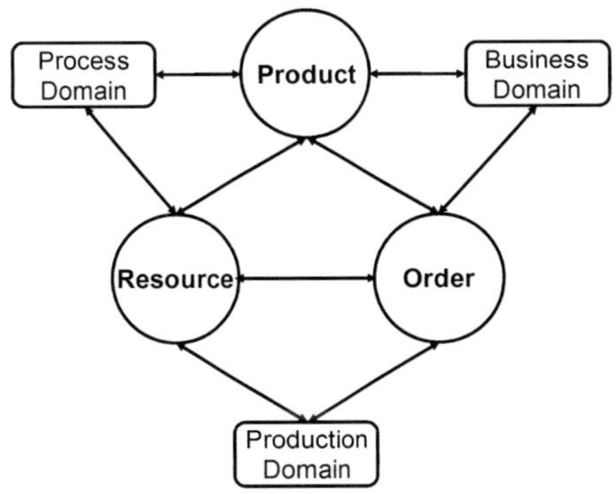

Fig. 1. Structure of manufacturing system entities and their related domains.

The manufacturing system exists in both a horizontal and vertical context. Horizontally, it is a part of a supply chain co-operating with suppliers, partners, and customers. Vertically, the manufacturing system is part of an enterprise network, a higher structuring level. At the same time it consists of several manufacturing units on a lower structuring level.

2.1 Entities as holons and activities as services

Manufacturing entities are described as holons, which are autonomous and co-operative building blocks of a manufacturing system (Valckenaers et al., 1994). They are capable of performing their own tasks without external help. They are also in co-operation with other holons to fulfil their mutual goals. Holons in a manufacturing system form a holarchy, which is a holon on an upper structuring level. The manufacturing activities are seen as services, holons requesting services and holons as service providers. The rules of co-operation, matching a service request with a provider, are the autonomy of the holarchy. The approach forms a network of manufacturing holons, which enables a manufacturing system to be distributed and dynamically changeable.

3. SIMULATION IN THE CONTEXT OF DISTRIBUTED MANUFACTURING SYSTEMS

Modelling is used to understand a system and its behaviour. It can be used to repeat or refine an activity to achieve a specific result, as well as to extract a system in order to apply it to a different content or context (Dilts, 2005). The simulation of manufacturing systems is used when the model evolves over time. The states of the manufacturing entities change at separate points in time. Simple models can be investigated analytically. Typically, in manufacturing systems the relations between the entities are too complex to solve without simulation (Law and Kelton, 2000).

The structure of a simulation model follows the structure of distributed manufacturing systems. The entities have digital and virtual parts, but lack a real part. The simulation model is constructed using digital information and knowledge which is the same as that used in the real manufacturing system.

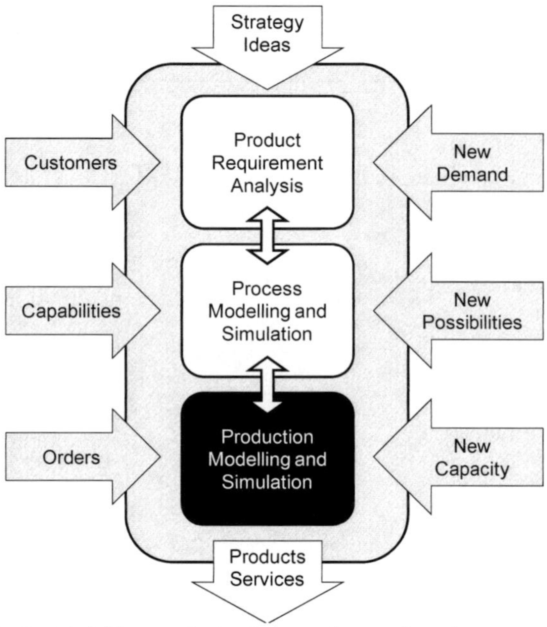

Fig. 2. Modelling and simulation of manufacturing systems from ideas to customer products and services.

Fig. 2 presents a process from strategy and ideas to products and services to customers. The process consists of product requirement analysis, as well as the modelling and simulation of processes and production systems. On the left-hand side in Fig. 2 are the activities that influence the existing manufacturing system, while the right-hand side illustrates examples of sources for change. The product requirements consist of existing customers, manufacturing strategy, and ideas combined with new possible product demand. This leads to new possible products. Using modelling and simulation, the capability and capacity of the system can be evaluated. The manufacturing system must be able both to manufacture the new products and handle the volume and variety of the existing orders.

In process modelling and simulation, the ability to manufacture the products can be verified. If the capabilities needed are lacking, applications such as machine tool and robot simulations can be used to digitally configure existing processes or to plan the implementation of new capabilities. As a result, the changes that are needed in the real system are known. When the capabilities needed are achieved, the capacity of the system must be verified. The system has an existing volume and variety of orders. By adding the changed capabilities and new products, the functioning of the new system can be evaluated. The need for new capacity can be found out and the possibilities of acquiring it can be planned.

3.1 Dynamic Nature of the Simulation Model

The simulation model is dynamic, as it is required to be able to communicate and adapt to changing states. The state of the manufacturing system is defined by the states of the units and the relations of the units. The units evolve over time by changing in structure and learning from their actions. Their information and knowledge, such as process times and requirements, are updated and available for the simulation model. Typically, the changing characteristics derive from the development of the manufacturing entities, such as continuous improvement, methods integration, and changes in the level of automation of the manufacturing processes.

The rules and goals for operation come from higher structuring levels and are derived from visions and ideas. The impacts from the higher levels usually also influence changes to the entities on the level of manufacturing units. Changes in the resources change their capabilities and in a similar fashion the requirements of product entities change. The product entities are aware of the features their manufacturing requires and the resource entities hold the information and knowledge of their capabilities of manufacturing methods. Matching the product features with manufacturing methods, the alternative chains of processes to manufacture the products can be defined.

3.2 Benefits of Simulation

Simulation is a key technology in an integrated approach to enhancing product and production engineering processes (Kühn, 2006). Typical areas of simulation are, for example: capacity requirements, throughput and bottleneck analysis, and the controlling and scheduling of manufacturing activities (Law and Kelton, 2000). Using modelling and simulation in the context of distributed manufacturing systems can be beneficial, for example in the following ways:

- Requirements of new products can be evaluated against manufacturing capabilities.
- If change is needed, different solutions can be simulated and the best solution selected.
- Solutions can be compared on the basis of requirements such as cost, quality, and time.

4. CONCLUSIONS

A distributed manufacturing system can improve manufacturing activities in a changing environment. Describing the system with entities and their related domains makes it possible to form a model of the manufacturing system. This model is a basis for modelling and simulation, from ideas to fully functional systems. Simulation can be used to evaluate solutions before they are actually executed. The dynamic nature of the simulation model enables it to adapt rapidly to the new state when the system changes.

5. ACKNOWLEDGEMENTS

This paper is part of the research project FMS 2010. It is co-financed by Tekes, the Finnish Funding Agency for Technology and Innovation, and several major companies in Finland.

6. REFERENCES

Dilts, R. (2005). Modeling, Robert Dilts NLP Home Page, *Available from:* http://www.nlpu.com/Articles/artic19.htm *Accessed: 2008-07-15*

Kühn, W. (2006). Digital Factory - Integration of simulation enchanging the product and production process towards operative control and optimization, *International Journal of Simulation*, Vol. 7, No. 7, pp. 27-39, ISSN 1473-8031

Law, AM. & Kelton, WD. (2000). *Simulation modeling and analysis*, 3rd ed. McGraw-Hill, ISBN 13: 978-0070592926, New York

Nylund, H.; Salminen, K. & Andersson, P. (2008). Digital Virtual Holons – An Approach to Digital Manufacturing Systems, In: *Manufacturing Systems and Technologies for the New Frontier, The 41st CIRP Conference on Manufacturing Systems*, Mitsuishi, M.; Ueda, K. & Kimura, F. (Ed), pp. 103-106, Springer-Verlag London Limited, ISBN-13: 9781848002661, London

Valckenaers, P.; Van Brussel, H.; Bongaerts, L. & Wyns, J. (1994) Results of the Holonic Control System Benchmark at the KULeuven, *Proceedings of the CIMAT Conference*, pp. 128-133, ISBN: 0-8186-6510, Troy, NY, USA,

Van Brussel, H.; Wyns, J.; Valckenaers, P.; Bongaerts, L. & Peeters, P. (1998). Reference architecture for holonic manufacturing systems: PROSA, *Computers in Industry*, Vol. 37, pp. 255-274, ISSN 0166-3615

Annals of DAAAM for 2008 & Proceedings of the 19th International DAAAM Symposium, ISSN 1726-9679
Editor B. Katalinic, Published by DAAAM International, Vienna, Austria 2008
Make Harmony Between Technology and Nature, and Your Mind will Fly Free as a Bird

SOFTWARE PACKAGE FOR EXTENDED DATA STORAGE OF AUTOCAD SOLIDS

OANCEA, G[heorghe]; CHICOS, L[ucia] A[ntoneta]; IVAN, N[icolae] V[alentin] & LANCEA, C[amil]

Abstract: *The paper presents a software package used for storing extended data related to AutoCAD solids. By means of a specialised language using its own syntax the software accepts and processes solid management operations, subsequently launching these for part model generating. After each command the data are processed and saved in the current drawing. When the current drawing is saved the extended data related to the part model are saved in the same file. The storaged data can be used later in the technological design by CAPP systems.*
Key words: *Software package, AutoCAD solids, Extended data, Visual LISP, CAPP systems.*

1. INTRODUCTION

It is well-known that the AutoCAD environment can be used both in constructive and technological design (Ivan et al., 2004). The efficiency of the AutoCAD system is significantly increased when it is combined with a programming language recognised by the environment (Ivan et al., 2004) allowing user interface for user customization in design process (Chung et al., 2006). Although AutoCAD is a popular 3-D CAD software (Chung et al., 2006), it has limitation regarding the access to the data associated to solids and the programmers can not add specific data to the solids (Stein, 2003).

The software package, presented in this paper, is used for storing extended data of AutoCAD solids necessary for CAPP systems. Its name is *SolidExtDATA* and it is compatible with *SolidDATA* module for data exchange between AutoCAD and virtual reality systems (Oancea et al., 2008). The software package is developed in AutoLISP/Visual LISP programming language and runs as an executable AutoCAD application. It works with AutoCAD Release from 2000 to 2008.

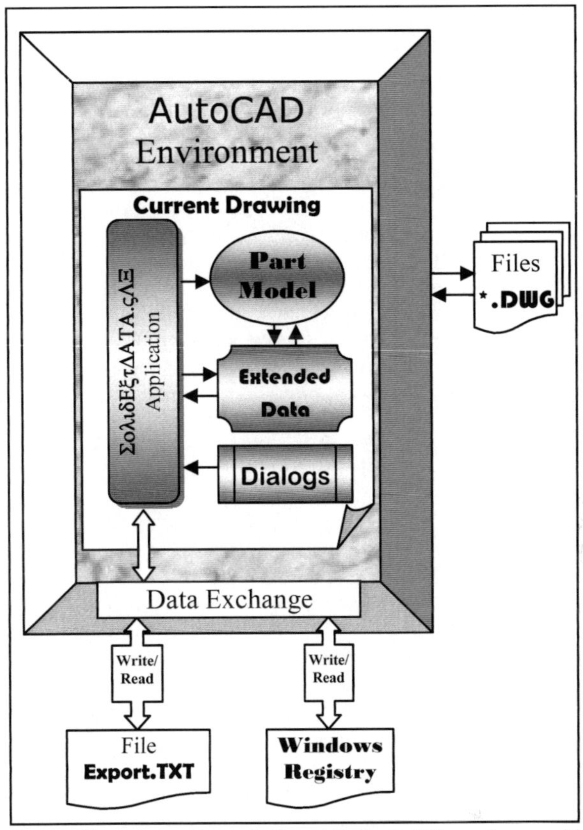

Fig. 1. The connection of the *SolidExtDATA* package with other software components.

2. THE SOFTWARE CONNECTION WITH AUTOCAD

According to the block diagram presented in the figure 1, the executable application has to be loaded in AutoCAD current drawing and it receives commands from other software components using one of the following three communication possibilities: file, Windows registry and variable.

In other software systems for data organising are used normal AutoLISP lists (Park et al., 2002), (Kim et al., 2002). In this software the commands are written in a specific syntax based on the principles of data organisation in lists associated to AutoCAD objects. When a command is received, the *SolidExtDATA* software identifies geometrical data, draws the solid, and then saves the extended data associated to the model in the current drawing. Thus saving the drawing file includes also saving the data, which can be easily accessed in another working session by a CAPP system for example.

The extended data associated to the solids are divided in two categories: geometrical data and accuracy data. The geometrical data like: diameters, radius, length, height and so on are necessary for generating part model. The accuracy data are dimensions deviations and surfaces roughness; they are used by CAPP systems and/or CAD/CAM systems.

For a more effective work in AutoCAD current drawing, the software package includes for launching the commands from the drawing, a set of dialog boxes likewise the box from the figure 2, which allows to the user, in a friendly dialogue manner, to specify the characteristic data.

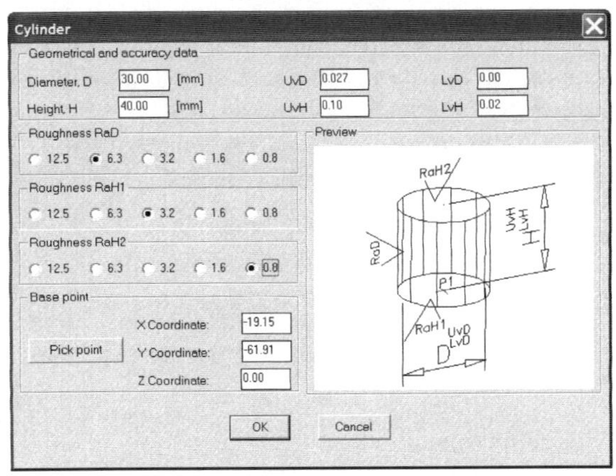

Fig. 2. The dialog box for specifying cylinder data.

3. THE COMMANDS USED BY THE *SOLIDEXTDATA* SOFTWARE PACKAGE

As it was mentioned before, the command received by the software *SolidExtDATA* running in AutoCAD has the form of an association list. It respects all the principles of the association list generated by AutoCAD/AutoLISP/Visual LISP for the management of the geometrical and non-geometrical data related to AutoCAD objects (Oancea, 2003).

An association list includes two types of data: dotted pairs and normal lists. The dotted pair contains as first element a key used as the significance of the second element that represents the useful value. The normal list includes as first element the code for the elements that follow, which typically represent the coordinates of a specific point (Oancea et al., 2008).

For generating part model the *SolidExtDATA* software package recognises the following commands categories: commands for generating primitives, commands for Boolean operations (union, subtract and intersect), and commands for chamfering and filleting. It also can process and manages all the commands used by the *SolidDATA* software module.

The following primitives can be generated based on a command received in the current drawing from other software system: box, wedge, cylinder, cone (including frustrum), torus and sphere. Due to the limited space, as an example, the syntax of a command used for generating a cylindrical surface (a cylinder solid primitive) has the following form:

$$((\text{TIP . "CYLINDER"})\ (\text{P1 x y z})\ (\text{D . Val}_D)\ (\text{DD Lv Uv})\ (\text{H . Val}_H)\ (\text{DH Lv Uv})\ (\text{RA Ra}_D\ \text{Ra}_{H1}\ \text{Ra}_{H2})) \quad (1)$$

where:

- *TIP* represents the key indicating the type of the primitive;
- "*CYLINDER*" represents the type of the primitive;
- *P1* materializes the key indicating the significance of the data succeeding it, i.e. the coordinates of the centre point for the base of the cylinder;
- *X, Y, Z* are the coordinates of the centre point;
- *D* is the key which specify the cylinder diameter;
- Val_D is the diameter value;
- *DD* is the key indicating the deviation of the diameter;
- L_v is the lower value of the diameter dimension;
- L_u is the upper value of the diameter dimension;
- *H* is the key specifying the height of the primitive;
- Val_H is the value of the height measured along the OZ axis;
- *DH* is the key indicating the deviation of the height;
- L_v is the lower value of the height dimension;
- L_u is the upper value of the height dimension;
- *RA* is the key indicating the roughness parameters;
- Ra_D is the value of the cylindrical surface roughness;
- Ra_{H1} is the value of the base surface roughness;
- Ra_{H2} is the value of the other base surface roughness.

The user has the possibility to launch the command, specifying the characteristic data in the current drawing using the dialog box from the figure 2.

Because in AutoCAD, the handle of any object is the unique identifier that does not change from one editing session to another, the association list presented in the relation (1), is completed by the software with the following dotted pair:

$$(\text{HANDLE . Val}_{Handle}). \quad (2)$$

The significance of the dotted pair elements from the relation (2) is:

- *HANDLE* represents the key which specify the identifier of the drawn object;
- Val_{Handle} is a hexadecimal number which represents the handle of the drawn object (cylinder).

After generating the cylinder, for a fully compatibility with *SolidDATA* software module (Oancea et al., 2008), the success of the operation is confirmed to the system which launches the command by the sequence of data as below:

$$(\text{CYLINDER Val}_{Handle}\ \text{X Y Z ValD Val}_L) \quad (3)$$

In the relation (3) the parameters have the significance presented above.

4. CONCLUSION

In contrast with AutoCAD the *SolidExtDATA* software package allows automatically storage of the solids data as extended data associated to the AutoCAD solids. The data saved in this manner can be any time easily accessed and used later in the CAPP and CAM systems.

The software package can be used with other software systems that accept one of the data transfer modalities: file, Windows registry, and variable in current drawing as well as recognise the syntax of the commands.

Future developments envisage the extension of the number of commands for generating model part using *SolidExtDATA* and a CAPP system which uses the extended data storaged in AutoCAD drawings. The CAPP system can add technological data to the part model in the same association lists.

5. ACKNOWLEDGEMENTS

This research has been supported by PN-II Program 71-133/18.09.2007. The authors express their gratitude to all partners for the fruitful collaboration.

6. REFERENCES

Chung, T.T.; Lee, C.C. & Fan, C.K., (2006). Optimum design of a 1×2 mechanical optical switch. *Structural and Multidisciplinary Optimization,* Vol. 31, No. 3, March, 2006, 229-240, ISSN 1615-1488

Kim, M.S.; Choi, J.C.; Kim, Y.H.; Huh G. J. et al., (2002). An Automated Process Planning and Die Design System for Quasi-Axisymmetric Cold Forging Products. *The International Journal of Advanced Manufacturing Technology,* Vol.20, No.3, August, 2002, 201-213, ISSN 1433-3015

Ivan, N.V.; Berce, P.; Drăgoi, M.V.; Oancea, Gh.; Ivan, M.C.; Bâlc, N. et al. (2004). *Sisteme CAD/CAPP/CAM, teorie şi practică (CAD/CAPP/CAM Systems, Theory and Practice),* Editura Tehnică, ISBN 973-31-1530-4, Bucureşti, Romania

Oancea, Gh. (2003). *Proiectare parametrizată asistată de calculator. VisualLISP/AutoLISP prezentare şi aplicaţii (Computer Aided Parameterised Design. VisualLISP/AutoLISP Presentation and Applications),* Editura Universităţii Transilvania din Braşov, ISBN 973-635-120-3, Braşov, Romania

Oancea, Gh.; Garbacia, F. & Nedelcu, A. (2008). Software Module for Data Exchange between AutoCAD and Virtual Reality Systems, In: *Product Engineering – Tools and Methods based on Virtual Reality,* Talabă D. & Amditis A. Editors, pp. 383-394, Published by Springer, ISBN 978-1-4020-8199-6 (HB), ISBN 978-1-4020-8200-9 (e-book), Dordrecht, The Netherlands

Park, D.H.; Kang1, S.S. & Park S.B. (2002). A Surface Area Calculation and CAPP System for Non-Axisymmetric Deep Drawing Products. *The International Journal of Advanced Manufacturing Technology,* Vol. 20, No. 1, July, 2002, 31-38, ISSN 1433-3015

Stein, D. (2003). *The Visual LISP Developers Bible, Available from:* http://www.steinvb.com/pub/vldb.asp *Accessed:* 2008-05-20

Annals of DAAAM for 2008 & Proceedings of the 19th International DAAAM Symposium, ISSN 1726-9679
Editor B. Katalinic, Published by DAAAM International, Vienna, Austria 2008
Make Harmony Between Technology and Nature, and Your Mind will Fly Free as a Bird

ESTIMINATION OF UNKNOWN PARAMETERS IN MECHANICAL MODELS FOR DETERMINATION VISCOELASTICITY PROPERTIES OF MATERIALS

OBUCINA, M[urco]; BAJRAMOVIC, R[asim] & DZAFEROVIC, E[jub]

Abstract: *Estimination of unkown parameters in mechanistic models for determination viscoelasticity properties of gluing laminate elements is given in this paper. Three parameters model (Standard Linear Model) and four parameters model (Burger model) are analyzed according to experimental data. The analysis showed that Burger model better approximates experimental data. Convergence improvement of nonlinear regression method for estimationa of unknown parameters in models can be obtained by parameters grouping.*

Key words: *viscoelasticity, models, nonlinear regression, laminate elements.*

1. INTRODUCTION

Wood and wood products have elastic and viscous characteristics and belong to viscoelasticity materials (Dinwoodie, 2000). Viscoelasticity wood behaviour is very complex can be described by nonlinear models for which the unknown parameters are determined according to the experimental data. Faster convergence of nonlinear regression can be obtained by parameter grouping (Christensen, 1971).

2. RHEOLOGICAL MODELS AND EXPERIMENT

Creep test of laminated elements was carried out on a device with four-wheel loading (Figure 1). The device was placed in a chamber where relative humidity was maintained constant (93%) with axial fans which forced air flow to move over a dish with saturated solution of water and K_2SO_4. The room temperature was 22.5 °C.

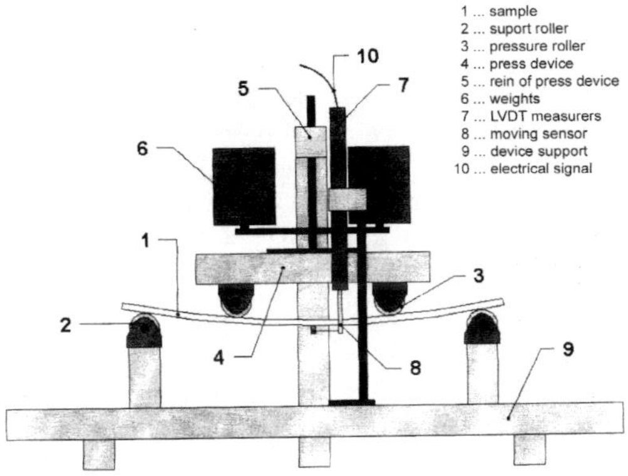

Fig 1. Device for creep test of laminated elemen*ts*

Constant loading was realized by weights (Figure 1 position 6). which is 17.5 % of the maximal loading.

At relatively low stress levels of humidity and temperature, wood may be considered as a linearly elastic material, and as a linearly viscoelastic material under others. Forms of the simplest viscoclastic models, which are used to represent the viscoelastic behavior of wood materials, are the Standard linear model Equation (1) and Burger model Equation (2) (Skrypek & Hetnarski 1993):

$$J(t)=\left[\frac{1}{E_1}+\frac{1}{E_2}-\frac{1}{E_2}\exp\left(-\frac{tE_2}{\mu}\right)\right] \tag{1}$$

$$J(t)=\left[\frac{1}{E_1}+\frac{1}{E_2}\left(1-\exp\left(-\frac{tE_2}{\mu_2}\right)\right)+\frac{t}{\mu_1}\right] \tag{2}$$

where $J(t)$ is the creep compliance; t is the time; E_i is elastic constant for springs; and μ_i is viscous constant for dashpots (i =1 and 2). According to Figure 1. the elastic deflection at middle point is (Obucina et al., 2006).

$$y=\frac{23(F/2)l^3}{648EI} \tag{3}$$

Where: F is the force ; l is the reference length; I is the moment of inertia of the beam's cross section and E is the elastic modulus.

If a viscoelastic beam is subjected to the same force, as in the elastic case, the deflection can be derived multiplying Equation (4) by $E(t)/E$, where $E(t)$ is the relaxation modulus. If $1/E(t)$ is replaced with $J(t)$, where $J(t)$ is the creep compliance, Equation (4) becomes

$$y(t)=K\cdot J(t)\,;\quad K=\frac{23}{648}\frac{(F/2)\cdot l^3}{I} \tag{4}$$

To determine the mechanism of any physical process it is necessary to establish a correspondence between experimental observations on the one hand and the mathematical model on the other. Any model can be written in the form

$$y_i=f(\mathbf{x}_i,\boldsymbol{\theta})+\varepsilon_i \quad i=1,2,\ldots,n \tag{5}$$

Where: -y_i is the value of the dependent variable at the i-th measurement

-$\mathbf{x}_i=(x_{1i},x_{2i},\ldots,x_{qi})'$ is a q-dimensional vector of independent variables at the i-th trial

-$\boldsymbol{\theta}=(\theta_1,\theta_2,\ldots,\theta_p)'$ is a p-dimensional vector of unknown parameters. - ε_i is an experimental error at the i-th trial.

Empirical models, which are usually polynomials in the independent variables but linear in the parameters, are usually used to predict for purposes of optimization and control of processes Another type of models is mechanistic model based on physical mechanism of the observed process. The models with four and three parameters used in this work are nonlinear.

In matrix notation the relationship between the dependent variable and the independent variables for models linear in parameters can be written as (Draper & Smith 1988).

$$\underset{nx1}{\mathbf{y}} = \underset{nxp}{\mathbf{X}}\ \underset{px1}{\boldsymbol{\theta}} + \underset{nx1}{\boldsymbol{\varepsilon}} \tag{6}$$

The vector of unknown parameters can be estimated by the method of least squares as

$$\hat{\boldsymbol{\theta}} = (\mathbf{X'X})^{-1}\mathbf{X'y} \tag{7}$$

Equation (7) can be solved if the matrix *X'X* is nonsingular. If this matrix is singular then vector of parameters is not uniquely defined.

With assumption that errors $\boldsymbol{\varepsilon}$ have normal distribution with constant variance –covariance matrix of unkown parameters is defined as

$$V(\hat{\boldsymbol{\theta}}) = (\mathbf{X'X})^{-1}\sigma^2 \tag{8}$$

It often may be assumed for nonlinear models that the dependent variable can be adequately represented over some interval in the parameters by a truncated (linear) Taylor series expansion about preliminary estimates of unknown parameters. With this assumption the nonlinear model can be written approximately in the form

$$\mathbf{z} = \mathbf{X}\left(\boldsymbol{\theta} - \tilde{\boldsymbol{\theta}}\right) + \boldsymbol{\varepsilon} \tag{9}$$

Where: $\mathbf{z} = \left[y_i - f\left(\mathbf{x}_i, \tilde{\boldsymbol{\theta}}\right)\right]$ *i*-th element (10)

$$\mathbf{X} = \left[\frac{\partial f(\mathbf{x}_i, \boldsymbol{\theta})}{\partial \theta_j}\right]_{\boldsymbol{\theta}=\tilde{\boldsymbol{\theta}}} \quad i,j\text{-th element} \tag{11}$$

$\tilde{\boldsymbol{\theta}}$ is a vector of the numerical values about which the linearization is made.

Equation (9) represents the approximate model linear in parameters and as before the unknown parameters can be estimated as

$$(\hat{\theta} - \tilde{\theta}) = (\mathbf{X'X})^{-1}\mathbf{X'z} \tag{12}$$

and residual sum of squares is given as $R = \mathbf{z}'\mathbf{z}$

The iterative procedure is terminated when the correction obtained by (12) becomes small enough; i.e. the change in parameters is insignificant. According to the given procedure software made in C^{++} computer language was made. For faster convergence "the doubling and halving technique" and optimization of parameter vector change were applied at each iteration step.

3. RESULTS

In order to compare two models F-test was carried out for the results given and the calculated results are given by curves in Figure 3. For 4- parameters model estimated variance is s_4^2 = 2.20360e-02 with 99 degrees of freedom and for 3-parameters model variance is $s_3^2 = 1.02391\text{e-}01$ with 100 degrees of freedom. The corresponding F value is $F=s_3^2/s_4^2=4{,}6465$. Since this value is greater than table values for both $\alpha = 0{,}05$ and $\alpha = 0{,}01$ ($F_{0.05,100,99}=1.39$ and $F_{0.01,100,99}=1.60$), it means that 4-parameters model is significantly better than 3 parameters model.

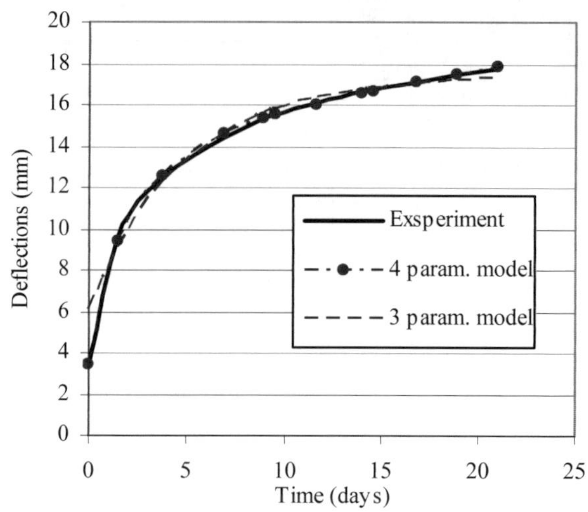

Fig. 3. Measurement and calculated creep values

In order to improve convergence of nonlinear regression method the grouping of parameters can be performed. For this case 4-parameters model was considered. This model can be transformed as

$$y_i = K\left(ae^{-bt_i} + ct_i + d\right)$$

Where $a = -1/E_2;\ b = E_2/\mu_2;\ c = 1/\mu_1;$ i $d = 1/E_1 + 1/E_2$

According to the nonlinear regression method the following results were obtained.

E_1=13701; E_2=11658; μ_1=420594; μ_2=28779; R=4.165
a=-8.57765x10^{-5}; b=0.405083; c=2.37754x10^{-6}; d=1.58763; R=4.165

Transforming *a*, *b*, *c* and *d* the next values are obtained
E_{1t}=13703; E_{2t}=11658; μ_{1t}=420602; μ_{2t}=28780; R=4.165.

4. CONCLUSION

According to the given F-test 4-parameters model is significantly better them 3-parameters model. The convergence was faster for transformed model obtained by parameter grouping. Obtained results are the same for both cases. The problem will appear if we want to determine confidence intervals for parameters. There is no problem for original model since parameters have physical meaning. In the case of transformed model confidence intervals for parameters *a, b, c* and *d* can not be transformed into confidence intervals for E_1; E_2; $\mu_1;\mu_2$. If we need models only for prediction then it is useful to use transformed model, but in the case that we need confidence intervals for real parameters then original model should be considered no matter how difficult is to estimate it.

5. REFERENCES

Christensen, R. M. (1971). *Theory of Viscoelasticity an Introduction*. Academic Press, 0-12-174250-4, New York

Dinwoodie J. M. 2000. *Timber its nature and behavior*, 0-419-25550-8, London and New York.

Draper, N.R. and Smith, H. (1988). *Applied Regression Analysis*. Wiley.

Obucina, M.; Dzaferovic, E.; Bajramovic, R. & Resnik, J. (2006). Influence of gluing technology on viscoelasticity propert of LVL,*Wood Research*,51(4), 11-22 1336-4561

Skrzypek J.J., Hetnarski R.B. (1993) *Plasticity and Creep, theory, examples, and Problems*, International Standard Book Number 0-8493-9936-X.

Annals of DAAAM for 2008 & Proceedings of the 19th International DAAAM Symposium, ISSN 1726-9679
Editor B. Katalinic, Published by DAAAM International, Vienna, Austria 2008
Make Harmony Between Technology and Nature, and Your Mind will Fly Free as a Bird
*Young Researcher and Scientist Paper / * Supervisor, Mentor*

BONDING QUALITY OF PANELS MADE OF CROSSCUT SOFTWOD BRANCHES

OLARESCU, A[lin] & BADESCU, L[oredana] A[nne]

Abstract: *In this paper there are presented results of bonding quality testing for new panel products made by progressive assembling of structural elements with crosscut texture obtained from softwood branches. Designing and manufacturing this new product is based on the concept of sustainable development and we intended to use these panels for manufacturing of furniture and for interior design.*
Key words: *wood, branches, panels, bonding.*

1. INTRODUCTION

Branch wood represents an important secondary wooden resource. For Romania the annual volume of exploited branches for the principal forest species is 3, 2268 millions cubic meters per year. This value represents 30% from the annually exploited volume of these species (Olarescu, 2007).

The annual volume exploited for dimensional assortment of branches with diameters above 3 cm is: 2004, 6 thou. m3/year for Fagus Sylvatica; 90, 32 thou. m3/year for Picea Abies; 25,48 thou. m3/year for Abies Alba (Olarescu, 2007).

Branch wood shows different properties compared to stem wood (Kollman, 1968; Hakilla, 1989) and the utilization of branch wood must be correlated with technological, economical and ecological constraints regarding to a satisfactory utilization of small-sized wood (Efthymiou, 2008).

Up to now the utilization of branch wood was conducted by the following criteria: forms of branches; some mechanical proprieties like elasticity; aesthetic qualities; the maximal utilization of valuable species; utilization to an industrial scale as a raw material for: cellulose, particle boards, fiber boards (Bowyer et al., 2003; Cionca, 2007; Lessard, 1999).

Today, the problem of the utilization of branch wood is going to be correlated with ecological aspects regarding sustainable development. These aspects start at forest level, methods for forest management, processing of raw material, life-cycle of products, possibility of repair, reintegration, recycling, biodegradation and transformation as well as the social impact produced by a factory in its community.

According to the concept of sustainable development, we have developed a tangible model for the utilization of branch wood. The new product is obtained by progressive assembling of structural elements with crosscut section made of softwood branches. We intended to use these panels for the manufacturing of furniture and for interior design.

Because the panels with crosscut texture made of softwood branches are new products, it is necessary to investigate their properties. The first property which was investigated was their bonding quality.

2. PANELS MADE OF CROSSCUT BRANCHES

The initial approach of the research project proposed and analyzed six technological variants for manufacturing this product. We were using two methods of assembling: green gluing and dry gluing.

In parallel with the realization of tangible models, the technological variants for manufacturing eco-panels with crosscut texture of branch wood were analyzed by using the

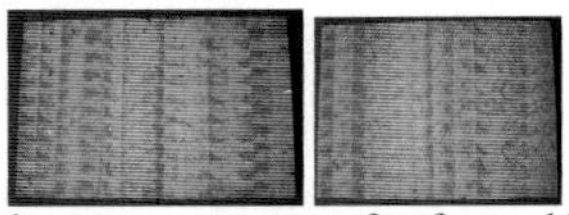

Fig.1. Panels with crosscut texture of softwood branches.

method of multi-criteria analysis. The following ecological, economical and aesthetical criteria were adopted: work consumption, energy consumption, aesthetic level of final product, possibility of manufacturing in small joiner's workshop, possibility of manufacturing at industrial scale, gluing time, costs of adhesive and control of clamping device. The results obtained by experimental research and by multi-criteria analysis conducted to the optimal technological variant for manufacturing the eco-panels with crosscut texture of softwood branches. The optimal technological variant supposes: drying the branches with bark with diameter over 3 cm, cutting the prisms, sorting the prisms, forming the linear block, gluing the prisms, clamping the linear block, conditioning the linear block, sanding to tolerance the linear block, cutting crosscut strips, sorting of the strips and forming the panels, gluing the strips and clamping the panel, conditioning of the panel, sanding to tolerance of the panel, sizing of the panel, packing and storage of panels. The panels obtained by this technology are shown in Figure 1.

The total efficiency of manufacturing eco-panels with cross-texture made of softwood branches with bark is 40 %, in the following working conditions: 11% of crosscut surface of branch wood is represented by bark and the efficiency of the manufacturing of prisms from branches is 53%.

3. BONDING QUALITY OF PANELS MADE OF CROSSCUT BRANCHES

3.1 Materials and methods

For the assembling of structural elements with crosscut texture made of softwood branches, a polyurethane adhesive Jowapur 687.40 was used. Specific consumption of adhesive was 282 g/m^{2}panels. A mechanical clamping device with screw was used for assembling the panel. The clamping pressure was 0, 02 N/mm^2. The time of clamping was 3 hours, at the temperature 20±2°C and relative humidity 40±5%.

A method for testing bonding quality of panels is described in *DD CEN/TS 13354:2003 – Solid wood panels – Bonding quality – Test method.* The principles of this method consist of the following: the test pieces are subjected to a pre-treatment, according to the service class as defined in ENV 1995-1-1, and then strained to failure by a compressive shear force. The shear strength is calculated with relation 1 and the cohesive wood failure of shear area is determined.

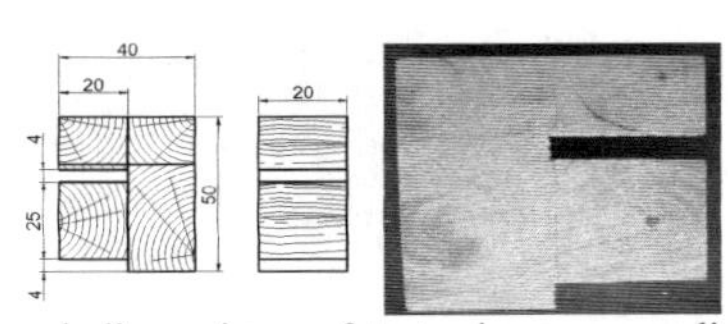

Fig.2. Shape and dimensions of test pieces according to DD CEN/TS 13354:2003.

$$f_v = \frac{F}{l * b}[N / mm^2] \quad (1)$$

Where:

f_v – shear strength, in Newtons per square millimetres;
F – maximum load of test pieces, in Newtons;
l – length of the shear area, in millimetres;
b – width of the shear area, in millimetres;

The minimum number of test pieces for single layer panels indicate to the *DD CEN/TS 13354:2003* is 10. We were using 16 test pieces. The moisture content of panels was 9,1%, according to EN 322, and the density of panels was 682,86 Kg/m^3, according to EN 323. Shape and dimensions of test pieces is presented in figure 2.

Because we intend to use these panels in dry conditions, the pre-treatment consists in 24 hours immersion of test pieces in cold water at 20°C.

According to the testing procedure the length and width of the shear area are determined before pre-treatment, with an accuracy of 0, 1 mm. For the determination of the length of the shear area, a digital Holzman caliper of 150 mm capacity and accuracy of 0, 01 mm was used. The caliper was checked with a 20 mm±µm setting rod. For the determination of the width of the shear area a Holzman Rotating Digital Dial Gage of 0 – 12, 5 mm and an accuracy of 0, 01 mm was used. The dial gage was checked with a 19 mm±µm setting rod (fig.3.a).

The test pieces were positioned in the shearing tool as in Figure 3b. The shearing force was applied parallel to the glue line with a constant rate of load, in such way that the failure occurs within approximately 60±30 s. We are using the universal testing machining ZDM 5t/51, calibrated annually by RENAR (Romanian Accreditation Association). The recording of the maximum load at failure was with accuracy to 10 N.

3.2 Results and discussions

Testing results are shown in Table 1, and the test pieces after testing are shown in Figure 3c.

According to *EN 13353:2003 Solid Wood Panels – Requirement,* for the single layer solid wood panel the lower 5-percentile of shear strength, calculated in accordance with EN 326-1, shall not be less than 2,5 N/mm^2 and the mean wood failure percentage shall be more than 40%.

In this case lower 5-percentile of shear strength is 4, 7072 N/mm^2 and the mean wood of failure percentage are 87, 7%.

The results obtained for test pieces number 1 and 12, marked with * in table, were eliminated by statistical interpretation of data, because failure occurs at 108,7 respectively 115 s.

For the test results marked in table with **, failure occurs in other zones of test pieces than zone of glue.

4. CONCLUSION

In accordance with the concept of sustainable development, we developed the tangible model for the utilization of branch wood. This product presents the following economical, ecological and esthetical advantages: increasing of the used biomass with 10 percent for softwood; affordable raw material and final products with higher added value; higher esthetical level for the final product; minimal cost for implementation in manufacture; visually attractive appearance of panels made of

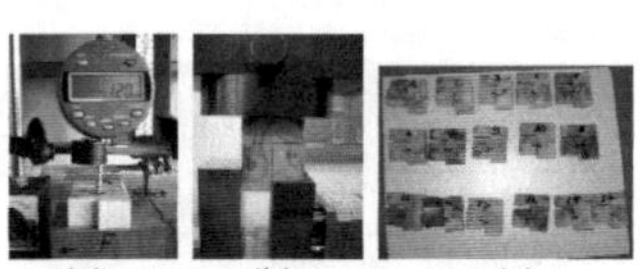

(a) (b) (c)

Fig.3. a – determination of width shear area; b – positioning of test pieces in the shearing tool; c – test pieces after testing

No.	Dimension		Maxi-mum load[N]	Shear strength , [N/mm^2]	Type of failure
	b [mm]	l [mm]			
1.	19,52	24,97	8300	17,0286*	100% W
2.	19,76	25,04	2380	4,810118	100%A
3.	19,72	24,85	3900	7,958502	75%W
4.	19,72	25,21	1710	3,439667	66% W
5.	19,67	24,9	2630	5,369725	100% A
6.	19,7	25,05	3400	6,889774	75% W
7.	19,72	24,96	1730	3,514751	75% W
8.	19,89	25,21	2400	4,786341	100%W
9.	19,8	24,86	1400	2,84421	100%W **
10.	19,84	25,11	2840	5,700723	75% W
11.	19,75	24,95	1620	3,287588	100%W **
12.	19,72	25,1	8720	17,61716*	100%W
13.	19,87	25,23	2860	5,704938	100%W
14.	19,78	24,95	2500	5,065743	100%W
15.	19,68	25	3000	6,097561	-
16.	19,81	24,98	2960	5,981565	100%A
Mean	**19,7468**	**25,0231**		**6,631062**	**89,7% W**
St. Dev.	**0,0891**	**0,1207**		**4,391188**	
5 %	**19,7078**	**24,9702**		**4,707722**	

Table 1. Results of bonding quality testing.

crosscut branches is encouraging and opens the perspective to eco-designers towards identifying alternative furniture components.

The bonding quality of eco-panels with crosscut texture made of softwood branches is within the limits of the specifications of *EN 13353:2003 Solid Wood Panels – Requirement* and the results is in according with the utilization of these panels.

The research regarding dimensional stability, mechanical and technological properties of these panels is in progress, but already confirms the hypothesis that making panels out of branch wood is feasible and rewarding.

5. REFERENCES

Bowyer, J.L.; Shmullsky, R. & HaygreenA, J.G., (2003). *Forest Products and Wood Science: an introduction*, Fourth Edition, Blackwell Publishing, ISBN 0-8138-2654-3 Ames, Iowa, USA.

Cionca, M., (2007). *History of Furniture. The Ancient Roman Lifestyle, House and Furniture*, Editura Universitatii Transilvania din Brasov, ISBN 973-635-835-6, Brasov.

Efthymiou, P.N., (2008), Main technological, ecological and economic constraints in utilizing small-sized wood, *Proceedings of 3rd International Scientific Conference Fortenchenvi,* pp. 18-25, ISBN 978-80-7375-182-1, Prague, may, 2008, Prague.

Hakkila, P., (1989). *Utilisation of Residual Forest Biomass*. Springer Verlag, ISBN 3-540-50299-8, Berlin, Heidelberg.

Kollmann F.F.P; Cote W.A., (1968) *Principles of Wood Science and Technology. Vol. I. Solid Wood*, Springer-Verlag, Berlin.

Lessard, M., (1999), Meubles anciens du Quebec : quatre siecle de creation, Edition de l'Homme, ISBN 2-7619-1524-0, Quebec, Canada.

Olarescu, A., (2007), *Lemnul din crengi. Structura, proprietati si mod de valorificare* (*Branches Wood. Structure, properties and utilization)*, Editura Universitatii Transilvania din Brasov, ISBN 978-973-998-157-0, Brasov.

EN 13353:2003 Solid Wood Panels – Requirement.

DD CEN/TS 13354:2003 – Solid wood panels – Bonding quality – Test method.

EN 325 – Wood –based panels- Determination of test pieces.

EN 323 – Wood –based panels- Determination of Density.

EN 322 – Wood –based panels- Determination of moisture content.

EN 326-1:1994 Wood based panels – Sampling, cutting and inspection - Part 1: Sampling and cutting of test pieces and expression of test results.

Annals of DAAAM for 2008 & Proceedings of the 19th International DAAAM Symposium, ISSN 1726-9679
Editor B. Katalinic, Published by DAAAM International, Vienna, Austria 2008
Make Harmony Between Technology and Nature, and Your Mind will Fly Free as a Bird
*Young Researcher and Scientist Paper / * Supervisor, Mentor*

LOCAL PHYSICAL ACTIONS ON SURFACE OF BILLET

OLT, J[ueri]; LEEMET, T[onu]; LAATSIT, T[oomas] & MAKSAROV, V[iacheslav]*

Abstract: *In the field of mechanical processing regular chip formation process is of great importance. It has beeb found that controlling chip formation process is the key for controlling stability of the technological system as a whole. The research paper proposes one of the most effective methods, which make it possible to reliably govern the process of splitting continuous chip, it appears with creation of preliminary local plastic action (LPA) on the external surface of the cut off layer, produced according to the specific laws. Also local thermal action (LTA) is introduced.*
Key words: *cutting, chip, LPA, engineering, control.*

1. INTRODUCTION

The process of cutting is one of the complex physical processes, with which appear elastic and plastic deformations; this process is accompanied by large friction, heat emission, scab formation, curling and shrinkage chip, by an increase in the hardness of deformed metal layer and by the wear of cutting tool.

At present in machine building with working of details made of corrosion-resistant high-temperature (strength) steels and alloys they encounter essential difficulties in connection with the formation of continuous chip, which leads to the ineffective operation of the highly productive automated technological equipment. One of the most effective methods, which make it possible to reliably govern the process of splitting continuous chip, is the method of preliminary local action by plastic deformation. The application of this method consists in the fact that the surface layers of metal, contacting with the tool of high hardness as a result of pressure, occur in the state of cubic compression and undergo plastic deformation. Because of this, under the tool is formed the pressure only in the contact zone, which creates in the zone of preliminary local action on the surface of billet internal structural changes.

2. RESEARCH OBJECT AND METHOD OF LPA

The preliminary local physical action on the external surface of the cut off layer produced according to the specific laws, makes it possible to change the conditions for the metal deformation with the cutting (Gulaev, 1960), (Maksarov & Timofejev 2003).

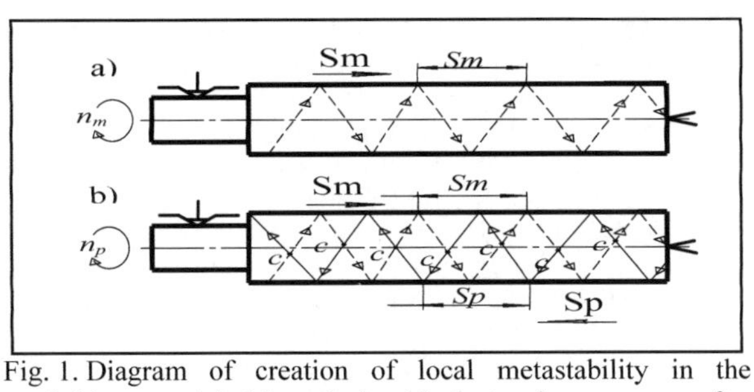

Fig. 1. Diagram of creation of local metastability in the workable material (a) and the blank turning process after preliminary plastic effect on the material (b).

Alternations in the conditions of cutting, in comparison with the source material, are the special feature of the process of grinding the billets, subjected to this action. Physical action on the surface of material in the local zone leads to a change in it of structure and mechanical properties of the workable metal.

In the process of working the zone of local action, find in the metastable state in comparison with the base metal, leads to an instantaneous change of the stress-strained state in the zone of chip formation (Maksarov & Timofejev 2003).

On the rheological parameters of the process of chip formation has an effect local the metastability, which is created in the field of the assumed allowance of the cut off layer of material on the external surface of billet along the specially predetermined trajectory by the point C (Fig. 1, A) which in the preparation stage is formed with the frequency of the rotation of billet n_m and with the supply S_m of device for the creation LPA (Gulaev, 1960), (Maksarov & Timofejev 2003).

Physical action on the surface of material in the local zone leads to a change in the density of the defects of crystal lattice, which form high-energy configurations, which leads to the appearance of that increased of the metastability of structure from this local region. Subsequently with cutting-edge mechanical processing with the frequency of the rotation n_p of billet and the supply S_m the tool cutting edge in the cutting plane intersects at point with the zone of the local physical action (Fig. 1, b). The zone of local action with distorted crystal lattice, which has other mechanical properties in comparison with the basic material, leads to an instantaneous change of the stress-strained state in the zone of chip formation (Fig. 2).

3. THEORETICAL RESULTS OF LPA

In the zone of preliminary local action on the surface of billet internal structural changes occur. It is known (Panin, 1990) that the plastic surface deformation in the local zone of the not heat-treated metal already at a temperature of T=+20°C leads to a change in structure and properties of material. Deformation is accompanied by origin, slip and accumulation of dislocations in the metal being deformed. With an increase in the dislocation density and imperfections of crystalline structure the free displacement of dislocations hinders.

Additional barriers for the dislocations are created due to the grain deformation and splitting of blocks. An increase in the quantity of point and linear defects of structure and volumes with the incoherent connection of crystals increases strength and hardness of material and is decreased its plasticity (i.e. capability for further deformation). Because of this, in the local zone of working by plastic deformation on the external surface of billet is formed the work hardening along the predetermined trajectory.

With the compression in the zone of local working in the region of the contact area within the limits of each grain many being intersected strips of shift in several parallel slip planes are formed. In this case occurs the turn of the disorderly oriented grains by the axes of the greatest strength along the direction of deformation - grain they are deformed and they flatten, being drawn out in the direction of deformation. Metal under the action of tool in the local zone forms the deformation texture of fibrous nature with preferred crystal orientation.

The specific volume of the riveted metal because of the increased quantity of defects of atomic-crystalline structure is more than annealed not riveted; therefore besides an increase in hardness and yield stress of metal in the surface layer of metal in the zone of local plastic action are created residual stresses. Increase in the number of defects of crystalline structure and appearance of internal stresses as a result of work hardening leads to the fact that free energy of metal will grow, and this in turn forms no equilibrium and unsteady state.

A rational change in the physical-mechanical properties of the material of the cut off layer in the zone of local plastic action (zone LPA) ensures an improvement in the conditions for chip formation with the subsequent machining, which is accomplished with a frequency of rotation n_p and supply S_p. With mechanical processing of plastic materials the basic portion of the work of cutting is expended on the plastic deformation of the removed metal. In the zone LPA the part of the work, spent on plastic deformation, it is already executed under the preliminary influence by contact tool. Consequently, in the process of cutting in the zone LPA cutting tool will be accomplished only the part of the work of cutting, spent on the plastic deformation of the basic cut off layer of metal. This it will give in the process of cutting in the zone LPA to a local variation in the volume of the plastic deformation of material, angle of displacement, shrinkage of shaving, force and temperature of cutting. The method indicated can be used for all steels, capable of plastically being deformed; however, the best results are obtained on steels with the hardness to HB 280.

Picture of work hardening with cold plastic deformation described below, Fig. 2, is observed in annealed carbon and alloy structural steels with the structure of ferrite, ferrite-pearlite and finely dispersed sorbite. In high-alloy corrosion-resistant (stainless), high-temperature (strength), nonmagnetic and other steels and alloys more complex picture is observed.

In special alloy steels because of the influence of the alloying elements on the expansion of γ - region (which in iron-carbide steels as stable structural constituting exists only at a temperature higher than Ac_3) (Panin, 1990), to an increase in the stability of super cooled austenite and to lowering the martensite point, austenite can be one of main structural components of steels in the state of their operation.

Alloyed austenite is subdivided into the stable and the unstable. Unstable austenite is capable of the phase transformation - the formation of martensite as a result of the application of external load (deformation) (Veitz & Maksarov 2000), (Veitz et al., 2001). Thus, the local surface deformation of steel with the structure of unstable austenite besides work hardening causes martensite transformation, which to an even greater degree strengthens difference in the structure and the properties of the base metal of work blank and zone LPA.

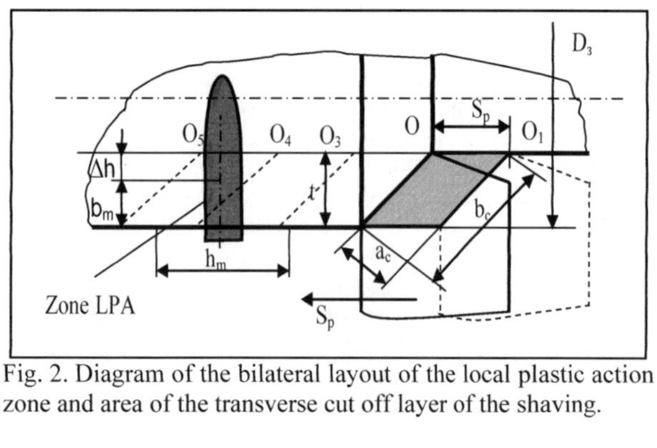

Fig. 2. Diagram of the bilateral layout of the local plastic action zone and area of the transverse cut off layer of the shaving.

Stable austenite does not undergo phase transformation under the effect of the deformation, which leads to a change only in its structure. In steels with the structure of stable austenite, just as in steels of ferrite and ferrite-pearlite classes, plastic action leads to an increase in the density of the defects of crystal lattice, which form high-energy configurations.

4. EXPERIMENTAL STUDY OF LTA

Experimental studies were conducted for different brand of steels - of 3X13, 08X18H10T and structural steel 45. For determining the area of continuous chip breaking under LTA.

Experiment was materialized at different depths, which enabled to conduct a study of the process of continuous chip breaking with different relations of the depth of action b and the depth of the cut off layer a (Fig.3).

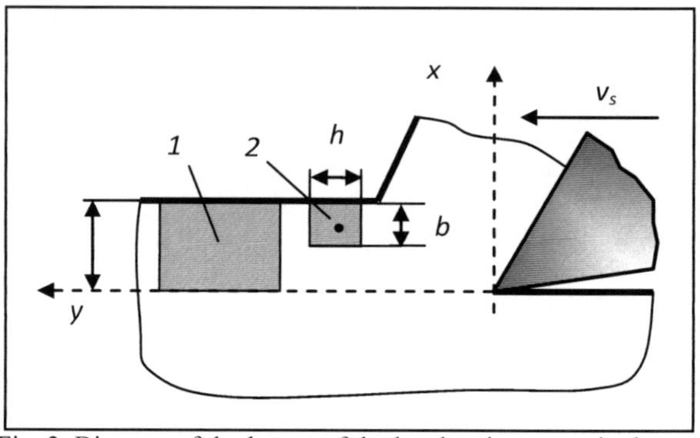

Fig. 3. Diagram of the layout of the local action zones in the cut off layer with different relations of the depth of action b and of the depth of the cut off layer a: 1–b≥ a; 2–b <a.

5. CONCLUSION

1. Chip breaking method is developed, based on the local plastic action on the surface of material, which leads to a change of the density of the defects of crystal lattice in the local zone. In this local region structure of the material has been changed. All this makes it possible to ensure an alternation in the conditions of cutting in comparison with the source material with the subsequent working.
2. It is established that in the zone of local plastic action occurs a change in structure and mechanical properties of the workable metal, also, in the process of working the zone LPA, found in the metastable state in comparison with the base metal, leads to an instantaneous change of the stress-strained state in the zone of chip formation. Also ensuring chip segmentation in the process of shaving.

6. REFERENCES

Gulaev, A.; (1960). *Thermal processing of steels,* Mashgis, Moskva

Maksarov, V. V.; Timofejev, D. (2003). Kinematical investication into process of ship formation with local physical actions given in the base material. *Problems in machinebuilding and usage,* (in Russian)

Panin, V. E.; (1990) *Structural changes in the plastic deformation,* Science, Novosibirsk

Veitz, V. L.; Maksarov V. V. (2000). *Dynamics and control of the process of chip formation with cutting-edge mechanical processing*, NWPI, Sankt-Peterburg (in Russian)

Veitz, V. L.; Maksarov V. V.; Lontsihh P. A. (2001). *Dynamics and the simulation of chip formation with cutting-edge mechanical processing*, RIO IGIUVA, Irkutsk (in Russian)

Annals of DAAAM for 2008 & Proceedings of the 19th International DAAAM Symposium, ISSN 1726-9679
Editor B. Katalinic, Published by DAAAM International, Vienna, Austria 2008
Make Harmony Between Technology and Nature, and Your Mind will Fly Free as a Bird

ESTIMATING THE PRECIPITABLE WATER VAPOUR USING ROMANIA'S PERMANENT GPS NETWORK

OLTEANU, V[lad]; BADEA, A[lexandru]; MOISE, C[ristian]; DANA, I[ulia] & JIVANESCU, I[ulia]

Abstract: *Initially developed by the U.S. Department of Defence for positioning and navigation for military forces, the GPS was afterwards discovered to be useful in other domains, one of them being meteorology. Due to the fact that the GPS signal travelling from the satellite to the receiver on the surface of the earth, which is a radio signal, propagates through atmospheric layers with different refractive indexes, a signal delay occurs. This paper aims to present the steps to follow in implementing software that estimates this delay and transforms it into PWV, value that may be later used in numerical weather prediction models. Our goal is to estimate these values for different EUREF and IGS permanent stations on Romania's territory using our developed software.*
Key words: *ZWD, PWV, GPS*

1. INTRODUCTION

The atmosphere influences the GPS signal when this passes through the ionosphere and the troposphere. The troposphere delay is divided into two parts based on the causes of the delay; therefore we have a wet delay caused by the presence of water vapour in the troposphere and a dry (or hydrostatic) part caused by the presence of other atmospheric constituents.

$$\delta S^i_{A[TROP]} = \delta S^i_{A[HYD]} + \delta S^i_{A[WET]} \quad (1)$$

We should mention that a more complex approach exists, approach that presumes the dividing of the wet and dry delays once more by taking into consideration the azimuth symmetric and asymmetric delay components (Schuler, 2001).

The hydrostatic delay is known to have values of about 230 cm for a ground pressure of 1000 mb (Rocken, 1995). This part of the delay may be computed with sufficient precision, if meteorological observations are available at the GPS site as well. The wet part of the delay takes values between 0 and about 40 cm. This is very difficult to model due to the irregular distribution of water vapour in the atmosphere. To estimate this value, another parameter will be introduced in the mathematical model of GPS observation processing, parameter that will represent the Zenith Wet Delay. To determine the delay for a certain elevation angle, the obtained parameter is multiplied by a "mapping function". The total delay given by (1) may now be expressed as:

$$\delta S^i_{A[TROP]} = m\left(\varepsilon^i_A\right)_{[HYD]} \cdot ZHD_A + m\left(\varepsilon^i_A\right)_{[WET]} \cdot ZWD_A \quad (2)$$

Here, ε denotes the satellite elevation angle.

Thus for each GPS receiver of a network, the Zenith Wet Delay (ZWD) may be computed, ZWD that may be further transformed into an estimate of the PWV, by multiplying it with a value determined as a function of constants related to the refractivity of moist air and the temperature. The computation of the ZHD may be done with the aid of the models developed so far; we mention here the Hopfield model, the Saastamoinen model, etc. We present here the Saastamoinen function for computing the ZHD:

$$ZHD = \frac{0.0022767\left[\frac{m}{hPa}\right] \cdot p_0}{1 - 0.00266 \cdot \cos 2\varphi - 0.00028\left[\frac{1}{km}\right] \cdot h} \quad (3)$$

The pressure, the temperature and the relative humidity values may be taken from the RINEX meteorological files.

This paper outlines the basics steps in estimating the zenith wet delays and in transforming them into PWV values. It also presents the way in which this process was implemented in our developed *"NWP Builder"* software.

2. ESTIMATION MODEL

The estimation model that we present here is the classic Least Squares method, with the mention that a Kalman filtering approach is more adequate. This kind of estimation is used by the *IfEN*'s *TropAC* software and by the JPL's GIPSY (but this is based on un-differenced observations) as well. The MIT's processing software GAMIT uses the Least Squares estimation that we also used.

To eliminate or diminish some of the errors, the functional model is based on the double difference equation and on a linear combination between the observations on L1 and L2. The linear combination is chosen in such a way to reduce or eliminate the first ionospheric effect (Teunissen, 1998).

The functional un-linear model is therefore given by:

$$v^{ij}_{AB[L1,2]} = \left[\frac{a_1}{\lambda_1}\nabla\Delta\varphi^{ij}_{AB[L1]} - \frac{a_2}{\lambda_2}\nabla\Delta\varphi^{ij}_{AB[L2]}\right] - \left(\frac{a_1}{\lambda_1} - \frac{a_2}{\lambda_2}\right)\nabla\Delta\rho^{ij}_{AB} -$$
$$-\left(\frac{a_1}{\lambda_1} - \frac{a_2}{\lambda_2}\right)\nabla\Delta\delta S^{ij}_{AB[TROP]} + \left(a_1\nabla\Delta N^{ij}_{AB[L1]} - a_2\nabla\Delta N^{ij}_{AB[L2]}\right) \quad (4)$$

where v represents the residual, $\Delta\varphi_{AB}{}^{ij}$ stands for the phase measurement (on the corresponding carrier frequency and expressed in cycles) between ground stations A and B and the satellites i and j, a_1 and a_2 are the linear combination's coefficients (in our case $a_1 = 1$ and $a_2 = f_2/f_1$).

Several parameters may be estimated depending on the user's necessities and preferences. In our case besides the zenith wet delays that we are interested in, the ground stations' coordinates and the ambiguities (actually the double differenced and linearly combined ambiguities) are estimated as well. For more complex approaches, but also more time and resource consuming, additional parameters may be estimated, such as orbit relaxations, azimuth dependent terms, etc (Hoffman – Wellenhof, 1997).

For the zenith wet delays, the corresponding coefficients that will be introduced in the design matrix may be computed with:

$$\omega_A^{ij} = \left[m(\varepsilon_A^j)_{[WET]} - m(\varepsilon_A^i)_{WET]}\right] \cdot \left(\frac{a_1}{\lambda_1} - \frac{a_2}{\lambda_2}\right) \quad (5)$$

$$\omega_B^{ij} = \left[m(\varepsilon_B^i)_{[WET]} - m(\varepsilon_B^j)_{[WET]}\right] \cdot \left(\frac{a_1}{\lambda_1} - \frac{a_2}{\lambda_2}\right) \quad (6)$$

Transforming the obtained zenith wet delays into precipitable water vapour may now be done if the weighted mean temperature of the atmosphere is known (T_m). This is done with the aid of the following relation:

$$PWV = \frac{10^6}{\rho \cdot R_\upsilon \cdot \left[\frac{k_3}{T_m} + k_2^{'}\right]} \cdot ZWD \quad (7)$$

An important problem that may appear in the estimation is the so called „convergence problem" (Schuler, 2001). For short baselines the satellites are viewed under almost equal elevation angles. Taking into consideration the relations (5) and (6) the values of the coefficients will be rather small causing the impossibility of sensing absolute values of the zenith wet delays. Only relative delays will be sensed. These may be afterwards transformed into absolute delays if at least one absolute value is known. In this case the model of estimation should be changed. The problem may be easily remarked in:

$$\Delta\delta S_{[WET]} = m(\varepsilon_B^i) \cdot ZWD_B - m(\varepsilon_A^i) \cdot ZWD_A \cong$$
$$\cong m(\varepsilon^i) \cdot (ZWD_B - ZWD_A) = m(\varepsilon^i) \cdot \Delta ZWD_{AB} \quad (8)$$

Therefore, the implemented software will need to go through the following steps, as they are presented in the scheme below (Fig. 1).

3. CONCLUSIONS

Our software, "NWP Builder", is an application that was developed based on the steps presented above. However, we have to underline that this application was especially created especially for processing baselines from EUREF and IGS permanent stations network, stations that are also equipped with meteorological sensors.

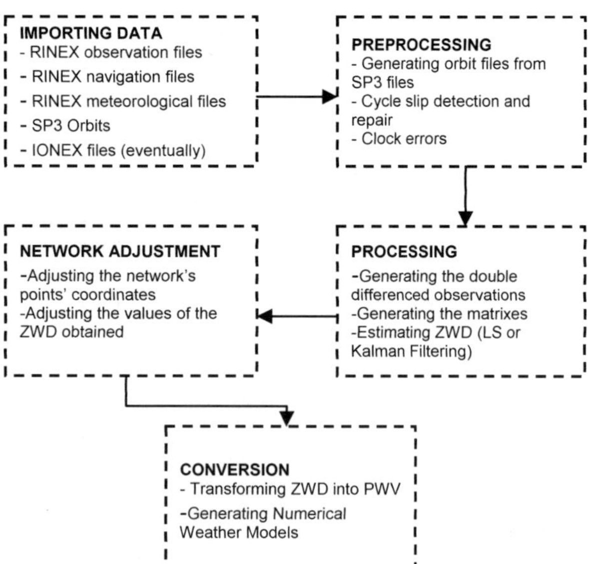

Fig. 1 – Basic steps of the implementation

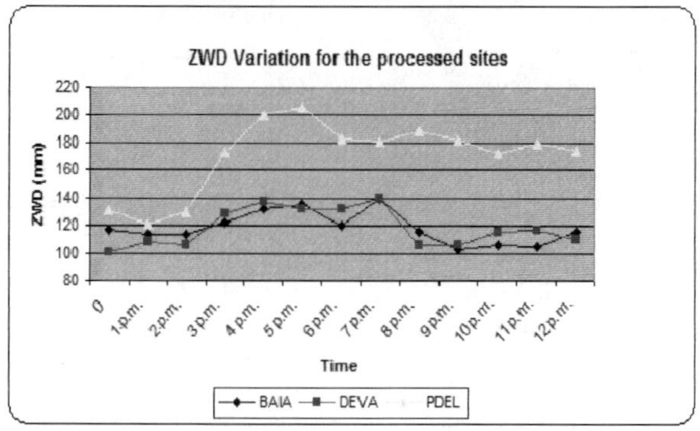

Fig. 2 – Zenith Wet Delay Variation over 12 hours for the studied sites

Thus, the parameters resulted from the estimation process are the zenith wet delays and the ambiguities.

In the tests that we are presenting here we used observations from 3 permanent stations: BAIA (Romania), DEVA (Romania) and PDEL (Portugal). The observation files were downloaded from the GNSS Data Center site (http://igs.bkg.bund.de/). The observations were made on the 29th of May 2008.

The BAIA – DEVA baseline has a length of about 200 km. In this case, we could only determine the relative zenith wet delay. This is why we used a third site which is further away from the other two. The BAIA - PDEL baseline has a length of about 4000 km. This baseline was processed first in order to obtain the absolute values for the zenith delay for the two other sites. With their help we obtained the absolute values for the DEVA site. The results over the 12 hours of observations that were processed are shown in the next figure (Fig. 2).

Today from all the available stations within Romania's permanent GPS network only 9 are equipped with meteorological sensors. In time their number will increase and therefore it will be useful to determine the PWV at all these stations and use it in numerical weather and climate predictions.

We should mention that we intend to further develop our software and somehow combine it with the Virtual GPS Station Technique in order to improve the possibilities of positioning.

4. REFERENCES

Hoffman – Wellenhof, B. et al. (1997). *GPS Theory and practice – fourth revised edition,* Springer, ISBN 3-211-82839-7, Vienna – New York

Rocken, C. et al (1995). GPS/STORM – GPS Sensing of Atmospheric Water Vapour for Meteorology, *Journal of Atmospheric and Oceanic Technology*, 12, 468-478

Schüler, T. et al. (2001) – GNSS Zenith Wet Delay Estimation Considering Their Stochastic Properties, *Available from:* http://forschung.unibw-muenchen.de/papers/khtesmcixcat5b37vhrgygjtenr6tm.pdf, *Accessed:* 2007-10-01

Schüler, T. (2001) On Ground - Based GPS Tropospheric Delay Estimation, *Available from:* http://137.193.200.177/ediss/schueler-torben/inhalt.pdf *Accessed*: 2008-02-01

Teunissen, P.J.G., Kleusberg, A. (1998) – *GPS for Geodesy*, Springer Verlag, ISBN 3-540-63661-7, Germany

Annals of DAAAM for 2008 & Proceedings of the 19th International DAAAM Symposium, ISSN 1726-9679
Editor B. Katalinic, Published by DAAAM International, Vienna, Austria 2008
Make Harmony Between Technology and Nature, and Your Mind will Fly Free as a Bird

THE APPLICATION OF GROUP TECHNOLOGY FOR THE PRODUCTION TIME ESTIMATION

OPETUK, T[ihomir] & COSIC, P[redrag]

Abstract: *The development of industry, transport, communications and technology results in tougher competition, i.e. the process of globalization is manifested. Small and medium enterprises (SME) which are greatly affected by this process are expected to accept or to refuse the received offer in a very short time. Consequently, it is impossible for a small enterprise to create a technological process in order to obtain production times. The idea of the authors of this paper is to develop a classification system of parts by means of group technologies. In the next step, an application to enable the production time estimation, and consequently the production costs estimation, for each family of parts will be developed.*

Key words: *Group technologies, estimation of production times, CAM, code classification.*

1. INTRODUCTION

At present, a lot of customers send a drawing or a 3D model to enterprises and expect to get the production costs and delivery times in answer to their enquiry. Due to a very short time at their disposal to provide answers, manufacturers cannot define the whole technological process with defined production times and the product price. In most cases, manufacturers give an approximate offer, with an assumed product price and the production times based on their experience and on what they can see on the drawing. The authors of this paper have an idea to develop a classification system of parts by means of group technologies. By using this system, the parts would be classified into families of parts. For each family, an application for the production time estimation, and consequently for the product cost prediction, will be developed. The paper presents application for the production time estimation developed here is intended only for shafts. The application only uses the example of the external turning, and it is oriented toward SMEs, i.e. toward private owners because small and medium enterprises have a large share in the total production in Croatia. The basic features of the single-item production and the small-scale production (Antolić, 2007) are:

- ➢A great number of demanding positions in small batches,
- ➢Short delivery times,
- ➢Tendency to approximate the prices achieved in mass-production.

2. CLASSIFIER OF PARTS

As there are a large number of different parts in production systems, the aspiration towards a classification system has led to the development of group technologies.
The group technology has two approaches. The first approach is to group parts according to the production plan so that the machine layout in a plant follows the technological process of the production of a particular group of products. This approach uses the visual method, production flow analysis and cluster analysis (Jurković & Tufekčić, 2000). The second approach is used for grouping parts according to their geometry and technological features. In this approach each part is given its code digit which is used to identify similar parts. Classification and codification systems are used in this approach. This paper deals with the latter approach in which all parts will be given a code digit which classifies them into a particular family of parts.

	1. DIGIT	
ROTATIONAL PARTS	0	symmetric
	1	nonsymmetric

2. CODE DIGIT FOR CATEGORY	
0	discs and rings
1	cylinders and shells
2	shafts
3	shorter rotational parts with eccentric surfaces
4	shafts with eccentric surfaces

3. DIGIT MATERIAL	
0	steel
1	aluminium
2	copper
3	castings

4. SURFACE ROUGHNESS DIGIT AND TOLERANCE CONVERTED INTO SURFACE ROUGHNESS	
0	$Ra \geq 0,2\ \mu m$ turning
1	$0,1\ \mu m \leq Ra < 0,2\ \mu m$ turning and grinding
2	$0,05\ \mu m \leq Ra < 0,1\ \mu m$ turning, grinding and honing
3	$0,025\ \mu m \leq Ra < 0,05\ \mu m$ turning, grinding, honing and superfinishing

5. L/D DIGIT RATIO	
0	$L/D \leq 3$ 2 clamping; face machining
1	$L/D > 3$ 2 clamping; face machining and machining of centre hole

Fig 1: Classifier of rotational parts (Opetuk, 2007)

3. APPLICATION FOR THE ESTIMATION OF PRODUCTION TIMES

In the application model, the authors of this paper have come up with the idea to divide the shaft into basic shapes (Opetuk, 2008.) as it is not easy to estimate the time required for the production of a shaft. This part of classification refers to the fine and rough profiling machining because the number of passes during turning is not the same; therefore, the division of the shaft into basic shapes makes this process of production time estimation easier. To estimate the production time it is necessary to calculate the time required for the production of each basic shape, and the sum of all these times will give the production time per piece. The application was tested by using the CAMWorks software to calculate the production time per piece required for the manufacture of 20 shafts. This production time per piece was then compared with the estimated production time per piece obtained by the Excel application.

3.1 Application for the calculation of base data

In addition to the above mentioned application for the calculation of the production time per piece, an application for the calculation of the following cutting conditions was developed using Microsoft Office Excel:

- Spindle speed – n [RPM],
- Cutting speed – v_c [m/min],
- Surface roughness – Ra [μm],
- Required machine power – P_c [kW],
- Metal removal – V [cm^3/min].

3.2 Application for the calculation of production times for shafts

The application for the calculation of the production time per piece for shafts is made by using Microsoft Office Excel. This application consists of the following 10 worksheets, i.e. the typical turning processes with respect to the surface:

- Machine data,
- Face machining,
- Machining of centre hole,
- Profiling – straight,
- Profiling – tilt,
- Slot machining,
- Radius machining,
- Threading,
- Support time,
- Production time per piece $\mathbf{t_1}$.

The order of worksheets in the application follows the order of operations in the technological process. The production time per piece comprises the following times:

- Machining time,
- Rapid time,
- Support time.

For the calculation of production times, one uses the formulae which are used in the standard process of creating a technological process (Gačnik & Vodenik, 1990). Figure 2 presents how the application works.

The application model has the following limitations:

- The model can be applied to rotational work pieces, i.e. shafts;
- The model is limited only to external turning;
- The operations of milling, grinding, etc., have not been taken into account in the model.
- Profile machining, or more precisely, radius machining is included into the application as part of the profiling – straight. Approximately, the error would be less than 0.1 % since it is a small segment of the shaft and its influence on the production time per piece is negligible.
- Another limitation is the x-axis machining which is not taken into consideration in this application. It can be accounted for by the fact that the sequence of basic shapes in fine and rough profiling should be defined.

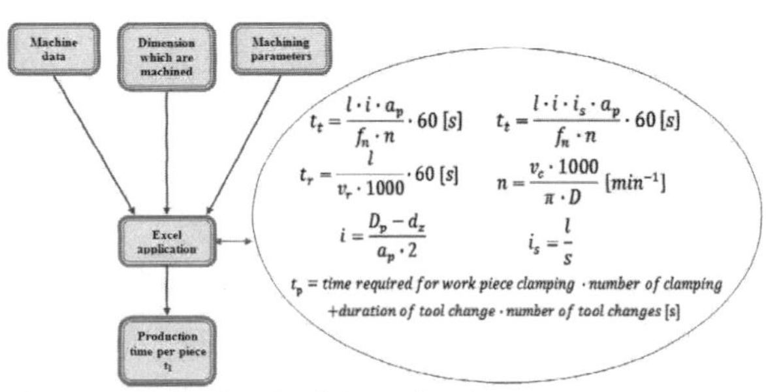

Figure 2. Mode of work of the application

4. CONCLUSION

This application gives good estimation of production times; therefore, it could find its application in the teaching process where students could use it for the estimation of production times of their work pieces. This application is not fully automatic and the user, i.e. the student, is expected to interact with this application while using it. In addition, the application can be used in industry, with the single-item and the small-scale production, for the estimation of the production time per piece of shafts. Figure 3 shows the differences between the production times per piece obtained by the Excel application and those obtained by the CAMWorks software.

The application could not be fully automated if one wanted to do it. The part that could be automated is the recognition of basic shapes, i.e. machining operations. In order to achieve this, a system that would automatically recognize these shapes, i.e. operations, should be developed on the basis of 2D and 3D models. The part that cannot be automated is the data input, i.e. the input of parameters related to machining and to the machine into the application because these parameters depend on the type of the machine applied, on the shaft material, on tools, etc. As far as the generalization of the model is concerned, i.e. the introduction of operations of milling, boring, grinding, etc., into the application, it can be achieved by adding worksheets referring to these operations. Using this application, one can achieve considerable savings when estimating the production times per piece. The estimation of production times per piece by using this application takes 15 minutes regardless of what is used as the basis: a 2D drawing or a 3D model. Savings in time are 4 times greater with respect to the CAMWorks when a 3D model is involved and 5 times greater when a 2D drawing is involved. If CATIA, less user-friendly and more complicated than CAMWorks, is used as the CAM software, savings in time are 6 times greater for the 3D model and 7 times greater for the 2D drawing. Thus, the time required for the estimation of production times is reduced; as a result, the price is also reduced.

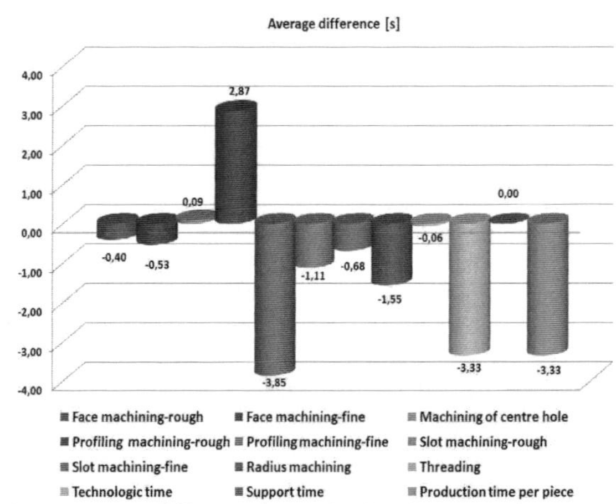

Fig 3. Average difference in production times per piece

5. REFERENCES

Antolić, D. (2007). *Estimation of the product production times by using regression models, (in Croatian),* Master's thesis, University of Zagreb, FAMENA, Zagreb, ISBN 86-7059-085-9

Gačnik, V. & Vodenik, F. (1990). *Design of process planning, (in Croatian)*, Tehnička knjiga, Zagreb

Jurković, M. & Tufekčić, Đ. (2000). *Technological processes – design and modelling, (in Bosnian),* Faculty of Mechanical Engineering, Tuzla

Opetuk, T. (2008). *Development and model application production time estimation for rotational parts, (in Croatian),* Master's thesis, FAMENA, Zagreb

Opetuk, T. (2007). *Application of group technology in the design of process planning, (in Croatian),* Graduation thesis, FAMENA, Zagreb

Annals of DAAAM for 2008 & Proceedings of the 19th International DAAAM Symposium, ISSN 1726-9679
Editor B. Katalinic, Published by DAAAM International, Vienna, Austria 2008
Make Harmony Between Technology and Nature, and Your Mind will Fly Free as a Bird

STRESS - RELATED ISSUES IN MODERN INDUSTRIAL ORGANIZATIONS IN ROMANIA

OPREAN, C[onstantin]; BUCUR, V[iorel] & TITU, M[ihail]

Abstract: *The paper sets forth some considerations regarding potential approaches to work-related stress in the knowledge-based modern industrial organizations, since they entail losses and negative consequences on the level of creativity, productivity and health condition of their employees, and especially the managers. The main goal is optimization of managers` activity, increasing stamina, as well as managing the ability of managers to identify stress-related sources in order to minimize its consequences and to improve the quality of the decision-making process.*
Key words: *stress, management, communication, performance, Selye-syndrome.*

1. INTRODUCTION

In countries like USA, Great Britain, France, Germany, a.o. stress-related issues are frequently approached in a number of research studies, scientific papers, where the authors attempt to identify the cause's for ever higher costs engendered by stress, which may reach an amount of hundreds or even billions of dollars per year.

The topic of the present paper is one of wide interest for specialists in the field, given the well-known stress-induced consequences on managers (and not only) – mention should be made of: discomfort, mental health, substance abuse, dissatisfaction, etc. – which decisively influence the quality of the decision-making process, work efficiency of employees.

More and more often, the complexity of production processes and the high degree of automation, associated with various other social, technical or economic-related issues, represent a major source for work-related stress in modern industrial organizations, entailing negative consequences for the employees.

Stress management should be a primary concern for organization managers in the larger context of E.U. organizations who pool their efforts towards a common strategy for occupational health and safety.

All studies that approach the issue of stress (i.e. occupational, professional, work-related, etc.) highlight significant losses on the labor market generated by stress.

Theoretical explanations help to identify stress-related sources for persons involved in the decision-making process and they will also provide the necessary scientific support for any legal regulations of stress-related problems in our country. (Bucur, 2002)

2. GOALS FOR OPTIMIZATION OF MANAGERS` ACTIVITY

Of all effects mentioned in specialized literature, the ones that strongly affect managerial activity within an organization hold a prominent place. (Goupil, 1991)

These effects can manifest on various levels – behavioral, physiological and psychological – and they are generated by certain responses of managers affected by professional stress. (Oprean & Ţîţu, 2008)

Their identification and understanding allow for the accomplishment of considered goals - part of the research program which includes the following activities:

- work-out a “methodology for identification of stress-related sources” and a “methodological guide”, which are currently missing;
- delineation of theoretical explanation and stress-induced consequences, for various types of organizations;
- improvement of decision-making process and implementation of efficient managerial strategies;
- training “stress resistant” managers;
- diminishing stress-related costs;
- work-out a “theoretical pattern” of guidelines for human interaction, meant to determine them to assume managerial responsibilities;
- set forth a strategy for the result assessment;
- build up a theoretical explanation of the notion of stress in view of modeling situations faced by a decision-making person, as well to simulate potential minimizing of consequences induced by stress-related sources;

Solving specific problems of managerial activities from the point of view of diminishing stress impact on managers.

3. CASE STUDY. APPROACHES TO STRESS WITHIN ORGANIZATIONS

The following categories of conflict have been taken into consideration: (Ţîţu & Oprean, 2007)

- family;
- professional;
- social;
- personal life.

The causes identified during a brainstorming session – the working group included 7 members and a chairperson (2 business people, 2 professors, 1 lawyer, 1 psychologist, 1 physician) have been classified according to “root” causes; (table 1). (Oprean & Ţîţu, 2008)

The hierarchy emphasizes major causes as well as secondary ones.

Their understanding enables the identification of a series of “root” causes underlying stress-generating events.

These conclusions should be confirmed by questionnaires whose results will be interpreted during research.

Taking into account the regulations issued by the European Commission, particularly the Health and Safety Framework Directive – 83/391 whose aim is to prevent, assess and avoid stress-induced injury, it is imperative that managers should increase their efforts in order to:

a) Identify the level of stress affecting managers before starting work;
b) Check and minimize the identified level of stress;
c) Identify the level of stress affecting managers at the end of the work.

FAMILY		PROFESSIONAL	
Children and parents	marriage	Potential promotion	Work acknowledgement
Siblings	In-laws Extra-marriage Relatives	Career	Excessive workload
		Level of payment	Managers
			Inappropriate work relations
			Employees
Living conditions		Inferiority complex	Dissatisfaction regarding work assessment
Material conditions			
Working conditions			Difficult socio-professional integration
Phonic pollution			
Cold environment			
SOCIAL		PERSONAL	

Table 1. Hierarchy of some stress-inducing causes.

In order to obtain information about the level of stress affecting managers before starting work, one can use (optionally) various methods of identification (personalized cards with PNC) while the results thus obtained will be "downloaded" on an information network. (Goupil, 1991)

The levels of stress – previously identified and distinguished by colors (red, black, blue, green, etc.) – may represent useful indicators for adopting proper measures.

It is a must to emphasize employers` responsibilities in order to take proper measures suitable for various identified levels of stress by means of specially designated areas for each level of stress. (Bucur, 2002)

Such areas can facilitate "relaxation" which induces an improved state for the whole body and positive detachment thus making the person able to start work over.

These relaxation areas may include:

- rooms for cardio fitness or medical therapy;
- aquarium, waterfalls;
- art room (it may include paintings of famous artists);
- room for meditation or recreation

Daily visits to these recreation areas or direct involvement in specific activities, with a time range of 2-5 minutes, may have positive effects aimed to minimize the level of stress.

The information thus obtained will allow for an increase of the "data bank" storage, its updating, and if the information is turned to good account it may render a picture of health condition as well as avoiding some stress-inducing agents, thus leading to the optimization of managerial activity and increase of "stress resistance". (Bucur, 2002)

4. CONCLUSION

Studies that approach stress analysis highlight the necessity of identifying stress-inducing causes.

Hierarchy of causes may also contribute to a better understanding of their role in generating various types of stress (i.e. professional, occupational, work-related, etc.), so that the multiple effects induced by stress can be checked.

Proper knowledge and understanding may lead to efficient dealing with stress-related issues (e.g. some causes mentioned in the present paper) as well as adopting preventive measures in order to do away with or minimize the effects with impact on certain activities (performing or managerial) within modern industrial organizations.

It is imperative that organisations insure training programmes for managers, and therefore convey specific management-related concepts allowing:

- an objective assessment of employees and their performances;
- identifying the causes of lower performances;
- supporting employees to cope with stress;
- encouraging continuous creative professional development;
- studying human qualities.

Managers should be aware that "emotions determine temperament", and intelligence control of our emotional life is a consequence of the knowledge and study of human qualities.

The word conflict can be interpreted - according to the Romanian Encyclopedic Dictionary - as "misunderstanding, conflict of interests, disagreement, quarrel, argument (violent, too)."

Therefore, conflict may be defined as "any form of socio-human relationships where the parts evince opposing, diverging interests." (Goupil, 1991)

Conflict is a ubiquitous notion in our daily lives, and it is a result of differences among people - their needs and perspectives being different from one another - however, to deny its existence or to disguise it on purpose will not lead to its alleviation, but to psychic disturbances, such as: diminishing concentration abilities, delayed answers, accompanied by the inability to make the right decisions, confusion, a feeling of uneasiness, etc. (Bucur, 2002)

Whenever two or more individuals or groups of people have diverging interests, aspirations, options, sets of values, information they acquired, understanding of events, etc. the outcome is disagreement - whether verbalized or not - which is the first, and easy, stage of conflict.

Wherever there are people, there are conflicts!

Conflict is an inherent element in the life of couples, the relationships between parents and children, siblings, professors and students, teachers and pupils, employers and employees, citizens and state authorities, ethnicities, parties or political supporters, races, etc. (Goupil, 1991)

So, in work teams, there may be task-related conflicts (which refer to ideas and methods used) and people-related conflicts (which refer to the competition for leadership, personality differences and unequally assigned tasks).

According to the effects they produce, we distinguish:

- destructive conflicts - characterized by total or partial destruction of competitors, while there is always a winner (e.g. wars);
- beneficial conflicts - facilitating changes and leading to increasing creativity and productivity (e.g. industrial revolution).

(Ţîţu & Oprean, 2008)

5. REFERENCES

Bucur, V. (2002). *Management. Management of Organisation,* Napoca Star Publishing House, Cluj Napoca

Goupil, G. (1991). *Stress Intelect,* Coresi Publishing House, Bucharest

Oprean, C. & Ţîţu, M. (2008). *Quality Management in the Economy and Organisation Based on Knowledge,* AGIR Publishing House, Bucharest

Ţîţu, M. & Oprean, C. (2007). *Quality Management*, University of Piteşti Publishing House, ISBN 978-973-690-646-6, Piteşti

Ţîţu, M. & Oprean, C. (2008). *Strategic Management,* University of Piteşti Publishing House, Piteşti

Annals of DAAAM for 2008 & Proceedings of the 19th International DAAAM Symposium, ISSN 1726-9679
Editor B. Katalinic, Published by DAAAM International, Vienna, Austria 2008
Make Harmony Between Technology and Nature, and Your Mind will Fly Free as a Bird

THE SIMULATION OF FLUID FLOW THROUGH A HYDRAULIC RESISTANCE

OPRUTA, D[an]; VAIDA, L[iviu] & PLESA, A[ngela]

***Abstract**: In this paper the authors present the mathematical method and the simulation program for the oil flow through hydraulic command resistances. Simulation software of the incompressible fluid flow was design in the Technical University of Cluj-Napoca. The authors used this software in order to simulate the oil flow through hydraulic command resistances. The mathematical model, based on Navier-Stokes, continuity and velocity equations is presented. There are also presented the steps followed for simulation and some final results.*

***Key words:** simulation, mathematical model, flow, resistance.*

1. INTRODUCTION

The main goal of thethis paper is the presentations of the mathematical model of the oil flow trough a hydraulic command resistance and also the simulation using a suitable program of simulation. Although there are many software programs involved in the simulation of the fluid flow (Susan-Resiga, et.al, 2003), we used CURGSIM, a software developed by the authors (Opruta, et al, 2001). The aim of the research is to find the proper geometry of the resistance. The validity of simulation will be realized by using an advanced method named PIV (Particle Image Velocimetry) (Adrian, 1988). The paper includes the equations of the mathematical model and the steps that were followed in the simulation.

The program displays the speed field, as well as the extreme values of pressure.

2. THE MATHEMATICAL MODEL

For a theoretical study over the quantitative and qualitative sides of fluid flow through the command resistances, the issue is to find the optimum mathematical model, to describe, as real as possible, the parameters and the associated phenomena that apear during the flow through hydraulic resistances: hydrodynamic forces, cavitation, flow coefficient (Backe, 1980).

We choosed the two-dimensional oil flow through hydraulic resistances, because the mathematical model is not very complex (Susan-Resiga, et al., 2003)

For determination of the mathematical model of a fluid flow through the hydraulic resistances we used MAC method (marker and cell). It involves the meshing of the fluid in cells (elementary domains). The accuracy of the results depends on the number of the cells.

The chosen mathematical model consists of:
Navier-Stokes equations:

$$\frac{\partial u}{\partial t}+\frac{\partial (u^2)}{\partial x}+\frac{\partial uv_x}{\partial y}=-\frac{1}{\rho}\frac{\partial p}{\partial x}+\gamma\left(\frac{\partial^2 u}{\partial x^2}+\frac{\partial^2 u}{\partial y^2}\right) \quad (1)$$

$$\frac{\partial v}{\partial t}+\frac{\partial (uv)}{\partial x}+\frac{\partial (v^2)}{\partial y}=-\frac{1}{\rho}\frac{\partial p}{\partial y}+\gamma\left(\frac{\partial^2 v}{\partial x^2}+\frac{\partial^2 v}{\partial y^2}\right)-g \quad (2)$$

Equation of continuity:

$$\frac{\partial u}{\partial x}+\frac{\partial v}{\partial y}=0 \quad (3)$$

Speed in the middle of the flow

$$w_x=\frac{w_0}{\sqrt{b}\left(1+\frac{2x}{D_0}tg\Theta\right)}\cdot\cos\alpha \quad (4)$$

$$w_y=\frac{w_0}{\sqrt{b}\left(1+\frac{2x}{D_0}tg\Theta\right)}\cdot\sin\alpha \quad (5)$$

Speed of a random point in the flow

$$w=w_0\left(\frac{r}{R}\right)^{\frac{1}{7}} \quad (6)$$

Equation for estimating the volume function

$$\frac{\partial F}{\partial t}+\frac{\partial (uF)}{\partial x}+\frac{\partial vF}{\partial y}=0 \quad (7)$$

The notations represent:

- u- speed on y direction,
- w_x- speed on x direction,
- w- mean speed in the flow,
- w_0- flow's initial speed ,
- x- flow's length,
- D- flow's diametre in the most contracted area,
- F- flow's volume function, $0 \le F \le 1$.

For two-dimensional flows, the volume function may be described as the area occupied by the fluid during the flow. The rest of the notations are the same as those used in fluid mechanics. To simplify the calculations, the density ρ and the cinematic viscosity ν are considered to be constants.

In the presence of some initial and boundary conditions, the equations system is capable of describing the isotherm flow of an incompressible fluid.

3. NUMERICAL METHOD

We consider D to be the rectangular cavity of a ω pattern. The D domain (field) is covered by a net of lines parallel with Ox and Oy, having the steps $\Delta x = h_1$ and $\Delta y = h_2$. The h_1 and h_2 values are chosen is such a manner that the lengths must be multiples of these numbers.

3.1. Organising stages for calculations

Cells that corespond to Ext(D) and doesn't belong to the entrance area are marked with A and the ones that do belong to this area are marked with B. The pressure in the B cells is calculated with:

$$p_{i,j}^0=\rho g\left(jh_2-\frac{h_2}{2}-d\right)+\rho\frac{u_0^2+v_0^2}{2}, \quad (8)$$

where the first term represents the static pressure and the second one the dinamic pressure. We have here the advantage of establishing the pressure in a certain point belonging to the flow domain, as well as pressure distribution on the hydraulic resistance's walls.(18)

1.3.) boundary conditions

For cells belonging to Ext(D) and those of Int (D) that have $F_{i,j}^{n+1} \in [0,1]$, boudary conditions are as follows:

a) if $j \neq 3$ and $I \notin \{3, l_x + 2\}$

where:

$$u_{i,j-1}^{n+1} = \begin{cases} u_{i,j-1}^{n+1}, if\, F_{i,j-1}^{n+1} = 1 \\ 0, if\, F_{i,j-1}^{n+1} \neq 1 \end{cases},$$

$$u_{i-1,j}^{n+1} = \begin{cases} u_{i-1,j}^{n+1}, if\, F_{i-1,j}^{n+1} = 1 \\ 0, if\, F_{i-1,j}^{n+1} \neq 1 \end{cases} \qquad (9)$$

$$v_{i,j-1}^{n+1} = \begin{cases} v_{i,j-1}^{n+1}, if\, F_{i,j-1}^{n+1} = 1 \\ 0, if\, F_{i,j-1}^{n+1} \neq 1 \end{cases},$$

b) if $j = 3$ then

$$u_{i,3}^{n+1} = 0 \text{ and } v_{i,3}^{n+1} = 0; \qquad (10)$$

c) if $i = 3$ and :

$$u_{3,j}^{n+1} = 0,\ v_{3,j}^{n+1} = \begin{cases} v_{3,j-1}^{n+1}, if\, F_{3,j-1}^{n+1} = 1 \\ 0, if\, F_{3,j-1}^{n+1} = 0 \end{cases} \qquad (11)$$

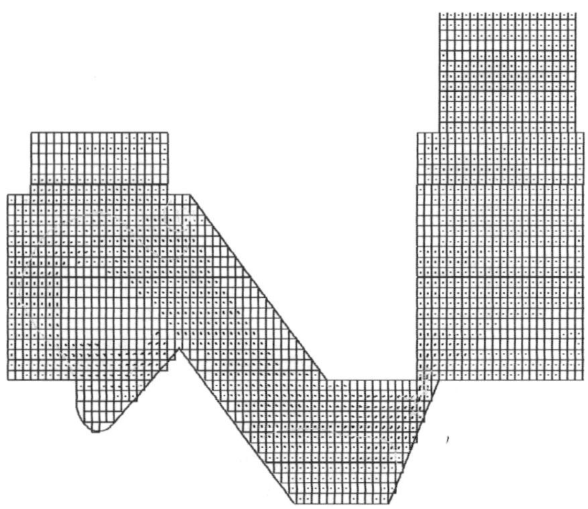

Fig.1. Flow simulation through a hydraulic resistance

d) if $i = 3$ and $j \in \{l_d + 1, \ldots, l_d + l_s\}$ then:

$$u_{3,j}^{n+1} = u_0 \text{ and } v_{3,j}^{n+1} = v_0 \qquad (12)$$

e) if $i = l_x + 2$ and $j \neq 3$ then:

$$u_{l_x+2,j}^{n+1} = 0\ ,\ v_{l_x+2,j}^{n+1} = \begin{cases} v_{l_x+2,j-1}^{n+1}, if\, F_{l_x+2,j-1}^{n+1} = 1 \\ 0, if\, F_{l_x+2,j-1}^{n+1} \neq 1 \end{cases} \qquad (13)$$

$$p_{i,j}^{n+1} = p_{atm} - \rho \frac{\left(u_{i,j}^{n+1}\right)^2 + \left(v_{i,j}^{n+1}\right)^2}{2} \qquad (14)$$

3.2. Details regarding the simulation

In order to verify the experimental data, a software application was made, developed from the mathematical model previously presented. The program is written in C programming language.

The application permits the simulation of laminar two-dimensional flow, of oil, for example, trough any geometrical configuration of hydraulic resistance.

The steps followed in using this program are:
the contour is generated using instructions like: LINE, ARC, CIRCLE or SPLINE; the meshing of the contour is made by function CDISCRET; definition of flow capacity using the instruction VITEZE; choosing of fluid material with CMATERIAL; preparation of simulation with CPREGCALC
starting the simulation with CALCCURG; display of the results with DES_PAS

The definition of geometrical patterns is realized in AutoCad version 12 (or higher).

The program has the advantage of being easy to use, owing to the explicit menus and also because of working with a widely used drawing software (AutoCad).

Another advantage is that it doesn't need special skills of programming.

A very important thing is choosing correctly the fluid for investigation, as well as adapting the number of iterations to the computer's capacity and speed of processing.

4. CONCLUSIONS

The paper deals with the mathematical model and simulation of fluid flow through a hydraulic resistance. Taking into account the processing capacity of the computers involved in data analysis, it was chosen a two-dimensional mathematical model of the flow.

The software application developed for fluid flow through hydraulic resistances, allows studying of this phenomenon for various geometrical patterns of the resistances.

An easy pattern description, the parameter's continuous display, as well as the results (shape of the flow lines, speed field, values and distribution of pressure), all these concluded that the application can be used in researching the phenomena associated with flow, regardless the shape of the hydraulic resistance.

5. REFERENCES

Adrian, R.J., (1988) "Statistical properties of particle image velocimetry measurements in turbulent flow", *Laser Anemometry in Fluid Mechanics*, Instituto Superior Tecnico,

Backé, W., (1982) Grundlagen der Ölhydraulik, RTWH Aachen;

Opruta, D., Vaida, L., Hedesiu H., (2001) „The influence of the geometric configuration of the hydraulic command resistances upon the cavitational phenomena", *ASME INTERNATIONAL, Proceedings of First National Conference on Recent Advances in Mechanical Engineering*, September 17-20, University of Patras, Greece;

Opruta, D, Vaida, L., (2001) A Method of Reducing the Hydrodymanic Forces in Hydraulic Ventiles, in *Acta Technica Napocensis of the Technical University of Cluj-Napoca*, pag. 13-17, vol. I,

Susan-Resiga, R., Muntean, S., Bernad, S., Balint, D., Balint, I., (2003) New methods in parallel calculus and simulation of the fluid flow, Ed. Orizonturi Universitare, Timisoa

Annals of DAAAM for 2008 & Proceedings of the 19th International DAAAM Symposium, ISSN 1726-9679
Editor B. Katalinic, Published by DAAAM International, Vienna, Austria 2008
Make Harmony Between Technology and Nature, and Your Mind will Fly Free as a Bird

CONSIDERATION ON THE KEY INDICATORS OF THE LABOUR MARKET FOR WORK ANALYSIS IN ROMANIA

OROIAN, M[aria]; FORGACIU, F[lavia] [L]elia & TOMULETIU, E[lena]-A[driana]

Abstract: *This paper tries to demonstrate the capabilities of a statistical informational system to analyse the labour market, as KILM - the fifth edition. For that we conducted an analysis of the labour market in Romania during 2000-2006 comparative with Bulgaria. Last but not least, the paper tries to stress the necessity of adapting the Romanian Statistical Information System to requirements of the Statistical Office of the European Union and of the Statistical Office of ILO, in order to ensure a higher degree of compatibility / comparability data.*

Key words: *statistical information system, KILM, labour market*

1. INTRODUCTION

Part of the market economy, in close liaison with other markets, the labour market is probably now one of the most dynamic segments of the economic mechanism, suffering profound mutations in the world, particularly in the context of globalisation, Romanian market is recording a series of fluctuations due primarily to the massive migration of labour, but also other causes, the system occupational being a homogeneous one, with many changes from one period to another.

Analysis of relevant labour market and developments, in order to develop policies and / or strategies - at the national, regional or global - is, for the reasons listed above, one approach is extremely difficult, given that it is hard to provided a perfect comparability of data collected nationally, with the aim of integration / comparison to their regional or global level. From this point of view we can find high enough variability data at national / regional / global perspective of harmonizing standards, design studies, implementation, analysis and statistical data sets of results for comparative research.

Relative recently, in this context what have gained an extremely important role, are statistical information systems of the labour market - structured on different levels of integration (for example, at the national level - Romania / INS, at regional level / European - European Union / Eurostat, at global level - International Labour Office / KILM) designed to ensure compatibility and comparability of data through a series of standardization indices / indicators of labour market.

2. INFORMATION

Unfortunately, after studying an important volume of publications, (scientific papers, doctoral theses and books) related to the recent years labour market in Romania, we could find that there is a certain retention and / or lack of information about the use of complex statistical informational systems, such as the International Labour Office, for a comparative analysis of the labour market in Romania.

It is emblematic that from Review of Theoretical and Applied Economics available online, in the period 2000-2007 only a single article (Vasile, 2007) mentions data from the indicators proposed by the statistical information system of labour KILM (Key Indicators of the Labour Market) developed by the International Labour Office, the majority preferring the calculating of certain indicators based on primary data to be provided either by the National Institute of Statistics of Romania, or by Eurostat - the Statistical Office of the European Union.

Starting from a series of articles (Cazes & Nesporova, 2004/5) and rules of good practice recommended by ILO in connection with the use KILM (ILO, 2005, 2007) for labour market analysis in different countries or regions (Global Employment Trends, 2008), we considered it appropriate to take into consideration the following aspects on labour market: labour force and participation indicators in the field of labour, indicators of employment and unemployment. These issues may well be emphasized with the help of 14 of the 20 indicators propose of KILM. Of course, for a more detailed analysis we would need all those 20 indicators.

For a more accurate reflection of the participation in the labour market, we used alongside the indicator KILM 1 - Labour force participation rate, a complementary indicator KILM 13 - Inactivity rate, an indicator reflecting the number of people who, despite located at the legal age for employment (15 + years) - are not part of the labour force (in other words, are neither employees nor unemployed). The two indicators mentioned give a percentage of 100%. Also, to illustrate the level of education of the workforce, we used indicator KILM 14 - Educational attainment and illiteracy, according to the rules contained in the standards of the International Classification of Education - 1997 edition (International Standard Classification of Education / ISCED 1997).

For the period 2000-2006, Romania had a relatively constant evolution, even if slightly downward of the KILM 1 - and a streak slight growth of inactivity rate, regardless of sex, with a good rate of participation in the case of the male population. Compared with Bulgaria the rate of participation in the labour market is significantly better (with more than 15% at the beginning of the period studied and more than 10% in the year 2006). At the beginning of the period studied (2000), Romania had a participation rate of workers above the average of the EU member countries, similar to the Nordic countries and Great Britain. At the end of the period (2006) due to the downturn of the indicator, we stand at the EU level (50-61%).

For KILM 14 we find out for period 2000 - 2005 a slight decrease in labour with the level of primary education for the benefit of employment with secondary education, the mainstream Romanian labour market. Unfortunately, the level of employment with high levels of education, although it suffered a series of mild fluctuations along the period studied, remained relatively constant. Comparative with Bulgaria, Romania has a much higher percentage of labour with primary education, over labour with tertiary education, where Bulgaria has a percentage almost double compared to Romania.

The next six indicators, KILM 2-7, relate to estimate the level of employment in the labour market. The importance accorded by BIM through KILM these indicators should not constitute a surprise for analysts of the labour market as employment growth in the labour market is, in itself, one of the main goals of all policies in the labour market.

KILM 2-7 offers a wide range of information, relating to employment-to-population ratio (KILM2), passing through a series of characteristics of taking up a job (KILM 3-6), to a special category of information, from a sector that is not always easy even to identify, engage in so-called informal sector of the economy (KILM 7). Altogether, these six indicators can provide a broad perspective on the degree of finding employment in a particular country. Unfortunately, for the period under study, the existing databases in KILM - the fifth edition, do not contain complete sets of data for Romania regarding the indicator KILM 5 (data only for the year 2000), or data is missing in the case of KILM 6, respectively KILM 7. For these reasons, the work we used to analyze the degree of find employment indicators KILM 2, KILM 3 - Status in employment, respectively KILM 4 - Employment by sector activity.

For KILM 2 (2000-2006) Romania had during studied period a relatively constant evolution, even if slightly downward the rate of employment-regardless of sex, with a better rate of employment for the male population. Compared with Bulgaria employment rate is more than 20 percent better at the beginning of the period studied, about 10 percent in the year 2006. It should be noted, however, that there was significant progress in Bulgaria in recent years, with a spectacular growth rate of employment for both sexes.

KILM 3 for the period 2000 – 2005 recorded a slight increase in the percentage of employed workers over the percentage of self-employment. Bulgaria has, compared with Romania, a much higher percentage of employed workers; the number of self-employed workers is approximately half of the level of Romania.

KILM 4 for the period 2000-2006 recorded a declining emphasis on the percentage of workers in agriculture (with over 10 percent), for the benefit of increasing the number of employees in industry, but especially in the services sector. Bulgaria has also a development similar to Romania, but the trends are more exacerbated in this case, the number of employees in agriculture has dropped to half, while the percentage has raised spectacularly employees in the sphere of services, the percentage found far beyond the level of Romania throughout this period.

The next indicators (KILM 8-13) trying to cover not only simplistic aspects of unemployment (lack of a job), but treats, for example, aspects of those groups of people who are engaged inappropriate - employment depending on time - but also those who, for one reason another strain, although at the age (15 + years) - is not registration in employment. We present here only KILM 8,9 and 11.

For KILM 8 - Unemployment rate (period 2000-2006) Romania has made a series of minor fluctuations, the level of this indicator stay relatively constant, with a higher rate of unemployment among the male population. Compared with Bulgaria, the unemployment rate is lower by over 10 percent to beginning of the period studied, about 3 percent in the years 2005 and 2006).

KILM 9 -Youth unemployment (15- 24 years) (percentage of total unemployed youth labour young): Romania has registered in the period 2000 - 2005 growth of the indicator light along the period studied. Regarding the percentage of unemployed youth of all young people (15-24 years), it is low (7-10%), probably because of the involvement of a large number of young people in secondary/ tertiary education. But the ratio: Unemployment young (15-24 years) / unemployment rate in adults (> 25 years) are worrying levels (average 3.5), which proves that Romania has problems with hiring young labour force. In Bulgaria, the ratio is one level lower than in Romania, due primarily to the large number of unemployed adults. Also the percentage of unemployed youth (15-24 years) in the total unemployed is high (over 30%). Unemployment among young people (15-24 years) (%) is higher by over 20 percent in Bulgaria in the first part of the range, decreasing to less than 5 percent in the last part of the period studied. The percentage of unemployed youth of all young people (15-24 years) has in Bulgaria a spectacular downward trend (from almost 12 percent in 2000 around 7 percent in 2005, which should be put on the involvement of a large percentage of young people in Bulgaria in the tertiary education level). The percentage of unemployed youth (15-24 years) in the total unemployed are much higher levels higher in Romania than in Bulgaria, because on the one hand lower rate of unemployment in our country, and on the other hand a high level the percentage of unemployed youth of all young people.

KILM 11 - Unemployment by educational attainment (period 2000-2006) is found a higher percentage of unemployed people with secondary level education, followed by those with primary level education. The percentage of unemployed is much lower among people with tertiary level education. Bulgaria has a much higher percentage of unemployed with basic education (approximately 1.5-2 times as high as in Romania), while Romania has a higher percentage (up to 20%) of the level of unemployed secondary education. As regards the number of unemployed with tertiary education, greater in Bulgaria, it must be in conjunction with the level much higher than in Romania labour with tertiary level education in existing neighbouring country.

3. CONCLUSIONS

The key indicators of labour market may serve as a tool for monitoring and evaluation of a whole series of issues related to the proper functioning of the labour market. Here is a few examples of the way in which they can be used for the analysis of the labour market, and for developing appropriate policies at national or regional level: promoting agenda BIM on "decent work"; monitoring developments on the labour market through the Millennium Development Goals proposed by United Nations; identify those policies that have led to positive developments of the labour market and promote these policies as possible "the rules of good practice" that can be implemented in other countries or economies.

REFERENCES

Cazes, S. & Nesporova, A. (2004/5). Labour markets in transition: balancing flexibility and security in Central and Eastern Europe, in *Revue de l'OFCE*, (no 91 bis), pp. 23-54, *Available from*: http://www.cairn.info/load_pdf.php?ID_ARTICLE=REOF_075_0023 *Accessed:* 2008-07-15

ILO. (2005, 2007). Guide to understanding KILM; *Available from*: http://kilm.ilo.org/2005/press/download/GuidEN.pdf *Accessed:* 2008-08-02

Global Employment Trends (2008); *Available from*: http://www.ilo.org/public/english/employment/strat/download/get08.pdf *Accessed:* 2008-08-20

UNESCO (1997). International Standard Classification of Education, *Available from*: ww.unesco.org/education/information/nfsunesco/doc/isced_1997.htm *Accessed:* 2008-03-12

Vasile, V. (2007). Efecte ale integrării şi globalizării asupra pieţei muncii (Effects of integration and globalization on the labor market), in *Integrarea României în Uniunea Europeană. Provocări şi soluţii*, supliment al Revistei de Economie Teoretică şi Aplicată *(Romania's integration into the European Union. Challenges and solutions*, supplement of the Review of Theoretical and Applied Economics), pp. 143 – 150.

Annals of DAAAM for 2008 & Proceedings of the 19th International DAAAM Symposium, ISSN 1726-9679
Editor B. Katalinic, Published by DAAAM International, Vienna, Austria 2008
Make Harmony Between Technology and Nature, and Your Mind will Fly Free as a Bird

THE WEARING OUT COMPENSATION OF DIESEL ENGINES COMBUSTION CHAMBERS BY MODERN SYSTEMS IMPLEMENTATION

OSTOIA, D[aniel]; NEGOITESCU, A[rina] & TOKAR, A[driana]

Abstract: *This paper presents a method of wearing out compensation in diesel engines combustion chambers by mechatronics systems implementation. The conection between geometrical combustion diesel engine parameters which also ship the level of wear and engine parameters can be realized with Electronic Central Unit (ECU). The Electronic Central Unit can corelate the existent changes by realizing the conection between combustion chamber pressure and geometrical changes due to wearing out and comanding engine actuators with new values describing the diesel engine working conditions.*

Key words*: wear, engine, pressure, mechatronic, pollution*

1. INTRODUCTION

The increase of the environment pollution generated by the internal combustion engines the vehicles are equipped with, represent a desideratum both for the manufacturers and for those who use them.

The problems generated by this pollution are complex and their causes are various (Negrea, 2001). Most researches refer to the new product's technology, but the influences of the wears that appear as a result of the exploitation are more difficult to underline and to mend.

This study approaches both the possibility to take over information regarding the wear of some engine devices and the possibility to mend the efects of these inerent wears by mending the engine's actuators' parameters. The modern engines cannot satisfy the requests imposed by pollution but by using the electronic and informatics elements, thus it results a mechatronic complex. The present work is a method for quantifying and reducing the wear effects in the combustion chamber by using the electronic and informational up-dating systems of some functional parameters. Thus, up-dating the new data of the mechanic parameters in the electronical administration of the engine (ECU), new configurations for the new functioning situations of the engine can be created.

2. THE RESEARCH METHOD

The experimental study of the pressure loss in the combustion chamber was achieved on a M 511 diesel engine. It is characterized by a power of 5 kW for 2500 rot/min speed and is switched on with the engine through the C coupling.

The engine parameters (the indicate pressure into the cylinder, the injection pressure) are measured with two transducers, a position photoelectric transducers, a Kistler type and an Elkon one. These transducer signals are introduced in a signal converter and than are captured in an acquisition device type Labview. The converter signals are processed by this device and with LabView software are saved in ASCII files for further computations.

The engine speed and smoke are measured with a Type AVL device. The sealing degree of combustion chamber is a result of wearing out compensation and the important parameters of degree comparison.

Fig. 1. The test stand

In order to simulate and control the sealing degree of combustion chamber, it was assembled a device that allows fitting gauged holes into the combustion chamber. Through these holes the motor fluid is lost and the performances of the combustion chambers decrease.

The tests on test stand were achieved in the following cases:

-engine under normal working conditions which means the hole diameter of chamber is equal to 0mm, so there are no loses in combustion chamber;

-engine under working conditions characterized by the hole diameter equal to 2 mm, so through this hole are supplementary losses. The test stand is presented in Figure 1.

The testing method purpose was to achieve verification on partial automated experimental installation that can be found in the Internal Combustion Engines lab of MMUT Department of University Politehnica of Timisoara (Ostoia, 2006; Ostoia et al., 2005). The stall land auxiliary installations were equipped as to obtain some parameters established at tests objectives.

The device has a good accuracy for a relevant experimental study duet o fact that convergent results were also experimental obtained for piston rings with different wear degrees.

The experimental program was achieved as every testing case was recurring for a convenient number of times and the resulted values are representing average values of these tests. The device was assembled into the combustion chamber and into the device can be achieved fitting gauged holes in order to simulate the losses through leakiness caused by piston rings wear.

In order to compare the simulation cases were achieved tests for the engine equipped as it follows:

-new piston rings

-weared piston rings

-new piston rings with fitting of 2mm device

To compensate the loss caused by the non-tighten of the combustion chamber due to the wears, it is proposed a change in the injection pressure from 175 bar standard to 200 bar. .

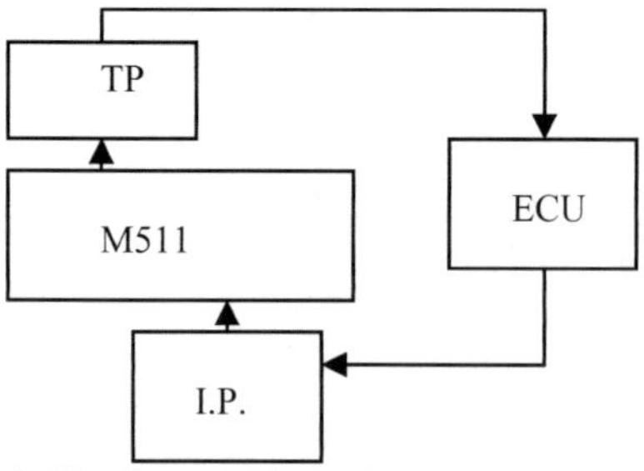

Fig.2. The diagram block of mechatronic system

Thus, together with the information received from the sensor in the combustion chamber, the engine's actuators' parameters can be corrected by changing the values of the Electronic Central Unit (ECU).

These changes must be corrected for each phase, together with their own diagnosis on board (OBD), but i tis preffered to be made together with the change of parameters in the combustion chamber (Madaras, et al., 2006).

The diagram block of connection between transducer pressure from combustion chamber TP, electronic control unit (ECU) and injection pressure actuators I.A. is shown in Figure2.

3. THE EXPERIMENTAL RESULTS AND THEIR INTERPRETATION

The main experimental data regarding the performance parameters of the engine (the effectiv power P_e, the engine speed n, the cyclic dosage d_c, the injection pressure p) are plotted in Figure3 and Figure 4 (Ostoia, et al., 2005).

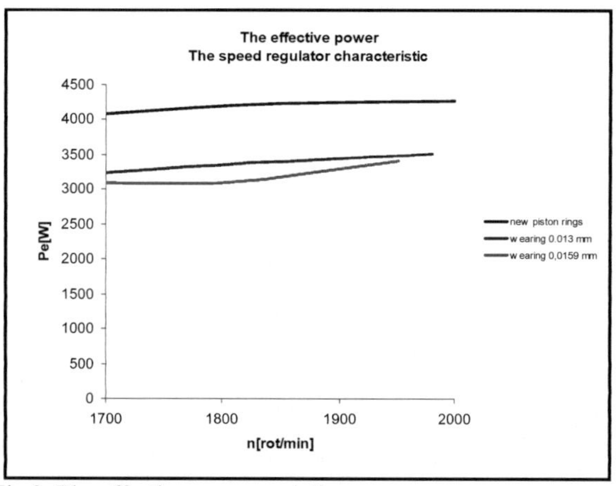

Fig.3. The effective power variation

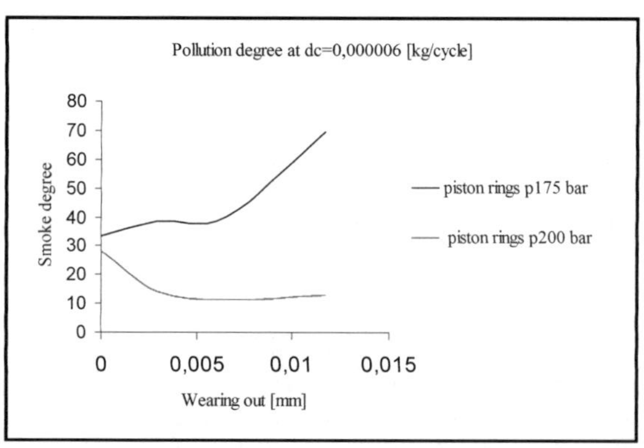

Fig.4. The influence of pressure injection about pollution degree

4. CONCLUSION

By analyzing the accomplished researches regarding the modification of conditions from the combustion chamber, results the following conclusions:

-The modulating methods allow the losses through leakiness evaluation of the engine piston rings

-By knowing the engine behavior in losses case, functional and constructive parameters can be adjusted as the engine working conditions to be optimum

-The tested engine and the ones of its family are energetically influenced by the sealing of the combustion chamber

-The compression and sealing degrees of the combustion chambers are corroborated with the wear of the sealing element. This fact was observed on the comsumption increasing together with the increase of leakinsess the combustion chamber

-The compensation of the leakeage due leakiness could be achieved through an electronic management of injection system until a certain value of these losses.

-The parameters correction of engines will be realized by unit electronic control (ECU) when the transducers of pressure indicate a different value of standard.

This study makes a connection between mechanics parameters (wearing out, effective power, pollution degree) and the electronics and informatics device which equipped the new generation of modern engine.

The study can be especially applied for high capacity diesel engines. These engines usually are related to heavy vehicles (trucks, buses) with an exploitation period larger than the cars one and they presume a high exploitation price.

The results for pollution and energetic performances improvements create the possibility of the combustion chambers wear degree quantification.

The experimental results of this study represent a research direction which must be in accordance with the studies regarding the engine operation optimization.

5. REFERENCES

COFRET, 2006

Mădăras L, Ostoia Daniel, Holotescu S., (2006). The influence of loses diesel engine combustion chambers to mechanical indicate work at idle characteristic, *The Third Edition of the French –Romanian Colloquim* Ostoia, D. (2006). Studies and researches of the formation control of mixture and burning and the influence of behaviour in work to the combustion chamber in correlation with pollution conditions to the diesel engines, *Ph.D. Thesis,* University Politehnica of Timisoara

Negrea, D. (2001). *Processes into internal combustion diesel engine. Economicity. Pollution control.* Vol I-II, Ed. Politehnica, 973-9389-89-9, Timişoara

Ostoia, D., Negoitescu, A.,S., (2005) A study regarding the combustion chambers behavior in corelation with emitted noxe sat diesel engines, *The 30th Annual Congres of the American Romanian Academy of Arts and Sciences, The Academy of Economic studies of Moldova,* pp. 576-578, 99-75-313-2, *july 200 ,Chsinau*

Ostoia,D., Negoitescu, A.,S., Tokar,A.,(2005). Improvemet possibilies of fuel pulverization at the multiple jet sprayer's by functional altering of the injection system, *Scientific reunion of the special program of the Alexander von Hhumboldt foundation concerning the reconstruction of the South Eastern Europe, Sustailnability for humanity & enviroment in the extended connection field science-economy-policy*, 1, pp. 131-134, 973-65-25-204-3, february 2005, Timisoara

Annals of DAAAM for 2008 & Proceedings of the 19th International DAAAM Symposium, ISSN 1726-9679
Editor B. Katalinic, Published by DAAAM International, Vienna, Austria 2008
Make Harmony Between Technology and Nature, and Your Mind will Fly Free as a Bird

A PROPOSITION FOR CLIENT DATA MODEL IN INSURANCE COMPANY

OSTOJIC, J[osip]; PAVLIC, M[ile] & MIKULIC, Z[elimir]

Abstract: *The paper suggests a client data model in an insurance company information system. Business processes in insurance company's sales and damages are key sources of client data as a basic entity type, i.e. the central table in the database. A proposition of client generalization is given, which includes memorizing, i.e. saving client attribute history. A basic conceptual client model is shown, which is the basis for the connection with data models in sales and damages of the insurance company. Entity-relationship method, i.e. an expanded variant of Chen's method, is used for modeling.*

Key words: *data model, insurance, client, E-R diagram.*

1. INTRODUCTION

In insurance, as a service operation, a business interaction between basic participants of business processes is observed and these are: insurer, policy holder, insured person, reinsurer, injurer and the damaged party. When today's level of IT development is considered, it is normal to expect that and insurance company information system supports business processes between participants. A basic assumption is that all who are a subject of an insurance contract are recorded in the information system with relevant data. A common name for all participants is usually client/party, as a term which is included in basic business processes.

Current, and especially future databases are required to be stable, semantically rich and consistent, active data, information i.e. knowledge repositories. In the traditional data modeling approach, before a relation logical scheme of database, comes a conceptual scheme or a conceptual specification which represents the result of conceptual modeling (Vuk, 2007).

How to fulfill all business demands and create a client data repository in an integral insurance company information system? A problem of frequent changes in the real world has been observed, and it reflects on the implementation of a client as an entity in business on information system. When modeling data it is good to have certain model flexibility which would, in future, be ready for various synchronizations and adjustments according to other data sources. Described problems are being explored for three years now, and will continue to be explored as a part of scientific project.

A method for modeling data is a defined procedure of finding and showing information objects and their interrelationships. The entity-relationship method is a method of conceptual and logical data modeling (Pavlić, 1996). The output result is a graphic illustration or a data model of an observed system. A data model is a diagram which uses text and symbols in order to present data grouping so the real data can be better understood (Hoberman, 2005).

Software producers have solved the client modeling problem in various ways. For example, it is often a separate application which controls all data on clients (SAP, 2007) or as a part of solution specialized for a certain industry, i.e. a key segment of a CRM solution (Oracle, 2002).

2. CLIENT GENERALIZATION

A client, as a business subject can appear in different roles in an insurance company information system. It is necessary to perform decomposition or division of client into types and kinds of persons. Figure 1 illustrates a generalization of entity type Client. The starting entity (Client) can be divided or grouped into different subtypes, depending on the point of view. Types of client: company, individual, trade or undefined, are, on a lower level, decomposed into entity subtypes or kinds of persons. With no regard to the shown hierarchy, the client is attributed all types and kinds of clients that they have on a lower level.

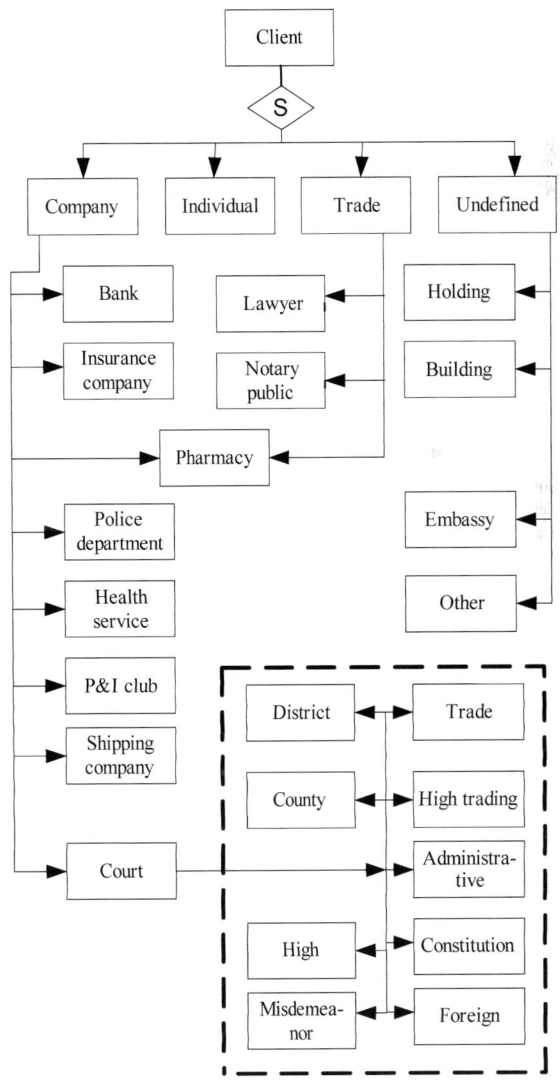

Fig. 1. Client generalization

Client attributes are: ID number for company and individual, that is a generated number for all other cases as a unique client identificator, type of person, kind of person, name and surname for individual, full and short name for company, client address, tax payer, bank account for home and foreign currency, the leading bank number and bank number in international business, court type, contact data (phone number, fax, e-mail, web address), date of birth and death, date of company founding and date of company closing, mark of application of a system from which the client was entered, etc.

3. BASIC CLIENT MODEL CONCEPTS

A basic client model concept assumes a set of entity types which define a client without connecting with the data models which describe information subsystems Sales and Damages. Figure 2 shows a ER diagram for client model. One can observe that the model has the following logical units: a client as a strong entity type and three weak entity types (address, account and contact), two aggregated entity types which relate to client attributes (current and preceding) and a territorial organization as a set of entity types which determine client address.

The address is an attribute of entity type Client. However, there can be more addresses and each can be associated to a different post or city. This is why exclusion of address as a different entity type (Address) is suggested, which is then connected to territorial organization and post. This way saving an unknown large number of addresses is enabled, and the null attribute value is avoided. Entity type Address, along with the current, can hold all preceding addresses. It is possible to have certain home clients and some foreigners from other countries. Each address is of one and just one Type of address.

A territorial organization represents the hierarchy of territory division according to the country organization. Four levels of hierarchy have been suggested. Country is an entity type which can consist of more Counties, and one County belongs to one and only one Country. A County consists of more Districts. Some cities have the status of a District. A Locality is the lowest form of people grouping in an urban area.
Two other weak entity types (Account and Contact) have joint entity types which relate to coding tables. Account is basically defined with Account type and connected to Bank in which it was open.

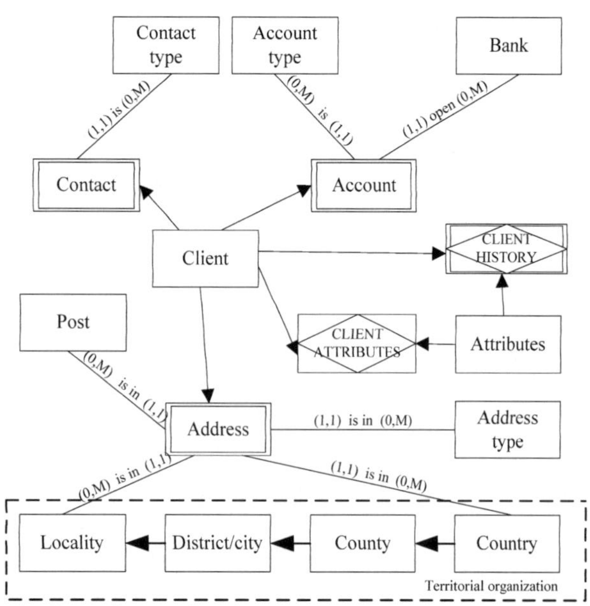

Fig. 2. ER diagram for client model

Contact contains attributes which describe preferred ways of communication with the client. Both entity types are important data classes from two different aspects of observing clients.

Entity type Attributes contains all possible client attributes, preceding and current. This way attributes are not saved in entity type Client because this enables model flexibility. During the information system life cycle it is possible to change client attributes, enter new attributes and remove old attributes. Attributes are never physically deleted; they are only marked as invalid. This way one can approach attribute history values in the database. Only the current valid attributes can be related in an aggregated entity type CLIENT ATTRIBUTES. This type preserves current values of all attributes of all clients. If a client value change happens, then data from CLIENT ATTRIBUTES is being copied into the aggregated entity type CLIENT HISTORY with the date marking when the values were valid. This way history of all values of all client attributes is preserved. This is not the case in many current databases. If the value in entity type Attribute should be deleted (i.e. setting the mark for an invalid attribute), then these values are deleted from CLIENT ATTRIBUTES for all clients and copied to CLIENT HISTORY. CLIENT HISTORY can have many value changes for the same client and the same attribute. In this case new changes would delete old ones, so it is suggested to use the concept of a weak aggregation type for CLIENT HISTORY which saves all changes of the same attribute belonging to the same client.

4. CONCLUSION

Insurance company information system should support business processes according to clients as key business subjects. Clients can appear in different roles, and often data on client is heterogeneous and placed in multiple systems. One of the prerequisites for client integration and reaching the whole picture is creating a robust and flexible data model.
A proposition for client generalization has been suggested and a client data model in an insurance company. The generalization hierarchy provides a view of types and kinds of persons (clients). Data model is shown by using the entity relationship method, which includes the client attribute history.
The following stages of exploring and applying this model can lead towards application (implementation) of the data model in a transaction system of an insurance company, which means it will be connected to data models describing information subsystems Sales and Damages.
The suggested model could be analyzed from the viewpoint of using or preparing client data for a commercial CRM solution. The analysis could especially focus on connecting clients on the level of important subjects, like business holdings, trades, families and their belonging data. This would further expand the already large number of attributes for each client.

5. REFERENCES

Hoberman, S.(2005), *Data Modeling Made Simple*, Technics Publications, LLC, ISBN 0-9771400-0-8, NJ 07720 USA

Oracle Corporation (2002), *Trading Community Architecture Concepts,* Oracle Corporation USA

Pavlić, M. (1996), *Information systems development,* Znak, ISBN 953-180-016-1, Zagreb

SAP AG, (2007), *SAP Business Partner for Financial Service, Available from*: http://help.sap.com, SAP AG Germany, *Accessed*: 2008-06-12

Vuk, D.(2007), *Practical approach to ER-data modeling*, Zbornik radova CASE 19 Metode i alati za projektiranje poslovnih i informacijskih sustava, CASE, Opatija, 18.6.-20.6.2007., ISBN 1334-448X

Annals of DAAAM for 2008 & Proceedings of the 19th International DAAAM Symposium, ISSN 1726-9679
Editor B. Katalinic, Published by DAAAM International, Vienna, Austria 2008
Make Harmony Between Technology and Nature, and Your Mind will Fly Free as a Bird
2nd DAAAM-ICAT

RESEARCH ON IMPROVING THE MECHANICAL PROPERTIES OF THE SLS METAL PARTS

PACURAR, R[azvan]; BALC, N[icolae]; BERCE, P[etru] & PREM, F[lorica]

Abstract: *The manufacturing of the injection moulding tools is one of the most important applications of Selective Laser Sintering (SLS) technology. The resulted porosity after post-processing stage in the oven has an important influence on the mechanical properties of the SLS metal parts. In order to reduce the porosity, a new post-processing method has been developed as compared to the classical one. The paper presents the new method and some results obtained by the authors.*

Key words: *Rapid Prototyping, Selective Laser Sintering, Injection Moulding, Rapid Tooling*

1. INTRODUCTION

Within the Selective Laser Sintering (SLS) process, there are a number of input parameters that can be controlled and modified to get different characteristics of the sintered parts (Childs et al, 1999). Some of this input factors pertain to the laser (e.g. laser power, laser scan spacing, etc.), while others refers to the metallic powder properties (e.g. particle size, percentage composition of the constituent materials, etc.) or to the sintering parameters (e.g. layer thickness, scanning speed, etc.) (Figure 1) Output parameters of interest might be hardness, density, strength, porosity, etc. (Chatterjee et al., 2003).

A considerable amount of work has been done and reported in this field, some of them mentioned bellow. Miler et al have carried out factorial experiments to express the strength of a sintered sample as a function of laser power, scanning speed and fill scan spacing and their respective interaction terms. The model developed takes into account the variation of small and large beam spot sizes and effect of heat loss on strength of sintered samples (Miler et al., 1997).

The Song paper states the influence of laser parameters like laser beam power and experimental parameters like scanning speed on various properties of a laser sintered bronze product. It is reported that density increases while surface roughness decreases with scanning speed decrease (Song, 1997)

Hardro et al determined the optimal process parameters for SLS of an elastometric polymer using an experimental design approach. Laser power, laser scan spacing and part bed temperature were the factors under consideration, while dimensional accuracy and material strength of the sintered samples were the response characteristics. It was concluded that all the factors as well as their interactions are statistically significant (Hadro et al., 1998).

The current paper is focusing mainly on the porosity issue. Theoretical and experimental methods for estimating the porosity of the SLS parts made by Laserform St-100 powder are revealed. A solution for decreasing the porosity of the SLS metal parts, after post-processing in the oven is also presented.

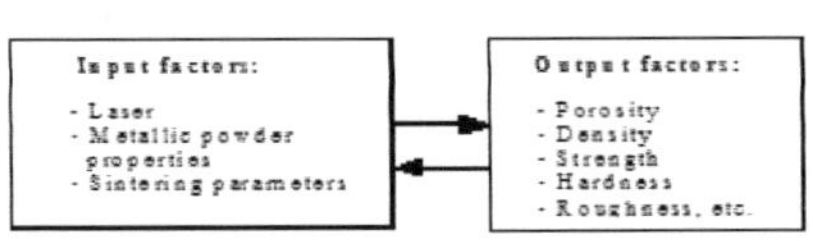

Fig.1. Input and output factors in selective laser sintering

2. POROSITY THEORETICAL APPROACH

A number of physical phenomena are possible for the formation of pores inside laser-sintered metal matrices. Three types of pores can be identified in the material structure, as illustrated in Figure 2 and presented below:

– large interconnected pores outside the balls caused by balling phenomenon
– relatively smaller pores inside the balls
– remained pores in the matrix after infiltration in the oven process.

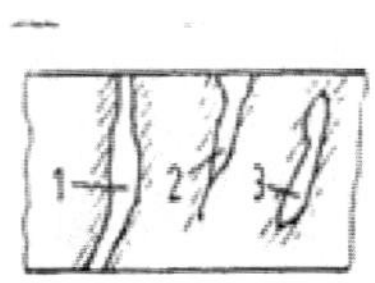

Fig.2. Types of pores – outside the balls, inside the balls and remained pores

Powder material always contains a considerable amount of air inside it in between the particles. The non-uniform porosity distribution of the porous material surface has an important influence over its mechanical and technological characteristics. In order to study the porosity, a schematic method is presented as illustrated in Figure 3. A vacuum impregnation equipment is presented. The pump PV that assures the suck of the resin from the porous probe 1, disposed into the pot 2, is creating the vacuum.

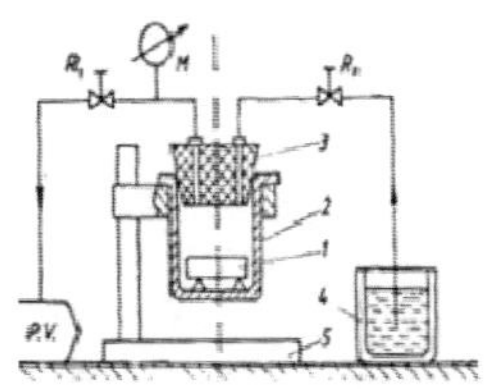

Fig.3. Impregnation equipment (under vacuum)

From the theoretical point of view, the porosity of the material can be calculated very simple using the formula presented below:

$$P := \frac{(m_2 - m_1)\cdot \rho_W}{(m_2 - m_3)\cdot \rho_L}\cdot 100 \quad (1)$$

where:

- P – is the part porosity
- m1 – is the sample mass, weighted in air before resin impregnation;
- m2 – is the sample mass, weighted in air after the impregnation
- m3 – is the sample mass weighted in water
- ρ_W is the water density
- ρ_L is the resin density (g/cm^3).

3. POROSITY EXPERIMENTAL APPROACH

One case study (a button for lever position adjusting of a grass cutting machine presented in Figure 4) has been developed within a project work, jointly at the Technical University of Cluj-Napoca and Plastor Company (RO).

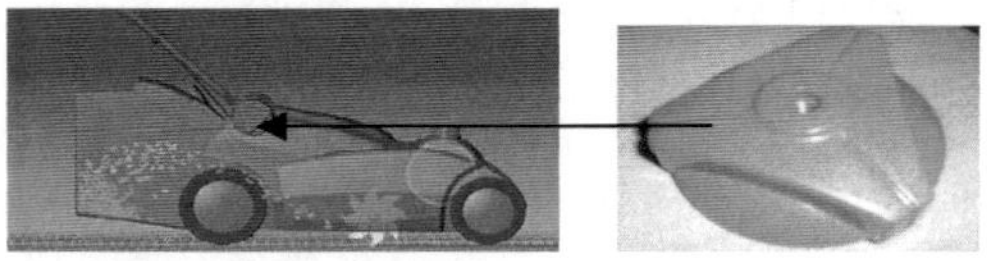

Fig.4. Case study- button for lever position adjusting

Two sets of active elements – punch (illustrated in Figure 5) – were manufactured using the Sinterstation 2000 equipment illustrated in Figure 6 (National Center of Rapid Prototyping – Technical University of Cluj-Napoca).

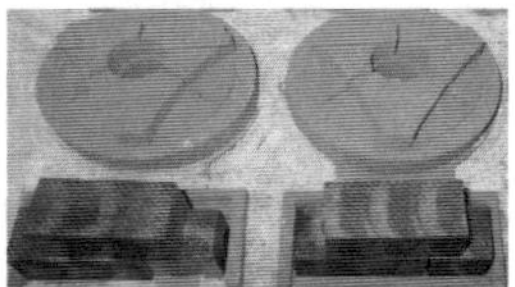

Fig. 5. Active tools manufactured using the Sinterstation 2000 equipment (TUCN)

The active tools made at the TUC-N on the Sinterstation 2000 machine have been post-processed (infiltrated and sinterization completed) at the TUC-N, as illustrated in Figure 5.

Fig. 6. Sinterstation 2000 equipment (TUCN)

One punch has been post-processed in classical way by infiltrating with bronze at 1070°C in the oven, the other one being supplementary impregnated with epoxy resins in vacuum, using the impregnation equipment and dry-oven illustrated in Figure 7.

Mixture Preparation:
½ liter of epoxy resin
5 % hardener

Mixing the resin and the hardener:
Boiling phenomena occurs after 8 min

Vacuum process cycle (Total time -56 min)
- Realizing the vacuum - 15 min
- Maintaining at 10^{-4} Torr - 10 min
- Releasing the vacuum - 3 min
- Vacuum process cycle repeat - 2 times

Drying in the oven - 72 hours

Tab. 1. Resin impregnation under vacuum

Both punches were weighed in air before and after the impregnation and in water before impregnation.

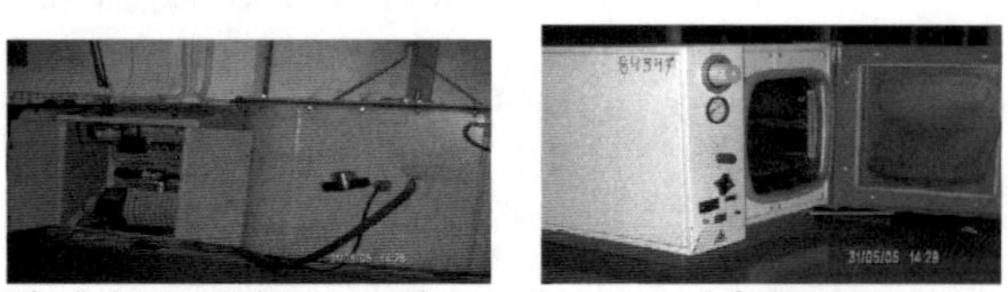

Fig.7. Vacuum impregnation equipment and dry-oven

The aim of resin impregnation under the vacuum was to fill all the existing blanks in the model structure. Several steps were followed, as illustrated in Table 1.
The practical tests for injection molding were made within the industrial company that we cooperated with.

4. RESULTS AND CONCLUSIONS

After finishing the experiment, theoretical porosity calculation was made, using formula 1. The obtained results are presented in Table 2.

	Punch 1			Punch 2		
	before Infiltration	after bronze infiltration	after resin impregnation	before infiltration	after bronze infiltration	after resin impregnation
Aparent density ρ	4,84	7,77	8,34	4,776	7,737	8,226
Porosity p	72,053	32,124	25,428	72,087	32,288	25,516

Tab. 2. Porosity calculation

By the experimental point of view, by using an electronic microscope from TUCN, some estimations were made regarding the porosity in all stages: before infiltration, after infiltration with bronze and after resin impregnation. (Fig. 8)

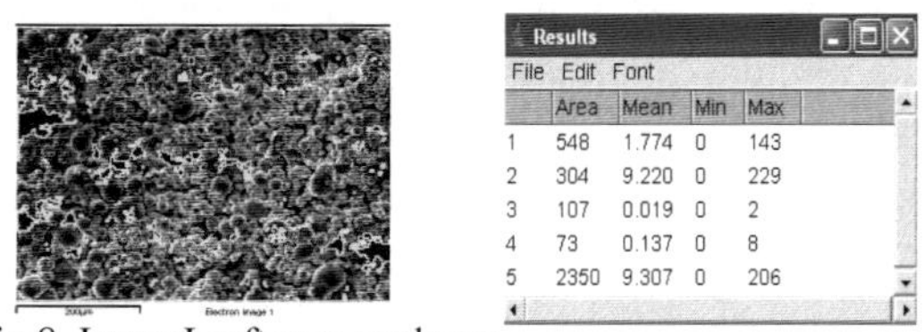

Fig.8. ImageJ software analyzer

ImageJ software was used to analyse the images. The porosity in this case was calculated by using the following formula:

$$p = \frac{\sum Ai}{Atot} \quad (2)$$

where A_i represents each granule area and Atot is the entire image area. Obtained results are comparable with the theoretical ones in all stages, revealing the fact that resin impregnation could be a reliable solution when trying to decrease the porosity of the SLS metal parts.

5. REFERENCES

Chatterjee, A. N. et al., (2003). An experimental design approach to selective laser sintering of low carbon steel. *Journal of Materials Processing Technology,* Vol. 136, No. 1 (2003) 151-157, ISSN 0924-0136.

Childs T. H. C. , et al., (1999). Selective laser sintering of an amorphous polymer – simulations and experiments. *Journal of Engineering manufacture,* Vol. 213, No. 4, (1999), 333-349, ISSN 0954-4054.

Hardro P. J. et al (1998). A design of experiment approach to determine the optimal process parameters for RP machines. *Proceedings of the 5th Int. Conference on Automation Technology*, July 1998, Chiao Tung University, Taipei.

Miller D. et al., (1997). Variable beam size SLS workstation and enhanced SLS model. *Rapid Prototyping Journal,* Vol. 3, No. 1 (1997), 4-11, ISSN 1355-2546.

Song Y. (1997). Experimental study of the basic process mechanism for direct selective laser sintering of low-melting powder. *CIRP Annals Manufacturing Technology* Vol. 46 No. 1 (1997), 127-130, ISSN 0007-8

Annals of DAAAM for 2008 & Proceedings of the 19th International DAAAM Symposium, ISSN 1726-9679
Editor B. Katalinic, Published by DAAAM International, Vienna, Austria 2008
Make Harmony Between Technology and Nature, and Your Mind will Fly Free as a Bird

MODEL IDENTIFICATION AND SIMULATION OF FUNCTION OF THE AMORTIZATION IN CONTEXT OF FUNCTIONING CHARACTERISTICS

PAJAZITI, A[gron]; LATIFI, A[hmet] & GECA, A[hmet]

***Abstract**: The construction of the amortization is done in several phases as in the geometrical, strength, stability and functionality aspects. The functionality aspect is in the adjustment of the extinguisher valve which actually still is done mainly by the driver activity. During the driver activity are tested several prototypes by the engineers that drove the cars through paths. Then, engineers evaluated the control of the extinguisher and also regulated the quicklime hydrators to drive better and better property for the vehicle. Nevertheless, whole process of regulation is the matter of subjective evaluation that can alter from a driver to the other as well as from time to time. Besides the subjective character of the actual quicklime hydrator, regulation of the second weakness is in the process of driver treatment.*

Further is given a brief description of the physical model. An elaborated presentation of the module can be found in the paper [1–3]. In the second section is given how can be found the parameters model through system identification. In the end in the third section it is confirmed the identification of the quicklime hydrator model.

***Key words**: absorber, model, pressure, simulation, valve.*

1. INTRODUCTION

In this paper is included development and identification of the extinguisher model that is foreseen the force of extinction as a function of speed displacement of the extinguisher for the given controlling parameters. In order to compare the coefficients of the model directly with controlling components of the extinguisher the approach of the model of black box, as neural nets (latifi, 2007) or force-state maps (Karnopp, 1991), is issued in the favor of so called models of the white box that offer contact to the physicality of the extinguisher. Even though the actual model is based in the previous physical models (Duym, 1997), it should be emphasized that are grouped in the global coefficients in that way that they can be identified only by dynamometer measurements. This is very important for many vehicle producers that want to have their identity that are obstructed by the other physical models where it is required detail measurements of the components but these are not always possible e.g. the measurement of the depressor, as well if the measurement is done only with dynamometer, it isn't necessary to avoid the extinguisher ant to obtain time and money.

The model of pressure entirely determines the conduct of the extinguisher dynamic through equations system that have physical adiabatic relations of the gas that is in the spare tube in low frequencies and in the oil pressure and in accordance to the cylinder faces in high frequencies. Undesired effects, as it is the mixture of gas with oil in pressure tube (Margolis, 1982), that can cause the increase of the pressure that usually refers if it isn't considered the ventilation. If it is considered the ventilation, the extinguisher efficacy can be obviously reduced because of the low module the effect produces impediments or membranes in order to separate the gas from fluid and as a consequence there is the avoidance of the oil and air mixture.

2. PHYSICAL MODEL OF THE ABSORBER

The quicklime hydrator (extinguisher) model can be classified in the model of pressure and in the model of bearing.

The model of pressure consists of the group of differential equations of indoor pressure in the chamber, although the bearing model calculates the bearing oil Q through the chambers in function of the pressure change Δp with help of nonlinear equations (statistic). The extinguishing force is calculated for the pressure chamber in the pressure and attraction, friction and possible striking force, i.e.,

$$F = (A_{pt} - A_{rod})\, P_{reb} - A_{pt}P_{com} + \text{Friction} + F_{bumper} \qquad (1)$$

To the absorber with two pipes, the geometrical change of the turning chamber volume because of the piston movement is equilibrated after the fluid bearing through piston valve as well with the pressure or attraction of oil in that chamber. On the other hand, the geometrical change of the volume of pressure chamber as well because of piston movement is equilibrated by geometrical amount with the quantity of flow through piston and basic valve and pressure or attraction of oil in pressure chamber. These preserving laws direct us to the differential equations system which expects the pressure increases in pressure pipe.

$$\dot{p}_{reb} = \frac{\left(\dot{x}\left(A_{pt} - A_{rod}\right) - Q_{pv}\right)\left(1 - \alpha p_{reb}\right)}{\left(L_{pt} - x - x_0\right)\left(A_{pt} - A_{rod}\right)\alpha} \qquad (2)$$

$$\dot{p}_{com} = \frac{\left(\dot{x}\, A_{pt} + Q_{bv} - Q_{pv}\right)\left(1 - \alpha\, p_{reb}\right)}{\left(x + x_0\right) A_{pt}\alpha} \qquad (3)$$

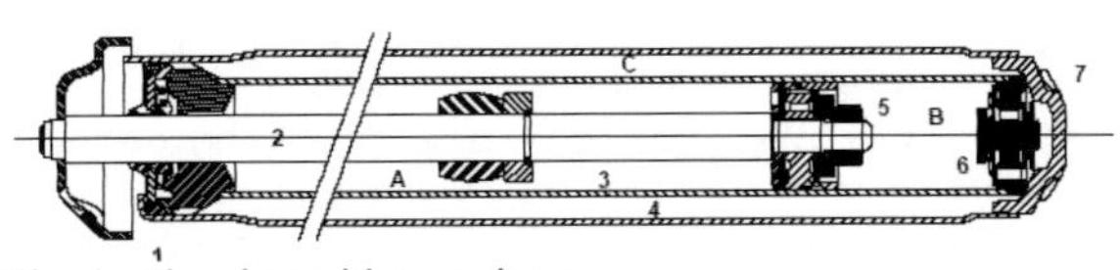

Fig. 1. Absorber with two pipes

Regarding to the spare pipe (tube), the pressure of oil is insignificant comparing to gas so the spare pressure can be expected by polytrophic pressure or attraction respectively by turning strike.

$$p_{rt} = p_{rt,0}\left(\frac{V_{rt,gas,0}}{V_{rt,gcs,0} + A_{rod}x}\right)^{\gamma} \qquad (4)$$

If the adiabatic condition is completed, i.e., it isn't transmitted the heating from the volume of gas, that is typical for high frequencies and is equal to Y = 1.4 for nitrogen gas. For volume change without temperature change that is usually for slowly changes, should be taken exponent Y = 1.

The bearing model has to do with bearing through valves for the appropriate pressure decline. In fig. 1 is showed a typical piston with valve. When the absorber moves to the width (turning), the oil will flow down through the indoor diameter of the entering valve, then through the port canal in the body of piston and then through the open valve determined by splitting and bending valve disks. When the flow meets the pressure extinction along the piston, so it is described as the bearing in pressure.
During the pressure the oil will begin to flow above through splitting and those are wholes in the body of piston with concentric rings around canals port.
At the end of splitting the oil will pressure and it will open the entering valve. The entering valve usually is control valve and it makes a slow resistance in oil when it moves up and it doesn't make any important pressure in order to reduce the relation of the valve. This bearing has to do with the bearing in entrance.

3. GEOMETRICAL CHARACTERISTICS

Geometrical parameters are simply determined by technical drawing. Except in the case if can't be found all geometrical parameters, the evaluation of valve parameters will compensate the appropriate no accuracy and it makes the model suitable for measurement. For BMW all relevant parameters of the model are given in the Table 1.

The length of pressure tube L_{PT} (mm)	380
The static length of piston x_0 (mm)	110
The length of piston (mm)	360
Diameter of piston $d_{pistoneta}$	25
Diameter of pressure tube ID d_{PT} (mm)	36
Diameter of spare tube OD d_{RT} (mm)	56
The length of low measurer up to the base d_{LMPT}	15
The static length of the extinguisher (mm)	485

Table 1. Geometrical parameters of the vehicle BMW.

4. THE MODEL EVALUATION OF THE EXTINGUISHER

Up to now, all the parameters of the vehicle BMW are determined except the entering parameters which are determined as well compressibility value (compression). It can be thought in the value identification of compressibility by the oil supplying data and these don't consider the suitability of cylinder faces (walls). Thus, it is better to determine compressibility of oil from measurements of extinction.
In this section, are used the value and compressibility parameters of $\alpha = 1e-9\ Pa^{-1}$ to anticipate the force of extinction for mature irritation. In fig. 2, together the curve force–displacement and force–speed are drawn for these selected rings of irritation for speed increased values. With the given curves are shown the measured values to those anticipated values by simulations of extinction complete model. As long as non linear nature of the extinguisher should be well devised, the dynamic bearing seems to be as well satisfactory. On the other hand, nonlinear nature of the extinguisher is in the aspect of nonlinearity of the curve force–speed. On the other hand, the dynamic bearing can be seen by hysteresis aspect of the similar drawing and the road of the curve force–displacement is steep. For the elaborated application, can be said that hysteresis is created by the gas compressing in the spare tube much more than oil compressing presented in (Duym, 1997). This comes out by the fact that hysteresis seems to be present always by low frequencies and it doesn't often increase. In order to qualify the shown evaluation , the value of RMS – is mature extinctive and the value of RMS is calculated the remain between the force which is shown in Table 1 and 4 rings of fig. 2 and the global sight between 1.23 and 3.08 s, considering high speeds. The general relative error of RMS is about 5 % that also means that the average square value is about 0.25 % ! The general value of RMS of the remain signal is 39 N for time unit it is comparative or better than the best quality which is obtained by using black boxes, especially with situation maps of the force where the extinction force is modulated as the function of speed and velocity (Margolis, 1982).
The actual model doesn't anticipate the effects of ventilation that are noticed as delays in the indoor pressure or of the force increase, particularly at the very beginning of the pressure (Karnopp, 1991).
As it is shown in fig. 2, there isn't an apparent delay presence (see curves force – displacement) that it means, that this absorber seems not to be sensitive in the ventilation during these tests and for this reason it is allowed to be used the model without ventilation effect.

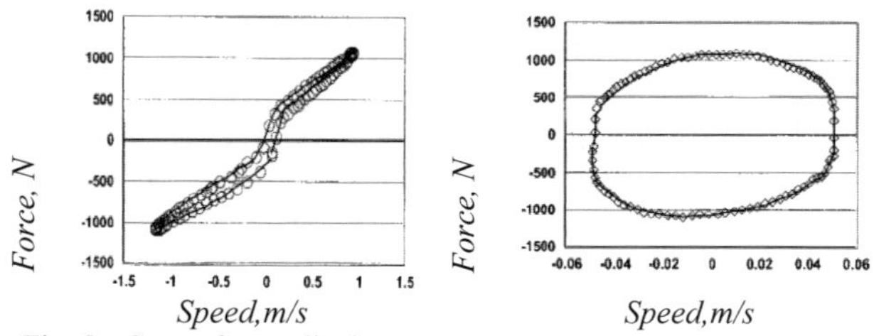

Fig. 2. Curve force–displacement

5. SUMMARY

In this paper is included the development and identification of the extinction model that anticipates the force of extinction as a displacement function of the extinction speed for the given controlling parameters. In purpose of the comparison of the model coefficients directly to the controlling components of the extinguisher the approach of black box models, as neural nets or maps of force – situation is issued in favor of so called models of white box that offers the approach in the physicality of the extinguisher.
Even though these models are based in the previous physical models, it should be emphasized that the coefficients of model are grouped in the global coefficients in that way that they can be identified only with measurement. It is very important for many producers of the vehicles that they wish to have their identity that is obstructed by other physical where are required detail measurements of the components which are not always possible.

6. REFERENCES

Latifi, A. (2007), *Modeling, Identification and Simulation of function of Shock Absorber in the connects of working diagrams*, Ph. D., FME, Prishtina, 2007.

Duym, S., Stiens, R., Baron, G. & Reybrouck, K. (1997), *Physical Modeling of the Hysteretic Behaviour of Automotive Shock Absorbers*, SAE Transactions – Journal of Passenger Cars SAE paper 970101, 1997.

Duym, S., Stiens, R. & Reybrouck, K. (1997): *Evaluation of Shock Absorber Semi – Vehicle System Dynamics* 27 (2), 109 – 127.

Karnopp, D., Crosby, M. & Harwood, R. A. (1991), *"Vibration Control Using Semi – active Force Generator,"* Jnl. Eng. for Industry, Vol. 20, pp. 207 – 218.

Margolis, D. C. (1982), *"The Response of Active and Semi–active Suspensions to Realistic Feedback Signal"*, Vehicle System Dynamics, Vol. 11, pp 267 – 282.

Annals of DAAAM for 2008 & Proceedings of the 19th International DAAAM Symposium, ISSN 1726-9679
Editor B. Katalinic, Published by DAAAM International, Vienna, Austria 2008
Make Harmony Between Technology and Nature, and Your Mind will Fly Free as a Bird

ORGANISATIONAL INNOVATION IN SLOVENIAN MANUFACTURING COMPANIES

PALCIC, I[ztok]; BUCHMEISTER, B[orut] & LEBER, M[arjan]

Abstract: *When we speak about innovation we usually think about product or service related innovation. On the other hand we forget about process innovation. This paper presents basic characteristics of organisational or non-technical process innovation. We present three organisational concepts and their adoption in Slovenian manufacturing companies: team work, continuous improvement processes and regular appraisal interviews. In the end we make a comparison with several European countries.*

Key words: *innovation, European Manufacturing Survey, Slovenia*

1. INTRODUCTION

Term "innovation" is still one of the hottest topics in current science. It is quite obvious that it is mostly linked to R&D of products (Bikfalvi, 2007). There are many studies on innovation revealing that increased R&D activities lead to innovative products which enable companies to achieve competitive advantages and to gain market shares (Freeman & Soete, 1997). The pioneer of innovation literature Schumpeter (1934) differentiates between: technical product innovation, technical process innovation, non-technical service innovation and non-technical process innovation, understood as organisational innovation. The first three groups were subject of many studies. On the other hand, there have been little conceptual and methodological contributions to monitoring of organisational innovation so far (Bikfalvi, 2007).

We have used the results of European Manufacturing Survey to find out about the extent of use of several organisational innovation concepts in Slovenian and some other European countries' manufacturing companies. We will present the basic characteristics of the selected organisational innovation concepts. In the future we want to extend our findings and look at correlations between specific organisational innovation concepts and the characteristics of the companies (size, field of activities, R&D expenditure etc.).

2. WHAT IS ORGANISATIONAL INNOVATION?

The existing literature on organisational innovation is diverse and scattered. There is no consensus on a definition of the term "organisational innovation", which remains ambiguous (Lam, 2005). Damanpour (1987) defines it as the use of new managerial and working concepts and practices. There are several ways to differentiate organisational innovation. The first possibility is into structural organisational innovations and procedural organisational innovations. Structural organisational innovations influence, change and improve responsibilities, accountability, command lines and information flows as well as the number of hierarchical levels, the divisional structure of functions (R&D, production, human resources, financing, etc.), or the separation between line and support functions. Procedural organisational innovations affect the routines, processes and operations of a company. Thus, these innovations change or implement new procedures and processes within the company, such as simultaneous engineering or zero buffer rules (Armbruster et al., 2008).

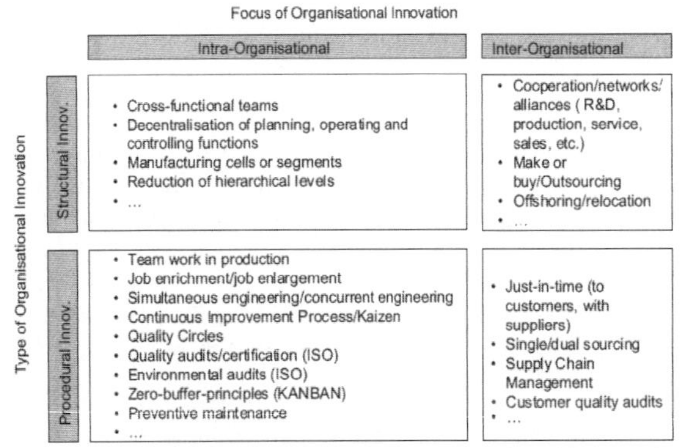

Fig. 1. An item-oriented typology of organisational innovations (Armbruster et al., 2008)

Organisational innovation can be further differentiated along an intra-organisational and inter-organisational dimension. While intra-organisational innovations occur within an organisation or company, inter-organisational innovations include new organisational structures or procedures beyond a company's boundaries. These comprise new organisational structures in an organisation's environment, such as R&D cooperation with customers, just-in-time processes with suppliers or customers or supply chain management practices with suppliers. Intra-organisational innovations may concern particular departments or functions or may affect the overall structure and strategy of the company as a whole. Examples for intra-organisational innovations include the implementation of teamwork, quality circles, continuous improvement processes or the certification of a company under ISO 9000 (Armbruster et al., 2008).

3. EUROPEAN MANUFACTURING SURVEY

The European Manufacturing Survey (EMS) was conducted in 2003/2004 as a pilot survey in nine European countries. The survey covers Austria, Croatia, France, Germany, Great Britain, Italy, Slovenia, Switzerland and Turkey. In total 2249 firms answered questions concerning manufacturing strategies, the application of innovative organisational and technological concepts in production and questions of personnel deployment and qualification. In addition, data on performance indicators such as productivity, flexibility, quality and returns was collected. The responding companies present a cross-section of the main manufacturing industries. Producers of rubber and plastics are represented by 11 percent, producers of metal works by 27 percent, mechanical engineering by 31 percent and electrical engineering by 10 percent.

In the year 2006 a new survey was conducted in even more European countries, where Greece, Netherlands and Spain joined the project. We will present results for Slovenia from 2004 and 2007 survey. We will compare results among European countries from the 2004 survey. In the year 2009 a new survey will take place in Europe. The authors of this paper were conducting a survey and analysis for Slovenian manufacturing companies.

4. SPECIFIC ORGANISATIONAL CONCEPTS

As we have mentioned we are going to present the use of three organisational concepts. The first is **Continuous improvement process (CIP)** as a management concept which tries to initiate favourable changes in companies by taking incremental, but continuous steps and avoiding quantum leaps. The second concept is **team work**. It is argued that the implementation of team work into the production process increases product and process flexibility as well as productivity (Womack et al., 1990). Team workers have a high variety of skills allowing for job rotation within the team so that they can fill in for one another. The enlargement of skills and responsibilities as well as the cooperation with other workers is supposed to have a positive impact on the worker's job satisfaction and task commitment which in turn positively supports the team's productivity. **Appraisal interviews**, as a third concept, are regular face-to-face meetings between employees and their managers and are one part of a human resources development concept. The appraisal presents an opportunity to give feedback on work content and work load as well as to review on what has been achieved during the reporting period and to agree on objectives for the following one (Armbruster et al, 2007).

Figure 2 presents the percentage of Slovenian manufacturing companies that have used all three mentioned organisational concepts in the year 2004 and 2007.

Slovenian companies are aware of the importance of introducing organisational concepts. Three out of four companies adopted CIP. There are many suppliers in Slovenia, who were forced to adapt their business processed to OEMs in order to remain competitive. In addition Ministry of Economy promoted one of CIP concepts in the beginning of this century – "20 keys". The situation in team work is even better, since the number of companies who are introducing team work in their production is rising. Appraisal interviews are less used in Slovenian companies but still present an important organisational concept.

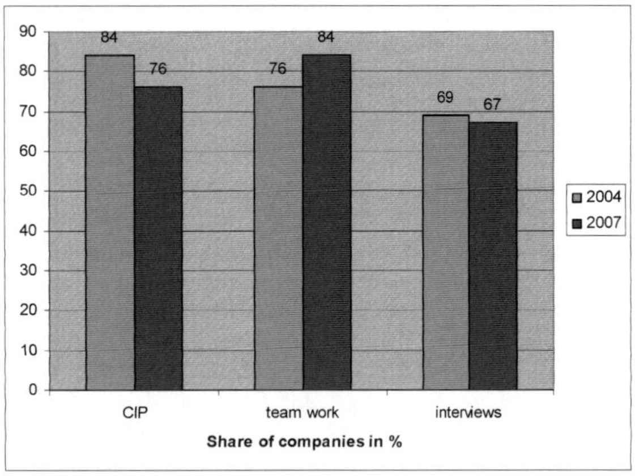

Fig. 2. Use of organisational concepts in Slovenian manufacturing companies

5. COUNTRIES' COMPARISON AS CONCLUSION

Figure 3 presents international comparison of organisational concepts between Slovenia, Germany, Austria, Switzerland and Croatia. Far more than half of the manufacturing companies in Slovenia, Austria, Switzerland and Germany indicate that they make use of CIP concept. The only exception is Croatia. Our findings suggest that supposedly the implementation of CIP occurs independently from sector structure and firm size.

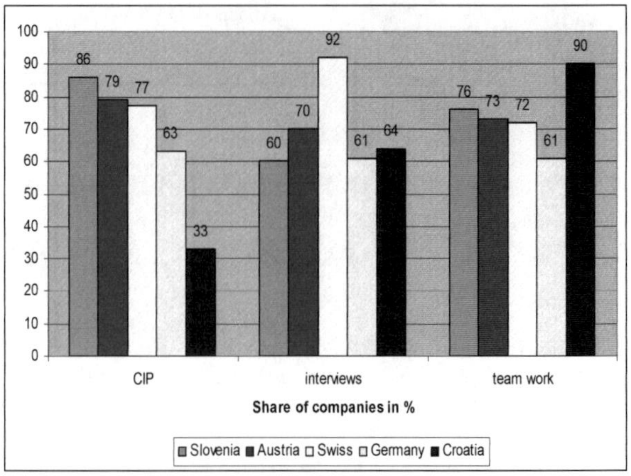

Fig. 3. International comparison of organisational concepts

In all surveyed European countries there is hardly any difference to be seen between the two values.

Appraisal interviews are highly popular in Switzerland, but are present in about 60 % of manufacturing companies in other countries as well. A possible explanation for the varying diffusion of appraisal interviews across countries might be found by taking a closer look at the cultural differences between the European countries. It is reasonable to assume that cultural aspects have a distinct influence on human resources management practices.

On average 75 percent of the surveyed manufacturing companies make use of team work in their production processes. It is surprisingly popular concept in Croatia; the lowest rate of companies using this concept is in Germany. The analysis points to two possible explanations: diffusion rates might vary due to national differences in work organisation across the investigated European countries and due to different business strategies of the examined companies (Armbruster et al., 2008).

6. REFERENCES

Armbruster, H., Kinkel, S, Lay, G. & Maloca, S. (2007). Techno-organisational innovation in the European manufacturing industry. Do European countries differ regarding the diffusion of technical and non-technical innovations in manufacturing companies?, *Proceeding of EUROMA 2007*, Acur, N., Erkip, N. K. & Gunes, E. D. (Eds), Ankara, June 2007, Bilkent University, Ankara

Armbruster, H., Bikfalvi, A., Kinkel, S. & Lay, G. (2008) Organizational innovation: The challenge of measuring non-technical innovation in large-scale surveys. *Technovation*, in Press

Bikfalvi, A. (2007). *Innovation, Entrepreneurship and outsourcing: essays on the use of knowledge in business environments*, doctoral dissertation, Girona

Damanpour, F. (1987). The adoption of technological, administrative and ancillary innovations: impact of organizational factors. *Journal of Management*, Vol. 13, No. (4), 675–688

Freeman, C. & Soete, L. (1997). *The Economics of Industrial Innovation*, Pinter Publisher, London, Washington

Lam, A. (2005). Organizational innovation. In: *The Oxford Handbook of Innovation*, Fagerberg, J., Mowery, D. C., Nelson, R. R. (Eds.), 115–147, Oxford University Press, Oxford

Schumpeter, J., 1934. *The Theory of Economic Development*. Harvard University Press, Cambridge, MA

Womack, J. P., Jones, D. T. & Roos, D. (1990). *The machine that changed the world*, New York

Annals of DAAAM for 2008 & Proceedings of the 19th International DAAAM Symposium, ISSN 1726-9679
Editor B. Katalinic, Published by DAAAM International, Vienna, Austria 2008
Make Harmony Between Technology and Nature, and Your Mind will Fly Free as a Bird

CORPORATE GOAL-ORIENTED FACTORY PLANNING CONSIDERING A LEAN PRODUCTION SYSTEM

PALLUCK, M[arkus]; DOMBROWSKI, U[we] & HENNERSDORF, S[ibylle]

Abstract: *Goals, methods and tools of an (Lean Production System) LPS are always chosen considering the corporate goals. At the same time goals for factory planning are defined. To guarantee an integrated approach the goals for factory planning have to be aligned with the corporate goals and the goals of an LPS. The following article presents an integrated model for a corporate goal-oriented factory planning. Thus the implementation of an LPS can be supported.*

Key words: *corporate goals, factory planning, factory reorganization, lean production*

1. INTRODUCTION

Factory planning has been developed from an unique and infrequently performed activity to a permanent management task with an immense complexity. Furthermore, the meaning of human-social and organizational aspects becomes more crucial, caused by the linkage of different processes (e.g. product development and order processing) (Dombrowski et. al., 2007). A new type of lean production system (LPS), which is currently implemented in many European enterprises, offers a holistic approach to arrange enterprise-specifically methods in order to design the production in a comprehensive and consistent way. In this context, the challenge of factory planning is to configure the elements of the factory in a way, that they support the fulfiment of the purposes of the production system in the best possible way. Based on the described challenge for factory planning, this paper offers an integrated model, which aligns the goals for factory planning with the goals of an LPS as well as with the corporate goals.

2. LITERATURE, CRITICAL OVERVIEW

Factory planning influences the implementation of an LPS on a major way. Caused by this, a coordinated approach aligning the different goals as well as the factory planning process with the implementation process of an LPS is necessary. However, in literature as well as in practice only a few approaches are addressing this fact. Stepping creates a methodology showing interrelations between LPS-elements and different steps of the factory planning process. Using this knowledge a methodology to configure factory planning projects for Value Added Networks is created (Stepping, 2007). An assessment for the strength of the interrelation is missing. A precise goal-oriented deriving of measures cannot be taken. Moreover, the approach by Oeltjenbruns defines interrelations between different elements of an LPS. On the other hand they do not align the process of factory planning with the implementation process of an LPS (Oeltjenbruns, 2000), (Matt, 2006).

Altogether, none of the given approaches consider that the goals given to the department which is responsible for the implementation of an LPS are aligned with the goals of factory planning. These presented facts often end in results of factory planning projects, which neither support the goals of the company-specific LPS nor the corporate goals. Furthermore, the available models do not consider the fact that the LPS-elements applied require a certain amount of qualification. Aside from the employees on the shop-floor especially the employees in the factory planning department are influenced by this fact. Only with necessary qualification-levels a goal-oriented planning, considering the LPS-elements, is effectively guaranteed.

3. RESEARCH COURSE

With the analysis of the existing solutions it turned out that a consisting set of goals aligning the goals of an LPS as well as the goals of factory planning is still missing. In order to correct these deficits, the Institute for Production Management and Operations Research (IFU) of the Technical University Carolo-Wilhelmina of Braunschweig developed a questionnaire to learn about the actual status of factory planning activities in combination with the implementation of an LPS. The questionnaire was sent to about 300 German companies in manufacturing industries. The response rate is about 4.6%.

The results of the survey show that in 69% of these companies an LPS according to the Toyota Production System exists, or is developed right now. At the same time, 87% of all companies recognized a change in the way how factory planning activities were handled (Fig. 1). On the other hand a shortening of planning cycles could be stated (19%). Consequently a rising amount of planning activities could be noticed (69%). 38% of all companies confirmed a decentralisation of factory planning activities - multiple choices are possible.

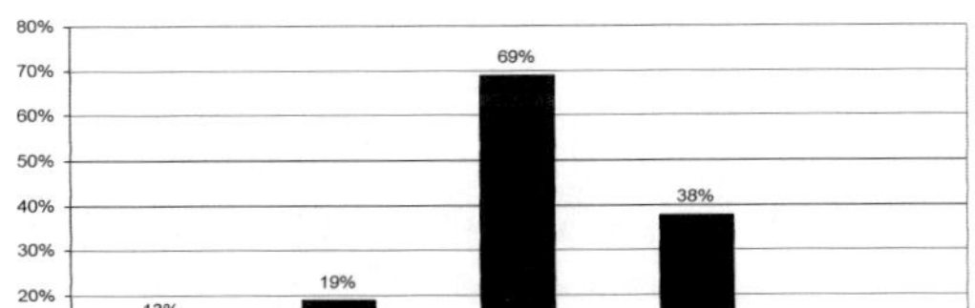

Fig. 1. Was there a change in factory planning activities (multiple choices are possible) (n=16)?

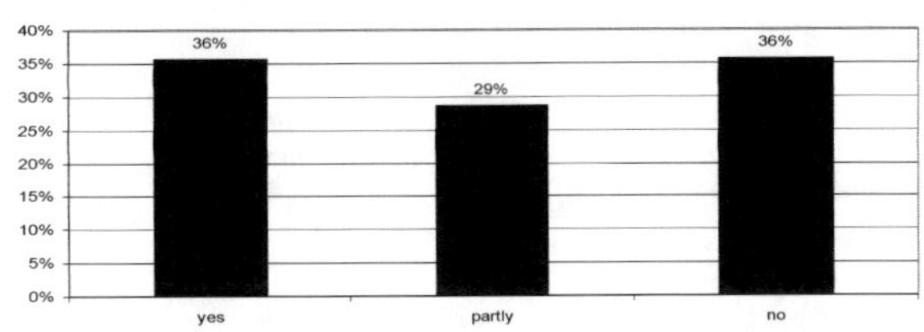

Fig. 2. Interrelations between change in factory planning activities and the implementation of an LPS (n=14)

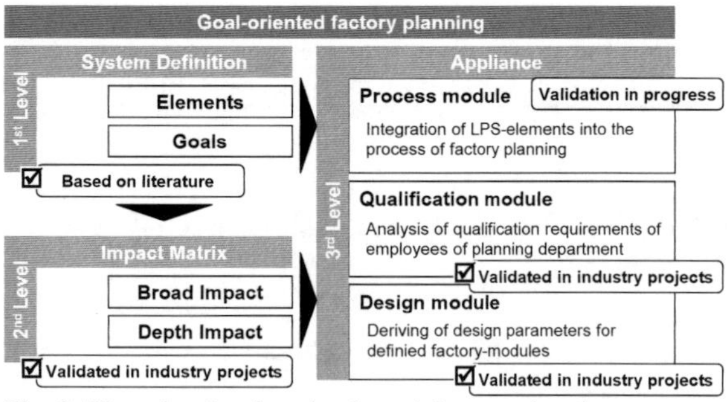

Fig. 3. Three levels of goal-oriented factory planning

Considering these results, the companies noticing a change in factory planning activities were asked if these changes were related to the implementation of an LPS, from their point of view. The results are shown in Fig. 2: Altogether 64% of all companies relate the change in factory planning activities fully (35%) or partly (29%) to the implementation of an LPS.

Based on these results, the IFU developed a three-level model aligning the goals of factory planning with the goals of an LPS as well as the corporate goals (Fig. 3). Hence, an essential contribution to a goal-oriented factory planning process is made. The first level gives a system definition based on literature. For this purpose the system theory, state-of-the-art of science and technology, is used. First, the elements of a factory and the elements of an LPS were defined. In case of a factory a differentiation of structural levels (location, general structure, building, area and workplace) were taken. The elements of an LPS were structured regarding different fields of activity (e.g. visual management, team work, continuous improvement, standardization).

The second level bases on an impact matrix to analyze the interdependencies between elements of an LPS and factory elements (Dombrowski & Palluck 2008). Therefore, the influences of these elements which are covered by the LPS interfere directly with the factory elements are evaluated. For the assessment a set of criteria is used which distinguishes the four impact coefficients: 0 (no influence), 1 (slight influence), 2 middle-sized influence) and 3 (strong influence). By analyzing the interdependencies between LPS-elements and factory elements, both the broad effect and the depth effect are calculated. For each LPS-element the following question is answered: "Which influence results from implementing the LPS-element?"

The **broad effect** of an LPS-element is measured in a percentage rate. It specifies the number of factory elements influenced from an LPS-element relating to the overall number of factory elements. Thus the design of the factory definitely is influenced by the implementation of these LPS-elements. On the other hand, LPS-elements defined by a low broad effect only have a slight influence on the design of factory elements. The **depth effect** of an LPS-element indicates the average intensity of the impact on a factory element. The depth effect can be calculated by the quantity of all ratings related to an LPS-element, whereas this result has to be divided by the overall number of factory elements, which have a rating > 0.

The third level comprises a process module (a) as well as a qualification module (b) and a design module (c). The process module describes the integration of LPS-elements into the factory planning process. Moreover, the connections between the factory planning process and the implementation process of an LPS are shown. To guarantee a lean factory planning process it is designed in a manner, that the three M's (muda: waste; mura: unevenness; muri: overburden) consequently are eliminated. The qualification module supports the analysis of the current qualification-status of employees in the factory planning department. Based on the previous knowledge of the employees, the approach of implementing the LPS (e.g. frontal experts-training or cascade-training) and the degree of complexity an approach for qualifying the employees is elaborated. The design module uses the results of the impact matrix (second level) to define factory modules for specific levels of structure of the factory.

4. RESULTS

The presented framework for corporate goal-oriented factory planning considering an LPS was validated in several industry projects in the commercial vehicle and automotive industry (series production). Each time the impact matrix was filled in with enterprise-specific data, accounting to the knowledge of about 5-8 company-intern experts (e.g. LPS-manager, employees of factory planning department). Considering the results of the analysis, LPS-elements with a strong impact on factory elements could be identified. Regarding this results in combination with the current status of qualification of the employees of the factory planning department (qualification module) factory modules supporting the corporate goals were defined (design module). Using this framework an alignment of the factory planning process within the implementation process of an LPS enables an adjustment of the factory as well as of the planning process relative to the corporate goals. Hence iteration loops during the planning process, caused by non-corporate goal-oriented planning, could be avoided. Furthermore, due to the alignment of the factory planning process within the implementation process of an LPS the continuous improvement of the factory could be fastened. Altogether the experience shows that the framework can be applied in different projects easily.

5. CONCLUSION

The goal-oriented approach to factory planning considering the methods and tools of an LPS and their impact on the factory elements supports the implementation of a LPS. The process module as well as the qualification and the design module help to align the factory planning process with the implementation process of an LPS.

6. REFERENCES

Dombrowski, U.; Hennersdorf, S. & Palluck M. (2007). System approach for plant design – development of a control loop for lean factory planning, *Annals of DAAAM for 2007 & Proceedings of the 18th InternationalDAAAM Symposium,* pp. 325-326, ISBN 3-901509-58-5, Zadar, Croatia, October 2007, Published by DAAAM, Vienna, Austria

Dombrowski, U.; Palluck M. (2008). Methodical approach to factory planning regarding a lean production system, *Proceedings of the 9th Biennial ASME Conference on Engineering Systems Design and Analysis,* ISBN 0-7918-3827-7, Haifa, Israel, July 2008, ASME Petroleum Division, New York.

Matt, D. T. (2006). Fast Configuration of Lean and Customer Driven Production Systems with Axiomatic Designed Production Module Templates, *Proceedings of the 5th CIRP International Seminar on Intelligent Computation in Manufacturing Engineering,* ISBN 88-950280-1-5, Ischia, Italy, July 2006

Oeltjenbruns, H. (2000). *Organisation der Produktion nach dem Vorbild Toyotas (Organizing the Production the Toyota Way),* Shaker Verlag, ISBN 3-8265-7966-6, Aachen

Stepping, A. (2007). *Fabrikplanung im Umfeld von Wertschöpfungsnetzwerken und Ganzheitlichen Produktionssystemen (Factory planning regarding Value Added Networks and Lean Production Systems),* Shaker Verlag, ISBN 978-3-8322-6275-4, Aachen

Annals of DAAAM for 2008 & Proceedings of the 19th International DAAAM Symposium, ISSN 1726-9679
Editor B. Katalinic, Published by DAAAM International, Vienna, Austria 2008
Make Harmony Between Technology and Nature, and Your Mind will Fly Free as a Bird

COMPUTER-AIDED DESIGN OF A DEVICE PROTOTYPE FOR SPECIFIC PURPOSE IN BIOMECHANICAL RESEARCH OF SOFT TISSUES

PALM, S[tjepan]; JELECANIN, J[elena] & MRCELA, T[omislav]

Abstract: *The design of an electropneumatic device for biomechanical research of tendons is presented in this paper. Device requirements are listed according to the expected elastic property of a tendon and the specific working conditions. The reasons for selecting pneumatic drive are explained and the device concept is laid out in form of pneumatic scheme. Component modelling using AutoCAD is presented, as well as the subsequent drawings for manufacturing of the specific component. Resulting device prototype is shown and the guidelines for further development are given. Finally, the importance and benefits of using CAD system in the design of a specific purpose device are discussed.*
Key words: *Design, CAD, biomechanics, pneumatics*

1. INTRODUCTION

Biomechanics of soft tissues, such as human tendon, includes research and analysis of its mechanical properties. Tendons, as most biological structures, have highly nonlinear stress-strain properties (Lieber & Burkholder, 2000). This elastic property is shown in Fig. 1. However, the property of tendon does not necessarily apply to tendon that has been transplanted and is yet to be tested.

In order to carry out these tests, the usage of especially designed device was necessary. Such devices are not commercially available because of their narrow filed of application. Thus, the idea of development of the device that would meet all the requirements was conceived. The cost of developing such a device is high considering the limited usage, therefore, the computer aided design is considered to be important tool in design process.

This paper addresses the problems in designing a device for biomechanical research according to specific requirements. These include the specific working conditions such as damp environment of a medical laboratory, expected force at which the tendons will be tested and the requirement on geometry of component for holding the tendon. Other important requirements are construction size, modularity, simplicity and the safety for use.

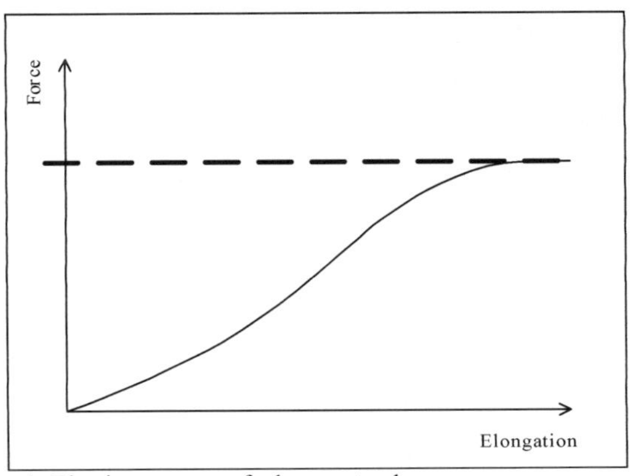

Fig. 1. Elastic property of a human tendon

2. DEVICE DESIGN

There are two types of criteria on device design; it has to meet requirement on its performance as well as requirements on its construction. The basic function of tendons is to transmit the force created in the muscle to the bone, thus making joint and limb movement possible. To do this effectively, tendons must be capable of resisting high tensile forces with limited elongation. (Josza & Kannus, 1997). Therefore, device components must be capable of producing such force. Requirements on working conditions led to selection of pneumatic system as the drive system. Pneumatic devices are characterized with especially robust and simple construction. These characteristics enabled the usage of pneumatic devices everywhere, where the working conditions are hard and where the device construction is required to be safe to operate. Thanks to the fact that they are always under the pressure of air higher than atmospheric, they can operate in aggressive chemical and damp atmosphere. (Zrnic, 1980).

Device layout is shown in Fig. 2. in form of pneumatic scheme. The device consists of pneumatic executive and control elements, and sensors for measuring system variables. Double-acting pneumatic cylinder was chosen for executive element because it can perform useful act in both motions. Relation for backwards force (Nikolic, 2002) in N is:

$$F_2 = A_2 \cdot p_{do} - A_1 \cdot p_1 - F_T \qquad (1)$$

where p_{do} is supply pressure in Pa, p_1 stagnated air pressure in an irradiated chamber in Pa, F_T force of friction in N, A_1 and A_2 back and front area of a piston in m^2 defined as:

$$A_1 = D_c^{\,2} \cdot \frac{\pi}{4}; \quad A_2 = \left(D_c^{\,2} - d_k^{\,2}\right) \cdot \frac{\pi}{4} \qquad (2)$$

where D_c is the diameter of cylinder in m and d_k the diameter of piston rod in m. Executive element is driven by mechanic direction control valve with proportional pressure actuator.

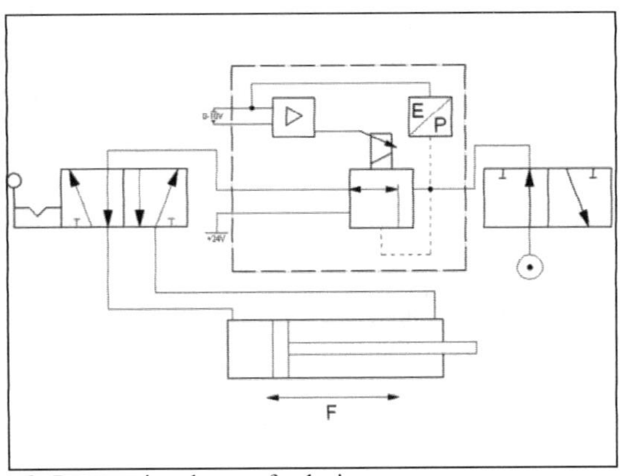

Fig. 2. Pneumatic scheme of a device

3. DETAILED PART MODELLING

Part modelling is based on the geometry of samples of human tendon whose elasticity is to be tested. The biomechanical research of human tendons includes tests on two types of transplanted tendons; so-called bon-tendon-bon and hamstrings. The first type of sample has of quadratic shaped bone at the ends of a tendon, while the other sample is made by folding only the soft part of a tendon. The requirement on the component for holding samples is therefore a construction that could manipulate both types of samples without changing the main holding part.

Based on these demands, the part for holding samples was conceived. In order to get a good sense of its appearance and functionality, 3D model (Fig. 3.) was made using AutoCAD because it makes it possible to make truly realistic model with solids and combine or subtract solids and get information about their physical properties (Finkelstein, 2005).

3D model of a component for holding samples showed in Fig. 3. consists of fixed and movable part. Elements for attaching ends of tendons are screwed on to the movable part. The fixed element is an integral part of the main case that has a joint for guideline for the other, movable part connected to pneumatic executive element. To ensure manipulation of both types of samples, two sets of movable parts are available. First part is for holding the bon-tendon-bon sample and it has the role of an adjustable jaw for holding the bone part of the sample. The other part is meant for holding the ends of hamstrings sample.

After all the modification and advancements were completed, and the final 3D model of a component for holding samples was available, drawings for manufacturing were made. Fig. 4. shows dimensional drawings from all the viewpoints necessary for manufacturing.

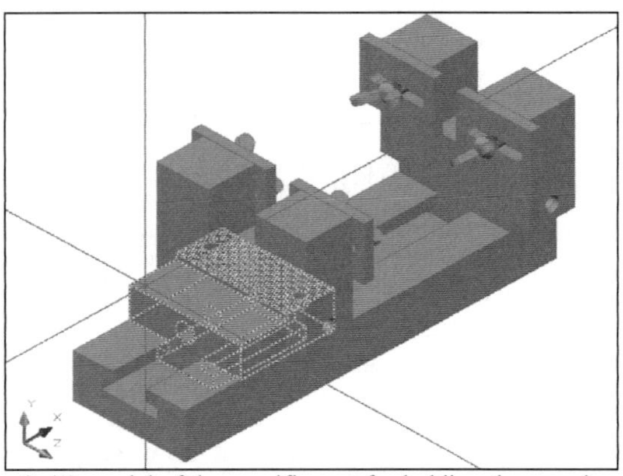

Fig. 3. 3D model of the specific part for holding the samples of tendons

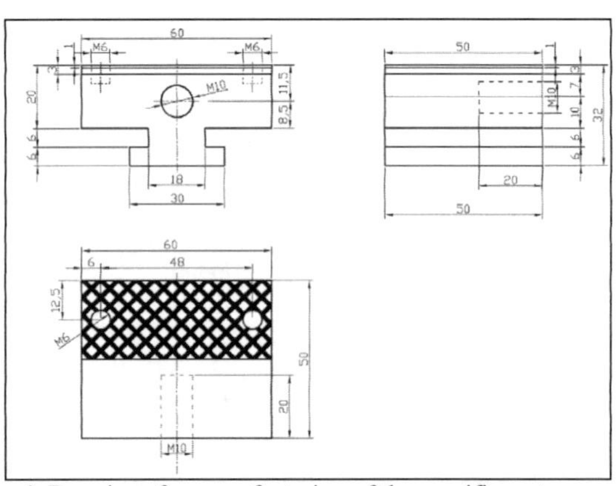

Fig. 4. Drawings for manufacturing of the specific part

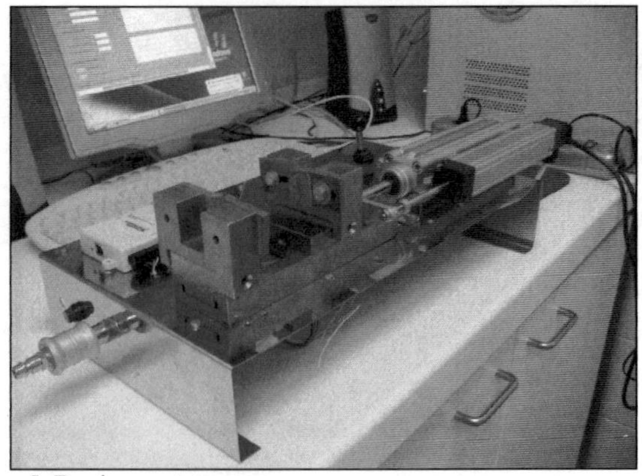

Fig. 5. Device prototype

4. DEVICE PROTOTYPE

The device prototype showed in Fig. 5. was manufactured directly from the 3D model and drawings for manufacturing. It consists of two types of systems; electropneumatic system for controlling the force applied to samples of tendons and mechanical part for holding samples of tendons. The electropneumatic system consists of executive and controlling elements for force control and sensors for measuring force and the displacement applied on samples. Further development of the device prototype includes implementing a third type of system; a computer supported system for force control and acquisition of values collected from sensors.

5. RESULTS

Mechanical properties of the device prototype were tested in biomechanical laboratory on both samples of tendons; bon-tendon-bon and hamstrings. It was demonstrated that the both types of samples can be easily attached by changing the movable parts without any additional tools. Final tests for functionality showed that the force for testing the elasticity of samples is adequate and the system is properly dimensioned.

6. CONCLUSION

In biomechanical research, the usage of especially designed test equipment is necessary. When developing the device prototype for biomechanical research of human tendons, the usage of computer aided design was an important stage in the design process that resulted with a successful device prototype. The time of the device development was minimized and overall costs of the device were decreased. Detailed engineering and analysis of 3D model lead to manufacturing the device prototype that needs no further modifications or advancements.

7. REFERENCES

Finkelstein, E. (2005). *AutoCAD 2006 and AutoCAD LT 2006 Bible*, Wiley Publishing, Inc., ISBN 0-7645-9675-6, Hoboken, USA

Josza, L. & Kannus, P. (1997). *Human Tendons: Anatomy, Physiology, and Pathology,* Human Kinetics, ISBN 0-8732-2484-1, Champaign, USA

Lieber, R. L. & Burkholder, T. J. (2000). Musculoskeletal Soft Tissue Mechanics, In: *The Biomedical Engineering Handbook*, Bronzino, J. D. (Ed.), Chapter 22, CRC Press, ISBN 3-5406-6351-7, Boca Raton, USA

Nikolic, G. (2002). *Pneumatika (Pneumatics)*, Skolske novine, ISBN 953-160-174-7, Zagreb

Zrnic, V. (1980). *Pneumatika (Pneumatics)*, Tehnicka knjiga, Beograd

Annals of DAAAM for 2008 & Proceedings of the 19th International DAAAM Symposium, ISSN 1726-9679
Editor B. Katalinic, Published by DAAAM International, Vienna, Austria 2008
Make Harmony Between Technology and Nature, and Your Mind will Fly Free as a Bird

THE MATHEMATICAL STUDY OF THE PROFILE OF TOOTHING KNIVES

PANTEA, I[oan]; STANASEL, I[ulian] & BLAGA, F[lorin]

Abstract: *Toothing heads are used while processing conical gears with curved teeth that have as their directing curve an elongated epicycloid or a circular arc. These tools are equipped with a great number of knives. The toothing knives have a profile made up of complex surfaces. The relieving method is used in order to rectify the profile but this method does not satisfy the imposed conditions for these tools. The paper presents, in this sense, the construction of a mathematical model for the rectification technology of the lateral positioning surfaces of knives used for toothing conical gears with curved teeth. It also serves the purpose of simulating the technology for determining the deviations of the chipping edge in different grinding sections of the knife.*
Key words: *Profile, relieving, lateral positioning.*
¶
1. INTRODUCTION
¶
The working profile of the knives used for toothing conical gears with curved teeth is made up of complex surfaces, surfaces that compose helixes.

A conical or cylindrical rectification tool cannot perfectly wrap around this kind of surface (Maroş, 1959).

The chipping edge of the knife is part of this surface, and the maximum permissible deviation is 0.08 mm. The large number of knives mounted on the head implies that the relieving process needs a large number of double strokes of the sled on one rotation of the toothing head

The paper presents the author's research regarding the execution of these surfaces through continuous processing following a cylindrical propeller and by using a conical abrasive tool (Pantea, 2004).

2. THE STUDY OF THE TECHNOLOGY OF TOOTHING KNIVES THROUGH THE SPATIAL GEARING METHOD

On profiling the toothing knives, the following issues must be addressed:

- determining the contour of the abrasive stone (the theoretical profile);
- determining the equations of the relieved surfaces.

Resolving these problems is done concomitantly in a calculus section using the spatial gearing laws and the coordinate system's transformation laws (Litvin, 1994). We go from the generating curve of the face gear, we determine the profile of the knife's chipping edge and the profile of the abrasive tool used for relieving. The chipping edge must have minimal errors after resharpening.
The generating curved is given under the form:

$$X_p = \begin{bmatrix} x_p(p) \\ y_p(p) \\ z_p(p) \end{bmatrix} = X_p(p) \qquad (1)$$

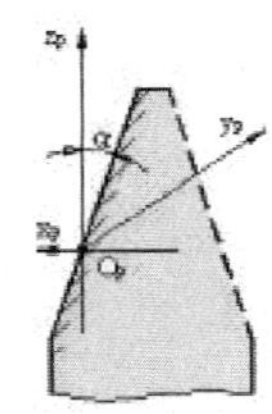

Fig.2 The generating curve

where p is the independent parameter of the curve.

The parametric equations and the limits of the independent parameter p used result from fig. 2
The parametric equations:

$$\begin{aligned} x_p &= p \cdot \sin(\alpha) \\ z_p &= p \cdot \cos(\alpha) \end{aligned} \qquad (2)$$

The limits of the independent parameters

$$\begin{aligned} p_s &= \frac{h' \cdot m}{\cos(\alpha)} \\ p_i &= \frac{-h' \cdot m}{\cos(\alpha)} \end{aligned} \qquad (3)$$

Particularizing (fig. 3) the arrangement of the systems for knives used for toothing conical gears with curved teeth arranged in a circular arc that have their directing curve a cylindrical propeller, the matrix product with the help of which the passing for the $x_iy_iz_i$ system to the $x_ky_kz_k$ system is achieved, is:

$$\begin{aligned} x_k &= x_i \cos(\theta)\cos(\varphi) + y_i \sin(\theta) - \\ &- z_i \cos(\theta)\sin(\varphi) - A_x \cos(\theta)\cos(\varphi) \\ y_k &= -x_i \sin(\theta)\cos(\varphi) + y_i \cos(\theta) + \\ &+ z_i \sin(\theta)\sin(\varphi) + A_x \sin(\theta)\cos(\varphi) \\ z_k &= x_i \sin(\varphi) + z_i \cos(\varphi) - \\ &- A_x \sin(\varphi) + A_z \end{aligned} \qquad (4)$$

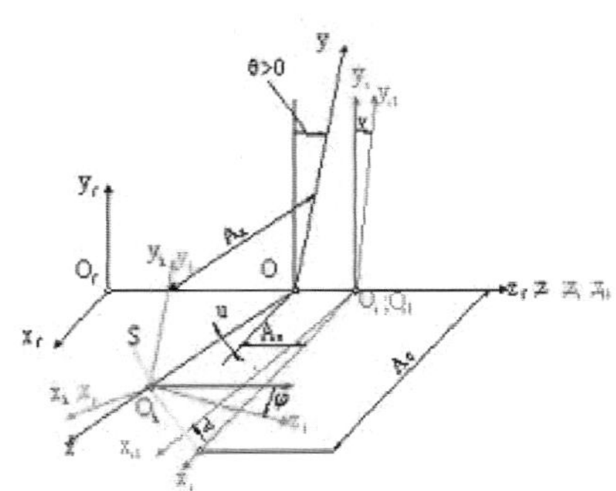

Fig. 3. The axis system.

Analogously, the passing formulas from the $x_k y_k z_k$ system to the $x_i y_i z_i$ system are deduced with the help of the reciprocal matrixes.

$$\begin{aligned} x_i &= x_k \cos(\varphi)\cos(\theta) - y_k \sin(\theta) + \\ &+ z_k \cos(\theta)\sin(\varphi) + A_x \cos(\theta) \\ y_i &= x_k \cos(\varphi)\sin(\theta) + y_k \cos(\theta) + \\ &+ z_k \sin(\theta)\sin(\varphi) + A_x \sin(\theta) \\ z_i &= -x_k \sin(\varphi) + z_k \cos(\varphi) - A_z \end{aligned} \quad (5)$$

The (5) equations express the trajectory of the relieving motion of the disk's surface points in relation to the $x_i y_i z_i$ system that is considered to be immobile.

The normal line to the chipping edge is determined from the condition of perpendicularity of vectors the tangent to the chipping edge and the relative speed in the relieving process (Litvin, 1994):

$$\overline{N} = \bar{t} \times \bar{v} \quad (6)$$

As for determining the chipping edge's profile in different degrees of resharpening, the speed vector of the relative motion's projections in the $x_k y_k z_k$ system must be determined (fig.4).

The normal lines N are described by following:

$$N = \begin{bmatrix} [\frac{\sin(u+\gamma_x)}{\tan(\alpha)} + \cos(u)\cos(\gamma_h)\tan(\gamma_v)]A_z + \\ + \cos(\gamma_x)x \\ \\ \cos(\gamma_x)y - [\frac{\cos(u+\gamma_x)}{\tan(\alpha)} + \\ + \sin(u)\cos(\gamma_h)\tan(\gamma_v)]A_z \\ \\ [\frac{\cos(u+\gamma_x)}{\tan(\alpha)} + \sin(u)\cos(\gamma_h)\tan(\gamma_v)]x + \\ + [\frac{\sin(u+\gamma_x)}{\tan(\alpha)} + \cos(u)\cos(\gamma_h)\tan(\gamma_v)]y \end{bmatrix} \quad (7)$$

The generated surface from the rectification with a conic disc will be:

$$X_g = \begin{bmatrix} [x(p)+r]\cos(v) \\ [x(p)+r]\sin(v) \\ z(p) \\ 1 \end{bmatrix} \quad (8)$$

The S_d relieved positioning surface is defined as the geometric location of the chipping edges M_a, ε positions the M_a chipping edge in relation to the technological device's $x_i y_i z_i$ system and for ε variable the X_{det} radius vector of the relieved surfaces is determined (Şteţiu et al., 1994).

$$X_{det} = \begin{bmatrix} x_{det}(p,\varepsilon) \\ y_{det}(p,\varepsilon) \\ z_{det}(p,\varepsilon) \\ 1 \end{bmatrix} = X_{det}(p,\varepsilon) \quad (9)$$

In order to determine the theoretical profile, the surface of the abrasive tool is considered the kinematic hull of the S_d relieved positioning surface.

The position vector's expression is:

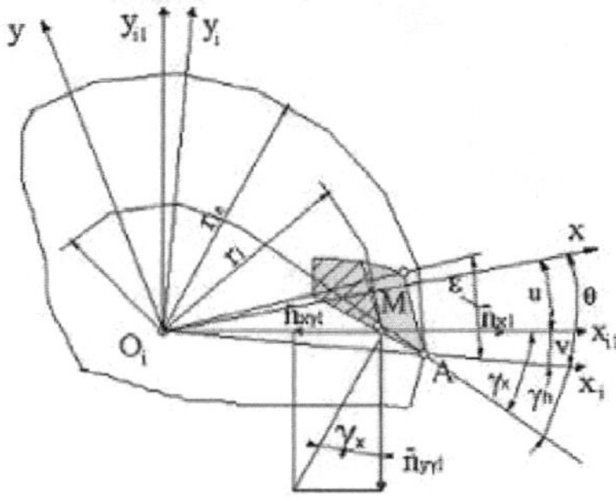

Fig. 4 The chipping edge.

$$X_k = M_{ki} X_d \quad (10)$$

If the p parameter is given discreet values the coordinates of the characteristic curves plotted on the abrasive tool can be determined.

$$X_k = \begin{bmatrix} x_k \\ y_k \\ z_k \\ 1 \end{bmatrix} = \begin{bmatrix} x_g \sin(\theta) + y_g \cos(\theta) - A_0 \\ x_g \cos(\theta) - y_g \sin(\theta) \\ z_g + A_2\theta \\ 1 \end{bmatrix} \quad (11)$$

By using the M_{ik} (5) transformation matrix, the real positioning surface is determined with the relationship that depends on θ.

$$X'_{det} = \begin{bmatrix} [x'_k \cos(\varphi) + A_x]\cos(\theta) - y'_k \sin(\theta) - \\ - z'_k \sin(\theta)\cos(\varphi) \\ [x'_k \cos(\varphi) + A_x]\sin(\theta) + y'_k s\cos(\theta) + \\ + z'_k \sin(\theta)\cos(\varphi) \\ - x'_k \sin(\varphi) + z'_k \cos(\varphi) - A_2\theta \end{bmatrix}$$

3. CONCLUSIONS

1. Using the algorithm, a research methodology of the geometry of toothing knives was constructed by using the spatial gearing method of research.
2. The mathematical modeling of the technology of rectification of toothing knives using the spatial gearing method implies establishing a response in regard to the optimization of the technology.
3. Through particularizing the research method for the I tool and the II tool and by using a Archimedes's spiral as a directing curve, the evaluation and optimization model of the profile of these knives is obtained through comparing the deviations in the grinding sections of the knives.

4. REFERENCES

Litvin, F.L. (1994). *Gear Geometry and Applied Theory.* University of Illinois, Chicago. PTR Prentice Hall, Englewood Cliffs, 0763. ISBN-13-211095-4, New Jersey.

Grămescu, Tr. ş.a. (1993). *Toothing technologies. Guide project book.* Ed.Universitas, ISBN 5-362-01009-3, Chişinău.

Şteţiu, G. ş.a. (1994). *Practice and theory of cutting tools, vol. I, II, III.* Editura Universităţii Sibiu, ISBN 973-95604-3-1, Sibiu.

Maroş, D. (1959). *Kinematics of tooth wheels.* Editura Tehnică Bucureşti.

Pantea, I. (2004). *Contributions regarding the technology of the tools used for bent teeth bevel gears teething.* Doctorate Thesis, University of Oradea.

Annals of DAAAM for 2008 & Proceedings of the 19th International DAAAM Symposium, ISSN 1726-9679
Editor B. Katalinic, Published by DAAAM International, Vienna, Austria 2008
Make Harmony Between Technology and Nature, and Your Mind will Fly Free as a Bird

THE SIMULATION OF THE PROFILING TECHNOLOGY OF TOOTHING KNIVES

PANTEA, I[oan]; STANASEL, I[ulian] & BLAGA, F[lorin]

Abstract: *Using a mathematical model that has as its study base the spatial gearing method of research, that is particularized for the technological gear formed at the rectification of toothing knives arranged in a circular arc pattern, calculus programs for optimizing the technology were drawn up in the programming environments of Visual C++ and MathCAD 2001. The chipping edge of the knife is in theory a line. The deviations of the chipping edge of the knife from a line that passes through the extreme points of the edge are determined through simulation. These deviations are evaluated in different sections of the resharpening domain of the knife.*
Key words: *Profile, relieving, lateral positioning surface.*

1. INTRODUCTION

Profiling knives used for toothing conical gears with curved teeth is difficult if the knives are not properly executed; after a small number of resharpenings profile aberrations appear, aberrations that negatively influence the flanks of the processed teeth. The relieving technology utilized by the firms that manufacture machine-tools and tools for toothing conical gears with curved teeth is unavailable. In our country attempts of profiling toothing knives through relieving were abandoned because after a small number of resharpenings the knife's profile does not continue to correspond to the imposed conditions (Grămescu et al., 1993). The large number of toothing machines existent in the country entails the identification of more economically viable and precise methods of relieving knives that equip the gear cutting ends. Assembling rectification devices on the cross-side rest of the lathe is difficult because of the large number of knives on the technological head and the confined space between them. The abrasive wheel, while processing a knife of the technological head, touches the adjacent knives therefore special heads should be used, implicitly increases the production costs (Pantea, 2004). The numeric research from the present paper uses an algorithm resulted from the geometric modeling of the technology of rectifying the profile of knives used for toothing conical gears with curved teeth through the particularization of the conical gear research method of spatial gearing (fig. 1).

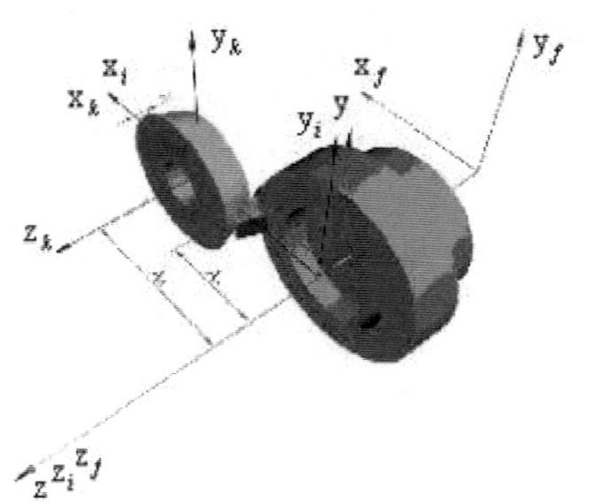

Fig. 1 The axis system.

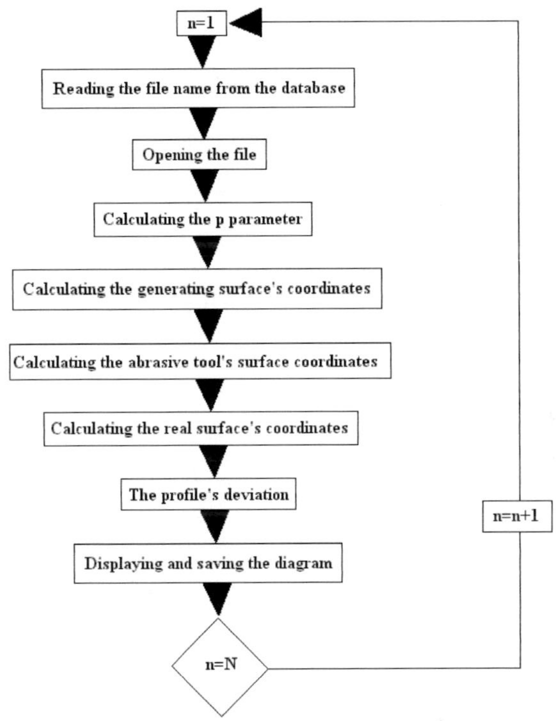

Fig. 2 The logical diagram of the program

If in the case of processing gears with curved teeth arranged in a circular arc there is the possibility of rectifying the gear's flanks, in the case of processing conical gears with teeth arranged in an epicycloid arc, there is none (Ștețiu, (1994).

The purpose of this numeric research is to determine the deviations of the rectified positioning surface of the knives used for toothing conical gears with curved teeth (Litvin, 1994).

The calculus programs are applied for the exterior toothing knives that have a straight chipping edge. The 5 mm module knives that equip the toothing machines Hardac-Gleason type 6'' were chosen. They are situated at the middle of the dimension range and are the most frequent utilized.

The finish rectification is accomplished through rectification with an abrasive conical wheel on the axes of the Niles type worm cutter-rectifying machine or on the thread rectifying machines (Pantea, 2003).

In order to attain the data presentation functions of the chipping edge's coordinates, of the abrasive tool's coordinates, of the deviations of the chipping edge from a line that passes through its extreme points and in order to display the diagram containing these deviations, the author has drawn up a program in the Visual C++ programming environment. The logical diagram of the data processing and diagram plotting program is shown in figure 2.

The program successively reads the names of the files and executes the calculation of each particular data set separately.

The program calculates using the algorithm:

- -the independent parameter p dependent on the knife's module;
- the generating surface's coordinates;
- the coordinates of the abrasive tool's surface;
- the real positioning surface's coordinates;
- the deviations of the chipping edge from a line that passes through its extreme points;
- the maximum deviation Δ.

2. NUMERICAL STUDY

The parameters employed while executing the pattern are:

- geometric elements of the knife:
- back angle α=19°
- angle of departure γ=20°
- knife module m=5 mm
- -elements of the tool assembly of the first/second order:
- exterior radius of the technological head r1 = 76.581 mm
- medium radius of the abrasive body r2 = 30 mm
- propeller pitch p = 100 mm
- pattern structure:
- number of points on the profile: 7
- number of evaluation sections: 7

Calculus program for the technology of toothing knives using the spatial gearing method (figure 3).

```
m_fR1 = 0.0f;
m_fR2 = 0.0f;
..................................................
a5 = m_fA * (m_fA2 * tan(RadianDe(m_iAlfa)) * sin(RadianDe(v0)) - cos(RadianDe(v0)) * tan(RadianDe(m_fGamma2))) + (m_fR1 - 1.104) * cos(RadianDe(v0));
b5 = -m_fA * m_fA2 * tan(RadianDe(m_iAlfa)) * cos(RadianDe(v0));
c5 = -m_fA2 * tan(RadianDe(m_iAlfa)) * cos(RadianDe(v0)) * (m_fR1 - 1.104);
```

The interface of the calculus program for the technology of toothing knives using the spatial gearing method is presented in fig.3.

The simulation results were obtained by taking into consideration the following parameters:

- number of sections of the real positioning surface:6
- number of points on the profile:7
- maximum permissible tolerance: 0.08 mm

The data obtained was processed and verified in MathCAD in order to plot the diagrams.

The deviations of the chipping edges in table 1.

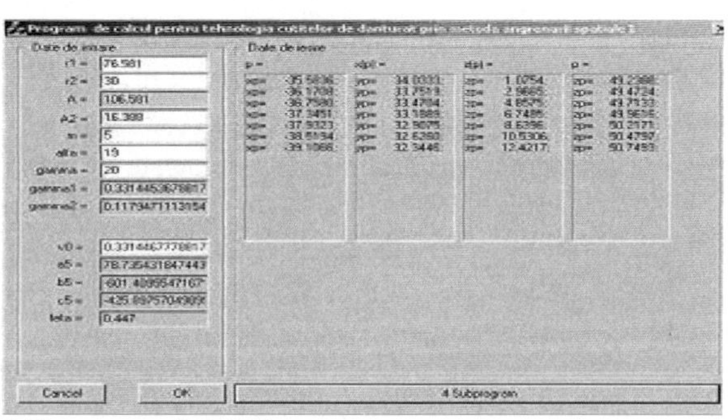

Fig. 3 The program's interface

The evaluated section	The deviation of the profile on rectification with a conical abrasive tool δ [mm]	The maximum deviation Δi [mm] Δi = max δ – min δ	The total deviation Δ [mm] Δi = maxΔmaxΔi
0	0.0047	-0.014	-0.014
1	-0.0021		
2	0.00447		
3	0.0012		
4	-0.014		
5	0.0038		

Tab 1. The deviations of the chipping edge

3. CONCLUSIONS

The "Calculus program for the technology of toothing knives using the spatial gearing method" program was drawn up. It allows the numeric research of the geometry of the positioning surface of knives used for toothing conical gears with curved teeth.

The advance relieving through rectification is a continuous process. It eliminates the negative dynamic phenomena that appear in the case of cam relieving because of the alternative axial movement.

Performing a study case relative to the application of the mathematical model presented, with reference to the I tool – II tool assembly tool on the Niles type worm cutter-rectifying machine that reproduces, on an experimental scale, the real processes of relieving for the case of rectification with a conical abrasive tool.

The numeric results obtained from the program for the toothing knives with teeth arranged in a circular arc emphasize the following:

- -the deviations of the chipping edge are set inside the permissible tolerance;
- -the working diameter of the toothing head does not suffer modification after the resharpening of the knives in the resistance limits of the knife (2/3 of the thickness).

The numeric results show that the profile obtained through the helical relieving with advance is situated within deviations under 0.05 μm (about 60% of the maximum permitted value)

4. REFERENCES

Grămescu, Tr. ş.a. (1993). *Toothing technologies. Guide project book.* Ed.Universitas, ISBN 5-362-01009-3, Chişinău.

Litvin, F.L. (1994). *Gear Geometry and Applied Theory.* University of Illinois, Chicago. PTR Prentice Hall, Englewood Cliffs, New Jersey 0763. ISBN-13-211095-4.

Pantea, I. (2003). *Numeric research regarding the tehnology of cutting tools with a cylindrical abrasive disc.* Annals of the Oradea University, pg. 215-218.

Pantea, I. (2004). *Contributions regarding the technology of the tools used for bent teeth bevel gears teething.* Doctorate Thesis, University of Oradea.

Ştețiu, G. ş.a. (1994). *Practice and theory of cutting tools, vol. I, II, III.* Editura Universității Sibiu, ISBN 973-95604-3-1, Sibiu

Annals of DAAAM for 2008 & Proceedings of the 19th International DAAAM Symposium, ISSN 1726-9679
Editor B. Katalinic, Published by DAAAM International, Vienna, Austria 2008
Make Harmony Between Technology and Nature, and Your Mind will Fly Free as a Bird

SCREEN ELEMENT SHAPE "R73" MUTATION WITH METHOD OF GROWTH FUNCTION PARAMETERIZATION

PAP, K[laudio]; VUJIC ZILJAK, J[ana] & ZILJAK, I[vana]

Abstract: *The paper gives research details on creating new shape of screening cell definition that have not even existed to day.Therefore they have not been researched nor applied yet. The need for new screening systems is evident when applying in security graphic product production, in large format designs, and when creating new graphic designs in general. The research goal is to stimulate the application of new ways of designing and reproduction, to research boundary areas that would not be applied solely in special security printing but in designing mass graphic products in general. Different in comparison to former research work where one screening shape was treated in the overall reproduction, this paper presents screening that applies an indefinite number of screening models and screening shapes in the same page. The «R73» mutating screen element shape (developed by the paper's author) is shown with a newly developed method of screen element shape growth function parameterization and stochastic functionality implementation. The research methods for testing growth function boundary conditions are 2D and 3D mathematical modelling. The mutation parameter has been designed in such a way that by its application there is altering of the screen element shape from symmetric shape to asymmetric shape. Due to this transformation of shape we have named them mutants.*

Key words: *Screening, Mutation, Function, Parameterization, Security*

1. INTRODUCTION

A screen is defined by conceptually laying a uniform rectangular grid of halftone cells over the device pixel array (Adobe Systems I., 1999). As the use of PostScript developed, there came a natural demand for the inclusion of halftones in the pages, first monochrome, around 1988, and then colour, around 1989/90, which brought a number of added difficulties to both rip and system designer (Brett, 2000). All kinds of methods have been used in the past to create this illusion of tone, including woodcuts, etchings, and engravings on copper and steel (Green, 1999). New halftone screening shapes have been created by us that alter their basic shape as their parameter alters. Due to this transformation of shape we have named them mutants. These shapes have a high dot gain in printing, so they are not to be recommended for high screen ruling. Therefore we recommend them for the area visible to the human eye. Such screen systems are applied in jumbo posters and in graphic work with exceptionally strong messages. Due to the fact that these designs are subject to pseudo-random generating where the initiator is known (congruent method seed), they are ideal for individualizing neutral surfaces in documents. The security is based on secrecy of parameters in the congruent generator algorithm and in parameters with which the screen shape is distorted.

It has been established during our research work on mutating screen element limitations with parameters that are found within the screen element mathematical definition that there is no general rule for free screening element altering. Mathematical modeling helps in setting the boundaries for permissible screen element shape shifting, although errors may occur here as well. The blow to the screening cell border can be set correctly only on basis of testing programs for carrying out graphic samples' screening.

2. SCREEN ELEMENT R73 AND ITS DEFINITION

The new complex stochastically modified screen mutant being the subject of this paper is given with mathematic equation for z growth function (1) .

$$z = 1 - Abs\,[\boldsymbol{Kory}\,x\,y + Abs(x) - Sin(\sqrt{x^4 + 4y^4})] \qquad (1)$$

Screen element growth function z is the function with the help of which the screening cell is built in the screening process for a set gray level and screen ruling (Foley et al., 1997). The 'kory' correction parameter has been incorporated into the mathematical formula program sequence with the help of which experiments are made as to generating the initial form mutating sequence.

We are using 2D and 3D mathematical models developed by ourselves (Žiljak at al., 2003) for experimental display of results during the development of this complex growth function. Model 2D (Fig. 1) shows the appearance of the screen element for 16 discrete gray levels for one and the same growth function. If the halftone screen that we are developing rapidly changes its shape through the grayscale, we can increase the number of discrete gray levels in this 2D simulation model in the course of experimenting. Changing of the shape throughout the grayscale should not be mixed up with halftone screen altering due to our mutation parameter. These are completely different issues.

The 2D model is not sufficient for determining the allowed 'kory' parameter values. This is the reason for using the 3D model (Fig. 2) for screen element growth. With this model we can research much better the growth function boundary conditions. The allowed values for 'kory' parameter within the [0, 1] interval have been determined on basis of experimental research of shapes with the 2D model and boundary conditions control in the 3D model.

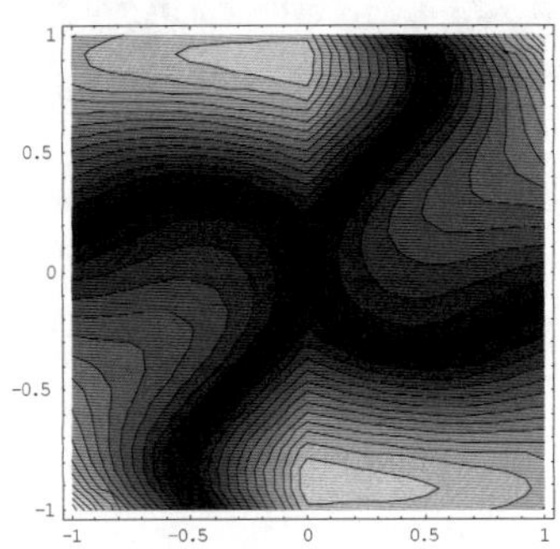

Fig. 1. The 2D model of screen mutant R73 with kory=1

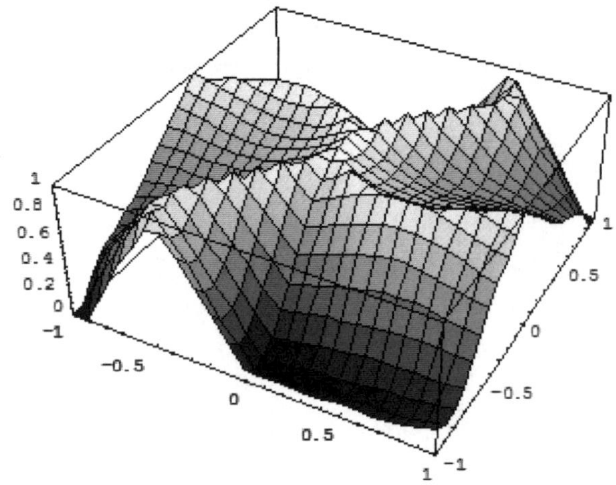

Fig. 2. The 3D model of screen mutant R73 with kory=1

3. PAREMETER OF MUTATION AND RESULT

The 'kory' parameter introduced here is included in the continuous function with values from 0 to 1 in the execution program. Figure 3 shows the mutant with the kory parameter that equals to =0.5, and figure 4 shows the mutant with the 'kory' parameter equalling to 0.

The mutation parameter has been designed in such a way that by its application there is altering of the screen element shape from symmetric shape (kory=0) to asymmetric shape (kory=1). Thus the 'kory' parameter is at the same time the shape deformation parameter for the screen shape so that the name mutant suits it even better.

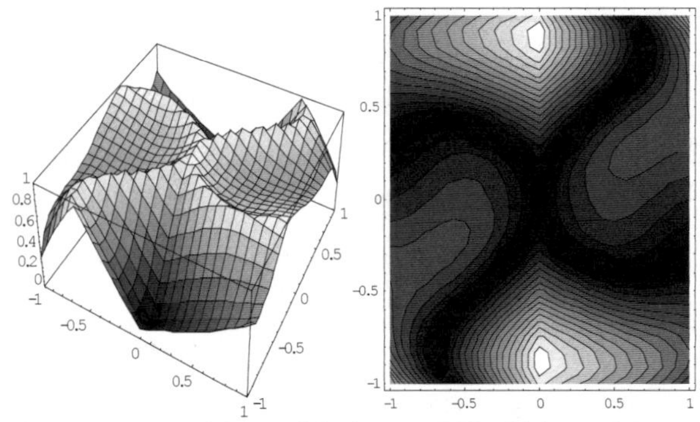

Fig. 3. The 3D and 2D model of screen R73 with kory=0.5

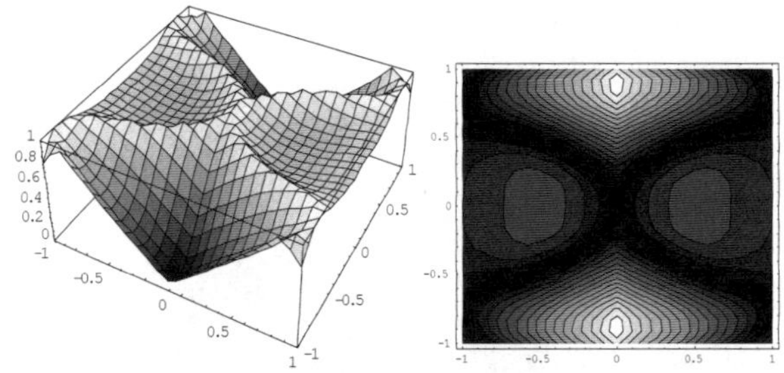

Fig. 4. The 3D and 2D model of screen R73 with kory=0

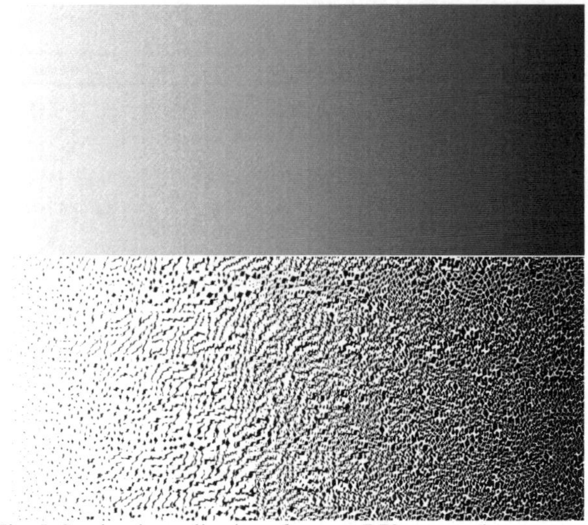

Fig. 5. Stochastic application of mutant R73 screen element

Figure 5 shows stochastic application of mutant R73 screen element on linear gradient from 0 to 88% grayscale. Parameter *Kory,* screen ruling and screen angle is stochasticly changed during screening phase.

4. CONCLUSION

There is no knowledge and experience on parameter screening. The original PostScript definition for pixel use allows pixel deformation, but the screen element itself does not change. Parameters screen element are a novelty in screening technology. The reason for studying them is the need for various different solutions in future stochastic screen element application used for special security design. A need for this type of application lies in the area of reproduction of graphics with a fi xed number of prints, reproduction of postal stamps, securities and other documents printed digitally in a single copy; an integrated passport photograph, badge or pass. This results are very excited for us because we can produce many experimental mutants in the future for different security applications.

5. REFERENCES

Adobe Systems I., (1999). *PostScript Language Reference, third edition,* Adobe Systems Incorporated, pp. 482, ISBN 0-201-37922-8,

Brett, G. (2000), *Digital Prepress Technologies*, Pira International Ltd, pp. 122, ISBN 1-85802-261-4

Foley, K.; van Dam; Feiner & Hughes (1997). *Computer Graphics Principles and Practice,* pp. 569, Addison-Wesley, ISBN 0-321-21056-5

Green, P. (1999), *Understanding Digital Color,* GATF 1999, pp. 224, ISBN 1-85802-450-1

Žiljak, J.; Vančina V.; Agić, D.;Žiljak, I. & Pap K. S. (2003). New screening elements in multi-colour printing for scecial purposes, *Advances in printing Science and Technolog,* pp. 261, Acta Graphica Publishers, ISBN: 953-9676-8-0

Annals of DAAAM for 2008 & Proceedings of the 19th International DAAAM Symposium, ISSN 1726-9679
Editor B. Katalinic, Published by DAAAM International, Vienna, Austria 2008
Make Harmony Between Technology and Nature, and Your Mind will Fly Free as a Bird

RESEARCH REGARDING SIMULATION AND OPTIMIZATION TRAJECTORY FOR A NACA PROFILE SURFACE

PARASCHIV, M[arius] D[aniel]; ISPAS, C[onstantin]; IVAN, I[oana] C[armen] & TILINA, D[ana] I[uliana]

Abstract: *This paper present a trajectory optimization for a NACA profile surface machining in CATIA V5R17 in Machining – Advanced Machining module, and correction of errors using a specialized software, trajectory correction and optimization for this type of surface.*
Key words: *trajectory optimization, simulation, machining, CAD, CAM.*

1. INTRODUCTION

This paper containes: simulation, corection of trajectory and generation of NC file for machining using CATIA V5R17.

By using certain commands for trajectory generation, it is possible to virtual machining a surface wich can have different degrees of difficulty.

This file will be imported into NCSIMUL software, used for simulation with respect of the real machining conditions for a milling machine tools.

2. GENERATION, SIMULATION AND ANALYS OF TRAJECTORY IN CATIA V5R17

CATIA V5R17 allows CAM simulation for different sceneries of machining. This can be used to generate and to optimize machining trajectories for different surface.

The « NC Manufacturing » software will generate a correct NC program for machining.

This software presents the following advantage:

- flexible administration of the programs, based on a simple and intuitive interface;
- operate with tools and tools library;
- can operate with machine tools with two up to five CNC axes.

Design of the blade is presented in Fig. 1. The surface of this piece is based on a NACA profile. The workpiece for the blade has following dimensions: 70x30x106 mm.

Due to the complex profile, it is difficult to machine this surface, but with the help of CATIA V5R17, we were able to machine it. We use Isoparametric machining, operation which can be found in module Machining – Advanced Machining.

The surplus material was removed in two cutting process. The first it was used a roughness machining in order to remove a lot of thickness from the half finished block, and the second was a finishing cutting process in order to obtain the necessary surface quality (Vlase, 1993).

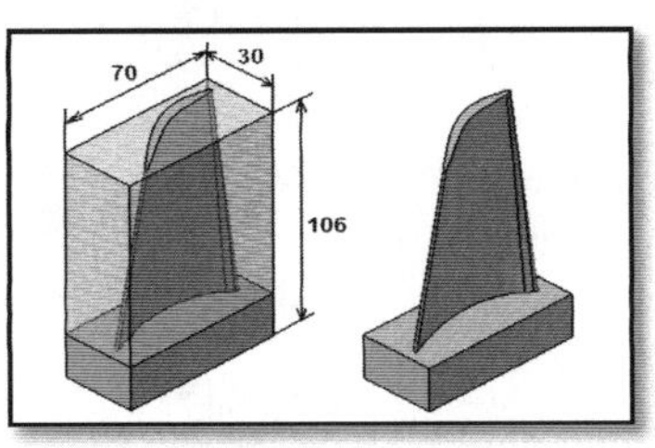

Fig. 1. Blade shape and workpiece.

Fig. 2. Principal settings of operation Isoparametric machining

After choosing the necessary operation, we have to specify some settings, for each operation. If the operation and details are different, we have five main categories (Fig. 2.).

In the first category we have to choose the trajectory settings such as: machining type, radial set-up and the inclination of tool axis. The second category is the machining pattern. Here we select the points for machining on the workpiece: part surface and corner used for machininig. In the third category we establish tool used for machining. In the fourth category, we set-up the necessary speed for the machine and tool. Here we choose also the approach and retract speeds, also, the spindle speeds. And in the last category are the finely set-up parameters the approach, retract clearance, linking retract and linking approach (Bernard, 2003).

After choosing the optimal cutting process, it was choose the parameters of machining: Machining, Radial and Tool Axis.

For machining strategy it was used a tolerance of 0,5 mm, because, the first operation is the roughing, and is no need for a very precise tolerance. In the second window, it was possible to choose the scallop height. This value represents errors which appear between real and theoretical surface of the blade. After that, it was set up the depth of cut. Number of steps represented the maximum number of crossing made by the tool in order to obtain the surface (Fig. 3.). It is not necessary to execute all the steps imposed. The program will execute the optimal number of steps (only the necessary steps). Also it was set up the start and the end extension.

In the third window it is the possibility of defining the trajectory of machining. It is very important because CATIA considers the machine table fixed. On the real machine tool used, the tool is able to move only in 3 translation axis on X, Y, and Z, and the machine table can move by rotation on two axis: A (around x direction) and C (around z direction).

After that, it was chosen the surface and the starting and the ending point of machining. We choose to have 2 mm of material, which represents the necessary surplus for the last operation, finishing (Ispas, 2008).

The cutting tools used, were:

-for roughing: a mill with 6 teeth and Ø28 mm diameter;

-for finishing: a ball mill carbide with 2 teeth and Ø20 mm diameter.

The value of machining speeds for approach and retract, and the tool speed were set up also.

In the last step it was chose the type of approaches and retract for the process of machining.

After that the trajectory of machining was checked, for roughing and the finishing.

After the cutting tool trajectory was computed, some visualization can be made. The second operation used to machine this blade is the finishing operation.

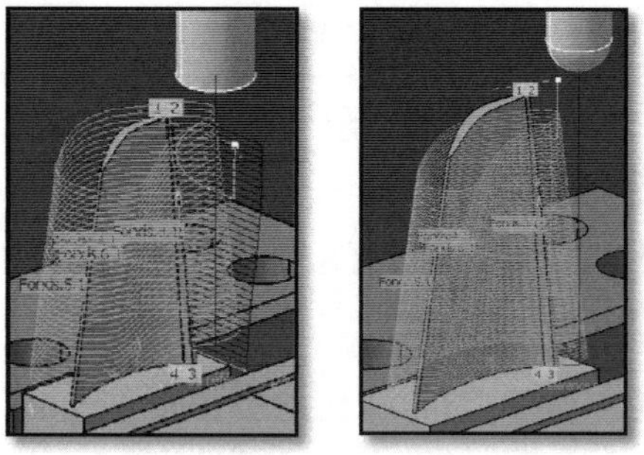

a b

Fig. 3. Trajectory generation for: a. roughing operation and b. finishing operation.

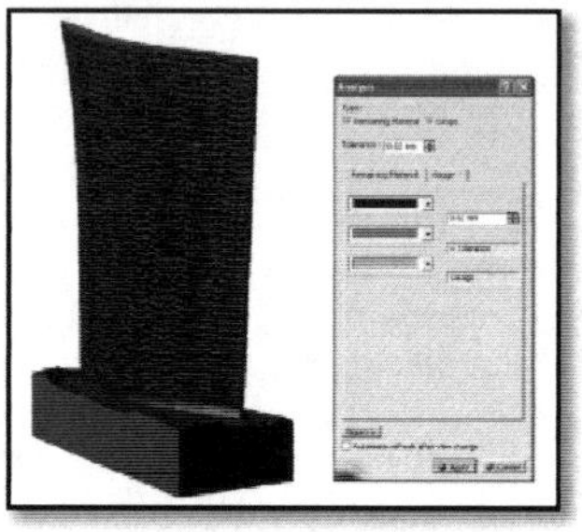

Fig. 4. Analys of dimensional and form deviation.

This operation is has almost the same set-up parameters as the precedent, but, the differences between them are the roughness obtain after finishing operation. After finishing operation exist the possibility to measure machining piece, in CATIA, by comparising with initial model, to identify form and dimensional deviation. To obtain a very good surface, we have to make different modification at the parameter described above.

3. GENERATION, SIMULATION AND ANALYS OF TRAJECTORY IN CATIA V5R17

After the simulation, an APT file was generated. The type of Catia output file was .apt file, because of high compatibility with other machining simulation software. The .cnc output file from Catia could contain errors because of Catia postprocessors and a limited simulation of the cutting process (according only with tool and pieces without machine tool)

So, to generate the program of machining it was used Generate NC Output in Batch Mode command, and the generated file, it is type .apt.

Another important aspect is the number of lines for the .apt file. For the machining with the tolerance of 0.02mm, the program has 4687 lines. This file was inserted in to another software for simulation in according with the machine tool structure and kinematics. This software has the possibility to create all the elements which take part in the machining process Fig. 5:

It has to be mentioned that it is absolutely necessary, a 3D model of the part in order to check the errors of machining and the resulting tolerance after machining.

In order to complete this simulation, it is necessary to make the following steps:

-uploading the necessary elements for machining (machine tool, tool, the fixed vice, the mobile vice, the workpiece and part).

-positioning of the parts on the machine table.

-uploading the APT file generated by CATIA.

-making necessary adjustments for simulation.

-setting up the collision detection.

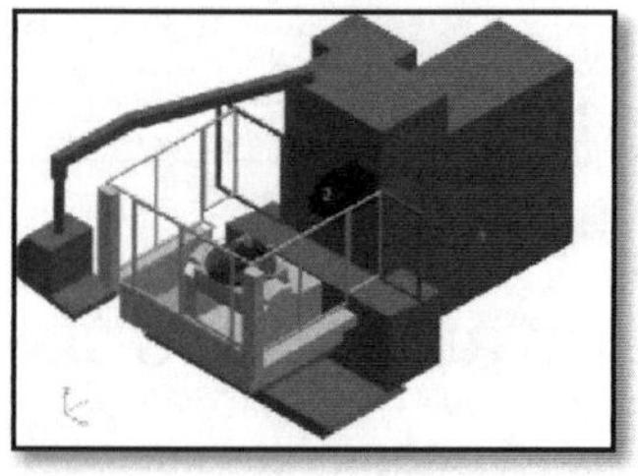

Fig. 5. Element wich determine the simulation.

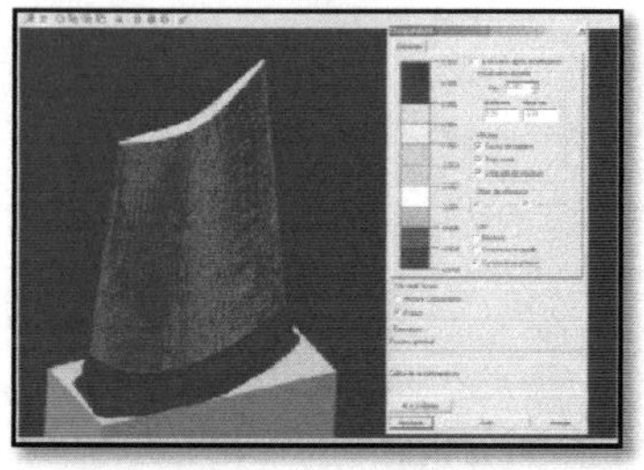

Fig. 6. Differences between machining simulated piece and the part imported from CATIA.

The element used for simulation are : 1-machine tool; 2-cutting tool; 3-fixed vice; 4-mobile vice; 5- workpiece; 6-piece

Is possible to make different set-up starting with conditions of contact between different elements up to the desired tolerances, and collision detection between different element wich compose the simulation.

And finnaly after the simulations is finished, we can analyze the difference between the obtained surface and the imposed surface Fig. 6.

4. CONCLUSION

By simulation it could be obtain the optimum machining parameters in order to assure the surface quality. Also, is recomendate to verify twice a machining operation in order to avoid errors or collision wich can often can appear in real machining process.

We used two simulation software, becose in the fisrt software we are able to generate the file for machininig operation, and in the secondary software we verify the correct solution offer by Catia simulation in order with machine cinematics.

In the conclusion, this paper offer the posibility of development of machining simulation in case of complex surface.

5. REFERENCES

Bernard, A. (2003), *Fabrication assistée par ordinateur* (Computer Aided Manufacturing), Traité IC2, série Productique, February 2003

Ispas, C., Paraschiv, M., Laboureau, L. & Anania, D. (2008). Research concerning the numerical errors correction for a naca profile surface on a 5 axis machine tool. *Academic Journal of Manufacturing Engineering,* Vol., No.6, (May, 2008) pp. 75-80, Timisoara, ISSN 1583-7904

Vlase, A., Sturzu, A., Stancescu, C. & Neagu, C. (1993). *Tehnologii de prelucrare pe masini de frezat* (Machining technologies for milling operations), Editura Tehnica, ISBN 973-31-0243-1, Bucharest

http://www.windmission.dk/workshop/BasicBladeDesign/bladedesign.html *Accessed:* 2008-02-27

http://www.ncsimul.tm.fr *Accessed:* 2007-11-12

Annals of DAAAM for 2008 & Proceedings of the 19th International DAAAM Symposium, ISSN 1726-9679
Editor B. Katalinic, Published by DAAAM International, Vienna, Austria 2008
Make Harmony Between Technology and Nature, and Your Mind will Fly Free as a Bird

DISTANCE LEARNING OF PROGRAMMING LANGUAGE PASCAL

PARDANJAC, M[arjana] & RADOSAV, D[ragica]

***Summary**: Research witch is a part of this work has empirical-theoretical character. It aims to provide answers about quality and possibilities of implementation of distance learning educational model of programming language PASCAL. Problem of this research is complex by its nature, and reflects in many processes and side effects witch take place in society, science, school and their interaction.*
Gathered research results should provide information if distance learning of programming language PASCAL has larger effect to user success versus the traditional way of learning.
***Key words:** Informatics, education, distance learning, high school, university education*

1. INTRODUCTION

The High school Curriculum is designed to enable the students to learn the application of techniques for solving particular problems (bloc of diagrams), elementary commands of programming language PASCAL, as well as writing of simple programs. Students who enroll in a major of Professor of Informatics and Graduate Engineer of Informatics at the Technical College "Mihajlo Pupin", show very weak knowledge in regards to solving problems, breaking down the problems into sub-problems, and the recognition of commands of the programming language PASCAL.

Thereupon such result, a question presents itself, what is the cause of that and how such issues can be avoided or reduced? Which methods and instruments can be used to influence and stimulate the changes of these conditions? Do the instructors have the required knowledge of informatics? Are they trained and qualified enough for the application of the information technology? Are the schools equipped with the adequate number of computers? Do the students have the optimal conditions for the study and practice? How the given requirements can be optimized? Can the motivation of the instructors and students be influenced upon and how?

One of the possible solutions is to present the courses through the distance learning (in further text DL). Consequently, universities started to offer more flexible courses, including technology-based distance education the use of which has grown steadily over the past decade, becoming widely used by universities around the world (Jara & Mellar, 2007). For that to be possible it is necessary for the instructors to be trained on how to present the study material in a different manner as well as to make the appropriate adjustments toward the change of the roll of the instructor in the educational/teaching process.

2. DISTANCE LEARNING

The improvement of teaching through the distance learning is focused on two different aspects of the educational process: the capability of the instructors to acknowledge how the distance learning can improve their presentation of the material and their ability to recognize the improvement in the success of their students as a positive outcome of the application of the DL. Many instructors believe that the DL can help in the improvement of studying and motivation of students, offering them diverse styles of studying, wide world of information, and the application of new types of learning.

In foreign countries many professors use DL to assist in the educational process. Some of them believe that this could be a possible solution toward the reformation of the educational system, through the stimulation of the new methods of teaching, and even the change of the role of the instructor in the teaching/learning process. Although some instructors prefer to determine their curriculum and activities as the course evolves, in this environment adult learners found it important that they be given a list of readings, assignments, and expectations at the beginning of the course (O'Lawrence 2007).

It fits the needs of a specific type of student, one who wants to choose when and where the learning will take place, rather than engaging in the ritual of spending 1 to 5 years sitting in class for 50 minutes twice a week for 30 weeks each year (Poteet, 1986). Students need guidance and support before and during their studies. This is especially true in the digital age when students need to be skilled computer users and able to deal with different types of media in order to study effectively. (Seppo, 2006). Also, it is very important to note the emphasis on influencing learner motivation, not controlling it. Systematic motivational design has been shown to be effective when used properly and within the boundaries of modifiable influences on learners (Keller & Suzuki, 2004).

3. EXPERIENCES IN DL OF THE PROGRAMMING LANGUAGE PASCAL

Significant numbers of foreign universities offer in their programs the possibility of attending a course in the programming language PASCAL. Among those are the following universities and schools: The University of York, Bond University, Oxford University Computing Services and University of Guyana.

All of the above universities offer one or several of the following links to the websites that contain the study material related to the programming language PASCAL.

http://library.thinkquest.org/11127/ – after introduction, the distance student can choose the Learning Center where they can select the level of courses and link to the given lessons. Lessons are classified in three levels: Class B – beginner level, Class I – intermediate class, Class A – advanced class.

http://www.geocities.com/hotdogcom/ptutor/paslist.html - The recommended method of study is to print the text for one or two chapters, download the example programs, and study the material by loading the example programs in the compiler's editor for viewing.

http://www.geocities.com/SiliconValley/Horizon/5444/pstart.htm – At the introduction page there can be found the main menu, from where the options can be selected introducing the user with the programming language PASCAL, and providing the basic and simple examples, followed by the writing of comments.

4. RESEARCH

4.1 Research results

Following the completion of the experiment, participants of the control and the experimental group, are put through the objective type of the final test of knowledge.

The difference between the experimental and the control group is statistically important, with certainty of 95%. That is why the null hypothesis is rejected, as it presumes that the differences are not important, while the main hypothesis is supported: that model of study of the programming language PASCAL based on the distance learning has a statistical importance in regards to the quality of the educational process.

Based on the presented information, it can be concluded that the suggested model of study of the programming language PASCAL based on the distance learning has an important effect contributing to the increasing level of knowledge of students in the topic of programming of the programming language PASCAL.

4.2 Results of the survey in regards to the opinion of the students about the DL after the research

Following the completion of the experiment and the final test of knowledge, the participants were surveyed again in regards to their opinion about the distance learning.

From the given results to the question "Would you like to use some form of system of distance learning (DL)?" it can be concluded that a significant number of users had changed their mind and now wanted to use some form of DL (number of participants that wanted to use the DL system prior to the experiment was at 20%, after the experiment that number has jumped to 72.82%.).

In response to question, in which form and in which way should the distance learning be applied following answers were obtained in Table 1.

In response to question in which way DL improves the results of learning (Table 2.), the participants were given the option to circle multiple answers, which they found to be of the greater importance. Following answers were obtained in Table 2 ((A – Provides more information, B – Provides only needed information, C – Not a loss of time, D – Work according to one's own pace, E – Learning is more interesting).

4.3 Conclusions of the Research

Based on the obtained and presented results the following can be concluded: *that the model of learning of the programming language PASCAL based on the distance learning has the same important influence/effect on the quality of the teaching process* – which supports the main hypothesis.

Answers	DL by itself	Addition (to the traditional system)	Should not be used
High school	23,08%	57,69%	19,23%
University	68,63%	21,57%	9,80%
Altogether	45,63%	39,81%	14,56%

Tab 1 Results of the answers to the question

Answers	A	B	C	D	E
High school	57,69%	42,31%	23,08%	57,69%	73,08%
University	45,10%	54,90%	29,41%	78,43%	82,35%
Altogether	51,46%	48,54%	26,21%	67,96%	77,67%

Tab 2 Results to the question "In which way DL improves the results of learning?"

The Experiment supports the following sub hypothesis as well:

- *Learning of the programming language PASCAL based on the model of distance learning is more economical in regards to time (minimized time needed for study) compared to the traditional classes;*
- *Learning of the programming language PASCAL based on the model of distance learning provides at the same time greater level and longevity of the obtained knowledge by the student;*
- *Learning of the programming language PASCAL based on the model of distance learning increases the motivation of students in the learning process;*
- *Learning of the programming language PASCAL based on the model of distance learning allows for the individualization of studying.*

5. CONCLUSION

Considering the role and the goal of the research, this research belongs to the group of verification type of researches. The results that were obtained with this research, have only assured and verified the fact that today's educational process, or rather the teaching cannot be imagined without the use of the computer.

From all the reasons mentioned above, this work presents intent to develop a model for the studying of the programming language PASCAL through the distance learning, with the goal of technological advancement of the lectures while at the same time allowing for the individualization of the educational process.

6. ACKNOWLEDGMENTS

This paper is part of research activities within preparation of national programmed for technology development Republic of Serbia for the project "On-line distance education system for continual education of ICT professionals". This project is a continuation of activities of implementing educational software and information systems in the field of distance learning developed as results of finalized projects where coauthors of this paper were participating: (1) "System of distance learning based on educational multimedia software" (2002/04) – financed by Ministry of science Republic of Serbia, (2) "E-learning: Data structures and algorithms" and "E – learning DMBS" – DL projects No.DL2002-03 – financed by WUS Austria – Distant Learning Program.

7. REFERENCES

Jara, M. & Mellar, H. (2007), Exploring the mechanisms for assuring quality of e-learning courses in UK higher education institutions, *Available from*: http://www.eurodl.org/materials/contrib/2007/Jara_Mellar.htm: 2008-07-15

Keller, J. M. & Suzuki, K. (2004), Learner motivation and E-learning design: a multinationals validated process, *Journal of Educational Media*, Vol. 29, No. 3, October 2004, page 229 –239, ISSN 1358-1651

O'Lawrence, H. (2007) An Overview of the Influences of Distance Learning on Adult Learners, *Journal of Education and Human Development*, Vol. 1, No. 1, 2007, page ISSN 1934-7200

Poteet, G. H. (1986), *We Succeeded through Home Study*, National Home Study Council, ISBN 0937471003, Washington

Sirkemaa, S. (2006), Information technology in developing a meta-learning environment, *Available from*: www.eurodl.org/materials/contrib/2006/Seppo_Sirkemaa.htm, *Accessed on*: 2008-07-15

Annals of DAAAM for 2008 & Proceedings of the 19th International DAAAM Symposium, ISSN 1726-9679
Editor B. Katalinic, Published by DAAAM International, Vienna, Austria 2008
Make Harmony Between Technology and Nature, and Your Mind will Fly Free as a Bird

IMPLEMENTATION OF AR BASED AUTOMOBILE ASSEMBLY SYSTEM

PARK, H[ong] S[eok]; CHOI, H[ung] W[on] & PARK, J[in] W[oo]

Abstract: *The AR technology as a new man-machine interface introduces a noteworthy perspective for a new manufacturing system design. In this paper, the architecture of AR browser is introduced and the optimal environment parameters for practical application of AR system are determined. Moreover, many methods are proposed in order to solve the problems suggested from the initial practical test. Based on these tests and results, operation programs for automobile assembly systems are generated using the AR system.*

Key words: *AR(Augmented reality) system, AR browser, Robust superimposition, Automobile assembly system*

1. INTRODUCTION

To survive in the turbulent and competitive market, automobile manufacturers must design and implement the manufacturing systems which respond rapidly to the market demands. Currently, the implementation of a manufacturing system for assembling new automobile body is done based on the prototype manufactured after the design of a new body model by changing a conventional system or newly designing. For this purpose, a system planner stops a current system and evaluates the manufacturability of a new model in that system as well as executes the work for changing it by putting a new model instead of the conventional automobile body to the current system. This traditional method leads to the critical problems such as reduction of system productivity and delay of the introduction of new product into market.

For these reasons, VR(Virtual Reality) based digital manufacturing technologies are used to analyze the static and dynamic behavior of system at all stages for configuration of manufacturing system(Günter et al., 2005; Park et al., 2006). However, most of methods and softwares for digital manufacturing require the perfect 3D models of the whole system in virtual environment to represent the target system and surrounding environment. That means this modeling work requires a lot of expenses and effort.

To solve these problems, the AR technology is applied in this paper. The AR technology can remarkably reduce modeling work because it uses the real manufacturing environment to design and plan manufacturing systems. It is also possible to obtain the information in real time from manufacturing field by superimposition of virtual objects to real scene(Wolfgang, 2004). Consequently, the application of AR technology is expected as an epochal method for implementing manufacturing systems in an efficient and user friendly manner.

For the development of an AR based assembly system, the architecture of AR browser is firstly introduced. Then, the optimal values of environment parameter for robust superimposition between virtual objects and real scene are proposed through lots of experiments. Through an initial test, the problems are derived for the application of AR system. For an implement of AR system through solving them, the methods for appropriate allocation of camera, avoidance of collision between virtual objects and marker structural configuration for recognizing in multi directions in consideration of robot behaviors are proposed. With these results and experiments, an operation program for cockpit assembly system is generated.

2. AUGMENTED REALITY SYSTEM

2.1 Architecture of Augmented reality browser

AR browser as the core component for the interaction and visualization in an augmented environment consists of 4 modules for digital image processing and interfaces for communicating between devices(Fig. 1). The modules of AR browser are classified to tracking module to track positions of target markers, rendering module for generation and handling of virtual objects and measurement module to calculate the distance between objects. Then, all cameras used in research are calibrated to compensate for radial and decentering distortion using calibration module.

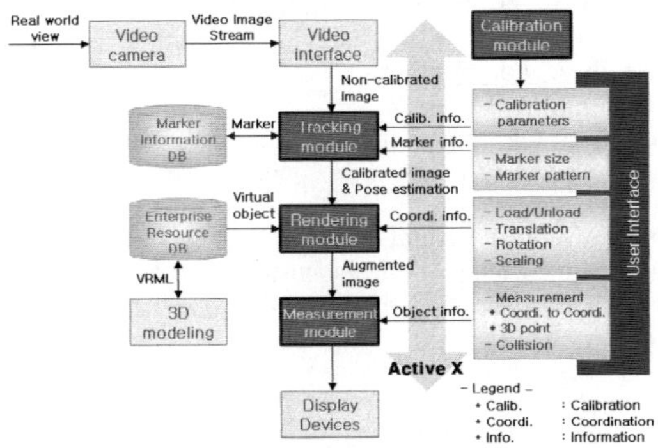

Fig. 1. S/W layout of AR browser

2.2 Environment variables for robust superimposition

While AR system internally executes the image processes for positioning virtual objects using a square marker, the chattering of virtual objects occures according to the environmental conditions of camera. In order to apply AR system to practice, it is important to find and maintain the condition to remove this chatter. To determine the appropriate conditions, a lot of tests were carried out with the environmental variables as below(Park et al., 2008);

- The angle between the marker and camera
- The distance between camera and marker
- The angle among the marker, camera and light
- The intensity of light
- The size of marker

Using the results of tests, the AR system is configured under appropriate conditions for robust superimposition between virtual objects and real scene.

3. PROBLEMS AND SOLUTIONS FOR IMPLEMENTATION OF AR BASED ASSEMBLY SYSTEM

To apply AR system to practice, there are some problems to be solved besides the environmental conditions for robust superimposition. For grasping them, a test bed of the H company's assembly station was realized with twelve times reduction model. Through lots of experiments with it, the problems occurring at implementing a AR based assembly system were examined(Fig. 2). To solve these problems, lots of methods are proposed as shown Fig. 3.

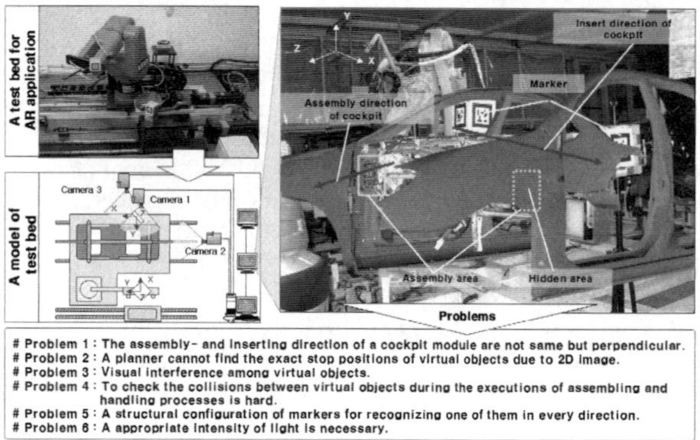

Fig. 2. Test-bed and problems for application of AR system

·Solutions for problem 1 & 2

- Because the image acquired from one camera deliver only 2D information, it is necessary to get more than two image information obtained from different places.
- In case of the cockpit module assembly, two cameras are required because this process should be done in two different directions such as assembly and insert direction.

·Solution for problem 3

- For viewing an assembly area, the virtual object was partitioned into several parts. Through that, the divided parts can be shown or hidden for supporting the execution of the assembly process.
- To increase the adjusting degree of two objects, the alignment line was modeled to guide the correct approach of one object to its counter part.

·Solution for problem 4

- The clipping plane was generated to express the only information of the sectional surface for showing the possibility of collision.

·Solutions for problem 5 & 6

- The marker system mounted on the robot gripper was manufactured in the structure of a box type, because the system should be seen in every direction due to free movement of robot.
- The light intensity should be adjusted to an environmental condition. In case of entering a robot gripper into the inside of automobile body, the light condition is so dark that a camera cannot catch the marker system. This problem was solved by setting up a light inside of the box type marker system.

With these methods, the system planner can generate operation program of assembly, i.e. a robot program.

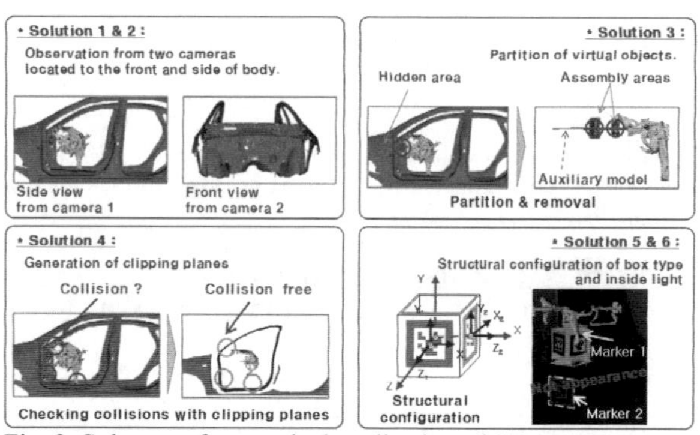

Fig. 3. Solutuons for practical application of AR system

4. GENERATION OF OPERATION PROGRAMS FOR ASSEMBLY SYSTEM OF NEW MODEL

For generating an operation program for assembling a new model of automobile in a conventional assembly station by using AR technology, the implemented test bed is presented. Base on the test bed, the camera and the marker systems were optimally allocated according to the results of the previous studies for applying AR technology to practical. With this test bed, the operation of the assembly station for a new model was planned and programmed by AR technology. The completeness of the generated operation program was proven by applying it to the conventional assembly station. Fig. 4 shows the pratical test for generating operation program for automobile chassis assembly. The boundary conditions such as the position tolerance range within 10 mm were fulfilled.

Fig. 4. Generation of opration program for chassis assembly by using AR technology

5. CONCLUSION

AR in manufacturing system applications faces several technical challenges. This paper outlined some of the major working areas and highlighted new approaches and developed solutions. These systematical solving approaches overcome the problems against the implement of an AR-based assembly system. With the experimental results of the developed system through the test bed, it is clear that AR system can help the system planer to analyze the actual assembly process in the early stages of the manufacturing process planning and save valuable costs and time for testing the real process with the prototype.

6. ACKNOWLEDGEMENT

This research was supported by the MKE(Ministry of Knowledge Economy), Korea, under the ITRC(Information Technology Research Center) support program supervised by the IITA (Institute of Information Technology Assessment).

7. REFERENCES

Günter, W. & Emmerich, S. (2005). Digital planning validation in automobile industry. *Computers in Industry,* Vol. 56, pp. 393-405.

Park, H. S. & Lee, G. B. (2007). Development of digital laser welding system for automobile side panels. *Int. Jr. of Automotive Technology,* Vol. 8, No. 1, 83-91.

Park, H. S.; Choi, H. W. & Park, J. W. (2008). Implementation of automobile cockpit module assembly system using augmented reality technology. *Transactions of NAMRI/SME,* Vol. 36, No. 1, 493-500.

Wolfgang, F. (2004). *ARVIKA: Augmented Reality für Entwicklung, Produktion und Service,* Publicis Corporate Publishing.

Annals of DAAAM for 2008 & Proceedings of the 19th International DAAAM Symposium, ISSN 1726-9679
Editor B. Katalinic, Published by DAAAM International, Vienna, Austria 2008
Make Harmony Between Technology and Nature, and Your Mind will Fly Free as a Bird

COLLABORATIVE DESIGN AND MANUFACTURING OF INDUSTRIAL PRODUCTS

PARPALA, L[idia] F[lorentina] & POPA, C[icerone] L[aurentiu]

Abstract: *In this paper is presented a case study on a virtual manufacturing of an industrial product. This case study is based on the membership to a European research network, VRL-KCiP trough a national network, INPRO. The major departments of a virtual enterprise were located to partners from these two networks. For example, the manufacturing department was located at the University of Ljubljana, Advanced Technologies Laboratory.*

Key words: *virtual, collaborative, manufacturing, design.*

1. INTRODUCTION

We have started this case study from the idea of customer demands propagation in industrial products manufacturing.

In this paper is presented a case study based on the membership to a European research network, VRL-KCiP (Figure 2) trough a national network, INPRO (Figure 3).

The connexion with the European network is University POLITEHNICA of Timişoara, which is the coordinator of the national network INPRO and is also representing Romania in the European Network.

2. CASE STUDY

Part of the VRL-KCiP network were taken into account for the manufacturing of the piece chosen for this case study: the first member of the VRL-KCiP network is University of Ljubljana, Advanced Technologies Laboratory and the second one is the 3S Laboratory, which is a mixed research institution having members from three institutions (INPG (Institut National Polytechnique de Grenoble), CNRS (Centre National de Recherche Scientifique) and UJF (Université Joseph Fourier)).

After an economical study of the technologies existing at the two partners, we have decided to localize the manufacturing department at the University of Ljubljana, Advanced Technologies Laboratory. The economical department is a team with representatives from University of Ljubljana, Advanced Technologies Laboratory and University Politehnica of Bucharest, Machines and Production Systems Department, PREMINV Laboratory.

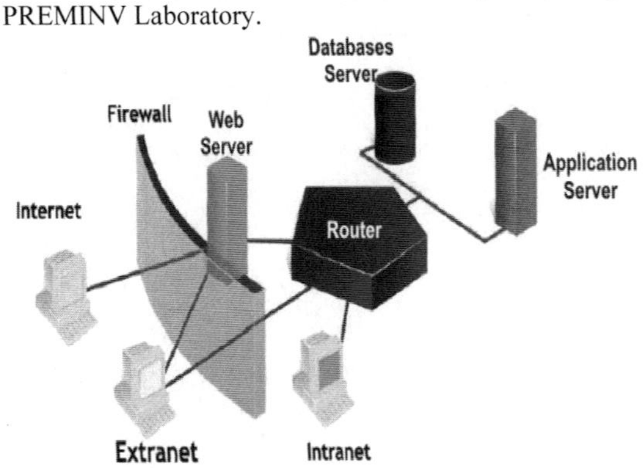

Fig. 1. INTERNET/INTRANET/EXTRANET platform architecture (Guran, 2005)

Fig. 2.VRL-KCiP Network

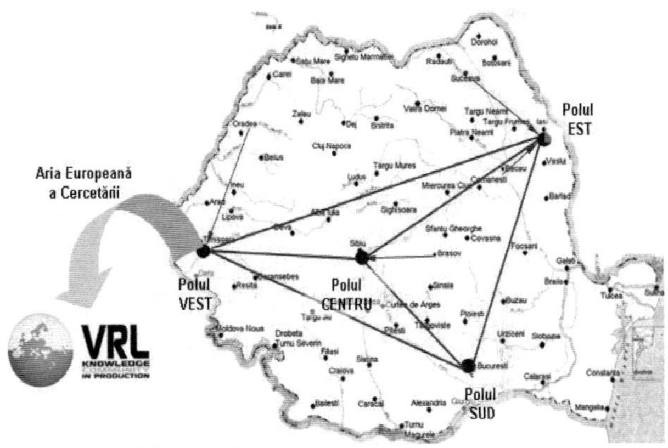

Fig. 3. INPRO Network

At the PREMINV laboratory were also located the design department. (Drăgoi,Guran&Coteţ, 2002)

Figure 1 is a schematic representation of the architecture of an INTERNET/INTRANET/EXTRANET platform for customer demands propagation which was the starting point for this case study. (Parpală & Popa, 2008)

Figure 4 represents the INTERNET/ INTRANET/ EXTRANET platform for collaborative design and manufacturing that we have developed in order to improve collaboration between the virtual enterprise departments. One can see that the customer may have an active role in product manufacturing trough the CRM application.

Figure 5 presents the way that customer demands are propagated in a virtual environment. There are also present the main virtual enterprise applications (CRM, SCM, ERP).

The design of the product was made using CAD/CAE software (CATIA V5). The chosen manufacturing technology was abrasive water jet cutting, so the software used was OMAX Make (delivered with the OMAX AWJC Machine).

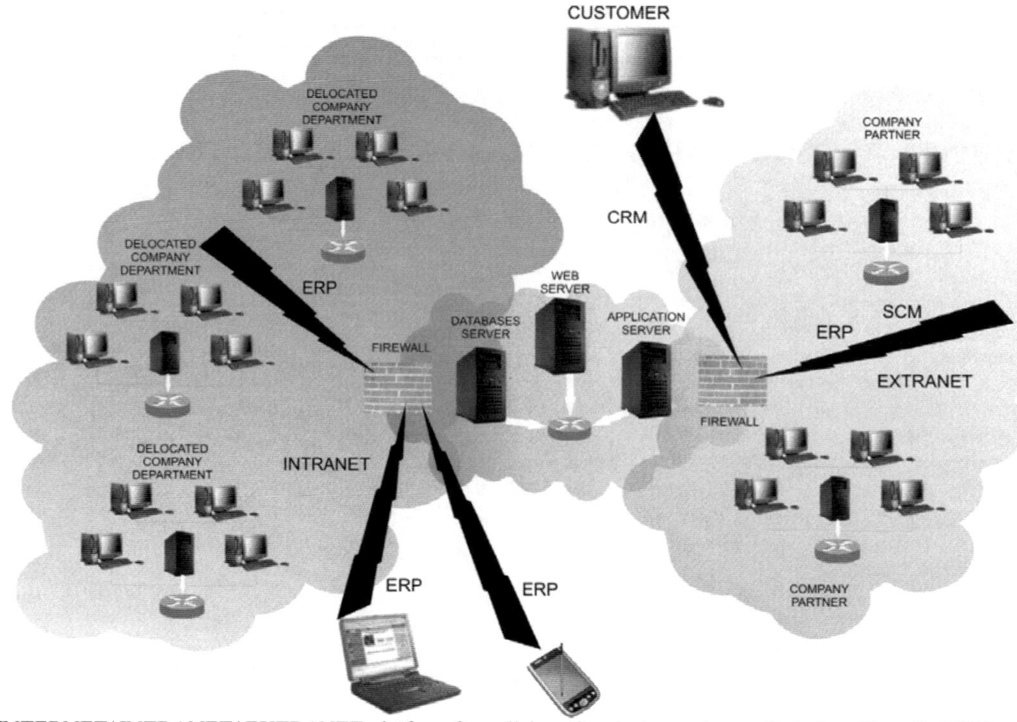

Fig. 4. INTERNET/ INTRANET/ EXTRANET platform for collaborative design and manufacturing (Parpală, 2007)

3. CONCLUSION

As opposed to the classical enterprise model, inside a Virtual Enterprise first we perceive the customers requirements and then the product design and manufacturing is started, keeping a continuous relation with the customer. (Camarinha-Mathos, 2002)

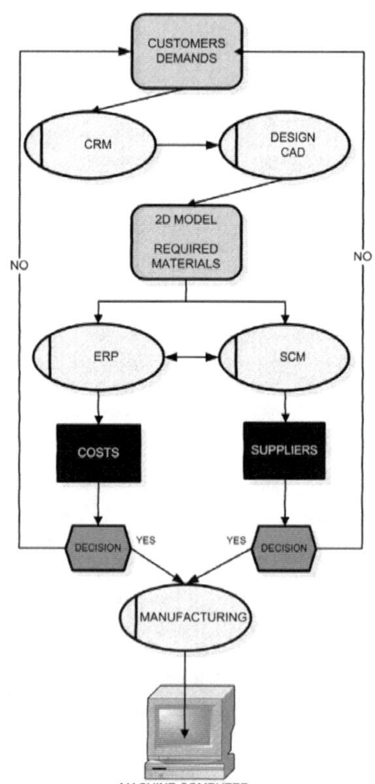

Fig. 5. Customer demands propagation in virtual environment (Parpală&Popa, 2008)

The technologies, techniques and methods used in Virtual Enterprises bring to cost, design and manufacturing time reduction, distribution time reduction and higher customers satisfaction.

The 3D modelling of the product was made in the design department. The following operation was the finite elements analysis. That helped as to make a topological optimization of the product in order to reduce its mass.

The design department delivers information also for the manufacturing department (2D model of the product) and for the economical department (bill of material).

4. REFERENCES

Camarinha-Mathos L.M. (2002). Virtual Organization in Manufacturing: Trends and Challenges in *Proceedings of International Conference on Flexible Automation and Intelligent Manufacturing (FAIM)*, Germania, 2002

Drăgoi G., Guran M., Coteţ C.E. (2002) –The PREMINV platform for training in management and engineering for virtual enterprise, *Colaborative Business Ecosystems and Virtual Enterprises*, Portugalia, 2002

Guran M. (2005). The Architecture of Virtual Industrial Enterprise - A New Paradigm in Distributed Data *Processing and communications in Proceedings of the 15-th International Conference on Control Systems and Computer Science*, Bucureşti, Mai, 2005.

Parpală L.F. (2007), *Contribuţii privind platforma INTERNET/INTRANET/EXTRANET pentru propagarea cerinţelor clienţilor în realizarea produselor industriale, Contributions on INTERNET/INTRANET/EXTRANET platform for customer demands propagation in industrial products realisation,* PhD Thesis, Bucureşti, 2007

Parpală L.F., Popa C.L. (2008), INTERNET/INTRANET/EXTRANET platform for customer demands propagation in products manufacturing, *ACADEMIC JOURNAL OF MANUFACTURING ENGINEERING, VOL. 6, NO.1/2008*

Annals of DAAAM for 2008 & Proceedings of the 19th International DAAAM Symposium, ISSN 1726-9679
Editor B. Katalinic, Published by DAAAM International, Vienna, Austria 2008
Make Harmony Between Technology and Nature, and Your Mind will Fly Free as a Bird

CAD – FEM IMPORT OF A MACHINE TOOLS FEED DRIVE SYSTEM MODEL

PARPALA, R[adu] C[onstantin]

Abstract: *Today manufacturers can no longer afford to consume time and money building and testing real prototypes of the large models, instead they use computers technologies Unfortunate at the moment there are specialized software packages so engineers are forced to use more than a software suit during the design phase. Mainly there are two categories of softwares that engineers use: CAD/CAM softwares and FEA softwares, so as a result there are many problems that appear when importing models from these two different software applications. The following steps are detailed in the paper: - Defeaturing of the 3D model and the setup of the FEA model for a machine tools feed drive system*

Key words: *import, preprocesing, mesh, design*

1. INTRODUCTION

Today manufacturers can no longer afford to consume time and money building and testing real prototypes of the large models, instead they use virtual prototypes.

In the virtual prototype approach it is possible to simulate the kinematic, static and dynamic behavior of the machine tools including all aspects of real life exploitation. Using different software packets one is able to simulate even the cutting process. This complex design concept was enabled by the use of high performance computer technology (Altintas et al., 2005).

Unfortunate at the moment there are specialized software packages so engineers are forced to use more than a software suit during the design phase. Mainly there are two categories of softwares that engineers use: CAD/CAM softwares and FEA softwares, so as a result there are many problems that appear when importing models from these two different software applications.

In this paper it's presented as a case study the model of a machine tools feed drive system. (Weck 2001)

2. COMPUTER AIDED DESIGN

The 3D model of the feed drive must be designed in order to be accepted as input by various software suits that will be used for further analyses.

Because it's hard to find an integrated environment in which to perform all analyses engineers must take into consideration software's compatibilities (Jeong Hoon Ko et al., 2003).

The 3D model of the feed drive was designed by using the CATIA V5 CAD software mainly because of his good integration with the ANSYS software which was used for statically and dynamical FEM analyses. All 3D part where fully parameterized in order to optimize needed parameters in FEM analyzes.

In order to generate all the contacts between surfaces it's necessary to correctly design the 3D assembly, it's also very usefully to check all the clashes and clearances within the CAD environment. Using the information provided by the CAD software we can set the correct tolerances for the automatic contact generation.

3. PREPROCESING THE MODEL

Because the model was not design in ANSYS Workbench native CAD system we must first check if all the model's features are imported before we proceed to further analyses.

Getting the geometry Into Design Modeler, simulation or advanced meshing is now much easier than in almost any other FEM software. The first thing to know about formats is that there are to classes readers and plug-ins. Readers simply translate the CAD format into workbench's internal format. A plug-in actually uses software from the CAD vendor and open up the geometry in the native format and gives the workbench all the information it needs in native format. Often the geometry provided by readers is referred as dumb and plug-in geometry as smart because this kind of geometry can be transferred back to the CAD files. It's a good practice to import all the parts into the Design Modeler before proceeding to simulation

After a successful import of the model it's very easy to automatically generate all the contacts between components, setting up the correct parameters for contact creation.

For a proper meshed structure it is necessary to repair the geometry before any further operations.

The main purpose of the CAD repair tool is to detect and close gaps between neighboring surfaces. Typically this procedure involve two steps:

- **Build topology** – build curves and points which will help to diagnose the model for geometrical problems. If the curves are within a geometric tolerance, they are merged together as one. The curves are then displayed in a specific color to illustrate their connectivity in the surface data (Table 1), which can be used to determine any gaps or holes in the geometry (**Fig. 2**).
- **Repair any gaps or hole in the topology** For any two faces (surfaces) that meet at a common edge (curve), there is typically a finite distance between the two edges. By default, a curve is associated with all the edges of each face.

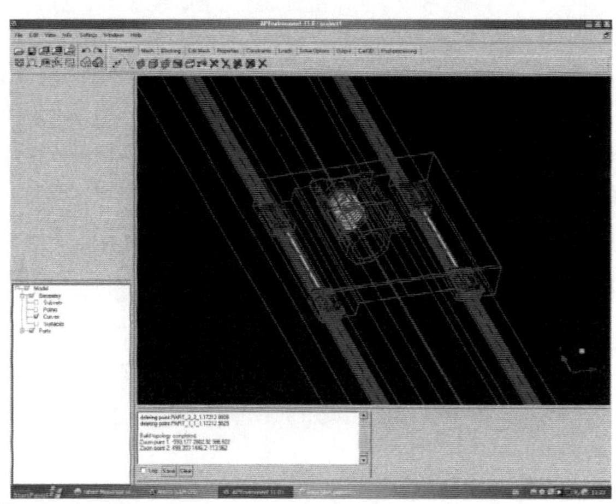

Fig. 1. Mesh quality

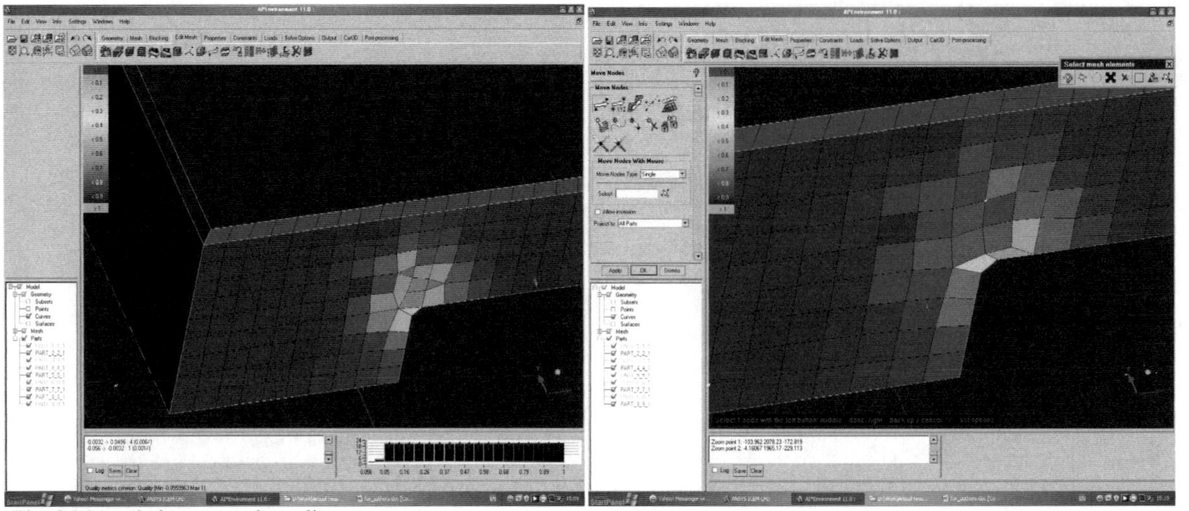

Fig. 2 Manual change mesh quality

This topology of the two surfaces would indicate a gap in the model. Typically, ANSYS meshers can handle this if the gap is smaller than the proposed element size on the surfaces or curves. Therefore, you would set a tolerance larger than the gap if you are using a large element size.

A tolerance smaller than the gap would create yellow curves which could be fixed. The recommended tolerance is approximately 1/10th the size of the average mesh size

The quality of the mesh needs to be checked before applying loads and constraints. It gives an idea of how close the mesh is to an ideal mesh. Mesh

Quality can be measured by the various criteria. For Hexa dominant meshes the quality is calculated as the determinant. The Determinant, more properly defined as the relative determinant, is the ratio of the smallest determinant of the Jacobian matrix divided by the largest determinant of the Jacobian matrix, where each determinant is computed at each node of the element. The Determinant can be found for all linear hexahedral, quadrahedral, and pyramidal elements.

Color	Semnification
Yellow	Single or free edge curves
Red	Double edge curves
Blue	Multiple edge curves
Green	Unattached curves

Tab 1 Build topology colors

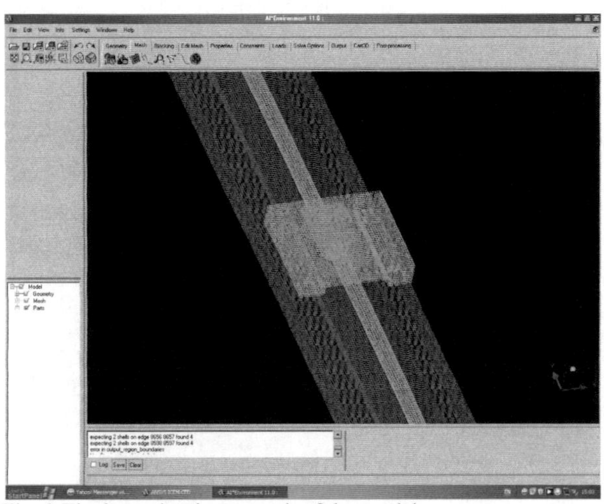

Fig. 3 The hexa dominant mesh of the model

A determinant value of 1 would indicate a perfectly regular mesh element, 0 would indicate an element degenerate in one or more edges, and negative values would indicate inverted elements.

To automatically improve the quality of the mesh elements different smoothing algorithms are available depending on which mesh type is loaded. Mesh can be smoothed with respect to a particular quality criterion and with a specified number of iterations to achieve a given quality level. A mesh containing tetras, pyramids, prisms and triangular and quad surface elements can be smoothed.

The mesh can be also improved manually by moving nodes. An example is shown in **Figure 2** some nodes where moved as a result we have in the left side of the figure a smother mesh this can be checked by visualizing element colors. Those colors change as the form of the elements changes.

4. CONCLUSION

Today the main problem in checking structures consists in importing and preprocessing the CAD model. It is well known that the geometry of the model can dramatically change FEM results (Zaeh & Oertli 2004)

Initiated mainly by the automotive and aircraft industry, the development of modern software tools for integrated simulation of products has been enchanted lately.

Unfortunate at the moment there are no integrated software platform for the virtual design and analyze of the machines tools.

5. REFERENCES

Altintas, Y.; Brecher C.; Weck M. & Witt S. (2005). Virtual machine tool, Annals of the CIRP , 54/2: 651 – 669.

Groß, H. & Hamann, J. (2001) Electrical Feed Drives in Automation. Basics, Computation, Dimensioning. Siemens, Publicis MCD Corporate Publishing, Erlangen and Munich

Jeong Hoon Ko; Won Soo Yun; Dong-Woo Cho; Kyung Gee Ahn & Seung Hyun Yun (2003). Development of a virtual Machine tool – Part 2, International Journal of the KSPE, vol. 4, No.3

Zaeh, M. & Oertli, Th. (2004) Finite Element Modeling of Ball Screw Feed Drive Systems, Annals of the CIRP, 53/1: 289-292.

Weck, M., (2001) Werkzeugmaschinen. Mechatronishce Systeme, Vorschube, Prozeßdiagnose (Mechatronical system, feed motion, Process diagnosis). Springer-Verlag, Berlin, Heidelberg, ISBN 3-540-67614

Annals of DAAAM for 2008 & Proceedings of the 19th International DAAAM Symposium, ISSN 1726-9679
Editor B. Katalinic, Published by DAAAM International, Vienna, Austria 2008
Make Harmony Between Technology and Nature, and Your Mind will Fly Free as a Bird

THE IMPLEMENTATION OF ABC METHODOLOGY IN CONCURRENT ENGINEERING

PARV, L[uminita] & LUPULESCU, N[ouras]

***Abstract**: Activity Based Costing (ABC) has received significant attention as means of identifying cost drivers that impact the facility output. This paper presents our approach to model product cost using this methodology. We address the support of information sharing and activity synchronization within a team and between teams. A database system for product cost modelling is presented in this paper. The developed system has the capability of selecting a material, as well as machining processes and parameters based on a set of design and production parameters and of estimating the product cost throughout the entire product development cycle. Generally speaking, building site industry and especially glass window heat insulated and window thermic proof can be new fields for the implementation of a concurrent engineering and of the ABC methodology for the costs determination.*

***Key words:** Activity Based Costing, heat insulator window, cost model.*

1. INTRODUCTION

The present work presents our results for the implementation of the concurrent engineering concept from the industry producing stained glass window insulated heat, an industry in full swing in our country. It has been started from the traditional organizing of such kind of factory and was implemented, after the shown model, a software responding to the methodology requirements. A special attention is paid to the estimation of the costs. In the attempt of defining a model of the costs, the researches proved that, especially the ABC methodology fit the best to this field (Laan, 2003). The implementation of this methodology allows the polishing of the cost model and the creation of a better distribution base of the indirect expenses on the product.

ABC provides more than a product pricing system. It improves the visibility of costs and shows how costs are passed down to products by activities. As a result, ABC is a valuable information tool that provides management with an unrivalled insight into the workings of the manufacturing system (Zhang & Liang, 2007).

2. DATA MODELLING DUE TO ABC METHODOLOGY

The cost estimation according ABC method assumes covering the main following stages (Thyssen et al., 2006):

- sharing the costs in direct and indirect costs;
- indirect costs separation on activities connected to survey centres;
- the indentification of cost determinants for each type of activity,
- activities realignment , based on cost determinants, in realignment centres and the unitary cost estimation/determinant;
- the total cost estimation on the product through totalizing the direct cost with those of the cost determinants deriving from the realignment centres necessary for the achievement of the product.

In conformity with the carried aut survey upon the costs situation on activities, we made the cost model shown in the following formula.

$$C = F + \sum_i b_i * x_i + \sum_{ij} D_i * B_{ij} \qquad (1)$$

where:
C - product cost;
b_i *- direct costs on the product unit (including human resource costs, direct costs with materials, etc);*
x_i *- number of type i of cost drivers;*
D_i *- cost driver rate of i activity;*
B_{ij} *- number of cost drivers on i activity used by j resource.*

The equation 1 will be:

$$C = F + VD + AD \qquad (2)$$

where:
VD - total direct variable cost;
AD - total indirect activity cost.

The previous shown model can be shown as a matrix in equation 2, so:

$$VD = \begin{bmatrix} b_1x_1 & b_2x_2 & b_3x_3 \ldots & \ldots & b_mx_m \end{bmatrix}^T \qquad (3)$$

where:
b_i *- direct costs on the product unit (including human resource costs, direct costs with materials etc.) ;*
x_i *– number of type i of cost drivers.*

The total indirect cost of activities will be presented in equations 4:

$$AD = \left[A_{ij} \Big/ \sum_{i=1}^{m} A_{ij} \right]_{mxn} \bullet \left[R_{jk} \Big/ \sum_{j=1}^{n} R_{jk} \right]_{nxs} \bullet p_k \qquad (4)$$

where:
A_{ij} *- number of activity j consumed of product;*
R_{jk} *- resources consumed by j activity;*
p_k *- price of type k .*

The adding of parameter time t to this formula leads to the determination of the costs drivers.

For the software implementation, the data were patterned using a relational model.

The stages for achieving this model are:

- achievement of the initial data dictionary;
- drawing up of conceptual data model;
- drawing up of logical model (relational data).

The following of these stages led to the achievement of relational model which respects the rules concerning the normal form FN3 defined by Codd.

COST TYPES	CONTENT
Operational costs	Costs generated by customers Costs generated by suppliers
Costs with the logistics	Typewriting of document costs Shipping costs and test Loading and unloading costs Stocking costs Packing costs
Typewriting costs	Costs related to the manufacturing process
Costs for raw materials	Costs for the raw materials and materials

Tab 1. Identification costs

For defining the specific production data of glass window heat insulated, we made an implementation project of ABC methodology, which referres to the defining of the costs determinants (table 1).

3. SOFTWARE IMPLEMENTATION

The model due to the cost estimation using ABC methodology will be integrated in the software destinated to the implementation of concurrent engineering. The architecture of the software system is shown in figure 1.

It is necessary to solve the problem between the programmed settings used. So, the relational model was implemented using ACCESS 2007, setting with multiple possibilities of sharing the data.

Of course, respecting the concurrent engineering principles of simultaneously solving of the emerged tasks during the process of making the system for the costs estimation are available even from the beginning of the defining of customres'requests.

The other modes by which the integration was done are:

- the mode which husbands the customer's requests, achieved in the same setting access, with a submenu choice of the manufacturing process due to the requested glass window heat insulated;
- CAD module, made in Solidworks which aims to the forms defining and of the glass window heat insulated dimensions;
- CAM module, represented by the X-cut application which allows the generation of the cutting codes.

This method was choosen from the specific conditions of the choosen product.

The detailed presentation of ABC methodology and of the modelling data due to this was made knowing that from the point of view of the management performances, the usefulness of ABC methodes can be revealed by the following aspects:

- the responsabilisation of those who decide regarding the level costs, especially for the supporting activities;
- a better handling of the costs for rolling aut of new products or of those referring to the changes of the existing products.

The advantages are reducing costs of manufacture preparation, either directly or indirectly by means of significantly, reducing the time necessary for the conceiving of the technological process, improving the quality of design activities by means of accelerating the informational flow and the elimination of routine.

The final goal is to determine the product cost in the early stage of development (Berling, 2008).

This field was choosen considering the extent of the building field. Modeling is forward looking, while ABC has a historical perspective (Ozbayrak et al., 2003). The result of the researches is integrated software. This software can be an assisting system of the decision, the principle of the costs optimization being the decisional one.

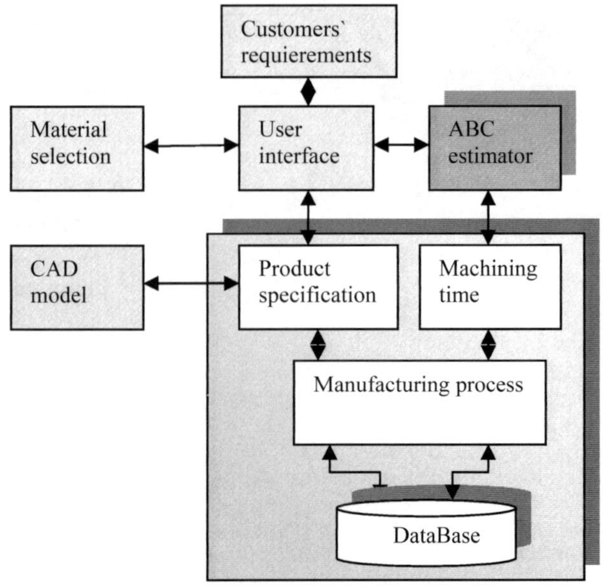

Fig. 1. The system`s architecture

4. CONCLUSION

This paper presents our own achievements regarding the modelling data due to the costs determination through ABC methodology. The model allows the integration of data regarding the costs in a software system for concurrent engineering. This software can be an assisting system of the decision, the principle of the costs optimization being the decisional one.

We address support of information sharing and activity synchronization within a team and between teams. We present an original methodology for ABC product model.

The paper has had detailed analysis and research of cost accounting and charging strategy by ABC in the glass window heat insulated industry. It also confirms different charging strategies according to the different customers, and then provides scientific reference for enterprise or department.

Once identified, corrective action can be taken. Combining the modeling tools and the ABC activity provides a method of identifying the areas that will have the largest impact in the future (Berling, 2008).

As a next trend development of this application they will try to extand this for other products with a high rank of diversification and personalization.

5. REFERENCES

Berling, P. (2008). Holding cost determination: An activity-based cost approach, *International Journal of Production Economics, Volume 112,* April 2008, Pages 829-840

Laan, E. (2003). An NPV and AC analysis of a stochastic inventory system with joint manufacturing and remanufacturing, *International Journal of Production Economics*, *Volumes 81-82*, January 2003, Pages 317-331

Ozbayrak, M.; Akgun, M. & Turkerc, A.K. (2003). Activity-based cost estimation in a push/pull advanced manufacturing system, *Available from:* http://www.sciencedirect.com *Accessed:* 2008-06-19

Thyssen, J.; Israelsen, P. & Jorgensen, P. (2006). Activity-based costing as a method for assessing the economics of modularization—A case study and beyond, *Available from:* http://www.sciencedirect.com *Accessed:* 2008-06-03

Zhang, B.; Xie, X. & Liang, H (2007). Activity cost based manufacturer cost and profit optimization, *Proceedings of the Sixth International Conference on Machine Learning and Cybernetics*, Hong Kong, 19-22 August 2007.

Annals of DAAAM for 2008 & Proceedings of the 19th International DAAAM Symposium, ISSN 1726-9679
Editor B. Katalinic, Published by DAAAM International, Vienna, Austria 2008
Make Harmony Between Technology and Nature, and Your Mind will Fly Free as a Bird

A COMPARISON OF SAME SLOPE AND EXPONENTIAL SMOOTHING FORECASTING MODELS

PASIC, M[ugdim]; BIJELONJA, I[zet] & BAJRIC, H[adis]

Abstract: *In this paper the same slope forecasting model is tested on real time series and compared with corresponding performances of fifteen exponential smoothing models. Tests are conducted using time series with different components, while standard deviation is used to measure the suitability of the model. Nonlinear programming is used for the estimation of parameters of the same slope forecasting model and exponential smoothing forecasting models. Developed model shows very good forecast results when trend component dominates.*

Key words: *forecasting, model, evaluation, smoothing, exponential.*

1. INTRODUCTION

Forecasting is the art and science of predicting future events, and very important factor that influences organizational performances, since forecasting enables efficient use of scarce organizational resources (Nikolopoulos & Assimakopoulos, 2003). Good forecasts are an essential part of efficient service and production and effective planning in both the short and long run depends on a forecast of demand for the company's services or products (Heizer & Render, 2006; Gardner, 2006).

There was a significant development of forecasting models in the last decade. Many different models are developed, and many science works and published papers are dedicated to this area (Sang at al., 2003). Particular development is achieved in the area of time series models (De Gooijer & Hyndman, 2006).

Models of exponential smoothing, which belong to the group of time series models, were introduced in the middle of recent century. From that time until now, many models are developed (Gardner, 2006; Gardner, 1985; Hyndman at al,. 2002).

Research (Hyndman at al., 2002) presents twelve exponential smoothing models with two types of error estimates for each model (additive and multiplicative), which gives total of twenty four different models. Study (Gardner, 2006) suggests fifteen basic exponential smoothing models, from which twelve are already presented (Hyndman at al., 2002). Some models are with additive and some with multiplicative error estimates, and three new models are introduced with dumped multiplicative trend.

Autoregressive Conditional Heteroskedasticity (ARCH) model describes the forecast variance in terms of current observations (Engle, 2004).

Research (Pašić at al., 2007) introduces basics of the same slope forecasting model, and the model is tested on the real time series. In general, results of the research shows better performances of the same slope model than moving average, weighted moving average and simple exponential smoothing models. These models are of the same complexity level as the same slope model.

The aim of this work is to compare performances of the same slope model with fifteen different exponential smoothing models using characteristic time series. Testing results show very good performances of the same slope model in comparison with the above mentioned exponential smoothing models, despite the fact that the same slope model is one-parameter model as compared to multi-parameter exponential smoothing models. Also, further improvement of the same slope model is suggested in this paper.

Trend	Seasonality		
	None	Additive	Multiplicative
None	N-N	N-A	N-M
Additive	A-N	A-A	A-M
Damped additive	DA-N	DA-A	DA-M
Multiplicative	M-N	M-A	M-M
Damped multiplicative	DM-N	DM-A	DM-M

Tab 1. Exponential smoothing models.

2. PROBLEM APPROACH AND METHODOLOGY

Mathematical formulation of the same slope forecasting model (Pašić at al., 2007) is expressed by:

$$F_t = D_{t-1} - \beta(D_{t-2} - D_{t-1}) \quad (1)$$

where:
F_t- Expected value of time series in period *t*,
D_{t-1}- Value of time series in period *t*-1,
D_{t-2}- Value of time series in period *t*-2,
β- Trend coefficient.

Depending on the value of the trend coefficient β, the following scenarios are possible:
$\beta < 1$- Forecasts smaller change rate in the period *t* to *t*-1 than in the prior period.
$\beta = 1$- Forecasts the same change rate in the period *t* to *t*-1 than in the prior period.
$\beta > 1$- Forecasts bigger change rate in the period *t* to *t*-1 than in the prior period.

The trend coefficient ß can be optimized by minimizing the forecast errors. In this research fifteen exponential smoothing models, (Granger, 2006), are used and listed in Table 1.

Models (Granger, 2006), treat trend component in five different ways and seasonal component in three different ways, which give fifteen different exponential smoothing models.

For an estimation of parameters of the same slope model and exponential smoothing models, nonlinear mathematical programming technique is used. The goal in nonlinear programming is to minimize standard deviation of a measure of an error. Standard deviation of the model is a measure of reliability of a forecasting model and can be calculated using the following equation:

$$\sigma = \sqrt{MSE} = \sqrt{\frac{\sum_{i=1}^{n}(D_t - F_t)^2}{n}} \quad (2)$$

where MSE denotes the mean square error or the model variance, and n stands for number of periods for which the model error is measured.

Standard deviation of the model will be used for the comparison of model performances.

3. RESULTS

All models are tested on ten time series as follows (source: Statistics Institute FB&H):

- TS1 - GDP of Federation of Bosnia and Herzegovina (FB&H). Time series from 1996 to 2004.
- TS2 - Total export of FB&H. Time series from 1996 to 2005.
- TS3 - Total import of FB&H. Time series from 1996 to 2005.
- TS4 - Realized investments in FB&H. Time series from 1996 to 2005.
- TS5 - Number of foreign tourists in FB&H. Time series from January 2002 to December 2004.
- TS6 - Number of arrivals of foreign tourists in FB&H. Time series from January 2002 to December 2004.
- TS7 - Live births in FB&H. Time series from January 2001 to December 2003.
- TS8 - Number of telephone units in FB&H. Time series from January 2002 to November 2005.
- TS9 - Number of minutes of telephone calls by mobile phones in FB&H. Time series from January 2004 to December 2006.
- TS10 - Number of calls by public telephones in FB&H. Time series from January 2004 to December 2006.

Time series TS1, TS2, TS3 and TS4 are with very dominant trend component. Time series from TS5 to TS10 are with very dominant seasonality, which means that time series values, in successive periods, constantly change, having values from local minimum to local maximum. The ranks of models are listed in Table 2. In this test, the same slope model is at the ninth position. Because of seasonality and continuous value changes, the same slope model does not fit the best to this kind of time series.

The final rank of the one-parameter same slope model, in comparison to other multi-parameter exponential smoothing models shows very good performances. These parameters present variables in a nonlinear programming model. The goal of this nonlinear programming model is to minimize forecast error of the model. Complexity of exponential smoothing models is not remarkable just because of number of parameters, but because of the fact that for every component of time series there is a different particular equation. So, all of the exponential smoothing models have two or three equations, and N-N model only has just one, as the same slope model has. Also, the same slope model uses a few historical records, precisely only two, while some of exponential smoothing models require more records in order to make good forecast. In some cases more than twenty historical records have to be memorized in order to make a forecast.

4. CONCLUSION

The same slope forecasting model is a very simple and easy to use model. It is very sensitive on trend changes and uses only two historical records no matter the type of time series.

Research results show that the same slope model is very suitable when dealing with time series where trend component dominates. However, the same slope model does not perform well when treating seasonal time series in a sense it uses only two last historical values of a time series. In general, this model is more suitable than simple exponential smoothing model, the only model of the same complexity used in this research.

Future research should be focused on improving the same slope model when dealing with seasonal time series. The model should be improved in a sense to recognize and eliminate nonstandard observations in time series which could significantly influence forecasting results. The model should also have capabilities to take into account non-equidistant time series data.

In the case that n-step forecast is needed the same slope model shows significant disadvantage due to the fact it forecasts time series values using the gradient of its previous time step.

Model developed in this paper could be successfully used in forecasting only one step in the future. Also this shortcoming will be overcome by taking into account seasonal time series effect in the model. This means that the gradients from the past seasons will be used to predict the future.

Model	TS1	TS2	TS3	TS4	TS5	TS6	TS7	TS8	TS9	TS10	Rank
N-N	9	10	9	9	11	13	9	5	5	7	10
N-A	5	9	10	7	9	7	5	4	11	3	7
N-M	8	11	11	8	3	2	1	11	3	1	5
A-N	7	1	3	5	10	11	8	8	4	8	6
A-A	3	3	5	2	7	6	6	3	8	5	3
A-M	2	4	6	3	1	1	2	12	2	4	1
DA-N	6	2	2	4	5	10	7	6	6	6	4
DA-A	1	5	1	1	6	8	4	2	10	2	2
DA-M	13	6	12	13	2	4	3	13	13	10	11
M-N	14	12	13	16	16	16	15	15	16	11	15
M-A	16	15	15	14	15	15	13	14	14	13	16
M-M	15	13	14	15	14	9	11	16	15	16	14
DM-N	12	16	16	12	12	12	16	7	7	12	13
DM-A	11	8	8	11	8	5	14	1	12	14	12
DM-M	10	14	7	10	4	3	12	9	1	15	8
SS	4	7	4	6	13	14	10	10	9	9	9

Tab 2. Models ranks.

5. REFERENCES

De Gooijer, J.G. & Hyndman, R.J. (2006). 25 years of time series forecasting, *International Journal of Forecasting*, Vol. 22, No. 3, (July-Sept. 2006) pp. 443-473, ISSN 0169-2070

Engle, R. (2004). Risk and Volatility: Econometric Models and Financial Practice, *The American Economic Review,* Vol. 94, No. 3, (June, 2004) pp. 405-420, ISSN 0002-8282

Gardner, E. (2006). Exponential smoothing: The state of the art – Part II, *International Journal of Forecasting*, Vol. 22, No. 4, (Oct.-Dec., 2006) pp. 637-666, ISSN 0169-2070

Gardner, E. (1985). Exponential smoothing: The state of the art – Part I, *International Journal of Forecasting*, Vol. 4, No. 1, (January-March, 1985) pp. 1-28, ISSN 0277-6693

Hyndman, R. J.; Koehler, A. B.; Snyder, R. D. & Grose, S., (2002). A state space framework for automatic forecasting using exponential smoothing methods, *International Journal Forecasting*, Vol. 18, No. 3, (July-September 2002) pp. 439-454, ISSN 0169-2070

Heizer, J. & Render, B. (2006). *Operations Management*, Pearson Education, Inc. 0-13-185755-X, New Jersey

Nikolopoulos, K. & Assimakopoulos, V. (2003), Theta intelligent forecasting information system, *Industrial Management & Data Systems*, Vol. 103, No. 9, pp. 711-726, ISSN 0263-5577

Pašić, M., Bijelonja, I., Šunje, A. & Bajrić, H. (2007). Same Slope Forecasting Method, *Proceedings of the 18th International DAAAM Symposium*, Katalinic, B. (Ed.), pp. 547-548, ISBN 3-901509-58-5, Vienna, Austria, October 2007, DAAAM International, Vienna

Sang, C. S.; Sam, I. S., Dan, L. & Jingmiao, G. (2004). A new Insight Into Prediction Modeling System, *Journal of Integrated Design and Process Science*, Vol. 8, No. 2, (April 2004) pp. 85 – 104, ISSN:1092-0617

Annals of DAAAM for 2008 & Proceedings of the 19th International DAAAM Symposium, ISSN 1726-9679
Editor B. Katalinic, Published by DAAAM International, Vienna, Austria 2008
Make Harmony Between Technology and Nature, and Your Mind will Fly Free as a Bird

PERIODIC INVENTORY SYSTEM CONTROL DECISIONS UNDER RISK

PASIC, M[ugdim]; BIJELONJA, I[zet] & KADRIC, E[din]

Abstract: *In this paper a mathematical model for periodic spare parts inventory management is developed. This model should ensure efficient inventory management, such that a balance between contradictory inventory management goals, low inventory cost and high service level is achieved. Software module for spare parts inventory management is developed according to the proposed mathematical model. Mathematical model and software module provide that calculation of relevant inventory management parameters depend not only on total costs of inventory, but on optimal balance between total costs and service level.*
Key words: *inventory, spare, parts, control, cost*

1. INTRODUCTION

Optimal inventory control is one of the crucial business functions, since business low-cost strategy can never be achieved without good inventory management (Ballou, 2004; Wild, 2002). Inventory is one of the most expensive assets of many companies, representing a significant percentage of total invested capital. High level service and smooth operations should be met at the minimum inventory cost. These goals are mutually contradictor, and should be balanced (Heizer & Render, 2006).

The most profitable policy does not consider individual optimization of one of these goals, but they must be jointly considered and optimized in order to achieve the optimal operations results. Those responsible for inventory decision making must take into account all these facts, in order to make decisions based on relevant evaluations of any possible alternative, and of course, consequences of applications of any of it. In general, inventory management functions are contained in the Enterprise Resource Planning (ERP) system, which provides ways to analyze the demand history, make forecasting evaluation, and suggest safety stock levels.

Today ERP systems are reasonable and sophisticated tools to forecast the demand of fast moving items, but most are ill-equipped to deal with the demand of slow moving items such as spare parts (Razi & Tarn, 2003). This is due to the fact that many inventory models available today show a number of difficulties in attempting to apply them to a spare part inventory management. Spare parts inventory management has some specific characteristics: high price, irregular demand hard to forecast, long and stochastic lead times and customers (internal or external) want those parts as soon as possible (Humphrey, 1998; Fortuin, 1999).

The model developed in this paper deals with spare parts inventory control and is generally based on previously developed inventory model (Razi & Tarn, 2003). In this model a specific cost of an item is used to determine the target stock level for that item. Items are grouped based on annual demand and lead time and a common group demand distribution is generated. The authors use a fixed number of review periods, and it was emphasized, that the number of review periods is the decision that must be made by an inventory manager.

In this paper, instead of using a fixed length of review period, developed model simulates a series of length of review periods. The length of review period is automatically chosen and depends on trade off between total inventory cost and the customer serves level. In this case a review period managerial decision making does not depend on manager's personal experience, intuition or approximate estimations. The model and software module are tested and verified on a real life example. Model developed in this paper shows excellent performances.

2. MATHEMATICAL MODEL

Mathematical model is defined by two parameters: number of review periods, T, and maximum inventory level, S. Total cost, TC, is composed of four components: total review cost, total ordering cost, total holding cost, and total penalty cost. Total cost, TC, can be calculated from the following equation:

$$TC(T,S) = \\ = C_r \cdot T + C_o \cdot T + I \cdot c \cdot \left[S - \frac{\overline{x}}{2} \right] + \pi \cdot T \cdot \sum_{x>S} (x-S) \cdot p(x) \tag{1}$$

where T is number of review periods, C_r is cost per review, C_o is cost per order, I is annual holding cost expressed as a percent of an item cost, c is an item cost, π is a penalty cost per item, and $p(x)$ is probability that demand is equal to x.

Any unsatisfied demand is assumed as lost sales, which means that backordering is not allowed, because lack of a spare part will surely result in a smaller production, and thus in a smaller revenue.

Number of review periods T represents total number of reviews that will be conducted in specified time period H. Number of review periods T is function of time period H, in which transactions are done, and length of review period r. Number of reviews T, within time period H, can be calculated from the equation:

$$T = \frac{H}{r} \tag{2}$$

For example, if time period H is two years, and review period r is one month, then the number of reviews T is 24. Length of review period r is a linear function of the lead time.

Mean demand, $\overline{x}$, over time period H, represents ratio of sum of all mean demands $\overline{x}_i$ of all reviews T_i and total number of reviews T, and can be calculated using equation:

$$\overline{x} = \frac{1}{T} \sum_{i=1}^{T} \overline{x}_i \tag{3}$$

It is assumed that demand is random, stochastic and that follows Poisson distribution. Poisson distribution is a discrete distribution, determined by one parameter only, its mean. The probability, $p(x)$, that demand is equal to x, if the mean demand over all review periods is equal to $\overline{x}$, can be calculated from equation:

$$p(x) = \frac{e^{-\bar{x}} \cdot \bar{x}^x}{x!} \qquad (4)$$

Service level represents probability that quantities on hand, during the lead time, will be sufficient to satisfy expected demands. Service level, *SL*, is calculated using following equation:

$$SL = \left[1 - \sum_{x>S} p(x)\right] \cdot 100 \qquad (5)$$

Figure 1 illustrates algorithm for periodic inventory control which is composed of four blocks and one loop.

Purpose of block 1 is to enable item selection, which will be analyzed, and to collect item data, and after that to define length of time period *H*, which serves as the basis for estimation of number of review periods *T* and mean demand $\bar{x}$.

In the preprocessing phase (block 2), periods are created, and then records about demand, in particular periods T_i, are collected. For given number of review periods *T* and known demands $\bar{x}_i$, mean demand $\bar{x}$ is estimated. When mean demand $\bar{x}$ is known, it is possible to create Poisson distribution with mean $\bar{x}$, and to estimate particular probabilities of random demand *x*.

As a result of execution of the processing phase (block 3), values of maximum inventory level *S*, total inventory cost *TC* and service level *SL* are estimated, for different lengths of review periods *r*, i.e. number of review periods *T*, within time period *H*.

Using loop, blocks 2 and 3 are executed for different lengths of review period *r*, enabling us to create a set of optimal solutions of total inventory cost *TC*, for different number of review periods *T* and maximum inventory levels *S*.

The main role of the postprocessing phase (block 4) is to show all results of analysis for different lengths of review periods *r*, i.e. number of periods *T*, and to select optimal solution for inventory control of selected spare part. Optimal solution represents optimal value of length of review period *r*, and maximum inventory level *S*, for which the relationship between total inventory cost, *TC*, and service level, *SL*, is acceptable.

3. RESULTS AND INTERPRETATIONS

Testing and verification of mathematical model and software module for periodic inventory control is done using

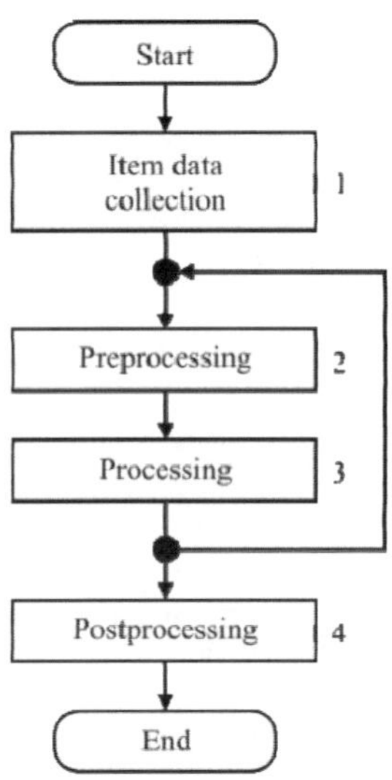

Fig. 1. Algorithm for periodic inventory control.

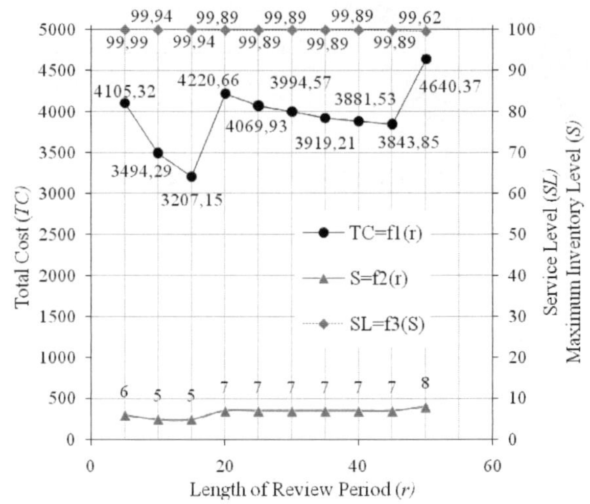

Fig. 2. Total inventory cost, maximum inventory level and service level for different lengths of review period.

demand records for a critical spare part of Sarajevo Public Transportation Company. This spare part is critical because if it is not available when it is needed, company suffers from unavailability of transportation vehicle. Demand records for this item were available for recent 50 weeks. Total demand for this item is 8 units, item price is 1.968,00 €, lead time is 5 days, annual holding cost is 30% of item value, ordering cost is 10,00 € and review cost is 1,00 €. Graphical representation of results is shown on Figure 2. It can be seen from Figure 2. that minimum inventory cost *TC* of 3.207,15 € and service level *SL* of 99,94 % are achieved for length of review period *r* of 15 days and maximum inventory level *S* of 5 item units.

4. CONCLUSION

Theoretical mathematical model and software module are tested and verified using real practical example. Results of tests show that theoretical model and software module are capable of ensuring efficient periodical control of spare parts inventory.

Length of review period, i.e. number of review periods, and maximum inventory level, determined for the case of optimal relationship between total inventory costs and service level, represent significant benefit to optimization of balance between three main aims of inventory control. Testing of the developed model showed excellent results.

5. REFERENCES

Ballou, R. H. (2004). *Business logistics/ Supply chain management and logware compact disc package*, Pearson Higher Education, ISBN-10: 0131492861, New Jersey

Fortuin, L. & Martin, H. (1999). Control of service parts, *International Journal of Operations & Production Management*, Vol.19 No.9, pp. 950-971, ISSN 0144-3577

Heizer, J. & Render, B., (2006). *Operations Management*, Pearson Education, Inc. 0-13-185755-X, New Jersey

Humphrey, A.S.; Taylor, G.D. & Landers, T.L. (1998). Stock level determination and sensitivity analysis in repair/rework operations, *International Journal of Operations & Production Management*, Vol. 18 No.6, 1998, pp.612-630, DOI: 10.1108/01443579810209566

Razi, M.A. & Tarn, J.M. (2003). An applied model for improving inventory management in ERP systems, *Logistics Information Management*, Vol. 16, No. 2, pp. 114-124, ISSN 0957-60-53

Wild, R. (2002). *Operations management*, Continuum International Publishing Group, ISBN-10: 0826449271, London

Annals of DAAAM for 2008 & Proceedings of the 19th International DAAAM Symposium, ISSN 1726-9679
Editor B. Katalinic, Published by DAAAM International, Vienna, Austria 2008
Make Harmony Between Technology and Nature, and Your Mind will Fly Free as a Bird

NUMERICAL STUDY OF APPEARENCE AND DEVELOPMENT OF DAMAGES IN COMPOSITE STRUCTURES

PASTRAMA, S[tefan] D[an]; HADAR, A[nton]; JIGA, G[abriel] & TABARA, I[ulian]

Abstract: *In this paper, a finite element code developed by the authors for the study of damages in composite materials is presented. Failure criteria are implemented and the equivalent stress in the laminae can be determined. The displacements are obtained in each node of the model and the stresses are calculated both in each lamina and interlaminar. Three criteria of stopping the post critical calculation are proposed and implemented in the program.*
Key words: *Finite element, damage, lamina, failure criterion.*

1. INTRODUCTION

Post-critical calculus of a laminate reinforced structure starts together with the appearance of the first delaminations in the composite structure. The main damages that could be produced in the composite material are matrix and fiber breaking and delaminations. In this case, the numerical methods are the only able to offer palpably results.

Post critical calculus is nonlinear and can emphasize both the failure mode of a composite structure and the mechanism of damage propagation (Hadar et al., 2007). In this way, the type of damage and the place where it appears can be detected.

After applying the failure criteria on each element, the finite element analysis can be continued in two ways: *i.* by keeping the initial load and performing the post critical calculus till the total failure of the structure and *ii.* by lowering the applied loads in order to determine the carrying capacity of the damaged structure.

2. IDENTIFICATION OF THE DAMAGES

The authors developed a finite element code in order to identify the damages and their propagation by an interactive calculus.

The application of failure criteria for plies is generally a difficult process. The authors show three possible failure modes of the fiber reinforced composite. The fibers are considered to be in a broken stage if (Fig. 1):

$$\sigma_l \geq \sigma_{af} \quad or \quad \sqrt{\tau_{lt}^2 + \tau_{lz}^2} \geq \tau_{af} \tag{1}$$

where σ_{af} and τ_{af} and are the normal and shear allowable stresses for the fibers. If one of the restrictions (1) is satisfied, the stiffness matrix of the element is eliminated.

A matrix breaking is considered to appear when the stresses in the center of each element satisfy the conditions (Fig. 1):

$$\sigma_t \geq \sigma_{am} \quad or \quad \sqrt{\tau_{tl}^2 + \tau_{tz}^2} \geq \tau_{am} \tag{2}$$

where σ_{am} and τ_{am} are the allowable stresses for the matrix.

The elasticity matrix $[\overline{D}]$ corresponding to the elements where one of the conditions (2) is satisfied, has another shape with respect to the equivalent one for transversely isotropic materials, (Wei, 1991).

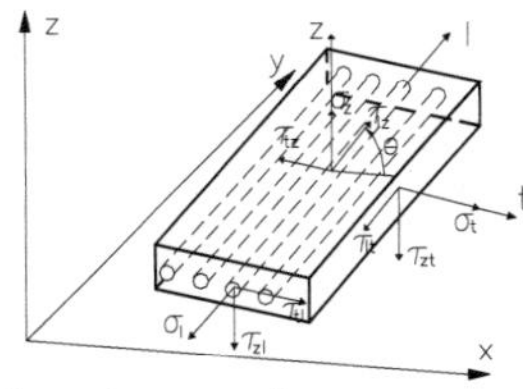

Fig. 1. Fiber reinforced composite

The finite element analysis is resumed using an iterative procedure; in this case, the elasticity matrix corresponding to the elements where matrix breakings appear, will modify its shape.

In order to discover the appearance of delaminations, the stresses in the center of the adjacent area between two layers are compared with the allowable contact stresses. The separation surface between two adjacent elements is considered to be a possible zone for delaminations, when one of the following relationships is satisfied (Fig. 1):

$$\sigma_z \geq \sigma_{ad} \quad or \quad \sqrt{\tau_{lz}^2 + \tau_{tz}^2} \geq \tau_{ad} \tag{3}$$

where σ_{ad} and τ_{ad} represent the allowable normal and shear stresses for delamination. If equation (3) shows the appearance of delamination, the calculus is continued by modifying the shape of the elasticity matrix $[\overline{D}]$ (Finn, 1993).

The stresses σ_{af} , τ_{af} , σ_{am} , τ_{am} , σ_{ad} and τ_{ad} can be experimentally determined and assigned as input data.

3. CALCULUS OF STRESS AND STRAIN STATE USING IMPOSED NODAL DISPLACEMENTS

Due to the fact that generally only the areas with high stress gradients are of interest, the study of a complex structure made from fiber reinforced layered composites is recommended to be done in two steps:

Step 1: **a.** Analysis of the stress and strain state for the whole structure using a commercial numerical code; **b.** Detection of areas with stress concentrators; **c.** Delimitation of these areas from the whole structure, retaining the displacements of the nodes situated on the contour of the isolated part, the displacements being obtained as a result of the sue of the numerical code (Flanagan, 1993);

Step 2: **a.** Analysis of the isolated areas with specialized finite elements destined to the study of local problems in multilayered composite materials (Hadar et al., 2006); in this case, the input data are the displacements obtained in the previous analysis; **b.** achievement of a post-critical calculus in order to determine the bearing capacity of the structure.

3.1 Methodology and algorithm for application of failure criteria and post-critical calculus

The flow chart of the algorithm is presented in Fig. 2 and is used for the application of the composite failure criteria, for the localization of the type of damage as well as for a post-critical calculus under an unmodified or modified (diminished) load, in order to pursue the way of damage propagation, up to the total failure of the structure.

In order to realize a post-critical calculus, the matrices of elasticity in a local system of coordinates, where the apparition of one or several damages has been observed, are modified:

- for the finite elements characterized by breaking of fibers, due to the fact that the strength element (in our case, the fiber) disappears, the corresponding matrices of elasticity are canceled; in order to continue the calculus with finite elements without any supplementary modeling of the structure, very small values will be assigned for these elements of the elasticity matrix ($10^4 - 10^5$ times less than for the initial ones);
- for the finite elements characterized by a matrix failure, the elasticity matrix form will be modified, corresponding to (Wei, 1991); for the disappearing elements of the elasticity matrix very small values will be also assigned;
- if between two adjacent layers a delamination will occur, the elasticity matrix corresponding to these elements will be also replaced (Finn, 1993), taking into consideration that the disappearing elements will acquire very small values;
- if in the structure one can find finite elements having multiple damages, the main flaw will be considered the fiber breaking, then the matrix failure and finally the delamination. For example, if in the same finite element delaminations and fiber breakings appear simultaneously, for this element, the matrix of elasticity will be replaced (Wei, 1991), the main flaw being considered to be the matrix failure.

3.2 Flow chart of the applicability of stop criteria for the post-critical calculus

In conformity with the code, the post-critical calculus continues till one of the following criteria is accomplished:

a) The criterion of the number of damaged elements.

In this case the program calculates the number of elements having damages as matrix breakings (N_{01}), fiber breakings (N_{02}) and delaminations (N_{03}) and compares this number with the total number of elements of the whole structure (N_0). According to Finn (1993), Hadar et al. (2006) and Flanagan (1993), the numerical code should warn that the structure fails if one of the following restrictions is accomplished:

$$\frac{N_{01}}{N_0} > 0.3 \; ; \; \frac{N_{02}}{N_0} > 0.15 \; ; \; \frac{N_{03}}{N_0} > 0.2 \qquad (4)$$

b) The criterion of the limit nodal displacements.

The displacement of each node of the structure is calculated with the equation $\delta = \sqrt{u^2 + v^2 + w^2}$, where *u*, *v* and *w* are the nodal displacements on the global system of axes *Oxyz*. The program considers that the structure fails when:

$$\delta > \delta_{lim} \qquad (5)$$

c) The criterion of the volume of damaged elements.

Let us consider V_{01}, V_{02}, V_{03} the volumes of the elements where damages (fiber or matrix breakings, delaminations) occur and let V_0 be total volume of the structure. The code proposed by the authors stops the post-critical calculus when one of the following restrictions is accomplished:

$$\frac{V_{01}}{V_0} > 0.3 \; ; \; \frac{V_{02}}{V_0} > 0.15 \; ; \; \frac{V_{03}}{V_0} > 0.2 \qquad (6)$$

All these criteria can be applied only if damages occur in the composite structure. The stopping criteria of post-critical analysis may be modified in function of the analyzed structure. In this case reference values can be established for the failure of the reinforced composite structure.

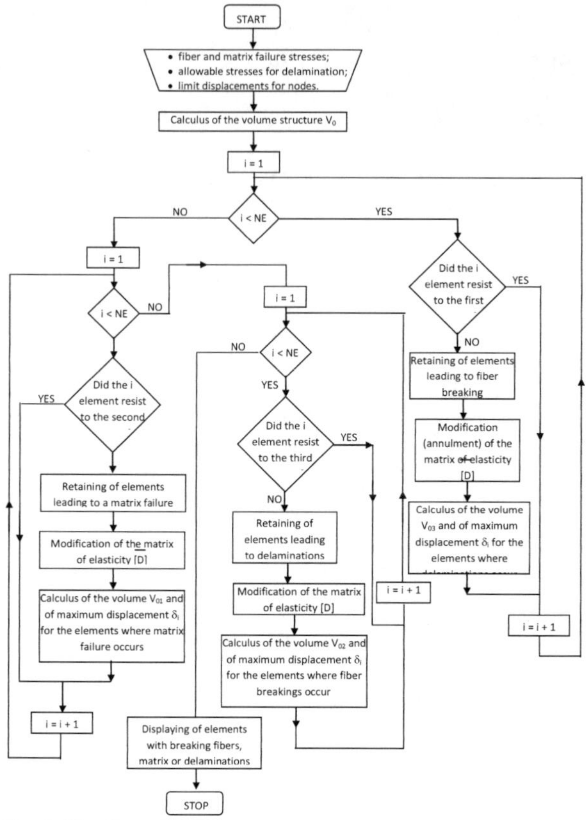

Fig. 2. The flow chart

4. CONCLUSIONS

The computer code calculates and displays the displacements u_x, u_y, u_z of each node of the whole structure. Through this program, the stresses in the center of each element can be calculated. The obtained stresses are related as well to the global system of axes *Oxyz* as to the local system *Oltz*. After applying the failure criteria, the computer code displays the number of elements where damages occur as well as the type of damage. The finite element calculus is resumed in a non-linear regime, in an iterative manner, the stiffness matrices of damaged finite elements being modified in function of the type of flaw. When one of the conditions referable to the structure failure is accomplished, the code displays it.

5. REFERENCES

Finn, S.C., Springer, G.S. (1993), Delaminations in Composite Plates Under Transverse Static or Impact Loads – a Model, *Composite Structures*, Vol.23.

Flanagan, G. (1993), A Sublaminate Analysis Method for Predicting Disband and Delamination Loads in Composite Structures", *Journal of Reinforced Plastics and Composites*, Vol.12.

Hadar, A., Constantinescu, I.N., Jiga, G., Ionescu, D. S., (2007), Some Local Problems in Laminated Composite Structures, *Materiale Plastice*, Vol. 44, nr 4, pp. 354-360.

Hadar, A., Nica, M.N., Constantinescu, I. N., Pastrama, S.D., (2006), The Constructive and Geometric Optimization of the Junctions in the Structures Made from Laminated Composite Materials, *Journal of Mechanical Engineering*, Vol. 52, No. 7-8/06, Ljubljana, p. 546-551.

Wei, J., Zhao, J.H. (1991), Three-dimensional Finite Element Analysis on Interlaminar Stresses of Symmetric Laminates, *Computers and Structures*, Vol.41, No.4.

Annals of DAAAM for 2008 & Proceedings of the 19th International DAAAM Symposium, ISSN 1726-9679
Editor B. Katalinic, Published by DAAAM International, Vienna, Austria 2008
Make Harmony Between Technology and Nature, and Your Mind will Fly Free as a Bird

ECOLOGICAL METHOD FOR SEPARATION OF INULIN FROM PHYTOEXTRACTS USING MOLECULARLY IMPRINTED POLY (VINYL ALCOHOL) FILMS

PATACHIA, S[ilvia]; CROITORU, C[atalin] & SCARNECIU, I[oan]

Abstract: *Molecular imprinting is a novel method for designing materials with molecular memory, which consists of cavities that bear the shape and size of a template molecule. The cavities are highly specific towards the molecule that imprints the polymer. In this work, a novel ecological method, having as aim the separation of inulin [IN] from aqueous extracts by using an imprinted poly (vinyl alcohol) [PVA] film has been proposed and tested. To demonstrate the obtaining of the molecular imprinted PVA and its selectivity, studies of IN sorption and desorption have been done.*

Key words: *poly (vinyl alcohol), inulin, hydrogel, selective absorption*

1. INTRODUCTION

Phytoextracts (aqueous or hydroalcoholic extracts from plants) are a natural source of a various class of compounds such as carbohydrates, alkaloids, steroids, flavonoids, terpenes and complex acids often used in traditional medicine as drugs, nutritional supplements, dyes and so forth. However, separation of a certain compound from the complex aqueous or hydroalcoholic mixture is a tedious task, often implying time and energy consuming procedures or exposing the biologically active substance to thermal degrading (Hart & Shea 2001).

Molecular imprinting is a novel method for designing materials with molecular memory, which consists of cavities that bear the shape and dimensions of a template molecule. The cavities are highly specific towards the molecule that imprints the polymer, making molecularly imprinted materials suitable for use in isomer separations, catalysis, biosensor assays or in controlled drug delivery (Rechichi et al. 2006; Dickert 2006). Recognition and separation of saccharides have been the focus of much recent attention. Boronic acid is a powerful tool for this purpose since it can form a stable complex (cyclic ester) with saccharides, especially with those comprising a cis-diol group. However; these supports have problems in mechanical and/or chemical stability. Hydrogels are reported to be rather weak under pressure, while glass beads exhibit poor stability under basic conditions (Hayden & Dickert 2002; Wulff & Karsten 2002). Such polymers obtained from bulk or solution polymerization, however, necessitate some tedious treatments, such as grinding and sieving before use (Bolman & Revsbech 2005). In this work, a new method of alternative molecular imprinting to design imprinted poly (vinyl alcohol) [PVA] stable hydrogelic films crosslinked with boric acid has been proposed (Patachia 2003; Patachia 2006). An oligomer of fructose with average molecular weight of 17000- inulin- has been used as template molecule. Inulin is widely used in diabetes treatment, as dietary fiber, blood pressure regulator and glomerular filtration rate measuring.

2. EXPERIMENTAL

2.1 Materials

PVA 120-98 (1200 polymerization degree and 98% hydrolysis degree) was purchased from Chemical Enterprises Rasnov, Romania. Inulin oligomer with average molecular mass of 17000, boric acid, vanillin and concentrated sulphuric acid have been purchased from Sigma-Aldrich, and were of reagent grade. All reagents have been used without further purification.

2.1 Imprinted films obtaining

PVA solution has been prepared by dissolving the polymer powder in Milli-Q deionized water, under magnetic stirring at room temperature, followed by heating at 80 °C for 4h. The solid content of the obtained solution was 10%.

PVA- IN mixtures have been prepared, by adding the weighted IN powder to 25 mL of PVA solutions with different solid contents.

Mixtures of IN to PVA ratios from 18 % to 38% (reported to 1g of dry PVA [xerogel]) have been obtained. The PVA films have been obtained by polymer solution (with the respective amount of IN added) casting and solvent evaporation in a vacuum drying stove (T=28^{0}C; p=120 mbar) for 24 hours.

The films have been immersed in a 0.8M boric acid solution for 3 hours to achieve crosslinking. Reference probe of PVA crosslinked film without IN has been prepared using the same procedure.

The template IN molecule has been removed from the polymer matrix in order to form the active cavities with the shape and size complementary with that of the template by curing the obtained films in distilled water for 3 days.

2.2 Imprinted films testing

The imprentation of the PVA films has been tested by comparing the amount of IN absorbed in the imprinted polymeric matrix with the amount of IN absorbed in the non-imprinted polymer.

IN absorption has been studied by immersing weighted film samples in a determined amount of 0.1 g/L IN solution.

At certain time intervals, 2 mL of IN solution were drawn and analyzed and the film samples have been immersed in a fresh IN solution of 0.1 g/L.

IN has been determined by the spectrophotometric method using a SPECOL 10 Karl-Zeiss Jena spectrophotometer, due to its ability to form a colored complex (absorption maximum at 520 nm) with vanillin in the presence of concentrated sulphuric acid.

IN desorption from the imprinted films has been studied by immersing the samples subjected earlier to absorption studies in a determined amount of distilled water. Solution samples have been drawn and analyzed as above, and after each determination the films have been reimersed in a fresh amount of distilled water.

3. RESULTS AND DISCUSION

Amount of IN absorbed in the imprinted PVA films (in terms of absorbed IN amount (g) reported to 1g of xerogel is plotted in Fig.1 (absorption kinetic). Amount of absorbed IN as a function of IN: PVA ratio is presented in fig.2:

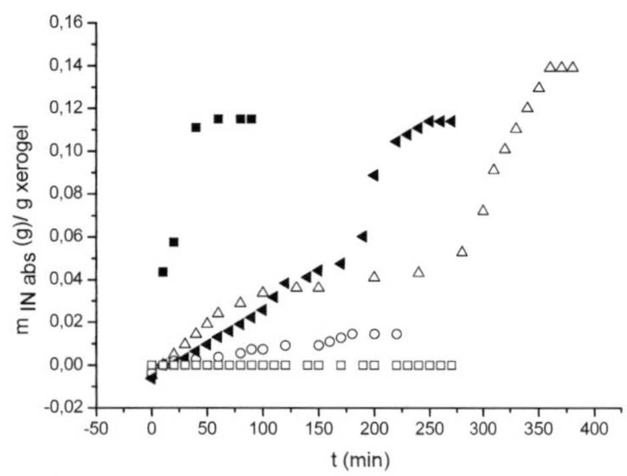

Fig.1: Absorption kinetic of IN from: □: PVA;
■:IN :PVA 18,04 % ; ▲ IN :PVA 24,33%;
◀: IN :PVA 30,46%; △: IN :PVA 38%

As it can be seen from Fig. 1, the imprinting of PVA with IN template was successful. All imprinted films absorb a higher amount of IN than the reference non-imprinted sample. A maximum of absorption is observed (Fig.2) at 30.46 % IN: PVA ratio. After this maximum, the absorbed IN amount decreases, probably due to the lower polymer content in the films, which could lead to insufficient formation of active cavities. The absorption rates are also higher for the imprinted samples, comparative with the non-imprinted one. The desorption kinetic is presented in Fig.3 in terms of desorbed IN amount/g xerogel:

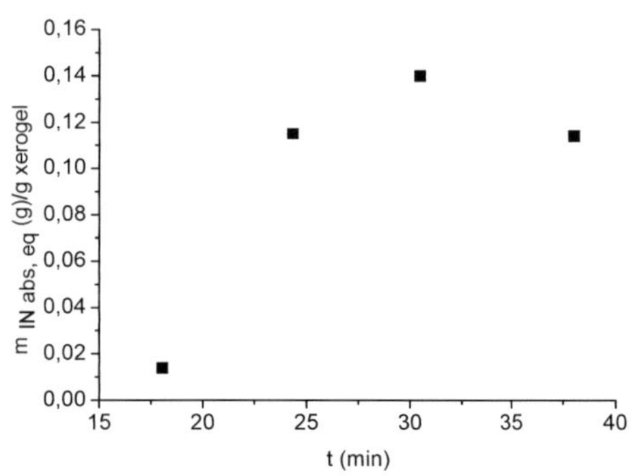

Fig.2. Absorbed amount of IN as a function of IN: PVA ratio

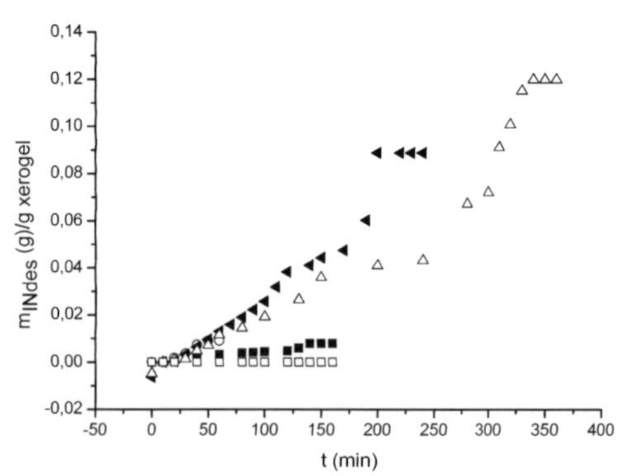

Fig.3. Desorption kinetic of IN from: □: PVA;
■:IN :PVA 18,04 % ; ▲ IN :PVA 24,33%;
◀: IN :PVA 30,46%; △: IN :PVA 38%

From the desorption kinetic (Fig.3) it could be seen that the desorption rate of IN from the polymeric matrix is higher in the films with optimum IN: PVA ratio, probably due to the presence of more non-deformed active cavities in the polymeric matrix, which leads to an increase in desorption. Also it could be seen that the IN molecule is retained in a higher amount in the imprinted PVA samples. Elimination of all IN amount from the polymeric matrix was not possible.

4. CONCLUSION

Imprinted PVA 120-98 films with inulin as template molecule have been obtained and characterized. Sorption and desorption analysis have been performed. The results indicated that the imprentation of PVA 120-98 with IN was successful.

All the imprinted films absorb a higher amount of IN than the non-imprinted PVA.

The optimum IN: PVA ratio has been found to be 30.46%. The imprinted films are transparent, homogenous and have good resistance to solvent action, due to crosslinking.

Taking into account the non-toxicity, biocompatibility, biodegradability of PVA and the possibility of obtaining molecular imprinted materials based on PVA/IN, new applications in special fields (medicine, pharmacy, separation processes) could be developed.

5. ACKNOWLEDGEMENTS

We would like to be thankful to The National University Council from Romania (CNCSIS) for funding of the research through the national grants CEEX10/2005, CEEX18/2005 and CEEX 148/2006.

6. REFERENCES

Bollmann, A., Revsbech, N. P. (2005). An NH_4^+ biosensor based on ammonia-oxidizing bacteria for use under anoxic conditions. *Sensors and Actuators B: Chemical*, Vol. 105, No. 2 (June 2005) pp. 412-418, ISSN 0925-4005.

Dickert, F. L., Hayden, O. (2002). Bioimprinting of polymers and sol-gel-phases. Selective detection of yeasts with imprinted polymers. *Anal. Chem.* Vol. 74, No.1 (February 2002) pp. 1302-1306, ISSN 0003-2700.

Hart, B.R., Shea, K.J. (2001). Synthetic peptide receptors: molecularly imprinted polymers for the recognition of peptides using peptide-metal interactions. *J. Am. Chem. Soc.*, Vol. 123, No. 3 (August 2001) pp. 2072–2073, ISSN 0002-7863.

Hayden, O., Dickert, F. (2001). Selective microorganism detection with cell surface imprinted polymers. *Adv. Mater.*, Vol. 13, No. 3 (February 2001) pp. 1480-1483, ISSN 0935-9648.

Patachia, S. (2003). Blends based on poly (vinyl alcohol) and the products based on this polymer, In: *Handbook of Polymer blends and composites*, Vasile, C, Kulshreshtha, A.K., (Ed), pp.230-256, RAPRA Technology Ltd, ISBN 1-85957-201-4, Shrewsbury, UK.

Patachia S., Valente A. J. M., Baciu, C. (2006). Effect of non-associated electrolyte solutions on the behaviour of poly (vinyl alcohol)-based hydrogels. *European Polymer Journal*, Vol. 43, No. 2 (July 2006) pp. 460-467, ISSN 0014-3057.

Rechichi, A., Cristallini, C., Vitale, U., Ciardelli, G., Barbani, N., Vozzi, G., Giusti, P. (2007). New biomedical devices with selective peptide recognition properties. Part 1: Characterization and cytotoxicity of molecularly imprinted polymers. *J. Cell. Mol. Med.* Vol 11,No.6 (May 2007), *pp. 1367-1376,* ISSN 0892-6638

Wulff, G., Karsten, K. (2002). Stoichiometric noncovalent interaction in molecular imprinting. *Bioseparation,* Vol.10, No. 2 (July 2002), pp.257–276, ISSN 1573-8272

Annals of DAAAM for 2008 & Proceedings of the 19th International DAAAM Symposium, ISSN 1726-9679
Editor B. Katalinic, Published by DAAAM International, Vienna, Austria 2008
Make Harmony Between Technology and Nature, and Your Mind will Fly Free as a Bird

AMMONIA SENSOR BASED ON POLY (VINYL ALCOHOL) CRYOGEL

PATACHIA, S[ilvia]; CROITORU, C[atalin]; FLOREA, C[laudia]; dumitrescu, l[ucia] & SCARNECIU, I[oan]

Abstract: *Polymeric materials have been extensively used as components of sensor assays due to their relatively fast response to various external stimuli. In this work, a potential ammonia sensing poly (vinyl alcohol) [PVA] ecological and fast responsive material has been obtained, having as aim developing new materials based on PVA for the selective detection of pollutants from wastewaters. Swelling degree variation with ammonium hydroxide concentration (from 0.6 to 6% wt) has been studied. The hydrogel response to NH_4OH concentration is linear in the studied concentration range.*

Key words: *poly (vinyl alcohol), cryogel, sensor, ammonia*

1. INTRODUCTION

PVA based polymers or PVA composite materials are often employed in capacitive humidity sensor assays due to their relative fast response (e.g. swelling degree modification) with relative humidity. Based on the fact that PVA has the ability to swell or collapse selectivity, depending on the electrolyte's nature (cation and anion type) and concentration, a novel type of electrolyte sensors could be developed (Patachia et al. 2006). The influence of inorganic aqueous salt solutions on synthetic or natural hydrophilic polymeric materials can be deduced from the Hofmeister series. According to Hofmeister series, ions are either kosmotropic (they tend to destabilize the hydrogen bonds between water molecules or between water and polymeric chains leading to a collapsing of the polymeric network) or chaotropic (they tend to stabilize the hydrogen bonds in hydrophilic polymers leading to polymeric matrix swelling) (Bagger et al. 2007). Although sensors can be outfitted today with biological molecules, sometimes those are not robust enough to survive in more rugged environments than laboratory confines (Bunde et al. 1998; Dickert & Hayden 2002; Hayden & Dickert 2001). NH_3 is the principal form of toxic ammonia and it can be found both freely dissolved or in the form of the NH_4^+ ion in wastewaters from chemical plants. It has been reported toxic to fresh water organisms at concentrations ranging from 0.53 to 22.8 mg/L. Hatching and growth rates of fishes may be affected. In the structural development, changes in tissues of gills, liver, and kidneys may also occur. Toxic concentrations of ammonia in humans body may cause loss of equilibrium; convulsions, coma, and death. That's why ammonia concentration level monitoring is an important task. Up to this extent there are some variants of ammonia sensors on the market, such as: electrochemical, solid state and capacitive sensors. All of them are suitable for determining the ammonia content from aqueous solutions, from approx. 2 to 20000 ppm (Bollman & Revsbech 2005). However they have certain drawbacks, such as: short sensor life, decreased sensitivity over time, increased calibration frequency, over 10% error in determining ammonia concentration bellow 200 ppm level, poor linearity of analytical response versus concentration, high cost and so forth (Hart & Shea 2001). In this work we have obtained a PVA based material with potential use as sensor. PVA is a non-toxic, non-carcinogenic, biocompatible, biodegradable, water-soluble polymer, in consequence easy to handle and friendly for the environment (Patachia 2003). Physical crosslinking using freezing-thawing cycles has been used for the PVA cryogel obtaining. The method of physical crosslinking of PVA is often employed in pharmacy and medicine (Patachia 2003), as an alternative to chemical crosslinking which uses potentially toxic reagents. One of our aims was to achieve good linearity of the cryogels response (e.g. swelling degree variation) as a function of the external stimulus (NH_4OH concentration variation).

2. EXPERIMENTAL

2.1 Materials

PVA 120-98 (1200 polymerization degree and 98% hydrolysis degree) was purchased from Chemical Enterprises Rasnov, Romania. Ammonium hydroxide (25% wt solution) has been purchased from Sigma.

2.2 PVA cryogel obtaining

PVA solution has been prepared by dissolving the polymer powder in Milli-Q water, under magnetic stirring at room temperature, followed by heating at 75 °C for 4h. The solid content of the obtained solution was 11%.wt to a certain amount of this solution, an aqueous NH_4OH solution has been added so as the PVA: NH_4OH ratio to be 12% wt. The PVA cryogel has been prepared by introducing a specific volume of PVA/ NH_4OH mixture in a PVC cylindrical recipient and submitting it to freezing at -20^{0}C for 12 hours, followed by thawing at room temperature (26^{0}C) for 12 hours. The above mentioned freezing-thawing procedure has been repeated four times. Alternative, a PVA cryogel without ammonia has been prepared following the same procedure.

2.3 Removal of NH_4OH molecule from the PVA cryogel

NH_4OH has been removed from the cryogel by washing with distilled water at room temperature. The ammonia removal has been monitored conductometrically, using a Radelkis OK-112 conductometer and by swelling studies. For the swelling studies, the cryogel samples have been soaked in distilled water at room temperature. At different time intervals, each sample was taken out, wiped with filter paper and weighted until a constant mass was observed. The degree of swelling (SD) was determined by using Eq. (1)

$$SD = \frac{m_s}{m_d} \cdot 100 \quad (1)$$

where m_s is the mass of the cryogel swollen in water and m_d the mass of the dried cryogel [xerogel].
The same procedure has been followed for the PVA/ NH_4OH cryogel [C_1] as well as for the reference [C_2].

2.4 Determination of cryogel sensitivity to NH_4OH solution

The weighted cryogel samples (after swelling equilibrium reaching) have been immersed in a determined volume of distilled water. To the amount of distilled water containing the sample, 25% wt ammonia solution has been added from a biurette, and the amount of NH_4OH in contact with the cryogel has been calculated. After 10 minutes, the sample was taken out, wiped with filter paper and weighted. Then, the cryogel sample has been reimersed in the ammonia solution, and a new volume of NH_4OH solution has been added from the biurette. The above procedure has been repeated for different NH_4OH concentrations (from 0.6 to 6% wt) for C_1 and C_2.

3. RESULTS AND DISCUSION

The elimination of NH_4OH from the polymeric matrix of the PVA/ NH_4OH cryogel and for the reference is plotted in Fig.1. The swelling kinetic of both cryogels is presented in Fig. 2:

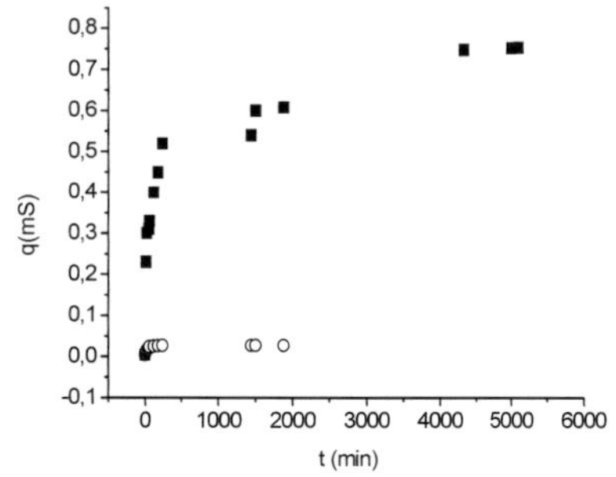

Fig. 1. Conductometric analysis of washing water from: (■): C_1; (○): C_2

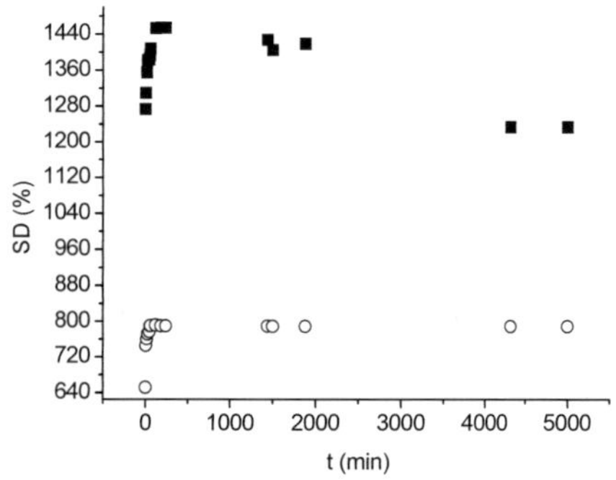

Fig. 2. Swelling degree of: (■): C_1; (○): C_2

The swelling degree of the cryogels as a function of ammonia concentration is plotted in Fig.3:

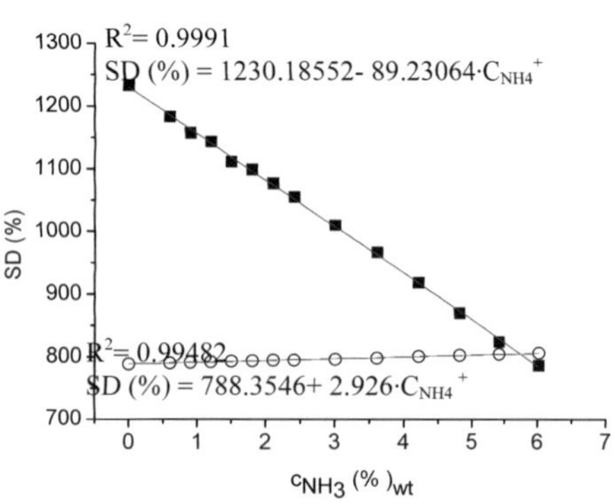

Fig. 3. Swelling degree of: (■): C_1; (○): C_2

As it can be seen from Fig. 1 and 2, elimination of NH_4^+ and OH^- occurred. The conductivity of the washing solution for C_1 is higher than that of the reference. The swelling degree of C_1 is higher, due to the kosmotropic effect of NH_4^+ ion (according to the Hofmeister series) which tends to break the hydrogen bonds between the water molecules, water and polymer molecules and between the polymeric chains, leading to an increased swelling of the cryogel.

Linear response of the swelling degree with increasing of ammonia concentration has been obtained for both cryogels (Fig. 3). The slope of the linear dependence in the case of the PVA/NH_4OH cryogel is higher, which means that the sensitivity of the PVA/NH_4OH sample is higher than that of the reference sample.

4. CONCLUSION

PVA 120-98 cryogels with or without NH_4OH in composition have been prepared by physical crosslinking and characterized, in terms of their swelling degree sensitivity to aqueous ammonia solution concentration range of 0.6 to 6 % wt.

The obtained cryogels are soft, opaque and have the property to swell selectively when immersed in aqueous media without dissolving.

Linear dependence of the swelling degree has been obtained. The response of the PVA/ NH_4OH cryogel to NH_4OH solution concentration is better than that of the cryogel reference, for the whole NH_4OH concentration range studied.

Use of the proposed PVA/NH_4OH material is limited to a temperature range of 4 to 35^{0}C

Based on the obtained results and on the fact that PVA is an environmental friendly material, frequently used in the field of pharmacy, medicine and advanced purifications, new sensor assay applications based on the mentioned material could be developed in our further studies.

5. REFERENCES

Bagger, L. H., Øgendal L.H., Westh P. (2007). Solute effects on the irreversible aggregation of serum albumin. *Biophysical Chemistry,* Vol. 130, No. 1-2 (July 2007) pp. 17-25, ISSN 0301-4622.

Bollmann, A., Revsbech, N. P. (2005). An NH_4^+ biosensor based on ammonia-oxidizing bacteria for use under anoxic conditions. *Sensors and Actuators B: Chemical*, Vol. 105, No. 2 (June 2005) pp. 412-418, ISSN 0925-4005.

Bunde, R. L.; Jarvi, E.; Rosentreter, J. (1998). Piezoelectric quartz crystal biosensors. *Talanta*, Vol. 46, No. 1 (September 1998) pp. 1223-1236, ISSN: 0039-9140.

Dickert, F. L., Hayden, O. (2002). Bioimprinting of polymers and sol-gel-phases. Selective detection of yeasts with imprinted polymers. *Anal. Chem*. Vol. 74, No.1 (February 2002) pp. 1302-1306, ISSN 0003-2700.

Hayden, O., Dickert, F. (2001). Selective microorganism detection with cell surface imprinted polymers. *Adv. Mater*., Vol. 13, No. 3 (February 2001) pp. 1480-1483, ISSN 0935-9648.

Hart, B.R., Shea, K.J. (2001). Synthetic peptide receptors: molecularly imprinted polymers for the recognition of peptides using peptide-metal interactions. *J. Am. Chem. Soc*., Vol. 123, No. 3 (August 2001) pp. 2072–2073, ISSN 0002-7863.

Patachia, S. (2003). Blends based on poly (vinyl alcohol) and the products based on this polymer, In: *Handbook of Polymer blends and composites*, Vasile, C, Kulshreshtha, A.K., (Ed), pp.230-256, RAPRA Technology Ltd, ISBN 1-85957-201-4, Shrewsbury, UK.

Patachia S., Valente A. J. M., Baciu, C. (2006). Effect of non-associated electrolyte solutions on the behaviour of poly (vinyl alcohol)-based hydrogels. *European Polymer Journal*, Vol. 43, No. 2 (July 2006) pp. 460-467, ISSN 0014-3057.

Annals of DAAAM for 2008 & Proceedings of the 19th International DAAAM Symposium, ISSN 1726-9679
Editor B. Katalinic, Published by DAAAM International, Vienna, Austria 2008
Make Harmony Between Technology and Nature, and Your Mind will Fly Free as a Bird

ANALYZE OF TRANSMISSION ERRORS IN GEARBOXES

PATER, S[orin]; FODOR, D[inu]; MITRAN, T[udor] & POLOJINTEF CORBU, N[icolae]

Abstract: *The static transmission error describes the composite effect of any deviation of the gear teeth from perfectly formed involute surfaces. The static transmission error is widely accepted as the principle source of vibration in gearboxes. The dynamic component of the static transmission error is a physical measure of any gear tooth surface deviation.*
Key words: *gear, frequency, bearing, transmission*

1. INTRODUCTION

Thus far, the discussion has been an idealization that has assumed that the gears are uniformly spaced, perfectly formed and completely rigid. Actual gears deviate slightly from perfect involute surfaces and elastically deform under load. These tiny surface deviations result in an unsteady force component of the torque. The unsteady forces are transmitted as vibration from the gear, through the shaft and bearings to the power transitions casing where it is measured. It is this unavoidable vibration that we exploit to no invasively investigate the operating condition of gearboxes while in operation.

Mathematically, this unsteady forcing function due to the pinion is best described by a complex Fourier series with a fundamental frequency equal to the pinion rotational rate, f_r .

$$s(t)=\sum_{n=0}^{\infty} c_n e^{jn(2\pi f_r)t} \tag{1}$$

The Fourier transform of the complex Fourier series is a one-sided pure line spectrum at multiples of the gear rotation rate.

$$S(f)=\sum_{n=0}^{\infty} c_n \delta(f-nf_r) \tag{2}$$

We can now develop a mathematical expression for the composite vibration due to both the pinion and the gear. This is accomplished by summing two infinite complex Fourier series. Therefore, if the pinion has N teeth and the gear has M teeth, then equation 3 describes the composite vibration due to both gears.

$$s(t)=\sum_{n=0}^{\infty} c_n e^{jn(2\pi f_r)t}+\sum_{m=0}^{\infty} c_m e^{jm\left(2\pi f_r\left(\frac{N}{M}\right)t\right.} \tag{3}$$

In equation 3, f_r is the pinion rotational frequency and $(N/M)f_r$ is the gear rotational frequency. The Fourier transform of the composite vibration is also a one-sided pure line spectrum.

$$S(f)=\sum_{n=0}^{\infty} c_n \delta(f-nf_r)+\sum_{m=0}^{\infty} c_m \delta\left(f-mf_r\left(\frac{N}{M}\right)\right) \tag{4}$$

It should be clear that the two summations share a common set of frequencies. In the second summation, whenever m equals any integer multiple of M, the summations share the same frequency component. Therefore, equation 4 can be further decomposed as equation 5.

$$S(f)=\sum_{l=0}^{\infty} c_l \delta(f-lNf_r)+\sum_{n=0,n\neq lN}^{\infty} c_n \delta(f-nf_r)+\sum_{m=0,m\neq lM}^{\infty} c_m \delta\left(f-mf_r\left(\frac{N}{M}\right)\right) \tag{5}$$

If we then transform back to the time domain, we arrive at the following equation.

$$S(f)=\sum_{l=0}^{\infty} c_l \delta(f-lNf_r)+\sum_{n=0,n\neq lN}^{\infty} c_n \delta(f-nf_r)+\sum_{m=0,m\neq lM}^{\infty} c_m \delta\left(f-mf_r\left(\frac{N}{M}\right)\right) \tag{6}$$

The first summation in equation 1-6 is composed of vibration components from both the pinion and the gear. The first summation's fundamental frequency (N f_r) is the gear mesh frequency. We have just shown that the components of vibration due to the gear mesh frequency and its harmonics are due to both the pinion and the gear. This is a fact that is neglected in the literature on gear diagnostics. In the next section, we will find that the summation over " in equation 6 is called the harmonic error signal. The second summation in equation 6 is then due solely to the pinion and the third summation is due solely to the gear. Also in the next section, we will find these two summations are called the residual error signals for the pinion and the gear. Thus, in the frequency domain, it should be possible to separate the vibration produced by different gears. The component of the static transmission error that occurs at multiples of the gear meshing frequency is caused by elastic tooth deformations and the mean deviations of the tooth faces from perfect involute surfaces (Zaveri 1985).The remaining components of the static transmission error that occur at multiples of the gear rotational frequency are caused by the dynamic components of the tooth face deviations. Thus, we have a concrete, physical justification for using the static transmission error for gear diagnostics. This includes, but is not limited to, heavily worn teeth, missing teeth, and cracked or chipped teeth. In the next section, we will show how the residual error signal is defined as the dynamic component of the static transmission error.

2. GEAR MOTION ERROR

The gear motion error is a real part of the static transmission error (Fhay & Perez 1990),. The gear motion error is a real signal, described by an infinite cosine series with fundamental period $f\,r$. The Fourier transform is then a two-sided pure line spectrum. Whereas the static transmission error was developed for predicting the amount of vibration produced by meshing gears, the gear motion error was developed for gearbox diagnostics. It should be noted that since we are dealing with real valued signals, the static transmission error and the gear motion error contain the same information and simply differ by a factor of 2. We will use the term gear motion error and its interpretation throughout the remainder of this thesis. The decomposition of the composite gear motion error has three components -- the harmonic error component □□□□$s_{eh}(t)$,

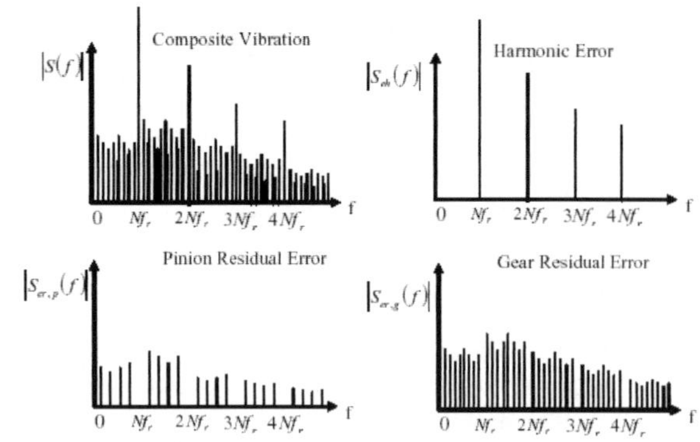

Fig 1. Graphical Representation of Equation 7

the residual error component due to the pinion $s_{er,p}(t)$, and the residual error component due to the gear □□□□ $s_{er,g}(t)$.

$$s(t) = s_{ch}(t) + s_{er,p}(t) + s_{er,g}(t) \quad (7)$$

Equation 7 is expressed graphically in figure 6. Notice how important good spectral resolution is when separating the vibration produced by the pinion and the gear.

The situation is made increasingly difficult when monitoring multiple gearsets. It should be apparent that several frequency components in the spectrum will be sufficiently close together so that it may not be practical to resolve them with reasonably sized FFTs. In this case, it is possible for information from another gear to leak into a different gear's residual error signal. There is no easy way of dealing with this situation when performing time domain averaging. This is part of the motivation behind developing the Comblet basis function (Dempsey & Zakrajsek 2001).

3. COMB FILTERS

We know that a comb filter is the most natural way to extract fault information from a composite vibration signal. Then we can do away with the adaptive sampling needed to compute the motion error signals. This is accomplished by giving the comb filter "teeth" a small but finite passband. Thus, even though we do not hit the shaft rate frequency bins exactly, we are still able to extract fault information. This approach also allows us to incorporate shaft speed measurement uncertainty into the filter design. This small passband can also capture variability in the shaft speed. Additionally, a modular wavelet basis will also allow us the flexibility to neglect "ghost" frequency components that mask early fault detection in multiple gear systems. Which is to say, there will be frequency bins that are contaminated by information from other vibration sources, which we can selectively neglect in our modular wavelet design? To my knowledge, the Comblet transform is the only signal processing technique for machinery diagnostics that addresses this issue.

An ideal comb filter is a unit amplitude periodically spaced impulse train in the frequency domain (Max 1981*)*. A finite impulse response (FIR) comb filter is designed to increase the signal to noise ratio for periodic signals in broadband noise. The z transform of a FIR comb filter with D unity gain peaks at $\omega_k = \frac{2k\pi}{D}$ and D zeros at $\omega_k = \frac{(2k+1)\pi}{D}$

$$H_{comb}(z) = b\left(\frac{1+z^{-D}}{1-az^{-D}}\right) \quad (8)$$

where the parameters 'a' and 'b' are determined by the half power bandwidth, Δω, from the following design equation

$$\beta = \tan\left(\frac{D\Delta\varpi}{4}\right) \quad (8)$$

4. CONCLUSION

In real functioning conditions, the gearing process has certain deviations versus the ideal conditions. These deviations are provoked both by the execution errors and the other transmission elements of the toothed wheels, and by the assembling errors.

The dynamic loads that appear in these conditions can be considerable, in comparison with the static forces, and their being token into consideration at the gearing planning is compulsory.

The toothed wheels transmission dynamic is influenced by the following facts:

- the rigidity variation of the gearing due to the variable deformations of the teeth in the process of gearing (the load is transmitted by a different number of teeth).
- the technological errors of the gearing
- the rotation speed, especially in those zones that correspond to the resonance phenomenon

The interior sources are represented by the deviations from the tooth-processing precision, especially the error of the measured step on the basis circle, that lends to the appearance of the periodical percussion between the teeth and creates a short term dynamic load and the profile error that creates a permanent dynamic load, as well as the periodic variation of the gearing rigidity, due to the periodic passing of the load from one tooth to two teeth. These sources are of a great interest for the gearing durability. The vibrations generated by these sources and together with them the dynamic forces and the noise become very strong high, especially when the frequency of the perturbation sources which is always in a relation determined by the gearing revolutions superposes on a frequency of it's own – the resonance phenomenon appears (Randall 1990).

The diagrams in fig.2, have been mode in order to diagrams the gearbox: - the diagrams of the signal acquired in time of the power spectrum in frequency and the cepstrum in time for the faultless gear box, considered as reference.

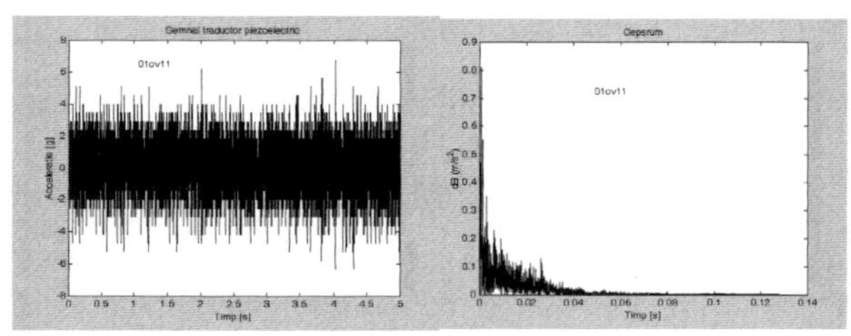

Fig.2. Diagrams of the signal

5. REFERENCES

Dempsey, P.J., & Zakrajsek, J.J., (2001) *Minimizing Loan Effects on NA4 Gear Vibration Diagnostic Parameter*, NASA/TM-2001-210671,

Fhay, K., &Perez, E., (1990), *Fast Fourier Transforms and Power Spectra in Lab. VIEW*, National Instruments Corporation, Application Note 040.

Max, J. (1981*). Méthodes et techniques de traitement du signal et applications aux mesures physiques.Methods and techniqes for signal analyse on phisical measurements.* Paris, Ed. Masson.

Randall, R.B., (1990), *Cepstrum Analysis and Gearbox Fault Diagnosis,* Bruel & Kjaer, Application Note, 1990.

Zaveri, K.(1985). *Modal Analysis of Large Structures Multiple Exciter Systems.* Bruel & Kjaer,

Annals of DAAAM for 2008 & Proceedings of the 19th International DAAAM Symposium, ISSN 1726-9679
Editor B. Katalinic, Published by DAAAM International, Vienna, Austria 2008
Make Harmony Between Technology and Nature, and Your Mind will Fly Free as a Bird

TOOL SHEAR STREES AND TEMPERATURE DISTRIBUTIONS PREDICTION USING FEM SIMULATION

PATRASCU, G[abriela] & CARUTASU, N[icoleta] L[uminita]

Abstract: *Metal cutting research improves the design of cutting tools, machine tool structures and the optimal planning of machining operation. Cutting forces, torque and power values are used in sizing machine tool structures. The stress and temperature field in the cutting tool edge, chip and finish work piece surface are used in designing cutting edge shape as well as optimizing feed, speed and depth of cut to avoid residual stresses on the finished surface. This paper shows the effect of cutting parameters on cutting forces in turning process with TNMG 332 insert simulation. TNMG insert was modelled using Catia V5 software and imported in FEM software. Orthogonal cutting was simulated using AdvantEdge 4.5 software developed for machining simulation.*

Key words: *shear stress distribution, temperature estimated distribution, FEM simulation, orthogonal machining model.*

1. INTRODUCTION

Computer simulation has been an expanding field being used in research studies and increasingly used in industrial applications for tool wear (Carutasu 2007). Many companies have software on the market to simulate tool wear for a wide variety of machining operations. These simulations provide information on how the cutting tool will react and respond to the workpiece properties. The simulation outcome depends largely on the model and laws it follows (Oxley 1998).

Many of the simulations use finite element analysis as a base to evaluate the phenomena (Jawahir 1991).

In the recent decades, with the emergency of more and more powerful computer and the development of numerical technique, numerical methods such as finite element method (FEM), finite difference method (FDM) and artificial Intelligence (AI) are widely used in machining industry (Kalhori 2001). Among them, FEM has become a powerful tool in the simulation of cutting process because various variables in the cutting process such as cutting force, cutting temperature, strain, strain rate, stress, etc can be predicted by performing chip formation and heat transfer analysis in metal cutting, including those very difficult to detect by experimental method (Huang et.al., 2003). Therefore a new tool wear prediction method may be developed by integrating FEM simulation of cutting process with tool wear model (Jawahir et.al., 1993).

Cutting tools are subjected to an extremely severe rubbing process. They are in metal-to-metal contact, between the chip and work piece, under conditions of very high stress at high temperature. The situation is further aggravated due to the existence of extreme stress and temperature gradients near the surface of the tool.

Interfacial friction on the tool rake face is not continuous and is a function of the normal and frictional stress distributions.

According to Zorev, the normal stress is greatest at the tool tip and gradually decreases to zero at the point where the chip separates from the rake face. The frictional shearing stress distribution is more complicated. Over the portion of the tool-chip contact area near the cutting edge, sticking friction occurs, and the frictional shearing stress is equal to the average shear flow stress at tool-chip interface in the chip. Over the remainder of the tool-chip contact area, sliding friction occurs, and the frictional shearing stress can be calculated using the coefficient of friction.

2. CASE STUDY

This paper presents the current modelling capabilities available in AdvantEdge 4.5 software to simulate metal cutting environment in turning process.

AdvantEdge machining modelling software is a central difference explicit finite element code using a Lagrangian mesh. The material model accounts for elastic-plastic strains and has an isotropic power law for strain hardening. The strain rate also affects the flow stress.

The material properties are temperature dependent and thereby it also accounts for thermal softening. A staggered method for coupled transient mechanical and heat transfer analysis is utilized. First an isothermal mechanical step is taken followed by a rigid transient heat transfer step with constant heating from plastic work and friction. Both steps have identical meshes. The central difference scheme is also used for the time integration in the thermal analysis. A six-node quadratic triangle element is used. The mesh, which becomes much distorted around the cutting edge, is periodically updated both refining large elements and coarsening small elements.

For turning process simulation it was used a plane strain deformation model. The insert and a part of work piece were meshed in order to have a practical number of elements for calculations. Work piece was made of Romanian OLC45 steel (AISI 1045). The TNMG 332 insert with PF chip breaker geometry (figure 1) were made available in STL form, generated from CATIA V5R8 system (Patrascu, 2007).

The cutting process parameters were:

- Cutting depth: 0.635 mm.
- Feed: 0.254 mm/rot.
- Cutting speed: 200 … 400 m/min.

For the proposed orthogonal machining model, cutting conditions and the material properties of the workpiece are the inputs. The outputs are process related variables, such as tool stress distributions, tool wear and temperature distribution in the chip and along the tool-chip interface (Patrascu, 2007).

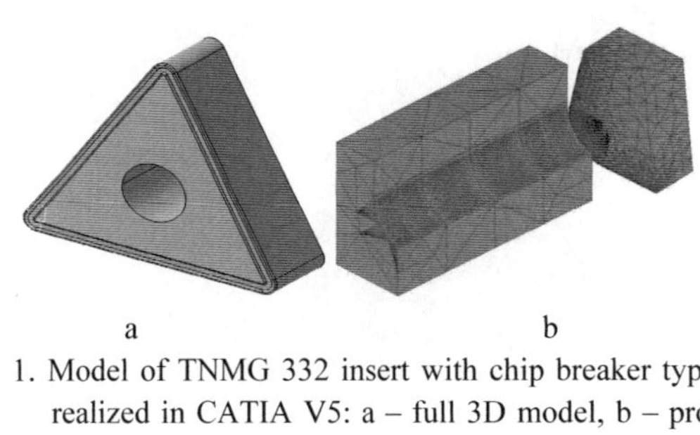

a b

Fig. 1. Model of TNMG 332 insert with chip breaker type PF realized in CATIA V5: a – full 3D model, b – process model in AdvantEdge.

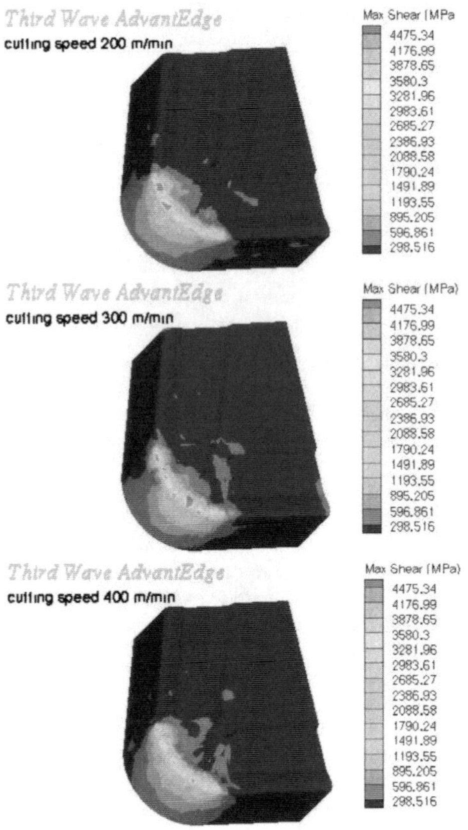

Fig. 2. Maximum shear stress distribution for 0.635 mm depth of cut.

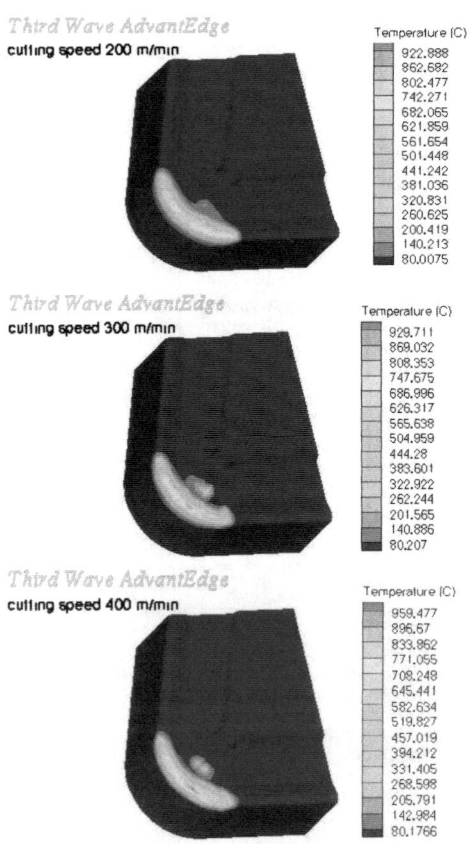

Fig. 3. Temperature distribution for 0.635 mm depth of cut.

Our FEM results provided a good estimation of temperature and shear stress values. The simulation is good for observing the influence of cutting conditions (speed, feed and depth of cut) and tool geometry. Comparison of computed values and the values from other studies on shear stress and temperature show a good agreement, even if FEM results overestimate the temperature levels.

The FEM results indicate a greater region of the tool to be above 1100°C, and maximum shear stress above 4900 MPa.

The results of our study prove the effectiveness of FEM simulation of turning process in order to select the optimum cutting process parameters.

3. CONCLUSION

This paper introduces a predictive modelling technique to determine forces, stresses, and temperature distributions in machining while considering influence of tool flank wear. The technique introduced in this paper combines oblique moving band heat source theory with nonuniform heat intensity at tool-chip interface and modified Oxley's parallel shear zone theory with ploughing effects due to tool flank wear to predict cutting forces, stress, and temperature distributions.

The proposed technique has been applied to machining of AISI-1045 steel using a carbide tool, and promising results have been obtained. The results have helped explain the heat partition behaviour of the tool-chip and tool-workpiece interfaces as width of flank wear increases.

Further studies are underway for improving the model to study the effects of chip-groove parameters on the natural contact length, fracture, the tool temperature and contact stress distribution, etc.

Acknowledgments: The authors would like to thank Prof. Jawahir I.S. from University of Kentucky for his support and Mr. Luis Zamorano - Applications Engineer from Third Wave Systems for the use of five months free evaluation license of AdvantEdge 4.5 software and for his support in using this license.

4. REFERENCES

Carutasu, N.L. (2007). *Contribution Regarding Designing Machines-Tools Structure Elements for High Speed Machining*, Ph.D. Thesis, University POLITEHNICA of Bucharest, Bucharest, Romania.

Huang, Y., &Liang, S.Y. (2003). *Modelling of the Cutting Temperature Distribution Under the Tool Flank Wear Effect*, Proc. Inst. Mech. Eng., Part C: J. Mech. Eng. Sci., 217, pp. 1195–1208.

Jawahir, I.S. (1991). *An Investigation of Three-Dimensional Chip Flow in Machining of Steels with Grooved Chip Forming Tool Inserts*, Transactions of NAMRI/SME, Vol. XIX, pp. 222-231, 1991.

Jawahir, I.S. & van Luttervelt, C.A. (1993). *Recent Developments in Chip Control Research and Applications*, Annals of the CIRP, Vol. 42 (2), 1993, pp. 659-693.

Kalhori, V. (2001). *Modelling and Simulation of Mechanical Cutting*, Ph.D. Thesis, *Luleå University of Technology, Sweeden.*

Oxley, P. (1998). *Development And Application Of A Predictive Machining theory*, Proc. CIRP International Workshop on Modeling of Machining Operations, Atlanta, GA, USA, May 1998, ISBN 0-9666706-0-4, University of Kentucky.

Patrascu, G. (2007). *Research Concerning the Optimization Through Simulation of the Cutting Process*, Ph.D. Thesis, University POLITEHNICA of Bucharest, Bucharest, Romania.

Annals of DAAAM for 2008 & Proceedings of the 19th International DAAAM Symposium, ISSN 1726-9679
Editor B. Katalinic, Published by DAAAM International, Vienna, Austria 2008
Make Harmony Between Technology and Nature, and Your Mind will Fly Free as a Bird

INFLUENCE OF THE LIGHTING AND THE ISO SPEED ON THE DIGITALLY SHOT BLACK AND WHITE PORTRAIT PHOTOGRAPHS

PAVLOVIC, I[vana]; MIKOTA, M[iroslav] & SKALA, T[ibor]

Abstract: *Gallery portrait art photographs because of its syntactic and semantic characteristics are often achieved in B&W technique. The aim of this paper is to contribute to the optimization of the digital shooting in the area of B&W portrait photography. In B&W shooting mode Munsell Color X-rite ColorChecker and face portraits were shot under the same condition ISO speeds 100/21, 400/27 and 3200/36 with flashlight, halogen, nitraphot and tungsten lighting were used. Reproduction curves and optical coverage for skin colours were defined and visual estimations, in conditions defined with standard ISO 9358, were done. Among tested lightings, flashlight gives the worst results while other give better providing more possibilities in light and shadow modulation. The evaluations of used ISO speeds show that digital photographic system enables using of the slightly higher speeds which enables working with the lighting of lower power but can be certain lack in photography syntaxes, as well.*

Key words: *portrait, lighting, ISO speed, tones.*

1. INTRODUCTION

From the very beginnings photography was developed as a specific portrait technique. With the arrival of the calotype process (patent from 1841) the development of black and white (B&W) negative-positive system began and, as a portrait technique, is still present, especially in the area of art gallery photography. The development of digital photography in the end of the 20th century and in the beginning of the 21st century imposes digital photographic system in this area. Most of the analysis and the development of digital photography are headed towards colour technique while portrait gallery photography, because of its semantic and syntactic characteristics, is often achieved in black and white technique.

The aim of this paper is to analyze influence of the lighting type and choosing ISO speed on the characteristics of the digitally shot portrait photographs and to contribute to the optimization of the digital shooting in the area of gallery black and white portrait photography.

2. THEORETICAL PART

Black and white photography is based on the effect of the modulation of the lights and the shadows and the impression of the volume and permeating of volume and space is created on that effect (Schnelle-Schneider, 2003.). With that, the scheme and the range of tones become main element of the photographic black and white syntax. Portrait is photographically very often treated as a black and white motive which uses associative character of the photographic medium more than iconic one. In the classic photography whole range of the different films for black and white photography is developed which with the suitable processing, give negative of desired characteristics and classic black and white photography beside semantic and syntactic standards has also put technical standards which digital photography, in the area of portrait photography, has to fulfil to become competitive in the area of the art gallery portrait photography (Hurter, 2005.). This is specially connected with the ISO speed and grained structure of the photograph with the wide range of creative demands – from fine-grained fine structure to stressed grains, tonal range (from high key to low key photography) and hardness (contrast) of the picture. Digital portrait shooting uses lighting which is also used in classical portrait photography and the effects which a photographer in classical photography defines with choosing the film type, are defined with settings on the digital photographic camera (Rice, 2005.). In the technical sense tones in black and white photography are traced through tracing of density. Maximal density shows how large is the entire range of tones achieved by the printing, and also shows the contrast and the dynamical range of the photograph. However, photography is experienced as a unique picture and an observer observes tones inside that particular photograph and very often without comparison with the tonal range of some other photograph. Because of that the optical coverage of the single field expressed in percentages is very important for the mark of the tones. It is calculated by the formula:

$$OC = ((1 - 10^{-D_f}) : (1 - 10^{-D_{max}}))\ 100 \qquad (1)$$

Where OC presents optical coverage expressed in percentages, D_f presents density of the field, and D_{max} presents density of the black field. This kind of determining of the optical coverage makes possible comparison of the tonal transmission through the single steps of certain photographic system (Mikota et al., 2005.). On the basis of determining the optical coverage the reproduction curve which represents dependence of the output optical coverage on the input optical coverage is also defined (Knešaurek & Brozović, 2006.).

3. EXPERIMENTAL PART

For experimental part in black and white shooting mode Munsell Color X-rite Color Checker and face portrait were shot. Colour chart and portrait were shot separately under the same conditions. For shooting leica digital photographic camera Canon EOS 5D in combination with Canon lens EF 70-300 mm F4-5,6 IS USM was used. For lighting flashlight Canon Speedlite 430 EX (6000 K), halogen reflector Kaiser 1000 (3400 K), nitraphot reflector Osram 500 (3400 K) and tungsten reflector Philips 300 (3200 K) were used.

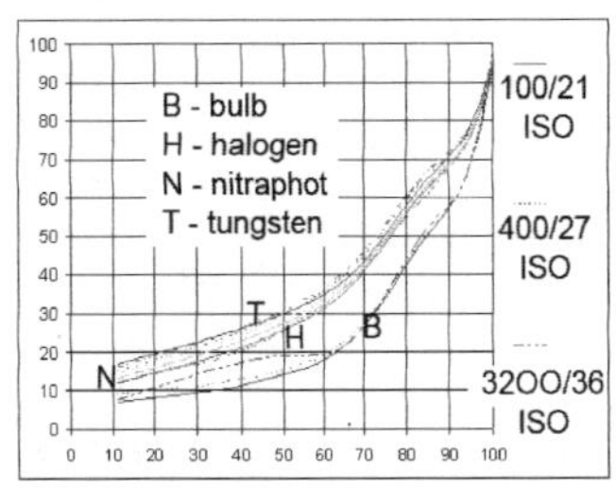

Fig. 1. Reproduction curves

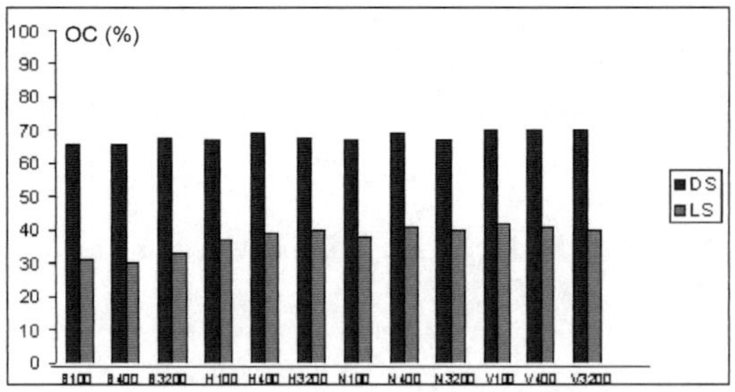

Fig. 2. Optical coverage for the light (LS) and the dark (DS) skin colour.

Lighting was put from the direction of the photographic camera. Light was measured with TTL system of the photographic camera with segment light metering on the standard colour of light and dark skin. With the "over all" integral light metering the other colours were also considered. Samples (colour chart and portrait photographs) were shot with the ISO speeds 100/21 ISO, 400/27 ISO and 3200/36 ISO. Shot photographs from the CF (Compact Flash) card were uploaded with the memory card reader after that. For shot photographs of the colour chart optical coverage for the fields from white to black and standard skin colours were determined by the Adobe PhotoShop CS3 programme and the reproduction curves were made. In the visual estimation 20 observers, who are professional users of the classical and the digital photographic system, participated. They were evaluating the quality of the shot portrait photographs by criterion of skin tones. Observers were estimating portraits (reproductions on the computer monitor) in conditions defined by standard ISO 9358 with the marks from 1-5. Observers also had to determine which ISO speed was used while shooting (by the experience of the gained effect in the classical photography systems).

4. RESULTS

Figure 1 shows reproduction curves for the different lighting types and different ISO speeds. Figure 2 shows optical coverage for the light and the dark skin colour. Table 1 shows results of the visual estimations of the shot black and white portraits and Table 2 shows evaluation of the ISO speeds that were used while shooting.

5. DISCUSSION

Reproduction curves of the black and white photographs (Figure 1) show that the flashlight shooting gives small contrast of the photograph in the area of low and middle optical coverage and higher contrast in the area of high optical coverage. The consequence of this is relatively small optical coverage for the light skin colour in the relation with the photographs, shot by the other light sources (Figure 2). Shooting by other light sources gives more similar and equalised reproduction curves (Figure 1) which results by "more pleasant" tones on the photographs in the area that suits the skin colour (Figure 2). In the visual estimation (Table 1) the best marks have the portraits shot with the nitraphot and the tungsten lighting. The worst marks had black and white portraits shot by the flashlight and that is obviously the consequence of worse possibility of the flashlight in the modulation of the light and the shadows. Generally, mark of the black and white portrait photographs wasn't in the essential dependence on the ISO speed used while shooting (Table 1).

Photograph	B 100	B 400	B 1200	H 100	H 400	H 1200
Mark	4	4	4,1	4,7	4,6	4,4
Photograph	N 100	N 400	N 1200	T 100	T 400	T 1200
Mark	4,9	4,9	4,9	4,8	4,7	4,8

Tab 1. Results of the visual estimation.

Real speed / Estimated speed	100/21 ISO	400/27 ISO	3200/36 ISO
50/18 – 200/24 ISO	20	8	0
200/24 – 400/27 ISO	0	12	0
400/27 – 800/30 ISO	0	0	9
800/30 – 1600/33 ISO	0	0	11
1600/33 ISO or more	0	0	0

Tab 2. Evaluation of ISO speeds.

Evaluation of the ISO speed (Table 2) shows that observers generally marked portraits shot with the ISO speed 400/27 ISO and 3200/36 ISO as photographs shot with lower ISO speed what is specially expressed at the portraits shot with 3200/36 ISO (neither one photographer-expert has evaluated that portraits were shot with ISO speed from 1600/33 ISO or more).

6. CONCLUSIONS

Among tested lightings, typical for portrait photography, in the area of black and white photography the worst results were given by flashlight. Other light sources give better results and make possible better modulation in the relation of lights and shadows in the area of skin colours which is the main expressive medium of the black and white portrait photography. Evaluations of used ISO speeds show that digital photographic system enables using of the slightly higher ISO speeds (in comparison with classical black and white negative-positive system) without visible grain structure of the picture what, in the portrait photography, enables working with the lighting of lower power which is more pleasant for the model. However, less visible grain picture structure with higher ISO speeds (for example 3200/36 ISO) can be certain lack in photography syntaxes, especially for the working in the low light photography technique which is very often in the black and white portrait photography.

The investigation pointed out that the digital shooting is competitive with the classical black and white negative-positive system also in its most demanding area – the area of gallery portrait art photography although in some elements it still falls behind. Further investigations of the phase of shooting in this area will potentially enable further optimization of shooting phase and defining of the suitable shooting stiles and professionally high quality shooting of portraits in JPEG format, that is, in the format with the compression. Also, further investigations will be headed on optimisation of system shooting - editing - printing to gain optimal results in the area of gallery black and white digital portrait photography.

7. REFERENCES

Hurter, B. (2005), *Portrait Photographer's Handbook*, Amherst Media, ISBN-10: 1584281405, Buffalo, New York

Knešaurek, N. & Brozović, M. (2005). The influence of substrate on tonal artwork reproduction, In: *Graphic Arts: Tecnology, Design, Communications – Scientific Book 2005*, Bolanča, Z. & Mikota, M., (Ed.), pp. 51-58, Faculty of Graphic Arts Zagreb, Ogranak Matice hrvatske Senj, Pulp and Paper Institute Ljubljana, ISBN 953-96020-3-3, Zagreb

Mikota, M.; Kropar-Vančina, V.; Mitrović Kukoč, B. (2005). Comparison of the Main Digital Photography Printing Techniqes on Glossy Photo Papers, In: *Graphic Arts - Technology, Design, Communication - Scientific Book 2005,* Bolanča, Z. & Mikota, M., (Ed.), pp. 37-49, Faculty of Graphic Arts, Ogranak Matice Hrvatske Senj, Pulp And Paper Institute, ISBN: 953-96020-3-3, Zagreb

Rice, R. (2005). *Professional Techniques for Black And White Digital Photography,* Amherst Media, ISBN: 1584281499, New York

Schnelle-Schneyder, M. (2003). *Sehen Und Photographieren: Von Der Aesthetik Zum Bild (Seeing and Photographing: From Estetic to Image),* Springer Verlag, ISBN: 978-3-540-43825-0, Berlin

Annals of DAAAM for 2008 & Proceedings of the 19th International DAAAM Symposium, ISSN 1726-9679
Editor B. Katalinic, Published by DAAAM International, Vienna, Austria 2008
Make Harmony Between Technology and Nature, and Your Mind will Fly Free as a Bird

CRITERIA DESIGN OF OPTIMALIZATION ULTRASONIC CONCENTRATORS

PECHACEK, F[frantisek]; CHARBULOVA, M[arcela] & JAVOROVA, A[ngela]

Abstract: *Paper deals about ultrasonic tool resonator. One part of this ultrasonic system is concentrator. The oscillations are amplified with this concentrator. Concentrator permits optimal matching of impedance between transducer and technological process. This is one of the most important concentrator attributes. Optimalization of ultrasonic tool resonator allows achieving required amplification of oscillations without endurance strength value excess. Endurance strength value depends to resonator material. Optimalization these resonators are based on laws and terms but on practice experience also.*
Key words: *Ultrasound, Machining, Concentrator, Ttransducer, Waveguide*

1. INTRODUCTION

Technology of ultrasonic machining requires use of ultrasonic tool resonator, which is the part of ultrasonic resonant system.

Term of ultrasonic system in machining includes whole set of components and parts, which are needed for realization, whereby those parts are energetic and mechanically connected. Ultrasonic system in machining performs transfer of incoming electric energy to outgoing mechanical energy of the tool, or to oscillating movement of the tool.

Ultrasonic resonant system consists of:

- ultrasonic transducer,
- ultrasonic concentrator,
- waveguide.

2. ULTRASONIC RESONANT SYSTEM PARTS CHARACTERISTICS

2.1 Ultrasonic transducer

Transducer is transforms electric energy of the generator to mechanical ultrasonic energy. Transducer generates harmonic deviation in longitudinal direction, as surface source of forced mechanical oscillation for ultrasonic resonant system.

In engineering praxis are used mostly piezoceramic transducers, which have lesser dimensions, larger use variability and higher efficiency and power stability, than magnetostrictive transducers. On the fig. 1 are schemes of piezoelectric ultrasonic transducers (a - symmetric arrangment of transducer‘s piezoceramics, b - asymmetric arrangment). They are characterized by lower material cost, undemanding technology of manufacturing and simplicity of design for various frequencies in used wave-length.

Nowadays all of the important manufacturers offer ultrasonic transducer on the basis of piezoceramic materials, where the transformation of the electric energy to mechanical energy depends on piezoelectric properties of the materials.

For power applications is necessary, that transducers transforms the energy under high deformations. By this reason must the piezoceramics have high toughness, quality and good piezoelectric properties. In term of process efficiency is necessary, that transducers are characterized by low electric losses under high electric intensities (Maňková, 2000).

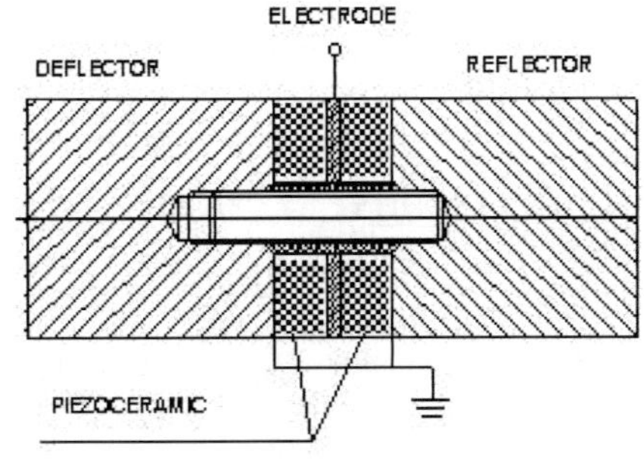

a

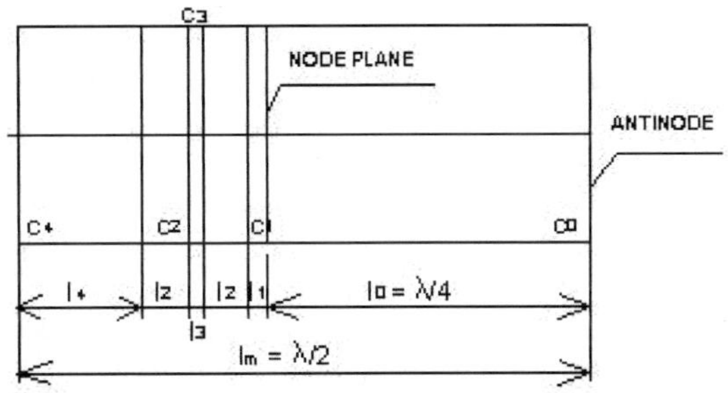

b

Fig.1 Model of the piezoceramic transducer with constant crosswise section
a - symmetric arrangment of transducer‘s piezoceramics
b - asymmetric arrangment

2.2 Ultrasonic concentrator

The most important property within ultrasonic resonant system, has the concentrator, whose task is concentration of mechanical oscillation energy on outgoing front, what makes the harmonic deviation stronger.

Among basic shapes of resonators belong circular resonators with following shapes: conic, exponential, catenoid, gradual. Besides those basic types we recognize combined resonators, which are combination of basic shapes. (e.g. conic - cylindrical, exponential - cylindrical, cylindrical - exponential - cylindrical etc.)

Important properties of resonators in ultrasonic resonant system are: amplitude of voltage course, amplitude of mechanical deviation course, amplitude of voltage peak, node plane placement etc.

For achievement of required results in ultrasonic machining, must be the resonator, transducer and other parts of system in frequency tune (Dillinger et al., 1997).

2.3 Waveguide

Waveguide is special example of ultrasonic concentrator with zero amplification of amplitude of mechanical deviation, with constant profile of the intersection. On the fig.2 is scheme of the ultrasonic waveguide. Major task of the waveguide is lengthening of ultrasonic resonant system. From several interesting and used properties in engineering praxis is for

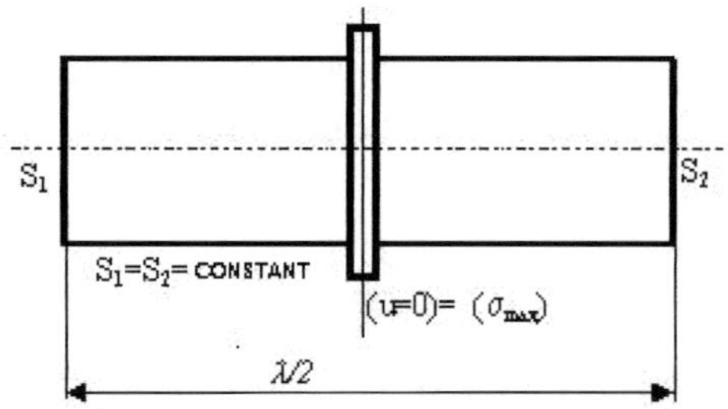

Fig.2 Scheme of the ultrasonic waveguide

example clear definition of node plane and amplitude of voltage peak, which is used for safe and clear gripping of the whole resonant system in technological application (Maňková, 2000).

3. OPTIMIZATION CRITERIA DESIGN

Optimization of ultrasonic tool resonators presents primarily achievement of required amplification of deviation amplitude, and not exceed allowed endurance strength, which depends on the material of the resonator. Important element of the optimization is choice of the material with good mechanical a physical properties, which depends on application. In choosing of the material is required to make provision for energy loss factor, endurance strength, abrasion resistance, machinability, availability and price.

To resonator optimization is required to know relations and patterns of calculation of particular types of resonators, but is also necessary to make provision for knowledge and results from the praxis (Tolnay, et al., 2000)..

According to that, practical criteria design is:

- Calculation of required values of particular parameters by means of known patterns for ultrasonic resonators calculation and substitution of the specific values.
- Comparison of calculated values and knowledge from praxis, to designation of advantages and disadvantages.
- Choosing the resonator for required application of ultrasonic machining.

Materials, which are used in ultrasonic tool resonators must have good mechanical properties, characterized by energy loss factor and longitudinal speed of wave propagation (Wilk &Tota, 2007).

Input parameters for resonator calculation without technological load:

- Amplitude of input harmonic deviation of the resonant system - u_x.
- Modulus of elasticity of concentrator material – E.
- Resonant frequency of the concentrator – f.
- Density of material of the concentrator - ρ.
- Dimensions of input and output diameter of the concentrator - D_1 and D_2.
- Function parameters of the geometric representation of the concentrator - S.

Besides of that, to calculation of the concentrator with technological load:

- Values and course of the technological load.

Output parameters of the concentrator without technological load:

- Placement of node plane on the concentrator.
- Placement of maximum amplitude of voltage on the concentrator.
- Value of the maximum amplitude of voltage on the concentrator.
- Acoustic amplification on the concentrator.
- Resonance length of the concentrator – l.
- Course of deviation amplitude on the whole length of the concentrator.
- Course of voltage amplitude on the whole length of the concentrator.

Besides of that outputs we use for concentrators with technological load:

- Formulation of single vibration work.
- Formulation of delivered technological input of the technological application process.

For achievement of equable distribution of mechanical tension along the axis and for larger amplification are used combined resonators. Those are combined of several parts of constant and variant intersection, whereby the basic shapes are used in combination.

Special solutions of the ultrasonic resonant system are combinations of the three basic parts, transducer, concentrator and waveguide integrated to one unit. By this design, the length of the ultrasonic resonator is reduced and there is a possibility to reduce spatial demandingness (Melo et al., 1997).

4. CONCLUSION

From calculated values, crafted graphic relations for particular shapes, data form literature and information form praxis is possible to evaluate the shapes of resonators for ultrasonic machining. In choosing of optimal concentrator shape is necessary to consider all advantages and disadvantages and choose the most favorable resonator for particular ultrasonic application.Way of designing and optimization of the ultrasonic resonators by means of mathematical patterns and graphic relations is suitable only for basic shapes of resonators. This way is not accurate for designing of combined resonators and is very work-intensive as well. More suitable is to use a software, which is designed for this task.

This paper was created thanks to the national grants:
VEGA 1/009/ 08
Optimalized systems and processes of performance ultrasound

5. REFERENCES

Dillinger, J.; Tolnay, M. & Mihalčák, P. (1997). Rezonančné systémy priečnej transformácie výkonového ultrazvuku. (Rezonance systems of cross transformation of power ultrasound.), *Proceedings of Zborník inžinierstvo 1997,* SjF STU Bratislava, 1997, Strojnícka fakulta STU, Bratislava

Maňková, I. (2000). *Progresívne metódy obrábania. (Progresive methods of machining)*, Vienala, ISBN 80–7099–430, Košice

Melo, Š.; Bigoš J. & Mihalčák, P. (1997). Analýza a optimalizácia ultrazvukových koncentrátorov. (Analysis and optimation of ultrasonic concentrators) *Proceedings of Zborník vedeckých prác 1997,* pp. 105–121, MtF STU Trnava, 1997, Strojnícka fakulta STU, Bratislava

Tolnay, M.; Greguš – Kollár, J. & Mihalčák, P. (2000). Konštrukcia nástrojov pre ultrazvukové obrábanie. (Ultrasonic tool design) *Proceedings of Zborník Náradie 2000*, pp. 185–188, Trenčín, 2000, SOPK, Trenčín

Wilk, W. & Tota, J. (2007). Modern ultrasonic machines in shaping of materials. *Proceedings of Zborník International Congress on Precision Machining 2007*, pp. 314-320, 2007, Praha

Annals of DAAAM for 2008 & Proceedings of the 19th International DAAAM Symposium, ISSN 1726-9679
Editor B. Katalinic, Published by DAAAM International, Vienna, Austria 2008
Make Harmony Between Technology and Nature, and Your Mind will Fly Free as a Bird

FINANCIAL STATEMENTS ANALYSIS WITH USAGE OF COMPUTER APPLICATION

PEJIC BACH, M[irjana]; JAKOVIC, B[ozidar] & MISULIC, M[arko]

Abstract: *Computer applications have a wide range of use in different areas of business. This paper presents an application that greatly assists with analyzing financial statements. The AFI application enables the user a clear, simple and quick view of the key business indicators of companies in order to make an informed and timely decision based on that data. The time needed to analyze financial statements is considerably reduced and the possibility of errors, that derives from applications that require copying data from the financial reports into their own modules, has been completely eliminated. With the AFI application it takes less than a second to complete an analysis of financial statements. Functionality, simplicity of use and easy access are the main characteristics of the existing AFI applications.*

Key words: *Computer application, financial statement analysis.*

1. INTRODUCTION

Financial statement analysis is a complicated process in itself. For one person to calculate fundamental and other financial ratios from a financial statement of a certain company it is necessary to copy all fields from the financial statement, which takes over an hour, and then make a calculation of the indicators from the data. The possibility of error during such of process of analyzing statements is considerable. Since having correct information at the right time is often a crucial factor in making investment decisions, especially in the dynamic capital market, procrastination can result in missing a great investment opportunity. By using the computer application, we can save a lot of time and get valuable information to our investors in time.

2. FINANCIAL STATEMENT ANALYSIS

Financial statement analysis is an integral part of investment analysis the goal of which is to provide information on financial situation, management, plans and strategies which the company is undertaking as well as the business environment so that business decisions can be made based on information gathered (Wild at al., 2001).

The basic groundwork of the analysis: balance sheet, income statement, cash flow statement, (Žager et al., 2007). The quality of information that the financial statements contain implies that the financial standing of the company and business success is presented in a truthful and fair manner pertaining to all materially important points (Sever, 2007). The process of analysis contains of the application of analytical rules and techniques with the intention of deriving important and useful data which will be used when making financial decisions. The beneficiaries of these information can be the owners, creditors, auditors, the state and many others. The analysis determines the current state of the company, real and objective information on liquidity, financial stability debt and profitability; it defines causes and symptoms of financial problems within the company (Žager & Žager, 1999).

Financial statement analysis is one of the main sources of information for investors because it provides insight into the business and financial standings of a certain company.

3. AFI – COMPUTER APPLICATION

AFI – „Analiza financijskih izvještaja" (Financial statement analysis) is a computer application that analyses financial statements of joint stock companies in Croatia that are available in Microsoft Excel format. MS Excel is a program that deals with data in tables. Data that is organized in tables ate possible to analyze and graphically present and thus generate estimates and various reports (Vlahović & Jaković, 2007). MS Excel is a very powerful tool that can be used for m any purposes. In his article, Abbott describes the application of MS Excel in resolving problems encountered when running a library (Abbott, 1990). It can be used for surveys (Foo, & Hepworth, 2000) or different data analysis (Granebring & Révay, 2007). In this article its use in financial statement analysis will be shown. The application is written in Visual Basic for Applications The AFI application has and empty module for entry of items from the financial statement which broadens the functionality of the program and enables the analysis of statements created outside Croatia. The main strength of the AFI application is that it is not necessary to copy data from the financial statement but only to upload the financial statement that you wish to analyze. During the analysis the data from the statement are automatically transferred and the result of the complete analysis is ready in less than a second. The user immediately receives an overview of the fundamental indicators and the financial standing of the company along with a graphic display of the key indicators.

4. APPLICATION PROCEDURES

The user interface enables the user easy access to program options; it is necessary to do only two things in order to analyze the uploaded financial statement. First the financial statement needs to be uploaded after which the user will see the fundamental indicators and the financial standing of the company for the current time period and for the same period the previous year. Based on the uploaded financial statement, the AFI application calculates the following indicators: Enterprise value, EBITDA, EBIT, BVPS, EPS, SPS, ROA, ROE, P/B, P/E, P/EBITDA, P/EBIT, P/S, EV/EBITDA, EV/EBIT, EBITDA margin, EBIT margin, gross profit margin, liquidity ratios, asset turnover ratios, financial leverage ratios, profitability ratios, activity ratios, debt ratios…

Since banks and insurance companies have different financial statement fields (Orsag, 2003), a special module is s provided for such companies within the application which computes the special indictors for such companies. Along with the calculation of said indicators, the application provides insight into the financial standing of the company (balance sheet, profit and loss statement, cash flow) as well as information on management, supervisory board and stock holders on the last day of reports. In order for the analysis to be complete it is only necessary to write the regular stock price in the field as well as the privileged stock price (if there is such a stock option) in order to get the accurate calculation of the indicators that use the market stock price as one of the variables.

5. TYPES OF ANALYSIS IN AFI APPLICATION

The AFI application enables you to make several types of financial statement analysis: (1) financial statement analysis of the company compared to the same period the previous year, (2) financial statement analysis compared to the last two periods – trend analysis, (3) comparative analysis – comparing the financial statement analysis of a certain company to financial statements of other companies in the same field of business which is especially practical when creating sector analysis, (4) *Trailing* analysis – indicators calculated on the basis of the last four quarter reports.

6. ADDITIONAL OPTIONS OF THE APPLICATION

In addition to the basic display of financial and fundamental indicators and the financial standing of the company, the user also has the graphic display of the information. The AFI application graphically shows the structures of the assets and liabilities, the ratio of the income and expenses and the net result. Furthermore, the AFI application enables the user to copy the results to Microsoft Word.

During the analysis of the financial statement of a certain company the user can change particular items of the financial standing of the company (balance sheet, profit and loss statement, cash flow report) and in that way see how the calculation of the indicators changes when they are altered. The user can predict and project certain items in the financial standing of the company in the future periods and observe the influence of his projections on the fundamental and financial indicators. The comparative analysis enabled the user to analyze financial statements of different companies and compare them. This option is particularly useful when making sector analysis and investigating and comparing companies that have the same or similar activities.

7. PROBLEMS ENCOUNTERED IN CREATING THE APPLICATION

One of the biggest challenges and problem in creating this application was limiting and preventing the actions of the user that could damage the functionality of the application and thus damage the regularity of the applications operations. The second problem was protecting the program code of the application which needed to be hidden so that it is not available to the public in order to prevent unauthorized duplicating and distributing of the application. The application is patented at the Croatian author agency, Centre for intellectual ownership.

8. LIMITATIONS AND POSSIBLE DEVELOPMENT OF THE APPLICATION IN THE FUTURE

The automated process of analysis of financial statements is limited on statements in Microsoft Excel format. If the proscribed format changes, the functionality of the application will be seriously undermined. The possibility of developing and upgrading the AFI application is extremely great. Development of the AFI application will enable the analysis of financial statements from other neighboring countries (Slovenia, Serbia, Montenegro, Bosnia and Herzegovina...) Before such a project can be realized, it is necessary to resolve a string of legal and technical issues – from different accounting standards to uneven structures and forms of financial statements in those countries. Our goal is to turn the existing AFI application into a Web service (Internet portal) that would offer its users the existing AFI applications and additional analytical tools used for stock analysis, technical studies, implications of stock price in comparison to similar stocks from the same sector, DCF model, automated information for users on the new financial statements in the base, daily capital market reports with added overview of the indicators for all stocks.

The other direction of AFI development relates to the modern portfolio theory and its application in practice. A tool would be made for selecting the optimal portfolio, calculating the risk index as well as many other models which are regularly used in developed markets.

9. CONCLUSION

Stock market development in Croatia and neighboring countries will result in a greater need for analytical tools when making investment decisions. Investigating companies and analyzing their financial statements is the only rational method and, of course, employing all other methods used in estimating fair value of stocks. Every device that will facilitate the process of conducting the analysis, such as the AFI application, will be an invaluable tool for any serious investor.

10. REFERENCES

Abbott, C. (1990). Program: electronic library and information systems. MCB UP Ltd, Volume: 24, Issue: 3, Page: 269 – 279, DOI: 10.1108/eb047062.

Brown, K. & Reilly, F. (2003). Investment analysis and portfolio management. Thomson – Southwestern, USA

Foo, A. & Hepworth, M. (2000). The implementation of an electronic survey tool to help determine the information needs of a knowledge-based organization. Information Management & Computer Security, Volume: 8 Issue: 2, Research paper.

Granebring, A. & Révay, P. (2007). Service-oriented architecture is a driver for daily decision support. Kybernetes, Volume: 36 Issue: 5/6, Case study.

Orsag, S. (2003). *"Vrijednosni papiri"*. Securities, Revicon, Sarajevo. ISBN 9958-757-15-X

Sever, S. (2007). *"Uloga financijskih i nefinancijskih pokazatelja u revizorovoj ocjeni kvalitete poslovanja, magistarski rad"* (*The role of financial and non-financial indicators in revisionist business quality estimation - Master's Thesis*). Faculty of Economics and Business – Zagreb.

Vlahović, N. & Jaković, B. (2007). *"Računalne vještine za ekonomiste: Priručnik za vježbe iz informatike"* (*Computer art for economists: Guidebook for practice in informatics*), 1st issue, Mikrorad, Zagreb. ISBN: 978-953-6286-83-6.

Wild, J. J.; Bernstein, L. A., Subramanyam, K. R. (2001). Financial Statement Analysis. Seventh Edition, McGraw Hill - Irwin, New York, str. 12.

Žager, K. & Žager, L. (1999). *„Analiza financijskih izvještaja". (Financial statement analysis*), Masmedia, Zagreb. ISBN 935-157-100-7.

Žager, K., Tušek, B., Vašiček, V. & Žager, L. (2007). *„Osnove računovodstva – računovodstvo za neračunovođe" (Basis of accounting – accounting for non-accounters*), HZRFD, Zagreb. ISBN 978-953-6480-93-7

Annals of DAAAM for 2008 & Proceedings of the 19th International DAAAM Symposium, ISSN 1726-9679
Editor B. Katalinic, Published by DAAAM International, Vienna, Austria 2008
Make Harmony Between Technology and Nature, and Your Mind will Fly Free as a Bird
*Young Researcher and Scientist Paper / * Supervisor, Mentor*

DEFINITION OF ELASTIC MEMBRANE PRESSURE IN HYDROELASTIC DRAWING OF THE PARTS OF CYLINDRICAL FORM

PERMINOVA, A[nastasia] & DRESVYANNIKOV, D[enis]

Abstract: *Elastic membrane pressure in hydroelastic drawing is determined. Stress-strained state of billet flange and energy equation are analyzed. A mathematical model defining a law of pressure change in hydroelastic drawing is built.*
Key words: *hydroelastic drawing, pressure, clamp, stresses, energy equation of stability*

1. INTRODUCTION

Hydroelastic drawing are progressive methods of sheet-metal stamping allowing to decrease different thickness of produced parts. Thinning of the wall of a half-pnished product produced by drawing is one of the main indexes of its quality: the less deviation of the wall thickness of the initial sheet billet, the more qualitative it is.

Hydroelastic drawing with high pressure requires only one rigid part of the stamp. A special diaphragm consisting of three layers simultaneously serves as a clamping of a sheet billet and a flexible element of the stamp. Under the influence of hydroelastic pressure of the fluid transferred by the diaphragm, the billet is deformed, taking the form of the rigid part of the die (Isachenkov E.I., 1976).

2. THE CLAMP FORCE AND MEMBRANE PRESSURE

In drawing of cylindrical parts with clamping proper choice of clamp force *Q* of the billet to the matrix is very important. Higher magnitude of this force leads to undesirable increase of radial tensile stresses and material thinning in a dangerous section of the part. With insufficient force of clamping the flange loses stability and obtained crimps could form folds that leads to decrease of the part quality or to unrecoverable rejects (Mihaylov Y.O., Dresvyannikov D.G., 2003).

Tangential stresses occurring in the flange due to plastic deformation are accordingly changed with the radius decrease of outer contour of the billet *Rx*, or, which is the same, with increase of drawing depth. Pressure of clamping *q* necessary for preventing of crimp-forming on the flange would also changed. It follows that dependence of clamping pressure on drawing depth could be found with joint consideration of stressed state of the flange and condition of its stability.

As a calculated scheme of definition of critical flange parameters and required clamping pressure let's consider slightly curved form of equilibrium lying on the rigid bottom of a ring plate fastened along inner contour and loaded by force Q (Fig. 1).

Radial tensile and tangential compression stresses are assumed to be uniformly distribution along ring and radial section according and be equal to average integral values σ_ρ, σ_θ.

To determinate clamping pressure *q*, per unit of flange area at the given moment of drawing we use energy equation of stability which for the ring plate set in curved orthogonal system of coordinates α_1, α_2, *z* would have view (Birukov N.M., 1974):

$$Y = \frac{1}{2}\int_0^{\alpha_1}\int_0^{\alpha_2}\left\{ t\left[\left(\sigma_\rho x_1 + \sigma_\theta x_2\right)\omega_f - \frac{\sigma_f^2}{E}\right] - 2q\omega_f^* + D'_c\left(P_{1f}x_1^2 + P_{2f}x_2^2 + P_{3f}x_2x_2 + \chi^2\right)\right\} A_1A_2 d\alpha_1 d\alpha_2 = 0\,, \quad (1)$$

where are: ω_f, ω_f^* – current and maximum values of flange deflection in the direction of normal to its inner surface.

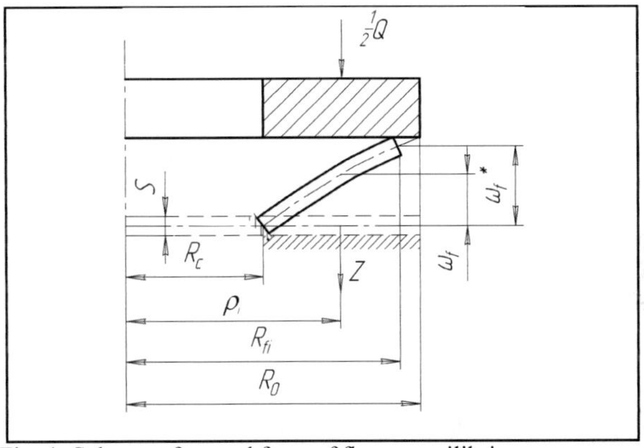

Fig. 1. Scheme of curved form of flange equilibrium.

$$P_{1f} = 1 - \frac{3}{4}\lambda^2(1-n)\overline{\sigma}_{1f}\,;\; P_{2f} = 1 - \frac{3}{4}\lambda^2(1-n)\overline{\sigma}_{3f}\,;$$
$$P_{3f} = 1 - \frac{3}{2}\lambda^2(1-n)\overline{\sigma}_{1f}\overline{\sigma}_{3f}\,. \quad (2)$$

For solving the equation (1) it is necessary for deflection ω_f to accept a suitable analytic expression that satisfies boundary conditions. Besides, this expression mustn't give positive values. In the case shown in Fig. 1, expression for ω_f may be taken in the form:

$$\omega_f = -a_f R_0\left(1 - \frac{R_c^2}{\rho_i^2}\right)\left[1 - \cos(n\varphi)\right], \quad (3)$$

where are: a_f – nondimensional coefficient; n_f – whole positive number (the number of waves to which the flange is divided in losing stability); φ – current angel calculated from the beginning of any wave (crimp).

When $\varphi=\pi$ and $\rho_i=R_{fi}$ deflection takes a minimum value

$$\omega_f^* = -2a_f R_0\left(1 - m_{cx}^2\right). \quad (4)$$

Taking into consideration equation 3 x1, x2, χ parameters characterizing change of curvature and twisting in buckling of the flange will be:

$$x_1 = -6a_f R_0 \frac{R_c^2}{\rho_i^4}\left[1-\cos(n_f\varphi)\right];$$

$$\chi = a_f R_0 n_f \left(\frac{3R_c^2}{\rho_i^4} - \frac{1}{\rho_i^2}\right)\sin(n_f\varphi);$$

$$x_2 = a_f R_0 \left\{ \left[\frac{n_f^2}{\rho_i^2} - \frac{R_c^2}{\rho_i^4}(n_f^2+2)\right]\cos(n_f\varphi) + \frac{2R_c^2}{\rho_i^4} \right\}. \quad (5)$$

After integration and transformation we receive

$$Y_f = \frac{2Ym_c^2}{\pi D'_c R_0^3 (1-m_{cx}^2)^2} = \quad ; \quad (6)$$
$$= \left(L_{1f} n_f^4 - L_{2f} n_f^2 + L_{3f}\right) a_f^2 + L_{4f} q a_f - L_{5f} = 0$$

$$L_{1f} = \frac{1}{6} Ð_{2f} (1-m_{cx}^2)^2; \quad (7)$$

$$L_{2f} = \frac{1}{2}\left[\frac{9\lambda}{2S^2}\bar{\sigma}_\theta \varepsilon_{if} m_c^2 \left(\frac{3-m_{cx}^2}{4} - \frac{\ln m_{cx}}{1-m_{cx}^2}\right) - \right.$$
$$\left. -\frac{1}{3}(1+m_{cx}^2 - 2m_{cx}^4)(3P_{3f} - 2P_{2f}) - 1 - 3m_{cx}^4\right]; \quad (8)$$

$$L_{3f} = \frac{27\lambda}{8\bar{S}^2} m_c^2 \varepsilon_{if} (1-m_{cx}^2)(3\bar{\sigma}_\rho - \bar{\sigma}_\theta) + \quad ; \quad (9)$$
$$+ 2(1+m_{cx}^2 + m_{cx}^4)(9P_{1f} + P_{2f} - 3P_{3f})$$

$$L_{4f} = \frac{9m_x^2\lambda}{2B\bar{S}^3} m_c^2 \varepsilon_{if}^{1-n} (1-m_{cx}^2); \quad (10)$$

$$L_{5f} = \frac{9B}{4E\bar{S}^2} m_x^2 m_c^2 \varepsilon_{if}^{1+n}. \quad (11)$$

Relative thickness of the billet $\bar{S}$ and intensity of deformation ε_{if} in (8)-(11) are defined from equations

$$\bar{S} = \frac{S}{2R_0}; \quad \varepsilon_{if} = \frac{2\varepsilon_{3f}}{\bar{\sigma}_{3f}-1}\sqrt{1-\bar{\sigma}_\rho\bar{\sigma}_\theta} \approx \mathrm{abs}(\varepsilon_{3f}). \quad (12)$$

Solving equations (6) we receive

$$q = \frac{1}{L_{4f}}\sqrt{L_{5f}\left(\frac{L_{2f}^2}{L_{1f}} - L_{3f}\right)}. \quad (13)$$

Meridian σ_ρ and tangential σ_θ stresses acting on the flange section (Fig.2) could be found in following dependencies is (Mihaylov Y.O., Dresvyannikov D.G., 2001)

$$\sigma_\rho = \sigma_t \frac{1}{1-\psi}\left(\frac{\varepsilon_\theta}{\psi}\right)^{\frac{\psi}{1-\psi}} \cdot \ln\frac{R_f}{R_0} + \quad ; \quad (14)$$
$$+ \frac{P_0(\mu-\mu_m)}{S_0}\cdot(R_f - R_0)$$

$$\sigma_\theta = \sigma_s - \sigma_\rho; \quad (15)$$

where is R_f – flange radius; P_0 – pressure in matrix; μ and μ_m – coefficients of friction of metal over metal and metal over elastic membrane; σ_t – tensile strength of the material.

Pressure in the matrix equals to clamp pressure q (Mihaylov Y.O., Dresvyannikov D.G., 2003) (Fig. 3):

$$P_0 = q \quad (16)$$

$$P_0 = \frac{1}{L_{4f}}\sqrt{L_{5f}\left(\frac{L_{2f}^2}{L_{1f}} - L_{3f}\right)}.$$

3. CONCLUSION

Stress-strained state of billet flange and energy equation are analyzed. A mathematical model defining a law of pressure change in hydroelastic drawing is built.

Drawing pressure at the initial stage is found from stability condition of billet flange; at the final stage the pressure is determined from the required force forming of a part.

The drawing pressure increases as the values of main parameters grow. The parameters are: tensile strength, relative thickness.

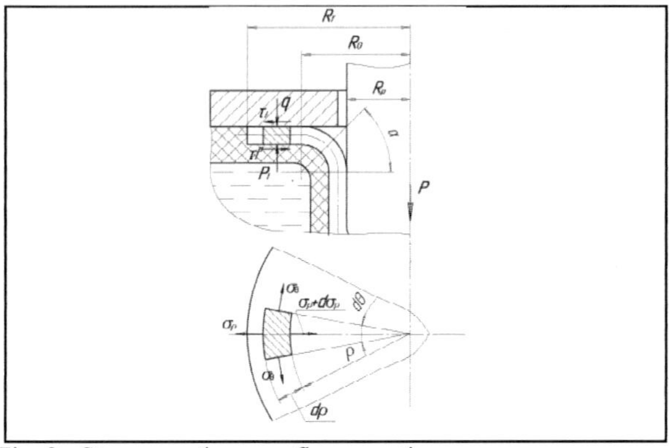

Fig. 2. Stresses acting on a flange section.

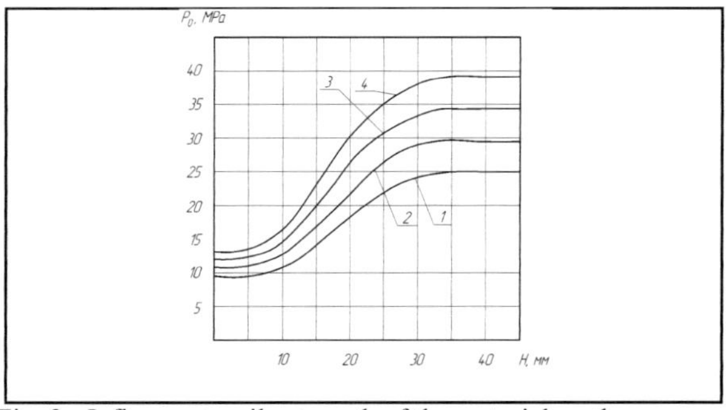

Fig. 3. Influence tensile strength of the material on the pressure in the matrix: 1 – σ_t = 300 MPa; σ_t = 350 MPa; σ_t = 400 MPa; σ_t = 450 MPa, H – depth of drawing, P^0 – pressure in matrix

4. REFERENCES

Birukov N.M. *Theoretical principles of obtain crimps in drawing pars from sheet billet.* Moscow, 1974, 291p.

Isachenkov E.I. Development of technology of punching by elastic liquid and gas environments. *Fogging and stamping manufacture*. No.7, 1976, pp. 2-5.

Mihaylov Y.O., Dresvyannikov D.G. Hydroelastic drawing of the parts of cylindrical form. *Vestnik of Izhevsk State Technical University*, No.2, 2001, pp. 60-63, Publisher ISTU, Izhevsk.

Mihaylov Y.O., Dresvyannikov D.G. Optimizing of pressure in the matrix in hydroelastic drawing. *Information technology in innovative projects.* pp. 94-95, Izhevsk, may 2003, Publisher ISTU, Izhevsk.

Mihaylov Y.O., Dresvyannikov D.G. Drawing by liquid and elastic medium as reduction thinning of the wall of parts. Higth technology. pp. 69-76, Izhevsk, may 2004, Publisher ISTU, Izhevsk.

Annals of DAAAM for 2008 & Proceedings of the 19th International DAAAM Symposium, ISSN 1726-9679
Editor B. Katalinic, Published by DAAAM International, Vienna, Austria 2008
Make Harmony Between Technology and Nature, and Your Mind will Fly Free as a Bird

THE PRESSURE CALCULATION FOR THIN CIRCULAR CLAMPED PLATES HYDROFORMING FROM ENERGETICAL REASONS

PERNEVAN, I[oan]

Abstract: *During the hydroforming of thin circular clamped plates the deformation is produced consequence to a work fluid acting, whereat theirs hydrodynamics parameters are known. From the theories of plasticity of the material subjected to deformation is known the minimum mechanical energy of deformation for an element of volume.*
Through the hydraulic energy assigned equalization with the mechanical energy necessary to deformation, the geometry of the deformed piece and the necessary pressure of deformation can be determined.
Key words: *pressure, hydroforming, energy, plasticity.*

1. INTRODUCTION

The aim of the work is to determine the expression of the hydroforming pressure for a thin circular clamped plate using the energetically criteria. The pressure determination problem is partially resolved by some complex analytical calculation procedures (Lee Wu, 1950; Timoshenko & Voinovski-Krieger, 1987; Mehdi BEN TAHAR, 2005). Hence i have proposed a more simple calculation procedure (method) that gives results with a good experimental validation.

The proposed method offers the possibility of development in after years for some theoretical and experimental researches regarding to the high plastic deformations.

2. THE NECESSARY HIDROFORMING PRESSURE CALCULATION

In concordance with the energetically plasticity hypothesis (Huber-Mises Hencky plasticity criteria) the plastic deformations arise in material when in the elementary volume is accumulate an amount of energy well determined. It has chosen the stress hardening σ_{ecr} as a characteristic dimension in the minimum deformation energy calculus because the plastic deformations are very big and the hardening phenomenon of the material is determinative in deformation forces calculus (Pernevan & Pernevan, 2005).The undeformed plate by circular disk shape with b radius and h_0 initial thickness is transforming into a spherical calotte by base radius b, poll height w and sphere radius of calotte ρ (Abrantes & Bataha, 2003), (Fig.1).

Employing the geometrical relationships for the circular disk and for spherical calotte by known geometries and employing the known relationships from plasticity theory for stresses and strains is determined the minimum energy U_d necessary to deform the disk.

$$U_d = \frac{-\pi \cdot b^2 h_0 \cdot \sigma_{ecr}}{2} \ln \frac{1}{e^{2\ln\left(1+\frac{w^2}{b^2}\right)}} \qquad (1)$$

In the hydroforming process the energy necessary for deformation is given by the hydraulic energy E_h of a work fluid placed into a V volume at p pressure:

$$E_h = p \cdot V \qquad (2)$$

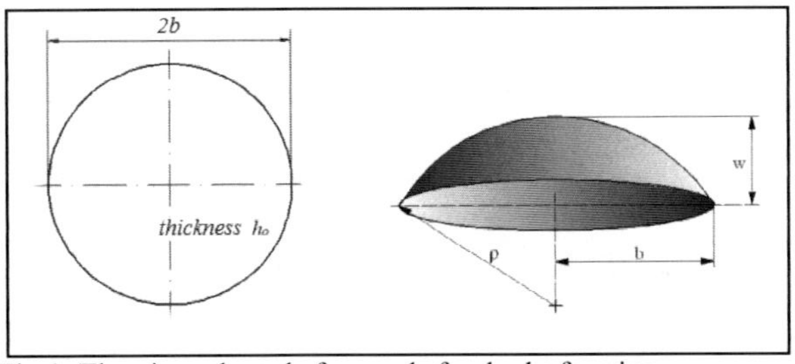

Fig.1. The piece shape before and after hydroforming

Equalizing the energy necessary to deformation exprimed by relationship (1) with the hydraulic energy exprimed by relationship (2) yields an expression of the hydraulic pressure necessary to deformation (3).

$$p = \frac{6 \cdot h_0 \cdot \sigma_{ecr} \cdot \ln\left(1+\frac{w^2}{b^2}\right)}{w \cdot \left(3+\frac{w^2}{b^2}\right)} \qquad (3)$$

3. THEORETICAL RESULTS

The calculus has performed for initial thickness $h_0=0.5mm$; $h_0=0.8mm$; $h_0=1.3mm$; $h_0=2.0mm$, the calotte base radius value $b=100mm$, and for the calotte poll heights $w=0 \div 50mm$ with step 5mm.

The theoretical results values are shown in Tab.1 and are graphically performed in Fig.3, 4, 5 and 6 with dots.

w	*p(w) [MPa]*			
[mm]	$h_0=2.0mm$	$h_0=1.3mm$	$h_0=0.8mm$	$h_0=0.5mm$
0	0.000	0.000	0.000	0.000
5	0.998	0.649	0.399	0.249
10	1.983	1.289	0.793	0.496
15	2.945	1.914	1.178	0.736
20	3.870	2.516	1.548	0.968
25	4.751	3.088	1.900	1.188
30	5.578	3.626	2.231	1.394
35	6.344	4.124	2.538	1.586
40	7.045	4.579	2.818	1.761
45	7.677	4.990	3.071	1.919
50	8.239	5.355	3.296	2.060

Tab.1. Theoretical results for *p(w)*

4. EXPERIMENTAL RESULTS

The geometry of the pieces subjected to the experimental testing is the same to the geometry of the pieces theoretical analysed.

The testing methodology was the following: the pressure values for a sequence of the calotte poll heights (*w=10mm, 15mm, 20mm....45mm, 50mm*) were measured and the experimental results are shown in Tab.2. and are represented in Fig.2 with line (Pernevan, 2007).

w [mm]	*p(w)* [MPa]			
	h_0=2.0mm	h_0=1.3mm	h_0=0.8mm	h_0=0.5mm
0	0.0	0.0	0.0	0.0
5	1.0	0.5	0.3	0.2
10	2.0	1.1	0.8	0.5
15	3.1	2.0	1.2	0.7
20	4.1	2.7	1.6	1.0
25	5.2	3.3	2.1	1.3
30	6.3	4.0	2.6	1.6
35	7.1	4.6	2.9	1.8
40	7.7	5.0	3.2	2.0
45	8.1	5.2	3.3	2.1
50	8.3	5.3	3.4	2.3

Tab.2. Experimental results for *p(w)*

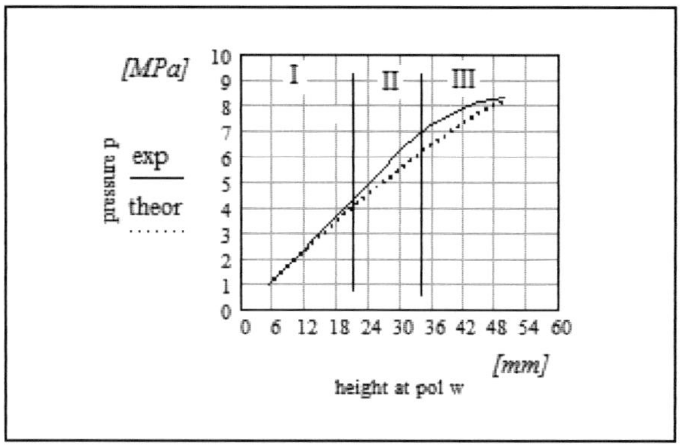

h_0=2mm.

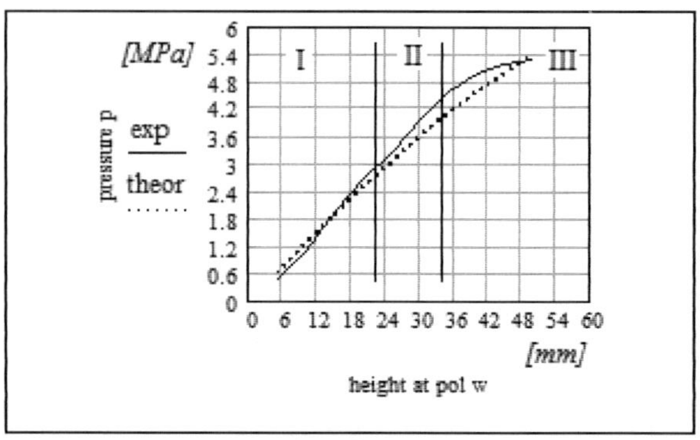

h_0=1.3mm.

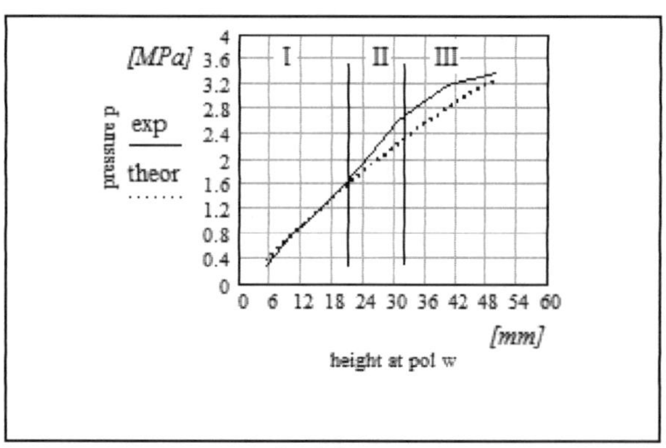

h_0=0.8mm.

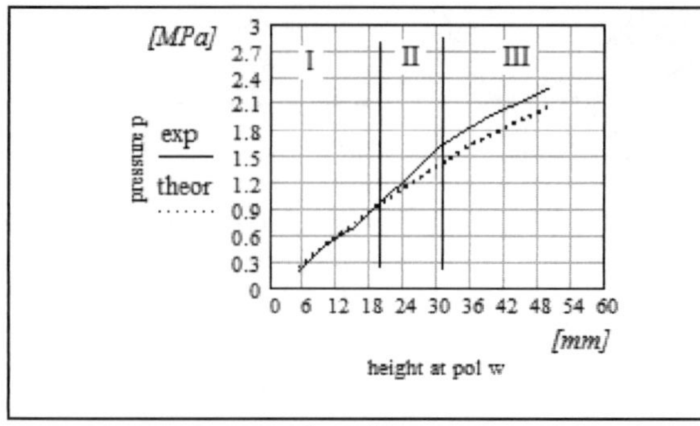

h_0=0.5mm.

Fig.2. The pressure *p* vs. poll heights *w* values (theoretical and experimental) for initial thickness h_0

5. CONCLUSIONS

For all the thicknesses of the pieces there can be distinguished three different zones for *p(w)* experimental curves: in the first zone *(I)* in which, in the piece material exist both elastic deformations and plastic deformations, the curve gradient is relatively small (Hook low) but slightly increased; in the second zone *(II)* the entire piece material is in the plastic deformation state with the initializing of the strain hardening process. The curve gradient is relatively bigger and is resting aproximative constant; in the third zone *(III)* the curve gradient is decreasing due an accentuate decreasing of the thickness and due the appearance of the micro breaks in the piece material. These micro breaks are preceding the catastrophically break of the piece. The shapes of the theoretical curves are different from the experimental curves shapes, but not in a significant mode, so as to offer indications regarding dependence *p(w).* These differences are explicated by that the theoretical pressure relationship *(11)* does not include elements that take account to the strain hardening process.

The originality of this paper consists in the fact that the *p(w)* calculation is based preponderant on the energetically transfer considerations that are available under an 30% deformation.

6. REFERENCES

Abrantes J. P. & Bataha G.F. (2003) „Simulaçă numérica via elementos finitos da conformação de chapas metálicas por pressão hidráulica (Bulge test)" ("The numerical simulation using Finite Element of the metal sheet hydroforming"), *Congresso Brasileiro de Engenharia de Fabricação 18 a 21 de Maio de 2003*

Lee Wu M.H.(1950) "Analysis of plane-plastic stress problems with axial symmetry in strain-hardening range", *Report 1021, NACA*

Mehdi BEN TAHAR M. (2005) *"Contribution a l'etude et la simulation du procede d'hydroformage"(Contributions to the study and the simulation of the hydroforming processes) ;* These pour obtenir le grade de Docteur de l'Ecole des Mines de Paris, 23 septembre

Pernevan I. (2007) *"Theoretical and experimental researching regarding the thin pieces hydroforming"* (in roumanian), Doctoral Thesis, Editura "Politehnica " Timişoara, ISSN 1842-7855, ISBN 978-973-625-430-7, seria 11: Ştiinta şi Ingineria Materialelor Nr.3

Pernevan I. & .Pernevan M. (2005) "Theoretical analysis of a circular thin plate under pressure clamped to rim in the plastic range deformation", *Modelling and optimization in the machine building field-MOCM* , Nov. 2005, pp.157-162, Bacău

Timoshenko S.P. & Woinovsky-Krieger S. (1987) *"Theory of Plates and Shells",* McGraw Hill Classic Textbook Reissue

Annals of DAAAM for 2008 & Proceedings of the 19th International DAAAM Symposium, ISSN 1726-9679
Editor B. Katalinic, Published by DAAAM International, Vienna, Austria 2008
Make Harmony Between Technology and Nature, and Your Mind will Fly Free as a Bird

MODELING A COMPRESSION VALVE IN A HYDRAULIC SHOCK ABSORBER

PERNEVAN, M[aria] & PERNEVAN, I[oan]

Abstract: *The goal of this work is to do a mathematic modeling of a hydraulic shock absorber valve, which, by his design, to render the feasibility of an assessment damper characteristic. This is achieved through a continuous variation of cross area valve orifice by which the work fluid is passed, as through the nonlinear characteristic of the valve spring.*
Keywords: shock absorber, valve, damping, velocity

1. INTRODUCTION

The moderns concerns of the specialists regarding to the shock absorbers designs consist in the achievement of somes shock absorbers that has an optim damping characteristic in conformity with the functioning conditions. This may be realised by many methods such: specials work fluids employment, valves exterior commands employment, hydraulic self adjusting valves employments. Obviously, the realizing of those obiectives involves supplementals shock costs comparative with the classic shock absorbers. From the methods exposed above the hydraulic self adjusting valve operates hydro-mechanically and requires no complex sensors or electronics (Fogiel, 2005).

The aim of this work is to realize the mathematical model of a valve that may to assure a cvasilineare damping force by self adjusting.

2. THE PHYSIC SHOCK ABSORBER AND VALVE MODEL

The shock absorber analized is a hydraulic twin tube shock absorber which has the physical model shown in Fig.1. The notation signification from Fig.1 is following: CV compression valve, Dr rod diameter, d valve diameter, D valve orifice diameter, b valve orifice length, Qv valve flowrate, x valve displacement

The analysed valve is a compression valve that by his geometrie allows the regulation of pressure inner the compression chamber and, implicit, from this, the regulation of the compression damping force (Pernevan, et. al, 1986).

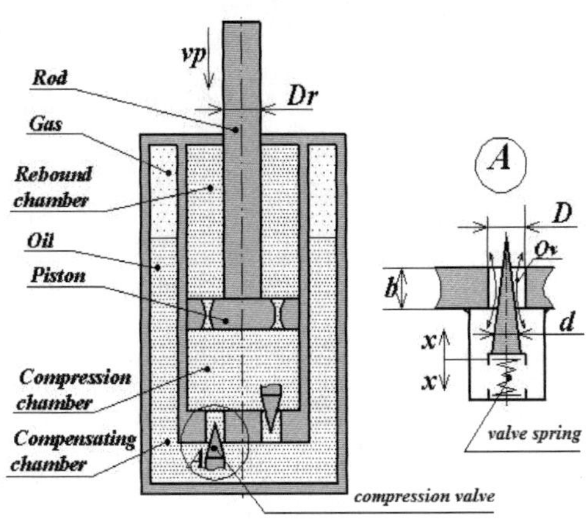

Fig.1. Shock absorber physical model

When the piston rod enters into cylinder the piston comprimes the oil from the compression chamber and the oil is forced to flow through the compression valve CV to the compensate chamber and by the piston orifices to the downside of the piston. It must be appreciated from the outset that adjustable dampers vary considerably in the way that the *p(Q)* curve of the valve and the associated *F(v)* of the damper are altered by the adjustement (Dixon, 1995).

By realizing an appropriate geometrie of the compression valve the damping force may be reglated to impose values.

3. MATHEMATICAL MODEL OF THE VALVE

In order to realize an mathematical model for the compression valve was imposed an predefined damping characteristic *Fd(vp)* (see Tab.1) with the display shown in Fig.2. (Pernevan et al., 2004).

Nr.crt	*Fd* [N]	*vp*[m/s]	*Qv*[m^3/s]	*p*[Mpa]
0	0	0	0	0
1	4010	0.1	0.00003142	4.168
2	4020	0.2	0.00006283	4.178
3	4030	0.3	0.00009425	4.189
4	4040	0.4	0.0001257	4.199
5	4050	0.5	0.0001571	4.209
6	4060	0.6	0.0001885	4.220
7	4070	0.7	0.0002199	4.230
8	4080	0.8	0.0002513	4.241
9	4090	0.9	0.0002827	4.251
10	4100	1	0.0003142	4.261

Tab.1. Damping forces, velocities, flowrates end pressure values

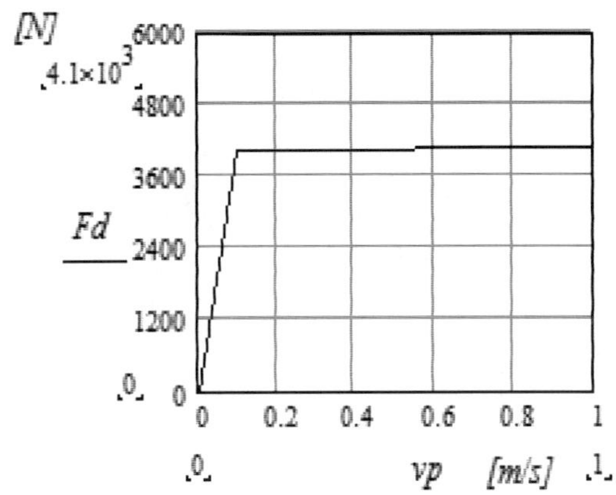

Fig.2. The damping characteristic *Fd(vp)*

The flow rate that passes through the compression valve eguals the flowrate due the piston moving when the piston rod enters inner cylinder. This flow rate has the following relationship:

$$Qp = Ar \cdot vp \quad (1)$$

Where:

$$Ar = \frac{\pi \cdot Dr^2}{4} \quad (2)$$

The drop pressure Δp across the compression valve is given by the compression chamber

$$\Delta p = \frac{Fd}{Ap} \quad (3)$$

Because:

$$Ap = \frac{\pi \cdot Dp^2}{4} \quad (4)$$

The oil flowrate that passes through the compression valve CV may be approximate with the oil flowrate that passes through annular interstices. (Douglas et al., 1996):

$$Qv = \frac{\pi \cdot d \cdot (D-d)^3 \cdot \Delta p}{96 \cdot \mu \cdot b} \quad (5)$$

According with the continuity equation theflowrate that passes through the compression valve equals the flowrate due the piston moving.

$$Qv = Qp \quad (6)$$

$$\frac{d \cdot (D-d)^3 \cdot \Delta p}{24 \cdot \mu \cdot b} = Dr^2 \cdot vp \quad (7)$$

From the expressions (7) may be determined the compression valve diameters *d* values function of: the compression pressure Δp, the oil viscosity μ, the rod diameter *Dr,* the piston diameter *Dp,* the valve orifice diameter *D,* the valve orifice length *b.*
The computation has made for the following geometrical parameters values of an usual shock absorber (Drăghici, 1974), employing the Mathcad program:

Dr=0.02m
Dp=0.035m
D=0.008m
$\mu=0.15Ns/m^2$
b=0.001m, b=0.002m, b=0.003m

The compression valve diameters *d* values are represented in Tab.2 and the graphical display of the valve diameters *d* versus shock absorver velocity *vp* for different values of the orifice length *b* are shown in Fig.3.

vp [m/s]	*d* [m]		
	b=0.001m	*b*=0.002m	*b*=0.003m
0	0.008	0.008	0.008
0.1	0.00784	0.00779	0.00776
0.2	0.00779	0.00774	0.00770
0.3	0.00776	0.00770	0.00766
0.4	0.00774	0.00767	0.00762
0.5	0.00772	0.00764	0.00759
0.6	0.00770	0.00762	0.00757
0.7	0.00769	0.00760	0.00754
0.8	0.00767	0.00758	0.00752
0.9	0.00766	0.00757	0.00750
1	0.00765	0.00755	0.00749

Tab.2. The valve diameters *d* versus shock absorver velocity *vp* for different values of the orifice length *b*

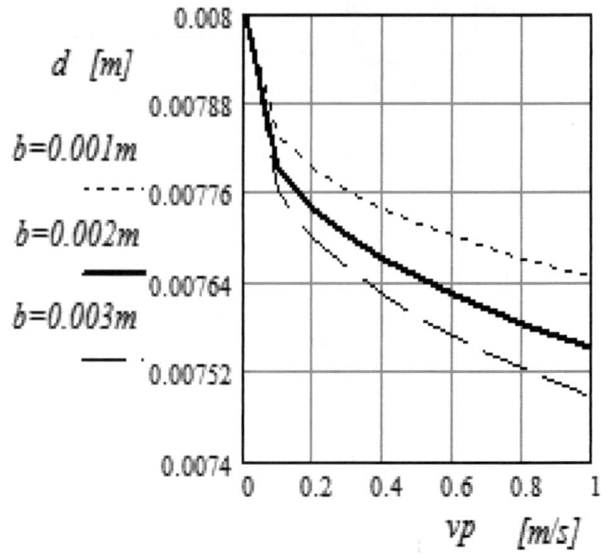

Fig.3. The valve diameters *d* variation graphical display versus shock absorver velocity *vp* for different values of the orifice length *b*

4. CONCLUSION

In order to determine the valve geometrie was imposed a predefined damping characteristic with a very small gradient. The damping force is quasiconstante.

The computations were made for an ussualy variante of shock absorber dimensions that may be extend also to other designs of shock absorbers.

Was imposed a pass orifice by annular cross area and was obtained a law of variation for the compression valve diameter and, implicity, a law of variation for the valve interstice cross area.

With the results obtained it follows to determine the deflection *x* of the valve metering element that realises an accurate deflection employing a spring with a nonlinear characteristic.

The originality of this paper consists in to realise a hydraulic valve matehematical model design that can to assure the shock absorber functioning at the imposed parameters.

5. REFERENCES.

Dixon, J. (1995). *The shock absorber handbook,* 1th Edition, Society of Automotive Engineers, ISBN 0-7680-0050-5, Warrendale, Pa

Douglas, J.F.; Gasiorek, J.M. & Swaffield, J. (1996), *Fluid Mechanics,* Longman Group Limited, ISBN 0-582-23408-5, Edinburg

Drăghici, I. (1974). *Suspensii şi amortizoare, (Suspensions and dampers),* Editura Tehnică, Bucureşti

Fogiel M. (2005), *Bypass Valve for a Hydraulic Dashpot (shock absorber),* Highland Park, US, Inventor: Hamers W Patent 2005012121273

Pernevan, M.; Bărglăzan, M.; & Pernevan, I. (2004), *Simple Action Shock Absorber Model,* Proceedings of the 6th International Conference on Hydraulic Machinery and Hydrodinamics, pp. 381-386, Scientific Bulletin of the "Politehnica" University of Timisoara, Romania, Transactions on Mechanics, Tom 49 (63), Special Issue, ISSN 1224-6027, Timişoara

Pernevan, I.; Gherasim, L. & Pernevan. M., (1986), *Amortizor Hidraulic, (Hydraulic Shock Absorber),* Brevet inventie nr. (Patent nr). RO 89976/25.03.1986

Annals of DAAAM for 2008 & Proceedings of the 19th International DAAAM Symposium, ISSN 1726-9679
Editor B. Katalinic, Published by DAAAM International, Vienna, Austria 2008
Make Harmony Between Technology and Nature, and Your Mind will Fly Free as a Bird

DEFORMATION HARDENING OF THE TITANIUM ALLOY OT4-1

PERNIS, R[udolf]; KASALA, J[ozef] & HAJDUCHOVA, L[uba]

Abstract: *The titanium alloy OT4-1 is used for the production of stationary blades of compressor rotors in aircraft engines. The blades are formed by a cold forming process during which the deformation hardening takes place. For the process of simulation of the production of blades from the alloy OT4-1, it is important to know certain material constants which make the simulation possible. The aim of this work was to determine functional mathematical model of the strain-hardening curve on the basis of physical modelling of deformation hardening of the titanium alloy OT4-1 as well as to process the strain-hardening diagram of the titanium alloy OT4-1 in a classic graphical form.*
Key words: *titanium, alloy OT4-1, hardening, strain-hardening diagram.*

1. INTRODUCTION

Titanium alloys are characterized by low density and therefore are very suitable for the aeronautical and space industry. Their high strength is another important feature of these alloys. Titanium alloys are increasingly finding their way into *aircraft construction (jet engines)*. This can be explained by indicator $R_{p0,2}/\rho$ expressing ratio of yield strength $R_{p0,2}$ and density ρ.
Titanium alloys can be used up to 600 °C, and around 0 °C they highly superpass steel materials (Ptáček et al., 1999). The presented titanium alloy OT4-1 is recommended for works up to 400 °C. This titanium alloy was used in an aircraft engine for fabrication of blades for the first and second stage of stationary blades of compressor. The reasons of the investigation of the titanium alloy OT4-1 were to find the possibilities of increase in mechanical properties by deformation hardening and process strain-hardening diagram of the titanium alloy. The next plan is to investigate forging of blades on computer by using Finite Element Method. For this purpose it was necessary to process mathematical model of yield strength in dependence on strain, either on strain ε or logarithmic strain φ.

2. BASIC CHARACTERISTICS OF THE TITANIUM ALLOY OT4-1

With respect to chemical composition, the titanium alloy OT4-1 belongs to the group of the Russian titanium alloys. Its basis is formed by the ternary system Ti-Al-M. The English titanium alloy IMI 315, with average composition corresponding to the alloy TiAl2Mn2, is considered to be its west-European equivalent. The chemical composition of alloy according to Russian standards is listed in Table 1. The content of elements is stated in weight percentage.

Element	Al	Mn	Impurities	Ti
Content %	1,0 - 2,5	0,7 – 2,0	max. 0,05	rest

Table 1. Chemical composition of the titanium alloy OT4-1.

Dependence of mechanical properties on temperature is shown in Fig 1. As can be seen from Fig 1, the alloy OT4-1 preserves its mechanical properties also below 0°C. The course of elongation A_5 indicates the forming property of the alloy OT4-1. Between 400 and 500°C, the alloy shows local minimum that should be avoided. Therefore, it is recommended to use cold forming of the alloy or forming at temperatures higher than 600°C.

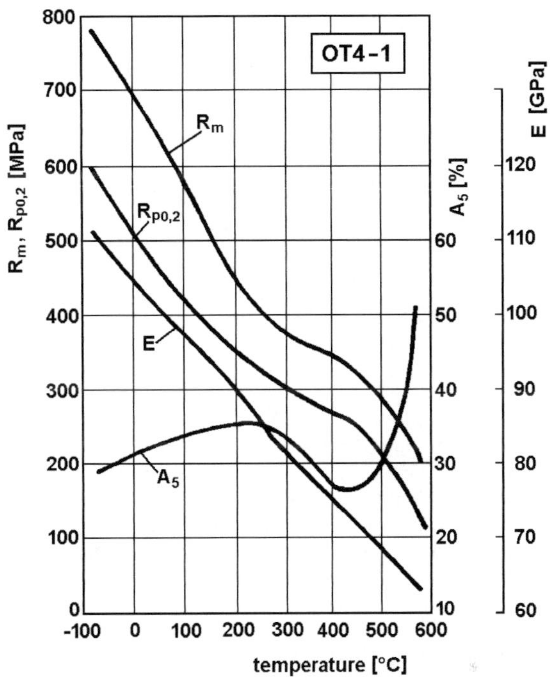

Fig. 1 Dependence of mechanical properties of the titanium alloy OT4-1 on temperature (Glazunov & Moiseev, 1974).

The 3.1mm thick sheet was used in the test experiments on deformation hardening. Samples with the width of 42mm and length of 300mm were prepared from this sheet. Chemical composition the titanium alloy OT4-1 samples is listed in Table 2. The samples of the sheet had these mechanical properties in the initial annealing state:

R_m = 632 MPa
$R_{p0,2}$ = 528 MPa
A_5 = 26,7 %
$HB_{1/30}$ = 235
E = 106 GPa

Element	Mn	Al	Fe	Mo
Content %	1,404	2,396	0,019	0,016
Element	Zr	Si	Ni	Ti
Content %	0,009	0,004	0,008	rest

Table 2. Chemical composition of the titanium alloy samples

3. COLD DEFORMATION OF THE ALLOY OT4-1

The test samples of the titanium alloy OT4-1 were submitted for cold deformation by rolling. Reduction plan of rolling was chosen in order to increase strain by 10 % on average. Strain ε_i after i-rolling is determined by the relation

$$\varepsilon_i = \frac{h_0 - h_i}{h_0} \cdot 100 \ [\%] \qquad (1)$$

where h_0 is the initial thickness of the sample, in the case of rolling h_0= 3,1mm. The thickness of the sample h_i is the thickness after i-rolling. Strain ε_i after i-rolling is stated in percents. The total number of rolling was n=7. Sample rolling was carried out in longitudinal direction. After the fourth rolling, cracks began to appear on edges of the samples. These were increasing in their size after another rolling. The middle of rolled flat samples was compact and without cracks. The part of sample was cut after each rolling with the aim to determine yield strength, ultimate tensile strength and hardness. Other samples were taken for the following metallographic analysis. From the measured thickness of sheets, the logarithmic strain φ_i after i-rolling was determined on the basis of the relation

$$\varphi_i = \ln \frac{h_0}{h_i} \ [-] \qquad (2)$$

Samples prepared by the process of cold rolling underwent the stress-strain testing and their Brinell hardness was measured.

4. EVALUATION OF MEASUREMENTS

Deformation hardening of the titanium alloy OT4-1 was evaluated by means of yield strength and ultimate tensile strength.

4.1 Strain-hardening diagram

Classically, dependence of mechanical properties on cold deformation was being stated graphically in the form of strain-hardening diagram. From measurements of yield strength $R_{p0,2}$ and ultimate tensile strength R_m, the strain-hardening diagram was made in relation to strain ε, as shown in Figure 2. In order to give more complex image of deformation hardening, strain-hardening diagram was completed by Brinell hardness. Measurements obtained from the stress-strain testing for full annealing state, which was considered to be as non-deformed state, were taken as the first initial point. That is, $R_{p0,2}$= 528 MPa for the yield strength, while the strain for this measurement is ε=0 %.

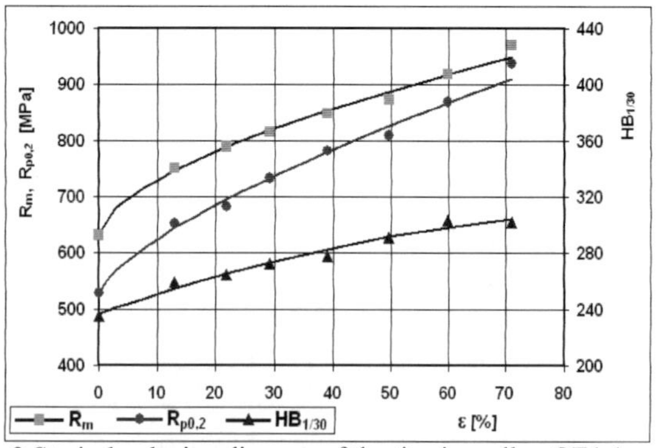

Fig. 2 Strain-hardening diagram of the titanium alloy OT4-1

4.2 Mathematical model of strain-hardening

There are many empiric and semiempiric relations in bibliographic reference materials. According to Talbert &Avitzur, the most often used are two- and three-parameter mathematical functions. Pernis & Jurenová use three-parameter function which is suitable for a large group of materials

$$R = R_0 + k \cdot \varphi^n \qquad (3)$$

where material characteristic R_0 is stress value measured in the annealing state. The yield strength of the titanium alloy OT4-1 has stress value R_0= $R_{p0,2}$=528 MPa. Material characteristics k and n for the previous equation (3) were determined by mathematical statistics, namely by Method of Least Squares. The following formation hardening equation for the yield strength of the titanium alloy OT4-1 was stated in dependence on logarithmic deformation

$$R_p = 528 + 359{,}1938 \cdot \varphi^{0{,}552555} \qquad (4)$$

The same type of mathematical function was used for the ultimate tensile strength R_m

$$R_m = 632 + 299{,}5676 \cdot \varphi^{0{,}470568} \qquad (5)$$

5. METALLOGRAPHIC AND FRACTOGRAPHIC ANALYSIS

The samples of the titanium alloy OT4-1 after cold deformation hardening underwent metallographic analysis. Fractographic analysis was carried out on fractures after the stress-strain testing. Fractures of the titanium alloy OT4-1 after uniaxial stress tests were submitted for fractographic analysis which was realized on electrical scanning microscope JSM 25 S (JOEL). Micrographic analyses were aimed at changes in the micromechanism of fracture in spreading of major fissure, in dependence on the grade of plastic deformation. Evaluation was realized in a similar way as states Horňak (Horňak & et al.). In all cases, fracture was formed by ductile transcrystal rupture.

6. CONCLUSION

From realized experimental tests on deformation hardening of the titanium alloy OT4-1, the following conclusions can be drawn:

- From the initial annealing state for the titanium alloy OT4-1, the yield strength increased from 528 MPa to 937 MPa by deformation hardening process.
- Mathematical model of strain-hardening curve with 3 parameters was proposed - the equation (3).
- The equation (4) enables to realize the simulation of the fabrication process of the stator blade by means of FEM
- The project supposes to optimize the size of forming tools used for the production of stator blades in the next stage.

7. REFERENCES

Glazunov, S. G. & Moiseev, V. N. (1974). *Konstrukcionnye titanovye splavy (Titanium alloys).* Metallurgia, Moskva

Horňak, P. & et al. (1997) Textúra a subštruktúra Si-ocele s orientovaným zrnom. *Acta Metallurgica Slovaca*, Vol., 3, (Special Issue 1/2), p. (33-39), ISSN 1335-1532

Pernis, R. & Jurenová, J. (2008). Brass Bullets of Improved Strength. *University Review*, Vol., 2., (2), p. (56-59), ISSN 1337-6047

Ptáček, L. & et al. (1999). *Nauka o materiálu I II (Theory of metals II).*. Akademické nakladatelství CERM, ISBN 80-7204-130-4, Brno

Talbert, S. H., Avitzur, B. (1996). *Elementary Mechanics of Plastic Flow in Metal Forming.* John Wiley & Sons, New York

Annals of DAAAM for 2008 & Proceedings of the 19th International DAAAM Symposium, ISSN 1726-9679
Editor B. Katalinic, Published by DAAAM International, Vienna, Austria 2008
Make Harmony Between Technology and Nature, and Your Mind will Fly Free as a Bird

SIMPLE NONLINEAR ADAPTIVE CONTROL OF HAMMERSTEIN MODEL

PERUTKA, K[arel]

Abstract: *The paper deals with the nonlinear control of Hammerstein model. The combination of simple nonlinear control together with the self-tuning control is the main idea of the paper. The used simple nonlinear control strategy can satisfactory control the controlled system even if the model does not exactly match the system. In purpose to get the exact model of the controlled system, the self-tuning control can be used. For the identification, the recursive instrumental variable method was used. The verification was realized by simulation in MATLAB, the Hammerstein model was used. Nonlinearity was modeled by the power function.*
Key words: *Hammerstein model, MATLAB, IV, self-tuning.*

1. INTRODUCTION

Most of physical, biological, chemical, economical and other systems are nonlinear in practice. Many nonlinear systems can be identified and controlled as linear systems around the steady state or working points. When nonlinear control is used, it possible to enlarge the working interval even in the case the linear control does not guarantee the sufficient quality of control. Moreover, some systems have nonlinearities, which cannot be linearly approximated, for instance friction, etc. Therefore, the necessity of nonlinear control occurs.

Nonlinear system is a set of elements of system, in which at least one of the elements is nonlinear. The elements are static and then they are described by algebraic equations or they are dynamic which are described by differential equations. Nonlinear systems differ from linear systems mainly in these properties (Modrlak, 2008):

- superposition principle does not hold on
- the steady states may appear beyond the coordinate origin
- the initial conditions influences reaching the steady states at autonomous systems
- the self excitation oscillation appears in the system, their frequency differs from the excitation frequency
- the step changes of amplitude appear during the changing frequency of excitation signal

Some nonlinear systems can be approximated by linear systems at a defined range and when specific conditions hold on.

In practice, such the systems exist as they can be divided into linear and nonlinear part. The dynamics of system can be approximated by linear model and its nonlinear part by the nonlinear characteristics. These systems can be modeled by two series connected blocks, nonlinear and linear block. Two possible models exist, Hammerstein and Weiner model. Hammerstein model represents the nonlinear system consisting of static nonlinearity followed by the system with the linear dynamics, see figure 1. The basic Wiener model consists of the system with the linear dynamics followed by the system with static nonlinearity. This is due to the fact that the superposition is not valid for the nonlinear systems, the output of Hammerstein model is different from the output of Weiner model. (Lin, 1994)

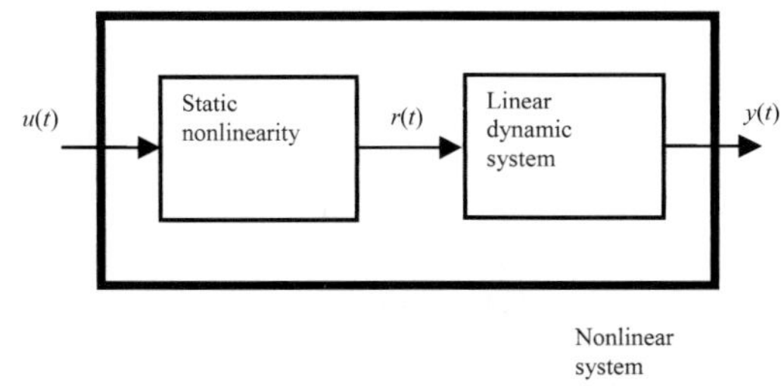

Fig. 1. Hammerstein model

This paper combines the simple nonlinear control introduced by Chen et al. (Chen et al., 2006) with the self-tuning control (Bobal et al., 2005).

The paper is organized as follows. Firstly, the introduction part is given. This followed by simple nonlinear controller description. After that, the results of simulation in MATLAB are given. The description of self-tuning control is not provided, but the used concept was taken from Perutka (Perutka, 2007).

3. SIMPLE NONLINEAR CONTROLLER

This method was introduced by Chen et al. (Chen, et al., 2006) and verified by Bubenikova (Bubenikova, 2007). The controller consists of three parts, the pure controller and generator giving together the nonlinear controller and the system model inversion. The pure controller is represented by the equation

$$u = y^{(n)} = \frac{a(1-\exp(-b\phi))}{1+\exp(-b\phi)}, \tag{1}$$

where $y^{(n)}$ is the nth-order derivation of the output y and the parameters a and b give the output level and slope of controller and they are chosen by the user.

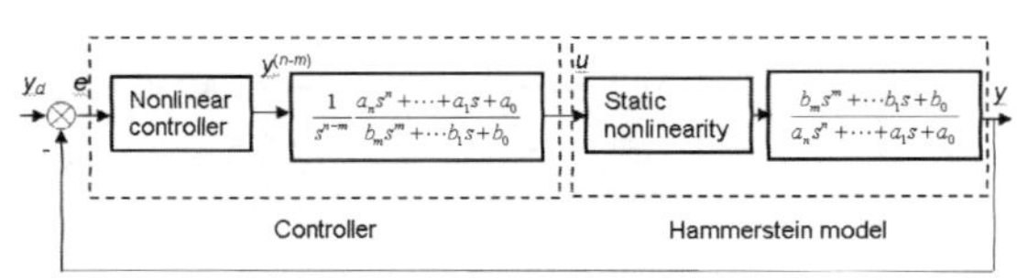

Fig. 2. Simple nonlinear controller of Hammerstein model

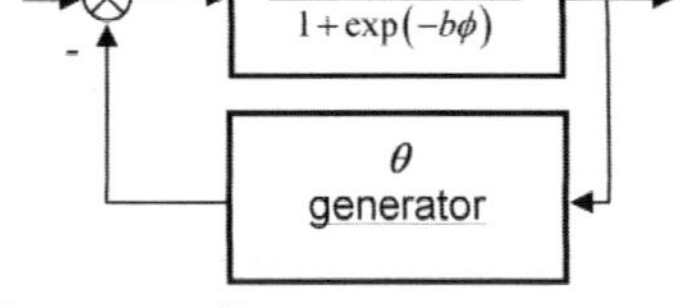

Fig. 3. Nonlinear controller

The ϕ parameter is counted according to $\phi = y_d - y - \theta$, where the generator is $\theta = \sum_{i=1}^{n-m-1} \lambda_i y^{(i)}$. where λ_i for i = 1, 2, …, n-m-1 are the controller parameters which have to be counted. Supposing

$$\frac{y(s)}{y_d(s)} = \frac{1}{\frac{2}{ab}s^{n-m} + \lambda_{n-1}s^{n-m-1} + \cdots + \lambda_1 s + 1} \tag{2}$$

and denoting $\varepsilon = \sqrt[n-m]{\frac{2}{ab}}$, the new form used to count the controller parameters is

$$\frac{y(s)}{y_d(s)} = \frac{1}{(\varepsilon s + 1)^{n-m}} \tag{3}$$

4. SIMULATION RESULTS

In this chapter, the results of control are shown. Figure 4 displays the history of control, the system output is displayed by solid line. In figure 5, the action signal displayed. Finally, there is the history of system parameters estimates in figure 6. The controlled system was in the form of Hammerstein model, the nonlinearity was in the form of the power function and the linear part of the controlled system was described by

$$G_L(s) = \frac{3}{s^2 + 12.5s + 22}$$

(4)

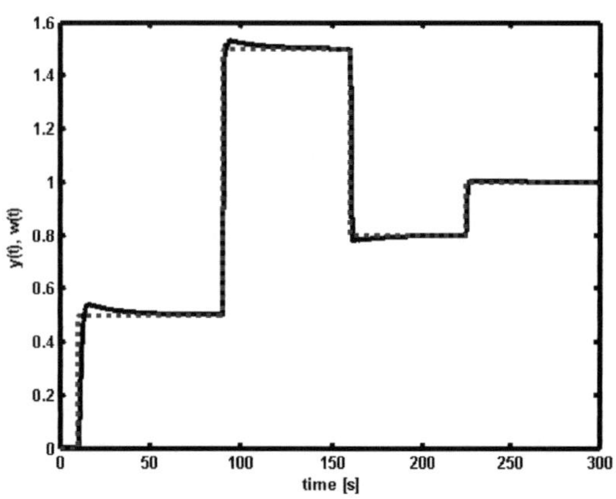

Fig. 4. History of control

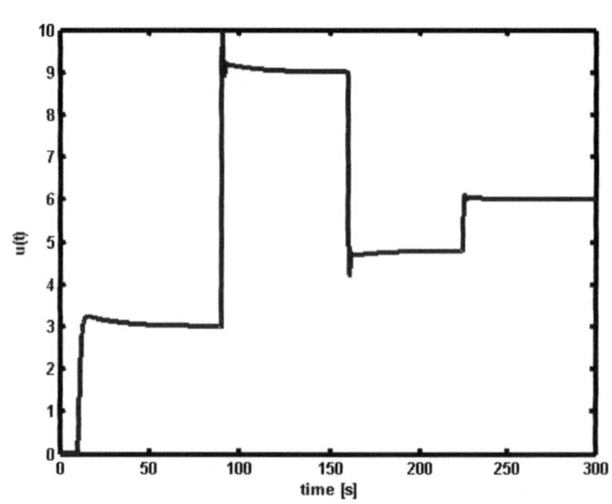

Fig. 5. Action signal

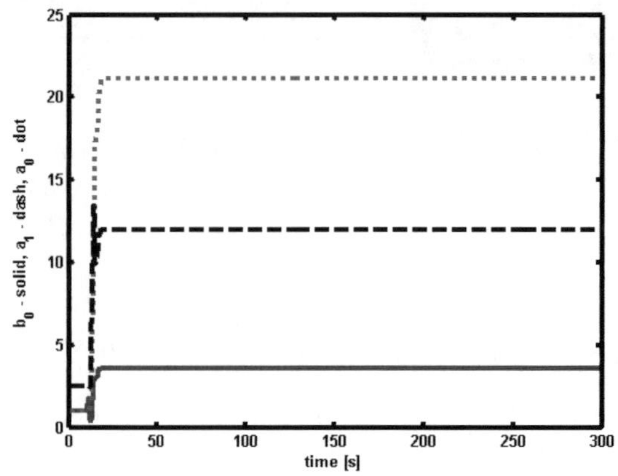

Fig. 6. Controlled system parameters estimation

The initial parameters estimates was in the form

$$G_E(s) = \frac{0.3}{s^2 + 0.5s + 0.1} \tag{5}$$

The self-tuning controller is a combination of online identification and controller correction according to the obtained system parameters. The feedback structure of closed circuited was used. Identification was done by recursive instrumental variable method. The optional parameters of the controller were: a = 100, b = 1, λ_1 = 0.282843.

5. CONCLUSION

In the paper, there was presented the simple adaptive nonlinear control of Hammerstein model. This approach was in the previous work non-adaptive. The paper deals with the implementation of online identification in this method for the first time. In future, the attention will be paid to implementation of adaptive version for real-time control in the laboratory.

6. ACKNOWLEDGEMENT

The paper was supported by the grant of the Ministry of Education, Youth and Sport of the Czech Republic, MSM 7088352101.

7. REFERENCES

Bobal, V.; Böhm, J.; Fessl; J. & Machacek, J. (2005). *Digital Self-tuning Controllers*. Springer-Verlag London Limited.

Bubenikova, M. (2007). Control of SISO systems with given type of nonlinearity. Thesis. Zlin: Tomas Bata University in Zlin, Faculty of Applied Informatics, 2007. 89 pp.

Chen, Ch.-T.; Chuang, Y.-Ch., Hwang, Ch. (2006). A Simple Nonlinear Control Strategy for Chemical Processes. In: Proc. of the 6th Asian Control Conference, Inna Grand Bali Beach Hotel, Bali, Indonesia, 18-21 July 2006. [CD-ROM]. pp. 64-66 ISBN 979-15017-0.

Lin, C.-F. (1994). Advanced Control System Design. New Jersey, USA: Prentice Hall, 1994. ISBN 0-13-006305-3.

Modrlak, O.: Theory of automatic control II: Nonlinear systems. Lectures notes. [on line] Liberec: Technical University of Liberec, [cit. 14.03.2008]. Available at <http://www.fm.vslib.cz/~krtsub/fm/modrlak/pdf/tar2_nel.pdf>.

Perutka, K. (2007). *Decentralized Adaptive Control*, Ph.D. thesis, Tomas Bata University in Zlin, Zlin, Czech Republic.

Annals of DAAAM for 2008 & Proceedings of the 19th International DAAAM Symposium, ISSN 1726-9679
Editor B. Katalinic, Published by DAAAM International, Vienna, Austria 2008
Make Harmony Between Technology and Nature, and Your Mind will Fly Free as a Bird

CAM STRATEGIES AND SURFACES ACCURACY

PETERKA, J[ozef]; POKORNY, P[eter] & VACLAV, S[tefan]

Abstract: *The article presents the impact of milling strategies on the accuracy of surfaces. Shaped surfaces may be created from simple shapes (cylinder, sphere, cone, etc.) or they are complex surfaces created in CAD systems. The surfaces were created in CAD software PowerSHAPE, which is also in possession of our department. CAM strategies were generated with software PowerMILL. Machining was realized by CNC milling centre EAGLE 1000 with control system Haidenhein i530. Applied finishing strategies were optimized constant Z, offset and spiral. The surfaces after milling were scanned by optical 3D scanner GOM ATOS I and then it was measured by CMM Contura G2 (Carl Zeiss). We have the surface deviation map (CAD Comparison in GOM ATOS I) and deviation from CMM measuring. These deviations show the influence of milling strategies on shaped surface accuracy.*
Key words: *technologies, machining, inspection, comparison, measurement*

1. INTRODUCTION

The basic sign of CAM is to minimize human intervention into the course of the production process by the exploitation of computer data processed in main elements of activities (Kuric et al., 2001).

The machining process is generally divided into roughing and finishing (Morovič, 2006).

Very important for CNC program generation is the choice of machining strategy. This strategy has influence on the final accuracy and final quality of machined part (Lipa & Moravčíková, 2007). Machining strategy is tool movement in machine space.

The machining strategy influences the surface accuracy and also herewith the technological properties of construction. It is very important for assembly (Senderská, 2006).

2. MODELLING AND PART MANUFACTURING

2.1 Part modelling

For research of influence of CAM strategies on surface accuracy was created the part (fig.1) in PowerShape system.

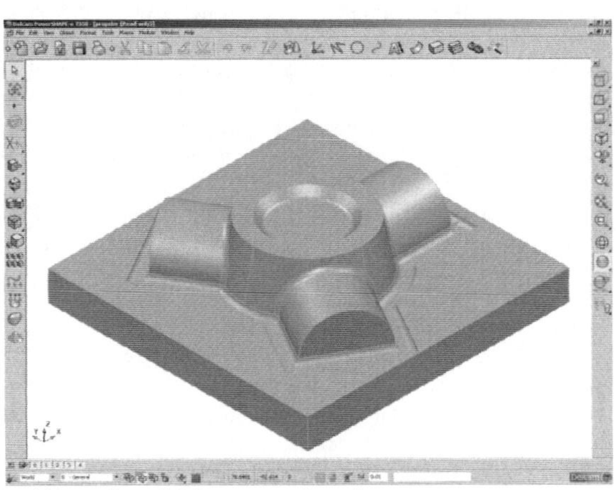

Fig. 1 Part of model for screw clamping

The model has three identical cylindrical surfaces. Three different finishing strategies will be applied on these surfaces.

2.2 CNC data generation

Model was loaded to CAM system PowerMill. There were (after block, tool and milling strategy choosing) generated CNC data for roughing, rest machining, shape machining and finishing.

For finishing of individual cylindrical surfaces these strategies choosen were:

- Optimized Constant Z (fig.2).
- Spiral Finishing (fig.3).
- Offset Finishing (fig.4).

Tool and cutting conditions were identical for all three choosen strategies:

- End mill - D = 6 mm, z = 2.
- v_f = 3000 mm.min^{-1}.
- n = 6000 min^{-1}.
- a_e = 0,305 mm.
- f_z = 0,25 mm.

Fig. 2 Toolpaths for strategy Optimized Constant Z

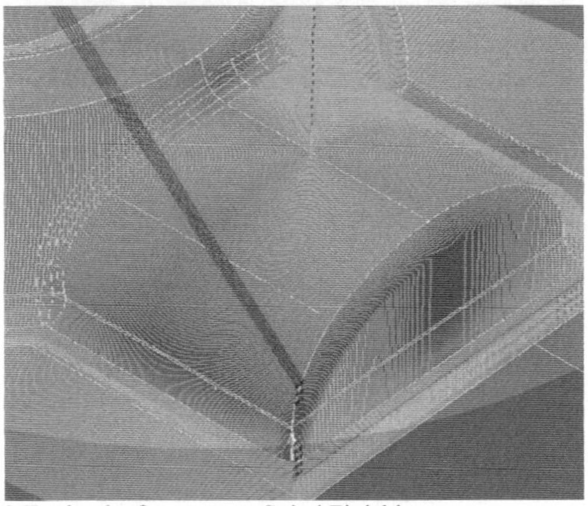

Fig. 3 Toolpaths for strategy Spiral Finishing

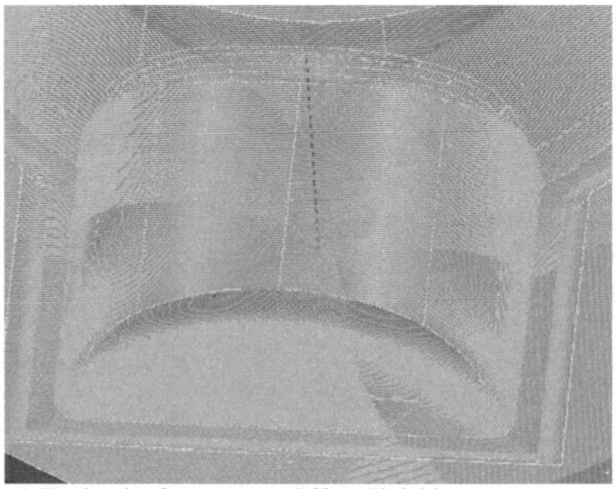

Fig. 4 Toolpaths for strategy Offset Finishing

Machining (roughing, rest machining, shape machining, and finishing) was realized by CNC milling centre EAGLE 1000 with control system Haidenhein i530.

3. MEASURING OF CYLINDRICAL SURFACES

Measuring of machined cylindrical surfaces was realized by optical 3D scanner GOM ATOS I - non contact measuring method (fig.5a) and after that the cylindrical surfaces was measured by Contura G2 - contact measuring method (fig.5b).

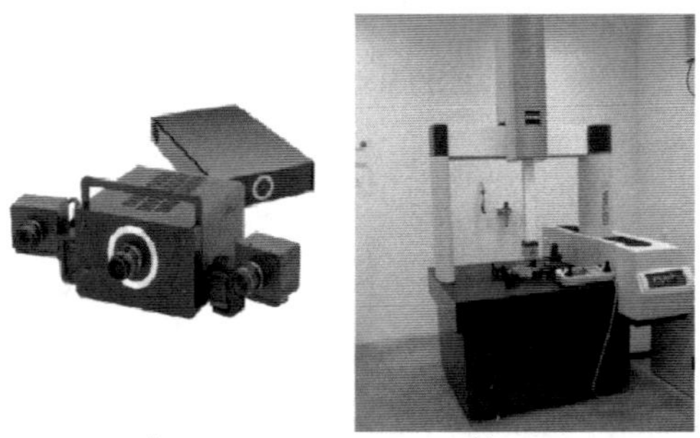

a) b)

Fig. 5. Machine for measuring of milling component

Firstly, we scanned milled model by measuring with machine GOM ATOS I. So we got computer model of produced component. We also compared this model with computer model which was created according to technical documentation. This method is called CAD Comparison. (Buranský, 2008) There is coloured map of deviations produced surfaces and computer surfaces (fig.6).

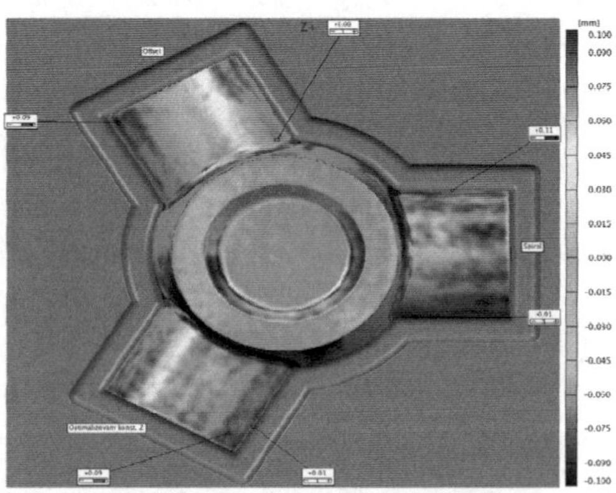

Fig. 6. Coloured surface deviation map of milled part

By measuring with CMM machine Contura G2 were all three cylindrical surfaces measured gradualy. Each measuring was realized three times. After the measuring the arithmetical means of measured values were calculated. We got the values of deviation for applied milling strategies (fig.7).

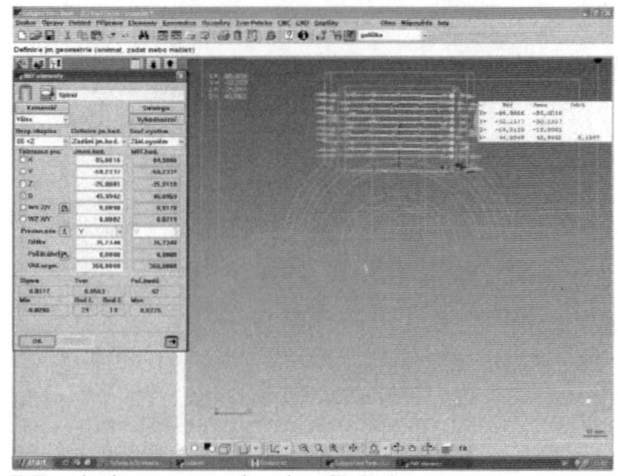

Fig. 7 Deviation measuring by CMM

4. CONCLUSION

The individual results of measuring are in (tab.1). Here are some used strategies of milling operation and appropriate deviations for non contact measuring and contact measuring.

STRATEGY	MEASURED VALUES [mm]	
	CMM Contura G2	GOM ATOS I
Optimized Constant Z	0,0356	0,08
Spiral	0,0442	0,12
Offset	0,0491	0,09

Tab 1. Results for milling strategies

It is liquet that for simple cylindrical surface the most suitable is the Optimized Constant Z milling strategy. The smallest deviations were recognized in both types of measuring.

According to the table it is possible to compare the differences between measuring of CMM machine and non contact machine. From CMM measuring we have expected the more correct results. And that was approved. The precision of results depends on the type of measuring.

In further research, we will mill more types of surfaces and also the free form surfaces. The various strategies of milling will be applied on these surfaces. The influence of milling strategies on precision of produced surfaces and the roughness of surface will be observed.

This article was written for grantproject VEGA 1/0130/08 - Research of influence of CAM strategies on achieved dimension accuracy and roughness of machined surface in conditions of university Hi-tech laboratory.

5. REFERENCES

Buranský, I.; (2008) Utilizing Optical Methods of Measuring in Machines Production Technologies, In: *International doctoral seminar:Proceedings*, AlumniPress, pp. 48-53 ISBN 978-80-8096-058-2, Smolenice

Kuric, I.; Košturiak, J.; Janáč, A.; Peterka, J. & Marcinčin, J. (2001). *Computer Aided systems in the Engineering*, Edis, ISBN 80-7100-984-2, Žilina

Lipa, Z.; Moravčíková, J.; (2007) The contribution of kinematic ratios in machining, In: *Development of metal cutting DMC 2007*, SjFTU pp. 225-22, ISBN 978-80-8073-858-7, Košice

Morovič, L.; (2006). Utilization of Reverse Engineering in the Design and Manufacturing of Sport Shoes, In: *International Doctoral Seminar*, STU MTF, pp. 178, ISBN 80-227-2387-8, Smolenice

Senderská, K.; (2006). Approaches and Tools to Evaluation Support of Product Construction for Assembly, In: *Inovations Transfer*, SjF TU, pp. 214-215, ISBN 80-8073-701-0, Košice

Annals of DAAAM for 2008 & Proceedings of the 19th International DAAAM Symposium, ISSN 1726-9679
Editor B. Katalinic, Published by DAAAM International, Vienna, Austria 2008
Make Harmony Between Technology and Nature, and Your Mind will Fly Free as a Bird

PRODUCT DESIGN SUITABLE FOR ASSEMBLY

PETERKA, J[ozef]; SENDERSKA, K[atarína]; VACLAV, S[tefan] & POKORNY, P[eter]

Abstract: *The paper deals with the product design from the point of view assembly. The application, created on the ground of known approaches to design for assembly is assigned to the product analysis and evaluation. For this purposes as a main method was selected the decision analysis, which combines the objective and subjective aspects. The proposed method was processed as an access application tool and was verified at some products.*

Key words: *design, assembly, manufacturing, analysis, method of assembly*

1. INTRODUCTION

The effective assembly assumes a product that responds to the requirements of manipulation and joining of separate parts into higher units (subassemblies and final products). In order to take in the consideration all aspects the designers need to understand to the assembly processes and procedures. The development of the complex products is teamwork. The basic problem, which we must to concern, is how the parts behave during the transport, manipulation, joining and testing. In this context are in the foreground different properties of the parts and components as by evaluation of their function.

2. PRODUCT DESIGN BACKGROUND

To ensure the required product function it is possible to design the product by various concepts often by using of entire different concepts and design principles (Václav 2005). By comparison of two products, that ensure the same function in the same range and in a comparable quality (so that they have a comparable utility value) can be observed great differences. It can be used not only other design principles but also other materials, forms and dimensions of the parts and components. The products have often other relative position of the parts and components, other tolerance limits and etc (Lotter & Wiendahl 2006). According to the design quality it is then possible to produce and to assemble the product relatively simply without serious technical problems and by relatively low costs or in the limit case is the production and also the assembly complicated or unacceptable from the point of view costs. A lot depends also on the production batch size. The design of a product with a great batch size should be differ from the design of the same product to the piece or robotized assembly (Boothroyd & Dewhurst 2006). The significance of the product design and their influence to production and assembly can be confirmed also by following data. Independently from the product kind are in the design process specified more than 70% manufacturing cots including costs of the assembly. 70-80% of the possible savings are depended on the product design, 10 –20 % are depended on the changes of technology, production means, fixtures and tools. 5-10% are connected with changes of the organization.

3. METHOD AND TOLL TO PRODUCT ANALYSIS FROM THE POINT OF VIEW ASSEMBLY

For the product design analysis from the point of view assembly was proposed a method issued from the decision analysis. The basis of this approach is a set of rules and recommendations for the assembly and the collection of questions. The analysis and evaluation is divided into four steps: product, parts, joints and assemblies. The method assumes to answer at five main questions in all four steps. The "answer" is a point evaluation and an own creative proposal that can be inspired so by existing stored proposal for similar products, parts, joints or assemblies as by the set of rules and recommendations (IPA 2007).

The developed access tool is dedicated to analysis of medium complex mechanical products and was tested at several products (Senderská 2006).

Main functions:

- Refill and modify the contained rules and recommendations.
- The evaluating questions/criteria selection and definition.
- Definition of the point values for decision analysis.
- Product/products analysis.
- Analysis of the parts, joint and subassemblies.

The conception enables also to modify the material, part symmetry, kind of the joint and other parameters assortment. The whole system is opened for further modification for instance as comparison of two or more variants of the products, evaluation from the point of view disassembly etc. For better information the system can include also drawings, 3D models or photographs of the products, parts, joints and subassemblies.

On the figure 1 is shown the main screen for the product data input.

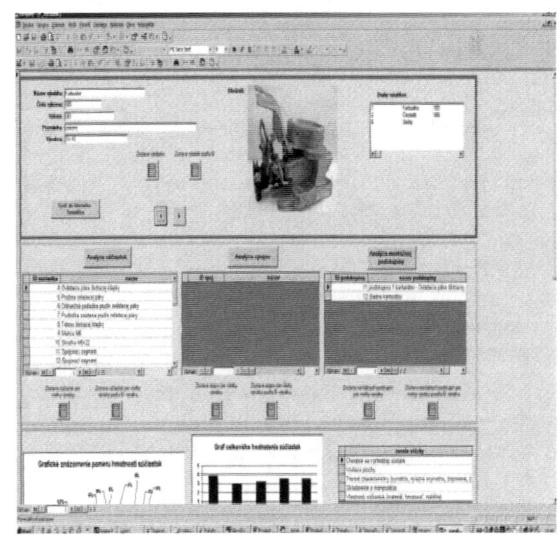

Fig. 1. Main screen – product data input.

The figure 2 shows the screen with processed data and recommendations to product change.

In this system was created some default reports. On the figure 3 is shown an output report of the product analysis. From the data and analysis results contained in this system can be created various reports according to the requirements.

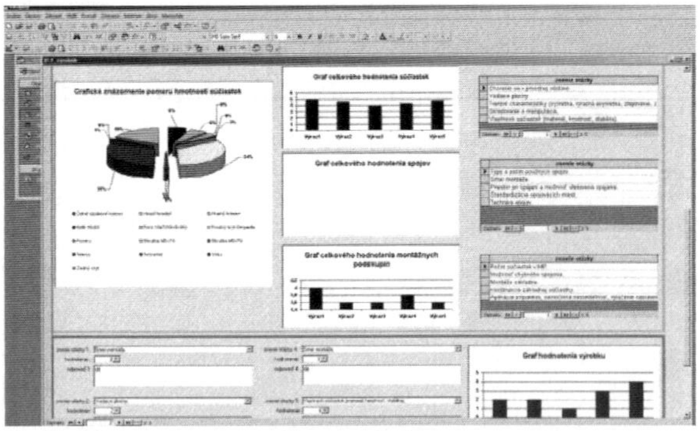

Fig. 2. Main screen – product analysis output.

Access aplikacia: Posudenie konstrukcie vyrobku z hladiska montaze

VÝROBKY

Názov výrobku:	Číslo výkresu
Čerpadlo	666

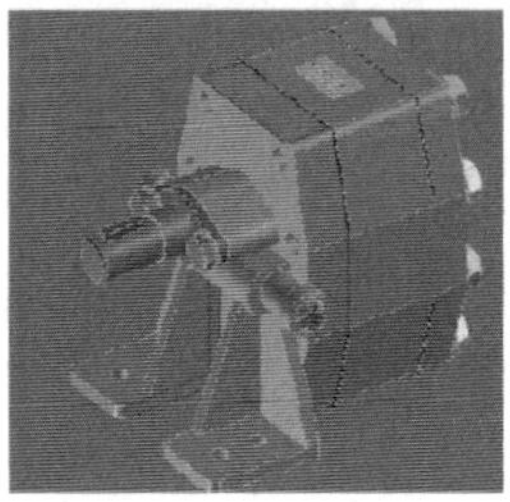

Identifikačné číslo výrobku v databáze :	2
Počet súčiastok vo výrobku:	13
Počet montážnych podskupín vo výrobku:	2
Počet spojov vo výrobku:	2
Nerotačné:	6
Rotačne symetrické:	3
Rotačne asymetrické I:	
Rotačne asymetrické II:	2
Rotacne asymetrické I a II:	2
Hmotnosť výrobku:	544

Zoznam súčiastok v výrobku:	Počet kusov vo výrobku [ks]	Číslo výkresu	Hmotnosť [kg]	Typ súčiastky symetria	Materiál
1 . Hranýhriadeľ	1	004	0,172	rotačná symetrická	kov, oceľ
2 . Zadný kryt	1	017	0,981	nerotačná	kov, oceľ
3 . Hnací hriadeľ	1	003	0,353	rotačná asymetrická I a II	kov, oceľ
4 . Kolík f6x65	2	005	0,015	rotačná symetrická	kov, oceľ
5 . Pero 10e7(h9x8x26)	1	007	0,005	nerotačná	kov, oceľ
6 . Predný kryt čerpadla	1	010	1,843	nerotačná	kov, oceľ
7 . Skrutka M8x16	2	012	0,016	rotačná asymetrická II	kov, oceľ
8 . Skrutka M8x70	8	013	0,065	rotačná asymetrická II	kov, oceľ
9 . Teleso	1	014	1,703	nerotačná	kov, oceľ
10 . Tesnenie	2	015	0,007	nerotačná	pružný materiál (guma, pr)
11 . Púzdro	4	011	0,006	rotačná symetrická	kov, oceľ
12 . Čelné ozubené koleso	2	001	0,333	rotačná asymetrická I a II	kov, oceľ
13 . Veko	1	016	0,045	nerotačná	kov, oceľ

Fig. 3. Example of the analysis output report.

The basic of this approach is a list of rules and recommendation for product design from the point of view assembly. The system is created so that in time of the system using can be the rules, recommendation and examples continuously refilled. On the figure 4 is shown the screen with the list of rules and appropriate examples.

The questions can also be modified or changed. On the figure 5 is shown the screen to creation of a relation between questions and defined rules. By changes in this filed must be clearly that the obtained results can be depended on the putted questions. This must be considered in the case of comparison of such results that was obtained by using different questions.

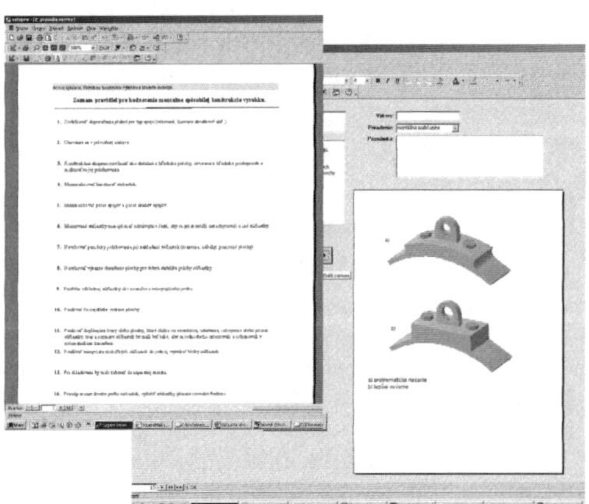

Fig. 4 List of the rules and recommendations

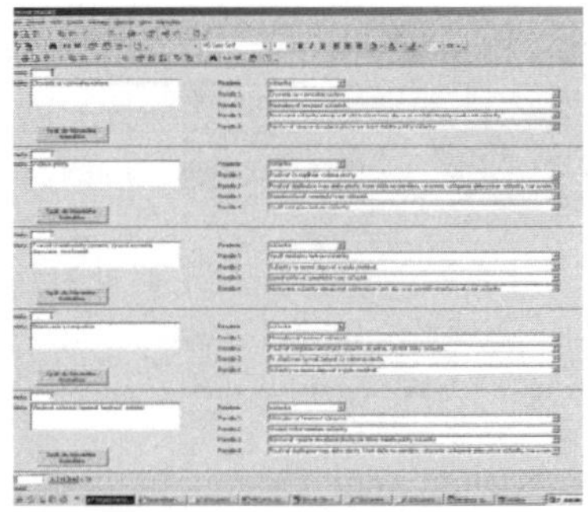

Fig. 5 Screen to creation of a relation between questions and defined rules

4. TOPICS TO FURTHER RESEARCH

In this phase of the development is performed an augmented verification in several directions. First augmented verification is oriented to the tool using for the analysis of various assembled products. The target is to verify the scope of the tool using from the point of view product variety. The second researched aspect is the possibility of the integration into the general procedure of the assembly process and system design and planning. The further considered characteristic is the product design change supporting. The main goal is to become the impulses and experiences from the industry. Important can be also the integration into the education. On the basis of the obtained answers to the above putted questions can be considered also the toll modification, selection of the other program tool or Internet publishing.

5. CONCLUSION

Experiences with creation and application of the developed tool to product design analysis from the point of view assembly show, that it is one of the possible directions in the design for assembly supporting. The tool can be easy used, modified and refilled. Important in also, that by this tool can be stored and managed any previous proposed solution of part, joint assemblies or product changes.

Acknowledgements

This paper was elaborated in the frame of the projects:
Objective method of assembled product design and assembly system VEGA No.: 1/3163/06 and
Innovative education materials for bachelor study line Automobile manufacturing KEGA No.: 3/6342/08
Supported by Ministry of Education of the Slovak Republic.
We are indebted for support.

6. REFERENCES

Boothroyd, G. & Dewhurst, Inc. (2006). Design For Assembly, *Available from:* http://www.dfma.com *Accessed:* 03-04

Fraunhofer institute (IPA) (2007). Design For Assembly and Manufacturing, Stuttgart, *Available from:* www.ipa.fhg.de *Accessed:* 2007-08-05

Lotter, B. & Wiendahl, H.P. (2006). Assembly in Production Industrial, ISBN 13978-3-540-21413-7 s.531, Springer, Berlin Heidelberg, New York

Senderská, K.; (2006). Approaches and Tools to Evaluation Support of Product Construction for Assembly, In: *Inovations Transfer*, SjF TU, pp. 214-215, ISBN 80-8073-701-0, Košice

Václav, Š. (2005). What is the Assembly? *Proceedings of 13th International Scientific Conference CO-MAT-TECH 2005*, pp. 1289-1294 ISBN 80-227-2286, Trnava

Annals of DAAAM for 2008 & Proceedings of the 19th International DAAAM Symposium, ISSN 1726-9679
Editor B. Katalinic, Published by DAAAM International, Vienna, Austria 2008
Make Harmony Between Technology and Nature, and Your Mind will Fly Free as a Bird

THE INVESTMENT RISK ASSESSMENT FOR A NEW PRODUCT LAUNCH

PETRE, M[ihai]; ISPAS, C[onstantin]; MOHORA, C[ristina] & LUNGU, B[ogdan]

Abstract: *This paper presents the importance of investment risk assessment analysis and simulation in the context of a new product launch. The risk analysis and the sensitivity emphasize the integrity and stability of the social economic analysis. According to that, it means to accept the work proposed hypothesis and also the fact that, in the conditions of unfavorable factors of influence, the investment will remain profitable.*

Key words: *risks, analysis, simulation, investment, new product*

1. INTRODUCTION

1.1 Investments Role in Social-Economic Environment

Investments play an important role for companies because it affect, at the same time, the demand and the offer.

The implementation of a project in products manufacturing and/or services area results in offer growth and diversification and, if the offer is validated on the market, additional income for the companies (Caouette,1998).

The projects and investment programs developement result in capital growth. So, investments are the main things for real economic restructuration, creating newer and more viable structures regarding the societies strategic targets. Investment projects are the main methods for obtaining newer technical and technological solutions provided by scientific research.

The investment concept is identical with: to transfer, to provide, to offer and it can be extended to efforts made in order to gain future benefits (by efforts, it does not mean only financial ones).

The share of **„intellectual" investments** becomes more important but, for present projects, it is very difficult to evaluate its value in money.

From the dedicated literature, the most pertinent definition, considered by a large group of specialists, is the one made by Pierre Masse („Les choix des investissements". Dunod, Paris) which defines **investment** as a **„change of an immediate and certain satisfaction, which is abandoned, for a future expectation that can be obtained using as prime element the invested goods"**, in short version, **„an uncertain spending for an uncertain future".**

The common point consists in the nature of engaged resources: material resources, human resources, technical resources, monetary resources, etc.

The concept of investment emphasis on the domain of application and through its content.

Starting from project manager "main paper", the investment risk is calculated. The first step in risk assessment, named TOP RANK, calculates the sensitivity analysis, choosing the significant inputs for which the selected output is more sensitive (Ispas, 2005).

The second step consists on the output variation risk and the final value in euros of the input significant variations.

2. INVESTMENT RISK ASSESSMENT FOR THE NEW PRODUCT LAUNCH

Sensitivity analysis shows the profitability indices distribution area, used for different evolution scenarios. The key factors are to test the project benefits and to order the indices regarding their impact on the outputs.

The purpose for a sensitivity analysis is to identify the model variables or to model critical parameters, of which variations, positive or negative, in comparison with the optimal case study values, leads to the most significant variations on profitability main indices, respectively RIR and VNP.

The identification criteria for these variables are changing in accord with the project characteristic and must be done with high accuracy.

The sensitivity analysis was made calculating in Excel the "main paper", with Palisade Decision Tools software, important software for sensitivity analysis and risk assessment case study.

The "What If" simulation has generated the results which proved that the most important share in risk assessment is "the exchange rate", the remaining parameters having insignificant influence on the output.

2.1. Risk analysis

2.1.1 Implementing the mathematical model in Palisade @Risk; Inputs

Starting from the sensitivity analysis, the mathematical model is made with Excel and it is correlated with the results obtained with TOP RANK. The economical and financial market information, the "exchange rate" input, in thousands of lei/thousands Euro at an exchange rate lei/Euro, was defined as the most significant variable that generates risk (Ispas, 2005).

Variable element modelling

From GENERAL DEVICE, the exchange rate in thousand of lei/ thousands of euros from 17.03.2008 was used. Considering the estimated forecast information from National Bank, triangular probability distributions were defined, concentrated around the medium value of 3,7538, with a maximum value of 3,9 and a minimum value of 3,5.

The program ran a Monte Carlo simulation with one hundred iterations. Each iteration acquired new values for "GENERAL DEVICE" (without VAT).

2.1.2. Monte Carlo simulation statistical results using 10 000 iterations

The results analysis

Detailed Statistics

Name	TOTAL GENERAL(fara TVA) / mii euro	In mii lei/mii euro la cursul lei/
Description	Output	RiskTriang(3,5; 3,75; 3,9)
Cell	D41	C5
Minimum	1238,175	3,517092
Maximum	1366,757	3,882334
Mean	1294,03	3,71661
Std Deviation	29,12673	8,311274E-02
Variance	848,3665	6,907728E-03
Skewness	0,3425137	-0,2498258
Kurtosis	2,519871	2,450162
Errors Calculated	0	0
Mode	1275,267	3,722806
5% Perc	1250,076	3,569106
10% Perc	1256,994	3,59666
15% Perc	1262,808	3,620731
20% Perc	1267,518	3,638352

Fig. 1. Statistical information for distribution results

Analyzing the table and variable statistical summary of the model it results that the **"GENERAL DEVICE"** value (without VAT) can vary between a minimum value of 1 250 thousands of € and a maximum value of 1 346 thousands of €, with a 90% probability of occurrence. The statistical average value is around 1 291 thousands of €, extremely close to the estimated value of 1 293 thousands of € obtained through direct calculation; this result places the project targeted value, in risk conditions, way below the +/- 10% value, accepted by UE standards for eligible reserve funds. Figure 1 presents the statistical information for the results.

The worst and best case scenarios

Sensitivity and risk analysis allows also a scenario analysis for the worst and the best case scenarios. From the values of statistical distributions obtained for **"GENERAL DEVICE"** (without VAT) result a probability value below 5% in order to obtain an amount over 1 350 thousands of €, also a probability value below 5% for an amount below 1240 thousands of €.

Conclusions for sensitivity and risk analysis

Sensitivity and risk analysis emphasize model integrity and stability for the social and economical analysis.

This leads in accepting the work examined hypothesis and the fact that under an unfavourable variation of influence factors, the investment will continue to be profitable (Chan, 2002).

Assumed risks are technical risk, financial risk, institutional risk, legal risks.

The project execution risks can be **internal type** or **external type**.

This is based on the three dedicated key systems of project management.

The supervising system

It consists in the permanent comparison of an actual situation with the project planning: natural evolution, financial expense, quality (the project objectives are identical with the created assets).

A suitable deviation from the supervising system (a programmed evolution) leads to a decision package from project managers who will decide if some recovering measures are possible or not.

The control system

This system will have to act fast and efficient when the supervising system indicates a deviation from its course.

Informational system

The informational system will support the control and supervising systems, providing the project team (in a short period of time) the necessary information.

For project supervision (the first key system of project management), the imperative information are the following:

- The measurement of physical evolution.
- The measurement of financial evolution.
- Quality control.
- Other particular information which offer specific interest.

Fig. 2. Iteration distribution histogram

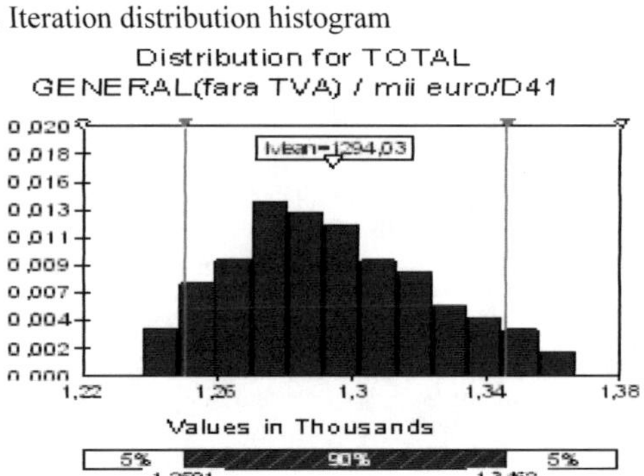

The financial control mechanism

A financial control mechanism, which assures an optimum found utilization, is a circular conditions system which helps on achieving project objectives, avoiding the unforeseen events and notify, in time, about the risks which demand corrective measures (Petit, 2000).

The main effective tools will rely on quantity and quality analysis of results.

Accountancy and financial management

The accountancy and financial management will be handled by a specialist accountant who will manage three major tasks:

1. Operations planning, control and registration.
2. Information presentation (the first two items are tasks of the specialist accountant).
3. Financial decisions (management competence).

3. CONCLUSIONS

The analysis for risk distribution is a common way for project manager, showing where the risk resources can be found. There were marked several imperative ideas:

•Through time, the project has to be adaptable to changes

•The information regarding risks represents critical elements for analysis.

In its evolution, the risk assessment for industrial companies' investments gained flexibility and adaptability through modelling and simulation of economic processes and also, the inter-disciplinary nature has been dramatically emphasized.

This paper presents an alternative in analyzing different economic processes with a simulation model made with a system of software internationally recognized. This software uses the specific Microsoft Excel routines, presenting a big degree of accessibility and comprehensively, creating a useful and easy to understand language used by a large number of experts from developed countries.

The modern management for investment projects leads in merging the real investments theories with financial theories in their superior form, enabling project management models to be made and to be optimized with simulation and risk analysis.

The risk analysis and the sensitivity emphasize the integrity and stability of the social economic analysis. According to that, it means to accept the work proposed hypothesis and also the fact that, in the conditions of unfavorable factors of influence, the investment will remain profitable.

4. REFERENCES

Caouette, J.B., Altman, E.I., Narayanan, P.(1998). *Managing Credit Risk, The Next Great Financial Challenge*, John Wiley&Sons.

Chan S., Ronald P., Kenneth, C., Ming. J. Zuo (2002) *Analyse economique en ingenierie* Ed. Pearson Education Canada Inc, Toronto, Ontario, ISBN 2 84211 198 2.

Damodaran A. (2002) *The Promise and Peril of Real Options,* Stern School of Business 44 West Fourth Street New York, NY 10012.

Ispas C., Petre M., Mohora C., Bălan E. (2005). *Hedging and others financial derivatives, new instruments of risk management in Romania-* Proceeding of the 7-th International Conference on: The Modern Information Technology in the Innovation process of the Industrial Enterprises pag. 64, Genoa, Italy September 8-9. ISBN: 88-7544-050

Petit J., *Evaluation,* Editura Stern Stewart & CO, Varşovia, 2000.

Annals of DAAAM for 2008 & Proceedings of the 19th International DAAAM Symposium, ISSN 1726-9679
Editor B. Katalinic, Published by DAAAM International, Vienna, Austria 2008
Make Harmony Between Technology and Nature, and Your Mind will Fly Free as a Bird

EXTRAJUDICIARY TECHNICAL EXPERTS AND CONSULTANTS' SOCIETY AND THE STANDARD ON WHICH IS BASED

PETRESCU, L[igia] & MARIN, G[heorghe]

Abstract: *Every activity concerning the engineering and technical field is organized in professional associations which promote practice professional and ethical standards. Extra Judiciary Technical Experts and Consultants' Society, member of The General Association of the Engineers of Romania and of The Association of European Recognized Experts of Paris, is a professional association of engineers and economists certified in conformity with the European Union standard No. EN 45013, and with the Law No. 156/2002, concerning the judiciary and extra judiciary expertise in Romania.*
Key words: *expert, expertise, society, engineer.*

1. INTRODUCTION

The Extra Judiciary Technical Experts and Consultants' Society from Romania has a great purpose according to the status of the Romanian technical expert and work this period to realize the Professional Standard for this activity. The term *expert* is frequently used but the Romanian legislation introduced the occupation of *extra judiciary technical expert* in the Occupations Cod of Romania only in 2005, position 241801, like answer to the efforts of our society and with the next definition: The extra judiciary expert is a specialist in technical field who draws up extra judiciary technical expertise reports with the purpose to establish and to specify some "fact circumstances", who analyses the technical and economical aspects of an organization, with regard to reorganize, to develop or to improve the production or commercial proceedings and who draws up extra judiciary technical and economical expertise reports for third person/party (Brown, S. 2001). The expertise activity, judiciary end extra judiciary, is materialized in the expert's report or report of expertise.

The expert is one who has acquired special skill in or knowledge of particular subjects through professional training and practical experience. The term expert is used to describe highly experienced professionals, but has been expanded to include any individual who attained their superior performance by instruction and extended practice.

2. THE EXTRAJUDICIARY EXPERTISE IN EUROPE

2.1 The expertise in Great Britain

- English dictionary: the expert is a person who has extended knowledge and abilities in a given field.
- The British legislation: the expert is the person whose personal practical experience in a given field allows him to be qualified to use proofs, to formulate conclusions and to give an impartial opinion.
- Experts' instruction begins from their base profession. The association Experts' Academy organizes courses for the expertise report drawing up, for the proofs collection and analysis, for the proofs and conclusions supporting.
- There is no recognized standards and formal attestation for the experts' activity.

2.2 The expertise in France

- The expert is a professional who is specialized in a certain technique.
- He is named by the instance.
- The Court of Cassation and the Court of Appeal establish every year the experts' list.
- The expert makes an oath for registration, accepts given missions and draws up reports of expertise.
- 25 years ago was created The France National Committee of Experts

2.3 The expertise in Germany

- There is no legal definition for the expert.
- Everyone who accomplishes the main conditions of knowledge, who proves personal integrity, who has financial and personal independence, can practice the expert profession.
- There is no law to establish the expert profession, but there are some authorized institutions to attest experts. These institutions are established by the industrial, or commercial sector, or by the professional associations.
- There are many types of experts in Germany:
  - ⇒ Experts certified by public institutions and attested by the Chamber of Commerce, by the Chambers of Artisans, and by the professional associations of engineers and architectures.
  - ⇒ Experts who realize inspections and supervisions workings requested by the law.
  - ⇒ Experts recognized by the private associations.
  - ⇒ Experts certified according to the European norm EN 45013.
- The newest system to attest experts is based on the European norm EN 45013.

2.4 The expertise in European Union

The organization EURO EXPERT is a non profit organization which objectives are:

- ⇒ To develop, to promote and to converge the professional and ethical standards for the European Union Experts, based on the next principles: height qualification, personal integrity, independence, impartiality, objectivity and the confidentiality respect.
- ⇒ A contact element between the experts and the European Union institutions.
- ⇒ Cooperation and relationship with governmental authorities, publics and private organizations.
- ⇒ A forum for the whole world experts.

The founding members are: France, Germany and Great Britain. The European Court of Justice can name like expert, every physical or juridical person, authority, comity or organization. The European Court of Justice names the expert, define the mission and establish the dead line for the expertise report.

3. THE CONTINUOUS EDUCATION FOR EXPERTS

It is important to know that *The General Association of the Engineers of Romania* (AGIR) is member of WFEO – World Federation of Engineers Organizations from 1990 and of IACEE – International Association of Continuous Engineering Education from 1992 (Greavu et al., 2005).

The activity of continuous engineering education by courses for extra judiciary technical experts and consultants is organized in the followings steaps (Fig. 1.):

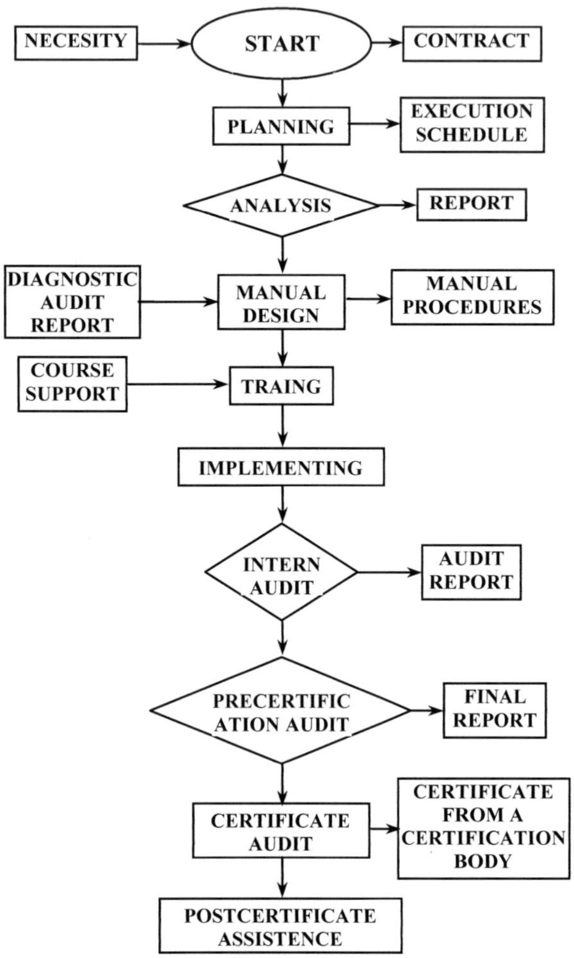

Fig. 1. The activity of continuous engineering education by courses for extra judiciary experts (Petrescu, L. 2005).

The activity to form new extra judiciary technical experts was assumed by our Extra Judiciary Technical Experts and Consultants' Society, member of AGIR, by organized courses for engineers with minimum 5 years of experience. The title of *extra judiciary technical expert* can be obtained by exam which verifies the professional qualification, the level of knowledge of the normative acts (Bolos, C. & Leaua, C. 2002) referring to his specialty and with some other conditions:

a. Is a Rumanian citizen and knows very well the Rumanian language;
b. Has a full capacity of discharge and judgment;
c. Is an university certificated graduate in the specialty in which wants to practice like expert and consultant;
d. Has minimum 5 years of experience in the specialty in which wants to practice like expert and consultant;
e. Is medically able to practice the expert activities;
f. Has not penal antecedents and has a good professional and social name;
g. Graduates with exam the course for extra judiciary technical experts and consultants.

All the graduated experts are accredited by the Office of Experts Certification – CERTEXPERT, member of The General Association of the Engineers of Romania.

4. THE OCCUPATIONAL STANDARD FOR EXTRAJUDICIARY TECHICAL EXPERTS AND CONSULTANTS

The realization of the Professional Standard for the activity of the extra judiciary technical experts was assumed by the Extra Judiciary Technical Experts and Consultants' Society from Romania (Greavu et al., 2005), (Radaceanu, E. 2001). According to the protocol to establish an occupational standard, such activity begins with an occupational analyze, what he represents, what he must know and what he must do. He must demonstrate fundamental, general and specific competences (Fig. 2).

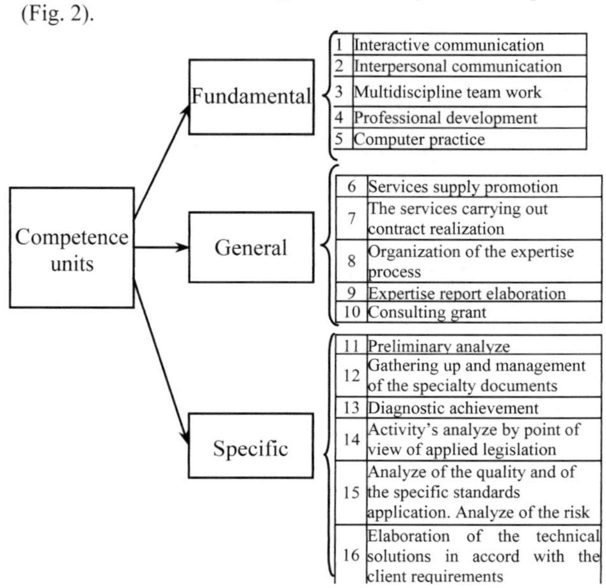

Fig. 2. Units of competence.

5. CONCLUSION

The expertise activity and the quality of the accepted experts and their reports are based on a sustained effort to improve the education, the continuous education, the standards and the laws which manage the domain. The international associations of experts try to generate a common program for all European countries to develop the expertise activity, and our society is part of it, and represents our future research plans.

6. REFERENCES

Bolos, C. & Leaua, C. (2002). *Calitatea – norme si legislatie (Quality – norms and legislation),* Petru Maior University Publisher, ISBN 973-8084-61-x, Targu Mures.

Brown, S. (2001),*La gestion de la relation client, CRM (Customer Relationship Management)*, Pricewaterhouse Coopers, village Mondial Press.

Greavu, V.; Popa, D. et al (2005). *Expert thnic extrajudiciar (Extrajudicial technical expert),* AGIR Publisher, ISBN 973-720-026-8, Bucuresti.

Petrescu, L. (2005), *Some modern ideas in sciences teaching,* Proceedings on The International Conference On Engineering Graphics And Design, ISBN 973-648-471-8, Bucuresti.

Radaceanu, E.(2001). *Management,* BREN Publisher, ISBN 973-8143-66-7, Bucuresti.

Annals of DAAAM for 2008 & Proceedings of the 19th International DAAAM Symposium, ISSN 1726-9679
Editor B. Katalinic, Published by DAAAM International, Vienna, Austria 2008
Make Harmony Between Technology and Nature, and Your Mind will Fly Free as a Bird

QUALITY OF EXPERTISE AND EXPERTISE OF QUALITY

PETRESCU, L[igia] & MARIN, G[heorghe]

Abstract: *The expertise activity, judiciary end extra judiciary, is materialized in the expert's report or report of expertise. The expertise is a product which quality is judged and is comply with all principles and standards of quality. A particular aspect in expertise achievement is involved with the quality of the organization which is analyzed and tested. That is the reason why the quality's aspects and peculiarities are essential in expertise approach.*
Key words: *quality, expert, expertise, management, customer*

1. INTRODUCTION

The most common interpretation of the term quality in engineering and manufacturing is the *non-inferiority*, *superiority* or *usefulness* of something. There are two common applications of the term quality as form of activity or function within a business. One is Quality Assurance which is the "prevention of defects", such as the deployment of a Quality Management System and preventative activities. The other is Quality Control which is the "detection of defects", most commonly associated with testing which takes place within a Quality Management System typically referred to as Verification and Validation. Taguchi's definition of quality is based on a more comprehensive view of the production system, and he relates Quality (or, more precisely, the lack of it) to "The loss a product imposes on society after it is shipped". Involving with expertise, quality can have two meanings: the characteristics of an expertise that bear on its ability to satisfy stated or implied needs and an expertise free of deficiencies. In expertise report, quality control and quality engineering are involved in developing systems to ensure expertise is designed and produced to meet or exceed customer requirements. These systems are often developed in conjunction with other business and engineering disciplines using a cross-functional approach.

2. THE EXPERTISE QUALITY ASSURANCE

The expertise quality assurance covers all activities from documentation, design, development, realization and consulting. This introduced the rules: "fit for purpose" and "do it right the first time".

The quality of the expertise is based on a special standard (Greavu et al., 2005). The realization of the Professional Standard for the activity of the extra judiciary technical experts was assumed by the Extra Judiciary Technical Experts and Consultants' Society from Romania. According to the protocol to establish an occupational standard, such activity begins with an occupational analyze, what he represents, what he must know and what he must do. He must demonstrate fundamental, general and specific competences, which are fundamental elements for a high quality expertise.

Critical to Quality (CTQ) is the key measurable characteristics of a product or process whose performance standards or specification limits must be met in order to satisfy the customer. CTQs align improvement or design efforts with customer requirements. CTQs represent the product or service, this time the expertise characteristics that are defined by the customer. They may include the upper and lower specification limits or any other factors related to the expertise. A CTQ usually must be interpreted from a qualitative customer statement to an actionable, quantitative business specification. To put it in layman's terms, CTQs are what the customer expects of an expertise the spoken needs of the customer.

If the quality of an expertise is to be evaluated, a quality model is needed, plus a method for evaluation and a quantitative scale of measures. A model is an abstract representation of an object: according to this very general definition, which implies the existence of several classes of models, it can be define a quality model as a structured set of quality requirements. The problem of defining a quality model to be used in the evaluation of an expertise is to obtain values that result from weighing the quality characteristics according to their criticality evidence, a quality profile which can be used both for evaluation and certification.

Generally speaking about quality characteristics, there are some external characteristics like: correctness, usability, efficiency, reliability, integrity, adaptability, accuracy and robustness. Like internal characteristics they can be mentioned: maintainability, flexibility, portability, reusability, readability, testability and understandability. It's important to work on a project what is the priority and it's also important not to underestimate the effect of all the unspoken tacit decisions made by developers in the course of their work: the unspoken assumptions. Setting clear goals is a pretty key management tenant, but one easily forgotten. Some of these characteristics are the most important:

- Customer Facing (most of the External Characteristics)
- Maintenance and Design (most of the Internal Characteristics)
- Defects.

Each review role had a cheat sheet which consisted of the relevant characteristics to look for, and some check lists. This helps ensure visibility of the quality priorities into the code review, which makes them doubly hard to ignore.

3. CUSTOMER RELATIONSHIP MANAGEMENT (CRM) CONCEPT IN EXPERTISE

Customer Relationship Management is often referred to the marketing. There is no complete agreement upon a single sense of the term. CRM can be considered from a number of perspectives. The customer "satisfaction" has a growing weight in the expertise success assurance. What is to do in this direction can be expressed in two strategies:

- Strategically vision of the organization which express orientation to the customer and planes to create new value for both customer and organization;
- Operational strategy which express the way to transpose in reality the planes of the customer relationship management through the customer needs and expectations knowledge and answering (Fig. 1).

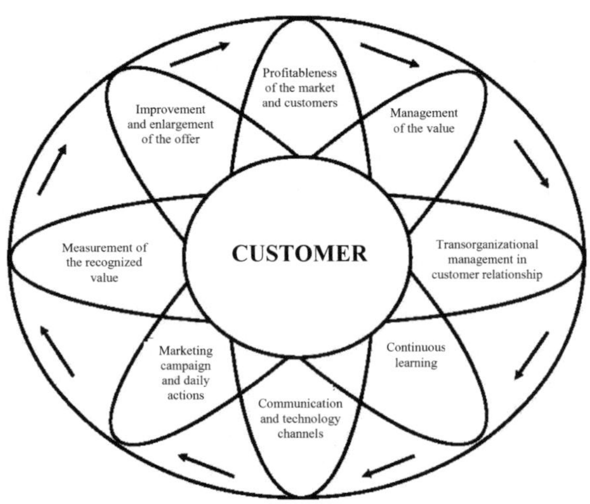

Fig. 1. Operational strategy for the CRM.

CRM is a combination of philosophies, policies and strategies connecting different players within an organization so as to coordinate their efforts in creating an overall valuable series of experiences, products and services for the customer.

The technical expert must adapt oneself to the market demand and supply. His service is accorded to the customer behaviors. The quality of expertise determines and is determined by the customer satisfaction / unsatisfaction. A dissatisfied customer imparts his unsatisfaction to minimum ten persons, but a satisfied customer imparts his satisfaction only to three persons. To gain a new customer is ten times expensive then to maintain an old one. Around 80-95% of the customers are disposed to spend more for a high quality of the needed expertise (Dragulanescu et al., 2003).

4. EXPERTISE OF THE QUALITY

Expertise of quality is similarly with the audit of quality. The quality's audit cans refer to an entire system of quality, to a process, or to a product. The audit of quality aims at the identification of the possible improvements of the system / process / product. Expertise of quality means in plus to co-operate to solve the problems using the knowledge, abilities and skills of an expert. The expertise gives the solution of the problems, and the expert grants consulting services.

The quality's management instruments (Fig. 2) are applied for three principal goals (Dragulanescu, N. et al., 2003):

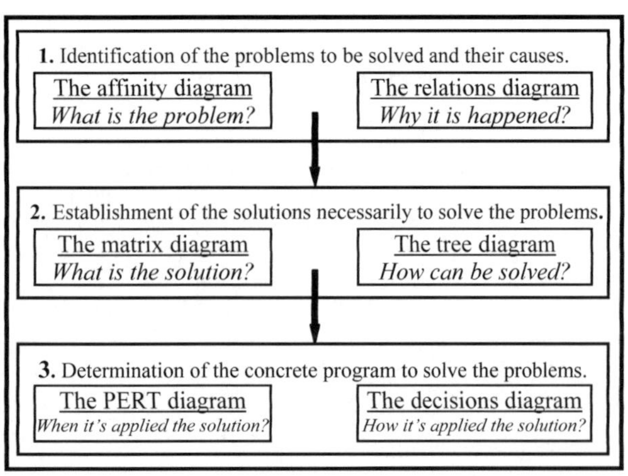

Fig. 2. The Quality's Management Instruments.

- Identification of the problems to be solved and their causes by The Affinity Diagram (Fig. 3) and The Relations Diagram (Cause-Effect) (Fig. 4);
- Establishment of the solutions necessarily to solve the problems by The Matrix Diagram and The Tree Diagram (Foster, T. 2001);
- Determination of the concrete program to solve the problems by The PERT Diagram and The Decisions Diagram (Radaceanu, E. 2001).

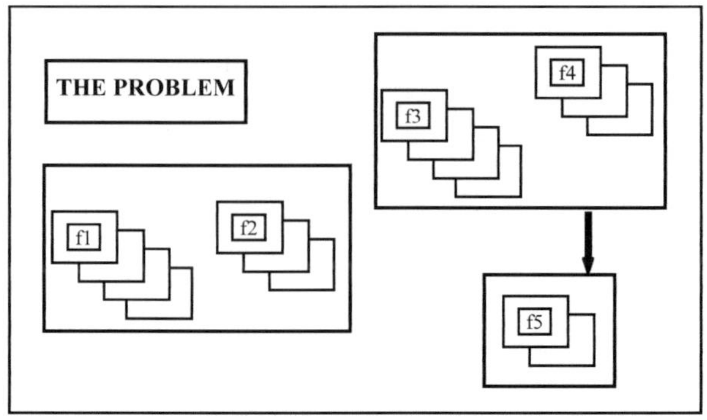

Fig. 3. The Affinity Diagram Application (Petrescu, L. 2004).

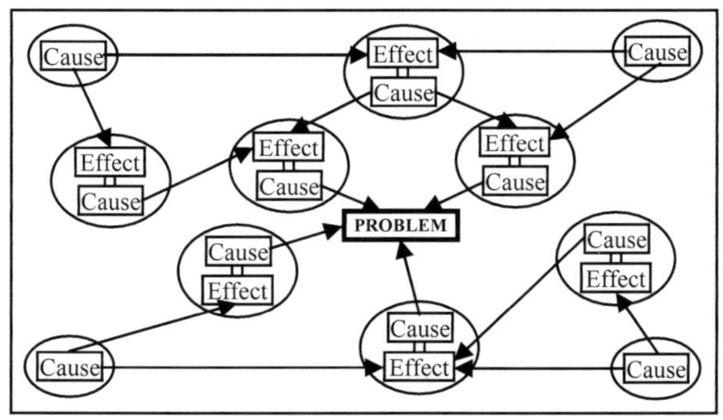

Fig. 4. The Relations Diagram (Cause-Effect).

A special instrument of the quality management is "benchmarking". It is used to evaluate the quality performances of an organization, with the purpose to make a comparison with an organization which is considerate a referential.

5. CONCLUSION

The expertise and the quality are the two elements on which the accepted experts and their reports are based on a sustained effort to improve their expertise activity.

6. REFERENCES

Dragulanescu, N. & Dragulanescu, M. (2003). *Managementul calitatii serviciilo, (Services Quality Management),* AGIR Publisher, ISBN 973-8466-33-4, Bucuresti.

Foster, T. (2001) *Managing Quality - an integrative approach,* Prentice Hall Publisher, New Jersey.

Greavu, V.; Popa, D. et al (2005). *Expert thnic extrajudiciar (Extrajudicial technical expert),* AGIR Publisher, ISBN 973-720-026-8, Bucuresti.

Petrescu, L. (2004), *The Engine System Study Of Reliability,* 4th International Conference "Research and Development in Mechanical Industry" RaDMI 2004 31 August - 04. September 2004, Zlatibor, Serbia and Montenegro, ISBN 86-83803-18X pag. 412-418

Radaceanu, E.(2001). *Management,* BREN Publisher, ISBN 973-8143-66-7, Bucuresti.

Annals of DAAAM for 2008 & Proceedings of the 19th International DAAAM Symposium, ISSN 1726-9679
Editor B. Katalinic, Published by DAAAM International, Vienna, Austria 2008
Make Harmony Between Technology and Nature, and Your Mind will Fly Free as a Bird

ANALYSIS OF THE CHARACTERISTICS OF METALLIC POWDERS REALISED FROM THE WASTE RESULTED FROM BALL BEARING MANUFACTURING

PETRESCU, V[alentin]; BIBU, M[arius]; NEMES, T[oderita]; POPESCU, F[lorian] & DEAC, C[ristian]

Abstract: *The experimental results and the method of elaborating powders from wastes of ball bearing steel of RUL type are presented. There are analyzed physical, chemical and technological characteristics of the powders as well as some properties of the sintered material. The studies show the parameter's variation function of compacting pressure.*
Key words: *powder metallurgy, ball bearing steel, technological characteristics, forging*

1. INTRODUCTION

After the finishing machining of bearing balls and ball tracks, important quantities of waste from RUL steel material may result, with a fine grain structure (Petrescu & Nemes, 2001; Amza, 2002). The reentry of the metallic powder into the economic circuit represents a superior exploitation of wastes of RUL steel, because the metallic powder prices at the global scale are higher than the prices of other types of metallic parts (obtained through casting, milling etc.

The studies presented in this paper represent a continuing preoccupation of the authors to recover the waste material as metallic powder (Urdas & Petrescu, 2001). Until the present time this technology was not exploited (Salak et al., 2001).

The waste resulted from the process contains, beside the RUL steel grains, particles of manganese, white cast iron detached from the finishing disks, emulsion oils, also iron oxides form the part's surface and from the hot pressing machining previously executed.

2. EXPERIMENTAL RESEARCHES

The experiments have been realized at temperature values of 800, 850 and 900°C, with time ranges of 30, 45 and 60 minutes.

The waste's purification can be realized by oxidizing burns (anneal), and the obtained reduced compounds have been treated to obtain some alloyed iron powders with low carbide ratio.

The metallic waste obtained in the above-mentioned conditions was annealed at 950°C, over a period of 60 minutes, was reduced at 950, 1000 and 1050°C temperatures with the reduction period of 60 minutes, in a dissociated ammonia atmosphere (160 l/h yield). The pressing curves, for the obtained powders in these three reduction regimes, have been raised for the compacting pressures of 2, 4, 6 and $8 \cdot 10^3$ daN/cm^2. In the sintered form, the hardness values have been determined on cylindrical plates with diameter d = 11.24 mm and height h = 20 mm, the increase of density through the sintering process on cylindrical plates with d = 16 mm, also tensile strength tests were carried out on pressed plates according to the MPA standards.

The sintering process was realized in a tubular sintering furnace, at a temperature of 1200°C, during a 90 minutes period, in a dissociated ammonia atmosphere (yield 150 l/h).

3. RESULTS AND DISCUSSIONS

The chemical composition of the recovered scrap is presented in table 1, the content of alloying elements matching the standard prescriptions for the RUL steels. It was determined that the carbide content concentration is higher than the standard concentration (0.95 - 1.10 %C), the initial scrap being unpurified with cast iron particles detached from the finishing disks and also with oils contained in the working emulsion.

The purpose of the annealing treatment is to obtain reduction compounds through the oxidation process, a decrease of the carbide concentration and a decrease of organic substances and oily impurities.

According to table 1, the total volume of carbide decreases with the increase of the annealing temperature, the lowest level was obtained at an annealing temperature of 950°C (0.08%).

This content may decrease with the decrease to 950°C of the reducing temperature and the spectral analysis shows only carbide residues in the reduced powder.

The amounts of the other elements maintain their values after the reduction and annealing processes. The reduction process at 950°C satisfies the necessities of elaborating the powder with a low carbide proportion. The reductions have been executed at higher temperatures in order to obtain discontinuous distributions adequate to sintering process conditions.

Table 2 shows the density and the fluidity of the obtained powders in the three reducing conditions. Those powders belong to the low compactness powders category, the values of the powder density being lower than iron powders, usually utilized at the sintering process, with a value of 2400 kg/m^3.

Because of the reduced powder fineness at 950°C, it was not possible to determine fluidity unless using a calibrated hole with a 4 mm diameter (the other two values of fluidity were obtained using a calibrated hole with a 2.54 mm diameter).

Figure 1 presents the pressing curves for the obtained powders in the three reducing conditions. With an increasing reducing temperature, the powder density increases in the pressed conditions, the powder reduced at 1050°C showing the highest density.

The increase of density through the sintering process as a function of compacting pressure is shown in figure 2.

Material type	Working regime		Chemical composition						
	Burning temp., °C	Reduction temp., °C	C	Si	Mn	Ca	Ni	Cu	Mo
Initial RUL steel scrap	-	-	1.89	0.28	0.31	1.51	0.14	0.18	0.06
Annealed RUL steel scrap	800	-	0.20	0.26	0.31	1.42	0.12	0.16	0.06
	850		0.14	0.27	0.30	1.39	0.11	0.15	0.06
	900		0.09	0.25	0.30	1.38	0.10	0.16	0.06
	950		0.08	0.26	0.30	1.35	0.10	0.15	0.06
RUL steel - reduced powder	950	950		0.25	0.30	1.35	0.10	0.15	0.06

Tab. 1. The chemical characteristics of the employed materials under different conditions

Reduced RUL powder	Density, kg/m^3	Fluidity, S/50g.
950 °C	1790	8.3*
1000°C	1950	11.6
1050°C	1660	121.6

Tab. 2. Density and fluidity of obtained powders

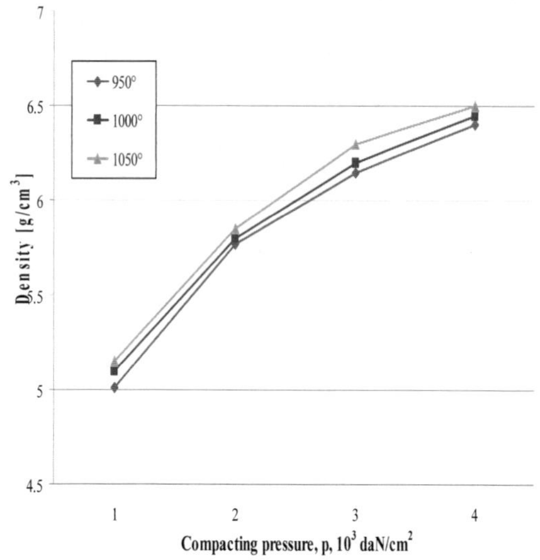

Fig. 1. Pressing curves for the steel powders

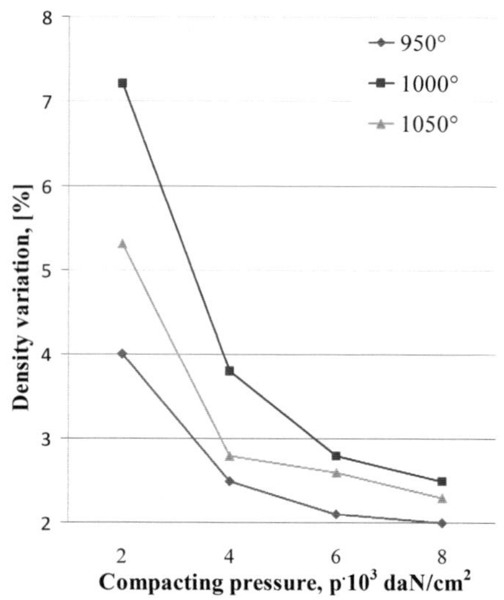

Fig. 2. Density increase as function of the compacting pressure

Analyzing the presented curves it can be concluded that the highest sintering density was obtained for the powder reduced at the temperature of 1000°C. The trend of the spheroid shape transformation of the powders, with the increase of the reducing temperature, explains the lower values for the density increasing for the sintering process of the powder reduced at 1050°C compared with the powder reduced at 1000°C. The mechanical characteristics of the sintered material using the three types of reduced powders are shown in fig. 3 and fig. 4.

Figure 3 presents the hardness variation as a function of the compacting pressure. The highest values of the hardness were obtained for the powder reduced at 1050°C with the trend of sintering hardness increasing with the increase of the reducing temperature.

Figure 4 presents the variation of the yield strength of the different sintered parts obtained from reduced RUL powders, in a close range, but superior to the commonly used iron powders (NC 100-24, Ph 160-27, FREM S 160-30).

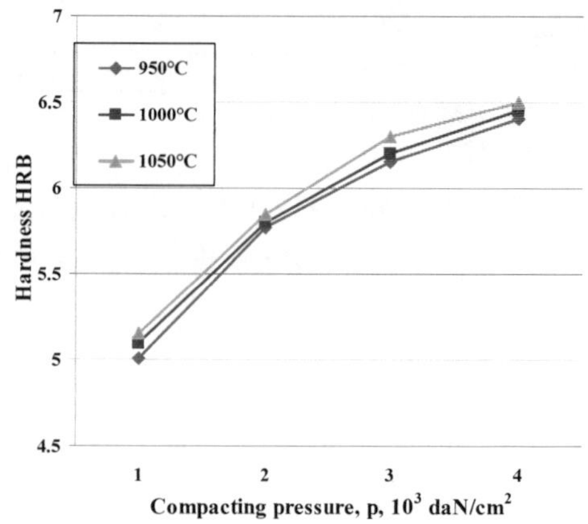

Fig. 3. Hardness variation as function of the compacting pressure

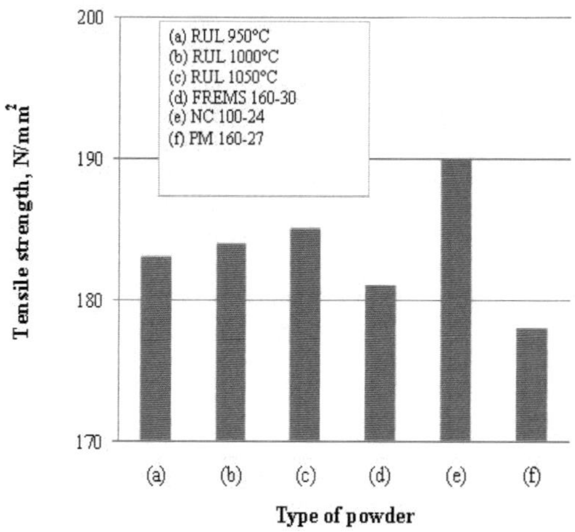

Fig. 4. Yield strength variation

4. CONCLUSIONS

The results of the presented research certify the possibility of elaborating alloyed powders with a low fraction of carbide from wastes of RUL steel, with good resistance characteristics after the sintering process, allowing the direct use of the powder mix recipes for the manufacturing of sintered parts with superior resistance characteristics, used in the machine manufacturing industry.

5. REFERENCES

Amza, Gh. (2002). *Tratat de Tehnologia Materialelor.* (Materials Technology Compendium). Romanian Academy House, Bucharest.

Petrescu, V., Nemes, T. (2001). *Tehnologia Materialelor 1. Elaborarea şi procesarea materialelor metalice* (Materials Technology – Elaboration and processing of metallic materials, Editura Universitatii "Lucian Blaga" din Sibiu, ISBN 973–651–250–9, Sibiu, 286 p.

Rud, V.D., Gal'chuk, T. & Povstyanoi, A.Y. (2005) *Powder Metallurgy Use of Waste from Bearing Production.* Powder Metallurgy and Metal Ceramics, Vol. 44, No.1-2, p. 106-112.

Salak, A., Selecka, M. & Danninger, H. (2001) *Machinability of Powder Metallurgy Steels.* Cambridge International Science Publishing, ISBN 1-8983268-2-7.

Urdas V. *et al.* (2001). *The study of materials used for the forging of metallic powders preforms Acta Universitatis Cibiniensis,* Vol. XLIV Technical Series, A. Materials Science and Technology, ISSN 1221-4949, pp. 51-54

Annals of DAAAM for 2008 & Proceedings of the 19th International DAAAM Symposium, ISSN 1726-9679
Editor B. Katalinic, Published by DAAAM International, Vienna, Austria 2008
Make Harmony Between Technology and Nature, and Your Mind will Fly Free as a Bird

EDUCATIONAL SYSTEM FOR WATER LEVEL REGULATION: DESIGN AND IDENTIFICATION

PETRLIC, D[ario]; MAJETIC, D[ubravko]; NOVAKOVIC, B[ranko] & BREZAK, D[anko]

Abstract: *This paper presents a student research work on modeling and identification of an experimental setup for educational purposes of control system demonstration and implementation. A brief description of the experimental setup is given with the emphasis on modeling and system identification. Both, mechanical and electrical parts of the setup have been realized in the Laboratory for automation and robotics. The communication with PC was established through serial port using MatLab software. The model parameters have been experimentally obtained based on various identification experiments which are in detail described in the paper.*

Key words: *Identification, Modeling, Liquid Level Control, Educational Experimental Setup*

1. INTRODUCTION

A laboratory setup for the purpose of education and analyzing of different control algorithms has been recently developed and implemented in the Laboratory for automation and robotics of the Faculty of Mechanical Engineering and Naval Architecture in Zagreb (Petrlić, 2005). The concept of the setup is selected in order to provide possibility of liquid level control. Generally, it comprises an open liquid container, a centrifugal pump, and various electronically controlled valves which provide possibility of disturbance simulation. This paper outlines the modeling and experimental identification of the setup. The results of the experimental validation are given. The work considered in the paper presents a basis for the future research which will include the methods of liquid level control.

2. SETUP DESCRIPTION

The experimental level regulation (LR) setup consists of eleven essential parts which are presented in the figure 1.

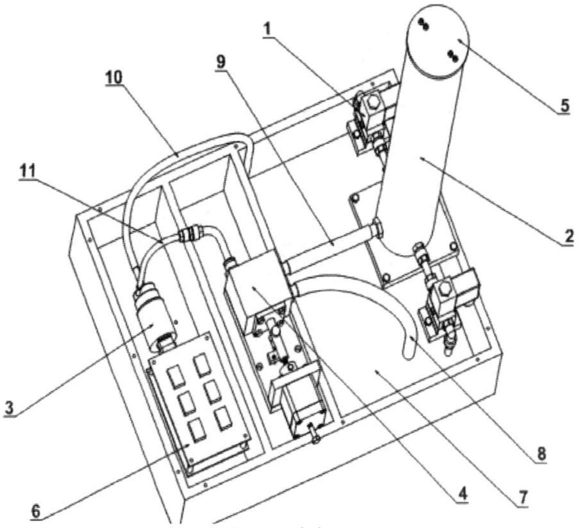

Fig. 1. Setup with marked essential parts

These parts are as follows: *1. solenoid 2/2 valve, 2. gauge glass, 3. the water pump which is activated by DC motor, 4. two-way valve with motor step drive, 5. sensor, 6. control unit, 7. water tank, 8. - 11. connective hoses*. Among all these parts, control unit and two-way value module were developed in the laboratory.

All these elements are located into the casing which is made from transparent plastic / perspex, so the system changes are easy to note. Construction of casing is realized as the union of three chambers. In first chamber the electronic elements are located. Function of second chamber is isolation of electronics from water. Finaly, all volume of liquid is stored in third chamber which represents the water tank from where the pump takes water and distributes it through the system (it is circular process). DC motor is used to drives the water pump and with pulse-width modulation (PWM) of voltage signal the pump-power is regulated. By changing the voltage signal at the DC motor, regulation or control of liquid-level in gauge glass is performed. Main characteristic of the two-way valve is that it has one input and two outputs as it shown in figure 2. The two-way valve module is made from polished transparent plastic / perspex, so that separation of flow is visible and easy to note.
By shifting the piston which is located inside the valve, the total volume of liquid that flows into the valve distributes in different rates at two outputs. The stepper motor is used for shifting the piston. The gauge glass (GG) represents the tank in which the fluid level is controlled. It is calibrated (from 100 mm to 300 mm with spacing of 5 mm) and it has four leaks: at the front of the GG there is a supply of liquid and at the bottom there is a drain which is always active, independently of the system state. On the left and right side, there are two leaks directly connected with solenoid valves. Solenoid valves are used to simulate disturbances (they have two states – on and off). Level of liquid is measured by sensor which is implemented into the gauge glass. The control unit enables the connection and control of main system parts (water pump, two-way valve, solenoid valves and sensor). By RS232 protocol control unit communicates with PC and the drivers are written as an M-functions using MatLab software.

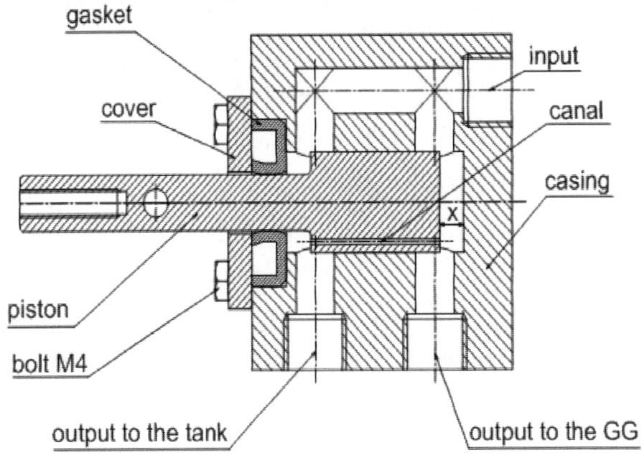

Fig. 2 Section view of two-way valve module

3. MATHEMATICAL MODEL

Existence of an accurate model of object dynamics is of crucial importance for the object control-related purposes (Isermann, 1996). In order to define the mathematical model of LR system, static and dynamic characteristics of system parts are defined (Lenart, 1995). On the basis of extensive measurements for every system part the mathematical equations are defined for each one. Structure of the mathematical model is given in figure 3, where the *Simulink* model is presented (Matlab help, 1999). It consists of power input, pump, two-way valve module, solenoid valves (left and right) and gauge glass with leak and height output. Two transformations of signal exist in this system. First, power input is transformed into the flow and second one, flow is transformed into the height.

This model presents the control of liquid level through the pump (control can be realized using the pump, two-way valve or in the combination). System input is power of the pump and the output is height of the liquid level in the gauge glass which is measured by sensor implemented into the gauge glass. Signal of the pump-power must also be sent as an input of the two-way valve module since it depends on the pump power and the piston shift. After the power signal is carried out through the pump and two-way valve, it is transformed into the flow-signal. Most system parts of LR setup can be identified as a simple dynamic elements (P0 or I0), as it shown in figure 3 (Kuo & Golnaraghi, 2003).

Identification of gauge glass required a different approach, because the parameters of constantly active drain could not be measured with existing sensor. Appling the Bernoulli's equation at the gauge glass the following relation is given:

$$q_{GG} = A_i \cdot \sqrt{\frac{1}{1+\xi_i}} \cdot \sqrt{2g} \cdot \sqrt{h(t)} + \left(\frac{dh}{dt}\right) A_m \qquad (1)$$

where q_{GG} is total GG flow, A_i is area of the GG drain cross-section, A_m is area of the GG cross-section, *h(t)* is height of liquid level in GG (from the drain), ξ_i is lost at the drain of the GG and *g* is gravitational constant. Coefficients of the first part of equation are measured with experimental method and approximate as a straight line ($q_{is}=0.0193*h+4.3771$, where q_{is} is leak flow from GG). The second part of equation can be expressed as:

$$h = \frac{1}{A_m} \int q_{mz}(t)dt \qquad (2)$$

where q_{mz} is input flow in GG. The GG without leak flow is determinate as an I0 dynamic element with gain $K_i=1/Am= 1/9*\pi = 0.0354\ cm^{-1}$, as it shown at figure 3. For more authenticity of the model, initial condition for gauge glass is set up (i.e. the height is measured from 100 mm up to 300 mm).

Mathematical description of two-way valve in *Simulink* is realized with the look-up table where different values of pump-power and piston shift are entered into the table.

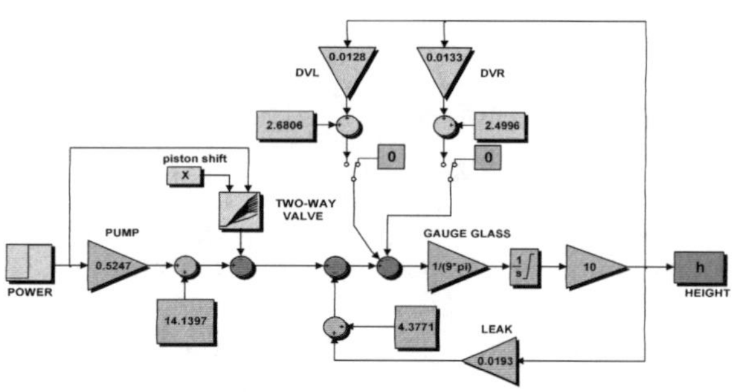

Fig. 3 Mathematical model described in *Simulink*

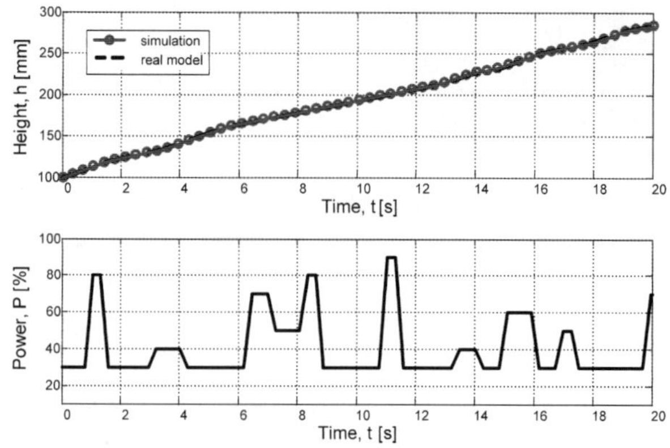

Fig. 4 Simulation results with variable power input

Thereby, the interpolation of output flow is accomplished. Solenoid 2/2 valves which simulate disturbances are P0 dynamic elements which are also approximated with straight line equation.

4. SIMULATION RESULTS

After the comparison of obtained results from the conducted simulations with the results obtained by measure at the LR setup, it turns out that mathematical model of LR system has shown a very good behavior and a minor deviation from the real model. Variation of system input (signal of pump-power) was the key-element for the conducted experiments. The results of one experiment are presented by the figure 4. In this case, power signal was variable in amplitude and the cycle aspect. Additional disturbance was the active left solenoid valve. From figure 4 high accuracy of mathematical model can be determinate regardless the irregularity of some objects in LR system (the pump show the most influence anomaly).

Besides, the other additional tests were made and all of them gave the similar results. Maximum height deviation between simulated and real LR model was *1.5 mm*. Obtained results tell us that the regulation of liquid level will be possible.

5. CONCLUSION

An experimentally supported work on identification and modeling of an educational level regulation system is given in the paper. A mathematical model of the system is derived. The model comprises dynamics of all crucial elements such as water pump, two-way valve, solenoid valves and gauge glass. The model has been experimentally identified and validated. A good correlation has been found. In the next phase of this project graphical user interface of the LR system is going to be developed in which basic PID and intelligent control algorithms will be integrated.

6. REFERENCE

Isermann, R. (1996) Modeling and Design Methodology for Mechatronic Systems, *IEEE/ASME Transaction on Mechatronics*, Vol. 1, No. 1, March 1996, pp. 16-28, ISSN 1083 – 4435

Kuo, C. B. & Golnaraghi, F. (2003) *Automatic Control System*, John Wiley & Sons, Inc., ISBN 0-471-13476-7, New York, USA

Lenart, Lj. (1995) *System identification Toolbox – For Use with MatLab*, The MathWorks, Inc., Natick, USA

Matlab help (1999) *Simulink – Dynamic System Simulation for MatLab*, The MathWorks, Inc., Natick, USA

Petrlić, D. (2005) Bsc. Thesis: *Level regulation of liquid in open container with disturbances*, Zagreb, Croatia

Annals of DAAAM for 2008 & Proceedings of the 19th International DAAAM Symposium, ISSN 1726-9679
Editor B. Katalinic, Published by DAAAM International, Vienna, Austria 2008
Make Harmony Between Technology and Nature, and Your Mind will Fly Free as a Bird

WELDING OF DUPLEX STAINLESS STEEL SAF 2507 BY LASER BEAM

PIKNA, L[ukas]; ULRICH, K[oloman] & KOLENIC, F[rantisek]

Abstract: *Modern duplex stainless steels have an optimized content of particular elements with assumption for desired ratio of austenite in HAZ and thereby also required mechanical and corrosion properties of weld joint.*
The research is focused on weld joint produced by laser beam with regard to anticorrosion properties, that are critical for exploitation of welded construction, mainly when the laser beam welding are not suitable methods of welding for these steels.
Key words: *Duplex, steel, laser, welding*

1. INTRODUCTION

Modern duplex (dual-phase) stainless steels have an optimized content of particular elements (50 % ferrite α, 50 % austenite γ) with assumption for desired ratio of austenite in HAZ and thereby also required mechanical and corrosion properties of weld joint. Finding the role of nitrogen, which is strongly austenite forming element and optimizes ratio of α/γ, is weldability of duplex steels comparable with weldability of austenitic steels.
Laser beam is a technology with a high concentration of energy. Laser welding may be used for almost all weldable materials The welding process is protected by inert gas.

We are focused on research of weld joint produced by laser beam with regard to anticorrosion properties, that are critical for pre exploitation of welded construction, mainly when the laser beam welding are not suitable methods of welding for these steels (1,2).

2. USED MATERIAL

Welding tests were performed on tube with diameter 42 mm, wall thickness 3,5 mm from duplex steel Sandvik SAF 2507.Chemical composition is in table 1. Steel have following properties : high resistance to uniform, pitting, intergranular corrosion, resistance to corrosion cracking, eroson and cavitation, good resistance to high temperatures and physical properties (1,2).

C max.	Si max.	Mn max.	P max.	S max.	Cr	Ni	Mo	N
0,030	0,8	1,2	0,035	0,015	25	7	4	0,3

Tab. 1. Chemical composition of duplex steel Sandvik SAF 2507

Application: material is suitable for in tube production, refinery, in solution contamined with chlorine. This steel is also partially sutiable for heat exchanger, where chloride or brackich water are used as coolant. Vessels for manipulation with organic acids and sulphuric acid and acid mixture. High strength makes this material an attractive alterante of austenitic steels (3,4,5).
Duplex Sandvik SAF 2507 have austenite - ferrite microstructure, which does not contain carbides in boundaries and intermetallic phases. Ferrite content is 40-60% .(fig. 1)

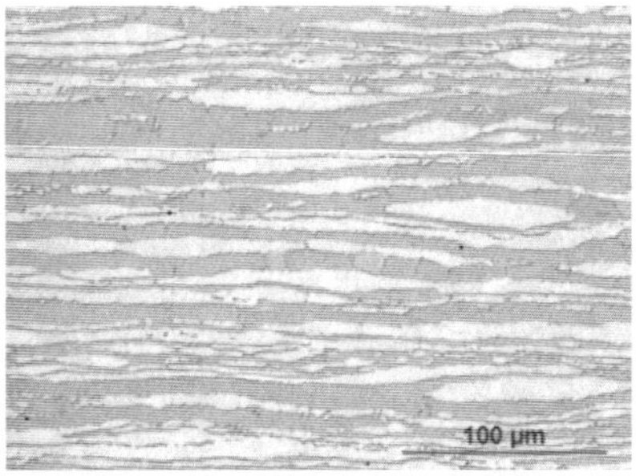

Fig. 1. Microstructure of duplex steel SAF 2205.

3.WELDING

Welding was performed on:
- Gas CO_2 laser machine Ferranti Photonics AF 8 with max. output 8 kW. With wave length 10,6µm.

On the base of effect of weld parameters to weld forming, its showed that the best weld joint are produced by parameters listed in the next table.

Specimen	29
Power	4,5 kW
Welding speed	15 mm.s^{-1}
Shielded gas	He
Welding Focus	on surface
Simultaneous Thermal Treatment (STT) Focus	+ 20 mm (above the surface)

Tab. 2. Parameters of welding

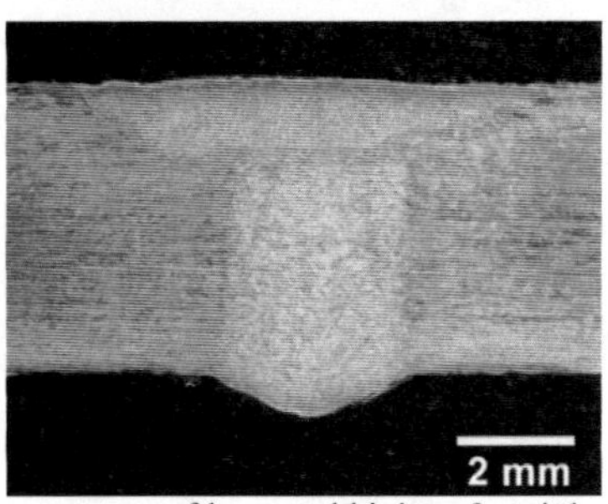

Fig. 2. Macrostructure of laser weld joint of stainless steel SAF 2507 with optimized formation by STT.

4. EVALUATION

The corrosion test ASTM A 923-01 and metallographic research were performed on specimens (laser-beam welding 15mm/s).

Metallographic analysis

Determination of ferrite content was performed in specimens in high, head part of joint. Both structural phases were specified by ASTM E 562, with these results:
Minimal ferrite content 39 %.

Corrosion test

For examination the corrosion resistance of joint of duplex steels we used ASTM G48 – in 6% $FeCl_3$, which serves to estimating the parameters to sensitivity to piting and crevice corrosion. In ASTM is specified the test temperature for duplex steel in testing solution to 40^0C for products with joint (for products without joint is specified to 50^0C).
Test result was evaluated by visual examinating under microscop with 20x zoom.
Surface of these specimens are showed on next figures:

Fig. 3 Pitting corrosion of specimen 29, at 45°C test temperature

Pitting attack only in place, when the joint was finished, on internal and external surface of tube. Base material is not attacked. Validation, that the pitting attack is not only the function of surface status of joint, is the weld etching on the side of treated specimen. This detection may be the reason for the discussion about the efficiency of chemical surfacing of joint surface.

5. RESULTS AND CONCLUSION

Specimen complied the corrosion test G48 after welding and STS at basic temperature 40°C, but at the temperature 45°C is unsufficient. Continuation in development of this type of welding may be suitable. We must focus on elimination of surface uncompactibilities of joint, finishing of joint without piping and elimination of spatter on internal side of tube.

Heat treatment, simultaneous thermal treatment on laser machine increase the austenite content. It´s application is practical and decreases costs on production of final product when the classical thermal treatment i sused, solution annealing.
It is possible predict, that corrosion resistance on external surface may be increased,

- post heat treatment 1050^0 C / 15 min. rapid cooling (best in water), which increase austenite content – unpractical for weldment – applicable in welding of tubes,
- suitable chemical treatment of external surface (elimination degenerated layer),
- preheating or heating of joint by defocusing beam,
- to achieving balanced content of austenite and ferite weld metal is required heat input 0,2 - 1,5 kJ/mm for SAF 2507
- improvement of austenitization of weld metal can be achieved using additional material with increased nickel content and using shielded gas with 2 - 4% Nitrogen content (both elements are autenitoforming).

Acknowledgements
The contribution was prepared under the support of APVV-0057-07, VMSP-P-0008-07 projects.

6.REFERENCES

Křeček, Ivan. *Duplex stainless steels, metalurgy and their welding.* In *ZVÁRAČ 1/2006.*

Pikna, Lukáš. *Welding of stainless duplex steel by laser beam.* In ZVÁRAČ 1/2008.

Pikna, Lukáš - Kolenič, František - Iždinská, Zita Fodrek, P. - Ulrich, Koloman - Emmer, Štefan - Kováč, Ľ.: *Influence of EB Welding Parameters on Properties of Welded Joints in Duplex steel type SAF 2205.*APVT-20-P01305. In: 60th Annual Assembly and International Conference : Proceedings of the IIW International Conference Welding and Materials. Technical, Economic and Ecological Aspects. 01-08 July 2007 Dubrovnik&Cavtat. Croatia. - Zagreb : Croatian welding society, 2007.

Pikna, Lukáš - Iždinská, Zita - Kolenič, František - Iždinský, Karol: *Welding parameters influence to the weld properties of duplex stainless steel created by electron beam.* In: Technológia 2007 : Zborník abstraktov. 10. medzinárodná konferencia. Bratislava, 19-20.9.2007. - Bratislava : STU v Bratislave, 2007

Pikna, Lukáš - Iždinská, Zita - Kolenič, František - Iždinský, Karol: *Welding of stainless duplex steel SAF 2205 by electron beam.* In: Technológia 2007 : Zborník abstraktov. 10. medzinárodná konferencia. Bratislava, 19-20.9.2007. - Bratislava : STU v Bratislave, 2007

Annals of DAAAM for 2008 & Proceedings of the 19th International DAAAM Symposium, ISSN 1726-9679
Editor B. Katalinic, Published by DAAAM International, Vienna, Austria 2008
Make Harmony Between Technology and Nature, and Your Mind will Fly Free as a Bird

LIFE TIME IMPROVEMENT OF A BUTTERFLY VALVE LAPPET BY SHAPE OPTIMIZATION

PINTILIE, G[heorghe] & ALBUT, A[urelian]

Abstract: *Butterfly lappets are parts for adjusting and control the fluid flow in hydraulic and pneumatic transportation network. Those parts are meant to assure a good and secure function of the valve for long time. Life time of a butterfly lappet is influenced by the external loads generated by the fluid pressure. This paper presents one method to decrease the loading level of fluid pressure and of the bending momentumum, the main solicitation, by optimization of the lappet shape. Mainly, the paper presents an analysis of the fatigue crack propagation in a butterfly lappet function to its initial and optimized shape.*

Key words: *butterfly lappet shape, fatigue crack*

1. INTRODUCTION

The butterfly lappet valves are used as parts for adjusting components of fluid flow in hydraulic and pneumatic transportation (fig. 1).

By rotating the lappet with an angle φ the valve decrease the fluid flow with a certain amount. It is well known that the lappet opening angle, φ, can have values till maxim 70°, going over this limit, cause a dramatic grow, followed by a reduction of the momentum generated by distribution of the fluid pressure on the valve lappet (Danbon & Solliec, 2000).

Therewith, is known that in 70° - 90° interval, in the butterfly lappet occur mechanical vibrations which disturb the adjusting process of fluid flow.

Due to the arguments presented above, results that function of butterfly lappet valves with angles over 70° is not recommended. Over this limit, the lappet plan is tilted with an angle between 0 and 20° with respect to the pipe axis, the lappet generate an important pressure drop (Eom, 1988).

Moreover, the forces that are taking place on the lappet can have higher values and can generate important stresses in the lappet, especially a substantial bending momentum.

The value of the bending momentum it is given by the pressure value and the way in which the pressure is distributed on the lappet surface (Leutwyler & Dalton, 2006).

Correlative with the fact that the bending momentum has a periodical variation, generated by its working condition, the butterfly lappet is loaded with fatigue bending that can lead to generation and propagation of fatigue crack.

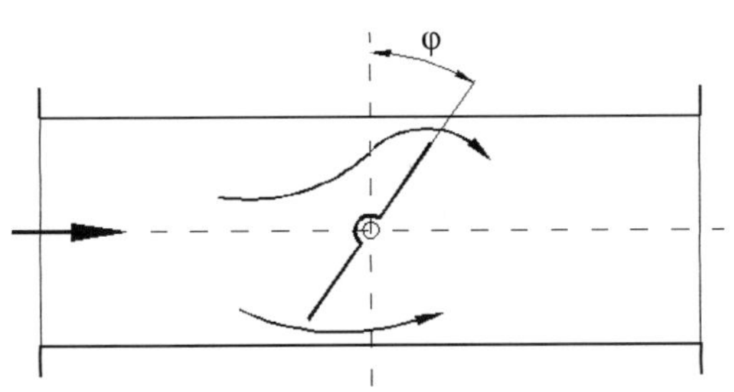

Fig. 1. Fluid flow around the butterfly lappet

One method to increase the life time of a butterfly lappet is to decrease the value of the bending momentum, implicit reduction of the maxim stress located in the transversal section, called critical solicitation area (Morris & Dutton, 1989).

Excepting the modification of fluid pressure and speed (both of them have great influence on the bending momentum) one method that lead to optimization of pressure distribution on the butterfly lappet, and simultaneously reduction of the bending momentum, is to optimize the lappet shape.

2. OPTIMIZATION OF THE LAPPET SHAPE

Usually, the butterfly lappets are made having a disk shape symmetric with respect to the longitudinal plan, with constant thickness and rounded edges (fig. 2.a). This lappet design do not generate a optimal fluid flow causing such a pressure distribution that the bending momentum is very high.

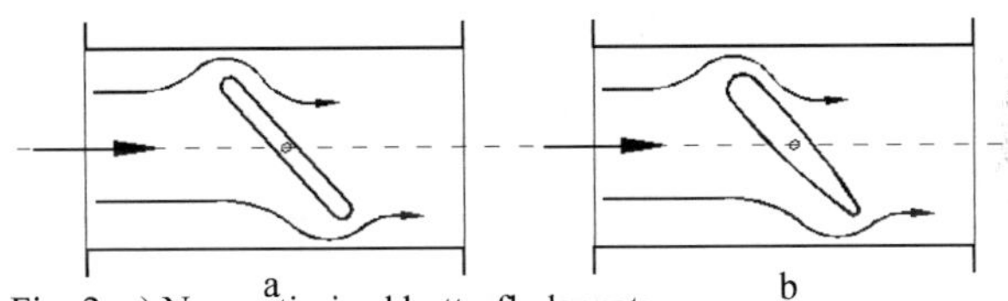

Fig. 2. a) Non optimized butterfly lappet;
b) Optimized butterfly lappet.

If the lappet shape is optimized, in such way that one section with a longitudinal plan to generate an aerodynamic profile, the fluid flow around the lappet and its pressure distribution will be optimal, resulting in a important reduction of the bending momentum value (Solliec & Danbon, 1999).

Due to complexity of the pipe flowing phenomenon, around the lappet, the optimal shape of the lappet section (profile) it was not determined by a theoretical approach, it was generated experimentally. The experimental tests have been made on butterfly lappets with diameters of , for which it was establish an optimal profile. Using this profile, the maximum bending momentum, generated in the critical section in the area of rotational axis, undergo a reduction of approximately 11.03%.

Total failure of the butterfly lappet by fatigue, is due to initiation and propagation of the fatigue crack. Generally, the crack initiate in the area with maximum load, an is generated by the bending momentum with respect to the lappet axis. Normally the crack initiation can happen in any point from that area.To determine the propagation speed of the fatigue crack, equation (1) is very often used:

$$\frac{da}{dN} = c \cdot \left[\frac{1-f}{1+R} \cdot \Delta K\right]^n \frac{\left(1-\frac{\Delta K_{th}}{\Delta K}\right)^p}{\left(1-\frac{K_{max}}{K_c}\right)^q} \qquad (1)$$

in which:

a – crack length;

c, n, p, q – the material coefficients, determined by experimental tests;

f – the crack opening function;
R – the asymmetry coefficient for the cyclic load;
$\Delta K = K_{max} - K_{min}$ – the stress intensity factor;
ΔK_{th} – the threshold value of the stress intensity factor;
K_c – the critical value of the stress intensity factor;
N – number of the loading cycles.

The bending momentum developed on the lappet varies between zero and maximal value. The maximal is determined experimentally, herby the asymmetry coefficient $R = 0$.

For this study case, equation (1) becomes:

$$\frac{da}{dN} = c \cdot \Delta K^n \frac{\left(1 - \frac{\Delta K_{th}}{\Delta K}\right)^p}{\left(1 - \frac{K_{max}}{K_c}\right)^q} \quad (2)$$

Using this equation and supposing that the crack grows with constant speed, it is possible to determine the cycles number between two different crack lengths, a_f and a_i:

$$N_{if} = N_f - N_i = \int_{N_i}^{N_f} dN = \int_{a_i}^{a_f} \frac{da}{f(\Delta K; R)} \quad (3)$$

The integration of equation (4) is the dependent equation of crack length function to the cycles number.

3. NUMERICAL RESULTS

To establish the lifetime of a butterfly lappet, have been determined the variation of crack length with respect to the loading cycles, $a = f(N)$, and speed variation of the crack growths function to the loading cycles, $\frac{da}{dN} = f(N)$.

Both butterfly lappets, disk shape (fig. 2.a) and optimized shape (fig. 2.b), were made from a material having the following mechanical properties: strength stress $\sigma_r = 310.3\,N/mm^2$; yield stress $\sigma_c = 172.4\,N/mm^2$; breaking tenacity $K_c = 2432\,MPa \cdot \sqrt{mm}$; the material coefficients for equation (1) $c = 0.5762$; $n = 3.6$; $p = 0.5$; $q = 0.5$;

For a certain fluid flow, for which experimental tests have been done, positioning the lappet in such way to obtain the maximum bending momentum, the maximal recorded stresses are: $\sigma_{max} = 159.7\,N/mm^2$ for disk shape and $\sigma_{max} = 142.1\,N/mm^2$ for optimized shape.

Considering that the crack initiate on the lappet surface, in the middle of the axis the following numerical results are obtained: fig. 3 a, b for disk shape and fig. 4 a, b for lappet optimized shape.

The numerical results have been obtained using the software NASGRO 5.01.

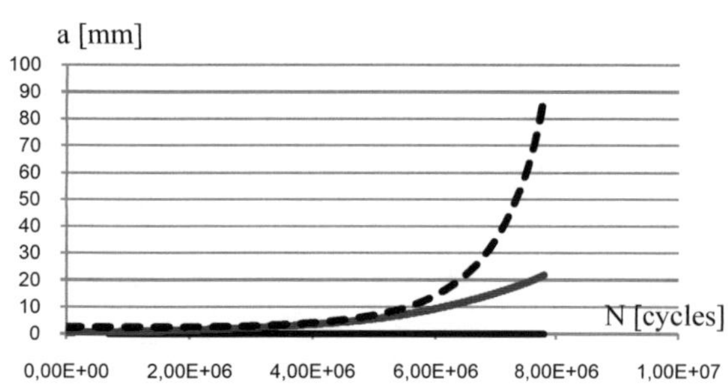

a. The $a-N$ function.

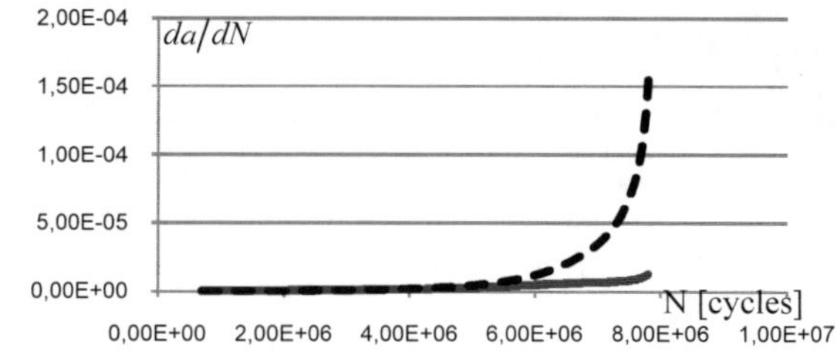

b. The da/dN function.

Fig. 3. Chart results for disk shape lappet $\sigma_{max} = 159.7\,N/mm^2$

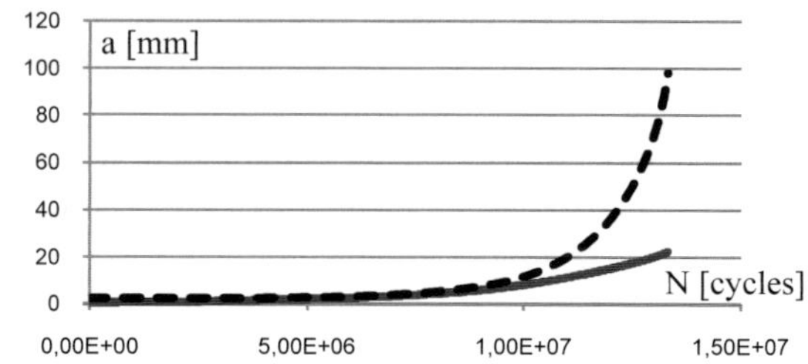

a. The $a-N$ function.

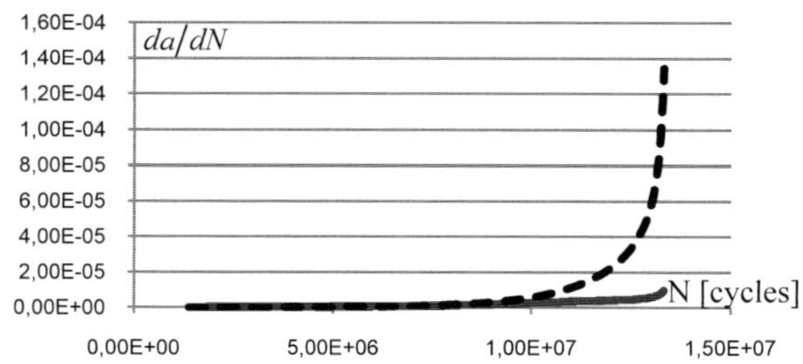

b. The da/dN function.

Fig. 4. Chart results for the optimized lappet $\sigma_{max} = 142.1\,N/mm^2$

4. CONCLUSION

The main conclusion which arises from the presented research work is that using a optimized shape of the butterfly lappet the maximal stress is considerably reduced and the durability of the lappet is increasing. Hence, for the initial lappet shape, the critical crack length $a_{cr} = 21.89\,mm$ and $c_{cr} = 89.74\,mm$ is reached after $N_c = 7.8 \cdot 10^6$ cycles, for the optimized shape lappet the critical crack length $a_{cr} = 22.562\,mm$ and $c_{cr} = 98.59\,mm$ is reached after $N_c = 13.32 \cdot 10^6$ cycles.

5. REFERENCES

Danbon F. & Solliec C., (2000), *Aerodynamic Torque of a Butterfly Valve-Influence of an Elbow on the Time-Mean and Instantaneous Aerodynamic Torque*, ASME J. Fluids Eng., 122, pp.337-344

Eom K., (1988), *Performance of Butterfly Valves as a Flow Controller*, ASME J. Fluids Eng., 110, pp.16-19.

Leutwyler Z. & Dalton C., (2006), *A Computational Study of Torque and Forces Due to Compressible Flow on a Butterfly Valve Disk in Mid-stroke Position*, ASME J. Fluids Eng., 128, pp.1074-1082

Morris M. J. & Dutton J. C., (1989), *Aerodynamic Torque Characteristics of Butterfly Valves in Compressible Flow*, ASME J. Fluids Eng., 111, pp.392-399

Solliec C. & Danbon F., (1999), *Aerodynamic Torque Acting on a Butterfly Valve. Comparison and Choice of a Torque Coefficient*, ASME J. Fluids Eng., 121, pp.914-917

Annals of DAAAM for 2008 & Proceedings of the 19th International DAAAM Symposium, ISSN 1726-9679
Editor B. Katalinic, Published by DAAAM International, Vienna, Austria 2008
Make Harmony Between Technology and Nature, and Your Mind will Fly Free as a Bird

THE MATERIAL TREATMENT INFLUENCE ON THE PROPAGATION SPEED OF A FATIGUE CRACK

PINTILIE, G[heorghe] & ALBUT, A[urelian]

Abstract: *The paper presents an experimental study made to demonstrate that the propagation speed of a fatigue crack is strongly affected by treatment of the material with a certain axial strain and with ultrasonic waves. In the paper are presented the experimental results obtained on a standard probe, for an steel having the ultimate tensile strength* $\sigma_r = 310.3\,N/mm^2$*, the yield strength* $\sigma_c = 172.4\,N/mm^2$ *and the breaking tenacity* $K_c = 2432\,MPa \cdot \sqrt{mm}$ *.*

Key words: *fatigue crack, propagation speed, ultrasonic*

1. INTRODUCTION

The mechanical properties of a material used to make certain parts or mechanical structures have a decisive effect on their strength. Especially, in case of variable solicitations, the structural integrity of a part or a mechanical structure is influenced by the propagation speed of the fatigue crack. The crack propagation speed depends on the mechanical properties of the material and the breaking occurs when a critical value is reached (Bao & McEvily, 1998; Baganoff & Kozin, 1984).

Taking in consideration that propagation of the fatigue crack happen without macroscopic plastic deformations or dimensional modification of the part or of the mechanical structure, determination of the required time period to reach the crack critical value becomes very important (Irwing, 1958).

2. METHOD TO MODIFY THE MECHANICAL PROPERTIES OF MATERIAL

The propagation speed of the fatigue crack for a certain material is affected by different factors as following: solicitation level, solicitation speed, asymmetry coefficient, stress concentrators, existence of residual stresses and so on.

The most influent factor for the crack propagation speed is the mechanical properties of the material, hereby the favourable modification of the material properties can lead to reduction of the crack propagation speed (Schijve & Broek, 1970).

One technical method that can lead to modification of the material mechanical properties, implicit to the modification of the fatigue crack propagation speed, consist in axial solicitation of the material until a certain level followed with treatment with ultrasonic waves.

As is showed in figure 1, the sample is loaded with axial force until $\sigma_1 < \sigma_e$, in which σ_e is the material elasticity limit.

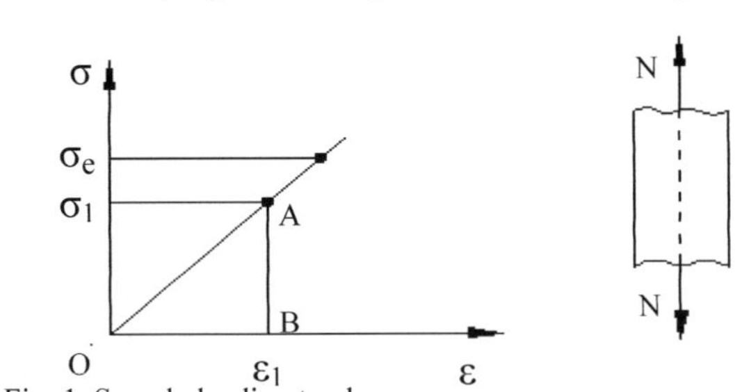

Fig. 1. Sample loading track

Maintaining constant the stress state corresponding to $\sigma 1$ value, the sample is treated with ultrasonic waves using an ultrasonic concentrator. Due to this ultrasonic treatment the material suffer structural modification which comes together with reduction of the stresses caused by axial load. After the ultrasonic treatment the sample has a residual deformation $\varepsilon 1$. It must be mentioned that, without the ultrasonic treatment, the come-back way of the sample is presented in figure 1 as AO segment.

Using the ultrasonic treatment while maintain constant the axial load, the come-back way of the sample is done following the AB segment. Is necessary to mention that the AB segment is approximately parallel with the σ axis.

The "residual" deformation ε_1 can be considered as a plastic deformation even that the loading stress did not exceed the material elasticity limit σ_e.

3. PROPAGATION SPEED OF FATIGUE CRACK

It is well known that the propagation speed of fatigue crack is function to variation of the stress intensity factor ΔK, as is shown in figure 2 (McEvily, 1977; Chen et al., 1991).

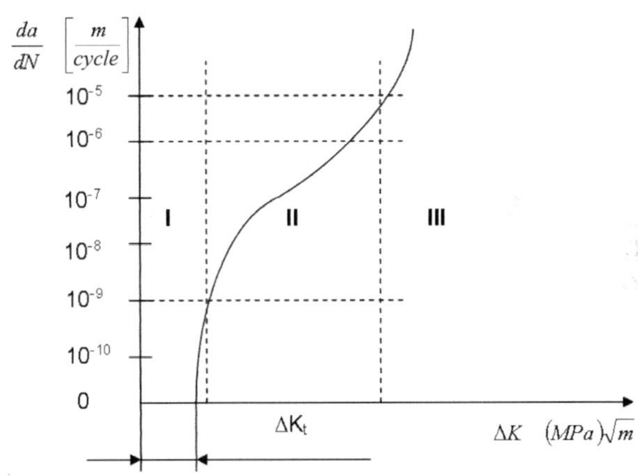

Fig. 2. Crack propagation domains

To define the curve $\frac{da}{dN} = f(\Delta K)$ for all three behaviours, it can be used the equation obtained my Forman, Newman, Shivakuman, Koning and Henriksen, this equation is the starting point of the working principle of NASGRO 5.0 software, for variable loads with constant amplitude:

$$\frac{da}{dN} = c \cdot \left[\frac{1-f}{1+R} \cdot \Delta K\right]^n \frac{\left(1 - \frac{\Delta K_{th}}{\Delta K}\right)^p}{\left(1 - \frac{K_{max}}{K_c}\right)^q} \qquad (1)$$

in which:

a – crack length;

c, n, p, q – the material coefficients, determined by experimental tests;

f – the crack opening function;

R – the asymmetry coefficient for the cyclic load;

$\Delta K = K_{max} - K_{min}$ – the stress intensity factor;

ΔK_{th} – the threshold value of the stress intensity factor;

K_c – the critical value of the stress intensity factor;

N – number of the loading cycles (Dugdale, 1960).

The crack opening function is calculated with Newman relation (Paris & Erdogan, 1963):

$$f = \frac{K_{op}}{K_{max}} = \max\left(R, A_0 + A_1 R + A_2 R^2 + A_3 R^3\right) \quad (2)$$

in the case in which $R \geq 0$ or:

$$f = \frac{K_{op}}{K_{max}} = A_0 + A_1 R \quad (3)$$

in the case in which $-2 \leq R < 0$.

In equation (2) and (3) K_{op} is the stress intensity factor for which the crack starts to open.

4. EXPERIMENTAL TESTS

The testing algorithm is composed by four individual stages:

a. Straining the sample till the σ_1 value is reached.

The sample has a stripe shape with thickness of 2 mm, and is made from steel having the following characteristics:

- strength stress: $\sigma_r = 310.3\,N/mm^2$;
- yield stress: $\sigma_c = 172.4\,N/mm^2$;
- breaking tenacity: $K_c = 2432\,MPa \cdot \sqrt{mm}$;
- the material coefficients for equation (1): $c = 0.5762$; $n = 3.6$; $p = 0.5$; $q = 0.5$;

The tensile test is done using a hydraulic driven machine.

b. Treatment of the sample with ultrasonic waves.

Maintaining the stripe sample under σ_1 stress, this is treated with ultrasonic waves, using a transductor able to generate ultrasonic waves with power of 150 W. The treatment is applied till the σ_1 stress is decreased almost to zero.

c. Fabrication of the sample for determination of the fatigue crack propagation speed.

The stripe used before is the base material for machining a sample having the shape and dimensions presented in figure 3.

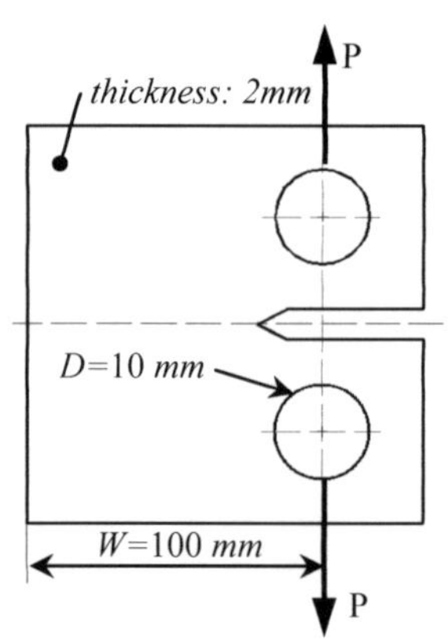

Fig. 3 Standard sample used for experimental tests

d. Determination of the fatigue crack propagation speed.

The propagation speed of the fatigue crack is calculated on two samples having the shape and dimensions presented in figure 3. The first sample is made form material without ultrasonic wave treatment and the second one is made from material treated with ultrasonic waves.

In fig. 4. is presented the variation of the crack length function to the cycles number N for both samples, treated and not treated with ultrasonic waves. Variation of the normal stress is periodical, between limits $\sigma_{max} = 160\,N/mm^2$ and $\sigma_{min} = 40\,N/mm^2$ $(R = 0.25)$.

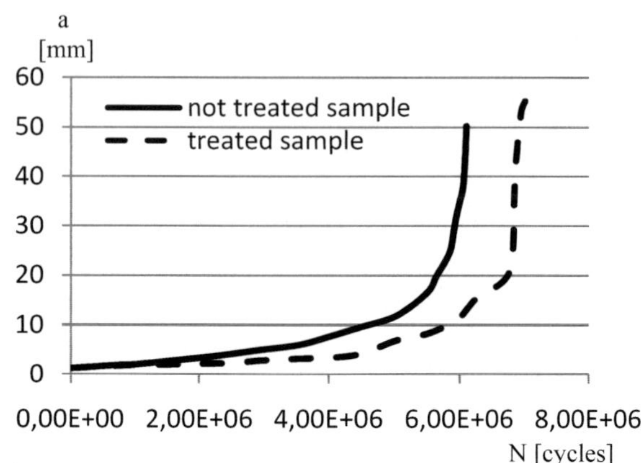

Fig. 4. The $a - N$ curve obtained experimentally

5. CONCLUSION

The structural modification which have been done to the material due to the treatment described in this paper, lead to modification of the mechanical characteristics of the material. These new mechanical characteristics influence the fatigue crack propagation speed.

Comparing the experimental results presented in fig. 4 it can be concluded that in the case of treated sample the crack length grows slower, hence the critical value $a_{cr} = 55.2\,mm$ is reached at approximately $7.01 \cdot 10^6$ cycles, while in the case of not treated sample the critical value $a_{cr} = 50.2\,mm$ is reached at approximately $6.10 \cdot 10^6$ cycles.

The way in which this treatment leads to the obtained results represents a new study direction.

6. REFERENCES

Baganoff, J.L. & Kozin, F., (1984), *Probabilistic Models of Fatigue Crack Growth - II*, Engineering Fracture Mechanics, Vol. 20, No. 2, pp. 255-270.

Bao, H. & McEvily, A.J., (1998), *On Plane Stress-Plane Strain Interactions in Fatigue Crack Growth*, International Journal of Fatigue, Vol. 20, No. 6, pp. 441-448.

Chen, Dou-Lun, Weiss, B. & Stricker, R., (1991), *A new Evaluation Procedure for Crack Closure*, International Journal of Fatigue, Vol. 13, No. 4, pp. 327-331.

Dugdale, D.S., (1960), *Welding of Steel Sheets Containing Slits*, Journal of Mechanics and Physics of Solids, Vol. 8.

Irwing, R.G., (1958), *Fracture*. Handbuch der Physik, Springer Verlag, Berlin.

McEvily, A.J., (1977), *Fatigue Fracture-Surface Roughness and the K-Opening Level*, International Journal of Fatigue, Vol. 19.

Paris, P. & Erdogan, F., (1963) *A critical Analysis of Crack Propagation Laws*, Transactions of ASME, Series D 85.

Schijve, J. & Broek, D., (1970), *Four lectures on Fatigue Crack Growth*, Engineering Fracture Mechanics, Vol. 2.

Annals of DAAAM for 2008 & Proceedings of the 19th International DAAAM Symposium, ISSN 1726-9679
Editor B. Katalinic, Published by DAAAM International, Vienna, Austria 2008
Make Harmony Between Technology and Nature, and Your Mind will Fly Free as a Bird

ERP IMPLEMENTATION IN A SMALL OR MEDIUM ENTERPRISE

PIRVU, M[arcel] & DUMITRESCU, C[onstantin]

Abstract: *The paper proposes to present, shortly, the concept of small and medium enterprise, to particularize the mode of ERP implementing inside of such kind of companies. It will be presented the benefices that the implementation bring along and in the end the conclusions which results after an ERP implementation in a small enterprise. The paper ends up with determination of the expectations we should have when it is implemented an ERP in a small / medium enterprise.*
Key words: *ERP, SAP, SME, Implementation, ASAP*

1. INTRODUCTION

Small and medium enterprises (SMEs) must make efforts for permanently improving and innovation for surviving in the present competitive environment. The managers' decisions take into consideration to keep the quantity, quality and delivery terms of the products/services, and also the real aspects regarding the development of internal activities according with the strategy initially decided. The assignment of this last desiderate enforces certain rules that have to be respected, enforces the use of an appropriate technology, in order to improve the efficiency, efficacy and adaptability to the requests of external environment. More, the programmes used inside the organisations must allow to combine the improving methodologies so that an individual optimum to result, and this optimum to be correlated with the general optimum imposed by the whole action system.

2. SMEs DEFINITION / FEATURES

An historical approach emphasize that the small enterprise, starting from the family association, is in fact the first and the oldest organisation structure, and represents the pregnant shape of the economical activity in the period before the industrial revolution. The small enterprise, the technical progress and the demand growth allowed the big structures development. Almost all of the production and administration models developed are for the big enterprises. During the '70s, the big enterprise, unique model for a long time, was reconsidered. (Burtica et al, 2006).

Some aspects, which redefine the small enterprise position in the framework of free market economy, are relevant:

1. Flexibility - the opening of a big enterprise directly depends on the number/availability of the small enterprises it is in contact with. 2. Small enterprise, being more flexible, more "human", bear up better a crisis, and it may be the solution until the crisis pass. 3. It can be relieved that the small enterprise has the ability to assimilate labour, to offer a diversified range of products/services, when the dismissals made by the big enterprises are great.

These abilities created the premises for small enterprise development and its transformation in a medium enterprise; so that in the last three decades in the free market economy sets in the collocation "small and medium enterprises" - SME; the medium enterprise take over, from the structural point of view, the features of the small enterprise, but its products/services offer, and especially its adaptability to the requirements of economical environment and big enterprises is much gradated, both quantitatively and mainly qualitatively. The dimensions, structure and economical indicators which characterise the SMEs activity are adapted at each state's legislation; in the specialty literature are emphasized the most relevant features:

- They have a reduced dimension, both as the aspect of human resources used and as the aspect of capital assets engaged;
- A centralised administration, coordinated by the leader, who is, same time, owner and entrepreneur; the costs shaping has to be a continue concernment for the entrepreneur;
- A reduced specialisation of managing/production jobs at enterprise's level and also a reduced specialisation of equipments;
- An intuitive strategy, slightly formalised, taking into consideration that the existence of the small enterprise is directly conditioned by the way the own strategy is correlated with strategies of the partners the enterprise is in contact with; the strategic vision regards different competitive advantages applying strategies specific for low costs, concentration strategies or differentiation strategies.
- An internal/external informational system less complex, adapted to the organisational structure which is specific for small enterprise, capitalising the opportunity of direct contact, and being based on the internal planning process.

During a period in which, due to the political, economical and social transformations, we assist to a great concentration in time and space regarding the economical phenomena development, concomitant with a concentration/internalisation of selling markets, is more obvious the fact that for a competitive SME its leader should react instantaneous, flexible and anticipated, and this implies the leader capacity to act, to know how, but especially when it needs to act. Here, the ERP planning systems role is very useful, if they are used in due time.

The clear finality which results is the next: SME is an ideal enterprise prototype, in an information based economy.

3. DESCRIPTION OF AN ERP SYSTEM

The integrated informational systems for the business administration – (Enterprise Resource Planning – ERP) appear in the enterprise's applications at the beginning of '90s. The efficient informational systems became the premise of the competitive growth, mainly through the costs reducing. The rule of availability of needed information, at the appropriate moment, is generally recognised, both for the big corporations and for the small enterprises.

ERP is considered the most precise expression of the interdependence between the economic environment and the informational technology. ERP has brought the integration advantage: a multi modular package which is adapted at enterprise's specificity and then can be extended with other modules. Enterprise Resource Planning represents a system based on client/server architecture, developed for transactions processing and for facilitation of all process integration, from production planning and developing phase, to the relations with suppliers, clients and other business partners. These integrated packages represent the base of e-business applications. (Fotache & Hurbean 2004). SAP offers SAP Business All-in-One

solution for SMEs, which includes modules such as Financial Accounting, Controlling, Materials Management, Sales & Distribution and Production Planning. The choosing of the appropriate ERP system, allows the beneficiary to implement a unique integrated system, replacing or redesigning of the existing functional system. Many organisations use powerful systems for enterprise resources planning in order to hold up a high competitiveness. A more efficient approach is to integrate these systems with production, engineering and with applications that exist in the enterprise to obtain in the end an enterprise resources planning system – ERP.
The main challenge consists of integration of all economic processes and available resources optimisation.

4. ERP IMPLEMENTATION

Accelerated SAP (ASAP) is the implementation solution provided by SAP. ASAP integrates several components that work in conjunction to support the rapid and efficient implementation of the SAP's ERP System.
The 5 steps of ASAP methodology is described below:
„**Phase 1: Project Preparation** - Provides initial planning and preparation for your SAP ERP project. In this step, we establish what are the resources that the client needs for SAP and he will purchase all of these.
Phase 2: Business Blueprint - In this phase, you create the Business Blueprint, which is detailed documentation of the results gathered during requirements workshops. Furthermore, the Business Blueprint serves to document the business process requirements of the company. On this basis, you will achieve a common understanding of how the company intends to run their business within the ERP System.
Phase 3: Realization - In this phase, you implement all the business and process requirements based on the Business Blueprint. You configure the system step by step in two work packages, Baseline and Final configuration.
Phase 4: Final Preparation - In this phase, you complete testing, end user training, system management and cutover activities to finalize your readiness to go live. Furthermore, the Final Preparation phase serves to resolve all critical open issues. Upon successful completion of this phase, you will be ready to run your business in your live ERP System.
Phase 5: Go Live and Support - Transition from a project-oriented, pre-production environment to successful and live production operation.
Continuous Improvement - Provide on-going support and assistance for post go-live. It covers solutions for those activities which are standard in a productive environment: business changes; technology changes; changes in the user community."(https://www.sdn.sap.com/irj/sdn/go/portal/prtroot/docs/library/uuid/501c66ff-c31c-2a10-62ad-9588691d78ae).
We have used ASAP methodology during the implementation of SAP Business All-in-One in a small enterprise called Electroconat. SAP implementation started due to the operations development, as volume and number of work centres, because the previous solution rapidly became outdated and was not trustful. In less then four month it was made the migration to the new system, the most difficult period being the one when both systems were parallel used.

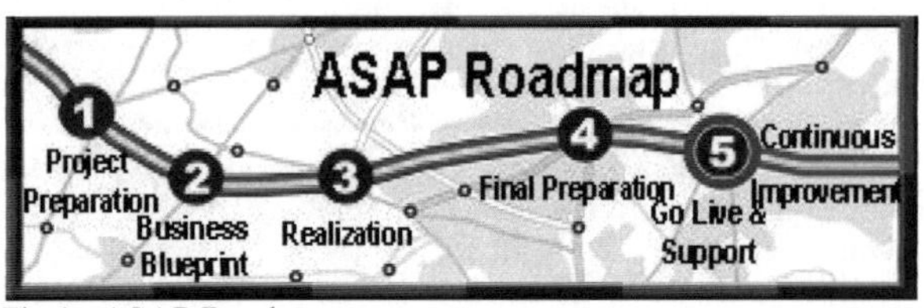

Fig.1. ASAP Roadmap
https://www.sdn.sap.com/irj/sdn/go/portal/prtroot/docs/library/uuid/501c66ff-c31c-2a10-62ad-9588691d78ae

In this way was made the control of procedures accuracy at figures level. The main objective was the integration of more departments at the level of informational system: the materials management, the one who administrate the stocks for the four stores Electroconat, selling department and the financial department.
The implementation objectives were:
• The integration of three departments from the informational flows point of view
• The functioning in online mode through which the data inserted in a point are updated inside the whole system
The implementation benefices were the next:
• The stocks and commands situation is centralised and updated any moment
• There is an information consistency during the processes, from offer to invoice.
• The costs reduction due to the process standardisation based on the best practices in industry.
The giving up of redundant processes and difficult procedures determined substantial costs reducing and these made the shareholders to see the SAP Business All-in-One implementation as strategic investment. Having the experience of the first seven months from the SAP Business All-in-One implementation, the company expects 100.000 de euro costs reducing, annually. "But maybe the most important advantage is that SAP offer a mirror of business, in which can be seen, every moment, how the company is functioning" (Electroconat Success Story). Change is an integral part of the business environment today. SAP supports and embraces the process of continuous improvement as new SAP functionality becomes available, as new business requirements are defined.

5. CONCLUSIONS

Do not forget that every improvement means change, but not every change means improvement. All changes interaction must be evaluated before initiating the change. It is very important to purchase a flexible system which can quickly adapt itself at changes which appear in business processes. Another important aspect for the implementation success is to establish, from the beginning, an influent insider coordinator. He must be able to take decisions about the system implementation, to assign tasks to the new users and to answer to the requests of persons who implement the system.
"The ERP system implementation must be finalised with efficiency growth for the enterprise's activities and processes, and also a best management of them. The production enterprises have procesual and managerial flows specific to their activity, which make them different from the commerce companies. Activity evaluation, redesigning of flows and the realistic assessment of the ERP implementation benefits are the needed steps for a successful implementation of an ERP." (http://www.ebsromania.ro/servicii/consultanta2.hm)

6. REFERENCES

Burtica, M., Taroata, A. & Tamasila, M. (2006), *Small and Medium Enterprises Management – Concepts, Evolution, Strategies*, Solness, ISBN 9737290690, Timisoara
Electroconat Succes Story – printed by SAP in 2008
Fotache, D. & Hurbean, L. (2004), *Informatics Solutions for Business Management – ERP*, Economica, ISBN 9735909847, Bucuresti
***http://www.ebsromania.ro/servicii/consultanta2.htm, *Accesed:* 2008-06-15
***https://www.sdn.sap.com/irj/sdn/go/portal/prtroot/docs/library/uuid/501c66ff-c31c-2a10-62ad-9588691d78ae, *Accesed:* 2008-06-17
***http://www.sap.com/solutions/sme/index.epx, *Accesed:* 2008-06-14

Annals of DAAAM for 2008 & Proceedings of the 19th International DAAAM Symposium, ISSN 1726-9679
Editor B. Katalinic, Published by DAAAM International, Vienna, Austria 2008
Make Harmony Between Technology and Nature, and Your Mind will Fly Free as a Bird
*Young Researcher and Scientist Paper / * Supervisor, Mentor*

FMEA ANALYSIS OF XEIKON ELECTROPHOTOGRAPHICS

PLEHATI, S[ilvio]; PAVLOVIC, T[omislav]; BOGOVIC, T[omislav] & PAP, K[laudio]*

Abstract: *By introducing FMEA analysis of Xeikon DCP32D electrophotographic printing model it is shown which components of the machine are weakest link in the printing. Each component is shown separately according to the modules and units of the machine, so that the critical component of each module can be read. Solution is in improving critical component and suggestion is to evaluate and validate this component using FMEA analysis.*
Key words: *Xeikon FMEA, Xeikon electrophotographic model, failures in digital printing, corona wire failure, scorotron failure*

1. INTRODUCTION

Today, construction and design of digital printing machines are known. To observe a specific digital printing technology and set forth corrective actions associated with potential faults in the model, it is necessary to analyze and evaluate each component of a digital printing machine. Evaluation of the failure is given critically: firstly we give the assessment of failure probability during the half-year work of the machine and afterwards we evaluate the consequences of error for that failure. For this kind of analysis, it is necessary to understand the construction and the design of the Xeikon digital printing machine, and to have the practical experience in operating the machine. FMEA (Failure Mode Effect Analysis) was taken as a widespread method of assessing the severity of specific modules and components of a digital printing machine (Stamatis, D. H., 2003).

1.1 Description of a model of digital electrophotographic printing process

Model of a digital printing process is taken from the Xeikon DCP32D (electrophotographic/xerographic model (Kipphan H. (Ed.), 2001)), and it is one of the most frequently used models in digital printing industry. This model is used in most digital printing machines, ranging from the photocopiers to home laser printers that have simplified and cheaper model. Our research is based on the most complex electrophotographic model, Xeikon DCP32D, whose functioning has been tracked and stored in the database during 12 years. All failures described and presented in this paper were analyzed by FMEA analysis.

1.2 Advantages and disadvantages of electrophotographic digital printing

Advantages and disadvantages are well known since the first implementation of electrophotographic models and we have extracted here advantages and disadvantages of basic model and of Xeikon DCP32 model.

The main advantages of electrophotographic printing model (Žiljak et al., 2007) are:

a) Ability to print one or more prints without making an offset plate, and to change them (variable press form)
b) The possibility of printing multipage books with a built-in postpress process.
c) Short start-up period for printing
d) Personalized printing (addresses, labels, personalized newspapers, etc.)
e) Quality reproduction for an acceptable price per print

The main disadvantages of electrophotogr. printing model are:

a) A slow speed of printing compared to offset press
b) Large life cycle cost of the machine, consisting of: a shorter working span, the costly parts and ecological disposal
c) Limited and expensive 'spot' colors
d) The problem with covering of uniform areas with 100% coverage when using the dry toner

The advantages of Xeikon electrophotographic digital printing model are:

a) Print from the roll (variable paper size on one axis depending on the RIP memory)
b) Simultaneous duplex color print
c) Possibility of fast individualized printing (depending on the RIP and PES configuration)

2. THE ROLE OF FMEA ANALYSIS IN IMPROVING THE PROCESS OF DIGITAL PRINT

To provide improvements of the Xeikon DCP32D digital printing machine we have taken data from its work database created during 12 years (Yang, K. & Basem El-H., 2003), and we have analyzed that data on the basis of half-year occurrence of certain problems. We used FMEA process is to: 1) assess the risk associated with potential errors, 2) rank the problems by relevance and 3) find most important problems of the product or the process. There are several types of FMEA analysis; we used Process FMEA analysis to describe, rank and find the failures that cause the greatest problems, and to assess the risks of all errors.

2.1 FMEA analysis of a complex system

To assess the risk associated with errors that happen on Xeikon DCP-32D, and to offer some solutions, we have introduced the division structure of the machine. We have divided the machine according to the basic structure:

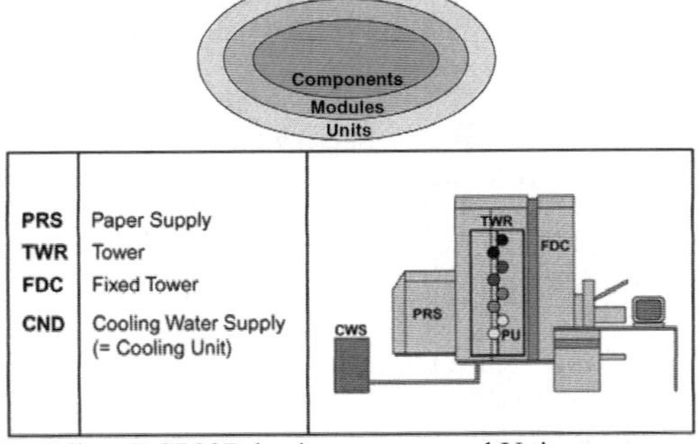

Fig. 1. Xeikon DCP32D basic structure and Unit structure

Fig. 2. Xeikon DCP32D

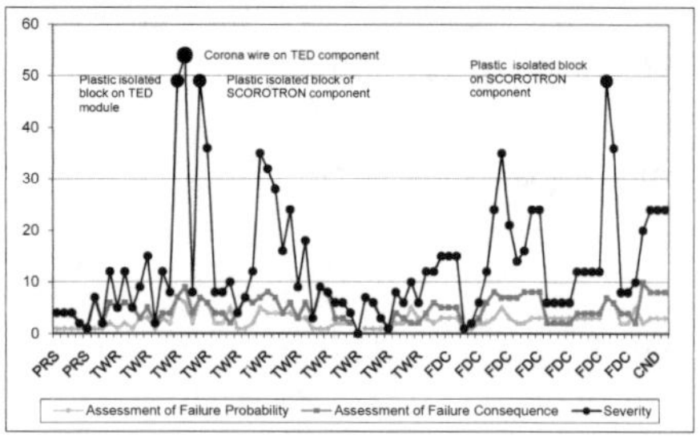

Chart 1. FMEA analysis of XeikonDCP32 electrophotographics

The basic structure is: Machine units, Machine modules and Machine components. Machine units are: PRS (Paper Supply), TWR(Tower), FDC(Fixed Tower) and CND-CWS(Cooling Unit) as shown Fig. 1. For individual machine units we have listed the machine modules and for each module we have extracted its components. When we put all this in Excel table, the FMEA analysis diagram is set. Then each component is critically assessed concerning the probability of failure in the ½ year span and the consequences of the failure for the system. By using a formula (1) we get the severity value. Formula (1) and Chart 1. is described as:

AFP = Assessment of Failure Probability
AFC = Assessment of Failure Consequence
SEV = Severity

Formula for FMEA analysis is:

$$AFP \times AFC = SEV \qquad (1)$$

3. RESULTS OF THE FMEA ANALYSIS

Critical component that came out of FMEA analysis is the TED unit component. It is a transfer corona, which is used to retain the applied toner on the opposite side of the print until the fusing. The problem is, when the corona breaks, it folds, falls down and touches the drum with the printing unit or touches the TED shield, bringing 650 V and thus creating a short circuit. The machine sanctions it and turns off the power supply of the TED driver board, but nevertheless the drum can be damaged. Also, the second failure is related to the TED unit module, which creates a similar problem (short circuit), but this time the plastic that holds the corona at the distance from the unit's shielding is the source of the problem. Since the corona is in a direct contact with the plastic, and the plastic is in contact with the shield, the plastic becomes fatigued, i.e. by the influence of toner dirt and the low quality plastic, the short circuit, alarm, and problems in the print occur. With this analysis we have defined the priorities and in that way we have found the critical points, we have gained an insight into the functioning of the machine, noticed which components we have to pay attention to during the work and which components need a possible change in design.

4. CONCLUSION

In this work we wanted to introduce FMEA analysis of Xeikon electrophotographic printing model for the first time and we wanted to find out which components of the machine are the weakest link in the printing (Fig.1., Fig. 2., Fig. 3. and Fig. 4.).

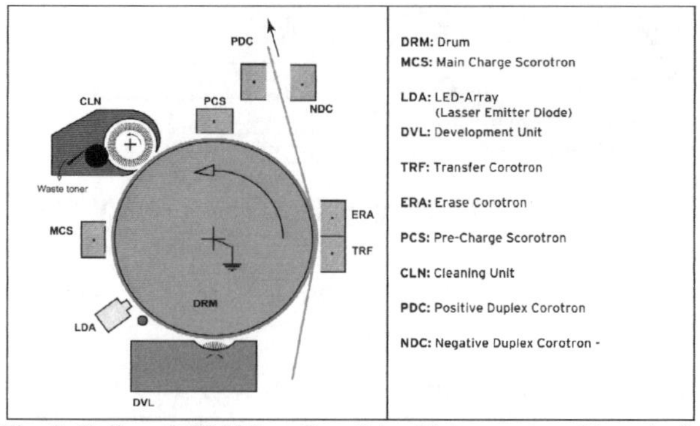

Fig. 3. Xeikon DCP32D – One Print Unit structure (there is 8 Units in Machine used for Duplex full colour printing (Xeikon NV, 1996))

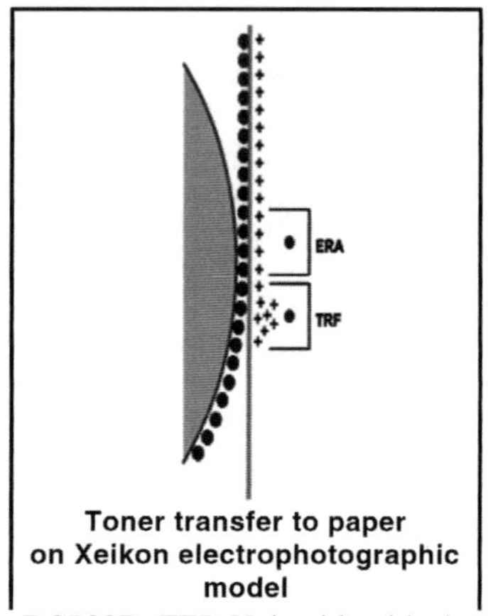

Fig. 4. Xeikon DCP32D TED Unit with critical component corona wire (used for electrostatic toner transfer on paper)

Each component is shown separately according to the modules and units of the machine, so that the critical component of each module/unit can be read (Chart 1). FMEA analysis has shown the critical component of the machine that causes the losses in production, delay and the additional costs. This component should certainly be improved and the further monitoring of the machine via FMEA analysis is suggested so the corrective actions can be validated (Yang G., 2007).

5. REFERENCES

Kipphan H. (Ed.) (2001). *Handbook of print media: technologies and production methods*, Springer, ISBN 3-540-67326-1, Berlin; Heidelberg; New York; Barcelona; Hong Kong; London; Milan; Paris; Singapore; Tokyo

Stamatis, D. H. (2003). *Failure mode and effect analysis: FMEA from theory's to execution* - 2nd ed., ASQ Quality press, ISBN 0-87389-598-3, Milwaukee

Xeikon NV (1996). *Xeikon DCP32D Service Manual*, Xeikon NV, Mortsel - Belgium

Yang G. (2007). *Life cycle reliability engineering*, John Wiley & Sons, ISBN 0-471-71529-8, New Jersey

Yang, K. & Basem El-H. (2003). *Design for Six Sigma: a roadmap for product development*, McGraw-Hill Professional, ISBN 0-07-141208-5, New York

Žiljak, V.; Pap, K. & Marciuš, D. (2007). Experimental Simulation Research of Digital Printing Cost-Efficiency in Comparison to Traditional Printing, PRE-CONFERENCE *Proceedings of the 1st Special Focus Symposium on Market Microstructure: From Orders to Prices - Best Execution in the Age of Algo Trading and Event Stream Processing*, Lasker, G.E. (Ed.), pp. 66-70, Zadar, October 2007, Faculty of Teacher Education of the University of Zagreb, Zagreb

Annals of DAAAM for 2008 & Proceedings of the 19th International DAAAM Symposium, ISSN 1726-9679
Editor B. Katalinic, Published by DAAAM International, Vienna, Austria 2008
Make Harmony Between Technology and Nature, and Your Mind will Fly Free as a Bird

E FLOW SIMULATION THROUGH A HEAT EXCHANGER CHANNEL

PLESA, A[ngela]; BODE, F[lorin] & OPRUTA, D[an]

Abstract: *The authors used FLUENT software in order to simulate the flow through the channels of a heat exchanger. LES method for swirl flow was applied to geometry of plate and fins aluminium heat exchanger channel, presented in figure 1. Those items were tested on an experimental stand developed by the authors. It gives the possibility to measure the parameters involved in the performances of the heat exchangers, such as: pressure, pressure drop, mass flow, temperature and humidity.*
Key words: *Heat exchanges, FLUENT software, flow simulation.*

1. INTRODUCTION

The authors used FLUENT software in order to simulate the flow through the channels of a heat exchanger. LES method for swirl flow was applied to geometry of plate and fins aluminium heat exchanger channel, presented in figure 1. Those items were tested on an experimental stand developed by the authors. It gives the possibility to measure the parameters involved in the performances of the heat exchangers, such as: pressure, pressure drop, mass flow, temperature and humidity. As it is known the flow through the channels of heat exchangers have to be turbulent. It is possible to analyze the entire turbulent field using Direct Numerical Simulation method (DNS). This method is inefficient for engineering problems due to the fact that considering the difference between maximum and minimum values of the Reynolds number, $Re_t^{3/4}$, the cells dimensions have to be proportional with $Re_t^{9/4}$. DSN method simulates the unsteady flow, with small time steps.These two aspects are more than DNS can support (Susan Resiga at al, 2003). There is another method, very easy to apply, based on Reynolds and Navier-Stokes, equations, known as RANS.

2. EXPERIMENTS

In order to simulate the flow through heat exchanger channels the authors used a method, named LES, placed between DNS and RANS.

2.1. Mathematical model

This method permits to solve the large vortices and model the small vortices. The main characteristics of ***LES*** method are:

- the momentum, mass and energy are scalars and they are carryed out by large vortices,
- large vortices are dependent by the geometry and frontier conditions,
- small vortices are not depending by the geometry and tend to be izotropic,
- there are better cahnces to find an universal model.

Solving only large vortices the regarding for meshing is less restrictive than for DNS. The time step is proportional with the turn over time of the vortices, which is also an unrestrictive condition.

LES method provides a good meshing in correlation with computer performances.

The equations characteristic for LES method are obtained either by filtering Navier-Stokes equations, by erasing the low energy vortices (Eq. 1)

$$\overline{\Phi}(x) = \int_D \Phi(x`) \cdot G(x, x`)dx` \tag{1}$$

where

G is the filtering function that gives the values of the vortices.

$$G(x,x`)=\begin{cases}\frac{1}{V}, & x`\in v\\ 0, & x`\notin v\end{cases} \tag{2}$$

The cell volume V, generates the filtering.

$$\overline{\Phi}(x) = \frac{1}{V}\int \Phi(x`)dx`, \quad x`\in v \tag{3}$$

After filtering Navier-Stokes equations are:

$$\frac{\partial p}{\partial t} + \frac{\partial}{\partial x_i}(\rho \overline{u}_i) = 0 \tag{4}$$

and
$$\frac{\partial}{\partial t}(\rho \overline{u}_i) + \frac{\partial}{\partial x_j}(\rho \overline{u}_i \overline{u}_j) = \frac{\partial}{\partial x_j}\left(\mu \frac{\partial \overline{u}_i}{\partial x_j}\right) - \frac{\partial \overline{p}}{\partial x_i} - \frac{\partial T_{ij}}{\partial x_j} \tag{5}$$

where

$$T_{ij} = \rho\overline{u_i u_j} - \rho\overline{u}_i\overline{u}_j \tag{6}$$

$$T_{ij} = \rho\overline{u_i u_j} - \rho\overline{u}_i\overline{u}_j \tag{7}$$

$$T_{ij} - \frac{1}{3}T_{kk}\delta_{ij} = -2\mu_t\overline{S}_{ij} \tag{8}$$

where μ_t is the turbulent viscozity and S_{ij} is:

$$\overline{S}_{ij} \equiv \frac{1}{2}\left(\frac{\partial \overline{u}_i}{\partial x_j} + \frac{\partial \overline{u}_j}{\partial x_i}\right) \tag{9}$$

μ_t can be derermined in FLUENT using two methods: Smagorinsky-Lilly and RNG.

Smagorinsky-Lilly model which was defined by Smagorinsky and improved by Lilly (Aziz & Fadehen, 2006), gave the equation:

$$\mu_t = \rho L_S^2 \left|\overline{S}\right| \tag{10}$$

where L_s is the medium lenght of the cells, expressed by:

$$L_S = \min\left(kd \cdot C_S V^{1/3}\right) \tag{11}$$

And: $$|\bar{S}| \equiv \sqrt{2\bar{S}_{ij}\bar{S}_{ij}} \quad (12)$$

where:

C_S is Smagorinsky constant, k – Kármán constant,

d – the distance to the nearest wall

RNG, model gave the equation:

$$\mu_{ef} = \mu\left(1 + H(x)\right)^{1/3} \quad (13)$$

where H(x) is Heaviside function:

$$H(x) = \begin{cases} x, & x > 0 \\ 0, & x \le 0 \end{cases} \quad (14)$$

where

$$x = \frac{\mu_S^2 \mu_{ef}}{\mu^3} - C \quad (15)$$

and:

$$\mu_S = \rho\left(C_{rng} V^{1/3}\right)^2 \sqrt{2\bar{S}_{ij}\bar{S}_{ij}} \quad (16)$$

2.2 Experimental results:

The simulation applied to geometry of the channel presented in figure 1 was made using FLUENT software.

The results, presented in figure 2, show the temperature in the channel.

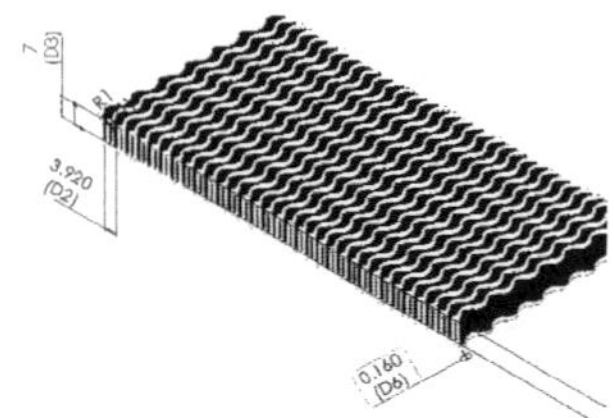

Fig. 1 Strip wings

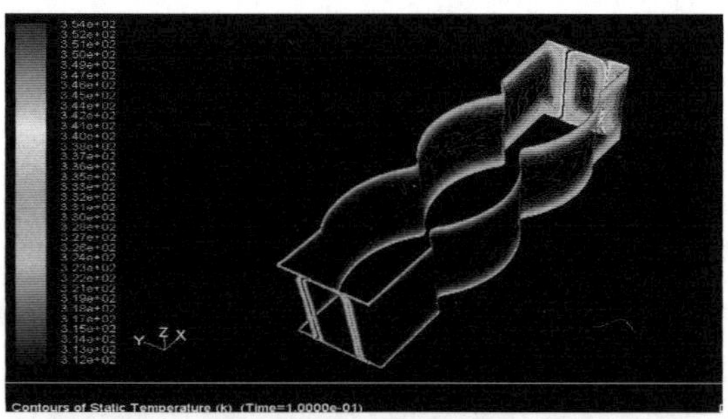

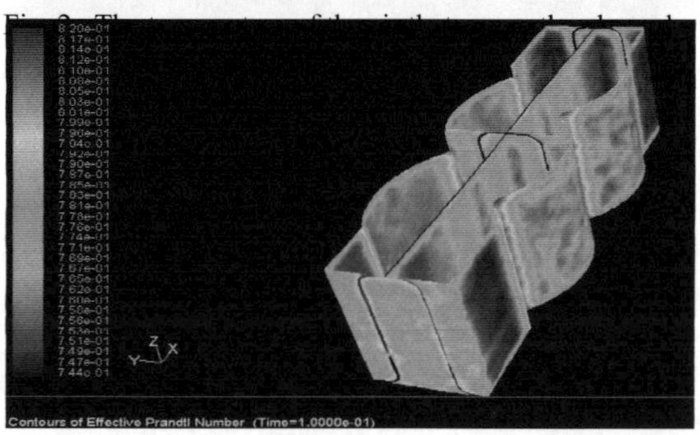

Fig.3 Nusselt number and vortices distribution

In figure 3 is presented the Nusselt criterium for the channel of the heat exchanger.

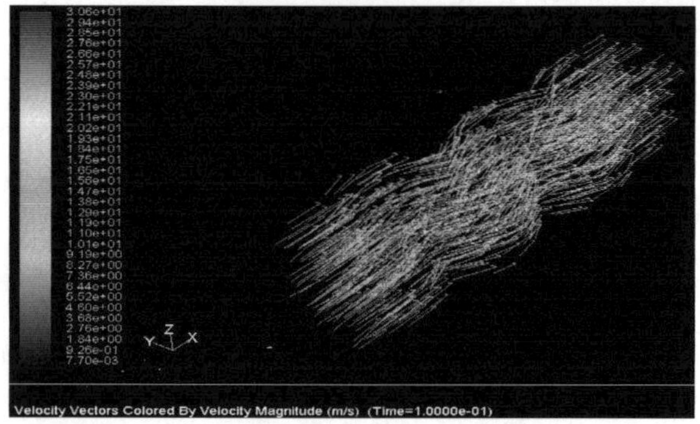

Fig.4. Velocity vectors distribution

Measured		Numerical simulation		Error
T_{e2}		T_{e2}*		E_{Te2}
[^{0}C]	[K]	[^{0}C]	[K]	[%]
53.37	326.52	53.51	326.66	0.0413
54.21	327.36	54.47	327.62	-0.1041
56.10	329.25	56.35	329.5	0.1518
57.95	331.1	57.6	330.75	-0.5460
60.30	333.45	60.6	333.75	-0.5122
62.88	336.03	62.47	335.62	-0.1229
56.99	330.14	57.25	330.4	0.3810
60.35	333.50	60.72	333.87	0.2698

Tab. 1 Measured and simulated values

In figure 4 is presented the velocity distribution and vectors for the flow through the heat exchanger channel. The authors design also an experimental stand, in order to validate the results of the simulation (Plesa at al 2006). Table 1 shows the comparison between the values obtained by simulation method to those obtained on the experimental stand

3. CONCLUSIONS

The results of the simulation are very close to the experimental values, as it can be seen in table 1. The simulation shows that in some regions of the channel the velocity increase and the vorticies appear. In these regions also the Prandtl number is high and the thermal convection is better. The aim of the research is to find the parameters that can provide turbulent flow and a good balance between the pressure drop and the global thermal coefficient.

4. REFERENCES

Azis, A, Mc. Fadehen, (2006) "Some new solution for extended surface heat transfer using symbolic algebra", *Heat Transfer Engineering* Vol.26, No. 8, oct.2006, pag. 15-23

Rush, T.A., Newell, T.A., Jacobi, A.M., (1999)"An Experimental study of flow and heat transfer in sinusoidal wavy passages", *Int. J. Heat Mass Transfer 42*,

Pleşa, A., Opruţa, D., Ungureşan, P., Teberean, I., (2006) „Some practical aspects regarding the flow in a plate and fins heat exchanger", *International Conference Experimental Fluid Mechanics*, Liberec, Czech Republic, pag. 157-160, ISBN 80-7372-141-4

Susan-Resiga, R., Muntean, S., Bernad, S., Balint, D., Balint, I., (2003) New methods in parallel calculus and simulation of the fluid flow, Ed. Orizonturi Universitare, Timisoara, Romania

Annals of DAAAM for 2008 & Proceedings of the 19th International DAAAM Symposium, ISSN 1726-9679
Editor B. Katalinic, Published by DAAAM International, Vienna, Austria 2008
Make Harmony Between Technology and Nature, and Your Mind will Fly Free as a Bird

EXAMINATION SOFTWARE FOR TECHNICAL SUBJECT EDUCATION

PODHORSKY, S[tefan] & MARONEK, M[ilan]

Abstract: *The aim of the paper is to outline developed program package for computer-based testing and assessments. The software is based on fully editable multimedia supported questions bank used for randomised test generation, allowing the test questions sequencing exactly according to examiner's needs. Several types of questions are supported with possibility of assigning audio, video or picture files and making variants of the same question. Special attention has been paid for the possibility of real usage of those types of question where student has to insert some text for the answer, not only to recognise an item from a given list. These question types are still not common in assessment software since automated marking of answers brings some difficulties.*

Key words: *education, examination, computer-based testing, computer-based assessment, software*

1. INTRODUCTION

An integral part of every education process, regardless of its form is assessment proceeding used to determine the reached study results. Beside traditional means of verification of acquired knowledge computer-based form can be used, offering higher level of efficiency, objectivity and reliability in the assessment process (Hrmo 2005). The computer-based assessment can be used not only for final tests, but also for gradual and formative assessment, with the major benefit of immediate feedback. In the case of computer-based testing the test questions must be circumscribed in the way to get definite answers, so the method of the correct answer selection from the given list, e.g. multiple-choice, multiple response, true or false test questions are most commonly used (JISC, 2007). Although this method of testing is quick and easy to implement, it is not able to verify the student's in-depth knowledge properly, especially in the case of technical education. This was the main reason for developing own system of computer-based testing and assessment in the Institute of Industrial Engineering at the Faculty of Materials Science and Technology.

2. THE *eTest* APPLICATION

eTest is a software package designed for computer aided preparation of didactic tests and for computer-based testing of student's knowledge. It consist of few modules. For item bank management the module *eTest Database* is used. The module *eTest Builder* is intended for generation of didactic tests, and in the case of paper form of testing it used to print test sheets. *eTest Student* and *eTest Supervisor* are the basic client applications for the computer-based form of testing. The next postulates had been defined before the start of the sofware developement:

- various type of test questions (considering the computer-based method of answers assessment),
- individual test building for every tested student,
- possible limitation to some part of the study subject (e.g. topic selection) for test generation,
- the count of generated test questions in every topic should be proportional to extension of those part in the curriculum,
- consideration of the questions difficulty level,
- possible usage of graphics (figures) into the test questions,
- integration of multimedia files into didactic tests,
- scoring of student's answers based on the weight of question signification and on the completeness of student's answer –balanced scoring (Huba et al., 2003).

Items bank data can be prepared using common text processor in rich text format (RTF) or a specialized tool can be used. Data is stored in own format, the IMS Question and Test Interoperability specification (QTI) is not internally supported, but data can be imported from and exported to other systems in the IMS QTI format. It seems to be a common practice in computer-based systems (Sclater & Low, 2002).

3. TOPICS

When a new test question is inserted into the item bank, it must be assigned to some topic of the study subject. Therefore, the structured list of topics is the integral part of each item bank of test questions. The list of topics has to respect the structure of the lectured matter. Each topics can be divided into smaller parts on the lower hierarchal level, similarly as the chapters of textbook are split into subheadings.

The assignation of test questions to topics is meaningful, it is used not only for easy categorisation of test questions, but the primary purpose is to guarantee uniform distribution of the generated test questions from every part of the subject matter. Therefore, the structured list of topics contains entries about the item's extensity in minutes. Based on these data and on the known summary count of the test questions, the software is able to determine how many questions should be picked from item bank in the frame of each topic.

4. TEST QUESTIONS

A test question is a record in the item bank database. To reach higher variability of the generated tests, the various variants of the same question should be inserted into the item bank. The term "*question variant*" means different formulation of the question, different way of the answer entering, different type of question and so on, but the object of the question stays the same. The software assures that only one variant of the question will be present each time in the generated test.

Beside the topic specification, after a new record is created, it is necessary to specify the *type of the question*. The type of the question determines the way the student will answer and in addition the question type determines the form of the answer entry into the appropriate field of the record. There are eight types of test questions in the present version of the software. Following the way of answer entering, the test questions type can be classified into three groups:

- student has to check the correct item/items in the list of provided items (answers),
- student has to enter a character string into input field (a word, few words, a sentence),
- student has to enter some numeric value.

Some fields in the item bank database can contain formatted text to enable complex formatting of the question's text in the case of technical subject education (special characters, Greek alphabet, superscript, subscript etc.). The form of the answer notation is specific for every type of question and consist a record for automatic assessment of the student's answer. A graphics can be attached to test questions – bitmaps or vector graphics. Beside graphics, multimedia files can be attached for the electronic form of testing (video or audio file). There are next numeric fields present in the question record:

- time in seconds, available for student to work out his answer – time limit,
- difficulty index of the question,
- rating in marks the student can gain for completely answered question.

4.1 Types of test question

The type of test question determines the way the student will answer to the question. Along with it, the construction and the form of text entry for automated answer assessment are defined by the question type – some syntax rules are valid for every type of test question. The algorithm of answer assessment and the way of mark assigning (weight scoring) also depends on the question's type. Multiple-choice test questions are most common ones. There are three types of test questions in this group, when the student is presented with a choice:

- Alternative – there is just one correct answer in the test question, – student gains full marks for correct choice, otherwise he gains no marks.
- Subset – there are any count of correct items in the test question, or none (multiple response question), – the count of the correct items is not given for student, – balanced scoring is used.
- Combination – there are at least two correct items in the test question, – the count of items the student has to check follows from the question, – balanced scoring is used.

Next group of the test question types are those ones that require some text input for the answer. In these cases, appropriate keywords have to be specified for every test question, which will be seeking in the student's answer. The student's answer is taken as correct only if all the specified keywords have been found within his answer. This method of automated assessment of the student's answer brings some complexity. One of them is the inflection of words in many of languages. The application solves this problem using only the base part of the keyword for seeking, not the whole form. Therefore, the keywords are wrote using tilde character "~" to separate invariable part of the keyword from the suffix. Another problem is the real possibility of student's misspell. The student would not be penalised for a mistake in entering the answer (Cook & Perli 2007). Therefore diacritic marks are omitted, when the keywords are sought in the student's answer, the difference between characters "*i*" and "*y*" is ignored as well. Furthermore, a student self-checking of his assessment is used. There are three types of test questions utilizing the method of keyword seeking:

- Keywords – one or a few keywords are seeking in the text of the student's answer, – student gains full marks if all the keywords are found in the text of his answer, otherwise he gains no marks.
- Enumeration – the task of student is to specify an enumeration of some items, – the order of the items in the answer is not relevant, – balanced scoring is used.
- Completion – a text is displayed with some words omitted, the task of student is to complete the text with appropriate words, – balanced scoring is used.

Beside of mentioned types of test questions the next ones can be used in the item bank as well:

- Number – a numeric value is required from student, – range of allowable values can be specified for the correct answer.
- Manual – the answer of student on test question is some text input and the software does not assess this answer, – the student's answer is assessed by the examiner.

5. THE TEST BUILDING MODULE

The test-building module is used to define a set of rules and conditions, which are used for test building process when tests are generated. The defined complex of rules and conditions are saved into a file separately for each of the subjects. Furthermore, a few such files can be created in the frame of one subject for continuous tests (credits tests) and for final test. Each of generated tests is unique from the view of answers combination, e.g. each student on examination will get different set of questions. It requires a few steps to create a setting configuration for a test building: 1. Selection of the desired item bank of test question; 2. Specification of a source set of test questions (database filtering); 3. Setting of preferences.

6. CONCLUSION

The main benefit of the *eTest* application is the high efficiency of assessment process of students' knowledge, especially in the case of high quantity of students. Notable time saving can be achieved without detriment to the quality and objectivity of the assessment process. If there are enough count of test question in the item bank the high variability of generated tests is guaranteed. Optimizing the needful time for every question in the test, the examination will be not only shorter, but in the same time students will not have enough time for prospective cheating. In addition, the system of electronic testing eliminates subjective effect of the assessing person.

When the project of software development started, it had been intended for technical education only. Considering the versatility of the designed concept, the software can be utilized in any field of education after building the appropriate item bank of questions. Item banks for subjects Technology of Foundry, Technology of Welding and Materials Jointing have been worked out at the Institute of Production Technologies. The server part of the application is being developed now, in the next academic term we plan to test the computer-based form of assessment in the education process. The paper was realized with the support of KEGA 3/4157/06 grant.

7. REFERENCES

Cook, J. & Perli, R. (2007). Writing assessment questions for online delivery: Principles and guidelines, *Available from:* http://www.ltss.bris.ac.uk/caa/writing_e-assessments/, *Accessed: 2008-06-05.*

Hrmo, R. (2005). Innovation of vocational technical subjects teachers pedagogical training, *Trendy technického vzdělávání 2005 (Trends in Technical Education 2005)*, pp. 7-16, ISBN 80-7220-227-8, Olomouc, june 2005, Votobia, Praha.

Huba, M.; Žáková, K. & Bisták P. (2003). *WWW a vzdelávanie (WWW and Education)*, STU, ISBN 80-227-1999-4, Bratislava.

Sclater, N. & Low, B. (2002). IMS Question and Test Interoperability: An Idiot's Guide, *Available from:* www.scrolla.ac.uk/resources/s2/idiots_guide.pdf.

The Joint Information Systems Committee (JISC) (2007), Effective Practice with e-Assessment, *Available from:* http://www.online-conference.net/jisc/content2007/JISC%20effective_e-assess.pdf, *Accessed: 2007-03-12.*

Annals of DAAAM for 2008 & Proceedings of the 19th International DAAAM Symposium, ISSN 1726-9679
Editor B. Katalinic, Published by DAAAM International, Vienna, Austria 2008
Make Harmony Between Technology and Nature, and Your Mind will Fly Free as a Bird
*Young Researcher and Scientist Paper / * Supervisor, Mentor*

SIMULATION CONCEPT FOR MACHINED SURFACE ROUGHNESS AND SHAPE DEVIATIONS PREDICTION

POLAKOVIC, M[ilos]; BURANSKY, I[van] & PETERKA, J[ozef]*

Abstract: *This paper deals with the problematic of machining simulation. Main aim is to show a concept for simulation algorithm which would be able to simulate a machining process, machined surface quality and possible machining errors due to the tool deflection. The building elements of models are described and the sub-algorithms which will be used in simulation development are mentioned as well.*
Key words: *simulation, voxel, roughness, deflection, deviation.*

1. INTRODUCTION

By machining process with geometrically defined cutting tools the deviations from designed CAD surfaces as well as dimension's deviations always arise. There are various factors such as machine-tool-clamping system stiffness, cutting force intensity and orientation, cutting edge geometry, tool wear, vibrations, used finishing strategy, CNC positioning precision, etc., which have different level of effect on the arising deviations.

Our goal is to develop a simulation algorithm, which would be able to simulate finishing process, specifically machining with ball end mill. Designed algorithm will take into account inputs such as tool geometry, stiffness, material properties, deflection, cutting forces and cutting parameters. The simulation output data will be used for graphical display of machined surfaces and for further analyses of shape deviations and surface roughness as well. Timeline of force intensity and tool load will also be analyzed and used for toolpath or cutting parameters optimization.

2. MACHINED PART VS. CAD MODEL

A low cost experiment for evaluation of machined surface deviations and roughness was prepared. According to information's from resources (Choi & Jerard, 1998), (Fornusek & Rybín, 2000) the appropriate model features and surface shapes were designed using the CAD software Delcam PowerSHAPE 5.8.21. The roughing and finishing strategies in CAM software PowerMILL 6.0.0.8 were applied for generating various roughing toolpaths as well as finishing toolpaths for ball end mill. In order to keep the machining costs as low as possible, the surfaces were machined with 3axis continuous milling strategy and only the shape deviations and surface roughness were evaluated. The test workpiece was machined by CNC milling machine VMC EAGLE1000 (positioning accuracy ± 0,005 mm, repeatable accuracy ± 0,003 mm).

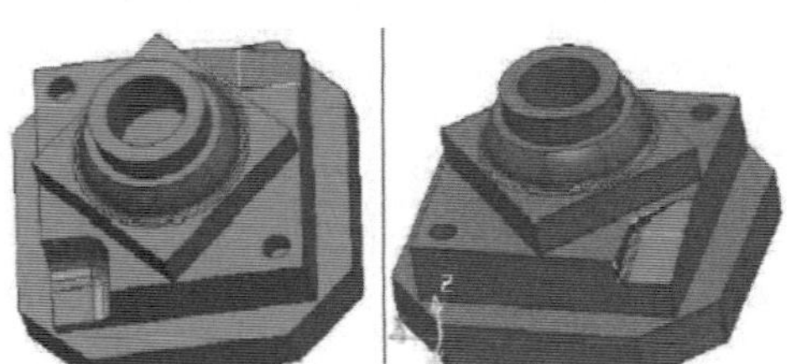

Fig. 1. The experimental 3D CAD model.

As a workpiece material the steel with hardness HB 264 was selected. For machining purposes the tool with PVD coated inserts and PVD coated cemented carbide ball end mills have been used.

2.1 Evaluation of machined workpiece

Machined workpiece has been digitized with non-contact optical 3D scanner GOM ATOS I 350 (scanning accuracy 0,02mm). Acquired digital data has been processed in software GOM ATOS v 6.0.2-5 and compared to original CAD model using the feature "CAD Comparison". After this procedure the deviation map has been obtained. This map indicates various areas on machined surface, where the tool deflection created significant deviations up to 0,2 mm.

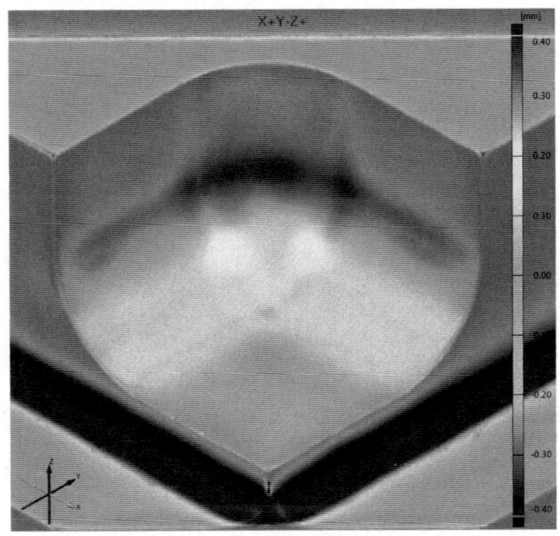

Fig. 2. Obtained deviation map on selected surface.

2.2 Roughness evaluation

In case of ball end mill finishing, the lowest roughness values (Ra = 0.8 μm) has been obtained on workpiece areas, where the cutting effective radius converged at nominal radius of used ball end mill. The highest values (Ra = 5 μm) has been detected on areas, where the effective radius was lowest, therefore the cutting velocity was insufficient. Various finishing strategies used by machining of this part show a different roughness pattern as well. For roughness evaluation the contact roughness measuring device Surtronick 3+ has been applied.

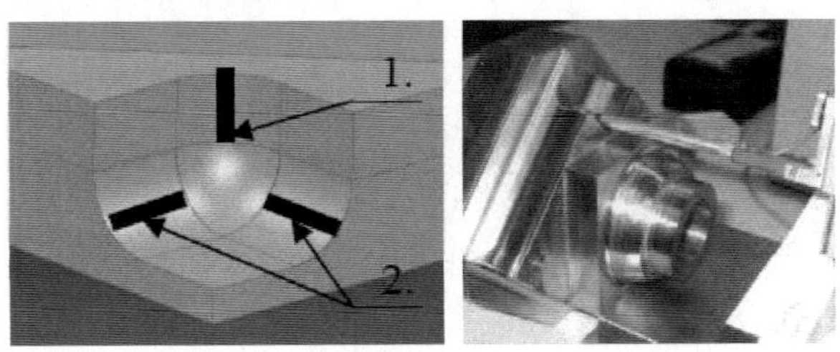

Fig. 3. Roughness measuring on selected surface of workpiece.

3. SIMULATION CONCEPT

The main aim is to simulate finishing process with ball end mill and determine suitability of applied strategy for toolpath generation as well as suitability of selected tool and cutting parameters, considering the arising shape deviations and surface roughness. Computational algorithm will simulate cutting movement of tool edge elements. This movement is a combination of translation and rotation of edge elements on the loaded tool path. The path is generated and saved into a CL data form in the CAM software. The interactions between cutting edge elements and workpiece elements will be detected and the corresponding chip cross sections, cutting forces and tool deflections will be calculated. The core of the algorithm development will be an appropriate implementation of tool, workpiece, cutting kinematics, cutting forces and tool deflection sub-models.

3.1 The smallest building element "The Voxel"

The Voxel will be used as a building element for physical representation of tool, cutting edge and workpiece geometry (Cohen-Or & Kaufman, 1995). Voxel is the smallest volume element, which can contain the information's about local volume (density, material, color, normal vector, forces, etc.). Voxels are usually distributed along the Cartesian's grid using the USD (Uniform Space Decomposition). Because the resolution needed for roughness simulation is very fine, the amount of needed Voxels would exceed any computational capability of common desktop computers. Therefore the algorithm called "Cube-Marching" (Lorensen & Cline, 1987) will be implemented. This algorithm is capable to convert Voxels to triangle mesh on the fly. Using the GPU hardware acceleration the triangle mesh will be recalculated and displayed much faster. Therefore the Voxel engine will be used only for physical representation of models and the triangles will be used for graphical representation of machined surfaces.

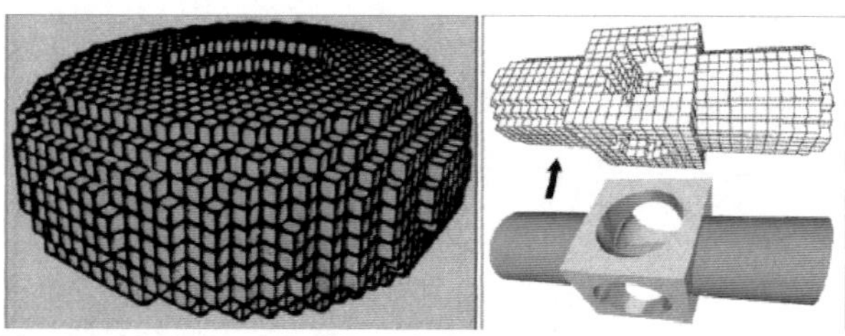

Fig. 4. Voxel decomposition of a CSG model.

3.2 Tool, cutting edge and workpiece models

Geometrical model of tool will be created in the CAD software, then converted to STL triangle mesh format and finally, using the USD decomposition, converted to Voxel model. The same procedure for acquiring workpiece blank voxel model will be applied as well.

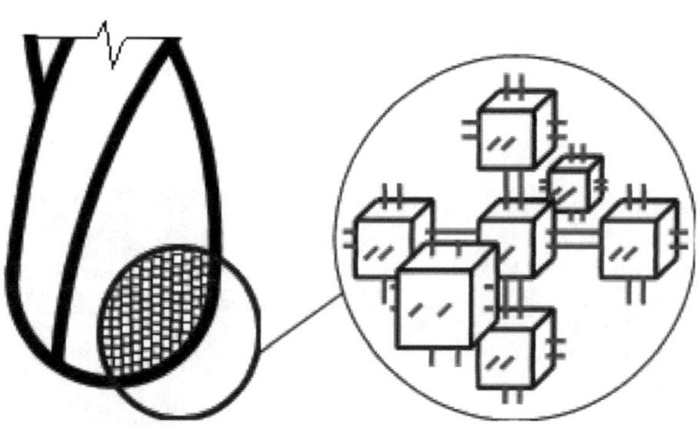

Fig. 5. The cutting edge model.

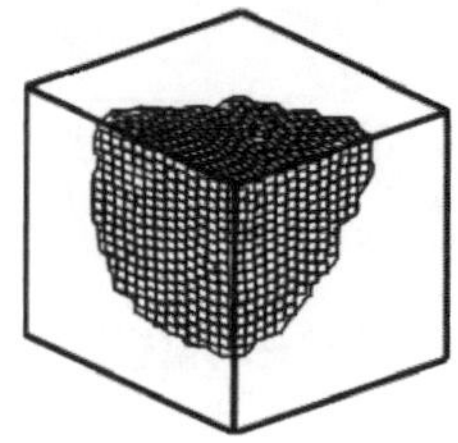

Fig. 6. The workpiece blank model.

3.3 Tool deflection model

Model for tool deflection will be based on results from work (Kim et al., 2003). Approximated equation of deviation between deflected and non deflected tool can be expressed as follows:

$$\delta = \delta_s + \delta_f + \phi_s (Lf - z) \tag{1}$$

-where δ_s is deflection of the shank, δ_f is deflection of the flute, ϕ_s is deflection angle of the shank, Lf is length of the flute and z is the coordinate where the deflection is being calculated.

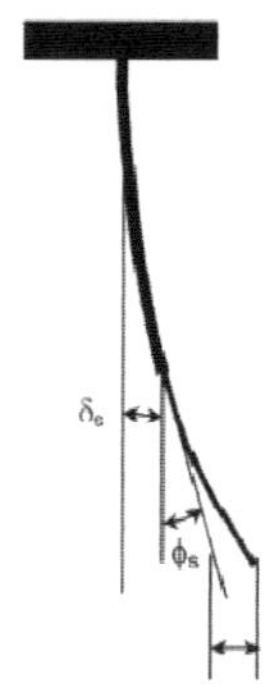

Fig. 7. *Simplified cutter deflection model.*

4. CONCLUSION

The CAM software nowadays provides possibilities of graphical cutting simulation. This type of simulation gives only an idea of cutter movement and material removal, other information's such as cutting forces, tool deflection, vibrations, surface quality after machining is unknown, respectively can not be determined. This paper tries to show a concept how to expand graphical simulation in a combination with physical cutting simulation, where the Voxel elements and triangle mesh are combined into a one cutting simulation packet.

5. REFERENCES

Choi, B.K. & Jerard, R.B. (1998). *Sculptured surface machining*, Kluwer Academic Publishers, ISBN 0 412 78020 8, U.S.A.

Cohen-Or, D. & Kaufman, A. (1995). Fundamentals of Surface Voxelization, *Available from:* http://www.sciencedirect.com/ *Accessed*: 2008-02-02

Fornusek, T. & Rybín, J. (2000). *Accuracy determining of multiaxis controlled milling machine, Available from:* http://www.fs.vsb.cz/ Accessed: 2007-12-10

Kim, G.M.; Kim, B.H. & Chu, C.N. (2003). Estimation of cutter deflection and form error in ball-end milling processes, *Available from:* http://www.sciencedirect.com/ *Accessed:* 2007-12-10

Lorensen, W.E. & Cline, H.E. (1987). Marching cubes: A high resolution 3D surface construction algorithm, *Available from:* http://portal.acm.org/citation.cfm?id=37401.37422 *Accessed:* 2007-10-10

Annals of DAAAM for 2008 & Proceedings of the 19th International DAAAM Symposium, ISSN 1726-9679
Editor B. Katalinic, Published by DAAAM International, Vienna, Austria 2008
Make Harmony Between Technology and Nature, and Your Mind will Fly Free as a Bird

CONSTRUCTIVE ASPECTS TO IMPROVE THE PERFORMANCE OF THE VIBRO-ROLLING DEVICE

POLOJINTEF CORBU, N[icolae]; PATER, S[orin]; FODOR, D[inu] & MITRAN, T[udor]

Abstract: *There are established methods for realizing manufacturing through vibro –rolling of the exterior surfaces. On the duration of the experiments, proposals of improving the vibro –rolling devices performances have been analyzed and formulated based on constructive new solutions. This method is intended primarily to reduce the friction forces on the direction of the roller oscillatory movement*
Key words: *rolling, tension, harshness, movement*

1. INTRODUCTION

There are already acknowledged procedures to process exterior surfaces by vibro-rolling.

During the experimental testing, several possible solutions have been expressed and analysed, to improve the performances of the vibro-rolling device based on new constructive methods.

2. CONSTRUCTIVE SOLUTIONS FOR THE ROLLING HEAD

The initial constructive solution for the rolling head during the experimental stage is presented in Figure 1.

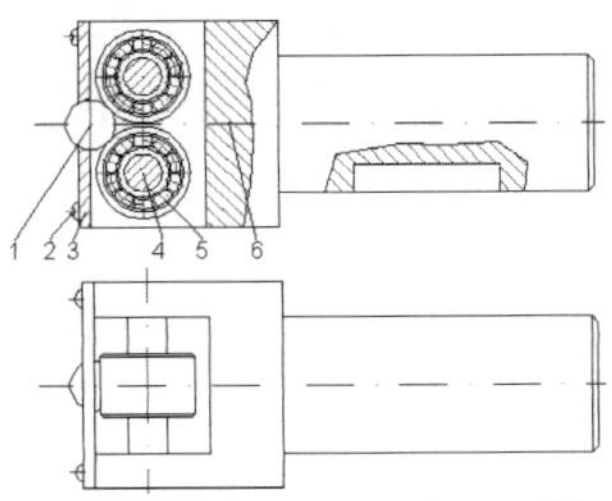

Fig.1. Rolling head used for the vibro-rolling device: 1. rolling ball; 2. screw for temple clamping; 3. temple; 4. shaft; 5. rolling bearing; 6. fork

In this case, the ball represents the pressure body pushing against the part surface. The backward push is taken by the radial bearings (5) mounted on the shafts (4), which are stiffly screwed, on their turn, in the fork bore holes (6). The fork tail is clamped into the frame bore holes, being protected against rolling by a parallel key. (Lee & Tarng 2001)

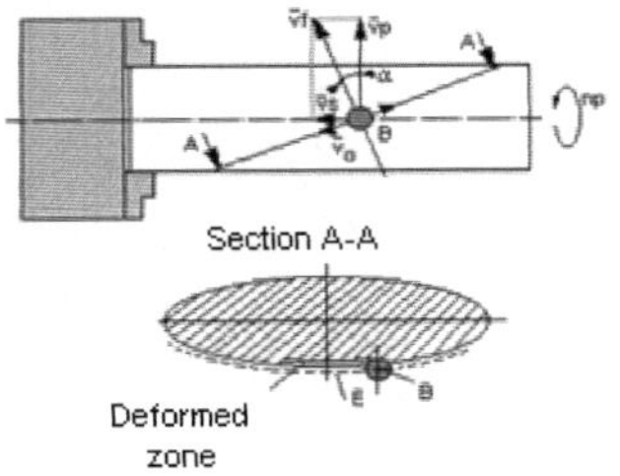

Fig.2.Chart presenting various speeds involved in vibro-rolling

The temple (3) is ment to maintain the ball in contact with the bearing exterior ring, in the absence of the pushing force.

The role of the bearings is to ensure a smooth functioning of the rolling head, without sliding friction. This can be achieved completely if the functioning does not involve feeding, but it doesn't determine the generation of a system of impressions, but only of a single sinusoidal impression.

The caption of Figure 2 is the following:

-v_o – speed of the oscillatory movement;
-v_p – tangential speed determined by the part rotation (n_p);
-v_s – feeding speed;
-v_f – speed resulting from the combination of v_p and v_s speeds;

If the oscillatory movement is determined on a parallel direction to the part axis, the rolling is achieved on v_p direction, also entailing a sliding movement on v_s direction.
If the oscillatory movement is determined on v_o direction perpendicular on v_f , the rolling will be achieved without sliding on v_s direction.

To obtain a movement without sliding on v_s direction, it is not enough to tilt the rolling head, but also to provide an oscillatory movement perpendicular on v_f direction. This implies tilting the entire vibro-rolling device, solution which entails constructive complications. Also, the device tilting must be calculated separately for every single change in the feeding value, which entails again operation complications. Another major impediment to a good rolling is the necessity to obtain a very subtle balance between the sliding forces. If the rolling friction between the ball and the part is higher than the sliding friction between the ball and the bearing exterior surface, the rolling on the feeding direction is replaced by sliding.

Next there are analysed two new methods used in vibro-rolling, which could replace the solution presented in Figure 1. These methods are only proposals, as they need further studies and development.

3. VIBRO-ROLLING DEVICE WITH CURVE CONVEX-PROFILE ROLLER

The convex-profile processing roller is fed against the processing surface by a concave-profile pressure roller (Figure 3).

The vibrating movement is an angular oscillatory movement having as oscillation centre the vertical axis which is parallel to z` - z axis. The vertical rotation axis is the same with that of the gear, which drives the gear rack. (Deacu & Pavel 1977)

The gear rack is parallel to the part axis and is rigidly mounted on the machine frame. The gear division radius must be the same as the profile radius of the processing roller. The main advantage of this method is that due to the combination (simultaneous operation) of movements “z” and “v”, the curve-profile roller will roll without sliding on the part surface in an angular oscillatory movement, thus preventing any possible friction between the roller and the part. (Keesen 1975)

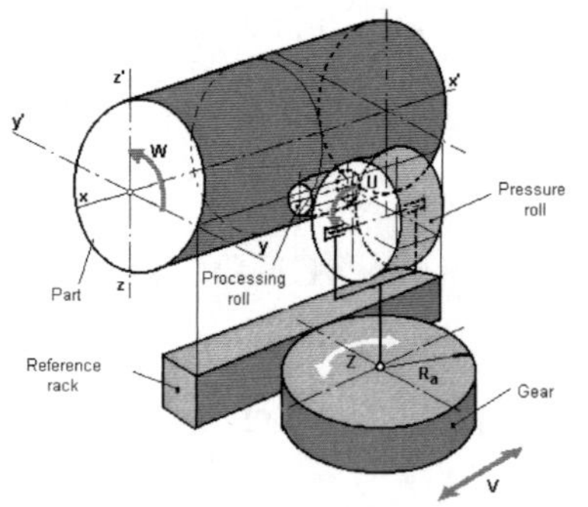

Fig. 3. Basic diagram of the vibro-rolling using a convex-profile roller

The gear – gear rack mechanism can be replaced by a band mechanism, just like that used on Maag grinding machines.

The drawback of this method is the fact that the roller impression on the part surface is much wider than with the ball processing, the resulting impression systems not being able to meet the requirements of the vibro-rolling processing.

4. VIBRO-ROLLING DEVICE WITH SELF-SUSTAINING OSCILLATIONS

The second scientific proposal is the use of an elastic system to obtain self-sustaining oscillation. This method could be used to remove the driving element of the vibration device. The following chart is meant to ensure the theoretical support for potential future construction of this device (Figure 4).

The processing roller is pressed down on the processing surface with an F force. It is rolled on the part surface and has the possibility to rotate around w – w` axis or to slide along the L-length guideway. The roller oscillation phases are presented in Figure 5. (Klocke.& Liermann 1996)

Once in A position, the roller will be rotated by the arm fitted with a spiral arc, as this one has the tendency to expand, and the end of the bar (N) will slide along the C-D curve guideway (β angle < angle of the friction cone), up to N` point, rotating the roller on w – w` axis (Figure 5).

At this point, the roller will be at an α angle relative to the perpendicular line on the part axis, determining a pushing force of the roller (of the M point) from A to B. During the movement from A point to B point, the arc will be compressed, so that when reaching point B, the movement can be reversed. N will move toward N` due to the arc expansion, modifying at the same time the position of the roller axis relative to the part axis. Thus, a transversal force is obtained, which is pushing the roller from point B to point A, compressing the arc at the same time. The cycle is repeated and a roller oscillatory movement is occurring, with a frequency and amplitude determined by the geometric parameters of the device.

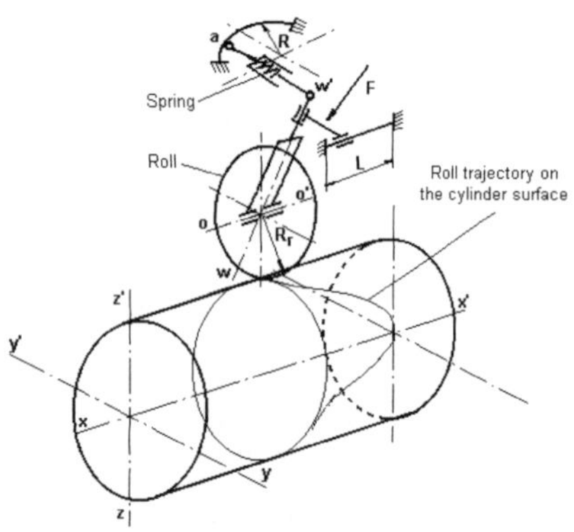

Fig. 4. Short presentation of the device; the roller impression on the part is a sine curve

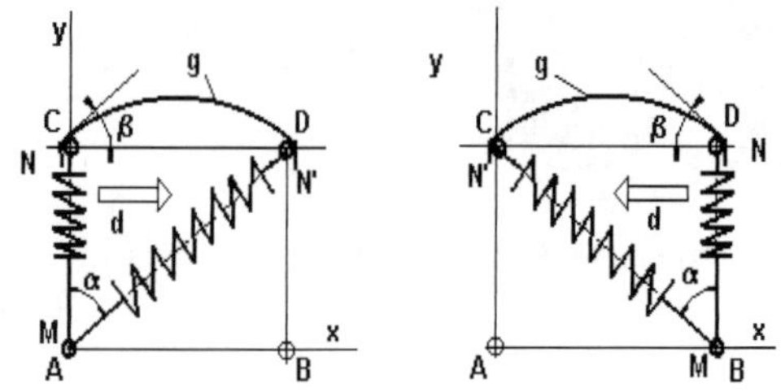

Figure 5. Representation of an oscillatory cycle

The disadvantage of the above-presented method is the difficult construction of this device.

5. CONCLUSION

The two "unconventional" solutions have both disadvantages related mainly to construction difficulties.

For the solution presented in Figure 1, the tensions needed to achieve a perfect rolling between the elements of the rolling head (ball, bearing) are also considerable, determining some constructive complications of the device.

A practical solution which allows the sliding friction between the ball and its frame, and facilitates a satisfying rolling between the ball and the part is presented in Figure 6. This variant has a simple and solid construction, the assembling of the parts 1 and 2, respectively 3 and 4 being realised by pressing. Part 3 is made of anti-friction bronze, to enable a slight ball rolling.

At the same time, the space between the ball and part 3 is filled with molybdenum disulphide (MoS_2), having superior anti-friction properties. (Wahl 1987)

This variant has been tested within the experimental stage of our doctoral studies, obtaining high-quality impression systems. What still remains unknown is the reaction of this type of head to long-term wearing, as it needs long-term experimental testing.

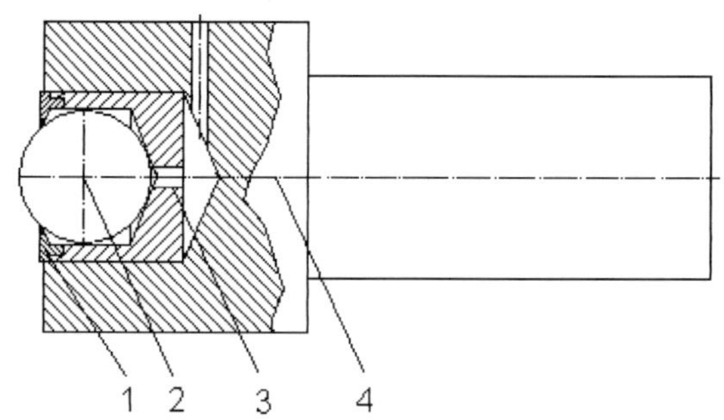

Fig. 6. Common rolling head. 1. retainer ring; 2. ball; 3. anti-friction part; 4. Frame

6. REFERENCES

B.Y. Lee, Y.S. Tarng (2001). *Surface roughness inspection by computer vision in turning operations*, International Journal of Machine Tools & Manufacture, 41 1251-1263;

Deacu L.& Pavel Gh.(1977) *Vibrații la mașinile – unelte, Vibration of manufacturing machines,* Editura Dacia, Cluj-Napoca,

Keesen G. (1975) *An in Depth Look at Roller Burnishing,* Cutting Tool Engineering.

Klocke F.& Liermann J.,(1996) - *Roller burnishing of hard turned surfaces,* Surface Conference, Goteborg, Sweeden,;

Wahl, F. M.(1987) *Digital Image Signal Processing*, Artech House, Boston,

Annals of DAAAM for 2008 & Proceedings of the 19th International DAAAM Symposium, ISSN 1726-9679
Editor B. Katalinic, Published by DAAAM International, Vienna, Austria 2008
Make Harmony Between Technology and Nature, and Your Mind will Fly Free as a Bird

GENERALIZED JOCOBIANS AND NEWTON METHOD FOR SOLVING THE FRICTIONAL CONTACT PROBLEMS

POP, N[icolae] & CIOBAN, H[oria]

Abstract: *This paper presents some results from nonsmooth analysis which leads to sufficient conditions such that the nonlinear and nondifferential system that modeling the contact problems with frictions, does have one solution, obtained by Newton's method using the generalized gradient.The generalizations by means of Newton's method can be applied to solve some aspect of friction contact problem.*
Key words: *frictional, nonsmooth analysis, 2D modelling*

1. INTRODUCTION

The theory of nonsmooth analysis shows that many results of standard analysis may be extended to a more general, nonsmooth framework. In this paper a new model based on nonsmooth equations is proposed for solving a nonlinear and nondifferential equation obtained by discretization of a quasivariational inequation that modeling the frictional contact problem. The main aim of this paper is to show that the Newton method based on the plenary hull of the Clarke generalized Jacobians (the nonsmooth damped Newton method) can be implemented for solving Lipschitz nonsmooth equation.

The theory and algorithms for solving nonsmooth equation have been developed by mathematicians such as Pang and Qi (1993), Xu (2001) etc. Recently, Hu et al. (2007) presented the employment of Newton's method to formulate B-differentiable equations involving projection for three dimensional contact problems. The contact constraints can be exposed as nonsmooth equation set in which the variables are related to the candidate contact nodes. For solving this nonsmooth equations can be propose the nonsmooth damped Newton method using the generalized Jacobians.

2. GENERALIZED GRADIENTS

Assume that $F{:}R^n \to R^n$ is locally Lipschitzian but not necessary differentiable and nonsmooth equation

$$F(x) = 0 \tag{1}$$

The most popular method for this solving equation (1) is the Newton method based on the Clarke generalization Jacobian,

$$x_{k+1} = x_k + (V(x_k))^{-1} F(x_k)$$

where $V(x_k) \in \partial F(x_k)$ is the Clarke generalized Jacobian of the F at x_k. The Generalized Jacobian of F is defined as a convex hull of all $n \times n$ matrices obtained as the limit of a sequence $(V(x_k))_l$, when $x_k \to x$, $k \to \infty$ and x_k is a point at which F is differentiable. In case when F is nonsmooth, $V(x_k)$ may not exist, and therefore, is should be replaced by a generealized derivative an using Rademarcher's Theorem, with F differentiable almost where.

Denote by D_F the set of points where F is differentiable, and by $\nabla F(x)$, a $n \times n$ Jacobian matrix of partial derivative whenever x_k is a point at which the partial derivative exist. By $\partial F(x)$ we denote the generalized derivative, in sense Clarke

$$\partial F(x) = \mathrm{co}\{V(x) = \lim_{k\to\infty} \nabla F(x_k) | \lim_{k\to\infty} x_k \to x,\ x_k \in D_F\} \tag{2}$$

If F is a locally Lipschitzian and nonsmooth function and for $h \in R^n - \{0\}$, $F'(x;h) = \lim_{t\downarrow 0} \dfrac{F(x+th) - F(x)}{t}$ exists, is said that $F'(x,h)$ is the directional derivative of F.

The one fundamental distinction between the smooth (F-differentiable) and nonsmooth (B-differentiable) functions is the absence of linearity in the directional derivative for nonsmooth functions. Generalization of the classical theorems on the existence of inverse or implicit functions of locally Lipschitz continuous functions are given in Clarke (1983) and Kummer (1991).

3. NEWTON'S METHOD WITH GENERALIZED JACOBIAN

Definition 1. Let $F{:}\ R^n \to R^n$ be a locally Lipschitz continuous function an the domain D, and let $x \in D$. The generalized Jacobian $\partial F(x)$ is said to be nonsingular (as of maximal rank), if all matrices $M \in \partial F(x)$ are nonsingulars in the usual sense, i.e. of maximal rank.

With the development of generalized derivatives one strives to extend the associated geometric concepts of tangents and normals to apply to nonsmooth sets. Under certain regularity conditions, these sets form locally Lipschitz continuous manifolds. Clarke's normal and tangential cones are only partly useful for geometrical characterization of such sets and of their images under smooth functions. The frictional contact problem, of an elastic body in contact with friction on a rigid obstacle, is discretized by means of the finite element method, and we obtain the following discrete problem

$$Ku + F(u) = Q \tag{3}$$

where u is the vector of the nodal displacement, K is the stiffness matrix of the elastic body, Q is the vector of the loads and $F(u)$ is the corresponding matrix which describes the influence of friction and the contact conditions. The next results leads to a sufficient condition such that the nonlinear and non-differentiable system (3) does have one solution, obtained by Newton's method using generalized gradient.

Theorem 1. (Clarke 1983) *Let the operator F: $R^n \to R^n$ be continuous, Lipschitz and affine on the number of the open cones with peak to u_0. Then $\det M \neq 0, (\forall)\ M \in \partial F(u_0)$ is a sufficient condition for F to be globally homeomorphism from R^n to R^n.*

The generalized Newton's algorithm for solving problem (3), is

$$M^k \Delta u^k = Q - Ku^k - F(u^k),\ M^k \in K + \partial F(u^k),\ k = 0,1,\ldots$$
$$\Delta u_k = u^{k+1} - u^k \tag{4}$$

or

$$u^{k+1} = u^k + (M^k)^{-1}(Q - Ku^k - F(u^k))$$

The Theorem 1 assures a sufficient condition for the existence of the solutions of problem (3) busing the generalized Newton's algorithm (4).

4. AN EXAMPLE

Let $\Omega \subset R^2$ be an elastic body discretized by finite element method, in contact with friction on a rigid obstacle. We suppose there are *NC* nodal points on the contact boundary, to be placed last in those *N* nodal points of the partition Ω^k. The vector *F* having the form

$$[\ 0 \ldots 0;\ F^l(u_l^{\,1});\ F1(u_l^{\,2});\ \ldots\ F^l(u_l^{\,1});\ F1(u_l^{\,2})\ ]^T \tag{5}$$

Because *F* is piecewise differentiable, at this singularities, the Jacobian *M* will be replaced with the generalized Jacobian ∂F. We consider the operator *F* only at the nodal contact point "*a*" and we denote it by F^a, and the operator F^a, for the all possible situations: open contact, closed contact, forward slide and back slide, having the form:

$$F^a(u) = \begin{cases} O & ,u \in Z_1 \\ P(u+D) & ,u \in Z_2 \\ J_+(u+D), & u \in Z_3 \\ J_-(u+D), & u \in Z_4 \end{cases} \tag{6}$$

where

$$O = \begin{bmatrix} 0 & 0 \\ 0 & 0 \end{bmatrix},\ P = \begin{bmatrix} p & 0 \\ 0 & p \end{bmatrix},\ J_+ = \begin{bmatrix} 0 & \mu p \\ 0 & p \end{bmatrix},\ J_- = \begin{bmatrix} 0 & -\mu p \\ 0 & p \end{bmatrix}$$

p is penalty coefficient which has the mechanical interpretation as contact force, μ is the friction coefficient, Z_i is open cone and *D* is the initial state.

An easy verification shows that F^a is piecewise linear, therefore *F* is continuous, Lipschitz and affine on a finite number of open cones with the peak to *D*. Consequently the generalized Jacobian ∂F, is the set of the matrices $2N \times 2N$ block-diagonal.

$$\partial F(u) = co\left\{ M = \begin{bmatrix} 0 & \ldots & \ldots & 0 \\ 0 & M^1 & 0 & 0 \\ 0 & 0 & M^2 & 0 \\ \ldots & \ldots & \ldots & \ldots \\ 0 & \ldots & 0 & M^F \end{bmatrix};\ M^a \in \partial F^a(u), a = 1,\ldots,NC \right\} \tag{7}$$

It remains to find out the condition on μ in order to guarantee the assumption from Theorem 1.

$$\det M \neq 0,\ (\forall) M \in \partial(K+F)(-D) \tag{8}$$

A simple example with only one nodal contact point will be given. The stiffness will be a matrix 2×2 symmetric and positive defined

$$K = \begin{bmatrix} a_{11} & a_{12} \\ a_{21} & a_{22} \end{bmatrix}. \tag{9}$$

For simplicity we consider $D = 0$, and ∂F becomes

$$\partial F(0) = co\left\{ \begin{bmatrix} 0 & 0 \\ 0 & 0 \end{bmatrix}, \begin{bmatrix} p & 0 \\ 0 & p \end{bmatrix}, \begin{bmatrix} 0 & +\mu p \\ 0 & p \end{bmatrix}, \begin{bmatrix} 0 & -\mu p \\ 0 & p \end{bmatrix} \right\}. \tag{10}$$

For $M \in \partial F(0)$ one obtain $M = \begin{bmatrix} \gamma p & (\alpha - \beta)\mu p \\ 0 & (\alpha + \beta + \gamma)p \end{bmatrix}$, with $\alpha \geq 0$, $\beta \geq 0$, $\gamma \geq 0$ and $0 \leq \alpha + \beta + \gamma \leq 1$, therefore all matrices $M \in K + \partial F(0)$ have the form

$$M = \begin{bmatrix} a_{11} + \gamma p & a_{12} + (\alpha - \beta)\mu p \\ a_{12} & a_{22}(\alpha + \beta + \gamma)p \end{bmatrix}$$

and

$$\det M = \det K + [a_{11}(\alpha+\beta+\gamma) + a_{22}\gamma - a_{22}(\alpha-\beta)\mu]p + \gamma(\alpha+\beta+\gamma)p^2 \tag{11}$$

Hence $\gamma(\alpha + \beta + \gamma) > 0$ and

$$\det K > 0 \Rightarrow \mu < \frac{a_{11}}{|a_{12}|}.$$

Consequently for the friction coefficient with $\mu < \frac{a_{11}}{|a_{12}|}$, the contact problem (3) has one solution and it can be obtained using the generalized Newton's method.

5. CONCLUSION

For solving systems of nonlinear equations, one usually preconditions of the linearized equations is an approximate inverse of the center of the interval Jacobian.

Further research in this area of interest is to develop the solution presented in this paper on friction contact applied on gears. Other developments can be in quasistatic frictional contact problems in 3D modelling

6. REFERENCES

Clarke, F.H.(1983). *Optimization and nonsmooth analysis*, Wiley and Sons, ISBN: 0-471-87504-X, New York

Hu, Z.Q., Soh, A.K., Chen, W.J., Li, X.W. & Lin, G.(2007). Nonsmooth nonlinear equations methods for solving 3D elastoplastic frictional contact problems, Computational Mechanics, Vol.39, No.6, 849-858, ISSN 0178-7675

Kummer, B.(1991). *Lipschitzian inverse functions, directional derivatives, and applications in $C^{1,1}$-optimizatin*, Journal of Optimization Theory and Applications Vol. 70, no. 3, 561-582, ISSN:0022-3239

Pang, J.S. & Qi, L.Q.(1993). *Nonsmooth equations: motivation and algorithms*, SIAM Journal of Optimization Vol.3, No.3, 443-465

Xu, H.(2001). *Adaptive Smooting Method, Deterministically Computable Generalized Jacobians and the Newton Method*, Journal of Optimization Theory and Applications, 109, no. 1, 215-224, ISSN:0022-3239

Annals of DAAAM for 2008 & Proceedings of the 19th International DAAAM Symposium, ISSN 1726-9679
Editor B. Katalinic, Published by DAAAM International, Vienna, Austria 2008
Make Harmony Between Technology and Nature, and Your Mind will Fly Free as a Bird

INTEGRATED CONTROLLING – A NEW PERSPECTIVE OF MANAGING A COMPANY

POP, N[icolae] A[lexandru]; PELAU, C[orina]; GHINEA, V[alentina] M[ihaela]; BENA, I[rina] & SANDRU, D[iana]

Abstract: *Only those companies that have the ability to adapt to the permanent changing conditions of our political, economical, social and technological environment, can survive and enjoy long term success. By permanently monitoring the company and its environment, as well as the knowledge on the performance of the company, controlling can improve this adaptability of the company and contribute by this to the long-term success of the company.*

This article concentrates on the analysis of the integrated controlling function and the controlling instruments of a company. It is not sufficient to know the sole instruments, but it is also required to integrate these instruments in the controlling system and thus improve the entire activity of the company and its performance.
Key words: *controlling, performance measurement, efficiency, efficacy*

1. INTRODUCTION

In a permanent changing environment, characterized by the well-known globalization phenomenon, the intensification of the competition process and fast developing technologies, every little detail can make the difference. It is not anymore about standardization, but only about emphasizing each particularity of the process, product or human capital of the company. It can be easily observed that only those companies which have the ability to adapt to these permanent changing conditions by implementing a permanent controlling system of its performances and results, can survive. Thus, a similarity can be observed between the business environment and nature. So, in the business world, as in nature, only those companies which have the capability to adapt to the business environment can survive and enjoy long term success. This statement relies on the theory „survival of the fittest“ of the socio-Darwinist philosopher Herbert Spencer. Under these conditions, controlling should be seen as a „helping hand“ of the management which contributes to the development of this adaptation capability of a company to the environment. So, the controlling department should set the informational basis for the management decisions and warn it in case of negative company developments.

Several decades ago, it was unheard of for a company to undertake contingency planning, critical incident analysis or implement a monitorization system of its performance. Nowadays, these activities have become a necessity, for companies focus their attention more and more on activities that minimize risk, avoid liabilities and ensure employees' safety and the company success.

2. INTEGRATED CONTROLLING

One of the main objectives of this paper is to define the concept and to make the delimitation among other disciplines and company departments. Ehrmann states the fact that there is no unanimously accepted definition of the concept controlling (Ehrmann 2004). Weber and Schäffer agree with him by stating that there are many definitions of controlling. Pemöller strengthens this idea by affirming that the term controlling is often used with different meaning in different situations (Peemöller 1992). Preißler affirms that even managers are not certain of what controlling is (Preißler 1995). Controlling is often compared with the simple "control", but many specialists agree that it is more than a simple control of the company activity (Ehrmann 2004, Peemöller 1992, Hungenberg & Wulf 2004). By analyzing the definitions and functions attributed to controlling a similarity can be observed regarding the management activity of the company. Thus Peemöller defines controlling as the management function which coordinates the activities of planning, control and collection of information in a company (Peemöller 1992). According to Ehrmann the main functions of controlling are planning, collecting and providing information, analysis, control and coordination of a company (Ehrmann 2004). Horvath combines these elements and defines controlling as a subsystem of the company management which systematically combines the planning and control of the company as well as the information system, and sustains by this, the adaptation and coordination of the whole system (Zerres 2000). On the other hand, if we analyze the definition of management, it refers to the target oriented planning, organization, leading and control of a company. By comparing these definitions one can observe that both management and controlling contain the activities of planning, control, and coordination. The conceptual question which has to be asked is where the limit between these two concepts is. Confusion exists regarding the financial, accounting orientation of controlling. American specialists perceive controlling as an activity of the financial department of a company (Zerres 2000). Starting from all these different definitions and functionalities of the controlling, there is a need to confer a certain meaning to controlling by making a delimitation of its objectives, function, roles, instruments and obligations in the company activity. It is also important to find the limit between controlling and the other departments of a company and to establish its clear position in the company organizational chart.

As presented in the existing literature in the field most of the instruments measure the performance of the activity of the company (Zerres 2000). So it can be stated that one of the functions of controlling is the performance measurement. There are presented several controlling instruments which measure the efficiency, efficacy and performance of the company's activity by which the adaptation capability of the company can be ensured. These elements are the internal processes, the customers, the human resources and the financial activity of the company, as mentioned in the balanced score card (fig. 1).

First of all, there are several instruments that measure the efficiency, profitability and also productivity of the internal process. An essential element for the performance measurement of the internal processes is the reduction of costs, which can lead to a better profitability of the company. In the marketing department of a company, there are controlling instruments which evaluate the company's customers and the four policies of the marketing mix. Each company has to analyze its customers in order to be able to invest in the "good customers" and reduce the costs with the "not that good customers" (Pop, Pelau, 2005, pg. 43-52). Besides this, for each component of the marketing mix there are instruments of efficiency and effectiveness evaluation. Another important department that uses controlling instruments is the human resources department. For an efficient and effective running of a company's activity, it is important to evaluate the performance and structure regarding the human resources both from a quantitative and qualitative point of view. More than that, an analysis of their degree of innovation provides a clear image about the company's development capability and its performance (Florescu, Mâlcomete, Pop, 2003). Last but not least, the financial department has also the duty to monitor and analyze all financial processes. This is done by using indicators like cash-flow, return-on-investment, etc.

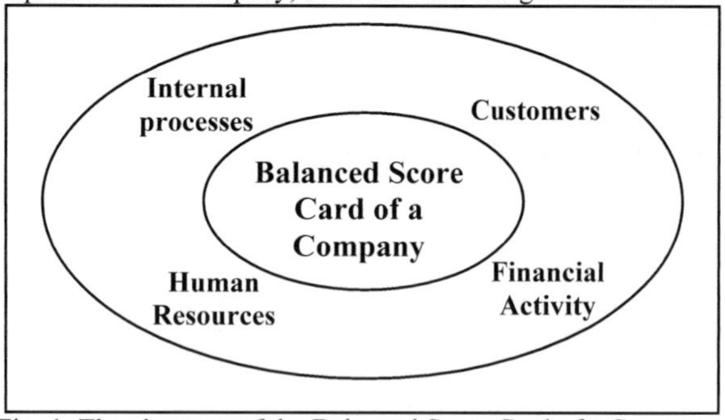

Fig. 1. The elements of the Balanced Score Card of a Company

In order to determine and analyze the existence and implementation of controlling in companies, a pilot study has been conducted in 23 Romanian companies from different fields. The research has consisted in interviews with the managers or controllers of the companies, and has had as main objective to determine if companies in Romania have a controlling department, if they use controlling or performance measurement instruments and which indicators do they use.

The analyses of the results have lead to the formulation of two hypotheses which should be proved within an extended research. The first hypothesis is that many companies use performance measurement instruments, even though they do not have an integrated controlling system. This is especially valid for small and medium companies. The second hypothesis is that there is a relation between the existence of a controlling system or the use of controlling instruments and the performance of the company. So the existence of an integrated controlling system influences in a positive way the performance of the company.

The limitations of this research consist in the subjective view and the limited information provided by the respondents and the quality of the application of a controlling system. According to the position of the company representatives, namely the interview respondents, the results might be influenced by his/her view and information of the controlling department or system. Besides this, the influence of the existence of a controlling department on the company performance depends on the quality of implementation and application. Regarding this issue, it is important to analyze the complexity of the controlling system and the quality of the communication between the controller and the manager. Even if a company has a complex controlling system, these reports can not have an influence on the company performance if the manager does not take the reports of the controller in consideration.

In order for a company to be able to make forecasts about future developments, it needs, first of all, an evaluation of the actual internal and external situation. Checking current performance against pre-established plans, with a view to ensure adequate progress and satisfactory performance has become compulsory for a company. It is not only about the performance measurement, but also about the correction of performance in order to ensure that the company's objectives are accomplished. Besides this, it has become compulsory to integrate all control activities into one well structured system that bridges gaps between departments.

3. CONCLUSION

In fast developing markets characterized by the intensification of competition and increased comparability between products, a company can gain a long term success by its capacity to adapt itself to the changing conditions of the markets. More "organic" forms of companies allow organizations to be more responsive and adaptable in today's rapidly changing world. These forms also cultivate empowerment among employees, much more than the hierarchical, rigidly structured organizations of the past, but they also provide a strong management control known as "controlling". In order to reach the ultimate company's goal, the profit maximization, this kind of ongoing communication between the internal processes, the marketing department, the human resources and also the financial activities tends to become one of the essence of the management science.

By permanent monitoring of the company and its environment as well as the knowledge about performance of the company, controlling can improve this adaptability of the company and contribute by this to the long-term success of the company.

4. REFERENCES

Baier, P. (1994). *Führen mit Controlling – So aktivieren Sie das Leistungspotential Ihres Unternehmens (Leading with controlling – That is how you activate the performance potential of your company)*, Walhalla, Berlin.

Ehrmann, H. (2004). *Marketing-Controlling*, Kiehl, Modernes Marketing für Studium und Praxis, 4th edition, Ludwigshafen.

Florescu, C.; Malcomete, P. & Pop, N. Al.(2003). *Marketing*, Dictionar explicativ, Editura economica, Bucuresti.

Geis, G.; Reinecke, S. & Tomczak, T. (2001). *Handbuch Marketingcontrolling – Marketing als Motor von Wachstum und Erfolg (Marketingcontrolling Manual – Marketing as a Engine for Growth)*, Wirtschaftsverlag Ueberreuter, Frankfurt/ Wien.

Hungenberg, H. & Wulf, T. (2004) *Grundlagen der Unternehmensführung (Basics of Company Leading)* , Springer, Berlin.

Peemöller, V. (1992) *Controlling – Grundlagen und Einsatzgebiete (Controlling – Basics and Application Fields)*, Neue Wirtschaftsbriefe, 2nd edition, Herne/ Berlin.

Pop, N. Al. & Pelau, C. (2005). *Elemente de Controlling în Marketing – Metode de evaluare a clienţilor (Controlling Elements in Marketing – Methods of evaluating customers)*, in: Revista de Management şi Inginerie Economică, Vol. 4, Nr. 3, pg. 43-52.

Preißler, P. R.(1995) *Controlling – Lehrbuch und Intensivkurs (Controlling – Manual and intensive course))*, Oldenburg, München.

Zentes, J. & Swoboda, B. (2001). *Grundbegriffe des Marketing – Marktorientiertes globales Management-Wissen (Basics of Marketing – Global Market Oriented Management Know-How)*, Schäffer-Poeschel, 5th edition, Stuttgart.

Zerres, M. & al. (2000). *Handbuch Marketing-Controlling (Manual Marketing-Controlling)*, Springer, Berlin.

Annals of DAAAM for 2008 & Proceedings of the 19th International DAAAM Symposium, ISSN 1726-9679
Editor B. Katalinic, Published by DAAAM International, Vienna, Austria 2008
Make Harmony Between Technology and Nature, and Your Mind will Fly Free as a Bird

AN APPLICATION OF NON-COMPRESSIBLE FLOWING IN THERMOPLASTIC PIPES IN INSTALLATIONS FOR FIRE PREVENTION

POP, P[etru]; LAZAR, L[iviu]; MARCU, F[lorin]; VERES, M[ircea] & UNGUR, P[etru]

Abstract: *This paper has presented a simple application of non-compressible fluids flowing in thermoplastic pipes from treated polyamide, thermocontractible used, in special for inner installations with cold water for fire prevention. These installations of cold water have a sprinkler, which is a special casing got in inner hydraulic circuit, that can automatic open at a predictable temperature get by fire, due to water pulverize under fired place. The paper has goal to substitute the complicated sprinklers with pipes sections from thermoplastic or thermocompressible material, and hydraulic resistance.*
Key words: *incompressible, thermocontractible, extinguishing*

1. INTRODUCTION

For fire prevention in special places, such as: medical stocks depot, warehouse, etc. as active medium is used cold water [Dubbel, 1988]. The cold water has capacity, under jet form or pulverization after warm bodies in burn, of heat absorption and evaporation, followed by cooling of bodies due to fire extinguishing. In general, the feed inner installations with cold water are equipped with special devices as sprinklers, for initiation and distribution of water for fire extinguisher. The sprinklers got in pipe section have fed with water, being special casings, which are automatic open at a predictable temperature of fire due to pulverization of water on burned place.

The constructions of sprinklers are complicate and expensive. A new, simple and efficient construction of hydrodynamic resistance this paper has presented. This construction has at base of polyamide property treated with gamma ray to be thermocontractible [Hutte, 1995]. Under pipe form, at flame temperature produced by fire the polyamide has contracted formed a necking, which under impulse force, pressure difference from pipe zone is broken assuring water circuit opening, which due to fire extinguishing.

2. THEORETICAL ASPECTS

Today, at fire extinguishing technique in water circuit have used sprinklers, which have to open themselves at predictable temperature by:

- Melting of low-fusible alloy, which closed sprinkler orifice [Carp&Ungur,2003;Dubbel,1988;Hutte,1995];
- Melting of low-fusible chemical composition, which supports of holding valve support [Ghimus,1984];
- Broken of glass bulb filled with a liquid, which has expanded at a higher temperature [Covrig,1996].

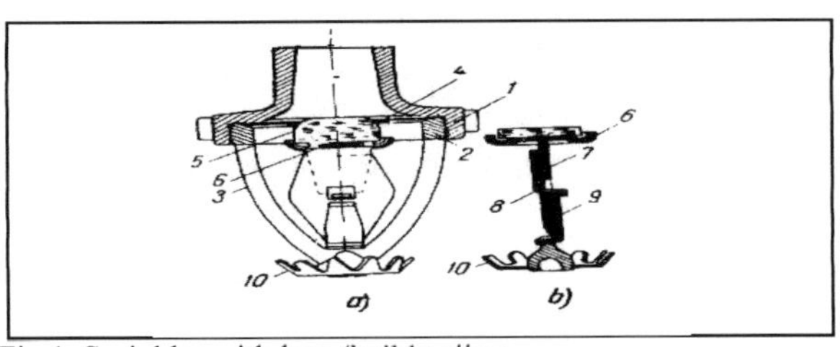

Fig.1. Sprinkler with low-fusible alloy.

The most used sprinkler is with low-fusible alloy (Fig.1-a: ensemble, Fig1-b: fusible support device). Its construction is complicated, has formed by: 1-sprinkler body, 2-bronze ring, 3-supported frame, 4-copper diaphragm, 5-valve, 6-shutting, 7,8,9-small plates of fusible alloy, 10-rosette. During fire braking out, shutting material is melting due to spurt open, on which the water get out under the pressure acting rosette and water pulverization is near uniform on desired surface.

A new, cheap and efficient device, which can change the sprinkle this paper has presented. At the base of new device construction is stand the continuity equation of Bernoulli for non-compressible fluids, impulse force in nozzles, and some thermoplastics and thermocontractible material's properties under pipes form narrow the section in heated portion, due to diminished hydrodynamic strength of pipe and its broken, and opening the water circuit which fire extinguishing [Dubbel,1988;Hutte,1995]. At water flowing through circular pipes with different sections, have different medium flow speeds, flow rates and pressures, the variations being function of section's size.

The main equations, which described the physical phenomena, have presented in following:

- *The continuity equation* [Hutte,1995]:

$$\dot{m} = \rho \cdot \dot{v} = \rho_1 v_1 A_1 = \rho_2 v_2 A_2 = \rho_0 v_0 A_0 \quad (1)$$

, or:

$$\dot{v} = \frac{\dot{m}}{\rho} = A_0 v_0 = A_1 v_1 = A_2 v_2 \quad (2)$$

, where: $\dot{v}$-is volumic flow rates, $\dot{m}$-masic flow rates, $\rho_1\rho_2\rho_0$-fluids densities, that for non-compressible mediums are constant ($\rho_1=\rho_2=\rho_0$), and $v_1v_2v_0$-fluid speeds within sections-A_1,A_2,A_0 of a narrow stream tube.

- *Bernoulli equation* [Hutte,1995] *for stationary flow* is:

$$\frac{v^2}{2} + \frac{p}{\rho} + g \cdot z = ct. \quad (3)$$

From continuity and Bernoulli's equations for flowing in narrow tubes have determined the medium flow speeds and flow rates in circular pipes. So, on (1-0)-way from Fig.2, it has:

$$\frac{v_1^2}{2} + \frac{p_1}{\rho} = \frac{v_0^2}{2} + \frac{p_0}{\rho} \quad (4)$$

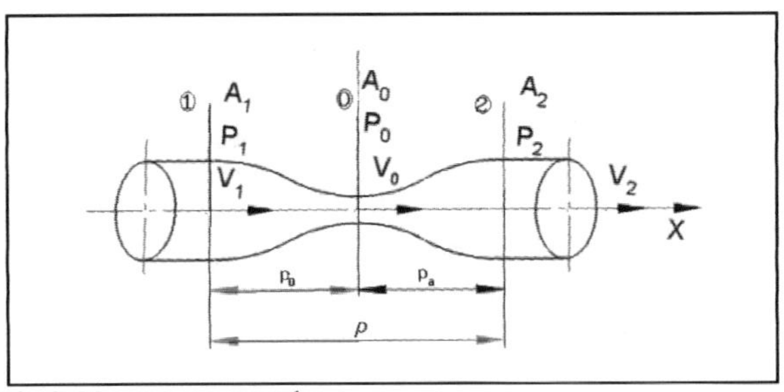

Fig.2. Narrow stream tube.

, which can obtain the speed-v_0 from section-0:

$$v_0 = \frac{1}{\sqrt{1-A_0/A_1}} \cdot \sqrt{\frac{2}{\rho}(p_1 - p_0)} = \alpha \cdot \sqrt{\frac{2}{\rho}(p_1 - p_0)} \qquad (5)$$

, where the constant-α depends only of surface's ratio-(A_0/A_1).

At section reducing from (1) to (0), the water speed-v_0 is raising significant, so and water flow rates due to broken of narrow thinner walls of pipe from themocontractible polyamide with small section-A_0. At suddenly decrease of section has produced the contraction of stream fluid.

- *The acting force of narrow water jet*:

From continuity and Bernoulli's equations of flow on (0-2)-way from Fig.2, for flow results:

$$p_0 - p_a = \frac{1}{2}\rho(v_2^2 - v_0^2) = \frac{1}{2}\rho v_2^2\left(1 - \frac{A_2^2}{A_0^2}\right) \qquad (6)$$

From which the action force of waterjet is:

$$F_A = -\frac{1}{2}\rho v_0^2 A_0\left(\frac{A_0}{A_2} - 1\right)^2 \qquad (7)$$

The force-F_A is orientated in negative direction of X-axis as $A_2 > A_0$, action on (0-2)-way to narrow tub's walls from thermoplastic and thermocontractible material going on tub's wall broken and opening the water circuit for fire extinguishing.

3. NEW DEVICE FOR FIRE PREVENTION

The sprinkler substitution with pipes from thermocontractible polyamide is easy and efficient, having the advantage of simple construction and easy fabrication [Argeseanu, 1999; Ghimus, 1984; Hornn, 1988].

In these installations, the thermocontractable pipe has the role of thermodynamic resistance, with a diameter-d_0 and thickness-a_o, the length-l_o is pre-established. At breaking of fire, as hot, the pipe has contracted altering the hydrodynamic flow parameters of water. Therefore, the narrow pipe zone has subjected of great speeds of water, an impulse force and a difference of pressure due to broken of wall's pipe and relieved the water from installation circuit of fire extinguishing. In Fig.3 has presented all phases process of thermocontractible pipe narrow and broken at hot temperature. The destroy phases of thermocontractable tub are: a-hydrostatic balance, b-narrow tub under adds heat, c-broken tub (hydrostatic unbalance).

The fast change of hydrostatic balance at thermocontractable pipe deformation due to gain of its hydrodynamic strength and its deterioration. At damaged tub under add heat, the hydrodynamic resistance from broken zone has narrowed, due to a hydraulic shock of tub.

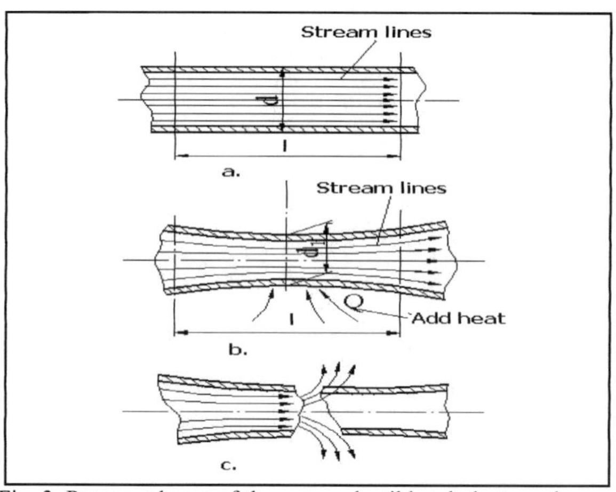

Fig. 3. Process phases of thermoconductible tub destroyed.

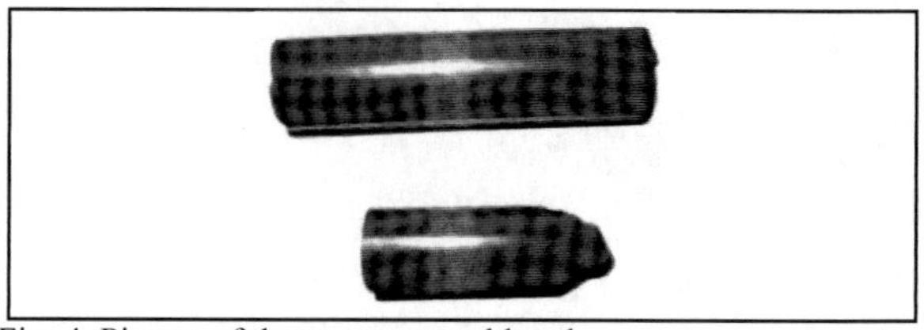
Fig. 4. Picture of thermocontractable tub.

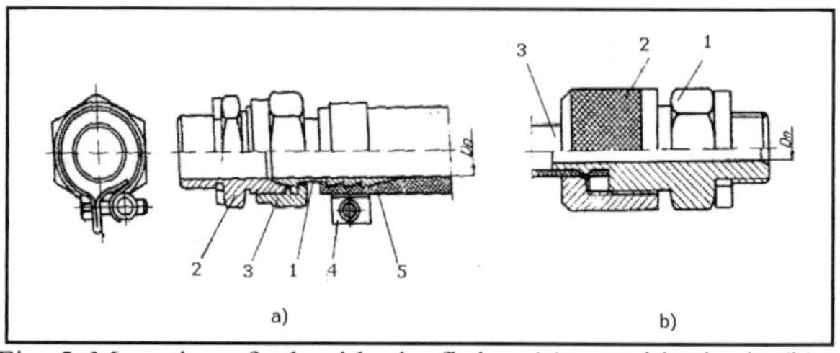

Fig. 5. Mounting of tub with pipefitting-(a), or with nipple-(b).

This phenomena is possible as a difference of pressure and contraction of stream fluid due to broken of pipe and relieved the waterjet, which fire extinguished [Dubbel,1988; Hutte, 1995].

In Fig.4 has depicted a partition from thermocontractable pipe, used as hydraulic resistance in installation of fire prevention.

In present, these tubs have used in decoration and anti-corrosive actions, fixed on pipes, rods, shank by contraction and heat contribution [Rufe, 2002].

In inner hydraulic circuit of prevention and extinguishing fire can be mounted in parallel the pipes by screw, joining type DIN 3930, or by joining with closed muffle or tapped muffle, metallic tapped snap rings, or stick by adhesives (fig.5). Where: (Fig.5-a) 1-is pipefitting, 2-junction, 3-nut, 4-snap ring, 5-thermoplatics tub, and (Fig.5-b) 1-nipple, 2-nut, 3-thermoplastics tub.

4. CONCLUSIONS

The solution of thermoplastic and thermoconductible resistance for substitution of sprinklers in installations of fire prevention and extinguishing from depots assures great economy and efficiency.

The construction of limitative hydraulic resistances from thermoplastic and thermoconductible tubs is simple, robust, a lightweight and low cost, with large perpective using in fire installations.

5. REFERENCES

Argesanu, V. et al. (1999), *Elements of Mechanical Engineering,* pp. 37-42, 173-193, Eurostampa Editor, Timisoara

Carp, V. & Ungur, P. (2003), *Study of Materials*, pp. 70-82, Didactical and Pedagogical Editor, Bucharest

Covrig, M. (1996), *Technology of Non-metallic Materials*, pp. 112-144, Lux Libris Editor, Brasov

Ghimus, D. & Ivanoff, M. (1984), Plastically Materials in Installations", Technical Editor, Bucharest

Hornn, S., et al. (1988), V*ade-mecum of Plastical Materials*, Tehnical Editor, Bucharest

Rufe, P.D. (2002), *Fundamentals of Manufacturing*, SME Editor, Dearborn, Michigan, USA

***Dubbel, (1988), *Handbook of Mechanical Enginnering, Fundamentals*, pp. D74-D99, G23-G24, N1-N38, Tehnical Editor, Bucharest

***Hutte, (1995), *Handbook of Mechanical Enginnering, Fundamentals*, 29th Edition, pp. B80-B84, D26-D27, E123-E125, Tehnical Editor, Bucharest.

Annals of DAAAM for 2008 & Proceedings of the 19th International DAAAM Symposium, ISSN 1726-9679
Editor B. Katalinic, Published by DAAAM International, Vienna, Austria 2008
Make Harmony Between Technology and Nature, and Your Mind will Fly Free as a Bird

HEATING INSTALLATION OF SEWAGE WATER WITH SOLAR CAPTOR

POP, P[etru]; VERES, M[ircea]; IANCU, C[armen]; MIHAILA, S[tefan] & UNGUR, P[etru]

Abstract: *The paper has presented a simple heating installation of sewage water with solar captor and boiler. Today, as rising combustible cost and electrical energy is looking for the alternative energy source, such as solar energy. The water has pre-heated through special pipes of solar captor and climbed up in an expansion vessel, from which is taken and going in inelar serpentine pipe into recuperate- storage boiler. Here the sewage water is heated and served out at domestic consumption points. The novel installation consists in a thermal resistance with porous core get in main closed circuit, substituted the boiler's serpentines, and heat storage endowed with tap, floating and separated rooms of cold and warm water. The installation is simple and easy to build up.*
Key words: *boiler, sewage water, serpentine, solar captor*

1. INTODUCTION

The installations of sewage water heating with boilers mounted in parallel have used in apartments, buildings and houses. The boilers have mounted so in primary thermal circuit, which circulated through boiler serpentine pipes and each on consumption water circuit (Ungur et al, 2008).

The circulation of primary agent by inner boiler serpentine's pipes has done by gravitation due to thermal pressure or with pumps. At circulation by gravitation, the boilers have mounted at predicted height, which at pumps this condition it's not necessary. The cold water has taken through a pipe and received heat from primary thermal agent by exchange surface of serpentine. The warm water of consumption, heated at 60-65^{0}C is served out at tap by a connection pipe. In our case, the heat flow has got by a solar captor and transmitted at water storage in boiler, which is working in storage-recuperate state (Chardonnet, 1973; Maghiar, 1995).

The heating installation of domestic water with thermal solar captor and storage-recuperate boiler is simple and easy to build up, great efficiency and large perspective for future.

2. THEORETICAL ASPECTS

In hydraulic circuit by thermal pressure is getting in recuperate-storage boiler, which is standard. The normal capacities of boilers are of 5, 10, 30, 50 and 100 l. The required volume for a boiler is getting by the relation (Ungur et al, 2008):

$$V = \frac{Q_{a\,max}}{c\cdot\rho(\theta_c-\theta_a)} \quad (1)$$

Where: V-is necessary volume [m^3];

$Q_{a\ max}$-total maxim quantity of transmitted heat by primary thermal agent of storage water in boiler on a consumption time [J] or [kcal];

c- specific heat of warm water for consumption [J/kg·K] or [kcal/kgf·^{0}C];

ρ- density of warm water for consumption [kg/m^3] or [daN/m^3];

θ_a- temperature of cold water for consumption [^{0}C];

θ_c- temperature of warm water for consumption [^{0}C].

If it has taken the temperature of cold water of θ_a=10^{0}C and temperature of warm water for consumption of θ_c=35-55^{0}C, for calculus of parameters-(ρ) and (c) has taken medium warm water for consumption:

$$\theta_m = \frac{\theta_a + \theta_c}{2} = \frac{10+55}{2} = 32.5°C$$

The determination of $Q_{a\ max}$ transmission of storage water in boiler has calculated with relation:

$$Q_{a\,max} = Q_s - Q_c \quad (2)$$

Where: Q_s-is heat loss of primary thermal agent (warm water by thermal captor) the water from inner installation-Q_c and heat flow loss of water from boiler-Q_a.

$$Q_s = Q_c + Q_a \quad (3)$$

The heat flow required for preparing of warm water is getting by relation:

$$Q_c = G_0 \cdot c(Q_c - Q_a) \quad (4)$$

Where: G_0-is hour's flow rate of warm water [l/h].

$$G_0 = \frac{G\cdot N\cdot p}{100} \quad (5)$$

Where: G-is daily consumption of warm water at 55-60^{0}C [l/day/person];

N-number of people from house;

p-variation percentage of water consumption.

For a consumption of normal water-G_0 of 10, 20, 30, 50 l/day, for a small kitchen is enough using as warm source a thermal solar captor of 1-2m^2. For specific weight of warm water for consumption γ=1000daN/m^3, at a volume of warm water of 50 l/day, is enough a heat quantity-$Q_{a\ max}$ of 3000 kcal/h, which can be assured by a thermal solar captor (Ardelean, 1978; Bute, 1987; Maghiar, 1995).

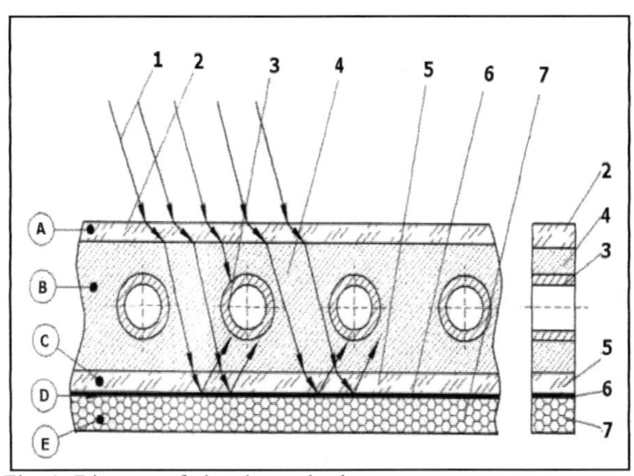

Fig. 1. Diagram of plan thermal solar captor

The scheme of thermal plan solar captor has presented in Fig. 1, where: 1-is light ray, 2,5-glass or Plexiglas, 3-capilar pipes from copper, 4-coloidal graphite, 6-aluminium foil or plate with reflexive paint, 7-plate from expandable polystyrene.

At construction of plan thermal solar stand the phenomena applications (Fig.1) of optics: A- reflection light, B-refraction and absorption light, C-reflection and refraction light, D-total reflection, E-thermo-isolate medium.

The working principal of plan solar captor consists in used of black body effect as absorbent of thermal radiation, conjugated with glasshouse effect of vacuum effect [Maghiar, 1995]. Between glass's plates of (2)-(5) from Fig.2, has exhibited intense glasshouse phenomena.

3. HEATING INSTALLATION OF SEWAGE WATER

The heating installation of sewage water with thermal solar captor and storage-recuperate boiler, with 50 l capacity, has composed from a thermal solar captor- A, and a heat storage battery –B presented in Fig.2. The thermal solar captor-A is getting in closed water circuit-1 equipped with taps, an expansion vessel-2 and a thermal resistor-4 (coil from cupper, or thermal register with porous core).

The primary circuit of agent has assured by thermal pressure. The heat storage battery-B is a simple water boiler, divided from a wall-5 in two partitions-c of cold water, and –d of warm water, being equipped with a valve and floating-3, the warm water from partition-d being heat by thermal register-4. The open storage battery-B is located up, for primary warm water permitted to rise up by thermal pressure and specific weight decreased. It is following a cold heated water circulation in adaptable boiler-B through the pipes-6 to tap, which assured the sewage warm water [Moraru&Popescu, 1978].

The novel characteristic elements consist in:

- heat storage battery-B, which is portioned and equipped with floating;
- thermal register-RT with porous core, located in storage battery, from which relieved heat from primary thermal agent at storage-recuperate boiler.

This porous thermal register substituted spirals from copper pipes used before, assured an improvement of thermal efficiency.

The thermal register is a device that realizes the heat transfer between fluid and environment, from which exist a temperature gradient. The function of thermal register with thermal resistance into a heat installation of warm water, without phase change with closed circuit is continuing.

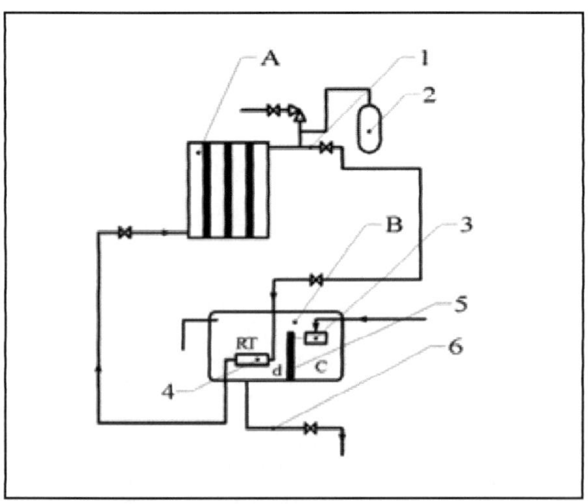

Fig. 2. Diagram of heating installation for sewage water, with thermal solar captor and adaptable boiler

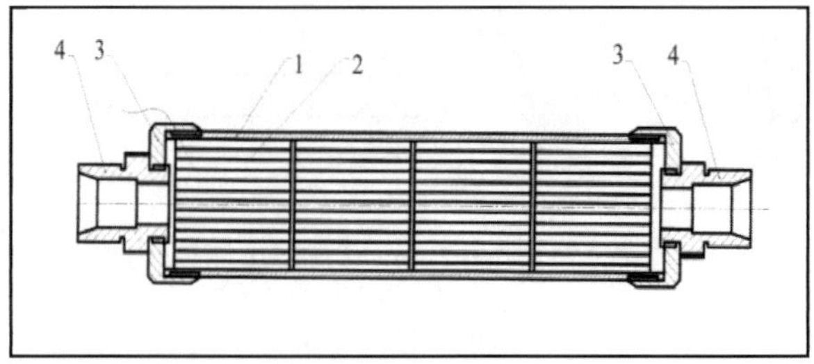

Fig. 3. Reinforced thermal register

In Fig.3 has presented a reinforced thermal register, which has composed from the following elements: 1 – is metallic pipe, 2 - porous ceramics parts, 3 - covers, 4 - couplings.

At thermal register, the heat transfer is indirect by a change surface (the wall of thermal register goes between fluid and environment).

These registers function for normal temperatures (50-150^{0}C), and water passing by installation with a low pressure. For realized a maxim heat transfer from environment to thermal register of steel or aluminum pipe, is necessary to grow up the transport section and hydraulic pressure, with decrease of water volume from installation [Nitu, 1982; Tomescu, 1975].

These requirements can be obtain by using of porous cores with adequate thermal conductivity and optimal specific heat, located within thermal register and get in small electro-pump of water, which assured a difference of pressure required fluid circulation through a heat installation with closed circuit [Danescu, 1980; Ungur, 2008].

For the future these installations with thermal solar captor and adaptable boiler will be used on large in heating installations of domestic water, because of them ecologic action and low cost of energy.

4. CONCLUSION

The installation presented in this paper is simple, ecological and easy to build up.

The installation can used any type of thermal solar captor, with large perspective in domestic houses and warmhouses.

The thermal solar captor is safety in working, can be assembly in any position, being ecologic.

Thermal register of heat exchanger is modern, equipped with porous core, having good thermal characteristics, and higher efficiency.

5. REFERENCES

Ardelean, I. & Folescu, G. (1978), *Solar Captor*, Scientifically and Encyclopaedically Editor, Bucharest

Bute, A. (1987), *General Energetic and Energy Conversion*, Traian Vuia Polytechnic Institute

Chardonnet, I. (1973), *Sources of Energy*, Sirsy Editor, Paris

Danescu, A., et al. (1980), *Using of Solar Energy*, Technical Editor, Bucharest

Maghiar, T. (1995), *New Source of Energy*, University of Oradea, Keysys Editor, Oradea

Moraru, D. & Popescu, C. (1978), *Solar Generators*, Scientifically and Encyclopaedically Editor, Bucharest

Nitu, V. et al. (1982), *General Energetic and Energy Conversion*, Didactical and Pedagogical Editor, Bucharest

Tomescu, F.M. (1975), *Energy Conversion and Sources*, Polytechnic Institute Bucharest

Ungur, P.; Pop, P.A.; Veres, M.; Iancu, C.; Mircea Gordan, M. & Craciun, D. (2008), Thermal Registers With Core For Heated Installations With Geothermal And Heat Water, *12th International Research/Expert Conference "Trends in the Development of Machinery and Associated Technology"* TMT 2008, Istanbul, Turkey, 26-30 August, 2008.

Annals of DAAAM for 2008 & Proceedings of the 19th International DAAAM Symposium, ISSN 1726-9679
Editor B. Katalinic, Published by DAAAM International, Vienna, Austria 2008
Make Harmony Between Technology and Nature, and Your Mind will Fly Free as a Bird

INTERNET BUSINESS REPORTING A CHALLENGE: THE CASE OF ROMANIAN MANUFACTURING ENTERPRISES

POPA, A[dina]; BLIDISEL, R[odica] & FARCANE, N[icoleta]

Abstract: *This paper contributes to the discussion of the present status of business reporting of the manufacturing enterprises from Romania. The internet represents a technological opportunity even in the field of business reporting. It facilitates the communication between managers and stakeholders and the decision making process. The paper posits some interesting views and our findings as regards the extent of internet business reporting of the manufacturing enterprises listed on Bucharest Stock Exchange (BSE).*

Key words: *business reporting, decision making, manufacturing*

1. INTRODUCTION

Due to the fact that manufacturing still represents an important part of the Romanian economy the objective of our paper is to measure the level of internet business reporting of Romanian enterprises from manufacturing sector listed on BSE and to test some variables that may influence the internet disclosure level.

After a relative deindustrialization period which Romania had passed during the '90, starting with the year 2000 it could be observed a tendency change and a gradual growth of the industry's role, sustained at the European level through the 2005 Lisbon Strategy. This tendency could be observed also within the number of manufacturing enterprises listed on BSE. In 2007 more than 70% of the BSE transacted companies are from manufacturing sector.

Communication by corporations of financial and other corporate performance data to stakeholders is a complex process of risk and information asymmetry reduction. The efficacy of this process is dependent upon the needs of the stakeholders and of the corporation management interests (Debreceny et al., 2001). Corporate disclosure is critical for the functioning of an efficient capital market (Healy & Palepu, 2001). The last decade of the 20th century brings an important scenery change in business communication: companies are attracted in their image's marketing game and trained to disseminate theirs information to the stakeholders through an environment which revolutionized the global communication and information technology – the internet.

International business environment and global capital markets impose new trends in corporate reporting; also Romanian companies that for the moment practice mostly mandatory financial reporting seem to be interested about voluntary internet reporting.

2. THEORETICAL FRAMEWORK AND METHODOLOGY

2.1 Literature review

A wide academic research exists in the field of the internet business reporting. In the early stages researchers had studied the internet reporting at country level, generally, using a descriptive or reflective research (Lymer, 1998; Deller *et al.*, 1999). Later by using different economic theories the study of internet reporting was oriented to find and explain the determinants that influence the internet business disclosure (Ettredge *et al.*, 2002; Debreceny *et al.*, 2002; Xiao *et al.*, 2004; Popa, 2008).

2.2 Methodology

Based on a quantitative content analysis the goal of our study is to find out if Romanian manufacturing companies provide the information contained in the business reports through an alternative media, the internet and not only by traditional hard copy. Because using internet for disclosing financial and non financial information facilitate stakeholders' access to information and decrease the costs with printing and disseminating the hard copy of companies' reports through the interested investors.

We investigate all manufacturing companies listed on BSE. Our sample consists of 61 companies listed on BSE at the end of year 2007 that are actively traded. The main reason to choose all the listed manufacturing companies is to explore the readiness of those companies to inform investors by providing information on their websites and to appreciate the conformity with the regulations of Romanian National Securities Commission (RNSC) regarding transparency of listed companies. This is due to the fact that RNSC and BSE are the Romanian governing bodies that encourage companies to disclose financial information on their website.

Our research instrument is based on the CIFAR study (CIFAR, 1995) and on the national and European regulations in the field of business reporting. As a research method we use content analyze for study the homepages of the sample companies developed from Weber model (Weber, 1985) and the hypothesis development.

After identifying the homepages of our sample companies we determine for each company the mean score for measuring internet business reporting, based on a list of variables predefined in the first stages of our study. To measure the disclosure level quantitatively in the present study, a disclosure index for internet business reporting (IBR) was developed:

$$ScorIBR = \sum_{i=1}^{47} IBR_i \qquad (1)$$

We build a composite index for each company by assigning value of "one" to every item disclosed on the company website and value "zero" otherwise. All online financial reporting items used in this study were considered equally important. The last stage in our framework was to develop a checklist instrument. By using statistic parametric test ANOVA we verify if the listed status and the ownership structure are influencing factors of the internet business reporting score (IBR). We develop two hypotheses:

H1: Company listing tier determines the level of information disclosed.

H2: Companies with more foreign shareholders disclose more information.

3. RESEARCH RESULTS

As we define, the objective of our study is the analysis of the level of business information currently provided on internet by Romanian manufacturing companies listed on BSE and to find out in an empirical way the factors that could influence the internet disclosure level. The lists of variables we look for on the companies' websites were divided in four groups: general information, business reporting fundamental items, information for shareholders and other users and other information.
The descriptive statistic for internet business reporting score is presented in table 1. ScorIBR that shows the level of internet reporting for Romanian manufacturing sector is quit low.

Companies	61
IBR Mean- *Scor IBR*	**12.23**
IBR Standard deviation	11.145
IBR Maximum	37

Tab. 1. Internet business reporting (IBR) statistics

Testing the first hypothesis H1, if internet business reporting score is influenced by the listing status we obtain the results presented in the table 2. The p-value in the ANOVA test (0.003) shows that hypothesis can be accepted. The information provided online by the companies varies by listing status.

	First tier companies		Second tier companies	Unlisted companies	
No. companies	6		29	26	
IBR Mean	21.67		16.62	5.15	
IBR Maximum	31		37	21	
IBR Std. dev.	8.571		11.785	5.662	
ANOVA					
	Sum of Squares	df	Mean Square	F	Sig.
Between Groups	16962	25	0. 678	2800	0.003
Within Groups	8481	35	0.242		
Total	25443	60			

Tab. 2. IBR by listing status and ANOVA results

Statistics and results of testing the second hypothesis H2, are presented in table 3. The enterprise ownership structure influences the level of information provided.

	Majority of foreign shareholders			Majority of domestic shareholders	
No. companies	15			46	
IBR Mean	22.60			8.85	
IBR Std. dev.	12.135			8.501	
IBR Maximum	37			31	
ANOVA					
	Sum of Squares	df	Mean Square	F	Sig.
Between Groups	7654	25	0.306	2930	0.002
Within Groups	3657	35	0.104		
Total	11311	60			

Tab. 3. IBR by ownership structure and ANOVA results

The p-value in the ANOVA (0.002) test shows that hypothesis can be accepted. The information provided on internet by the manufacturing companies is influenced by the majority of foreign shareholders.

4. CONCLUSIONS

The internet like a disclosure media has proved its potential for business reporting. Companies listed on BSE provide their financial statements and other additional reports by using online disclosure. The level of internet reporting obtained by us for Romanian manufacturing sector is low, but it can be appreciate as a good one, taking in consideration that BSE is a relatively new and small stock exchange and the Romanian business environment can not be compared, yet with the western developed economies. According to the results of our hypotheses: manufacturing companies listed in the first and second tier provide more complete financial information and have a high level of internet reporting, manufacturing companies that had a majority of foreign shareholders seems to use more the internet to reduce the information asymmetry between the firm and investors, this is reflected in the high IBR Score obtained by these enterprises. We can outline that company's performance and foreign stakeholders are important factors that may contribute to the improvement of the level of internet business reporting of Romanian listed companies.

The study shows that Romanian companies listed on BSE have accepted the importance of the internet as a corporate reporting media and are interested in online reporting of their business. A better understanding of these practices is an important issue for future research, identification and analyzes of all the other factors that could influence the level of internet disclosure of Romanian companies is our future plan and will give us the possibility to form a complete image about the internet reporting on BSE and also to make international comparisons.

5. REFERENCES

Centre for International Financial Analysis & Research (1995). *International Accounting and Auditing Trends*. 4th Edition, CIFAR, Princeton New Jersey

Debreceny, R.; Gray, G.L.& Mock, T.J. (2001). Financial Reporting Web Sites: What Users Want in Terms of Form and Content. *The International Journal of Digital Accounting Research*. Vol.1, No.1, 1-23, ISSN 1577-8517

Debreceny, R.; Gray, G. L.& Rahman, A. (2002). The Determinants of Internet Financial Reporting. *Journal of Accounting and Public Policy*, Vol. 21, No. 4, pp. 371-394, ISSN 0278-4254

Deller, D.; Stubenrath, M.& Weber, C. (1999). A survey on the use of the Internet for investor relations in the USA, the UK and Germany. *European Accounting Review*, Vol. 8, No. 2, pp. 351-364, ISSN 0963-8180

Ettredge, M.; Richardson, V.J. & Scolz, S. (2002), Dissemination of Information for Investors at Corporate Web Sites, *Journal of Accounting and Public Policy*, Vol. 21, No. 4, pp. 357-369, ISSN 0278-4254

Healy, P. & Palepu, K. (2001). Information asymmetry, corporate disclosure, and the capital markets: A review of the empirical disclosure literature. *Journal of Accounting & Economics*, Vol. 31, No. 1-3, pp. 405-440, ISSN 0165-4101

Lymer, A. (1998). The use of the Internet for Corporate Reporting - a discussion of the issues and survey of current usage in the UK, paper presented at the *21st Annual Congress of the European Accounting Association*, Belgium, April 1998

Popa, A. (2008). Online Financial Reporting: an empirical study of companies listed on BVB., In: *Managementul calității informației contabile*, Georgescu I., Tugui A., pp. 239-247, Al. I.Cuza University Publishing, ISBN 978-973-703—299-3, Iași

Weber, R.P. (1985) *Basic Content Analysis*, Sage Publications, ISBN: 0803924488, Beverly Hills

Xiao, Z.; Yang, H. & Chow, C. (2004). Patterns and Determinants of Internet-Based Corporate Disclosure in China. *Journal of Accounting and Public Policy*, Vol. 23, No. 3, pp. 191-225, ISSN 0278-4254

Annals of DAAAM for 2008 & Proceedings of the 19th International DAAAM Symposium, ISSN 1726-9679
Editor B. Katalinic, Published by DAAAM International, Vienna, Austria 2008
Make Harmony Between Technology and Nature, and Your Mind will Fly Free as a Bird
*Young Researcher and Scientist Paper / * Supervisor, Mentor*

MULTIPOLAR DISTRIBUTED MATERIAL FLOW SIMULATION OF A VIRTUAL ENTERPRISE

POPA, C[icerone] L[aurentiu] & COTET, C[ostel] E[mil]*

Abstract: *In this paper we will present a multipolar distributed material flow simulation of a virtual enterprise. In our case study three partners are found, organized under the shape of a virtual organization. Two of them represent product suppliers, and the third represents the enterprise where the assembling will take place, after which the final product will result. In order to evaluate the architecture of the system, we used simulation models of the material flow for the three FMS, using the Witness software. The purpose of the simulation is the identification of the ways to improve the configuration of the system in order to raise productivity.*

Key words: *multipolar, simulation, Witness, FMS, material flow.*

1. INTRODUCTION

The virtual enterprise represents a temporary alliance of enterprises who wish to share resources and aptitudes for the purpose of making a product in the shortest time possible, at the smallesc price possible, and with the maximum satisfaction of the client (Camarinha-Matos et al., 1997). It is based on a technical infrastructure, represented by the informational technologies and by the computer communication networks (Dragoi, 2005).

Thus, the projection, planning of production and marketing, supplying, fabrication, services etc. can be realized anywhere in the country, the continent or the world, because of the facilities offered by the infrastructures which allow the change of information, goods and services (Ciobanu & Popa, 2004).

2. MULTIPOLAR DISTRIBUTED SIMULATION

The multipolar distributed simulation can be defined as an integrated system of monitoring more than two material flow simulations, interconnected inside the architecture of a virtual enterprise (Cotet & Dragoi, 2003).

In order to realize a simulation model of the material flow in the system, the first step is to define the necessary machine-tools, the work-pieces and the parts involved in the process on the level of each FMS (Flexible Manufacturing Systems). The inputs in the system are represented by the parameters of the machine-tools, of the work-pieces and of the parts involved, and the output of the system is represented by the optimized architecture of the system. The main limitations of the system are: the realization plans, the available realization stages, and the capacity of the CAD-CAM-CAE system to analyze the behavior of the machine-tools, the work-pieces and the parts in the work regime.

In the following study case three partners are found, organized under the shape of a virtual organization. Two of them represent product suppliers, and the third represents the enterprise where the assembling will take place, after which the final product will result.

Using specific solutions, each partner can realize a simulation in order to check the capability of its own system to fulfill the function traced in the virtual enterprise. After this, all these simulations can be integrated in a model of multipolar simulation in order to optimize the material flow in the entire system.

In the following we will define the three partners implicated in the system as being the three flexible fabrication systems, and thus FMS1 and FMS2 will represent the components supplier partners and FMS3 will be the system that makes their assembly.

3. CASE STUDY

In order to evaluate the architecture of the system, we used simulation models of the material flow for the three FMS, using the Witness software. The simulation of this project was realized in order to demonstrate and confirm the productivity of a fabrication process, based on the presentation of the proposed design and the operational data. Another purpose of the simulation was the identification of the ways to improve the configuration of the system in order to raise productivity.

The improving of productivity means the identification of the material flow concentrator for each FMS and for the virtual enterprise as a system, and the proposal of another manufacturing architecture that eliminates the concentrators.

According to the previous statements, FMS1 and FMS2 are diffused systems, and FMS3 is a concentrated system. Therefore, a concentrated FMS can be defined as a manufacturing architecture based on a single work point surrounded & assisted by transport, transfer & deposit facilities. A diffused FMS system can be defined as a manufacturing architecture with more than two work points connected by transport & transfer systems and using deposits at local or system level. (Cotet et al., 2004)

In figure 1 the FMS2 is presented during simulation. The parameters for each machine and conveyor were defined, and following the simulation, the place where the concentrators of the respective systems were identified.

The rapport regarding the activity of conveyor C2_1 is found in figure 2. We can observe the blocking of activity at this unit for 45% of the time, the conveyor is stopped for the intervention of the operator 5% of the time, and the machine works normally just 50% of the total time.

Fig. 1. The FMS2 during simulation.

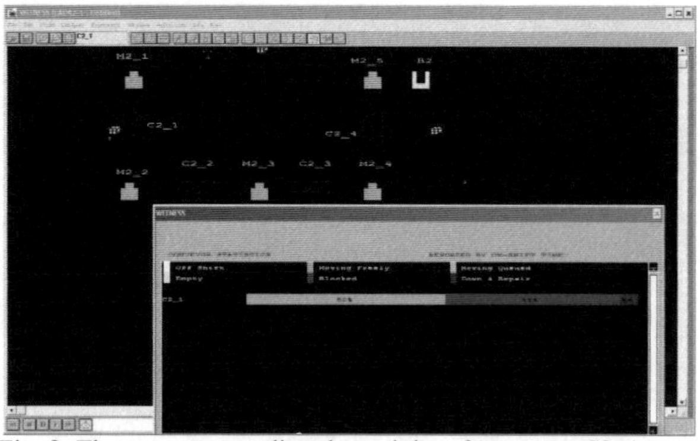

Fig. 2. The rapport regarding the activity of conveyor C2_1

The FMS2 system was remodeled and a new simulation of the material flow in the system was made.In figure 3 the rapport regarding the activity of conveyor C2_1 after the remodeling is presented and we can observe the total disappearance of blockages found in the previous simulation. Thus, the conveyor is in function 55% of the time, 34% of the time it waits for parts arrival, and the time necessary for the intervention of the human resource is of 11%. In figure 4 the third manufacturing system is presented during the simulation. We can observe: the B1 and B2 buffers that are the inputs in system 3 and also the outputs of system 1 and 2; the two conveyors and the machine on which the final assembly is done (M3_1). Also, the two parts (P and P2) can be found, which come from the previous systems FMS1 and FMS2, the operator and the number of pieces being worked at the time of reporting. From the study of the rapports and from the localization of the concentrators, especially in conveyor C3_2, results the necessity of remodeling the system for eliminating the concentrators. This will be done for system 3, and if concentrators will be maintained it results the necessity of intervention over the inputs in the system, and therefore over systems 1 and 2, which through the results of their own outputs determine the two inputs of system 3.

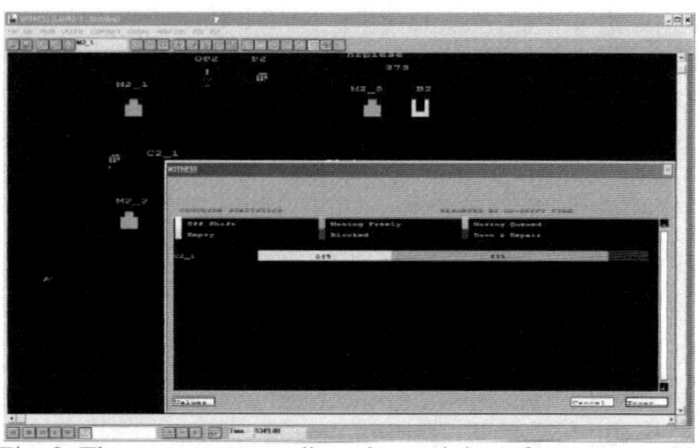

Fig. 3. The rapport regarding the activity of conveyor C2_1 after system remodeling

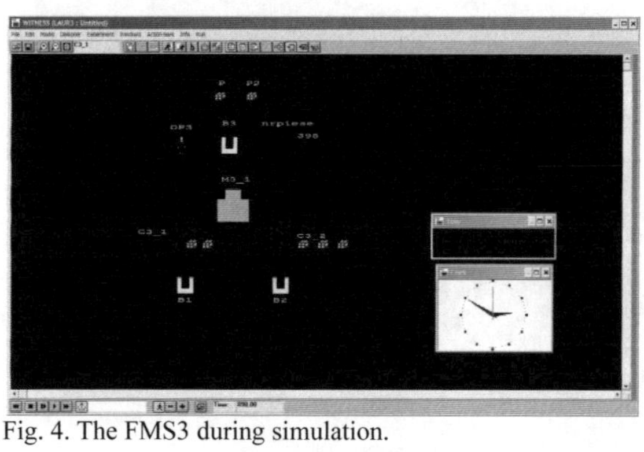

Fig. 4. The FMS3 during simulation.

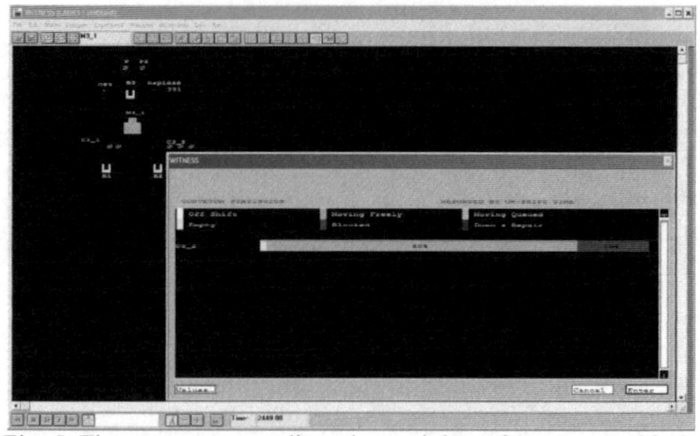

Fig. 5. The rapport regarding the activity of conveyor C3_2 after system remodeling

Following the remodeling of system 3, according to the demands, the elimination of the concentrator in the case of conveyor C3_2. In the case of conveyor C3_2, according to the rapport presented in figure 5, the existing blockage is reduced from 84% to 19% of time. Thus, it results that it works normally 80% of the time, which indicates a major improvement of its activity and implicitly of that of the system.

4. CONCLUSION

If we want to optimize the architecture of a usual material flow inside a virtual enterprise, a simple simulation is not enough. This is the reason why we proposed as a solution, the multipolar simulation, capable of evaluating the performances on a virtual enterprise as an integrated system. The analysis of the multipolar simulation does much more than just concatenates the results of isolated simulations. The flow concentrator which results from the multipolar simulation can be any of the concentrators of isolated simulations, but it can also be a totally different one. Some of the main causes of this particularity are the difference in algorithms for diffuse, isolated algorithms. For this case study, the elimination of the material flow concentrator of FMS3 gives important improvements of production at the level of the entire virtual enterprise, but the most influent concentrator of the entire system is localized in FMS2.

5. REFERENCES

Camarinha-Matos, L.M.; Carelli, R.; Pellicer, J.; Martín, M. (1997). Towards the virtual enterprise in food industry, *Proceedings of the ISIP'97 OE/IFIP/IEEE International. Conference on Integrated and Sustainable Industrial Production*, ISBN 0-412-79950-2, Portugal, May 1997, Chapman & Hall, Lisboa

Ciobanu, L.F.; Popa, C.L. (2004). Simulation and CAD/CAM/CAE cooperative systems integration in virtual environments, *Proceedings of the 7th Conference on Management of Innovative Technologies (MIT'2004)*, pp. 15-18, ISBN 973-700- 028- 5, Constanta, Romania, October 2004, Ed. AIUS, Craiova

Cotet, C.E.; Abaza, B.F.; Carutasu, N.L. (2004). Multipolar distributed simulation for concentrate and difussed FMS, *Proceedings of the International Conference on Manufacturing Systems (ICMaS 2004)*, pp. 647-650, ISBN 973-27-1102-7, Ed. Academiei Romane, Bucharest

Cotet, C.E.; Dragoi, G. (2003). Material flow management in validating concentrate and diffused FMS architectures. *International Journal of Simulation Modelling IJSIMM*, Vol. 2, No. 4, December 2003, pp.109-120, ISSN 1726-4529

Dragoi, G. (2005). Informational Infrastructure of modern enterprise (*Infrastructura informatională si de comunicatii a intreprinderii moderne)*, Ed. Politehnica Press, ISBN 973-8449-73-1, Bucharest

Annals of DAAAM for 2008 & Proceedings of the 19th International DAAAM Symposium, ISSN 1726-9679
Editor B. Katalinic, Published by DAAAM International, Vienna, Austria 2008
Make Harmony Between Technology and Nature, and Your Mind will Fly Free as a Bird

USING AGENT TECHOLOGY TO MANAGE ACTIVITIES IN THE MANUFACTURING DEPARTMENT

POPA, C[icerone] L[aurentiu] & PARPALA, L[idia] F[lorentina]

Abstract: *Agent technologies are suitable for application like manufacturing: (planning highly complex production, diagnostics, repair, reconfiguration and re-planning, workflow simulation) and virtual enterprises (forming business alliances, managing supply chains, using distributed systems). In this paper we will preset a case study about the way that agent technology can be used in order to manage and solve problems in a manufacturing department. In our example we will use a very simplified communication model between agents and we will present a possible dialogue between the agents that manage the activities of the machines from the manufacturing department.*

Key words: *agent, distributed, system, algorithm, manufacturing*

1. INTRODUCTION

Agent technologies are suitable for domains where:

- highly complex problems need to be solved or highly complex systems to be controlled;
- the information required for solving problems or controlling systems is distributed and is not available centrally, because: it is distributed geographically and the agents do not want to share it.
- in domains with dynamically changing environment and problem specification.
- different pieces of software need to be integrated in a run time and in the design times. (Vokrinek & Rollo, 2004)

In agent-based programming, the agent is the basic element of distribution. Each agent serves as an independent component with its own local state and execution model. The agent designer can choose to assign a particular set of functionalities to an agent, specify the types of events and messages the agent may invoke or respond to, and implement those triggers and/or responses. In building an agent-based system, each agent behaves independently, interacts with other agents through the events and messages that are communicated from one agent to another. (Weiss, 1999). The agents involved must be able to take decisions, to offer possibilities of cooperation and to negotiate relations between different domains involved in the project. (Popa&Hadar 2006). Agents have the folowing characteristics: autonomy, reactivity, pro-activeness, social abilities, mobility, coordination, communication, and cooperation. (Wooldrige, 2002).

2. USING AGENTS TO MANAGE THE ACTIVITIES

Manufacturing planning represents a matter difficult to achieve, especially if it's part of an open and distributed application, from a dynamic environment. In such an application rarely will happen that planning will be up to date with the schedule, the system may require fulfillment of tasks that were not originally foreseen, and sometimes it may be decided to omit certain tasks or unexpected tasks may occur. Also, some resources may not be available at the appropriate time or other types of resources may be needed. Differences may occur concerning execution times, some activities may take longer than was expected, and these delays could affect other activities in progress. In such a system an agent can be responsible for managing a machine, a tool or a human resource.

The agent must have all the information concerning the machine: all the technical data, processing systems, types of tools that can be used, types of tools available in the store, types of workpieces that can be processed. This agent must permanently interact with other agents from the system in order to obtain information regarding workpieces to be processed, the desired shape and characteristics of the processed part etc.

The agent must also obtain information about the activity of the other machines in real time, especially when certain processing operations of the part should initially be made on another machine or even after. The agent must be able to manage a complex database where information is permanently recorded. For example, the agent will know when a tool is used and replaced because the database contains information about the work time of that tool. The supervising agent is in permanent connection with the agent that manages the conveyor or the robot that handles the workpieces and the parts. He should know all these times precisely, in order to avoid a pieces blockage. In fact, all these agents are supervised at the level of the multiagent system by a "supervising agent" who has the role to manage the entire data flow and the agent's activities in the system that is organized hierarchically.

The supervising agent must manage efficiently the crisis; he will communicate to the other agents involved in the process that a problem occurred in the system. The supervising agent can decide: to interrupt all the activities from the system, or to overlying the operations. The supervising agent will offer a solution by planning-programming-reprogramming the activities affected. (Popa et al, 2007)

3. CASE STUDY

Two languages were imposed as the most used: KQML and FIPA-ACL, but generally each creator of a multiagent system uses its own language of communication. FIPA language describes in detail each participant's: name, address, its class, and presents in detail the message itself that is meant to be sent. In conclusion, it gives details of the language used, protocols etc. In our example we will use a very simplified communication model between agents. It will contain only the receiver agent, the sender agent, commands and information defined by the user etc. In this case study will use the system architecture described in Figure 1, in which a possible dialogue between the agents that manage the activities of the machines from the manufacturing department will be exemplified.

The adapted system for this case study is composed of: store for workpieces and parts; MS32 CNC multi-axis turning machine, AFC 105S CNC boring and milling machine, Blohm Precimat CNC grinding machine, four conveyors for parts transport; five robots for the transfer of parts between machines and conveyors.

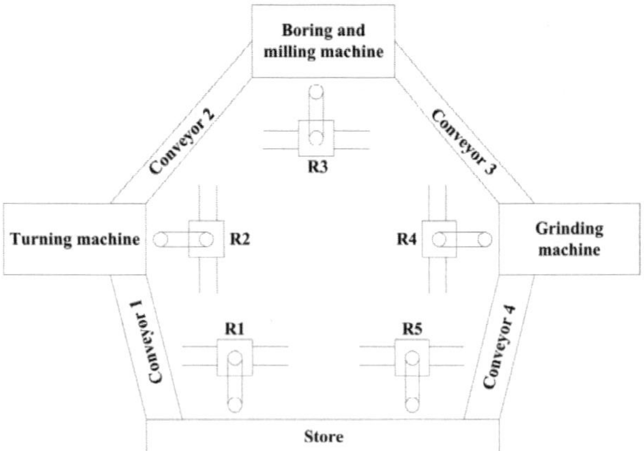

Fig.1. The system arhitecture in our case study

The agents defined in this case study are:

- "superviz" is the system's supervising agent in the manufacturing department;
- "turning_machine" the agent that manages the multi-axis turning machine activity in the department;
- "machine_AF" the agent that manages the boring and milling machine in the department;
- "grinding_machine" the agent that manages the grinding machine in the department;
- "store" is the agent who manages the entrances, exits of the store. This agent owns a database that contains all existing tools, workpieces and parts;
- "depan" is the agent who manages the human resource for troubleshooting;
- "conv1" is the agent that manages the activity of the first conveyor of the system; this conveyor is the link between the store and turning machine and so on;
- "rob_transf1" is the agent that manages the activity of robot 1, which makes the transfer between the store and conveyor 1 and so on;

The supervising agent announces the start of dialogue with all the agents in the system:

To *all* From *superviz* Hello

It also ensures that each agent of the system is online and ready to begin the activity:

To *conv1* From *superviz* Ping

To *rob_transf1* From *superviz* Ping

To *turning_machine* From *superviz* Ping

Such verification is done for each agent part of the system, if the agent is "online" it will reply:

To *superviz* From *store* Online

To *superviz* From *turning_machine* Online

We can have a situation in which an agent does not respond, in our case the agent of conveyor 1. The supervising agent will repeat the message and if in a given period of time he will not receive a response he will assume that there is a problem with that machine or with the way the agent that manages that machine was programmed. The supervising agent will notify the agent that manages the human resource to intervene on the machine and the application that sustains that agent in order to identify the error and the activity of that agent:

To *depan* From *superviz* conv1 goto

To depan From *superviz* conv1 71f390b problem

The "depan" agent receives the code of the problem, in our case the conveyor's agent does not respond to the supervising agent's request and the application will automatically contact a qualified human resource for this kind of problem and to which it will submit a report on the previous activities of agent and the machine managed by him. Also, time estimation for solving the problem will be required; when it will be possible the "depan" agent will send this estimation:

To *depan* From *superviz* conv1 time estimation

To superviz From *depan* 0.15.21 time

The supervisor announces the agent that manages the store to prepare 50 parts to be transported to the turning machine in order to be machined:

To *store* From *superviz* 50 part1 prepare

Once the parts are ready, the agents that manage the robots which make the transfer are notified, as well as the conveyor's agent who will carry the part to its destination. They will confirm that they are ready to begin the activity:

To *rob_transf1* From *store* part1 ready

To *store* From *rob_transf1* begin activity

To *conv1* From *rob_transf1* part1 ready

To *rob_transf1* From *conv1* part1 begin activity

The supervising agent announces the agent that manages the turning machine that part 1 must be processed and the number of such parts that will arrive:

To *turning_machine* From *superviz* part1 arrive

To *turning_machine* From *superviz* 50 parts arrive

It requires an estimation of the time needed for processing a part:

To *turning_machine* From *superviz* part1 time estimation

In order to make time estimation, the machine's agent requests the part's data and full information about it. The agent has access to the complete database of the system, it extracts comprehensive information regarding the part: types of surfaces, the material of the part, the operation plan, necessary tools and delivers them to the requesting agent.

To *turning_machine* From *superviz* part1 information

Based on the obtained information, the agent makes a list of the order of necessary operations, necessary tools. Based on this analysis he makes an estimation of the time necessary for processing and transmits it to the supervising agent.

To *superviz* From *turning_machine* 0.10.30 time

The supervising agent examines this time, compares it with the initially estimated time in the design stage of the part and of its material flow. If the processing time falls within the limits accepted by him he forwards the approval for starting the processing.

To *turning_machine* From *superviz* time ok

4. CONCLUSIONS

This simple system model can be extended towards a distributed architecture, because the agent oriented technology facilitates a good administration of the system. Thus, the presented system can be integrated in a much more complex one. Our further work will concentrate on the analysis and the simulation of a distributed system, based on agent orientated technology.

5. REFERENCES

Popa, C.L. & Hadar, A (2006). A cooperative system design based on agent technology, *Proceedings of 17th International DAAAM Symposium*, Katalinic, B. (Ed.), pp. 311-312, ISBN 3-901509-57-7, Vienna-Austria, Published by DAAAM International, Vienna

Popa, C.L., Parpala, L.F. & Aurite, T (2007). Algortihm for problems rezolving in distributed system using agent technology, *Proceedings of 18th International DAAAM Symposium*, Katalinic, B. (Ed.), pp. 575-576, ISBN 3-901509-58-5, Zadar-Croatia, Published by DAAAM International, Vienna

Vokrinek, J.; Rollo, M. (2004). *Agent System Design. Available from: http://agents.felk.cvut.cz/ Accessed:2008*-04-20

Weiss G. (1999). *Multiagent Systems: A Modern Approach to Distributed Artificial Intelligence*, The MIT Press, ISBN 0-26223-203-6, Cambridge

Wooldrige, M. (2002). *An Introduction to MultiAgent Systems*, John Wiley&Sons

Annals of DAAAM for 2008 & Proceedings of the 19th International DAAAM Symposium, ISSN 1726-9679
Editor B. Katalinic, Published by DAAAM International, Vienna, Austria 2008
Make Harmony Between Technology and Nature, and Your Mind will Fly Free as a Bird

DEVICE AND EXPERIMENTS FOR COLOURING THE ALDER WOOD THROUGH IMMERSION

POPA, E[lena] & POPA, V[iorel]

Abstract: *The paper studies the importance of the colouring parameter control with a view to obtaining surfaces of an improved aesthetic quality. This is entailed by the necessity of finding ways of using the fast growing wood species for manufacturing highly aesthetic pieces of furniture. The paper also presents a device which has been designed and made to enable the immersion colouring of the wood surfaces in laboratory conditions. By means of this device, which is operated through a PC compatible computer, the parameters of the colouring process can be controlled and maintained constant.*
Key words: *device, control, parameters, wood colouring*

1. INTRODUCTION

Safeguarding wood resources is a consistent concern throughout the world. This means not only using rational quantities or incrementing the wood usage efficiency, but also finding new solutions for including fast growing wood species (such as the alder, poplar and acacia) in the production of furniture pieces. One solution is to ennoble the wood surface through colouring.

The colouring process, performed chiefly manually in laboratory conditions, has the disadvantage that the obtained surface is not uniform on account of the human error and the uneven wood structure. This has triggered difficulties in the colouring process and intolerable dispersions of the colour measurement values.

Literature research has lead to no references regarding the use of devices for avoiding the human error as far as wood colouring in laboratory conditions is concerned.

Consequently, three original devices have been conceived to enable the control of a number of parameters of the colouring process through immersion, rolling and pulverization.

This paper presents the device designed and created by the authors for the immersion colouring, in laboratory conditions, of the alder wood, as well as the results obtained after the experiments on the measurement of the colour and colouring process parameters.

2. RESEARCH METHODOLOGY

2.1. Theoretical Background

The present paper is part of a larger research program focusing on the colouring processes of the fast growing wood species with a view to their ennoblement.

For this purpose, it has been necessary to study the wood features: macromolecular structure, wood surface roughness, liquid permeability on the main wood sawing directions, and wood colour (measuring method, expression values, differences determined by wood structure) (Popa & Popa, 2004). These features have an important role in the superficial colouring process of the wood.

The wood colour can be measured by means of three methods:

- The spectral method (McCurdy et al., 2003);
- The method of the trichromatic colorimeter (Stenudd, 2003);
- The computerized measurement; a colour is expressed through values varying from 0 to 255 (Mitrofan, 1983).

This study has employed the computerized method for measuring the wood colour, through an original procedure aided by CorelDraw (Popa & Popa, 2004).

With CorelDraw one can measure the colour of the scanned surfaces in different colorimetric systems, such as:

RGB: Red (R); Green (G); Blue (B).

CMY: Cyan (C); Magenta (M); Yellow (Y).

CIELAB, defined by the International Commission of Illumination (CIE) in 1976, is widely used in colorimetry and is a part of the CorelDraw program. Within this system, a colour is featured through the L*a*b* rectangular proportions where:

- L* represents the brightness of the colour;
- a* represents the red-green degree;
- b* represents the yellow-blue degree.

Using the L*, a* and b* rectangular coordinates, the chrome (satiation) can be determined, according to the relation 1(Camargos et al., 2005):

$$\text{Chrome ab CIE 1976} \qquad C^*ab = (a^{*2} + b^{*2})^{1/2} \qquad (1)$$

In literature and practice there are devices that commonly use one of the following colouring procedures: immersion, rolling or pulverisation; however, the control of the colour parameters has often been neglected. Starting from these procedures, followed in laboratory conditions, the authors have created and used original simplified devices, which mechanise the application of the colouring solution in order to eliminate the errors of manual colouring.

2.2. Device Design and Functioning

The immersion colouring device has been conceived to allow for the modification of the following: the immersion speed of the piece in the immersion bath, the immersion time, and the return speed.

The device shown in the figure 1a consists of a frame (2) sustaining a double-acting pneumatic motor (1), to whose bar has been attached a fitting element (3) of the piece (4) to be coloured through immersion. The frame is fixed on a 400 x 400mm board (6), which constitutes the support of the device. On the board, between the two columns of the frame, is placed the vessel (5) containing the colouring solution.

The functioning of the installation is controlled through a PC compatible computer. An original program has been made in Delphi to determine the advance motion of the piece for its immersion, the maintenance in the colouring solution for a certain time and the return to the initial position. The immersion time can be regulated within large limits, starting from a second to tens of minutes.

2.3. Sample Presentation

For the immersion operation, the experiments have been conducted on 80x50x10 mm alder wood samples (fig.1.b).

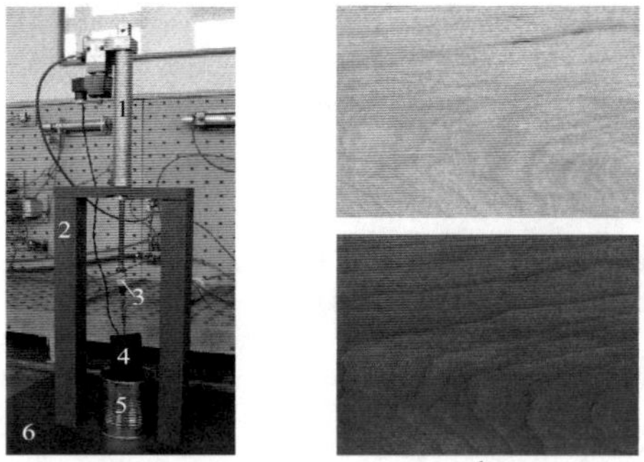

a b

Fig. 1 a. The immersion colouring device; b. Natural and coloured alder sample

Since wood as a material is inhomogeneously structured, absorbing the colouring solution according to the sawing direction, the study has been conducted on samples sawed tangentially (Tg), radially (Rd) and transversally (Tr).

For the same reason, it has been necessary to measure the same surface several times and calculate the average of the performed measurements.

3. RESULTS AND COMMENTS

The experimental determinations have tracked the variation of the colouring solution consumption, as well as the variation of two of the colour parameters, to wit the L* luminosity and the C*ab satiation. The measure unit for the colour parameters is [u.CIE], a unit which was adopted by the International Illumination Commission in 1976.

The immersion time has been regulated with the help of the designed and created device at the 2, 8, 14, 20 and 25 sec values. For these immersion times the colouring solution consumption has been calculated. The obtained results are presented in the table and diagram in the figure 2. It is noticeable from the analysis of the experimental results that the immersion colouring procedure has the advantage of performing the wood colouring operation in a shorter time. Furthermore, the procedure can be easily automated. Conversely, it is disadvantageous from the viewpoint of the colouring solution consumption. Even in terms of a minimum immersion time, the consumption is high, reaching ~ $125 g/m^2$. This is due to the high permeability of the alder wood.

By using CorelDraw the L*, a* and b* rectangular measures have been determined and the C*ab colour satiation has been calculated with the relation (1). The results are shown in the table and diagram in the figure 3. The presented values for the colour parameters are the average values for the three sawing directions.

The experiments have shown that the speed of the introduction and taking out of the sample from the immersion bath does not influence the colour parameter values. The two measured colour parameters, the L* luminosity and the C*ab satiation, are influenced by the immersion time. An observation of the diagram in the figure 4 leads to the conclusion that while the L* luminosity of the colour tends to decrease along with the increase of the immersion time, the C*ab satiation is relatively constant. As a result of the colouring process with the created device, due to the possibility of controlling the immersion time, the colour reproducibility can be assured.

4. CONCLUSIONS

The created device, controlled by means of a PC compatible computer, has removed the human error and has assured the colouring uniformity and reproducibility, which represents an important advantage of the presented solution.

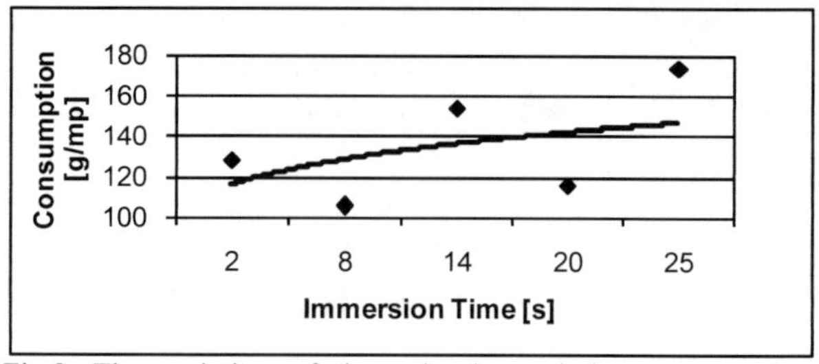

Fig.2. The variation of the colouring solution consumption depending on the immersion time

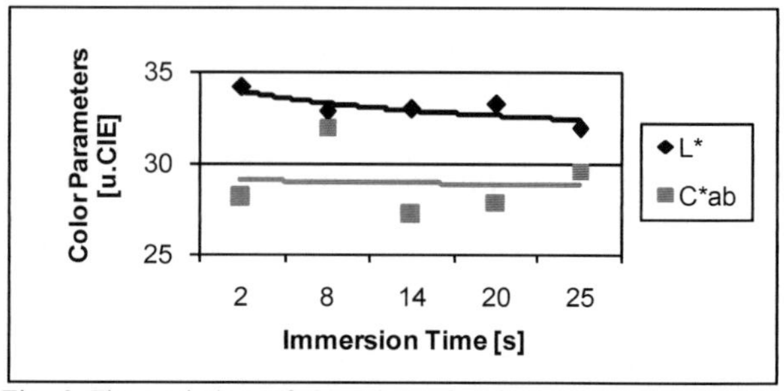

Fig. 3 The variation of the L* luminosity and C*ab satiation depending on the immersion time

Colouring with this device has resulted in beautiful colours, which have covered the faults of the alder wood (such as the colour stains), has emphasized the fibre drawing (fig.1b) and has enhanced the wood value. The industrial implementation of the colouring and measuring device and method for the colour parameters proves to be a possible solution for using the fast growing species in the production of highly aesthetic furniture pieces.

Through colouring, the more valuable species can be replaced successfully, which will eventually lead to a more diversified wood exploitation and assure a natural and durable renewal of the wood species.

5. REFERENCES

Camargos, J.; Goncalez, J.; Janin, G.; Bakour, R. & Ananias, R. (2005). Colorimétrie quantitative appliquée à l'élaboration d'un classement numérique et aussi visuel, en un catalogue de 25 groupes de couleurs pour 400 espèces de bois tropicaux brésiliens de l'Amazonie (Quantitative colorimetry applied to the drawing up of a numerical and visual classification in a catalogue of 25 colour groups for 400 Brazilian species of Amazonian tropical wood). In: *Les cahiers scientifiques du bois*, vol 3, ISBN 2-907086-11-1, Nancy, France

McCurdy, M.; Pang, S. & Keey, R. (2003). Measurement of Colour Development in Pinus radiata Sapwood Boards During Drying at Various Schedules, *Proceedings of the "8-th International IUFRO Wood Drying Conference"* pp. 445-448, ISBN 973-635-198-X, "Transilvania" University Publishing House, august 2003, Braşov, Romania

Mitrofan, Gh. & Pflanzer, G. (1983). *An introduction to Colour Television,* Technical Publishing House, Bucureşti, Romania

Popa, E. & Popa, V. (2004). *Lemnul de anin, plop şi salcâm. Proprietăţi, colorare (Alder, Poplar, Robinia Wood. Properties, Colouring),*"Transilvania" University Publishing House, ISBN 973-635-279-X, Braşov, Romania

Stenudd, S. (2003). Colour Response of Silver Birch to Press Drying, *Proceedings of the "8-th International IUFRO Wood Drying Conference"* pp. 449-454, ISBN 973-635-198-X, "Transilvania" University Publishing House, august 2003, Braşov, Romania

Annals of DAAAM for 2008 & Proceedings of the 19th International DAAAM Symposium, ISSN 1726-9679
Editor B. Katalinic, Published by DAAAM International, Vienna, Austria 2008
Make Harmony Between Technology and Nature, and Your Mind will Fly Free as a Bird

THE PROGRESS THROUGH CLUSTERISATION IN EUROPEAN UNION

POPA, H[oria] L[iviu]; IZVERCIANU, M[onica]; MOCAN, M[arian] & PATER, L[iana] R[odica]

Abstract: *The persistent negative competitiveness discrepancies in the European Union as compared to those in North America and in the Asia-Pacific Area require the elaboration of new actions, structured on categories of member countries. The concepts of cluster and clusterisation are defined in a systemic way. The clusterisation of the EU 27 countries is being analysed and three directions of new actions meant for increasing sustainable competitiveness in the European Union will be submitted for approval.*
Key words: *progress, EU 27, competitiveness, clusterisation*

1. INTRODUCTION

The enlarged European Union is taking new steps in enforcing the 2000-2010 Lisbon Strategy, updating it three times a year but without succeeding in eliminating the negative competitiveness discrepancies as compared to North America (3 countries) and the Asia-Pacific Area (13 countries). The various competiveness classifications realized in the last 10 years confirm this reality (GCR, -2007; WCY, -2008) and require integrative systemic innovation, adjusted to the various categories of EU 27 member countries. One of the means to increase sustainable competitiveness in the European Union is to improve clusterisation and cluster performances in all domains of activity.

In the present work, the concepts of cluster and clusterisation are defined in a systemic way. The evolution of clusterisation and the enforcement of competitiveness strategies are analysed from an objective perspective, new integrative patterns for achieving progress through clusterisation are submitted, based on total competitiveness cycle (Popa, 2007/1).

Future research will develop new models and methods for clusterisation optimisation.

2. SYSTEMIC DEFINITION OF CLUSTERS AND CLUSTERISATION

In our Universe, clusters are the systems of systems $\{S_S\}$ which generate becoming (progress / stagnation / regress) in all natural and / or artificial environments, from macro-cosmos to micro-cosmos.

In economy, **clusters C** are **geographical agglomerations / concentrations of organizations** which consist of four categories of sub-systems:

[1] Enterprise networks (EN), product / service providers, which offer competitive assortments $\{K_i\}$ to the consumers in the target markets and consist of: **[1.1]** Enterprises / organization groups "leader" within EN; **[1.2]** Innovation, production companies / service providers, selling, human resources training etc; **[1.3]** Research-development-innovation companies (Universities, institutes and R&D companies,….); **[1.4]** Financing companies (banks, refundable financing funds etc.),

[2] Facilitators (Fa) such as institutions / organizations / networks for cooperation and competitiveness which include: **[2.1]** Competitiveness Councils and Centres, **[2.2]** Development Agencies (regional, county, local), **[2.3]** Chambers of Commerce, Industry and Agriculture,

$N_{Copt} \leftrightarrow K_{ismax}$		innovation	
		partial	integrative
clusterisation	innovative	K_{idmax}	$\mathbf{K_{ismax}}$
	trade	K_{idmin}	K_{id}

Tab. 1 The optimal clusterisation level N_{Copt}

[2.4] Other facilitators: employers, professional organizations, unions etc.

[3] Territorial public authorities (TPA): **[3.1]** local; **[3.2]** regional; **[3.3]** euro-regional; **[3.4]** governmental

[4] Target markets / segments / niches with specific consumers and clients under continuous behavioural changes (**Consumers - Co**).

Based on strong public-private partnerships, competitive clusters are: a) potentially lasting competitive on the national and global market; b) in competition and internal / external cooperation relations (coopetition); c) determinant at a large extent in the 21st century of the evolution of sustainable competitiveness at regional, national and global level; d) decisively contributors to increasing life standards in the countries and regions where they operate, according to a particular stage of their life cycle.

Clusterisation is a continuous process of creating, functioning with regular restructuration and liquidation of successive-parallel cluster generations g = 1, 2, …., n, n+1, … / clusters of clusters in a D_{str} space-time-resources domain which create the conditions for the existence of $\{S_S\}$ systems of systems. Natural clusterisation is spontaneous and it can take decades when operating in the business community; clusterisation based on strong public-private partnerships is planned and, at present, it lasts for 1-2 years. The competitive integrative clusterisation and innovation within the clusters generate in successive-parallel manner the lasting integrative competitiveness K_{id} (long period of time) / sustainable integrative competitiveness K_{is} (unlimited period of time) in external environments M_{ext}.

The optimal clusterisation level N_{Copt} depends on a great number of factors [natural, demographical, psychological, social, cultural, political, juridical, administrative, socio-economical (business), technological-managerial, military], on the nature of the factor innovation and it is defined by the maximal integrative competitiveness K_{imax}.(Table 1).

3. SUSTAINABLE COMPETITIVENESS AND OPTIMAL CLUSTERISATION IN EU 27

After the acceleration of globalisation, started in 1960, the concept of cluster and the priority of competitive clusterisation in all domains of activity have been understood in the European countries with great delay. The clusterisation process has evolved differently in EU 27 (Table 2), thus contributing to the increase of regional, national and global competitiveness of member countries. The 15 European countries (+) / 12 EU countries (+) in the world 27 top, according to „Global Competitiveness Index GCI", have superior performance (economic performance, government efficiency, business efficiency and infrastructure) which explain the

Nr.	Competitiveness [rank **GCI 2007-2008**] for **EU 27** and *nonEU* countries (from 131 countries) (GCR, -2007)	The clusterisation level (Gallup, 2006)
1	**England [9]** +	**99 – 50%**
2	**Ireland [22]** +	
3	**Latvia [45]**	
4	**Austria [15]** +	**49 – 25%**
5	**Italy [46]**	
1	*Turkey [53]*	
6	**Bulgaria [79]**	
7	**Sweden [4]** +	**24 – 20%**
8	**Finland [6]** +	
2	*Norway [16]* +	
9	**Belgium [20]** +	
10	**Portugal [40]**	
3	*Croatia [57]*	
11	**Germany [5]** +	**19 – 10%**
12	**Netherlands [10]** +	
13	**France [18]** +	
4	*Iceland [23]* +	
14	**Estonia [27]** +	
15	**Lithuania [38]**	
16	**Slovenia [39]**	
17	**Slovak Republik [41]**	
18	**Hungary [47]**	
19	**Malta [56]**	
20	**Romania [74]**	
5	*Serbia [91]*	
6	*Switzerland [2]* +	**9 – 1%**
21	**Denmark [3]** +	
22	**Luxembourg [25]** +	
23	**Spain [29]**	
24	**Czech Republic [33]**	
25	**Poland [51]**	
26	**Cyprus [55]**	
27	**Greece [65]**	

Tab. 2. The nowadays correlation competitiveness - clusterisation level in Europe

competitiveness – clusterisation correlations as shown in Table 2. In all these countries, clusterisation has been playing a very important role in increasing sustainable competitiveness; the pioneer in generating the natural / spontaneous economic clusters is (EC, 2006): England (1780), France (1800), Germany (1804), Switzerland (1820), Italy (1823).

In Europe, the sustainable competitiveness growth and the clusterisation are still approached in a traditional way (EC, 2006; Gallup, 2006; CLUNET, 2008; HLAGC, 2008): non-systemic (concepts, patterns, quantification, methods) and non-integrative (innovation in policies, management, culture of competitiveness, resources, technologies, critical masses / thresholds for competitive clusters, optimal clusterisation levels).

Sustainable clusterisation is the key of the progress in any D_{str} space-time-resources domain and it relies upon two categories of systemic integrative processes : (1) clusterisation of entities in the domain, while generating efficient critical masses M_{crp} / thresholds able to produce lasting progress; (2) cluster-based integrative innovation (in resources, processes and executive structures / programmes, self-control structures and processes, radical and continuous improving programmes), capable for generating sustainable competitiveness in external environments M_{ext}. This pattern was fully confirmed after 1960 by the USA, Canada, Asia-Pacific Area (the 13 developed countries), by the European developed countries, respectively (+).

At present and in the years to come, EU 27 shall reconsider Lisbon 2000 strategy in a systemic / integrative manner and urgently orient its efficient and effective enforcement, in order to attain as soon as possible the target envisaged: "the most dynamic and competitive knowledge-based economy in the world capable of sustainable economic growth with more and better jobs and greater social cohesion, and respect for the environment by 2010". **Three categories of new DAK inter-related actions are necessary for increasing sustainable competitiveness**, as follows:

-DAK for the entire EU 27, focused on clusterisation optimisation,

-DAK for the 12 EU countries (+) identified in Table 1, focused on cluster-based integrative innovation (in resources, processes and executive structures / programmes, self-control structures and processes, radical and continuous improving programmes),

-DAK for the other 15 EU countries in Table 1, focused on centres, clusters and competitiveness programmes in the enlarged Europe (Popa, Mocan & Pater, 2007/2).

4. CONCLUSIONS

The persistent negative competitiveness discrepancies in the European Union as compared to those in North America and in the Asia-Pacific Area claim the elaboration of new actions, structured on categories of member countries. The necessary DAK actions for increasing sustainable competitiveness are: (1) DAK for the entire EU 27, focused on clusterisation optimisation in all domains; (2) DAK for the 12 EU countries (+) identified in Table 1, focused on cluster-based integrative innovation; (3) DAK for the other 15 EU countries in Table 1, focused on centres, clusters and competitiveness programmes in the enlarged Europe.

5. REFERENCES

CLUNET, (2008). Cluster Policy Guidelines Report, PRO-INNO Europe, *Available from:* www.proinno-europe.eu/admin/uploaded_documents/CLUNET_policy_Guidelines_080108.pdf *Accessed:* 2008-05-02

EC, (2006). *Innovation clusters in Europe – A statistical analysis and overview of current policy support*, European Commission, Office for Official Publications of the European Communities, ISBN 978-92-79-07289-5, Luxembourg

Gallup, (2006). The Gallup Organization Flash EB No 187 – 2006 Innobarometer on cluster's role in facilitating innovation in Europe. Analytical Report, *Available from:* ftp://ftp.cordis.europa.eu/pub/innovation/docs/innobarometer_2006.pdf *Accessed:* 2008-05-02

GCR, (-2007): *Global Competitiveness Report*, World Economic Forum, Geneva, 1997 / .... / 2007, Palgrave Macmillan, ISBN 1-4039-9636-9, New York, USA

HLAGC, (2008). *The European Cluster Memorandum*, The High Level Advisory Group on Clusters, *Available from:* http://www.vinnova.se/upload/dokument/VINNOVA_gemensam/ Kalender/2008.pdf *Accessed:* 2008-05-02

Popa, H.L., (2007/1). The integrative competitiveness. A systemic approach, *Scientific Bulletin of the "POLITEHNICA" University of Timisoara, Romania, Transactions on Management. Engineering Economy. Transportation Engineering*, Tom 52 (66), 2007, Fasc. 1, pp. 5-18, ISSN 1224-6050

Popa, H.L.; Mocan, M.L. & Pater, L.R. (2007/2). Competitiveness Strategies for the International Scientific - Academic Associations and Clusters, *Annals of DAAAM for 2007 & Proceedings of the 18th International DAAAM Symposium*, Katalinic, B. (Ed.), pp. 579-580, ISBN 3-901509-58-5, ISSN 1726-9679, Zadar, October 2007, Published by DAAAM International, Vienna, Austria

WCY, (-2008). *World Competitiveness Yearbook*, International Institute for Management Development, 1997 / .... / 2008, ISBN: 185-743-451-X, Lausanne, Switzerland

Annals of DAAAM for 2008 & Proceedings of the 19th International DAAAM Symposium, ISSN 1726-9679
Editor B. Katalinic, Published by DAAAM International, Vienna, Austria 2008
Make Harmony Between Technology and Nature, and Your Mind will Fly Free as a Bird

NEW TRENDS IN NON-CONVENTIONAL TECHNOLOGIES AND ELECTRIC DISCHARGE MACHINING

POPA, M[arcel]; CONTIU, G[lad] & PRECUP, M[ircea]

Abstract: *The paper wants to emphasize the importance of the unconventional technology and more precisely the importance of knowing the parameters of a process before planning a new thechnology. The new claims require organisatorical methods and management of fabrication, respeotively of development of more complex and more flexible products with high chances of success. In the technology, it is necessary to find optimal solutions which should not conditionals directly anymore the three factors: time, quality and costs. By knowing a technology and optimising it so that the prduct should be concurential with the market. The nonconventional technologies are taking a serios position in the industry. Practicaly it is imposible in some cases to process a workpiece without these technologies. Researches of unconventional manufacturing methods and technologies, mainly electric discharge machining process, have been performed for over 20 years at the Faculty of machine Building from the Techncal University of CLuj-Napoca.*

Key words: *nonconventional, electroerosion, engineering, machining*

1. INTRODUCTION

In 1942 the Lazarenko couple discovered the EDM(Electric Discharge Machining) process. At the beginning it was used only in the military industry. EDM performed a long way till to its present high performances. EDM has appeared as a necessity to manufacture materials with better mechanic and thermal characteristics. The major advantage of EDM in comparison with other manufacturing processes is represented by the fact that the hardness of material is not important, the only condition being that the processed material must be electro conductive.

We are participating today to the most extraordinary growth of electronic and automatic, aeronautic and spatial industry, nuclear industry or micro technologies. It is noted also, a sharp rise of unconventional techniques and technologies. In this field we find today from the unconventional "economic war" and "unconventional terrorism" to the "clasic"unconventional technologies - laser, EDM, etc.- from unconventional forms of informational or economical war and "exotic" techniques as those para psychological, satellite telepathy, to the unconventional weapons.

The most important means of development, but also the principal financial support of the unconventional technologies is represented by the military industry and techniques. We note also, that some unconventional technologies are still in different stages of investigation or improvement. All the new technological applications have an unconventional character, but after it became known and applied, there are rapidly turned into conventional(Popa et. al., 2008).

2. GENERAL ASPECTS OF EDM PROCESS

The phenomena of electro erosion of metals and their alloys were well known in the electro technology industry, because switches, contactors, micro switches etc. are destroyed by the sparks that appear during functioning or when they are switched on or off.

The discharge in the electric sparks last a very short time, producing high energy on the very small surfaces and punctiform a very high temperature instead of discharges through electric arcs which are very powerful on the more bigger surfaces and can last longer(welding, shortcutting).

Today, the machining by electro erosion can practically realize any kind of manufacturing that is made with classic methods: drill, saw, turn, mill, ream, grind, hone, etc.

This machining can be classified in two groups: EDM (machining with massive electrode), WEDM (machining with wire electrode)(Westkämpfer & Warnecke, 2001).

The EDM machines can be separated in two types: universal and special machines. The physic process of material machining and the evacuation of the processed material from the spark take place in the interstice between work piece and tool. In the process of electro erosion, the electric parameters have the main importance.

In EDM, conductive workpiece materials are removed for the purpose of machining in a dielectric by electrical discharge. The material removal results from the erosive effect of subsequent, time wise separated, no stationary or quasistationary discharges between electrodes, i.e., between tool and work piece. Each discharge generates a microscopic removal on the two electrode surface. In principle, the process is based on thermal erosion. Hence, an efficient EDM process can only be realized by a purposefully uneven material removal on the two electrodes. Wire – EDM as kinematical variant of EDM allows hereby the machining or respectively, production of complex geometrical contours.

3. ELECTRICAL DISCHARGE MACHINING. EXPERIMENTAL RESULTS.

3.1 Wire Electrical Discharge Machining

The process can be compared with the classic process of sawing or cutting with diamond wire, but the particular characteristic of not stressing the work piece and the wire that don't get in contact with. Comparing with the classic machines, the work piece flows forward and the wire rolls only.

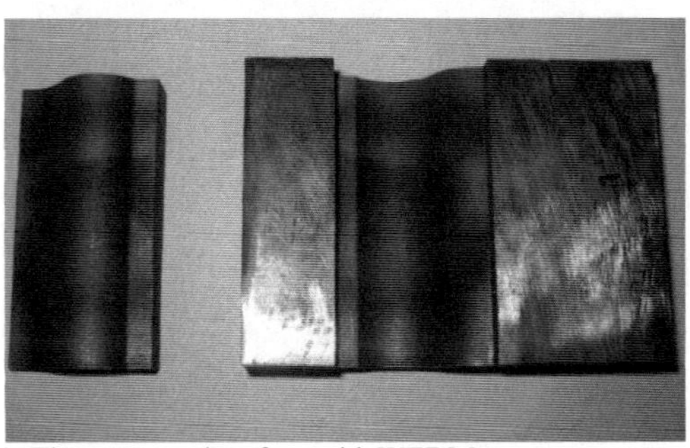

Fig. 1. The processed surface with WEDM

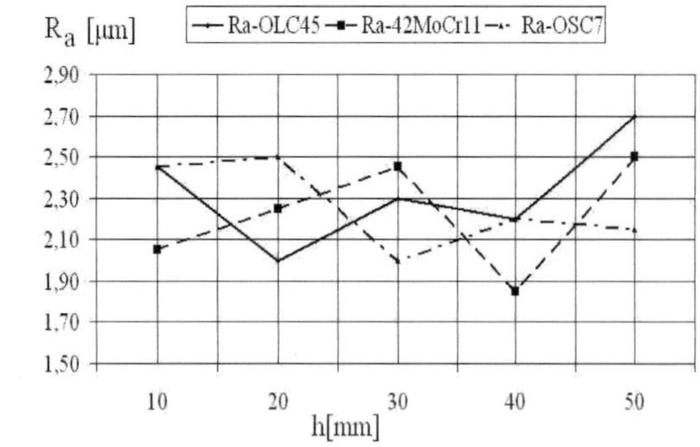

Fig.2. Roughness variations depending on the material

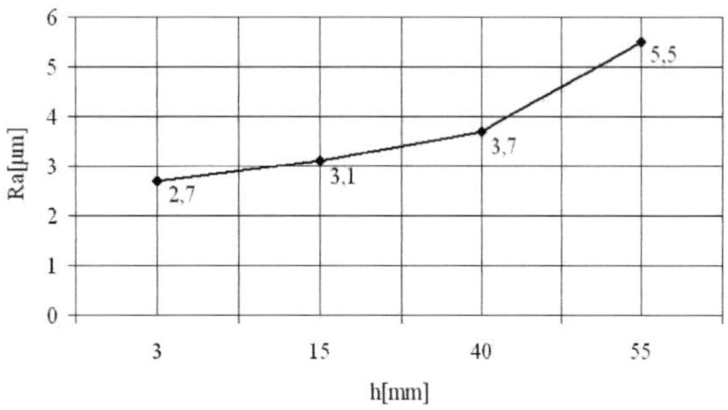

Fig. 3. The diagram of roughness variation depending on the thickness of the work piece

One of the parameters that are followed to be obtained after machining is the roughness of the surface. By machining with this method can be obtained very good surfaces with roughness that can arrive till 0,2 µm, but only in finishing mode. In the industry is not always necessarily such a good surface. So in the industry, it is important to know the roughness of the surface to be obtained in order to prepare a technology as economic as possible. Obtaining a roughness that is wanted by the client, it avoids an additional machining which would require additional time and cost. The autors realised some experiments to obtain results regarding the time of machining and the quality of the processed surface. The roughness will increase when the intensity of the current grows and when the frequency decreases.The process can be compared with sawing or cutting with diamond wire. As it can be observed in the fig. 1, the texture of the surface is not the same on the entire surface of the work piece. This phenomenon is due to the dielectric, that doesn't wash equable the processed work piece. As in the conventional processes, the EDM can also do rough, finishing or super finishing works.With the new machines of EDM, the roughness R_a that is obtained on the existing machines that work in the industry is between 0, 2 ÷ 6,3 µm.

In other cases, due to the aspect that must be obtained on the surface, the work pieces are processed with very high Ra value, over 12, 5 µm. The productivity that is obtained in this case is very high due to the very high energy that is used. For the very high precision processes, it can be obtained in special conditions a roughness Ra of 0, 05 ÷ 0,1 µm. (Dodun Oana, 2001).

Because usually, the process is used in industry after a thermal treatment of the work piece, we used hard steel.

- OLC 45; 42MoCr11; OSC7.

For a higher hardness, the test pieces were thermal treated. (Table 1.)

The dimensions of the test pieces are:
- length L= 80mm;
- width l= 20mm;
- height h= 10÷50mm;

The roughness of the surface was measured on the middle of the height. For the same thickness of the work pieces the value of the roughness are oscillating between 1, 85 µm and 2, 70 µm as it can be seen in the figure 2.

For the work pieces that have thicknesses over 70 mm, it can be observed the difference between the roughness from the bottom and the middle of the work piece – figure 3.

This phenomenon takes place because the nozzles for dielectric are placed at the top and at the bottom of the work piece and the pressure is not big enough to be able to wash all the particles.

In the industry the roughness that is obtained is very important. On this parameter depends the quality and the aspect of the product.

Today the new machines with the command and control system that assist the process make the work of the operator easier.

4. CONCLUSIONS

In the industry, this technological method of machining is expanding more and more. To be mentioned is that it begins to be used in micromachining. It's important to be able to establish a technological itinerary and after this a price for a work piece or for a large series of production.

The operator is using the CAM programs in order to write the CNC program, but the parameters of the machine must be set concerning the material, the thickness of the work piece, the roughness that must be obtained and the tool material. For a better planning of the technology, it's necessary to build a bibliotheca with most common materials that are to be used in a factory. For the same kind of machining (roughing or finishing), the values of parameters can be obtained from the tables of the machine-tools. Although the parameters are optimized, the quality of surface is different for each machining because external conditions are deferent for each machining.

But for industry it is important to be prepared a bibliotheca with specified parameters for machining different materials. For example for the same kind of machining (roughing or finishing), in the tables of the machine-tools, the parameters are optimized but it is not specified the quality of the surface that can be obtained.

Our researches have shown the knowledge importance of all parameters that can influence the quality of the process and in the future we will prepare a bibliotheca that will complete the CAM programs by having a larger table of materials and machining parameters.

5. REFERENCES

Ceausescu, Nicu, Popescu, I., 1982 – *Tehnologii neconventionale, "Unconventional Technlogies"* Ed. Scrisul Romanesc, Craiova.

Dodun Oana. *Tehnologii neconventionale. Prelucrari cu scule materializate. "Unconventional Technologies. Machining with materialized tools".* Editura Tehnica Info Chisinau 2001.

Popa, M., Contiu, G. Precup, M., Preja, D., Gaina, O., Fagarasan, C.. *Unconventional Technologies and competitive engineering in the 21st century,* Proceeding of the TMCE 2008, Kusadasy Turkey,

Popa M., Contiu G., Precup M., Preja D., Gaina O., *Mathematical model and simulation of the heat transfer in the EDM process*, IEEE International Conference of Automation , Quality and Testing, Robots AQTR 2008, May 22-25, 2008.

Westkämpfer, E., Warnecke, H.J. *Einführung in die Fertigungstechnick, „Introduction in the machininng technology",*Teubner Verlag, Stuttgart 2001.

Annals of DAAAM for 2008 & Proceedings of the 19th International DAAAM Symposium, ISSN 1726-9679
Editor B. Katalinic, Published by DAAAM International, Vienna, Austria 2008
Make Harmony Between Technology and Nature, and Your Mind will Fly Free as a Bird

LASER WELDING AUTOMATION

POPA, M[arcel]; PRECUP, M[ircea] & CONTIU, G[lad]

Abstract: *The article presents a method about the automation for welding with a laser device, HL54P Trumpf. In some cases the edge of the metal sheet is not linear, it can take different shapes: circular, elliptic, and other shapes more complex. The goal of the authors is to find a solution to determine the position of the edge and a path which the processing optics to follow. The solution found by the authors is to read the edge with a video camera, identifies the shape edge in a known scale image, and generate the path for processing. This method is necessary in the case of non-CNC (Computer Numerical Control) machine or the CAD (Computer Aided Design) files and CAM (Computer Aided Manufacturing) software for CNC programming are missing.*
Key words: *Laser, Nd:YAG , Automation, Welding*

1. INTRODUCTION

The goal of the team was the automation of the HL54P laser device for the overlap welding, in the case of non-CNC (Computer Numerical Control) machine or in the case of missing CAD (Computer Aided Design) file and CAM (Computer Aided Manufacturing) software for CNC programming. In our case the laser device had the laser beam generator, the processing optics and optical cable. The optical cable guides the laser light from the laser device to the processing optics which focuses the laser light onto the surface of the work piece.

The HL54P was obtained after an international collaboration program with University of Stuttgart. It is a solid-state laser with an Nd: YAG rod (Neodymium-doped Yttrium Aluminium Granat).

The advantages of laser beam machining compared to other methods are: greater precision, increasing work speed, allowing the point-like processing of the piece - this is a process that does not imply touching the piece with an instrument (beam) (Hugel, 1992).

The use of this procedure reduces the heating of the working area because the energy is introduced faster and on a small surface. This way, large power densities can be reached. Other advantages are flexibility and accessibility (Popa, 2005).

2. GENERAL ASPECTS

Nd:YAG lasers emit light in the near infrared range, at a wavelength of 1.06μm. This means that the light emitted by Nd:YAG lasers is almost in the visible range (Koechner, 2006).

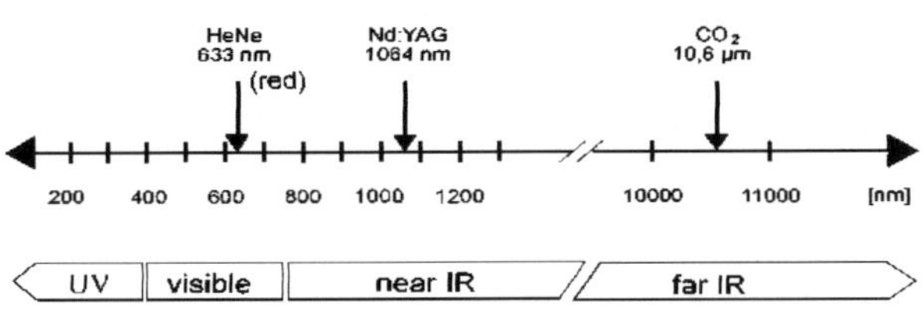

Fig. 1. Light spectrum (Trumpf Laser, 2005)

The laser light of an Nd: YAG laser can be routed through glass optics and optical fibbers.

The positioning system has the following configuration: two axes for moving the working table in xOy plane and one axis for moving the processing optics Oz axis in order to focus the laser beam on the work piece.

The precision of the positioning system depends on the accuracy of the axis. In our case we use DGE -25-200-SP for the Oz axis and two DMES-25-200 for the positioning in the xOy plane.

The repeatability of the focusing axis, Oz, is 0.02 mm and for the other axes, Ox and Oy is 0.05 mm. The stepping motor, MTR-ST-57-48S, has the full step angle of 1.8° with a 5% maximum error. It is possible to supply two coils simultaneously with different current share. The result is a 1/2, 1/4, 1/5, 1/8, 1/10, 1/32 of a step, depending on the current share. These allow considerable refinement in the maximum resolution of positions to which a stepping motor can run.

The smallest incremental path (resolution) on a positioning axis is determined by the motor's step angle (number of steps per revolution) and the feed constant of the positioning axis (determined by the diameter of the input pinion or the slope of the spindle).

This can be calculated as follow:

$$n_s = \frac{360°}{\alpha_s} \tag{1}$$

$$d_r = \frac{d_f \cdot i}{n_s} \tag{2}$$

In the equations above n_s represents number of steps per revolution, α_s motor angle, d_r resolution, d_f feed constant of the axis, i gear multiplication.

For the axis of the work table, in case of the full step, the resolution is 0.012 mm per step and for the focusing axis is a 0.05 mm resolution per step. From this point the position of the work piece can be calculated. Some error can occur if the stepping motors lose steps. This happens only when the load exceeds the maxim load on the axis.

3. RESEARCH COURSE

The connection between positioning system and laser device was made by programming an Atmel microcontroller (Atmel ATMEGA8535). For machining was necessary to control the laser device at the same time with the positioning system. With the help of the PC interface made in Delphi software was possible to synchronize the laser device and the positioning system for machining. The laser pulses were generated when the work piece was in the processing position.

For path identification we used the monitoring system of laser optics and an adapted web camera. The web camera had a manual focus which allows calibrating the focus distance to 182mm. After this calibration the working distance can be adjusted by moving the optics on the Oz axis.

Fig. 2. The edge of the overlap work piece

For material processing the laser beam has to be focused to the necessary power density. A convex lens is used for this.

A collimation unit is used to focus the laser beam onto the work piece. The collimation lens converts this beam into a parallel beam. A mirror is inserted into the beam guidance to monitor and adjust the work piece. The mirror allows the transmission light having the wavelength of the laser beam. It reflects visible light which comes from the work piece through the lens into the beam guidance.

Most of the existing path following solutions performs complex analysis on the images of the path using curve matching or Kalman filters (Ma et al., 1999). These methods produce accurate information about the curvature of the path and allow a positioning system, typically moving at slow speed, to accurately drive the path. However these methods require a relatively large amount of processor time and often require considerable modification if the position of the camera is altered.

The colour image proved too difficult to work with. Fortunately with the PC interface provides an efficient routine to convert the 24-bit colour image to a 8-bit grey scale. A threshold, chosen to select the maximum amount of track and a minimum amount of the surroundings, is then applied to the grey scale image to produce a black and white image that is further processed. We extract two pieces of information from the image: the position of the path, offset from the centre of the image, and the gradient of the path.

Rather than processing the entire region, which would require too much computing time, only the outer edges of the processing region are checked to locate the path.

These edges are offset from the edges of the image due to the presence of a one pixel black border and the presence of noise caused by synchronisation problems if we use the 30 frames per second camera driver.

Fig. 3. The edge after conversion

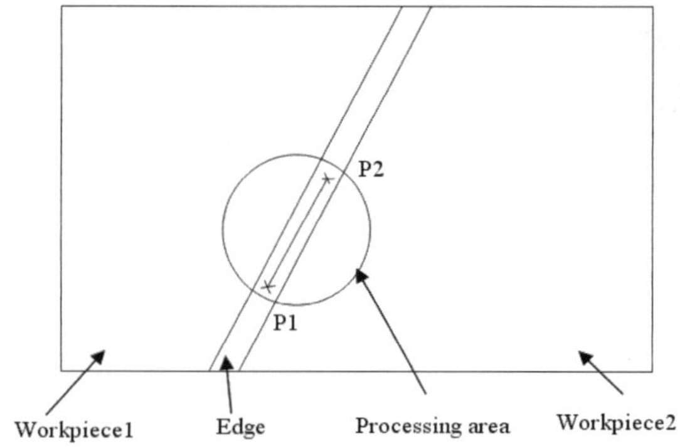

Fig. 4. Image processing detail

The point P1 is tested at first, across the lower edge, then up the left edge and finally up the right. If P1 does not exist the track will be assumed to be absent. Point P2 is located by checking the top edge, then the left and right edges. If P1 is found on one of the side edges, that edge will not be checked for P2. The positions of P1 and P2 are computed by taking the average position of the white pixels on the edge. One result of this is that if two parallel lines are present P1 and P2 will be between the lines, which would allow the positioning system to perform simple autonomous road driving if the field of view from the camera is wide enough. The line between P1 and P2 gives the trend of the path in the region of interest.

The offset of the path is computed by averaging the differences between the horizontal positions of P1 and P2 and normalising the result so that it is positioned between –1, for the left-hand side of the image, and 1, for the right hand side.

The gradient is computed and then scaled so that a gradient of zero indicates a vertical line (in the image), a negative gradient indicating that the line slopes to the left and a positive gradient indicating a slope to the right. A horizontal line is represented by a gradient of 1 (the maximum positive gradient) as there is no way of knowing which direction it is leaning in.

With the processing path obtained we can start the welding process by setting the laser parameters. The laser parameters are specific for different types of material, in our case metal sheets.

4. CONCLUSION

With this method the team proposes the optimisation of the welding process by decreasing the process time. This can be obtained by automatically identifying the working zone, with a minimal intervention from the operator. Certain issues need further research: false edge, discontinued edge, the overlap of the paths for the proximate edges, start and end points.

¶

5. REFERENCES

¶

Hugel, H. (1992). *Strahlwerkzeug Laser (Laser Beam Tool)* Teubner Verlag, ISBN 3-519-06134-1, Stuttgart

Koechner, W. (2006), *Solid-State Laser Engineering,* Springer Science and Business Media, Inc., e-ISBN: 0-387-29338-8

Ma, Y., Kosecká, J. and Sastry, S. (1999). Vision Guided Navigation for a Nonholonomic Mobile Robot, *IEEE Transactions on Robotics and Automation*, vol. 15, no. 3, (june 1999) page numbers (521-536), ISSN: 1042-296X

Popa, M. S. (2005). *Tehnologii şi maşini neconvenţionale, pentru mecanică fină şi mecatronică (Unconventional Technologies and machines for Fine Mechanics and Mechatronics)*, U.T. Pres, ISBN 973-662-148-0, Cluj-Napoca, Romania

Trumpf Laser (2005). *LCB Laser Devices. Basic Training*, Trumpf Laser GmbH+Co.KG, Germany

Annals of DAAAM for 2008 & Proceedings of the 19th International DAAAM Symposium, ISSN 1726-9679
Editor B. Katalinic, Published by DAAAM International, Vienna, Austria 2008
Make Harmony Between Technology and Nature, and Your Mind will Fly Free as a Bird

STATISTICAL MODELLING OF MEASURES OF THERMAL COMFORT

POPA, M[onica]; SIRBU, D[ana]; CURSEU, D[aniela] & POPA, M[arcel] S[abin]

Abstract: *The correlation coefficient of subjective experienced PMV and calculated PMV (ISP 7730) was 0.25, which is considered weak. This means that the calculated PMV measured poorly the workers' degree of discomfort. The linear regression model was used to get a better fit between the subjective estimated PMV and the measured environmental thermal variables. The correlation was now 0.52, being two times higher than the obtained by the method given in ISO 7730. A logistic regression model was used to fit the actual feeling of draught to the environmental thermal variables. The estimated logistic regression model fits the data set better than the deterministic model (PD). These models seem to explain the actual sensations of the workers much better than the complicated indices.*

Key words: *comfort, thermal, regression, model, questionnaire*

1. INTRODUCTION

The first requirement for an acceptable thermal environment is that a person feels thermally neutral as regards the whole body, i.e. the worker does not know whether he would prefer a higher or lower ambient temperature level. This is usually evaluated by the PMV index (ISO7730). The index is a function of the activity level, clothing insulation, air speed operative temperature and humidity. Thermal neutrality as described by the PMV index is not the only condition for thermal comfort. A person may feel thermally neutral as regards the body as a whole, but may not be comfortable if one part of the body is warm and other cold. It is therefore a further requirement for thermal comfort that no local warm or cold discomfort exists in any part of the body. Such local discomfort may be caused by an asymmetric radiant field, by local convective cooling (draught) and by a vertical air temperature gradient. The percentage of those who were dissatisfied (PD) due to draught in ventilated working areas is described by the PD index (Melikov et al, 1989).

The logical way of investigating how the thermal climate is perceived in buildings is to ask people to describe the thermal climate with the use of questionnaires and rating scales. Seven-point scales, such as Bedford or PMV scale have often been used. In addition, questions on air motion and indoor air quality are usually asked. Great individual differences were found in the thermal work environment between subjective estimated and calculated PMV values (Popa, 2000).When the temperature at the work site rose, the workers graded the thermal comfort according to the PMV scales as "too warm", although the air temperature was estimated to be lower than the actual value. In offices, the calculated PMV values were lower than the estimated PMV values. Then observation is in agreement with the findings of other studies (Croome et al., 2000; De Dear et al., 2001). The PD index predicts that usually considerably less than 20% of the workers should feel the velocity-related discomfort. The results of the questionnaire concerning draught indicated that there was a markedly great difference between the subjectively estimated and calculated values. The aim of the study was to develop a thermal comfort model by using statistical modeling of the measured environmental thermal variables and subjective ratings of the thermal environment given by the workers in a moderate climate.

2. MATERIAL

A thermal questionnaire was used in 80 working locations. The workers were asked to fill out at least one self-administered questionnaire inquiring thermal sensation, ventilation and whether they felt air movement, and if yes, whether the air movement felt uncomfortable. The workers also graded their thermal comfort according to the PMV scale and estimated the air temperature of the working areas. Altogether 600 workers evaluated their degree of comfort by using the PMV scale. The thermal insulation of a whole clothing ensemble (I_{cl}) was estimated by summing up the values for individual garments. Activity level (M) was estimated by means of tables. The predicted mean vote (PMV index) was determined from the measured values of thermal variables and the values of M and I_{cl}. At the same time, continuous of the thermal variables at the work sites was conducted using indoor climate analysers. The Brüel & Kjaer 1213 Indoor Climatic Analyser measured all the thermal variables including radiant temperature asymmetry. One of the faces of the asymmetry transducer was directed towards the window. The second thermal analyzer was constructed using an HP portable computer, an HP 3421 data acquisition unit, YSI 401 thermistors, Disa 54N50 low velocity flow analyzer and Humicap humidity sensors. Air temperatures were measured at the ankle and neck levels and air velocities at the neck level only. The indoor thermal climate measurements were made according to the ISO 7730 standard. The sensors were fixed on a tripod placed near the worker.

3. METHODS

We consider three response variables which measure the experience of thermal comfort of a person. The first is the subjective experience Y_0. The response Y_0 is a random variable depending on environmental variables including biological and psychological factors of a person. The second measure Y_1 is the PMV index or PD index. In both cases the model of estimated response Y_2 of the fitted statistical model of Y_0. The predictors $x = (x_1, x_2, \dots x_k)$ of model are the measured environmental variables such as air temperature, air velocity, relative humidity, etc. The selected model is a regression model which is an ordinary linear one (Weissberg, 1985) if Y_0 is continuous (PMV) and logistic (Hosmer et al., 1989) if Y_0 is a dichotomous (PD) variable. The statistical analyses have been performed by the SAS software system. The ordinary regression model of the continuous response Y_0 is:

$$Y_0 = \beta_0 + \beta_1 x_1 + \beta_2 x_2 + \dots + \beta_k x_k + \epsilon \qquad (1)$$

The coefficients β_i, I = 1, … , k are unknown parameters. The estimates b_i of the coefficients β_i are computed from the data set by the regression analysis. The statistical error ϵ is supposed to be normally distributed. The fitted model of Y_0 is now defined as:

$$Y_2 = b_0 + b_1x_1 + b_2x_2 + \ldots + b_kx_k \quad (2)$$

Comparison of the measures Y_1 and Y_2 can be made by the correlation coefficient R. If R (Y_2, Y_0) > R (Y_1, Y_0) then Y_2 will fit the data set better than PMV index Y_1. If the response Y_0 is a dichotomous variable with two possible values: 1 = satisfied or 0 = unsatisfied, the measure Y_2 is constructed by a logistic regression model. In this case we are modeling the probability p(x) that a person is satisfied (=1) in a thermal environment $x = (x_1, x_2, \ldots, x_k)$:

$$p(x) = P(Y_0 = 1 \mid x) \quad (3)$$

Thus the probability that a person is unsatisfied in the environment is:

$$P(Y_0=0 \mid x) = 1 - p(x) \quad (4)$$

In a logistic regression model the logit transformation of the probability p(x) is modeled as follows:

$$g(x;\beta) = \log\left[\frac{p(x)}{1-p(x)}\right] = \beta_0 + \beta_1x_1 + \beta_2x_2 + \ldots + \beta_kx_k \quad (5)$$

The subjective experience Y_0 is the response of the model and the estimates b_i of the coefficients β_i are computed from the data set by the logistic regression analysis. The model of Y_2 will be the estimated logistic regression model of Y_0:

$$p(x,b) = \frac{e^{g(x;b)}}{1-e^{g(x;b)}}, g(x;b) = b_0 + b_1x_1 + b_2x_2 + \ldots + b_kx_k \quad (6)$$

Thus $P(Y_2=1 \mid x) = p(x;b)$. As a measure of difference between the observed value of Y_0 and the fitted value of either Y_1 or Y_2 is the Pearson residual:

$$r(y_j, p_j) = \frac{y_j - n_jp_j}{\sqrt{n_jp_j(1-p_j)}} \quad (7)$$

Here y_j is the observed value of Y_0, $p_j = P(Y_1 = 1 \mid x_j)$ (PD index) or $p_j = P(Y_2 = 1 \mid x_j)$, n_j is the number of observations classified according to x. In unclassified data $n_j = 1$, j = 1, 2,...n.

The measure Y_1 or Y_2 which has the smallest X^2 is to be considered as the best estimate of Y_0. The Pearson chi-square statistic is the summary statistics:

$$X^2 = \sum_{j=1}^{n} r(y_j, p_j)^2 \quad (8)$$

4. RESULTS

The response variable Y_0 is subjective experienced PMV. A stepwise procedure has selected the following environmental variables to the estimated regression model for the measure Y_2:

Environmental variable	Parameter estimate b	Standard error of b	P-value
Constant	-6.5054	0.7808	0.0001
Winter/summer	0.5184	0.1349	0.0001
Morning/afternoon	-0.3456	0.0981	0.0005
Man / woman	0.2768	0.1052	0.0088
Turbulence intensity	-0.0055	0.0025	0.0307
Metabolic rate	0.0063	0.0018	0.0007
Air temperature, ankle	-0.0972	0.0465	0.0373
Air temperature, neck	0.3608	0.0435	0.0001
Relative humidity	0.0128	0.0060	0.0332
Temperature asymmetry	-0.0347	0.0197	0.0789

The correlation coefficients are R (Y_2, Y_0) = 0.52 and R (Y_1,Y_0) = 0.25. This indicates that the estimated statistical regression model of Y_2 (PMV) fits the data set better than the deterministic model of Y_1 (PMV). When the variable Y_0 is the subjective experienced PD, a stepwise procedure has selected the following environmental variable to the estimated logistic regression model for the measure Y_2

Environmental variable	Parameter estimate b	Standard error of b	P-value
Constant	5.4216	1.6977	0.0014
Metabolic rate	-0.0124	0.0039	0.0014
Morning/afternoon	-1.2140	0.6027	0.0440
Air temperature, neck	-0.1959	0.0588	0.0009
Air velocity	1.7300	0.7988	0.0303

The Pearson chi-square statistic for the logistic regression model of Y_2 was 592 and for the deterministic model of Y_1 was 3199.This indicates that the estimated logistic regression model of Y_2 fits the data set better then the deterministic model of Y_1(PD).

5. DISCUSSION

The correlation coefficient of subjective experienced PMV and calculated PMV (IS0 7730) was 0.25, which was considered weak. This indicates that the calculated PMV measured poorly the workers' degree of discomfort. The linear regression model was used to get a better fit between the estimated PMV and the measured environmental thermal variables. The correlation coefficient was now 0.52, being two times higher than that obtained by the method given in ISO 7730. The workers' subjective estimation correlated with the percentage of people feeling draught to the measured environmental thermal variables. The estimated logistic regression model fits the data set better than the deterministic model of the PD according to the Pearson chi-square-statistic. These models seem to explain the actual sensations of the workers much better than the complicated index.

6. REFERENCES

Croome, D.C. & Baizhan, L (2000) Productivity and indoor environment, *Proceedings of Healthy Buildings 2000*, vol.1, pp 629 -634.

De Dear R. & Brager S.G. (2001) The adaptive model of thermal comfort and energy conservation in the built environment, *International Journal of Biometeorology*, vol.45, nr.2, pp. 100-108.

Hosmer, D.& Lemeshow, S.(1989) *Applied logistic regression*. John Wiley & Sons, New York.

ISO 7730 (1996) Moderate thermal environments. Determination of the PMV and PPD indices and specification of the conditions for thermal comfort. International Organization for Standardization, Geneva.

Melikov, A.K.; Fanger P.O. (1989) Draught risk in practice, *Proceedings of 2nd World Congress on Heating, Ventilating, Refrigerating and Air-conditioning,* vol. 3, pp. 121 – 126.

Popa, M. & Popa, M.S. (2000). Indoor air quality problems in Romania: monitoring and control, *Proceedings of the International Conference on Advanced Metrology „Metrologia 2000*, vol.1, pp. 476-482.

Weissberg, S. (1985) *Applied linear regression*, second edition, John Wiley & Sons, New York.

Annals of DAAAM for 2008 & Proceedings of the 19th International DAAAM Symposium, ISSN 1726-9679
Editor B. Katalinic, Published by DAAAM International, Vienna, Austria 2008
Make Harmony Between Technology and Nature, and Your Mind will Fly Free as a Bird

NEURO-FUZZY CONTROL FOR KHEPERA III MOBILE ROBOT

POPESCU, C[ristina]; BUCUR, G[abriela] & POPESCU, C[atalin]

Abstract: *Navigation of mobile robots in unstructured environments is an actual problem for scientists and researchers. In this paper, a neuro-fuzzy based approach is proposed for mobile robot control, which co-ordinates the sensor information and robot motion together. A fuzzy logic system is designed to implement obstacle avoidance behaviour for mobile robot. It is used a Takagi-Sugeno controller which can be named a neuro-fuzzy controller. Simulation studies and experiments on a real robot demonstrate the efficiency of the proposed approach.*

Key words: *neuro-fuzzy, control, mobile robot*

1. INTRODUCTION

Since 1985, many researchers investigated the application of fuzzy controllers in mobile robotics. Mostly, the results are obtained by simulation.

In this paper we will describe experiments with the Khepera III mobile robot, developed in 2007 in the Microcomputing Laboratory at the Swiss Federal Institute of technology. The earliest versions of the Khepera robot were available since 1992.

This problem was similarly studied for Khepera II robot (Godjevac, 1997), but for this robot the sensorial system was change, so we adapted the control structure. We also developed Matlab functions to communicate with the robot, via bluetooth, using a Windows platform.

2. THE KHEPERA III MOBILE ROBOT

Our work is based on the experiments with a robot named Khepera III. It is cylindrical in shape, measuring 130 mm in diameter and 70 mm in height and its weight is 690 g.

The basic configuration of Khepera is composed of DsPIC 30F5011 at 60 MHz processor, system and user memory, extension busses and a serial link. The microcontroller includes all the features needed for interfacing with memories, with I/O ports and with external interruptions.

The sensory/motor board includes two DC motors with incremental encoders, 9 infrared proximity and ambient light sensors with up to 25 cm range, 2 infrared ground proximity sensors for line following applications and 5 ultrasonic sensors with range 20 cm to 4 meters.

3. NEURO-FUZZY CONTROLLER FOR OBSTACLE AVOIDANCE

Our experiment is controller design for obstacle avoidance of the Khepera III with a minimal number of linguistic rules. The robot has 9 infrared proximity sensors and 2 motors. The inputs are the linguistic variables: *distances between the robot and the obstacle* and the outputs are the linguistic variables: *motor speeds*. We work with four input linguistic variables: *distance to the left* D_s, *to the front* D_f, *to the right* D_d and on *the back* D_{sp}, like in figure 1. If we note with S_1, …, S_{11} the sensor values normalised within [0,1], then the input variables are calculated in the following way (Godjevac&Steele,2001):

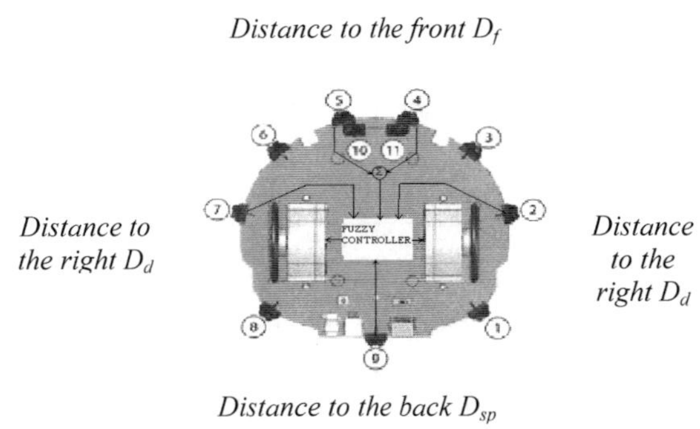

Fig.1. Inputs and outputs variables of Khepera III.

$$D_s = S_7 \quad (1)$$

$$D_f = (S_4+S_5)/2 \quad (2)$$

$$D_d = S_2 \quad (3)$$

$$D_{sp} = S_9 \quad (4)$$

For each input linguistic variable, we define 3 linguistic values: *small*, *medium*, *big* , and for each output we define 7 linguistic values: *backward fast*, *backward medium*, *backward slow*, *stop*, *forward slow*, *forward medium*, *forward fast* (Popescu, 2006).

The first step is up to set the initial base of the rules and the parameters of the controller. We set 16 rules, which constitute the initial base rules, to implement the obstacles avoidance behaviour. The second step is to implement the controller. The Takagi Sugeno controller's parameters will be adapt by a supervised learning method based on gradient descent, which consists of modifying parameters in order to obtain the desired output in response to given inputs (Sugeno&Murakami,1985). We choose an on-line identification method because it uses less computational time and less memory space than an off-line algorithm (Popescu, 2007). After learning, linguistic rules are extracted from the system parameters. They may be different than the initial rules. The parameters of membership functions are initialised to uniformly cover the input space. Each membership function for input variables are Gaussian functions and for output variables are singleton functions, like in figure 2.

The initial base rules describing obstacle avoidance behaviour are presented in the table 1.

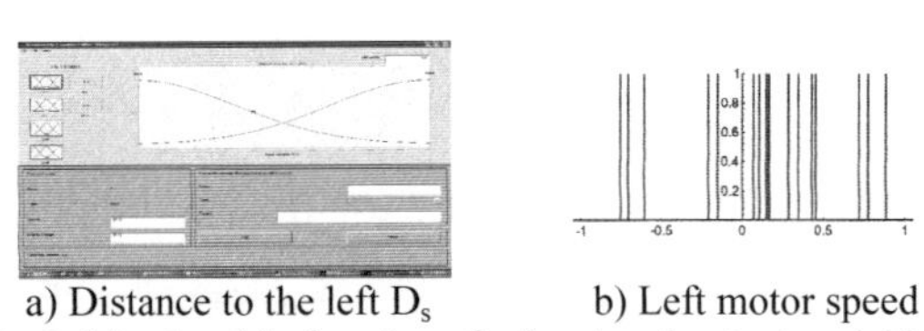

a) Distance to the left D_s b) Left motor speed v_s

Fig. 2. Membership functions for input and output variables.

Rule	D_s	D_f	D_d	D_{sp}	v_s	v_d
R1	Small	Small	Small	Small	Stop	Stop
R2	Small	Small	Small	Big	Backward fast	Backward slow
R3	Small	Small	Big	Small	Backward slow	Forward fast
R4	Small	Small	Big	Big	Forward medium	Backward medium
R5	Small	Big	Small	Small	Backward fast	Backward fast
R6	Small	Big	Small	Big	Forward medium	Stop
R7	Small	Big	Big	Small	Backward slow	Stop
R8	Small	Big	Big	Big	Forward medium	Forward medium
R9	Big	Small	Small	Small	Backward fast	Forward fast
R10	Big	Small	Small	Big	Forward medium	Forward fast
R11	Big	Small	Big	Small	Forward slow	Backward medium
R12	Big	Small	Big	Big	Backward slow	Backward slow
R13	Big	Big	Small	Small	Backward slow	Forward medium
R14	Big	Big	Small	Big	Backward medium	Backward slow
R15	Big	Big	Big	Small	Stop	Backward slow
R16	Big	Big	Big	Big	Backward medium	Forward medium

Tab. 1. The initial base rules of the controller.

The particularity of the Takagi and Sugeno's controller is that the consequent parts of linguistic rules are expressed as functions of linguistic variables, and in this case we will consider only the special type of Takagi and Sugeno's controller were these functions are constants.

The Takagi and Sugeno's fuzzy controller has three types of parameters to adapt: centre values $a=(a_{11},...,a_{nm},...,a_{NM})^T$, width values $b=(b_{11},...,b_{nm},...,b_{NM})^T$ and consequent values $c=(c_{11},...,c_{nk},...,c_{NK})^T$, where M is the number of input, K is the number of outputs and N is the number of linguistic rules.

For Gaussian membership functions, the equation for the adaptation of the parameter a, b and c is like following:

$$a_{nm}(t+1)=a_{nm}(t)-\Gamma_a \frac{u_n}{\sum_{n=1}^{N} u_n} \frac{x_m(t)-a_{nm}(t)}{b_{nm}(t)^2} \sum_{k=1}^{K} \big(y_k(t)-y_{d_k}(t)\big)\big(c_{nk}(t)-y_k(t)\big) \qquad (5)$$

$$b_{nm}(t+1)=b_{nm}(t)-\Gamma_b \frac{u_n}{\sum_{n=1}^{N} u_n} \frac{\big(x_m(t)-a_{nm}(t)\big)^2}{b_{nm}^3} \sum_{k=1}^{K} \big(y_k(t)-y_{d_k}(t)\big)\big(c_{nk}(t)-y_{d_k}(t)\big)$$

$$c_{nk}(t+1)=c_{nk}(t)-\Gamma_w \frac{u_n}{\sum_{n=1}^{N} u_n} \big(y_k(t)-y_{d_k}(t)\big)$$

During the learning process the rules base was changed and we observed that there are no contradictions – two rules with the same antecedent parts and different consequent parts. For inference process we used Min-Max method and Centre of Gravity method for defuzzification.

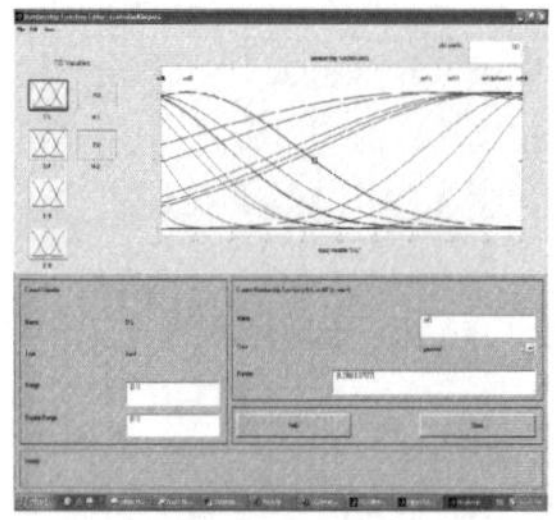

Fig. 3. Modified membership functions after learning for D_s.

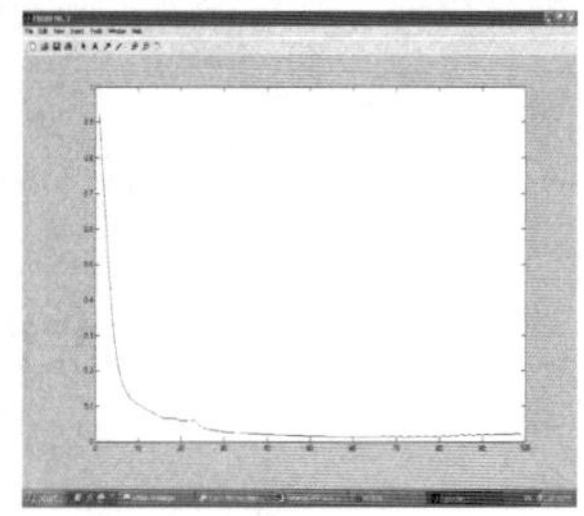

Fig.4. Evolution of error in learning process.

The membership functions are modified after learning process like in figure 3. It is observed that after 200 iterations Khepera learned the desired behaviour and the evolution of the error is very good, as presented in figure 4.

When implementing this controller in the real robot Khepera, we observe that with 16 rules the robot avoid obstacle with success and if we change the rule base, we can implement another behaviour, like wall following or reached a target.

4. CONCLUSION

In this paper we studied the design and the implementation of neuro-fuzzy controller on a mobile robot. Godjevac (1997) was first studied the principle, method and algorithm and we adapt them for the new Khepera III robot. The neuro-fuzzy controller was first validated in simulation, then applied to control a real mobile robot.

The first step is to build an initial set of linguistic rules, which is set up to run its assigned task. For adjust the parameters of the fuzzy controller we used an adaptive algorithm similar to a neural network learning. The learned knowledge can be represented in the form of linguistic rules, which can be extracted from the adapted parameters of the fuzzy controller and this is considered as the biggest advantage of a neuro-fuzzy network compared to a neural network.

For the future, we want to resolve the "dead cycle" problem that can appear on navigation of mobile robot and we want to implement another behaviour for the Khepera, like reach a target.

5. REFERENCES

Godjevac, J. (1997). *Neuro-Fuzzy Controllers Design and Application*, Presses Polytechniques et Universitaires Romandes, ISBN 2-88074-355-9, Lausanne.

Godjevac, J. & Steele, N. (2001). Neuro-Fuzzy Control for Basic Mobile Robot Behaviours, In: *Fuzzy Logic Techniques for Autonomous Vehicle Navigation*, Driankov, D. & Saffiotti, A. (Ed.), pp. 97-117, Physica-Verlag Heidelberg, ISBN 1434-9922, Heidelberg.

Popescu C. (2006). Mobile Robot Navigation Using Fuzzy Logic For Obstacle Avoidance, *Bulletin of "Petroleum-Gas" University of Ploiesti* , 5th International Symposium on Process Control SPC 2006, Ploiesti, ISSN 1224 – 8495.

Popescu C. (2007). Neural Network Techniques For Mobile Robot Navigation, *Annals of the University of the Petrosani*, Electrical Engineering, vol 9, pp. 332-338, ISSN 1454-8518.

Sugeno, M. & Murakami, K. (1985). An Experimental Study on Fuzzy Parking Control Using a Model Car, In: *Industrial Applications of Fuzzy Control*, Sugeno M. (Ed.), pp.125-138, North-Holland.

Annals of DAAAM for 2008 & Proceedings of the 19th International DAAAM Symposium, ISSN 1726-9679
Editor B. Katalinic, Published by DAAAM International, Vienna, Austria 2008
Make Harmony Between Technology and Nature, and Your Mind will Fly Free as a Bird

STABILITY OF TRANSVERSAL VIBRATIONS IN HIGH-SPEED DRILLING AND THE INFLUENCE ON THE PROCESSING PRECISION

POPESCU, D[aniel] & CHERCIU, M[irela]

Abstract: *The paper presents a general mathematical model for gyroscopic systems. This is presented in operator form, in case of transversal vibrations of a high-speed drill, for which is determined the critical rotation speed at which the system stability is lost. Together with the drill deformations it is also computed the maximum transversal deformation. It is shown the influence of transversal vibrations on the processing precision.*
Key words: *gyroscopic systems, high-speed drilling, transversal vibrations, stability, processing precision*

1. INTRODUCTION

Gyroscopic systems include as examples translating and rotating strings, beams, membranes and plates. These systems posses combinations of system parameters that produce vanishing eigenvalues. Such combination of system parameters are designated herein as critical system parameters and their vanishing eigenvalues and the corresponding eigenfunctions as critical eigenvalues and critical eigenfunctions.

One interesting phenomenon for gyroscopic dynamic systems is that unstable conservative systems can be stabilized by increasing the gyroscopic forces. Criteria to determine reasonably small gyroscopic forces for stabilization are important for practical problems. Stability criteria for gyroscopic systems have been formulated for both discrete and continuous systems. For discrete systems, the criteria usually relate the definiteness of a combination of the coefficient matrices in the equation of motion to the stability of the system (Huseyin et al., 1983; Huseyin, 1991). The extension of these criteria to continuous can only ensure stability when the system stiffness operator is positive definite (Shieh, 1971; Wickert & Mate, 1990).

Other stability criteria have been developed to examine continuous systems whose operator is not definite, but these criteria usually require knowledge of the system eigenfunctions (Shieh, 1971), which is normally as difficult to obtain as the eigenvalues themselves.

2. THE MATHEMATICAL MODEL

It is considered the general equation of continuous gyroscopic systems (Renshaw, 1996):

$$M\ddot{u} + G\dot{u} + Ku = f \tag{1}$$

where: M, G, K – differential linear matrices in spatial domain
u – displacements
f – forcing term

In order to obtain the mathematical model for the main spindle vibrations, the following assumptions are made:

a. there are no distributed superficial forces on the external spindle surface
b. there are no additional fulcrums or joints that can produce shocks
c. the initial state of the spindle is unstrained
d. a plane section normal on the spindle axis before deformation remains plane without keeping perpendicularity

Under these assumptions, for the mathematical model of transversal vibrations given by (1):

$$M = \begin{pmatrix} 1 - p\dfrac{\partial^2}{\partial x^2} & 0 \\ 0 & 1 - p\dfrac{\partial^2}{\partial x^2} \end{pmatrix};$$
$$G = \begin{pmatrix} 0 & 2\Omega \\ -2\Omega & 0 \end{pmatrix};$$
$$K = \begin{pmatrix} c^2\dfrac{\partial^4}{\partial x^4} - \Omega^2 & 0 \\ 0 & c^2\dfrac{\partial^4}{\partial x^4} - \Omega^2 \end{pmatrix}; \tag{2}$$
$$u = \begin{pmatrix} u_1 \\ u_2 \end{pmatrix};$$
$$f = \begin{pmatrix} f_1 \\ f_2 \end{pmatrix}$$

where: $u_1(x)$, $u_2(x)$ – transversal deformation of the spindle on two perpendicular axes; $f_1(x)$, $f_2(x)$ – external load; Ω – spindle angular speed; $p(x)$, $c(x)$ – coefficients depending on section characteristics and spindle material.

We have:

$$p(x) = \frac{I(x)}{A(x)}; c^2(x) = \frac{EI(x)}{\rho A(x)} \tag{3}$$

The limit conditions in this case are:

$$\begin{gathered} u_1(0;t) = u_1(l;t) = 0 \\ \frac{\partial^2 u_1}{\partial x^2}(0;t) = \frac{\partial^2 u_1}{\partial x^2}(l;t) = 0 \end{gathered} \tag{4}$$

$$\begin{gathered} u_2(0;t) = u_2(l;t) = 0 \\ \frac{\partial^2 u_2}{\partial x^2}(0;t) = \frac{\partial^2 u_2}{\partial x^2}(l;t) = 0 \end{gathered} \tag{5}$$

In this case larger deformations are allowed and applying the finite Fourier transform gives:

$$M = \begin{pmatrix} 1 + p\dfrac{n^2u^{-2}}{l^2} & 0 \\ 0 & 1 + p\dfrac{n^2u^{-2}}{l^2} \end{pmatrix};$$

$$K = \begin{pmatrix} c^2 \dfrac{n^4 u^{-4}}{l^4} - \Omega^2 & 0 \\ 0 & c^2 \dfrac{n^4 u^{-4}}{l^4} - \Omega^2 \end{pmatrix} \quad (6)$$

If the angular speed of the spindle is smaller than the critical value:

$$\Omega_{cr} = \frac{\pi^2}{4} \cdot \sqrt{\frac{E}{\rho}} \cdot \frac{d}{l^2} \quad (7)$$

then the spindle vibrations remain stable.

3. CASE STUDY

Is considered a steel drill with $d = 4 \cdot 10^{-3}$ m, $l = 0.1$ m; $\Omega_{cr} = 5120\ s^{-1}$ corresponding to $n_{cr} = 48895$ rpm.

The system is stable and initially the transversal vibrations on the two axes present maximum values:

$$u_{1\max}(l_0) = \frac{2F}{l_0}\left(\frac{c^2\pi^{-4}}{l^4} - \Omega^2\right)^{-1} \cdot \left(1 + \cos\frac{\Omega l^2}{c\pi}\right) \cdot \left(1 - \cos\frac{\pi d_0}{l}\right) \cdot \sin\frac{\pi d_0}{l}$$
$$u_{2\max}(l_0) = \frac{2F}{l_0}\left(\frac{c^2\pi^{-4}}{l^4} - \Omega^2\right)^{-1} \cdot \sin\frac{\Omega l^2}{c\pi} \cdot \left(1 - \cos\frac{\pi d_0}{l}\right) \cdot \sin\frac{\pi d_0}{l} \quad (8)$$

where:

$$f_1(x;t) = \begin{cases} \dfrac{F}{l_0}; & x \in [0, l_0] \\ 0; & x \in [l_0, l] \end{cases}$$
$$f_2(x;t) = 0 \quad (9)$$

4. RESULTS AND CONCLUSIONS

In the example presented it was determined the critical angular speed for a $d = 4 \cdot 10^{-3}$ m drill. The next figure presents the variation of the critical angular speed as function of the diameter and length of the drill.

The dependency relation is:

$$\Omega_{cr} = 12803 \cdot \frac{d}{l^2} \quad (10)$$

where $d \in (0.001 \div 0.01)$m ; $l \in (0.05 \div 0.15)$m

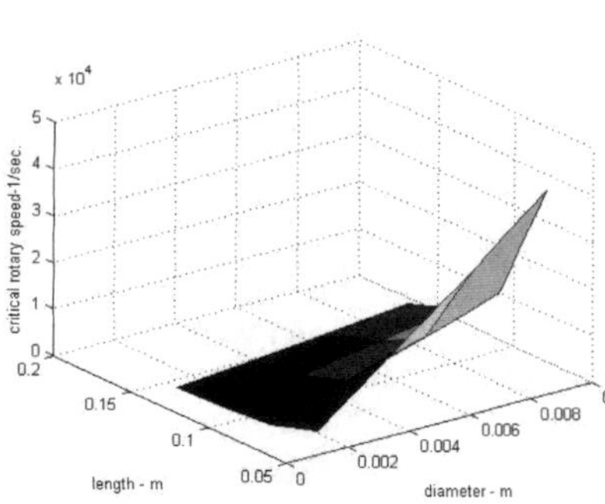

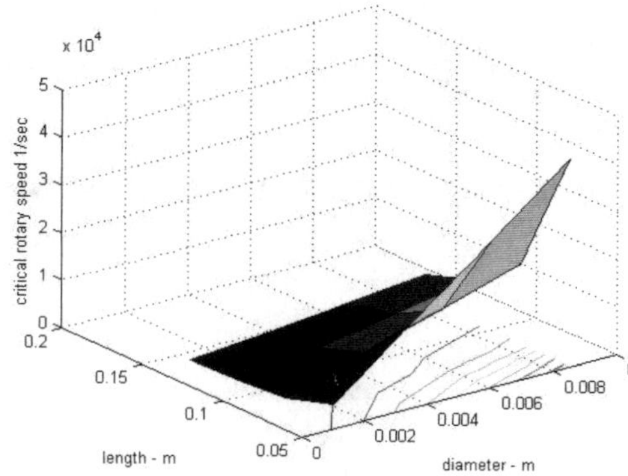

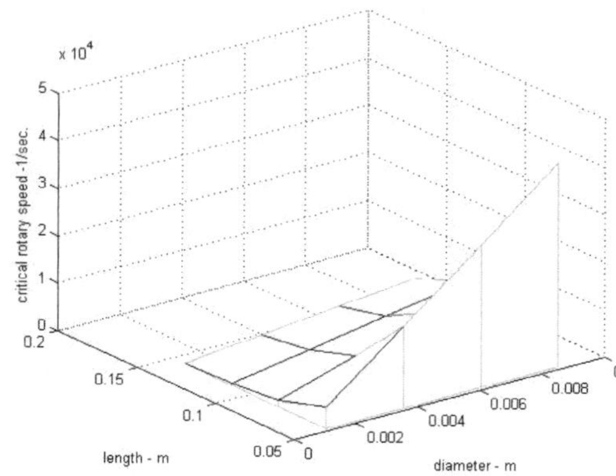

Fig. 1. Variation of the angular speed as function of the drill diameter and length

It can be observed that the critical speed increases proportionally with the drill diameter and decreases with its length square.

Using (8), (9) we can determine the amplitudes of transversal vibrations at the surface of the work material. For example, for a drilling process with $l = 0.1$ m, $l_0 = 0.01$ m, $d = 0.01$ m, $F = 1550$ N it is obtained:

$$u_{\max}(l_0) = \sqrt{u_{1\max}^2 + u_{2\max}^2} = 0.554 \cdot 10^{-6}\,m \quad (11)$$

Consequently, for this process the influence of transversal vibrations on the processing precision is insignificant.

5. REFERENCES

Huseyin, K.; Hagedorn, P. & Teschner, W. (1983). On the Stability of Linear conservative Gyroscopic Systems, In: *Journal of Applied Mathematics and Physics*, vol. 34, pp. 807-815.

Huseyin, K. (1991). On the Stability Criteria for Conservative Gyroscopic Systems, In: *Journal of Vibration and Acoustic*, vol 113, pp. 58-61.

Renshaw, A.A. & Mote, C.D. (1996). Local Stability of Gyroscopic Systems Near Vanishing Eigenvalues, In: *ASME Journal of Applied Mechanics*, vol. 63, pp. 116-120.

Shieh, R.C. (1971). Energy and Variational Principles for Generalized (Gyroscopic) Conservative problems, In: *International Journal of Non-Linear Mechanics*, vol. 5, pp. 495-509.

Wickert, J.A. & Mote, C.D. (1990). Classical Vibration Analysis of Axially Moving Continua, In: *ASME Journal of Applied Mechanics*, vol. 57, pp. 738-744.

Annals of DAAAM for 2008 & Proceedings of the 19th International DAAAM Symposium, ISSN 1726-9679
Editor B. Katalinic, Published by DAAAM International, Vienna, Austria 2008
Make Harmony Between Technology and Nature, and Your Mind will Fly Free as a Bird

ASPECTS ABOUT ADDERS PSEUDORANDOM TESTING TECHNIQUES

POPESCU, D[aniela] & PATER, M[irela]

Abstract: *The exhaustive testing is not efficient for adders having the dimension greater than 8. They can be tested by using the logic partitioning technique. By applying the sensitized-path logic partitioning technique, which is presented in the first part of the paper, we present comparative analysis of testing possibilities for different kinds – with various sizes – of adders.*
Key words: *Testability, adders, pseudorandom testing*

1. INTRODUCTION IN PSEUDO--EXHAUSTIVE TESTING

Pseudoexhaustive testing achieves many of the benefits of exhaustive testing but usually requires far fewer test patterns (Abramovici et al., 1996). It relies on various forms of circuit segmentation and attempts to test each segment exhaustively (Wu Woan Kim, 2004).

There are several forms of segmentation:

1. Logical segmentation
   a. Cone segmentation (verification testing) (Abramovici et al., 1996)
   b. Sensitized path segmentation (Abramovici et al., 1996)
2. Physical segmentation

The verification testing for adders will degenerate in exhaustive testing, which is impracticable for large adders.

We will determine, based on this sensitized-path segmentation method of logic partitioning, the tests needed for detecting all single s-a-0 (or s-a-1) faults for the adders that are structured in blocks.

2. THE TESTING OF THE CARRY SKIP ADDER (CSKA) BASED ON THE SENSITIZED-PATH SEGMENTATION

For testing the *n* bit CSkA (Omondi, 1996, Hayes, 1998) with this technique we need to test each m_i block individual, by applying all the $2^{2 \cdot m_i}$ input patterns ($i = 1, M$ - where M is the number of CSkA's Blocks), separately with cu C_{-1}=0, and with C_{-1}=1.

During the testing of one block, the other blocks will have as inputs $A_i B_i = 01$; this input patterns provides the conditions for sensitized- path segmentation.

Supplementary, in order to realize the complete testing for detection the single s-a-0 (and s-a-1) faults (Vlăduţiu & Crişan, 1989), we need to test the adder with the input pattern: 00 00. 00 with C_{-1}=0.

In principle, for implementing the Test Patterns Generator (TPG) there are two methods: one based on PROM utilization and the other based on using the synchronous sequential circuits implemented with counters (or LFSR) and multiplexors.

The first method supposes to use a NUMPROM count for addressing the PROM and some multiplexors for selecting the adder's input.

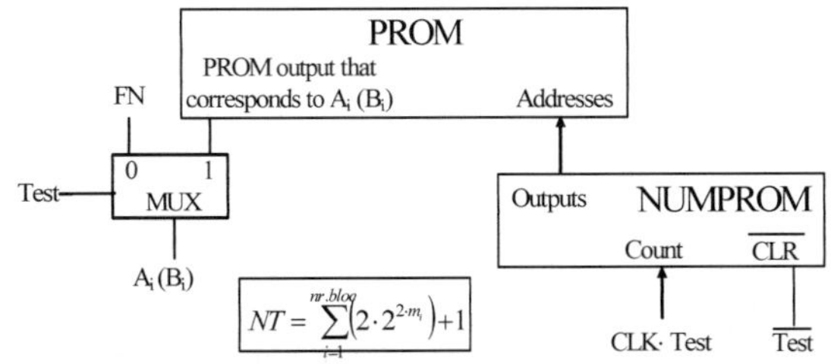

Fig. 1 PROM utilization method

NUMPROM counts from 0 to NT, where NT represents the total number of tests for complete testing the adder. *n* circuits like Fig. 1 will be connected to the inputs of the *n* bit adder. FN denotes the normal functioning of the adder.

The second method supposes the use of the following elements:

- The NB counter which identifies the block which is tested; it has $\log_2 M$ stages (where M represents the number of adder's blocks) and the count condition for numbering from 0 to $(M-1)$ is: $Count = CLK \cdot EndNUM_i \cdot Test \cdot$ The End NUM_i signal is used for signalizing the end of counting the states for testing the block i. It is obtained by delaying with one clock the $FEndNUM_i$ signal. So,

$$FEndNUM_i = \sum_{i=1}^{M} (NB_i \cdot (\text{binary comb. which ident. the last test for bl. i}))$$

This delay is necessary for providing the test 00 00 ... 00 with C_{-1}=0; it will be generated after each complete testing of a block (in order to obtain an important simplification of the TPG, the number of tests was grown with M-1).

The counter initializing is made by the $\overline{Test}$ signal.

- The NUM counter needed for generating, under the condition $Test = 1$, the tests for the block being in test. It is initialized by the signal:

$$\overline{Test} + \sum_{i=1}^{M} (NB_i \cdot \text{binary comb. for the last test for block i})$$

When the tested adder is with variable dimension for blocks, the NUM counter must function as a *modulo-t sequential counter* (Hayes, 1998); so, the design is much more complicated. The NUM counter has $2 \cdot m_{max}$ stages. The C_{-1} signal is given by the counter's most significant stage that corresponds for each block test counting, that is:

$$\sum_{i=1}^{M} (NUM_j \cdot NB_i)^{\cdot}$$

- The MUX_{Ai} and MUX_{Bi} multiplexors for controlling the adder's inputs. Let A_i and B_i be the stage *i* of block *j*. The multiplexors associated with this stage, together with their commands are shown in Fig. 2. FN denotes the connection for normal working and NUM_{2i}, the connection to the corresponding stage of the NUM counter.

In order to increase the operating speed of these adders we must modify the structure of the multiplexors by using controllable gates.

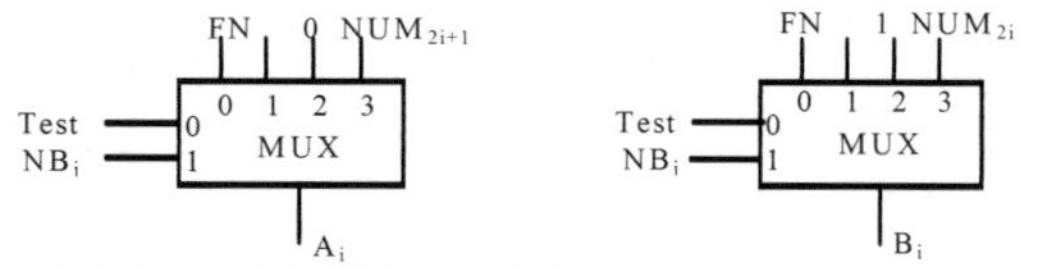

Fig. 2 MUX_{Ai} and MUX_{Bi} multiplexors

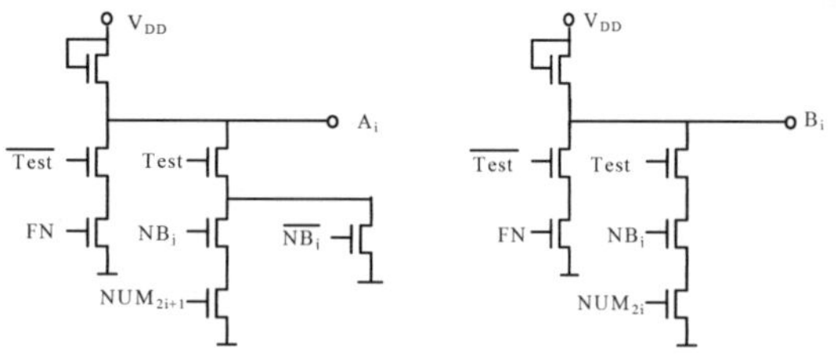

Fig. 3 n-MOS implementations of MUX_{Ai}

The n-MOS implementations of these MUX_{Ai} and MUX_{Bi} multiplexors are shown in Fig. 3. Obviously, the concept of a controllable gate can be applied to other types of gates and to different technologies. The same tests are suitable for multilevel CSkA

In conclusion, Table 1 presents the number of tests needed for testing the CSkA (with various sizes) based on this technique. The logic partitioning of the pure Carry-Lookahead Adder (CLA) is not possible due to its complex and irregular structure. So, it can be tested only either in an exhaustive manner, or in a deterministic manner. The exhaustive testing of a pure CLA with *m* stages is realized by $2^m \cdot 2$ tests. Half of them are applied with C_{-1}=0 and half with C_{-1}=1.

Consider a RCLA (Vlăduţiu & Crişan, 1989) for *n*-bits operands, organized in *N* blocks, each of them having *m* stages $(N = n/m)$. The exhaustive testing of each block is realized by applying the all 2^{2m} input tests with C_{-1}=1 and with C_{-1}=0. For sensitized-path segmentation, the other blocks beside the tested block will have as input pattern: 01 01 ... 01. To these $N \cdot 2 \cdot 2^{2m}$ tests it must be added the supplementary test: 00 00 .. 00 with C_{-1}=0. So, for the complete testing of this adder it results $\left(N \cdot 2 \cdot 2^{2m}+1\right)$ tests needed.

The BCLA testing is realized in the same manner.

Consider a SBCLA for *n*-bits operands, organized in *N* superblocks, each of them having M blocks with *m* stages $(N = n/m)$. The exhaustive testing of each block is realized by applying the all 2^{2m} input tests with C_{-1}=1 and with C_{-1}=0. For each of these tests, the blocks which are most significant than the tested block will have as input patterns those of Table 1, and the less significant blocks will have the input patterns 01 01 .. 01.

To these tests, for each of the $M \cdot N$ blocks, it must be added the supplementary test: 00 00 ... 00 with C_{-1}=0. In this way is realized sensitized-path for the adder's untested blocks, as well as the complete testing of the superblocks.
So, for the complete testing of this adder, it results a number of tests.

$$\left((n/m)+(n/m)^2-\sum_{k=1}^{n/m}k\right)\cdot 2\cdot 2^{2m}+1=\left(\frac{(n/m)\cdot(n/m+1)}{2}\right)\cdot 2\cdot 2^{2m}+1$$

The first SBCLA block	The second SBCLA block	...	The second block that follows the tested block	The first block that follows the tested block
01 01 ... 01	01 01 ... 01	...	01 01 ... 01	01 01 ... 01
01 01 ... 01	01 01 ... 01	...	01 01 ... 01	11 11 ... 11
01 01 ... 01	01 01 ... 01	...	11 11 ... 11	01 01 ... 01
...				
01 01 ... 01	11 11 ... 11	...	01 01 ... 01	01 01 ... 01
11 11 ... 11	01 01 ... 01	...	01 01 ... 01	01 01 ... 01

Tab. 1 Total number of tests

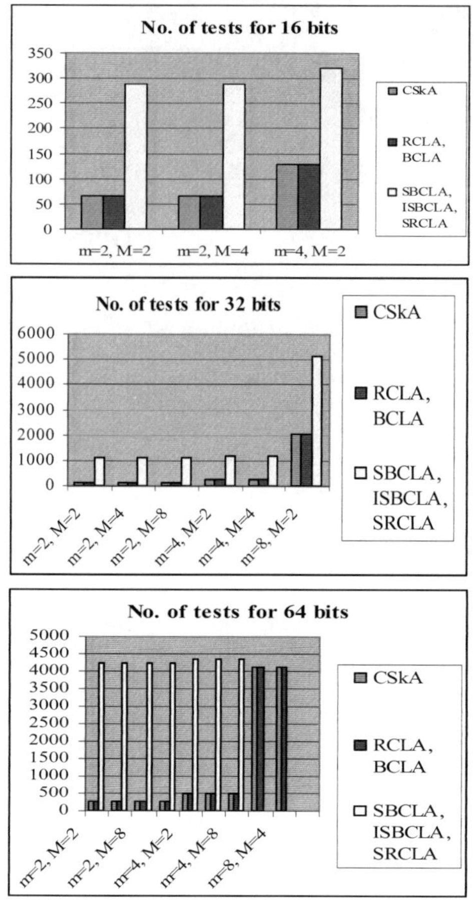

Fig. 4 Performance study

The ISBCLA and SRCLA testing is similar; it results the same number of test patterns (Fig.4.). Table 1 shown the total number of tests which are needed for complete testing for detection the single s-a-0 (and s-a-1) faults for various kind of adders with various sizes (obtained by this method).

3. CONCLUSIONS

The conclusions of our analysis show that the sensitized-path partitioning is efficient only for small dimensions of logic blocks (implicit for the adders' blocks) and it is not dependent on the dimension of superblocks. This is due on the fact that the logic segmentation at the supeblocks level is not efficient. The paper is the result of the author's activities concerning the testing and the design for testability for Computer arithmetic systems.

4. REFERENCES

Abramovici, M., Breuer, M.A., & Friedman, A.D., (1996), *Digital Systems and Testable Design,* Computer Science Press

Hayes, J. P., (1998), *Computer Architecture and Organization*, McGraw-Hill Book Company

Omondi, A.R., (1996), *Computer Arithmetic Systems Algorithms, Architecture and Implementation*, Prentice Hall International,

Vlăduţiu, M., & Crişan, M., (1989), *Testing techniques for automated data sistems*, Editura Facla, Timişoara

Wu Woan Kim (2004), *Parallel Testing Method by Partitioning Circuit Based on the Exhaustive Test*, Lecture Notes in Computer Science, Computational Science and Its Applications – ICCSA 2004, Springer Berlin / Heidelberg

Annals of DAAAM for 2008 & Proceedings of the 19th International DAAAM Symposium, ISSN 1726-9679
Editor B. Katalinic, Published by DAAAM International, Vienna, Austria 2008
Make Harmony Between Technology and Nature, and Your Mind will Fly Free as a Bird

AN ASSESSMENT OF GEOMETRICAL FEATURES FOR A NEW RAPID PROTOTYPING BENCHMARK PART

POPESCU, D[iana]; GHINEA, M[ihalache] & CHIRCOR, L[idia]

Abstract: *The assessment of Rapid Prototyping processes capabilities are made based on benchmark parts, but the lack of standardization in the field determined the existence of a relatively large variety of such parts, some of them addressing specific issues for particular processes. The current paper makes an analysis of different independent benchmarks in order to determine the main geometrical features considered, and then presents a new test part which gathers other geometrical elements useful for an objective evaluation of RP processes capabilities in terms of position and form accuracy. The importance of such an approach resides offering the necessary tools for assessing the capability of RP processes to build features which can be encountered in functional or medical prototypes.*

Key words: *Rapid Prototyping, benchmarking, accuracy, features*

1. INTRODUCTION

Due to the existence of a relative large variety of test parts for evaluating RP processes, the literature is offering information about the prototypes different geometrical features and their manufacturing resolution. Each of the test parts presented in the literature contains a certain number of features (holes, slots, threads, chamfers, slope walls, overlap features, very thin walls, freeform features, etc.) of different sizes and placed at different locations, which can be used by a designer (both in technical and medical field) in defining the geometrical form of a product. But there are also geometric features that have not been studied yet and which should be identified and evaluated, thus offering useful information for an objective evaluation of RP processes capabilities, in terms of position and form accuracy.

In order to complete the data set gather from literature, the first step is to consider the independent benchmark parts which were used to test more than one RP process. Further a broad list of geometrical features is consider, each with the corresponding evaluating purpose. Based on this information, a benchmark part containing new geometrical features was designed.

2. RP GEOMETRICAL BENCHMARK PARTS

An extensive study of the RP benchmarking and comparative studies helped distinguish the following important issues (Ispas et al. 2002), (Gatto & Iuliano, 1998), (Childs & Juster, 1994), (Makesh et al., 2004):

- There are more than 20 benchmark parts available in RP industry, designed by both RP machines producers and researchers in the field. Benchmarks developed by RP systems' producers tend to have a design which underlines the advantages of a particular process and machine. Most of the time these are user-dependent, system-dependent or process-dependent benchmarks;
- There is a lack of standardization regarding both the design of benchmark part and measurement/test methods, which makes difficult the evaluation of the results of the comparative studies;
- There are three types of benchmark parts: geometric (used for evaluating dimensional and geometric accuracy, parallelism, symmetry, concentricity, flatness, repeatability, etc.), process (for assessing aspects like parameter settings, building orientation, build styles, etc.) and mechanical (used for evaluating strengths, curling, shrinkage, swelling, etc).
- In the literature, comparative analyses have been made considering mostly the most common RP processes (SLA, 3DP, FDM, LOM, SLS). The purposes of benchmarking these RP processes are to determine feature size limitations, build orientation issues or surface and dimensional accuracy capabilities, and to make comparison between different processes. Also benchmark studies are very useful for determining the optimal RP process for a certain application;
- The benchmark studies performed so far involves building of only one sample for each process and material, when there are other factors to be considered: machines parameters settings or building styles;
- Geometric benchmark parts are designed to include many different geometric features (spheres, cones, parallelepipeds, cylinders, cubs, slots, slope and thin walls, draft angles, freeform surfaces, etc.). These features are built in different positions and number in order to check characteristics such as: repeatability, relative position, parallelism, symmetry, etc.

In this context, in order to avoid the use of system, user or process-dependent benchmarks, rules for building an "ideal" benchmark part for RP were developed (Gatto & Iuliano 1998):

- Its dimensions must be proportionally distributed in the modeling envelope of the RP machine, in order to accept the performances in the central and extreme zones;
- Small, medium and large dimensions uniformly distributed; building time not too big;
- Easy to measure using a coordinate measuring machine;
- Not be specific to a certain RP process.

For the present study, based on the above mentioned observations and developed criteria for designing a benchmark part, five independent test parts were considered as relevant because they gather the biggest number and variety of geometrical features.

Benchmark parts presented in the analyzed studies were used for assessing different aspects regarding RP processes:

- Linear accuracy and feature repeatability for: SLA, FDM, SLS, EOS, SGC and LOM (Childs & Juster, 1994). The benchmark has a plane base on which are placed different geometric features: parallelepipeds, cylindrical holes, free form features, slope and thin walls. Also, the test part has small, medium and large size features;
- In (Ippolito et al., 1995) on the based of the 3D Systems benchmark parts, the study was also focused on analyzing the surface accuracy for different RP systems;
- Accuracy of curved surfaces and surfaces roughness. The benchmark part studied in (Gatto & Iuliano 1998) consists in a cylinder combined with a sphere through sloping and planar surfaces;
- Accuracy and surface quality. The test part contains only planar surfaces, oriented at different angles to the vertical

building direction: 0°, 10°, 40°, 80° and 90°. 44 prototypes were built using the following RP processes: SLA, SLS, LOM, EOS, 3DP and FDM. An important particularity of this test part is that different materials were used with the same process, i.e. for SLA – resins: SL5185, SL5195, for SLS – resins: Somos6120, Somos7120, Somos8100, for FDM: ABS P400, ABS401, wax ICS400, for EOS: PA2200, PA3200GF, PS2500, etc. (Shellabear, 1999).

- In the test part presented in (Kruth et al. 2005), the researchers introduced the slope walls for evaluating the staircase effect, sharp edges to check the heat accumulation at the angle tips, small dimensions holes and thin walls for determining processes accuracy and resolution on x, y, z axes.
- The benchmark part in (Ispas et al. 2002) has a geometrical form obtained by intersecting a cylinder and a cone. The dimensions are grouped in small, medium and large, a comparative analysis of dimensional and form accuracy in vertical and horizontal planes being performed.

3. DESIGNING A NEW GEOMETRICAL BENCHMARK PART

Analyzing the information collected in table 1, the following geometrical features were not considered so far to the best of our knowledge:

- Inclined cylindrical surfaces for studying the relative position of inclined axis
- Hexagonal hole with inclined axis for evaluating the need for support structures and form tolerances such us angularity
- Torus for assessing the ability to build this type of feature and other similar features like helicoidally channel with circular profile (for ball screws)
- Pyramid for measuring the flatness and straightness of sloping planes
- Shoulder hole for assessing for the same feature coaxiality, concentricity and perpendicularity
- Completely enclosed spherical holes

Also, the slope plane and the edge fillet are considered for comparatively assessing the staircase effect innate to all layered manufacturing processes. All the geometrical features can be used for measuring the position accuracy on x, y, z directions.

Geometrical feature	Form and position tolerances - objective
Plane in parallelepipeds and cubes	Flatness, parallelism, linear accuracy, straightness, repeatability, perpendicity,
Cylindrical surfaces in solid or cylindrical holes with horizontal and vertical axes	Concentricity, roundness, coaxiality
Thin walls	Capability to build this type of features
Spherical surfaces in sphere or hemispheres	Roundness, symmetry
Overhang features such us in horizontal holes etc.	Capability to build this type of features
Sloping surfaces at different angles	Linear accuracy, relative position, angularity
Completely enclosed features such as spherical or cubic voids	Capability to build this type of features
Circular holes	Cylindricity, roundness, repeatability, symmetry, coaxiality
Hollow parts: cylinders, squares	Straightness, ability to build thin walls, concentricity
Cones	Sloping profile
Mechanical features: chamfer, blending or fillet	Capability to build this type of features

Tab. 1. List of geometrical features and the form and position tolerance objectives envisioned from their built

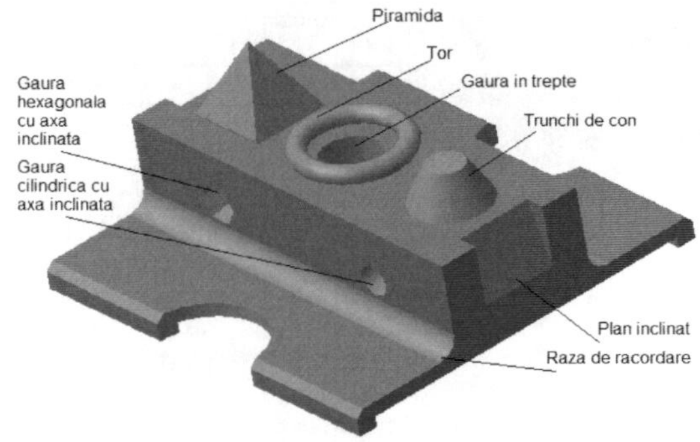

Fig.1. Proposed benchmark part

4. CONCLUSION AND FURTHER WORK

The research presented in this paper focused on designing a new benchmark part for assessing the form and position tolerances of different geometrical features not analyzed in the RP literature. This benchmark part can be used for all processes, its features not being particular to a certain technology or RP process. By using this type of benchmark parts and this approach, geometrical features which can be encountered in functional parts can be assessed.

The test part proposed contains features as completely enclosed spherical voids, torus, pyramid or shoulder holes, which can be used for assessing on one hand the ability of RP processes to build these features, and on the other hand evaluating the form and position tolerances.

Further work will consider building the benchmark part using 3DP, FDM and SLA processes and comparatively analyzing the results. Moreover, an analysis of the causes determining the defects could be significant.

5. REFERENCES

Childs, T.H.C. & Juster, N.P., (1994), Linear and Geometric Accuracies from Layer Manufacturing, *Annals of CIRP, vol.43/1/1994, pp.163-166*

Gatto, A. & Iuliano, L., (1998), *Prototipazione rapida. La tecnologia per la competizione globale (Rapid Prototyping. Technology for global competition.)*, Tecniche Nuove, Milano ISBN 88-481-0294-8

Ippolito, R., et al. (1995), Benchmarking of Rapid Prototyping Techniques in Terms of Dimensional Accuracy and Surface Finish, *Annals of CIRP vol.44/1/1995, pp.157-160*

Ispas, C., Popescu, D., Rigal, J.F. (2002), Research on the dimensional and form accuracy of a FDM1650 system, IDMME 2002, CD

Jurens, K.K., (1999), Standard for the rapid prototyping industry, *Rapid Prototyping Journal*, vol.5, no.4, pp.169-178, ISSN 1355-2546

Kruth, J.P. et al., (2005), Benchmarking of different SLS/SLM processes as Rapid Manufacturing Techniques, *Int. Conf. Polymers&Moulds Innovations, Gent Belgium, April 20-23, pp.1/7-7/7*

Makesh, M., et al., (2004), Benchmarking for comparative evaluation of RP systems and processes, *Rapid Prototyping Journal,* vol.10, no.2, pp.123-135, ISSN 1355-2546

Popescu, D. (2007), *Design for Rapid Prototyping: Implementation of Design Rules Regarding the Form and Dimensional Accuracy of RP Prototypes*, Annals of DAAAM for 2007 & Proceedings of The 18th International DAAAM SYMPOSIUM, pp.591-592, ISSN 1726-9679, ISBN 3-901509-57-7, Zadar, Croatia

Shellabear, M., (1999), Benchmark study of accuracy and surface quality in RP&M models, *RAPTEC*, Task 4.2, Report 2

Annals of DAAAM for 2008 & Proceedings of the 19th International DAAAM Symposium, ISSN 1726-9679
Editor B. Katalinic, Published by DAAAM International, Vienna, Austria 2008
Make Harmony Between Technology and Nature, and Your Mind will Fly Free as a Bird
*Young Researcher and Scientist Paper / * Supervisor, Mentor*

AERO-DERIVATIVE GAS TURBINES FUELLED BY BIOGAS

POPESCU, J[eni]; VILAG, V[aleriu]; PETCU, R[omulus]; SILIVESTRU, V[alentin] & STANCIU, V[irgil]*

Abstract: *The aero-derivative gas turbines represent an advanced solution for technologic transfer from aeronautic field to ground applications, along with an increase in economic efficiency. The utilization of aviation gas turbines with finished flying resources in cogenerative plants constitutes an advantage for valorizing renewable energy sources such as landfill gas. The paper presents the theoretical research, the results of CFD numerical applications and preliminary results for the injection and combustion of a biogas, in equal molar proportion of CH_4 and CO_2, in an aviation turbo-shaft.*
Key words: *biogas, gas turbine, combustion, CFD.*

1. INTRODUCTION

Environmental laws adopted at Brussels require a stabilized level of waste production until 2012. Moreover, the member states confronts to new limitations in the field of waste management. Studies are made and solutions are demanded in many countries. The use of waste materials is not only excellent suitability for alternative fuel production – biogas, it also creates some additional benefits by reducing large polluting biomass quantities. The biogas is produced by means of anaerobic digestion its main components being CH_4 and CO_2 in different proportions (Nikolić, 2005).

2. BIOGAS COMBUSTION IN GAS TURBINES

2.1 Background

The use of biogas in gas turbines has gained interest in recent years because of its CO_2 neutral aspect. However, only two ground applications, cogenerative plants, are known in this direction: first developed in 1992 in U.K. by SITA Packington, using a Rolls-Royce TM 501-KB5, and the second one in Sweden, by TURBEC, using a T100 micro gas turbine. Both gas turbines were primarily designed for gaseous fuels and industrial use.

2.2 Theoretical considerations on combustion process

Considering known the chemical composition of a system, the calculation of its thermodynamic properties are permitted, proprieties that can be used in a large variety of applications in chemistry and chemical engineering. The combustion process was analyzed through the Gordon program (Gordon, 2002) allowing

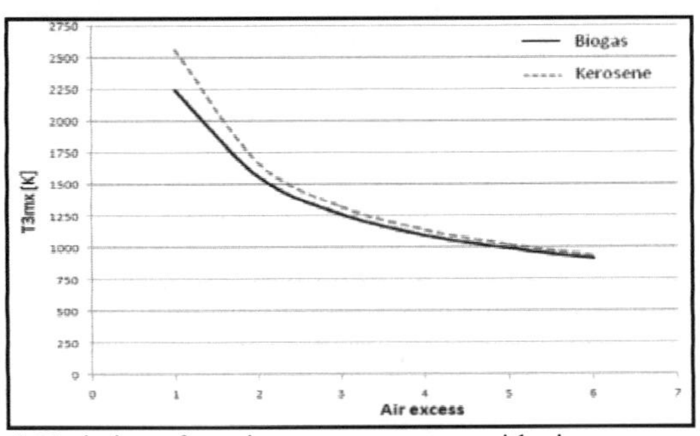

Fig. 1 Variation of maximum temperature with air excess

Fuel	minL	λ	$\dot{m}_{fuel}$ [kg/s]
Kerosene	14.6	4.6	0.0983
Biogas	4.57	4.3	0.3355

Table 1. Air quantity, air excess and mass flow rate for the two studied fuels

the obtaining of the chemical equilibrium of the composition for imposed thermodynamic states by defining the parameters. The considered gas turbine is an aviation turbo-shaft of 882 KW, with known performances and design, originally equipping a transport helicopter.

Knowing the admission temperature in the compressor the compression ratio for a considered turbo-shaft, of 6.6, and the adiabatic coefficient for air, 1.4, the admission temperature in the combustion chamber can be calculated as input data for the calculus program.

It is of interest for us to compare the combustion process parameters for the classic jet fuel, kerosene, and the bio-gas.

The maximum temperature for kerosene is calculated from the general combustion reaction, taking into consideration an air excess of 1÷6, is:

$$C_{12}H_{23}+a\cdot(O_2+3.76N_2)\rightarrow b\cdot H_2O+c\cdot CO_2+d\cdot N_2+e\cdot O_2 \quad (1)$$

The bio-gas composition is 50% methane and 50% carbon dioxide. The general combustion reaction is applied in the same manner as for kerosene and helps determining the variation of the maximum temperature (Fig.1).

$$(CH_4+CO_2)+a\cdot(O_2+3.76\cdot N_2)\rightarrow b\cdot H_2O+c\cdot CO_2+d\cdot N_2+e\cdot O_2 \quad (2)$$

From the obtained diagrams, the air excess results for the maximum temperature reached in the gas turbine, 1063 K. Knowing (KLIMOV CORPORATION, 1973) the fuel mass flow rate, the quantity of air necessary for the combustion process and the air excess, we can calculate the air mass flow rate and the fuel mass flow rate (Pimsner, 1964).

2.3 Numerical applications for biogas combustion

Computational Fluid Dynamics (CFD) is a computer-based tool for simulating the behaviour of systems involving fluid flow, heat transfer and other related physical processes. It works by solving the equations of a fluid flow over a region of interest, with specified conditions on the boundary of that region.

Different calculus techniques for the thermodynamic phenomenon inside the combustion chamber of the gas turbines were mostly based on experiments (Cârlănescu, 1997). The high cost of experimentation determined its reduction by using virtual experiments with the help of numerical simulations.

The geometry of the combustion chamber was considered well known. The fuel to be burned with the air inside this chamber is a biogas in equal volumetric proportions of methane and carbon dioxide. The other contained species are neglected, considered as impurities. The combustion chamber geometry was modelled in CAD software and processed in a CFD environment.

The first numerical study consists in the determination of

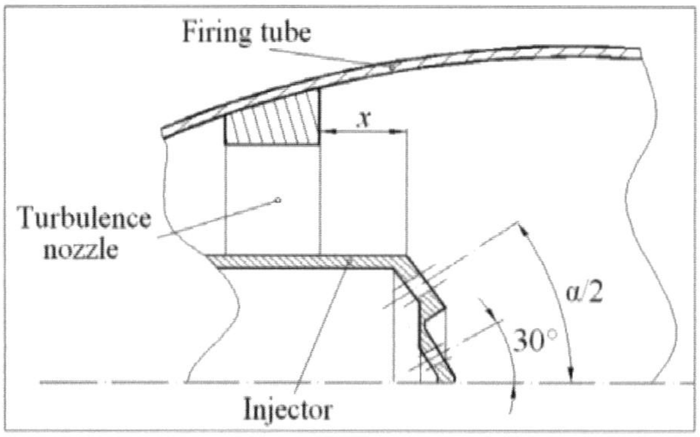

Fig. 2 The studied configuration of the injection system

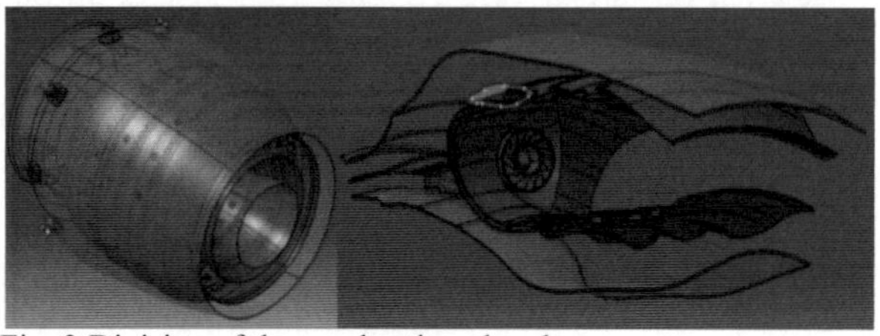
Fig. 3 Division of the combustion chamber

the geometry of the injector such as to obtain a good injection jet. The injector is considered as in Fig. 2, the simulation consisting in the variation of x, but mainly the angle a.

For the numerical simulation some hypotheses have been proposed in order to reduce the calculation domain and, consequently, the necessary time to obtain the solution. Due to the fact that the original combustion chamber has eight identical injectors positioned in the same plane at the same radius from the longitudinal axis of the gas turbine, the main simplifying hypothesis is that the calculation domain can be reduced to a single sector from the eight identical ones formed, Fig. 3. This hypothesis implies that the air mass flow rate is equally divided among the eight "pies" and the fuel mass flow rate is equally divided among the eight injectors.

The second hypothesis is that the walls of the combustion chamber adiabatically isolate the internal flow, meaning that are no loses of heat produced during the combustion.

The third important hypothesis, which reduces the number of equations to be solved, is that the thermal conductivity of the wall is neglected, their only purpose being to direct the flow. This is close to reality since our purpose is to study steady states only.

The boundary conditions used for the simulations have been calculated from theory or have been taken from the given data from the constructor of the engine. The main goal of the CFD application is to define the injector geometrical configuration to transform the combustor chamber for bio gas so that the new gas turbine engine functions in safety conditions.

The air pumped into the combustion chamber is formed by oxygen O_2 and nitrogen N_2 in 1:3.76 molar ratios. The mass flow is 6.75 kg/s = 8 x 0.84375 kg/s and the working pressure is 6.4 bars. The mass flows of the fuels were determined from thermodynamics after selecting the working regime from the comparison shown before between the original kerosene fuel and the biogas.

2.4 Numerical results

Considering the previously described hypotheses, the injection angle and the temperatures in the combustion chamber were determined. From the temperature and its variation with the air excess for biogas it was determined the necessary biogas mass flow to be injected. The value was used to define the geometry of the injector, length x and angle a. The results for different angles are presented in Fig. 4.

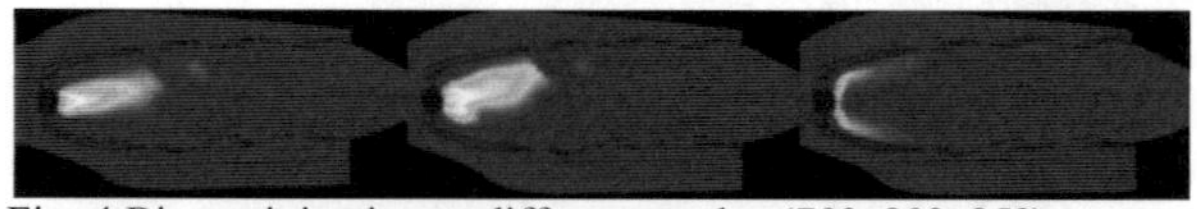
Fig. 4 Biogas injection at different angles (70°, 80°, 85°)

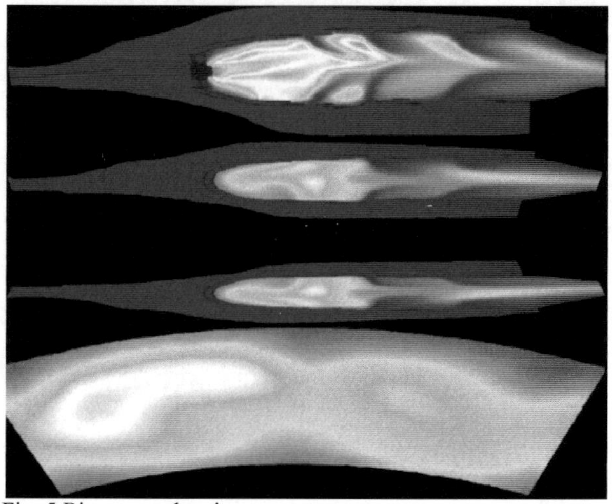
Fig. 5 Biogas combustion

The most important numerical result is the temperature in of the combustor chamber and at the exit of the combustion chamber, in front of the turbine inlet, and it was used like a constriction for configuration and dimensioning of the burner. Actually, by its help and by the given mass fuel flux it was verified that a certain working regime was obtained. The obtained average temperatures at the exit of the chamber are close enough to the ones given by the constructors.

3. CONCLUSIONS

The bio-gas was analyzed from the combustion point of view compared to kerosene, for an aviation gas turbine and diagrams were realized tracking the variation of the maximum temperature with the air excess.

CFD simulations are used for design purposes in many activity fields from the engineering domain. For the current work CFD was used for redesign the combustion chamber of an aviation gas-turbine in order to be able to work on bio-gas instead of the original fuel, kerosene. First it was studied the geometry of the injector by varying two of its parameters, second the necessary injection velocity and third the combustion itself for the desired working regime. All the simulations were conducted with respect to reliability and safety since the gas-turbine shall be use for industrial plants.

4. REFERENCES

Cârlănescu, C. (1997), *Turbomotoare de aviaţie. Aplicaţii industrial (Aviation Gas-Turbines. Industrial Applications)*, Editura Didactică şi Pedagogică, Bucureşti

Gordon, S. (2002), *NASA Glenn Coefficients for Calculating Thermodynamic Properties of Individual Species*, TP-2002-211556

KLIMOV CORPORATION (1973), *Motorul de aviaţie turbopropulsor TV2-117-A şi reductorul VR-8 (TV2-117-A Aviation Turboprop and VR-8 Gear Box)*,

Nikolić, V. (2005), *Producerea şi utilizarea biogazului (Producing and Using the Biogas)*, Chiminform Data, Bucureşti

Pimsner, V. (1964), *Energetica turbomotoarelor cu ardere internă (Energetics of Internal Combustion Gas-Turbines)*, Editura Academiei Republicii Populare Române

Annals of DAAAM for 2008 & Proceedings of the 19th International DAAAM Symposium, ISSN 1726-9679
Editor B. Katalinic, Published by DAAAM International, Vienna, Austria 2008
Make Harmony Between Technology and Nature, and Your Mind will Fly Free as a Bird

E.R.P. SYSTEMS AND ORGANIZATION'S PERFORMANCE

POPESCU, L[uminita]; POPESCU, V[irgil] & IANCU, A[nica]

Abstract: *The faster and faster changes in the business environment and the increase of an activities' complexity within a firm need permanent adapting, in a vivacious rhythm which put at trial in many occasions the human capacities of effort and analysis. The E.R.P. Systems are able to process a very high volume of data and information aggregated in order to make processes more optimal and efficient.*
The E.R.P. Systems allow the synchronization of all enterprise functions by making fluid the data exchange between departments because they represent a very good mean for information integration and ordering.
From the functionality point of view, the system covers all the domains of interest for an organization: production management, stocks management, relations with suppliers, management of the relations with clients, orders' monitoring, financial management, human resources management.
Key words: *E.R..P. Systems, performance, management, organization, development*

1. INTRODUCTION

The Enterprise Resource Planning (E.R.P.) Systems have appeared in mid '60, at the same time the development of the M.R.P. (Material Requirements Planning) technique. These systems, MRP1, used the following parameters: the volume required to be made, based on the clients' firm orders; technological charts; the stocks situation on each article/good; the medium duration of a supplying. Then, at the begging of the '80, the MRP2 (Manufacturer Resource Planning) Systems have emerged, integrating the planning instruments for sales, production, supplies and the necessary of production units, as well as creating some informational situations of economic-financial nature (Popescu, 2007).
Though they come from the production environment, the ERP systems try in the present to cover all the basic functions of an organization, no matter its type.
The ERP represents in essence (Radu et al., 2005):

- an assembly intercorrelated by informatics subsystems meant for the human resources, materials and finance management within an organization;
- an informatics system oriented towards the identification and planning of the necessary resources in order to satisfy the requests of the organization's clients;
- an informatics system with many modules designed to make the main activities in a commercial firm more efficient.

The development process of an Enterprise Resource Planning system is a complex one and is usually accompanied by a re-design of the entire management process, therefore implying significant financial resources and long period of implementation.
The ERP systems are designed on modular concept and facilitate the view of the enterprise as a whole (Brehm & Gomez 2007), covering the next categories of activities:

- *the financial-accounting management*: automation of the financial-accounting information registration taken from the primary documents, achievement of a complete financial-accounting evidence, the operative monitoring of the enterprises' financial-accounting situation, the automatic print of the informative and synthesis documents.
- *the production management*: identification of the principal performance indicators and transmitting them to the decision taking factors; the production models' flexibility; information given in real time about the operational indicators; creation of an assisting instrument for the production's planning; calculation of the ante and post production costs.
- *the human resources management*: the management of the organization's employees personal data; the employees management within the organizational structure; simulation of variants for salary modifications in order to correctly allocating the budget; monitoring the employees' efficiency and evaluating their performances.
- *the stocks management*: daily actualization of the stocks situation in order to eliminate the phenomena of over stocking or lack of stock; establishing the request for supplying and making the supplies on the base of an optimal model.
- *the fix assets management*: management of fix assets during the entire utilization period; generating the situations specific for the fix assets; making the history of the fix asset from the chronological point of view from the entry in the patrimony until the exit from it.

2. INFORMATION

By definition, an ERP system is a complex software solution, its elements being integrated in a general platform, for managing the company's resources so to achieve its objectives. These systems achieve in present days the integration of all leadership functions within the organization, starting from the planning, assuring the stock of materials and resources, defining technologies, coordinating the production processes, managing the financial-accounting administration, the human resources, the stocks of finite products but also maintaining the relationships with clients and business partners. Such an ERP system permits to the decision taking factors to make a complete analysis on the achievement of the business plan. Prediction plans, evaluations, presumptions on the industry's evolution tendencies to which the company belongs, qualitative analyses, integration of the new e-business technologies and on-line communications can be made by the simulation options and the flexible and dynamic character of applications.
A major role of the ERP systems is to assist the management in the planning, budgeting and monitoring the company's financial performance, by the mean of control function. Control allows to the management to adjust processes so all things to happen according to the preset scenario and given rules and dispositions (Burdus & Carparescu 1999).
the control process is made of three major elements, coordination, communication and information and implies three stages of execution:

1) *Planning and Budgeting*. This stage means the alignment of the strategical and operational objectives and generated the plans of action.
2) *Monitoring and Analysis*. the performance monitoring, the achieved performances analysis and reporting them to the executive management team are included in this stage.
3) *Conclusions and Recommendations*. This stage include measures for the profit growth, costs decrease and recommendations for strategy adjustments.

The execution of the control function means the processing of very large data volumes, obtaining certain reports and determining some performance measuring indicators or criteria. The processing of of this significant data volume can be done much more easily if the company has implemented an ERP type system.

A control solution signifies the strategical combination of three elements: informatics systems, business processes and human resources. Each of them has an essential role in executing the control because the strategical decisions for the company can be taken based on them.

The organization objectives are evolving from the extension of the market rate to the increase of profitability and services' quality at the same time with the growth of competitiveness and diversity, but also with the increasing exigencies of clients. The decisions to change objectives and evaluate the degree in which they are achieved can be done only with the help of control processes and through the existence of an informational infrastructure able to offer the decisional support based on information and analysis, not on intuition.

Targeted objectives' defining and the results' measurement methods are essential in defining and implementing a control solution. Measurement of an organization performance and establishing the future strategies implies the defining of some synthetic performance indicators to which be able to compare the company's evolution.

3. CONCLUSION

An ERP system is made of a series of modules meant for the various departments of an enterprise: production, finances and accountancy, acquisitions, sales, human resources etc. but the implementation of the system is a difficult process because brings along major changes in the employees' activity, its cost can not be predicted (it depends on the number of subunits that will be administrated by the system, on the number of modules to be installed, on the project's complexity). Besides those expenses, there also the so called hidden costs, that might raise the project's implementation cost:

- *The employees' professional training* is the most important factor that is underestimated when the ERP system implementation cost is calculated. These expenses have high values, because employees must learn new procedures and processes (Bingi et al., 1999).
- *Integration and Testing.* The creation and testing of the links between the ERP system and other software components implemented inside the organization is a really slow process, especially when the already existing components are old and there is no support for them from the firm that had produced them.
- *Customizing* must be avoided as much as possible, because, besides the fact it is expensive, makes more difficult the upgrading of the ERP center when a new version of it is available.
- It has been noticed consequent to the studies made on the organizations that have implemented at ERP type solution that the benefits have not appeared immediately after the system was introduced. In fact, during that period, a decrease of the firm's performances was observed, due especially to the numerous changes that the implementation of an ERP system brings up.

The principal benefits identified by the organizations that have implemented such systems are:

- *Reducing the operational costs*. Though it has an extremely various nature, the information is introduced only once and is accessible to any module that needs it. Thus, it saves time, resources, the probability to make operational errors is lower and gives access to most recent information etc.
- *Facilitating the management of daily operations.* The daily decisions are taken on the ground of the information available at that moment. By using the standard reports given by the ERP system, the information are new, coherent and presented intuitively.
- *Stocks optimization.* There have been signaled major cut backs regarding the fix assets from the stocks in many cases, while the clients satisfaction has grown. This was due to the new instruments of operational and decisional management which allow the control of supplying, delivery, picking, production processes etc. At the same time, the stocks with short expiration terms and slow rotation are kept under control etc.
- *Increase the level of respecting orders.* By the automation of the chain of taking orders, acquisitions, stocks management, storage management, logistics, the great majority of potential defects that lead to not respecting the orders or delivering unordered goods are eliminated.
- *Cash flow improvement.* The company controls in a stricter manner the cash flows and the periods of regaining the money by implementing coherent commercial policies and by the easiness with which the specific indicators are obtained, thus imposing a financial discipline to all the factors involved.
- *integration of financial information* Since there is only a system, there is a single version of the results. It is eliminated therefore the "multiple truths" problem, when each department has its own set of results, analyses and conclusions, that are most often contradictory.

We can say that an ERP system represents a combination between management practices and the technology of information, through which the organization's business processes are integrated within the informatics system in order to achieve some specific objectives.

4. REFERENCES

Burdus, E. & Caprarescu, Gh. (1999). *Fundamentele managementului organizatiei* (*The Basics of Organizational Management)*, Editura Economica, ISBN 973-590-209-5, Bucuresti

Bingi, P.; Sharma, M. & Godla, J.K. (1999). Critical Issues Affecting an ERP Implementation, *Available from*: http://carl.sandiego.edu/gba573/critical_issues_affecting_an_erp.htm *Accessed*: 2008-04-10

Brehm, N. & Gomez, J.M. (2007). Service-oriented Development of Federated ERP Systems, *Available from*: www.se.uni-hannover.de/semsoa-2007/2007/proceedings/paper6.pdf *Accessed*: 2008-04-03

Popescu, L. (2007). Schimbari organizationale determinate de tehnologia informaţiei (Organizational Changes brought up by the Information Technology), *Gândirea Militară Românească,* Nr.1, Ianuarie/februarie 2007, pp. 116 – 120, ISSN 1454-0460, 1842-8231

Radu, I.; Ursaceanu, M.; Vladeanu, D.; Cioc, M. & Burlacu, S. (2005). *Informatica si Management. O cale spre performanta* (*Informatics and Management. A Path toward Performance*), Editura Universitara, ISBN 973-7787-59-5, Bucuresti

Annals of DAAAM for 2008 & Proceedings of the 19th International DAAAM Symposium, ISSN 1726-9679
Editor B. Katalinic, Published by DAAAM International, Vienna, Austria 2008
Make Harmony Between Technology and Nature, and Your Mind will Fly Free as a Bird
*Young Researcher and Scientist Paper / * Supervisor, Mentor*

A NEW ENHANCED PRODUCTION SYSTEM AS CFS-ERP TECHNOLOGICAL-INFORMATICS SYSTEM

POPESCU, S[tefan] E[ugen]; NICOLESCU, A[drian] F[lorin] & MINCIU, C[onstantin]*

Abstract: *Associating a technological system CFS (Cellular Fabrication System) type with a specific software system ERP for technological resources planning is formed a new enhanced technological-informatics system which highly increased the manufacturing efficiency. In description of the new identified composed system are presented the block diagram of CFS-ERP system, the visual diagram of ERP modules and the work flow of the ERP software system for technological resources planning in CFS.*
Key words: *System, Manufacturing, ERP, CFS, Management.*

1. INTRODUCTION

The paper come to present the new enhanced technological-informatics system composed from ERP software system for production planning and CFS technological system. Implementing the specific ERP software system together with technological systems the efficiency in technological resources planning is highly increased and the production management is more efficient leading to performance in parts production.

2. INFORMATION

It was developed a new software system composed from an ERP web application for technological resource management and the database. The software system and the referential technological system as flexible manufacturing cell (FMC) are forming a new technological-informatics system having specific interconnected practical features.(Popescu et al., 2007)

2.1 The block diagram of CFS-ERP system

In Fig. 1. it can be observed the block components of the technological-informatics system consisting in CFS technological system and ERP software system. The ERP software system, specifically design for production programming in flexible machining cells is formed from .NET web application and SQL database. The two main components of the software system are running on hardware servers, recommended separately, on web server respectively on database server. The ERP software system is interconnected with the CFS technological system in the Enterprise intranet. Accessing the specific features of the ERP software system, it become possible to generate the needed reports for launching the parts production. When the production is started the CFS technicians, monitoring the production status can update the number of manufactured parts by accessing a specific feature of the Production Planning module. In the diagram it can be observed the main possibility of accessing the ERP software system by Client PCs, equipped with Web Browser, used for data entries and reports generating. Also in the intranet is assured the IT&C development and maintenance service for the ERP web application and the specific database. The access to ERP software system is possible also from internet connecting to the ERP for production planning modules by authentication and authorization corresponding to an username and a password assuring confidentiality of production data & controlled access.

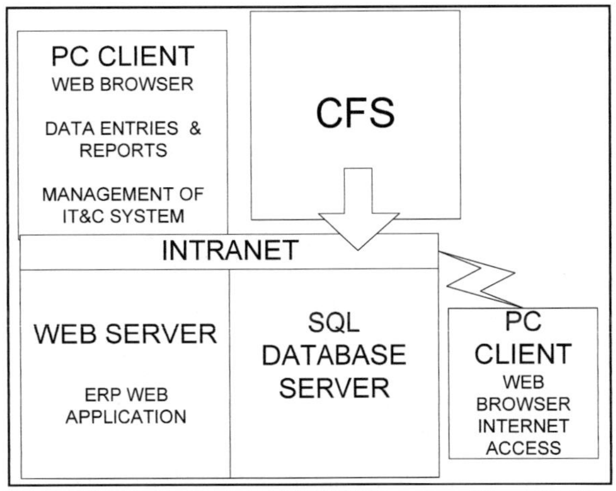

Fig. 1. The block diagram of CFS-ERP system. (Leon, 2005)

The activities in part manufacturing from CFS level are bring together with the preparing production activities by the same ERP modules generating efficiency and fluency in work.

2.2 The visual diagram of ERP modules

The new enhanced technological-informatics system as CFS-ERP can be observed as visual diagram of ERP modules, hardware disposal and allocated human resources in Fig. 2. The CFS technological system component consist mainly from CNC Tool Center, Industrial Robot, Conveyor and Command and Control Unit.(Nicolescu, 2007) The CFS was studied for manufacturing parts with revolution surfaces starting from bars. The ERP software system consist in a distributed structure specific to the manufacturing activities and also consist in a main core formed from the Web Server for the .NET application and database server for the SQL Server database.

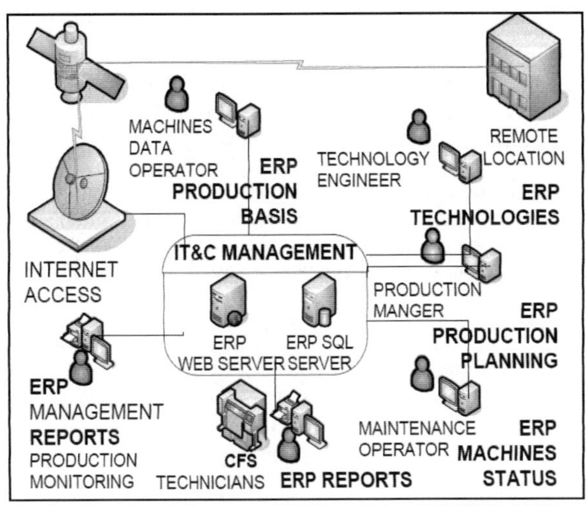

Fig. 2. The visual diagram of ERP modules.(ADCOS, 2007)

The software requirements for running the ERP software system for technological resources planning in CFS are:
- Microsoft Internet Explorer (or other web browser)
- Operating System with IIS - Msft. Windows 2000 or better
- .NET Framework 2.0
- RDBMS Microsoft SQL Server 2000 or 2005
- Microsoft Excel for reports generated as excel file
- Microsoft Project for viewing the production project
- For updating and maintenance of ERP as ASP.NET web application is needed Visual Studio 2005 the Microsoft tool for .NET development.(Negara, 2004). The hardware requirements for running the ERP software system for production programming in CFS are identified according to Fig. 2.:
- User PCs as Work station terminals connected in intranet
- Internet Server (with OS Windows Server 2000 or better)
- Data Base Server for RDBMS Microsoft SQL Server 2005

2.3 The work flow of the ERP software system for CFS

The ERP software system for technological resources planning in CFS supposes a specific work flow in preparing, launching and monitoring the production. Initially in the work flow, in the Production Basis Module, is assured by the Machines Data Operator the data management for production components identified as CNC Tool Centers, Industrial Robots, Parts, Semi Parts, Machining and Tools needed in fabrication. The following step is in the Technology Module for the Technology Engineer to elaborate the part fabrication technology by selecting the specific machines, machining and tools and completing the machining parameters and uploading the machining drawing for each technological step and the manufacturing drawing for part. The Production Manager check the Machine Maintenance Operator input in Machines Status module and in function of Machines availability, elaborate and update the Production Project for specific ordered part and introduce the Start Date, the Finish Date and the number of parts to be manufactured, in Production Planning Module. The CFS Technicians accessing the Reports Module and generate the needed reports for Launching the production. The main reports are consisting in Part Manufacturing Technology, Part Technological Film, Tool Sets and Machines, Production Planned as time period and number of parts. Once the right information is available at the CFS level, the Technicians can proceed to CNC Programming, select and solicit the right tools and piece bars. The CFS Technicians are assuring the Operative Management of technological system reporting at the end of the working day the production level as number of pieces manufactured and updating in the Production Planning Module as chronologically records. If are appearing unpredicted interruptions of fabrication the initial planning has to be updated by the Production Manager noticed by the CFS Technicians who are reporting the cause and time period estimated for remediate the flow of production, as Machines functional status to the Maintenance Operator. After the production is Launched, in the Management Department it can be tracked the Production Status as level of parts manufactured and CFS Functional Status as Part Production Plan updated by accessing the Reports Module of the ERP software system for CFS becoming possible the estimation in programming new parts production. The work flow described above is containing the integration of software activities with specific technological activities in parts manufacturing in CFS systems as one expression for the new enchased technological-informatics system. The software system and the referential flexible manufacturing cell (FMC) are forming a new technological-informatics system having specific interconnected practical features. (Popescu et al., 2007) The technological-informatics integration conduct to optimizing the utilization of technological systems by improving data flow communications, data management and decreasing the time for transmitting the information to FMC production level. (Popescu et al., 2007).

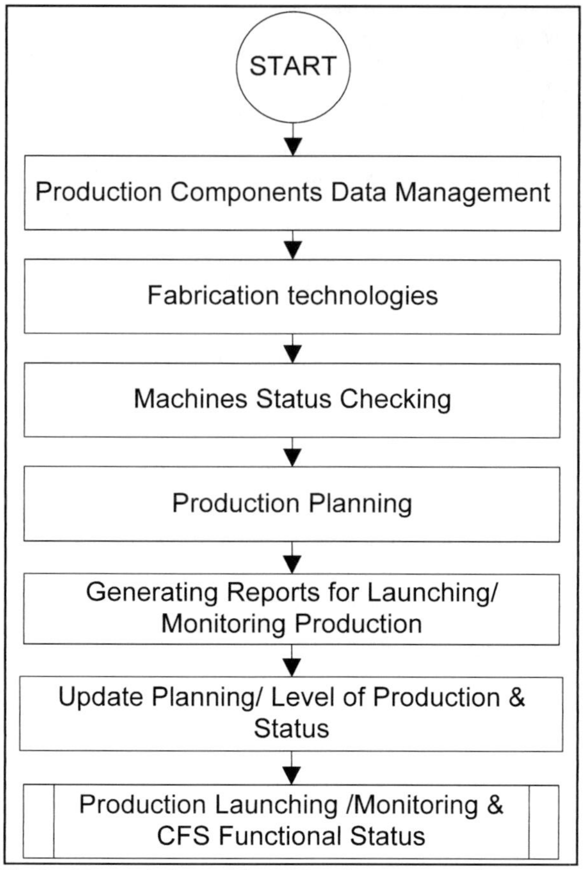

Fig. 3. The work flow of the ERP software system for CFS.

The utilization of the new integrated system with specific features and emergent properties offer a clear evidence of the resources, production and machines and an increase speed of making the parts and diffusing the information flow in the system and to FMC personnel. (Popescu et al., 2007).

3. CONCLUSION

The paper presents the integration of software systems ERP and technological systems CFS as a new enhanced production system CFS-ERP technological-informatics system. The integration was presented as specific block diagram and visual diagram together with the specific work flow of the ERP software system for CFS. Implementing the new ERP-CFS technological-informatics system the resources planning efficiency in CFS is highly increased leading to performance in part production activities. Overall there were made original contributions by theorizing the new enhanced CFS-ERP system according to practical developed ERP and database.

4. REFERENCES

Leon, F. (2005). *Programming Enginery-Course material*, Faculty of Automatics and Calculators, Iasi, 2005

Negara, G. (2004) *Application Development in Visual Studio .NET-Course material*, Faculty of Informatics, Iasi, 2004

Nicolescu, A. (2007). *Industrial Robot Implementing into Manufacturing Systems - Course material*, PUB, Bucharest

Popescu, S.E.; Minciu, C. & Nicolescu, A.F. (2007). *Basics of New ERP Software Package and Data Base for Technological Resources Management in FMC* (2007). ISBN 3-901509-58-5, ISSN 1726-9679, B. Katalinic (Ed.), pp 301, DAAAM International, Vienna, 2007

ADCOS Romania (2007). AIS - *Application documentation.*

Annals of DAAAM for 2008 & Proceedings of the 19th International DAAAM Symposium, ISSN 1726-9679
Editor B. Katalinic, Published by DAAAM International, Vienna, Austria 2008
Make Harmony Between Technology and Nature, and Your Mind will Fly Free as a Bird
*Young Researcher and Scientist Paper / * Supervisor, Mentor*

INTEGRATING MANUFACTURING ENTERPRISES FUNCTIONS WITH NEW ERP FOR CFS SPECIFIC MODULES

POPESCU, S[tefan] E[ugen]; NICOLESCU, A[drian] F[lorin] & MINCIU, C[onstantin]*

Abstract: *Initially it was designed and developed a new Production ERP for technological resources planning in FMC (Flexible Manufacturing Cells). Analyzing the Enterprise functions were identified the Enterprise ERP modules similar to general modules of an ERP. In the analysis of utilization structure of production planning ERP for CFS (Cellular Fabrication Systems) in the paper are presented the modular structure of an Enterprise ERP, the components of the Production ERP for CFS, the general work flow in production planning in CFS and the work flow from top management level.*
Key words: *Enterprise, ERP, CFS, Management, Manufacturing.*

1. INTRODUCTION

The paper come to present the association between production Enterprise activities and specific ERP for CFS modules and work flow. Integrating the ERP for CFS software system in the Enterprise production activities the efficiency in part production is highly increased by having a main MRP core.

2. INFORMATION

The Enterprise functions can be described as ERP general modules. The production function of the manufacturing Enterprise is composed from specific activities according to specific production planning ERP modules. The utilization of specific Production ERP for CFS it supposes an specific work flow and an adequate top management work flow. The specific work flow in production planning is described according to technological resources management activities from CFS. The specific work flow structured the management activities and increase the Enterprise production planning efficiency in CFS.

2.1 The Enterprise and Production ERP modules structure

The identified Enterprise functions are Management, Development & Research, Production, Personnel, Commercial and Financial Services. Each main module is composed from adjacent sub modules as it can be observed in Fig. 1. where the Management module has the Reporting as sub module, the Development & Research has the IT&C sub module, the Production module has the CFS, CNC Programming, Resources Planning and Technologies sub modules, the Personnel module has the Salaries and Human Resources sub modules, the Commercial module has the Suppliers, Stocks and Delivery sub modules and Contracts, Orders and Bills sub module adjacent to Financial module whit the Accountancy sub module. In the image shown in the Fig. 1. it can be observed also the main core of the ERP software system consisting of Web Server and SQL Server as hardware components according to main software components as ERP Web applications and SQL database. The Enterprise ERP modules structure is showing the usefulness of an ERP software system in addition with the Enterprise functions. The New ERP for FMC was developed as .NET web Application in Visual Studio 2005 and database in RDBMS SQL Server 2005 according to requirements of technological resources management&planning (Negara,2004).

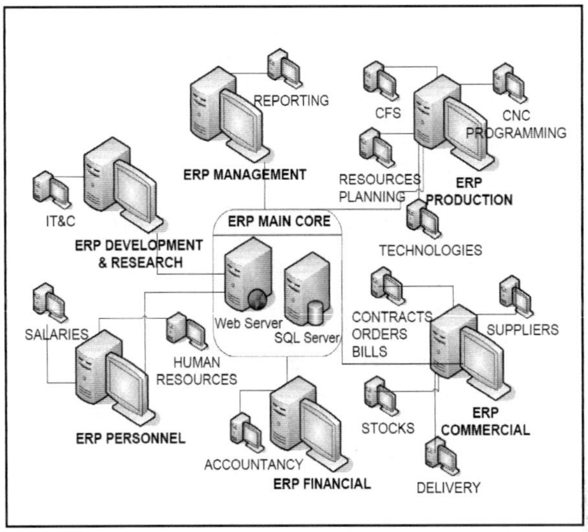

Fig. 1. The Enterprise ERP modules structure.(EBS, 2005)

In the Fig. 2. the Resources Planning sub module cover the Production Basis, Production Planning, Machine Status and Reporting data management together with Technologies sub module.(Popescu et al., 2007). In addition to the production ERP module was considered the CFS technological system and the CNC Programming as sub modules. Considering both systems, will result the specific work flow shown in Fig. 3.

2.2 Specific work flows in CFS-ERP Production Planning

The specific work flow in CFS manufacturing is describing the activities starting with Order Receiving, Top Management Activities, Order Acceptance, continuing with Preparing Production and ERP Planning, Flexible Configuring and Testing of CFS, Launching and Monitoring the Production and finalizing with CFS Operative Management, Production Delivery and Production Finish. The CNC Programming is developed in the ERP Planning when the Part Technology is ready. (Nicolescu, 2007)

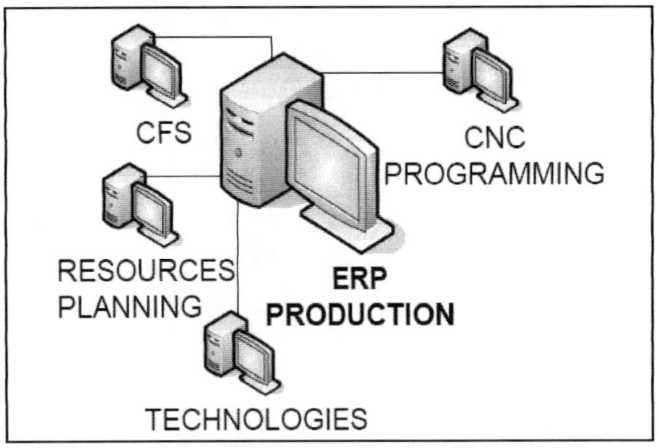

Fig. 2. The components of the Production ERP for CFS.

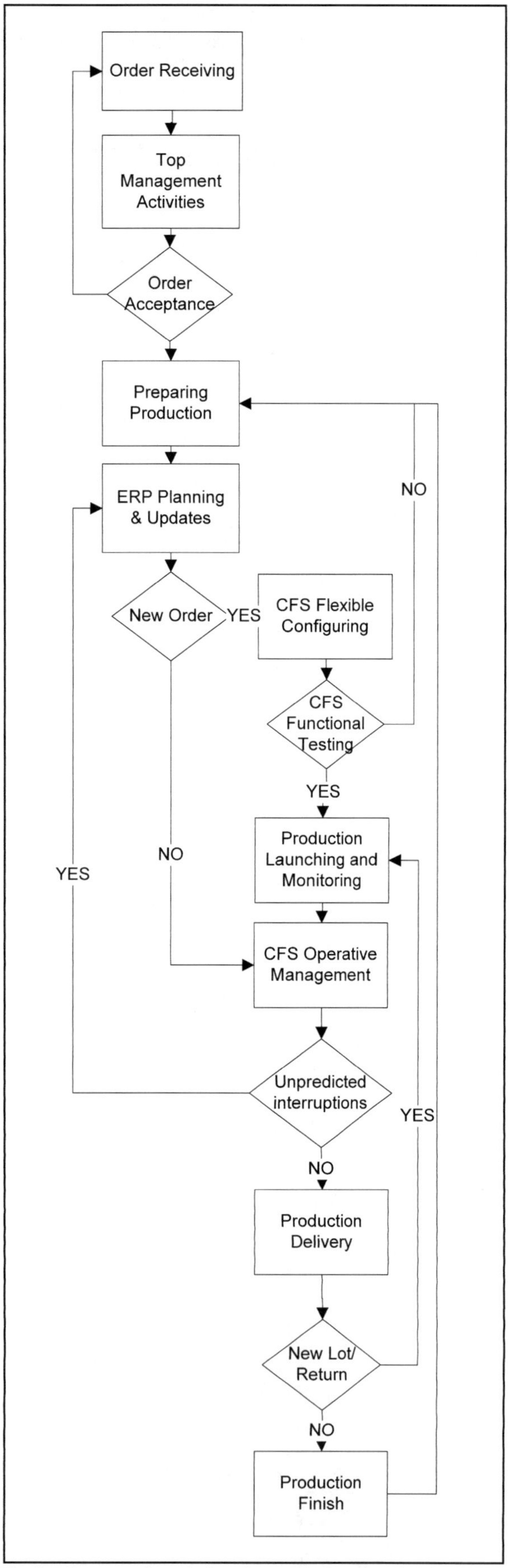

Fig. 3. The general work flow in production planning in CFS.

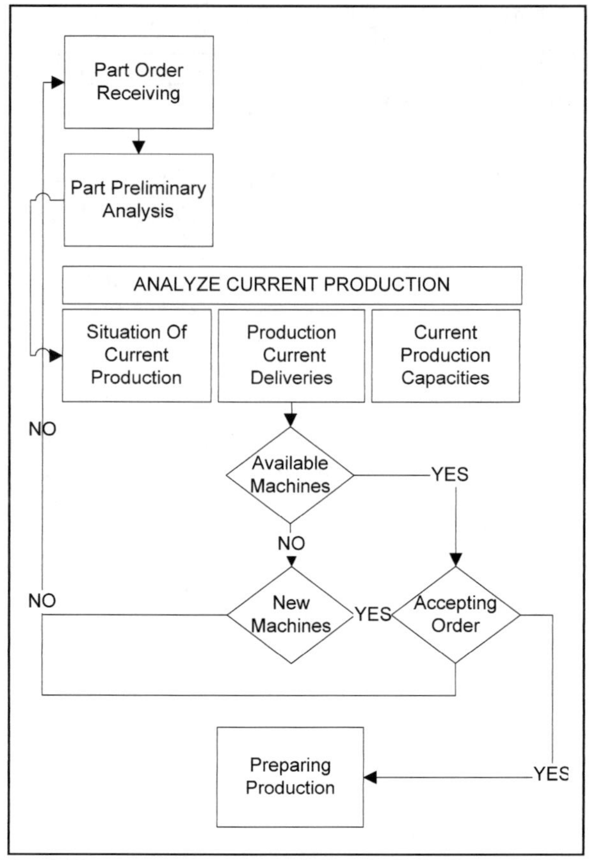

Fig. 4. The top management work flow in CFS. (Leon, 2005)

The top management work flow is showing the analyze of Current Production according to Enterprise Deliveries and Capacities and the decision to accept or not the part order.

3. CONCLUSION

The paper presents the integration of manufacturing Enterprises functions with new ERP for CFS specific modules. The integration was presented as specific ERP Enterprise modules together with Production Planning components in CFS and specific work flows according to Enterprise functions. Implementing the ERP for technological resources planning in CFS, all the specific Enterprise functions has to be adequately adapted into efficient work flow leading to performance in part manufacturing and production management. Overall there were made original contributions by theorizing the CFS specific work-flow according to Enterprise functions and ERP modules.

4. REFERENCES

Leon, F. (2005). *Programming Enginery-Course material*, Faculty of Automatics and Calculators, Iasi, 2005

Negara, G. (2004) *Application Development in Visual Studio .NET-Course material*, Faculty of Informatics, Iasi, 2004

Nicolescu, A. (2007). *Industrial Robot Implementing into Manufacturing Systems - Course material*, PUB, Bucharest, 2006 - 2007

Popescu, S.E.; Minciu, C. & Nicolescu, A.F. (2007). *Basics of New ERP Software Package and Data Base for Technological Resources Management in FMC* (2007). Proceedings of the 18th International DAAAM Symposium, ISBN 3-901509-58-5, ISSN 1726-9679, B. Katalinic (Ed.), pp 301, DAAAM International, Vienna, 2007

EBS Romania (2005). *Clarvision ERP - Application documentation*, 2005-11-12

Annals of DAAAM for 2008 & Proceedings of the 19th International DAAAM Symposium, ISSN 1726-9679
Editor B. Katalinic, Published by DAAAM International, Vienna, Austria 2008
Make Harmony Between Technology and Nature, and Your Mind will Fly Free as a Bird

THE DEVELOPMENT OF HEXAPOD KINEMATIC STRUCTURE

POPPEOVA, V[iera]; URICEK, J[uraj]; ZAHORANSKY, R[obert]; REJDA, R[udolf]; BULEJ, V[ladimir] & DURICA, J[an]

Abstract: *This paper describes parallel kinematic structures used in the area of machine tools and robots, history of their application in production systems. There is information about design of hexapod kinematic structure with six degrees of freedom at University of Žilina. There were designed hexapod mechanical elements, inverse kinematic analyse, 3D model and simulation software was used for simulation of hexapod moving platform motion. Algorithm was assembled for calculation of the length of active legs during moving in real time. There was designed also control system of hexapod and control of drives.*
Key words: *parallel kinematic structure, hexapod, simulation software, control system*

1 INTRODUCTION

Parallel kinematics structure (PKS) is mechanism with closed kinematical chain, which consists of the base, platform and at least two reciprocally independent leading legs. Parallel links (struts, legs) are joining between the base and the movable platform. Motion of the platform relatively to the base is controlled by a change of linear or angular parameters of several prismatic joints or legs acting in parallel system (kinematic chains). The most common configuration comprises six active legs and it is named Hexapod. These legs are realized as the linear actuators, most frequently by electromechanical cylinders. In this case the movable platform has six degrees of freedom (DOF) - moving in three linear directions and three angular directions independently or in any combination (Merlet, 2000; O'Connor, 1998).

This type of structure has been known for long time. Around 1800 the mathematician Cauchy studied the stiffness of the so-called "articulated octahedron" (Knoflíček & Kolíbal, 2000). More recently in 1949, Gough used similar mechanism for the test of tyres. Then later in 1965, these mechanisms rediscovered and used very widely in the flight simulator by an engineer called D. Stewart. Since that time, any parallel - linkage mechanisms are referred to as "Stewart platforms", although Gough discovered this mechanism before him.

The main fields where are applied this kind of mechanisms, are design of machine tools and robotics. Although the speed, accuracy, and flexibility of conventional machine tools have improved over the years, the basic design remains unchanged-tool and workpiece are still constrained to move along linear axes. Now, this long-standing configuration has competition from machines based on parallel or hybrid kinematics.

Since with the hexapod design the only forces on the machine's structure are tension and compression, it is less susceptible to bending than a conventional machine design.

Starting with the public presentation of the first parallel kinematic machines (PKM) on the IMTS fair in Chicago (1994), extensive research has been conducted university laboratories as well as in machine tool industry. During the last decade, more than 200 different parallel mechanisms have been built, mostly as prototypes or academic studies.

A first parallel mechanism device was used in a robotics assembly cell by McCallion in 1979. Then, Parallel manipulators have been under increasing development over the last few decades, so they are considered to be attractive alternative to the serial linkage devices, such as the conventional robotics arms (Marek, 2006; Chren, 2007).

2 DESIGN OF THE SCHOOL HEXAPOD

The school hexapod and its control system were developed at the University of Žilina at Department of Machining and Automation in last few years.

Mechanical construction

The mechanical construction consists of base, platform and six identical linear actuators with trapez system screw-nut. For the conection to the base and also to the platform are used the cardan joints (Fig. 1).

Parameters of school hexapod:

- DOF : 6 °
- Dimensions a x b x c : 700x600x600 mm
- Weight: 40 kg
- Max. angle of platform rotation Ux, Uy, Uz: 30 °
- Max. operating space X-axis, Y-axis: -50 to +50 mm
- Max. operating space Z-axis : 0 to +100mm

Mathematical model

To fully describe the position and orientation of the movable platform in the workspace, six coordinates are needed. Three of them are positional displacements that define the actual position of tool centre point (TCP) fixed on the platform with respect to global coordinate system (fixed with base). The other three coordinates are angular displacements that define the orientation of the platform. We define generalised Cartesian space coordinates **p**, whose elements are the six variables chosen to describe the position and orientation of the platform, as $\mathbf{p} = f(x, y, z, \varphi, \theta, \psi)$.

The position of the platform frame is specified with respect to the base frame by a vector $\vec{P}_p = (x, y, z)$, which gives the coordinates of point P_0 with respect to the base frame. The orientation of the platform frame is described with respect to the base frame by a matrix of rotation.

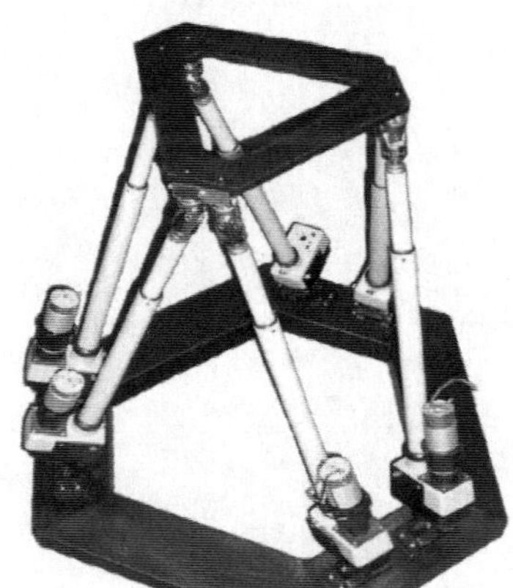

Fig. 1 Design of school hexapod

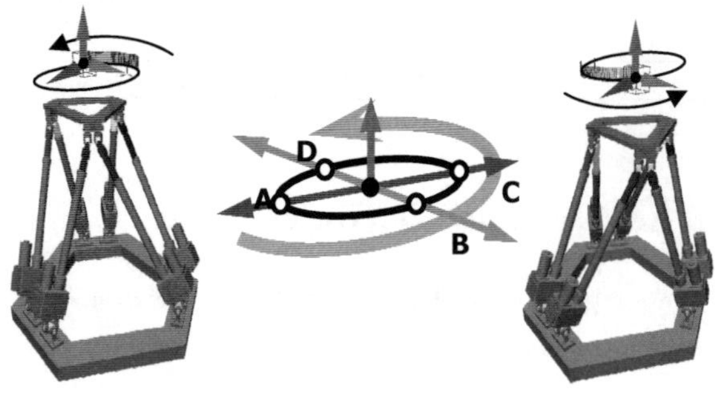

Fig. 2 Hexapod circle interpolation calculation

3 SIMULATION SOFTWARE

The simulation software for school hexapod was created in the same time than the engineering design of mechanical structure. It was designed in environment of program language Delphi 7 with the support of library Open GL.

Computer simulation of school hexapod structures is used for system study on its mathematical model. System's action is defined as relation system for transformation of parameters from imput and status space to parameters in its output and status space. It is possible to use the data from simulation software for control of real hexapod (Fig. 3).

For mechanism simulation is important information about dimensions and forms its kinematic chains and joins, presentations of position its elements.

4 CONTROL SYSTEM

It was designed one of possible architecture of control system for designed PKS. Proposed scheme of control system is based on master PC (on this PC is running a simulation/control software), which generates the data necessary for control systems of each of drives. The communication between the master PC and each control moduls is realized by converter USB - RS232 (Fig. 4).

The control system of drives is designed for DC motor with feedback included incremental or Hall sensor. DC motor is controled by the pulse width modulation (PWM). Current protection is realized by comparator, which swich off the H-bridge in case when become overload (Fig. 5).

5 CONCLUSION

PKS development is short – run compared with long – run machinery development with serial kinematic structure.

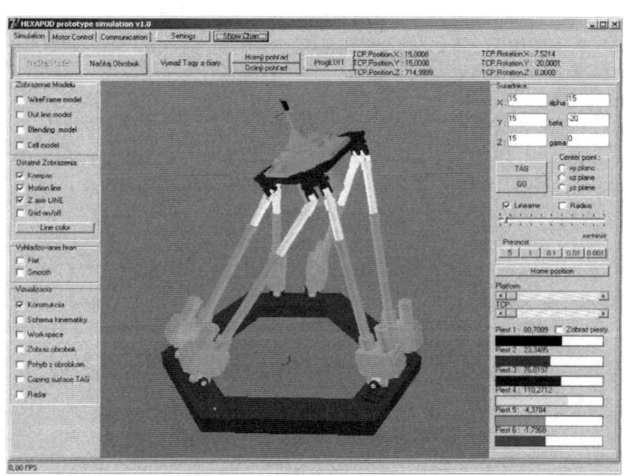

Fig. 3 Computer simulation of hexapod

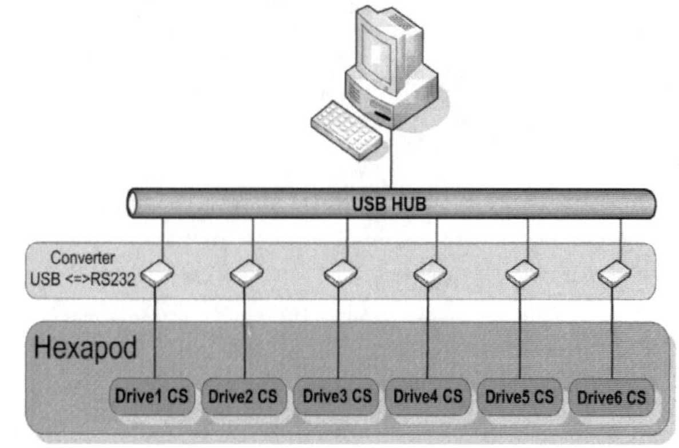

Fig. 4 Architecture of hexapod control system

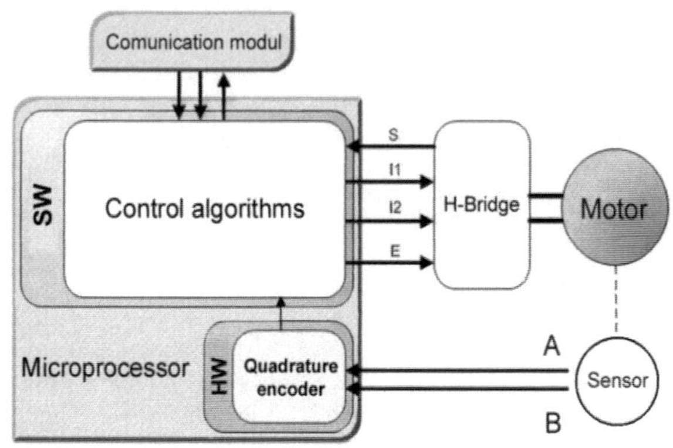

Fig. 5 Schema of the drives control system

Whereupon it is possible to expect, that in the future will arise more and more projects supporting PKS research (Knoflíček & Kolíbal, 2000, Marek, 2006).

At University of Žilina (Slovakia) was designed school hexapod with six degrees of freedom. There were designed hexapod mechanical elements, inverse kinematic analyse, 3D model, simulation software and hexapod control system.

Designed hexapod can be used as teaching aid for better understanding of parallel structure behavior and also as basis for next development in the area PKS.

6 REFERENCES

Chren, O. (2007). A kinematical variation of the tricept, *ERIN2007 International Conference*, pp. 14-17, ISBN 978-80-227-2636-8, STU Bratislava, April 2007, Slovakia

Knoflíček, R. & Kolíbal, Z. (2000). Morphological analyse of Industrial Robots Design (in Czech), VIENALA Košice, ISBN 80-88922-27-5, Košice, Slovakia

Marek, J. (2006). Machining centres with unconventional structure (in Czech). *MM Průmyslové spektrum, specialedition – Design of CNC machine tool. September 2006*, Praha, MM publishing, p. 234-243, pp. 282, ISSN 1212-2572

Merlet, J. - P. (2000). *Parallel robots,* Kluwer Academic Publisher, Dordrecht

O'Connor, R. (1998). Design News - Inverted struts improve hexapod rigidity. *Accessed:* 1998-02-16. *Available from:* http://www.designnews.com/article/CA112337.html

www.parallemic.org

www-sop.inria.fr/members/Jean-Pierre.Merlet//merlet_eng.html

Acknowledgement

The article was made under support of Slovak Grant Agency KEGA – project No 3/4098/06.

Annals of DAAAM for 2008 & Proceedings of the 19th International DAAAM Symposium, ISSN 1726-9679
Editor B. Katalinic, Published by DAAAM International, Vienna, Austria 2008
Make Harmony Between Technology and Nature, and Your Mind will Fly Free as a Bird

AVOIDANCE TRAJECTORY DESIGN FOR MOBILE ROBOTS

POZNA, C[laudiu] & ALEXANDRU, C[atalin]

Abstract: *The aim of this paper is to present a new methodology for avoidance trajectory design. More exactly the mobile robot follows an of line defined trajectory, during this task, initial unknown obstacles must be avoided. This maneuver generates the following problems: designing the avoidance trajectory and controlling the mobile robot on this trajectory. Present work focuses on the first problem: designing the avoidance trajectory. After an introduction in which are discussed the know solutions of this problem, we propose our solution based on the road potential concept. In the end we simulate this solution.*

Key words: *potential, avoidance trajectory, robot, obstacles*

1. INTRODUCTION

After we have solved the trajectory definition and control problems (Pozna & Alexandru 2008) we are confronted with the avoiding obstacle problem. In (Brandt 2005) and (Hrich 2005) two solutions are presented: the first is based on the mathematical theory of differential games, and the second, which is called "the elastic bands", was developed by Quinlan and Khatib for mobile robots path planning. In (Brandt 2005) the two methods are compared and it is proved that the "elastic band" method is more suitable for path planning. This method is based on repulsive potential field generation. More precisely the road (the right and left kerb) and the obstacles are modeled by an elastic network which is composed by several nodes linked by springs. The nodes are disposed on the right and left kerbs. The spring's stiffness is related to the desired trajectory which is the equilibrium position of the network in the absence of the obstacle. The presence of the obstacle (a new point and new springs) will perturb the initial position of the network and will generate a new equilibrium position. Using this equilibrium position a smooth avoiding trajectory is designed.

From our point of view this method has the following inconvenient: first, because of the nonlinear stiffness definition, the equilibrium position computation is a time consuming process, and second, in practice, the shape and the dimensions of the obstacle are discovered during the avoiding maneuver.

2. MINIMUM POTENTIAL FIELD CONCEPT

The car is following a trajectory and in a certain moment, discovers an obstacle which makes the initial trajectory impossible to pursuit. The car control system must react and perform the following on line tasks: discover the shape and dimensions of the obstacle; compute the avoiding trajectory; control the robot on the trajectory and, in the end, return to the initial trajectory. Some comments are necessary: because the obstacle is discovered on line, during the avoiding manoeuvre, the avoiding trajectory will be composed by several parts; this means that, on line, the control system must perform a loop which is composed by: discovering the obstacle and the road (the universe), designing a part of the avoiding trajectory and controlling the robot (car) on this trajectory part.

Because this is an on line process we must avoid time consuming computation. For this reason we must link together, from mathematical point of view, the universe knowledge with the trajectory generation. This means that it is important to consider the sensors behavior when we design the avoiding trajectory. In fact, if we analyze the "elastic bands" solution we can see that the escaping trajectory is the static deformation of a hypothetically elastic network associated with the car, road and obstacle (Brock 1999; 2002).

It is also know that the equilibrium of these kinds of structures can be obtained by imposing a minimum potential energy condition. Our idea starts from this point: it is not necessary to construct such a complicated potential field and to compute the nonlinear solution of equilibrium; it is more suitable to construct a simple potential field which can be used very quickly.

The construction of the potential field is related on finding a mathematical function which has the following properties: in absence of the obstacle, the minimum of this function is the initial trajectory (see figure 1); the presence of the obstacle will generate a new minimum which will avoid the obstacle; the computation of minimum must be a non time consuming process; the function definition can be linked to the relative speed between the obstacle and the car; the function can be easily constructed, based on the data sensors.

The mathematical expression of this function (in the car referential system) is:

$$z_R = \frac{1}{2}\left(m_2\left(sign\left(y-e_y\right)+1\right)-m_1\left(sign\left(y-e_y\right)-1\right)\right)\cdot\left|y-e_y\right| \quad (1)$$

$$e_y = \frac{X_O - e}{\cos(\psi)} - x\tan(\psi),\; m_1 = \frac{1}{L-e},\; m_2 = \frac{1}{e} \quad (2)$$

where e is the desired position of the car in the road referential frame, X_O & Y_O - the car position on the road referential frame, ψ - the car orientation, L - the wide of the road.

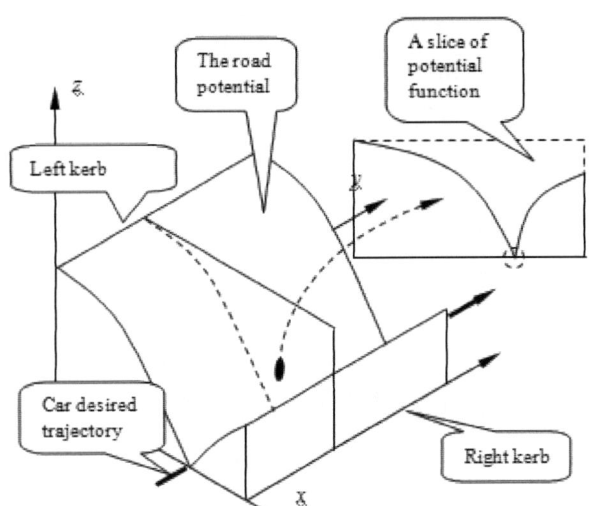

Fig. 1. The road potential function (no obstacles).

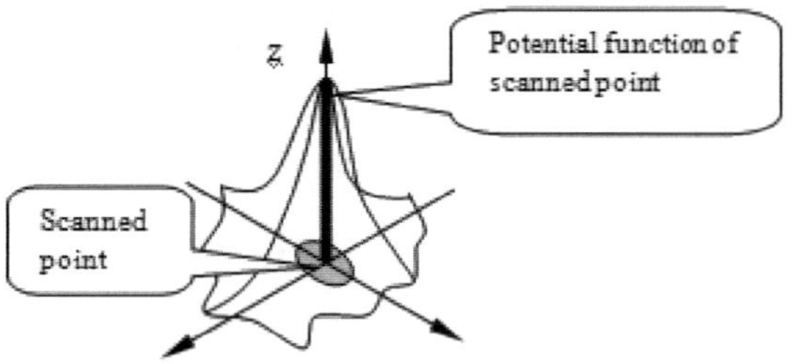

Fig. 2.The one of the obstacle scanned points potential function.

Because we intend to use a laser scanner the obstacle(s) is a collection of scanned points. The idea is to generate a potential function for each point, so in the end the obstacle potential function will be a sum.

$$z_O = \sum_{i=1}^{n} \exp\left(-\frac{c_1(y-y_1)^2 + c_2(x-x_1)^2}{2\sigma^2}\right) \qquad (3)$$

where c_1, c_2 & σ are parameters linked to the steering motor performance and the relative speed between the car and the obstacle, x_i & y_i - the position of the scanned point $i = \overline{1,n}$.

To obtain the minimum points we must solve the following equation:

$$\begin{Bmatrix} x_i \\ y_i \end{Bmatrix} = \begin{Bmatrix} \min(z_R + z_O) \\ y_i \end{Bmatrix}\Bigg|_{y_i \in (y_{i-1}-I,\, y_{i+1}+I)} \qquad (4)$$

where x_i=0:delta:Lr, Lr - the road length in the xoyz referential frame, I - the parameter which restrict the search area in order to avoid shaded minimum points.

3. AVOIDING TRAJECTORY – SIMULATION RESULTS

With equation (4) we have obtained a collection of minimum potential points. Using this result we must design a smooth C2 curve which will be the avoiding trajectory of the mobile robot. Because the obstacle is discovered during the avoiding maneuver the avoiding trajectory will be composed by several parts, for each part definition we imagine a strategy composed by two steps: first, recognize the environment obtain the minimum potential points, use the first k points and define the trajectory, and second, control the car (robot) on this trajectory.

In order to simulate this strategy we have designed in Matlab a program structure able to reproduce the car behavior during the avoiding maneuver. The road and the obstacles potential function is presented in figure 3. During the avoiding maneuver this function will change for each scanning step.

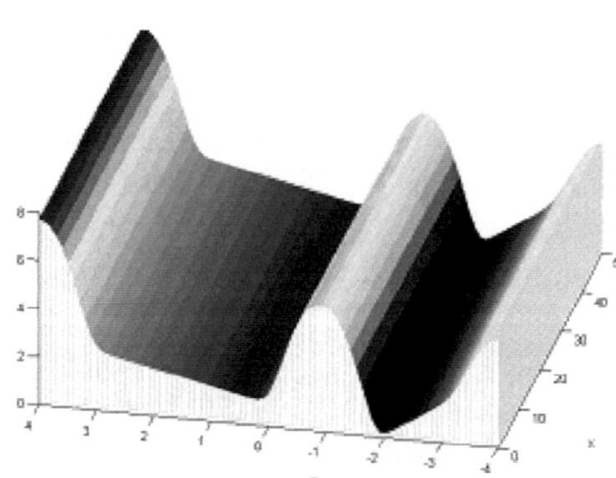

Fig. 3. The road & obstacle potential function.

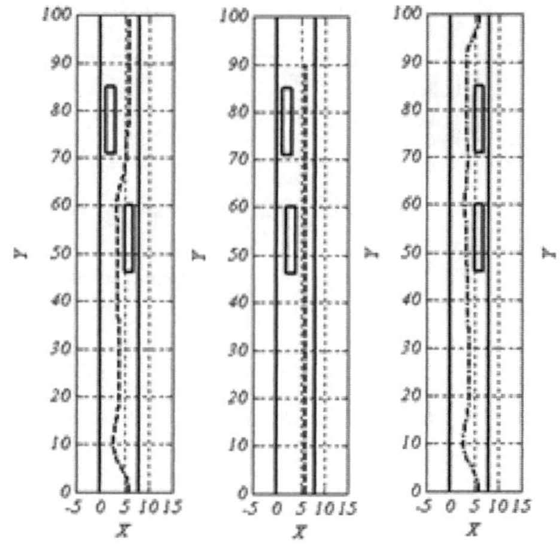

Fig. 4. The avoiding trajectories.

From several simulations the avoiding trajectories, which are constructed step by step, are presented in the diagrams shown in figure 4.

4. CONCLUSIONS

Avoiding obstacle means first of all, to design a trajectory and after this to control the car on the trajectory. This paper focuses on the trajectory design problem, where a new concept: the potential field of the road is proposed. The designing of this trajectory is an "on-line process" because the obstacles forms and the obstacles dimensions are discovered during the avoiding maneuver.

Using the concept of the potential field we have integrated the sensors output in a mathematical function. The minimum of this function, which is the avoiding trajectory, can be easily computed.

Even the simulation results confirm our idea to obtain all the parameters of the function, needs experimental tries. In order to pursuit these try we must design the controller which allows following the designed trajectory.

5. REFERENCES

Pozna, C. & Alexandru, C. (2007). Mobile Robot Control by Learned Behaviour, *Annals of DAAAM for 2007 & Proceedings of the 18 th International DAAAM Symposium: "Inteligent Manufacturing & Automation: Focus on Creativity, Responsability and Ethics of Engineers"*, Katalinic, B., pp. 605-606, ISBN 3-901509-58-5, October 2007, DAAAM International, Zadar

Brandt, T.; Sattel, T. & Wallaschek, J. (2005). On Automatic Collision Avoidance Systems, *Proceedings of SP-1920 SAE World Congress*, pp. 286-291, ISBN 978-0-7680-1565-2, April 2005, SAE International, Detroit

Hrich, K. (2005). Optimization of Emergency Trajectories of Autonomous Vehicles with Respect to Linear Vehicles Dynamic, *Proceedings of IEEE/ASME 2005 International Conference on Advance Intelligent Mechatronic*, pp. 135-144, ISBN 0-7803-9047-4, July 2005, IEEE Xplore, Monterey, CA

Brock, O. (1999). Generating Robot Motion the Integration of Planning and Execution, *PhD thesis*, Stanford University

Brock, O.; Khatib, O. & Viji, S. (2002). Task - Consistent Obstacle Avoidance and Motion Behaviour for Mobile Manipulation, *Proceedings of the 2002 IEEE International Conference on Robotics & Automation - ICRA*, pp. 388-393, ISBN 0-7803-7273-5, May 2002, IEEE Xplore, Washington DC

Annals of DAAAM for 2008 & Proceedings of the 19th International DAAAM Symposium, ISSN 1726-9679
Editor B. Katalinic, Published by DAAAM International, Vienna, Austria 2008
Make Harmony Between Technology and Nature, and Your Mind will Fly Free as a Bird

RIGID AND FLEXIBLE ASSOCIATION OF KINEMATIC CHAINS IN GEAR PROCESSING MACHINES

PREDINCEA, N[icolae]; CONSTANTIN, G[eorge] & GHIONEA, A[drian]

Abstract: *The article establishes the theory of complex surface generation specific to the processing of tooth gear processing with the purpose of associating rigidly or flexibly two or more closed kinematic chains. It has been proposed the foundation of the mixed association of the complex generating kinematic chains through the definition of the adjusting functions for defining the kinematic structure of the tooth processing machine tools. In perspective, the authors have in view the study of the kinematic and dynamic accuracy using the properties of the Botez mechanism.*

Key words: *closed kinematic chain, open kinematic chain, botez mechanism, association, adjusting function*

1. INTRODUCTION

For the generating of a surface on machine tools, the generatrix *G* or the director curve *D* are flat analytical curves (involute, ellipse, Archimedean spiral, polygonal curves, closed contour, resulting from combinations of simple flat curves, etc.) or three-dimensional curves (cylindrical helix, cone-shaped helix, combinations of these with certain flat curves, etc.). The generation of these curves requires the quantitative composing of at least two simple generating motions, of rotation (*R*) or of translation (*T*); the composition laws are expressed by kinematic conditioning function resulting from the kinematics of generating the paths *G* or *D* (Predincea & Constantin, 1993). These functions, defined in relation to the parameters of the simple generating motions (y_s, y_p – tool *S* or workpiece *P* speeds) express, at the same time, the kinematic connection of the output values y_{e1} and y_{e2} of the two kinematic chains.

The kinematic connection as rigid kinematic link imposed to the outputs y_{e1} and y_{e2}, can be obtained by means of a fictitious mechanism *W* (Fig. 1) that connects the two output ends of the two associated chains with each other (kinematic loop or closed kinematic chain). The transfer ratio of the fictitious mechanism is expressed by the kinematic conditioning function itself that is essential for achieving the paths *G* or *D*.

$$i_w = \frac{w_e}{w_i} = \frac{y_{e1}}{y_{e2}} = \frac{Y_i}{Y_e} = \frac{1}{i_T} = f_{12}(y_s, y_p) = f_{12}^{*}(x_s, x_p), \qquad (1)$$

where w_i and w_e are the inputs/outputs of the Botez mechanism (Predincea et al., 1995); Y_i, Y_e – inputs/outputs of the closed kinematic chain; i_T – total transfer ratio of the closed kinematic chain.

The Botez mechanism is also an ideal mechanism characterized by the following features: ideal accuracy, infinite rigidity, neglecting the weight and friction between its elements; the geometrical and motion parameters represent theoretic values for comparison purposes, having in view the study of the cinematic and dynamic accuracy.

For closed kinematic chain, the adjusting equation derives from the transfer equation (Predincea & Constantin, 1993):

$$i_R = A_i / B_i = K_i (1/i_w)(i_{D1}/i_{D2}), \qquad (2)$$

where $K_i = i_{11} i_{12} / i_{21} i_{22}$ is the kinematic chain constant; *i* becomes *F* for treading, *R* for rolling and *d* for relieving.

2. RIGID ASSOCIATION OF CLOSED KINEMATIC CHAINS

In many cases, the generating of complex-shaped surfaces requires the kinematic obtaining of both generating curves, *G* and *D*. For each of these there are kinematic conditioning functions of the form (1). Sometimes, however, the three-dimensional director curves and the complex shaped curves are obtained kinematically, as resultants of two or more paths of the simple generating motions.

The same degree of complexity is also specific to the situation when the kinematic generating conditions of rigid character are added conditions of technological (worm gear processing with tangential feed) and/or economic nature (spur and helical cylindrical gear processing).

Consequently, the achievement of the paths *G* and *D* with the above mentioned features requires either the existence of two or more closed kinematic chains in mixed association, or that of a single kinematic chain of closed type associated with one or several technological kinematic chains. Frequently, closed kinematic chains are mixed associated by means of a differential mechanism M_Σ (Figs. 1 and 2).

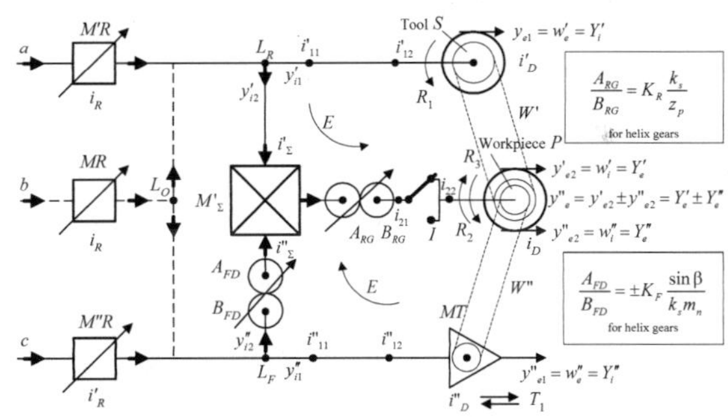

Fig. 1. Rigid mixed association of closed kinematic chains for cylindrical gear tooth processing.

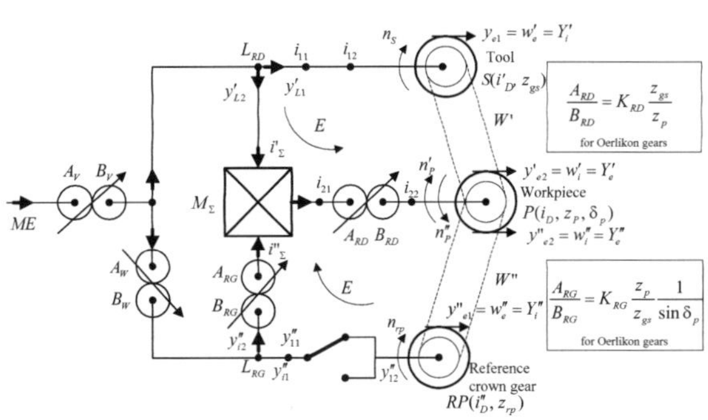

Fig. 2. Rigid mixed association of closed kinematic chains for bevel gear processing with continuous dividing.

The general association structure is characterized by the following features:

- the output values of the two kinematic chains in mixed association in parallel-series are summed up algebraically by a differential mechanism M;

- there is a branch common to both closed kinematic chains comprised between the differential mechanism and workpiece; at the output end of this branch there are the values w_i, and w_e of the fictitious mechanisms W' and W'', identical to the generating parameters of the curves G and D;

- each Botez mechanism correlates the motion parameters that are necessary for achieving the curves G and D, so that the adjustment function of the closed kinematic chains becomes

$$\frac{A}{B} = f'_{1,2}(i'_w, i_D, i'_D) = K \cdot f_{1',2}(x_s, x_p); \quad (3)$$

$$\frac{A''}{B''} = f_{1''}(i''_w, i_D, i''_D) = \pm K'' \cdot f^{*}_{1''2}(x''_s, x''_p, x^{*}_p). \quad (4)$$

- equation (3) and (4) allow for an easy and highly accurate determination of the change gears; calculation difficulties appear in case of the geometric parameters x, resulting from the closing condition of the closed kinematic chain, is a trigonometric function (e.g. x^{*}_p);

- the existence of the two closed kinematic chains requires in their structure a relatively large number of mechanisms, which contributes to the substantial increasing of the errors of kinematic and dynamic nature, the differential gear and change gears having one of the most influential contributions.

3. FLEXIBLE ASSOCIATION OF KINEMATIC CHAINS

The flexible CNC association (Figs. 3 and 4) refers to:

- the kinematic structure of the machine tool is constituted only by independent generating kinematic chains (main and feed chains) (Kief & Roschiwal, 2007);

- simple structures of the kinematic chains due to the adjustment achieved only electromechanically or electrically,

- the kinematic closing condition imposed by the generation of the curves G and D is easily achieved through CNC;

- if in rigid association the inputs Y_i of the two associated kinematic chains are subject of a rigid connection (L_R, L_F) in association with flexible program, the connection is pseudo-elastic (Weck & Brecher, 2005);

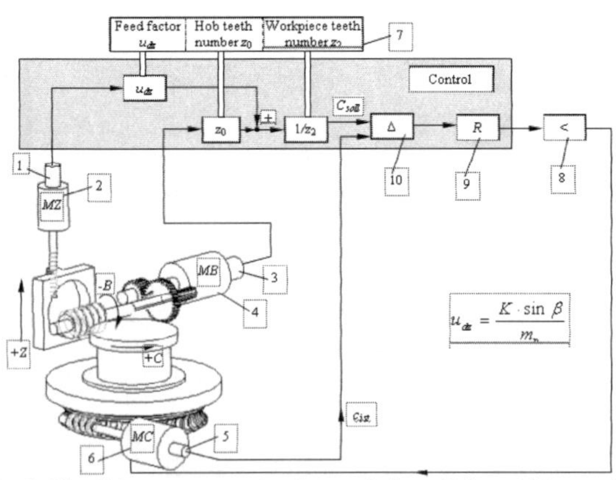

Fig. 3. Flexible program mixed association of closed kinematic chains for cylindrical gear tooth processing: 1 – transducer on Z axis; 2 – driving system along Z axis; 3 – transducer on B axis; 4 – main kinematic chain driving; 5 – transducer on C axis; 6 – circular feed kinematic chain (C); 7 – block for data introducing; 8 – amplifier; 9 – regulator, 10 – summator; K – kinematic chain constant; β – workpiece helix angle; m_n – normal modulus.

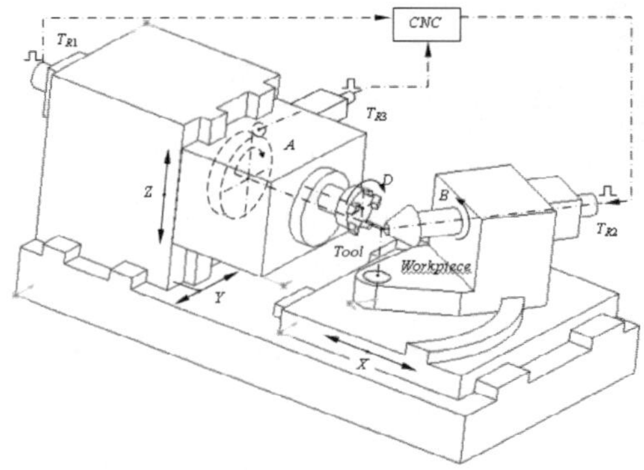

Fig. 4. Flexible program mixed association of closed kinematic chains for bevel gear processing with continuous dividing (Sandu et al., 2000): B – workpiece gear continuous rotation, D – mill rotation, A – reference crown gear rotation alternating with feed motion X, T_R – rotary transducers.

- possibility of identification, reducing and compensating for in real time the errors caused by static, dynamic and thermal behavior of the mechanisms in the kinematic chain structure.

4. CONCLUSIONS

The conditions imposed by Botez mechanism are available both for rigid and flexible association. Although the flexible type programming, by NC, has penetrated in this field of the machine-tools generating surfaces by thread cutting and. rolling, the rigid program-carrier of the machine-tools in this field is still present.

The advantages of the machines with rigid program-carrier are known, i.e.: accuracy, reliability, availability to any producer, low costs, absence of difficulties with respect to the supply and operating of certin components, that are specific to the equipment required for the flexible program-carrier.

Those with flexible program-carrier have the following advantages: simple structures of the kinematic chains, and consequently a high accuracy, possibility of easy integration in flexible production cells and systems, reduced costs in fabrication preparing and maintenance, easiness in adaptation to any production type. But acquiring the processing precision (profile and pitch errors) is limited due to the errors resulting in gear wheel calculation.

5. REFERENCES

Kief, H., Roschiwal, H. (2007). *NC/CNC Handbuch 2007/2008* (NC/CNC Handbook 2007/2008), Hanser Verlag, ISBN-10: 3-446-40943-2.

Predincea, N. & Constantin, G. (1993). Asocierea mixtă a lanţurilor cinematice închise la maşini de danturat (Mixed association of the closed kinematic chains in gear processing machines), *Scientific Bulletin of University Cluj-Napoca*, Series Machine Building, pp. 315-318.

Predincea, N.; Ispas, C., Minciu, C. & Ghionea, A. (1995). Teoria asocierii lanţurilor cinematice închise (Theory of Association of Clsoed Kinematic Chains), *T.C.M.M.*, No. 11, Edit. Tehnică, Bucharest, pp. 24-36.

Sandu, C.; Predincea, N. & Bălan, E. (2000). CNC-Freiformmaschinen zum Zyklo-palloid-verzahn von Kegelrädern (CNC Machines for processing the cyclo-palloid teeth of bevel gears), *Scientific Bulletin of North University of Baia Mare,* Series C, Vol. XIV, Fascicle: Tribology, Machine Manufacturing Technology, pp. 234-244.

Weck, M., Brecher, C. (2005). *Werkzeugmaschinen 1-Maschinenarten und Anwendungsbereiche* (Machine tools 1 - Machine types and fields of use), Springer-Verlag, Berlin, Heidelberg, ISBN: 978-3-540-22504-1.

Annals of DAAAM for 2008 & Proceedings of the 19th International DAAAM Symposium, ISSN 1726-9679
Editor B. Katalinic, Published by DAAAM International, Vienna, Austria 2008
Make Harmony Between Technology and Nature, and Your Mind will Fly Free as a Bird

THE USE OF MARKOV MODELS FOR IMPROVING CUSTOMER RELATIONSHIP MANAGEMENT IN ROMANIAN BANKS

PREJMEREAN, M[ihaela] & VASILACHE, S[imona]

Abstract: *The paper analyses one of the components of customer satisfaction in the Romanian banking system, the waiting time in call centres and the retrials in order to get in contact with an available server. The process is modelled in the form of a Markovian queue, taking as parameters the size of the call centre and the its corresponding load. The results of the study serve as a guide for Romanian banks in better dimensioning their call centres, in order to better serve their clients.*

Key words: *Markov, models, customers, Romania, banks*

1. INTRODUCTION

Customer relationship management is technology-driven, (Peelen, 2005) relying on the organizational capacity to get fast in contact with the clients and to use these efficient contacts as loyalty-building instruments. The fluctuations in trust which are characteristic to the passage from the face-to-face relationship with the bank officer to the automation of the service are mitigated by means of technology-based offerings (Greenberg, 2002) which allow the bank to provide its clients with faster and safer ways to find a solution to their requests. A very common instrument in rapidly transmitting information and solutions to customers, on demand, and also getting client information is the call center, which is widely used in the Romanian banking system (Popa and Popa, 2008).

The call centers of the Romanian banks have limited capacity, usually below the number of clients which simultaneously ask for the service at a time t. This results in some clients being directed to the servers, until their capacity is exceeded, at t_1, and the rest of the customers already in the system to be put on hold, or asked to call back at a later moment. The customer satisfaction decreases with the increase of the waiting time. In order to avoid this situation, clients are asked to come back later whenever all the servers are blocked. Those clients coming back at random intervals interfere with the new calls. This makes it difficult for the bank to approximate the actual number of clients soliciting its call centre in a certain period of time (Aguir et al., 2007).

Following Altman and Borovkov (1997), Templeton (1999), Aguir et al. (2003, 2007), Artalejo and Resing (2006), Shin (2006), Gimenes-Guzman et al. (2008), who have studied the retrial queue phenomenon, we adapt the Markov model to the particular situation of the Romanian call centres market, in the banking sector.

By applying this model, banks can get a realistic dimension of the retrial volume, and thus project their call centres capacity accordingly, by discriminating between new and repeated calls, and limiting the number of coming back clients, which leads to ensuring better levels of customer satisfaction. In this way, automation forecasting contributes to improvements in customer relationship management, by making banking services more flexible and user friendly for the clients of the bank. Our purpose here is to present an overview of the model, which can be further developed in the best interest of the bank, and of its clients.

2. METHODOLOGY

2.1 Premises

Let us take a call centre having c available servers, and C customer arrivals, in the form of a Poisson process with rate λ. Some of these clients meet an available server, the rest, $C - c$, occupy the orbit, and reattempt after an exponential time, distributed by ε. The probability that the *jth* client will retry, when signalled to do so, is P, while the probability that the same client abandons and exits the orbit is 1-P. In the probability space, we assume that clients on the orbit change places with clients who are being answered by means of a Γ process, $\Gamma = \Gamma(t)_{0 \le t \le n}$. If this process takes a number of finite states (two, in our case), and the jumping from one state to another can be attributed a stochastic intensity dependent upon some given variables (e.g., the patience of the client), then Γ is a Markov process.

2.2. Model

If we take the (i, j) state, where $i \le c$, in which i clients are being taken by the servers and j clients are waiting on the orbit, and associate to it a probability p_{ij}, then the Markov equations, at the equilibrium, are:

- Case I – all the clients are being covered by the servers, assuming that the distribution of the real queue for the (0,0) state is the same as the distribution of the virtual queue for the (1,0) state

$$(\lambda + \varepsilon i)\, p_{i,0} = \lambda p_{i-1,0} + \varepsilon p_{i-1,1} + (i+1)\, \varepsilon p_{i+1,0} \qquad (1)$$

where i takes values from 1 to c-1

$$\lambda p_{i,0} + c\varepsilon p_{i,0} + (i-c)\,(1-P)\, p_{i,0} = \lambda p_{i-1,0} + \varepsilon p_{i-1,1} + \varepsilon c p_{i+1,0} + (i-c+1)\,(1-P)_{i+1,0} \qquad (2)$$

where i takes values from c to C-1

$$\lambda P p_{C,0} + \varepsilon c p_{C,0} + (C-c)\,(1-P)\, p_{C,0} = \lambda\, p_{C,0} + (1-P)\, p_{C-1,1} + (1-P)^2\, p_{C,1} \qquad (3)$$

- Case II – there is at least one client on the orbit, assuming that the distribution for the real queue, for state (0,j), plus the delay for the jth client on the orbit, is the same as the distribution of the virtual queue, for the (1,j) state

$$\lambda p_{i,j} + \varepsilon i\, p_{i,j} + j p_{i,j}\,(1-P) = \lambda p_{i-1,j} + (j+1)\,(1-P)\, p_{i-1,j+1} + (i+1)\, \varepsilon\, p_{i+1,j} \qquad (4)$$

where i takes values from 1 to c-1

$$\lambda p_{i,j} + c\varepsilon\, p_{i,j} + (i-c)\,(1-P)\, p_{i,j} + j\,(1-P)\, p_{i,j} = \lambda\, p_{i-1,j} + (j+1)\,(1-P)\, p_{i-1,j+1} + \varepsilon c\, p_{i+1,j} + (i-c+1)\,(1-P)\, p_{i+1,j} \qquad (5)$$

where i takes values from c to C-1

$$\lambda P p_{C,j} + \varepsilon c p_{C,j} + (C-c)\,(1-P)\, p_{C,j} + j\,(1-P)^2\, p_{C,j} = \lambda p_{C-1,j} + (j+1)\,(1-P)\, p_{C-1,j+1} + (j+1)\,(1-P)^2\, p_{C,j+1} + \lambda P p_{C,j-1} \qquad (6)$$

These equations are solved using a method by Tijms and van de Coevering (1990), for exploiting the geometric tail of the state probabilities, and thus reducing the Markov chain.

3. RESULTS AND DISCUSSIONS

Following Aguir et al. (2007), we computed the retrial rate, λ', as:

$$\lambda' = \sum_{j=1}^{C'} j\,(1-P) \sum_{i=0}^{C} p_{i,j}$$

We have based our estimations on a call centre with 30 available servers and 5 clients waiting. Varying the system load, l_s, from 50% to 150%, we obtained the data presented in Table 1. The dependence of the retrial rate on new calls rate is modelled in Figure 1, taking into account the system load.

The bank should operate with two types of constraints, on the one hand, the number of servers (because of cost constraints, it can't afford having more servers than necessary), and the ratio attended calls to non-attended calls, in other words, clients who are served versus clients who are waiting. In doing the computation, the bank should take into account that, for a high level of system load, only approximately half of the observed calls volume is represented by new calls, the rest belonging to clients on the orbit. Knowing the number of servers and the total call volume, the new calls rate is found out. Based on Markov equations, the relationship between the service time and the waiting time can be also modelled, for obtaining the best equilibrium between serving clients well and not making them wait. Markov modelling is useful in HR projections, by estimating the servers needed for the customer volume.

System load	*New calls rate*	*Retrial rate*
50%	4	0.08
60%	4.5	0.18
70%	5	0.4
80%	5.5	0.77
90%	6	1.26
100%	6.5	2.34
110%	7	3.71
120%	7.5	5.4
130%	8	7.38
140%	8.5	9.53
150%	9	11.97

Table 1. Retrial rate as a function of the new calls rate

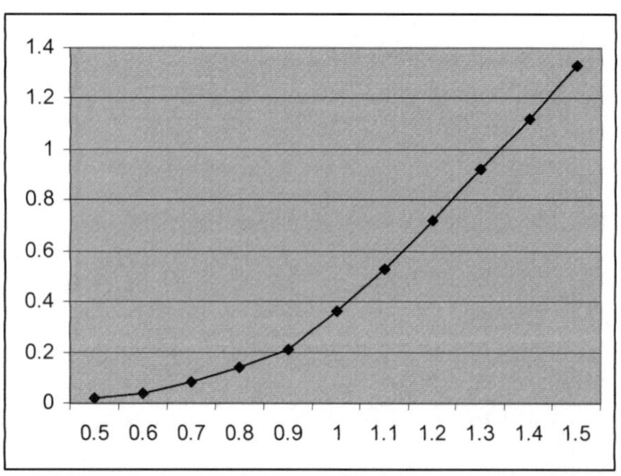

Fig. 1. The ratio between new calls rate and retrial rate, based on system load

As it can be seen, the evolution of the retrials depending on the new calls is almost exponential, for large system loads the number of retrials exceeding the number of new calls. This means that, usually, clients who are asked to come back later do it immediately, which burdens the system and makes it difficult to discriminate between calls which are entering the system for the first time and repeated calls.

4. CONCLUSION

If in the banking system retrials are not taken into account, estimations of the actual volume of customers which are accessing the bank call centre lead to significant errors. As proven by the application presented, for higher degrees of system load, the rate of the retrial is larger than the rate of the new calls. Knowing which is the retrials rate and modelling its relationship with the new calls rate helps the bank, on the one hand, dimension correctly the amount of staff needed for this service which is expanding on the Romanian market and, on the other hand, keep an account of the retrial phenomenon. Busy servers can scary customers away, so the bank should estimate the response time as compared with the waiting time, in order to meet its quality thresholds, but it is nonetheless true that clients who feel they get the information they want from the call centre and are able to solve their problems this way will insist. A high rate of retrials may signal the bank that the call centre is working properly, that the system load is optimum, and that clients will be stimulated to try to get in touch with the servers, on the condition that the limit of nerves abuse, if clients are too often and for a too long time put on hold or asked to call back later, is observed.

5. REFERENCES

Aguir, M.S., Aksin, O.Z., Karaesmen, F., Dalley, Y. (2007) On the interaction between retrials and sizing of call centers, *European Journal of Operational Research*, doi: 10.1011/j.ejor.2007.06.051, accessed July, 20th 2008

Aguir, M.S., Karaesmen, F., Aksin, K., Chauvet, F. (2003) Analyse du problème des rappels et le dimensionnement dans un centre d'appels, *Proceedings of MOSIM '03*, Toulouse, France

Altman, E., Borovkov, A.A. (1997) On the stability of retrial queues, *Queuing Systems*, 26: 343-363

Artalejo, J., Resing, J.A.C. (2006) *Mean value analysis of single server retrial queues*, SPOR Report No. 2006-13, Eindhoven: Technische Universiteit Eindhoven

Gimenez-Guzman, J.M., Domenech-Benloch, M.J., Pla, V., Casares-Giner, V., Martinez-Bauset, J. (2008) Value Extrapolation Technique to solve retrial queues: A Comparative Perspective, *ETRI Journal*, 30 (3): 492-494

Greenberg, P. (2002) *CRM at the speed of light: Capturing and keeping customers in internet real time* (2nd ed.). Berkeley and London: McGraw-Hill

Peelen, E. (2005) *Customer Relationship Management*, Prentice Hall

Popa, L., Popa, A. (2008) Bancile la examenul "call center", *Banii nostri*, http://www.baniinostri.ro/stiri/stiri.php?ContentID=8986, accessed on: 2008-07-20

Shin, Y.W. (2006) Monotonicity properties in various retrial queues and their applications, *Queuing Systems*, 53 (3): 147-157

Templeton, J.G.C. (1999) Retrial queues, *TOP*, 7 (2): 351-353

Tijmes, H.C., van de Coevering, M.C.T. (1990) *How to solve numerically the equilibrium equations of a Markov chain with infinitely many states*, Research Memoranda, Free University Amsterdam, ftp://zappa.ubvu.vu.nl/19900046.pdf, accessed on: 2008-07-05

Annals of DAAAM for 2008 & Proceedings of the 19th International DAAAM Symposium, ISSN 1726-9679
Editor B. Katalinic, Published by DAAAM International, Vienna, Austria 2008
Make Harmony Between Technology and Nature, and Your Mind will Fly Free as a Bird

ALGORITHMS AND SOFWARE FOR VISION SYSTEMS OF MOBILE ROBOTS

PRIANICHNIKOV, V[alentin]; KIRSANOV, K[irill]; KII, K[onstantin] & LEVINSKY, B[oris]

Abstract: *The applications of remote sensors in the robotics technology is connected with the appearance of delays in the measuring channels that are significant in control, as well as with distortions caused by the medium of the signal propagation. To overcome these problems, a system of mathematical models of ultrasonic and optical sensors in the control systems of mobile robots was developed.*
Key words: *Robotics, computer vision, remote sensors*

1. INTRODUCTION

To provide qualitative solution of the problems arising in mobile robotics, it is necessary to improve the navigation aids and the control of an MR (mobile robots) and to develop new software that will be able to realize one of the central goals of programmed control.The crucial problem is to receive and process data on the environment for both adaptive control and the supervisory mode. These data are predominantly provided by remote sensors (RS). Under difficult conditions, the set of effective sensors becomes significantly narrower; therefore ultrasonic and optical methods of forming a feedback come in the forefront. As a result, the need in designing intelligent software and developing new special models, methods and means for logical filtration and processing complex information about real world arises.

2. PROCESSING SCHEME

Let us present in more detail algorithms of the upper level from the point of view of the functions realized by them. We also determine the interrelations of the algorithms and data flows between them.Denote by *X(t)* the coordinate system of the operating space of a mobile robot, which is a data set determining the functioning of a particular robot with remote sensors. Let us call the operating environment one-, two- or three-dimensional depending on what robot motions are prevailing, linear, or motions on a plane adequately represented in the form of a flat map, or motions in 3D terrain (for example, as for manipulators or underwater robots). Let us note that the dimension of the space of the generalized coordinates *X(t)*, defining the motion of robots as an electromechanical system, may be higher than 1, 2 or 3, depending on the design of these robots. However, in the coordinate system *X(t)* of the mobile robot, the locomotion problem is posed on a certain subspace, whose dimension is 1, 2 or 3 (we will call it the subspace of the locomotion task or the navigation subspace). Let *Au* be the algorithm of forming control commands *U(t)* on the basis of an internal representation *Mr* (model of the dynamic capabilities of the robot and its parameters). The dimension of the control vector *U(t)* corresponds to the dimension of the vector of the generalized coordinates *X(t)*, i.e., it is supposed that a part of the generalized coordinates of robot are controlled, and the remaining coordinates are either calculated in the virtual model of robot *Vr* or changed in the process of the motion of the real robot in accordance to its kinematic constraints. Let *As* be the algorithm of processing the measurements and formation of the results of determination of *Xs(t)* from the primary measurements *Ls*, (accumulated into a train or an image), which has been filtered by using the model of sensors *Ms*. Denote by *An* the navigation algorithm for determining the position *Xm(t)* and formation of the target function *Xp(t)* on the basis of *Xs(t)* and the map of external environment *Me* (stored in the robot memory). *Xp(t)* can be input to the algorithm *Au* through the interface *Vi* from the operator *Mu*. Let *Vs*, *Vr*, and *Ve* be the virtual representation of the sensors, the robot and the environment for the substitution of real devices during testing the algorithms of the robot within the framework of mathematical simulation or in the course of operation of a trainer. In the case of work in the real world, *Vs* and *Vr* are the drivers of the remote sensors and robot, respectively, and *Ve* is directly the real environment. Real location *X* of the robot in the space of its generalized coordinates is not directly accessible to control system. Let *Xp* be a purposeful program trajectory formed either through the interface of integrated software *Vi* with the operator of supervisory control *Mu* or calculated by the algorithm *An* according to the data *Xm* (for example, we form it based on the rules) (Pryanichnikov 2008).
Assume that *t is* the current time, changed with a step of *dt : tk = t - k dt, k=0,1,...,* are the moments of the time for observation and formation of digital control commands. Let *Ms*, *Mr*, and *Me* be data (structure of data, model, and parameters) that provide the input of the corresponding algorithms.
The interrelations of the developed algorithms and programs and data flows can be presented more clearly in the form of the following diagram:

$$X(t) = Vr\,(Ve, U(t)\,);\ Ls(t) = Vs\,(Ve, X(t)\,)$$
$$Xs(t) = As\,(Ms, Ls(t, t\text{-}dt, \ldots, t\text{-}k\,dt)\,)$$
$$Xm(t) = An\,(Me, Xs(t)\,),\quad Xp(t) = Vi\,(Mu, Xm\,)$$
$$U(t) = Au\,(Mr, Xm(t) - Xp(t)\,)\ \text{go to } X.$$

It contains 3 cycles - (*X,...,U,X,...*) and 2 asynchronies in *Vs, Vi.* complex information about real world arises.

3. INSTRUMENTAL SOFTWARE

The proposed software (including the subsystems of development, fixing and profiling) makes it possible to create recognition and controlling software within the framework described above. The final recognition and/or controlling program is implemented in the form of a directed graph, where particular nodes and edges correspond to program modules (subprograms) and data flows between them, respectively. A similar organization of the process of development can be found in the professional program packages, such as Mathlab: Simulink, Microsoft Robotics Studio, etc. The main distinction of the proposed software is that in contrast to the mentioned packages, which are aimed at operation under laboratory conditions, it is mainly oriented to the implementation on a real robot. This becomes possible because of the following specific features:

1. Two programming languages for the development were employed. Each module of the system can be realized either in a binary form (Windows\Linux x86, and ARM are supported) or in the form of programs in the Python language, whose interpreter is embedded into the system. This organization provides a high performance (optimized modules in C++, and a high rate of the development or the prototyping algorithms in Python).
2. The built-in systems for parallel dispatching and memory management was implemented. The proposed parallel dispatcher (Kirsanov 2007) was implemented to ensure the parallel processing of the program graph without blockings (except for the cases, when the programmer explicitly specifies this behavior) providing in this case the integrity of data. The use of non-blocking smart pointers (Edelson 1991) reduces considerably the problem with memory release. For this system, a set of the modules was developed, which ensure the base functionality, namely, the video input from files, from databases of images, video cameras or framegrabbers connected directly to the onboard computer. In addition, the capture of different types of data from ultrasonic and range-finders is supported. The set of modules which ensures the distributed synchronous operation of several programs of this type on a network and makes it possible to move modules executed on the onboard computer to the remote computer and in the opposite direction were designed. This simplifies considerably the process of debugging and developing programs for processing sensor reading.

4. COMPUTER VISION MODELS AND COMPUTER VISION SUBSYSTEM

One of the units of the proposed system is a subsystem for processing and interpreting visual data. This subsystem is based on the approach to processing grayscale and color images proposed in (Kiy 2004). In contrast to the existing systems of color segmentation (Chang 2007) in the proposed approach, the main emphasis is placed on producing in real-time a brief geometric scene description convenient for solving image recognition problems arising in robotics, rather than on selecting a rational palette and indexing image points. The image processing in the developed subsystem is based on the representation of color images in the ($G/(G+B)$, $G/(G+R)$, I) coordinate system, where I is the grayscale intensity), which allows us to recover the complete color information and is convenient for reducing the original 3-D problem of color vision to problems of lower dimensions. The vision system consists of subsystems with the following basic functions: early vision (selection of local objects, intermediate vision (construction of global objects), and approximate qualitative classification of global objects selected at the intermediate stage. At the stage of local (early) vision, the system has no definite ideas what a problem is going to be solved. It does not concentrate its efforts on finding predetermined features (for instance, straight segments), but prepares a general-purpose local image description and classification. This description simultaneously comprises intensity, texture, color, and shape characteristics of the local objects. In the course of processing, we divide images into several parallel vertical or horizontal strips of a fixed width. The results of the early vision stage are represented in the form of collections of intervals on the vertical or horizontal axis of the image plane (this depends on the employed type of strips (vertical or horizontal)). Each of these intervals has color (hue and saturation) and grayscale intensity characteristics. To select colored objects, we group colored intervals in the strip with close color characteristics. As a result of grouping, we obtain color bunches in the considered strip. Each color bunch in the strip is a set of intervals with close color characteristics and a range of grayscale intensity. At the stage of intermediate processing, we construct from local objects in strips parts of global objects joining local objects in adjacent strips. The shape of these global objects is characterized by a continuous system of collections of intervals in strips.

Based on the developed technique a software package has been developed. The package can deal both with a single image and video sequences. This package has an interface for working through a network (including the work via the Internet). For a standard Pentium IV processor, 3 GHz, color images of resolution 640x480 are processed at 10 fps to 8 fps. The processing rate for images of resolution 320x240 is four times greater. The developed technique has been successfully applied in driving autonomous vehicles on roads and is being applied to autonomous navigation of a mobile robot in offices. Figure presents an example of processing results of the subsystem. A real-world image is reduced to a finite number of color intervals describing the scene, which are superimposed on the grayscale image corresponding to the original color image.

Fig. 1. The results of segmentation of a real-world image.

5. CONCLUSION

We proposed a computational scheme for the solution of the problems of controlling data flows (including visual information) from the sensor system to the navigation modules of a mobile robot and for designing software tools for development and fixing of the algorithms. The proposed software reduces considerably the development and debugging of algorithms for sensor processing and recognition algorithms. The presence of the built-in means of recording, parallel performance, dynamic updating, and memory management distinguishes the proposed system from the tools available on the market. We also integrated a new technique for segmenting real-world color images in real time with the system.

6. REFERENCES

Chang, Y. C., Archibald, J. K, Wang, Y.-G &. Lee, D. J. (2007a). "Texture-based color image segmentation using local contrast information" *Int. J. of Inform. Technol. Intel. Computing,* Vol. No. 2, (May 2007).

Edelson, D. R. & Pohl, I. (1991), Smart pointers: They're smart but they're not pointers. P*roceedings of the 1991 Usenix C++ Conference*, p. 20, April, 1991.

Kiy, K.I. & Dickmanns E.D., (2004), A color vision system for real-time analysis of road scenes, *Proceedings of IEEE Intelligent Vehicle'04 Symposium*, pp. 54-59, ISBN 0-7803-8311-7, Parma, Italy, June, 2004.

Kirsanov K.B. (2007) "Development of a distributed information-control system and its application in identification of images by a mobile robot," *Proceedings of Workshop in Mechatronics and Robotics.* pp. 112-116, TSNIII. St. Petersburg, March, 2007.

Pryanichnikov V.E. (2008), "Remote sensors in the composition of motion control systems of mobile robots," - Informational--Measuring and Control Systems, № 1, Vol..No. 6,(January, 2008), pp 5-18.

Annals of DAAAM for 2008 & Proceedings of the 19th International DAAAM Symposium, ISSN 1726-9679
Editor B. Katalinic, Published by DAAAM International, Vienna, Austria 2008
Make Harmony Between Technology and Nature, and Your Mind will Fly Free as a Bird

THE CALCULUS OF THE VIBRATION AMPLITUDE WITH THE MECHANIC MODEL APPROAPCH FOR VIBRATING TUBE MILLS

PRICHICI, M[ariana] A[driana]

Abstract: *This paper attempt to show ways to determine the operating range where the most effective grinding effected can be anticipated using simple theoretical consideration of the energy input for vibrating tub. The results will be used for determine the optimal filling ratio of grinding elementsfor optimizing the energy consumption of the mills. This research was implemented already in a cement factory near Oradea.*
Key words: *grinding tube, vibration amplitude, frequency, filling ratio*

1. INTRODUCTION

The mechanic model approach of "dividing mass in two cylinders" (fig.1), the entire grinding element bulk is divided in two compact cylinders with a diameter d, length L and density ρ_{mc}. The two cylinders move up and down on Δ axes and have a circular movement with frequency ω of the vibrating grinding tube. The grinding tube has a harmonic oscillatory movement with vibration amplitude A and frequency ω.

$$f = \frac{\omega}{2\pi}\ perioadT = \frac{2\pi}{\omega} \tag{1}$$

2. REQUIREMENT AND OBJECTIVE

A prerequisite for the following model observation is the assumption that the attainable grinding results are in direct relationship to the energy input (Hengel,Staudinger 1992),. Thus those parameters that have considerable influence on the input energy should be investigated. The parameters which primarily influence the vibration state and thus the input energy are the vibration amplitude of grinding tube, the size of the grinding chamber and the filling ratio of grinding elements.

The objective is the study of the parameters mentioned before so that a maximum energy can be transferred to the grinding bulk. The investigation of these elements will be based on the model of "dividing mass in two cylinders".

The model of "dividing mass in two cylinders "will be used to describe the movement behaviors of different extreme cases.

3. GEOMETRICAL CONDITIONS FOR THE "DIVIDING MASS IN TWO CYLINDERS" APPROACH.

The energy transferred by the vibrating grinding tube wall should be so big, that, the entire "dividing mass in two cylinders" converts twice the distance *s* within an oscillation period T (Prichici,Indrie&Gherghel, 2008). With this, the distance s covered during the up and down movement is given by

$$s = \omega A t - \frac{g}{2}t^2 \tag{2}$$

for the inferior cylinder and for the superior cylinder:

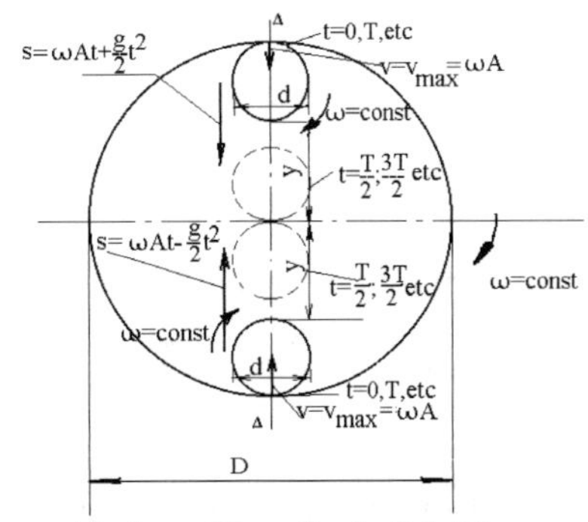

Fig.1 Geometrical conditions for the "dividing mass in two cylinders"mechanic model (Prichici,2005)

$$s = \omega A t + \frac{g}{2}t^2 \tag{3}$$

For the second part of the period in the two expressions we change the sign of the term $\frac{g}{2}t^2$. The peripheral speed *v* represents the product of vibration amplitude and the frequency ω of the vibrating grinding tube given by the formula (4)

$$v = \omega A \tag{4}$$

is changing with the modiffing of vibration amplitude *z* , of harmonically up and down movement of the grinding tube and is given by the next formula, and you can see the correlation between z and the speed v from formula (6).

$$z = A \sin \omega t \tag{5}$$

$$v = \dot{z} = \omega A \cos \omega t \tag{6}$$

The maximum speed of the inferior cylinder was achieved at the moment: t=0,t=T;t=2T,etc, when the grinding tube is perfect cylinder (fig.2) :

$$v_{\max} = \omega \cdot A \tag{7}$$

The tow cylinders are dropt on vertical in opposite ways with this maximum speed for the inferior cylinder. In this moment they have a circulary movement with frequency ω of the vibrating grinding tube.

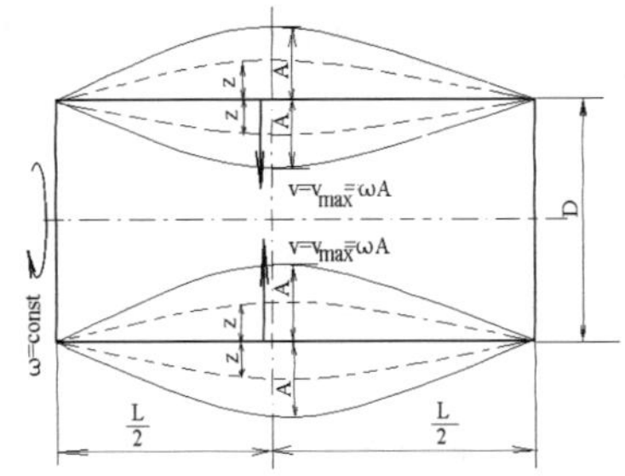

Fig.2 Calculus scheme of mechanic model

The low of movement of the tow cylinder is given by:

$$s = \omega \cdot A \cdot t \mp \frac{g}{2} t^2 \quad (8)$$

The vibration input should take place after exactly half an oscillation period (t=T/2). In figure 1, at the moment $t = \frac{T}{2}, s = y = \frac{D}{2} - d$, we consider the space of the two cylinders in the time interval T/2 as being equal, and equal with y , so , we can write:

$$y = \frac{D}{2} - d = \omega \cdot A \cdot \frac{T}{2} \pm \frac{g}{2} \frac{T^2}{4} == \pi \cdot A \pm \frac{g}{2} \frac{\pi^2}{\omega^2} \text{ and:}$$

$$y = \frac{D}{2} - d = \pi \cdot A \pm \frac{g}{2} \frac{\pi^2}{\omega^2} \quad (9)$$

The "minus" sign is for the ascendant movement of the inferior cylinder, which is smaller than the movement of superior cylinder , in its descendent movement. To complete the space: $y = \frac{D}{2} - d$ of the inferior cylinder, we adopt "plus" sign in relation (9), and we obtain a maximum value for the amplitude A.

$$A = \frac{y}{\pi} + \frac{g\pi}{2\ \omega^2} \quad (10)$$

The two cylinder will run, the distance $y = \frac{D}{2} - d$, (in the period T/2), the term $\frac{g}{2}\left(\frac{T}{2}\right)^2$ will be smaller then $\omega \cdot A \frac{T}{2}$ to y. So, for D=0,6 m , η=0,4 si ω=20π rad/sec, we obtain : $y \cong 0{,}0316m, \frac{g}{2}\left(\frac{T}{2}\right)^2 \cong 0{,}0122m$ and $\frac{g}{2}\left(\frac{T}{2}\right)^2$ cannot be neglected in front of the value of y . With a *decreased of* frequency ω and diameter D of the vibrating grinding tube, term: $\frac{g}{2}\left(\frac{T}{2}\right)^2$ has smaller value in coherency with y. So, for the diameter D=1m of grinding tube, η=0,4, ω=40π rad/sec, we obtain the following values:

$$y \cong 0{,}057m; \frac{g}{2}\left(\frac{T}{2}\right)^2 \cong 0{,}00306m.$$

The movement of grinding tube mills is imaginary represent. The dynamic process is very complex. The mechanic model gives a mathematical schema and a qualitatively and a quantitatively method for the study of dynamic process. The harmonically vibration amplitude is maximum at the middle of cylinder, and zero at the beginning of the it. We can simplify in the following way; we consider that the harmonically vibration amplitude of grinding tube is constant.

4. DETERMINATION OF VIBRATION AMPLITUDE

For the characterization of the movement of the " divided mass in two cylinder "it is necessary to describe the geometrical relationship in the grinding tube (Prichici,Indrie&Gherghel, 2007). Here we start out with the definition of the filling ratio η:

$$\eta = \frac{2\frac{\pi \cdot d^2}{4} L}{\frac{\pi \cdot D^2}{4} L} = \frac{2d^2}{D^2} = 2\left(\frac{\frac{D}{2} - y}{D}\right)^2 \quad (11)$$

By transpose, one obtains the following for the distance y:

$$y^2 - Dy + D^2\left(\frac{1}{4} - \frac{\eta}{2}\right) = 0 \quad (12)$$

with:

$$y_{1,2} = \frac{D}{2}\left(1 \pm \sqrt{2 \cdot \eta}\right) \quad (13)$$

Because $y < \frac{D}{2}$, we adopt for the distance y:

$$y = \frac{D}{2}\left(1 - \sqrt{2 \cdot \eta}\right) \quad (14)$$

Substituting (14) in (10) results in the required vibration amplitude :

$$A = \frac{D}{2\pi}\left(1 - \sqrt{2\eta}\right) + \frac{\pi \cdot g}{2\omega^2} \quad (15)$$

5. CONCLUSIONS

From the graphical evaluation of equation (15), it can be recognized that:
- With an increased of filling ratio, only low amplitudes are required for obtaining the described sequence of movement ;
- An increase in diameter of the grinding tube requires a proportional increase in the oscillation amplitude;

Mills with large grinding tube diameters should be operated with a high filling ratio, since oscillating amplitudes greater than 10 mm can only be achieved at great tehnical effort (Prichici,2005).

However, in order to make statements of the filling ratios which should be selected for a specified grinding tube diameter, the energy input must be examined as a function of filling ratio

6. REFERENCES

Hengel, M.; Staudinger G.(1992),*Calibration of the photo - sedimentometer lumosed with a powder of known particle size distribution*- Tehniche Universitat Graz , Austria

Prichici, A.(2005), *The calculus of optimum filling ratio of grinding elements in vibration bolls mills with a mechanical model of dividing mass in two half cylinders*, Oradea University, ISSN 1583-0691, Oradea

Prichici, A.(2005), *Tratat de mori cu bile*, Oradea University ed., ISSN 973-613-047-9, Oradea

Prichici, A.; Indrie, L. & Gherghel, S. (2007). *The dynamic study of vibrating tube mills. The calculus of vibration amplitude with the model mechanic approach of dividing mass,* Oradea University ed., ISBN 1583-0691, Oradea

Prichici, A.; Indrie, L.; Gherghel, S.(2008),*Vibrating tube mills of construction materials, industry model mechanic approach of divided mass in N bar for the calculus of vibration amptitude fascicle of management and technological engineering,* Oradea University, ISSN 1583 – 0691, Oradea

Annals of DAAAM for 2008 & Proceedings of the 19th International DAAAM Symposium, ISSN 1726-9679
Editor B. Katalinic, Published by DAAAM International, Vienna, Austria 2008
Make Harmony Between Technology and Nature, and Your Mind will Fly Free as a Bird

INTERCONNECTED POLY(DIMETHYLDIPHENYLSILOXANE)/SILICA NETWORKS FOR ELECTRICAL ACTUATION

PRISACARU, G[heorghe]; DIACONU, I[lie]; MARTA, C[onstantin]; DOROFTEI, I[oan]; CAZACU, M[aria] & VLAD, A[ngelica]

Abstract: *The electrostatic field induced strain of a series of dimethyldiphenylsiloxane copolymer/silica networks differing by diphenylsiloxane unit's content and copolymer/silica ratio was investigated at ambient conditions.*
At fields up to about 30 MV/m, thickness induced strain of all films showed a quadratic dependence suggesting that the electrostatic effect is involved. Under the actual experimental conditions (rigid electrodes and unprestrained films) induced strains up to 25 % were obtained at fields of about 50 MV/m. These values are remarkable when compared to those found in prestrained (Pelrine et al., 2000) or unprestrained (Pelrine et al., 1998) films with compliant electrodes at very high fields (> 110 MV/m). High electrostrictive coefficients (up to about 10^{-14} m^2/V^2) and small response times (~ 1s) were found. The electromechanical parameters depend both on the copolymer composition and copolymer/silica ratio.
Key words: *networks, copolymer, films, actuation, strain.*

1. INTRODUCTION

Some technologies of actuators require electroactive materials capable to operate sometimes in special environmental conditions such as low temperatures and high humidity.

On the other hand, there is a growing demand in mechatronics, robotics and bioengineering, regarding polymer elastomers with high actuation strains and stresses, fast response time, good flexibility, high efficiency, stability and durability.

Silicone elastomers seem to be very promising in meeting these properties (Pelrine et al., 1998, 2000; Kornbluch et al., 2001; Trujillo et al., 2004). In order to achieve these requirements, high driving electric fields are needed for silicone actuators which may be an important disadvantage in certain applications.

The actuation electric field can be reduced by preparing new compounds based silicone with improved electromechanical properties.

First time, some attempts in this direction has been focused on the development of composites (Mazzoldi et al., 2004; Carpi & De Rossi, 2005; Gallone et al., 2007), which preserve, to some extent, the mechanical properties of silicone and increase the dielectric constant of the material. However, this approach led to limited results.

Recently, a blend composed of silicone and a high polarizable conjugated polymer showed remarkable increase of the electromechanical response (Carpi et al., 2008).

The goal of our paper is to determine some electromechanical parameters of a different type of electroactive dielectric material consisting of a series of new networks based on dimethyldiphenylsiloxane copolymer and tetraethylorthosilicats (TEOS) differing by copolymer composition and TEOS content. Electromechanical parameters, strain, apparent electrostrictive coefficient and response time, were determined from the thickness deformation of polymer films induced by d.c. driving electric field.

In our knowledge there are no reports in literature concerning the synthesis and electromechanical properties of these dimethyldiphenylsiloxane copolymer/silica networks.

2. EXPERIMENTAL

The synthesized dimethyldiphenylsiloxane copolymers having different contents in diphenylsiloxane units (between 6.7 – 21.6 % moles) were mixed in different ratios with tetraethylorthosilicate as crosslinker in presence of dibutyltindilaurate as catalyst. TEOS was used in large excess, thus in situ formation of silica network occurs concomitantly with crosslinking (Figure 1).

Interpenetrated silica networks were processed as films cast before crosslinking. Thickness induced strain of films (0.02 – 0.2 mm thick) was measured under ambient conditions using d.c. driving electric fields up to 60 MV/m. The measuring devices was composed of a micro Box Data Acquisition System, a MTN/EP080 Monitran displacement sensor and an electrode-film unit presented elsewhere (Diaconu et al., 2006).

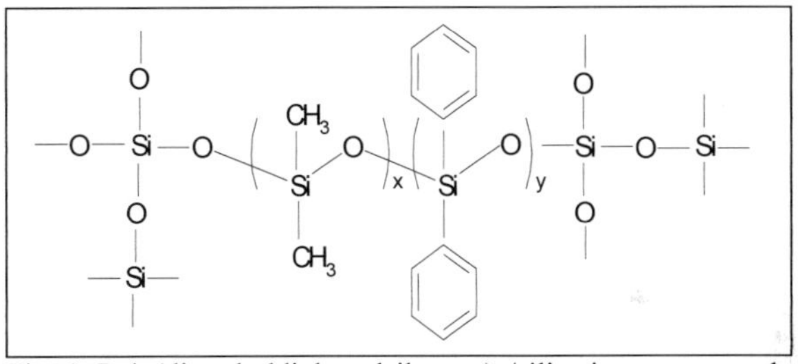

Fig. 1. Poly(dimethyldiphenylsiloxane)s/silica interpenetrated networks

3. RESULTS

All films showed compression in thickness direction under applied static electric field irrespectively of its sign (Figure 2).

For lower electric fields up to about 30 MV/m, the induced strains showed a quadratic dependence (Figure 3), whereas at higher values of the electric fields the strain presented a lower increase.

The quadratic dependences suggest the electrostatic nature of the strain responses. For the highest electric fields remarkable strains up to 25 % were found. From the slope of straight lines quite high values of the apparent electrostrictive coefficient up to about 10^{-14} m^2/V^2 were determined (Table 1).

Contracting responses are quite quick and showed a steady strain for all films (Figure 4).

Electromechanical parameters depend both on the copolymer composition and the copolymer/silica ratio (Table 1). The induced strain and electrostrictive coefficient (M) decrease with diphenylsiloxane unit's content and increase with the copolymer/silica ratio decrease.

The response time (τ) does not depend significantly on the copolymer composition and the copolymer/silica ratio.

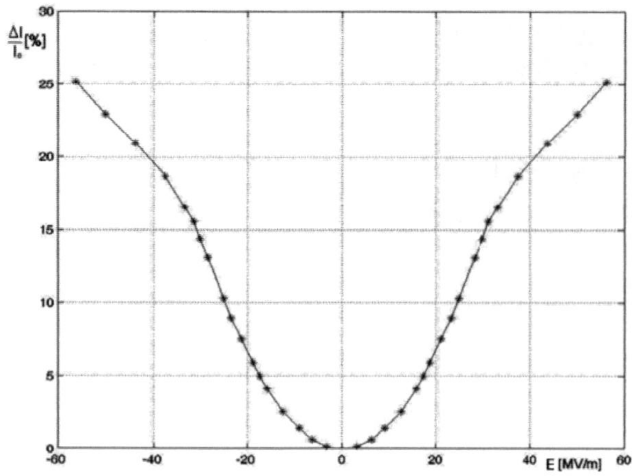

Fig. 2. Dependence of the strain on electric field for FSG19 film

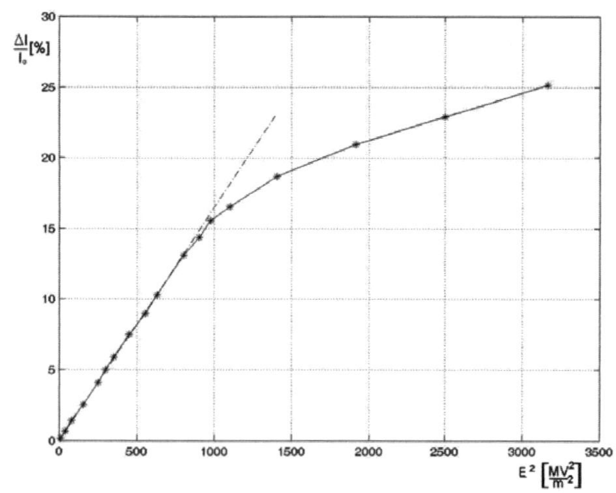

Fig. 3. Dependence of the strain on the square field for FSG19 film

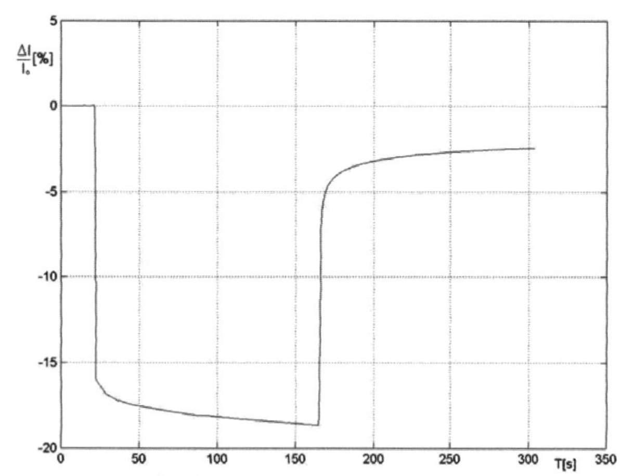

Fig. 4 Time response of contraction and relaxation processes for FSG19 film at 600 MV/m

These dependences may be explained by the higher elastic stiffness of diphenylsiloxane units as compared to that of the dimethylsiloxane ones and by the higher positive effect of dielectric constant of silica than that of the negative effect of silica elastic stiffness respectively.

Sample code	MR*	y(%)**	$\Delta l/l_0$ (%)	E (MV/m)	τ (s)	M (m^2/V^2)
FSG16	0.64	6.7	6.0	3.5	1.0	3.99×10^{-15}
FSG17	0.64	18.8	5.0	3.5	1.25	8.62×10^{-15}
FSG20	0.64	21.6	0.268	50	1.0	1.07×10^{-18}
FSG19	1.07	18.8	0.244	3.5	1.25	1.64×10^{-16}

Tab 1. Electromechanical parameters for some films with different copolymer compositions and copolymer/silica ratios

** Copolymer/TEOS masic ratio;*

*** Relative content of diphenylsiloxane: $100\cdot y/(x+y)$.*

4. CONCLUSIONS

The investigated siloxane copolymer/silica networks show remarkable electromechanical parameters such as thickness strain, apparent electrostrictive coefficient and response time. These parameters depend both on the copolymer composition and copolymer/silica ratio. Further research is now developing to determine other parameters (effective compressive pressure, mechanical energy density and efficiency) and to synthesize new siloxane/silica networks in order to establish correlations between structure and properties so as to produce actuators with higher forces, motions, energy density, speed of response and stability.

5. REFERENCES

Carpi, F. & De Rossi, D. (2005). Improvement of electromechanical actuating performances of a silicone dielectric elastomer by dispersion of titanium dioxide powder, *IEEE Transactions on Dielectrics and Electrical Insulation,* 12, 4, August, 2005, pp.835- 843, ISSN: 1070-9878

Carpi, F.; Gallone, G.; Galantini, F. & De Rossi, D. (2008). Silicone-Poly(hexylthiophene) Blends as Elastomers with Enhanced Electromechanical Transduction Properties. *Advanced Functional Materials*, 18, 2, January, 2008, pp.235-241, ISSN: 1616-301X

Diaconu, I.; Dorohoi, D. O. & Topoliceanu, F. (2006). Electrostriction of a Polyurethane Elastomer-Based Polyester. *IEEE Sensors Journal* 6, 4, August, 2006, pp. 876-880, ISSN: 1530-437X

Gallone, G.; Carpi, F.; De Rossi, D.; Levita, G & Marchetti, A. (2007). Dielectric constant enhancement in a silicone elastomer filled with lead magnesium niobate-lead titanati. *Materials Science and Engineering:C*, 27, 1, January, 2007, pp.110-116, ISSN: 0928-4931

Kornbluch, R.; Pelrine, R.; Pei, Q. & Shastri, S. (2001), Aplication of Dielectric Elastomer EAP Actuators, In: *Electrostrictive polymer (ESP) Actuators as Artificial Muscles*, Yoseph Bar-Cohen, pp. 457-495, Spie Press, ISBN 0-8194-4054-X, Bellingham, Washington 98227-0010

Mazzoldi, M.; Carpi, F.; Pioggia, G.; Ferro, M. & De Rossi, D. (2004). EAP Activity in Italy, *WW-EAP Newsletter*, 6, 2, 2004, 7-9, *Available from:* http://eap.jpl.nasa.gov

Pelrine, R. E.; Kornbluh, R. D. & Joseph, J. P. (1998). Electrostriction of Polymer Dielectrics with Compliant Electrodes as Means of Actuation. *Sensors and Actuators A*, 64, 1, January, 1998, pp. 77-85, ISSN: 0924-4247

Pelrine, R. E.; Kornbluh, R. D.; Pei, Q. & Joseph, J. P. (2000). High-Speed Electrically Actuated Elastomers with Strain Greater than 100%. *Science*, 287, 5454, February, 2000, pp.836-839, ISSN: 0036-8075.

Trujillo, R.; Mou, J.; Phelan, P. & Chan, D. (2004). Investigation of electrostrictive polymers as actuators for mesoscale devices. *The International Journal of Advanced Manufacturing Technology*, 23, 3-4, February, 2004, pp.176-182, ISSN: 0268-3768

Annals of DAAAM for 2008 & Proceedings of the 19th International DAAAM Symposium, ISSN 1726-9679
Editor B. Katalinic, Published by DAAAM International, Vienna, Austria 2008
Make Harmony Between Technology and Nature, and Your Mind will Fly Free as a Bird

TORQUE MULTIPLIERS FOR MACHINE-TOOLS

PRODAN, D[an]

Abstract: *This paperwork presents a variant of torque multiplier that can be used together with two-speed gearboxes with electric switching that equips Heavy Vertical Turning and Boring Mills and Horizontal Boring and Milling Machines. The two-range torque multiplier is simple and has no gears and no backlash. It can be also use within feed kinematic chains of CNC machine tools.*

Key words: *machine tools, torque multiplier, belt transmission.*

1. INTRODUCTION

Using two-speed gearboxes with electric switching on some machine-tools such as Heavy Vertical Lathes does not ensure all the speed-torque ranges that can be achieved with classic gearboxes with sliding gears (Prodan, 2008). Most of the Vertical Lathes with classic gearboxes ensure the transmission ratios $i_1 = 1$, $i_2 = 1/3$ and $i_3 = 1/9$. Figure 1 shows torque-speed characteristics as obtained at the pinion on the crown gear level, when machining with constant torque (Botez, 1977) – condition specific to heavy machine tools – and the main motor has the following parameters: rated power $P_N = 100$ KW, $n_{nom} = 1500$ rpm, $n_{max} = 4500$ rpm. When a two-speed gearbox is used for a motor with the same parameters as above, then the torque-speed characteristics as shown in Figure 2 are obtained. By comparing the two characteristics one may notice that a lower torque is obtained within the range of 166-373 rpm when using a two-speed gearbox. Also, the maximum possible developed torque is 2.25 times smaller. Many machine-tools manufacturers prefer this type of gearbox due to its compactness and lack of backlash. However, if the characteristics obtained with a classic gearbox are needed, then a more powerful motor should be used or, for lower powers, a couple of two-speed gearboxes with electric switching can be connected. A more powerful motor increase both the size and the cost. The second option increases the cost, too, and it is limited by the value of the maximum input torque of the second gearbox, depending on its technical parameters. The performances of the classic three-speed gearbox can be reached by using a classic gear reducer or a classic gearbox with sliding gear and two speed ranges, but the backlashes make this system improper for the feed kinematic chains (Perovic, 2006).

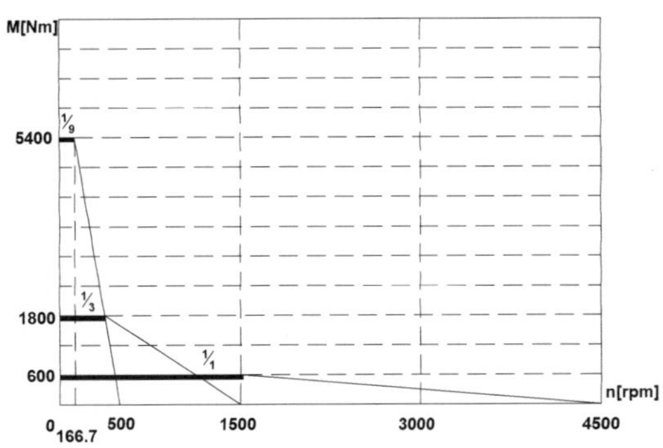

Fig. 1. The torque-speed characteristics when machining with constant torque.

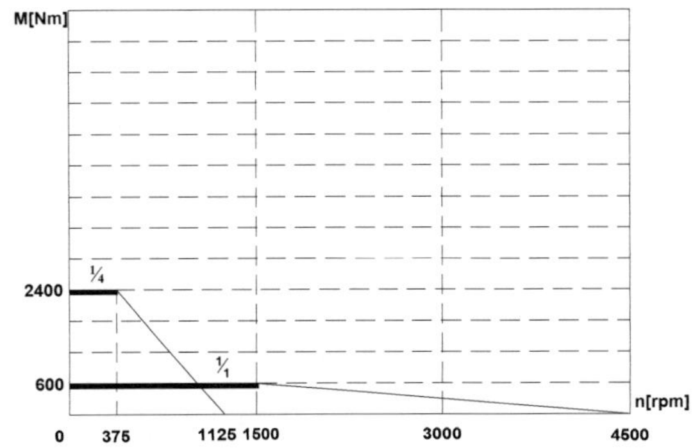

Fig. 2. The torque-speed characteristics when a two-speed gearbox is used.

2. TORQUE MULTIPLIER WITH TIMING BELTS

Nowadays timing belt transmission is frequently utilized due to high powers and torques that can be transmitted, but also for other advantages such as: simple construction, backlash-free, low noise, lubrication-free and lower costs than those of gear transmission. Usually, one step torque multipliers are used. A problem specific to these transmissions is the need of pulling systems. Figure 3 presents the proposed system. The electric motor ME is coupled with the two-speed gearbox CV. The transmission ratios of the gearbox are 1/1 and 1/4. The motion is brought to the shaft I that drives the clutch C. If the clutch is on position I, then the motion is transferred to the shaft II by means of the transmission 1D1/1D2 = 2/3. The belts 2D1/2D2 = 2/3 transfer the motion from the shaft II to the shaft III. Then, the belt transmission 1D3/2D3 = 1/1 transfer the motion to the shaft IV. Finally, the pinion Z_1 engages with the crown gear Z_2 rotating the table. If the clutch C is switched on position II, the motion from the shaft I reaches directly the shaft III and then the table through 1D3/2D3 and Z_1/Z_2. Therefore the torque multiplier by itself ensures two ratios: 1 /2.25 and 1/1.

The formulas below describe the system working:

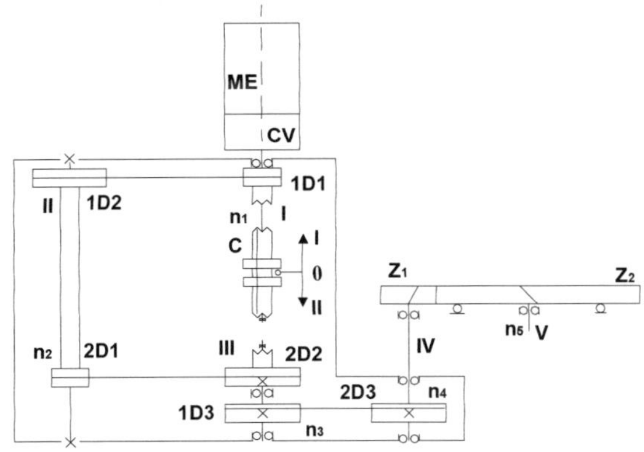

Fig. 3. The proposed system.

$$C \to I$$

$$n_5 = n_1 \times \frac{1D1}{1D2} \times \frac{2D1}{2D2} \times \frac{1D3}{2D3} \times \frac{Z_1}{Z_2} \quad (1)$$

$$C \to II$$

$$n_5 = n_1 \times \frac{1D3}{2D3} \times \frac{Z_1}{Z_2} \quad (2)$$

$$C \to 0$$

$$n_5 = 0 \quad (3)$$

Figure 4 shows system characteristics. This variant contains four speed ranges that include those three obtained with the classic gearbox. The system is backlash-free and the torque multiplier requires no lubrication. This constructive solution reduces the overall size and uses only one pulling system although there are two belt transmissions. To make such a system suitable for the feed kinematic chains of the CNC machine tools, it is necessary to compensate the backlash occurring at the level of the final transmission pinion Z_1/crown gear Z_2 (Catrina et al., 2005). Depending on the machine-tool type, mechanical or hydraulic systems can be used for backlash compensation (Prodan & Marinescu, 2005). The use of a pneumatic system for actuating the clutch C is recommended by the small forces required and by the fact that no oil should contaminate the belts (Bucuresteanu & Isar, 2007). Figure 5 presents the pneumatic diagram of the actuating system of the clutch C.

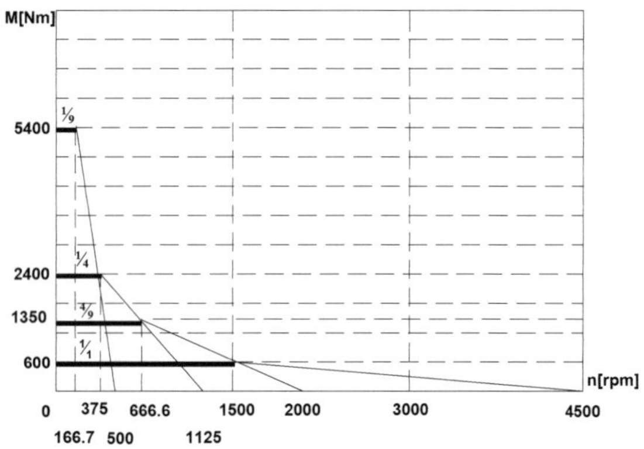

Fig. 4. The proposed system characteristics.

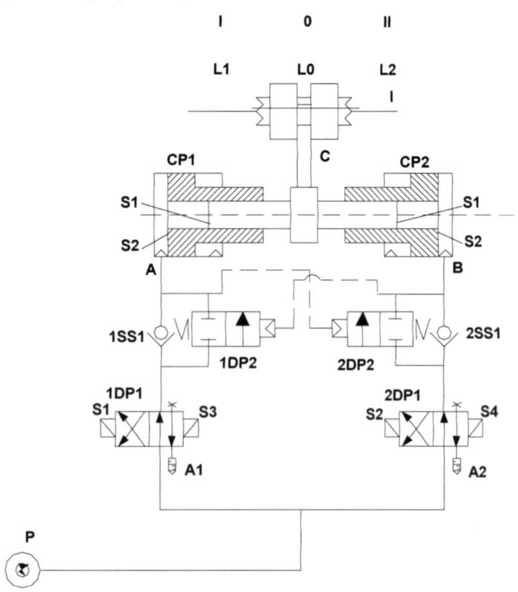

Fig. 5. The pneumatic diagram.

The pneumatic linear motors CP1 and CP2 move the clutch C to those three positions I, 0 and II. The motors are telescopic with the active surfaces S1 and S2.

The inductive micro-switches - L1, L0 and L2 - confirm those three positions. The clutch C can slide along the shaft I. The directional control valves 1DP1 and 2DP1 ensure the selection of the position. The pneumatically controlled valves 1DP2 and 2DP2 and the check valves 1SS1 and 2SS1 help maintaining the selected position. Noise dampers A1 and A2 reduce the noise.

The cyclogram shown by Table 1 describes the working of the system.

	SOLEINOIDS				SWITCHS		
	S1	S2	S3	S4	L1	L0	L2
0			+	+		+	
I	+			+	+		
II		+	+				+

Tab 1. The cyclogram of the system.

3. CONCLUSION

When using integrated gearboxes with electrical switching of the speed, especially on heavy machine-tools, torque multipliers should be provided for obtaining the torque required to rough machining.

The classic torque multipliers – with gears – are expensive, noisy, require tooth machining and have backlash that might compromise the precision of the operated system. In such situations torque multipliers with belt transmissions are recommended.

The advantages specific to belt transmissions are:

- no backlash
- low noise
- no complicated machining is required
- lubrication-free
- and, nevertheless, they can be procured from specialized producers.

When the gearboxes operate as feed boxes as well – see the Vertical Turning and Boring Mills – the belt transmissions represent a simple, cheapper and more liable solution. If there are special requirements with regard to the precision of the total transmission, then mechanical, hydraulic, electric systems for backlash compensation must be provide.

4. REFERENCES

Botez, E. (1977). *Masini-unelte. Bazele teoretice ale proiectarii (Machine-Tools. Theoretical Fundamentals in Designing),* Technical Publishing House, ISBN C.Z.621.9, Bucharest

Bucuresteanu, A. & Isar, D. (2007). *Elemente si sisteme pneumatice. Aplicatii Industriale (Pneumatic Elements and Systems. Industrial Applications),* Certex Publishing House, ISBN 978-973-1716-06-0, Bucharest

Catrina, D.; Totu, A.; Croitoru, S.; Carutasu, G.; Carutasu, N. & Dorin, A. (2005). *Sisteme flexibile de prelucrare (Flexible Systems for Machining),* Matrix Rom Publishing House, ISBN 973-685-981-9, Bucharest

Perovic, B. (2006). *Handbuch Werkzeug-Maschinen,* Carl Hanser Verlag, ISBN 10:3-446-40602-6, Munchen, Wien

Prodan, D. & Marinescu, S. (2005). *Refabricarea masinilor-unelte. Sisteme hidraulice (Re-designing of Machine-Tools. Hydraulic Systems),* Technical Publishing House, ISBN 973-31-2255-6, Bucharest

Prodan, D. (2008). *Masini-unelte grele (Heavy Machine-Tools),* Printech Publishing House, ISBN 978-973-718-892-2, Bucharest

Annals of DAAAM for 2008 & Proceedings of the 19th International DAAAM Symposium, ISSN 1726-9679
Editor B. Katalinic, Published by DAAAM International, Vienna, Austria 2008
Make Harmony Between Technology and Nature, and Your Mind will Fly Free as a Bird

HYDRAULIC DEVICE FOR RELIEVING THE ROTARY TABLE BEARINGS ON HEAVY VERTICAL LATHES

PRODAN, D[an] & MOTOMANCEA, A[drian]

Abstract: *This work aims to present an efficient solution for relieving the loads appearing on the rotary table bearing system of the heavy vertical lathes during some specific cutting procedures. The technical problem solved by this solution consists in taking up the additional axial force that exceeds the rated capacity of the axial bearing provided for the bearing assembly of the rotary table, by a hydraulic system with a rotary and linear hydraulic engine as its central element.*
Key words: *Rotary table, relieving, hydraulic engine*

1. INTRODUCTION

The work refers to a hydraulic system of relieving the axial loads appearing during the machining process on the rotary tables of heavy vertical lathes (turning and boring lathes).
For the time being, as far as the authors know, such types of relieving processes are not mentioned anywhere and therefore on heavy vertical lathes with rotary table diameter between 1700 mm and 4300 mm, the parts with weight exceeding the maximum capacity imposed by the manufacturer of the axial bearings fitting the bearing assembly of these rotary tables cannot be machined. This fact obviously restricts the capacity of machining some heavy parts, even if their diameters allow the parts to be turned on such machines.
The solution proposed and presented by the authors is an extremely efficient way to solve a highly stringent technical problem in the metal cutting area.

2. DEVICE DESCRIPTION

For the machines currently manufactured, the solution of taking up the weight of the part is of the type shown in Fig.1. (Botez 1977) The part 1, with weight G, is clamped on the rotary table 2. This table is rotated by the toothed rim 3. The weight is completely relieved by the bearing 4 onto the frame 5.
The proposed solution is shown in Fig.2. Besides the elements 1-5, also used for the solution in Fig.1, there are also: the bearing assembly 6 and the rotary and linear hydraulic engine 7. Its piston, with diameter d and working surface S, can rotate simultaneously with the table 2. The piston can also perform a vertical translation movement. The weight G decomposed into

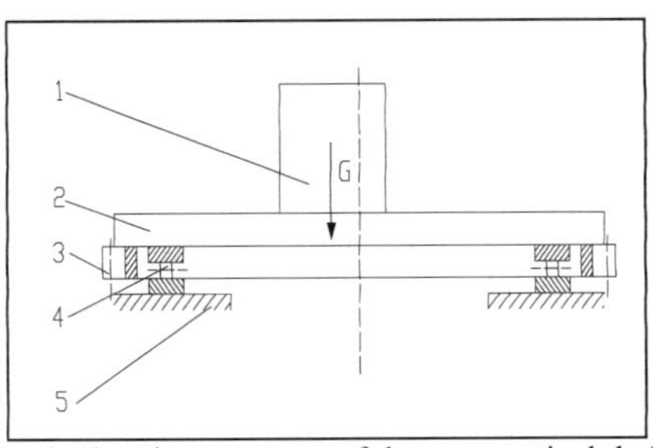

Fig.1. Classic bearing system of heavy vertical lathe rotary table.

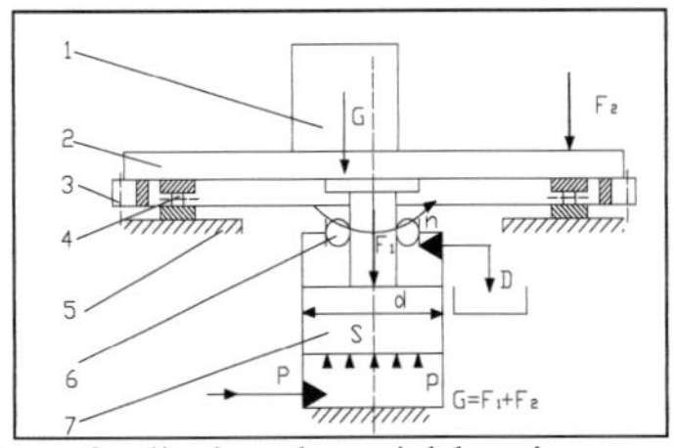

Fig.2. Solution of relieving the axial bearing, proposed by the authors

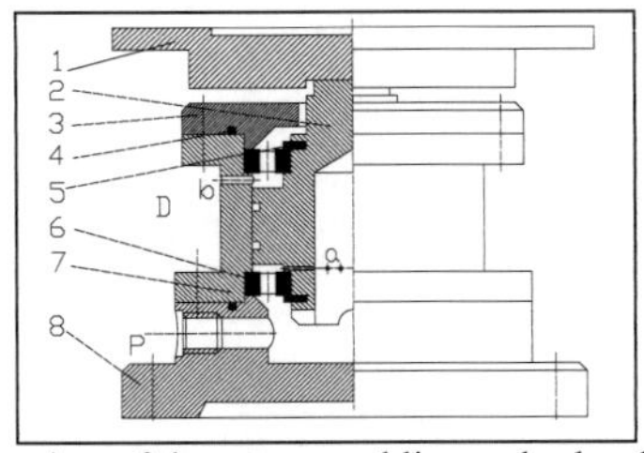

Fig.3 Construction of the rotary and linear hydraulic engine.

the forces F_1 and F_2. The force F_1 is taken up by the hydraulic engine 7, and the force F_2 is relieved through the bearing 4 onto the frame 5. The force F_1 can be controlled by means of the pressure p generated by the hydraulic driving installation that shall provide the possibility of adjusting the pressure according to the machined part, so that the capacity of the bearing should not be exceeded. The hydraulic engine used, as it has already specified, allows the rotation of the piston and also its vertical translation. Losses are drained onto the circuit D.
The construction of such an engine is shown in Fig.3. Part 1 is the one that leans onto the rotary table. It is caught by the piston 2. The piston can rotate, but also can perform a vertical movement, due to the roller bearings 6. These bearings are secured with the caps 3 and 8 and with the flexible rings 5. The eventual losses between the body 7 and the caps 3 and 8 are stopped by the sealing elements 4. The pressure oil feeding is made through the path P. The pressures on the two sides of the lower bearing are balanced through the holes a. The inevitable losses between the piston and the body are drained through the holes b towards the path D. The piston is pushed upwards by the oil pressure and it is driven into rotation by the rotary table. The piston can move vertically due to the fact that there is the possibility of relative movement, on small strokes, between the rings of this type of bearing.
Fig. 4 shows the location of such a system in a vertical lathe currently manufactured. (Prodan&Marinescu 2005). The rotary table 1 leans on the bearing 2 and the hydraulic engine 4. The stresses taken up by the hydraulic engine are also relieved on the frame, by means of the construction 5. This is a schematic diagram meant to explain how the relieving system operates. In practice the rotation transducer is also mounted in the central area.
The bearing load is relieved directly on the frame 3.

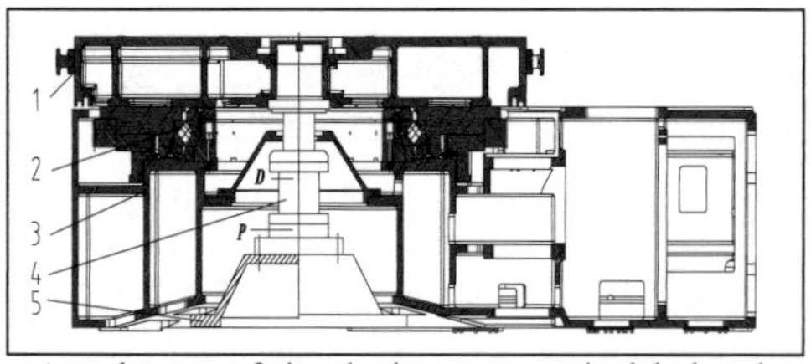

Fig.4. Attachment of the device to a vertical lathe already in operation.

The hydraulic schematic diagram is shown in Fig.5. The pump 1P1 driven by the electric engine 1ME1 sends oil from the basin T, through the check valve 1Ss1, by the relieving system on the path P. The maximum operating pressure is adjusted by means of the pressure valve 1Ssig.1. The pressure is detected and confirmed by the pressure relay 1Rp1.

The accumulator 1Ac1 shall provide a minimum pressure in case of accidental shutdown of the pump during operation. The accumulator is fitted with the security group 1Rb1., 1Rb2 and 1Ssec.1. The gauges and valves 1M,2M and respectively 1RM and 2RM are used for reading the pressures at the pump and the accumulator. The oil filtration is made in the return circuit, by means of the filter 1F1.

3. MATHEMATICAL MODELS USED

3.1.Static calculations

For the operating pressure, at an imposed load F_1, the following relationship shall be used (Prodan 2004):

$$p = 1{,}1 \cdot \frac{F_1}{\frac{\pi \cdot d^2}{4}} \tag{1}$$

In the relationship (1) d stands for the diameter of the piston. The output of the feeding pump can be calculated using the expression:

$$Q_P = 2{,}5 \cdot \frac{\pi \cdot d \cdot p \cdot j^3}{96 \cdot L \cdot \mu} \tag{2}$$

In the relationship (2) there are the following notations: j-piston-to-cylinder clearance, L-piston width, μ- viscosity of the oil used. The power of the electric engine for pump driving shall be determined with the relationship:

$$P_{ME} \geq \frac{Q_P \cdot p}{450} \tag{3}$$

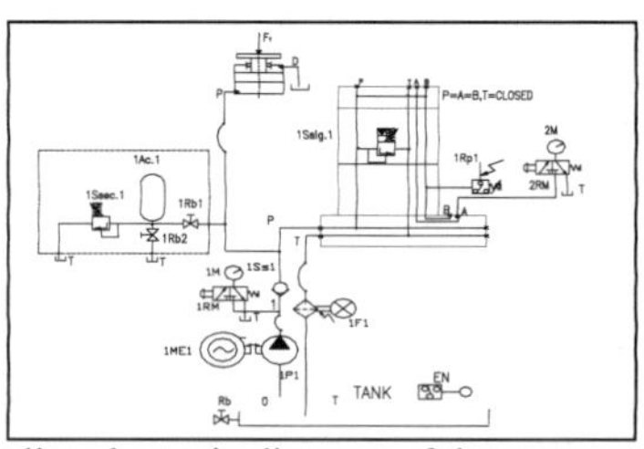

Fig.5. Hydraulic schematic diagram of the proposed system.

In the above mentioned relationship, the pressure is expressed in bar., the pump output in L/min. and the resulted power in KW.

The accumulator shall be charged with nitrogen under a pressure p_0 :

$$p_0 = 0.9 \cdot p_{min} \tag{4}$$

where p_{min} is the minimum operating pressure, determined according to the machine type.

If it is considered that the installation operates with the maximum weight part for which the corresponding pressure is p_{max} , the minimum volume recommended for the accumulator V_0 is (Prodan 2002):

$$V_0 = \frac{Q_P \cdot p_{max} \cdot t}{0.9 \cdot (p_{max} - p_{min})} \tag{5}$$

In the above mentioned relationship, t stands for the period of time when a proper relieving is provided for the load corresponding to pressure p_{min}. This time shall be longer than the time required for stopping the rotary table. If a single accumulator cannot cope with the output requirements, accumulator packs can be used.

3.2. Dynamic calculations

According to the hydraulic diagram shown in Fig.4, the pump output Q_P, assumed to be constant, is found in the flows passing through the pressure valve Q_1 , through the functional clearance Q_2 and is dissipated by compression Q_3, within the hydraulic engine:

$$Q_P = Q_1 + Q_2 + Q_3 \tag{6}$$

The expressions of these flow rates, as well as the calculation of other hydraulic parameters under dynamic conditions are found in (Prodan&Dobrescu 2000).

4. CONCLUSIONS

The proposed system allows the relieving of the bearing assembly, mainly of its axial component. It is recommended to be used on vertical lathes using bearing assemblies with axial-radial bearings. The load taking up is performed according to the machined part and to the recommendations of the bearing manufacturer. The hydraulic engine, with possibilities of translation and rotation, is actually a more complicated construction that also allows the connection of the rotational transducer for the C-axis at the CNC machines. The functional piston-to-cylinder clearance depends on the construction overall size and the technological possibilities of manufacturing. The actual hydraulic installation allows the continuous adjustment of the operating pressure according to the machined part. The pressure value is permanently monitored by the machine equipment. It is noticed that the time of maintaining the minimum pressure in case of emergency is directly proportional with the accumulator volume. This time should be longer than the time necessary for the complete stop of the rotary table. The system cannot be used in case of the machines with hydraulic lifting force.

5. REFERENCES

Botez, E. (1977) *Machine-Tools, Fundamentals Theory of Design*, Editura Tehnică,1977, Bucharest.

Prodan, D.& Dobrescu, T. (2000*), Hydraulic systems of hydraulic relieving*, Hydraulics, Hydraulics & Pneumatics Research Institute, Publishing House, December 2000/3-4, Bucharest.

Prodan, D. (2002) *Modernization of hydraulic installations on vertical lathes,* Hydraulics & Pneumatics Research Institute, Publishing House, June 2002/2, Bucharest

Prodan, D. (2004) *Machine-tool Hydraulics*, Printech Publishing House, Bucharest.

Prodan, D. & Marinescu, S. (2005) R*efabricarea maşinilor-unelte, sisteme hidraulice*, Editura Tehnică, 2005, Bucharest.

Annals of DAAAM for 2008 & Proceedings of the 19th International DAAAM Symposium, ISSN 1726-9679
Editor B. Katalinic, Published by DAAAM International, Vienna, Austria 2008
Make Harmony Between Technology and Nature, and Your Mind will Fly Free as a Bird

PLANNING ALGORITHM FOR DEVELOPMENT OF AN INTELLIGENT SCHEDULING SYSTEM

PROSTEAN, G[abriela]; TAROATA, A[nghel]; TAUCEAN, I[lie] & TAMASILA, M[atei]

Abstract: *Process planning is one of the most important activities in a company. Much research effort has been devoted in the development of knowledge based process planning systems, which have proved the large potential of computer applications in all the types of project activities. The aim of this paper is to present the planning algorithm of the activities of a project which forms the basis of the development of a planning intelligent system (PIS). PIS optimize the production operations planning.*
Key words: *planner, critical, algorithm, states.*

1. INTRODUCTION

The efficient and effective manufacturing operations planning require the assistance of methodologies and information systems meant for decision support. The PIS is designed first to decide for the best planner that determines what sequence of different operations should pass through the production facilities (Beerel, 1993).

To build a planning intelligent system today, a knowledge engineer performs different types of activities such as: knowledge-processing that refers to extracting knowledge, organizing knowledge, assembling knowledge and refining knowledge; planning algorithms, expertises and reports (Prostean, 2007). These are the primary tasks in the process of building an intelligent system in operation planning. Each primary task involves some specific knowledge engineering activities (Prostean & Prostean, 2002).

At the basis of building the model of the knowledge base there have been the pieces of knowledge furnished by the following 3 algorithms also called "computation procedures applied to knowledge", representing the assembly of the representation formalism associated to the inference strategy of the Planning Intelligent System (Cantor, 1998):

1) The Computation of the Critical Path,
2) The Project Synchronization,
3) The algorithm of the diagnosis analysis of the evolution of the project activities.

This paper presents the essential aspects referring to the implementation of the computation algorithm of the Critical Path.

The Critical Path represents the sequence of the activities of maximum time length between the initial event and the final event of the project.

This algorithm has been inspired by Floyd's algorithm (Crețu, 1992) for the computation of the minimum path in a graph, being reconceived for the computation of the maximum path in a graph.

The algorithm uses a "work" adjacency matrix. On account of the fact that the lengths of the activities are greater or equal to "0", there has been conventionally established that:

- In the initial state (before a project specific data are activated) all the elements of "Work" matrix are equal to "-1";
- If there is a path between knot i and knot j, then work[i][j] is equal to the average optimistic/probable/pessimistic value of the given activity.

2. THE SEARCHING ALGORITHM IN THE SPACE OF THE STATES FOR THE IDENTIFICATION OF THE CRITICAL PATH

The algorithm can be defined as follows: *there shall be executed maximum iterations on "work" matrix, and after k iteration, each location from "work" matrix will contain the maximum length of any path to knot **i** to knot **j**, a path that shall not pass through any knot with the index greater than **k**.*

The knots **i** and **j** can be any two knots of the graph which mark the beginning and the end of the path, and **k** marks any intermediate smaller knot. For the computation of work matrix, there shall be used the following relation (1), (figure 1).

$Work_k[i][j]=maxim(work_{k-1}[i][j],work_{k-1}[i][k]+work_{k-1}[k][j])$(1)

where k represents the moment for which there is computed **work** matrix. **Work** matrix is unique. The interpretation of relation (1) for the computation of the path from knot **i** to knot **j** (**$work_k$**[i][j]) shall be made as follows: There shall be compared **$work_k$**[i][j] – the value of the path from **i** to **j** without passing through knot **k**, with the sum **$work_{k-1}$**[i][k]+ **$work_{k-1}$**[k][j] – the value of the path from **i** to **k** plus the value of the path from **k** to **j**.

If the path is longer, there shall be attributed to **$work_k$**[i][j] the value **$work_{k-1}$**[i][k] + **$work_{k-1}$**[k][j], on the contrary, the value **$work_k$**[i][j]=**$work_{k-1}$**[i][j] remains unchanged.

In figure 2 there is presented the activities network for a project.

The system follows 6 steps within the strategy of identifying the Critical Path with the help of work matrix, as follows:

1. The system verifies the maximum value of the activities path. work[i][k] + work[k][,j] > work[i][j] . For: i = 0, k = 0, j = 0 ⇒ work[0][0] + work[0][0]>work[0][0]. For this situation, the system has not identified any links between the events.

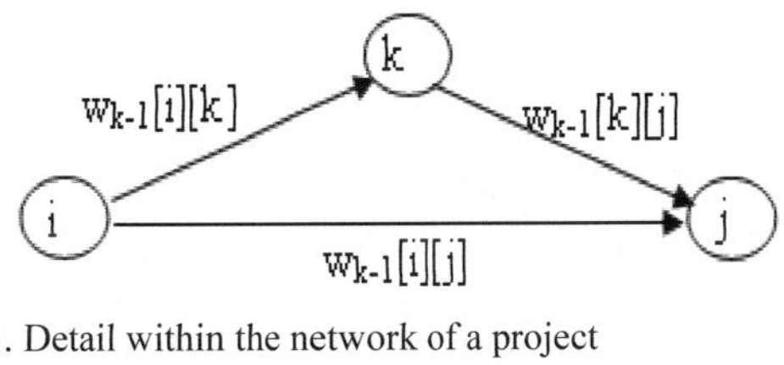

Fig. 1. Detail within the network of a project

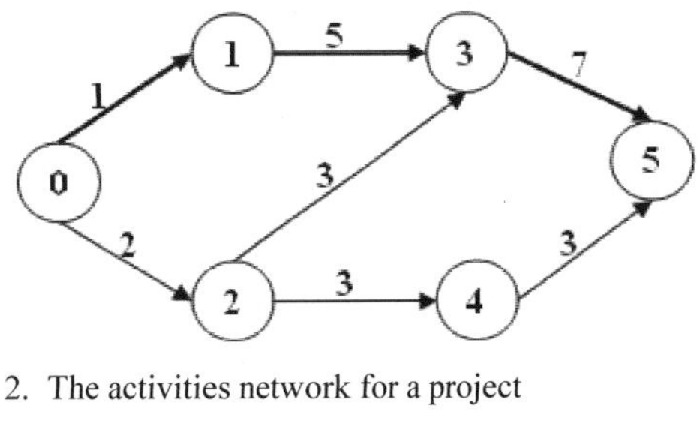

Fig. 2. The activities network for a project

$$\xrightarrow{k,\,j} \qquad i, k \downarrow \qquad work = \begin{pmatrix} -1 & 1 & 2 & -1 & -1 & -1 \\ -1 & -1 & -1 & 5 & -1 & -1 \\ -1 & -1 & -1 & 3 & 3 & -1 \\ -1 & -1 & -1 & -1 & -1 & 7 \\ -1 & -1 & -1 & -1 & -1 & 3 \end{pmatrix}$$

2. For the values i=0, k=0, j=1,…5, there shall be incremented "j ", and the same situation can be observed up to j = 5, following which there shall be incremented "k".
3. For the values i=0, k=1, j=0,…3, there shall be incremented "j" , and the same situation will result up to j = 3, for which there will be identified paths between knots (a sequence of activities): 0 → 1 → 3. By applying condition (1), there results the following:

work[i][k] + work[k][j] > work[i][j] ⇒
work[0][1] + work[1][3] > work[0][3] ⇒1 + 5 > - 1 ⇒ 6 > -1

$$work = \begin{array}{c|cccccc} & 0 & 1 & 2 & 3 & 4 & 5 \\ \hline 0 & -1 & 1 & 2 & -1 & -1 & -1 \\ 1 & -1 & -1 & -1 & 5 & -1 & -1 \\ 2 & -1 & -1 & -1 & 3 & 3 & -1 \\ 3 & -1 & -1 & -1 & -1 & -1 & 7 \\ 4 & -1 & -1 & -1 & -1 & -1 & 3 \end{array}$$

Element **work**[0][3] = -1 of **work**[i][j] matrix will be replaced by **work**[0][3] = 6, and there results the following $\mathbf{work_1}$[i,j] matrix.

$$work_1 = \begin{array}{c|cccccc} & 0 & 1 & 2 & 3 & 4 & 5 \\ \hline 0 & -1 & 1 & 2 & 6 & -1 & -1 \\ 1 & -1 & -1 & -1 & 5 & -1 & -1 \\ 2 & -1 & -1 & -1 & 3 & 3 & -1 \\ 3 & -1 & -1 & -1 & -1 & -1 & 7 \\ 4 & -1 & -1 & -1 & -1 & -1 & 3 \end{array}$$

4. For the values i=0, k=1,j=1,...5, there shall be incremented "j " up to j = 5, these being situations for which there have not been found paths between knots, following which there shall be incremented k=2 and j= 0,…3, a situation for which there will be identified the paths (a sequence of activities): 0 → 2 → 3. Condition (1) shall be verified in order to establish the possible passage to the following state, and thus there results that:

work[i][k] + work[k][j] > work[i][j] ⇒
$work_1$[0][2] + $work_1$[2][3] > w_1[0][3] ⇒ 2 + 3 > 6

Element $work_1$[0][3] =6 of $work_1$[i][j] matrix is greater than the sum $work_1$[0][2] + $work_1$[2][3] = 5, consequently, there shall be selected $work_1$[0][3] =6, and thus there can be got on to the following state.

5. For the values i=0, k=2, j=4, there shall be incremented j = 4, for which the following paths will be identified (a sequence of activities): 0 → 2 → 4. By applying condition (1), there results the following situation:

$work_1$[0][2] + $work_1$[2][3] = 5, $work_1$[i,j] matrix remaining unchanged.
work[i][k] + work[k][j] > work[i][j] ⇒
$work_1$[0][2] + $work_1$[2][4] > $work_1$[0][4] ⇒2 + 3 > -1 ⇒ 5 > -1

$$work_2 = \begin{array}{c|cccccc} & 0 & 1 & 2 & 3 & 4 & 5 \\ \hline 0 & -1 & 1 & 2 & 6 & 5 & -1 \\ 1 & -1 & -1 & -1 & 5 & -1 & -1 \\ 2 & -1 & -1 & -1 & 3 & 3 & -1 \\ 3 & -1 & -1 & -1 & -1 & -1 & 7 \\ 4 & -1 & -1 & -1 & -1 & -1 & 3 \end{array}$$

6. For i=0, k=2, j=5, and for i=0, k=3, j=0,...1, respectively, the situation of the evolution within the space of the states is as follows: there shall be incremented j = 5, a situation for which there have not been identified paths between knots, after which there shall be incremented k=3 and j= 0,…5, the last situation being the one for which the following sequence of activities will be identified: 0 → 1 → 3 → 5 . Condition (1) shall be verified in order to establish the possible passage to the next state.

work[i][k] + work[k][j] > w[i,j] ⇒
$work_2$[0,3] + $work_2$[3,5] > $work_2$[0,5] ⇒
6 + 7 > -1 ⇒ 13 > -1

The maximum value attributed to the critical path is 13, it being retrieved in $work_3$[0][5] matrix.

$$work_3 = \begin{array}{c|cccccc} & 0 & 1 & 2 & 3 & 4 & 5 \\ \hline 0 & -1 & 1 & 2 & 6 & 5 & \mathbf{13} \\ 1 & -1 & -1 & -1 & 5 & -1 & -1 \\ 2 & -1 & -1 & -1 & 3 & 3 & -1 \\ 3 & -1 & -1 & -1 & -1 & -1 & 7 \\ 4 & -1 & -1 & -1 & -1 & -1 & 3 \end{array}$$

3. CONCLUSIONS AND FUTURE WORK

The highlighting, by means of the states space representation of the computation process of the Critical Path by using the procedure of "The Critical Path Computation", enables to easily establish the acquisition forms of the knowledge for the PIS. The use of these operation scheduling techniques can be accepted to be successfully applied in manufacturing, in building industry, and as well in software development projects.

The main contribution of the planning intelligent system is the capacity to drastically shorten the project duration in real time, having the updated reports for the project and each specific task. The algorithm has been developed for a suitable embedding into a PIS, that is an object oriented planning intelligent system, where the computation is distributed in time.

4. REFERENCES

Beerel A. (1993). *Expert System in Business, real world applications*, Ellis Horwood Limited, ISBN:0-13-296633-6.

Cantor M., R. (1998). *Object-Oriented Project Management with UML*, Published by John Wiley & Sons, Inc, USA, ISBN: 0-471-25303-0.

Crețu V. I.(1992), *Data structures and advanced programming techniques,* Timişoara, Centrul de multiplicare al UTT.

Prostean G., Prostean O. (2002). Intelligent project manager using a plannifying, supervising and diagnosing integrated process, *International Conference of Artificial Intelligence*, Las Vegas, Nevada, USA, June 24-27, CSREA Press

Prostean G. (2007*).* Considerations above Operations Scheduling Intelligent System, *Proceedings of the 5th IEEE International Conference on Computational Cybernetics*, Gammarth, Tunisia, Oct. 19-21, 2007, pag.173-178

Annals of DAAAM for 2008 & Proceedings of the 19th International DAAAM Symposium, ISSN 1726-9679
Editor B. Katalinic, Published by DAAAM International, Vienna, Austria 2008
Make Harmony Between Technology and Nature, and Your Mind will Fly Free as a Bird

IMAGE TREATMENT PROCESS TO OBTAIN THE KINEMATIC PARAMETERS DURING HUMAN LOCOMOTION

RADU, C[iprian] & ROSCA, I[leana]

Abstract: *To determine the kinematic parameters of human locomotion, first of all is necessary to choose the right measuring method and to design an experimental installation. In this way, in the first scientific paper, called: On the Design of an Experimental Installation to Determine the Kinematic Parameters During Human Locomotion, published in the same proceeding, we have chosen the adequate measuring method and we have measured the kinematic parameters of human locomotion. Thus, the aim of this scientific paper is to process all the measured data by using editor software called Adobe After Effects. Using this software we have determined the trajectories of the light markers and also the angles between anatomical segments. The results are displayed graphically and numerically.*
Key words: *angle, kinematic, markers, software, trajectories.*

1. INTRODUCTION

The systematic construction of biomechanical models of bipedal locomotion and corresponding stability criteria in the sense of gait synthesis took place in the 1970s in the works of Stepanjenko and Vukobratonic. The biomechanical model consists of few segments, the lower extremities being represented quite faithfully (the upper leg, the lower leg and the foot), while the total upper body torso is be represented by just one segment (Medved, 2001).

Our solution to construct a biomechanical model of lower extremities consists in representing the upper leg and lower leg as a straight beam and the foot as a triangle (the ankle joint, the heel and second metatarsal). Our model is simple and is very useful to determine the most important kinematic parameters during normal locomotion, such as: the angles between anatomical segments, respectively the angle between foot and ground (Rosca & Radu, 2008).

2. METHODOLOGY AND RESULTS

In order to determine the position of a particular observed point experimentally, a stereometric procedure has to be performed. With the help of 3D positions of a one point on the rigid body, the position of this body is determinated. In applying this principle, markers are attached to the subject's body and the positions of the characteristic body landmarks are extracted later from recordings obtained by using a high-speed video camera (Panagiotakis & Tziritas, 2004). Video recording of the light markers attached to the lower right limb (hip, knee and ankle joints heel and the second metatarsal's pick), of a human subject during locomotion, is possible by using a high speed video camera, in special conditions (1/250 frames/second and an adjustable depth recording field of 3500 mm). The recorded images are transmitted to the acquisition system and saved as *.avi* format, by CamLink software, which is the interface software for high-speed video camera.

For our experiment, the movie length is 8.24 seconds that means 247-recorded frames (for a displayed rate of 30 frames/second in normal conditions). The experimental solution to record the human locomotion was presented in the first scientific paper, called: *On the Design of an Experimental Installation to Determine the Kinematic Parameters During Human Locomotion,* published in the same proceeding.

The recorded images of human locomotion event in *.avi* format are imported in Adobe After Effects 6.5 software for image treatment technique, to determine some kinematic parameters.

The most important module of editing software is tracking module. This module has been used by as to track the light markers from the right foot of human subject. In this way, for each light marker we have generated a virtual tracking marker (track1, track2 … track5) (see figure 1). This virtual tracking marker is positioned in light marker's centre. For an accurate measure, both centers of light marker and virtual tracking marker have to coincide (James, 2001).

Once they are generated and virtual tracking markers have been edited, the next step is to establishing the tracking mode. For our case we have used the color-tracking mode. In this way, we have established that the white color of the light markers to be tracked by virtual tracking markers, during all length movie.

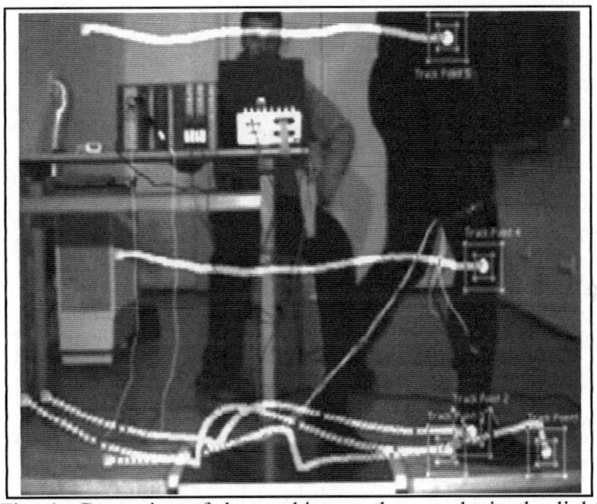

Fig. 1. Generating of the tracking marker to obtain the light marker's trajectories.

As it can be seen in figure 1, the tracking markers will follow the light markers, finally obtaining the trajectories of the most important landmarks of lower right limb. Also, as it can be seen in the same figure, there are some major differences regarding the tracking points appearing in the trajectories. The different velocity of the light markers attached to the anatomical parts can explain that.

The velocities of the light markers attached on the foot segment (on the heel, ankle joint and second metatarsal) are higher then velocities of the light markers attached to the knee and hip joint during swing phase of lower limb during locomotion (Radu et al., 2008).Because there are five distinct phases of human normal locomotion, we have determinated for all five of them the angles between anatomical segments, respectively between foot and ground (see figure 3), as follow:

- Phase one (initial contact): α = 180°, β = 143°, φ = 16°, δ = 94°;
- Phase two (loading): α = 166°, β = 138°, φ = 0°, δ = 95°;
- Phase three (midstance): α = 169°, β = 144°, φ = 0°, δ = 91°;
- Phase four (heel off): α = 166°, β = 132°, φ = 10°, δ = 87°;
- Phase five (toe off): α = 161°, β = 120°, φ = 24°, δ = 87°;

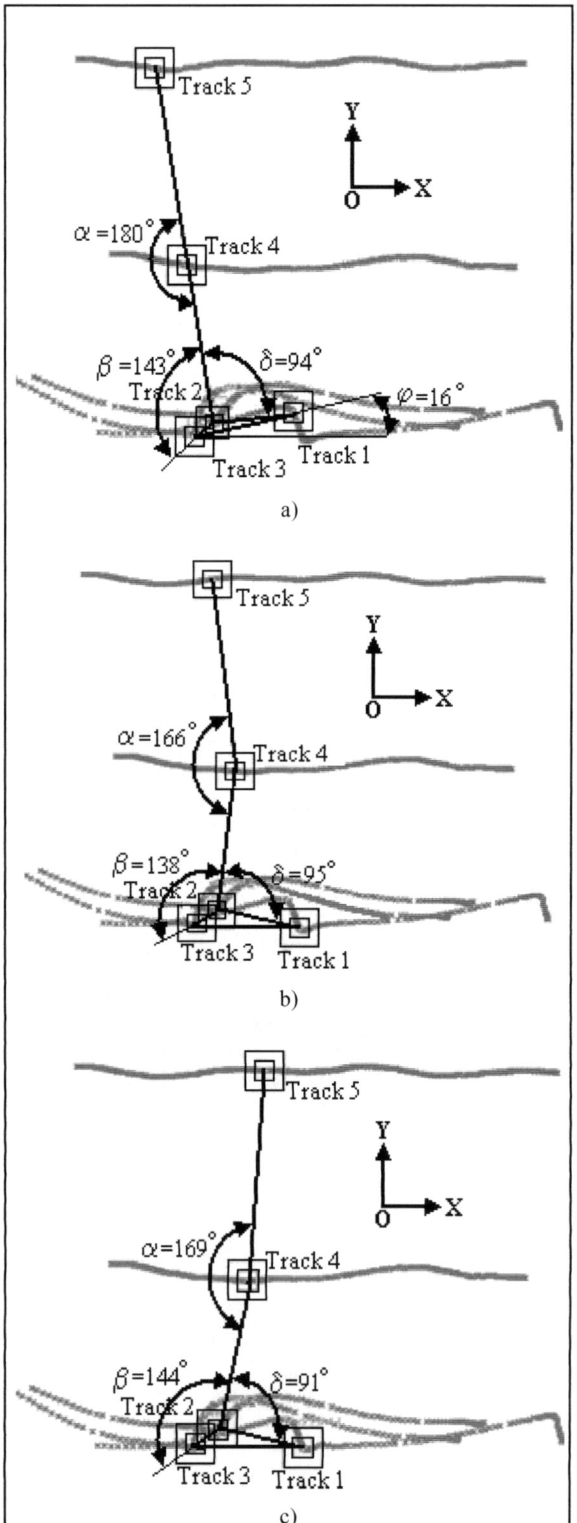

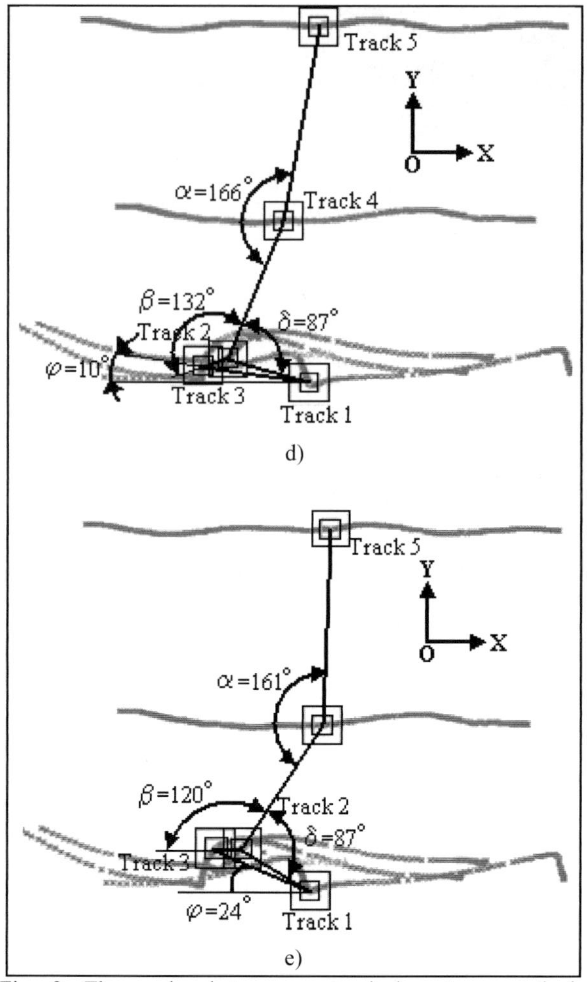

Fig. 2. The angles between anatomical parts respectively between foot and ground, during normal locomotion for: a) initial contact phase; b) loading phase; c) midstance phase; d) heel off phase; e) toe off phase.

3. CONCLUSIONS

Our solution to construct the biomechanical model of lower extremity, respectively to obtain some kinematic parameters during human locomotion, is a very simple and accurate method, that because the recorded images are affected by deformation errors in proportion of 0.857%, which is suitable for the error limits admissible for this measuring method.

4. REFERENCES

James, R.G. (2001). *Gait Analysis in Cerebral Palsy*. Mac Keith Press, ISBN 0 90126 90 8, London, U.K.

Medved, V. (2001). *Measurement of Human Locomotion*, Congress Library, ISBN 0-8493-7675-0, USA.

Panagiotakis, C. & Tziritas G. (2004). *Recognition and Tracking of the Members of a Moving Human Body*, Springer Publisher, ISSN 0302-9743, Berlin, Germany.

Radu, M.I, Atanasiu, V., Budescu, E & Ibanescu, R. (2008). *Aspects Concerning the Biomechanics of Ankle Joint*. Published by Gheorghe Asachi Technical University from Iasi, ISSN 1011-2855, Romania.

Rosca, I.C & Radu, C. (2008). *On a Mechatronic System to Determine Dynamic Parameters of the Human Limb During Locomotion Used For Artificial Muscles and Implant Design*, 6th International DAAAM Baltic Conference, ISBN 978-9985-59-783-5, Tallin, Estonia.

Annals of DAAAM for 2008 & Proceedings of the 19th International DAAAM Symposium, ISSN 1726-9679
Editor B. Katalinic, Published by DAAAM International, Vienna, Austria 2008
Make Harmony Between Technology and Nature, and Your Mind will Fly Free as a Bird

TRIBOLOGICAL PERFORMANCE OF BIPHASIC LUBRICATION

RADULESCU, A[lexandru] & STANESCU, T[itu]

Abstract: *The present paper presents a study for the evaluation of the tribological properties of the biphasic lubrication system, by comparison with the classical pump-driven liquid lubrication solution. An experimental stand was design and realized, in order to test a standard gearbox, alternatively lubricated with both systems. The influence of the two lubrication systems on the performance of the gears and ball bearings was investigated, with variations of load and speed. Statistical techniques were used to design an efficient experimental plan. The dependent variables chosen to be monitored in this experiment were the operating temperature and the level of noise. The extended tests show that the biphasic lubrication system offers better performance than the classical lubrication system.*

Key words: *Lubrication, Biphasic, Experiment, Temperature, Noise.*

1. INTRODUCTION

The present paper presents a study for the evaluation of the tribological properties of the biphasic lubrication system, by comparison with the classical pump-driven liquid lubrication solution.

The liquid-oil circulating systems, while providing advantages, were nevertheless found to impose additional handicaps upon machine design. They can be expensive, machine assembly is more difficult and the pump itself becomes an extra maintenance item. Moreover, the development of extremely high-speed grinders and spindles demonstrated that the liquid-oil environment can cause certain bearings to run hot (Pinkus et al., 1977), (***, 1985).

Concerning the biphasic lubrication system, it was observed that the distribution of lubricant by conveying oil particles in compressed air overcomes many of the drawbacks of a circulating liquid system. Using the energy of the compressed air stream allows the generation of a fine, dry, smoke-like fog of oil particles, which can be conveyed over distances up to 150 m and it is possible to be applied directly to the machine element to be lubricated (Anand & McFarland, 1989), (Shamim & Kettleborough, 1997), (Wame & Halder, 1984).

The most important characteristic of biphasic lubrication which influences the operating temperature, friction and vibration of rolling element bearings is that biphasic lubrication does not require an oil sump in the bearing housing. With biphasic lubrication, the rotating bearing does not churn oil; this results in lower friction, due to the lower energy loss, and lower operating temperature (Graham & Klaus, 1986). Also, as the biphasic lubricant at room temperature flows through the narrow passages and over the hot operating surfaces of the bearings, some heat is removed by forced convection from the bearing housing (Erdemir et al., 1991), (Lauer & Dwyer, 1990).

In the following study, the influence of two lubrication systems on the performance of the test bearings was investigated, with variations of load and speed. The dependent variables chosen to be monitored in this experiment were the operating temperature and the level of noise. The extended tests show that the biphasic lubrication system offers better performance than the classical lubrication system.

2. EXPERIMENTAL STAND

The experimental stand used for the determination of the tribological performance was a modified FZG stand, with two different lubrication systems: a classical pump-driven liquid lubrication system and a biphasic lubrication system (Figure 1). The main technical parameters of the experimental stand are presented in Table 1.

Statistical techniques were used to design an efficient experimental plan. The experiment was statistically balanced, randomized and blocked. For each experimental run, after applying load and speed, the test gear boxes were allowed to achieve steady state temperature. At steady state condition, the temperature and the level of noise were measured. The lubricating oil level in the oil sump was maintained at one half the height of the lowest ball in the test bearing at stand still condition as recommended by the bearing manufacturer (***, 1985).

The load range was chosen to approximate industrial application conditions. The maximum and minimum test speeds at the pitch diameter were chosen as 5.65 m/s (1000 rpm) and 2.83 m/s (500 rpm) respectively. The maximum speed and load combination was limited by the requirement that a minimum oil viscosity has to be maintained at the operating temperature as recommended by the gear box manufacturer.

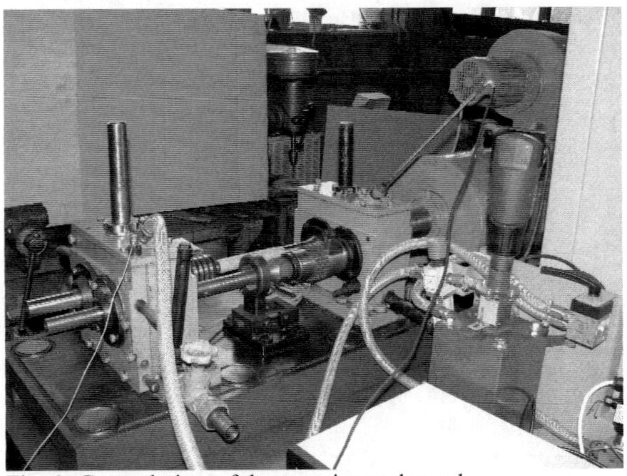

Fig. 1. General view of the experimental stand

Technical parameter	Reference value	
Mechanical power	16 kW	
Load	10; 30; 50 daNm	
Speed	500; 750; 1000 rpm	
Temperature	20 ^{0}C	
Time	60 min	
Type of the oil	pump-driven liquid lubrication	TIN 125 EP
	biphasic lubrication	HP38EP
Volume of the oil	pump-driven liquid lubrication	1,25 l
	biphasic lubrication	0,75 l

Tab 1. Main parameters of the experimental stand

3. RESULTS

Analysis of Variance (ANOVA) tests were run on the data using Statistical Analysis Software (SAS). It was found that in these tests the bearings and the shafts had negligible influence on both operating temperature and level of noise. Also as expected, the type of lubrication system, load and speed had strong influences on the operating temperature and noise.

Next, General Linear Model (GLM) procedure, which estimates and tests hypotheses about linear models, was used to model operating temperature and level of noise as functions of the independent variables which have strong influence on the operating temperature as identified in the ANOVA test.

For the operating temperature, the log-log model gave the highest regression coefficient (0.93). This model is plotted in Figure 2 and Figure 3 to show the variation of the operating temperature in bearing and liquid bath with variation of load and speed.

For level of noise, the linear model gave the best R-square value (0.84). This model is plotted in Figure 4 to show the variation of level of noise with variation of load and speed.

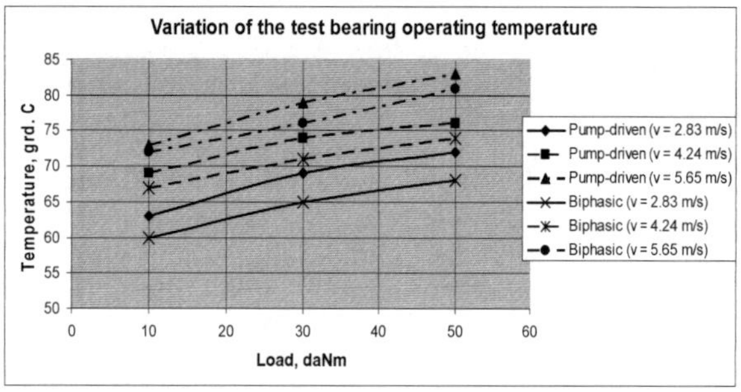

Fig.2. Graphical representation of the statistical model describing the variation of the test bearing operating temperature with load and speed

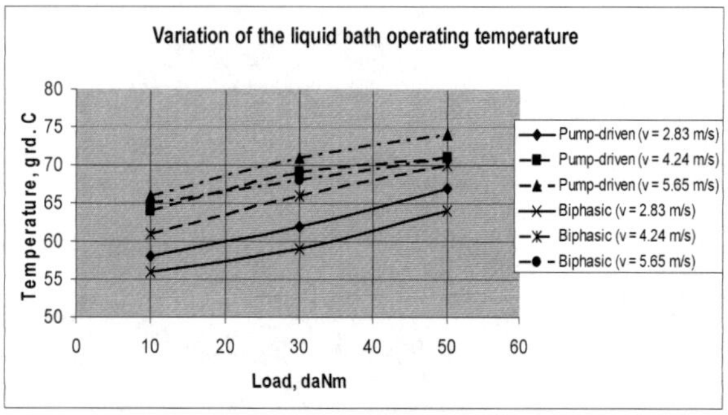

Fig.3. Graphical representation of the statistical model describing the variation of the liquid bath operating temperature with load and speed

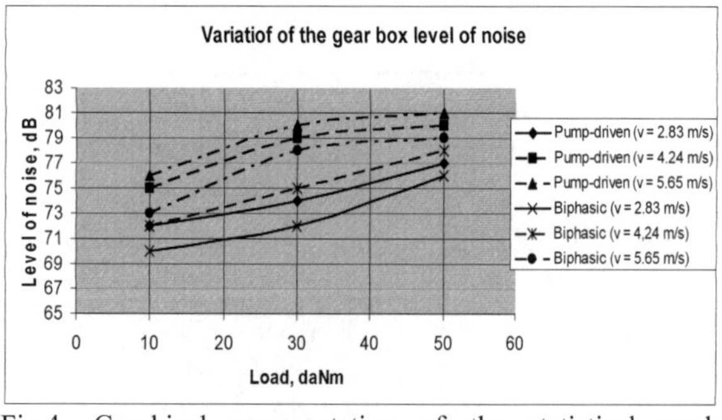

Fig.4. Graphical representation of the statistical model describing the variation of the gear box level of noise with load and speed

4. DISCUSSIONS

It can be seen from the plots for the models of temperature and noise that in general the biphasic lubricated bearings operate at lower temperature (5 to 8 ^{0}C) and the level of noise for the corresponding gearboxes decrease with 10-15 % by comparison with pump-driven liquid lubrication system.

It has also been found that the difference in operating temperature and the level of noise between biphasic and pump-driven lubrication systems remain approximately constant with variation of load and speed. Moreover, they do not change appreciably with time.

The reduction of the temperature for the bearings can lead to significant savings in energy when elements of a total petrochemicalcomplex are considered. Also, through better wear and fatigue protection, biphasic lubrication can reduce maintenance requirements and downtime of expensive equipment, leading to saving in resources.

5. CONCLUSIONS

Analysing the previous results, it can be observed that the biphasic lubrication has significant advantages over classical pump-driven liquid lubrication solution:

- the reduction of the bearing temperature, liquid bath operating temperature and level of noise for the entire gear box;
- considering the difference between the temperatures for pump-driven and biphasic lubrication systems, it must be precised that the load and the speed of the bearing have no influence;
- the low rate of lubricant application, which is made possible with biphasic lubrication, provides continuous lubrication without the requirement of a circulating system design;
- the continuous application of the oil reduces the manufacturing and installation costs and the oil consumption;
- the biphasic lubrication is ideally in chemically corrosive environment or in environment where hard particles may invade the bearing cage and cause wear to the load bearing surfaces.

6. REFERENCES

Anand, N. K. & McFarland, A.R. (1989). Particle deposition in aerosol sampling lines caused by turbulent diffusion and gravitational settling, *American Industrial Hygiene Association Journal*, Vol. 50, No. 6, pp. 133-140

Erdemir, A.et al. (1991). The synergistic effects of solid and liquid lubrication on the tribological behavior of transformation-toughened ZrO_2 ceramics, *STLE-ASME Conference on Tribology*, Report No. DE91-018617, St. Louis

Graham, E. E. & Klaus, E. E. (1986). Lubrication from the vapor phase at high temperatures, *ASLE Transactions*, Vol. 22, No. 2, pp. 229-234

Lauer, J. L. & Dwyer, S. R. (1990). Tribochemical lubrication of ceramics by carbonaceous vapors, *Tribology Transactions*, Vol. 33, No. 1, pp. 25-32

Pinkus, O., Decker, O. & Wilcock, D. F. (1977). How to save 5% of our energy, *Mechanical Engineering*, Vol. 32, pp. 32-39

Shamin, A. & Kettleborough, C. F. (1996). Aerosol aspects of oil mist lubrication – reclassification and deposition of oil mist in bearings, *Energy Week Conference and Exhibition*, Vol. IV, NewYork, pp. 380-385

Wame, T. M. & Halder, C. A., (1984). Toxicity of lubricating oils, *Journal of the American Society of Lubrication Engineers*, Vol. 42, No. 2, pp. 97-103

*** (1985). Bearing installation and maintenance guide, *SKF Industries Inc.*, Catalog No. 140-710, Pennsylvania

Annals of DAAAM for 2008 & Proceedings of the 19th International DAAAM Symposium, ISSN 1726-9679
Editor B. Katalinic, Published by DAAAM International, Vienna, Austria 2008
Make Harmony Between Technology and Nature, and Your Mind will Fly Free as a Bird

EXPERIMENTAL RESEARCHES CONCERNING THE LUBRICANTS DEGRADATION PROCESS

RADULESCU, A[lexandru]; RADULESCU, I[rina] & BALAN, C[orneliu]

Abstract: *The present paper describes experimental researches concerning the lubricants degradation process, for their diagnosis and performance monitoring. The paper is focused on the idea that the lubricant have to be changed when it is really used and not according to the producers recommendations. For the appreciation of the degradation stage and of the changing moment, there are made periodically physic-chemical oil analyses. The paper proposes a rheometrical method for the evaluation of the lubricants degradation degree, based on periodically rheological measurements. The experimental tests show that the viscosity downs by the number of the working hours, for hydraulic lubricants and motor oils. An exponential relation is proposed for the modelling of the law variation of the viscosity versus working time.*

Key words: *Degradation, Monitoring, Diagnosis, Lubricant, Rheology.*

1. INTRODUCTION

The present paper describes experimental researches concerning the lubricants degradation process, for their diagnosis and performance monitoring. The beginning of the studies was given by the actual oil change method, which is before the completely use of lubricant; this is out-of-date, it has big economical waste and applying ecologically effects (Balan et al., 2000), (Lukas & Anderson, 1992). In Western Europe, USA or Japan, the oil is changed based on the instructions written on the oilcans, but at the minimal recommended covered distance.

The determination of the oil changing moment is different because it depends on difficult to control factors. According some cases the oil for combustion engine is changed more frequently that is necessary, in order to be sure that its properties were not all vanished. This type of procedure is not dangerous, from theoretical point of view, but it is expensive. The degradation of "working" lubricant oil depends on many factors, but most important ones are the oxidation and the contamination (Nadkarni, 1991), (Marshall, 1993). The contamination level and the "lifetime" of used oil depend on the exploitation conditions, the oil quality, the construction and the technical state of engine (Sieber &Salmon, 1994).

The paper proposes a rheometrical method for the evaluation of the lubricants degradation degree, based on periodically rheological measurements. The experimental tests show that the viscosity downs by the number of the working hours, for hydraulic lubricants and motor oils. An exponential relation is proposed for the modeling of the law variation of the viscosity versus working time.

2. EXPERIMENTAL PROCEDURES

The rheological model proposed for the industrial lubricant is the Newtonian behavior, expressed as:

$$\tau = \eta \frac{du}{dy} \tag{1}$$

The experimental stand used for measuring the rheological parameters of the lubricants is a cone and plate viscometer, which offers absolute viscosity determination with precise shear rate and shear stress information.

In order to estimate the wear degree of the used oils, a theoretical relation is proposed, which established the variation of the viscosity versus the equivalent distance covered by the motor vehicle (Czarny, 1989):

$$\eta = \eta_0 e^{-Kd} \tag{2}$$

The two parameters characteristics for Eq. 2 are the initial viscosity η_0 for the fresh oil and the wear intensity coefficient K. These values are determined using the regression analysis method.

Experimental investigations were carried out at the ambient temperature of 20 ^{0}C, using a Physica MCR 301 rheometer (***, 2007). They offer absolute viscosity determinations with precise shear rate and shear stress information readily available. The sample volumes required are extremely small and temperature control is easily accomplished.

In order to obtain practical informations concerning the durability of the lubricants, three types of oils have been tested, comming from motor vehicles with different wear degrees (***,1993):

– M30 from a Diesel motor vehicle with 87000 km way;
– 10W40 from an essence motor vehicle with 17000 km way;
– 5W30 from an essence motor vehicle with 80000 km way.

3. RESULTS

As results, there are obtained characteristic rheograms, using the Physica MCR 301 rheometer, for all the tested oils, in fresh and used state. Figure 1 shows a tipical diagram, where four curves are presented, corresponding for different wear degrees for the tested oil.

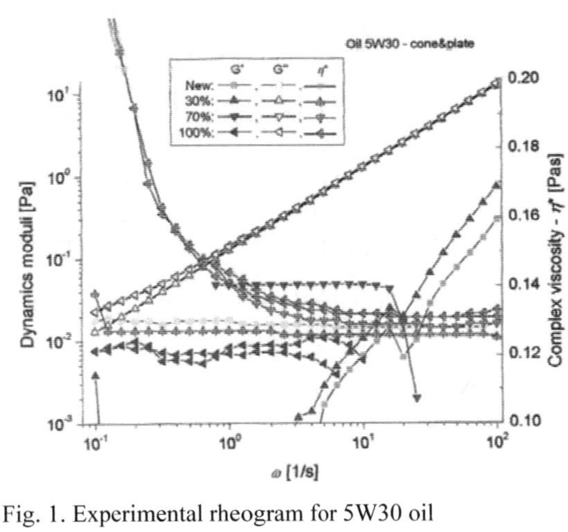

Fig. 1. Experimental rheogram for 5W30 oil

Measurements results show that the viscosity downs by the equivalent distance of the motor vehicle, and it cleary depends of the oil type.

Experimental results confirmed previous observations: oil degradation generally leads to the decrease of the viscosity and the elasticity (if the oil presents this property). This phenomenon is due to the molecular forces modifications induced by the fluid wear; the solid particles presence in used oils cannot determine the viscosity increasing, typical for the fresh fluids.

The rheological measurements results put in evidence remarkable structural modifications in tested oils, which prove that the fluid structure has been affected in quantitative and qualitative ways, by the mechanical wear.

4. DISCUSSIONS

The experimental results give information about the normal rheological behavior of the used fluids: the decreasing of the viscosity is proportional with the wear degree of oil. Also it is important to emphasize that the con and plate tests give better results for the evaluation of ellasticity of tested oils.

Even if the measured values of the dynamic modulus G'are small (generally $G' < 10^{-2}$ Pa), measurements prove that the oils have as component molecular additives. This phenomenon is observed especially in small oscillations frequency domain, where the dynamic modulus has small variations with the frequency.

The experimental data have been numerical analized with the regression analysis method, in order to obtain the mean values of the viscosity, for fresh and used oils (see Table 1). The same table presents the values of the correlation coefficient, which is an indicator of the correlation level between the theoretical Newtonian model and the experimental data.

In order to obtain the main values of the initial viscosity η_0 for the fresh oil and the wear intensity coefficient K (see Eq. 2), the data from Table 1 are numerically treated and the results are presented in Table 2.

The wear characteristics of the tested oils, which include the experimental data and the theoretical fitted curves, are plotted in Figure 2.

Type of oil	Wear degree	Equivalent distance, km	Viscosity, Pa.s	Correlation coefficient
5W30	Fresh oil	0	0.128	0.963
	Used oil	2920	0.119	0.982
		7090	0.103	0.991
		9930	0.094	0.994
10W40	Fresh oil	0	0.253	0.957
	Used oil	3160	0.228	0.978
		6900	0.212	0.989
		10240	0.178	0.995
M30	Fresh oil	0	0.548	0.968
	Used oil	2850	0.511	0.975
		7100	0.492	0.988
		9870	0.463	0.990

Tab 1. Regression parameters for tested oils.

Type of oil	Initial viscosity, Pa.s	Wear intensity coefficient, km^{-1}	Correlation coefficient
5W30	0.278	0.0199	0.991
10W40	0.549	0.0394	0.854
M30	0.791	0.0348	0.859

Tab 2. Main values of the initial viscosity and wear intensity coefficient

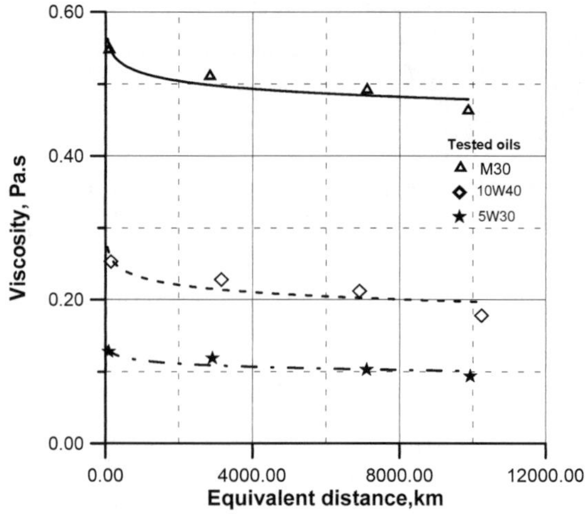

Fig. 2. Wear curves for the tested lubricants

5. CONCLUSIONS

Comparative tests for the fresh and used lubricants have been realized, which point out the dependence between the degradation stage and the fluid viscosity.

The principal results of these researches are to obtain new criteria for the "life reserve" diagnosis of lubricants oils, to assure a modern laboratory and to create new premise to develop new directions of approach for the lubricants durability problems.

Experimental researches show significant differences concerning the viscosity of lubricants for different state of wear; they are due to the changing of physical – chemical properties. It is an obvious conclusion and a valid assumption that lubricants' aging is directly connected to lubricants viscosity decrease.

Practically application of tested method offers important data concerning wear process evolution and about lubricants influence on durability, at high precision. The method may be applied in many practical situations, having economical conditions.

6. REFERENCES

Balan, C., et al. (2000). The Rheology of Lubricating Grease, *European Lubricating Grease Institute Publications*, Amsterdam

Czarny, R. (1989). A study of thixotropy phenomena in lubricating greases, *EUROTRIB' 89*, Vol. IV, Helsinki, pp. 198-203

Lukas, A. & Anderson, D. P. (1992). Machine and lubricant condition monitoring for extended equipment lifetimes and predictive maintenance, *Int. Symp. Mining*, September 15 to 17, Pretoria, South Africa, pp. 145-149

Marshall, E. R. (1993). Used Oil Analysis-A Vital Part of Maintenance, *Lubrication*, Vol. 79, No. 2, pp. 120-126

Nadkarni, R. A. (1991). A review of modern instrumental methods of elemental analysis of petroleum related material, *Modern Instrumental Methods of Elemental Analysis of Petroleum Products and Lubricants, ASTM STP 1109*, pp. 51-57

Sieber, J. R. & Salmon, S. G. (1994). Elemental analysis of lubricating oils and greases, *Lubrication*, Vol. 80. No. 1, pp.66-72

*** (1993). Catalog de Produse Petroliere, PECO, (in Romanian)

*** (2007). Physica MCR 301 rheometer catalog

Annals of DAAAM for 2008 & Proceedings of the 19th International DAAAM Symposium, ISSN 1726-9679
Editor B. Katalinic, Published by DAAAM International, Vienna, Austria 2008
Make Harmony Between Technology and Nature, and Your Mind will Fly Free as a Bird

CODEMO - A COLLABORATIVE DESIGN MODELER FOR COMPLEX MULTIDISCIPLINARY OPTIMIZATION PRODUCTS

RADULESCU, B[runo]; RADULESCU, M[ara] & DUMITRAS, C[atalin]

Abstract: *Integrated product and process development is accomplished by multidisciplinary teams. To support the team approach we have developed an inter-disciplinary communication medium. This paper describes the application of a Co-operative Design System to optimize the design of a product from a multidisciplinary perspective. Also, demonstrates an approach to solve some problems of interoperability, heterogeneity of platforms, and data sharing. One of the goals of CoDeMo is to bring engineers from the various disciplines, servicemen, and even customers early in the product development process to access design of the product concurrently. On the technical level, achieving these goals depends on integrating the modeling and simulation tools used by the different engineering disciplines in the areas of design, analysis, manufacturing, etc.*

Key words: co-operative system, multidisciplinary team, integrated design,

1. INTRODUCTION

This paper describes the application of the Cooperative Design System to optimize the design of the product from a multidisciplinary perspective.

The new methodologies, the new technologies and the new advanced tools must be employed to design products of high quality at the low price.

In this economic context at the beginning of 21st century, the evolution of the software tools, allowed to propose a space representation of the objects, which facilitates the study and the simulation of their behaviours, (Gardarin, 1999).

2. CODEMO FOR A COMPLEX PRODUCT

The concept of product model, (Roucoules, 1999), results from the will to substitute the real product by a model of information (virtual product) which not limiting itself to purely geometrical information.

This concept finds it place in the integrated design because it aims is to handling information concerning all the cycle of life of the product. (Haug et al., 1995)

The product model is a virtual model of product, which allows to different trades to work on the same product at the same time. The product model responds to the needs of functionalities of the management of the database, to the commands and to the notifications of the various actors.

As support of the design activity, the Co-operative Design Modeler (CoDeMo) uses a client-server structure based on Ilog product libraries with C++ object classes. The product model is composed of a knowledge model (entity and production rules) and of data model (components, links and relations). A main interest in using entity is to be able to declare the instance of these entities without to be immediately obliged to define the values of their characteristics.

The design parameters are considered as attributes of entities in the base definition and remain with the entities when

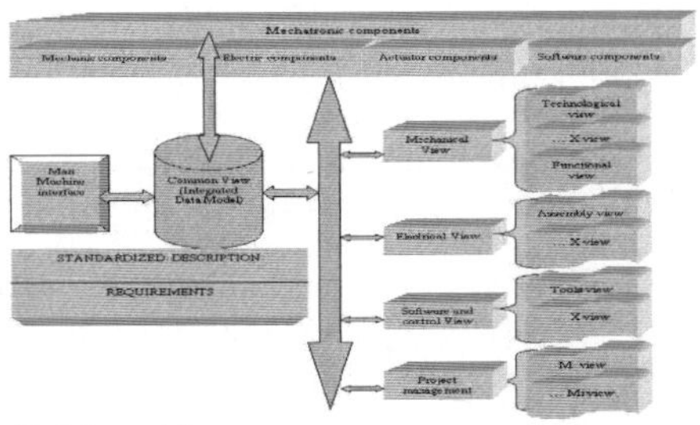

Fig.1. CoDeMo architecture

they are regrouped in engineering views, (Tichkievitch & Veron, 1997).

The framework proposed for interdisciplinary communication between different engineers views for a multidisciplinary optimisation of a mechatronic product; illustrate how across disciplines this system is able to facilitate the communication and also to share the information concerning all the actors that take part to the design process. Other team members, thus allowing for a collaborative design of the mechatronic system can use the simulation results stored in the data base (figure 1).

In the context of a mechatronic system, the idea is to find the best mechatronic solution of the system, that isn't the best mechanical solution coupled with the best electrical solution and the best software solution. The mechanical characteristics of these modules, the electric systems and the software, must be integrated to form a co-operative system of concurrent engineering. The objective of this system is to analyze and propose a solution to improve these problems, (Weston 1996).

3. APPLICATION OVERVIEW. CASE STUDY: A MOTION CONTROL UNIT

If we take an example with a motion unit control made up from an electrical motor, shafts, a wheal and a charge, where the electrical engine can be a direct current motor, we can make the optimizations to achieve the best mechatronic solution.

In our example we shall define in the Technological trade view (figure 2) the entities that we need. The entity is an object which is manipulated by the actors and it represents the class of the types of the elements. Every entity has a giving name and an series of attributes which makes it dependent of the trade view and editable or modifiable.

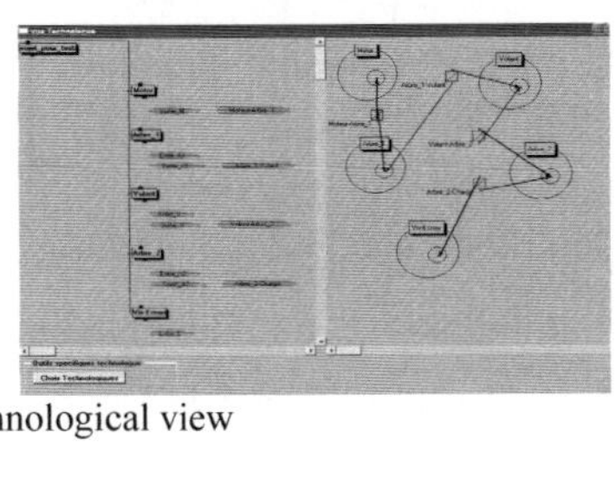

Fig. 2. Technological view

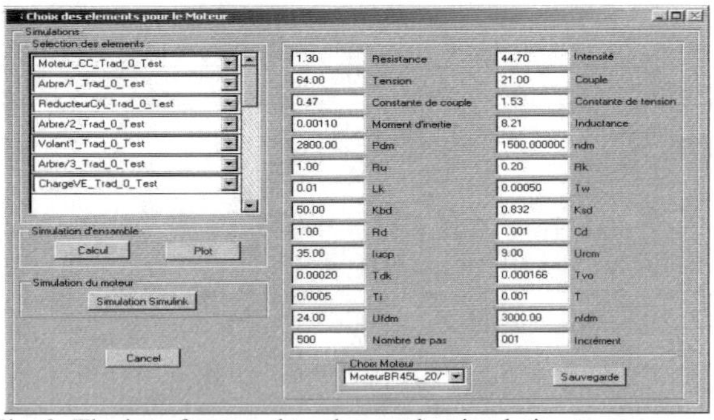

Fig. 3. The interface used to choose the simulation type

To have a cooperative system of conception of mechatronic products, we should integrate the mechanical module, electrical module and the software module. Each actor thus will use the data model for inserts his knowledge and will build the views of product.

The Technological view is built by the technological trades actors according to the technological choices realized in the phase of the initial design. The Mechatronic view is then built by translating the Technological view.

Also with the Mechatonical view are translated the attributes of the entities from the Technological view. The entities translation is one of the automatic tasks managed by the internal actor.

For example, if we have a entity shaft and it's attributes in the Technological view are the elasticity, the length and the diameter, for this entity the internal actor will make a translation in the Mechatronical view if in the neutral file it will find this information. This neutral file can be modified without modifying the code. The line that makes the translation is:

{Component_Name
Shaft Technolog name
Traduction
Component Shaft_Mech Mechatronic name_Trad_0_Mech}

So, in the Mechatronic view we will have for the entity Shaft_Mecha an attribute which is a property of it. This is stiffness. The value for this attribute is compute after a formula which is also in a neutral file. The formula which is loaded from this file is read by a parser.

We did a simulation of the direct current motor using Simulink directly linked to CoDeMo by its API's functions. We also have the possibility to test and to simulate the influence of the other elements (mechanical or command) to the system using Matlab API functions (Figure 3).

There are many industrial types of equipment that require high-order errorless systems. In this case, Symmetric Optimal Method can be employed. The model for determining speed controller Rω is shown as figure 4.

For necessary calculations we can obtain the speed controller with corresponding transfer function (1):

$$R_\omega = \frac{T_0.s+1}{K_{PI}.T_0.s} = \frac{K_I.J.}{8T_{\Sigma}.K\Phi.K_\omega.4T_{\Sigma}} \cdot \frac{(8T_{\Sigma}.s+1)}{s} \tag{1}$$

by using Simulink from MATLAB to simulate this controller we have the diagram (Figure 5).

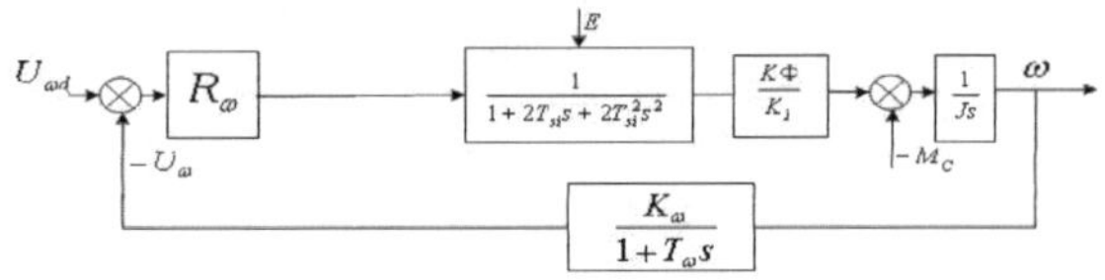
Fig. 4. The speed controller

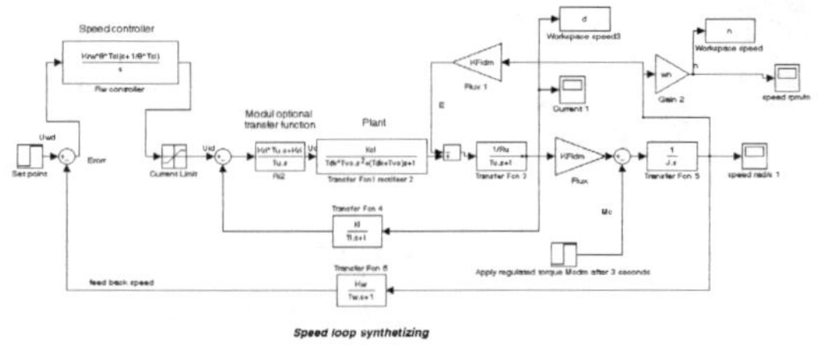
Fig. 5. Speed simulation controller

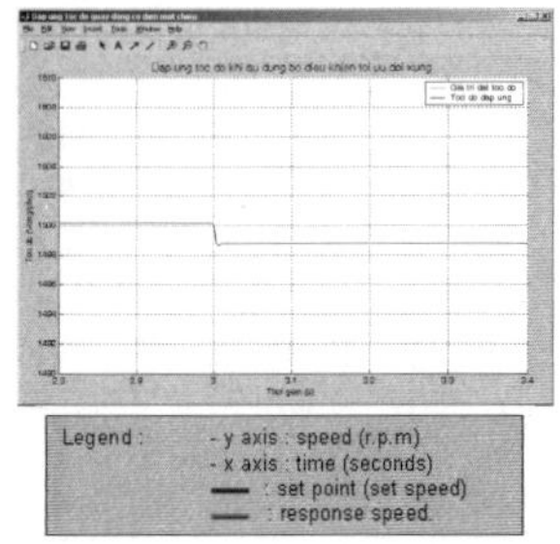

Fig. 6. Speed simulation

All the parameters used for this simulation, are exported from CoDeMo like a *.m file. The result of simulation is presented in the figure 6.

4. CONCLUSIONS

The speed controlling loop is second-order errorless with control signal and first-order errorless with disturbance signal. Therefore, in stable condition speed error will be zero.

We can see that when load torque is equal to regulated torque then after 3 seconds the speed response is reduced slightly compared to the set signal.

Using CoDeMo we can simulate and test the role of the others elements of the system because during the functioning of the system the apparition of the vibrations is inevitable. These vibrations are periodical displacements of the system around the equilibrium position. The vibrations amplitude depends to the system elements and to the interactions between those elements.

5. PERSPECTIVES

The results presented in this paper had validated the concept employing a simple example. Concerning the perspectives, a few short time perspectives can be the implementation of others trade views in CoDeMo. Also, the mechanism of translation and substitution of different system's elements having attached equations can be improved.

6. REFERENCES

Gardarin, G. (1999) *Internet/Intranet and data bases*, Edition Eyrolles

Haug, E.J., Choi, K.K. & Kuhl, J.G. (1995) *Virtual Prototyping Simulation for Design of Mechanical Systems*, ASME Journal of Mechanical Design, Vol. 117 B, pp. 63-70.

Roucoules, L. (1999) *Knowledge and methods: Development contributions of a integrated design environment* phd., I.N.P. Grenoble.

Tichkievitch, S., Veron, M. (1997) *Methodology and product model for integrated design suing a multi-view system*, , No 46/1, 1997, Annals of the CIRP

Weston, R.H. (1996) *Model-Driven Configuration and Information Sharing in Concurrent Engineering*, , Vol.45/1, pp. 449-454 Annals of CIRP

Annals of DAAAM for 2008 & Proceedings of the 19th International DAAAM Symposium, ISSN 1726-9679
Editor B. Katalinic, Published by DAAAM International, Vienna, Austria 2008
Make Harmony Between Technology and Nature, and Your Mind will Fly Free as a Bird

A MODIFIED FASTMAP ALGORITHM

RADULESCU, F[lorin] & BOICEA, A[lexandru]

Abstract: This paper investigates the possibilities to use a modified version of Fastmap for the nodes of a graph, creating an alternative Euclidian space in order to quickly find an approximation of the distance between nodes. The algorithm may be used to estimate distances between streets in a simplified map for a city or town (a dual graph of the location). A search engine may use this estimated distance for ordering the search results from 'near' to 'far'.
Key words: *Multidimensional Scaling, Fastmap, graph, distance.*

1. INTRODUCTION

Local searches involve finding a resource in a specified geographical area. This kind of searches account already more than 10 percent of all Internet queries and are expected to grow (Helft, 2006).

These approaches are almost all based on real maps. In (Radulescu, 2007) we described a simple alternative method for local searches. This method uses a graph as a simplified map for a city or town and a simple distance estimation method from a specified point to a resource for ordering the search result from 'near' to 'far' without using real maps and visual interfaces.

A real map is actually like a graph and there are two ways to code a map in a graph:

1) Street segments as graph edges
2) Streets as nodes: every street is a node and every intersection between any two streets is an edge.

The second modeling approach was selected in this paper for the simplified map of a city because it leads to a much simple representation.

In such a graph the distance between any two nodes may be defined as the length of the minimal path between them. For equal-weighted edges (the length of each edge is 1), the meaning is "how many streets away is one street from another".

Storing the graph or all distances between any two streets is either slow or memory consuming. The idea investigated in this paper was to create a Euclidian space knowing only the distances between any two points. This operation is called multidimensional scaling (Torgerson, 1952). In that case every street may be represented by few coordinates and so the distance between any two streets can be easily computed.

There are many algorithms for multidimensional scaling: FastMap (Faloutsos & Lin, 1995; Yang et al., 2006), MetricMap and Landmark MDS (LMDS). These algorithms approximate classical MDS using a subset of the data and fitting the remainder to the solution.

2. FASTMAP

2.1 Classical Fastmap

For our approach FastMap was selected to be used for multidimensional scaling of the simplified map. It is a recursive algorithm and at each step the following operations are done:

1) Two distant points (a, b) - nodes in the graph in our case - are selected as an axis.
2) For every point c compute the coordinate x for this axis using the generalized Pythagoras theorem (also known as the law of cosine):

$$x = (D^2(a, c) + D^2(a, b) - D^2(b, c)) / (2 * D(a, b)) \qquad (1)$$

3) Use for further axis not the original distances D between points but the remainder D' after subtracting the distance with respect with the coordinates already computed.

In Fig. 1, (a, b) are the two points selected as axis, (x, y) are the coordinates of c and d for this axis (the first axis point a being the origin) and D and D' are the distances between c and d before and after computing the current axis coordinates:

$$D'^2 = D^2 - (x - y)^2 \qquad (2)$$

This process stops after computing the desired number of coordinates for every point or no more axes can be found.

2.2 Problems with real graphs

For real graphs the problem is that the distance matrix is not a Euclidian one, so when we use (2) for computing D' we obtain for some pairs of nodes a negative value for square D'. In that case the only way to continue is to assume D' is 0.

But this assumption leads to propagated errors in computing real good coordinates for our purpose. The algorithm was run with this assumption on a 2000 nodes matrix and the optimal number of axes was 18 (too much) and also the average differences between real distances and computed distances (computed from the resulting coordinates) was high: 0.78 after 18 steps, as shown in Fig. 2.

2.3 Modified Fastmap

For solving this problem we tried different types of adjustments, modifying the original FastMap algorithm. One of them had very good results.

The modified algorithm is the following. The third step was introduced in order to recalculate the distance matrix and have integer distances between nodes.

Step 1: Search for a pair of points (a, b) with maximum distance between them. These points are selected as an axis.
Step 2: For every point c compute the coordinate x for this axis using the generalized Pythagoras theorem (1).
Step 3: For every pair of points c and d (Fig. 3.) recalculate the distance between c and d using:

$$D^2(c, d) = INT\,(D^2(c, e) + D^2(e, d)) \qquad (3)$$

Step 4: Use for further axis use D' as in the original algorithm, using (2), but D' is now computed starting from the modified distance matrix.

D(c, e) can be obtained directly from the coordinates of c and d for the current axis.

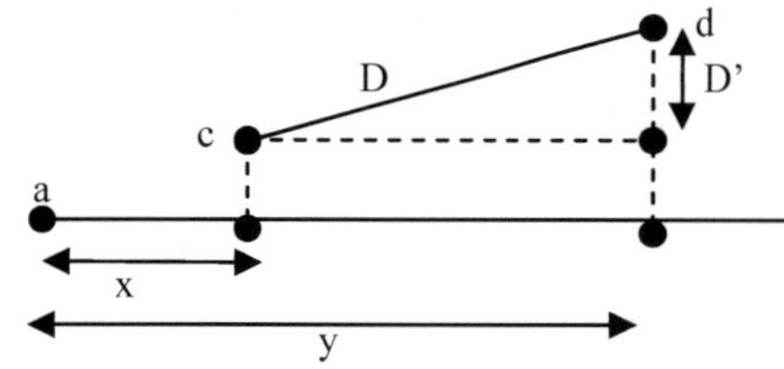

Fig. 1. Computing D'

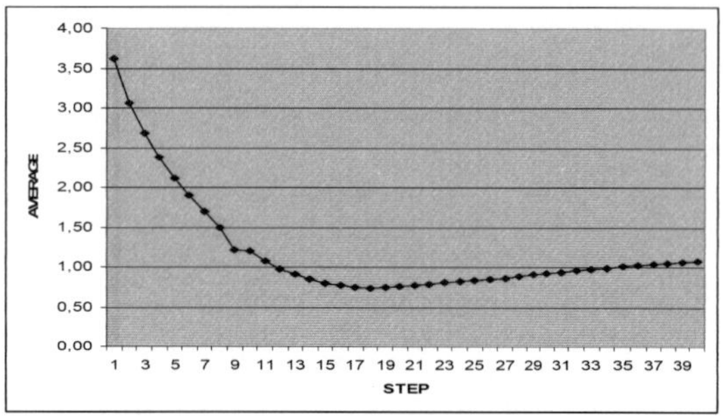

Fig. 2. Average of | Dreal – Dcomputed | over all pairs of nodes

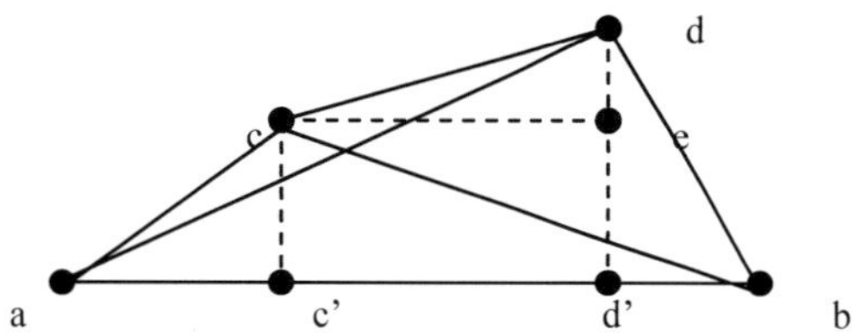

Fig. 3. Adjusting D(c, d)

D(e, d) = |D(d, d') – D(c, c')| and the two distances from the right side can be obtained using (1) for the two square triangles (a, c, c') and (b, d, d') where all two other edges are known.

3. EXPERIMENTAL RESULTS

This proposed algorithm was used on 12 distance matrices containing random generated data:

- 4 matrices for a graph with 1000 nodes and 4 edges for every node. In Table 1 the files are labeled F1-F4.
- 4 matrices for a graph with 2000 nodes and 4 edges for every node. In Table 1 the files are labeled F5-F8.
- 4 matrices for a graph with 2000 nodes and 3 edges for every node in Table 1 the files are labeled F9-F12.

Table 1 contains the results of running the algorithm for all 12 matrices and for only two steps (first two coordinates).

The table columns are the following: **File**: The file with the distance matrix. **AvgErr**: The average of Abs(Dreal(x, y) – Dcomputed(x, y)) over all pairs of nodes (x, y). Dcomputed(x, y) is the distance between nodes x and y resulting from their coordinates using the Euclidian distance (1). **Diff 0**: The number of matrix positions where | Dreal(x, y) – Dcomputed(x, y) | = 0. **Diff 1**: The number of matrix positions where | Dreal(x, y) – Dcomputed(x, y) | = 1. **Diff >1**: The number of matrix positions where | Dreal(x, y) – Dcomputed(x, y) | > 1. **%0-1**: The percentage of 0 and 1 difference values from the total: (Diff0 + Diff1) / (Diff0 + Diff1 + Diff>1).

The results are clear: for over 95% of the graph node pairs, if the described algorithm is used for finding the first two coordinates and then this coordinates are used for computing the distance between the two nodes, the distance is the same or almost the same (0 or 1) with the original (real) distance. These first two coordinates can be pre-computed and stored. More than that, for some graphs this percentage is over 98-99%.

File	AvgErr	Diff 0	Diff 1	Diff > 1	% 0-1
F1	0.17	419.680	77.024	3.796	99.24
F2	0.16	423.536	74.221	2.743	99.45
F3	0.17	416.155	82.031	2.314	99.54
F4	0.16	423.264	74.831	2.405	99.52
F5	0.20	1.604.398	384.910	11.692	99.42
F6	0.34	1.431.513	472.224	97.263	95.14
F7	0.18	1.645.942	344.927	10.131	99.49
F8	0.37	1.387.147	493.416	120.437	93.98
F9	0.29	1.440.510	538.753	21.737	98.91
F10	0.28	1.495.317	457.619	48.064	97.60
F11	0.28	1.456.455	527.476	17.069	99.15
F12	0.30	1.419.806	559.414	21.780	98.91

Tab 1. Experimental results

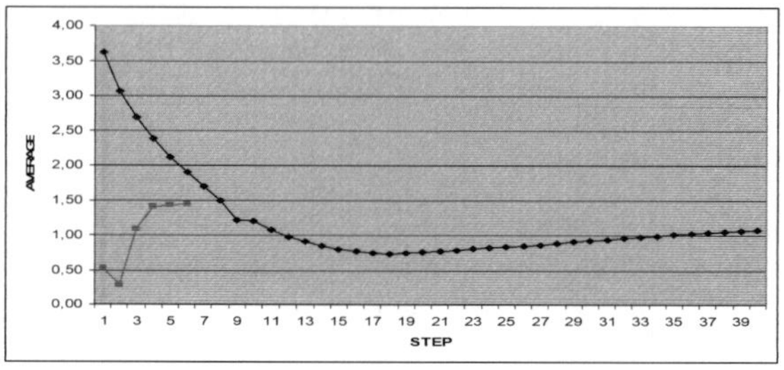

Fig. 4. Average of | Dreal – Dcomputed | over all pairs of nodes for both implementations: original and modified algorithm.

For a comparison between the original and modified algorithm, Fig. 4 presents the graphic for both (update for Fig. 2, for the same dataset).

The modified algorithm stops after only 6 steps but only first two coordinates are enough (and optimal). The original algorithm has a minimum after 18 steps and this is minimum is greater.

4. CONCLUSIONS

The results show that using this modified Fastmap algorithm we can associate with every node in a graph two coordinates that leads to a good approximation of the real distance between them. For many applications (including search engines that order the results form 'near' to 'far') it is then not necessary to store an entire graph distance matrix but only these few coordinates because the time for calculating any distance using Euclidian distance formula is reasonable short.

5. REFERENCES

Helft, M. (2006), The Retooling of a Search Engine, *New York Times*, December 4th 2006.

Radulescu, F. (2007). "Locality Sensitive Search: A Fast Approach", *Proceedings of the CSCS16*, Bucharest, May 2007, vol. 2, pp. 192 – 195.

Torgerson, W.S. (1952). Multidimensional Scaling: Theory and Method, *Psychometrika*, vol 17, 1952, pp. 401-419.

Christos F. & Lin K.I. (1995). FastMap: A Fast Algorithm for Indexing, Data-Mining and Visualization of Traditional and Multimedia Datasets, *Proceedings of the 1995 ACM SIGMOD International Conference on Management of Data*, San Jose, California, May, 1995.

Yang, T., Liu, J., McMillan, L., Wang, W. (2006). A Fast Approximation to Multidimensional Scaling, *Proceedings of the ECCV Workshop on Computation Intensive Methods for Computer Vision,* Graz, Austria, May 2006.

Annals of DAAAM for 2008 & Proceedings of the 19th International DAAAM Symposium, ISSN 1726-9679
Editor B. Katalinic, Published by DAAAM International, Vienna, Austria 2008
Make Harmony Between Technology and Nature, and Your Mind will Fly Free as a Bird

SELF-SUSTAINED VIBRATIONS

RADULESCU, M[ara]; RADULESCU, B[runo] & COZMINCA, I[rina]

***Abstract**: To stay in competition on the world market, enterprises must design innovative products in a shorter interval of time and at lower prices. In this context, the phase of product designing has an enormous importance on the economic success of the enterprise. For this reason enterprises focus more and more on the optimization of the designing phase. By approaching the themes proposed in this paper, we touch a domain of high importance and in continuous development. The studies and researches that preoccupied us are the ones focused on the self-sustained vibration cutting especially the self-sustained vibrating drilling.*

***Key words:** self-sustained, vibration, cutting, chips*

1. INTRODUCTION

Vibrating drilling exists and has been used for more than 50 years more in the lab than in industrial applications.

The first studies were made in *Russia* by Professor V. N. Poduraev, who had the idea to use axial vibration for drilling. He established for the first time the interest presented by this processing technology (Poduraev, 1977).

The second wave of researchers appeared in *Japan* and focus on Kumabe's papers (Kumabe, 1979). The researches effectuated by him especially focused on drilling with vibrations generated by a mechanism producing ultrasounds. As for the processing techniques by low frequency vibration, the rigid excitation mechanism has always been used.

In *China*, they started to make researches on this topic. Deyuan and Lijiang (Deyuan & Lijiang, 1998) made investigations related to the chip formations in the moment of drilling by rigid excitation vibration taking also into account beading phenomenon.

Here we should mention that in (Wang et al., 1998) are making references to studies asserting a melioration of the life span of the tools used up to 400%, and also of the axial precision from 0,028 to 0,074 mm.

At the Purdue University from USA, in 1996, a 3 year project started having as objective the study of influence of controlled vibration necessary for drilling. The excitations used were of a rigid nature and had as generation mechanism the piezoelectric excitation. The results of this project were published in 1998 (Toews et al., 1998).

The explanations given in the conclusions of this report are similar to those presented in (Wang et al., 1998), but they do not present sufficient arguments. The particularity of these researches refers especially to the amplitude used which is very small that is 2,5-3 μm.

2. SELF-SUSTAINED VIBRATINONS CUTTING

The disadvantages of forced vibration presence prevented the application of this processing procedure. Most of the essays effectuated in the specialized literature started from test floors intended especially for the lab study of problems. Vibrations, in the case of cutting by self-sustained vibrations, damp and sustain themselves allowing the system dynamics to adapt by

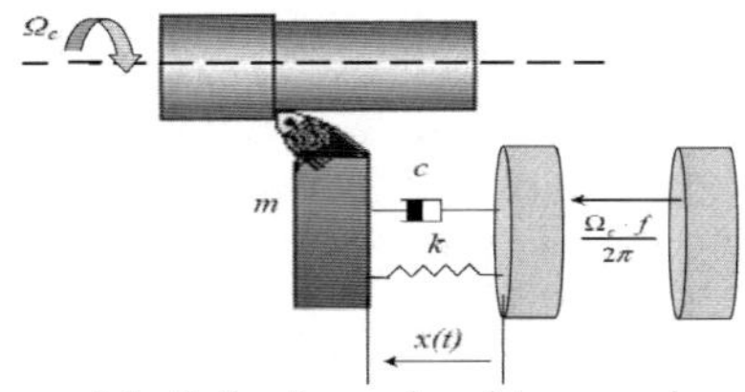

Fig. 1. The model of vibration reduced to one axis

itself to the cutting conditions. In this case we must search the instability domain and favor the vibrating phenomenon that might allow the tool to vibrate with sufficiently high amplitude so that the chips might fragment (Gouskov & Brun-Picard, 1996).

The cutting by self-sustained vibration process using a self-vibrating head is that to favor and encourage the appearance of the self-sustained vibration on the axial direction.

The models allowing the simulation of this cutting procedure should contain that mode of vibration too. In the specialized literature we did not find studies focusing on the phenomena that appear in cutting in the conditions of vibration use. The researchers' interest concerns more the predictive methods capable to avoid and control vibration and less the aspects regarding the frequencies and amplitudes appearing in the system tool-machine – systems – devices – parts. This phenomenon is seen as an expression of some negative effects on the quality of the processed surface, the tools or tools-machine life span. Thus, this phenomenon of appearance of vibrations was and is seen by some researchers as a problem that must be avoided.

The model took into account (Figure 1):

- a constant rigidity of the machine tool,
- a linear characteristic of the cutting process that came from a simple model of the Merchant type in which force depends directly proportional to the chip dimensions (feed rate, cutting depth)
- the classical dependence between the cut layer and the previous passage (what brings a delay in the system).

The mathematic model contained an equation of the type:

$$m\ddot{x}(t)+c\dot{x}(t)+kx(t)=-C_F\cdot b\cdot\left(f+x(t)-x(t-r)\right) \qquad (1)$$

where:

- $r=\frac{\frac{2\pi}{\Omega_c}}{z}=\frac{\frac{60}{N}}{z}$ represents the delay between the two consecutive passages [s].
- N represents the relative rotation speed between tool and part in [r/min]
- Ωc represents the same rotation speed but in [rad/s]
- z represents the number of tool teeth
- f represents the feed programmed in [mm/rot]
- CF represents the specific pressure of the material under study (the force necessary to detach a 1 mm² chip)
- b represents the chip width [mm]. The cutting rigidity is considered linear.

So, eliminating the trivial solutions we have:

$$\overline{x}(t) = \frac{C_F \cdot b \cdot f}{k} = -\frac{F_p}{k} \qquad (2)$$

where :

- Fp represents the penetration force on a perpendicular direction to the processed surface.

We have the following equation:

$$\ddot{x}(t) + 2\varsigma\omega_0\dot{x}(t) + \omega^2{}_0 x(t) = -\frac{C_F \cdot b}{m} \cdot \left(x(t) - x(t-r)\right) \qquad (3)$$

The asymptotic stability of solution x(t) for equation 1 corresponds to the back stroke generated by the cutting force that is equivalent to the stability of fixed point x(t) for equation 3. This model uses delayed differential equations.

The main characteristic of this system is that it does not have a finite dimension. There are several possibilities to study this type of system, but the one that interests us in this case is the one that was used to approach the stability problem in metal cutting. This method is based on a series of particular generalizations.

The study of system stability may be also approached using the Lyapunov method accepted for the solving of the determinist dynamic systems. This method has a more general character but unfortunately it is not easy to use.

The general method that can be used for a large class of differential equations is named the method of D partitions. In fact it is not something else but a methodology of application for Lyapunov's method for the systems presenting infinity of solutions of the characteristic equations attached to the differential equations of the system.

The stability of a system modeled with equations of the 3 type will have a characteristic D function:

$$D(s) = s^2 + 2\varsigma\omega_0 s + \omega_0^2 + \frac{C_F \cdot b}{m} \cdot \left(1 - e^{-sr}\right) \qquad (4)$$

Where D curves represent the stability frontier corresponding to a solution of characteristic equations D(s), crossing the imaginary axis of the complex plane. These curves are defined by:

$$\begin{cases} \mathrm{Re}\left(D\left(j(\omega)\right)\right) = 0 \\ \mathrm{Im}\left(D\left(j(\omega)\right)\right) = 0 \end{cases} \qquad (5)$$

The D variants from the general case represent surfaces of dimensions 1 in the parameter space but they belong to the curves for a plane space defined by two parameters.

The D curves build the stability frontier of the system by the reunion of the instable areas:

$$\begin{cases} \mathrm{Re}\left(D\left(j(\omega)\right)\right) = -\omega^2 + \omega_0^2 + \frac{C_F \cdot b}{m} \cdot \left(1 - \cos(\omega \cdot r)\right) = 0 \\ \mathrm{Im}\left(D\left(j(\omega)\right)\right) = -2\varsigma\omega_0\omega + \frac{C_F \cdot b}{m} \cdot \sin(\omega \cdot r) = 0 \end{cases} \qquad (6)$$

These equations give a stability frontier that the field researchers call stability lobes (Figure 2).

The phenomenons of no linearity by determinations made on the milling machine introduce only quantitative changes at the level of expression of the cutting law or the transfer function from machine-tool.

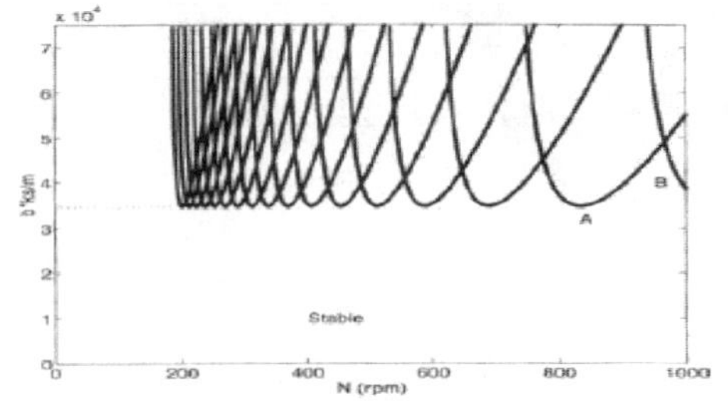

Fig. 2. Stability lobes in physical parameters

The nonlinear rigidity of the machine on a perpendicular direction of the processed surface a damping depends on the oscillation frequency on the same direction and a cutting law also nonlinear.

The equations obtained allowed, by a convenient choice of the parameters considered, the deduction of a stable limited cycle highlighted in determinations.

3. CONCLUSIONS

The explanations are still superficial. For example the considered damping is dependent on the tool oscillation frequency. What happens at the contact level is much more complex, this damping coming from a continuous variation of cutting conditions lengthways the tool trajectory. In the same time the tool-machine had a nonlinear rigidity whereas the great majority of the tool-machine' structures present own frequencies with values quite different from one another.

The determination of amplitudes, frequencies and other phenomena could not find an explanation in the linear theory of vibrations of the machine-tools. This phenomenon refers to the fact that these vibrations could be excited even in stable cutting conditions and once installed the vibration phenomenon persisted.

4. PERSPECTIVES

Starting from the results presented in this paper we are presuming that conceiving a more reliable model of the self-sustained chipping process in the case of the drilling operation could solve some of the problems of the complexity, modeling the phenomena composing the tool-machine, tools, devices and parts in the vibrator mode in the case of the drilling operation with self-sustained vibrations and designing the device to make the drilling operation with self-sustained vibrations could be the right perspective for this field.

5. REFERENCES

Deyuan Z. & Lijiang W., (1998) *Investigation of Chip in Vibration Drilling*, Elsevier, Int J Mach Tool Manu, vol. 38, no. 3, pp. 165-176.

Gouskov A. & Brun-Picard D., (1996) *Instability of vibrating drilling*, Colloque PRIMECA, Nantes, France.

Kumabe J.,(1979) *Fundamentals and Application of Vibration Cutting*, Jikkyo Publishing Co.,Japan.

Poduraev, V. N., (1977) *Cutting with Vibrations,* Mashinostroenie, Moscow, Russia.

Toews H. G., Compton W. D. & Chandrasekar S., (1998) *A Study of the Superimposed Low-Frequency Modulation on the Drilling Process*, Elsevier, Precision Engineering, vol. 22, pp. 1-9.

Wang L.-P., Wang L.-J., He Y.-H. & Yang Z.-J., (1998) *Prediction and Computer Simulation of Dynamic Thrust and Torque in Vibration Drilling*, ImechE Procedure vol. 212 part B, pp. 489-497.

Annals of DAAAM for 2008 & Proceedings of the 19th International DAAAM Symposium, ISSN 1726-9679
Editor B. Katalinic, Published by DAAAM International, Vienna, Austria 2008
Make Harmony Between Technology and Nature, and Your Mind will Fly Free as a Bird

QUESTIONS ON THE SALES MANAGEMENT IN INSURANCE INDUSTRY

RAHAU, L[oredana]; CIOTEA, V[alentin] F[lorin] & CIOTEA, M[argareta]

Abstract: *Insurance developing and improving the ratio between insurance premiums and gross national product is the result of the action of objective and subjective factors, the increase of the average income per capita being the most important. Insurance companies are the main providers of insurance market with an offering increasingly varied and adapted to the needs of individuals and companies. Competition shown in this market requires insurance companies to formulate management strategies appropriate in the circumstances in which the management of insurance cannot ignore the organizational and functional specificity and activity purpose of these companies. Such an element of specificity is the management of insurance sales.*

Key words: *insurance, sales, management system, insurance agent*

1. INTRODUCTION

Insurance present a great importance for the economy of a country because it participates to the process of creating added value and because it provides jobs. It also participates in the bid of lending capital on the financial market intermediary between those insured and the applicants for loaning resources. We are talking in general about a process of nonmonetary financial intermediation with positive effects on the stability of the national currency, because the flows of financial resources that start from insurance companies to various beneficiaries do not increase the monetary mass in circulation, they only redistribute it. Insurance Companies influence (positive or negative) the payment balances of the country when making, in addition to direct insurance, reinsurance operations because of occurring foreign exchange flows expressed in the shape of denominated premiums, commissions, compensation, participation in benefits received or paid, all reflected in balance of payments. The insurance business contributes to reducing economic uncertainty and creates the possibility to resume unwanted interrupted activity as a result of sinister product.

2. BODY OF PAPER

On the world market there is a trend of globalization of the insurance business, three regions (North America, Europe and Asia) contributing each about one third of the total gross premiums collected. On the American insurance market, North America holds first place, but in South America the insurance market knows a slow development, as it is strongly affected by economic crisis. The European insurance market holds the western European and eastern European Markets. The first is dominated by the markets of industrialized states while the markets of Eastern Europe holding several sufficiently financial solid competitors are barely consolidating. Asia is dominated by the Japanese market, Japanese insurers being considered the most important in Asia, despite their diminishing market share, as a consequence of American and European expansion (Cistelecan, 2006).

In the developed countries, the insurance is a market of buyers with a wide offer of insurance products and heavy competition. (Petrescu, 2005). Increasing life expectancy of people, the emergence of web portals that enables comparison of the insured's offers, efficient risk management of insurance companies are several factors that contributed to reducing the premium rates and increased interest in insurance products.

In Romania, the insurance market is in training and constantly increasing. This is demonstrated by the favorable evolution of indicators such as the degree of insurance penetration of the economy and insurance density. The degree of penetration has seen a slow and constant growth, but is situated at a low level compared with that recorded in the states of the European Union. Insurance density has increased every year, also. The insurance is a competitive market, but most of the specific activities (over 80%) is undertaken by a limited number of companies (the first top 10 companies). In terms of geographical distribution, but also the types of insurance products required, the insurance demand is not uniform. The difference is manifested between effective demand and potential demand of insurance. Reviewing offer insurance, we can observe the trend of diversification of the insurance portfolio, the emergence of specific insurance products, the increase of quality insurance products offered, the increase of the professional level of insurance staff.

The products offered by insurance companies must be sold to potential consumers of insurance (individuals or legal persons). Purchasers are seeking insurance protection, saving and fruitfulness through insurance, and the insurance company wants to make a higher profit. We believe that the satisfaction of the buyer of insurance products and the insurance company`s gain depend both on the preparation and training of personnel involved in the sale of insurance and the coordination of insurance. This means that sales management is vital for the sales activity insurance.

Management of sales is regarded as a process by which the planning, organizing, directing, endowing with personal and the control of operations and sales made by sales agents of a company to achieve its objectives are accomplished (Donaldson, 2001). In insurance, sales management represents the application of the management process in sales and takes into account the sales strategy, the specific objectives of the sale, the sales budget, sales policy and procedures used in these domains (Cistelecan, 2006).

The hypothesis from we started in this research carried out is that, in the field of insurance in Romania, sales management features are compared with other industries.

Primarily, the main factors that determine specificity of the sales management process in the insurance field are (Vorzsak, 2004):

- the insurance products are non-sought products, which the consumer avoids to buy, although he is aware that they are needed;
- insurance is a specific non-corporal product and the risk is the object of any insurance contract;
- differences between the insurance products are hardly notice to the ordinary consumer;
- the insurance products are too complex to be explained by advertisements in various media of communication.

Secondly, the management of sales activities in insurance agencies or other subsidiaries in the territory of an insurance company (offices or work points) highlights the role of insurance agent in managing its business. Insurance agencies meet two main functions: the sale of insurance products and providing services for the insured. If the insurance contract is not closed, the customer can no longer be served and it is possible that he may seek the services of another insurance agency. Therefore, the success of the insurance agency depends on the efficient completion of the two fore mentioned functions.

Thirdly, sales staff (insurance agents at the various steps of their career) is the direct link of the insurance company to the insurance consumers. There are situations when the customer signs a contract of insurance due to the confidence given by insurance agent. Why? Because offering insurance products are actually offering promises.

Fourth, if is the case of non-life insurance the classic steps of the process of selling personal life insurance are being taken in the private life insurance market several stages are being taken: prospecting, contact, preliminary discussion, interview and follow-sale, the insuring of the insurance policy and post-sales contact. If not so long ago the only concern of the insurance agent was selling, now this is not enough. Competition in the insurance market, the variety of insurance products, consumer reluctance for the insurance are some causes for which the sale activity is only one of the tasks accomplished by the sales agent now, who is also undertaking the following activities: prospecting (assumed to discover and contact new customers), communications (assumed to transmit customer information relative to products offered by insurance), services (counseling, financial assistance); collection of information (the agent is collecting information useful to the society for which he works, draws up at regular's intervals reports on the visits made and results).

Fifth, in the Romanian insurance practice, we can identify two sales management systems:

- a top-down, case is which, the management sets long term as well as short term objectives for the activity of sales, then determines the ways in which they can be achieved. For each insurance agent is established, annual and monthly minimum number of insurance contracts to be concluded and the minimum total annualized insurance premium;

- a bottom-up, when each insurance agent will mention the results that are expected to be achieved, then these results will be examined, discussed and reformed at each hierarchical level, until a single global sales plan is achieved. The aim of the insurance agent is a minimum number of insurance contracts to be made monthly and annual, increase and a minimum of the first annualized insurance.

These two sales management systems are similar because the business planning and long-term measurement results are made at top management level, but differ in the way they set the targets of the insurance agent at the individual general and level.

We appreciate that whatever system used, the management of insurance agency must ensure the operational frame for the process of management sales, must weigh the weaknesses and strengths, should provide tools for overcoming the problems that arise, whether they would mean a more numerous, better prepared personnel. Insurance companies must carefully determine the specific objectives to be achieved by their sales agents. These objectives are represented by the number of annualized gross premium contracts, the value of the contract.

Sixth, training and improving staff involved in the insurance sale is achieved through participation in courses organized by the insurance companies with their own trainers. Recruitment and retention of officers insurance is more difficult than in other areas. To attract professionals they need, the insurance companies must rethink the system of recruitment, payroll, training of employees.

In the countries of Central and Eastern Europe we can find that a growing number of people involved in selling insurance, but this increase is lower than the revenue growth of insurers' gross premiums received, which demonstrates a high productivity in the field.

Insurance companies are investing in training of insurance agents through programs aiming several objectives: knowing the company for which they work for and identify with it; knowledge of company products; knowledge of the customer`s defining traits, knowing how to conduct effective commercial presentations, understanding the methods and responsibilities involved in field work. For the improvement of insurance personnel, insurance companies may opt for staff training system (mostly used) or staff training in specialized units. Companies seek the achievement of efficient management`s sales activities. The effectiveness of insurance products sales may be measured by the ratio between the number of insurance agents, the number of insurance contracts sold and the annual average premium of a contract.

The Annual Reports of the Commission of Insurance Supervision published in Romania until 2006 find that in 2004 the number of contracts of insurance in force was 9,030,820 in the year 2005 we see a growth of 17.71% and in 2006 the growth was 20.45% compared to the situation in 2005. The number of insurance agents (working with a working card and collaborators) rose from 36,130 in 2004 to 43,260 in 2005 and 45,462 in 2006, namely an increase of 25.82%. If we refer to the earnings from subscribed gross premiums, then we find a real annual growth of 17.01% in 2005 versus 2004 and 23.68% in 2006 compared to the situation in 2005 (CSA Annual Report 2004 - 2006). These recorded increases; we believe are a result of improving management of insurance sales, other factors of influence being the increasing confidence of the public in insurance and measures to provide information about insurance.

3. CONCLUSIONS

Research carried out on the insurance market in Romania shows that the management of sales activities is important for the insurance company, pursues large underwritten gross premiums, insurance benefits in the smallest amount possible and a profit as large as possible.

The undertaking and correct application of the personal sale process is vital because insurance has certain characteristics: intangibility, loss of value through the inflation effect, negative association, lack of awareness of the need. Achieving proper management in sales activity involves security: human resources (insurance agents) appropriate numerically in terms of competence, motivation, hierarchical control, recognizing that insurance agents constitute the basic operators in promoting insurance products, business insurance agents activity is reflected in the management, training and improvement of insurance agents through specific programmers in the insurance field. The continuation of this research will be a comparative analysis of insurance sales management life and nonlife.

4. REFERENCES

Annual Report of the Commission of Insurance Supervision, 2004-2006 collection, ISSN 1841-7256, Bucuresti

Cistelecan, L. (2006). *Insurance,* Petru Maior University Publisher House, C.Z.U. 368 (075.8), page 160-165, 191-192, Targu-Mures

Donaldson, B. (2001), *Sales management*, Codecs Publisher House, ISBN 973-8060-37-0, Bucuresti

Petrescu, E.C. (2005). *Marketing insurance*, Uranus Publisher House, ISBN 973-7765-10-9, Bucuresti

Vorzask, A.(coord.) (2004), *Marketing services*, Presa Universitara Clujeana Publisher House, ISBN 973-610-305-6, Cluj Napoca

Annals of DAAAM for 2008 & Proceedings of the 19th International DAAAM Symposium, ISSN 1726-9679
Editor B. Katalinic, Published by DAAAM International, Vienna, Austria 2008
Make Harmony Between Technology and Nature, and Your Mind will Fly Free as a Bird

IMPROVING BUSINESS PROCESSES WITH RFID TECHNOLOGY

RAKIC-SKOKOVIC, M[arija]; OSTOJIC, G[ordana]; LAZAREVIC, M[ilovan] & STANKOVSKI, S[tevan]

Abstract: *The actual and potential uses of RFID are wide-ranging and diverse. While many organizations are adopting RFID by force of mandate, even more of them are seeking ways to use the technology to improve their operations. Implementation of the RFID technology, in an existing parking lot access control system in city of Novi Sad, has given benefits to all interest parts (the Parking Operator, parking place users and parking collectors).*
Key words: *RFID technology, business processes, parking lot.*

1. INTRODUCTION

In the last two decades, business process improvement has become a widely accepted path to greater efficiency and profitability in operations. Oriented around outcomes instead of routines, business process improvement aims to provide significant change in organizational performance. With the proper solution companies can develop, debug, and run sophisticated business processes that bridge distributed organizations and their suppliers, partners, and customers and meet large-scale information requirements. Implementation of RFID (Radio Frequency Identification) technology can be a potential way to meet these demands.

RFID technology allows a machine-readable and thereby rapid and efficient identification of tagged items. Data obtained is immediately passed on to the overall management information system, thus providing crucial assistance for the exact and up to date IT-supported tracking of movements taken by items within a business. The potential offered by RFID for operative management improvements is enormous and it has become indispensable in the automobile and consumer goods industries, as well as for providers of logistical services.

The use of RFID is particularly beneficial in closed loops requiring the highest level of process security. The examination of RFID deployment must first identify process innovation and improvements. Only then can the selection of suitable technology components and the system integration take place. A structured approach helps to both minimize any potential start-up problems and to fully exploit the potential of RFID.

In order to organize the deployed RFID technology so that it is as effective as possible, the carrying out of a material flow analysis for the business is essential. A resulting draft solution should explain the organizational, process-related and technical changes to be made, should contain a cost-benefit analysis and also provide an initial project plan for the implementation, (Gizanis, 2006).

2. LITERATURE REVIEW

A survey of 275 manufacturing companies across a wide range of enterprise sizes and industries found that two-thirds of the companies' RFID implementations were driven in part or entirely by process improvement goals, rather than just by customer mandates. The report found that 67% of smaller companies (under $25 million) are implementing or planning to implement RFID. And these smaller firms are the most likely to be driven by process improvement goals (63%) or both process improvement and mandates (27%), rather than by customer mandates alone (10%), (McBeath, 2006).

By applying RFID technology on the plant floor, manufacturers can seamlessly integrate the newly captured information into the existing information and control infrastructure, thereby using the RFID tag as a unique identifier and minimizing capital equipment costs and investment risk. On the plant floor, RFID will provide the greatest impact in the areas of inventory visibility, labour efficiency and tracking and genealogy. The better a manufacturer is able to collect, manage and use information to drive production assets and processes, the more visibility (and value) it can provide to its trading partners. Increasingly demanding FDA requirements are forcing consumer goods manufacturers of all types to more effectively manage product information, including lot tracking and related quality data across their entire supply chain. In the event of a product recall, access to reliable, accurate and real-time information is absolutely critical. RFID can also complement existing manufacturing execution system efforts in genealogy tracking, (Chand, 2007).

Scientific literature has widely debated the reasons and the current limits to RFID adoption and related benefits on supply-chain processes. Jones et al. (2004) argue that a main reason for RFID diffusion is the capability of tags to provide more information about products than traditional barcodes. Manufacturing site, production lot, expiry date and components type are among information that can be stored into the tag chip.

Prater et al. (2005) have discussed the main benefits of RFID adoption for supply-chain processes, for the specific case of the grocery retailing. The availability of real-time information is regarded as the main benefit, although additional outcomes can be found in increased inventory visibility, stock-out reduction, real-time access and update of current store inventory levels, etc (Bushnell, 2000). Despite the achievable benefits, several authors agree that the main limit to a wide use of RFID technology has to be found in its cost (Prater et al., 2005; Karkkainen & Holmstrom 2002; Burnell, 1999). Consequently, critics to RFID argue that investments in tags and readers, as well as in the related informatics infrastructure, are still not profitable.

Not every conceivable use of RFID makes sense, at this time. The cost of tags, maturity of solutions, and availability of skilled implementers, is improving dramatically over a short period of time. Ever-increasing numbers of RFID-enabled business process improvements are becoming feasible and cost-justified. Selecting the right approach and timing requires knowledge and creativity.

3. EXAMPLE OF IMPLEMENTATION OF RFID TECHNOLOGY IN CITY OF NOVI SAD

One of the business processes which can be improved by implementation of RFID technology is parking revenue. Parking plays an important role in the traffic (and logistic) system since all vehicles require a storage location when they are not being used to transport passengers. Whether it is a parking lot or on-street parking there is a problem of parking revenue convenience.

There are different types of parking system in city of Novi Sad: on-street parking, parking lots and garages. Parking revenue is done by a Parkomat, SMS (Short Message Service) message over mobile phone and by parking ticket. Instead of parking revenue collectors there are parking checkers (checking parking price payment). Parking revenue for parking lots and garages is done by parking collectors, because the parking price varies with the time spent on a parking lot or in a garage.

We have implemented RFID technology in parking system of city of Novi Sad, meaning both on-street parking and parking lots and garages, thus creating intelligent parking control management system which integrates RFID technology, automatic control technology and applications software.

This parking system could not be fully automated because of limitations like: lack of needed infrastructure, cost of system implementation, security and privacy concerns.

Implemented parking system expects that parking place user stop a vehicle on inductive loop, and then reader query's a RFID tag. The RFID tag detects the interrogating signal and transmits a response signal containing encoded data back to the receiver. This data is: UID - Unique Identification - encode check, validity period - locked data for date, time, and year, so that the tag can be used only in defined period (after locking this data no one can unlock these blocks for reuse), check bit (one block memory) to determine is the user's vehicle already is in the parking lot or not. This check is practiced for prevention of malversation. PLC software analyses gathered data and if it is correct, sends a signal to barrier to lift an arm and sends a command to RFID reader to write on RFID tag information that the vehicle is entering/exiting parking lot. While PLC is processing information, green led diode is blinking, thus informing a user of activity of checking and writing information on tag process. After writing data to RFID tag green led diode stops blinking and user can drive in/out a parking lot (or garage).

If one or all of the named data are not correct, system will inform a user by lighting red led diode. An arm of a barrier will not lift a user can't enter/exit a parking lot. If a user still wants to enter a parking lot (or a garage) he or she must press the pushbutton, placed on a control centre side and follow the procedure already mentioned above. In case of error during exiting parking lot, a parking collector checks an RFID tag to determine the problem. If a problem is related with the validity period parking collector calculates revenue for the time spent from the end of the validity period till now. Calculated time and a price for the parking service are displayed on the PC monitor. Those data could be seen by both a parking collector and a parking place user. After paying for the parking service, parking collector prints a receipt and exit barrier arm lifts, so the user can leave a parking lot or garage. If the problem relates to the value of the check bit then a user of a RFID tag an authorized institution deals with this problem.

Parking collectors at parking lots and garages are authorized to visually check, every now and then, validity period and quality of graphics printed on a RFID tag (in case of a forgery).

4. CONCLUSION

Implementation of the RFID technology, in an existing parking lot access control system in city of Novi Sad, has given benefits to all interest parts (the Parking Operator, parking place users and parking collectors).

Parking Operator has gain robust system, easy to operate, easy for maintenance, with the reliable RFID tag-ticket check for the prevention of malversation. Also, RFID reader is placed in a way that it doesn't violate exterior. Till now collection of the parking service revenue has been increased for 17%.

People using parking place are spending much less time in waiting in line to buy the tickets for the on-street parking and much less time waiting at the entry and exit barriers of a parking lot or garage.

Parking collectors that are working at parking lots and garages are much less involved in collecting revenue. They are only active when dealing with parking place users who don't have correct RFID tag-ticket and, of course, when they have to charge for the parking place users who don't have RFID tag-ticket.

Conclusion is that implementation of RFID technology in parking system leads to process improvement related to:

- Reduced cash handling and streamlined back-office operations
- Scalable system that can fit current and future needs
- Automatic data capture and detailed reporting
- Improvement of traffic flow at peak hours
- Improvement of customer service
- Cash-free convenience
- Quick online access to parking users data related to revenue, thus enabling giving penalty tickets for users.

RFID technology will be more widely used in the next period, when RFID infrastructure will be installed in number of object (facilities, roads, parking lots, etc).

Future research will be directed towards the designing a parking system so that a parking place user will be able just to drive thru and enter/exit parking lot or garage. Also we will try to automate the in and out privileges of the subscriber and then transfer this data to the enterprise software for the traffic analysis that will allow optimization of the human resources needed for traffic flow in and out. For customer payment, the RFID tag could be read to debit a pre-pay system or charge the parking services against a credit card. All of this will facilitate customers entering and leaving and this improves service levels and increases capacity in the parking lot. These benefits will drive higher revenues.

5. REFERENCES

Burnell, J. (1999). Users will overcome many obstacles and implement RFID, study predicts. *Automatic ID News* Vol. 15 No.5, pp. 26.

Bushnell, R. (2000). RFID's wide range of possibilities. *Modern Materials Handling* Vol. 55 No.1, pp. 37.

Chand, S. (2007). Embracing RFID technology drives process improvements, *Available from:* http://www.plantengineering.com/article/CA6414253.html *Accessed:* 2008-06-27

Gizanis, D. (2006). Process Improvement with RFID, *Available from:* http://www.snt-world.com/Content.Node/boxcontent/process_improvement_with_rfid.doc *Accessed:* 2008-06-27

Jones, P., Clarke-Hill, C., Shears, P., Comfort, D., Hillier, D., 2004. Radio frequency identification in the UK: Opportunities and challenges. *International Journal of Retail and Distribution Management* Vol. 32 No. 3, pp. 164–171.

Karkkainen, M., Holmstrom, J. (2002). Wireless product identification: Enabler for handling efficiency, customization and information sharing. *Supply Chain Management: An International Journal* Vol. 7 No. 4, pp. 242–252.

McBeath, B. (2006). How Manufacturers are Improving Processes by Using RFID, *Available from:* http://www.chainlinkresearch.com/research/detail.cfm?guid=823E0D5A-9F41-5EBA-2428-6BAF7BFD1737 *Accessed:* 2008-06-27

Prater, E., Frazier, G.V., Reyes, P.M. (2005). Future impacts of RFID on e-supply chains in grocery retailing. *Supply Chain Management: An International Journal* Vol. 10 No. 2, pp. 134–142.

Annals of DAAAM for 2008 & Proceedings of the 19th International DAAAM Symposium, ISSN 1726-9679
Editor B. Katalinic, Published by DAAAM International, Vienna, Austria 2008
Make Harmony Between Technology and Nature, and Your Mind will Fly Free as a Bird

REMOTE MONITORING AND DATA ACQUISITION OF INDUSTRIAL PROCESS PARAMETERS THROUGH INTERNET

RANCEA, I[rina]; SGARCIU, V[alentin] & DICHIU, D[aniel]

Abstract: *In this paper we used the concepts of remote control of sensors for automated monitoring and data acquisition of industrial processes. The application can monitor and command using the same acquisition server, different industrial processes. Our case study is using a multifunctional board from National Instruments and the National Instruments LabView framework for acquisition of the temperature parameter and bi-positional command of a heating oven.*
Key words: *acquisition, remote command, remote control*

1. INTRODUCTION

Technological process supervision involves acquisition, processing and returning the commands for a large number of parameters. Using the Internet connection it is allowed to acquire data from long distances and to be concentrated in high performance servers.

The paper is focused on a solution for data acquisition/distribution using LabView framework, designed for long distance (Ahmad *et. al,* 2003). The advantages of control at distance of the industrial equipments are that people should not work in risk area, the possibility to control more than one equipment in the same time from different hosts, the possibility of making statistics in real time.

2. ACQUISITION SYSTEM

2.1 Hardware Resources for Data Acquisition

The system uses the multifunctional board of analogical and numerical inputs-outputs NI PCI-6014E from National Instruments (National Instruments Data Sheet).

I/O Connector – the connector is prolonged through a ribbon cable next to a connection board where, through screw terminals, electrical links with real world are made

- Analog Input – contains the analogical multiplexer MUX, the instrumental amplifier NI-PGIA with programmable amplification factor, the numerical-analogical converter ADC, the temporary memory for numerical equivalents of input signals AI FIFO and the trigger modes for input signals Analog Trigger

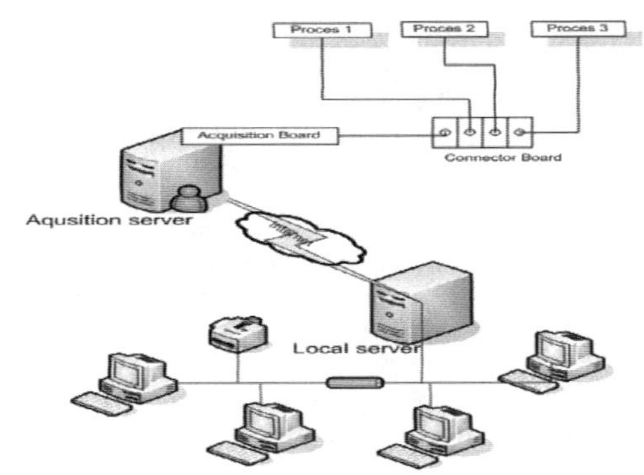

Fig. 1. The application architecture

- Analog Output – contains two numerical-analogical converters (DAC0 and DAC1) and a memory (AO FIFO) where the numerical equivalents of analogical output signals are brought
- Digital I/O – numerical inputs/outputs block – 8 pins for general usage that can be individually software configured
- Two counters (24 bits), each one having two inputs (Source and Gate), one output (Out) and two software registers
- PFI – programmable Function Interface – 10 connections with the outside of the board through which one can collect temporary signals from outside and can generate different kinds of internal board's signals
- Digital Routing – responsible with the management of data flow between PC's pipe-line and the acquisition systems. The numeric routing circuits use FIFO memory in each subsystem to ensure efficient flow of data.
- RTSI – Real Time System Integration Bus – an additional pipeline through which one can achieve synchronization for specific measuring functions,
- Bus Interface – contains the hardware components through which the acquisition board connects to the PC's pipe-line

2.2 Application Architecture

The monitoring and command configuration is installed on a local workspace; the access to the application server is done through an Internet connection (Shirer, 2001; Matousek, 2006) (Fig. 1.) For compatibility both application server and the workspaces have installed the LabView 6.1. framework. (National Instruments) The acquisition board NI-PCi-6014E allows the measuring of continuous tension voltages and TTL numeric signals, while the commands could be analogical or TTL numeric.

3. CASE STUDY

3.1 Monitoring the temperature of the heating oven

The temperature acquisition and bi-positional command of the heating oven layout schema is presented in Fig. 2.

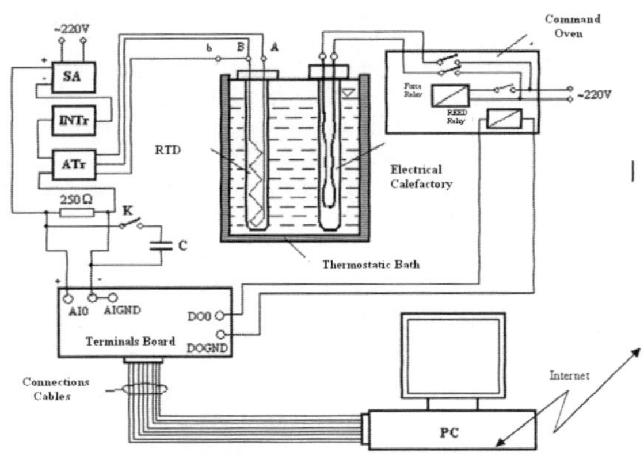

Fig. 2. The layout schema for the heating oven

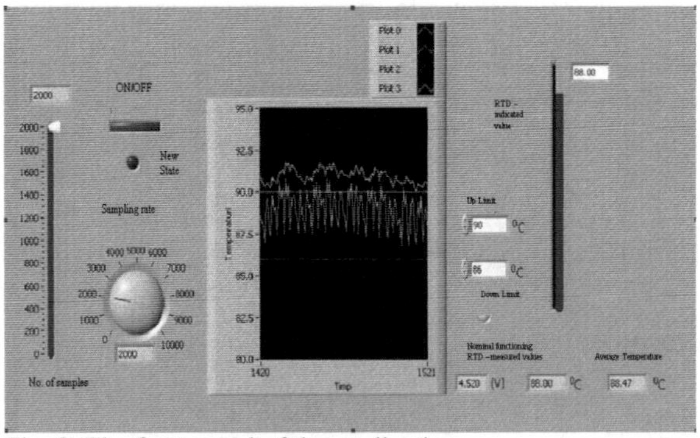

Fig. 3. The front panel of the application

A part of the application goal is to acquire the analogical signal provided by the RTD and also generating a stop command when the superior temperature is reached. When the oven temperature reaches its inferior limit, the heating process is restarted. (The goal is to obtain a positional temperature regulation). So, the two sensors have the role to maintain the oven temperature between certain limits. Because of the constructive differences between them, one will effectively command the oven starting/stopping, while the second will have a protection role in the installation – it will stay between the limits and in case of the first sensor's malfunction, it will take over the tasks of the first one. The logic of the application is based on RS two-state. In the right hand side of the panel from Fig. 3 there are provided the commands for setting the optimum working values for each thermo element. As long as the current value is out of the fixed values the bare graph indicator will be doubled by a different coloured line. The wave graph allows a permanent monitoring of the current values.

We tested the behaviour when having passive filtration (Fig. 6), numerical filtration and the case of no filtration (Fig. 3).

The application is made up of several modules:

- Data acquisition module – deals with starting, configuration and initialization of reading and also with shutdown of the connection with the data acquisition board. At this moment it should be set which channels are used for data acquisition. The number of samples and the rate of sampling. After these, the data acquired are processed by the LabView.
- Conversion module – converts the voltage into temperature so that the user can monitor and take decision based on "human readable data"
- Communication module – responsible with the remote data transmission. It was developed using NI DataSocket Server-Client communication. The server module writes data in the server, and every client can read/write data due to his rights. Also, the server application can read command data from the server and apply it to the sensors (Fig.4 and Fig. 5).
- Command module – transmits on/off signals based on the information received from the logical module
- Logical module – it has to keep the temperature between the boundaries set by the user. One characteristic of the equipment should be taken into consideration – the thermal inertia due to the resistance of the system.

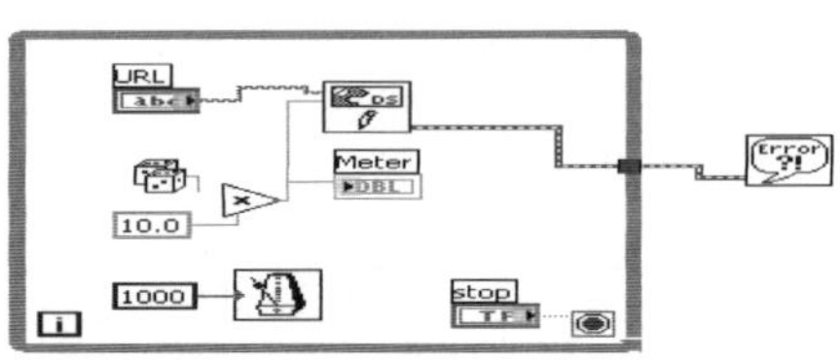

Fig. 4. DataSocket Server

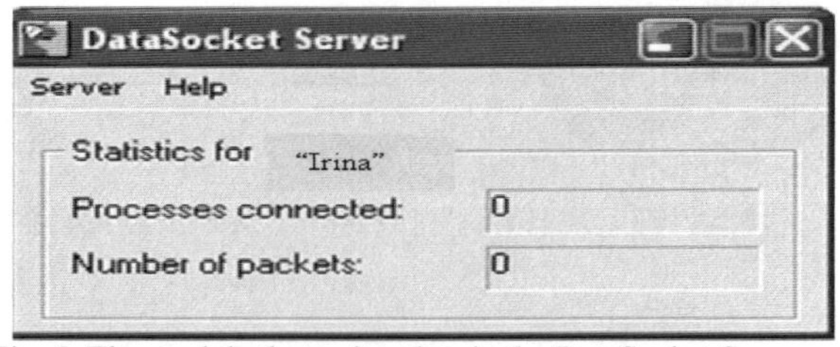

Fig. 5. The module that writes data in the DataSocket Server

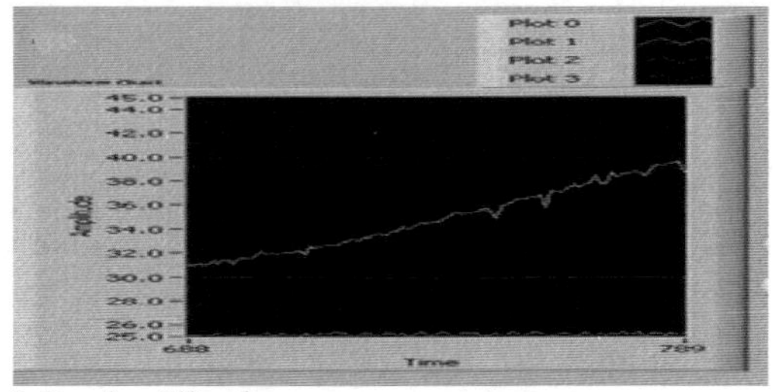

Fig. 6. Passive Filtration

- One can notice that when the command sensor has reached the inferior boundary due to the sensor's protection elements' resistance, certain inertia takes place and the temperature of the precincts drops by at most 1 Celsius degree below the inferior boundary. This problem will be solved in the future by using better sensitive elements.

4. CONCLUSIONS

The application was designed and developed to test the possibility of adding remote controlling/monitoring. It was used for a series of experiments between several laboratories.

As main part of this paper we presented the process of remote controlling of the precinct temperature, but the server can deal with more than one application in the same time - it can monitor parameters provided by 16 sensors and it can lead till four industrial equipments in the same time, using the data acquisition board NI PCI-6014E.

5. REFERENCES

Ahmad, M.R.; Khan, N.; Hooong, C.C.; Abas, N. (2003). Remote data acquisition through Internet based telemetry, *Research and Development. 2003. Scored 2003. Proceedings. Student Conference,* pp. 237-244, ISBN 0-7803-8173-4, August 2003

Matousek, M. (2006). Internet Data Acquisition, *Space Mission Challenges for Information Technology 2006.SMC-IT 2006, Second IEEE International Conference,* ISBN 0-7695-2644-6, July 2006

National Instruments Resources, LabView Graphical Programming, *Available from:* http://www.rmuti.ac.th/~nopparat/document/pdf/Labview%20Graphical%20Programming.pdf *Accessed:* 2008-05-25

National Instruments, NI PCI-6014E Data Sheet, *Available from:* http://sine.ni.com/nips/cds/view/p/lang/en/nid/11442 *Accessed:* 2008-05-25

Shirer, D. (2001). Labview 6i Adds Internet Features to Data Acquisition Environment. *Computing in Science and Engineering,* Vol. 3, No. 4, (July/August 2001), pp. 8-11, ISSN 152109615

Vlad, M.; Sgarciu, V.; Rancea, I. (2005). Acquisition and monitoring of process parameters using Internet, *CSCS15 - 15th International Conference on Control Systems and Computer Science*, Politehnica University Bucharest, Vol. 1, pp. 8-14

Annals of DAAAM for 2008 & Proceedings of the 19th International DAAAM Symposium, ISSN 1726-9679
Editor B. Katalinic, Published by DAAAM International, Vienna, Austria 2008
Make Harmony Between Technology and Nature, and Your Mind will Fly Free as a Bird

REMOTE MONITORING OF PARAMETERS FOR A PRESSURE TRANSDUCER THROUGH HART PROTOCOL AND LABVIEW ENVIRONMENT

RANCEA, I[rina]; SGARCIU, V[alentin] & STAMATESCU, G[rigore]

Abstract: *In this paper we used the concepts of remote control for automated monitoring of parameters for a pressure transducer with sensitive piezorezistive element. The application can monitor the parameters of different industrial processes using the same server. Our case study is using a transducer provided by Siemens – SITRANS P, MS series, 7MF4013 equipped with a HART modem and a LabView environment*

Key words: *pressure transducer, HART protocol, remote monitoring, LabView*

1. INTRODUCTION

Technological process supervision involves acquisition and monitoring of a large number or parameters. Using the Internet connection one can store those parameters in high performance servers and data bases. The paper is focused on a solution for monitoring parameters for a pressure transducer with sensitive piezorezistive element through the HART protocol, designed for long distances.

2. SYSTEM ARCHITECTURE

2.1 Hardware Resources

The system uses the relative pressure transducer SITANS P, MS series, 7MF4013 manufactured by Siemens. Its schema is presented in Fig.1. The transducer measures the relative pressure (the difference between process pressure and atmospheric pressure); this pressure is then converted in unified electric signal, scale 4 ÷ 20 mA cc. The transformations flow is : pressure difference → force → modification of electric resistivity → electric current. The pressure p_e is passed through process connector to the measure cell. The pressure is sent to a silicone piezorezistive sensor through the diaphragm and the filling liquid that takes over the distorsions of the measure diaphragm. As a result of pressure variations the resistance of the four sensistive pizorezistive elements connected in a bridge is changed. This generates an output tension from the bridge balanced with the input pressure. The tension is amplified by the measure amplifier and converted in digital signal that is evaluated by the microcontroller and then converted through the Digital-Analog convertor. The analog output has a variation between 4mA and 20 mA for a pressure variation between p_{min} and p_{max}. The transducer can be set to a conditional point of working through the two buttons or through the HART modem.

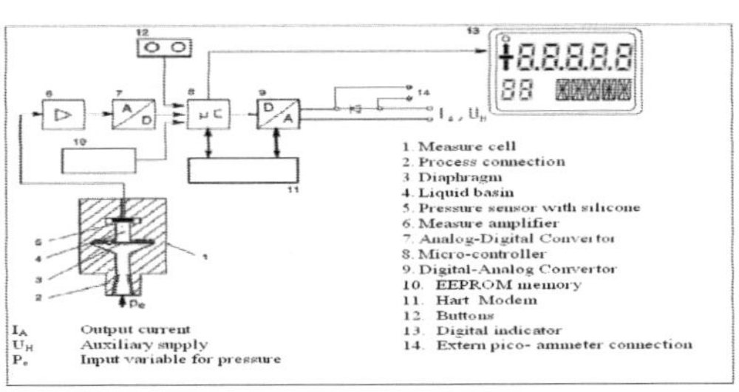

Fig. 1. The principle schema of the SITANS P

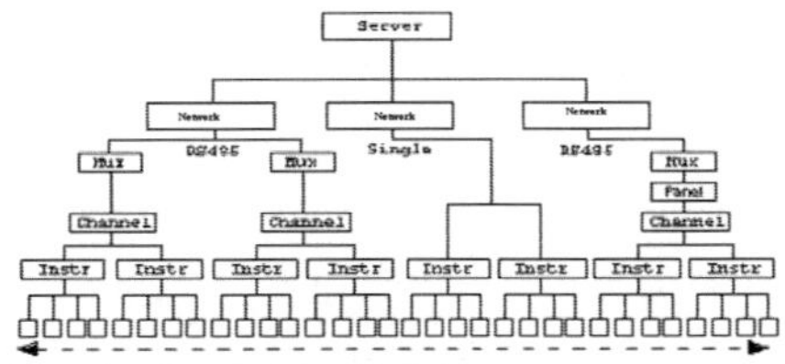

Fig. 2. Hierarchy for HART OPC Server

2.2 HART Communication Protocol

The HART communication is realized between two devices: a field device and a control or monitoring system. HART provides two simultaneous communication channels: analog signal 4-20mA and digital signal. The 4-20mA signal communicates the primary measured value using a current loop of 4-20mA – it is one of the fastest and reliable industrial standards. (Merget, 2003)

HART is a master-slave communication protocol – every slave communication (field device) is initiated a communication master device. On every HART loop two masters can be connected. The primary master is, in general, a distributed control system (DCS), programmable logic controller (PLC) or another PC. The slave devices can be transducers, executive elements and control equipments. (Lu *et. al,* 2007) (Fig. 2) The HART communication protocol is based on the Bell 202 communication standard and operates using the FSK principle. (Pereira *et. al,* 2003) The devices that can communicate through the HART protocol can work in two network configurations: point-to-point and multipoint.

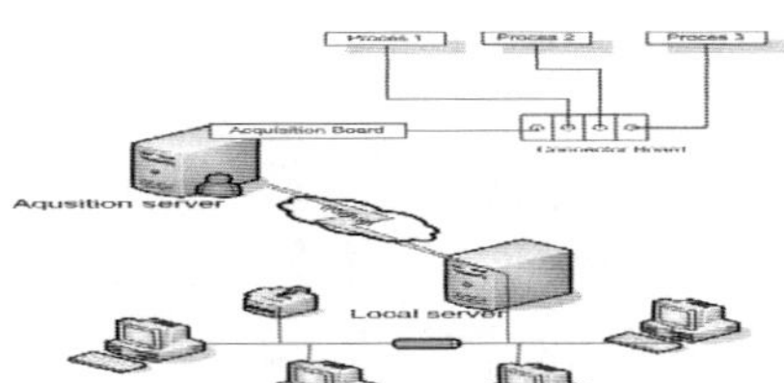

Fig. 3. Application Architecture

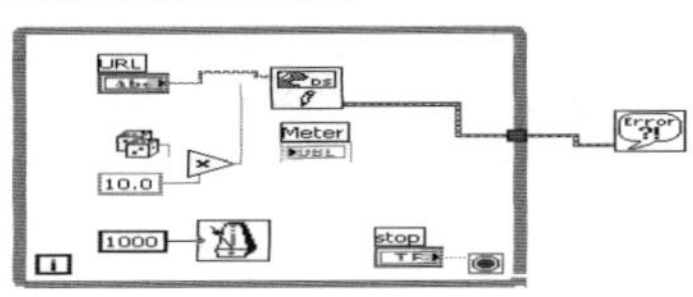

Fig. 4. DataSocket Server

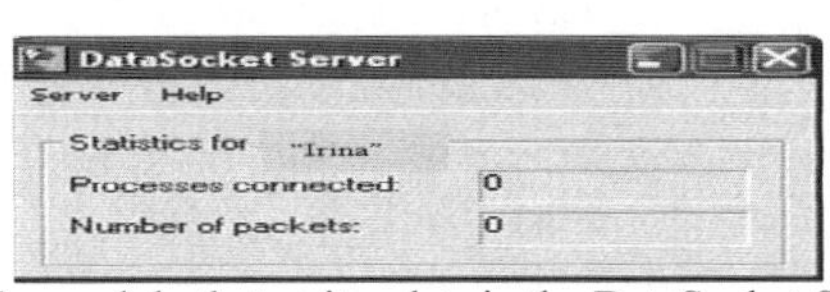

Fig. 5. The module that writes data in the DataSocket Server

HART Server is a specialized software that allows building hierarchically extended configurations and facilitates the data access from HART devices and associated processes in a format that can be used in different management applications. (Fig. 4) (Wang *et. al,* 2007)

2.2 Application Architecture

The monitoring system is installed and configured on a local workspace; the access to the application server is done through an Internet connection, as in Fig. 5., Fig. 6. and Fig. 10. (Vlad *et. al,* 2005)

The application has a communication module that is responsible with the remote data transmission. It was developed using NI DataSocket Server-Client communication. The server module writes data in the server, and every client can read/write data due to his rights. Also, the server application can read command data from the server and apply it to the sensors (Fig.4 and Fig. 5).

3. CASE STUDY

The assembly contains the following modules (Fig. 9):
- Compressor - power supply with compressed air able to maintain o pressure of 8 bars. When the pressure is under 6 bars it starts automatically.
- RP – pressure reducer
- ME - Manometer – pressure analog indicator of high precision, HEISE manufactured, class 0.1, domain 0-4 bars
- Basin – container that simulates a tehnological installation working with pressure till 10 bars
- TP - SITRANS P DM – pressure transducer, configurable through HART protocol
- RE – evacuation faucet and simulate disturbances
- SA – power supply of 24 V cc
- SV – servo-valve 8288200.9650.02400
- DI-158U – acquisition/distribution of numerical and analog data module connected to PC through USB (DATAQ manufactured)

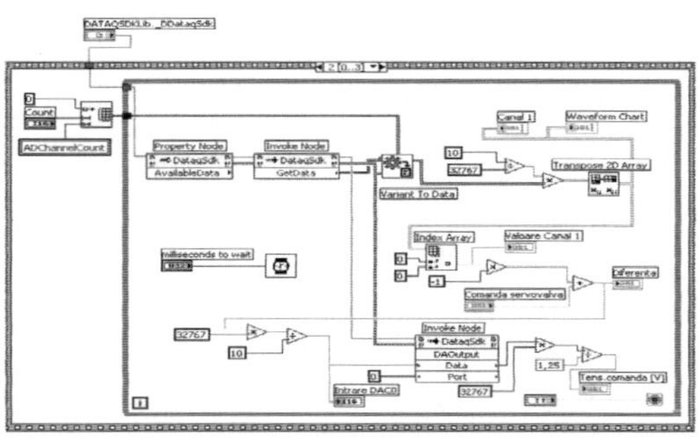

Fig. 6. The block diagram of the application

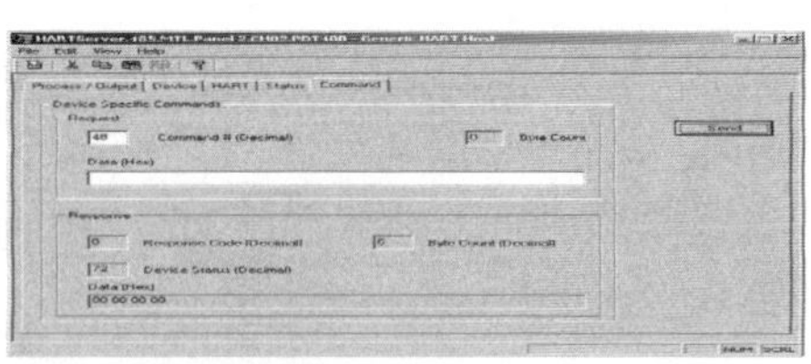

Fig. 7 HART Server application

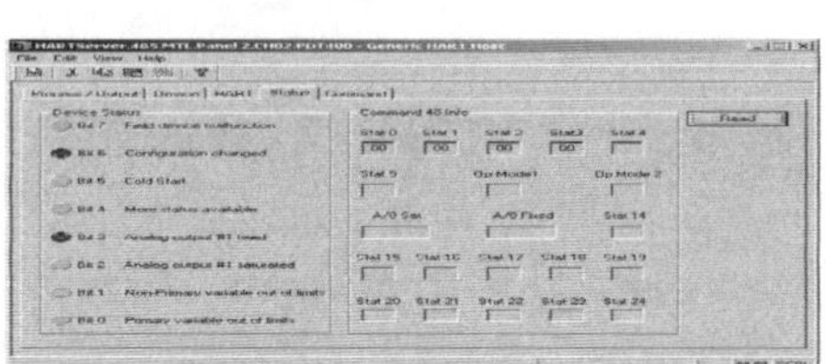

Fig. 8. HART Server Status window

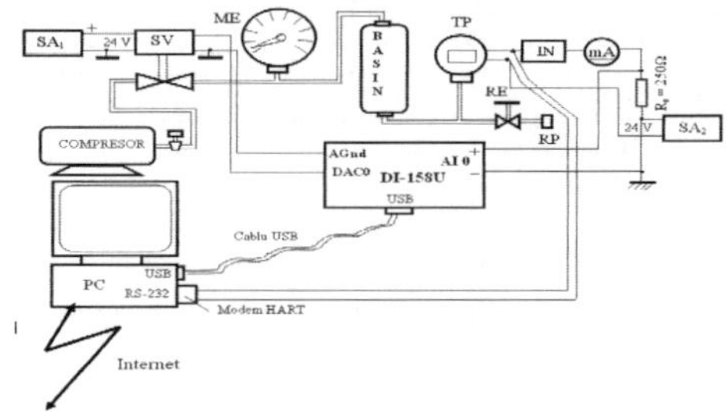

Fig. 9. Assembly Schema

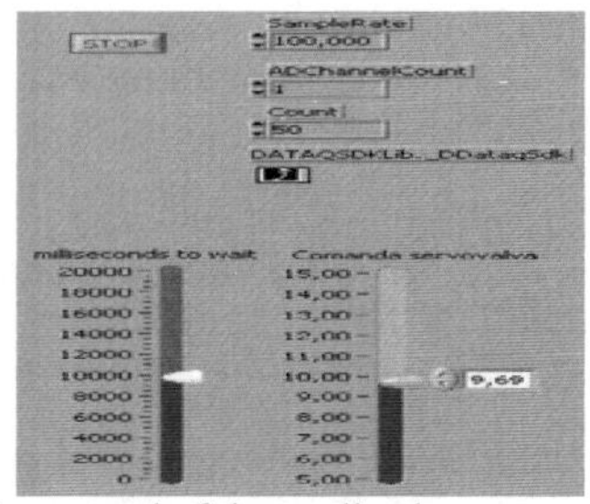

Fig. 10. The front panel of the application

The operating mode of the transducer through HART protocol can be configured wit the HART Server application (Fig. 7 and Fig. 8). OPC (OLE for Process Control) has a communication interface between various sources of data and applications that provides the information from those sources. The main target of the OPC Server is to transfer data through unshared routes of communication to physical devices. HART Server can obtain information from the configured HART devices such as their state, their configuration parameters and their tolerance.

4. CONCLUSIONS

The application was designed and developed to test the possibility of adding remote monitoring. It was used for a series of experiments between several laboratories. As main part of this paper we presented the process of remote monitoring of the pressure transducer with sensitive piezorezistive element, but the server can deal with more than one application in the same time. In the future we will implement it in an industrial factory.

5. REFERENCES

Lu X.; Guo W. (2007). The Application of HART protocol in the Convertor of Magnetic Flowmeter, *Proceedings of the World Academy of Science, Engineering and Tehnology*, Vol. 23, august 2007, ISSN 1307-6884

Merget O. (2003). The Electronic Integration of Field Devices, *IEEE Computing & Control Engineering Journal*, Vol. 14, october-november 2003, pp. 22-23

Pereira J.; Postolache O.; Girao P.S. (2003). HART Protocol Analyser based in LabView, *Intelligent Data Acquisition and Advanced Computing Systems: Technology and Applications, proceedings of the second IEEE International Workshop,* september 2003, pp. 174-176

Vlad, M.; Sgarciu, V.; Rancea, I. (2005). Acquisition and monitoring of process parameters using Internet, *CSCS15 - 15th International Conference on Control Systems and Computer Science*, Politehnica University Bucharest, Vol. 1, pp. 8-14

Wang H.; Shi L.; Yank X.; Xu S. (2007). Research and Development of On-line Monitoring and Management System to Hart Field Devices, *IEEE Electronic Measurement and Instrument, ICEMI, 8th International Conference,* pp. 3-410 – 3-402

Annals of DAAAM for 2008 & Proceedings of the 19th International DAAAM Symposium, ISSN 1726-9679
Editor B. Katalinic, Published by DAAAM International, Vienna, Austria 2008
Make Harmony Between Technology and Nature, and Your Mind will Fly Free as a Bird

VIRTUAL REALITY TOOL FOR ORTHOGLIDE PARALLEL ROBOT

RAT, N[adia] R[amona]; GOGU, G[rigore]; STAN, S[ergiu] D[an] & BALAN, R[adu]

Abstract: *The paper presents a 3DOF Virtual Reality (VR) application for the kinematic simulation of Orthoglide parallel robot. The interface was developed in MATLAB/Simulink and SimMechanics environment, based on the CAD model realized in SolidWorks environment.*
Key words: *parallel robots; orthoglide; kinematics simulation; virtual reality.*

1. INTRODUCTION

The parallel robots are mechanisms with kinematics closed chains, composed by an end-effector (the mobile platform) with n degrees of freedom, connected to the fixe base by two or more kinematical chains called limbs or legs. A simple or a complex kinematic chain can be associated with each limb (Gogu, 2008).

The Virtual Reality (VR) immerses the user in a three-dimensional (3D) environment that can be actively interacted and explored (Halliday, 1994).

Virtual reality environment tool is used by many researchers in design, development and manufacturing of the robotic industry (Stan et al., 2008).

Using the virtual reality simulation with a virtual robot, we can observe a three dimensional design and the real-time behavior of the robot; that fact is relatively new and allows testing the robot before accomplishment a physical implementation. In this way, we can save important resources (money and time) and we can adjust the various problems without affecting the physical robot.

This paper presents the necessary steps to develop the virtual environment for kinematic simulation, starting from the SolidWorks model of Orthoglide parallel robot.

Orthoglide is a 3DOF parallel robot (Wenger and Chablat 2000) with three coupled motions of type 3PRPaR (3-Prismatic-Revolute-Parallelogram-Revolute), composed by a mobile platform connected to the fixed frame by three kinematic chains presented in figure 1 (Gogu, 2004).

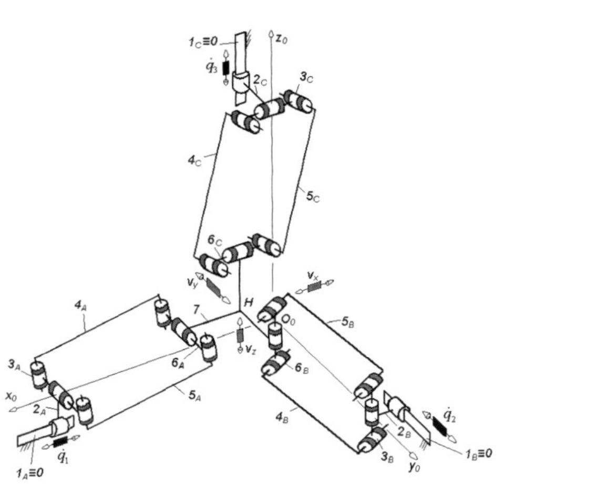

Fig. 1. Kinematic structure of Orthoglide parallel robot (Gogu, 2004)

2. ALGORITHM FOR DEVELOPMING A VR APLICATION BASED ON SOLIDWORKS CAD MODEL

This approach has a great advantage to making possible to observe the behavior of the robot in virtual medium using the interface SolidWorks – SimMechanics – MATLAB/Simulink – Virtual Reality.

The first step is to make the CAD model of Orhoglide parallel robot in SolidWorks environment.

Once the CAD model is done, the next step is to export the model in SimMechanics environment for transposition in MATLAB/Simulink environment.

For a better comprehension of the needed steps for obtaining the VR model an algorithm is presented in figure 2.

The inertia proprieties and the coordinates of the joints were determined automatically when the CAD model was imported from SolidWorks in Matlab/Simulink environment. Inputs for the model can be: the position, velocity or acceleration of actuated joints or the generalized forces.

In this paper, the input was chosen to be the speed of all three actuator joints of the Orthoglide parallel robot.

In Matlab Simulink environment, the three parallel links connected at the mobile platform can be seen in figure 3.

Figure 3a presents the SolidWorks model of the Orthoglide parallel robot. SimMechanics model of Orthoglide is presented in figure 3b.

The dynamic model of Orthoglide parallel robot is presented in figure 3c. We can see the three links (noted with arm 1 to 3) connecting the base (noted with batiu) and the mobile platform (noted platforma). A parallelogram loop can be observed in each limb (figure 3d).

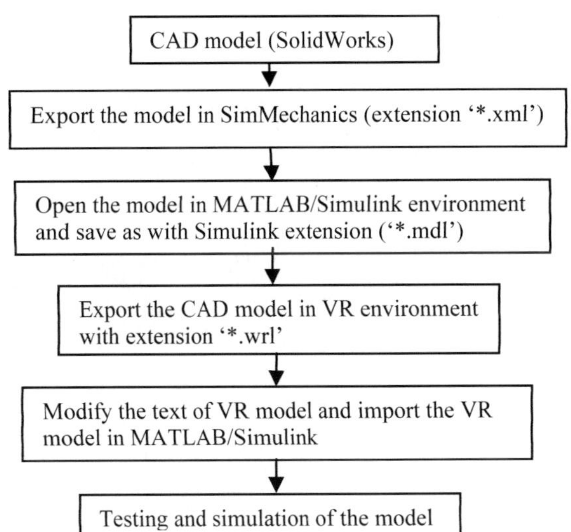

Fig. 2. The algorithm for the VR model obtained from the SolidWorks CAD model of the robot

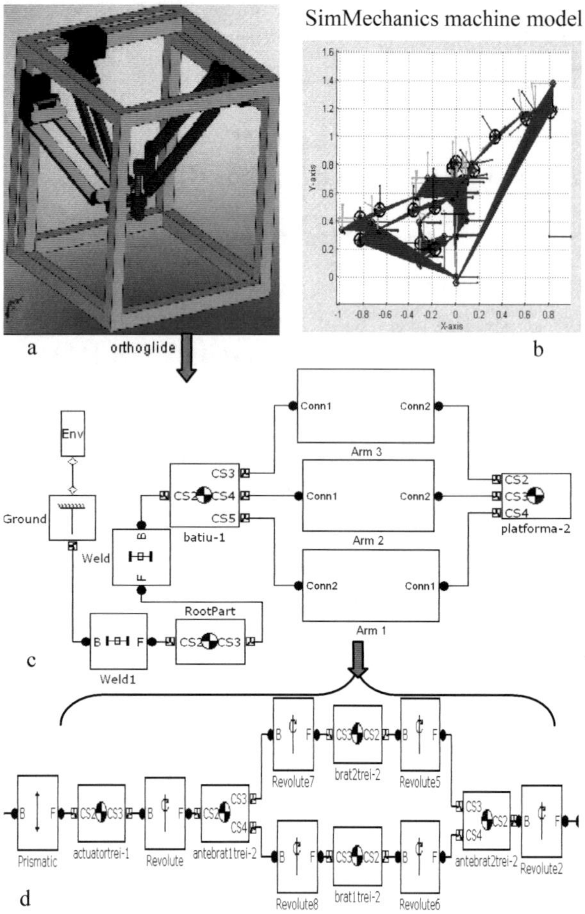

Fig. 3 Dynamic model of the Orthoglide parallel robot in MATLAB/Simulink

In this step we can verify the validity of the SolidWorks model. This model is valid if a direct correspondence exists between the links and the kinematic joints in figures 1 and 3d (ex: a revolute joint from figure 1 need to be also a revolute joint in figure3d). If the SimMechanics machine model is set up correctly, then we can go to the next step and create the virtual reality environment.

3. VIRTUAL REALITY TOOL

Once the dynamic model is set up in MATLAB/Simulink environment, the next step is to make the Virtual Reality environment imported from SolidWorks model. To do this, we need to export the SolidWorks model like VRML files and with few modifications the Virtual Reality environment for Orthoglide parallel robot can be obtained (figure 4).

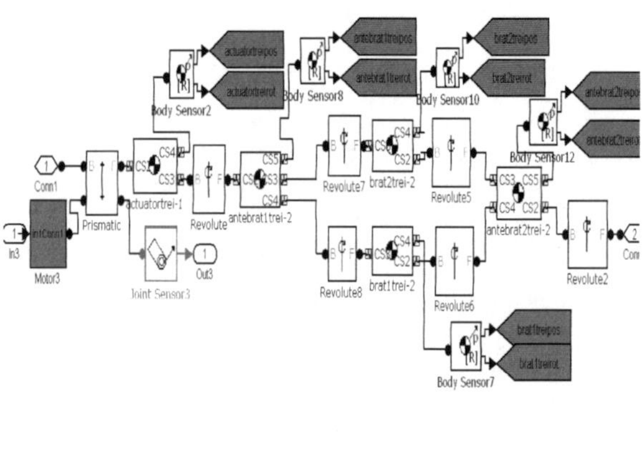

Fig. 4 Dynamic model of the Orthoglide parallel robot (completed for VR).

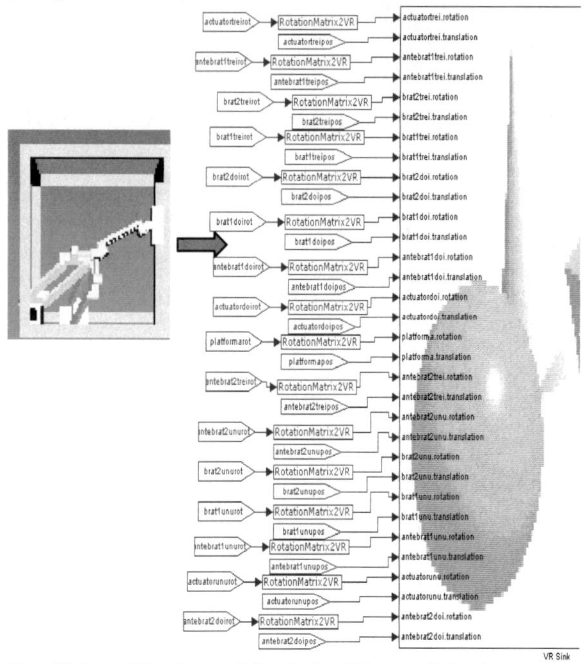

Fig. 5 Virtual Reality tool for Orthoglide parallel robot

In figure 4, we can see how the dynamic model from figure 3 was modified to obtain the connection with the Virtual Reality environment. A special attention must be paid to use the same notations in the dynamic and Virtual Reality models (figure 4 and figure 5).

4. CONCLUSIONS

A Virtual Reality (VR) interface for the control of Orthoglide 3PRPaR parallel robot has been presented in this paper. The interface for high level control presented in this paper was verified and tested for Orthoglide parallel robot and results were presented in MATLAB/Simulink and SimMechanics environment. If we need to feel the behaviour of the robot, a Joystick can be easily integrated in the model. A control model can also be added to get a complete view of robot behavior.

5. REFERENCES

Halliday, S., Green, M., (1994) *A Geometric Modeling and Animation System for Virtual Reality*, Virtual Reality Software and Technology (VRST 94), 71-84, Singapore, August 1994.

Gogu, G. (2008) *Structural Synthesis of Parallel Robots, Part 1: Methodology*, Springer.

Gogu, G., (2004). *Structural synthesis of fully-isotropic translational parallel robots via theory of linear transformations*, European Journal of Mechanics / A – Solids, vol. 23, pp. 1021-1039.

Wenger, P.; Chablat, D. (2000). *Kinematic analysis of a new parallel machine tool: the Orthoglide*. In: Lenarčič J, Stanišić ML (eds) Advances in Robot Kinematics. Kluwer Academic Publishers, pp 305-314.

Stan, S.-D.; Manic, M.; Bălan, R.; Lăpuşan, C. (2008) *IG3pR-VRI, a Novel Virtual Reality Robot Interface for Isoglide3 Parallel Robot,* 13th IEEE International Conference on Emerging Technologies and Factory Automation, Hamburg, Germany.

Annals of DAAAM for 2008 & Proceedings of the 19th International DAAAM Symposium, ISSN 1726-9679
Editor B. Katalinic, Published by DAAAM International, Vienna, Austria 2008
Make Harmony Between Technology and Nature, and Your Mind will Fly Free as a Bird

SERVO-HYDRAULIC LINEAR AXES BEHAVIOUR MODELING AND SIMULATION

RATIU, C[laudiu] I[oan]

Abstract: *The tendencies obvious today in robot construction relate to flexible automation and mechatronics, i.e. increasing their intelligence level to compensate for nonlinearities and to adapt to distortions. The applied research programme developed by the author was directed towards optimizing the static and dynamic performance of the linear servo-hydraulic axes with the purpose of investigating nonlinear phenomena occurring while in operation and towards producing a package of numerical simulation software that provide useful information to their design to make them operate at the level of performance that is today required. A test stand was also produced and an experimental research programme ran to correct and the mathematical model and numerical simulation software of an adaptive controller*
Key words: *servo-hydraulic axis, nonlinear system, oil compressibility, adaptive controller, numerical simulation,*

1. INTRODUCTION

An automated control hydraulic drive provided with a linear hydraulic motor has the composition presented in Figure 1,
That is: (a) the linear hydraulic motor, (b) the load itself, (c) elements to measure the controlled physical magnitudes, (d) an acquisition processing and control element, (e) the hydraulic energy control element. The linear hydraulic motor, (Fig.2) in whose chambers the oil volumes V_A and V_B are closed is a damped oscillating system, with own frequency ω_P, defined by relationship (1).

$$\omega_{P(x)} = \sqrt{\frac{A \cdot E'_{ULEI}}{M_{tot}} \cdot \left(\frac{1}{x_A + x} + \frac{\alpha}{x_B - x} \right)} \qquad (1)$$

The hydraulic stiffness of the oil in volumes A and B in the hydraulic motor modifies dependent on piston position and the own frequency of the hydraulic motor also changes.
In Figure 3, the own frequency variation curves are represented as effective values for various loads determined in the hydraulic cylinder used for the experiment (Ratiu, 1998).
Due to this particularity of the cylinders in the structure of the linear servo-hydraulic axes, the electronic controller, that is constants P, I and D are optimally tuned in only one position where the positioning accuracy and stability are optimal. For the rest of positions, the properties mentioned above are damaged.

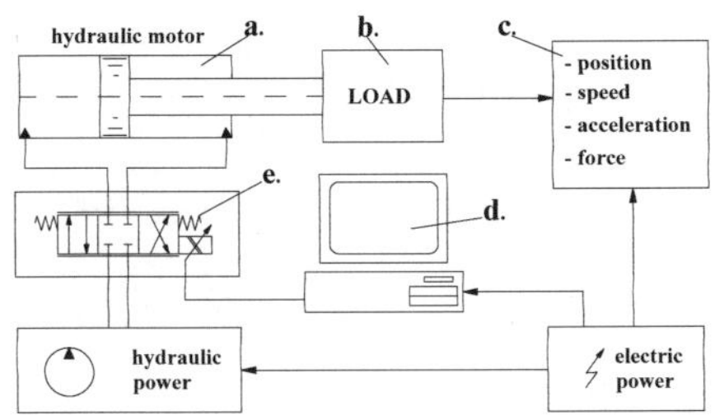

Fig.1. the block diagram of a servo-hydraulic application

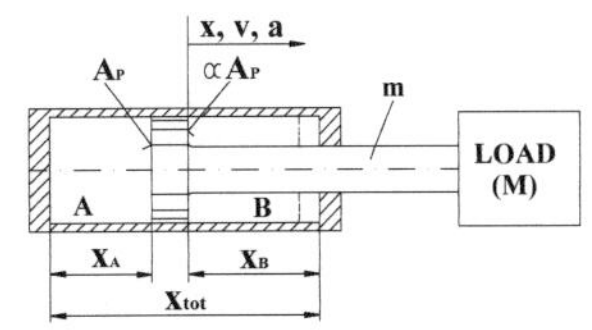

Fig. 2. Linear hydraulic motor

2. PROBLEM FORMULATION

The elaboration of the nonlinear model aimed at defining the static and dynamic transfer property for the system made up of the actuator, valve, linear hydraulic motor and load. The elaboration of the nonlinear model requires the determination of the motion differential equations for the valve and engine as well as of the differential equation for the pressure formed in the motor chambers (Ratiu & Nascutiu, 2003), described by the differential equation system (2):

$$\begin{aligned}
& m_v \cdot \ddot{y}_v = -c_{EP} \cdot I_{EP} + c_a \cdot (y_{0arc} - y_V) - c_f \cdot \dot{y}_v + 2\alpha_D [S_2 \cdot (p_0 - p_A) - S_1 \cdot p_A] \cdot \cos(\varphi) \\
& \dot{p}_A = \frac{E'_{ULEI}}{A_P \cdot x} \cdot \left[Q_A + c_{LIP} \cdot (p_0 - p_A) + c_{LIV} \cdot \dot{x} - A_P \cdot \dot{x} \right] \\
& M_{tot} \cdot \ddot{x} = A_P \cdot p_A - \alpha \cdot A_P \cdot p_0 - c_{fM} \cdot \dot{x} - F_S \\
& Q_A = \alpha_D \cdot \sqrt{\frac{2}{\rho}} \cdot \left(S_2 \cdot \sqrt{p_0 - p_A} - S_1 \cdot \sqrt{p_A} \right) \qquad (2) \\
& S_1 = n \cdot (a_1 + a_2 \cdot y_V + a_3 \cdot y_V^2 - a_4 \cdot y_V^3) \ , \ -y_{OV} \le y_V \le y_{OV} \\
& S_2 = n \cdot (a_1 - a_2 \cdot y_V + a_3 \cdot y_V^2 + a_4 \cdot y_V^3) \ , \ -y_{OV} \le y_V \le y_{OV} \\
& S_1 = S_2 = 0, y_V < -y_{OV} \ \textit{and} \ y_V > y_{OV}
\end{aligned}$$

3. PROBLEM SOLUTION

In order to solve the system, it should be converted into an equivalent first order differential equations system.

3.1. Matlab transposition of the differential equation system

The new system contains equations where unknowns are found as state variables (Barkana, 2005). They are physical magnitudes describing the system state at a certain moment where: v_1 = the valve speed, v_2 = the valve position, $v_3 = p_A$ (pressure in motor chamber A), v_4 = motor speed, $v_5 = x$ (motor position).

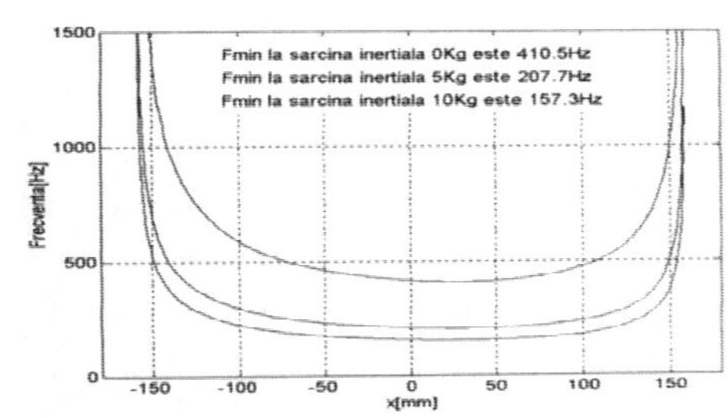

Fig.3. The hydraulic motor frequency variation diagram

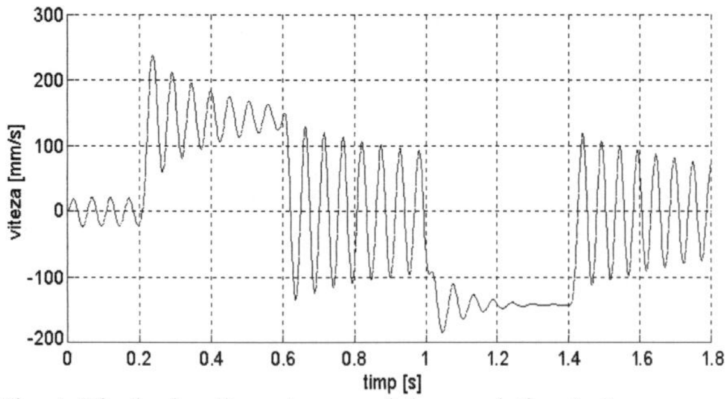

Fig. 4. The hydraulic motor speed time variation (v_4)

$$
\begin{aligned}
&\dot{v}_1 = \frac{c_{EP}}{m_V} \cdot I_{EP} + \frac{c_a}{m_V} \cdot (y_{0arc} - v_2) - \frac{c_f}{m_V} \cdot v_1 + \frac{2\alpha_D \cdot \cos\varphi}{m_V} \cdot [S_2 \cdot (p_0 - v_3) - S_1 \cdot v_3] \\
&\dot{v}_2 = v_1 \\
&\dot{v}_3 = \frac{E_U'}{A_P \cdot v_5} \cdot [Q_A + c_{LIP} \cdot (p_0 - v_3) + c_{LIV} \cdot v_4 - A_P \cdot v_4] \\
&\dot{v}_4 = \frac{A_P}{M_{tot}} \cdot (v_3 - \alpha \cdot p_0) - \frac{c_{fM}}{M_{tot}} \cdot v_4 - \frac{F_S}{M_{tot}} \\
&\dot{v}_5 = v_4 \\
&Q_A = \alpha_D \cdot \sqrt{\frac{2}{\rho}} \cdot \left(S_2 \cdot \sqrt{p_0 - v_3} - S_1 \cdot \sqrt{v_3}\right) \qquad (3) \\
&S_1 = n \cdot \left(a_1 + a_2 \cdot v_2 + a_3 \cdot v_2^2 - a_4 \cdot v_2^3\right) \;,\; -y_{OV} \le v_2 \le y_{OV} \\
&S_2 = n \cdot \left(a_1 - a_2 \cdot v_2 + a_3 \cdot v_2^2 + a_4 \cdot v_2^3\right) \;,\; -y_{OV} \le v_2 \le y_{OV} \\
&S_1 = S_2 = 0 \;,\; v_2 < -y_{OV} \;\; \text{si} \;\; v_2 > y_{OV}
\end{aligned}
$$

3.2. Numerical solutions for the differential equations system

There was been analysed the variation of the both five state variables.In Figure 4 given, for example, the state variable variations v_4, for the system starting behavior corresponding to the minimum current at the electromagnet, I_{0EP}=0,6 A, and for a current step I_{EP}=0,3 A. It is evident that the damping is low both in the start and one control step after the system stabilizes. The speed of the motor also has a proportional behavior of second order delay, and a low damping.

3.3 The adjustment of the controller for the speed automated control

The controller for the speed automated control is tuned by establishing the weight of its characteristic effects, so that the entire system reaches the required speed static and dynamic performance (Tao, 2003). For this purpose, it is aimed to minimize the control stationary deviation, the response time and overadjustment of indicial response. In Table 1, the adjustment values for the adaptive controller parameters are given in the case of the speed automated control.

The calculation programme for the controller constants is achieved in the MATLAB – SIMULINK environment.

Numerical simulations in two hydraulic motor piston positions were performed in this circumstances.

CURENT	0.15 A			0.30 A			0.50 A		
POSITION	P	T_I	T_D	P	T_I	T_D	P	T_I	T_D
0-32 mm	0.62	0.009	0.151	0.62	0.009	0.167	0.81	0.009	0.217
96-128 mm	0.52	0.009	0.149	0.54	0.009	0.175	0.68	0.009	0.220
192-224 mm	0.30	0.009	0.144	0.35	0.009	0.169	0.53	0.009	0.221
288-320 mm	0.13	0.009	0.165	0.18	0.009	0.185	0.35	0.009	0.232

Tab. 1. The adaptive controller parameters adjustment values

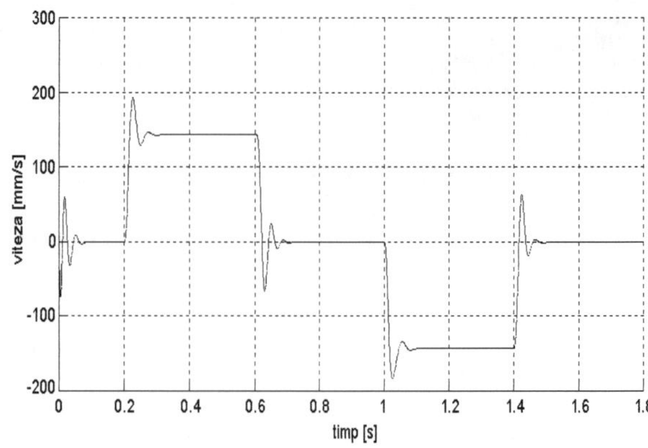

Fig.5. The indicial response in speed for the piston position at 50 mm

Thus, for an optimal adjustment in position *x=50 mm*, with speed control succesive stepf of: *v=±150 mm/s,* with PID controller constants : $P = 0,6$, $T_I = 0,0094$ *s*, $T_d = 0,015$ *s*, the indicial response in speed is given in Figure 5.

The analysis of the diagrams produced showed that the parameters of the PID controller do not satisfy the entire range of motor operation. This is due to the variable hydraulic capacity of the chambers of the hydraulic motor.

4. CONCLUSION

Another experimental program that confirmed the mathematical model validity in the case of the servo-hydraulic axis was performed, too. The tests aimed at finding out the behavior of the axis to run under automated control conditions of speed and position. The adaptive control systems that can solve the issues that occurred in the simulation programs contain complex adjustment elements, circuits identifying operational points and the adaptive logic of the controller constants.

The solution for this kind of applications is that of including an electronic controller where constants P, I and D will be permanently tuned dependent on the motor position. The conventional adaptive systems are extremely complex and the processing time is relatively large, so that it can affect the quality of real time process adjustment. A modern concept that has recently been used for adaptation purposes is based on Fuzzy logic use for hydraulic control circuits (Jang et al., 1997). This solution is adopted by author and the experimental research is on progress.

5. REFERENCES

Barkana, I., *Output Feedback Stability and Passivity in Nonstationary and Nonlinear Systems,* International Conference on Control and Automation, June 27-29, (2005), Budapest, Hungary

Jang, J-S.R., C-T.E. Sun and E. Mizutani, *Neuro-fuzzy and soft computing: a computational approach to learning and machine intelligence,* Prentice Hall, (1997).

Ratiu, C., *Linear electro-hydraulic axis with NC adaptive control*, PhD Thesis, TUCN, Cluj-Napoca, (1998).

Ratiu, C., Nascutiu, L., *Linear hydraulic axes with multiple reactions*, in Jurnal of: Machine tools and Hydrodinamics Conference, Timisoara, (2003), ISSN-1224-6077, 201-208

Tao, G., *Adaptive Control Design and Analysis*, John Wiley & Sons, 2003 (ISBN 0-471-27452-6)

Annals of DAAAM for 2008 & Proceedings of the 19th International DAAAM Symposium, ISSN 1726-9679
Editor B. Katalinic, Published by DAAAM International, Vienna, Austria 2008
Make Harmony Between Technology and Nature, and Your Mind will Fly Free as a Bird

MEASUREMENT LINE FOR STATIC TORQUE

RISOVIC, S[tjepan]; ZEZELJ, D[ragan] & PANIC, N[enad]

Abstract: *This work is driven by the idea to create a design solution of semi-automatic measurement line that would provide more accurate results and increase the speed of the recalibrating process of torque tools. Tools considered here are operated by hand, therefore subjected to the performance (skill, strength and knowledge) of the operator. Since the pulling speed and the force are difficult to maintain or repeat results differ from expected values. The analysis filtered out the solution with gearbox and trapezoidal spindle which is taken as the base for prototype line to be produced. Testing has shown enhanced repeatability of the measurement, ease of operation and suitability for implementing automation features and thus fulfilling of initial requests.*

Key words: *torque, tools, measurement*

1. INTRODUCTION

Torque tools are used in everyday life by machinists, auto mechanics, steel construction workers, dentists, etc. They differ in the working principles and load ranges and therefore, in design, size, accuracy and so on. They are briefly divided into two groups: indicating and setting torque tools (ISO 6789, 2003). While indicating tool is intended to indicate torque exerted by the tool at the output drive, the purpose of setting torque tool is to exert the prescribed value of the torque followed by appropriate signal. General use of the setting tool is to set the desired amount of axial force into a screw or bolt joint. The connection between the introduced torque and the achieved force is complex though often reduced to only a few parameters on account of simplicity of calculation (Decker, 1975). However, the basic assumption that must be met to even approximately achieve the desired force is that the torque tool exerts the prescribed value of torque within the limits of acceptable tolerance (ISO 6789, 2003). In order to provide such performance, torque tools must be recalibrated periodically. The issue of this work is the recalibration procedure of the setting torque tool that should be conducted by a certified calibration laboratory and the appropriate equipment that should be used.

2. EXISTING MEASUREMENT LINE

The measurement line (Fig. 2) consists of the tool driving square adapter (2) (ISO 1174-1, 1996) mounted on the top flange of the measurement shaft (1). Four strain gages forming full bridge are attached to the outside diameter in the middle of the shaft and connected to the digital measurement acquisition system (3). The last item in the line is a PC with software compatible to the acquisition system (4).

2.1 Measurement procedure

The measurement procedure (Schicker & Wegener, 2002; ISO 6789, 2003) is conducted by hand pulling the torque tool mounted on a driving square attached to the top of the measuring shaft (Fig. 1). Due to angular deformation of the shaft electrical resistance of gages is changed and picked up by the acquisition system in the form of electrical potential change. Those changes are recorded continuously by PC software as an array of discrete points (Figures 2a, b and c).

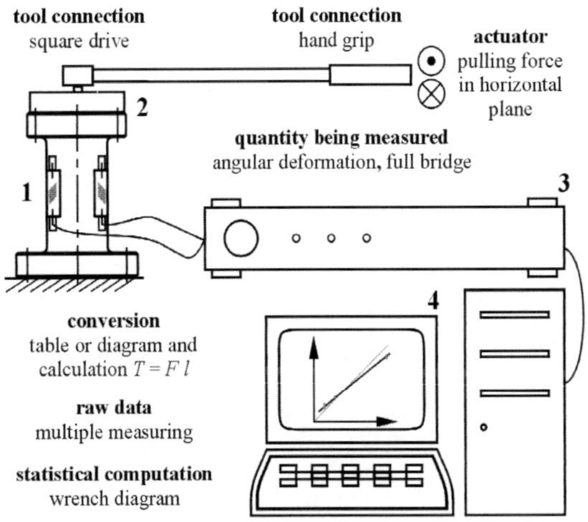

Fig. 1 Existing measurement line

Fig. 2a shows the diagram obtained by usual operation, while 2c is achieved by smooth pulling with fine force increment. Fig. 2b is the most important because of the lack of indication of tool triggering and also another peak that occurs when the tool mechanism hits the wrench casing allowing overload. Since the curves are not smooth and in cases 2b and 2c neither are they characteristic, they are not suitable for automated software processing.

It should be pointed out that the operating torque was set at 70 Nm which should be considered as a small value, thus easy to handle. Wrenches designed for higher amounts of torque are usually robust, equipped with appropriate grip extension in order to provide less force on a larger distance for the operator to have better control over the force and the speed. However, the extension requires more space for manipulation forcing the operator to maintain control over a longer path leading back to measurement inaccuracy.

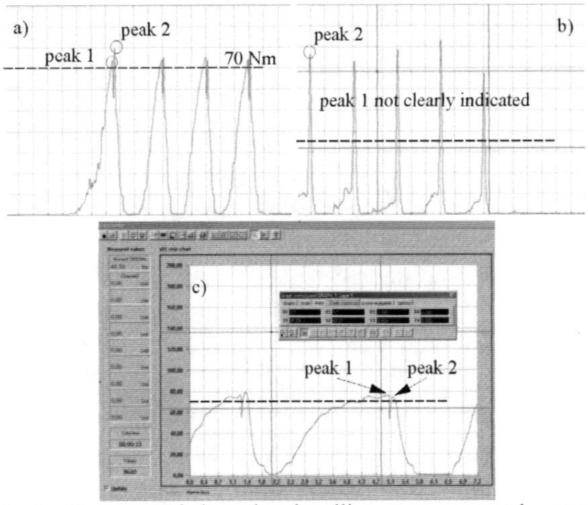

Fig. 2 Data recorded on hand pulling torque wrench set to 70 Nm: a) normally, b) aggressively, c) smoothly

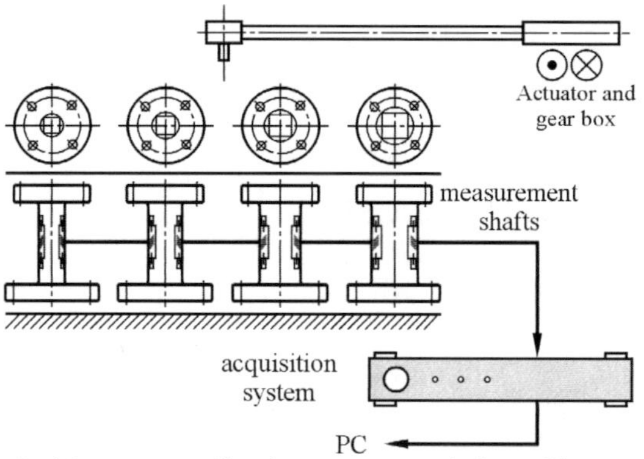

Fig. 3 Measurement line improvement solution with actuator and transmission, additional shafts and software procedure (not shown)

3. IMPROVEMENT REQUESTS

One can conclude that the operating performance has significant influence on the measuring accuracy as well as on the measuring repetition. In order to provide both, a new transmission device is considered, preferably electrically controlled with fine gradient adjustment. Instead of hand pulling, the wrench would be driven by a new actuator while the measurement shaft would stay still (Fig. 3) (Opalić et al., 2004). The equipment must be suitable for operation in both directions to meet various tools capabilities. Additionally, since the measurement shaft is optimally designed for particular torque range, several shafts should be produced to cover at least the most usual ranges. The procedure should be kept simple without the need of changing measurement shafts, thus saving the time in preparation of measurement. Finally, because of better control of the input, smooth output diagrams could be created, suitable for automated processing. Therefore, a new procedure for peak detection should be created and implemented in the existing software. On that basis, the coupling between software and the actuator could be created allowing semi-automatic operation within one measurement procedure.

4. RESULTS

Based on the requested properties a prototype of new, low-cost measurement line is produced (Fig. 4). Among the analysed design solutions the one with gearbox and trapezoidal spindle was chosen. The actuator and the transmission are chosen as compact product capable of both electrical and hand operation. The force is provided by transversal movement while adjustment in the longitudinal direction gives capability for testing various sizes of torque tools. The limiting load and the operating dimensions correspond to the tool with maximum torque value of 3000 Nm.

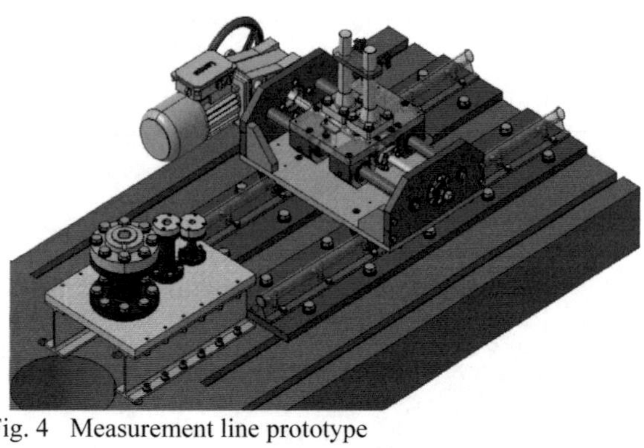

Fig. 4 Measurement line prototype

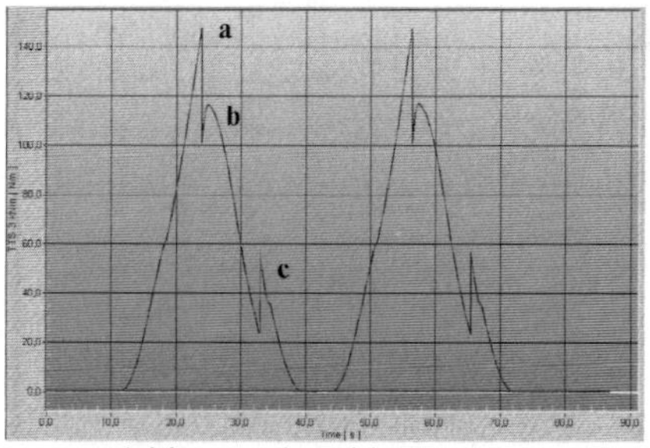

Fig. 5 Data acquisition results with new measurement line

Testing of the prototype line was conducted for three different setting torque tools with nominal torque values of 200, 760 and 2000 Nm. The equipment was operated manually by wire since the automation part has not been implemented yet. Nevertheless, the measurement results (Fig. 5) are practically identical, independent of the operator's physical performance and suitable for automated peak detection. Three significant moments could be clearly seen in the measurement cycle: a) torque peak, b) secondary load due to mechanism impact with tool casing and c) releasing of the mechanism allowing another operating cycle.

5. CONCLUSION

The undertaken analysis of recorded data (Fig. 2) outlined the weaknesses of the old measurement line as follows:

- accuracy of the measurement depends on the operator's physical performance,
- output data do not form smooth curves, and thus are not suitable for automated peak detection,

As opposed, the new line, not even finished, offers benefits in both areas. Further work on the improvement will be focused on the development of the procedure for peak detection and automated operation within the measurement cycle. For this purpose the loop back control must be created between the actuator and the acquisition system or PC. When a peak point is reached the operation of the actuator should be stopped and reversed until the measured quantity reaches zero. The measurement cycle could be repeated as many times as needed, potentially determined by the data stored about a specific torque tool. If so, the operator's concern would be to mount the tool, set the torque to be tested and introduce the tool type to the software. The rest of the work would be done by the system.

This work is partially funded by the Croatian Ministry of Science, Education and Sports.

6. REFERENCES

Decker, K.-H. (2007). *Machinenelemente: Gestaltung und Berechnung, 16. Auflage*, C. Hanser Verlag, ISBN 3-446-19194-1, München,

ISO 1174-1:1996, Assembly tools for screws and nuts - Driving squares - Part 1: Driving squares for hand socket tools, ISO, 1996

ISO 6789:2003, Assembly tools for screws and nuts - Hand torque tools - Requirements and test methods for design conformance testing, quality conformance testing and recalibration procedure, ISO, 2003

Opalić, M.; Vereš, M. & Žeželj, D. (2004). Measurement line design, *Proceedings of 45th International Conference of Machine Design Departments,* Mazal, P. (Ed.), pp. 78-83, ISBN 80-214-2702-7, Blansko, Czech Republic, September 2004, Brno University of Technology, Brno

Schicker, R. & Wegener, G. (2002). *Measuring torque correctly*, Hottinger Baldwin Messtechnik, ISBN 3000089454, Bielefeld

Annals of DAAAM for 2008 & Proceedings of the 19th International DAAAM Symposium, ISSN 1726-9679
Editor B. Katalinic, Published by DAAAM International, Vienna, Austria 2008
Make Harmony Between Technology and Nature, and Your Mind will Fly Free as a Bird

PRACTICAL APPROACH OF USING EMBEDDED CONTROLLERS FOR VIRTUAL INSTRUMENT OVER TCP/IP AND 802.11 B/G PROTOCOLS

RISTEIU, M[ircea]; ILEANA, I[oan]; TULBURE, A[drian]; MARC, G[heorghe] & CROITORU, B[ogdan]

Abstract: *The paper describe a real time remote measuring system based on a reconfigurable controller system for environment parameters measurements. We are integrating embedded controllers for measurements into TCP/IP infrastructure as a remote measurement implementation, and we collect many heterogeneous types of parameters, as a measurement implementation. We are evaluating the performances of high speed FPGA and/or real time implementation via TCP/IP - 802.11g. The quality of the evaluation will be described by the time delay of the measurements. Based on these aspects we have developed virtual instruments and we have built up applications module to optimize measurements performances taking into account the sampling time, measurements delay, time delay processing, and applications behavior. Running VI-s for each criterion, the main related expectations have been fulfilled. For combined criteria, some improvements have to be developed into the software side, and some actual facilities of LabVIEW must be combined with operating system programming.*
Keywords: *remote monitoring, virtual instrumentation, measurement delay, 802.11b/g, embedded controllers.*

1. INTRODUCTION

We have started using LabVIEW implementation for wind energy evaluation taking into account the power of G-based programming language. Today's measurements for this kind of implementation are based on a data logger with GSM-based gateway for downloading data from the site measurements. This implementation is required for real time measurements.

The main requirement for using real time measurement around LabVIEW implementation was the demand for adding some environment monitoring parameters to the main wind energy parameters list (T. Savu, G. Savu, 2000). This aspect requested to mix together some heterogeneous parameters. The entire in-site implementation is powered by an optimized solar panel module. In our approach we use some sensors for typical measurements, as follow: temperature, wind speed; wind direction, noise level, stroboscopic effect and 2 axes vibrations; wind direction; wind speed sensor – anemometer; atmospheric pressure; humidity; temperature sensor attached on the metallic wind direction support; tilt sensor (this sensor includes vibration and temperature measurement parameters); solar radiation sensor.. In our approach, the data is sent to a computer through a wireless connection. The main advantage of this implementation is based on the fact that the TCP/IP- based reconfigurable controller allows us to organize a mobile and real time measuring system with remote control facilities. For such issue, we have access to buss controller and monitor, and remote terminal. A brief description of the reconfigurable controller we use here is focused on: embedded controller use 40MHz processor; 512 MB of Nonvolatile memory; real-time data transmission capabilities; local and remote working capability; configuration and programming accessibility by using the FPGA and Real-Time; the TCP/IP Ethernet interface.

For this embedded controller, there is used a measuring chassis with 2 x cRIO-9201/9221 modules. The cRIO-9201/9221 returns uncalibrated, as binary data. It has to be calibrated and scaled in software side. The accuracy of the measurements with cRIO 9201 is proved in lab experiments, and is resumed in table 1.

Error	Percent of Reading	Percent of Range*
cRIO-9201/9221		
Calibrated max (–40 to 70 °C)	±0.25%	±0.25%
Calibrated type (25 °C, ±5 °C)	±0.04%	±0.07%
Uncalibrated max (–40, 70 °C)	±0.67%	±1.25%
Uncalibrated typ (25 °C, ±5 °C)	±0.26%	±0.46%
* Range equals 10.53 V for the cRIO-9201, 62.50 V for the cRIO-9221		

Tab. 1. Accuracy of measurements with cRIO 9201/9221

To transmit all recorded environmental data from controller to a high level processing unit (PC-based) over a large distance (remotely) we implemented a wireless network for this purpose. The controller was directly connected to a Wireless Acces Point (WAP) and a 26dB antenna, via TCP/IP port. For extending wireless range we attached external antennas to WAP. With both 26 dB antennas we have got stable connection at 30 km distance. For the proposed application we use Wired Equivalency Privacy (WEP) general encryption protocol, that easily exploited by software that can break the encryption after capturing traffic and recognizing encryption patterns. We also used WAP 2 encryption method for dedicated situations, like local configuring session.

Hardware configuration (UTP port, measuring module, and other local facilities) is done through software assistance. We used MAX- controller dedicated software- to configure the Ethernet interface of controller, to assign an IP and a name to the device, in order to be able to work in the network. Then measuring modules are settled up with particular properties. For local configuring procedures, a 232 port with TPC device is also enabled. After this step is done we tested the connectivity and the link between all devices in the network (PING command). Then we are setting up first communication parameters (time delay of the measurements, level of radio signal, antennas positions) with Windows- based software on the host computer.

2. LABVIEW CONTROLLING APPLICATIONS

The most interesting and challenging part of the project was the designing of an application module in LabVIEW, for effectively display and manipulate the read data from all connected sensors. To do this, we needed to create a new Lab View project on a target device (in our case, cRIO named as "*myRioK*"). We created two VIs: one on FPGA Target and second on the REAL-TIME part. The first VI called "*FPGA buf read 2.vi*" on FPGA Target is reading the data from the channels of NI9221 and stores it into a FIFO for each channel (buffer). We used a FPGA I/O node to be able to read data (signals) from NI9221 – cRIO.

Each "FIFO – read" has a numeric indicator where data is displayed, and the indicators are invoked in the second VI to manipulate the contained data (for the real-time mode). The second VI, named "*Real time read-scalare 2.vi*", is realized on the REAL-TIME part of the project and invokes the first VI and to be more precise, the four indicators (*Element4, Element5, Element6, and Element7*) are used, and invoked as a VI reference object on FPGA Target. In this VI we made few mathematic calculations that are called calibrations which mean we used few linear transformations to adjust the displayed value to the real value. These kinds of sensors that we have used are not factory default scaled, so the values recorded and transmitted needs a linear transformation to display the real value. The general formula used (Zhang, W.; Branicky, M.S.;& Phillips, S.M., 2001) to scale the results is:

$$x = \frac{a * abs\ (S)}{2^k} * \varepsilon \qquad (1)$$

Where:

x- Real displayed value on the application side;
a- Data read from remote channel;
+/- t - superior/ inferior module voltage limit;
S- Unsigned sum of 't';
ε- Error coefficient used for accuracy of scaling process

Depending on different used sensors and different signals, each channel might be scaled with different parameters.

It was attached to this VI (inside it) a loop timer to read data from cRIO only in specified time intervals (dynamic control). The reason of this approach is based on the fact that the speed of the FPGA is 40 MHz (250 ns) and it is useless to display so fast changes at the client side. An important aspect of measuring stage is the oscillation of the displayed value. That value is permanently changing in a small interval. The problem of the chaotic oscillation depends in high percent on what hardware is used. This means the FPGA controller frequency is responsible for this kind of oscillations and the reason is very simple; the FPGA makes a reading at 40 MHz that means at every 250 ns a reading is done and the human eye cannot distinguish the displayed values. The specified reading has certain accuracy and of course the error level of reading is included in an interval. The general idea is to create a VI on the FPGA part where the read data are put (or memorized) into an array using an I/O FPGA node and an array object (exactly how the data is read at 40 MHz), then in a VI on the Real-Time part the data stored in the array is taken, and using a "for" or "do...while" loop for every element in the array is calculated it's average value. The loop counts only 4000 elements, this represents all the readings made in 250 ms by FPGA controller. This algorithm can be changed or easily modified depending on what the user wants to do or to be displayed on the front panel.

3. TESTING DELAY MEASUREMENT IN SITUU

Next step of our implementation is focused on developing delay measuring setup for evaluation of the effects of wireless transmission for real time measurements. The block diagram of this experiment is shown in figure 1 (Risteiu, M.; Ileana, I.; &Tulbure, A., 2007). The assumption that the network can be described by a delay distribution implies that there is no correlation between consecutive packets sent over the network. The assumption holds if the packets are transmitted sufficiently far apart, so that the queue lengths have time to change between the packets. This is not the only requirement since the delay is usually also dependent on the utilization of the network and the utilization changes. To model a network with correlation, a Markov chain can be used. For this reason the repeatability of the data has to be ensured. For the proposed measurements we are using Visual Route (Windows based software that access DataLink layer on OSI model). Based on fixed conditions (Zhang, W.; Branicky, M.S.;& Phillips, S.M., 2001), first records with packet losses are shown in figure 2. The accuracy of the measurements is summarized in table 2. For our purpose, the measurements are valid (Zhang, W.; Branicky, M.S.;& Phillips, S.M. 2001), (Cabulea, L., 2003).

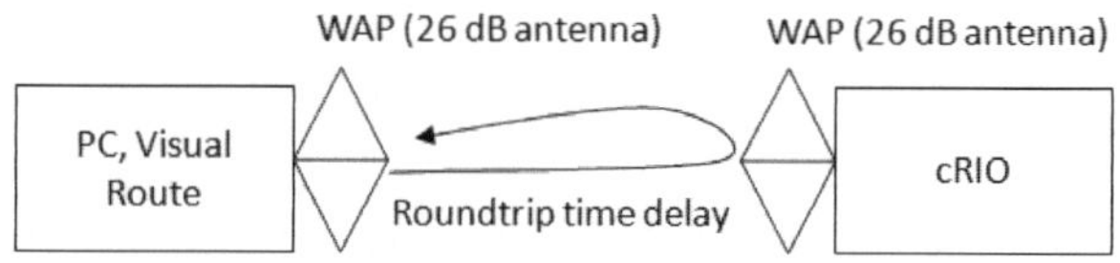

Fig. 1 Block diagram of round trip time measurement setup (Zhang, W.; Branicky, M.S.;& Phillips, S.M. ,2001)

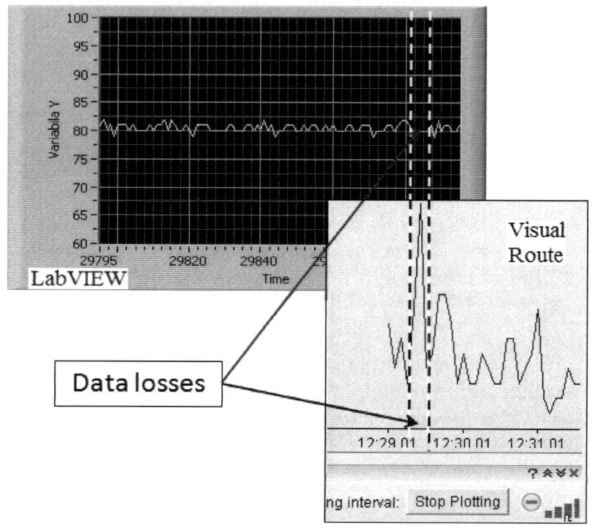

Fig. 2 Measurement losses in "*FPGA buf read 2.vi*" Block Diagram on FPGA Target

4. CONCLUSION

Real time measurements are challenging because of time delay and packet losses. It is necessary to built up on host computer individual configuration process. Using 802.11b/g protocol for UTP port asks for testing the delay of data sending session. We proved that 20 ms is actual limit delay, and over this limit another sending data session has to be designed.

5. REFERENCES

Cabulea, L., (2003), Procedee de aproximare in statistica si probabilitati (Approximation approaches in statistics and probabilities), Aeternitas Publishing House, 2003, ISBN 978-973-7942-84-2

Mikael Pohjola, (2006). PID Controller Design in Networked Control Systems, Helsinki University Of Technology Report, Issue 2, Department of Automation and Systems Technology, January 2006.

Risteiu, M.; Ileana, I.; &Tulbure, A. (2007). Integration of the Zigbee Smart Sensors in TCP/IP Infrastructure for Fast Enabling Remote Data Acquisition, Process Control 2007

T. Savu, G. Savu (2000). Tehnologii Asistate de Calculator (Computer Aided Technologies), Editura ALL, Bucureşti,

Zhang, W.; Branicky, M.S.;& Phillips, S.M. (2001), Stability of Networked Control Systems. IEEE Control Systems Magazine. Vol. 21, Issue 1, February 2001. ISSN: 0272-1708.

Annals of DAAAM for 2008 & Proceedings of the 19th International DAAAM Symposium, ISSN 1726-9679
Editor B. Katalinic, Published by DAAAM International, Vienna, Austria 2008
Make Harmony Between Technology and Nature, and Your Mind will Fly Free as a Bird

ISSUES REGARDING DIGITAL SIGNATURE INFRASTRUCTURE AND DIGITAL CERTIFICATE MANAGEMENT

ROBU, R[aul] & FILIP, I[oan]

Abstract: *The paper presents an original systematization of main elements of digital signature infrastructure, and some aspects regarding the management of digital certificates, highlighted by practical experience of authors in implementing of digital signature in Romania, in hospitals and in territorial work inspectorates from several counties. The goal of the paper is to clarify some problems concerning the digital certificates, server certificates, certification authorities, respectively regarding registration in order to obtain a digital certificate, certificate issuing, certificate searching and digital certificate renewal and revocation.*

Key words: *cryptography, signature, authentication, certificate, key.*

1. INTRODUCTION

The high performances reached by the information technology during the last years and its decreasing costs made it possible to present, manipulate and archive documents more and more in an electronic format. In this context, an essential role is played by the digital signature, the instrument through which the content of an electronic document and its issuer are authenticated in an almost infallible manner.

The digital signature is implemented using cryptographic methods based on two keys: a private key and a public key (Diffie & Hellman, 1976). The base principle of these methods is: a hash function -such as MD5, SHA1 (NIST, 1995), SHA2 (NIST, 2002)- is applied on document, in order to extract from the document a binary code, known as hash value (we note it H1), representing essential data. The hash value is encrypted (Rivest et al., 1978) with the private key of the signer, obtaining a cryptogram which represents it self the signature and which is attached to the original document. Signature validation procedure is achieved by extracting the hash value from the signed document (we note it H2), decrypting the cryptogram which is attached to it by using the public key, in order to obtain the initial hash value (H1). If the two hash values are identical (H1 and H2), the signature is valid.

2. THE DIGITAL SIGNATURE INFRASTRUCTURE

The required infrastructure to implement a digital signature system mainly consists from: digital certificates, server certificates, certification authorities (CA), facilities for the certificate management

2.1 Digital certificates

The digital certificate represents a collection of data in electronic format which confirm the connection between the validation data of the digital signature and a person, authenticating the identity of that person (Romanian Parliament, 2001). The most important data contained by a digital certificate are: public key, the name and the signature of the certification authority that issued the certificate, the holder's name, the cryptographic algorithm and the hash function used.

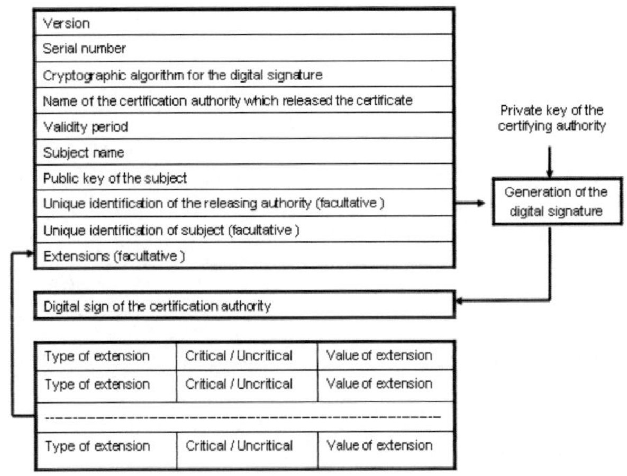

Fig. 1. Certificate format - X. 509

The most widely used digital certificate is the one with the number X.509, standardized ISO/IEC/ITU -version 3 (figure 1).

A system based on public key certificates also involves the existence of a Certification Authority, which releases certificates for a certain group of a key pair's owners (public and private). Each certificate contains the value of the public key and information that identifies in a unique mode the subject of the certificate (which can be a person, an application, a device or another entity which has the private key corresponding to the public key integrated in the certificate). The certificate represents a connection, impossible to fake, between a public key and a certain attribute of its possessor. The certificate is digitally signed by a Certification Authority, which confirms the identity of the subject. Once the certification infrastructure is established, a user of that public keys' infrastructure can obtain a public key for any of the users certified by that Certification Authority, simply by obtaining the certificate for that user and extracting the required public key (Patriciu et al., 2001).

2.2 Certificate servers

A certificate server, also known as a Cert-server or a Key-server, is in fact a database which allows the users to search or generate digital certificates. A Cert-server usually provides some administrative components which allow certification companies to maintain the security policies – for example, to permit storage only for the keys that fulfill certain conditions.

The most well known certificate servers are (Patriciu et al., 2001):

- Netscape Certificate Server: interoperates only between Netscape clients.
- Microsoft Certificate Server (IIS): interoperates between Netscape clients, as well as Entrust clients.
- Entrust Web/AC Server: permits SSL, S/MIME certificates to be released and published as well as the signing of objects for Netscape Enterprise Server, for MS IIS and Apache Stronghold servers.

2.3 Certification authorities

Because it is not practical for all the users in the world to trust a single organization or company concerning their secret communications, there are more Certification Authorities. They mutually detain certificates – not necessarily each one to the other one–, with the help of which an user having a certificate released by a certain authority – such as CA1 – may obtain the public key of an user possessing a certificate released by another authority – such as CA3, for example -. Therefore, a so called certification chain or path is constituted –see figure 2-, in which, at the present time, all the great systems that distribute public keys are registered in.

A credible certification authority which does not have a digital certificate issued by another certification authority stands at the highest level of the previously mentioned hierarchy. The public key released by such a certification authority must be recognized in an independent manner, i.e. published on a large scale (DigiSign, 2006).

For example, in Romania, there are two certification authorities, DigiSign and e-Sign both headquartered in Bucharest. These companies are certified by ADACOM, and VERISIGN is the root of this certification tree. A certificate released by the e-Sign company is presented in figure 3.

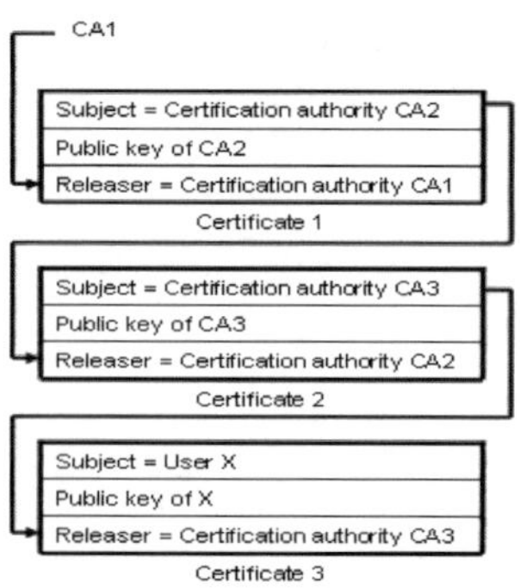

Fig. 2. A chain of certification

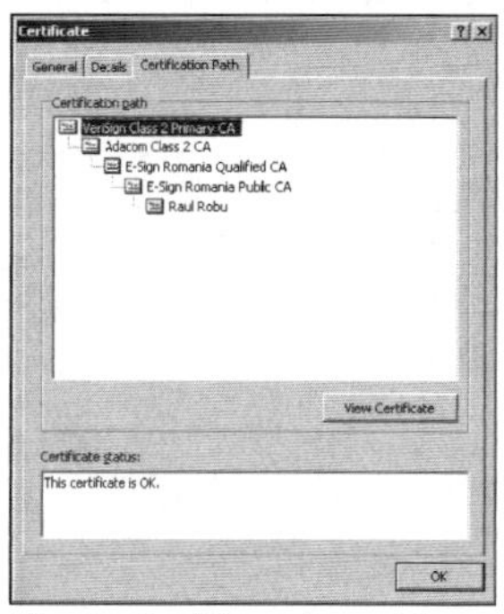

Fig. 3. Certificate issued by e-sign

3. THE MANAGEMENT OF DIGITAL CERTIFICATES

The certification authorities usually have a secure site that allows the management of certificates. The management of digital certificates comprises (DigiSign, 2007):

a) Registration in order to obtain a digital certificate:

A person who buys a digital signature from a certification authority receives a cryptographic secure device (like e-token or smart card). Any use of the secure device involves entering the pin code. At three wrong introductions of the pin code, the device locks and can be unlocked by entering the puk code.

In order to obtain a digital certificate, one has to register on a secure page of the certification authority's website. Before registration, the future signatory must send certain documents to the certification authority. These documents consist of copies of personal identification documents, as well as a statement certified by the public notary through which one declares on his own risk that he learned about the legal provisions regarding the digital signature. The next step involves the installation of secured device drivers. Once these steps are taken, the registration process can be carried out only using the cryptographic device connected to the computer. The person fills in the form in which he enters personal data. Some of these personal data, such as one's name, email address, workplace, phone number, will appear in the certificate. The future signatory must also enter a password, which is needed in case of the certificate's revocation.

b) Receiving the certificate: Within 24 hours of the registration time, the administrator of the application who manages the certificates verifies the accuracy of the registration data, comparing them with the ones contained in the papers sent by the person. In case the registration data are valid, the administrator sends to the applicant, an email which contains a link and a password that the applicant can use to download the certificate. The applicant must introduce the cryptographic device in the computer and then he can download the certificate on the secure device. After this operation, the applicant can digitally sign any document.

c) Searching the certificates: all digital certificates generated by certification authorities are public and accessible to everyone. The confidentiality of a document or email message can be assured by encrypting the email message or document with the recipient's public key. This way only the recipient can decrypt the email or document with his private key located on his secure device. That is why certification authorities offer the opportunity to search and download any certificate generated by them, to anyone who is interested.

d) Renewal of the certificate: digital certificates have a period of validity of one year. They must be renewed annually, the expired certificates having no legal power.

e) Revocation of the certificate: if a person has lost possession of the secured device and believes that the pin code necessary in order to be able to sign is compromised, he may revoke the certificate. The digital signature done with a revoked certificate no longer has any legal value.

4. CONCLUSIONS

Digital signature offers many advantages and can be used in a wide range of applications, being able to substitute physical signature in any domain.

This technology assures important time reductions, allowing the achievement of remote transactions between individuals and / or institutions and eliminates paper work.

The authors offer in this paper a systematic presentation of the infrastructure of the digital signature and of the management of digital certificates, founded on their practical experience in implementing digital signature in hospitals and territorial work inspectorates from several counties of Romania.

5. REFERENCES

Diffie, W. & Hellman, M. (1976). New directions in cryptography, *IEEE Transactions on Information Theory,* Vol. 22 (Nov. 1976), pp. 644-654, ISSN 0018-9448

DigiSign (2006). The trust chain, *Available from:* http://www.digisign.ro/products-and-services/security-services/qualified-certificates/lant.html, *Accesses:* 2008-06-14

DigiSign (2007). Registration, Searching, Renewal and Revocation of certificates, The homepage of Center of Digital Certification, *Available from:* https://ca-services.digisign.ro/public/digitalidCenter.htm, *Accessed:* 2008-06-15

NIST -National Institute of Standards and Technology (1995). Announcing the Secure Hash Standard, *FIPS 180-1* (1995)

NIST -National Institute of Standards and Technology (2002). Announcing the Secure Hash Standard, *FIPS 180-2* (2002)

Patriciu, V.V.; Ene-Pietroşanu, M.; Bica, I.; Vaduva, C. & Voicu, N. (2001). *Security of electronic commerce,* Bic All, ISBN 973-571-325-X, Romania

Rivest, R.; Shamir, A. & Adleman, L. (1978). A method for obtaining digital signatures and public key cryptosystems, *Communications of the ACM,* Vol. 21 (Feb. 1978), pp. 120-126, ISSN 0001-0782

Romanian Parliament (2001). Law nr. 455 about digital signature, *Official Gazette of Romania,* No. 429 (Jul. 2001), *Available from:* http://www.mcti.ro/index.php?L=0&id=20&lege=3, *Accessed:* 2008-06-14

Annals of DAAAM for 2008 & Proceedings of the 19th International DAAAM Symposium, ISSN 1726-9679
Editor B. Katalinic, Published by DAAAM International, Vienna, Austria 2008
Make Harmony Between Technology and Nature, and Your Mind will Fly Free as a Bird

TECHNICAL AND LEGISLATIVE ASPECTS REGARDING THE DIGITAL SIGNATURE

ROBU, R[aul]; VOISAN, E[mil] & UNGUREANU, D[an]

Abstract: *This paper represents a synthesis of the key aspects regarding the digital signature, based on the personal expertise and experience of the authors, acquired from their position of pioneers as effective promoters of the digital signature in Romania. Legislative aspects, including definitions introduced by laws, the principle of the digital signature, its mathematical base, the manner in which the documents are signed, the verification, respectively the validation of signature and the way in which the confidentiality is assured are presented.*
Key words: *cryptography, signature, authentication, certificate, key.*

1. INTRODUCTION

The high performances reached by the information technology during the last years and its decreasing costs made it possible to present, manipulate and archive documents more and more in an electronic format. In this process, an essential role is played by the digital signature (NIST, 1994), the instrument through which the content of an electronic document and its issuer are authenticated in an almost infallible manner.

2. THE PRINCIPLE OF THE DIGITAL SIGNATURE

The digital signature is implemented using cryptographic methods with two keys: a private key and a public key, in the concept introduced by Diffie and Hellman, from Stanford University (Diffie & Hellman, 1976). The first key must have a maximum secrecy regime and that is why it is known only by its owner, being named the private key (PRIV). The second key, on the other hand, is made public, that is the reason it is called a public key (PUB). Both keys are in fact string of bits, provided by a special program. While the public key can be issued anywhere in the world, the private key must be kept in a safe place. That is why the private key is usually stored on a secured cryptographic device (etoken or smart card, which can be accessed only by entering a password, the PIN code) and cannot be extracted from this device in a comprehensible form, but only as a cryptogram. The public key is included on a digital certificate which is also stored on the secured cryptographic device. The certificate may be exported and published on the Internet because it does not contain confidential information.

3. THE MATHEMATICAL FUNDAMENTALS OF THE DIGITAL SIGNATURE

The existing implementations of the digital signature systems, most often use the following cryptographic algorithms:

- MD2, MD5 -"Message Digest"- (Wang & Yu, 2005), created by Ronald Rivest and SHA -"Secure Hash Algorithm"-, created by the USA Standards Institute for digital signature (NIST, 2002), all used for digest's extraction;
- RSA, created by Rivest, Shamir and Adleman (Rivest et al., 1978), El Gamal (El Gamal, 1985) and DSA -"Digital Signature Algorithm"- (NIST, 1994), all used for digest's encryption / decryption.

4. SIGNING AND CHECKING THE SIGNATURE OF THE DOCUMENTS

In **practical implementations,** the public key algorithms are often inefficient and slow. In order to decrease the time, a digest of the document is realized based on a hash function (Fig. 1).

Signing document M (Fig 1.) determines the following actions (Patriciu et al., 2001):

- the digest D of the document M is done with the help of a hash function (SHA1, SHA2, MD2, MD5);
- the document digest is ciphered with the private key of the releaser (which is stored on a secured device) and its obtained a cryptogram S which represents the digital signature, which is attached to the document M;
- the digital certificate which contains the public key is attached to document M, operation which allows the verification, respectively the validation of the signature.

The encrypting algorithms have a great crypto-complexity, generally based on complex mathematical operations, with huge numbers (hundreds of decimal numbers or thousands of bits).

The digital signature protocol guarantees that (Patriciu & & Ene-Pietrosanu, 2001):

a. the signature is **authentic**, because it can be verified only with the public key of the releaser;
b. the signature **cannot be forged**, because only the releaser knows his own secret key;
c. the signature is **not reusable**, because it depends on the content of the document which is encrypted;
d. the signature is **not alterable**, because any alteration of the content of the document results in the fact that the signature cannot be validated with the public key of the releaser;
e. the signature is **not reputable**, because the only one who had the necessary instruments to sign is the releaser and that is why the verification / validation process which takes place at the receptor, does not even involve the releaser.

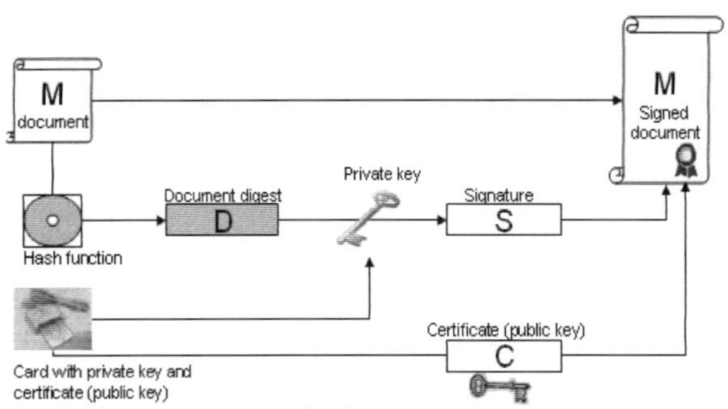

Fig. 1. Signing an electronic document

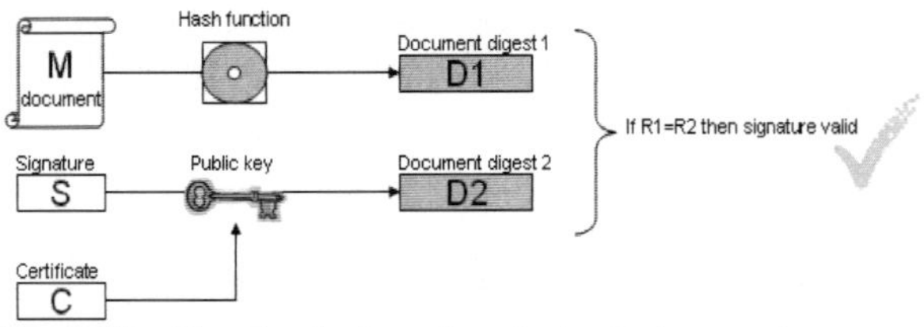

Fig. 2. Checking the signing of an electronic document

The signature is checked through 3 steps (fig. 2) (Patriciu et al., 2001):

- a new digest of the alleged signed document is created
- the signed digest is decrypted with the public key of the releaser
- the 2 digests are compared and if there is coincidence the signature is validated.

5. INSURING CONFIDENTIALITY

The pair of keys can be used not only to sign and check the signature, but also in order to insure the confidentiality of the documents. If a signatory X desires to send a confidential document to a signatory Y, signatory X may encrypt the document using the public key of signatory Y. In this way, only signatory Y will be able to decrypt the document with the help of his private key.

6. LEGISLATIVE ASPECTS

A series of documents settle, from a legislative perspective, the digital signature concept and the way it is used in EU (EP & EC, 2000).

The Directive 1999/93/EC of the European Parliament and of the Council of 13 December 1999 on a Community framework for electronic signatures (EP & EC, 2000) introduces the following fundamental definitions:

The digital signature represents data in an electronic format, which are attached or logically associated with other electronic data and which serve as an authentication method.

The extended digital signature represents the digital signature which fulfills all of the following conditions:

a) it is linked in an unique manner to the signatory
b) it insures the identification of the signatory
c) it is created through means exclusively controlled by the signatory
d) it is linked to the electronic data that it is related to, in such a manner that any ulterior modification of these data can be identified.

A signatory is a person who possesses a device which creates the digital signature and who acts for himself or as a representative of a third party.

Digital signature creation data are any electronic data with a unique character, such as codes, or private cryptographic keys, which are used by the signatory to create a digital signature.

A digital signature creation device is a configured software and / or hardware, used for the implementation of the data necessary for the creation of the digital signature.

A secured digital signature creation device is the device that creates the digital signature which fulfills all the following conditions:

a) the data for the creation of the signature, used to generate it, must appear only once and their confidentiality must be insured
b) the data necessary for the creation of the digital signature, used to generate it, must not be inferred
c) the signature must be protected against forgery by the technical instruments available at the time it is generated
d) the data for the creation of the digital signature must be effectively protected by the signatory against being used by unauthorized persons
e) it must not alter the data to be signed or prevent such data from being presented to the signatory prior to the signature process.

Digital signature verification data are electronic data, such as codes or public cryptographic keys, which are used with the purpose of checking/verifying a digital signature.

A digital signature verification device is a configured software and / or hardware, used for the implementation of the data needed for the checking of the digital signature.

A digital certificate is a collection of electronic data which attest the link between the data used for the verification of the digital signature and a person confirming the identity of that person.

7. CONCLUSIONS

The digital signature has the following attributes: it is authentic, it cannot be forged, it is not reusable, it is not alterable and it is not reputable.

Digital signature offers many advantages and can be used in a wide range of applications, being able to substitute hand signature in any domain. This solution eliminates hard copies and assures important time reductions, allowing the achievement of remote transactions between individuals and / or institutions.

The digital signature also guarantees that there were no text adding or changing operations in the document after the document was signed. This cannot be guaranteed while using the hand signature.

This paper presents a systematic description of the technical and legislative aspects regarding the digital signature, based on the authors' practical experience in implementing digital signature in hospitals and territorial work inspectorates from several counties of Romania.

8. REFERENCES

Diffie, W. & Hellman, M. (1976). New directions in cryptography, *IEEE Transactions on Information Theory,* Vol. 22 (Nov. 1976), pp. 644-654, ISSN 0018-9448

El Gamal, T. (1985), A public key cryptosystem and a signature scheme based on discrete logarithms, *IEEE Transactions on Information Theory*, Vol. 31 (Jul. 1985), pp.469-472, ISSN 0018-9448

EP -The European Parliament and the EC -European Council (2000). Directive 1999/93/EC, a community framework for electronic signatures, *Official Journal of the European Communities* (Jan. 2000), pp. 12-20

NIST -National Institute of Standards and Technology (1994). Announcing the Standard for Digital Signature Standard (DSS), *FIPS 186 (1994)*

NIST -National Institute of Standards and Technology (2002). Announcing the Secure Hash Standard, *FIPS 180-2 (2002)*

Patriciu, V.V.; Ene-Pietroşanu, M.; Bica, I.; Vaduva, C. & Voicu, N. (2001). *Security of electronic commerce,* Bic All, ISBN 973-571-325-X, Romania

Patriciu, V.V. & Ene-Pietrosanu, M. (2001). New technologies regarding the Internet's security, *Revista Informatica Economica*, Nr 1 (17), (2001), pp. 60-68

Rivest, R.; Shamir, A. & Adleman, L. (1978). A method for obtaining digital signatures and public key cryptosystems, *Communications of the ACM,* Vol. 21 (Feb. 1978), pp. 120-126, ISSN 0001-0782

Wang, X. & Yu, H.; (2005). How to Break MD5 and Other Hash Functions, *Advances in Cryptology – EUROCRYPT,* Vol. 3494/2005, pp. 19-35, ISSN 0302-9743

Annals of DAAAM for 2008 & Proceedings of the 19th International DAAAM Symposium, ISSN 1726-9679
Editor B. Katalinic, Published by DAAAM International, Vienna, Austria 2008
Make Harmony Between Technology and Nature, and Your Mind will Fly Free as a Bird

PRESSING AND SINTERING OF TiO_2 POWDER CATHODE USED TO OBTAIN TITANIUM BY ELECTROCHEMICAL REDUCTION

ROMAN, C[ostel]; CARCEA, I[oan]; CHELARIU, R[omeu] & SOARE, V[asile]

Abstract: *The physico-mechanical properties of sintered cathode influence the efficiency of subsequent electrochemical reduction process as well as the titanium quality. The paper presents the results of experimental studies concerning the influence of pressing/sintering processing parameters on both the apparent density and electrical resistivity of TiO_2 sintered cathode. Correlations between processing parameters and sintered cathode properties have been established. These correlations will be used to increase the efficiency of process.*
Keywords: *electroreduction, cathode, sintering, titanium*

1. INTRODUCTION

Titanium and especially titanium alloys find usage in different industries such as aeronautics, chemical, medical, metallurgical, naval etc. At the present days, the applications development of titanium-based metallic materials is limited due to its high cost (Nagesh& Ramachandran 2007, Nie et al. 2006).

Now, the most used commercial process to produce titanium is Kroll process (Qiu et al. 2008), which is very expensive one with low efficiency. Thus, the long-term success of the titanium-based materials remains to find new efficient solutions to produce titanium with a minimum impact on the environment. Taking into consideration this, in the last years there have been tested, especially in laboratories, new extraction processes of the metallic titanium such as: $TiCl_4$ electrolysis in alkaline chloride mixtures or directly from TiO_2 through the electrolysis in melted salts, usually calcium chloride (Liu et al. 2007). The most important variants of the electrochemical reduction are FCC, OS, QIT, MER Corporation, BHP Billiton process, respectively. Among these, FCC process seems to have the most success chances at commercial level (Wang&Li 2004).

This paper describes the pressing/sintering process of TiO_2 powder to make the cathode. Afterwards, the cathode will be used into an electrochemical reduction process that is presented hereinafter.

The experimental procedure for titanium extraction uses an installation whose principle scheme is shown in fig. 1. It consists of electrolysis cell, resistance electric furnace, DC source, inert gas injection system, evacuation device of the melt salt. The resistance electric furnace (l) with a power of 4,5 kW and the chamber made of ceramic provides a maximum heating temperature of 1200°C. The electrolysis cell is mounted inside the heating furnace. The crucible is made of silicon and aluminum oxinitride, a resistant material to high temperatures and chemical degradation of the melted electrolyte. It has the inner diameter of 68 mm and a height of 100 mm. At the bottom of the ceramic crucible was made a circular hole for mounting cathode's support (7) made of refractory steel. In order to insulate, the space delimited by metallic shell-ceramic crucible-cathode support (2) it was filled with a mixture made of oxides and silicon nitride.

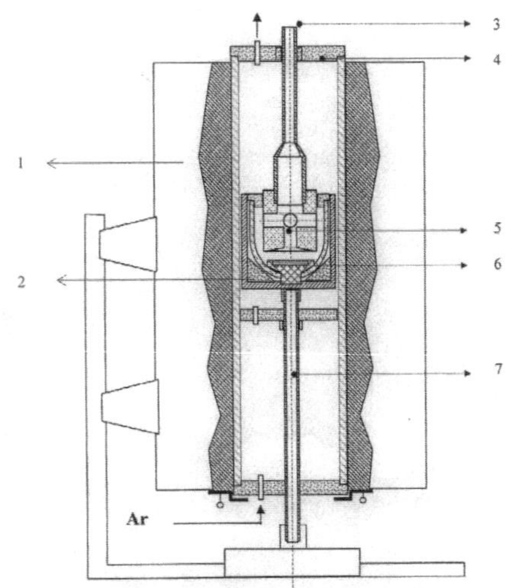

Fig. 1. The experimental unit to produce titanium by electrochemical reduction (laboratory scale).

Cell's anode (5) made of high density graphite has a cylindrical shape with the dimensions of Φ 54 mm and height of 80 mm. Into the anode there are two horizontal and one vertical holes in order to assure a good eviction of the anodic gases and to improve electrolyte's circulation. Anode bar (3) is made from two stainless steel concentric tubes jointed by a tronconic bush. Thus, it allows adjusting anode-cathode distance and to collect and evict anodic gases and the heat insulating material. The used electrolyte is a mixture of $CaCl_2$+NaCl+CaO. Before to be used, the TiO_2 powder cathode have been passed through a pressing/sintering process. It has a cylindrical shape with Φ 30 mm and height between 5-10 mm. Dimensions are adequate to the used electrochemical cell.

Beside electrolyte's properties, we consider that the geometrical shape and cathode's characteristics are equally important for the success of the electrochemical reduction process. That is why, in the paper, we present some aspects regarding the implications of the pressing/sintering process on apparent density and electric resistivity of the TiO_2 powder cathode.

2. MATERIALS AND METHODS

The cathodes were made of TiO_2 powder (Merck, Germany) having chemical reagent purity and the granulation between 0.1-10.0 μm. The stages of pressing/sintering process are given in the following. In order to eliminate the possible humidity, TiO_2 powder was dried in air at a temperature of 200°C for 2 hours. After weighting the necessary mass, the TiO_2 powder was pressed. The die used to make the cathodes with the desired dimensions and shape, was designed to bear forces till 500 kN and to remove cathode after pressing operation.

A press with maximum pressing force of 750 kN assured the pressing force. The press is drove with a hydraulic driving system having the maxim pressure of 1000 mbar, and it is equipped with force and displacement transducers making possible pressing process monitoring.

To record parameters values of the pressing process one computer with interface for data acquisition was used.

Pressing time was 5 minutes for each cathode. After die removing, each cathode supported sintering process in two stages (T = 400°C, t =2 h, T = 900°C, t = 8 h) by using Vulcan 550 resistance electric furnace. After furnace cooling at room temperature, the cathodes were weighted and dimensionally verified in order to determine apparent density. Apparent density was determined by calculation of the ratio between cathode's weight and the volume that corresponds to measured dimensions.

The resistivity was determined with relation $\rho = RA/l$ where R is electric resistance (Ω) of cathode, A is contact area (cm^2), l is the length between measurement points of the resistance (cm).

3. RESULTS AND DISCUSSIONS

The pressing force, pressing time, cathode's weight, dimensions and apparent density are indicated in the table 1, and the dependency D_a=f(F,m) are graphically showed in fig. 2.

The values of electric resistivity of the experimentally achieved cathodes at room temperature are indicated in table 2.

The dependence of electric resistivity of cathodes (6g weight) on apparent density both before and after sintering is graphically shown in fig. 3.

4. CONCLUSION

To keep the integrity of cathode to dry the TiO_2 powder before pressing is necessary. Also, for the same reason two stages of the sintering process must be used. The pressing force of 75 kN is adequate to obtain minimum electric resistivity of sintered TiO_2 powder, and a reasonable value of apparent density. The validation of this choice remains to be demonstrated by subsequent electroreduction experiments.

Pressing force, F [kN]	Cathode's weight, m [g]	Cathode's diameter, [cm]	Cathode's height, [cm]	Apparent density, D_a [g/cm³]
50	6	2.95	0.49	1.77
50	8	2.98	0.63	1.81
75	6	2.98	0.45	1.92
75	8	2.96	0.62	1.85
100	6	2.97	0.45	1.95
100	8	2.98	0.61	1.87

Table 1. Variation of apparent density in terms of pressing force and cathode's weight.

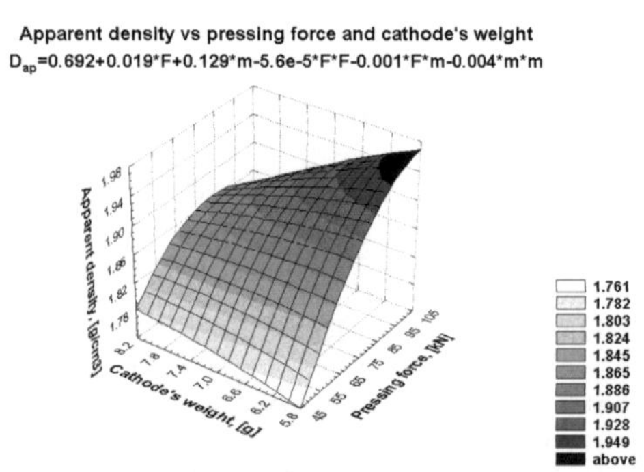

Fig. 2. Force pressing and weight dependency of the apparent density.

Pressing force, F [kN]	Cathode's weight, m [g]	Apparent density, D_a [g/cm³]	Electric resistivity, [MΩ·cm] Before sintering	After sintering
50	6	1.77	77±7	18±2
50	8	1.81	86±10	20±3
75	6	1.92	43±7	14±1
75	8	1.85	68±5	78±12
100	6	1.95	51±12	15±2
100	8	1.87	76±7	209±50

Table 2. Electric resistivity values of TiO_2 cathodes

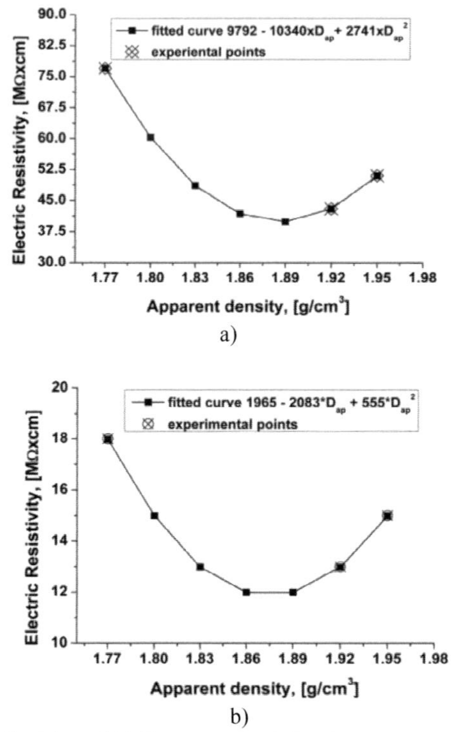

Fig. 3. Electric resistivity vs. apparent density:
a) before sintering; b) after sintering.

5. REFERENCES

Liu, M., Lu, S., Kan, S., Li, G. (2007) Effect of electrolysis voltage on electrochemical reduction of titanium oxide to titanium in molten calcium chloride, *Rare Metal,* Vol. 26, no. 6, 547-551, ISSN 1001-0521.

Nagesh, CR., Ramachandran, CS. (2007). Electrochemical process of titanium extraction, *Trans. Nonferrous Met. SOC. China,* Vol. 17, no. 2, 429- 433, ISSN 1003-6326.

Nie, X., Dong, L., Bai C., Chen D., Qiu G. (2006). Preparation of Ti by direct electrochemical reduction of solid TiO_2 and its reaction mechanism, *Trans. Nonferrous Met. SOC. China,* Vol. 16, no. 3, 723-727, ISSN 1003-6326.

Qiu, G., Feng, X., Liu, M., Tan, W., Liu, F. (2008). Investigation on electrochemical reduction process of Nb_2O_5 powder in molten $CaCl_2$ with metallic cavity electrode, *Electrochimica Acta* Vol. 53, no. 12, 4074–4081, ISSN 0013-4686.

Wang, S., Li, Y. (2004). Reaction mechanism of direct electro-reduction of titanium dioxide in molten calcium chloride, *Journal of Electroanalytical Chemistry,* Vol. 571, no. 2, 37–42, ISSN 0022-0728.

Annals of DAAAM for 2008 & Proceedings of the 19th International DAAAM Symposium, ISSN 1726-9679
Editor B. Katalinic, Published by DAAAM International, Vienna, Austria 2008
Make Harmony Between Technology and Nature, and Your Mind will Fly Free as a Bird

KNOWLEDGE REPRESENTATION DATABASE FOR THE DEVELOPMENT OF TOLERANCES AND FITS IN DESIGN

ROMANESCU, M[aria] C[amelia]; PRUNA, L[iviu]; ANTONESCU, I[on] & ANGHEL, A[lina]

Abstract: *This paper describes the architecture of TFDES (Tolerances and Fits in Design Expert System), an expert system for fit choosing in a Computer Aided Design environment and presents how recent improvements in the openness and functionality of tolerances and fits in design system have made a major contribution for a notable expert system.*
A logical algorithm is developed on a computer software with user-friendly interface. The algorithm generates an optimum layout considering factors such as the relevant ISO tolerance and fits on surface-finish information in the decision process.
Key words: *tolerance, fits, knowledge based system*

1. INTRODUCTION

The research presented in this paper allows the authors to bring theoretical contributions on the rules specific for structuring data bases characteristic to fits, tolerances fields and ISO limit deviances values from the perspective of using them, to create software programs able to run under CAD systems.

From a practical perspective, we have created a program able to run under CAD systems, that allow the user to choose and automatically mark on technical drawings the data related to the tolerances fields and the limit deviances values.

2. RELATED WORK

Unfortunately, design references do not give clear information of principles involved in fits and tolerance field establishing. Instead, papers present general design criteria and rules of thumb that should be used in design process. Parkinson (Parkinson, 1982) views tolerances and basic dimensions as parameters that affect the overall cost of machining and assembly. He does not address the parts dimensions and tolerances required for its function, but only those needed for assembly. M. Anjanappa (Anjanappa et al., 1991) discusses the identification, development, and operation of an approach that uses a global data base of flexible manufacturing.

A. Desrochers (Desrochers & Rivière, 1997) presents a matrix-based method for defining the constraints for tolerance representation function of the imposed rules for computer aided design and computer assisted manufacturing. Tarek M. Sobh (Tarek et al., 1999) addresses the problems related to the representation and analysis of tolerances in the fields of quality control, computer aided design and production.

M. Kalta (Kalta & Davies, 2001) describes the EXCAP program for cast parts that interacts with the CAD system. The program extracts the geometrical data and the dimension devices, necessary for the parts manufacturing.

In the field of mechanical design, a module developed in the Czech Republic is available, for the automatic computation of tolerances and fits incorporated in the MITCalc program.

The research status reveals that currently in the field of workpiece dimensional precision there is no program for establishing the fits type, the tolerance fields or the values of limit deviances that can run within a CAD software.

3. OVERVIEW OF CURRENT WORK

The expert program TFDES aids the designer by automating a specific design task while having also the capability to store new knowledge, as a result of experience that can be used latter (Romanescu et al., 2007). The authors' experience in both production and CAD domain will lead to the development of new solutions for structuring the data base such that it can be used by a program able to run on 2D CAD systems.

The successful planning requires the following activities:
1) Creating a data base incorporating the fits, tolerance fields and limit deviances values and structuring it in a quantifiable format that works in 2D CAD system;
2) Creating a base of knowledge in a quantifiable format;
3) Developing of expert system computer code;
4) Testing the computer program.

The authors describe developments in the first three activities, which have been made possible by their experience.

4. SYSTEM ARCHITECTURE

Block diagrams for incorporating the fits, tolerance fields and limit deviances values strategy can be divided into three sections: the input stage, the knowledge base, and the output stage. An algorithm developed is depicted in a flow chart, as shown in figure 1.

Input stage

Several important factors that may affect a blank layout in choosing the tolerances field and fits are first discussed:
- Establishing a nominal dimension;
- Choosing the fitting system;
- Choosing the precision class;
- Choosing the type and group of the fitting;
- Marking on the drawing tolerance fields and values of the limit devices.

Knowledge representation and decomposition

The base structure of these systems includes the following:
- a component that contributes to the development in the computer science field;
- a component constituting the knowledge base.

The requirements imposed on the system are the following: the utilization of the real knowledge about the tolerances and fits system, the extension ability, the transparent characteristic of the system, the user-friendly representation of the knowledge, the capacity for user-computer integration.

As an user-aiding mechanism, the system simplifies the designer reasoning - the design project is reduced to selecting automatically without operator intervention and placing the tolerances, the times for detailing are reduced, the tolerances can be efficiently reused and the constructions can be CAD systems.

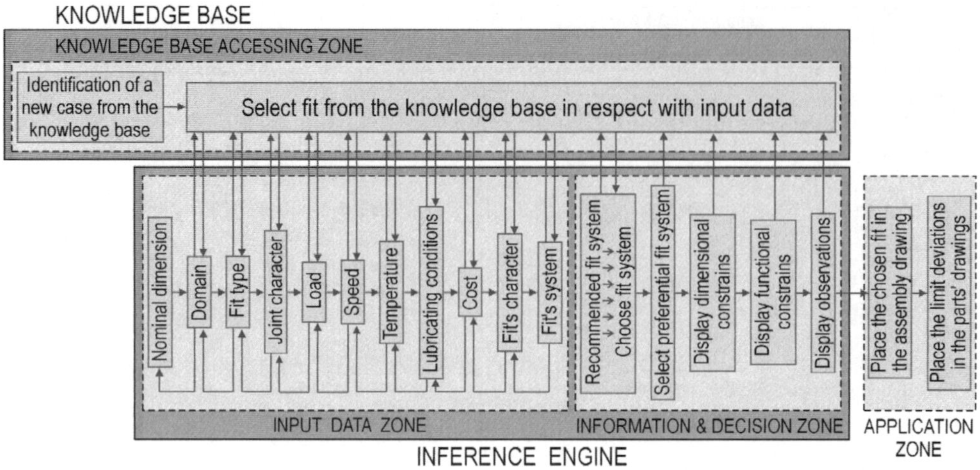

Fig. 1. Flow chart depicting the main program

System formulation

The user calls the tolerance input program, which assign default tolerances. Each module interacts with another to arrive at the final design. Tools to assist this process were written in the AutoLISP programming language. Figure 2 presents a fragment of the index_filter function that establishes the index that must be saved after the user selection:

```
(defun index_filter( / parameter searchList
indexes positions)
[...]
(if (= step 2)    ; after we know the Domain, prepare to
save the corresponding indexes
 (progn             ; for the item selected by the user
  (setq searchList domList) ; determine in what list to
search for the parameter
  (setq parameter (cadr (assoc 'Dom state))) ;
save the selected parameter
  (setq positions (cdr (assoc 'Diameter
indexes_list))) ; save in <<positions>> the indexes that
we are searching for in <<searchList>>, to check if they match
the ones selected by the user
  (filter parameter)
  (if (/= indexes nil)
   (progn
    (if (> (nth 0 indexes) (nth 1 indexes)) ;
order ascending the indexes that match the user selection
[...]
  (if (not (assoc 'Dom indexes_list))
   (progn        ; if the article is not <<indexes_list>>
    (setq indexes_list (cons (cons 'Dom indexes)
indexes_list)) ; save it in <<indexes_list>>
   )             ; end progn
   (progn
    (save_indexes 'Dom)
   )             ; end progn
  )              ; end if
 )               ; end progn
)                ; end if step 2
[...]
```

Figure 2. The index_filter function

The output stage

The selection of the fits function of the conditions set (economic, manufacturing and functioning) by TFDES for the assemblies made as CAD applications will determine the choosing and subsequent positioning of the tolerances and limit devices corresponding to the ISO standards values for shop drawings components. Determining the algorithms needed, the structure of the program and developing an expert type program at code level led to finalizing the representation knowledge in an expert system building process in CAD system.

5. CONCLUSION AND FUTURE WORK

The work described in this paper has been also implemented in educational process of our students. The major conclusions are:

a) The program TFDES developed considers most of the requirements needed in a selection of tolerances and fits in design.

b) Solid groups and subgroups information with open access to its data structure and functionality is necessary for an expert system used for ISO tolerances field and jigs.

The following future work is recommended:

a) Developing a user manual for the software.

b) Turning the expert program commercial into a commercial product that can be subsequently integrated into the complex CAD systems.

c) Developing a manual containing the expert system for dimensional tolerances in AutoCAD.

6. REFERENCES

Anjanappa, M.; Courtright, M. J. & Anand, D. K. (1991). Manufacturability Analysis for a Flexible Manufacturing Cell, *Journal of Mechanical Design*, Vol. 113, December 1991, pp. 372-376

Desrochers, A. & Rivière, A. (1997). A Matrix Approach to the Representation of Tolerance Zones and Clearances, *The International Journal of Advanced Manufacturing Technology*, Springer-Verlag London Ltd, Volume 13, Number 9, September 1997, pp. 630, ISSN: 0268-3768

Kalta, M. and Davies, B. J. (2001). Integration of 2D CAD Models of Turned Components with CAPP, *Engineering with Computers*, Springer-Verlag London Ltd, Volume 17, Number 3, October 2001, pp. 315-325, ISSN: 0177-0667

Parkinson, D. B. (1982). Assessment and Optimization of Dimensional Tolerances, *Computer Aided Design*, Vol.17, No.4, May 1982, pp. 191-199

Romanescu, C. & Pruna, L. (2007). *Expert Systems Methodology*, The 2nd International Conference on Engineering Graphics and Design 2007, June 2007, Galati

Sobh, T. M. & Henderson, T. C. (1999). Tolerance Representation and Analysis in Industrial Inspection, *Journal of Intelligent and Robotic Systems*, Springer Science+Business Media B.V., Formerly Kluwer Academic Publishers B.V., Volume 24, Number 4, 1999, pp. 387-401, ISSN: 0921-0296

Annals of DAAAM for 2008 & Proceedings of the 19th International DAAAM Symposium, ISSN 1726-9679
Editor B. Katalinic, Published by DAAAM International, Vienna, Austria 2008
Make Harmony Between Technology and Nature, and Your Mind will Fly Free as a Bird

ON THE DESIGN OF AN EXPERIMENTAL INSTALLATION TO DETERMINE THE KINEMATIC PARAMETERS DURING HUMAN LOCOMOTION

ROSCA, I[leana] & RADU, C[iprian]

Abstract: *Human locomotion is a phenomenon that progresses in real time, so to determine the kinematic parameters during human gait it's necessary to choose a right method to be concluding in this way. Considering these, we have chosen the high-speed photography method, which consists in continuously video recording of the human subjects during locomotion (4.5 ÷ 5.6 km/h) on the force plate, followed afterwards by image discrete quantity, which succeed in 1/250 frames/second. The discrete frames can be edited using image edit software, finally to obtain the successive positions of the markers, which are positioned in the lower limb's joints (hip, knee and ankle joints), heel and the second metatarsal's pick.*
Key words: *camera, kinematic, marker, trajectory*

1. INTRODUCTION

The design of external prosthetic elements of locomotor system involves previously some complex kinematic and dynamic studies of the interested anatomical system.

Kinematics is part of mechanics concerned with the description of movement. The kinematics description of movement encompasses positions and angular positions of body segments. To determine these unknowns there are o lot of methods, such as: exoskeletal method, accelerometry method and stereometric method (Medved, 2001).

The stereometric method is the largest and the most diverse group used to determine the kinematic parameters of human movement. This measuring method may be classified in several ways. This classification includes: close range analytical photogrammetry fundamentals method and the high-speed photography method. The last method is the most used that because is the simplest and the less affected by errors method. Besides attaining a very high spatial resolution and accuracy of less then 1%, the basic quality of the photographic method is its non-invasiveness, not disturbing the subject and no expensive costs (Vaughan et al., 1999).

A further, important comparative advantage of this method is the permanent storage of the complete photographic record of the movement structure recorded. This offers a rich informational background during analysis, as well as a possibility to redefine characteristic body points in later analysis. This is further suitable for combined qualitative and quantitative analysis (Vaughan et al, 1999).

2. EXPERIMENTAL DESIGN SOLUTION

2.1 Experimental objectives

The experimental objectives consist in determining of kinematic parameters of lower right limb, more exactly the angles between anatomical segments during locomotion. These angles are represented by: the angle between thigh and shank (α angle), the angle between and shank and posterior side of foot (β angle), the angle between shank and frontal side of leg (δ angle) and the angle between foot and ground (φ angle), during stance phases of normal locomotion (see figure 1).

Also, for a complete biomechanical analyze we have decided to determine the trajectories of lower limb's joints during normal locomotion.

2.2 Measuring method

Biped human locomotion is a phenomenon that progresses in real time, so it's necessary to choose a right method to be concluding in this way. Since we try to analyze qualitatively some continuously and successive positions of lower limb's joints during normal locomotion, it's necessary a high precision visual analyze.

Once, human eye is perceptible for a distinct number of 24 different frames per second (for 24 frames per second the images will be in continuously movement), it's necessary that the tracking to be made to a high speed frequency that is comparable with locomotion speed and then to bring it close to the normal observation frequency of human eye (Hunt & Lataman, 2006).

Considering these, we have chosen the high-speed photography method, which consists in continuously video recording of the human subjects during normal locomotion (4.5 ÷ 5.6 km/h), followed afterwards by image discrete quantity, which succeed in 1/250 frames per second.

The discrete frames can be edited using image edit software, finally to obtain the successive positions of the markers, which are positioned in the lower limb's joints (hip, knee and ankle joints), heel and the second metatarsal's pick.

2.3 Experimental installation description

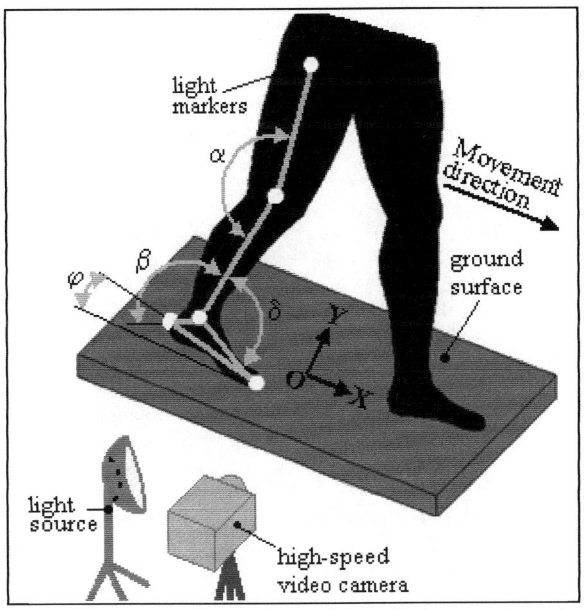

Fig. 1. The proposed experimental installation to determine kinematic parameters during normal walking.

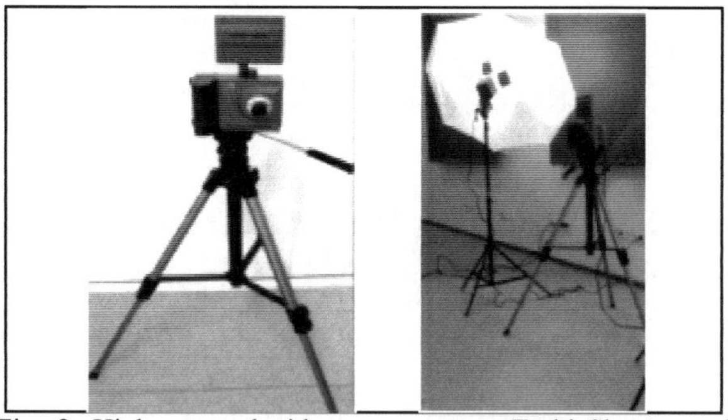

Fig. 2. High – speed video camera type TrobleShooter and illumination system used for this experiment.

To determine the kinematic parameters of human normal movement is necessary to design an experimental installation. The proposed experimental installation, presented in figure 1, is compound from: a high speed video camera TroubleShooter, three light sources, five light markers and a central acquisition system (Dell laptop).

High-speed video camera is a device that allows analyzing video processes in a very short time. This video camera records an image clip with a higher speed then eye can do and afterward plays it back in a very short time, allowing to observe, to measure and to analyze the events (Panagiotakis & Tziritas, 2004). TroubleShooter video camera is produced by Fastec Images Company, and has a maximum video resolution of 640 x 480 pixels and a record speed of 25 to 500 frames/second in normal conditions (see figure 2).

Video camera was embedded on a tripod, perpendicularly on the human subject's movement direction, on an optimal distance of 3500 mm. This distance permits a good visualization of the human subject from down chest's level and implicit correct visualization of all markers mounted on the lower limb.

Minimum illumination for a recording rate of 25 frames/second and a maximum resolution of 640 x 480 pixels is 10 lux. The higher number of frames the higher illumination needed (Rosca & Radu, 2008).

For our experiment we have used a normal optical objective with a focal distance of 12 mm and a recording rate of 250 frames/second with a maximum resolution of 640 x 480 pixels. Because of that, we have used three light sources (halogen light) with a total illumination of 17548.4 lux. Using this video camera and an image software editor (Adobe After Effects), it helps as to determine the lower limb's joints trajectories and the angle between human anatomical segments, during human locomotion.

The *light markers* are used in combination with a high-speed camera, to determine kinematic parameters of lower limb's joints. There are five markers, which are placed on the hip, knee and ankle joint, on the heel and the second metatarsal's pick. They have a spherical shape, with a 30 mm diameter and they were painted with a special paint, which is highly sensible to the halogen light (see figure 3) (Roşca & Radu, 2008).

Also, the lower limbs of the human subject are dressed up in black-moulded pants because the marker's colour is in highly contrast to the pant's colour. For concluding results, the markers have to be placed exactly on the interest areas, more precisely the marker's centre has to coincide to the joint's centre off rotation, and heel and second metatarsal's markers have to be on the same axis. In advance, to determine the right position of each marker, we have made repetitive palpations of the anatomic areas. Anyway, the marker's placement is relative, that because it's impossible to determine the right position of the joint's centre of rotation and on the other hand because of the relative sliding between moulded pants and skin tegument.

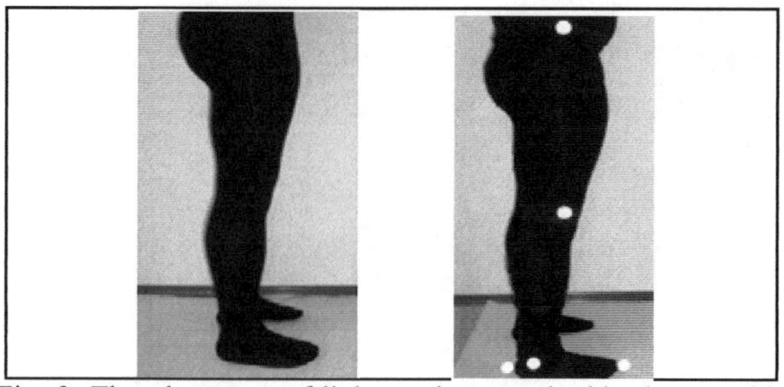

Fig. 3. The placement of light markers on the hip, knee, ankle joint, on the heel and second metatarsal area.

For a correct interpretation of the recorded movie data, by using geometrical correction equation, we have determinated the maximum error of image deformation, on the vertical and horizontal directions (Roşca & Radu, 2008):

$$\Delta m = \frac{100 \cdot 1}{L \sin \alpha}, \qquad (1)$$

where: Δm is maximum error of image deformation [%], l is the length of light marker [mm] (l = 30 mm), L is the distance between marker centre and objective centre of video camera [mm] and α is the angle between recording axe and light marker direction movement [°].

Thus, for a video camera height of approximately 600 mm to the ground, a recording maximum distance of 2130 mm and a distance of 3500 mm between camera central optical axe and the symmetrical axe of light markers, we have determinate that for each marker, for both recording axes, the maximum error of image deformations is 0.857%.

3. CONCLUSIONS

The proposed measuring method is a very simple and efficient method and also quite accurate, that because the recorded images are affected by deformation errors in proportion of 0.857%, which is suitable for the error limits admissible for this measuring method.

The aim of this paper is to design an experimental installation, based on some theoretical information about measuring of locomotion kinematics and the next step of our study is to process all recorded data obtained by using this experimental installation, to determine the kinematic parameters during human locomotion. This is done by using editing software, named Adobe After Effects 6.5 software, in the following scientific paper, called: *Image Treatment Process to obtain Kinematic Parameters During Human Locomotion*, published in the same proceeding.

4. REFERENCES

Hunt, J.L. & Latam, V. (2006). *Control of Biped Locomotion*, Springer Publisher, ISSN 0140-0118, Berlin, Germany.

Medved, V. (2001). *Measurement of Human Locomotion*, Congress Library, ISBN 0-8493-7675-0, USA.

Panagiotakis, C. & Tziritas G. (2004). *Recognition and Tracking of the Members of a Moving Human Body*, Springer Publisher, ISSN 0302-9743, Berlin, Germany.

Rosca, I.C & Radu, C. (2008). *On a Mechatronic System to Determine Dynamic Parameters of the Human Limb During Locomotion Used For Artificial Muscles and Implant Design*, 6th International DAAAM Baltic Conference, ISBN 978-9985-59-783-5, Tallin, Estonia.

Vaughan, C.L.; Davis, B.L. & O'Connor, J.C. (1999). *Dynamics of Human Gait*, Kiboho Publisher, ISBN 0-620-23558-6, Cape Town, South Africa.

Annals of DAAAM for 2008 & Proceedings of the 19th International DAAAM Symposium, ISSN 1726-9679
Editor B. Katalinic, Published by DAAAM International, Vienna, Austria 2008
Make Harmony Between Technology and Nature, and Your Mind will Fly Free as a Bird

EXPERT SYSTEM FOR INFORMATION SYSTEM RISK ESTIMATE AT THE ENTERPRISE LEVEL

ROSU, S[ebastian] M[arius]; GURAN, M[arius] & DRAGOMIRESCU, C[ristian]

Abstract: *As a result of a new product development paradigm, there is a greater need for software tools to risk estimation and to effectively support the formal representation, collect and exchange of product information, during the product development stage. In this paper we describe the basic constrains and components for a project management process (PMP) and risk management process (RMP) as part of project management process and we present a model for information system risk estimate using an expert system.*
Key words: *Risk management, expert system, knowledge base.*

1. INTRODUCTION

In order to survive, whether organizations are composed of one or many enterprise (holding) or organizations are small, medium or large, it is necessary to learn from the past, supervise the present and plan the future. In this regard, the enterprises continue to implement IT strategies & architectures, to improve manufacture, research, products quality, sales, and services and to control costs. All enterprises have a local area network, an intranet or/and Internet, servers and workstations for operations, administration and management that work together for the same objective: profits. The enterprises uses Internet or Internet technologies to attract, retain and cultivate relationships with customers, streamline supply-chain, manufacturing, and procurement systems and automate corporate processes to deliver the right products and services to customers quickly and cost-effectively (Rosu et al., 2007), also to capture, explore, analyze, and automate corporate processes information on customers and company operations in order to provide better business decisions. As a result of a new product development paradigm, there is a greater need for software tools to effectively support the formal representation, collect and exchange product information during the product development stage (Dragoi et al., 2005). In a real meaning, an e-business is any business that uses Internet or Internet technologies (Dragoi, 2003) to attract, retain and cultivate relationships with customers, streamline supply-chain, manufacturing, and procurement systems and automate corporate processes to deliver the right products and services to customers quickly and cost-effectively, also to capture, explore, analyze, and automate corporate processes information on customers and company operations in order to provide better business decisions.

2. RMP AS PART OF PMP

Project Management is a method of organizing and managing resources (data, information, knowledge, skills, tools, techniques, etc.) so that the project is completed within defined destination, quality, time and cost constraints. Generally, projects need to be performed and delivered under certain constraints. Traditionally, these constraints have been listed as destination, time, and quality and cost (see figure 1 where each side represents a constraint).

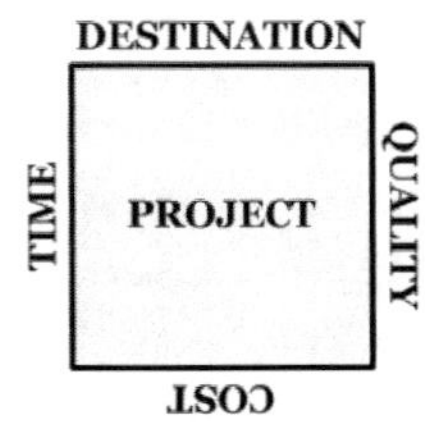

Fig. 1. The project constrains.

In the enterprise, we can distinguish some basic project components (steps) in the development of a project (Rosu & Draghici, 2007): start of project study, start project planning, start project execution, basic project design ready, project introduction, project monitoring and controlling, release project product, project product acceptance, close project and retrospective investigation of project. Enterprise project management is all that mix of components of control, communications, leadership, and teamwork, resource management, which goes into a successful project basis on complete quality process components. Good project management deals with three factors: time, cost and performance. In order to bring the many components of a large project into control there is a large toolkit of techniques, methodologies, and tools. These techniques provide the tools for managing different components involved in a project: planning and scheduling, developing a product, managing financial and capital resources, and monitoring progress - see the PMP model in figure 2 (Rosu et al., 2007).

Enterprise project development presupposes knowledge and assumption of multiple risks. The risk management process includes three phases:

- Risk identification;
- Risk analyses;
- Risk feedback.

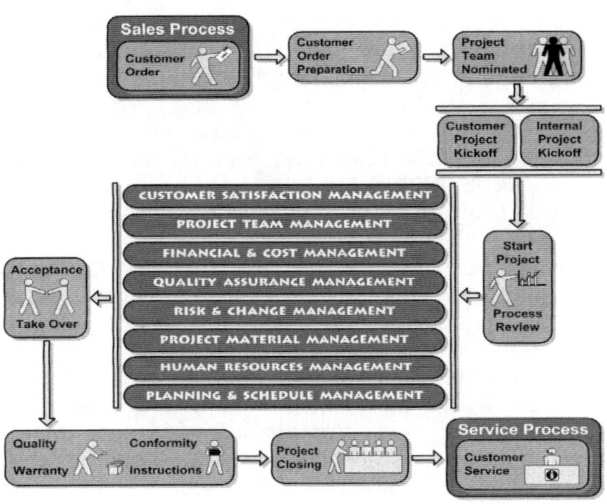

Fig. 2. The Project Management (PMP) – Process Overview.

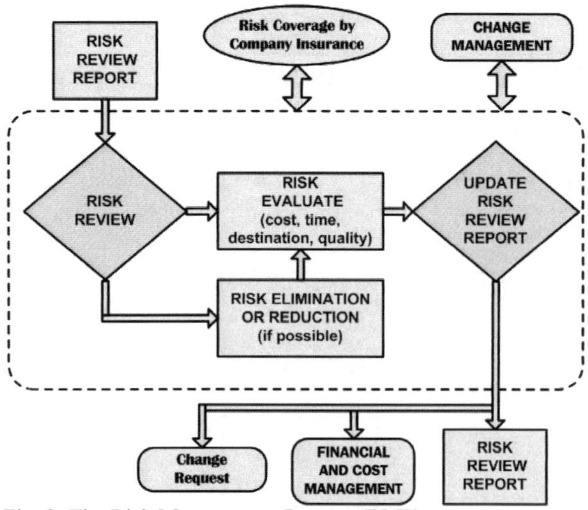

Fig. 3. The Risk Management Process (RMP).

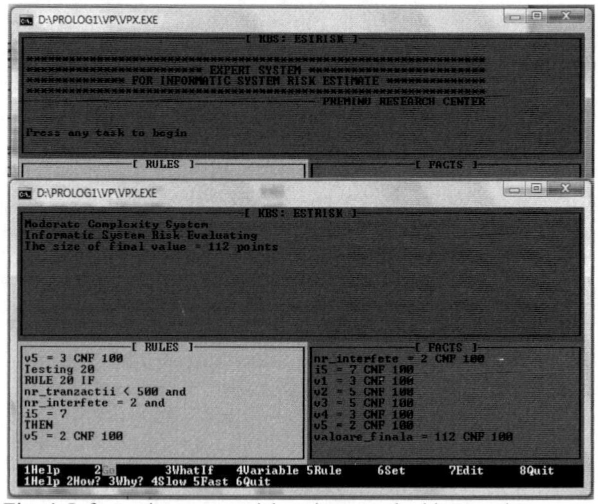

Fig. 4. Informatics system risk estimate – the kb consults.

In the enterprise activities risk eludes probability to not perform the establish objectives such as:

- Performance – quality standards failure;
- Schedule – execution terms failure;
- Costs – budget exceeding.

Risk factors are all factors that can have probability to deviate a plan. Risk management process is an important component of a successful project development process with informational system support. Risk is the net negative impact of the exercise of vulnerability, considering both the probability and the impact of occurrence. Risk management is the process of identifying risk, assessing risk, and taking steps to reduce risk to an acceptable level (see the RMP model in figure 3).

The risk management process should not be treated primarily as a technical function carried out by the IT experts who operate and manage the IT system, but as an essential management function of the organization.

3. AN EXPERT SYSTEM FOR RISK ESTIMATION

Information systems evaluation problem become important by dimension, cost, strategic place occupied by the system advance in time. For projects development in good conditions, the information systems must be assured and surveyed from reliability, quality and maintenance viewpoint. All information system has effects at project development activity, effects that make up in risk factors.

We analyze all these factors and elaborate an expert system implemented in VP-Expert (we used expert system generator - VP-Expert version 2.1, by Brian Sawyer, Educational Version, distributed by Paperback Software International), which effectuates risk evaluation for an information system, emerged in harnessed in an enterprise pursuant to proper rules. Production rules form the knowledge representation model used in this work. In *ESTRISK.KBS* knowledge base there are *if-then* structure rules excluding the rules for inference engine operations.

Before make the knowledge base, it was establishing the code variables: activity character, utility for project team and for management team, revenues, cost, users number, interfaces project tem proportion, etc.

For all variables, we assign a value (between 1 up to 5) adequate to minimum and maximum risk level. Each value has an importance expressed by a factor with predetermined values (between 1 up to 10). For all variables, the pondered factors must be introduced from keyboard.

According to final score, the evaluated informatics system receives a qualifying – *very complex*, *moderate complexity* or *simple system* (see figure 4 the evaluation results).

4. CONCLUSION

In this new digital era, enterprises use automated information technology (IT) systems to process their information for better support of their missions, risk management plays a critical role in protecting an organization's information assets, and therefore its mission, from IT - related risk. Risk management process is an important component of a successful project development process with informational system support.

Here, it was describe the risk management as part of project management process, and was present a model for information system risk estimate using an expert system realized at PREMINV Research Center (Polytechnic University of Bucharest) in a university and Romanian SME partnership.

In Romania just few companies developed proper risk measure and cover mechanisms. In the future, we intend to realize other similar expert systems to evaluate all enterprise activities who involve risks.

5. REFERENCES

Rosu, S. M.; Draghici, A. & Guran, M. (2007). Knowledge Transfer in the Enterprise Business Intelligence, *Annals of DAAAM for 2007 & Proceedings of the 18th International DAAAM Symposium, October 24th - 27th*, B. Katalinic (Ed.), published by DAAAM International, Vienna, Austria, pp. 647-648, ISBN 3-901509-58-5

Dragoi, G.; Cotet, C. E.; Patrascu, G.; Rosu, L. & Rosu, S. M. (2005). Internet/Intranet/Extranet-based systems for virtual product development environment in the CESICED platform, *Proceedings of the 8th Conference on Management of Innovative Technologies (MIT'2005)*, 24th - 26th September 2005, pp. 67-72, ISBN 961-6238-96-5, Fiesa - Piran, Slovenia

Dragoi G. (2003). *Intreprinderea Integrata: metodele, modele, tehnici si instrumente de dezvoltare si realizare a produselor,* published by PolitehnicaPress, Bucharest

Roşu, S. M. & Draghici, A. (2008). Information system risk estimate as part of project development by virtual teams using expert systems, *Academic Journal of Manufacturing Engineering*, no. 2/2008, pp. 135-142, ISSN 1583-7904

Roşu, S. M., Drăgoi, G., Guran, M., Roşu, L. (2007). Material management process with database support for virtual project development by virtual teams at the enterprise level, *Proceedings of the 9th International Conference on Management of Innovative Technologies - MIT'2007*, October 9th - 10th, pp. 99-106, ISBN 978-961-6536-19-6, Fiesa - Piran, Slovenia

Annals of DAAAM for 2008 & Proceedings of the 19th International DAAAM Symposium, ISSN 1726-9679
Editor B. Katalinic, Published by DAAAM International, Vienna, Austria 2008
Make Harmony Between Technology and Nature, and Your Mind will Fly Free as a Bird

ENTERPRISE E-LEARNING SITES DEVELOPMENT USING OMT METHODOLOGY

ROSU, S[ebastian] M[arius]; ROSU, L[uminita] & GURAN, M[arius]

Abstract: *In order to keep pace with competitor companies it is imperative for the enterprises to ensure a permanent training to their employees, training adapted to market changes or trends and enterprise specifically. In that direction, e-learning represents today the best alternative solution to traditional methods for organizations considering that the employees have all time access to the enterprise resources, knowledge and information. In this paper, there will be provided a model for e-learning sites built in the enterprises with open source software (OSS) support.*
Key words: *Enterprise, site architecture, OMT methodology.*

1. INTRODUCTION

E-learning is an efficient method to put at employees disposal needed the knowledge and information's for to keep pace with quick market changes and furnish the possibility to check their assimilated knowledge level. People, companies and countries which have the most information's, knowledge, abilities and the biggest capacity to create, to take and efficient change the information's had the biggest success chance in the new Internet economy, based on information's. E-learning begins to be use in the training departments of companies, schools and universities like an alternative or complementary to traditional training methods. This training is process assist by Internet or by means of Internet/Intranet/Extranet platform. Actually, e-learning integrate many solutions which can be used at organization level. By e-learning it is created for employees a possibility to exploit the existing knowledge resources by knowledge presentation in a more accessible format, individualized and pointed to training beneficiary. Nowadays, the enterprises manage the professionals re-conversion for their people based on e-learning products or start to development properly e-learning trainings systems.

2. AN E-LEARNING SITE ARCHITECTURE

A method for developing an e-learning site is using OMT (Object Modelling Technique) methodology. Object model built is one of the biggest OMT methodology fazes. An object is an abstract concept and it is built for problem understanding before solution implementation. Generally, an object is an entity that can be distinguished among other entities and have significance in modeling application context. Each object has one specific identity even if it is similar to other objects. An object class describes a multitude of objects with similar properties, similar behaviors and similar relations regard others objects. For a site designing, OMT method can be particularized (Rosu & Guran, 2007) as well as figure 1. The Web pages modeling such as an object have proper characteristics. It is easy to observe that an object model shows us the real static system structure. The object model is represented by charts, graphs which bows are object and arches representing the relations between objects. They are also representing the relation between web pages and databases. For this site there were identified and created the following classes: *index*, *main_page*, *main_theory*, *main_solving_problems*, etc.

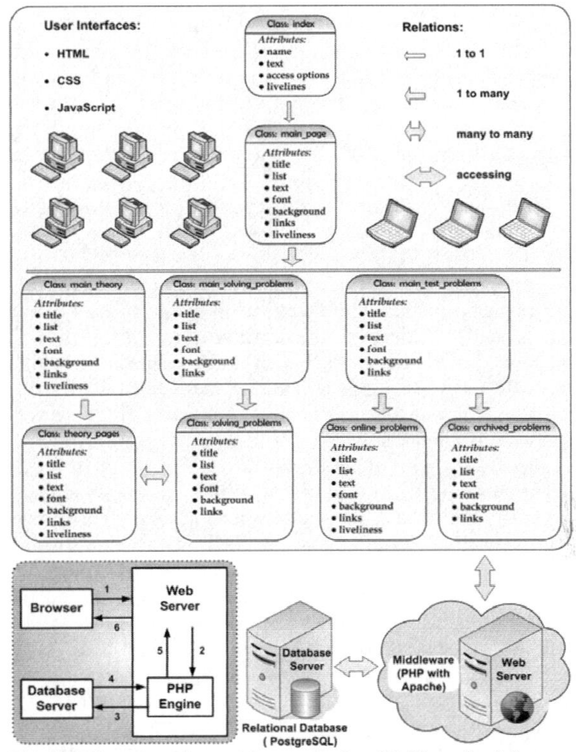

Fig. 1. E-learning site architecture using OMT methodology

After that, there were identified associations between objects, relations, attributes and there was created the database. The database development using this model requests an application server and a database server. Application servers are responsible for communications between users. Usual steps in users working process with databases are the following:

- The user makes a request for certain dates (e.g. a special form, a rapport or a situation).
- The application server receives the request, locates the files and gives request to PHP for processing.
- The PHP starts processing, makes the connection with database and sends the request.
- The Database server receives the request, processing and sent to PHP the result.
- PHP returns the result to application server.
- Server applications send the result to the user.

Regarding software, our option was to use open source software with free sources, these products having a remarkable quality because these projects often get together distinguished programmers and because everything is public, open and anybody can come. Sources access and modified possibility is a great advantage even if very rarely somebody makes these modifications. Certainly, in the actual market the price makes the rules. Enterprise software solution is influence by shrift, medium or long-term calculation. Decision can be take consequence a comparative analysis (Rosu et al., 2007).

3. SOFTWARE SUPPORT

For added, accessed and processed database information is necessary to have a database administration system and for this site was used PostgreSQL, an object-relational database management system. A Relational Database Management System (RDBMS) is a complex software program (Darie et al., 2006), which purpose is to store, manage and retrieve data as quickly and reliable as possible. PostgreSQL use the SQL language - Structured Query Language, the most prevalent standardized language for database interrogation (The PostgreSQL Global Development Group, 1996-2003). There were created *beginner*, *medium*, *advanced* and *master* tables for tests evidence (see figure 2) in the e-learning database. The corresponding files (with download tests) can be put on server in a specific folder (e.g. Pdf for *files.pdf*, Doc for *files.doc*, etc.). From the downloaded page (e.g. archieved.html) the database is accessed and after that the selected test can be downloaded.

All PostgreSQL features are support by graphical interface of PgAdmin III, interface who makes administration to be very easy.

A language that can translate the browser request, process these request, interact with software on server to satisfy the requests and to indicate to the server exactly which need to send back to user's browser is request. These languages, which work with the server and can execute these, are middleware. Such as middleware languages class was used PHP, a server-side, web scripting software. PHP is a scripting language specialized for the applications developing through code integration in the HTML document. PHP syntax derived from C, Java and Perl. The difference between PHP and other similar products, such as JavaScript is, that PHP is executing on the server when JavaScript is execute on the user computer. A HTTP (HyperText Transport Protocol) server is requested for application rolling. A good choice is Apache because of its flexibility, portability, safety and extensibility.

The interfaces with users are indicating to be simply used, friendly, and attractive, without many graphical pictures or animations because they can make loading operation slower. For the forms there can be used different language as HTML (HyperText Markup Language), CSS (Cascading Style Sheets) or JavaScript.

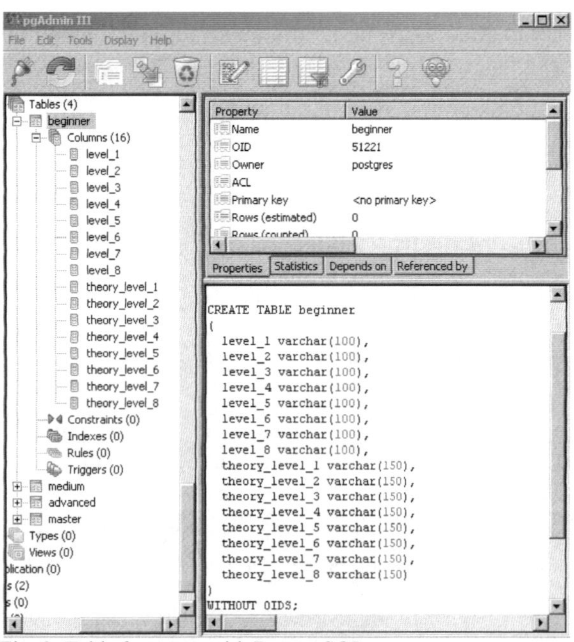

Fig. 2. Table *beginner* with PostgreSQL

Easy to used is HTML because this language can be edited in any text editor software and integrated the PHP script. The most used HTML editors are Notepad for Windows and Pico for Linux. JavaScript gives us the capability to make dynamic Web pages by integrating a script in the HTML structure. In this mode it can put the process behind the button or can make actions when users get over the mouse cursor over an HTML element or a *document* object. JavaScript can make Web page to be alive, more interesting, more fun and isn't very difficult to learn.

4. CONCLUSION

We present here a work method for an e-learning site realization at the enterprise level according to OMT methodology and using OSS. Our option was to use open source software with free sources, these products having a remarkable quality because these projects often get together distinguished programmers and because everything is public, open and anybody can come. In his multiple forms, e-learning offers a lot of advantages for enterprise and for employees such as: a personalized training experience, low cost (e.g. travel cost for training elimination), access from many parts (it is necessary just a Intranet connection), collaboration learning based on mutual reliance and exchange of information between participants, total accessibility to courses and complementary materials (e-library), access to all resources at the same time, learning in the proper rhythm, in a personal manner, interactive technologies utilization, better understood courses presented by means of multimedia, storage rooms and numerous training personnel elimination, the employees' ability to choose the course materials for printing and the materials for electronic format keeping, shipping low cost, courses dispensed in modules, each module having at the end an evaluating test – at the beginning and the end of the course provide tests to see the employees' knowledge level, diversity learning mediums integration (liveliness, movies, sound, etc.) (Rosu et al., 2007).

The validation and development of e-learning sites using OMT methodology will be carried out based on a practical application in a university and Romanian SME partnership. In the future, we intend to apply this model for medical sites development such as support for Romanian e-Health activities.

5. REFERENCES

Darie, C.; Balanescu, E. & Bucica, M. (2006). *Beginning PHP and PostgreSQL E-Commerce: From Novice to Professional*, ISBN 1-59059-648-x, published by Apress, December

Rosu, S. M. & Guran, M. (2007). E-learning sites development in the enterprise using open source software, *Academic Journal of Manufacturing Engineering*, supplement, number 3/2007, pp. 136-142, ISSN 1583-7904

Roşu, S. M.; Guran, M. & Roşu, L. (2007). Enterprise Databases Support for Product Development Using Open Source Software, *Annals of DAAAM for 2007 & Proceedings of the 18th International DAAAM Symposium*, October 24th - 27th, B. Katalinic (Ed.), published by DAAAM International, Vienna, Austria, pp. 649-650, ISSN 1726-9679, ISBN 3-901509-58-5

Roşu, S. M.; Guran, M. & Mateescu, L. M. (2007). E-learning sites development using Open Source Software support in the enterprise, *Proceedings of the 4th Balkan Region Conference on Engineering Education*, July 12th – 14th, Lucian Blaga University Press, Sibiu, pp. 98-103, ISSN 1843-6730.

The PostgreSQL Global Development Group (1996-2003). PostgreSQL 7.4.2 Documentation, *Download from: http://www.postgresql.org/docs/manuals*

Annals of DAAAM for 2008 & Proceedings of the 19th International DAAAM Symposium, ISSN 1726-9679
Editor B. Katalinic, Published by DAAAM International, Vienna, Austria 2008
Make Harmony Between Technology and Nature, and Your Mind will Fly Free as a Bird
*Young Researcher and Scientist Paper / * Supervisor, Mentor*

AUDIT RISK AND INITIATIVES OF IMPROVEMENT

ROTARU, H[oratiu]

Abstract: *In this work, the author describes the audit risk and ways of it's improvement that represent interest to the users of financial statement. One aspect of importance is represented by the materiality threshold, whose use arises from the auditor's judgement. Moreover the author presents the opinion reflected in the Big 4 and non Big 4 literature, not just the recommendation of the audit standards. Another aspect of this work I would like to underling are the proposed initiatives suggested by the author that will surely improve the audit quality. The author considers that the audit literature gained a developed opinion considering the proposed initiatives, that follow as a result of an attentive study of audit risk literature and the author's practical experience.*

Key words: *audit, risk, model, opinion, users, improvement.*

1. INTRODUCTION

The audit risk, as defined by SAS 107 „Audit Risk and Materiality in Conducting an Audit", represents the positive opinion of an auditor of financial statements that are materially misstated, opinion that unknowingly failed to be appropriately modified.

As the title of this work suggests, there is a correlation between audit risk, materiality (that must be considered together) and the result of an audit that will influence the users of the financial statements, the audit opinion.

2. USERS OF FINANCIAL STATEMENTS

In order to establish the audit risk and assess materiality the auditor must take in concern the recommendation of the audit standards (SAS no. 107, 2007), witch refers to user needs. SAS no. 107 specifies that the user must have knowledge of business activities, of limitations of an audit because of estimation and materiality and to be willing to study the financial statements that represent interest to him/her. The auditor shouldn't consider just the needs of specific individuals, according to SAS 107, but the needs of users as a group.

One problem that arises speaking of users, in the audit field, is the possible confusion between audit risk, economic risk and audit failure. If the characteristics of the users specified by SAS no. 107 are accomplished by one user, these should understand that an auditor should be considered guilty only in the case of audit failure, this meaning that the audit report contains an opinion where significant misstatements were not detected and „the judgement of a reasonable person relying on the information would have changed or influenced by the omission or misstatement" (ICAEW, Accounting Recommendation 2.301, 1967). The economic risk shouldn't make an auditor responsible, from the perspective of user that has an „appropriate knowledge of business and economic activities and accounting" as the standards suggest.

3. AUDIT RISK AT THE ACCOUNT BALANCE

The audit risk (AR) consists of inherent (IR) and control risk (CR) and detection risk (DR) (Graham & Messier, 2006).

The inherent risk (IR) represents a misstatement that could be material, individually or aggregated, when there are no related controls. An example is that cash is more likely to be stolen than a building. Another example is represented by the technological developments that make a product more susceptible to overstatement (William & Lizabeth 2005).

The control risk (CR) represents a misstatement that could be material, individually or aggregated, and not be prevented or detected by the entity's internal control.

These risks exist independently of the audit of the financial statements. Inherent risk and control risk may be assessed separately, but if combined they describe the risk of material misstatement (RMM). It is recommended that the auditor assesses RMM throughout tests of controls to prove how effective the internal control is and to obtain audit evidence, although it represents a professional judgement rather that a precise measurement.

The detection risk (DR) represents the misstatement that could be material, individually or aggregate that the auditor will not detect. This is possible because the auditor uses sampling and does not verify 100 percent of a class of transactions, for example. Other factors are the selection of an inappropriate audit procedure or misapplying an appropriate audit procedure. In order to reduce the level of detection risk the author suggests proper assignment of personnel, the use of professional skepticism, supervision of audit stages performed.

The detection risk consists of substantive analytical procedures (AP) and tests of details (TD).

The components of audit risk can be assessed in nonquantative terms (low, medium, high) but the author suggests the use of quantative terms, percentages. In addition to this SAS no. 107 presents an audit model that shows the relation between audit risk components, that isn't meant to be applied as a mathematic formula, but a lot of professionals use it as a mathematic formula. The audit risk model is:

$$AR = RMM \times DR \quad \text{Or} \quad AR = IR \times CR \times TD \times AP \qquad (1)$$

The audit risk model can be also represented as:

$$DR = AR/(IR \times CR) \qquad (2)$$

This is how a risk model can be used for detection risk of an account (Rotaru 2008):

1. The asses of detection risk at 5 percents, due to auditor's professional judgement, possible because the audit risk is assessed also at 5 percents;

2. The assessment of inherent risk at 60 percents, due to the auditor's professional judgement; this account is semnificative, the calculation is complex, a significant number of transactions are recorded in this account yearly;

3. The assessment of control risk at 30 percentage because the control structure proved effective in prior years and few misstatements were detected with tests of control.

The detection risk is assessed at:

$$0.05\%/(0.6\% \times 0.3\%) = 0.28\% \qquad (3)$$

4. THE ASSESMENT OF MATERIALITY

The assessment of materiality represents the professional judgement of the auditor and often represents a percentage applied to a benchmark; when choosing a benchmark he/she should have in mind one of the following elements:

- Assets, liabilities, equity, income, expenses or financial position, performance and cash flows;
- The nature of the entity;
- The size of the entity, nature of ownership and he sources of financing (gross profit, profit before tax).

The auditor takes in concern, for example, the profit before tax for a profit oriented entity, but this wouldn't be a suitable benchmark for a non-for-profit entity; for an asset based entity the benchmark should be the assets (Graham, L.; Messier, W.).

In the professional literature, Big 4 and non-Big four manual, an example of materiality threshold is represented, in the opinion of multiple authors, by:

0.2% to 10%	of turnover	Plumhoff [1952]; Anderson [1977]; Towers [1986]; Woolf [1994], Turley and Cooper [1991]
0.5% to 5%	of gross profit	Carmichael [1969]
0.5% to 36%	of net profits	Bernstein [1967,1970]; Copeland and Frederick [1968]; Neumann [1968]; Thomas [1978];Turley and Cooper [1991]
0.1% to 5%	of total assets	Woolf[1994]; Turley and Cooper [1991]
10% to 20%	of related total	Plumhoff [1952]; Mitchell [1972]; Towers [1986]
0.5% to 5%	of gross profit	Carmichael [1969]
0.1% to 10%	of total assets	Mitchell [1972]; Woolf[1994]; Turley and Cooper [1991]
10%	of total liabilities	Mitchell [1972]
10%	of equity	Mitchell [1972]
0.2% to 10%	of turnover	Woolf [1994]; Turley and Cooper [1991]
3.3% to 36%	of net profit	Turley and Cooper [1991]; Chong [1992, 1993]

Table 1. Opinion reflected in the professional literature

The auditor should take in concern not only quantative considerations (Fogarty et al., 2006), but also qualitative materiality considerations, qualitive factors that the auditor may consider relevant include:

a) Potential effect on trends, especially profitably trends;
b) A misstatement that changes a loss into an income, and vice-versa;
c) A misstatement that has as effect the increasing management's compensation (for example the award of bonuses)
d) The circumstances of surrounding effects for misstatements involving fraud, illegal acts, conflicts of interest;
e) The significance of misstatements to reasonable users, for example earnings to investors and equity amounts to creditors;
f) The existence of offsetting effects of individual different misstatements;
g) The possibility that an immaterial misstatement could gain an material effect in the future periods;
h) The cost of making the correction, it may not be beneficial to correct the immaterial misstatements;
i) The risk that possible undetected misstatements would affect the auditor's evaluation.

5. INITIATIVES THAT IMPROVE AUDIT QUALITY

Initiatives proposed by the author considered to be able to help firms improve audit quality, effectiveness and efficiency as they enter the stage of the risk assessment are:

1. Assure that all members of the audit team understand the procedures assigned to them, this is important because of certain audit procedures used;
2. Encourage active participation of partners and managers who have a complete understanding of the risk assessment process, thus there will result a more effective, efficient and higher application of the standards and risk assessment.
3. Tailor the risk assessment process by customizing internal control tools and templates for certain industries, the industry tailoring being a good implementation examples of risk assessment from prior audits;

   Tailoring tools and templates:

- Customized audit work area programs
- Planning documents, memos and risk assessment summary forms;
- Internal control templates with control objectives, examples and scenarios, the following should be included:
  - ✓ Information technology control assessment;
  - ✓ Entity's level of control assessment;
  - ✓ Level of control assessment of the activity

4. Collaboration with experienced risk-assessment based auditors in audit engagements, in this way the audit team can be trained as the audit process of an engagement develops, communicating audit results, both positive and negative across the audit firm as the audit process develops;
5. Encourage industry team meetings in order to share ideas, questions and best practices and the result of these meetings to be further communicated to the audit team;
6. Collaboration with internal control an information technology experts in order to develop templates, tools and training which help the implementation of information technology in risk assessment;
7. Enable a internal control evaluation process in such a manner that identifies the controls that prevent and detect material misstatement, this way the audit team should be focused on the true risks of material misstatement;
8. Ensure that the risk assessment standards have been successfully implemented in the audit team in order to provide a more effective risk assessment result; one significant factor is the personnel of the audit firm and also the creativity of the implementation process.

6. CONCLUSION

Audit failure has cost large amounts of money and for many of them meant bankruptcy because of negative advertising. An auditor or audit team should thing twice when assessing the audit risk that it is very important for risk standards (104-111) to be implemented by an audit firm and should take in concern the opinion presented in this article that has as a result the improvement of audit quality, effectiveness and efficiency.

The author considers that the subject treated brings new insights to audit risk mainly by the methods proposed to be used by the auditor and will still be of interest to the public in the future when new ways of improvement are implemented in the field of audit risk.

7. REFERENCES

Fogarty, J.; Graham, L. & Schubert, D. (Jul 2006). *Assessing and Responding to Risks in a Financial Statement Audit*, Journal of Accountancy, Vol. 202, Iss. 1, p.43, New York;

Graham, L. & Messier, W. (May 2006) *Audit Risk and Materiality in Conducting an Audit*, Journal of Accountancy, Vol. 201, Iss. 5, p. 116-119, New York;

Rotaru, H. (May 2008) *Audit risk model - actual status and possibilities of improvement,* The International Economic Conference, University Publishing House, ISBN 978-973-739-594-8, Sibiu;

William M. Jr & Lizabeth A. (Fall 2005) *Inherent risk and control risk assessments*, Auditing, Vol. 19, Iss. 2, US;

SAS no. 107 „Audit Risk and Materiality in Conducting an Audit"

Annals of DAAAM for 2008 & Proceedings of the 19th International DAAAM Symposium, ISSN 1726-9679
Editor B. Katalinic, Published by DAAAM International, Vienna, Austria 2008
Make Harmony Between Technology and Nature, and Your Mind will Fly Free as a Bird
*Young Researcher and Scientist Paper / * Supervisor, Mentor*

UNIVERSE RECESSION POSSIBLY EXPLANED BY INVARIANTIVE MECHANICS

RUGESCU, D[ragos] R[onald]

Abstract: *Mathematical principles of the continuum mechanics are extended to the general relative motion. Four different types of relative time derivatives are proven. The derivative transport theorem of Reynolds proves being independent of the referential. The same approach is introduced in the Invariantive Mechanics of Onicescu. It shows that intrinsically the gravity possesses a repulsive term, responsible for the recession of the universe.*
Key words: *Continuum, mechanics, motion, cosmology*

1. INTRODUCTION

The sound mathematical principles of absolute motion of continua, like those previously introduced (Hamel, 1956; Iacob 1959; Onicescu, 1972; Liggett, 1994) are a proper basis for an extension to relative motion in large spaces. That means a motion is referred to a non-inertial, moving and high-order referential. Such principles are targeted at least for rigorously defining the motion of distant space systems, for the general treatment of fast developing flow fields in a vacuum and simulation of very large-scale motion of space matter (Rugescu 2000), usually related to star formation (Osterbrock, 2006). A step by step development is required.

2. EXTENTION OF EULER FIELD DESCRIPTION

As the fluid occupies entirely its given territory, the position vector may not be attached to a specific particle. One can imagine an *Eulerian* observer flatting free while inspecting the fluid territory from the tip of its absolute position vector $\mathbf{x}_A$ (the local one $\mathbf{x}$) and watching the fluid like in front of a small "window". This is the well-known *Eulerian field* description of a fluid. While the variables ($\mathbf{X}_A$, t) and ($\mathbf{X}$, t) are called *material*, the *Eulerian field description* is based on the independent variables ($\mathbf{x}_A$, t) and ($\mathbf{x}$, t) which are called *spatial variables*, namely the *absolute spatial variables* ($\mathbf{x}_A$, t) and the *local spatial variables* ($\mathbf{x}$, t).

A *scalar property* F of the fluid may then be described either as a function of the spatial variables ($\mathbf{x}_A$, t) and ($\mathbf{x}$, t), or of the material variables ($\mathbf{X}_A$, t) or ($\mathbf{X}$, t). The *absolute and local field descriptions* of the property are distinct but give the same value for F,

$$F = F(\mathbf{x}_A, t) = F(\mathbf{x}, t). \tag{1}$$

The *absolute material description* and the *local material description*, give also the same value, which reads respectively

$$F = F(\mathbf{X}_A, t) = F(\mathbf{X}, t). \tag{2}$$

The Lagrangean description may be seen as a peculiar case of the Eulerian one, when the observer decides to move attached to a specific particle. Although the algebraic expression (2) of the function F may be very different, the value of F in one point and at one moment is unique,

$$F(x_{1A}, x_{2A}, x_{3A}, t) = F(x_1, x_2, x_3, t).$$

Things are going a completely different way when the rate of change of property F is considered. The four representations of the property lead to at least *four types of time derivatives*, possibly with distinct numerical values, for which distinct symbols are to be used. These four distinct derivatives are defined by the following partial time derivatives

$$\frac{\partial F}{\partial t} = \frac{\partial F(\mathbf{x}, t)}{\partial t} \quad \text{local, partial;} \tag{3}$$

$$\frac{\partial_A F}{\partial t} = \frac{\partial F(\mathbf{x}_A, t)}{\partial t} \quad \text{absolute, partial;} \tag{4}$$

$$\frac{dF}{dt} = \frac{\partial F(\mathbf{X}, t)}{\partial t} \quad \text{local, material;} \tag{5}$$

$$\frac{d_A F}{dt} = \frac{\partial F(\mathbf{X}_A, t)}{\partial t} \quad \text{absolute, material.} \tag{6}$$

In fact, the last two derivatives are nothing but the *total time derivative of F*, an invariant for the Euclidean transform as F is a scalar. They need no different notation and the index for the differential operator may be suppressed,

$$\frac{d_A F}{dt} = \frac{dF}{dt}. \tag{7}$$

Relations of equivalency between these derivatives may be produced on that basis. This particularly refers to the derivatives (3), (4) and (7).

Important is to note that an *arbitrarily* moving (neither with the fluid or continuum, nor with the frame) Eulerian observer feels a different rate of change of the property F and this derivative requires a specific notation. It is clear that neither $\mathbf{x}$ nor $\mathbf{x}_A$ are fixed for a general Eulerian observer. When the "floating" of the observer through the fluid is specified in the absolute referential, it writes

$$\frac{DF}{Dt} = \frac{\partial_A F}{\partial t} + \mathbf{w}_a \cdot \nabla F, \tag{8}$$

while the expression must be used

$$\frac{DF}{Dt} = \frac{\partial_A F}{\partial t} + \mathbf{w} \cdot \nabla F, \tag{9}$$

when the motion of the Eulerian observer is surveyed from the local referential. Here $\mathbf{w}_a$ is the absolute eigen-velocity of the observer, $\mathbf{w}$ is the relative or local value of that speed and the invariant value for the spatial gradient of F is

$$\nabla F = \frac{\partial F}{\partial x_{A,i}} \mathbf{e}_A^i = \frac{\partial F}{\partial x_i} \mathbf{e}^i. \tag{10}$$

Transfer between these derivatives was proven (Iacob, 1959; Rugescu, 2000).

3. THE INVARIANTIVE PRINCIPLE OF CARTAN

The Levi-Civita rigid referential, associated with the formula $\mathbf{v} = \mathbf{v}_M + \boldsymbol{\omega}\wedge\mathbf{x}$, is extended at astronomical distances by the seven-dimensional *phase-time* space $\mathbf{U}=\mathbf{S}_N \times\mathbf{E}_3\times T$. The basic invariantive mechanics, limited to small particles and velocities, is extended to large spaces by the Cartan derivative, or the *external derivative* of a scalar form Ω_δ, denoting a variation of the *impulse-energy* vector $(\mathbf{p},E)$. It should be expressed in the subspaces $\mathbf{S}_N$ and $\mathbf{E}_3$. A Cartesian T space is adopted and the scalar form, in the $\mathbf{S}_N$ sub-space, becomes

$$\Omega^{(i)}_{\delta,\mathrm{A}} = \int_m \mathrm{d}\Omega^{(i)}_{\delta,\mathrm{A}} = \int_m \left(\frac{\mathrm{d}_\mathrm{A}\xi}{dt}\cdot\delta_\mathrm{A}\xi - c^2\delta t\right)\mathrm{d}m_\mathrm{A} \,. \quad (11)$$

In the $\mathbf{E}_3$ sub-space it reads

$$\Omega^{(i)}_{\delta} = \int_m \left(\frac{\mathrm{d}\xi}{\mathrm{d}t}\cdot\delta\xi - c^2\delta t\right)dm \,.$$

The absolute position of the particle with the dynamic mass $\mathrm{d}m_\mathrm{A}$ is defined by the vector $\xi=\mathbf{x}+\mathbf{r}$, while the local position in $\mathbf{E}_3$, where the dynamic mass is $\mathrm{d}m$, is defined by $\mathbf{r}$. When an absolute displacement occurs, this is described by the vector $\delta_\mathrm{A}\xi(\delta_\mathrm{A}\xi_1,\delta_\mathrm{A}\xi_2,\delta_\mathrm{A}\xi_3)$. This is related to the displacement $\delta\mathbf{r}$ and the angular displacement $\delta_\mathrm{A}\boldsymbol{\alpha}$ by

$$\delta_\mathrm{A}\xi = \delta_\mathrm{A}\mathbf{x} + \delta_\mathrm{A}\boldsymbol{\alpha}\wedge\mathbf{r} + \delta\mathbf{r}$$

Here $\mathbf{x}$ is the position of the adopted origin of the moving non-inertial rigid referential for $\mathbf{E}_3$. This way the Onicescu operator, extended to the absolute movement in $\mathbf{S}_N$ stands in

$$\mathrm{D}_\mathrm{A}\omega_{\mathrm{A},\delta} = \mathrm{d}_\mathrm{A}\mathbf{p}_\mathrm{A}\cdot\delta_\mathrm{A}\xi + (E'(\alpha)\mathbf{p}_\mathrm{A}\,\mathrm{d}t - \mathrm{d}_\mathrm{A}\xi) - \mathrm{d}_\mathrm{A}E_\mathrm{A}\,\delta t$$

and for $\mathbf{E}_3$ in the formula

$$\mathrm{D}\omega_\delta = \mathrm{d}\mathbf{p}\cdot\delta\xi + (E'(\alpha)\mathbf{p}\mathrm{d}t - \mathrm{d}\xi) - \mathrm{d}E\,\delta t \,.$$

Computations are performed when constitutive non-variants for the *instant-position* and *impulse-energy* vectors are introduced. Derivatives stay unchanged while T space remains Cartesian. The dynamic mass is related to speed through

$$\frac{\partial m}{\partial \mathbf{p}^2} = \frac{1}{2}m\cdot c^2 \,.$$

This is an intrinsic inertial feature of all moving masses, a main proof of the invariantive mechanics. Extending to the case of two interacting particles, conservation of the impulse results

$$m_1\mathbf{v}_1 + m_2\mathbf{v}_2 + \mu(\mathbf{v}_1+\mathbf{v}_2) + \kappa(h_1+h_2)\mathbf{r} = \mathbf{C} \quad (12)$$

Denoting through ξ_i the projections of speeds on the distance vector $\mathbf{r}=\mathbf{r}_2-\mathbf{r}_1$, the law of motion returns to

$$1 + \frac{1}{c^2 r}M + \frac{1}{c^4}Nr^2 = 0 \,, \quad (13)$$

with μ the *gravity* and κ the *dilatational interaction mass*,

$$M = -\frac{1}{2}f\frac{m_1^0 m_2^0}{\sqrt{1-\mathbf{v}_1\mathbf{v}_2/c^2}}\cdot\frac{\xi_1+\xi_2}{m_1\xi_1+m_2\xi_2-C\cos(\mathbf{C},\mathbf{r})}\left(1+\frac{k}{c^2 r}\right)$$

$$N = -\frac{1}{2}g\frac{l_{12}}{m_1^0+m_2^0}\cdot\frac{m_1^0\xi_1+m_2^0\xi_2}{m_1\xi_1+m_2\xi_2-C\cos(\mathbf{C},\mathbf{r})}\cdot\frac{1}{\sqrt{1-u}} \,.$$

The dilatational mass is part of the elastic interaction that naturally appears in the general law of motion (13),

$$v_2\cos(\mathbf{v}_2,\mathbf{r}) - v_1\cos(\mathbf{v}_1,\mathbf{r}) = \frac{1}{2}\left|\frac{\dot N}{N}\right|\cdot\left|\frac{1+\dfrac{c^2}{r^3}\dfrac{\dot M}{\dot N}}{1-\dfrac{c^2}{2r^3}\dfrac{M}{N}}\right| = 0 \,. \quad (14)$$

We recognize here a value for the relative velocity between the bodies. Extension is made to a number of particles, associated to distant galaxies in the universe (Onicescu, 1974)

$$w = \frac{1}{2}\left|\frac{\dfrac{\dot N}{N}+\dfrac{c^2}{r^3}\dfrac{\dot M}{N}}{1-\dfrac{c^2}{2r^3}\dfrac{M}{N}}\right| r \,. \quad (15)$$

It shows that the relative velocity depends on the sign of the derivative of N. For very large spaces ($r\to\infty$) we have indeed

$$w = \frac{1}{2}\left|\frac{\dot N}{N}\right| r \,,$$

where $\dot N/N$ is not depending on r. Examining the relation, a distance may be found where the effect of the two forces cancels. This is a neutral gravisphere and its order of magnitude results from the value of the dilatational mass. The only determination to date (Onicescu, 1974) is found as

$$\kappa = \frac{9}{2}f\frac{m_1^0+m_2^0}{c^2r^2\sqrt{1-\mathbf{v}_1\mathbf{v}_2/c^2}}$$

4. CONCLUSION

The invariantive mechanics renders new inertial interaction properties, without the assumptions of the theory of relativity and this can be a start in developing the laws for the evolution of the universe. It is suggested that the properties of the so-called dark matter are explained (Rugescu, 2008) by the invariantive mechanics as stated above. New developments are necessary however to adapt the given theory to the physics of large spaces in general, where the relaxation of time is sometimes inevitable. The problem of large space structures may well be served by such developments.

4. REFERENCES

Hamel, G. (1956), *Mechanik der Kontinua* (*Mechanics of Continua*), Stuttgart

Iacob, C. (1959), *Mathematical Introduction to Fluid Mechanics*, Gauthier-Vilars, Paris

Liggett, J.A. (1994). *Fluid Mechanics*, McGraw-Hill, NY, USA

Onicescu, O. (1974), *Invariantive mechanics and the cosmology*, Ed. Rom. Acad., Bucharest

Osterbrock, D.E. & Ferland, G.J. (2006), *Astrophysics of Gaseous Nebulae and Active Galactic Nuclei,* 2-Ed. USB

Rugescu, R. D. (2000), On the Principles of Relative Motion of Continua, *Sci. Bulletin U.P.B.*, series A, **62**(2) p. 97-108

Rugescu, D. R. (2008), Gravity and black holes, *Paper on High School Symposium in Physics*, April 19

Annals of DAAAM for 2008 & Proceedings of the 19th International DAAAM Symposium, ISSN 1726-9679
Editor B. Katalinic, Published by DAAAM International, Vienna, Austria 2008
Make Harmony Between Technology and Nature, and Your Mind will Fly Free as a Bird

DESIGNING THE CHEMICAL KINETICS CALORIMETER AS A TOOL FOR ADVANCED THERMOCHEMICAL RESEARCH

RUGESCU, R[adu]; SILIVESTRU, V[alentin]; ALDEA, S[orin] & IONESCU, M[ircea] D[an]

Abstract: *The chemical kinetics calorimeter (CKC) is developed as a low cost and highly efficient tool for advanced thermochemical research. The research team had found that during isochoric combustion of solid fuel samples the equilibrium among the combustion products is interrupted at a high temperature of more than 1500 degrees Kelvin. The calorimeter is consequently able to act as a research tool in chemical kinetics in a much easier way than an actual motor. The technological design of such a combustor is presented, with emphasis on the pressure transducer positioning for fast unsteady measurements during combustion, visualization means, sealing and surface finishing for catalytic study.*
Key words: *Manufacturing, calorimetry, chemical kinetics.*

1. INTRODUCTION

The process of fast burning solid propellant samples within the limited closed space of a calorimeter is a typically static process. No chemical relaxation phenomena are expected and the equilibrium between the gaseous combustion products has no reason to be unfulfilled (Beckstead 2006; Schuricht 2001). Except for the fast self-burning of solids at the beginning of the process, when incomplete combustion sometimes occurs (Kubota 2007; Kuhl et al, 2006), the residing time of the chemical species considerably surpasses the time of chemical reactions (Westenberg & Favin 1962). It was therefore a surprise to find that the residual gas within the calorimeter, although slowly cooled down to the room temperature, presents a chemical composition of a high temperature gas (Rugescu 2005). The processes within the static calorimeters are further investigated, which involve a complex chemical behavior. Currently considered for high speed flows only (Bray 1959; Williams 1965; Hill et al. 1967), the chemical relaxation also occurred in these small and static devices, offering a simple and cheep means to investigate these processes in the laboratory. Considering the complex and very fast initial flow development within the combustor, we propose a multiple point measuring system with adequate pressure transducers and visualization means. This type of combustor is called *Chemical Kinetics Calorimeter* and such a precursor, simplified device was designed and manufactured by the COMOTI-UPB team for experimental investigation of the process.

The following criteria and principles were adopted for designing the CKC device:

1. Closed, high pressure vessel with V=283 cmc;
2. Easy and fast dismounting of the hood;
3. Secured sealing for pressures up to p_{max}=200 bar;
4. Electrical connector for sample ignition;
5. Overpressure release membrane security valve;
6. Manual valve for vacuum pump and gas discharge;
7. High rate variable pressure transducer mounting;
8. High strength, stainless steel vessel construction;
9. Inner surface covering for ceramic catalyst studies;
10. Electric remote operation of the combustor.

These design conditions were embedded into the device manufactured by the *Works for Powders* in Fagaras, Romania.

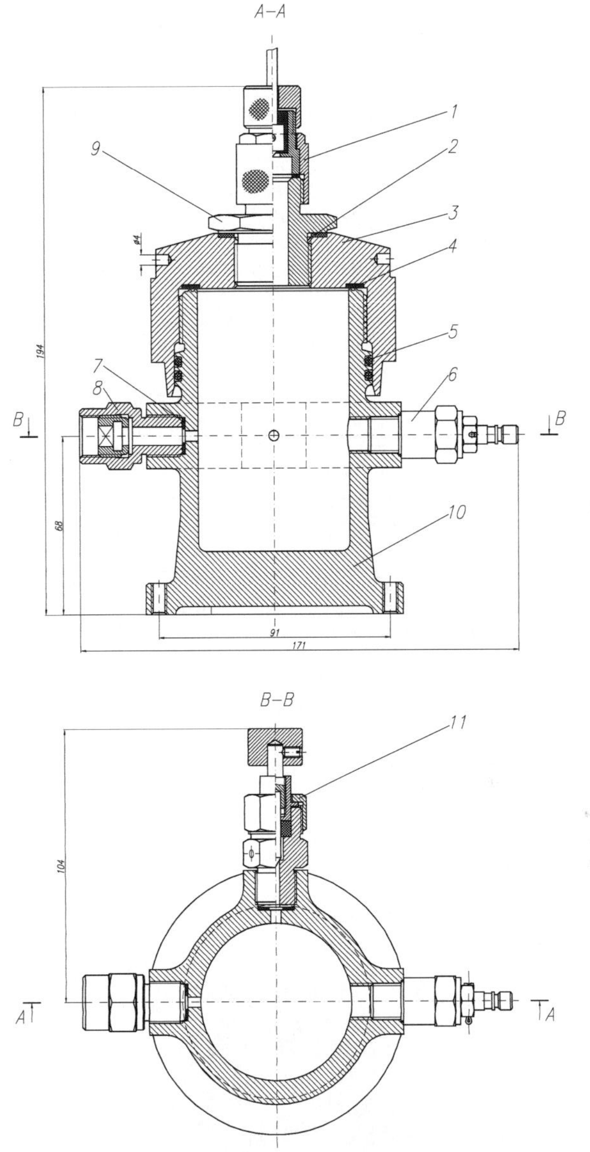

Fig. 1. Assembly of the UPB isochoric CKC combustor.

The combustor assembly includes the independent valves and igniter mounting that actually are not parts of the set.

2. TECHNOLOGY OF THE ISOCHORIC COMBUSTOR

The general draft of the set is given in fig. 1. The set consists of two main parts, connected through a screw mounting, secured with two standard "o-ring" seals beneath the upper hood. This highly simplifies the combustor's block design. The sub-assemblies are numbered and described below.

The numbered items in Fig. 1 represent the following parts:

1. Electrical fuse igniter mounting (aerospace standard);
2. Dome nipple elastic seal;
3. Dome (hood) with mounting lateral Ø4 mm holes;
4. Hood high-pressure elastic seals;
5. Overpressure double o-ring seals;
6. Manual valve for vacuum pump and gas discharge;
7. Elastic seal for the overpressure release security valve;
8. Overpressure explosive membrane security valve;
9. Dome nipple for electrical igniter mounting;
10. Main body of the isochoric combustor;
11. Fast pressure variation transducer mounting.

In the improved version of the device the upper electric connector in the hood is replaced with a sealed, quartz translucent visualization port, where a video CCD camera is fitted, while the igniter itself is re-located to the fourth lateral position within the enlarged flange from the middle body of the combustor, now free.

The material used tor manufacture the body and hood is a high strength, stainless C-120 Chromium steel, equivalent to the 205 Cr 115 alloy. The entire set is fixed with screws on the working table, as seen in Fig. 2 from below. The device is intended for use on the fast variable inner pressure recordings but not intended in the measurement of heat released through isochoric combustion. These constant volume heat measurements are performed in separate, standard calorimetric devices, through known standard procedures.

The electric fuse igniter deserves a special attention. It comprises a standard detonating primer, based on 2 grams of colloidal double base powder (uppermost piece in Fig. 2). The effect of this quantity of powder is adding it to the combustion effect of the main propellant cartridge and is consequently taken into account while performing the pressure variation measurements and its comparison with the numerical simulation. The heating effect released by the electric fuse, previously determined, is also considered, as a heat added to the heat released by combustion. The melted fuse, consisting of a coper alloy, does not add gaseous species in the mixture however and the presence of the mass and volume of melted fuse metal in the vessel is consequently neglected.

The explosive membrane safety valve is a standard type also, involving a calibrated coper membrane that covers an orifice with a diameter of 6 mm, just large enough to assure a safe gas discharge when an overpressure arises during propellant sample combustion. Between combustion runs, the safety valve is covered with a protection cap which is removed before any run of the CKC. The check-in list includes the removal of this protection cap. This enters routine safety procedures.

Fig. 2.The MRE-1B engine block drawing with parts.

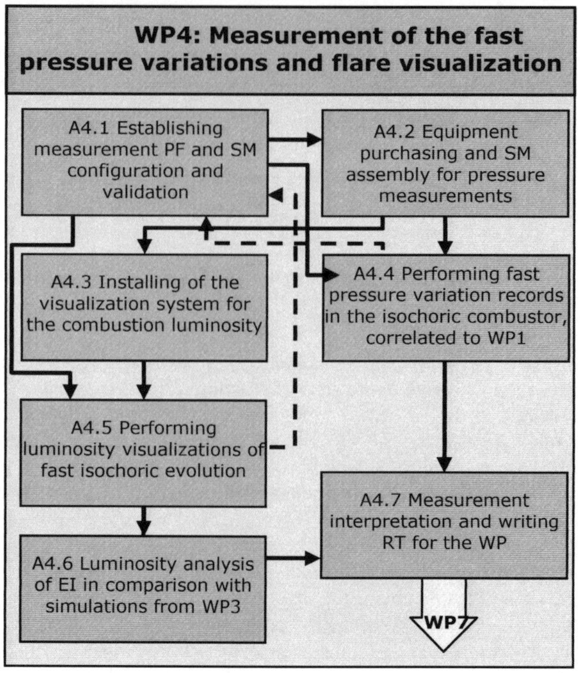

Fig. 3 Management plan of the CKC experimental exploitation.

3. EXPERIMENTAL TECHNOLOGY

The management of experimental measurements with the CKC device is developed under a national CNCSIS grant proposal. The flow chart in Fig. 3 (PF is for the factorial research plan, SM for the measuring system, EI for the isochoric evolution and RT for the technical report) shows that the PF is targeted on proving with high confidence the temperature of chemical freezing, as a basis for its explanation.

4. CONCLUSIONS

Technology of the CKC for simple and cheap chemical kinetics research is presented. The device may easily be used in the laboratory, replacing high costs reaction engines experiments for propellant combustion studies. It may open new avenues for accessible chemical kinetics studies, with major consequences for an improved combustion control with reduced emissions.

5. REFERENCES

Beckstead, M. (2006), Recent progress in modeling solid propellant combustion, *Combustion, Explosion, and Shock Waves* (*CESW*), Vol. 42, Nr. 6, pp. 623-641 (19).

Bray, K. N. C. (1959), Atomic recombination in a hypersonic wind-tunnel nozzle, J. Fluid Mech., **6**(1), p. 1-32.

Hill F. K.; Unger H. F. & Dickens W. (1967), *AIAA J.* 5(5) 54.

Kubota N. (2007), Propellants and Explosives-Thermochemical Aspects of Combustion, 2nd Ed., Wiley, N. York.

Kuhl, A., Neuwald, P., & Reichenbach, H. (2006), Effectiveness of combustion of shock-dispersed fuels in calorimeters of various volumes, *CESW*, **42**(6), p. 731.

Rugescu, R. D. (2005), Thermal Freezing in the Technology of Aerospace Propellants, *Eng. Meridian*, TUM Chisinau, **4**(4)

Schuricht, S. R. (2001), *Numerical simulation of high speed chemically reacting flows*, Purdue University, 162 pages; AAT 3075724.

Westenberg ,A. A. & Favin,S. (1962), *Nozzle Flow with Complex Chemical Reaction*, JHU, Appl. Phys. Lab, AD0275464.

Williams, J. C., III (1965), Correlation of the sudden freezing point in nonequilibrium nozzle flows, *AIAA J.* **3**(6), p. 1169.

Annals of DAAAM for 2008 & Proceedings of the 19th International DAAAM Symposium, ISSN 1726-9679
Editor B. Katalinic, Published by DAAAM International, Vienna, Austria 2008
Make Harmony Between Technology and Nature, and Your Mind will Fly Free as a Bird

DESIGNING OF AUTOMATED MANUFACTURING AND ASSEMBLY SYSTEMS

RUZAROVSKY, R[oman]; HORVATH, S[tefan] & VELISEK, K[arol]

Abstract: *Automation of manufacturing and assembly processes is the actual trend of machine industry. In this paper is described the proposition about design and planning of manufacturing and assembly devices and systems. This paper gives some preliminary methods of automated system design. The important step of automated manufacturing process design is the analysis of manufacturing and assembly process, the request designation of control system realization and also the several components definition of technical control system. It is necessary to create the control algorithm and to define the technical system for design of automated system. It is needed to find the reply to three basic questions for the right control system realization.*

Key words: *automation, assembly, manufacturing, designing, control system*

1. INTRODUCTION

Automation of manufacturing and assembly processes is the actual trend of machine industry (Velíšek et al., 2004). Supposition of automation process is the using of available technical devices to several operations realization of technological and helpful process. This main manufacturing process should be realized automatically by predefined program. The knowledge about automated technical devices is more important by realization of automated manufacturing and assembly. The important segment in automated manufacturing system is the realization of automatic process control system and the coordination of concrete control device.

2. REALIZATION REQUIREMENTS

The automation of manufacturing and assembly process can be run in term of the manufacturing profile, material flow or information flow. These basic elements of the manufacturing process are usually automated together in praxis, but it is not the condition (Velíšek et al., 2005). The process control is characteristic as the process organization in system that provides the achievement of the required final state; assembly or manufacturing process. The necessary element of control system is the back-connection information. The utilization of this information for the manufacturing process control could be only then if it is known the noted model of system behavior by means of that it is possible to create the control algorithm and technical system for its realization by defined aim.

3. ANALYSIS OF PROCESS CONTROL

It is needed to know the reply to three basic questions for the right control system realization (Fig.1). Those are the questions:

- **What has to be controlled?**

In this case it is necessary to define the mathematical models of controlled system by physical basis and identification. At the same time it is required to analysis the manufacturing process in term of the operative motion sequence. Also it is necessary to know the character of manufacturing object before the definition of automated control of the manufacturing process. At once also to analysis the manufacturing process in term of several in sequence connected operative motions. The important step of analysis of manufacturing process is the dividing into partial operations. The basic element of the manufacturing process is the operative motion

- **What shall we to use for control?**

In this case is necessary to choose the right devices of automatic control, the type of energy in the automated manufacturing process. The devices of the automatic control is the group of tools, devices, instruments a knowledge by these it could be realized the practical control of the manufacturing process type. The devices of the automatic control are in general all technical devices, these could enable to achieve, transfer, save, process and utilize the information and also all helpful devices, and these could enable the activity of those named technical devices.

- **What way shall we choose for control?**

The manufacturing process control is often the integral member of the manufacturing process automation. The control is understood as the technical advance by that the machines and devices in the limited system are influenced in its physical and technical values in terms of required method by the valid regularity.

The manufacturing process is created with more quantity of elements, operations and motions. It is existed the more quantity of interacted connections. The control system that is used in the manufacturing process has the hierarchical structure. The hierarchical structure is the one of the manufacturing process basic feature. It is necessary to give one's time to the control of the individual motions and operations and also to interaction cooperation and communication between them and with the environment by control realization. It is used more method, systemic technologies and programming instruments.

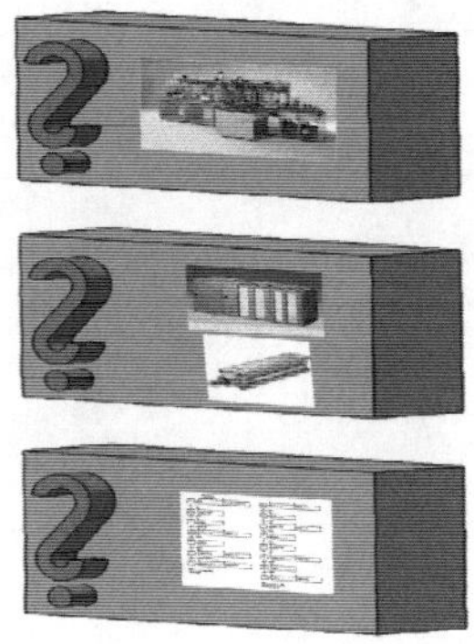

Fig. 1. Manufacturing process control analysis

4. MANUFACTURING PROCESS DEFINITION

In the fist time it is necessary to analysis the manufacturing and assembly process by the several criteria, in term of the technological advance. The model of the manufacturing and assembly process has the task to describe the solved problem. The model describes the relationship between the manufacturing process and the user and is divided into (Fig.2):

- topological model,
- physical model,
- mathematical model,
- numerical model.

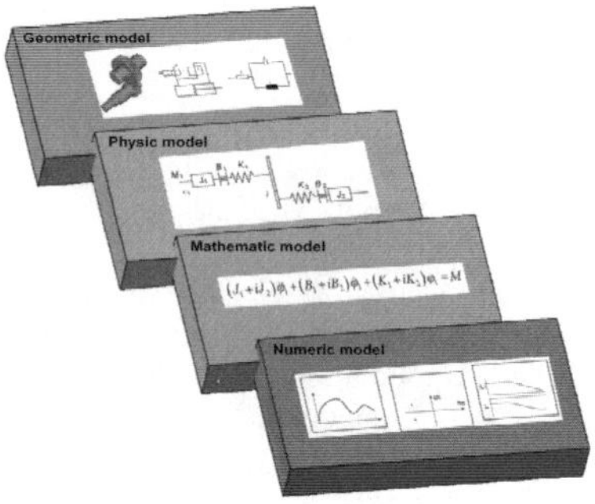

Fig. 2. Manufacturing process model

5. TECHNICAL CONTROL STRUCTURE OF AUTOMATED SYSTEMS

The manufacturing and assembly process is realized effectively by digital and numerical control. It is existed a new functions, new control elements, diagnostic e.g. (Zvolenský, 2007). It is realized the control of manufacturing assembly process by integrated control system. The basic control structure of manufacturing system is designed out of five basic elements:

- mechanical systems (mechanics)
- sensor systems (sensors)
- motion systems (actuators)
- communication and information systems (informatics)
- process systems (processors).

The control system of manufacturing and assembly systems is possible to understand as integration of combined elements of control system (Fig.3). The control system is realized as mechanical, pneumatic, hydraulic, electro-pneumatic or electro-hydraulic control structure. That means the control system could be to accept as an individual system that is represented by hierarchical ordered control system of manufacturing and assembly process.

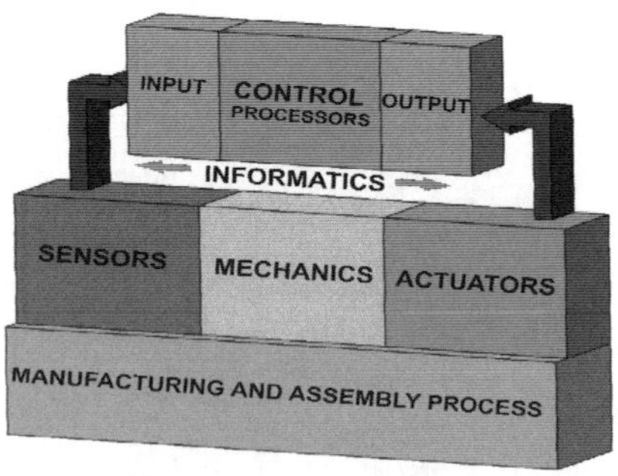

Fig. 3. Schematic description of controlled manufacturing process

5. FURTHER RESEARCH

The field of research is the definition of methods of design and realization the automated assembly and manufacturing system. The method makes the provision for optimal choice of technical devices and tools that are needed for realization of manufacturing and assembly systems and devices. The technical tools and devices are defined on base of input parameters. At the same time the method of design will make the provision for choice of using the technical devices and tools by using of signal energy. At the second time the choice of control system in term of input request of manufacturing process. The result of research is the method of design and planning of automated manufacturing and assembly system and its verification by using of the technological devices and tools. With help of assembly and manufacturing device design methodology we would be able to create peripheral assembly and manufacturing device, which will operate in automated manufacturing and assembly cell (Javorová, 2007). (Fig.4).

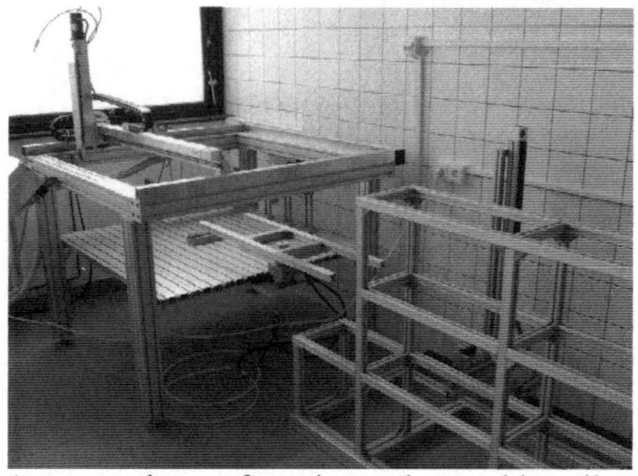

Fig. 4. Automated manufacturing and assembly cell

6. CONCLUSION

It is important to define the manufacturing process and process control before the realization of designing and planning method and design of automated manufacturing system. The designing method includes the analysis of manufacturing process and its partition to minimal defined operation motion, the definition of control algorithm and also sequence of separate motions of manufacturing process. The result of analysis is the method of design and planning of automated device and machine and also choice of optimal technical devices and tools of control. The method of design as aim of research would be realized and verified at the Institute of manufacturing systems and applied mechanics.

7. REFERENCES

Javorová, A. (2007): Assembly and manufacturing cell, *Research and Development in Mechanical Industry - RaDMI 2007*, p. 599-602, ISBN 86-83803-22-4, Belgrade/Serbia/, September 2007, High Technical Mechanical School of Trstenik, Trstenik

Velíšek, K. & Katalinič, B. (2004). *Manufacturing systems I.*, Vydavateľstvo STU, ISBN 80-227-2009-7, Bratislava

Velíšek, K., Pecháček, F., Košťál, P. & Štefánek, M. (2005) *Assembly machines and devices,* Vydavateľstvo STU, ISBN 80-227-2187-5, Bratislava

Velíšek, K., Javorová, A. & Košťál, P. (2007). Flexible assembly and manufacturing cell, *Academic Journal of Manufacturing Engineering*. Vol. 5, No. 2, p. 141-144, ISSN 1583-7904

Zvolenský, R. (2007). Design of automated disassembly devices. *Research and Development in Mechanical Industry - RaDMI 2007*, p. 697-700, ISBN 86-83803-22-4, Belgrade/Serbia/, September 2007, High Technical Mechanical School of Trstenik, Trstenik

ACKNOWLEDGEMENT

This paper was created thanks to the national grant: VEGA 1/3193/06 - ***Multifunctional manufacturing and assembly cell.***

Annals of DAAAM for 2008 & Proceedings of the 19th International DAAAM Symposium, ISSN 1726-9679
Editor B. Katalinic, Published by DAAAM International, Vienna, Austria 2008
Make Harmony Between Technology and Nature, and Your Mind will Fly Free as a Bird
2nd DAAAM-ICAT

MACHINING SIMULATION DESIGN BY SOFTWARE CATIA

RUZAROVSKY, R[oman]; MATUSOVA, M[iriam] & ZVOLENSKY, R[adovan]

Abstract: *The goal of this paper is a description about machining simulation. The machining simulation is represented by selected non-rotary parts. It is necessary to choose the specification of parts, clamping fixtures, methods of setting the place and clamping. Individual part models, clamping fixtures and also machining simulations are realized by software CATIA VR15. The paper is described the individual machining operations that are used by machining simulation. It is created the manual of one part machining. The manual includes the part creation, combination creation to definition of data that are needed to machining simulation. The data are: mainlining reduction, tool axis, type of tool, motion direction of tool, machining tolerance e.g. The program will be generated NC program after the ending of machining simulation. The NC program will be used for next machining without order modification. In conclusion the paper is about machining simulation and also about design and procedure to realization of machining simulation.*

Key words: *design, machining simulation, drilling simulation, software Catia, clamping fixture*

1. INTRODUCTION

The market of today is more and more dynamical and the market development is managed by customers. It is not only needed mass manufacturing of products. Manufacturing company has to compete by broad range of products in present. Manufacturing programme variety expressively raises complexity of manufacturing design and control. The most important presumption of company good result is the request achievement to general production cycle reduction, from product design to its customer delivery with minimum production cost (Kuric, 2002). The other requirements that are asked to the manufacturing system and modern machine industry in present are oriented to:

- high productivity in short production load and production assortment flexibility,
- through time shortening of manufacturing,
- storage and reserve state reduction,
- production costs depression,
- increase of time and performance utilization of machines,
- increase of production and products quality.

2. SIMULATION

It is possible to characterize the term simulation as automatic software control of total blocks format (programme sentences – lines). It is normally synchronous displayed the tool motion in simulation with text file control (or binary file of CAD/CAM systems). This motion characterizes the next motion of cutting tool by machining and it is showed the curved line of tool axis position change. It is removed the material through the cutting tool on tool machine which is displayed by most of graphic simulation systems. It is important to have the product documentation before and after tool machining in machining simulation creation. Software has to save the work-piece design and also semi-finished product in the memory. It is needed to define the semi-finished product in term of material type. It is qualified the physical, chemical and physic-chemical factors that could be problematic. It is possible to choose the machining method on the ground of the defined factors. (Hrušková, 2007).

It is determined the choice of clamping fixture by design of machining method. Eventually it could be:

- Universal clamping fixtures,
- Grouped clamping fixtures,
- Modular clamping fixtures,
- Special clamping fixtures.

2.1 Modular clamping fixtures

Elements of the modular clamping fixture system are the one of the right selected choice of clamping fixtures. The flexibility of modular clamping fixtures enables to adapt its position to the work-piece (Javorová, 2004). It is possible to create the database of modular fixture clamping elements by software Catia. The database elements are deigned as parametric. It is possible to change the dimension of clamping fixture in the work-piece clamping design process (Charbulová, 2007). The selection of the concrete modular clamping fixtures is conditioned by the type of production. The costs of the clamping fixture preparation could rise by the piece production.

3. THE SIMULATION DESIGN PROCESS

It is realized the machining simulation in one of the working environment in Catia software. The name of working environment module is "Surface Machining" (Start => Machining => Surface Machining). After the activation of relevant icon will be divided the tree into three parts:

- Process List – technology information,
- Product List – before designed model geometry information,
- Resources List – information about the machine and tools.
-

By design of simulation is it possible to select the machining such as roughing, milling, drilling e.g. It has to set the basic parameters of machining after the selection of concrete machining method. It is possible to select from more alternatives by hole drilling design (Fig. 1).

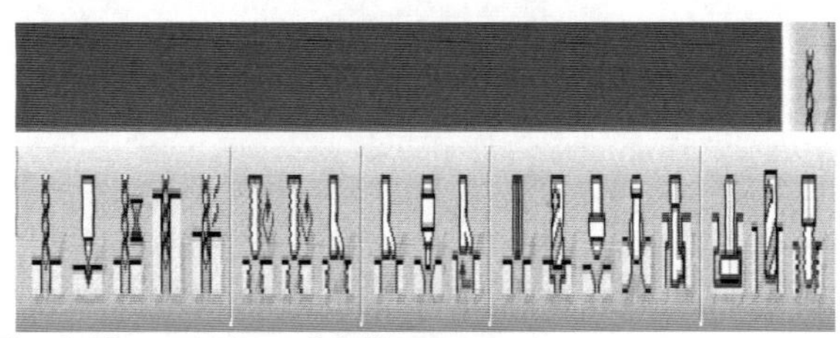

Fig. 1. The variations of drilling process.

It is activated the table for drilling parameters definition after selection of concrete drilling method. It is going to be set in colour change of drill tool after the confirmation of basic data drilling as are machining surface, breakthrough of tool, drilling holes e.g. (Fig. 2).

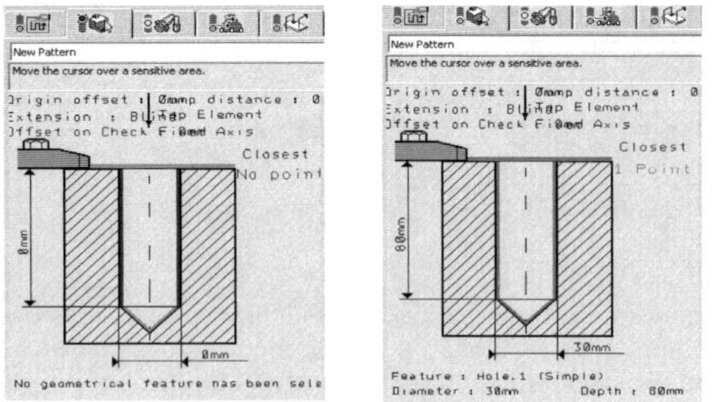

Fig. 2 The basic parameters of drilling process.

After the click on fist mark on the top of table will be appeared the table (Fig. 3). In this table it is necessary to assign the distance of tool breaking into the work-piece (Approach clearance “A”), the cut depth before break (Max. depth of cut “Dc”) and the cut breaking with drive up (Retract offset “Or”). The distance “A” will be choosing to stop the tool crash in the work-piece. The distance “Dc” will be choosing in 2 mm and the distance “Or” in 1 mm e.g. So the drill will be drilled into the depth in 2 mm and then will be coming back in 1 mm and the whole process will be repeated to requested depth. At the end it will be changed in depth mode “By tip (Dt)” to “By shoulder (Ds)”.

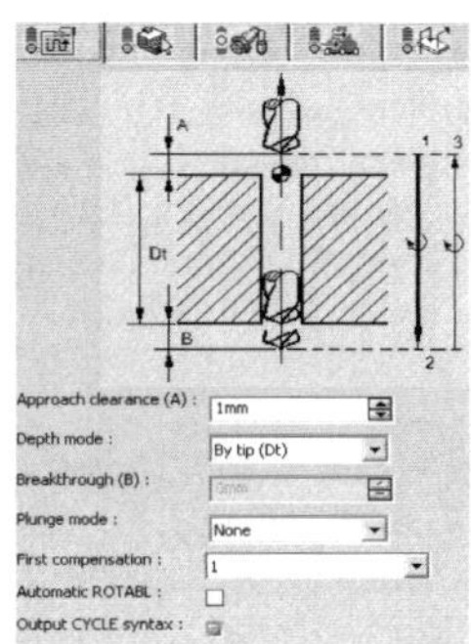

Fig.3 The parameters of the first mark.

It will be appeared the table with definition of select drill parameters after the click on third mark on the top of table (Fig.4). In this table will be defined the length of tool (the length of cutting edge, the total length e.g.). These lengths it is possible to adapt into tool which can be choosing for drilling process. In the next step it will be also choosing tools, drill, milling cutter e.g.

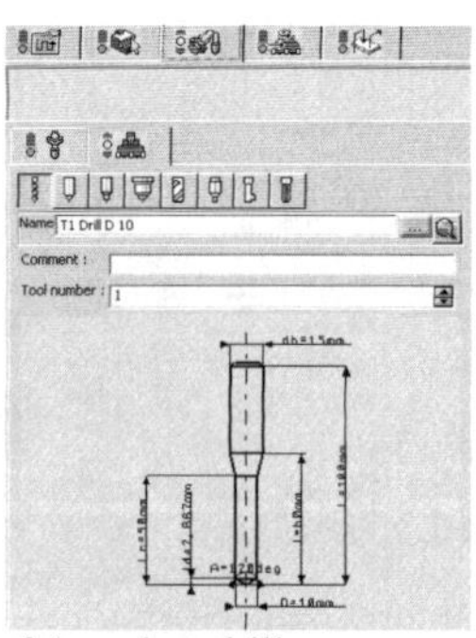

Fig.4 Parameters of the select drill.

It is needed to define the clamping elements, the machine and the coordinate system following the definition of basic drilling parameters. The simulation will be realized in the defined coordinate system (Fig. 3).

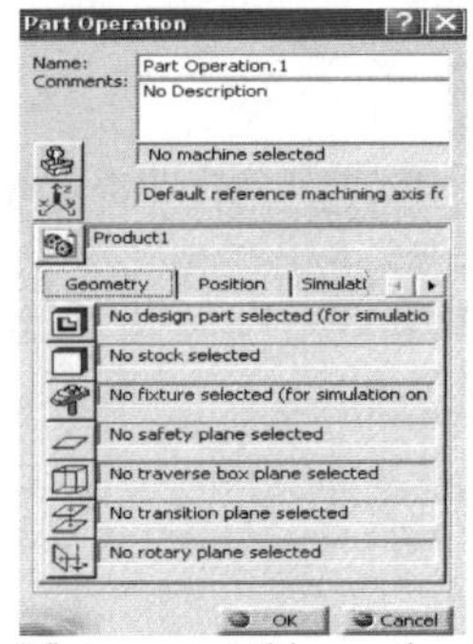

Fig. 5 Definition of fixtures, machine and coordinate system

Following is integrated the drilling operation with right defined drill into the sequential graph by the software CATIA V5.After the confirmation of the simulation running it came to the graphics simulation, Fig.6.

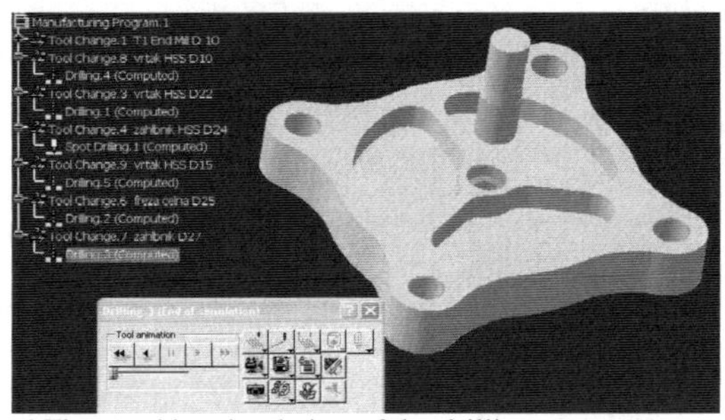

Fig. 6 The graphics simulation of the drilling process

The advantage of the software CATIA is the possibility to choice of machining method. The advance of the milling simulation design is the same as by the drilling operation. It is possible to generate NC program thereafter. The disadvantage of the simulation design is the necessity to leave the exact course training of software by the operator.

4. CONCLUSION

The creation of the database of modular clamping fixture elements is able to provide for effective and flexible selection and the following application of the clamping fixtures by work solutions in the praxis. It is enabled to reduce costs to repeating creations of the same or the design similar clamping fixtures.

On the model example it is possible to verify the designed methodology of clamping fixture creation right in the software. In this way it is increased the productivity by design and also it is removed the potential lost time. The lost time is caused by the incorrect assembly and formation of the fixture clamping.

5. REFERENCES

Charbulová,M.;Pecháček, F.(2007).*Modular clamping systems.* pp. 149-153, ISBN 86-83803-22-4,7th International Conference RaDMI 2007,Trstenik

Hrušková,E. (2007). *Selection efficiency of parametrical designed elements.* pp .191-194, ISBN 973-9087-83-3, 8th International Conference Modern Technologies in Manufacturing 2007, Cluj-Napoca

Javorová,A. (2004).*Modular fixtures analysis.* pp. 215-218, ISBN80-227-2076-3, Akademická Dubnica 2004

Kuric, I.; and kol. (2002). *Počítačom podporované systémy v strojárstve*, EDIS, ISBN Žilina 351s.

Annals of DAAAM for 2008 & Proceedings of the 19th International DAAAM Symposium, ISSN 1726-9679
Editor B. Katalinic, Published by DAAAM International, Vienna, Austria 2008
Make Harmony Between Technology and Nature, and Your Mind will Fly Free as a Bird

THE MAXIMUM DELAMINATION FORCE AT DIFFERENT TYPES OF COMPOSITE STRUCTURES

SABAU, E[milia]; IANCAU, H[oratiu]; HANCU, L[iana]; BERE, P[aul] & POPESCU, C[onstantin]

Abstract: *The paper presents some results of analytical-numerical-experimental study regarding delamination at composite materials structures. For the "delamination" test we realized an experimental installation to can determine de maximum delamination force hereupon the composite is submitted. Propagation of interlaminar cracks is one of the main damage mechanisms in composite materials. Propagation of delamination mainly depends on the initial crack extension and its loading mode. We developed a comparative study of the maximum delamination force obtained for different types of materials.*
Key words: *delamination force, composite materials, fibres glass, polyester resin.*

1. INTRODUCTION

The composite materials represent arrangements of reinforce materials in continuous or discontinuous fibers, covered by a matrix having a lower mechanical resistance, (Zgura & Moga, 1999). From technical point of view, the notion of composite materials reports to materials which possess the following properties:
- they are created artificial, through the combination of different components;
- represent a combination of at least two special materials from chemical point of view, between which exists a distinct separation surface;
- presents properties which no other component taken separately could have.

The interface includes the contact surface between matrix and surfaces of the fibers, as well as the next area. The nature and the interfacial force connection depend on the structure, on the surface characteristics of the fibers (roughness, determined surface, porosity, crystal dimension, the presence of functional classifications chemical active) and on structural characteristics of the matrix (chemical composition, macromolecules conformation). The maximum performances of the composite materials are obtained in the case of realization o optimum adhesion between matrix and fibers, (Misca, 2001), (Iancau & Nemes, 2005). Shocks, impact, loadings or repeated cyclic stresses can cause the laminate to separate at the interface between two layers, a condition known as delamination. This phenomenon can be produced locally or can cover a large area which leads in the end at the breaking up of the composite structure, (Sabau, 2006). Delamination is interlaminar damage; it is the separation on a certain length of the sheets from the interior composite material, (Kachanov, 1990). Delamination can be produced during production or exploitation of the composite structure and can have a great variety of causes (the unsuitable choice of the component materials, technological imperfections, stress solicitation, and so on).

2. EXPERIMENTAL INSTALLATION

For the "delamination" test we realized an installation, which has the components represented in figure 1.

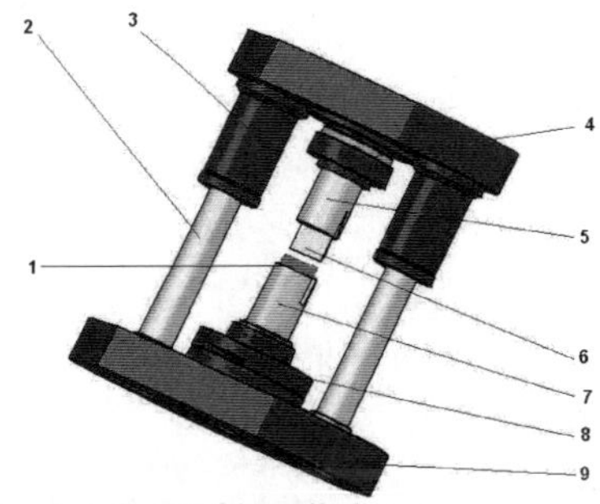

Fig. 1. The experimental installation

Fig. 2. Instron 1196 pressing machine

The components of installation are: composite sample (1), guiding columns (2), guiding socket (3), clamping screw (4), knife holder (5), slice knife – role of sample slicing (composite structure) (6); sample holder – fixes the sample (7); holder – fixes the sample holder (8), foot plate (9).

The experimental installation is actuated with the help of the Instron 1196 pressing machine, figure 2. The method consists of a force application along the main axle of the sample, with a constant speed, until the delamination sample. The operation is realized through the drop displacement on the vertical direction until the knife is penetrating the sample, moment that the delamination takes place. The sample is fixed on the holder (7) through the agency of the clamping screw. The values of the forces can be read on the diagrams delivered of the press.

3. THE DELAMINATION TEST AT DIFFERENT TYPES OF SAMPLES

The used samples are of three types:

A: Mat 270 g/m^2, 60% reinforcing grade, with polyester resin Lerpol TIX 3603/R type, 4 layers;

B: Fibers glass tissue 450 g/m^2, 60% reinforcing grade, with polyester resin Lerpol TIX 3603/R type, 4 layers;

C: Cotton tissue, 60% reinforcing grade, with epoxy resin Ephiphen RE 4020/DE 420 type, 4 layers.

Cr. no.	Samples	Maximum Delamination Force [N]	Specific force [N/m²]
1.	A	1	21277
2.	B	2.4	79365
3.	C	4.56	86805

Tab. 1. The experimental results

Five test pieces have been taken from each plate, by cutting out with a cutter. The first phase consists in the sample penetration by the knife. The speed of application was of 10 mm/s. For the determining of the specific force we related the maximum delamination force of the test piece that's been arisen at the strained area.

$$F_{spf} = F_{del} / S \quad (1)$$

$$S = b \times h \quad (2)$$

The average of the experimentally obtained results were processed and inserted in table 1.

We can notice that the maximum delamination force for cotton tissue is more resistant then glass or Mat tissue. Diagrams were assigned to each test pieces depending on the maximum delamination force, figure3, 4 and 5. The mechanical behavior of polymeric composite materials depends in principal on nature and architecture of the reinforced material, reinforced grade and the technological process. The architecture of the reinforced grade influence meaningful the mechanical characteristics.

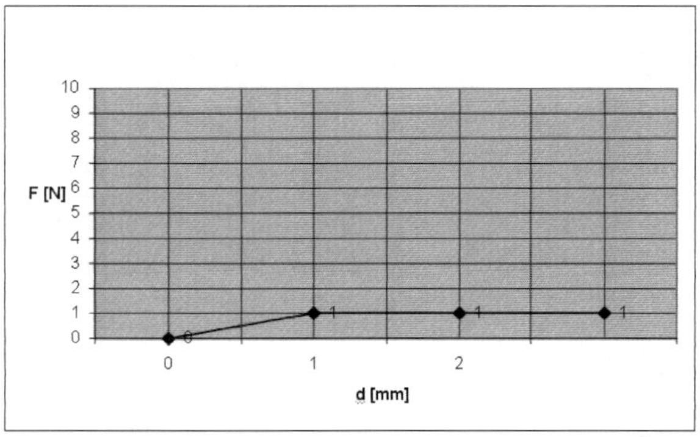

Fig. 3. The maximum delamination force for Mat 270 g/m^2 with polyester resin Lerpol TIX 3603/R type

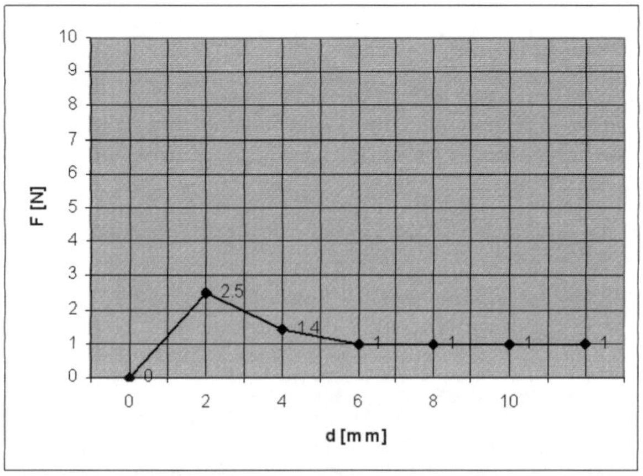

Fig. 4. The maximum delamination force for glass tissue 450 g/m^2 with polyester resin Lerpol TIX 3603/R type

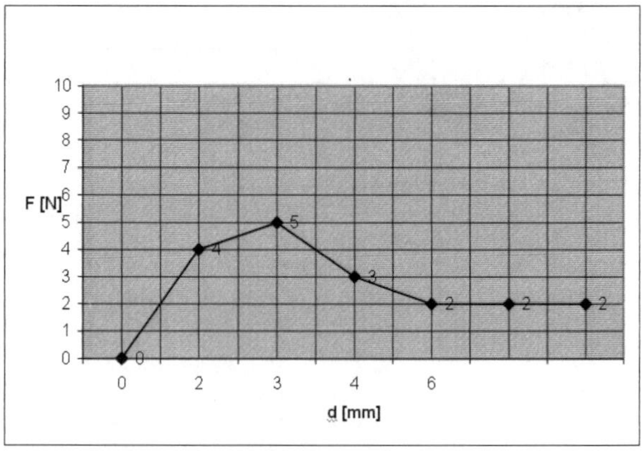

Fig. 5. The maximum delamination force for cotton tissue with epoxy resin Ephiphen RE 4020/DE 420 type

The specific force gives as indications about strength and quality of the realized composite structure. This installation is precise because of guiding columns and of possibility to obtain diagrams where we can read the forces that actuate on the sample and the moment of the sample delamination.

4. CONCLUSION

The mechanical behavior of polymeric composite materials depends in principal on the nature and the architecture of the reinforced material, reinforced grade and the technological process. The degree of reinforcement also depends by the used fiber.

Delamination is interlaminar damage and can be produced during production or exploitation of the composite structure and can have a great variety of causes. It is very important in industrial practice to know the principal causes of delamination appearance, to can be eliminated.

This installation is precise because of guiding columns and of possibility to obtain diagrams where we can read the forces that actuate on the sample and the moment of the sample delamination.

The specific force gives as indications about strength and quality of the realized composite structure.

In the future, we will have in view, the extension of the experimental investigations with the help of suggested installation and verification in the practice of mathematical model.

5. REFERENCES

Inacau, H. & Nemes, O. (2003), *Materiale compozite. Conceptie si fabricatie, Composite materials. Manufacture and conception.* Ed. Mediamira, ISBN 973-9357-24-5, Cluj-Napoca, Romania

Kachanov, L.M. (1990), *Delamination Buckling of Composite Materials*, Library of Congress in Publication Data, ISBN 90-247-3770-2, Brookline, Massachusetts, USA

Misca, B.R.H. (2001), *Materiale compozite polimerice, Polymer Composite Materials*, Presa Universitara Clujeana, ISBN 973-8095-13-1, Cluj-Napoca, Romania

Sabau, E; Iancau, H. &. Crai, A. (2006), *Experimental researches regarding the delaminating process at composite materials with organic matrix, Mechanical Engineering 2006, The 10th international conference*, pag. 612-616, ISBN 80-227-2513-7, Bratislava, Slovakia

Zgura, G. & Moga, V. (1999), *Bazele proiectarii materialelor composite, Designing basements of composite materials*, Ed. BREN, ISBN 973-9493-01-7, Bucuresti, Romania

Annals of DAAAM for 2008 & Proceedings of the 19th International DAAAM Symposium, ISSN 1726-9679
Editor B. Katalinic, Published by DAAAM International, Vienna, Austria 2008
Make Harmony Between Technology and Nature, and Your Mind will Fly Free as a Bird

INFLUENCE OF CUTTING CONDITIONS AND TOOL COATINGS ON THE SURFACE FINISH OF WORKPIECES OF MAGNESIUM OBTAINED BY DRY TURNING

SAENZ DE PIPAON, J[ose]; RUBIO, E[va]; VILLETA, M[aria] & SEBASTIAN, M[iguel]

Abstract: *The main reason for carrying out this work is that in industries, such as the aeronautical and motor, it is usually necessary to make repair operations on pieces of light alloys with inserts of other types of metals such as steel or stainless steel and that have stringent requirements of surface finish. In this work, the goal is to find the relationship between the different combinations of cutting parameters and tool coatings, and the quality of the surface finish of the workpieces made up of magnesium UNS M11311. The followed methodology consists of a series of tests of dry horizontally turning carried out with tools of different coatings and under cutting conditions usually used in repair operations. The principal result is that the best surface finishes are obtained for low feeds. Once the lower feed value is fixed the minimum roughness value is obtained for certain combinations of cutting speed and coating. As a first conclusion, it is possible to affirm that for repairing operations the machine tool is not a critical issue.*

Key words: *Dry machining, Dry turning, Magnesium UNS M11311, Auto ignition, Repair operations.*

1. INTRODUCTION

The magnesium is the lightest metallic material that can be used in industry. Namely, it is ideal for applications in industries such as aeronautical, aerospace, automotive, medical, electronic or sports, where the ratio density to resistance must be low (Caton, 1991, Gil et al., 2001, Tönshoff et al., 1997).

However, this material presents problems with the heat generated in the machining process, since it has a tendency to be flammable. Autoignition temperature (430ºC) is lower than the melting temperature (650ºC), so any spark can cause the ignition of chips or dust. Besides, the use of water based coolants in the machining of magnesium alloys is dangerous in case of chip ignition; burning magnesium will decompose water to form hydrogen atmospheres which are highly explosive. If a fire occurs, water should never be used to extinguish magnesium fires, dry sand or a suitable extinguisher for fire involving metals should be used.

On the other hand, the use of lubricants or coolants during the machining constitutes an undesirable factor owing to economical and environmental factors, being necessary to develop cleaner manufacturing technologies such as the dry machining. Then, the machining of the magnesium should be under dry conditions, not only because of the economical and environmental factors but, as well, of the security of the process.

In such situation, it is necessary to combine the cutting parameters and types of tools to optimize machining processes that involves magnesium. This allows obtaining pieces with a good dimensional precision and a high quality of the surface finish with a cost as low as possible, and of course with security conditions for workers and equipment.

In industries such as aeronautical and motor because of the high cost of manufacturing components, repair and maintenance operations are usually required. In such operations are used low values of cutting speeds, feeds and depths of cut, maintaining stringent requirements of surface roughness (usually $0.8 < Ra < 1.6\mu m$).

This study is focused on the measure of the surface finish (in terms of Ra) of bars of magnesium UNS M11311 obtained by dry turning varying the cutting parameters and type of tools used. Low cutting conditions have been used as it is usually in repair operations. Besides, three different types of tool with identical geometry and different coatings have been used. Concretely, one specifically for non ferrous metals and two for steels since magnesium normally forms part of hybrid materials (magnesium-aluminium and magnesium-steel).

2. METHODOLOGY

This work is framed within a study involving different light alloys, types of cutting tools (geometries and coatings) and cutting conditions. The methodology used is as follows (Rubio *et al.*, 2005):

- Previous activities to machining process. These activities consist of the design of experiments, namely a Taguchi L_{27} collected in Table 1 (Taguchi, 1987), the preparation of the test material and the protocols to calculate the cutting parameters values and to registry data and observations of the machining process.
- Turning tests. In these tests workpieces of different light alloys are mechanized under certain conditions of feed, cutting speed, depth of cut and types of tools.

N. Test	v(m/min)	f(mm/rev)	Tool	Ra (μm)			
				L1	L2	L3	L4
1	v3	f2	T1	G1	G3	G2	G1
2	v1	f3	T2	G2	G2	G1	G1
3	v3	f3	T1	G2	G1	G1	G3
4	v1	f1	T3	G2	G2	G2	G2
5	v3	f3	T2	G1	G3	G3	G2
6	v2	f1	T3	G3	G1	G3	G1
7	v1	f1	T1	G1	G1	G1	G1
8	v3	f2	T2	G3	G2	G1	G3
9	v3	f1	T3	G1	G3	G1	G3
10	v3	f2	T3	G2	G1	G3	G2
11	v2	f2	T1	G3	G1	G1	G2
12	v2	f3	T2	G3	G1	G2	G3
13	v2	f2	T3	G1	G2	G2	G3
14	v3	f3	T3	G3	G2	G2	G1
15	v1	f3	T1	G3	G3	G2	G2
16	v3	f1	T2	G2	G1	G2	G1
17	v2	f3	T1	G1	G2	G3	G1
18	v2	f1	T1	G2	G3	G2	G3
19	v1	f2	T2	G1	G1	G2	G2
20	v3	f1	T1	G3	G2	G3	G2
21	v1	f2	T3	G3	G3	G1	G1
22	v2	f3	T3	G2	G3	G1	G2
23	v1	f3	T3	G1	G1	G3	G3
24	v2	f1	T2	G1	G2	G1	G2
25	v2	f2	T2	G2	G3	G3	G1
26	v1	f1	T2	G3	G3	G3	G3
27	v1	f2	T1	G2	G2	G3	G3

Tab. 1. Tests conditions

Tool coating	HX	TP200	TK2000
v(m/min)	75	150	225
f(mm/rev)	0.05	0.10	0.15
d(mm)	0.25		

Tab. 2. Cutting conditions

- Monitoring processes. In order to get graphics documents that can be analyzed after the process, all the turning tests described before have been photographed and recorded by video and both the chips obtained and tools used have been photographed with a camera of high resolution.
- Roughness measurement. Measurements of the surface roughness have been made using a surface roughness tester in three generatrices separated one from each other 120° and denoted by G1, G2, G3. In each one of them, the roughness, in terms of *Ra*, has been measured in four different sections of the length of the workpiece denoted by L1, L2, L3, L4.
- Data processing and analysis of results. The data thus obtained have been treated with mathematical techniques according to the Taguchi L_{27} design of experiments.

3. APPLICATIONS

For this study, the workpieces used in the turning tests were cylindrical bars with a diameter of 40mm and length of 125mm (useful 100 mm) of magnesium alloy UNS M11311.

The cylindrical bars were dry turned on an EMCO Turn 120 CNC lathe equipped with an EMCO Turn 242 numerical control. The cutting conditions, collected in Table 2, have been expressed in units usually employed in manufacturing workshop. Although they are not System International units (S.I.), they give a more intuitive idea of the values used.

Three different types of tool, from SECO manufacturer, with identical geometry and different coatings have been used. Concretely, one specifically for non ferrous metals and two for steels with a coating of $Ti(C,N)+Al_2O_3+TiN$. The manufacturer references are: HX, TP200 and TK2000 respectively. They have been denoted by T1, T2, T3 in Table 1.

To observe the machining tests carried out, videos and photographs of the tools were systematically taken during the tests using a Sony Cibershot DSC-P100 digital camera of high resolution. To measure the roughness of the workpiece were used a surface roughness tester Mitutoyo Surftest SJ401. The roughness was measured on three generatrices separated 120º in four sections L1=0-25mm, L2=25-50mm, L3=50-75mm, L4=75-100mm (Table 3).

4. RESULTS

Once the results of the machining tests have been analysed, it can be seen that the best surface finish depend crucially on the feed and, as well although to a lesser extent, on the tool, the cutting speed and especially on its relationship. The best surface finish has been obtained for a feed of 0.05 mm/rev.

Taking only into account the coating factor but no its interactions, the tool shows slightly differences in the surface finish depending on the type used, so for the coatings TP200 and TK2000, similar average values have been obtained while for the HX the surface finish is of a slightly better quality. However, the coating has a little influence at feed level. The effect of the coating increases as the feed grows.

5. CONCLUSIONS

The best surface finishes are obtained for low feeds. Once the lower feed value is fixed the minimum roughness value is obtained for certain combinations of cutting speed and coating. This allows affirming that for repairing operations the machine tool is not a critical issue.

N.Test	Ra (μm)			
	L1 0-25 mm	L2 25-50 mm	L3 50-75 mm	L4 75-100 mm
1	0.563	0.522	0.510	0.462
2	1.425	1.162	1.212	1.176
3	0.966	0.838	0.898	0.885
4	0.415	0.455	0.390	0.420
5	1.340	1.316	1.295	1.287
6	0.392	0.390	0.396	0.525
7	0.474	0.475	0.507	0.374
8	1.041	1.119	1.075	1.110
9	0.309	0.255	0.412	0.422
10	0.574	0.807	0.725	0.726
11	0.516	0.631	0.444	0.676
12	0.985	1.223	1.020	1.032
13	0.534	0.465	0.526	0.577
14	0.852	0.927	0.891	0.961
15	1.099	1.193	0.958	0.988
16	0.460	0.514	0.475	0.540
17	0.724	0.762	0.901	0.681
18	0.376	0.279	0.454	0.323
19	0.622	0.785	0.702	0.689
20	0.361	0.324	0.342	0.437
21	0.844	0.821	0.781	0.804
22	1.287	1.296	1.197	1.237
23	1.103	1.009	1.397	1.334
24	0.330	0.322	0.350	0.378
25	0.712	0.571	0.668	0.498
26	0.420	0.398	0.390	0.329
27	0.679	0.497	0.773	0.561

Tab. 3. Roughness obtained in each test in terms of *Ra* (μm).

Cutting tools used in the machining of other types of materials (aluminium, steel) can be used obtaining a quality of the surface finish similar to that obtained with tools for specific use of non-ferrous metals. This is important if hybrid materials of magnesium alloys with inserts of other materials such as steel are machined.

Finally, it is important to optimize the cutting parameters to obtain a good quality of the surface finishes, but also to get a chip that does not present high risk of ignition.

6. ACKNOWLEDGMENTS

Funding for this work was provided in part by the Spanish Ministry of Education and Science (Directorate General of Research), Project DPI2005-09325-CO02-02 and the material used in conducting the machining tests by Engine Overhaul Shop of the company IBERIA L.A.E.S.A.

7. REFERENCES

Caton, P.D., Magnesium: an old material with new applications (1991). Materials & Design, 12(6), 309-316.

Gil, F.J.; Manero, J.M.; Rodríguez, D.; Aparicio, C. (2001). *Light Alloys,* Edciones de la Universidad Politécnica de Cataluña S.L., ISBN: 84-8301-480-7, Barcelona

Taguchi, G. (1987). *System of experimental design*, American Supplier Institute. Vol 2, ISBN: 0-527-91621-8, New York.

Rubio, E.M; Camacho A.M.; Sánchez-Sola, J.M.; Marcos, M. (2005). Surface roughness of AA7050 alloy turned bars. Analysis of the influence of the length of machining. *Journal of Materials Processing and Technology*, 162-163C May 2005, 682-689.

Tönshoff, H.K.; Winkler, J., The influence of tool coatings in machining of magnesium (1997). Surface and Coatings Technology, 94-95, 610-616.

Annals of DAAAM for 2008 & Proceedings of the 19th International DAAAM Symposium, ISSN 1726-9679
Editor B. Katalinic, Published by DAAAM International, Vienna, Austria 2008
Make Harmony Between Technology and Nature, and Your Mind will Fly Free as a Bird

IMPROVED NEGOTIATION PROCESS WITHIN DISTRIBUTED MANUFACTURING SYSTEMS

SALMINEN, K[ai]; NYLUND, H[asse] & ANDERSSON, P[aul]

Abstract: *A modern distributed manufacturing system (DiMS) operates on intensive internal negotiation that basically has three levels: the human interaction level, based on human cognitve capabilities, and the machines and processes levels, based on machine cognition and communication with non-intelligent devices. In this paper a logical level enhanced negotiation process is suggested that would enhance self-organisation within a DiMS. This research belongs to the FMS 2010 project studying next-generation distributed manufacturing systems at Tampere University of Technology.*
Key words: *Holonic Systems, Self-Organising Systems, Human-Computer Interaction, Intelligent Systems*

1. INTRODUCTION

The core of a self-configuring manufacturing system is in its service negotiation process. Within the Distributed Manufacturing Systems (DiMs) concept (Salminen et al. 2006), service holons are used as the basic value offerings given by a holon or a holarchy for a service broker, either as a service request or as a service offering. The role of service holons within DiMS is central. It represents the value offering of a manufacturing holon and, at the same time, it has the vital manufacturing knowledge embedded in it. The XML-based holon and the related ontologies thus form the basis of the internal communication and thus make self-organisation possible (Van Brussel 1998). As DiMS operates on the basis of the information embedded in services, it is also vital to ensure the existence and quality of the services offered. As service holons are communicating entities in a constantly evolving self-organising system, their formal structure has to support emergent properties, i.e. to have internal change logic. Important research has been done within the "Methodology of Emergent Synthesis" programme in Kobe, where the central question is realising an artifactual system that achieves its purpose in unpredictable conditions (Ueda et al. 2001). The action-based learning approach (Kolb 1984) aims at a similar outcome.

2. LOGICAL LEVEL SYSTEM

A framework of neurological levels was developed by Robert Dilts (Dilts 1990). Dilts attributed this system to the anthropologist Gregory Bateson, who introduced the notion of logical levels (1972, 1979). It was then adapted for Business Process Re-engineering and change management.

The highest level in the logical level system is the vision level, which gives the purpose of the holon. The second level is the strategic level, defining the goal, the third level is the value and paradigm level guiding the use of strategy for alternative changes, the fourth level is the identity level, on which the holon gets its identity within a holarchy, the fifth level is the resource and skills level, describing the resources and skills available for executing the task, the sixth level describes the management of the holon – it relates, for example, to scheduling, maintenance, and co-operation with other holons – and finally the seventh level describes the operative part and context of the service. Each level represents a different conceptual negotiation process with its related ontology. The system supports the decision-making process by allowing internal coherence verifications. Within goal-oriented self-organised systems changes on a higher level can be used to initiate corresponding changes on lower levels in a fractal manner. Change demands on a lower level can also challenge higher-level logic. By addressing test changes to a virtual system their effects on coherence can be evaluated.

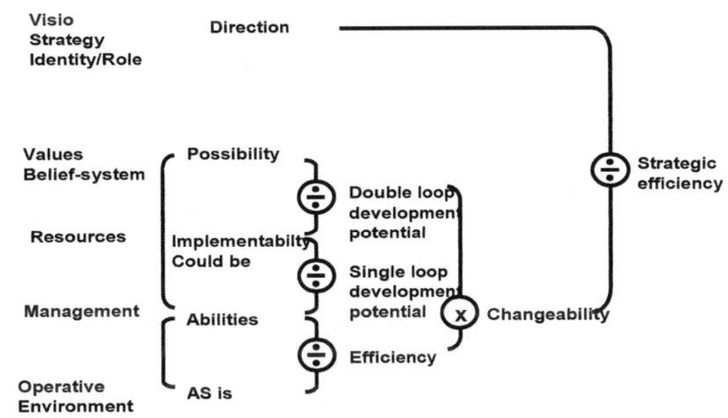

Fig. 1. Logical level system and change.

2.1 Logical Level system in context of a service holon

Within Within DiMS all service holons have a similar internal structure that consists of a resource holon, product holon, and order holon fractals. After creation they have a passive status and are activated in reponse to a service request as a corresponding service offering. After acceptance they become operative service holons and after their execution history data holons.

Each service holon structure is pre-created and formalised for executing a specific request. As the service holons are used as a means of communication they must have a suitable internal logic. For this purpose the use of the logical level system is proposed. As a DiMS is in constant evolution, each holon is also bound to develop with time. Therefore a holon is communicating constantly on each conceptual level within a holarchy as each level is experiencing constant changes.

As various issues affecting the quality and availability of a service appear within the holarchy, they are always addressed from a higher level than the one where they exist

.

3. COMMUNICATION AND FORMAL REPRESENTATION OF A SERVICE HOLON

The service holons are central communication and knowledge entities within DiMS, They are similar in structure, consisting of digital, virtual and real parts as well as a communication part. From process control From the process control point of view their communication part understands the digital past, present, and future. Thus the communication within the holarchy is related to all of these three levels of existence within three time concepts. (Nylund 2008). As the service holon is designed it is first conceived, then modelled and tested in a real context. During tests and use their quality gets ensured. Each holon thus has a specific context in which it is valid. This context can be defined with a fuzzy set of rules that can be verified before execution..

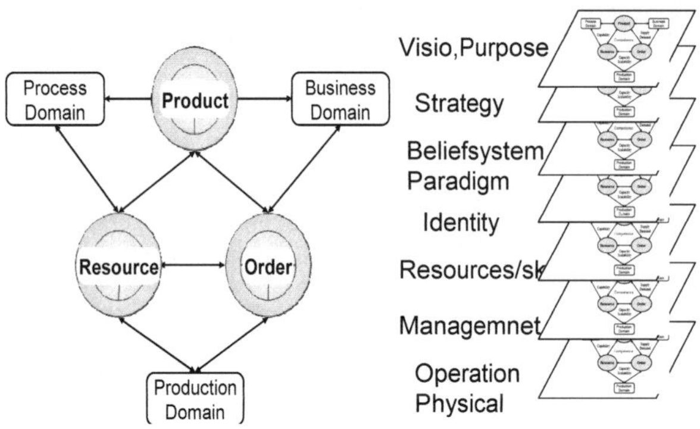

Fig. 2. Holon and Fraktal Logical Level view of a service holon.

3.1 Complexity of service holons

The complexity of individual service holons varies according to the individual case. A DiMS has basically four different types of resources: first, intelligent resources, i.e. human resources; second, machine cognition types of resources, i.e. service robots with machine intelligence; third, machine tools with machine intelligence, and fourth, resources such as tools and material. Only intelligent resources have a holonic appearance and have a communication part and therefore have an equivalent structure. Human to uman communication is most complex, partner robots need to co-operate with humans and each other to perform demanding tasks requiring cognition (Zäh 2007), whereas machine tools perform simple predefined tasks.

3.2 Creation process of a service holon within DiMS

EachEach service holon is created during the emergent planning process of a new value offering by human-holarchy interaction. As a service holon has product, order, and resource parts, the creation process is also performed by a multidisciplinary team.

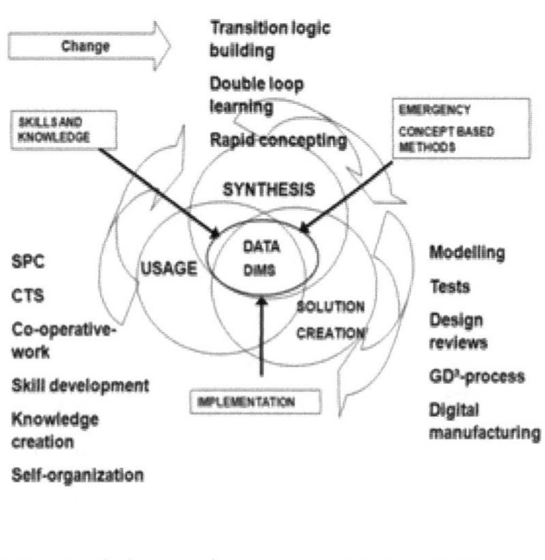

Fig. 3 .Service holon creation process.(Salminen 2008)

As As constant change is allowed, there has to be a logical change process guided by the logical level system to ensure the coherence of the decisions. One important notion is the usage of the value level for sustainable solutions. At the heart of the process lies meta-competence, or new learning that can be transferred to new situations, thus forming the basis for self-organisation abilities (Kolb 1984).

The constant self-organisation of a distributed manufacturing system needs a supporting continuous multi-path interdisciplinary change process. For DiMS a methodology was created that consists of three phases: first, a hybrid emergence process, then a virtual-digital-based implementation and usage process supported by service-based knowledge management.

This approach allows the efficient re-use of services that have been created and tested and thus the rapid ramp-up of new cases. The formal description allows their usage as early as from the early concept phase

4. CONCLUSION

This paper explains the use of the logical level concept to support the self-organisation of service holons within a DiMS. Service holons within a distributed manufacturing environment clearly benefit from a logical level structure that allows negotiation processes to take place on different conceptual levels as they manufacture themselves through a DiMS. Thus self-organising is coherent with the logical levels of the holarchy. Especially in the case of humans communicating with patner robot types of holons, the enhanced negotiation process is justified. The efficient usage of the concept calls for a good and coherent ontology and an efficient service holon creation system.

5. REFERENCES

Bateson, G. (1979). “Mind and Nature: A Necessary Unity (Advances in Systems Theory, Complexity, and the Human Sciences)”. Hampton Press. ISBN 1-57273-434-5.

Dilts, R (1990). “Changing Belief Systems with NLP”, Meta Publications, Capitola, CA.

Kolb, D. Experiential Learning, Englewood Cliffs, New Jersey: Prentice Hall (1984)

Nylund, H., Salminen, K., & Andersson, P. (2008). Digital Virtual Holons – An Approach to Digital Manufacturing Systems, In: Manufacturing Systems and Technologies for the New Frontier, The 41st CIRP Conference on Manufacturing Systems, Mitsuishi, M.; Ueda, K. & Kimura, F. (Ed), pp. 103-106, Springer-Verlag London Limited, ISBN-13: 9781848002661, London

Salminen K., Nylund, H., & Andersson, P.H. (2008) “The emergence process of an adaptable manufacturing system based on evolution paradigm” Proceedings of The 18th International Conference on Flexible Automation and Intelligent Manufacturing (FAIM 2008), de Vin Leo J. (Ed), pp. 1011-1019, ISBN: 978-91-633-2757-5 University of Skövde, June 30th – July 2nd, 2008 Runit AB, Skövde

Ueda, K. Markus, A., Monostori L., Kals H.J.J., & Arai, T. (2001). Emergent Synthesis Methodologies for Manufacturing, CIRP Annals-Manufacturing Technology, Elsevier

Van Brussel, H., Wyns, J., Valckenaers, P., Bongaerts, L. & Peeters, P. (1998). Reference architecture for holonic manufacturing systems: PROSA, Computers in Industry, Vol. 37, pp. 255-274, ISSN 0166-3615

Zäh, M. & Holst, F. (2007). “Towards the Cognitive Factory”, CARV 2007 Keynote Proc.of the 2nd Int.Conf.on Changeable, Agile, reconfigurable and Virtual production. (2007), Toronto, Canada,pp.2-17, ISBN 978-0-9783187-0-3

Annals of DAAAM for 2008 & Proceedings of the 19th International DAAAM Symposium, ISSN 1726-9679
Editor B. Katalinic, Published by DAAAM International, Vienna, Austria 2008
Make Harmony Between Technology and Nature, and Your Mind will Fly Free as a Bird

EXPLOITING CONCURRENCY IN MOTION AND PROCESS CONTROL

SAMAK, S[amoil]; SAMAKOSKI, B[lagoja] & SOKOLOSKI, Z[latko]

Abstract: *This paper explores the recent change in processing power increase where the execution power per unit is peaking, while the number of execution units on a chip are increasing, and how they can be used to the advantage of motion and process control.There are increasing demands placed on these control systems, and they can be handled through this increase in total processing power, but they will require new applications algorithms to exploit the increased concurrency.*
Key words: *motion & process control, concurrency, multi-core*

1. INTRODUCTION

Using a PC-based solution with off-the-shelf components has been a long-time trend in the Industrial Automation (IA) industry, and the benefits that it has brought have been numerous. There is a crucial trend in the computing industry that has to be embraced to make full use of PC-based control solutions. The trend is the peaking of the processing power per execution unit, and the rise of multi-core computing. For example, looking at AMD's server roadmap (Ferguson, 2008), one can see that at the end of their 2 year projection they expect to have 12-cores per CPU. Total processing power will continue to increase, but it will be in the form of closesly packed multiple execution units.

This means that software developed today should not be expected to benefit from the increase in processing power per unit as before, in other words, 'The Free-Lunch is Over' as a popular software writer Herb Sutter wrote (Sutter, 2005). In his article he discusses the major implications this has on the way we write software. This means that the motion and process control software developers have to look for ways in which the current tasks that need to be improved can be broken down and parallelized, and look for features which previously were considered costly in processing time, for ways to exploit the currently numerous available cores.

In this paper we take a look at some techniques used today in process and motion control that can be parallelized and new operational demands that can be fulfilled by these systems. The areas that can be improved are in the consolidation of process control processes, NURBS interpolation and adaptive control. The increased processing capacity can be used to add features for easier machine to enterprise systems integration, better machine diagnostics and maintenance, and closer integration to CAD/CAM systems.

2. AREAS OF PARALLELIZATION

Motion controllers are being placed into more sophisticated machines where motion is just a part of the overall process that needs to be controlled. A prime example is controlling Advanced Fiber Placement machines, where the motion control needs to be accompanied with precise process control of 32 tow fibers length, motion, tension and temperature. Multiple cores will allow the process control to run alongside the motion control, and each independent process control parameter can run on its own core. Processes can share memory for internal communication, and on multi-core CPUs, this can even be the L2 or L3 cache, which allows for the lowest latency communication, possibly increasing response times of critical tasks.

In cases where the necessary I/Os for process control have to be performed at a farther distance from the motion controller, a communication bus based on the standard Ethernet can be used. Since Ethernet has become extremely popular in most communications not involved with real-time behavior, a number of proprietary protocols such as EtherCAT have sprung up to deal with the non-deterministic behavior of the collision response. These proprietary protocols deliver real-time behavior in the order of 1ms or much less response times, but at the cost of not being able to use standards based components. However, a review of these protocols (Zwerina, 2005) has shown that even using the standard Ethernet/IP together with QoS, UDP and IEEE1588 can still deliver deterministic behavior, not as high as the proprietary ones, but still within the range of most control applications. Integrating this as the main communication bus of the motion controller can allow it to exploit further concurrency that lies in distributed high-performing FPGA, DSP and specialized application processors.

The most computationally intensive process of the motion controller is the computation of the interpolation points for the motor drives, taking into account numerous other static and dynamic parameters. Ideally, the motion control should be interpolating Non-Rational Uniform B-Splines (NURBS) based paths, as today all the popular CAD/CAM systems have moved into using curves to store the geometry of shapes. Storing paths in NURBS allows for consistency between the model and the paths by maintaining geometric precision up to the point of axis interpolation. NURBS interpolation is a costly operation, and there is already research (Yau et al., 2005) for real-time, accurate and parallel algorithms using FPGA boards.

Multiple cores will allow the creation of separate parallel tasks to perform the lookahead calculations. Before the motion control starts, multiple processes can be started that will perform precise interpolation calculations, of different segments of the path. Once they are done with a segment, their results will be buffered, and can be immediately used to drive the motors. The advantage of running multiple such tasks is that each one can take a longer time than if it were meant to run in stringent real-time, and once the results are done, the buffer will contain data for as many segments as the number of tasks that were running. This allows for more complex and futher-in-time lookahead methods to be implemented.

The motion controller has an ever expanding role as it becomes a focal point of connection between a machine's operation and the overall plant and company operations. As there is an increasing pressure for manufacturing companies to integrate their factory floor systems with their enterprise planning and operation systems (Brousell, 2008), the motion controller will need to be interconnected to these systems. The requirements of such integration on the controller will be to collect operational data, and respond to informational requests, probably from higher level management systems. This has to be done without interfering with the primary function of the controller – process and machine control.

Tying together manufacturing operations and enterprise planning systems is crucial for companies that want to have an integrated operating model. To achieve this, the controller will have to provide data to already existing IT systems, by connecting to the company's middleware, usually through Web services such as SOAP. To provide a custom and rapid development platform, this would have to be based on a popular OS, such as Windows or Linux, as those platforms already provide numerous Rapid Application Development (RAD) tools such as .NET Framework and Sun's Java to easily interconnect to the existing IT infrastructure. To have access to the machine's data, this OS can run on its own cores, and be able to read and query, with a lower priority from the main real-time OS that controls the machine. Both systems can be run on multiple-cores, on the same machine, providing extremely fast and reliable communication. To achieve the different priorities expected from this system, where a real-time and a non-realtime OS are competing for peripheral resources, a real-time hypervisor can be used, such as the RTS Real-Time Hypervisor by Real-Time Systems GmbH. This will fulfill the necessary real-time requirements of the process and motion control processes, while allowing the integration system to be running in parallel.

Using a real-time hypervisor, the controller can also contain virtual machines that provide Human-Machine Interface (HMI), and CAD/CAM integration. The HMI is a common part of the control system, and in this case, the real-time OS gets to not be disturbed by failures caused in the HMI component, but gets almost the same speed and reliability of communication through shared memory. The CAD/CAM integration is a relatively new concept, that has been researched more in-depth only recently due to the availability of multiple execution units and larger processing power. This research comes from the Brigham Young University where the Direct Machining And Control (DMAC) and the Reconfigurable Mechanism for Application Control (RMAC) (Lin, 2005) concepts have been developed. These concepts try to remove the M&G codes that are so prevalent, and also modularize a machine motion calculation from a given path, to allow machine operation directly from a CAD/CAM system in order to increase machining production rate, quality and flexibility.

The motion controllers of the future will be strained to contain more and more control and measurement points. Machines will need to collect serviceability data, such as vibrations, precision feedback, and other parameters that can determine the degradation of the performance characteristics of the machine. Such degradation could be caused either by improper usage, incorrect design or normal wear and tear. This data can allow the manufacturers to diagnose issues faster and make better decisions in future design. To relate to the software industry, companies such as Microsoft are already using so-called Software Quality Metrics (SQM) data in their popular software products, to generate diagnostics data. This has been used numerous times by the product maintenance teams to diagnose recurrent and popular failures, and by the product development teams to make better designs in the future.

3. ISSUES

The hardware side has already promised the delivery of ever increasing number of cores, and even though they will eventually hit certain chip-die limits for the maximum cores per chip, software will probably be the lagging factor to exploit those gains. Writing reliable concurrent software is still very hard work, and concurrent software development is plagued with bugs that are hard to reproduce, which removes even more determinism during the quality assurance process. This has partly been due to the primitive model of using threads and synchronization elements established a while ago (such as the standard POSIX APIs), which proved to be very error-prone for complex projects. Race-conditions and dead-locks appear in software long after those same scenarios have been tested, and even worse, they cause data corruption that would be hard to reproduce. There is an understandable aversion towards concurrent programming in reliable systems.

All is not lost though, as more and more software corporations and research institutions such as University of California at Berkley (Merritt, 2008) and Stanford University are concentrating efforts and funds into developing concurrency models that will make development easier and less error prone. There are also efforts by programming language committees like the C++ Standard Committee to standardize on a viable memory model that will abstract the hardware and provide a foundation for developing parallel code (Boehm, 2008) that runs consistently on different architectures. The debates are quite heated, as there are many competing models being presented.

Besides the issues in creating parallel code, there is another limit to multiple core exploitation – non-parallel algorithms. There are certain types of problems that do not have parallel algorithms developed, and at the current state, they cannot be improved through multi-core use.

4. CONCLUSION

The rise in available concurrent processing power will be the next challenge for motion and process controllers to exploit. There is a demand for better machines that can peform faster to increase production throughput, that can be managed better, providing more diagnostics, and that can integrate better with the factory and enterprise systems. Creating a robust system that manages processes for reliable and scalable operation will take time, as concurrency issues are still being ironed out, but it is the future for the modern motion and process controller. Exploiting concurrency will provide immense benefits to consumers of such systems. This type of controller is already in development at the research laboratories of MIKROSAM, where it will be used for efficient control of the next-generation wind-blade production and in their new Advanced Fiber Placement (AFP) machines.

5. REFERENCES

Boehm, H. (2008). Threads and memory model for C++. *Available from:* http://www.hpl.hp.com/personal/Hans_Boehm/c++mm/. *Accessed:* 2008-05-17.

Brousell, D.R. (2008). The Long Climb. *Manufacturing Executive*. April 2008. 8-13.

Ferguson, S. (2008). AMD Roadmap Calls for 45nm Chips in Late 2008. *Available from:* http://www.eweek.com/c/a/Infrastructure/AMD-Roadmap-Calls-for-45nm-Chips-in-Late-2008/ *Accessed:* 2008-05-15.

Lin, W. (2005). Dynamic Reconfigurable Machine Tool Controller. *Available from:* http://contentdm.lib.byu.edu/ETD/image/etd675.pdf. *Accessed:* 2008-05-10.

Merritt, R. (2008). CS gets with parallel program. *Available:* http://www.eetimes.com/showArticle.jhtml;?articleID=206905134. *Accessed:* 2008-05-20.

Sutter, H. (2005). The Free Lunch is Over: A Fundamental Turn Toward Concurrency in Software. *Dr. Dobb's Journal*. March 2005. 30-33.

Yau, H.; Lin, M. & Chan, Y. (2005). Design and implementation of real-time NURBS interpolator using a FPGA-based motion controller. *IEEE International Conference on Mechatronics*. July 2005. 56-61.

Zwerina, K. (2005). Standards-based real time Ethernet now off-the-shelf. *Available from:* http://ethernet.industrial-networking.com/articles/articledisplay.asp?id=942. *Accessed:* 2008-05-23.

Annals of DAAAM for 2008 & Proceedings of the 19th International DAAAM Symposium, ISSN 1726-9679
Editor B. Katalinic, Published by DAAAM International, Vienna, Austria 2008
Make Harmony Between Technology and Nature, and Your Mind will Fly Free as a Bird

BLACK SEA WAVES, ENERGY SOURCE

SAMOILESCU, G[heorghe]; SERGIU, N[icolae]; DORIAN, M[arin]; BEJAN, M[ihai]; MOROIANU, C[orneliu] & BANDRABUR, D[iana] L[ilia]

Abstract: *The paper reveals research results of a hydro pneumatic conversion system using sea wave energy. We take into consideration the Black Sea storm and wave regimes followed by an irregular wave analysis. We are also presenting a micro power station fully functional model. Its predicted location is Black Sea waters, near Constanta. The proposed micro power station produces clean energy. It has a positive economical impact and is also protecting the shore.*
Key words*: Energy, conversion, hydro pneumatic.*

1. INTRODUCTION

West side of the Black Sea is baric disturbance more subjected than other areas as the meteorological data reveals. The cyclone and anticyclone genesis high frequency is the real reason (Bologa 1998).

The storm conditions are frequently created in the cold season when low pressure atmosphere mass is coming over worm waters. Air stratification is not stable and this generates vertical and horizontal high wind gradients.

Wave height increasing is also the consequence of constant wind flow, stable in time and direction. Six hours at a 9 m/s wind speed produces a 0.9 to 1.2 meters height increase and from 15 to 28 meters in length. In areas more than 40 meters water depth creating waves time is 20 – 200 minutes depending on wind speed. The extinguish time in this case is between 2 and 5 hours. (Iulian 1990)

Black Sea currents regime also depends on river water volume contribution, water density gradient and sea floor. Measurements from Constanta reveals: 36.6% S currents, 22.6% N and 40.8% SV-SSE.

2. IRREGULAR WAVE CHARACTERISTICS

Natural waves are irregular, with different periods and amplitudes (Bondar 1998), (Bondar1992). The analysis will use statistics and probabilistic techniques. Wave parameters are presented in figures 1 and 2:

– ξ – sea surface instant displacement from reference position;
– ξ_a – relative wave amplitude (distance from reference position to wave comb);
– h_w – relative wave height (vertical distance between two combs or two gaps);
– T_4 – relative period at zero intersection (time between two consecutive zero displacement $\xi(t)$ with ascendant trend);
– T_3 – relative period between two wave combs;

If ξ displacement depends on *x* we shall define:

– $(L_w)_4$ – relative wave length at zero level considering a positive trend (distance between two zero values in movement way);
– $(L_w)_3$ – relative wave length between two successive combs (horizontal distance between two combs in movement way);
– $\bar{h}_w$ – average irregular wave height (arithmetic average in a sea specific point, excepting values lower than 0.1 m).

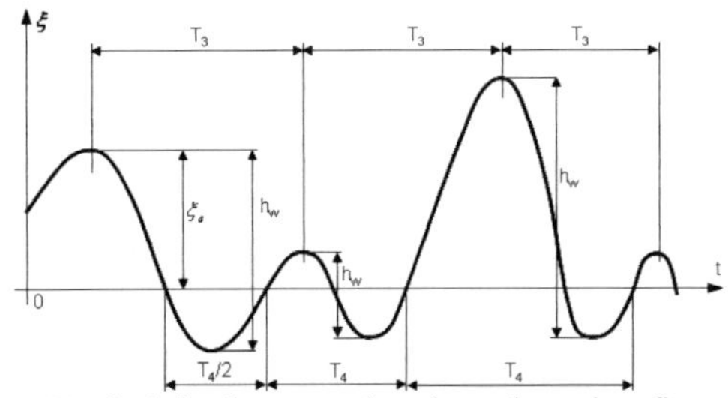

Fig. 1. Vertical displacement time dependence in a fix point

The results are:

– average wave period, corresponding to $\xi(t)=0$

$$\overline{T}_4 = \frac{T_{44}}{N_{10}} = \frac{86}{125} = 0.688s; \quad (1)$$

– average wave period, corresponding to combs

$$\overline{T}_3 = \frac{T_{53}}{N_{20}} = \frac{120.2}{127} = 0.94s; \quad (2)$$

– average wave period

$$\overline{T}_0 = \frac{T_t}{N_{01}+N_{02}} = \frac{206.2}{252} = 0.81s. \quad (3)$$

For average wave length the same pattern must be used. Calculus must be done for all sea levels from table 1.

Processing a great number of measured values in a long time period is the key for right results.

Measurements revealed a stationary character of irregular wave. This opens the way for statistic processing of measurements.

We offer a calculus sample in tables 1 and 2.

In order to obtain the characteristics of wave, short time measurements are the advantage of statistic processing.

Elaborated by Meteorology and Hydrology Institute of Bucharest, Black Sea wave specter has the general form

$$E(\omega)=2.202\cdot\left(\frac{E}{\bar{\omega}}\right)\cdot\left(\frac{\omega}{\bar{\omega}}\right)^{6.9285} exp\left[-0.3883\cdot\left(\frac{\omega}{\bar{\omega}}\right)^{5.9285}\right] \quad (4)$$

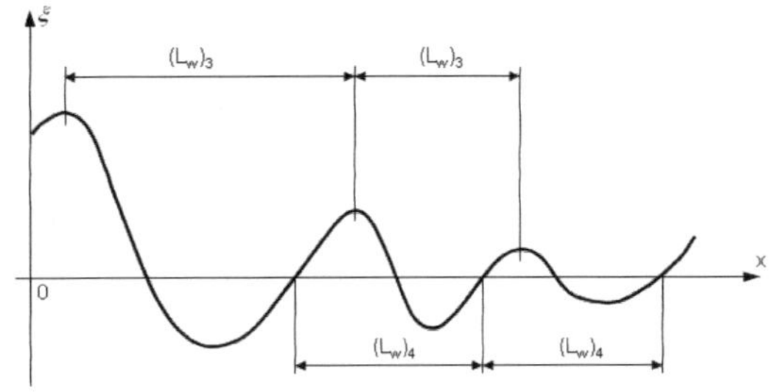

Fig. 2. Irregular wave vertical displacement at *t* moment

h_w (m)	No. of waves	%	Cumulated wave numbers	1 × 2	Average wave height 1 × 2			
					$\bar{h}_w$	$(\bar{h}_w)_{\frac{1}{3}}$	$(\bar{h}_w)_{\frac{}{1}}$	$(\bar{h}_w)_{\frac{}{1}}$
1	2	3	4	5	6	7	8	9
2.0	4	4	4	8	8	–	–	–
2.25	40	39	44	90	90	–	–	–
2.5	31	30	75	77.5	77.5	17.5 =(2.5×	–	–
2.75	25	25	150	68.5	68.75	68.75	22 = 2.7 × 8	–
3.0	2	2	102	6	6	6	6	3
Σ	102	100	–	250.25	250.25	92.25	28	3

Tab. 1. Wave characteristics

In equation (4)

– $E=\rho gH^2/8$ is the average specific energy of waves, in $[kJ/m^2]$,

–$\bar{\omega}=2\pi/\bar{T}$ is the average pulse, in$[rad/s]$,

– ω is the wind pulse, in $[rad/s]$,

– ρ is the water density and

– H is the height of the wave, in$[m]$.

Specific energy from 1994 to 2007 had values between 0.05 to 4.46 $[kJ/m^2]$, 88.9% from it being in 0.7 – 3.7 $[kJ/m^2]$ limits.

Wave characteristics are subjected to a significantly change over a year and from one year to another due to a great gradient of wind.

We will consider a Gauss distribution of the relative height and a Rayleigh one for amplitudes. Five energy values will be taken into consideration for the above mentioned interval 0.7 – 1.4 – 2.1 – 2.8 – 3.6 $[kJ/m^2]$.

Average heights will be calculated for this energy values.

A micro power station is proposed to work with Black Sea waves. It will be designed according to 3.3 s wave period and 0.65 m average height.

T_4 between two values $\zeta(t)=0$, (s)	Periods number from column 1	T_3 between two combs, (s)	Periods number from column 3	1 × 2	3 × 4	5 + 6
1	**2**	**3**	**4**	**5**	**6**	**7**
0.5	20	0.6	21	10	12.6	22.6
0.7	40	0.8	42	28	33.6	61.6
0.9	30	1.0	28	27	28	55.0
1.1	10	1.2	12	11	14.4	35.4
1.2	25	1.4	24	30	33.6	63.6
Σ	$N_{10}=125$	–	$N_{20}=127$	$T_{44}=8$	$T_{33}=120.2$	$T_0=206.2$

Tab. 2. Wave characteristics

3. HYDRO PNEUMATIC POWER STATION MODEL

A micro power station using wave energy is the aim of present research. It will work in an insular regime using the hydro pneumatic principle.

The caisson is made from steel and orientated to open sea. This particular orientation was chosen in order to use all types of wave. The generator module is placed into a cylindrical column fixed in caisson's ceiling. One way turbine and an electric generator are generator's main components. Cylindrical column is vertically fixed over the caisson. At a safe level, where waves cannot reach, it has a hole for air circulation.

One way turbine requires a special attention. The concept takes into consideration a variable geometry turbine. Overload turbine protection use a centrifugal device.

The whole installation is sustained by concrete embedded columns. This is able to ensure the necessary stability when the installation is placed on the sea floor.

4. CONVERSION SYSTEM CHARACTERISTICS

We will present assemble main components most important characteristics.

Micro power station characteristics are:

- nominal power of micro station 3 kW
- system length 5 m
- system height 7.4 m
- system width 5 m
- micro power station global output 0.329

The turbine characteristics are:

- type one-way
- pale aerodynamic profile NCA 0012
- nominal power 3.5 kW
- spinning 3000 rot/min
- air speed at turbine nominal power 53.7 m/s
- turbine diameter 0.42 m
- estimated output 0.5

Generator characteristics are:

- type vertical, synchronic
- nominal power3 kW
- spinning 3000 rot/min
- estimated output 0.91

5. CONCLUSIONS

Waves are wind generated. Black Sea wave regime is highly unpredictable regarding direction, height and period.

The most important aspect in wave creation is the relation between wind speed and wave dimensions.

The micro power station will be designed according to particular characteristics of the Black Sea.

6. REFERENCES

Bologa, A.S. (1998). *I.R.C.M. Contract no. 31/1998*, Regional Research and Management Development in the Black Sea, Constanta

Bondar, C. (1992). *Contribution at Profile, Deformation and Breaking of the Waves in the Coastal Region*, Meteorology Institute of Bucharest, Bucharest.

Bondar, C. (1998). *Regarding of Energetic Specter Aspects of Waves in Romanian Sea Shore of the Black Sea*, Meteorology Institute of Bucharest, Bucharest.

Iulian, C. (1990). Waves Energy Exploit, , Technical Publishing House, Bucharest

*** (2007). *Hydro pneumatic power plant using waves, Contract 21-001,* National Center for Programs Management CNMP, Bucharest.

Annals of DAAAM for 2008 & Proceedings of the 19th International DAAAM Symposium, ISSN 1726-9679
Editor B. Katalinic, Published by DAAAM International, Vienna, Austria 2008
Make Harmony Between Technology and Nature, and Your Mind will Fly Free as a Bird

HYDRO PNEUMATIC MICRO POWER STATION USING SEA WAVES

SAMOILESCU, G[heorghe]; SERGIU, N[icolae]; DORIAN, M[arin]; BEJAN, M[ihai]; MOROIANU, C[orneliu] & BANDRABUR, D[iana] L[ilia]

Abstract: *The paper reveals research study in wave energy conversion into electric energy. Calculus pattern for the 5 kW hydro pneumatic power station using sea waves is the actual target. There are also presented station working mode and none polluting, renewable energy advantages. Micro power station working principle is based on hydro pneumatically transformation into electrical energy. The main micro power station components are: the* caisson, *the cylindrical column, the one-way turbine and three phased generator.*

Key words*: Hydro pneumatic, micro power station.*

1. INTRODUCTION

This paper will focus onto wave and wind regimes, from all range of intense meteorological phenomenon (Iulian 1990), (Demirbilek & Panchang 1998), (Trusca 2005).

Using statistics offered by Hydrological and Meteorological Institute of Bucharest, the annual average height of the waves in Black Sea is 0.46m (Holthuijsen & Booij 2004). This happens between constant pressure curves of 5-9m.

In order to design the hydro electric micro power station using waves the energetic balance is taken into consideration. It contains wave energy, waste energy produced by pressure over oscillating column and turbine and generator outputs.

Station's generator module has a tri phased generator excited by rare soil permanent magnets, in order to achieve a higher output. (Olaru & Lazar 1990)

Elements of micro station working regime: turbine start at $\pm 40^0$ pale angle in order to ensure a quick start using low speed air; waves propagation direction is also the entrance one, in this way avoiding changes in energy flux direction; two systems possible to adopt for generator module over spinning protection – one is breaking the turbine over 50Hz and the other one changes the pale angle using a centrifugal device.

The pale angle changes automatically depending on air flux speed and loading torque. (Desire 1982)

Turbine has three possible working regimes:

– starting phase – the intake air flux creates a pressure over turbine pales until they reach maximum angle; the turbine starts spinning;

– low wave regime – the air flux increase the turbine spinning; pale angle changes depending on air speed and loading torque;

– height wave regime – turbine spinning overcome nominal value; at 30% overcome the system allows intake air to bypass the turbine; this design protects turbine structure and generator module.

2. POWER STATION CALCULUS

The calculus pattern was designed for this particular power station placed on Constanta level of the Black Sea. Starting parameters are:

– P_n theoretical nominal power 5000 kW;
– n turbine spinning speed n = 3000 rot/min;
– U_n nominal voltage 220 V;
– η_t turbine output 0.5;
– η_g generator output 0.92;
– φ diameter ratio – ratio between pale outer and inner diameter 0.7;
– λ speed modulus – ratio between laminar turbine speed and air speed through turbine 1.25;
– L length of caisson's precincts 5 m;
– B length of caisson's precincts 3 m;
– ρ density 1230 kg/m^3;
– γ specific weight 9926.

We calculated:

– P_t turbine nominal power

$$P_t=\frac{P_N}{\eta_g}[W]\ P_t=\frac{5000}{0.92}=5435[W] \tag{1}$$

– P_{ta} turbine consumed power

$$P_{ta}=\frac{P_N}{\eta_g\cdot\eta_t}[W]\ P_{ta}=\frac{5000}{0.92\cdot 0.5}=10869.5[W] \tag{2}$$

– air speed inside turbine

$$v_t=\left[\frac{P^t\cdot n^2}{143.2\cdot\rho\cdot(1-\varphi^2)\cdot\lambda^2\cdot\eta^t}\right]^{\frac{1}{5}}[m/s] \tag{3}$$

$$v_t=\left[\frac{5.435\cdot 3000^2}{176\cdot(1-0.7^2)\cdot 1.25\cdot 0.5}\right]^{\frac{1}{5}}[m/s]$$

$$v_t=58.7[m/s]$$

– turbine inner section

$$S_t=\frac{2\cdot P_t}{\rho\cdot v_t^3\cdot\eta_t}[m^2] \tag{4}$$

$$S_t=\frac{2\cdot 5435}{1.23\cdot 58.7^3\cdot 0.5}=0.0868m^2$$

– precincts inner section

$$S_i=L\cdot B=5\cdot 3=15m^2[m^2] \tag{5}$$

– air speed in turbine channel

$$c=\frac{2\cdot P_t}{\rho\cdot S_t\cdot v_t\cdot\eta_t}=\frac{2\cdot 5435}{1.23\cdot 0.0868\cdot 58.7\cdot 0.5}=3469\,m/s \tag{6}$$

$$v_2=\frac{4\cdot v_t^2-c}{4\cdot v_t}=\frac{4\cdot 58.7^2-3469}{4\cdot 58.7}=43.9\,m/s \tag{7}$$

$$v=2\cdot v_t-v_2=2\cdot 58.7-43.9=73.5\,m/s \tag{8}$$

– aerodynamic pressure

$$\Delta P=\rho\frac{v^2-v_2^2}{2}=1.23\frac{73.5^2-43.9^2}{2}=2137\,N/m^2 \tag{9}$$

– volume decrease due to aerodynamic pressure

$$v_p=S_i\cdot h_p=15\cdot 0.214=3.21m^3 \tag{10}$$

– air column height

$$h_p=\frac{\Delta P}{\gamma}=\frac{2137}{9986}=0.214\text{m} \quad (11)$$

– volume decrease debit due to aerodynamic pressure (equivalent debit of pressure gradient)

$$Q_p=\frac{v_P}{T}=\frac{3.21}{T}[m^3/s] \quad (12)$$

– requested turbine air debit

$$Q=S_t\cdot v=0.0868\cdot 73.5=6.379\ m^3/s \quad (13)$$

Wave period dependence of wave height $T=f(h)$ and wave length dependence of the same wave height $\lambda=f(h)$ are known for Constanta Black Sea area. Wave parameters for nominal power of system are determined.

The following values are considered:
– wave height – $h=1.1$m;
– wave length – $\lambda_v = 17m$;
– wave period – $T=4.2$s;
– wave power

$$P_v=P_{ta}+P_p=624\cdot h^2\cdot\frac{\lambda_v}{T}\cdot L=624\cdot 1.1^2\cdot\frac{17}{4.2}\cdot 5 \quad (14)$$

– wave volume

$$V_v=\frac{1}{\pi}\cdot L\cdot h\cdot\lambda_v=1.59\cdot h\cdot\lambda[m^3] \quad (15)$$

– wave debit

$$Q_v=Q+Q_p=\frac{31.1}{4.2}=7.4\ m^3/s \quad (16)$$

– equivalent debit of pressure gradient

$$Q_p=\frac{v_p}{T}=\frac{3.21}{4.2}=0.764\ m^3/s \quad (17)$$

– global output

$$\eta=\frac{P_N}{P_v}=\frac{5000}{15.280}=0.327 \quad (18)$$

For turbine calculus v_t, n, P_t, S_t, λ and φ are used. Other requested characteristics are calculated using the following relations:
– pale outer diameter

$$D_{p1}=\sqrt{\frac{S_t}{0.785\cdot(1-\varphi^2)}}[m] \quad (19)$$

$$D_{p1}=\sqrt{\frac{0.0868}{0.785\cdot(1-0.7^2)}}=0.467\text{m}$$

– pale inner diameter

$$D_{p2}=\varphi\cdot D_{p1}=0.7\cdot 0.467=0.327\ [m] \quad (20)$$

– turbine diameters

$$D_1=D_{p1}+2\Delta=467+3=470\text{mm} \quad (21)$$

$$D_2=D_{p2}-2\Delta=327-3=324\text{mm} \quad (22)$$

where Δ is distance between pale and turbine carcass;
– pale length

$$l_p=\frac{D_{p1}-D_{p2}}{2}=\frac{467-327}{2}=70\text{mm} \quad (23)$$

– pale number

$$N_p=12 \quad (24)$$

– pale cover coefficient

$$\beta=0.85 \quad (25)$$

– pale angular opening

$$\alpha=\beta\frac{360}{N_p}=0.85\cdot\frac{360}{12}=25.5° \quad (26)$$

– two consecutive pale angle (Desire 1982)

$$\phi=\frac{360}{N_p}=\frac{360}{12}=30° \quad (27)$$

– pale angles

$$I=25°\ i=7°\ \alpha=18° \quad (28)$$

– section ratio

$$R_S=\frac{S_i}{S_t}=\frac{15}{0.0868}=173 \quad (29)$$

3. CONCLUSIONS

Hydro pneumatic system advantages are:
– very simple kinematic scheme;
– no gear box for turbine is needed; the turbine is directly coupled with turbine;
– energy transmission does not use liquid and the consequence is no water-tight system required;
– the system does not have kinematical structures in contact with the sea water; the corrosion is reduced.

The micro power station working is based on following principles:
– the system has a variable geometry turbine which allows starting at low air speed;
– the waves are coming into system on their propagation direction which is distinct from other solutions; this approach avoids energy flux direction changing;
– the system is able to capture and convert waves with different height;
– generator module protection against over spinning is realized through turbine breaking and pale angle changing by an centrifugal device.

Efficiency of implementing non polluting energy will be proved. In order to protect the environment, important knowledge and technology will be transferred into renewable resource conversion field.

The turbine will have a completely new design, having a variable geometry and over spinning protection.

4. REFERENCES

Demirbilek, Z. & Panchang, V. (1998). *A Coastal Surface Water Wave Model of the Mild Slope Equation*, Technical Report CHL – 98 – 26, U.S. Army Engineer Research and Development Center, Vicksburg
Desire, L. G. (1982). *Wind Energy*, Eyrolles, Paris
Holthuijsen, L. H., Booij, N. et al (2004). *User Manual for SWAN version 40.31*, Delft University of Technology, Delft
Iulian, C. (1990). *Waves Energy Exploit*, Technical Publishing House, Bucharest
Olaru, G. & Lazar, P. D. (1990). *Contributions Regarding Implementing of Energetic Dams in Black Sea"*, Marth – April, Energetic Magazine, Bucharest
Trusca, C. V. (2005). Research and *Contributions Regarding Dynamics of Wave's Associate Phenomenons Near Sea Coast*, PhD Thesis, Galati
***. *Wavegen a world leader in marine renewable energy*
***. *WEC survey of energy resources*

Annals of DAAAM for 2008 & Proceedings of the 19th International DAAAM Symposium, ISSN 1726-9679
Editor B. Katalinic, Published by DAAAM International, Vienna, Austria 2008
Make Harmony Between Technology and Nature, and Your Mind will Fly Free as a Bird

AGGREGATION OF AUTOMATED INSTRUMENTATION VIA INTERNET USING REMOTE VISA CALLS

SANDU, F[lorin]; NEDELCU, A[drian] V[alentin]; MORARU, S[orin] A[urel] & BORZA, P[aul] N[icolae]

Abstract: *The paper is addressing remote experiments in a client-server architecture based on web-servers (with forwarding to dedicated ports) and high-tech workbench servers that are compliant with the Virtual Instrument System Architecture (VISA) and can be accessed as stand-alone (grouped) or distributed configurations (aggregated by VISA Remote Servers). Most implementations are done in the National Instruments software environment.*

Key words: *Instruments, VISA, Internet, Remote experiment.*

1. INTRODUCTION

The problem addressed by the authors is the discovery, aggregation and invoking (in automated test sequences) of remote instrumentation resources (mainly equipment but also interfaces and software objects). In a larger context, such capabilities should be advertised, searched and probed (assessed) by controllers that build a distributed test & measurement (T&M) configurations based on "ontologies". The work aims to enlarge the access to existing high-tech laboratory infrastructure for distance-learning purpose. It is included in an E.U. "Leonardo da Vinci" Transnational Network that should integrate distributed experiments as educational SCO (Sharable Content Objects)

2. STATE OF THE ART

Most of the outstanding remote laboratories (with real and/or virtual experiments) are presented in published surveys (e.g. of Gomes and Garcia Zubia, 2007) that include also best-practice and case-studies.

One of the best approaches belongs to researchers from the University of Genova (Bagnasco et al, 2006) who manage instruments as web-services. This is a top-down approach of the OSI (Open System Interconnect) stack. Compared to this, we would rather approach the problem bottom-up, using the Virtual Instrument System Architecture ("VISA").

3. RESEARCH COURSE

3.1 Publishing the full-capability remote GPIB control

A high-tech automated "tower" of instruments can be (almost) fully controlled remotely by using the complete GPIB (General Purpose Interface Bus – IEEE488) drivers offered by the equipment manufacturers, in order to be used with National Instruments (NI) "LabVIEW" (LV). The integrative LV Virtual Instrument (VI) developed at "Transilvania" University of Brasov – Romania (Fig.1) groups the front panels of 4 instruments (3 with full controllability and one – the HM1007 oscilloscope – limited to waveform visualization). The VI-s offered by NI, HP (for HP34401A) and Hameg – for HM8142 and for HM8131 were integrated with authentication blocks, instruments' reset, stimuli annulling and outputs' disabling. For remote control (via Internet) of a LV VI panel, the built-in LV web-server must be activated and configured on the workbench-server by the administrator.

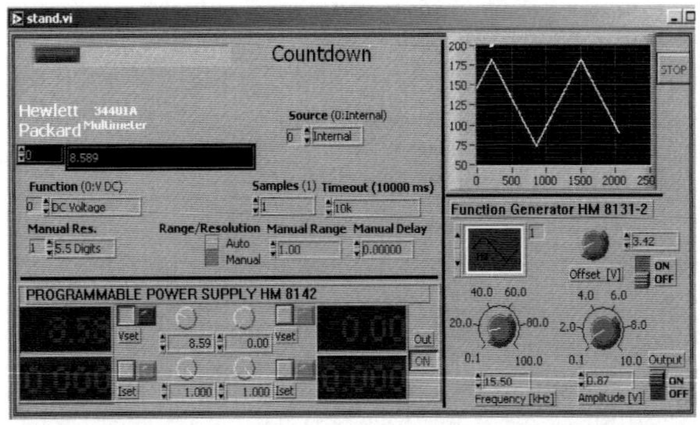

Fig. 1. Integrated panel for remote access to a GPIB workbench

Users should address *http:// <workbench-server name> : <port> / <application name>.htm.* Out of security reasons only the web-server is directly visible on the Internet, but not the workbench-server (it was connected – together with the workbench-servers of other networked lab-s – in the *Intranet* of the web-server, through a switch connected to its second network card). For this reason *port forwarding* was used to route the remote users' requests directly to the workbench-server (addressed with its Intranet address, 192.168.200.x).

3.2 Controlling remote instruments with NI-VISA

VISA is the origin of the well-known plug-and-play concept in modern operating systems. Session-orientation of VISA (based on interrupts' servicing) encouraged us to use the most recent virtual instrumentation software – like NI LV – together with NI Measurement and Automation Explorer (MAX) in order to integrate remote and local resources and to build test sequences.

Using NI Remote VISA Servers, the authors implemented a method of sharing instruments and even interfaces to remote users that can integrate them in their experiments (if they are only didactical, even Units-Under-Test, UUT, can be far from the user, if not, at least UUT and intelligent transducers should be local and other equipment can be remotely).

The configuration implemented consisted of a instrumentation (workbench) PC, running NI Remote VISA Server, to which all the instruments are connected using different interfaces (such as GPIB, USB, serial and parallel). The workbench server is also connected in the Intranet of the web-server which is "visible" in the Internet.

Port forwarding has also been configured in order to route Remote VISA data (coming through the dedicated port 3537) from the web-server to the workbench server. Any security requirements can thus be fulfilled by configuring access lists for port 3537. A client who has access to the system will use only the IP address of the web-server for remote access to the "published" experimental resources. The resources' "offer" of a certain server can be viewed in a tree-like manner in the Remote Systems window of NI MAX (Fig. 2). The availability of these resources can also be tested in MAX, using a graphical user interface.

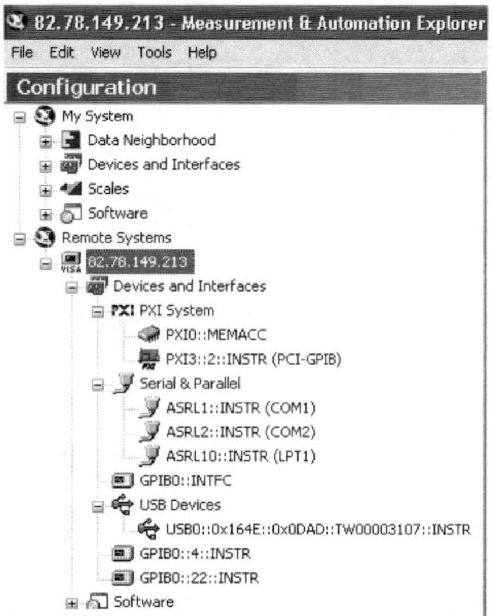

Fig. 2. The Remote Systems "explorer" window of NI MAX

3.3 Implementing remote laboratories with NI-VISA

The authors have used the Remote VISA technology for the implementation of remote experiments. The workbench consists usually of the following typical hardware parts:

- Experiment board(s) – the UUT
- Stimuli generators (e.g. programmable power supplies and/or function generators)
- Measurement instruments (e.g. digital multimeters, DMMs)
- Switches (in relay matrix and/or multiplexer configuration); a robust implementation is suggested by the authors in Fig.3.

Mechanical relays were chosen because of their larger bandwidth, of their better separation between the control circuit and the controlled one; the relays are controlled via parallel port of the workbench. "Test Points" are connected to common relay terminals; in order to multiplex them, "normal opened" terminals are connected to DMM input by appropriate DIP switches. Because the parallel port, LPT (which has also the advantage to be latched) is a resource which is "visible" through VISA, the relays can be controlled using Remote VISA. In order to do that, pins 11 (Busy) and 12 (Error) had to be tied to GND. Otherwise, the hardware driver will think the peripheral (it is talking to) is busy or wrong and will not output any data.

The software (see also Cheij, 2002) is built in NI LabVIEW (LV). A typical remote experiment (e.g. with an electronic UUT) can have a diagram like in Fig. 4.

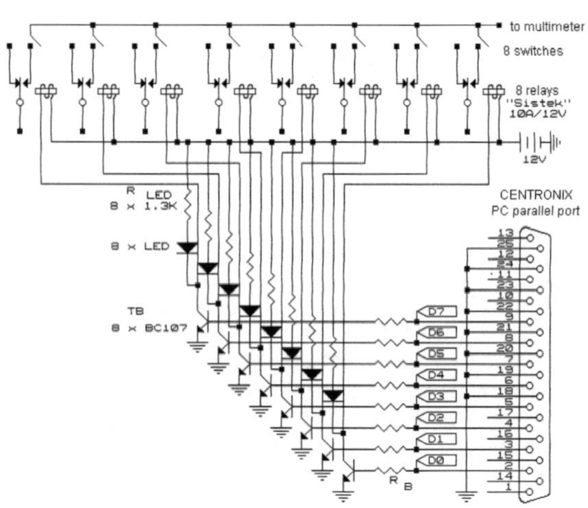

Fig. 3. Relay bench controlled by the (latched) LPT port

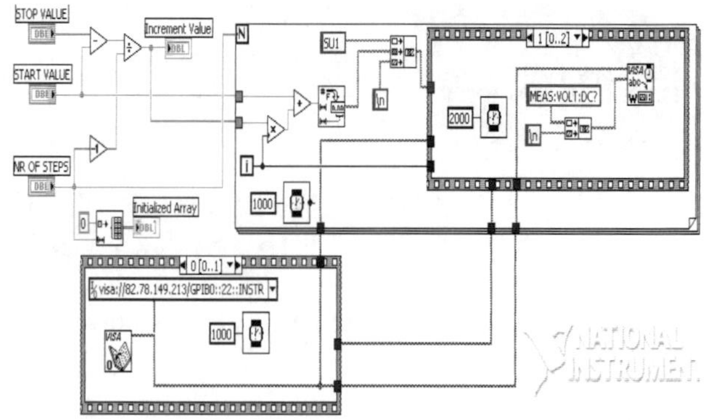

Fig. 4. LV measurement with remote DMM (GPIB address 22)

It sweeps a given range of input stimuli (e.g. between a start and a stop voltage, with a user-defined number of steps) and performs certain measurements (e.g. of voltages or currents).

A measurement cycle involves opening a VISA session to each of the instruments used in the laboratory. The syntax for identifying a remote instrument is the following: *visa://ip_adrress_of_the_webserver/instrument_ID.*

After opening a VISA session, an instrument can be controlled using standard SCPI - Standard Commands for Programmable Instruments (Pieper, 1998) commands (e.g. the syntax for measuring a DC voltage is: *MEAS:Voltage:DC:?*). Such commands are sent to the instrument using the VISAWrite.vi LV function. The measured value can be viewed by calling the VISARead.vi function, whose output is converted to double format and placed in an array.

4. CONCLUSIONS

Successful integration of local and remote instrumentation into automatic test sequences was demonstrated. Experiments were centered not on UUT but on instruments. Full access to their capabilities implies some specific security issues that are challenging further research.

5. ACKNOWLEDGEMENT

The research was financed by the European Union, inside the „Leonardo da Vinci" Trans-national Network RO / 06 / B / F / NT175014 "VET-TREND" (Valorization of an Experiment-based Training System through a Transnational Educational Network Development).

6. REFERENCES

Bagnasco, A., et al (2006). *Exposing Measurement Instruments as Grid Services in Distributed Cooperative Laboratories*, Networking Instrumentation, and Measurements, Davoli, F., et al (Ed.), Springer Science and Business Media, Inc., ISBN 0387298118

Cheij, D. (2002). *Software Architecture for Building Interchangeable Test Systems*, IEEE Aerospace and Electronic Systems Magazine, Vol. 17, N°1, pp 27-30, ISSN 0885-8985

Gomes, L. ; Garcia Zubia, J. (Ed.), (2007), *Advances on Remote Laboratories and e-Learning Experiences*, Published by University of Deusto, ISBN 978-84-9830-077-2

National Instruments Corp., U.S.A. (2003), *Remote Instrument Control with NI-VISA,* accessed on the following website: http://zone.ni.com/devzone/cda/tut/p/id/4789; date of access 19.02.2008

Pieper, J. M., (1998). *Standard Commands for Programmable Instruments* ,SCPI Consortium, ACEA, Wierden, 1998, *Available on*: http://www.scpiconsortium.org, *Accessed on*: 2007-09-17

Annals of DAAAM for 2008 & Proceedings of the 19th International DAAAM Symposium, ISSN 1726-9679
Editor B. Katalinic, Published by DAAAM International, Vienna, Austria 2008
Make Harmony Between Technology and Nature, and Your Mind will Fly Free as a Bird

THE EFFICIENCY ENHANCEMENT OF SHELL AND TUBE HEAT EXCHANGER BY MAGNETIC FIELD

SANSOMBOONSUK, S[iriluk]

Abstract: *Mathematical model has produced to study the efficiency enhancement of shell and tube heat exchanger by magnetic field. This model owing to the fact that magnetic field has direction crosses fluid movement, both internal and external of tube. For analyzing the results used mathematical method by solve differential equation and numerical method. Including examine theoretical accuracy by producing experimental equipment. The experiment fluids are water and salt solution. Magnetic flux has been compared from 0-0.23 Tesla. Indicate that, quantity of heat transfer is increased, temperature efficiency increased about 5 percents, and heat transfer coefficients increased were found. These increasingly affect the heat efficiency of heat exchanger by 7 percents. Moreover, results of experiment and simulation were good similary.*
Key words: *Heat Exchanger, Magnetic Field, Heat Transfer, Temperature Efficiency*

1. INTRODUCTION

Heat exchanger is equipment which widely use in industry. Therefore, the efficiency increasable of heat exchanger is very interesting. Generally, there are two techniques to increase heat transfer in heat exchanger. The first is passive technique which the structure inside the equipment is changed, such as increasing heat transfer area, changing the flow profile of working fluid. The second is active technique, which uses the external energy to enhance heat transfer rate. The examples of external energy are magnetic energy, electrical energy and mechanical energy (Genin, 1996). From reviewable the past research, it can be found that many research studied the effect of magnetic field to temperature and velocity of fluid in heat exchanger. The study applied for vertical and horizontal slab both in force and free convection (Ali, 2000, Uda, 2002, Nasreen, 1998, Ralphl, 1996). Almost of study was focused on heat transfer only one direction. Working fluid flows though heat exchanger within tube. However, the efficiently increasing are still not satisfying results. The goal of this study is to utilize the principle of magnetrohydrodynamics (MHD) for enhancing efficiency of shell and tube heat exchanger. The efficiency of heat exchanger is pointed by temperature efficiency. In this study, there are two working fluids which flow inside both shell and tube. For investigating the effect of magnetic field on temperature profile mathematical model at steady state is developed and solved by analytical technique. The equipment is set up and compared results with the simulation results in order to examine the accuracy of model.

2. MATHEMATICAL MODEL

It will be valuable thing to know the characteristics of heat exchanger with magnetic field. Hence, development of mathematical model is necessary to predict and investigate the effects of magnetic field on heat transfer rate.

2.1 Magnetrohydrodynamics principle

Magnetic force incurs on working fluid when it move pass magnetic field strength with high electric conductivity. This force can increase heat transfer of working fluid.

The current density, that acting on working fluid, is

$$J = \sigma(uB) \tag{1}$$

The magnetic force, acting on working fluid, is

$$F_m = JB = \sigma u B^2 \tag{2}$$

Energy from magnetic field, generated on working fluid, is

$$W_m = \frac{JJ}{\sigma} = \sigma u^2 B^2 \tag{3}$$

where u is fluid velocity (m/s)
σ is electric conductivity (semen/m)
B is magnetic field strength (N/ A.m^2)

2.2 Mathematical model of shell and tube heat exchanger

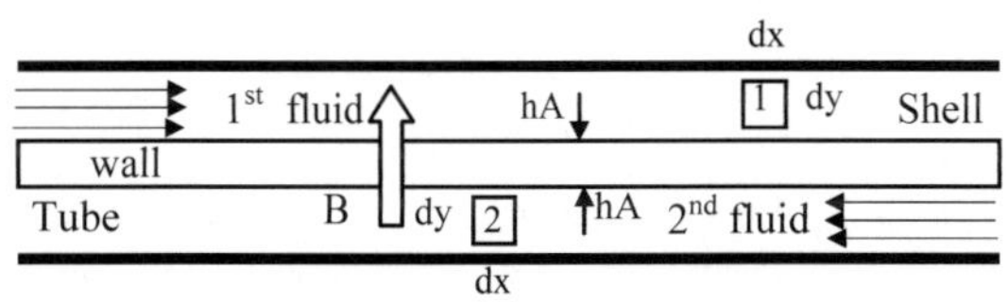

Fig.1.Working fluid in shell and tube with magnetic field.

Fig. 1 shows working fluid flowed though shell and tube.As shown in Figure, there are two counter flow working fluid. The first fluid flows inside shell whereas another flows inside tube. Magnetic field has transverse direction on both working fluids.

The assumptions developing mathematical model, are written by following:

1) temperature is changed only on horizontal direction,
2) velocity is changed on horizontal direction,
3) thermal capacity(c_s, c_t) is constant,
4) convective reductance $(hA)_s, (hA)_t$ is constant,
5) heat conduction is invisible,
6) thermal resistance at wall is zero,
7) magnetic field strength along flow path is constant.

From considering fluids, which flow inside and outside control volume based on law of mass conservation, theory of momentum, law of energy conservation, therefore mathematical model can be written below.

Equation of working fluid inside tube, is

$$h_t A_t (T_w - T_t) = \rho_t c_t A_t L (u \frac{\partial T_t}{\partial x}) + \sigma B_t^2 u_t^2 A_t L \tag{4}$$

Equation of working fluid inside shell, is

$$h_s A_s (T_w - T_s) = \rho_s c_s A_s L (u \frac{\partial T_s}{\partial x}) + \sigma B_s^2 u_s^2 A_s L \tag{5}$$

Equation of wall of tube, is

$$h_t A_t (T_t - T_w) + h_s A_s (T_s - T_w) = 0 \tag{6}$$

3. MATHEMATICAL MODEL ANALYSED BY ANALYTICAL TECHNIQUE

By analytical technique, temperature of working fluid inside shell and tube from equation (4), (5), (6) can be illustrated as equation (7), (8).

Temperature of fluid inside shell is

$$T_s = (T_{si} + \frac{M_s}{K_1})e^{k_1 x} - \frac{M_s}{K_1} \quad (7)$$

Temperature of fluid inside tube is

$$T_t = (T_{ti} + \frac{M_t}{K_2})e^{-k_1 x} - \frac{M_t}{K_2} \quad (8)$$

The temperature efficiency is calculated by equation (9).

$$\eta_{Temp} = \frac{T_{s,in} - T_{s,out}}{T_{s,in} - T_{t,in}} \quad (9)$$

where ρ is working fluid density (kg/m^3)

L is length of shell and tube (m)

$$K_1 = \left(\frac{B_s(B_s D_t - D_s B_t}{D_s D_t (B_t + B_s)} - \frac{B_s}{D_s} \right)$$

$$K_2 = \left(\frac{B_s B_t M_t - B_s B_t M_s}{D_t (B_t + B_s)} - \frac{B_t B_s M_s}{D_t (B_t + B_s)} \right)$$

$$M_t = \left(\frac{\sigma_t B_t^2 u_t}{L \rho_t c_t A_t} \right) \quad , \quad M_s = \left(\frac{\sigma_s B_s^2 u_s}{L \rho_s c_s A_s} \right)$$

4. EXPERIMENT SETTING

The equipment contains two water tank of cold and hot water, heater, two centrifugal pump, and ball valve for adjust flow rate. Working fluids used in the experiment are water and salt solution. Permanent magnetics are installed on top and bottom of tube bundle. Cold water is pumped to tube whereas hot water is pumped to shell. Flow meter is used to measured flow rate, thermocouple type K is selected to measure temperature at inside and outside of shell and tube. The experimental equipment is presented in Fig.2.

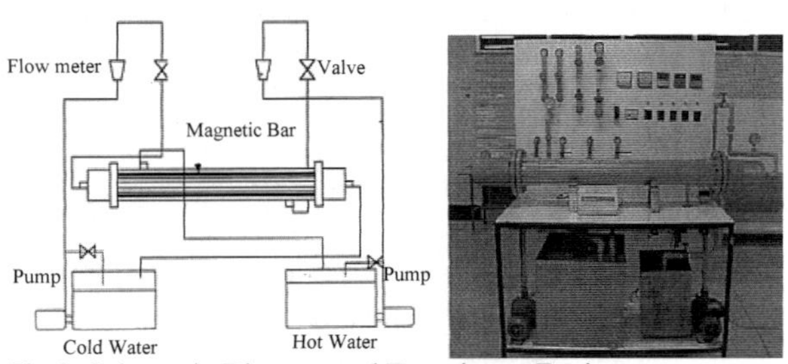

Fig 2. Schematic Diagram and Experiment Equipment

5. RESULTS AND DISCUSSION

In the simulation and experiment, magnetic field strength was slightly increased from 0 to 0.23 Tesla. This paper shows the effect of magnetic field and type of working fluid to temperature efficiency (η_{Temp}) of working fluid in heat exchanger when the concentration of working fluid (salt solution) was varied from 5% to 15%. The results (as shown in Fig. 3.) indicated that temperature efficiency of working fluid both flow through shell and tube increased 5%. Heat transfer efficiency was increased of 7% when compare in case of without magnetic field as well as overall heat transfer coefficient was increase. Temperature efficiency of tube increased more than shell because tube was installed closer with magnetic field than shell. In addition, working fluid which had more concentration provided higher efficiency. The results of experiment and simulation had good agreement. Results from simulation were however higher than experiment.

Heat transfer of shell and tube heat exchanger is increased because acceleration of working fluid is delay and its distribution is growth by magnetic field. Thermal efficiency of heat exchanger with magnetic field is also heightened. Therefore, the parameters affected to the increasing of heat transfer are the magnetic field strength and conductivity of working fluid. Magnetic field with crowded flux and intensity can produce more force and energy to working fluid. There are fluid with higher conductivity can get more influence form magnetic filed.

6. CONCLUSIONS

From the study, it can be concluded that magnetrohydrodynamics can enhance the efficiency of shell and tube heat exchanger. The rising of efficiency is depened on magnetic field strength, the concentration of working fluid and flow rate of working fluid. This study could be the way for efficiency enhancement of shell and tube heat exchanger in another way.

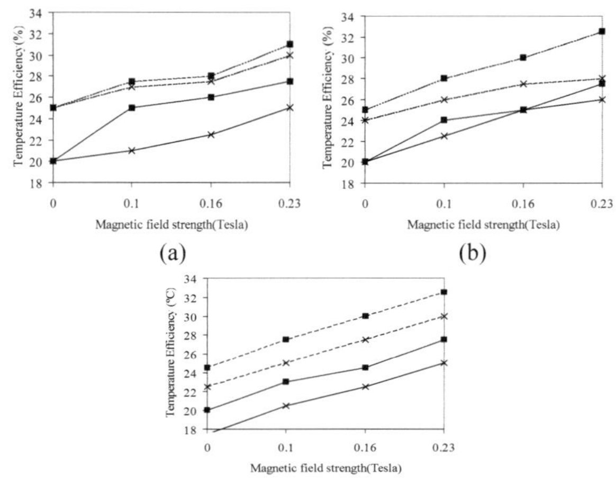

Fig.3.Temperature efficiency of heat exchanger when flow rate of working fluid 300 l/hr.

(a) salt solution 5% concentration
(b) salt solution 10% concentration
(c) salt solution 10% concentration

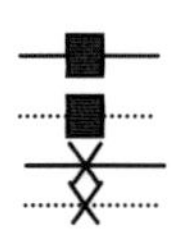
$\eta_{Temp}$of working fluid in tube, from experiment
$\eta_{Temp}$of working fluid in tube, from simulation
$\eta_{Temp}$of working fluid in shell, from experiment
$\eta_{Temp}$of working fluid in shell, from simulation

7. REFERENCES

Ali J. Chamka (2000). *Unsteady Laminar hydromagnetic fluid-particle flow and heat transfer in channels and circular pipes,* International Journal of Heat and Fluid Flow

Genin LG (1996). *Turbulent flow in a longitudinal magnetic field,* Heat Transfer Research No.1-4

N.Uda, N.Yamoda, H.Horiike, K. Mitazaki (2002). *Heat transfer in Lithium annular flow under transverse magnetic field,* Journal of Energy Conservation and Management

Nesreen K. Ghaddar (1998). *Numerical simulation of a vertical thermosyphonic loop placed in transverse magnetic field,* Journal of Numerical Heat Transfer, Part B: Fundamental

Ralphl W.(1996). *Principle of Enhance Heat Transfer,* John Wiley and Son, Inc

Annals of DAAAM for 2008 & Proceedings of the 19th International DAAAM Symposium, ISSN 1726-9679
Editor B. Katalinic, Published by DAAAM International, Vienna, Austria 2008
Make Harmony Between Technology and Nature, and Your Mind will Fly Free as a Bird

TWO AXLE ORIENTATION DEVICE, USED IN POSITIONING A MOTION PICTURE CAMERA

SARBU, F[lavius]; CALEFARIU, G[avrila]; POPESCU, R[adu] & TOMA, V[ictoria]

Abstract: *The paper achieves dynamic modeling of a mechanical belt drive, by using a camera positioned on one of the axes. A step by step engine is used for actuation. The dynamic model is solved with the help of program conceived in Matlab.*
By keeping enter data, the limit driving torque, where step loss starts, correlated with the according maximal frequencies, may be finding. In such a way, a drive is achieved, which has an optimal ratio of transmitted power and its global mass.
Key words: *modelling, transmission, mechanical, system, dynamic.*

1. INTRODUCTION

Robots endowing with artificial seeing systems, image acquiring and checking, presume a fine positioning of the camera, for being able to comprise the whole available eyeshot. In movable robots, inclusive removing the camera, angular distances and positions of the camera towards the environmental objects are variable.

Objects identification by the system needs image acquiring from an optimal angular position. The positioning fineness is determined by the algorithms used in the image analysis and by the building up manner of the objects data base being in the system's memory. These elements determine the maximal angular increment of the angular positions.

A second issue is represented by the angular positioning speed. This is determined by the robot's moving speed, together with the camera, as well as by the calculation time needed to the image analysis. Image acquiring and analyzing is completed by a decision concerning the robot's moving. The robot's decision taking frequency concerning its movement determines also the camera's positioning frequency. The angular positioning time has to be significantly lower than the minimal time between two decisions taken within the calculation system, with regard to the robot's displacement.

The third issue is represented by the specification of the camera's position towards the main pivots of the orienting device. From this point of view, determining factor is the calculation time needed for the image analysis. Certainly, this time is intended to be as less as possible. The calculation time is determined by factors, such as: computer performances, type and structure of the algorithm being on the grounds of the image analysis, as well as the number of algorithmic and logic operations carried out.

The last of the previous factors is influenced, among others, also by the camera's position towards the main pivots (Angeles2007).

The number of algorithmic, logic and arithmetic operation will be the less, if the inner area of the camera, where the light flux is converted into electric signal, is fixed during the orienting rotations.

This is possible only if the main pivots of the orienting device are crossing and if the above mentioned area is in the main pivots' crossing point (Marsden & Ratiu 1999; Dumitru & Cherciu 2002).

2. CINEMATIC AND CONSTRUCTIVE STRUCTURE OF THE ORIENTING DEVICE

The above formulated requirements led to the specification of the following performances of the orienting device:

- positioning angular increment: max. 0,4°;
- angular rotation speed: min. 45°/s;
- rotation angle on axis: min. 120°;
- rotation angle on axis: min. 270°.

For driving, step by step motors are used, because of the calculation exactness of the positioning angular increment it executes. Mechanical adapting between the engine performances and orienting device performances is realized by bending elastic wire drives and reduced elasticity of extension.

The main organological elements of the device and the calculation elements of the dynamic behavior on axis are presented within figure 1.

Dynamic parameters are: J_1 – rotational inertia of the step by step motor's rotor and of the driving roll fixed on its axis, J_2 – moment of inertia of the camera and of its driving roll, R_1 , R_2 – radii of the rolls performing the elastic wires drive, F_1 , F_2 – stresses within the elastic wires, F_i – tensile force , M_e – driving moment of the electric engine, φ_1 – rotational radius of the motor, φ_2 – rotational radius of the camera. Friction within bearings and roll inertia are neglected, having very low values.

3. DYNAMIC MODEL

The dynamic model is achieved under the following hypotheses:

- the electric engine's inertia, the camera's inertia, the belt's elasticity, the belt's pre-tensioning are taken into account;
- all the other mechanical elements are considered perfectly rigid;
- the tensioning roll's inertia, frictions within roll bearings and friction of the belt drive are neglected.

Dynamic model (Calefariu 2002; Dumitru & Cherciu 2002) is represented by the system (1).

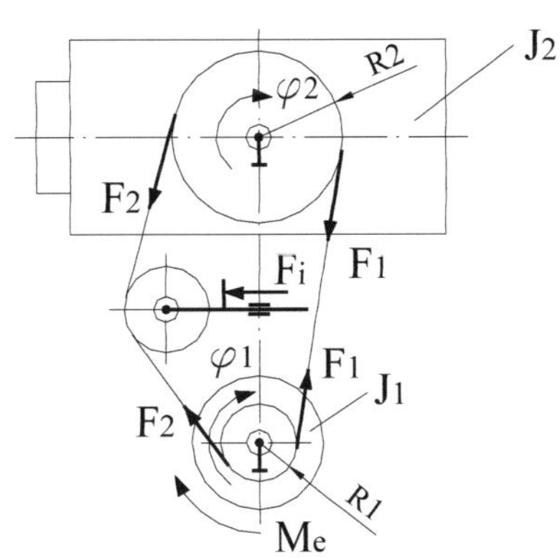

Fig. 1. Organological elements and the dynamic parameters

$$
\begin{aligned}
F_1 &= (\varphi_1 R_1 - \varphi_2 R_2 + e_1)c_1 \\
F_2 &= (-\varphi_1 R_1 + \varphi_2 R_2 + e_2)c_2 \\
J_1 \ddot{\varphi}_1 &= M_e + F_1 R_1 - F_2 R_1 \\
J_2 \ddot{\varphi}_2 &= F_1 R_2 - F_2 R_2
\end{aligned} \quad (1)
$$

In the relations (1), there appear also the following sizes: e_1, e_2 – initial pre-tensions on the two sides of the elastic wire, c_1, c_2 – coefficient of rigidity of the elastic wire. The coefficients of rigidity are depending on each branch and on an experimental determined parameter.

The previous model allows theoretic testing of different motor types, different performances, for an adequate choice of the driving solution. As a result of testing more versions, the step by step motor drive has been selected.

4. NUMERICAL RESULTS

The expression of the electromagnetic torque in dynamic drive, for step by step motors is complicated enough. Dynamic researching is made for the case of a two phased step by step motor.

The relation for calculating the electromagnetic couple of a two phases and 50 teeth step by step motor with variable reluctance (magnetic resistance) is given by (Kuo et al., 2006). The shape of controlling tensions is given by the same paper and looks like in figure 2.

System (1) has been solved with the help of a program written in MATLAB language, for the following values of the dynamic parameters: J_1=1,6 [Nmm^2], J_2=220 [Nmm^2], R_1=7 [mm], R_2=21 [mm], e_1=0,2 [mm], e_2=0,2143 [mm], c_1=450[N/mm], c_2=420 [N/mm] and electromagnetic moment M_e= 0,1 :Nm].

The results obtained are presented within the diagrams of figure 3.

The continuous maximum amplitude curve represents the motor rotation (φ_1), the continuous less amplitude curve represents the camera's rotation angle (φ_2), and the dotted line represents the electromagnetic moment applied on the electric engine's rotor.

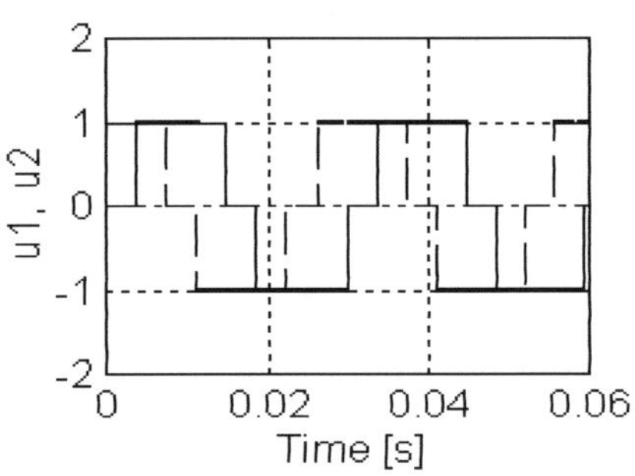

Fig. 2. The form of command tension

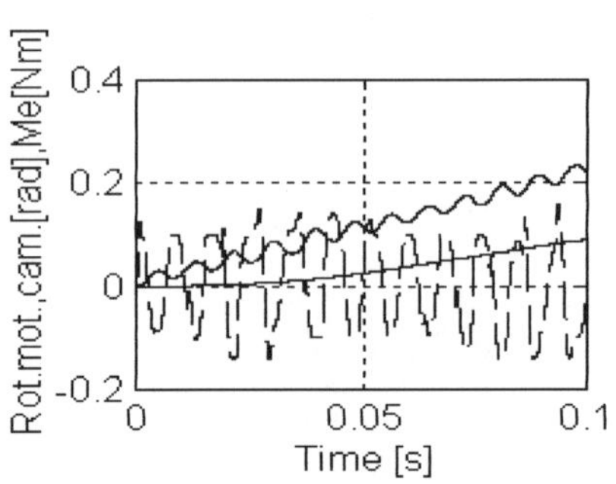

Fig. 3. The rotation of motor and camera, torque 100 %

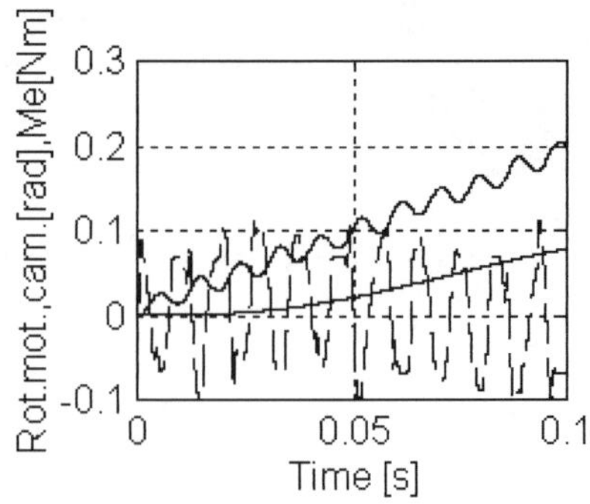

Fig. 4. The rotation of motor and camera, torque 70 %

The dynamic model may be used in various manners. With its help, the optimal electric engine from the point of view of the electromagnetic torque and the rotor's inertia may be chosen. Also the perfect longitudinal elasticity of the wire used in the mechanical drive, may be studied and specified.

A very important element for the step by step motors used in open systems (without reaction) is represented by the steps loosing phenomenon. Applying different frequencies and electromagnetic moments, with the help of the dynamic model, the study of this phenomenon is completely approachable.

If the rated couple decreases by 30%, the motor is at limit of the step loosing phenomenon (figure 4).

For a 50% couple decrease, the number of loosed steps is higher than that of executed steps. By increasing rated couple, for example by 100%, as expected, a much better control of the rotor is achieved.

5. CONCLUSIONS

The researches presented within this paper offer a study instrument for dynamic performances of a dynamic belt drive. They have been applied for researching the process of positioning a camera acted by a step-by-step motor.

For the above system the dynamic performances are: positioning angular increment 0,36° and angular rotation speed 57°/s. The achieved dynamic model allows the optimization of the presented camera. The applied model may be applied to any belt drive.

By an adequate modeling of pre-tension, this model may be applied to any mechanical transmission/drive. By using different parameters and solving the dynamic model under different conditions, an optimal structure for the mechanical drive may be obtained.

6. REFERENCES

Angeles, J. (2007). *Fundamentals of Robotic Mechanical Systems: Theory, Methods, and Algorithms,* Springer Science, ISBN 0-387-29412-0, NY

Calefariu, G. (2002). *Dynamics of mechanical systems*, Transilvania University of Brasov, ISBN 973-635-010-X, Brasov

Dumitru, N. & Cherciu, M. (2002). The dynamics of the kinematical chains considering the friction in the joints, *Proceedings of The International Conference on Manufacturing Systems ICMaS 2002,* Romanian Academy, pp. 121-124, ISBN 973-27-0932-4, Politehnica University of Bucharest, October 2002, Bucharest

Kuo, B., Kelemen, A., Crivii, M. & Trifa, V. (1981). Command systems for incremental control of position, Technical Editor, ISBN – No-, Bucharest

Marsden, J. & Ratiu, T. (1999). *Introduction to Mechanics and Symmetry: A Basic Exposition of Classical Mechanical Systems*, Springer Verlag, ISBN 0-387-98643-X, NY

Annals of DAAAM for 2008 & Proceedings of the 19th International DAAAM Symposium, ISSN 1726-9679
Editor B. Katalinic, Published by DAAAM International, Vienna, Austria 2008
Make Harmony Between Technology and Nature, and Your Mind will Fly Free as a Bird

INFLUENCE OF THE WORK PARAMETERS ON THE MICROHARDNESS OF THE COLD PLASTIC DEFORMED SURFACES

SARBU, I[onel]; CARAUSU, C[onstantin]; PARASCHIV, D[ragos]; PRUTEANU, O[ctavian] & TABACARU, L[ucian]

Abstract: *In this paper we present the results of the research regarding the micro hardness of the bearing races of the type 6210 radial ball bearings'rings processed through cold plastic deformation. We have traced variation charts and we have determined the dependency relations between the work parameters and the depth of the deformed ring.*
Key words: *microhardness, plastic deformation, bearing, ring, empirical relation.*

1. INTRODUCTION

The experimental trials have been done at SC Rulmenţi SA Bârlad pursuing the influence of the technological elements (the force of deformation, the deformation feed and piece's speed of rotation) on the microhardness of the bearing races of rings on the cold plastic deformed bearings (Wagoner, 2001). The measurements of the microhardness have been done following the diagram in fig. 1 (Pruteanu et al., 2006).

2. EXPERIMENTATION CONDITIONS

2.1. Material and semi-manufactured

We used the special material for the manufacturing of the bearings rings, 100Cr6 and the ring shape forged and hot rolled semi-manufactured.

2.2. Measurement machines and devices

For the plastic deformation of the bearing races of the bearing rings we have used special machines type CRF-120 OR for the ring 6210-10 (external ring) and type CRF-70 IR for the ring 6210-20 (interior ring). For the measurement of the micro hardness we have used the device Akashi MVK-D (Pruteanu et al., 2007).

3. MATHEMATICAL PROCESSING OF THE EXPERIMENTAL RESULTS

For the deduction of the empirical relations and the representation of the working parameters'influences we have used the following functions: F1(x) = a.x + b; F2(x) = a.eb.x; F3(x) = a + b.ln.x; F4(x) = a = b/x; F5(x) = a + b.x + c.x2.

We deducted the coefficient of correlation R, the estimated error, we applied the test F of significance and through comparison we established the most suitable model for the data in the experiment.

After the cold plastic deformation we have done the measurements of micro hardness HV0.5, depending on the working parameters, mentioned in table 1 for the external ring 6210-10. Each time we varied one of the parameters deformation force, F, deformation feed A and rotation speed of the piece V and we maintained constant the other two.

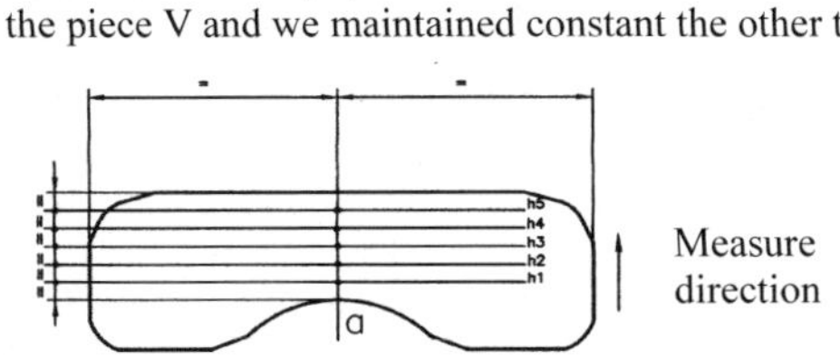

Fig. 1. The diagram of the microhardness measurements

For exemplification we present the diagram of the empirical modules, the tables with the estimated parameters and the relations of the functions which describe best the variation of the micro-hardness (values of the relations "bold" in table 1), are presented in figures 2, 3 and 4. One can notice that for the external ring the micro-hardness of the bearing races for all the working parameters vary depending on the linear function and very closely on the exponential function.

In table 2 we present the values of the micro-hardness of the bearing race for the internal ring 6210-20, as well as the empirical relations of the functions.

Parameters		Micro hardness $HV_{0.5}$				
		h_1	h_2	h_3	h_4	h_5
P = 8,0 MPa	A=34mm/min V=68 rpm	302	309	297	283	281
P = 7,0 MPa		302	306	302	295	289
P = 6,0 MPa		298	303	295	378	246
P = 5,5 MPa		309	303	295	250	207
A=34,0 mm/min	P = 5,5 MPa V= 68 rpm	315	309	309	266	269
A=30,0 mm/min		302	301	302	301	286
A=27,5 mm/min		302	309	303	295	287
A=25,0 mm/min		312	309	295	287	265
V=82,0 rpm	P = 5,5 MPa A=34mm/min	297	299	287	280	260
V=76 rpm		301	309	303	282	287
V=68 rpm		311	303	308	289	265

Parameters		Dependency relation
P = 8,0 MPa	A=34mm/min V=68 rpm	$HV_{0.5}$= 315,4.$e^{-0,046.h}$
P = 7,0 MPa		$HV_{0.5}$= 310,1.$e^{-0,025.h}$
P = 6,0 MPa		$HV_{0.5}$=281+45,4.h-23,7.h^2
P = 5,5 MPa		**$HV_{0.5}$= 350 – 51,4.h**
A=34,0 mm/min	P = 5,5 MPa V= 68 rpm	**$HV_{0.5}$= 334,1 – 27.h**
A=30,0 mm/min		$HV_{0.5}$= 308 – 6,4.hy
A=27,5 mm/min		$HV_{0.5}$= 312,7.$e^{-0,03.h}$
A=25,0 mm/min		$HV_{0.5}$= 328,4 – 23,2.h
V=82,0 rpm	P = 5,5 MPa A=34mm/min	$HV_{0.5}$= 312,5 – 18,6.h
V=76 rpm		$HV_{0.5}$= 312,9 – 11.h
V=68 rpm		**$HV_{0.5}$= 327 – 21,2.h**

Tab. 1. Micro-hardness of the 6210-10 external ring's bearing races

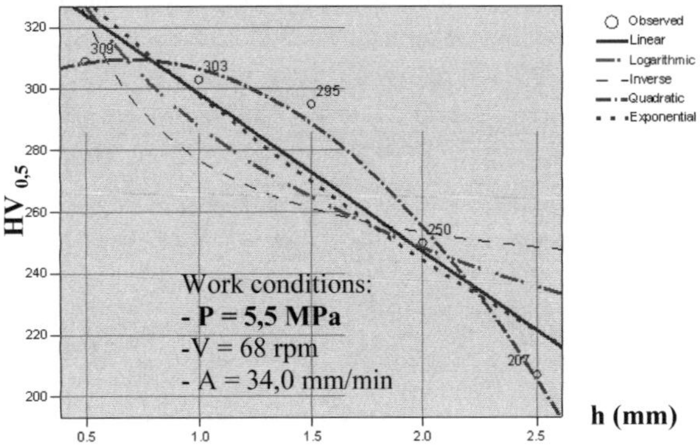

Fig.2. Empirical models of the variation of hardness on the depth of the deformed ring

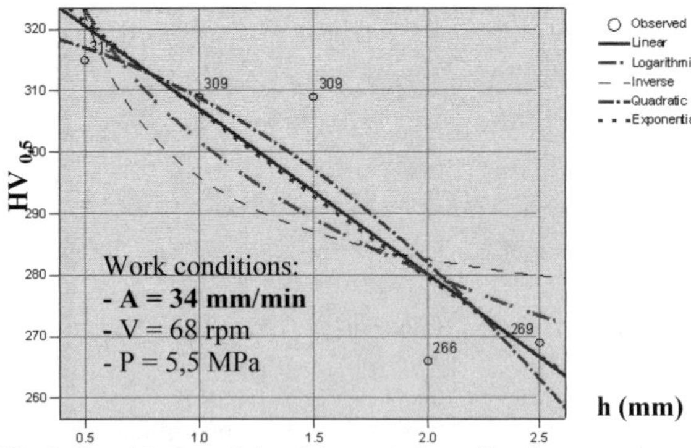

Fig.3. Empirical models of the variation of hardness on the depth of the deformed ring

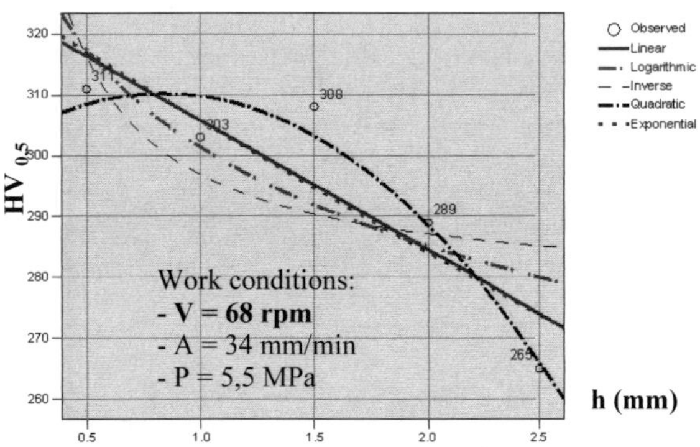

Fig.4. Empirical models of the variation of hardness on the depth of the deformed ring

To exemplify we present the diagram of empirical models, tables with estimated parameters and the relations of the functions which describe best the variation of micro-hardness (values of the relations "bold" in table 1), are presented in figures 5 and6.

One can notice that in the case of internal ring the micro-hardness of the bearing races for all the working parameters vary depending on the inverted function.

Parameters		Micro hardness $HV_{0,5}$				
		h_1	h_2	h_3	h_4	h_5
P = 4,5 MPa	A=25mm/min V=118 rpm	270	280	289	291	281
P = 4,0 MPa		283	295	301	301	294
P = 3,5 MPa		274	289	289	290	292
P = 3,0 MPa		272	286	291	281	263
A=27,5 mm/min	P = 4,0 MPa V=118 rpm	280	283	291	289	268
A=25,0 mm/min		271	281	287	294	289
A=22,5 mm/min		270	280	290	291	286
V=118 rpm	P =4,0 MPa A=25mm/min	272	298	302	295	290
V=98 rpm		285	306	294	298	312
V=84 rpm		285	291	290	306	298

Parameters		Dependency relation
P = 4,5 MPa	A=25mm/min V=118 rpm	$HV_{0.5}$= 292 – 10,8/h
P = 4,0 MPa		**$HV_{0.5}$= 304,6 – 10,3/h**
P = 3,5 MPa		$HV_{0.5}$= 296,7 – 10,8/h
P = 3,0 MPa		$HV_{0.5}$= 246+63.h-22,6.h^2
A=27,5 mm/min	P = 4,0 MPa V=118 rpm	$HV_{0.5}$=258,6+46.h-16,6.h^2
A=25,0 mm/min		**$HV_{0.5}$= 296,2 – 12,9/h**
A=22,5 mm/min		$HV_{0.5}$= 294,7 – 12,45/h
V=118 rpm	P =4,0 MPa A=25mm/min	**$HV_{0.5}$= 304 – 14,2/h**
V=98 rpm		$HV_{0.5}$= 309,65 – 11,7/h
V=84 rpm		$HV_{0.5}$=291,3 + 10,3ln(h)

Tab. 2. Micro-hardness of the 6210-20 internal ring's bearing races

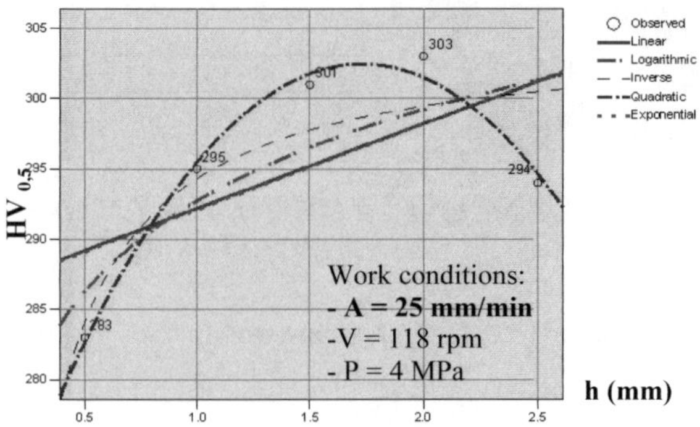

Fig.5. Empirical models of the variation of hardness on the depth of the deformed ring

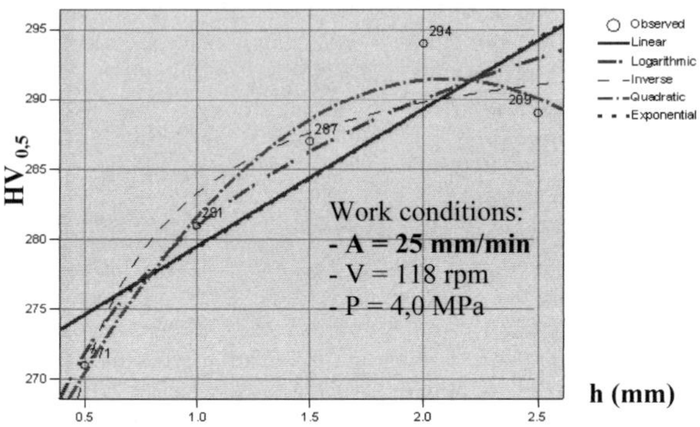

Fig.6. Empirical models of the variation of hardness on the depth of the deformed ring

4. CONCLUSIONS

When processing through cold plastic deformation through rolling of the bearing rings they get superficially hardened.

The experimental research regarding the variation of micro-hardness with the depth of the hardened layer when modifying the values of the work parameters show that:

- in the case of the external ring the micro-hardness vary depending on a linear function, very closely being the exponential one.
- in the case of the internal ring the micro-hardness varies depending on the inverted function.

5. REFERENCES

Pruteanu ,O.V., Carausu, C. Paraschiv, Sarbu, I., (2006). *The influence of the working parameters on the parameters of the superficial layer when procesing the steels through grinding in the rolling's technology*. Мировая зкономика и бизнес – администрирование, Minsk, Belarus, may, 186-197.

Pruteanu, O.V., Paraschiv, Carausu, C., Sarbu, I., Bunea, M., Bardac, D. (2007). *Processing through plastic deformation of the bearing rings.* MicroCAD 2007, International Scientific Conference, Miskolc, 143-148.

Pruteanu, O.V., Paraschiv, Pricope, Bunea, C.M., Sarbu, I., Cărăuşu, C. (2007). *Considerations Regarding the 100Cr6 Material's Conduct through the cold Rolling of the Bearing Rings,* International Journal for Manufacturing Science & Technology, U.S. Library of Congress, vol. 9, no.1, p.51-57, ISSN 1524-1548.

Pruteanu, O.V., Paraschiv, Carausu, C., Sarbu, I., Bunea, M., Bardac, D. (2007). *Processing through plastic deformation of the bearing rings.* MicroCAD 2007, International Scientific Conference, Miskolc, 143-148.

Wagoner, R.H., Chenot, J.L. (2001). *Metal forming analysis.* Cambridge University Press. New York.

Annals of DAAAM for 2008 & Proceedings of the 19th International DAAAM Symposium, ISSN 1726-9679
Editor B. Katalinic, Published by DAAAM International, Vienna, Austria 2008
Make Harmony Between Technology and Nature, and Your Mind will Fly Free as a Bird

MASTERING IT SECURITY ISSUES IN OUTSOURCING

SAVII, G[eorge]; NEIDENBACH, N[orbert] & WOLF, E[dwin]

Abstract: *- In order to lower the costs many companies choose for the translocation of business processes within outsourcing. The present article highlights the most important inherent security aspects and proposes some measures for their implementation. It presents the risks related to outsourced IT security, the integration of outsourced IT security in the business processes and control means within an outsourced IT security.*

Key words: *outsourcing, ITIL, Managed Security Services.*

1. INTRODUCTION

During the past years the demand for IT (Information Technology) security has been continuously growing given the more complex applications as well as the inherent risks (Cisco 2007). At the very same time the business processes are depending more than ever on an efficient IT (Phifer 2000). Regarding this very need for action, the IT decision-makers must also take into consideration some aspects of IT security related to the question "make or buy". The present study aimed to develop a method for mastering IT security issues in outsourcing

2. PARTIAL OUTSOURCING

2.1 Characteristics

The partial outsourcing centers on IT security cost reduction and control. By translocating some parts of IT security, one can reach greater efficiency and transparency. By means of partial outsourcing the IT security personnel-related costs can be avoided. Due to the IT security outsourcing, the outsourcing service provider can concentrate more on the central processes of its own company.

All these advantages are possible by using a Managed Security Service Provider (MSSP) (Cisco 2007, Choi & Seo 2005). Within Managed Security Services (MSS) one can initially mention the firewall and Intrusion Prevention Technology operation as well as various antivirus measures (Ott 2001). The service performance, the quality, the service availability as well as the support are controlled in the so-called Service Level Agreement (SLA) which has been previously agreed upon (Brittain & Matlus 2002).

The services imply activities which go from the sheer client system monitoring to service operation as well as to service outsourcing towards dedicated-systems or towards shared-systems. On the European market, the MSSP is in the initial stage; nevertheless it never seized to make its presence felt on the American market.

The purpose of outsourcing should harmonize with the general strategy of the company as well as with the aims, the established Return of Investment (ROI) and with the Total Cost of Outsourcing (Ott 2001).

2.2 Risks of IT-security outsourcing

The introduction and the implementation of MSSP may have as a result failure of IT security functions (Alner 2001, Phifer 2000).

A major risk is represented by the fact that one company entrusts the IT infrastructure protection to another company and thus depends on an outsourced service provider. In this way expert knowledge and competence are lost in the process and cannot be recovered again.

The company providing the outsourcing services has a considerable insight in the company security system. A risk assessment is therefore a foundation of each and every outsourcing.

At the very beginning of the outsourcing the responsibilities, the contact persons, the processes, information chains as well as technical and organizational connections must be harmonized, defined and documented (Pescatore 2001).

The Information Technology Infrastructure Library (ITIL) is a framework of best practice approaches intended to facilitate the delivery of high quality IT services (Van Schaik 2006, van Bon 2002). In order to avoid risks, all the ITIL processes, especially those processes in relation with the service-support must be observed. Furthermore there are response and maintenance intervals, monitoring, reporting which must be established.

An in-depth analysis of the reported security problems has led to the conclusion that one also must define the way of handling the demand as well as the way of harmonizing the knowledge transfer between the outsourcing provider and the internal IT.

2.3 Security measures

By means of IT security functions outsourcing the costs and the internal capacities can be saved by the outsourcing provider and in this way one can achieve a greater quality and a better control.

In this respect one must also asses the risks and the internal expenses regarding the control of the MSSP and of the necessary in-house processes. It is therefore of utmost importance to include the results of the above assessment in the decision.

Higher costs as well as outsourcing inefficiency are the result of neglecting of these aspects.

In case of a contract negotiation one must:

- observe the security measures
- observe the service providing and SLA provisions
- observe the agreement upon regular auditing reports
- take into consideration the risk minimizing measures

3. TOTAL OUTSOURCING

3.1 Characteristics

In most cases one may choose the outsourcing of the whole IT security as a consequence of the outsourcing of the operative IT and this should be distinguished from the outsourcing of individual IT security functions.

All the operative activities of the IT security are indicated below and the outsourcing provider is limited to:

- the strategic orientation within the field of IT security
- the service provider control

- to the definition of the security demand on the basis of the business processes
- to the monitoring of the legal requirements observation (e.g. the person in charge with data security or revision)

In this respect one must also take into consideration the internal expenses as well as the tasks resources and the functions within the general IT outsourcing costs.

3.2 Risks

A total IT outsourcing and implicitly the IT security to a MSSP is known to present both risks and advantages (Alner 2001, Phifer 2000).

It is of utmost importance to create a universal security management in collaboration with the IT provider and to establish this contractually in the SLAs. Under these circumstances the feasibility of such a security management is also very important. The risks presented by IT outsourcing can be reduced by means of evaluation and efficiency which also guarantees a strong bond between the outsourcing provider and the client.

3.3 Managed Security Services Process (MSS) in case of the client

The Managed Security Services (MSS) are based on the security requirements of the business processes, on the general conditions of the IT strategy as well as on the risk management (among others general security directives) and legal prescriptions.

The identification of business process requirements regarding the availability, the privacy and the integrity is to be carried out by means of a risk assessment which must be in conformity with the general risk management of the client.

Figure 1 presents an example of the risk classes' representation of an IT service, based on the risk assessment.

The IT service classification in risk classes and a risk potential assessment of the business processes can occur by means of IT service client consultation.

Of utmost importance is the risk and expenses analysis by means of which the measures for risk minimization with regard to the expenses can be assessed.

The expenses inherent to the security measures must be assessed in relation to the prejudices in case of a security event.

In order to verify the measures implementation a client reporting form (for the revision and the as well as for the data security in charge personnel) must be drawn up. By automating the reporting the expenses are reduced and the relevance is ensured.

The security issues are thus transparently rendered and can also be efficiently highlighted.

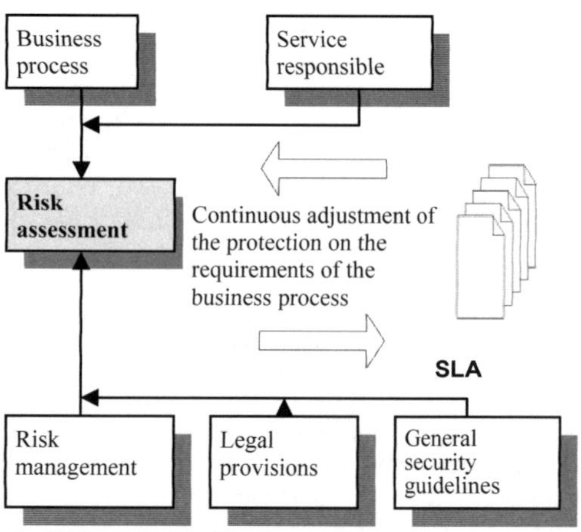

Fig. 1. Risk analysis of business process.

3.4 Managed Security Services Process at the Service Provider

The IT-Service provider must comply to the requests of the client by IT-security measures and persevering processes.

To allow standardization and a practical transition, it is a good choice to implement organizational, technical and physical measures for the adopted protection class in the form of guidelines or process directives. These will be later tailored on the individual systems.

The full implementation of the measures of a security norm can later lead to the certification of the IT infrastructure.

To prove the implemented measures, a reporting must be done, for internal use, for the customer (IT-department, data security responsible, etc.) and also for the possible revisions. This reporting must be automated as much as possible, so the costs will be minimized, the updating and the necessary transparency assured.

4. CONCLUSIONS

The analysis of the IT outsourcing processes and markets allowed the highlighting of the most important inherent security aspects related to IT outsourcing and to propose some measures for their implementation. As important aspects, the risks related to outsourced IT security, the integration of outsourced IT security in the business processes and control means within an outsourced IT security were presented and analyzed.

The universal and unitary process starting with the risk evaluation and up to the risk reduction measures presented in the paper makes possible to avoid the process gaps, to minimize the losses and to economically and efficiently operate in the IT security domain.

By means of the proposed methods, the IT security performance can be fully controlled even if this occurs within an outsourcing project. Moreover the risks can be held at bay.

5. REFERENCES

Alner, M. (2001) The Effects of Outsourcing on Information Security. *Information Systems Security*, Auerbach Publications, CRC Press LLC, May/June 2001, ISSN 1065-898X.

Brittain, K. & Matlus, R. Road Map for IT Service-Level Management. *Gartner Article Top View*, 28 January 2002.

Cisco (2007) *Managed Security for the Small and Medium-Sized Business – The Benefits of Out-Tasking Security Services to a Qualified Service Provider*, White Paper, Cisco Systems, Inc.

Choi, Y.S. & Seo, D.I. (2005) An analysis of ISP's role as managed security service providers (MSSPs), *The 7th International Conference on Advanced Communication Technology, ICACT 2005*, Volume 1, pp. 624 – 626, ISBN: 8955191235, Korea, February 2005, Phoenix Park.

Ott, J. L. (2001) Managed Security Services, *Information Systems Security*, Vol 10, No 4, (September/October 2001), 299-321, ISSN 1065-898X.

Pescatore, J. (2001) Choosing a Managed Security Services Provider. *Gartner Research Note*, 31 August 2001.

Phifer, L. (2000) Outsourcing Security Needs to a Managed Security Service Provider. *Available from*: http:// searchsecurity.techtarget.com/onlineEventsTranscript/0,289 691,sid14_gci511332,00.html. *Accessed*: 2002-02-15.

van Bon, J. (Ed.) (2002), *The guide to IT service management*, Addison Wesley, ISBN 0-201-73792-2.

Van Schaik, E. A. (2006) A Management system for the Information Business, *Red Swan Publishing*, ISBN: 1-933703-03-2.

Annals of DAAAM for 2008 & Proceedings of the 19th International DAAAM Symposium, ISSN 1726-9679
Editor B. Katalinic, Published by DAAAM International, Vienna, Austria 2008
Make Harmony Between Technology and Nature, and Your Mind will Fly Free as a Bird

SYSTEM AND METHODOLOGY FOR ANALYSING, EVALUATING AND CORRECTING HUMAN BIPED LOCOMOTION

SCHOR, V[ladimir]; SINDILA, G[heorghe]; POPESCU, D[iana] & CARUTASU, N[icoleta]

Abstract: *The difficulty in establishing a mathematical model of human motion and in developing a methodology for gait analysis resides in the complexity of assessing and controlling the parameters involved. The solution proposed in this paper considers the use of a new device called Ergo control for evaluating, correcting and optimizing human gait, by measuring in real time all the parameters contributing to propulsion of the center of mass of a human subject. This device can be use with success for a faster diagnosis and recovery of persons with walking problems and for correcting pathological human gait pattern.*
Key words: *biomechanics, gait, Ergocontrol, diagnosis, rehabilitation*

1. INTRODUCTION

There is a worldwide interest for a deeper understanding of a very complex process such us human gait, which implies a synchronization of the skeletal, neurological and muscular systems of the body. This interest especially regards the clinical assessment of pathological gait and rehabilitation of people who, for a different medical reason (stroke, amputation, neurological impairment, muscle disease, etc.), suffered a modification of the normal walking pattern. Human gait analysis constitutes a research subject for both medicine doctors (for diagnosis and rehabilitation, prosthesis design) and engineers (for developing new methods of visually investigating human movements, mathematical modeling of human gait and implementing the gait pattern in robots, human gait recognition, etc.).

The first researches in the field date from 1975 when Mann summarized in a chart the kinematics interactions that occur during the gait cycle, and made the distinction between stance and swing phases. The work continued over the years with studies regarding the dynamic of human gait, biomechanics of walking and running, movement tracking and recognition, study regarding the locomotion models and the brain activity during the foot movements, study regarding the reflexes involved in muscles activity during walking, balance and control of walking, etc.

In Romania, the researches have been conducted in different hospitals and research centers (Orthopedic Department of St. Pantelimon Hospital – Bucharest, University Politehnica of Timisoara – BioArt CEEX Project, Recovery Department of University of Medicine and Pharmacology Carol Davila – Bucharest, etc.) especially for and diagnosis and movement corrections.

2. METHODS FOR GAIT ANALYSIS

Due to the large number of persons having walking problems and due to an increasing interest of the health and insurance companies in a faster diagnosis and rehabilitation, independent centers for study the human gait have been founded for the study of all the processes involved in this natural and complex activity - walking. In this context, there are laboratories focused on using markers for studying different anatomical segments involved in locomotion (Gavrila, 1999). The inconvenient of such an approach resides in the small number of steps considered and the difficulty of experimentally studying the dynamic of gait.

Another type of kinematical analysis is goniometry, which even if it gives the possibility to study a bigger number of steps (walking cycles), does not offer information regarding the dynamical aspects. The dynamic analysis depends of forces measuring platforms on three directions, or baropodometers which principal disadvantage is the small number of recorded steps not enough for evaluating the human subject movement stereotype.

Other methods, more complex, combine in different manners the above-mentioned methods, adding also biological investigations (electromyography, electroencephalography) needed for evaluating the human gait control mechanisms.

The study of the literature in the field showed that the devices used in the gait analysis labs and the correspondent methods of assessment, are limited when it comes to evaluate the movement stereotype of a human being, due to the small number of cycles, which can be used to record the gait dynamic.

In the last couple of year, the research focused also on modeling and simulating human locomotion, in order to have a global and integrated view on dynamical and kinematical aspects. However, not even this approach can not properly describe the human gait stereotype due to its complexity and to the fact that each simplification leads to modified and irrelevant results. (Cappozzo, 1984), (Ackermann& Schiehlen, 2006), (Kale et al., 2002).

Devices such us Electronic Baropodometer, Body Analysis Kapture, Biomechanics, etc, are mostly used in diagnosis, not considering the movement pattern and thus not offering the possibility to vary in a controlled manner the conditions in which the gait is taking place. Moreover, these devices do not have the capability to correct the gait stereotype. In this context, another device called Ergocontrol was developed by S.C. Medisport S.R.L., a Romanian company headquartered in Bucharest. This device is using a new approach in evaluating, correcting and optimizing human biped locomotion, by measuring in real time all the parameters contributing to propulsion of the center of mass of a human subject.

Avoid leaving a heading at the bottom of a column, with the subsequent text starting at the top of the next page/column. Use extra spacing (between earlier figures or sections) to push the headings up to the top of the same column as its text. In view of the tight page constrains, however, do please make the fullest possible use of the text area.

3. ERGOCONTROL – SYSTEM FOR EVALUATING, CORRECTING AND OPTIMIZING HUMAN GAIT

Ergocontrol device (fig.1-2) offers the capability of characterizing the human gait stereotype, and then to correct and optimize it in order to increase the physical performances of sportsmen or to restore the normal walking ability of those who suffered different trauma.

As mentioned before, all the parameters involved in human locomotion are measured in real time and the results are represented as curves describing the dependency between the force of propulsion and the position of different elements of the human locomotor chain (fig.3). This information is directly linked to prioreceptive information coming from the specialized anatomical elements from muscles, tendons and articulations used in the natural subcortical gait control.

Another novelty introduce by Ergocontrol is that the graphical representation allows the patient to correct the movement pattern in real time, attempting to follow the shape of the curves presented on the screen.

An electromechanical system and a force transducer are used for collecting the information regarding the force of propulsion, and sent to specialized software for a real time analysis and evaluation. The basic scheme of the device is presented in figure 4.

The electrodynamical break **FE** and the electrical motors **ME1** and **ME2** allow controlling the conditions in which the movement is taking place, by varying the load depending on the diagnosis and the methodology used for correcting the gait pattern. Motors **ME1** and **ME2** have digital signal processors, position transducers (which allow a precise position evaluation both in relative ad absolute movement), and analog digital converters for force measurement. The information referring to the force of propulsion developed by the human subject **S** are collected by a computer assisted electromechanical system and its corresponding force and position transducers, and then are used as input in an evaluation software. The results are presented in real time as curves on the screen **M**, both to the patient and the medical assistant for correcting the movement pattern. The use of visual feedback constitutes the interface between the subcortial movement control and the voluntary movement control.

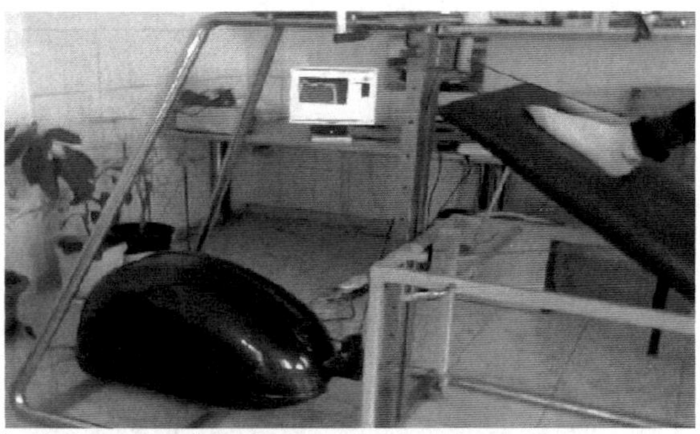

Fig. 1. Ergocontrol device.

Fig.2. Ergocontrol – device interior view.

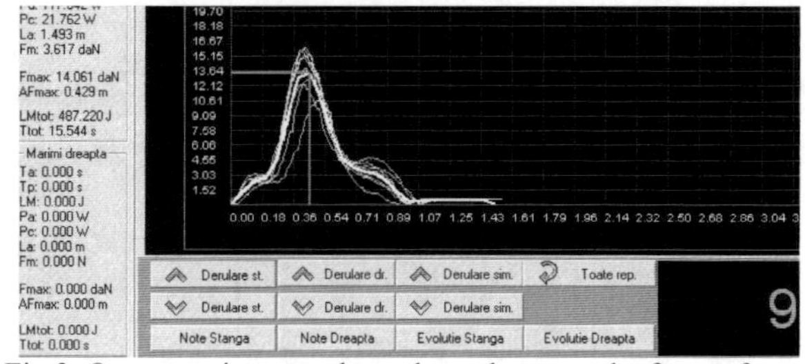

Fig.3. On screen image – dependency between the force of propulsion and position of locomotor chain elements.

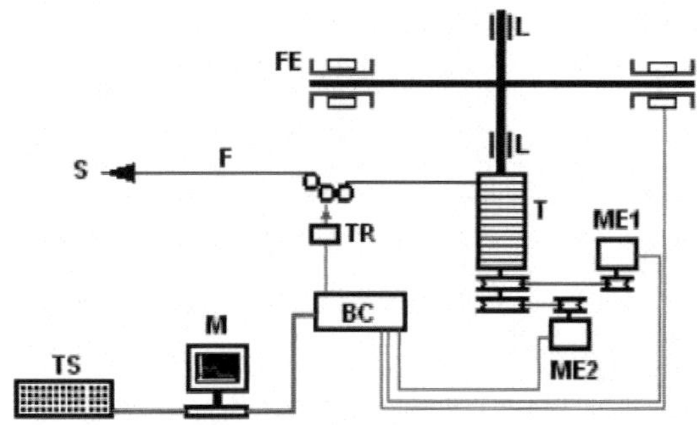

Fig.4. Basic scheme of Ergocontrol device.

4. CONCLUSION

Ergocontrol device described in this paper represents a new electromechanical system which can be used for human gait diagnosis and correction. The functioning of the device is based on a new approach, which consists in real time measurement of parameters involved in the gait and controlling the parameters characterizing the external conditions in which the gait is taking place (initial speed, moving forward friction coefficient, etc.).

Moreover, the approach combines the classical methods used in gait analysis (digital image analysis, goniometry and podometry) with the capabilities of the device to record a large number of gait cycles for properly characterizing the movement stereotype and to use corrective working models.

Further research will focus on improving the electromechanical subsystem of the device and implementing new corrective working models for movements of arms.

5. REFERENCES

Ackermann, M., Schiehlen, W. (2006), *Dynamic Analysis of Human Gait Disorder and Metabolical Cost Estimation*, Archive of Applied Mechanics, Volume 75, Numbers 10-12/October

Cappozzo, A. (1984), *Gait Analysis Methodology*, Hum Movt Sc 1984; 3: 27–50

Gavrila, D.M. (1999), *The Visual Analysis of Human Movement: A survey*, Computer Vision and Image Understanding, vol.73, no.1, January, pp.82-98

Kale, A., et al., (2002), *Gait analysis for human identification*, Proceedings of 3rd International Conference on Audio and Video Based Person Authentication, June 2003, ttp://citeseer.ist.psu.edu/kale03gait.html

Mann, R.A. (1975), *Biomechanics of the foot*, American Academy of Orthopaedic Surgeons (Ed.) Atlas of orthotics: Biomechanical principles and application, pp. 257-266, St. Louis: C.V. Mosby

Schor, V. (1983), *A multifunctional Modular Ergometer for Simulating the Specific Dynamics in Swimming*, International series on sport science, vol.14, Biomechanics and Medicine in Swimming, Human Kinetics Publisher Inc.

Annals of DAAAM for 2008 & Proceedings of the 19th International DAAAM Symposium, ISSN 1726-9679
Editor B. Katalinic, Published by DAAAM International, Vienna, Austria 2008
Make Harmony Between Technology and Nature, and Your Mind will Fly Free as a Bird

PRODUCTION OPTIMIZATION BY USING GENETIC ALGORITHMS AND SIMULATION MODEL OF PRODUCTION SYSTEM

SCHREIBER, P[eter]; VAZAN, P[avel] & TANUSKA, P[avol]

Abstract: *The paper presents the way how it is possible to combine the behaviour of simulation model of production system and its optimization by genetic algorithms considering the fulfilment of quantitative production goals in defined conditions. The objective function is mathematical formula for production costs (it is additive nonlinear function with more variables). The constrains are: flow time, capacity utilization and number of finished parts. Mathematical model is very complicated for analytical solution. Although optimizing methods are implemented in simulators, often they are not effective enough. The genetic algorithms seem as simply alternative for problem solution.*
Key words: *production system, simulation, optimization, genetic algorithm.*

1. INTRODUCTION

The lot size is one of the directions of production which markedly influences production costs. The lot size influences production flexibility, amount of parts in process, flow time, capacity utilization etc. The goal is to determine the lot size so that production costs will be minimal.

There are several known methods for determination of lot size in the world. One of them is called economically optimal lot size. This method defines the mathematical formula which minimizes the set-up costs and storage costs. This lot size is expressed by mathematical model that solves the compromise between the reduction of fixed costs per piece and increasing of lot size, and on the other side increasing of the storage costs.

The optimal lot size is determined as follows:

$$D_o = \sqrt{\frac{2 * Q_p * N_{pz}}{N_j * n_s * t}} \quad (1)$$

D_o - optimal lot size in pieces, Q_p – planned number of parts in pieces, N_{pz} – batch set-up costs, N_j – costs per one piece, n_s – annual storage costs, including credit interest, t – fraction – period of the year.

Another approach is based on search of minimal lot size which is needed for effective utilization, usually as bottleneck or capital-intensive capacity unit. The lot size, defined in this way, provides economical utilization of chosen capacity units (bottlenecks).

The minimal lot size is determined as:

$$a = \frac{t_{pz}}{t_K * D} \Rightarrow D = \frac{t_{pz}}{t_K * a} \quad (2)$$

t_{pz} – time for set-up in min., t_k – part time in min., D – lot size in pieces, a – coefficient. a = 0,04 for complicated parts and a = 0,1 for production with automatic machines.

The other methods are based on empirically found variables with many defects. There are used different approaches, e.g. optimal lot size is defined at many possibilities and chosen is that one with its own minimal costs. This includes e.g. optimal lot size according to Teplov or Bankovsky (Gregor, 2000).

These classic calculation methods require accurate input values (set-up costs, storage costs, etc.). In fact these values are qualified approximations but not exact values. The analytical solution is more complicated when the model concerns constrains (Ramaswamy, 2006).

2. SIMULATION

Simulation of real production system behaviour by software tools is one of the possible approaches to determination of optimal lot size. It is possible to optimize the system through evaluation of system behaviour with different input parameters. The most appropriate simulation method is the discrete-event simulation for manufacturing model building. Rapid expansion of simulation tools for manufacturing allowed the usage of this procedure very effectively. The model building takes a short time and the model is very detailed. The authors used the simulation package Witness of the company Lanner group Ltd.

The objective function had to be defined for optimising of production cost. The production cost is computed according to following function in the realised model.

$$SumCosts = \sum_{i=1}^{P} In\_part\_costs_i + \sum_{i=1}^{P} Added\_value_i \quad (3)$$

where In_part_costs are initial costs for each entered parts, Added_value is gained amount in production system, P is number of entered parts. Total added value is calculated:

$$Sum\_Added\_value = \sum_{i=1}^{P} OC_i + SC_i + LC_i + TC_i + STC_i \quad (4)$$

where OC – operation costs, SC – setup costs, LC – labour costs, TC – transport costs and STC – storage costs.

$$operation\_costs = \sum_{j=1}^{m} t_{Aj} * unit\_costs_j \quad (5)$$

$$setup\_costs = \sum_{x=1}^{s} t_{Bx} * unit\_costs_x \quad (6)$$

where m is number of machines, t_{Aj} is processing time, unit_costs is rate per unit of time, s – number of set up operations, t_{Bx} – set up time, unit_costs is rate per unit of time.

$$labour\_costs = \sum_{k=1}^{op} t_{LBk} * unit\_L\_costs2_k \quad (7)$$

where op is number of operations with labour, t_{LBk} is operation time with specific labour, unit_costs2 are rates per unit of time.

$$transport\_costs = \sum_{l=1}^{top} transport\_operation\_costs_l \quad (8)$$

$$storage\_costs = \sum_{n=1}^{NS} t_{Sn} * unit\_costs_n \qquad (9)$$

where top is number of transport operations and NS is number of storages.

All functions are calculated in elements of simulation model of production system. Partial values of objective function are always calculated when specific element of production system finishes its activity. SumCosts is calculated at the same time. Discrete-event simulation allows this process.

The function SumCost is growing up with rising of number of finished parts. Therefore it is not proper to use it as objective function. The function Unit_Cost is designed by the authors as objective function. This function calculates the production cost per finished part.

Subsequently the objective function is revised so that the production costs are optimized, but other production goals have to reach defined values (Važan & Tanuška, 2003).

3. PRODUCTION SYSTEM OPTIMIZATION BY GENETIC ALGORITHM

Let suppose that production costs depend on the lot sizes and their input intervals will be minimized. The constraints are flow time, capacity utilization and the number of finished parts. Then :

$$N(d_1, t_1, d_2, t_2, ..., d_n, t_n) \rightarrow \min \qquad (11)$$
$$T_i(d_1, t_1, d_2, t_2, ..., d_n, t_n) \leq T_{imax}, i = 1..n \qquad (12)$$
$$C_j(d_1, t_1, d_2, t_2, ..., d_n, t_n) \geq C_{jmin}, j = 1..m \qquad (13)$$
$$P_i(d_1, t_1, d_2, t_2, ..., d_n, t_n) \geq P_{imin}, i = 1..n \qquad (14)$$
$$d_{imin} \leq d_i \leq d_{imax}, i = 1 .. n \qquad (15)$$
$$t_{imin} \leq t_i \leq t_{imax}, i = 1 .. n \qquad (16)$$

N – production costs, n – number of parts, d_i – lot size of the ith part, t_i – input interval of batch of the ith part, T_i – flow time of ith part, C_j – capacity utilization of jth part, M – number of equipment, P_i – number of finished i^{th} part, T_{imax} – maximally acceptable flow time of ith part, C_{jmin} – minimally acceptable capacity utilization of jth equipment, P_{imin} – minimally acceptable number of finished i^{th} parts, d_{imin}, d_{imax}, t_{imin}, t_{imax}, - limits of scanned space.

The objective function (1) and constraints (2) – (4) cannot be expressed analytically but the simulation model determines values N, Ti, Cj, Pi (Duzinkiewicz & Kwiesielewicz, 1998)

It is necessary to associate production system and its parts together with terms used in genetic algorithms for production system optimization by means of genetic algorithm (Sekaj, 2001).

Chromosome (the solution of optimising problem) will be a vector of numbers which represents input parameters of the system (d1, t1, d2, t2, ...,dn, tn). Its elements are genes.

Population: Let there are 40 individuals (solutions) in one generation.

Objective function is given by relation (1) and evaluates each solution. The value N is determined by simulation.

Selection: Genetic algorithm requires surviving "good solutions. They are those with small value N. The fitness of individual solutions is inversely related to their costs. If we use roulette selection, the probability of survival is

$$p_k = \frac{F_k}{\sum_{k=1}^{g} F_k} = \frac{\frac{1}{N_k}}{\sum_{k=1}^{g} \frac{1}{N_k}} = \frac{1}{N_k \sum_{k=1}^{g} \frac{1}{N_k}} \qquad (17)$$

where p_k is selection probability of kth solution, F_k is fitness of kth solution, N_k are costs of kth solution obtained by simulation, g is the number of solutions in population (40).

Elitisms: The best 4 solutions turn to the next generation unchanged to keep the best solutions. The other 36 solutions are crossovered and mutated.

Crossover operation: one-point crossover of two individuals in randomly chosen position. Let $p_c = 1$.

Mutation: Mutation is realized with probability $p_m = 0.05$.

Solution ending: Number of generations or calculation time is an ending condition.

The form of genetic algorithm is following:

1. Random solutions - vectors (d_1, t_1, d_2, t_2, ...,d_n, t_n) are generated, the generated values should fulfil conditions (12) - (16).
2. The solutions are evaluated according to relation (11).
3. If the ending condition is fulfilled, the best individual is the desired solution.
4. The best 4 solutions are chosen into new generation. The serial numbers 1 – 4 will be assigned.
5. Remaining 36 individuals in new generation will be obtained in the following way:
    a. Roulette selection (according to relation (17)). 36 individuals are given to the positions 5 – 40.
    b. Neighbouring pairs crossover. Conditions (12) – (14) are checked by simulation. If the solution is inadmissible, another random crossover point is used.
    c. Some genes are chosen for mutation in solutions 5 – 40. New genes have to fulfil conditions (15) - (16) and also new solutions have to fulfil conditions (12) – (14).
6. New generation is formed by solutions gained in steps 4 and 5. Algorithm continues by the step 2.

The results of this optimization are comparable or better than the values obtained by embedded optimizers of Witness.

4 CONCLUSION

The simulation optimization is a proper method for the lot size determination. It is able to respect more factors, which determine the lot size, than classic optimizing methods. It requires the existence of simulation model. On the other side classic methods are faster and simpler.

The authors want to verify this method by various types of production systems, especially by flexible manufacturing systems. The next goal is to compare systematically the results obtained by various optimising methods.

This paper has been supported as a part of solution of projects VEGA 1/0368/08 and VEGA 1/0170/08.

5. REFERENCES

Duzinkiewicz, K. & Kwiesielewicz, M. (1998). Multicriteria approach to production tasks planning in systems with continuous switchable production processes. In: *Proceedings of 13th International Conference on Systems Science*, pp.25-31, Wrocław, Poland, II

Gregor, M., Košturiak, J., Mičieta, B., Bubeník, P., Růžička, J. (2000) *Dynamické plánovanie a riadenie výroby (Dynamic planning and control of production)*, KPI - ŽU EDIS, ISBN 80-7100-607-6, Žilina, Slovakia

Ramaswamy K.V.(2006) Optimal lot sizing in manufacturing revisited. *Journal of Information and Optimization Sciences*. Vol. 27 no.1, (2006) pp 97-105, ISSN: 0252-2667

Sekaj, I. (2001) Genetic algorithms. *AT&T Journal*, vol.8 (11, 2001) pp.46-48, ISSN 1335-2237

Važan, P. & Tanuška, P. (2003) Optimization of control goals of flexible manufacturing system, In: *CIM: Computer Integrated Manufacturing : Advanced Design and Management,* Wydawnictwa Naukowo-Techniczne, pp. (586-591), ISBN 83-204-2830-5, Warszawa

Annals of DAAAM for 2008 & Proceedings of the 19th International DAAAM Symposium, ISSN 1726-9679
Editor B. Katalinic, Published by DAAAM International, Vienna, Austria 2008
Make Harmony Between Technology and Nature, and Your Mind will Fly Free as a Bird

AN EVALUATION OF EXISTING STEP SOLUTIONS FOR AUTOMATED PROCESS PLANNING

SCHUMANN, C[hristian]; TEICH, T[obias]; MILITZER, J[oerg] & UNGER, K[atja]

Abstract: *The response time to customer requests, next to costs or delivery reliability, in a B2B environment is one key factor for gaining purchase orders. Therefore, the authors develop a new model to assist the process of offer preparation. Besides a detailed cost and time planning, the approach takes a new description of company resources into consideration. This description is assigning features to machines and production techniques. The resulting specification represents the supply of a company, the ability to produce single features.*
In the developed approach, the description of the demand is a model created by a CAD System. In regards to the functional description of resources by features only, a requirement is the interoperability of the data format produced by several CAD systems. Taking these two points into account, the authors have chosen STEP (STandard for the Exchange of Product model data) as the most advanced standard. This paper will deal with the usability of STEP in the context indicated above.
Key words: *CAD, STEP, Process Planning, CAM*

1. INTRODUCTION

The concept of which this paper is about was discussed in different papers on various conferences like the 18th International DAAAM Symposium (Teich et-al. 2007) or recently the German-Russian Logistics Workshop in Moscow (Teich et-al. 2008). As a short summary, the goal is the automated generation of offers to a customer in a B2B (Business to Business) environment. Therefore various theoretical and practical concepts have been revised, especially the results of the SFB 457 and resulting doctoral thesis's and habilitation treatise (Teich 2003).

One major problem is the description of the demand in a functional way, which means that the single elements of a single piece part can be identified and related to a meaning, the function of the feature. The paper is structured into three parts,

- Detailed description of requirements for data model
- Evaluation of implementation of STEP in market-leading CAD/PLM solutions (Autodesk Inventor, CATIA, PARTSolutions)
- Review of the new developed AP 224
- Conclusion derived from results

2. DETAILED DESCRIPTION OF REQUIREMENTS FOR DATA MODEL

For the automated preparation of offers, a very detailed demand is required. To accomplish our model, the supplier needs to gain all possible information for matching customer's needs. This results from the requirements determined by the concept:

- Work Scheduling has to be performed automatically as far as possible. Therefore the product has to be decomposed in its intermediate products. Intermediated products can be derived by feature separation. Therefore the product has to be described by features only.
- For determination of the raw material, the description needs to cover information about the initial shape, where all features are subtracted from.
- The description of the features has to be accomplished by a relation to other features (this doesn't apply to the initial shape).

As we talk about an ideal situation, the supplier can take for granted that he will get a 3D-drawing of the demand, enriched with data that enables him to give a detailed answer to the request.

Furthermore, the 3D-drawing has to be accomplished by using features, which is a need for the automated process planning and work scheduling later on (See also Amaitik & Kilkic 2005; 2007). Features in this context are defined as an accumulation of information which can be used for process planning. Examples are counter holes, filets etc.

These features need to be enriched by physical properties which are always related to a specific feature or a feature face. Furthermore, tolerances and feature type related attributes have to be added. Features, which include these needed information have a higher level of value according to the advanced reason they are used to.

The method we are using is FBD (Feature based design). We are not realizing Feature Recognition because of the additional complexity this will result in. The authors clearly understand that Feature Recognition and FBD need to be accomplished both to make automated offerings and especially the integration of CAD and CAP resulting from it, feasible.

As a last requirement, the language needs to be producer-independent since dependency will lead to the characteristic interface problem or to technology dependencies for companies which try to adopt our approach.

3. EVALUATION OF IMPLEMENTATION OF STEP IN MARKET-LEADING CAD/PLM SOLUTIONS

In a first face, the most reasonable standards for our model seemed to be IGES and STEP. Since IGES is only for the exchange of data between CAD (Computer Aided Design) – Systems, the second face brought our attention to STEP.

The overall objective of STEP is to provide a mechanism that describes a complete and unambiguous product definition throughout the life cycle of a product, independent of any computer system (See SCRA 2006).

The main implemented Application protocols in industry are AP 203 and 214, which are supported by the main CAD vendors for the mechanical engineering industry like CATIA, Solid Edge, Autodek Inventor or UGS (See SCRA 2006 p 142).

For our solution we decided to take a deeper look into the implementation of an independent data standard into proprietary CAD and PLM (Product Lifecycle Management) systems, especially in Autodesk Inventor, CATIA and PartSolutions.

The implementation study can be described as follows:

- All descriptions are done for a prototypical product.

- Different methods are used for describing this product
- The export is tested by interpreting the resulting STEP-file

The analysis shows the main disadvantages of STEP as it is implemented today. The evaluation is demonstrating the effect of different drawing methods (only subtractive from the raw material feature, additive with defined features only and rotation of a base shape) and compares the output in STEP-Files afterwards.

An Overview of the research work done, is shown in figure one. As stated in the requirements in chapter two, the subtractive feature drawing method is preferred by the authors. Since we also wanted to see the consequences for the output depending on the method a designer is drawing his part.

Results can be summarized shortly. The translator of each program produces boundary representation models only. Here are the elements needed for our approach, which get lost:

- material properties
- tolerances
- Feature based design history.

Especially the features used for drawing while using the subtractive drawing approach are very important for the automated process planning. Also tolerances need to be in the model of the demand to determine the resource to accomplish a process step.

It can be stated that the existing implementations are not useful for our approach. Reasons for not using the possibilities of AP 203 and 214 can be seen in the company principles of these CAD-solutions. We assume that they are not really interested in giving the drawing history in a neutral format to the outside of the system.

As a result of this, our team needed to look ahead and search for a more convenient solution, which was found by STEP AP 224.

4. REVIEW OF THE NEW PASSED ISO 10303 AP 224

The STEP AP we decided to go for is AP 224. It is called "Mechanical product definition for process planning using machining feature". A detailed analysis of the AP brought us to this decision, leaded by the following points:

- The determination of the raw material gets possible by the definition of so called base shapes. They can be a block, a cylinder or a polygon with any number of sides. The base shape can also be defined explicit by declaring a boundary represented volume.
- The features which then can be removed by the designer are separated into machining features (for example a hole or a pocket), transition features (such as chamfers or filets) or compound features, which are machining features taken together by a path, on which the same features in terms of size and sort are arranged on the final product.
- Tolerances can be assigned to measurements (the definitions are based on ISO 1101 and ANSI Y14.5M (See SCRA 2006)).

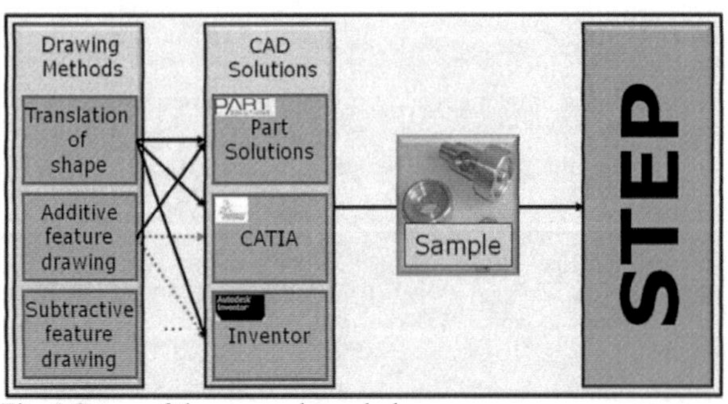

Fig. 1.Setup of the research work done.

One disadvantage can be seen in the description of dependent features, which is realized by appointing shapes which are not used to describe the features they lie inside. This disadvantage can be eliminated by generating the shapes of all features and then determine the location of the shapes described explicit by such a definition. This AP is an ISO-Standard since late 2007. Therefore implementations can only be found in some tools. Mainly Universities (See Amaitik & Kilic 2005; Amaitik & Kilic 2007; Liu et-al. 2004; Sharma et-al. 2002) and some companies have a solution based on STEP AP 224 in place (LOCAM, STEP Tools, SCRA (See SCRA 2006)).

Since we don't have access to these solutions, we developed a workaround to have testing data for process planning in place. Basically we take the EXPRESS (which is the language of STEP) –schema of AP 224 and derive our instances.

5. CONCLUSION

The results from analyzing the existing implementation of STEP were not promising. Developing our own design guide and extracting information using proprietary API (Application Programming Interface) would be the result. However, using STEP AP 224 enables our approach to be realistic and achievable for the machine building branch. For other sectors the application seems easier, the definition of features not so complex (for example textile industry).

6. REFERENCES

Amaitik, S.M. & Kilic, S.E. (2005) STEP-based feature modeler for computer –aided process planning. In: *International Journal of Production Research*, Vol. 43, No.15, p. 3087-3101, Taylor & Francis Group Ltd., London.

Amaitik, S.M. & Kilic, S.E. (2007) An intelligent process planning system for prismatic parts using STEP features. In: *International Journal of Advanced Manufacturing Technology*, Vol. 31, p. 978-993, Springer-Verlag Ltd., London.

Liu, H.; Maekawa, T.; Patrikalakis, N. M.; Sachs, E. M. & Cho, W. (2004) Methods for feature-based design of heterogeneous solids. In: *Computer-Aided Design*, Volume 36, 2004, p. 1141-1159, Elsevier Ltd.

SCRA (2006) STEP Application Handbook, pdesinc.aticorp.org/downloadable_files/ STEPapplicationhandbook63006BF.pdf, last visited 05.28.2008.

Sharma, R. & Gao, J. X. (2002) IMPLEMENTATION OF STEP 224 IN AN AUTOMATED MANUFACTURING PLANNING SYSTEM. In: *Proceedings of the Institution of Mechanical Engineers*, Part B: Journal of Engineering Manufacture, Volume 216, Number 9, 2002, Professional Engineering Publishing, London.

Teich, T.; Mildenberger, U.; Richter, M.; Militzer, J. & Unger, K. (2007) Functional Characterization of Supply and Demand for Automatic Responses of ERP-Systems to Consumer Requests. In: Proceedings of the 18th DAAAM International Symposium 2007, Katalinic, B. (Ed.), Zadar, Croatia, October 24-27, 2007, pp. 745 f.

Teich, T.; Militzer, J.; Unger, K.; Gaese, T. & Winkler, S. (2008) Ein formalisierter Ansatz zur Generierung von Kundenangeboten auf Basis heuristischer Verfahren. In: Logistics and Supply Chain Management: Trends in Germany and Russia: *Proceedings of the German-Russian Logistics Workshop*, Publishing House of the Saint Petersburg State Polytechnical University, pp. 272-282

Teich, T. (2003) Extended Value Chain Management - ein Konzept zur Koordination von Wertschöpfungsnetzen, Chemnitz: Verlag der GUC 2003, ISBN 3-934235-23-9, 690 S.

Annals of DAAAM for 2008 & Proceedings of the 19th International DAAAM Symposium, ISSN 1726-9679
Editor B. Katalinic, Published by DAAAM International, Vienna, Austria 2008
Make Harmony Between Technology and Nature, and Your Mind will Fly Free as a Bird

WINGS WITH IMPROVED AERODYNAMIC CHARACTERISTICS

SCURTU, D[an] & CIOBANU, B[ogdan]

Abstract: *The paper proposes a method to increase the lift capacity of wings by using a jet launched through the trailing edge. Theoretical model is based on potential movement theory. Fluid emission through trailing edge is assimilated with a line-form surge with uniform flow distribution. The lift coefficients determined for the prototype wing are bigger with 25% related to the ones obtained for classic wing. We observe an increase of the lift coefficient in the same time with a decrease of the drag coefficient. For incident angles between 0 and 14 degrees, we have obtained negative values for the drag coefficient (the drag force becomes propulsive force). In addition, the boundary layer is positively affected.*

Key words: *Wing; jet; lift; drag; coefficient.*

1. INTRODUCTION

The paper aims at establishing, in a theoretical and experimental manner, the influence of a q controlled debit jet on its aerodynamic features. The jet is launched through the trailing edge of the lift wing.

The issue was already approached at a theoretical level at the half of the 20th century. (Davidson I.M. 1956) theoretically studies the jet-flaps lift wing. In 1955, Kadasch M. sets a theoretical form of a complex function that describes the fluid movement around a fluid emission lift wing through the trailing edge. (Malavard L. 1956) sets a linear form of functions for the disturbing potential determined by the jet. (Nagy A.B. 1969) extends the theory Kutta-Jukovsky to jet air foils. (Roy M. 1956) sets an expression for the disturbing potential generated by the jet from the trailing edge. (Dumitrescu L. 1957 and Patraulea N. 1961) theoretically set the effect of fluid emission on the lift force. All this research aimed at the plane wing and became less effective along with the development of engines used in aviation engines.

In the context of present energy deficit, we believe it is useful to continue the research applying to the singular wing and the network working wing that applies to axial aeraulic generators.

2. GENERAL CONSIDERATIONS

The lift wing put in a real current fluid generates a lift force $F_L = \frac{\rho}{2} \cdot b \cdot c \cdot C_L \cdot W_\infty^2$ and a drag force $F_D = \frac{\rho}{2} \cdot b \cdot c \cdot C_D \cdot W_\infty^2$ in which: ρ - the density of the fluid current moving around the wing, C_L - lift coefficient, C_D - the draft coefficient, W_∞ is the speed of the general current, b the span and c the air foil cord. In practical applications, the lift force F_L must be as big as possible, and the drag force F_D, as small as possible.

In order to achieve this, the lift coefficient values C_L must be as big as possible and the values of the drag coefficient C_D must be as small as possible.

The paper has in view a physical method that allows the increase of the lift coefficient, at the same time with the decrease of the drag coefficient.

The method consists of the creation of a controlled emission of fluid in the trailing edge of an aerodynamic air foil. The fluid emission is assimilated to an equally distributed source on the span, figure 1.

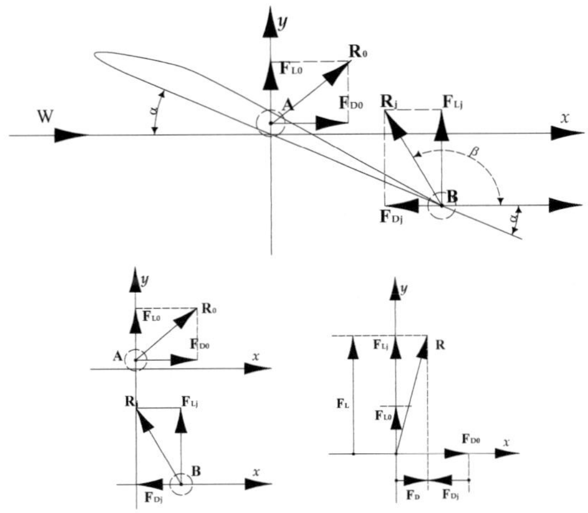

Fig. 1. Forces on the jet air foil.

3. MATHEMATICAL MODELLING OF THE MOVEMENT

The study of fluid flow around the solid bodies is based on the theory of potential movements and the use of the custom transformation theory.

The general movement of the fluid around the circle of custom transformation with fluid jet is described by the global complex potential function:

$$h(z) = -w_\infty \left[z \cdot e^{i\alpha} + \frac{a^2}{z} \cdot e^{-i\alpha} \right] - i\frac{\Gamma}{2\pi} \ln z + \frac{q}{2\pi b} \ln(z-a) \quad (1)$$

The lift theoretical coefficient C_{LT} is given by the relation:

$$C_{LT} = \frac{2\pi c W_\infty \sin(\alpha+\sigma) + \frac{q}{b}\sin\beta}{cW_\infty} = 2\pi \sin(\alpha+\sigma) + \frac{q}{cbW_\infty}\sin\beta \quad (2)$$

The theoretical modelling of the flow was made in the following conditions:
- The chosen aerodynamic air foil is type Gö593.
- The values of the incidence angle α [deg] = -6°, -1°, 2°, 4°, 6°, 8°, 10°, 12°, 14°, 16°.
- The q debit [m^3/s] of the jet $q_0 = 0$; $q_1 = 0.001$; $q_2 = 0.002$; $q_3 = 0.003$; $q_4 = 0.004$; $q_5 = 0.005$; $q_6 = 0.006$.
- $q_0 = 0$ stands for the classic aerodynamic air foil (no jet in the trailing edge).
- The speed of the general current $W_\infty = 20$ [m/s]; cord $c = 0.1$ [m]; span $b = 0.5$ [m].

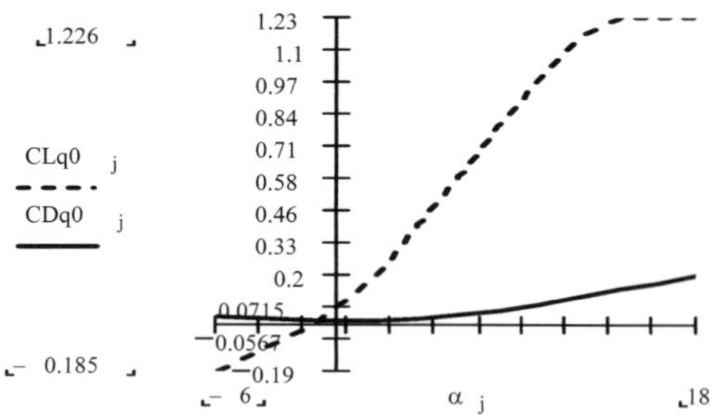

Fig. 3. Aerodynamic characteristics for Gö593 (Re=1.4·10^5)

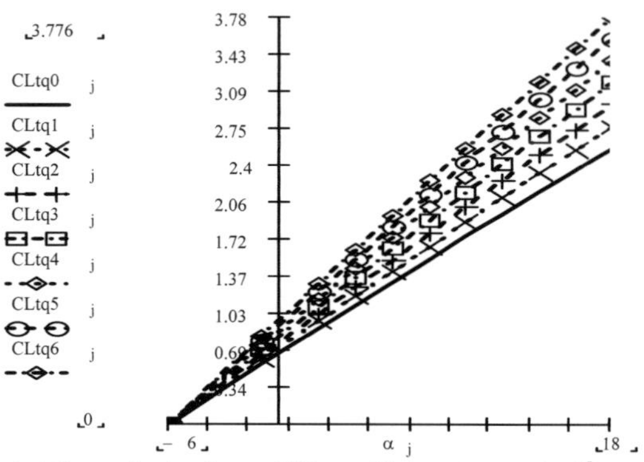

Fig. 4. Theoretical values of lift coefficient (Re=1.4·10^5)

Figure 3 present the lift and drag coefficients for Gö593 airfoil, without fluid emission, as given in catalogues.

In figure 4, the variations of the theoretical coefficient C_L are presented for all the values of q, taking into consideration all the values of the incidence angle α.

4. THE EXPERIMENTAL STUDY OF THE PHENOMENA

We aimed at assessing the lift coefficient C_L and the drag coefficient C_D for a controlled fluid jet lift wing emitted by the trailing edge, as compared to the classic wing in the same conditions as in theoretical modelling.

The tests were performed on an open circuit aerodynamic tunnel, with continuous speed adjustment in the working chamber and a turbulence degree of 0.006; the aerodynamic balance used for measurements is made up of type FUTEK FBB350 force transducers.

The results of experimental determinations referring to the C_L lift coefficient and the drag coefficient C_D are shown in figures 5 and 6.

5. CONCLUSIONS ON EXPERIMENTAL DETERMINATIONS

- The increase of the lift coefficient depends on the q debit jet evacuated through the trailing edge.
- The values of lift coefficient for the wing with jets emitted through the trailing edge are bigger than those for the classic wing up to 40%.
- The increase of the lift coefficient for the jet wing is determined by circulation Γ growth.
- The values of drag coefficient for the fluid jet wing are smaller than the values determined for the classic wing.
- For values of the incidence angle α from 0° to 14°, the drag coefficient values are negative. In this area of incidence angles, the lift wing incidence shows a propulsive character.

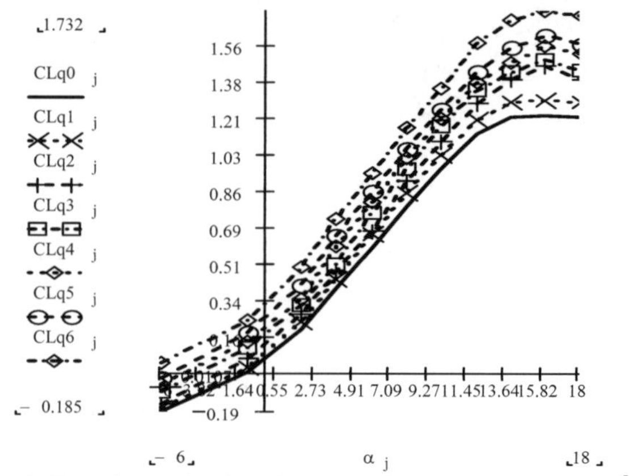

Fig. 5. Experimental values for lift coefficient C_L - Re=1.4·10^5

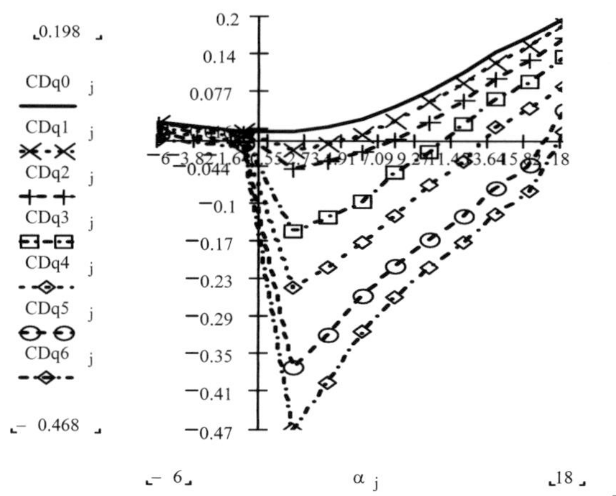

Fig. 6. Experimental values for drag coefficient C_D Re=1.4·10^5

The look forward of presented studies refers at:
- Theoretical and experimental studies about flow in laminar layer.
- Establish of a method to draw an air foil for an imposed pressure (velocity) distribution.
- Experimental studies of an air foils network with fluid emission through trailing edge.

6. REFERENCES

Davidson, I.M. (1956). *The Jet-Flap*, Journal of the Royal Aerodynamic Society, London

Dumitrescu, L. (1957). *Determinarea efectului de ejectie la profilele cu volet fluid (Determination of emission effect for airfoils with fluid volet)*, Revista studii si Cercetari de Mecanica aplicata, tom VIII

Kadasch, M. (1955). *Theorie des profiles avec soufflage au bord de fuitte (Theory of airfoils with fluid emission through trailing edge)*, Bul. De la Soc. Francaise des Mecaniciens, nr. 18

Malavard, L. (1956). *Sur une theorie lineare du soufflage au bord de fuitte d'un profile d'aille (About a lineare theory of emission through the trailing edge of an airfoil)*, Comptes Rendus nr. 242, Paris

Nagy, A.B. (1969). *Asupra teoriei Kutta-Jukowski la profilele cu jet (About Kutta-Jukowski theory of airfoils with jet)*, Studii matematice, Ed. Acad. Romane

Patraulea, N. (1961). *Aripi cu jet in bordul de fuga (Wings with jets through trailing edge)*, Studii si cercetari de mecanica aplicata nr. 5

Roy, M. (1956). *Theorie d'un profil d'aille a jet (Theory of a jet-wing)*, Comptes Rendus nr. 244, Paris

Annals of DAAAM for 2008 & Proceedings of the 19th International DAAAM Symposium, ISSN 1726-9679
Editor B. Katalinic, Published by DAAAM International, Vienna, Austria 2008
Make Harmony Between Technology and Nature, and Your Mind will Fly Free as a Bird

TEMPERATURE FIELD AT DISSIMILAR METALS WELDING

SCUTELNICU, E[lena]; IORDACHESCU, M[ihaela] & IORDACHESCU, D[anut]

Abstract: *FEM models of heat flow were created for the case of thin plates TIG welded joints of low carbon steel and stainless steel, and low carbon steel and copper, respectively. These models have used a double ellipsoidal asymmetric heat source. Discussion on the graphs of temperatures variation enables conclusions on the possible undesired structural changes and related preventing measures. The asymmetric position of the heat source closer to the material having a higher thermal conductivity is an effective method for controlling the temperature field.*
Key words: *welding, steel, copper, modelling, temperature*

1. INTRODUCTION

The temperature history of the welded components has a significant influence on the structural changes, residual stresses, distortions and, therefore, on the fatigue behaviour of the welded structures. Classical solutions for the transient temperature field such as Rosenthal's equations dealt the semi-infinite and infinite bodies subjected to an instant point heat source, line heat source, or surface heat source. These solutions can be used to predict the temperature field far from the heat source, but fail to predict the temperatures near the heat source. An advance in modelling was introducing the three-dimensional double ellipsoidal heat source to predict the temperature field of semi-infinite body (Goldak et al., 1985). Following this idea, analytical solutions for the transient temperature of a semi-infinite body subjected to a 3D moving heat source were further reported (Nguyen et. al., 1999). In case of thin plates the source influence in the thickness direction could be negligible; thus, the heat source becomes a Gaussian surface-distributed one (Mackerle, 2002), or an elliptical disk heat source (Scutelnicu et al., 2007).

Recently, results on temperature field simulation in case of TIG welding of copper were also reported (Lei et al., 2006). Other research has investigated the main metallurgical and mechanical issues related to low carbon steel-stainless steel welding (Kou, 2003).

Based on the state-of-the-art, but considering the knowledge gaps identified in the field, a FEM model based on double ellipsoidal heat source was developed to predict the heat flow in case of low carbon steel-stainless steel and low carbon steel–copper plates welding. This investigation approach is based on the new technological variant of moving the heat source closer to the material having a higher thermal conductivity. Further research envisages the simulation of the whole process, after appropriately coupling the modules for strain-stress field and microstructure prediction, to supporting the optimization of the welding process.

2. DISSIMILAR JOINT MODEL DESCRIPTION

The mathematical model considers the isotropy of the base metals, the materials thermal properties, the heat loss by convection and radiation, and a double ellipsoidal heat source; the thermal source influence in the plate's thickness direction was neglected.

The process model features are:

- Two dissimilar metals (1, 2) having the same thickness g, initial temperature T_0 and different thermal properties ($c_{1,2}$ – specific heat, $\rho_{1,2}$ – density, $\lambda_{1,2}$ – thermal conductivity);
- The time function of the heat input, welding time dependent as a triangle functions sum; it was optimised the number of these triangle functions simulating the welding source movement.
- A constant value of the maximum heat input.

The heat transfer in the plates was modelled as 2D heat transfer problem using SHELL element group type. Because of different thermal properties, both plates were discretized using 4-noded quadrilateral elements. To obtain accurate temperature field distribution in the region of high temperature gradients, a refined mesh was used in the vicinity of the contact line of the parts. The heat source was moved closer to the material having a higher thermal conductivity, low carbon steel in the first analysed case, and copper in the second, respectively.

3. THERMAL ANALYSIS RESULTS

3.1 Low carbon steel – stainless steel case

The main issues of welding low carbon steel with stainless steel are related to the formation of different composition transition layers and diffusion zones in the weld. These increase the weld susceptibility to intercrystalline corrosion and cracking, and formation of the undesired σ phase. Finally, unfavourable tensile residual stresses may develop, as too. Therefore, in deep thermal analysis of the process is a useful instrument in understanding these.

The analysis results correspond to the simulation of TIG butt welding of low Carbon Steel-CS, with Stainless Steel–SS; the dimensions of the plates were 10x200x300mm; the process parameters, namely the heat source power, P, was 1760W, and the welding speed, v=3.8mm/s.

Graphs illustrating the FEA-obtained temperature distribution on transverse section, corresponding to CS-SS welding process after 8s, and 70s, respectively, are presented in Fig. 1 and 2. The peak temperature is achieved on the weld line.

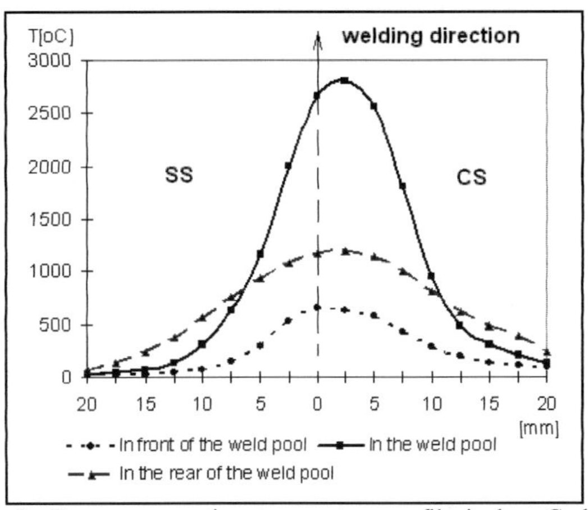

Fig. 1. Transverse section temperature profile in low Carbon Steel (CS)–Stainless steel (SS) welded joint (t=8s).

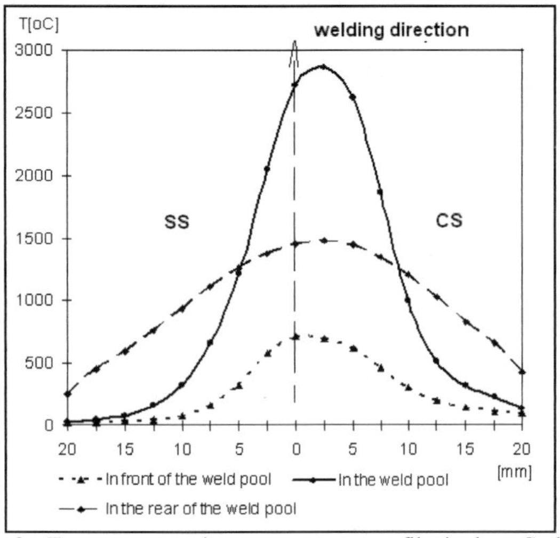

Fig. 2. Transverse section temperature profile in low Carbon Steel (CS)–Stainless steel (SS) welded joint (t=70s).

Due to CS thermo-physical properties, a larger HAZ results in its case. Both graphs confirm the higher concentration of the heat in SS and its larger dissipation in CS.

3.2 Low carbon steel – copper case

The main issues of welding copper with low carbon steel are related to the weldability of these metals; cracks may occur in HAZ of the low carbon steel, and fragile intermediate layers may form in the carbon steel. The undesired thermal effects of the welding process on the low carbon steel, and the related issues are reduced by asymmetrically positioning (closer to the copper) the heating source moving along the gap.

Moreover, a smaller thermal conductivity causes heat concentration and less extent of HAZ such is, for instance, the CS case, requiring a smaller heat input per weld length unit. A larger conductivity requires a large heat input per weld length unit such is the copper case. Thus, the relationship between the welding parameters and the thermo-physical properties of the base metals and the correspondent dimension of the HAZ must be considered.

The analysis results correspond to the model of TIG butt welding of copper–Cu with low carbon steel; the dimensions of the plates were 10x200x300mm; the process parameters, namely the heat source power, P, was 2400W and the welding speed *v*=5mm/s. Graphs illustrating the FEA-obtained temperature distribution on transverse section corresponding to the Cu-CS welding process after 6s, and 54s, respectively, are presented in Fig. 3 and 4.

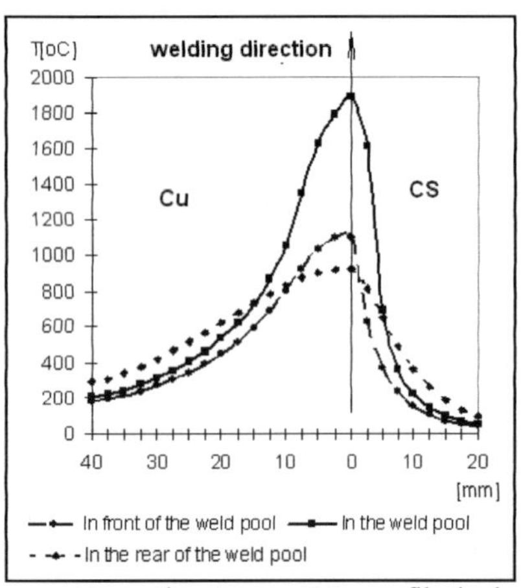

Fig. 3. Transverse section temperature profile in low Carbon Steel (CS)–copper (Cu) welded joint (t=6s).

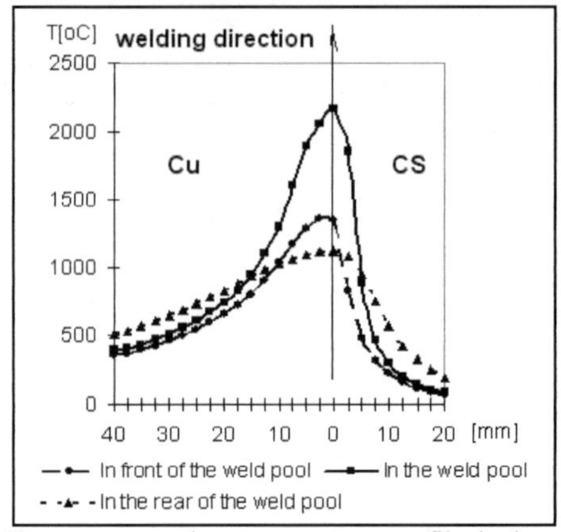

Fig. 4. Transverse section temperature profile in low Carbon Steel (CS)–copper (Cu) welded joint (t=54s).

Peak temperature is achieved in carbon steel, but with increasing the distance from the weld, the temperature values are higher in the non-ferrous metal, as a result of higher thermal conductivity. In both analysed cases (Fig. 3-4), due to materials different thermal conductivity, the asymmetry of HAZ is visible, namely its larger extension on the copper component.

4. CONCLUSION

This model based on FEM has used a double ellipsoidal asymmetric heat source moving closer to the material having a higher thermal conductivity. Larger copper HAZ results for Cu-CS welding, whilst HAZ extension in the CS plate features the CS-SS case. Peak temperature is achieved in the less conductive material, but with increasing the distance from the weld, the temperature values are higher in the higher thermal conductivity material. Discussion on the graphs of temperatures variation enables conclusions on the possible undesired structural changes and related preventing measures. The asymmetric position of the heat source is an effective method for controlling the temperature field in arc welding of dissimilar plates.

5. REFERENCES

Goldak, J.; Chakravarti, A. & Bibby, M. (1985). A Double Ellipsoid Finite Element Model for Welding Heat Sources, *International Institute of Welding,* Doc. SG-212-603-85

Kou, S. (2003). *Welding Metallurgy*, 2nd edition, John Wiley and Sons, ISBN 978-0-471-43491-7, New York

Lei Y.; Yu W.; Li C. & Cheng X. (2006). Simulation on temperature field of TIG welding of copper without preheating, *Transactions of Nonferrous Metals Society of China,* Vol.16, No.4, August 2006, pp. 838-842, ISSN 1003-6326

Mackerle, J. (2002). Finite element analysis and simulation of welding - an addendum: a bibliography (1996–2001), *Modelling and Simulation in Materials Science and Engineering*, Vol.10, No.3, May 2002, pp. 295–318, ISSN 0965-0393

Nguyen, N. T.; Ohta, A.; Matsuoka, K.; Suzuki, N. & Maeda, Y. (1999). Analytical Solutions for Transient Temperature of Semi-Infinite Body Subjected to 3-D Moving Heat Sources, *Welding Journal*, Vol.78, No.8, August 1999, pp. 265 – 274, ISSN 0043-2296

Scutelnicu, E.; Iordachescu, M. & Iordachescu, D. (2007). Dissimilar metals welding: mathematical modelling and experiments, *Proceedings of 14th International Conference on the Joining of Materials*, CD-ROM, ISBN 87-89582-15-2, April 29-May 2, 2007, JOM Institute, Helsingor, Denmark

Annals of DAAAM for 2008 & Proceedings of the 19th International DAAAM Symposium, ISSN 1726-9679
Editor B. Katalinic, Published by DAAAM International, Vienna, Austria 2008
Make Harmony Between Technology and Nature, and Your Mind will Fly Free as a Bird

FERMENTER VERIFICATION: MICROBIOLOGY AND GAS PRODUCTION MONITORING

SEDLACIK, R[adek]; RUZICKA, J[an] & BIENERT, J[an]

Abstract: *The main purpose of this work was to determine the influence of bioreactor running and construction on microbial characteristic of activated sludge. Four experiments under different conditions were carried out. Results of these tests have shown importance of regulation and negative impact of specific bioreactor construction on sludge flocs concentration. The second goal of the work was a creating of new formulas for computing of carbon dioxide (CO_2) and methane (CH_4) production, as the indicators of microbial activity.*

Key words: *bioreactor, image analyse, biogas production, sludge flocks.*

1. INTRODUCTION

Laboratory fermenter is a part of LABI (Laboratories of integrated automation) project (Dvorackova et al., 2005). This device was tested from automation and measurement point of view (Dvorackova et al., 2006). However, when complex microbiological community (activated sludge) is used as functionary process agent, effect of reactor´s equipment on its base morphological structures – flocs – must be thoroughly proved. Furthermore, composition and abundance of microbial population strongly depend on available substrate, temperature, level of dissolved oxygen, pH and on sludge retention time (Yuan & Blackall, 2002; Li & Yang, 2007). Small change of one parameter could have a significant impact on microbes' community. Nowadays, microscopic evaluation of microbial structures (flocs or bio-film) for quick checking of proper process is very common. Modern and objective methods of digital image analysis have been for that purpose developed (Liwarska-Bizukojc, 2005). This paper describes results of both observation of flocs behaviour (depending on substrate character and mode of bioreactor running) and creating of new formulas for software computing of biogas productions for quantifying microbial processes.

2. EXPERIMENTAL

2.1 Materials

A mixture of aerobic activated sludge and a mineral-medium ($MgSO_4$; $CaCl_2$; $FeCl_3$; $(NH_4)SO_4$; phosphoric buffer; trace elements: B^{3+}, Fe^{2+}, Zn^{2+}, Mn^{2+}, Cu^{2+}, Co^{2+}, Mo^{6+}) was utilized in the first, second and third experiment. In the fourth experiment, aerobic activated sludge was replaced by anaerobic sludge. Glycerol (12g) / protein hydrolysate Hycol E (0,5 g/l) and starch (0,5 g/l) were added as a substrate in experiments.

2.2 Methods

The first experiment was carried out in six glass separators (1 l volume – laboratory scale). After one week of resting period, 0,5 g/l of Hycol with minerals in two separators, 0,5 g/l of starch with minerals (without nitrogen) into two next separators were added at day 7. Two separators without substrate served as blank. Air bubbling provided good aeration and perfect stirring of employed mixtures.

Figure 1 shows the scheme of the fermenter linkage in the second, third and fourth experiment. The procedure under aerobic conditions (A) was follows: Firstly, the inlet air was divested of CO_2 by absorbing this gas in 5M NaOH solution (1). Then, 0.05M $Ba(OH)_2$ solution (2) was used for indication of residual CO_2. Secondly, the air passed through the flow-meter (3), fermenter (4) and drying jars (5). Finally, the concentrations of gases were measured in CO_2 and CH_4 sensors (GASCARD II, Edinburgh Instruments, UK) (6).

The third experiment was similar to the second, but the substrate was continuously dosed via peristaltic pumps (12,5 mg per litre per day). In addition, the temperature was rapidly changed (from 22°C to 45°C) during the fourth experiment.

Arrangement for anaerobic experiment (B) was different. Biogas (mixture of CO_2 and CH_4) was produced by microbes in bioreactor. The operating staff opened outlet valve (O) via the Internet interface and the biogas flowed through drying jars (5) into the gas sensors (6). Then, it streamed through flow-meter (Bronkhorst, Holland) with small range (Q).

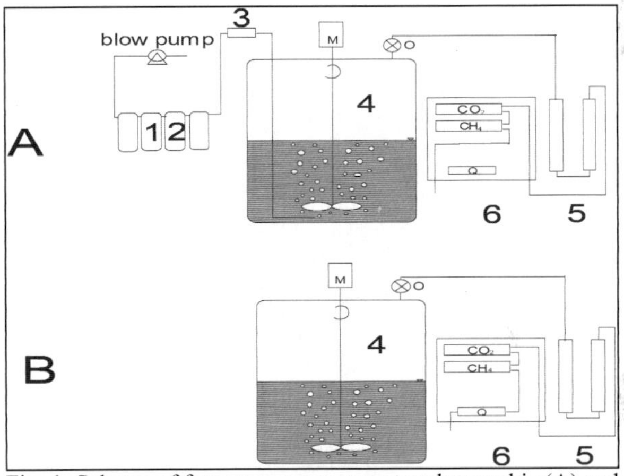

Fig. 1. Scheme of fermenter arrangement under aerobic (A) and anaerobic (B) conditions

3. RESULTS

3.1 Series no. 1 – laboratory testing

Table 1 presents flocs counts in aerobic activated sludge during the first experiment. The flocks larger than 100 x 100 µm were counted manually and "digitally" by using ImageJ software. It was found that data from digital image analyse were more accurate and objective, especially when the floc size was close to intended size. After Hycol and starch solutions addition, number of flocs increased by 50 % (Hycol) and 40 % (starch). The flocks were more compact, larger and population of other microbes (protozoa) was more abundant than in blank glass separators.

3.2 Test No. 2 – bioreactor experiment AE1

The addition of Hycol E at 7th day to sludge microorganisms in reactor was tested. As can be seen in Table 2, flocs numbers in bioreactor was relatively stable (AE1 column), although a slight increase after Hycol addition was observed. However, this rise was negated by biomass sedimentation - larger flocs settled to the bottom of fermenter vessel and clung to mantle and sensors inside the bioreactor. Clunning was partially reduced by increasing experiments stirring speed from 115 rpm to 140 rpm. The flocs were generally smaller than in glass separators and their structure was more open, probably due to mechanical effect of reactor running.

	Blank		Hycol		Starch	
Day	M	D	M	D	M	D
1	72,5	48,0	79,5	56,0	73,5	60,0
4	95,5	60,0	96,0	64,0	95,0	68,0
8	93,0	72,0	118,5	68,0	111,5	72,0
10	86,5	68,0	128,5	80,0	133,5	76,0
15	85,5	52,0	176,5	112,0	142,5	96,0

Tab. 1. Number of flocks in glass separators counted manually (M) and digitally (D – ImageJ software)

	AE1		AE2		AN	
Day	M	D	M	D	M	D
1	90,7	68,0	84,5	84,0	152,5	96,0
4	91,7	72,0	78,0	64,0	128,0	80,0
8	88,7	60,0	84,0	60,0	99,5	56,0
10	97,3	64,0	62,0	56,0	---	---
15	85,7	72,0	53,5	36,0	---	---

Tab. 2. Number of flocs in bioreactor counted manually (M) and digitally (D) during aerobic (AE) and anaerobic (AN) tests

3.3 Test No. 3 – bioreactor experiment AE2

Experiment was run in another mode than the previous ones – substrate additions were done every second day and one week of sludge cultivation was performed at 22°C, whilst the second week the temperature was increased to 45°C (imitation of external intervention). Significant decrease of flocs quantity was found (more than 50 %) in experiment, especially in the latter period. Although pH, level of dissolved oxygen and substrate concentrations were stayed stable, change of temperature caused destruction of large flocs. Furthermore, higher temperature depressed protozoa and other free microbes in the sludge. Generally, all these changes have negative impact on sludge applicability. On the other hand, this step change and its repercussions clearly illustrate sensitivity of microbial processes and could be conveniently used in microbiological lessons.

3.3 Test No. 4 – bioreactor experiment AN

Anaerobic experiment proceeded one week. Glycerol was added at the beginning of the test, but number and size of flocs decreased continually, due to biomass sedimentation and clunning to sensors.

3.4 Formulas for computing of CO_2 and CH_4 production

Determination of gas production in bioreactor is necessary for biochemical balances. However, sensors installed on fermenter measure only gas flow and concentrations of carbon dioxide and methane. Gas production has to be computed by specific software. Nevertheless, students or colleges from foreign universities probably do not have this software. For this case, a set of simple formulas for MS Excel® was prepared.

A gas volume passing through sensors during experiment could be computed as a bounded integral. Time is on x-axis and flow rate is on y-axis. Unfortunately, aerobic experiments have a long data archiving period and flow rate can be very variable during this long time interval. To solve this problem, polynomial function of degree four was instituted as an equation of integrated function. Coefficients of this polynomial (*a,b,c,d,e*) were in general found out by the Newton's interpolation polynomial:

$$P_4(x) = k_0 + k_1(x - x_0) + k_2(x - x_0)(x - x_1) + k_3(x - x_0)(x - x_1)(x - x_2) + k_4(x - x_0)(x - x_1)(x - x_2)(x - x_3) \qquad (1)$$

Where:

$$k_0 = f(x_0) \qquad k_1 = \frac{f(x_1)-f(x_0)}{x_1-x_0} \qquad k_2 = \frac{\frac{f(x_2)-f(x_1)}{x_2-x_1} - k_1}{x_2-x_0}$$

$$k_3 = \frac{\frac{\frac{f(x_3)-f(x_2)}{x_3-x_2} - \frac{f(x_2)-f(x_1)}{x_2-x_1}}{x_3-x_1} - k_2}{x_3-x_0}$$

$$k_4 = \frac{\frac{\frac{\frac{f(x_4)-f(x_3)}{x_4-x_3} - \frac{f(x_3)-f(x_2)}{x_3-x_2}}{x_4-x_2} - \frac{\frac{f(x_3)-f(x_2)}{x_3-x_2} - \frac{f(x_2)-f(x_1)}{x_2-x_1}}{x_3-x_1}}{x_4-x_1} - k_3}{x_4-x_0} \qquad (2)$$

Formulas for coefficients *a,b,c,d,e* were obtained after substitution in equations (1) and (2):

$$\boldsymbol{a} = k_4 \qquad \boldsymbol{b} = k_3 - k_4(x_3 + x_2 + x_1 + x_0)$$

$$\boldsymbol{c} = k_2 - k_3(x_0 + x_1 + x_2) + k_4[x_0(x_1 + x_2 + x_3) + x_1(x_2 + x_3) + x_2 * x_3]$$

$$\boldsymbol{d} = k_1 - k_2(x_0 + x_1) + k_3[x_0(x_1 + x_2) + x_1 * x_1] - k_4[x_0(x_1 * x_2 + x_1 * x_3 + x_2 * x_3) + x_1 * x_2 * x_3]$$

$$\boldsymbol{e} = f(x_0) - k_1 * x_0 + k_2 * x_0 * x_1 - k_3 * x_0 * x_1 * x_2 + k_4 * x_0 * x_1 * x_2 * x_3 \qquad (3)$$

These coefficients were integrated in MS Excel® formulas for computing gas volume, CO_2 and CH_4 production (in ml, mg and mmol) by integration of bounded integrals.

4. CONCLUSION

The results of experiments have indicated negative impact of used bioreactor construction on large flocs numbers. In addition, Test No. 3 has demonstrated importance of a stability of regulated parameters during whole experiment. Finally, new formulas for data sheet programs facilitate simpler gas balancing. Further research will be focused on use of pure microbial cultures which are more sensitive and less stable during biotechnological processes.

5. ACKNOWLEDGEMENTS

This work was supported in part by the Ministry of Education, Youth and Sports of the Czech Republic under grant FRVS, No. MSM 281100001 and MSM 7088352101.

6. REFERENCES

Dvorackova, M.; Hruska F. & Hoffmann J. (2005). Laboratory biochemical model and WEB communication, *Proceedings of the 16th International DAAAM Symposium,* Opatia, Croatia, pp. 105-106. ISBN: 3-901509-46-1.

Dvorackova, M.; Hruska, F. & Sedlacik, R. (2006). Technological and metrological verification of DE5 model, *Annals of DAAAM for 2006 & Proceedings of the 17th International DAAAM Symposium*, Katalinic, B. (Ed.), pp 113-114, ISBN 3-901509-57-7, ISSN 1726-9679, Vienna, Austria 2006, Published by DAAAM International,

Li, X.Y. & Yang, S.F. (2007). Influence of loosely bound extracellular polymeric substances (EPS) on the flocculation, sedimentation and dewaterability of activated sludge. *Water Research* Vol. 41, No.5, (March, 2007), p. 1022-1030, ISSN: 0043-1354

Liwarska-Bizukojc, E. (2005) Application of image analysis techniques in activated sludge wastewater treatment processes. *Biotechnology Letters* Vol. 27, No. 19, (July, 2005), p. 1427-1433, ISSN: 0141-5492

Yuan, Z. & Blackall, L.L. (2002). Sludge population optimisation: a new dimension for the control of biological wastewater treatment systems. *Water Research* Vol. 36, No.2, (January, 2002) p. 482-490, ISSN: 0043-1354

Annals of DAAAM for 2008 & Proceedings of the 19th International DAAAM Symposium, ISSN 1726-9679
Editor B. Katalinic, Published by DAAAM International, Vienna, Austria 2008
Make Harmony Between Technology and Nature, and Your Mind will Fly Free as a Bird

INSTALLATION AND METHODOLOGY FOR THE STUDY OF THE STABILITY OF IMPLANTS USED IN TREATING FEMORAL NECK FRACTURES

SEGHEDIN, N[eculai]; CIONCA, D[an] &DROSESCU, R[adu]

Abstract: *In the case of femoral neck fracture, according to the viability of the femoral head, one can use two therapeutic methods: internal fixation and arthroplasty. The paper presents an experimental installation and a methodology for the study of the stability of various types of implants used in the femoral neck fractures treatment. The offered installation allows the simulation of the actions that solicit the femoral head while walking. Thus, various actions can be applied to the femoral head, in various angles, according to the various positions of the thigh-bone during walking. The installation also allows applying to the femoral head forces of up to 200 daN, with a frequency of about 1 Hz. With the aid of a force transductor, measurements can be made regarding the forces that acted upon the femoral head. Besides that, one can determine the plastic and elastic displacements that occur between the fractured area and the main material. Various comparative studies can be made between the various types of implants that are used for the internal fixation of the femoral neck fracture.*

Key words: *stability, implants, forces, displacement, measurement.*

1. INTRODUCTION

In treating femoral neck fractures through the method of internal fixation, one of the most important problems that surgeons must overcome is choosing the right type of implant (Brandt et al., 2006), (Chen et al., 2004), (Heini et al., 2004).

This is the more difficult considering that the current number of implants used in achieving femoral neck osteosynthesis is very large. The ideal implant has to meet two contradictory conditions. On the one hand it has to insure the stability of the fracture focal point, and, on the other hand, it must have a dynamic behaviour that allows the axial compression of the fragments. With this aim in mind, various installations were created, installations which could allow an investigation of the static and dynamic behaviour of the implants used in the femoral neck fracture osteosynthesis. Therefore, the paper (Ehmke et al., 2005) presents a stall that has the purpose to determine the deformation of the screws used in hip implants. The stall offers a two-direction distribution of the forces pressing against the femoral head. The system which allows the simulation of the loadings occurring during walking consists of a biaxial actuator. The variable force is generated by an axial cam with an angle of 23°. The implant screw is placed in a main body made of polyurethane. The paper (Bonnaire & Weber, 2002) presents a system of determining the static and dynamic stability of implants used in treating femoral neck fracture. Also, the paper proposes an original transducer for determining the displacements in the area of the fracture. The paper (Cionca, 2008) aims at using osteosynthesis in treating femoral neck fractures, presenting, among others, a comparative study concerning the elastic and plastic deformations that occur when using two types of implants. The methodology and experimental installation presented have as a main advantage the high degree of universality. The final structure of the stall resulted after systematic and scientific studies of numerous cases that can be encountered in the medical practice, thus permitting the performance of a great number of experiments involving the stability of implants used in treating femoral neck fracture.

2. THE PROPOSED METHODOLOGY

The methodology proposed for determining the static and dynamic behaviour of implants used in femoral neck fracture osteosynthesis presents the following stages: 1. Establishing the type of fracture tract which the implant is used for; 2. Establishing the type of bone that is going to be used in researches. One can use bones of corpses, animal bones, or synthetic femoral models made of composite materials; 3. Establishing the type of implant that will be subject to experiments; 4. Creating the fracture tracts. This stage can be achieved by cutting, on condition that one removes as little material as possible, so that the shape of the femoral bone is not deformed; 5. Implant placement. Implants should be placed identically into the bone in order to reduce the variability of the studied phenomenon. In the case of screw implants, it is necessary that they are placed aided by a dynamometric key, so that all screws are clamped at the same time as the action force; 6. Establishing the dynamic character of external forces. External forces can simulate walking or they can have an evolution imposed by the experiment requirements. When walking, the loadings are of the type: periodical variable loadings or tiredness loadings which occur when there is a periodical variation of the efforts, repeated several times. The efforts the implant is subjected to have a variation of the *pulsating cycle* type to which the external force (the loading) varies from 0 to a maximum value. It is thought that the frequency of loading appliance to a normal walk is 1Hz. 7. Establishing the force number that solicits the femoral head. Only one resulting force can be applied – the one which actions vertically, or several other forces can be predicted, in various directions; 8. Establishing the application points of the external forces; 9. Establishing the direction of the external forces. The femoral bone can be placed into ortostatic position or in various other walking positions; 10. Establishing the size of the forces. The forces the implant is subjected to can be of maximum four times the body weight. Thus, these forces can reach 240... 300 daN; 11. Establishing the experimenting strategy. This is done according to the number of available models and implants, the chosen variable parameters, the time available for doing the experiments, etc.

3. THE EXPERMENTAL INSTALLATION

To study the deformation state of the implants used in femoral neck fractures, an experimental stall was created which allows the simulation of implant behaviour (fig. 1).The stall is made of a basic disc (1), on which a frame is placed (2). A worm restrictor is placed on the frame (2) and it is actioned by a triphase asynchronous engine (4). The restrictor (3) rotates a cam with variable cam action throw (5).

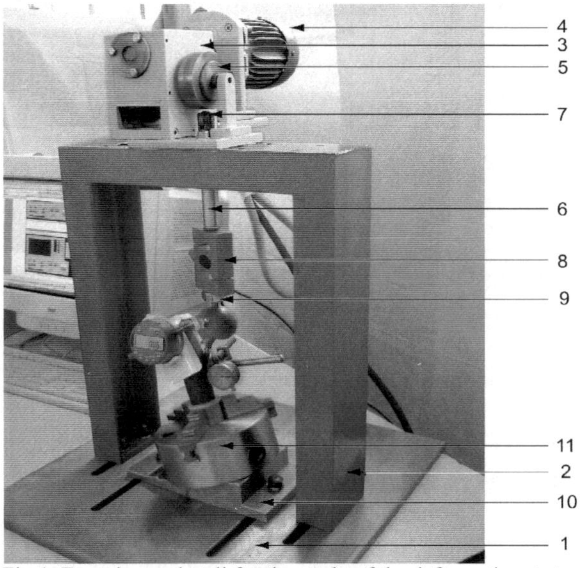

Fig.1. Experimental stall for the study of the deformation state of the implants used in femoral neck fractures

The cam clamps on a tappet (6) which has a roller (7) on the upper part. The tappet (6) is connected to a resistive transducer (8).

The transducer has a pressure head (9) attached. On the basic disc, a supporter (10) is placed and on it the centring box chuck (11) is placed. The chuck clamps the superior part of the femoral bone that is to be subjected to the deformation state research. The supporter (10) allows the bevelled positioning of the femoral bone in an angle of 16°. The ensemble electric engine-restrictor-cam allows applying a frequency force of about 1Hz. The electrical engine's rev is about 1300 rot/ min, and the worm restrictor permits a rev reducing gear of about 26 times. This way, a cam rev of 52 rot/ min. (0,86 Hz) is obtained. The cam allows obtaining variable processes and implicitly variable forces, which are applied to the femoral head. These forces are applied through a tensometric transducer of the type Höttinger Baldwin Messtechnik, S9, which permits the measurement of forces of the maximum value of 500 daN. The transducer has an entry resistance of 505,64 ohm and an exit resistance of 350,43 ohm. The transducer is calibrated so that the connection between the measured force and the tension is 1 daN=0,0001 V. The transducer signal is taken up by a signal conditioner of the type SC-2043-SG National Instruments, with 8 entries. The analogical signal is taken up by the signal conditioner and is transmitted to the data acquisition disc of the type 6023E National Instruments, placed on the basic disc of a computer. The acquisition disc turns the analogical signal into a digital signal, having a sampling rate of 200,000 samples per second. The resolution is 12 bit and 16 analogical entries. To operate, visualise and store data, a virtual measurement instrument of the type Continuous Strain Measurement was used and created in the LabView programming medium of the National Instruments trade mark (fig. 2). This instrument allows a periodical visualisation of the force developed with the help of the tensometric transducer.

The displacements that occur at the level of the fracture focal point are measured in two directions, aided by two comparative, digital and mechanic instruments of the precision 0,01 mm. To determine the absolute displacements of the femoral head from the residual femoral neck, the comparative instruments were placed in such a way that they were unitive with the residual femoral neck. The displacements were measured in two directions.

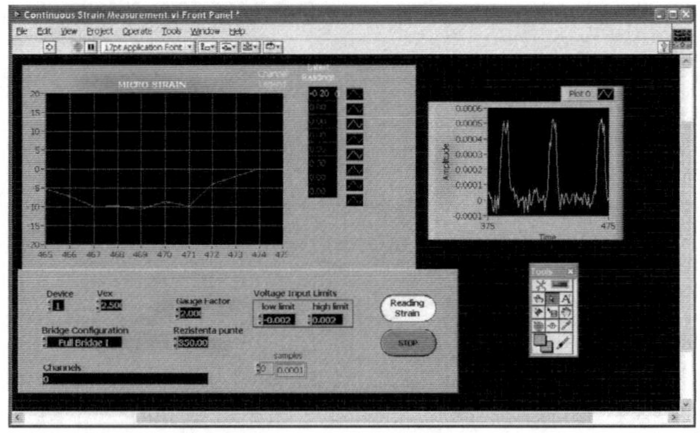

Fig.2. The Continuous Strain Measurement virtual measurement instrument

According to the literature of the field, one displacement measurement focused on the direction perpendicular on the fracture tract, at its superior limit. The other displacement measurement consisted of measuring the displacement of the femoral head on a vertical direction, along the fracture tract.

4. CONCLUSIONS

The stall was used to study the stability of two types of implants: the AO screw system and the Hansson pin hook system. Various comparative studies can be made between the various types of implants that are used for the internal fixation of the femoral neck fracture.

These researches proved the universality, reliability and precision of the proposed installation. Further activities connect to the development of the stall by including a hydraulic installation which could permit enlarging the range of forces to be applied.

5. REFERENCES

Bonnaire, F.A. & Weber A.T. (2002). Analysis of fracture gap changes, dynamic and static stability of different osteosynthetic procedures in the femoral neck. *Injury, International Journal of the care Injured*, **33**, *Suppl. 3*, pp. S-C24-S-C32.

Brandt, E.; Verdonchot N.; Vugt, A van & Kampen, A van (2006). Biomechanical analisys of the percutaneous compression plate and sliding hip screw in intracapsular hip fractures: Experimental assesment using synthetic and cadaver bones. *Injury, International Journal of the care Injured*, **37**, pp. 979-983.

Chen, W.P.; Tai C. L.; Shih C. H.; Hsieh P. H.; Leou M. C. & Lee M. S. (2004). Selection of fixation devices in proximal femur rotational osteotomy: clinical complications anf finite element analysis. *Clinical Biomechanical*, **19**, pp. 255-262.

Cionca, D., (2008). *Contemporaneousness of Osteosynthesis in Treating Femoral Neck Fractures.* Doctoral Thesis, "Grigore T. Popa" Medicine and Pharmacy University, Iaşi, Romania.

Ehmke, L.W.; Fitzpatrik, D. C.; Krieg, J. C.; Madey, S. M. & Bottlang M. (2005). Lag screws for hip fracture fixation: Evaluation of migration resistance under simulated walking. *Journal of Orthopaedic Research*, **23**, *6*, pp. 1329-1335.

Heini, P.F.; Franz, T.; Fankhauser, C.; Gasser B. & Ganz R. (2004). Femoroplasty-augmentation of mechanical properties in the osteoporotic proximal femur: a biomechanical investigation of PMMA reinforcement in cadaver bones. *Clinical Biomechanics*, **19**, pp. 506-512.

Annals of DAAAM for 2008 & Proceedings of the 19th International DAAAM Symposium, ISSN 1726-9679
Editor B. Katalinic, Published by DAAAM International, Vienna, Austria 2008
Make Harmony Between Technology and Nature, and Your Mind will Fly Free as a Bird

NEW DESIGN ASPECTS OF A LOCOMOTORY REHABILITATION MECHATRONIC SYSTEM

SEICIU, P[etre] L[ucian]; LAURIAN, T[iberiu]; FILIPOIU, I[oan] D[an] & PARVU, F[elix]

Abstract: *The design of a mechatronic system for the locomotory rehabilitation of disabled persons is a very difficult task, due to the complexity of walking. There are a lot of parameters to be considered, but the most important is to achieve mechanically a foot movement as close as possible to the real one. In this respect, the feet driving mechanism is the most important mechanism and, for its simpler design, we propose a similar pedalling new model.*
Key words: *mechatronic, locomotion, design.*

1. INTRODUCTION

Walking is one of the most important features of humans. Walking failure leads to serious problems in fulfilling the most common daily activities such as stair climbing, sidewalk walking and street crossing leading in time to total loss of the limb control. Therefore the scientific study of the human movement is mandatory both for its theoretical understanding and for medical purposes. The mechatronic technology and virtual reality in rehabilitation proved to be promising research directions. The modern rehabilitation methods were used only in the last period at a small level, due to a lack of equipments. The use of the gait trainers, as a new method appeared like an alternative to the classical rehabilitation. The results are similar to the classical methods. A consistent difference appeared with the use of the mechatronic gait simulation systems. In this moment there are only a few systems operating (in Germany, Switzerland and USA).

The designing and assembling of the state-of-art mechatronic system (MS) presented in this paper lead the authors to the conclusion that a more thorough scientific study of the walking is needed in order to improve MS functioning (Filipoiu et al., 2007). The main goal is a new design of the foot driving system (FDS), since this is the most complex feature of the MS, mainly due to the foot complex movement.

2. WALKING – PEDALING ANALOGY

The most difficult task in the design of the FDS (figure 1) is to replicate human walking as close to reality as possible.

That is why walking is studied intensely by scientists world-wide.

There are several theories that study walking (six determinants of gait, inverted pendulum, dynamic walking etc.) but none of them is fully applicable for design purposes.

Figure 2 presents a comparison of kinetic parameters from a dynamical simulation with the experiments (Zajac et al., 2003). The black lines (kinetic trajectories) are plotted from the simulation and the grey lines (average kinetic trajectories) are plotted from experimental subjects. VGF and HGF are the vertical and horizontal ground reaction forces from subject measurements. AM and AP are the ankle moment and power computed from inverse dynamics by using subjects' measurements of ground reaction forces and kinematics.

The forces are normalized by body weight; moment and power by body mass.

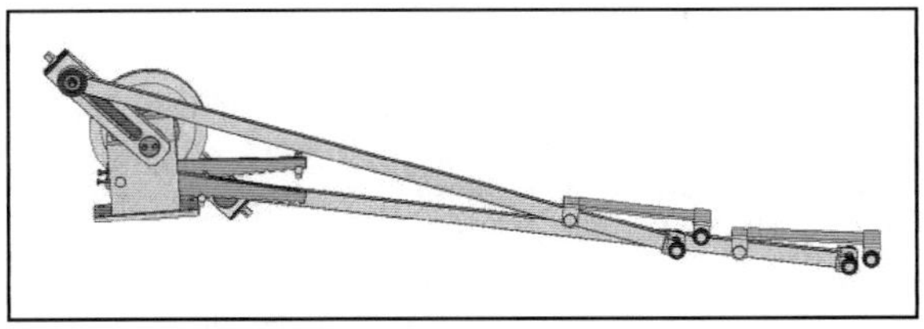

Fig. 1. Feet driving mechanism.

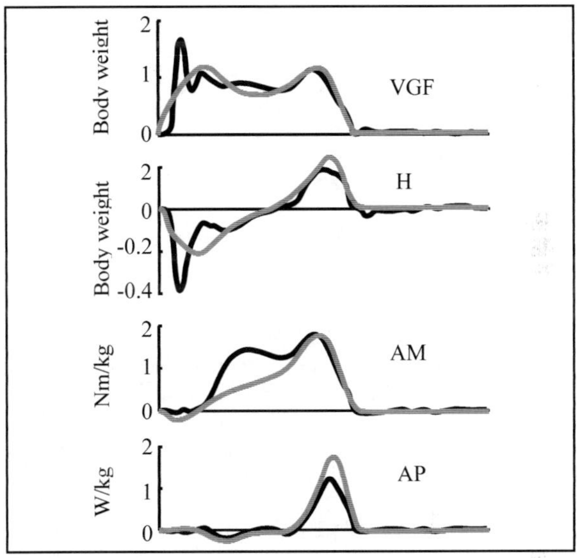

Fig. 2. Kinetic parameters per gait cycle.

In pedalling, muscles have to produce energy over the crank cycle and deliver the energy to the crank to overcome crank resistance and inertia. Muscles deliver power to the crank by developing a tangential crank force. Power delivered to the crank is calculated by (Zajak et al., 2002, 2003).

$$P = \omega \cdot T = \omega \cdot r \cdot F_t = \omega \cdot r \cdot F \cdot \cos\alpha \quad (1)$$

where ω is the crank angular velocity, T is the crank torque, r is the crank arm length, F_t is the tangential crank force, F is the normal force on the pedal and α is the instant angle. All the terms are determined or measurable. The energy delivered to the crank over the crank cycle (0 – 360°, figure 3), which is the external work done on the environment in a cycle, can be calculated by integrating crank power over the cycle.

A comparison between simulated and experimental hip, knee and ankle powers, during the crank cycle, is presented in figure 4 (Zajak et al., 2002). Experimental data are plotted as grey lines and simulation data as solid lines. Propulsive energy to the crank occurs during leg extension (the area above 0 line, during 0÷180° crank angle) is positive.

If we consider the ankle diagrams in figure 3 and 4 we can observe easily that the 2 lines are almost similar.

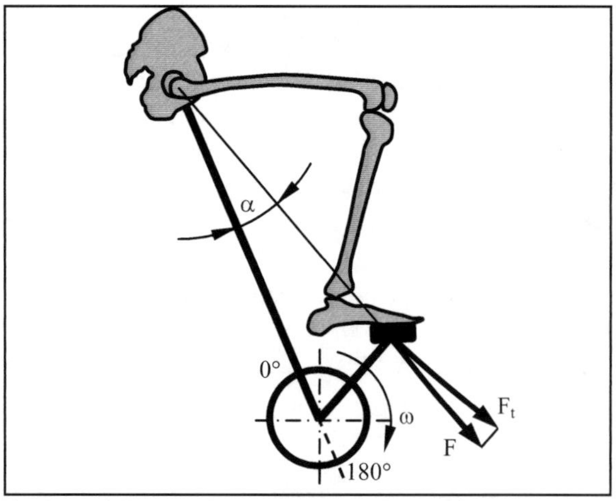

Fig. 3. Pedalling analogy.

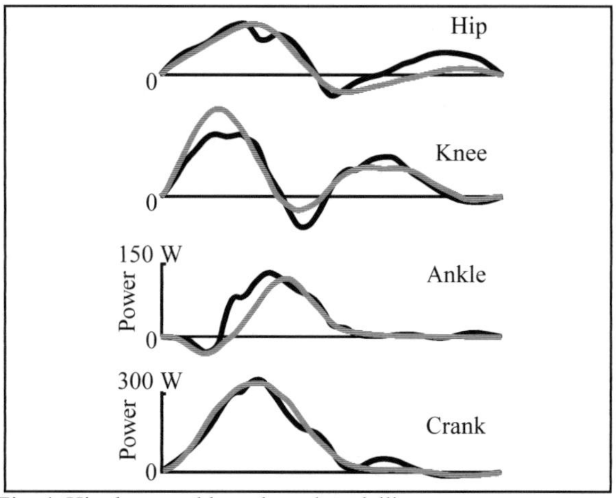

Fig. 4. Hip, knee, ankle and crank pedalling power.

This can lead to the conclusion that the study of pedalling, under some conditions is similar to the study of walking.

3. MECHANICAL WALKING ENERGY ASPECTS

Human walking is theoretically explained by two prevailing, but contradictory theories. The first theory called "The Six Determinants of Gait" (SDG) aims to minimize the energetic cost of locomotion by reducing the vertical displacement of the body centre of mass (COM). The second theory called "The Inverted Pendulum Analogy" (IPA) suggests that walking is a movement combination of two pendulums: the inverted pendulum for the stance leg and a direct pendulum for the swing leg. Walking efficiency is considered for these two theories in order to design properly the mechatronic system. According to the SDG theory, the cost of transport depends on the relative proportions of single and double support. The most economical flat COM trajectory is achieved with single support dominating the step and with an instantaneous double support phase. The dimensionless work rate is (Kuo et al., 2007)

$$WR = \frac{s^2 \cdot f}{4 \cdot \sqrt{4 - s^2}} \tag{2}$$

The IPA theory states that the dimensionless positive work rate is (Kuo et al., 2007)

$$WR = \frac{1}{8} \cdot f^2 \cdot s^2 \tag{3}$$

where s is the step length and f is step frequency, normalized by the natural frequency of the leg.

The theoretical lower limit for the cost of transport (Collins et al., 2005) in walking models is $c_{mt} = 0$. This can be achieved by swaying the upper body with springs in such a manner as to totally eliminate the collision losses. Without swaying the upper body, a rough lower bound on energetic cost can be estimated from the point-mass small-angle model

$$c_{et} \geq c_{mt} \geq J \frac{(s - s_f)^2 \cdot v^2}{2 \cdot g \cdot d \cdot l^2} \approx 0.0003 \tag{4}$$

where J is the collision reduction factor, which is 1/4 for push-off before heel-strike, $s \approx 0.4$ m is the step length , $s_f \approx 0.2$ m is the foot length, $\ell \approx 0.8$ m is the leg length, $v = 0.4$ m/s is the average velocity, and $g \approx 10$ m/s2 is the gravity constant.

Another dynamical model (Kuo et al., 2002) find out that $c_{et} = 0.003$. Other values of c_{et} for real models (Collins et al., 2005) are presented in table 1.

Cornel biped	Deft robot	MIT learning biped
0.2	5.3	10.5

Table 1. Estimated cost of transportation for various robots.

4. CONCLUSIONS

The most significant conclusion is that the analogy between walking and seated pedalling is very useful for FDS design, due to the simplicity of the pedalling model.

A total new conclusion is that the SDG theory applies best for the new MS design, since it uses a flat COM trajectory.

The cost of transport can be estimated easily, but the results vary largely from one application to another. We aim to reach low cost of transport values with the above conclusions applied in the newly designed MS.

All these conclusions will lead to a newly designed MS which will simulate walking closer to reality and, hence, a better and more efficient recovery of the persons with locomotory disabilities.

5. REFERENCES

Collins, S.; Ruina, A.; Tedrake, R. & Wisse M. (2005). Efficient Bipedal Based on Passive-Dynamic Walkers, *Science Magazine*, Vol. 307, no. 5712, February 2005 p. 1082-1805, ISSN: 0036-8075

Filipoiu, I. D.; Seiciu P. L.; Laurian, T. & Carutasu, N. (2007). Mechatronic System for Neuro-Motor Disabled Persons: Computer Simulation, *Ann. DAAM proc. Int. DAAM Symposium*, Katalinic, B. (Ed.), pp. 281-282, ISBN 3-901509-58-5, Zadar, Croatia, $24\text{-}27^{th}$. October 2007, DAAAM International Vienna, Vienna

Filipoiu, I. D.; Seiciu P. L.; Laurian, T. & Carutasu, N. (2007). SIMESIM – The Mechatronic System For Neuro-Motor Disabled Persons, In.: *DAAM International Scientific Book 2007,* Katalinic, B. (Ed.), pp. 387-398, DAAAM International Publishing, ISBN: 3-901509-60-7, Vienna

Kuo, A. D. (2002). Energetics Of Actively Powered Locomotion Using the Simplest Walking Model, *Journal Of Biomechanical Engineering*, ASME, Vol. 124, Issue 1, February 2002, pp. 113-120, ISSN: 0148-0731

Zajak, F. E.; Neptune, R. R. & Kautz, S. A. (2002). Biomechanics and Muscle Coordination of Human Walking. Part I. *Gait and Posture*, Vol.16, Issue 3, December 2002, pp. 215-232, Elsevier, ISSN: 0966-6362

Zajak, F. E.; Neptune, R. R. & Kautz, S. A. (2003). Biomechanics and Muscle Coordination of Human Walking. Part II. *Gait and Posture*, Vol.17, Issue 1, February 2003, pp. 1-17, Elsevier, ISSN: 0966-6362

Annals of DAAAM for 2008 & Proceedings of the 19th International DAAAM Symposium, ISSN 1726-9679
Editor B. Katalinic, Published by DAAAM International, Vienna, Austria 2008
Make Harmony Between Technology and Nature, and Your Mind will Fly Free as a Bird

APPROACHES OF LAYOUT DESIGN SYNTHESIS FOR INTELLIGENT MANUFACTURING SYSTEMS

SEMINSKY, J[aroslav] & WESSELY, E[mil]

Abstract: *Paper is focused to methodology approaches of layout design problems in mechatronic facilities such as IMS. For a long time, engineering design research has been focused on the development of various design theories, methodologies, methods, tools, and procedures. Engineers to more efficiently design artefacts have subsequently used that design methods. However, as the artefacts have grown in complexity, the need for new methods has become obvious. Also, in a nowadays world, increased competition and globalisation require organizations to re-examine traditional product development strategies. While the difficulties in design synthesis are caused by a wide variety of issues, the complicatedness under problem size is so essential that it make procedural design knowledge insufficient to generate superior design solutions.*
Key words: *IMS, Design Methodology, Design Engineering, Complexity, CAD.*

1. INTRODUCTION

Environment around us has from greater part artificial origin – it was created by activity of human. Objects made like product of industrial production (Kusiak 1990), (Warnecke 1992) create the great part of human environment (Košturiak 2000). Character of these objects is casual (Gregor at al. 1997), they were made with some goal to fulfil some before stated conditions (functional, aesthetic, technologic, ...), . Solving so problem is identified as constructive problem.

2. CONSTRUCTIVE PROBLEM

On the one hand the constructive problem can be characterised like definition needs of human to technical object so way that it would have defined parameters in near future. So that is not chaotic, random activity but it is planed systematic activity of human. Solving of so problem is very strong as at beginning it is not clear how way we can fulfil defined parameters. On the other hand it is significant for advance technical objects that their structure is from many heterogeneous subsystems. Present industrial product is made from combination of mechanical, electrical, electronic (hardware and software) components. Products have more or less smart function (Madarász et al., 2006). This smart function is result of synergetic effect of above-mentioned components. Functional and structural diversity of design components results to request of another like mechanistic access. Conceptions of so-called mechatronic access (Isermann 2005) passed over certain development to present time. Historical oldest access use solving of constructive problem so way that function and structural subsystems are solved individually at serial engineering conditions (Janíček & Ondráček 1998). This means that design is made via solving functional, geometrical or power characteristics and later manufacturing, assembly or operation problems are solved (Armeanau & Doicin 2002). As for mechatronic objects the complex function is characteristic achieve by integrated innovative and coordinated design of elements and relations of mechanical, electrotechnical and control subsystems that have to be integrated also at horizontal plane (Tichkiewitch & Brissaud (2004).

For a long time, engineering design research has been focused on the development of various design theories, methodologies, methods, tools, and procedures. The design methods have been subsequently used by engineers for more efficiently design of artefacts. However, as the artefacts have grown in complexity, the need for new methods has become obvious. Also, in a nowadays world, increased competition and globalisation require organizations to re-examine traditional product development strategies. Traditional methods focused exclusively on the numerical optimality of produced artefacts, or their manufacturing processes, are no longer adequate. Creativity and innovation of designed artefacts provide organizations not only with a competitive advantage but are, in fact, a matter of their survival.

3. COMPLEXITY AND INTEGRITY

The increasing of complexity and uncertainty (Peklenik 2001) brings a lot of practical and theoretical difficulties in all domains of designing process. Three major design objectives are usually solved: namely optimality, creativity and robustness. It is not easy to choose approach of solving such problems only by existing principles as analysis and determinism (Hubka, V. 1987). Instead of traditional analytic, deterministic approaches based on for example top-down problem decomposition, TRIZ, evolutionary computation, cellular automata, knowledge based engineering etc., the more smart approaches are being developed with both the bottom-up and top-down features. There many theoretical works were made based on emergent synthesis (Ueda 2001).

New approach is inspired by biology and called like emergent synthesis (Arciszewski & Kicinger 2005). The definition of "emergence" varies in such fields as biology, physics, philosophy, etc. From the system-theoretical view point, however, authors use here the "emergence" in the following sense: "a global order which expresses new function, structure, and action is formed through bi-directional dynamic processes where a local interaction of elements reveals a global behaviour and the global behaviour feedback to the local elements as certain constraints". Moreover, this definition implies the property that implicit globality emerges from explicit locality. As for the characteristics of an emergent system, such key words as evolution, adaptation, learning, co-ordination and interactivity are mentioned. The ways to utilise these concepts as basic mechanisms for solving synthesis problems like synthesis of systems with artificial intelligence, complex technical systems, intelligent manufacturing systems, etc.

4. DESIGN AND MAN-MACHINE INTERFACE

Emergent design tool based on bottom-up and top-down features is good platform from methodology point of view (presently based on multiagent approach frequently); but simulation and visualisation for designer is necessary for decision making in development of solution. Up to this time computers in designing of automated manufacturing systems were used mostly or only for drawing and visualisation of layout (CAD) (Kuric, at al., 2002) and various analytical steps conducted to planning and scheduling. Nowadays we are acting witness of a new class of design tools for mechatronic systems, like IMS are, integrated layout and structural design with simulation and control program generation.

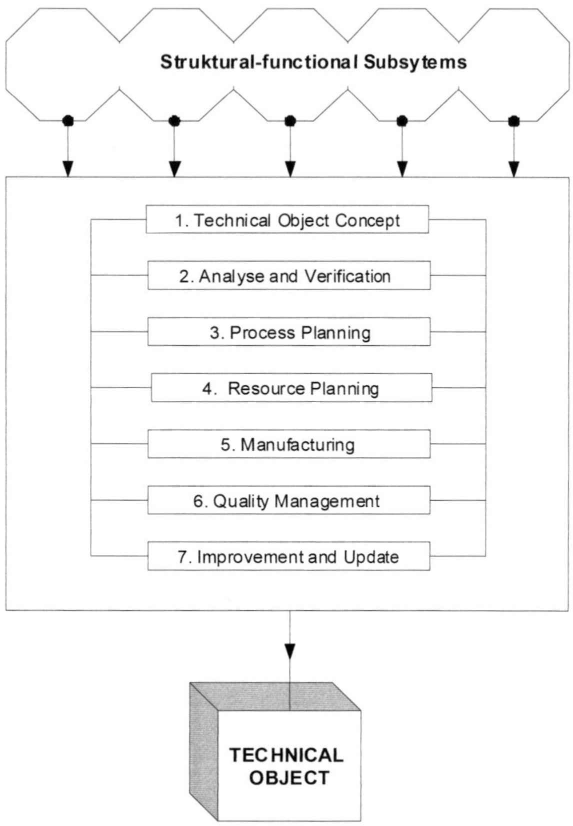

Fig. 1. Designing of FMS as complex and integrated process of analyse and synthesis of technical object

5. CONCLUSION

With regard to design of complex (mechatronic) engineering/technical systems the necessity of knowledge about designing theory is more and more important from these minimal points of view:

- quick changes on the market and therefore of production conditions resulting in systematic product and production systems innovation. So innovation requires adequate concepts and new engineering concepts supported by corresponding abstract basis,
- new engineering complexes are build like integrated from various components: mechanical, electrical and electronic, optical, maybe bionic sometime in future), what brink new quality but new problems in designing of integrated systems also,
- computer aided design of new products with simple connection to computer aided manufacturing and computer aided process planning are standard in modern production. Implementation of information systems supported production in frame of all product life cycle needs adequate abstract basis of designing.

Therefore new generation of smart design computer aided tools for intelligent manufacturing systems have to be enhanced in the bottom-up and top-down features. Research has to consider for need of more complexity and integrity to be applied

6. ACKNOWLEDGEMENT

Paper was made under grant support of Scientific Grant Agency of Ministry of Education and Slovak Academy of Sciences: VEGA 1/3145/06 “Solving of Constructive Problem, Classical versus Evolutionary Access” and VEGA 1/0559/08 “Virtual designing of mechatronic systems”.

7. REFERENCES

Arciszewski, T. and Kicinger, R. (2005). "*Structural design inspired by nature.*" Innovation in Civil and Structural Engineering Computing, B. H. V. Topping, ed., Saxe-Coburg Publications, Stirling, Scotland, 25-48.

Armeanau A. & Doicin C. (2002). Integration of Product Engineering Methods within Functional-Technological Engineering FTE, *Open and Global Manufacturing Design in Machine Engineering,* vol. 2, No. 1-2, p. 233-240, ISSN 1642-6568

Gregor, M. ; Košturiak, J. & Halušková, M (1997). *Priemyslové inžinierstvo. (Industrial Engineering).* BLAHA, , ISBN 80-966996-8-7, Žilina

Hubka, V. (1987) *Teorija techničeskich sistem. (Theory of Technical Systems).* Moskva

Isermann, R. (2005). *Mechatronic systems: fundamental.* Springer, ISBN 1-85233-930-6, London

Janíček, P. & Ondráček, E. (1998). *Řešení problému modelováním. (Problem solving through the modelling).* PC-DIR Real, s.r.o., ISBN 80-214-1233-X, Brno

Kicinger, R., Arciszewski, T., & De Jong, K. A. (2005). "Emergent Designer: An integrated research and design support tool based on models of complex systems." *International Journal of Information Technology in Construction,* 10, 329-347. Available from: http://www.kicinger.com/publications/pdf/KicingerITCon2005.pdf. Accessed: 2008-07-25

Košturiak, J. (2000). *Projektovanie výrobných systémov pre 21. storočie. (Designing of Manufacturing Systems for 21. Century).* Žilinská univerzita. ISBN 80-7100-675-0, Žilina

Kuric, I.; Košturiak, J.; Janáč, A. & Peterka, J. (2002). *Počítačom podporované systémy v strojárstve. (Computer aided Systems for Manufacturing).* University of Zilina / EDIS, ISBN 80-7100-948-2, Žilina

Kusiak, A. (1990). *Intelligent Manufacuring Systems,* Prentice Hall, New Jersey

Madarász, L.; Bučko, M. & Andoga, R. (2006). *Integračné aspekty tvorby a prevádzky systémov CIM (Integration Aspects of CIM Systems Creating and Operation).* TU V Košiciach. ISBN 80-8086-043-2. Košice

Peklenik, J. (2001). Complexity and Emergent Properties of the Work systems in Manufacturing. *3rd International Workshop on Emergent Synthesis* – IWES’01, p.9-23, ISBN 961-6238-49-3, , Bled, Slovenia, March, 12 – 13, 2001, University of Ljubljana, Slovenia

Tichkiewitch, S. & Brissaud, D. (Eds.) (2004). *Methods and Tools for Co-operative and Integrated Design.* Springer Verlag ISBN: 978-1-4020-1889-3

Ueda, K. (2001). Emergent synthesis based approaches to complexity and uncertainty. *3rd International Workshop on Emergent Synthesis* – IWES’01, p.25-30, ISBN 961-6238-49-3, Bled, Slovenia, March, 12 – 13, 2001, University of Ljubljana, Slovenia

Warnecke, H. J. (1992). *Fraktale Fabrik. Revolution der Unternehmenskulur.* Springer Verlag, Berlin

Annals of DAAAM for 2008 & Proceedings of the 19th International DAAAM Symposium, ISSN 1726-9679
Editor B. Katalinic, Published by DAAAM International, Vienna, Austria 2008
Make Harmony Between Technology and Nature, and Your Mind will Fly Free as a Bird
2nd DAAAM-ICAT

ADDITIVE LAYER MANUFACTURE OF TENSILE TEST SPECIMENS IN STAINLESS STEEL 316L BY LASER CONSOLIDATION

SEWELL, N[eil] T[homas]; BASSOLI, E[lena]; GATTO, A[ndrea] & EVANS, K[enneth]

Abstract: *Laser Consolidation (LC) is an Additive Layer Manufacturing (ALM) process which produces net-shape fully dense metal parts. LC uses a synchronised pulsed Nd:YAG laser and a powder delivery unit to build parts which exhibit excellent mechanical properties and impact resistance. These properties may be tailored at build time by modifying build parameters but the non-coaxial nature of the LC head assembly causes direction dependant features. This paper investigates the mechanical properties and microstructure of AISI 316l parts built by LC. Tensile test specimens were constructed in different orientations and underwent tensile testing and SEM observation. All the specimens exhibited high elongation at break and Ultimate Tensile Strength (UTS) but parts properties were significantly affected by build direction. Parts built by relative head movement along the plane of LC head assembly (X direction) were more ductile and stronger compared with parts constructed orthogonally. SEM analysis revealed a higher presence of voids in the latter. It was speculated that the voids were the major factor in the reduction in strength and ductility of the parts.*
Key words: *Laser Consolidation Stainless Steel Testing*

1. INTRODUCTION

As Additive Layer Manufacturing (ALM) systems mature, more is expected of the functional parts produced using the technology. For ALM to be accepted as an industrial manufacturing technique, parts need to be produced reliably to high tolerances and with good mechanical properties. Currently, one of the major limitations for ALM is the difficulty in predicting part characteristics to ensure their robustness and repeatability.

Laser Consolidation (LC) is a powder-fed based technology which produces net-shape parts without the need for additional finishing. Powder-fed ALM offers promising advantages: metallurgical soundness, high strength and ductility of parts; ease of change of the powder supplied to enable multi-material or graded parts (Domack & Baughman, 2005); operation using standard CNC know-how enabling acceptance in most industries. As in every ALM system, the building process in LC is inherently anisotropic. In particular, the non-coaxial nature of the LC head assembly and its path are likely to cause direction dependant features varying not only along and perpendicular to the build direction, but also within each layer.

Previous work has suggested that parts produced using LC exhibit excellent mechanical properties (Xue & Islam, 2000) but often leave out important information regarding the experimental procedure and processes used (Toyserkani & Khajepour, 2006). More information is available for a similar powder-fed system known as Laser Direct Deposition (LDD), using a continuous CO2 laser coaxial with the powder nozzle where fully dense specimens were made in a nickel based superalloy resulting in high strength and remarkable ductility (Zhang et al., 2007). Studies on Titanium alloys (Gao et al., 2007) proved that parts can be successfully built by LDD with higher mechanical properties than obtained by casting and equal to those obtained by wrought annealed parts. These authors also investigated and proved part anisotropy.

With LC, the direction of build, either axially aligned with the head or orthogonal to it, will affect the characteristics of the part due to the non-coaxial head configuration. This paper examines the use of LC to produce some simple dog-bone samples in stainless steel 316L using two different orientations, parallel and orthogonal to the build head. The results illustrate that although the mechanical and microstructural properties of the parts are excellent, there are some differences due to part orientation. Explanations for the differences are investigated.

2. EXPERIMENTAL PROCEDURE

2.1 Part Manufacture

Parts for testing were manufactured using the Accufusion Laser Consolidation System, owned by Airbus and based at the University of Exeter. This system is currently involved in a UK Government Technology Strategy Board sponsored project, named DAMASCUS, to investigate the ALM of parts for aerospace and automotive. The LC system head assembly consists of a non-coaxial powder feeder nozzle and a pulsed Nd:YAG laser. Powder is fed into the laser beam at its focal point and consolidates as the laser source is removed. Building starts on a substrate that is moved beneath the head assembly to produce a consolidated bead. The head unit is raised incrementally to build up the part focusing the beam on the previously built section. In this experiment, beads of material were deposited one on top of another at 0.1mm increments to create rectangular parts just over 20mm high (200 layers) by 180mm long. Bead width depends on the amount of powder that consolidates within the laser spot: in this experiment parts were 1mm thick.

Effectively LC is a head assembly mounted on the Z-axis of a CNC system with the X and Y axis moving in the horizontal plane beneath. Additional rotational axes are available. A simple CNC program was designed to deposit the 180mm long beads of material on the substrate, a 30mm thick steel plate. For the specimens with the axis parallel to X direction the head scan during build was parallel to the direction of the laser beam and the powder fed (Figure 1). Reciprocating axial passes were used, i.e. the first part being built from -X to +X (X+ve), the second from +X to –X (X-ve) and repeated. After manufacturing eight test samples along the X axis (4 for each vector), parts were made along the Y axis in a similar manner. For Y specimens the head unit moved orthogonally to the head assembly plane. Parts were then removed from the substrate along the first few beads as allowed for by the 2mm overbuild added. The parts then had waists added by spark erosion obtaining geometry consistent with standard BS EN ISO 527-2 (reduced section width 10mm, length 75mm).

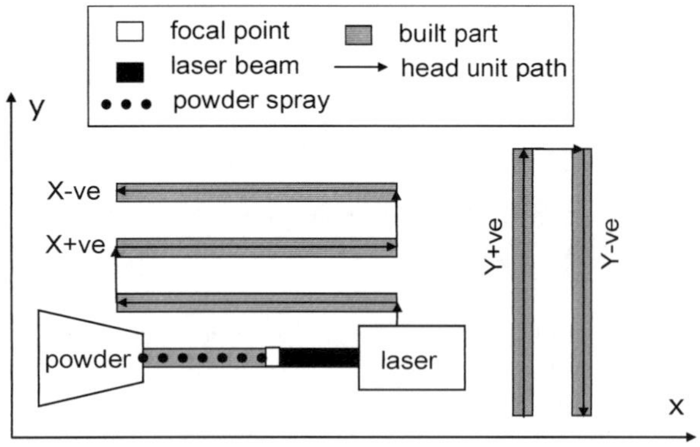

Fig. 1 – Building strategy.

2.2 Processing Parameters

Parts were produced using previously derived settings for processing stainless steel 316L by LC. 10ms pulses at 30Hz with 5J per pulse were used with a 5g per minute 316L powder feed rate. An initial build speed of 300mm per minute was used, deliberately lower that usual, to allow for the heat absorbance of the substrate affecting build height for the first 10 layers (Toyserkani & Khajepour, 2006). After this a rate of 375mm per minute was used, derived from a desire to overlap laser spots by a minimum of 50%. Given that the laser spot diameter is 0.5mm, the laser needs to fire once every 0.25mm. At 30Hz with a 50% overlap, substrate movement would be 450mm per minute, however, preliminary experiments showed that inconsistent build-up occurred at rates higher than 375mm per minute in the axis perpendicular to the laser head assembly.

2.3 Tests

Ultimate Tensile Strength (UTS) tests were carried out on at least 5 parts in each orientation at a speed of 2mm per minute with strain measured using an extensometer acting on a gage length of 50mm. After testing, rupture surfaces and un-melted powder were observed using the scanning electron microscope (SEM) to investigate failure mechanisms and joining phenomena between the particles.

3. RESULTS AND DISCUSSION

Table 1 shows the main results of the UTS tests. The mean and Standard Deviation (SD) were calculated for the positive and negative vectors of X and Y directions separately and together. The separate build vectors are inconclusive as there were too few parts for a meaningful SD. Yet, no remarkable differences can be seen and the stress-strain graphs are very similar. The X direction results show high consistency in their behaviour in both directions, whereas the Y direction data is less consistent. UTS of parts built in the Y direction is on average 40MPa lower than that of the X specimens and the maximum strain is almost half. All LC parts show high strength when compared with sheet stainless steel 316L values. Elongation at break is only slightly smaller for X parts. Typical rupture surfaces of X and Y direction parts are shown in Figure 2.

Sheet AISI 316L		UTS = 485MPa		□$_b$□□□□□□	
	No. parts	UTS (MPa)		□$_b$ (%)	
		mean	SD	mean	SD
X +ve	4	589	5.6	31.0	3.4
X -ve	3	582	7.9	34.7	1.0
X all	7	586	8.1	32.8	2.8
Y +ve	3	537	34.8	15.6	8.1
Y -ve	2	555	0.0	18.6	0.2
Y all	5	544	28.4	16.6	6.4

Tab. 1. Tensile tests results.

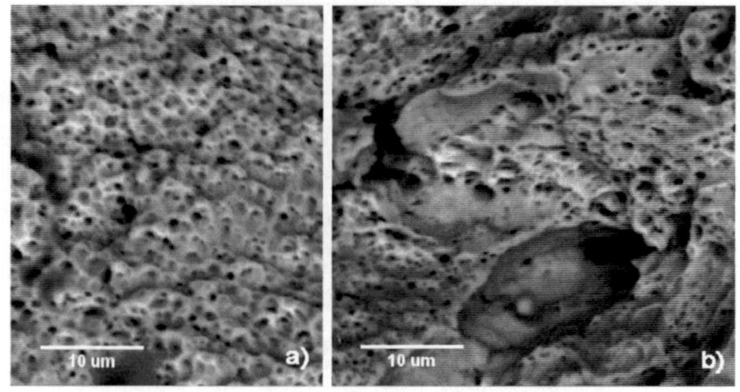

Fig. 2. Rupture surface: a) X sample; b) Y sample.

X specimens (Fig. 2a) exhibit uniform ductile failure morphology based on micro voids forming under strain; porosity previous to rupture is almost absent and joining between the layers is not visible. For Y parts, instead, many large voids are evident throughout the surface (Fig.2b), up to 100μm wide. The smooth inner surface suggests they were formed during laser consolidation. In Y direction the powder is sprayed across and not along the section buing built, which can likely be the cause for less material consolidated in the building area.

4. CONCLUSION

In conclusion, this study confirmed that LC parts show high mechanical properties, equal to or better than parts produced traditionally. In particular, ductility is remarkably high compared with other additive processes. Anisotropy within the building plane was analysed, finding that the specimens created in the X direction exhibit higher strength and much higher ductility when compared with parts created in the Y direction. The parts in the X direction demonstrate consistent results throughout the test samples, where as the parts in the Y direction displayed greater variation between the results. SEM observation of the rupture surfaces revealed numerous voids in the structure of the parts in the Y direction; it was proposed that these void could be a dominant factor in tensile properties. Further study will be carried out to clarify this speculation and to investigate the factors contributing to void formation. Polished sections will be observed to calculate the percentage porosity in X and Y directions. No definitive result has been obtained as to positive and negative scan vectors, that will be further investigated. As a direct consequence of this research, a specific CNC programming employing the rotation axis of the worktable could ensure maximum mechanical response in the load direction.

5. REFERENCES

Domack, M.S. & Baughman, J.M. (2005). Development of nickel-titanium graded composition components. *Rapid Prototyping J.*, Vol. 11, No. 1, 41 - 51, ISSN: 1355-2546.

Gao, S.Y.; Zhang, Y.Z.; Shi, L.K.; Du, B.L.; Xi, M.Z. & Ji, H.Z. (2007). Research on Laser Direct Deposition Process of Ti-6Al-4V Alloy. *Acta Metall. Sinica (English Letters),* Vol. 20, No. 3, (June 2007) 171-180, ISSN: 1006-7191

Toyserkani, E & Khajepour, A. (2006). A mechatronics approach to laser powder deposition process. *Mechatronics,* Vol. 16, No. 10, (Dec. 2006) 631-641, ISSN: 0957-4158

Xue, L. & Islam, M.U. (2000). Free-form laser consolidation for producing metallurgically sound and functional components. *J. of Laser Applications,* Vol.12, Num.4, (Aug 2000) 160-165, ISSN: 1042-346X

Zhang, K.; Liu, W. & Shang, X. (2007). Research on the processing experiments of laser metal deposition shaping. *Optics & Laser Technology,* Vol. 39, 549–557, ISSN: 0030-3992.

Annals of DAAAM for 2008 & Proceedings of the 19th International DAAAM Symposium, ISSN 1726-9679
Editor B. Katalinic, Published by DAAAM International, Vienna, Austria 2008
Make Harmony Between Technology and Nature, and Your Mind will Fly Free as a Bird

A NOVEL INTUITIVE DYNAMIC MODELLING METHOD FOR DIGITAL I&C SYSTEMS IN NUCLEAR POWER PLANTS

SHIN, S[eung] K[i]; GOH, G[young] T[ae] & SEONG, P[oong] H[yun]

Abstract: *Conventional static reliability analysis methods are inadequate for modelling dynamic interactions between components of a system. Various techniques have been proposed for modelling dynamic systems based on improvement of the conventional modelling methods. However those methods are inadequate to apply to digital instrumentation & control (I&C) systems in nuclear power plants from the viewpoint of intuitiveness. In this paper we introduce dynamic nodes to the existing reliability graph with general gates (RGGG) as an intuitive modelling method to model dynamic systems including the digital I&C systems.*

Key words: *reliability analysis, dynamic systems, digital I&C systems, reliability graph with general gates*

1. INTRODUCTION

Reliability of systems in a nuclear power plant is analyzed to compare with regulatory criteria designed to ensure the availability. There are several reliability analysis methods: fault tree and event tree method, reliability graphs, reliability block diagrams, Markov chains, and Monte Carlo simulations.

As digital systems are introduced to nuclear power plants, issues related with reliability analyses of these digital systems are being generated. The conventional static modelling methods, such as the event tree and fault tree methods, present significant shortcomings when used in the reliability modelling of digital I&C systems in nuclear power plants, as they cannot properly account for dynamic interactions between the digital systems and components (Aldemir et al., 2006).

To overcome the limitation of the conventional static modelling methods, dynamic modelling methods should be adopted for the reliability analyses of digital systems in nuclear power plants. A dynamic fault tree method is a powerful method that can model dynamic systems; however, the dynamic fault tree cannot intuitive model a system and experiences a state space explosion problem in the phase of quantitative analysis (Amari et al., 2003).

Our objective is to improve the existing RGGG method for the purpose of developing an intuitive method that can capture the dynamic behaviour of a system failure.

2. VARIOUS DYNAMIC MODELLING METHODS

Most research has involved existing modelling methods in order to integrate easily with static methods, which have already been widely implemented in the system modelling field. In this chapter we review those approaches.

2.1 Dynamic Fault Tree

A dynamic fault tree modelling technique was developed to handle the difficulties that arise in the reliability analysis of fault-tolerant computer systems when critical applications are complicated by several factors (Dugan et al., 1992). Four dynamic gates (FDEP, SPARE, SEQ, PAND gate) were newly adopted in the conventional fault tree method. Each dynamic gate can express each dynamic failure process that is related to the failure sequence of the component parts.

Dynamic fault trees are solved by conversion to equivalent Markov models. However it should be noted that the conversion to a Markov chain generally induces a state space explosion problem as the number of basic events increases (Amari et al., 2003). If the number of basic events in n and each event has two conditions, then the total number of states in the Markov chain is 2^n.

2.2 Dynamic Modelling Based on a Petri-Net

Generalized stochastic Petri nets (GSPN) are a performance analysis tool based on a graphical system representation typical of Petri nets, where some transitions are *timed* while others are *immediate* and techniques to generate a continuous time Markov chain (CTMC) from the GSPN and solve the CTMC are available and implemented in several software tools, such as *GreatSPN* (Chiola et al., 1995). However, since this approach also uses the Markov chain, it is hard to avoid the problem raised from the size of states.

To model dynamic performances of systems using Petri nets, a firing delay concept is used in association with transitions. Memory can be introduced to transitions using general distributions for firing delays. Several transitional memory policies were proposed to model the dynamic flexibility as memory is associated with statically positioned transitions. To overcome this shortcoming, memory was assigned to tokens (aging tokens) rather than to transitions (Volovoi, 2004). Aging tokens provide dependability modelling flexibility and compact representation of complicated dynamic interactions among multiple entities.

2.3 Dynamic Reliability Block Diagrams

Dynamic reliability block diagrams (DRBD) was proposed based on an extension of the extending the existing reliability block diagrams formalism (Distefano & Xing, 2006). In a DRBD model, each component is characterized by a variable *state*: *active*, *failed*, and *standby*. Reliability relations between components are expressed by a specific connection representing the logical condition applied by the *driver* component to the *target* component and the dependency is represented using a dash line with a circle in the target component termination that represents the connection between the related components.

The main advantage of DRBD is the capability to model dependencies among subsystems or components concerning their reliability interactions. A quantitative analysis of DRBD can be conducted using the existing methods such as Markov chains and Monte Carlo Simulation.

3. DYNAMIC RGGG

Although the fault tree method is the most frequently used method, because it focuses on the causal relations of the failure between components, it cannot intuitively model a system. In order to model the system intuitively, a reliability graph with general gates (RGGG) method was proposed; it can make a

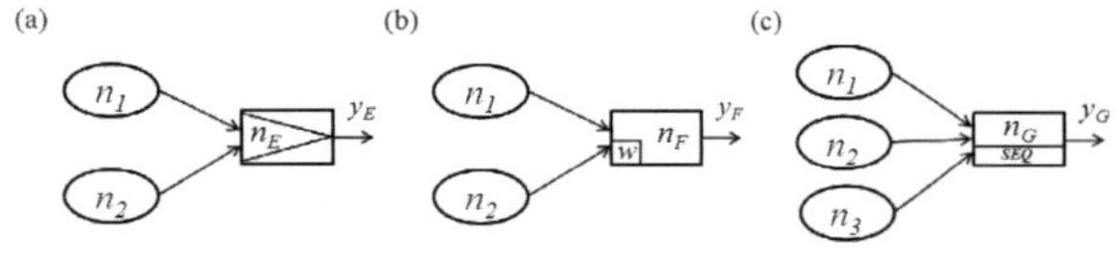

Fig. 1. The dynamic nodes of a dynamic RGGG:
(a) a PAND; (b) a WSP; and (c) an SEQ node

one-to-one match from the actual structure of a system to the reliability graph of the system (Kim & Seong, 2002). However, the existing RGGG method is inadequate if the failure of the system is related to the sequence of component failures. Therefore, in this section we improve the existing RGGG method for the purpose of developing an intuitive method that can capture the dynamic behaviors of a system failure.

3.1 Addition of Dynamic Nodes to the RGGG

To enable the RGGG to model dynamic systems, we introduce additional nodes that can express dynamic processes based on the dynamic gates of a dynamic fault tree. Figure 1 shows the proposed dynamic nodes; PAND, SPARE, and SEQ node.

A novel FDEP node is not required in the RGGG. The existing RGGG can illustrate the property of an FDEP gate by using only the OR nodes.

3.2 Quantitative Analysis of Dynamic Nodes

To evaluate an RGGG, we convert the RGGG into an equivalent Bayesian network by determining the probability tables of all the nodes. We use the discrete-time method to determine the probability tables.

We divide the line of the total process time (T) into n equal intervals. The time of one interval then becomes t. The output of each node is one of $\{I_1, I_2, ..., I_n, I_\infty\}$. If the output of a certain node is I_k, the node fails in the kth interval; and if a node has output I_∞, the node never fails during the total process time. If P_{ij}^k denotes the probability that an arc (a_{ij}) from node i (n_i) to node j (n_j) fails in the kth interval, and if $F_{ij}(t)$ denotes the cumulative failure distribution function of a_{ij}, we can derive P_{ij}^k as follows:

$$P_{ij}^k = \int_{(k-1)t}^{kt} \frac{dF_{ij}(t)}{dt} dt \qquad (1)$$

To obtain accurate reliability, we should increase the value of n; however, as n increases, the probability table of each node becomes more complex. The number of blanks that should be filled in for a probability table of a node with two inputs is $(n+1)^3$. Without rules, the task of making the tables is difficult and prone to errors. We therefore propose a set of rules to establish the probability table of each dynamic node. First, let the outputs of n_1, n_2, n_3, n_E, n_F, and n_G be I_x, I_y, I_z, I_e, I_f, and I_g, respectively ($x, y, z, e, f, g \in 1, 2, ..., n, \infty$).

Rules for making the probability table of a PAND node shown in Figure 1(a) are as follows:

i. If $e > y$,
0.

ii. If $e = y \leq x$,
$Pr\{a_{1E}$ fails before the eth interval$\} \cdot (1 - Pr\{a_{2E}$ fails before the eth interval$\})$.

iii. If $e \leq x$, $e < y$,
$Pr\{a_{1E}$ fails before the eth interval$\} \cdot Pr\{a_{2E}$ fails at the eth interval$\}$.

iv. If $e = y > x$,
$1 - Pr\{a_{2E}$ fails before the eth interval$\}$.

v. If $x < e < y$,
$Pr\{a_{2E}$ fails at the eth interval$\}$.

		n_E			
n_1	n_2	I_1	I_2	I_3	I_∞
I_1	I_1	0	0	0	1
	I_2	0	$1 - P_{2E}^1$	0	$1-\Sigma$
	I_3	0	P_{2E}^2	$1 - P_{2E}^1 - P_{2E}^2$	$1-\Sigma$
	I_∞	0	P_{2E}^2	P_{2E}^3	$1-\Sigma$
I_2	I_1	0	0	0	1
	I_2	0	$P_{1E}^1(1 - P_{2E}^1)$	0	$1-\Sigma$
	I_3	0	$P_{1E}^1 P_{2E}^2$	$1 - P_{2E}^1 - P_{2E}^2$	$1-\Sigma$
	I_∞	0	$P_{1E}^1 P_{2E}^2$	P_{2E}^3	$1-\Sigma$
I_3	I_1	0	0	0	1
	I_2	0	$P_{1E}^1(1 - P_{2E}^1)$	0	$1-\Sigma$
	I_3	0	$P_{1E}^1 P_{2E}^2$	$(P_{1E}^1 + P_{1E}^2)(1 - P_{2E}^1 - P_{2E}^2)$	$1-\Sigma$
	I_∞	0	$P_{1E}^1 P_{2E}^2$	$(P_{1E}^1 + P_{1E}^2)P_{2E}^3$	$1-\Sigma$
I_∞	I_1	0	0	0	1
	I_2	0	$P_{1E}^1(1 - P_{2E}^1)$	0	$1-\Sigma$
	I_3	0	$P_{1E}^1 P_{2E}^2$	$(P_{1E}^1 + P_{1E}^2)(1 - P_{2E}^1 - P_{2E}^2)$	$1-\Sigma$
	I_∞	0	$P_{1E}^1 P_{2E}^2$	$(P_{1E}^1 + P_{1E}^2)P_{2E}^3$	$1-\Sigma$

Tab. 1. Probability table for a PAND node for $n=3$

vi. If $e = \infty$,
1 – (sum of the other probabilities in the same row).

When we fill in the probability table by using rules 1 to 5, we exclude the case where the value of e is ∞. Furthermore, because the sum of the probabilities in each row should be 1, we apply rule 6. Table 1 shows a probability table for a PAND node which is made by above six rules. The rules for SPARE and SEQ nodes are not illustrated in this paper.

4. SUMMARY & FURTHER WORKS

We proposed an intuitive modelling method for dynamic systems. We upgraded the capability of the RGGG so that it can be used to model dynamic systems. However it is not easy to make each probability table as n increases. A software tool for generation of probability tables was developed on the basis of the proposed rules and we are currently developing a tool to connect the generated probability tables with an existing Bayesian software tool.

5. REFERENCES

Aldemir, T.; Miller, D. W.; Stovsky, M.; Kirschenbaum, J.; Fentiman, A. W. & Mangan, L. T. (2006). Current State of Reliability Modeling Methodologies for Digital Systems and Their Acceptance Criteria for Nuclear Power Plant Assessments, *NUREG/CR-6901,* U. S. NRC

Amari, S.; Dill, G. & Howald, E. (2003). Dynamic fault-tree models for fault-tolerant computer systems, *Proceedings of the Annual Reliability and Maintainability Symposium,* Amari, S. (Ed.), pp. 374-379, 0-7803-7717-6, USA, Jan. 2003, IEEE Computer Society, Washington, DC USA

Chiola, G.; Franceschinis, G.; Gaeta, R. & Ribaudo, M. (1995). GreatSPN 1.7: Graphical Editor and Analyzer for Timed and Stochastics Petri Nets. *Performance Evaluation,* Vol. 24, No. 1-2, (Nov. 1995) 22 (47-68), 0166-5316

Dugan, J. B.; Bavuso, S. J. & Boyd, M. A. (1992). Dynamic fault-tree models for fault-tolerant computer systems. *IEEE Transactions on Reliability,* Vol. 41, No. 3, (Sep. 1992) 15 (363-377), 0018-9529

Kim, M. C. & Seong P. H. (2002). Reliability graph with general gates: an intuitive and practical method for system reliability analysis. *Reliability Engineering & System Safety,* Vol. 78, No. 3, (Dec. 2002) 8 (239-246), 0951-8320

Volovoi, V. (2004). Modeling of system reliability using Petri nets with aging tokens. *Reliability Engineering & System Safety,* Vol. 84, No. 2, (May 2004) 13 (149-161), 0951-8320

Annals of DAAAM for 2008 & Proceedings of the 19th International DAAAM Symposium, ISSN 1726-9679
Editor B. Katalinic, Published by DAAAM International, Vienna, Austria 2008
Make Harmony Between Technology and Nature, and Your Mind will Fly Free as a Bird

TECHNOLOGY OF FIBROUS MATERIALS PRODUCTION FROM THE MELT OF THERMOPLASTIC MATERIALS BY MEANS OF BLOW WITH AIR

SHIROBOKOV, K[onstantin]

Abstract: *The technological process of fibrous materials production from the melt of thermoplastic materials by means of vertical blow with air is researched. It is shown that reception fibrous material by way vertical blow allows greatly to reduce the prime cost got product under get prettier the quality ready filament. The experimental – industrial facility is given, the results of studies are presented, for shaping fibrous material from melt thermoplastic, and further directions of the work are intended.*
Key words: *facility, thermoplastic, melt, blow, fibrous*

1. INTRODUCTION

The technical progress in different branches of national economy is determined by the quality of machines and units realizing different technological processes and is characherized by the quality of output production. The quality and the prime cost of goods production firstly depend on the materials which they are made of (Leskov, 2001).

Fibrous materials are widely used in different spheres of person activites. The main function of such materials is the heat insulation of different kinds of power and transport equipment in technical industry and the heat insulation of buildings and constructions in industrial and civil engineering. Fibrous materials are often used for sound insulation of premises and equipment (Sentyakov & Timofeev 2004). Basalt and glass fibres based on mineral materials are the most extended in these industries. Only synthetic staple fibre (for example, polyethylene terephthalate) which is used in light industry can compete with these materials. However, the use of synthetic staple fibre in construction is considered to be not rational because of its high cost of production. That is why the aim of this work is improving the quality of fibrous materials from the melt of thermoplastic materials by means of vertical blow with air.

2. INFORMATION

To achieve this aim it is necessary to solve the following problems:

- To analyse modern technologies and methods of producing fibrous materials such as wind, duplexing, mechanical, and combined ones with the purpose of finding out advantages and disadvantages of each method to come to a right decision of their reasonable realization in specific production conditions from necessity of fibrous production of desired quality and productivity;
- To work out the mathematical model of producing fibrous materials from the melt of thermoplastic materials by means of vertical blow with air;
- To work out and manufacture the experimental – industrial facility for fibrous materials production from the melt of thermoplastic materials by means of vertical blow with air;
- To study the influence of technological and constructive parameters of the process to produce fibrous materials of the desired quality from the melt of thermoplastic materials by means of vertical blow with air;
- To work out the method of the experimental determination of the main geometric and physical parameters of the process for producing fibrous materials from the melt of the thermoplastic materials by means of vertical blow with air;
- To work out the estimating method of qualitative characteristics of fibrous materials.

A well known technological process of chemical fibres production consists of three main stages. The first stage of any chemical fibre production includes the preparation of the spinning paste. This viscous liquid is carefully cleaned by frequent filtration. Then hard particles and air bubbles are deleted. The second stage includes the fibres forming through the spinneret feeder. Spinneret is a small vessel made of strong heat and chemical resistant material with a flat bottom having a lot of (up to 25 thousand) small holes. Diameter of each hole can vary from 0,04 to 1,0 mm.

Fine streams of melt from the holes of the spinneret get into space forming fibres from melt polymer. Then they are cooled and they solidify. After forming fibres are collected in bunches or ropes consisting of many fine fibres. Then threads are washed and subjected to special processing (soaping or lubrication) to facilitate textile conversion or dried. Ready made threads are wound on reels or spools. Threads are split to staples producing staple fibre. Then staple fibre is collected in stacks (Papkov, 1990).

Having analyzed the methods of fibrous materials production we came to the conclusion that the most interesting presents the technological process blow expiring from the melting unit stream melted source cheese complying with direction of its current by flow of the power system, for instance, pair or compressed air (Dzhigiris & Mahova 2002).

In comparison with the well-known method of synthetic fibre production, the method under our consideration provides unite the operations forming and exhaustion elementary threads, reduction of the number labour-consuming technological stage. Follows to note that unlike known technological process, way vertical blow air is the most perspective, on its technical-econmic factor, but in the same way allows solves the problem of utilization of plastics made of polyethylene terephthalate, expands the range of physical characteristics of fibres and creates the premises of using fibres in mechanical ingineering, heat power engineering, and civil engineering. Only duplexing method in quality of fibres is better (Sentyakov & Timofeev 2004).

The practical realization of this method in industry is based on experimental data being got by manufacturers. The difficulty of the physical processes interacting the jet of melted material with the flow of power carrier hasn't allowed to give the theoretical description of this process yet. Many authors set only different hypothesises of the process of fibre forming. After analysing the hypothesises of fibre forming the author offers the original conceptual model of fibre materials production by means of vertical blow with air. The main idea of this model is that the quality of producing fibre depends on the cross coordinate of the jet of melted materials with the flow of compressed air. The experimental data corroborates the model and the recommendations of setting the parameters of the technological process for fibrous materials production from mineral melts by means of vertical blow with air propely are given.

Scheme of the experimental installation submit for fig. 1. In this facility the outflow of melted material occures through the action of hydrostatic pressure which is maintained constantly due to the periodical material supply in equal amount to the melted material expiring through the nozzle in the lower part of the facility. This is determined in the period of facility adjustment. The experimental methods proved the possibility of fibrous materials production from the melt of thermoplastic materials by means of vertical blow with air. The original technical devices for realization this technology are elaborated, namely the construction of blow head for splitting the jet of melted material into elementary fibres, the oscillation mechanism of the blow head for forming linen from this fibre, the device for dosed supply of raw materials in melting unit. The possibility of fibrous materials production from secondary raw materials, for example, crashed plastic bottles, is confirmed and data of physical characteristics of this fibre are received.

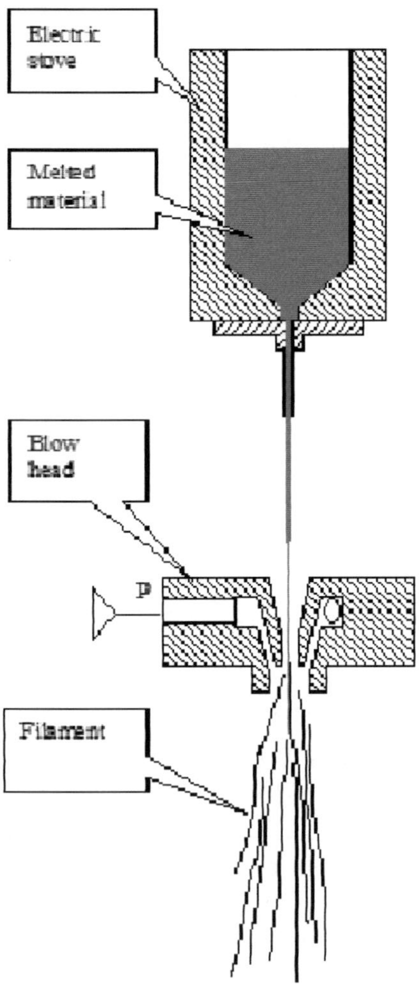

Fig.1. Scheme of the process of the reception filament from melt thermoplastic

The results of the studies of the processes of fibre production from materials of different quality are shown that the fibre of the best quality is produced from the original polyethylene terephthalate. The physical characteristics of the samples of polyethylene terephthalate fibres of different qualities in comparison with the charachteristics of superfine basalt fibre are determined. They are water absorption, compressibility, elasticity and ultimate tensile strength according to international standart GOST 17177-94 "Materials and civil heat insulating products. Methods of testing."

The result of the studies of heat insulating characteristics of the stamples of polyethylene terephthalate staple fibre shows that their heat-conductivity coefficient is practically identical to heat-conductivity coefficient of superfine basalt fibre and staple glassfibre URSA and heat-conductivity coefficient is 0.037...0.040 W/m°K. In the same way noted following advantage of polyethylene terephthalate filament, got by new way, in contrast with similar fibrous basalt material. First, this half smaller absorption of water and incomparable more intensive drying that can be important in determined condition to usages. Secondly, the best springy characteristic that allows the polyethylene terephthalate a filament in greater depth to restore its form and sizes after deforming. And, finally, in 5 - 10 once big tensile strength. All this allows to speak of polyethylene terephthalate filament, as about competitive material on the market of the modern building materials.

3. CONCLUSION

Thereby, as a result called on studies are received following scientific results:

1. The experimental data characterising the process of the staple fibre forming from the melt of thermoplastic materials by means of vertical blow with air are received;
2. The rated dependence for calculating the average diameter of elementary fibres taking into account the main geometric and physical parameters of the process is received;
3. The method of experimental-rated determination of the main geometric and physical parameters of the process for the staple fibre production from the melt of thermoplastic materials by means of vertical blow with air is developed and confirmed by the results of industrial tests of new constructions of blow heads.
4. Hereinafter necessary to develop the method of experimental determination of dencity, viscosity, surface tension of low-melt simulating medium and to improve separate parts of the facility for the fibrous materials production from the melt of thermoplastic materials by means of vertical blow with air.

4. REFERENCES

Dzhigiris D.D. & Mahova M.F. (2002). Bases production basalt filaments and product, Moscow, Teploenergetik

Leskov S.P. The Mini-plants for production of basalt fibres.// Building materials, 2001. № 4 , p. 25 - 26.

Papkov S.P. (1990). The Theoretical bases production chemical filaments. Moscow: Himia

Ponomarev V. B. (2003). Gabbro-basalt raw material for production mineral filament. Moscow - Permi

Sentyakov B.A. & Timofeev L.V. (2004). Technology production heat insulating material on base basalt filament. Izhevsk: Publishers IZHGTU

Annals of DAAAM for 2008 & Proceedings of the 19th International DAAAM Symposium, ISSN 1726-9679
Editor B. Katalinic, Published by DAAAM International, Vienna, Austria 2008
Make Harmony Between Technology and Nature, and Your Mind will Fly Free as a Bird

PRELIMINARY INVESTIGATION ON PERFORMANCES OF A BERNOULLI GRIPPER IN GRASPING LEATHER PRODUCTS

SIKETOVA, K[atarina]; LISKA, J[an]; HORVATH, S[tefan]; RUZAROVSKY, R[oman]; DINI, G[ino] & FAILLI, F[ranco]

Abstract: *The manipulation of leather products (such as plies, bags, wallets, etc.) presents many difficulties mainly due to the high flexibility of the objects, their irregular shapes and, last but not least, the possibility of producing imprints on their delicate surfaces during grasping. To overcome these problems, the use of non-contact grippers can be considered instead of more traditional suction cups or fingered grippers. This paper gives some preliminary results on the use of a Bernoulli gripper for grasping leather objects. In particular, it investigates on the grasping force that this gripper can be exerted on the leather surface, varying some design parameters for different values of air flow. The results have been compared with those ones obtained grasping a rigid and non-porous material.*

Key words:*Leather products, Bernoulli gripper, grasping force, Venturi channels*

1. INTRODUCTION

In leather industry there are many problems related to labour: the environment is critical because of the presence of different chemical products, the leather plies could be heavy and their manipulation for loading the machines could be tiring for operators.

Nevertheless, the automation of these processes presents many difficulties. In this phase the leather plies are very flexible, delicate and their surface is liable to imprint, if grasped by unappropriatetools. In particular, the irregularity of their borders makes impossible the grasping of each single ply from the top of the stack. All these characteristics in practice avoids the use of standard grippers to automate the manipulation of leather plies by robots.

Interesting contributions in the field of flexible object manipulation is given in [Seliger, 2003]. Furthermore, in [Failli, 2004], a specific gripper based on special vacuum cups disposed as a grid has been developed. This technique assures the required grasping forces, their correct distribution on the surface of the ply, and a sufficient delicacy in approaching the object, due to the cup material and to its special internal shape.

However, in some conditions (e.g.: very delicate leather surfaces, wet conditions, etc.) some imprints remain on the surface after manipulation.

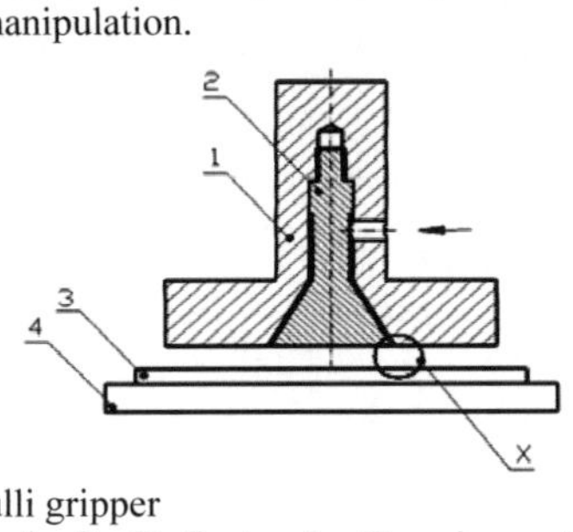

Fig. 1 Bernoulli gripper
1 – Gripper body, 2 – Deflector, 3 – Grasping material, 4 – Work plate, X – Turbulent flow point

Therefore, the study of a non-contact gripper based on the Bernoulli effect seems to be very attractive for this kind of problem.

2. BERNOULLI GRIPPER

Bernoulli grippers (or air flow grippers) are typical non-contact grasping devices [Monkman, 2007]. Some applications have been investigated in literature, such as in [Erzincanli, 1996] and [Davis, 2006], but the use of such devices in grasping leather has to be deeply demonstrated.

2.1 Working principle

The working principle is based on the well-known Bernoulli's law. The air flow moves between the internal gripper ring surface and the deflector. The vacuum for grasping the object is created by both the turbulent flow in point X (Fig. 1) and the increasing of air speed in the gap between the gripper and the object.

2.2 Bernoulli gripper configurations

The design proposal of the Bernoulli gripper includes two basic parts: the gripper body and the deflector. Three air inputs on the external cylinder surface have been realized by reason of obtaining a stable and constant air flow.The aim of the design is to investigate on the influence of: i) different air flow angles; ii) different disk surfaces (Fig. 2). In particular, two different deflectors having an air flow angle respectively of 30° and 60° have been designed and realized. As far as the disk surfaces, two different types have been considered: a simply flat surface and a radially slotted surface, including eight Venturi channel used to increase the vacuum effect (a detail of these channel is given in the lower part of Fig. 2).

In this way, four different gripper configurations result: i) 30°-flat, ii) 60°-flat, iii) 30° slotted, iv) 60°-slotted.

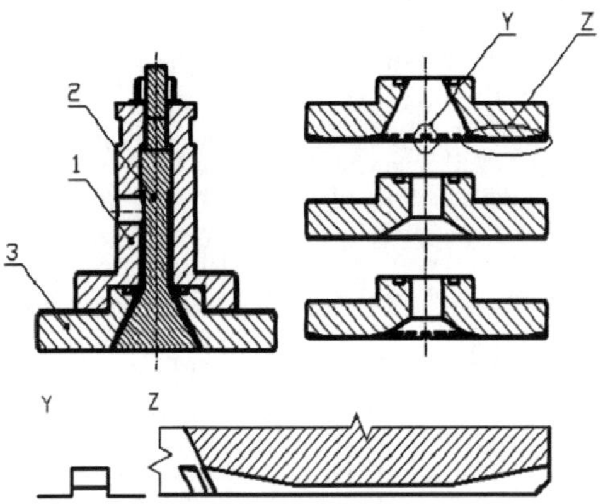

Fig. 2 The design proposal of Bernoulli gripper with interchangeable elements
1 – Gripper body, 2 – Deflector, 3 – Disk

3. EXPERIMENTAL TESTS

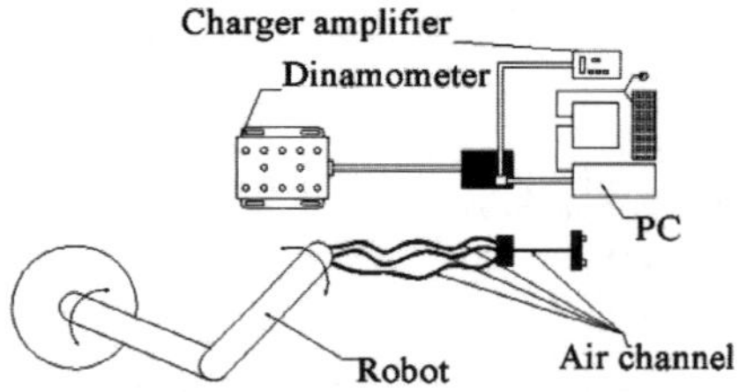

Fig. 3. Experimental setup of the measuring

The aim of the experimental tests is to identify the influence of different process and design parameters on the measurement of the maximal suction force exerted by the gripper.

It is very important to emphasize that these tests are preliminary and concern only the evaluation of lifting force on leather objects in some specific conditions.

3.1 Experimental setup

Experimental tests have beencarried out in the laboratory of Department of Mechanical, Nuclear and Production Engineering of the University of Pisa. The experimental setup is schematically illustrated in Fig. 3 and is mainly formed by a Scara robot, a Kistler dynamometer model 9257B, a charge amplifier and a PC.

Three different objects have been grasped:

- leather A: leather with a thickness of 1 mm (very flexible material);
- leather B: leather with a thickness of 2 mm (flexible material);
- plastic (rigid and non-porous material).

The lifting force has been measured using 3 different values ofair flows (100, 200 and 250 LPM), using the following repetitive cycle:

1. the gripper goes down at a given distance from the object surface (0.5, 1 and 1.5 mm);
2. itstops at the previous distance for 5 s;
3. it comes back to the start position.

3.2 Results

Fig.4 to 7 show some results obtained during experimental tests. Each graph reports the average value of the lifting force (*F-average*) measured during the working cycle in function of the distance between the gripper and the object (*gap*) for 3 different materials.

The following preliminary considerations can be made:

- generally, lifting force increases with gap, with the exception of using the 60° deflector in grasping rigid material;
- the maximum lifting forces measured in grasping leather object are lower than those ones obtained for rigid material (from 25 to 50%);
- the best performances in grasping leather have been obtained using the 60° deflector with a maximum value of about 5 N using a flat disk and about 7 N using a slotted disk. Unsatisfactory values have been obtained using a 30° deflector.

4. CONCLUSIONS

The preliminary experimental tests performed in this study have shown that the performances of the grippers are different using different design parameters and different materials. The presence of the Venturi channels and a 60° deflector gives a positive effect, but further investigations should be made in the future.

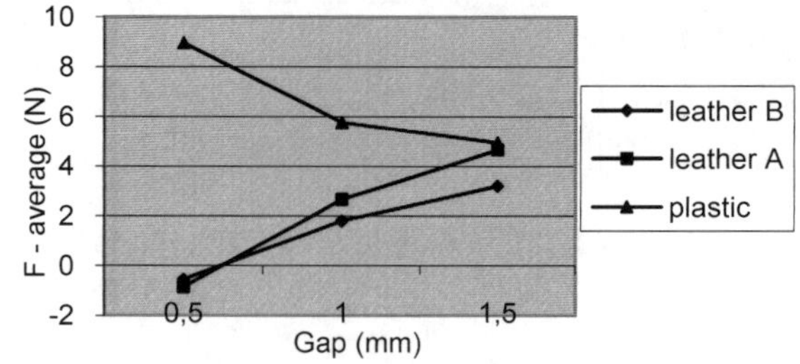

Fig.4 – Lifting force measured using "60°-flat" gripper.

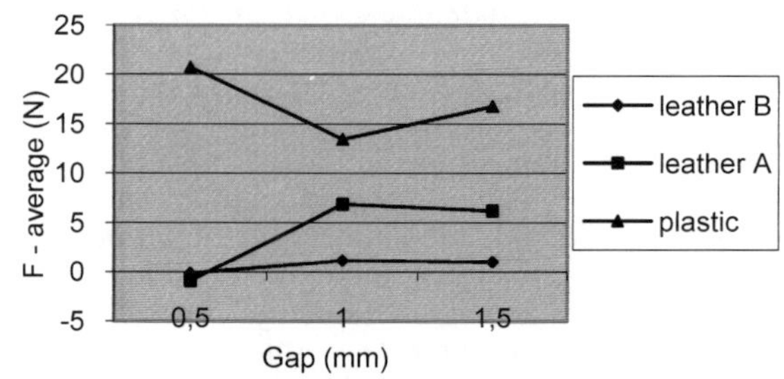

Fig.5 – Lifting force measured using "60°-slotted" gripper.

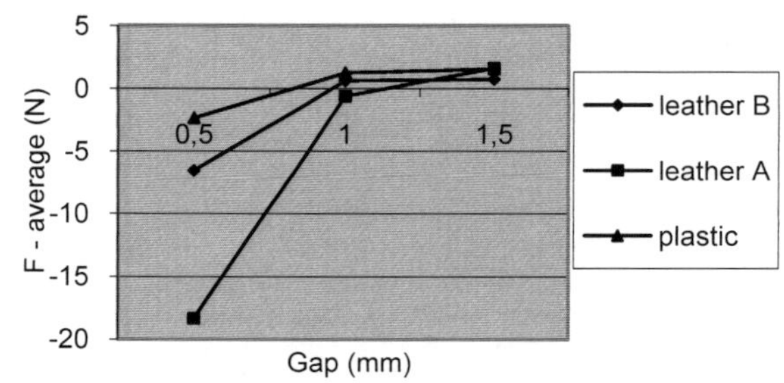

Fig.6 – Lifting force measured using "30°-flat" gripper.

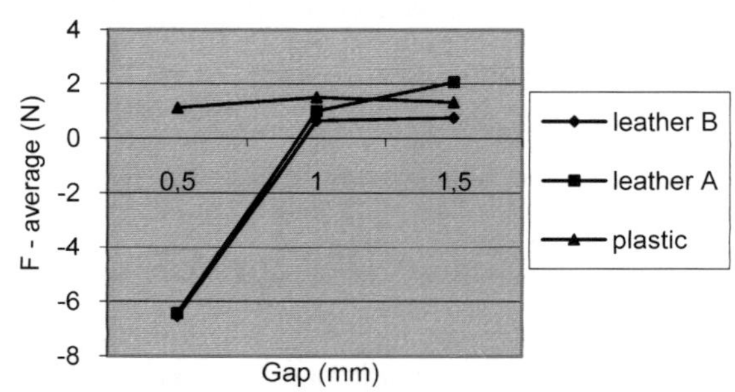

Fig.7 – Lifting force measured using "30°-slotted" gripper.

5. REFERENCES

Davis, S.; Gray, J.O.; Caldwell, G. (2006) An end effector based on the Bernoulli principle for handling sliced fruit and vegetables, *Robotics and Computer Integrated Manufacturing*.

Erzincanli, F.; Sharp, J.M. (1996) Development of a non-contact endeffector for robotic handling of non-rigid materials, *Robotica,* Vol.15, pp 331-335.

Failli, F.; Dini, G. (2004), An innovative approach to the automated stacking and grasping of leather plies, *Annals of the CIRP,* Vol.53/1, pp 31-34.

Monkman, G.J.;Hesse, S.; Steinmann, R.; Schunk, H., (2007), *Robot grippers,* Wiley-Vch.

Seliger, G.; Szimmat, F.; Niemier, J.; Sephan, J. (2003), Automated handling of non-rigid parts, *Annals of the CIRP,* Vol.52/1, pp 21-24.

Annals of DAAAM for 2008 & Proceedings of the 19th International DAAAM Symposium, ISSN 1726-9679
Editor B. Katalinic, Published by DAAAM International, Vienna, Austria 2008
Make Harmony Between Technology and Nature, and Your Mind will Fly Free as a Bird

STATISTICAL APPROACH TO THE DIFFERENCE BETWEEN PUBLIC AND PRIVATE TELEVISIONS IN PRIME TIME TV COMMERCIALS

SIMICEVIC, V[anja] & RACIC, I[figenija]

Abstract: *The purpose of this study is to present media advertising situation in a transitional environment. This paper is focused to statistical analysis of the TV commercials content on the Croatian television channels. The original empirics survey research was conducted at the random sample, including both, public and private television models, in order to analyse TV commercials, as a form of modern economic promoting component. The paper put special emphasis on significant differences of the commercials distribution in prime time, between these two television models.*
Key words: *statistical analysis, advertising, TV commercials*

1. INTRODUCTION

The modern company's marketing environment includes various types of public, where the greatest importance is attached to media public. Most often used promotional tool, for the purpose of achieving the greatest possible communication effect, is the mass media advertising as an important marketing channel through which companies can have a direct influence on their own image and sales (Kotler et al., 2006). As media usage increases month over month, marketers have found a way to use the popular television programs for promotion. The TV commercial has become the dominant one-way communication form that surrounds us, in almost every moment of our lives. The aim of this study was to research the contents and characteristics of TV commercials on the Croatian television channels, considering the differences between two television models (public and private) and apply statistical analysis of TV commercials. The original empirics survey research was conducted on the random sample of Croatian TV channels, in order to realize the aims of this study. The results of the research, as well as suggestions how to involve it in the future media researches, are the scientific contribution of this paper.

2. RECENT RESEARCH

The quantitative analysis of TV commercials has been limited. Some reasons for their limited usage were a lack of archives and because of this a difficulty in searching and browsing them. The research papers that analyse methodological tools are mainly focused on typology of commercial message types and content analysis. Crask (1998) developed a typology of commercial message types involved coding. Coding necessitates a detailed description of the variables and the development of categories that are exhaustive and mutually exclusive (Neuendorf, K. A., 2002). Content analysis (Pegoraro, A. & O'Reilly, N. 2008.) relies on coding of data classified in a conceptual framework, driven by the objectives of the research questions. Intermedia Advertising Group (2001.) has launched the first ever ad effectiveness system, which measures the performance of all prime time advertising on the major broadcast networks on a daily basis.

3. THE TELEVISION MARKET SITUATION

The media industry worldwide is experiencing a significant change in major business direction, from the acknowledged servant to public interest, towards the ever-increasing commercialisation of the sector, where media organizations become first and foremost a profit generator for its owner. The transitional context makes this process more complex because of the undeveloped media regulation, up to the recent political heritage and influences, changes in cultural values, weak economic situation and lower buying power than in developed countries, poor accompanying infrastructure of media researches (Tomašević Lišanin, et al., 2004). Television advertising is a highly significant service industry in the world, and thus also in Croatia. Republic of Croatia, as a country, is highly oriented towards media advertising. The Croatian TV advertising market is to large extent connected to the situation in its broader business and social environment. Market fragmentation has resulted in media fragmentation, in explosion of more focused media that better match today's targeting strategies. The media landscape is constantly evolving. The fragmentation of Croatian media market resulted in the significant increase the number of TV stations. A consequence of the growing dynamics of the media market is the increased influence of advertisers. For this moment, three dominant television media houses exist in Croatia: (1) Croatian (public) television: The First channel (HTV1) and the Second channel (HTV2); (2) Private televisions: NovaTV (the member of Central European Media Enterprises Ltd.) and RTL (formed by Bertelsmann AG group, a media company that originated in Germany). The main difference between these two television models is their financial source and the profit gain.

4. THE METHODOLOGY OF RESEARCH

The aim of this study was to research the advertising frequency and content of TV commercials, as one of modern forms of economic propaganda on the TV channels in Croatia. In order to realize the aims of this study, the original empirical research was conducted at the random sample of Croatian televisions: private and public, by survey research method including aforementioned three dominant television media houses, in prime time. The data were collected by recording method, on a sample size of 1233 TV commercials, which is completely acceptable for this type of research (Kish, 1995). 84-hours of recorded television programming was realized, in prime time, from 19h to 22h, the interval when the most TV audience usually sit in front of TV screens. Advertising variables are defined: (1) Topic of the TV commercial, (2) Target advertising (commercial) group or audience with respect to gender and age reference, (3) Main protagonist or character of the TV commercial, (4) Message appeal (5) Presentation procedure of the TV commercial, i.e. humorous/non-humorous messages. After the classification of commercial main messages, TV commercials characteristics (categories and counties) are testing in this research. The hypothesis of equal distribution of TV commercials is tested, on the public and private television in prime time, in order to research the influence of financial source to TV commercials broadcasting. The descriptive statistics methods and techniques, correlation analysis and hi-square test, were used. All the methods are processed using Statistica software.

5. RESEARCH RESULTS

Three televisions exist on Croatian national level media market: one public television HRT, and two new private, NovaTV and RTL. The survey results based on descriptive statistics methods indicate the most frequently used statistics: the average duration of TV commercials in prime time, is 19 seconds. The longest duration of the message is 120 seconds (HTV1), and the shortest is 3.39 seconds (NovaTV). Most advertising is intended to adults, men and women, equally. Great deal of advertising does not use humour, as a process of exposure, meaning, most of them are non-humorous messages. The analysis of the main area of interest, variable topic of the TV commercial, shows that NovaTV advertises mostly hygienic and cosmetic products (45 %) and the RTL products related to the consumption of food and beverages (35%). On public television, advertising of products is related to food and drink is 25 per cent (HTV1), or 27 per cent (HTV2). Distribution of advertisements on the topic, according to public television and TV commercial, shows a high positive correlation. The communicator has to figure out an appeal (rational, emotional or moral) theme that will produce the desired response. The commercials focused on health (36%), dominated on NovaTV, and the RTL-to those that appeal on youth and beauty (22%). The commercials that appeal to social success and prestige dominate on public television (27%). According to the research results, there is a strong (Reinard, J.C. 2006.) positive correlation between commercial's distribution by topic on the public and private television channels (r = 0,680472). A Chi-square test of the equal probability hypothesis was conducted in the purpose of which showed that there was statistically significant relation between the television models and TV commercials distribution. Results of the analysis showed that null hypotheses is rejected, what indicates the statistically significant (with an alpha risk of 0.05) relation between the television models (public and private) and TV commercials distribution in prime time, meaning that private televisions, NovaTV and RTL, broadcast, significantly more TV commercials in prime time than the public channels. The researched results, as well as, confirmation of alternative hypothesis are expected, for the reason that TV commercials are the main source of the profit gain. TV commercial statistics and data on insertion frequency, time length in seconds provide with a comprehensive grasp of the competitors and the advertising environment.

6. THE INFLUENCE OF TV COMMERCIALS

In the new Millennium, the consumer material fascination for materials goals is progressing. Commercials, as well as other media messages, are only instrument, and can be used in a good or a harmful way. Any restrictions and warning consumers will not cancel advertising appeal. Three essential elements are required for successful advertising: (1) information (uniqueness and innovativeness of the products that are sold), (2) reasoning (encouragement to think about advertisers message in the direction of interest and estimation of products) and (3) emphasis (highlighting the most attractive topics of the messages (Sheehan, K. B., 2004). People can vary according to the orientation values. The companies make great efforts in researching characteristics of the pioneers of consumption products (the first 2.5 per cent of the buyers to adopt a new idea; those beyond two standard deviations from mean adoption time) and the early adopters (the next 13.5 per cent between one and two standard deviations), and direct even aggressive communication by them, while other individuals accept new products much later. After a slow start, an increasing number of people adopt the new product (Rogers, M. E., 2004). In spite of aggressive advertising process, people fortunately, have strong defence mechanisms against advertising and other marketing tools.

7. CONCLUSIONS

Modern, large companies make contemporary efforts in researching TV commercial's characteristics, especially because the TV commercials are generally considered the most effective mass-market advertising format. For that purpose, the theoretical and methodological basis of TV advertising situation is researched in this article, which put special emphasis on significant differences of the TV commercials content distribution in prime time. Results of the analysis confirmed: (a) a strong positive correlation between TV commercials' distribution by topic on the public and private television channels; (b) the private channels, as a profit generator for its owners, broadcast significantly more TV commercials in prime time than the public channels. Research in the applied statistical methods in Croatian media has been rare, i.e., only few have been published (Jurišić et al., 2007). In this context, this study provides a preliminary platform for future media research.

8. REFERENCES

Crask, M. R. (1989) Typology of Main Message Strategies for Television Commercials. *Available from*: Journal of Advertising, *Accessed*: 1989-01-01.

IAG - Intermedia Advertising Group.(2001). IAG Launches TV Ad Effectiveness System. *Available from*: Business Wire *Accessed*: 2001-12-10.

Jurišić, J.; Kanižaj, I.; Jokoš, I.; Janeš, S. & Jurić, J. (2007). Manipulating Readers: Disguised Advertising in Croatian Newspapers, *Politička misao*, Vol. 19, No. 1, (3, 44) pp. 117-137, ISSN: 0032-3241.

Kish, L. (1995). *Survey Sampling,* John Wiley & Sons, ISBN: 978-0-471-10949-5, New York, USA

Kotler, P.; Wong, W.; Saunders, J. & Amstrong, G. (2006). *Principles of Marketing*, FT Prentice Hall, ISBN: 0273684566, London, UK

Neuendorf, K. A. (2002). *The Content Analysis Guidebook*. Sage Publications, Inc. ISBN: 0-7619-1978-5, London, UK

Reinard, J. C. (2006). *Communication Research Statistics*, Sage Publications, Inc. ISBN: 0-7619-2987-8, London, UK

Rogers, M. E (2004). *Diffusion of Innovations,* The Free Press, ISBN: 0-02-926671-8, New York, USA

Sheehan, K. B. (2004). *Controversies in Contemporary Advertising,* Sage Publications, Inc. ISBN: 13-978-0761926351, London, UK

Pegoraro, A. & O'Reilly, N. (2008). An analysis of television commercials during tennis grand slam finals: Does the equal prize money gender equality in advertising? *Available from:*http://www.nassm.com/files/conf_abstracts/2008-320.pdf *Accessed*: 2008-05-31.

Tomašević Lišanin, M.; Vranesevic, T.& Palic, M. (2004). Acquisition in the Print Media Sector: The Case of Austrian "Styria Medien" and Croatian "Vecernji list", *International Journal of Management Cases,* Vol.7, No.2, pp. 15-29.(6/7), ISSN: 1741-6264

Annals of DAAAM for 2008 & Proceedings of the 19th International DAAAM Symposium, ISSN 1726-9679
Editor B. Katalinic, Published by DAAAM International, Vienna, Austria 2008
Make Harmony Between Technology and Nature, and Your Mind will Fly Free as a Bird

CAD SYSTEM FOR CALCULATING ORIENTATION ERRORS

SIMION, I[onel]; DOBRE, D[aniel] & MARIN, D[umitru]

Abstract: *The inaccurate placement of the workpiece relative to the cutting tool can negatively affect the quality of the processed part. An original system for computer aided determination of the orientation error is presented in this paper. The system is built starting from an original mathematical model and allows the analysis of complex errors. The model can be used inside a CAD system for fixture design.*
Key words: *CAD, orientation error, fixture design.*

1. INTRODUCTION

In most cases, the determination of errors that appear to orientation process of parts in devices for processing presumes laborious calculations and assumes a good experience for designers. This process is often made by approximation, reffering at samples existing in literature which are more or less appropriate with the concrete design case.

We started from the typology of real surfaces from the workpiece and from the locator element. We analyzed the relative position of bases (geometrical elements defined as plans, lines and points associated with the surfaces). Following these, we built a database containing graphic patterns used as a library schemes targeting, by encoding the information.

The library scheme presented in this work is built using a mathematical model based on the theory of transformation of coordinates (Simion, 1995). The results from our earlier work show that the proposed math model is compatible with the standards and that it provides three-dimensional relations for orientation error. The project includes the implementation of the error analysis system into a CAD system.

2 LITERATURE REVIEW

Various computer-aided fixture design methods have been developed through the years to assist the fixture designer. One of the approaches consists in using relevant design experience from a design library and adapting it to provide a new fixture design solution. Another approach is based on simulating the orientation process, in order to select the optimal scheme.

From another point of view, some of the approaches can be based on geometric patterns. Other approaches are based on cinematic models. Asada and By (1985) created the Jacobian Matrix to model the 3D fixture-workpiece relationship. Xiong (1993) applied the kinetic model from multi-fingered robot hand grasping problem to the fixture configuration. Rong et al. (1996) have a series of studies on tolerance and stability analysis. Zhang et al. (2001) analyzed the locating error and tolerance assignment for computer-aided fixture design. Bragaru (1998) introduced a formula for calculating the vector error guidance, based on the relative position of bases.

3 THE BASIC MODEL

In order to cover possible situations, the database has been structured for the following levels of diversification:

- level 1 - depending on the number of orientation surfaces;
- level 2 - depending on the orientation surfaces type: plane, cylinder, cone, sphere;

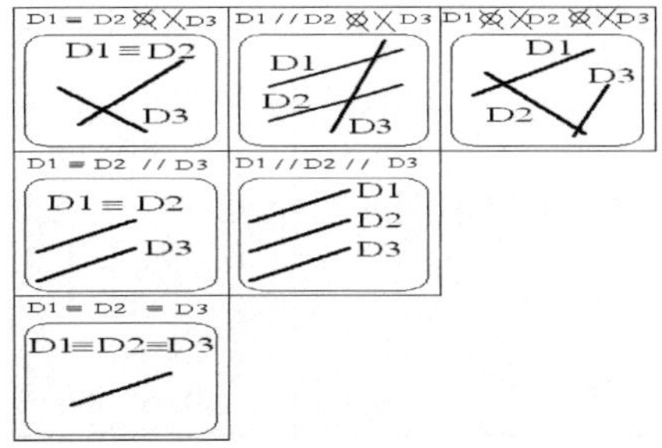

Fig. 1. Three bases of the same type (line).

- level 3 - according to the typology and the relative position of associated bases: plane (Γ), line (D), point (P); for example, Fig.1 shows the analysis of the case of three bases of the same type (line);
- level 4 - in a graphic model there are more technically possible oriented schemes, differentiated by the type of the dimension that is analyzed in order to determine the orientation error and by the symbols of locators.

In Fig. 2 there is presented a sample model from a database built as such.. The orientation schemes shown in this figure is modelled on an inside or outside cylinder surface (the orientation base is an axis, meaning a fictional line) and on a flat surface (the orientation base is a plane which can be real or fictional).

Table 1 is associated to the model in Fig. 2. The components of error orientation are defined as variations of the relative positions between the dimension bases (DB) and the active bases (AB), projected on the direction of the dimension for which the orientation error is determined (Simion, 1995).

We used the following notations: TDp is the tolerance for the diameter of the dimension surface; TA is the tolerance for dimension A; Ai is the minimum effective dimension A; J1max is the maximum clearance between the workpiece and the locator; γ is the half-angle of the V-block; δ is the angle between measuring direction and the bisector plane of the V-block.

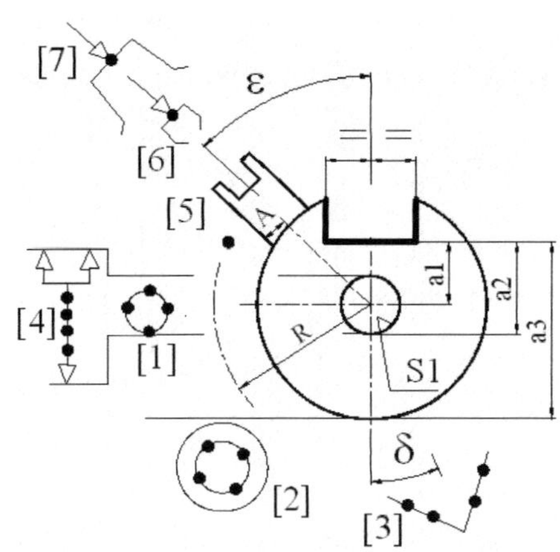

Fig. 2. Orientation on two surfaces: cylinder and plane.

Orientation schemes	DB component	AB component
Dimension type: a1, a2, a3		
[1]+[5],[6],[7]; [2]+[5],[6],[7]	0 for dimension **a1**, TDp/2 for dimensions **a2, a3**	$J1 \max \cos \varepsilon$
[3]+[5],[6],[7]		$TDp \cos d / 2 \sin \gamma - Dp \cos d \cos gT\gamma / 2 \sin^2 \gamma$
[4]+[5],[6],[7]		0
Dimension type: ε		
[1]+[5]; [2]+[5]	$\frac{TA}{R + Ai}$	$\frac{J1 \max}{R + Ai}$
[1]+[6],[9]; [2]+[6],[7]	0	$\frac{J1 \max}{R + Ai}$
[3]+[5]	$\frac{TA}{R + Ai}$	$\mathrm{arctg}(\frac{TDp \cos \delta}{2 \sin^2 \gamma R} - \frac{Dp \cos gT\gamma}{2 \sin^2 \gamma R})$
[4]+[5]	$\frac{TA}{R + Ai}$	0
[3]+[6],[7]; [3]+[6],[7]	0	0
Dimension type: =		
[1],[2]+[5],[6],[7]	0	$J1 \max \cos \varepsilon$
[3],[4]+[5],[6],[7]	0	0

Tab. 1. The orientation error components associated to the model in Fig. 2.

4. THE ERCAD SYSTEM

The original ERCAD software for automate determination of the orientation error is designed for schemes involving one, two or three orientation surfaces and includes a database comprising over 5000 orientation schemes type.

The program operates for dimensions of parts between 0 and 500 mm and for precision between IT5 and IT12. The system operates in an AutoCAD environment and has its own menus, which are based on the types of schemes. Depending on the data explicitly introduced by the user or taken from the formal description of the workpiece (dimensions, tolerances etc.), the system automatically calculates and displays the specific orientation error.

Logical sequence of steps is shown for the model in Fig. 3, referring to the determination of orientation error for dimension $t = 90°$ (the workpiece is oriented on two cylindrical surfaces).

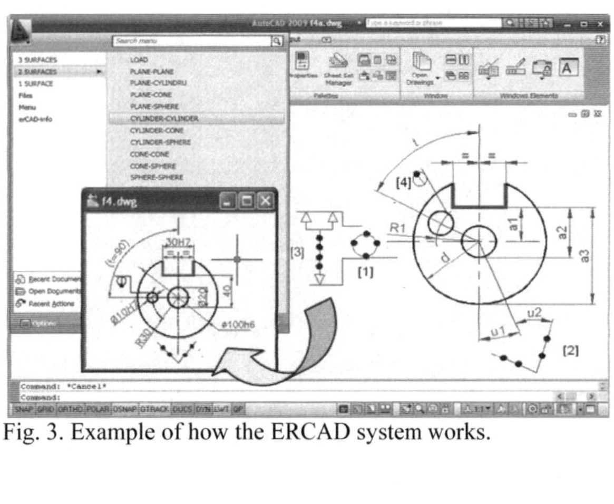

Fig. 3. Example of how the ERCAD system works.

Step 1: The user decides the number of surfaces on which the workpiece is oriented, by selecting the "2 SURFACES" menu.

Step 2: The user decides the type of orientation surfaces. In this case the appropriate option couple "CYILINDER-CYLINDER" must be selected.

Step 3: Depending on the relative position of the two orientation surfaces (parallelism in this case) the user chooses the appropriate graphical model. As a result, the model is displayed, as in Fig. 3.

Step 4: The user is required to confirm the choice of the appropriate model. Assuming an affirmative answer the user will proceed to the next step.

Step 5: As the model usually contains several possible schemes, the user is required to choose the scheme for which the error is calculated. The prompter indicates possible schemes to avoid choosing an impossible scheme: "*Choose scheme targeting (1+7/2+7/3+7/4+7):*". In this case, "3 +7" is to be written usig the keyboard.

Step 6: It is required to choose the dimension type (linear or angular and the direction) for which the error is determined. For the presented case, as a response to the prompter "*Choose dimension (a1/a2/a3 / = / t):*" the option "t" is to be typed.

Step 7: The system requires the introduction of user data regarding size, shape and relative position, needed for calculating the specific orientation error, according to the associated database.

Step 8: The prompter displays the calculated value of orientation error: "*The error is 0.1215 degrees*".

5. CONCLUSION

This paper presents an original system for computer aided determination of the orientation error. Note that if the presented system is integrated within a comprehensive fixtures design system, all the values introduced by the user (the type of orientation scheme, the nominal size, precision etc.) are automatically taken from previous design activities.

A second remark refers to the correctness of values calculated using the described software. Relations obtained by the method of coordinates processing and placed in the database were verified by data from literature, where they existed, or by applying other methods of calculating the orientation errors. The results were validated.

We anticipate that successful completion of the project will provide the designers a helpful tool for making better decisions about optimizing the fixture design.

6. REFERENCES

Asada, H. & By, A. (1985). Kinematic Analysis of Workpart Fixturing for Flexible Assembly with Automatically Reconfigurable Fixtures, *IEEE J. of Robotics and Automation*, Vol. 1, pp. 86-94.

Brăgaru, A. (1998). *Proiectarea dispozitivelor (Fixture Design)*, Editura Tehnică, ISBN 973-31-0717-4, Bucharest.

Rong, Y. & Bai, Y. (1996). Machining Accuracy Analysis for Computer- Aided Fixture Design, J. *of Manufacturing Science and Engineering*, Vol. 118, pp. 289-300.

Simion, I. (1995). *Research concerning the precision of the orientation schemes*, Ph.D.Thesis, University "Politehnica" from Bucharest.

Xiong, Y. L. (1993). Theory and Methodology for Concurrent Design and Planning of Reconfiguration Fixture, *Proceedings - IEEE* International Conference on Robotics and Automation, Vol. 3, May 2-6, pp. 305-311.

Zhang,Y.; Hu, W.; Kang, Y.; Rong, Y. & Yen, D. W. (2001). Locating error analysis and tolerance assignment for computer-aided fixture design, *International Journal of Production Research*, Vol. 39, No. 15, pp. 3529-3545.

Annals of DAAAM for 2008 & Proceedings of the 19th International DAAAM Symposium, ISSN 1726-9679
Editor B. Katalinic, Published by DAAAM International, Vienna, Austria 2008
Make Harmony Between Technology and Nature, and Your Mind will Fly Free as a Bird

FORMAL MODEL FOR DESCRIBING ORIENTATION ERRORS

SIMION, I[onel]; RAICU, L[ucian] & ARION, A[urel]

Abstract: *Orientation error specifies the ranges of permissible variations of the relative position between the basis systems so that a manufactured part will be correctly oriented. In this research, mathematical model based on coordinates transformation theory is developed to analyze locating variations and their influences on orientation error. The model can be used inside a CAD system for fixture design.*
Key words: *fixture design, orientation error.*

1. INTRODUCTION

One of the most critical tasks in fixture design is to determine the orientation error of the workpiece within the locators system. As one of the deterministic components of a machining error, orientation error is primarily caused by size and position variations both from locating elements on a fixture side and from locating features on a workpiece side. Because of the diversity of part shape, locating feature form and tolerance specification, there is no comprehensive solution to estimate orientation error.

According to Bragaru (1998) the accuracy of the piece orientation into the device is determined by the variation of the relative position of the basis systems belonging to the workpiece and to the support element, as in Fig.1. The basis are reference points, lines or planes which are assumed to be exact. They are established based on real features. For example, Fig.2 shows the basis system for a cylindrical workpiece oriented on V-block.

Our earlier works (Simion, 1995) presented a model based on coordinates transformation in order to automate the calculation of the orientation error. This model is based on the mathematical coordinates transformation theory and can be used into a CAD system.

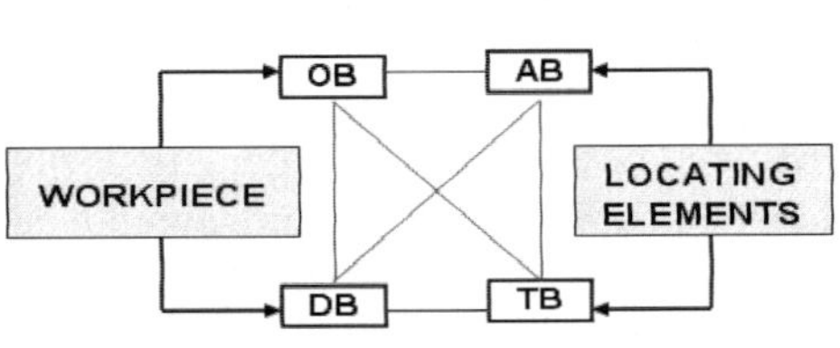

Fig. 1. Basis system. OB – Orientation Base, DB – Dimensioning Base, AB – Active Base, TB – Tuning Base, GS – Generating Surface.

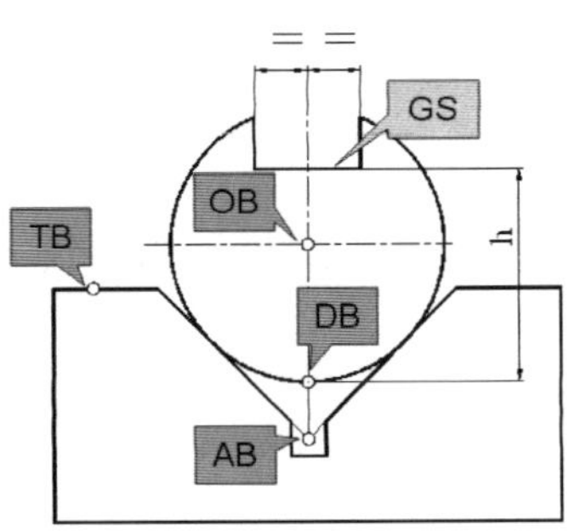

Fig. 2. Workpiece oriented on V-block.

2. LITERATURE REVIEW

Many works in the field of fixture design dealed with determining the precision of workpiece location. Various methods for the determination of workpiece location have been developed. Cai et al (1997) used variational methods to model the workpiece resultant error. Kumar et al. (2000) used a neural network approach to conceptually design complete fixture units. Kang et al. (2003) used two models (geometric and kinetic) to verify the fixture design. Wang (2002) developed a tolerance analysis method to assist fixture layout design for 2D workpieces. Bragaru (1998) used a formula for calculating the vector guidance error, based on the relative position of bases. Zhang et al. (2001) analyzed the locating error for computer aided design.

3. THE MATHEMATICAL MODEL

Let $(Oxyz)$ be the reference system, connected to the locating elements that determine the piece orientation and let $(O'x'y'z')$ be the reference system connected to the workpiece, as in Fig.3.

The relationship between the rectangular coordinates (a,b,c) and $(a'b'c')$ of a point in the two systems, is established by the formula:

$$\begin{bmatrix} a' \\ b' \\ c' \end{bmatrix} = \begin{bmatrix} c_{11} & c_{12} & c_{13} \\ c_{21} & c_{22} & c_{23} \\ c_{31} & c_{32} & c_{33} \end{bmatrix} \begin{bmatrix} a \\ b \\ c \end{bmatrix} + \begin{bmatrix} a_0' \\ b_0 \\ c_0 \end{bmatrix} \qquad (1)$$

with (a_0', b_0, c_0) being the coordinates of the origin of the $(O'x'y'z')$ system in the $(Oxyz)$ system.

The orientation error represents the extreme values of the coordinates variations v_a, v_b and v_c, considering that:

$$v_a = i(a'-a), v_b = i(b'-b), v_c = i(c'-c). \qquad (2)$$

The value of the "i" coefficient is 1 or 2, corresponding to the situation when the workpiece can move on the support elements in one, respectively two senses, in the error's direction.

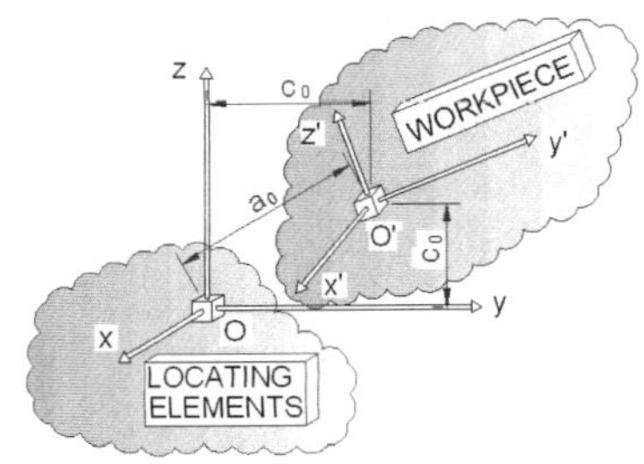

Fig. 3. The coordinates transformation method for rectangular dimensions.

From the analytical equations of the functions (2) there results the orientation error formula.

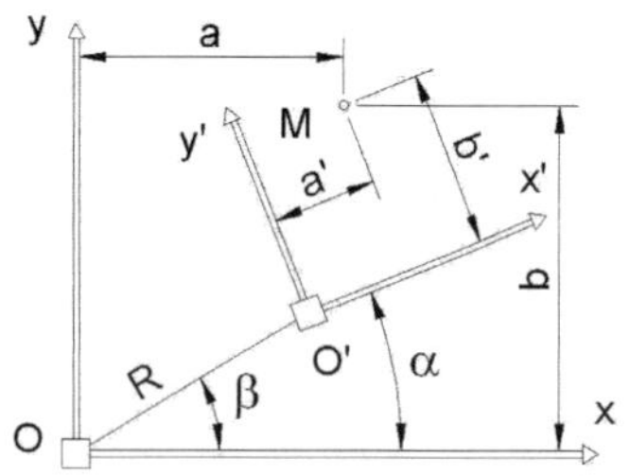

Fig. 4. The coordinates transformation method for rectangular dimensions.

For example, in 2D, for rectangular coordinates, (1) can be written as:

$$a' = a\cos\alpha + b\sin\alpha - r\cos(\beta - \alpha)$$
$$b' = -a\sin\alpha + b\cos\alpha - r\sin(\beta - \alpha) \quad (3)$$

Where: (a,b) respectively (a',b') are the coordinates of a point M; α is the angle between Ox and O'x' axis; (R, β) define the position of the origin of the $(O'x'y')$ system in the (Oxy) system and r is the variation of the R variable – Fig.4.

By substituting (3) in (2) we finally obtain the orientation errors for the dimensions a and b:

$$\varepsilon_o(a) = \max\{i[a(\cos\alpha - 1) + b\sin\alpha - r\cos(\beta - \alpha)]\}$$
$$\varepsilon_o(b) = \max\{i[-a\sin\alpha + b(\cos\alpha - 1) - r\sin(\beta - \alpha)]\} \quad (4)$$

Fig. 5 shows two examples of orientation error determined by using the coordinates transformation method. The schemes are build on a single orientation surface.

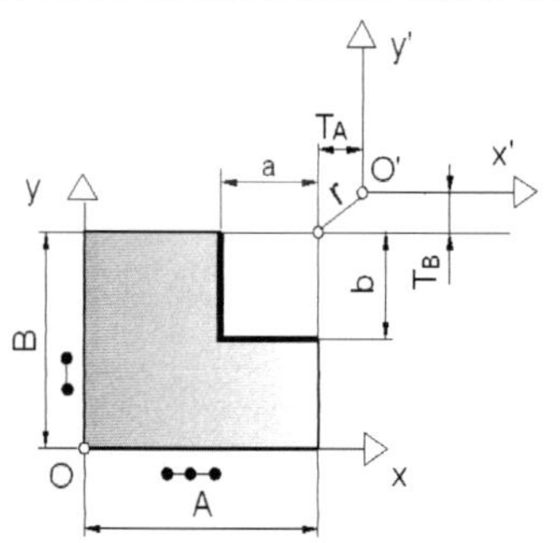

$r = \sqrt{T_A^2 + T_B^2}$,
$\alpha = 0$, $i = 1$
$\beta = \text{arctg}(T_B / T_A)$,
Coordinates variation:
$v_a = -r\cos\beta = -T_A$,
$v_b = -r\sin\beta = -T_B$.
Orientation error:
$\varepsilon_o(a) = T_A$,
$\varepsilon_o(b) = T_B$.

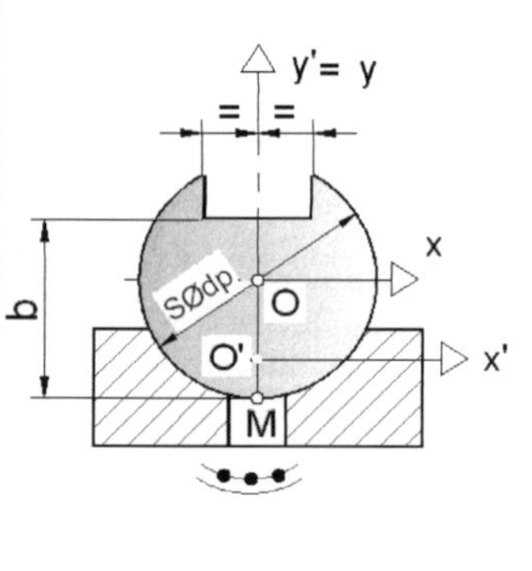

$r = T_{MO}$, $\alpha = 0$,
$\beta = \pi/2$, $i = 1$.
Coordinates variation:
$v_= = 0, v_b = -r$,
where: $r = T_{dp}/2$.
Orientation error:
$\varepsilon_o(=) = 0$;
$\varepsilon_o(b) = \frac{T_{dp}}{2}$.

Fig. 5. Orientation schemes based on single surfaces.

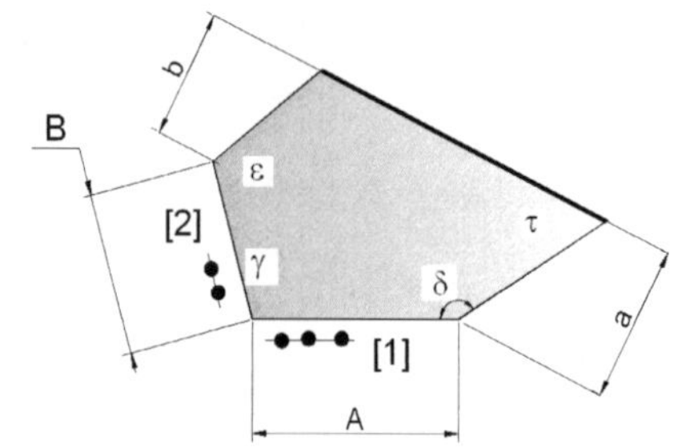

i=1, $\alpha = 0$, $\beta = 0$
$r_1 = T_A\sin(\delta + \tau) + A\cos(\delta + \tau)(T_\delta + T_\tau)$
$r_2 = T_B\sin(\gamma + \delta + \tau) + B\cos(\gamma + \delta + \tau)(T_\gamma + T_\delta + T_\tau)$.
Orientation error:
$\varepsilon_o(a) = T_A\sin(\delta + \tau) + A\cos(\delta + \tau)(T_\delta + T_{\tau\_})$
$\varepsilon_0(b) = T_B\sin(\gamma + \delta + \tau) + B\cos(\gamma + \delta + \tau)(T_\gamma + T_\delta + T_\tau)$.

Fig. 6. Orientation schemes based on two surfaces.

Fig. 6 illustrates the orientation error for a complex scheme, based on two orientation surfaces. The error is also determined using the proposed model.

4. CONCLUSION

This paper presented an original model describing the workpiece location error caused by fixture geometric errors.

The designed model was used to create a database and a specialized software for analyzing the orientation error. This software vas verified on particular design situations. The results were confirmed by comparison with the existing data in literature.

However, this work has not reached it's limits and can be further exploited. The model was used inside a specialized CAD system for fixture design (Simion, 1995). Future work will address further experimental validation of the model by applying it to different fixture layouts and varying part geometries.

5. REFERENCES

Brăgaru, A. (1998). *Proiectarea dispozitivelor (Fixture Design)*, Editura Tehnică, ISBN 973-31-0717-4, Bucharest.

Cai, W.; Hu, S.J. & Yuan, J.X. (1997). A Variational Method of Robust Fixture Configuration Design for 3-D Workpiece, *Journal of Manufacturing Science and Engineering*, Vol. 199, pp. 593-602.

Kumar, A.S.; Subramaniam, V. & Teck, T.B. (2000). Conceptual design of fixtures using machine learning techniques, *International journal of Advanced Manufacturing Technology*, Vol. 16, pp. 176-181.

Kang, Y.; Rong, Y. & Yang, J-C. (2003). Computer-aided fixture design verification, *International Journal of Advanced Manufacturing Technology*, Vol. 21(10-11), pp. 827-849.

Simion, I. (1995). *Research concerning the precision of the orientation schemes*, Ph.D.Thesis, University "Politehnica" from Bucharest.

Wang, M. (2002).Tolerance analysis for fixture layout design, *Assembly Automation*, Vol. 22, pp. 153-162.

Zhang,Y.; Hu, W.; Kang, Y.; Rong, Y. & Yen, D. W. (2001). Locating error analysis and tolerance assignment for computer-aided fixture design, *International Journal of Production Research*, Vol. 39, No. 15, pp. 3529-3545.

Annals of DAAAM for 2008 & Proceedings of the 19th International DAAAM Symposium, ISSN 1726-9679
Editor B. Katalinic, Published by DAAAM International, Vienna, Austria 2008
Make Harmony Between Technology and Nature, and Your Mind will Fly Free as a Bird

AN ANALYTICAL APPROACH OF THE APPEARING OF THE BLOWHOLES IN THE CASTS

SIMIONESCU, G[heorghe] & CIUCESCU, D[oru]

Abstract: *The blowholes represent discontinuities in the walls of the casts and, for this reason, conduct to the diminishing of the mechanical properties, wherefrom the importance of these defects. The appearing of the blowholes depends on many factors. This paper tries to analyze some main factors influencing the appearing of the blowholes in the casts.*
Key words: *gas solubility, partial pressure, viscosity.*

1. INTRODUCTION

All types of discontinuities in the walls of the casts, including the blowholes, conduct to the diminishing of the mechanical properties. When the blowholes have a prolonged or an angular form, the negative effect is greater (Facy & Pompidou, 1997).

The blowholes may be provoked both by the alloy as well by the mould. Those blowholes due to the alloy are distributed either in whole volume or in the superficial layers, when those blowholes due the mould are distributed in only in the superficial zones (Askeland, 1984; John, 1972).

In this paper are analyzed some main factors influencing the appearing of the blowholes in metallic casts (Ciucescu, 2001).

2. THE FORMING MECHANISM OF THE BLOWHOLES

The forming mechanism of the blowholes consists in forming and raising up of the bubbles in the liquid alloy. The germination takes place, usually, in the pores of the mould sand, of the inclusions from the liquid alloy, as is shown in figure 1 (Simionescu et al., 2001).

In the proximity of a pore appears a cupola, which begins to grow. This cupola elongates vertically where the metal – static pressure is less. At the base the cupola begins to get thinner till its detachment in a form of a bubble. Then this process is repeated (Ciucescu, 2007).

If the viscosity is not great, the bubbles achieve to leave the liquid alloy otherwise, the bubbles remain in the alloy forming blowholes (Ciucescu, 2001).

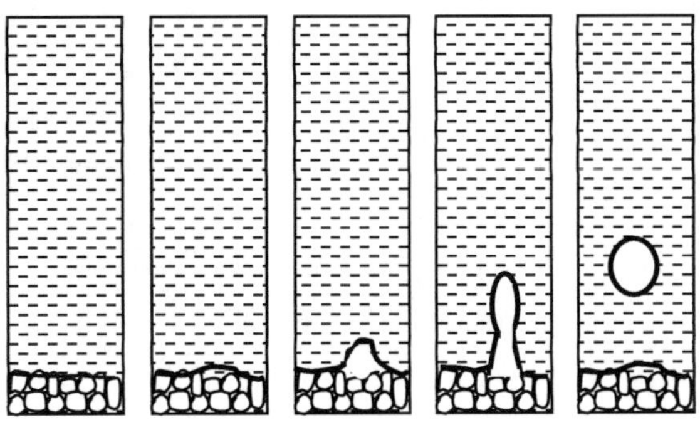

Fig. 1. Stages in the forming mechanism of the blowholes

Firstly, the forming of the blowholes depends on the partial pressure of the gases. The total pressure (p_t) of the gases is the sum of the partial pressures of each gas:

$$p_t = p_{H_2} + p_{N_2} + p_{CO} + p_{CO_2} + \dots \quad (1)$$

where: p_{H_2} is the partial pressure of hydrogen;
p_{N_2} - the partial pressure of nitrogen;
p_{CO} - the partial pressure of carbon oxide;
p_{CO_2} - the partial pressure of carbon dioxide.

To this pressure is opposed the sum of atmosphere pressure (p_{at}), metal-static pressure (p_m) and the superficial pressure of the alloy (p_{ts}):

$$p_{ext} = p_{at} + p_m + p_{ts} = p_{at} + \rho_m \cdot g \cdot h_m + \frac{2\sigma_m}{r} \quad (2)$$

where: p_{ext} is the external pressure;
ρ_m - the density of the alloy;
g - the gravity acceleration;
h_m - the height of the column of the liquid alloy;
σ_m - the superficial tension of the alloy.

If:

$$p_{gas} \rangle p_{ext} \quad (3)$$

the bubbles will go out of the liquid alloy and, sometimes, the metallic surface is agitated, giving the impression of "boiling".
If:

$$P_{gas} \langle p_{ext} \quad (4)$$

the bubbles are do not form and the gases remain in liquid alloy.

If the bubbles are forming in the liquid alloy, the capillar pressure (p_c) which is necessary to be surpassed to form and to increase these bubbles, may be determined by the relation:

$$p_c = \frac{2\sigma_m \cdot \cos\theta}{r_{cr}} \quad (5)$$

where: θ is the humectation angle of the bubble by the alloy;
r_{cr} - the critical ray of the germ of the bubble.

The mechanical work (L) to form the bubbles is given by the relation:

$$L = \frac{4\sigma_m r_{cr} \cdot \sigma_{gm}}{3} \quad (6)$$

where: σ_{gm} is the superficial tension at the surface gas - metal.

The forming intensity of the bubbles is given by the relation:

$$I = C \cdot e^{-\frac{4\pi \cdot r_{cr}^2 \cdot \sigma_{gm}}{3R \cdot T}} \quad (7)$$

where: C is the proportionality coefficient;
R - the constant of Boltzman;
T - the temperature of the alloy.

In the case of a little viscosity of the alloy and the lack of convection currents, the critical ray of the bubble in the moment of its detaching is given by the relation:

$$r_{cr} = \sqrt{\frac{3r_p \cdot \sigma_m}{2g \cdot (\rho_m - \rho_g)}} \quad (8)$$

where: ρ_g is the density of the gas.
r_p - the ray of the pore.

The raising speed of the bubble (V_B) is given by the relation:

$$V_b = \frac{2}{9} g \frac{1}{\eta_m} r \cdot (\rho_m - \rho_g) \quad (9)$$

where: r is the ray of the bubble;
η_m - the dynamic viscosity of the alloy.

The condition that the bubble can detach itself from the germination place is given by the relation:

$$\tau_{cr} = \frac{h_m}{V_b} \quad (10)$$

where: τ_{cr} is the making time of the metallic layer whose the thickness is enough great not to be penetrated by the bubble.

In the figure 2 it is represented the forming of the sieve – type bubbles in an ingot of steel.

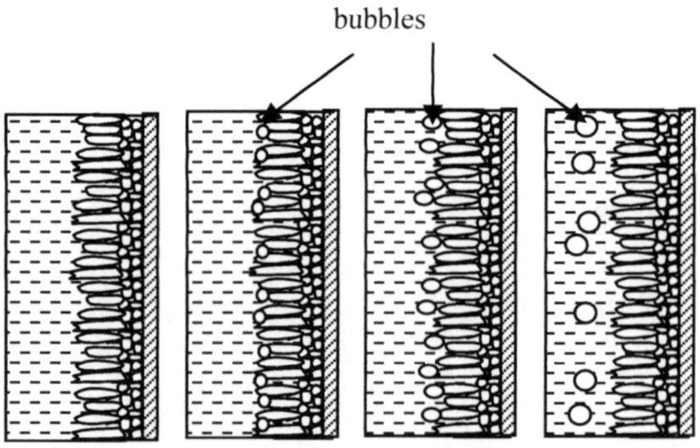

Fig. 2. The forming of the sieve – type bubbles in an ingot of steel (Simionescu et al., 2001).

The proper dimension of the diameter of the whistler (d) is given by the relation:

$$d = \frac{s}{2} \quad (11)$$

where: d is the thickness of wall of the cast;
s - the diameter of the whistler.

In the figure 3 is represented the best dimension of the whistler.

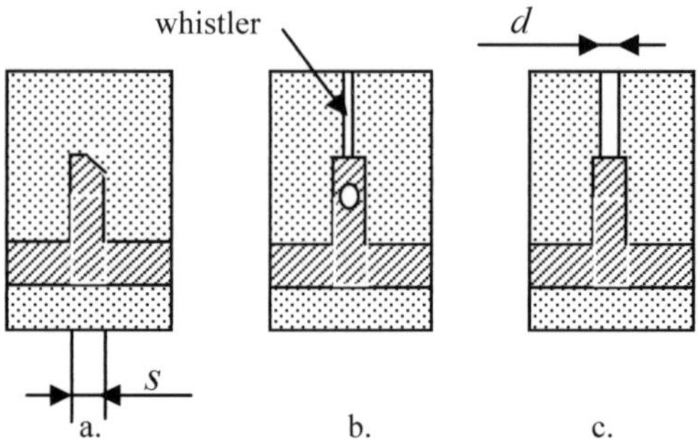

Fig. 3.The representation of the proper dimension of the whistler: a. without the whistler; b. with an under-dimensioned whistler; c. with a proper whistler.

3. CONCLUSIONS

The appearing of the blowholes is a very stringent problem in the casts (Ciucescu, 2003).

The germination and the growing up of the bubbles in the liquid alloy is a very complex phenomenon. The available relations for calculation to avoid blowholes are not always used by lack of data concerning the exact values of the viscosity and the superficial tension (Petrescu et al., 2000).

In the most cases are taken experimental relations as the relation (11).

4. REFERENCES

Askeland, R.D. (1984), *The Science and Engineering of Materials,* ISBN 0-534- 029757-4, PWS-KENT Publishing Company, Boston, Massachussetts

Ciucescu, D. (2007), *The Science and Engineering of Materials,* ISBN 973-30-1465-6, E.D.P., București

Ciucescu, D. (2001), *The Study of Materials * Metallic Materials,* ISBN 973-9362-68-2, Plumb Publishing Company, Bacau

Ciucescu, D. (2001), *The Study of Materials * Non-metallic Materials,* ISBN 973–9362–70-2, Plumb Publishing Company, Bacau

Ciucescu, D. (2003), *Elements of Physical Metallurgy,* ISBN 973– 634–015-5, Plumb Publishing Company, Bacău

Facy, G. & Pompidou, M. (1997), *Fondery Handbook*, ISBN 2-12-263221-6, AFNOR, Paris

John, V.B. (1972), *Introduction to Engineering Materials,* Mac-Millan Publishing Company, ISBN 0-02-949140-1, New-York

Petrescu, M.; Petrescu, I.; Calin, M. & Petrescu, N. (2000), *Metals, Ceramics, Polymers,* ISBN973-98857-0-5, "Tipografia" Publishing Company, Bucharest

Simionescu, Gh.; Cernat, C.; Bratu, C. & Pană, M. (2001), *The interaction liquid alloy – mould,* ISBN 973–96398–1-X, "Elvarom" Publishing Company, Bucharest

Annals of DAAAM for 2008 & Proceedings of the 19th International DAAAM Symposium, ISSN 1726-9679
Editor B. Katalinic, Published by DAAAM International, Vienna, Austria 2008
Make Harmony Between Technology and Nature, and Your Mind will Fly Free as a Bird

EXPERIMENTAL RESEARCH REGARDING ELECTROMAGNETIC FIELD DEFORMATION

SINDILA, G[heorghe] & OCNARESCU, C[onstantin]

Abstract: *The electromagnetic field deformation enlarges the range of manufacturing possibilities offered by classical cold deformation procedures. One of the main advantages of manufacturing by means of electromagnetic field bending is the highly precise control over the energy level required for the deformation to take place. This involves a deep knowledge of the main parameters that are involved in the process, as well as of their influence upon the deformation degree. This paper establishes, based on experimental research, dependence relations between the diameter of the deformed part, the energy level charged in the condensers battery and the main variables that come up in the deformation procedure of cylindrical parts.*

Key words: *electromagnetic deformation, cylindrical parts, deformation procedure, energy level*

1. INTRODUCTION

The electromagnetic field deformation, as a non-conventional process of manufacturing, enables the practical achievement of all classical cold-working procedures. Taking into account its relatively low efficiency (< 30%), the process applies only when the classical cold deformation procedures do not meet certain technical and economical requirements. Threading (fig. 1) and even assembling metallic parts with non-metallic ones (fig.2) represent the domains that the electromagnetic field deformation procedure is mostly applied to. In order for a good assembling to take place, it should be performed a rigorous check of the battery's of condensers charging degree, according to the required level of deformation (Balaban, 1993).
This is the reason why this paper establishes the influence of the main constructional and functional parameters (which characterize the deformation process in electromagnetic field), upon cylindrical parts' degree of deformation. (Ciocârdia &al., 1992)

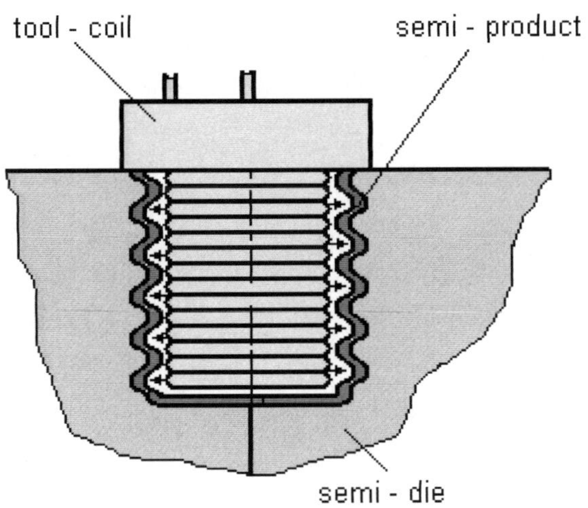

Fig. 1. Threading in electromagnetic field

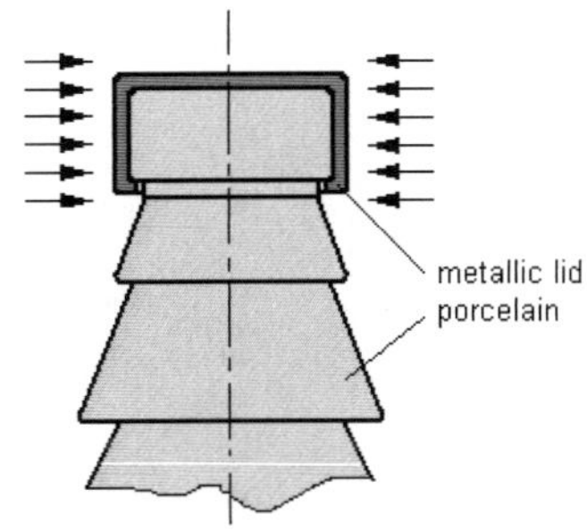

Fig. 2. Assembling metallic parts with non-metallic ones

2. EXPERIMENTAL ESTABLISHING OF THE DEFORMATION DEGREE

The experience in applying the deformation process in electromagnetic field emphasizes that, for the deformation of the same material by means of the same deformation device, the main parameters that influence the deformation degree are: the energy level charged in the condensers battery **U**, the diameter of the part subject to deformation **D**, the thickness of the material **g**, the diameter of the coil's wire **d**, the total number of coil's wires **N**, the clearance between the coil and the blank **j**. The determination of a dependency function between the diameter of the deformed part $\mathbf{D_f}$ and the above-mentioned parameters is experimentally performed by employing the method of response surfaces:

$$D_f = f(U, D, g, d, N, j) \quad (1)$$

Because most of the multi-variable dependency functions (determined practically by means of different experimental methods) are found in a polytrophic form, , we propose the following dependency function (Sindila & al., 2003):

$$D_f = a_0 \cdot U^{a_1} \cdot N^{a_2} \cdot d^{a_3} \cdot D^{a_4} \cdot j^{a_5} \cdot g^{a_6} \quad (2)$$

Choosing the domains where the parameters vary is performed according to their frequency of appearance in experimental research studies, the technological limitations of the material, the constructional characteristics of the experimental device, etc. Thus, relation (2) leads to:

$$\begin{aligned} \ln D_f &= \ln a_0 + a_1 \cdot \ln U + a_2 \cdot \ln N + \\ &+ a_3 \cdot \ln d + a_4 \cdot \ln D + a_5 \cdot \ln j + a_6 \cdot \ln g \end{aligned} \quad (3)$$

In order to simplify the computations, the parameters' values were established in geometrical progression and were encrypted in accordance with those displayed in tabel 1. There were manufactured tool-coils and tubular parts whose diameters,

Natural variables	Encrypted variables	Encrypted level values		
		- 1	0	+ 1
		Natural level values		
Tension, U [V]	x_1	4000	4475	5000
No. of wires, N	x_2	5	8	13
Coil's wire diameter, d [mm]	x_3	1,1	1,6	2,3
Conductor's diameter, D [mm]	x_4	30	40	60
Clearance, j [mm]	x_5	0,7	1,0	1,5
Material's thickness, g [mm]	x_6	0,5	0,7	1,0

Table 1

respectively thickness of material, have values according to those stated in table 1.
There were also observed the values of the charging tension of the condensers battery as well as the values of the clearance between the tool-coil and the tubular part (Mihail, 1976).

Employing the established methodology for an entire experimental factorial program with central points, there were performed 16 different experiments and 4 within the centre of the experiment, as a result of which there were determined the a_{ir} values of the parameters under study (Sindila, 1976).

Fig. 3. Set of coils

The experiments were run on a DEMARO-type installation of deformation in electromagnetic field, employing the set of coils displayed in fig.3 and tubular parts made of aluminium alloys (fig.4).

Fig. 4. Set of obtained parts

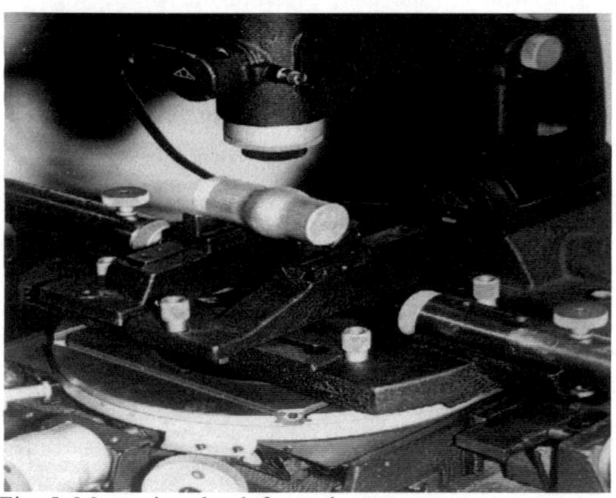

Fig. 5. Measuring the deformed parts

The analysis of the experimentally-obtained values revealed the model's adequacy and enabled the determination of the independent variables' significance, as well as the importance of reliance/trust intervals, statistic errors and the required dependency function. It was concluded that not all the variables under study (such as the number of wires and the clearance between the coil and the part) have an important significance on the deformation and thus, they were dropped out (table 2).
The statistical errors lower than 5% emphasize the fact that the theoretical results and the practical ones (obtained in a real case of deformation) are conveniently close.
Replacing the values thus determined in relation 2, it was established the computation formula of the deformed part's diameter, taking into account only the significant parameters:

Independent variable	Fisher function	Meaning
Coefficient a_0	46376,45 >> 1	significant
Tension, U	1,53 > 1	significant
No. of wires, N	0,00	insignificant
Conductor's diameter, D	4,40 >1	significant
Coil's wire diameter, d	316,21 >1	significant
Clearance, j	0,64 < 1	insignificant

$$D_f = 0{,}345 \cdot \frac{U^{0,130} \cdot D^{1,018}}{d^{0,113} \cdot g^{0,091}} \quad (4)$$

In order to determine the energy level charged in the condensers battery needed to obtain a certain diameter D_f of the deformed final part, it's recommended to use the following relation:

$$U = 3582 \cdot \frac{D_f^{7,692} \cdot d^{0,869} \cdot g^{0,699}}{D^{7,83}} \quad (5)$$

3. CONCLUSIONS

The specialty literature offers, in general, few data regarding the deformation process in electromagnetic field and the main parameters that are involved in the process. The data derived from the experimental method offer the opportunity of estimating the way the main constructional (D, d, n, U) and functional (j) parameters influence the degree of deformation of cylindrical aluminium parts under the bloating process in electromagnetic field. The energy level charged in the condensers battery in order to obtain a certain diameter of the deformed part, can also be estimated. The same methodology can be extended as well to parts made of different materials. The graphic representation of the experimentally-obtained relations enables, in practical situations, the rapid determination of the way in which the parameters under consideration influence the purpose function.

4. REFERENCES

Balaban, C. – *Strategia experimentării şi analiza datelor experimentale*, E.A.R. Bucureşti, 1993.

Ciocârdia, C., Drăgănescu, Fl ., Sindilă, Gh., Carp C ., Pârvu C . - *Tehnologia presării la rece*, E. D. P., Bucureşti , 1992 .

Mihail, R. – *Introducere în strategia experimentării, cu aplicaţii din tehnologia chimică*, E.S.E., Bucureşti, 1976.

Sindilă, Gh. - *Cercetări teoretice şi experimentale privind deformarea plastică în câmp electromagnetic*. Teză de doctorat. Bucureşti, 1985.

Sindilă Gh., Rohan R., Sindilă G. – *Deformation of metals in electromagnetic field.* Proceedings. International conference on manufacturing science and education challenges of the European integration. 6-7 Nov., 2003, Sibiu.

Annals of DAAAM for 2008 & Proceedings of the 19th International DAAAM Symposium, ISSN 1726-9679
Editor B. Katalinic, Published by DAAAM International, Vienna, Austria 2008
Make Harmony Between Technology and Nature, and Your Mind will Fly Free as a Bird

COMPUTER - AIDED METAL STRIP LAYOUT

SINDILA, G[heorghe]; LUPEANU, M[ihaela] & ROHAN, R[odica]

Abstract: *The benefits provided by the procedures of cold plastic deformation process lead to the transformation of an increasing amount of metal into pieces obtained by cold compressing processes. Taking into consideration that from the cost of the parts processed by this method, the material cost is involved with a very large share (which can reach up to 90% of the cost of the processed part), the problem of determining an optimal metal strip layout becomes a main issue in designing technological cold compressing processes.*
Key words: *optimum scheme, layout, metal strip, cold compressing*

1. INTRODUCTION

The paper brings forward a programme that determines the optimal layout for the outline of the parts that are punched out from a metal strip.

The material saving in the cold compressing process can be achieved through: designing a technological form of the outline leading to a layout without stamping trims (fig.1), establishing the correct cell connector size (fig. 6), designing the optimum metal strip layout for punching.

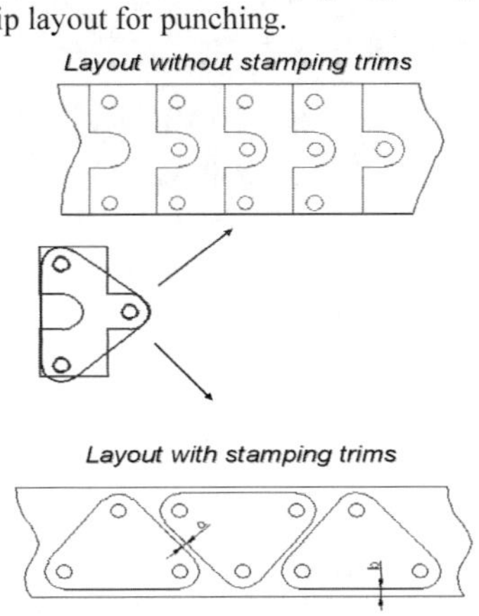

Fig. 1. Designing a technological form of the outline

To decrease the necessary time to highlight the multiple variants of cutting-on, the efficiency calculation and determination of each optimal variant, the programme "Optimum layout" was designed.

2. OPTIMUM LAYOUT

The suggested program, "Optimum layout", is quite different from those produced by similar specialized companies, in the first place because it is a particular application to the cutting-on on semis in the shape of metal strips with standard widths. The program executes on command whichever among the acknowledged types of layouts (on a straight line, tilted, on several rows, etc.) without running through all the intermediate variants. As perspective, the programme will be able to fully accomplish a combined layout. This implies the possibility to dispose two or more pieces on the metal strip aiming to a rational use of the material.

This programme is equipped with a library of data on standardized thicknesses and widths of metal strips depending on the value of the cell connectors for multiple variants of layout. Each activity of the algorithm shown in Figure 2, programmed in LabView (fig.3, 4), finally allowed through the cutting – on coefficients and use of material, Kc, respectively Ku (fig.5), to emphasize the optimal metal strip layout scheme (fig.5).

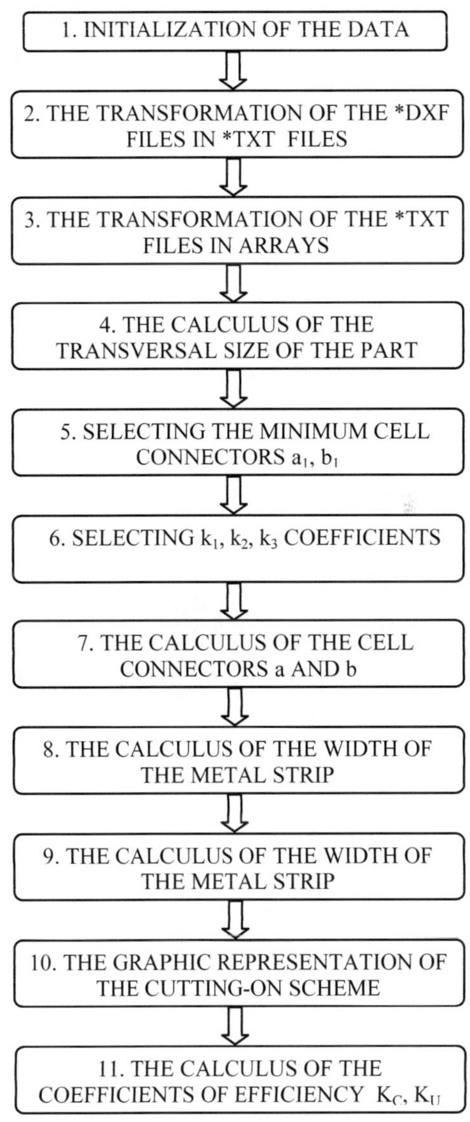

Fig. 2. The algorithm of the program

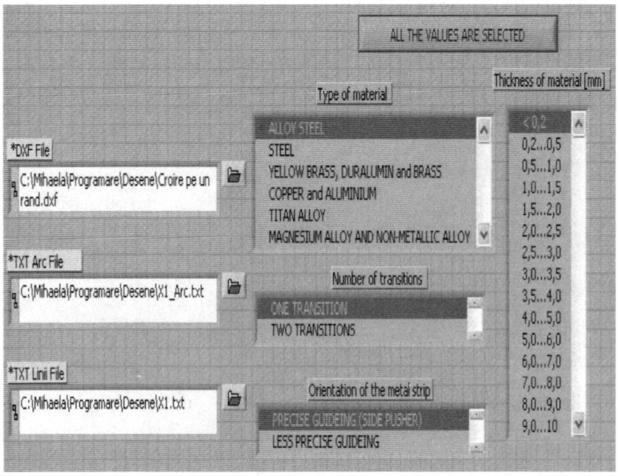

Fig. 3. Initialization of the data

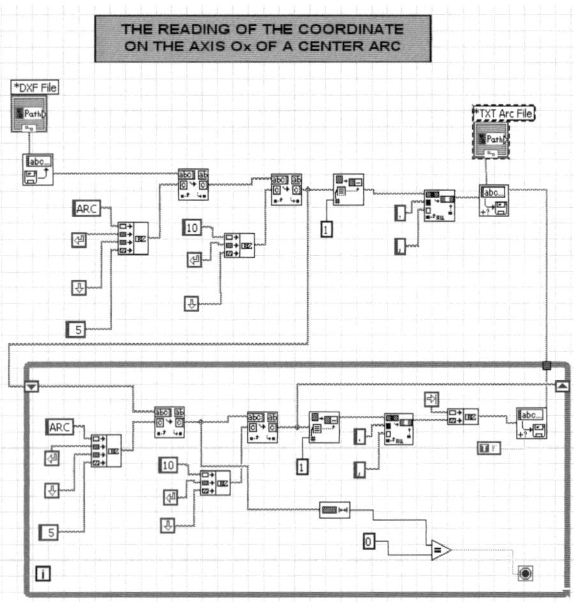

Fig. 4. The transoramtion of the *dxf files in *txt files

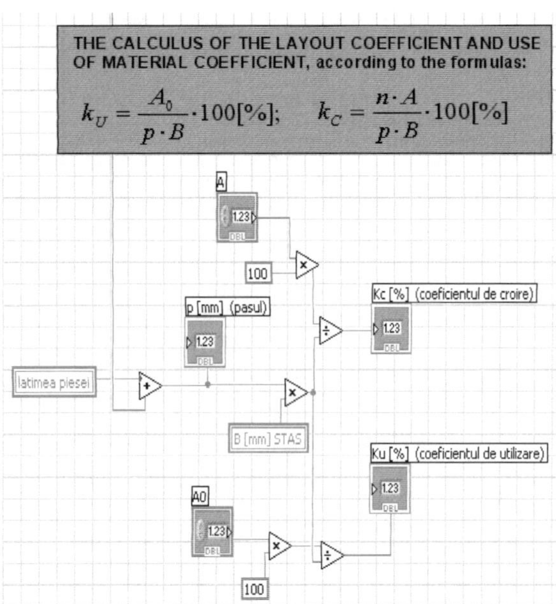

Fig. 5. The calculus of the coefficients of efficiency k_c, k_u

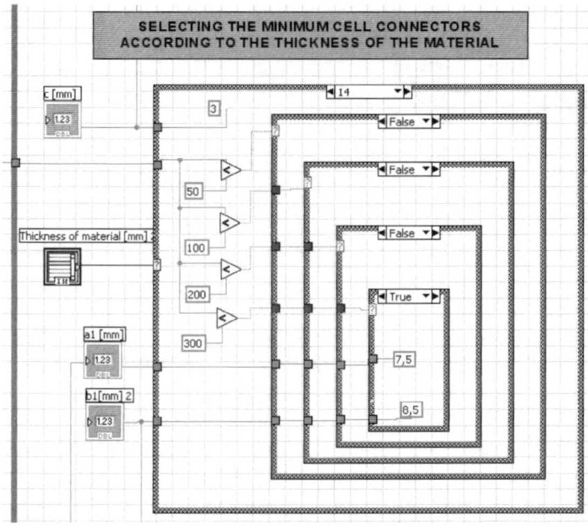

Fig. 6. Selecting the cell connectors

Depending on the type of layout studied (on a straight line, end-to-end, tilted, coalesced, on several rows) the programme automatically shows optimal metal strip layout (Fig. 7).

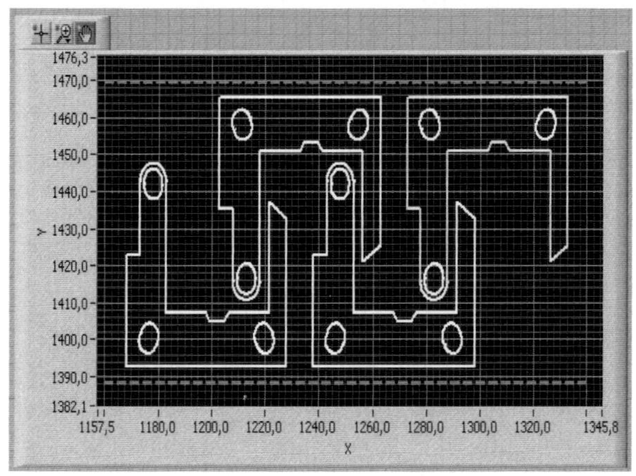

Fig. 7. Optimal metal strip layout

3. CONCLUSION

Computer-aided cutting – on allows taking into consideration numerous variants of cutting – on and the rapid and objective determination of the optimum scheme of cutting – on. The designed programme can make available to a user a ranking of analizated layout schemes, according to decreasing values of the coefficients of efficiency.

4. REFERENCES

Brabie, Gh. (2004). *Deformarea plastică la rece a tablelor metalice (Sheet metal cold plastic strain),* Bucharest, 2004.

Sindilă, Gh. (2008). *Tehnologii de fabricare prin presare la rece (Cold compressing manufacturing techonologies),* BREN, Bucharest, 2008.

Ungureanu, Ion (2004). *Tehnologii de prelucrare a profilelor prin deformare plastică volumică la rece (Profile manufacturing technologies using cold compressing),* Bucharest, 2004.

National Instruments Corporate Headquarters (2008), *NI Ultiboard User Manual, Available from:* http://digital.ni.com/manuals.nsf/websearch/8C021D4E2540C2208625742400646DA8 *Accessed:* 2008-01-12.

Braha, V.; Nagat, Gh., Negoescu,F. (2003), *Tehnologia presarii la rece (Cold compressing Techonology)*, Iasi, 2003.

Annals of DAAAM for 2008 & Proceedings of the 19th International DAAAM Symposium, ISSN 1726-9679
Editor B. Katalinic, Published by DAAAM International, Vienna, Austria 2008
Make Harmony Between Technology and Nature, and Your Mind will Fly Free as a Bird

METHOD FOR WELDING WIRE CONSUMPTION OPTIMIZATION

SINTEA, S[orin]; COOPER, E[dward]; GREBENISAN, D[an] & BORDEA, N[icolae]

Abstract: *In modern automated welding plants a big problem is to monitor online wire welding consumption. Main process of the welding automated machines and robots are designed to optimize the welding quality by monitoring arc welding parameters. Our concern is to monitor the welding wire speed and wire quantity consumed. Monitoring these parameters we are able to warn operator the reaching of wire drum low level. The system we present is able to communicate at the central point this information.*

Key words: *wire welding automation for automotive industry.*

1. INTRODUCTION

In welding plants were welding process is made using automated welding machines or welding robots, the main concerning of operators is focused in monitoring quality of welding. For this they monitor arc welding parameters (welding voltage, welding current, welding gas pressure). Using these measured parameters the welding controller trying to keep welding current into preset domain of values (Joni N. & Trif N.2005).

We propose to monitor the wire parameters (wire speed and wire consumed quantity) and include them automation controlled process. Using wire speed control we can include a new major loop in welding process control. This new control loop can inform process about:

- Status of welding wire motor load;
- Status of welding wire mechanism;
- Any jam or obstruction of welding wire.

Monitoring this parameter we can warn operator about problems with welding wire, and stopping in time of a poor welding. Another important parameter is warning of welding wire drum limit. This parameter is important for inform welding engineers and supply department to prepare a new drum unit for welding unit (Schmidt, J. S. 2002).

All these features are provided by Wizard View 1 controller, produced by ELCo Enterprises Inc, USA (ELCo Enterprises Inc. 2008).

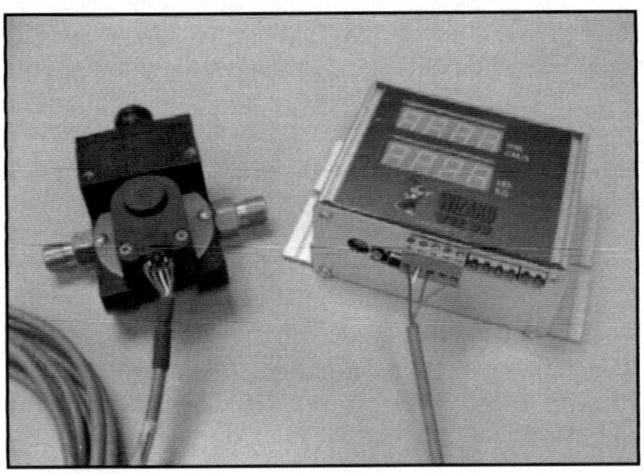

Fig. 1. ELCo Wizard View 1 Controller

2. CONTROLLER

This controller is a new entry in welding automation line controllers. The controller measuring wire speed using ELCo's HHSS wire speed sensor and local display the current wire speed and consumed wire quantity (see Fig. 1.) (ELCo Enterprises Inc. 2008). The controller is able to measure values all the time or only in active welding interval, for this the controller can be enabled or disabled using a signal connected to automated welding machine (ELCo Enterprises Inc. 2008). The controller has two operating modes:

- Simply counter – in this mode the controller counts the total weight of wire consumed in welding process;
- Controller mode – in this mode the unit outputs through two relays. One relay indicates when wire speed exceeds a preset low limit value measured. The other one relay indicates when counted welding wire is overtaking the drum warning level.

All internal data of Wizard View Controller can be accessed through serial port.

We are using this Wizard View 1 controller in controller mode and both outputs connected with welding programmable logic controller (PLC).

3. SOLUTION

In most solutions we have a welding PLC associated with a welding unit (Joni, N. & Trif, N. 2005). This PLC is responsible with welding process. One solution is to add the Wizard View 1 controller outputs to the PLC inputs and to update the PLC software to monitor and get the right decisions in according with these inputs. In big welding plants it is not a problem, because there is "Software for Automation Department" and they can make this update.

In small plants, where we do not have possibility to change the software in controllers, we can addition a PLC to the welding unit, to monitor welding parameters (Margineanu, I. 2005). On this PLC we can add the normal welding parameters (welding voltage, welding current, welding gas and a signal from welder named RUN. The "RUN" signal is active when welding cell is welding or can be active when operator or robot is in welding interval.

3.1 Implementation

We can choose a PLC (see Fig. 2.) with one CPU, three modules (one 8 optoisolated digital inputs, one with four 0-10 Volts (or 0-5 Volts) isolated analogue inputs and one with 8 relays output digital module) and optional an Ethernet module (if CPU does not have Ethernet capabilities on board) (Direct Logic. 1998). For implementation we take Koyo PLC model DL-205 with DL-240 CPU unit.

To the PLC we add the following signals:

- digital inputs: RUN is the running signal which indicate that operator or robot is in welding process; SPD_OK is the "Normal Speed" digital output of Wizard View 1 controller

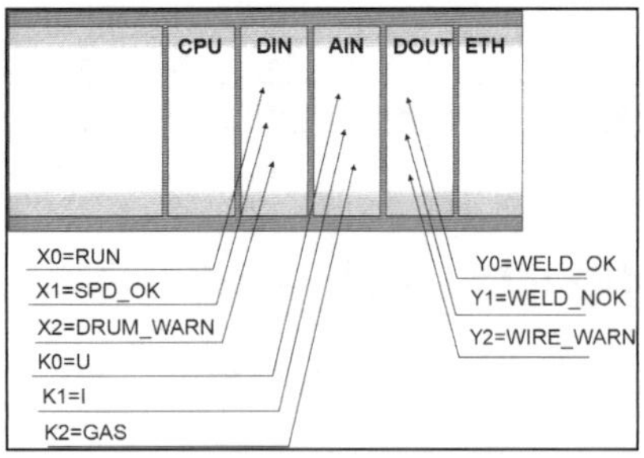

Fig. 2. PLC

and DRUM_WARN is the "Drum Warning" signal generate by Wizard View 1 controller

- analogue inputs: U is the analogue input connected to welding voltage through resistive divider (100:1), I is analogue input connected to welding current using a current transformer and GAS is analogue input connected to the gas pressure sensor.
- digital outputs: WELD_OK indicates that all welding parameters are in normal limits, WELD_NOK signal indicates that at least one parameter is out of limits and WIRE_WARN signal indicates that welding wire reach the preset value in Wizard View 1 controller.

3.2 Programming

In PLC software we implement the algorithm for welding monitoring. Our implementation is made on KOYO Direct Logic-205 PLC with DL240CPU (Direct Logic, 2008). For implementation we declare the following constants in CPU memory:

- voltage sensor constant in location V2100,
- current sensor constant in location V2101,
- gas pressure sensor constant in location V2101,
- lower preset value for welding current in location V2103,
- lower and upper value of welding voltage in locations V2104 and V2105,
- lower level of gas pressure in location V2106.

The algorithm is based on following mathematical and logic relations:

- converting the input analogue signals in decimal floating point format (Zaciu, R. 2002):

$$V2110 = K0 \times V2100 \tag{1}$$

$$V2111 = K1 \times V2101 \tag{2}$$

$$V2112 = K2 \times V2102 \tag{3}$$

where, relations describe the conversions for voltage, current and gas pressure measures.

- comparing of analogue signals with presets and establish the values for internal Boolean variables which describe the status of welding:

$$C1 = V2111 > V2102 \tag{4}$$

$$C2 = V2110 > V2104 \tag{5}$$

$$C3 = V2110 < V2105 \tag{6}$$

$$C4 = V2112 > V2106 \tag{7}$$

$$C5 = X0 \,\&\, C1 \,\&\, C2 \,\&\, C3 \,\&\, C4 \,\&\, X1 \tag{8}$$

where, C5 variable describe the correct status of welding.

- generating normal and correct welding output signal:

$$Y0 = X0 \,\&\, C5 \tag{9}$$

- generating out of limits welding output signal:

$$Y1 = X0 \,\&\, !\,C5 \tag{10}$$

- generating drum warning signal in time of welding:

$$Y2 = X0 \,\&\, X2 \tag{11}$$

and optional, we can generate a drum warning signal

$$Y3 = X2 \tag{12}$$

The PLC can be equipped with an Ethernet module, to be able to send an e-mail when quantity of consumed wire reaches the preset value. The e-mail is send to a central server or direct to "Service Team" or to the "Supply Department" to minimize the time request for supplying with welding wire drum.

4. RESULTS

By introducing the wire speed factor in welding analysis we can monitor the status of wire supplying mechanism; and we can detect problems quickly. Detecting wire supplying mechanism problems in time, we can avoid other unexpected stopped time in welding process and eliminated death time.

By adding this new feature to welding units we can short the time with supplying with wire of welding automated machines. This can increase the productivity and can automate the supplying process in the entire plant.

5. CONCLUSIONS

This system presents a new vision in optimising the welding wire consumption and minimizing the time request for supplying with wire welding in automated plants.

The system can be developed in future by adding new features in PLC and introducing the fuzzy logic algorithms in detecting of the correct generating of WELD_OK signal. Also the PLC can be programmed to send data over Ethernet port to a central server, for creating of central welding monitor system. All data centralized on the server level can be accessed by anyone who is involved with welding process from the plant.

Another changing in presented system can be obtained by replacing the DL240CPU with WinPLC CPU for Direct Logic 205 PLC. The WinPLC CPU is based on WindowsCE operating system. All software can be made using Microsoft Visual Basic or Microsoft Visual C and the most important advantage is possibility to present all data and welding status on internal web site. The WinPLC system has internal web server integrated.

6. REFERENCES

Direct Logic. (1998) *DL205 User Manual,* Koyo, PLC Direct Inc, D2-USER-M, USA

ELCo Enterprises Inc. (2008). *Wizard View 1. User Manual, rev. D.,* ELCo Enterprises Inc, -, Jackson, Michigan, USA,

Joni, N. & Trif, N. (2005). *Automated Electrical Arc Welding,* LUX Libris, ISBN 973-9495-34-3, Brasov, Romania,

Margineanu, I. (2005). *Programmable Controllers,* Blue Editor, ISBN 973-650-156-6, Cluj Napoca, Romania

Schmidt, J. S. (2002). *Protocols for interconnecting the components of a welding system,* -, IIW DOC XII-1697-02, Lisboa

Zaciu, R. (2002). *Signals Digital Editing,* Blue Editor, ISBN 973-650-095-0, Cluj Napoca, Romania

Annals of DAAAM for 2008 & Proceedings of the 19th International DAAAM Symposium, ISSN 1726-9679
Editor B. Katalinic, Published by DAAAM International, Vienna, Austria 2008
Make Harmony Between Technology and Nature, and Your Mind will Fly Free as a Bird

ASSESEMENT FOR MANAGEMENT OF WASTEWATER QUALITY FROM A HEALTHCARE UNIT - CASE STUDY

SIRBU, D[ana]; CURSEU, D[aniela]; POPA, M[onica] & POPA, M[arcel] S[abin]

Abstract: *Hospitals give significant volumes of wastewater loaded with micro-organisms, heavy metals, toxic chemicals and radioactive elements which may have an impact on the environment and human health. One of the main environmental problems caused by the hospital effluents is their discharge in the urban sewer network without preliminary treatment. The aim of this paper is to presents the basic characteristics and hazards of wastewater from one Romanian health-care establishment with 175 beds. All the results illustrate the load of pollutants that the hospital effluents can bring to the municipal sewage system and hence the need to elaborate guideline management in order to have a healthier environment.*

Key words: *hospital, wastewater, pollution, disinfectants, quality.*

1. INTRODUCTION

Hospital complexes use for their activities and their hygiene, great volumes of water which are then rejected, fill of micro-organisms of which some are multiple resistant, often toxic and sometimes radioactive chemicals. The hospital is a heavy water consumer, while in medium the consumption of the households is 150 to 200 liters per capita and by day, the average value passes from 500 to 1200 liters/bed/day in the hospitals. Quality and quantity of wastewater produced in every hospital depends on different factors, such as: bed number, accessibility to water, kind of services, number of units/departments, climate situation, people's culture and geographical situation (Sabzali & Shivaii, 2006). All the new technologies which answer progress as regards care (hemodialysis, radiology, laboratories of high technologies, etc), should not compromise the watery ecosystem of our rivers and seas. The harmful effects related to their wastewater should be a subject to consider for all hospitals managers and it is necessary to assess the quality of their hospital effluent.

The aims of this study are: to characterize the effluents from the chemical and microbiological point of view and to compare with the national and international recommendations in order to develop efficient projects to improve the quality of these rejections.

2. MATERIALS AND METHODS

A hospital wastewater from Cluj County, with a capacity of 175 beds, which discharges the total volume of effluents into the urban wastewater network without pre-treatment, has been selected. The investigated hospital has following departments: internal medicine, pediatrics, surgery, psychiatry, gynecology and obstetrics, laboratory and pharmacy. In order to characterize the pollutant composition of the wastewater discharged, a questionnaire concerning susceptible matters of trapping various pollutants, were applied and evaluated.

To define quality parameter, sampling from out coming sewages of the hospital and from the municipal network system, was conducted. The following chemical and microbiologic analyses were carried out: pH, COD (chemical oxygen demand), BOD_5 (5-day biological oxygen demand), chlorides, TSS (total suspended solids), total phosphorous, MPNTC (most probable number of total coliforms) and MPNFC (most probable number of fecal bacteria). The samples were collected on three days of the week: Monday, Wednesday and Sunday, two samples per day at 10 a.m. and 3 p.m., 4 weeks. All the analyses were carried out in conformity with the national standards. The results were compared with the Romanian and European threshold values, and the Pc/Vt ratio (Pc – pollutant concentration in the hospital effluent, Vt – threshold values) was calculated (RD, 2002; Sirbu et al., 2005).

3. RESULTS AND DISCUSION

Investigation of hospital daily water consumption shows that the average value is 165 m^3/day which is equal to 942 L/day/bed. Quantitative amount of daily wastewater production, considering water to wastewater conversion coefficient of 0.8 was estimated at 132m^3/day (754 L of wastewater day/bed). On the basis of questionnaire, we found several types of hospital rejections which are shown in Table 1.

COD and BOD_5 were determined for the measurement of total organic load. In the hospital effluent sample, COD concentration (535 mg/l), and BOD_5 (368 mg/l) exceeded the discharge national standards ($P_c/V_t > 1$), showing that hospital wastewater were heavily loaded with organic and mineral substances (Table 2).

The biodegradability of organic substances is a measure of the speed and completeness of their degradations by microorganisms and therefore the BOD_5/COD ratio could be used to analyze the degree of degradation.

Rejections	Types and sources
Rejects of domestic nature	rejections of the kitchens, detergent products, rejections of the garages and workshops, those of the laundry, and the boiler room
Rejects of specific nature	***Chemical pollutants:*** - cleaners and disinfectants which are often highly complex products or mixtures of active substances. The most often used products are mainly chlorinated derivatives, the products containing aldehydes (formaldehyde and glutaraldehyde: toxic molecule both for man and the environment), the iodized compound (betadine). - drugs (analgesics, antibiotics, radio elements, and cytostatic agents) - chemical agents from the laboratories (acids, strong bases, solvents) - heavy metals such as mercury and the silver resulting, for one of breakings of the mercury thermometers and for the other of the service of radiology. ***Biological pollutants:*** - the pathogenic rejections of germs resulting from the people's diseases (*Pseudomonas aeruginosa, Staphylococcus aureus, E. coli, Streptococcus pneumoniae, Klebseilla pneumoniae*)

Tab. 1. General types of hospital rejections

Parameters	Concentrations in: hospital effluents	urban effluents	Threshold values
Temperature (^{0}C)	26.8	19.8	40
pH (U)	8.7* (P_c/V_t> 1)	7.5	6.5 – 8.5
TSS mg / l	335* (P_c/V_t> 1)	282	300
BOD_5 mg/l	368* (P_c/V_t> 1)	280	300
COD mg/l	535* (P_c/V_t> 1)	498	500
BOD_5 / COD	0.68	0.56	0.5 - 0.6
Total Phosphorous mg/l	7.6 (P_c/V_t> 1)	6	5
Chlorides mg/l	287	60	400

Tab. 2. Comparison of the average concentrations in pollutants of hospital and urban effluents

A BOD_5/COD ratio 0.5 – 0.6 could be considered as threshold value to study the biodegradability of organic substances into the hospital wastewater (Seiss et al., 2001). In our study the BOD_5/COD ratio in hospital wastewater is 0.68, which indicate that the pollutants would be difficult to degrade, which describes a potential impact on the WWTP (wastewater treatment plant) efficiency. However, COD might give false results in the sample contained organic or inorganic elements that were refractory, i.e., resistant to oxidation. For the accuracy of biochemical knowledge of hospital wastewater, according to biodegradability, it is necessary to introduce in the monitoring of hospital effluents the TOC (total organic carbon) determination and the COD/TOC ratio. TOC degradation rate depends on the chloride concentration of the wastewater. The different organochlorine compounds resulting from disinfections could contribute to high TOC concentrations in the hospital effluents. The results of bacteriological characterization are summarized in Table 3. Accordance with previous studies on microbiological characterization of hospital wastewater the concentrations in germs are weaker in the hospital effluents than in the urban effluents, which is probably related to the higher concentration of the disinfectants and the antibiotics (Kummerer, 2001; Sirbu et al., 2007). The level of contamination is variable according to the hour, the day or the flow at the time of the taking away. The hospital stocks are characterized by their resistance to antibiotics. These results showed that the application of important quantity of disinfectants in hospital can reduce bacteria in wastewater but give rise to toxicity effects on aquatic organisms (Popa et al., 2005). So, the parameters used for the microbiological characterization was also considered in this study like an indirect detection of the massive presence of disinfectants and/or antibiotics. Total/fecal coliform populations of hospital wastewater were affected because chloride values of these effluents are greater than values generally uncounted in domestic wastewater. Many studies conclude that chloride could be used as indicator of the toxicity of hospital wastewater on aquatic organisms when NaOCl was used (Emmanuel et al., 2001). Chloramines and other agents, such as sodium hypochlorite and some iodized substances are responsible for the disinfectant effect. Organic halogen compounds are formed by reaction of chlorine with organic wastewater compounds. An additional AOX in the wastewater will result. The AOX have a bad biodegradability and several toxic effects on aquatic organisms. Hospitals cannot be neglected as contributors to AOX in urban wastewater (Kümmerer, 2001). For more accuracy it seams to be important to introduce the determination of AOX (organohalogen compounds absorbable) in hospital wastewater monitoring in our country.

Parameters (average values)	Threshold values*	Concentration in : Hospital effluent	Urban effluent
MPNTC germs/100 ml		4.8 x 10^4	3.9 x 10^6
MPNFC germs/100 ml	1 x 10^8	3.9 x 10^4 P_c/V_t< 1	2.8 x 10^6

Tab. 3. Microbiological characterization of hospital effluent

These substances, which are not biodegradable, may finally enter surface water by WWTP effluents, and enter groundwater after the application of sewage sludge as fertilizers (Gartisser et al., 1996). Among other substances, various pharmaceuticals made the object of ecological research, such as: the sex hormone, the radio elements, antibiotics and cytostatic agents (Sirbu et al., 2007).

4. CONCLUSION

The problems of wastewater in health care units have not been monitored in Romania so far. All the results confirm the existence of hazardous substances in the studied hospital effluents, and hence the need for continuing the approach by establishing the ecotoxicological risk assessment of hospital wastewater for the aquatic ecosystems. It should also be noted that the effluent of the hospital undergoes major dilution in the communal sewage system. To align with the European Commission Directives for each medical unit, it would be important to be able to evaluate their effluents and to develop efficient projects of management them.

5. REFERENCES

Emmanuel, E.; Blanchard, J.; Keck, G. & Perrodin, Y. (2001). Caractérisation chimique, biologique et ecotoxicologique des effluents hospitaliers (Chemical, biological and ecotoxicological characterisation of hospital wastewater). *Sci. Techn.*, 22: 31-33.

Gartisser, S.; Brinkler, L., Erbe, T; Kummerer, K & Willmund, R. (1996). Contamination of hospital wastewater with hazardous compounds. *Acta Hydrochim*, 24: 90-97.

Kümmerer, K. (2001). Drugs in the environment: emission of drugs, diagnostic aids and disinfectants into wastewater by hospitals. *Chemosphere* 45:957-969.

Popa, M.; Curseu, D.; Sîrbu, D. & Chakirou, C. (2005) Toxicological Effects of Some Chemicals from Hospital Wastewaters. *Proceeding of Sustainability for humanity & environment in the extended connection field Science – Economy-Policy,* vol.II, pp. 308-312, ISBN 973-625-206-X, Timisoara, febr. 2005, Ed. Politehnica Timişoara.

RD (2002). Guidelines for Wastewater Evacuation in Urban Sewer Network (In Romanian). Romanian Directive 188/2002, Bucharest, Romania.

Sabzali, A. & Shivaii, H. (2006). Hospital wastewater treatment. *Adv. Res. Magazine*. Tehran, 1:44-50.

Seiss, M.; Gahr, A. & Niessner R. (2001). Improved AOX degradation in UV oxidative wastewater treatment by dialysis with nanofiltration membrabe. *Wat. Res.*, 13:3242-3248.

Sirbu, D.; Curseu, D.; Popa, M. & Popa, M.S. (2007). Approaches on Management for Pharmaceutical and Personal Care Products in Water. *Proceedings of the 18th International DAAAM Symposium*, Katalinic, B. (Ed.) vol 18, no1, pp. 693 – 694, ISBN 3-901509-58-5, Zadar, Oct. 2007, Croatia.

Sirbu, D.; Popa, M.; Curşeu, D.; Achimaş, A. & Hegedüs, L. (2005). Environmental Risk of Hospital Wastewater, *Central European Journal of Occupational and Environmental Medicine,* vol.11, nr.4, pp.281-287, ISSN 1219-1221.

Annals of DAAAM for 2008 & Proceedings of the 19th International DAAAM Symposium, ISSN 1726-9679
Editor B. Katalinic, Published by DAAAM International, Vienna, Austria 2008
Make Harmony Between Technology and Nature, and Your Mind will Fly Free as a Bird

EFFICIENCY OF FRICTIONAL AND MR DAMPERS IN MODEL OF THE ROBOT LANCE

SKOCZYNSKI, W[aclaw]; ROSZKOWSKI, A[ndrzej] & MAREK, B[artosz]

Abstract: *An excessive robots' arm vibration occurrent during realization of technological cycle that produced some structural defects in lead crystal articles was the reason to engage in this study. The aim of the study was to develop methods for passive and semi-active damping of the glass pick-up robot lance vibration. Two kinds of dampers were developed and built. The passive dampers made possible to change frictional elements preload. The semi-active damper enabled alteration of magnetorheological (MR) modules properties using current control devices. Dynamic characteristics of the lance with each type of damper were experimentally determined. Logarithmic decrements of the system were estimated using half-power bandwidth method. It can be concluded that the most effective solution for damping of the robots' lance remained the MR dampers with active control system.*

Key words: *robot, damper, dynamic characteristics, vibration, testing*

1. INTRODUCTION

Glass pick up robots (Fig. 1) are employed in glass-works to deliver of molten glass gob to forming devices. Robot units vibrations have an effect formation of produced lead crystal articles. Preliminary experiments showed that in the free vibration spectrum dominates one low frequency. It is related to the most flexible element of this robot - i.e. the lance equipped with gathering ball. Application of dampers fixed to the lance was accepted as the most effective way to decrease this vibration.

The aim of the study was to develop methods for passive and semi-active damping of the glass pick-up robot lance vibration. Two kinds of dampers were developed and built: frictional and semiactive. They were tested and the most efficient damper's solution was recommended as a practical industrial application.

2. SYSTEM WITH PASSIVE FRICTIONAL DAMPER

Two types of frictional dampers were developed and built. The first of them consisted of overlapped conical rings (Fig. 2). The second one (inertial) was equipped with adjoining planar rings. Preload of the rings was adjusted using prestressed spring. Frictional behavior occurred between these rings. Both dampers employed friction power to absorb the vibration

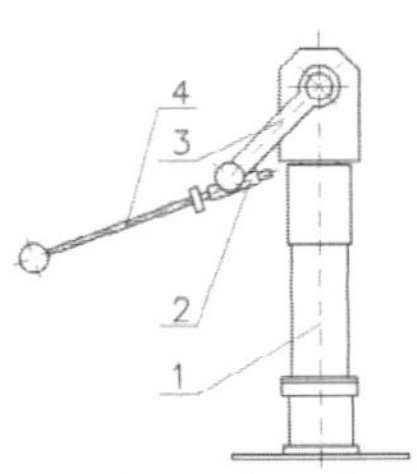

Fig. 1. Scheme of glass pick up robot: 1) column, 2) and 3) robots arms, 4) lance with gathering ball

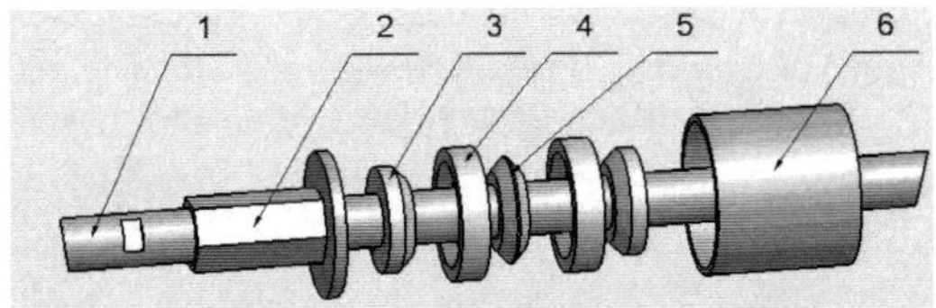

Fig. 2. Scheme of the passive frictional damper elements on physical model of the lance: 1) lance, 2) tightening element, 3) utmost internal conical ring, 4) external conical ring, 5) internal conical ring, 6) casing (clamp nut)

energy (Popp et al., 2003). They were fixed directly on the lance in some adjusted locations.

3. SYSTEM WITH SEMI-ACTIVE MR DAMPERS

The prototype of system for lance vibration damping of glazier robot was built (Fig. 3). Two linear semi-active MR dampers type was RD-1005-03 from Lord corporation (Carlson, 2002).They were fixed to the lance and applied in the test stand. Based on earlier investigations of dynamic properties of the glazier's robot lance the design assumptions were accepted in relation to the dampers. System should provide damp vibrations in the frequency range of 1 - 25 Hz in two reciprocally perpendicular planes in the lance longitudinal section. Moreover it was assumed that the lance with fixed dampers set require the possibility of free rotation (similarly to the real system). The rotation was assured by fixing the linear MR dampers to the lance with bearing application.

The MR dampers required the use of control system and two independent power supply circuits. For used dampers the maximum power voltage was 12 V RMS and current intensity was limited to 2A RMS. There were impose limitations by manufacturer of dampers with regard to limited possibilities of heat sink (Sapinski, 2004).

Damper control signals come from non-contact inductive sensors MGB 102 of vibration velocity. The sensors were fastened to the body of the damper. Fixing made possible the measurement of the piston rod velocity in relation to the dampers body. A damper control algorithm was developed for above mentioned technical parameters of dampers and velocity sensors. For MR dampers two kinds of power supply were used both direct and impulse current under active control.

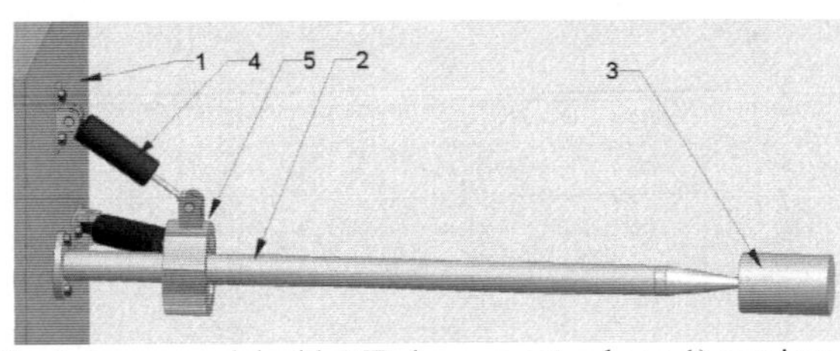

Fig. 3. Lance model with MR dampers set, where: 1) cast iron block, 2) lance, 3) supplementary ending of lance, 4) MR dampers, 5) element (bearing mounting) transferring of lance vibrations to dampers

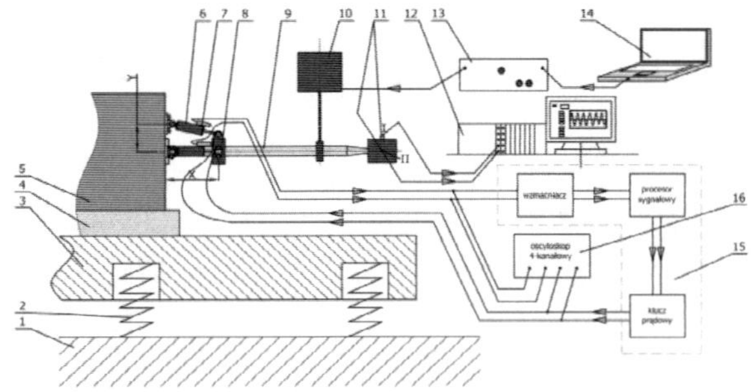

Fig. 4. Schematic diagram of the test stand: 1) foundation, 2) springs, 3) concrete plate, 4) wooden logs, 5) cast- iron block, 6) MR dampers, 7) inductive sensors of vibration velocity, 8) double-row ball bearing, 9) lance, 10) electrodynamical exciter, 11) accelerometers, 12) PXI measuring computer, 13) power amplifier, 14) computer with audio card, 15) control system of MR dampers, 16) 4-channel digital oscilloscope

4. TEST STAND

The test stand with the lance model and each damper type was designed and made (Fig. 4). Elements of the stand were placed on the concrete plate of nearly 20-ton mass lying on 10 springs. It should provide isolation between the test stand and outer vibration. The lance model was made according to the original engineering design. It was fixed to the cast- iron block of 1200 kg mass. Equivalent element of mass equal to the largest ball applied in glass-works and molten glass gob that is gathered by the ball during realization of technological cycle is fixed at the end of the lance. Each type of damper was separatelly installed on the lance.

The lance was subjected to the harmonic excitations using electrodynamical exciter. The exciter was applied with signal, that was initially modelled in MATLAB environment. Each signal was an audio file of defined frequency parameters. The above file was reproduced in Winamp program. Signal from the sound card was amplified and used to supply electrodynamical exciter. In MATLAB program, the file of sweep type of length 300 seconds about frequency changing with step 0.1 Hz in the range of 5 - 16 Hz, was generated.

For MR dampers, the inductive velocity sensors, attached to them, were employed to measure relative vibration velocity between the piston rod and the body of the damper. The signal from dampers reached the dampers control system in real time criteria. The system based on DSP. The task of the card with DSP controled the level of signal coming from the velocity sensors. If the absolute value of voltage signal exceeded the set threshold level then the control system supplied dampers MR. Supply current of dampers was switched off at the moment of falling down the absolute value of this signal below the threshold level. The control system made possible adjusting the triggering level of the dampers power supply. The power supply allowed to set current level.

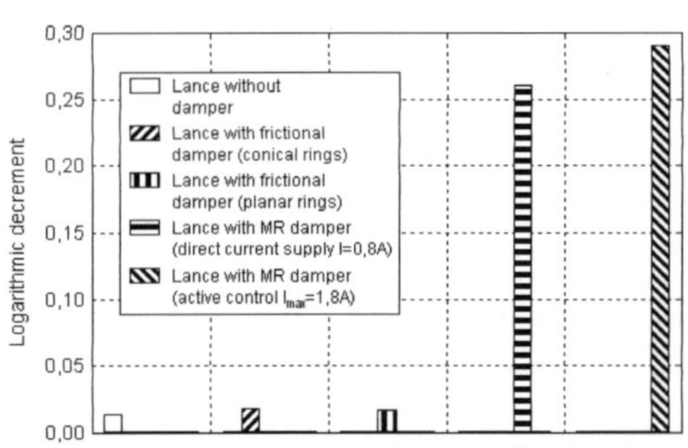

Fig. 5. Logarithmic decrement for the lance vibration for the best settings of dampers

The test stand was equipped with measuring computer with data acquisition card. Two accelerometers placed on the equivalent mass served a measurement of the lance vibration in two mutually perpendicular directions. The signals acquisition card type was PXI-4462. It was placed in measuring computer type of National Instruments PXI. Signals from the accelerometers were processed offline in MATLAB software environment. Baseed on the accelerations signals, amplitude-frequency characteristics were determined. They corresponded with the first resonant mode. These characteristics were used for calculation of the logarythmic decrement of the lance using half-power bandwidth method (Gade et al., 2002).

5. RESULTS

For various settings of both types of dampers, i.e. different locations on the lance, variable preloads (in frictional dampers) and varied kind of current supply (in MR damper), dynamic characteristics of the lance were determined. The tests pointed out that MR dampers were much more effective then frictional ones (Fig. 5). Especially the system with active control proved the best damping properties. The most advantageous location of MR damper was also determined.

Unfortunatelly, the proposed frictional dampers demonstrated complitelly useless in the tested structure, in spite of adjusting the best their damping parameters.

6. CONCLUSION

The work presents two original solutions with dampers attended to use in glass pick up robot lance. They were developed, built and tested in the laboratory field. Dynamic characteristics of the lance with each type of damper were experimentally determined. The first resonant vibration mode of the lance was taken into account because of its domination in the spectrum. Logarithmic decrements of the system were estimated using half-power bandwidth method.

Application of frictional damper attached to the tested robot lance requires a ground revision. It is possible that a general construction of this damper was inadequate to the relative microslips occurring in the contact surface of frictional rings. It is also to consider application for the lance a semiactive system using controllable friction damper.

It may be concluded that the most effective solution for damping of the robots' lance usage of the MR dampers with active control system. The following studies should concentrate on instalation of MR dampers in the real robot structure and performing industrial tests.

7. REFERENCES

Carlson, J.D. (2002). *Controlling Vibration with Magneto-rheological Fluid Damping*. Sensor Technology and Design , February 2002

Gade, S. Herlufsen, H. & Konstantin-Hansen, H. (2002). How to determine the modal parameters of simple structures. *Journal of Sound and Vibration*, Vol. 36, No. 1, (January 2002) 72-73, ISSN 0038-1810

Popp, K.; Panning, L. & Sextro, W. (2003). Vibration damping by friction forces: Theory and applications. *Journal of Vibration and Control*, Vol. 9, No. 3-4 (March 2003) 419-448, ISSN: 1077-5463

Sapinski, B. (2004) *Linear Magnetorheological Fluid Dampers for Vibration Mitigation: Modelling, Control and Experimental Testing,* Uczelniane Wydawnictwa Naukowo-Dydaktyczne AGH, ISSN 0867-6631, Krakow

A scientific project founded from the budget resources appropriated to science in the years 2005 to 2008 as a research project.

Annals of DAAAM for 2008 & Proceedings of the 19th International DAAAM Symposium, ISSN 1726-9679
Editor B. Katalinic, Published by DAAAM International, Vienna, Austria 2008
Make Harmony Between Technology and Nature, and Your Mind will Fly Free as a Bird

APLICATION OF DIGITALIZATION AND SIMULATION TO COINING

SKUNCA, M[arko]; LONCAR, D[amir] & MATH, M[iljenko]

Abstract: *Coining as metalforming technology is known for its technological complexity and phenomenological problems. Practitioned technicians and engineers in the field of coining are the key to the solution of those problems. Development of 3D scanners and FE codes, have enabled the simulation of coining. This includes simulation of details up to scale where flow related phenomenon is tracable. Modification of coin/medal geometry can be made before tooling has been made. Results is significant savings in tool production, allowing technology approved geometry modification, at the moment of artist creation of coin's vers and avers. Paper demonstrates the simulation of coining, using the appropriate FE numerical background that enables the completion of virtual coining.*
Key words: *Coining, 3D scan, FEM, Metalforming, Metal flow*

1. INTRODUCTION

Papers regarding particular technology like coining or minting are scarce (Ike & Plancak, 1998), (Ike, 2005). Reason to this is highly commercial nature of latter technology. Moreover commemorative medals and coins tend to exhibit increase in price (Ladany, 1981). Therefore monetary institutes keep there technology as secret as possible.

High complexity of medal surface within the hight span of ± 0.05 mm in present case, represents a great challenge to FEM modelling. Therefore a 3D FE coining simulation of the golden medal shown in was performed. (Fig. 1)

In order to reduce details (Buffa et al., 2007) have created a model of simplified coin surface geometry. In this paper no geometry simplification was made and simulation was performed over 1/12 of the coin showed as marked area. (Fig. 2)

2. DIGITALIZATION

Digitalization of coin geometry was performed at FSB Zagreb, using ATOS Standard 3D digitizer manufactured by GOM mbh (http://www.gom.com). ATOS measuring head with two CCIR-50Hz cameras 0,8 MPixel, was used with retro reflective illumination. Medal of 37 mm diameter was digitized using available measuring volume of 50x40 mm (one set of lenses: 35 mm projector lens, 50 mm camera lenses, calibration object 50x40 mm).

Fig. 1. Croatian red cross medal. Damir Mataušić - Academic sculptor & medallier.

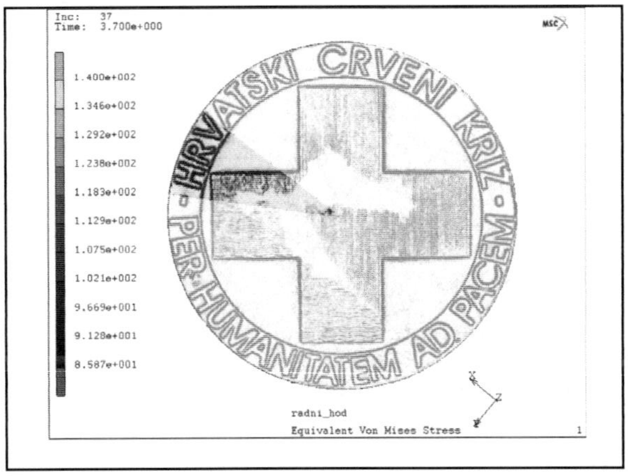

Fig. 2. Digitized medal with 1/12 marked for performing FE simnulation.

Implemented photogrametric triangulation technique resulted in high detail scan. Basic steps of digitalisation were as follows; After fixture on base plate, medal was sprayed with penetrant in order to neutralise surface reflection. Calibration and digitalisation were performed in short time. Once point cloud was recorded, ATOS software was used to digitize, process, visualize and export the measured data to Mentat for preprocessing and FE model preparation.

3. PREPROCESSING

3.1 Geometry

Large number of triangles were converted to the geometry of 7e5 surfaces modelling desired geometry. From this number, 3e4 surfaces was chosen to model one twelvth of the coin vers, shown shadowed in Fig. 2. Minimum surface sizes of 0.05 mm were taken as satisfactory modelling of the surface geometry. With a details up to 0.02 mm.

One twelvth of the rigid punch was modelled using 3e4 surfaces. Underneath the punch FE mesh was generated as $1/12^{th}$ of the cylinder of radius 18.3 mm and height 0.68 mm. Divided into mesh of 0.17 mm element size, cylinder is made of 1e5 finite elements. Tip of the cylinder near axis of symmetry was cut off in order to avoid (well known bug) of node penetration trough wedge formed by two intersecting surfaces.

Two wedge plains were used to impose boundary conditions upon 1/12 of the coin.

Flow of the material at the outer edge was restricted by the cylinder surface intersecting wedge planes.

Base plane was modelled as a flat plane, in order to keep the model as small as possible.

In every case of defining rigid surfaces, non-uniform rational B-splines were avoided.

3.2 Material

Rigid plastic formulation of material was used in order to make model numerically as simple as possible. Stress-strain curve for gold was assigned to material after an internal data sheet not intended for publishing.

3.3 Contact bodies

As previously mentioned, besides deforming FE billet, there are six contact bodies in numerical simulation. All those six bodies were modelled as rigid bodies. For the practical reasons of limited calculation time, none of the bodies was divided into FE elements.

3.4 Boundary conditions

Posed by the use of the symmetric wedge planes enabled the simulation to run to its completion. Moreover, likewise posed BC's enable 3D remeshing that was excecluded from the first processing.

3.5 Loadcases

Only one, static, time independent loadcase was considered. Basic constant time step loading procedure was used for the same reason of simplicity. Regarding global stiffness matrix positive definiteness was required. Number of recycles was increased to ensure completion of the simulation. Convergence criterion was set using relative residual force.

3.5 Job

Job was set upon one and single loadcase using linear tetrahedral elements written for the updated lagrangian framework.

4. PROCESSING

Numerical simulation lasted for 60 hours on average desktop PC of the year 2004. It was performed in MSC Marc via input text file created in MSC Mentat. (Buffa et al., 2007) used DEFORM-3D^{TM}, but no FE simulation processing time was given.

5. POSTPROCESSING

MSC Mentat was use to perform psotprocessing. Sueccesful simulation was obtained using the simplest allowaable element type and avoiding B-spline surface description. FE simulation has pointed to critical stress areas. (Fig. 3) Qualitatively these areas coincide with the critical points identified by a engineers involved in technology. Although not included in simulation, part of the geometry thet caused the most problems during coining is identified by highly stressed area denoted by the arrow. (Fig. 3) Indded this area is a bit to the right at the right angled tip.

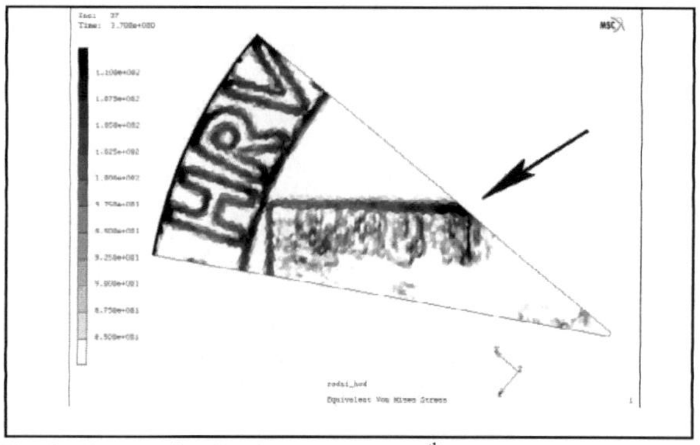

Fig. 3. Stress distribution along 1/12th of the medal. Darker color shows highly stressed areas.

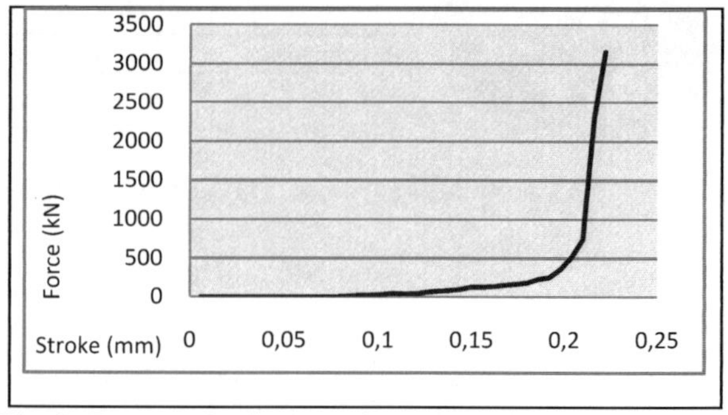

Fig. 4. Force-stroke diagram of medal minting

Force stroke diagram shows adequate calibration steep force rise at punch displacement of 0.22 mm. The order of magnitude is appropriate to the force used in a workshop to mint a medal.

6. CONCLUSION

As shown in this paper it is possible to create and perform 3D FE medal minting simulation. Besides the primarily qualitative accordance of numerical and simulation data, one has to be aware of the limits of the simulation. Capturing the fine details emerging in minting operation requires extremely large number of finite elements. Therefore only critical areas should be modelled, i.e. 'virtual reality' should be only partially applied.

More quantitative interconnections between experiment and simulation should be established. Force-stroke diagram should be recorded, surface hardnesses or even grain texture via MLI (mean linear intercept) should be introduced. Aim of all mentioned is creation of experiment / simulation / production integration, neccesarry for any serious technological advancement.

When considering large models, simple numerical methods regarding element types, surface representation and integration models should be used (MSC.Marc – Vol. A).

7. ACKNOWLEDGMENTS

The present work has been supported by TheMinistry of Science, Education and Sports of Republic of Croatia.

7. REFERENCES

Buffa G., Fratini L. and Micari F. (2007).: The Relevance of the Preform Design in Coining Processes of Cupronickel Alloy, *Proceedings of the 9th International Conference on Numerical Methods in Industrial Forming Processes,* J.M.A. César de Sá and A. D. Santos (Ed.), pp. 1005-1010, ISBN 978-0-7354-0415-1, Porto, Portugal, 17 - 21 June 2007, American Institute of Physics, New York

Ike H., Plancak M. (1998).: *Coining process as a means of controlling surface microgeometry,* Journal of Materials Processing Technology, Volumes 80-81, 1 August 1998, pp. 101-107, ISSN: 0924-0136

Ike H. (2005).: *Nanoscopic surface texture formed by indentation and sliding of a smooth wedge tool,* Wear, Volume 258, Issue 9, April 2005, pp. 1404-1410, ISSN: 0043-1648

MSC.Marc – Volume A Theory and user information (Version 2003) 2003 MSC.Software Corporation, 2 MacArthur Place, Santa Ana, CA 92707

Shaul P. Ladany (1981). : A minting policy for commemorative coins, *European Journal of Operational Research,* Vol. 8, No. 2, (October 1981), pp. 130-138, ISSN: 0377-2217

Annals of DAAAM for 2008 & Proceedings of the 19th International DAAAM Symposium, ISSN 1726-9679
Editor B. Katalinic, Published by DAAAM International, Vienna, Austria 2008
Make Harmony Between Technology and Nature, and Your Mind will Fly Free as a Bird

THEORETICAL MODEL OF THE SURFACE ROUGHNESS AT THE END MILLING WITH CIRCULAR TIPS

SLATINEANU, L[aurentiu]; TOCA, A[lexei]; MAZURU, S[ergiu]; DODUN, O[ana] & COTEATA, M[argareta]

Abstract: *The end milling by the using of the circular inserts is applied to obtain plane surfaces of reduced roughness. The inclination of the tool axis can influence the shape of the asperities generated on the machined surface. An interesting problem concerning the machined surface could be the variation of the surface roughness parameter* R_a *along different directions existing on the machined surface. The paper presents a theoretical model able to emphasize the influence exerted by the radius of the circular inserts, by the feed, by the radius at which the cutting edges are placed and by the distance between the plan defined by the tool axis and the feed direction and the direction where the surface roughness must be evaluated. The mathematical model shows that the surface roughness has approximately the same size on the entire machined surface*
Key words: *milling, inserts, roughness, influences*

1. INTRODUCTION

The milling is the cutting process which uses rotating tools having many cutting teeth placed on the revolution or plane surface; there is a feed motion achieved usually in a plan perpendicular on the tool axis. There are milling techniques which use cylindrical mills, end mills, conical mils etc. If in the case of the rough milling the interest is focused on the material removal rate, on the tool wear, on the sizes of the cutting forces, at the finish milling the main aspects are the surface roughness and the machining accuracy. To obtain a diminished surface roughness, there are used cutting teeth with secondary approach angle κ_1=0 or teeth with circular inserts; the main advantage of the last solution is the possibility to re-establish the cutting properties of the worn tool by the rotation of the circular inserts.

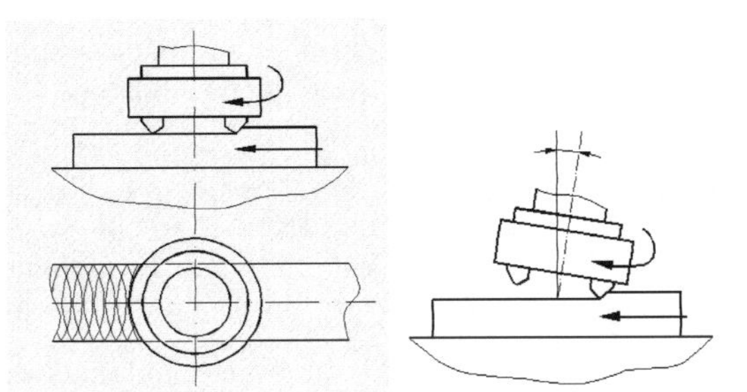

a *b*

Fig. 1. Influence exerted by the inclination of the tool axis on the asperities shape existing on the surface machined by the using of mills with circular inserts.

At the plane surfaces milling by the using of the end mills with circular inserts, there are (fig. 1) two situations: *a)* when the tool axis is perpendicular on the plane of the feed motion, the asperities generated by each tool tooth are intersected (fig. 1, *a*); *b)* if the tool axis is a little inclined (practically, it is difficult to ensure the absolute perpendicularity of the tool axis on the machined surface), the asperities generated by each tooth do not intersect (fig. 1, *b*). In the last case, the machined surface roughness is usually appreciated only along the direction corresponding to the tool axis motion. The problem is to establish if the surface roughness parameter R_a is constant or variable along other directions existing on the machined surface.

As above mentioned, different researchers measured the surface roughness especially along the direction defined by the tool axis motion.

Thus, Lu et al. (Lu et al., 1998) established a multiple regression equation corresponding to the CNC end milling; they considered that the model predicts the size of the surface roughness parameter R_a with about 90% accuracy. In the case of the theoretical profile of the machined surface consisting of elliptical or circular arcs,

Jun Qu and Albert J. Shih developed (Qu and Shih, 2003) some dimensionless form expressions for three surface roughness parameters (R_t, R_a, and R_q). Amin et al. (Amin et al., 2007) studied the influence exerted by the preheating on the surface roughness, in the case of the end milling of carbon steel by the using of the circular carbide inserts. There is not clear information concerning the variation of the surface roughness parameters in different directions included in the plan of the machined surface.

2. THEORETICAL MODELLING

A theoretical model could emphasize the eventual variation of surface roughness parameters along the surface machined by the using of the end mills with circular inserts. To establish such a model, the information included in the figure 2 can be taken into consideration. One can notice that if the rake angle has the value γ=0° and the face of the circular insert is placed in the axial plan, the profile of the machined surface includes a concatenation of arcs of circle along the direction corresponding to the tool axis motion (fig. 3). For any other direction, the surface profile includes arcs of curves different of the arcs of circles.

Fig. 2. Distinct shapes of the machined surface profile in

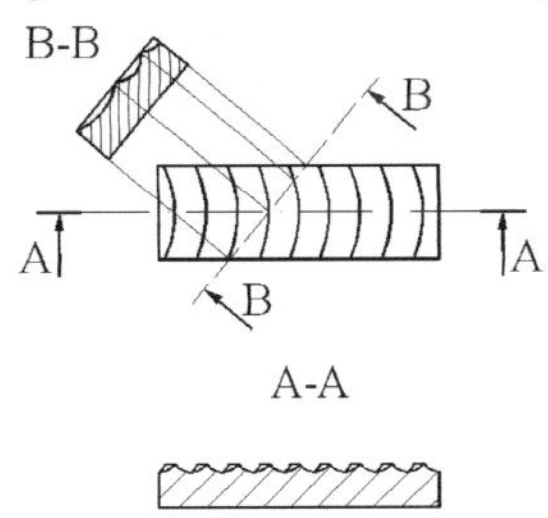

different sections.

The machining scheme is presented in the figure 1; the end mill achieves the rotating motion and the feed motion. The radius of the circular inserts is symbolized by R; the cutting edges of the circular inserts are placed at the radius a.

If the circle corresponding to the insert active edge is placed in the plan yOz, its defining relations are the following:

$$(x-a)^2 + z^2 - R^2 = 0 \qquad y = 0 \tag{1}$$

At the distance b from the plan zOy (this plan is defined by the mill axis and by the direction of the feed motion), the profile of the surface is given (Slătineanu, 2004) by the relation:

$$z = \sqrt{R^2 - \sqrt{y^2 + b^2} - a)^2} . \tag{2}$$

On the other hand, the surface roughness parameter R_a is generally given by the relation:

$$R_a = \frac{1}{n}\sum_1^n |y_i| , \tag{3}$$

where y_i is the distance from a point of the profile to the mean line and n is the number of the points on the segment of the abscissa semiaxis where the roughness parameter R_a must be calculated. Because the surface roughness parameter is defined by taking into consideration the position of the so-called *mean line,* defined by the parameter m (the distance from the mean line to the axis Ox), in the presented case the surface roughness parameter R_a corresponds to the relation:

$$R_a = \frac{\int_{i=1}^{n} |z_i - m|}{n} . \tag{4}$$

A numerical method was used to determine the position of the mean line. It is known (Surface, 2003) that the mean line must have such a position that the sum of the squares of the deviation z_i of the profile from the mean line be minim:

$$S = \int_{i=1}^{n} (z_i - m)^2 = min . \tag{5}$$

Adequate software (Creţu, 1992) was used to determine the size of the parameter m. This software is based on the following considerations (Slătineanu et al., 2004); one can consider the initial position of the mean line just near the axis Oy_1. By taking into consideration an increment Δy, the first sum S_1 of the squares of the profile deviations $(z\text{-}m)^2$ (for a pre established number n of the points of the profile) can be determined.

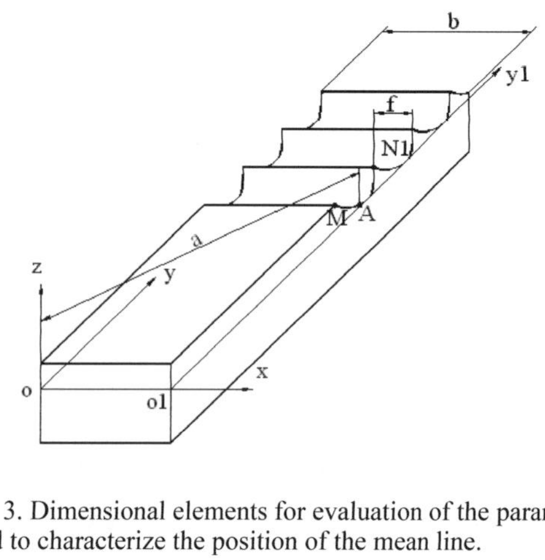

Fig. 3. Dimensional elements for evaluation of the parameter m, used to characterize the position of the mean line.

Within the next step, a new line placed at the distance from the axis Oy increased with the increment Δm can be taken into consideration and a new sum S_2 is calculated. The process is retaken; initially, to series of the sum S_1, S_2, S_3 etc. decreases; when the series begins to increase, the penultimate size of m can be appreciated as offering information about the position of the mean line. The limits of the interval used for the calculus (fig. 3) are

$$\sqrt{\left(a-\frac{f}{2}\right)^2 - b^2} , \sqrt{\left(a+\frac{f}{2}\right)^2 - b^2} .$$

Knowing the size of the parameter m, the size of the surface roughness parameter R_a can be determined, by the using of the relation (4). This relation was used to calculate the size of the parameter R_a, by taking into consideration two levels of the radius a, the feed f, the distance b and the radius R of the circular insert. The mathematical processing of the results thus obtained (by means of software based essentially on the method of the smallest squares) led to the relation:

$$R_a = 20.51 R^{-1.068} a^{0.1214} f^{1.977} b^{-0.0001579} . \tag{6}$$

The analysis of this relation (based on a simplified geometrical model) shown that the size of the surface roughness parameter R_a does not depend on the size of the distance b. On the other hand, the main influence on the size of the parameter R_a is exerted by de feed f, the absolute size of the exponent attached to this size having the biggest value.

The established model offers an image concerning the influence exerted by different factors on the surface roughness at the end milling. In the future, one will try to use other methods to study the surface roughness within a surface obtained by end milling with circular inserts and practically verify the validity of such mathematical models.

3. CONCLUSION

A theoretical model able to emphasize the influence exerted by the feed f, the radius R of the circular inserts, the radius where the cutting edges are placed and by the position of the plane where the surface roughness parameter R_a is measured was proposed.

4. REFERENCES

Amin, K.M.N.; Abraham, I.; Khairusshima, N. & Mirghani I. Ahmed, M.I. (2007). Influence of preheating on performance of circular carbide inserts in end milling of carbon steel. *Journal of Materials Processing Technology* Vol. 185, No. 1-3, 2007-04-30, 97-105, ISSN 0924-0136

Creţu, G. (1992). *Fundamentals of experimental research. Laboratory handbook* (in Romanian). Technical University "Gh. Asachi" of Iaşi, Romania

Lou, M.S.; Chen, J.C. & Li, C.M. (1998). Surface Roughness Prediction Technique for CNC End-Milling. *Journal of Industrial Technology*. Vol. 15, No. 1, November 1998 - January 1999. *Available from:* www.nait.org/jit/Articles/ lou1198.pdf. *Accessed:* 2008-06-23

Qu, J. & Shih, A.J. (2003) Analytical Surface Roughness Parameters of a Theoretical Profile Consisting of Elliptical Arcs. *Machining science and technology*, Vol. 7, No. 2, 281–294, ISSN 1091-0344

Slătineanu, L.; Dodun, O.; Coteaţă, M. & Santo, L. (2004). Study of the surface roughness at the machining of the plane surfaces with end mills. *Progressivnye technologhii i sistemy machinostroenia. Mejdunarodnyi sbornik nauchinyh trudov.* ISBN 966-7907-16-3, Donetskii Natsionalnyi Tekhnicheskii Universitet, Ukraine, 28, 265-270

* * * (2003). *Surfaces and Profiles* (2003). *Available from:* http://www.predev.com/smg/intro.htm. *Accessed:* 2005-03-02

Annals of DAAAM for 2008 & Proceedings of the 19th International DAAAM Symposium, ISSN 1726-9679
Editor B. Katalinic, Published by DAAAM International, Vienna, Austria 2008
Make Harmony Between Technology and Nature, and Your Mind will Fly Free as a Bird
2nd DAAAM-ICAT

SOME ASPECT OF ACCURACY AND QUALITY OF THE SMALL HOLES IN STEREOLITOGRAPHY AND 3DP MODELS

SOBOLAK, M[ariusz]; BUDZIK, G[rzegorz] & BORAWSKA, P[atrycja]

Abstract: *Stereolitography (SL) and 3DP are methods of rapid prototyping (RP). Models are made adequately from the photohardening resin (SL) and from the special powder (3DP). When small holes are created in RP models some errors arise. For example small holes are plugging. The reason for plugging holes in stereolitography is twice or three times highlighting the same point on the surface of resin. The reason for plugging holes in 3DP is, however, the structure of the material.*
Test holes were set perpendicular to the working platform. Holes were tested with rectangular and round cross section.
The results of the research on models made by listed methods of RP are presented.
Key words: *Stereolitography, 3DP, small holes, plugging*

1. INTRODUCTION

Nowadays requirements for rapid prototyping techniques are growth. Precise details from using these methods are maked often. If only about resosultion of RP models (Sobolak & Rejman, 2004; Sobolak & Budzik, 2005) and possibility of its enlargement (Sobolak et al., 2006) was written a lot, problems with small elements were omitted very often. In case of making small holes using RP techniques, technological problems step out. In the article is presented area of applicability selected methods RP in making the small holes.

2. SMALL HOLES IN STEREOLITHOGRAPHY

Stereolithography is the most widely used rapid prototyping technology. It is additive technique. Models were made from layers about settled thickness. The hardening of the resin is made would by the laser light or ultraviolet light exposure with the proper wavelength (Richter & Jacobs, 1992; Kochan & Chua, 1995).
The technology is possible as there is and special device thanks this which models have texture and accurate surface (Stereolithography Buildstation, 2001). It was made layer by layer, surroundings had been hardened, and after that the cut layer is filled. Errors are occurred in holes with small diameters. Picture of model experiments are presented on Fig. 1. The row of round holes was made from ø0,1mm to ø1mm every 0,05mm and square holes from ø0,1mm to ø1mm every 0,05mm.

Fig. 1. Stereolithographic models: round and square holes.

Round and square holes are being made, there is a phenomenon of plugging holes below nominal dimension (Fig. 2). It can be accepted that both round and square holes have border nominal dimension as 0,7mm. However cause of plugging holes are different. Fig. 3 has shown the schedule of laser bundle speed when round hole is being hardened. The internal side of the laser bundle has lower speed than external side. For this reason the internal side has wider hardened (inward). In case of square holes, the internal part of surroundings is being hardened longer. It causes that smaller holes can be plug (Fig. 4).

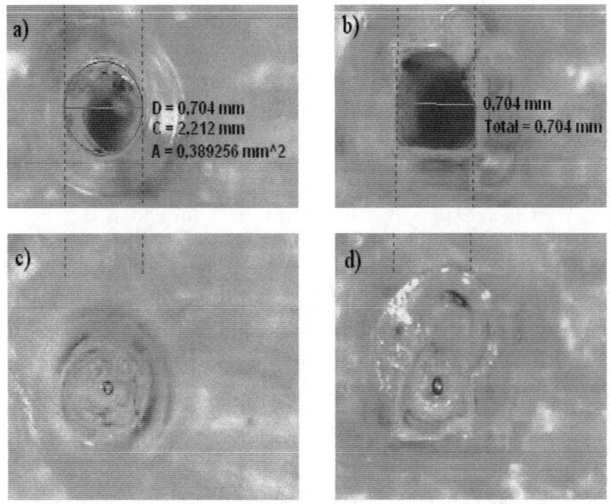

Fig. 2. Holes making by stereolithography; a,b – holes (ø0,7mm, 0,7×0,7mm), c,d – filled holes (ø0,65mm, 0,65×0,65mm).

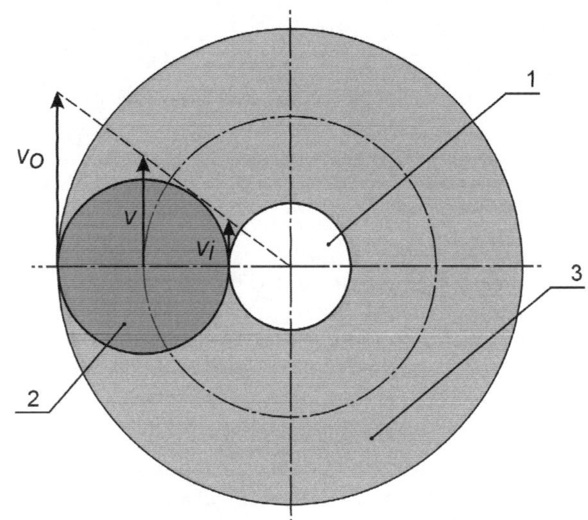

Fig. 3. Harden of round hole's surroundings; 1 – hole, 2 – laser beam, 3 – harden surroundings; v – speed of laser beam, v_o – outer speed, v_i – internal speed.

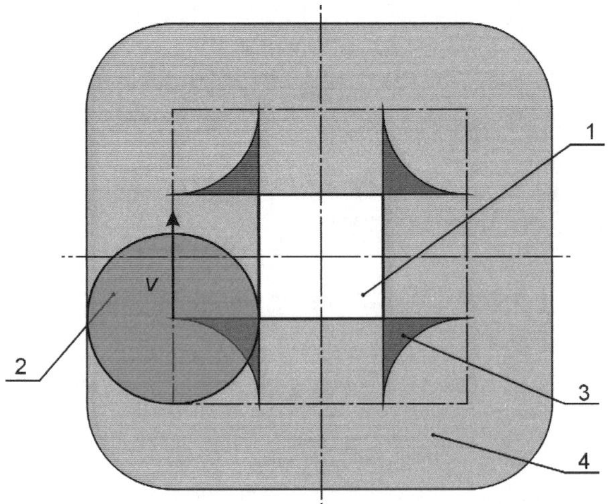

Fig. 4. Harden of square hole's surroundings; 1 – hole, 2 – laser beam, 3 – twice hardened fields, 4 – harden surroundings, v – speed of laser beam.

3. SMALL HOLES USING THREE-DIMENSIONAL PRINTING (3DP)

3DP is another method of RP. Using standard inkjet printing technology, Z Corp. 3D printers create models layer-by-layer, depositing a liquid binder into thin layers of special powder (Budzik, 2008).

Fig. 5 has shown models which was used to testing the correctness of making small holes. The row of round holes was made from ø0,1mm to ø1mm every 0,05mm and square holes from ø0,1mm to ø1mm every 0,05mm.

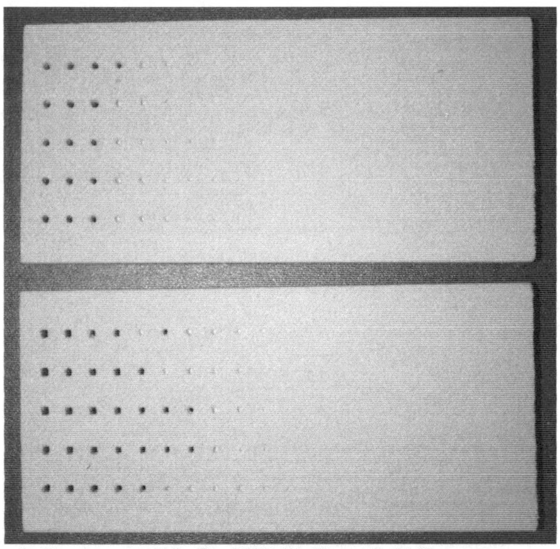

Fig. 5. Testing models for 3DP (Z-Corp, Z-510).

Small holes are filled or it is unfeasible to remove powder from inside of hole. Removing powder couses that there is spalling from holes with bigger dimension on their edges.

External elements about section both round and square holes are made correctly about 0,7mm. Internal, round holes are made from nominal diameter 0,9mm, and square holes – about 0,85mm.

3DP images flat geometric of layer on numerical mesh (just like using ink printer or computer screen). Granularity powder and viscosity binder are responsible for surface structure.

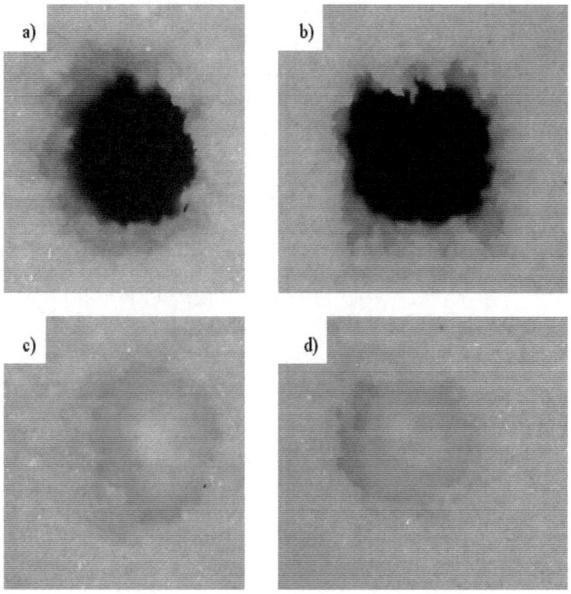

Fig. 6. Holes 3DP with nominal dimension: a – ø0,9mm, b – 0,8×0,8 mm, c – ø0,85mm, d –0,75×0,75mm.

4. CONCLUSIONS

Ranges of technological possibilities of making small holes using RP techniques were presented in this article. Directions of further research will show how to make small diameter holes as well as how to make them in different settings perpendicular to hardened layer.

This research work is financed from the funds allotted for science in 2007-2009, as research and development project (R03 021 02).

5. REFERENCES

Budzik, G. (2008). *Wytwarzanie i analiza modeli wyrobów z wykorzystaniem technologii trójwymiarowego druku (3DP)*, Oficyna Wydawnicza Politechniki Rzeszowskiej, Rzeszów 2008

Kochan, D., Chua, C.K. (1995). *State-of-the-art and future trends in rapid prototyping and manufacturing* . International Journal of Information Technology. 1/1995, p.173-184.

Richter, J., Jacobs, P. (1992). *Rapid prototyping and manufacturing fundamentals of stereolitography SME*, Dearbon, MJ, 1992, p.287-319.

Sobolak, M., Rejman, E. (2004). Some aspects of accuracy and quality of the model surfaces produced In the stereolitography process, *Proceedings of Fourth International Congress: Mechanical Engineering Technologies'04*, Vol 4/72,Varna, Bułgaria, s. 152÷155.

Sobolak M., Budzik G. (2005). Dokładność geometryczna metody stereolitografii. Western Scientific Centre of Ukrainian Transport Academy, Lviv 2005.

Sobolak, M., Budzik, G., Markowski, T., (2006). Metody zwiększenia dokładności prototypów wykonywanych wybranymi technikami RP, *Proceedings of Projektowanie procesów technologicznych TPP 2006*, Komisja Budowy Maszyn PAN O/Poznań, Poznań 2006, s. 65÷70.

Stereolithography Buildstation (2001). 3D Systems Publications.

Annals of DAAAM for 2008 & Proceedings of the 19th International DAAAM Symposium, ISSN 1726-9679
Editor B. Katalinic, Published by DAAAM International, Vienna, Austria 2008
Make Harmony Between Technology and Nature, and Your Mind will Fly Free as a Bird

CAD METHOD USED FOR IMPROVING THE PERFORMANCES OF LIFTING SYSTEMS USED FOR VEHICLES

SPANU, A[lina]; STOENESCU, A[drian] & ANGHEL, F[lorina]

Abstract: *The paper is focused on the problems of studying the dynamic conditions as well as the efforts and linear displacement for the lifting system used for vehicle services. We have analysed the kinematics parameters and the forces acting on the entire scissor linkage during the lifting process. In order to do this task we have designed a 3D model of the assembly and we have made an assembly analyse. Finally we have designed a supplementary subassembly for improving the linear displacement of the platform.*

Key words: *lifting linkage,3D model, assembly analyse.*

1. INTRODUCTION

The paper is focused on the study of the lifting assembly used for a great variety of automotive services in order to achieve a high positioning accuracy as well as high technical performances for the dynamic process. Such lifting systems are used for motocycles, cars, vans and trucks, so that we have to compute the optimal dimensions and actuating system in order to deliver a product which will meet the customer requirements. Much more, the range of products must be reliable and it must be built in compliance with regulations in force. The state of art takes mainly into account two kinds of lifting systems: the first one - a platform hydraulically actuated using a central hydraulic cylinder (Multi Ram, 2008); the second one – a scissor linkage hydraulically actuated too (Scissor Car Lift, 2008). We consider that the last one is more reliable due to its linkage and its safe system in case of cutting-out the hydraulic supplier system. Furthermore the system could be improved regarding the dynamic process by doing some research into the field of actuating system as a feed-back one.

2. DYNAMIC STUDY OF THE LINKAGE

The linkage system was designed as a scissor one (Fig.1) and we have been focused on two main aims: the study of the mechanical forces and the actuating system; the second one – the study of efforts and mechanical deformation during the loading process.

Fig. 1. The assembly of the lifting system.

The kinematical study of motion for the assembly shown in Fig. 1, have been made using the scheme represented in Fig. 2 which are given below.

The kinematical study of motion has been made using the following nonlinear system of six equations:

$$JD * \sin(\varphi 1) + DE * \sin(\varphi 2) = 0 \quad (1)$$

$$JD * \cos(\varphi 1) + DE * \cos(\varphi 2) = s1 \quad (2)$$

$$DI * \sin(\varphi 2 - \pi) + s3 * \sin(\varphi 4) = DG * \sin(\varphi 1) \quad (3)$$

$$DI * \cos(\varphi 2 - \pi) + s3 * \cos(\varphi 4) = DG * \cos(\varphi 1) \quad (4)$$

$$s2 * \sin(\varphi 3) = AD * \sin(\varphi 1) + DC * \sin(\varphi 2 - \pi) \quad (5)$$

$$s2 * \cos(\varphi 3) = AD * \cos(\varphi 1) + DC * \cos(\varphi 2 - \pi) \quad (6)$$

where φ_4 is the angular value between IG and JX axis.

We have computed it by using numerical method Newton-Raphson, so that we could determine the values for the six unknown variables: φ_1, φ_2, φ_3, φ_4 and s_1, s_2. The independent variable is s_2 the displacement of the hydraulic piston of the cylinder for which we have given a constant increment. In Fig. 2 P_1 is the force acting on the platform during the process of lifting a vehicle as we have specified above.

Finally, we have written a twenty one linear equation system for computing the forces acting in each kinematical joint, in order to determine the deformation and the force acting on the hydraulic piston used as actuator. We have written for each kinematical element three equations, two equations for forces and one equation for the values of mechanical couple.

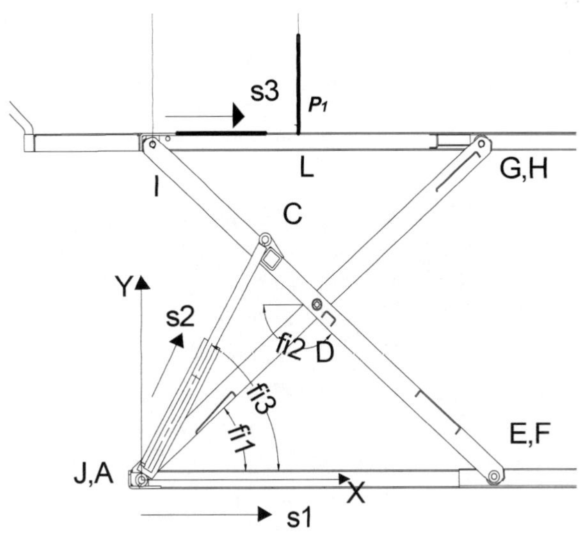

Fig.2. The lifting linkage.

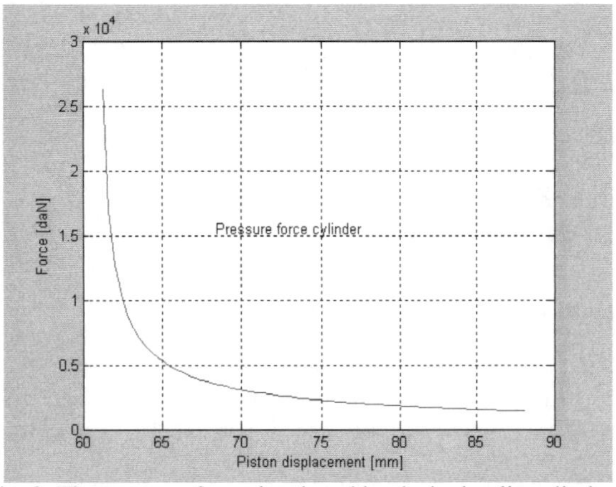

Fig. 3. The pressure force developed by the hydraulic cylinder.

The final result for the force developed by the hydraulic system is given in Fig. 3. We may infer that the value for this force is absolutely great at the beginning of the motion, due to the influence of pressure angle established during the kinematical analysis. (Hong, 2002; Weisheng, 2002).

Based on the results of this study we have turned to the other aim of our work.

3. THE STUDY OF EFFORTS AND DEFORMATIONS FOR THE ENTIRE ASSEMBLY

Taking into account the value for the force acting on the platform, we have made the analyses for the entire assembly using CATIA V5 R16. First of all we have designed the assembly using the soft mentioned above. (Ghionea, 2007).

For the best result of such analyse we have to specify the kind of restraints for each joint as well as the type of connections.

We have considered for the computation the value of pressure force needed for hydraulically actuation, which was established during the above theoretical study.

The load on the platform may vary depending on the vehicle type – car, van, truck and so on. The load was considered distributed along a limited surface and following a percentage of the vehicle heavy: 40% for the front of the platform and 60% for the back of the platform.

As we may infer from the Fig. 4 and 5 the biggest values for efforts and linear displacement was achieved for the platform. Another critical zone was that of G and H joints (Fig. 2 and Fig. 4) due to their type especially regarding the use of this kind of ball bearings. The main idea was to find out a technical solution for increasing the rigidity of these two kinematic elements.

Fig. 4. The efforts developed in joints and element for the entire assembly.

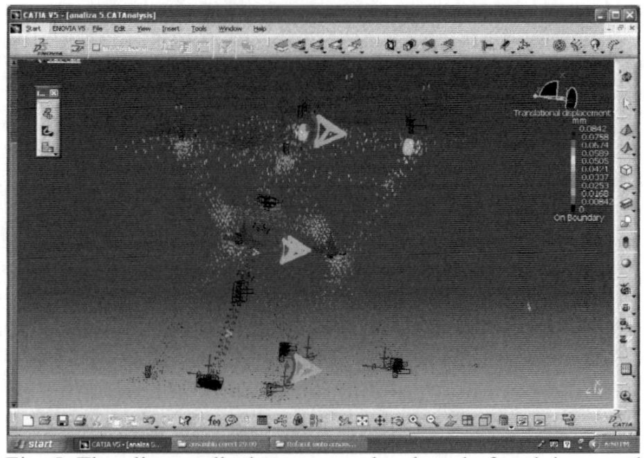

Fig. 5. The linear displacements developed for joints and elements for the entire assembly.

Finally we have designed and we have analysed a new subassembly having the aim of improving the dynamic process. This subassembly could be attached to the sides of the platform. We have made the analyses again with these new parts and the results show us a considerable decreasing of efforts.

All the results of our research were applied by a company which produces such lifting systems and they have achieved expected functional parameters.

Future works will be focused on the improving the hydraulic supplier system especially regarding its dynamics.

4. CONCLUSION

The lifting linkage used for vehicles during their service activities has to be very reliable and compliant with regulations in force. A major problem is the computation of deformation and linear displacement of the entire linkage during the lifting process. First we have made a kinematic analyse which helped us to determine the values for speed and acceleration.

The second step has been regarding the computation of force acting on each kinematic element taking into account the dynamics for the hydraulic supplier system.

The final aspect was the 3D analyse using the 3D model of the entire assembly. We have studied the linear deformations taking into account the entire linkage, so that we could determine the direct influence between them. The critical zones were the platform as well as the upper side of the kinematic elements of the linkage used to sustain the platform.

Consequently, the future works would study technical solutions for the upper side elements and for the dynamics of hydraulic supplier system for a better lifting process using this kind of linkage.

5. REFERENCES

Ghionea, I. (2007). Proiectare asistata in CATIA V5. Elemente teoretice si aplicatii (Computer Aided Assisted in CATIA V5. Theory and applications), *Editura Bren*, ISBN 978-973-648-654-8, Bucuresti, Romania.

Hong, S., Chiu, G., T. (2002). Motion Synchronization for Dual-Cylinder Electrohydraulic Lift Systems, *Mechatronics IEEE/ASME Transaction*, Volume 7, Issue 2, June 2002, Pg. 171-181.

MultiRam Lifting System, *Available from*: http://www.slift.de *Accessed*: 2008-03-21.

Scissor Car Lift Systems, *Available from*: http://www.cartuningcentral.com *Accessed*: 2008-03-21

Weisheng, Z. (2002). Automatic Control of the Hydraulic Lifting System, *Proceedings of the Fifth (2002) ISOPE Pacific/Asia Offshore Mechanics Symposium*, Daejeon, Korea, November 17-20, 2002.

Annals of DAAAM for 2008 & Proceedings of the 19th International DAAAM Symposium, ISSN 1726-9679
Editor B. Katalinic, Published by DAAAM International, Vienna, Austria 2008
Make Harmony Between Technology and Nature, and Your Mind will Fly Free as a Bird

ACTIVE STEERING CONTROL OF A RAIL VEHICLE: A NEW APPROACH

SPIRYAGIN, M[aksym]; SPIRYAGIN, V[alentyn] & Ulshin, V[italy]

Abstract: *The object of this investigation was to design, develop and simulate a fuzzy logic controller for an active steering control system of a rail vehicle. The active steering railway bogie is built upon the application of axle boxes with links and actuators inclined to the longitudinal axle of bogie in the horizontal plane. Actuator actions depend on analysis of noise in the wheel-rail contact. The obtained results of performed simulations with a locomotive model running in a curve show the possibility of the application of the proposed design for new rail vehicles and locomotives produced earlier, which have running gear based on Alstom designed constraints.*

Key words: *Bogie, control system, actuator, rail vehicle, link.*

1. INTRODUCTION

The dynamics of rail vehicles has been significantly improved during the last years by means of implementation of complex mechatronic systems. The behavior of locomotive running in a curve is more complicated because the lateral forces appear between a flange of the wheel and the head of rail in the contact zone, and the values of wheel slip are higher than for straight parts of track. These two factors are the main reasons for wheel wear. To solve this problem and to improve vehicle dynamics, the steering systems have found wide application.

An active steering system for rail vehicles, where mechanical components are replaced by the use of control systems and actuators, is one of effective solution (Spiryagin, 2004). A very good review and analysis in the field of active steering control systems for railway vehicles was made by R.M. Goodall and T.X. Mei. A number of steering strategies are possible for solid-axle wheelsets, but the perfect steering can be achieved if the angle of attack for two wheelsets can be controlled to be equal, and the bogie is in line with the track on curves (Iwnicki, 2006). However, in the real work conditions, it is quite difficult to reach the required equality. In the existing common strategies, the significant performance improvements can be obtained by means of measuring the relative yaw angle between the individual wheelset and the bogie and compensating for the primary forces (Iwnicki, 2006). Nevertheless, it is better to make a direct measurement of the angle of attack because this parameter is a one of the main characteristics, which has a big influence on wear process (Masliev, 2002). Therefore, the steering system can be designed if the angle of attack is taken into account as one of the important input signals. The angle of attack can be obtained based upon noise spectrum analysis in the wheel-rail contact.

This paper considers the design of active control for the wheelset steering, which includes a fuzzy logic controller for getting a desired performance. The proposed design is validated through simulation using a model of vehicle and a control system.

2. MECHANICAL SYSTEM

The mechanical system based on Alstom designed constraints, where the radius links are positioned on different levels in an anti-parallelogram application (Iwnicki, 2006), is used for our investigation.

The theoretical and experimental investigation described in (Spiryagin, 2004) shows that more effective results can be reached in case of the use of inclined links to the longitudinal axle of bogie in the horizontal plane. However, to enhance the steering ability of a railway vehicle, it is better to install dual direction actuators on a bogie, because if the angle of steering is not enough then actuators could transfer additional displacements to axle boxes in two directions.

Based upon the above, we proposed a decision for the active steering of wheelsets on curved track, which is shown in Fig. 1. Several design ideas are being explored for the development of a bogie with better performance. One of the similar design constructions was applied for the three axle bogie (Gorbunov et al., 2003).

3. WHEEL-RAIL NOISE

The situation with noise analysis for vehicle's movement on the curved part of track is more complicated. Recently, the papers in this research area were published. The publication (Hsu et al., 2007) describes theoretical and experimental investigation performed with the twin disc rig. The analysis confirms that noise levels are different and depend on yaw angle for different adhesion coefficients. Also, the obtained results of experimental investigation in publication (Kocha et al., 2006), show that squeal sound pressure amplitudes emitted by undamped wheels increase in direct proportion to the rolling speed and to the angle of attack, and curve squeal disappears when the wheel flange is in contact with the rail.

Based upon what is written above, it is possible to develop an active steering control that allows for the accounting for the contact characteristics depending on noise analysis in the contact.

4. CONTROL SYSTEM

The block scheme of the proposed microprocessor control system for a railway vehicle is shown in Fig. 2.

The detection of the angle of attack is based upon a noise spectrum analysis in the wheel-rail contact and the use of GPRS and GPS technologies. GPS satellite system is used for obtaining the position of a railway vehicle at a specific moment of time. After receiving the current position on the curve, the track characteristics for the current position can be obtained by means of the GPRS from the station computer.

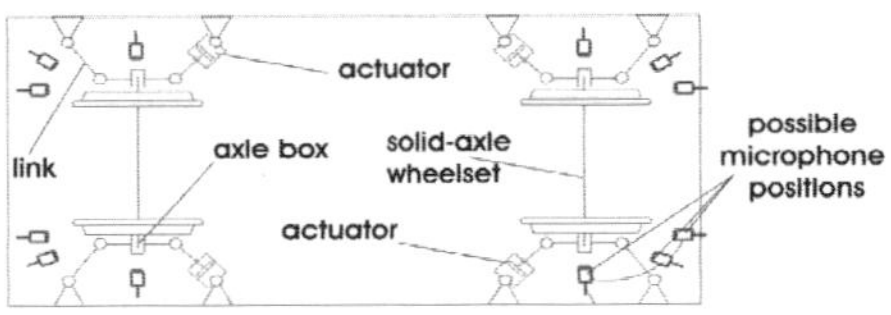

Fig. 1. View from above of the bogie with actuators for active steering.

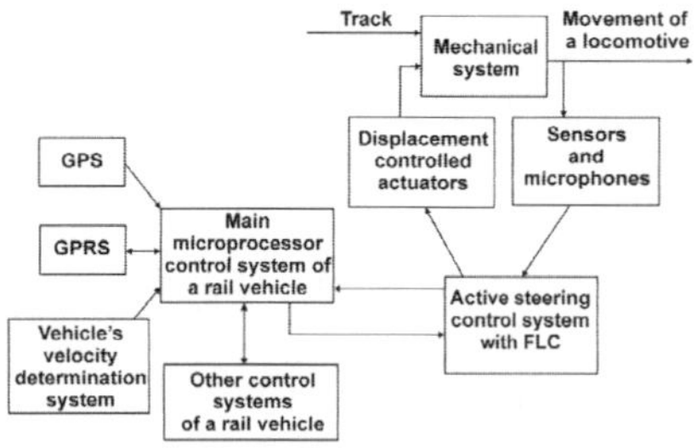

Fig. 2. Microprocessor control system of a rail vehicle.

The proposed active steering control system is presented in Fig. 3. First, the observer collects and processes all information from sensors and microphones, and provides the optimal and the estimated steering angles as outputs. After that, the values of angles are received by the feedback controller.

The radius of the track curvature R is known now by means of using GPS/GPRS technologies and the ideal steering angle γ_{IDEAL} (Fig. 4) for the wheelset can be calculated with the following equation

$$\gamma_{IDEAL} = \arcsin\left(\frac{b}{2R}\right) \tag{1}$$

where b is the distance between the leading ant trailing axles.

The noise in the contact zone are picked up by using a microphones provided near each wheels of bogies. The obtained noises are processed by a special algorithm to obtain a sound pressure level of a certain frequency band. By looking up a special database data, received from experimental and theoretical research, the dependence between the sound pressure level, vehicle velocity V, relative slip, the lateral displacement and the angle of attack ψ can be obtained.

The estimated steering angle γ_{est} (in our case, γ_1 and γ_2 for each wheelset) can be calculated by means of the following equation obtained from (Zharov & Markov, 2002):

$$\gamma_{est} = \psi - \frac{y_1 - y_2}{b} + i^* \cdot \frac{b}{2R} \tag{2}$$

where y_1, y_2 are the lateral displacements of wheelsets; $i^* = 2i - 3$ (for the leading wheel set i=1 and for the trailing one i=2).

The control law is based on a simple feed-back control and can be represented as the following equitation:

$$F_{in} = F + \Delta F \tag{3}$$

where F is the initial control force given to the wheelset from the actuator. The value of the initial control force F at the starting moment of locomotive movement on a curved part of track can be obtained by looking up a special database data, received from theoretical research. The corrective actuator force ΔF can be obtained by means of fuzzy logic rules as shown in Table 1. We chose triangle and trapezium shapes for membership functions in our controller because they are very basic and widely applied.

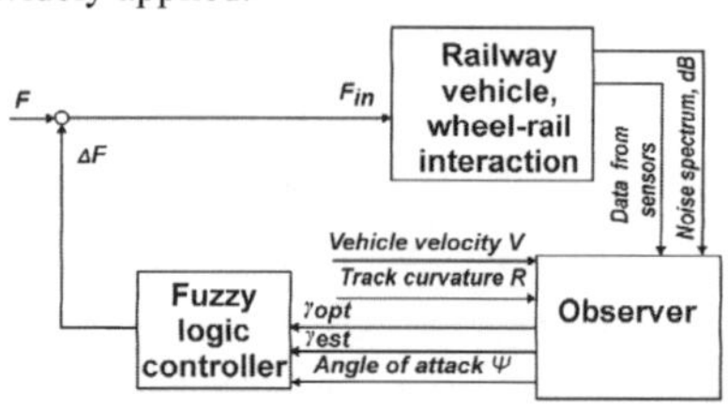

Fig. 3. Block diagram of the active steering control system.

The group of linguistics variables was used for the both input and output variables. This group includes N and P stand for negative and positive values, S, M and B means small, medium and large values, and ZE means zero. The range of values for difference of angles for membership functions were obtained from analysis of mechanical design constraints.

	$\gamma_{IDEAL} - \gamma_{est}$						
	NB	NM	NS	ZE	PS	PM	PB
ΔF	NB	NB	NM	ZE	PM	PB	PB

Tab. 1. The set of rules for the controller.

5. SIMULATION

For the generation of the simulation model the non-linear MBS software package called Simpack has been used. The proposed control strategy was modeled in Matlab/Simulink. The Simpack model of the vehicle was linked with the control unit in Matlab/Simulink by means of the SIMAT-interface. Based on the software packages described above the system was investigated by using co-simulation.

As a result, we achieved a satisfactory control system.

6. CONCLUSION

The active steering control based on noise analysis with the application of fuzzy logic controller has been examined. It has been confirmed by means of simulations that the control system can offer good curving performance of the actively steered vehicle. However one of the main objectives of the proposed active control is the detection of the angle of attack. As a sequence, it is necessary to do further research in this field for a more detailed study on the dependence of the angle of attack from noise in rail-wheel contact for different values of slip, velocity, wheel load and wheel's positions on track. The obtained results can give the further development of active primary suspensions.

7. REFERENCES

Spiryagin V. (2004). *Improvement of dynamic interaction between the locomotive and railway track*, Ph.D. dissertation, East Ukrainian National University named after Volodymyr Dal, Lugansk, Ukraine.

Iwnicki S. (2006). Handbook of Railway Vehicle Dynamics, CRC Press, UK.

Masliev V. G. (2002). *Scientific fundamentals of a choice of design-engineering parameters of devices for locomotive sprockets bindings wear reduction*, D.Sc. dissertation, Ukrainian State Academy of Railway Transport, Kharkov, Ukraine.

Gorbunov, M.; Spiryagin, V.; Spiryagin, M. & Lapin, D.; *Triaxial truck of a railway vehicle*, Patent UA62900, 15 December, 2003.

Hsu, S.S.; Huang, Z.; Iwnicki, S.; Thompson, D.J.; Jones, C.J.C.; Xie, G. & Allen, P.D. (2007). Experimental and theoretical investigation of railway wheel squeal, In: *Proc. IMechE Part F: J. Rail and Rapid Transit*, vol. 221, pp. 59-73.

Kocha, J.R.; Vincenta, N.; Chollet, H. & Chielloc, O. (2006). Curve squeal of urban rolling stock—Part 2: Parametric study on a 1/4 scale test rig, In: *J. of Sound and Vibration*, vol. 293, pp. 701–709.

Zharov, I.A. & Markov, M.A. (2002). Vliyanie radiusa krivoj i smazyvaniya na soprotivlenie dvizheniya telezhki pri kvazistaticheskom dvizhenii, Vesnik VNIIZHT, vol. 3. *Available from:* http://www.css-rzd.ru/vestnik-vniizht/v2002-3/v2-9_1.htm *Accessed:* 2008-05-26

Annals of DAAAM for 2008 & Proceedings of the 19th International DAAAM Symposium, ISSN 1726-9679
Editor B. Katalinic, Published by DAAAM International, Vienna, Austria 2008
Make Harmony Between Technology and Nature, and Your Mind will Fly Free as a Bird

KINEMATICS OF THE SPHERICAL PARALLEL ROBOT

STAICU, S[tefan] & RUGESCU, R[adu]

Abstract: *Matrix recursive relations (MRR) are established for the inverse kinematics (IK) of the Agile Wrist spherical parallel robot and presented in this paper. The three actuators of the robot produce three independent rotations, controlled through torque. Assuming a given position and a rotational motion of the platform, the IK problem is solved through MRR.*
Key words: *Matrix, kinematics, parallel, robot, platform*

1. INTRODUCTION

When compared with serial mechanisms, the parallel robot is a spatial mechanical structure with specific features: smaller in weight and with improved stiffness, good positioning accuracy, suitable arrangement of actuators and a low workspace volume for a similar target area.

Considerable effort has been devoted to the kinematics of full parallel manipulators. Among these robots, the class of manipulators known as Stewart-Gough platforms had focused special attention. The Star parallel manipulator (Herve & Sparacino, 1992) and the Delta parallel robot (Clavel, 1988; Tsai & Stamper, 1996) are equipped with three motors that drive the moving platform in a general translational motion. The general direct kinematics of the Agile Wrist spherical parallel robot was previously developed in a tedious manner (Gosselin & Angeles, 1989).

Spherical parallel manipulators provide high stiffness and accuracy due to their mechanical arrangement in which all the actuators are anchored to the base. Three kinematical chains support the end-effector. Accuracy and precision in the execution of the task of the wrist are essential since the robot is intended to operate on fragile objects; where positioning errors of the tool could end in costly damages.

The matrix approach here presented largely simplifies the computations and transforms them into a highly reliable tool.

2. INVERSE GEOMETRIC MODEL

Some recursive matrix relations for the kinematics of a three-degree-of-freedom spherical parallel robot are first established. This mechanism presents a structure with nine axes, concurrent in a common *centre of rotation*, namely the fixed point O (Fig. 1). Let $Ox_0y_0z_0(T_0)$ be a fixed Cartesian frame, about which the manipulator moves. It possesses three legs of known size and mass. The architecture consists of two main elements, the base and the moving platform, which is free to undergo arbitrary rotations around its centre and three identical legs, each of these composed of two links coupled by revolute joints.

The first element T_1 of leg A is called *proximal link* and is one of the three driving parts of the robot. It is a homogenous rod of radius r, rotating around the Oz_1^A axis with the angular velocity $\omega_{10}^A = \dot{\varphi}_{10}^A$ and the angular acceleration $\varepsilon_{10}^A = \ddot{\varphi}_{10}^A$. The *distal link* T_2 of the same radius r is connected to the $A_2x_2^Ay_2^Az_2^A$ frame and manifests a relative rotation with the angle φ_{21}^A and angular velocity $\omega_{21}^A = \dot{\varphi}_{21}^A$.

The platform of the robot is a three arms star with length equal to $l = r\sqrt{(2+\sin^2\delta)/3}$, which rotates with an angular velocity $\omega_{32}^A = \dot{\varphi}_{32}^A$ with respect to the neighboring body. The link angles of the manipulator are securely the same on each of the legs connecting the end-effectors to the base. They are denoted by $\beta = \pi/2$ for the proximal link and $\gamma = \pi/2$, for the distal link. The angles $\alpha_A = 0$, $\alpha_B = 2\pi/3$, $\alpha_C = -2\pi/3$, $\delta = \pi/6$ and $\sin\theta = 1/\sqrt{3}$, $\cos\theta = \sqrt{2/3}$ give the geometry of the base and the initial position of the manipulator's arm. Each leg of the robot is a serial spherical wrist: the first joint couples the base with the proximal link, the second joint couples the proximal link with the distal link and the last one couples the distal link with the platform (see Fig. 1).

Let us consider the rotation angles $\varphi_{10}^A, \varphi_{10}^B, \varphi_{10}^C$ of the three actuators A_1, B_1, C_1 as the variables that give the instantaneous position of the mechanism. In the inverse geometric problem however, it can be considered that the three Euler angles $\alpha_1, \alpha_2, \alpha_3$ give the absolute orientation of the mobile platform. Pursuing the leg A along the path $OA_1A_2A_3$, we obtain the successive transformation matrices (Staicu, 2007; Rugescu, 2003)

$$a_{10} = a_{10}^{\varphi}\theta_1 a_\beta a_\alpha^A,\ a_{21} = a_{21}^{\varphi} a_\delta \theta_1 \theta_2,\ a_{32} = a_{32}^{\varphi}\theta_1, \quad (1)$$

with the displacement matrices given in (2).

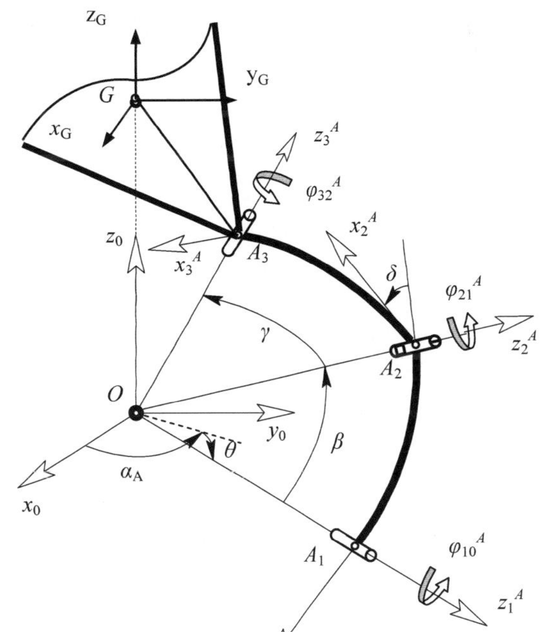

Fig. 1 Kinematical scheme of first leg A of the mechanism

$$\theta_1=\begin{bmatrix}0&0&-1\\0&1&0\\1&0&0\end{bmatrix},\ \theta_2=\begin{bmatrix}-1&0&0\\0&-1&0\\0&0&1\end{bmatrix},\ a_\theta=\begin{bmatrix}\cos\theta&0&-\sin\theta\\0&1&0\\\sin\theta&0&\cos\theta\end{bmatrix}$$

$$a_\alpha^A=\begin{bmatrix}\cos\alpha_A&\sin\alpha_A&0\\-\sin\alpha_A&\cos\alpha_A&0\\0&0&1\end{bmatrix},\ a_\delta=\begin{bmatrix}\cos\delta&\sin\delta&0\\-\sin\delta&\cos\delta&0\\0&0&1\end{bmatrix},\qquad(2)$$

$$a_{k,k-1}^{\varphi}=\begin{bmatrix}\cos\varphi_{k,k-1}^A&\sin\varphi_{k,k-1}^A&0\\-\sin\varphi_{k,k-1}^A&\cos\varphi_{k,k-1}^A&0\\0&0&1\end{bmatrix},\ a_{k0}=\prod_{j=1}^{k}a_{k-j+1,k-j}$$

Let us suppose for example, that the absolute motion of the platform is a 3-D rotation around the fixed point O, defined by the Euler angles

$$\alpha_l=\alpha_l^*(1-\cos\frac{\pi}{3}t),\quad (l=1,2,3)\,. \qquad(3)$$

From the orientation of the platform in the fixed frame, successive relative rotations around axes Ox,Oy,Oz lead to the rotation matrix $a=a_3a_2a_1$, with the following identities

$$a_{30}^{\circ T}a_{30}=b_{30}^{\circ T}b_{30}=c_{30}^{\circ T}c_{30}=a\,, \qquad(4)$$

$$a_{30}^{\circ}=a_\alpha^{AT}a_\theta^T\theta_1^T\theta_2^T\theta_1^Ta_\delta^T\theta_1^T. \qquad(5)$$

From these relations we obtain the real time evolution of nine characteristic angles $^{j}_{10},\ ^{j}_{21},\ ^{j}_{32}\ (j=A,B,C)$.

3. VELOCITIES AND ACCELERATIONS

The kinematics of the elements of leg A, for example, is characterized by the absolute angular velocities given by the recursive relations (6) and the velocity the joint A_k by (7)

$$\vec\omega_{k0}^A=a_{k,k-1}\vec\omega_{k-1,0}^A+\omega_{k,k-1}^A\vec u_3,\omega_{k,k-1}^A=\dot{}^A_{k,k-1}\,. \qquad(6)$$

$$\vec v_{k0}^A=a_{k,k-1}\vec v_{k-1,0}^A+a_{k,k-1}\tilde\omega_{k-1,0}^A\vec r_{k,k-1}^A,\ \vec v_{k,k-1}^A=\vec 0. \qquad(7)$$

the the velocity the joint A_k. Knowing the rotation of the platform through relations (3), we develop the *inverse kinematics model* (Staicu, 2007) and determine the velocities $\vec v_{k0}^A,\vec\omega_{k0}^A$ and accelerations $\vec\gamma_{k0}^A,\ \vec\varepsilon_{k0}^A$ of each moving joint A_k by differentiating (6) and (7). The *matrix conditions of connectivity* is:

$$\omega_{10}^A\vec u_i^Ta_{10}^T\vec u_3+\omega_{21}^A\vec u_i^Ta_{20}^T\vec u_3+\omega_{32}^A\vec u_i^Ta_{30}^T\vec u_3=$$
$$=\vec u_i^T\{\dot\alpha_1a_1^T\vec u_1+\dot\alpha_2a_1^Ta_2^T\vec u_2+\dot\alpha_3a_1^Ta_2^Ta_3^T\vec u_3\}, \qquad(8)$$

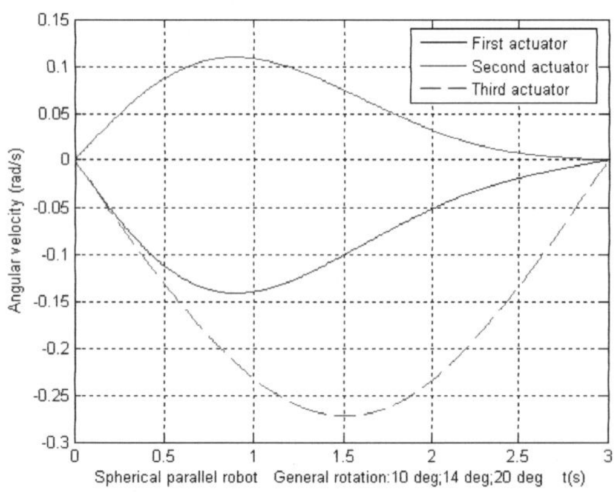

Fig. 2 Angular velocities ω_{10}^j of all three robot actuators

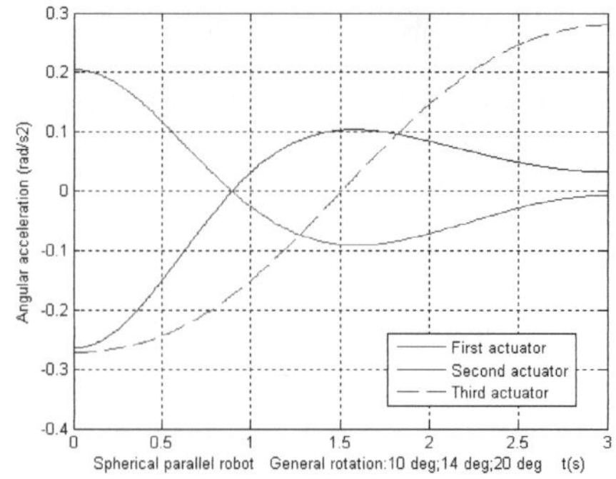

Fig. 3 Angular accelerations ε_{10}^j of the three actuators

with

$$\vec u_1=[1\ 0\ 0]^T,\ \vec u_2=[0\ 1\ 0]^T,\ \vec u_3=[0\ 0\ 1]^T \qquad(9)$$

These relations give the relative angular velocities ω_{10}^A, $\omega_{21}^A,\omega_{32}^A$ as a function of the angular velocities $\dot\alpha_1,\dot\alpha_2,\dot\alpha_3$ of the platform. The relative angular accelerations $\varepsilon_{10}^A,\varepsilon_{21}^A,\ \varepsilon_{32}^A$ are given by the *conditions of connectivity* resulting from (8):

$$\varepsilon_{10}^A\vec u_i^Ta_{10}^T\vec u_3+\varepsilon_{21}^A\vec u_i^Ta_{20}^T\vec u_3+\varepsilon_{54}^A\vec u_i^Ta_{30}^T\vec u_3=$$
$$=\vec u_i^T\{\ddot\alpha_1a_1^T\vec u_1+\ddot\alpha_2a_1^Ta_2^T\vec u_2+\ddot\alpha_3a_1^Ta_2^Ta_3^T\vec u_3+$$
$$+\dot\alpha_1\dot\alpha_2a_1^T\tilde u_1a_2^T\vec u_2+\dot\alpha_2\dot\alpha_3a_1^Ta_2^T\tilde u_2a_3^T\vec u_3+ \qquad(10)$$
$$+\dot\alpha_3\dot\alpha_1a_1^T\tilde u_1a_2^Ta_3^T\vec u_3-\omega_{10}^A\omega_{21}^Aa_{10}^T\tilde u_3a_{21}^T\vec u_3-$$
$$-\omega_{21}^A\omega_{32}^Aa_{20}^T\tilde u_3a_{32}^T\vec u_3-\omega_{32}^A\omega_{10}^Aa_{10}^T\tilde u_3a_{21}^Ta_{32}^T\vec u_3\},$$

where $\tilde u_1,\tilde u_2,\tilde u_3$ are three skew symmetric matrices.

4. RESULTS AND CONCLUSIONS

Consider, for example, a spherical robot that has the following characteristics: $\alpha_1^*=\pi/36,\alpha_2^*=\pi/18,\alpha_3^*=\pi/12$, $r=0.074$ m, $\Delta t=3$ s. The graphs for angular velocities ω_{10}^j (Fig. 2) and angular accelerations ε_{10}^j (Fig. 3) of the three revolute actuators result. The recursive matrix method proves superior, avoiding any analytical resolution of tedious equations. It gives directly the motion of any point of the robot, easy, accurate, free of algebraic errors and fast. The computational program is elementary and the method perfectly suited for a real time automatic control of robots.

5. REFERENCES

Clavel, R. (1988) Delta: a fast robot with parallel geometry, *Proceed. 18th Int. Symp. Industrial Robots*, Lausanne

Gosselin, C. & Angeles, J. (1989), The optimum kinematic design of a planar three-degree-of-freedom parallel manipulator, *ASME J. Mech. Trans. Aut. Des.* 110(1)p.35-41

Hervé, J-M. & Sparacino, F. (1992) Star. A New Concept in Robotics, *Proceedings of the Third International Workshop on Advances in Robot Kinematics*, Ferrara

Rugescu, R. (2003), Bound Vectors & Spectra of Rigid Rotation Matrices, *Proc. SISOM-2003*, Bucharest, Romania, pp.134-139.

Staicu, S., Liu, X-J. & Wang, J. (2007) Inverse dynamics of the HALF parallel manipulator with revolute actuators, *Nonlinear Dynamics*, Springer, 50, 1-2, pp. 1-12

Tsai, L-W. & Stamper, R. (1996) A parallel manipulator with only translational degrees of freedom, *ASME Design Engineering Technical Conferences*, Irvine, CA.

Annals of DAAAM for 2008 & Proceedings of the 19th International DAAAM Symposium, ISSN 1726-9679
Editor B. Katalinic, Published by DAAAM International, Vienna, Austria 2008
Make Harmony Between Technology and Nature, and Your Mind will Fly Free as a Bird

KINEMATICS OF A 3-*PRR* PLANAR PARALLEL MANIPULATOR

STAICU, S[tefan]; MAGHETI, I[oan] & CARP-CIOCARDIA, D[aniela] C[raita]

Abstract: *Recursive modelling for the kinematics of the well known 3-PRR planar parallel manipulator with prismatic actuators is established in this paper. Three identical planar legs connecting to the moving platform are located in a vertical plane. Knowing the motion of the platform, some matrix equations offer recursive expressions and graphs for the displacements, velocities and accelerations of the three actuators.*

Key words: *kinematics, planar parallel manipulator, platform*

1. INTRODUCTION

The parallel architectures have the following potential advantages in comparison with serial robots: higher kinematical precision, lighter weight and better stiffness, stabile capacity and suitable position of arrangement of actuators.

For a *planar mechanism*, the loci of all points of its bodies can be drawn conveniently in a same plane.

Aradyfio and Qiao (1985) examined the inverse kinematics solution for the three different 3-DOF planar parallel robots. Gosselin and Angeles (1988) as well as Pennock and Kassner (1990) present a kinematical study of a planar parallel manipulator, where a moving platform is connected to a fixed base by three legs consisting of two binary bodies and three parallel revolute joints. Williams et al. (1988) analysed at Ohio University the control of a planar parallel manipulator.

A recursive method is introduced in the present paper, with the purpose to reduce significantly the number of equations and computation operations. This method uses a set of matrices for the kinematics of the 3-*PRR* planar parallel manipulator (Fig.1).

2. KINEMATICS ANALYSIS

Having a closed-loop structure, the planar parallel robot is a special symmetric mechanism composed of three planar kinematical chains of variable length, with identical topology, all connecting the fixed base to the moving platform. Each leg consists of two bodies, with one prismatic joint and two revolute joints. The prismatic joints of the manipulator ($\underline{P}RR$) are in fact, three actively controlled prismatic cylinders, which can be installed on the fixed base. The whole mechanism consists of seven moving bodies, six revolute joints and three prismatic joints.

Let us attach a Cartesian frame $x_0y_0z_0$ to the fixed base

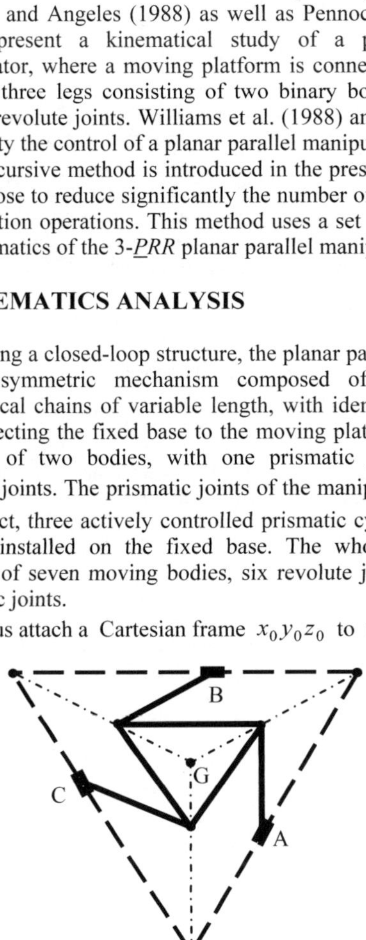

Fig. 1 The 3-*PRR* planar parallel manipulator

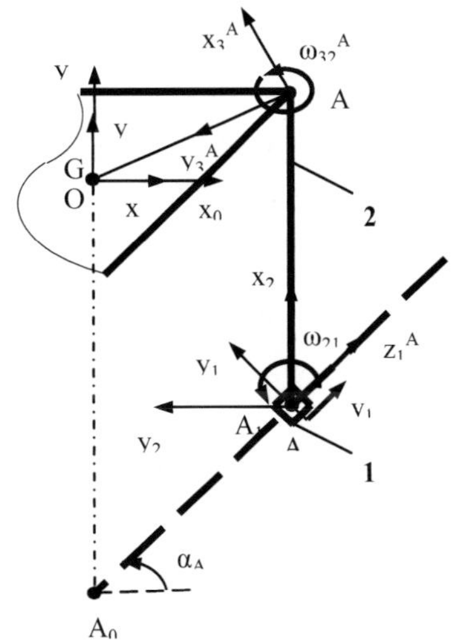

Fig. 2 Kinematical scheme of first leg A of the mechanism

with its origin located at the centre O and the z_0 axis perpendicular to the base. The origin of the central coordinate system $x_Gy_Gz_G$ is located at the centre G of the moving platform (Fig. 2).

One of the three active legs consists of a prismatic joint, which is like a piston **1** linked at the $x_1^Ay_1^Az_1^A$ frame, and has a displacement λ_{10}^A, a velocity $v_{10}^A = \dot{\lambda}_{10}^A$ and an acceleration $\ddot{\lambda}_{10}^A$. The second element of the leg is a rigid rod **2** of length l_2 linked at the $x_2^Ay_2^Az_2^A$ frame, having a relative rotation about z_2^A axis with the angle φ_{21}^A, the velocity $\omega_{21}^A = \dot{\varphi}_{21}^A$ and the acceleration $\ddot{\varphi}_{21}^A$. Finally, a revolute joint is connected to the equilateral moving platform with the edge $l = r\sqrt{3}$, which rotates with the angle φ_{32}^A and the angular velocity $\omega_{32}^A = \dot{\varphi}_{32}^A$ about z_3^A axis.

Pursuing the first leg A in the $OA_0A_1A_2A_3$ way, we obtain the following matrices of transformation:

$$a_{10} = \theta_1 a_\alpha^A, \quad a_{21} = a_{21}^\varphi \theta_2 \theta_1^T, \quad a_{32} = a_{32}^\varphi \theta_2, \tag{1}$$

where a_α^A, θ_1, θ_2 are three constant matrices while $a_{k,k-1}^\varphi$ is an orthogonal rotation matrix (Staicu et al., 2007).

In the inverse geometric problem, we can consider that the position of the mechanism is completely given by the coordinates x_0^G, y_0^G of the mass centre G and the orientation angle ϕ, which are expressed by the analytical functions

$$\frac{x_0^G}{x_0^{G*}} = \frac{y_0^G}{y_0^{G*}} = \frac{\phi}{\phi^*} = 1 - \cos\frac{\pi}{3}t \tag{2}$$

We obtain the following relations between angles, from the rotation conditions of the moving platform:

$$\varphi_{21}^{A}+\varphi_{32}^{A}=\varphi_{21}^{B}+\varphi_{32}^{B}=\varphi_{21}^{C}+\varphi_{32}^{C}=\phi\,. \qquad (3)$$

Other six variables $\lambda_{10}^{A}, \varphi_{21}^{A}, \lambda_{10}^{B}, \varphi_{21}^{B}, \lambda_{10}^{C}, \varphi_{21}^{C}$ will be determined by several vector-loop equations, as follows:

$$\vec{r}_{10}^{\,i}+\sum_{k=1}^{2} q_{k0}^{T}\vec{r}_{k+1,k}^{\,i}+q_{30}^{T}\vec{r}_{3}^{\,Gi}=\vec{r}_{0}^{\,G}\,, \qquad (4)$$

where one denoted:

$$\begin{aligned}&\vec{r}_{10}^{\,i}=\vec{r}_{00}^{\,i}+(l_1+\lambda_{10}^{i})q_{10}^{T}\vec{u}_3\,,\ \vec{r}_{00}^{\,A}=l_0[0\quad -1\quad 0]^{T}\\&\vec{r}_{00}^{\,B}=0.5l_0[\sqrt{3}\quad 1\quad 0]^{T},\ \vec{r}_{00}^{\,C}=0.5l_0[-\sqrt{3}\quad 1\quad 0]^{T}\\&\vec{r}_{21}^{\,i}=\vec{0},\ \vec{r}_{32}^{\,i}=l_2\vec{u}_1,\ \vec{r}_{3}^{\,Gi}=r\vec{u}_2\quad (i=A, B, C).\end{aligned} \qquad (5)$$

Recursive relations express the absolute angular velocities $\vec{\omega}_{k0}^{A}$ and the velocities $\vec{v}_{k0}^{A}$ of the joints A_k :

$$\begin{aligned}&\vec{\omega}_{k0}^{A}=a_{k,k-1}\vec{\omega}_{k-1,0}^{A}+\omega_{k,k-1}^{A}\vec{u}_3\,,\quad (k=1, 2, 3)\,.\\&\vec{v}_{k0}^{A}=a_{k,k-1}\vec{v}_{k-1,0}^{A}+a_{k,k-1}\tilde{\omega}_{k-1,0}^{A}\vec{r}_{k,k-1}^{\,A}+v_{k,k-1}^{A}\vec{u}_3\,.\end{aligned} \qquad (6)$$

Equations of geometrical constraints (3) and (4) can be derivate with respect to the time to obtain the following *matrix conditions of connectivity* (Staicu et al., 2008)

$$\begin{aligned}&v_{10}^{A}\vec{u}_j^{T}a_{10}^{T}\vec{u}_3+\omega_{21}^{A}\vec{u}_j^{T}\{a_{20}^{T}\tilde{u}_3\vec{r}_{32}^{\,A}+a_{20}^{T}\tilde{u}_3a_{32}^{T}\vec{r}_{3}^{\,GA}\}+\\&+\omega_{32}^{A}\vec{u}_j^{T}a_{30}^{T}\tilde{u}_3\vec{r}_{3}^{\,GA}=\vec{u}_j^{T}\dot{\vec{r}}_{0}^{\,G}\,,\ \omega_{21}^{A}+\omega_{32}^{A}=\dot{\phi}\,,\end{aligned} \qquad (7)$$

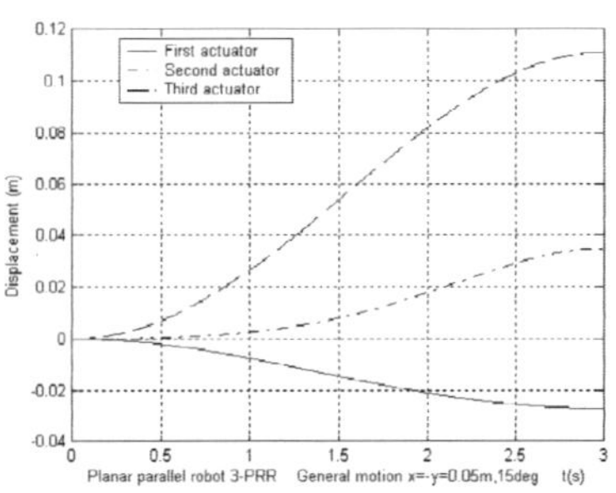

Fig. 3 Displacements of the three actuators

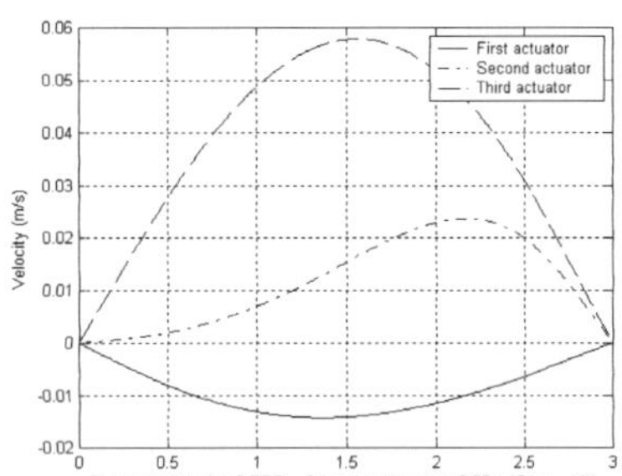

Fig. 4 Velocities of the three actuators

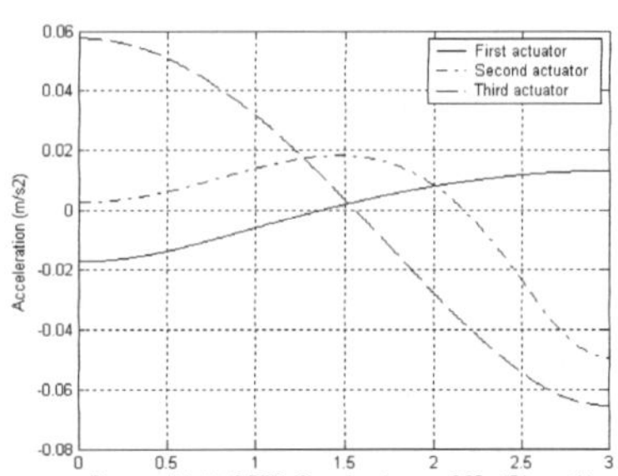

Fig. 5 Accelerations of the three actuators

where $\tilde{u}_3$ is a skew-symmetric matrix associated to the unit vector $\vec{u}_3$ directed to the positive direction of z_k axis.

As for the relative accelerations $\varepsilon_{10}^{A}, \gamma_{21}^{A}, \varepsilon_{32}^{A}$ of the manipulator, the derivatives with respect to the time of the equations (7) give other following conditions of connectivity

$$\begin{aligned}&\gamma_{10}^{A}\vec{u}_j^{T}a_{10}^{T}\vec{u}_3+\varepsilon_{21}^{A}\vec{u}_j^{T}\{a_{20}^{T}\tilde{u}_3\vec{r}_{32}^{\,A}+a_{20}^{T}\tilde{u}_3a_{32}^{T}\vec{r}_{3}^{\,GA}\}+\\&+\varepsilon_{32}^{A}\vec{u}_j^{T}a_{30}^{T}\tilde{u}_3\vec{r}_{3}^{\,GA}=\vec{u}_j^{T}\ddot{\vec{r}}_{0}^{\,G}-\\&-\omega_{21}^{A}\omega_{21}^{A}\vec{u}_j^{T}\{a_{20}^{T}\tilde{u}_3\tilde{u}_3\vec{r}_{32}^{\,A}+a_{20}^{T}\tilde{u}_3\tilde{u}_3a_{32}^{T}\vec{r}_{3}^{\,GA}\}-\\&-\omega_{32}^{A}\omega_{32}^{A}\vec{u}_j^{T}a_{30}^{T}\tilde{u}_3\tilde{u}_3\vec{r}_{3}^{\,GA}-2\omega_{21}^{A}\omega_{32}^{A}\vec{u}_j^{T}a_{20}^{T}\tilde{u}_3a_{32}^{T}\tilde{u}_3\vec{r}_{3}^{\,GA}\,,\\&\varepsilon_{21}^{A}+\varepsilon_{32}^{A}=\dot{\phi}\,,(j=1, 2)\end{aligned} \qquad (8)$$

The relationships (7) and (8) represent the *inverse kinematical model* of the planar parallel manipulator.

As application let us consider a mechanism having the following characteristics:

$$\begin{aligned}&x_0^{G*}=-y_0^{G*}=0.025\,m\,,\quad \phi^{*}=\frac{\pi}{24}\,,\quad \Delta t=3\,s\\&r=0.1\,m\,,\quad l_0=0.3\,m\,,\quad l_1=l=r\sqrt{3}\,,\quad l_2=0.2\,m\,.\end{aligned} \qquad (9)$$

Using the MATLAB software, a computer program was developed to solve the studied inverse kinematics problem. To illustrate the algorithm, we assume that for a period of three seconds the platform starts from a central configuration, rotates and moves along a rectilinear trajectory.

Based on the computational program, the displacements (Fig. 3), the velocities (Fig. 4) and the accelerations (Fig. 5) of the actuators are plotted versus time.

3. CONCLUSIONS

Within the inverse kinematical analysis, some exact relations that give the time-history evolution of the displacements, velocities and accelerations of each element of the parallel robot have been established in the present paper.

The simulation by the presented program certifies that one of the major advantages of the current matrix recursive formulation is a reduced number of additions or multiplications and consequently a smaller processing time of numerical computation. Also, the proposed method can be applied to various types of complex robots when the number of components of the mechanism is increased.

4. REFERENCES

Aradyfio, D.D. & Qiao, D. (1985) Kinematic Simulation of Novel Robotic Mechanisms Having Closed Chains, *ASME Mechanisms Conference*, Paper 85-DET-81

Gosselin, C. & Angeles, J. (1988) The optimum kinematic design of a planar three-degree-of-freedom parallel manipulator, *ASME Journal of Mechanisms, Trans. and Automation in Design*, 110, 1, pp. 35-41

Pennock, G.R. & Kassner, D.J. (1990) Kinematic Analysis of a Planar Eight-Bar Linkage: Application to a Platform-type Robot, *ASME Mechanisms Conference*, pp. 37-43

Williams II, R.I. & Reinholtz, C.F. (1988) Closed-Form Workspace Determination and Optimisation for Parallel Mechanisms, *The 20th Biennial ASME Mechanisms Conference*, Kissimmee, Florida, DE, Vol. 5-3, pp. 341-351

Staicu, S., Liu, X-J. & Wang, J. (2007) Inverse dynamics of the HALF parallel manipulator with revolute actuators, *Nonlinear Dynamics*, Springer, 50, 1-2, pp. 1-12

Staicu, S. & Zhang, D. (2008) A novel dynamic modelling approach for parallel mechanisms analysis, *Robotics and Computer-Integrated Manufacturing*, Pergamon-Elsevier, 24, 1, pp. 167-172

Annals of DAAAM for 2008 & Proceedings of the 19th International DAAAM Symposium, ISSN 1726-9679
Editor B. Katalinic, Published by DAAAM International, Vienna, Austria 2008
Make Harmony Between Technology and Nature, and Your Mind will Fly Free as a Bird

BAROTHERMOPOLYMERIZATION APPARATUS WITH ELECTROMAGNETIC INDUCTION

STAMATE, V[alentin]; CHICOS, L[ucia] A[ntoneta]; OANCEA, G[heorghe] & LANCEA, C[amil]

Abstract: *The paper presents a barothermopolymerization apparatus used for polymerization of dental prosthetic works as well as for polymerization of different types of composites. For obtaining the polymerization temperature the apparatus uses a heating system based on the electromagnetic induction system. The entire polymerization process is controlled automatically the process taking place in conditions of pressure and temperature required by the composite materials.*
Key words: *Barothermopolymerization apparatus, Dental technique, Electromagnetic induction, Composites.*

1. INTRODUCTION

Polymerization of certain composite materials used in the dental technique imposes a controlled medium of temperature, time and pressure (Bechir et al., 2006). For obtaining the polymerization temperature, various types of apparatus used in the dental laboratories, use different external heating sources (Mestra, 2008) or an electric resistance submerged in a liquid bath within the working vessel (Dreve, 2008), (Westdental, 2008) or winding the vessel on the outside (Caloris, 2005).

The main drawback of these apparatus is the heating source. In case of the resistance on the outside it will heat first the vessel and then the polymerization liquid and it requests a supplementary cooling system using water or air as cooling fluid.

As for the electric resistance submerged in a polymerization liquid because of the limestone deposition, the metal corrosion, the starts/stops consecutively of the apparatus determine a decrease of the resistance reliability and implicitly of the apparatus.

The apparatus presented in this paper and named CIF A-01 (fig. 5) is a barothermopolymerization apparatus that for obtaining the polymerization temperature uses the electromagnetic induction principle. The apparatus can use various polymerization mediums such as a liquid bath or air.

Using of this method determines only heating the bottom side of the vessel and then the liquid/air within the working vessel.

2. *CIF A-01* BAROTHERMOPOLYMERIZATION APPARATUS

The apparatus comprises, as main components, a working recipient, a thermoregulator, a timer, a heating system based on the electromagnetic induction principle and an electronic system that controls the apparatus operation.

2.1 Working recipient

The working recipient of the barothermopolymerization apparatus, presented in the figure 1, comprises a working vessel 1 and the lid 2. The vessel has a cylindrical shape and at the upper side has a collar provided with a slot 3 for the silicone gasket 4.

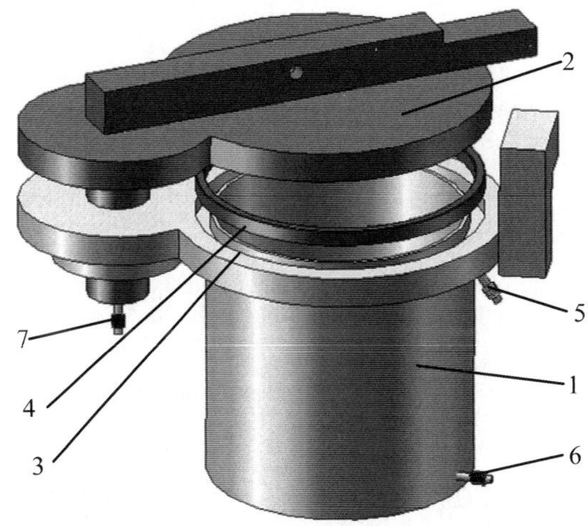

Fig.1. The working recipient

The wall of the working vessel is made of stainless steel without magnetic properties. The bottom side of the apparatus is made of stainless steel with ferromagnetic proprieties. The lid is made of stainless steel and it can be opened or closed by pivoting.

The upper nozzle 5 allows passing the air that action upon the silicone gasket 4 and the bottom nozzle 6 allows the discharge of water under pressure. The air under pressure is introduced in the vessel through nozzle 7.

Sealing up between the vessel and lid is done by means of a gasket specifically designed with cross-section in the shape of the U letter (fig. 2).

In contrast with the gaskets with circular section, this shape allows a hermetic sealing of the vessel because the more pressure will increase the more silicone gasket is pressed on the vessel walls and its lid thus optimizing the sealing. The gasket is made of a silicone resistant to wear and temperature.

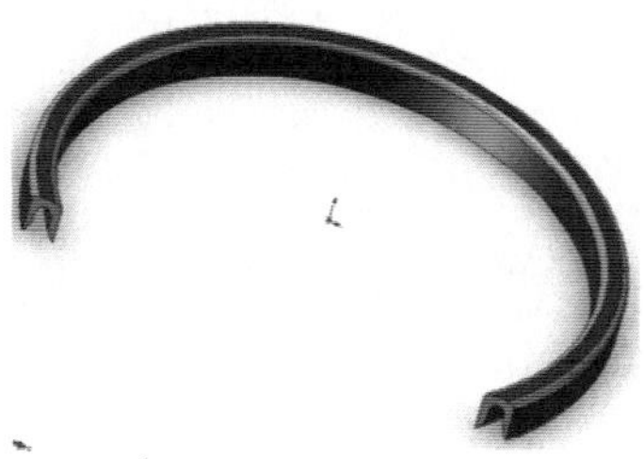

Fig. 2. The silicone gasket

2.2. Heating system

As mention before CIF A-01 apparatus uses for heating polymerization liquid the electromagnetic induction principle.

For this propose it uses an electronic circuit 1 presented in the figure 3 that amplify the electric current frequency from

Fig. 3. Electronic circuit of the heating system

50 Hz to about 30 kHz.

This high-frequency current passes through the inductor 2 from the figure 3/figure 4 and makes around it a magnetic field 3 (fig. 4).

The vessel is placed at a certain distance above the inductor 2 thus the magnetic flux which penetrates the bottom side of the vessel induces an electromotive force that determines the creation of the eddy-currents (Foucault) (Morega, 2005), (Anghel, 2006). These currents generate heat by Joule effect and first will heat the vessel bottom 1 and then the medium used for polymerization.

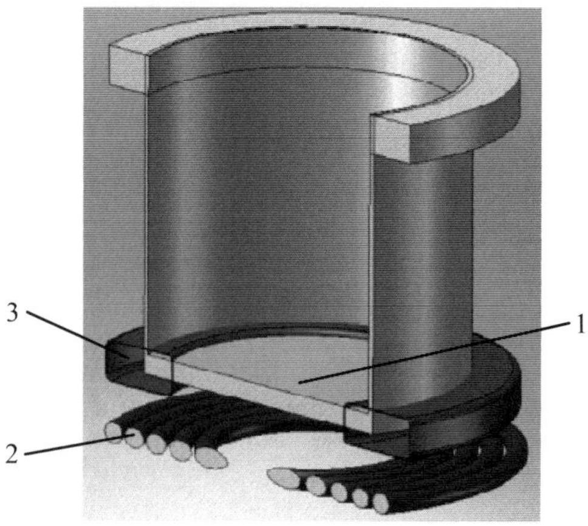

Fig. 4. Heating principle

3. OPERATING PRINCIPLE

The barothermopolimerization apparatus CIF A-01 (fig. 5) is able to generate a temperature between 20 - 150°C and an adjustable pressure up to 10 atm.

The dental work that shall be polymerized is inserted into the working vessel and will be covered with water or glycerine. After closing the lid of the vessel the compressed air is introduced at the desired pressure by pressing button 1. Air will enter first in the opening of the silicon joint achieving sealing and then the air is pumped into the working vessel.

The air pressure gauge 2, fixed on the apparatus casing, is able to give information about the air pressure value at any moment. The thermoregulator 3 is used for setting the polymerization temperature of the material and the timer 4 is used for programming the polymerization time for each dental work.

To turn on the apparatus the buttons 5, 6 and 7 shall be pressed successively. From this point begins warming of the vessel bottom then the heat is transmitted to the inside liquid.

The polymerization process starts when the specific temperature of each material is reached. The polymerization time depend on the material properties which will be polymerized.

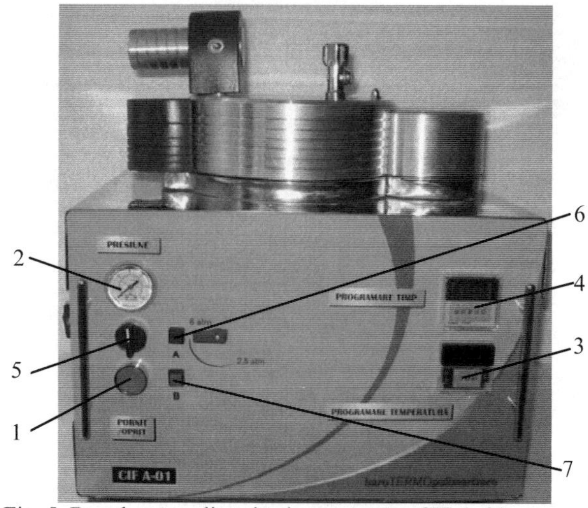

Fig. 5. Barothermopolimerization apparatus CIF A-01

4. CONCLUSION

Upon expiry time polymerization the electronic circuit of the apparatus commands the following:

- turning off the heating system;
- opening an electrovalve for evacuating the pressurized liquid from the vessel.

Finally, the lid is opens and the dental work can be recovered.

Barothermopolymerization apparatus CIF A-01 (fig. 5) by using the electromagnetic induction principle allows increasing heating efficiency, the reliability and security operation and removes the drawbacks of the apparatus that use other heating sources.

The apparatus can be used in polymerization of various types of dental works and a wide range of composite materials used in various fields.

In the future developments we intend to use a light source in order to polymerize the photopolymerizable composite materials, as well.

5. REFERENCES

Anghel, F.S.; Anghel, D. (2006). Metode si procedee tehnologice (Technological Methods and Procedures), *Available from:* http://www.amotion.pub.ro/oferta_educationala/download/capitolul16.pdf *Accessed:* 2008-08-25

Bechir, A.; Pancă, A.; Toma, I.; Nicolau, A. (2006). *Materiale dentare utilizate în laboratorul de tehnică dentară (Dental Materials Used in the Dental Technique Laboratory),* Editura Leda & Muntenia, ISBN 973-692-100-X, Constanța, Romania

Caloris. (2005). Termobaropolimerizator tip BARO 2000 (Thermobaropolymerizator BARO 2000), *Available from:* http://www.caloris.ro/content/medicalstomatologic.htm *Accessed:* 2008-05-16

Dreve. (2008). Polymax1 (120°C), *Available from:* http://www.dreve.de/dreve_neu/dentamid_gb/labor_polymax1120_dent_gb.htm *Accessed:* 2008-07-28

Mestra. (2008). Poly-Drucktopf, *Available from:* http://www.mestradeutschland.de/pdf/bedienungsanleitungen/BEDDrucktopf.pdf *Accessed:* 2008-06-23

Morega, M. (2005). Campul electromagnetic Herzian (Hertzian Electromagnetic Field), *Available from:* http://www.iem.pub.ro/~mihaela/cursuri/bioelectromagnetism/BEM_curs/C11_campul%20electromagnetic.pdf *Accessed:* 2008-04-15

Westdental. (2008). Multicure, *Available from:* http://www.westdental.ro/index.php?pag=detalii&prod_id=3307&nod_id=135 *Accessed:* 2008-07-02

Annals of DAAAM for 2008 & Proceedings of the 19th International DAAAM Symposium, ISSN 1726-9679
Editor B. Katalinic, Published by DAAAM International, Vienna, Austria 2008
Make Harmony Between Technology and Nature, and Your Mind will Fly Free as a Bird

LOW POWER DEVICES FOR GESTURAL HUMAN - COMPUTER INTERACTION

STAN, A[ndrei]; PANDURU, L[ucian] & LUPU, R[obert]

Abstract: *This paper presents a wireless system for remote human-computer interaction involving low power devices. The system is based on the user's hand movement recognition in order to control remotely the software applications running on a PC. The hand movement is detected by a three axes accelerometer integrated in a digital embedded system placed on hand's wrist, representing the endpoint of the wireless network. A wireless communication interface allows data exchange between the endpoint and the network coordinator that is connected to the PC via RS232. Some aspects concerning software and digital signal processing are presented.*
Keywords: *accelerometer, microcontroller, firmware, embedded, wireless*

1. INTRODUCTION

The common interaction between humans makes use of hands movement to indicate, represent or express certain object properties or attitudes about the subject of conversation. The main goal of the system described in this paper is to control remotely the software applications that run on PC machines or digital embedded systems by using a set of intuitive movements of the user's hands.

The gestural control system for software applications running on a PC is based on the user's hands movement recognition. The system behaves like an alternate input device for digital systems that needs an input interface for the human user. The main modules of the proposed system are presented in Figure 1:

- The endpoint or bracelet is a digital embedded system placed on a hand's wrist and its function is to detect the hand movements and to transmit to the coordinator the digital information. This module is the main control element of the system. The analog signals from accelerometer MMA7260QT are filtered and converted in digital values using the analog to digital converter of the MSP430F1611IRDT microcontroller (Texas Instruments, 2006). The main challenge is to properly interpret the accelerometer output and translate acquired data into meaningful commands to the PC applications. The data packets are sent to the coordinator using a XBee modem;
- The coordinator is a bridge between the wireless medium and the PC that hosts the controlled application. This embedded device contains a XBee modem for receiving messages, a MSP430 microcontroller for processing the information and a RS232 transceiver for communication with PC;
- The PC hosts the controlled application, receives the digital movement information and processes it in order to obtain specific commands for the controlled software application. A friendly graphical user interface is needed.

The wireless communication between the endpoints and the coordinator is implemented using the ZigBee stack of protocols (MaxStream, 2006).

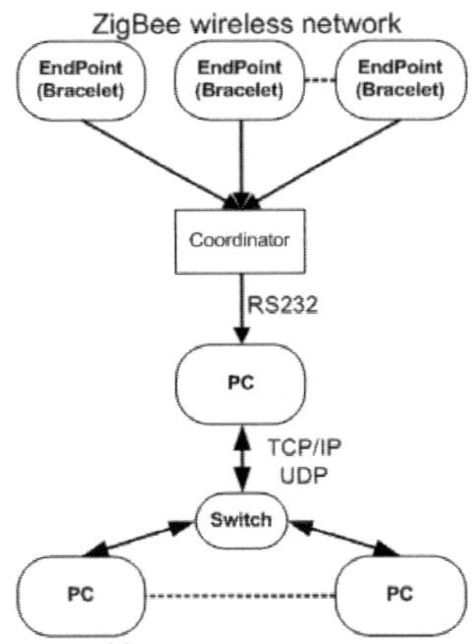

Fig. 1. The system architecture

This solution is low power and low cost and it is in concordance with the data bandwidth required by application. It also supports various topologies with multiple endpoints.

The coordinator is the management unit of the ZigBee network: it allows the endpoints to be associated to a network. The PC is the main data processing module in the system and it may be connected in a network to further enhance the functionality of the application.

The system has to face the challenge implied/involved by a low power consumption demand.

The devices we have proposed could be involved in e-health applications and patient assistance or social integration.

2. THEORETICAL ASPECTS

The most important physical parameter that changes and that can be measured when the hand moves starting from stillness is the acceleration along a certain axis. The hand movement detection is based on physical phenomena of acceleration appearance (Freescale, 2008). The acceleration (***a***) represents the rate of velocity (***v***) change of an object. Consequently, the velocity represents the rate of position (***s***) change of an object. In other words, the velocity is the position derivative and acceleration is velocity derivative.

It is supposed that the acceleration is known as a function of time and by applying a double integration over this function of time the position of the object can be find, assuming initial zero conditions:

$$\vec{v} = \int(\vec{a})dt, \vec{s} = \int(\vec{v})dt \Leftrightarrow \int(\int(\vec{a})dt)dt \qquad (1)$$

The computing algorithm for these two integrals uses the trapezes approximation method, as follows:

$$\begin{aligned} v_{current} &= v_{previous} + (a_{current} + a_{previous})\frac{T}{2} \\ s_{current} &= s_{previous} + (v_{current} + v_{previous})\frac{T}{2} \end{aligned} \qquad (2)$$

In conclusion, the hand velocity is computed by integrating the hand acceleration signal and the hand position is computed by integrating the hand velocity.

3. DIGITAL SIGNAL PROCESSING

The endpoint has various software modules that make data conversion and processing for hand acceleration, data message construction and transmission to the coordinator using ZigBee wireless connection (Shorey et al., 2006).

The conversion cycle is performed once at 10 ms. Because this is a low power system, the accelerometer is disabled between two conversion cycles. Each conversion cycle has a 2 ms startup period for accelerometers to recover from sleep mode. After data acquisition, a message is built with acquired data and transmitted to the coordinator.

Because of the sensitivity of the accelerometer and the high resolution of analog to digital converter, the useful signal (the output voltages of accelerometer) has a large amount of superposed noise with possible sources in handshaking or even heart beating. In Figure 2, the acquired signal affected by noise is presented. There can be seen the initial rise of voltage corresponding to the start of movement. This is explained by the fact that the movable central mass of g cell moves in opposite direction of the movement. As the hand movement velocity becomes constant, the acceleration decreases and movable central mass comes back to initial position. This settling process implies acceleration opposite to the direction of movement: the latter half of the signal presented in Figure 2.

Noise filtering is achieved by using a 64 point rolling average filter. Mathlab simulation was used with real datasets for filter calibration. Without filtering, the signal is almost unusable for further processing. After filtering, the noise is substantially reduced.

To easily distinguish between positive and negative accelerations (e.g. to determine the sense of movement) the accelerometer has an offset of half of supply voltage for 0g acceleration. So, for positive or negative accelerations the corresponding output voltage is positive in all cases.

To make the difference between positive and negative accelerations, it is necessary to fix the reference value corresponding for 0g. This is done at the application startup, when an average with 512 points is made. During this process, the hand must be held in a precise and fix position.

In Figure 3, the results of the algorithm are plotted. The first plot is the raw unfiltered data from accelerometers outputs. The second plot represents the velocity computed by integrating first filtered signal and the third plot represents the distance computed by integrating the velocity.

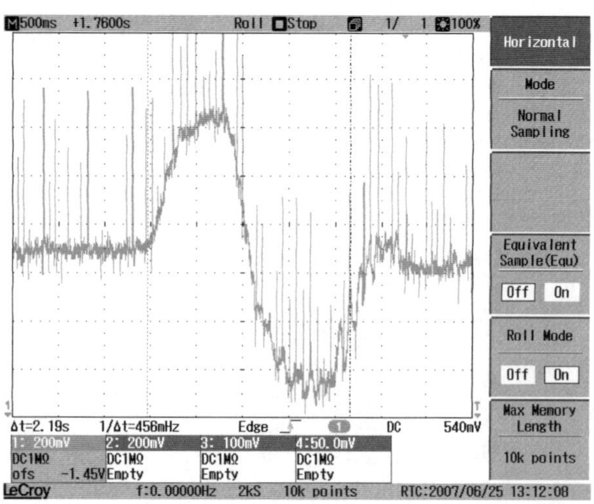

Fig. 2. The accelerometer time response

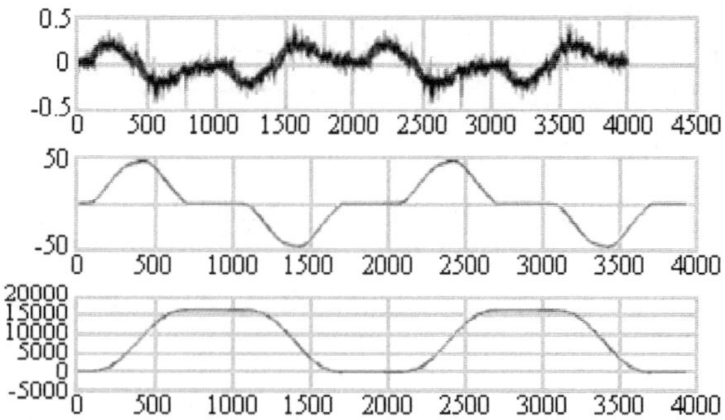

Fig. 3. The processed signals

4. FIRMWARE ISSUES

The developed firmware consists of several tasks and each of them manages a particular resource. The communication between tasks is implemented with semaphores and waiting queues allowing a high level of parallelism between processes. Each process may be individually enabled or disabled. This feature is very important in increasing the flexibility of the application and also to optimize the overall power consumption of the system (Phoha at al., 2006).

To address power consumption issues several techniques are used. Power gating involves temporarily shutting down blocks in the design when the blocks are not in use. One special aspect of power gating is managing the in-rush current when the power is reconnected. All this issues are implemented in this design. The microcontroller offers different power saving modes providing the mobile device the ability to trade-off between power consumption in standby and recovery times.

5. CONCLUSIONS

This paper presents some challenges of hardware and software design of a low-power gestural control system.

The system successfully interprets the accelerometer data and computes the coarse hand displacement based on acceleration. Using displacement information, various hand trajectories are decoded and translated to commands which are then issued to the PC applications.

The implemented solution is original and all the devices features ultra-low power operation modes controlled by firmware.

The most consistent part of the work was focused on digital signal processing of accelerometer output, on wireless network development and on efficient microcontrollers programming.

Besides PC application control, the proposed devices could be also used in patient monitoring systems.

6. REFERENCES

Freescale, MMA7260QT accelerometer datasheet, *Available from:*http://www.freescale.com/files/sensors/doc/data_sheet /MMA7260QT.pdf?pspll=1,*Accessed:* 2008-03-08

MaxStream, ZigBee module datasheet, *Available from:* http://ftp1.digi.com/support/documentation/manual_xb_oe mrfmodules_802.15.4.pdf, *Accessed:* 2008-03-28

Phoha S., La Porta T., Griffin C., (2006). *"Sensor Network Operations"*, John Wiley & Sons, Inc., ISBN 0-471-71976-5, New Jersey

Shorey R., Ananda A., Chan M. C., Ooi W. T. (2006). *"Mobile, wireless, sensor networks"*, John Wiley & Sons, Inc., Hoboken, ISBN 0-471-71816-5, New Jersey

Texas Instruments, MSP430F1611 microcontroller datasheet, *Available from:* http://focus.ti.com/lit/ds/symlink/ msp430f1611.pdf, *Accessed:* 2008-04-30

Annals of DAAAM for 2008 & Proceedings of the 19th International DAAAM Symposium, ISSN 1726-9679
Editor B. Katalinic, Published by DAAAM International, Vienna, Austria 2008
Make Harmony Between Technology and Nature, and Your Mind will Fly Free as a Bird

CONSTRUCTIVE MEASURES FOR INCREASING THE ACCURACY OF THE MACHINE TOOLS HAVING VERTICALLY MOVING ELEMENTS

STAN, G[heorghe]; FLORESCU, I[ulian] & FLORESCU, D[aniela]

Abstract: *Currently, in order to minimize the errors met during the vertical motions of the moving elements in the structure of a machine tool; various systems are used that are relatively complex and whose adjustments are difficult to do. This work is presenting a balancing – compensation system of the elements moving vertically; it has a simple structure that is easily adaptable to various types of machine tools. The new balancing-compensation system has been applied, manufactured and tested on three types of machine tools. The related results have been reified in increasing by approximately 10 times the geometrical accuracy and, by default, the positioning accuracy of the element moving in a vertical plane.*
Key words: *accuracy, balancing, compensation, machine tools*

1. INTRODUCTION

The structure of a numerical control (CNC) machine tool includes elements moving in a vertical plane that are numerically controlled axes of the machine and directly taking part to the machining process. In order to move the element vertically and to obtain an optimization of the kinematical linkage as well as of the driving servomotor it is necessary the respective moving element to be balanced, fig.1. Thus, the moving element 1 moves vertically along the direction Y being guided by the elements 2 fixed to the column 3. There is a component 4 (spindle) into the structure of the moving element 1 that moves horizontally along, Z axis, such a situation can be met on horizontal spindle boring and milling machine. The presence of the moving element balancing requires the introduction of a balancing force F_e, equal and of contrary sense to the vector G_{EM} that means the weight of the vertically moving element.

$$F_e = G_c + G_b = G_{EM} \tag{1}$$

Where the symbols mean: G_c – the weight of the moving element 1; G_b – the weight of the spindle 4. In this manner, during the up and down motion at no load of the moving element 1 the resisting force met by the servomotor will be the same. The presence of unbalance between F_e and G_{EM} requires a different resistant force on the servomotor that drives the vertical motion along both senses. Another negative effect caused by unbalance consists of the elastic deformation of the kinematical linkage that is different while moving along the two senses, with negative effects on the positioning accuracy (Bell, 2005). The vertically moving element can be balanced either mechanically or hydraulically. The mechanical version is based on a counterweight and a hoist system. This solution has a good reliability but the transient time of the vertically moving element is negatively affected. The response time of the axis would increase by 40 – 50% that leads to an increase of the path error in case when the respective axis is taking part to contouring through interpolation. The hydraulic balancing version of the moving element is more suitable because it does not affect the response time of the vertically moving element. A correct balancing imposes the two forces F_e and G_{EM} be equal, of contrary sense and on the same direction (axis); otherwise a bending torque will occur that acts on the moving element 1. There are cases when the weight of the moving element 1 becomes variable by adding the various attachments or tools to the spindle. In case of the bridge type milling machines (plano millers), boring and milling machines and vertical lathes the vertically moving element has a variable gravity center. The example shown at figure 1 shows that while the spindle is horizontally moving along Z axis, the gravity center of G_{EM} will move horizontally as well, generating a bending torque that acts on the moving element 1. The repercussions will consist of modifying the geometrical position of the moving element 1 towards the guideways 2 and increasing the elastic deformation of the kinematical linkage during the vertical motion because of the increase of the friction force in the guideways of the moving element 1 (Bezier, 2006).

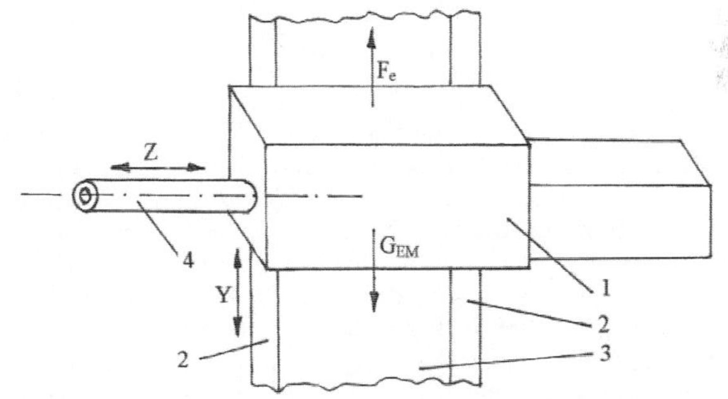

Fig. 1. The vertically moving element of a machine tool.

2. SYSTEM OF ELASTIC DEFORMATION COMPENSATION APPLIED TO A SINGLE COLUMN VERTICAL LATHE

The example of a vertical lathe, fig. 2, having as structure elements the column 13, the cross rail 2 moving along the W axis that performs positioning only afterwards being clamped to

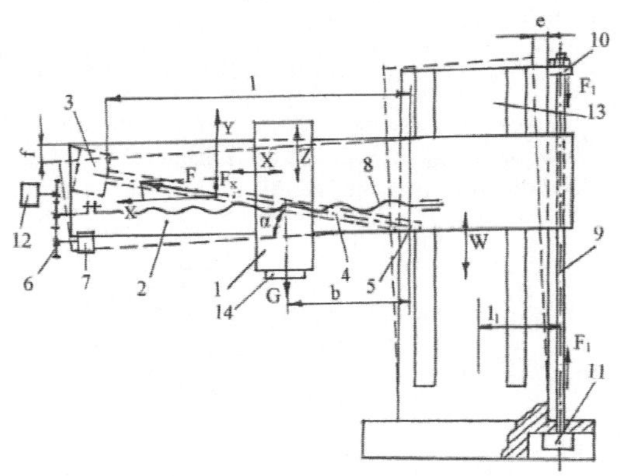

Fig. 2. System of elastic deformation compensation applied to a single column vertical lathe.

the column 13, the ram slide 1 moving horizontally along the X axis. On the ram slide 1 the ram 14 is vertically moving along Z axis. The ram 14 is taking part to machining, because the cross rail 2 is clamped to the column. While the ram slide 1 is moving along the cross rail, X axis, because of its weight G and the console distance b a bending torque will occur "$M = G \odot b$" that acts on the cross rail and produces an elastic deformation f of the cross rail 2. The same torque M acts on the column 13 giving the deformation e. The maximum rate of the elastic deformation f will occur when the ram slide 1 is at the maximum travel b. At the same time, after due researches it results that the elastic deformation e has the maximum rate when the bending torque M is maximal and the cross rail 2 is located at the upper end of the W axis (Patent No. 119131). For compensating the two elastic deformations the following solutions having as main element the pressure regulator, are used: a) For the elastic deformation f of the cross rail the rod 4 is used that is supported at one end to the support 5 and the other end to the support 3, fig. 3. At the end to the support 3 of the rod, the compensation hydraulic cylinder with multiple areas is located. When supplying with oil the compensation hydraulic cylinder a force F will be developed that requires the rod 4 for compression and the cross rail to stretching. The vertical component of the force F_x has the rate:

$$F_x = F \sin\alpha \tag{2}$$

The angle α is the angle of the rod 4, fig. 2. The compensation effect of the elastic deformation f will take place when there will be the equality (Zetu, 2002):

$$F_x \cdot l = G \cdot b \tag{3}$$

The symbols mean: l – the distance from the rod 4 end to the column 13; G – the weight of the ram slide; b – the distance from G gravity center to the column 13. Because the distance b is variable, for maintaining the equality as per the relation (3) it is imposed that the pressure into the compensation hydraulic cylinder to be variable in order to modify the component of the force F_x in function of the position of the ram slide 1 on the cross rail 2. The pressure will be modified by the pressure regulator 7 that picks up its motion from the ball screw 8 through the gearing 6. The ball screw 8 belongs to the kinematical linkage that provides the ram slide 1 motion along the X axis.

b) For compensating the elastic deformation e another rod 9 is used that is supported at one end to the support 10, fig. 2. At the end of the rod towards the lower support the compensation hydraulic cylinder 11 is located. It is supplied at the pressure P and develops a force F_1, stretching the rod 9. For compensating the deflection e, the equality presented below should be achieved (Patent No. 119132):

$$F_1 \cdot l_1 = G \cdot b \tag{4}$$

where l_1 means the distance from the axis of the rod 9 to the middle area of eth column. Because the distance b is variable, it is imposed the force F_1 to be variable too, for the condition of the relation (4) to be achieved. The variation of the force F_1 is performed by modifying the pressure with the help of the pressure regulator 12 that receives its motion from the ball screw 8 through the gearing. The diagrams of the pressure variation for compensating the two elastic deformations f and e are different; this requires each compensation to have its own pressure regulator. This solution has been applied manufactured and tested on a single column vertical lathe having the table diameter 2500 mm, the ram slide travel 2000 mm and the cross rail vertical travel 1900 mm. The test outputs are the following:

•The case when the deformation compensations have not been corrected and the cross rail has been positioned and clamped at 800 mm from the W axis origin. The rate of the horizontality deviation of the cross rail while the ram slide was moving has been 0.042/1000 mm.

•The case when the f and e elastic deformation compensations have been corrected. The rate of the horizontality deviation of the cross rail while the ram slide was moving has been 0.003/1000 mm.

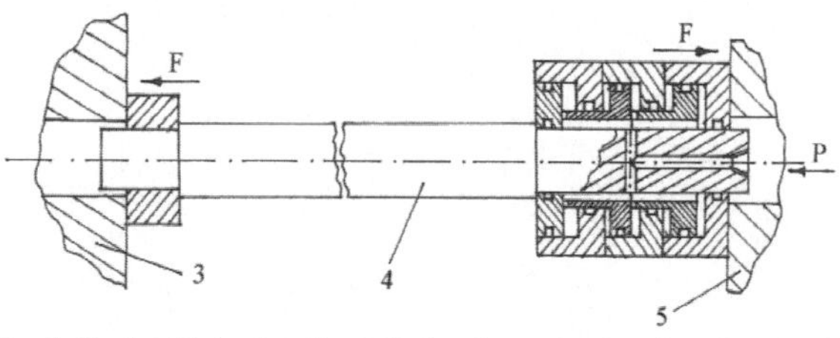

Fig. 3. Rod with hydraulic cylinder for compensating the elastic deformation of the cross rail.

3. CONCLUSIONS

The presence of the compensation systems for the vertically moving elements, especially on the middle and large size machine tools, is strictly necessary both for the reason of duly sizing the kinematical linkages and geometrical and positioning accuracy. The enhancement of the geometrical accuracy of the numerical control machine tools requires an attentive research of the elastic deformations caused by the modification of the gravity center. The phenomena presented above can be counteracted through intervening either on the effect or on the cause. Interventions on the effect are performed through "duly" sizing the components in order to decrease the deformation rates or through the intervention of some compensation elements, as for instance: increasing the guideway rigidity; enlarging the distance between the moving element guideways; compensating the elastic deformations of the elements supported in console (overhang) etc. The interventions on the cause refer to removing the bending torque caused by the modification of the gravity center of the vertically moving element. According to the results obtained on the machine tools where these new balancing-compensation systems have been applied, the geometrical accuracy of the machine increased by 10 times approximately. At the same time, on the machines using the vertical axis for machining through interpolation, in order to avoid a negative influence the transient duty of the vertically moving element, it is necessary to use hydraulic balancing, not counterweights. In general, when selecting the balancing-compensation system for the vertically moving elements, the weight of the respective element will be considered, as well as the admissible elastic deformation rates, positioning accuracy rates, guiding rigidity, constructive possibilities and the cost of the solution.

4. REFERENCES

Bell, Y. (2005). Compensation system of backlash and pitch errors of the feed kinematical linkages of the numerical control machine tools, *Progresivie Tehnologii i Sistemi Masinoctroenia*, Donet, ISBN 966-7907-00-7.

Bezier, P. (2006). Optimizing the transmission ratio on mechanical systems for decreasing the response time", *BIP Iasi,* Tome XLVII(LI), Department-Modelling in the machines building field, ISSN *1012*-2944, pp. 555-565.

Stan, G. (2004). *Balancing System*, Patent No. 119131, http://www.osim.ro/publicatii/brevete

Stan, G. (2005). *Balancing-Compensation System*, Patent No. 119132, http://www.osim.ro/publicatii/brevete.

Zetu, D. (2002). Machine Tools with Numerical Control, Publisher Junimea, Iasi, ISBN 973-37-0667-1.

Annals of DAAAM for 2008 & Proceedings of the 19th International DAAAM Symposium, ISSN 1726-9679
Editor B. Katalinic, Published by DAAAM International, Vienna, Austria 2008
Make Harmony Between Technology and Nature, and Your Mind will Fly Free as a Bird

MATHEMATICAL MODEL FOR CURVATURE STUDIES OF THE FLANKS OF CYLINDRICAL GEAR WITH CYCLOIDAL TEETH

STANASEL, I[ulian]; BLAGA, F[lorin]; PANTEA, I[oan] & BUIDOS, T[raian]

Abstract: *The curved cylindrical cycloidal gear is generated by rolling with a mobile straight line, using a milling cutter with several units of knives mounted on the tangential slide of the toothing machine. In order to locate the contact point, the aim of this paper is to determine the flanks curvature resulted after generation. Based on the determined analytical relationships it was done a numerical calculation program. The results being used to study the parameters which influence the size of the flanks curvature and also to establish the type of the knife necessary to be used in order to get a right position of the contact point.*

Key words: *curved teeth, cyclidal flank, milling cutter.*

¶

1. INTRODUCTION

¶

Knowing the values of the curvature radius, especially in contact zone, is necessary for the strength calculations regarding the stress contact and the stress inflection of the teeth and the thickness of lubricant film between the flanks as well.

A cycloid is the geometric place of a fixed point on a circle that is rolling without sliding on a straight line. Depending on the position of the considered fix point on the circle O_s, the resulted trajectory can be:

The cycloid is generated in the Γ_D plan as a trajectory of a point which is fixed by the rolling circle O_s, that rolls on a fixed straight line (fig.1).

The cycloidal director is transposed by rolling on surface of wheel part that has radius. The generation hook that defines the cycloidal teeth has the flanks defined by straight cutting edges that are attached on the fix point on the circle O_s.

The reference profile of generation hook is settled in its median plan and the reference angle α_0 is 20 .

The flanks of the wheel part are simultaneously generated by rolling with straight line and continuous division. The two cycloidal directories of the flanks must be generated with

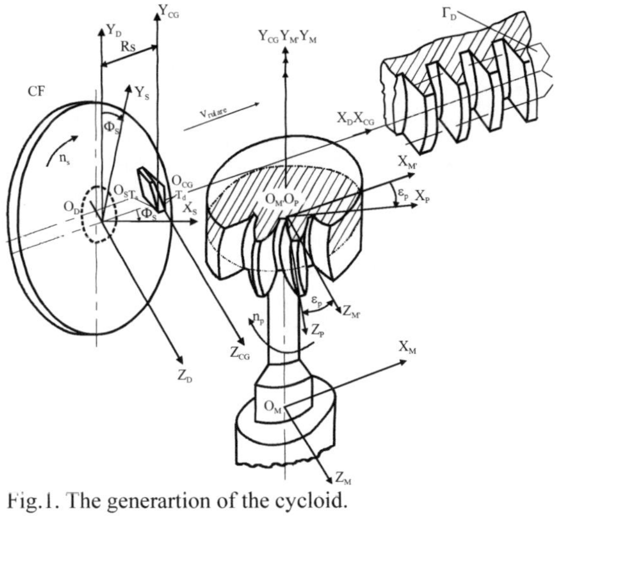

Fig.1. The generartion of the cycloid.

different curvature radius in order to realize the camber of the teeth. The curves which define the flanks of tooth are kinematic generated simultaneously by correlated motions.

2. THE FLANKS OF THE GENERATING HOOK

For the purpose to generate the cycloid of the rolling circle it can be attached to the circle a disc, where are fixed a few equidistantly knives on, resulting a head milling tool. In order to perform the analytical study of generation motions and the geometry of the flanks generated on the wheel part you consider several coordonate systems (fig.1). The considered systems are triortogonale, each of them being attached to an element that participates at generation process: S_S – the reference system of cutting edges T_s and T_d. Its origin is in the centre of the head milling tool. It is rotated simultaneously with the tools; S_{CG} – the reference system of generating hook. It is a mobile one, which moves with the rolling speed correlated with rotation motion of wheel part. Its origin is in the centre of the knife located in the middle plan of generating hook (Stănăşel, I., Mihăilă, I., Ghionea, A. 2003). The parametric equations of the flanks of the generation hook are obtained by changing the reference system of the tool edges from S_S to S_{CG} system.

$$\begin{cases} x_{CG} = (R_S + k_{1,2}\cdot\dfrac{m\cdot\pi}{4} + k_{1,2}\cdot u\cdot\sin\alpha_0)\cdot\cos\Phi_S - R_S - R_R\cdot\Phi_S \\ y_{CG} = -(R_S + k_{1,2}\cdot\dfrac{m\cdot\pi}{4} + k_{1,2}\cdot u\cdot\sin\alpha_0)\cdot\sin\Phi_S \\ z_{CG} = -u\cdot\cos\alpha_0 \end{cases} \tag{1}$$

3.THE CURVATURE RADIUS OF THE FLANKS

For the curves indicated by the parametric equations, the calculation of the curvature radius is made by the relation:

$$\rho = \frac{\sqrt{(x'^2 + y'^2)^3}}{x'y'' - y'x''} \tag{2}$$

In order to determine the curvature radius of the generating hook flanks the relationship (2) will be applied for (1). The derivates of equations are settled related to parameter Φ_S. After doing the calculations you obtain (Ionescu, Gh., D 1984).:

$$\rho = \frac{A}{B} \tag{3}$$

$$A = \left[\begin{array}{l} \left(R_S + k_{1,2}\dfrac{m\cdot\pi}{4} + k_{1,2}\cdot u\cdot\sin\alpha_0 \right)^2 + \\ + 2R_R\cdot\left(\begin{array}{l} R_S + k_{1,2}\dfrac{m\cdot\pi}{4} + \\ + k_{1,2}\cdot u\cdot\sin\alpha_0 \end{array} \right)\cdot\sin\Phi_S + R_R^2 \end{array} \right]^{\frac{3}{2}} \tag{4}$$

$$B = -\left(R_S + k_{1,2}\frac{m\cdot\pi}{4} + k_{1,2}\cdot u\cdot\sin\alpha_0\right)^2 - \\ -\left(R_S + k_{1,2}\frac{m\cdot\pi}{4} + k_{1,2}\cdot u\cdot\sin\alpha_0\right)\cdot R_R\cdot\sin\Phi_S \qquad (5)$$

This relationship can be used in a calculation program made in Matlab (Ghinea, M., Fireţanu V., 1998) to determine the curvature radius for the two points M_s and M_d which belong to the two flanks generated by the cutting edges

In tables 1 are summarized data obtained by the calculations mentioned before.

The geometric place of the curbure centers of a plan curve indicated parametric is exprimed by:

$$\begin{cases} X = x - y'\dfrac{x'^2+y'^2}{x'y''-y'x''} \\ Y = y + x'\dfrac{x'^2+y'^2}{x'y''-y'x''} \end{cases} \qquad (6)$$

The parametric equations of the centers of the curvature radius can be obtained by applying the relationship (5) for (1). After doing the calculations are obtained

$$X = -\cos\Phi_S R_R\left[\frac{\left(R_S + k_{1,2}\frac{m\pi}{4} + k_{1,2}u\sin\alpha_0\right)\sin\Phi_S + R_R}{\left(R_S + k_{1,2}\frac{m\pi}{4} + k_{1,2}u\sin\alpha_0\right) + R_R\sin\Phi_S}\right] - \\ - R_R\Phi_S - R_S$$

$$Y = \sin\Phi_S\left[\frac{R_R\left(R_S + k_{1,2}\frac{m\pi}{4} + k_{1,2}u\sin\alpha_0\right)\sin\Phi_S + R_R^2}{\left(R_S + k_{1,2}\frac{m\pi}{4} + k_{1,2}u\sin\alpha_0\right) + R_R\sin\Phi_S}\right] - \\ -\frac{R_R}{\left(R_S + k_{1,2}\frac{m\pi}{4} + k_{1,2}u\sin\alpha_0\right)}\times \\ \times\left[\frac{\left(R_S + k_{1,2}\frac{m\pi}{4} + k_{1,2}u\sin\alpha_0\right)^2}{\left(R_S + k_{1,2}\frac{m\pi}{4} + k_{1,2}u\sin\alpha_0\right) + R_R\sin\Phi_S} + \\ + \frac{2R_R\sin\Phi_S\left(R_S + k_{1,2}\frac{m\pi}{4} + k_{1,2}u\sin\alpha_0\right) + R_R^2}{\left(R_S + k_{1,2}\frac{m\pi}{4} + k_{1,2}u\sin\alpha_0\right) + R_R\sin\Phi_S}\right] \qquad (6)$$

The program allows determining the coordonates of the curvature radius centers for any point that belongs to the cycloidal directory by using the relationship (6).

y_{CG} mm	m=2,5 mm, Rs=50 mm, i_c=1		
	$\lvert\rho_s\rvert$ mm	$\lvert\rho_d\rvert$ mm	$\lvert\rho_d\rvert-\lvert\rho_s\rvert$ mm
15	48,864	52,729	3,8646
-10	48,605	52,489	3,8841
-5	48,345	52,249	3,9037
0	48,085	52,009	3,9233
5	47,825	51,768	3,9430
10	47,564	51,527	3,9628
15	47,303	51,286	3,9826

Tab. 1. The values of the curvature radius for i_c=1.

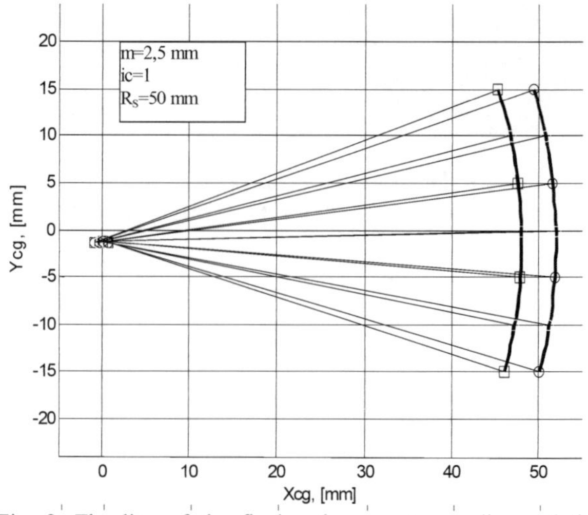

Fig. 2. The line of the flanks, the curvature radius and the geometrical place of the radius centers

The data obtained after using the program were used for graphical representations from fig.2, which illustrates the line of the flanks of the generating hook, the curvature radius and its centre for different points.

In the relations below were used the following notations:

R_S – the head milling tool radius;

i_c – the number of the groups of knives;

R_R – the rolling circle radius;

Φ_S – the rotation angle of head milling tool;

m – module of the gear

α_0 – presure angle of the gear

u- parameter that pointed the points on the cutting edges of the tool;

$k_{1,2}$ –tell the right or the left cutting edge

4. CONCLUSIONS

Based on the coordinating systems there were established the analytic relationships of the cycloidal trajectory of the tool.

There were determined the analytical relationships and it was elaborated a performed calculation program for the curvature radius of the flanks generated by the cutting edges.

The analytical study shows that by the proposed generating procedure is possible to obtain cylindrical gears with curved teeth which assure the localization of contact zone in the centre of the teeth.

5. REFERENCES

Ghinea, M., Fireţanu V., (1998) *Matlab, numerical calculus, graphics, applications,* ISBN 973-601-275-1, Editura Teora, Bucuresti.

Ghionea, A., Constantin, G., Stănăşel I., Ghionea, I., (2008) Milling heads for processing curved teeth in cylindrical gears, *Proceedings of the Oradea University* CD Rom edition, pp. 1407-1412, ISSN 1583-0691, may 2008 Editura Universităţii din Oradea, Oradea.

Ionescu, Gh., D. (1984). *The differential theory of the curves and surfaces with technical applications,* Editura Dacia, Cluj-Napoca.

Stănăşel, I., Mihăilă, I., Ghionea, A. (2003)- Contributions to the study of generation of the flanks of the generating hook of cylindrical curved gear in oblong cycloidal arc, DETC'03 ASME 2003 *Design Engineering Technical Conferences and Computers and Information in Engineering Conference* pp. 785-789, ISBN 0791837025, Chicago, Illinois, USA.

Annals of DAAAM for 2008 & Proceedings of the 19th International DAAAM Symposium, ISSN 1726-9679
Editor B. Katalinic, Published by DAAAM International, Vienna, Austria 2008
Make Harmony Between Technology and Nature, and Your Mind will Fly Free as a Bird

INTEGRATED SYSTEM FOR MODELING AND MANUFACTURING THE CAMS

STANASEL, I[ulian]; DRAGOMIR, G[eorge]; BLAGA, F[lorin]; PANTEA, I[oan] & STANASEL, O[ana]

Abstract: *The paper presents a study on the Kurz cam profiles, ended with the creation of a software which enables the statement of a complex motion law allowing the good functioning, without shocks or detachment, of the cam-displacing shoe tappet mechanism, and a study regarding the 3D model rendering and the creation of a CNC program for processing a cam whole profile.*
Keywords: *Mathematical model, optimizing, simulation, procesing.*

1. INTRODUCTION

The cam mechanisms are used for an accurate construction of some complex functions for the motion transmission.

The Kurz profiles ensure a smooth, shock-free variation of the acceleration which determines a decrease in mechanical stress, noise and vibration level, along with an increase in reliability and the improvement of the intake and exhaust processes (Antonescu P., et al., 2000).The Kurz cam is divided into 10 areas. Within the paper the symmetrical profile of the cam was considered, so that the laws determining the tappet upward motion in the four elevation areas are similar to these defining its downward motion.

The upward motion in the first area represents a cosine curve, has the size of a quarter of wave and corresponds to an interval marked as Φ_0, (Fig. 1).

It is expressed by the relation:

$$h_0 = h_p \left(1 - \cos\frac{\pi}{2\Phi_0}\alpha_0^*\right) \tag{1}$$

The elevation in the second area represents half a sine curve and corresponds to an interval noted with Φ_1. It provides positive acceleration and is expressed by :

$$h_1 = h_p + c_{11}\alpha_1^* - c_{12}\sin\frac{\pi}{\Phi_1}\alpha_1^* \tag{2}$$

The upward movement in the third area is a quarter of a sine curve corresponds to an interval marked as Φ_2 and enables the transition from positive to negative acceleration.

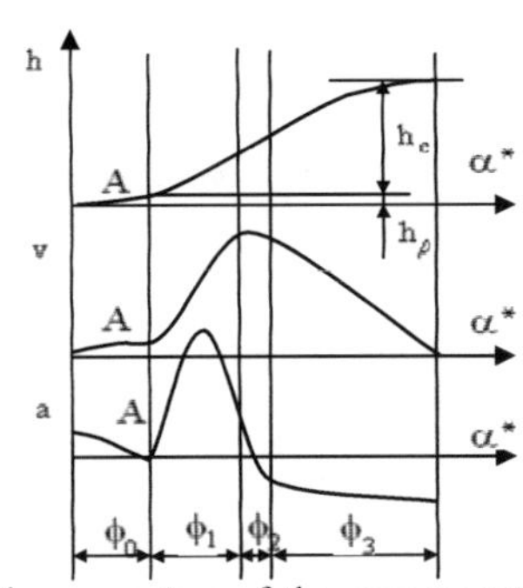

Fig.1. Kinematic parameters of the cam-tappet mechanism for Kurz profiles.

The mathematical relation of the tappet movement in this area is the following:

$$h_2 = h_{1f} + c_{21}\alpha_2^* + c_{22}\sin\frac{\pi}{\Phi_2}\alpha_2^* \tag{3}$$

The tappet elevation in the last area corresponds to an interval noted with Φ_3 and is defined by a fourth-degree polynom according to the relation:

$$h_3 = h_{2f} + c_{31}(\Phi_3 - \alpha_3^*)4 - c_{32}(\Phi_3 - \alpha_3^*)2 + c_{33} \tag{4}$$

In order to determine the c_{jk} values, the determination by differentiation was required of the mathematical functions of the acceleration and speed levels during the tappet elevation.

After applying the minimal conditions which provide the functions continuity at the starting and final moment of the interval, a system of six equations are obtained:

$$\begin{gathered} c_{11}\Phi_1 + c_{21}\Phi_2 + c_{22} + c_{33} - h_c = 0 \\ c_{31}\Phi_3^4 - c_{32}\Phi_3^2 + c_{33} = 0 \\ c_{11} - c_{12}\frac{\pi}{\Phi_1} - \tilde{v}_{0f} = 0 \\ c_{11} + c_{12}\frac{\pi}{\Phi_1} - c_{21} - c_{22}\frac{\pi}{2\Phi_2} = 0 \\ c_{21} + 4c_{31}\Phi_3^3 - 2c_{32}\Phi_3 = 0 \\ c_{22}\left(\frac{\pi}{2\Phi_2}\right)^2 + 12c_{31}\Phi_3^2 - 2c_{32} = 0 \end{gathered} \tag{5}$$

where i and f values represent the starting and the final moments of the interval.

The cam profiles were determined starting from the valve elevation laws of the distribution system. In ordere to represent these laws a software programme was made up in Matlab. The data offered by this programme were used to calculate the polar and carthesian coordinates and to mark points on the cam profile (Ghinea, M., Fireţanu V., 1998).

2 THE MODELING AND MANUFACTURING THE CAM

The cam solid 3D model was designed in SOLID EDGE using the Cartesian coordinates of the points defining its profile, obtained as a result of optimizing the functional performances. In order to use these matrix elements in the SOLID EDGE program, it was required to imput data from Matlab to EXCEL using the "Excel Link"for MATLAB (Stănăşel, I., et al. 2008). The acquired matrix is presented in Fig.2. Next step in order to obtain the cam solid model, is to use the "curve by table" comand from the "Part" menu of SOLID EDGE. This command enables the curves spatial representation using their coordinates from EXCEL. In fig 3 are represented the curve that defines the cam profile in Cartesian coordinates and in SOLID EDGE program. The solid model of cam is obtained by usind the "Protrusion" command (fig. 4).

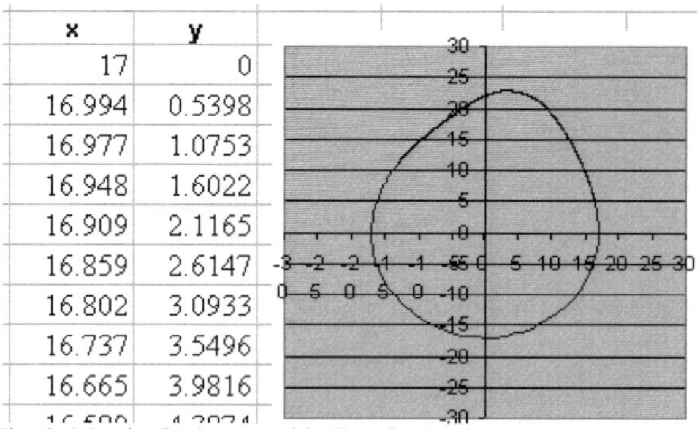

x	y
17	0
16.994	0.5398
16.977	1.0753
16.948	1.6022
16.909	2.1165
16.859	2.6147
16.802	3.0933
16.737	3.5496
16.665	3.9816

Fig. 2. Matrix G1 imported in Excel program

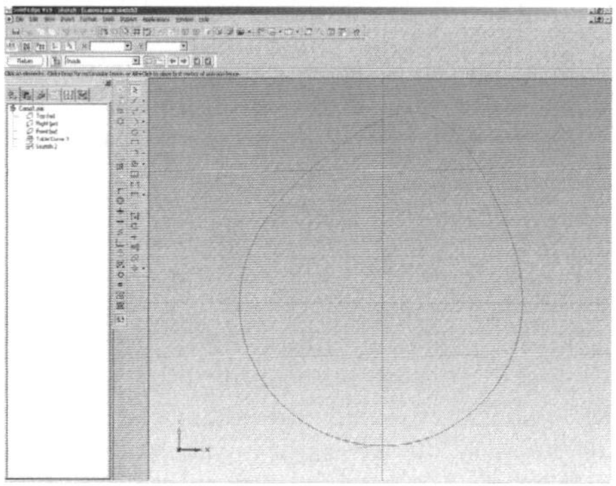

Fig. 3. Cam sketch in SOLID EDGE

In order to realise the CNC programme it is used NX machining from Solid Edge (fig.4). The application can be used only if is is later installed, being not included in Solid Edge.

It is established then the size of semi-manufactured goods it is chosen the tool, it is established the milling method and the programme performs the path of the tool and it simulates the manufacturing (fig.5).

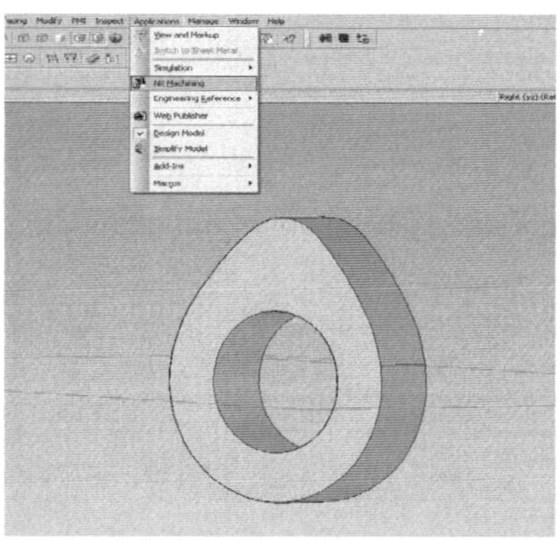

Fig. 4. The 3D model of the cam

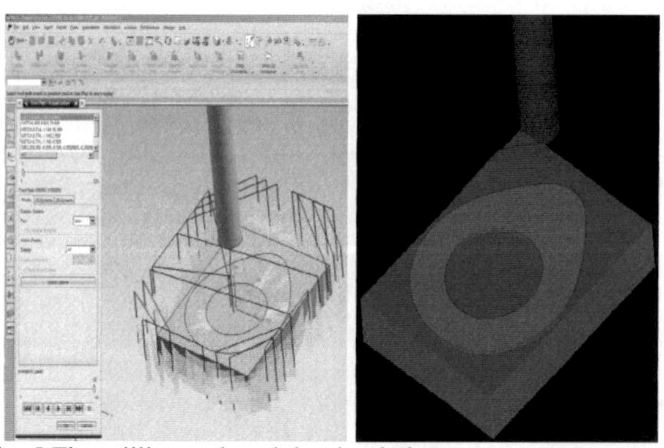

Fig. 5 The milling path and the simulation.

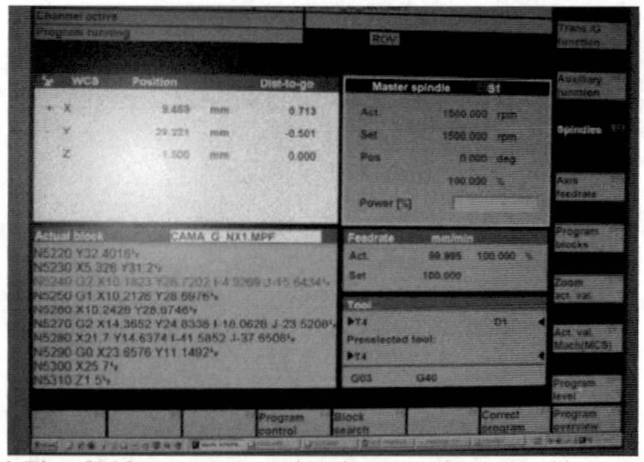

Fig. 6 The CNC programme implemented on the Siemens 840 D equipment.

Fig.7. The manufacturing process of the cam.

After post processed, we obtain the CNC programme. This programme is implied on Siemens 840 D equipment of the machine tool which is done the manufacturing on (fig. 6).

In fig. 7 are presented aspects within the manufacturing with the elaborated programme and the unfinished cam in a manufacturing phase.

3. CONCLUSION

Knowing the laws of movements of the cam it can be written a programme that outputs the coordinated points from the cam profile edge with an enough manufacturing precision.

These points can be exported in Excel and then to SE for the purpose of performing the solid model of cam.

By the use of installed NX machine application it is simulated the cam manufacturing; the CNC programme being post processed for equipments specified to machine tools where you prelucrate on.

The paper presents an integrated method Matlab-Excel-Solid Edge-Unigraphics NX for cam design and manufacturing, the way of technique documentation and how was elaborated the manufacturing programme.

The proposed method can be extended.

4. REFERENCES

Antonescu P., et all (2000)., *Mechanisms,* Editura Printech Bucureşti.

Ghinea, M., Fireţanu V., (1998) *Matlab, numerical calculus, graphics, applications,* ISBN 973-601-275-1, Editura Teora, Bucuresti.

Muscă G. (2007) *Computer aided design using Solid Edge* ISBN 973-37-1172, Editura Junimea Iaşi.

Stănăşel, I., Blaga, F., S., Stănăşel, O., (2008) Integrated system for constructive optimization of recipients mate by sheet metal, *Proceedings of the Oradea University* CD Rom edition, pp. 1791-1798, ISSN 1583-0691, may 2008 Editura Universităţii din Oradea, Oradea

*** (2004) *Sinumerik 840 D Programming Guide*

Annals of DAAAM for 2008 & Proceedings of the 19th International DAAAM Symposium, ISSN 1726-9679
Editor B. Katalinic, Published by DAAAM International, Vienna, Austria 2008
Make Harmony Between Technology and Nature, and Your Mind will Fly Free as a Bird
*Young Researcher and Scientist Paper / * Supervisor, Mentor*

INFLUENCE OF THE STRENGTHENING BARS OF GUITAR'S PLATES ON THE NORMAL MODES OF VIBRATIONS USING FINITE ELEMENT METHOD

STANCIU, M[ariana]; CURTU, I[oan] & ITU, C[alin]

Abstract: *The paper presents the results of the modal analysis of six types of structures made from plates. Firstly, was done geometrical modeling of structures, after which they were numerical modeled using shell and solid finite elements. The next step consisted in simulating the structures behavior to free vibration for different thicknesses and materials. The results were processed and compared in order to establish the influence of various factors on their natural frequencies, modal shapes and amplitudes. This study is a part of a wider research whose purpose is optimizing structures in order to improve the acoustic performance of classical guitar.*
Key words: *stiffness, plate, vibration, frequency, guitar, FEM*

1. INTRODUCTION

The guitar represents a complex vibrating system. One of the most important parts of guitar is the body. This is made up of ligno-cellulose plates (top plate, back plate, ribs) which vibrate, radiate and amplify the sounds. The acoustic box have a low stiffness compare with other parts of the guitar.

The plates must have a thin thickness in order to vibrate under the exciting forces of strings. In the same time, these structures must resist to the cyclic stresses (Curtu, 2007). Both requirements are fulfilled by the stiffening braces glued on the top plates.

The paper aims to analyze the aim in which number and pattern of strutting system influence the natural frequency and the modal shapes (Curtu, 2008). First, it has been modeled the six types of plates structures with FEM. Next it was performed the dynamical analysis and finally, the results were compared and discussed.

2. LITERATURE REVIEW

The natural frequency and modal shape of the plates of the guitar has been studied by many researchers, but each approach differs by the objectives of analyses, the method, types of guitar or the features of materials.

Elejabarrieta, Ezcurra and Santamaria (2007) are the authors of numerous articles regarding to the modal analysis and vibration behavior of the classical guitar in different construction stages using FEM. Serghei Vladimirovici (2006) researched with FEM and experimental methods the influence of the Russian strutting systems on the dynamical behaviour of the plates.

Paul M. Shaheen (2004) focused on the sensitivity analyses of the folk guitar simulating the building process. Becache, Chaigne, Derveaux and Joly (2004) studied the time-domain numerical modelling of the fluid – structure interaction of guitar.

Bernard Richardson with colleagues at Cardiff University performed studies on classical guitar to establish the influence of individual features on the modal shapes and frequency (Wrigth, 1996). They provided simple methods for luthiers to control the natural frequency and the modal effective weight in order to improve sound quality.

3. COMPUTATIONAL MODEL

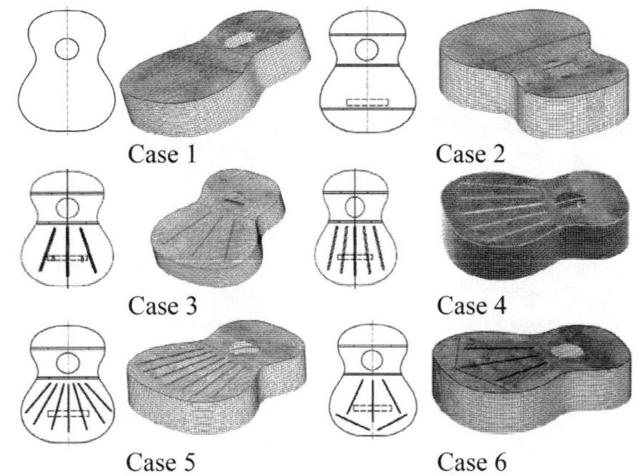

Fig.1. Types of strutting system of guitar plates

According with the types of guitars made up of S.C. Hora S. A. Reghin Romania, it was modeled six types of top plates different from the point of view of stiffening braces pattern namely: case 1– box with simple plate, case 2– plate with reinforced strips disposed transversal to the axis, case 3 – plate with 3 braces, case 4– plate with 5 braces, case 5– plate with 7 braces and case 6– plate with 3 radial braces and 2 in V pattern (Fig. 1). The neck and bridge were neglected. The plates were modeled with shell type elements with 4 nodes. The constraints in X, Y, Z axis were applied in area of the neck and guitar body jointing. For the dynamic analysis with FEM it was used Patran-Nastran 2004 package. It was used the following parameters: thicknesses (*h= 2.5 mm*), density (*ρ= 450 kg/m^3*), the Young's modulus (*E= 14000 MPa and G=5000 MPa*) and the Poisson's coefficient (υ=0.36) (Curtu, 2007).

4. RESULTS AND DISCUSSIONS

A normal mode analysis was run for the whole structures and the first ten modes were extracted. The mode shapes were selected based on visual similarity comparison to the mode shapes from literature. The shape of vibration patterns is almost equal: the number and emplacement of braces does not imply significant change in the form of vibration modes, especially for first five modes (Fig.2).

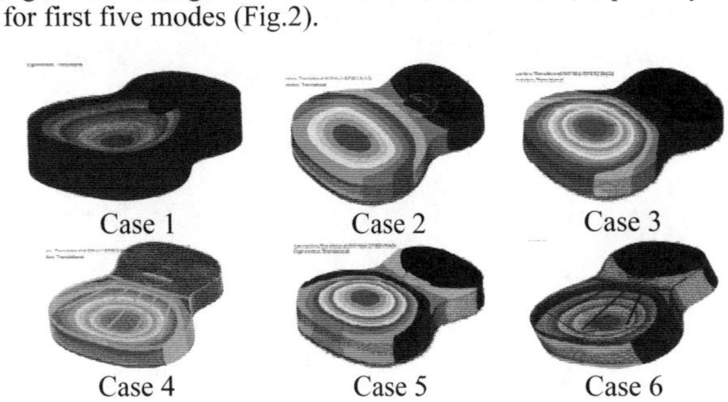

Fig.2. The modal shapes for first normal modes (0,0)

		Case n / Case 1	Case n / Case 2	Case n / Case 3	Case n / Case 4	Case n / Case 5	Case n / Case 6
	F [Hz]	187	216	210	209	208	209
Case 1/ Case n	187	0 %	-13.4%	-10.9%	10.5%	-10.1%	10.5%
Case 2/ Case n	216	15.5%	0%	2.85%	3.3%	3.8%	3.3%
Case 3/ Case n	210	12.3%	-2.7%	0%	0.48%	0.96%	0.48%
Case 4 / Case n	209	11.8%	-3.2%	-0.48%	0%	0.48%	0%
Case 5 / Case n	208	11.2%	-3.8%	-0.95%	-0.48%	0%	-0.48%
Case 6 / Case n	209	11.8%	-3.2%	-0.48%	0%	0.48%	0%

Table 1. Normal frequencies for different types of guitar plates

Table 1 shows the first normal frequency and the percentage difference of the each studied cases, for first normal mode. The dynamic behavior of the soundboard was substantially modified in the case 4, 5, 6) (top plates with five and seven internal rods). The normal frequency decrease with 3% compared to the case 2 (plate with transversal struts).

In Fig. 3 it can be noticed big differences between the plates with stiffening braces disposed radial and the other ones (simple and with transversal strips). A high fundamental frequency of 216.39 Hz was obtained for plates with reinforced braces (case 2).

The close values were obtained for case 4 (plate with 5 radial braces) and case 6 (plate with 5 braces disposed 3 radial and 2 in V pattern) (Fig. 3 and 4). The explanation for this similarity is the approximate effective mass of structures and the equal mass distribution between the upper and the lower zones.

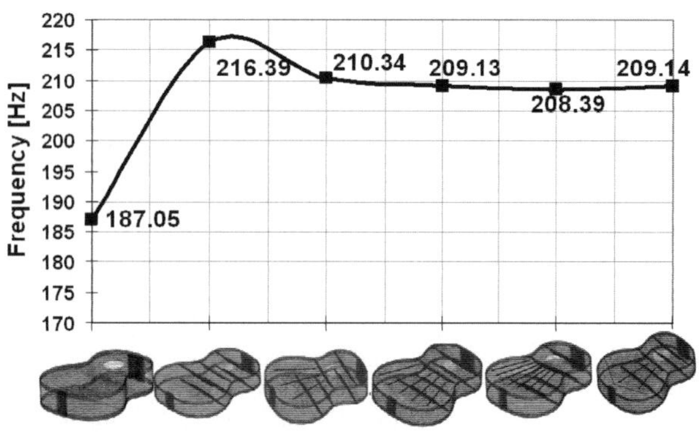

Fig.3. Frequency of guitar plates mode (0,0) by different strutting systems

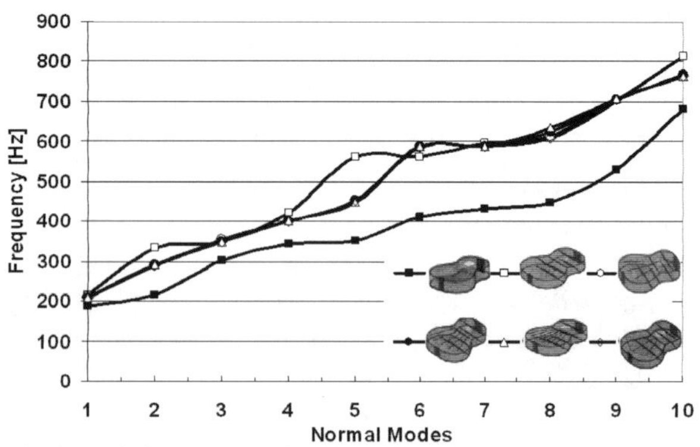

Fig.4. Variation curves of overtones

5. CONCLUSION

This study analyzed the dynamical behavior of the guitar body as a whole, and not as a individual parts of instruments. So the spectrum of the normal frequencies were identified for all structure, which assume that can be some deviations from the eigenmodes and eigenfrequency of the individual parts (top plate, back plate, ribs), similar of Bader's conclusion (Bader, 2005).

The modes are different for each type of guitar due to the physical, mechanical and dynamical features. The normal frequency influences the quality of sound in terms of the balance in timbre. Therefore one of the acoustic criteria for guitar quality is to be able to resonate in a rich range of harmonics and to assure an unequal filter of tones. The sensitivity analysis of the natural frequency and modal shapes obtained in six cases of strutting systems of the plates revealed that variations are relatively small but from the musical point of view, they can influence the quality of the acoustic guitar.

The objectives of the further work are to analyze the frequency response of the guitar body in the each six cases, to compare the results of the dynamical analysis of the individual plates to the behavior of the guitar body, to performed real test.

6. REFERENCES

Bader, R. (2005). *Computational Mechanics of the Classical Guitar,* Springer-Verlag Berlin Heidelberg N.Y. , ISBN 3-540-25136-7, in Netherland

Bécache, E., Chaigne, A., Derveaux, G., Joly, P. (2004). Numerical simulation of a guitar. In: *Computers and Structures* , **83**,. 107–126.

Curtu, I., Stanciu, M., BABA, M. (2007). Aspects regarding to the influence of the anizotropy on the statically modeling of the guitar. In: *Proc. of the 2nd Int. Conf. „Computational mechanics and virtual engineering",* pp. 87., Ed. Universitatii Transilvania din Brasov, Romania.

Curtu, I., Stanciu, Mariana, Itu, C., Grimberg, R., (2008) Numerical Modeling of the Acoustic Plates as Constituents of Stringed Instruments, in *Proceedings of the 6th International Conference of DAAAM Baltic Industrial Engineering, 24-26th April 2008,* ISBN 978-9985-59-783-5, p. 53-58, Tallinn, Estonia.

Elejabarrieta, M., Ezcurra, A., and Santamaria, C., (2007), Resonance box of guitar: structure and fluid interaction. *19th International Congress on acoustic, Madrid, 2-7 September 2007,* Spain

Shaheen, P. M., (2004). Sensitivity analysis of the natural frequency and modal effective weight of mode (0,0) of the top and back using FEM, PhD Thesis.

Vladimirovici, S. (2004). *The Computational Method of The Musical Instruments Body Parts.* PhD Thesis, The State Thechnical University of Marii, Russian Federation, in Russian.

Wright, H., (1996). *The Acoustics and psychoacoustics of the guitar.* PhD Thesis, University of Wales, 1996.

ACKNOWLEDGEMENT

This paper is supported by Romanian Ministry of Education and Research under Research of Excellence Program – Contract No.49 RoLight and CNCSIS Bucuresti TD cod 182, no. 222/2007. This work was accomplished under the tehnical support of Transilvania University of Brasov, Romania.

Annals of DAAAM for 2008 & Proceedings of the 19th International DAAAM Symposium, ISSN 1726-9679
Editor B. Katalinic, Published by DAAAM International, Vienna, Austria 2008
Make Harmony Between Technology and Nature, and Your Mind will Fly Free as a Bird

INTELLIGENT VISION SYSTEM FOR GRASPING ERROR COMPENSATION OF A WAFER MANIPULATION ROBOT

STANCIU, M[ihai]; NICOLESCU, A[drian]; ENCIU, G[eorge] & GHINEA, M[ihalache]

Abstract: *Industrial robots working in the semiconductor industry (for silicon wafers manipulation) operates at high speed. The robot effectors use a vacuum system to hold the wafer but very often due to high inertial forces the wafer position is changed during the manipulation cycle. The paper proposes an intelligent vision system integrating an auto calibration procedure that is used to evaluate on the fly the real position of the wafer in the robot hand. The system is based on an edge recognition procedure able to give an evaluation of the positioning error as a feedback for the robot controller. Using the error value, the robot controller can adjust in real time the target point coordinates avoiding wafer collision with the flexible cell components.*

Key words: *vision, scara, wafer, error, compensation.*

1. INTRODUCTION

Most of the industrial robots used in silicon wafer manipulation are SCARA type articulated arms with 3 or 4 degrees of freedom. Generally the silicon wafer manipulation process is performed in vacuum or in a clean room environment. In present the semiconductor industry try to optimize the manufacturing process using 200mm and 300mm diameter wafers and the average manipulation speed is 1.2m/s. The silicon wafer weight is in the range 0.5-0.8kg but considering the very high value of the manipulation speed involving too a high acceleration value on the trajectory, usually in 5-8% of the cases a positioning error of the wafer in the grasping mechanism occurs (***,2001). This error is affecting the final positioning of the wafer in the dropping point (that can be a place in a wafer cassette or processing station).

The authors of this paper developed a specific vision system able to detect the positioning error and to provide the correction parameters trough a feedback loop to the robot controller (Nicolescu et al., 2002). The error evaluation process and correction parameters calculations are performed in the final stage of the trajectory trough the dropping point and the robot controller compensate this error by changing the coordinates of programmed endpoint (Stanciu et al., 1997).

2. VISION SYSTEM ARCHITECTURE

The vision system includes (fig.1):

- 1 or 2 video cameras installed on the bottom side of the robot end effector (for the 3 degrees of freedom version of the robot is required only one video camera due to the robot kinematics allowing the error compensation only on one direction – radial). For the 4 degrees of freedom robot 2 cameras are used to detect two components of the error in an horizontal plane;
- computer having installed the image processing software and two usb free ports;
- 1 or 2 usb cables connecting the cameras with the computer;
- RS232 serial cable connecting the computer and the robot controller.

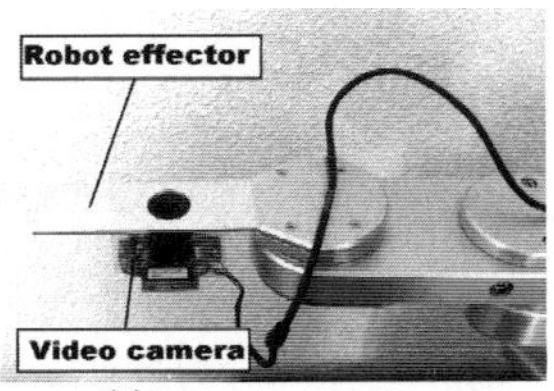

Fig. 1. Vision system architecture

Video cameras positions on the robot end effector are correlated with the wafer diameter. If the wafer is right positioned his edge will cover half of video cameras visual field in order to allows the detection of a positive / negative error value. This set-up increases the accuracy of the edge detection routine maximizing the image contrast. In the same time to increase the error detection software speed will be used gray scale images having a reduced size and less color information.

3. VISION SYSTEM CALIBRATION

The calibration process of the above presented vision system is performed automatically when the working cycle of the robot is started. In the first step the robot will pick up the reference wafer from a wafer holder. The vision system will detect the wafer edge and will set this edge (Guerra & Vlsi, 2000) as a calibration reference line (line position is expressed by pr). In the second step the robot will pick up the reference wafer from a wafer holder with an offset of 10mm. The vision system will detect the wafer edge (p10) and will set the vision system conversion parameter (cp) as the difference between this edge and the calibration reference line (fig.2). In the last step is calculated the vision system aspect ratio parameter (ar):

$$ar = \frac{10}{|cp|} \quad \text{[mm/pixels]} \tag{1}$$

$$cp = p10 - pr \quad \text{[pixels]} \tag{2}$$

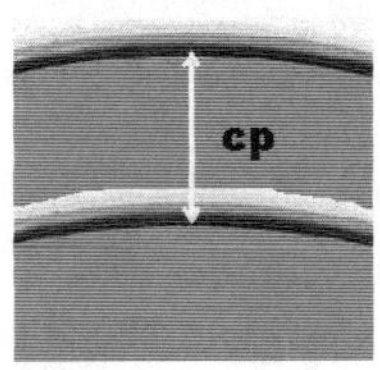

Fig. 2. Calibration image

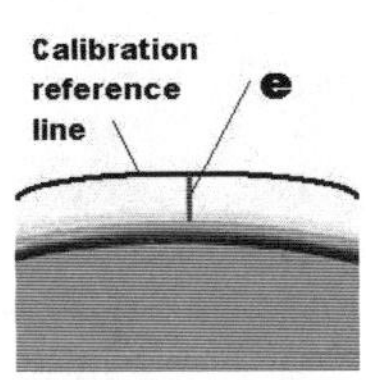

Fig. 3. Error detection image

Concluding, during the calibration stage is detected the reference line of a zero error positioned wafer and is also calculated the vision system aspect ratio parameter that will be used in error evaluation (fig.3). To optimize the error detection process will be used the reference wafer holder as a working station in the application architecture. In the same time periodically after 50-60 robot working cycles the vision system will be calibrated to increase the accuracy of the error evaluation process.

4. ERROR MEASUREMENT

The positioning error evaluation process is performed in two different ways depending on the number of robot DOF. A 3-axis robot can compensate the positioning error only on one direction. In this case will be used one video camera positioned on the longitudinal symmetry axis of the robot effector. The edge detection algorithm will select the 2 points defining the intersection between the wafer edge and the left and respectively right image limits (Spacek, 1986). To determine the error value is required only one point but using a second one will increase the accuracy of the calculation routine (fig.4).

A four axis robot can compensate both components of the positioning error. In this case will be used two video cameras positioned on the sides of the robot effector. The edge detection algorithm will select the two points defining the intersection between the wafer edge and the bottom and respectively right or left image limits for both images. To determine the error value are required only two points but using a second set will result an increased accuracy of the calculation routine (fig.5). Usually the aspect ratio parameter corresponding to a vision system resolution of about 3-3.2 pixels / mm and the overall accuracy of the positioning error routine will be in the range 0.3-0.35 mm. This value is very rough considering an usual 0.1 mm accuracy of this type of robots. To increase this accuracy can be used a more advanced video camera but the vision system resolution will be limited to 5-6 pixels / mm. To solve this problem the authors developed a mathematical method to increase the resolution by "devised pixel" technique.

In fig.6 is presented an enhanced image of a detected point with his 8 neighbors. On the figure are also marked the light intensity values for every pixel. With a standard resolution algorithm the detected point on the edge of the wafer will be the pixel marked on fig.6 with the light intensity value = 155. Assuming too that the vision system resolution is 3 pixels / mm resulting the reference point position as in fig.7.

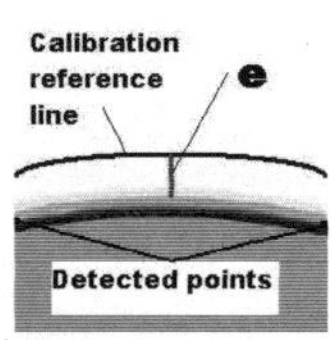

Fig. 4. Edge detection image

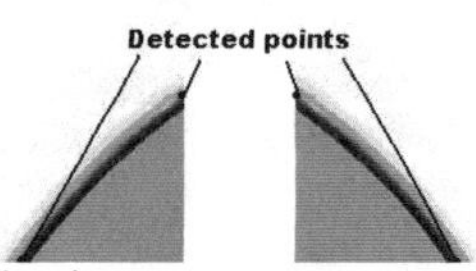

Fig. 5. Edge detection image

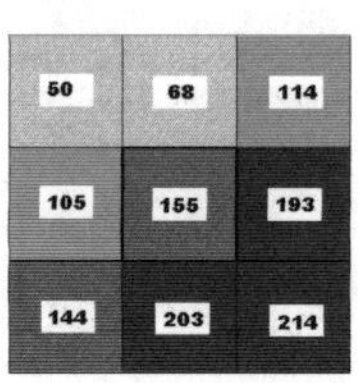

Fig. 6. Edge detection image

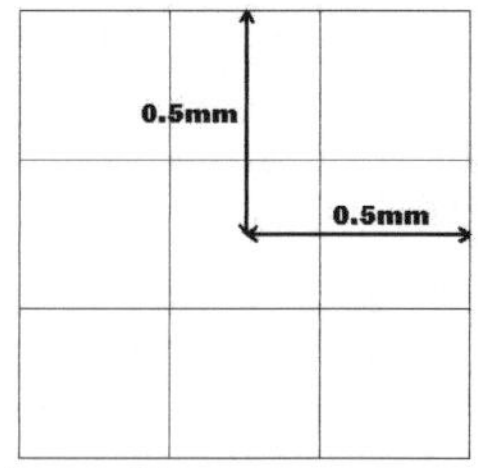

Fig. 7. Detected point position

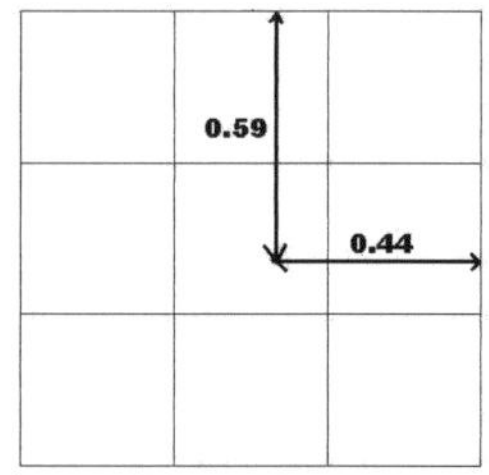

Fig. 8. Detected point position using "devised pixel" technique

By applying by "devised pixel" technique is assumed that every pixel weight is proportional with the light intensity as defined in fig.6. In this case the position of the reference point will be the position of the center of gravity of the fig.6. The equations of the center of gravity are presented in (3) and (4).

$$x = \frac{\sum_{i=1}^{9} x_i \cdot I_i}{\sum_{i=1}^{9} I_i} \qquad (3)$$

$$y = \frac{\sum_{i=1}^{9} y_i \cdot I_i}{\sum_{i=1}^{9} I_i} \qquad (4)$$

The results for the image presented in fig.6 are: x=0.44mm and y=0.59mm. Assuming that the vision system resolution is the same (3 pixels / mm) resulting the reference point position determined with the "devised pixel" technique as in fig.8.

Experimental research revealed that using "devised pixel" technique is possible to achieve an accuracy of 0.05 mm with a standard video camera (resolution 800x600 pixels).

5. CONCLUSION

Based on the requirement to minimize the positioning error resulting in wafer damage, the authors developed the vision system presented in the paper. The error detection model is based on an original "devised pixel" technique that allows increasing significantly the accuracy with standard hardware and low processing resources.

6. REFERENCES

* * *. *SHR3000 - Manual*, JEL Corporation, 2001.

Guerra, C.; A Vlsi. *Algorithm For The Optimal Detection Of A Curve*, CAPAIDM, No.86 /2000, pp. 197-202.

Nicolescu, A.; Popescu, D.; Costian, D. *Industrial robot assisted design using Catia P3 V5R7 environment,* Construcţia de maşini, 2002 (54), nr. 6/ 2002, ISSN 0573-7419, OID-ICM, Bucuresti,

Spacek, L.A. *Edge Detection And Motion Detection*, IVC, No.4, 1986, pp. 43-56.

Stanciu M. D.; Nicolescu A.; Ispas C. *Identification of Joint's Elastic Behavior Influence on Volumetric Accuracy of Industrial Robots.*, "IEEE's 6th International Workshop RAAD'97", pag.163...pag.168, STUDIO 22 Edizioni, ISBN 88-87054-00-2, 1997, Cassino, Italy.

Annals of DAAAM for 2008 & Proceedings of the 19th International DAAAM Symposium, ISSN 1726-9679
Editor B. Katalinic, Published by DAAAM International, Vienna, Austria 2008
Make Harmony Between Technology and Nature, and Your Mind will Fly Free as a Bird

AUTOMATIC ALIGNMENT SYSTEM FOR COORDINATES MEASURING MACHINES

STANCIU, V[alerian]; DAN, S[tefan]; STANCIU, M[ariana] & CURTU, I[oan]

Abstract: *In nowadays industry, due to rising pressure from the market in terms of quality, productivity and cost reduction the inspection process becomes one of the main focuses for engineering studies. It is well known that the quality inspections are time-consuming processes and wherever involve manual setups or direct measurements will always be a potential failure risk. One of the main disadvantages of the actual CMM is the manual alignment sequence that act for highly automated production flows as bottle necks and could lead to alignment errors and consequently to false inspection reports. This paper proposes an automated solution instead of manual alignments based on laser sensors and shape recognition algorithm.*

Key words: *Moiré algorithm, Triangulation, Laser sensors, Shape recognition, Alignment sequence.*

1. INTRODUCTION

The measurement systems for dimensional measurements that are on the market nowadays rely on the experience and the precision of the manual setup of the operator. A complex 3D measurement conducted with a Coordinate Measuring Machine is subject the errors inducted by misalignments during the manual alignment of the part in the working volume of the CMM can lead to false measurements and, consequently, to misjudgments of the production process involving unnecessary costs. These kinds of misalignments are often done by inexperienced operators and the training costs of the personnel for operating/ programming the CMM's are high. For mass production, the vendors offer dedicated solution for each production line, which can run semiautomatic in conjunction with unique fixtures for each type of product. The aim of this paper is to present an automatic alternative alignment system to the actual CMM architecture that assumes manual alignment. Using simple laser system built on the portal of the CMM, that can measure the distance from the emitter to any solid obstacle in the working volume and return this data as 3D point, the machine can execute at the beginning of the inspection sequence an automatic alignment routine.

2. ACTUAL CMM STRUCTURE

The actual commercialized CMM's are based on two major methods of point acquisition.

First class, based on the direct contact of the touch-probe with the part to be measured is the most used solution in steel and plastic industry and basically in each application for solids with a reasonable elasticity. The principle is to transform the deflection on the solid in reading triggers of the optical rulers and of the normal vector of the deflecting surface into vector components on each axis depending on the coordinates system (Pettersson, 2004). The alignment sequence is manually performed by the operator by indicating with successive deflections (points) the areas that give the coordinate system of the part: plan – three points; line – two points on a normal direction to the defined plan's vector; one point with the normal vector perpendicular to the plan created by the two vectors defined before (initial plan and the line). The errors of the manual alignment are propagated throughout the entire measurement with very low possibility of detection. The main producers of contact CMMs are Brown & Sharpe Inc., Carl Zeiss and Mitutoyo.

The second class of coordinate measuring machines is so-called non-contact CMM. Using an optical projector and a CCD camera to acquire information regarding dimensions and position of the part to be inspected, these measuring devices are the top of the arrow of the dimensional measurements industry.

There are several methods to identify the position of a spatial point in the working volume without mechanical contact.

The first is to send a laser beam from the projector (optical spot) and with a CCD camera to read, based on the triangulation principle the spatial coordinates of the measured point. This technology is used by the Carl Zeiss CMM with a VEST non-contact probe.

The second solution (optical scanning) provided by Perceptron (US) is to project a slit lighting beam onto the surface to be measured and the beam image is recorded to calculate the x, y, z coordinates from the beam central line on the surface.

Probably the most revolutionary non contact measuring solution is offered by OPTON using the Moiré algorithm. A grating pattern is projected on the surface to be measured. The projection lighting is provided by the light source of the projector (OPTON, 2008). The original grating pattern is deformed along the curved surface. (Figure 1) Then this deformed grating pattern is taken into a computer via a CCD camera and saved as a digital image.

Regardless the sensor system and the algorithm chosen for calculation a manual alignment sequence is need it to indicate the CMM exactly where is the part to be inspected in the entire volume of the machine. Another inconvenient of the non-contact CMM is related to the material of the part. The transparent or translucent surfaces like glass or plastic, the mirror surfaces – high finished metal surfaces, cast iron surface in black color or even opaque materials that have a color similar to the grating pattern can not provide reliable measurements data.

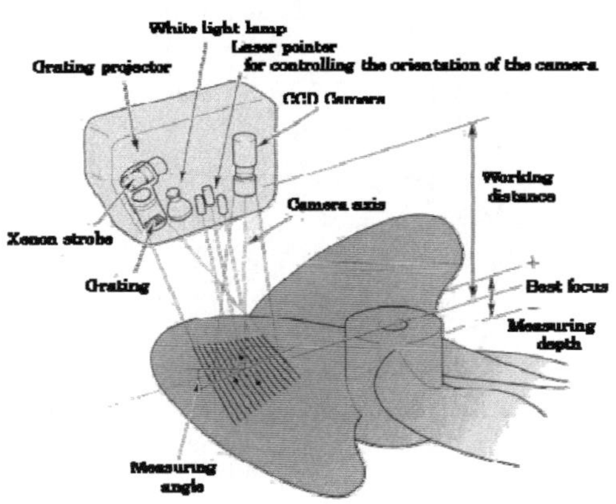

Fig. 1. OPTON 3D Moiré CMM

3. SELF ALIGNMENT CMM ARCHITECTURE

The system proposed to eliminate manual alignments that could lead to false measurements is simple and can be used by both types of coordinate measuring machines without major changes of the architecture.

A bridge/ portal CMM consist of three major elements: a solid table (preferable made of granite), a light metal structure in shape of a portal that travels along the length of the table and a beam that slides on the portal left-right, up and down (Fan et al., 1998).

Laser pointers that are able to measure the distance between the emitter point and the focus point are common tools used in construction or any other application that require straight measurements. The costs of this kind of laser pointers is relatively low and for short distance (up to 3 meters) the precision is acceptable (about 2-3 mm).

In the figure 2 is presented a CMM with set of laser pointers built in each foot of the portal. The pre-hit (safety distance between surface of the part and the point where the machine changes from traveling speed to measurement speed) is normally defined at 5 mm and consequently a precision of 5 mm is enough for a rough estimation of the part position on the table, before the automatic alignment sequence is performed. The sensors are evenly distributed at each 5 mm, vertically on the foot detecting the presence of any object by measuring the distance to the object. If there is no object, the sensors measure the distance between the feet. To increase the vertical resolution from 5 to 2.5 mm the sets of sensors can be shifted with 2.5 mm as is shown in the picture 2.

The pre-alignment sequence (identification of an object position in the working volume) is a simple algorithm that is performed at once by traveling the portal along the entire X-axis of the CMM. The position of the object on the X-axis is given by memorizing the coordinates from the optical ruler when the sensors detect first and last an obstacle. To avoid overcharging of the lasers, overheating of the part or of the environment the lasers are inactive by default and trigger every 5 mm a reading; at a speed of 5mm/s this means every second a measurement and for a standard table with 1500 mm length the entire pre-alignment sequence is finished in about 5 minutes.

The position and the boundaries of the part on the Y-axis are determined directly from the measurements of the lasers and the height of the part (Z-axis) is given by the sensors vertical pattern. Each time the sensors trigger a reading two sets of points are created in the computer. These points can be void if the sensors measured the standard distance between the portal feet or data points containing (x,y,z) coordinates of the surface detected.

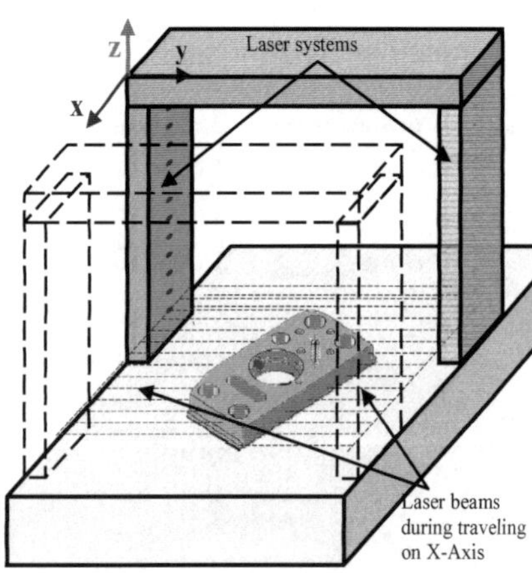

Fig. 2. Self alignment CMM structure

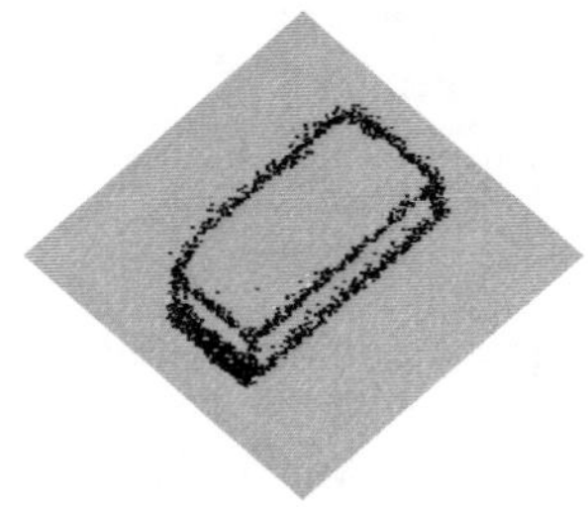

Fig. 3. Cloud of points stored after pre-alignment sequence

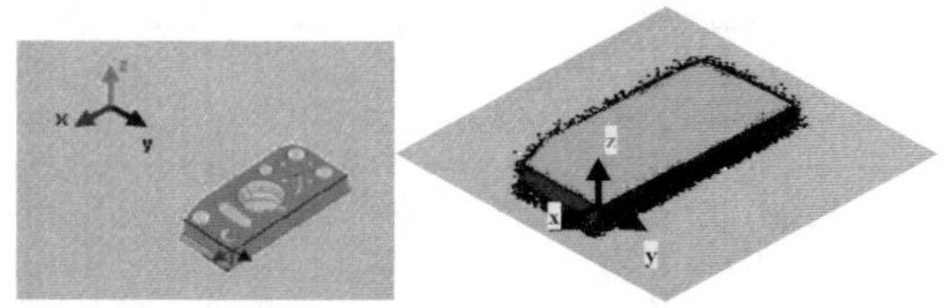

Fig. 4. 3D model lapped over the cloud of points

After a complete pre-alignment cycle a cloud of points determining the boundaries surfaces of the object is stored in the computer with a minimum resolution of 5 mm on each direction – figure 3.

A shape recognition algorithm (Hofer et al., 2005) overlaps the points cloud on the 3-D model and determines automatically the set of rotations and translations between a local coordinate system and an absolute system – figure 4.

The machine is ready in less then 7 minutes to run the standard automatic alignment.

4. CONCLUSION

Automatic alignment feature gives the opportunity to eliminate manual operation on a CMM, this way the possible false readings of the operator are eliminated as well. Additionally, the supervision of the CMM can be done from distance (e.g. operating room) given the possibility to perform measurements in unfriendly environments. Combining the automatic alignment with features like fuzzy path planning algorithm (Fiorentini et al., 1992), that calculates the optimum path between given inspection areas/ points and with automatic loading/ unloading conveyor systems several production lines can be served only by one supervisor. The mechanics of the CMM need no change to run the automatic alignment. Minimum electronics must be implemented (laser system) supported by an intelligent software, transforming the CMM into an semi-independent machine.

5. REFERENCES

Fan, K. C.; Chen, M. J.; Huang, W. M. (1998). *A Six-degree-of-freedom Measuring System for the Motion Accuracy of Linear Stages*, Inst. J. Mach. Tools and Manufact., Vol. 38, No. 3,

Fiorentini, F.; Moroni, G.; Palezzato, P. & Semerano, Q. (1992). *Probe path generation for an automatic inspection system,* 8th International Conference of CAPE

Hofer, M.; Odehnal, B.; Pottmann, H.; Steiner, T. & Wallner, J. (2005). *3D shape recognition and reconstruction based on line element geometry,* Institute of Discrete Mathematics and Geometry, TU Wien

OPTON, *OPTON's non contact 3D Moiré measurement machine Available from:* http://www.opton.co.jp *Accessed:* 2008-07-15

Pettersson U. (2004). *Identification and Adaptive Control of a Coordinate Measuring Machine,* Linköping

Annals of DAAAM for 2008 & Proceedings of the 19th International DAAAM Symposium, ISSN 1726-9679
Editor B. Katalinic, Published by DAAAM International, Vienna, Austria 2008
Make Harmony Between Technology and Nature, and Your Mind will Fly Free as a Bird
*Young Researcher and Scientist Paper / * Supervisor, Mentor*

TECHNOLOGY OF ELECTRONIC-BEAM WELDING OF BERYLLIUM BRONZE WITH COPPER.

STARKO, K[irill]

Abstract: *Non-ferrous metals and alloys on their base are applied more and more in different branches of national industry, especially in those ones featured by high requirements stipulated by special operation conditions (temperature, aggressive environment, high electrical, thermal conductivity). When using electronic-beam welding, which possesses such advantages, as compared to other methods, as high productivity, reliable protection, small diameter of the heating spot and etc., it makes it possible to obtain high-precision and important welded joints (Volchenko & Yampolsky, 1988).*
Key words: *electronic-beam welding, technology of welding.*

1. INTRODUCTION

This paper covers the development of a technological process of welding of cooper-based heterogeneous metals. We shall consider some specific aspects of welding of copper and copper-based alloys, describe various physicochemical processes taking place during heating, melting and evaporation, and give some recommendations on overcoming various difficulties.

2. INFORMATION

The proposed construction consists of 2 parts (Fig. 1). The first part is a copper cylinder made of M0 material with 10 mm thick wall (Fig. 2), and another part is a thin bronze foil (Material – BrB2) with 0.3 mm thick wall (Fig. 3).

In chemical and energy machinery construction the copper is used thanks to its high electrical and thermal conductivity, high corrosion resistance in some aggressive environments. All these features will be better, if the purity of metal is higher, that imposes special requirements on welding products made of pure copper. The total concentration of impurities in M0 material does not exceed 0,05%.

Beryllium bronze is a sort of alloy of copper with beryllium, with percentage ratio of beryllium being restricted by 2%. Application of beryllium bronzes is justified in situations requiring high electrical conductivity, high thermal conductivity, high mechanical and elastic properties, high corrosion resistance, absence of any ferromagnetic properties of material, and application of spark-free material.

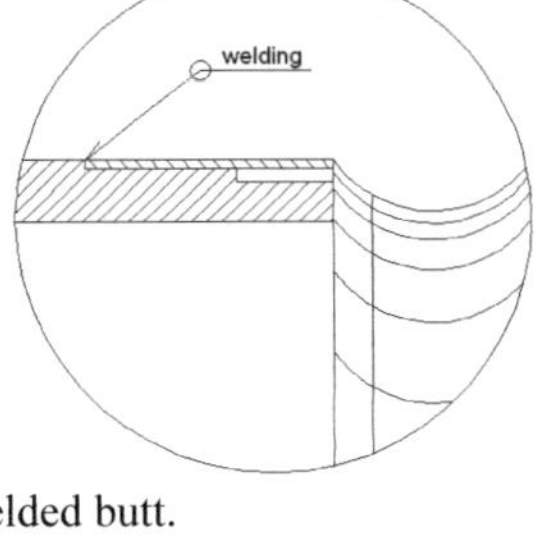

Fig. 1. The welded butt.

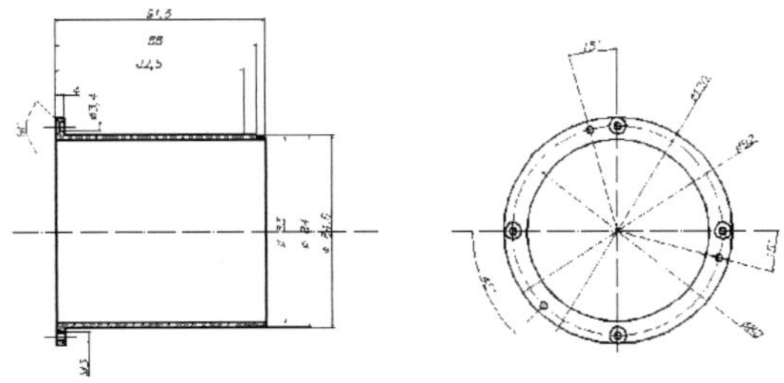

Fig. 2. The cylinder of M0 materials

When developing welding technology, the following peculiarities of the welded materials were taken into consideration:

- In connection with high temperature and thermal conductivity that make local heating difficult, more concentrated heating sources and heavier duty modes of welding are required. However, in connection with susceptibility of copper to grain growth in welding laminated seams, the metal of each pass is subjected to forging under temperature of 550-800ºC in order to achieve grain size refinement.
- Slight oxidability of copper at high temperatures leads to contamination of the seam metal with high-melting oxides. Thus, the welding bath needs to be deoxidized.
- High linear expansion coefficient (1.5 times more than that for steel) may lead to raised temperature and residual weld stresses and deformations. High temperature stresses together with degradation of mechanical properties may lead to formation of cracks. To reduce deformation of the construction, welding process shall be performed using rigid fixation along the tack welds. In case of raised metal thickness, the size of clearance is adjusted.
- Copper in melted state absorbs hydrogen significantly, which leads to formation of pores and cracks in the seam. It is necessary to reduce hydrogen content in the area around the welding bath.
- Raised fluidity of melted copper and alloys on its base (especially bronze) makes it difficult to perform welding in vertical and overhead positions, so, welding is often performed in lower position.

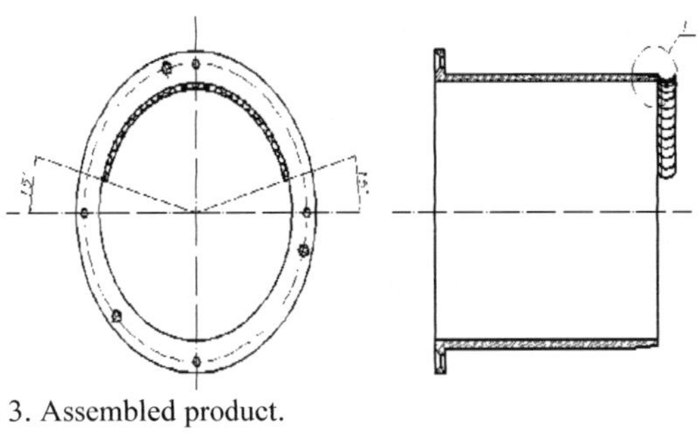

Fig. 3. Assembled product.

Development of technology for welding these items encounters with a number of problems, namely: different thickness of welded parts, influence of steam tension and process of bronze wetting on the surface of the copper cylinder.

Different thickness of the welded parts. When coupling two metals by means of welding, propagation of heat and propagation of temperature have their own traits (Kaidalov, 2004). This process is touched in more details in thermal calculation.

Steam tension. Reviewing scientific source (Babichev & Babushkina, 1991) and diagram for dependency of saturated copper and beryllium vapor pressure on temperature (Fig. 4), one can draw a conclusion that at high temperatures steam tension of beryllium is greater than that of copper. In addition to fact that bronze will be depleted by beryllium and enriched with copper, beryllium vapors may so affect the contact area of the welded parts, that being acted on by the response forces it will get loose that may lead to various defects.

Wetting processes. With this technological process of welding one may encounter with the following unpleasant phenomenon: melted bronze, getting onto copper, forms independent drops, and this is inadmissible when welding so precise and important parts. According to recommendations, when welding such parts, copper must either be pretreated with a dope including silver micro particles, fluorine and chlorine salts, or it is necessary to apply some other surface activation ways in order to improve wetting properties, for example, etching.

Taking into account thermo physical properties of the materials and geometry of the parts, when being guided by reference item (Borovik, 1998), there was performed a thermal calculation, based on which there were built temperature fields for the welded parts.

For the reviewed technological process the width of the welding bath (and, consequently, the width of the seam) is of great value. Its dimensions may be defined relying on analysis of the available isotherms. Melting temperature for M0 material is T_1=1356 K, and that for BrB2 material –T_2=1228K. It is quite logically that the area confined between these isotherms is the welding bath (Fig. 5). Thus the width of the welding seam makes up about 1.5 mm. When performing the thermal calculation, we used certain values for such parameters as power of heating source and speed of welding (based on which it is possible to calculate other welding parameters).

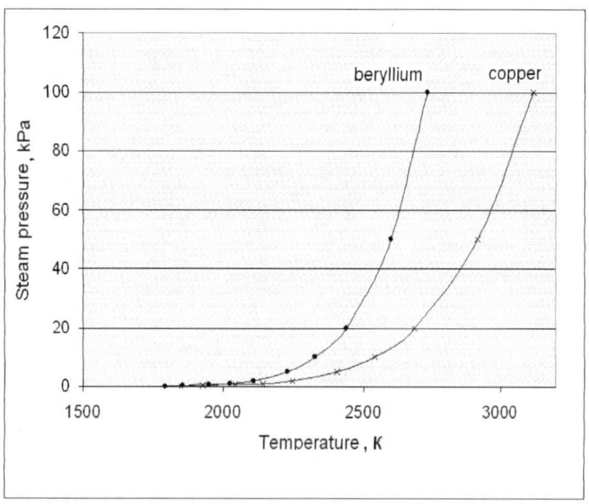

Fig. 4. Dependency of saturated copper and beryllium vapors pressure on temperature

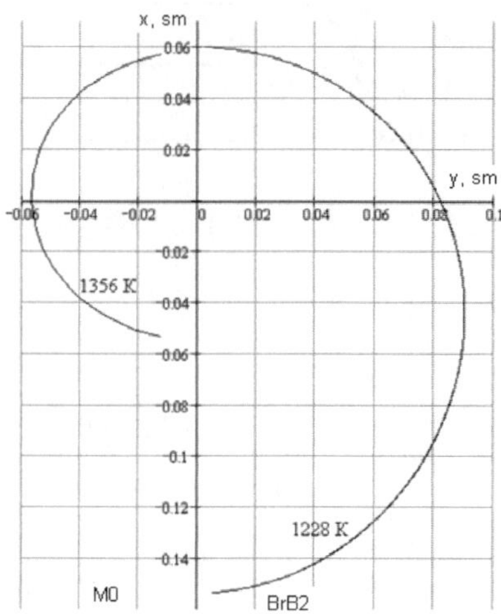

Fig. 5. The isotherms of melting for the welded parts

Technological process of welding is performed in several stages:

- Pretreatment of the parts. The given parts must be treated mechanically or chemically to achieve clean surface (absence of dirt, grease, foreign substances, and the like).
- Preparation for welding. Just for improvement of weld ability and through electrolytic method, the bronze foil is coated with a small layer of the copper with thickness of δ=0,025 mm at the distance of 2-3 mm from the welding place.
- Welding. Welding may be performed either over the entire perimeter in 1 turn of the part (in order to give the electron beam to reach the specified operation mode), or the welding may begin 10-15 mm before the place where the bronze foil is attached, so that the beam also may be able to reach the specified operation mode.
- Control of the ready product. Non-destructive control targeting to reveal various defects by using capillary method. The method and materials were selected based on the reference item (Gurvich, 1992).

3. CONCLUSION

As a result of the work I've done: the technological process of welding of products of heterogeneous metals was developed though widths of items differ from each other for several orders. Also physicochemical processes taking place during heating, melting and evaporation were described in details; and recommendations on overcoming various difficulties were given.

4. REFERENCES

Babichev, A. & Babushkina, N. (1991). *Physical quantities,* Energoatomizdat, ISBN 5-283-04013-5, Moscow.

Borovik, V. (1998). *Theory of welding processes*, MEI, ISBN 5-02-001508-3, Moscow.

Gurvich A. (1992). Non-destructive control, Visshaya Shkola, ISBN 5-06-002166-1, Moscow.

Kaidalov A. (2004). *electronic-beam welding and adjoining technologys* , Ekotehnologiya, ISBN 966-8409-03-05, Kiev.

Volchenko, V. & Yampolsky, V. (1988). *Theory of welding processes,* Visshaya Shkola, ISBN 5-06-001473-8, Moscow.

Annals of DAAAM for 2008 & Proceedings of the 19th International DAAAM Symposium, ISSN 1726-9679
Editor B. Katalinic, Published by DAAAM International, Vienna, Austria 2008
Make Harmony Between Technology and Nature, and Your Mind will Fly Free as a Bird

MACHINE VISION SYSTEM FOR SEAM WELD DETECTION IN LONGITUDINALLY WELDED PIPES

STAROVESKI, T[omislav]; MAJETIC, D[ubravko]; UDILJAK, T[oma] & MIHLJEVIC, V[iktor]

Abstract: *This paper presents the development of machine vision system for seam weld detection in longitudinally welded pipes. The system is to be used for positioning operations prior to mechanical working of such parts. Application which uses several different approaches is developed for this purpose. Experimental results show that robust seam weld detection system can be achieved using simple algorithms with low computational cost.*
Key words: *machine vision, seam weld detection, edge detection, Hough transform*

1. INTRODUCTION

Prior to mechanical working of longitudinally welded pipes (LWP), it is often necessary to determine exact position of seam weld. Reason for this is mostly due to inadequate mechanical properties of weld zone. Pipes must thus often be properly oriented with respect to seam weld position, in order to reduce strain during bending or avoid machining of weld zone.

Various seam weld detection systems, based on X-ray imaging, ultrasonic testing, or eddy currents exist today in industry, but these systems are primarily designed for non-destructive testing. Although most of these systems can be applied for indexing operations, their cost, processing and setup time, as well as often required human supervision are usually limiting factors.

Based on material, making technology and various other parameters, LWPs can have different properties. In order to develop practical machine vision (MV) based solution, which could be used regardless of material or making technology used, this paper is focused on common properties of LWPs such as changes in geometrical or optical surface characteristics of weld zone. Robustness to various forms of noise, such as corrosion of surfaces, presence of holes and other features is considered as most important factor, as these features can significantly influence success rate of detection algorithms.

Main goal behind this work is to develop compact and reliable MV based indexing system for orienting these parts. Motivation for this work is found in fact that MV approach has advantages as easily configurable versatile non-contact method for this task.

2. TASK ANALYSIS

Indexing system is briefly described as follows: Pipe is first placed in clamping device of indexing system by robot or some other machine attendance system. After clamping, camera is placed in the center axis of the pipe, which is then continuously rotated in steps for a given rotation angle.

After each rotation step, MV system is triggered, which executes weld detection algorithms within region of interest (RoI). Process completes upon successful detection of seam, with confirmation signal sent back.

Several samples of LWP with different characteristics were obtained, as shown in Fig.1. In some samples, seam weld is optically visible. Some of selected pipes also have holes present. In second case, there is very low contrast between weld and rest of material due to corrosion, but seam weld is still detectable as geometric feature.

Fig. 1. Two sets of LWPs obtained for testing

3. RECOGNITION ALGORITHMS

Approach used in case of optically visible longitudinal seam welds, consists of two steps: Canny's algorithm followed by Hough transform (HT).

Using Canny's algorithm (Canny, 1986), edge extraction is performed on greyscale image of RoI for each rotation step. In this technique, greyscale images are sequentially transformed into edge images using Gaussian smoothing, Sobel edge detection, thinning (non-maxima suppression) and finally, hysteresis thresholding. End result of this algorithm is binary image with edges extracted as single pixel thickness lines. Edge detection parameters can easily be adjusted in order to match sensitivity for specific pipe being indexed. These parameters include low and high hysteresis thresholds, as well as size of Gaussian convolution mask.

If seam weld exists for the given rotation step, it will be visible as straight vertical line in binary edge image. Such lines are easily detected using the simplest form of HT (Shapiro and Stockman, 2001), which is the second step of detection process.

This algorithm can be described as evidence gathering process in which edge pixels of image space are mapped to Hough space (accumulator array) and vice versa. In this case, polar representation of lines i.e. $r(\theta)=x0\bullet\cos(\theta)+y0\bullet\sin(\theta)$ was used for transformation. Each collinear point increases accumulator value at corresponding coordinates (r,θ). Maximum values of accumulator space after transformation indicate edges with maximum lengths. Sensitivity is again adjusted to match specific pipe being indexed, by taking into account only values exceeding adjusted threshold of accumulator array. Main advantages of this algorithm are robustness to noise with relatively low computational cost.

For detection of seam weld in second case, previously described approach is not suitable, since heavy surface corrosion causes low contrast between weld zone and base material. Even tough, seam weld is still possible to detect as there are changes in surface topology of weld zone.

These changes are easily expressed by illuminating seam topology using LED laser with line generating optics, oriented at low axial angle and normal to inner surface of LWP.

Orienting Laser in such way, beam will be projected as straight line on smooth surfaces, while on the other hand, broken line will be projected collinear with weld face reinforcement. It was experimentally determined that Hough transform is not suitable with this type of illumination, since it is difficult to adjust accumulator threshold for different batches and thus, achieve sufficient robustness. Number of visible pixels within RoI is therefore used to determine weather the seam weld is present for the given pipe orientation.

4. RESULTS

Two experimental systems were used for testing different sets of samples (Fig. 1). Mimas MV library is used as basis for application development in both setups. In this paper, only two images are shown for each setup. Results of more extensive testing can be found in (Mihljevic, 2008).

In the first case, a small mirror is placed in the center of the pipe, which is used to reflect inner surface of LWPs to camera placed above, as shown in Fig.2. Seam weld is visible within RoI of the second image, obtained after rotation step.

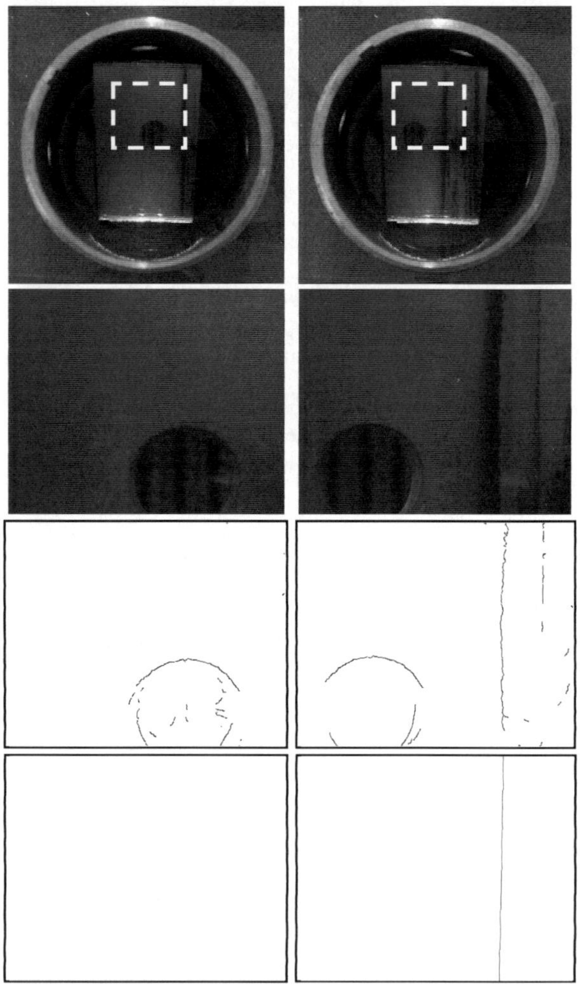

Fig. 2. Results of weld detection for two sequential rotation steps. Original images are shown in first row, followed by grayscale images of RoI, results of edge detection and finaly, Hough transform.

LWPs from second group are illuminated using LED Laser with line generating optics, as described in previous section. Images in Fig.3 show inner surface of LWP illuminated in such way. In the first case, smooth surface is illuminated, while second case shows illumination of weld face reinforcement. It can be seen that there is lesser number of pixels present in second case, indicating presence of seam weld at given rotation step.

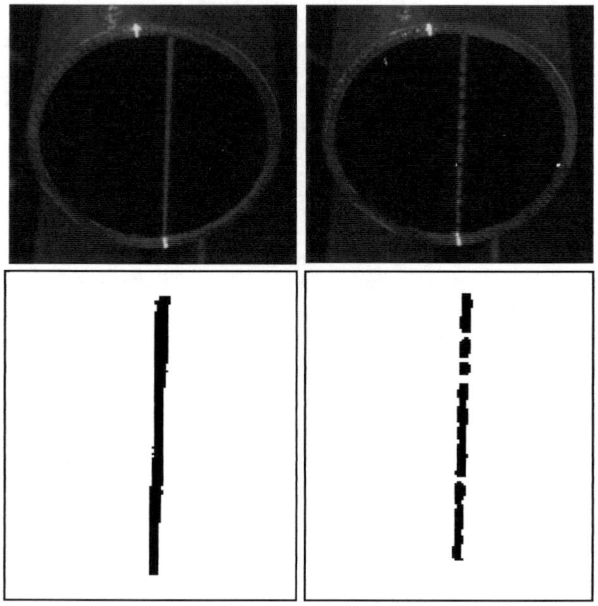

Fig. 3. Results of weld detection for two sequential rotation steps. Bitmap images of RoI after thresholding are shown in second row.

5. CONCLUSION

This paper presents practical MV based solution for indexing of LWPs, which is a common problem in modern production systems.

In this paper, only longitudinal seam welds are considered as features, but technique presented here can be easily expanded for detection of arbitrary shapes, i.e. by using generalized form of Hough transform algorithm (Ballard, 1981).

Although this work is still in early stage, overall repeatability of the process as well as robustness to noise shows promising results.

Future work will be conducted on expanding currently implemented algorithms for detection of objects with different features and will include investigations of other methods and algorithms for similar applications.

6. REFERENCES

Canny, J., (1986). A computational approach to edge detection. *IEEE Transactions on Pattern Analysis and Machine Intelligence* 8(6), pp. 679–698.

Shapiro, L & Stockman, G. (2001). *Computer Vision*, Prentice-Hall, Inc., 0130307963, Upper Saddle River, NJ, USA.

Mimas C++ Real-time Computer Vision Library, *Available from:* http://vision.eng.shu.ac.uk/mmvlwiki/index.php *Accessed:* 2008-07-01

Mihljevic, V. (2007). Methods for seam weld detection in longitudinally welded pipes, *Diploma thesis*, University of Zagreb

Ballard, D. (1981). Generalizing the Hough transform to detect arbitrary shapes. *Pattern Recognition* 13(2), pp. 111–122.

Annals of DAAAM for 2008 & Proceedings of the 19th International DAAAM Symposium, ISSN 1726-9679
Editor B. Katalinic, Published by DAAAM International, Vienna, Austria 2008
Make Harmony Between Technology and Nature, and Your Mind will Fly Free as a Bird

INFLUENCE OF THE MODE OF THE COOLING UNDER GRINDING ON QUALITY OF SURFACE LAYER

STARSHEV, D[enis]

Abstract: *It is shown that under grinding on characteristic surrface layer essential influence renders the nature of the cooling. The broughted graphs to dependencies to hardness surrface layer from velocity of the cooling. The graph of the experimental studies of the change the temperature in grinding process is presented. The offered method of management factors quality of the processed surfaces.*
Key words: *Grinding, cooling, temperature, hardness*

1. INTRODUCTION

The grinding process is the high temperature characterized by arising in zone of the cutting, promoting appearance heat defect on processed surface, reducing working factors of the details (Yaschericyn & Eremenko 1990).

One of the methods of the reduction of the temperature of the grinding process is used the cooling (Nosenko, 2000). However cooling capable not only to reduce the temperature in zone of the cutting, but also influence upon physico-mechanical characteristic of the surrface layers. In this connection are organized theoretical and experimental studies of the influence of the nature of the cooling on characteristic of the processed surfaces (Orobinskiy, 2000).

2. INFORMATION

In process of the cooling under grinding coolling ambiences, first, change its condition in consequence of increasing of the temperature. Secondly, exists passing physico-chemical processes, in accordance with evaporation of the ambience and change the mechanism of the process of the cooling. In accordance with this coolling ambiences can be divided into two groups.

For the first group process cooling turns out to be unchangeable from begin before the end. By cooling are change only temperature to surfaces of the stocking up and ambiences and, hereupon, heat and physical of the characteristic of the process. The process of the cooling is described fluent crooked (Fig. 1) and will comply with to equation:

$$t = t_{н} \cdot e^{-k\tau} \qquad (1)$$

where t - a temperature at moment τ; $t_{н}$ - a difference between the temperature of the stocking up and ambiences at initial moment; k - constant, definied size and the form of the stocking up, characteristic of the metal and coolling ambiences.

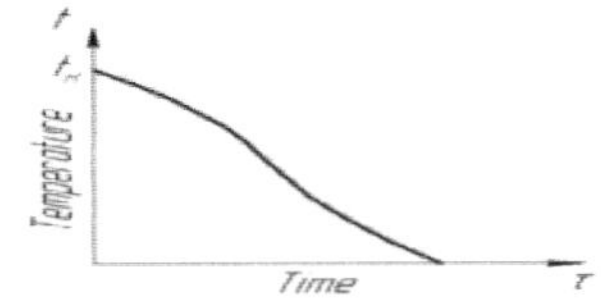

Fig. 1. Cooling in ambiences of the first group

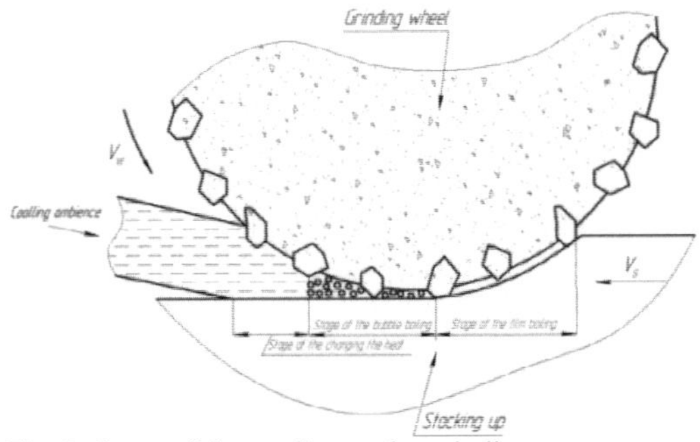

Fig. 2. Stage of the cooling under grinding

Curve (Fig. 1), described by equation (1), pertains to cooling in gas ambience (on air), and in all events, when at contact (the ambience-stocking up) does not change the condition of the ambience.

The second group is characterized by change conditions of the ambience in connection with boiling (Fig. 2).

Forming the steam film at moment of the interaction coolling ambience with surface of the cutting causes slow cooling in consequence of small velocity of the tap of the heat (the stage of the film boiling). The destruction of the steam film brings about contact of the separate volumes coolling liquids with heated surface of the stocking up; at liquid is vaporized. The evaporation continuously enterring to coolled surfaces of the volumes to liquids is connected with absorption of the hidden heat of the evaporation and brings about intensive cooling (the stage of the bubble boiling).

At reduction of the temperature coolled surfaces of the stocking up begins the third stage - changing the heat, which is also characterized decelerated by cooling.

The classical description of the processes of the cooling is realized by means of equations mathematical physicists (Lahtin & Rahshtadt 1980). In general type change the warm-up field is described by differential equation thermal conductivity (the equation Furie):

$$\frac{dt}{d\tau} = a\left(\frac{d^2t}{dx^2} + \frac{d^2t}{dy^2} + \frac{d^2t}{dz^2}\right) \qquad (2)$$

Where t - a temperature; τ - time; $a = \frac{\lambda}{C_p \cdot \gamma}$ - a factor to conductivities of the temperature (λ - a factor to conductivities of the heat; C_p – heat capacity; γ - specific density).

For decision this equation necessary to assign the marginal conditions: initial sharing the temperature inwardly stocking up and law heat exchange between surface of the stocking up and coolling ambience.

Then for processed stocking up shall get:

$$\theta = 1 - \sum_{n=1}^{\infty}(-1)^{n+1}\left[\operatorname{erfc}\frac{(2n-1)-\xi}{2\sqrt{Fo}} + \operatorname{erfc}\frac{(2n-1)+\xi}{2\sqrt{Fo}}\right] \qquad (3)$$

Here - θ warm-up criterion (the relative temperature in given to point and in determined time); ξ = x/s (the zone of the

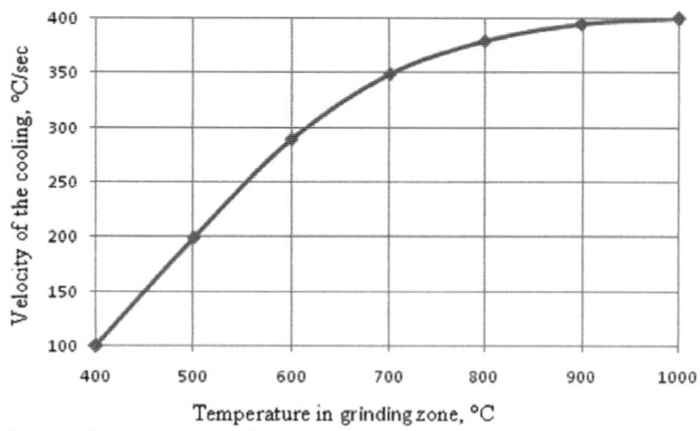

Fig. 3. The influence of the temperature of the heating on velocity of the cooling in coolling ambience (2% solution of the soda in water)

contact), s - a thickness under investigation surrface layer; $Fo = \frac{a\tau}{R^2}$ - a criterion Furie (Samohockiy & Parfenovskaya 1976).

On velocity of the cooling essential influence renders the temperature of the heating processed surfaces (Fig. 3).

In connection with small duration of the influence of the grinding wheel with unit area to surfaces of the stocking up and sharp cooling, heat does not have time to to extend to greater depth, in connection with than, depth layer with changed a structure of the material turns out to be very small. However its follows to take into account, since it renders the essential influence upon longevity made to product.

The most remote from processed surfaces layers are warmed before smaller importances of the temperature, in connection with than, decreases and velocity of their cooling (Fig. 4). Hardness falls with reduction of the velocities of the cooling considered layer a material of the stocking up.

The mode of the cooling under grinding must, first of all, provide the necessary depth raise to hardness. On the other hand, mode of the cooling must be such to do not appear the strong heat voltage, bring about warping the product and formation heat rifts.

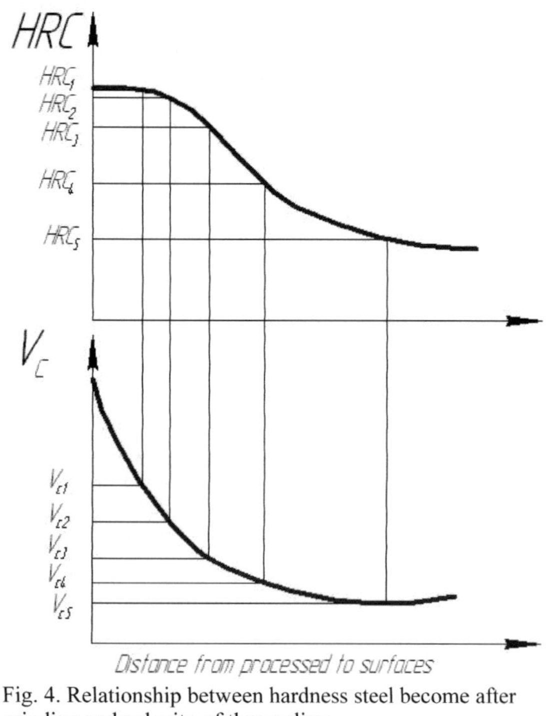

Fig. 4. Relationship between hardness steel become after grinding and velocity of the cooling

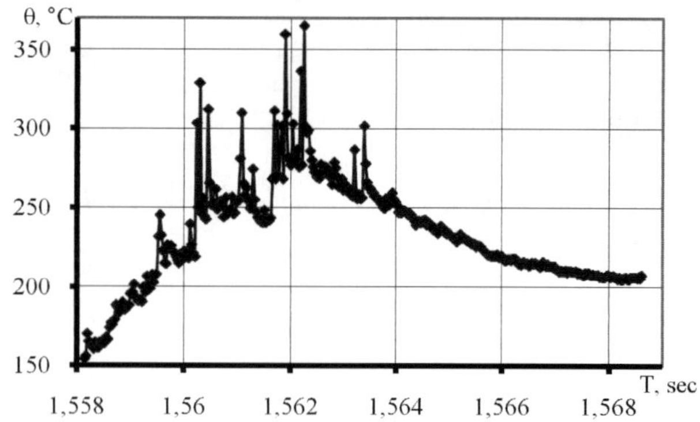

Fig. 5. Graph change the temperature in grinding process

The total heat voltage grow with increase the temperature of the heating and with increasing of the velocities of the cooling, since in both these events increases the swing of the temperature on section of the stocking up. Increase the swing of the temperature brings about growing termal and structured voltages.

From wrong passing of the termal processes under grinding in stocking up can appear the different heat defects.

Define the temperature of the heating, time of the endurance and velocity of the cooling in grinding process possible with the help of thermocouples. The under investigation thermocouple was fixed in stranglehold tool, was included presenting the table. For time of the passing of the thermocouple through zone of the cutting at miscellaneous to value of the presenting were measured parameters of the temperature such as time of the heating, endurance, velocity of the cooling after leaving the thermocouple from zone of the cutting. Got at results are of the form of, submitted for Fig. 5.

On the temperature of the heating, time of the endurance and velocities of the cooling for any processed material on reference table possible to define got in process of the processing hardness, and depth of the spreading the remaining voltages. Changing modes of the cutting are possible to control the velocity and nature of the heat processes, accompanying grinding.

3. CONCLUSION

Thereby, knowing law of the change the temperature processed surfaces at time, to account of the regulation mode cuttings, on the grounds of graphics, got at thermocouples possible to control physico-mechanical characteristic surrface layer processed stocking up. On result of the studies was installed that with reduction of the velocities of the cooling hardness and toughness surrface layer fall and increases probability of the origin heat defect.

4. REFERENCES

Lahtin, J.M. & Rahshtadt, A.G. (1980). *Termal processing in machine building: Reference book,* Moscow, «Machinostroenie»

Nosenko, V.A. (2000). *Grinding active metals.* - Moscow, «Machinostroenie»

Orobinskiy, V.M. (2000). *The methods of the abrasive grinding, and they optimization.* Moscow, «Machinostroenie»

Samohockiy, A.I. & Parfenovskaya, N.G. (1976). *Technology of the termal processing metal,* Moscow, «Machinostroenie»

Yaschericyn, P.I. & Eremenko, M.L. (1990) *Theory cuttings. Physical and heat processes in technological system,* Minsk "Vysheyshaya School"

Annals of DAAAM for 2008 & Proceedings of the 19th International DAAAM Symposium, ISSN 1726-9679
Editor B. Katalinic, Published by DAAAM International, Vienna, Austria 2008
Make Harmony Between Technology and Nature, and Your Mind will Fly Free as a Bird

STUDIES REGARDING MANUFACTURING TECHNOLOGIES OF NATURAL BASALT SINTERING PARTS

STEFANESCU, W[erner]; PUGNA, A[drian] & PLENICEANU, A[ristica]

Abstract:) *The present paper highlights the experimental results obtained for a series of basaltic samples elaborated through sintering and also some aspects regarding the technologies involved. The identification of the components substances by X-rays diffraction method had the purpose to identify the structure of this material and also the possibility of optimizing the technology of the sintering basalt. The quality improving of the sintering basalt parts depends by the forming regime, calcinations process and sintering regime.*
Key words: *sintering, diffraction, structure, X- rays.*

1. INTRODUCTION

In the actual stage of the society development, characterized by the acute demanding of materials and energy, the natural resources are not unlimited; in this case recycling it's not a feasible option on the long term.
Basalt parts cam be obtained through casting, followed by a recrytalization heat treatment, or by the method of powder sintering. The basic raw material is the natural basalt rock. The basalt improves its main chemical, physical and mechanical characteristics through melting and recrystallization. Molten basaly cam be cast in various moulds, resulting in polyfunctionnal and efficient parts for different industrial fiels, machine building included (Apostolescu, 1982)
The elaboration of modern making techonologies of basalt parts having superior physical and mechanical qualities implies study and research on substances the basalt rock is made of. After having apllied the heat treatments, the structure parameters and Miller indices of parts are measured. Heat treatments result in structural changes and the occurance of new substances as an outcome of sintering chemical reaction.
Following chemical analyzing, it was ascertain that basaltic materials are mixtures of binary oxide with a variant perceptual structure containing between a maximum 52% for Silicon oxide and a minimum of 2% for Titanium oxide, as shown in Table 1, similar values than pit his estimations by Chais (Becherescu, 1982; 1987), and a diversified of structural links of covalent - ionic type for Si and Al elements, of ionic type for alkaline metals (Na; K) and alkaline (Mg; Ca), and metallic type (Fe; Ti).
The presence of the three types of chemical links is influencing the structure and the properties of the chemical compounds.

Nr. Crt.	Chemical composition	Extreme values (%)	
		Romania	Optimal
1	SiO_2	44-52	43,5-47
2	Al_2O_3	14-16	11-13
3	Fe_2O_3+FeO	9-14	5-8
4	CaO	9-12	10-12
5	MgO	7-10	8-11
6	Na_2O+K_2O	3-8	3-5
7	Ti_2O	2-3	2-3,5

Tab.1. Chemical composition oxides of the basalt

During the experimental processes of sintering depending on the included substances in recipe and the temperature, in the weigh of the samples has took place some structural changes, determined by the chemical reactions, and the results have not coincide with the one determined theoretical.
Because of this the results obtained thru chemical and theoretical method, have required an investigation of the basalt material structure, thru a highly resolution method – X ray diffraction.

2. MATERIALS AND METHOD

X rays diffraction method is one of the most modern and safe means of determining the structural characteristics of basaltic materials (Gheorghies, 1990); (Ekstrom, 1986).
Considering the diffraction patterns recorded with the rocks collected from the four representatives basins in Romania, presented in figure 1, it resulted that they consist of the following substances: anorthite ($CaOAl_2O_32SiO_2$), magnetite (Fe_3O_4), olivine $(MgFe)_2SiO_4$, dolomite ($CaCO_3$-$MgCO_3$) and ilmenite ($FeTiO_3$) (Stefanescu, 2000).
From the analysis of the basaltic rocks diffraction patters, resulted a homogenous structure with very little variation. Which denotes that the structural homogeneity of the basaltic rocks cannot generate physical-mechanical properties well differentiated.
In this case the main role of obtaining basaltic pieces with high physical-mechanical properties is due to elaboration technology and thermal treatment

3. EXPERIMENTAL RESULTS

Turning basalt as raw material, into a product or finished part, implies several operations: choosing the raw material and its processing till it results into powder (washing, drying, sorting, crushing, screeminig, preparation and screeming of powder-binder mixture, shaping and pressing of parts), drying and calcinations, machining and sintering. When elaborating the execution technologies for the parts to be used in machine-building, the parameters of the shaping, calcinations and sintering regime were established experimentally

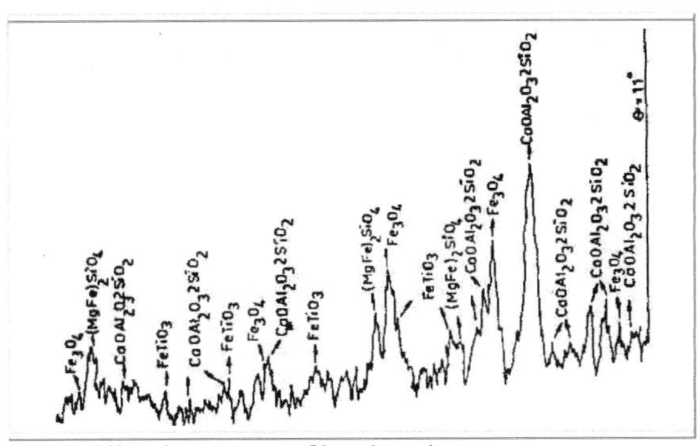

Fig. 1. Diffraction pattern of basalt rock

(Stefanescu & Iancului 2004) considering the shape, dimensions and weigh of parts.
Technological experiments (Stefanescu & Sandru, 1995) on test pieces of various shapes and dimensions demonstradet that when marking basalt parts the following aspects should be considered:
1.The compactness, structure, cohesion of basalt - binder mixture and the compacting during sintering process are influenced by the force employed to press the material into the dies.
2.The dimensional differences and the dispersions, between the test pieces formed at low pressures (0,5-1tf) as compared to the test pieces achieved at high pressures (1,5-2tf) are important
3.The quality of the parts made of basalt through sintering is influenced by the pressure they have been shaped at and the sintering temperature. The parts made at pressures of 1.5-2tf are more compact, the dimensional deviations are smaller, and as the shaping pressure in dies rises the contraction coefficient decresses, also with the improvement of the parts quality.
The sintering technologies (Stefanescu, 2000); (Stefanescu & Sandru, 1999) are conditioned by the shaping regime, the calcinations regime and the sintering regime:
1.The shaping regim is established according to the proportion of mixture component and the parts shaping pressures.
2.The calcinations regime depends on the heating speed and on the calcinating temperature.
3.The sinterimng regime as well as the calcinations regime depends on the heating speed and on the sintering temperature.
The sintering regime of differently shaped and dimensioned parts has been established on the basic of sintering parameters with speeds of 85^0C/h, 125^0C/h and 140^0C/h: the heating time, the duration of one impulse, the number of impulses, the time afforded to the keeping of parts in the sintering process have been calculated for the specified speeds and the cooling time, as shown in Table 2.
To emphasize the changes of crystalline phases evident, during the elaboration process of parts destined to machines building, substances where identified through X – ray diffractometric methods and sintered basalt parts.
The comparative study of rocks diffractograms for parts manufacture and those of the basalt parts figure 2 proved that, during sintering process, there appeared no new substances, and no significant structural changes have been produced at the level of elementary cell so as to determine changes of physical – mechanical properties.
Analyzing each diffractogram it was ascertain a big anorthit presence and frequency in the mass of basalt rocks and sintered basalt parts, which confer them an under – cooling stable and rigid state (lack of elasticity).
The under – cooling state it could be also the result of the presence of olivine and magnetite; their presence being more than 10%, which would favor the crystallization process and from here the vitreous state in the parts mass.

Sintering parameters	Speeds values		
	85^0C/h,	125^0C/h,	140^0C/h,
Heating time [h]	13,2	8,9	8,0
Duration of 1 impulse [s]	43	29	26
Number of impulses [-]	1093	1093	1093
Maintaining time [h]	1,7	1,2	1,0
Cooling time [h]	9,1	6,1	5,5

Tab. 2. The basic parameters of sintering

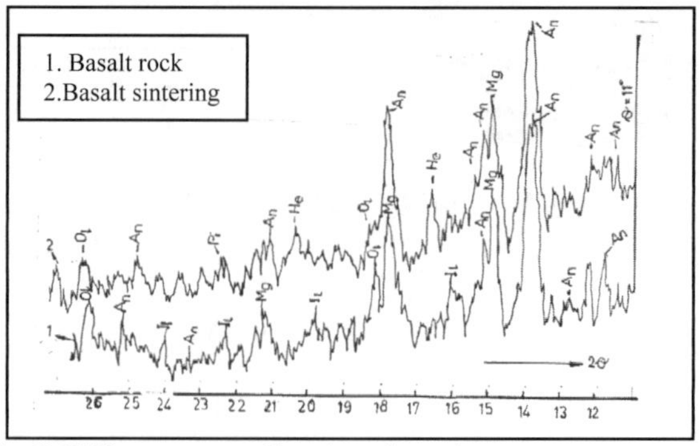

Fig. 2. Diffractograms of basalt rocks and parts

4. CONCLUSION

The studies and the experimental results revealed the following conclusions:
-regarding the basaltic rocks collecting place from the chemical composition point of view there is no significant differences which confer homogeneity and identical properties to basalt sintered parts;
-by comparison analysis of the component substances from basalt rocks and sintered basalt parts resulted that no new substances appeared;
-high-frequency of anorthite in basalt rocks and sintered basalt parts offers them an under – cooling state characterized by a total lack of elasticity;
-further studies should concentrate on trying to obtain the so called “complete sintering” with positive effects on sintered basalt parts by using new thermical treatments in order to modify the crystalline structural arrangement, or by using mineralizers in order to inhibit granules growth and pore transfer to the granules limits.

5. REFERENCES

Apostolescu,.D.(1982),*Cristallography, Mineralogy*, Ed. Didactica si Pedagogica, Bucuresti;
Becherescu, D.(1987), *Solid State chemistry*, Ed. Stiintifica si Enciclopedica, Bucuresti;
Becherescu, D.(1982), *Physical methods used in chemistry*, Ed. Stiintifica si Enciclopedica, Bucuresti;
Ekstrom, H. (1986) *X-ray powder diffraction techniques in the studies of high performances ceramics,* Chem. Scr. 26. A;
Gheorghies, C. (1990), *Diffraction spectrum and investigation techniques,* Ed. Tehnica, Bucuresti;
Stefanescu, W. & Iancului, D. (2004), Studies concerning of basaltic materials in agricultural machinery building, in *USAB University from Timisoara, Scientifical Research & Horticultural*, pp. 325-329, ISSN 1453-1402;
Stefanescu, W.& Sandru A. (1995), Utilization of some basalt parts for sprinklers, in *U.P.T University from Timisoara, Bul. St. Mecanica Agricola & Metalurgie*, nr.40(54), pp 5-8, ISSN1224-6050;
Stefanescu, W.& Sandru A. (1999), The utility of basaltic materials usage with mineralizers input in machine building, in *U.P.T University from Timisoara, Bul. St. Mecanic*, nr.44(58), pp 333-338, ISSN1224-6077;
Stefanescu, W. (2000), *Contribution to the study of structure influence and physical-mechanic properties of basalt based components over the system operational reliability of farm equipments*. PhD; Timisoara.

Annals of DAAAM for 2008 & Proceedings of the 19th International DAAAM Symposium, ISSN 1726-9679
Editor B. Katalinic, Published by DAAAM International, Vienna, Austria 2008
Make Harmony Between Technology and Nature, and Your Mind will Fly Free as a Bird

STUDIES CONCERNING THE ADHERENCE STRENGTH OF SOME THERMOGLUED TEXTILE ENSEMBLES – RESULTS AND INTERPRETATIONS

STEFANUTA, I[oan]; VLAD, D[orin] & VRINCEANU, N[arcisa]

Abstract: *The paper consists of experiments performed on seven different basic woollen and cotton textile materials and two types of representative insertions with adhesive. The obtained results leaded to the adherence values calculus for these underensembles. The interpretations provided by the researchers will constitutes an analyse fundamentals for the specialists of manufacturing field.*
Key words: *traction force, adherence, extension, insertion*

1. INTRODUCTION

The underensembles have been made at some comercial companies (in manufacturing regime), and the experimental tests were performed at ``Lucian Blaga`` University laboratories, within some doctoral research studies.

The researches in this field, performed up to now have not had a unitary character. ''Lucian Blaga`` University research team have proposed an approaching of the topic in a wider frame, consisting parts of doctoral thesis and disertations.

This leads to the finishing off procedures, as well as young researchers forming, besides practical results achievement, usefull in manufacturing field.

2. EXPERIMENTAL WORK

For adherence determination the procedure will be as follows (DIN 53859 Part 2):

-8 samples for each type of material cut on the longitudinal direction have been made, so that 4 samples for each type of material with one of two types of insertion, with 160 mm x 50 mm size, should be;

-for each sample, 40 mm remain unglued or ithout being glued;

-the samples will be conditioned;

-the distance between the dynamometer jaws should be 50 mm, and the speed of lifting the jaw C_1, up to 100 mm/min, according to the DIN 53859 Part 2;

-the data necessary for the test, the test's name, the number of samples, the type of basic material and the type of insertion are introduced on the computer;

-the unpasted layers will be fixed between the dinamometer jaws, and subsequently, by means of computer control, the mechanism of C_1 jaw movement is running;

-after the glued layers separation, the C_1 jaw lifting is stopped; in that moment on the computer screen some testing data will appear (tension – elongation plot/diagrame, forces necessary for the unglueing);

-on the basis of data provided by the computer, the adherence is calculated (Cioara and all).

Within the framework of the thermogluing process, it is noticed that, while the underensemble advances through the machine, its pressure remains constant, and it appears as its passing among the heating modulus, and the temperature shows first a sudden increasing, and subsequently a smaller one in a longer duration (Stefanuta, I.-2001). In the calender cyllinders area, the pressure reaches very quickly the highest and at the outlet, both temperature and pressure decrease to values near to environment. After such a work regime, an intimate and effiecient contact between the adhesive and basic textile material is provided (Preda et all, 1996).

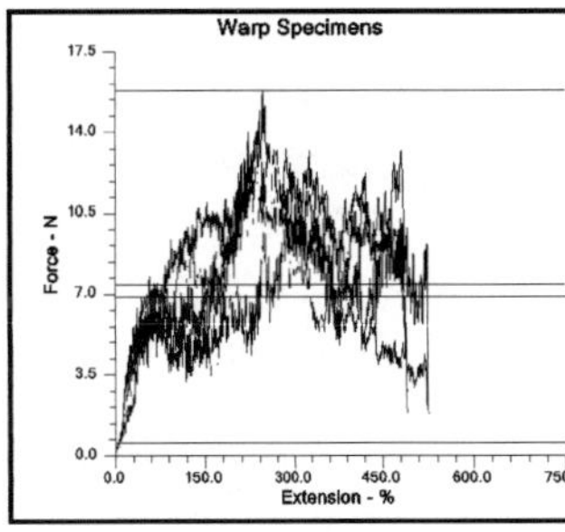

Fig. 1. 100% Cotton

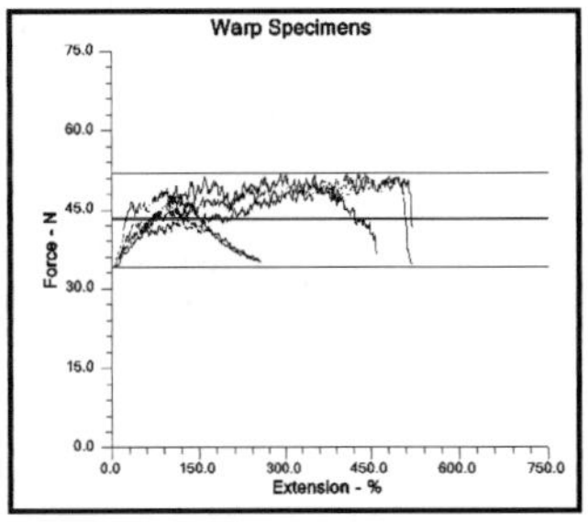

Fig. 2. 80% Cotton - 18% PA -2%

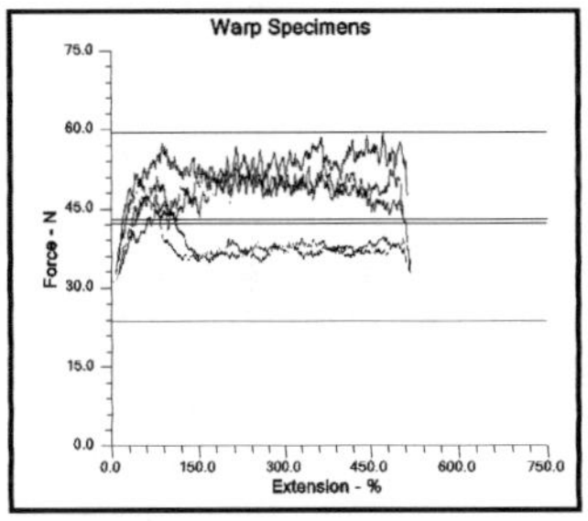

Fig. 3. 98% Cotton -2% Ellastan

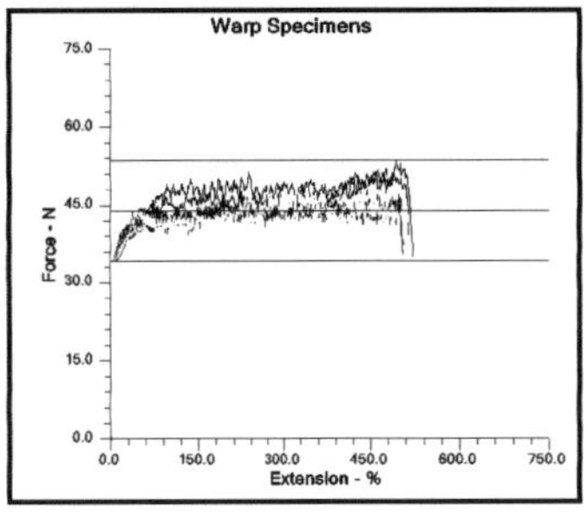

Fig. 4. 53% Cotton -43% PES – 4% Ellastan

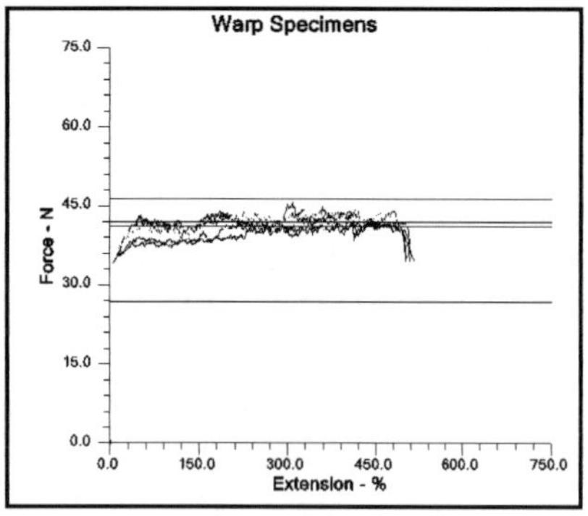

Fig. 5. 60% Wool -38% PES -2% Ellastan

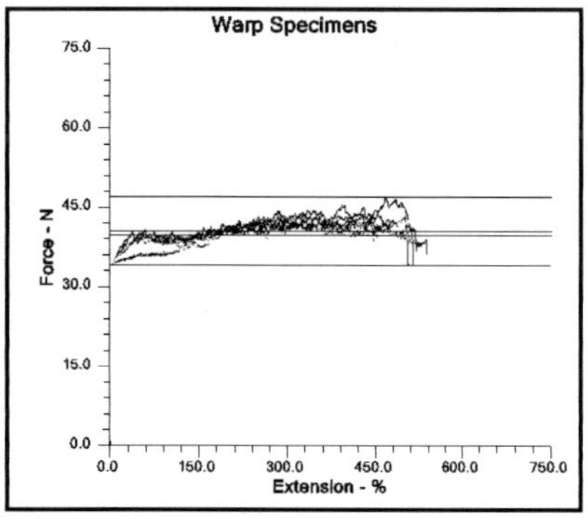

Fig. 6. Wool 96%-Elastan 4%

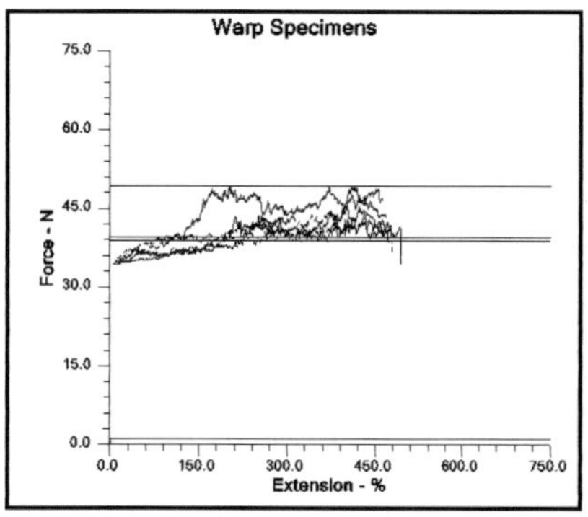

Fig. 7. 44% Wool -54% PES -2% Ellastan

3. THE ADHERENCE CALCULUS

Tabel 1 contains the results obtained for each type of sample subjcted to analyses and the average (mean) for the determinations where adhesives A and B were used, respectively (Technical book of Titan[2]).

Specimen	Force peak [N]	Force peak [N]	Average force peak [N]
1	9.16	13.30	9.39
2	8.73	12.36	9.78
3	7.27	10.93	8.00
4	8.48	13.62	8.96
Mean for adhesive A			9.03
5	6.46	15.78	5.97
6	6.82	11.56	6.81
7	4.32	13.52	2.83
8	3.93	12.93	2.65
Mean for adhesive B			4.56
Statistics			
Mean	6.90	13.00	6.80
Maximum	9.16	15.78	9.78
Minimum	3.93	10.93	2.65
Coeff. of Var	28.35%	11.35%	41.37%

Tab: 1. Results for material 1 – Cotton 100%

By the means of averages obtained in Tabel 1, in Tabel 2 the adherence for each type of material will be calculated, taking into account the adhesive used (Preda et all, 1996).

Nr. crt.	Basic material	Adherence type A insertion [N/cm]	Adherence type B insertion [N/cm]
1	cotton – 100%	1.806	0,912
2	cotton – 80% PA – 18% Elastan – 2%	9.42	8.112
3	cotton – 98% Ellastan – 2%	9.892	7.418
4	cotton – 53% PES – 43% Ellastan – 4%	9.03	8.652
5	Wool – 60% PES – 38% Ellastan – 2%	8.402	8.172
6	Wool – 96% Ellastan – 4%	8.21	7.934
7	Wool – 44% PES – 54% Ellastan – 2%	8.116	7.704

Tab: 2. The adherence calculus

4. CONCLUSION

In case of underensembles with type A insertion, a better adherence at cotton type textile materials is found, unlike wool type textile materials. The difference is provided by the high affinity of the adhesive at cotton fibre and the better denseness of made bond points, (the cotton fabric has a more compact and thinner structure);

In case of underensembles with type B insertion, the adherence is aproximately the same, at both types of basic materials.

If the strength at traction solicitation is studied, one concludes that forces values are aproximately in the same range for mixture basic materials, unlike 100% cotton type textile materials.

These tests highlighted the investigation method and procedure correctness. In the future, the above mentioned tests will also be performed on any other types of textile materials.

5. REFERENCES

Cioară, L., Cristian, I., Onofrei E., (2000) *Textile tissues structure characteristics and properties.*, Performantica, ISBN 973-730-029-7

DIN 53859 Part 2 – Tear growth testing of textile fabrics.

Preda, C.; Preda, Cr. (1996). *Methods and apparatus for Quality Control of textiles designed for wearing clothes manufacturing, BIT, ISBN 973-97907-2-0, Iasi;*

Stefanuta, I. (2001). *Processes and machines for textile clothing industry,* ULB, ISBN: 973-651-229-0, Sibiu;

Technical book of Titan[2] – *Universal Strength Tester,* Model710

Annals of DAAAM for 2008 & Proceedings of the 19th International DAAAM Symposium, ISSN 1726-9679
Editor B. Katalinic, Published by DAAAM International, Vienna, Austria 2008
Make Harmony Between Technology and Nature, and Your Mind will Fly Free as a Bird

STUDIES ON STRESS AND STRAIN STATE OF A HIP JOINT ENDOPROSTHESIS

STICLARU, C[armen]; DAVIDESCU, A[rjana]; CRAINIC, N[icolae] & FAUR, C[osmin]

Abstract: *The paper presents a study on a hip oint enoprosthesis. The paper deals with some stress and strain aspects for hip joint endoprosthesis. These aspects are depicted from a complex finite element analyze in Ansys. This study has two stages: obtaining the 3D model for the bone and endoprosthesis, analysing the model in Ansys – to obtain stress and stree state informations. With these informations the orthopedist takes the good decision for a proper hip joint endoprosthesis used for a named patient.*
Key words: *total hip arthroplasty, computational model, hip prosthesis, stress and straint state of a hip joint*

1. INTRODUCTION

Arthroplasty is the reconstruction or reshaping of a damaged or diseased joint. This elective surgery most often involves joint replacement, the implantation of an artificial joint (*prosthesis*). In addition to osteoarthritis, arthroplasty can be a treatment for conditions including hip fractures, other source of acute trauma and rheumatoid arthritis.

Arthroplasty may be used to:

- Replace all or part of a joint with a prosthesis
- Resurface a joint with the patient's own tissue
- Reshape the bone and cartilage that make up the joint

The purpose of this procedure is to relieve pain, to restore range of motion and to improve walking ability, thus leading to the improvement of muscle strength. Indications for arthroplasty include: osteoarthritis (OA), rheumatoid arthritis (RA), avascular necrosis (AVN) or osteonecrosis (ON), congenital dislocation of the hip joint (CDH), acetabular dysplasia (shallow hip socket), frozen shoulder, loose shoulder, traumatized and misaligned joint, joint stiffness.

The people may benefit from hip replacement surgery if: hip pain limits everyday activities such as walking or bending; hip pain continues while resting, either day or night; stiffness in a hip limits the ability to move or lift your leg; the patients have little pain relief from anti-inflammatory drugs or glucosamine sulfate; the patient's harmful or unpleasant side effects from your hip medications; other treatments such as physical therapy or the use of a gait aid such as a cane do not relieve hip pain.

The hip joint supports most of the upper body weight. As a person ages the bone become thinner and more brittle, increasing the risk for injury. The hip joints are connecting the torso to the legs and support the upper body weight (fig. 1 a). The bones of the pelvis, the pubis, the ischium and the illium form a ball-socket joint together with the head of the femur, the log thigh bone (fig. 1 b).

A total hip prosthesis is composed of two components: the femoral (thighbone) component and the cup component (fig. 2).

During the procedure, the joint is fully exposed and the damaged bone and cartilage are cut away or reshaped. A plastic cup is placed in the enlarged hip socket (fig. 1c). Then, the top of the femur is removed and a metal ball is inserted into the top of the femur (fig. 1c). Also a metal stem is also inserted into the femur to add stability to the prosthesis (fig. 1 d). The joint is tested before the incision is closed. The whole stem component is there only to keep the relatively small ball component fixed to the skeleton. The stem component is big, it engages about one third of the whole thighbone. It is ballast. When the total hip fails, the one third of the thighbone skeleton round this big ballast suffers, it is damaged or destroyed. In the healthy hip joint the femoral head is continually in close and stabile contact with the socket during all movements. during total hip replacement a portion of these supporting structures (muscles, ligaments, capsule) is cut (divided) for easier access to the hip joint. Even if the surgeon tries to restore muscle and soft tissue balance by suturing together the cut ligaments, muscles, and joint capsule after the total hip replacement, there is usually some imbalance of soft tissues left. The total hip prosthesis must be anchored securely within the skeleton for good function.

There are two methods how to secure the fixation of a total hip prosthesis to the skeleton:

1. The cemented total hip - the surgeon uses bone cement for fixation of the prosthesis to the skeleton
2. The cement less total hip - the surgeon impacts the total hip directly into the bed prepared in the skeleton.

When prostheses are used, they may be made of polyethylene, metal, ceramics or silicone. The most common design is metal-on-polyethylene, although metal-on-metal designs have become more popular in recent years.

2. THE SOLID MODEL FOR THE FEMUR AND HIP JOINT PROSTHESIS

The solid model for the femur was obtained using computer tomography images. These images were imported in MIMICs and then exported in ProEngineer to make the assembly with the endoprosthesis. The assembly was created using cut operations to obtain the form for the femoral head. This form is obtained by the medical doctor during the surgery (fig. 3). In fig. 3 are presented 3 cases: a. correct position for the femur; b. the position for leg hitting ground; c. leg swinging free.

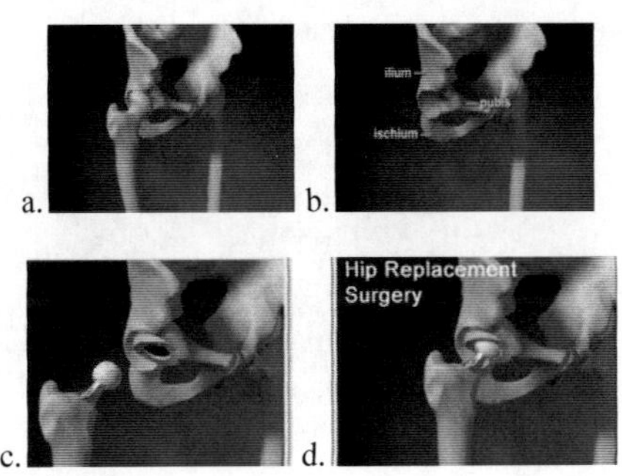

Fig. 1 The hip joint

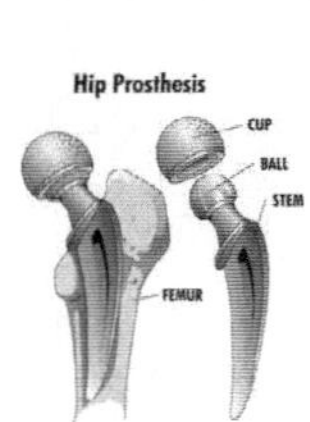

Fig. 2 Hip prosthesis

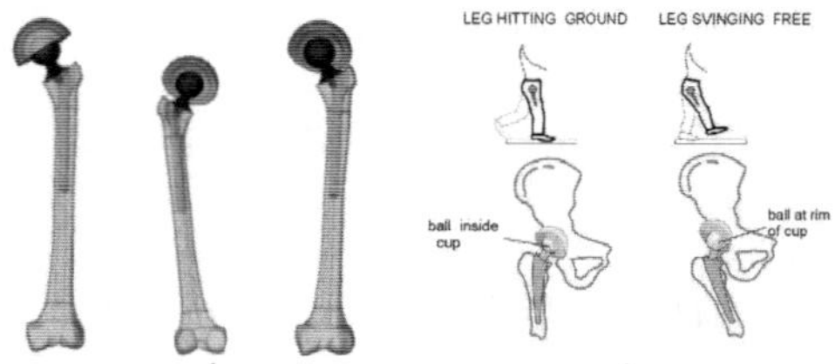

a. b. c. d.

Fig. 3 The solid models for the femur with endoprosthesis

3. THE FINITE ELEMENT MODEL DEVELOPED IN ANSYS

The steps for solving the finite element analysis are:

- import the CAD model from proEngineer (Sticlaru, 2007);
- assigning the materials for the femur and endoprosthesis elements (stainless steel for the stem, ball and exterior cup, polyethylene for interior cup, for the femur (Weinans H., 1992), bone properties (Gahr R.H., 1999));
- creating the mesh for the model (fig. 4) (a. the femur, b. the stem, c. the cup, d. interior cup);
- creating the environment for the assembly;
- run analyze - for different body position (fig. 3 a, b, c).

The environment for the model looks like that presented in fig. 5 (a. the forces of the hip in single leg stance b. x-ray image of a normal hip showing the compression trabeculae oriented parallel to the resultant compressive load on the femoral head (Bibb 2006) c. the environment for the solid model). The simulations were performed with the distal end of the femur rigidly constrained. The load for the assembly was obtained by studying the influence of different positions for the human body (Davidescu, 2007). In normal walking the hip is subjected to wide swings of compressive loading from one-third of body weight in the double support phase of gait to 4 times body weight during the single leg support phase.

In normal walking the hip is subjected to wide swings of compressive loading from one-third of body weight in the double support phase of gait to 4 times body weight during the single leg support phase.

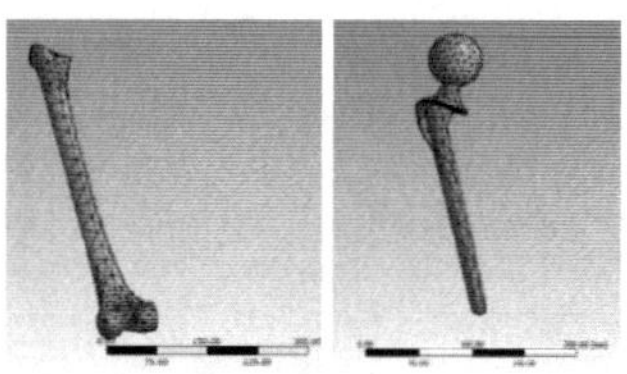

Fig. 4 The mesh for the solid model

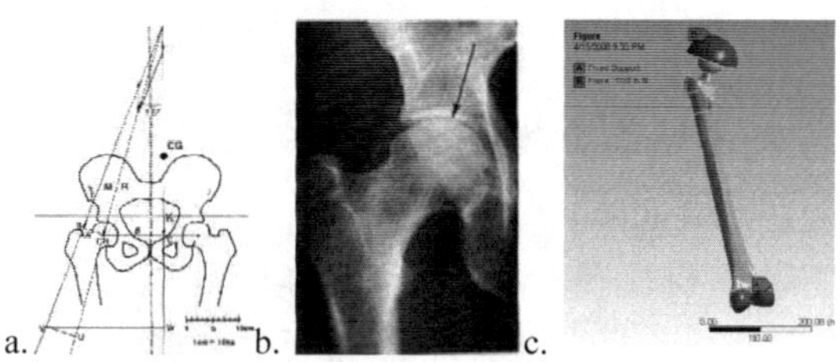

a. b. c.

Fig. 5 The forces on the hip jont

4. RESULTS AND CONCLUSIONS

From the obtained results there can be depicted some aspects:

• total deformations are presented in fig. 6; it can be observed that the distribution is asymmetrical distributed and the major values appear in the cup;

• directional deformation along the femoral axis is resented in fig. 7 – the deformation of the bone is different and the values are greater for the ball at rim of the cup;

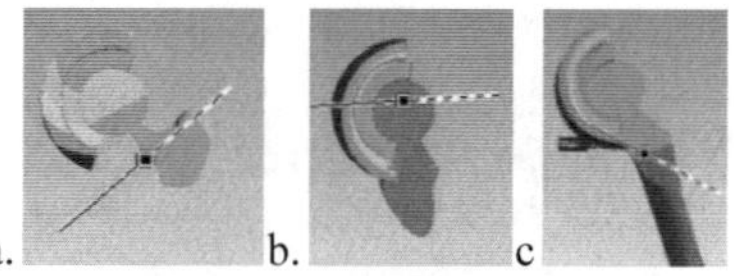

a. b. c

Fig. 6. The total deformation for the model

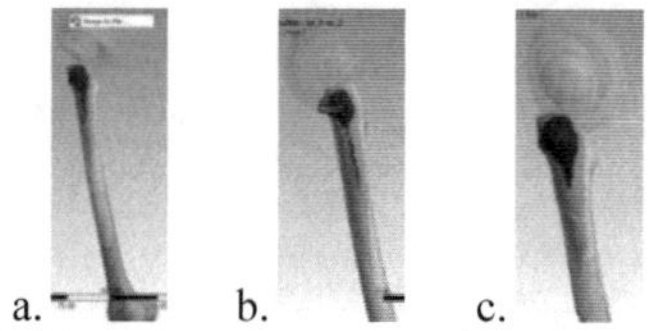

a. b. c.

Fig. 7 The deformation alog the bone axes

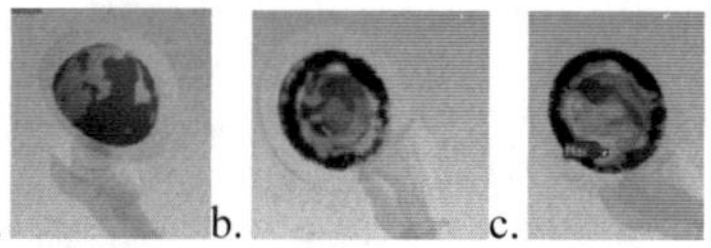

a. b. c.

Fig. 8 The strain state for the cup

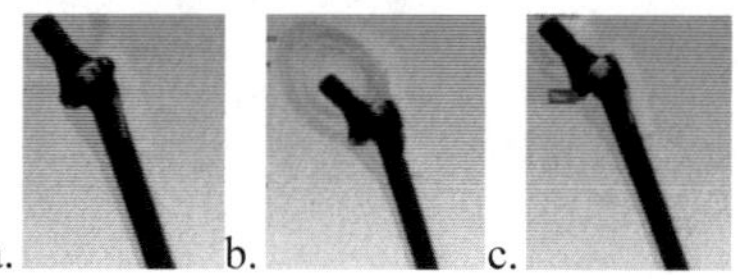

a. b. c.

Fig. 9 The stress state for the stem

• strain state for the cup is presented in fig 8 – it is an assymetrical distribution, with greater values for swinging up the leg (c);

• stress state for the stem is presented in fig. 9 – the distribution for the stress state is simmilar, but the values are very different – the greater values are for the swinging up the leg position;

The presented aspects depicted from FEM analyze for a hip joint endoprosthesis are very useful for the orthopedist, because that can be taken better decisions for the patients who need total hip arthroplasty. Using the parameterized capabilities of the proEngineer and the link between ProE and Ansys, many models can be developed easy starting from this one. This study is useful for obtaining personalized hip joint endoprosthesis for patients that have skeleton deformation. For this kind of patients to realize a total hip arthroplasty is difficult. In these cases after a computer tomography, the model of the bone is imported in proE – the stem and the cup are designed using the bone form, than in Ansys and it can be decided if the endoprosthesis is a good one.

5. REFERRING

Bibb R. (2006) *Medical modelling - The application of advanced design and development techniques in medicine* Woodhead Publishing Limited and CRC Press LLC © 2006, Woodhead Publishing Limited;

Davidescu A., Sticlaru C. (2007) Studies by Finite Element Method of some Devices for Treatment of Intertrochanteric Fractures, *The 12th IFToMM World Congress*, Besançon (France), vol 1, pg.112-117, www.iftomm.org;

Gahr R.H., Leung K.S., Rosenwasser M.P., Roth W.(1999) *The Gamma Locking Nail*, Einhorn-Press Verlag GmbH Reinbek, ISBN 3- 88756-808-7.

Sticlaru C., Davidescu A. (2007), Comparative Study of Fixation Devices for Intertrochanteric Fractures *The 12th IFToMM World Congress*, Besançon (France), vol 1, pg.118-123, www.iftomm.org

Weinans H., Huiskes R., Grootenboer H. (1992) - Effects of Material Oroperties of Femoral Hip Components on Bone Remodeling, *Orthopaedic Research Society.*

Annals of DAAAM for 2008 & Proceedings of the 19th International DAAAM Symposium, ISSN 1726-9679
Editor B. Katalinic, Published by DAAAM International, Vienna, Austria 2008
Make Harmony Between Technology and Nature, and Your Mind will Fly Free as a Bird

UNICORTICAL - BICORTICAL BEHAVIOR OF A HYBRID BONE SCREW

STOIA, D[an] I[oan] & TOTH-TASCAU, M[irela]

Abstract: *The goal of the paper is to determine both the unicortical and bicortical behavior of a hybrid bone screw when threading through the vertebral body and pedicle. The behaviour (unicortical and bicortical) refers to the moment of torsion needed to be applied on the screw at a specific axial load, in order to self–thread in the bone structure. This moment of torsion have to be applied by the orthopedist, during the surgical implantation procedure.*
Key words: *bone screw, unicortical, bicortical, bone threading.*

1. INTRODUCTION

Bone is a complex, highly organized and specialized connective tissue. The vertebra consists of a vertebral body anteriorly resembling a tubular bone with a central portion surrounded by a thin cortical shell. The posterior vertebral arch consists of the pedicles with dense cortices and relatively little intervening spongy bone and laminae which are flattened bones with also relatively dense cortex and less dense cancellous bone (Carter, 2002), (Tan, 2003).

Basically, there are two types of bone screws designed to be assembled with cortical or cancellous bone structures. Taking into account that the vertebral bone is a composite structure composed by a cancellous core and a cortical shell, the screw must proper threading the both bone structures (Butscher, 2003).

The bone screws are available in a variety of designs and materials. For instance, the treads may be vee, square or buttress shaped and the major and minor diameters may be straight or tapered. The helix pitch plays also an important role. These features affect the pull-out strength, the stiffness, and the holding power of the screw. It is assumed that a well-fixed screw will help maintain a surgical reduction until fusion occurs (Dawson, 2003).

In the framework of the project *Researches Regarding the Improvement of the Modeling and Manufacturing Techniques for the Human Spine Implants*, were designed, developed and tested three types of screws according to the bone structure: one screw type for cancellous vertebral bone, a second screw for cortical vertebral bone and a third screw, designed as a combination of the first two, named hybrid. All of the screws are destined to fix a plate implant in the cervical bone structure. This paper will present the torque function of angular displacement during the threading/screwing tests of the hybrid screw.

Due to the self-threading requirements of the hybrid screw, it is necessary to measure the torque needed to be applied on the screw (for a certain axial loading value), in order to realize the threading and screwing in the composite bone structure. The tests were performed for two cases: unicortical and bicortical threading (Stoia, 2008).

2. TESTING EQUIPMENTS AND METHODS

The tests were realized in the frame of CIDUCOS Testing Laboratory, accreditated by RENAR– Romanian Accreditation Association, on Vortex-*i,* testing equipment (Fig. 1.). The testing conditions:

- Axial load on the screw P=35 N;
- Angular velocity 2.5 rot/min;
- Maximum torque 1Nm;
- Angular displacement 4500 deg - equivalent of 12.5 rotations.

In order to measure the unicortical and bicortical screw behaviour, five cadeveric specimens of lumbar L4 vertebras were used. The vertebral body specimens were coded CV1, CV2 and CV3, while the vertebral pedicles were coded PV1 and PV2.

At the places of the screw insertions, in all the specimens, 2.5 mm holes were drilled. After that, these were fixed on the rotational table of the testing machine. The hybrid titanium screw was fixed in the grip of the torque cell, being aligned with the bone hole. The machine control was made by computer, using the dedicated Emperor software.

The type of the control program used for tests was *run*, who involves all the parameters presented before. The maximum angular displacement (θ) was calculated using the following relation:

$$\theta = l / p = 7.5 / 0.6 = 12.5 \text{ rot} \quad (1)$$

where: l represents the thread length of the hybrid screw and p, the pitch of the screw.
An important issue for the tests stability and validity is to have a good alignment between the bone holes and the screw.

3. RESULTS

The results express the mechanical behaviour of the hybrid screw when threading the two structures. The intermediary positions for threading the vertebral bodies and pedicles are presented in the Fig. 2.

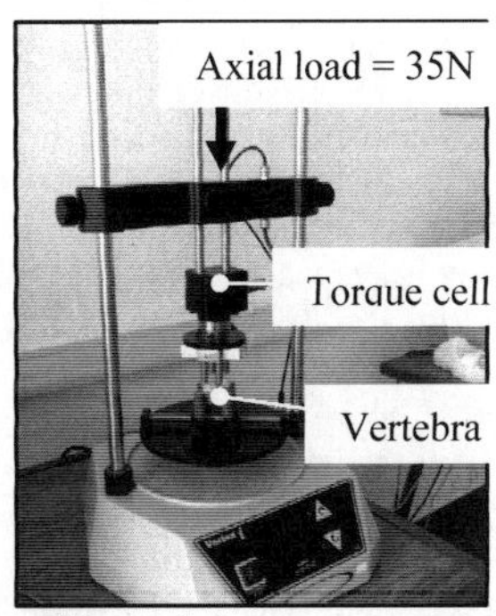

Fig. 1. Vortex-*i* testing equipment.

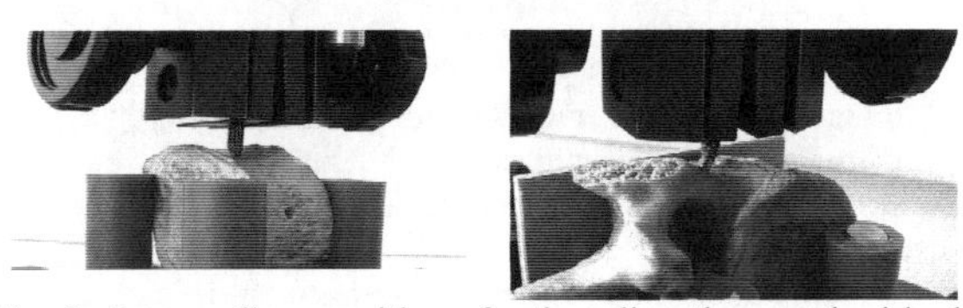

Fig. 2. Intermediary positions for threading the vertebral body (left) and pedicle (right).

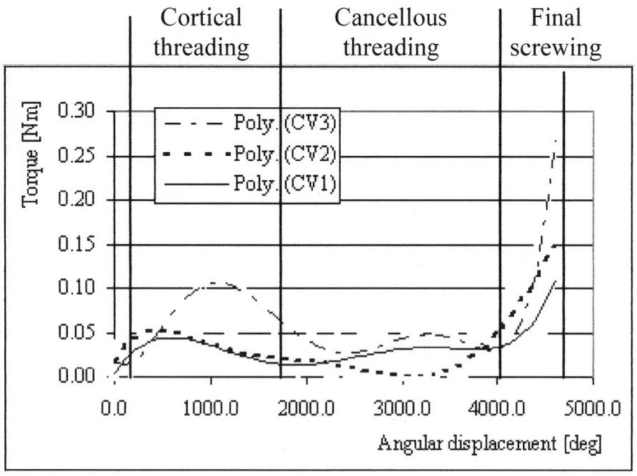

Fig. 3. Unicortical threading behaviour. Six-order polynomial tendency curves of the measurements.

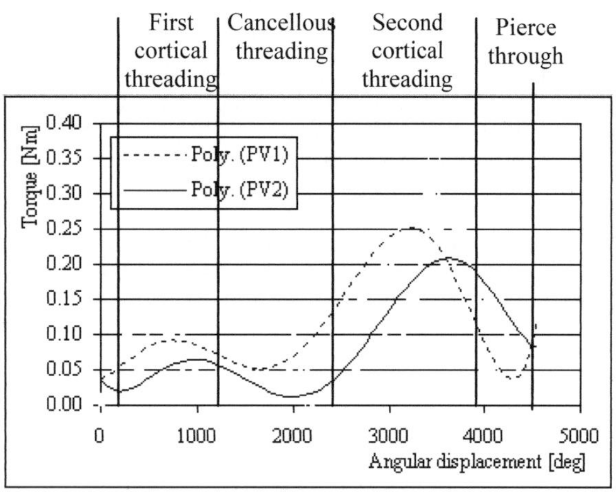

Fig. 4. Bicortical threading behaviour. Six-order polynomial tendency curves of the measurements.

Because of the large sampling rate of the measured values, only the tendency curves of the torques function of angular displacements will be presented.

In the Fig. 3. are presented the tendency curves of the torques function of angular displacements for the three vertebral specimens CV1, CV2 and CV3. The shape differences recorded for the three curves are generated by two factors: the first one is generated by the differences in the bone density between each specimen; the second factor is cause both by the differences in the angular starting position of the screw and the thickness of the bone layers (cortical and cancellous).

The first factor is responsible by the amplitude variation while the second generates a phase difference.

In the case of unicortical threading three characteristic intervals can be identified:

- Cortical threading – from 0 to 3.5 rot and a maximum torque of 0.11 Nm.
- Cancellous threading – from 3.5 to 11.5 rot with a relative constant torque. The spontaneous variations are determined by the variations in the bone density.
- Final screwing – from the 11.5 to 12.5 rot is characterized by the sudden increasing of the torque.

In the Fig. 4. is presented the bicortical threading behaviour of the PV1 and PV2 specimens. In the same manner as in the first case, the torques, function of angular displacements are presented..

The threading and screwing characteristic intervals for bicortical structures cases are:

- First cortical threading – form 0 to 3.5 rot – corresponding to the penetration of the first bone layer.
- Cancellous threading – from 3.5 to 6.6 rot – corresponds to the second bone layer and is characterized by the decreasing of the torque due to the reduced density of the layer.
- Second cortical threading – from 6.6 to 10.8 rot – record a maximum torque value of 0.25 Nm; 2.5 higher than in the first cortical threading.
- Pierce through – form 10.8 to 12.5 is the interval where the torque decreases due to the pierce of the second cortical layer.

The maximum torque values recorded in all the testing cases are very much dependent on the cortical layer thickness.

4. CONCLUSIONS

The performed tests prove the hybrid screw capacity to self-threading in both structures: unicortical and bicortical bone.

The maximum measured torques in both cases for an axial force of 35 N, are 0.11Nm for the first cortical threading and 0.25mm for the second cortical threading. These values indicate the surgeon effort, required during the implantation procedure.

It must be mentioned that, when threading and screwing in the living bone, the maximum torque values can be different due to the viscoelasticity of the living bone toward cadaveric bone.

Also, the torque values required for self-threading of the hybrid screw are dependent by the following factors: anisotropy of the bone structure (Dong, 2004), the age and the sex of the human subjects.

One can say that the measurements reveal some reference values for the torque, in the case of self-threading in the unicortical and bicortical bone structure.

5. REFERENCES

Butscher, A.; Schneider, R.; Wahl, D.; Gasser, B.; Linke, B. (2002). Holding strength of conventional and locked plate screws, ACTA of Bioengineering and Biomechanics, Vol.4, No.1, pp.782-783, 83-7085-639-x, Wroclaw, Poland.

Carter, D.R. & al. (2002). *Bone in Clinical Orthopedics,* Thieme, 3-13-125721-0, Stuttgart-New York.

Dawson, J.M.; Boschert, P.; Macenski M.M. & Rand, N. (2003). Clinical Relevance of Pull-out Strength Testing of Pedicle Screws, In: *Spinal Implants: Are We Evaluating Them Appropriately?, Melkerson, M.N., Kirkpatrick, J.S., Griffith S.L.,(ASTM International), pp.68-77,0-8031-3463-0, Dallas, Texas.*

Dong, X.N.; Guo X.E. (2004). The dependence of transversely isotropic elasticity of human femoral cortical bone on porosity, Journal of Biomechanics, Vol.37, No.8, pp.1281-1287, 0021-9290, Elsevier.

Stoia, D.I.; *Researches Regarding the Improvement of the Modeling and Manufacturing Techniques for the Human Spine Implants*, Type TD, Code 95, Contract GR98/11.06.2008, Financed by the Romanian National University Research Council.

Tan, J.S.; Kwon, B.K.; Samarasekera, D.; Dvorak, M.F.; Fisher, C.G. & Oxland, T.R. (2003). Vertebral Bone Density-A Critical Element in the Performance of Spinal Implants, In: *Spinal Implants: Are We Evaluating Them Appropriately?, Melkerson, M.N., Kirkpatrick, J.S., Griffith S.L.,(ASTM International),68-77,0-8031-3463-0, Dallas, Texas.*

Annals of DAAAM for 2008 & Proceedings of the 19th International DAAAM Symposium, ISSN 1726-9679
Editor B. Katalinic, Published by DAAAM International, Vienna, Austria 2008
Make Harmony Between Technology and Nature, and Your Mind will Fly Free as a Bird

USE OF CROSS-SPECTRUM MEASUREMENT TECHNIQUE IN FINDING DAMPING AND STIFFNESS PROPERTIES OF POLYMERS

STOICA, G[ina] F[lorica]; IORDAN, A[ndrei]; CANANAU, S[orin] & GRIGORAS, S[tefan]

Abstract: *In the following, a experimental method for obtaining the stiffness and damping material properties of polymers will be presented, using the Virtual Field Method for identification and deflectometry for measurement. The importance of correct measurements for the phase lag will be shown, applying the cross-spectrum phase measurement technique. The experimental arrangement will be revealed and also the methods used for obtaining Poisson's ratio. The improvements to the experimental set-up are then developed by introducing the phase lag measurement by the cross-spectrum technique. The results of this measurement technique are presented and comments upon their validity have been made. Several experiments have been made, by taking into account several frequencies of excitation and several apparatus for measurements. Noise sensitivity has also been studied.*
Key words: *Poisson Ratio, Deflectometry, Cross Spectrum, Virtual Field Method, Polymers.*

1. INTRODUCTION

The materials possessing both elastic and viscous properties are called viscoelastic, the polymers being the typical example. Poisson's ratio is defined as being the ratio between the lateral contraction and the elongation of an isotropic homogenous material in an infinitesimally small extension. For those viscoelastic materials, Poisson's ratio is a function of time, frequency or temperature (Tschoegl, 2002). In the following, we will present the ameliorations to a method of obtaining the frequency depending Poisson ratio. Due to experimental difficulties, the literature on this subject is rather scarce. Despite this, we have identified to ways to determine the Poisson ratio: an indirect and a direct way (Tschoegl, 2002).

For most frequency dependent measurements, a beam-like specimen is submitted to sinusoidal charges, the ratio being measured with strain gauges. For identification, the least square or the reduction of variables method is used. The method presented here, gathers full-field information by aid of an optical measurement method (deflectometry) and uses the Virtual Fields Method for identification. Also, the specimen used is not a beam, but a plate in vibration.

2. EXPERIMENTAL ARRANGEMENT

The experimental set-up was introduced by Giraudeau and Pierron (Giradeau, 2005). It consists of an arrangement is presented in the Fig. 1. The tested material was a PMMA type polymer cut into thin plates size 200x 160x3 mm and a mirror treatment was applied on one of their faces. This mirror treatment was required by the optical method required for the measurement. The plate is submitted to a sine driving motion, leading to the bending of the plate. For the full description of the sine driving movement two images are taken at different moments in time with a CCD camera and a flash.

As it is shown, the plate is being moved by an electrodynamical shaker and driven by a drivin device introduced to guide the movement and limit it's the amplitude.

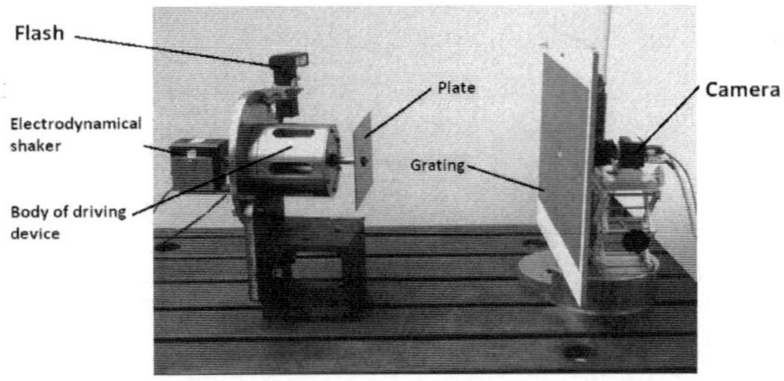

Fig. 1. The experimental set-up

The driving device consists of a series of metal membranes mounted perpendicular to an axle.

After these precautions, we can assure that the movement has small amplitude and that bending vibrations are preponderant. An accelerometer is mounted on the axle holding the plate in order to describe its movement. Facing the mirror treated plate is a grating and behind the grating a CCD camera used for picture taking. Near to the plate we have a flash used to trigger the camera. In order to set the electrodynamic shaker in motion we use a National Instruments signal generating and measurement card. With the help of this card, we generate two signals, a sine one, set to generate the movement and a square one for triggering the flash. The method we use for measurement (deflectometry) requires 2 pictures taken at two precise instants in time, therefore the two signals (sine and square) must be synchronized. We measure the amplitude of the sine movement with the accelerometer and we impose a lag of 0 or pi/2 between the measured signal and the generated square.

3. IMPROVEMENTS–USE OF THE CROSS-SPECTRUM

In order to introduce the improvements we added to this experimental set-up, we will talk about the synchronization of the two signals present in the set-up: the sine and the square. This synchronization is done according to the following graph (Surrel, 2003) (Fig. 2).

Where, the phase lag was measured with a phasemeter (case A.) or by means of the cross-spectrum (case B.).

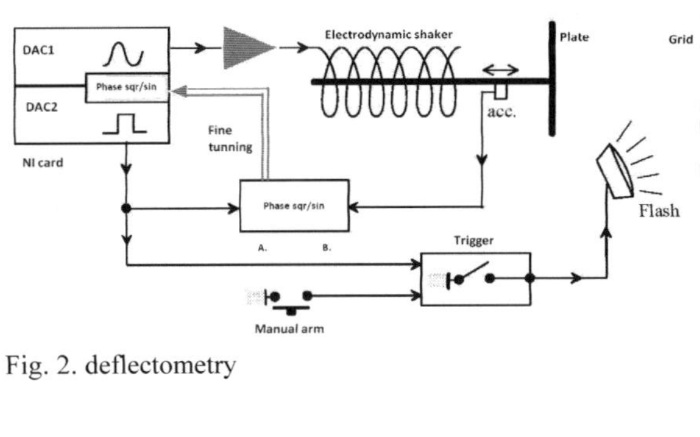

Fig. 2. deflectometry

The reason we introduced the cross-spectrum measurement technique was that the phasemeter could only allow us to impose a phase lag between the two signals with 0.5 degrees of precision maximum. We obtained 10^{-3} degrees of precision. The cross spectrum method of analysis is based on the Fourier transform, which serves in the conversion of information from the time domain to the frequency domain and vice-versa. No information concerning the signal is lost by doing this transformation. For a time-domain signal, a(t), the Fourier transform is: (Herlufsen, 1984)

$$A(F) = \mathcal{F}\{a(t)\} = \int_{-\infty}^{+\infty} a(t) e^{-j2\pi f t} dt \qquad (1)$$

Likewise, for another signal *b*(*t*):

$$B(F) = \mathcal{F}\{b(t)\} = \int_{-\infty}^{+\infty} b(t) e^{-j2\pi f t} dt \qquad (2)$$

The cross-spectrum is defined like: (/cite)

$$S_{AB} = E\left[\hat{A}_i^*(f) \cdot \hat{B}_i(f)\right] = \lim_{n_d \to \infty} \frac{1}{n_d} \sum_{i=1}^{n_d} \hat{A}_i^*(f) \cdot \hat{B}_i(f) \qquad (3)$$

So we can see the phase lag clearly:

$$\left|A_i(f)\right| \cdot \left|B_i(f) \exp^{j\Delta\Phi(f)}\right| = \left|A_i(f)\right| \cdot \left|B_i(f) \exp^{j[\Phi B(f) - \Phi A(f)]}\right| \qquad (4)$$

By averaging, for we apply a FFT (fast Fourier transform), we can obtain the phase lag between the two signals a(t) and b(t). In the following we will present the steps took for validating this technique for the experimental set-up and the results obtained by doing so.

There have been more stages in the validation of this technique. Firstly, we have used a DAC by DSP Technology – called SIGLAB[i] to generate two cosines; the same apparatus was used for measurement. The second series of tests consisted in adding a second SIGLAB for measurement to prove the fact that the method works, not indifferent to the apparatus used. To finalize this stage, we have crossed the input and output channels of the two machines following this diagram: Input 1→ → Output 2; Input 2 → Output 1.

The imposed phase lags, frequencies and sample rates of the signals are presented below (Table 1 to Table 3).

In order to fully emulate real life conditions, we finished the set of experiments by simulating noise into our signals. The results were also very impressing, a precision of 10-2 degrees exists at high levels of noise, and only at 30% level of noise precision sadly diminishes. The table 4 present the results obtained during the study of the influence on noise levels.

Imposed lag	0	π/2	π/4	π/6	π/10
Measured lag	10^{-4}	89,998	44,998	29,997	17,997

Table 1. Same SIGLAB for generation and measurement; frequency 200Hz, sample rate 1024

Imposed lag	π/2	π/4	π/6	π/10	π/100	π/1000
Measured lag	89,9 9994	44,9 9977	29,9 9910	18,0 2220	1,798 277	0,178782

Table 2. 2 SIGLABS;
bandwidth 500Hz, frequency 200Hz, sample rate 1024

Imposed lag	π/2	π/4	π/6	π/10	π/100	π/1000
Measured lag	89,998 72	45,0 0014	30,0 0054	18,0 0128	1,798 127	0,1784 44

Table 3. For 50 Hz

	100 Hz		150 Hz	
Noise level	Imposed lag	Measured lag (degrees)	Imposed lag	Mesured lag (degrees)
3%	π/4	44,995	π/4	44,997
	π/6	30,006	π/6	30,002
	π/10	17,999	π/10	17,997
	π/100	1,8039	π/100	1,8008
	π/1000	0,1799	π/1000	0,1809
5%	π/4	45,004	π/4	44,996
	π/6	30,004	π/6	29,996
	π/10	18,001	π/10	18,007
	π/100	1,8050	π/100	1,7952
	π/1000	0,1838	π/1000	0,1727
10%	π/4	45,002	π/4	44,998
	π/6	30,026	π/6	29,993
	π/10	18,005	π/10	18,001
	π/100	1,7958	π/100	1,7973
	π/1000	0,1723	π/1000	0,1876
30%	π/4	45,014	π/4	44,984
	π/6	30,015	π/6	30,021
	π/10	17,968	π/10	17,988
	π/100	1,8536	π/100	1,8233
	π/1000	0,1837	π/1000	0,2093

Table 4.

4. CONCLUSION

A technique for measurement of the Poisson coefficient was discussed here and the improvements to the existing method (Giraudeau, 2006) were underlined. The cross-spectrum technique for phase lag measurement was presented and also its results. Sadly, due to the lack of stability of the acquired signals, this technique did not produce the break-through it was expected, because the results of the measurements were practically the same as in the previous work (Giraudeau, 2006) when a phasemeter was used to measure phase lag.

The reasons this technique did not worked as expected are, in our opinion, due to the fact that the plate's driving movement is assured by an electrodynamical shaker, therefore its amplitude is not constant in time. Since the cross-spectrum calculus of the phase lag is an average of the signal, it is obvious that on the experimental set-up we didn't obtained the same results as we did in the presented tests. Due to this factor and of course of random errors, the precision of measurement of the phase lag diminished to 0.4 degrees, comparable with that of a phasemeter.

5. REFERENCES

Giradeau, A. (2005), Identification of Stifness and Damping Properties of Thin Isotropic Vibrating Plates Using the Virtual Fields Method: Theory and Simulations, *Journal of Sound and Vibration,* 284, 757-781

Giradeau, A. (2006), Stiffness and Damping Identification from Full Field Measurements on Vibrating Plates, *Experimental Mechanics*

Herlufsen, H. (1984), Dual Channel FFT Analysis (part I), *Technical Review*

Surrel, Y. (cours CNAM 2003-2004), *Images optiques; mesures 2D et 3D*

Tschoegl, N.W. (2002), Poissons Ratio in Linear Viscoelasticity. A Critical Review, *Mechanics of Time Dependent Materials,* 6, 3-51.

Annals of DAAAM for 2008 & Proceedings of the 19th International DAAAM Symposium, ISSN 1726-9679
Editor B. Katalinic, Published by DAAAM International, Vienna, Austria 2008
Make Harmony Between Technology and Nature, and Your Mind will Fly Free as a Bird

ANALYTICAL DETERMINATION OF POSITION OF THE DELTA CABLE SYSTEM

STOLLMANN, V[ladimir] & VACEK, V[ladimir]

Abstract: *The article is devoted to solution of basic navigation task of the delta cable systems. The delta cable systems represent a prospective locomotive mechanism of forest robots, so called the deltastats, investigated at the Technical University in Zvolen. In the article is derived an analytical procedure of determination the actual position by odometrical method.*

Key words: *forest, robot, cableway, deltastat.*

1. INTRODUCTION

Conception solution of Delta cable systems was created at the Faculty of Forestry of the Technical University in Zvolen. It represents the locomotive mechanism of the deltastats, which are a new technical means intended for transport and handling of bulky and heavy objects on large distances, mainly for automization and robotization of forest operations (Stollmann & Belanova 2006). A typical solution of the Delta cable system to application in forest area is depicted in the Fig 1, where: – poles with firmly fasten cables, 2 – cable carriage, 3 – hauling road.

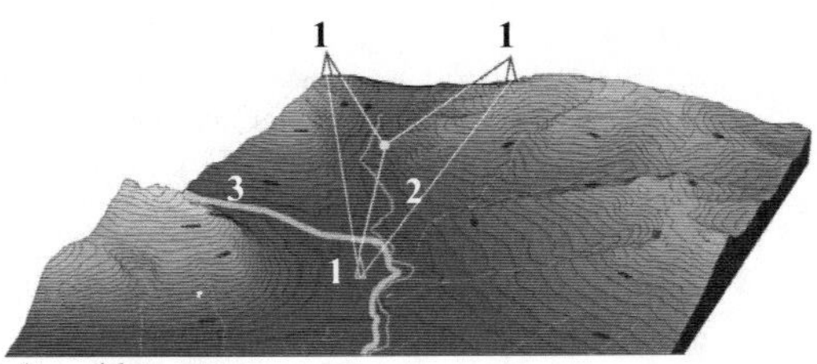

Fig1 Delta cable system

2. KINEMATIC SCHEME

The Delta cable systems are characterized by parallel kinematical structure, which is in nowadays robotization technique considered a modern structural element (Smrcek & Hajtinger 1996, Barborak et al. 2006). The Delta systems enable navigation and at the same time an orientation of working floor in space and require the use of six cables, which are firmly fasten on three posts – see Fig 2, where 1 – working floor.

For applying of the Delta cable systems in the sets of robots (deltastats) it is typical the gravitational establishment of the robotic working unit to vertical position. This enables to reduce the number of cables of the Delta cable system to there pieces – see Fig 4.

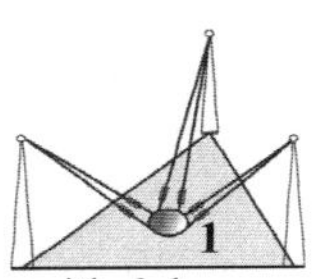

Fig 2 Delta cable system with 6 degree of freedom

3. STRUCTURE OF DRIVE

The Delta cable system represents a locomotive mechanism of the deltastats, which in its substance are airborne robotic systems. The structure of the Delta cable system drive is given in Fig 3, where: 1 – control unit, 2 – drive unit, 3 – brake, 4 – clutch, 5 – drum, 6 – length of cable sensing unit, 7 – cable carriage.

Its characteristic sign is location of driving units and cable drums directly in the carriage.

The movement of cable carriage 2 (Fig 1) is performed by means of steel cables. The steel cables are firmly fastened to posts and they are wound on the drums, which are located in the cable carriage 7 (Fig 3).

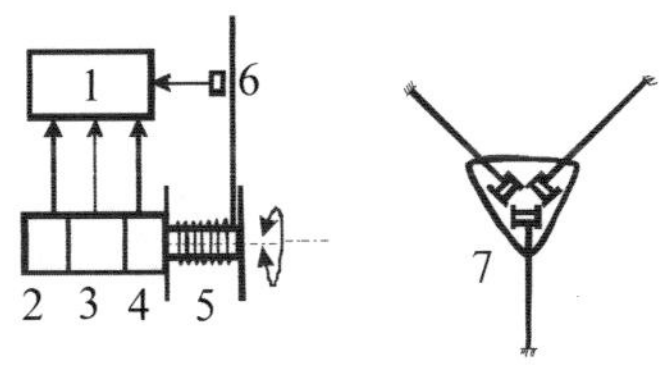

Fig 3 Structure of drive

4. DEFINING OF NAVIGATION TASK

Determining of the carriage position 2 (Fig 1) in the space is a key task for the need of navigation of a carriage to the target point. We can consider various methods of the task solution, out of which we are choosing the odometric method for its simple technical realization. In the case of the Delta cable systems, this method requires the measuring the length of cables, where we can use the measuring system consisting of measuring pulley and incremental readers (Sakal et al. 1999).

5. ANALYTICAL SOLUTION OF NAVIGATION TASK

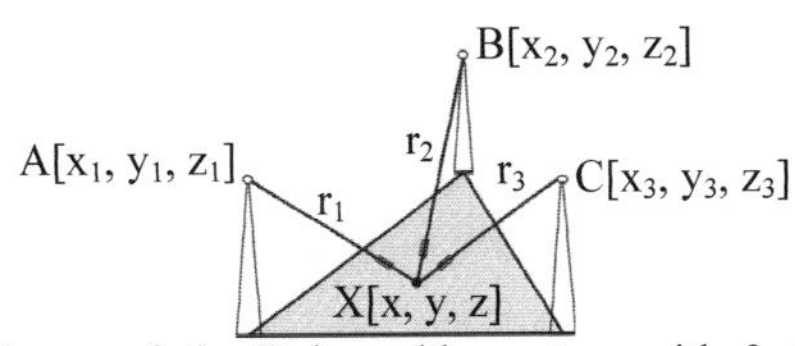

Fig 4 Scheme of the Delta cable system with 3 degrees of freedom.

Let presuppose that we known the lengths of cables r_1, r_2, r_3 at the point X and the coordinates of the points A, B, C – see fig 4. Our task is to determine the position of the carriage i.e. the coordinates of the points X. We are obtaining the searched solution as an intersection of three spherical surfaces.

The equations of spherical surfaces are:

$$(x - x_1)^2 + (y - y_1)^2 + (z - z_1)^2 = {r_1}^2 \qquad (1)$$
$$(x - x_2)^2 + (y - y_2)^2 + (z - z_2)^2 = {r_2}^2 \qquad (2)$$
$$(x - x_3)^2 + (y - y_3)^2 + (z - z_3)^2 = {r_3}^2 \qquad (3)$$

By powering the equations (1), (2) and (3) we get:

$$x^2 - 2xx_1 + x_1^2 + y^2 - 2yy_1 + y_1^2 + z^2 - 2zz_1 + z_1^2 = r_1^2$$
$$(4)$$
$$x^2 - 2xx_2 + x_2^2 + y^2 - 2yy_2 + y_2^2 + z^2 - 2zz_2 + z_2^2 = r_2^2$$

$$x^2 - 2xx_3 + x_1^2 + y^2 - 2yy_3 + y_1^2 + z^2 - 2zz_3 + z_1^2 = r_3^2 \quad (5)\ (6)$$

By subtracting of the equations (4), (5) and (4), (6) we get:

$$2x(x_2 - x_1) + x_1^2 - x_2^2 + 2y(y_2 - y_1) + y_1^2 - y_2^2 + \\ +2z(z_2 - z_1) + z_1^2 - z_2^2 = r_1^2 - r_2^2 \quad (7)$$

$$2x(x_3 - x_1) + x_1^2 - x_3^2 + 2y(y_3 - y_1) + y_1^2 - y_3^2 + \\ +2z(z_3 - z_1) + z_1^2 - z_3^2 = r_1^2 - r_3^2 \quad (8)$$

Let substitute:

$$a_1 = 2(x_2 - x_1) \qquad a_2 = 2(x_3 - x_1) \quad (9)$$
$$b_1 = 2(y_2 - y_1) \qquad b_2 = 2(y_3 - y_1) \quad (10)$$
$$c_1 = 2(z_2 - z_1) \qquad c_2 = 2(z_3 - z_1) \quad (11)$$
$$d_1 = r_1^2 - r_2^2 - x_1^2 + x_2^2 - y_1^2 + y_2^2 - z_1^2 - z_2^2 \quad (12)$$
$$d_2 = r_1^2 - r_3^2 - x_1^2 + x_3^2 - y_1^2 + y_3^2 - z_1^2 - z_3^2 \quad (13)$$

Equations (7), (8) can be written as:

$$a_1x + b_1y + c_1z = d_1 \quad (14)$$
$$a_2x + b_2y + c_2z = d_2 \quad (15)$$

If we expressed the variables y, *z in the equations* (14), (15) by the variable *x then we have:*

$$b_1y + c_1z = d_1 - a_1x \quad (16)$$
$$b_2y + c_2z = d_2 - a_2x \quad (17)$$

This system of linear equations (16), (17) can be solved by the Cramer's rule where D, D_y, D_z are determinants (Matejdes.2005):

$$y = \frac{D_y}{D} = \frac{\begin{vmatrix} d_1 - a_1x & c_1 \\ d_2 - a_2x & c_2 \end{vmatrix}}{\begin{vmatrix} b_1 & c_1 \\ b_2 & c_2 \end{vmatrix}} = \frac{c_2d_1 - a_1c_2x - c_1d_2 + a_2c_1x}{b_1c_2 - b_2c_1} \\ = \frac{c_2d_1 - c_1d_2}{b_1c_2 - b_2c_1} - \frac{a_1c_2 - a_2c_1}{b_1c_2 - b_2c_1}x \quad (18)$$

Let denote:

$$q_2 = \frac{c_2d_1 - c_1d_2}{b_1c_2 - b_2c_1} \qquad k_2 = \frac{a_1c_2 - a_2c_1}{b_1c_2 - b_2c_1} \quad (19)$$

Then equation (18)can be rewritten as:

$$y = q_2 - k_2x \quad (20)$$

Similarly:

$$z = \frac{D_z}{D} = \frac{\begin{vmatrix} b_1 & d_1 - a_1x \\ b_2 & d_2 - a_2x \end{vmatrix}}{\begin{vmatrix} b_1 & c_1 \\ b_2 & c_2 \end{vmatrix}} = \frac{b_1d_2 - a_2b_1x - b_2d_1 + a_1b_2x}{b_1c_2 - b_2c_1} = \\ = \frac{b_1d_2 - b_2d_1}{b_1c_2 - b_2c_1} - \frac{a_2b_1 - a_1b_2}{b_1c_2 - b_2c_1}x \quad (21)$$

$$q_3 = \frac{b_1d_2 - b_2d_1}{b_1c_2 - b_2c_1} \qquad k_3 = \frac{a_2b_1 - a_1b_2}{b_1c_2 - b_2c_1} \quad (22)$$
$$z = q_3 - k_3x \quad (23)$$

By substituting relations (20) and (23) into the equation (1) we obtain following quadratic equation with one variable *x*:

$$(x - x_1)^2 + (q_2 - k_2x - y_1)^2 + (q_3 - k_3x - z_1)^2 = r_1^2 \quad (24)$$

$$x^2 - 2xx_1 + x_1^2 + \\ +k_2^2x^2 - 2k_2q_2x + 2k_2y_1x + q_2^2 - 2q_2y_1 + y_1^2 + \\ +k_3^2x^2 - 2k_3q_3x + 2k_3z_1x + q_3^2 - 2q_3z_1 + z_1^2 = r_1^2 \quad (25)$$

after modification:

$$(1 + k_2^2 + k_3^2)x^2 + \\ +(-2x_1 - 2k_2q_2 + 2k_2y_1 - 2k_3q_3 + 2k_3z_1)x + \\ +x_1^2 + q_2^2 - 2q_2y_1 + y_1^2 + q_3^2 - 2q_3z_1 + z_1^2 - r_1^2 = 0 \quad (26)$$

Let denote:

$$a = 1 + k_2^2 + k_3^2 \quad (27)$$
$$b = -2x_1 - 2k_2q_2 + 2k_2y_1 - 2k_3q_3 + 2k_3z_1 \quad (28)$$
$$c = x_1^2 + q_2^2 - 2q_2y_1 + y_1^2 + q_3^2 - 2q_3z_1 + z_1^2 - r_1^2 \quad (29)$$

Then the equation (26) can be expressed in the form:

$$ax^2 + bx + c = 0 \quad (30)$$

Two solutions of this equation are given by the formula:

$$x_{1,2} = \frac{-b \pm \sqrt{b^2 - 4ac}}{2a} \quad (31)$$

The answer to the navigation task has two solutions. The x coordinate of these points is given by the relation (31) and their relating coordinates y, z are determined from the equations (20) and (23).

6. CONSULSION

In the article is described an analytical solution of navigation task of the Delta cable systems by odometric method. The merit of navigation by this method is its simplicity, reliability, including its independence from satellite systems. The results are two solutions. The both are real in the practice and have a practical meaning at the deltastats operations. We suppose that certain disadvantage of their use can be their lower accuracy. Therefore, in the future we would like to focus on its evaluation.

The contribution was elaborated within the research project VEGA No. 1/3523/06 *The research of new technical and technological principles for wood concentration* and research project KEGA No. 3/6448/08.
Translation: PhDr. Olga Lejsalova, CSc.

7. REFERENCES

Barborak O. ; Liptak P., & Jozefek M. (2006). *Robotization – one of the key trends of expert preparation in specialized machime amd production technology.* TU Kosice, ISBN 80-8073-560-3, pp. 35 – 98

Matejdes, M. (2005): *Applied Mathematics.* Zvolen: Matcentrum, ISBN 80-89077-01-3, pp. 558

Sakal, P.; Bozek, P.; Nemlaha, E. (1999): *Numeric control systems (in Slovak).* Tripsoft Trnava, ISBN 80-968294-1-6,. pp.154

Smrcek, J.; Hajtinger, K. (1996). The analysis of the robots implementation possibilities in the non-standard environment. In: Int. Cong. *„MATAR 96 – Machine Tools, Automation and Robotics in Mechanical Engineering“,* Praha, ISBN 80-364-2568-6, pp. 185-189

Stollmann,V.; Belanova,K. (2006). Deltastats (in Slovak). Patent application PP 0079/2006, Urad priemyselneho vlastnictva Banska Bystrica

Annals of DAAAM for 2008 & Proceedings of the 19th International DAAAM Symposium, ISSN 1726-9679
Editor B. Katalinic, Published by DAAAM International, Vienna, Austria 2008
Make Harmony Between Technology and Nature, and Your Mind will Fly Free as a Bird

VIRTUAL TECHNOLOGIES FOR SPECIALIZED ROBOTIC WORKPLACES

STOLLMANN, V[ladimir]; BOZEK, P[avol]; BARBORAK, O[to]; NASCAK, L[ubomir] & WALEKOVA, G[abriela]

Absract: *Specialized robotic workplaces or systems are not only a complex issue, but at the same time they are financially demanding in the field of production, especially, when the prototypes essential for mechanical engineering or forestry are concerned. Such a robotic workplace calls for an original project of a small series production and the proposed virtual environment meets the requirements for the verification of the technological properties, reliability and construct possibilities of the prototypes. The specialized robotic workplace is an interactively described 3D object and is programmable in accordance with the real environment requirements. Virtual technologies represent an convenient solution for the preparation of a safe ergonomic, economic and environmental workplace.*

Key words: *robot, virtual, workplace, animation*

1. INTRODUCTION

Economic and global trends in mechanical engineering mamangement in the Slovak Republic are heading to maintain the related EU conditions aimed at the sustainable development of the machine engineering production. These technologies related to the effectiveness of national economic development in accordance with the EU countries´ machine production management are accepted in the project phase of the new specialized robotic workplace. (Lustig et al., 2007). Authors focus on the implementation of the results in forestry.

2. VIRTUAL CONTROL SYSTEMS

It is obvious, that the information acquisition at real technical means of the control systems is financially demanding. The basic training for the plant and its structure´s project engineers is available at the control system supplier, and limited renewing courses at the customer´s control system. However, the principle problem lies in the operators´ preparation for emergency situations of the plant equipment. Virtual control systems connected with virtual models of control systems mean an effective solution of the abovementioned problems. Current research results and further research development in the field of virtual control systems implementation are supported by VRML /Virtual Reality Modeling Language/, designed for interactive description of 3D objects and worlds (Barborák et al., 2006).

2.1 VRML for virtual concepts

VRML /Virtual Reality Modeling Language/ is designed for interactive description of 3D objects and worlds. It is also a universal variable format for 3D graphics and multimedia. The use of VRML can vary and comprises also the possibility of technical and scientific visualization, multimedia presentations, entertainment, computer-aided education, www pages and virtual worlds.

It is a standardized file system defined by ISO/IEC 14772. VRML is capable to represent static and dynamic (animated) 3D objects, multimedia objects with hyperlinks for individual components of multimedia, such as a text, sound, picture, animation and film. VRML was designed to meet the following requirements:

a. The possibility to create automated scripts. It allows for the development of computer programs for the VRML creation, editing and operation based on automated translation programs for the conversion of other common 3D formats into VRML files.
b. The arrangement provides the ability to combine dynamic 3D objects and VRML worlds.
c. The spreadability allows for the addition of new objects not explicitly defined in VRML
d. Performance
e. The scale allows for the development of arbitrary large dynamic 3D worlds.

The standard ISO/IEC 14772 does not need to define the physical equipment nor other concepts depending on the implementation, e.g. screen contrast or input devices. On the contrary, this standard is for wide range of equipment and implementations which do not include even a monitor /display/ or a mouse (Smrček & Hajtinger 1996).

3. THE DECISION PROCESS FOR SIMULATION ALGORITHM DEFINITION

By the CNC program design as well as by the robots programming on the virtual scene, there are situations when for some machines it is necessary to wait for a defined time period. The movement speed, either angle or translation speed, represents the time function. It is essential to use the simulation algorithm controlled by events. For example : figure 1 shows the real scene of a robotic technological workplace with the possibility to program the technological process of materials manipulation.

3.1 Some advantages of virtual control systems

The training in control system real technical means is financially demanding. The basic training for the plant and its structure´s project engineer is available at the control system supplier, and limited renewing courses at the customer´s control system. However, the principle problem is represented by the operators´ preparation for emergency situations of the plant equipment. Virtual control systems connected with virtual models of control systems mean an effective solution of the abovementioned problems.

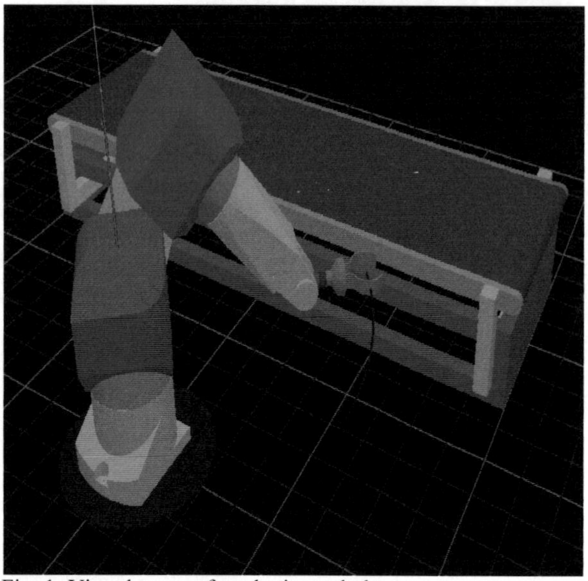

Fig. 1: Virtual scene of a robotic workplace

The development process based on virtuality is at present a prerequisite for the successful process of a new product or workplace creation. In the phase of project, there exists the concept of a virtual robotic workplace model with the possibility do design real technology and define the basic principles of the technological process control system. The implementation of virtual methods in this paradigm is highly effective and leads to early decisions of real workplace effectiveness from various aspects (Javorek et al., 2000). The original idea of a virtual technological workplace project was based on a know-how workplace. It is similar to the constant education level increase not only at universities - where it is necessary to implement new technologies into education.
Plant information environment as well as its system integration with the use of new IT and economic tools of IS organization and control, provides the transparency of information and knowledge processes and allows for the integration, consolidation and restructuralization of necessary information for the needs of individual plant management levels.
New trends in the IS organization and control as well as in the information and knowledge support present the creation of the whole range of new specialized functions. They are especially informatic, economic and manager positions focused on the organizational, technical, economic and technological part of the information processes, whereas the positions focused on information processes content are missing. Drucker, a management expert, describes the lack of information responsibility. Some experts relate this field to the information and knowledge managers and comment on the difficulties with the combination of IT initiatives and knowledge management content.
The practice shows that the knowledge assurance of corporation processes calls for the need of workers interested in the information and communication processes with the emphasis on managing positions. This proves these workplaces stayed out of the system, though the work with the contents has been always the strength of IT professionals in the information databases focused on the economic growth.
Professional operator profile focused on economic and knowledge information represents an interdisciplinary problem, e.g. in the competitive intelligence or in the competences for special librarians of the 21 century for the Special library association. From the point of needs analysis, the necessary knowledge, abilities and skills are not incorporated and the information professional is missing. In the system of positions it is necessary to create the place for an information specialist. The professional assurance of individual IS dimensions should be enhanced with the computer science, informatics, theory of information systems, information and knowledge management and information science.
Economic relations of the mentioned fields of science overlap in the professional area, though their common subject provides specialization in the assurance of plant strategic processes . It is necessary to involve the long neglected information professionals, who are responsible for the information content, into the information and knowledge processes and consider them the key personalities of the processes (Štollmann, 2006).

4. CONCLUSIONS

The new possibilities of specialized robotic workplaces design, project, implementation and management allow for the simulation of the use in the project phase while meeting economic, ergonomic and sensitive environmental requirements accepted in EU countries at the same time.
It is still clear that the information acquisition in real technical means is financially demanding. The basic training for the plant and its structure´s project engineer is available at the control system supplier, and limited renewing courses at the customer´s control system. However, the principle problem is represented by the operators´ preparation for emergency situations of the plant equipment. Virtual control systems connected with virtual models of control systems offer a convenient solution of the abovementioned problems.

The contribution was elaborated within the research project VEGA No. 1/3523/06 *The research of new technical and technological principles for wood concentration* together with the project KEGA No. 3/6448/08.
The contribution was elaborated within VEGA No.1/0421/08 "Research of possible applications of non-metal and composite materials by the action members production in the production technology" carried out at Trenčín Univerzity of A. Dubček in Trenčín together with related university workplaces.

5. REFERENCES

Barborák O. ; Lipták P., & Jozefek M. (2006). *Robotization – one of the key trends of expert preparation in specialized machime amd production technology,* TU Košice, ISBN 80-8073-560-3, pp. 35 - 98

Javorek, Ľ. ; Naščák, Ľ., & Oswald, J. (2000). *Monitoring proceeding digging*, Starý Smokovec, ISBN 458 – 235 – 36 – 5 page 87 – 92

Lustig, F. ; Schauer, F. & Ožvoldová, M. (2007). *Physics Labs for Integrated E-learning Classical Real, Real Remote and Virtual* In: *Proceeding* of *ICTE Information and Communication Technology in Education* 2007, Ostrava, ISBN 978-80-7368-388-7, pp. 107-115

Smrček,J. & Hajtinger, K. (1996). *The analysis of the robots implementation possibilities in the non-standard environment. In: Int. Cong. „MATAR 96 – Machine Tools, Automation and Robotics in Mechanical Engineering“, Praha,* ISBN 80-364-2568-6, *pp. 185-189*

Štollmann, V. ; (2006). *Progress forest techniques*, In: OPTIROB 2006 : first international conference Optimization of the robots and manipulators / ed. Adrian Olaru, Mihai Stanciu, Editura Bren, Bucuresti, ISBN 973-648-572-2. - page 249-255.

Annals of DAAAM for 2008 & Proceedings of the 19th International DAAAM Symposium, ISSN 1726-9679
Editor B. Katalinic, Published by DAAAM International, Vienna, Austria 2008
Make Harmony Between Technology and Nature, and Your Mind will Fly Free as a Bird

DESIGN ASPECTS OF THE OPC UNIFIED ARCHITECTURE

STOPPER, M[arkus] & KATALINIC, B[ranko]

Abstract: *OPC Unified Architecture is the most recent OPC specification from the OPC Foundation and differs significantly from its predecessors. The OPC Foundation's main goals with this development was to provide a path forward from the original OPC communications model (COM/DCOM) to a current communications model (SOA) and introduce a cross-platform architecture for process control, while enhancing security and providing an information model. By defining abstract services, OPC UA provides this service-oriented architecture (SOA) for industrial applications, starting from factory floor devices and ending with enterprise applications. OPC UA integrates the different aspects of the former OPC specifications into a unified address space accessible with a single set of services. This paper gives an overview over some edge design aspects of OPC UA regarding the underlying SOA-conform architecture.*
Key words: *service oriented architecture, ole for process control, opc unified architecture, industrial web services*

1. INTRODUCTION

The OPC Foundation is approaching completion of the first stage of its new Unified Architecture (UA). The new Unified Architecture is meant to unify all of the existing OPC technology (OPCOLE, 2003). Importantly it means open standards and no longer just the proprietary Microsoft standard (OPCDCOM, 2000). OPC vendors are already working through the design and implementation phase of their OPC UA technology products. After three years of specification work and a further year of prototyping the first version of Unified Architecture is now being released considering the following characteristics for the communication stack (OPCUA, 2008):

- multiple platform implementation including portable ANSI C, Java and .NET implementations
- scalability from embedded controllers up to Mainframes
- the stack can support multi-threaded as well as single-threaded/single-task operation, which is necessary for porting the stack onto embedded devices
- an own security implementation, based on new standards, realizes true security (e.g. compare with [OPCSEC, 1998])
- configurable time-outs for each service
- chunking of big datagram's

Consequently perhaps the most obvious change compared to the legacy OPC standard that users will notice right away is that OPC-UA is even not based on Microsoft COM/DCOM. Everyone on the market is embracing open, internet-based communication standards. Naturally, OPC has done the same by basing UA on TCP/IP, HTTP, SOAP, and XML. Not being tied solely to Microsoft platforms makes OPC-UA highly scalable. It is expected to see OPC-UA implemented on everything from embedded field devices through to Unix or mainframe-based enterprise applications (OPCUA, 2008).
All of the base services defined by OPC are abstract method descriptions which are protocol independent and found the basis for the whole OPC UA functionality. The transport layer puts these methods into a protocol, which means it serializes/de-serializes the data and transmits it over the network. Currently there are two protocols specified for this purpose. One is a binary, towards high performance optimized TCP protocol and as second, a Webservice-based one. The OPC information model isn't just a hierarchy based on folders, items and properties any more, but a so-called Full Mesh Network based on nodes instead. These nodes can additionally transmit all varieties of meta information. Additional new and important features of OPC UA are redundancy support, heartbeat for connections in both directions, which means that the server as well as the client recognizes interruptions, and buffering of data and acknowledgements of transmitted data, so lost connections don't lead to lost data anymore. Lost datagram's can get fetched repeatedly (OPCUA, 2008).

2. ARCHITECTURE BASE DESIGN

The OPC Unified Architecture blurs the distinction between Data Access, Historical Data Access or Alarms & Events in a way, that existing OPC applications are largely broken down on functional lines based on which of the OPC specifications are supported. Clients and Servers based on the Unified Architecture will blur this distinction since the transfer of all data, no matter of what its type or purpose, is transferred using the same base UA services.
This mechanism coupled with the support of complex data (arrays, binary structures, XML documents, etc.) leads to a situation where server vendors expose data with complete trustworthiness as it naturally exists in the system. They presently expose only a subset of data forced into an OPC-defined data format. Forward thinking vendors will realize that there will no longer be a need for any proprietary vendor defined interfaces with this capability and they will only expose the data via UA (OPCUA, 2008).
The OPC Foundation has been working with other standards organisations to gear UA as a premier vehicle for conveying and transporting data defined by these other specifications. With the acceptance of UML based models and XML Schema to define and render complex data, many information model specifications are emerging.
It can be expected to see companion specifications emerging that bind the data models described by these and other specifications to OPC UA, thus providing for complete information level interoperability from one application to another. Note also that applications that use this type of data are not the current round of typical factory floor applications that traditionally implement OPC. Instead they are enterprise applications for use at MES and ERP level. It can be expected to see OPC-UA implemented in certain applications far outside the scope of nowadays OPC applications (OPCUA, 2008).
Further robustness and reliability play an important role. Rather than live at the kindness of the underlying internet protocols, the Unified Architecture takes direct responsibility for all of the necessary failure detection and recovery logic making UA viable for mission-critical communication. Beyond robustness and reliability, several redundancy schemes are defined such that vendors may provide this capability in a standard way. It is expected to see OPC-UA used as the native communication

method by applications, usurping proprietary vendor protocols and possibly some existing field bus protocols (OPCUA, 2008). Existing OPC applications vary greatly in their compatibility and interoperability with other OPC applications. Although OPC is currently stressing the need for products to be both certified and tested at interoperability workshops, the reality is that only a small percentage of OPC products on the market presently meet that mark. With OPC Unified Architecture, the certification tests cover both servers and clients and will be administered by sanctioned testing labs. These facilities are expected to be available in time to test the very first UA products coming to the market. The certification status of any UA component will be available in real time via electronic certificates issued by OPC directly though the UA interface. OPC-UA applications which are properly certified thus greatly increase the likelihood of true plug-n-play (OPCUA, 2008).
Backward compatibility will ensure quick adoption. The OPC Foundation is providing its members with several code-based deliverables to simplify vendor development and aid with backward compatibility. First, the OPC-UA API provides much of the plumbing and low level generic serialisation and wire support so that vendors need only to code the high level aspects of UA into their applications. The API is expected to be available in several languages supporting multiple popular computing platforms.
In addition to the API, the OPC Foundation is also providing its members with shippable binary adapter modules which allow new OPC-UA clients direct access to all of the legacy COM-based OPC servers, and legacy COM-based OPC clients access to a subset of the features of new OPC-UA servers. The two adapters can even be used together to connect a legacy client to a legacy server, replacing the standard DCOM communication with the new OPC-UA internet-based protocols. With complete backward compatibility ensured, vendors will have much to gain and nothing to loose by adopting the new OPC Unified Architecture (OPCUA, 2008).
Summarized can be said that OPC UA specifies an abstract set of services and the mapping to a concrete technology. OPC UA does not specify an API but only the message formats for data exchanged on the wire. A communication stack is used on client- and server-side to encode and decode message requests and responses. Different communication stacks can work together as long as they use the same technology mapping. The following section gives an overview over the different components of OPC UA clients and servers followed by a section of technology mappings and a section of possible interactions between OPC UA servers (OPCUA, 2008).

3. CLIENT AND SERVER PARTS

An OPC UA client consists of a client implementation using an OPC UA communication stack. The client implementation accesses the communication stack using the OPC UA API. Note that the API is not standardized. It may vary for different programming languages and potentially for different communication stacks. Several communication stacks may exist for different operating systems, programming languages and mappings. The client-side communication stack allows the client to create request messages based on the service definitions. The client-side communication stack communicates with a server-side communication stack. The OPC Foundation standardized only this communication. Thus, everybody can develop his or her own communication stack with its own API as well. The server-side communication stack delivers the request messages to the server implementation via the OPC UA API. Since the OPC UA API realizes the abstract service specifications, it may be the same as on the client-side. The server implementation implements the logic needed to return the appropriate response message. The OPC UA server implementation gets its data from some underlying system. For example, this can be a configuration database, a set of devices or some OPC server (OPCUA, 2008).

4. TECHNOLOGY MAPPING

The OPC UA mapping specification currently defines two mappings: UA Web Services and UA Native. The first mapping uses SOAP and the various WS-* specifications. The second mapping uses only a simple binary network protocol and integrates certain security mechanisms.
The encoding of the data can be done in XML or UA Binary. UA Binary specifies the serialization of data into a byte string. The UA Binary encoding is faster than the XML encoding since the message size is smaller than for the XML encoding. On the other hand, the XML encoding allows generic SOAP-clients to interpret the data in the SOAP message, while they would only get a binary string using the UA Binary encoding. In theory, the encoding of the data is independent of the mapping. However, the XML encoding will typically only be used in the UA Web Service mapping in combination with the various WS*-specifications. The protocol of the UA Web Service mapping is SOAP/HTTP(S) while the UA Native mapping typically runs directly on TCP/IP. For bypassing firewalls, the UA Native mapping also allows putting the binary encoded messages into SOAP messages using HTTP(S). UA messages can also be encoded and transported with other protocols, for example using WCF Binary. However, by using this technology interoperability is lost since only communication with WCF clients is then possible (Resnick, 2008).

5. AGGREGATING SERVERS

Build-into the OPC UA specification is the concept of aggregating servers. An aggregating server aggregates one or more OPC UA server and provides the information of those servers – or an excerpt of the information – in its address space. Thus, a client does not have to access several servers but only one server. This mechanism allows a flexible architecture by chaining several OPC UA servers for different clients with different requirements (OPCUA, 2008).
For example, several OPC UA servers running on small devices will be aggregated by one OPC UA server. Several clients of the distributed computing system may access this server. Another OPC UA server aggregates this server and provides part of the information to the MES system. The MES system could work as an aggregating server, too. OPC UA supports aggregating servers by allowing marking the origin of data.

6. CONCLUSION

OPC UA is an important step to integrate new technologies and concepts into industrial applications. The strong meta model, the open and standardized service-oriented architecture (SOA) and the ability of unifying different kind of OPC servers opens the door for new areas of application. Especially in the field of process automation, it will help to simplify the integration of other applications like MES, HMI or SCADA systems.

6. REFERENCES

OPCDCOM (2000). *DCOM on Non-Microsoft Platforms*, Software AG, http://www.opcfoundation.org/, 2000

OPCOLE (2003). *OPC Specification: Data Access, Alarms & Events, Historical Data Access*, OPC Foundation, http://www.opcfoundation.org/, 2003

OPCSEC (1998). *OPC, DCOM and Security*, Intellution Inc., http://www.opcfoundation.org/, 1998

OPCUA (2008). *OPC UA Specification: Parts 1–11*, OPC Foundation, http://www.opcfoundation.org/, 2008

Resnick, S., Crane, R. & Bowen C. (2008). *Essential Windows Communication Foundation (WCF): for .NET Framework 3.5*, ISBN 978-0321440064, March 2008, Addison-Wesley Longman, Amsterdam, Netherlands, 2008

Annals of DAAAM for 2008 & Proceedings of the 19th International DAAAM Symposium, ISSN 1726-9679
Editor B. Katalinic, Published by DAAAM International, Vienna, Austria 2008
Make Harmony Between Technology and Nature, and Your Mind will Fly Free as a Bird

THE KINEMATIC GENERATION OF THE DIRECTRIX CURVES OF THE GEOMETRICAL SURFACES

STRAJESCU, E[ugen] & SANDU, I[on]

Abstract: *The paper presents a theoretical study of the principle of the kinematics generation of the directrix **D** curves of the geometric complex surfaces using the composition of the simple movements. In that sense, as a new original application, it is presented the cinematic generation of largely used curves **D** of the real complex surfaces generated on machine tools, respectively the generation of the circular helix at the generation of the cylindrical helical surfaces.*

Key words*: surface generation, directrix, circular helix, cylindrical helical surfaces.*

1. INTRODUCTION

The complex surfaces are these geometric surfaces of that theoretical curves D or/and G has a such geometric form that for their generation are necessary compositions of sample movements cinematic correlated by a cinematic correlation ratio, RCCIN+ (Sandu & Strajescu, 2007). According to the cinematic principle of the generation of the real surfaces on machine tools, for the generation of complex surface in a general case, the generating element must effectuate a movement along the theoretic generatrix G and a movement along the theoretical directrix D (Sandu & Strajescu, 2006), (Strajescu & Sandu, 2006). At many type of complex surfaces, the movement along the curve D is a movement composed resulted from the composition of single movements, cinematic correlated. In that sense, at the generation of the helical cylindrical surfaces where the theoretic directrix curve D is a circular helix, it is used the composition of a rotation movement with the rectilinear translation movement of the generating element in order to generate a helical circular trajectory in conditions of cinematic correlation (Sandu & Strajescu, 2004), (Botez, 1967).

2. THE COMPOSITION OF A ROTATION MOVEMENT WITH A TRANSLATION RECTILINEAR MOVEMENT MADE IN AN AXIAL DIRECTION, NORMAL ON THE ROTATION PLANE

We consider a rotation right cylinder, with the radius R and a circular base in the plane YOZ of the coordinate system. To this cylinder we impart a rotation movement around the OX axe with the frequency n (rot/min). In the same time, an initial generating point M_o of the cylinder makes a rectilinear translation movement on the axial direction of the OX axis, with the speed $\vec{v}_A$, and, after a certain moment, come up in M in its resulting movement, but relative to the cylinder. In order to obtain a relative movement, we consider the cylinder fixed by its rotation in an opposite sense with the frequency $n^* = -n$, on the rotation angle Φ (fig. 1). By consequence, the relative speed's movement of the generating point M is the vector $\vec{v}_M$, given by the vectorial relationship:

$$\vec{v}_M = \vec{v}_T + \vec{v}_A \quad (1)$$

where: $\vec{v}_T^*$ is the vector of the tangent speed from the rotation movement with the $|n^*|$ frequency, having the size:

$$\left|v_T^*\right| = 2\pi Rn \quad \text{[rot/min]} \quad (2)$$

The right linear translation movement is an uniform movement made by the generating point M along the generatrix G of the cylinder, on cyclical cinematic distances with the size p.

A cinematic cycle is a rotation of the cylinder and is made in the time T_{CIC} having the size:

$$T_{CIC} = 1/n \quad \text{[min]} \quad (3)$$

In this case, the size of the axial speed $\vec{v}_A$ from the right linear uniform movement is given by the relation:

$$v_A = \frac{p_E}{T_{CIC}} = p_E n \quad \text{[m/min]} \quad (4)$$

In this way it results:

$$R_{CCIN} = \frac{v_A}{n} = p_E \quad \text{[mm]} \quad (5)$$

That imposes to the two corposants movements the cinematic coordination in the purpose that the generating point M cross cyclically along the generatrix G of the cylinder constant distances p_E.

If the vectorial relation (1) is projected on the coordinated system's axes $OXYZ$, we obtain the sizes of the components of the $\vec{v}_M$ speed on these axes, having the next expressions:

$$\begin{cases} v_{MX} = v_A = p_E \cdot n \\ v_{MY} = -\left|v_{TY}^*\right| = -\left|v_T^*\right| \sin\Phi = -2\pi Rn \sin\Phi \\ v_{MZ} = -\left|v_{TZ}^*\right| = -\left|v_T^*\right| \cos\Phi = 2\pi Rn \cos\Phi \end{cases} \quad (6)$$

The differential coordinates dX, dY, dZ of the generating point M are differential spaces crossed respectively with the speeds v_{MX}, v_{MY} and v_{MZ}, in a differential time dT:

$$\begin{cases} dX = v_{MX} \cdot dT; \\ dY = v_{MY} \cdot dT; \\ dZ = v_{MZ} \cdot dT \end{cases} \quad (7)$$

where:

$$dT = d\Phi / 2\pi \cdot n \quad (8)$$

We replace now the relations (6) and (8) in the relation (7) and we obtain the expressions of the instantaneous components X, Y and Z of the generating point M.(9).

Effectuating the integrals we obtain the coordinates X, Y and Z of the point M that represent the parametric equations of the cinematic curve C, described by the generating point M as consequence of the composition of the two movements (10).

$$\begin{aligned} X &= \int v_{MX} dT = \frac{1}{2\pi n}\int v_{MX} d\Phi = \frac{p_E}{2\pi}\int d\Phi \\ Y &= \int v_{MY} dT = \frac{1}{2\pi n}\int v_{MY} d\Phi = R\int -\sin\Phi d\Phi \\ Z &= \int v_{MZ} dT = \frac{1}{2\pi n}\int v_{MZ} d\Phi = R\int \cos\Phi d\Phi \end{aligned} \tag{9}$$

$$\begin{cases} X = \frac{p_E}{2\pi}\Phi + C_X \\ Y = R\cos\Phi + C_Y \\ Z = R\sin\Phi + C_Z \end{cases} \tag{10}$$

where integration constants C_X, C_Y and C_Z are determinate knowing that at $\Phi = 0$, $X = 0$, $Y = R$ and $Z = 0$, resulting $C_X = 0$ $C_Y = 0$ and $C_Z = 0$.

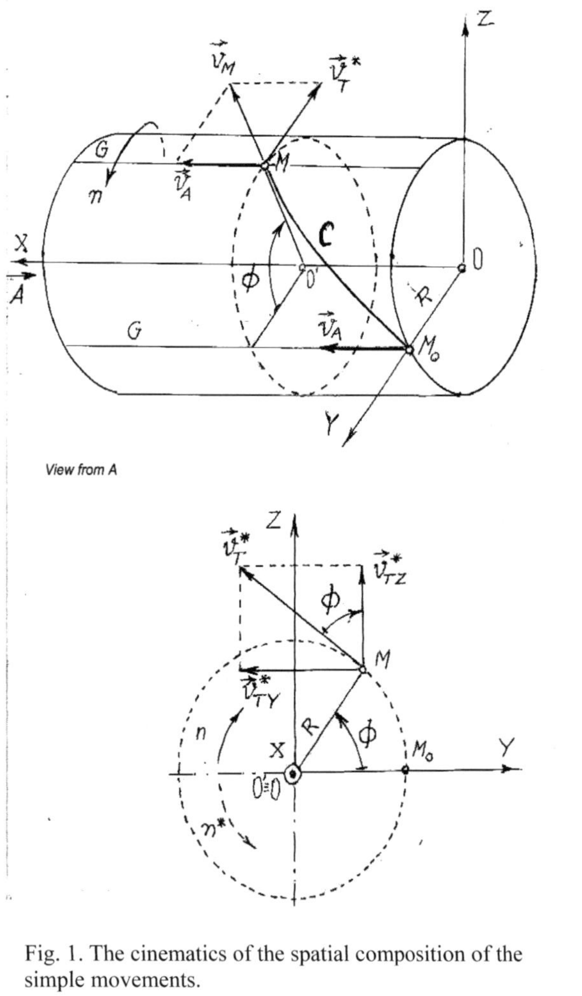

Fig. 1. The cinematics of the spatial composition of the simple movements.

By consequence, the parametric equations of the cinematic curve C are:

$$X = p_0\Phi; Y = R\cos\Phi; Z = R\sin\Phi. \tag{11}$$

that represent the known parametric equations of the circular helix and in which

$$p_0 = p_E / 2\pi \tag{12}$$

is the parameter and p_E is the step of the circular helix.

So, by composing a rotation movement with a right linear translation movement made on an axial direction, normal on the rotation plane, a generating point M generates as a cinematic curve C a circular helix, with the condition to respect R_{CCIN} given by the relation (5) that impose to the two movements the cinematic correlation with the view to the generating point M generate the helical circular trajectory with the pitch p_E and with the radius R given.

Practically, at the generation of the complex real surfaces the curve C as circular helix is generated as directrix D of the surface.

The most extended application of this case of simple movements composition is represented the generation of the directrix D at the generation of the main extended complex surfaces in techniques, the helical cylindrical surfaces with all kind of profiles.

3. CONCLUSION

The aspects shown in this paper represent original contribution concerning the cinematic generation of the circular helix used as directrix D at the generation of real complex surfaces of the type cylindrical helical surfaces.

The treatment of these aspects by original and unique mathematical demonstrations contributes to the improvement of the cinematic generation theory by simple movement composition of the plane and space curves used in practice as real complex curves D

This theory of the cinematic generation from composition of single movements cinematically correlated by ratios R_{CCIN}, for the directrix curves D and for the generatrix curves G of the geometrical surfaces, permits the cinematic correlation of the mobile parts of the machine-tools with NC at the generation of the complex surfaces on them, and, in the same time, the creation of the cinematic algorithms available for the generation of the most complex forms of the real surfaces on these machine tools.

We must precise that this area of researches is largely covered by the activities of the Department of Machines and Production Systems from the "Politehnica" University of Bucharest, in correlation with the problems of the machine-tools and their command (NC)

4. REFERENCES

Botez, E., (1967). *Bazele generarii suprafetelor pe masini-unelte. (Basis of the surface generation on machine-tools).* Editura Tehnica, Bucharest.

Sandu, I., Gh., Strajescu, E., (2007) *New Theoretical Aspects Concerning the Generation of the Complex Surfaces.* ICMaS 2007, Bucharest, Editura Academiei Romane, ISSN 1482 – 3183, pag. 181-184.

Sandu, I., Gh., Strajescu, E., (2006). *Contributions at the theory of the Generation of the Directrix and Generatrix Curves of the Geometrical Surfaces.* ICMaS 2006, Bucharest, Editura Academiei Romane, ISSN 1482 – 3183, pag. 327-330.

Sandu, I., Gh., Strajescu, E., (2004). *The Cinematic Generation of the Directrix and Generatrix Curves of the Geometrical Surfaces.* ICMaS 2004, Bucharest, Editura Academiei Romane, ISSN 1421- 2943, pag. 171 - 173.

Strajescu, E., Sandu I., Gh., (2006) *Theoretical Studies about the Generation of the complex Surfaces* ICMaS 2006, Bucharest, Editura Academiei Romane, ISSN 1482 – 3183, pag. 327-330.

Annals of DAAAM for 2008 & Proceedings of the 19th International DAAAM Symposium, ISSN 1726-9679
Editor B. Katalinic, Published by DAAAM International, Vienna, Austria 2008
Make Harmony Between Technology and Nature, and Your Mind will Fly Free as a Bird

DOUBLE-FLUX WATER TURBINE DYNAMICS

STROITA, D[aniel] C[atalin]; BARGLAZAN, M[ircea]; MANEA, A[driana] S[ida] & BALASOIU, V[ictor]

Abstract: *This article presents the Double-flux water turbine and a method for calculating the characteristic times. Those times are the coefficients of the machine's dynamic equation and they would be obtained through a specific method. The dynamic equation permits the study of hydraulic turbines which are working in unsteady regimes. For all this, we will use the stereo-dynamic method (Barglazan, 2001) for a turbine with the rotor dimensions: diameter D=250 mm and width B=200 mm. The hill chart is known from the experimental part and is given in common coordinates Q11 and n11with the efficiency given as a relative efficiency. The new part of the method is that the derivates are calculate with computer programs through a graph - analytical procedure.*

Keywords: *double-flux turbine, hill chart, velocity triangles, characteristic times, dynamic equation,*

1. INTRODUCTION

The Double Flux turbine belongs to action type water turbines and it was discovered by Antony Michell in 1903. Professor Donat Banki introduced the evelventic channel for water admission improving the running of the turbine.

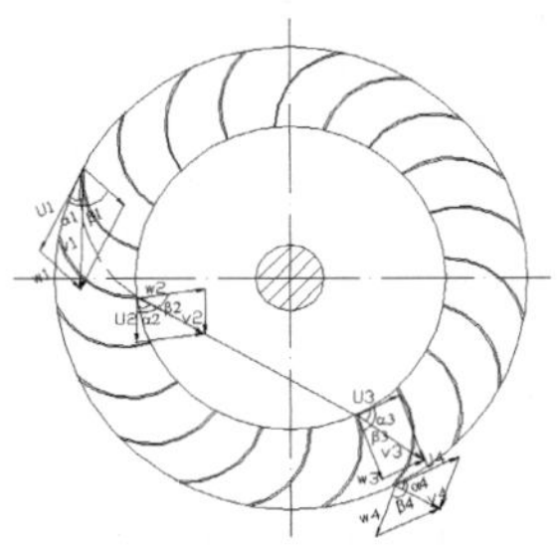

Fig. 1. Velocity triangle for Double-Flux water turbine

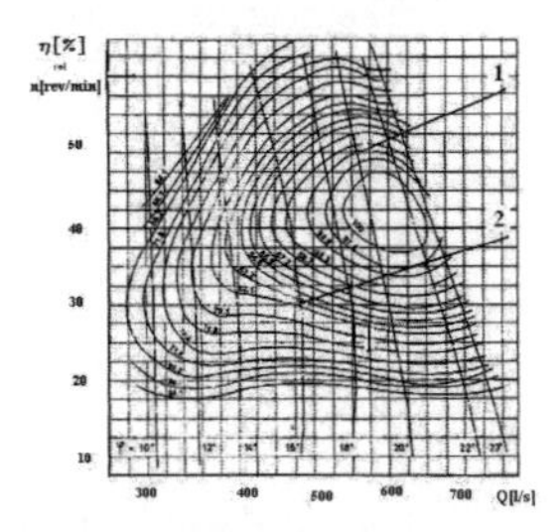

Fig. 2. The hill chart for double flux turbine with the runner D=250mm and B=200mm

The double-flux name comes from the double passing of the water in the turbine runner like this the kinetic energy from the water is absolved by the runner twice, we can see this from the velocity triangle in the figure 1.

Further we will present the most important equations of this kind of turbine.

$$P = \rho Q u_1 (v_1 \cos \alpha_1 - v_4 \cos \alpha_4) \quad (1)$$

Equation (1) represents the hydraulic power of the turbine, this equation comes from Euler equation adapted for the double-flux turbine. Turbine's efficiency is presented in equation 2, and is estimated like a ratio between runner's diameter and turbine's head.

$$\eta = 0.863 - 0.264 \frac{D}{H} \quad (2)$$

Although the efficiency is not very high the great advantage of this kind of micro-hydro turbine is the simplicity of the construction and the low cost. The double flux water turbine covers a great running domain and it can be compared with very popular water turbines like Francis, Kaplan or Pelton.

2. HYDRAULIC TURBINE DYNAMIC EQUATION APPLIED FOR DOUBLE FLUX TURBINE

We will start by presenting the general dynamic equation of a hydraulic turbine (Barglazan, 2001):

$$\frac{d\omega}{dt} = \frac{\Delta a}{a_0} \cdot \frac{1}{T_a} + \frac{\Delta \varphi}{\varphi_0} \cdot \frac{1}{T_\varphi} + \frac{\Delta \omega}{\omega_0} \cdot \frac{1}{T_\omega} + \frac{\Delta H}{H_0} \cdot \frac{1}{T_H} + \frac{\Delta \rho}{\rho_0} \cdot \frac{1}{T_\rho} = \quad (3)$$

$$= \frac{\alpha}{T_a} + \frac{\varphi}{T_\varphi} + \frac{\Omega}{T_\omega} + \frac{h}{T_H} + \frac{r}{T_\rho}$$

This general equation will be particularized for the double-flux turbine and it will be:

$$\frac{d\omega}{dt} = \frac{\Delta \varphi}{\varphi_0} \cdot \frac{1}{T_\varphi} + \frac{\Delta \omega}{\omega_0} \cdot \frac{1}{T_\omega} = \frac{\varphi}{T_\varphi} + \frac{\Omega}{T_\omega} \quad (4)$$

The general formula is simplified because for the double-flux turbine we have only one element for adjusting the flow-rate, which is the nozzle's vane and the density of the liquid ρ and the head of the turbine H are considered constant. The characteristic times for this type of turbine are T_φ and T_ω the times for the variation of the flow-rate and of rotor's angular speed.

In this new method we will use the classical method (Barglazan, 1979), improved by computer programs (Stroita 2007). We consider two points on the hill chart (1, 2) corresponding to 2 working regimes of the hydraulic turbine. The method is presented step by step and is exemplified for the working regime (point 1), for the other regime (2) the results are only presented, the steps used being similar as for the first.

The formulas for T_φ and T_ω are given below:

$$T_\varphi = \frac{J \cdot \omega_0}{M_{a0}} \cdot \frac{1}{\frac{\varphi_0}{Q_0}\left(\frac{\partial Q}{\partial \varphi}\right)_0 + \frac{\varphi_0}{\eta_0}\left(\frac{\partial \eta}{\partial \varphi}\right)_0} \quad (5)$$

$$T_\omega = \frac{J \cdot \omega_0}{M_{a0}} \cdot \frac{1}{\frac{\omega_0}{Q_0}\left(\frac{\partial Q}{\partial \omega}\right)_0 + \frac{\omega_0}{\eta_0}\left(\frac{\partial \eta}{\partial \omega}\right)_0 - 1} \quad (6)$$

where J is the inertial momentum of the hydraulic turbine, electric generator rotor and shaft and Ma_0 is the active momentum.

$$M_{a0} = \frac{\rho \cdot g \cdot Q_0 \cdot H_0}{\omega_0} \quad (7)$$

Through similitude relations we can calculate the flow-rate Q and the speed n, using the Q_{11} and n_{11} from the hill chart. Also from the hill chart we can obtain the partial derivatives through a grapho-analytical method.

$$Q = Q11 \cdot D^2 \cdot \sqrt{H} \quad (8)$$

$$n = \frac{n11 \cdot \sqrt{H}}{D} \quad (9)$$

$$\omega = \frac{\pi \cdot n}{30} \quad (10)$$

The hill chart is presented in the figure 2, with the coordinates Q_{11} and n_{11} and with the parameters φ, which is nozzle's vane opening and η_{rel} which is the relative efficiency.

In order to obtain the partial derivatives $(\partial Q / \partial \varphi_0)_0$ and $(\partial \eta / \partial \varphi_0)_0$, we extract from the hill chart (Barglazan, 1974) the dependence Q=Q(φ) and η=η(φ) presented in the figures 3 and 4.

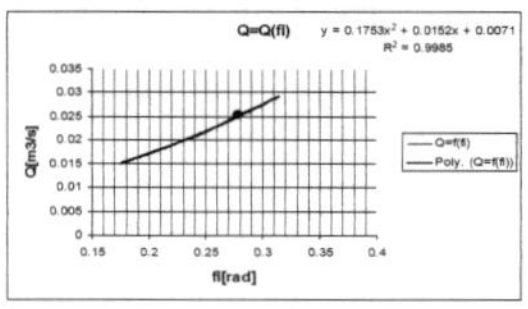

Fig.3 The variation of the flow-rate with the opening of the nozzle's vane, regime 1

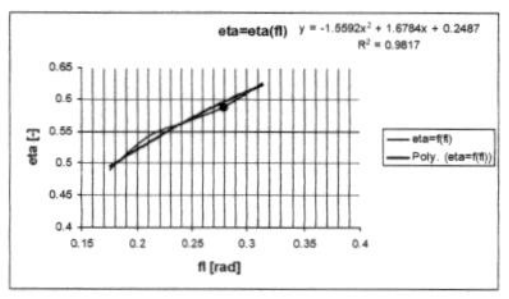

Fig.4 The variation of the efficiency with the opening of the nozzle's vane, regime 1.

In those graphs the working regime 1 is presented with the big dot. In order to obtain the equation of the variation we added a trendline which corresponds to a polynomial with the order 2, it seems from practice that in this case the polynomial with the order 2 gives the best approximation for the variation. Now, with the equation known we used a MathCad program for obtaining the partial derivate in the specified point. Like this we have the value for the partial derivatives. The same method is used for the other two variations Q=Q(ω) and η=η(ω) presented in the figures 5 and 6.

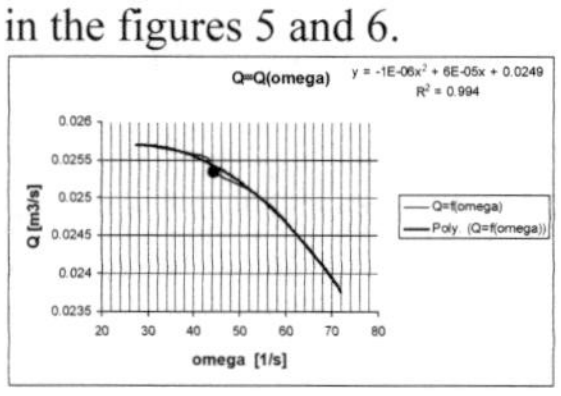

Fig. 5 The variation of the flow-rate with the angular speed, regime 1

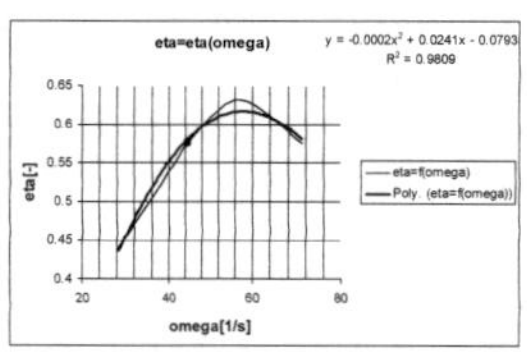

Fig. 6 The variation of the efficiency with the angular speed, regime 1

In the MathCad program is needed to enter the dependence along the line n11=const and along the curve φ=const, the parameters of the rotor, the operating regime and it will calculate automatically the characteristic times. For this operating point the results were:

Tφ=0.0138800 and Tω=0.002421 seconds, and now we can write the dynamic equation of the double-flux water turbine as:

$$\frac{dx}{dt} = \frac{\varphi}{0.0138800} + \frac{\Omega}{0.002421} \quad (11)$$

As the same is done for the other two working regimes, the graphs and the characteristic times are presented below:

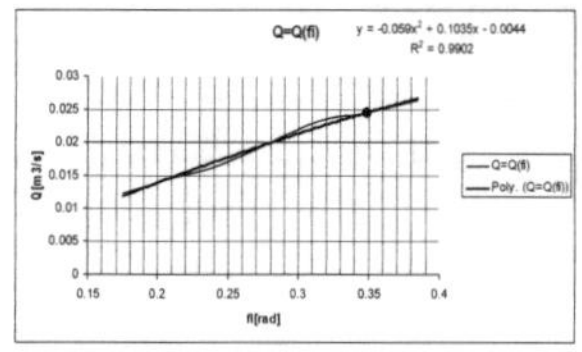

Fig.7 The variation of the flow-rate with the opening of the nozzle's vane, regime 2

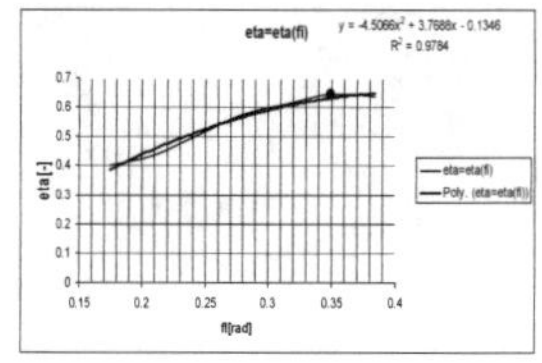

Fig.8 The variation of the efficiency with the opening of the nozzle's vane, regime 2

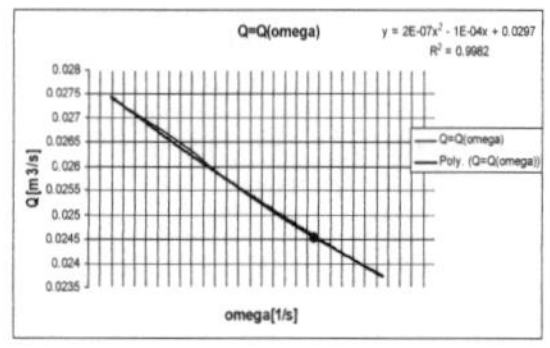

Fig. 9 The variation of the flow-rate with the angular speed, regime 2

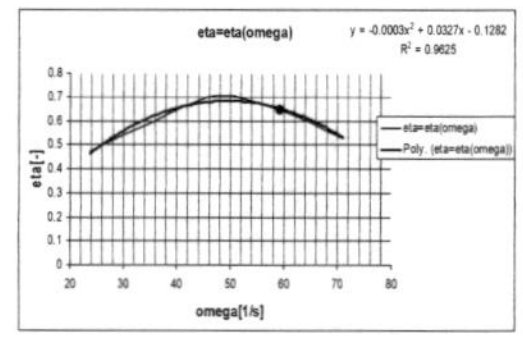

Fig. 10 The variation of the efficiency with the angular speed, regime 2

At this operating point the characteristic times are T_φ=-0.0000831 and T_ω=-0.058970 and the dynamic equation can be write

$$\frac{dx}{dt} = \frac{-\varphi}{0.0000831} - \frac{\Omega}{0.058970} \quad (12)$$

3. CONCLUSIONS

- ✓ A presentation of Double Flux water turbine has been made
- ✓ It was presented the calculus of the characteristic times for two working regimes (1, 2).
- ✓ The presented method is an improved one and it can be applied to all types of water turbine.
- ✓ With the help of computer programs the method is more precise that the classic one.
- ✓ The dynamic equation can predict the running of the turbine in unsteady regimes

4. FURTHER WORK

As further work in the Phd. Thesis called "Dynamic Identification of Double Flux turbines" we will solve the derivative dynamic equation, identifying the dynamics of this kind of water turbine through experimental results.

5. REFERENCES

Barglazan, M. (2001) *Hydraulic turbines and hydrodynamic transmissions*, Ed. Politehnica, ISBN 973-9389-39-2, 482 p, Timişoara.

Barglazan, M. (1974) *Automation control of hydraulic machinery, laboratory works*, Institutul Politehnic "Traian Vuia" Timişoara.

Barglazan, M. (1979) *Automation control of hydraulic systems*, Institutul Politehnic "Traian Vuia" Timişoara

Stroita D. C., Barglazan M. (2007) *A Method for Calculating the Characteristic Times for Double-Flux Water Turbines,* Scientific Bulletin of The „Politehnica" University of Timisoara, Transaction on Mechanic, Tom52(66), Fasc 4.

AKNOWLEDGEMENT This paper was possible through the CNCSIS Grant IDEI cod 35/nr. 68/01.10.2007, director Prof. Dr. Ing. Victor BALASOIU

Annals of DAAAM for 2008 & Proceedings of the 19th International DAAAM Symposium, ISSN 1726-9679
Editor B. Katalinic, Published by DAAAM International, Vienna, Austria 2008
Make Harmony Between Technology and Nature, and Your Mind will Fly Free as a Bird
*Young Researcher and Scientist Paper / * Supervisor, Mentor*

NUMERICAL INVESTIGATION OF THE FLOW IN A STRAIGHT AND BENDED HYDROFOIL CASCADE OF AN INDUCER

STUPARU, A[drian]; MUNTEAN S[ebastian]; BAYA A[lexandru] & ANTON L[iviu]*

Abstract: *This paper presents the numerical investigation of the 2D flow in a straight and bended hydrofoil cascade of an inducer by using commercial code FLUENT 6.3. First the investigated inducer is described and then the equations that govern the flow and the boundary conditions imposed on the computational domain. The results of the flow simulation with different types of turbulence models for the two types of hydrofoil cascade are compared with the experimental results, the streamlines distribution for the two hydrofoil cascade and a set of conclusions are presented.*

Key words: *pump inducer, straight hydrofoil, bended hydrofoil, numerical analyse of the flow, streamlines distribution*

1. INTRODUCTION

The inducer is an axial impeller which is mounted on the same shaft with the impeller of a pump. By producing a supplementary specific energy on the inlet of the impeller of the pump, because of the rising of the pressure over the vaporising pressure, it prevents or reduces the cavitation. The inducer is used for the pumps with severe suction condition and it has a "sacrificial role". It will be replaced after a certain operating time and that is why the technology for its execution has to be simple and economical. To use straight and bended hydrofoil cascade for the construction of the blades of the inducer satisfies these requirements. We investigated two inducers, one with the blades constructed from straight hydrofoil cascade and one with the blades manufactured from bended hydrofoil cascade. Each inducer has two blades.

In order to validate the numerical investigations of the flow, the results obtained are compared with experimental results from investigations performed in a wind tunnel upon two inducers which respect the geometrical similitude laws.

2. COMPUTATIONAL DOMAIN, FLOW EQUATIONS AND BOUNDARY CONDITIONS

The computational domain, figure 1 and figure 2, was generated using the pre-processor GAMBIT from FLUENT, based on the existing geometry. The geometrical characteristic of the two types of hydrofoil corresponding to the middle radius of the blade and the investigated operating condition are given in table 1.

The generated mesh for the computational domains is structured and has 60,000 cells each. A boundary layer was attached to the two hydrofoils in order to be able to compute the flow near a solid wall, (Thomson et al., 1997).

The periodic boundaries of the domains are positioned at a distance, regarding the chord.

Hydrofoil	Q [m^3/h]	D [mm]	d [mm]	s [mm]	β_s [°]	β_0 [°]	t/l [-]
Straight	100	300	100	3	36.5	20.52	0.804
Bended	100	300	100	3	31.5	20.52	0.804

Table 1. Geometrical characteristics and operating conditions of the investigated hydrofoils

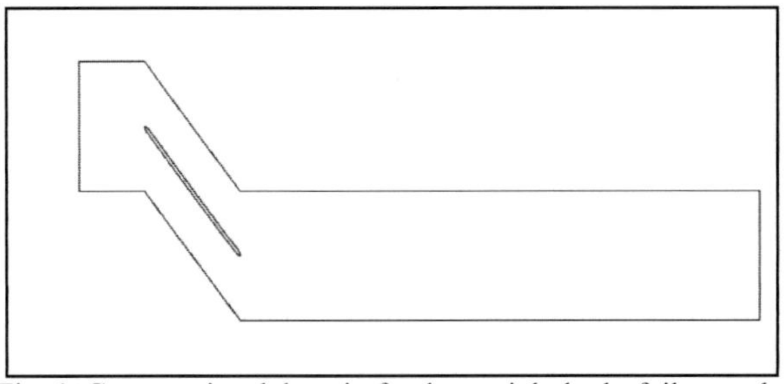

Fig. 1. Computational domain for the straight hydrofoil cascade

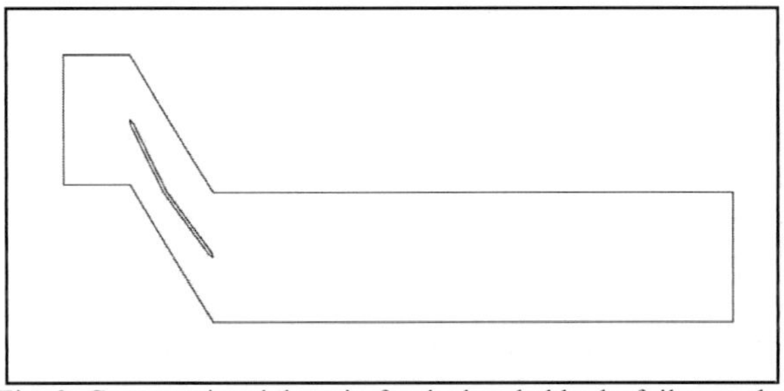

Fig. 2. Computational domain for the bended hydrofoil cascade

This distance is equal with the space of the cascade, while the inlet is positioned at a distance equal with half of the space of the cascade and the outlet at a distance equal with four times the space of the cascade, (Susan-Resiga&Muntean, 1999)

A steady relative 2D flow is computed in the computational domain:

$$\nabla \cdot \vec{V} = 0 \tag{1}$$

$$\rho \frac{d\vec{V}}{dt} = \rho g - \nabla p + \mu \Delta \vec{V} \tag{2}$$

The numerical solution of flow equations (1) and (2) is obtained with the expert code FLUENT 6.3, using a Reynolds-averaged Navier-Stokes (RANS) solver.

First the inviscid flow is calculated, that means that the viscosity of the water is not take into account.

After that, the flow is calculated using two models of turbulence: standard k-ε and Reynolds Stress Model.

The standard k-ε model is a two-equation model in which the solution of two separate transport equations allows the turbulent velocity and length scales to be independently determined.

The Reynolds Stress Model (RSM) is the most elaborate turbulence model that FLUENT provides. Abandoning the isotropic eddy-viscosity hypothesis, the RSM closes the Reynolds-averaged Navier-Stokes equations by solving transport equations for the Reynolds stresses, together with an equation for the dissipation rate. This means that five additional transport equations are required in 2D flows, (Fluent, 2001).

We imposed on the inlet section the two components of the velocity, corresponding to the prescribed flow rate and flow angle, together with the turbulence parameters,

$$w_x = \frac{Q}{S_{IN}} \tag{3}$$

$$w_y = \frac{w_x}{tg\beta_0} \tag{4}$$

On the outlet section of the computational domain and of the computational domain a radial equilibrium condition is chosen, (Gostelow, 1984).

On the periodic surfaces of the domain the periodicity of the velocity, pressure and turbulence parameters were imposed.

The remaining boundary conditions for the domain correspond to zero relative velocity on the blade,

3. NUMERICAL RESULTS

The pressure coefficient is defined by the following equation:

$$c_p = \frac{p - p_{IN}}{\frac{\rho}{2} w_{IN}^2} \tag{5}$$

A comparison between the pressure coefficient obtained from the numerical investigation of the flow with different types of turbulence and from experimental investigation (Anton, 1994) is made, in order to validate the numerical results. It can be observed that the pressure coefficient distribution obtained with the RSM turbulence model is most similar to the distribution obtained from measurements, for both hydrofoil cascades, as presented in figure 3 and 4.

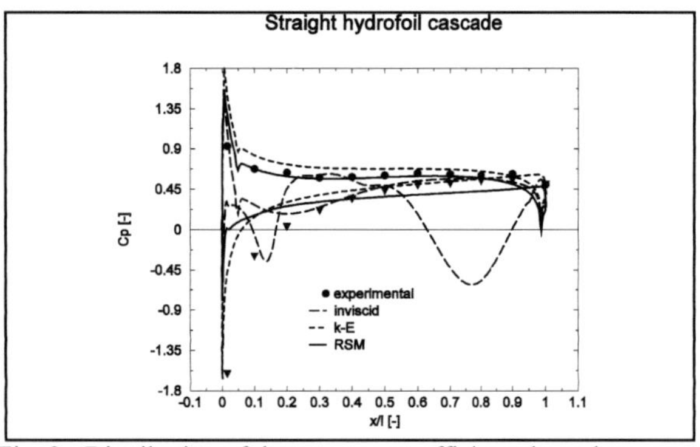

Fig. 3. Distribution of the pressure coefficient along the straight hydrofoil.

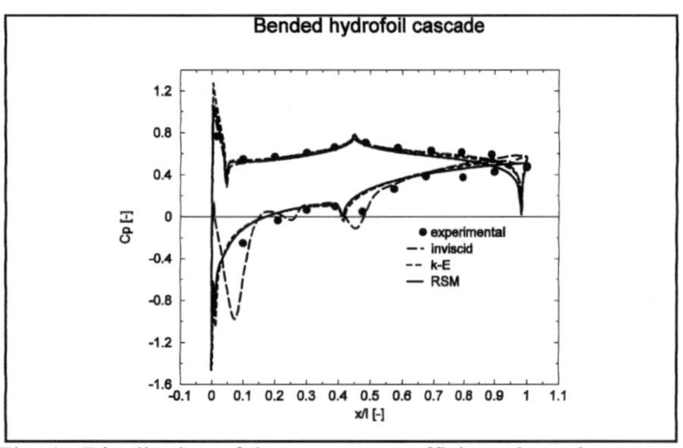

Fig. 4. Distribution of the pressure coefficient along the bended hydrofoil.

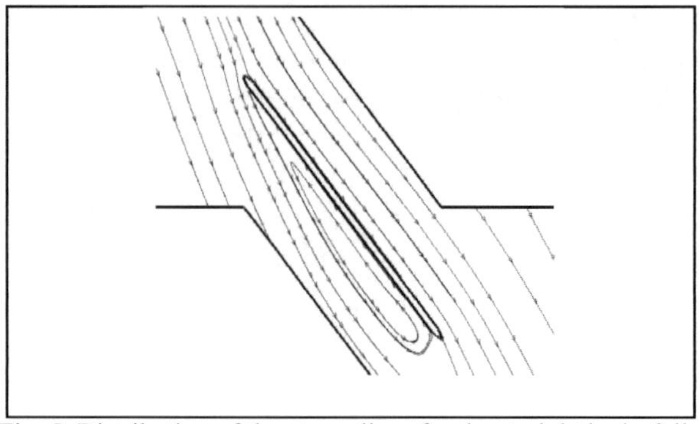

Fig. 5. Distribution of the streamlines for the straight hydrofoil

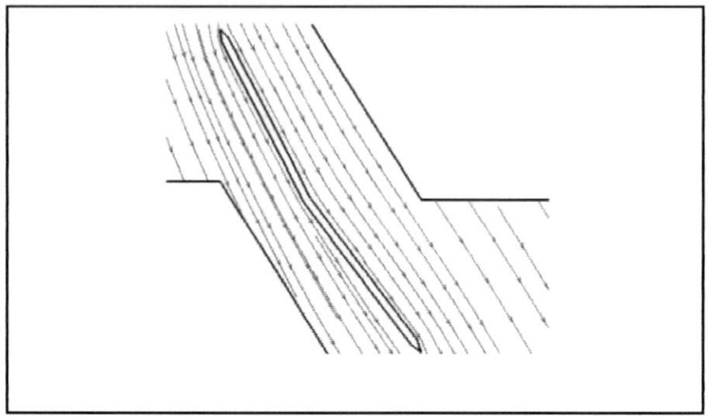

Fig. 6. Distribution of the streamlines for the bended hydrofoil

In figure 5 and 6 the streamlines for the two hydrofoil cascades are presented. It can be observed that for the straight hydrofoil appears a recirculation on the suction side which leads to the partial blockage of the flow channel. This phenomenon is eliminated by using a bended hydrofoil cascade, as shown in figure 6.

4. CONCLUSIONS

From the comparison of the numerical data with the experimental data regarding the pressure coefficient distribution along the two hydrofoils it results a good agreement for the RSM turbulence model. The numerical investigation of the flow predicts the presence of a dead zone of the flow on the suction side of the straight hydrofoil. This dead zone is not present at the bended hydrofoil.

This leads to the conclusion that the bended hydrofoil is suited for the construction of the blade of the inducer and that the use of the RSM turbulence model for numerical investigation is recommended.

5. REFERENCES

Anton, L. E. (1994). Determination of pressure distribution on the blades of an inducer, *Proceedings of XVII IAHR Symposium*, pp. 321-328, China, September, 1994, Beijing

Fluent Inc., (2001). *FLUENT 6.3 User's Guide*, Fluent Incorporated, Lebanon

Gostelow, J.P.(1984). *Cascade Aerodynamics*, Pergamin Press, Oxford

Susan-Resiga, R.; (1999). Periodic boundary conditions implementation for the Finite Element Analysis of cascade flows, *Scientific bulletin of Politehnica University Timisoara*, Vol. 44(58), pp. 151-160

Thomson, J.F.; Warsi, Z.U.A.; Mastin, C.W. (1997). *Numerical Grid Generation,* Elsevier Science Publishing Co.

Annals of DAAAM for 2008 & Proceedings of the 19th International DAAAM Symposium, ISSN 1726-9679
Editor B. Katalinic, Published by DAAAM International, Vienna, Austria 2008
Make Harmony Between Technology and Nature, and Your Mind will Fly Free as a Bird

DETERMINATION OF QUALITY CHANGES DURING THE PRODUCTION OF DRIED FOOD PRODUCTS USING COLORIMETRY

STURM, B[arbara] & HOFACKER, W[erner]

Abstract: *Colour is one of the most important criteria for the quality of food products. It does not only influence it's subjective appreciation by the consumer but it can also be used for the determination of quantitative quality criteria which depend on the drying process. The paper gives a synopsis of the techniques of measurement and quantification methods for colour. Special emphasis is given to the use of low cost hardware available as digital equipment and to the possibility of continuous measurement. The different methods of colour designation in terms of RGB, XYZ or L,a,b –values are discussed. It is shown, that information relevant for the control of the product quality can be derived from the image data.*
Key words: *Colorimetry, Imaging, Fruit, Vegetables*

1. INTRODUCTION

A lot of effort has been made to use machine vision systems when processing agricultural products. Generally the surface colour or the dimensions of the product particles are determined and evaluated. It is assumed, that these information allow a quality assessment.

A comprehensive study of activities in this fields is given by (Brosnan & Sun, 2004) and (Cheng-Jin Du & Sun, 2004).

New developments in hardware, particularly the availability of cheap and robust imaging systems make it possible to develop systems which derive quality information from the images or from the video stream.

2. OBJECTIVES

For the purpose of quality assurance, but also for process control, the quality changes can be determined by continuously measuring the surface colour of the product. However, some problems have to be resolved:

- The functional link between the information derived from the image data, the quality criterion to be controlled and the external process parameters (e.g. temperature, humidity) must be known.
- The system hardware must provide for the images in requested rate and (constant) quality and must be able to perform the image analysis and information extraction in an acceptable delay.

Finally, the system should be simple, robust and reliable. The investment for the system should be as low as possible and comparable to a conventional system controller.

3. EQUIPMENT AND PROCEDURES

3.1 Hardware¶

The experiments are carried out with the experimental setup described in (Sturm & Hofacker, 2008). This equipment can be used for drying experiments in the range of 5°C to 200 °C air temperature, and 4°C to 80 °C dew point temperature. It can be used as an overflow dryer as well as a single layer dryer. Weight of the sample, pressure drop, surface colour and temperature are measured continuously. A digital camera (IC-Imaging, Model DFK21F04) supplies digital images to a PC running under Windows XP. The visual field of the camera covers one or several product particles. Special efforts were made to protect the camera from heat and humidity to guarantee for constant image quality. Dryer and optical system were protected from daylight to avoid interference with the artificial illumination (D65) of the product samples. The images are taken continuously and are sent to the computer in intervals which can be set in a wide range; it was set to 5 minutes for experiments described below. The images are stored for later analysis or can be analysed immediately after exposure for the purpose of automatic control.

3.2 Image Analysis

The analysis of the image data is carried out by a program specially written for this purpose. The region of the image to be analysed is defined by mouse operation. Information about the background colour and a white standard is put in. All pixels in the analysis region are evaluated: if they are part of the product surface, the RGB-values are extracted and stored; if they are of the background colour, they are discarded. Glossy regions are identified by their luminosity. The minimum- maximum- and mean RGB values are determined. Pixels which undoubtedly are part of the particle surface are counted to get information of the particle size and the change of the particle size due to shrinkage.

The RGB-values are used to calculate the X,Y,Z-Values and then the L,a,b values of the pixels by the procedures given by (Hunt, R. W. G. et al., 1991) and (León, K. et al., 2006):
The first step is to transform the RGB-values in XYZ-(tristimulus)-values:

$$\begin{vmatrix} X \\ Y \\ Z \end{vmatrix} = \begin{vmatrix} M_{11} & M_{12} & M_{13} & M_{14} \\ M_{21} & M_{22} & M_{23} & M_{24} \\ M_{31} & M_{32} & M_{33} & M_{34} \end{vmatrix} * \begin{vmatrix} R \\ G \\ B \\ 1 \end{vmatrix} \quad (1)$$

The second step is to transform the X,Y,Z values in CIE-L*a*b* values (CIE, 1932):

$$L^* = \begin{cases} 116 * \left(\frac{Y}{Y_n}\right)^{1/3} - 16 & \text{if } \left(\frac{Y}{Y_n}\right) > 0.008856 \\ 116 * \left(\frac{Y}{Y_n}\right) & \text{if } \left(\frac{Y}{Y_n}\right) < 0.008856 \end{cases}$$
$$a^* = 500 * \left[\left(\frac{X}{X_n}\right)^{1/3} - \left(\frac{Y}{Y_n}\right)^{1/3}\right] \quad (2)$$
$$b^* = 200 * \left[\left(\frac{Y}{Y_n}\right)^{1/3} - \left(\frac{Z}{Z_n}\right)^{1/3}\right]$$

The values given by (CIE, 1932) for the matrix coefficients and the constants in the transformation equation for a 2° observer, D65 illumination and reference RAL white standard are used.

4. RESULTS

Tomatoes were dried at temperatures of 40°C, 55°C and 70°C with a dew point temperature of 10°C. In Figure 1 the variation of the moisture content is shown. The dring process accelerates, as already known, with higher air temperatures.

In Figure 2, the Total Colour Change TCD during drying is shown. The TCD Values is calculated from the L*,a*,b* Values with the function (León K., et.al., 2006)

$$TCD = \sqrt{\Delta L^2 + \Delta a^2 + \Delta b^2} \tag{3}$$

The colour changes proceed very fast at high temperatures. It seems, that the colour change is due to a decomposition reaction. Further investigation is necessary in this field, notably if colour changes should be avoided during processing.

In Figure 3 the visible cross section of the product particle is shown. This parameter indicates the shrinkage of the product, as the surface of the particle visible to the camera can be determined by counting the pixels which belong to it. The change of the number of these pixels corresponds to the change in visible cross section. The correlation between this value and shrinkage of the particle volume depends on the particle's shape and can be expressed for simple geometries as spheres, cubes etc. only.

Figure 3 leads to the assumption, that the product shrinks more and faster if dried at high temperatures. However, the shrinkage process is rather complicated due to the inhomogenous moister distribution during the drying process. Fast surface drying and – if present the transition of the materials from rubbery in glassy state – may provoke case hardening and thus surprising effects: As shown in Figure 4 shrinkage rate is lower if the product is dried with higher temperature.

This effect may be due to the rapidly drying and hardening of the surface of the particles. During the remaining drying process, these parts of the particle can not change their size and shape anymore, even if moisture is removed from the particle.

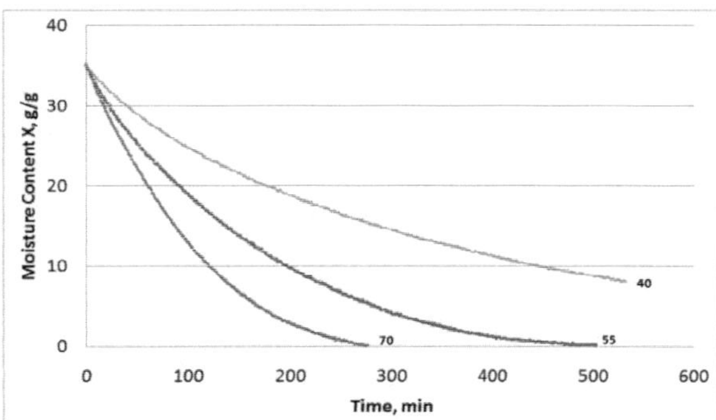

Fig. 1. Drying courves of tomatoes at different temperatures (overflow, dew point temperature 10°C, air velocity 4 m/s)

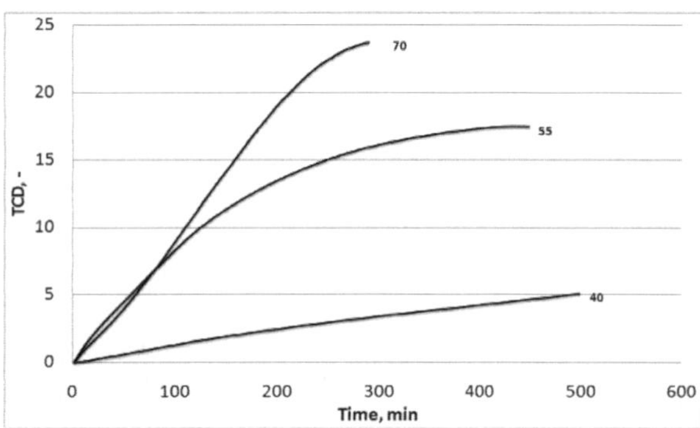

Fig. 2. Total Colour Difference, continuously measured during drying, drying parameters as in Fig. 1

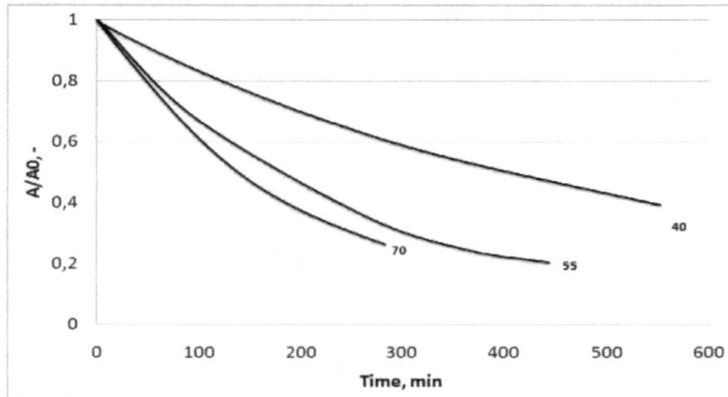

Fig. 3. Visible Cross section as a measure for the particle dimension and for the shrinkage as a function of time

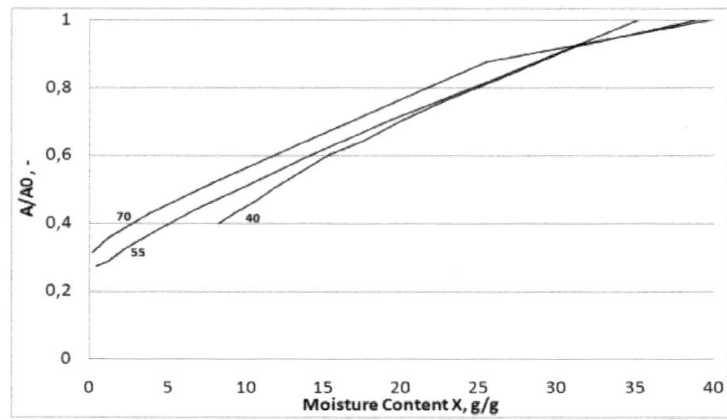

Fig. 4. Visible Cross section as a function of moisture content

5. CONCLUSIONS

First experiments have shown that the continuous measurement of colour and size of particles during drying give valuable information on the changes, a product undergoes during drying. Further work is to be done in the field of the determination of the functional links between colour changes and quality changes. A first step will be, to determine more clearly the correlation between the quality parameters and the colour changes.

6. ACKNOWLEDGEMENTS

The authors acknowledge the German Federal Ministry of Economy and the University of Applied Science Konstanz for financial support.

7. REFERENCES

Brosnan, T, Sun Da.-Wen (2004). Improving quality inspection of food products by computer vision -a review, *Journal of Food Engineering,* Vol. 61, No. 1 (January 2004), pp. 3–16

CIE (1932). *Commission Internationale de l'Eclairage proceedings, 1931.* Cambridge University Press, ISBN 9780470049044, Cambridge.

Du, J.-W. Sun, D.-W.(2004). Recent developments in the applications of image processing techniques for food quality evaluation. *Trends in Food Science & Technology.* Vol. 15 No 5 (May 2004) pp. 230–249

Hunt, R. W. G., Chichester, E. H. (1991). *Measuring colour* (2nd ed.), Wiley, ISBN 0135676789, England.

León, K.; Pedreschi F. & León J.: Color measurement in L*a*b* units from RGB digital images. *Food Research International,* Vol. 39 No 10 (Dec. 2006), pp. 1084-1091

Sturm, B. & Hofacker, W. (2008) *Farbveränderungen während des Trocknungsprozesses von komplexen biologischen Materialien(Colour Changes during Drying of complex biologic Materials)* , ProcessNet-Fachausschuss „Trocknungstechnik", Halle / Saale 5.-6.3.2008

Annals of DAAAM for 2008 & Proceedings of the 19th International DAAAM Symposium, ISSN 1726-9679
Editor B. Katalinic, Published by DAAAM International, Vienna, Austria 2008
Make Harmony Between Technology and Nature, and Your Mind will Fly Free as a Bird

APPLICATION OF 3D OPTICAL SCANNING FOR THE SHAPE ACCURACY ANALYSIS OF MACHINE PARTS PRODUCED BY MULTI-PASS METAL SPINNING

SUGAR, P[eter]; SUGAROVA, J[ana] & MOROVIC, L[adislav]

Abstract: *The paper brings the partial results of the shape accuracy analysis of machine parts produced by multi-pass metal spinning. The thin sheets with 1 mm thickness, materials EN 30-69 (STN 41 1375.21) and EN 10088/1-3-95 (STN 41 7240) were used for experimental study. The shape accuracy was analysed by the optical 3D measuring machine based on the principle of triangulation (3D coordinates for each camera pixel are calculated with high precision and a polygon mesh of the object's surface is generated). The 3D digitilizing with the mobile GOM ATOS I 350 system was used. For the visualization of shape deviations of final parts compared to nominal shape was realised by using colour plots.*

Key words: *forming, spinning, measurement, shape.*

1. INTRODUCTION

Metal spinning is relatively old, but now high-tech computerized forming technology, which produces axisymmetrical hollow parts with advantageous surface layer properties. The theoretical strains are considered under two idealised models. The first process, in which the thickness of workpiece is almost unchanged, is called conventional spinning and the second process where the hoop strain is zero, is named pure shear forming, also named spin forging (Quigley &Monaghan, 2000). In conventional spinning, it is assumed that the sheet thickness is constant and the hoop strain is compressive, the radial strain is tensile and the thickness strain is zero. In shear forming the radial position of any element is unchanged and a single pass is usually all that is require to produce the desired shape (Razavi, 2005).

Spinning, compared with deep drawing, involves lower forming forces and less power. As a result, equipment is cost-effective engineered and process is very flexible. Other benefits are: increased level of automation, high mechanical streght and hardness of part material, cost reduction throught high material yield and favorable cycle time.

The nature of conventional metal spinning proces is evident from the Fig. 1.

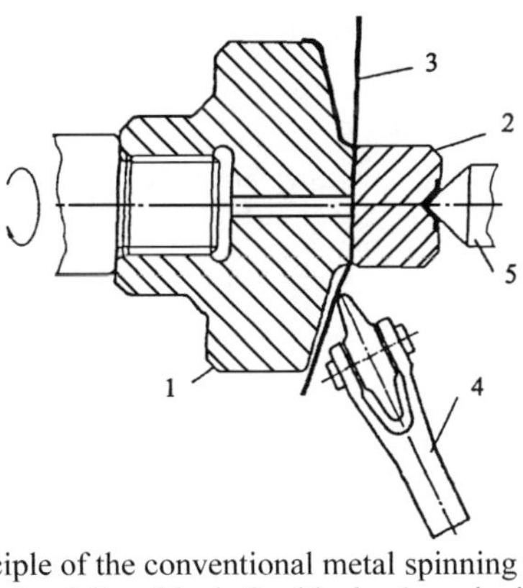

Fig. 1. Principle of the conventional metal spinning process 1 – mandrel, 2 – follow block, 3 – blank, 4 – spinning tool, 5 – tailstock centre

Nowadays the process is getting increasing importance in many fields of industrial applications, as for example in the manufacturing of process equipment such as centrifuges, funnels, tanks, medical and gastronomy equipments as well as drive technologies. It also has the possibility of producing parts that could not be deep drawn (Palten &Palten, 2002).

Benefits available to users of spinning technology include: high flexibility, increase level of automation, very low risk of crack propagation, high mechanical strenght and hardness of spun parts, cost reduction through high material yield, variety of operations possible in one setup and favorable cycle time.

These benefits of the metal spinning technology apply to production of one-off parts and prototypes as well as to small to medium runs.

2. SHAPE ACCURACY MEASUREMENT

The experimental investigation of spinning parameters on the shape accuracy of part formed by conventional (multi-pass) spinning technology was realised at the Department of Manufacturing Technology and Material Science, Faculty of Environmental and Manufacturing Technology, Technical University in Zvolen (Gašparová, 2006), (Šugárová & Šugár, 2002), (Šugár & Šugárová, 2007). To achieve knowledge about the coherence of the metal spinning input parameters and the formed part shape accuracy the experimental measures were caried out. The methodology and results of the spun parts shape accuracy measure by optical 3D measuring machine, are presented in the paper.

2.1. Methodology of experiments

A spherical shape of radius 107,5 mm blending via a 25 mm radius into a 168 mm diameter cylinder was used for these experiments (Fig. 2). Parts were made from thin steel sheets EN 30-69 (STN 41 1375.21) and EN 10088/1-3-95 (STN 41 7240), thickness s_0 = 1 mm. Mechanical properties of the material shows Table 1. The manual spinning lathe was used. Variable parameters were spindle revolution (n = 315, 490 a 890 min^{-1}) and number of passes of spinning tool (spinning without sizing p = 3 and spinning with sizing p = 4).

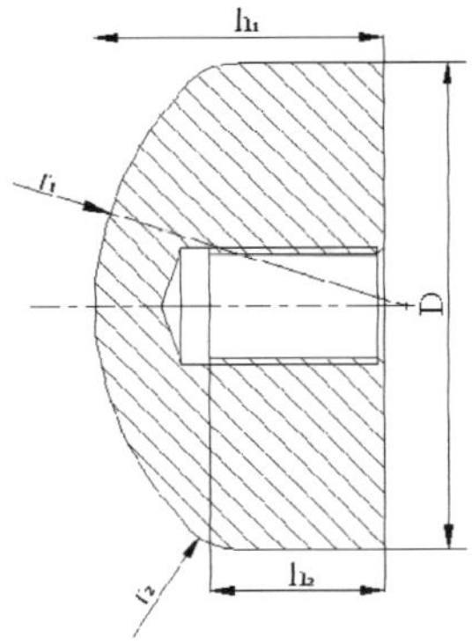

Mandrel's dimensions [mm]	
D	168 ± 0,3
h_1	100
h_2	60
r_1	107,5
r_2	25

Fig. 2. Mandrel's dimensions

Mechanical properties	Material	
	EN 30-69	EN 10088/1-3-95
R_m [MPa]	389	604
$R_{p0,2}$ [MPa]	269	235
$R_{p0,2} / R_m$	0,69	0,39
A [%]	32,3	59,7

Tab. 1. Mechanical properties of experimental materials

The experimental samples were analysed from the shape accuracy point of view. 3D digitalization of the metal spinning part shape was realized by the system of optical scanning GOM ATOS I 350 (Advanced Topometric Sensor), consisting of two CCD cameras (they are located in the way that objective lens form an angle of 29 degrees) and projector. The evaluation of computer models was realized by using of software GOM ATOS v. 6.0.2-5, which allows dimensional and shape deviations analysis of scanned forms in relation to nominal data.

The basic aim of the experimental works was to consider the possibilities of the 3D optical scanning application to handle the shape accuracy analysis of the manufactured parts. The results of these works were processed in the visualization form of the part shape deviation and in the form of graphical surface deviation plots in a 3D presentation.

2.2. Results and discussion

For illustration, in the article, the result of shape accuracy measurement of the part made from material EN 100088/1-3-95 (STN 41 7240) is presented. The thickness of sheet was 1 mm, spindle revolution n = 890 min^{-1} and the conventional, multi-pass spinning without sizing was realised. Fig. 3 shows the comparison of the final part profile curve to the CAD data, represented the nominal profile curve.

In the second case, the comparison of surface deviations to the reference data for multi-pass conventional spinning with and without sizing is presented. The material EN 30-69 (STN 41 1375.21) with thickness 1 mm was used. The deviations are sumarized in Table 2.

All obtained graphical and numerical outputs showed a high degree of shape deviation in the places, which are the furthest from the symmetry axis. In this case, the resulted state was caused by the material properties and by the fact, that the manufacturing of the parts was manual. By the using of a spinning roller operated by a hydraulic feeding mechanism (CNC spinforming machines), here is not expected such a wide difference between real and nominal shape of the manufactured part.

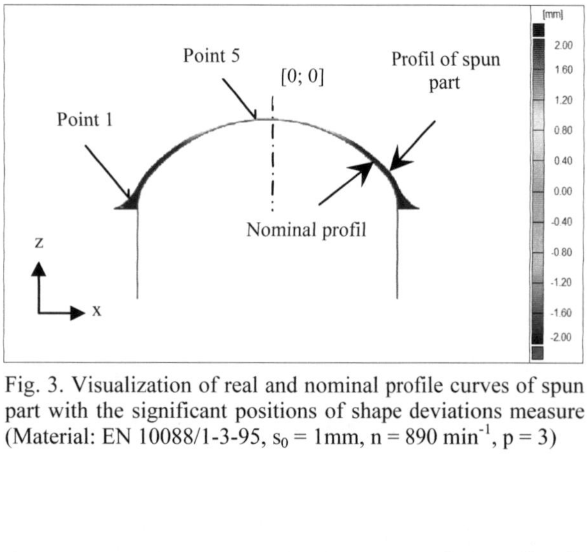

Fig. 3. Visualization of real and nominal profile curves of spun part with the significant positions of shape deviations measure (Material: EN 10088/1-3-95, s_0 = 1mm, n = 890 min^{-1}, p = 3)

Position of measure (x, z) [mm]	Surface deviation [mm]	
	Number of passes	
	p = 3 (spinning without sizing)	p = 4 (spinning with sizing)
Point 1 (- 86,7; - 60,1)	+ 2,97	+ 1,81
Point 2 (- 79,7; - 34,9)	+ 0,53	+ 0,30
Point 3 (- 68,9; - 24,3)	+ 0,62	+ 0,53
Point 4 (- 51,2; - 12,5)	+ 0,58	+ 0,25
Point 5 (- 1,7; - 0,62)	+ 0,03	+ 0,00

Tab. 2. Comparison of surface deviations for spinning with and without sizing (Material: EN 30-69), s_0 = 1mm, n = 890 min^{-1})

3. CONCLUSIONS

Presented methodology and results are partial task of the research aimed at complex evaluation of qualitative parameters of parts manufactured by metal spinning. The results of the shape accuracy analysis, obtained by 3D optical scanning method, show the factuality of the monitoring of large number of input process parameters impact (e.g. springback error) on the spun parts shape accuracy. In the future the methodology will be used for comparison of shape accuracy of parts produced by metal spinning and the same parts produced by drawing. The obtained results supposes development of computer aided system for the process design optimization.

4. ACKNOWLEDGMENTS

The results presented in the paper are based on investigation of research project "Expert system for technological parameters optimization of metal spinning process" – project No. 1/4097/07, funded by the Slovak Ministry of Education.

5. REFERENCES

Gašparová, J., Šugárová, J. & Kalincová, D. (2006). Study of microstructure and formed part wall thickness changes after metal spinning. *Development of Mechanical Engineering as a Tool for Enterprise Logistic Progress*. Legutko, S. (Ed.), pp. 187 – 192, ZPW M-DRUK, ISBN 83-89873-28-1, Poznan.

Palten H. & Palten, D. (2002). Metal spinning – From Ancient Art to High-Tech Industry. *Metal forming*, Vol. 36, No. 9, pp. 30 – 34.

Quigley, E. & Monaghan, J. (2000). Metal forming: an analysis of spinning processes. *Journal of Materials Processing Technology*, Vol. 103, No.1, pp. 114 – 119, ISSN 0924-0136.

Razavi, H.; Biglari, F,. R. & Torabkhani, A. (2005). Study of Strains Distribution in Spinning Process Using FE Simulation and Experimental Work. *Available from:* http://me.aut.ac.ir/frbiglari.htm *Accessed:* 2008-06-24.

Šugár, P. & Šugárová, J. (2007). Metal spinning force analysis. *International Tool Conference 2007*, pp. 123 – 127, ISBN 978-80-7318-572-5. Zlín, May 2007, University of Tomas Bata Zlín.

Šugárová, J. & Šugár, P. (2002). Research study of strain distribution throughout the part after metal spinning operations. *Annals of DAAAM for 2002 & Proceedings of the 13th International DAAAM Symposium,* Katalinic, B. (Ed.), pp. 545-546, ISBN 3-901509-29-1. Vienna, October 2002, DAAAM International Viena.

Annals of DAAAM for 2008 & Proceedings of the 19th International DAAAM Symposium, ISSN 1726-9679
Editor B. Katalinic, Published by DAAAM International, Vienna, Austria 2008
Make Harmony Between Technology and Nature, and Your Mind will Fly Free as a Bird

THE USE OF TRACE TRANSFORM FOR CLASSIFICATION OF VISUALIZED TRANSPARENT POLYMERIC FOILS

SYROVA, L[ivia] & RAVAS, R[udolf]

Abstract: *The contribution treats the topic of processing the experimentally gained information on the properties of optically transparent polymeric foils. Schlieren optical system was used to examine and visualize refractive index distribution. Trace transform method enables classification of different types of foils into classes. One of the possibilities how to gain invariant features is to use trace transform based on the triple features that are combination of three funcionals (trace, circus and diametric). The suitability of trace transform was tested on the images of visualized samples of polymeric foils. Experimental results are presented as well.*
Keywords: *visualization, schlieren, foil, trace, classification.*

1. INTRODUCTION

Properties of optical transparent polymeric foils are related to their production and therefore the evaluation of the quality of the foils is an important part of their production (Bolf et al., 1993). Optical visualization methods and especially schlieren method (Smith & Lim, 2000) allows visual quality assessment, because it permits to locate on a pre-processed picture the places with elastic strains that are marked by variations in the level brightness (Caastle et al., 1994). By the image processing it is possible to evaluate quantitative properties as well. Properties of visualized images of the foils are similar to the properties of textures. In order to recognize images of the foils methods suitable for textures recognition such as statistical, frequency methods and wavelet transform can be used. The completion of the optical visualization method with suitable characteristics calculated from the pre-processed image has found the application in the field of defectoscopy, identification and classification of polymeric foil. Numerical characteristics calculated from the image function of the region called features correspond to the object. Elements of the signature vector are formed by the aligned set of features. Position of ending point of signature vector in the n-dimensional space, where n means number of elements of signature vector, represents pattern of the object. The main problem is to create such a description of object in ideal case whose patterns belonging to the same class can be grouped into the subspace in that no pattern of object belonging to the different class is situated. Very important group of methods of region description enables to calculate characteristics that are invariant to translation, rotation and scaling of the image. Classifiers utilizing the decision rules make a decision upon the classification of the image to the corresponding class. In the simplest case it is possible to construct the decision rules analytically. In the completion case classifiers based on learning, e.g. on the artificial neural network are used (Grman, 2000). The contribution treats the principle of description of images by the trace transform method (Kadyrov & Petrou, 2001).

2. TRACE TRANSFORM

We can regard trace transform as a generalization of the principle of Radon transform. Radon transform is useful when to calculate the characteristic numbers of image especially an integral calculated along the lines crossing the image. Trace transform uses in addition to the integrals calculated along lines also other functionals calculated along lines crossing the image. It uses combination of triple different or the same functionals along the lines in the image. The advantage of transform represents the possibility to create many characteristic numbers – features, whose choice depends on the character of classification problem.

An arbitrary line crossing the image can be expressed by the function of three parameters $l(r,\theta,t)$, t is position of point in the line, θ is angle between line and x axis and r represents distance of the line from the origin of coordinate system.

A functional F is real-valued function on a vector space, usually of functions. Image is described by a triple features. Procedure of the triple features construction can be described by the steps on:

- Produce the trace transform of the image by applying a Trace functional T along lines crossing the image. As a result there is the matrix with the columns containing values of functionals for the different values of line distance from the origin of coordinates. Rows contain functionals values in dependence of the rotation of lines.
- Produce the "circus function" of the image by applying a diametric functional R along the columns (along values of distances) of the trace transform,
- Produce triple feature by applying a circus functional θ along the string (along values of rotation) of numbers produced in preceding step.

The triple features can be defined by the system of functionals

$$G(f)=\Phi[R[T[f(r,\theta,t)]]] \tag{1}$$

f() is image (brightness) function.

There are some rules based on functionals with specific properties to triple features calculation used in (Turán et al., 2006). The first is the shift property. If

$$F[f(x+a)]=F[f(x)] \quad \forall a\in R \tag{2}$$

then functional is called shift invariant functional, if

$$H[f(x+a)]=H[f(x)]-a \quad \forall a\in R \tag{3}$$

then functional is called sensitive functional. The second is specific property to the scaling. For shift invariant functionals it is the abscissa homogeneity property

$$F[f(bx)]=b^{\alpha_F}F[f(x)] \quad \forall b>0 \tag{4}$$

and ordinate homogeneity property

$$F[bf(x)]=b^{\beta_F}F[f(x)] \quad \forall b>0 \tag{5}$$

shift sensitive functionals booth condition (6) should be fulfilled

$$H[f(bx)]=\frac{1}{b}H[f(x)]\wedge H[bf(x)]=H[f(x)] \quad \forall b>0 \quad \alpha_H=-1\wedge\beta_H=0 \tag{6}$$

It is possible to create triple features invariant to the translation, rotation and scaling by two combinations of transforms T, R, Φ:

Combination 1:

- Functional T is shift invariant (2) with abscissa homogeneity property (4)
- Functional R is shift invariant (2) with abscissa homogeneity property (4) and ordinate homogeneity property (5)
- Functional Φ is shift invariant (2) with abscissa homogeneity property (4) and ordinate homogeneity property (5)

Combination 2:

- Functional T has the same properties such as in case 1.
- Functional R is sensitive (3) with properties (6)
- Functional Φ is shift invariant (2) with ordinate homogeneity property constant (5), and is not sensitive to the first harmonic of the function f.

Triple features computed for original f and geometrically modified image f_M are related by:

$$G[f_M]=b^{\delta}G[f] \quad \delta=\begin{cases}-\beta_\Phi(\alpha_T\beta_R+\alpha_R) & \text{for combination 1.}\\ \beta_\Phi & \text{for combination 2.}\end{cases} \tag{7}$$

The invariance is fulfilled for $\delta = 0$ or when to put instead triple features ratio of two normalized triple features

$$G_r=\frac{G_{1N}(f)}{G_{2N}(f)} \quad \text{where} \quad G_N(f)=|G(f)|^{\frac{1}{\delta}}.sign(G(f)) \tag{8}$$

3. EXPERIMENTAL RESULTS

In order to verify applicability of trace transform for classification of polymeric foils we have used six different types of foils that are in fig. 1 presented. From each type of the foil we captured 16 images of different samples. We used a digital camera with resolution 5 megapixels. Triple features were calculated from the part of images with dimensions 1400x1400 pixels.

From images it is seen, that different foils have different characteristic features that are well observed, images of different foils can be distinguished and on the contrary images of different samples of the same foil show some similarity. Features show some regularity of stochastic character.

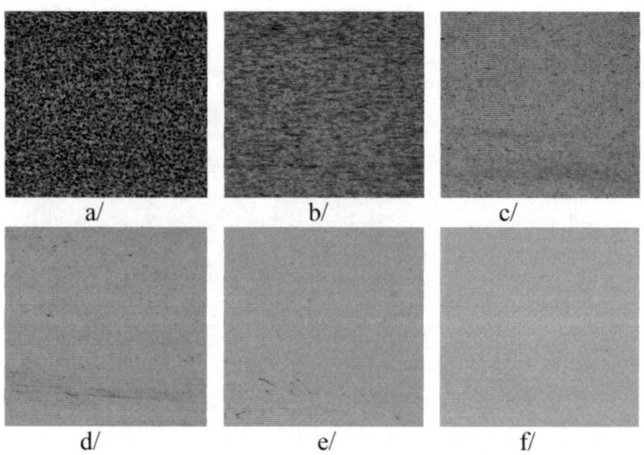

Fig. 1. Images of foils a/OOL30, b/C2, c/C1, d/C6, e/ KXE20, f/ON25

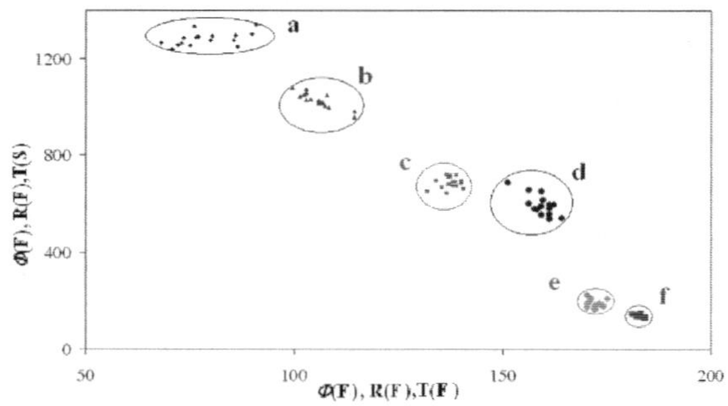

Fig. 2. The tendency of clustering of selected signature vectors for images of samples of polymerics foils from fig. 2

In fig. 2 is the tendency of clustering of signature vectors corresponding to the same types of foils and the tendency of clustering of signature vectors corresponding to the different types of foils presented. Functionals F and S were used.

The signature vector consists of triple features Φ□=F, R_1=F, T_1=F and Φ_2=F, R_2=F, T_2=S

$$F=median\{f(x)\} \tag{9}$$

$$S=\int|f'(x)|dx \tag{10}$$

4. CONCLUSIONS

Experimental results present suitability of the trace transform for calculation of features of images of visualized polymeric foils that are necessary for their recognition and for the detection of different foils damages. In general it is important that features would be invariant to the rotation and scaling. In case of visualized images of transparent polymeric foils the requirement of invariance to the scaling is not necessary to be fulfilled in regard to the technique of their images scanning. On contrary this method is not sensitive to the rotation of foil sample that can be the most advantage.

5. REFERENCES

Bolf, J.; Bajcsy, J. & Bolf, P. (1993). Methods of the Properties Testing of Optically Transparent Materials. *Journal of Electrical Engineering,* Vol. 44, No. 8, pp. 244-247

Castle, D. A.; Gibbins, B. & Hamer, P. S. (1994). Physical Methods for Examining and Comparing Transparent Plastic Bags and Cling Films, *Journal of Forensic Science Society,* Vol. 34, pp. 61-68

Grman, J. (2000). Neural Network Application in the Defectoscopy. *Proceedings of XVI. IMEKO World Congress IMEKO 2000*, pp. 205-210, Vienna, 25. - 28.9.2000.

Kadyrov, A. & Petrou, M. (2001). The Trace transform and its applications. IEEE *Transactions on Pattern Analysis and Machine Intelligence,* Vol. 23, No. 8, pp. 811-828.

Smith, A.J. & Lim, T.T. (2000). *Flow visualization,* Imperial College Press, London

Turán, J.; et al. (2006). Trace Transform and KLT Based Invariant Features and Image Recognition System. *Acta Electrotechnica et Informatica*, Vol. 6, No. 3, pp. 1- 11

ACKNOWLEDGMENT

We gratefully acknowledge that the results presented here were solved as partial goals of the research task VEGA No 1/3099/06.

Annals of DAAAM for 2008 & Proceedings of the 19th International DAAAM Symposium, ISSN 1726-9679
Editor B. Katalinic, Published by DAAAM International, Vienna, Austria 2008
Make Harmony Between Technology and Nature, and Your Mind will Fly Free as a Bird

WINDMILL'S DESIGN AND IMPLEMENTATION ASPECTS

SZEIDERT, I[osif]; PROSTEAN, O[ctavian]; ROBU, A[ndreea] & JURCA, L[ucian]

Abstract: *The paper presents main problems that appear and must be solved in the case of on-site wind aggregates farm's design and implementation. These issues regarding the analysis and design of windmill farms are focused on resources and constraints. The main objective for a windmill's farm designer is obviously to maximize the energy captured from the wind considering certain constraints, such as financial and environmental issues. This paper analyses the main aspects regarding the windmill's farms.*
Key words: *wind, farms, integrated, design*

1. INTRODUCTION

The task of development and implementation of a certain wind farm site is an extremely complex job, which appeals interdisciplinary domains of engineering. (Heier, 1998) (Szeidert et. al., 2005) In technical literature, the analysis and design aspects of a wind farm are divided into two main separate parts: the resources and the constraints.

The main objective for the wind farm designer is to maximize the energy captured within the bounds placed on it by constraints, such as environmental and financial issues.

The main problem of using wind energy is the fact that this energy introduces significant uncertainty in operating of an energetic power system: it is variable and unpredictable (practically the energy presents a pumping character).

In this order of idea, the must be considered a suitable management for the uncertainties, there is required a higher flexibility in the energetic power system, either in the form of more flexible generation, demand or transmission between areas.

Nowadays, researchers are investigating the energetic power system impacts of wind power. Thus, the results on the costs of integration differ and comparisons are difficult to be conducted due to different methodology, data and tools used, as well as terminology and metrics in representing the results. The wind power must be considered always in relation with the other elements of the power system.

An in-depth review of the studies is needed to be able to draw realistic conclusions on the range of integration costs for wind power.

The design process for a wind farm must be iterative.

The design algorithm must establish at an early stage that there is a sufficient resource to make the project viable. Simultaneously, with this it is necessary to assess:

- Planning issues: ecology, landscape designations, visual influence, electromagnetic interference, sound (noise);
- Technical issues: grid connection, ground conditions and access which may affect the development and implementation of wind farms.

All previously mentioned issues may constrain the design of a wind farm and therefore must be early considered. After setting up the terms and constraints on the development, many aspects of the design process are solvable by using usual computational methods. In technical literature, a comprehensive methodology has been developed over the last few years using a number of sophisticated software packages.

These are mainly used to:

- Optimize the design of the wind farm,
- Calculate objective characteristics of the wind farm for use in the environmental assessment process,
- Predict the wind farm's yearly energy production.

Practically, the methodology links the various site, turbine, resources, environmental and economic issues together to produce an optimum and well defined solution under specified constraints.

Wind power energetic is one of the fastest growing electric power industries in the world. Current global installed capacity exceeds 25.000 MW with a projected growth rate of about 10.000 MW/year for the next five years. The phenomenal growth of this industry can be attributed to the fast progress made in wind turbine technology.

The actual technological progress has led to the development of turbines with high power capture efficiency. Nowadays, many wind farms produce electrical power at a Cost-of-Energy almost comparable to the one of coal and natural gas based power plants.

The studies on wind turbines development has been mainly focused on large wind turbines, due to economic reasons. Typical large wind turbines of today are massive structures with enormous blade spans (70-120m in diameter), tall towers (80-150m in height) with power ratings up to 5 MW.

Modern large wind turbines are endowed with sophisticated control systems which are organized to support several modes of operation such as start-up. shut-down, power production etc. During low wind speed operation, the goal is to maximize energy capture. However, in conditions where wind speeds are greater than the turbine's "rated" wind speed (usually around 10-13 m/s) the primary objective is to minimize fatigue loading of the turbine structure.

2. ISSUES REGARDING ENVIRONMENTAL IMPACT

One of the main issue regarding the environmental impact is the sound (noise) produced by the wind farms while functioning at site. Nowadays, it has become a standard procedure and requirement for manufacturers to quote the sound source power of their turbines together with the associated octave band analysis.

This approach is rapid and simple and can be used effectively for wind farm design. Unfortunately, in the case of sound (functioning noise), the precise values at any location are dependent on: the sound source power, the topography, ground cover, background sound levels and on air's quality.

In the analysis procedure, there must be adopted the sound envelope estimation.

Additionally to visual and sound implications of wind farms there are a number of other environmental, planning and energy issues that must be addressed, such as: EMI (electromagnetic interference problems), fauna/flora disturbances and other planning issues.

On national level, relevant issues should be national energy objectives such as the reduction of CO_2 emissions problem, security of energy supply, local manufacturing potential and other related issues.

3. RESOURCE PLANNING - WIND SPEED

The energy production of a wind farm is very sensitive to wind speed. Therefore, an accurate estimate (as much as possible) of the long-term mean wind speed at a site must represent an important input to the financial assessment of a project. This fact can only be accurately achieved by making on-site wind speed measurements at a level close to the hub height of the wind turbine.

As there can be large seasonal variations in mean wind speed and annual variations in the order of up to 20%, data recorded over a period of several years is required to define the long-term values. An accurate estimate of long-term mean wind speed can, however, can be obtained from a short-term data set recorded on the site by additionally using concurrent and long-term data recorded at a nearby meteorological station.

The long-term mean wind speed will vary over the site and these variations will impose the design, implementation and the energy production of the wind farm (the economic issue). (Zinger & Muljadi, 1997)

4. ENERGY PRODUCTION (OUTPUT) OF THE WIND FARMS

The yearly energy production of a proposed wind farm can be forecasted by combining the wind farm power curve, net of the effect of topography, wake interaction and electrical losses, with the wind speed direction frequency table. Availability in excess of 90% is considered to be typical for a modern wind farm. There are continuously developed prediction techniques for the wind speed on site, for energy consumption, etc. Some, of them were studied also by the paper's authors, such as modern techniques based on the usage of recurrent neural networks. (Vasar et. al., 2007) (Heier, 1998)

5. ISSUES REGARDING THE COST MODELING

The cost-modeling task studies the costs of energy production of the wind farm that results from the design activities in several cases, as depicted in Fig.1. (Zinger & Muljadi, 1997)

The cost analysis procedure takes into consideration the economic parameters, environmental conditions, performance parameters and cost components in order to elaborate the afferent cost model.

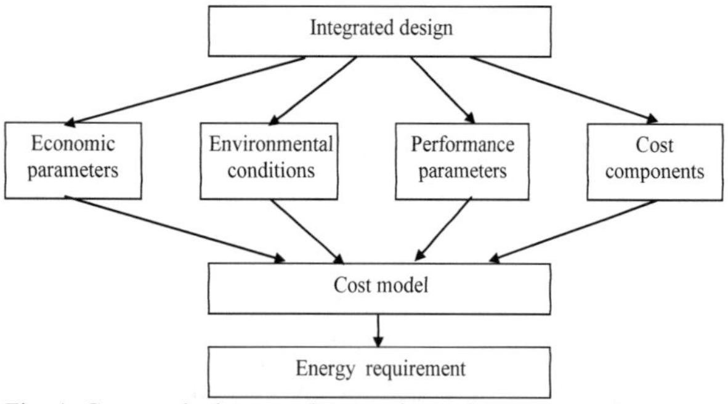

Fig. 1. Cost analysis procedure - schematic representation.

6. INTEGRATED APPROACH.

In present, the design of wind farms is conducted by using, an integrated approach regarding the electrical and civil engineering permits the site design.

In the case of wind farms, the management software should take into consideration and solve the following issues:
- The EC (electromagnetic compatibility), the IEC standards;
- Switching and operating strategies;
- Protection strategies;
- Monitoring and diagnosis issues;
- The wind farm's effects over the national energy power networks. (Lewis & Garrad, 1994) (Reinebach, 2007)

7. CONCLUSION

Because of the complexity of wind farms there is quite difficult to define cost functions for the automatic optimization of concept or design parameters. Therefore, an optimization process of wind farm concepts should be considered to be solved by multidisciplinary design teams. The optimization procedure should start with a baseline, followed by an objective learning phase and a consistent combining phase. However, in the case of differences between concepts, the selection could be performed by considering other criteria. (Bauer et al., 2003) (Garrad et al., 1993)

The wind energy market is moving so fast that construction has outpaced the availability of construction-ready sites. This has forced owners with fixed turbine deliveries to develop, design and build often at the same time. A coordinated and experienced team of development, engineering and construction professionals under central leadership is critical to having these sites completed on time and on budget. There are many decisions that have to be taken in designing a modern wind turbine.

As a conclusion, the optimal wind turbine design for one specific site is not necessarily the optimal design for another site because the wind speed distribution varies between the sites. Similarly, the turbine with the highest efficiency is not mandatory to be the optimal turbine. It is possible for a less efficient wind turbine to have a lower cost of energy. (Szeidert et. al., 2005)

In conclusion, the paper deals with the main issues and aspects that occur in the design and implementation of wind farms.

8. REFERENCES

Bauer P., De Haan S.W.H., Dubois M.R. (2003) Introduction to wind energy and offshore problematic, PCIM 2003, Nuremberg, Germany, 20-22 May.

Garrad, A. D., Mercer, A. S., and Adams, B. M. (1993) An analytical approach to wind farm design, *Report of Garrad Hassan & Partners Ltd*, with Jenkins, N. R., University of Manchester Institute of Science and Technology.

Heier S. (1998) *Wind energy conversion systems*, John Wiley & Sons Inc., New York, 1998.

Lewis, P. E. and Garrad, A. D. (1994) Wind farm optimization, *Proceedings, 16th British Wind Energy Association Conference*, Sterling, 1994.

Reinebach D. (2007) A Coordinated Approach To Wind Farm Construction, *North American Windpower Magazine*, November 2007, www.nawindpower.com,*Accessed*: 2008-02-05

Szeidert I., Budisan N., Filip I., Szeidert R. (2005) Consideration and Perspectives about Wind Energy Usage in Romania, *Simpozion Electrotehnica si Energetica, Zilele Academice Timisene*, 26-27 mai 2005, ISBN: 973-625-253-3, CD-ROM.

Vasar C., Szeidert I., Filip I., Prostean G. (2007). Short Term Electric Load Forecast with Artificial Neural Networks, *Preprints of the 4th IFAC Conference on Management and Control of Production and Logistics (MCPL 2007)*, Sibiu, Sept. 27-30, 2007, ISBN 978-976-739-481-1, pag.443-450.

Zinger D.S., Muljadi E. (1997) Annualized wind energy improvement using variable speeds, *Industrial & Commercial Power Systems technical Conference, 1997.* Conference Records, Papers presented at the 1997 Annual Meeting, IEEE 1997, 11-16 May 1997, pp. 80-83.

Annals of DAAAM for 2008 & Proceedings of the 19th International DAAAM Symposium, ISSN 1726-9679
Editor B. Katalinic, Published by DAAAM International, Vienna, Austria 2008
Make Harmony Between Technology and Nature, and Your Mind will Fly Free as a Bird
*Young Researcher and Scientist Paper / * Supervisor, Mentor*

MANAGEMENT ASPECTS OF E-BUSINESS – AN EMPIRICAL RESEARCH

SZENTES, B[alazs] & GAAL, Z[oltan]*

Abstract: *The use of various new technologies (first of all ICT) changes fundamental aspects of organizations: new business processes are developed and organizational structure and cultures change in order to manage these new challenges. To leverage the new technologies, managers need to do their job in a different way: rather than concentrating on day-to-day operation and administrative tasks, they can focus more on the future of their organization, on business opportunities and their leadership role. They must transform the increased efficiency of the new technologies into higher effectiveness of the organization.*

Key words: *e-Business, leadership, organizational culture.*

1. INTRODUCTION

Many books or studies are written on how to operate a good e-business system. The authors describe how to build up the system, what are their advantages, but they usually hardly deal with the core importance: how to implement or how it should operate in the right way. How does it change the life of an organization? How does it change the role of managers? What kind of organizational culture is needed to be able to use all the benefits offered by the various e-business methods and technologies?

In this paper we want to describe a possible research method (a Ph.D. study carried out at the University of Pannonia) for understanding the management changes caused by the use of various e-business solutions. In the first part of the paper we define the framework of the research, how we planned the research, after that we introduce our hypothesis, at the end we describe our research, through which we can examine the above mentioned questions, and show some of the preliminary results.

2. THEORETICAL BACKGROUND

2.1 e-business

Narrowly e-business is the transformation of the key business processes through the use of Internet technologies. But in a broader view it is "all electronically mediated information exchanges, both within an organization and with external stakeholders supporting the range of business processes" (Chaffey, 2004). This definition shows, that e-business is not just a commercial solution, but it needs a revision of all the thoughts of the organizations own business.

If an organization decides to employ various e-business solutions, among others it will face the following challenges:

- organizational changes required by e-business;
- management attitud changes;
- organizational flexibility;
- etc. (Papazoglou & Ribbers, 2006)

2.2 e-business and organization

e-business technologies and solutions have great impact on the organization. According to Jackson and Harris the greatest challenge is to fullfill the sometimes great organizational changes (Jackson & Harris, 2003). The changing process sometimes generates radical changes. e-business could make organizations flatter, but it increases the need of coordination.

2.3 Organizational culture

As Schein says the culture of a group/organization can be defined as a pattern of shared basic assumptions that was learned by a group as it solved its problems of external adaptation and internal integration, that has worked well enough to be considered valid and, therefore, to be taught to new members as the correct way to perceive, think, and feel in relation to those problems. (Schein, 2004)

To diagnose an organizations culture, Cameron and Quinn suggest the Organizational Culture Assessment Instrument (OCAI). (Cameron & Quinn, 2006) The instrument is in the form of a questionnaire that requires individuals to respond to just six items:

- dominant characteristics;
- organizationals leadership;
- management of employees;
- organization glue;
- strategic emphases;
- criteria of success.

After the scoring and evaluating of the questionnaire, the culture profile of the organization can be drawn. The method of Cameron and Quinn for examining the organizational culture is a well described, easy to fill and evaluate method.

In Fig 1. the four culture types of Cameron and Quinn can be seen. But which one of the types supports the most the use of e-business solutions? Which one suites the most for e-business solutions?

3. MODEL OF THE RESEARCH

As we see, e-business has impact on four dimensions of the organizations: structure, management style, culture and efficiency.

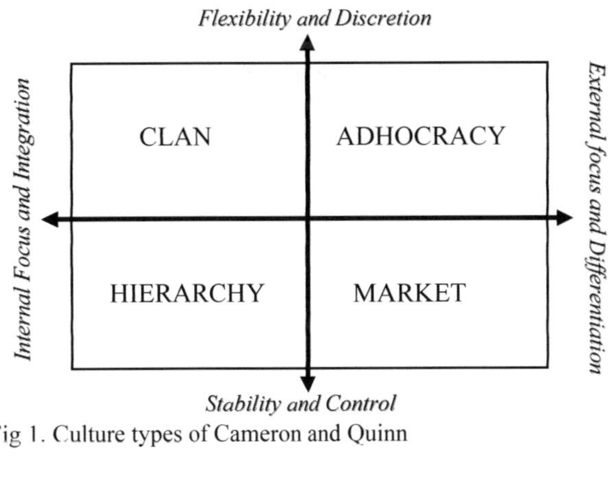

Fig 1. Culture types of Cameron and Quinn

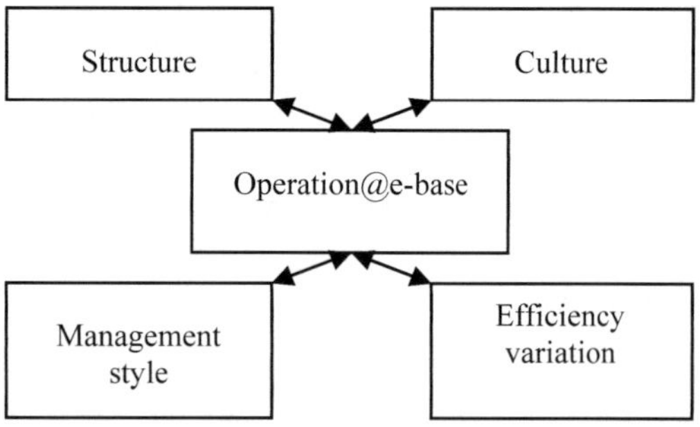

Fig. 2. Model of thoughts

Our hypotheses are:

- the structure of organisations will be flatter due to the adoption of e-Business: virtual organisations will emerge, the number of decision making levels will be lower but therefore the need for coordination will increase.
- The management style of managers will change: their role as a leader will strengthen, they will focus much more on the future of their organisation and less on day-to-day operation.
- The culture of organizations will be more external focused, and the role of flexibility and discretion will be higher.
- The above mentioned changes will raise the efficiency of organisations, due to more efficient decision making processes, quicker customer responses, etc.

The model of thoughts for the research is shown in Fig.2.

4. STANDING OF RESEARCH

To verify our hytothesis we assembled a questionnaire. After expert interviews and literature research, we made up the questionnaire, which is available online at www.gtk.uni-pannon.hu/eBusiness. In order to get the needed variety of datas, an offline version of the questionnaire was formed as well, and posted. The language of the questionnaire is Hungarian.

The aim of the questionnaire is to identify, wether the organization is well "e-supported" or not. We expect, that there are sygnificant differences between highly "e-supported" organizations and organizations, with lower "e-support".

In our target group there are companies operating in Hungary, regardless of size, or industry. In the following of the research we plan to make case studies in western countries (for example in Germany) as well, in order to be able to make comparison of the used best practices.

The preliminary results of the questionnaire show the following.

The majority (about 50%) of the filled questionnaires came from bigger organizations (nr. of employees over 250, the annual revenue over 50 million Euro), the rest is equally shared between small and middle sized organizations.

The majority of those who filled in the questionnaires are middle managers.

Counting the number and maturity of the applied e-business methods and solutions half of the examined organizations can be called "e-supported" organization.

As the result shows, according to the managers' judgement their organization got flatter (the average number of hierarchy levels: 4) due to the use of the e-business solutions.

The managers said that they can arrange their day to day jobs easier with the help of e-business solutions. They spend less time on routine tasks, have much time for "planning the future" – their leader role enforced.

The culture of organizations is not jet to be typed because of the low number of the filled questionnaires (40 pieces, the questionnaire of Cameron and Qiunn is used to identify the organizational culture.).

As the preliminary result shows, the questionnaire is able to fullfill it's task: the above mentioned questions will be answered. We hope, that the increase of the number of the filled in questionnaires will make our results more appropriate.

5. FOLLOWING STEPS OF THE RESEARCH

In the near future we will gather the needed amount of datas to be able to fullfill the planned quantitative analyses.

To get even more proper results, case studies will be created as well. These case studies will represent some well formed e-supported organizations and leaders, who have successfully overcome with the challenges of e-business.

6. SUMMARY

In this paper we summarized the various management aspects of e-business. We showed, e-business could influence the life of organizations, it alters the organizational structure, culture, and the role of managers.

We introduced a possible research model which is capable of examining the changes caused by the implementation of e-business methods and technologies.

The research is on its way, and we hope, that we can show the freshest results at the 2nd European DAAAM International Young Researchers´ and Scientists´ Conference in October, 2008.

7. REFERENCES

Cameron, K.S. & Quinn, R.E. (2006). *Diagnosing and Changing Organizational Culture – Based on the Competing Values Framework, Revised ed.,* Jossey-Bass, A Wiley Imprint, ISBN: 978-0-7879-8283-6, San Francisco

Chaffey, D. (2004). *E-Business end E-commerce Management (2nd. ed.),* Prentice Hall FT, ISBN: 0 273 68378 0, Harlow, England

Jackson, P. – Harris, L. (2003): E-business and organisational change – Reconciling traditional values with business transformation, *Journal of Organizational Change Management,* Vol. 16 No. 5, pp. 497-511, ISSN: 0953-4814

Papazoglou, M.P.; Ribbers, P.M.A. (2006). *e-Business – Organizational and technical foundations,* John Wiley and Sons, ISBN: 978-0-470-84376-5, Chiceshester, England

Schein, E.H. (2004). *Organizational Culture and Leadership, (3rd. ed.),* ISBN: 0-7879-6845-5, Jossey-Bass, A Wiley Imprint, San Francisco

Annals of DAAAM for 2008 & Proceedings of the 19th International DAAAM Symposium, ISSN 1726-9679
Editor B. Katalinic, Published by DAAAM International, Vienna, Austria 2008
Make Harmony Between Technology and Nature, and Your Mind will Fly Free as a Bird

THE INFLUENCE OF WORKING PARAMETERS ON PRECISION IN PROCESSING SHORT, CIRCULAR HONING, PROCESSING OF PIECES MADE OF CAST IRON 200

TABACARU, L[ucian]

Abstract: *This technical paper presents a study concerning the working parameters influence upon precision in processing short, interior, circular honing, processing of pieces made of cast iron 200. The standard of iron 200 is SR ISO 185-94. It was taken into the consideration the honing machine, the honing head, abrasive bars and working material. There are also presented the influence of the longitudinal and tangential speed upon circularity deviation for two different sections.*
Key words: *honing, parameters, precision, processing*

1. INTRODUCTION

One of the main issues of the technical progress in Machines design is the mechanism and device reliability of the machine – tool. In order to satisfy these increased requirements the hole accuracy most be enhanced. Based on this assumption there is a need to improve the machining technology. Honing is a technological process which provides high precision. Thus the study of the influence of working parameters on surface quality will lead to a better accuracy of short interior circular surfaces.

2. TECHNOLOGY USE

Honing is a surface finishing process using a set of abrasive bars or diamonds, interlinked, and assembled in a honing head. The hydraulic pressure ensures the permanent contact between the head and the surface to be machined. There are two main movements of the honing head: spinning at the speed v_t and rectilinear alternative movement at the speed v_L. Appropriate lubrication is used for cooling and oiling. There are several important characteristics of the honing process (Yokoyamak, I., 1983 & Yokoyamak, I., 1989).

- A large surface between the surface and honing bars;
- Low pressure upon the surface to be machined,
- Low cutting speeds.

These characteristics lead to:

- Low surface roughness;
- Increased dimensional accuracy;
- Correction of initial geometrical errors;
- Relative high productivity.

1.1 *The honing machine.* For testing, it was used a single axis, vertical honing machine, model ZG 833 (figure 1) design for diameters range 30 to 165 mm of the pieces which could be assembled on the machine table

1.2 *The honing head.* Figure 2 shows a 6 bars honing head actually used in the research, covering a diameter range between 95 and 110 mm, its main elements are:

1- double cone action rod;
2- honing head;
3- driving bolt;
4- abrasive bar holder;
5- abrasive bar;
6- spring for continuous contact between the bar support and the double cane plunger pin;
7- elastic element for plunger return to initial position during the time when it does not work;

Fig.1. Vertical honing head ZG 833. General assembly.

1.3 *Abrasive bars.* The bars are produced by S.C. ABROM Bîrlad and have the following technical specifications: $22C180J_6V_{18}T_3$, and the size 100x13x13.

1.4 *Working material.* The piece is made of cast iron: 200 SR ISO 185 – 94 – ϕ_{int}= 95 mm., ϕ_{ext}=115 mm, L=140 mm.

Before honing, the piece had been fine turned using the following working parameters: n=63 rev/min. s=0.1 mm/rev., t=0.25 mm tip radius r=2 mm. The roughness obtained after lapping, as an average of 10 units varies R_a=4.83 – 4.63 µm.

1.5 *The working parameters* (Tabacaru, L., 1998 & Tabacaru, L., Pruteanu, O.V., 2004)

a. Tangential speed
Considering the number of rotations/ min available on the honing machine (155, 280, 400 rot/min) and the working diameter three values had been obtained for tangential speed: 46.25, 83.56 and 119.37m/min.

b. Longitudinal speed
There are three speeds available on the honing machine ZG 833 (8,11.5 and 18 m/min).

c. Pressure – can be computed using the equations:
$p = Fn/A = S/A$; $S = Q/ tg (\alpha 1 + \varphi 1) + tg (\alpha 2 + \varphi 2) = 43/ 0,826 = 52$ daN
$p = 52/ 10x1,3 = 53/ 13 = 4$ daN /cm²
For all experiments the pressure had a constant value.

d. Machining time
Machining times used was 1; 2; 3; 4; 5 min.

e. Cooling liquid

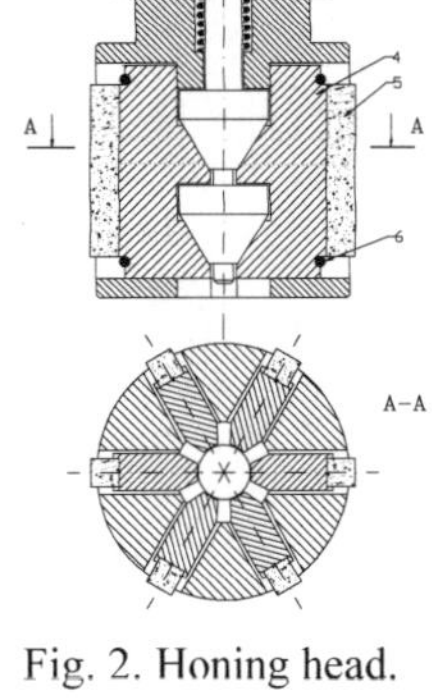

Fig. 2. Honing head.

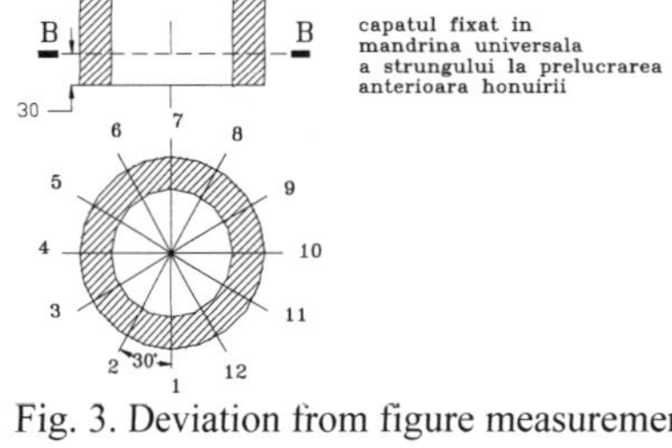

Fig. 3. Deviation from figure measurement.

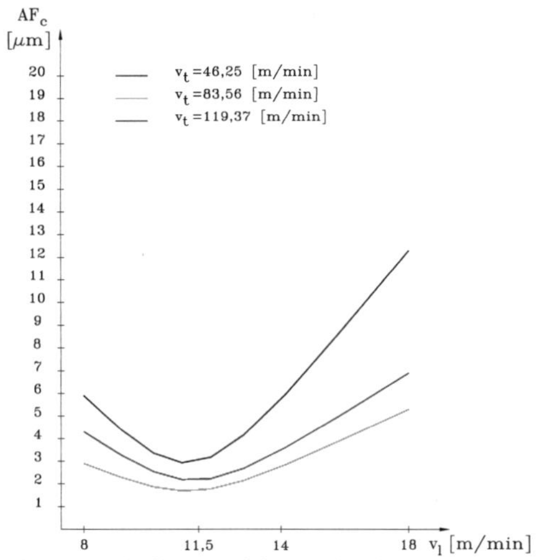

Fig. 4. The influence of the longitudinal speed v_l, upon circularity deviation AF_c, for A-A measurement section, at iron 200 honing.

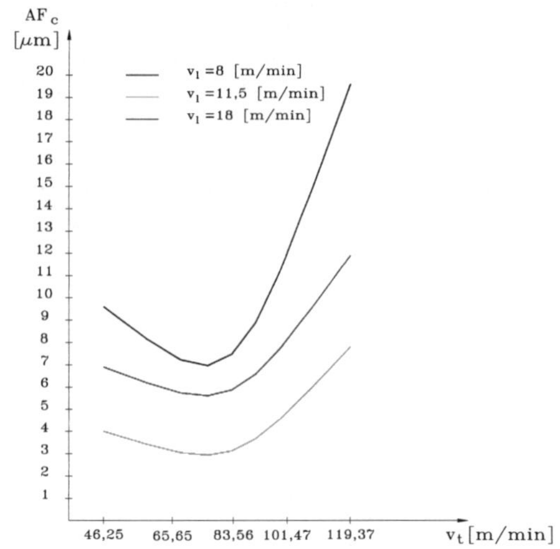

Fig. 5. The influence of tangential speed v_t, upon circularity deviation AF_c, for B-B measurement, at iron 200 honing.

It had been used a mixture of kerosene (60:1) and oil (8:1) in a proportion of 86.6% kerosene.

1.6 *Experimental research on the influence of working parameters on precision in process.*

To study the deviation from circularity AF_c in processing by honing of cylinders using the vertical honing machine (Pruteanu, O.V., Nedelcu, D., 1996)., model ZG 833 (figure 1), it was used a Russian production coordinate
holing machine, model 2D-450 and a measurement machine including three main elements: the palpation element, the dial comparative with a precision of 0,001 mm and the element to catch hold of tool machine. It was used the coordinate holing machine model 2D-450 because it was possible the adjustment on the two directions *0x* and *0y,* with a precision of $1\mu m$, and the superposition of the revolution axis of the piece over the rotation axis of measurement machine.
The measurements were effected at every 30^0, for the two sections A and B (figure 3).
The influences of the longitudinal speed upon circularity in two different sections are presented in figure 4 and 5.

$$AF_C = 30{,}9205 - 0{,}2652574\,V_t + 1{,}816299E\text{-}03\,V_t^2 - 3{,}540013\,V_l + 0{,}1507697\,V_l^2 \quad (1)$$
$$s_{min} = 1{,}300001$$

$$AF_l = 42{,}84639 - 0{,}3685379\,V_t + 2{,}660099E\text{-}03\,V_t^2 - 4{,}887184\,V_l + 0{,}2037378\,V_l^2 \quad (2)$$
$$s_{min} = 2{,}091666$$

3. CONCLUSIONS

1. Referring to the honing operation, the values of longitudinal speed parameter are greatest than the values of tangential speed parameter and reveals that longitudinal speed has a great influence on deviation from circularity and cylindricality than the tangential speed parameter. We could affirm that longitudinal speed influences 13-14 fold the deviations from circularity and 10-11 fold the deviation from cylindricality than the tangential speed.
2. An increase of longitudinal speed leads to a decrease of deviation from circularity to a certain value $V_l = 11{,}5\,\mathrm{m/min}$. An increase over this value leads to an increase of deviations from circularity.
3. An increase of tangential speed leads to a slowly decrease of deviations from circularity values to the value $V_t = 83{,}56\,\mathrm{m/min}$ A continuous increase of tangential speed determines a faster increase of deviations from circularity.
4. The lowest values of the deviations from circularity were obtained for the following values of the working parameters: tangential speed: $V_t = 83{,}56\,\mathrm{m/min}$ Longitudinal speed: $V_l = 11{,}5\,\mathrm{m/min}$ The minimal deviation from circularity: $AF_c = 1{,}4 \div 2\,\mu m$

On the future I want to develop this research for others metals like steel with industrial implementation on machine manufacturing.

5. REFERENCES

Pruteanu, O.V., Nedelcu, D., (1996). *The stress analysis on deformation roll*, Proceedings of 7th International DAAAM Symposium, Katalinik, B. (Ed.), pp. 361-362, ISBN 3-901509-02-X, Vienna, Austria, October 1996, Vienna

Tabacaru, L., (1998) *Studiul comparative intre honuirea si vibrohonuirea suprafetelor interioare de revolutie, scurte, The comparative study between the honing and vibro-honing of interior surfaces, short,* Ph.D. Thesis, Iasi

Tabacaru, L., Pruteanu, O.V., (2004), *The influences of cutiing speed on the efficency of superfinishing,* Bulletin of Politechnical Institute, Fasc. 1-2, pp. 34-38, Iasi

Yokoyamak, I., (1983*) Analyses of Thermal Deformation of Workpiece in Honing Process (2nd Report)* "Bulletin of the Japan Society of Precision Engineering" V17, No 4, pp. 56-60, Publishing Japan Society, Japan

Yokoyamak, I., (1989) *Analyses of Thermal Deformation of Workpiece in Honing Process (5th Report) "Bulletin of the* Japan Society of Precision Engineering", V22, No 1, pp. 60-64, Publishing Japan Society, Japan

Annals of DAAAM for 2008 & Proceedings of the 19th International DAAAM Symposium, ISSN 1726-9679
Editor B. Katalinic, Published by DAAAM International, Vienna, Austria 2008
Make Harmony Between Technology and Nature, and Your Mind will Fly Free as a Bird

THE INFLUENCE OF WORKING PARAMETERS ON PRECISION IN PROCESSING BY CIRCULAR VIBRO-HONING, PROCESSING OF PIECES MADE OF CAST IRON 200

TABACARU, L[ucian] & NEDELCU, D[umitru]

Abstract: *This paper presents a point of view concerning the influence of working parameters on precision in processing in vibro-honing of circular short surfaces for pieces made of cast iron 200 (SR ISO 185-94). Based on the analysis of experimental values, the authors have obtained the relationship* $AF_c=f(v_t,v_l,v_0)$ *and* $AF_l=f(v_t, v_l, v_0)$*. There is also presented the influence of longitudinal speed upon circularity deviation for two different oscilatorie speed.*
Key words: *vibro-honing, surface, roughness, parameters.*

1. INTRODUCTION

Vibro-honing is a technological finishing which uses a set of abrasive or diamond bars interlinked, mounted in a vibro-honing head. This device (the head) ensures continuous contact between the set of bars and the piece surface. It has several working speeds (equation 1).
Two components of the cutting speed, the longitudinal speed v_l and the tangential speed v_t;
A high frequency oscillatory movement, of low amplitude, with oscillatory speed v_0. This supplementary movement could be either the piece or the vibro-honing head.
Equation (1) represents the optimal ratio which should exist between the three working speeds (Yokoyamak.,1983,1989).

$$v_t : v_o : v_l = 3 : 2 : 1 \quad (1)$$

2. TECHNOLOGY USE

During the experiments the high oscillations were made by the piece. Figure 1 depicts the mechanical device designed and used during the experiments. Its functional scheme is represented in Figure 2.
The permanent contact between the pick-up roll and frontal cam 3 is made through a set of spiral springs.
A change in the frontal cam speed will modify the frequency. The amplitude could be changed by mounting or descending the roll on the cam and can be set through a radial movement of the pick-up device.
Figure 2 presents the formulae for the calculus of oscillatory speed.

$$v_o = a\omega = a\pi n_{camă}/30 = A\pi n_{camă}/60 \quad [m/min.] \quad (2)$$

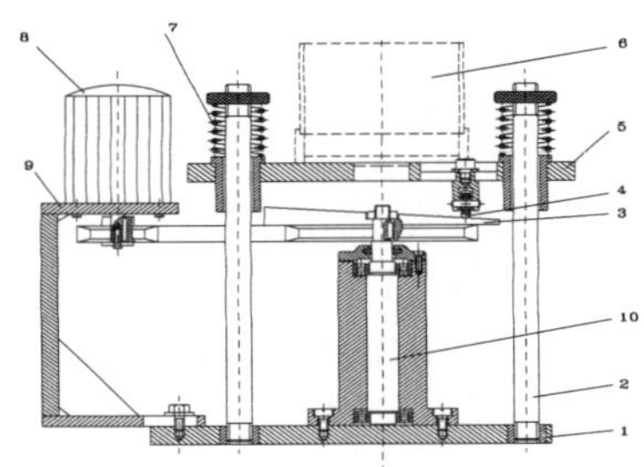

Fig.1. Oscilatory mechanical device.

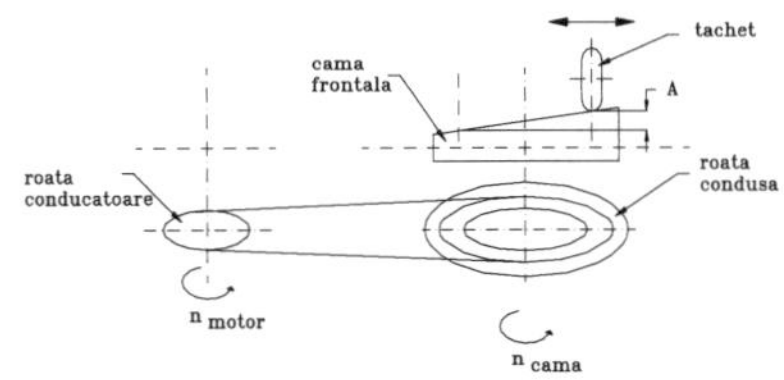

Fig. 2. Functional design of the oscillatory mechanical device.

1.1 *The honing machine.* For testing it, we used a single axis, vertical honing machine, model ZG 833 (figure 3) design for diameters range 30 to 165 mm of the pieces which could be assembled on the machine table
1.2 *The honing head.* Figure 4 shows a 6 bars honing head actually used in the research, covering a diameter range between 95 and 110 mm; its main elements are:
1.3 *Abrasive bars.* The bars were produced by S.C. ABROM Bîrlad and have the following technical specifications: $22C180J_6V_{18}T_3$, and the size 100x13x13.
1.4 *Working material.* The piece is made of cast iron: 200 SR ISO 1083 – 93 – ϕ_{int}= 95 mm., ϕ_{ext}=115mm, L=140 mm.

Before vibro-honing, the piece had been fine turned using the following working parameters: n=63 rev/min. s=0.1 mm/rev., t=0.25 mm tip radius r=2 mm. The roughnesses obtained after lapping, as an average of 10 measurements varies R_a=4.83 – 4.63 μm.
1.5 *The working parameters* (Tabacaru, 1998, Tabacaru & Pruteanu, 2004).
a. Tangential speed

Fig.3. Vertical honing head ZG 833. General view.

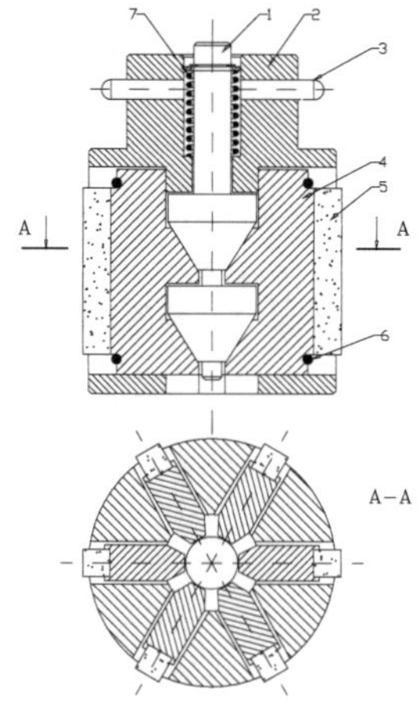

Fig. 4. Honing head.

Considering the number of rotations/ min available on the honing machine (155, 280, 400 rot/min) and the working diameter, there had been obtained three values for tangential speed: 46.25, 83.56 and 119.37m/min.

b. Longitudinal speed

There are three speeds available on the honing machine ZG 833 (8,11.5 and 18 m/min).

c. Oscillatory speed. During the research there had been used the following values for oscillatory speed: v_0 = 30.84; 55.70; 79.58 [m/min]

d.Pressure – can be computed using the equations:
p=Fn/A=S/A;

S=Q/tg(α1+φ1)+tg(α2+φ2)=43/0,826=52daN,
p=52/10x1,3=53/13=4 daN/cm²

For all experiments the pressure had a constant value.

d. Machining time
Machining times used was 1; 2; 3; 4; 5 min.

e. Cooling liquid
A mixture of kerosene (60 l) and oil (8 l) in a proportion of 86.6% kerosene was used.

3. **Data analysis.** In order to obtain a general equation $AF_c=f(v_t, v_l, v_0)$ and $AF_l=f(v_t, v_l, v_0)$ a computer software was developed using TURBO PASCAL. The following equation was found using the software (Pruteanu & Nedelcu, 1996):

$$AF_C = 8,126581 \square 2,015869E-02\,V_t - 6,811509E-05\,V_t^2 - 1,304274V_l \square 0,060 - 2,40585E-03V_o \square 2,602441E-04V_o^2 \quad (3)$$

$$s_{min} = 0,5933334$$

$$AF_l = 9,030414 \square 2,290881E-02\,V_t \square 2,081946\,E\text{-}05\,V_t^2 - 1,570633V_l \square 7,175521E-02V_l^2 \square 2,365186E-02V_o \square 6,761353E-05V_o^2 \quad (4)$$

$$s_{min} = 0,8386108$$

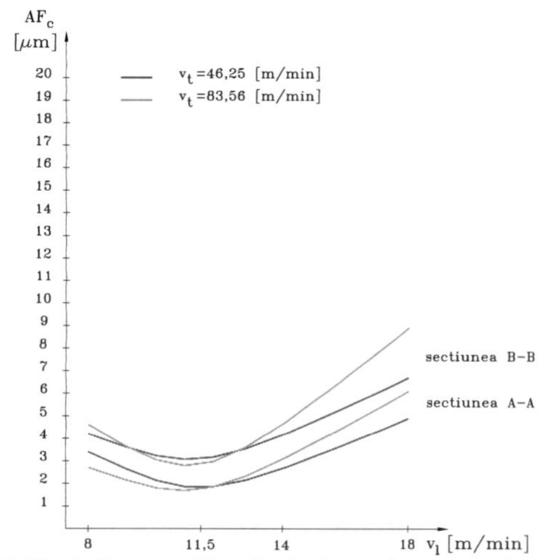

Fig. 5. The influence of longitudinal speed v_l, upon circularity deviation AF_c, for oscillatory speed v_o=30,84 m/min, at vibro-honing of iron 200.

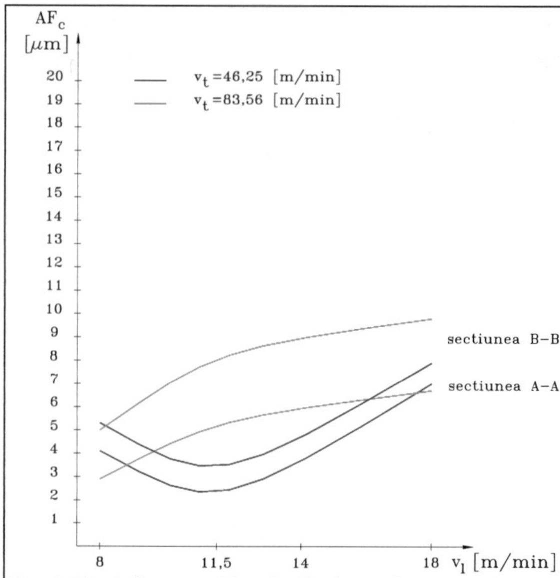

Fig. 6. The influence of longitudinal speed v_l, upon circularity deviation AF_c, for oscillatory speed v_o=79,58 m/min, at vibro-honing of iron 200.

On the future we want to develop this research for others metals like steel with industrial implementation on machine manufacturing.

3. CONCLUSIONS

The longitudinal speed and the tangential speed influence the circularity deviation the same way the honing process does for small speed values. For oscillatory big values of speed and tangential big values of tangential speed, the curve is changeable, the increasing of the longitudinal speed leads to increasing the circularity deviation. Around the values of longitudinal speed about Vl=11,5m/min when the deviation value remains constant.

The smallest values of circularity deviation were obtained taking into account the following parameters: tangential speed 83,56 m/min; longitudinal speed 11,5 m/min; oscilatory speed 30,84m/min; the minimum value of circularity deviation obtained (1,7-2) μm.

4. REFERENCES

Pruteanu, O.V., Nedelcu, D., (1996). *The stress analysis on deformation roll*, Proceedings of 7th International DAAAM Symposium, Katalinik, B. (Ed.), pp. 361-362, ISBN 3-901509-02-X, Vienna, Austria, October 1996, Vienna

Tabacaru, L., (1998) *Studiul comparative intre honuirea si vibrohonuirea suprafetelor interioare de revolutie, scurte, The comparative study between the honing and vibro-honing of interior surfaces, short,* Ph.D. Thesis, Iasi

Tabacaru, L., Pruteanu, O.V., (2004), *The influences of cutiing speed on the efficency of superfinishing,* Bulletin of Politechnical Institute, Fasc. 1-2, pp. 34-38, Iasi

Yokoyamak, I., (1983*) Analyses of Thermal Deformation of Workpiece in Honing Process (2nd Report)* "Bulletin of the Japan Society of Precision Engineering" V17, No 4, pp. 56-60, Publishing Japan Society, Japan

Yokoyamak, I., (1989) *Analyses of Thermal Deformation of Workpiece in Honing Process (5th Report) "Bulletin of the* Japan Society of Precision Engineering", V22, No 1, pp. 60-64, Publishing Japan Society, Japan

Annals of DAAAM for 2008 & Proceedings of the 19th International DAAAM Symposium, ISSN 1726-9679
Editor B. Katalinic, Published by DAAAM International, Vienna, Austria 2008
Make Harmony Between Technology and Nature, and Your Mind will Fly Free as a Bird

ENHANCED HYBRID COMPOSITE FOR ADVANCED AEROSPACE APPLICATIONS

TACHE, F[lorin]; DOBRE, T[anase] & CHIRILUS, A[lina] A[lexandra]

Abstract: *Current research thoroughly investigates the design, production and testing of a new composite material that combines the properties of homogeneous materials with those of carbon fibers, thus resulting a lightweight structure with both mechanical and thermal resistance while having a price comparable to that of existing products and complying with all applicable environmental requirements. CFRAl has proven great potential so far, consequently giving reason for continuing the work started within the last two years at UPB.*
Key words: *carbon fibers, explosion, bullet-proof, composite.*

1. INTRODUCTION

In the last few decades, composite materials have been used more and more often wherever lightweight, yet strong structures are needed (Beukers & van Hinte, 2005). There are numerous types of composite materials, differing by type itself, production method, uses etc. (Beukers, 2005). Regardless of type, composites have induced a major change in the classical way of parts design, mainly due to their heterogeneity and anisotropy (Zgura & Moga, 1999).

The material being analysed in this paper, namely *CFRAl* (Carbon Fibers Reinforced Aluminum), involves reinforcing an aluminum matrix with carbon fibers. In order to make the two constituents work properly with one another, a fundamental issue must be surpassed, which is the electro-chemical incompatibility between aluminum and carbon. A handy solution has been employed with respect to this matter. Even in the early stages of a composite material development, its final destination must be considered. Aspects like what kind of loads will act on the composite structure, their order of magnitude etc. should be taken into account and thoroughly examined. The material considered herein is supposed to withstand the impact of small-dimension high-velocity objects and to have a good behaviour in the event of an explosion in its vicinity. Practical examples regarding its application would be bulletproof plates and any kind of containers that would successfully endure the explosion of a bomb inside them.

2. MAKING CFRAl

For making the material, two types of carbon fibers fabrics have been chosen. They differentiate from one another by having different fiber orientations, $0/90^o$ and $\pm 45^o$, and by fabrics type, *plain weave* and *satin weave* (Wikipedia, 2008). By combining the two, a better resistance to an explosion is conferred to the composite material, the carbon fibers actually covering four different orientation directions. The matrix involved herein consists of an epoxy resin having aluminum powder inside it. The metal particles are completely enveloped by the resin, thus avoiding a direct contact between carbon and aluminum, and leading to a delamination-free hybrid composite material that benefits from all of its constituents properties. And by a carefully studied manufacturing technology, CFRAl develops properties well above the sum of its constituents properties. It takes tenacity from aluminum, strength from carbon fibers and ease of making from the epoxy resin.

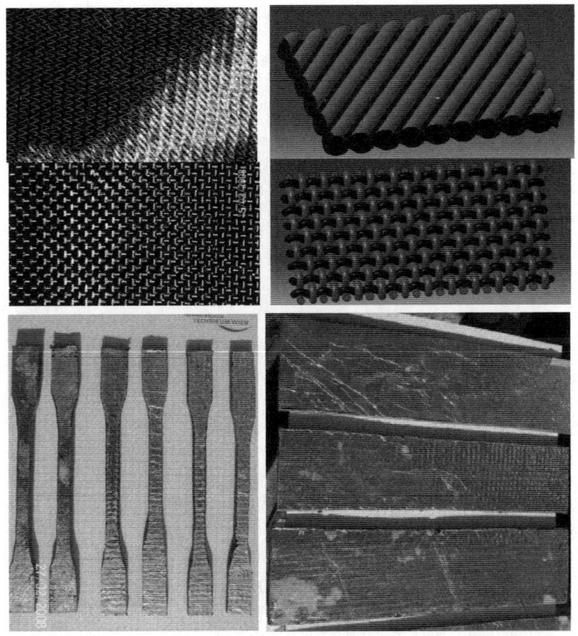

Fig. 1. CFRAl – from raw materials to finite test samples

The images in figure 1 show the two different types of carbon fibers fabrics used in making CFRAl (real material and symbolically depicted) and different test specimens for conducting standard tensile and bullet impact tests.

3. MECHANICAL PROPERTIES ASSESSMENT

Having in mind CFRAl's aforementioned purpose, a proper way to evaluate its behaviour in the event of an explosion is to consider a spherical container made of CFRAl, with a high pressure gradient in its center. In this model, the thermal and short fragments impact effects are neglected and will be dealt with later. Hence, figure 2 shows a symbolic representation of the forces being developed and the tensions induced by these forces within the container wall. Considering a thin wall with respect to the sphere radius, the only true loads acting on it are tension loads uniform in all directions, called *membrane tension loads* (Gere, 2002).

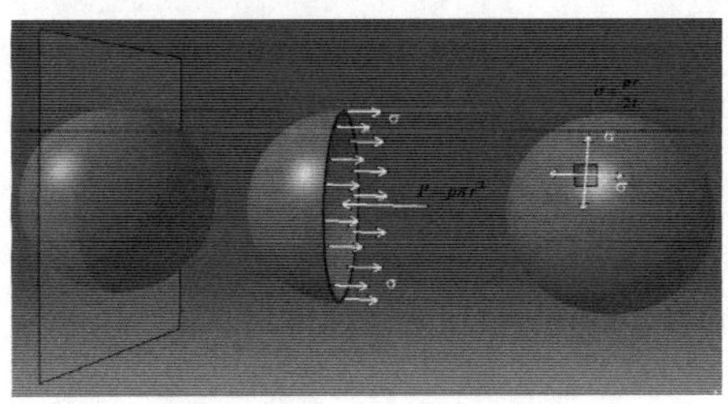

Fig. 2. Forces acting on the sphere wall

The tension load within the sphere wall, acting in all directions, is:

$$\sigma = \frac{p\,r}{2t}, \qquad (1)$$

where p is the uniform pressure generated by the explosion, r is the sphere radius and t the wall thickness.

Extrapolating and generalizing the model, by considering a sphere with a very long radius, the spherical shape can be approximated with a planar surface. Consequently, a good indication of CFRAl resistance to an explosion is its tensile strength, much easier to be evaluated by conducting standard tensile tests. Several tensile test specimens have been tested and the results will be thoroughly analysed. Preliminary data reveal tensile strength of a 10-layer CFRAl plate above *300 MPa*.

With an excellent behaviour of carbon fibers at high temperatures, only one aspect remains to be assessed, namely the resistance to short fragments impact during the explosion. For this, bullet impact tests have been and are to be conducted on different CFRAl plates.

Ballistic tests have revealed an outstanding CFRAl resistance against a *7.65 mm* calibre bullet shot from a standard 10-metre distance (figure 3) and room for improvement when it comes to a *9 mm* calibre bullet. For the latter one, new samples have been prepared, being made up of 14 carbon fibers layers and arranged in such a manner that the shock waves produced by the impact with the speeding bullet are better dissipated in the entire composite mass. Later tests with these new improved samples have also proven excellent resistance to a *9 mm* calibre bullet. As it can be seen from the picture above, the bullet only affects the first layers it encounters in its path and, as it penetrates them, it eventually breaks apart inside CFRAl, leaving the last layers intact. The picture on the right shows the last 4 layers with no delamination, similar behaviour being observed in the *9 mm* test as well, but with a CFRAl plate made of 14 carbon fibers layers.

While the thickness of the material has grown, its density has been actually reduced, now being roughly *1650 kg/m³*, by adopting a higher volume fraction of the fibers (above *50%*) inside the composite material. Nevertheless, a cross-section view of a CFRAl specimen, magnified 200 times using a BX 51 Olympus microscope (figure 4), shows the matrix completely enveloping the carbon fibers.

Fig. 3. *7.65 mm* bullet stopped by a 10-layer CFRAl plate

Fig. 4. CFRAl cross-section views (200 x magnification rate)

4. FINANCIAL ASPECTS REGARDING CFRAl

It is a long way from the raw materials (resin and fibers) to the 10-layer test specimen shown in figure 5, employing resources which are not always easy to quantify. Some of these are: materials for the moulds, screws and nuts, tools for the lay-up process and preparation for curing in the oven, cutting gross samples into standard test specimens, time etc.

Considering that at least some of the abovementioned resources costs would be reduced in an eventual CFRAl mass production, the price of this material is given below, for different lay-ups, in Euros per mass unit and per area unit.

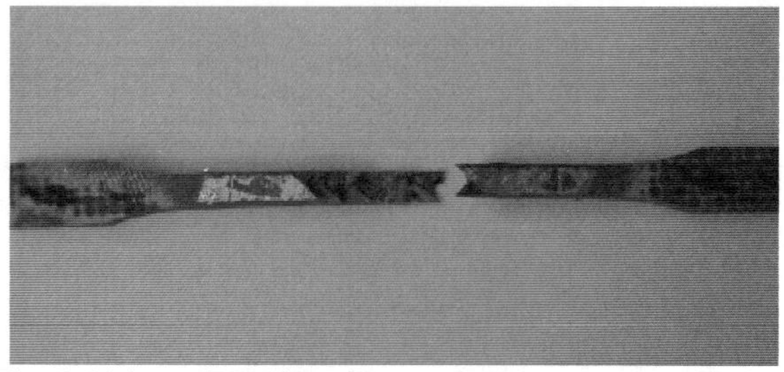
Fig. 5. 10-layer CFRAl tensile test specimen

CFRAl type	Price [Euro/kg]	Price [Euro/m^2]
CFRAl 5 x ±45°	35.98	331.84
CFRAl 5 x 0/90°	30.85	289.83
CFRAl 10 x ±45°	52.22	536.84
CFRAl 10 x 0/90°	43.32	452.80
CFRAl 5 x ±45° + 5 x 0/90°	36.09	494.82

Table 1. CFRAl price for different lay-ups

5. CONCLUSIONS

Though it is a daring challenge, Carbon Fibers Reinforced Aluminum is a new type of composite material being developed by the Chemical Engineering Department in cooperation with the Faculty of Aerospace Engineering at University 'POLITEHNICA' of Bucharest, Romania. The research so far shows promising results, CFRAl having foreseeable excellent behaviour in applications requiring a reliable, yet lightweight, reinforcement material that can also withstand powerful thermal shocks. Current study represents a continuation of the work started within the last two years at UPB and is an integral part of the first author's PhD study.

6. REFERENCES

Beukers, A. & van Hinte, Ed. (2005). *Flying Lightness*, 010 Publishers, ISBN 90-6450-538-1, Rotterdam

Beukers, A. (2005). *Engineering with Composites*, Lecture Notes, Delft University of Technology, Faculty of Aerospace Engineering

Gere, J. M. (2002). *Mechanics of Materials*, 5th SI Edition, Nelson Thornes Ltd., ISBN 0-7487-6675-8, Cheltenham, UK,

Zgură Gh., Moga V. (1999). *Bazele proiectării materialelor compozite (Basics of Composite Materials Design)*, Editura Bren, ISBN 973-9493-01-7, Bucuresti.

***http://en.wikipedia.org/wiki/Plain_weave, Plain weave – Wikipedia, the free encyclopedia, accessed 2008-06-03

Annals of DAAAM for 2008 & Proceedings of the 19th International DAAAM Symposium, ISSN 1726-9679
Editor B. Katalinic, Published by DAAAM International, Vienna, Austria 2008
Make Harmony Between Technology and Nature, and Your Mind will Fly Free as a Bird

COMPOSITES FUEL TANK TECHNOLOGY FOR THE NERVA LAUNCHER

TACHE, F[lorin]; SILIVESTRU, V[alentin]; DOBRE, T[anase] & RUGESCU, R[adu] D[an]

Abstract: *The Romanian project NERVA for a small size space launcher is considered for using improved technologies of that type for the manufacturing of structural elements and especially of the fuel and oxidizer tanks. Problems that stand for the basis of the research are related to the unusual use of this technology to fit with cryogenic liquids of liquid oxygen type. As a modeling environment, the liquid nitrogen was used and physical-mechanical properties of the composites at low temperatures were investigated. It was shown that, under such conditions, a good behavior may be accomplished with a technology developed by the U. P. B. team.*
Key words: *composites, rockets, fuel, tanks*

1. INTRODUCTION

The NERVA is a typical Romanian project in the family of small orbital launch vehicles (Rugescu, 2008), largely considered today as an alternative to tenders on orbital launch capabilities from Russia. They answer the inability of reusable systems, like the space shuttle, to acquire a low cost capability of launching scientific and commercial 1payloads in low Earth orbit (LEO). Many other similar projects like the Spanish AQUARIUS air-launch project (Simon, 2006) are known. Some private companies with the capability of developing cheep space transporters are raising in this market, to only mention the international "Sea Launch" (Intl. Launch Services, 2005), the US "SpaceX" (SpaceX, 2006), "Orbital" & "t/Space" or the Russian "Air Launch Corporation" (Sarigul-Klijn et al., 2006).

The recent successes of University "Politehnica" of Bucharest (UPB) into FP7 European Space priority projects ORPHEE and NANOPROP bolsters the development of the *NERVA* launcher for *PUBSAT* nanosatellite (Rugescu *et al.*, 2008). NERVA will be a readily available, low cost rocket launcher for achieving half of the local orbital velocity at 100 miles altitude. Enhanced with a high efficiency, dual mode combustion third stage, it will be able to inject the three-axes controlled (3AC) PUBSAT nano-satellite into LEO. The rocket transporter is accessible to Romanian current aerospace technology. The foreground of NERVA is the military soil-air SA-2 *Guideline* weapon, now obsolete in Romania, subjected to a precise re-conversion to scientific application. The source weapon is a derivation of the famous *Rheintochter* missile, successfully developed at Peenemunde in the early 1945 by the team of Wernher von Braun. The new NERVA is a peaceful renovation of that brilliant work, transforming the soil-to-air missile into a high performance orbital vehicle, perhaps the first orbital project developed from such a low performance, conventional rocket system.

2. COMPOSITE TANKS TECHNOLOGY

A desired improvement is the composite tanks technology, aimed to add further confidence in this project. The upper stage with 5000 m/s ideal velocity performance is at the limit of the current Romanian rocket technology. All align into building a cheap, ground launch vehicle by minimally modifying the basic SA-2 rocket system. In order to accommodate the high thrust

Container	σ
Hen's egg	8.9
Railway oil car	13.0
Coke bottle, half gallon (2 liters)	33.5

Table 1. Natural and ordinary containers

enhancement of the solid motor (SM) booster, thrust vectorization of the liquid motor (LM) and lightweight structure for the second stage, with highly extended propellant tanks and lightweight guidance, a drastic improvement of the structural efficiency of the SA-2 Guideline system is however required.

Very light structures are required for the launch platforms that mainly depend on the manufacturing technology of the fuel tanks, as more then 90% of the vehicle mass concerns the consumable propellant. The tank structure must be as light as possible to allow a convenient payload to be added for injection into the LEO. Estimating the vehicle performance that a good tank technology induces, in terms of mass ratio, only proves the efficiency of a given manufacturing technology,

$$\mu_0 = \frac{M_0}{M_0 - M_f} = \frac{M_0}{M_f} \quad (1)$$

Here, the total launching mass of the orbital rocket M_0 is the sum of the structure and payload mass M_f and the propellant content M_p. The same mass ratio is also defined for the liquid propellant tanks alone and is consequently called the *structural factor* σ of the tanks,

$$\varepsilon = \frac{M_t + M_p}{M_t} \quad (2)$$

It is suggestive to compare the technological figures of the structural factor with some natural or well-known liquid containers (Table 1), from where the limits of the present technology are clearly emphasized.

It is really impressive that the structural factor 28.7 of the Shuttle ET lies very close to the limit of a simple Coke bottle. These antipodal structures make profit of the inner pressure stiffening effect, or the *balloon* effect. The Shuttle ET must also withstand high concentrated loads from the two solid boosters, from the main propulsion system of the orbiter and the hydrostatic pressure from the large mass of liquid propellant during the boost. The solution was found to distribute the loads, which are incoming through the dual strut joints (or *bipod fitting*) on the rear of the tank, along the thin structure itself (Fig. 1). The two bolt catchers fixed to the forward, or top area of the External Tank at the Solid Rocket Booster/External Tank forward attach point are only provisioned to prop laterally the SRB-s, not transmitting any axial thrust (Nemeth et al., 1996).

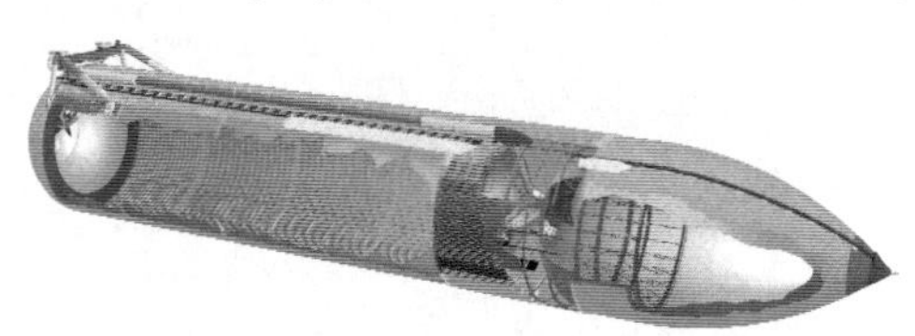

Fig. 1. Shuttle external tank with bipod fitting (left, up)

For the NERVA vehicle no such problems of uneven loads distribution into the tank walls are present, because almost entire loading goes into the longitudinal direction. Only small dynamic loads due to pitch and conning motions of the rocket in response to commands from the autopilot, inducing some sloshing in the propellant, create unimportant asymmetry in the axial loading, far below their level into the ET structure however. The knowledge and management of the bending moments and transverse forces during the NERVA ascent are still a main part of the design study to cover all unexpected occurrences.

3. COMPOSITES TECHNOLOGY ASSESSMENT

Extensive use of magnesium alloys matrix is envisaged, in combination with either boron fibers or carbon fibers fabric (Volkova, 2006). The following criteria were used to assess the technology of propellant tanks:

- preserve fairly unchanged the basic second stage diameter;
- preserve the second stage liquid engine construction equally unchanged;
- accept a sensible extension in combustion time of the second stage liquid motor;
- allow for a sharp lengthening of propellant tanks of the second stage;
- remove all military equipment and replace it with extra tankage, engine gimbals bearing and a much smaller navigation hardware;
- relocate the pressure gas vessel in front of the propellant tank.

One possible candidate solution for the NERVA launcher fuel tanks is a new composite material being developed within UPB. Carbon Fibers Reinforced Aluminum (CFRAl) is a new type of composite material being developed by the Chemical Engineering Department in cooperation with the Faculty of Aerospace Engineering at University 'POLITEHNICA' of Bucharest, Romania. The research so far shows promising results, CFRAl having foreseeable excellent behavior in applications requiring a reliable, yet lightweight, reinforcement material that can also withstand powerful thermal shocks without excessive load concentration.

Consequently the material would have to possess good mechanical properties regarding, but not limited to, its behaviour under internal pressure loads and cryogenic temperatures of the propellant components (liquid oxygen or LOX at -183°C and liquid methane or LME -162°C).

Internal pressure is equivalent, considering a thin wall classical problem, with tension loads acting isotropic in all directions into the wall. To experimentally proof this assumption, standard tensile tests have been performed, revealing that *CFRAl* has a very good specific strength (ratio of tensile strength over specific weight), as compared to other homogeneous or composite materials (Tache *et al*, 2007). *CFRAl* fibres are used for tank construction, with expected good cryogenic resistance. Extensive tests are still under development, with emphasis on load concentrators created at sharp profile changes, around openings in the structure and around other discontinuous geometries of the tank shell.

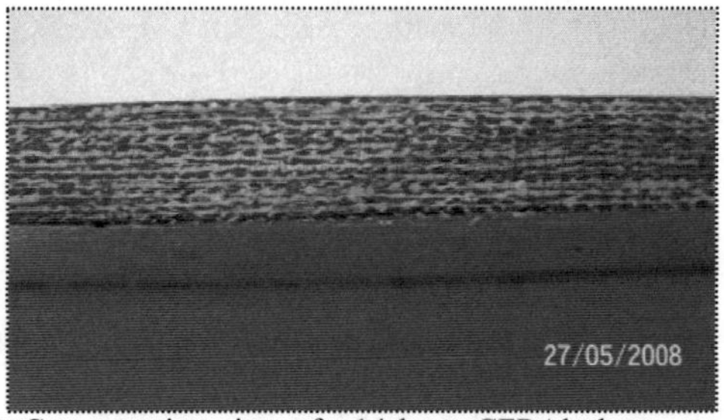

Fig. 2. Cross-section view of a 14-layer CFRAl plate

Fig. 3. A 10-layer CFRAl plate subjected to bullet impact tests

As an extension, ballistic tests have proved that *CFRAl* is bullet-proof with respect to a *7.65 mm* calibre bullet shot from a 10-metre distance. It needs some improvement when faced with a *9 mm* calibre bullet, shot from similar distance. For the latter one, new samples made up of 14 carbon fibre layers and arranged in such a way that the impact shock waves are better supported by the entire composite mass are prepared.

4. CONCLUSIONS

The development of reliable technologies for composite tanks for cryogenic propellants is still incipient, on the international scale, due to low temperature cracks and difficult manufacturing of fittings for plumbing. Recent advances in some new magnesium alloys have re-launched this technology for lightweight magnesium-matrix composites.

For the specific application into the NERVA small orbital vehicle, the prediction of the available performance is performed by computer simulations, proving the feasibility of the NERVA vehicle as an orbital system when lightweight propellant tanks are used. Investigations have shown up to now that for storable propellants the MMC formulation with magnesium matrix is highly promising. Nevertheless, a lot of development is required for achieving acceptable reliability and load capacity within the end joints of the tanks.

The target of the described research is to advance in improving of *CFRAl* material up to potential magnesium or beryllium alloys, continuing the work of the first author's PhD study, started at UPB in the last two years.

5. REFERENCES

International Launch Services Press Release (2005), Tuesday, September 6

Nemeth, M. P., Britt, V. O., Collins, T. J. & Starnes, J. H., Jr. (1996), *Nonlinear Analysis of the Space Shuttle Superlightweight External Fuel Tank*, NASA TP-3616

Rugescu, R. D. (2008), *NERVA Vehicles, Romania's Access to Space*, Scientific Bulletin of U. P. B., Series D in Mechanics, 70, no. 3, p. 31-44

Rugescu, R. D., Predoiu, I. & Aldea, S. (2008), *PUBSAT and NERVA Launcher Fuel Sloshing Dynamics*, Proceedings of ICAI-2008 Conference, June 24-26, Bucharest, Romania, p. 200-203

Sarigul-Klijn, M., Sarigul-Klijn, N., Morgan, B., Tighe, J., Leon, A., Hudson, G., McKinney, B. & Gump, D., *Flight Testing of a New Air Launch Method for Safely Launching Personnel and Cargo into LEO*, AIAA-2006-1040

Simon, J. (2006), *The Aquarius*, a proposal for a nano-satellites launcher vehicle, INTA, Torrej´on de Ardoz, Paper IAC-06-B5.5.05, Proceedings of the 57th IAC of the IAF, IAA and IISL, Valencia, Spain, Oct. 02-06

SpaceX (2006), *Excellent Engineers Wanted*, Aviation Week and Space Technology, Sept. 25, p. 67

Tache, F., Stanciu, V., Chiciudean, T.G., Toma, A.C., Stoica, A., Dobre, T. & Fuiorea, I. (2006), IAC-06-C2.8.04, 57th IAC, Valencia, Spain

Volkova, E. F. (2006), *Modern magnesium-base deformable alloys and composite materials (a review)*, Metal Science and Heat Treatment, Springer, New York, Vol. 48, No. 11-12/Nov, DOI 10.1007/s11041-006-0120-0, pg. 473-478.

Annals of DAAAM for 2008 & Proceedings of the 19th International DAAAM Symposium, ISSN 1726-9679
Editor B. Katalinic, Published by DAAAM International, Vienna, Austria 2008
Make Harmony Between Technology and Nature, and Your Mind will Fly Free as a Bird
2nd DAAAM-ICAT

RAPID MANUFACTURING OF BIOCERAMIC IMPLANTS BY DIRECT WRITING

TALLIS, A[ndrea]; JONES, J[ason] & WIMPENNY, D[avid]

Abstract: *A new process for the manufacture of customised implants for replacing damaged or diseased bone is being developed at DMU under a UK Government funded project – TSB Direct Writing of Bioceramic Implants. The new process is based on the extrusion of a fine bead of colloidal slurry which is then rapidly cooled to below -40°C to initiate an irreversible freeze casting reaction. When the slurry is returned to room temperature it remains a brittle solid which can be fired at elevated temperatures to generate the required mechanical properties. By sequentially depositing layers of material it is possible to generate a complex 3D object which has an internal lattice structure which enables rapid integration of the implant with the surrounding natural bone to produce a secure, long lasting repair. In this paper the results extrusion and freeze casting trials will be presented and discussed.*

Key words: *Direct Writing, Freeze Casting, Shear Thinning, Extrusion*

1. INTRODUCTION

The complexity of geometry found in the maxillofacial area means that off the shelf prosthetics are an impractical solution to meeting the needs of surgeons when tackling bone loss in the skull. A UK government funded research project seeks to create a viable process for manufacturing custom made ceramic based prostheses for the maxillofacial region using the rapid prototyping technique of Direct Writing, sometimes known as Solid Freeform Fabrication. In this process a ceramic slurry mixture is extruded to create a thin filament of material which can be built up layer by layer and 'freeze cast' by rapid cooling to create a complex 3D geometry specifically designed to match the needs of the patient.

This work seeks to carry out the initial optimisation of the slurry formulation in order to obtain a mix that is capable of both freeze casting and accurate extrusion.

2. LITERATURE

Rapid prototyping of ceramics has previously been confined to Selective Laser Sintering techniques, however the Direct Writing of ceramics has recently received considerable research interest, e.g. Wang and Shaw (2005) looked at using dental porcelains in RP applications to directly fabricate dental implants. Morrissette and Lewis (2000) carried out work on aqueous alumina and PVA gelcasting suspensions, this technique however requires large quantities of organic additives to promote cross linking in the green body making burn out more complex.

Laurie et al (1992) showed how a modified sol-gel processing technique could be used to make bulk castings of ceramic material when an aqueous solution of colloidal silica is used as the liquid phase of the slurry. Rapid freezing of parts causes an irreversible gelling reaction to take place and near net shape castings can be made with reduced levels of both cracking and shrinkage compared with the conventional sol-get processing route. Use of this gelation method would therefore reduce the need for organics, simplifying the slurry mixture.

However, when freeze casting aqueous alumina slurries in bulk the addition of small amounts of glycerol to the slurry mix serves as a cryoprotectant, effectively disrupting the crystalline behaviour of water contained in the slurry (Sofie and Dogan, 2001). As water crystals in the freeze cast body translate into microstructural defects in the dried sample this homogenisation of crystal growth leads to a much more even porosity without the large scale defects that can be associated with the process.

Substrates suitable for 3D scaffolds must exhibit both microporosity to aid cell attachment and an interconnected macroporosity to enable vascularisation to take place (Hench, 1991). The even microporosity generated by the processing conditions of the freeze cast slurry, coupled with the interconnected macroporosity that the rapid prototyping technique allows us to 'write' into the structure makes this an ideal substrate for cell growth and proliferation.

3. METHOD

In the trials conducted at DeMontfort University the ceramic slurry was made from a mixture of mono-modal alumina powder with a sub-micron particle size (CT3000SG, Almatis, Germany) and a sol made up of an aqueous suspension of colloidal silica, a dispersant and glycerol. The slurry was prepared by mixing the sol and powder in a ball mill (Retsch PM100, Germany) to ensure homogeneity before being transferred to either the rheometer or extrusion apparatus. Four different slurry formulations were prepared, A – D, these consisted of identical quantities of powder mixed with varying quantities of sol to give a range of material viscosities.

Trials to determine the viscosity of each of the slurry formulations were carried out using a Bohlin Instruments CVO 120 HR rheometer (Malvern, England) fitted with a 4° / 40 mm cone and plate geometry. Isothermal controlled shear ramps from 0.1/s to 100/s were conducted at 21°C with a solvent trap fitted.

A PVM syringe pump (Sapphire Engineering, UK) fitted to an XYZ robot was used as the extrusion apparatus. The 5 ml extrusion syringe is fitted with a syringe tip with an internal diameter of 900 µm. This system was programmed to write a freestanding triangular structure with wall length of 22 mm composed of 10 layers. Extrusion started with a tip height of 1 mm above the glass substrate and was raised by 1 mm for each subsequent layer. The rate of travel of the extrusion head was set to 900 mm min^{-1} and the syringe piston was set to dispense 5 $mm^3 s^{-1}$.

Immediately after extrusion, samples were sealed in a polypropylene box which was then placed in a bath filled with

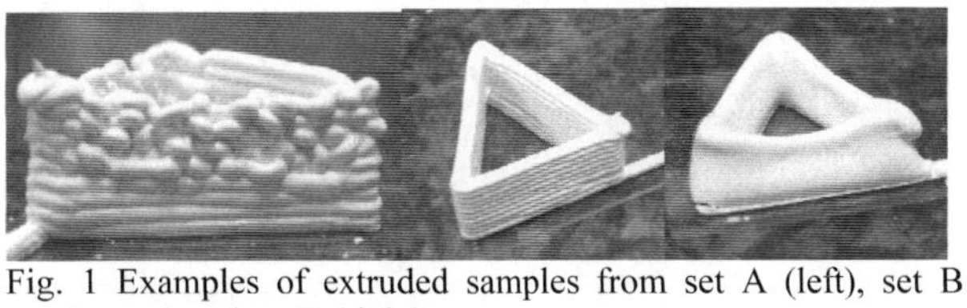

Fig. 1 Examples of extruded samples from set A (left), set B (centre) and set D (right)

Sample Set	Mean (µm)	Standard Deviaiton (µm)
A	10 239	127
B	10 126	169
C	10 269	185

Tab. 1. Mean wall height of 10 layer structures from sample sets A – C

an isopropanol / dry ice slurry with a temperature ~ -70°C. After one hour the samples were removed and allowed to dry for several days on the bench top before measuring.

Samples that did not have uniform extrusion due to trapped air within the syringe were discarded, these only occurred in sample set A. Wall height was measured to the nearest 10 µm on each of the three walls of the samples apart from sample set D where excessive slumping made measurement impossible.

4. RESULTS

Figure 1 shows examples of extruded structures. The left hand image shows a discarded sample of material A where the extrusion has been affected by trapped air within the syringe body, one third of samples constructed from slurry formulation A had to be discarded due to this effect. The central image is from sample set B, formulation C produced similar parts to B. The right hand image is from sample set D where it is possible to see that excessive slumping of the material means that control of part geometry is impossible.

In order to determine the effect of viscosity on the slumping behaviour of the different slurries the height of the ten layer wall was measured with the exception of sample set D where slumping was extreme. The mean and standard deviation of these sample sets is set out in Table 1 above.

The slumping behaviour of slurries is determined by their rheological characteristics. The viscosity of each of the samples A – D was measured whilst the rate of shear they were subjected to was linearly increased. Figure 2 shows a graph plotting sample viscosities as a function of shear rate.

The degree of shear experienced by the slurry as it passes through the extrusion tip can be calculated using

$$\text{Shear rate} = \frac{4Q}{\pi r^2} \qquad (1)$$

where Q is the volume flow rate and r is the radius of the extrusion orifice. This gives us an applied shear rate of 70 s^{-1}, at this level of shear the viscosity range of the formulations is very narrow ranging from 6.8 Pa s for A down to 1.6 Pa s for D.

5. DISCUSSION

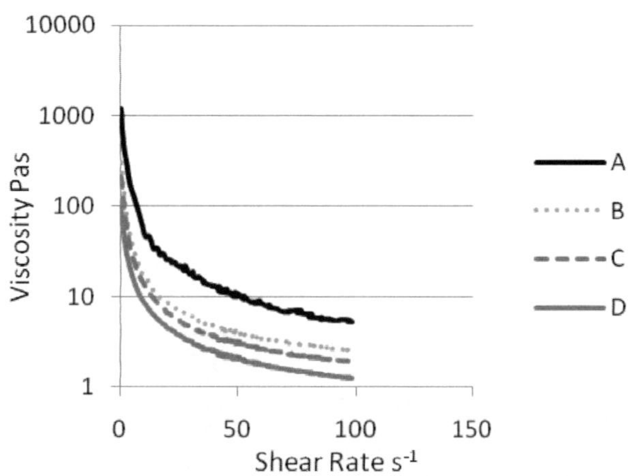

Fig. 2 Viscosity of alumina slurry samples A – D as a function of shear rate

The rheological plots in Fig 2 show that all of the slurry formulations exhibit shear thinning (pseudoplastic) behaviour. This means that as the level of shear that the material is exposed to is increased, its viscosity decreases. This characteristic makes these slurries suitable for extrusion as the action of passing through the syringe orifice subjects the slurry material to considerable shear stresses. Once the shear is removed the viscosity of the slurry increases once again, resisting further flow of the material and allowing the construction of self supporting 3D structures.

The results of this trial show us that there is a viscosity 'window of opportunity', the viscosity of sample set D is insufficient to prevent a degree of slumping that makes precise control of geometry impossible. Whereas the increased viscosity of sample set A means that air pockets can become trapped within the slurry in the syringe body of the extrusion head. As a consequence, when these voids reach the syringe tip no material is extruded until the trapped volume of air has been expelled. This leaves a gap in the wall which on the next layer of the program effectively means that the tip height above the substrate has been increased and is too far above the substrate, allowing the extrudate to distort before settling on the next previous layer. Thus irregular parts like those illustrated in the left hand image of Figure 1 become more likely and part reproducibility decreases. When formulation A extrudes successfully, uniform parts are produced, giving the lowest standard deviation value from each of the sample sets, however the high degree of wastage from this group makes this too an undesirable slurry mix.

Both formulations B and C yield parts with good reproducibility without the problem of trapped air volumes within the syringe. This would indicate that both of these formulations fall within the desirable viscosity 'window' for this process. Further, the slightly lower standard deviation of sample set B would indicate that this formulation is closer to the optimum and should be adopted for further trials.

6. FURTHER RESEARCH

Further work includes sintering these samples to determine whether firing shrinkage is influenced by changes in formulation. In addition, tensile testing of samples made with each of the slurry mixes should be carried out in order to determine whether the strength of the finished artefact is affected by changes in formulation.

7. REFERENCES

Hench, L. (1991) Bioceramics: From Concept to Clinic, *Journal of the American Ceramic Society*, 74, 4, (March 2005) 1487-510, ISSN 1551-2916

Laurie, J, Bagnall, C, Harris, B, Jones, R, Cooke, R, Russell-Floyd, R, Wang, T, Hammett, F. (1992) Colloidal suspensions for the preparation of ceramics by a freeze casting route, *Journal of Non-Crystalline Solids*, 147 & 148, (August 1992) 320-325, ISSN 0022-3093

Morissette, S. & Lewis, J. (2000) Solid Freeform Fabrication of Aqueous Alumina – Poly(vinyl alcohol) Gelcasting Suspensions, *Journal of the American Ceramic Society,* 83, 10, (October 2000) 2409-16, ISSN 1551-2916

Sofie, S. & Dogan, F. (2001) Freeze Casting of Aqueous Alumina Slurries with Glycerol, *Journal of the American Ceramic Society*, 84, 7, (July 2001) 1459-64, ISSN 1551-2916

Wang, J. & Shaw, L. (2005) Rheological and extrusion behaviour of dental porcelain slurries for rapid prototyping applications, *Materials Science and Engineering A*, 397, (April 2005) 314-321, ISSN 0921 5093

Annals of DAAAM for 2008 & Proceedings of the 19th International DAAAM Symposium, ISSN 1726-9679
I! litor B. Katalinic, Published by DAAAM International, Vienna, Austria 2008
Make Harmony Between Technology and Nature, and Your Mind will Fly Free as a Bird
*Young Researcher and Scientist Paper / * Supervisor, Mentor*

A DYNAMIC PORTFOLIO OPTIMIZATION FRAMEWORK WITH DISTINCTION BETWEEN "RISK" AND "UNCERTAINTY"

TALPOS, I[oan]; PIRTEA, M[arilen] & DIMA, B[ogdan]

Abstract: *The current theory of financial assets' portfolio management is dominated by the Sortino's critical concept of "downside risk". Still, there is no clear distinction between the components of "the downside risk" ("loss" or "unrealized return") and between "risk" and "uncertainty" - which are, on different extent, critical variables of portfolio optimization. Thus, the objective of the present paper is to provide a simple and broad analytical framework of portfolio management able to address these issues. The methodology is based on a dynamic optimization framework and is applied on 5 series of simulated random walks. A 46% level of return-to-risk ratio is obtained for these data. Such results offer an empirical support for the thesis that the advanced analysis could provide a better explanation of the investors' behavior in the portfolio management processes.*
Key words: *portfolio, management, "risk", "uncertainty", optimization*

1. INTRODUCTION

The actual theory of financial assets' portfolio management is a refinement of the "modern portfolio theory" (Mean-*Variance Analysis*) initiated by Harry Markowitz. Along with the establishing works of Frank Sortino and The Pension Research Institute at San Francisco State University, new models have been developed that recognize investors' preferences for upside over downside volatility of their returns. Still, such models do not differentiate in terms of the downside volatility ("loss" or "unrealized return" and do not clearly distinguish between "risk" and "uncertainty" which are, on a different degree, critical variables for the portfolio optimization. The objective of this paper is to provide a simple and broad analytical framework of portfolio management in the light of the *Post-modern portfolio theory* (**PMPT** - Rom & Ferguson, 1993) with two major features: (1) a dynamic and systematic choice of the portfolio's "optimal" structure and (2) a particular definition of "risk".

2. LITERATURE REVIEW

Resumed by a poetical definition: "Risk, like beauty, is in the eye of the beholder" (Balzer, 1994), the **PMPT** is currently based on the *Sortino ratio*, which measures returns adjusted for the target and downside risk (Sortino & Horsey 1996; Sortino, 2001). Simultaneously, a more robust model for investment returns' pattern has been introduced: the three-parameter lognormal distribution one, which permits both positive and negative skewness in return distributions (Forsey, 2001). Rom & Ferguson (1993) have developed the analysis based on *volatility skewness.* This measures the ratio of a distribution's percentage of total variance from returns above the mean, to the percentage of the distribution's total returns from returns below the mean. These methods are not explicitly taking into account: the issues of the "risk/uncertainty" distinction and some of the possible mechanisms for their incorporation into a general model of portfolio' construction and management.

3. THE ANALYTICAL FRAMEWORK

A successful model could be represented by the framework of the *multi-periodic optimization of the patrimonial structures* model. More exactly, it supposes that the economic system is formed by N groups of identical agents, each group having its individual utility function. Each of these is chosen based on a specific structure of their wealth by incorporating both M monetary and Q non monetary assets. The aim is to try to balance the group's return to risk ratio and preserve an "optimal" structure of the wealth for a certain number of successive periods in order to minimize the *adjustments costs* for suboptimal structures by considering the *budgetary restriction.* So, the current and expected values for the non monetary assets costs, returns and risks as well as the incomes from labor and capital are all involved in the optimization process. If the information is "imperfect" (is incomplete, unequal distributed and there are costs of obtaining, updating and using it), a *bounded rationality* anticipation mechanism will be involved (the anticipation will be formed based on a *mixed* mechanism which will incorporate "all" the available information from the current and past periods). The differences between diverse agents are reflected by their individual utility functions, whereas the return to risk ratio is particularly weighted to reflect the "risk aversion".

At a *global* level, the formal description of the optimization problem looks like an aggregation of S individual problems as:

$$x_{ijt} \geq 0, \quad \sum_{j=1}^{S}\sum_{i=1}^{N} x_{ijt} = 1, \forall (j,t) \tag{1}$$

$$\sum_{j=1}^{S}\sum_{i=1}^{N}\sum_{k=1}^{t}\left(c_{ijk} + c^{*}{}_{ijk+l}\right)x_{ijk} + \sum_{j=1}^{S}\sum_{p=1}^{M}\sum_{k=1}^{t}\left(L_{pk} + L^{*}{}_{pk+l}\right) = \sum_{j=1}^{S}\sum_{z=1}^{Q}\sum_{k=1}^{t}\left(Y_{zk} + Y^{*}{}_{zk+l}\right) \tag{2}$$

$$\frac{\sum_{j=1}^{S}\sum_{i=1}^{N}\sum_{k=1}^{t}\left(\eta_{ijk} + \eta^{*}{}_{ijk+l}\right)x_{ijk}}{\sum_{j=1}^{S}\sum_{i=1}^{N}\sum_{k=1}^{t}\left(R_{ijk} + R^{*}{}_{ijk+l}\right)x_{ijk}} \rightarrow MAX \tag{3}$$

$$\sum_{j=1}^{S} E\left(*_{k+1}\right) = \sum_{j=1}^{S}\left(\sum_{k=1}^{t} \alpha_{kj} *_{k} + \beta_{tj} *_{t}\right) \tag{4}$$

where $\boldsymbol{x}$ is the weight of a non monetary asset $\boldsymbol{i}$ in the wealth structure in the current period $\boldsymbol{t}$; $\boldsymbol{N}$ is the total of non monetary assets from the *selection universe*; $\boldsymbol{c}$ are the costs associated with the buy and hold non monetary assets; $\boldsymbol{L}$ are the monetary and quasi monetary assets with a high degree of liquidity; $\boldsymbol{Y}$ are the incomes from labor and capital obtained in the current period and / or accumulated from previous periods; $\boldsymbol{\eta}$ are the returns of non monetary wealth components consisting of monetary flows generated by their utilization and by their prices variations; $\boldsymbol{R}$ are their associated risks and * indicates the anticipated values of the involved variables formed in the current period for $\boldsymbol{l}$ future periods. *Relation (1)* is a **logical restriction:** the weights of a particular non monetary asset could be only positive or null and their sum could not exceed "1". *Relation (2)* is a **budgetary restriction.** *Relation (3)* is the

objective function: each group of agents is trying to maximize not only the individual level of return or risk but their ratio. *Relation (4)* is the **anticipation mechanism**.

This description of the optimization problem implies that:

- Each group of agent is looking to systematically preserve an "optimal" patrimonial structure.
- In order to reduce the *transactional costs*, a certain adopted structure should be kept at least for some future periods, thus it is necessary to include the anticipated values of the variables.
- The "optimal" level of *monetary balances* (the stocks of different means of exchange and payment) is obtained simultaneous with the level of the non monetary assets by "solving" the optimization problem, so that there is no "residual" accumulation for *prudential* or *speculative* reasons.
- The objective function implies a "balance" in return to risk ratio's elements, thus the agents could be described as "risk neutral" (with different degrees of risk tolerance). In other words, they accept to assume a higher risk that the "perfect risk aversion" agents and a lower level of return that the "perfect risk takers", in order to reach a better correlation between these variables. At "aggregate" level, there are neither "casino" economical systems, nor "old granny" ones.
- The *bounded rationality* model implies that all the information which can be obtained at an "efficient level of implied costs" is used both from previous as well as from current periods. The goal is to adopt the "second best" decisions with "incomplete information".

A specific issue concerns the definition of the "risk" concept. The key distinction in defining and measuring a risk is similar to the one between "risk" and "uncertainty". *"Risk" is the probability to obtain an unfavorable result of an economic decision. "Unfavorable" means that the result is "positive" but lower that the expected one or that the result is "negative".* Therefore the risk concept incorporates both the situations of "unrealized" results and "losses". *"Uncertainty" means that the observable result deviates (in a "positive" or "negative" sense) from the expected one.* "Uncertainty" reflects both the situation of "unfavorable" results as well as that of "excess results". Suppose for instance that the returns of a project are "normally" distributed around a certain "objective" or "subjective" target value. In such case, three main areas could be delimited: *Area 1* where the returns are positive but lower that the target value, that can be settled based on the average of the previous values, the average of the sectors' returns, the "concurrencies average", the interest or inflation rate, the growth rate of the financial markets etc. or can be a pure subjective value; *Area 2* where the returns are negative and respectively *Area 3* where the returns are positive and higher that the target value. *Area 1* and *Area 2* form together the **risk zone,** while all three areas reflect the uncertainty **zone**. However, the relative importance of *Area 1* and *Area 2* for the risk definition is not the same: the agent will perceive a higher risk associated with losses that with values of returns, which are lower that the target but still positive.

Thus, an implementation methodology of such risk definition, at the operational level, consists of the next steps:

1. The construction of a risk values' set r_{jt} according to the following rules:

$$r_{jt} = \begin{cases} m_1 * \left(target_j - \eta_{jt}\right) & \text{if } target_j > \eta_{jt} \text{ and } \eta_{jt} > 0 \\ m_2 * \left(target_j - \eta_{jt}\right) & \text{if } \eta_{jt} < 0 \\ m_3 * \left(\eta_{jt} - target_j\right) & \text{if } target_j < \eta_{jt} \\ \text{with } m_2 > m_1 > m_3 \end{cases} \tag{5}$$

2. The construction of a global risk measure. If r_{jt} *values are* ***iid*** *processes then for this measure could be utilized the so-called* ***normalized Euclidian norm*** *of the risk values set components*:

$$R_{jt} = \sqrt{\sum_{k=1}^{t} \frac{\left(m_m \left|target_j - \eta_{jk}\right|\right)^2}{\sigma^2_{jk}}}, \quad m = 1,2,3 \tag{6}$$

The ***m*** parameters could be seen as measures of *risk aversion* specific to each group ***j***. With these characteristics, for the optimization problem, formulas (1), (2) and (4) remain as above and formula (3) becomes:

$$\frac{\sum_{j=1}^{S}\sum_{i=1}^{N}\sum_{k=1}^{t}\left(\eta_{ijk} + \eta^*_{ijk+1}\right)x_{ijk}}{\sum_{j=1}^{S}\sum_{i=1}^{N}\sum_{k=1}^{t}\left(\frac{\left(m_m\left|target_j - \eta_{jk}\right|\right)}{\sigma^2_{jk}} + m_m\frac{\left(m_m\left|target_j - \eta^*_{jk}\right|\right)}{\sigma^{*2}_{jk}}\right)x_{ijk}} \rightarrow MAX \tag{7}$$

For instance, in order to illustrate the potential of this analytical framework, we had run a ***simulation over 5 random walk series of returns with null mean and unitary variance with a length of data equal with 1000 values***. The target value for the return is equal with 30% and m_2=3, m_1=2, m_3=1.

	RETURN	RISK
Mean	1.171783	2.528298
Median	0.529424	2.461687
Maximum	29.15225	9.817127
Minimum	-0.445627	-6.545283
Std. Dev.	2.391756	0.680791
Skewness	6.710659	0.324029
Kurtosis	62.99006	56.37651
Jarque-Bera	155566.3	117303.3
Probability	0.000000	0.000000
Sum	1157.721	2497.958
Sum Sq. Dev.	5646.130	457.4507
Observations	990	990

Tab. 1. The main features of the simulation's results

4. COMMENTS AND FURTHER RESEARCH

The simulation's results suggest that, even for *random walk* data, it is possible to obtain with this methodology a "decent" level of **return to risk ratio** (46% in this case). Despite some important *caveats*, we consider that the advanced analysis could provide a better explanation of the investors' behavior in the portfolio management processes. In order to fully achieve this, we further intend to develop the analysis by: (1) building-up a "non-neutral" model of risk selection; (2) constructing a target selection methodology; (3) extending the concept of "bounded rationality" by incorporating the time factor. Our final goal is to prove that any approach of portfolio management which does not take into account a more complex view of risk is like "talking about Hamlet without the prince of Denmark".

5. REFERENCES

Balzer, L.A. (1994), Measuring Investment Risk: a Review. *Journal of Investing*, Fall

Forsey, H. (2001) The Mathematician's View: Modelling Uncertainty with the Three Parameter Lognormal, In: *Managing downside risk in financial markets*, Sortino, F. & Satchell, S., (Ed.), 51-59, Elsevier Science, ISBN 0 7506 4863 5, Oxford

Rom, B. M. & Ferguson K. (1993). Post-Modern Portfolio Theory Comes of Age. *Journal of Investing*, Winter

Sortino, F. & Horsey.F (1996). On the Use and Misuse of Downside Risk. *The Journal of Portfolio Management*, Winter

Sortino, F.(2001). From alpha to omega, In: *Managing downside risk in financial markets*, Sortino, F.& Satchell, S., (Ed.), 3-26, Elsevier Science, ISBN 0 7506 4863 5,Oxford

Annals of DAAAM for 2008 & Proceedings of the 19th International DAAAM Symposium, ISSN 1726-9679
Editor B. Katalinic, Published by DAAAM International, Vienna, Austria 2008
Make Harmony Between Technology and Nature, and Your Mind will Fly Free as a Bird

MOBILE ROBOT POSITION ESTIMATION BASED ON MULTI SENSOR FUSION

TAMAS, L[evente]; LAZEA, G[heorghe]; ROBOTIN, R[adu]; MARCU, C[osmin] & MAJDIK, A[ndras]

Abstract: *This paper gives an overview about the state estimation techniques based on Kalman filters in the robot navigation field. It deals with the problem of the position estimation based on odometric and ultrasonic measurements. Its main aim is to improve the position estimation in navigation. There are presented the results from applying the Kalman filter for odometric and ultrasonic measurement techniques.*
Key words: *robot navigation, Kalman filter, position estimation, sensor fusion.*

1. INTRODUCTION

The localization algorithms are important components of control software of the autonomous systems. This is usually based on the information from the localization sensors as a feedback about the driving action that was performed in the surrounding world. Based on this information, the position has to be estimated as accurately as possible.

The difficulty of the localization problem is due to the existence of the uncertainty in both the model and the sensing of the robot. However, the uncertainty in information can be reduced by applying an optimal filtering algorithm (Welch & Bishop, 2003)

The Kalman filter is a well known estimation technique that combines the information of different uncertain sources in order to estimate the state of the robot.

The Kalman filters covered in this paper are the Linear Kalman Filter (LKF) for LTI systems, the Extended Kalman Filter (EKF) which is suitable also for nonlinear systems. The Unscented Kalman Filter (UKF) is based on the Unscented transformation and it can be also applied for systems with discontinuities and is directly comparable to the second order Gaussian filter (Julier et al., 1995).

2. KALMAN FILTERS

2.1 Linear Kalman Filter

The LKF addresses the general problem of trying to estimate the state of a discrete-time controlled process which is governed by the general stochastic difference equation given by (1) and (2).

$$x_{k+1} = F\left(x_k, u_k, w_k\right) \quad (1)$$

$$y_k = H\left(x_k, v_k\right) \quad (2)$$

The linear state space representation of the system is given by (3) and (4).

$$x_k = Ax_{k-1} + Bu_{k-1} + w_{k-1} \quad (3)$$

$$z_k = Hx_k + v_k \quad (4)$$

The random variables w and v represent the process and measurement noise and it is supposed that they are white, independent noises with a Gaussian distribution.

The vector x represents the states of the system while z relates the state of the measurements and u is the input vector for the system. The matrices A, B, H are assumed to be constant, although they may change from step to step in real life applications.

The prior estimate is based on the current state and the process model, while the posterior state estimate is computed taken into account also the measurement from the step k.

The two kinds of estimates are in fact related to the two phases of the Kalman filter algorithm: the time update and the measurement update (Welch & Bishop, 2003).

The first phase of the Kalman filter can be summarized in the equations described in (5) and (6)

$$\hat{x}_k^- = A\hat{x}_{k-1} + Bu_{k-1} \quad (5)$$

$$P_k^- = AP_{k-1}A^T + Q \quad (6)$$

The Q represents the covariance of the process noise v while with R is denoted the covariance of the measurement noise w. These values also might change with time but in this algorithm, they are supposed to be constant.

After the time update step in the correction step there are updated the priory estimates as shown in (7), (8) and (9)

$$K_k = P_k^- H^T \left(HP_k^- H^T + R\right)^{-1} \quad (7)$$

$$\hat{x}_k = \hat{x}_k^- + K\left(z_k - H\hat{x}_k^-\right) \quad (8)$$

$$P_k = \left(I - K_k H\right) P_k^- \quad (9)$$

2.2 Extended Kalman Filter

The EKF considers the linear system model around the current value of the state vector (Mayback, 1979).

The linearization is done by computing the Jacobian of the matrices A, B, C, D as follows

$$A_k = \left.\frac{\partial F\left(x, u_k, \overline{v}\right)}{\partial x}\right|_{\hat{x}_k} \qquad B_k = \left.\frac{\partial F\left(\hat{x}_k^-, u_k, v\right)}{\partial v}\right|_{\overline{v}} \quad (10)$$

$$C_k = \left.\frac{\partial H\left(x, \overline{n}\right)}{\partial x}\right|_{\hat{x}_k} \qquad \mathrm{D}_k = \left.\frac{\partial H\left(\hat{x}_k^-, n\right)}{\partial n}\right|_{\overline{n}} \quad (11)$$

Although the EKF is widely used in the control community it is difficult to implement, difficult to tune, and only stable for systems which are almost linear on the time scale of the update interval (Julier et al., 1995).

3. MODELLING

3.1 Odometer Positioning and Heading Estimation

The simulations and experiments were carried out on a skied steered mobile robot platform. The position $x_{odo,k}$, $y_{odo,k}$ and heading $\omega_{odo,k}$ at time step k are calculated using the equations (12), (13), (14), where $\Delta d_{R,k}$ and $\Delta d_{L,k}$ denote the distances travelled at time step k by the right and respectively left side wheels.

$$x_{odo,k} = x_{odo,k-1} + \left(\frac{\Delta d_{R,k} + \Delta d_{L,k}}{2} \right) \cos \omega_{odo,k} \qquad (12)$$

$$y_{odo,k} = y_{odo,k-1} + \left(\frac{\Delta d_{R,k} + \Delta d_{L,k}}{2} \right) \sin \omega_{odo,k} \qquad (13)$$

$$\omega_{odo,k} = \omega_{odo,k-1} + \tan^{-1} \left(\frac{\Delta d_{R,k} - \Delta d_{L,k}}{W} \right) \qquad (14)$$

The W is the distance between the centers of the right and left wheels (Kyriakopoulos, 2004).

4. EXPERIMENTAL SETUP

In the following it will be described the steps that were performed in a real life application in order to measure and estimate the position of a robot.

Both the infrared and ultrasonic measurements were carried out on a microcontroller based data acquisition board. The details of the experiments and the architecture of such a measurement system are described in (Sabatini, 1995).

4.1 Low Level Ultrasonic Measurement

The main difficulty in the ultrasonic measurement is the detection of the first echo in a noisy measurement environment with multiple echoes.

A possible solution was proposed in (Agrisani, 1995) by computing the envelope of the received echo and estimating the time of flight (TOF) from this envelope. A pair of emitter-receiver has a specific echo envelop for which discrete time modeling be given by.

$$A(kt_s) = A_0 \left(\frac{kt_s - \tau}{T} \right)^{\alpha} e^{\left(\frac{kt_s - \tau}{T} \right)} \qquad (15)$$

where the A_0 is the echo of the amplitude, α and T are specific parameters for the emitter-receiver pair; τ is the expected time of flight and t_s is the sampling time. A typical experiment result is shown on Fig. 1.

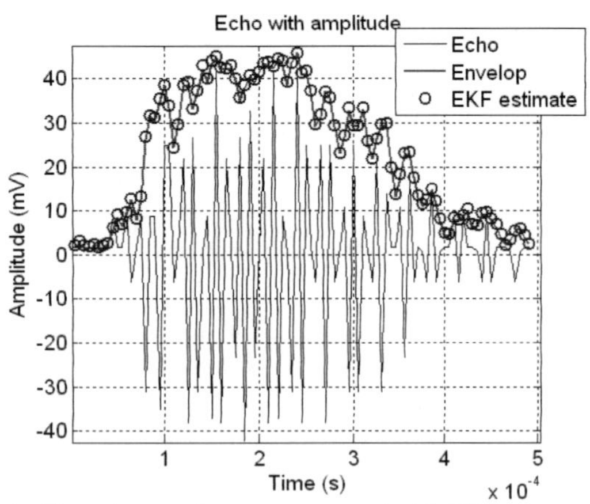

Fig. 1. The echo envelope reconstructed with EKF.

The estimation of the TOF consists of the following:

- acquisition of the echo;
- applying the Hilbert transformation on the signal
- based on the echo envelope, applying the EKF the parameters of the can be determined (dotted line)
- based on the determined parameters, the TOF it can be computed with a better accuracy.

4.2 Odometric Measurements

The odometric measurements were carried out using the dead-reckoning system of the mobile robot platform. During the test procedure for the state estimation problem a 2x2 meter square was chosen as a reference path for the robot.

The main difficulties with the dead-reckoning systems are the accumulation of the systematic errors, which may corrupt the position estimation.

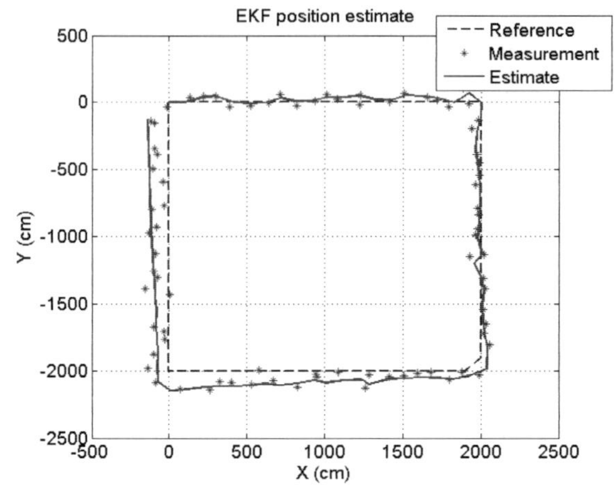

Fig. 2. Odometric measurement and EKF estimation

On Fig. 2 are shown the obtained measurement and the EKF estimate of the state for the robot that has a rather noisy trajectory with certain drift in the orientation.

5. CONCLUSION

Based on the results it can be concluded that the Kalman filter and its extensions are important tools for mobile robot state estimation enhancement. The main advantages of the different extensions of the Kalman filter can be seen especially in the case of the nonlinear estimations. By applying these algorithms, it was possible to enhance the performance of the position estimation in case of the robot navigation.

As further development it can be taken the multiple data fusion: based on the information from several sensors, it can be performed the state estimation fusing the information using different extensions of the Kalman filter, like the UKF.

6. REFERENCES

Angrisani, L. (2004). Ultrasonic Time-of-Flight Estimation through Unscented Kalman Filter, *IEEE Transactions on Instrumentation and Measurement*, pp 211-217.

Julier, S. J.; Uhlmann, J. K. & Durrant-Whyte, H. F. (1995), A New Approach for Filtering Nonlinear Systems, *The Proc. of the American Control Conf.*, pp 1628–1632, Seattle USA

Kyriakopoulos, K.J. (2004). A dead-reckoning scheme for skid-steered vehicles in outdoor environments, *Proc. Int. Conf. Robotics Automation*, pp 580-585, Los Angeles

Mayback, P. S. (1979) *Stochastic models, estimation and controls,* Academic Press, New York, USA

Sabatini, A.M. (1995) A Low-Cost, Composite Sensor Array Combining Ultrasonic and Infrared Proximity Sensors, *IEEE Int. Conf. on Intelligent Robots and Systems,* pp 120-126, Pittsburgh, USA

Welch, G. & Bishop, G. (2003). *An Introduction to the Kalman Filter,* UNC, California

Annals of DAAAM for 2008 & Proceedings of the 19th International DAAAM Symposium, ISSN 1726-9679
Editor B. Katalinic, Published by DAAAM International, Vienna, Austria 2008
Make Harmony Between Technology and Nature, and Your Mind will Fly Free as a Bird

MODEL FOR THE ANALYSIS OF THE INNOVATION LEVEL FOR IT&C ENTREPRENEURIAL ACTIVITY IN ROMANIA

TAMASILA, M[atei] & TAUCEAN, I[lie] M[ihai]

Abstract: *The present paper proposes a methodology for the assessment of the innovation level, in general, and of the entrepreneurial activity, in particular. This methodology uses Kano's model of satisfaction, the five levels of inventiveness proposed by Altshuller, and the activities and products degree of ideality. In order to sustain the practicability of the proposed methodology, the research includes a case study about Romanian IT&C entrepreneurial activity innovation level, witch turn out relatively modest in this moment (6.92 out of 10).*

Key words: *model, entrepreneurship, innovation, IT&C*

1. INTRODUCTION

The methodology for the assessment of the innovation level of the proposed entrepreneurial activity is based on a global indicator achieved through the level-headed summing up of three indicators: the indication of the entrepreneur satisfaction I_{ES}, corresponding to the charming characteristics from Kano's model (Kano, 1984), the indicator of inventiveness, I_{IN}, associated to Altshuller's five levels of inventiveness (Altshuller, 1999), and the indicator of ideality, I_{ID}, which reflects the degree of ideality.

Thus, in order to assess the level of innovation, we propose a global indicator of the innovation level, I_I, which can be determined by means of the following relation:

$$\begin{aligned} I_I &= I_{ES} \cdot p_{ES} + I_{IN} \cdot p_{IN} + I_{ID} \cdot p_{ID} = \\ &I_{ES} \cdot p_{ES} + I_{IN} \cdot p_{IN} + \left(\sum_{k=1}^{9} i_k \cdot q_k \right) \cdot p_{ID} \end{aligned} \quad (1)$$

where p_{ES}, p_{IN} and p_{ID} represent the weights corresponding to the three indicators, $i_1, i_2, \ldots, i_9$ are the indicators of the degree of ideality, and $q_1, q_2 \ldots, q_9$ are the weights of the indicators of ideality. The sum of the weights from each category is equal to 1, i.e. $p_{ES} + p_{IN} + p_{ID} = 1$ and $q_1 + q_2 + \ldots + q_9 = 1$. Some weights can be null, as function of the field where the methodology is applied.

2. THE INDICATORS LEVEL ESTABLISHING

Generally, the characteristics of a product, according to Noriaki Kano's model, from the point of view of customers' satisfaction, are as follows: non-satisfying characteristics, satisfying characteristics, and charming characteristics.

In the proposed methodology, we have considered that it is necessary that the indicator of the entrepreneur's satisfaction, I_{ES}, refer only to the charming characteristics from Kano's model, because, on the one side, their achievement implies an innovation effort from the part of the entrepreneurs, and, on the other side, they represent, most often, the decisive element for the success of a business (see table 1).

Grades	1	2	3	4	5	6	7	8	9	10
Characteristics	1		2		3		4		≥ 5	

Tab. 1. Scale for entrepreneur's satisfaction level.

For the assessment purposes, there has been proposed that I_{ES} indicator be attributed grades on a scale from 1 to 10, as function of the number of charming characteristics, as follows: grades 1 and 2 for one charming characteristic, grades 3 and 4 for two charming characteristics, grades 5 and 6 for three charming characteristics, grades 7 or 8 for four charming characteristics, and grades 9 or 10 for five or more charming characteristics.

In order to take into account the degree of inventiveness, there has been proposed in the paper to make an adjustment on the basis of the five levels of the inventiveness solutions and of the required inspirations sources established by Altshuller, with the grades corresponding to the inventiveness indicator, I_{IN} (see table 2) as follows:

Level one activities – which do not imply any invention, can be achieved by routine improvements of some existing products through well-known methods in the field, and for which the inspiration source comes from own knowledge (grade 2).

Level two activities – which imply minor improvements of some existing products for which the inspiration source comes from the scientific field where the designing engineers work (grade 4).

Level three activities – which imply fundamental improvements of some existing products through known methods, for which the solutions shall be looked for in related fields or in other fields (grade 6).

Level four activities – which principally imply new products or new generation products for which there are used new principles, and the solutions come from the clearing up of some phenomena from various fields, phenomena that have been less understood until that moment (grade 8).

Level five activities – which imply rare scientific discoveries, whole new products, whose solutions can be found by exceeding the borders of the scientific knowledge known at a certain moment (grade 10).

The ideality indicator, I_{ID}, takes into account the ideality of the product through nine indicators. These indicators are as follows: The indicator of the system dimensionality degree (i_1); The indicator of the aggregation status (i_2); The indicator of the type, nature, and frequency of the actions applied to the system (i_3); The indicator of the degree of the system "porosity" (i_4); The indicator of the degree of the system dynamic capability (i_5); The indicator of the degree of human involvement (i_6); The indicator of the degree of the system multiplicity (i_7); The indicator of the nature, type and dimensionality of the system functions and features (i_8); The indicator of the degree of the system convolution (i_9).

Inventions/ improvements	Grades
0	2
Minor	4
Fundamental	6
New	8
Completely new	10

Tab. 2. Scale for inventiveness level.

Grades	1	2	3	4	5	6	7	8	9	10
i_1		0		1		2		3		≥4
i_2	AP		micro		small		medium		big	large
i_5	large		big		medium		small		micro	AP
i_6			reduced			medium			big	intense
i_7		1		2		3		4		5
i_8		1		2		3		4		5

Tab. 3. Scale for ideality level (AP = authorized person)

The indicator of ideality, I_{ID}, can be determined with the following relation:

$$I_{ID} = \sum_{k=1}^{9} i_k \cdot q_k \quad (2)$$

where i_k represents the nine indicators of the degree of ideality, and q_k – their weights, with the feature that their sum is 1. The indicators weights are established as function of the object of assessment.

3. CASE STUDY

In order to highlight the use of the developed researches, here we will present below the application of the proposed methodology for the very assessment of the level of innovation of the IT&C entrepreneurial activities from Romania.

3.1 The determining of the indicator of the entrepreneur satisfaction (I_{ES})

It has been considered that there are at least 5 charming characteristics, such as (Burtica et al., 2006): the possibility to apply own ideas, the motivation to obtain higher income, the advantage to be one's own chief and to have a flexible work schedule, a way to obtain a better social position. According to the grades scale, the indicator I_{ES} is given grade 10.

3.2 The determining of the inventiveness indicator (I_{IN})

For this case study, to the indicator I_{IN} there was given grade 4, because as per the classification established by Altshuller it has been deemed that the entrepreneurship is of level two, on account of the fact that following a study carried out on the Small and Medium Size Enterprise (SME) from Romania (Tamasila et al., 2008): only 40% are concerned with innovation, 65% do not have a compartment dedicated to innovation, in 54% of the cases there is only one person in charge with innovation, and 61% have declared that the new products contribute with 30% to the obtained turnover, at the most.

3.3 The determining of the indicator of ideality (I_{ID})

This indicator reflects the degree of ideality of the entrepreneurial activity. In order to determine the level of ideality, there have been taken into account only the indicators i_1, i_2, i_5, i_6, i_7 and i_8 because it has been considered that only these ones are relevant for the analysed case study. Thus, indicator i_1, which gives the system degree of dimensionality, was given grade 8, according to the presented classification, because the entrepreneurial activity involves three planes, as per STEP model: the social, the technical and the economic ones (and less the political plane). Indicator i_2, which estimates the size of the system where the entrepreneurial activities are carried out, was given grade 7, because the entrepreneurial activity is being developed in smaller enterprises, almost all of them SMEs. Indicator i_5, which shows the degree of flexibility, was given grade 9 because the system dynamic capability develops inversely proportional to its dimensions. Indicator i_6, namely the degree of human involvement, has been given grade 10, because in this case the human involvement is intense and decisive for the start and for the success of the business. The degree of the system multiplicity, taken into account through indicator i_7, was given grade 6, because it can be achieved with three functions of the enterprise (production, trading, and financial-accounting). Indicator i_8, which estimates the dimensionality of the entrepreneur managerial functions (Taucean et al., 2008), was given grade 8, because the entrepreneur's activity implies at least 4 functions (organizing, decision, coordination and control).

The five indicators taken into account were given the following weights: $q_1=q_2=q_7= 0{,}1$; $q_5=q_8= 0{,}2$; $q_6= 0{,}3$. The other weights corresponding to the eliminated indicators have been considered to be null.

3.4 The calculation of the global indicator of the innovation level I_I

In order to calculate the global indicator, there have been established the following weights associated to the three indicators: the weight of the indicator of the entrepreneur's satisfaction $p_{ES} = 0.30$, the weight of the indicator of inventiveness $p_{IN} = 0.45$ and the weight of the indicator of ideality $p_{ID} = 0.25$.

Finally, through the use of relation (1), the global indicator of the innovation level of the methodology proposed in this paper came to be equal to 6.92.

4. CONCLUSION

By applying the methodology proposed, subjectivity can be eliminated in the assessment of the activities & products innovation level. The weights can be modified according to the type of the assessed entity and by taking into account the importance of the indicators in the given assessment.

The innovation level of the Romanian IT&C entrepreneurial activity, in accordance by the proposed evaluation model, it is relatively modest in this moment (scored 6.92 out of 10). In our opinion, using this model for innovation level establishing is a very good way to improve the inventiveness score by increasing the number of firms that are concerned with innovation, increase also the firms that have a compartment dedicated to innovation and those with more persons in charge with innovation.

5. REFERENCES

Altshuller, G. (1999). *The Innovation Algorithm: TRIZ, Systematic Innovation and Technical Creativity*, Technical Innovation Ctr, ISBN 978-0964074040, Worcester

Burtica, M.; Taroata, A. & Tamasila, M. (2006). *The management of Small and Medium size Enterprises*, Solness, ISBN 978-973-729-069-4, Timisoara

Kano, N. (1984). Attractive quality and must-be quality, *The Journal of the Japanese Society for Quality Control*, April, 1984, pp. 39-48

Tamasila, M.; Mocan, M.L. & Taucean, I.M. (2008), *Proceedings of The 6th International Conference of DAAAM Baltic Industrial Engineering,* Kuttner, R.(Ed.), pp. 369-373, ISBN 978-9985-59-783-5, Estonia, April 2008, DAAAM Baltic University of Technology, Tallinn

Taucean, I.M.; Taroata, A. & Tamasila, M. (2008), *Proceedings of The 6th International Conference of DAAAM Baltic Industrial Engineering,* Kuttner, R.(Ed.), pp. 381-386, ISBN 978-9985-59-783-5, Estonia, April 2008, DAAAM Baltic University of Technology, Tallinn

Annals of DAAAM for 2008 & Proceedings of the 19th International DAAAM Symposium, ISSN 1726-9679
Editor B. Katalinic, Published by DAAAM International, Vienna, Austria 2008
Make Harmony Between Technology and Nature, and Your Mind will Fly Free as a Bird

THE INFLUENCE OF INPUT PARAMETERSIN CASE OF COLD PLASTIC DEFORMATION

TANASA, R[aluca]; IURAS, E[milia] & CARP, I[ulia]

Abstract: *A theoretical analysis based on the effects observed in the final stage of mechanical shocks machining considering the influence of different factors characterizing the cutting conditions (the work feed f, the cutting speed v and the ball radius r_{ball}) is highlighted in this paper. Systematic analysis on the influence of the working conditions is presented obtaining a precise classification of those factors able to affect the final surfaces of a product.*

Key words: *Surface roughness, Mechanical shocks, Plastic deformation, Systemic analysis.*

1. INTRODUCTION

The smoothening through plastic deforming using centrifuged balls is considered a simple and economic method of fine processing and also of hardening for the metal surfaces; is a complex phenomenon that includes dynamic interactions between the metallic structure of the tool and the work piece (Stepanenko- Aizenberg M.V., 2007). The method can be successfully applied for cylindrical surfaces (internal and external) and for plane surfaces, as well.

Refering at cold deformation smoothing by mechanical shocks, one regards as an easy and efficient method of the smoothing and hardening surfaces (Shiou F-J. et al., 2003).

Among many advantages of this method (Drăghici G., 1984) we could mention the most important ones a) the possibility to be applied on common machine-tools usually used for turning, grinding etc, b) the possibility to be applied by the operators that do not possess a high qualification; c) applying for a large set of external and internal cylindrical surfaces; d) the significant increasing of the fatigue strength etc. As disadvantages, one can consider the lack of the surface continuity, the possibility that the results be influenced by the variation of the initial hardness of the workpiece material etc.

2. THE TECHNOLOGICAL EQUIPMENT

The simplest technological equipment which can be successfully used in superficial plastic deformation processing by percussion, is constituted of a device which can be mounted on a universal machine-tool, lathe or milling-machine, and depending on the shape of the surface to be processed could be revolved or plane, respectively. To process external revolved surfaces, a device as the one from fig.1 can be used, being mounted on a universal lathe's carriage.

The device has an electric motor which, through a cinematic transmission induces the rotary motion n_s of the port-balls disc. The tool-disc, mounted on the lathe's carriage can be displace along the generatrix of the workpiece, assuring the feed motion s_l.

The revolved surface is assured by rotating the workpiece with a certain number of rotations per minute n_p, which is reffered as the circular feed motion s_c.

During smoothing procedure, the tool-disc is placed very close to the processed surface for the balls to hit the workpiece when it rotates.

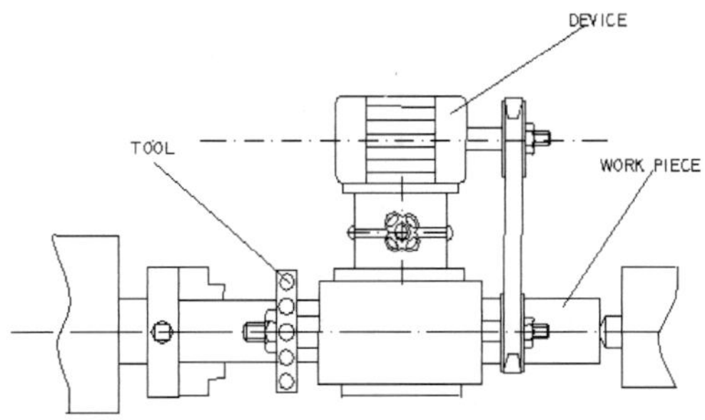

Fig.1. Schematical representation of experimental setup.

The frequency of the percussions will be dependent on the disc number of rotations per minute and on the number of balls on the disc.

3. WORK HYPOTHESIS

The studies made till now consider that the surface smoothening can be realized in good conditions also by cold plastic deforming. After the collision of some balls with a surface, this is locally deformed and its micro-relief modified. During processing micro-craters appear on the surface, having the shape of spherical calotte. In the theory of the plastic deformations that consider the collision of two bodies, the local deformation is evaluated as the distance between the two bodies after collision, on the direction of their displacement. To observ the modification of the height of micro-irregularities we are using the mathematical relations for local plastic deformations (Ponomariov S.D. et al., 1964) and the equations from the contact theory. We can calculate the elements of the trace left by the deforming ball, i.e. the depth of the micro-crater (the maximum deformation) and its diameter as well. These parameters have been compared with the micro-irregularities of a surface finished through other methods.

3.1. Theoretical considerations

The irregularity assemblies represent the real surface relief. These irregularities appear due to oscillation movements of the tool tip, friction of the tool edge on the surface, high frequencies vibrations of the tool or of the machine-tool. The presence of irregularities on the surface has, in hard work conditions, some disadvantages, as narrowing the effective contact surfaces, worsening the friction and functioning conditions of the part, decrease the resistance to alternate efforts of the material by tensions summing, decreasing the tightness and variation in the effective dimensions of the piece. On the other hand, the absence of these elements makes impossible maintaining the oil film on the contact surface by normal oiling.

The parameters that influence the roughness are: the geometric parts of the tool (by the beam of the pin tool, tool clearance, rake angle, angle of inclination), the cutting edge micro geometric (by usage tool, the roughness of the back edge and front face), the cutting conditions parameters (speed, feed, depth of cut), the physics-mechanic properties of the material piece, cutting oil and the technological system hardness.

4. THE EXPERIMENTAL RESULT

We considered different initial roughness obtained with a lathe tool reinforced with ceramics plate, having radius angle 0, 8 mm, and different working regime for steel OLC45, according to STAS 791-80.

In order to establish an empirical model of the surface roughnesses obtained during the plastic deformation process the experiments were made at different working parameters, presented in table 1. The parameters are: n_p-piece revolution, f_l-longitudinal feed, d_b-balls diameter, R_{ai} [µm] – initial surface roughness parameter. In figure 2 we indicate the profile of the initial roughness, and figure 3 (a and b) presents the final surface roughness surface after machining using different working conditions. The influence of the initial surface roughness on the final surface roughness parameter is shown in figure 4; one can observe the influence of the initial roughness; if the initial roughness (marked II in graph) is important after manufacture the roughness surface improvs as indicated by dependence (I).

Nr. Crt.	n_p [rot/min]	f [mm/s]	d_{ball} [mm]	$R_{a\ init}$ [µm]	$R_{a\ final}$ [µm]
1	20	0,05	6,78	7.73	1.77
2	160	0,05	6,78	4.23	1,4
3	20	0,224	6,78	4.94	1,8
4	160	0,224	6,78	5.45	2,13
5	20	0,05	10	5.46	1,0
6	160	0,05	10	4.15	0,98
7	20	0,224	10	3.83	1.02
8	160	0,224	10	4.82	3,10

Tab. 1. Input parameters values

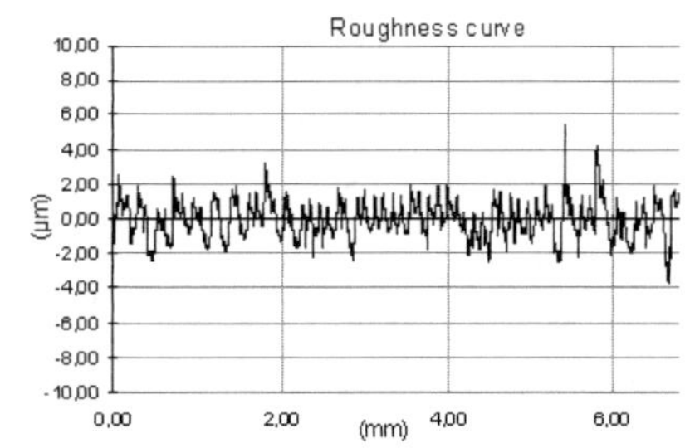

Fig. 2. Initial roughness curve

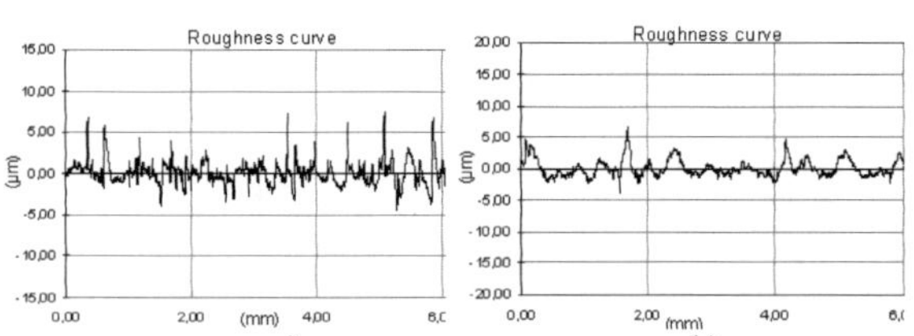

a) b)

Fig.3. Roughness curve final. a) n_p=20[rot/min], f_l=0, 05[mm/s], d_b=10[mm]; b) n_p=160[rot/min], f_l=0,05[mm/s], d_b=10[mm].

The analysis of the experimental data on the surfaces roughness, obtained after the manufacturing process was realized using home-made computer software written in GWBASIC language.

Using an alghoritm based on the least square method (Creţu, Gh. 1992), from experimental result we have determined the following empirical relation, that describe in the most adequate way the studied model:

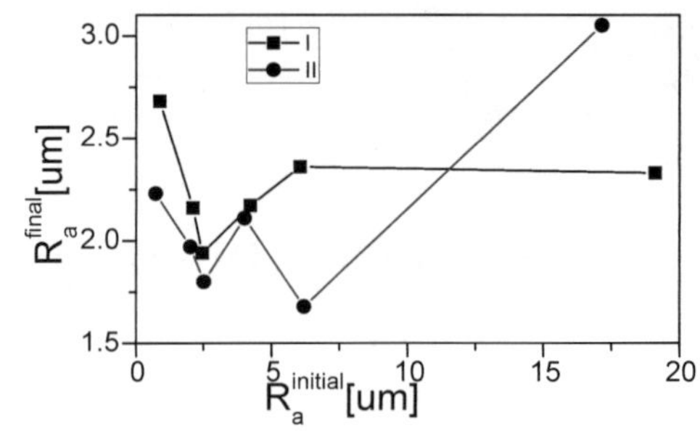

Fig. 4. Influence of the initial roughness on the surface final roughness parameter

$$R_{af} = 1,008391 + 0,005464 n_p - 0,14367 f + 0,02795 d_{ball} \quad (1)$$
$$s(1) = 0,2869167$$

In order to determine what element has the highest influence on the R_a surfaces roughness parameters, we have established a function type power:

$$R_a = 5,953731 - \frac{17,93975}{n_p} - \frac{0,05896}{f} - \frac{3,791355}{d_{ball}} - \frac{13,15997}{R_{ain}} \quad (2)$$
$$s(5) = 0,1735373$$

From the empirical model obtained for the R_a – roughness parameter, and studying the size of the exponents (2), it can be observed that the initial roughness of the workpiece R_{ainit} **[µm]** has the most influence on the R_{afinal} parameter followed by the n_p – piece revolution [rot/min], f_l-longitudinal feed [mm/s] and d_{ball} -ball diameter [mm].

When the ball diameter and the speed of work piece are increased, the surface roughness value decreases. These experimental observations confirm our initial hypothesis.

5. CONCLUSION

In general, it is indicated to have an initial high roughness, obtained with a low cost procedure in order to obtain in the end a material with excelent roughness properties.

We could remark that taking into consideration the absolute sizes of the exponents belonging to different factors; we can put in order these factors on the base of their importance. Some simplifying hypothesises were accepted to develop theoretical models for the surface roughness parameter Rz which characterizes the finished surface. In the future, are planned to use autocorellation equation to verify the validity of the theoretical models.

6. REFERENCES

Creţu, Gh. (1992). *Fundamentals of the experimental research. Handbook for the laboratory activities* (in Romanian). Institutul Politehnic, Iaşi

Drăghici G. (1984), *Manufacturing machines technologies* (in Romanian), Editura Didactică şi Pedagogică, Bucureşti

Ponomariov S.D. et al. (1964) Resistance *calculation in machine manufacturing* (in Romanian), Editura Tehnică, Bucureşti

Shiou F-J. & Chen C-H. (2003), Freeform surface finish of plastic injection mold by using ball-burnishing process, *Journal of Materials Processing Technology*, Vol. 140, pp. 248-254

Stepanenko- Aizenberg M.V. (2007), Model of high velocity penetration of an indenter into a medium. *Journal of Minning Science*, Vol. 43, pp. 254-272

Annals of DAAAM for 2008 & Proceedings of the 19th International DAAAM Symposium, ISSN 1726-9679
Editor B. Katalinic, Published by DAAAM International, Vienna, Austria 2008
Make Harmony Between Technology and Nature, and Your Mind will Fly Free as a Bird
*Young Researcher and Scientist Paper / * Supervisor, Mentor*

USING FUZZY LOGIG TO CONTROL THE POSITION OF A TWO-COORDINATE SYSTEM AND ITS APPLICATION TO PNEUMATIC DRIVES

TARATINSKY, I[lya] & MOGILNIKOV, P[avel]

Abstract: *Modelling of systems with uncertain nonlinearities and little physical insight such as pneumatic drives is a domain of those black box models showing universal approximating capabilities such as fuzzy models. In this article, a fuzzy controller, which normalizes uncertain nonlinearities, was developed and implemented in an industrial controller (Siemens S7-300) to control the position of a two-coordinate system of pneumatic drives, which was used for playing checkers.*
Key words: *Fuzzy, modelling, nonlinear, identification, pneumatic.*

1. INTRODUCTION

Since the mid 20th century, pneumatic control technology has been widely applied in the industry automation; in recent years, the technology is even applied in the different precision control machines, e.g., in the electronic industry. Owing to the compressibility characteristic of the air, the pneumatic control system has a very high level of nonlinearity and the system parameters are time variant with environmental change, therefore, the mathematical model isn't easy to derive and to obtain the accurate values of the coefficients of the model (Aldridge, 1992). The fuzzy theory and the application to the control technology have been successfully developed in recent years. The uncertainties affecting control problems can be solved through the fuzzy logic thinking and decision method.

For the pneumatic cylinder position control system, the dry friction forces, the compressibility of air, the pressure and the temperature are time variant uncertainties, affect the performance of position control. In different articles the authors (Kroll, 1996, Daugherity; Rathakrishnan & Yen, 1992) obtain the performance indices with rise-time, overshoot, settling time and steady state error; besides, with different searching methods to find the optimal control parameters and get the better performance. Adjusting the normalization factors for the optimal performance through trial and error is time-consuming and difficult.

Identification means computing the parameters of a given model structure assessing input/output data of the system considered. Various model structures are known. In general, linear models are easy to handle but will not yield satisfactory performance if the model validity is not restricted to small deviations from a fixed operating point. Nonlinear conventional models such as state-polynomial (e.g. bilinear or quadratic) ones require apriori knowledge for the choice of an appropriate model structure. In contrast, fuzzy models represent a rather general and flexible approach. They describe the input/output behaviour of a system using a set of IF—THEN-production rules of the form: IF *condition* THEN *action.* Such models may be interpreted as a weighted combination of several local models resulting in a nonlinear global model.

They describe the input/output behavior of a system using a set of If—THEN-production rules of the form: If condition Then action. Such models may be interpreted as a weighted combination of several local models resulting in a nonlinear global model. Hence a mismatch between the nonlinearities of local models and process is less significant compared with a single nonlinear model. Therefore fuzzy modelling has been applied especially to modelling tasks with uncertain nonlinearities, uncertain parameters and/or high complexity.

This fuzzy controller was applied to the pneumatic two-coordinate system that was used for playing checkers.

2. SYSTEM DESCRIPTION

The experimental layout of the pneumatic cylinder position control system is shown in figure 1. The control signal is sent from the industrial controller to the servo valve. The air flow rate into the cylinder can be regulated through the valve. The pressure difference between the chambers of the cylinder is built up and then the motion of the cylinder can be controlled. The position of the cylinder is measured by a pulse scale and the load of the cylinder is controlled by a proportional pressure control valve.

1. Compressed air source
2. F-R-L unit
3. Servo valve 1
4. Pneumatic cylinder 1
5. Pulse scale 1
6. Pulse scale 2
7. Pneumatic cylinder 2
8. Servo valve 2
9. Controller (Siemens S7-300)
10. PC

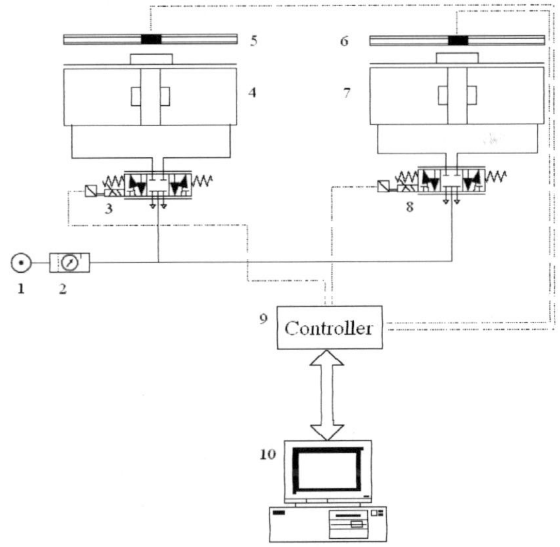

Fig. 1. Scheme of a two-coordinate pneumatic drive system.

3. DYNAMICAL FUZZY CONTROLLER

In this article functional fuzzy models (often referred to as models of the Sugeno type (Terano; Asai & Sugeno, 1992)) with multi-dimensional reference fuzzy sets as are described and used. Compared with the commonly used one-dimensional reference fuzzy sets the multi-dimensional approach is more flexible and promises higher model quality with less parameters.

A functional fuzzy model with M inputs X_j and one output y is described by a set of c rules such as (Zadeh, 1965):

$$R_i : IF(x(k-1)\ Is\ v_i)\Big|_{v_1,\dots,v_c}\ THEN\ \hat{y}_i(k) = f_i(x(k-1)) \quad (1)$$

R1: if e_k is P_1 then Δu_k is P_{u1}
R2: if e_k is N_1 then Δu_k is N_{u1}
R3: if Δe_k is P_2 then Δu_k is P_{u2}
R4: if Δe_k is N_2 then Δu_k is N_{u2}
R5: if $\Delta^2 e_k$ is P_3 then Δu_k is P_{u3}
R6: if $\Delta^2 e_k$ is N_3 then Δu_k is N_{u3}

where: $e_k = r - y_k$ is the error; $\Delta e_k = e_k - e_{k-1}$ is the error, multiple of 1st; $\Delta^2 e_k$ is the error, multiple of 2nd; $\Delta u_k = u_k - u_{k-1}$; r is the reference input, y_k is the output.

Comparing to PID, rules R1 and R2 correspond to integral action; R3 and R4 correspond to proportional action; R5 and R6 correspond to differential action based on fuzzy rules.

$$\mu_{P_i}(e_i) = \frac{1}{\pi}\cdot arctg(d_i \cdot e_i) + 0.5\,, \quad (2)$$

$$\mu_{N_i}(e_i) = \frac{1}{\pi}\cdot arctg(-d_i \cdot e_i) + 0.5\,, \quad (3)$$

$$\mu_{P_{ui}}(\Delta u) = \frac{1}{2\cdot b_i}\cdot \Delta u + 0.5\,, \quad (4)$$

$$\mu_{Nui}(\Delta u) = -\frac{1}{2\cdot b_i}\cdot \Delta u + 0.5\,,\ i = 1,2,3\,. \quad (5)$$

Membership functions $\mu_{P_i}, \mu_{N_i}, \mu_{P_{ui}}, \mu_{Nui}$ of fuzzy numbers P_i, N_i, P_{ui}, N_{ui} are shown on Fig. 2.

$$\Delta u_1 = \frac{1}{\pi}\cdot\frac{(arctg(d_1\cdot e_k) + arctg(d_2\cdot \Delta e_k))}{(g_1 + g_2)}\,, \quad (6)$$

$$\Delta u_2 = \frac{1}{\pi}\cdot\frac{(arctg(d_2\cdot \Delta e_k) + arctg(d_3\cdot \Delta^2 e_k))}{(g_2 + g_3)}\,, \quad (7)$$

$$\Delta u_3 = \frac{1}{\pi}\cdot\frac{(arctg(d_3\cdot \Delta^2 e_k) + arctg(d_1\cdot e_k))}{(g_3 + g_1)}\,, \quad (8)$$

$$d_i = \frac{tg(0.45\cdot\pi)}{a_i}\,,\ g_i = \frac{1}{2\cdot b_i}\,. \quad (9)$$

The fuzzy controller can be expressed as following equations: $\Delta u_k^* = medium\{\Delta u_1, \Delta u_2, \Delta u_3\}$. The membership function of Δu_k^*. can be defined as follows (Fig. 3):

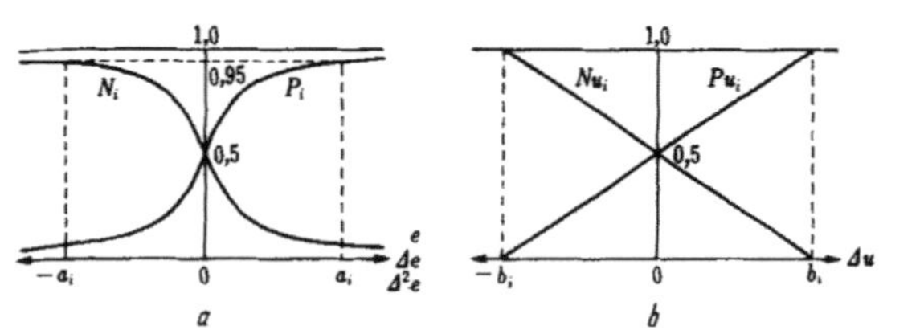

Fig. 2. Membership functions of fuzzy numbers P_i, N_i, P_{ui}, N_{ui}.

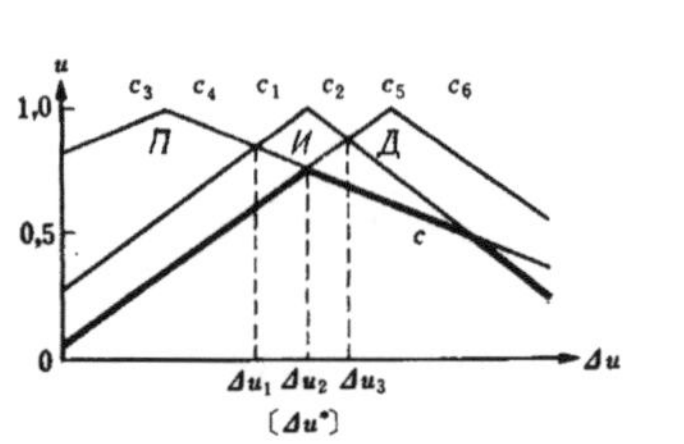

Fig.3. Membership function of Δu_k^*.

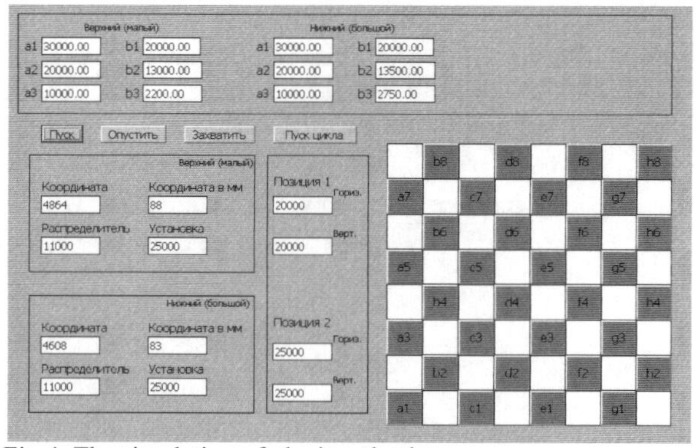

Fig.4. The simulation of playing checkers.

Fig.5. Two-coordinate robot system, which intended for playing checkers.

4. CONCLUSIONS

Modelling of systems with uncertain nonlinearities and little physical insight is a domain of those black box models showing universal approximating capabilities such as fuzzy models. Pneumatic drives belong to this kind of systems. Through the above explained controller design method and the experimental results, one can make the following conclusions:

1. The effect of uncertain factors, e.g., the dry friction force, the time response and the steady state error of the pneumatic cylinder position control with the designed controller are minimal. The experimental results have shown the advantages of such self tuning controllers.

2. The adaptivity of the external disturbance effect with the designed self tuning neural fuzzy controller is much better than that of controller without tuning capabilities. The fuzzy controller, which normalizes uncertain nonlinearities, was developed and implemented in industrial controller Siemens S7-300.

5. REFERENCE

Aldridge J. (1992). *Automated tuning and generation of fuzzy control system*, ISATrans.31, pp. 15-17.

Daugherity W.; Rathakrishnan B. & Yen J. (1992), *Performance evaluation of a self -tuning fuzzy controller*, IEEE Internet Conf., pp. 389 -397.

Kroll A. (1996). *Fuzzy-modelling of systems with uncertain nonlinearities and its application on pneumatic drives*, in IEEE *Int. Conf. on System, Man and Cybernetics*, Peking, 14.-17.10.1996.

Terano T.; Asai K. & Sugeno M. (1992). *Fuzzy systems theory and its applications*, Academic Press, INC., San Diego.

Zadeh L.A. (1965). *Fuzzy sets*, Information and Control, Vol. 8, pp. 338-353.

Annals of DAAAM for 2008 & Proceedings of the 19th International DAAAM Symposium, ISSN 1726-9679
Editor B. Katalinic, Published by DAAAM International, Vienna, Austria 2008
Make Harmony Between Technology and Nature, and Your Mind will Fly Free as a Bird

VIRTUAL AND REMOTE CONTROL LABORATORY USING MATLAB

TARCA, R[adu]; TARCA, I[oan] & POPENTIU-VLADICESU, F[lorin]

Abstract: *The work presented in this paper pretends to provide a general architecture for a virtual and remote control laboratory for a physical system using the Matlab platform. The virtual and remote control laboratory was developed at the University of Oradea and will be use by our students.*
Key words: *e-learning, virtual, remote control.*

1. INTRODUCTION

The accessibility of Information and Communication Technology (ICT) allowed the enhancement of the traditional learning methods. Nowadays, several universities have material and software facilities allowing students to perform laboratory experiments by simulation or with real equipment without any geographical or temporal limitations (Amadou, 2006).

The relationship between ICT and process control has reached a new stage, encouraging the creation of applications such as monitoring and control through the Internet, as well as teleworking, telemedicine, and telerobotics.

At this time, several e-learning laboratories have been developed. Two categories can be distinguished:

- remote distance-learning laboratories, which offer remote access to real laboratory equipment and instruments – figure 1;

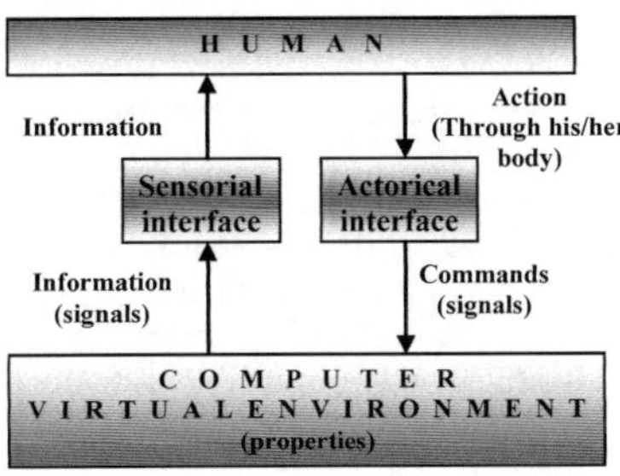

Fig.1. Remote distance-learning laboratory scheme

- virtual learning laboratories, which offer access to an virtual environment using for this simulation software;

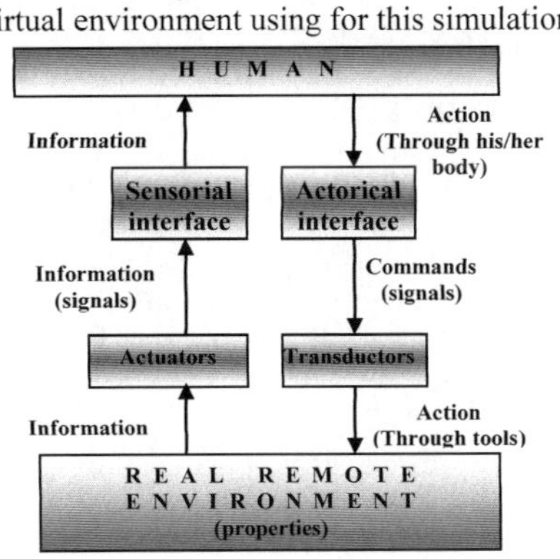

Fig.2. Virtual learning laboratory scheme

Based on the Internet, either a virtual learning laboratory (Schmid, 2001) or a remote distance-learning laboratory (Baccigalupi, 2006), (Callaghan, 2005), (Corter, 2004), (Ferrero, 2003), (Jimenez, 2007), (Khamis, 2003), (Sicker, 2005) are available for setting up a laboratory in a learning environment. A virtual laboratory allows continuous access to a simulated process on a computer. Remote distance-learning laboratories are an option halfway between traditional and virtual laboratories, allowing remote users to perform real experiments (Valera, 2005).

To implement virtual and remote laboratories, we have used one MATLAB-based software package, namely, MATLAB Web Server (by Mathworks).

2. THE VIRTUAL AND REMOTE CONTROL LABORATORY ARCHITECTURES

The scheme of the virtual and remote control architecture is shown in figure 3.

In the picture two main areas can be seen:

- local area in which the user works, and
- remote area where the whole physical system and control elements are located.

The elements of local and remote area are the following:

For the local area:

- Computer with Internet connection and an HTTP 4.0 client application. The application is optimized for Internet Explorer 6 and Netscape 7.

For the remote area:

- High speed Internet Connection.
- Computer Server.
- Data acquisition system – the control unit of the physical system.
- Physical system to control.
- Images capture system and web video server: a CCD camera with MPEG-4 video compression streams.
- Http Server. This server allows the communication of the computers using the http protocol.
- MATLAB R2007 with SIMULINK.
- Real - Time Windows Target Toolbox V.2.1: this toolbox allows Simulink schemes to be executed in real time. For this purpose, it provides the necessary blocks for the interaction with the data acquisition system.

The Software part of the system consists of two modules:

1. Web application: this includes client-server communication using HTTP/HTML protocol, the user interface, user's access control, and the main CGI application. The Common Gateway Interface (CGI) is a standard for interfacing external applications with information servers, such as HTTP or Web servers.
2. Real-time application: this is a set of predefined Simulink control schemes and Matlab code, based on Real time Windows Target toolbox, which implements the real time execution of Simulink schemes over a specific physical system.

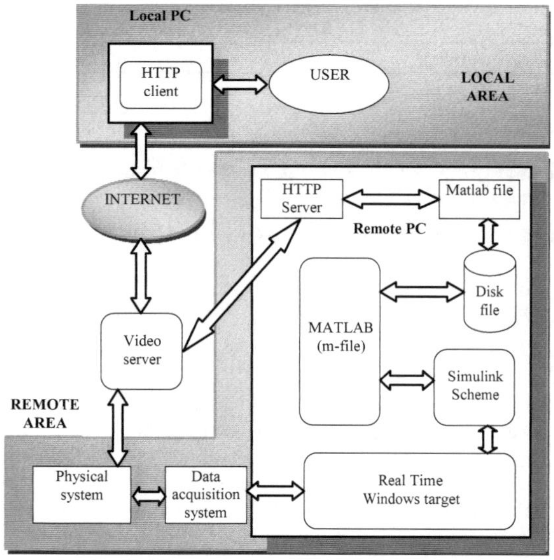

Fig.3 General architecture of the system

PHP is a popular script language that has been chosen as far as it is an open language widely supported by most web servers and O.S. platforms, and with an extensive library that supports every network protocol. PHP code runs on the web server so it shows a controlled environment for the programmer and can communicate with any other process running in the server (Matlab application in our case).

3. EXPERIMENTAL SYSTEMS

The experimental system is presented in figure 4.

The system consists in:

- one DC servomotor;
- worm-gear transmission;
- two synchronous belt transmissions;
- an incremental rotation transducer.

Once the user has accessed the system, a page appears in which all needed data to perform the real-time execution is requested. User can choose the physical system (DC motor, worm gear transmission), the control model (speed or position feedback, space state feedback, identification), the type of execution (Simulation or Real-time execution) and Regulator type (PID, algebraic regulator, etc). After introducing all data, the experiment can be performed; when is finished, the output signal is shown on the screen. Moreover, the application allows to download a "*.mat" file with the values of the most significant signals (output, control action, etc.) in order to be analyzed by the user.

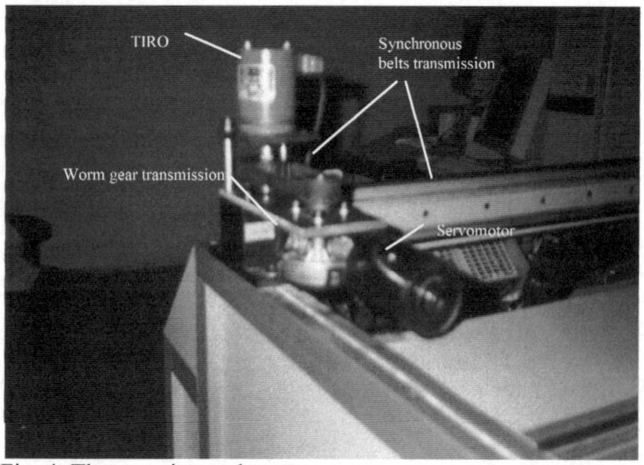

Fig. 4. The experimental system

After the execution, a web page with the graph of the output signal is presented to the student (in this case the engine velocity). This page shows that the system allows the download of all the signals involved in the execution in order to be analyzed by the students.

During real-time execution the user has access to a compressed video stream showing the experiment. For this, a high bandwidth internet access is required.

4. CONCLUSION

This paper presents an experiment realised by our research team in the field of virtual and remote control laboratories. The advantage of the proposed system is that it helps the student to perform practice experiments remotely without a strict timetable. The tool developed, presented in this paper can also be used to test new control schemes over different physical equipments.

5. REFERENCES

Amadou, M. M. D.; Saad M.; Kenné, J. P.; Nerguizian V.; *Virtual and remote laboratories*, Proceedings of the 1st IEEE International Conference on E-Learning in Industrial Electronics, pp. 176-173, Dec. 2006, ISBN 1-4244-0324-3.

Baccigalupi, A.; De Capua, C.; Liccardo, A.; (2006) *Overview on Development of Remote Teaching Laboratories: from LabVIEW to Web Services,* IMTC 2006 - Proceedings of the 23rd IEEE Instrumentation and Measurement Technology Conference, Sorrento, Italy 24-27 April 2006, pp. 992-997, ISBN 0-7803-9360-0.

Callaghan, J.M.; Harkin, J.; El Gueddari, M.; McGinnity, T.M.; Maguire, P.L.; (2005) *Client-server architecture for collaborative remote experimentation*, Proceedings of the Third International Conference on Information Technology and Applications (ICITA'05), Volume 2, 4-7 July 2005 pp. 125 - 129 vol.2, ISBN 0-7695-2316-1.

Corter, J. E.; Nickerson, J. V.; Esche, S. K.; Chassapis, C.; (2004) *Remote versus hands-on labs: A comparative study.* In Proceedings of the 34th ASEE/IEEE Frontiers in Education Conference. October 20 – 23, 2004, Savannah, GA, pp. F1G.17-F1G.21, ISBN 0-7803-8552-7.

Ferrero, A.; Salicone, S.; Bonora, C.; Parmigiani, M. (2003) *ReMLab: a Java-based remote, didactic measurement laboratory,* Instrumentation and Measurement, IEEE Transactions on Volume 52, Issue 3, June 2003, pp. 710 – 715, ISSN 0018-9456.

Jimenez, L.M.; Puerto, R.; Reinoso, O.; Neco, R.P.; Fernandez, C.; (2007) *Remote Control Laboratory Using Matlab and Simulink,* IEEE International Symposium on Industrial Electronics, 2007. ISIE 2007., 4-7 June 2007, pp. 2963-2967, Vigo, ISBN: 978-1-4244-0755-2.

Khamis, A.; Rivero, D.M.; Rodriguez, F.; Salichs, M.; (2003) *Pattern-based Architecture for Building Mobile Robotics Remote Laboratories* Proceedings of the 2003 IEEE Intsmatiooai Conrerenee on Robotics & Avtomntion Taipei, Taiwan, Sept. 2003, pp.3284-3289, ISBN 0-7803-7736-2.

Schmid, C. (2001) *Virtual control laboratories and remote experimentation in control engineering*, Proc 11th Annual Conference on Innovations in Education for Electrical and Information Engineering, pp. 213-218, 2001.

Sicker, D.C.; Lookabaugh, T.; Santos, J.; Barnes, F.; (2005) *Assessing the Effectiveness of Remote Networking Laboratories,* Proceedings 35th Annual Conference: Frontiers in Education, 2005. 19-22 Oct. 2005 pp. S3F-7 - S3F-12, ISBN 0-7803-9077-6.

Valera, A.; Díez, J. L.; Vallés, M.; Albertos P.; (2005) *Virtual and Remote Control Laboratory Development,* IEEE Control Systems Magazine, february 2005, pp. 35-39, ISSN 0272-1708.

Annals of DAAAM for 2008 & Proceedings of the 19th International DAAAM Symposium, ISSN 1726-9679
Editor B. Katalinic, Published by DAAAM International, Vienna, Austria 2008
Make Harmony Between Technology and Nature, and Your Mind will Fly Free as a Bird

TITANIUM OBTAINING THROUGH ELECTROCHEMICAL REDUCTION OF TITANIUM DIOXIDE IN CHLORINATED MEDIA

TARCOLEA, M[ihail]; COTRUT, M[ihai] C[osmin]; CIUCA, S[orin]; SOARE, V[asile] & SURCEL, I[oan]

Abstract: *A variant of Fray-Farthing-Chen (FFC) Cambridge process for direct extraction of metallic Titanium in molten salts is presented. In our solution, the cathode of electrochemical cell (TiO_2 powder) is pressed and sintered as a disc and is reduced in a molten mixture of Calcium and Sodium chlorides with a graphite anode. Electrochemical process develops in two stages:* ***ionization*** *of oxygen from TiO_2 and its transit into electrolyte bath, phenomenon accelerated by the large solubility of the Calcium oxide in the chlorinated mixture, and* ***calciothermic*** *reduction of TiO_2 with the Calcium provided by electrolysis of Calcium oxide that is solubilised in molten electrolyte. After process, at 850°C and 2÷3V, the purity of obtained metal was about 95 %.*

Key words: *Titanium, reduction, molten salt, FFC process*

1. INTRODUCTION

A variant of FFC Cambridge process (Fray, 1998) for direct production of Titanium is presented. The declared intention is to replace Kroll process (Kroll, 1940).

The method consists in reduction of Ti oxide through an electro-deoxidation and electro-reduction process, with molten Calcium chloride as electrolyte.

The cathode is a pressed-sintered disc, consisting in Ti dioxide powder.

The established method is based on the following physical and chemical considerations:

- CaO solubility in Calcium chloride is very high (about 20% moles), making possible the electrolysis process and cathode deposited Calcium reduces the metal oxide that constitutes the cathode (Fray et al., 1998).
- decomposition voltage of CaO is less than that of $CaCl_2$, hereby only CaO electrolysis determines oxygen liberation in anode (eventually with CO_2 and CO forming), avoiding gaseous Chlorine liberation. Concomitantly, by Oxygen consuming during electrolysis, through dissolved CaO re-establishment (Ca solubility in $CaCl_2$ is about 4% mol), the chemical composition of the electrolyte maintains quasi-constant, with a favorably influence on the electrochemical process.
- high affinity of Ca for Oxygen allows the advanced reduction of the Oxygen content in cathode, towards level of ppm, enabling high quality alloys obtaining.

The procedure bases are the flowing electrochemical processes, expressed as characteristic reactions (Chen et al., 2000; Chen & Fray, 2000; J.F Suzuki et al., 2003, Suzuki, 2005):

- deoxidation of TiO_2 through contained Oxygen ionization as a result of applied voltage on the electrodes.

$$\tfrac{1}{2}O_2 + 2e^- = O^{2-} \quad (1)$$

$$2CaCl_2 + O^{2-} = 2\underline{CaO} + 2Cl^- \quad (2)$$

$$Me_xO_y + 2e^- = xM + yO^{2-} \quad (3)$$

- calciothermic reduction of the oxides through the action of deposited Ca on cathode, as a result of the dissolved CaO electrolysis in $CaCl_2$ electrolyte.

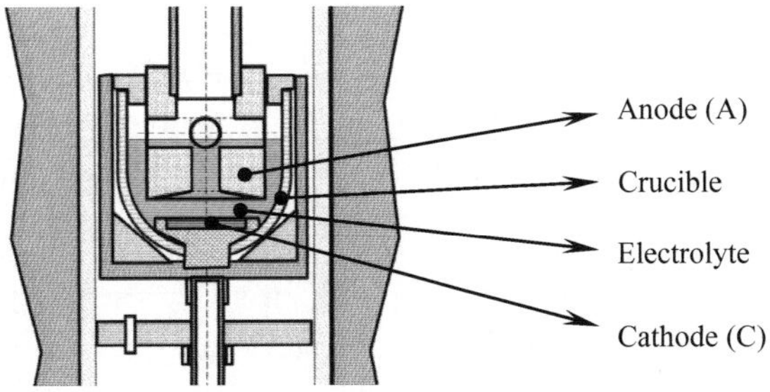

Fig. 1. Detail of the installation for electro-reduction process.

$$CaO = Ca^{2+} + O^{2-} \quad (4)$$

$$2O^{2-} \rightarrow O_2\,(gas) + 2e^- \quad (5)$$

$$O_2 + C = CO_2\,(or\ CO) \quad (6)$$

$$yCa + Me_xO_y = x\underline{Me} + yCaO \quad (7)$$

2. EXPERIMENTAL PROCESS EVOLUTION

The experimental installation for electro-reduction process (fig. 1) consists mainly in a crucible (Silicon and Aluminum oxy-nitride) that contains the electrolyte (Calcium and Sodium chloride), having an anode (super dense graphite) and a cathode (pressed/sintered disc of Titanium dioxide to be reduced). All of them are included in a heating furnace to assure the necessary thermal conditions.

The electrochemical process develops in two stages (fig. 2). In the first stage the electric potential is 2.6 V (less than CaO decomposition voltage); during this stage current intensity constantly decayed (fig. 2). In the second stage the voltage was increased to 3 V (less than $CaCl_2$ decomposition voltage, to avoid gaseous Chlorine appearance), with an initial current intensity of about 3.7 A. The time current intensity decay was more abrupt in the second stage (fig. 2). Entire process develops at 850°C. The main process parameters are listed in Table 1.

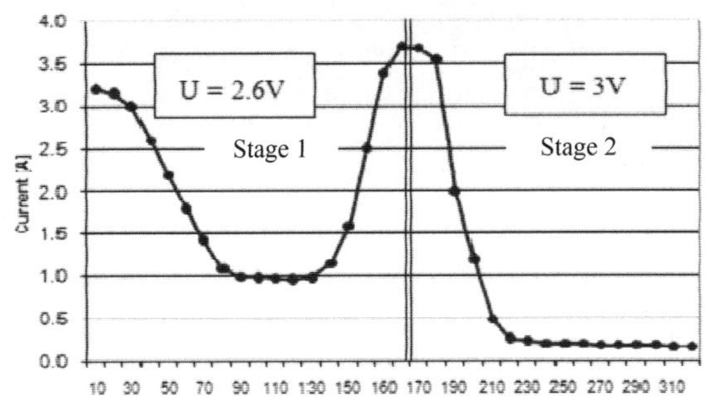

Fig. 2. Variation of current intensity during electrochemical process.

Sample	Stage	Potential, V	Current intensity, A	Distance A-C, mm	Duration, min
A	I	2.6	3.20 ÷ 0.20	20	160
	II	3.0	3.75 ÷ 0.30		200
B	I	2.6	3.30 ÷ 0.25	15	160
	II	3.0	3.80 ÷ 0.35		190
C	I	2.6	3.45 ÷ 0.25	10	160
	II	3.0	3.95 ÷ 0.35		180

Tab. 1. Process parameters for samples A, B, and C.

Cathode (pressed/sintered TiO_2)	Surface, cm^2	Density, g/cm^3	Porosity, %
A	6.25	2.33	44.20
B	6.25	2.34	44.05
C	6.20	2.38	42.85

Tab. 2. Characteristic measurements of cathode discs.

Sample	Chemical composition, % weight				
	Ti	O	Ca	Cl	Fe
A	95,00	0,70	1,49	2,61	0,20
B	95,30	0,75	1,23	2,46	0,26
C	94,43	0,74	1,56	2,98	0,29

Tab. 3. Chemical composition of cathodes after calciothermic reduction process.

3. RESULTS

After electrolyte removal and disc extraction from cathode holder, disc was washed in a slightly acid solution (pH = 2÷4) and physical, chemical and structural analyzed. *Dimensional measurements* of discs before the electrochemical reduction process are presented in Table 2.

Chemical analysis on samples assayed from plane surfaces of discs, in depth towards 1 mm, (Table 3), demonstrates the success of this direct obtaining method; the Titanium content can be raised further, by raising the cathode discs porosity.

SEM and EDAX analysis of the sintered cathode after electrolysis was performed to reveal the structural aspects (fig. 3) and to determine phase composition on the analyzed surface (fig. 4) for the sample C. It is also obvious the advanced reduction of Titanium oxide to metallic Titanium.

5. DISCUTION

In the first stage of the electrochemical process, the current intensity decay can be explained through the corresponding decay of dissolved Oxygen in melted electrolyte by its discharging at anode. The same phenomenon, in the second stage of the electrochemical process, can be explained through the raise of the ions quantity in the melted electrolyte, as a result of dissolved CaO decomposition.

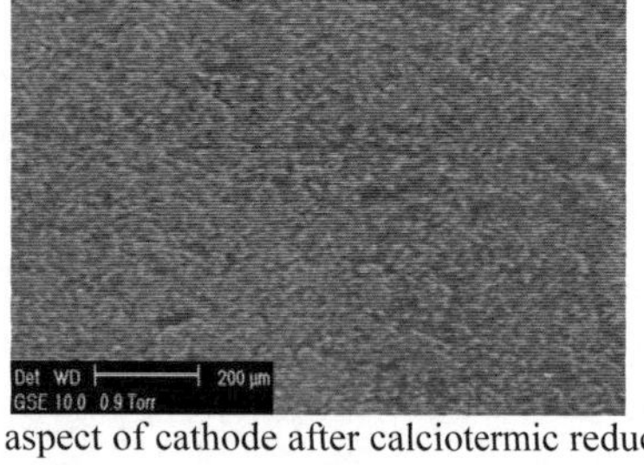

Fig. 3 SEM aspect of cathode after calciotermic reduction.

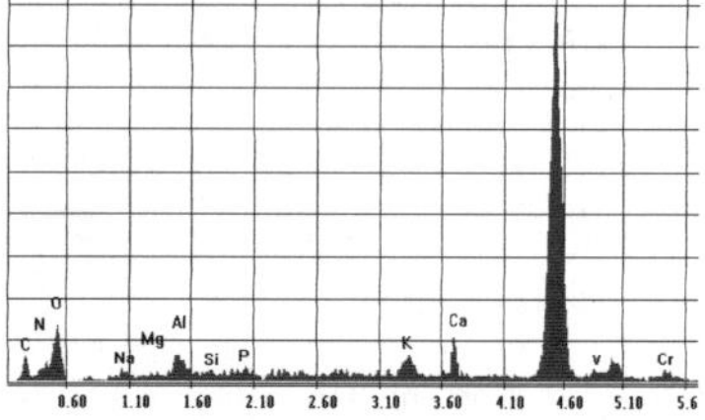

Fig. 4. EDAX spectrum of cathode after calciotermic reduction.

The experimental data demonstrate the possibility to obtain high purity Ti through an electrochemical process of deoxidation – reduction of Titanium oxide as follows:

- deoxidation of metal oxide through Oxygen ionization, developed at voltages about 2.0 ÷ 2.5 V,

- calciothermic reduction of the oxide through the action of Ca deposited at the cathode as a result of electrolysis of the dissolved CaO into electrolyte, at voltages of 2.7 ÷ 3.0V.

Through successive deoxidation processes, starting from Oxygen ionization from oxide, followed by its reduction with Calcium originated from the electrolysis of dissolved CaO in the electrolyte, the duration of alloy obtaining is reduced to about 4 hours, comparing with minimum 15 hours necessary to obtain the alloy only by deoxidation (Chen et al., 2000).

Oxygen and Calcium diffusion into electrolyte bath and sintered oxides, respectively, can be accelerated by decreasing height of sintered cathode and increasing cathode porosity. These precautions can improve the reduction process kinetics, with its duration decreasing.

4. CONCLUSIONS

In contrast to classical technologies for Titanium (Kroll, 1940), this experimental process presents spectacular advantages: drastic adjustment of ample operations number, of process duration, and of production costs through decrease of energetic and raw materials consumption.

This new technology is ecological (the electrochemical process develops at voltages lower than 3.2V, corresponding to $CaCl_2$ decomposition potential), without insult emission (Chlorine and chlorinated compounds, dust, etc).

Our method is based on cathode porosity control. This is the major difference from other authors' experiments: the use of porous sintered oxide cathode allows improving Titanium purity, process duration minimizing and a better process parameters control.

This procedure can be used other refractory metals and alloys direct obtaining. We already produced Ti6Al4V, following a similar procedure starting form Ti, Al, and V oxides.

5. REFERENCES

Chen, G.Z., Fray, J.F., (2000), Novel Direct Electrochemical Reduction of Solid Metal Oxides to Metal using Molten Calcium Chloride as the Electrolyte, *Euchem 2000 Proceeding*, Denmark, pp 157-161

Chen, G.Z., Fray, D.J., Farthing, T.W., (2000), Direct electrochemical Reduction of Titanium Dioxide to Titanium in Molten Calcium Chloride, *Nature*, vol. 407, pp. 361–364

Fray, D.J., Farthing, T.W., Chen, G.Z., (1998), *Removal of Oxygen from Metal Oxides and Solid Solutions by Electrolysis in a Fused Salt*, UK Patent, PCT/GB 99/01781, Inter. Pub. No. WO 9964638

Kroll, W.J., (1940), The Production of Ductile Titanium, *Trans. Am. Electrochemical Society*, vol. 78, pp. 35-47

Suzuki, R., (2005), Calciothermic Reduction of TiO2 and in situ Electrolysis of CaO in Molten CaCl2, *Journal of Physics and Chemistry of Solids*, volume 66, pp. 461-465

Suzuki, R., Teranuma, K., Ono, K., (2003), Calciothermic Reduction of Titanium Oxide and in-situ Electrolysis in Molten $CaCl_2$, *Metallurgical and Materials Transactions B*, volume 34 B, pp. 277-285

Annals of DAAAM for 2008 & Proceedings of the 19th International DAAAM Symposium, ISSN 1726-9679
Editor B. Katalinic, Published by DAAAM International, Vienna, Austria 2008
Make Harmony Between Technology and Nature, and Your Mind will Fly Free as a Bird

MODULAR ORTHOPEDIC IMPLANTS FOR FOREARM BONES BASED ON SHAPE MEMORY ALLOYS

TARNITA, D[aniela]; TARNITA, D[an]; BIZDOACA, N[icu]; NEGRU, M[ihai] & COPILUS, C[ristian]

Abstract: *The present paper describes the application of a shape memory alloy in the construction of an orthopedic implant. The main theme of this paper is the design of modular adaptive implants for fractured bones. After a particular stage of the healing period has passed, using implant modularity, the load is gradually transferred to bone, assuring in this manner a gradual recovery of bone function. The adaptability of this design allows the doctor to make the implant to correspond to the specific anatomy of the patient. Using 3D virtual models of the forearm bones (radius and ulna), we have undergone the mechanical simulation of different types of loading on the bones.*
Keywords: *numerical simulations, modularity, orthopedic implants, Nitinol.*

1. INTRODUCTION

Bionics or Biomechatronics is a fusion science which is interrelated with medicine, mechanics, electronics, control and computers. The finite products of research in this field are implants and prostheses for humans and animals. The roll of the implants and prostheses is to interact with muscles, the skeleton, and the nervous system to assist with or enhance motor control lost by trauma, disease, or defect. Prostheses/implants are typically used to replace parts lost by injury (traumatic) or missing from birth (congenital) or to supplement defective body parts. In addition to the standard artificial limb for every-day use, many amputees have special limbs and devices to aid with practicing sports or recreational activities.

1.1 Shape memory alloy

Shape memory alloys (SMA) constitute a group of metallic materials with the ability to recover to a previously defined length or shape when subject to an appropriate thermo-mechanical load. When there is a limitation of shape recovery, these alloys promote high restitution forces. Because of these properties, there is great technological interest in the use of SMA for different applications. Super-elastic NiTi has become a material of strategic importance as it allows specialists to overcome a wide range of technical and design issues like the miniaturization of medical devices or the increasing trend for less invasive and therefore less traumatic procedures. Essentially nitinol is an alloy containing approximately 50 at.% nickel and 50 at.% titanium. The shape memory alloy Nitinol has several advantages such as greater ductility, more recoverable motion, excellent resistance to corrosion, stable transformation temperatures, high biocompatibility, good kink resistance, less sensitivity to magnetic resonance imaging, fatigue life and the ability to be electrically heated to recover shape and this properties are presented in different papers (Pelton et al., 2000; Duering et al., 1999). The biocompatibility of these alloys is one of the important points related to their biomedical applications as orthopedic implants, cardiovascular devices, and surgical instruments. Studies about the medical applications of the shape memory alloys have been made (Friend & Morgan, 1999).

2.METHOD FOR VIRTUAL 3D BONES MODELS

The natural variability of the geometry and mechanical properties among bones from one to the other is a crucial problem that creates real difficulties in the field of biomechanical research. The dimensions, form, mechanical properties, elastic constants and physical constants of the bone are different among bones of different types and even among different bones of the same type. They depend on: age, sex, height, profession etc. The geometrical aspects of modelling the bone are dominated by the necessity of using spatial models because most of the bone elements have complicated spatial geometrical forms. The ulna and the radius are long bones of the forearm, broader proximally and narrower distally.

To obtain the cross sections of the bones, a PHILIPS AURA CT tomograph installed in the Emergency Hospital of Craiova was used. To obtain the tomography of the two bones of human forearm (radius and ulna) we used scanning schemes: for the ends of the bones the scanning operation was made at distances of 1 mm and for the medial areas at distances of 3 mm. In Figure 1 we present main images of the medial ulna, which show the changes in the shape of the bone. The obtained images were re-drawn in AutoCad over the real tomographies and the drawings were imported in SolidWorks (a parametrical CAD software), section by section, in parallel planes. The sections of the ulnar bone are presented in figure 1. Solidworks allows the yield of a solid by "unifying" the sections drawn in parallel planes. The command which makes these sections appear as a solid was Loft Shape. For the next step we used the ANSYS program for the discretisation by means of the finite elements method of the spatial structure of the ulna and radius bones (Tarnita et al, 2006).

2.1.Virtual simulations of bones stresses

In order to identify the target shape of modular implants, different studies were made (Tarniţă et al., 2007). The studies identify the stresses and deformations maps developed by radius and ulna for different external statical solicitations. The geometrical models were subjected to different solicitations to allow the study of the most solicited parts of the bone which will be the likely fracture sites. In figure 2 we exemplified two simulation results, for the forearm bones: radius bone and ulna bone, important for identifying stresses distribution and fracture areas.

The clinical observations and the laboratory tests confirm the results obtained by virtual simulation.

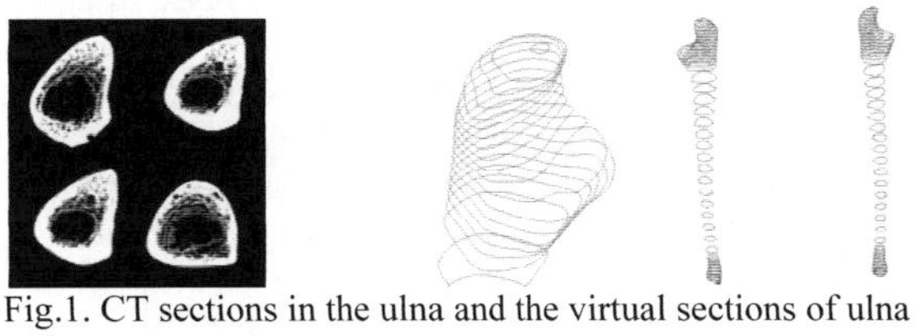

Fig.1. CT sections in the ulna and the virtual sections of ulna

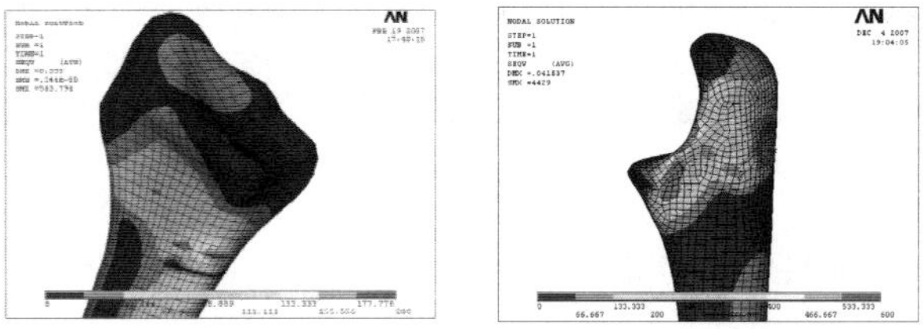

Fig.2. Stresses distributions for radius bone and ulna bone

3. MODULAR ADAPTIVE IMPLANT

The design idea of modular adaptive implants results from the following observations:

-doctors have limited degrees of freedom in selecting the proper dimensional apparatus for bone fractures;

-the current mechanical devices used in orthopedics lose some of their mechanical characteristics after some time (especially elasticity, which should ensure a constant tension that is mandatory for the correct anatomical healing of the fractured bones);

-the process of fracture healing has a particular dynamic, which imposes the necessity of particular progressive tension or discharge to improve the recovery time, depending on the normal structure and function of the bone;

-to improve the healing process, the fractured parts have to be in permanent contact in order to ensure the proper conditions to develop bone calluses. The actual or external fixator has to be manually adjusted with respect to the main axis of the bone. Unfortunately, the degrees of freedom of current devices are limited to 3 or 4 screws;

-a minimally invasive surgery ensures protection from blood edema and improves bone recovery and vascularization of the region.

The solution to these problems is the **M**odular **A**daptive **I**mplant - MAI. The proper shape of MAI is related to the microscopic structure of the bone (Bizdoaca et al., 2008). As one can observe, comparing the structure of a healthy bone with that of an osteoporotic bone (figure 3), the internal architecture of the healthy bone has a regular modular structure.

A modular net, identical in structure with the bone and locally configurable in terms of tension and release, is best design solution in terms of biocompatibility (Bizdoaca et al., 2008). The identification of the mechanical solicitation of the particular bone structure leads to the concept of the practical implementation of a feasible device able to undertake the functionality of normal bones. This device will partially discharge the tensions in the fractured bones (the fractured parts still need to be tensioned to allow the formation of the callus) improving the recovery time and the healing conditions. The proposed intelligent device has a network structure, with modules made out of Nitinol, especially designed in order to ensure a rapid connection and/or extraction of one or more MAI modules. The binding of the SMA modules ensures the same function as other immobilization devices, but also respects additional conditions concerning variable tension and its discharge. Moreover, these modules allow little movement in the alignment of the fractured parts, reducing the risks of wrong orientation or additional only bones callus. We suggest the design shown in Figure 4 for the unitary SMA module structure

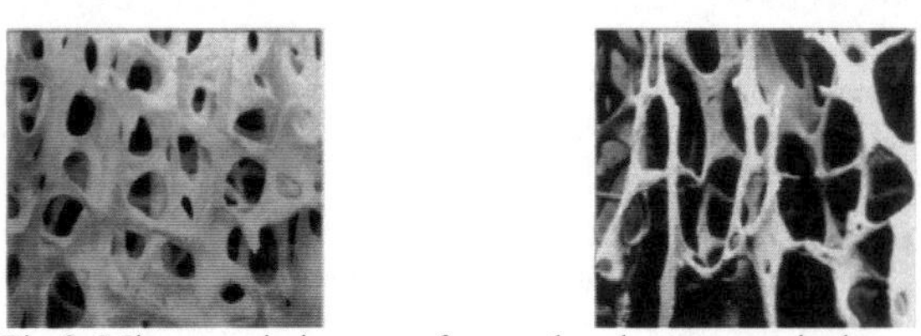

Fig.3. Microscopic images of normal and osteoporotic bone

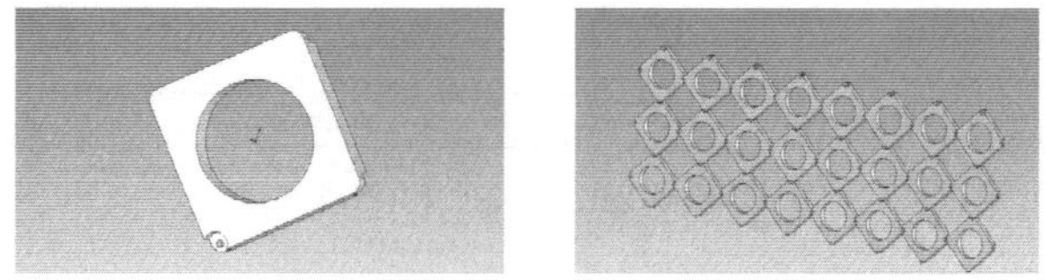

Fig.4. The SMA module and the SMA connection network

The design of the SMA module ensures not only the stability of the super-elastic network and constant force requirements, but also a rapid coupling/decoupling procedure.

Doctors can use SMA modules with different internal reaction tension, but all the modules will have same shape and dimension (Figure 4). The connection with affected bones and the support for this net are similar to those of a classic external fixator, but allowing for the advantages of minimal invasive techniques. The new device leads to a simple post-operatory training program of the patient.

4.CONCLUSIONS

The study of this technique offers a feasible direction. In the future, we want to realize a different type of SMA module and to experiment with them on real bones. At the same time, the studies will be developed in the direction of numerical simulation of the complex ensemble made up of the bone and the MAI network for different functional regimes, for different weight, temperature and physical-chemical condition and especially for different types of bone fractures. A very promising direction is the design and implementation of an independent and permanent functional adaptive MAI network, which can ensure a very rapid reintegration of the patient in the social life with minimal costs of the treatment.

Acknowledgement

Research activity supported by Ministry of Education, Research and Youth, programs PNCDI2 Id_92/2007 and CEEX 259/2006

5.REFERENCES

Bizdoaca, N.; Tarnita, D.; Tarnita, D. N.; Popa, D. & Bizdoaca, E. (2008). Shape Memory Alloy based Modular Adaptive Ortophedic Impants. *Proceedings of The International Conference on biomedical electronics and biomedical informatics, BEBI'08,* pp 123-128, Rhodos, August, 2008.

Bizdoaca, N.; Tarnita, D.; Tarnita, D. N.& Bizdoaca, E. (2008). Application of smart materials-bionics modular adaptive implant, *Proceedings of The 11th International Conference on Climbing and Walking Robots, CLAWAR'08*, pp.190-197, Coimbra, Portugal, September, 2008.

Duering, T.; Pelton, A.R. & Stöckel, D. (1999). An overview of nitinol medical applications. Material Science and Engineering, A273-275, pp. 149-160, ISSN 0921-5093.

Friend, C.M., & Morgan, N.B. (1999). Medical Applications for Shape Memory Alloys. *Professional Engineering Publishing Ltd.*, UK, pp.1-16.

Pelton, A.R.; Stöckel, D. & Duerig, T.W. (2000). Medical uses of nitinol, *Materials Science Forum*, 327-328, pp. 63-70, ISSN 0255-5476.

Tarnita, D., Popa, D., Tarnita, D. N., Grecu, D. (2006). CAD method for 3D model of the tibia bone and study of torsion and compression stress using the finite element method, *Romanian Journal of Morphology and embriology,* Vol. 47. No.2, (october, 2006), pp.181-186, ISSN 1220-0522.

Tarnita, D.; Negru, M.; Tarnita D.N. & Grecu, D. (2007). Stresses and Displacements of Radius Bone Using Finite Element Method. *Proceedings of The 8th EUROPEAN Congress on Ortopedics, EFORT*, pp.354, Firenzze, Italy, May, 2007.

Annals of DAAAM for 2008 & Proceedings of the 19th International DAAAM Symposium, ISSN 1726-9679
Editor B. Katalinic, Published by DAAAM International, Vienna, Austria 2008
Make Harmony Between Technology and Nature, and Your Mind will Fly Free as a Bird

MINIROBOTS WITH ADAPTABLE STRUCTURE

TATAR, O[limpiu]; MANDRU, D[an]; ALUŢEI, A[drian] & LUNGU, I[on]

Abstract: *The possibilities to use robots for inspection, exploration and maintenance of the pipes are highlighted. In this paper, the authors' contribution in this field is discussed. The minirobot presented in this paper is composed of slider-crank mechanisms and uses three geared mechanisms which are driven by a DC geared motor placed in the central region of the module.*

Key words: *in pipe minirobot, structure, adaptable, mechanism*

1. INTRODUCTION

The studied minirobots are characterized by an adaptable structure, based on linkages mechanisms. A very important design goal of these minirobots is the adaptability to the inner diameters of the pipes. Pipe diameter, which is one of the important size parameters, limits the working space occupied by the inspection robot. Therefore, it is necessary to be considered that a robot is designed for a certain size of pipe diameter.

Various pipe robots are developed for the inspection of pipes. For example, it was developed a three-wheeled pipe robot that uses scissors-like structure (Okada & Sanemori, 1987), a multi-joint pipe robot that uses the active universal joints as the steering mechanism (Choi & Ryew, 2002) or a wheel type pipe robot based on the differential drive principle (Roh & Choi, 2005).

2. THE DEVELOPED IN-PIPE MINIROBOTS

The minirobot that is presented is composed of slider-crank mechanisms placed at 120° angles around the central axle. This structure can adapt more easily to the variation of the pipe's diameter (Fig. 1).

The mass of the minirobot, including the power wires, is of 630 [g]. The wheels have a radius r = 25 mm, a length of 7 mm and the component elements have the lengths: h_1 = 95 mm, h_2 = 58 mm, h_3= 53 mm (Tătar et al., 2008).

The minirobot's propulsion is achieved by using three drive wheels. The drive wheels are placed into motion using three geared mechanisms which are driven by a DC geared motor placed in the central region of the module.

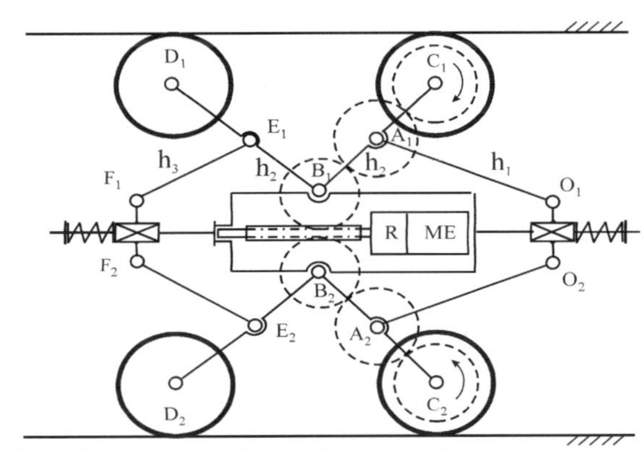

Fig. 1 The structural scheme of the minirobot

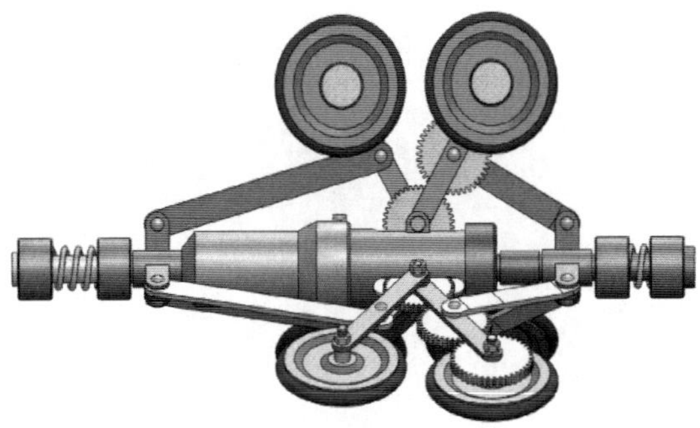
b)

c)

Fig. 2 The 3D model of the minirobot

The minirobot also has in its structure two sliding elements and two helicoidal springs which generate the force needed for the wheel to press against the inner surface of the pipe.

The movement transmission from the motor to the drive wheels is achieved by using three chains placed on gears and driven by the worm gear placed on the motor's axle (Fig. 3).

In figure 3 we have the following annotations: ME - DC motor with gear reducer (1/53 ratio motor/gearbox drives, 6V), *1* – worm, *2, 3, 4* – gears, z_1= 1 one thread, z_2 = 42, z_3 = 38, z_4 = 38 teeth, n_M the motor rotation frequency and n_R motor wheel rotation frequency (Tătar & Mândru, 2008).

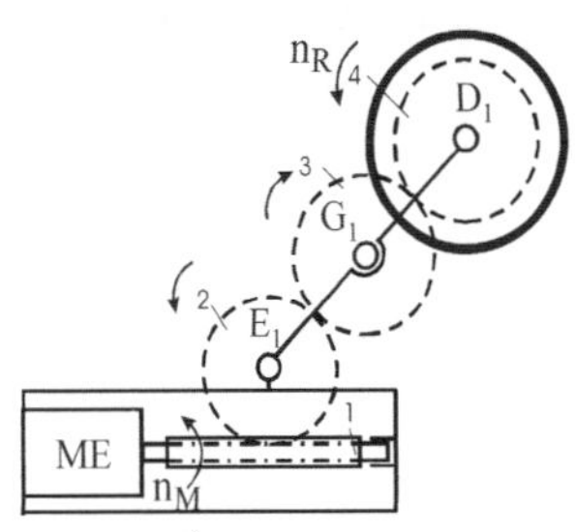

a)

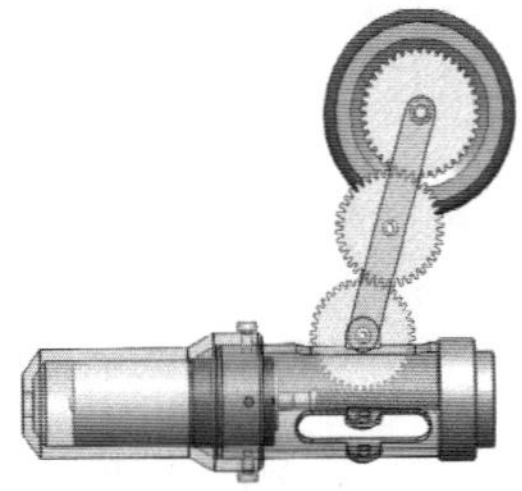
b)

Fig. 3 Transmission with gears in the structure of the minirobot
a) The structural scheme b) The 3D model

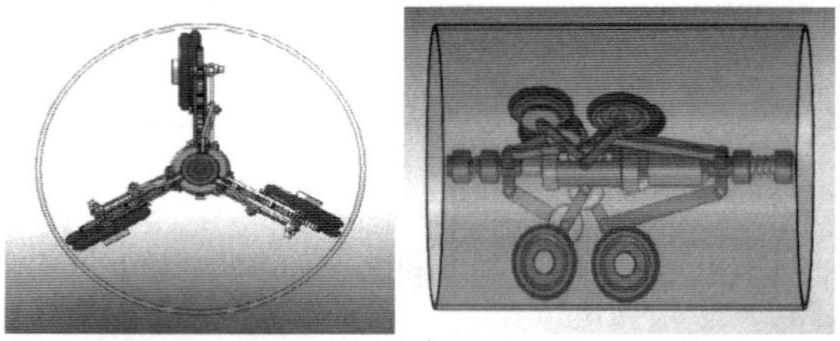

a)

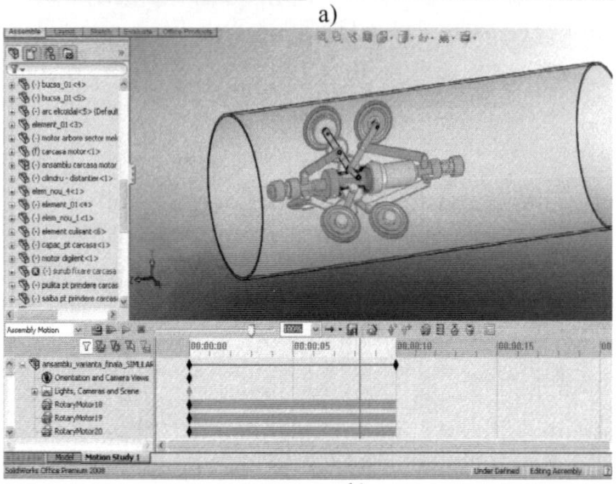

b)

Fig. 4 The model of the minirobot inside the pipe and the simulation of its locomotion

Fig. 5 The photography of the minirobot

These minirobots have movement *capacities for inspection in (140 – 180) mm diameter* pipes.

3. THE POWER AND CONTROL SYSTEM

The hardware component consists of a CEREBOT II motherboard (Fig. 6) and a few additional peripheral modules (Fig. 7), all developed by Digilent (http://www.digilentinc.com/).

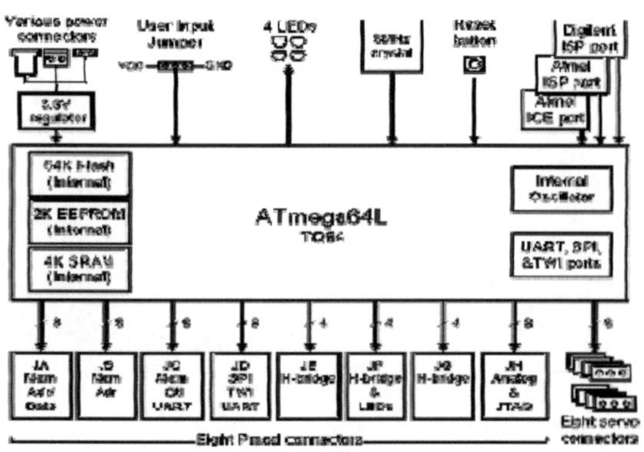

a)

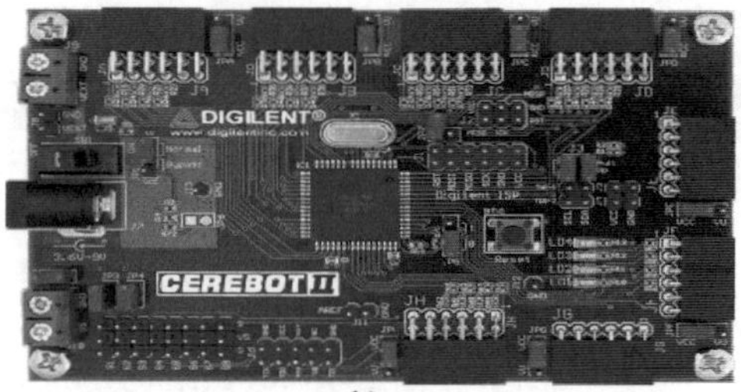

b)

Fig. 6 The CEREBOT II platform: a) block sketch; b) Photo of the circuit board

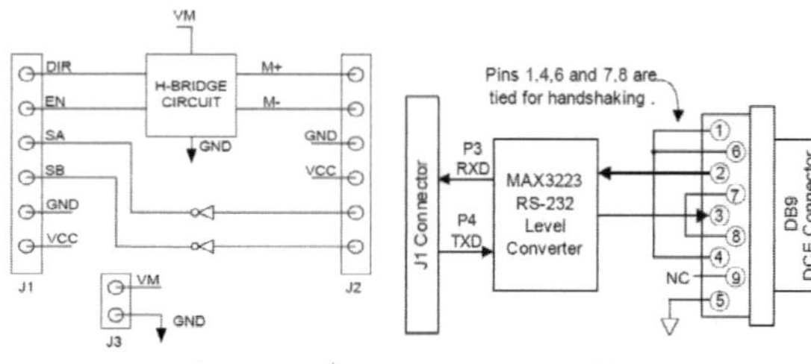

a) b)

Fig. 7 a) H-Bridge module for powering the motors
b) RS232 module for PC communication

For powering the DC motors we used a PMOD HB5 peripheral module with a H bridge (Fig. 7 a). For communicating with the PC a PMOD RS232 communication module (Fig. 7 b) was used, which represents the hardware support needed by the two software components for communication. For signalling certain states and parameters of the system a four PMOD LED module was used.

The embedded software component was developed in the BASCOM-AVR environment and the PC component was developed in the DELPHI programming environment and it accomplishes both the control of the robotic system and the user interface.

4. CONCLUSION

A very important design goal of the mobile in-pipe robotic minirobot is the adaptability to the inner diameters of the pipes. In this paper, we proposed the wheel type in-pipe minirobot, defined by an adaptable structure, based on linkages mechanisms. The prototype was designed in order to inspect pipes with variable diameters within 140 and 180 mm. It is the active module of a modular robotic system which is still in the phase of developing.

5. REFERENCES

Choi, H.R.; Ryew, S.M. (2002). Robotic system with active steering capability for internal inspection of urban gas pipelines, *Mechatronics*, Vol. 12, pp. 713-736.

Okada, T.; Sanemori, T. (1987). MOGRER: A vehicle study and realization for in-pipe inspection tasks, *IEEE Journal Robotics, Automation*, Vol. 3, No. 6, pp. 573-582.

Roh, S.G.; Choi, H.R. (2005). Differential-drive in-pipe robot for moving inside urban gas pipelines, *IEEE Transactions Robotics*, Vol. 21, No. 1, pp. 1-17.

Tătar, O.; Stan S.; Mândru, D. (2008). The modular robotic systems, *The 79th Annual Meeting of the International Association of Applied Mathematics and Mechanics*, GAAM 2008, 31 March– 4 April, 2008, Bremen, Germany.

Tătar, O.; Mândru, D. (2008). Design of In-Pipe Modular Robotic Systems, *The 4th International Conference Mechatronic Systems and Materials*, MSM 2008, Bialystok, Poland, 14-17 Iuly, 2008, pp. 20-30.

http://www.digilentinc.com/ *Accessed:* 2008-09-04

Annals of DAAAM for 2008 & Proceedings of the 19th International DAAAM Symposium, ISSN 1726-9679
Editor B. Katalinic, Published by DAAAM International, Vienna, Austria 2008
Make Harmony Between Technology and Nature, and Your Mind will Fly Free as a Bird

STUDIES ON THE CORROSION BEHAVIOR OF THE DENTAL ALLOYS

TATARCIUC, M[onica] & VITALARIU, A[nca]

Abstract: *A wide variety of dental devices are applied in oral cavity for different clinical aims. Restorative dental materials are required to have long-term durability in the oral cavity, which is a complex environment where dental materials are in contact with many different chemical media. The corrosion resistance and biocompatibility of commercial dental alloys have been investigated. Five dental alloys: Gold, Palliag, Gaudent, Verasoft, and Amalgam were surveyed. We appreciate the dental alloys behaviour by measuring the corrosion potential. The experimental medium was represented by three different artificial saliva: Afnor, Fusayama Meyer and Rondelli. The results showed that Amalgam had the highest corrosion potential and Rondelli saliva represented the highest corrosive medium.*

Key words: *alloy, corrosion, biocompatibility, saliva*

1. INTRODUCTION

Corrosion behaviour of dental alloys is one factor used to determine the material degradation and biocompatibility. It can be assumed that saliva, food components and beverages can degrade and age dental restoratives under in vivo conditions. Many researches have studied the corrosion behaviour of dental alloys in physiological solutions (Aeimbhu et al, 2005). To measure the corrosion potential is a relatively simple concept used to evaluate the behaviour of dental alloys into the oral cavity. Measuring the corrosion potential we can appreciate the active or passive behaviour of the dental alloys (Kobayashi et al, 2005). Using the Evans diagram we can evaluate the corrosion from the thermodynamic point of view.

2. MATERIAL AND METHOD

Materials used in this study were five commercially available alloys: Gold, Palliag, Gaudent, Verasoft and Amalgam with different contents described in Table 1.

The potential of corrosion E_{cor} is a measure that expresses the tendency of corrosion of a metal or alloy introduced into an electrolytic fluid. The potential of corrosion can be directly measured in report to a reference electrode that is characterised by a potential half-cell very stable. A reference electrode is introduced into a corrosive medium with the studied alloy and using a milivoltmeter with high impedance we can directly measure the potential. The potential of corrosion can be indirectly evaluated using the curves of linear polarisation, the diagram Evans. In the diagrams it is represented the logarithm of the density of the electric power depending on the potential of the electrode corresponding to a limit of over potential equal to ±50……60Mv. The crossing of the linear portion of the anodic and catholic lines of the polarization curve indicate on the potential axe the value of the corrosion potential E_{ctor}.

The electrodes were realised from dental alloys used in oral prosthesis removed from some of our treated patients. The surface of the electrodes was planed and ground on 320, 500, 800, 2400 silicon carbide abrasive papers followed by 1μm alpha alumina polishing and 0.3 μm finishing. After that, the electrodes were sheathed with epoxy resin. The electrode surface was planted with a grinding paper, cleaned and ultrasound treated in a solution of ethylic acid, distilled water and acetone.

Three types of electrolytes fluid were studied, represented by three artificial saliva: Aynor, Fusayama Meyer and Rondelli. The composition and the pH are described in Table 2.

Sample	Alloy	Composition	Density (g/cm^3)
1	GOLD	Au-77%, Ag-13,5%, Pt-1,5%	16.5
2	PALLIAG	Ag-58,5%, Pd-27,4%, Cu10,5%, Au2%, Zn1,5%, Ir-0,1%	11,1
3	GAUDENT	Cu-82%, Al-9,97%, Ni-4,35, Fe-1,32%, Mn-2,04%	7,8
4	VERASOFT	Ni-53,6%, Mn-19,5%, Cr-14,5%, Cu-9,5%, Al-1,6%, Si-1,5%	7,7
5	AMALGAM	Alloy (Ag-69,7%, Sn17,7%, Cu-12%, Zn-0,9%)(50%)+Hg50%	11,6

Tab. 1.The composition of the studied alloys

Corrosion medium	Composition	pH
Afnor saliva	Na Cl-0,7 g/L, KCl-1,2g/L, Na_2HPO_4 H_2O-0,26g/L, $NaHCO_3$-1,5g/L, KSCN-0,33g/L, urea-1,35g/L	6,78
Fusayama saliva	Na Cl- 0,40 g/L, KCl-0,40 g/L, Na_2HPO_4 H_2O-0,69 g/L, urea-1,00g/L, $Na_2S.9H_2O$-0,005 g/L, $CaCl_2$-0,79 g/L	6,24
Rondelli saliva	KCl-1,47g/L, $NaHCO_3$_1,25g/L, KSCN-0,520g/L, KH_2PO_4H $_2O$-0,190g/L	7,52

Tab. 2. The pH and composition of the corrosion medium

3. EQUIPMENT AND DEVICES

To evaluate the corrosion potential and to register the potentiodynamic polarisation we used the potentiostat PGP201, from the Economic Electrochemical Lab: VoltaLab 21 (Radelkis Copenhagen).

VoltaLab 21 is a compact potentiostat/galvanostat with built-in signal generator that can be used as a stand-alone instrument when it is programmed through its front panel. In this manual mode, the potentiostat scan rate can be selected up to 2.5 V/s. With VoltaMaster 4, the maximum scan rate is 10 mV/s. Ideal for corrosion studies, VoltaLab 21 records the polarization resistance and the corrosion potential over very long periods. The resolution of the electric power was 100Pa. The electrochemical cell also contained a saturated calomel electrode as reference. Then the alloys were anodically polarized. We registered the polarisation curve for a opened potential circuit (OPC) between -300 Mv and +300 Mv with a scanning rate of the potential equal to 0,5 mV/s (Kim et al, 2005). The potential into an opened circuit was evaluated into the Evans diagram. The examination of the surfaces of the electrodes was performed with an optical microscope.

4. RESULTS

The crossing point of the anodic and cathodic curves indicates the corrosion potential. All the values registered for the OPC and for the corrosion potential are showed in Table 3.

With only one exception (Palliag into the Afnor solution), the corrosion potential appreciated by the Evans diagram was smaller in comparison to the potential into the open circuit. That can be explained because the potential into the open circuit polarisation is measured into statically conditions and in the Evans method into dynamic conditions (the potential applied on the electrode being different to the equilibrium potential). The Evans diagram reflecting the potential in opened circuit was registered for each alloy. In figures 1-3 are shown the Evans diagrams registered for gold, palliag and amalgam.

The highest corrosion probability was registered for the amalgam and the lowest for Palliag. The corrosion of dental amalgam is thus a complex process, which involves contributions from each of the phase present as well as intergranular corrosion (Brett et al, 2002).

The artificial medium Fusayama Meyer saliva was the lowest corrosive medium and the Rondelli saliva had the highest corrosive potential.

The choice of a dental alloy became a real challenge because of wide variety of products offered by the dental market. The results of our study are useful for the practitioners (the dentist and dental technician) representing a guide in the selection of the appropriate dental alloy for each patient.

Because of the increased aesthetic demands of the patients, the orthodontic fixed appliances became very frequent among the teenagers and young patients. This is the reason why our future researches will be focused on the evaluation of the corrosion behaviour of dental alloys used in orthodontic treatments.

ALLOY	Artificial saliva	OPC	E_{cor} mV (ESC) (Evans)
Gold	Afnor	-137	-201
	Fusayama-Meyer	-152	-181
	Rondelli	-249	-287
Palliag	Afnor	-3	+2,3
	Fusayama-Meyer	-22	-117
	Rondelli	-65	-194
Gaudent	Afnor	-323	-353
	Fusayama-Meyer	-170	-174
	Rondelli	-268	-356
Verasoft	Afnor	-430	-599
	Fusayama-Meyer	-143	-181
	Rondelli	-237	-302
Amalgam	Afnor	-480	-668
	Fusayama-Meyer	-322	-388
	Rondelli	-449	-654

Tab. 3. The potential in open circuit and the corrosion potential for the studied alloys

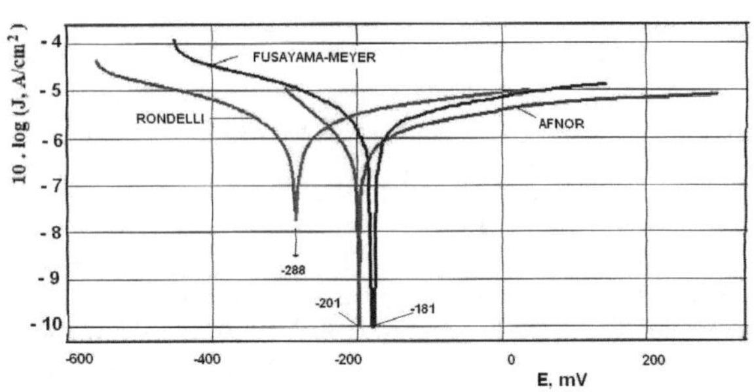

Fig.1. Evans diagram for gold alloy in different artificial saliva

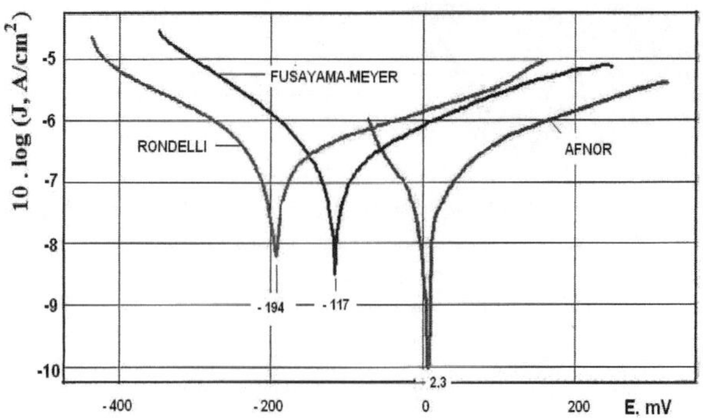

Fig.2. Evans diagram for Palliag in different artificial saliva

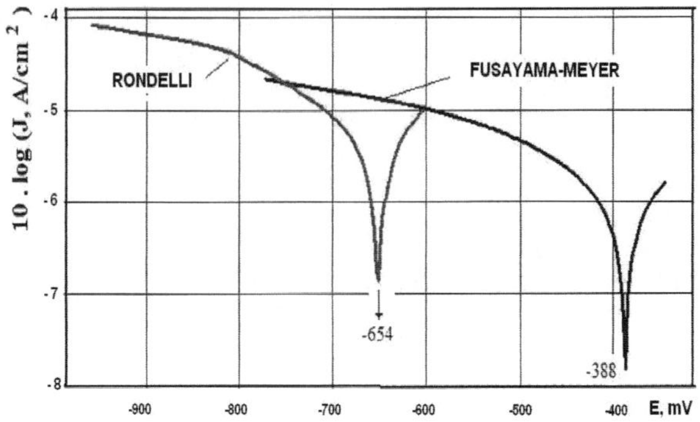

Fig.3. Evans diagram for Amalgam in different artificial saliva

5. CONCLUSION

The influence of different artificial saliva on metallic corrosion was investigated by measurements of open circuit potential, polarisation curves, chosen owing to its high rate of corrosion. The corrosion of the dental alloys in different artificial saliva brought clearly into evidence the importance of the contacting solutions on the rate of corrosion (Brett &Muresan, 2002).

Our results are clinically significant because demonstrate that the individually oral conditions, the specific parameters and composition of the oral fluids are responsible for the corrosion behaviour and the biocompatibility of the precious and non-precious dental alloys. The selection of the dental alloy should be based on the salivary pH evaluation of each patient.

6. REFERENCES

Aeimbhu, A.; Castle, J. & Singjai, P. (2005). The influence of albumin on the electrochemical behaviour of commercial titanium. *Key Engineering Materials*, Vol.288-289, (2005), pp. 615-618

Brett, C.; Acciari, H.& Guastaldi, A.(2002). Corrosion of dental amalgams-studies of individual phases. *Key Engineering Materials*, Vol.230-232, (2002), pp. 463-466

Brett, C. & Muresan, I. (2002). The influence of artificial body fluids on metallic corrosion. *Key Engineering Materials*, Vol.230-232 (2002), pp. 459-462

Kim, Y.S.; Yoo, R.Y. & Sohn, C.G. (2005). Role of alloying elements on the cytotoxic behavior and corrosion of austenitic stainless steels. *Materials Science Forum,* Vol. 475-479, (2005), pp. 2295-2298

Kobayashi, S.; Ohgoe, Y. & Ozeki, K. (2005). Biocompatibility of diamond-like carbon coated Ni-Ti ortodontic wirw and acrylic resin teeth. *Key Engineering Materials*, Vol. 284-286,(2005), pp. 783-786

Annals of DAAAM for 2008 & Proceedings of the 19th International DAAAM Symposium, ISSN 1726-9679
Editor B. Katalinic, Published by DAAAM International, Vienna, Austria 2008
Make Harmony Between Technology and Nature, and Your Mind will Fly Free as a Bird

MODELING AND SIMULATION OF A VIBRATORY MACHINE MECHANICAL SYSTEM WITH ELASTIC-DAMPED SUPPORTING ELEMENTS

TATARU, M[ircea] B[ogdan]; RUS, A[lexandru] & ABRUDAN, G[heorghe]

Abstract: *This paper presents some aspects concerning modeling and simulation of mechanical systems with elastic motions. Presentation is concentrated on the general aspect of modeling these systems considering also the particular case of elastic-damped motions. Problem of modeling was studied in the case of a one mass (one degree of freedom) vibratory machine elastic-damped supported. A six-degree of freedom joint with elasticity and damping properties was used to model this supports. Specific in this case is the feed-back connection between elementary subsystems of the system structure.*

Key words: *modeling, systems, elasticity, feed-back, joint*

1. INTRODUCTION

Mathematical modeling for different kind of mechanical systems is already well-known. This paper presents a specific way of using theory, mathematical calculus, systems theory, numerical calculus and soft techniques for a vibratory machine modeling

2. GENERAL MODEL

Calculus model for this machine is obtained using the schematic showed in the in Fig. 1 (see Munteanu, M., 1986, 1974)).

In order to determine the mathematical model of this model of this systems many authors (Buzdugan, Gh. et. al., 1975), (Munteanu, M., 1986,) are using the well-known fundamental equation of dynamic equilibrium between inertia forces on one part and resultant between applied forces and elastic-damped supports reaction forces on the other side. It may be obtained,

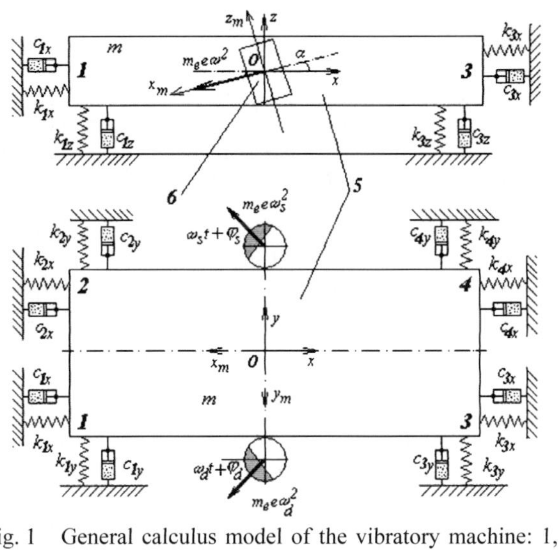

Fig. 1 General calculus model of the vibratory machine: 1, 2, 3, 4 - the four elastic damped supports; 5 - working element; 6 - non - equilibrated masses generators;

in this way, the following system of three equations on x, y, and z:

$$\begin{aligned} m\cdot\ddot{x} &= F_{xgs}\cdot\cos(\alpha_s)+F_{xgd}\cdot\cos(\alpha_d)-k_x\cdot x-c_x\cdot\dot{x};\\ m\cdot\ddot{y} &= F_{ygs}+F_{ygd}-k_y\cdot y-c_y\cdot\dot{y};\\ m\cdot\ddot{z} &= F_{xgs}\cdot\sin(\alpha_s)+F_{xgd}\cdot\sin(\alpha_d)-k_z\cdot z-c_z\cdot\dot{z} \end{aligned} \quad (1)$$

where:

$$\begin{aligned} F_{xgs} &= -m_{es}\cdot e_s\cdot\omega_s{}^2\cdot\cos(\omega_s\cdot t+\varphi_s);\\ F_{xgd} &= -m_{ed}\cdot e_d\cdot\omega_d{}^2\cdot\cos(-\omega_d\cdot t+\varphi_d);\\ F_{ygs} &= -m_{es}\cdot e_s\cdot\omega_s{}^2\cdot\sin(\omega_s\cdot t+\varphi_s);\\ F_{ygd} &= -m_{ed}\cdot e_d\cdot\omega_d{}^2\cdot\sin(-\omega_d\cdot t+\varphi_d) \end{aligned} \quad (2)$$

and: m – is the inertia mass of all the system, $k_{x1,2,3,4}$, $k_{y1,2,3,4}$, $k_{z1,2,3,4}$, - are the rigidity constants of the four supports and, k_x , k_y , k_z – are their equivalent rigidity constants for each direction, $c_{x1,2,3,4}$, $c_{y1,2,3,4}$, $c_{z1,2,3,4}$, -are the damping constants of the four supports and, c_x , c_y , c_z – are their equivalent damping constants for each direction, m_{es}, m_{ed} – is the non-equilibrated mass of the generators left and right side respectively, e_s, e_d – is the distance between weight centre and rotation axis of the non - equilibrated masses left and right side respectively, ω_s, ω_d – is the angular velocity of the generators, left-side and right-side, respectively, φ_s, φ_d – is the initial angular position of the weight centre of the non-equilibrated masses, left-side and right-side respectively.

It may be obtained the differential system of equations for this model based on (1), (2) and expressed it in matrix forms (see (Buzdugan, Gh. et.al.1975) and (Ionescu, V, 1986)):

$$\{r''\}=[H_i]\cdot(\{F_s\}+\{F_d\}-[H_a]\{r'\}-[H_e]\{r\}) \quad (3)$$

where:

$$\begin{aligned} \{F_s\} &= [H_{\alpha s}][H_\pi]H_{mes}.H_{acfs}[H_{\varphi s}]\{u_{gs}\};\\ \{F_d\} &= [H_{\alpha d}][H_\pi]H_{med}.H_{acfd}[H_{\varphi d}]\{u_{gd}\} \end{aligned} \quad (4)$$

in which: $\{r''\}=\{x''\ \ y''\ \ z''\}^t$ - -is the accelerations vector; $\{r'\}=\{x'\ y'\ z'\}^t$ -is the speeds vector; $\{r\}=\{x\ \ y\ \ z\}^t$ - is the displacements vector; $\{u_{gs}\}$, $\{u_{gd}\}$ – are motion unit module vectors for the left and right generator; $[H_i]$ – is the inertia transfer matrix; $[H_e]$ – is the rigidity transfer matrix; $[H_{as}]$, $[H_{ad}]$ – is the damping transfer matrix; $[H_\pi]$, $[H_{\varphi s}]$, $[H_{\varphi d}]$ - and geometrical transfer matrices; H_{mes} = m_{es}, H_{med} = m_{ed} - inertia transfer matrices for the left and right generators; H_{acfs}=e_s. $\omega_s{}^2$,

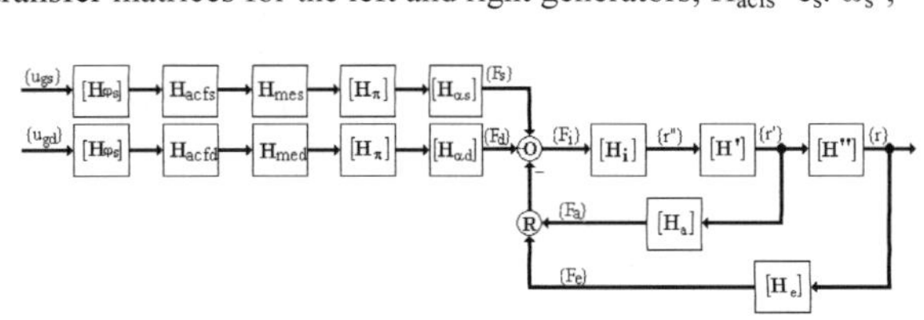

Fig. 2 Structural block diagram of the vibratory machine

$H_{acfd}=e_d\cdot\omega_d^2$ - centrifugal acceleration transfer matrices

In the end, based on the equations written above it may be obtained a structural scheme for the analyzed system (Fig. 2).

3. SIMULATION MODELS.

All the models provide the same pattern, because are following the same or same like steps (Bernhard P. & Algis Džiugys, 2002), (*** *Matlab with Simulink®*R13, 2003). Blocks presented in the simulation scheme in the next figures are the correspondent of the general one in the Fig. 2. Diagrams of the models provide trajectory information shown in Fig. 4 and kinematic parameters like displacements and their time derivative up to acceleration. A more detailed model for these mechanical systems is given in Fig. 3. This provides cinematic detailed aspects concerning measurements points and detailed manner of evaluating motion in the inertia center of the working element.

It may also be mentioned that modeling elastic-damped supports is similarly to the model presented in Fig. 2 (the same feedback connection aspect), as it may be seen in Fig. 4. In modeling elastic-damped mechanical systems a six degree of freedom joint with elastic and damping properties was used to model.

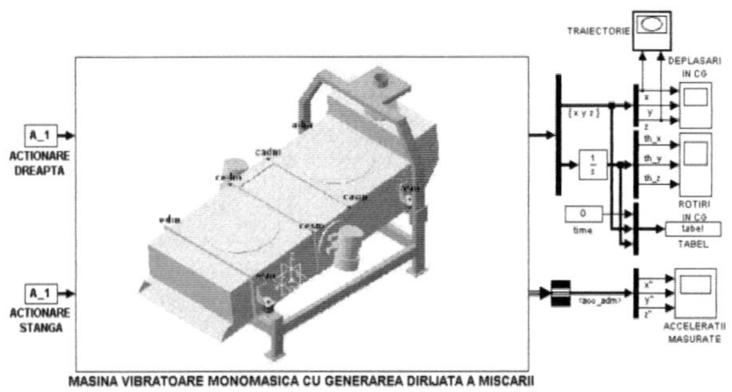

Fig. 3 Matlab with Simulink®R13 diagram of the vibratory machine – with measurement points.

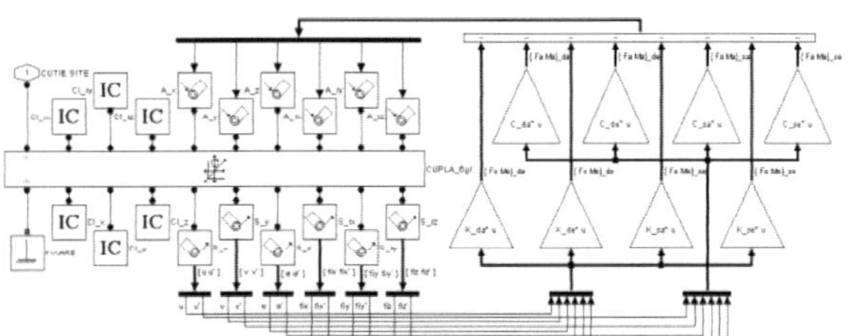

Fig. 4 Elastic-damped supports model diagram in Matlab with Simulink® R13 SimMechanics Toolbox was used

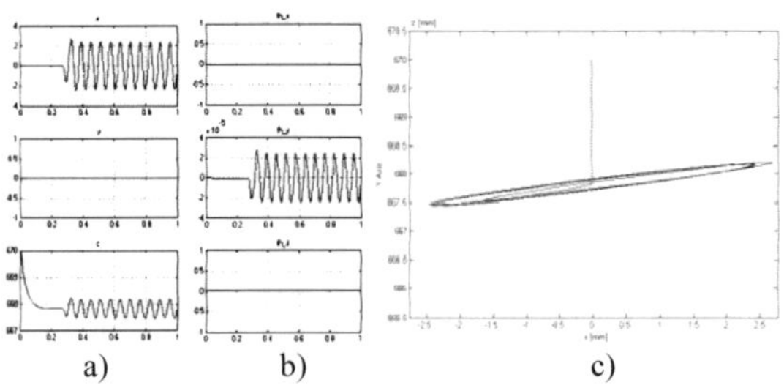

Fig. 5 a) – displacements; b) – rotations and c) - trajectory of the mass center in xOz coordinate plane in mass center of the working element of the vibratory machine (15 , optimum declination angle of the generators in order to give a good throwing angle for 6 inclination angle of the working element)

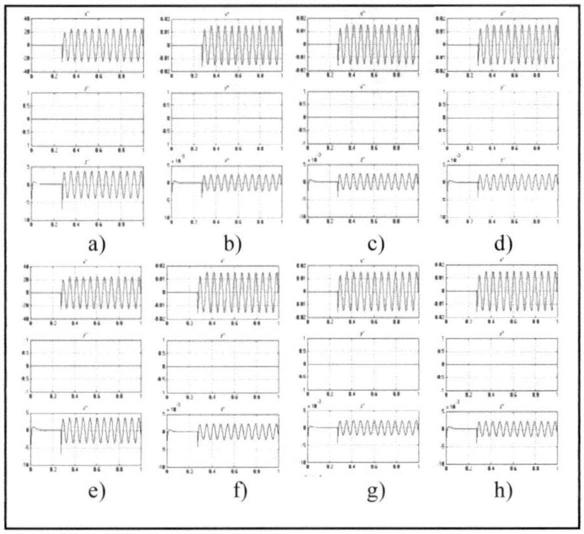

Fig. 6 Accelerations measured in different points of the working element of the vibratory machine: a) "adm"-admission-right-motion part; b) "asm" – admission-left-motion part; : c) "cadm" – sieves box-admission-right-motion part; d) "casm" – sieves box-admission-left-motion part; e) "cedm" sieves box-evacuation-right-motion part; f) "cesm" – sieves box-evacuation-left-motion part; g) "edm" – evacuation-right-motion part; h) "esm" – evacuation-left-motion part

Displacements, rotations and trajectory of the working element are given in Fig. 5:

Accelerations for distinctive different points of the working element of the vibratory machine are given in Fig. 6.

This paper does not treat these cases. Studies for these malfunctioning cases make the object of more detailed working papers. Here is only the "perfect" example of the system, with an accent on modeling method and ways of realization.

4. CONCLUSIONS

As one may observe this work has developed such an understanding that creates a possibility to directly model and simulate in virtual form mechanical motion of vibratory machines.

5. REFERENCES

Bernhard Peters, Algis Dziugys, (2002), *Numerical simulation of the motion of granular material using object-oriented techniques*, Computer methods in applied mechanics and engineering, pg.1984 Elsevier Press Release

Buzdugan, Gh. et. al., (1975), *Vibratiile sistemelor mecanice (Vibrations of Mechanical Systems)*, Editura Academiei Romane, Bucuresti (Romanian edition)

Ionescu, V., (1986), *Teoria sistemelor (Systems Theory), vol.1*, Editura Didactica si Pedagogica, Bucuresti (Romanian edition)

Munteanu, M., (1986), *Introducere în dinamica masinilor vibratoare (Introduction to vibratory Machines Dynamics)*, Editura Academiei R.S. Romania, Bucuresti (Romanian edition)

Munteanu, M., (1974), *Dinamica masinilor vibratoare suspendate pe elemente elastic (Dynamics of vibratory machines, elastic elements supported)*, Constructia de masini no 25, (Romanian edition)

***, (2003) *Matlab with Simulink®*R13, *Help Documentation*, SimMechanics Toolbox

Annals of DAAAM for 2008 & Proceedings of the 19th International DAAAM Symposium, ISSN 1726-9679
Editor B. Katalinic, Published by DAAAM International, Vienna, Austria 2008
Make Harmony Between Technology and Nature, and Your Mind will Fly Free as a Bird

STRATEGY DEVELOPMENT MODELFOR INNOVATION IN ENTREPRENEURSHIP

TAUCEAN, I[lie] M[ihai] & TAMASILA, M[atei]

***Abstract**: The Ansoff's Matrix is a model for strategy development that is use with proved success and which correlates existing/new products with existing/new markets. To this model we add the third element: the existing/new business, in a new model we called the PMB Model (Product/Market/ Business Model). This third key element we consider that complete the bi-dimension model and can give more ways for strategy analysis. We based the model on the dynamism and many and necessary movement in entrepreneurship activity, regarding products, markets and business. The model can give the risk awareness concerning innovation, defined by the each possible strategy position in the model (in the strategy "cube").*
***Key words:** model, strategy, entrepreneurship, innovation, Ansoff.*

1. INTRODUCTION

Entrepreneurs are dynamic persons (Dynamicbusiness.com, 2007) that must make many strategic innovative movements to be successful. They can be innovative regarding products and services that they sell, they can be innovative in markets that they decide to be in, or they can be innovative in their business that they runs.

Innovation here means that there is innovation in existing product or new product development, innovation in existing market or development of new market, innovation in existing used business model or new business model development. We based on that our proposed model and on the dynamism and many and necessary movement in entrepreneurship.

2. ANSOFF'S PRODUCT/MARKET MATRIX

The Ansoff Growth matrix is a tool that helps businesses decides their product and market growth strategy. Ansoff's product/market growth matrix suggests that a business' attempts to grow depend on whether it markets **new or existing** products in **new or existing markets** (Ansoff, 1957).

The output from the Ansoff product/market matrix is a series of suggested growth strategies that set the direction for business strategy. These are described below (tutor2u, 2008):

Market penetration refers to a growth business strategy focused on selling existing products into existing markets. Market penetration seeks to achieve four main objectives:

• maintain/increase the market share of current products can be achieved by a combination of competitive marketing mix;
• secure dominance of growth markets
• restructure a mature market by driving out competitors;
• increase usage by existing customers

Product / Market	**Existing products**	**New products**
Existing markets	Market penetration	Product development
New markets	Market development	Diversification

Table 1. Ansoff's Product/Market Matrix.

A market penetration marketing strategy is very much about "business as usual". The business is focusing on markets and products it knows well. It is likely to have good information on competitors and on customer needs.

Market development is the name given to a growth strategy where the business seeks to sell its existing products into new markets. There are many possible ways of approaching this strategy, including:

• New geographical markets (for example exporting the product to a new country)
• New distribution channels
• Different pricing policies to attract different customers or create new market segments

Product development is the name given to a growth strategy where a business aims to introduce new products into existing markets. This strategy may require the development of new competencies and requires the business to develop modified products which can appeal to existing markets.

Diversification is the name given to the growth strategy where a business markets new products in new markets. This is an inherently more risk strategy because the business is moving into markets in which it has little or no experience. For a business to adopt a diversification strategy, therefore, it must have a clear idea about what it expects to gain from the strategy and an honest assessment of the risks.

3. NEW MODEL FOR ENTREPRENEURSHIP

Starting from the Ansoff's Matrix, we can develop two new matrixes using "business" as a third element (see table 2 and 3).

Here business means new model development of existing business or new model of business that it's imperative to be successful in entrepreneurship activity. "New" means new management model, new management method and techniques, new management functions (new organization, new decision model, new motivation and coordination of the business etc.).

New product means innovation in existing product (product modernizations, less or more functions, level of complexity or simplicity of the product etc.), or new product development (revolutionary new materials/technologies, new functions etc.).

Product / Business	**Existing products**	**New products**
Existing business	Business penetration	Product development
New business	Business development	Diversification

Table 2. Product/Business Matrix

Market / Business	**Existing market**	**New market**
Existing business	Business penetration	Market development
New business	Business development	Diversification

Table 3. Market/Business Matrix

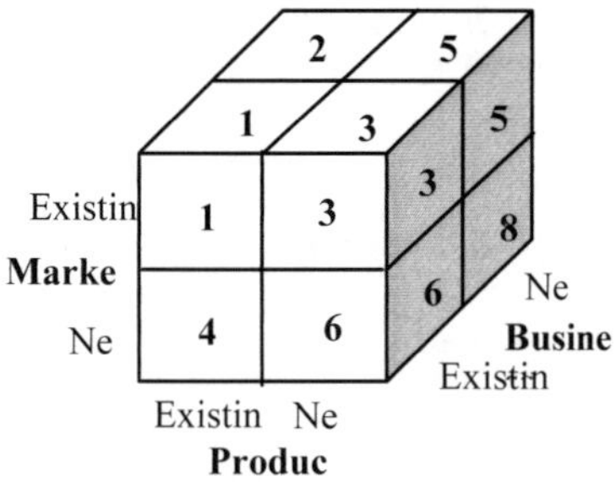

Fig. 1. Product/Markets/Business Model

Strategies	Products	Markets	Business
1	Existing	Existing	Existing
2	Existing	Existing	New
3	New	Existing	Existing
4	Existing	New	Existing
5	New	Existing	New
6	New	New	Existing
7	Existing	New	New
8	New	New	New

Table 4. New model strategies

Strategies	Innovation	Risk degree
1	No innovation	No risk
2	Business innovation	Low risk
3	Product innovation	Low risk
4	Markets innovation	Low risk
5	Product & business innovation	Medium risk
6	Product & market innovation	Medium risk
7	Market & business innovation	Medium risk
8	Total innovation	High risk

Table 5. Innovation and risk degree for new model strategies

New markets means innovation in existing market segments (development of the existing market segment, new combination of existing segment characteristics etc.) or new market creation (creation new demands, new wishes, new needs for and from the clients).

Figure 1 present the "cube" of the new model, that we called the PMB Model (Product/Markets/Business Model), by adding the business element to the Ansoff Model. The eight strategies resulting from the new models are presented in the table 4. Also we can evaluate the degree of risk for the strategies according to the innovation degree of the three elements: product, market and business innovation (see table 5).

Business risk depends on the nature of business and on managers/entrepreneurs. A business investment is usually riskier for a new business or for a start-up enterprise versus a business with a long history. But also an old and obsolete business or in decline stage business, from a point of view of product/markets/business matrix, could be riskier versus a new business or a start-up enterprise.

We should take into account the entrepreneur's attitude to risk and its influence on the final business decision. Entrepreneurs differ in their attitudes to risk. For serious/strategic business decision, entrepreneurs are usually defined as "risk seekers" (prefers a risky solution, see figure 2, curve *c*). This is an assumption about the utility (U) or satisfaction derived for money. A "risk seeker" entrepreneur is a person for whom utility function increases as his level of income (I) increases. The utility function of a "risk averter" entrepreneur declines as the level of income rises (see curve *a*). A "risk neutral" entrepreneur regard each increment of income as having the same value (curve *b*).

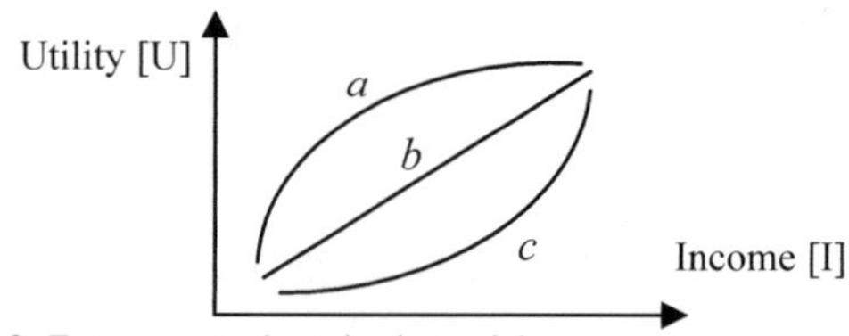

Fig. 2. Entrepreneur's attitude to risk

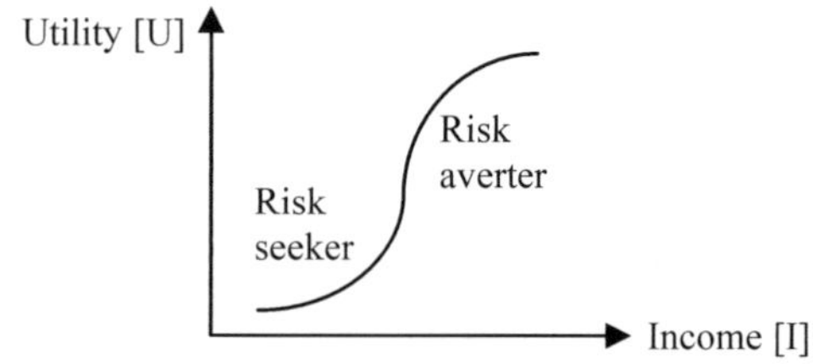

Fig. 3. Entrepreneur's combine attitude to risk

Friedman and Savage propose a utility function that combines attitudes to risk (Friedman & Savage, 1948), in order to explain a comportment that is at first a risk seeker and next is risk averter and so on (figure 3). Entrepreneurship is about taking risk (Drucker, 1970). The behavior of the entrepreneur reflects a kind of person willing to put his or her career and financial security on the line and take risks in the name of an idea, idea for a product, for a market or for a business.

From the nature of the entrepreneurship, we can say that entrepreneurs are innovators, but they have different degree of risk (as we saw in table 5). They persist on discovering new products and services, they create and developed new markets, and they run their business in a new way to reach the success.

4. CONCLUSIONS

The new model of strategy development for entrepreneurship involves key innovation indicators for entrepreneurs. The proposed PMB Model can be use in successful strategy positioning and development. We based that on the dynamism in entrepreneurship regarding products, markets and business. The model can give the risk awareness concerning innovation, defined by the each possible strategy position in the model (in the strategy "cube").

The model can be develop by a more detailed analysis of the three dimensions (product, market and business indicators which are very complex) and by adding new indicators that can be considered keys indicator for entrepreneurship success. In the last case, when using more than three dimensions, the model can lose the visual analysis advantages that are given by the three dimensional model (or the two dimensional model).

5. REFERENCES

Ansoff, I. (1957). Strategies for Diversification. *Harvard Business Review*, Vol. 35, Issue 5, Sep-Oct 1957, pp. 113-124, ISSN 0017-8012

Drucker, P. (1970). Entrepreneurship in Business Enterprise, *Journal of Business Policy*, vol. 1, 1970, pp. 3-12

Dynamicbusiness.com (2007). 50 Favourite Entrepreneurs, 28 June, *Available from*: http://www.dynamicbusiness.com/articles/articles-entrepreneur-profiles/50-favourite-entrepre.html, *Accessed* 2008-05-28

Friedman, M. & Savage, L.P. (1948). The Utility Analysis of Choices Involving Risk, *Journal of Political Economy*, Vol. 56, No. 4, January 1948, pp.279-304, ISSN 0022-3808

tutor2u (2008). Ansoff's Product/Market Matrix, Strategy Revision Notes, *Available from*: http://www.tutor2u.net/business/ strategy/ansoff_matrix.htm, *Accessed* 2008-05-28

Annals of DAAAM for 2008 & Proceedings of the 19th International DAAAM Symposium, ISSN 1726-9679
Editor B. Katalinic, Published by DAAAM International, Vienna, Austria 2008
Make Harmony Between Technology and Nature, and Your Mind will Fly Free as a Bird

THE INFLUENCE OF THE POST-CURING TREATMENT ON THE MECHANICAL PROPERTIES OF THE GLASS FIBERS MAT POLYMERIC COMPOSITE MATERIALS

TEODORESCU, H[oratiu] & MOTOC LUCA, D[ana]

Abstract: *The paper focuses on the influence of the post-curing thermal cycles on the mechanical performance of polymeric/glass fibre composites. The values are compared with those retrieved in case of the composite structures tested at room temperature in order to size the mechanical properties variation. The results indicate that if the composite structures are aimed to be used in low-cost engineering applications a single thermal cycle is enough to improve their mechanical properties. Results also indicate that the amount of fillers embedded into the structure has its influence on the mechanical characteristics of the polymeric composites.*

Key words: *post-curing, polymeric, composite, mechanical, properties.*

1. INTRODUCTION

It is well acknowledged and proven by a wide spectrum of scientific research that the performance of composite structures depends on the constitutive elements characteristics (e.g. material types, quantity, architecture within the structure, etc.), manufacturing technology, working conditions, etc. Temperature influence on the mechanical characteristics it is one of the subjected that was approached quite extensively in the literature for different types of material combinations and in all cases material degradation were reported (Dimitrienko, 1999).

With respect to the subject of the paper, flexural modulus is usually reported into the literature, for various reasons that will not mentioned in here and temperature influence upon it comes usually as being an external factor with an increasing trend (Cerny & Glogar, 2004), (John & Venkata Naidu, 2004).

The influences of the constitutive elements on the elastic properties of the composite materials are another trendy subject in the domain of composite materials characterization, usually reported from fibre length, distribution, and volume fraction and architecture point of view (Xu, Eichmiller & Barndt, 2001).

One of the author of the herein paper reported the influence of the thermal treatment on the electrical characteristics of particle reinforced polymeric composite materials, in the sense of improvements on these characteristics. This idea was extended to other types of composite materials, such the ones approached in this paper (Oltean & Motoc Luca, 2008).

The present work focuses on the mechanical properties of the random E-glass fibres composites, reinforced with different volume fraction of mineral fillers (used to lower the costs of the materials, improve the UV resistance and electromagnetic characteristics). More specifically, the paper focuses on the experimental values retrieved from flexural testing as a part of a more extensive work focused on this class of composite materials. The samples tested were subjected to thermal cycles, simple and consecutively, with the aim of sizing the influence of the post-curing treatment on the mechanical characteristics of the composite structures implied into the studies. Particular to this paper, different from other and in the same time new is the fact that it shows that a single relatively long post-curing thermal treatment it is enough to lead to mechanical characteristics improvements.

The experimental results retrieved provide useful information for design purpose regarding to the improvements that can be done especially when the structure is being employed in aggressive media environments.

2. EXPERIMENTAL ANALYSIS

2.1 Materials

In this study were used random, long E-glass fibre mats in a combination with a polymeric matrix from DSM Composite Resins (Switzerland), the latter being choose due to its availability and very good fibre wetting and impregnation properties. The fibres - MultiStrat^{TM} Mat ES 33-0-25 supplied by Johns Manville, USA, were made up from multidirectional continuous E glass fibres, leading to a 60% volume fraction in the composites' panels. The fillers considered were $CaCO_3$ particles embedded into different volume fraction into the composite's matrix material (5 %, 10 % respectively).

2.2 Testing methods

The samples' mechanical properties were evaluated at room temperature, after a single thermal cycle and 3 consecutive thermal cycles, respectively. One thermal cycle applied, by aid of a controlled temperature oven, had the following variation: 1 hour of heating up (5 min. until the temperature was reached) and maintaining at 120^0 C followed by 1 hour of cooling down at room temperature. The samples were shaped using a standardized form (SR EN ISO 178) and subjected to flexural loading conditions (3 points), all of them with a crosshead speed of 1 mm/min using a LR 5K Plus device from Lloyd Instruments Ltd. The results obtained applying a statistical analysis was considered as corresponding to the mean values.

3. RESULTS AND DISCUSSION

The experimental results show the influence of the post-curing thermal treatments on the mechanical characteristics of the composite samples investigated. In figure 1 was plotted the Young modulus of bending retrieved for samples without thermal treatment and samples subjected to a single thermal cycle, respectively.

As can be seen, in case of the structure without mineral fillers or with a 10% of fillers content an improvement of the flexural modulus with 34% and 46 %, respectively, was reached. The 5% content of mineral fillers seems to lead to an opposite effect in the modulus variation. This behaviour was found for the values retrieved from tension and compression testing and may help us to conclude that this percentage represent a critical amount that has to be avoided during the composites manufacturing.

In case of applying consecutively thermal cycles (3 in this case) the experimental values do not reveal huge modifications with respect to the values retrieved after one thermal cycle applied as part of a post-curing treatment. This behaviour can be regarded to the composite material structure, namely to the fact the previous thermal loading was long enough to ensure the

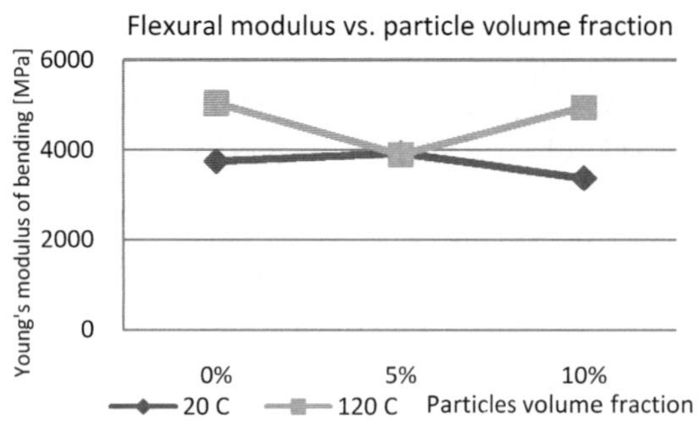

Fig. 1 Flexural modulus variations with fillers volume fraction, at room temperature and after a single thermal cycle

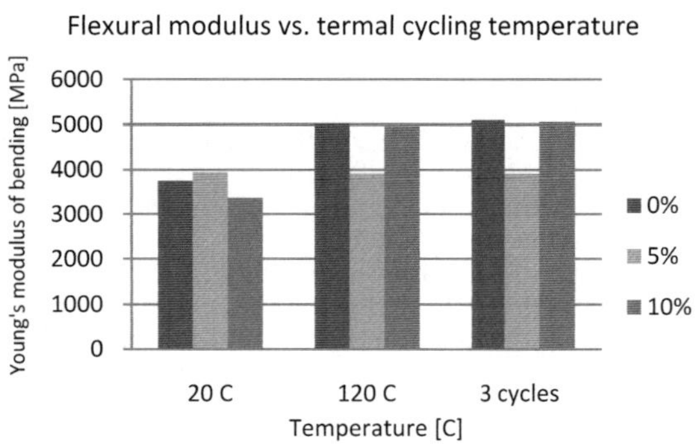

Fig. 2 Flexural modulus variations for different thermal cycles

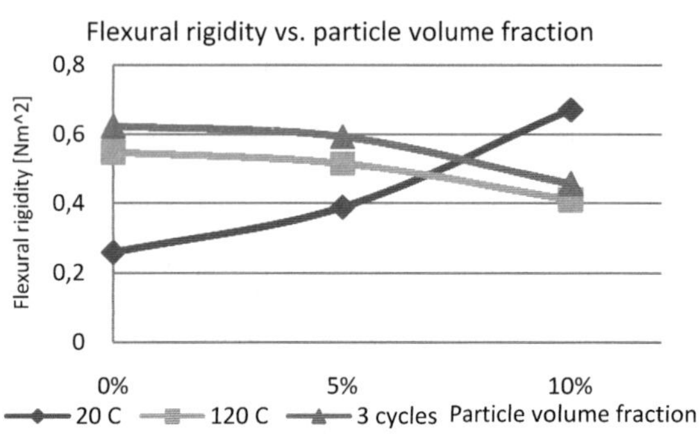

Fig. 3 Flexural rigidity variations for different fillers volume fraction, at room temperature and after the thermal cycles

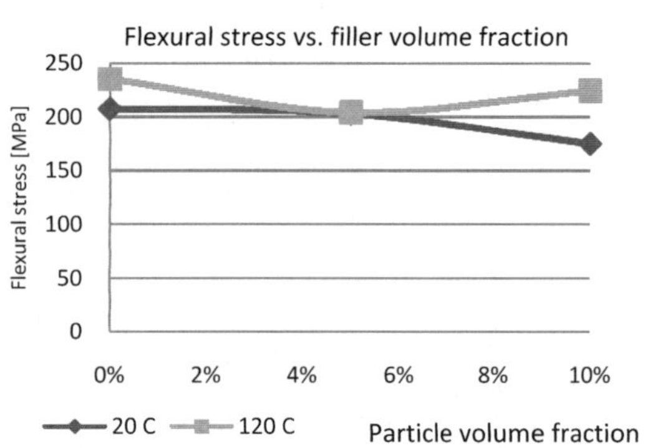

Fig. 4 Flexural stress variations with fillers volume fraction, at room temperature and after the first thermal cycling

Composite type glass fibre & polymeric matrix	1 thermal cycle	3 thermal cycles
0 % mineral filler	6.0	0.01
5 % mineral filler	2.4	0.12
10 % mineral filler	4.4	0.01

Tab. 1. Relatively weight loss after the thermal cycles

volatile compounds evaporation and polymerization process ending.

The same findings can be applied to the variations of the samples' flexural rigidity, were again there are no huge changes due to an extra thermal treatment. The flexural stress (used in structure designs) experience the behaviour as the flexural modulus along with the thermal treatments applied upon the composite samples.

A weigh reduction (see table 1) and dimensional shrinkage was recorded in case of the samples subjected to thermal treatments. No further thermal treatments were applied in order to avoid samples degradation or maturation during the experimental trials.

4. CONCLUSIONS

The variations of the flexural modulus and flexural rigidity of glass fiber mat reinforced polymeric composites reinforced by various volume fractions of mineral filler, without and after subjected to a single thermal cycle at a temperature of 120^0 C, or 3 consecutively thermal cycles with the same parameters, were investigated. The thermal influence was considered as a natural consequence of the manufacturing process, and can be viewed as post-curing treatment. The improvements of the flexural modulus sized after the first thermal cycle were sufficiently to help authors to conclude that a single relatively long thermal cycle will be enough to improve the mechanical characteristics of any type of composite structures, especially in industry environments were the manufacturing and materials costs are of high concerns along with improvements on their mechanical characteristics.

5. REFERENCES

Cerny, M. & Glogar, P. (2004).Young's modulus of the ceramic matrix composites with polysiloxane based matrix at elevated temperature. *Journal of Material Science,* Vol. 39, 2239-2242.

Dimitrienko, I. (1999). Modelling of the mechanical properties of the composite materials at high temperatures: Part 1. Matrix and fiber. *Applied Composite Materials*, Vol. 4, 219-237.

John, K. & Venkata Naidu, S. (2004). Effect of fibre content and fibre treatment on flexural properties of sisal fibre/glass fibre hybrid composites. *Journal of Reinforced Plastics and Composites*, Vol. 23, 1601-1606.

Oltean, I. D. & Motoc Luca, D. (2008). Conductive polymeric composites behaviour under various loading conditions. *Proceedings of the 6th International Conference of DAAAM Baltic Industrial Engineering*, Editor R.Kyttner, Tallin, Estonia, ISBN 978-9985-59-783-5, 513-518.

Xu, K., Eichmiller, F. & Barndt, P. (2001). Effects of fibre length and volume fraction of calcium phosphate cement. *Journal of Materials Science: Materials in medicine*, Vol. 12, 57-65.

Acknowledgement
The research was supported from grant ID_135, 108/1/10/2007, CNCSIS, Romania.

Annals of DAAAM for 2008 & Proceedings of the 19th International DAAAM Symposium, ISSN 1726-9679
Editor B. Katalinic, Published by DAAAM International, Vienna, Austria 2008
Make Harmony Between Technology and Nature, and Your Mind will Fly Free as a Bird

NEW ULTRA LIGHTWEIGHT AND EXTREME STIFF SANDWICH COMPOSITE STRUCTURE FOR MULTIPLE APPLICATIONS

TEODORESCU-DRAGHICESCU, H[oratiu]; VLASE, S[orin]; MOTOC LUCA, D[ana]; SCUTARU, M[aria] L[uminita]; STANCIU, A[nka] & ROSU, D[orin]

Abstract: *The paper presents an ultra lightweight sandwich composite structure with extreme stiffness that can replace the classic stiffening solution with ribs of a panel with imposed weight and dimensions. The structure features two carbon/epoxy skins reinforced with twill weave fabric (0.3 kg/m²) and an expanded polystyrene (EPS) core. To accomplish a theoretical approach, the structure is subjected to a biaxial field of static normal loads combined with a shear load. A comparison between sandwich structure's stiffness's with carbon- and glass fibers skins plies has been accomplished.*

Key words: *sandwich, stiffness, core, fabric, fibres.*

1. INTRODUCTION

Sandwich composite structures have been used increasingly in applications in aeronautics, transportations, in automotive industry, in machine-tools construction, robotics, etc., where high rigidity parts are needed. In most cases, these structures exhibit failures consisting in the connection detachment between core and skin due to impact applied loads. The main objective of the authors' researches is to obtain a sandwich composite structure that can replace the classic stiffening solution with ribs of a panel with imposed weight and dimensions.

2. CRITICAL OVERVIEW

There are a lot of improvements in the construction of sandwich composites especially to improve their impact resistance. For instance, foam filled 3D integrated core sandwich composite laminates with and without additional face sheets have been manufactured using vacuum assisted resin infusion moulding process in multiple steps (Hosur et al., 2007). In order to handle large LCD panels, robot structures have been produced using carbon fibers/epoxy composite material with a polyurethane foam core to increase the stiffness of the wrist blocks (Lee et al., 2002). Developments of sandwich structures have been made reinforcing cores by way of three-dimensional Z-pins embedded into foam, honeycomb cells filled with foam and hollow/space accessible Z-pins acting as core reinforcement (Vaidya et al., 2001). Tooling and manufacturing methods, designing, stress, loads and load testing, vacuum bagging, autoclaves, etc., are presented in three basic references (Marshall, 2006; Smith, 2005; Wiedemann, 1989). Materials described are fibers and fabrics based on E-glass, S-glass, pan based carbon, pitch based carbon, and several Kevlars.

3. THE SANDWICH STRUCTURE

The sandwich composite structure taken into account presents two skins based on epoxy resin reinforced with a 0.3 kg/m² twill weave fabric and an expanded polystyrene (EPS) 9 mm thick core with a density of 30 kg/m³. The final thickness of the structure is 10 mm (fig. 1). The carbon fibers fabric used in this structure is a high stiffness one, that presents so called twill weave. The main feature of this weave is that the warp and the weft threads are crossed in a programmed order and frequency, to obtain a flat appearance. In order to accomplish a mechanical analysis, an equivalence model of the twill weave fabric is presented in fig. 2. The skins have been impregnated under vacuum with epoxy resin and glued to the core. Data regarding the sandwich architecture are presented in table 1.

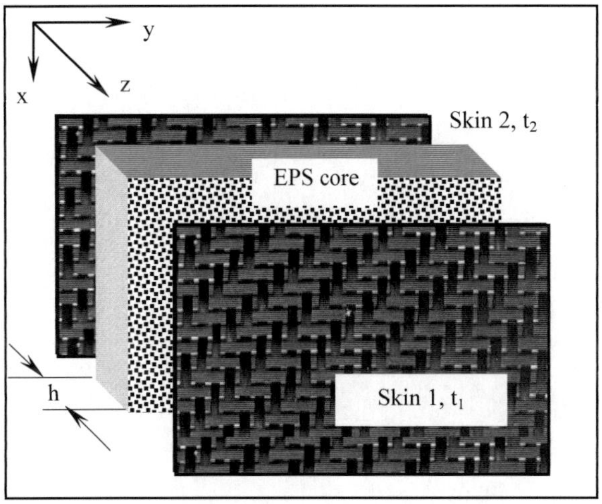

Fig. 1. The sandwich structure.

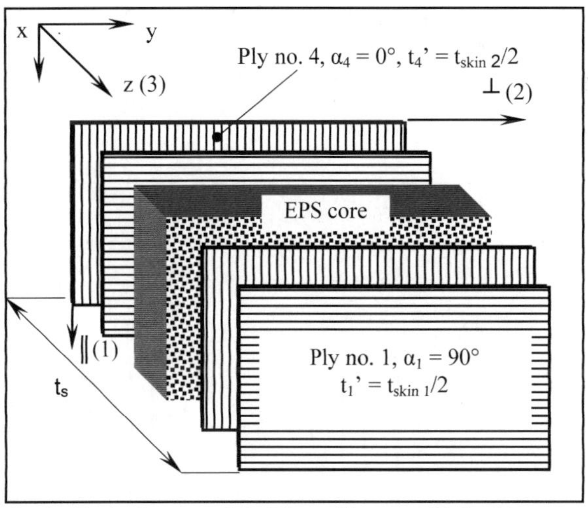

Fig. 2. Equivalence model of the twill weave fabric skins.

Structure thickness	$t_s = 10$ mm
Thickness of each ply	$t'_{1\text{-}4} = 0.175$ mm
Skins thickness	$t_{skin} = 0.35$ mm
Core thickness	$h = 9$ mm
Plies fibers disposal angle	$\alpha_{1,3} = 90°$, $\alpha_{2,4} = 0°$
Plies fibers volume fraction	$\varphi_{1...4} = 56\%$

Tab. 1. Sandwich architecture

The core stiffness's are (Wiedemann, 1986; Jones, 1999):

$$r_{core\,11} = r_{core\,22} = \frac{E_{core}}{1-\upsilon_{core}^{2}} \qquad (1)$$

$$r_{core\,12} = \frac{E_{core}\cdot\upsilon_{core}}{1-\upsilon_{core}^{2}};\; r_{core\,33} = G_{core} \qquad (2)$$

Then, sandwich structure stiffness's can be computed:

$$\underline{r}_{ij} = \sum_{K=1}^{N}\left(r_{ijK}\cdot\frac{t'_{K}}{t_{skin}}\right) + r_{core\,ij}\cdot\frac{h}{t_{s}}. \qquad (3)$$

Sandwich structure compliances are:

$$\underline{c}_{ij} = \frac{1}{\underline{r}_{ij}}. \qquad (4)$$

4. RESULTS

A comparison between sandwich structure's stiffness's with carbon- and glass fibers skins plies as well as bending strengths are presented in figs. 3 and 4.

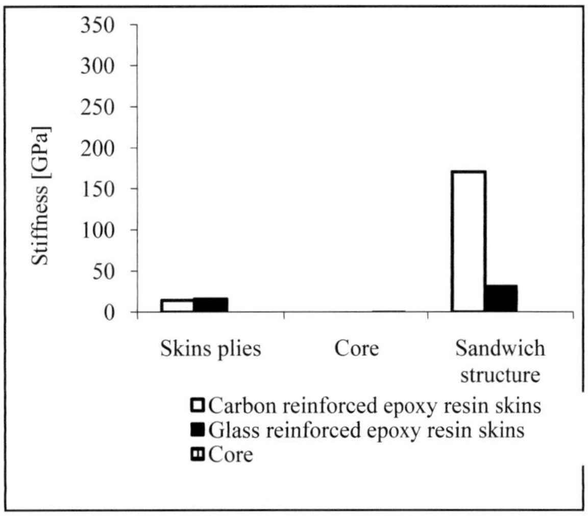

Fig. 3. Stiffness's of the structure's components.

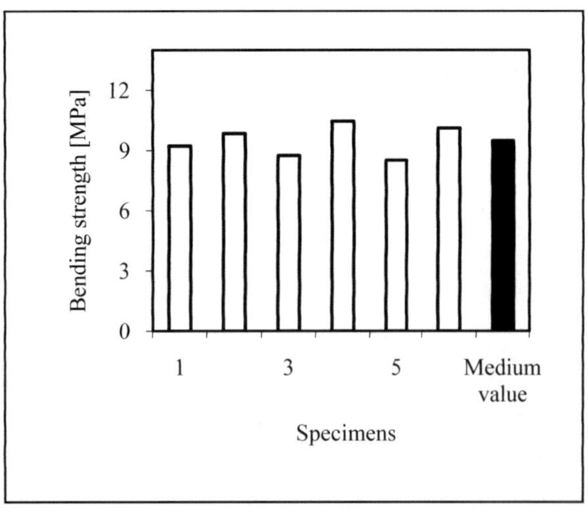

Fig. 4. Bending strengths of six specimens.

5. CONCLUSIONS AND FURTHER RESEARCH

The sandwich structure with two carbon/epoxy skins reinforced with a 0.3 kg/m^2 twill weave fabric and an expanded polystyrene (EPS) 9 mm thick core with a density of 30 kg/m^3 fulfils following special requirements: panel dimensions 10 x 2350 x 4070 mm and overall weight, maximum 10 kg. This sandwich structure is currently used to sustain large solar cells panels but applications can be in the aeronautic and automotive fields, transportations, naval industry and military domain. Sandwich structure's strains with skins based on twill weave carbon fabric reinforced epoxy resin are comparable with those of the structure with skins based on EWR-300 glass fabric/epoxy resin. Stresses in fibers direction in case of the sandwich structure with carbon fabric/epoxy resin reinforced skins, are up to six times higher than those existent in EWR-300 glass fabric/epoxy resin skins. Stresses transverse to the fibers direction in case of the sandwich structure with carbon fabric/epoxy resin reinforced skins are 20% lower than those existent in EWR-300 glass fabric/epoxy resin skins.

The shear stresses in carbon fabric/epoxy resin reinforced skins plies are almost identical with those existent in EWR-300 glass fabric/epoxy resin skins plies. Both sandwich structures core stresses can be neglected, the loading is taken-over exclusively by the two structures skins. Using a 9 mm thick expanded polystyrene core (EPS), according to fig. 3, the stiffness of the sandwich structure with carbon fibers reinforced epoxy resin skins is more than ten times higher than the skins plies stiffness, which is an outstanding achievement.

Further researches will be accomplished in the following directions:

- Determination of the influence of the adhesive type on the sandwich structure stiffness and bending strength.
- Determination of extensions field at bending using optical methods.
- Structure damping analysis computing the most important damping features starting from the dampings, dynamic Young moduli and Poisson ratio for every lamina using the so called correspondence principle of linear viscoelasticity theory.
- Structure sound damping analysis using various sound pressure waves.

6. REFERENCES

Chang Sup Lee; Dai Gil Lee; Je Hoon Oh & Hyun Surk Kim (2002). Composite wrist blocks for double arm type robots for handling large LCD glass panels. *Composite Structures,* Vol. 57, Issues 1-4, pp. 345-355, ISSN 0263-8223

Hosur, M.V.; Abdullah, M. & Jeelani, S. (2007). Dynamic compression behavior of integrated core sandwich composites. *Materials Science and Engineering:A,* Vol. 445-446, pp. 54-64, ISSN 0921-5093

Jones, R.M. (1999). *Mechanics of Composite Materials,* Taylor & Francis, Inc., ISBN 1-56032-712-X, Philadelphia

Marshall, A.C. (2006). *Composite Basics,* Aircraft Technical Book Co., ISBN 978-0966454048, Tabernash, Colorado

Smith, Z. (2005). *Advanced Composite Techniques,* Aeronaut Press, ISBN 978-0964282841, Napa, California

Vaidya, U.K.; Nelson, S.; Sinn, B. & Mathew, B. (2001). Processing and high strain rate impact response of multi-functional sandwich composites. *Composite Structures,* Vol. 52, Issues 3-4, pp. 429-440, ISSN 0263-8223

Wiedemann, J. (1986). *Leichtbau. Band 1: Elemente, (Lightweight constructions. Vol. 1: Elements),* Springer, ISBN 3-540-16404-9, Berlin, Heidelberg, New York, Tokyo

Wiedemann, J. (1989). *Leichtbau. Band 2: Konstruktion, (Lightweight constructions. Vol. 2: Construction),* Springer, ISBN 3-540-50027-8, Berlin, Heidelberg, New York

Annals of DAAAM for 2008 & Proceedings of the 19th International DAAAM Symposium, ISSN 1726-9679
Editor B. Katalinic, Published by DAAAM International, Vienna, Austria 2008
Make Harmony Between Technology and Nature, and Your Mind will Fly Free as a Bird

FINGERPRINT RECOGNITION WITH PHASE-ONLY CORRELATION

TEUSDEA, A[lin] C[ristian] & GABOR, G[ianina]

Abstract: *This paper presents fingerprint registration in a fingerprint database using phase-only correlation (POC) techniques. There are presented two modified phase-only correlation techniques with efficient registration results: a rectangle band limited phase-only correlation, BPOC, and an elliptic limited phase-only correlation, EPOC – proposed by the authors. The results of the computer simulations of these two POC methods are analyzed on the same fingerprint database. This analysis denotes that the proposed method EPOC has better performances than BPOC method for fingerprint registration in a database.*
Keywords: *fingerprint recognition, fingerprint registration, phase-only correlation, biometrics.*

1. INTRODUCTION

Nowadays security systems are a generic issue. Biometrics is a popular security criterion to restrict access to some systems and preserve their security. A part of biometrics is fingerprint recognition or matching. From the several methods developed in the past years, the phase-only correlation (POC) (Chien, 2004; Ito et al., 2005; Ito et al., 2004; Miyazawa et al., 2005; Takita, 2003) is important because its sub-pixel image translation capability. The past experiments developed a modified POC by rectangular band filtering the cross-spectrum of the POC function (BPOC) (Ito et al., 2005; Ito et al., 2004) in order to improve the genuine-impostor rejection.

This paper presents a theoretical introduction and some experiments for evaluating recognition performances of the proposed method and the dedicated ones.

2. METHODS AND SAMPLES

2.1 Phase only cross-correlation

The recognition process is used for object registration which means that one object is "compared" with several objects. Comparison criteria concludes if the compared objects are or not similar with other objects. The comparison process basically works with two objects. In our case the comparison method is the cross-correlation while the objects are the fingerprints.

In a single cross-correlation process the two objects are denoted as reference and non-reference. That means that from the cross-correlation process we obtain the information if the reference is similar or not with the non-reference object.

The cross-correlation considers two (N×M) images, $ref(x,y)$ as reference image and $nref(x,y)$ as non-reference image. The 2D discrete Fourier transforms of these images, denoted as $Ref(u,v)$ and $NRef(u,v)$, are given by (Ito et al., 2005; Ito et al., 2004)

$$Ref(u,v) = REF(u,v)\cdot \exp\left[i\cdot \varphi_{ref}(u,v)\right], \quad (1)$$

$$NRef(u,v) = NREF(u,v)\cdot \exp\left[i\cdot \varphi_{nref}(u,v)\right], \quad (2)$$

where $REF(u,v)$ and $NREF(u,v)$ are the amplitude parts and $\varphi_{ref}(u,v)$ and $\varphi_{nref}(u,v)$ are the phase parts of the 2D discrete Fourier transforms.

The phase-only cross-spectrum (Ito et al., 2005; Ito et al., 2004; Takita, 2003) is defined by

$$PO\_CS(u,v) = \frac{CIC(u,v)}{|CIC(u,v)|} = e^{i\cdot\Delta\varphi_{rn}(u,v)} \quad (3)$$

and thus the phase-only cross-correlation is given by

$$POC(x,y) = \frac{1}{N\cdot M}\sum_{u=1}^{N}\sum_{v=1}^{M}\left[e^{i\cdot\Delta\varphi_{rn}(u,v)}\cdot e^{-\frac{2\pi i}{N}ux}e^{-\frac{2\pi i}{M}vy}\right]\cdot \quad (4)$$

If $ref(x,y) = nref(x,y)$ then $\Delta\varphi_{rn}(u,v) \equiv 0$ and the phase-only correlation is given by

$$POC(x,y) = \frac{1}{N\cdot M}\sum_{u=1}^{N}\sum_{v=1}^{M}\left[1\cdot e^{-\frac{2\pi i}{N}ux}e^{-\frac{2\pi i}{M}vy}\right] = \delta(x,y)\cdot \quad (5)$$

This means that if the two images are identical then the POC gives a highly sharp peak so the matching accuracy is higher than in the classical method.

2.2 Band limited phase-only cross-correlation

The cross-correlation process on a database is basically characterized by two quantities: the autocorrelation peak intensity (API) and the cross-correlation peak intensity (CPI). The API value is obtained from all "autocorrelations" between the new fingerprint and the genuine class fingerprints as witness fingerprints. More precisely the minimum value from this set of "autocorrelations" peaks is denoted as API. In the same manner the CPI value is the maximum value of all the other cross-correlation peaks generated with the impostor classes fingerprints.

Phase-only correlation is a very precise matching method and effective for the verification process. This is done by fine "comparing" of the high frequencies in the Fourier transforms of the fingerprints.

When one has to register a new fingerprint then the correlation process must match it with some deformed representations of it. The deformations alter exactly the high frequencies. The reference-witness correlation can have a lower API value than the reference-nonreference correlation CPI value. This means that the involved matching process fails (figure 1 a, b, c; figure 2 a, b).

Fingerprint database registration matching process has to correlate only frequencies that are common to all fingerprints from the same class. This is the reason why the band limited phase only correlation (BPOC) (Ito et al., 2005; Ito et al., 2004) was introduced. This correlation uses a 2D band filter on the phase-only cross-correlation spectrum. The band filter is defined with two sub-unitary valued coefficients: over the rows direction, cL, and over the columns direction, cC.

The BPOC results as in figure 3 a, b, show that the

matching process is successful as the *API* value (figure 3a) is greater than the *CPI* value (figure 3b).

a b c

Fig. 1. Fingerprint examples: a – reference; b – witness (genuine class); c – non-reference (impostor class).

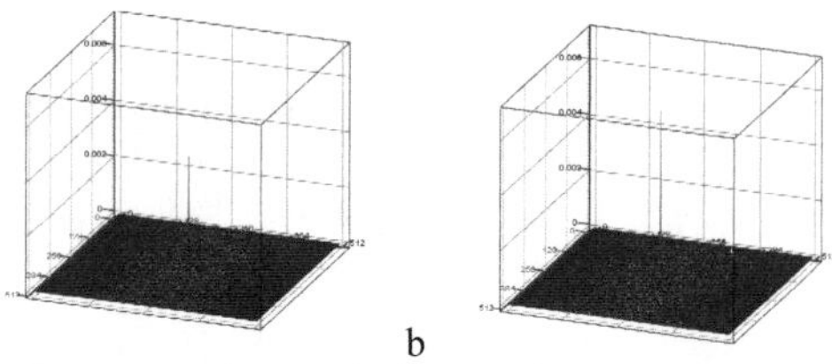

a b

Fig. 2. Phase-only correlation results between: a – reference and witness (same genuine class); b – reference and non-reference class (different classes).

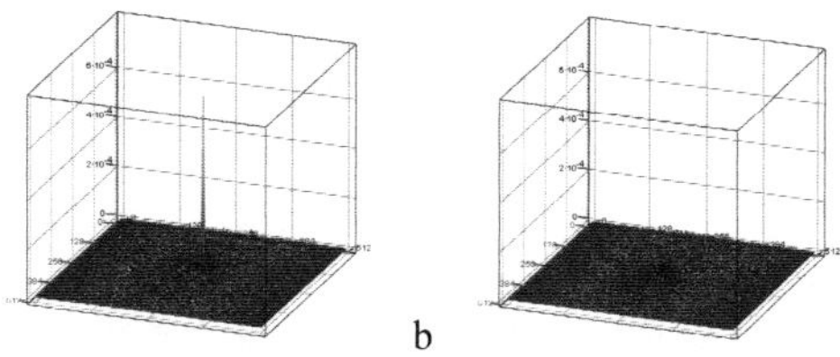

a b

Fig. 3. Band limited phase-only correlation results between: a – reference and witness (same genuine class); b – reference and non-reference class (different classes).

2.3 Elliptic band limited phase-only cross-correlation

In this paper the authors proposes an elliptic band phase-only correlation, (EPOC). In this method, there is used an elliptic band filter with the same cL and cC parameters instead of a rectangle band filter with cL and cC parameters. The reason of this choice is that the power spectrum of the fingerprints usually presents the highest density of the information in a centered elliptic form. As mentioned before, this centered ellipse contains that kind of spatial frequencies that can accommodate the database fingerprint registration.

The POC, BPOC and EPOC, phase-only correlations comparative results are presented in the next section.

3. RESULTS AND DISCUSSIONS

In this paper, a fingerprint database was used which was scanned with Cross Match Verifier 300 Classic (USB) at 500dpi in 30.5 × 30.5 mm image size (http://www.neuro technologija.com/download/CrossMatch_Sample_DB.zip).

The database contains fingerprint classes with 6 fingertips × 8 different scanned fingerprints for each of the 76 persons. The fingerprint index denotes "*ppp_ff_s.tif*", as *ppp* is the person number, *ff* is the fingertip number and *s* is the number of the fingerprint version.

The experimental results from the BPOC and EPOC are presented in figure 4, person with index 12 and fingertip index 3, 4, 5, 6, 7, 8, over the same set of fingertips of persons with indexes 13, 14, 17, 22, 27, 45, 47, 76. The band limited POC parameters were selected with the values: $cL = 0.80$ and $cC = 0.45$. There are presented the Genuine Acceptance Rate, GAR, and the False Rejection Rate, FRR. The GAR is calculated when the False Acceptance Rate is $FRR = 0$. This means that there was selected a threshold value (Th_C), which is the highest value from all the maximum intensity of the cross-correlation peaks (CPI). Genuine class POC values higher than Th_C were considered in GAR calculation while the smaller ones were considered in FRR calculation.

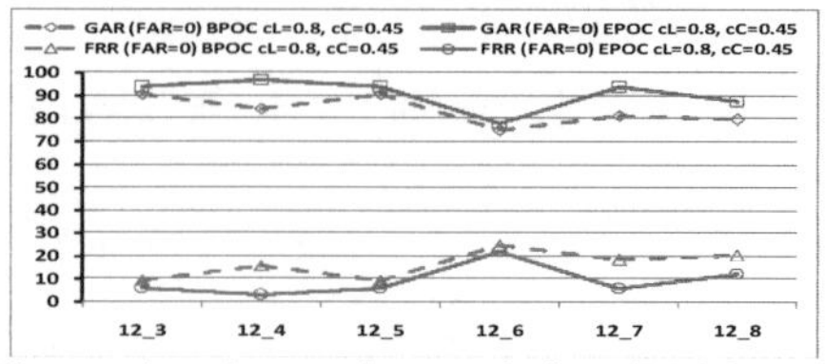

Fig. 4. Fingerprint phase-only correlation (BPOC and EPOC) results for the person with index 12.

The EPOC technique has the highest GAR value for all six fingertips of the person with index 12, in an experiment with 336×48= 16,128 POC correlations. In this experiment the differences of GAR coefficients performed with EPOC and BPOC techniques for the same fingertip lies between 3.44% and 15.38%.

4. CONCLUSIONS

In this paper, there are presented two modified phase-only correlation methods: the rectangle band limited, BPOC, and the proposed elliptic band limited, EPOC. These matching methods are very efficient for fingerprint recognition (single correlation experiment).

The results in figure 4 emphasize that elliptic band limited phase-only correlation, EPOC, has better performances (higher GAR values) than the rectangle band limited phase-only correlation, BPOC. Thus, for fingerprint database registration, the EPOC method is more efficient than the BPOC method.

Our future research will develop a more robust EPOC method to geometrical deformations of the fingerprints, with a Log-Polar transform. Another future plan is to work with a much larger fingerprint database to ensure statistical significance so as to be able to use it in biometrics technology.

5. REFERENCES

Chien, L. H. & Aoki T. (2004). Robust motion estimation for video sequences based on phase-only correlation, *Proceedings of the 6th IASTED SIP 2004*, Hamza M.H. (Ed), pp. 441-446, ISBN 0-88986-434-9, Honolulu, USA, Aug. 2004, ACTA Press, Canada.

Ito ,K.; Nakajima H.; Kobayashi K. & Aoki T., Higuchi T. (2004). A fingerprint matching algorithm using phase-only correlation, *IEICE Transactions. Fundamentals*, E87-A, No. 3, March 2004, pp. 682-691, ISSN 1745-1337.

Ito K.; Morita A.; Aoki T.; Higuchi T.; Nakajima H. & Kobayashi K. (2005). *A fingerprint recognition algorithm using phase-based image matching for low-quality Fingerprints*, Proceedings of IEEE Int. Conf. on Image Processing, pp. II-33 - II-36, ISBN: 0-7803-9134-9, September 2005, Genova, Italy, IEEE, NJ USA.

Miyazawa, K.; Ito K.; Aoki T.; Kobayashi K. & Nakajima H. (2005). *An efficient iris recognition algorithm using phase-based image matching*, Proceedings of IEEE Int. Conf. on Image Processing, pp. II-33--II-36, ISBN: 0-7803-9134-9, September 2005, Genova, Italy, IEEE, NJ USA.

Takita, K.; Aoki T.; Sasaki Y.; Higuchi T. & Kobayashi K. (2003). *High-accuracy subpixel image registration based on phase-only correlation*, IEICE Transactions. Fundamentals, E86-A, No. 8, August 2003, pp. 1925-1934, ISSN 1745-1337.

Annals of DAAAM for 2008 & Proceedings of the 19th International DAAAM Symposium, ISSN 1726-9679
Editor B. Katalinic, Published by DAAAM International, Vienna, Austria 2008
Make Harmony Between Technology and Nature, and Your Mind will Fly Free as a Bird

SUBSURFACE DRAINAGE AND ITS REVERSABLE FACILITIES IN SUBIRRIGATION

TEUSDEA, A[lin] C[ristian]; DAVID, I[oan] & MANCIA, A[urora]

Abstract: *The paper work focuses on subsurface drainage system design and its reversible use in subirrigation. It emphasizes David I. equations which completes Ernst equations for subsurface drainage design with the term for the load loss at the filter drain entrance. There are also provided the David I. equations for subirrigation design. These equations are the missing link to a complete numerical design algorithm. The computer simulations analysis accomplished can be used as a design tool for subsurface drainage systems which can work reversible too, as subirrigation systems.*
Keywords: *subsurface drainage, subirrigation, drain spacing, real drain.*

1. INTRODUCTION

Subsurface drainage structures may be also employed in the case of subirrigations. The double use reduces the costs of these structures. For this purpose, the underground drainage structure must be firstly designed and then verified as a subirrigation system.

2. METHODS AND SAMPLES

Two-layered soil subsurface drainage structure is designed using Ernst relationship (Sutton, 1971; Beers, 1976; Martinez Beltran 1978; Wehry et al., 1982; Kroes & Van Dam, 2003) completed by David I. for a real drain with filter (Wehry et al., 1982, David, 1983).

The basic scheme of this structure is depicted in figure 1a where: z - is the drainage norm, h - is the hydraulic head, K_1, K_2 - are the hydraulic conductivities of the two soil layers, D_0 - is the distance between the drain separation layer (the distance corresponds to the radial flow), $D_1 = D_0 + 0{,}5 \cdot h$ - is the water level above the drain (the distance corresponds to the horizontal flow in K_1 layer), D_2 - is the thickness of the layer below the drain, characterized by the hydraulic conductivity K_2, $D_v = h$ - the vertical distance of the vertical flow, d_0 - is the diameter of the drain tube and L - is the drain distance.

Ernst relationship for the calculation of the distance between ideal drains is (Sutton, 1971; Beers, 1976; Martinez Beltran 1978; Wehry et al., 1982; Kroes & Van Dam, 2003)

a 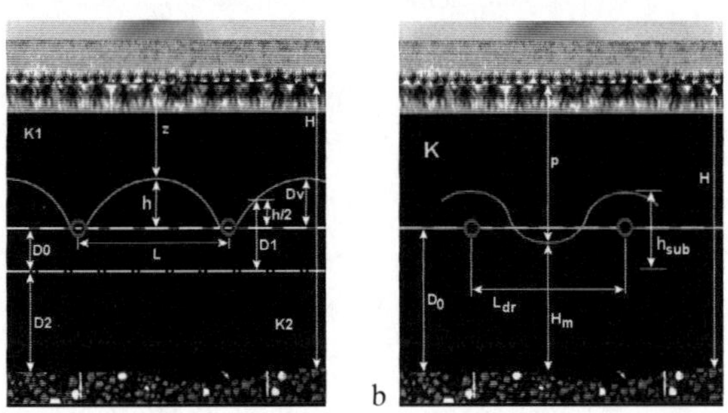 b

Fig. 1. a- design of the subsurface drainage structure (scheme); b - design of the subirrigation structure (scheme).

$$h = h_v + h_o + h_r = \frac{q \cdot L}{K} \cdot \left(\zeta_o + \zeta_v + \zeta_r\right) = \frac{q \cdot L}{K} \cdot \zeta_0, \qquad (1)$$

where h_o, h_v, h_r are the hydraulic head losses due to the horizontal, vertical and radial flow, respectively, and ζ_0 the total loss coefficient with an ideal drain. The resistance coefficients, ζ_o, ζ_v, ζ_r, corresponding to three flow types are calculated according to the relationships below (Sutton, 1971; Beers, 1976; Martinez Beltran 1978; Wehry et al., 1982; Kroes & Van Dam, 2003)

$$\zeta_o = \frac{L}{8 \cdot \left(D_0 + 0{,}5 \cdot h\right)}; \quad \zeta_v = \frac{d_0 + h}{L}; \quad \zeta_r = \frac{1}{\pi} \cdot \ln \frac{2}{\pi} \cdot \frac{D_0}{d_0}. \qquad (2)$$

If the drains are real - with discontinuous slits - and overcastted with a filter, then David I. (David, 1983) proposes the addition to the Ernst relationship with a term for the load loss at the filter drain entrance, δh_{if},

$$h = h_0 + \delta h_{if} = \frac{q \cdot L}{K} \cdot \left(\zeta_0 + \zeta_{if}\right). \qquad (3)$$

The resistance coefficient at the filter drain entrance is given by the relationship (David, 1983)

$$\zeta_{if} = \alpha \cdot \left[\ln\left(\frac{1}{\sin\left(\frac{nb}{2d_0}\right)} \right) + \frac{1-\chi}{2\chi} \cdot \ln\left[\left(A_1 + \sqrt{A_1^2 + 1}\right) \cdot \left(A_2 + \sqrt{A_2^2 + 1}\right) \right] \right] + $$
$$ + \beta \cdot \left[\ln\left(\frac{1}{\sin\left(\frac{\pi \cdot l}{2B}\right)} \right) + \frac{1-\chi}{2\chi} \cdot \ln\left[\left(B_1 + \sqrt{B_1^2 + 1}\right) \cdot \left(B_2 + \sqrt{B_2^2 + 1}\right) \right] \right] \qquad (4)$$

with

$$A_1 = \frac{d_0}{nb} \cdot \frac{\left(\frac{d_f}{d_0}\right)^n - 1}{\left(\frac{d_f}{d_0}\right)^{n/2}}; \quad A_2 = \frac{d_0}{nb} \cdot \frac{\sqrt{\left(\frac{d_f}{d_0}\right)^{2n} - 1}}{\left(\frac{d_f}{d_0}\right)^{n/2}}; \quad B_1 = \frac{\sinh \frac{\pi\left(d_f - d_0\right)}{2B}}{\sin \frac{\pi \cdot l}{2B}};$$
$$B_2 = \frac{1}{\sqrt{2}} \cdot \sqrt{\frac{\sqrt{\sin^4 \frac{\pi \cdot l}{2B} + \sinh^2 \frac{\pi \cdot \left(d_f - d_0\right)}{B}}}{\sin^2 \frac{\pi \cdot l}{2B}} - 1}, \qquad (5)$$

where: n - is the number of slits situated on the circumference of the drain, b - is the width of the slit, l - is the length of the slit, B - is the distance between slits and d_f - is the filter diameter. The coefficients α and β are calculated in different ways: in the case of longitudinal slit ($l > b$) and in the case of transversal slit ($l < b$) (David, 1983).

The coefficient $\chi = K_f / K_0$ represents the proportion between the hydraulic conductivity coefficient of the filter, K_f, and of the soil around the drain, K_0 respectively.

The distance between the drains, L, results from the second order equation in L, which is given by the relationship (3) (only the positive term is retained). After the calculation of this distance, the design of the drainage structure is practically finished. The next step consists in a verification if the structure is fit for a subirrigation system (figure 1b), meaning that it may be used reversibly.

The goal of the subirrigation verification is to determine the level difference $h_{sub} = H_0 - H_m$. The difference represents the total load loss which secures the water reserve at the root level. The basic scheme of this structure is depicted in figure 1b. The notations are as it follows: p - the width of phreatic water, H_0 - the width of the saturated soil zone at the drain, H_m - the width of the saturated soil zone midway between the drains, D_0 - the distance between drains and the impermeable layer, $L_{dr} = L$ - the drain distance, H - the height of the impermeable layer and K - the hydraulic conductivity of the soil. The total hydraulic head loss is composed of two components

$$h_{sub} = h_o + h_r \tag{6}$$

where: h_o - is the hydraulic head loss due to the horizontal movement and h_r - is the hydraulic head loss due to radial movement around the drain. These losses are defined by the relationships below (David, 1983)

$$h_o = \frac{\varepsilon L}{8KT_e}, T_e = \frac{H_0 + H_m}{2}, H_0 = H - z - h_{if} - h_{ld}, \tag{7}$$

$$H_m = H - p, h_{ld} = \frac{8(\varepsilon L_{dr})^2}{\pi^2 g d_0^4} \cdot \left[\left(\frac{\lambda}{3} \cdot \frac{X_{dr}}{d_0} \right) - 1 \right], \tag{8}$$

$$h_{if} = \frac{\varepsilon L_{dr}}{2K} \cdot \zeta_{if}, \tag{9}$$

$$h_r = \frac{\varepsilon L^2}{4KH_m} \left[\frac{\alpha - 1}{\alpha + 1} - \right. \\ \left. - \frac{2H_m}{\pi L_0} \cdot \ln \left[2^\alpha \cdot \left[\left(1 + \cos\left(\frac{\pi D_0}{H_0} \right) \right) + \frac{\pi r_0}{H_0} \cdot \sin\left(\frac{\pi D_0}{H_0} \right) \right]^\alpha + \left(1 + \cos\left(\frac{\pi D_0}{H_0} \right) \right)^\alpha \right] \right] \tag{10}$$

with $\alpha = H_0 / H_m$.

From these relationships, h_{if} is the hydraulic head at real drain with filter entrance, h_{ld} is the hydraulic head loss along the drain of $X_{dr} = H/i$ with i slope, $g = 9.81(m/s^2)$ is the gravitational acceleration and $\lambda = 0.04$ is the roughness coefficient.

The validation criterion of the drainage structure for the reversible usage (meaning, subirrigation) is described by the following relationship

$$(H_c + z) < H, H_c = H_m + h_{sub} + h_{if} + h_{ld} \tag{11}$$

where H_c represents the water height in the collecting channel.

3. RESULTS AND DISCUSSIONS

Computer simulations were done involving the equation (1-5) for subsurface drainage, and (6-11) for subirrigation validation. Two examples were taken into consideration: the first with hydraulic head $h = 1.1$ *m* and the second with $h = 0.6$ *m*, with two values for $H = 3.00$ *m* and $H = 4.75$ *m*. The initial parameters were $q = 0.01$ *mm/day*, $z = 0.6$ *m*, $D_0 = 0.5$ *m*, $D_1 = 0.8$ *m*, $D_2 = 2.0$ *m*, $d_0 = 0.065$ *m*, $d_f = 0.073$ *m*, $l = 0.001$ *m*, $b = 0.005$ *m*, $B = 0.012$ *m*, $p = 1.6$ and $i = 0.02$. The first situation considers an ideal drain, which means that there is no ζ_{if} coefficient involved. The second situation considers the real drain with filter which means that there is a real non-zero value of the ζ_{if} coefficient involved.

The drain distance, L, decreases with *4.16 %* in the first example, from the ideal drain situation to the real drain with filter situation. In the second example, the drain distance, L, decreases with *13.74 %* in the first example, from the ideal drain situation to the real drain with filter situation. This means that the real drain with filter needs a smaller drain distance.

Subirrigation validation fails in the first three cases of the first example because the $(H_c + z)$ values are greater than the H values. The forth case generates a subsurface drainage structure that can be used in subirrigation, as the validation criterion $(H_c + z) < H$ is satisfied.

In the second example, all the four cases generate subsurface drainage structures that can be used in subirrigation, as the validation criterion $(H_c + z) < H$ is satisfied. Regardless of the H values, the $(H_c + z)$ values are smaller in the situation of real drain with filter than in the ideal drain situation. These results show that those subsurface drainage structures are more stable to subirrigation usages.

4. CONCLUSIONS

This paper studies the effect of using a real drain with filter in subsurface drainage design and its subirrigation validation. In the subsurface drainage design, the Ernst equations completed by David I. are used with a term for the load loss at the filter drain entrance. The subirrigation design is also accomplished with David I.'s equations. Computer simulation results point out that the subirrigation validation criterion is agreed upon only by the subsurface drainage system with real drain with the selected input parameters.

In the future research, we plan to verify if, according to this paper's analysis, the already designed subsurface drainage systems in the Western part of Romania are suitable for subirrigation.

5. REFERENCES

Beers, W.F.J. (1976). *Computing Drain Spacings – A generalized method with special reference to sensitivity analysis and geo-hydrological investigations*, pp. 14-19, IRLI Bulletin 15, Wageningen, The Netherlands.

David I. (1983). Uber ein berechnungsverfahren der verlusthohe beim eintrittswiderstand des wassers in die drainrohre / Calculation method of the head loss by the ground water entrance in drainage, pp. 3-20, *Revue Roumaine des Sciences Techniques – Serie de Mecanique appliqué*, Tome 28, Nr, 1, Bucharest.

Kroes J.G. & Van Dam, J.C. (2003). *Reference Manual SWAP version 3.0.3*, pp. 52-58, Alterra Report 773, Alterra, Green world Research, Wageningen.

Martinez Beltran, J. (1978), *Drainage and Reclamation of Salt-Affected Soil – Bardenas Area Spain*, pp. 276-278, ILRI, Publication 24, Wageningen.

Sutton J. (1971). *Section 16 Drainage of agricultural land*, USDA – Soil conservation Service, National Engeneering Handbook, pp. 4-1 – 4-122, Washington DC.

Wehry A.; David I. & Man E., (1982), *Probleme actuale în tehnica drenajului / The present state of drainage technique*, pp. 67-85, Editura Facla, Timişoara.

Annals of DAAAM for 2008 & Proceedings of the 19th International DAAAM Symposium, ISSN 1726-9679
Editor B. Katalinic, Published by DAAAM International, Vienna, Austria 2008
Make Harmony Between Technology and Nature, and Your Mind will Fly Free as a Bird

DRAGAN DAM DEFORMATIONS ANALISYS WITH FOURIER CORRELATION

TEUSDEA, A[lin] C[ristian]; MODOG, T[raian] & MANCIA, A[urora]

Abstract: *The monitoring of Drăgan Dam's structural deformations in time is carried out through physical methods (inverse pendulum) and topographical methods (optoelectronic). The different conclusions drawn from the deformations measured with these two systems reveal the importance of the correlation analysis between them. If there is a significant correlation, then the differences between the values of those deformations will be provided by measuring errors. If the correlation is not present, then the differences between the values of those deformations will be provided by coarse errors. This paper intends to identify and exclude the problems of the presented systems, which provide wrong deformations. The Fourier analysis is used to generate the correlations between the deformations measured by these two different methods.*
Keywords*: structural deformation, spectral analysis, Fourier transform, dam monitoring.*

1. INTRODUCTION

Drăgan Dam presents a double arch concrete structure featuring 120 m height and 450 m length at the crest. It has 33 vertical plots and generates a basin of about 120 million m^3 of water. Monitoring the deformations of large concrete dams is important to prevent fatal accidents of dam cracking. The deformations of the dam crust are measured physically with an inverse pendulum with a very good precision (10^{-2} mm) given by an optical coordiscope. The surveying method readings of dam crust deformations are done with an optoelectronic device called total surveying station. This method involves building a surveying network of reference points, from which sets of readings are measured for the same deformations (Behr, 1998).

For plots 7, 12, 19, 24 and 29, the time series provided from inverse pendulums consist in 2010 readings, from May 2000 until November 2005. This is the reason why the time series provided from the surveying targets (i.e. the control points) consist only in the readings of deformations at control points placed nearest to the measuring points of the inverse pendulum.

This paper presents the time series correlation for five measuring points and their nearest control points, done only for plot 19, which is the middle vertical axis of the dam.

2. METHODS AND SAMPLES

There are two ways to get the correlation information between two time series that have different numbers of readings, i.e. 2010 readings for the inverse pendulum and just 12 seasonal readings for the surveying method. The first way is to select only the corresponding 12 dates, from the 2010 dates, which match the dates for the surveying method (figure 1). The second way is to interpolate the 12 dates from the surveying method and to obtain 2010 readings dates, which match the inverse pendulum time series dates.

In this paper, we chose the first way which correlates these two different time series. As mentioned before, the correlation process involves two time series over 12 dates. The inverse pendulum time series (1D+t) are denoted by X (figure 1 – the circle doted line) and the surveying time series are denoted by XT (figure 2 – the diamond doted line).

Furthermore, we consider horizontal deformations (X and Y) of plot 19 vertical axis. The vertical axis of plot 19 consists of five measuring/control points (figure 3) spatially distributed along the plot height (2D information) – i.e. one vertical bolded line from figure 3a, 3b.

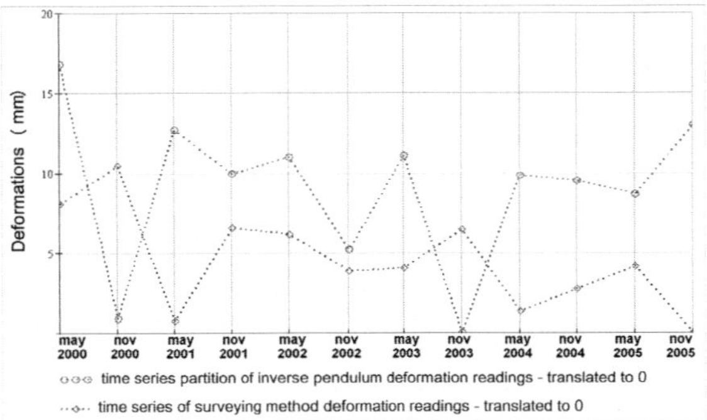

Fig. 1. Upstream-downstream direction inverse pendulum deformations (X) and surveying deformations (XT) readings for one measuring/control point of plot 19 for 12 dates.

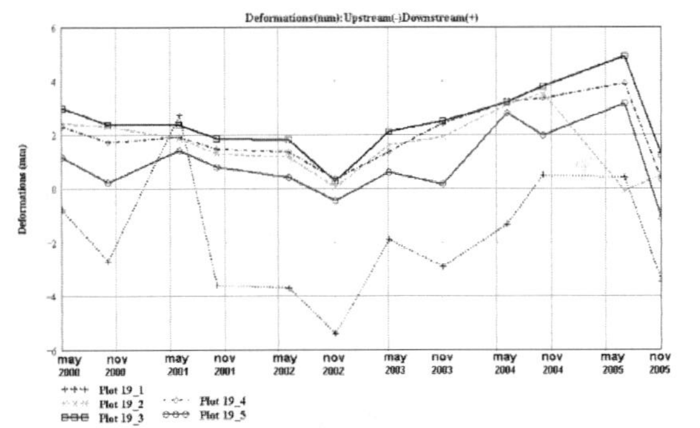

Fig. 2. Surveying deformations readings in five control points of plot 19 for 12 dates.

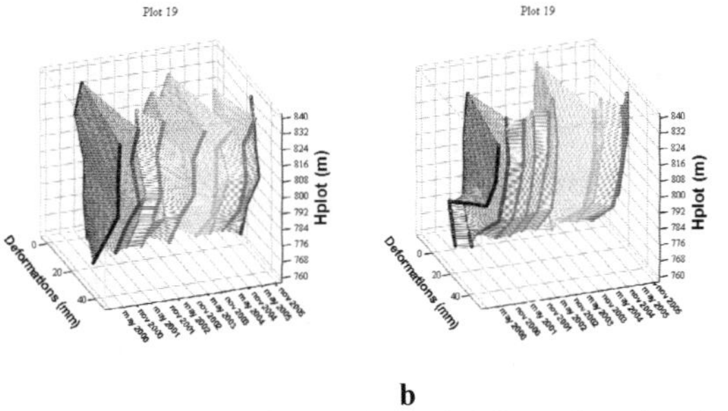

a **b**

Fig. 3. (2D+t) surfaces of the horizontal deformations for plot 19 vertical axis: a - for upstream-downstream deformations HX, b - for upstream-downstream deformations HXT.

Time series of the vertical axis horizontal deformations give the (2D+t) surfaces: first upstream-downstream deformations, denoted by HX, for the inverse pendulum readings and second upstream-downstream deformations, denoted by HXT, for the surveying readings.

The correlation process may be a statistical or a Fourier spectral analysis one. The normalized Fourier correlation coefficient, $NFCC$, can be built from the Fourier analysis, described (Gârlaşu et al., 1982; Grierson, 1982) by

$$
\begin{aligned}
NFCC(f(x),g(x)) &= \max_x \left[\frac{\left[\left| f(x) \otimes g(x) \right|^2 \right]}{\max_x \left[\left| f(x) \otimes f(x) \right| \right]^{0.5} \cdot \max_x \left[\left| g(x) \otimes g(x) \right| \right]^{0.5}} \right] \qquad (1) \\
&= \max_x \left[\frac{\mathcal{F}^{-1}\left[\overline{F(-k)} \cdot G(k) \right](x)}{\max_x \left[\mathcal{F}^{-1}\left[\overline{F(-k)} \cdot F(k) \right](x) \right]^{0.5} \cdot \max_x \left[\mathcal{F}^{-1}\left[\overline{G(-k)} \cdot G(k) \right](x) \right]^{0.5}} \right] \\
&= \max_x \left[\frac{\mathcal{F}^{-1}\left[\overline{F(-k)} \cdot G(k) \right](x)}{\max_x \left[\mathcal{F}^{-1}\left[\left| F(k) \right|^2 \right](x) \right]^{0.5} \cdot \max_x \left[\mathcal{F}^{-1}\left[\left| G(k) \right|^2 \right](x) \right]^{0.5}} \right]
\end{aligned}
$$

where $f(x), g(x)$ are two functions, $F(k), G(k)$ are their Fourier transforms, x and k are two Fourier-conjugate variables (i.e. t as time and ν as frequency) and $\mathcal{F}^{-1}$ is the inverse Fourier Transform. From the recommendations provided by literature the best way to correlate pure time series is the Fourier analysis method (Pytharouli, S, 2004; Pytharouliand & Stiros, 2005). When the information is time-spatially distributed, the only way the correlation process can achieve consequent results is by Fourier correlation and not by statistical correlation – as the Pearson coefficient – despite the symmetry of the two definitions of the correlation coefficients.

Statistical significance of the correlation coefficient values are: 0.10 – 0.29 for weak; 0.30 – 0.49 for average, 0.50 – 1.00 for strong.

3. RESULTS AND DISCUSSIONS

Fourier correlations were done in the first step for the (2D) case (i.e. 1D data of five control points of one horizontal dimension distributed as 1D along H) between the HX, HY and HXT, HYT series (table , 1lines 2-13, column 3 and 4). In the second step the Fourier correlation were done for the (2D+t) case (i.e. 1D data of five control points of one horizontal dimension distributed as 1D along H, over the 12 dates as t) between HX, HY and HXT, HYT (table 1, line 14 and 15, column 3 and 4).

1.	**Date**	**HX vs HXT**	**HY vs HYT**
2.	may.2000	*0,928**	*0,895 **
3.	october.2000	*0,601**	*0,950 **
4.	july.2001	*0,976**	*0,857 **
5.	october.2001	*0,909 **	*0,892 **
6.	june.2002	*0,988 **	*0,891 **
7.	november.2002	*0,777 **	*0,414 **
8.	may.2003	*0,944 **	*0,812 **
9.	november.2003	*0,642 **	*0,753 **
10.	june.2004	*0,964 **	*0,939 **
11.	october.2004	*0,841 **	*0,504 **
12.	july.2005	*0,988 **	*0,865 **
13.	november.2005	*0,955 **	*0,607**
14.	***Lp* (2D+t)**	***0,886***	***0,800***
15.	**NFCC (2D+t)**	***0,841***	***0,535***

Tab. 1. The results of the Fourier correlation of plot 19 time series (* have the significance of (1D data of one horizontal dimension + 1D along H) Fourier correlation, where H is the plot height).

The Fourier correlation of the (2D) series has *NFCC* minimum values: *0.601* for the HX and HXT pair (upstream – downstream deformations) and *0.414* for HY and HYT pair (left side – right side deformations). This means that the horizontal deformations measured in upstream – downstream direction (HX, HXT time series) are strongly correlated and measured in left side – right side direction (HY, HYT time series) are just averagely correlated.

The (2D+t) correlation was done in two ways. In the first way, the Lp norm (*p=2.5*) of the (2D) results of correlations is calculated providing the Lp score over all 12 dates. In the second way the pure (2D+t) Fourier correlations of surfaces from figure 3a, 3b are done. The Lp score denotes highly strong correlations between all horizontal deformations measured by the inverse pendulum (HX, HY time series) and by the surveying method (HXT, HYT time series). Instead, the results of Fourier correlation of (2D+t) time series denote that the horizontal deformations measured by the inverse pendulum and by the surveying method are highly strong correlated for (HX vs. HXT) and just strongly correlated for (HY vs. HYT).

4. CONCLUSIONS

Spectral correlation analysis (via Fourier transform) of the structural dam horizontal deformations measured by physical method and by surveying method is presented. As correlation inputs were used: the (2D) time series of the deformations at control points and the (2D+t) time series of the deformations of entire vertical axis of the median plot of the dam. Despite that the (2D) correlation results show an overall averagely correlation, the (2D+t) correlation results show a strongly correlation for both horizontal deformations. This means that the (2D+t) Fourier correlation analysis is more suitable to diagnose the dam's status.

Correlation analysis accomplished in this paper reveals that the surveying method is reliable for periodical monitoring of the dam's deformations and can be used to make accurate diagnosis of the dam's status. In the future research, we plan to make a (3D+t) analysis of all the dam's plots in order to provide a more accurate dam's status diagnosis.

5. REFERENCES

Behr, J. A.; Hudnut, K. & King, N. (1998). *Monitoring structural deformation at Pacoima Dam, California, using continuous GPS*, Proceedings of the 11th International Technical Meeting of the Satellite Division of the Institute of Navigation ION GPS-98, pp. 59-68, September 1998, Nashville, TN. *Available on*: http://pasadena.wr.usgs.gov/scign/group/pacoima_dam/ION/PacoimaGPS.pdf.

Gârlaşu, St.; Popp, C. & Ionel, S. (1982). *Introducere în analiza spectrală şi de corelaţie / Introduction to spectral and correlation analysis*, Ed. Facla, Timişoara.

Grierson, B.A. (2006). *FFT's, Ensembles and Correlations*, *Available from:* http://www.ap.columbia.edu/ctx/ctx.html, *Accessed:* 2007-08-12.

Pytharouli, S; Kontogianni, V.; Psimoulis P. & Stiros, S. (2004). *Spectral Analysis Techniques in Deformation Analysis Studies,* International Conference on Engineering Surveying and FIG Regional Conference for Central and Eastern Europe (INGEO 2004), Bratislava, November 2004, Proceedings-CD, 10 pp.

Pytharouli, S. I. & Stiros, S. C. (2005). Ladon dam (Greece) deformation and reservoir level fluctuations: evidence for a causative relationship from the spectral analysis of a geodetic monitoring record, *Engineering Structures*, Volume 27, Issue 3, February 2005, Pages 361-370, ISSN 0141-0296.

Annals of DAAAM for 2008 & Proceedings of the 19th International DAAAM Symposium, ISSN 1726-9679
Editor B. Katalinic, Published by DAAAM International, Vienna, Austria 2008
Make Harmony Between Technology and Nature, and Your Mind will Fly Free as a Bird

POSSIBILITIES THE ROLLING REDUCTION BY REFERENCING DIRECTION WITH SUSPENSION CORRELATION

THIERHEIMER, W[alter]; TANE, N[icolae] & THIERHEIMER, D[iana]

Abstract: *In this particular case we considered the elements that have big influence in dynamic behavior of the automotive, such as: steering geometry, the particular aspects regarding constructive solution. The problem of kinematical, dynamical and mathematical models is considered to characterize the dynamical behaviour of vehicles on curvilinear movement. The differential equation described in here was solved using the Laplace's method. The theoretical and experimental data comparison confirms with good accuracy the hypothesis made on the study of vehicle's stabilization system.*
Key words: *maniability, control, suspension, kinematics*

1. INTRODUCTION

The optimisation of the correlation between the suspension and steering systems with the aim to increase the stability and maniability of automotive vehicles, assumes a complex simulation completed by rigorous experimental research. Over the past decades research have been done on improving systems like the one under discussion without detailed analysis of the stabilizer rod influence. The goal here is to provide certain theoretical models and practical solutions to correlate the steering system with the suspension one to generally increase traffic safety in case the vehicle enters and exit turnings. Detailed analyses in case of numerous degrees of freedom are taken into account. In these cases the vehicle's suspended and unsuspended masses were influenced through the elastic (springs and tires) and damping elements. In case the vehicle exits a turn a new method of stabilizer rod positioning was proposed, the piston rods being coupled with the toothed rack of the steering module (Oţăt & Thierheimer 2007).

2. MATHEMATICAL MODEL OF VEHICLE

The mathematical model that comprises the dynamical behaviour of a vehicle is a very important assumption for its stability characterization. In the study of phenomenon two dynamical equivalent methods having seven degree of freedom were proposed, the difference between them consisting of stabilizer rod location. Therefore, the modified solution for the stabilizer rod placement consists of placing it behind the steering wheels axle and has the extremities coupled through the piston rod on the steering box levers. The basic movement of the vehicle is a linear and/or a curvilinear one, having imposed on it the vertical and rotational oscillations. The main disturbance factor is the road micro-profile. On such various surfaces vehicle's velocity is different in both cases on discussion. As it is shown in Figure 1, the most important dynamical variables are: the roll angle φ_1, the pitch angle φ_2, z_i (i=1,2,3) the vertical displacements of the left and right front wheels, respectively of the rigid rear bridge. In addition, there are the following constants: vehicle suspended mass m and unsuspended m_i (i=1,2,3) masses of left and right steering wheels, respectively mass of rigid rear bridge, elasticity and amortization constants, and the distance a (b) between the front (rear) wheels and the centre of gravity, (Pereş et all., 1998).

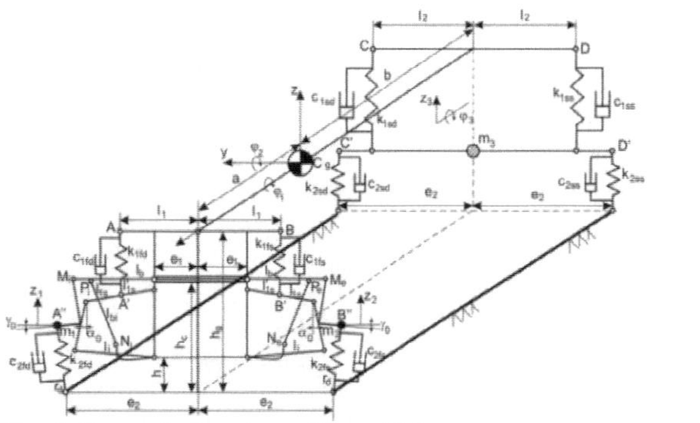

Fig. 1. Dynamical system of the vehicle

Based on the balance of the forces acting on the vehicle, the torques and kinematical conditions, the following represents the matrix form of the differential equation characterizing vehicle's curvilinear behaviour, valid also in case of linear movement:

$$[M]\cdot[\ddot{q}]+[C]\cdot[\dot{q}]+[K]\cdot[q]+[P]=0, \qquad (1)$$

where: $[M]$ - represents the mass matrix, $[C]$ - matrix of amortization coefficients, $[K]$ - matrix of elastic constants, $[P]$ - matrix of disturbance elements, $[\ddot{q}]$, $[\dot{q}]$ and $[q]$ acceleration, velocity and displacements vectors of the system ($[q]$ = $[z\ \varphi_1\ \varphi_2\ z_1\ z_2\ z_3\ \varphi_3]^T$).

The use of the Maple V computer program to solve the differential equation written previously by Laplace's method.

In the matrix of disturbance elements, in case of curvilinear behaviour of the vehicle, were included the rotational angles of the front stabilizer rod extremities. These angles are directly proportional with the vertical displacements of the pivots at the moment of wheels levelling.

These raising can be determined with the following equations:

$$h_e = \rho\cdot(1/\cos\theta_e)\cdot\sin\alpha_0 - \rho\cdot\sin\theta_e\cdot\sin\beta_0, \qquad (2)$$

respectively

$$h_i = \rho\cdot(1/\cos\theta_i)\cdot\sin\alpha_0 - \rho\cdot\sin\theta_i\cdot\sin\beta_0, \qquad (3)$$

where ρ is the radius of the circle generated by the wheel contact point rotation around the pivot axis (the circle lay into a perpendicular plane on the pivot axis), α_0 is the pivot transversal inclination angle, β_0 is the pivot longitudinal inclination angle, θ_i and θ_e the levelling angles of the wheel from the inside, respectively outside the curve on which the vehicle is moving. By means of the Eqns. (2) and (3) the angular displacements of the front stabilizer rod can be determined together with:

$$\psi_e = arctg\left[h_e\cdot l_s/(l_s+l_s^{'})\right],\ \psi_i = arctg\left[h_i\cdot l_s/(l_s+l_s^{'})\right]. \qquad (4)$$

3. KINEMATICAL ASPECTS REGARDING THE PROPOSED SYSTEM

Both classical solution and that proposed Figs. 1 and 2 the steering swivel has the angle of longitudinal inclination β_0 and transversal α_0, the camber angle γ_0 and steering angle δ_0.

In the straight line displacement of the vehicle this angles provides the stability of movement without produce perturbations. In the curve displacement especially at the beginning and at the end, the angles α_0, β_0 produces displacements, in the vertical plane, of the steering swivel and pivots axis (Thierheimer 1998).

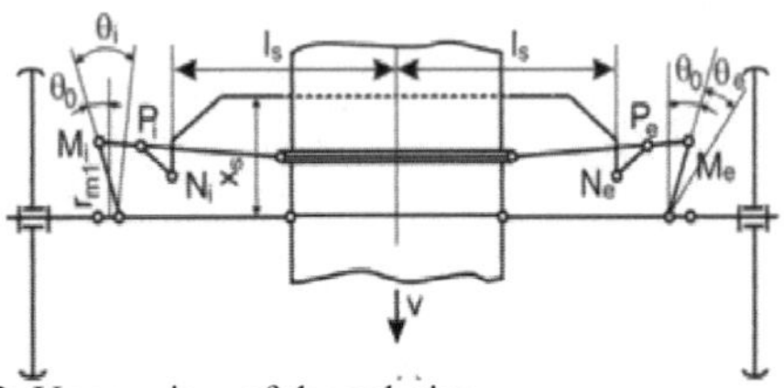

Fig. 2. Upper view of the solution

These displacements aren't conditioned by the position of the stabilizer crossbar, but they are transmitted to the suspended and non-suspended masses of the vehicle by the wheel and springs.

P_i, N_i refers to the inner wheel of the trajectory respectively P_e, N_e to the outer wheel.

Analyzing the kinematics schema, Figure 2, and take into account all the possible movements of the system, we may consider that in the curve displacement the points P_i and P_e moves on the spherical surfaces. In this situation the points N_i and N_e describe circles with radius r_{m2} – who has the center on the geometrical axis of the stabilizer steering. These circles are in two parallel planes and these are parallel with rolling plane.

In the curve displacement P_i and P_e describe different trajectories, the points N_i and N_e describe two arcs of a circle that has different length. As a result the stabilizer crossbar is spinning. In this way result an elastic torque. The determination of the movements of the points P_i, N_i, P_e and N_e, respectively angular displacements ψ_i and ψ_e is realized in relation with the mass center of the vehicle (Thierheimer 2004).

This coordinates are:

- for P_e

$$
\begin{aligned}
x_{P_e} &= a - r_{m1}\cdot\cos(\theta_e - \theta_0);\\
y_{P_e} &= (e_1 + l_{1s}) + r_{m1}\cdot\sin(\theta_e - \theta_0)\\
z_{P_e} &= -(h_g - h_c) - r_{m1}\cdot[1-\cos(\theta_e + \theta_0)]\cdot\\
&\cdot\frac{l_{1s}}{l_b}\cdot\sin\beta_0 + r_{m1}\cdot\sin(\theta_e + \theta_0)\cdot\frac{l_{1s}}{l_b}\cdot\sin\alpha_0
\end{aligned} \tag{5}
$$

- for N_e

$$
\begin{aligned}
x_{N_e} &= a - x_s + r_{m2}\cdot\cos\psi_e;\\
y_{N_e} &= l_s\\
z_{N_e} &= -h_g + h + h_a + r_{m2}\cdot\sin\psi_e.
\end{aligned} \tag{6}
$$

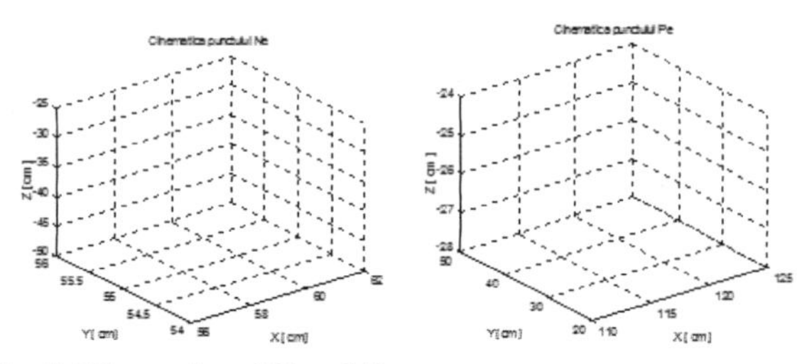

Fig. 3. Kinematics of N_e and P_e

By processing the results we obtained the following graphics for the kinematics and movements for the points P_e and N_e, Figure 3.

The kinematics of these points is studied because the calculation of the angular displacements ψ_i and ψ_e is relative to the position of the points.

4. EXPERIMENTAL ANALYSIS

The experimental tests were carried out using a MC-32 system that represents an acquisition measurement apparatus controlled by a PC computer (Thierheimer 2004).

Figure 4 are shown the theoretical and experimental curves that were obtained for vehicle's centre of gravity displacements and roll angle φ_l in case the speed of the vehicle moving on curvilinear path having a radius of 15 *m*, was 14 *m/s*. The results obtained after the theoretical and experimental data comparison can be used to modify the kinematical, dynamical and mathematical models. In this way the possible results can describe with accuracy vehicle's movement parameters no matter the external influence.

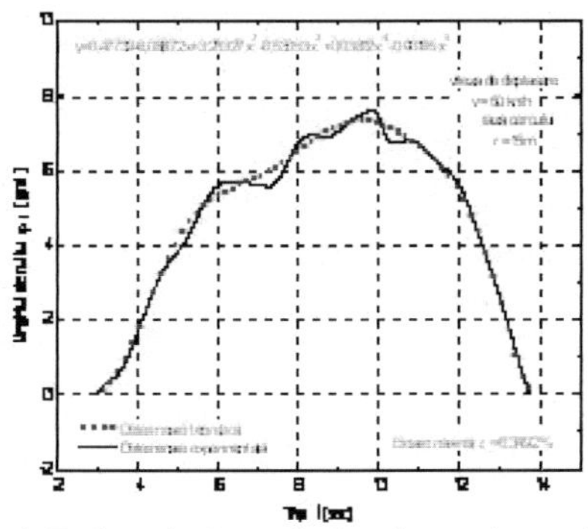

Fig. 4. Roll angle theoretical and experimental curves

5. CONCLUSION

This solution is conceived, designed and realised to improve the re-establish of the steering mechanism, especially in the cornering displacement. By taking account to the objective of stabilisation of the vehicle movement at the beginning and at the end of the curve displacement, we consider that dynamic model (and the vehicle also) is on the influence of the perturbation factors determined by the steering geometry (the suspension is an independent one) when the displacement is realised on straight, dry end even road. As a result of these excitations the suspended and non-suspended masses of the vehicle realise rotation and translation moves related to the mass centre of the vehicle.

6. REFERENCES

Otat, V. & Thierheimer, W. (2007), ş.a. *Dynamics of vehicles*, Editura Universitaria, Craiova, ISBN 973-742-023-3, Romania.

Peres, Gh.; Thierheimer, W.; Luca, D. & Zamfira, S. (1998) Contributions to the study of vehicle's dynamic behaviour in curvilinear movement, In: *Proceedings of the 9th International DAAAM Symposium*, Katalinic, B. (Ed.), pp. 381-382, ISBN 3-901509-08-9, Wien.

Thierheimer, W. (1998), Research concerning the optimisation of the correlation between the suspension and the steering systems in case of vehicles with front drive, Ph.D. thesis, May, Transilvania University Brasov.

Thierheimer, W. (2004) *Research corelation with the direction of suspension from the front cars involved*, Editura Universitatii Transilvania Brasov, ISBN 973-635-420-2.

Annals of DAAAM for 2008 & Proceedings of the 19th International DAAAM Symposium, ISSN 1726-9679
Editor B. Katalinic, Published by DAAAM International, Vienna, Austria 2008
Make Harmony Between Technology and Nature, and Your Mind will Fly Free as a Bird

STUDIES CONCERNING THE ROLLING MOTION REDUCTION BY CONTROLLING THE STEERING AND GEAR AND THE SPRING RIGGING CORRELATION

THIERHEIMER, W[alter]; TANE, N[icolae] & THIERHEIMER, D[iana]

Abstract: *The study features a working solution to optimize the correlation between the steering gear and the spring rigging, thus improving vehicle's dynamic output. Performance streamlining, materialized in the quadratic linear hydraulic regulator synthesis QLHR, was a result of a complex correlation analysis between Safety and Convenience indicators. To control the release of the half decks' spring rigging, one uses the pressure of the working liquid in the steering box. The theoretical and experimental data comparison confirms with good accuracy the hypothesis made on the study of vehicle's stabilization system.*
Key words: *hydraulic regulator, stability, semi-active control*

1. INTRODUCTION

Usually, stabilizers are solid torsion rod springs, linked to the wheels' spring rigging by articulated steering rods. This precaution allowed stabilizers to react only to those disturbances coming from the rolling track and yielded by the uneven deck rigging compression and expansion.

Accordingly, stabilizers' active influence, which must be compatible with this driving principle, should be achieved in the stabilizer – steering mechanism unit or in an equivalent constructive solution (Oţăt & Thierheimer, 2007).

The most convenient solution for vehicle's movement security would be to transmit a $\Delta\triangleleft$ angular rotation lag between the two working positions, compression and expansion, for the steering gear and the torsion rod, respectively.

When the system collapses, the stabilizer continues to function as a passive elastic system. If certain forces are exerted, such fail-safe functions would be more difficult to achieve. One must also take into consideration the fact that the output regulator used to correct angular lags based on constructive achievement conventional elements is achieved quite easy and cost-effectively, with the use of the electro-mechanical and mecatronic devices (Thierheimer, 1998).

Inducing $\Delta\triangleleft$ angular lags caused by the rod torsion leads to an angular shift that equals the stabilizing rod tips' $\Delta\triangleleft$ coefficients, namely $\Delta c = c_s \, \Delta\phi$, as related to vehicle's symmetry longitudinal axis.

Apart from the reduction in the rolling oscillations, active stabilizers offer the possibility to permanently correlate the spring rigging with the steering gear, thus eliminating the unwanted induced turning phenomenon, related to the rolling track quality.

2. THE MATHEMATICAL MODEL OF VEHICLE SYSTEM

The aforementioned limitations concerning the two degrees range pattern, although not as important as it was suggested, are eliminated for the seven range degrees pattern suggested and presented as follows. Far from being exhaustive – and in this case, one ignores vehicle's flexible body movement, the pattern allows the illustration within reasonable complexity limits of the calculation for different control synthesis procedures, including the preview and apriori information on the rolling track profile (Pereş et all., 1998).

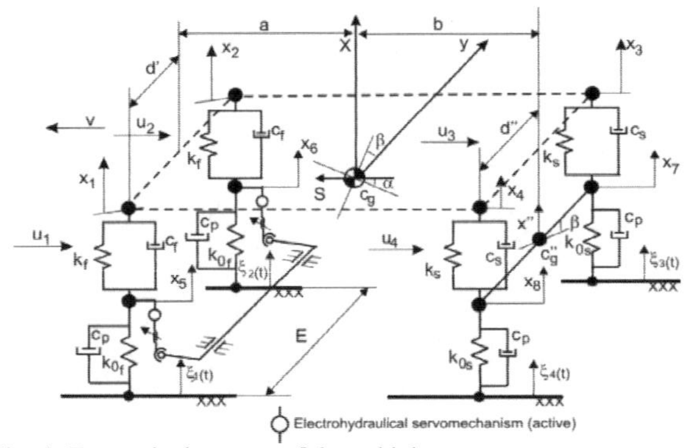

Fig. 1. Dynamical system of the vehicle

The seven range degrees are identifiable as follows: the range degrees, Figure 1, for the suspended volume (the x gravity center vertical movement, the α tilting angle and the φ rolling angle) and two range degrees for each unsprung mass (x_5 and x_6 vertical movements for each front wheel's independent spring rigging and x'' and the β'' rolling angle for the rear deck, respectively).

Finally, the movement equation for the seven range degrees pattern, with included control, can be written in a template-assembled form:

$$\dot{x} = Ax + Bu + Dw \tag{1}$$

The semi-active control application, employing stabilizer's variable elasticity for the two range degrees pattern, is based on the moving equations system

$$\begin{aligned} & M\ddot{x}_M + c\left(\dot{x}_M - \dot{x}_m\right) + u\left(x_M, x_m, \dot{x}_M, \dot{x}_m\right) + \\ & + k_1\left(x_M - x_m\right) = 0, \\ & m\ddot{x}_m - c\left(\dot{x}_M - \dot{x}_m\right) - u\left(x_M, x_m, \dot{x}_M, \dot{x}_m\right) - \\ & - k_1\left(x_M - x_m\right) + k_2\left(x_m - \xi\right) = 0, \\ & \dot{\xi} + \alpha_a \xi = w \end{aligned} \tag{2}$$

and the control method based on the forces balancing logic.

$$u = \begin{cases} k_1 \left|x_M - x_m\right| \operatorname{sgn}\left(\dot{x}_M - \dot{x}_m\right), \\ \qquad for\left(x_M - x_m\right)\left(\dot{x}_M - \dot{x}_m\right) < 0 \\ 0, \quad diferent \end{cases} \tag{3}$$

The simplified model of achieving the control pattern (2) is featured in Figure 2. The hardware includes the hydraulic pump *PH*, the pressure regulator R_p, the p_0 constant pressure *T* tank, actuators (valves) with W_{S1} and W_{S2} on and off options (when one is closed, the other is open) included in the steering gear, a unitary device made up of the relative movement-pressure T_{px} transducer and the *EEH* hydraulic execution element, the *CRH* hydraulic regulator cylinder, the device allowing to introduce in the system the force controlling the torsion angles lag for the $\Delta\Psi_{sd}$ stabilizing rod tips. Such system is able to introduce a force so as to position the stabilizing rod tips according to the movement $(x_M - x_m)$ or to cancel this force according to the sign of the (3) conditioning function; there is of course the need to provide apart from the described hardware, the assistance of certain transducers to monitor the sign of the conditioning force and of the central unit (the processor) to control the *HEE* hydraulic execution elements with the help of the W_{S1} and W_{S2} valves. To be more rigorous, in order to perfectly compensate the suspended elastic force in the time sequences when the force regulating stabilizer's tips positioning is active, the following physical parameters must be correlated so that: the product between the surface friction coefficient F_U and the relation between this surface and the T_U surface associated to the T_{px} transducer must be unitary.

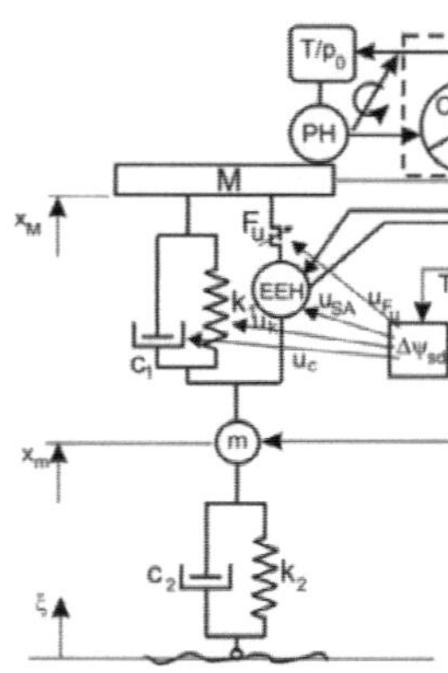

Fig. 2. A possible reification of the semi-active control principle (3)

3. EXPERIMENTAL ANALYSIS

The experimental tests were carried out using a MC-32 system that represents an acquisition measurement apparatus controlled by a PC computer. The system comprises interface circuits proper to each subsystem-measurement, acquisition, memory and the processing one (Thierheimer, 2001).

A compelling representation of the suggested solution concerning the seven range degrees pattern is offered in Figure 3 *a* and *b* respectively, by comparison to both the classical system and the QLHR quadratic linear hydraulic regulator adaptive system, together with the theoretical measurements and those resulting from experimental trials. The unevenness, the cosine curve mentioned in chapter 6 and 8 of (Thierheimer, 2005) respectively, having 6 *cm* amplitude and a 0.1 *s* width, is perceived in the right side of the vehicle with an 0.18 *s* phase shifting between the two decks, the vehicle's speed for this trial being estimated of at least 20 *m/s*. The comparison favors the active control strategy; the comparison criterion is the acceleration level for the suspended mass and the relating movements of the wheel center – the rolling track during the transitory periods perceived as a mechanical impulse and caused by an evident unevenness such as the simulated one. The two sets of weights mentioned - the control weight being equal to the unit each time - illustrate the flexibility if the focus is on improving Safety.

4. CONCLUSION

As there is no relation between the measured variables when the immobile vehicle is no longer in equilibrium, by simply manually modifying the rolling values, the automated regulator is activated, which imposes a thorough study, as there is an important disadvantage if there is a lateral wind - the vehicle rises on the front side.

Thus, one achieves a certain control by volumetric accumulation in the HEE hydraulic execution elements, an accumulation exerting on the non-turning wheels and causing a rectilinear course on crossing deep hindrances (holes).

One can assert that the present study opens the path to new research topics concerning the suggested solution, such a s the fuzzy logic or even the neuronal networks in controlling the course influence on the spring rigging and implicitly on vehicles' safety in traffic.

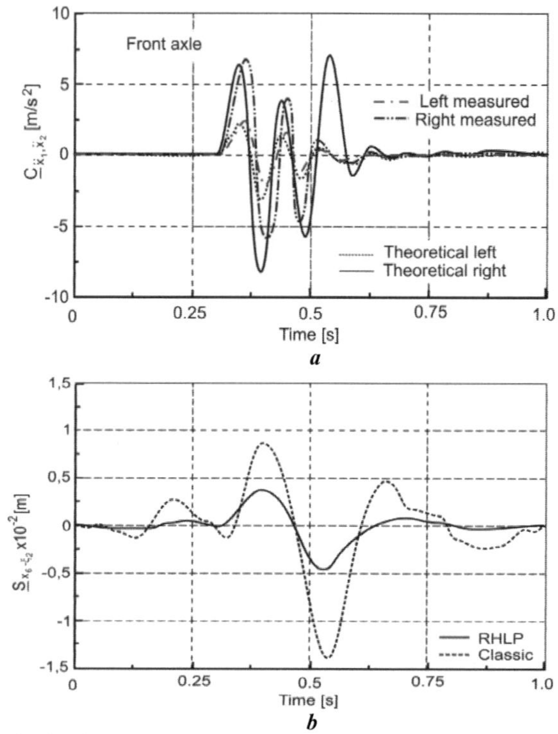

Fig. 3. The theoretical and experimental analysis of system's reaction to an unevenness unfolding with a 0.18 phase shifting on wheel 1 and 4

5. REFERENCES

Otat, V. & Thierheimer, W. (2007) *Dynamics of vehicles*, Editura Universitaria, Craiova, ISBN 973-742-023-3, Romania.

Peres, Gh.; Thierheimer, W.; Luca, D. & Zamfira, S. (1998) Contributions to the study of vehicle's dynamic behaviour in curvilinear movement, In: *Proceedings of the 9th International DAAAM Symposium*, Katalinic, B. (Ed.), pp. 381-382, ISBN 3-901509-08-9, Wien.

Thierheimer, W. (1998), Research concerning the optimisation of the correlation between the suspension and the steering systems in case of vehicles with front drive, Ph.D. thesis, May, Transilvania University Brasov.

Thierheimer, W. (2001) *The theory of tehnical systems of agriculture and food industry*, Editura Universitatii Transilvania Brasov, ISBN 973-8124-76-X.

Thierheimer, W. (2005) Aspects regarding the active steering for vehicle dynamics control, in *Proceedings of the Third International Conference "Challenges in Higher Education and Research in the 21st Century"* Lubomir Dimitrov (Ed.), Heron Press Sofia, Bulgaria, pp. 116-119, ISBN 954-580-178-6, 1-3 June, Sozopol.

Annals of DAAAM for 2008 & Proceedings of the 19th International DAAAM Symposium, ISSN 1726-9679
Editor B. Katalinic, Published by DAAAM International, Vienna, Austria 2008
Make Harmony Between Technology and Nature, and Your Mind will Fly Free as a Bird
2nd DAAAM-ICAT

APPLICATION OF THE KINEMATIC MODEL CONSTRUCTION IN THE MEDICAL FIELD

TIBA, Z[solt]

Abstract: *The human skeleton from mechanical point of view is a multi degree of freedom system supporting structure. Depending upon the fact whether muscle groups are tightened, the supporting structure can be either determinate structure or statically indeterminate structure. The tightened muscles turn into the part of the supporting structure, the slack muscles make it possible the structure operates as a mechanism. In this paper the simplified kinematics model of the leg and its analysis are presented.*

1. INTRODUCTION

The human leg made up of bones, joints and muscles is a multi degree of freedom system construction. The muscles yield a determinate construction in a tightened state. In this case all of the degrees of freedom of the construction are fastened by supporting. An slackened musculature limb is an unstable construction, acts as a mechanism, which moves due to the acting force and moment in the particular direction. Due to this construction which is an unstable from mechanical point of view, a motion can be carried out. It can be achieved by tightening certain muscles and slacking others. Tightened muscles can be modelled with springs and act as rods in a construction.

The purpose of the paper is to subject the leg to kinematics analysis. On the base of fundamental mechanical theory I elaborated a very simple kinematics model of the leg. In the knowledge of basic principle, more sophisticated model can be constructed which takes more muscles and joints into account, It facilitates the more accurate modeling of operating of the leg. In the following I outline the engineering fundamental concepts regarding the determinate and statically indeterminate structure, and then I build up and analyze the kinematics model of the leg.

The kinds of supporting		The number of unknowns	
		Planar structure	Spatial structure
Carrier idler set		1 (F_y)	1 (F_y)
Pinned support		2 (F_x, F_y)	3 (F_x, F_y, F_z)
Fixed end		3 (F_x, F_y, M_z)	6 (F_x, F_y, F_z, M_x, M_y, M_z)
Joint guide		1 (F_n)	1 (F_n)
Rodded needling		1 (F_n)	1 (F_n)

Tab. 1. Analysis of supporting

Degree of freedom: **s**
The number of the scalar data determining the position of the formation Ugural, (2004):

	planar case	spatial case
mass point:	2 (x, y)	3 (x, y, z)
figure:	3 (x, y, φ_z)	6 (x, y, z, φ_x, φ_y, φ_z)

Determinate structure: $s=n_{supporting}$
s: number of the degree of freedom of the structure
$n_{support}$:number of the reactions of the structure

Unstable structure:$s> n_{supporting}$

Statically indeterminate structure:$s< n_{supporting}$

2. ANALYSIS OF THE JOINTED STRUCTURES

The number of the equations: 3 x the number of the rods
The number of unknowns: the number of reactions of the whole structure: depending upon the supporting and the number of reactions between the rods: 2 x the number of pins (Shigley et al, 2004) (the number of pins: the number of the rods connecting with the pin, minus 1)

2.1. Application of the analysis in the case of structure containing 3 joints

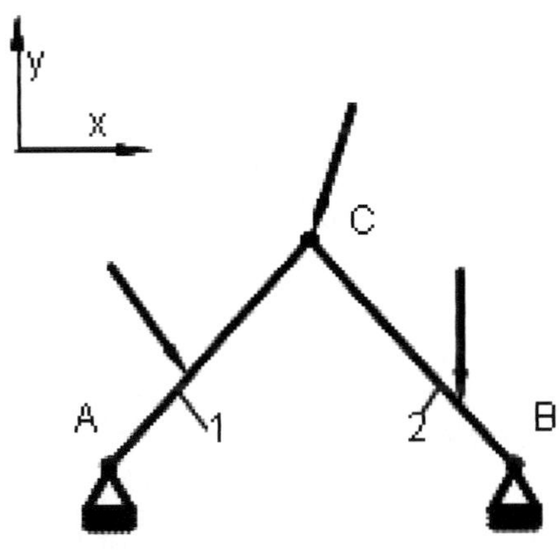

Fig. 1. Structure containing 3 joints

E: the number of equations
U: the number of unknowns

E = 2 x 3 = 6
U = 2 + 2 + 2 =6
E = U, consequently all the reactions can be calculated

The equations of static

- spatial: 6
- planar: 3 (generalized sum of F_x, F_y, M_z are 0), $\sum F_x = 0$, $\sum F_y = 0$, $\sum M_i = 0$

For the whole structure: $M_a = 0 \Rightarrow Y_B$ $M_b = 0 \Rightarrow Y_A$

For the rod 1

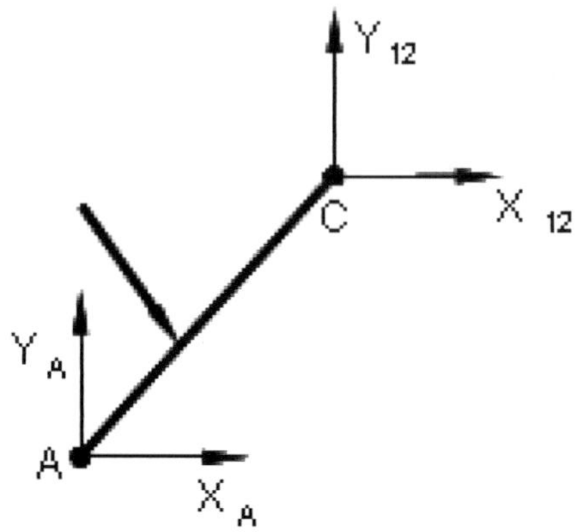

Fig. 2. Analysis of Rod 1

$$M_c = 0 \Rightarrow X_A$$
$$\sum X = 0 \Rightarrow X_{12}$$
$$\sum Y = 0 \Rightarrow Y_{12}$$

For the rod 2

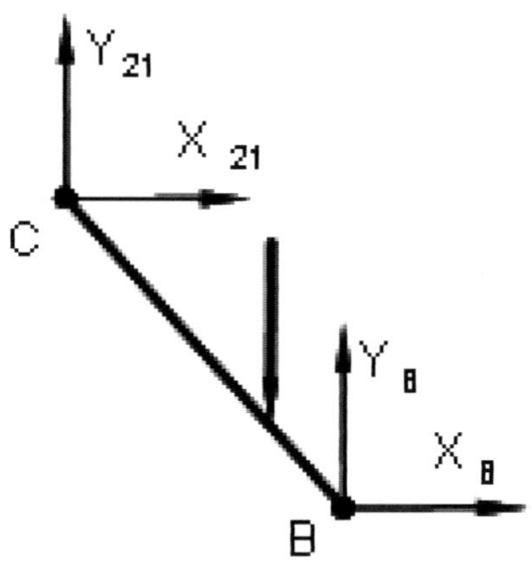

Fig. 3. Analysis of rod 2

$$\sum X = 0 \Rightarrow X_B$$

3. STEPS OF THE KINEMATIC MODEL CONSTRUCTION OF THE LEG

- substitution of the shin and thigh-bone by rods
- substitution of the ankle, knee, hip joint by joint
- consideration of muscles as springs and damping
- construction of the model as a mechanism

Approximate kinematic model of the leg

Further simplifications

- substitution of the muscle groups only by one spring
- consideration of springs tightened as rods

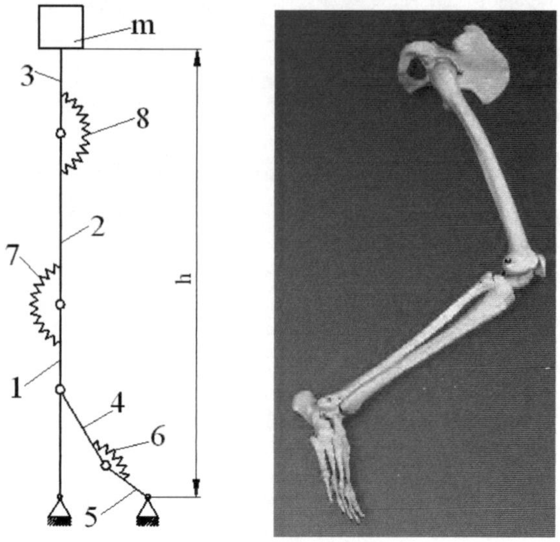

Fig. 3.. Build-up and analysis of the mechanism (Home-position)

Equations = 8 x 3 = 24
Unknowns = 12 x 2 = 24, consequently all the reactions can be calculated

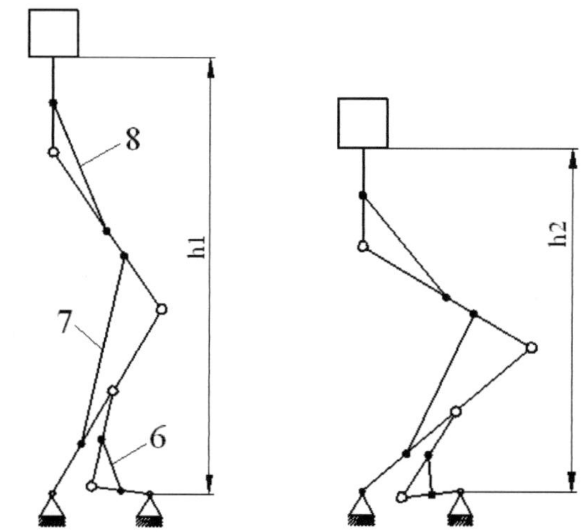

a. muscles are flabby b. muscles are tightened

Fig. 4. Positions of the mechanism during the step

4. CONCLUSION

In possession of the expertise relating to the kinematics modeling of leg and limb, it becomes possible to design such artificial limbs which motion form is similar to the original one. The loads and forces acting on the skeleton made statically determinate by tightened muscles can be calculated, in this way the stressing of artificial limbs can be carried out.

5. REFERENCES

Ugural, A.(2004). *Mechanical Design,* McGraw-Hill, ISBN 0-07-242155-X, New York

Shigley, J.; Mischke C.; & Budynas R.; (2004). *Mechanical Engineering Design,* McGraw-Hill, ISBN 007-252036-1, New York

Annals of DAAAM for 2008 & Proceedings of the 19th International DAAAM Symposium, ISSN 1726-9679
Editor B. Katalinic, Published by DAAAM International, Vienna, Austria 2008
Make Harmony Between Technology and Nature, and Your Mind will Fly Free as a Bird

DESIGN OF THE BRIDGES STRUCTURES USING MATHCAD

TIEREAN, M[ircea]; BALTES, L[iana] & EFTIMIE, L[ucian]

Abstract: *The paper will present two F.E.M. applications, developed for truss and frame structures, using MathCAD. There are presented the basic elements of the programs and the drawings of distorted structures. The structure bars are checked at traction, bending (in the case of frame structures) and buckling, considering the stiffness of the nodes. The attained results were verified with ANSYS.*
Key words: *bridge, FEA, simulation, structures.*

1. INTRODUCTION

There are thousand of papers that describe F.E.A. applications, one of there realised with individual soft, others with commercial programs. We consider useful to present two simple F.E.A. applications, realised with MathCAD, for truss and frame bridge structures.

Because there are simple structures realised from straight bars, the meshing is easy, the element boundaries being the successive junctions.

2. APLICATION FOR TRUSS STRUCTURES

We considered as application the parabolic truss structure of a bridge (fig. 1). The forces are $Q_1 = 2\cdot10^5$N and $Q_2 = 10^5$N.

The application starts with the drawing of the structure. The input data are the number of bars, the horizontal dimension "b" and height of first column "c". The structure drawing is generating by successive movement of the position vector across the nodes (fig. 2) (Tierean et al., 1999, 2006).

Next, a nodal junction matrix will be built, which defines the elements of the structure. In this matrix the column number is the same with the element number, the first line shows the start node and the second line the arrival node. Thus the column 0 defines the 0 bar, the start node being 0 and the arrival node 1. The last column defines the 24 bar the start node being 7 and the arrival node 10.

In the next step is generate the expansive incidence matrix, which has the same number for the lines and for the structure nodes; the number of columns is the same with the number of its elements.

Additional, with these definitions (nodal junction and expansive incidence matrix) the elements are toggled in plan by polar co-ordinates. In order to assembly the stiffness matrix (K), the computing of the rotated matrix of incidence (SER) is necessary,

$$SER_{2\cdot i+u,2\cdot z+v} := SE_{i,z}\cdot ROT(z)_{u,v} \qquad (1)$$

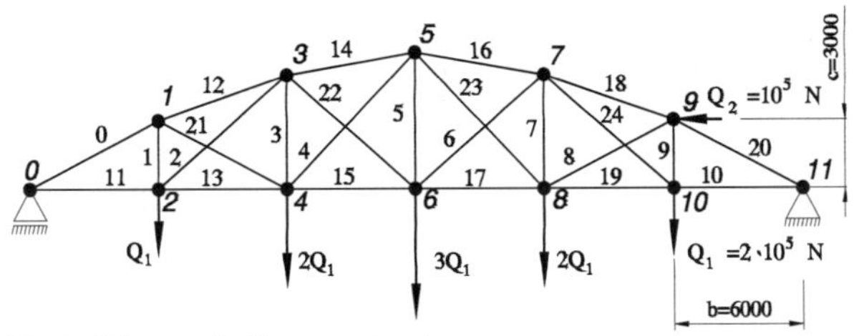

Fig. 1. The parabolic truss structure.

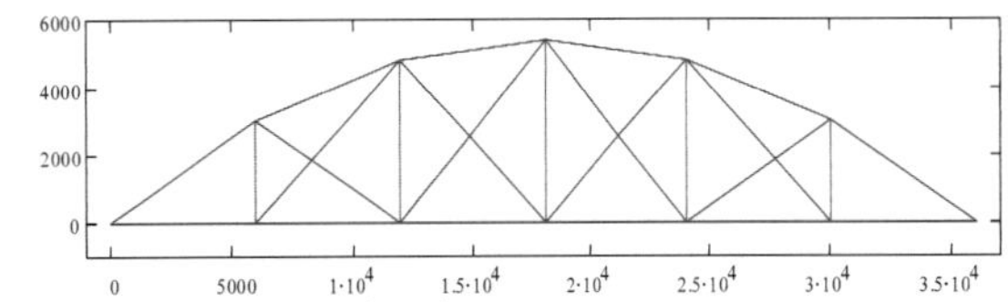

Fig. 2. The structure drawing.

trough the multiplication of the incidence matrix (SE) and the rotation matrix (ROT), and for the local stiffness matrix (KL) also.

$$KL_{2\cdot z+u,2\cdot z+v} := KLOC_{u,v}\cdot\left(\frac{E\cdot A_z}{l_z}\right) \quad K := SER\cdot KL\cdot SER^T \qquad (2)$$

After this step, it is necessary to explain the boundaries of structure; also the loadings are introduced in nodes. Every node has two values: the first refers on horizontal force and the second refers on vertical force. In the next step there are computed the nodal displacements (two for each node) by multiplying the inverse of stiffness matrix with the loadings vector.

These nodal displacements draw the distorted structure (fig. 3), amplified by 100.

The forces on the bar direction, in the starting node, are obtained by projecting the nodal distortions in the bar local base, amplified by their stiffness.

$$DEF^{<z>} := \begin{pmatrix} \delta_{2\cdot I_{1,z}} - \delta_{2\cdot I_{0,z}} \\ \delta_{2\cdot I_{1,z}+1} - \delta_{2\cdot I_{0,z}+1} \end{pmatrix}\cdot E\cdot\frac{A_z}{l_z} \qquad (3)$$

$$F^{<z>} := -ROT(z)\cdot KLOC\cdot ROT(z)^T\cdot DEF^{<z>} \qquad (4)$$

After the nodal equilibrium and the calculation error verification, (in the first occurrence the error is less than 10^{-6}, and in the second one the error is less than 10^{-7}), the structure bars are checking at traction and buckling. The results were checked with ANSYS.

3. APLICATION FOR FRAME STRUCTURES

As application is considered the frame structure (fig. 4), loaded by the forces $Q_2 = Q_6 = 25\cdot10^3$N, $Q_3 = Q_4 = 10^4$N and the bending moments $M_3 = M_4 = 10^4$Nm. As in the previous chapter, the calculation program starts with the drawing of the structure, the input data being the number of bars, the horizontal "b" and the vertical dimension "c".

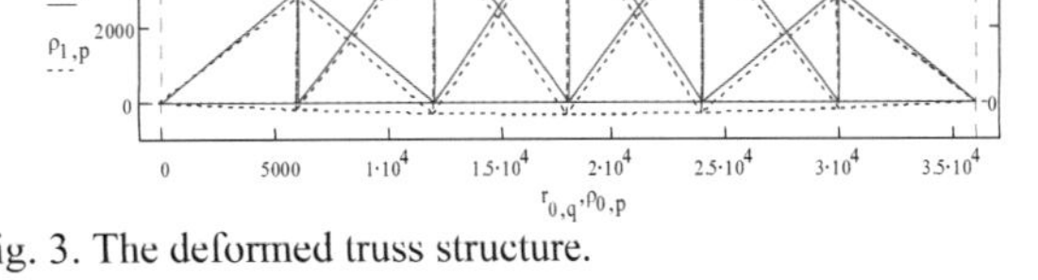

Fig. 3. The deformed truss structure.

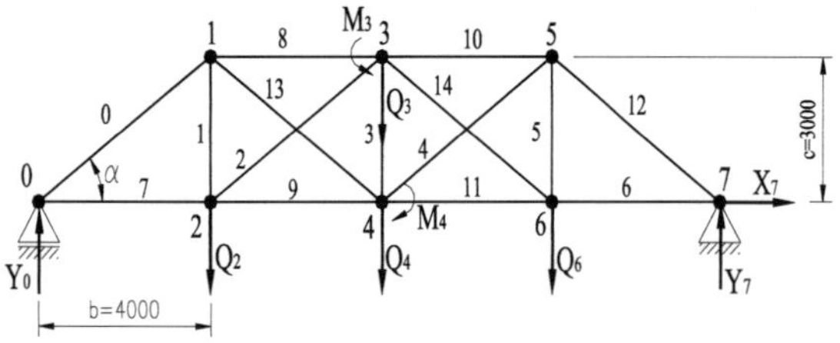

Fig. 4. The frame structure.

Because in this case are three displacements the rotation matrix is:

$$ROT(z) := \begin{pmatrix} \cos(\alpha_z) & -\sin(\alpha_z) & 0 \\ \sin(\alpha_z) & \cos(\alpha_z) & 0 \\ 0 & 0 & 1 \end{pmatrix} \quad (5)$$

and by multiplying with the translation matrix (because of bending) result the roto-translation matrix.

$$RT(z) := ROT(z)\cdot Td(z)^T \quad (6)$$

The components of rotated matrix of incidence (SER) are the same as the elements of rotation matrix, for the starting node and same as roto-translation matrix, in the case of arrival node.

$$SER_{3\cdot i+u,3\cdot z+v} := if(SE_{i,z} > 0, SE_{i,z}\cdot ROT(z)_{u,v}, SE_{i,z}\cdot RT(z)_{u,v}) \quad (7)$$

The stiffness matrix results by projecting the free elements in the incidence space of the structure (transformation of the transformation).

$$K11(z) := \begin{bmatrix} \frac{E\cdot A_z}{l_z} & 0 & 0 \\ 0 & \frac{12\cdot E\cdot Iz_z}{(l_z)^3} & \frac{6\cdot E\cdot Iz_z}{(l_z)^2} \\ 0 & \frac{6\cdot E\cdot Iz_z}{(l_z)^2} & \frac{4\cdot E\cdot Iz_z}{l_z} \end{bmatrix} \quad (8)$$

After the fixing of the structure, the loadings are introduced in nodes. Every node corresponds with three values: the first refers on elongation force; the second refers on shearing force and the third to the bending moment.

After that step the nodal displacements are calculated (three for each node) by multiplication of the inverse stiffness matrix with the loadings vector. These nodal displacements draw the distorted structure, using cubic Hermite interpolation.

$$H(x) := \begin{pmatrix} 1 & 0 & -3 & 2 \\ 0 & 1 & -2 & 1 \\ 0 & 0 & 3 & -2 \\ 0 & 0 & -1 & 1 \end{pmatrix}\cdot\begin{pmatrix} 1 \\ x \\ x^2 \\ x^3 \end{pmatrix} \quad (9)$$

In the fig. 5 is presented the deformed frame structure considering the nodes stiffness.

To obtain the forces is necessary the displacements imagine in the global reference system.

$$TD(z) := ROT(z)\cdot Td(z)\cdot ROT(z)^T \quad (10)$$

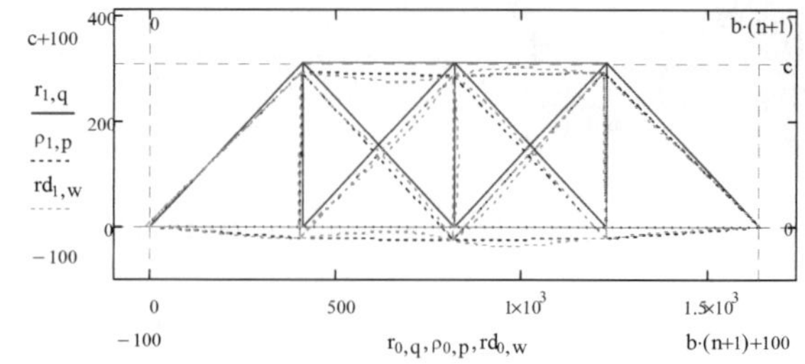

Fig. 5. The deformed frame structure.

It is also necessary to calculate the strain matrix in the start (δEF) and in the arrival node (ΔEF). With these matrix are calculated the force matrix in the local (FL) and in the global reference system (F), in the start (i) and in the arrival node (j) (Tofan et al., 1995).

$$\delta EF^{\langle z\rangle} := TD(z)\cdot\left[\begin{pmatrix} \beta_{3\cdot I_{1,z}} \\ \beta_{3\cdot I_{1,z}+1} \\ \beta_{3\cdot I_{1,z}+2} \end{pmatrix} - \begin{pmatrix} \beta_{3\cdot I_{0,z}} \\ \beta_{3\cdot I_{0,z}+1} \\ \beta_{3\cdot I_{0,z}+2} \end{pmatrix}\right] \quad (11)$$

$$\Delta EF^{\langle z\rangle} := -TD(z)^{(-1)}\cdot\delta EF^{\langle z\rangle} \quad (12)$$

$$FLi^{\langle z\rangle} := -K11(z)\cdot ROT(z)^T\cdot\delta EF^{\langle z\rangle} \quad (13)$$

$$Fi^{\langle z\rangle} := ROT(z)\cdot FLi^{\langle z\rangle} \quad (14)$$

After the bars' equilibrium checking (less than 10^{-6}), nodal equilibrium (less than 10^{-8}) and the computing of error (less than 10^{-9}), the structure bars are checking at traction, bending and buckling, considering the nodes stiffness. The results were checked with ANSYS.

4. CONCLUSION

The presented programs give the deformed structures, the displacement values, and the bars stresses, in the local and in the global reference system. Using a medium equipped Windows XP PC configuration, the computing time is 4s for the first program and 9s in the second case.

Due to the simplicity and the higher velocity, the program is easily to implement in any type of calculation for the structure bars avoiding the laboriously methodology of the drawing, introducing of the data and calculation, claimed by commercial programs for F.E.A.

5. REFERENCES

Mathsoft, Inc. (2001) *MathCAD User's Guide*, Cambridge, MA

Ţierean, M., Tofan, M., Goia, I., F.E.A. of bars structures using MathCAD, *NAFEMS World Congress '99 on "Effective Engineering Analisys"*, Newport, USA, 25-28, April, 1999, pag. 947-952.

Tierean, M.H., Baltes, L.S., Mirza Rosca, J., Santana Lopez, A., Applications for finite element analysis of strength structures using Mathcad, *Third international conference „Mechanics and Machine Elements"*, 2-4 Nov. 2006, Sofia, Bulgaria, 2006, ISBN 10: 954-438-587-8, pag. 240-245.

Tofan, M.C.; Goia, I.; Tierean, M.H. & Ulea, M. (1995). *Deformatele structurilor (Structure's strains)*, Editura Lux Libris, ISBN 973-96854-2-0, Brasov, Romania

http://www.ptc.com/go/mathsoft/mathcad/

Annals of DAAAM for 2008 & Proceedings of the 19th International DAAAM Symposium, ISSN 1726-9679
Editor B. Katalinic, Published by DAAAM International, Vienna, Austria 2008
Make Harmony Between Technology and Nature, and Your Mind will Fly Free as a Bird

OPTIMISATION OF STRUCTURES BASED ON RELIABILITY

TIEREAN, M[ircea]; BALTES, L[iana] & EFTIMIE, L[ucian]

Abstract: *This paper presents a method for the optimisation of strength structures based on reliability. The problem's variables are considered as random values, having the advantage of considering some accidental strains, which can predictable load the structure, but having a low appearance probability. There are presented the calculus hypothesis and the problem's general wording. With a view of this purpose, is used a structure composed of 10 bars, discussed in two cases: with jointed junctions and with fixed junctions. For both cases there were done finite element programs using MathCAD. For buckling, the frame structure is considered with elastic joints. There are presented the bars' area variations for the last iteration of frame and truss structures.*

Key words: *FEA, optimisation, reliability, simulation, structures.*

1. INTRODUCTION

In the case of deterministic optimisation, the input values of the problem (admitted stress and strains, geometrical dimensions, forces, strain, and boarding conditions) are considered univocally as known. The no statistic approaches to those values impose the choice of a safety factor, which covers every possible breaking case. The choosing of this safety factor has a great influence regarding the cost and structure mass, sometimes bigger than the results of the mathematics program. In the case of probabilities optimisation, these problem variables are considered as random values, here existing the advantage of considering some accidental strains, which can predictably load the structure, but they have a small appearance probability.

The purpose of the optimisation based on reliability is to find the geometrical dimensions of structure elements, by imposing the failure probability.

2. HYPOTHESIS

1. The probability concept is used not just with a relative frequency, but as a measure for a certain hypothesis trust grade, which has been used in computing without being confirmed by experiments.
2. The structures loading and bearing capacity are random values, with normal distribution.
3. The uncertainty that affects the safety is considered in calculus by scattering parameters of the random variables.
4. The structure's geometry and loading positions are considered as deterministic.
5. There are neglected the shearing forces and torsion moments affect regarding truss and frame structures.
6. The loading action is independent.
7. The bars' cross sections are constants, related to their longitudinal axes; they are completely characterised by a single parameter.
8. The loading is statically applied.
9. The temperature and the structure's environment do not influence the resistance.
10. The destruction of every element brings about the destruction of the structure (Tierean et al., 2006).

3. THE PROBLEM'S GENERAL WORDING

There is considered a structure consisting of "n" bars, traction-compression strained for the first case (truss), and bending-traction-compression strained for the second case (frame). The structure is loaded by "l" forces. There are considered as initial data: the geometrical dimensions, material and loading conditions. The shape of the bars' cross section is specified, so it is considered that only one dimension (g) defines univocally all other geometrical characteristics that may appear into calculus. The optimisation problem can be phrased this way: determine the geometrical characteristics (g), so that the elementary failure probability $F_i(g)$ is equal with the admitted value (Thoft-Christensen et al., 1986):

$$F_i(\bar{g}) = P[L_i(\bar{g}) \le 0] = F_{ai},\ (i = 1,2,...,n). \tag{1}$$

For the truss structure, the safety limit that corresponds to the "m" breaking case is:

$$L_j = \sum_{j=1}^{n} a_{ij} T_j - \sum_{j=1}^{l} b_{ij} S_j,\ (i=1, 2,..., m), \tag{2}$$

where:

l = the number of loading cases;
T_j = the admissible loading of the "j" element;
S_j = the loading in the "j" case;
a_{ij} = the "j" element resistance factor;
b_{ij} = the "j" element loading factor, for the "i" breaking type.

In the case of frame structure, the "i" element safety limit is:

$$L_i = X_{1i}^2 A_j W_j - (M_j A_j + N_j W_j) X_{1i}, \tag{3}$$

where:

X_{1i} = the admissible strength of the "i" bar material;
M_j = the bending moment, for the critical section "j";
N_j = the axial force, for the critical section "j";
m = the number of critical sections;
$W_j = W(g_j)$ = the strength modulus of the "j" section;
$A_j = A(g_j)$ = the critical section "j" area.

For a Gaussian random design variable's distribution, the elementary reliability is:

$$F_i = P[L_i \le 0] = \phi\left(\frac{0-\bar{L}_i}{\sigma_{Li}}\right) = \phi\left(-\frac{\bar{L}_i}{\sigma_{Li}}\right) \le F_{ai}, \tag{4}$$

where: $\phi(x) = \frac{1}{\sqrt{2\pi}} \int_{-\infty}^{x} \exp(-\frac{t^2}{2}) dt;$

L_i = the safety limit mean value;
σ_{Li} = the safety limit mean square error.

Let $F_{ai} = \phi(-f_i) \Rightarrow f_i = -\phi^{-1}(F_{ai})$ be the optimisation problem restriction; it results:

$$\phi\left(-\frac{\bar{L}_i}{\sigma_{Li}}\right) = \phi(-f_i) \Rightarrow f_i = \frac{\bar{L}_i}{\sigma_{Li}} \tag{5}$$

or to avoid the square root: $f_i^2 = (\frac{\overline{L}_i}{\sigma_{Li}})^2$. (6)

In each iteration (q) the Eq. (6) is solved and the obtained vector is the optimum solution for that iteration. The algorithm has the following steps:

Step 1. There are established the elementary failure probabilities (F_{ai}), the elementary diameters' initial values $g_i^{(0)} = (g_1^{(0)}, g_2^{(0)}, ..., g_n^{(0)})$, q=0 and the admitted error limit ε.

Step 2. There are calculated the forces (and moments) corresponding to the $g^{(q)}$ vector.

Step 3. There are calculated the elementary failure probabilities.

Step 4. If the relation

$$\left|\frac{F_i(\overline{g}^{(q)}) - F_{ai}}{F_{ai}}\right| \le \varepsilon, \quad (i = 1,2,...,n), \tag{7}$$

is accomplished, it results that the optimum solution is $g^{(q)}$. For the opposite case it will be made the jump to step 5.

Step 5. There is modified the "g" vector, which will have the components $g_i^{(q+1)}$ resulted by solving the equation (4).

Step 6. For compression strained bars it will be made the buckling check. If the buckling exists, the optimum solution is $g^{(q)}$. For the opposite case it will be made a jump at step 2, with q=q+1.

4. CALCULUS EXAMPLE

The presented method was applied to a 10 bar structure (Fig. 1). It is considered that the bars are made of steel with $X_1 = 273$ MPa and $\sigma_{X1} = 19.492$ MPa. For the calculus, there were used the loading variation factor $V_Q = 0.2$; the admitted failure probability $F_i = 10^{-6}$ and the admitted range $\varepsilon = 10^{-3}$ [2].

The loadings are $Q_1 = 5 \cdot 10^5$N, $Q_2 = 10^5$N and the structure's dimensions are b=6000 mm, c=3000 mm.

The calculus program starts with the structure drawing section, valid for a 5·n bars structure generating process. In the input data block there are introduced the bar's number and the horizontal dimensions "b", respectively the vertical ones "c". In the first stage there are generated the left diagonals and the vertical beams, by the successive traversing of all junctions by the position vector "r". In the second stage there are generated the right diagonals and the inferior sole, by returning to the initial junction and in the contrary sense traversing the previously generated vertical beams. In the last stage it results the structure's final draw, by joining the second before the last junction with the second one (Tofan et al., 1995).

In the next step, by using the finite element method, there are computed the internal forces and moments. After the initial structure check, the bars are dimensioned again, by using the Eq. (6). The compressed bars must be checked at buckling and will be verified the Eq. (7) accomplishment. Finally it is graphically represented the bar's section variation into the current iteration, the total area variation factor and the structure's strain. With a view for reaching the optimum solution there are 5 iterations necessary, finally obtaining a $\mu = 77.6\%$ total area decrease for truss and 4 iterations with $\mu = 57.2\%$ for frame.

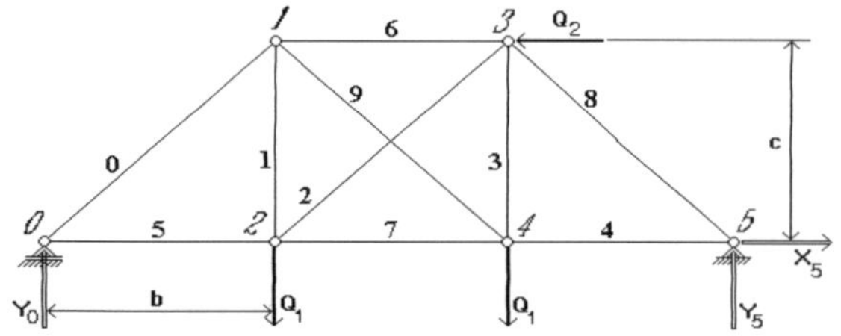

Fig. 1. The 10 bars structure.

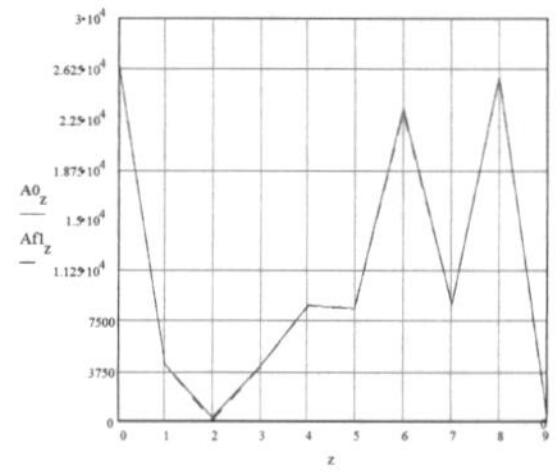

Fig. 2. Last iteration of the truss structure.

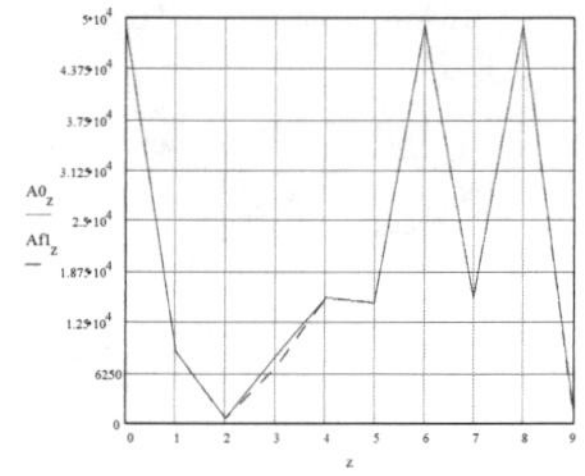

Fig. 3. Last iteration of the frame structure.

Fig. 2 presents the bar's area variation for the 5th iteration (truss) and fig. 3 presents the bar's area variation for the 4th iteration (frame). In the case of the truss structure, in the last iteration, the bar's reliability spectrum is $[10^{-6}; 5.577 \cdot 10^{-6}]$. In the case of the frame structure, in the last iteration, the bar's reliability spectrum is $[9.976 \cdot 10^{-7}; 3.539 \cdot 10^{-6}]$.

In the case of the frame structure, for the buckling length determination, the junctions are considered as elastic-fixed junctions.

5. CONCLUSIONS

Comparing the results obtained at optimisation of the 10 bars structures it can be concluded:

- the method ensures a fast convergence to the optimum value;
- in the case of the truss structure's optimisation, a number of bars do not change their section, other bars decrease it, but there are some bars which after decrease in the first iteration, present in the following ones some increases;
- in the last step of the optimisation process, there can be very easily found the elementary reliability lowest value; this way it is diagnosed which element will break first;
- for frame structure's optimisation, almost all bars change their section during the iterations; the optimisation time is short;
- optimising the same structure, initially considered with fixed junctions, and then with joint junctions, it is established that, in the case of the frame structure, the optimisation process is performed in a less number of iterations; it is obtained a shorter area decrease than in the case of the truss structure;
- the optimisation programs are done in such a way, that for every step there can be closely watched the work parameters; there is easy service here.
- From these conclusions, it can be deduced that the optimisation on the basis of reliability is an efficient designing instrument; this way it can be ensured a low value for the bar's areas, but only respecting the reliability requirements.

6. REFERENCES

Tierean, M.H.; Baltes, L.S.; Mirza Rosca, J. & Santana Lopez, A. (2006). Reliability, a useful way for the strength structures' optimisation, *Third international conference „Mechanics and Machine Elements"*, 2-4 Nov. 2006, Sofia, Bulgaria, ISBN 10: 954-438-587-8, pag. 97-103

Tofan, M.C.; Goia, I.; Tierean, M.H. & Ulea, M. (1995). *Deformatele structurilor (Structure's strains)*, Editura Lux Libris, ISBN 973-96854-2-0, Brasov, Romania

Thoft-Christensen, P.; Murotsu, Y. (1986). *Application of structural systems reliability theory*, Springer-Verlag, Berlin

Mathsoft, Inc. (2001) *MathCAD User's Guide*, Cambridge, MA http://www.ptc.com/go/mathsoft/mathcad/

Annals of DAAAM for 2008 & Proceedings of the 19th International DAAAM Symposium, ISSN 1726-9679
Editor B. Katalinic, Published by DAAAM International, Vienna, Austria 2008
Make Harmony Between Technology and Nature, and Your Mind will Fly Free as a Bird

A MODEL ABOUT THE ORGANIZATIONAL STRESS MANAGEMENT

TITU, M[ihail]; BUCUR, V[iorel] & GHITESCU, T[eodor]

Abstract: *In this paper the authors aim is to draw attention to certain aspects regarding stress - related influence on managerial performance, emphasizing the need of organisational decisions for the training of managers so that they become aware and minimize stress-related effects. We have also underlined the relevance of acknowledging "post-shock" traumatic experience that the body is undergoing and, consequently, helping the body to get to the initial state in order to overcome any pressures.*

Key words: *stress, management, communication, performance, quality.*

1. INTRODUCTION

Position holders in any organization are specialists in various fields who, by means of specific technical methods, perform activities in the area of production, services, research, etc. in view of accomplishing organizational objectives. Each position holder is a well-educated and professional bio-individual, interacting with other job holders, its constituent elements as well as its environment, receiving and transmitting "stimuli", thus requiring permanent biological and behavioral self-adjustment. The emergence of stress is connected with the understanding of this complex process of human self-adjustment to internal and especially external stimuli. There are at least two study patterns of self-adjustment: cybernetic pattern, exclusively based on the interactions due to natural energies as well as on their analysis by formal mathematics methods; the present paper sets forth the study of human self-adjustment (especially position holders in organizations) based on the Jay Forrester pattern, to which one can add the indirect self-adjustment loop, based on intellectual energy and adapted in keeping with figure 1. Figure 1 exemplifies the self-adjustment process of bioenergy recovery, consumed in any kind of activity, by means of human feeding flow, which is different to any other animal, due to the existence of two categories of self-adjustment loops: direct self-adjustment loop, due to natural stimuli (under genetic supervision), similar to any other animal; indirect self-adjustment loop, due to intellectual energy (employing the degree of flow knowledge and its dependence on other personality flows). To the decision element of the loop we have added self-adjustment parameters required to the feeding process by this form of particularly human energy: the adequate level of food in the stomach determined by intellectual methods, NNHI and feeding time selected by intellectual methods as well, TAI. Obviously, the decision center on the intellectual adjustment loop is placed on the area controlling human thinking and not within the hypothalamus, where the reflex feeding center is placed. The descriptive pattern explains the fact that the feeding flow (consumed bioenergy recovery) can be shaped on: an input (feeding) rhythm, RH, which is accumulated in the real level of stomach food NRH. NRH value is compared on the natural negative reverse connection to the necessary value NNH, genetically designed in the hypothalamus for each bio-individual. Whenever there is a significant difference between NNH – NRH, the bio-individual is urged by means of the feeding reflex arch to look for food. After getting it, the RH feeding rhythm depends on the available time for TA feeding, subconsciously determined and which can be mathematically expressed as: RH = (NNH-NRH)?TA, [food units/time unit]. The **rhythm** expresses the **dynamics** of feeding flow, while the **real level** of NRH stomach food, the **condition** of the feeding flow. (Forrester, 1979), (Ghiţescu, 2006)

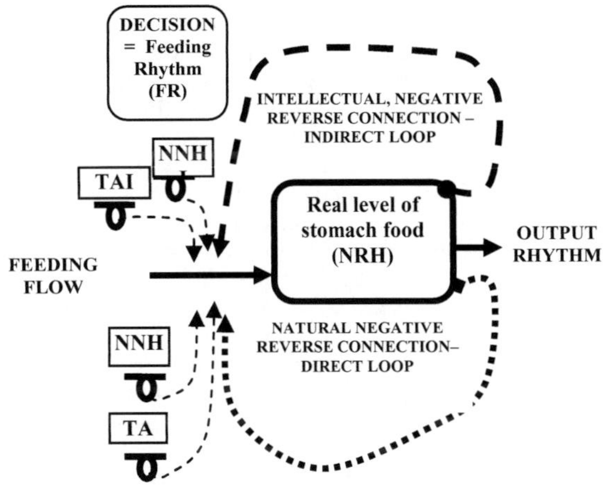

Fig. 1. Descriptive chart of self-adjustment feeding flow, by means of two categories of negative reverse connection loops, specific to human individuals.

2. SELF-ADJUSTMENT OF POSITION HOLDERS IN ORGANIZATIONS

Unlike any other bio-individual and due to "knowledge" – the synthetic expression of human education, the human being benefits of an additional negative reverse connection self-adjustment loop called in *indirect loop.* This loop makes all feeding processes aware by means of maternal language, at a different knowledge level (intellectual energy) form one individual to another, in keeping with the available information about healthy food and the influence of feeding on other subsystems of personality. The educated human being in his relation with food, by means of the indirect loop, has a different feeding rhythm than the uneducated human being (specific to the animal bio-individual). The difference is required by the following information: Good manners require that the food is placed on the plate, all the necessary cutlery will be prepared, thus modifying the TA feeding time into TAI; If the food contains elements contrary to medical specifications, then the NNH necessary level of food changes into NNHI; If the food is contrary to moral principles (in keeping with specific religious beliefs and dogmas), then RH=0, therefore the individual will look for adequate food and will change the corresponding TAI into NNHI; The feeding rhythm, RHI= (NNHI-NRHI)/TAI [food units/time units] due to the loop influence, is different to the rhythm determined by the natural loop, the direct one, and this will be the real rhythm; The output rhythm is not determined by the input rhythm (as in the cybernetic pattern), but it is influenced by another real level of the same flow, or on a different flow, and it might have the two categories of self-adjusting loops: *natural and intellectual.* (Bucur, 2002). The above-mentioned pattern, called realist pattern explains in a more accurate manner than the cybernetic pattern, the self-adjustment difference between the educated individual and animal, by the existence of two categories of self-adjustment reverse connection loops: a direct one, based on natural energies (bioenergies), and the other one, indirect, exclusively based on intellectual energy, artificial energy, obtained by means of human education (knowledge). (Forrester, 1979), (Ghiţescu, 2006), (Goupil, 1991)

3. STRESS SOURCES AND THEIR INFLUENCE ON THE SELF - ADJUSTMENT OF MANAGEMENT AND EXECUTIVE POSITION HOLDERS

Should we accept the fact that this pattern expresses more accurately the differences between self-adjusting processes of human beings and animals, then we can also understand the self-adjusting differences among various position holders, having only executive or leadership positions, or mixed positions. Due to increased specialization (acquiring knowledge about job-related technologies and processes), the position holder becomes a self-adjusting "energetic amplifier", with two other self-adjusting loops presented in figure 2: ***Direct loop, specific to executive positions,*** by which the position holder amplifies his own bioenergetic component, by means of natural energy resources provided by employed technologies in order to produce goods in a greater quantity and of a better quality compared to the situation of using only bioenergetic resources acquired genetically. Bioenergetic self-adjustment on each flow involved in the position responsibilities can be explained by means of the pattern in figure 1; ***Indirect loop, specific to management positions,*** by which position holders can influence subordinated organizational sub-systems, *employing exclusively intellectual energy.* Therefore, any manager can be considered an "amplifier of amplifiers", and each stimulus conveyed to subordinates will be further amplified in keeping with the amplification coefficient of each executive position holder. The manager, as a personality, will of course be subject to self-adjustment loops explained in figure 1.

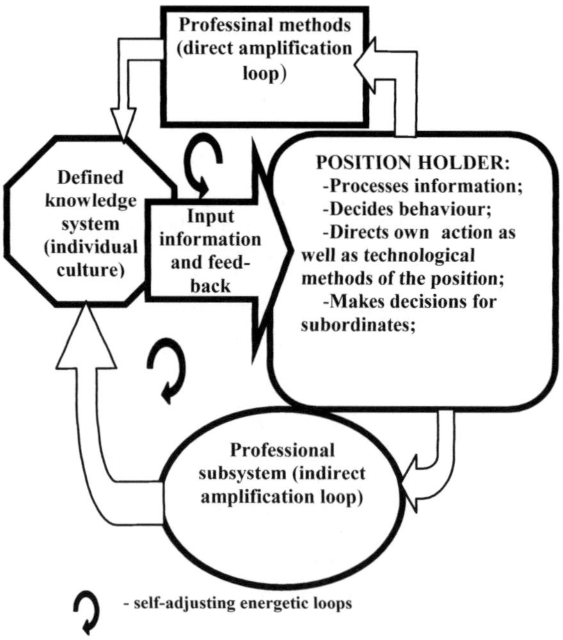

Fig. 2. Position holder self-adjustment within an organization

Figure 2 may represent a starting point in the creation of new and more accurate analysis patterns of organizational stress. (Forrester, 1979), (Ghiţescu, 2006)

4. CONCLUSIONS

Stress factors are different for managers and executive staff especially because of the two categories of self-adjustment loops explained in figure 2, but they are also similar due to the self-adjustment loops explained in figure 1. The present study aims to accomplish some instruments for the assessment of stress factors and their influence on organizational processes in keeping with the patterns in references and books relying on the patterns presented in figures 1 and 2 on a minimum of two categories of loops for the two extreme types of organizational positions: *self-adjustment loops in keeping with bioenergetic flows (natural energies); self-adjustment loops in keeping with intellectual energy flows (artificial energy).* (Titu & Oprean, 2008).

The figures evince the following: self-adjustment is a bioenergetic process where the carriers of adjustment energy forms are components of the human personality. Therefore organizational stress can be referred to exclusively at the component of human resources of the organization; stress can be explained by an imbalance between consumed and regenerated energies or each flow of the personality subsystems due to different categories of stress factors. Therefore, the definition of assessment instruments for the stress level is directly related to the description of complex personality patterns which should evince, in a descriptive and assessing manner, the interconnection of natural energetic flows with intellectual ones. Such patterns can be quite accurate to the real ones, by means of multidisciplinary research: medicine, labor psychology, ergonomics, ecology, management; the general stress level of a position holder is the result of an interaction among stress factors level (usually external ones) with the self-adjustment rhythms for each personality flow (internal processes), where each flow has two categories of loops: *direct* and *indirect;* there are *several similar categories of stressors* between executive and management positions, there are also *different categories of stressors* due to the predominant form of energy employed in the accomplishment of job objectives: *natural energies* for executive positions, and *intellectual energy* for management positions; even within the same category of stressors, there is a difference in the action of stressors, between executive and management position holders, due to the statistical average value of sensitivity thresholds on each self-adjustment loop; regeneration of each category of consumed energy during self-adjustment processes has *different conditions* for the two categories of position holders: for the executive, the conditions are predominantly determined by management quality; for the management, the conditions are predominantly defined by the external environment.

The above-mentioned statements lead to the following conclusions: Predominant internal stress sources for executive position holders; Predominant external stress sources for management position holders. (Forrester, 1979)

5. REFERENCES

Bucur, V. (2002). *Management. Management of Organisation,* Napoca Star Publishing House, Cluj Napoca

Forrester, J. (1979). The Principles of Systems, Technical Publishing House, Bucharest

Goupil, G. (1991). *Stress Intelect,* Coresi Publishing House, Bucharest

Ghiţescu, T. (2006). Educational Management of Systems, Matrix Rom Publishing House, Bucharest

Ţîţu, M. & Oprean, C. (2008). *Strategic Management,* University of Piteşti Publishing House, Piteşti

Annals of DAAAM for 2008 & Proceedings of the 19th International DAAAM Symposium, ISSN 1726-9679
Editor B. Katalinic, Published by DAAAM International, Vienna, Austria 2008
Make Harmony Between Technology and Nature, and Your Mind will Fly Free as a Bird

CONTRIBUTION ABOUT THE QUALITY MANAGEMENT IN THE HIGHER EDUCATION

TODERICIU, R[amona]; MUSCALU, E[manoil] & TITU, M[ihail]

Abstract: *In any organization, for having a high quality activity, maintaining and developing it, had to be used adequate means and developed specific actions, in a coordinated way. These are arranged under some well known concepts like: the quality control, to certify the quality, to improve the quality. We will try to present few of these concepts, considered to be essential for understanding how to get organised for quality in the higher education institutions.*

Key words: *organisation, quality management, higher education, organizational culture.*

1. INTRODUCTION

It is said that for the Romanian universities implementing the quality management it represents a very important step in changing the institutional culture, what means to create a culture based on good management and excellence. We emphasize the fact that all the concepts and the basic principles of the management of quality can be adapted for the higher education but do not have to be applied like for a firm but have to be adapted, shaped according to the higher education needs. For ISO 9000 the management of quality represents "the ensemble of the activities of the general function of management, which determines the policy in the field of quality, the objectives and responsibilities which are implemented in the system of quality through the means like quality planning, insuring the quality and improving the quality". Starting from this definition we can establish at the university's level elements contained by the orientation and the control which refers to the quality as follows: establishing the policy which refers to quality and the quality's objectives, quality planning, the quality control, insuring the quality and improving the quality. We present as follows some key aspects of the educational process's quality: (Muscalu, 2001)

- "to establish a coherent profile of training for each faculty and to harmonise these profiles inside the university's educational offer;
- to establish the study program's opportunities (majors) and to adapt the university's offer;
- to determine the requests and real demands of the social and economic environment regarding the graduates competences after each study program (major), to correlate them with the university's experience and the international practice (European);
- to elaborate and implement some adequate curricula;
- to implement some criteria and quality evaluation procedures on every segments of the educational process;
- introducing a *feedback* from the students, alumni and employees, having in view the structure and the quality of training and its improvement".

The management of quality supports the universities to learn there were it is necessary to learn about themselves, to improve and change where it is needed, to interact efficiently with the external environment, national and international. It supposes at the institution's level of higher education the existence of some adequate organizational structures which to favour developing and applying the quality system. The quality management became in the past few years an essential element of development and success, the price being replaced by the quality, from the consumer's point of view. The management of quality essence principles are presented as follows:

- oriented toward the client – the organization has to have as main objective to satisfy the clients and to overcome the client's expectations;
- the leadership – the leaders (clear vision over the future, to establish the objectives, to sustain the values) ascertain the unity between the purpose and the organization's objectives. They are preoccupied with the creation and maintenance inside the internal environment of the employees implication in fulfilling the organization's objectives;
- the employees implication - (the use of the employees talents and skills, the motivation of innovation, creativity) the essence of an organization it is the personnel which through its total devotion allows the skills to be used in the organization's benefit;
- the process approach – the activities and the resources need to be lead as a process for the obtained results to be efficient;
- the systemic approach – in order that an organization to be efficient in reaching its objectives the management, the identification and understanding of the process need to be correlated as a system;
- continuous improvement – (to establish the objectives of change, of improvement, the measurement of the mechanisms efficiency which determine and implement these changes) the organization's objective has to be represented by the continuous improvement of the global performances;
- to base the decision on deeds – the efficient decisions are based on the analyses of the data and information;
- the relationships out of which both parties can benefit – must cultivate a reciprocal relationship of the organization with the suppliers in order to grow the abilities of creating value.

2. GENERAL CONSIDERATION

The specialists say that a quality management system implies in an compolsory way: a politic focused on the client; an adequate organization structure; an efficient documentation; an informational system which can allow to detect and to eliminate the dysfunctions, presented in figure 1. The management system of quality according to the ISO standards "ensure the university's academic management, to understand the task and responsibilities by all the employees and the continuous improvement of the entire activity. The quality systems built like that are created to contribute to the control maintenance, stability and university's capability, to determine them to have a higher level". (Todericiu, 2007)

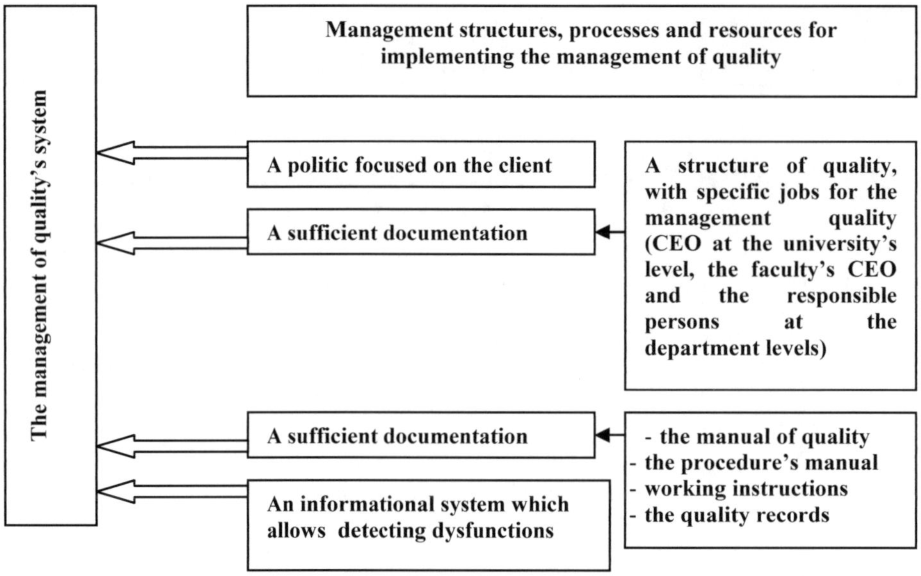

Fig.1 The university's quality system.

3. THE TOTAL QUALITY MANAGEMENT IN HIGHER EDUCATION

SMQ supposes the same fundamental elements it doesn't matter of the chosen pattern: (Oprean & Ţîţu, 2008)

- to focus on satisfying the demands and the student's expectations, the employer and the stakeholders;
- in the quality problem, the proactive attitude of the institution's management which it is expressed through creating an adequate environment for performance for all its actions;
- the approach of the quality problem in strategic terms: mission, value, principles, politics, strategy, objectives, etc.
- maintaining under control and improving the institution's processes - to implicate the staff;
- to identify some relevant items of quality and to introduce the mechanisms of internal evaluation of them, the system's documentation in order to provide some objective proves which can create trust.

The specialists in academic management say that, the existence and the development at normal parameters of a system of quality management in a university leads to:

- the efficient and transparent use of the resources which came from the state budget but the extra budgetary resources allocated to the higher education;
- to make a high quality training to be able to respond to the Romanian market's demands;
- to align at the standards and the practice of the European higher education institutions in the field of management of quality, helping directly to the university's prestige and to the development of international partnerships;
- to increase the trust of all the stakeholders in the educational process that it will be satisfied all the explicit and implicit expectations of the educational process, offering credibility and diploma recognition, recognising the students and teachers activity from that higher education institution;
- to determine all the weak spots and to eliminate the errors which can appear in the academic process; (Ţîţu & Oprean, 2007)
- to become more efficient in all the activities to create the academic process and the continuous improvement of the quality of the academic process.

4. CONCLUSIONS

The university's system for quality insurance fulfil two important functions. ***The function of quality management:*** Has as main purpose the focus on the institution's performance at all levels of its activity and supposes an adequate system of organization for developing the function of quality. The quality management system refers to: the university's management responsibility; the way in which they assume this responsibility through defining and following the strategy, politics and its objectives regarding quality, through insuring the necessary resourced for realising it; to focus all the organization's activities on satisfying the demands and the needs of the external and internal clients; to create inside the institution the environment where the entire staff to be encouraged to performance, to assume the individual responsibilities in order to fulfil its mission and its objectives. ***The quality insurance function***: It purpose is to create the client's trust and to the other external interested parties of the university, having in view the capacity and the organization's openness to satisfying the demands and the expectations. *The trust it is based* "on one hand evaluating the graduates skills, the research results, the direct assistance for the economic and social environment, reported to the existent standards or with the similar products of other organizations from the field (benchmarking), on the other hand evaluating the quality's internal system from the institution of higher education reported to the existent standards and accepted by the stakeholders".

5. REFERENCES

Muscalu, E. (2001). *Universitary Strategies*, Eficient Publishing House, Bucharest

Oprean, C. & Ţîţu, M. (2008). *Quality Management in the Economy and Organisation Based on Knowledge,* AGIR Publishing House, Bucharest

Ţîţu, M. & Oprean, C. (2007). *Quality Management*, University of Piteşti Publishing House, ISBN 978-973-690-646-6, Piteşti

Ţîţu, M. & Oprean, C. (2007). *Strategic Management,* University of Piteşti Publishing House, Piteşti

Todericiu, R. (2007). *Universitary Management from the European and Globalising Perspective*, ULBS Publishing House, Sibiu

Annals of DAAAM for 2008 & Proceedings of the 19th International DAAAM Symposium, ISSN 1726-9679
Editor B. Katalinic, Published by DAAAM International, Vienna, Austria 2008
Make Harmony Between Technology and Nature, and Your Mind will Fly Free as a Bird

METHOD OF SETTING UP MAINTENANCE STRATEGY

TOMA, V[ictoria] & MARASCU KLEIN, V[ladimir]

Abstract: *This paper presents an efficient using method of data generated by failure of production equipment, with regard the identifying critical elements (sub-systems, wearing items), setting and optimizing maintenance strategy. The method includes several stages, such as: collecting and primary data processing provided by the running behavior of equipment, their analysis and statistical processing. Finally maintenance strategy adopting is achieved depending on the evolution of failure rate and minimal average maintenance cost of equipment under the given running conditions.*
Key words: *maintenance, strategy, modeling.*

1. INTRODUCTION

The existence of spoiling and failing processes of production equipment imposes adoption of an ensemble of maintenance activities for reducing their effects. Maintenance represents a combination of technical, administrative and management activities, with regard to keep and restore the operating state of technical equipment under secure operating conditions, (Monchy, 2003). Key words: *keeping* and *restoring* of operating state of the equipment, designate the manner of carrying out maintenance. So, according to the moment of intervention, maintenance may be:

- *preventive* – carried out before occurring the failure;
- *corrective* – carried out after failure has occurred.

Setting up corrective or preventive maintenance strategy, may be based on different criteria, such as (Monchy, 2003):

- impact of failure on the production or environment security;
- costs of failure;
- flexibility of using survey techniques;
- cost of technical surveying equipment;
- variation of the production equipment's failure rate.

2. METHOD OF SETTING UP MAINTENANCE STRATEGY

Manufacturer companies have maintenance departments, using traditional working methods, consisting in recording and transmitting information on paper support: registers, files, etc. Generally, this information is incomplete, difficult to access for the maintenance staff and their valuation on long terms is excluded and so, individual specialists have no opportunity to take benefit from the experience of the group.

Because, in the case of corrective maintenance, removal of failure represents almost an urgent task, setting up of accessible working instruments, for facilitating identification, diagnoses and removal of failures are of a huge importance for the production system. With this purpose, an information product, based on a method of acquiring, analyzing and processing of data generated by failures of equipment has been conceived, for identifying critical elements and optimizing their maintenance.

Figure 1. shows the proposed method, its essential stages and supporting instruments. The arrows are indicating the flow of the method.

The study hypotheses are:

- failures of elements are independent events;

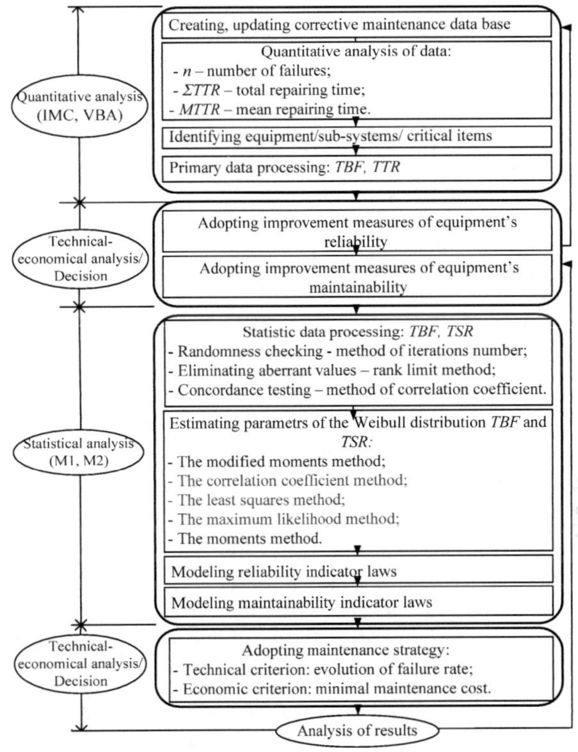

Fig. 1. Diagram of the method.

- all failures needing a time of restarting higher than 5 minutes are taken into account;
- several simultaneous maintenance interventions are considered as a single one;
- systematic preventive maintenance interventions are not taken into account in determining times of good working.

2.1. Data base of corrective maintenance

The first stage of the method consists in *creating the information data base*, proper to the company operating the equipment. The data base shall represent the operating instrument for informing, recording of the failures and corrective maintenance interventions. In this way the history of corrective maintenance interventions for all equipment shall result, featured by the following data:

- failing component, nature of failure, failure cause;
- operating, repairing, standing of production;
- replaced (modified) spare parts, manpower, materials;
- report concerning the executants' activity;
- cost of the corrective intervention.

Accordingly, the information system shall provide both the management of corrective maintenance activities and primary processing of data concerning equipment failure and maintenance interventions, due to them. Also within this stage, a quantitative analysis of the collected data is carried out, by using the Pareto diagram for identifying equipment, subsystems, and critical spare parts, presenting a high number of failures, failures and repeating failing causes. The analyses shall have on their grounds the following criteria: number of failures n, average repairing time t and total repairing time for a given period.

Statistic reports, histograms, supplied by the informational program substantiate identification of improvement directions (reliability, maintenance, availability), representing decisional instruments for the maintenance manager.

For critical elements the time evolution of failing rate is surveyed, with regard to conceive an adequate maintenance strategy. Accordingly, for the critical item, the informatic program supplies the sequence of operating times between successive failures (TBF) and the sequence of repairing times (TTR). Lifetimes represent the modelling base of the operating time between successive failures distribution law and, consequently, of the time between maintenance actions and repairing times provide modelling of maintainability.

2.2. Statistic data modeling

The next stage of this method consists in the selection of operating time without failures and of repairing time. The Weibull model has been adopted, because it covers random, aging and spoiling processes (Monchy, 2003), (Boucly, 1998). The Weibull distribution also provides a good concordance with experimental results (Lyonnet, 2007), (Kececioglu, 1993) because of its flexibility, given by the presence of the three parameters β, γ, η.

Estimation of Weibull distribution's parameters needs to pass some under/stages, as follows:

- checking randomness of experimental data;
- testing and eliminating aberrant values;
- testing concordance of experimental values with the three-parametric Weibull distribution.

Checking randomness of experimental data is achieved by using iterations method. Statistic decision concerning randomness is taken on grounds of the existing iterations' maximal length.

Eliminating aberrant values is achieved by applying a specific test to the Weibull repartition, i.e. the rank limit method (Kececioglu, 1993), which tests the values of the given sequence, not only its extreme values.

Testing concordance between the empiric distribution of experimental data and the Weibull distribution represents a fundamental stage, because all maintenance forecasts have on their bases, the adopted model. Therefore the hypothesis formulated with regard to the theoretic model is checked for its correspondence with experimental data, by using as concordance indicator, the correlation coefficient. The values of correlation coefficient $Ccor\ (\gamma) \geq 0{,}90$, indicate a good correlation (Kececioglu, 1993).

Estimating parameters of Weibull distribution is achieved by using: the modified moments method, the correlation coefficient method for estimating the location parameters of the Weibull distribution, the least squares method, the maximum likelihood method *(MVM)* and the moments method. According to the recommendations of the specialty literature (Kececioglu, 1993) the calculated parameters have been adopted by *MVM.*

Synthesizing, within this stage, the method provides the determination of following sizes:

- distribution function of operating time without failures $F(t)$, respectively the maintainability function, $M(t)$;
- failure rate, $z(t)$ / repairing rate $\mu(t)$;
- mean operating time, *MTBF* / mean repairing time, *MTTR.*

2.3. Setting up maintenance strategy

In the final stage, *adopting maintenance strategy* is made according to the evolution of the equipment's failing rate and the maintenance costs as per time unit.

For equipment with operating times without failures modeled by the Weibull distribution, the possible variants of maintenance strategy are set up according to the shape parameter β. Thus, from the expression of the failing rate function, characteristic to the Weibull model:

$$z(t) = \frac{\beta}{\eta}\left(\frac{t-\gamma}{\eta}\right)^{\beta-1} \qquad (1)$$

it is to be observed that:

- $\beta = 1$, indicates a timely constant failing rate, recommending corrective maintenance, failing risk being all the time the same, even if preventive replacements are or aren't carried out;
- $\beta < 1$, defines a decreasing failing rate, justifying also a corrective maintenance, because after a preventive replacement, "failure risk" due to early failures, will be higher than former.

$\beta > 1$, indicates a timely increasing failing rate. In this case, for a fixed applying period the cost of the two strategies has to be calculated and compared. Preventive maintenance is adopted only if cost on timely unit is smaller than the cost of corrective maintenance.

3. CONCLUSIONS

On grounds of the hypotheses formulated, this paper develops a method of acquisition and processing date from failures of production equipment, with regard to set and optimize the maintenance strategy.

Quantitative analysis of failing data provides an identification of critical elements prom the reliability's, maintenance's or availability's point of view.

Estimation of the failure model's parameters and that of the repair's ones, is needed in forecasting reliability, respectively maintenance indicators, which ground the maintenance strategy setting up. Thus, the decision as to when to repair or replace components are made by the owner using actual data instead of having to rely on manufacturer's recommendations which tend to be conservative.

Effective use of the method of setting up maintenance strategy is supported by informatics programs (fig.1.), developed in Microsoft Access (IMC), Visual Basic for Application (VBA) and Mathcad (M1, M2).

The results of the case studies carried out with this method have been published within scientific papers (Toma et al., 2008), (Morariu et al., 2008).

4. REFERENCES

Boucly, F. (1998). *Maintenance management*, AFNOR, ISBN 2-12-466 721-1, Paris

Kececioglu, D.B. (1993). *Reliability & Life Testing Handbook*, PTR Prentice-Hall, ISBN 0-13-772377-6, Englewood Cliffs

Lyonnet, P. (2007). *Maintenance: Mathematics and Methods*, Tec Et Doc, ISBN 2-85206-745-5, Paris

Morariu, C.; Toma, V. & Mărăscu-Klein, V. (2008). Mathcad program for estimating the Weibull distribution parameters, *Annals of the Oradea University, Fascicle of Management and Technological Engineering*, Volume VII (XVII), pp 1619-1628, ISSN 1583 - 0691

Monchy, F. (2003). *Maintenance – Methods and organisation*, Edition Dunod, ISBN 2-7430-0731-1, Paris

Toma, V., Morariu, C. & Klein-Mărăscu, V. (2008). Mathcad program for checking randomness of samples, eliminating outlier data and testing the concordance with the Weibull distribution, *Annals of the Oradea University, Fascicle of Management and Technological Engineering*, Volume VII (XVII), pp 2804- 2811, ISSN 1583 - 0691

Annals of DAAAM for 2008 & Proceedings of the 19th International DAAAM Symposium, ISSN 1726-9679
Editor B. Katalinic, Published by DAAAM International, Vienna, Austria 2008
Make Harmony Between Technology and Nature, and Your Mind will Fly Free as a Bird

IMPROVING DENOISING PERFORMANCE WITH ADAPTIVE WAVELET TRANSFORMS

TOMIC, M[laden]

Abstract: *Introducing adaptivity into traditional wavelet transform further improves the quality of denoised signal. Common approach is to choose wavelets such as to match the properties of the analyzed signal. The lifting scheme made it possible to easily introduce such adaptations. This paper provides an overview of adaptive wavelet transforms in signal denoising. Non-wavelet domain denoising method based on the intersection of confidence intervals (ICI) rule is also briefly investigated.*

Key words: *adaptive, wavelets, lifting, denoising, ICI*

1. INTRODUCTION

Wavelet transforms, in general, perform very well in signal denoising. Since there is an indefinite number of wavelet transform basis functions (wavelets) possible, the efficiency of noise removal is greatly influenced by the choice of the wavelet. Shorter wavelets perform better for signals with many discontinuities or lots of high frequencies. On the other hand, longer and smoother wavelets perform better for smoother signals consisting of primarily lower frequency components. Ideally, shorter wavelets should be used for higher frequency parts of the signal, while longer wavelets should be used for lower frequency parts of the signal. To achieve this, different adaptation techniques are used widely with varying success.

This paper will provide a brief overview of current research achievements in signal denoising using adaptive wavelet transforms.

2. LIFTING SCHEME

Traditional wavelet transforms implemented with filter banks were widely accepted and used in many applications, however, introducing adaptivity to such transforms is not always an easy task. A new framework for the wavelets construction was proposed in (Sweldens, 1996; Sweldens, 1997) - the lifting scheme. It is a computationally efficient way of implementing the wavelet transforms and it makes it easy to introduce adaptivity to transform. Several detailed case study examples were given in (Sweldens & Shcroder, 1997).

A simple two-band decimated filter bank based on the lifting scheme is shown in Fig. 1. Before filtering, the signal *X(z)* is split into even-indexed and odd-indexed components. There are two different types of lifting steps – the dual lifting step and the primal lifting step, which are commonly referred to as the predict and the update step, respectively. The filter *P(z)* is used in the predict step and it predicts the value of one of the odd-indexed components, based on the values of the even-indexed components. The prediction error of the prediction will represent the values in the high-pass channel of the filter bank, i.e. the wavelet coefficients. The filter *U(z)* is used in the update step. Its purpose is to update the even-indexed components, based on the wavelet coefficients. The net effect will be the low-pass filtering of the even-indexed components. Output of the filter bank are coarse approximation, or average *A(z)* and signal details *D(z)*. If signal denoising is a target application of

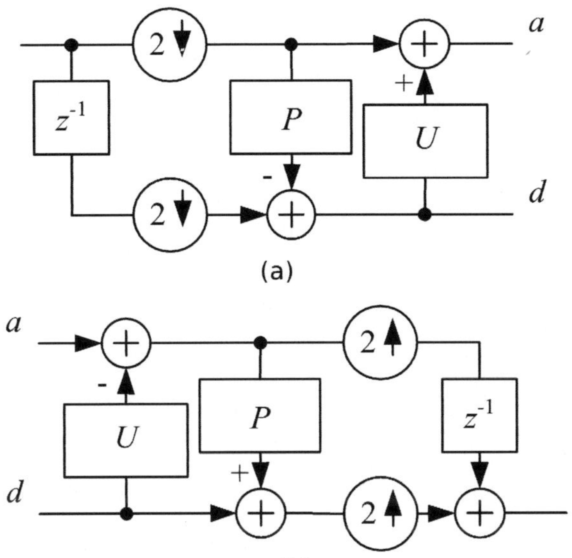

Fig. 1. (a) Analysis and (b) synthesis wavelet filter banks based on the lifting scheme

the filter bank then the decimators can be removed and filters *P(z)* and *U(z)* upsampled, resulting in an undecimated transform.

In (Daubechies and Sweldens, 1998) it was proved that any discrete wavelet transform can be decomposed into a finite sequence of lifting steps. Also, in (Kovacevic & Sweldens, 2000) generalisation of the lifting scheme approach to wavelet transforms to arbitrary dimensions was presented.

3. ADAPTIVE WAVELET TRANSFORMS USING LIFTING SCHEME

Since all calculations in the lifting framework are done in the spatial domain, it is easy to introduce adaptivity into transform. The idea of adaptation is commonly focused on modifying the transform properties based on a certain signal features or local properties. Typically, shorter wavelets should be used in the neighbourhood of discontinuities or generally higher frequencies in a signal, while longer wavelets should be used for the signal regions where lower frequencies prevail.

Two adaptive transforms were proposed in (Claypoole et al, 1997; Claypoole et al, 1998; Claypoole et al. 2003) - the scale-adaptive transform (ScAT) and the space-adaptive transform (SpAT). In the ScAT, an adaptation is performed on a scale-by-scale basis. From a set of predictors they choose the one which better fits the prevailing signal properties. The objective measure of the quality of fit is a simple sum of squared prediction errors. The predictor which yields the smallest sum of squared errors is chosen as the predictor for the entire signal at a given scale. The approach in the SpAT transform is to perform adaptation on a point-by-point basis. The predictor is chosen for each data point as the one which gives the smallest prediction error. It is to be expected that such predictor will often fit well to the local signal properties and make it possible for the wavelets to contract in the neighbourhood of signal discontinuities. The shortcoming of SpAT is that it makes the predictor spacevarying.

Since the update step depends on the predictor it may complicate its design or introduce the necessity of bookkeeping of the predictor selection. To resolve this, Claypoole proposed the update first framework where the update step is performed before the predict step.

Unlike Claypoole et al. who adapt the predictor, Piella in (Piella & Heijmans, 2002) proposed an adaptive update transform where only the update filter is adapted, while the predictor is fixed. In the general form presented, choice of the update filter depends on the decision maps which can be based on any meaningful criteria. In the original work, an example of a gradient based decision maps is suggested while in (Piella et al., 2002) derivative based decision maps are proposed. However, the method is only briefly investigated and no comparison to other methods is provided.

Algorithm called switch thresholding was proposed in (Wu et al., 2004). It is a very simple adaptation scheme where two conventional wavelets are used - Haar and CDF(2,2). The adaptation is performed on a point-by-point basis by choosing the appropriate wavelet for the current point. For the step edges, Haar transform is to be used, while for the other parts of the signal CDF(2,2) is chosen. Whether the current point is a step edge is determined by the difference to its neighbouring point. If the difference is larger than the threshold it is considered to be an step edge. Although some coarse limits for the threshold are suggested to follow from the statistics theory, the actual threshold value is determined empirically.

Instead of modifying the transform, an opposite approach is proposed in (Chan & Zhou, 1999). It was showed that the signal itself can be modified in the neighborhood of discontinuities. After the discontinuity is detected, signal values from one side of the discontinuity is used to extrapolate its values on the other side of the discontinuity. Traditional wavelets may now be applied to such a signal region because the discontinuity is eliminated and the whole region is smooth. If we record how the changes are made to the signal, it should be possible to recover the discontinuities at synthesis, using the inverse filters.

4. INTERSECTION OF CONFIDENCE INTERVALS

In a separate field, local polynomial approximation was used in (Katkovnik, 1999) as a filter design tool for denoising. Katkovnik uses kernel estimators to estimate the true value of the noisy signal. Again, the idea is to use larger estimator support for the smooth signal regions and to contract the support for regions which contain high frequencies. To achieve adaptive support selection, he proposed the intersection of confidence intervals (ICI) method. The algorithm calculates confidence intervals (CI) for estimators of growing support. If the order of estimation is kept constant and estimator support grows, the CI gets smaller. The ICI rule states that as long as the support spans data points with similar statistics the CIs should at least partially intersect. Ones there is no more intersection, it is likely that a change in a signal statistics occurred and estimator support should not include these points. For smaller supports estimation bias is low but the variance is high. As the support grows, estimation bias is getting larger and the variance is getting smaller. It was proved in (Stankovic, 2004) that the ICI rule gives an optimal bias-to-variance tradeoff.

5. CONCLUSION

Real world signals are often corrupted by noise which may severely limit their usefulness. For this reason, signal denoising is a topic that continually draws great interest. Wavelets perform very well in the area but the denoising efficiency can be significantly improved by introducing adaptivity to traditional wavelet transforms. The lifting framework allows researcher to do it in a straightforward way and many adaptive algorithms were developed to explore this capability. Commonly, the adaptation algorithm tries to contract the wavelet support in the neighbourhood of discontinuities in the signal in order to avoid visual artifacts in the denoised signal, such as Gibbs phenomena. For smooth signal regions longer wavelets are used.

There are many ways to further improve the transform efficiency and the focus of our current research is on connecting the ICI rule with the wavelet transforms.

6. REFERENCES

Chan, T. F. and Zhou, H. M. (1999). Adaptive ENO-wavelet transforms for discontinuous functions, Technical Report 99-21

Claypoole, R.L.; Davis, G.; Sweldens, W. and Baraniuk, R. (1997). Nonlinear wavelet transforms for image coding, *Conference Record of the Thirty-First Asilomar Conference on Signals, Systems & Computers*, volume 1, pages 662–667 vol.1

Claypoole, R.L.; Baraniuk, R.G. and Nowak, R.D. (1998). Adaptive wavelet transforms via lifting. In *Proc. IEEE International Conference on Acoustics, Speech and Signal Processing*, volume 3, pages 1513–1516 vol.3

Claypoole, R.L.; Davis, G.M.; Sweldens, W. and Baraniuk, R.G. (2003). Nonlinear wavelet transforms for image coding via lifting, *IEEE Transactions on Image Processing*, Volume 12, Issue 12, Dec. 2003 Page(s):1449 - 1459

Daubechies, I. and Sweldens, W. (1998). Factoring wavelet transforms into lifting steps. *J. Fourier Anal. Appl.*, 4(3):245–267, 1998.

Katkovnik, V. (1999). A new method for varying adaptive bandwidth selection. *IEEE Transactions on Signal Processing*, 47(9):2567–2571

Kovacevic, J. and Sweldens, W. (2000). Wavelet families of increasing order in arbitrary dimensions, *IEEE Transactions on Image Processing*, 9(3):480–496

Piella, G.; and Heijmans, H. (2002). Adaptive lifting schemes with perfect reconstruction, *IEEE Transactions on Signal Processing*, vol. 50, no. 7, pp. 1620--1630

Piella, G.; Pesquet-Popescu, B. and Heijmans, H. (2002). Adaptive update lifting with a decision rule based on derivative filters. *IEEE Signal Processing Letters*, 9(10):329–332, 2002.

Stankovic, L. (2004). Performance analysis of the adaptive algorithm for bias-to-variance tradeoff, *IEEE Transactions on Signal Processing*, 52(5):1228–1234

Sweldens, W. (1996). The lifting scheme: A custom-design construction of biorthogonal wavelets, *Appl. Comput. Harmon. Anal.*, 3(2):186–200, 1996.

Sweldens, W. and Schroder, P. (1997). Building your own wavelets at home, *Wavelets in Computer Graphics*, SigGraph

Sweldens, W. (1997). The lifting scheme: A construction of second generation wavelets. *SIAM J. Math. Anal.*, 29(2):511–546

Wu, Y.; Pan, Q.; Zhang, H. and Zhang, S. (2004). Adaptive denoising based on lifting scheme, In *Proc. 7th International Conference on Signal Processing ICSP '04*, volume 1, pages 352–355 vol.1

Annals of DAAAM for 2008 & Proceedings of the 19th International DAAAM Symposium, ISSN 1726-9679
Editor B. Katalinic, Published by DAAAM International, Vienna, Austria 2008
Make Harmony Between Technology and Nature, and Your Mind will Fly Free as a Bird

WAVELET TRANSFORMS WITH APPLICATION IN SIGNAL DENOISING

TOMIC, M[laden]

Abstract: *This paper is focused on application of the wavelet transforms in signal denoising. Wavelet transforms have ability to localize the signal properties in both the frequency and the time. Signal decomposition results in packing the main signal properties in a small number of large wavelet coefficients and constraining noise to small coefficients. To remove noise from the signal, wavelet shrinkage technique is used – thresholding is applied to wavelet coefficients at each scale. This way, only larger coefficients are retained, while the coefficients representing noise are eliminated before the reconstruction. An example of signal decomposition and denoising is provided, in which these effects can be easily observed.*

Key words: *wavelets, denoising, shrinkage, thresholding*

1. INTRODUCTION

Wavelet transforms have gained wide acceptance as a valuable tool for common signal processing tasks. The most important difference between the wavelet transforms and transforms such as a Fourier transform is that the wavelets are localized in both the frequency and the time. This makes it possible to better localize properties of the analyzed signal. The result is a well known ability of the wavelet transforms to pack the main signal information into a very small number of large wavelet coefficients. Another advantage of wavelet transforms is that there is an indefinite number of basis functions. An appropriate wavelet can then be chosen for a specific signal which makes the transform adjustable and adaptable. Because of these excellent properties, wavelet transforms have been used with great success in many different applications, such as signal denoising and compression or feature detection. In this paper the focus will be on the application in signal denoising, where wavelets are used extensively.

2. WAVELET TRANSFORMS IN SIGNAL PROCESSING

For efficient application of the wavelet theory in signal processing it was necessary to simplify the calculations and allow the adaptivity to be introduced into the transform. In (Mallat, 1988; Mallat, 1989a; Mallat 1989b; Mallat 1989c) the theory was successfully connected with the concept of the multiresolution analysis. He also showed that the wavelet analysis can be performed using iterated filter banks of certain properties. Calculating the discrete wavelet transform (DWT) became as easy as performing a simple signal filtering. In a

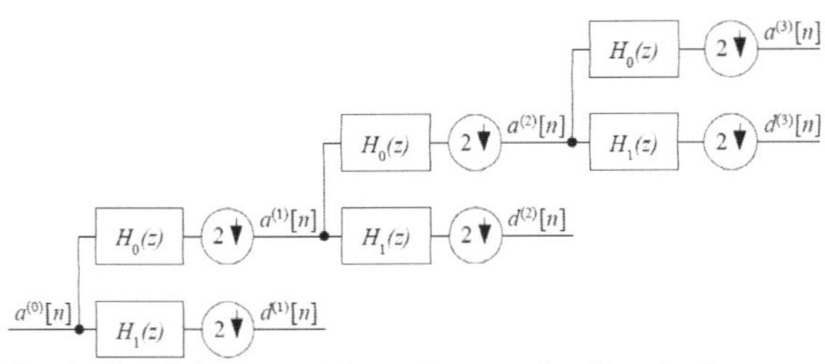

Fig. 1. Signal decomposition using wavelet filter bank

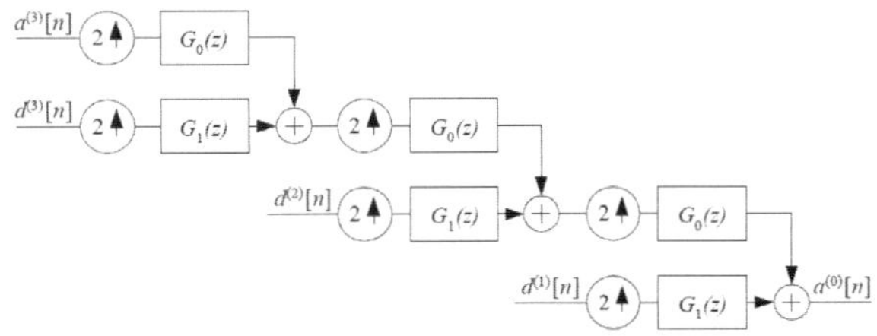

Fig. 2. Signal reconstruction using wavelet filter bank

typical scenario, the analysis filter bank consists of a low-pass and a high-pass channel. Output from the low-pass channel represents the coarse approximation of the signal, while the output from the high-pass channel contains the fine signal details. Both, the approximation and the details, are decimated by a factor of 2 and low-pass channel output is iteratively filtered with the same filter bank until the desired level of decomposition is reached. This is illustrated in Fig. 1. Details coefficients at each decomposition level are also called the wavelet coefficients. To reconstruct the original signal, synthesis filter bank is used, with its inputs upsampled by a factor of 2, as shown in Fig. 2.

3. DENOISING WITH WAVELET SHRINKAGE

The above procedure covers the basics of the wavelet analysis of the given signal but it does not provide a tool to remove a noise from the noisy signal. For denoising purposes, the wavelet shrinkage technique is used. It is based on the fact that the magnitude of the wavelet coefficients is important. Coefficients of higher magnitude are more likely to represent the main signal properties, while the coefficients of smaller magnitudes are more likely to be caused by the noise, as shown in Fig. 3 and Fig. 4. To remove the noise, a thresholding is applied to wavelet coefficients before signal reconstruction. A work was done in this area by Donoho and Johnstone (Donoho & Johnstone, 1994; Donoho & Johnstone, 1995; Donoho et al., 1995; Donoho & Johnstone, 1998). They successfully used the wavelet shrinkage technique for signal denoising and done significant research of the topic.

Although it has been showed that the wavelet transforms perform well in signal denoising they still exhibit unwanted visual artifacts, such as a Gibbs phenomena around signal discontinuities. Since the traditional wavelet transform is not translation invariant the size and the impact of the artifacts might heavily depend on the position of the discontinuity within the signal. To reduce the artifacts regardless of the discontinuity position, the translation-invariant denoising by means of the undecimated wavelet transform was proposed (Coifman and Donoho, 1996). In this transform the decimation step is not used, which results in an overcomplete representation of the signal. This was proved to both, significantly lower the overall root mean square error (RMSE) of the denoised signal, and also improve its visual appearance - subjective quality. The undecimated wavelet transform is now a common approach to a signal denoising problem. An example of denoising is shown in the Fig. 5. Additive white Gaussian

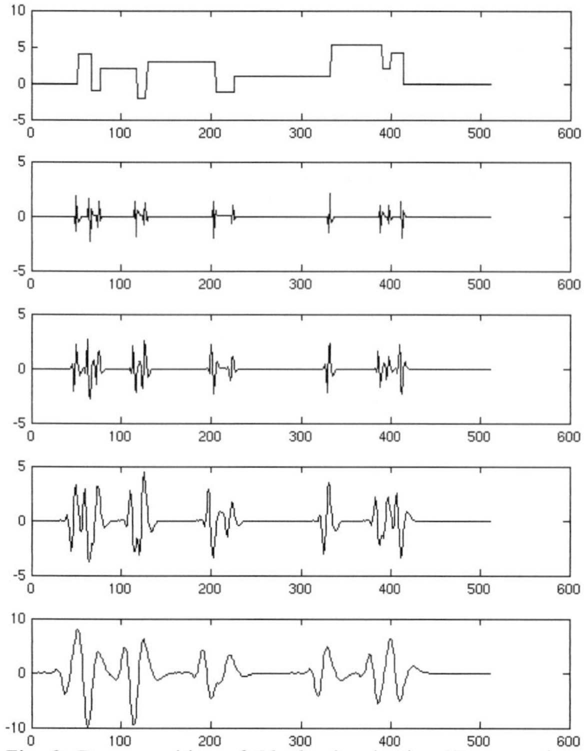

Fig. 3. Decomposition of *Blocks* signal using Haar wavelet– original signal and wavelet coefficients at each level

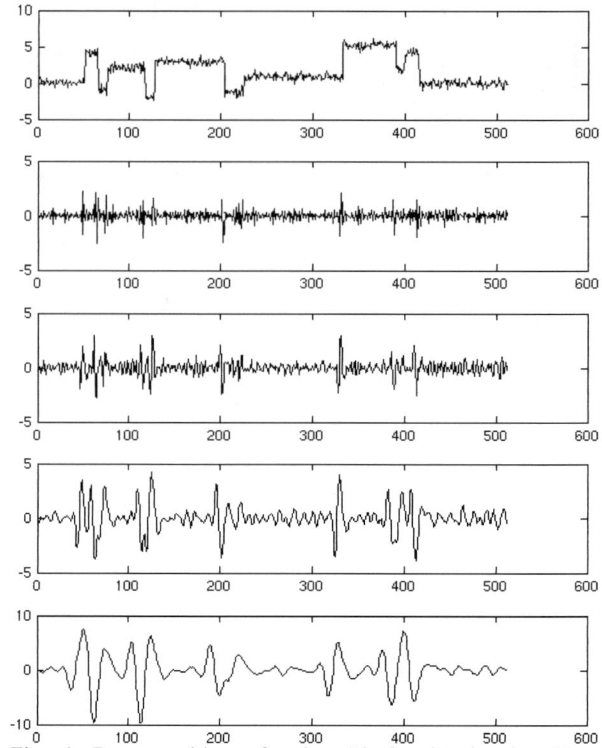

Fig. 4. Decomposition of noisy *Blocks* signal using Haar wavelet – noisy signal and wavelet coefficients at each level

noise with standard deviation equal to 5% of maximum signal magnitude is added to "Blocks" signal and Haar wavelet is used to construct a 4-level decomposition tree, as in Fig. 1. and 2.

4. CONCLUSION

Since many real-world signals are susceptible to noise, signal denoising is a very important topic in a wide range of applications. Efficient denoising algorithms can improve otherwise unusable noisy signals to an acceptable level. The advantage of the wavelet transform when compared to traditional transforms such as a Fourier transform is its ability to localize the signal properties in both the frequency and the time. Because of this, wavelet transform packs the main signal properties in a small number of higher magnitude coefficients, while less important properties and noise get constrained to smaller coefficients, which can easily be observed in a provided denoising example. Denoising is performed by thresholding the wavelet coefficients at each decomposition level before the reconstruction. This way only main signal properties are reconstructed during the synthesis while the noise is attenuated.

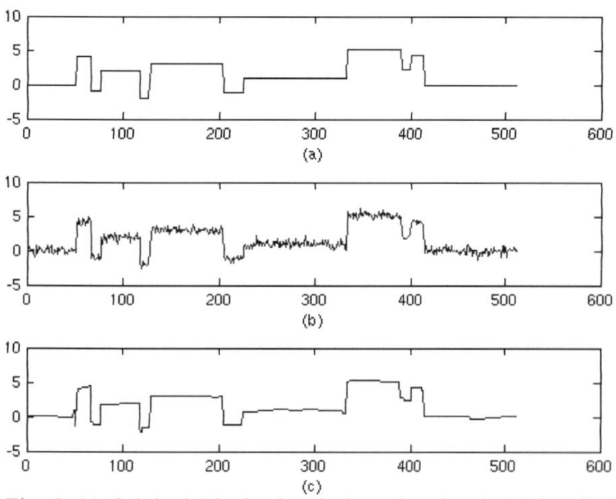

Fig. 5. (a) Original *Blocks* signal, (b) noisy signal, (c) denoised signal

Wavelet transforms proved to perform very well in signal denoising. Further improvements are achieved with introducing adaptivity into wavelet transform, where research is commonly focused on choosing different wavelets for a different class of signals.

5. REFERENCES

Coifman R. R. and Donoho, D. L. (1996) Translation-invariant denoising, *Lecture Notes in Statistics: Wavelets and Statistics*, New York: Springer-Verlag, pp. 125—150

Donoho, D. L. and Johnstone, I. M. (1994). Ideal spatial adaptation by wavelet shrinkage, *Biometrika*, 81(3):425–455

Donoho, D. L. and Johnstone, I. M. (1995). Adapting to unknown smoothness via wavelet shrinkage, *Journal of the American Statistical Association*, 90(432):1200–1224

Donoho, D. L. et al. (1995). Wavelet shrinkage: Asymptopia?, *J. R. Statist. Soc. B.*, 57(2):301–337

Donoho, D. L. and Johnstone, I. M. (1998). Minimax estimation via wavelet shrinkage, *Annals of Statistics*, 26(3):879–921

Mallat, S. G. (1988). *Multiresolution representations and wavelets*, University of Pennsylvania, Philadelphia, PA, USA

Mallat, S. G. (1989a). Multiresolution approximation and wavelet orthogonal bases of l2(R), *Transactions of the American Mathematical Society*, Vol. 315, No. 1, pp. 69-87

Mallat, S. G. (1989b). Multifrequency channel decompositions of images and wavelet models, *IEEE Transactions on Acoustics, Speech, and Signal Processing*, Volume 37, Issue 12, Page(s):2091 - 2110

Mallat, S. G. (1989c). A theory for multiresolution signal decomposition: The wavelet representation, *IEEE Trans. Pattern Anal. Mach. Intell.*, 11(7):674–693

Annals of DAAAM for 2008 & Proceedings of the 19th International DAAAM Symposium, ISSN 1726-9679
Editor B. Katalinic, Published by DAAAM International, Vienna, Austria 2008
Make Harmony Between Technology and Nature, and Your Mind will Fly Free as a Bird

COORDINATE CONVERSIONS AND TRANSFORMATIONS

TOMOIAGA, T[iberius]; ALEXEI, A[drian]; MARINESCU, M[irel] & BADEA, A[lexandru]

Abstract: *In this paper is described a software product realized by authors in order to compute coordinate conversions and transformations from/to coordinate systems and projection used in Romania. This software can be used by both military and civilian specialists to work out dedicated problems*
Key words: *coordinates, datums, parameters, transformation.*
¶

1. INTRODUCTION

¶

The problem of coordinate conversions and transformations has continued to be one of the most important preoccupations of worldwide geodetic specialists.

In the past, the positioning specific problems were solved at regional or local level. This fact leaded to the existence of many coordinate and projection systems. Recent studies prove, from geodetic point of view, that the globalization phenomenon leaded to some discontinuities of geographic data at the boundary of certain countries.

While new GPS based data are geocentric referred (Fig. 1, left), local data are referred to local or regional datums (Fig. 1, right).

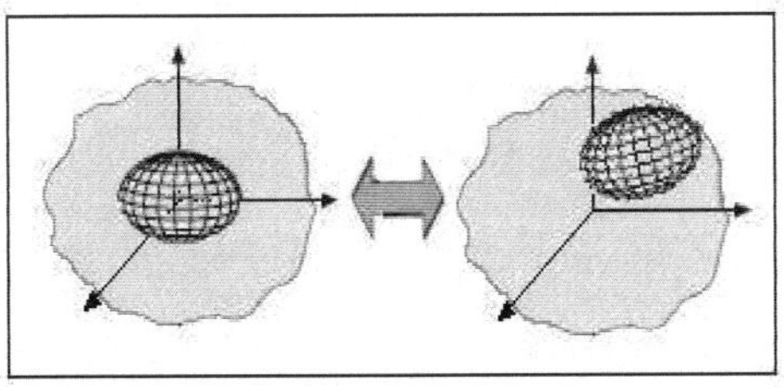

Fig. 1 Geocentric and local coordinate systems

Nowadays, taking into account the huge complexity of the problems revealed after our country acceptance in NATO and European Union, we consider very important to be respected by Romanian specialists the existent standards elaborated by both National Geospatial-Intelligence Agency (NGA) from USA (DMA, 1987a,b; DMA, 1991) and EuroGeographics from Germany. In this way we could eliminate the matters mentioned above.

To work out all the problems revealed at coordinate conversion/transformation from/to coordinate systems and projections used in our country, the authors of this paper come to support military and civilian specialists by realizing software structured on two modules.

2. SOFTWARE DESCRIPTION

The software package was developed using Borland Delphi 7.0 programming language and several open source components for tables management, reporting and for geodetic network graphical representation.

By choosing this solution, the authors can have total control of the software, having the possibilities of modifying, updating, improving and adding new modules anytime found suitable or to derive new customized versions.

The software package developed follows the general procedure described in fig. 2.

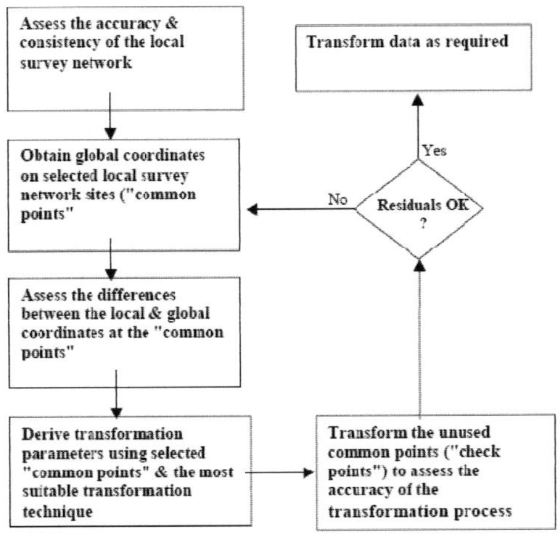

Fig. 2: General transformation procedure

The first module realizes 2D and 3D transformation parameters computation (Fig. 3) between different datums in the following cases:

- 4 parameters (with/ without baricenter);
- 7 and 12 parameters (with/ without baricenter) in the following situations:
  - without weights;
  - with individual weights;
  - with global weights.

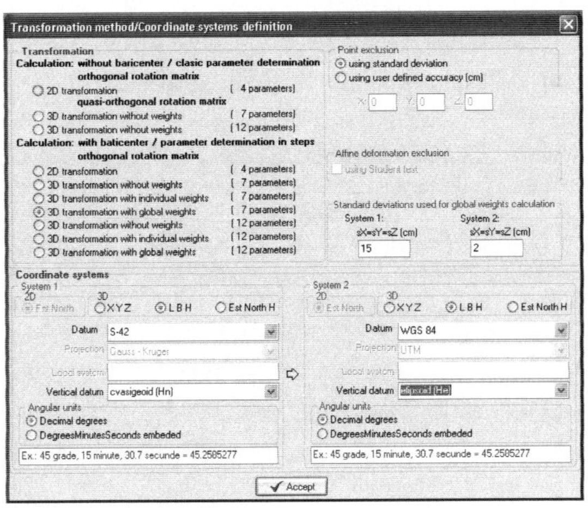

Fig. 3: The interface of the first module (computation of transformation parameters).

The model used to compute the coordinate transformations with 4 parameters is given by the following relations (Moritz, 1980):

$$\begin{bmatrix} x_2 \\ y_2 \end{bmatrix} = \begin{bmatrix} \Delta X_0 \\ \Delta Y_0 \end{bmatrix} + k \begin{bmatrix} \cos\theta & -\sin\theta \\ \sin\theta & \cos\theta \end{bmatrix} \begin{bmatrix} x_1 \\ y_1 \end{bmatrix} \tag{1}$$

The model used to compute the coordinate transformations with 7 parameters is described by the following relations (Moritz, 1980):

$$\overline{X}_2 = (1+k)R\overline{X}_1 + \bar{t}_x \tag{2}$$

In case of transformation without baricenter this is given by relation:

$$\mathrm{R} = \begin{pmatrix} 1 & \mathrm{d}\gamma & -\mathrm{d}\beta \\ -\mathrm{d}\gamma & 1 & \mathrm{d}\alpha \\ \mathrm{d}\beta & -\mathrm{d}\alpha & 1 \end{pmatrix}. \tag{3}$$

The difference between transformation with and without baricenter is given by rotation matrix conformation.

In case of transformation with baricenter the rotation matrix is expressed like a function of Hamilton normats quaternions (Grafarend & Richter, 1977):

$$R = (I_3 + S)(I_3 - S)^{-1}, \tag{4}$$

where:

$$\mathrm{S} = \begin{pmatrix} 0 & \mathrm{c} & -\mathrm{b} \\ -\mathrm{c} & 0 & \mathrm{a} \\ \mathrm{b} & -\mathrm{a} & 0 \end{pmatrix}. \tag{5}$$

The second module is designed to coordinate conversion respectively transformation (Fig. 4 & 5):

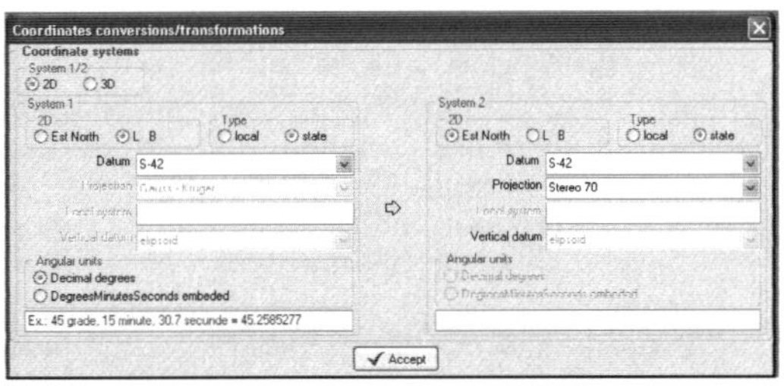

Fig. 4: The interface of the second module (coordinate conversions).

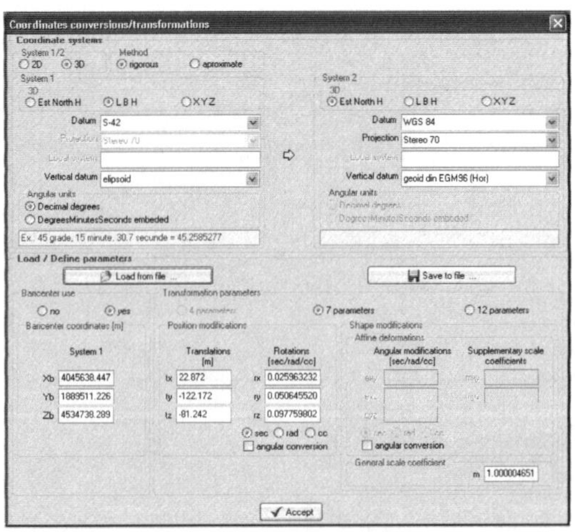

Fig. 5: The interface of the second module (coordinate transformations).

This module allows realizing the following operations:

- coordinate conversion (Fig. 4) between:
  - usual projections used in our country: UTM, Gauss-Kruger and Stereo70;
  - geocentric coordinates;
- coordinate transformation (Fig. 5) between:
  - S-42 and WGS84 datums using the projection mentioned above;
  - ETRSYY and ETRS89.

This module offers the possibility of loading the parameters from a file generated with the first module or to be typed.

Both modules uses like reference surfaces the ellipsoid, the European Gravimetric (Quasi)geoid EGG97, the geoid computed from Global Geopotential Model EGM96 and the Romanian quasigeoid.

3. CONCLUSIONS

The described software is still under development. Upon request, the finished modules can be added with supplementary coordinate systems and projections.

Particularities of the current version, by modules, are:

- module 1:
  - global and individual weighting;
- module 2:
  - residual mean square error calculation for transformed points;
  - transformation distortions (module and direction for error bias of transformed point).

Besides perfecting the already finished module, authors wishes to add some new ones. The final version of this software will allow the following operations:

- global transformation parameters determination between two coordinate systems, using S42, WGS84 and ETRS89 datums and Stereo70, Gauss-Kruger and UTM projections;
- computation and graphical representation of the vector of the horizontal and vertical errors;
- computation and graphical representation of Delaunay triangulation of the network;
- adding new points in a network by simulation;
- network scale variation determination.

Using the orientation and the module of the horizontal error vector and Thiessen proximity polygons different regions can be drawn for regional transformation parameters determination.

4. REFERENCES

DMA, (1987a). Supplement to Department of Defense World Geodetic System 1984 Technical Report: Part I - Methods, Techniques, and Data used in WGS84 development. *DMA TR 8350.2-A*, first edition, December 1.

DMA, (1987b). Supplement to Department of Defense World Geodetic System 1984 Technical Report: Part II - Parameters, Formulas, and Graphics for practical application of WGS84. *DMA TR 8350.2-B*, first edition, December 1.

DMA, (1991). Department of Defense World Geodetic System 1984: Its definition and Relationship with Local Geodetic Systems. *DMA TR 8350.2*, second edition, September 1.

Grafarend, E. and B. Richter, (1977). The generalized Laplace condition. *Bulletin Géodesique*, 51, 4, pp. 287-293.

Moritz, H, (1980). Geodetic reference system 1980. *Bulletin Geodesique*, Vol. 54, No. 3, pp. 395- 405.

Annals of DAAAM for 2008 & Proceedings of the 19th International DAAAM Symposium, ISSN 1726-9679
Editor B. Katalinic, Published by DAAAM International, Vienna, Austria 2008
Make Harmony Between Technology and Nature, and Your Mind will Fly Free as a Bird

BUILDING HUMAN RESOURCES IN ROMANIA: PRACTICES AND OPPORTUNITIES

TOMULETIU, A[driana]; OROIAN, M[aria] & POP LUPU, A[nca]

Abstract: *This paper brings to attention a very important aspect required by an increasingly competitive labour market, namely the need for developing human capital through increased investment in human resources. To be as clear as possible, we have presented a best practice model, that is, a grant financing project in which we were directly involved. This Phare project was won at regional level by a Romanian manufacturing company from Tirgu-Mures. Its aim was to develop a set of training services for the staff of the company and to initiate a network providing consultancy in the human resources field, for other similar companies and beyond.*
Key words: *training, human resource, labour market*

1. INTRODUCTION

A priority objective of socio-economic development in the current Romanian and European context, as it emerges from a series of national and international documents, is the investment in human capital. This objective is stipulated both in the Lisbon Strategy (Commission of the European Communities, 2005), and in the National Development Plan of Romania 2007 - 2013 which states as a priority "The development of human resources, increase in employment and combat of social exclusion" (Government of Romania, 2005).

Europe has moved to a knowledge-based economy and society, in which access to information and knowledge, together with the motivation and ability to use them intelligently become keys to strengthening the competitiveness of Europe, improving the level of employment of the human capital and its degree of adaptability (Cimpean, 2006).

Human resources development and promotion of human capital by increasing adaptability and acquiring special skills required in the workplace are requirements arising from changes that are taking place very quickly. This is partly due to the industrial revolution in the present society which is governed by new means of communication and biotechnology with a strong impact on the economic and cultural relationships, on people's life style, and especially on the working environment. More investments in human resources represent, for our ongoing integration into the European Union society, the key for adaptability to the structural changes on the labour market, and also reduce the gap that exists between Romania and the other EU developed countries, regarding the level of development of human capital.

2. INFORMATION

A series of studies conducted at several organizations show that Romania is facing low investment in continuous professional training (Chisu, 2007; Sandi & Moarcas, 2007). Many companies perceive training more as an expense rather than an investment, which triggers a large competency deficit on the work force market. On the background of the stringent need for coverage of this deficit, for the development of human capital and increase of its competitiveness in the labour market, the European Union will give Romania grants of 3.47 billion euros by 2013. The aim will be to promote equal opportunities regarding lifelong learning and the development of a modern, flexible and attractive labour market, which would lead by 2015, to a sustainable integration of 900,000 people into the labour market (Ministry of Labor, Social Solidarity and Family, 2006).

Funds will be allocated on 7 priority axes, namely:

- Axis I. "Education and training in support of growth and development of knowledge-based society" - 797 million euros.
- Axis II. "Connecting learning with the labour market" - 911 million euros.
- Axis III. "Increasing the adaptability of the workforce and companies - 450 million euros.
- Axis IV. "The modernization of public service employment" - 176 million euros.
- Axis V. "Promoting active employment measures" - 476 million euros.
- Axis VI. "Promoting social inclusion" - 524 million euros
- Axis VII. "Technical assistance" - 139 million euros.

EU Grants Funds for the development of human resources will be granted for the following types of investments:

- Organizing seminars for training employees in order to increase the adaptability to advanced technologies.
- Preparation of awareness campaigns regarding corporate social responsibility and programmes designed to transform undeclared labour into legal employment.
- Implementation of programs and support services for the development of entrepreneurial culture, and for starting a business.
- Development and promotion of modern managerial skills, especially for micro, small and medium-sized enterprises.

Such funds can be accessed by schools, universities and research centers, accredited training suppliers, chambers of commerce and industry, trade unions and employers unions, micro, small and medium-sized enterprises.

The best-known program of financing by European Union funds so far is the PHARE program - the first non-refundable financial instrument designed by the European Union to support Central and Eastern Europe in the progress towards a democratic society and a free market economy.

Phare program - Economic and Social Cohesion - Development of Human Resources - Promoting Human Capital has aimed, years in a row, at investing in human resources so that they would become more adaptable to structural changes in the labour market, but also at reducing the competency deficit found between the work force from Romania and the one from European Union.

One of the local initiatives to meet the needs for continuous training of human resources is the project called *Center for Training and Development of Human Resources*. The project was won at regional level by an average size manufacturing company from Tirgu-Mures, in partnership with People's University of Tirgu-Mures (as a training adults provider), being funded by the European Union and the Government of Romania, entered in the Ministry of European Integration program - Economic and Social Cohesion, Human Resource

Development, Promoting Human Capital, the Phare budget line /2003/005-551.05.03.02.

The aim of this project was the development of an overall service training for staff of the company and the initiation of a consultancy providing network for promoting best practices in human resources for other similar companies and beyond.

Project activities were justified by the needs identified in the organization and in other similar organizations:

- The need to carry out a professional selection for hiring specialized human resources most appropriate to the job description.
- The imperative need to correlate the demand on the market labour with an appropriate training offer.
- The need to develop professional skills in order to avoid redundancies and fluctuation of the work force.

Thus, this centre has been implemented within the organization which was equipped with advanced training materials, appropriate furniture, library, licenses for batteries of psychological evaluation and selection of personnel tests, all with the support of the granted funds.

The basic activities of the project focused on providing training programmes which included not only specialized courses, but also courses for the development of social competencies, courses based on a curriculum appropriate to each course session, as follows:

- Efficient communication and organizational interrelation courses addressed to all employees, with the purpose of ensuring efficient circulation of information at all company levels, and of preventing work conflicts.
- PC training courses for those lacking such competences and IT systems implementation courses inside the company.
- Leadership and diversity management courses addresses to the staff in leading positions, in order to create premises for a more efficient cooperation between them and their subordinates.

The assessment of staff with the acquired psychological tests was designed to identify any inconsistencies of the characteristics identified with the existing job profile and vocational shifting, where appropriate. Also, psychological testing aimed at the development of a more efficient professional selection, leading to the hiring of the most suitable people for the positions in the organization, in agreement with the profiles of those positions, so that in the future this would contribute to reducing staff fluctuation, at least on an average duration of time.

A national conference with the participation of specialists in the human resources, communication and leadership fields was also organized as part of the project, with the aim of promoting it and its results. The results were disseminated through the media, but also through a conference volume.

The project activities have had a major impact on employees whom had never before benefited from training programmes, although this is specified in the law, and is an educational imperative of the new era of knowledge. Through the courses the target group situation has improved, not only in terms of cognitive or social skills, but also in financial terms, because employees have become able to respond more effectively to professional activities. Indirectly, the work motivation has increased, the communicational environment within the organization has improved, which resulted in a more sustained and interested participation in all business activities.

The training had a direct impact on adults who have become aware of the need to continuously develop social and professional skills to be able to deal with the structural changes that occur almost overnight.

The project also resulted indirectly in positive consequences on the quality of family life. Those who have benefited from the advantages of the project and have made at the same time, efforts in this direction, have increased their chances of reintegration into the labour market in crisis situations and also their career promotion chances; they became more flexible and open to change and to the need for development and permanent training. The direct beneficiaries have also acquired a sense of social security as a result of increasing confidence in their ability to cope with all requests of the socio-economic life.

All measures developed by us by means of this project have been designed to contribute to improving the quality of human resources working in this company, measures and practices that will have positive consequences for the long-term, by lowering unemployment and staff fluctuation and also by improving communication and interpersonal/intergroups relations as a direct consequence of a change of attitude. As a result of access to lifelong learning at work, the foundations will be set for the quality development of people's life, for economic performance criteria within companies, elements that will lead to social-economical development of the community.

3. CONCLUSION

The findings of this project as well as the elements we have decoded during its implementation have determined us to become supporters of professional training, by carrying out professional qualification and re-qualification programs. We would like to conclude, by emphasizing the fact that Romanian organizations of all types should become increasingly aware of the importance of granting credit to factors interrelated to human resources development and continuous professional training, with the purpose of increasing performance and economic competitiveness. Thus, small and large-sized Romanian entrepreneurs, and not only them, should become increasingly preoccupied with the people that are directly responsible for production activity, that is, their own staff.

Our theoretical and practical approach is intended as an example, from which one can learn, one can reflect upon and which can be developed into research at a much higher scale.

4. REFERENCES

Chisu, V. (2007). Educational Marketing. Development and Integration Strategies for Romanian Educational Services in the Context of the Labour Market and Economic Globalization, *Available from*: http://www.biblioteca.ase.ro/downres.php?tc=9020 *Accessed*: 2008-08-10

Cimpean, E.A. (2006). *The Lifelong Learning – Significations and Content*, pp. 15, University Press Publishing House, ISBN (10)973-7665-37-6, ISBN (13)978973-7665-37-9, Tîrgu-Mureș

Commission of the European Communities (2005), Communication to the Spring European Council. Working Together for Growth and Jobs. A New Start for the Lisbon Strategy, pp. 26-28, *Available from:* http://eur-lex.europa.eu/lexuriserv/site/en/com/2005/com2005_0024en01.pdf *Accessed*: 2008-07-30

Government of Romania (2005). The National Plan of Development, 2007-2013, *Available from:* http://eufinantare.info/Documente/Sinteza_PND2007_2013.pdf *Accessed:* 2008-08-25

Ministry of Labor, Social Solidarity and Family (2006). Human Resources Development Divisional Operational Program 2007-2013, *Available from*: http://www.mmssf.ro/website/ro/autoritate/200406pdsdru.pdf *Accessed*: 2008-08-11

Sandi, A. M. & Moarcas, M. (2007). Romania Education Policy Note, *Available from:* http://siteresources.worldbank.org/INTROMANIA/Resources/EducationPolicyNote.pdf *Accessed:* 2008-09-15

Annals of DAAAM for 2008 & Proceedings of the 19th International DAAAM Symposium, ISSN 1726-9679
Editor B. Katalinic, Published by DAAAM International, Vienna, Austria 2008
Make Harmony Between Technology and Nature, and Your Mind will Fly Free as a Bird

FUNDAMENTAL FEATURES OF TECHNOLOGICAL SYSTEMS RIGIDITY

TONOIU, S[ergiu] & CATANA, M[adalin]

Abstract: *Because no definition and no determination method are formally recognized for the rigidity, characteristics of rigidity and some entities associated to technological system rigidity, these will be defined to acquire a maximum degree of generality. So, there were defined with a high level of generality essential elements for the study of technological systems rigidity: structure, state, references, deformations, rigidity and compliance.*
Keywords: *system, loading, deformation, rigidity, compliance.*

1. INTRODUCTION

Because the rigidity, characteristics of rigidity and some entities related to technological system rigidity are not unitary and generally defined in the literature (see Tonoiu, 1999, and Weck et al., 1989), several proposals concerning these topics were presented by Tonoiu & Doicin, 2002; Tonoiu et al., 1997a; Tonoiu et al., 1997b. This paper develops the previous researches of the authors.

2. PRELIMINARIES

Study method characteristics. Besides the essential phenomenological parameters, the definition and theoretical or/and experimental determination of an entity are influenced by the study method characteristics, such as: calculus /modeling hypotheses, measurement instrumentation accuracy, etc.

Technological system structure. A technological system, TS, is defined as a reunion of physical entities, E_i, constructive-functional interdependent, i.e.: $TS = \bigcup E_i$.

Let D(STR) represent the set of the defining characteristics of the technological system structure, i.e. characteristics referring to the system type, entity type, etc.

Technological system state. It is being considered that the state of a technological system or of its component entities is type of: rest; quasi-rest; passive functioning; active functioning.

Let D(STA) represent the set of the defining characteristics of the technological system state, i.e. characteristics referring to the state type, working parameters, etc.

References. It is being considered that the references associated to technological systems are of type of: physical or geometrical reference; absolute or relative reference. Physical references are of type of guide, fixed centre, etc. Geometrical references are commonly of type of Cartesian coordinate system.

Let RPR and RGR represent a relative physical reference and a relative geometrical reference associated to a given technological system (Fig. 1), so that: $RPR \equiv E_r$, $RGR \equiv Oxyz$.

Let D(REF) represent the set of defining characteristics of the references of a technological system, i.e. characteristics referring to the reference type, physical entity being reference, etc.

3. LOADINGS AND DEFORMATIONS

Loadings. Loading is a quantity of type of force, moment or pressure. The forces are of fixing, of machining, of excitation etc.; the moments are of type of bending moment or torsion moment; the pressure is acting because of a corresponding force etc. Let $\overline{F}$ represent one loading or a resultant of more loadings.

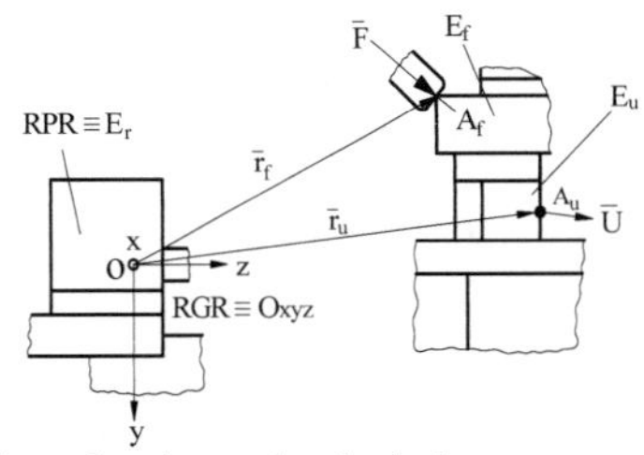

Fig. 1. Entities related to technological systems.

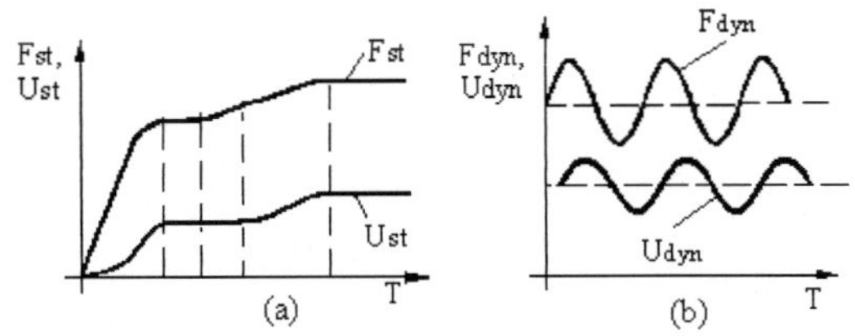

Fig. 2. Variation of loading and deformation: (a) static case; (b) dynamic case.

The $\overline{F}$ loading is applying on an element, E_f, $E_f \in TS$ (Fig. 1). Let A_f represent a point, $A_f \in E_f$, which is identical with the application point of a force or is participating to define the acting space of a moment or pressure; the position of A_f is defined e.g. by its positional vector, $\overline{r}_f$, (Fig. 1).

The direction of $\overline{F}$ can be indicated in a pre-established manner. Let F represent magnitude of the $\overline{F}$ loading, i.e. $F = |\overline{F}|$.

It is defined as being loading – time function the relationship between the loading magnitude, F, and the time, T, i.e.: F=F(T).

Dependent on the intensity of the F variation in time, the loading is static, $\overline{F}$st, or dynamic, $\overline{F}$dyn, if the F variation in time is "slow" (Fig. 2a) or "rapid" (Fig. 2b), respectively; a dynamic loading is occurring with a certain "frequency". Dependent on the monotony type of the F variation in time, a loading is of type of charging, constant or of discharging, if the F(T) function is increasing, constant or decreasing, respectively (see Fig. 2).

Let D($\overline{F}$) represent the set of the defining characteristics of the $\overline{F}$ loading, i.e. characteristics referring to the application point, direction, etc.

Deformations. It is considered as being deformation a quantity of type of: elastic deformation, plastic deformation or displacement; linear deformation or angular deformation. Let $\overline{U}$ represent one deformation or a resultant of more deformations. The $\overline{U}$ deformation is considered to be of a point A_u, belonging to an entity, E_u, of the considered technological system. The position of A_u is defined e.g. by its positional vector, $\overline{r}_u$, (Fig.1). Thus: $E_u \in TS$, $A_u \in E_u$. It is to be noticed that: $E_u \equiv E_f$ or $E_u \neq E_f$; $A_u \equiv A_f$ or $A_u \neq A_f$.

The direction of deformation can be indicated in a pre-established manner.

Let U be the magnitude of the $\overline{U}$ deformation, i.e.: $U = |\overline{U}|$.

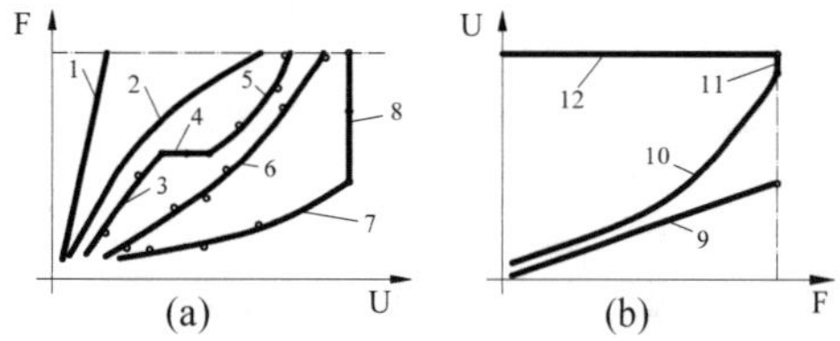

Fig. 3. Functions F(U), section (a), and U(F), section (b).

Deformation-time function is the relationship between the deformation magnitude, U, and the time, T, i.e.: U=U(T).

Dependent on the intensity of the U variation in time, the deformation is static, $\overline{U}$ st, or dynamic, $\overline{U}$ dyn, if the U variation in time is "slow" (Fig. 2 a) or "rapid" (Fig. 2 b), respectively; a dynamic deformation is producing with a certain "frequency". Dependent on the monotony type of the U variation in time, a deformation is of type of charging, constant or discharging, if the U(T) function is increasing, constant or decreasing (see Fig. 2).

Let D($\overline{U}$) represent the set of defining characteristics of $\overline{U}$.

4. LOADINGS-DEFORMATIONS RELATIONSHIPS

The useful relationships are of type of loading–deformation function/curve and deformation–loading function/curve, as follows.

The loading–deformation function is the dependence relationship between the loading magnitude, F, and the deformation magnitude, U, i.e.: F=F(U).

The deformation–loading function is the dependence relationship between the deformation magnitude, U, and the loading magnitude, F, i.e.: U=U(F).

In the case of static loadings and deformations, $\overline{F}$st and $\overline{U}$st, respectively, the F(U) and U(F) functions could be nonlinear or linear (Fig. 3), in correspondence with the deformation type, as follows: the F(U) and U(F) functions which are nonlinear (Fig. 3, curves 2, 5, 6, 7, 10) or linear not constant (Fig. 3, curves 1, 3, 9) correspond to different deformation types – elastic, plastic, displacement (see § 3.2); a constant F(U) function and a variable U(F) function (Fig. 3, curves 4, 11) correspond to deformations of type of displacement caused by clearances from joints; a constant U(F) function and a variable F(U) function (Fig. 3, curves 8, 12) correspond to "not deformable" structure – in the considered conditions.

5. RIGIDITIES AND COMPLIANCES

In numerous theoretical and applicative developments, two important specific system characteristics are of major interest: rigidity and compliance.

Rigidity is the characteristic that expresses the intensity of loading – deformation relationship. Thus, quantitatively, the rigidity, K, is the ratio between the loading magnitude variation, ΔF, and the deformation magnitude variation, ΔU, i.e.: $K=\Delta F/\Delta U$, $\Delta U \neq 0$.

Compliance is the characteristic that expresses the intensity of deformation – loading relationship. Thus, quantitatively, the compliance, C, is the ratio between the deformation magnitude variation, ΔU, and the loading magnitude variation, ΔF, i.e.: $C=\Delta U/\Delta F$, $\Delta F \neq 0$.

In the case of loadings and deformations for which F(U) and U(F) are nonlinear, let $(F_j;U_j)$ and $(F_l; U_l)$ represent two different couples/points associated to different states, j and l, j≠l, of the system.

There are being defined the relative rigidity, K_{jl}, and the relative compliance, C_{jl}, as:

$$K_{jl} = \frac{F_l - F_j}{U_l - U_j},\ C_{jl} = \frac{U_l - U_j}{F_l - F_j},\ F_l \neq F_j,\ U_l \neq U_j \qquad (1)$$

In a point (F_l, U_l) where F(U) and U(F) are differentiable, there may be defined the instantaneous rigidity, K_l, and the instantaneous compliance, C_l, as: $K_l = (dF/dU)_l$, $C_l = (dU/dF)_l$.

An important particular case is when, for a well defined $\overline{F}_\varphi$ loading – where φ is associated to the $\overline{F}$ direction, the $\overline{U}$ deformation is of unknown direction. In such of case, let $\overline{U}_{\lambda\varphi}$ represent the components of $\overline{U}$ parallel to the axes of Oxyz reference, φ and λ= x, y, z. Thus, $K_{\lambda\varphi}$ and $C_{\lambda\varphi}$, expressed on the basis of the definition relationships, are:

$$K_{\lambda\varphi} = \frac{\Delta F_\varphi}{\Delta U_{\lambda\varphi}},\ C_{\lambda\varphi} = \frac{\Delta U_{\lambda\varphi}}{\Delta F_\varphi},\ \varphi \text{ and } \lambda = x, y, z \qquad (2)$$

6. MATRICES OF RIGIDITIES AND COMPLIANCES

For each representative couple of points (A_f, A_u), $\overline{F}_\varphi$ loading and $\overline{U}_{\lambda\varphi}$ deformation there may be defined the matrix of fundamental rigidities, $[K]_{(A_f,\ A_u)}$, and the matrix of fundamental compliances, $[C]_{(A_f,\ A_u)}$. For example, the matrix $[K]_{(A_f,\ A_u)}$ is as follows:

$$[K]_{(A_f,\ A_u)} = \begin{bmatrix} K_{xx} & K_{xy} & K_{xz} \\ K_{yx} & K_{yy} & K_{yz} \\ K_{zx} & K_{zy} & K_{zz} \end{bmatrix}_{(A_f, A_u)} \qquad (3)$$

7. CONCLUSION

Defining fundamental elements have been introduced referring to study method, structure, state, references, deformations, rigidity and compliance associated to technological system. The considered elements have been developed until a level that assures a high degree of generality.

The assembly of entities proposed by this paper represents an important theoretical basis for new interpretations and developments of standards, determination methodologies and of the other matters linked to the technological system rigidity or compliance.

8. REFERENCES

Tonoiu, S. & Doicin, C.V. (2002). Technological manufacturing systems' (TMS) rigidity, *Proceedings of the 4th workshop "Human Factor and Environmentalist"*, Katalinic, B. (Ed.), pp. 101-102, ISBN 3-901509-37-2, Kosice-Slovakia, December 2002, DAAAM International, Vienna

Tonoiu, S. (1999). *Contributions on the study of machining technological systems rigidity*, Ph.D. Thesis, POLITEHNICA University of Bucharest, 1999, Romania (in Romanian)

Tonoiu, S.; Dulgheru, L.; Catană, M. & Purcărea, M. (1997a). Methods for experimental determination of static rigidity for machining technological systems, *Proceedings of the 9th International Conference on Machine Tools*, pp. 501-508, ISBN 973-31-1139-2, Bucharest-Romania, 1997, Ed. Tehnica, Bucharest (in Romanian)

Tonoiu, S.; Purcărea, M. & Catană, M. (1997b). Considerations on contact rigidity and dumping in machining technological systems, *Proceedings of the 9th International Conference on Machine Tools*, pp. 509-516, ISBN 973-31-1139-2, Bucharest-Romania, 1997, Ed. Tehnica, Bucharest (in Romanian)

Weck, M.; Eckstein, R. & Schafer, W. (1989). Methods for determination of machine tool static rigidity, *Mechanik*, No. 4, 1989, pp. 125-129 (in Polish)

Annals of DAAAM for 2008 & Proceedings of the 19th International DAAAM Symposium, ISSN 1726-9679
Editor B. Katalinic, Published by DAAAM International, Vienna, Austria 2008
Make Harmony Between Technology and Nature, and Your Mind will Fly Free as a Bird

METHODOLOGY FOR EXPERIMENTAL DETERMINATION OF STATIC RIGIDITY FOR NORMAL LATHES

TONOIU, S[ergiu] & CATANA, M[adalin]

Abstract: *The paper presents a unitary methodology for experimental determination of static rigidity for normal lathes. An example application of this methodology for the main spindle's assembly (MSA) of normal lathes is presented also. There are presented several schemes for loading the main's spindle assembly and for determining the corresponding deformations. Relevant experimental data are presented also.*
Keywords: *lathe, shaft, loading, measuring, rigidity.*

1. INTRODUCTION

For experimental determination of static rigidity for one system's element, this element will be submitted to a progressive increasing load, whose value is measured at each level. The corresponding deformation is measured with an adequate instrument, as shown by: Tonoiu et al., 1997a; Tonoiu et al., 1997b; Weck et al., 1989.

In order to determine the element's rigidity, there must be stated some things regarding: its structure, references, loading, strains. Measuring of loading and deformation is performed according to certain loading – measuring schemes. For loading measuring there is used a dynamometric device, which provides 3D forces. For the case of MSA of normal lathes, the forces are usually applied onto a rigid shaft assembled with the spindle. The shaft materializes the application points for force, A_F, and deformation, A_U (see Tonoiu, 1999 and Tonoiu & Iliescu, 2002).

2. LOADING – MEASURING SCHEMES

A loading scheme provides characteristics such as: loading type, direction, application point, sense, size (see Tonoiu, 1999).

The loading – measuring scheme also refers to the materialization of application points for force, A_F, and deformation, A_U.

It is necessary that the loading scheme, and therefore the loading device, to allow the variation of loading force, F, with respect of two axes of the geometric reference, such as the variation of angular coordinates φ_y and φ_z (see fig. 1, a).

The scheme and the corresponding measuring instruments of a deformation, U, have to allow the measuring of deformation components, X, Y, and Z, with respect to geometric reference (see fig. 1, b). In the case of an angular deformation, these components have to be obtained for at least two points (see fig. 1, c).

For example, for the case of MSA of normal lathes, the measuring scheme can be without pieces attached to main spindle (see fig. 2, a), with shaft 1 (see fig. 2, b), or with chuck 2 and shaft 1 (see fig. 2, c). Application points of force, A_F, and deformation, A_U, can be different (see fig. 2) or identical (see fig. 3), and may be considered in different positions.

Two rigid shafts that may be used for MSA loading are shown in fig. 3. The rigid shafts are composed of two assembled parts: body, 1, and spherical part, 2. Body 1 has a spherical surface, *a*, for deformation measuring, and part 2 has a spherical surface, *b*, for applying the loading force. 3 holes, *c*, in part 2 permit the access of the gauges used for deformation measuring. Surfaces *d* and *e* are used for assembling the two parts.

A loading-measuring scheme for MSA is shown in fig. 4.

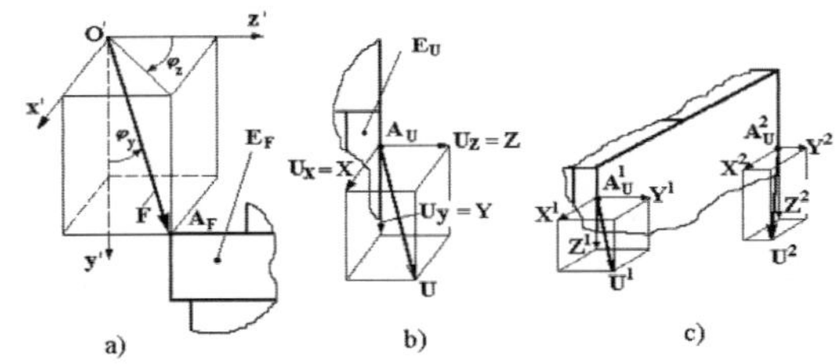

Fig. 1. Schemes for measuring loading and deformation.

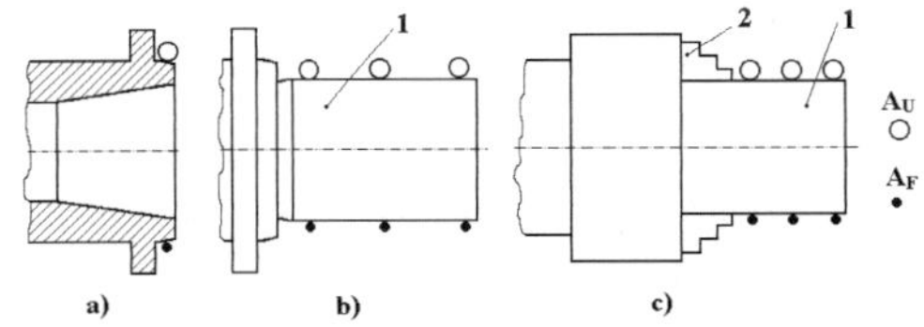

Fig. 2. Loading schemes of MSA.

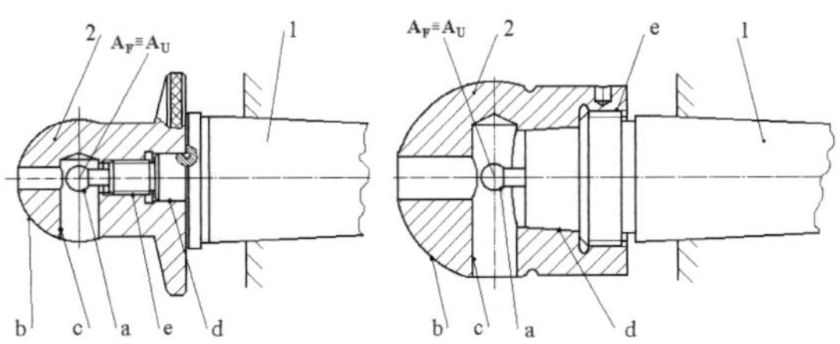

Fig. 3. Rigid shafts for loading MSA.

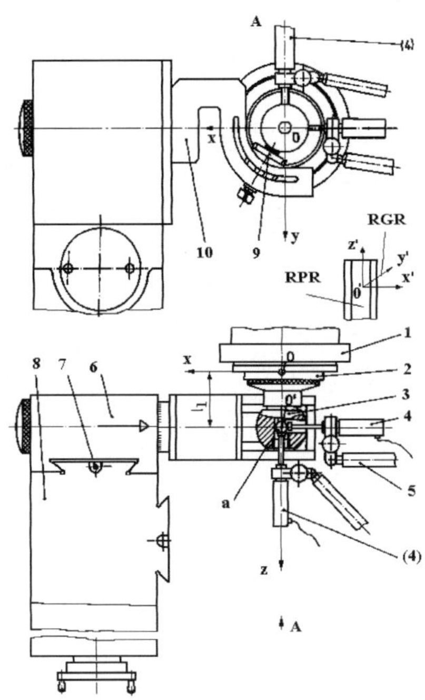

Fig. 4. A practical loading-measuring scheme for MSA.

The scheme in fig. 4 includes the MSA 1 and a rigid shaft composed of body 2 and the spherical element 3. MSA can be loaded on different directions. The deformation is measured for the force's application point ($A_F = A_U$) because of the spherical head of the body that transmits its deformation to three inductive gauges 4. The inductive gauges are fixed upon the normal lathe's bed with the magnetic holders 5. The loading force is realized by a dynamometric device that includes a prismatic part 6 and a rotating part 10. Part 6 is fixed to the cutting tool's support 8 of the lathe. In order to have a correct position on *y*-axis for the dynamometric device, there is used an adjusting screw 7. The rotating part 10 supports the force detecting element 9. The loading of MSA can be realized at different axial positions, l_1. RPR and RGR are relative physical reference and relative geometrical reference associated to the normal lathe. RPR is the lathe's bed and RGR is placed on the lathe's bed.

3. EXPERIMENTAL DATA

The experimental conditions are as follows:
- Normal lathe: SNA500;
- System's state: repose;
- Main spindle's assembly is equipped with the right hand side shaft in fig. 3 ($A_F = A_U$);
- RPR: lathe's bed;
- RGR: O'x'y'z';
- Axial position: $l_1 = 65$ mm;
- $F = F_{st}$: F_x, F_y, F_{xy}, F_{xyz} [daN];
- $U = U_{st}$: X, Y [μm].

Experimental data are presented in tables 1, 2, and 3.

F_x [daN]	X [μm]		F_y [daN]	Y[μm]	
	incr.	decr.		incr.	decr.
0	0	0.9	0	0	2.3
50	10.5	11.1	50	10	15.5
100	19.2	20.1	100	19.5	24.7
150	25.8	27.1	150	24	28.6
200	33.2	33.7	200	26.9	31.7
250	38.1	39.4	250	29.9	34.1
300	45.1	45.9	300	33	36.2
350	49.9	50.5	350	34.3	36.3
400	52.1	52.1	400	36.3	36.3

Tab. 1. F_x and F_y loadings.

F_{xy} [daN]	X [μm]		Y[μm]	
	incr.	decr.	incr.	decr.
0	0	2	0	2.7
50	7.4	10	11.1	11.6
100	15.3	20.1	20.5	28.6
150	22.7	28.4	29.1	37
200	30.6	34.1	38.7	44.9
250	37.2	38.5	47.5	50.2
300	40.5	41.1	53.2	55
350	41.3	42	55.4	57
400	42.4	42.4	57.2	57.2

Tab. 2. F_{xy} loading ($\varphi_y = 30°$).

F_{xyz} [daN]	X [μm]		Y[μm]		Z[μm]	
	incr.	decr.	incr.	decr.	incr.	decr.
0	0	0.9	0	2.7	0	1.7
50	6.9	7.8	6.6	11.9	0.5	3.6
100	16.4	18.4	14.3	21.6	2.1	5.3
150	24	27.1	23.5	31.7	3.6	7.2
200	32.8	36.5	33	41.1	4.5	7.8
250	40.9	45.9	42	50.6	5.8	8.2
300	48.1	53.4	50.2	58.4	6.8	8.7
350	55.1	60	58.1	65.1	7.8	9
400	62.3	62.3	70	70	9.1	9.1

Tab. 3. F_{xyz} loading ($\varphi_y = 30°$ and $\varphi_z = 75°$).

In table 1, $F_{st} = F_x$ and $F_{st} = F_y$, respectively. In table 2, $F_{st} = F_{xy}$ and $\varphi_y = 30°$. In table 3, $F_{st} = F_{xyz}$, $\varphi_y = 30°$, and $\varphi_z = 75°$.

Using the data in table 1, MSP rigidities in daN/mm are as follows:

$$K_{xx} = \frac{F_x}{X} = \frac{400}{0,0521} = 7678\ ;\ K_{yy} = \frac{F_y}{Y} = \frac{400}{0,0363} = 11019 \quad (1)$$

Using the data in table 2, MSP rigidities in daN/mm result as follows:

$$K_{x30} = \frac{F_{xy}}{X} = \frac{400}{0,0424} = 9434\ ;\ K_{y30} = \frac{F_{xy}}{Y} = \frac{400}{0,0572} = 6933 \quad (2)$$

Using the data in table 3, MSP rigidities in daN/mm are:

$$K_{Sx} = \frac{F_{xyz}}{X} = \frac{400}{0,0623} = 6420\ ;\ K_{Sy} = \frac{F_{xyz}}{Y} = \frac{400}{0,0669} = 5979 \quad (3)$$

$$K_{Sz} = \frac{F_{xyz}}{Z} = \frac{400}{0,0091} = 43956\ ; \quad (4)$$

where S stands for the direction of spatial force F_{xyz}.

4. CONCLUSION

Determination of the rigidity of normal lathes and of the main spindle's assembly (MSA) of lathes claims for a particular work environment, which refers to the structure of technological manufacturing systems, references, states, loading, and strains.

The paper presents loading - measuring schemes for main spindle's assembly of normal lathes and some research devices. So, a dynamometric device for 3D loading has been realized, and is described in the paper. In order to materialize the application points of forces and deformations, two rigid shafts with spherical heads were also realized and presented in the paper.

The paper presents a unitary methodology for experimental determination of static rigidity for main spindle's assembly of normal lathes. The methodology can be extended to other components of normal lathes or to other technological manufacturing systems.

5. REFERENCES

Tonoiu, S. & Iliescu, M. (2002). Determination of static rigidity for main spindle's assembly of normal lathe, *Proceedings of the 4th workshop "Human Factor and Environmentalist"*, Katalinic, B. (Ed.), pp. 105-106, ISBN 3-901509-37-2, Kosice-Slovakia, December 2002, DAAAM International, Vienna

Tonoiu, S. (1999). *Contributions on the study of machining technological systems rigidity*, Ph.D. Thesis, POLITEHNICA University of Bucharest, 1999, Romania (in Romanian)

Tonoiu, S.; Dulgheru, L.; Catană, M. & Purcărea, M. (1997a). Methods for experimental determination of static rigidity for machining technological systems, *Proceedings of the 9th International Conference on Machine Tools*, pp. 501-508, ISBN 973-31-1139-2, Bucharest-Romania, 1997, Ed. Tehnica, Bucharest (in Romanian)

Tonoiu, S.; Purcărea, M. & Catană, M. (1997b). Considerations on contact rigidity and dumping in machining technological systems, *Proceedings of the 9th International Conference on Machine Tools*, pp. 509-516, ISBN 973-31-1139-2, Bucharest-Romania, 1997, Ed. Tehnica, Bucharest (in Romanian)

Weck, M.; Eckstein, R. & Schafer, W. (1989). Methods for determination of machine tool static rigidity, *Mechanik*, No. 4, 1989, pp. 125-129 (in Polish)

Annals of DAAAM for 2008 & Proceedings of the 19th International DAAAM Symposium, ISSN 1726-9679
Editor B. Katalinic, Published by DAAAM International, Vienna, Austria 2008
Make Harmony Between Technology and Nature, and Your Mind will Fly Free as a Bird

THE ANALYZE OF THE LIFE CYCLE FOR A PNEUMATIC CONNECTOR

TOPLICEANU, L[iliana]

Abstract: *The selection of the technology and materials to make any product is an important decision and is important to consider, in this choice, all the environment aspects. The Life Cycle Assessment is the most comprehensive methodology to find out about the environmental impact. Sometimes, for a faster answer a simplified LCA tool is useful to use. The paper presents this type of methodology in the case of the pneumatic connector.*

Key words: *connector, impact, environment, eco-indicator*

1. INTRODUCTION

Life Cycle Assessment (LCA) is a tool for the systematic evaluation of the environmental aspects of a product or service system through all stages of its life cycle, "from cradle to grave". LCA provides an adequate instrument for environmental decision support. The International Organization for Standardization (ISO) has standardized this framework within the ISO 14040.

2. LCA FOR THE EVALUATION OF THE PRODUCT ENVIRONMENTAL IMPACT

A product's life cycle starts when raw materials are extracted from the earth, followed by manufacturing, transport and use, and ends with waste management including recycling and final disposal (SETAC, 1993). At every stage of the life cycle there are emissions and consumption of resources. The environmental impacts from the entire life cycle of products and services need to be addressed. To do this, life cycle thinking is required.

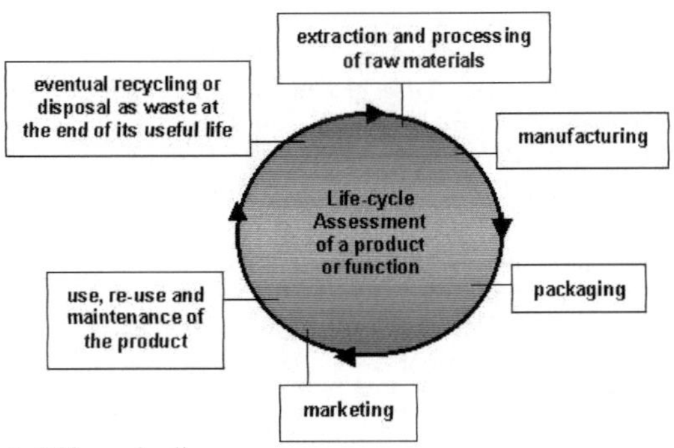

Fig. 1. Life cycle diagram

By definition, LCA is a technique, which assists in assessing the environmental aspects and potential impacts associated with a product, by:

- Compiling an inventory of relevant inputs and outputs of a product system.
- Evaluating the potential environmental impacts associated with the selected inputs and outputs.
- Interpreting the results of the inventory analysis and impact assessment phases in relation to the objectives of the study (ISO, 1997).

3. SIMPLIFIED LCA

LCA is a holistic and iterative process and needs time to be realized, so, in some situation it is recommended to use a Simplified LCA. The SETAC EUROPE LCA Screening and Streamlining Working Group defines simplified LCA as: **Simplified LCA** is an application of the LCA methodology for a comprehensive screening assessment i.e. covering the whole life cycle but superficial e.g. using generic data (qualitative and/or quantitative), standard modules for transportation or energy production, followed by a simplified assessment i.e. focusing on the most important environmental aspects and/or potential environmental impacts and/or stages of the life cycle and/or phases of the LCA and a thorough assessment of the reliability of the results.

The aim of simplifying LCA is to provide essentially the same results as a detailed LCA, but with a significant reduction in expenses and time used. Simplification presents a dilemma, however, since it is likely to affect the accuracy and reliability of the results of the LCA. Thus, the primary object of simplification is to identify the areas within the LCA which can be omitted or simplified without significantly compromising the overall result.

Simplification of LCA consists of three stages (Vigon& Jensen, 1995) which are iteratively interlinked:

- Screening: Identifying those parts of the system (life cycle) or of the elementary flows that are either important or have data gaps
- Simplifying: Using the findings of the screening in order to focus further work on the important parts of the system or the elementary flows.
- Assessing reliability: Checking that simplifying does not significantly reduce the reliability of the overall result.

The terms "Screening LCA" and "Streamlined LCA" are often used as synonyms for a simplified LCA. However, a clear distinction should be made. Screening as a part of the simplification procedure can help to identify the parts (or life cycle stages) of a product system that can be left out in a simplified LCA.

In principle, a screening LCA which already has certain parts missing would not be capable of identifying all key issues, as it does not cover the full life cycle or all environmentally important aspects (Curran, 1996). In other words, the screening step in simplified LCA should be comprehensive in coverage, but may be superficial in detail. Screening LCA is for instance used in environmental labeling to identify the environmental "hot spots", i.e. the areas where labeling criteria are assumed to have the greatest effects. Another use of screening LCAs is to identify the processes where emissions of particular interest occur in the life cycle. This procedure can be applied for many other environmental management tools, e.g. risk assessment, to assess whether unwanted effects actually will occur.

4. THE SIMPLIFIED LCA IN A CASE OF A PNEUMATIC CONNECTOR

We exemplify the use of the simplified LCA method in the case of a pneumatic connector and we want to study the environmental impacts of production and utilization for 10 years of 1000 pieces. One connector is composed of nipple, bolt nut and lock nuts.

Among the large numbers of software tools (approximately 37) it was choused SimaPro 7.1 and the BUWAL 250 database. Corresponding of these processes and using the Eco-indicator 99 method, attached of the SimaPro, results the environmental impacts. The Eco-indicators 99 project was commissioned by the Dutch Ministry of Housing, Spatial Planning and the Environment (VROM) in order to update the Eco-indicators 95 methodology. The primary differences of the Eco-indicators 99 method compared to the previous version lie in the characterisation and weighting steps. The effects are allocated to three endpoints categories and these are: human health, ecosystem quality and resources, with units of measurements directly indicating the damages to these endpoints.

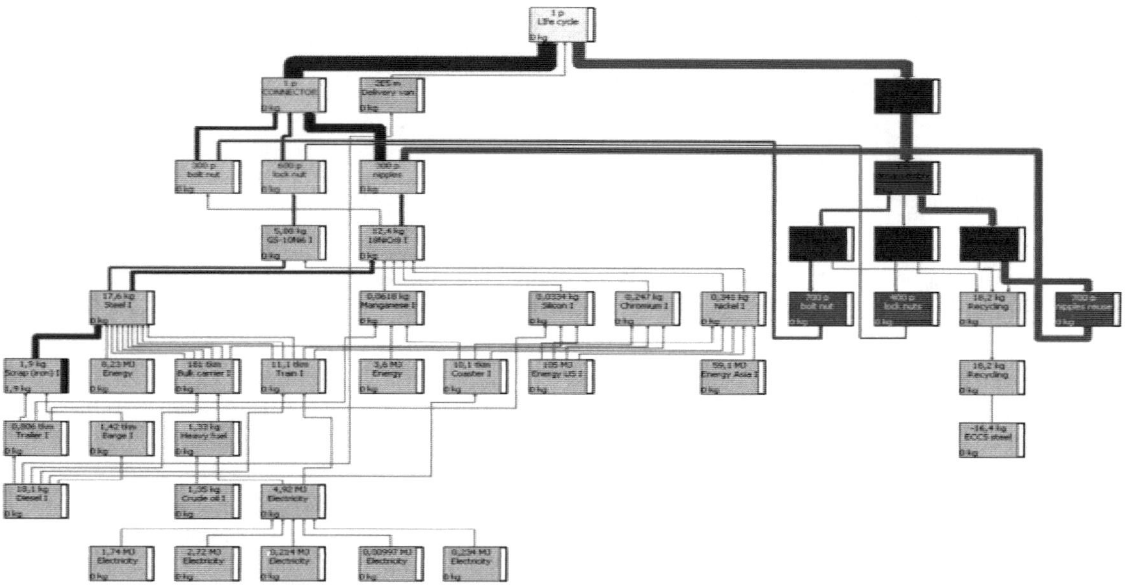

Fig. 2. Connector life cycle tree.

The utilization of a simplified LCA imposes a careful selection of the boundaries and inputs of the system (Blair &Wyckoff 1989). The first step of the method consists in the creation of the life cycle of connectors choosing the materials for which of them. With the help of SimaPro we obtain the environment diagram's impact of the connector assembly. The life cycle of a product can't be discussed without to consider the end life. In the same time it is important to reuse the different components of assembly to diminish the consumption of raw material and energy and the environmental impact. Considering all of this we estimate 10 years long life for the connectors and we can create the life cycle tree. We take nipples and consider that after 10 years 30% of them are recycled and 70% are reutilized and the same for bolt nut- made from the same material. The screw nut, created from a different material, is recycled 60% and reutilized 40%. Now, the life cycle tree can be created (Fig.2). The study was made considered that the pieces are transported with a delivery fan on 100 kilometers. With all of these data, and using Eco-Indicator 99 method the impacts of the connectors' life cycle is shown (Figure 3).

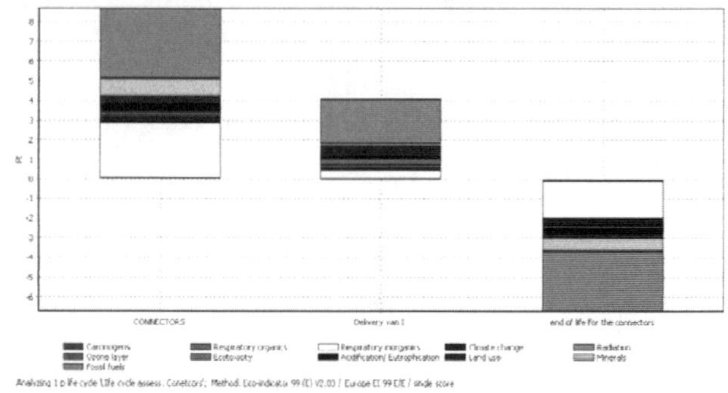

Fig. 3. Life cycle assessment impacts diagram

5. CONCLUTIONS

The simplified LCA can give important information about the environmental impacts of a products or processes and help to choose the right material, technology or transportation way (Arthur, 1989). Very important is to take care of the inputs and outputs considered and the boundary in time and space of the system. Even if a great number of elements can be directly observed and measured in the plants, developing a direct analysis of all elements is not always suitable or possible and the usages of the Simplified LCA is recommended.

6. REFERENCES

Arthur, W. (1989) *Competing Technologies, Increasing Returns and Lock-in by Historical Events*. The Economic Journal

Blair, P., Wyckoff, A. (1989) *The Changing Structure of the U.S. Economy: An Input-Output Analysis.* In Frontiers of Input-Output Analysis, eds. R. Miller, K. Polenske, and A. Rose. New York: Oxford University Press.

Curran, M. A. (1996). *Environmental life-cycle assessment.* New York, USA: McGraw-Hill.

SETAC. (1993). *A Conceptual framework for life cycle impact assessment.* Pensacola, USA: SETAC

Vigon, B. W., & Jensen, A. A. (1995). *Life cycle assessment: data quality and database practitioner survey.* Journal of Cleaner Production, 3, 135–141.

Annals of DAAAM for 2008 & Proceedings of the 19th International DAAAM Symposium, ISSN 1726-9679
Editor B. Katalinic, Published by DAAAM International, Vienna, Austria 2008
Make Harmony Between Technology and Nature, and Your Mind will Fly Free as a Bird
*Young Researcher and Scientist Paper / * Supervisor, Mentor*

EXPLOSION SURFACING OF Mg ALLOYS WITH ALUMINIUM AND CrNi AUSTENITIC STEEL

TRSO, M[artin]; TURNOVA, Z[uzana]; NESVADBA, P[etr] ; TURNA, M[ilan]*

Abstract: *The article is devoted to explosion welding of combined metals by explosion. Within experimental activity, technical aluminium was deposited on Mg alloy of AZ 63 type and CrNi austenitic steel (18Cr8Ni type) was deposited on Mg alloy type. Parallel situation of welded materials was employed in welding operation. Welding parameters and conditions were determined respecting chemical composition and condition of the above-mentioned metals. Explosive type of S 30 Semtex was applied for welding.*
Bimetals were fabricated in co-operation with VÚPCH Explosia Inc. Pardubice-Semtín. Quality of joints was evaluated by ultrasonic defectoscopy, optical microscopy, and microhardness measurement through the joint boundary and by X-ray microanalysis.
Key words: *Explosion welding, Mg alloys, Optical microscopy, Microhardness measurement, X-ray microanalysis.*

1. INTRODUCTION

Present technical practice nowadays often employs joining of materials with different chemical and physical properties (Nesvadba, 2007). There is in many cases endeavour to attain, that one of materials (metals) would in optimum measure meet the requirements laid on chemical, physical and mechanical characteristics desired for a certain application (Turňa, 2007). The proportion of application of diverse technical materials is shown in Fig. 1.

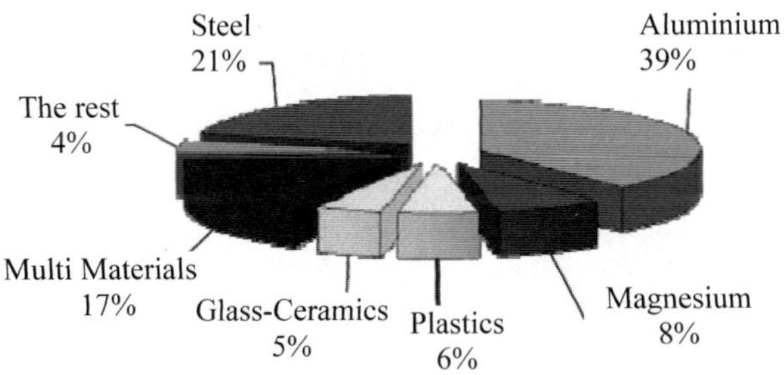

Fig. 1 Percentage proportion of technical materials used in practice (Middeldorf, et al., 2005).

2. MAGNESIUM ALLOYS

Magnesium and its alloys are very interesting materials for technical practice, mainly for the field of transport technology. Commercially used magnesium alloys have specific weigh around 1 700 kg.m^{-3}, what is by about 35 % lower value than in case of Al alloys and by 75 % lower than in case of steels. Chemical composition of AZ 91 alloy is shown in Table 1.

Chemical composition (wt.%)							
Mg	Al	Zn	Mn	Si	Cu	Ni	balance
base	8.3-9.7	0.35-1.0	Min. 0.13	0.50	0.10	0.03	0.30

Table 1. Chemical composition of AZ 91alloy.

3. EXPERIMENTAL

Used materials: AZ 91, AZ 63, technical Al (99.5 %), CrNi austenitic steel (18Cr8Ni). Accelerated materials were Al and CrNi austenitic steel.Parallel location of welded materials and Semtex explosive was used in welding. Weld edges were mechanically machined and degreased prior to welding. The quality of welded joints was assessed by:

- ultrasonic defectoscopy
- optical microscopy (by the study of macrostructure and microstructure)
- microhardness measurement across the bimetal boundary
- X-ray microanalysis of the zone of bimetal boundary.

Figs. 2 and 3 show the macrostructures of 18Cr8Ni-AZ 91 (Belokostolský, 2007) and Al-AZ 63 (Demianová, 2007) bimetals and Figs. 4 and 5 show the microstructure of mentioned bimetals.

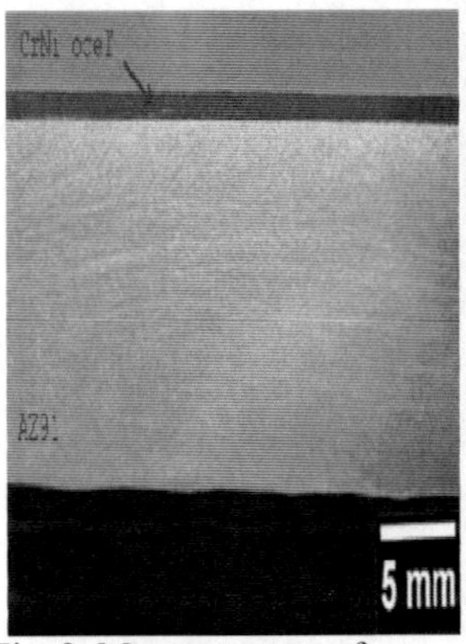

Fig. 2, Macrostructure of bimetal18Cr8Ni-AZ 91 in detonation direction (v_d = 1956.35 m/s).

Fig. 3, Macrostructure of bimetal Al-AZ 63 (v_d = 2027 m/s).

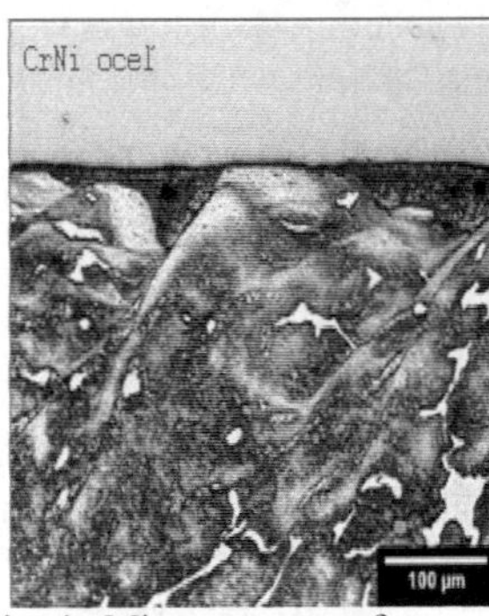

Fig. 4, Microstructure of welded joint of 18Cr8Ni-AZ 91 in detonation direction.

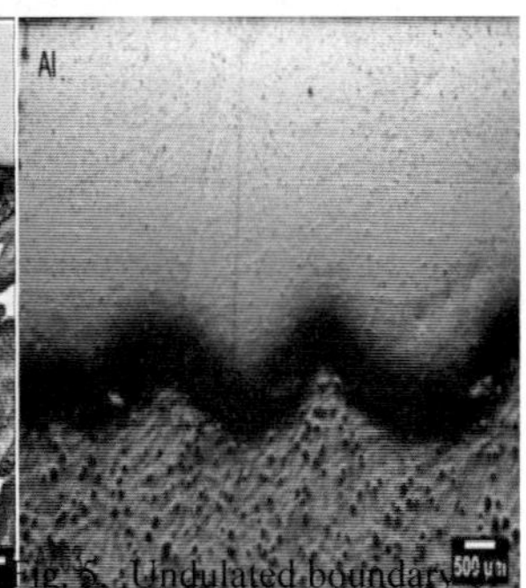

Fig. 5, Undulated boundary of Al-AZ 63 bimetal.

Microhardness measurement was performed across the boundary of combined joint Al -Mg alloy type AZ 63. Location of indents in microhardness measurement is marked in Fig. 6. The values of measured hardness are plotted in graph (Fig. 7).

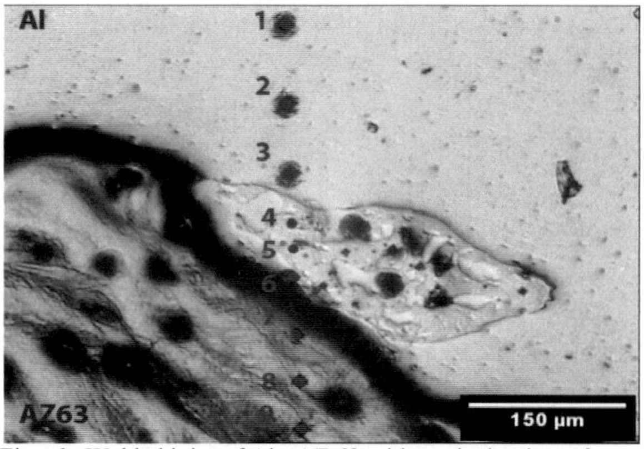

Fig. 6, Welded joint of Al - AZ 63 with marked points of microhardness measurement (Demianová, 2007).

X-ray microanalysis was performed on energy-dispersion microanalyser type JEOL JXA-80 at FCHPT STU Bratislava.

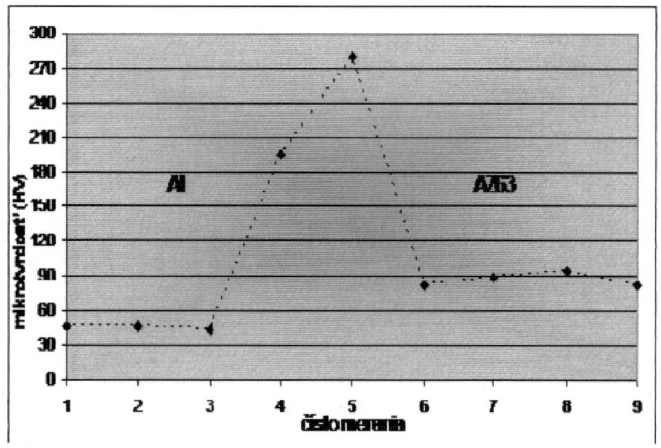

Structural situation in the boundary zone of bimetal was studied in details. Figs. 8 and 9 show: situation of line microanalysis and the courses of concentration change of Al and Mg over the bimetal boundary.

Fig. 7, Course of microhardness in normal direction to boundary of Al - AZ 63 joint (Demianová, 2007)..

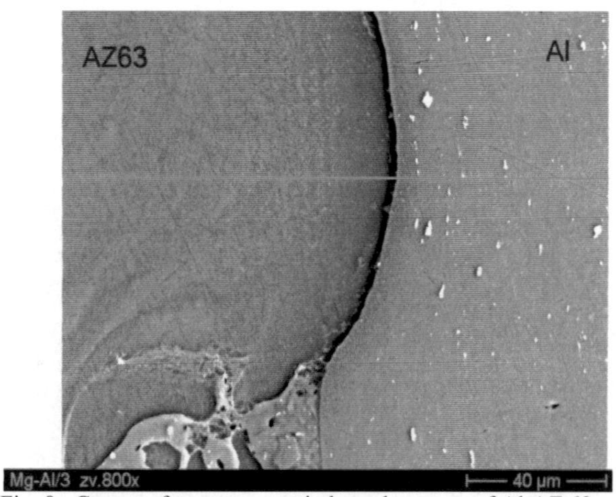

Fig. 8, Course of measurement in boundary zone of Al-AZ 63 bimetal, by X-ray microanalysis (Demianová, 2007)..

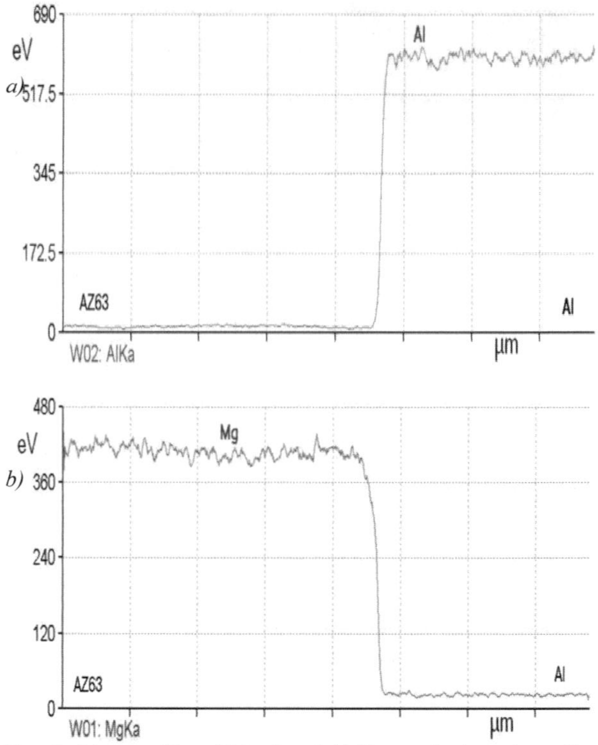

Fig. 9, Line profiles of Mg a) and Al b) over the boundary of welded joint of Al - AZ 63 (Demianová, 2007).

4. CONCLUSIONS

The aim of work was to suggest the technology, parameters and conditions of explosion welding Mg alloy type AZ 63 with aluminium and austenitic steel type (18Cr8Ni) with magnesium alloy type *AZ 91*.

Quality of joints was assessed by defectoscopy and metallographically. Macrostucture and microstructure was studied and microhardness measurement in joint boundary was also performed. Very good quality of joints was attained in case of Al - AZ 63 bimetal. An undulated boundary in direction of detonation speed $v_d = v_k$ (welding speed) was observed, which is the guarantee of quality joint.

Boundary quality was in details assessed and proved by X-ray microanalysis.

In case of austenitic CrNi steel - AZ 91 alloy combination it is still necessary to optimise the parameters and conditions of welding.

This work is a part of VEGA *no. 1/3191/06* project.

5. REFERENCES

Belokostolský, T. ; (2007) *Design of welding technology for joining Mg alloy with selected metals. Thesis*, Welding Department at ÚVTE , MTF STU, Trnava.

Demianová, K. ; (2007) *Design of welding technology for joining Mg alloy with aluminium.* Thesis, Welding Department at ÚVTE , MTF STU, Trnava.

Middeldorf, K. ; Herold, H. ; von Hofe, D. ; (2005) *Trends in Joining. IIW International Conference Benefits of New Methods and Trends in Welding to Economy, Productivity and Quality*, Czech Welding Society, Prague.

Nesvadba, P. ; (2007) *Explosion welding.* Lectures presented at FS VUT Brno.

Turňa, M. ; (2007) *Special methods of solid state welding.* Lectures presented at IWE. FS CVUT Prague.

Annals of DAAAM for 2008 & Proceedings of the 19th International DAAAM Symposium, ISSN 1726-9679
Editor B. Katalinic, Published by DAAAM International, Vienna, Austria 2008
Make Harmony Between Technology and Nature, and Your Mind will Fly Free as a Bird

APPLYING SIGNALS OF CONTROL SYSTEM IN TOOL WEAR MONITORING

UDILJAK, T[oma]; MULC, T[ihomir] & CIGLAR, D[amir]

Abstract: *Open structure of modern digital controls open up new possibilities and prospects in monitoring of machining systems and processes. In many cases, the combination of digital plant signals and internal data of the machine control system, along with advanced methods of signal analysis can replace the external control systems. The integration of process control software module into the machine control system allows fast reactions should there be any process disturbances, without any additional hardware expansion. This paper studies the sensitivity of signals contained in the control system to the cutting tools wear processes in face turning.*

Key words: *monitoring, control, forces, turning*

1. INTRODUCTION

Over the recent years, machine tools and production systems have gone through dramatic changes caused to the greatest extent by the development of information technology and flexible automation. Control of high-speed machines is a very demanding task which requires powerful and efficient systems of process monitoring and diagnostics. Basic conditions for good management of machining monitoring include knowledge about the process state and undertaking of adequate actions (Isermann, 1994, Stute, 1981, Mulc et al. 2004). The diversity of input parameters, constant development of new materials, geometry and new tool materials, as well as higher machining speeds, with simultaneous setting of increasingly strict standards regarding safety, complicate the control process monitoring, so that process monitoring remains one of the most demanding tasks in further development of machining devices. Controller significantly affects the capabilities of machining systems. It offers some possibilities for establishing simple, inexpensive and easy-to-manage monitoring systems. Thus, standard functions library can be supplemented by specific modules for tool monitoring in order to provide the users with new possibilities in the field of "on-line" process monitoring with regard to avoiding collision, breakdown, overload and monitoring of tool wear. However, the sensitivity and applicability of such systems in various processing conditions need to be checked for every individual case.

2. ESTIMATION OF THE FEED CUTTING FORCE

2.1 Modeling of the Feed and Main Drive System

Reliability of monitoring process is strongly dependent on quality of information extracted from the measuring signals. With adequate procedure it is possible to extract the influence of inertial forces, influence of friction of moving components (eq. guideways, bearings, spindles), and influence of static coefficient of friction. Mechanical chain of servo axis consists of slider, transmission and electro motor, Fig. 1.

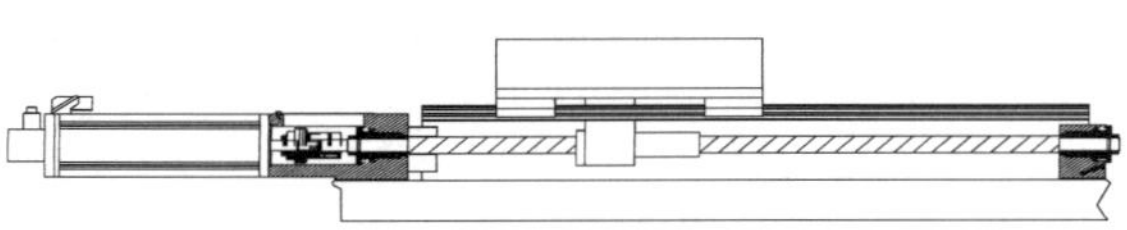

Fig.1. Mechanical chain of the servo axis

Taking in consideration the momentum of inertia, momentum of friction on the motor side, and momentum of load, the mechanical equation for the i-th axis could be as follows:

$$T_{mi} = J_{mi}\ddot{q}_{mi} + T_{Tmi} + T_{pmi} \qquad (1)$$

The momentum of load, reduced to motor axis, is expressed as:

$$T_{pmi} = \frac{1}{N_i} T_{poi} \qquad (2)$$

T_{Tmi} represents momentum of friction on the motor side, T_{poi} momentum of load, and N_i transmission ratio. Momentum of load consists of inertial part, friction resistance T_{Ti}, gravitational influence G, and cutting resistance force T_{ri}:

$$T_{poi} = J_{oi}\ddot{q}_{oi} + T_{Ti} + T_{ri} + G \qquad (3)$$

For horizontally arranged feed drives, the influence of gravitation could be neglected, G=0. The same could be done for vertically arranged feed drives with compensation (electrical or mechanical) of slider weight.

The friction is very complex phenomena and it is difficult to express it mathematically. According to [Isermann, 1994], the losses caused by friction could be presented as follows:

$$T_{Ti} = T_{Toi} sign(\dot{q}_{0i}) + T_{T1i}\dot{q}_{0i} + T_{T3i}\dot{q}_{0i}^3 \qquad (4)$$

$T_{T0i} sign(\dot{q}_{0i})$ - dry friction, (Coulomb's friction)

$T_{T1i}\dot{q}_{0i}$ - viscous friction depending on velocity and temperature

$T_{T3i}\dot{q}_{0i}^3$ - friction in guideways

Motor torque must overcome the resistance cutting forces, inertial forces and friction forces. The resistance cutting forces are:

$$T_{ri} = K_{fi} F_{ri} \qquad (5)$$

where coefficient K_{fi} depends on transmission. Having in mind the transmission ratio it could be written:

$$q_{mi} = N_i q_{oi}, \; \dot{q}_{mi} = N_i \dot{q}_{oi}, \; \ddot{q}_{mi} = N_i \ddot{q}_{oi} \qquad (6)$$

By including the equations (2), (3), (4), (5) and (6) in equation (1) we obtain:

$$T_{mi} = J_{ei}\ddot{q}_{mi} + D_{ei}\dot{q}_{mi} + T_{Te0i} sign(\dot{q}_{mi}) + \frac{1}{N_i} T_{T3i}\dot{q}^3 + \frac{1}{N_i} T_{ri} \qquad (7)$$

Motor torque is proportional to current:

$$T_{mi} = K_{mi} I_{ai} \tag{8}$$

Including equation (8) in equation (7) results in mathematical model of servo axis, (9):

$$K_{mi} I_{ai} = J_{ei}\ddot{q}_{mi} + D_{ei}\dot{q}_{mi} + T_{Teoi} sign(\dot{q}_{mi}) + \frac{1}{N_i} T_{T3i}\dot{q}^3 + \frac{K_{fi}}{N_i} F_{ri} \tag{9}$$

Each particular realization of servo axis needs estimation of influence of individual load component.

2.2 Estimation of the Cutting Force

For proper estimation of parameters by using control system signals, equation (9) should be modified in matrix form:

$$y_i(t) = \Psi_i^T(t)\theta_i \tag{10}$$

$$y_i(t) = I_{ai}(t) \text{ - output vector} \tag{11}$$

$$\Psi_i^T(t) = \left[\dot{\omega}_{mi}, \omega_{mi}, sign(\omega_{mi}), \omega_{mi}^3, 1\right] \text{ - measuring vector} \tag{12}$$

$$\theta_i^T = \left[\frac{J_{ei}}{K_{mi}}, \frac{D_{ei}}{K_{mi}}, \frac{T_{Teo1}}{K_{mi}}, \frac{T_{T3i}}{N_i K_{mi}}, \frac{F_{ri}}{K_{mi} N_i / K_{fi}}\right] \tag{13}$$

-vector of parameters for i-th joint

Estimation of n unknown in parameters vector for i-th joint demands acquisition of at least n measuring values in various measuring points: t=kT_0, k=1, 2,.....4,. By applying least square method, the equation (10) gives following solution:

$$\hat{\Delta} = \left[\Phi^T \Phi\right]^{-1} \Phi^T Y \tag{14}$$

Equation (14) is suitable for on-line estimation of the dynamic parameters. The estimated parameters are consecutively compared with previous values by applying equation 15:

$$d\hat{\Delta} = \frac{(|\hat{\Delta}_n| - |\hat{\Delta}_{n0}|)}{\hat{\Delta}_{n0}} \tag{15}$$

The process could be monitored by analyzing the magnitude and direction, $sign(d\hat{\Delta}_n)$, of deviation of the estimated parameter. The change of parameter is a change generated in the observed system which does not cause immediate system failure, but has negative impact on system behavior.

3. EXPERIMENT PLANNING

The aim of the experiment is to determine the sensitivity of drive system parameters to tool edge wear in process of fine turning. The turning unit was fitted within the unit of special machine tool controlled by Siemens digital control system (Mulc et al., 2004), Sinumerik 840D, Fig. 2.

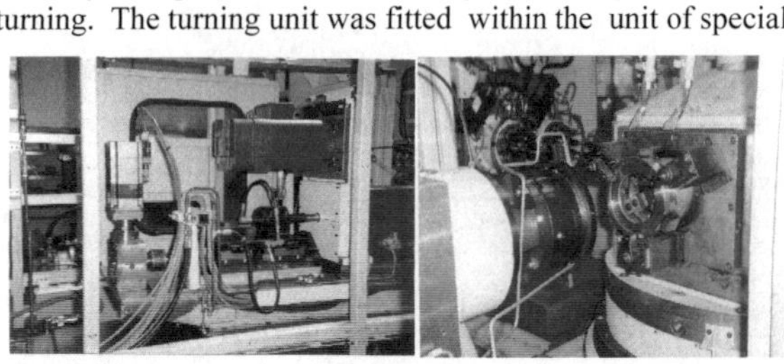

Fig. 2. Unit for fine turning (SAS-Zadar)

4. RESULT ANALYSIS

During the period of automatic working of the system (till the tool wear out) the system stores the correction values, i.e. tool wear values suitable for wear curve, Fig. 3.

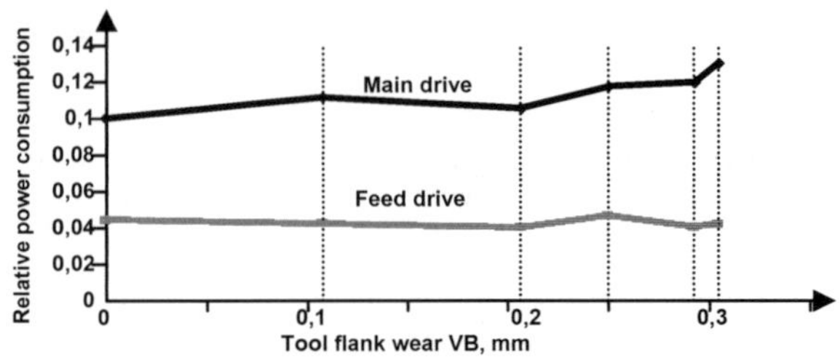

Fig.3. Dependence of relative power consumption on tool wear

It has been shown that tool wear mostly influence main spindle, i.e. main drive (Cuppini et al., 1990). Current signal of the main drive shows increase of approx. 30% during increase of tool wear. It is a significant increase and could be used for judging on tool condition. The experimental results confirm that feed drive signal is not suitable for the judging on tool condition in fine turning. Because the share of power necessary to prevail friction and mechanical loses in feed drive is very high, it is not possible to isolate the power changes in feed drive that are consequence of increase in tool wear.

5. CONCLUSION

Open control with digital drive system open up new possibilities and prospects in "on-line" monitoring of the machining systems (Shi & Gindy, 2007). By combination of digital drive systems with additional information from the control system, methods of isolating characteristic features from the signal and sophisticated data processing technologies, high reliability and safety of signal analysis is achieved. Further development of such systems, and the methods of isolating characteristic features, together with technologies of artificial intelligence (Brezak et al., 2004), presents a significant step towards realizing a simple, reliable, and user friendly way of monitoring of cutting tools and machining processes.

6. REFERENCES

Isermann R.(1994). *Monitoring and Diagnostics*, VDI-Verlag, Dusseldorf

Stute G.(1981). *Control of Machine Tools*, Carl Hanser Verlag Munchen Wien

Brezak, D., Udiljak, T. Mihoci, K., Majetic, T., Novakovic, B., Kasac, J.(2004). Tool Wear Monitoring Using Radial Basis Function Neural Network, *International Joint Conference on Neural Networks & IEEE International Conference on Fuzzy Systems*, Budapest 2004

Cuppini D., D'Errico G., Rutelli G.(1990). Tool wear monitoring based on cutting power measurement, *Wear*, 139(1990), 303-311.

Damodarasamy S., Raman S.(1993). An inexpensive system for classifying tool wear states using pattern recognition, *Wear*, 170(1993), 149-160

Mulc, T., Udiljak, T., Čuš, F.(2004). Milfelner, M. Monitoring Cutting Tool Wear Using Signals from the Control System, *Strojniški vestnik*, 50(2004)12, ISSN 0039-2480, p. 568-579

Shi, D., N.N. Gindy. (2007). Industrial Applications of Online Machining Process Monitoring System, *IEEE/ASME Transactions on Mechatronics*, Vol. 12, No. 5, October 2007, 561-564

Annals of DAAAM for 2008 & Proceedings of the 19th International DAAAM Symposium, ISSN 1726-9679
Editor B. Katalinic, Published by DAAAM International, Vienna, Austria 2008
Make Harmony Between Technology and Nature, and Your Mind will Fly Free as a Bird

INTEGRATED DESIGN AND MANUFACTURING SYSTEM FOR BLADES MOULD

UDROIU, R[azvan]

Abstract: *The aim of this research is to develop an integrated design and manufacturing system for composite blades moulds. Feature technology can facilitate the integration of design and the subsequent applications such as engineering analysis, process planning, machining and inspection. In this work it is proposed a helicopter blade mould model based on constructive-technological features (CTF). CTF wants to help the CAD/CAPP/CAM integration in an innovative way. Using design by features and features recognition techniques it is proposed a software package that achieve the integrated design and manufacturing of the blades mould. This contribute to the automation of the design-manufacturing process of the blades mould and to the reduction of product development cycle time.*
Key words: *CAD/CAM, mould, feature, software, milling.*

1. INTRODUCTION

Most discrete parts are manufacturing by using a near net shape process NNS (Machover, 1996) such as forging, die casting, stamping, injection moulding or autoclave moulding. These processes use die and moulds to impart the NNS geometry on an incoming shapeless material or a preformed work piece. The material that the parts are from can be metal, polymeric or composite. The production of complex parts from composites materials corresponds to a long manufacturing process compared to a mechanical part obtained by chip removing machining. The manufacturing process of complex parts from composites materials is characterized by three design functions: the design of external shape of finished part, the design of internal structure of composite part and the definition of bond tool or mould, and two manufacturing functions: the mould machining and the production composite parts. A helicopter blade is a complex part made of composite materials. In this work the focus is on the design of external shape of the helicopter blade, definition of the mould and mould machining.

2. FEATURES AS INOVATIVE ELEMENTS LINKING DESIGN AND MANUFACTURING

Actual production of mechanical parts is diversified the slogan being higher quality product obtains cheaper and without pollution the nature. In the same time the concurrence at the market place is better. These things claim with acuteness to enhancing the companies' competitiveness by cutting the product-development time, reducing the time-to-market and improving the quality of design using CAD/CAM/CAPP/CAE systems integrated in concurrent engineering philosophy. Concurrent engineering helps bring higher-quality products to market faster and a less cost by involving all the players, including customers and suppliers, at the earliest stage possible.

During the design process, different engineers take information about the design, and manipulate it in order to generate the new information required for the development of the new product. Various engineering representations of the designed product are created and used for different engineering tasks. These represent, for example, a manufacturing engineering viewpoint, a structural analysis viewpoint, a process engineering viewpoint, and so on (Lee et al., 2003).

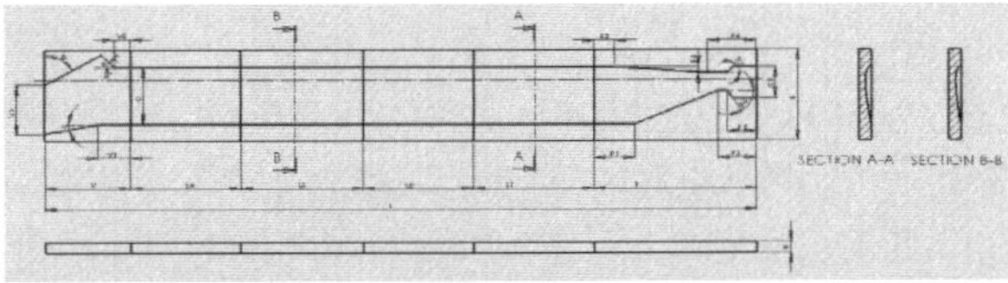

Fig. 1. Generic model of helicopter blade mould

Feature technology, is expected to be able to provide for an adequate basis for the integration of design and the subsequent applications such as engineering analysis, process planning, machining and inspection. Features may be generated in three distinct ways (Udroiu, 2003), (Ibrahim & McCormack, 2005), known as "feature recognition", "design with features" and "interactive feature definition/ identification".

Currently, three main views are discerned on how to obtain application features, such as manufacturing features, analysis and inspection features, from a product model. Today, the focus is on the integration of feature based design with feature based manufacturing. In (Chen et al., 2007) the machining feature model of a design part, needs to be built from the design feature model of the part during its design process and automatically adjustable to maintain the consistency of these two models.

In this work I proposed two features models to design a helicopter blade mould: the model of the helicopter blade based on design features and the model of helicopter blade mould based on "constructive-technological features" (CTF), fig.1. CTF wants to help the CAD/ CAPP/ CAM integration in an innovative way. It is proposed the following definition for CTF: "a constructive-technological feature is a geometric shape that has attached several minimal geometrical configurations for milling (MGCM) for each manufacturing operation (roughing, semi-finishing, finishing, rest-material) and a set of information regarding the cutting process, like as cutting strategy, NC files, etc". The concept of minimal machining configuration (Mawussi et al., 2000) for prismatic machining (MMC) and later MGCM (Udroiu, 2003) for complexes parts allow in the context of integrated design, to inform the designer about the machining ability of the part from a cutting tool point of view or allow to inform automatically the NC engineer about the optimal cutting tools use for machining a feature of a part. In the same way the fact that a feature can be machined only with one MGCM whose cutting tool is relatively expensive makes it possible draw the attention of the designer to the portions of the part which increase the total machining cost.

3. INTEGRATED CAD/CAM SYSTEM THE CORE OF CONCURRENT ENGINEERING

The work presented in this paper falls under the step of integrated engineering. The purpose is principally the development of an integrated design and manufacturing system for helicopter blade.

The software resources of system proposed is composed from the software packages realized by the author, some CAD/CAM commercial software and software resource of NovaMill CNC machine.

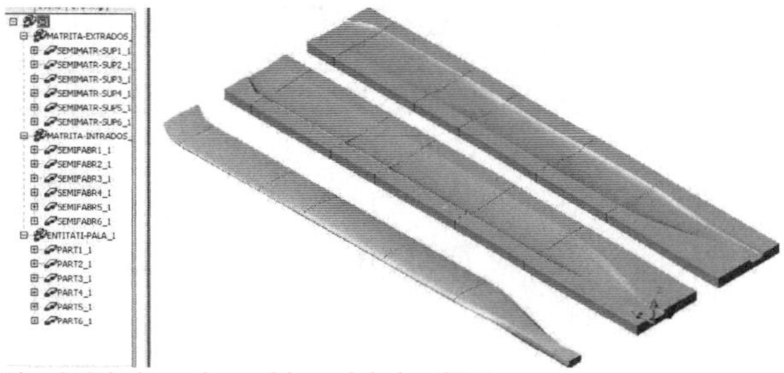

Fig. 2. Blade and mould models by CTF

The innovative software packages proposes in this paper are the following: PACPE (Computer Aided Design of Helicopter Blade), PACAPE (Computer Aided Design of Helicopter Blade Mould), DACRAF (Computer aided establish of cutting parameters concerning complex surface milling) and ECU-Virtual (Complex Milling Features – Virtual). PACPE and PACAPE are written in Visual Lisp and Dialog Control Language programming language and are implemented in Autodesk Mechanical Desktop. DACRAF and ECU-Virtual are written in Visual Basic programming language and can be added like add-in application to commercial CAD/ CAM system PowerShape/ PowerMill of Delcam. The software resource of milling system NovaMill is called Virtual Reality CNC Milling of Denford.

PACPE allows the automatic and parametric design of a helicopter blade, using features modelling technique. Thus, it was defined a hierarchic system of aerodynamics complex macro-features (ACMF), features and an aerodynamic airfoils database. The PACPE interface between computer and user is constituted from a pull down menu, an ensemble of dialogue boxes and an own captor of errors in view to avoid the wrong data introduction.

Starting with the 3D virtual model of the blade, obtained by PACPE, the software system PACAPE (Udroiu, 2007) generates the 3D model of the blade mould. This mould is realised using design by features and a hybrid modelling (solid modelling combining with surface modelling). The mould is composed from two parts (fig. 2): the upper mould part and the bottom mould part. The system of features is composed from constructive-technological macro-features (CTMF) and features. It is defined four categories of CTMF: CTMF of root, CTMF with constant airfoil, CTMF with variable airfoil and CTMF of tip. In the process of blade mould modelling it was defined three 3D digital models such as: billet mould model, raw upper-lower mould model and mould model.

An example of experimental blade and its mould obtained by PACPE and PACAPE is presented in figure 1. The upper and the lower part of the mould are composed from six CTMF.

ECU-Virtual and DACRAF solve the main issues of the manufacturing process of helicopter blade mould. ECU-Virtual is articulated around three following modules: VTOOL (Virtual specification of cutting tools), AsistGPM (Assistance to the development of machining process) and StrategEnt (Machining features strategies and generation of CN file). The input data for ECU-Virtual is a CAD neutral file, such as IGES or STEP. All CTMF of the mould can be exported from DWG parametric file format in a neutral file format.

The purpose of VTOOL module is the virtual and automatically identification of optimal cutting tools for each CTMF of the mould and determination of their main parameters. After that, VTOOL allows the automatic generation of these cutting tools in PowerMill commercial CAM system. VTOOL use features recognition technique.

The next module of ECU-Virtual, named AsistGPM provide assistance to the development of machining process. This module automatically determines the size of raw material for each CTMF, generate the block of raw material and check in if the CTMF can be manufactured with the selected CNC machine. It can be specify, that AsistGPM has incorporate the main specifications (spindle speed, travels, table size etc) about a lot of CNC machine. The establishing of cutting parameters concerning complex surface milling (CTMF of the mould) it is obtained using the software system DACRAF. These modules, VTOOL, AsistGPM, DACRAF provide information necessary to the forth module named StrategEnt.

StrategEnt establish in automatic way the machining strategies for each CTMF of helicopter blade mould and generate automatically the tool paths and NC files. The optimum strategies proposed within StrategEnt software are focused on 2 ½ and 3 axis milling machine.

4. CONCLUSION

This research is focused on the integration of design, process planning and manufacturing process of moulds for composite blades, using like tool the concept of constructive-technological feature. Thus, it was development an integrated system for helicopter blades. Briefly, this system allows the automatic design of a helicopter blade and its mould, virtual specification of the optimal cutting tools, the establishing of cutting parameters concerning complex surface of the mould, choosing in automatic way the machining strategies for each CTMF of helicopter blade mould and generate automatically the tool paths and NC files. The mains benefits of the proposed system are the automation of the process and the reduction of product development cycle time. The validation of this system was done using the 2 ½ axis NovaMill CNC milling machine from Denford. The further researches will be focus on the development of a knowledge based system for assembling and manufacturing of internal structure of the composite blade.

5. ACKNOLEDGEMENTS

This study was supported by a National Research and Development Program of Romania, CEEX Grant 41/ 07.10.2005. The author expresses his gratitude to all partners for the fruitful collaboration.

6. REFERENCES

Chen, Z., M.; Gao, S., M.; Li, W., D. (2007). An approach to incremental feature model conversion, *International Journal of Advanced Manufacturing Technology*, Vol. 32, No.1-2, pp.99–108, ISSN 0268-3768

Ibrahim, R., N.; McCormack, A., D. (2005). Robustness and generality issues of feature recognition for CNC machining, *International Journal of Advanced Manufacturing Technology*, Vol. 25, No.7-8, pp. 705–713, ISSN 0268-3768

Lee, K., H.; McMahon, C., A.; Lee, K., H. (2003), Design of a feature-based multi-viewpoint design automation system, International Journal of CAD/CAM, Vol.3, No.2, pp. 67-75, ISSN 1598-1800

Machover, C. (1996). *The CAD/CAM handbook*, McGraw Hill, ISBN 0-07-039375-3, USA

Mawussi, K., Duong, V., H., Kassegne, K. (2000). Determination of the parameters of cutting tools in integrated design of products, *CD-ROM Proceedings of the IDMME International Conference*, Canada

Udroiu, R. (2003). Design and manufacturing of complex shape parts, PhD Dissertation, Transilvania University of Braşov, Romania

Udroiu, R. (2007). Computer aided design of tooling for aerospace composite parts, *Proceedings of the 8th international conference "Modern Technologies in Manufacturing"*, Technical University of Cluj-Napoca, 4-5th October, ISBN 973-9087-83-3, pp. 449-452, Cluj Napoca, Romania.

Annals of DAAAM for 2008 & Proceedings of the 19th International DAAAM Symposium, ISSN 1726-9679
Editor B. Katalinic, Published by DAAAM International, Vienna, Austria 2008
Make Harmony Between Technology and Nature, and Your Mind will Fly Free as a Bird

REQUIREMENTS TO PRECISION LASER CUTTING PROCESSING OF REFRACTORY METALS

UEBEL, M[artin]; BUERGER, W[olfgang]; SCHOELE, H[olger]; STOERZNER, F[alko]; MEUDTNER, A[ndreas] & STIBRITZ, G[uido]

Abstract: *Precision laser cutting of tungsten composite materials is an important problem in the production of refractory metallic products. To meet the demanding requirements of the users, an improvement on the surface quality as well as the reduction of structure size and production costs are necessary.*
Processing of test parts is performed on systems with pulsed Nd:YAG rod lasers. Main emphases of the examinations are on the observation of the suitability of different cutting gases as well as on the optimization of technological parameters of the process of laser cutting, of the laser beam parameters and the gas current.
To enable statements about the suitability of other laser plants, examinations were conducted at a fiber laser system and a laser system of waterjet leaded ray.
Key words: *tungsten, refractory metals, laser, precision, cutting*

1. INTRODUCTION

The work examination serves the method development of the precision laser cutting of refractory metallic materials; in particular of tungsten composite materials.

In previous use of these methods the cut area qualities, structure size and production costs did not suffice for many applications. Due to bad cut edge quality, rip danger as well as large warmth influence zones caused by high process energies, tolerances and other requirements of the user are not or only conditionally met.

The market of refractory metals is constantly growing. However, the potential of this market cannot be exhausted due to the insufficiently developed technologies available today.

Essential improvements of the project are the attainable cutting quality and the rise of the process safety compared to the level of technology and to the competitors.

The task in this work examination consists of planning cutting attempts for the parameter optimization, of conducting laser cutting experiments and evaluating the cut tests. Furthermore material tests are processed by outside companies. These have to be evaluated in the context of the work examination and to be compared to each other.

2. EXPERIMENTAL SETUP

The materials to be examined in the project are metal based alloys with a very large amount of tungsten. A very high temperature is characteristic for these alloys. The materials are processed with a thickness of 0.1 mm to 2.3 mm as metal sheets. The products manufactured of the semi-finished products find new possibilities for the application in different industrial areas, e.g. in medical engineering, the beam technology, the automobile industry and the aerospace.

In this work examination pulsed Nd:YAG laser plants were used. The laser cutting systems are optimized especially for fine cutting use. High-precision work pieces are mainly manufactured of metal based alloys like high-grade steel, aluminium and copper-based alloys; beside those, some non-metals are used as well. In addition to the available lasers at the work place, material tests were processed by outside companies at other laser plants.

The features of the cut opening to be determined were carried out with the VDI guideline 2906 (cut opening quality when cutting, cut and perforate work pieces made of metal). Of special significance for the work examination are conicity, roughness, the warmth influence zone, cut opening breadth and burr formation. For measuring the roughness a stylus instrument is at disposal. Interesting characteristic quantities are Ra, Rz and Rt. Also a variety of further surface identification values can be determined next to this one for the work examination according to DIN 4287.

For the visual evaluation stereo microscopes are used. It is of great importance for the regulation of the cut breadth both on the beam admission side and on the beam leaving side and the burr height. Material ejection and warmth influence zones can be swiftly recognized on the top side of the test. The micro examination views the cross section of the cut opening. The cross section test of the cuts will be poured in curable plastics and then polished. The test can be examined now with regard to the geometric qualities via microscope.

3. EXPERIMENTS AND RESULTS

Using special cutting gases different surface results can be achieved. The cutting gases which are of special interest are air, nitrogen and oxygen.

The quantitative evaluation of the cutting area of the cut tests is carried out with the stylus instrument. When compared, oxygen cut shows the lowest roughness (figure 1). Air and nitrogen produce a larger roughness. Under the reaction with oxygen the surface profile is burned off or oxidized and therefore it is smoothed. At the air gas flow the effect, however, is considerably lower due to the smaller amount of oxygen, and at the nitrogen cut it does not occur. While the oxidation of the metal also brings in an additional thermal energy which contributes to melting the tungsten alloy with the oxygen, the melt is made stiffen and cools down fast using the nitrogen gas current.

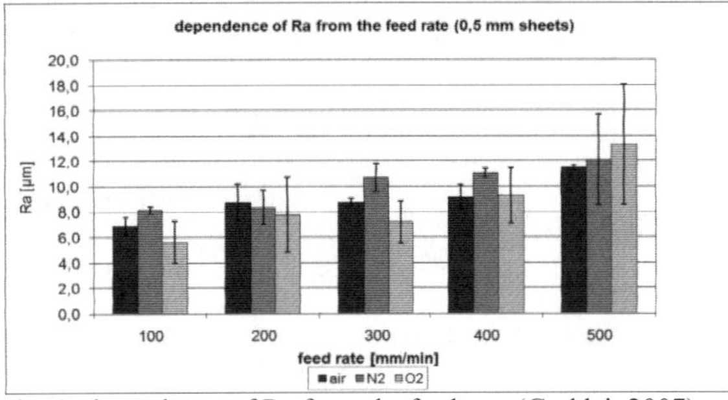

Fig. 1: dependence of Ra from the feed rate (Guddei, 2007)

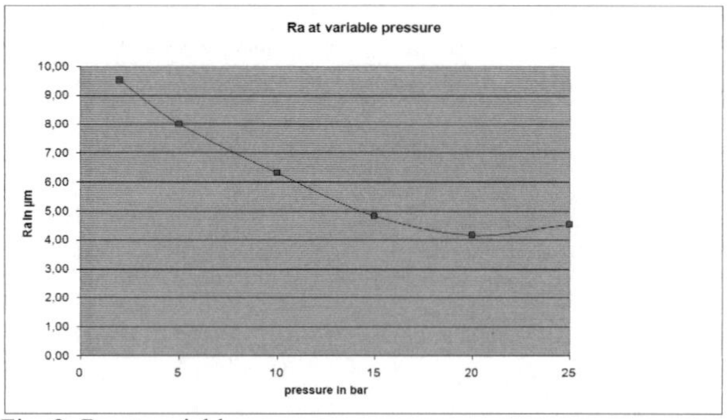

Fig. 2: Ra at variable pressure

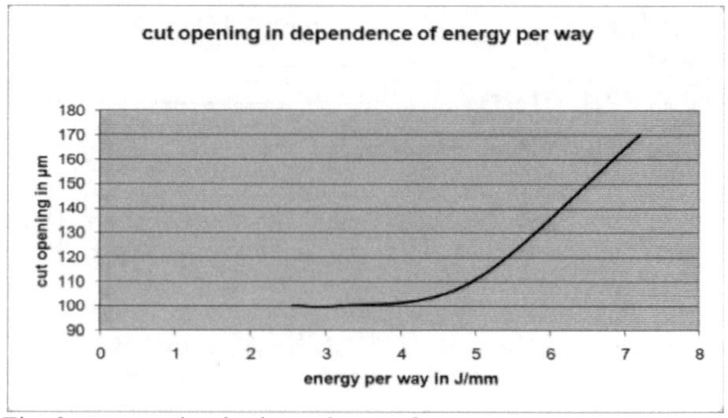

Fig. 3: cut opening in dependence of energy per way

While the nitrogen gas current has developed an oxide free surface and a low discoloration, the oxygen gas current surface is strongly oxidized and discolored by the warmth entry into the surrounding structure. The tempering tarnishes are obvious there considerably stronger and penetrate into the material further than at the nitrogen cut.

The feed rate is varied for this examination from 100 mm/min up to 500 mm/min. Apperently, the roughness increases with an increasing feed rate. As seen in figure 1, this statement applies to the three different cutting gases similarly. To be able to determine the cut surface geometry, micro examination specimen were made. While the melt is blown clean with oxygen at the slow cut tests below and the burr accumulates with a narrow, long form at the underside, the melt of the fast cut tests is deposited mainly at the top side and in the cut opening. The burr matured at the top side of the test is distributed by the cutting gas flow conditionally very broadly and can be removed only with difficulty. Therefore it is not favourable under the given conditions to cut with high speed, although this leads to higher production costs. Rather a melt which produces a narrow, well removable burr has to be striven for.

The examinations to the gas pressure were carried out at the metal sheets with a thickness of 1 mm. Air served as cutting gas. The pressure varied from 2 bar to 25 bar in steps of five. During the tests it became obvious that it is difficult for a low pressure to separate the tests from the metal sheet. While at a low pressure a lot of melt is blown up, at a high pressure no melt is taken up on the top side or in the cut opening. The rough value Ra is compared in illustration 2 with the cutting gas pressure for the measuring of the metal with a thickness of 1.0 mm. The most favourable pressure is between 15 to 20 bar. The possibility of increasing the amount of air continues to exist by the use of a 1.5 mm nozzle. For later use in the batch production such a high cutting gas consumption is not economical

Another aim of this work was to reduce the energy per way as much as possible to achieve the minimal amount of energy necessary for cutting 0.1 mm thick material. To do this the parameters pulse frequency, charging voltage and pulse duration were adapted. It was observable that the cut opening was narrower with relieving energy per way. The pulse frequency had to be increased simultaneously, selected so that the pulses overlap sufficiently to produce a continuous opening. The dependence of the cut opening breadth of the way energy is represented in figure 3. The experiment shows that it is important to reduce the energy per way at a high pulse frequency as far as possible. On the one hand, the cut opening gets narrower thereby; on the other hand, the interaction zone can be reduced.

In contrast to experiments before, a fiber laser system of the company IPG Laser GmbH and a water jet conducted laser system of the company Synova S.A. are used here now. The two essential advantages of the fiber laser are the high feed rate and the almost inexistent melt deposit in the cut opening. A very big warmth influence zone is to be mentioned as an essential disadvantage. The water jet conducted method of laser cutting is based on the principle of the total reflection of the laser beam in a jet of water. When comparing the method of the company Synova with other methods, it achieves the highest cut edge qualities (Saint-Ghislain, 2007). Melt deposits are not available at all, the cut surfaces are parallel and the opening is very narrow. Deficits can be seen in the low processing speed and the complicated handling.

4. CONCLUSION

The knowledge of the process of cutting tungsten alloys won by the experiments and their evaluation is suitable to considerably improve the cut area quality in comparison to the qualities obtained before the work examination.

A rise of the cutting gas pressure up to 20 bar is of special significance for the improvements mentioned. Economic considerations must be included since such a rise of the pressure is achieved only with larger technical and financial effort. Broader possibilities for improvement of the cutting area quality for the thicker metals (= 0.5 mm) lie in the combination of high pulse frequency and simultaneous maximization of the application of energy per way. By contrast, it is favourable for thin metals (0.1 mm and 0.2 mm) to choose the energy as small as possible to ensure a narrow cut opening. Regarding the feed rate it can be noticed that this depends substantially on the material strength. The focus situation must be selected in order to produce low tapering at the cut opening. This ensures an optimal removal of the melt from the opening and provides a straight cutting edge simultaneously. The evaluation of the cut samples of the two outside companies has shown that other laser systems are in principle suitable as well to cut tungsten alloys with considerably higher feed speeds or with substantially better cutting area qualities. A decision on the actual efficiency of these plants can only be taken after further examinations which take the economic parameters into account.

5. REFERENCES

Forschungsprojekt der Thüringer Aufbaubank (2007): *Rationalisierungsmöglichkeiten bei der Laserpräzisionsbearbeitung von refraktären Metallen. Vorhabenskurzbeschreibung PräMet.* University of Applied Sciences Jena

Guddei, Yvonne (2007*): Zwischenbericht PräMet.* University of Applied Sciences Jena

Norm DIN EN ISO 4287 (1998). *Oberflächenbeschaffenheit: Tastschnittverfahren*, Deutsches Institut für Normung e. V., Berlin

Richtlinie VDI 2906 (1994). *Schnittflächenqualität beim Schneiden von Metallen,* Verein Deutscher Ingenieure, Düsseldorf

Saint-Ghislain, Michael (2007): *Thungsten plate cutting by Laser-MicroJet®*, Synova SA, Ecublens

Annals of DAAAM for 2008 & Proceedings of the 19th International DAAAM Symposium, ISSN 1726-9679
Editor B. Katalinic, Published by DAAAM International, Vienna, Austria 2008
Make Harmony Between Technology and Nature, and Your Mind will Fly Free as a Bird

FLEXIBLY PREPARING OFFERS BY ADAPTING FUNCTIONAL RESOURCES

UNGER, K[atja]; MILITZER, J[oerg]; ZIMMERMANN, M[atthias]; GAESE, T[homas]; WINKLER, S[ebastian] & TEICH, C[ornelia]

Abstract: *To cope with intensified competition along global production networks, prompt and precise offers concerning technologic or economic feasibility for an individual product demand is needed. Therefore, this paper will first briefly present the three-columned research project: the customer's performance request, the functional description of company resources, and enhanced ERP modules to match both sides. The structure and the functional adaptation of the resource model is considered more detailed in this paper, because it enables the scalability of manufacturing resources by the use of technical or economic production functions (PF). This will help implementing a firm's real technological capability, being able to answer product requests and transfer them into expected planning environment automatically.*
Keywords: *features, resources, functions, manufacturing, capability*

1. INTRODUCTION

Not only for bigger companies but also for small and medium enterprises (SMEs) the attraction of new business is an essential success factor in highly competitive value-added production networks. In recent years mass customization (MC) has advanced to a key strategy to offer customers more choice (Ismail et al., 2007). The present trend to individualization means matching customer's individual needs, forces especially SME's to improve their manufacturing agility and product flexibility finally to benefit from cross-company cooperation and planning.

This paper deals first with a short delimination of the project and the methodical framework for an automated offer preparation process with short response times. To improve the needed manufacturing agility, the structure and functionality of the functional resource model is outlined in the second part.

2. RESEARCH PLACEMENT

Some publications refer to manufacturing capability for computer aided engineering (CAE) support and differentiate between manufacturing resources, processes and strategies (Molina et al., 1995). This context is similar to ours, but strategies are outlined because of the operative focus of the offer preparation process.

As one of the key components CAD/CAM integration, CAPP systems, and feature technology were already contemplated (Gao & Huang 1996). The latter is an important element for the presented functional resource model the authors are developing within a research project between the University of Applied Sciences Zwickau and the Chemnitz University of Technology.

The following drawing illustrates the tripartite focus of research (figure 1), whereas the functional resources implementation is embedded on the right side. Together with the description of a product demand on the left side ERP modules are enhanced for generating, scheduling and evaluating process variants to benefit from adapting functional resources.

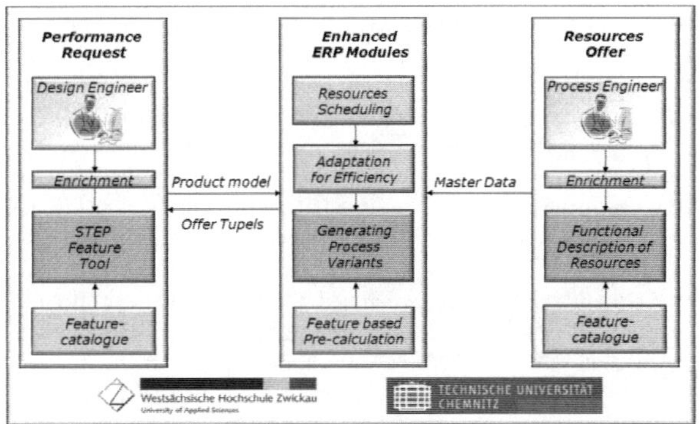

Fig. 1. Three-columned research project.

2.1 Performance Request

A customer's performance request consists of general customer data, a CAD model with required product characteristics, and order-specific information, e.g. price expectations and quantity. Already done is product model design based on STEP, enabling standardized product model data (Amaitik et al., 2005). The named project uses features to limit the design engineer according to the abilities of manufacturing resources, already published by the concept of SBDCR (Teich et al., 2008a).

2.2 Resources Offer

The right pillar is the functional description of company resources, where the available operating facilities and equipment with their capabilities are represented in an innovative manner – as character-changing competence cells. An evaluation of technical and economic PFs will be provided in this paper and helps to integrate static and dynamic influences in offering.

2.3 Enhanced Planning Environment

To make a use of modeling resources and customer requests in a functional way, enhanced ERP modules are being created to present concrete offers to customers.

First module is the generation of process variants (PVP) and production plans, combined with an initial measurement by a matrix of economical requirements, done by ant colony algorithms for multi-criteria optimization problems.

The module pre-calculation calculates the costs feature-based. Each generated variant is scheduled with extracted orders from an existing ERP/SCM system (Teich et al., 2008b) to evaluate capacity restrictions of the resources.

Finally, the customer gets a representation of possible solutions as concrete offers, varying by price, date and delivery conditions as well as product implementations.

The greatest benefit for a flexible preparation of offers comes from the adaptation of PFs, because benchmarked process variants will be rescheduled after time, factor and parameter-oriented scalability. This adaptation is considered more deeply in the next paragraph.

3. ADAPTIVE RESOURCE MODEL

To explain the structure and functionality of the adaptive resource model, the authors refer to Competence Cells (CC) which is an educational term for representing abilities (Erpenbeck & Heyse 1999). These abilities have to be transferred to a manufacturing context, where PFs are used to describe the scalable behavior of CCs to flexibilize the preparation of offers.

3.1 Structural Assignment

The company resources – so called character-changing competence cells (4Cs) – have an indirect impact on the products' attributes. More precisely, the resources are bundled in Competence Cells and characterised by their parameters (general and specific ones) according to the manufacturing method or to its physical properties and conditions. These parameters with interdependences are encapsulated in so called manufacturing features. A design feature in its product-related view mainly refers only to its geometric and technological attributes, not to adaptive parameters of the production process.

As mentioned before, enhanced ERP modules generate and schedule PVPs. To break this alternative plan down, intermediate products (IPs) with Features are connected by different process categories. They are to disaggregate into manufacturing stages and those again to process elements characterized by manufacturing methods that constitute a granular operation of a work schedule (figure 2).

3.2 Functional Adaptation

Prior research of the theory of production already located the field of applicable PFs. Most relevant is the technical funding of GUTENBERG that evolves technical set values as adaptable parameters (Sonntag & Steven 2005).

Based on the structural assignment, technical PFs define each process step and cover consumption factors as input, a process step itself as throughput as and the production rate as output factor (figure 3).

Each process step is characterized by only technical efficient possibilities. However, a technical less efficient variant can be more economical. Evaluating this, it requires adapting the production rate, technical set values, or time to influence the output factors of a process step. In our model the technical PF are to transform into economic ones for re-scheduling and re-evaluating process variants.

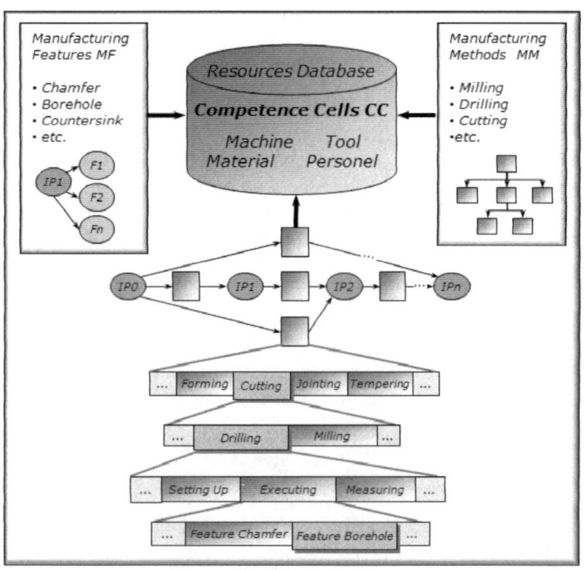

Fig. 2. Assigning CC to MF and to MM

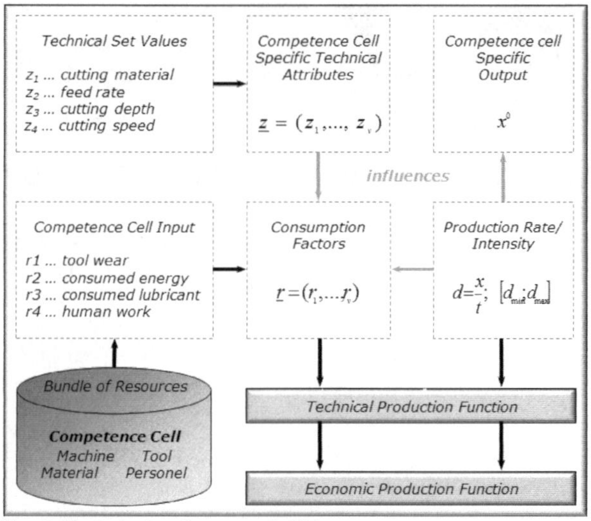

Fig. 3. Technical and economic PF

4. CONCLUSION

The functional resource model should enable an efficient and dynamic use of manufacturing capabilities, embedded in the framework for generating additional orders with reduced response times in a value-added production network. Present work themes include a prototypic implementation of the framework, the preparation of an UML resource model and filling a database for turning operations/ competences. For further research, the scalability of the resources model to other fields of application will be analyzed and evaluated.

5. REFERENCES

Amaitik, S.M.; Kilic, S.E. (2005). STEP-based feature modeler for computer –aided process planning. In: *International Journal of Production Research*, Vol. 43, No.15, Taylor & Francis Group Ltd., London, p. 3087-3101

Erpenbeck, J.; V. Heyse (1999). The biography of competences. Waxmann Publishing, New York, Munich

Gao, J. X.; Huang, X. X. (1996). Product and manufacturing capability modeling in an integrated CAD/process planning environment, In: *The International Journal of Advanced Manufacturing Technology*, Vol. 11, pp. 43-51

Ismail, H.; Reid, I. R.; Mooney, J.; Poolton, J.; Arokiam, I. (2007). How Small and Medium Enterprises Effectively Participate in the Mass Customization Game, In: *Engineering Management, IEEE Transactions on*, Vol. 54, pp. 86-97

Molina, A.; Ellis, T.; Young, R.; Bell, R. (1995). Modelling Manufacturing Capability to Support Concurrent Engineering, In: *Concurrent Engineering*, Vol. 3, No. 1, SAGE Publications, pp. 29-42

Sonntag, S.; Steven, M. (2005). Technical Fundings of Gutenberg's production function, In: *Quantitative business management*, Physica Publishing, pp. 159-183

Teich, T.; Militzer, J.; Zimmermann, M.; Unger, K. (2008a): SBDCR for Standardization of Customer Requests, In: *Proceedings of the Fifteenth International Working Seminar on Production Economics*, Vol. 1, pp. 495-503

Teich, T.; Militzer, J.; Unger, K.; Gaese, T.; Winkler, S. (2008b): A formalized approach for generating customer offers based on heuristic methods, In: *Logistics and Supply Chain Management: Trends in Germany and Russia,* Proceedings of the German-Russian Logistics Workshop, Publishing House of the Saint, pp. 272-282, Petersburg State Polytechnical University

Annals of DAAAM for 2008 & Proceedings of the 19th International DAAAM Symposium, ISSN 1726-9679
Editor B. Katalinic, Published by DAAAM International, Vienna, Austria 2008
Make Harmony Between Technology and Nature, and Your Mind will Fly Free as a Bird

MATHEMATICAL MODEL FOR STUDY AND ASSESSING OF ADHESIVE ATTRITION PROCESS OF COUPLING ELEMENTS IN SLIDING FRICTION

UNGUR, P[etru]; POP, P[etru]; GORDAN, M[ircea]; GORDAN, C[ornelia] & POP, M[ircea]

Abstract: *In paper has presented certain opinions and appreciations about of quadratic mathematical functions of friction, derivate from classical concept of Lorentz attrition curve, for adhesion attrition in time of coupling elements sliding friction. The Lorentz wear curve W=f(t) presents the evolution of third equation on range-OABC. The new concept given other interpretations of wear functions W=f(B,S) and W=f(t), being description as parabolic functions in time and a parabolic function of wear in time for running in stability wear.*
Key words: *attrition, contact, friction, parabola, sliding.*

1. INTRODUCTION

In specialty papers [Balasiu,1990;Grama,2007; Bowden, 1956;Ceausu,1980], the wear is defined as a destroyed process of surfaces in contact of parts, in relative friction moving one to another, followed by a change of quality geometry and superficial layer properties of coupling materials.

The wear is a process with evolution in time and due to occurring of failure in working of gages, machine tools, installations, limited work precision [Kragelsky,1978]. At friction coupling, as altering of contact surface coupling elements state in relative sliding moving has produced dislocation and removing of material from bodies surfaces by interaction of mechanical, thermal and chemical action [Pavlescu,1983;Tudor,1984]. The wear of friction coupling surfaces can be quantification by static characteristics:

- wear speed-v_w, expressed loss of mass, altering volume or of a dimension in time, being reported of wear at function time-t;
- wear intensity-I_w, expressed by wear to unite sliding length, knowing as non-dimensional linear intensity wear.

The destroy wear due to excessive material removal from surfaces of friction coupling get to increase roughness and its destroyed. This evolution in time of wear it has known as Lorentz wear curve [Shrader et al, 2000].

This curve get by a plotting of wear function W=f(t), has an inflexion point at maxim wear in working, it's complicate used in practice, liked better a line equation of form W_m=f(t) [Ceausu,1980].

The paper has presented the plotting of normal wear curve, accidental or damage wear curve and of wear line, following by a new mathematical model, derivates from them by using correlation of mathematical model and analytical approximation.

The paper's novel consists in new mathematical wear model defined.

2. ASPECTS ABOUT OF LORENTZ WEAR CURVE

After Kragelski, the normal wear of couple in friction has considered normal until the coupling and equipment became inadequate. Continuous operating of device or installation due to passing from normal wear to accidental or damage wear.

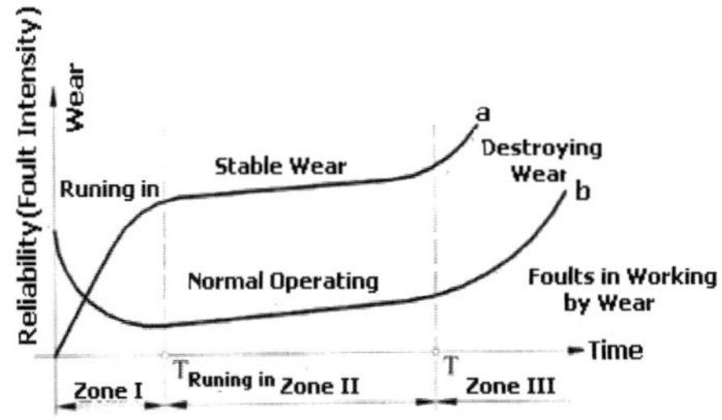

Fig. 1. Mechanical wear of Lorentz curve in time.

Normal wear of operating and damage wear of couplings in sliding friction has depicted in Fig.1, by well-known Lorentz curve W= f(t) [Georgescu,1984].

The limit of attrition or operating of installation or machine has determined by promptitude of wear process deployment of machine's components. The wear of elements in moving depends by: type of wear and moving, cycle moving, normal force-F_N, wear speed-v_w, loading-B, time of loading-t_B, temperature of working-T, etc. In tribo-technical system [Balasiu, 1990], friction and wear has determined by interaction between characteristic elements of friction state and wear device. A process of friction and wear has dependent of structure system-S and load force's group-B, defined as:

$$F_N = f(B,S) \tag{1}$$

$$W = f(B,S) \tag{2}$$

Where, F_N is similar with F_R-friction force from coupling; B-load group; S-structure system and W-wear measurement. In Fig.2 has presented wear curve by loading time and in Fig 3 in detail of Lorentz's wear diagram in time. The normal wear is getting by the relation:

$$W_2 = W - W_1 \tag{3}$$

Where, W-is admissible maxim wear value in working; W_1-running in wear value; W_2-normal wear value in working.

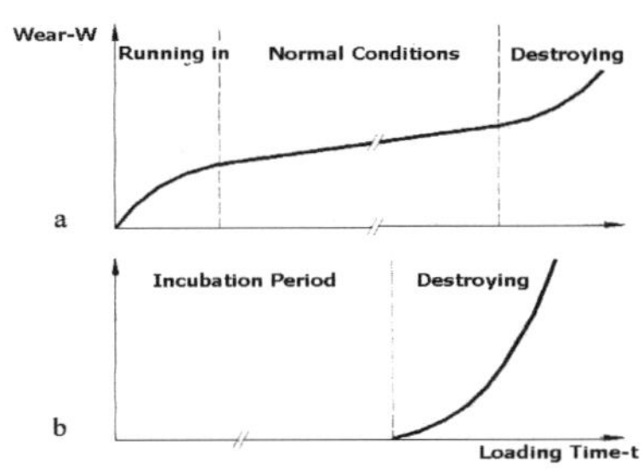

Fig. 2. Wear in function of loading time: a.-normal wear, b.-destroy wear.

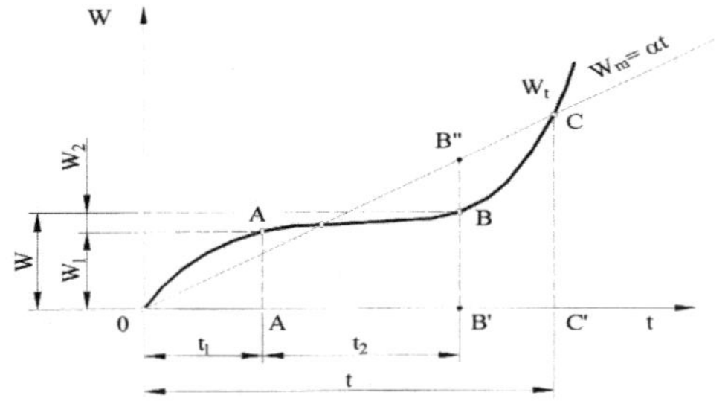

Fig. 3 Detail of Lorentz wear diagram in time.

The wear intensity (Fig.3) has determined with relation:

$$\tan\alpha = \frac{W_2}{W_1} \tag{4}$$

, or:

$$\tan\alpha = \frac{W - W_1}{t - t_1} \tag{5}$$

Where: t-is maxim time in operating, t_1-time in running in, t_2-time in normal operating of equipment. Intensity of wear-tanα depends of surfaces roughness, group of pressing forces-B, wear speed-v_w, material characteristics, lubrication, etc. In calculus, the wear curve W=f(t) is more less used, preferred by line $W_m = \alpha \cdot t$, as:

$$t = \frac{W}{\tan\alpha} + \left(t_1 - \frac{W_1}{\tan\alpha}\right) \tag{6}$$

By analysing of Fig.3 and equation (6), it has resulted:

- if angle-α is small, that wear time-t will be great;
- term $\left(t_1 - \frac{W_1}{\tan\alpha}\right)$ characterized the running in probe;
- if time of running in-t_1 has great value, and wear after finished running in-W_2 is small, that maxim time of operating-t will grow.

Because of complexity wear functions W= f(B,S) and W=f(t), these can be described more easy by quadratic complex functions. A mathematic model of wear as a parabolic, with a low slope has presented in following.

3. NEW MATHEMATICAL MODELS FOR ATTRITION

The wear workpieces depend by quality of cutting surface, materials used, lubrications, and of speed and specific pressure of parts in moving, etc. The attrition goes to altering of shape, volume and weight of parts, without modification of physics-chemical properties of material in friction, exhibited by adhesion, abrasion and erosion. After [Bowden &Tabor, 1956], the abrasion wear is characterising by micro-plastics deformations occurring and by cut of thin metallic layers from abrasive hard particles from surfaces.

Another variant of a wear function of quadratic equation (Fig.4) has given by:

$$W = \frac{1}{2p} t^2 \tag{7}$$

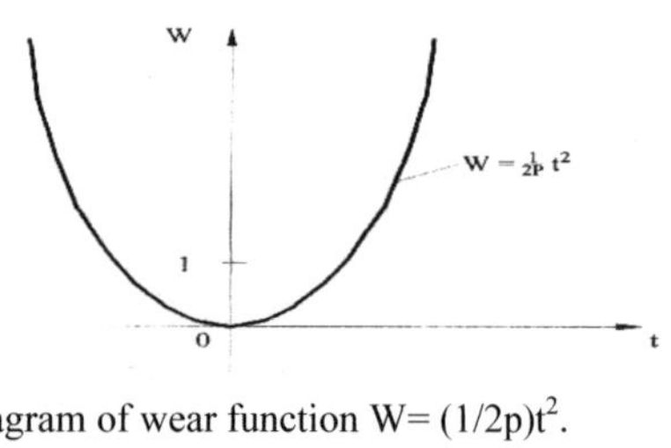

Fig. 4 Diagram of wear function $W = (1/2p)t^2$.

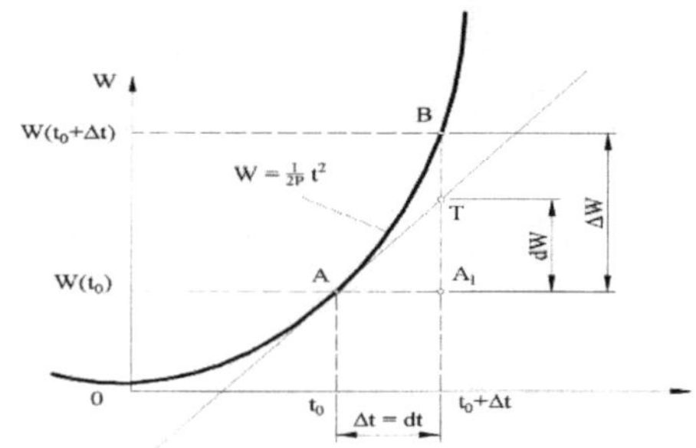

Fig. 5 The differentiation of wear function $W=(1/2p)t^2$.

Equation (7) is a symmetric parabola in report with axe-OW, with a flatten curve. Where: p-is parameter of wear parabola, t-durability of coupling in sliding friction.

The differentiation of function $W= (1/2p)t^2$ in point-t_0 is:

$$dW = \frac{1}{\rho} dt \tag{8}$$

In Fig.5, the size dt= Δt is differentiation of independent variable. The wear study has goal approximation of $W= (1/2p)t^2$ in proximity of point-A. Diagram from Fig.5 represents wear function, presumed differential in time interval and tangent at this curve in point A [t_0, f(t_0)], curve-OA represents running in zone. The ordinate of proximity point-B [t_0+Δt, f(t_0+Δt] is intersected by tangent in point-T. The portion of curve-AB corresponds of stability wear, and OB of normal operating period. The raising of function $\overline{A_1T} = \Delta W = f(t_0 + \Delta t) - f(t_0)$, corresponded of abscissa-Δt growing, has decomposed in two terms $\overline{A_1T}$ and$\overline{TB}$. The raising of function $\overline{A_1T}$ is differentiation of wear function W(t), notated with dW. The size of dW= (1/p)dt is differentiation wear in point t_0.

4. CONCLUSIONS

The mathematical model proposed for analysis of adhesive mechanical wear reflected physical process of fast wear of friction couplings with high sliding speeds, used in modern machine tools with high speed machining.

At high speed machining, it has required a new mathematical model on which wear speed, volume of wear, intensity wear and its process has a parabolic evolution.

5. REFERENCES

Balasiu, D. (1990), *Techniques of Investigation of Damage Process*, Technical Editor, Bucharest

Bowden, F.P. & Tabor, D. (1956), *Friction and Lubrification*, Methuen et 6, London

Ceausu, I, et al. (1980), *Organize and Leading of Activities of Maintenance and Reparation,* Technical Editor, Bucharest

Georgescu, A. et al, (1984), Lubrification Practice in Industry, Vol. I-II, Technical Editor, Bucharest

Grama, L. (2007), *Experimental Forecast in Construction of Machines, Methodology, Applications and Problems*, Veritas Editor, Tirgu-Mures

Kragelsky, I.V. & Alisin, V.V. (1978), *Friction Wear Lubrication. Tribology Handbook,* Vol. I-II, Mir Publishers, Moscow

Pavelescu, D. (1983), *Tribotechnica*, Technical Editor, Bucharest

Schrader, G.F., Elshennawy, A.K. (2000), *Manufacturing Processes & Materials,* SME Editor, ISBN-087263-517-1, Dearborn, Michigan, USA

Tudor, A. (1984), Some Aspects about Reliability Indicators of Friction Couplings, Tribotechnica 84, Iasi 28-29 Sept.1984

Annals of DAAAM for 2008 & Proceedings of the 19th International DAAAM Symposium, ISSN 1726-9679
Editor B. Katalinic, Published by DAAAM International, Vienna, Austria 2008
Make Harmony Between Technology and Nature, and Your Mind will Fly Free as a Bird

THE EFFICIENT MEASUREMENT METHOD OF PHONIC-ABSORBENT COEFFICIENT FOR CONSTRUCTION MATERIALS

UNGUR, P[etru]; POP, P[etru]; UNGUR, P[atricia]; GORDAN, M[ircea] & VERES, M[ircea]

Abstract: *This paper has presented an efficient and simple method of acoustical characteristics determination of phonic-absorbent and phonic-insulating construction materials. From these materials are: mineral and glass wadding, panels and subassemblies of false ceilings, for base walls from modeling alpha plaster with cellulose fiber, granules and powders polymeric reinforcement. For phonic-absorbent and phonic-insulating construction materials are necessary imposing for manufactures the sonic absorbent coefficient-*α_w *to estimate of required absorbent sound. Today, it has used for determination of* α_w *many methods, which did not so accuracy and more expensive, every method get different* α_w*-value for the same material type. This work presents an acceptable method for determination of sonic absorbent coefficient-*α_w*, has at base the using of stationary sounds waves get in by predictable sounds generator within Kundt tub, with uniform transversal section, which has at the end an sample of tested material.*
Key words: *insulating, noise, phonon-absorbent, sound.*

1. INTRODUCTION

The noise is an uncomfortable sound for human body, products by different noise sources by mechanical, hydro and aerodynamic, and electromagnetic action, machine tools, pneumatic installations, cars [Darabont, 1983, Pupazan, 1970].

The nocive action of sounds and noise have diminished by using of phonon-absorbent panels fabricated from composite materials reinforcement with fibres, and modelling α-plaster materials reinforcement with breakage granules of polystyrene from expandable pearlite. In general, these materials have designed and manufacturing with improved quality and quantity indicators of noise level. The phonon-absorbent materials for construction dissipated absorbent sound energy of phonon-absorbent elements, because of small size of transportable sound energy and get in thermal energy by friction and incident acoustic processes on phonon-absorbent element due to oscillations in porous medium [Pop,et al,2008].

In practical, the capacity of phonic insulating for construction materials has determined by absorbent coefficient-α_w. This paper has goal determination of absorbent coefficient.

2. THEORETICAL ASPECTS

Determination of insulation capacity of construction elements at sounds and noise has basic on sound transmission analysis through different physical medium, by air elements of walls and floor, which partitioned a building [ISO2923-1996].

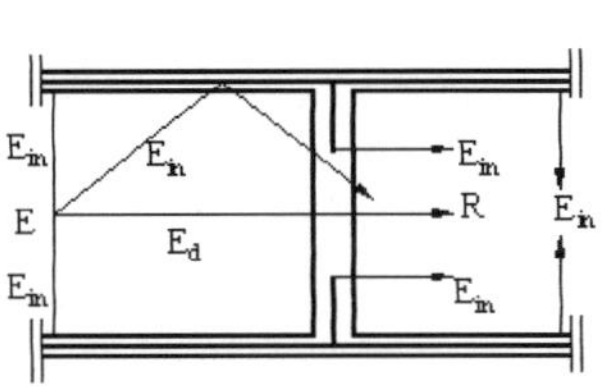

Fig. 1. Propagation of sound diagram emitted into a room.

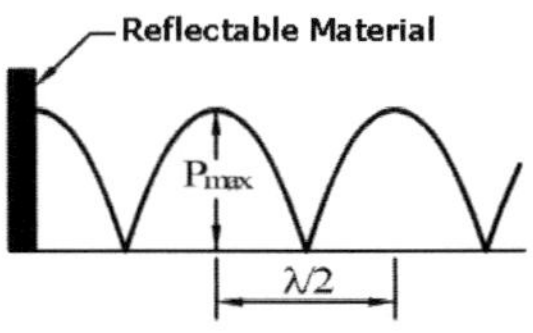

Fig. 2. Stationary waveform into a tub, with acoustic reflection wall.

The transmission of emitted sound in a room (Fig.1), named of emission-(E) by a sound source-(S) and received in reception room-(R) has realized from two ways [Darabont,1974,1983]:

-direct way, from which the energy has transmitted by vibrant element from rooms, formed direct energy-E_d;

-indirect way, from which the energy has transmitted by collateral ways, formed indirect energy-E_{ind}.

The phonic insulation grade between rooms isfunction of:

-itself phonic insulation capacity of partition element;

-sound level from inner and external room;

-transmissions, that of structure and linkage's partition elements.

For construction materials, that covered phonon-absorbent walls and ceilings has measured phonic absorption coefficient-α_w. A simple measurement method is stationary wave method. In this case, it has used a loudspeaker, as sound source for production of stationary acoustic waves into a tub with uniform transversal section, which has determined with a sample of tested material.

Used a steel plate with perfect reflection wall as master, the stationary acoustic waves within a tub have the form presented in Fig.2 [Darabont, 1983].

When the tub has finished with a phonon-absorbent material, the stationary waves from tub have the shape from Fig.3. For determination of absorbent coefficient of tested material, have to measure the minim and maxim of static pressure wave. The quantitative measures got the name of stationary wave ratios (RUS), and defined by relation [Darabont, 1983]:

$$RUS = \frac{p_{max}}{p_{min}} = \frac{A+B}{A-B} \quad (1)$$

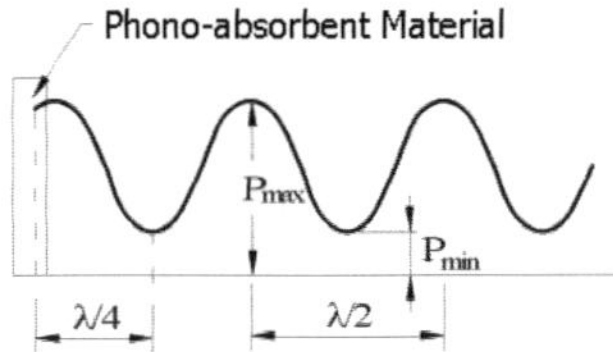

Fig. 3. Waveform of stationary waves in tub, with a phonic-absorbent end wall.

Where: RUS- is stationary wave coefficient;

p_{max}- maxim sonic pressure;

p_{min}- minim sonic pressure;

A- amplitude of incident wave;

B- amplitude of reflected wave.

The absorbent coefficient-α_w has calculated with relation:

$$\alpha_w = 1 - \left(\frac{RUS-1}{RUS+1}\right)^2 \qquad (2)$$

For calculus, the relation (2) can be reducing as:

$$\alpha_w = 1 - \left(\frac{A}{B}\right)^2 \qquad (3)$$

The absorbent coefficient of sample has defined as a ratio between incident total energy and square of sonic pressure. The gage for absorbent coefficient determination has in design phase and execution, as bilateral collaboration between Oradea University and Congips Co. from Oradea [Pop et al, 2008].

3. GAGE FOR DETERMINATION OF PHONIC-ABSORBENT COEFFICIENT

The gage for phonic-absorbent coefficient determination of construction materials it is a lab sound gage, composed in principal, from a long tub with circular section and a loudspeaker at the end. At the other end has mounted a plate of reflected material, or a circular sample from tested phonon-absorbent material. The maxim and minim of acoustic pressure within tub can be received by a microphone type probe, which is get in from a carriage by an axial orifice made in loudspeaker.

In fig.4 has depicted the block diagram for phonic-absorbent measurement by stationary waves with Kundt tub. Where: G-is sound generator, 1,2-filters, 3-sound tub, 4-sample tested, 5-probe, 6-osciloscope, 7-electronic interface, 8-PC computer.

For equipped of installation is necessary a Bruel&Kjaer gage. The instrument of phonic-absorbent coefficient measurement has made from two waves tubs, one with diameter of 100mm for frequencies range of 90-1800Hz, and the second with diameter of 30mm for range of 800-6500Hz [Broch,1975;Bruel&Kjaer2007].

Such as orientation, in Tab.1 has presented the frequencies [Hz] and reference level [dB], in conformity with ISO 717 [Darabont, 1983;STAS6161/1-89; STAS6161/3-82].

For tests and determination of phonic-absorbent coefficient would be use a generator of reference sound level of 70dB. The gage has executed in a bilateral accord between Oradea University and Congips Co from Oradea [Pop et al, 2008].

The novel of gage consists in construction, sound generator and inner cylindrical cavity with reflexion surfaces.

The qualitative graphic of phonic-absorbent curve has depicted in fig.5.

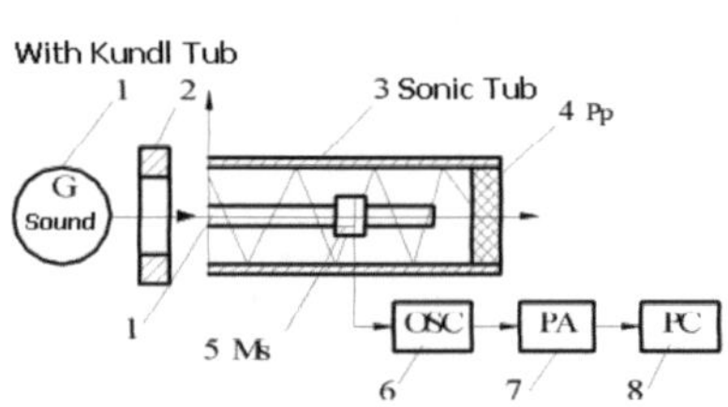

Fig. 4. Block diagram of phonic-absorbent coefficient measurement by stationary waves method.

	Frequency [Hz]						
Reference Level	100	125	160	200	250	400	500
Admissible L_n [dB]	67	67	67	67	67	66	65
Reference Level	630	800	1000	1250	1600	2000	2150
Admissible L_n [dB]	64	63	62	59	53	50	47

Tab. 1. Reference level values of noise in room by ISO 717.

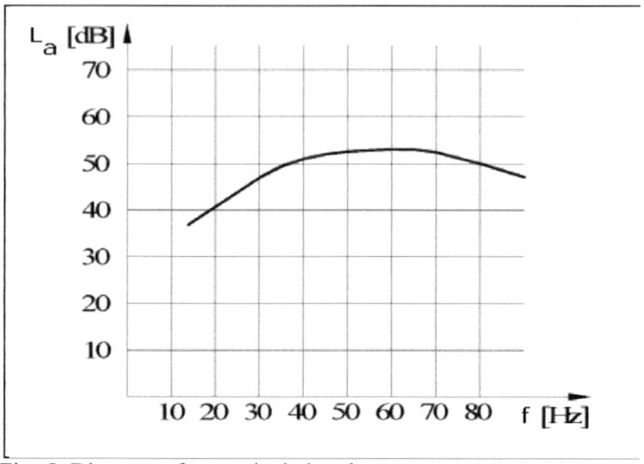

Fig. 5. Diagram of acoustical absorbent curve.

4. CONCLUSIONS

The method of phonic-absorbent coefficient-α_w used stationary waves and Kundt tub is approximately, such as other methods, but has the advantage of small size.

This method of measurement of phonic-absorbent coefficient with stationary waves and Kundt tub is simple and of lab technique.

The installation and gage is not difficult to build up, has great capacity of sound reflection by Kundt tub, assured by special reflection paint, applied on inner surface of cylindrical cavity.

5. REFERENCES

Broch, J.T. (1975), *Acoustique Noise Measurement*, Bruel&Kjaer

Darabont, A, et al, (1983), *Sound Measurement and vibration in Techniques*, Technical Editor, Bucharest

Darabont, A. & Vaiteanu, D. (1974), *Prevention of Noise Pollution and Vibrations*, Technical Editor, Bucharest

Pop, P.A.; Ungur, P.A.; Patcas, E. & Ungur, P. (2008), Recovering, Treating And Using Waste From Expandable Polystyrene By Mechanical Brakeage Process Of Manufacturing Composite Materials, *The 9th ASME Conference ESDA2008 July 7-9, 2008, Haifa, Israel, Proceedings of ESDA2008*, Paper No. ESDA2008-59176, pp. 1-8, ASME International

Pupazan, C. (1970), *Acoustics in Construction. Noise Propagation and Phonic Insulating*, Romanian Academy Editor, Bucharest

*** (2007), Measures Aquistiques, Bruel&Kjaer

*** (1996), *Acoustique. Mesurage du Bruit a Bord des Bateaux,* ISO 2923-1996)

*** (1989), *Acoustical in Construction, Methods of Noise Level Measurement in Civil Constructions*, STAS 6161/1-89

*** (1982), *Urban Acoustic, Methods of Noise Level Determination in Towns*, STAS 6161/3-82

Annals of DAAAM for 2008 & Proceedings of the 19th International DAAAM Symposium, ISSN 1726-9679
Editor B. Katalinic, Published by DAAAM International, Vienna, Austria 2008
Make Harmony Between Technology and Nature, and Your Mind will Fly Free as a Bird

VALVE WITH UMBILICUS FOR PAINTING BY HYDRAULIC SPRAYING OF CONSTRUCTION MATERIALS

UNGUR, P[etru]; POP, P[etru]; UNGUR, P[atricia]; VERES, M[ircea] & CRACIUN, D[an]

Abstract: *The painting is a cover process with one or more films of paint due to desire colour and aspects of object. The thickness of film and numbers of layers are condition by a certain factors as: content of paint binding material, capacity of paint covering, property of paint adhesion and uniform drying. At special material's construction, as cement, plated plaster of panels and ceilings have applied the hydraulic spaying to avoid the exfoliation and exfloration phenomena. For realized a good work and minim paint at this modern process has used a hydraulic pump at high pressure, with a special valve, presented in this paper. The novel consists in valve construction, which relieves the paint with high speed and very fine covering of paint on desire surface of part.*
Key words: *covering, paint, spray, umbilicus, valve.*

1. INTRODUCTION

The thermo and phonon-absorbent construction materials has used in construction industry and installations for obtained thermal insulations, due to reducer of fuel and electrical energy, by diminished heat loss, vibration and noise.

The construction materials as plated α-plaster of panels and ceilings, or aerial bindings of α-plaster, with great absorbent capacity of sound, because of have many openings and great pores. By passing the sound wave through pores, it has subjected more repeated reflections with walls due to dumping.

A main problem of these products of gyps, regyps, etc. is them painting. It has well known that at these materials normal spraying with air jet did not assure a good adhesion of film, due to exfoliation and exfloration of paint. A way to avoid these defects is using painting by hydraulic spraying, characterized as a large deposition. In paper has presented a special valve construction, which favoured the smooth mixture of pigments with fillers and binding materials, eliminated separation in layers, because of density difference for every pigment [Constantinescu, 1979;Veteleanu, 2000].

The advantage of this valve consists in that did not change the paint colour, because of smooth mixture of paint's components into a special mixed room with core of rotated hyperboloid or frustum of a cone.

2. THEORETICAL FUNDAMENTALS

Application of film forming materials from hydraulic spraying consists in them dispersion in minute particle with a paint installation by spraying without air and them engaged on surfaces of painting [Math Encyclopedia, 1980].

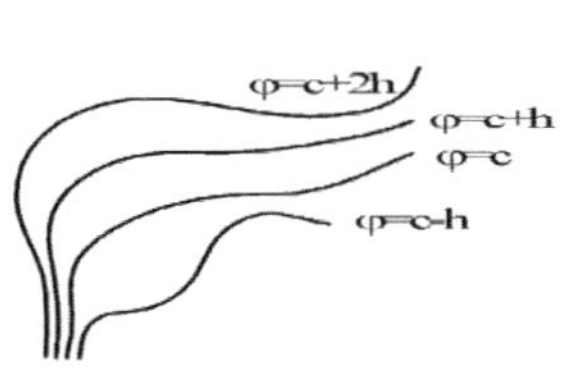

Fig. 1. Level lines of scalar field.

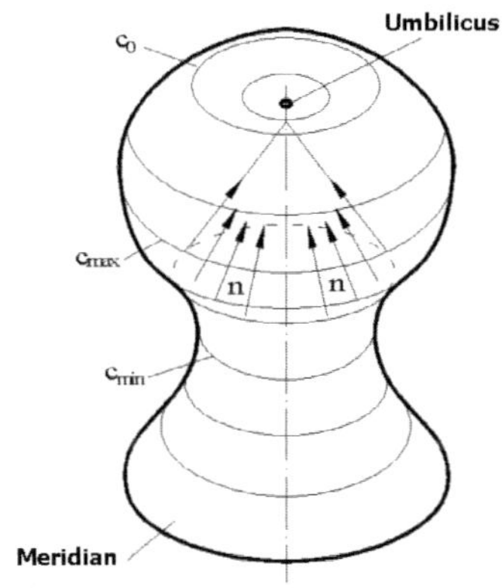

Fig. 2. Surface with level lines, umbilicus.

In hydraulic spraying process a great importance, have scalar fields as temperature and density of paint, represented by level surfaces. The scalar field or scalar function of time represented a correspondence between every point P(x,y,z), respectively every position vector from certain range its conferred a scalar φ(x,y,z)=φ(Γ). The fields have represented intuitive by surfaces φ(x,y,z)=ct., named level surfaces in space or plan by level lines- φ(x,y)=ct.

The scalar functions, as same lines of altitude and density are level lines [Math Encyclopedia, 1980]:

$$\varphi(x,y,z) = \varphi(r) \tag{1}$$

$$\varphi(x,y) = ct. \tag{2}$$

In construction of valve spraying under hydraulic pressure, for obtaining a smooth mixture of paint before spraying has used level surfaces and gradient. Ones of them are revolution surface, which has named umbilicus, presented in Fig.2. The umbilicus is a surface of revolution level, which has covered with an orthogonal network of meridians and parallel circles. The normal's-n of surfaces, along a parallel circle formed by regular points created a frustum of a cone. At a special valve the rotation cone has degenerated in a rotated hyperboloid or frustum of a cone, around of them by revolution in a special room, smoothing the spray paint, and get out by a special clearance-umbilicus [Math Encyclopedia, 1980].

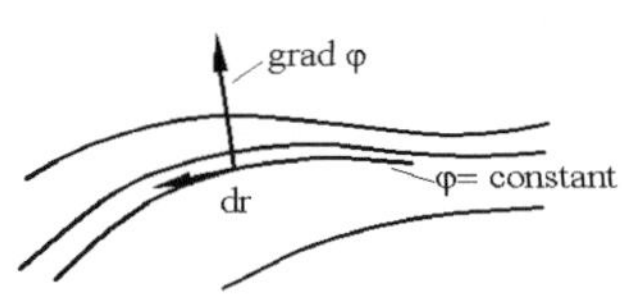

Fig. 3. Surface of level φ=ct and gradient.

The gradient of level surfaces (Fig.3) is getting by relation:

$$grad\ \varphi = \frac{\partial\varphi}{\partial x}i + \frac{\partial\varphi}{\partial y}j + \frac{\partial\varphi}{\partial z}k \quad (3)$$

For the function φ=φ(r)=φ(x,y,z) the variation-dφ a function-φ, for:

$$dr = dxi + dyj + dzk \quad (4)$$

, is:

$$d\varphi = \frac{\partial\varphi}{\partial x}dx + \frac{\partial\varphi}{\partial y}dy + \frac{\partial\varphi}{\partial z}dz \quad (5)$$

This expression can be sight as a dot product between vector-dr and vector-gradφ, given by relation (3). The function variation by position vector variation-dr would be:

$$d\varphi = grad\ \varphi \cdot dr \quad (6)$$

If it is choose direction-dr, so that dr to be level surface φ=ct, than φ did not change dφ = 0, or vector-gradφ is perpendicular on level surfaces φ = ct.

An application of level surfaces and gradient represents a construction of valve with hydraulic spraying, used at spraying without air of construction materials under form of panels, plates and ceilings from plaster with add of cellulose, expandable polystyrene, pearlite, phonon-absorbents. The smooth paint in special room of valve has made without mass loss, divergence of scalar field, named solenoid being zero.

$$div\ a = \frac{\partial u}{\partial x} + \frac{\partial v}{\partial y} + \frac{\partial w}{\partial z} = 0 \quad (7)$$

The total mass loss for a finite range-G has calculated by volume integration-$\iiint_G divad\tau$. This mass has flow by boundary-S of range-G. The expression in Gauss theorem components is:

$$\iiint_G \left(\frac{\partial u}{\partial x} + \frac{\partial v}{\partial y} + \frac{\partial w}{\partial z}\right) d\tau = \iint_S [ucos(x,n) + v\ cos(y,n) + w\ cos(z,n)] \quad (8)$$

This theorem is available for example of revolution body, type umbilicus, used at paint valve, valve body being with surface-G simple connected and Gauss boundary-S with normal-n, continuing on surface. Thus, this valve is an application of transformation's volume integration in a surface integration. It has resulted that for a range with div a=0 the total flux is zero on boundary range.

3. VALVE FOR SPRAY PAINTING WITHOUT AIR

The proceeding of construction materials painting by spray without air is modern film forming materials has done by a hydraulic pump at high pressure [Secasiu, 1970; Tarina, 1982].

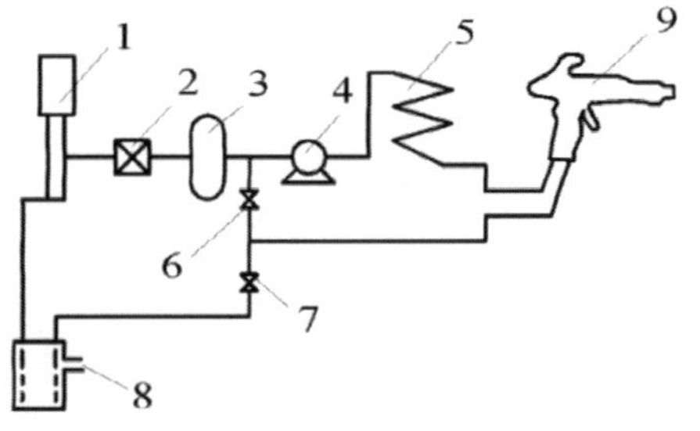

Fig. 4. Flow spraying installation diagram without air.

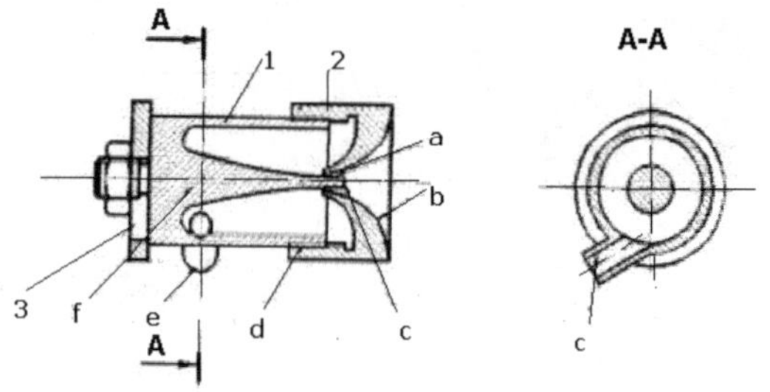

Fig. 5. Valve of spaying with umbilicus.

Behind friction with air, the speed of minute particles paint is decrease, due to a slow lay of them on painted part, avoided the possibility of rebound and forming of oil mist as spraying with compressed air. At this proceeding the productivity is higher, realized a paint economy. The quality painting is higher, avoided exfolier and exflorer at panels, ceilings of phonon-absorbent materials by α-plaster [Gordan, 2006; Ungur, 2007].

In Fig.4 has presented a diagram flow of spaying without air, where: 1,4-are pumps, 2-tap, 3-filter, 5-heating element, 6-tap, 7-valve of air exhausted, 8-vessel with paint, 9-spray gun with special valve The tests had done at Congips Co. for painting of tiles and phonon-absorbent panels, which it was a success.

The valve spraying with umbilicus has presented in Fig.5, where: 1-body with tubular cavity and central tapering core, 2-cover with umbilicus, 3-support of fixing, a-umbilicus, b-curved surface, c-pin, d-thread, e-tub, and f-core.

The concept of valve with umbilicus can be used more that at painting, at sprinkle by hydraulic pump.

4. CONCLUSIONS

From special construction of valve of painting by hydraulic spray with room of mixed has assured a great smooth of paint.

Spraying delicacy is emphasized by select form of fit through is get out the liquid (umbilicus).

The valve presented its assured the tendency of paint jet transformation in a fascicle of minute sprayed particles.

The construction of valve is easy to build up, assured a higher quality of paint and large using in future.

5. REFERENCES

Constantinescu, M. (1979), *Anticorrosive Protection of Metals*, pp. 106-109, Technical Editor, Bucharest

Gordan, M.; Ungur, P. & Fodor, M.D. (2006), An Application of the Gauss Divergent them to the Exhusted Gates, *13th Conference on Applied and Industrial Mathematics, KAIM 2006*, Oct.14-16, pp.49-50, ISBN 9975-7859-1 Pitesti, Romania

Secasiu, I. et al. (1970), *Paint Handbook*, pp. 245-247, Didactical and Pedagogical Editor, Bucharest

Tarina, A. et al. (1982), *Book Operator in Lacquer and Paint Industry*, pp. 286-291, Technical Editor, Bucharest

Ungur, P; Pop, P.A. Gordan, M. & Muresan, N. (2007), Optimal Construction of Hydraulic Pump with Elliptic Rotative Piston by Kakeya-Besikovitch Application, *ASME International Design Enginnering Technical Conference, Computers and Information Engineering, Sept.4-7, 2007, Las Vegas, Proceeding of DETC2007*, pp.1-8, ISBN 0-7918-3806-4, ASME International, USA

Veteleanu, A. & Olaru, S. (2000), *Anticorrosive Protection of Metals*, pp. 142-144, Expires Editor, Brasov

*** (1980), *Small Mathematical Encyclopedia*, Translated by Ostelnicu, V & Coatu after Kleine Enzyklopadie Der Mathematic, 6th Edition, pp. 245-246, 590-595, Technical Editor, Bucharest.

Annals of DAAAM for 2008 & Proceedings of the 19th International DAAAM Symposium, ISSN 1726-9679
Editor B. Katalinic, Published by DAAAM International, Vienna, Austria 2008
Make Harmony Between Technology and Nature, and Your Mind will Fly Free as a Bird

ASPECTS REGARDING THE PROGRAMMING OF CONTROL SYSTEMS FOR INDUSTRIAL PROCESSES

UNGUREANU, D[elia]; KRISTALY, D[ominic]; SINTEA, S[orin]; VULPE, A[nca] & PERNIU, L[iviu]

Abstract: *This paper contains some aspects of application programming realized using control systems of industrial processes. The aspects on which we are focused are to define specific types of data and functions. In order to define an automation strategy, we will add new functions that transpose algorithms for specifically needs. There are two different implementations: using structured programming languages, like C, and using programmable controller programming languages, like Function Block Diagram (FBD) and it is made a comparison between them. The modules of programs obtained are used as material for the specific topics in learning activities of departments of the universities. For dissemination, analyzes, discussions and tests, there will be used the facilities for distance learning offered by the transnational network "VET-TREND – Valorisation of an Experiment-based Training System through a Transnational Educational Network Development".*
Key words: *programming, data types, structure of data, control systems.*

1. INTRODUCTION

In this paper we discuss about some aspects of application programming of the automation systems belonging to two categories: real-time controls and automation systems. These systems are designed to aid facilities managers and owners to reduce operating, maintenance, and management costs. To efficiently use the hardware components, is very important how the programming of applications is done, so we analyze some aspects of programming a control system for industrial processes.

We use as support a Freelance 2000 system offered by ABB Automation Systems GmbH. The problem on which we are focused is the manner of the building of the program for different applications.

The application programs can be built with graphic configuration with high-performance editors in programming languages according to IEC 61131-3: function block diagram (FBD), ladder diagram (LD), instruction list (IL), sequential function chart (SFC).

The program modules obtained will be used later to built an expert system for managing and monitoring equipments in thermo-electric power stations. They have two different implementations: using structured programming languages, like C, or using programmable controller programming languages, like Function Block Diagram (FBD) (Ungureanu, et al., 2005).

2. USE OF STRUCTURED DATA TYPES

When we design a program for a complex application it is useful to define structured data types to process jointly several variables even if they have different data types. In this manner we can manage information easier. Variables are used for storing and processing information. Various different data types are available in the system, e.g. byte, word, integer, real, data&time. To enable several variables to be processed jointly even if they have different data types.

Name	Type	Comment	Initial value
val	REAL	agent thermic capacity	
aux	REAL	auxiliar	
validare	BOOL	validation	0
mark	BOOL	mark selection	1
index	INT	index	

Table 1. Creating data type structure.

It is possible to define new *structured data types.*

System variables are created every time a new resource like a process station, Field Controller or gateway is added. The status details for the resource are stored in these variables.

Default values can be assigned to each variable and to the separate elements of a structured variable. These values are assumed after a cold start, or when a station is initialized.

Application-specific data types can be created, i.e. defined in addition to the structured ones, with the aid of the specific editor (Table 1.). For every component of the structure we must specify the name and the type. A default initial value may be entered. When a warm start is made, this value is used as the structured data type for all variables which have that data type.

If we use use a variable of defined type, named BB, we can describe the state of one component of the equipment (eg. a boiler). The variable is represented in Fig. 1.

3. IMPLEMENTING THE SEARCH ALGORITHM

A case study is executed for an installation for the management of the equipments in thermo-electric power stations. There are suggested several program-modules for implementation. In the station there are more boilers, noted with B1, ...B7, with different capacities of thermic agent. To obtain the requeisted quantity of thermic agent is necessary to select the boilers that must to be in function.

It is proposed as search algorithm the more general class of search: search for solution in the state space search (Chen, 2000).

A state space can be viewed as a graph and a state in the state space can be viewed as a vertex in the graph. However, unlike the graph search where the graph is presented in a static fashion, the states and their connections considered in a state space search are usually constructed in a dynamic manner on a need base. This is to prevent the huge number of states to be presented. Searching a graph may require visiting all the elements in that data structure, in this case the search operation actually has the same effect of the traversal operation.

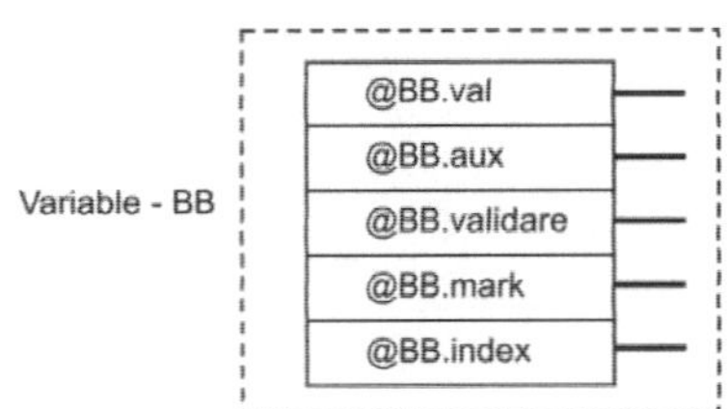

Fig. 1. A variable of structured data type

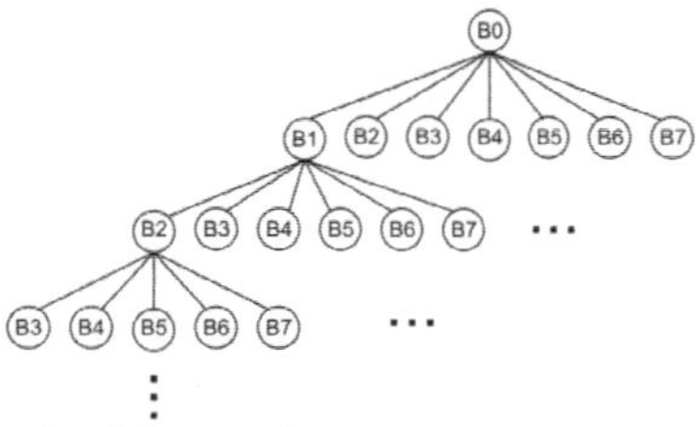

Fig. 2. Example of the search tree

We use the Breadth-first Search (BFS). It can be considered as an extension of level order traversal for the trees represented in Fig. 2 (Ungureanu, 2005).

To visit all elements of data structure, we use a stack where are put temporary the elements of the tree. We verify the relation between the tree elements until we find the better solution. The algorithm is presented in short in Fig.3.

4. ADDING USER FUNCTIONS

Among the facilities offered in the programming of the application, which brig in a lot of flexibility, there are the possibility of defining in completion of the standard function set particulary user functions, that is a reuse of the written code.

The scope of functions and function blocks provided by the Freelance 2000 process station corresponds to the basic supply defined in IEC 61131-3, in addition to numerous other tested functions and function blocks.

Available function categories are: analog value processing, digital value processing, loop control, logic control, logic functions, monitoring, acquisition, arithmetic functions, Modbus functions.

While designing the station and during configuration, the processing capacity and speed of the process station can be easily adapted to the demands of the automation task (Margineanu, 2004). They are accommodated in a function block library and can be supplemented by user-specific function blocks.

In order to define an automation strategy, we will add new functions for specifically needs. The specification of a *user-created function block class* will determine the functionality and the appearance of a function block.

In a function block instance, all local and output variable values are retained from one execution to the next. This means that the function block instance has an internal state, with the result that the same inputs need not always result in the same outputs. A user function block class is made up of the following components: interface, parameter list, text list, program, faceplate.

To implement the BFS algorithm is necessary to use a stack structure. So, it is useful to define a function that implament the stack. We defined a function bloc that offers the specifical type of operations for a stack (Boed, 1999).

It has the next inputs:

```
Select start node
repeat
        calculate: diff =quantity – ∑ node
        if (diff==0) then
                found solution
        else
        for node = 1 ... number_of_nodes
                pop node
                calculate: diff =quantity – ∑ node
                if( min>diff)
                        min =diff
```

Fig. 3 Breadth-first Search algorithm

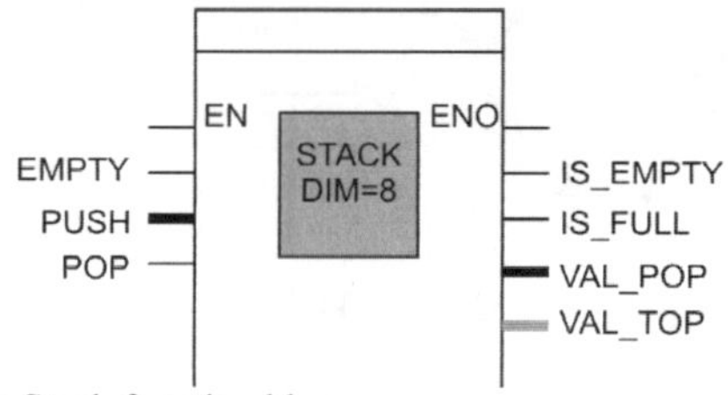

Fig. 4 The Stack function bloc

- EN (enable) – enable the use of the bloc – type: BOOL
- EMPTY – empty the stack – type: BOOL
- PUSH – put a value in the stack – type: REAL
- POP – extract a value from the stack – type: BOOL

The outputs are:

- ENO (enable out) (0 = disable, 1 = enable) – type: BOOL
- IS_EMPTY-tests if the stack is empty – type: BOOL
- IS_FULL – tests if the stack is full – type: BOOL
- VAL_POP – return a the extract – type: REAL
- VAL_TOP – return the index of the top (the number of elements of the stack) – type: INT

All standardized function blocks and all functions are available in the configuration of a user function blocks.

A function block instance is created by choosing a class out of a list of user function block classes (fig. 4). Instances can only be created from classes which have passed the plausibility check.

5. CONCLUSION

In this paper we discussed about some aspects of programming a process control system. We used the facilities offered by the system to define new data types and new functions specifically to our applications. We defined some simple functions and we try so to define complex functions, such as the implementation of the Breadth-first Search algorithm. The implementation using FBD language is more difficult than a implementation using a structurated programming language like C.

The final goal is to create an expert system to manage the equipment in a thermo-electric power station. The programmer must decide what version of implementation is used, such as to use efficiently the hardware components.

6. REFERENCES

Boed, V. (1999). *Networking and Integretion of Facilities Automation Systems*, CRC Press LLC, ISBN 0-8493-0699-X, United States of America.

Chen, Z. (2000). *Computational Intelligence for Decision Support.* CRC Press LLC, ISBN 0-8493-1788-1, United States of America.

Mărgineanu, I.(2004). *Automate Programabile*, Editura Albastră, ISBN: 973-650-195-7, Cluj-Napoca, Romania.

Ungureanu, D. (2005). Computational Intelligence - Problem Solving by Intelligent Search, *Proceedings of The 31-st Internationally Attended Scientific Conference of the Military Technical Academy*, Bucharest, november, 2005, Section 11, ISBN 973-640-074-3, Military Technical Academy of Bucharest.

Ungureanu, D.; Sisak, Fr. & Truican, I. (2006) Programming of control systems for industrial processes, *Proceedings of The 15-th International Scientific and Applied Science Conference ELECTRONICS-ET'2006*, Sozopol, september 2006, pp. 35-42, ISBN 954-438-518-5, Technical University of Sofia.

Annals of DAAAM for 2008 & Proceedings of the 19th International DAAAM Symposium, ISSN 1726-9679
Editor B. Katalinic, Published by DAAAM International, Vienna, Austria 2008
Make Harmony Between Technology and Nature, and Your Mind will Fly Free as a Bird

EVALUATION OF MAINTENANCE SYSTEMS

UNGUREANU, N[icolae]; UNGUREANU, M[iorita] & COTETIU, R[adu]

Abstract: *The paper present a point of view about the evaluation of maintenance of mechanical devices used in industry. The evaluation of maintenance is necessary for measure the efficiency of maintenance. We talk about technical efficiency, economical efficiency, etc. The result of evaluation is a helpful tool in management of industrial activities.*
Key words: *Maintenance, reliability, indicators.*

1. INTRODUCTION

Maintenance is defined as all actions which have the objective of retaining or restoring an item in or to a state in which it can perform its required function. The actions include the combination of all technical and corresponding administrative, managerial, and supervision actions.

The maintenance activities are influenced by many factors. The most important are presented in figure 1. According with influences of these factors in present in Romania are applied a few systems of maintenance (fig.2) (Antonescu 1988).

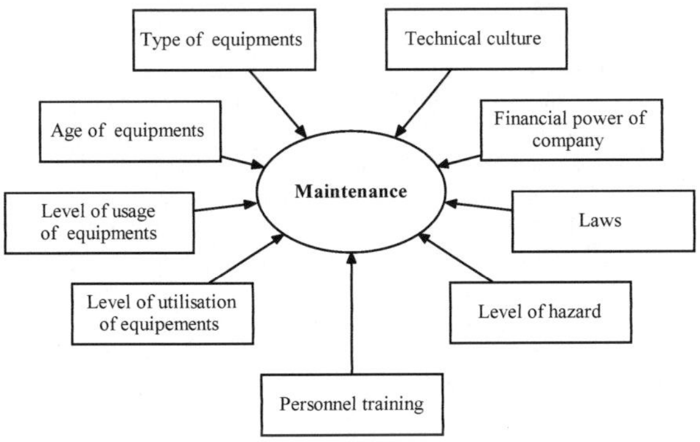

Fig. 1 Factors of influence on maintenance

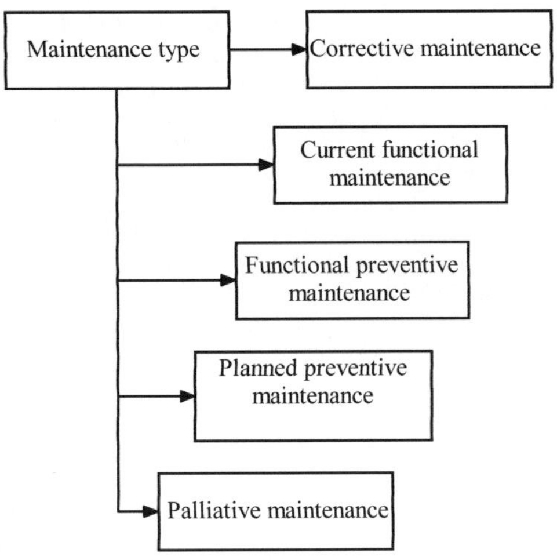

Fig. 2. Maintenance systems applied in Romania

The parallel between Romanian maintenance systems (Ungureanu 2003) and maintenance system applied in other countries in European Union (Smith 2008) is easy. Thus corrective maintenance correspond to discretionary maintenance where is possible to make rearrangements and modifications of construction, current functional maintenance, functional preventive maintenance and planned preventive maintenance correspond to proactive maintenance with scheduled operations. The palliative maintenance is as a matter of fact a reactive maintenance used in case of breakdowns and emergencies.

Beside this systems are used and increasing more and more the modern maintenance systems especially the predictive maintenance system in all forms: vibration analysis, infrared, ultrasonic and machine history.

2. EVALUATION OF MAINTENANCE

The biggest problem for the management is the selection of the best maintenance system for concrete situation. To solve this problem is necessary to have a set of tools to evaluate the efficiency of one or other of maintenance systems. At this moment for evaluate the efficiency of maintenance systems in Romania are used the indicators of maintainability and availability (Ungureanu, 2003):

- **Mean time to repair (mean time to restore), MTTR**. For calculus of MTTR in the phase of design is used the formula:

$$MTTR = \frac{n_1\lambda_1 t_1' + n_2\lambda_2 t_2' + \cdots + n_k\lambda_k t_k'}{n_1\lambda_1 + n_2\lambda_2 + \cdots + n_k\lambda_k} = \frac{\sum_{i=1}^{k} n_i\lambda_i t_i'}{\sum_{i=1}^{k} n_i\lambda_i} \quad (1)$$

where

- n is number of similar components;
- λ – failure rate for same type components;
- $n_i\lambda_i$ – average number of failures for group of elements n_i;
- t'_i – average time estimated for discharge a failure for one element of group n_i;
- k – number of distinct group with component elements with know reability.

In the case of experiments or on base of observations collected from exploitation the MTTR can be calculating with formula:

$$MTTR = \frac{t'_1 + t'_2 + \cdots + t'_r}{r} = \frac{\sum_{i=1}^{r} t'_r}{r} \quad (2)$$

where

- r is number of maintenance actions;
- t'_r – time used for discharge the failure r.

- **Availability coefficient** can be calculate with formula:

$$K_D = \frac{MTBF}{MTBF + MTTR} \quad (3)$$

where

- MTBF is mean time between failure;
- MTTR – mean time to restore.

- **Percentage of inactive time**:

$$K_{IN} = \frac{MTTR}{MTBF+MTTR} \quad (4)$$

- **Percentage of availability**:

$$K_{DI} = \frac{MTTR}{MTBF} \quad (5)$$

- **Coefficient of use**:

$$K_U = \frac{MTBF}{T_E} \quad (6)$$

where T_E is the calendar time of use, including time of effective use of machines, time for maintenance and time of interruption of use.

At the beginning of 2007 was approved a new European standard entitled "Maintenance – Key Performance Indicators". This standard use a system of indicators structured in three categories: economic, technical and organizational. (figure 3)(European Committee for Standardization 2007)(Grencik &Legat 2007)

The economic indicators are 24 divided in 4 levels and use as unit of measurement the money. In the most cases the reference for indicators is "total maintenance cost". (13 indicators from 24). For example E14 is defining as:

$$E14 = \frac{Preventive\ maintenance\ cost}{Total\ maintenance\ cost} x100 \quad (7)$$

The technical indicators are 19 divided in 4 level and use as unit of measurement the time. Same as in the case of economic indicator we have a reference "number of failure". In the list of indicator we meet "classic" indicators as men time between failure, MTBF – reliability indicator and mean time to failure, MTTF – reliability indicator.

The organizational indicators are 34 divided in 4 level and use as measurement unit the time and in some cases number of persons.

The E.U. standards "Maintenance – Key Performance Indicators" present also the definition of indicators and a methodology to use the Key indicators. The methodology follows the logical diagram presented in figure 4.

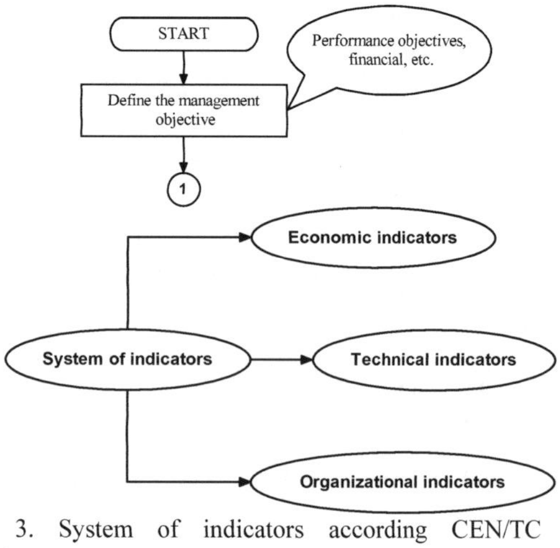

Fig. 3. System of indicators according CEN/TC 319 Maintenance

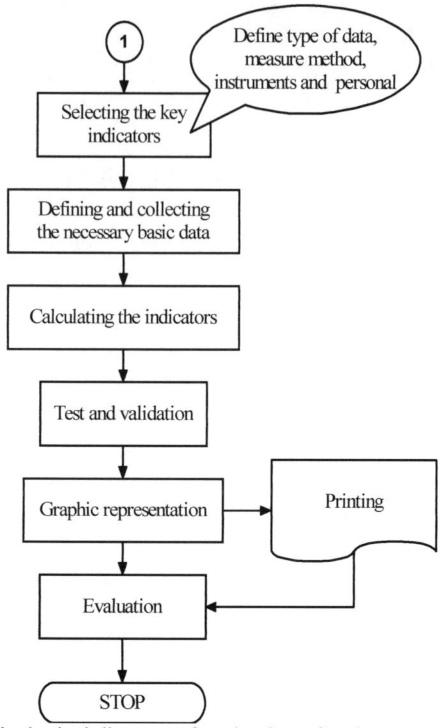

Fig. 4. The logical diagram of evaluation of maintenance according European standard entitled Maintenance – Key Performance Indicators

3. CONCLUSION

The process of evaluation of maintenance systems is not an easy job because are many factors involved and is not possible to make a pattern ready to use. Each time before evaluation is necessary to analyze the concrete situation and choose from the list of indicators the agreed indicators. In the same time is possible, in particular cases, to not find in list of EU standard or in national standards an indicator that uses the avaible data. In this case is possible to create a specific indicator, but only with condition to define coherent the terms.

The recommendation for evaluation is to choose a small number of indicators and the calculus of these indicators must respect the principles of statistics mathematic.

4. REFERENCES

Antonescu, V., Stoichitoiu, D., *Elemente de teorie si culegere de probleme de fiabilitate, mentenabilitate, disponibilitate, vol I, II*, (*Elements of theorie and practical aspects of reliability, maintainability and availability*) Institutul central pentru industria electrotehnica, Oficiul de informare documentara, Bucuresti, 1988

Grencik, J., Legat, V., Maintenance audit and benchmarking – search for evaluation criteria on global scale, *Eksploatacja i niezawodnosc, Polska Akademia Nauk Oddzial w Lublinie* nr 3/2007, pag 34-38.

Smith, M., Planned Maintenance Techniques, *Available from:* http://elsmar.com/Planned_Maintenance/index.htm, *Accessed:*2008-02-17

Smith R., Mobley K., *Industrial Machinery Repair: Best Maintenance Practices Pocket Guide*, ISBN 0-7506-7621-3, 2003, Elsevier Science

Ungureanu N., *Fiabilitate si diagnoza, (Reliability and diagnosis)* ,Editura Risoprint Cluj Napoca, 2003

*** "Maintenance – Key Performance Indicators" European Committee for Standardization, 2007

Annals of DAAAM for 2008 & Proceedings of the 19th International DAAAM Symposium, ISSN 1726-9679
Editor B. Katalinic, Published by DAAAM International, Vienna, Austria 2008
Make Harmony Between Technology and Nature, and Your Mind will Fly Free as a Bird

THE MODELLING OF END-EFFECTOR'S POSE ERRORS FOR THE EQUIVALENT KINEMATICAL CHAIN OF THE UPPER HUMAN LIMB

VACARESCU, V[aleria]; LOVASZ, E[rwin] C[hristian] &VACARESCU, C[ella] F[lavia]

Abstract: *Based on the modelling of the upper human limb movements through opened kinematical chain with seven freedom degrees, the present paper proposes the differential modelling of its end-effector's pose errors, for develop an arm prosthesis and improve its performance by geometrical calibration. In the specialty literature, is appreciated the contribution of different sources to these errors, concluding the fact that errors due to the geometrical factors represent 90-95% of the prosthesis arm's pose errors.*
Key words: *modelling, structure, prosthesis, errors, calibration.*

1. INTRODUCTION

A study upon the structure of the upper human limb is also used for acknowledging the morph-functional characteristics of the body (the interdependency between anatomy and biomechanics) and for obtaining important data in the "humanoid" modelling process for functional prosthesis. (Moeslung et al., 2005) proposed the modelling of the 3D position of a human arm only with two parameters to express the motion coursed by the anatomical joints. (Reinbolt et al., 2005) used the multi-joint kinematical model for the development of dynamics patient's specific model for the clinical problems in rehabilitation. A model of the human arm as a 7 DOF system is used by (Rosen et al., 2005) to resolve the equations of motion. (Bronzino et al., 2000) defines the geometry of joint's surfaces of upper human limb and the reference system attached to each joint, for studying upper human limb's biomechanics. (Vacarescu et al., 2005) models the upper human limb like an opened kinematical chain with 7 DOF and establishes the Jacobian's matrix for its analyses.

In this paper, it is proposed the differential modelling of 3D pose errors of the arm's kinematical chain with 7 DOF. Based on the above mentioned, the authors of this paper developed a model of human arm prosthesis, following the improvement of its performances through geometrical calibration using the proposed mathematical model.

2. THE MODELLING OF THE UPPER ARM

Structural, the upper human limb can be associated with an open kinematical chain with 7 DOF and eight links. In figure 1 is presented the equivalent kinematical chain. Because the shoulder's blades and the clavicles moves are too hard to model, in prosthesis fabrication, these links together with the afferent joints are included in the fixed link. So, the first kinematical joint of the equivalent kinematical chain is the shoulder's joint, respectively the spherical scapula-humeral joint. In kinematical studying, this joint can be replaced with three rotation joints, their axes having one common point. The three moves in this joint are maid along the axes perpendicular on the characteristic plans (sagittal, frontal, transversal). In the elbow joint, the main move is the flexion-extension move that is associated with a rotation kinematical joint. The wrists moves can be done, each, by a rotation kinematical joint. So, the equivalent kinematical chain is continued with three lower kinematical joints with perpendicular axes.

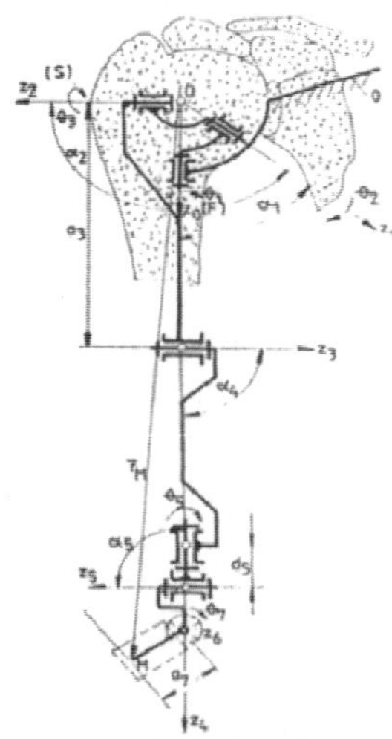

Fig.1 Equivalent kinematical chain with seven freedom degrees

The directions of the three axes of the kinematical joints form a dextral trihedral and they are oriented upon the frontal, transversal and sagittal direction of the shoulder's joint. Properly choosing the other joints axes, the transform matrixes can be written, between the system's axes, for determining the position vector $\overrightarrow{r_M}$, of the characteristic point situated on the arm's end-effector. To do this, are used the Hartenberg-Denavit parameters. For fig.1, the values of the kinematical chains parameters Hartenberg Denavit are shown in the table 1.

Joint i	q_i	α_i	a_i	d_i
0	θ_1	0^0	0	0
1	θ_2	90^0	0	0
2	θ_3	90^0	0	0
3	θ_4	-90^0	a_3	0
4	θ_5	90^0	0	0
5	θ_6	-90^0	0	d_5
6	θ_7	90^0	a_7	0

Tab. 1. The values of Hartenberg-Denavit parameters

3. THE DIFFERENTIAL MODELLING OF ERRORS

The major problem is the difference between the nominal geometry of the arm's kinematical chain, obtained by designing, in concordance with its functions and its real geometry, affected by manufacturing tolerances, assembling errors occurred during arm's assembling. Generally, the nominal geometrical models are simple, based on some presumptions like the parallelism and perpendicularity of axis. Are used Hartenberg-Denavit parameters ($\theta_i, d_i, a_i, \alpha_i$) in

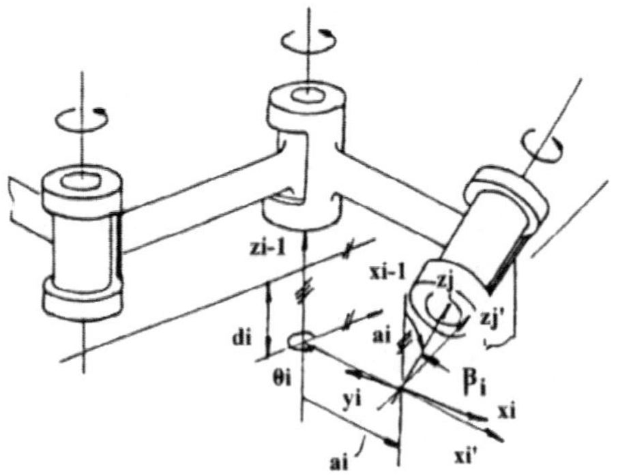

Fig.2. The geometrical parameters of errors

modelling of geometrical errors of the relative position parameters between the chains elements.It is known the fact that the small geometrical errors of these four parameters can lead to important variations of the pose parameters. It is necessary an external parameter, the β_i angle, which defines a rotation system (i) as a new system (i'), around the y_i axis (fig. 2). In the absence of β_i angle (called twist angle), the non-parallelism of the axis must be set off by an artificial correction of lenghts a_i and d_i, even if these parameters were initially correct. The β_i rotation angle is used only for the parallel axis of the successive rotation joints. For the nominal geometrical model, $\beta_i = 0$. The geometrical model allows to establish the pose vector of the arm's end-effector, as function of Hartenberg-Denavit geometrical parameters:

$$x = f(\theta(\alpha, a, d, \beta) \quad (1)$$

where: θ,α,a,d,β are R^n vectors, for the arm's "n" joints. This relation between geometrical parameters is, generally, nonlinear. In their identification order, the geometrical model is linearised in around an initial estimation, $\hat{\theta},\hat{\alpha},\hat{a},\hat{d},\hat{\beta}$, (nominal parameters) of the real parameters θ,α,a,d,β. Since two axis of some successive rotation joints are (in the nominal geometrical model) initialy supposed to be parallel, it is chosen $\hat{\beta}_i = 0$, to each element "i" of the arm. A differential model of the errors is obtaines such as:

$$\Delta x = J_\theta \cdot \Delta\theta + J_\alpha \cdot \Delta\alpha + J_a \cdot \Delta a + J_d \cdot \Delta d + J_\beta \cdot \Delta\beta \quad (2)$$

where: $\Delta\theta$ is the error vector of the angle θ_i (measured in the rotation joints with transducers); $\Delta\alpha,\Delta a,\Delta d,\Delta\beta$ are the error vectors of the torsion angle α_i, the length a_i, d_i and angle β_i (in the case of translation joints, d_i is measured with transducers), and:

$$J_\theta = \frac{\partial f}{\partial \theta};\; J_\alpha = \frac{\partial f}{\partial \alpha};\; J_a = \frac{\partial f}{\partial a};\; J_d = \frac{\partial f}{\partial d};\; J_\beta = \frac{\partial f}{\partial \beta} \quad (3)$$

Each one of these jacobian matrix is a sensitive matrix of end-effector's position and orientation, considering the variation of the Hartenberg-Denavit geometrical parameters (moderating coefficient). These matrix are calculated using the $\hat{\theta},\hat{\alpha},\hat{a},\hat{d},\hat{\beta}$ geometrical parameters nominal values. Generally, the vector $J_{\theta i}$ can be written such as:

$$\left\{J_{\theta_i}\right\} = \left\{\begin{matrix} \left[{}^0R_{i-1}\right] \left(\left\{{}^1z_{i-1}\right\} \times \left\{{}^{i-1}P_N\right\}\right) \\ \left[{}^0R_{i-1}\right] \left\{{}^1z_{i-1}\right\} \end{matrix}\right\} \quad (4)$$

where: "[]", "{}", and "×" mean: matrix, vector and vector product; $\left[{}^0R_{i-1}\right]$ is the rotation matrix of the (i-1) coordinate axis system related to the base (0) coordinate system; $\left\{{}^{i-1}P_N\right\}$ is the end-effector's position vector related to system (i-1) and $\left\{{}^1z_{i-1}\right\} = \left\{0 \quad 0 \quad 1\right\}^T$. A similar expression can be used for each jacobian matrix. The differential model of the errors, expressed in the relation (2) can be used for a geometrical calibration of the arm, in biomechanical applications. In this situation, is estabilshed the end-effector's pose error, expressed by Δx in the relation (2), by measuring with any system (with ultrasound measurement module or optical system, in our research project). Also are measured, with transducers, $\Delta\theta_i$ or Δd_i in joints and using as initial dates in the relation (2), Δx, $\Delta\theta_i$ (or Δd_i), such as the nominal values $\hat{\theta}, \hat{\alpha}, \hat{a}, \hat{d}, \hat{\beta}$ of the Hartenberg-Denavit geometrical parameters, are calculeted the errors $\Delta\theta, \Delta\alpha, \Delta a, \Delta d, \Delta\beta$, used for obtaining real, corrected Hartenberg-Denavit geometrical parameters, $\theta^*, \alpha^*, a^*, d^*, \beta^*$:

$$\theta^* = \hat{\theta} + \Delta\theta, \alpha^* = \hat{\alpha} + \Delta\alpha, a^* = \hat{a} + \Delta a, d^* = \hat{d} + \Delta d, \beta^* = \hat{\beta} + \Delta\beta \quad (5)$$

The computation is made iterative until it riches the minimum of errors, by introducing in the relation (2) the new estimated values of the Hartenberg-Denavit geometrical parameters, $\theta^*, \alpha^*, a^*, d^*, \beta^*$.

4. CONCLUSION

The method presented in this paper it is developed in the CEEX 88 research project, for biometrical measurements. The present algorithm is applied in programming the control system of the human arm prosthesis, for the improvement of position and orientation accuracy of the end-effector. In the future work, the authors of the paper propose the developing of the human arm model, for the improvement of the prosthesis as well as for the use of a robot arm on a wheelchair.

5. REFERENCES

Bronzino, J. (2000). *The biomedical engineering handbook,* CRC Press, IEEE Press, IEEE Order No: PC5788, USA

Moeslund, T.;Granum, E.; (2005). *Modeling the 3D pose of a human arm and shoulder complex using only two parameters,* Integrated Computer Aided Engineering, vol.12, issue 2, 159-175, ISSN 1069-2509

Reinbolt, J; Schutte,J et al. (2005). *Determination of the patient-specific multi-joint kinematic models through two-level optimization,* Journal of biomechanics, vol.38, issue 3, 622-626, Elsevier, Florida

Rosen,J; Perry, J.C.; Manning, N. et al. (2005). *The human arm kinematics and dynamics during daily activities – toward a 7DOF upper limb powered exoskeleton,* Advanced Robotics ICAR'05 Proc.,ISBN:0-7803-9178-0, Seattle

Vacarescu, V.; Vacarescu, C.F.(2005). *The kinematics analysis of the equivalent kinematical chain of the upper human limb,* 5th International Conference RaDMI 2005 Proceedings, Serbia & Montenegro, section E25, ISBN 86-83803-20-1.

Annals of DAAAM for 2008 & Proceedings of the 19th International DAAAM Symposium, ISSN 1726-9679
Editor B. Katalinic, Published by DAAAM International, Vienna, Austria 2008
Make Harmony Between Technology and Nature, and Your Mind will Fly Free as a Bird

THE METHOD FOR THE EVALUATION OF SPINAL COLUMN POSTURE

VACARESCU, V[aleria]; LOVASZ, E[rwin] C[hristian]; VACARESCU, C[ella] F[lavia] & MARGINEANU, D[an]

Abstract: *Idiopathic scoliosis generates complex deformation on spinal column and rib-cage. The clinicians used 2D Cobb angle calculated from X-ray images to assess the severity of spine deformation. In this paper, the evaluation of spinal column posture is made with the 3D Measurement Ultrasound Module. It is assumed that every vertebra is associated with a precision point. The coordinates of the points allow the determination of the parameters of spinal column position and orientation. The paper propose a new method for establishing the position and orientation matrix of spinal column and the correction matrix of spinal column resulted from medical therapy. Treatment efficiency and the correction of the spine posture can be evaluated.*
Key words: *spine, idiopathic, scoliosis, correction, matrix.*

1. INTRODUCTION

A fundamental issue related to our research is scoliosis. Scoliosis is defined as abnormal lateral curvature of the human spine. It is usually considered as a three dimensional deformity, because axial rotation will always accompany the lateral curvature. Severe scoliosis curves may alter balance and coordination, interfere with the function of internal organs, determines premature degeneration of the spinal column, and cause deterioration of neurological function. To understand the disease and to document the morphology of the scoliotic spine, actual studies (Boisvert et al.,2006) were made about the positions and orientation of the vertebrae in scoliotic patients, using a set of rigid inter-vertebral transforms on the radiographic imaging. (Ghanem et.al.,1997) uses the optic measurements, during surgery, to determine the vertebrae position. (Chi et al., 2006) presents a new method for measuring axial rotation of vertebra. (Bronzino et al, 2000) makes the gait evaluation by comparing the technical reference system attached to a cluster with 3 markers, with an anatomical reference system with z axis along each foot bone. In [5], (Vacarescu et al,. 1998), applies the rigid transforms between reference systems for analyzing the robots errors, respectively for their calibration. Based on the above mentioned, the authors of this paper propose the application of the methods used in analyzing the robots errors for evaluating the correction of the spine's position. In this paper, the evaluation of spinal column posture is made with the 3D Measuring Module with ultrasounds. It is assumed that every vertebra is associated with a precision point on the crest of the vertebrae. It is established the position and orientation matrix of spinal column and correction matrix of spinal column, correction which resulted from medical therapy. So, it can be evaluated the efficiency and evolution in the treatment of the spine posture.

2. MATRIX OF HUMAN SPINE POSTURE

The report of measurements shows the spinal crest line in sagittal, frontal and transversal projections. This is subdivided automatically into the number of vertebrae. The rotation angles are shown in addition to the representation in sagittal and frontal projections (angles φ and ψ). The angles in transversal

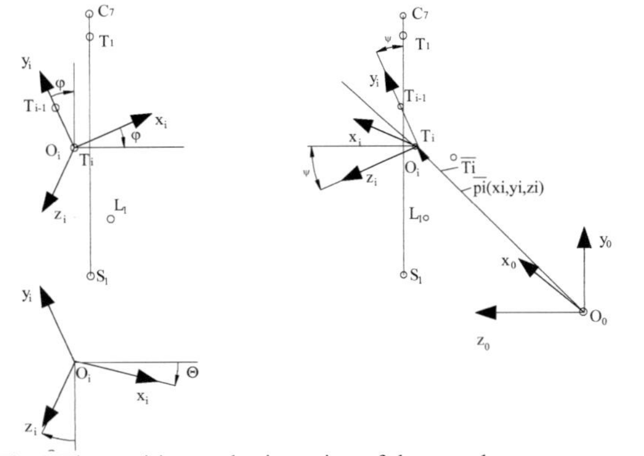

Fig.1.The position and orientation of the vertebrae

projection may be determined for each vertebrae (angle θ); the coordinates x_i, y_i, z_i of the vertebrae "i" are shown in the measurements report. For analyzing the 3D human spine scoliotic deformations, this paper proposes the use of rigid transforms between references systems (figure 1). For this purpose, reference $O_0\, x_0\, y_0\, z_0$, having the $O_0\, y_0$ axis in vertical position, will be attached to the reference marker (fixed on the patient's body). Each vertebrae "i" will have attached a reference system $O_i\, x_i\, y_i\, z_i$, having the $O_i\, y_i$ axis in direction i, i-1 (figure 1). The position of $O_i\, x_i\, y_i\, z_i$ reference system, reporting to the $O_0\, x_0\, y_0\, z_0$ reference, is defined by the following transform:

$$O_0x_0y_0z_0 \xrightarrow{\ {}^0T_i\ } Ox_iy_iz_i$$

Shown through the relation:

$$^0T_i = Transl(x_i, y_i, z_i) * RPY(\varphi, \theta, \psi) \qquad (1)$$

3. CORRECTION MATRIX

In relation (1), the x_i, y_i, z_i, φ, ψ, θ parameters are determined by measurements using the Measuring Module in examining the human spine posture of patient "x". In order to estimate the human spine correction, obtained by applying the therapeutic techniques, are compared two series of measurements, obtained after the human spine posture investigation of the patient "x" before and after applying the therapy (figure2). In the figure 2, the $O_i\, x_i\, y_i\, z_i$ reference system defines the vertebrae "i" position, reporting to the $O_0\, x_0\, y_0\, z_0$ reference marker, corresponding to the first investigation. The $O_c\, x_c\, y_c\, z_c$ reference system defines the corrected vertebrae "i_c" position and orientation), corresponding to the human spine investigation after applying the medical therapy. The position of corrected "i_c" vertebrae, defined by the $O_c\, x_c\, y_c\, z_c$ reference is described by the following transforms:

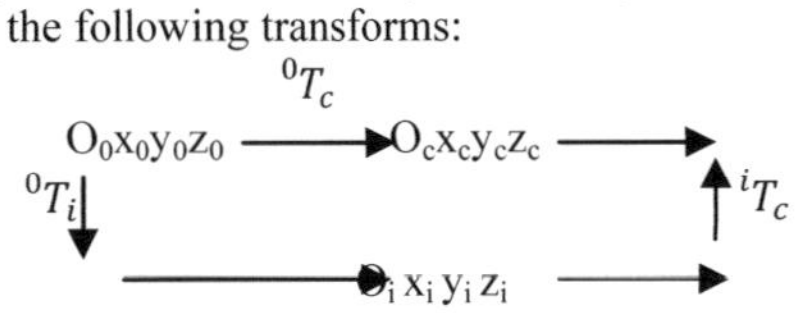

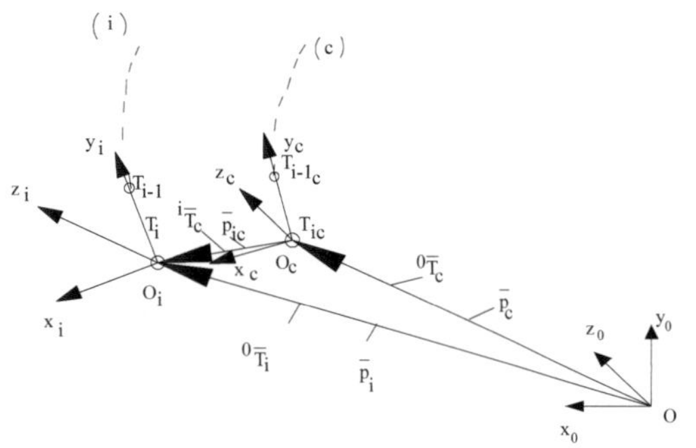

Fig. 2 The rigid transforms corresponding to the correction of the spine

The ${}^{i}T_c$ transform expresses the size of the position and orientation correction of "i_c" vertebrae, resulted after the rehabilitation therapy. The ${}^{0}T_i$ transform expresses the corrected "i_c" vertebrae position and orientation, reported to the $O_0\,x_0\,y_0\,z_0$ reference:

$$ {}^{0}T_c = \mathrm{Transl}\,(x_c, y_c, z_c) * \mathrm{RPY}\,(\varphi_c, \theta_c, \psi_c) \quad (2)$$

$$ {}^{i}T_c = \mathrm{Transl}\,(x_{ic}, y_{ic}, z_{ic}) * \mathrm{RPY}\,(\varphi_{ic}, \theta_{ic}, \psi_{ic}) \quad (3)$$

In the figure 2:

$$\overrightarrow{p_{ic}} = \overrightarrow{p_i} - \overrightarrow{p_c} \quad (4)$$

In the relations (2), (3), (4):

$$x_{ic} = x_i - x_c;\ y_{ic} = y_i - y_c;\ z_{ic} = z_i - z_c$$

$$\varphi_{ic} = \phi_i - \phi_c;\ \theta_{ic} = \theta_i - \theta_c;\ \psi_{ic} = \psi_i - \psi_c \quad (5)$$

in which:

${}^{i}R_c$ –is the correction matrix of the orientation of "i" vertebrae;

${}^{i}p_c$- is the correction matrix of the position of "i" vertebrae.

4. CASE STUDY

The described method was applied to a group of 15 scoliotic patients from the City Hospital of Timisoara, partner of the Politehnica University of Timisoara, in the Research Project CEEX 88/2006. For one of the patients in the group, the initial geometrical parameters of spine position and orientation were measured with the Measuring Module and are presented in figure 3. It can be observed a small scoliosis deviation to the 6th vertebrae, T_6 (fig. 3). The shape and the position geometrical parameters of the correction spine, resulted from the application of the therapeutic procedures, are presented in the fig. 4. It can be observed the T6 position's correction. The measured dates were processed in Matlab. Comparing the values from fig. 3 with the values from fig. 4, for the T6 vertebrae, results:

x_{ic}=0.640038; y_{ic}=0.409; z_{ic}=1.02709; ψ_{ic}=2.8; ϕ_{ic}=-1.3; θ_{ic}=0

The spine correction matrix of the T6 vertebrae is:

$$ {}^{i}T_c = \begin{vmatrix} 0.2675 & -0.8183 & -0.3228 & -0.4950 \\ -0.9636 & -0.2520 & -0.0896 & -0.8118 \\ 0 & 0.3350 & -0.9422 & -0.8307 \\ 0 & 0 & 0 & 1.0000 \end{vmatrix} \quad (6)$$

5. CONCLUSION

The obtained data was used to estimate the size of the spine's correction for the patients included in the project and which benefited of medical rehabilitation therapy.

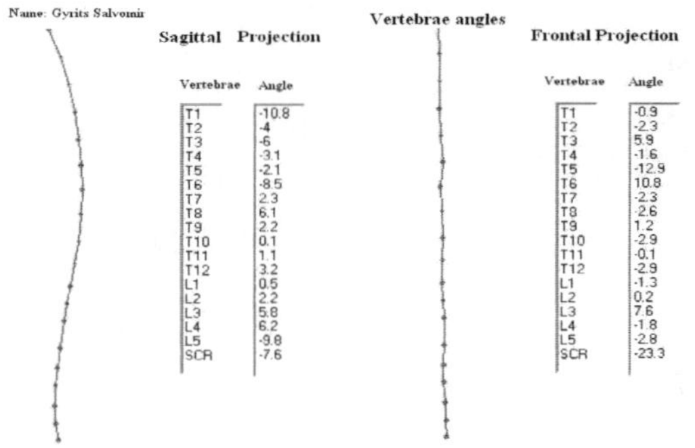

Fig. 3 The vertebrae form in before the therapy

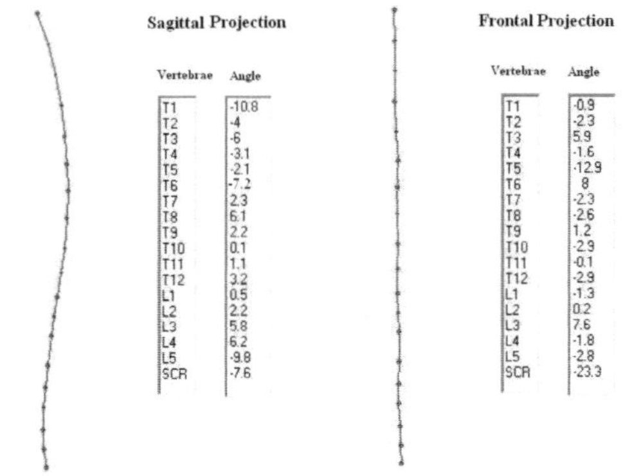

Fig. 4 The vertebrae form of the correction spine after the therapy

The proposed method offers relevant information about the variability of the human spine, using rigid transforms. In the future research, we will use the proposed method for the situations presented below:

– for an initial analysis, comparing the geometrical parameters of the patient's spine obtained by measurements with geometrical parameters of the "ideal" spine of the same patient, having as a result the maximum values of correction needed by the patient;

– for analyzing the efficiency of the therapeutic technique applied in order to correct the posture of the spine, comparing the geometrical parameters of patient's spine before and after the therapy;

– for the improvement of the intelligent brace developed in the research project, by using the proposed method in programming its control system, in order to have progressive adjustments of the spine, proposed by the medical specialist.

6. REFERENCE

Bronzino, J.(2000). *The biomedical engineering handbook,* CRC Press, IEEE Press, IEEE Order No: PC5788, USA

Boisvert,J.et al. (2006). *3D anatomical variability assessment of the scoliotic spine using statistics on Lie groups,* Proc. 3rd IEEE International Symposium, pp.750- 753, ISBN: 0-7803-9576-X, USA

Chi,W.M.et al (2006). *Vertebral axial rotation measurement method,* Computer methods and programs in biomedicine, vol. 81, no1, pp. 8-17, ISSN 0169-2607, Elsevier Science, Ireland

Ganem,I.P et.al (1997). *Intraoperative optoelectronic analysis of three-dimensional vertebral displacement after Cotrel-Dubousset rod rotation,* Journal Spine, vol. 22, no16, pp. 1913-1921, ISSN 0362-2436, USA

Vacarescu, V.; Vacarescu, I.N.(1998). *Industrial robots: performances and testing,* Mirton Publisher, ISBN: 973-578-590-0, Timisoara, Romania

Annals of DAAAM for 2008 & Proceedings of the 19th International DAAAM Symposium, ISSN 1726-9679
Editor B. Katalinic, Published by DAAAM International, Vienna, Austria 2008
Make Harmony Between Technology and Nature, and Your Mind will Fly Free as a Bird
2nd DAAAM-ICAT

INVESTMENT CASTING WAX PART COOLING TIME DETERMINATION

VALENTAN, B[ogdan]; DRSTVENSEK, I[gor]; BRAJLIH, T[omaz]; HRIBERNIK, A[les] & BALIC, J[oze]

Abstract: *Investment casting of metal parts is a rapid tooling/manufacturing method for small series production, where wax core is melted out of plaster mould. Wax parts are obtained from wax casting in to silicone rubber moulds. After wax is casted some time is needed for wax to harden enough, that wax part can be removed. That time is critical, since if is to short, wax part will be destroyed, if it's to long, time is wasted and in production cycle this can not be tolerated. Simplified experimental method for wax cooling time determination is presented in this paper.*

Key words: *Investment casting, wax casting, silicone rubber moulding, cooling time*

1. INTRODUCTION

In a case of small series production (Buchmeister et al., 2008) Silicone Rubber Moulding (SRM) is the most cost efficient method. SRM are commonly used for castings of Polyurethanes, Wax, and Low Melting Point Alloys and even in food industry (Marzipan sweets). In case of Polyurethanes we exactly know, when our product is ready (polymerization finished). In case of Wax parts there is no simple and accurate method to determinate time, when our Wax part is finished (hardened). Results are damaged parts, that in some cases can be repaired but often they need to be cast again. This leads to excessive waiting for the wax part to cool down in order to be sure that it is properly hardened. Consequently a lot of time is wasted, reducing the efficiency of investment casting as a rapid tooling technology.

To determinate time needed for Wax to harden in SRM we need to take in consideration some parameters. From thermodynamics we can find some useful parameters and equations. From there we have two options:

1. Use one of commercially available software used for casting simulations (Moldflow, Moldex, Cosmos/Flow or similar more advanced methods (Kovacic et al., 2005). In this case we need to adjust parameters and get the proper material characteristics. Casting can then be simulated. This procedure can give us very good results, but it takes a lot of time, since we need exact CAD model of our mould and proper parameter preparation.
2. Use of simplified equations with experimentally chosen parameters.

Paper presents experiments needed to determinate time needed for Wax parts to cool down and derivation of simplified equation to determinate cooling time from basic mould and cast part parameters.

2. SRM AND COOLING TIME

Cooling time - time from casting till the time we open our mould, depends from several major influences. Usually we need pre-heated moulds, since wax needs slower cooling curves to obtain his shape. Cracks and too big material shrinkages are the most common defects caused by rapid cooling. Moulds used in experiment were pre-heated to 70 °C. Wax used in our experiment was industry wax from company MCP-Hek type WA-70 (MCP-Hek homepage) with melting point at 68-70°C and recommended casting temperature at 80°C. For cooling a temperature of 30 °C was sustained in oven. Cooling time is time from end of casting until our wax product in a core has reached 35 °C.

3. SRM MOULDS

For experiment several different moulds were used but for basic principle presentation only one will be presented. SRM moulds are usually block shaped (Fig. 1), this shape gives us uniform heat distribution from wax part through SRM to the surface.

Fig. 1. SRM mould for wax medal production.

4. MEASURING AND CASTING EQUIPMENT

Temperature measurements were taken with Metrix M-4560A digital multimeter (Fig. 2) equipped with thermocouple type **K**. Thermocouple has a class 1 tolerance of ±1.5 between −40 *°C* and 375 *°C* (Wikipedia - Thermocouple). Measuring accuracy was previously checked with ice-water mixture (0 *°C*) and boiling water (100 *°C*) to get maximal possible accuracy of the measurements. Wax was cast in Vacuum-chamber MCP 4/01 equipped with wax heater. Vacuum-chamber is capable of absolute pressure down to 5 *mbar*, where wax is cast to prevent non-homogeny structure caused by remaining air or water steam. To cool down the mould in a controlled environment electronically regulated oven was used with preset temperature of 30 *°C* (with tolerance ±1 *°C*). TFD 128 temperature and air moisture data logger was used to assure a control on mould surrounding air.

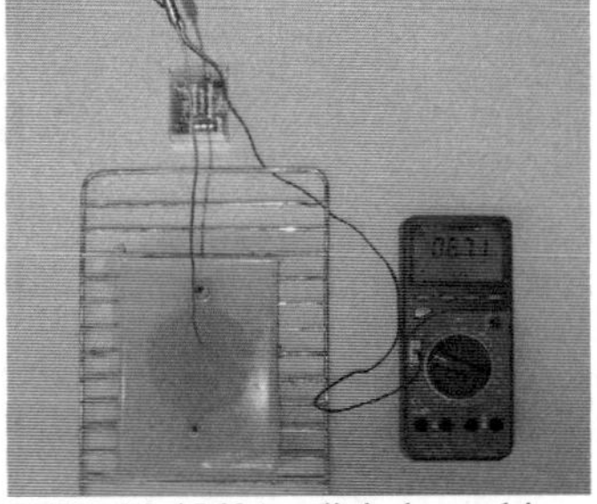

Fig. 2. METEX M-4560A digital multimeter used for temperature measurements. Multimeter measures momentarily

temperature, so measurements in known time intervals were needed. Data logger could be used to simplify data acquisition.

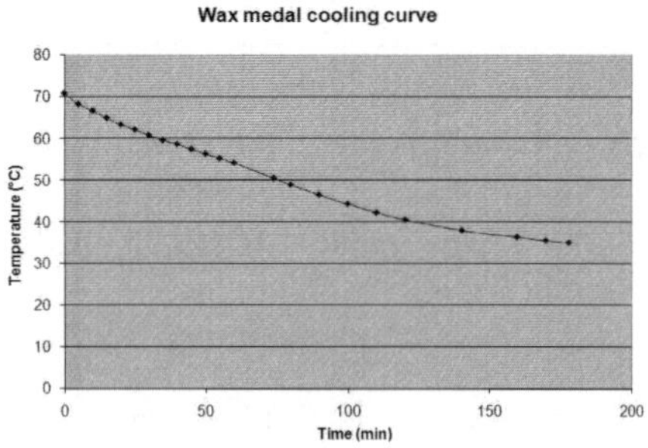

Figure 3: Cooling curve of medal mould. Cooling curves of other Wax parts are similar; cooling time depends from wax part and silicone mould mass and shape.

5. JOINT QUANTITI OF HEAT – Q

Joint quantity of heat from wax part and silicone rubber mould is heat quantity caused by temperature change and heat of fusion:

$$Q = Q_{wax} + Q_{silicone} \tag{1}$$

Wax heat quantity depends from wax mass, temperature change (wax casting temperature is 80 *°C* and end temperature of wax core is 35 *°C*)

$$\Delta T_v = T_{zv} - T_{kv} = 80°C - 35°C = 45°C \tag{2}$$

and Wax specific gravity and in our case also a heat of fusion must be taken into a consideration (Q_{sv} = 160 [*J/kg*]).

$$Q_{wax} = Q_{\Delta t} + Q\Delta = m_v C_{pv} \Delta T_v + m_v \cdot 160 \left[\frac{J}{kg} \right] \tag{3}$$

Silicone rubber mould heat quantity is calculated on the same principle:

$$Q_{silicone} = m_s C_{ps} \Delta T \tag{4}$$

Since silicone is preheated to 70 *°C* and end temperature of silicone is approximate 35 *°C*:

$$\Delta T_s = T_{zs} - T_{kv} = 70°C - 35°C = 35°C \tag{5}$$

This is joint quantity of heat that we need to lead out of the system in order to cool down wax part to 35 *°C*. Heat is delivered by conduction since wax is not moving in a mould after casting procedure is finished. Temperature of surrounding air is constant (or we presume that in our equations).

$$\Delta T = \frac{T_{zv} + T_{zs}}{2} - T_z = \frac{80°C + 70°C}{2} - 30°C = 45°C \tag{6}$$

6. THERMAL ENERGY TRANSFER THRUGH TIME

Our starting point is a simplified equation for thermal energy transfer:

$$\frac{\Delta Q}{\Delta t} = -K \cdot A \cdot \frac{\Delta T}{\Delta x} \Rightarrow \Delta t = \frac{Q}{K \cdot A \cdot \left(\frac{\Delta T}{\Delta x} \right)} \tag{7}$$

Because of simplification and some unknown parameters (material and geometrical parameters) we cannot calculate exact time and parameter *K* so approximation is needed:

$$K \Rightarrow k_B \tag{8}$$

Parameter k_B is experimentally set by measuring cooling time. k_B presents all unknown parameters, influence of all parameters are presented as linear.

Therfore, the required cooling time *t* is:

$$t = \frac{Q}{k_B \cdot A \cdot \left(\frac{\Delta T}{\Delta x} \right)} \tag{9}$$

Where is:
t – time of cooling for Wax part center,
Q – joint heat energy of our system,
k_B – correction factor,
A – outside area of silicone mould,
ΔT – average temperature difference of Wax/Mould/Surrounding system,
Δx – distance from center of wax part (on the thickest area) and mould outside shell.

From figure 3 we can see, that 35 *°C* is reached in 178 *min*. Parameter k_B derivate from equation 9:

$$k_B = \frac{Q}{t \cdot A \cdot \left(\frac{\Delta T}{\Delta x} \right)} \tag{10}$$

Where t is our cooling time (178 min); k_B is:

$$k_B = \frac{47285{,}16 \ J}{10680 \ s \cdot 0{,}0658 \ m^2 \cdot \left(\frac{45 \ °C}{0{,}0548 \ m} \right)} = 0{,}082 \left[\frac{J}{ms \ °C} \right] \tag{11}$$

7. CONCLUSION

Investment casting is one of indirect rapid manufacturing/tooling procedures (Badida et al., 2006) and wax core is there our key part. Wax parts must be absolutely perfect, since this is the only way to get parts of satisfying accuracy (Brezocnik et al., 2004) and shape. Besides wax vacuumation and vacuum casting, we need to establish slow and even cooling (in order to prevent micro cracks and other specific cooling defects). On the other side this is a small serial production procedure so we want to shorten the cooling time, to speed up the production. Cooling time is difficult to predict based only on experience, and since commercially available software is to expensive and time consuming, that leads to relatively large amount of unsuitable wax parts. These parts can be reused as a raw material, but a lot of time is wasted that way.

Presented calculations and measurements are only a part of research, since there are still a couple of unused factors that need to be taken in to a consideration to get better and more accurate end results. The largest error is caused by the shape of the wax part, resulting in an error of around ±15% between simple and complex shape (Valentan et al., 2008). For more accurate time calculations additional experiments are in progress and accuracy of ± 5 to 7% is expected.

8. REFERENCES

Badida, M.; Hricová, B. & Vargová, J. (2006). *Ecodesign and new product development*, 10th International Research/Expert Conference: "Trends in the Development of Machinery and Associated Technology", TMT 2006, Barcelona-Lloret de Mar, Spain, 11-15 September, 2006.

Buchmeister, B.; Pavlinjek, J.; Palcic, I. & Polajnar, A (2008). *Bullwhip effect problem in supply chains;* Advances in production engineering & management, Volume: 3 Issue: 1 Pages: 45-55, Mar. 2008.

Brezocnik, M., Kovacic, M. & Ficko, M. (2004). *Prediction of surface roughness with genetic programming* Journal of materials processing technology, Volume: 157 Special Issue: Sp. Iss. SI Pages: 28-36, Dec. 2004.

Kovacic M, Brezocnik M, Pahole I, et al (2005). *Evolutionary programming of CNC machines*; Journal of materials processing technology Volume: 164 Pages: 1379-1387, May 2005.

Valentan, B.; Brajlih, T.; Drstvensek, I. & Balic, J. (2008). *Basic solutions on shape complexity evaluation of STL data.* J. Achiev. Mater. Manuf. Eng., Volume: 26, iss. 1, Pages: 73-80, Jan. 2008.

Annals of DAAAM for 2008 & Proceedings of the 19th International DAAAM Symposium, ISSN 1726-9679
Editor B. Katalinic, Published by DAAAM International, Vienna, Austria 2008
Make Harmony Between Technology and Nature, and Your Mind will Fly Free as a Bird

THE FUTURE TREND OF THE OPTION MARKET IN ROMANIA

VANCEA, S[maranda]; HORJA, M[onica] I[oana]; AVRAM, E[leonora] L[aura] & IGNAT, A[ndreea]

***Abstract**: The present paper tries to bring into light particular aspect regarding the Romanian Market dedicated for option on futures trade. In order to realize a viable analyze about Option Trading Volumes in Romanian Capital Market is necessary to make an analytic overview of the past options trading volume. The importance of such study consists in revealing the option market sentiment, useful in pointing the option trading place in Romania. The anticipated trend of option trading volume is based on real data analyze realized in 1998 – 2007. The mathematical calculation reveals a linear trend for the next 3 years, in case of the same influences factors constancy previously considered.*
***Key words:** option market, trading trade.*

1. INTRODUCTION

The paper brings into focus the futures option trend at the Sibiu Monetary–Financial and Commodities Exchange (SMFCEX). Within this market there are traded options that have as underlying asset future contracts traded within the same market. What kind of market is this? A futures or an option market? A correct answer to such a question can be made only appealing to the specialist's definitions. We remark the opinions of two specialists in the domain. On one hand there is the opinion of Dragoescu E. who says that "in Romania there isn't a distinct options trade market, because until now the options trades resumed at options having as underlying asset futures contracts"(Dragoescu, 2001). On the other hand there is also the opinion of the specialists from SMFCEx, who are saying that "options that have as underlying assets futures contracts [...] are being traded exclusively at futures markets"(Ancuta, 1999).

2. OPTION MARKET IN ROMANIA

As an emergent capital market, "the growing potential of the exchange market from Romania is very high" (Corduneanu & Iovu, 2008), but also in direct correlation with reaction on important capital market in the word (US, Frankfurt, Tokyo).There are studies (Cerny, Koglas, 2008) that suggest that the emergent markets, but not only them, react very quickly to the information revealed in the prices from other markets. The derived financial instruments market organized by the Sibiu Monetary-Financial and Commodities Exchange (SMFCEx) is the first market regulated in term established in Romania. Within it where established the bases of futures contracts transactions, in the first instance and subsequent options contracts transactions. As it is well known, options are derived financial contracts, that offer the buyer the right, without obliging him, to choose in the matter of exercise the contract stipulations, about an underling asset. The reservation of this right realizes through the payment of a sum of money negotiated with the partner. Since 1998, SMFCEx is the only regulated option market, having as underling assets the futures contracts(options on stocks futures –29 instruments (80% of total options / the support stocks form futures contracts are the most liquid units traded at Bucharest Stock Exchange); options on currency futures – 3 instruments (10% of total options); options on SIBEX futures – 2 instruments (4% of total options); options on interest rate futures – 1 instruments (3% of total options); options on gold futures – 1 instruments (3% of total options). Today the market of option contracts negotiations is a regulated market, organized according to the stipulations of the capital market law, by the Monetary–Financial and Commodities Corporate Sibiu, which administrates the derived financial instruments market, SMFCEx, so that it functions regularly, by the regulations that are emitted and subjected to the approval of the market surveillance authority, The National Commission of Securities (CNVM) and which also respects the requirements needed to relate to it.

3. THE OPTIONS TRADING VOLUMES IN ROMANIAN ANALYZE

The assessment of an activity within an option market is based on the trading volumes indicator, meaning on the number of closed deals and on the number of opened positions on the market, at a given time. The value of the options transactions is not a relevant indicator for the option market, because the transactions value is related to the value of the bonuses negotiated at the market, and they only represent the prices of some contracts, not the prices of the underlying asset to which the option contracts report to. What information are the trading volumes within an option market offering us? This type of information is useful to evaluate the market's "sentiment" (Summa, 2004). A series of investors are using this information to take trading decisions. In the situation when an increase of the trading volumes is being estimated, a speculative fervor is suggested. If this volume is based on PUT contracts, the tendency of the underlying asset's quotation evolution is bearish. If this volume is based on CALL contracts, it signifies a bullish tendency of the underlying asset's market.

The Analyse Methodology Basis: the problematic of the perspectives of the Option Market in Romania is investigated appealing to the extrapolation methodologhy, as a research method adequate in the Romanian capital market context. The research is based on option contracts number evolution trend between 1998 and 2007 within SMFCEX. The analyse methodology is based on the following: every influence factor of the traded contracts volumes are known and are contained by the method; these factors will manifest influences in the future time; the random factors that can appear in the future can't be estimated

Year	Option	Futures	Total
2007	34.900	3.456.023	3.490.923
2006	36.651	4.232.059	4.268.710
2005	11.629	696.109	707.738
2004	2.273	72.901	75.174
2003	19.369	168.545	187.914
2002	67.300	225.069	292.369
2001	53.731	135.242	188.973
2000	46.752	158.536	205.288
1999	15.215	159.927	175.336
1998	133	193.203	193.336
Total	287.953	9.575.491	9.863.444

Tab. 1 the Trading Volume in SMFCEX

Souce:http://www.sibex.ro/?sibex=raport/2007

The dates from table 1 werw used in order to realized the specified analize.

The graphic presentation of the number of option contracts traded within SMFCEX between 1998 and 2007 is not suggestive to determine the tendency manifested within the market, which imposes multiple variants testing.

Testing the Parabolic Trend: the adusting function for a parabolic trend has the mathematical form as the equation (1) show us, with the a,b,c real parameters which are determined by imposing the conditions (2) and the time has no influence (3)

$$\hat{Y}= a+b*t+c*t^2 \quad (1)$$

$$(Y_t-\hat{Y}_t)^2 \text{ is minimum} \quad (2)$$

$$(\Sigma t = 0) \quad (3)$$

That means that a ,b ,c will be the sistem's solutions :

$$\begin{cases} n*a+c*\Sigma t^2 = \Sigma y \\ b*\Sigma t^2 = \Sigma t*Y_t \\ a*\Sigma t^2 + c*\Sigma t^4 = \Sigma t^2 * Y_t \end{cases} \Leftrightarrow \begin{cases} 10a+330c = 253.063 \\ 330b = 276.951 \\ 330a+19.338c = 4.899.199 \end{cases}$$

The adjusting function will be:

$$\hat{Y}_t=395.623+839*t-432*t^2 \quad (4)$$

The adjusted values are: $\hat{Y}_1$=353.080, $\hat{Y}_2$=368.582, $\hat{Y}_3$ =380.628, $\hat{Y}_4$=389.218, $\hat{Y}_5$=394.352, $\hat{Y}_6$=396.030, $\hat{Y}_7$ =384.252, $\hat{Y}_8$=389.018, $\hat{Y}_9$=380.328, $\hat{Y}_{10}$=368.182.

The large adjusting values resulted from the calculation are imposing the testing of new evolution variants of the option contracts number.

Testing the Exponential Trend The exponential adjusting function has the form as the equation (5) show us, with a,b real parameters which determine the condition (2) and (3)

$$\overline{Y}=a*b^t \quad (5)$$

The conditions imposed lead to the system's solutions :

$$\begin{cases} n*\lg a = \Sigma \lg Y_t \\ \Sigma t^2 * \lg b = \Sigma t * \lg Y_t \end{cases} \Leftrightarrow \begin{cases} 10 \lg a = 37,80766 \\ 330 \lg b = 40,48954 \end{cases} \Rightarrow$$

a=6.036 and b=1.327 ⇒ the adjust function is (6):

$$\overline{Y_t}=6.036*1.327^t \quad (6)$$

The adjusted values will be: $\overline{Y_1}$=473, $\overline{Y_2}$=833, $\overline{Y_3}$=1.467, $\overline{Y_4}$=2.583, $\overline{Y_5}$=4.549, $\overline{Y_6}$=8.010, $\overline{Y_7}$=14.106, $\overline{Y_8}$=24.838, $\overline{Y_9}$ =43.743, $\overline{Y_{10}}$=77.019.

The high adjusting values resulted, do not lead, for now, to the conclusion that the exponential trend variant of the option contracts number evolution is the most designated for the realization of stipulations related to future possible evolutions.

Testing the Linear Trend: the linearity in the evolution of the number of traded option contracts signifies a relative constant evolution. The linear adjusting function is presented in equation (7) where a and b are real parameters which can be determined from conditions like (2) and (3):

$$\tilde{Y}=a+b*t \quad (7)$$

It results that a and b are the system's solutions:

$$\begin{cases} n*a = \Sigma Y_t \\ b*\Sigma t^2 = \Sigma t * Y_t \end{cases} \Leftrightarrow \begin{cases} 10a = 253.063 \\ 330b = 276.951 \end{cases} \Rightarrow a=25.306,3 \; b=839$$

b>0 means an incerasing tendency. As follows, the adjusting function is (8):

$$\tilde{Y}=25306+839*t \quad (8)$$

The adjusting values will be: $\tilde{Y}_1$=17.759, $\tilde{Y}_2$=19.433, $\tilde{Y}_3$ =21.107, $\tilde{Y}_4$=22.789, $\tilde{Y}_5$=24.467, $\tilde{Y}_6$=26.145, $\tilde{Y}_7$=27.823, $\tilde{Y}_8$=29.505, $\tilde{Y}_9$=31.179, $\tilde{Y}_{10}$=32.857

4. CONCLUSION

In conclusion: $\Sigma (Y_t-\hat{Y}_t)^2 > \Sigma (Y_t-\overline{Y_t})^2$ and $\Sigma (Y_t-\hat{Y}_t)^2 > \Sigma (Y_t-\tilde{Y}_t)^2$ which suggests that the parabolic trend is not proper to aproximate the option trading's evolution at SMFCEX. The results with the smallest adjustments correspond the best to the linear trend.

Such a solution determines us to be pesimists relating the evolution of the number of option contracts that will be traded within SMFCEX.

An explanation related to the slow growth of the number of option contracts in comparison with the one of the futures contracts could be the low notoriousness of the product between the effective investors from the market, and probably the low tolerance relating the acceptance of the trading risks of a sofisticated market product, topics that deserved to be investigated for the next researches.

Obviously there are limitations of the present research consisting in low relating options trading iphotesis whit macroechonomics indicators

5. REFERENCES

Ancuta, T.(1999). *Everything about Futures and Options*, Publisher Tipotrib, ISBN 973-98325-6-3, Sibiu,

Cerny, A., Koglas, M. (2008). Stock Market Integration and the Speed of Information Transmission, *Czech Journal of Economics and Finance*, Vol.58/2008, page 2-20, ISSN 0015-1920

Corduneanu, C, Iovu, L.R (2008). Steps in the Development of the Romanian Financial System and the Correlation with the Level of Economic Growth,*Timisoara Journal of Economics*, No.1/2008, ISSN 1844-7139

Dragoescu, E. (2001). *Commodities Exchanges,* Publisher Risoprint, ISBN: 973-656-093-7, Cluj-Napoca

Summa, J.F. (2004). *Trading Against The Crowd,* Publisher Wiley Trading, ISBN: 0-471-47121-6, New Jersey

http://www.sibex.ro/?sibex=raport/2007/, Accessed on: 2008-07-12

Annals of DAAAM for 2008 & Proceedings of the 19th International DAAAM Symposium, ISSN 1726-9679
Editor B. Katalinic, Published by DAAAM International, Vienna, Austria 2008
Make Harmony Between Technology and Nature, and Your Mind will Fly Free as a Bird
*Young Researcher and Scientist Paper / * Supervisor, Mentor*

DYNAMIC AND THERMAL ANALYSIS OF THE FEED DRIVE SYSTEM

VARBAN, R[amona] C[ristina]; PARPALA, R[adu] C[onstantin] & PREDINCEA, N[icolae]

Abstract: *Using a set of dedicated software's engineers are able to analyze all aspects of real life usage of machine tools elements without spending money on real prototypes. A high-speed machine tools drive system generates high frictions at contact areas, into the screw-ball mechanism and bearings, thereby causing thermal expansion of the screw which adversely affects machining accuracy. In the dynamic analysis the position of all bodies are determinate as a result of time dependent forces applied to the mechanism. The finite element method is used to determinate dynamic characteristics of the feed drive system, e.g. natural-frequencies and mode shapes. The structural behaviour under dynamic and thermal loads is evaluated in order to obtain an optimized design of the feed drive.*

Key words: *thermal, FEM, thermocouple, natural-frequencies*

1. INTRODUCTION

In a feed drive system, the ball screw plays an important role as a power transmission unit and a linear scale. When a large speed of motion is demanded, the rate of feed of the ball screw transmission is much increased. Much heat is produced at both ends of the bearing support and the nut because of a greater speed of rotation (Kim & Cho, April 1997).

The accumulated heat causes the temperature to rise in these areas. Then the ball screw deforms thermally and has a seriously negative effect on machine accuracy. Several types of analysis must be performed for a complete study of machine tools behavior. The structural behavior under thermal and dynamic loads is evaluated in order to obtain an optimized design of the feed drive.

Because of the impact that feed drives imply over the finite piece quality the design demands are very high so accurate analysis must be performed in order to assure a very good behavior of the whole machine tools.

2. COMPUTER AIDED DESIGN

One of the first steps in virtual prototyping is building the CAD model. The 3D model of the feed drive must be designed in order to be accepted as input by various software suits that will be used for further analyses (Altintas et al., 2005).

During the design phase simplified simulation models are used to estimate the impact of design parameters over the machine performance. These simplified models are also used in order to improve the time needed for calculation and to eliminate the computing errors. It is well known that a very complex model can generate errors during the FEM analysis; because of this many features from the 3D model are eliminated. Also some details that are essential for FEM analysis are not so important for Kinematical analysis and must be ignored. A CAD model must be easily redesigned so it's important to use parameters to define all the key elements of the model. The 3D model of the feed drive was designed by using the CATIA V5 CAD software mainly because of his good integration with the ANSYS software which was used for statically and dynamical FEM analyses.

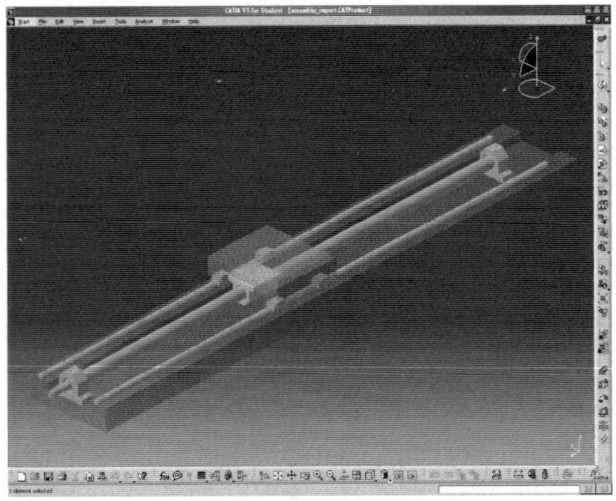

Fig. 1. 3D CAD model of the feed drive.

All 3D part where fully parameterized in order to optimize needed parameters in FEM analysis.

In order to generate all the contacts between surfaces it's necessary to correctly design the 3D assembly, it's also very usefully to check all the clashes and clearances within the CAD environment. Using the information provided by the CAD software we can set the correct tolerances for the automatic contact generation.

The CATIA software provides some very usefully tools for space analysis. In fig. 2 it's an example of a misplaced component. In this case the error is detectable by a visual inspection of the model. Because we need to use the 3D model for further analyses it's a good practice to define also material properties by using predefined model from the CATIA library or by defining new materials. By defining material properties we can also check important aspects of the assembly like volume, mass, moment of inertia.

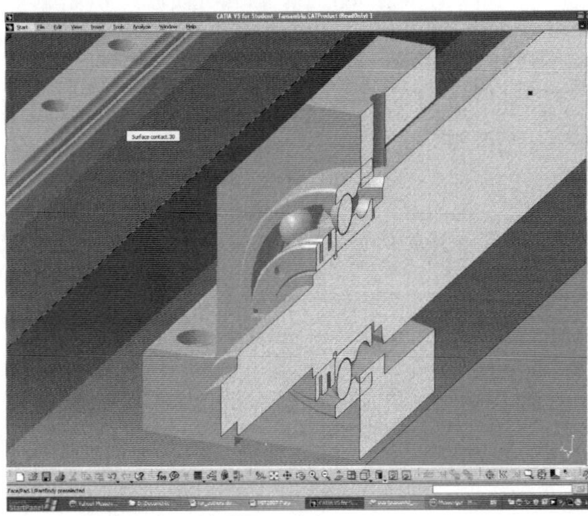

Fig. 2. Section through 3D model.

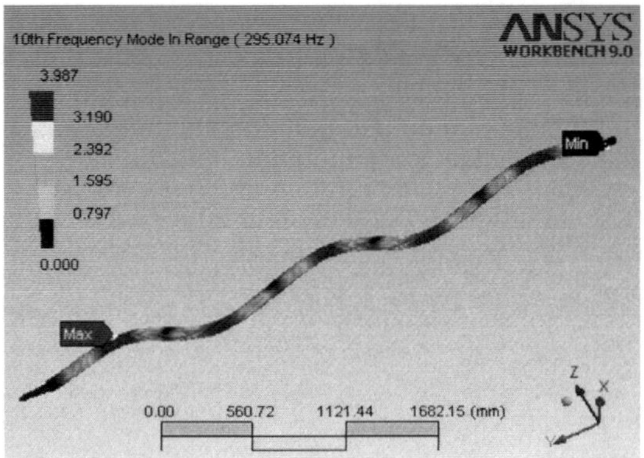

Fig. 3. 10^{-th} Frequency mode for the screw-drive

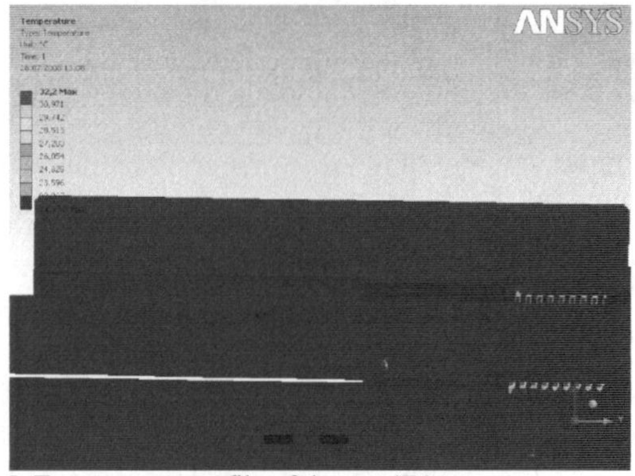

Fig. 5. Temperature profile of the nut (T_2).

3. ANSYS FEM ANALYSIS

3.1. Dynamic analysis for the feed drive system

The dynamic response of a machine, structure, or system can be determined by superposing it's natural modes of vibration when the amplitudes of motion are small. Thus a complete dynamic description of the machine requires the determination of the modal frequencies, mode shapes, and the system parameters-equivalent mass, stiffness, and damping ratio. The procedure determining this information of a system is called Modal Analysis (Zaeh & Oertli, 2004).

The Finite Element Method (FEM) is widely used to perform a Modal Analysis. FEM is extremely useful for complicated devices and structures with unusual geometric shapes. The frequencies that are calculated by the program (Table 1) can be further used for other verifications.

Mode	Frequency
1	22,29 Hz
2	22,3 Hz
3	61,41 Hz
4	61,43 Hz
5	120,18 Hz
6	120,23 Hz
7	198,15 Hz
8	198,23 Hz
9	294,94 Hz
10	295,07 Hz

Table 1. Natural frequency.

3.2. Thermal analysis for the feed drive system

The thermo-elastic machine behavior, i.e. the load dependent temperature distribution and the resulting deformation of the machine tool, is influenced by a variety of constructional and thermotechnical parameters (Schmitt, 1996).

The legend of figure 4: 1 - uniaxial feed drive system; 2 - table; 3 - servomotor; 4 - ballscrew; 5 - thermocouple; 6 - displacement gauge; 7 - data processing software; 8 - data collection system.

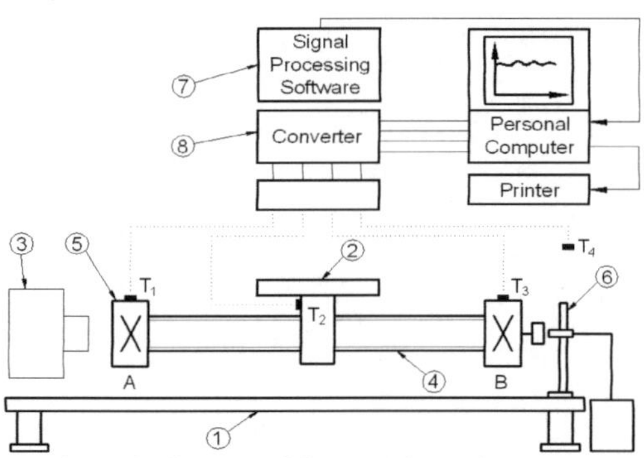

Fig. 4. Schematic diagram of the experimental set-up.

The objectives of the experiments were to establish a system to measure the ball screw transmission system thermal errors, to determine a suitable model for the thermal error data, and to predict (and to compensate) thermal error according to the determined model.

The thermocouples were attached to positions T_1, T_2 and T_3 as shown in fig. 4 to measure temperature increases of the front bearing, nut and rear bearing as key heat sources, respectively. The fourth thermocouple was used to measure the environmental temperature (Predincea et al., 28-30 May 1998).

The variation of temperature rises of positions T_2 and the thermal error is shown in fig. 5.

4. CONCLUSION

¶

By using ANSYS Workbench one is able to optimize the design process by changing one or more of the initial parameters; those parameters are automatically updated into the CATIA 3D CAD model. By analyzing the calculation result in the post-processing program the designers can evaluate the machine properties during the design stage.

Today the main problem in checking structures consists in importing and preprocessing the CAD model. It is well known that the geometry of the model can dramatically change FEM results. The multiple regression analysis is adequate to predict the ball screw thermal errors with variation of table speed and temperature history. The predicted thermal error data can be used to correct the error with a suitable numerical control route.

The principle and the method used in this work can be used to solve a system of multiaxis feed drives.

¶

5. REFERENCES

Altintas, Y.; Brecher C.; Weck M. & Witt S. (2005). *Virtual machine tool*, Annals of the CIRP , 54/2: 651-669.

Kim, S.K. & Cho, D.W. (April 1997). *Real-time estimation of temperature distribution in a ball-screw system*, International Journal of Machine Tools and manufacture, Vol. 37, No. 4, pp. 451-464.

Predincea, N.; Pupăză, C. & Toma, O. (28-30 May 1998). *Analiza deformatiilor termice ale mecanismului surub-piulita*, (*Analysis of thermal deformations of ball-screw system*), The VIII^{-th} Conference of Managerial and Technological Engineering, pp. 499-506, Timişoara, România.

Schmitt, T. (1996). *Modell der Wärmeübertra-gungsvorgänge in der mechanischen Struktur von CNC-gesteuerten Vorschubsystemen*, (*Model of the heat transfer in the mechanical structure of a CNC feed drive system*), ISBN 3-8265-1476-9.

Zaeh, M. & Oertli, Th. (2004). *Finite Element Modeling of Ball Screw Feed Drive Systems*, Annals of the CIRP, 53/1: 289-292.

Annals of DAAAM for 2008 & Proceedings of the 19th International DAAAM Symposium, ISSN 1726-9679
Editor B. Katalinic, Published by DAAAM International, Vienna, Austria 2008
Make Harmony Between Technology and Nature, and Your Mind will Fly Free as a Bird

PRODUCT DESIGN AND SUSTAINABLE DEVELOPMENT IN ENGINEERING PROCESS

VARGOVA, J[ana]; BADIDA, M[iroslav] & HRICOVA, B[eata]

Abstract: *Engineers are able to optimise the design using their knowledge of scientific principles. Decisions are made in the design process that have a considerable effect on the sustainability of the product, process or system. Engineers can also be involved in the design process as a representative of the organisation that commissions the design or the organisation that will use it. The core of the design process is the function of the product that is to be designed. Function does not only mean the technical function, but also any social, cultural, psychological and economic functions the product will perform.*

Key words: *environment, design, ecodesign, sustainable development, Life cycle analysis*

1. INTRODUCTION

Traditionally, the design process is seen as one link in a chain of steps. An organisation commissions a new process or product to help it reach its goals. The engineer that accepts the task of delivering the design is given a programme of requirements, which the product should ultimately adhere to. This list of requirements is also used subsequently to evaluate the design that the engineer has produced.

Once the designer has accepted the programme of requirements, the next step is to obtain a solution that fits into all requirements completely (even if the demands seem to be inconsistent). This lack of critique depends on the engineer's culture.

2. THE DESIGN PROCESS

Most engineers concern themselves with physical products in the broadest sense, i.e. every physical system designed with a certain purpose is a product. A bicycle is a product, but so is a factory, a water treatment facility, and a new area of a city. Designing is the developing and planning of such a product.

Different disciplines have different design processes, though they share some common characteristics. The design process can be characterized by the cycle of design, which describes certain steps present in each design process (Fig. 1). When designing for sustainable development, designers should bear it in mind throughout the design process.

The most important decisions concerning sustainable development take place in the initial phase of the design process—analysis; the earlier sustainable development plays a role in the process, the larger its influence. It is much easier to alter the assignment to improve sustainability than trying to increase the sustainability of an already finished detailed design. (Mihok & Liberkova, 2005)

2.1 Analysis

The core of the design process is the function of the product that is to be designed. 'Function' does not only mean the technical function, but also any social, cultural, psychological and economic functions the product will per-form. Although the programme of requirements will contain the main functions, these need to be analysed for further requirements.

Every design process follows from a problem that needs to be solved. It is the designer's job to identify the actual problem, clarify it and express it in a problem statement, i.e. the designer analyses what the real problem is. This often turns out to be a different problem from the one expressed by the commissioner. Only when the problem is defined clearly and in 'do-able' terms can the designer search for the most sustainable way of solving it. (Rusko & Voláková, 2004)

The designer may conclude that major changes need to be made for a sustainable solution. But the commissioner will frequently not allow the designer to bring about such a large change. The programme of requirements is generally so restrictive that only small product improvements are possible. The engineer should therefore review the commissioner's programme of requirements critically before accepting a design assignment.

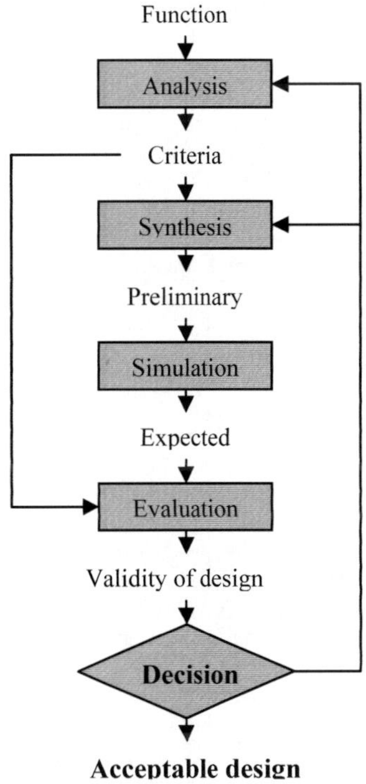

Fig. 1. The cycle of designing

3. LIFE-CYCLE ANALYSIS (LCA)

LCA is a tool that allows the total environmental impact of a design or a product to be analysed. It can be used during different phases of the design process. It can also be used to optimise the environmental performance of a design.

LCA quantifies the environmental impact of a certain product-system. The LCA of an existing product or system can set the bottom line for a new design. The product system encompasses all phases of the product life, i.e.

- Raw materials acquisition and refining (e.g. mining,

drilling, agriculture, forestry, fisheries)
- Processing and production of product and production equipment
- Distribution and transport
- Use, re-use and maintenance
- End-of-life landfilling, incineration, litter and recycling

In all these phases, the contribution of the product to different forms of pollution (e.g. the greenhouse effect, ozone layer depletion and acidification) is calculated.

However, these different forms of environmental impact cannot be added together. In order to calculate one single number as the result of the LCA, weight factors have to be introduced that set the relative priority for each environmental problem. Weight factors can be derived from the relative distance of the current situation in regard to the goals set out in policy documents. (Muranský & Badida, 2005)

Alternative designs and materials can thus be compared. If priorities change, the LCA score will change too.

4. DESIGN TOOLS AND STRATEGIES

There is a large number of strategies that a designer can follow to design eco-efficiently. In the LiDS wheel (Lifecycle Design Strategies), these strategies are clustered and visualised. Each strategy contains a number of basic rules. (Mulder, 2006)

Strategy 1: choose materials with low environmental impact

Chose materials that are:
- Clean materials. Choose non-toxic and harmless materials. For example, avoid materials containing heavy metals, asbestos, chlorofluorocarbons (CFCs) and endocrine disruptors such as phthalates.
- Renewable materials. Avoid scarce materials, i.e. materials from a non-renewable or slowly renewing sources such as fossil fuels, copper, tin, zinc and platinum. Plastics too are made from fossil fuels and are counted as scarce materials.

Strategy 2: dematerialise

The principles of dematerialisation are:
- Reduction of weight. Using less material often reduces the product's environmental impact. Less material means less resource consumption, less waste and a lower environmental impact during transportation.
- Reduction in volume. When the product and its packaging are reduced in size and volume, more products can be transported in a given transport facility, making transportation more efficient. Another solution is to make the product foldable or 'nest-able'.

Strategy 3: select environmentally efficient production techniques
- Environmentally sound production processes. For example, DSM-Cist in Delft replaced 15 chemical production steps in pharmaceutical production with five bio technological ones. This drastically reduced the consumption of water and harmful emissions.
- Fewer production steps. For example, combining various parts into one cast item can save assembly steps and simplify the design. Chose materials that require no finishing touches or only apply finishing to parts that absolutely require it (e.g. because they are visible).
- Lower/cleaner energy consumption. Make efficient use of (sustainable) energy sources. Use sustainable energy sources such as windmills and solar cells in the production process. Also use waste heat from other companies situated nearby.

Strategy 4: select an environmentally sound distribution system
- Less and cleaner packaging. Use less packaging material or packaging with a lower environmental impact, e.g. replace polystyrene packaging with cardboard. Bioplastics made from starch, for example, can be a good replacement for plastic foams in some applications. A reduction in packaging can also be obtained by designing the product in such a way that there is no need for packaging or less packaging is needed.
- Energy-efficient transport mode. Some modes of transport are more polluting than others. Freight transport by train or boat is preferable to transport by road or plane.
- Energy-efficient logistics. The cleanest transport is no transport. Nowadays, production sites in a supply chain are often far apart. Reducing transport distances can result in significant energy savings.

Strategy 5: reduce environmental impact in the use phase

A product also has an environmental impact during the use phase. This should be considered in the design process. The following principles hold:
- Reduce energy consumption during use. For example, energy consumption by a vehicle is generally far greater than the energy consumed in its production.

Strategy 6: optimise the life-span

A product with a long life-span will have a lower environmental impact because fewer materials are used overall. Prolongation of the technical lifespan can be obtained by making:
- The product more reliable and easier to repair.
- Maintenance easier.

Strategy 7: optimise end-of-life system
- Stimulate re-use of the entire product.
- Remanufacturing/refurbishing. Stimulate re-use of parts. But to enable the re-use of parts, it is necessary that the product can be dismantled easily.
- Recycling of materials the less variation in materials used, the more efficiently materials can be recycled.
- Safe incineration. Avoid the use of substances that generate hazardous fumes when incinerated. (Mulder, 2006)

5. CONCLUSION

A company´s societal an economical responsibilities are more than just making a profit. The role of companies is to create value by using scarce goods in an effective and efficient manner to produce goods and services. Sustainable entrepreneurship generally implies that corporations are willing and able to behave responsibly towards society and environment.

The paper was elaborated in connection with the projects VEGA No. 1/3232/06, solved at the Technical University in Košice.

6. REFERENCES

Hauschild, M., Jeswiet, J. & Alting, L.(2004): "*Design for the Environment – Do we get the focus right*". 2004 Annals of CIRP Vol. 53/1/2004; pp 1 – 4

Mihok, J.- Liberkova, L.(2005): *Vyhodnocovanie efektívnosti znižovania miery zaťaženosti ŽP. Evaluation of efficiency of environmental load elimination.* In: Moderné prístupy k manažmentu podniku, Bratislava, STU, 2005, s. 360-363,ISBN80-227-2284-7

Muranský, J.- Badida, M. (2005): *Ekodizajn v strojárstve. Základy metodiky. Ecodesign in Mechanical Engineering.* Technická univerzita v Košiciach, Vydavateľstvo Michala Vaška, Prešov, 2005, 304 s., ISBN 80-8073-119-5

Mulder, K (2006): *Sustainable development for engineers.* Greenleaf publishing, Sheffield, UK. 288p. ISBN-10: 874719-19-5.

Rusko, M., Voláková, M.(2004): *Prieskum využívania environmentálneho označovania a LCA na Slovensku. Review of utilisation of environmetal labelling and LCA in Slovakia.* Environmentální značení 2004/2, Praha, 2004, s. 6-7, ISSN 1212-4761

Annals of DAAAM for 2008 & Proceedings of the 19th International DAAAM Symposium, ISSN 1726-9679
Editor B. Katalinic, Published by DAAAM International, Vienna, Austria 2008
Make Harmony Between Technology and Nature, and Your Mind will Fly Free as a Bird

QUALITY AND COST IMPROVEMENT IN SPECIFIC VEHICLES MANUFACTURING

VARTOLOMEI-M, M[ihael]; JADANEANT, M[ihai]; VARTOLOMEI, M[ihaela]; POPESCU, M[ihaela] & DRAGHIA, M[arius]

Abstract: *The on-track vehicles don't have surplus of suspension or shock absorber, so the shock and vibration effects are more aggressive for the man inside. In this paper, starting from the tank real model we drew the equivalent model: a metal bar with two degree of freedom, six weights coupled with the metal bar representing the cobbles, the dumping constants, and the elastic constants of the torsion metal bars and helical spring, the elastic constants and shock-absorber constants of the track and cobble rubber rim. Than we calculated noise level reduction by using noises absorber materials, and we propose intelligent materials for the noises absorber room manufacturing in order to damp the shocks and vibration in specific vehicles and airplanes. All these are arguments for the designers and manufacturers to ameliorate the life and fight conditions in military technique for the success of the security and military missions.*

Key words: *intelligent materials, on-track vehicle, vibration effects, noises amelioration, quality and cost management.*

1. INTRODUCTION

In the activity of tank soldier, the main problem is to be insured some specific conditions in order to eliminate or reduce the vibrations and their effects (Krause, 1995). In this millennium the process of substitution the classic materials with superior technical-functioning and economic features is very necessary, especially in armored transporters engineering realm, in naval and aerospace industry, for reduce armored vehicles' vibrations and harmful effects. The response is simply: „new materials", "intelligent materials" (Vartolomei-M & Jadaneant, 2007b), composite based on boron filament, glass, asbestos, carbon and Kevlar, nano-materials, materials with form's memory, amorphous metallic materials (Popescu, 2004).

2. ON-TRACK VEHICULES AND HARMFUL EFFECTS

Caterpillar transporters, as any land motor, don't have surplus of suspension and own absorption, so the effects of the shocks and vibrations are more aggressive for tank man than other vehicle's driver. The experimental studies regarding the influence of the vibrations on human body have been made generally in environment of exciting with sinusoidal vibrations. But, during driving a tank in hilly land, the vibrations are certainly different than the lab experiences. The land where the tanks are used is so hilled that the multidirectional requesting leads to resultant forces much bigger than those who action in the lab experimental situations (Vartolomei-M & Jadaneant, 2007a). The research results emphasized a negative impact for the driver ability, especially regarding to maintain the human equilibrium and a constant pressure of the foot on the pedals (Darabont et al., 1983). In this study, the real model (figure 1 a) is replaced with its equivalent model (figure 1 *b).*

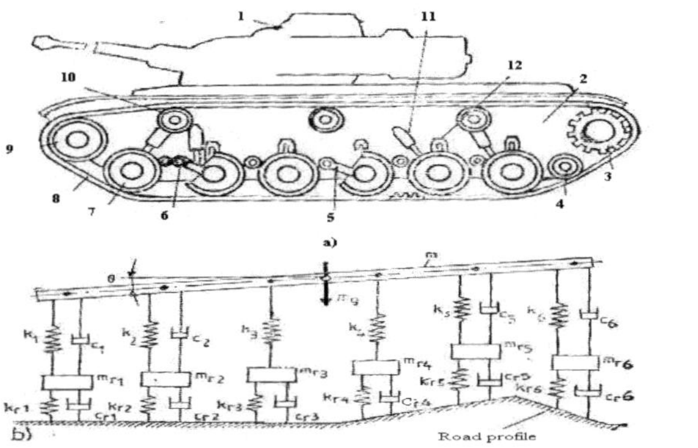

Fig. 1. The tank – real model (a) and equivalent model (b): 1 – turret; 2 – tank's body; 3 – drum for training; 4 – wheel for guiding and expending; 5 – oscillatory arm; 6 – torsion bar; 7 – cobble wheel; 8 – track; 9 – wheel for track guidance and expending; 10 – support roller; 11 – shock-absorber; 12 – helical spring.

In the equivalent scheme, the body of the tank is drawn as a metal bar with two degree of freedom (it can make vibrations on vertical and balance in rocking). Six weights are coupled by this metal bar (from m_{r1}, to m_{r6}, figure. 1 *b),* representing the cobbles. The torsion metal bars and helical spring have the elastic constants $k_1,...,k_6$, and the dumping constants $C_1,...C_6$. The track and cobble rubber rim features are represented by the elastic constants $k_{r1},...,k_{r6}$ and shock-absorber constants $C_{r1},...C_{r6}$.

3. NOISES AMELIORATION IN ROMANIAN ON-TRACK VEHICULES TECHNIQUE

Noise reduction in a room treated with noises absorber materials can be obtained by reduction the reverberation time.

Starting with the work of Darabont et al. (1983), we calculated the reduction of noises level by covering noises sources with noises-absorber materials, using relation (1):

$$L = 10 \lg \frac{A_1}{A_2} [\text{dB}] \qquad (1)$$

Where: A_1 is the equivalent absorption surface of the room after noises-absorption treatment, in m^2 AU (squares meters acoustic unities),

A_2 is the equivalent absorption surface of the room before noises-absorption treatment, in m^2 AU.

In the case of rooms untreated with noises absorbers the value of the equivalent absorption surface, A_2, is determined with relation (2):

$$A_2 = 0.35 V^{3/2} [m^2 \text{AU}] , \qquad (2)$$

Where V is the volume of the room, in m^3. The equivalent absorption surface A_1, is calculated with relation (3):

$$A_1 = 0,35\, V^{3/2} + \sum_{i=1}^{n} A_i [\mathrm{m}^2 \mathrm{AU}] \quad (3)$$

where $A_1 \ldots A_i$ are the equivalent absorption surfaces of some acoustic treatments, in m^2AU.

Every surface is calculated with the relation (4):

$$A_i = \alpha_i S_i (i = 1,2,\ldots,n) \quad (4)$$

where α_i is the acoustic absorption coefficient of some construction elements of the room.

S_i is that surface with the acoustic absorption coefficient α_i, in m^2.

Absorption coefficient of noises isolation and absorber materials α (%) are given in special tables in Romania.

For reverberation time T we can use the relation (5):

$$T = -\frac{0,163V}{S(1-\alpha_{med})} [\mathrm{s}] \quad (5)$$

Where

$$\alpha_{med} = \sum_{i=1}^{n} \frac{\alpha_i S_i}{S} \quad (6)$$

As we can see in figure 2, the noise can be propagated by such elements: the room's walls (1), leaky spaces and inevitable orifices (2,3), elements for linking the frame and the machine (4), vehicles' components passing that are rigidly linked by the room's walls (5), elements of the machine, leaky passage of some elements of the machine through the room's walls (pipes, tubes, axles etc), neighboring elements of the construction rigidly linked by the room's walls (7,8), construction elements the room and the machine are neighbors (9,10). In order to determine the attenuation degree of the noise level achieved by using a room for a noisy source (if we consider that there is a diffuse field inside this room), we use relation (7):

$$\Delta L = R - 10\lg\frac{S_C}{A_C} = R - 10\lg\frac{S_C}{\sum_{i=1}^{n}\alpha_i S_i}[dB], \quad (7)$$

where R is the sonorous reduction of the room's wall, in dB;

S_C is the surface of room's wall, in m^2;

A_C is the equivalent noise absorption surface of the interior of the room's wall, in m^2AU;

α_i is the noises absorption coefficient of the elements from the room's interior;

S_i is the corresponded inner surfaces, in m^2;

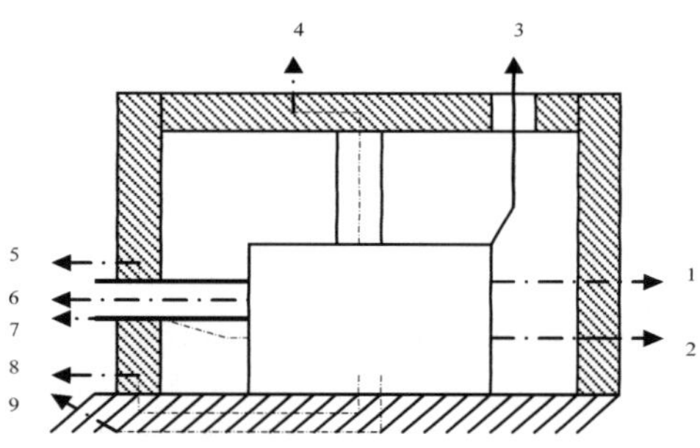

Fig. 2. Noise propagation in the case of noise source shielding
Source: Darabont et al., 1983, pg.77

We noticed that from expression (7) results the attenuation achieved by the room is a function of sonorous reduction R, of the room's wall, respectively a function of wall's weight and of absorption coefficient. Using the weight law, sonorous reduction can be calculated with relation (8):

$$R = 20\lg pf - 47,5[dB], \quad (8)$$

where p is the weight of a separating element on square meter (m^2) in da N/m^2,

f is the frequency in Hz.

The value of frequency bands is $\Delta f = 2,5 f_{\lim}$ [Hz], where $f_{\lim}$ represents the frequency when sonorous reduction R starts to decrease.

Furthermore, we stated that the room's wall must be covered with special materials (Popescu, 2004) and we first propose to replace the classic materials with modern ones with superior technical, functional and economic properties (cost reason), as a real necessity for military transporter production.

4. CONCLUSIONS

This working paper is a part of research in doctoral study, and it is in the experimental stage. Here we studied another possibility to increase the quality of military technology manufacturing through noises amelioration. The methods we presented in this paper (advanced material use in isolation the noisy sources) can be applied to all caterpillar transport vehicles, heavy technique „on track", fighting vehicles, but also for air or naval technique. The measures proposed are needed to be taken in order to prevent and combat the noises during the functioning of some equipments and vehicles in military technique, in order to ameliorate the conditions and the comfort of the soldiers. Thus we calculated the noise level reduction. Furthermore, we took into account the possibility to entirely lock the noises source in a noises-separator room, to ensure phonic isolation of these noises sources, using advanced materials. The great technological progress in advanced materials applied in military technique leads to a new perception of military interventions. This is possible only if the manufacturing system is well thought out and well worked. New solution in the management of specific transport manufacturing (Lean Manufacturing, cost reduction, quality management increase) is necessary as military endowment must be prepared to cope with third millennium challenges.

5. REFERENCES

Darabonţ, A.; Costin, A.; Vaiteanu, D. (1983). *Combaterea zgomotului si vibratiilor in tehnica militara (Noises and Vibrations Combating in Military Technique)*, Military Publishing House, Bucuresti

Krause, K. (1995). *Arms and the State: Patterns of Military Production and Trade*, Cambridge: Cambridge University Press, ISBN 0-521-55866-2

Popescu, M. (2004). *Tehnici de imbinare a materialelor plastice (Techniques of Plastic Materials Joining)*, Politehnica Publishing, ISBN 973-625-165-9, Timisoara

Vartolomei-M, M.S; Jadaneant, M. (2007a). Aspects Regarding Quality Improvement in Romanian Military Technique in NATO Framework, *AGIR Bulletin,* No.2, (april-june 2007), p.86-90, ISSN 1224-7928

Vartolomei-M, M.S., Jadaneant, M. (2007b). The Management of Military Technical, Technological and Manufacturing Systems, *4-th International Conference on Economic Engineering and Manufacturing Systems, ICEEMS 2007 Brasov*, 25th-26th October 2007, p.371-374, ISSN 1582-0246, RECENT Journal, vol.8, nr.3a(21a), November 2007

Annals of DAAAM for 2008 & Proceedings of the 19th International DAAAM Symposium, ISSN 1726-9679
Editor B. Katalinic, Published by DAAAM International, Vienna, Austria 2008
Make Harmony Between Technology and Nature, and Your Mind will Fly Free as a Bird

ECONOMIC ENGINEERING STUDY REGARDING AEOLIAN INSTALLATION MANUFACTURING SYSTEM USING CPM AND PERT METHODS

VARTOLOMEI, M[ihaela]; MILOS, T[eodor] & VARTOLOMEI, M[ihael]

Abstract: *This paper started from a Partnership Project research began in 2007 at Politehnica University of Timisoara. It is another approach for the financial and manufacturing parts of Aeolian investment as ERS alternative. The methods we used, in order to calculate the term for planning and designing the Aeolian investment, provides from investment theory, especially investment strategy in manufacturing: the CPM and PERT Methods. The goals of these methods use in the investment process are: the achievement of the project objectives in stipulated terms and conditions. First the specific terms of the paper are defined: "work", "activity", "activity time", execution time. Then the draft with activity succession and activity time is drawn. We appreciated investment manufacturing term using probability theory and Laplace function. After inputs are established and investment graph are built, it is create Gantt graphic. Any change in the system and timing leads to a new program re-optimization.*

Key words: *CPM, PERT, investment graph, Average Squared Deviation, Dispersion*

1. INTRODUCTION

Solar energy is the first energy source used by humans. It represents essential element in mankind's development and there is no way to be life on Earth without sun energy (Lazarescu, 2003). The total solar energy intercepted by the Earth is $178 \cdot 10^9 MW$, by 15.000 times more than mankind's current necessity. Medium density of radiant flux on horizontal area is between 250 W/m^2, in Sahara and 100W/m^2, in Central Europe (Ambros et al, 1999). The Partnership Project this paper has started is called "Energetic Supply for a Local Community Using Draughts". It will choose the optimum solution for Aeolian turbine which will be manufactured by a partner of the project. The peculiar of the project is the possibility to choose the configuration of the turbine palette with the calibration and simulation of the flow with the peculiar soft existent in the University. In this paper, our attention is fixed on planning and designing period of the Aeolian investment.

Non-fossil energy sources or energy renewable sources (ERS) represents the environment resources, that are continuously regenerating or in a certain periodicity and their consumption don't lead to a possible depletion. These kinds of ERS are constantly renewed or restored and include: sun (solar), internal heat of the earth (geothermal energy), wind (wind power), vegetation (biomass), falling water, tides, and wave motion (hydropower). A clear feature of ERS is their independent existence by any man's consistently activity (Ambros et al., 1999).

2. INVESTMENT DESIGNING PROCESS USING CRITICAL PATH METHODS (CPM)

2.1 CPM Principles

In this paper we call "work" the whole process of work and manufacturing activities targeted towards execution and setting the Aeolian investment.

Activity symbol	Start-end events	Activity name	Activity time (in days)	Symbol of right preceding activity
a_1	1,2	Execution project elaboration	180	-
a_2	2,3	Work space organization	30	a_1
a_3	3,5	Building execution	360	a_2
a_4	2,4	Equipments contraction	30	a_1
a_5	4,5	Equipment acquisition	300	a_4
a_6	5,7	Equipment assembling	60	a_3
a_7	2,6	Prepare for manufacturing	360	a_1

Tab. 1. Investment activities and execution time.

The work has more activities. An activity represents the sum of rather homogeneous operations, which participate to the achievement of a part of the work.

The program is referring to tidy up activities executions, so that the work frames in planned terms, without surpass resources allocated. Every activity is formed by a throng of events (event is the physical stage an activity can be at a certain moment). Any activity is bordered upon two events: starting one and ending one. If the investment's work (W) has n activities then:

$$A=(a_1, a_2, \dots, a_i);\ i=1\dots n \qquad (1)$$

Where: A=throng of activities
a_i=activity i

In this investment situation, we had identified 7 activities, from a_1 to a_7 (table 1). Furthermore, it is necessary to introduce a succession relation between these activities. Thus, the investment activities and events' graph (networks system of oriented arcs, bordered by knots) is showed in figure 1. The arcs represent the activities and the knots represent the start-end events. Looking figure 1, we can notice that there are more paths from event 1 to event 7: P_{12357}, P_{12457} and P_{1267} (so called full paths), the longest full path is called critic path:

P_{12357}=180+30+360+60=630 days
P_{12457}=180+30+300+60=570 days
P_{1267}=180+360+0=540 days
P_{cr} =max (P_{12357}, P_{12457}, P_{1267})= P_{12357}=630 days

P_{12357} is the critical path (P_{cr}) because it has the longest length. Thus, the uncritical path has time reserves (represents the difference between maximum terms till activity execution is accepted and possible minimum terms). There are four terms: minimum term of activity start (t_i), maximum term of activity start (T_i), minimum term of activity end (t_j) and maximum term of activity end (T_j).

The minimum and maximum terms are established using optimality principle (relation 2 and 3).

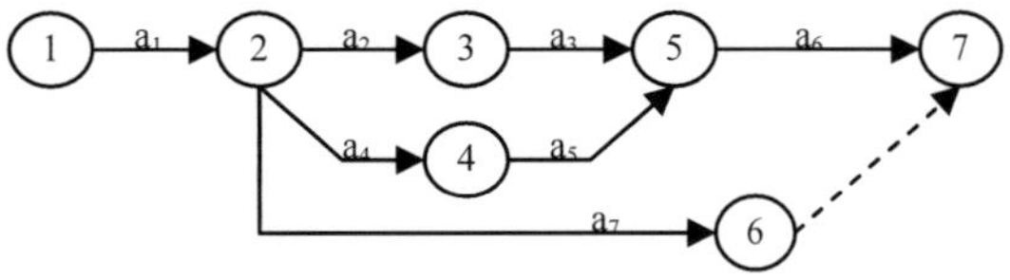

Fig. 1. Investment work graph

$$t_i = \max P_{1i} \text{ and } t_j = \max P_{1j} \quad (2)$$

$$T_i = P_{cr} - \max P_{in} \text{ and } T_j = P_{cr} - \max P_{jn} \quad (3)$$

Where n represents the final event of the work

Using minimum and maximum term of the events we can calculate time reserve for each activity. So we have:

$t_1=0$;
$t_2=\max P_{12}=P_{12}=180$ days;
$t_3=\max P_{13}=P_{123}=180+30=210$ days;
$t_4=\max P_{14}=P_{124}=180+30=210$ days
$t_5=\max P_{15}=\max\ (P_{1235};\ P_{1245})=\max(180+30+360;\ 180+30+ +300)=\max(570;510)=570$ days
$t_6=\max P_{16}=P_{126}=180+360=540$ days
$t_7=\max P_{17}=P_{15}+P_{57}=570+60=630$ days
$T_7=P_{cr}-0=630$ days
$T_6=P_{cr}-P_{67}=630-0=630$ days
$T_5=P_{cr}-P_{57}=630-60=570$ days
$T_4=T_5-P_{45}=570-300=270$ days
$T_3=T_5-P_{35}=570-360=210$ days
$T_2=\min(T_3-P_{23};\ T_4-P_{24};\ T_6-P_{26}) = \min\ (310-30;\ 270-30;\ 630-360) = \min\ (180;\ 240;\ 270) = 180$ days
$T_1=T_2-P_{12}=180-180=0$ days

2.2 PERT Principle

The essential difference between CPM and PERT (Program Evaluation and Review Technique Methods) is the execution period evaluation: CPM the period is evaluated in a certain number of days, PERT the period is appreciated in three situations: optimistic (the shortest, "a"), pessimistic (the longest, "l") and probable ("b"). The average value is calculated with relation 4 (Dogaru, 2004):

$$\bar{d}_{ij} = \frac{a+4b+l}{6} \quad (4)$$

$$\sqrt{\sigma^2} = \frac{l-a}{6} \quad (5)$$

$$z = \frac{d_p - \bar{d}_n}{\sigma^2} \quad (6)$$

Where: $\sqrt{\sigma^2}$ - represents Average Squared Deviation
σ^2 - represents Dispersion
z – represents probability factor
d_p – represents planned duration of investment
$\bar{d}_n$ - investment average duration

If planned term of investment execution is 366 days, the probability factor is 0.127 (using relation 6).

Act. symb	Activity duration				$\sqrt{\sigma^2}$	σ^2
	a	b	l	$\bar{d}_n$		
1	2	3	4	5	6	7
a_1	160	175	220	180	10	100
a_2	20	30	40	30	3.33	11
a_3	300	360	420	360	20	400
a_4	20	30	40	30	3.33	11
a_5	250	290	390	300	23.33	544
a_6	40	60	80	60	6.67	44
a_7	300	360	420	360	20	400
Tot.	520	625	760	630	x	555

Tab. 2. Average Squared Deviation and Dispersion

Then using Laplace function and interpolation methods we obtain the probability of 54.09% that indicate a proper programming (under 25% it means very short term and above 60% indicates many time reserves, so it will be redacted).

2.3 PERT/time

After program finalizing in its optimum variant, we'll elaborate Gantt graphic.

2.4 PERT/cost

A very important facility of this method is that prolonging execution dates of some activities, in the framework of time reserves, it is possible to obtain cost reduction linked by the work deployment (Buglea, 2002). This is why this model is known in scientific literature as cost-time function optimization. In some condition, the investment execution period can be shorted, but this supposes cost increase (urgency cost). But also, an exaggerated prolonging can produce certainly cost increases (figure 2) (Popa, 2005).

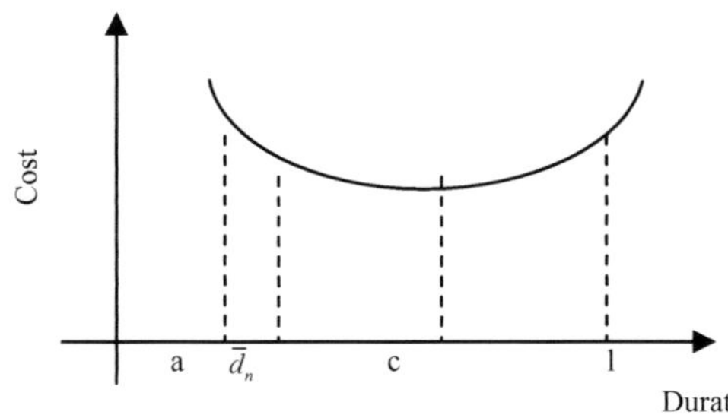

Fig. 2. Cost/duration evolution

2.5 Work launching and execution

After the convenient variant is chosen, and after all terms was established, we go to the investment launching process, re-drawing Gantt graphic depending by chosen variant, needed resources and mean for achieve any activity in established terms, passing, step by step Gantt graphic. Any modification in the system leads to the program re-optimization.

3. CONCLUSSIONS

In planning and designing process of our project Aeolian investment in good conditions, we use Critical Path Methods. First we ordered different activities so that the final results to correspond with the main target of the project, from terms and resources allocation points of view (Anton et al., 2007). The designing process supplies data about economic and technical optimum and the planning process bunches economic and technical optimum with social optimum.

4. REFERENCES

Ambros, T.; Arion, V.; Gutu, A.; Sobor, I.; Todos, P.; Ungureanu, D. (1999). *Energy Renewable Sources*, Tehnica-Info Publishing House, Chişinău, ISBN 9975-910-79-3

Anton, L.E.; Baya, A.; Milos, T.; Stuparu, A. (2007), *Experimental Hydrodynamics*, "Orizonturi Universitare" Publishing House , ISBN 978-973-638-330-4.

Buglea, A. (2002). *Investments and Their Financing*, West University Publishing, Timişoara, ISBN 973-8433-13-4

Dogaru, V. (2004) *Statistics in Economy,* Eurostampa Publishing House, Timişoara, ISBN 973-687-269-6

Lazarescu, S. (2003). *Energetic Concepts and Wind Energy Utilization*, Orizonturi Universitare Publishing House, Timisoara, ISBN 973-638-039-4

Popa, A. (2005). *Investments Efficiency*, Sitech Craiova, ISBN 973-746-059-6

Annals of DAAAM for 2008 & Proceedings of the 19th International DAAAM Symposium, ISSN 1726-9679
Editor B. Katalinic, Published by DAAAM International, Vienna, Austria 2008
Make Harmony Between Technology and Nature, and Your Mind will Fly Free as a Bird

IN-NETWORK AGGREGATION WITH SIZE REDUCTION FOR WIRELESS SENSOR NETWORKS – QUANTITATIVE ANALYSIS

VASAR, C[ristian]; BIRIESCU, M[arius] & MIHET POPA, L[ucian]

Abstract: *In this paper is presented a study case considering a wireless sensor network (WSN) with no data aggregation and then using in-network data aggregation in the idea of reducing in-network communication and implicit energy consumption, and propose a solution for optimizing the energetic autonomy. This issue is very important for increasing the wireless sensor autonomy and reducing the lifecycle cost extending the battery life. The power constraint has a great impact on the wireless sensor networks operation; the battery life makes the data communication very high-priced and restricts the amount of transferred data.*
Key words: *Wireless sensor network, data aggregation, energy optimisation*

1. INTRODUCTION

A WSN is set of spatially distributed autonomous unwired sensors (devices) that monitor at different site locations, various physical or environmental conditions, such as: physical properties (temperature, pressure), motion properties (velocity, position), contact properties (force, torque), presence (on), biochemical etc. The development of wireless sensor networks was initially motivated by military applications such as the case of battlefield surveillance. However, in many non-military applications, related to environment and habitat monitoring, healthcare applications, home automation or traffic control, the usage of wireless sensor networks become very broaden. (Cook & Das 2004)

Wired connections have important disadvantages: limit units mobility, does not allow wired devices to be closed to the monitored phenomenon, and not in the last considering the large number of sensors or actuators from a real environment, the wires implicate important maintenance problems, with high costs (Rabaey et al. 2000). These facts make wireless communication between such devices, an inevitable necessity in many applications. However, the wireless communication can be extremely costly in comparison with local processing of data, in the specialized literature it is state that sending a single bit over radio is at least three orders of magnitude more expensive than executing a single instruction locally (Shrivastava et al.2004).

2. WIRELESS SENSOR NETWORK

In wireless sensor networks, each sensor collects information from the surrounding environment, perform basic processing and transmit the data to user using the network infrastructure. A simple approach is based on periodically communication between sensors and a base station, where data is stored and processed. The main inconvenience of this basic method is the significant amount of transmitted data. The communication optimization becomes a demand, considering the "energy efficiency vs. data accuracy" trade-off. Usual means to improve these issues are based on multi-hop routing protocols applied to WSN. Data from multiple sensor nodes is merged together and then rerouted to base station to reduce the number of packets that propagates through the network. The base station becomes the root of the routing tree and the sensor nodes are organized in branches. Data flow from the sensors toward base station along this routing tree. The number of messages propagated through the network is reduced at the cost of rising the size of the messages (Heinzelman et al. 2000). Also the benefit of in-network data aggregation over significant energy savings within WSNs has been confirmed theoretically (Krishnamachari et al. 2002) by reducing the in-network communication, even if this is achieved by increasing the local processing at mote level.

In the simulations performed in the paper the WSN topology is according to Fig. 1 (but different WSN sizes were considered). Each mote has sensors, limited processing and radio capabilities. All the measurements performed inside WSN are required at the base station level, which has limited radio capability but can be integrated into a powerful computational system for complex processing of data provided by WSN.

Due to the power limitation of radio communication results a limitation on the feasible distance between the sender and the receiver. Because of this restriction, the simple and direct communication between source and destination is not always possible, specially in WSNs, that are designed to cover large areas (environmental applications) or that operate in difficult radio environments with strong attenuation (in buildings). In order to solve the issue of limited distances, an usual approach is the usage of relay units, where the packets take multi hops from source to destination (Karl & Willig 2005).

If the data is communicated using minimized routes this energetic costs are optimized. This can be achieved by defining routing trees, which will route all communication to motes located on the first diagonal and then along it.

In our case, the distances between sensor nodes were maximized, but obviously they cannot exceed the radio-communication range. Each node can communicate directly only with nearing neighbours. For example, the covered radio-communication area for node 22 is marked with grey colour. In this circumstances, the transmission from other node (except node 36) to base station will involve other nodes that will be used as repeaters, and the communication will require more receptions and transmissions that will increase the energetic consume as the distance between the node and the base station grows.

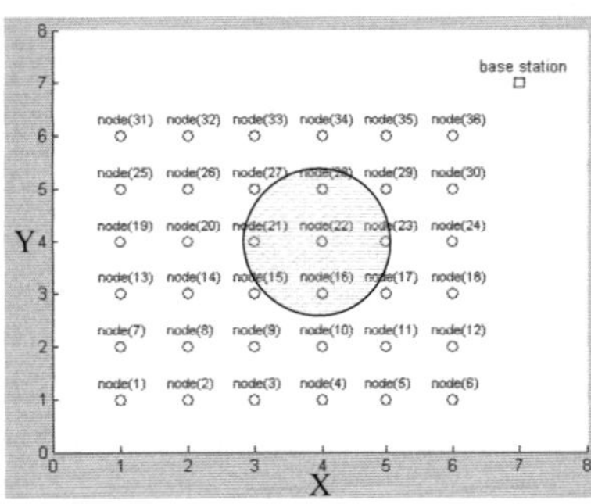

Fig. 1. Wireless Sensor Network

The main objective of the wireless sensor network is to extract the global maximum and the secondary objectives are to find local maximums from the monitored area.

3. SIMULATION STUDIES

To determine the quantitative impact of in-network data aggregation on the total amount of network and implicit over the energy consume of a wireless sensor network, a number of simulations had been performed, considering networks of different size and different distributions of measured values.

For example, in Fig.2 is depicted the case of a WSN containing 625 sensors, and the measured values are uniform distributed among the network. In this case, using a WSN with no aggregation and extracting all the measured values from it requires a total number of 10725 communications at each sample time.

In order to reduce this amount of communications but not loose information regarding global and local maximum values, since each sensor receive directly the communicated values from its nearing neighbours (maximum four neighbours), and if its own value is smaller than the maximum value of its neighbours it will not communicate its value at the present sample time. The communication start from the mote situated at the longest distance from base station and is performed row by row, and at one sample time all measured values are evaluated and/or communicated.

Taking into all above considerations, for uniform distributed values of measurements, the number of total communications is decreased at half (an average of 5650 instead of 10725) – extending with about 90% the average battery life. In Fig. 3 is presented the evolution of total communications number during an interval of 40 sample time. The simulated networks had 625 motes (placed in a 25 X 25 matrix). The number can be decreased even more by aggregate the maximums over clusters defined over the routing tree, but in that case the local maximums are not detected.

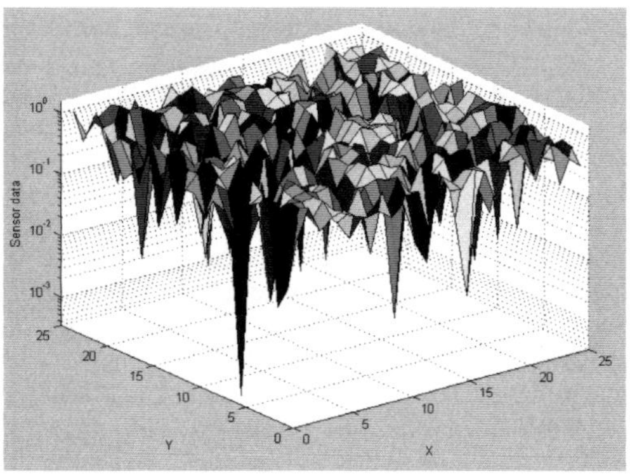

Fig. 2. Wireless Sensor Network data (25x25)

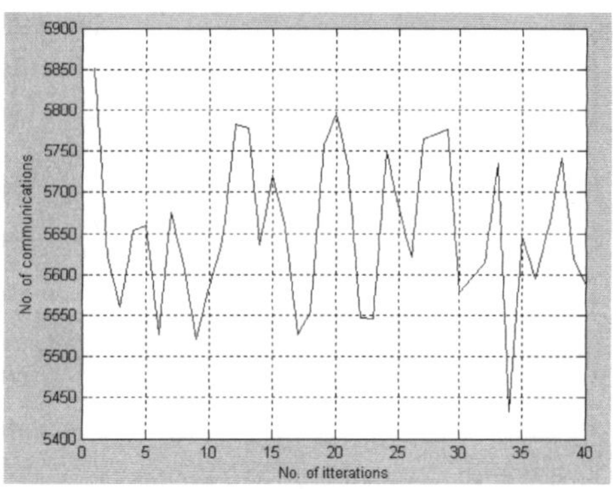

Fig. 3. Number of communications

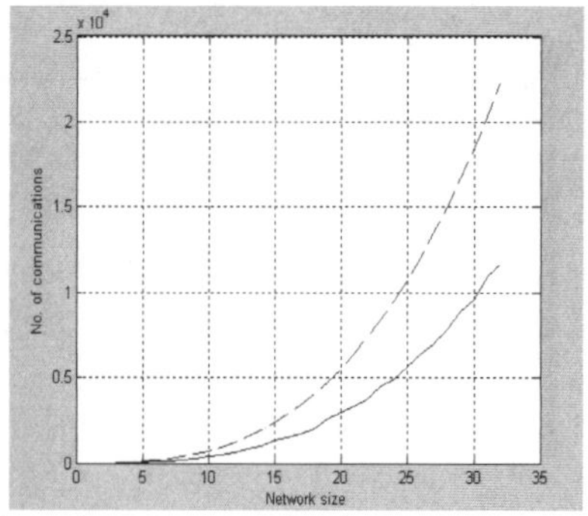

Fig. 4. Total number of communications

To study the influence of data aggregation on different network size, another simulation study has been performed, considered wireless sensor networks having 9 to 1024 sensors and a number of 20 iterations for each network case. The results are depicted in Fig. 4, with solid - the network with data aggregation and with dotted - the case of network without aggregation. The network size *n* is on the abscise, considering that all network have n elements, placed in a n X n matrix.

4. CONCLUSION

The presented aggregation method decreases at half the number of communication at each sample time, but this can be further lowered combining this method with other techniques (using clusters for example). Performing in-network data aggregation in case of wireless sensor networks with limited resources (as bandwidth or energy) can optimize the communication process, without loosing sensitive information. This improves the trade-off between battery life and data accuracy. One disadvantage of the presented method is the relative reduced tolerance to node malfunctions, in case if one node fails, the entire communication suffers. This can be overcome by using dynamical routing trees.

Future research directions will consider the possibility to use a priori known information about the monitored process to optimize the energy consumption of the entire WSN (for example defining the routing tree according with the probability to occur the maximum for different monitored area, or using multiple base stations).

5. REFERENCES

Cook D.J. & Das S.K. (2004). "*Smart Environments: Technologies, Protocols, Applications*", Chapter 2 "Wireless Sensor Networks- F.Lewis", Wiley-Interscience, New York.

Heinzelman W., Chandrakasan A., & Balakrishnan H. (2000). „Energy-efficient Communication Protocols for Wireless Microsensor Networks", In *Proc. Hawaaian International Conf. on Systems Science*, January 2000.

Karl H. & Willig A. (2005). "*Protocols and Architectures for Wireless Sensor Networks*", John Wiley & Sons, Ltd. ISBN: 0-470-09510-5, pp 60-64.

Krishnamachari B., Estrin D., & Wicker S. (2002). Impact of data aggregation in WSN. *International Workshop on Distributed Event-Based Sytems*, Vienna,Austria,July 2002.

Rabaey J. M., Ammer M. J., da Silva J. L., Patel D., & Roundy S. (2000). "PicoRadio Supports Ad Hoc Ultra-Low Power Wireless Networking" *IEEE Computer*, 33(7): 42–48.

Shrivastava N., Buragohain C., Agrawal D. & Suri S. (2004). "Medians and Beyond: New Aggregation Techniques for Sensor Networks," in *Proceedings of the Second ACM Conference on Embedded Networked Sensor Systems (SenSys 2004)*, August 2004.

Annals of DAAAM for 2008 & Proceedings of the 19th International DAAAM Symposium, ISSN 1726-9679
Editor B. Katalinic, Published by DAAAM International, Vienna, Austria 2008
Make Harmony Between Technology and Nature, and Your Mind will Fly Free as a Bird

PERFORMANCE INDICATORS IN PUBLIC SECTOR ENTITIES MANAGEMENT

VASICEK, V[esna]; BUDIMIR, V[erica] & LETINIC, S[vjetlana]

Abstract: *At a time when resources are getting more and more limited, it is necessary for spending public funds to be economic and rational. There is a visible need for economic, efficient and effective decision - making and management in the public sector. Information from external financial reports are not sufficient for public sector managers to make management decisions and therefore is necessary to develop cost and management accounting systems, as well as performance indicators which would be used as basis for quality business activity management in public sector entities. By using scientific methods of interview, analysis and synthesis, the intention of this work is to establish the performance measurement's importance for management decision - making at public sector entities.*
Key words: *Public sector entities, performance indicators, quality, management*

1. INTRODUCTION

Today's public sector is characterized by constant changes, uncertainty and complexity. Citizen's demand for faster and better service gets greater, whereupon appears the need for public sector reform. The routine way of conduct of public sector managers is changing. Public expenditure management becomes a central issue when making economic policies at most countries. The need for responsible, appropriate and efficient state administration, which is better at performing their duties and costs less, imposes the need to monitor efficiency of plan performance, as well as the performance of public sector as a whole. Despite the above, public sector entities still mostly use information from external financial reports as a basis for decision-making. The intention of this work is to show on the example of a public sector entity, that the information obtained that way is not sufficient for quality decision – making.

2. SIGNIFICANCE OF ACCOUNTANCY DATA FOR DECISION – MAKING IN PUBLIC SECTOR ENTITIES

An indispensable basis for effective decision – making in all subjects, including the public sector entities, is appropriate information. The information needed for decision – making partly comes from the business environment, but some of its most significant part results from the organization itself, and this from the accounting information system.
Accountancy information presented in a standard format and form of financial reports are primarily intended for external users. The goal of external financial reports is to give information to users, about the financial position and cash flow of business entity and the performance in fulfilling the objectives of business activity.
Accountancy information presented that way includes most, but not all the necessary information to users when making quality management decisions. Therefore, public sector managers, in order to increase the quality of decision – making, reach for information from other sources. We are primarily thinking of internal financial reports, which content and form is adapted to their needs, i.e. they are not standardized. For the purposes of management decision – making it is necessary to process accountancy information in a way to get targeted focused information. Instrument for such data processing consists of cost and management accounting (Anthony & Young, 2003). Cost management and internal cost accounting exist in all forms of business activity, including the public sector, but are not formal and binding. They are implied as a "good practice". It is therefore important in reform processes of the public sector, to develop cost and management accounting which are becoming the tools of the new public management and enable performance measurement in the public sector. In this way they contribute to public liability of management (Vašiček, 2006).

3. PERFORMANCE INDICATORS AS BASIS FOR QUALITY DECISION – MAKING IN PUBLIC SECTOR ENTITIES

Considering that public services' users pay in funds as taxpayers to finance them, it is important for them to be satisfied with the quality and efficiency of the public sector. For quality and efficiency evaluation of the public sector is not possible to use methods of analysis appropriated to enterprises, and is essential to define the methods adjusted by its definition and structure. In most countries are defined, for the stated need, performance indicators that measure and value information on functionality and quality of the service provider and of the public sector itself. (Rowe, 2004).
Performance indicators put in relation different financial and/or nonfinancial data in order to collect information about the efficiency of business activity. If the performance indicators were to be significant basis for quality business activity management in public sector entities, it is necessary for the public sector management to define them. Public sector managers, when defining performance indicators, determine the volume of the desired financial and nonfinancial information, so that the results obtained by their use enable performance evaluation to users, in relation to:

- default goals,
- previously achieved performance and/or
- similar institution's performance.

Following the aforementioned, performance indicators represent the basis for making quality decisions since they enable public sector entities to (Bolton, 2003):

- make savings,
- use the insured assets in the most appropriate way,
- improve decision – making in planning and management process,
- increase transparency and reliability,
- increase efficiency, and ultimately
- increase the quality of services provided.

It is visible how information obtained through performance indicators can be used for quality management, as well as external or internal assessment and evaluation of public sector entities' business activity.
Economic, efficient and effective management in the public sector requires, alongside with performance indicators, clear

definition of institution's mission and vision, as well as strategic and operative objectives and their connection with concrete measures and tasks. One of the high – quality models that would meet these needs is Balanced Scorecard model, as a balanced system of results measuring, which connects achieved business events with future planed quantities that are defined by strategy. BSC's importance in the function of public entities management will be the topic of future research.

4. PERFORMANCE INDICATORS IN FUNCTION OF HIGHER EDUCATION INSTITUTIONS' MANAGEMENT IN CROATIA

Among the programs that are funded as part of the public sector, are also programs in the field of higher education. Because of the level and significance of public funds invested in the specified programs, economic and rational spending and transparent public reporting on the achieved program's results is important (Dražić Lutilsky, 2006).
From the previous consideration in the framework of this work, is evident how public sector entities' financial reports, and thus financial reports of institutions of higher education as well, represent adequate, but not sufficient basis for making quality management decisions. It is therefore necessary, alongside with the existing financial reports, to design sets of performance indicators and thus to significantly improve business activity management in institutions of higher education. Information obtained from performance indicators could contribute to the development of higher education institution’s policies, as well as to their responsibility for received public funds.
Considering the structure and functioning of higher education institutions in Croatia, as part of the scientific research conducted for the period 2002.-2004., there has been suggested a model of measuring and reporting through four groups of indicators (Budimir, 2006):

a. Students performance indicators,
b. Finance performance indicators,
c. Research performance indicators and
d. Staff performance indicators.

The intention of this work is to identify the significance of performance indicator measurement in public sector entities management. For the purpose of argumentation on significance of performance indicators for making quality decisions in higher education institutions, the measurements results of individual indicators within the above groups will be shown in the continuation of this work. Indicators will be based on financial and nonfinancial data of a higher education institution, monitored through three years.

1. The ratio of enrolled and graduated students (the number of graduated students / the number of enrolled students)			
Year	2004/2005	2005/2006	2006/2007
Ratio	15,1 %	18,5 %	16,6 %

Tab. 1. Students performance indicator

2. The share of tuition income in the total income (tuition income and enrolment income / total income)			
Year	2005	2006	2007
Ratio	38,9 %	42,8 %	44,2 %

Tab. 2. Finance performance indicator

3. The number of published scientific and expert papers per teaching staff employed (the number of published scientific and expert papers and projects / total number of employed teaching staff)			
Year	2005	2006	2007
Number	1,47	1,95	2,50

Tab. 3. Research performance indicator

4. The ratio of the institution’s own staff and total staff number (the number of employed staff / total staff number)			
Year	2005	2006	2007
Ratio	30,9 %	38,8 %	45,5 %

Tab. 4. Staff performance indicator

The results obtained through the above indicators provide quality information necessary for analyzing business activity, and efficient public sector entity's business management. Based on the analysis performed, it is visible how information from basic financial reports constitutes only a part of data necessary to calculate performance indicator. Thus confirms that the external financial reports are not sufficient basis for making quality management decisions, and they need to supplement with qualitative data as a part of notes.
In institutions of higher education, as well as in the rest of the public sector in Croatia, legal acts do not regulate obligation of reporting on performance indicators. The specified indicators do not officially exist as an analytical tool of public sector's business activity analysis. However, with regard to all the previously mentioned reasons and facts, their forming is required and necessary.
The principle that is stressed out is public sector efficiency, because it concerns disposal and expenditure of public funds. However, the methodology of monitoring the effectiveness and efficiency is still not sufficiently developed and is reduced only to financial data displays and comparison with results achieved in previous periods or with the set plans, or to financial monitoring of accomplished results (Kaplan & Norton, 1996).

5. CONCLUSION

Based on the analysis carried out, the increase in business activity performance of public sector entity is visible through period. By comparing peer indicators on business activities of other institutions of higher education, we could rate their performance monitored in their interaction. By comparing performance indicators and set goals, it is possible to evaluate public entities' business activity in relation to the adopted plans. Through conducted research and analysis, it can be concluded that measurement and reporting on performance are significant for decision – making, or for quality management and leadership in public sector entities.

6. REFERENCES

Anthony, R. N. & Young, D. W. (2003). *Management Control in Nonprofit Organizations,* McGraw-Hill Higher Education, ISBN 0-07-250825-6, New York

Bolton, M. (2003). Public sector performance measurement: delivering greater accountability. *Work Study,* Vol. 52, No. 1, pp. 20-24, ISSN 0043-8022

Budimir, V. (2006). *Significance of accounting information for performance analysis of higher education institutions in the Republic of Croatia, master thesis,* Faculty of Economics and Business, Zagreb, title retained on 2006-07-04

Dražić Lutilsky, I. (2006). *Cost information in the function of public sector management,* doctoral thesis, Faculty of Economics and Business, Zagreb, title retained on 2006-12-21

Kaplan, R. S. & Norton, D. P. (1996). *The Balanced Scorecard: Translating Strategy into Action,* Harvard Business School Press, ISBN 0-87584-651-3, Boston

Rowe, K. J. (2004). Analysing and Reporting Performance Indicator Data : “Caress” the data and user beware!, *Available from:* http://acer.edu.au/ *Accessed* 2008-08-01

Vašiček, V. (2006). Status and directions of cost-accounting application in the public sector, *Accounting and Finance, Treasury,* Vol. 52, No. 4, pp. 2-7, ISSN 1845-8696

Annals of DAAAM for 2008 & Proceedings of the 19th International DAAAM Symposium, ISSN 1726-9679
Editor B. Katalinic, Published by DAAAM International, Vienna, Austria 2008
Make Harmony Between Technology and Nature, and Your Mind will Fly Free as a Bird

GRAPHICAL USER INTERFACE FOR QUADRUPED'S LEG ROBOT

VATAU, S[teliana] & CIOI, D[aniel]

Abstract: *In this paper is presented a Graphical User Interface for a custom quadruped leg's robot and the main features of the interface. This Graphical User Interface (GUI) is developed in Java and is essential in a development motion for legged robot. The friendly interface, allows any user to define and test movements for this robot. The cross-platform capability was the first reason to choose Java language for developing this application.*
Key words: *GUI, leg, robot, JAVA*

1. INTRODUCTION

With advances in science and technology, the interest to study the animals walking has developed the demand for building the legged robots. The development of legged robot involves research in heterogeneous areas. A Graphical User Interfaces (GUI) for monitoring and controlling a custom mobile robot named jRobo are presented in this paper. There are three components that form the basis of the design approach: visibility of system status; matching system and real world; user control and freedom. This approach requires that a lot of information be displayed on the user interface. But this is not practical, since with this amount of data on the screen the user is sure to get confused. So at any given time only a few details about the remote unit are to be supplied to the user. But the details should include the critical information regarding the status for the control of robots. A user can interact with the robot through one of the three types of graphical user interfaces: observer, commander, and a super user. The observer can only observe the system. The commander can send commands to the robots. The super user has the ability to control and modify not only the robot but also the environment of the system (Dixon et al., 1999).

2. LEG DESIGN

The aim has been to do research that would enable the design of an autonomous mobile robot, where the main direction, that has been taken, is to build a self-contained four-legged walking robot, capable of dynamic walking. We selected a design where each leg is composed of three links and has three degrees of freedom. The first joint of each leg is an abduction/adduction in the hip, i.e. a rotation around the longitudinal axis of the robot. The second joint is a flexion/extension joint also in the hip, i.e. a rotation around the lateral axis of the body. The axes of rotation for the first two joints in the hip are set such that the axes of rotation intersect orthogonally. Finally, the third joint is a flexion/extension joint in the knee. The total number of D.O.F. is three of the leg. Figure 1 shows the real leg of robot and the CAD model.

3. GRAPHICAL USER INTERFACE

A GUI is a rich environment for real-time two-way communication between the remote unit and the user (Sushila et al., 2002).

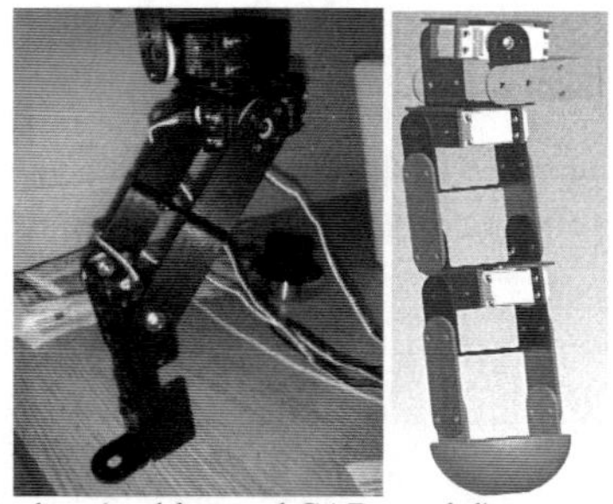

Fig. 1. Leg of robot (real leg and CAD model)

The main features of the interfaces are described in the following. Figure 2 shows a screenshot of GUI for the leg control made in Java (Sun, 2008).

Using this GUI, the user can control the robot by two methods. Using first method, in the top part of the GUI (see figure 3), the user can add (by select one of the button of the left part) or remove (by select one of the button of the right part) a one of the motors, which correspond of the robot.

When the user adds a motor, it can be possible to set the rotation (in degree) and how long time (in ms) must perform that rotation. The position in diagram specifies the start and stop time of that joint rotation. The motors' time of operation is show by a colour rectangle. In this way it can be possible to define a gait diagram (i.e. pace, walk, trot, or gallop). After that user define a gait diagram, it is possible to save all information in a file with xml extension using a "Save As" button.

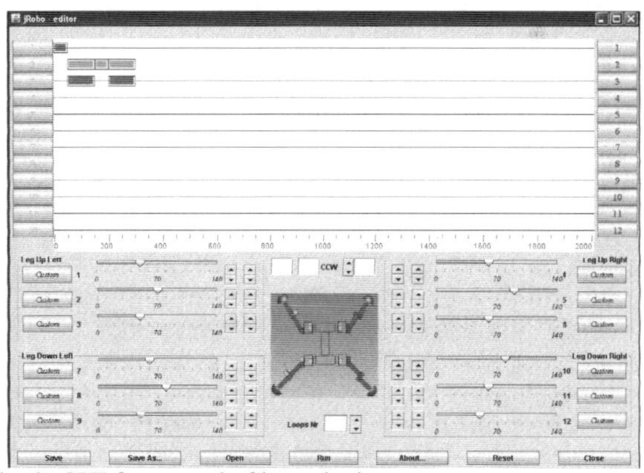

Fig. 2. GUI for control of legged robot

Fig. 3. The top part of GUI for control of legged robot

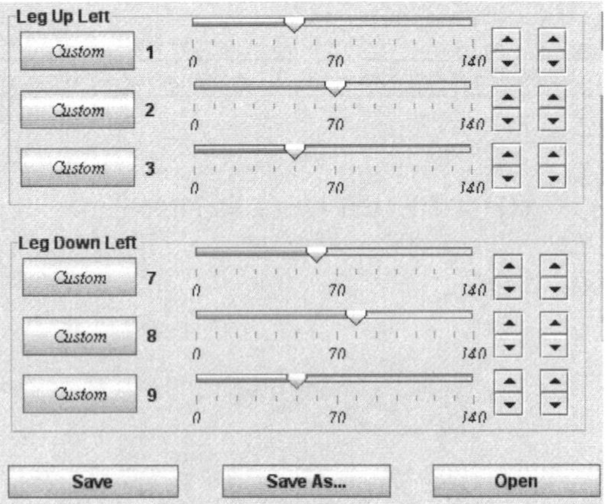

Fig. 4. The left bottom part of GUI for control of legged robot

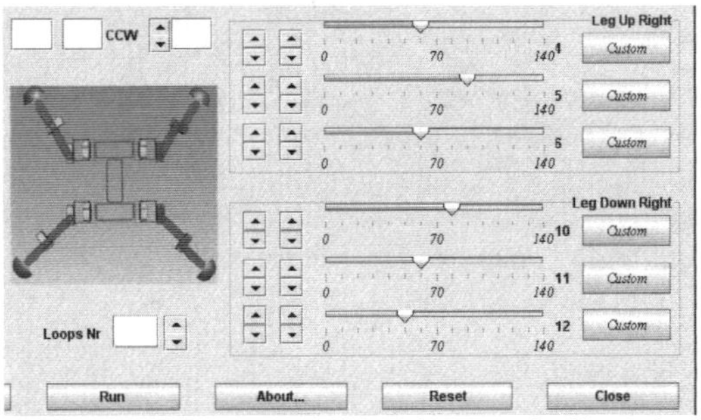

Fig. 5. The right bottom part of GUI for control of legged robot

The "Save As" button can be find in the bottom part of the GUI (see figure 4). In this mod, the user can define the forward, backward, turn left and turn right walking, concordant a different gait.

The "Open" button is in the bottom part also, and permitted of the user to open an existing file.

The GUI has a "Run" function, which gives the user possibility to move the robot in concordance with its settings.

With a "Reset" function, the user set the default value. Another method to move the legged robot is by using the slider for rotation and two slide bars to control the time (start and stop) correspond to each motor marked with a number from 1 to 12. The user can set the sense of motors rotation (CCW or CW) and how many step execute the robot using "Loops Nr" (see figure 5).

The "Custom" button allows customize the angular velocity, represented by a graph. Using the last method any adjustment is sent in real time to the robot, which performs the commands.

4. CONTROLLING OF LEGGED ROBOT

Generally any robot has a combination of motors and sensors, which are controlled by microcontrollers (Koditschek et al., 2004). There are wide varieties of motors, sensors and microcontrollers available. In this project for the quatruped robot named jRobo, low cost controller board and actuators are used. There are 12 D.O.F., each D.O.F. has one HS-5955TG servomotor and it is controlled by SSC-32 controller board.

The robot assemble controller board with servomotor is shown in figure 6. At this time, the communication between computer and robor is made using a serial cable and a USB to serial adapter.

The SSC-32 (serial servo controller) is a small preassembled servo controller with some big features. It has high resolution (1μs) for accurate positioning, and extremely smooth moves. The range is 0.50ms to 2.50ms for a range of about 180°.

Fig. 6. SSC-32 Controller Board and HS-5955TG servomotor

The motion control can be immediate response, speed controlled, timed motion, or a combination. A unique "Group Move" allows any combination of servos to begin and end motion at the same time, even if the servos have to move different distances. This is a very powerful feature for creating complex walking gaits for multi servo walking robots (Lynxmotion, 2008).

HS-5955TG servomotors are basically geared DC motors with positional feedback control, which can accurately position the shaft. The motor shaft of HS-5955TGs servomotor is positioned by Pulse Width Modulation (PWM).

Generally angles are coded as pulse width, so based on the pulse width duration the motor rotates. Algorithm: all the 12 motors are controlled and actuated simultaneously while maintaining the previous positional values. Initially, the first motor will be serviced with on-time pulse period and during the off-time pulse period of the motor, second motor will be serviced with on-time pulse period. This type of actuation is continued till all the 12 motors are serviced. Positional values loaded in the file and are retrieved and pulses are sent to the motors accordingly.

5. CONCLUSION

Quadruped robots are the fundamental block of any advanced walking robots. By making the quadruped robots fully autonomous, it can be used in environment where human cannot enter.

This paper presents an easy way to control a quadruped robot using a Graphical User Interface.

On the next step the control communication will be done through wireless Ethernet devices. Also, the robot will be equipped with proximity sensors and the information from these sensors will be represented on this Graphical User Interface.

6. REFERENCES

Dixon, K.; Dolan, J; Huang, W.; Paredis, C. & Khosla, P. (1999). RAVE, a real and virtual environment for multiple mobile robot systems. *Available from:* http://www.cs.cmu.edu/afs/cs.cmu.edu/user/cjp/www/pubs/IROS99_RAVE.pdf *Accessed:* 2007-10-10

Koditschek, D. E.; Full R. J. & Buehler M. (2004). Mechanical aspects of legged locomotion control. *Available from:* http://polypedal.berkeley.edu/twiki/pub/PolyPEDAL/PolypedalPublications/Mechanical_2004_ASD.pdf *Accessed:* 2007-12-10

Lynxmotion Inc., SSC-32 controller board, (2008).*Available from:*http://www.lynxmotion.com/images/data/ssc-32.pdf *Accessed:* 2007-06-24

Sun Microsystem Inc., Java tutorial, (2008). *Available from:* http://java.sun.com/docs/books/tutorial/ *Accessed:* 2008-04-12

Sushila, S.; Lakshmi, V. & Arvin A. (2002). Graphical user interfaces for mobile robots. *Available from:* http://citeseer.ist.psu.edu/612670.html *Accessed:* 2008-02-10

Annals of DAAAM for 2008 & Proceedings of the 19th International DAAAM Symposium, ISSN 1726-9679
Editor B. Katalinic, Published by DAAAM International, Vienna, Austria 2008
Make Harmony Between Technology and Nature, and Your Mind will Fly Free as a Bird

JROBO: SIMULATION AND CONTROL WITH JAVA FOR QUADRUPED ROBOT

VATAU, S[teliana] & CIOI, D[aniel]

Abstract: *Physics-based simulation and control of quadruped locomotion is difficult because quadrupeds are unstable, under actuated, high-dimensional dynamical systems. We develop a simple control strategy that can be used to generate a large variety of gaits and styles in real-time, including walking in all directions (forwards, backwards, sideways, turning). The application named "JRobo" is developed in Java and Java3D API.*
Key words: *robot, simulation, control, java, java3d*

1. INTRODUCTION

In the past few decades, the development of walking or running machine has drawn significant attention in the field of robotics. Many previous studies of legged robots have been performed. About walking on terrain, monopod biped and quadruped robots have been studied. Most of these earlier studies employed precise models of a robot and an environment and also involved planning foot trajectories as well as controlling joint motions on the basis of an analysis of the models (Hong et al., 1999).

With advances in science and technology, the interest to study the animals walking has developed the demand for building the legged robots. The development of legged robot involves research in heterogeneous areas. Design of legged robot involves equal amount of mechanical and electronics considerations. There are many factors which are to be considered are cost, actuator, size, weight and controlling of actuators. All these factors have been considered and designed.

2. ROBOT MODEL

The JRobo robot has 12 degrees of freedom, with three degrees of freedom per leg. Each leg has hip, knee and ankle. The hip joint is actuated in vertical plane (Pitch) and horizontal plane (Roll), knee joint is actuated in vertical plane (Pitch) and ankle is not actuated. Figure 1 shows the quadruped robot model. The mechanical design is divided into four phases: determining the mechanical constraints, conceptual design, building the prototype model, specification and fabrication of the model. There are various design considerations when designing a quadruped robot. Among them, the major factors that have to be considered are robot's size selection, degrees of freedom (D.O.F) selection, link design, stability and foot pad design. Robot size plays a major role. Based on this the cost of the project, materials required for fabrication and the number of actuators required can be determined. In this project miniature size of the robot is preferred so a height of 500mm is decided which includes mounting of the control circuits, but the actual size of the robot is 400mm without controlling circuits. The leg has got six degrees of freedom (Hip–3 D.O.F, Knee–1 D.O.F, Ankle–2 D.O.F), but implementing all the six D.O.F is difficult due to increase in cost of the project and controlling of the actuators which become complex, so in this project reduced degrees of freedom is aimed so 3 D.O.F per leg has been finalized.

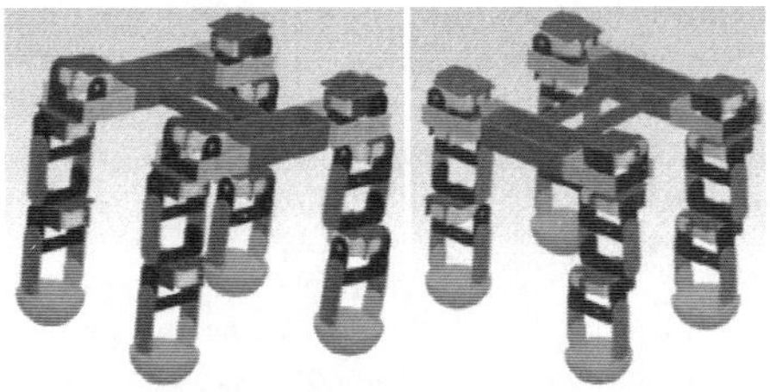

Fig. 1. CAD model of the robot

Fig. 2. Brackets A and B

In this project U shaped bracket like arrangement is used for joints formation. The bracket consists of two parts namely servomotor bracket A and B (figure 2). Servomotor are fixed in the bracket A and the bracket B is used to transmit the output of the servomotor. Bracket B and servomotor are coupled using servomotor horn. By using the brackets there is a greater flexibility and individual joint can be actuated without disturbing the other joints. The servomotor brackets are designed in accordance with the motor size.

With quadruped mechanism, three points will be in contact with the ground surface. The stability of the robot is determined by the foot pad. In a statically stable gait, the vertical projection of the center of gravity (G) onto a horizontal plane is kept within the support area at all times, as shown in figure 3. In the absence of any inertial or external forces and if the ground is sufficiently rigid, the robot can remain stable as long as the G is within the support area (Fukuoka et al., 2003). For robots with point feet, a necessary condition for static stability is that the robot has at least three legs on the ground at all times. This is necessary in order to form an area of support that can contain the projection of G within its borders.

In figure 4, in the left part of the figure, three legs provide support and the projection of center of gravity is located inside the support area such that the robot is statically stable. The foot placement in the right part projects the center of gravity outside the support area, which leads to instability due to a tipping moment caused by gravity.

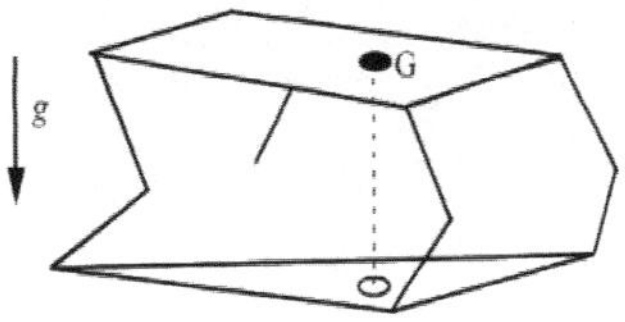

Fig. 3. Vertical projections of feet contact points and center of gravity (G) on a horizontal plane

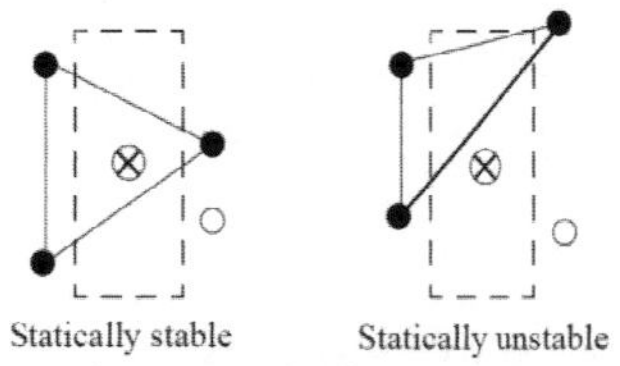

Fig. 4. Support polygon, statically stable and unstable cases

Generally there is a concept that over sized and heavy foot pad will have more stability due to more contact area. But there is a disadvantage in using the oversized and heavy foot pad, because the torque requirement of the motor is more and lifting the leg against the gravity becomes difficult. By considering this disadvantage an optimal sized foot pad was used.

Stable walking pattern can be obtained only if the centre of mass and centre of pressure are within the supporting (Buehler et al., 1998). Generally walking cycle consists of two steps namely "initialization" and "walking". In the "initialization" step the robot will be in balanced condition and in this step the servomotors are made to return to home position. This will certainly help the robot to advance into the next step.

The support diagrams of four typical quadruped robotics' gait (trot, gallop, pace, walk) (Schmiedeler & Waldron, 1999) are shown in figure 5. A solid circle indicates that the leg is in contact with the ground providing support, and an open circle indicates that the leg is in the air returning to the appropriate position for its next period of support.

3. ROBOTIC CONTROL SYSTEM

The quadruped robot is controlled using a GUI (Graphical User Interface), which implementing different diagram gait. In figure 6 is show a screenshot. This GUI is made in Java and Java 3DAPI (Sun, 2008). In the buttom part of interface it is displayed the result of positional values combinations, which are possible to implementing a certain style for walk. In this mode, the users can choice the optimal positional value for locomotion. In right part of interface, in Keys control panel, there is four control buttons (Forward, Left, Right, and Back).

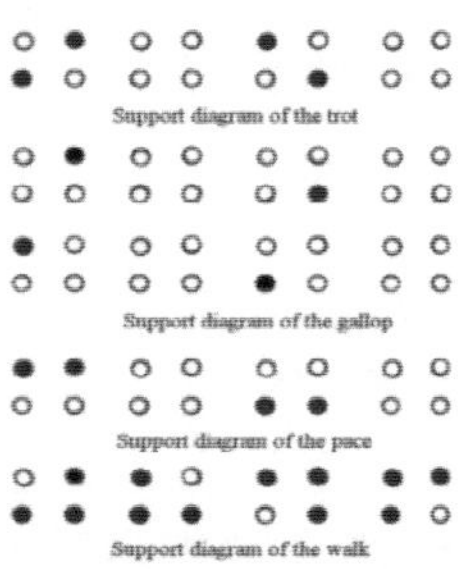

Fig. 5. Quadruped gait diagrams

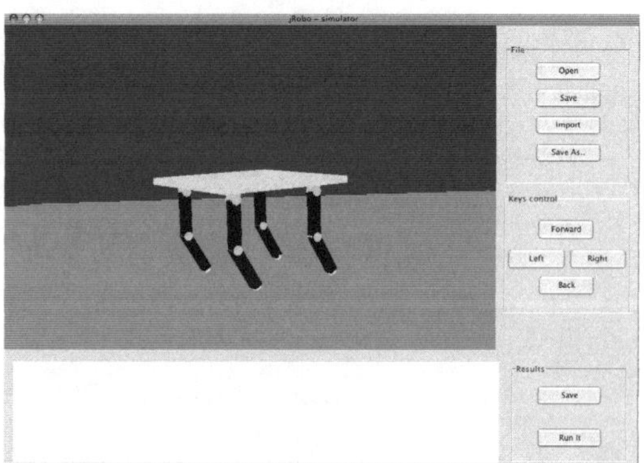

Fig. 6. A command interface screenshot

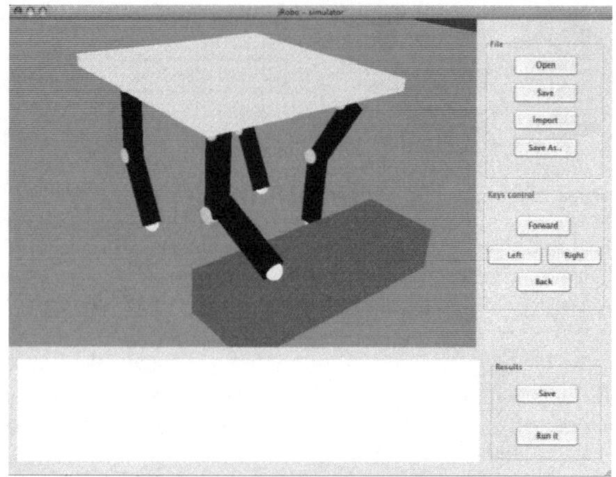

Fig. 7. The robot crossing an obstacol - screenshot

With these buttons the robot can be controlled manual. In Results panel there are two buttons. With "Save" button can be saved the optimal positional value for all motors, which are corresponding to a style gait.

When the user click on the "Run it" button, the robot walk in concordance with the style gait choiced from the bottom part of interface. The buttons from File panel perform the following operation: permitte to user to open an existing file, save the result in an xml file, or import a compatible file.

The robot has four sensors for collision detection. Figure 7, shows the robot crossing an obstacole. On the real robot there are four micro-switches for collision detection with obstacles (whitch are implementing in foot leg's robot).

The algorithm for crossing the obstacoles is simple. The robot up his leg, and beat the obstacol. If is obstacol, the robot increase up leg and verified again is obstacol in face. When is no more obstacol in face, then robot move in concordance with a strategy to displacement over obstacle.

4. CONCLUSION

In this paper, the authors present a simulator for study different walking gait for a quadruped robot. Using this simulator, the user obtains the positional values for implementing in a microcontroller.

Complex movements can be achieved by increasing the D.O.F. Proximity sensors (which are able to detect the presence of nearby objects without any physical contact) and remote control through wireless ethernet mode will be considered.

5. REFERENCES

Buehler, M.; Battaglia, R.; Cocosco, A.; Hawker, G.; Sarkis, J. & Yamazaki, K. (1998). Scout: A simple quadruped that walks, climbs and runs. *Proceedings of ICRA98*, 1707-1712, (1998).

Fukuoka, Y.; Kimura, H. & Cohen, A. H. (2003). Adaptive Dynamic Walking of a Quadruped Robot on Irregular Terrain Based on Biological Concepts. *In The International Journal of Robotics Research, Vol. 22, No. 2, February 2003, pp. 187-202.*

Hong, Y.S.; Lee, H.K.; Yi, S.Y. & Lee C.W. (1999). The design and control of a jointed-leg type of a quadrupedal robot for locomotion on irregular ground. *In Robotica vol.17, 1999, pp. 383–389.*

Schmiedeler, J. & Waldron, K. J. (1999). The Mechanics of Quadrupedal Galloping and the Future of Legged Vehicles. *In The International Journal of Robotics Research, Vol. 18, No. 12, December 1999, pp. 1224-1234*

Sun Microsystem Inc., Java tutorial, (2008). *Available from:* http://java.sun.com/docs/books/tutorial/ *Accessed:* 2008-04-12

Annals of DAAAM for 2008 & Proceedings of the 19th International DAAAM Symposium, ISSN 1726-9679
Editor B. Katalinic, Published by DAAAM International, Vienna, Austria 2008
Make Harmony Between Technology and Nature, and Your Mind will Fly Free as a Bird
2nd DAAAM-ICAT

RAPID PROTOTYPING OF CONTROL DEVICE FOR PREVENTIVE CONTROL OF CHILDREN PLAYTHING

VAUPOTIC, B[ostjan] & PAHOLE, I[vo]

Abstract: *The market requirements have changed very much. One of the principal reasons is the market globalization and saturation. Thus, now the products have to be adapted to buyers' needs. The products become more and more complex, more beautiful, flawless and safe. Safety is of particular importance for children's toys, where it is in the first place. The European standard for the toy safety EN 71.1 specifies the devices for establishing the proper safety of future toys. After adoption of that law in Slovenia the demand for those devices emerged. The manufacture of the devices was taken up by means of rapid prototyping according to PolyJet process offering verification of functionality, putting together and aesthetic acceptability on a 3D CAD model. The 3D model in hand gives a more complete feeling of size, shape and other aspects of the object than the best 3D graphics. The model made can be used also as the final product provided that the mechanical properties are low.*

Key words: *Rapid prototyping, PolyJet process, 3D CAD model, prototypes, final products*

1. INTRODUCTION

The philosophy of the common European market is based, in addition to the free flow of goods, personnel and capital, particularly on assurance of safety of products entering the market. The consumers are more and more self-confident, therefore, when buying a product they want to have the guarantee that it works well and that it is safe.

On the admittance of Slovenia to EU the Europe standard for toy safely EN 71.1, specifying the requirements and the test methods for establishing mechanical and physical properties of toys, came into force. That standard is concerned about the children's' toys, the toy meaning any product or material intended to be used in the plays of children younger than 14 years. The standard refers to new toys and considers the period of the anticipated and normal use and the usual behaviour of children provided that the toys are used as planned and in the specified way. It includes special requirements for toys intended to be used by children younger than 36 months and by children, too small to sit independently (Urad RS, 1999). The Health Care Institute in Maribor will implement the test methods for establishing whether the mechanical and physical properties of toy comply with the European standard EN 71.1. To that end it was necessary to make proper devices, produced at the Faculty of Mechanical Engineering Maribor in the Laboratory for intelligent manufacturing systems, for implementing the test methods. The device for rapid prototyping was of great assistance for that purpose. It helped to make real prototypes in scale.

2. MANUFACTURE OF THE TEST DEVICES ACCORDING TO EUROPEAN STANDARD 71.1

The first stage covered the manufacture of 3D CAD (Jaballi et al., 2007) models according to the requirements of standard EN 71.1.

A roll for small parts (simulating the child's esophagus), two templates A and B (for checking adequacy of geometrical shape of certain toys), two link probes A and B (for checking accessibility of part or component), a device for measuring the edge sharpness and a device for measuring the tip sharpness were designed. Designing was effected by programme package SolidWorks assuring simple adaptation and corrections of 3D models (Fig. 1).

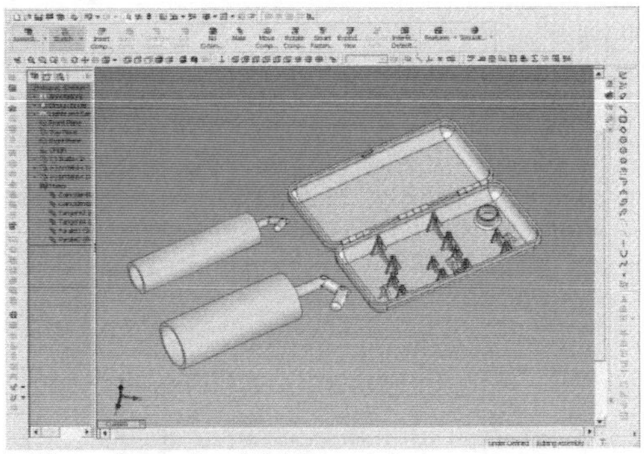

Fig. 1. Three dimensional CAD model.

The geometrical and functional requirements, specified in standard EN 71.1., were taken into consideration; in cases when products were made in composed form attention was paid to proper clearance on hinges and moving parts. Also the costs, the exactingness and the time periods of delivery were considered. A good compromise was the 3D printing technology as one of the rapid prototyping technologies assuring the manufacture of prototypes and products (Pham & Dimov, 2002). 3D models in stereo lithographic format (STL file) were used as the basis for printing. During the export from SolidWorks into the STL file the proper size of triangles must be set so that the made surface is flawless (Drstvenšek & Balc, 2004). In the graphic environment attention is paid to optimal setting of the model on the working tray (Fig. 2).

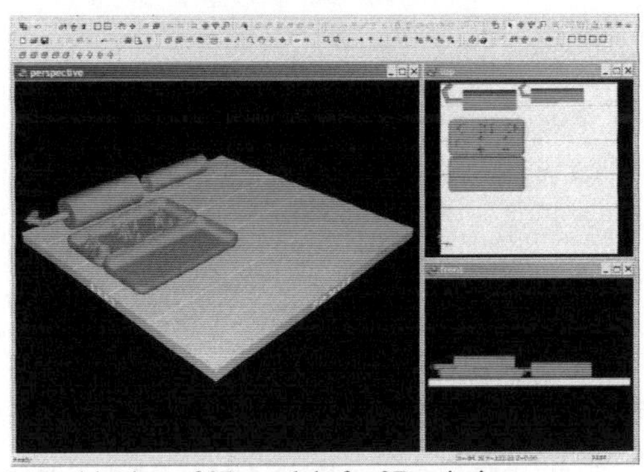

Fig. 2. Positioning of 3D models for 3D printing.

Then the STL file is imported into the specific purpose control program determining the delivery time and the consumption of the bonding agent for full parts and the supporting material for hollow model parts on the basis of the model volume and position. The mass and the time frame serve for the financial estimate. Automatic determination of supports and layering of the model in vertical direction (Z axis) follow. Afterwards, individual layers (bit maps) are sent to the printer which takes them over and deposits them successively into the tray (Drstvenšek, 2004).

When the product has been printed by the machine, the product is ready for immediate use and does not require any subsequent treatment, except removal of supporting material. The supporting material looks like "marmalade", although it is not sweet and sticky. It can be removed by means of a high pressure pump and water. The supporting material is simply "washed away" by water jet under the pressure of about 40 bar. The water jet pressure mainly depends on the model to be cleaned. Models with thin walls and very small details were cleaned with lower pressure, while more robust models were cleaned with higher pressure so that the time of cleaning was shortened.

The first prototypes ensured the tests of correctness of the product design (Fig. 3). It was found out that on the link probes for testing of accessibility (they were called little finger by us, since they simulate the child's little finger) the fits were slightly too large, therefore they were reduced on the final products.

Also the design of the box for containing the device for measuring the tip sharpness, printed in assembled form, underwent corrections on the hinges and fixing clamps. Further, the shape of openings for installation of keys and diode lights on the casing for the device for measuring of edge sharpness was slightly changed.

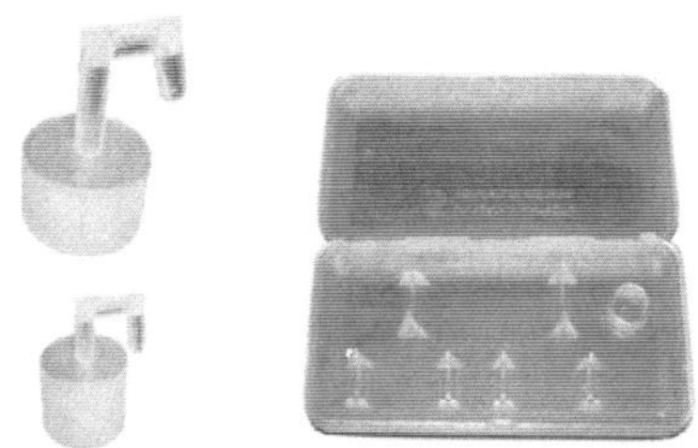

Fig. 3. Prototypes made according to PolyJet process.

The 3D CAD models, adapted to new requirements were then printed and functioning, shape and mountability were tested. Prior to manufacture of final products all prototypes were delivered to the client who confirmed correctness and adequacy of prototypes.

The last stage comprised the manufacture of final product.

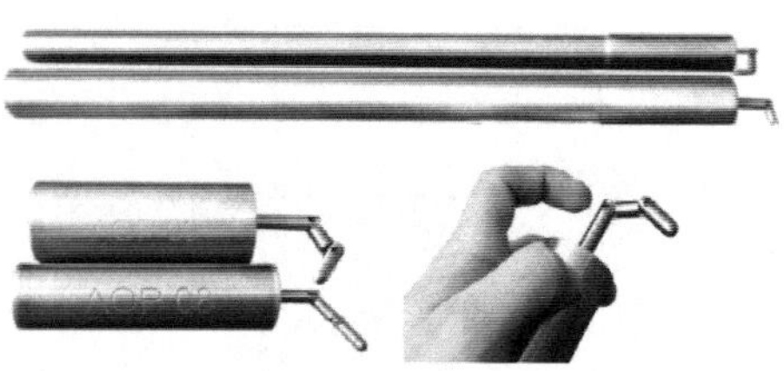

Fig. 4. Final products.

On the basis of findings obtained from prototypes made by rapid prototyping according to the PolyJet process the final products were made and calibrated according to the specified standard (Fig. 4). On the other hand, the models (housing of the device measuring the tip sharpness and the box for its accommodation and housing of the device for measuring the edge sharpness) made by layer technology were used as final products (Fig. 5).

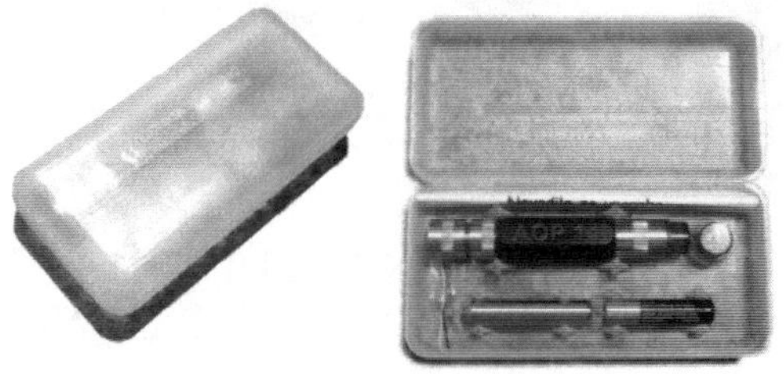

Fig. 5. Final products made by RP.

The models made by rapid prototyping allow also painting; our models were coated with a black mat paint coat additionally protecting against UV rays and minor mechanical damages.

3. CONCLUSION

The manufacture of testing devices for verifying the children's toy safety according to the PolyJet process of rapid prototyping has proved to be a good approach since all requirements were complied with. A minor quantity of products of rather complicated shape was concerned. We must be aware that the price of rapid prototyping is influenced only by the size of the product. The shape complexity is quite unimportant. Due to the high number of different requirements and the introduction of new products the rapid prototyping technology according to the PolyJet process has enabled us to establish by testing the adequacy of shape, functions and mountability of the devices and to adapt it accordingly. As in case of rapid prototyping the data on the shape of the product are in the form of 3D CAD model, the changes on the product are made very simply.

On the basis of the experience in rapid prototyping, the applicability of prototype models and the responses from users we conclude that the PolyJet technology for rapid manufacture of prototype model is adequate with respect to shape as well as dimensional requirements. With rapid prototyping it is possible to correct the deficiencies of the shape, before the final products are made. The advantages are obvious particularly in case of exacting complicated shapes which would be hardly manufactured by conventional processes or the price would be too high and the manufacturing time too long. In case only a few products, not exposed to major mechanical loadings, are needed, the models made by PolyJet process are usable also as final products. With wider choice of materials the applicability of this technology will increase.

4. REFERENCES

Drstvenšek, I. & Balc, N. (2004). *Layered Technologies*, Faculty of mechanical engineering, University of Maribor, Maribor

Drstvenšek, I. (2004). Rapid prototyping, *Available from:* http://maja.uni-mb.si /slo/index.htm, *Accessed:* 2008-04-13

Jaballi, K.; Bellacicco, A.; Louati, J.; Riviere, A. & Haddar, M. (2007). A computer aided toleracncing: algorithm for 3D manufacturing tolerancing, *APEM journal,* Production Engineering Institute (PEI), University of Maribor, ISSN 1854-6250, Maribor

Pham, D.T. & Dimov, S.S. (2002). *Rapid manufacturing: The technologies and applications of rapid prototyping and rapid tooling*, Springer, London

Urad RS: za standardizacijo in meroslovje, Ministrstvu za visoko šolstvo, znanost in tehnologijo (1999). *Slovenian standard DSIST EN 71-1: Safety of Toy-Part1: Mechanical and physical properties*, Slovenia

Annals of DAAAM for 2008 & Proceedings of the 19th International DAAAM Symposium, ISSN 1726-9679
Editor B. Katalinic, Published by DAAAM International, Vienna, Austria 2008
Make Harmony Between Technology and Nature, and Your Mind will Fly Free as a Bird

NEURAL NETWORK MODEL FOR TURNING OPERATION

VAUPOTIC, B[ostjan]; BALIC, J[oze] & CUS, F[ranci]

Abstract: *In the paper an intelligent system, selecting the best set of tools on the basis of the 3D CAD model and the other relevant selection factor, has been conceived. For solving the complex classification problem the artificial intelligence method has been used; the neural networks.*
This concept has been used for the most widespread cutting process, i.e. turning. The results reached are in conforming with the expectations. A high degree of classification has been reached. The constant increase of the knowledge which the system takes from the growing data base is considered to be a great benefit of the proposed system. Due to robustness and universality it is proper for classifying the cutting tools also in other cutting processes. The resulting solutions are comparable with the solutions given by experts. The system can be quite practically used with minor corrections adapted to the user.
Key words: *turning operation, intelligent automatic selection, intelligent systems, 3D CAD model, neural networks*

1. INTRODUCTION

Today, many different computer-aided CNC machine programming systems are available on the market offering different solutions for selection of cutting tools, how-ever, still always the programmer himself must select the optimum set of tools and their correct sequence on the basis of his knowledge and experience. To that end, he applies most frequently the recommendations of the tool makers in their catalogs and the general instructions of CAM systems (Balič, 2004). Such selection method requires the user to coordinate the recommendations with his knowledge, experience and intuition. In practice that means that still always the human's decision is necessary for selection of the cutting tool. The use of CAM systems requires well qualified users and a lot of time, since they are exacting for programming which is very time-consuming, too. The CAM systems do not allow capturing of the feed-back information, reached in the process of selection, and repeated preparation of those data in similar cases, therefore they can be considered to be inflexible.

The proposed intelligent system summarizes all the advantages and does away with the disadvantages of such systems and can work independently or it can be integrated into the CAM system.

2. AI AND NN IN MANUFACTURING SYSTEMS

The term "artificial intelligence" appeared in the eighteenth century in science fiction. In the fifties it was made a scientific discipline with the birth of "electronic brain" (Kokol et al., 2001). Today, the artificial intelligence deals, particularly, with the development of the techniques, methods and arhitectures for solving logically complicated problems which would be difficult or even impossible to solve by conventional methods.

The hitherto researches and findings in the area of the use of the artificial intelligence techniques in the modern manufacturing system show that the latter can considerably contribute to the improvement of the necessary flexibility and efficiency required by modern manufacturing systems. Since its appearance the artificial intelligence has broken through into many areas in the manufacturing systems. While during the early years the knowledge-based systems attracted the highest attention, recently the soft systems have been in the foreground. Among them the genetic algorithms, the case-based decisions, the fuzzy logic and the neural network are most widely used. The neural networks were formed in the early forties of the previous century as one of the branches of artificial intelligence. Haykin (Haykin, 1994) defines the neural network as a mass parallel distributed processor that stores the experimental knowledge and ensures its use . Thus, the neural network is very similar to the brain, since the knowledge is gathered by the network through the learning process and the interneural connections known as synaptic weights are used for storing of that knowledge.

3. INTELLIGENT SELECTION OF CUTTING TOOL IN TURNING BY THE USE OF ANN

The proposed intelligent system for automatic selection of the cutting tool for the turning process is so conceived that it can select independently, without the human's interference into the system, the optimum set of cutting tools on the basis of the 3D CAD model and important parameters. The system must be capable to learn in order to work intelligently. However, learning only is not sufficient. To be capable to learn at all, the system must have certain capabilities such as sufficient memorizing capacities (memory), concluding capacity (processor capacities), perceptive capacities (input and output) etc.

In addition, learning requires certain initial previous knowledge which is inherited in the living systems. By learning the capability of the system increases, consequently, its intelligence increases (Hopfield, 1982). The artificial neural networks, used by our system, have such capabilities.

3.1 Creation of data base for learning

For the needs of learning of the neural network the data base was created. This is the model of the environment serving as a basis for the control, decision making and execution of action. The data bases as such are a kind of the driving medium for operation of intelligent systems and/or a kind of "reservoirs" ensuring the existence of intelligent systems (taking of data) at all.

The data base consists of four units. In each unit the relations, on the one hand, between the geometrical features, workpiece material, quality of the require surface and other parameters and, on the other hand, between the cutting tools are defined.

In the unit one there are eight cutting tools for roughing, i.e. four left and four right tools. The unit two contains twenty cutting tools for finishing, out of which eight left, eight right and four neutral cutting tools. The third unit contains four grooving tools for different size of depth cutting and cut width. The fourth unit contains four threading tools with two different thread pitches. The cutting tools differ in geometry, material type to be machined, surface quality after cutting and other tool parameters (Sandvik Coromant, 2008).

Figure 1 shows the maximum and/or minimum entering and/or outlet angle on one workpiece for a specific tool.

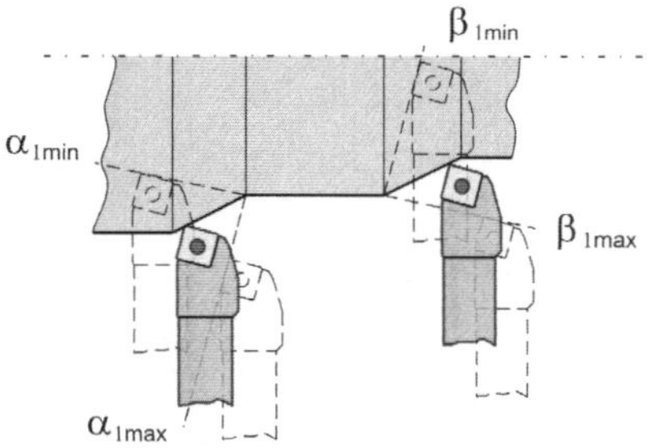

Fig. 1. Limitations of entering and /or outlet angle on workpiece for one of cutting tools.

Identically, also other cutting tools are limited. In the data base each geometrical feature, with required surface quality and by considering also other selection parameters, has the relevant cutting tools which can machine it. To this end, the tool entry into the workpiece and iths exit are taken into account.

3.2 3D CAD model

The starting point for operation of our system is the 3D CAD model, which is an advantage nowadays. When designing the products, the users more and more frequently use three-dimensional CAD systems because of higher flexibility, accuracy and intuitiveness of three-dimensional modeling.

The selection of appropriate cutting tools for the manufacture of products requires the familiarization with all geometrical features of the product. As, usually, the turning process takes place in two stages, i.e.rough and fine turning, the geometrical features must be known for the final as well as semi-finished product.

From the basic 3D CAD model, representing in the same time the final product and/or containing the geometrical features for fine turning the semi-finished product, after rough turning, as well as the blank necessary for the manufacture of such product are automatically created.

3.3 Operation of the system

The system works in several stages. The first stage covers the preparation of the 3D CAD model. The blank, the workpiece for rough turning and the workpiece for fine machining representing the final product are automatically formed. The next step is the recognition of geometrical features and/or higher geometrical elements, first on the workpiece for rough and then on the workpiece for fine machining. The features thus recognized are then coded and, together with the material and the workpiece surface quality and other selection parameters, they represent the input data in the neural network in the second stage of the system. The second stage comprises the selection of the most suitable sets of tools for the concerned workpiece, which totally machine the product. In the meantime, two neural networks have learned one to turn roughly and the other to turn finely on the basis of known cutting tools, required surface quality, workpiece material and features occuring on the workpieces. For easier work and more clear presentation of operation our research was limited to workpieces with specific features, which, however, can be randomly superstructed and/or extended. It must be pointed out that the two neural networks can learn again new rules and laws by adding new cutting tools and features into the base and, thus, the capacity of networks increases with the extension of the base. After completion of the second stage the output data from neural networks serve as input data into the other two neural network (third stage).

Like the human the two neural networks have learnt to propose solutions on the basis of solutions of similar cases. The two networks have learnt one to turn roughly and the other to turn finely with the examples of workpieces, where all the limiting features and tool sets for those workpieces occur.

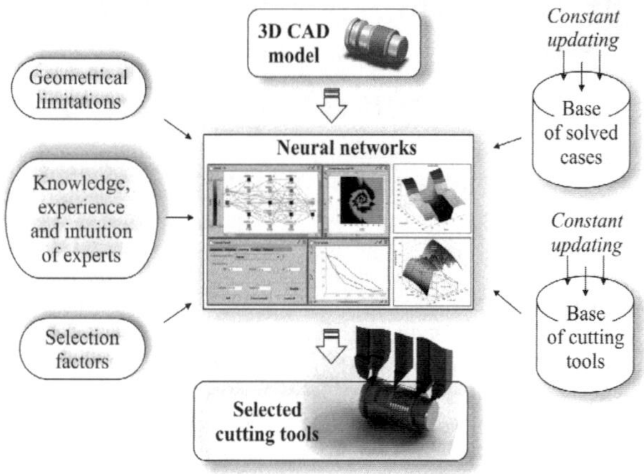

Fig. 2. Intelligent automatic selection of cutting tool.

On the basis of their knowledge and experience the experts have selected the best tool set. Those decisions were taken into account in learning of those networks. Each example newly solved is stored in the base, so that the two networks permanently learn and gain knowledge and experience similarly as the human expert. Still other selection parameters can be included into learning of the networks and, thus, still more accurate results can be reached. The last stage (the fourth stage) covers the selection of the best set of cutting tools, decoding and graphic representation of selection (Figure 2).

4. CONCLUSION

The model developed is simple, robust, user simple and cheap at the price. The organization and the determination of parameters of the neural network take place in the NeuroSolutions environment with simplified transition of data from and into that environment through data tables and graphs in the Excel programme which can be used by the operator without deep knowledge about the neural network. The system works independetly of the human, completely autonomoustly. The system does away with the inconveniences which are a consequence of the human factor (monotony, repeating operations, inaccuracy, slowness). If such a system is included into other intelligent solutions which have been or will be developed (such as intelligent automatic systems for generation of the tool path, for the determination of optimum cutting conditions, for selection of optimum fixing device, for generation of the NC code etc.), the "intelligent machine tool" will be able to make the desired product on the basis of the 3D CAD model autonomously, independently of the human, quickly, accurately, safely and reliably.

5. REFERENCES

Balič J. (2004). *Inteligentni obdelovalni sistemi*, Faculty of mechanical engineering, Maribor

Kokol, P.; Hleb Babič, Š.; Podgorelec, V.; Zorman, M. (2001) *Inteligentni sistemi,* Faculty of electrical engineering and Computer Science, Maribor

Haykin, S. (1994) *Neural Networks.* A Comprehensive Foundation, Macmillan college publishing company, New York

Hopfield, J. J. (1982). Neural networks and physical systems with emergent collective computational abilities, *National Academy of Sciences*, vol. 79, pp. 2554-2558.

Sandvik Coromant (2008). *Available from:* http://www.coromant.sandvik.com/ [20.2.2008].

Annals of DAAAM for 2008 & Proceedings of the 19th International DAAAM Symposium, ISSN 1726-9679
Editor B. Katalinic, Published by DAAAM International, Vienna, Austria 2008
Make Harmony Between Technology and Nature, and Your Mind will Fly Free as a Bird

THE COMPARISON OF SELECTED ALGORITHMS OF SIMULATION OPTIMIZATION

VAZAN, P[avel]; MORAVCIK, O[liver] & KRIZANOVA, G[abriela]

Abstract: *The paper presents the methods of simulation optimization. The authors point out the problems of simulation optimization. They compare some typical algorithms on the real example. The main goal is to give the general procedure for effective usage of simulation optimization. This procedure involves the process of selection of algorithm, input variables, their set up of range, step selection. The proposal of the authors has already been realized and has brought significant time reduction.*
Key words: *simulation optimization, method, algorithm*

1. INTRODUCTION

The simulation optimization is the most significant simulation technology in the last years according to many authors. It eliminates one of disadvantages of simulation and it is used to find the best solution from many simulation experiments.

The combination of simulation and optimization has already been expected for a long time, but only in the last decade it has achieved real development.

Today, leading simulation software vendors have introduced optimizers that are fully integrated into their simulation packages. Now simulation practitioners have access to robust optimization algorithms and they are using them to solve a variety of "real world" simulation optimization problems (Waller, 2006).

There also exist many barriers which have to be over-came for broader simulation optimization using. Great scepticism predominates to the results of simulation optimization in concrete applications (Boesel, 2001).

2. SIMULATION OPTIMIZATION METHODS

Understandably there are a lot of methods that could be used for simulation optimization. The major simulation optimization methods are displayed on Fig. 1. However, most developers have involved heuristic search methods into the software packages for simulation optimization. The heuristic search algorithms provide good, reasonably fast results on a wide variety of problems.

We want to mention a few important heuristic algorithms. Here belong genetics algorithms, evolutionary strategies, simulated annealing, simplex search, tabu search (Tanuška & Kopček, 2007).

The computational demands of simulation optimization cause, that the practical usage of simulation optimization is possible without software support. The software packages are solved as plug-in modules which are added to the basic simulation platform. The approach to simulation optimization is based on viewing of the simulation model as a black box function evaluator. The optimizer chooses a set of values for the input parameters and uses the responses generated by the simulation model to make decisions regarding the selection of the next trial solution (April, 2003).

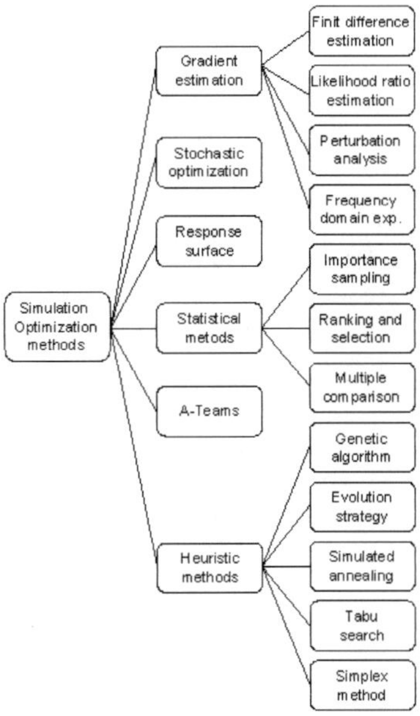

Fig. 1. The methods of simulation optimization.

3. COMPARISON OF SELECTED ALGORITHMS

In spite of the progress it is necessary to emphasize that the realization of simulation optimization will be a compromise between acceptable time and accuracy of found solution.

The authors realized the research oriented to comparison of selected algorithms of simulation optimization. The following criteria were compared:

- Time for optimal solution finding
- Success of optimal value finding
- Number of runs for evaluation
- The influence of change of parameters of algorithms for optimal solution finding

The authors have tested several algorithms on various simulation models. The limited range of this paper does not allow present the whole extensive work. This is why we show the problem of successful usage of simulation optimization only on chosen example. This process was used for searching of optimal values of lot size in the real production system.

The simulation model of system was created in simulator Witness. The basic advantage of simulation optimization is that the objective function can be simply defined and it does not need to contain input values. The objective function is defined inside the simulation model. In this case if quantitative values of other defined manufacturing goals are fulfilled , the objective function will return the value of costs per finished part. (Važan, 2008). The realization of simulation optimization has been realized in plug-in module Optimizer.

These algorithms were included into the comparison:

- All combinations – brute force algorithm.

- Random solutions – to enable an appreciation of the shape of the solution space
- Min/Mid/Max – tests the extremes and mid points of all parameter settings.
- Hill Climb – a simple algorithm. Fast but prone to get stuck in local optima.
- Adaptive Thermostatistical simulated annealing – the main algorithm, a variant of simulated annealing with extra adaptive nature. Developed by Lanner.

Lot size is an input parameter (LS1, LS2 in tables) in our solving procedure. The right values of input intervals to the lot sizes also have to be connected, because lot sizes and their input interval hang together. It is very important to constrain the input parameters meaningfully.

The number of optimization variables and their ranges is defined by the following table 1. It is not possible to use the algorithm All combination because of long time, this arises from the results. Other algorithms were only approximated to searching optimum. See table 2. In the next experiments we have tried to reduce scanned set and then bigger step 3 has been chosen for variables P1 and P2 and step 2 for variables LS1 and LS2. The number of possible combinations has been decreased to 8280. The results of these experiments are presented in table 3. The dramatic reduction of the number of combinations allowed to use algorithm All combinations as is shown by the results of experiment. Then better value of objective function has been found in previous experiment. The found value need not to be the searched minimum. This has been caused by the reduction of the number of combinations.

The found better value allows the range reduction of input variables. If the range of input variables is reduced enough, it will be possible to use step 1. If the number of combinations allows using of algorithm All combinations then it will be possible to find the global extreme of the objective function in relatively short time. It is possible to declare that the solution is really the searched global minimum of objective function. The tables 4 and 5 document the last step of this procedure.

Var. name	Range of var.s	No values	No combinations
P1	5 až 30	26	26
P2	5 až 70	66	1716
LS1	1 až 10	10	17160
LS2	1 až 10	10	171600

Tab. 1. The number and range of input variables.

Algorithm	Found optimum	Estimated time	Real time	No evaluations
Adaptive T.SA	202	0:16:23	0:05:15	658
All Comb.	----------	2days23h	-------	-----------
Random Sol.	216	0:17:29	0:17:51	800
Min/Mid/Max	271	0:01:48	0:01:48	81
Hill Climb	239	0:18:24	0:00:26	312

Tab. 2. Results of experiment1.

Algorithms	Found optimum	Estimated time	Real time	No evaluations
Adaptive T.SA	189	0:27:51	0:06:59	627
All Combin.	189	2:58:17	2:56:48	8280
Random Sol.	249	0:24:16	0:23:56	800
Min/Mid/Max	271	0:01:50	0:01:40	81
Hill Climb	252	0:25:00	0:00:36	317

Tab. 3. Results of experiment2.

Var. name	Old range of var.	New range of var.
P1	5 - 30	8 - 14
P2	5 - 70	7 - 13
LS1	1 - 10	1 - 3
LS2	1 - 10	1 - 3

Tab. 4. New ranges of variables

Algorithm	Found optimum	Estimated time	Real time	No evaluations
Adaptive T.SA	162	0:18:09	0:01:34	321
All Comb.	155	0:25:00	0:22:28	990
Random S.	155	0:17:48	0:12:39	800
M/M/M	170	0:01:49	0:01:48	81
Hill Climb	1114	0:27:07	0:00:23	311

Tab. 5. Final results

4. THE PROPOSAL OF FINAL PROCEDURE

The selection of the proper algorithm is the important part of this procedure. We can notice that the selection of algorithm has to respect mainly two basic factors:

1. What data will include individual sets of variables – it includes whether we know all data files of given range or whether we only know the min., mean and max. value.
2. Time of optimization process.

The authors recommend the following procedure (simplified) for algorithm selection and realization of optimization process based on extensive research:

- Reduce the range of inputs variables by specially designed preparatory experiments. The right range represents such states of the system that will be explored. The constraints of inputs variables represent upper and lower limits for system loading in the presented example.
- Use algorithm Random solutions or Adaptive Thermostatistical SA with bigger step (2 and more).
- Reduce again the range of input variables and repeat experiment by using the Adaptive Thermostatistical SA algorithm.
- If it is possible to reduce again the range of input parameters or if time of obtained result is acceptable, and then repeat the experiment by using All combinations algorithm or Hill Climb algorithm, else repeat the experiment by using the Adaptive T. SA algorithm.

Although simulation optimization represents powerful tool, its real usage needs complex knowledge and appropriate procedure. This procedure is only a part of bigger project which is realized by authors. The proposal of the authors has already been realized and has brought significant time reduction. This result will contribute to more effective using of simulation optimization.

This paper has been supported as a part of a solution of projects VEGA 1/0170/08.

5. REFERENCES

April, J.; Glover, F.; Kelly, J.P. & Laguna M.(2003). Practical introduction to simulation optimization, *Proceedings of the WSC 2003.*, Chick, P. J. Sánchez, D. (Eds.), pp. 71-77, ISBN: 0-7803-8131-9 New Orleans, 12. 2003, WSC.

Boesel J., Glover F., Bowden O.R., Kelly J. P. & Westvig E. (2001). Future of simulation optimization. *Proceedings of the 2001 Winter Simulation Conference*, pp. 1466-1469, ISBN 0-7803-7309-X 12,2001, ACM, New York.

Tanuška P., & Kopček M. (2007). The technological process control through the use of virtual controllers. *Proceedings of the XIV Meždunarodnja naučno techničeskaja konferencia*. ISBN 966-7907-22-8. Sevastopol. Ukraine.

Važan, P., Schreiber, P. & Križanová G. (2008). The Simulation Optimization in Simulator Witness, *Proceedings of the Third Conference: Ecumict* pp. 461-469, ISBN 9-78908082-553-6 Gent, 3, 2008. Nevelland

Waller A.P. (2006). Optimization of simulation experiments, *Available from :* http://www.solventure.be/web/Jahia/pid/9 *Accessed:* 2008-06-02

Annals of DAAAM for 2008 & Proceedings of the 19th International DAAAM Symposium, ISSN 1726-9679
Editor B. Katalinic, Published by DAAAM International, Vienna, Austria 2008
Make Harmony Between Technology and Nature, and Your Mind will Fly Free as a Bird

VIRTUAL LABORATORY IN PNEUMATIC SYSTEMS EDUCATION

VELISEK, K[arol]; KOSTAL, P[eter] & MUDRIKOVA, A[ndrea]

Abstract: *Labour power is an important development aspect of present day informatics society. This importance can be found on firm level as well as in global economic development use. Labours, which are suitable and high, qualified for these conditions can be found in groups of university-educated graduates. Especially we are talking about graduates of economy and technical universities. This kind of graduates is very required on the labour market.*
Researches operating about employers requirements and also operating about student opinions to the university training are indicating that teaching schedules and education process are not creating for praxis requirements. That mean, that universities have to teach special practical knowledge and skills re-bound to the praxis and its needs.
Key words: *laboratory, education, experiences, competencies.*

1. INTRODUCTION

In these days social situation, political situation, economical situation and also globalization are creating special environment in which can survive only the best personalities.

●Important requirement of potential employers is an "correct" relation between professional competences (high level of professional knowledge a knowledge of present branch stay) and key competences. (Mudriková 2006)

●Key competences can be represented as universal skills complex, which overhung level of specific professional knowledge and skills. Key competences are skills, which are telling something about that, how will some person behave accordingly to the situation. Professional competences are of course very important and key competences are helping for better use of professional competences in praxis.

●Professional competences, which ensure good managing of working tasks.

Key competences, are widely usable skills such as decide ability, solving problems ability, tem working ability, communication in foreign language, learning ability, ability to work with information, and so on.

2. LABORATORY OF PNEUMATICS SYSTEMS

In these days, at our department we have the possibility to work in laboratory of pneumatic system. The laboratory includes two workplaces, which means that four student are able to work in the laboratory.

Student can work in the laboratory and so they are able to improve their practical skills and they also reach new skills from studied information. Pneumatic mechanisms can be found in many forms in number of specialized subjects. The ambition of the lector is to show all these mechanisms in form of practical exercises, to show its function in some practical form. (Fig. 1)

In many cases the mechanism is not working exactly, so student have to rebuild the exercise, what leads to the increase of manual skills and this activity increase failure searching skill. Many mechanisms are not working, because there is not enough hardware, and building elements replacement is also thing of money.

Fig. *1*: Students in laboratory of pneumatics

The aim of whole project is creating of virtual pneumatic and electro-pneumatic control system laboratory. This laboratory will be used for learning of designing principles deals about creation of pneumatic and electro-pneumatic control system and its simulation. (Zvolenský 2007)

Creation of such laboratory will lead to the quality increase of teaching process of some subjects which belongs to our study branch called manufacturing devices and systems (automation and mechanization, automates theory, and so on…) Student which will go through courses in these kind of laboratory will get needed professional skills and competences.

These competences and skills will strengthen their skills, which are very important for designing process of such devices. The students will also know the methodology how to create several types of systems, which will lead to their labor market value increase.

3. VIRTUAL LABORATORY

New created virtual laboratory of pneumatic system will be used for teaching of pneumatic and electro-pneumatic systems principles. Various control systems can be realized with help of pneumatic and electro-pneumatic systems.

Pneumatic and electro-pneumatic can also simulate some real industrial systems. By creating of virtual laboratory we would like to increase the quality level of some our subjects. Student with studying and working in this laboratory will increase their skill needed for design of such devices. Following to these new information and skill students will be also more attractive for its possible employers. (Danišová et al 2007)

Student will in virtual laboratory create pneumatic or electro-pneumatic models of control systems and they will be also able to simulate the function of such device. Student will be able to create its own virtual pneumatic or electro-pneumatic control system, and they will be also able to test this system. So they will be able to verify the proper function of the system by using of simulation.

Fig. 2: Virtual laboratory of pneumatics systems

Virtual laboratory equipment will be composed of 10 PC with actual hardware parameters and they will have the functions of single working units. (Fig. 2) Then network server will be installed in the laboratory. The function of such server will be internet access and other one function will be the function of licence server. The environment of this laboratory will also include data projector of high quality with silver screen, interactive board. All this equipment will lead to the using of modern teaching methods.

3.1 Manuals, studying literature and E-learning

Within the frame of this project will be created manuals which will be accessible through internet access. The manuals and materials will deal about connections of chosen pneumatic and electro-pneumatic control systems. Solved tasks will start with simple layouts and will end with more complex tasks.

Every such task will include detail device function description and standard connection scheme. These schedules will serve as supplementary study literature for students of internal and also distant study form. Experiences connected with creating of such teaching system will be implemented in to the distant study system by help of e-learning.

If we are looking to the e-learning as to the effective using of information technologies in education process, we can say that we are talking about two new possibilities, which can be used in education process. (Velíšek et al 2007)

E-learning is created to improvement of these disadvantages and it is an optimal solution for complex education. To create interesting, address, individual and interactive education process, which can be integrated in to the every day live, we have to combine proper methods of classical and e-learning education style. Also in areas in which is personal contact with teacher necessary, can be students learned before through electronic. Students are then coming to the lectures with new amount of information and questions. After such process teacher is able to speak only about interesting or difficult fields of study. Also the time of lectures can be reduces. (Mudriková et al 2007)

E-learning brings many communication ways such as emails, video conferences, which allows teacher to individual access to individual students. Information such as, how good was the students in single lectures, how many points reach student in single fields of study or haw many time it takes for each student. Thanks to the number of communication ways is teacher able to communicate with his students more intensive as before by typical classes of classical education system.

E-learning is a solution designated for education process, but the education is showed in other context. E-learning is not restricted only for student education, but in some view it is a method of information passing or sharing. Such sharing and passing information is important in field of whole life education process, which is very important for technician. Classical information technologies and systems are handling about information searching and searching in the exact time. E-learning gives emphasis on kind of information passing.

E-education as a new progressive way of education brings new possibilities. This education form can be used in all degrees of formal education and can be also used in whole life education process.

4. CONCLUSION

To the most important requirements influencing to the future employees – technical engineers belongs:

- Creativity,
- Foreign language skills,
- Independence,
- Computer working skill,
- Responsibility,
- To be flexible sill,
- Knowledge of actual trends in the branch,
- Willingness of other education,
- Communication skills. (Mudriková 2006)

Students can create all these competences by the work in virtual and real pneumatic and electro-pneumatic laboratory.

Systems and its functions, which are learned in virtual laboratory, can be found in real industrial environment. Knowledge and skills with those systems are highly reputable and are also searched by wide number or employers, such as huge international technical industries and concerns such as (PSA, KIA, Volkswagen, and others.)

ACKNOWLEDGEMENT

The research has been supported by national grant: **KEGA 3/5211/07 - Building of virtual laboratory of pneumatic systems**

5. REFERENCES

Danišová, N.; Zvolenský, R. & Ružarovský, R. (2007) Inteligent manufacturing systems. Inteligentné výrobné systémy. (Intelligent manufacturing systems), *Proceedings of: ERIN 2007. Education, Research, Innovation* : 1. medzinárodná konferencia mladých výskumníkov a doktorandov. Bratislava 25.-26.4. 2007. Zborník abstraktov. - Bratislava : STU v Bratislave, 2007. - ISBN 978-80-227-2636-8

Mudriková, A. (2006) Celoživotné vzdelávanie ako jedna z podmienok profesnej úspešnosti inžiniera – technika. *Proceedings of SCHOLA 2006*. 7. medzinárodná vedecká konferencia KIPP : Kvalita výchovy a vzdelávania. - Bratislava : STU v Bratislave, 2006. - ISBN 80-227-2389-4. - S. 274-279

Mudriková, A.; Charbulová, et al. (2007): Virtual laboratory of pneumatics. *Proceedings of 25 jubilejna naučna konferencija s meždunarodno učastie MTF 2007.*, Sbornik dokladi. Tom II. - Sozopol, 14.-16. 9. 2007. - Sofia : Techničeskij universitet, 2007. - ISBN 978-954-438-624-5. - 10-15, Tom 1

Velíšek K.; Košťál, P. & Mudriková, A. (2007) TVORBA A BUDOVANIE VIRTUÁLNEHO LABORATÓRIA PNEUMATICKÝCH A ELEKTROPNEUMATICKÝCH SYSTÉMOV RIADENIA, (Building of virtual laboratory of pneumatic systems) Projekt KEGA.2007

Zvolenský, R(2007): Design of automated disassembly devices. *Proceedings of RaDMI 2007 7th International Conference "Research and Development in Mechanical Industy - RaDMI 2007"*, Belgrade/Serbia/, 16-20 September 2007. - Trstenik : High Technical Mechanical School of Trstenik, 2007. - ISBN 86-83803-22-4. - S. 697-700

Annals of DAAAM for 2008 & Proceedings of the 19th International DAAAM Symposium, ISSN 1726-9679
Editor B. Katalinic, Published by DAAAM International, Vienna, Austria 2008
Make Harmony Between Technology and Nature, and Your Mind will Fly Free as a Bird

SENSORLESS CONTROL FOR INTERIOR PERMANENT MAGNET SYNCHRONOUS MOTOR WITH EXTENDED KALMAN FILTER

VESELY, L[ibor] & ZBRANEK, P[avel]

Abstract: *Permanent magnet synchronous motors have been used in many industrial applications because they have several inherent advantages. Extended Kalman filter with model of interior permanent magnet synchronous motor (IPMSM) is described in this paper. Model of IPMSM ($L_d \neq L_q \neq L_s$) makes it possible for EKF operates in the low speed, because information about rotor position is contained in changes of stator inductance.*
Key words: *sensorless, control, IPMSM, EKF*

1. INTRODUCTION

Permanent magnet synchronous motors have been used in many industrial applications because they have several inherent advantages e.g. rugged construction, easy maintenance, high power factor, high efficiency and suitability for wide speed ranges of constant power operation. However, rotor position information is necessary even to be able to control drive speed. Conventional speed and position sensing was obtained using encoder, resolver. These sensors add cost, weight, and degrade reliability.

Many authors publish paper about algorithms for rotor position and speed estimation. Most of these algorithms are based on state estimation. Once of state estimation possibility is usage extended Kalman filter (EKF) with α-β stationary axis model. Conventional algorithms base on EKF are developed with simple model of permanent magnet synchronous motor ($L_d = L_q = L_s$), thereby provide bad performance in the low speed range (Fu & Xu 1997). Extended Kalman filter with model of interior permanent magnet synchronous motor (IPMSM) is described in this paper (Julier & Uhlman 2004). Model of IPMSM ($L_d \neq L_q \neq L_s$) makes it possible for EKF operates in the low speed, because information about rotor position is contained in changes of stator inductance (Kosaka & Uda 2003).

2. MODEL OF PMSM

Mathematical model of interior permanent magnet synchronous motors in the α-β stationary system describes equations (1). Fig. 1. shows interior permanent magnet motor with salient rotor type. Fig. 2. shows surface permanent magnet motor with non-salient rotor type (Bowen et al., 2001) (Pillay & Krishnan 1989).

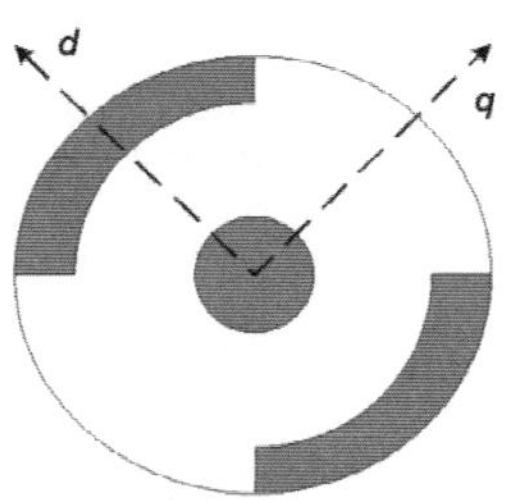

Fig. 1. IPMSM ($L_d \neq L_q$)

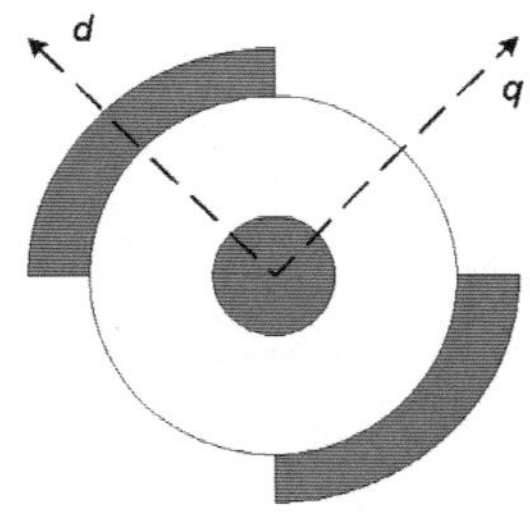

Fig. 2. SPMSM ($L_d = L_q$)

$$\frac{d}{dt}\begin{bmatrix} i_\alpha \\ i_\beta \\ \omega_s \\ \theta \end{bmatrix} = \begin{bmatrix} -\frac{RL_\beta}{L_\alpha L_\beta - L_{\alpha\beta}^2} & \frac{RL_{\alpha\beta}}{L_\alpha L_\beta - L_{\alpha\beta}^2} & \frac{L_\beta \psi_{af} \sin\theta}{L_\alpha L_\beta - L_{\alpha\beta}^2} & \frac{L_{\alpha\beta} \psi_{af} \sin\theta}{L_\alpha L_\beta - L_{\alpha\beta}^2} \\ \frac{RL_{\alpha\beta}}{L_\alpha L_\beta - L_{\alpha\beta}^2} & -\frac{RL_\alpha}{L_\alpha L_\beta - L_{\alpha\beta}^2} & -\frac{L_{\alpha\beta} \psi_{af} \cos\theta}{L_\alpha L_\beta - L_{\alpha\beta}^2} & -\frac{L_\alpha \psi_{af} \cos\theta}{L_\alpha L_\beta - L_{\alpha\beta}^2} \\ 0 & 0 & 0 & 0 \\ 0 & 0 & 1 & 0 \end{bmatrix} \cdot \begin{bmatrix} i_\alpha \\ i_\beta \\ \omega_s \\ \theta \end{bmatrix} + \begin{bmatrix} \frac{L_\beta}{L_\alpha L_\beta - L_{\alpha\beta}^2} & \frac{L_{\alpha\beta}}{L_\alpha L_\beta - L_{\alpha\beta}^2} \\ \frac{L_{\alpha\beta}}{L_\alpha L_\beta - L_{\alpha\beta}^2} & \frac{L_\alpha}{L_\alpha L_\beta - L_{\alpha\beta}^2} \\ 0 & 0 \\ 0 & 0 \end{bmatrix} \cdot \begin{bmatrix} u_\alpha \\ u_\beta \end{bmatrix} \tag{1}$$

where

$$L_\alpha = L_0 + L_1 \cos 2\theta \tag{2}$$

$$L_\beta = L_0 - L_1 \cos 2\theta \tag{3}$$

$$L_{\alpha\beta} = L_1 \sin 2\theta \tag{4}$$

and

$$L_0 = \frac{L_d + L_q}{2} \tag{5}$$

$$L_1 = \frac{L_d - L_q}{2} \tag{6}$$

u_α, u_β stator voltage components in stationary frame;
i_α, i_β stator current components in stationary frame;
ω_s rotor speed (electrical speed);
θ rotor position (electrical angle);
L_d, L_q stator inductances
Ψ_{af} denote the flux linkage of the rotor magnet without stator current

3. EXTENDED KALMAN FILTER (EKF)

Let us assume that our process has a state vector $x \in R^n$ with a measurement $y \in R^n$ and then the process is governed by the non-linear stochastic difference equation:

$$x_k = f\left(x_{k-1}, u_k, \omega_{k-1}\right) \quad (7)$$

$$y_k = h\left(x_k, v_k\right) \quad (8)$$

ω_k, v_k random variables represent the process and measurement noise

f non-linear function relates the state at the previous step *(k-1)* to the state at current time step *k*

The equations for the extended Kalman filter fall into two groups: time update and measurement update equations. The time update equations are responsible for projecting forward (in time) the current state and error covariance estimates to obtain the priori estimates for the next step. The measurement update equations are responsible for incorporating a new measurement into priori estimate to obtain an improved a posteriori estimate (Welch & Bishop 1995)(Kalman, 1960).

1. Time update equations

$$\hat{x}_k^- = f\left(\hat{x}_{k-1}, u_k, 0\right) \quad (9)$$

$$P_k^- = M_k P_{k-1} M_k^T + W_k Q_{k-1} W_k^T \quad (10)$$

2. Measurement update equations

$$L_k = P_k^- C_k^T (C_k P_k^- C_k^T + V_k R_k V_k^T)^{-1} \quad (11)$$

$$\hat{x}_k = \hat{x}_k^- + L_k (y_k - h(\hat{x}_k, 0)) \quad (12)$$

$$P_k = (1 - L_k C_k) P_k^- \quad (13)$$

$\hat{x}_k$ posteriori state estimate;

$\hat{x}_k^-$ priori state stimate;

Q_k process noise covariance matrix;

R_k measurement noise covariance matrix;

P_k estimate error covariance matrix;

L_k Kalman gain that minimisez the posteriori error covariance matrix;

Linear stochastic difference equations derive from mathematical model (1) of IPMSM:

$$x(k+1) = \mathbf{M}x(k) + \mathbf{N}u(k) + \omega(k) \quad (14)$$

$$y(k) = \mathbf{C}x(k) + v(k) \quad (15)$$

where

$$\mathbf{M} = (I + J \cdot T_{vz}) \quad (16)$$

$$\mathbf{N} = N \cdot T_{vz} u_{k-1} \quad (17)$$

I ones metrix;

T_{VZ} sampling period;

J Jacobian matrix of partial derivation of non-linear function *f;*

and

$$J = \begin{bmatrix} \frac{\partial f_1}{\partial i_\alpha} & \frac{\partial f_1}{\partial i_\beta} & \frac{\partial f_1}{\partial \omega_s} & \frac{\partial f_1}{\partial \theta} \\ \frac{\partial f_2}{\partial i_\alpha} & \frac{\partial f_2}{\partial i_\beta} & \frac{\partial f_2}{\partial \omega_s} & \frac{\partial f_2}{\partial \theta} \\ \frac{\partial f_3}{\partial i_\alpha} & \frac{\partial f_3}{\partial i_\beta} & \frac{\partial f_3}{\partial \omega_s} & \frac{\partial f_3}{\partial \theta} \\ \frac{\partial f_4}{\partial i_\alpha} & \frac{\partial f_4}{\partial i_\beta} & \frac{\partial f_4}{\partial \omega_s} & \frac{\partial f_4}{\partial \theta} \end{bmatrix} \quad (18)$$

4. CONCLUSIONS

This paper has introduced algorithm for rotor position and speed estimation based on state estimation idea. The sensorless algorithm use extended Kalman filter with model of interior permanent synchronous motor. Model of IPMSM ($L_d \neq L_q \neq L_s$) makes it possible for EKF operates in the low speed, because information about rotor position is contained in changes of stator inductance. Designed algorithm was tested in Matlab Simulink and attained very good results also in low speed region. The future research will be aimed to implementation algorithm to hybrid controller and testing on real electromechanical system with IPMSM.

5. ACKNOWLEDGEMENTS

The research has been supported by Czech Science Foundation under the project GA 102/06/0949 "Intelligent Control of Induction Machine and Synchronous Machine Electrical Drives", the Ministry of Education of the Czech Republic under the project 1M0567 "Centre for Applied Cybernetics"

6. REFERENCES

Bowen, C., Jihua, Z. & Zhang, R. (2001) Modeling and Simulation of Permanent Magnet Synchronous *Motor Drives*, Northwestern Polytechnical University, pp. 905-908

Fu, M. & Xu, L . (1997) A novel sensorless control technique for permanent magnet synchronous motor (PMSM) using digital signal processor (DSP) Aerospace and Electronics Conference, 1997. NAECON 1997., *Proceedings of the IEEE 1997* National Volume 1, Issue , 14-18 Jul 1997 Page(s):403 - 408 vol.1

Julier, S. & Uhlman, J. (2004) Unscented filtering and nonlinear estimation. *Proceedings of the IEEE*, vol. 92, no. 3, pp. 401–422, March

Kalman, E. R. (1960) A new approach to linear filtering and prediction problems, *Transactions of the ASME–Journal of Basic Engineering*, vol. 82, no. Series D, pp. 35–45.

Kosaka, M. & Uda, H. (2003) Sensorless ipmsm drive with ekf estimation of speed and rotor position, *Proceedings of 42nd IEEE Conference on Decision and Control*, vol. 6, December 2003, pp. 5915–5920.

Pillay, P. & Krishnan, R. (1989). Modeling, Simulation, and Analysis of Permanent Magnet Motor Drives, Part I: The Permanent Magnet Synchronous Motor Drive, 0093-9994/89/0300-0265

Wang, L. & Loreny, D. R. (2000) Rotor positron estimation for permanent magnet synchrounous motor using saliency-tracking sefl-sensing metod, 0-7803-6401-5/00, IEEE 2000

Welch, G. & Bishop, G.(1995) An introduction to the Kalman filter, Technical Report 95-041, Univ. of North Carolina at Chapel Hill, Dept. of Computer Science

Annals of DAAAM for 2008 & Proceedings of the 19th International DAAAM Symposium, ISSN 1726-9679
Editor B. Katalinic, Published by DAAAM International, Vienna, Austria 2008
Make Harmony Between Technology and Nature, and Your Mind will Fly Free as a Bird

MOBILE ROBOTIC TESTING PLATFORM

VESELY, M[ilos]

Abstract: *This paper presents "Mobile robotic testing platform" project. The project is aimed at development and implementation of mobile testing robots. The robots will be used for testing autonomous and cooperative algorithms for mobile robots. The robots will be used especially for testing algorithms for multi-robot cooperative area exploration. These robots will be also used for education. Students can be involved in the project. They can develop hardware and software for these robots. The paper describes briefly rescue robotics, tepresence, multi-robot rescue systems and autonomous robots. Construction, differential chassis and hardware of the testing robots are described. Communication bus specially developed for the robots is mentioned at the end of the paper.*
Key words: *Robot, Reconnaissance, Autonomous, Cooperation.*

1. INTRODUCTION

Rescue and reconnaissance robots are important part of robotics. These robots are used by rescue teams, firemen and army in dangerous and hardly accessible areas. Robot's main mission objectives mostly are:

- exploration of area
- creating map of environment
- finding victims
- finding dangerous zones

Using these robots can avoid human injuries and rescuers do not need go in to dangerous areas. In addition performance of the rescue team is increased by using these robots and their sensors. Robots with properly chosen sensors can perform in exploration even better than humans. Especially robots equipped with thermal camera perform very well exploration of ruins and finding victims. There are many research teams working on rescue and reconnaissance robotics. Nowadays there are many sources of information about and navigation of mobile robots (Siegwart & Nourbakhsh, 2004; Borenstein et al., 1996), robotic hardware, software and sensors (Solc & Zalud, 2002) as well as many valuable instructions and recommendations about developing mobile robots (Jones et al., 1999). An extensive amount of research has been carried out in the area of localization and mapping (Thrun, 2002). My research is aimed at multi-robot cooperative area exploration.

2. TELEPRESENCE ROBOT

Most of present rescue and reconnaissance robots are telepresence robots. That means that robot is remotely controlled by operator. Information from robot sensors and video camera are transmitted to the operator. This allows the operator to feel as if he was present.

3. RESEARCH TREND

Performance of rescue and reconnaissance systems can be increased by autonomous and cooperative elements (Birk & Caprin, 2006). Cooperation of robots lowers time needed for mission tasks. Typical example of such task is exploring an area with multi-robot system. Autonomous elements in the system allow the operator to control whole group of robots. General trend is to develop more autonomous and cooperative robots. Field of distributed autonomous mobile robotics is wide and many areas have not been fully studied (Parker, 2000).

4. MULTI-ROBOT RESCUE SYSTEMS

Importance of autonomous robots will rise in the future. In the present the operator is still irreplaceable. Hence semi-autonomous systems are developed in which robots autonomously perform individual tasks and human operator is responsible for whole mission's progress and important decisions. Proposal of such system follows. A group of autonomous cooperative robots is incorporated in rescue system. The robots autonomously explore selected area and create map. The map contains important information like positions of obstacles, dangerous zones and positions of victims. Autonomous activity of robots will be interrupted when important event happen for example victim is found. In that case the operator is noticed and he can make a decision about mission's proceeding. The operator can also assign tasks to individual robot or he can control one of the robots telepresently. The robots have collective communication system and share information. Behavior of the system is based on state and position of each robot and on state of the mission. The basic element of this multi-robot system is a single autonomous robot.

5. AUTONOMOUS ROBOT

Autonomous robots are robots which can perform desired tasks in unstructured environments without continuous human guidance. Many kinds of robots have some degree of autonomy. A high degree of autonomy is particularly desirable in fields such as space exploration, cleaning floors and mowing lawns.

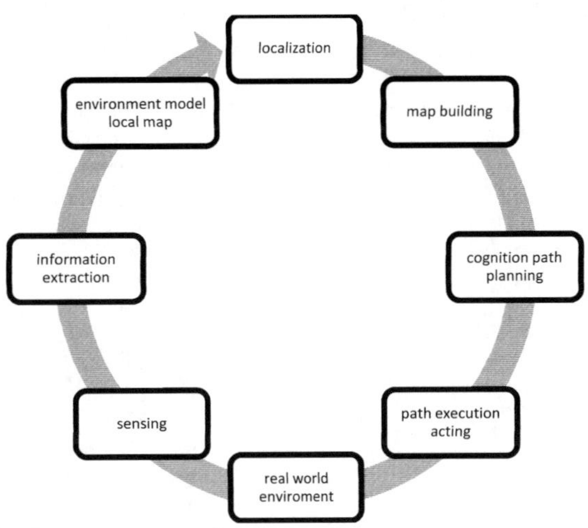

Fig. 1. Autonomous robot operation.

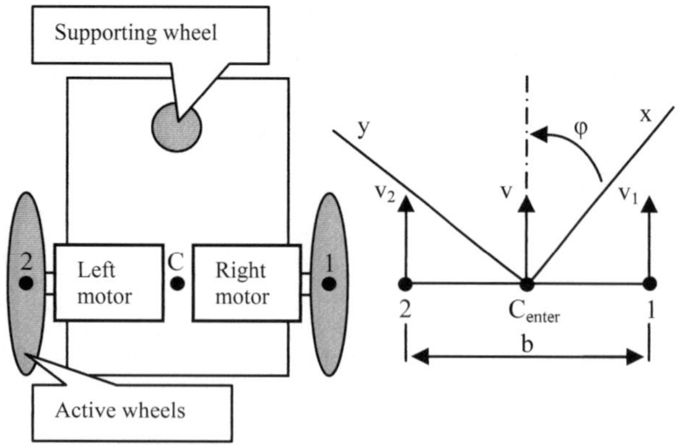

Fig. 2. Differential chassis diagram.

Autonomous robot operation in general is shown in the figure 1. Diagram description starts at the bottom block "real world environment" and continues clockwise. Robot gathers information about environment with various types of sensors. Important data from sensors are extracted and passed in to environment model and local map. Navigation system localizes robot in the map and updates this map with newly acquired information. The "Cognition path planning" block controls behavior of the robot and plans robot path according to mission's objectives. The path is executed by power drivers connected to motors and manipulators.

6. MOBILE ROBOT CONSTRUCTION

Mobile robotic testing platform is constructed for indoor environment. It is made of aluminium construction profile and aluminium plate. Construction is developed to be reliable simple and cheap. Robot can be easily demounted or modified. Most of joints are detachable. The robot is built on differential chassis. This type of chassis is commonly used in robotics. Chassis has two active wheels. Each active wheel has one degree of freedom. The third wheel is freely revolving and acts as a supporting point. This solution is not suitable for hard terrain but it can turn around on one place and it can be controlled easily. Velocity (v) and turning (φ) of the robot can be controlled only by velocity of each active wheel (v_1, v_2). Block diagram of differential chassis is in the figure 2.

7. MOBILE ROBOT HARDWARE

Computation system of the robot can be divided in to three layers (see figure 3). All decision making, path planning and algorithms responsible for robot's behavior are executed in the first layer. This layer is equipped with mATX PC motherboard with Intel Celeron processor and wifi communication with other robots. Hardware parts are common and can be easily updated from the actual market. The second layer contains sensors, power drivers, electronics needed for gathering data from sensors and other. The last hardware part of the robot is mechanic construction, motors and manipulators.

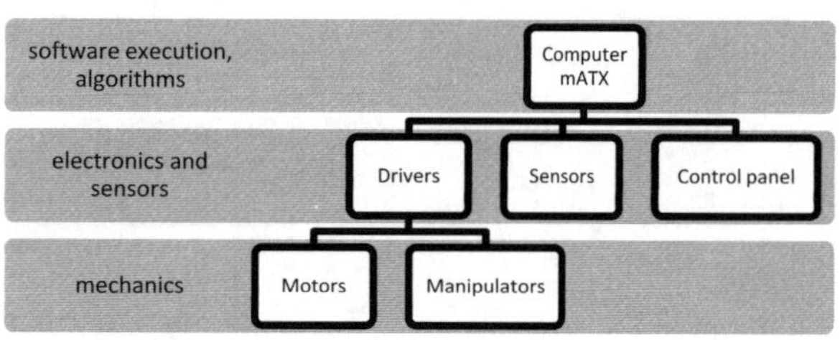

Fig. 3. Mobile robot hardware.

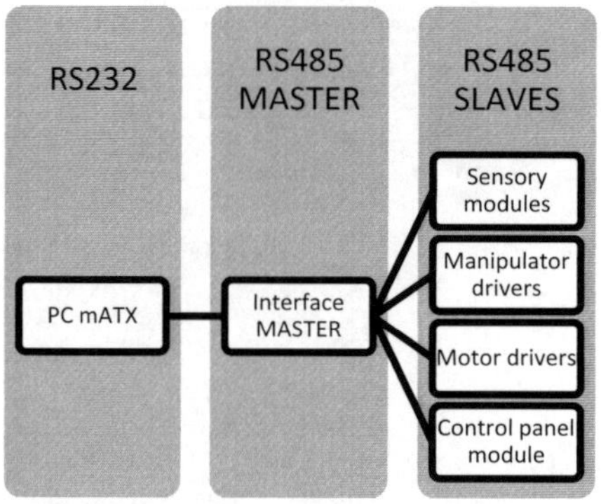

Fig. 4. Mobile robot communication block diagram.

8. MOBILE ROBOT COMMUNICATION BUS

Communication system bus for testing robot has been developed (see figure 4). This bus works in Master/Slave mode and has its own data protocol. The protocol was designed for robot sensory system. The protocol can address 32 slaves and is simple and reliable. Physical layer of the bus is industrial standard RS485. Master module is an interface between the bus and superior computer system. Master module is connected with computer system by RS232 link.

9. CONCLUSION

Most of present rescue and reconnaissance robotic systems use telepresence robots. Performance of these systems can be increased by autonomous and cooperative elements. Presented project is aimed at development and implementation of mobile testing robots. The robots will be used for testing autonomous and cooperative algorithms for mobile robots. The results of the project are two prototypes of the testing robots. Next steps are development of the control software for single autonomous robot and then development of the cooperative robotic system and incorporation of individual robots.

10. ACKNOWLEDGEMENT

The paper has been prepared with the support of the research plan MSM 0021630529 "Intelligent systems in automation"
Applicant: Brno University of Technology
Administrator: Faculty of Electrical Eng. and Communication

11. REFERENCES

Birk, A. & Caprin, S. (2006). Rescue Robotics - a crucial milestone on the road to autonomous systems. *Advanced Robotics Journal,* Vol. 20, No. 5, p. 11.

Borenstein, J.; Everett, B. & Feng, L. (1996). *Navigating Mobile Robots: Systems and Techniques*, A. K. Peters, Ltd., MA, ISBN 1-56881-058-X, Wellesley

Jones, Joseph L.; Flynn, Anita M. & Seiger, Bruce A. (1999). *Mobile Robots, Inspiration to Implementation*, A K Peters, Ltd., ISBN 1-56881-097, Natick, MA, USA

Parker, L. E. (2000). Current State of the Art in Distributed Autonomous Mobile Robotics, In: *Distributed Autonomous Robotic System 4,* p. 3-12, Springer-Verlag, ISBN: 4-431-70295-4, Tokyo

Siegwart, R. & Nourbakhsh, I. R. (2004). *Introduction to Autonomous Mobile Robots,* A Bradford Book, The MIT Press, ISBN 0-262-19502-X, London, England

Solc, F. & Zalud, L. (2002). *Robotics*, lecture notes, Brno University of Technology (BUT), Brno

Thrun, S. (2002). Robotic mapping: A survey, In: *Exploring Artifical Intelligence in the New Millenium,* p. 1-35, Morgan Kaufmann, ISBN: 1-55860-811-7, San Francisco

Annals of DAAAM for 2008 & Proceedings of the 19th International DAAAM Symposium, ISSN 1726-9679
Editor B. Katalinic, Published by DAAAM International, Vienna, Austria 2008
Make Harmony Between Technology and Nature, and Your Mind will Fly Free as a Bird

FLEXIBLE SYSTEM FOR METALLOGRAPHY ANALYSIS

VESSELENYI, T[iberiu]; MOGA, I[oan] & MUDURA, P[avel]

Abstract: *A laboratory for automated preparation and analysis of metallographic probes would improve the quality of the products at sustainable costs and would offer services of analysis and verification for clients. The flexible system should integrate the subsystem for microstructure and the subsystem for chemical investigations creating a distributed structure. This paper outlines the advantages of a project to develop a flexible system for metallography analysis.*
Key words: *metallography, flexible systems, intelligent agents, laboratory automation.*

1. INTRODUCTION.

The idea of an automated metallographic laboratory is sustained by an increasing need for an efficient quality control of products in the context of a higher competition level on the global market (Vesselenyi, Mudura,2001), (Vesselenyi, 2005).

A laboratory for automated preparation and analysis of metallographic probes would improve the quality of the products at sustainable costs and would offer services of analysis and verification for an important number of industrial companies, because small and medium companies can hardly afford necessary investments in laboratories of their own, and in the big companies, more frequently the strategy of services externalization is used. In this way potential beneficiaries are all the companies that use, in their products, metallic materials which have to fulfill proven quality materials.

The microstructures database of the system can be used by the clients not only for production connected activities but in the R&D activities, developed for elaboration of new materials and as an information centre in this field. Beside industrial companies, the flexible system and its database can be used by universities and other institutions for educational purposes.

This paper highlights the advantages of a project which would aim to develop a flexible system for metallography analysis.

2. FLEXIBLE SYSTEM STRUCTURE

In metallographic domain, there are systems that have in their structure a microscope, storage device microscope, and a service robot („CLEMEX Robotic Imaging Analysis System", all together with a software for image processing Clemex Vision), but without automated probe preparation integration, this task being fulfilled by human operators. Automated metallographic image analysis also remains an open field, not fully investigated yet, especially in what concerns classification, interpretation and automated recognition of microstructure (MetLab Corporation), (VISIONx Inc.), (ASM Materials Information), (Alicona Imaging), (METADEX).

According to the studied documentations, it is estimated that a flexible system which can solve the problems both for automated preparation of probes and automated metallographic analysis, (even for a small group of microstructures), was not yet developed.

The purpose of a project to develop such system, on short term, is the development of a distributed flexible system which has to be able to provide services in metallic materials verification and identification and a database containing microstructures, with Internet accessibility which may be consulted by industrial clients, universities and research institutes. The strategic purpose is to initiate a platform in order to support product quality maintenance in which the partners represent a core of expertise.

Command and control of the flexible system, using intelligent agents and the scattered nature of the flexible system are representing new elements in the field.

The flexible system should integrate the subsystem for microstructure investigations (METSYS) and the subsystem for chemical investigations (CHEMSYS) creating a distributed structure, connected through Internet. This flexible system can solve complex problems, which today can be solved only by human operators. During flexible system construction the mechanisms for innovative solution implementation are created.

The flexible system for metallographic analysis represents a new concept, namely that of an expert flexible system, in which the final products are not physical objects but the information about materials obtained after their processing. The system involve innovative construction and information technology solutions: a new command language of the system based on intelligent agents; construction solutions for the flexible system mechanics of preparing metallographic probes, etching, washing and drying of probes, positioning the probes on the microscopes table and also microstructure image processing algorithms (Vesselenyi, 2005).

The proposed metallographic analysis distributed flexible system is structured from two subsystems with a permanent Internet connection, one dealing with chemical analysis (CHEMSYS) and one dealing with preparing probes and metallographic analysis (METSYS). The command program of the flexible system is based on decisions and actions of four intelligent agents (I.A.) generically denominated: "Coordinator" (it coordinates actions, assists decisions, solve conflicts between the other agents), "Metallographer" (metallography analysis), "Chemist" (chemistry analysis) and "Assistant" (figure 1). Intelligent agents are in a permanent cooperation, are accessible to the human operator, demands and give information, deliver warnings and messages to the operator, take decisions on their own action domain and launch different tasks to be executed, coordinates actions for common use of system's resources. Intelligent agents are manipulating information objects in a direct connection with physical objects: system components (robots, magazines, modules for probe processing) or processed probes.

Real objects (system's components) are materialized by probe grinding machine, laboratory industrial robots (with max. manipulated weight of 0,2 kg), magazines, motorized stage microscopes, washing and drying modules, etching modules.

The other goal of this project would be an Internet accessible database which has (in addition from a keyword

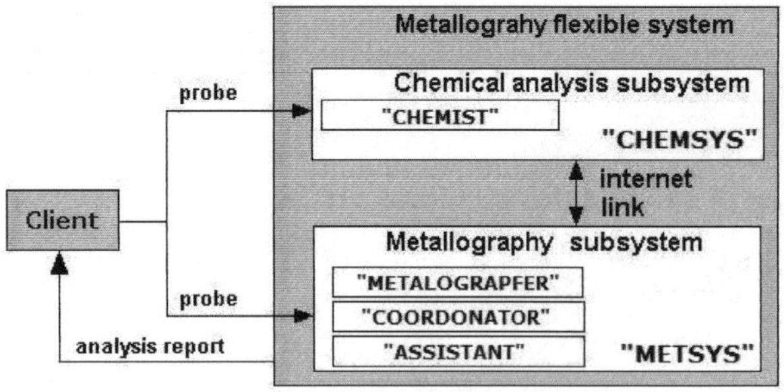

Fig.1. Diagram of the distriduted flexible system.

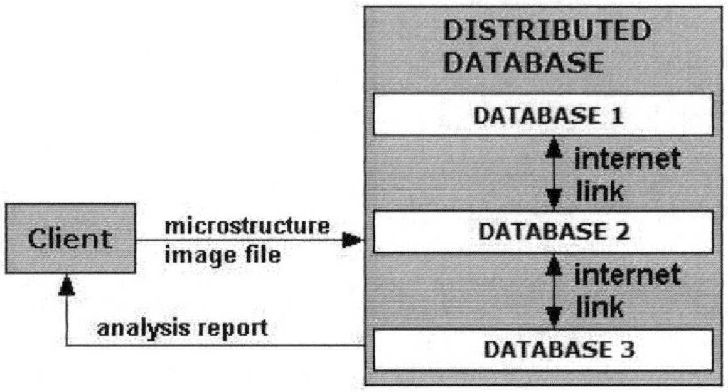

Fig.2. Diagram of microstructure database .

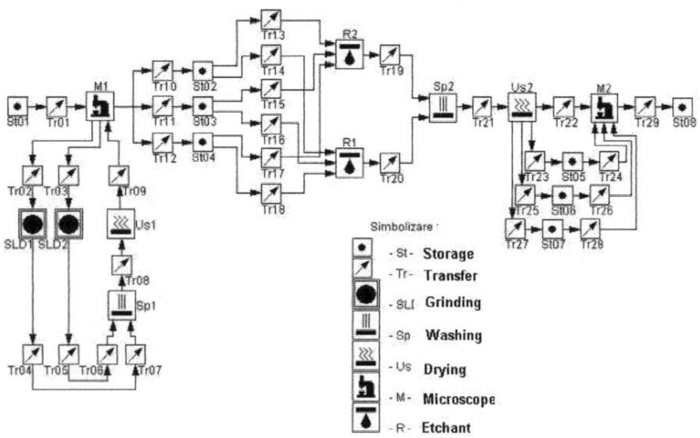

Fig.3. Probe flux schematics.

based browsing engine), a "searching by content" service centre, in which the client transmits an image file and different information about an analyzed material and the system, based on image processing algorithms and developed decision making algorithms, is capable to return the microstructure, that is best related to the presented one (figure 2). In this case the same algorithms are used as in the probe's case.

A generic technological flux for probe preparation and analysis is shown in figure 3.

In what concerns programming media that are used, software applications can be firstly developed in MATLAB package (Image Processing Toolbox, Neural Network Toolbox, Fuzzy logic Toolbox, Control System Toolbox, MATLAB Compiler), and then in the operational stage every application is to be implemented in Visual C++ language.

3. ADVANTAGES OF THE PROPOSED SYSTEM

In contrast to a classical laboratory the flexible system for metallographic analysis is distinguished by a higher capacity of preparation and analysis (productivity augmentation) and by a higher and constant quality of provided information. Metallographic probes preparation and analysis capacity can be augmented up to 10 times compared to a classic metallographic laboratory (compared to the case in which all the operations are made by a human operator). Accuracy of the resulted information and constant quality are also advantages which produce long term economical benefits.

Another major advantage of the analysis flexible system, that produces immediate economic effects, is the promptitude of deliverance to client of the information that is needed to guarantee a higher.

Generally speaking, lack of information about materials blocks the beginning of some manufacturing processes, but in the case of obtaining rapid information these blockages are diminished or eliminated. We may also remark that an analysis flexible system with a higher productivity offers a support for industrial companies that cannot invest in a metallographic laboratory of their own.

Chemical substances automated management (as it is foreseen in the flexible system) imposes strict and clear rules in order to avoid environment contamination risks. A clear evidence is kept concerning quantity, origin and using of chemical substances. In a classical metallographic and chemistry laboratory management, using of chemical substances is provided totally by a human operator which is submitted to human errors.

4. CONCLUSIONS

Estimated scientific and technical results of the proposed project, are as follows:

- development of a metallographic analysis flexible system and a microstructure database, accessible by Internet, which offer services to clients. Based on the developed techniques similar flexible systems can be realized (including other investigation domains too – ecology, biomedical);
- command program of the system, based on intelligent agents, represents a new information technology which may be applied in different domains;
- by developing of automated roughness analysis and microstructures analysis there are created a set of image processing methods useful on a large scale of these domains;
- Internet accessible microstructure database, with a "search after content" browser, constitutes a development core of an information system in the field of material science.

In the near future, the authors plan to develop the described flexible system as part of a national project, in the benefit of universities and education institutions and also of manufacturing companies.

5. REFERENCES

Alicona Imaging (TEX 3D Professional), (2008), *Available from:* http://www.alicona.com/cms/front_content.php, *Accessed*: 2008-06-27

ASM Materials Information (Micrograph Center), (2008) *Availablefrom*:http://products.asminternational.org/mgo/index.jsp, *Accessed:2008-06-27*

METADEX (2007), CSA, *Available from:* http: www.csa.com *Accessed*: 2007-03-12

MetalloROM (2007), HDH Thermal, Dr. Sommer Werkstofftechnik, *Available from:* http: www.werkstofftechnik.com, *Accessed: 2007-03-12*

MetLab Corporation - ImagePro Plus(2008); *Available from:* http://www.mediacy.com, *Accessed*: 2008-06-27

Vesselenyi, T.; Mudura, P. (2001) Metallography image, FFT characteristics analysis, (in Romanian) Simpozion materiale avansate, tratamente termice şi calitatea managementului, Zilele Academice Timişene, ISBN 973–8247–32–2.

Vesselenyi, T., (2005) Metallography image analysis automation , (in Romanian) PHD Thesis, University "Politehnica" Timisoara.

VISIONx Inc. - Vision Gauge; Leco - IA32) (2008).

Annals of DAAAM for 2008 & Proceedings of the 19th International DAAAM Symposium, ISSN 1726-9679
Editor B. Katalinic, Published by DAAAM International, Vienna, Austria 2008
Make Harmony Between Technology and Nature, and Your Mind will Fly Free as a Bird

REDUCTION OF OSCILLATING DEMAND MAGNIFICATION EFFECT IN SUPPLY CHAIN

VEZA, I[vica]; GJELDUM, N[ikola]; BILIC, B[ozenko] & BAJIC, D[razen]

Abstract: *The changing business environment in which supply chain is acting, needs a more effective production process planning and control method that are able to deal with uncertain product demand and forecast from downstream supply chain participant. Distorted information along the supply chain can lead to tremendous inefficiencies like excessive inventory, misguided capacity plans, missed production schedules, poor costumer services and ineffective transportation planning. Because costumer demand is rarely perfect stable, business must forecast its own demand in order to properly position inventory. The aim of this paper is to detect the main prerequisite for demand magnification effect in supply chain and to introduce the simple approach by linear mathematical model for production planning and inventory control in all supply chain stages.*

Key words: *Production, Bullwhip, inventory, planning.*

1. INTRODUCTION

In a supply chain, for a typical final product consumer, even when consumer sales do not seem to vary much, there is pronounced variability in the retailers' orders to the wholesalers. Orders to the manufacturer, and to the manufacturers' supplier, oscillate even more. According to the lean production principles, excessive production and high inventory level are the biggest waste in production process (Rother, 2003). To solve the problem of distorted information, companies need to first understand what creates the oscillating demand magnification effect, so called Bullwhip effect so they can counteract it (Tempelmeier, 2006). Innovative companies in different industries have found that they can control the Bullwhip effect and improve their supply chain performance by coordinating information and planning along the supply chain.

2. CAUSES OF THE BULLWHIP EFFECT

Because costumer demand is rarely perfect stable, business must forecast its own demand in order to properly position inventory. Demand forecasting is frequently different from the actual production plan (Nishioka, 2003). Because forecast errors occurs, companies need to have an inventory buffer called safety stock. Some of the main causes for Bullwhip effect are (Lee, 1997):

- Demand forecast mistakes
- Order batching
- Price and season fluctuation
- Costumer order reductions or cancellations
- Adjustment of inventory control parameters with each demand observation
- Misperceptions of feedback and time delays
- Panic ordering reactions after unmet demands

The described effect can lead to either inefficient production or excessive inventory as the producer needs to fulfil the demand of its predecessor in the supply chain. This also leads to a low utilization of the distribution channels. It is important to use techniques and tools that can control the Bullwhip effect, that is, to control the increase in variability in the supply chain (Simchi-Levi 2000).

3. CHANGING DEMANDS IN A SUPPLY CHAIN

In this paper, a three-stage supply chain is presented, where manufacturer, responsible for consumer order demands fulfilling, is supplied with raw materials and semi products by two tiers of suppliers. The supply chain is modelled in ProModel software, to acquire data and graphs about intermediate orders and inventory levels of supply chain participants by its statistical module.

The most common changing demand cases as input data for simulation are described together with direct consequences on supply chain material flow. All participants in the supply chain change the order demands for resumption of inventory level in one order period. The demands quantity changes as follows:

1) The order demand is constant at initial value of 100 items, up to period 4. Intermediate order quantities are constant at all three stages.
2) In period 4 the order demand is increased for 10%, and then returned to initial value in. The demands in intermediate participants are respectively magnified in period 4, and the last participant, Supplier 2 increases production for 80 %. The next period, results with cancellation of Supplier 2 production. This variation is caused by Bullwhip effect.
3) For next two periods, 6 and 7, of same effect like in point 2), the magnification is even larger, and results with 120 % higher Supplier 2 production.
4) Periods 7 to 13 are set at constant initial value to present speed of stabilization for intermediate demand quantities.
5) Periods 14 to 19 show the reaction on 20 % demand reduction. The magnification is strong enough to cause cancellation of Supply 2 production in period 15.
6) Periods 20 to 31 are set to change order from 80 items to 250 items in 10% incremental increases in every period. Intermediate orders are unstable and magnified in start of increasing period, but stabilized as order quantity increasing continues.

The conclusion can be made that the Bullwhip effect occurs only in case of demand quantities change after periods of constant orders quantity, or after periods of constantly increasing or decreasing orders quantities. The bigger deviation from previous order quantity trend results with stronger Bullwhip effect.

The strong Bullwhip effect can be recognized in periods 4 to 8, and than in period 14. In periods 20 and 32, the effect is weak, but also influential on warehouse level quantity.

In the simulation model, the inventory level of 500 units is set in every supply chain participant warehouse. The data about sequence of released market orders, the intermediate order quantities along the supply chain and the contents of every participant warehouse are shown separately for every participant of a supply chain on Fig 1. and in Table 1.

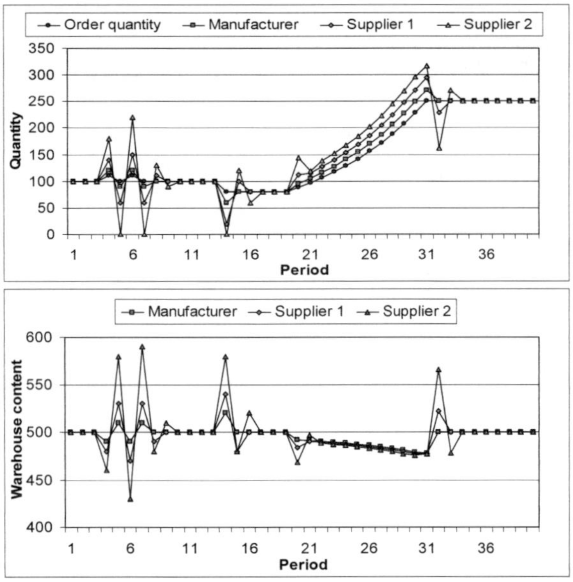

Fig. 1. Order quantities and warehouse levels of supply chain participants for resumption of inventory level in one period.

Per.	Mark. Order	Manu. order	Supp.1 order	Supp.2 order	Per.	Mark. Order	Manu. order	Supp.1 order	Supp. 2 order
1	100	100	100	100	21	97	94	87	74
2	100	100	100	100	22	106	116	139	190
3	100	100	100	100	23	117	128	139	140
4	110	120	140	180	24	129	141	153	167
5	100	90	60	0	25	142	155	169	184
6	110	120	150	220	26	156	170	186	202
7	100	90	60	0	27	171	187	204	223
8	100	100	110	130	28	189	206	224	245
9	100	100	100	90	29	207	226	247	269
10	100	100	100	100	30	228	249	272	296
11	100	100	100	100	31	250	272	295	317
12	100	100	100	100	32	250	250	228	162
13	100	100	100	100	33	250	250	250	272
14	80	60	20	0	34	250	250	250	250
15	80	80	100	120	35	250	250	250	250
16	80	80	80	60	36	250	250	250	250
17	80	80	80	80	37	250	250	250	250
18	80	80	80	80	38	250	250	250	250
19	80	80	80	80	39	250	250	250	250
20	88	96	112	144	40	250	250	250	250

Tab. 1. Order quantities of supply chain participants

In order to reduce oscillating demand magnification effect in those critical periods, the mathematical model is presented. The production demand can be expressed by a linear equation:

$$Q = x_0 + f_1x_1 + f_2x_2 \quad (1)$$

The value x_0 is production demand, and gives a production plan according to the produced quantity in the previous period of simulation. The first parameter x_1 is the difference between quantity in current order from upstream participant and produced quantity in last period. The second parameter x_2 is difference between current inventory level and inventory level which has to be maintained.

In order to achieve the smallest range of order variation, and smallest range of inventory level during simulation, factors of signification f_1 and f_2 are optimized. For this particular order sequence factors of signification are $f_1=0.75$ and $f_2=0.62$, so the optimal mathematical model is:

$$Q = x_0 + 0.75x_1 + 0.62x_2 \quad (2)$$

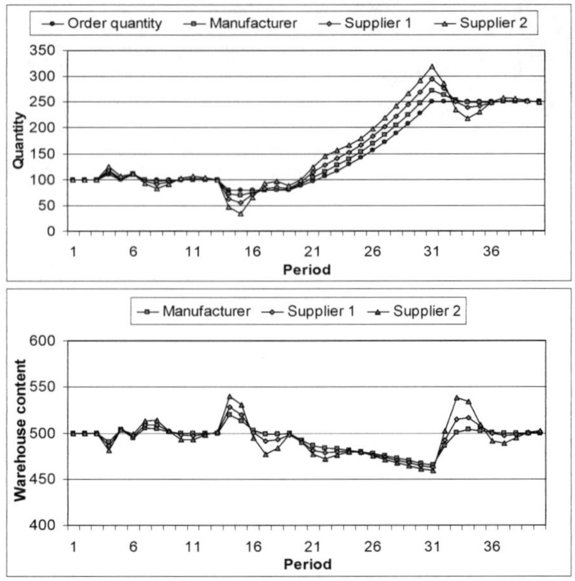

Fig. 2. Order quantities and warehouse levels of supply chain participants for resumption of inventory level according the mathematical model.

The graphical results of improved behaviour of supply chain are shown on Fig 2.

4. CONCLUSION

The Bullwhip effect occurs in case of demand quantity change from achieved routine in previous period. The routine can be constant order quantity and increasing or decreasing order quantity during previous period. The second main prerequisite for Bullwhip effect is rapid response on order quantity demand change with intention for resumption of inventory level in one period assuming the new order quantity will maintain in next periods. This results with progressive increase or decrease order quantity in upstream supply chain stages. The aim of this paper was to define approach with mathematical model used for defining intermediate order quantities in supply chain which reduce Bullwhip effect by changing the factors of signification in order to achieve satisfying speed of response on changing order quantities. The research described in this paper was necessary preliminary work before modelling of virtual production network without significant influence of Bullwhip effect in order to reduce excessive inventory, improve costumer services and achieve effective transportation planning.

5. REFERENCES

Lee, H. L.; Padmanabhan, V. & Whang, S. (1997). The bullwhip Effect in Supply Chains, *Sloan Menagement, Review 38,* Spring 1997, 93-102

Nishioka, Y. (2003). Collaborative Agents for Production Planning and Scheduling, Available from: http://www.pslx.org/en/doc/TR-001.pdf, *Accessed:* 2008-06-18

Rother, M. & Shook, J.(2003), *Learning to See*, The lean enterprise institute, ISBN: 0-9667843-0-8, Brookline

Simchi-Levi, D.; Kaminski, P. & Simchi-Levi, E. (2000). *Designing and managing the supply chain*, McGraw-Hill Higher education, ISBN:0-0256-26168-7, New York

Tempelmeier, H. (2006) *Inventory menagement in supply networks – problems, models, solutions*, ISBN 3-8334-5373-7, New York

Annals of DAAAM for 2008 & Proceedings of the 19th International DAAAM Symposium, ISSN 1726-9679
Editor B. Katalinic, Published by DAAAM International, Vienna, Austria 2008
Make Harmony Between Technology and Nature, and Your Mind will Fly Free as a Bird

A BINARY ENERGY SOURCE OF A TIMBER-DRIER

VIGLASKY, J[ozef]; LANGOVA, N[ada]; DANIHELOVA, A[nna]; SUCHOMEL, J[ozef] & ANDREJCAK, I[mrich]

Abstract: *The research and development described in the paper was focused on the design of an efficient heating system for an application in a commercial timber-drier. With regard to the fact that the period of solar radiation is determined by natural laws, it is necessary to provide apart from the solar energy, the heat supply from another energy source in a continual drying kiln. The problem of an additional heating system used in the solar kiln to improve its efficiency was successfully solved in an original way of regulating the heating system with a binary source of heat at the TU in Zvolen .*

Key words: *solar timber-drier, solar air collector, drying quality, energy savings*

1. INTRODUCTION

The Slovak Republic similarly as other countries with the well-developed industry including wood processing industry can easily spend five or more per cent of its national energy consumption on drying processes (Strnád, 2007). Therefore, in order to be able to indicate and determine ways on how to improve operation efficiency, decrease overall costs of drying, achieve satisfactory quality of dried timber, and at the same time implement the above technology in an environmental-friendly way, it is necessary to specify both in a complex way and in detail the currently highly topical process of timber-drying.

2. OBJECTIVES

The research was focused basically on the design of the kilns, and the aspects of regulation and control have been studied with regard to the fact that the period of solar radiation is determined by natural laws, i.e. its range varies greatly during the day and even season, therefore it is necessary to provide, apart from the solar energy, the heat supply from another energy source in a continual drying kiln to optimise the co-operation of two heat energy sources.

3. THE SOLAR TIMBER DRYING KILN

The prototype of the solar kiln placed in the furniture factory is schematically shown in Figure 1. The test runs described in this contribution were carried out in the town of Skalica, 48° 51` Northern latitude, 17° 14` Eastern meridian, 178 m a.s.l. and with approximately 2,100 hours of sunshine per year.

Construction details of the kiln / Structure of the kiln. We used a commercial drier slightly accommodated to be joined to a solar air collector.

The single-row low-capacity kiln has a 7 m^3 sawn timber volume. The timber stack measures are: the width of 1,500 mm, the height of 1,750 mm, and the length of 6,000 mm. The kiln is designed to dry coniferous and broad-leaved timber at 100 °C max. drying temperature.

Solar air collector. As specific conditions occur during the drying process using solar energy by means of a solar air collector, the first part of our investigation was aimed at a theoretical analysis followed by an experimental analysis of the solar air collector work. An experimental system was designed at the TUZVO to test the characteristics of solar air collectors.

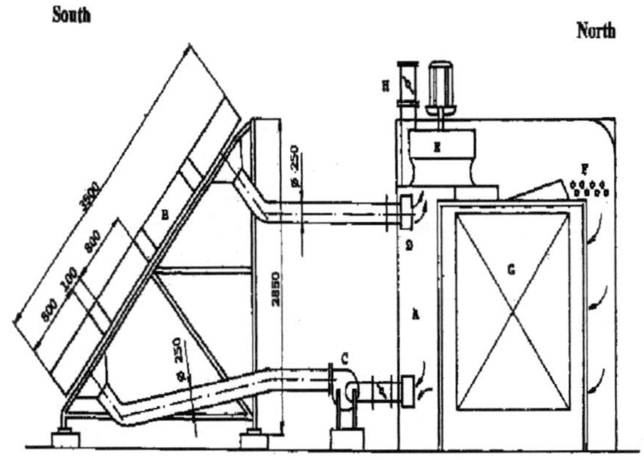

Fig. 1. The Solar Timber Drying Kiln

The Solar Kiln Scheme: (A) Drying chamber; (B) Solar collector; (C) External blower to induce air flow through the collector; (D) Hot air discharge to internal fans; (E) Internal fans; (F) Heaters; (G) Wood stack; (H) Aeration stacks.

The heat for the kiln is supplied by the 20 m^2 external air heating solar collector facing south and inclined at 45° to the horizontal for optimum year round performance. Air was circulated over the collector and through the load at 3 $m.s^{-1}$. Vents and a spray system were used to control the relative humidity in the kiln. The whole control system can be manually checked by switches, but a sensitive and effective equilibrium moisture content (EMC) control system based on control logic is required for performance in the solar kiln to protect wood quality as well as to maintain optimum drying rate and also to prevent energy loss through useless venting. All measured variables inside and outside the kiln were recorded at 15 min. intervals. The ratio of collector area to loading capacity is 2.86 $m^2.m^{-3}$ which agree with the recommended solar collector ratio in respect to the drying rate and lies in the range between 2 and 7 $m^2.m^{-3}$ run (Steinmann, 1992).

4. EXPERIMENTAL PART

Solar drying. Test drying runs were carried out using wood species of Pine Monterey (Pinus radiata D. Den.), thickness dimension of 24 mm to assess the kiln by the following criteria:

- Drying time
- Final wood moisture content
- Drying quality
- Supplemental energy
- Temperature and relative humidity conditions in- and outside the kiln
- Efficiency of the drying process.

Solar radiation. Solar radiation recorded during runs was done by the radiation pyrometer HEMI (Sensor 118 SN-8580, Swiss Instrument - HEMI, 100 V = 1 000 W.m^{-2}). Furthermore, the average initial moisture content of the sawn timber to be solar-dried was 60 % in the evaluated run. The species dried was *Pine radiata*, sawn in planks measuring 3,000 mm in length, 100 mm in width and 24 mm in thickness. In the evaluated run the 7.0 m^3 kiln load was piled using 18 mm square piling sticks and placing the boards edge to edge in each layer. The intensity of casehardening was assessed after allowing the prongs to dry indoors for 8 hours. After 80-hours or 5-day solar drying period, the average moisture content of sawn timber was 20 % in a representative evaluated run. During the run the drying chamber temperature was varying in range of 30-40 °C, but the drying condition expressed by the equilibrium moisture content (EMC) was constant, in the mean average of 10 %.

Air-drying. Simultaneously with the solar drying and using the same species and sizes, air-drying test runs were carried out. The drying quality was evaluated, and both drying time and final wood MC were recorded (Bakoš et al., 2000).

Results and discussion. The variables were selected according to a preliminary plan, which would allow conclusions about the two different drying methods, by comparing the results. The variables considered were:

- Species of wood
- Thickness
- Time of year
- Initial and final moisture content.

Comparing the drying time of runs (same thickness - e.g. 24 mm, same species - e.g. Pine, solar kiln and air drying), the conclusion is that solar drying is 6 times faster (80 hours) than air-drying (480-510 hours). Since solar drying is only faster than air-drying below 70 to 80 % wood MC, the initial moisture content is of importance. The influence of the final moisture content is particularly marked below 20 %, as the solar method becomes more profitable the lower the desired final wood MC is. With regard to cracks, splits, honeycombing and final wood MC, the quality of solar dried timber is better than for air-dried stock. The quality is comparable or even superior to that obtained by the conventional kiln drying method. As expected, the efficiency of solar drying improves with an increase of the average external temperature and is more effective during the first stage of the drying process than in the second stage when the wood MC is already below 40 %. Regarding economics it must be pointed out that the supplemental energy consumed (kWh) per kg of evaporated water is only approx. one-tenth to the amount used in a conventional kiln. As the efficiency of the tested solar timber-drying kiln depends strongly on the environmental conditions and mainly on the intensity of the solar radiation collected, the results are only transferable directly to locations of similar latitude.

Solar timber drying kiln control. In a solar kiln, the maximum temperatures are approximately 50 °C, which lasts only for a very short period of time each day, and is only achieved when the wood is partially dry and can endure these temperatures. Temperature control is not necessary in solar wood drying kilns and, therefore, only the relative humidity of the kiln air has to be controlled according to the given drying schedule. The relative humidity inside the solar kiln should be computer controlled in such a way that venting will take place if it rises above the target value as given by the drying schedule and if, at the same time, the absolute humidity of the outside air is less than the absolute humidity of the air inside the kiln. Unless both conditions are satisfied simultaneously, venting will not occur. This will be to prevent air with too high moisture content from entering the kiln. With this control, venting will be optimised preventing loss of energy (Horbaj et al., 2000). An efficient solar kiln control system giving the best possible quantitative and qualitative results for the final product should ensure that the drying operation is always modulated to match a current timber condition at any moment of a drying period.

Additional heating system. Regarding that the period of solar radiation is determined by natural laws, it is necessary to provide, apart from the solar energy, the heat supply from another energy source in a continual drying kiln. The problem of an additional heating system used in the solar kiln to improve its efficiency was successfully solved at the TU in Zvolen. Several control aspects were investigated. This problem has been solved in an original way of regulating the heating system with a binary source of heat (Viglasky & Novak, 1989). The above regulation makes it possible to optimise the co-operation of two heat energy sources in the kiln in such a way that it maximises the utilisation of solar energy and minimises the heat consumption from conventional energy source, e.g. steam or hot water.

5. CONCLUSIONS

1 The original way of regulating the heating system enables us to optimise the cooperation of two energy sources in kiln in such a way that it maximizes the utilization of solar energy and minimizes heat consumption from conventional energy source.
2 Solar drying is 6 times faster than air-drying.
3 The application of solar energy to wood processing holds an exciting promise of energy savings up to 34 %.

Acknowledgements: This research was partly supported by VEGA grant - the Contract No. 1/0511/08/5 *Use of a neuron net technology in diagnostic and reliability of technical systems used in an energy sector.* The authors are indebted to the mentioned institution for sponsoring this research work.

6. REFERENCES

Bakoš, M. ; Zolotová, I. & Sarnovský, J. (2000). *Remote Labs – Web Based System to Support Education,* 6th International Conference on Virtual University, Bratislava, ISBN 80-227-2336-3, pp. 233 - 237

Horbaj, P. ; Barborák, O. & Andrejčák I. (2006). Transfér energie vetra ako ďalší podnet pre strojársku prax, 8rd Conference TRANSFÉR 06, Trenčín, ISBN 80-8075-154-4), pp 165 – 169

Steinmann, D. E. (1992). The effect of collector area and solar tracking on the performance of a solar lumber drying kiln. In: Proceedings of the 3rd IUFRO International Wood Drying Conference. Vienna, Austria. pp. 283-291.

Strnád O. (2007) *ISMS implementation methodology*, International Experience Exchange TUV Nord, Essen, , ISBN 80-86229-79-3, pp. 236-238

Viglasky, J. & Novak, V. (1989). Technological equipment for optimal exploitation of solar energy by combined heating system. The Invention - Patent No. 255 339, Czechoslovakia (in Slovak).

Annals of DAAAM for 2008 & Proceedings of the 19th International DAAAM Symposium, ISSN 1726-9679
Editor B. Katalinic, Published by DAAAM International, Vienna, Austria 2008
Make Harmony Between Technology and Nature, and Your Mind will Fly Free as a Bird
*Young Researcher and Scientist Paper / * Supervisor, Mentor*

NUMERICAL STUDY OF AN AXIAL GAS-TURBINE STAGE

VILAG, V[aleriu]; POPESCU, J[eni]; PETCU, R[omulus]; BOGOI, A[lina] & SILIVESTRU, V[alentin]

Abstract: *Gas-turbines engines are very largely used in power generating systems due to advantages related to small dimensions to power ratio and despite great efforts in design, production and test phases. In the paper is presented a way to virtually test an axial gas-turbine resulting data to be used in improving the design and data to be compared with other analytical or experimental studies. The study begins with a short theoretical part to present the way the profiles of the studied axial turbine were obtained. It is shown the design of the geometry and the corresponding 3D mesh to be used as computational domain. There are also presented the conditions, parameters and models that are used as input data for the CFD code and then the numerical results. The numerical results are interpreted in terms of power and efficiency versus the rotational speed of the gas-turbine and they are used for taking decision in the overall design process.*
Key words: *gas-turbines, numerical experiments, CFD.*

1. INTRODUCTION

Gas turbines engines are very largely used in power generating systems such as aircraft, energy and industry. Their main advantages are related to small dimensions to power ratio when speaking about aviation and relatively small fuel consumption and high reliability when speaking about ground applications (Cârlănescu, 1997).

The advance in computer technology made possible the virtual tests in form of Computational Fluid Dynamics simulations for many thermodynamic and flow applications. These simulations have advantages related to lower costs and shorter time to market in comparison to classical analytical and experimental methods. This relatively new tool allows us to validate geometries from gas-turbines and to predict the performances for new or improved products. It offers a better perspective on parameter variation helping us to better understand the phenomena by conducting numerical experiments.

2. BLADE GEOMETRY DESIGN

2.1 Symetrical profile

The following parameters are used for the base profile:

$\bar{y}_{Gmax}$: maximum thickness (% chord)

$\bar{R}_{ba}$: radius of curvature of the leading edge (% maximum thickness)

$\bar{R}_{bf}$: radius of trailing edge (%maximum thickness)

ω: half of the angle between the tangents at the trailing edge (degrees)

For simplifying formulas some additional notations are required, resulting the following formula for thikness distribution along the symmetrical profile (Ainley 1951):

$$\bar{y}(\bar{x}) = \begin{cases} 0.05 \cdot \bar{y}_{Gmax} \cdot s\left(a \cdot 0.5\sqrt{\bar{x}} + b\bar{x} + c\bar{x}^2 + d\bar{x}^3\right), \\ \quad for\ 0 \le \bar{x} \le 0.01 \cdot \bar{x}_{Gmax} \cdot s \\ 0.05 \cdot \bar{y}_{Gmax} \cdot s \cdot \begin{bmatrix} e + f(1-\bar{x}) + g(1-\bar{x})^3 + \\ +h(1-\bar{x})^4 \end{bmatrix}, \\ \quad for\ 0.01 \cdot \bar{x}_{Gmax} \cdot s < \bar{x} \le 1 \end{cases} \quad (1)$$

The graphical expression of that is displayed in Fig.1

2.2 Actual profile

This base profile must be curved and aligned considering the desired working regime. The working regime is translated into necessary angles at the inlet and the outlet of the row blade and they are function of the radius and the type of row blade: stator or rotor. The profile camber is curved along a parabola tangent to these necessary angles (Novak 1967), as shown in Fig. 2:

α_1- inlet flow angle

α_2- outlet flow angle

β_1- inlet blade angle

β_2- outlet blade angle

In order to obtain the curved profile we need to find the arc length, to be able to position the resulted thickness of the base profile.

The parabola is now written parametrically:

$$\begin{cases} x(t) = a_0 t^2 \\ y(t) = 2a_0 t \end{cases} \quad (2)$$

and the arc length is given by:

$$\bar{s}(t) = a_0\left[t\sqrt{1+t^2} + \ln\left(t + \sqrt{1-t^2}\right)\right] \quad (3)$$

If we apply one more rotation we obtain the profile in one section, aligned with the gas-engine axis.

3 CFD CASE

3.1 Geometry and mesh description

The profiles with respect to the radius of the axial-turbine, for the stator and the rotor blade rows, were given by analytical methods in form of sets of points (Sellers 1975). Using this series of points and the radiuses of the flow canal into CAD software the following geometry was obtained. Due to the fact that we are studying the first turbine stage, the stator blades are cylindrical. In order to obtain a predominant axial velocity at

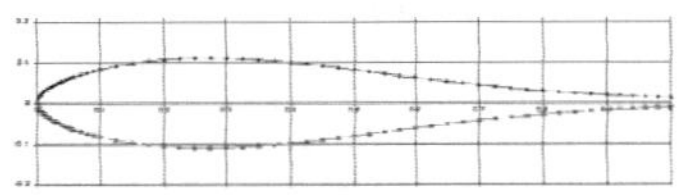

Fig.1 The base symmetrical profile

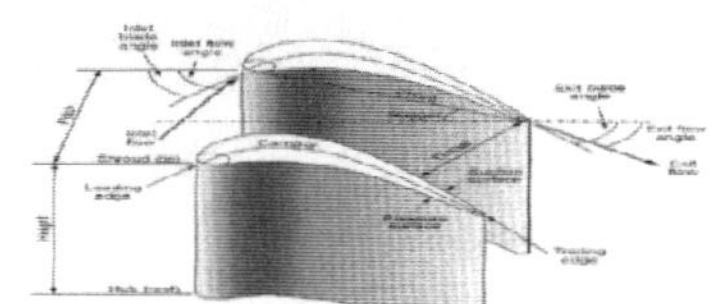

Fig.2 Blade geometry and angles

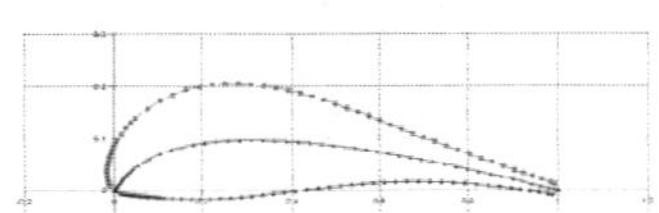

Fig.3 Actual profile

the outlet of the rotor row blade, the rotor blades are twisted. The inner and outer diameters of the flow canal are constant and are taken from the preliminary design calculus.

The necessary CFD mesh for such a complex geometry would be too large to be use on desktop computers or even on dedicated calculus machines, so we need mesh reduction. The reduction of the mesh is done by using the assumption that the stator blades are identical and the rotor blades are identical and also that they respectively behave identical. This results in that we can use only a smaller part of the geometry but with some constrains:

- the angle of the used sector must contain integer number of sections corresponding to one blade; this is valid both for stator and rotor
- the ratio between the angle used for the stator and the angle used for the rotor must be close to unity.

The first constrain is obvious, and the second contains relation to mass flow through the interface between the stator and the rotor.

In our case study we used 3 stator blades giving an angle equal to $\alpha_s = 3x\frac{360^0}{20} = 54^0$ and 7 rotor blades giving an angle equal to $r = 7x\frac{360^0}{47} = 53.61^0$. The ratio between them is $\frac{\alpha_s}{\alpha_r} = 1.0071$ that is close enough to unity, Fig. 4.

3.2 Working conditions

Thermodynamic and mechanic imposed conditions are the user input for the CFD code concluded into inlet and outlet conditions and rotational speed for the rotor.

It was imposed on the stator inlet the total pressure at the value of 9.12 bars and the total temperature equal to 1300 K, and on the outlet of the rotor the mass flow per machine equal to 8.1 kg/s. All imposed values are taken from the thermodynamic cycle (Pimsner 1988) proposed for the gas-turbine to be equipped with this axial turbine, MTI- 1500. Each sector is limited by two periodic boundaries, by two walls, the hub and the shroud, and by the inlet and the outlet surfaces.

The working fluid is assumed to be air ideal gas that is close to reality at these relatively reduced pressures.

The rotational speed of the rotor has been varied between 14000 and 28000 rpm with a step equal to 2000 rpm. At every rotational speed, the regime was considered stationary.

3.3 Results

The major advantages of CFD is that it can give parameters variation in space, and time if appropriate, in form of contours plot. This means that one can easily identify problems related to flow continuity, backflow or other types of losses. We present in the following figures contour plots of static pressure and static temperature, Fig. 5, at half span of the axial-turbine. The pictures are taken at the nominal working regime, at 22000 rpm.

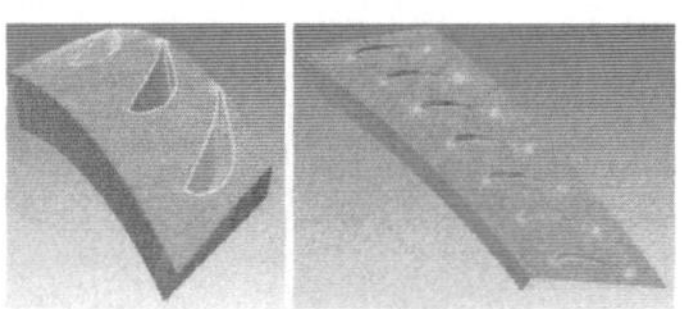

Fig.4 Mesh sectors for 3 stators and 7 rotors

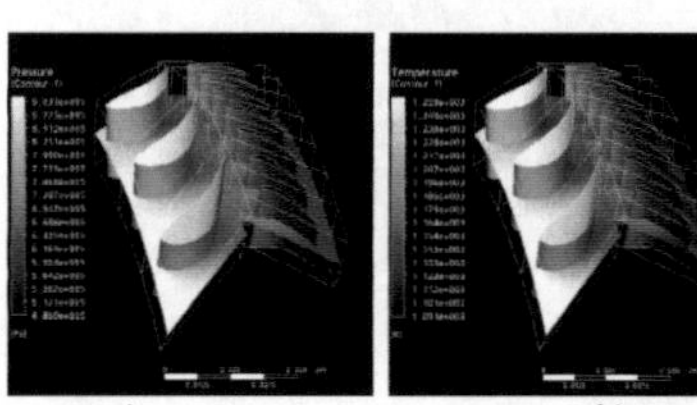

a) b)

Fig. 5 Contour plots of: a) static pressure; b) static temperature

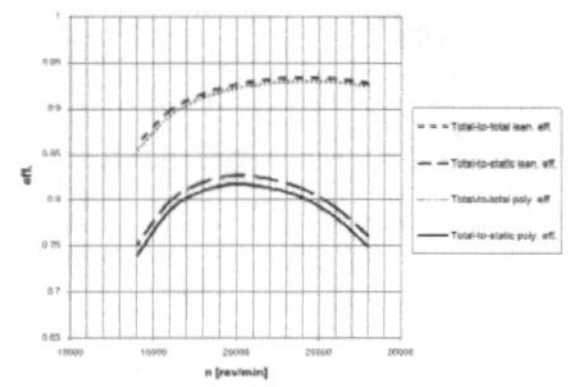

Fig. 6 Efficiency variation

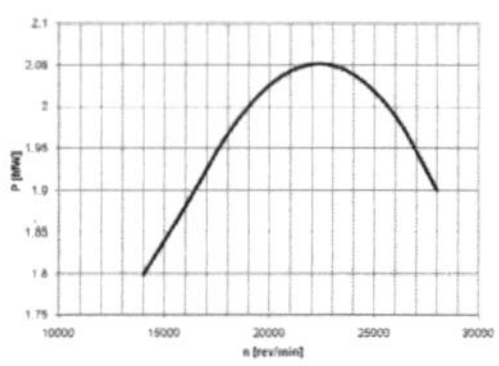

Fig. 7 Power variation

One other important result is the efficiency variation with respect to the rotational speed of the rotor. We present the variation of four types of efficiency in Fig. 6.

The most important parameter in this stage of development is the useful power of this axial turbine stage. We present in Fig. 7 the variation of the power with the rotational speed of the rotor row blade.

We can observe that according to CFD results the maximum power is obtained at 22000 rpm which is the nominal proposed working point. The value of the power is around 2050kW which is higher than expected, $\frac{1}{3}4500kW = 1500kW$. This is mainly due to the higher temperature imposed at the inlet of the stator row blade.

4. CONCLUSIONS

The overall scope of the research is to have a good algorithm using analytical and numerical methods for designing axial turbines. The main conclusion is that we have obtained a good correlation between the analytical way to draw turbine profiles and CFD numerical simulations. The maximum efficiency and the maximum power are obtained at the same rotational speed which was imposed for the analytical method.

Another conclusion is that by combining the two methods we can improve the efficiency, into an iterative cycle, by modifying parameters of the analytical method and verifying the corresponding changes using CFD simulations. The efforts of doing that are considerable lower comparing to experiments for which the production stage takes time and costs a lot more money. The experiments are not excluded, but their time should arrive into a further development stage, after the best geometry had been obtained through the proposed combined method.

Future work consists in CFD simulations of the entire axial turbine for the MTI 1500 industrial turbo-engine.

5. REFERENCES

Cârlănescu C. (1997), *Turbomotoare de Aviaţie. Aplicaţii Industriale (Aviation gas turbines. Industrial Applications)*, Editura Didactică şi Pedagogică, Bucureşti

Ainley D.G. (1951), *A Method of Performance Estimation for Axial-Flow Turbines*, Reports and Memoranda No.2974, A.R.C. Technical Report, Decembre

Novak R.A. (1967), *Streamline Curvature Computing Procedures for Fluid-Flow Problems*, Transaction of the ASME Journal of Engineering for Power, A Series

Sellers J.F. (1975), *DYNGEN – A Program for Calculating Steady-State and Transient Performance of Turbojet and Turbofan Engines*, Lewis Research Center, NASA TN D-7901

Pimsner V. (1988), *Maşini cu Palete (Bladed machines)*, Editura Tehnică, Bucureşti

Annals of DAAAM for 2008 & Proceedings of the 19th International DAAAM Symposium, ISSN 1726-9679
Editor B. Katalinic, Published by DAAAM International, Vienna, Austria 2008
Make Harmony Between Technology and Nature, and Your Mind will Fly Free as a Bird

CONSTRUCTION FOR ANAEROBIC DIGESTION AND BIOGAS PRODUCTION

VINDIS, P[eter] & MURSEC, B[ogomir]

Abstract: *Construction for anaerobic digestion is a system that harnesses natural process to treat waste, produce biogas that can be converted to heat and electricity and anaerobic digestate, a soil improving material. The paper presents the construction for anaerobic digestion and biogas production from energy plants. The amount of biogas production (methane) is observed. The basic structure is made of stainless steel, on which other components are fixed. The experiment is made according to DIN 38 414, part 8. The highest biogas yield gives maize 327 Nl kg VS^{-1}, then sorghum 312 Nl kg VS^{-1} and amaranth 197 Nl kg VS^{-1}. The results of tests show as most suitable energy plant maize.*

Key words: *Construction, biogas, methane, anaerobic digestion.*

1. INTRODUCTION

Microbial conversion of energy crops and organic wastes to biogas has become one of the most attractive technologies for energy production, resource recovery, and waste treatment. It creates a wide breadth of positive environmental impacts because it reduces emissions of greenhouse gases, improves the management of manure and organic wastes, and replaces mineral fertilizer. Biogas is used today mainly for electricity and heat production, but it can also be applied as a vehicle fuel or for the production of hydrogen which is necessary for fuel cells. Biogas production in the agricultural sector is a very fast growing market in many European countries (Weiland, 2006).

Biogas is produced in different environments, e.g., in landfills, sewage sludge and biowaste digesters during anaerobic degradation of organic material (Wellinger & Linberg 2000). Anaerobic digestion is becoming more and more attractive for the treatment of high strength organic wastes such as swine manure, since it produces renewable energy, methane, and valuable digested residues, liquid fertilizer and soil conditioner (Bonmati et al., 2001).

Sustainable biogas production from energy crops must not be based on maximum yields from single crops, but on maximum methane yield from sustainable and environmentally friendly crop rotations (Amon et al., 2007).

2. CONSTRUCTION OF MINI DIGESTER FOR BIOGAS PRODUCTION

The mini digester consists of twelve units and serves to produce the biogas from various energy crops and other organic waste material. During the test the biogas production must be read daily. Figure 1 shows the entire structure (mini digester) for anaerobic digestion.

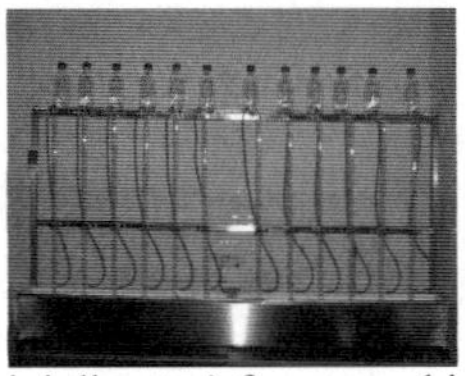

Fig. 1. Structure (mini digester) for anaerobic digestion.

Three tests with three repetitions simultaneously are possible, whereas three units serve for the control (inoculum). The volume produced is let out in case of each reading, each day at the beginning of test, later on every two or three days, when the gas formation diminishes

The mini digester comprises twelve gas cells. Each cell consists of a reaction vessel (500 ml fermenter) and a well closed gas pipe. The gas pipe - eudiometer is of 350 ml size and contains the confining liquid. It is connected to the levelling vessel with solution. The biogas produced in fermenters supplants the confining liquid in the gas pipe into the outside levelling vessel of 750 ml volume. The gas produced is read on the gas pipe. The fermenters are connected with the glass gas pipe and submerged into water with constant temperature 35+/-1 degrees C.

The basic structure of the mini digester is 2500 mm long, 1000 mm high and 350 mm wide and is made of stainless steel (inox). At the top a shelf is provided on which there are the levelling vessels for surplus confining liquid. At the bottom, a trough 2500 x 200 x 200 mm lined with insulating material is provided to prevent excessive heat losses and to enable the fermenters to be in the dark. Heating pump ensures constant temperature and water circulation. The eudiometers are fixed to a metal beam above the structure. A thermometer and a barometer measuring, through a sensor, the water temperature in the trough and separately the adjacent air temperature are fixed on the left side of the steel structure. The barometer serves for measuring air pressure.

The biogas produced contains 50 - 75% of methane, 10 - 40% of carbon dioxide and other matters (O_2, H_2, H_2S, N_2, NH_4,...). Oxygen is an indicator of anaerobic fermentation and the level of oxygen must be under 1%.

2.1 Measuring biogas production and composition

Measurements were conducted according to DIN 38 414, part 8. Biogas composition (CH_4, CO_2, O_2) was analysed 10 times in course of the 5- week digestion. The exact composition of biogas is measured by gas detector Geotechnical Instruments GA 45 (Figure 2).

Biogas production is given in norm litre per kg of volatile solids (Nl (kg VS)$^{-1}$), i.e. the volume of biogas production is based on norm conditions: 273 K, and 1013 mbar (DIN 38 414, 1985). The biogas produced is monitored every day and the substrates are mixed every day for 10 min. A water bath tempers the digesters.

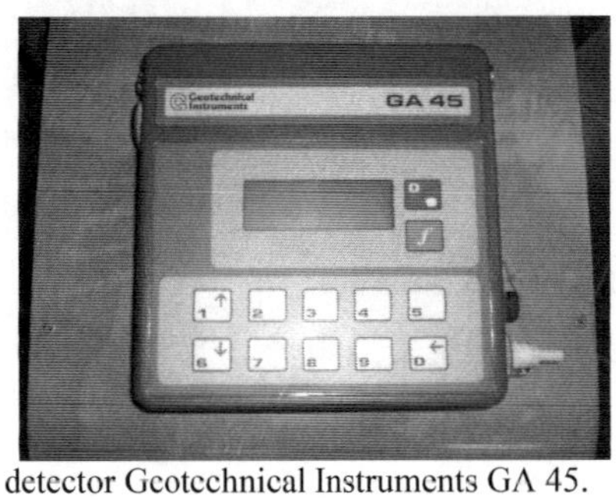

Fig. 2. Gas detector Geotechnical Instruments GA 45.

3. ANAEROBIC DIGESTION OF ENERGY PLANTS

Maize, sorghum and amaranth was chopped after harvest. Whole plants were anaerobically digested and biogas yields and biogas quality were measured and compared. Actively digested pig manure slurry was collected from biogas plant, filtered and used as inoculum to prepare substrate/inoculum ratios.

Biogas production from inoculum alone was measured as well and subtracted from the biogas production that was measured in the digesters that contained inoculum and biomass.

3.1 Biogas production

Biogas production was measured during 5 – week digestion. The lowest biogas and methane yield was in case of amaranth (197 Nl/kg VS^{-1}). The highest biogas and methane yield was in case of maize (327 Nl/kg VS^{-1}). Sorghum has a biogas yield 312 Nl/kg VS^{-1}.

Average biogas and methane yield in norm litre per kg of volatile solids (Nl/kg VS^{-1}) from three replicates per treatment during 5 – week digestion is shown in figure 3.

The predominant crop for biogas production is maize. Maize is considered to have the highest yield potential of field crops grown in Central Europe. In our experiment maize have the highest biogas yield. Biogas production of maize in norm litre per kg of volatile solids during 5- week digestion is shown in Figure 4.

Most of the biogas is produced in the first week of the experiment, after twenty days the anaerobic digestion is mostly finished. After 35 days the amount of biogas is very low.

Methane production from organic substrates mainly depends on their content of substances that can be degraded to CH_4 and CO_2. Composition and biodegradability are key factors for the methane yield from energy crops.

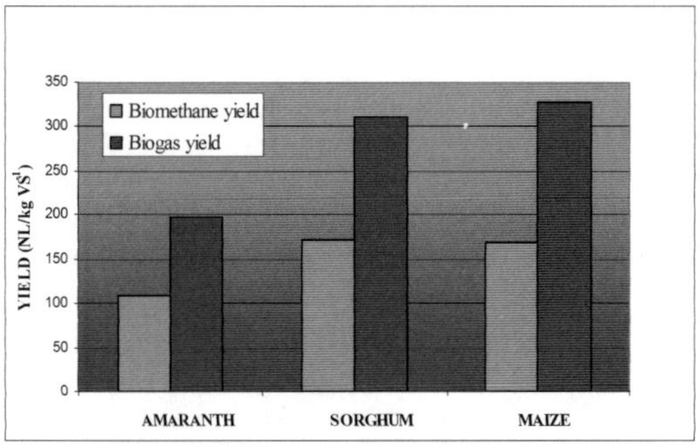

Fig. 3. Average biogas and methane yield.

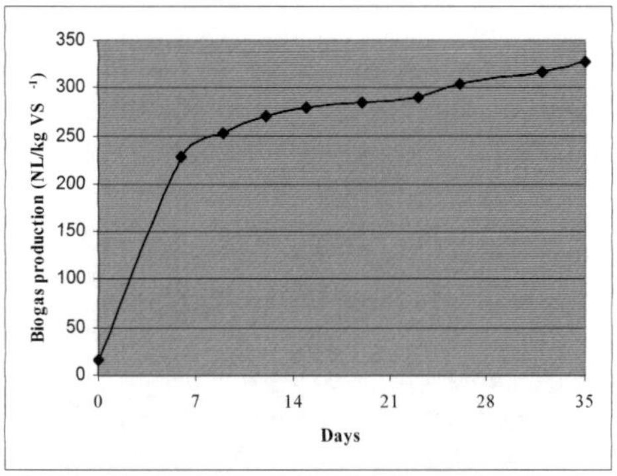

Fig. 4. Biogas production of maize.

3.2 Biogas composition

Biogas composition (CH_4, CO_2 and O_2) was analysed 10 times in course of the 5 - week digestion with gas detector Geotechnical Instruments GA 45.

The methane content ranged from 21,8 to 60,4% (mean 51,8%). The maximal level of CO_2 was on the second day of the experiment, 48,8%, and then decreases to 37% and stays stable. The average content of CO_2 was 40% during 5 – week digestion. That the digestion was anaerobic shows oxygen content in the biogas, which was under 1%.

The biggest differences in biogas composition occur in the first week of the digestion and then the gas content is more or less stable.

Figure 5 shows the biogas composition of biogas produced from maize.

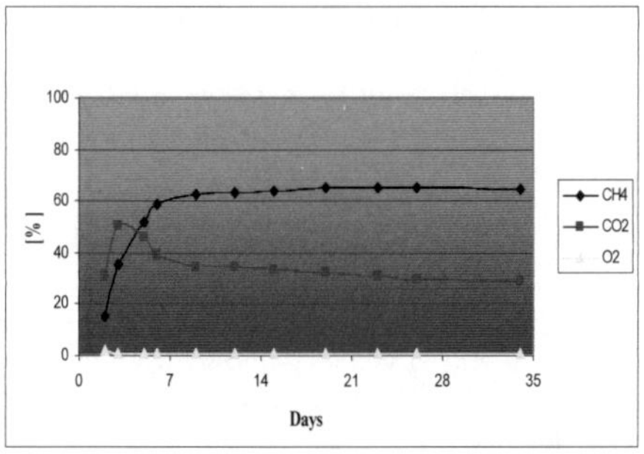

Fig. 5. Biogas composition of biogas produced from maize.

4. CONCLUSIONS

Maize, sorghum and amaranth silage are very suitable substrates for anaerobic digestion. Economic biogas production requires the methane yield from organic substances to be calculable. Methane yield depends on the nutrient composition of organic substrates. Energy plants should be conserved as silage prior to anaerobic digestion as this increases the methane yield.

In the experiment three energy plants were used: amaranth (Amaranthus sp.), sorghum (Sorghum halepense) and maize (Zea mays). Each experiment was per-formed in three repetitions; meanwhile three units serve for inoculum. The highest biogas yield gives maize 327 Nl kg VS^{-1}, then sorghum 312 Nl kg VS^{-1} and amaranth 197 Nl kg VS^{-1}. The average content of methane in biogas from amaranth is 55,5%, from sorghum is 55,4% and from maize is 51,8%. The content of oxygen was under 1%.

The results of tests show as most suitable energy plant maize, meanwhile other energy plants should be tested also (sunflower, clover grass, sudan grass, …). Maize should be conserved as silage prior to anaerobic digestion as this increases the methane yield.

5. REFERENCES

Amon, T.; Amon, B.; Kryvoruchko, V.; Zollitsch, W.; Mayer K. & Gruber. L. (2007). Biogas production from maize and dairy cattle manure - Influence of biomass composition on the methane yield. *Agriculture, Ecosystems and Environment,* Vol. 118, (173–182)

Bonmati, A.; Flotats, X.; Mateu, L. & Campos, E. (2001). Study of thermal hydrolysis as a pretreatment to mesophilic anaerobic digestion of pig slurry. *Water Science Technology*, Vol. 44, No. 4, (109–116)

DIN 38 414 (1985). Determination of digestion behavior ''sludge and sediments''. Beuth Verlag, Berlin (in German)

Weiland, P. (2006). Biomass digestion in agriculture: A successful pathway for the energy production and waste treatment in Germany. *Engineering in life sciences,* Vol. 6, No. 3, (302-309)

Wellinger, A. & Linberg, A. (2000). Biogas upgrading and utilization. *IEA Bioenergy Task*, Vol. 24. Paris, France: International Energy Association

Annals of DAAAM for 2008 & Proceedings of the 19th International DAAAM Symposium, ISSN 1726-9679
Editor B. Katalinic, Published by DAAAM International, Vienna, Austria 2008
Make Harmony Between Technology and Nature, and Your Mind will Fly Free as a Bird

BUSINESS STRATEGIES IN THE ROMANIAN BANKING SYSTEM

VINTILA, G[eorgeta]; BARBU, T[eodora]; NEDELESCU, M[ihai]; PREDA, O[ana] & STANESCU, C[ristina]

Abstract: *The evaluation of the present offer of banking products and services in the developed countries as well in the emerging ones shows the extent to which they fulfill or not the principles specific to social responsibility and ethics in economics. The Romanian banking system, currently one of the markets with the lowest penetration in EU has posted sustained growth for the past years. Client confidence in financial institutions is increasing, and the system's dynamics is attracting new foreign players, which have an easier market access since Romania's EU entry. Competition is expected to heat up, resulting in pressure on margins and diversified product portfolios, which ultimately benefit the end consumer.*
Key words: *banking system, banking strategies, distribution channels, banking products.*

1. INTRODUCTION

The main objective of the research is to identify the new distribution channels and to increase the efficiency of the existing ones, in order to improve the relation between bank and customer, as a starting point to increase profitability of banks. In our economy are many different ways for a bank to make money. Some of them are focused on the traditional banking strategies such as: attracting house-hold deposits in exchange for interest payments and transaction services and lending those funds to business customers at higher interest rates and other are focused on the non-traditional strategies such as mortgage banks. These approaches to banking business strategies are focusing on local markets or serving customers nationwide; catering to house hold customers or business clients; using an internet delivery system. These business strategies are relatively new development in the Romanian banking system. The studies which were made in this area have performed on systematic analysis of the profitability, risks or long term viability of these banking models.

2. BANKING PRODUCTS AND DISTRIBUTION OVERVIEW

Competition is heating up in the Romanian banking system. A simple method for comparing the performance of different banking strategies is to separate banking companies by size. Scale is clearly important (DeYoung et al., 2004) a large bank company give it access to low-unit-cost marketing and production techniques;

- a small bank company allows it to build person-to-person relationships with its customers.

The 40 banks registered in Romania compete for a highly fragmented market with modest financial insight which supports the on going consolidation trend. The top 10 banks accounted for over 80% of the banking assets at the end of 2007 (http://rbd.doingbusiness.ro/ 2008-07-21).

The concentration ratio for top 5 exceeded 62% already fairly high and expected to increase further, as the main banks will continue to strengthen their position. The stability of the economic environment, the downward inflation trend, a higher average salary and increased business opportunities in various industry sectors, together with the fall of the BNR interest rates, have all contributed in the last years to shift the banks focus from a cautionary approach, targeting operations, towards financing the real economy. As Romania is still an emerging financial services market, banks tailor their offer to a less sophisticated demand for retail and corporate products using a mass market approach, as opposed to mature markets where banks differentiate their offer and customize per specific customer segments. Most Romanian banks are clustered in the services area that means broad products range distributed via various channels. The major banks in this area have capitalized on various strengths. (Financial Stability Report 2007)

- BCRŠ and BRD-GSG are examples of the universal bank concept, with clear scale benefits and cross-selling potential;
- Raiffeisen Bank developed an aggressive corporate and retail acquisition strategy;
- UniCredit has emerged as a strong universal bank, combining the strong Tiriac retail operations and HVB's corporate profile.
- In the opposition Transilvania Bank and OTP Bank are mostly oriented towards the retail segment while ABN AMRO's traditional focus is on the corporate segment.
- Porsche Bank, Raiffeisen Bank and HVB Bank, are examples of niche players that focus on specialized products, distributed via different channels including own agents, branches, car dealerships, brokerage companies.

At retail banking level, products are developed for retirees, young people, and affluent clients. The drop in interest rates in the course of a few years has opened up the opportunity of refinancing loans, now offered by almost all banks and used especially for previously contracted long-term loans, such as mortgage. On the other hand, SME clients are now in focus of many banks, such as Transilvania Bank or BRD-GSG, those are actively developing products for this particular segment. The bank's corporate portfolio is also being extended – for example derivatives, currently offered by banks such as ABN AMRO, UniCredit Bank or BRD-GSG. The accumulation of wealth has also prompted the development of private banking services for high net worth individuals, offering special relationships and investment services. ABN AMRO has been one of the pioneers of private banking in Romania; several other international banks have subsequently entered this segment.

3. DISTRIBUTION CHANNELS OF THE BANKING PRODUCTS

The development and expansion of electronic payment channels and instruments have permitted banks to offer their customers a higher level of convenience, often at lower costs (about 34% of household pavements are made using electronic channels like debit cards, credit cards and automated bill pay) (http:// rbd.doingbusiness.ro/ 2008-07-21). For those who wish to make classical cash transactions, ATM networks has made access to cash more convenient, while generating fee income for banks and creating an new financial service sector for owners of ATMs. Banks have invested heavily into the expansion of their branch networks, which is still the major distribution channel, and plan to do so even more aggressively. Overall, banks increased their branch networks by 35% in 2006 compared to 2005, subsequently reaching more than 4,850 units by the end of 2007.(http://rbd.doingbusiness.ro/ 2008-07-21).

Several top 10 banks invested heavily in expanding their networks, such as BRD-GSG is now second in terms of number of units. The largest branch network belongs to CEC. Further growth prospects remain solid, although the gap to CEE branch density has shrunk: the current average in Romania is approximately 22 bank units per 100,000 inhabitants. However, severe discrepancies remain between branch density in Bucharest and the rest of the country, especially in the rural areas. These are signs that banks are prepared to invest in developing all distribution channels in an attempt to foster penetration and improve customer services and convenience. In this direction, banks support their investments with specific products to encourage their clients to use cards and to reduce the number of cash payments. The traditional distribution channels seem to be challenged by a few players which adopted different distribution strategies that are meant to maximize the efficiency of the customer interfaces:

- ING Bank started to address the retail segment, via an innovative "Self Bank" concept that is further supported by a franchising network of over 120 franchisees;
- ABN AMRO Bank pioneered a retail banking practice consisting of mobile banking agents, quickly adopted by a large number of players for selected products;
- Citibank started in 2004 the implementation of its retail arm - CitiFinancial, via a network of retail agencies;
- Volksbank set up its first franchising outlet in 2005, and is now offering basic bank products, as well as insurance and leasing services, in 75 franchised units.

Large banks, such as BCRŠ, BRD-GSG, Raiffeisen, and UniCredit have developed their own leasing, insurance, asset management or consumer finance companies, benefiting from group synergies and leveraging on their extensive branch networks. Cross-selling is an efficient way to leverage the traditional banking network. To mention just a few examples of financial entities that promotes and sells, besides banking products, various other financial products: (Annual Report 2006).

- Insurance products: BCRŠ, ING, UniCredit, Volksbank;
- Mutual fund products: BCRŠ, BRD-GSG;
- Leasing: BCRŠ, BRD-GSG, UniCredit, Raiffeisen.

Competition will prompt banks to focus even further on innovative sales techniques and look at long-term strategic planning trends. Overall, banks will have to decide whether they will develop their products in-house or will acquire from third parties. Mature market distribution models rely on the in-sourcing/outsourcing of products with direct benefits on the cost side. In addition, along with an increasingly sophisticated demand, banks will have to decide on how much of non-banking products the customer expects to be offered by banks. Things may work the other way too, with banking products that are sold via non-banking financial entities. For the time being, the proportion of loans intermediated by brokerage companies is insignificant. However, as competition on the maturing retail segment intensifies and banks begin to pay more attention to the costs entailed by an extensive branch network, financial intermediation will increase. In addition to credit brokers, some banks have already tapped into partnerships with retailers, real estate developers, real estate agents, car dealers etc. Banks such as OTP, Raiffeisen or Alpha Bank have partnerships with credit brokers; BCRŠ, BRD-GSG and Volksbank are examples of banks that use real estate agents to attract mortgage clients. One main challenge consists in the need for accelerated expansion of the bank's distribution network in order to tap into the full potential of different geographic areas, but with increasing efficiency. Therefore, channels such as e-banking, m-banking, are expected to become standard distribution channels in the future. All these alternative distribution channels will lower the expansion and operational costs of banks.

4. CONCLUSION

We can define a variety of banking business strategies that are based on differences in product mix, funding sources, geographic focus and other dimensions. In conclusion we can make some notice (DeYoung et al., (2004).

- In the studies of banking business strategies it was find substantial differences in profitability and risk across the various banking strategies. Low profitability does not necessarily doom a banking business strategy. High average strategies like corporate banking generate high amounts of risk, while low average return strategies like community banking generate less risk. However both banking strategies may be financially viable.
- During the studies it was find some evidence that banks without a competitive strategy perform poorly, as do banks which are focused on the traditional banking strategies without embracing efficient new banking products and distribution channels.

Beside the strong growth in mortgage loans, Romanian financial institutions should look at an entire spectrum of attractive growth options such as new products or new distribution channels. The long term driver for the Romanian banking system is the low financial penetration that should continue to fuel double digit growth over at least the medium term. The average Cost Income Ratio for the banking system, calculated by BNR at 67.7% is well above the best practice ratios for banks in mature banking environments – 45%-50% – indicating opportunity for cost improvement (Barbu & Vintilă 2007). However, for the moment the focus is still on growth. The 2008 is expected to bring further pressure on margins, driven by intensifying competition from both banks and non-bank financial institutions and possibly by BNR's policies. Thus profitability will be closely watched in order to survive as competition becomes ever more intense. On the other hand, a favorable cost base compared to other countries could be used by international banking groups to relocate regional back office processing to Romania. To sum up, the banks' market positioning is the key for success. It makes sense to position as a universal bank if the bank has a significant market share, while it may be more appropriate for smaller banks to pursue a niche strategy. The increasing competition and sophistication of the market, along with this pricing pressure should gradually result in cheaper, more convenient and more diversified banking services for Romanian consumers, as well as alternative delivery channels.

5. REFERENCES

Barbu T.; Vintilă G. (2007) *The Emergence of Ethic Banks and Social Responsibility in Financing Local Development*, Theoretical and Applied Economics nr.12, ISSN 1844-0029.

DeYoung, R.; Hunter, C.; Udell, F. (2004)*The past, present and probable future for community banks*, Journal of Financial Services Research, Vol. 25, No. 2/3, pp 85-133.

DeYoung, R.; Rice, T. (2004) *How do banks make money? A variety of business strategies*, Economic Perspectives, Federal Reserve Bank of Chicago, pp 52-67.

http://rbd.doingbusiness.ro/ (2008-07-21).

Annual Report 2006 http://www.bnro.ro/ (2008-07-21).

Financial Stability Report 2007 http://www.bnro.ro/ (2008-07-21).

Annals of DAAAM for 2008 & Proceedings of the 19th International DAAAM Symposium, ISSN 1726-9679
Editor B. Katalinic, Published by DAAAM International, Vienna, Austria 2008
Make Harmony Between Technology and Nature, and Your Mind will Fly Free as a Bird

NEW TRENDS AND ORIENTATIONS IN INDIAN CONSUMER BEHAVIOUR

VINTILA, G[eorgeta]; PREDA, O[ana]; NEDELESCU, M[ihai] & STANESCU, C[ristina]

Abstract: *From the market perspective, people of India comprise different segments of consumers, based on class, status, and income. An important and recent development in India's consumerism is the emergence of the rural market for several consumer goods. Three-fourths of India's population lives in rural areas, and contribute one-third of the national income. India is a lucrative market even though the per capita income is low and it remains a huge market, even for costly products.*
Key words: *Rural consumer behaviour, Marketing strategies, Online marketing, Spending behaviour.*

1. INTRODUCTION

Consumer behaviour will change world over and more so in developing countries like India where social behaviour and literacy levels too are changing. As people migrate from the villages to the cities in search of employment, so will their buying patterns albeit slowly. The main objective of the paper is to provide a real image of the Indian world, and is the beginning of a larger research on the Asian consumer behaviour. The interest on this part of the world's consumers is given by the specific cultures and habits. For those companies who want to be successful in this part of the world consumer behaviour must be of major concern. The interest on this topic is also given by the lack of data in the international literature. The huge Indian market is a target for a lot of companies. In order to be successful on this market one should understand the Indian people's social life, personal tastes and preferences, way of life, how they identify an effective product and what makes them get attracted towards a product.

2. CHARACTERISTICS OF THE INDIAN CONSUMER BEHAVIOUR

The Indian consumers are noted for the high degree of value orientation. Such orientation to value has labeled Indians as one of the most discerning consumers in the world. Even, luxury brands have to design a unique pricing strategy in order to get a foothold in the Indian market. Indian consumers have a high degree of family orientation. This orientation in fact, extends to the extended family and friends as well. Brands with identities that support family values tend to be popular and accepted easily in the Indian market. Indian consumers are also associated with values of nurturing, care and affection. These values are far more dominant that values of ambition and achievement. Product which communicate feelings and emotions gel with the Indian consumers.

Apart from psychology and economics, the role of history and tradition in shaping the Indian consumer behaviour is quite unique. (http:// www. knowledgestorm. com). Perhaps, only in India, one sees traditional products along side modern products. For example, hair oils and tooth powder existing with shampoos and toothpaste.

About three quarters of the Indian population are in the rural areas and with the growing middle class, specially in the Indian cities, the spill over effect of the growing urban middle class is also felt in the rural areas.

The Indian rural market has been growing at 3-4% per annum, adding more than 1 million new consumers every year and now accounts for close to 50% of the volume consumption of fast-moving consumer goods in India. The market size of the fast moving consumer goods sector is projected to more than double to US$ 23.25 billion by 2010 from the present US$ 11.16 billion.

3. INCREASING AWARENESS OF INDIAN CONSUMERS

In a fast paced marketing environment mangers in the present day are facing a constant challenge- adapting western marketing strategies to the Indian context. (http:// www. iimb. ernet.in).

The complexities of the Indian marketing context are vastly different from those in the west- ranging from daily paid consumers to those who can afford designer cars, from consumers who have biscuits for breakfast to those customize their diet for fitness and beauty reasons and from those who buy unbranded apparel to those who buy heritage labels, to name a few marketing situations. Over the years, as a result of the exposure to the west, satellite television, foreign magazines and newspapers, there is a significant increase of consumer awareness among the Indians.

Today more and more consumers are selective on the quality of the products/services. This awareness has made the Indian consumers seek more and more reliable sources for purchases such as organized retail chains. The consumer also seeks to purchase from a place where his/her feedback is more valued. Indian consumers are now more aware and discerning, and are knowledgeable about technology, products and the market and are beginning to demand benefits beyond just availability of a range of products that came from 'trusted' manufacturers.

The Indian consumers are price sensitive and prefer to buy value for money products. (http:// edms. matrade. gov.my). Even it is seen as an impenetrable market many companies such as Ford, Reebok, Nokia, had a great success. What was the key? Let's take the example of Ford. Before establishing their base in India, they engaged in a lot of researches. (http://www.123eng.com). Their researches were made on the Indian people's social life, personal tastes and preferences, way of life, how they identify an effective product and what makes them get attracted towards a product. The social and economic conditions were analyzed. The general economy of India was also researched on. They had modified their product to suit the Indian conditions. Their technology had to be adjusted and suited to such an extent that their car is adaptable to Indian conditions. Indians are especially taking into account the road conditions and other social factors they designed the product in such a way that it's best suited to the conditions and it's received by the target customers.

4. MARKETING STRATEGIES

Online Marketing: A study by the Confederation of Indian Industry and the International Trade Centre predicts that e-commerce activity in India will rise from US$ 0.10 million in 2000-2001 to US$ 5.8 billion in 2008-09, of which the business to business segment will account for US$ 5.41 billion. Currently, the products Indian consumers are buying through online are greeting cards, clothes, CDs, cassettes, books, magazines, medicine and educational material.

Celebrity Influence: In India, celebrities are being increasingly used in marketing communication by marketers to lend personality to their products. With the visual media becoming more popular the use of celebrities in the TV media has increased. Celebrities create headlines. Their activities and movements are being closely watched and imitated. What they endorse sell like hot cakes. In India, it is not difficult to look for the reasons why companies are increasingly using celebrities. Indians love their heroes. When a consumer likes the celebrity in the advertisement, he or she is more likely to accept what the celebrity says about the advertised product and therefore will develop more positive feelings toward the advertisement and the brand itself.

Quality Oriented Outlets: Indian consumers looking for quality choose expensive brands as they feel that price is an indicator of quality. However, in the absence of well known brands in selected product range, consumers are likely to take cues from well established retail outlets hoping that these outlets carry quality products.

Freebies: Freebies are consumer products given free of charge as gifts to purchases of selected products above a certain value. TVs, washing machines, refrigerators, and ready made clothes are some of the product categories in which freebies are given to Indian consumers. Freebies generally comprise tooth paste, soaps, detergent, cooking oil etc.

Eco-Friendly Products: In general, Indian consumers are likely to buy environmentally responsible products and packs. The future key for marketing could be to select more ethical and ecological responsible products and packaging, which is also convenient for consumers, thus, balancing environmental concerns with commercial considerations. Consumers in India are taking lead in prompting manufacturers to adopt technologies to produce eco-friendly products.

5. CHANGING TRENDS IN INDIAN CONSUMER BEHAVIOUR

Bulk Purchasing: Urbanization is taking place in India at a dramatic pace and is influencing the life style and buying behaviour of the consumers. The working urbanites are depending more on fast and ready-to-serve food, they take less pain in traditional method of cooking and cleaning. Bulk purchases from hyper stores seems to be the trend these days with purchasing becoming more of a once-a-week affair, rather than frequent visits to the neighborhood market/store/vendor. The popular growing shopping trend among urbanities is purchasing from super markets to hyper stores.

Trendy Lifestyles: The current urban middle and upper class Indian consumer buying behaviour to a large extent has western influence. There is an increase in positive attitude towards western trends. The Indian consumer has become much more open-minded and experimental in his/her perspective. There is now an exponential growth of western trend reaching the Indian consumer by way of the media and Indians working abroad. Foreign brands have gained wide consumer acceptance in India.

Buyers' Market In The Making: The sellers' market is slowly moving towards becoming the buyers' market. Since, India's economic liberalization policies were initiated in 1991, many new product offerings have entered the Indian market and product variety has also increased manifold. Import licensing restrictions are being eliminated and tariffs significantly reduced and this has led to large range of consumer goods made available in India. Indian consumers have always preferred foreign goods and with the liberalization, they now have a choice of foreign products vis-à-vis the local products.

6. CONCLUSION

Culture is the most fundamental determinant of a person's wants and behaviour. A child growing up in India is exposed to the following values: achievement and success, activity, efficiency and practicality, progress, material comfort, individualism, freedom, external comfort, humanitarism, youthfulness (Agrawal, 2006).A consumer's behaviour is also influenced by social factors, such as the consumer's reference group, family, and social roles and statuses. A person's behaviour is strongly influenced by many groups .A persons reference group are those groups that have a direct (face to face) or indirect influence on the person's attitudes or behaviour. Members of the buyer's family can exercise a strong influence on the buyer's behaviour. A traditional family in India is different from a western family in that the "empty nest" stage of the family life cycle is not very common. The decision making process for the purchasing of consumer durables is quite elaborate, as many members with their diverse view are involved in it. At the same time, the characteristics of Indian families are changing under the influence of external factors such as liberalization and media explosion. Indian families are presently in a state of flux, shifting from being strictly hierarchical to more egalitarian in character. We can distinguish between two families in the buyer's life. The family of orientation consists of one's parents. From parents a persons acquires an orientation towards religious, politics, and economics and a sense of personal ambitions, self –worth, and love. Even if the buyer no longer interacts very much with his or her parents, the parents influence on the unconscious behaviour of the buyer can be significant. In countries where parents continue to live with their children, their influence can be substantial like India. In case of expensive products and services, husband and wives engage in more joint decision making. The market needs to determine which member normally has the greater influence in the purchase of a particular products or services. either the husband or the wife, or they have equal influence. The following products and services fall under such: Husband – dominant: life insurance, automobiles, television Wife – dominant: washing machines, carpeting, non –living – room furniture, kitchenware Equal: Living – room furniture, vacation, Housing, outside entertainment.

Adapting to social conditions play the most important role in establishing your brand in the market. (http:// www.consumerpsychologist.com). This also means that customers are open to new and different products from time to time. It's just that they want the product to be flexible and adaptable to their needs and preferences. People are changing from time to time, so do their tastes and preferences. Identifying those is the first step towards achieving success and the rest depends on the performance of the product.

7. REFERENCES

Agrawal, Meenu. Consumer Behaviour and Consumer Protection in India, New Delhi, New Century Pub, 2006 ISBN 81-7708-107-1

(http:// www.consumerpsychologist.com/ 2007-05-03)

(http:// edms. matrade. gov.my 2007-05-03)

http://www.knowledgestorm.com/search (2007-05-03)

http://edms.matrade.gov.my (2007-05-03)

http://www.123eng.com/ (2007-05-03)

http://www.iimb.ernet.in/iimb/docs/eep07-08/28.htm (2008-02-10)

Annals of DAAAM for 2008 & Proceedings of the 19th International DAAAM Symposium, ISSN 1726-9679
Editor B. Katalinic, Published by DAAAM International, Vienna, Austria 2008
Make Harmony Between Technology and Nature, and Your Mind will Fly Free as a Bird

FEA OF ALUMINIUM FSW: TEMPERATURE FIELD AND STRESSES

VISAN, D[aniel]; BIRSAN, D[an]; MIRCEA, O[ctavian] & CONSTANTIN, E[mil]

Abstract: *In this paper, we presented the temperatures distribution, Von Mises stresses distribution and the Cauchy stresses distribution for the FSW welding process of the AA6061 T4 aluminium alloy. We took into account the case in which the tool is mobile. To stimulate displacement or movement of the thermal source it used 200 functions, corresponds to the number of the welding tools positions that move along the longitudinal axis of the joint.*
Key words: *stir, aluminium, welding, modelling.*

1. INTRODUCTION

The aluminium alloys are using in the aero-spatial industry from the year 1930. In the aero-spatial industry, the basic alloys are 2xxx and 7xxx, but, nowadays, the 6xxx aluminium alloys present a particular interest.

Troeger states that the 6xxx alloys have a series of advantages like: medium resistance, plasticity, good welding characteristics, corrosion resistance and a low cost (Troeger & Starke 2000). Nowadays, after applying thermal treatments, the 6xxx aluminium alloys are used in many applications (the exterior of the planes fuselages, panels and even in automotive shock absorbers), in the detriment of the 2xxx and 7xxx alloys. On the other hand, in practice there can appear some issues such as the low wear resistance and casting defects (Iordachescu et al., 2006). In order to improve the low resistance a number of methods applied, including the melting-welding processes, laser, plasma – but the porosities and cracks maintained near the processed areas.

In 1991, Thomas Wayne from the Welding Institute in UK (TWI) has patented the active rotary element friction welding, FSW, extending the method of using heat generated by friction in order to obtain some joints in the solid state (Colligan, 1999). The mechanical and metallurgical researchers are investigating the processing procedures with active rotary element (FSP), a procedure derived from FSW, in order to improve, by using this technique, the surface properties of the aluminium alloys. Certain properties improved using this technique, in precise areas, at a considerable depth (Vilaça et al., 2005).

The paper presents the fundamental parameters of the FSW process (the advance and the rotational speed of the processing device), as well as the influence of those upon the characteristics of the processed area, in the case of the cast aluminium alloy AA6061. Applying the FSW procedure, the analysis performed for a single layer.

2. HYPOTHESIS AND SIMULATION CONDITIONS

The finite elements analysis applied for the rotary active element welding procedure case for plates having the dimensions 10x100x300 [mm]. The simulation of the heat transfer and the visualization of the thermal fields from the joint applied for the mobile thermal source, analysis that is valid in the case of the real FSW welding. The calculus method proposed for the simulation of the thermal-mechanical fields from the joints welded using this procedure based upon the following hypothesis:

- The isotropy of the base metal;
- The thermo-physical properties of the materials are introduced according to the temperature;
- The base metals are in perfect contact;
- The external heat losses are made through convection and radiation;
- The yield strength of the base metal varies with the temperature;
- The external heat losses considered through conduction, convection and radiation.
- In order to make the model there were used 540 elements for the modelling of the processing device and 14400 elements for the modelling of the base metal;

The process parameters used for the modelling were:

- The rotation speed: $\Omega = 1180$ rpm
- The advance speed: $v = 150$ mm/min = 2,5 mm/s
- The pressure force: $F_z = 25$ kN
- The tilt angle of the processing device: $\alpha = 2.50$

It used an element SOLID 3D in the variant "coupled" with thermal and mechanical characteristics. The behaviour model of the material considered Von Mises elastic-plastic (bilinear model) with the tangent modulus different from zero.

In order to stimulate displacement or movement of the thermal source, it is necessary to define some time functions associated to the displacement or movement of the welding tool, functions that should be able to simulate a continuous welding process during the whole time interval starting from the moment 0 to the moment t_n (the total welding time). Thus, the number of finite functions, in total 200 functions, corresponds to the number of the welding tools positions that move along the longitudinal axis of the joint. The thermal field is the result of the summing up the thermal effects during the completely welding period $0...t_n$.

The influence of the deformation temperature upon the value of the friction coefficient on the contact surface between the deformation tool and the metal manifested using the temperature variation, the composition and the properties of the oxide layer formed on the surface of the piece that will be deformed. The value of the friction coefficient increases with the increase in temperature, and after reaching a maximum, the friction coefficient is decreasing once with the further increase of the temperature. This variation manner was explained in the case of the aluminium in this way: the increase of the temperature to values around 200-250°C leads to the increase of the friction coefficient due to the appearance of the primary oxides on the surface of the piece, oxides that have abrasive properties (Birsan et al., 2007).

Regarding the influence of the deformation speed as well as the relative sliding speed that appears between the two pieces that are in contact with each other in the deformation process, upon the value of the friction coefficient, was noticed the fact that with the increase of the deformation speed, the value of the friction coefficient decreases.

Hereupon of FEA of the FSW aluminium welding process, some images that simulate its expansion or unwinding in time and data concerning the tensions, deformations and material displacement on the process duration, obtained.

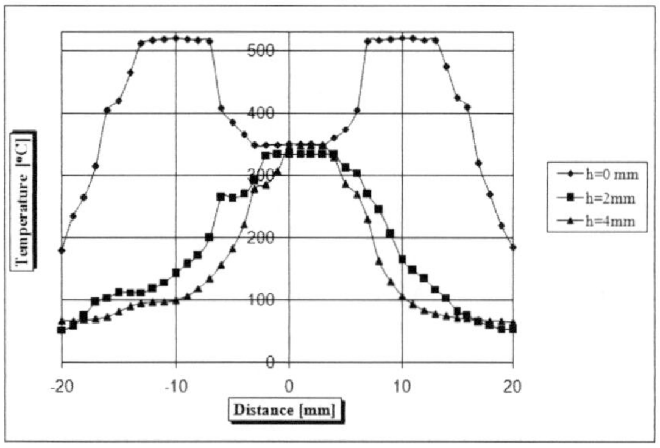

Fig. 1. Temperature distribution in the base metal.

3. DISTRIBUTION'S ANALYSIS FOR THE MOBILE THERMAL SOURCE CASE

3.1 The analysis of the temperature distribution in the base material

In the case of the mobile thermal source, the simulation of the thermal transfer in the FSW welded joints requires the fine meshing on the entire length of the axis of the two base metals are welded.

In Fig. 1 the ascending slope, as well as the descending slope of the temperature variation chart with time, in the chosen measurement point, for a single run, can have different values, and there are present a series of influencing factors. Important in this chart is the maximum value that the temperature reaches in that point. On the plate on which the thermal source is moving parallel and symmetrical with the movement axis of the source, we will find the points that have reached successively the same maximum temperature.

3.2 The distribution and the variation of the stresses

Figure 2 presents the global distributions of the Von Mises stresses at a certain moment in time. The Von Mises stresses represent the equivalent tension calculated according to the main stresses P1, P2 and P3. During the process the Von Mises stresses has a value that can reach up to 180 N/mm^2, the minimum values are registered in the area where the friction is maximum, due to the influence of the thermal field. In the cooling phase, once with the increase of the cooling time and of the levelling of the internal stresses it can be noticed a levelling of these stresses from the point of view of the values and a time distribution. Maximum values of the Von Mises stresses obtained within the interval 162 - 180 N/mm^2, values that exceed the yield strength of the base material.

In Fig. 3 there are presented the global distributions of the Cauchy stresses in the cross section at different time steps, from the moment in which the tool penetrated the base material until the time step t=18s. The maximum value of the compression Cauchy stresses in the FSW welding is of 365 N/mm^2.

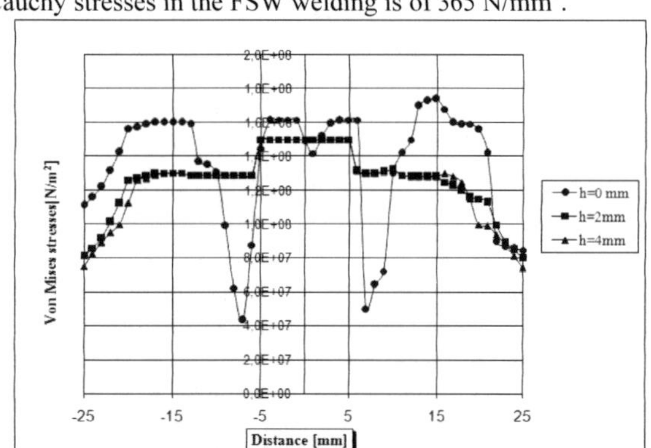

Fig. 2. Von Mises distribution in cross section.

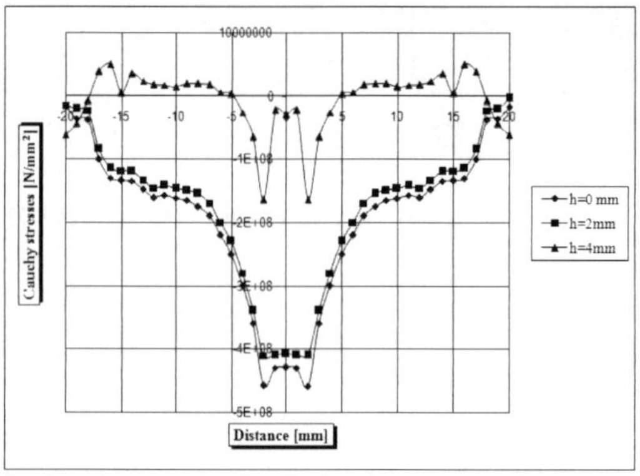

Fig. 3. Cauchy stresses distribution in cross section.

The maximum values of the Cauchy stresses registered at the base of the device, where the pressing and the friction of the material are maxim.

4. CONCLUSION

After the analysis made, some conclusions regarding the processing of the cast aluminium alloy AA 6061, are:

- The finite elements analysis of the FSW process allowed the estimation of the maximum process temperature; the value of this temperature is around 550°C, temperature that is inferior to the solidus temperature of the AA6061-T4;
- The maximum frictions due to the rotation movement as well as the pressure force required for the penetration of pin of the processing device leads to maximum stresses in the contact area with the base material.
- The maximum values of the Cauchy stresses registered at the base of the device.

It must be noticed that for now, worldwide, there are no communicates regarding to the distribution of the stresses and deformations of the FSW process. The model made using the MARC MENTAT specialised code allows the extraction of a series of results of the stresses and deformations field that subsequently will have to test through practice.

New relevant data for the processing of different materials and the elaboration of new types of devices will obtain based on the experimental research extent made for a series of other alloys and refinements of the analytical analysis model of the FSW process that will allow the welding by friction with active rotary element of carbon steels or of titanium.

5. REFERENCES

Birsan, D., Visan, D., Iordachescu, M. (2007) 3D Thermo-Mechanical Model of Fusion Welding, *Welding in the World*, , vol. 51, no. SPI, Ed. Elsevier, ISSN 0043-2288, pp. 657-666.

Colligan, K. (1999) Material Flow behaviour during Friction Stir Welding of Aluminium, *Welding Journal*, 75(7), 229s-237s.

Iordachescu, M.; Scutelnicu, E. & Iordachescu, D. (2006) Fundamentals of the Process and Tools Design: Friction Stir Processing of Materials, *The Annals of Dunarea de Jos University of Galati, Fascicle XII, Welding Equipment and Technology, Year XVII*, 63-72, ISSN 1221-4639.

Troeger, L.P. & Starke, E.A., (2000) Microstructural and Mechanical Characterization of a Superplastic 6xxx Aluminium Alloy, *Materials Science and Engineering, A277*, 102-113, doi:10.1016/S0921-5093(99)00543-2.

Vilaça, P., Santos, J. P., Góis, A., Quintino, L. (2005) Joining Aluminium Alloys Dissimilar in Thickness by Friction Stir Welding and Fusion Processes, *Welding in the World*, Vol. 49, No. 3/4, 56-62.

Annals of DAAAM for 2008 & Proceedings of the 19th International DAAAM Symposium, ISSN 1726-9679
Editor B. Katalinic, Published by DAAAM International, Vienna, Austria 2008
Make Harmony Between Technology and Nature, and Your Mind will Fly Free as a Bird

EVALUATION OF MODERN POST AND CORE SYSTEMS BY TESTING THE MECHANICAL FATIGUE RESISTANCE

VITALARIU, A[nca] & COMANECI, R[adu]

Abstract: *The aim of the present study was to evaluate the fatigue resistance of the teeth restored with different posts (carbon fibre post, zirconia post, titanium post) when subjected to a cyclic loading test. The teeth were intermittently loaded at an angle of 45 degrees to the long axis of the tooth, at a frequency of 1,3 loads per second until failure occur. The results show that the survival rate of the teeth reconstructed with carbon fibre post was higher than that of the teeth reconstructed with titanium and zirconia posts. In conclusion, the post reinforced with fibres reduces to a minimum the risk of root fractures of teeth restored with post and core systems.*
Keywords: *post, fibre, zirconia, fatigue*

1. INTRODUCTION

The post and core systems (PCS) represent an important issue in dental practice because they are the only chance to save the teeth with large coronal destructions and to avoid their extraction. Treatment failures caused by mechanical deficiencies of post and cores represent a problem of clinical significance and justify the researches orientated through the increasing of their performances and clinical longevity.

In the latest years there have been developed new techniques and materials for PCS: ceramic posts (cast or prefabricated), and composite posts reinforced with different kind of fibers: carbon, glass, quartz, silica, in order to replace the metallic posts wich have a lot of disadvances: poor esthetics and biocompatibility (the migration of metallic ions from the post's alloy in the oral tissues is well known, high Young modulus that increase the risk of catastrophic (irreversible) root fractures, followed by the tooth's extraction (Rosentrit et al., 2000).

Ceramic posts offer mechanical properties very close to those of dental alloys, an excellent esthetics and biocompatibility, because eliminate the risk of corrosion, bimetallism and allergic reactions. Ceramic posts are made of zirconium dioxide partially stabilized by the addition of yttrium (Y_2O_3) (Hochman & Zalkind, 1999). The usage of fiber reinforced post is based on harmony between Young modulus of the dentinal root, the post, luting cement and composite core material, and on the chemical connection between them. This result in a system that behaves like a mono-block under the occlusal forces (Bateman et al.,2003). Even the fiber reinforced posts have many advantages: proven biocompatibility, non corrosive, an equilibrate transmission of occlusal forces, no interferences with modern techniques of investigation (RMN, scanner), easily removable to permit endodontic re-treatment, their clinical reliability must be demonstrate by passing time. Even this "in vivo" tests are the most reliable, they have the disadvantage that are time consuming because are requesting long clinical observation periods. To overcome the difficulties of clinical studies, numerous "in vitro" (in laboratory) methods were developed to address specific properties of post and core restorations. Most of them, have been focused on the determination of the mechanical resistance to a simple stress, being static tests (Maccari et al.,2003). But the experience proved that the fatigue of the restorative materials is a primordial factor in clinical failures. The fatigue resistance tests represent an essential research instrument because they simulate the repetitive cyclic pattern of the occlusal forces and replace the clinical tests that are time consuming and thus contribute to decrease of clinical evaluation costs and time (Heydecke et al., 2002).

2. MATERIAL AND METHOD

2.1 Preparing of the samples for the test

For this experiment we used forty human maxillary incisors having similar dimensions and without cracks, decay, or abrasion. The teeth were endondontically treated and mounted in acrylic resin blocks with a simulated periodontal ligament. The samples were divided into three experimental groups and one control group:

- group C – teeth restored with prefabricated carbon fiber reinforced posts
- group Z– teeth restored with prefabricated zirconia posts
- group T – teeth restored with prefabricated titanium posts
- group M (control) – endodontically treated teeth but with no posts

The teeth from control group were covered with cast metallic crowns. In experimental groups *(C, Z and T)* the anatomic crowns of the teeth were sectioned perpendicular to the long axis, and the root canals were prepared for the posts. Gutta-percha was removed from the roots leaving 3 mm of root canal filling in the apical portion to ensure the apical seal of the endodontic space.

The posts were cemented with a dual-cure resin cement (Panavia F) and the core build-up was performed with a light curing composite resin. For the mechanical test the teeth were mounted in stainless steel cylinders filled with self-polymerizing resin. In order to simulate the periodontal ligament, a thin layer of silicone (0,2 mm) was applied along to the root surfaces. All teeth were covered with cast metallic crowns. On the palatal surface of each metal crown, right above the cingulum, a 0.3mm deep and 1mm-wide notch was made, in order to place there the compressive load during the test. The load was applied using a chisel-shaped steel pin, corresponding to the shape and dimension of the incisal edge of a lower incisor.

2.2 Experimental fatigue tests of post and core systems

Each test specimen was intermittently loaded to a maximum level of 25 daN at an angulation of 45 degrees to the long axis of the tooth (Figure 1), at a frequency of 1.3 Hz until the failure.

The testing device was stopped after 400 000 cyclic loads if no failure had occurred. Root fractures, post fractures and post and crown decementation were considered causes of failure. The survival rates of the groups were statistically compared with a Kaplan-Meier analysis.

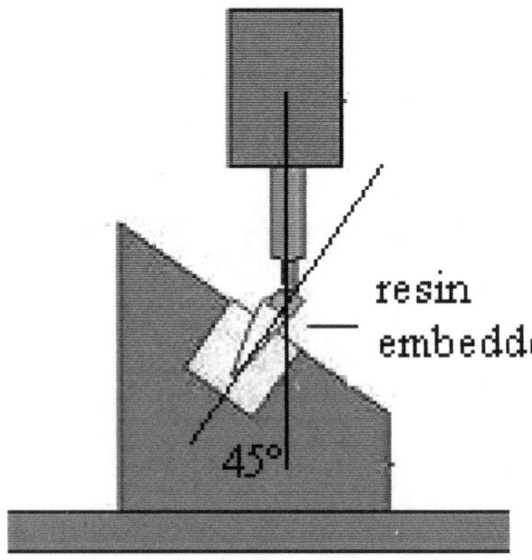

Fig.1. The sample in the testing device

3. DEVICES AND EQUIPMENT

Measurement system assisted by computer consists of load cell with power supply, signal conditioner, data acquisition (DAQ) device and virtual instrument software, (Figure 2).

- **Load cell** type Gefran Cu1KM – Italy with nominal load of 10kN and 350Ωm strain gauges in full bridge configuration with 0.2% accuracy is connected to a stabilized 10VDC power supply;
- **Signal Conditioner** type SC 2043 SG National Instruments - USA provides both the nominal excitation voltage for load cell and the router of the conditioned strain gauge bridge signal to DAQ card;
- **DAQ Card** type PCI 6023E National Instruments - USA take the load cell signal (with a sampling rate of 200 kS/s) via signal conditioner and displays both the graphic and numerical evolution in real time;
- **Virtual instrument VI** is the LabView application which provides the communication between DAQ device and the computer.

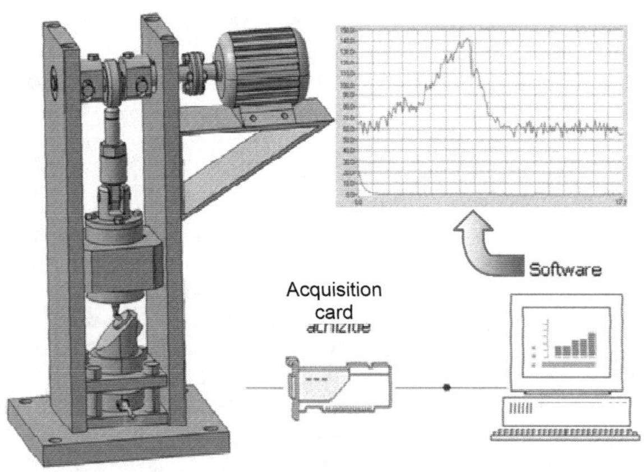

Fig.2. The testing device and the data acquisition system

4. RESULTS

The results obtained after 400.000 impacts are shown in Table 1 and Figure 3. Only one fracture was observed in the control group (teeth with no post) and three in carbon fiber group while in zirconia and titanium groups were recorded six fractures in each group. The survival rates recorded in the experimental groups (C, Z and T) were lower than that of the control group. The differences between the experimental groups were significant, the teeth reconstructed with posts reinforced with carbon fibers had a higher survival rates than the teeth reconstructed with titanium and zirconia posts.

M (control)	Z (zirconia)	C (carbon)	T (titanium)
92.794 ✖	26.171	✖ 251.123 ✖	18.763 ✖
400.000 ⚠	28.419 ✖	258.434 ✖	81.109 ✖
> 400.000 ✓	149.762 ✖	262.257 ✖	90.981 ✖
> 400.000 ✓	182.143 ✖	> 400.000 ✓	95.012 ✖
> 400.000 ✓	241.112 ✖	> 400.000 ✓	101.448 ✖
> 400.000 ✓	272.018 ✖	> 400.000 ✓	117.674 ✖
> 400.000 ✓	> 400.000 ✓	> 400.000 ✓	> 400.000 ✓
> 400.000 ✓	> 400.000 ✓	> 400.000 ✓	> 400.000 ✓
> 400.000 ✓	> 400.000 ✓	> 400.000 ✓	> 400.000 ✓
> 400.000 ✓	> 400.000 ✓	> 400.000 ✓	> 400.000 ✓

Legend ✖ - fracture; ✓ - intact; ⚠ - crack

Tab. 1. The results after 400.000 cyclic loads

Fig.3. Kaplan Meier curves of the experimental groups

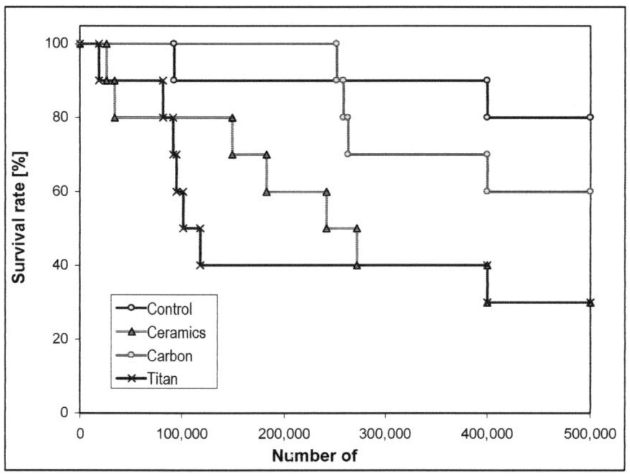

5. CONCLUSION

Teeth reconstructed with fiber reinforced posts exhibit a higher fatigue resistance than teeth reconstructed with rigid posts (titanium or zirconia posts). Fiber posts reduce the risk of root fracture and are the best choice in reconstruction of endodontically treated teeth. Knowing the advantages and disadvantages of different type of post, the dentist will be able to avoid a post that predispose to irreparable fractures.

6. REFERENCES

Bateman, G.; Ricketts, D.N.J.& Saunders, W.P. (2003). Fibre-based post systems: a review of the literature. *Br Dent J*, No.195, (2003), pp.43-48

Heydecke, G.; Butz, F.; Hussein, A.& Strub, J.R. (2002). Fracture strength after dynamic loading of endodontically treated teeth restored with different post-and-core systems. *J Prosthet Dent* , No.87, (2002), pp. 438-445

Hochman, N.& Zalkind, M.(1999). New all-ceramic indirect post-and-core system. *J Prosthet Dent*, Vol.81, 1999, pp. 625-629

Maccari, P.C.; Conceicao, E.N.& Nunes, M.F. (2003). Fracture resistance of endodontically treated teeth restored with three different prefabricated esthetic posts. *J Esthet Restor Dent*, Vol.15, No.1, (2003), pp.25-30

Rosentritt, M.; Furer, C.; Behr, M.; Lang, R.& Handel, G. (2000). Comparison of in vitro fracture strength of metallic and tooth-coloured posts and cores. *J Oral Rehabil*, Vol.27, No.7, (2000), pp.595-601

Annals of DAAAM for 2008 & Proceedings of the 19th International DAAAM Symposium, ISSN 1726-9679
Editor B. Katalinic, Published by DAAAM International, Vienna, Austria 2008
Make Harmony Between Technology and Nature, and Your Mind will Fly Free as a Bird

INVESTIGATION OF THE HARD FILMS DEPOSITED BY PVD-MAGNETRON SPUTTERING ON THE CUTTING TOOLS

VLASCEANU, D[aniel]; COTRUT, C[osmin]; TARCOLEA, M[ihai]; BACIU, F[lorin] & NICODIM, M[ariana]

Abstract: *Hard coatings based on hard materials are especially suitable for the purposes of protection against metallurgical tools wear. The properties involving high hardness and thermal stability with low wear resistance and friction coefficient are the main goal to achieve. Hard materials such as WC are widely used as cutting tools material due to their high hardness, strength and fracture toughness. Elemental composition and topography, of the coatings, it was determined by the SEM method. Phase composition and texture has studied by X-ray diffraction technique. The mechanical property of coatings has investigated by microhardness measurement and scratch test. The results of experimental research proved that the investigated coatings exhibited a high microhardness having good adhesion on the substrate (> 50 N).*
Key words: *sputtering, hard coating, SEM, XRD, microhardness*

1. INTRODUCTION

Generally, coatings have a major influence on the performance of cutting tools. The insufficient adhesion is still one of the most important aspects in the development of new coating technologies for hard metal cutting tools. Particularly, the principal purpose was to find coatings that could be deposit at lower temperatures, in order to allow to sharper edges tools to be coat without any embrittlement effect (Brookes, 2001).

The solution was PVD (physical vapor deposition), where deposition temperature normally can be kept about 500°C.

A particular aspect of PVD technology it is linked to the coating thickness; one can say that this quantity is strongly dependent on the loading of the reactor and the supporting device. The main purpose of PVD is to obtain a straight line of deposition (Wissmann, 1972).

For hard metal cutting tools, thickness control is one of the key factors in order to optimize the properties and to facilitate the PVD coating techniques, with a well-defined processing way, the fully control of charges offering a clear competitive advantage. The cutting tools are obtained from hard metals (made from WC powders), coated with TiN through PVD-magnetron sputtering technology (Ruset et al., 2007).

The PVD methods have been preferred for a wide range of materials and applications due to their relative simplicity, versatility, and the general high quality of the resultant coatings (Kelly, et al., 1993).

Physical vapor deposition is a generic term for a number of processes, which are use to deposit solid coatings, or films, from the vapor phase. Processes currently included in this definition are vacuum evaporation, ion plating and sputtering (Mateescu, G., 1998). The most widely used material for wear resistant coatings is titanium nitride (TiN).

2. EXPERIMENTAL PVD-MAGNETRON SPUTTERING DEPOSITION

In TiN deposition process, Ti layer is deposit on the surface of the wafer by creating a high partial pressure of N near the wafer surface and by creating a high partial pressure of Ar at or near the target. This minimizes nitrogen concentrations at or near the target, thus minimizing any nitriding of the target. To accomplish this, the Ar gas is introduced through gas inlet located near the target and the N gas is introduced using a lower gas inlet located at or near the wafer surface or otherwise located near the bottom of the chamber.

In order to investigate the effect of the angle between magnetron axis and sample surface, the depositions at two different inclinations (15° and 45°) was made.

3. INVESTIGATIONS AND RESULTS

3.1 Microhardness

Code	Conditions of deposition	$HV_{0.1}$
P1	Inclined to 45°/2 hours	2587
P2	Inclined to 15°/3 hours	2587
P3	Inclined to 15°/5 hours	2850

Tab. 1. Micro hardness values

The hardness-testing machine was loaded with a force of 1 N. Several tests have been made on three different samples (Table 1), having different thicknesses on the deposited layer.

Although the principal parameters of deposition process have been the same, the microhardness it is influenced by the sample geometry.

One of the principal factors taken into account is the angle between the plane area and the treatment chamber axis.

This angle depends in a direct manner on the incidence of ions energy of the substratum area.

The microhardness obtained is superior by respect to the microhardness obtained by usual deposition techniques (CVD, mechanical deposition).

A possible explanation for large values of microhardness obtained through PVD – magnetron-sputtering method is due to the big compressible tensions induced between the substratum and plasma.

3.2 Layer growth

After preparation of the specimen for metallographic analysis, the following aspects have been taken into account:

- thickness determination after the deposition;
- thickness determination, where there is possible, of the titanium intermediate layer;
- deposition inspection on the cutting edges;
- interface inspection between deposited layer and substratum.

Code	Conditions of deposition	Thickness (μm)
P1	Inclined to 45°/2 hours	9
P2	Inclined to 15°/3 hours	15
P3	Inclined to 15°/5 hours	21

Tab. 2. Layer growth of the cutting tools

3.3 SEM investigations

Fig. 1. Layer aspect (×1250)

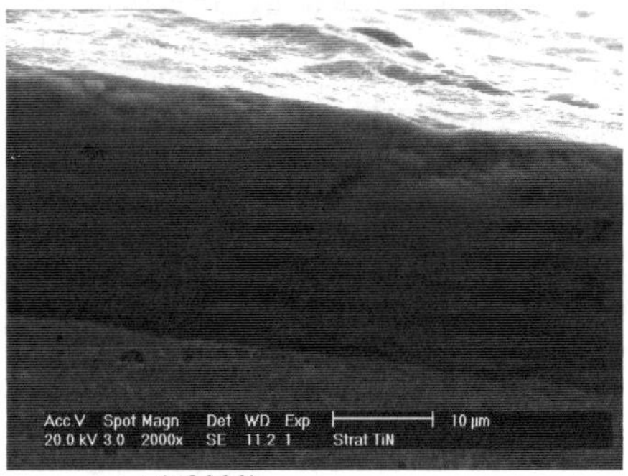

Fig. 2 Layer aspect (×2000)

3.4 XRD analysis

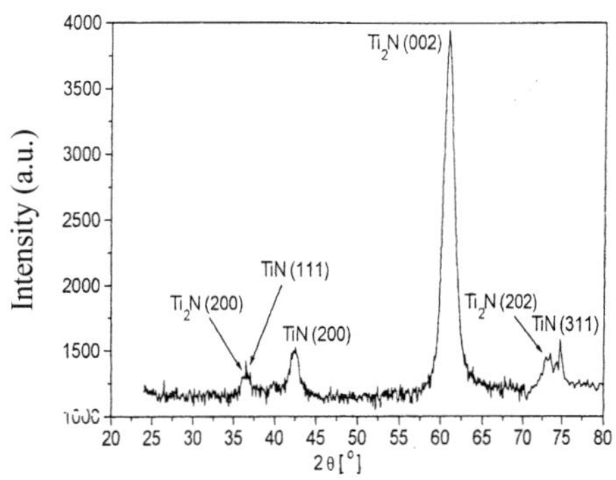

Fig. 3 XRD spectrum for P2 (layer thickness 15 μm)

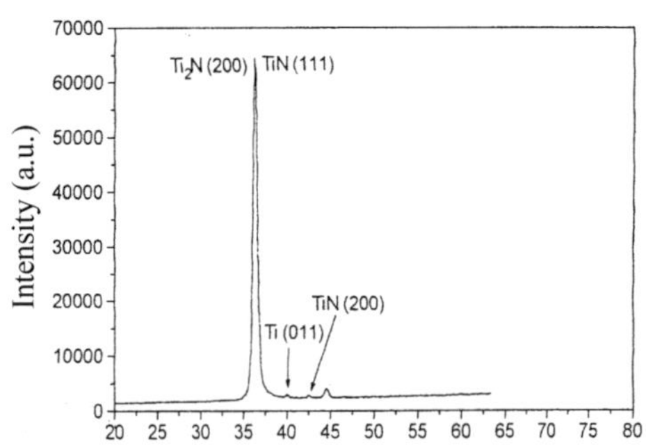

Fig. 4 XRD spectrum for P1 (layer thickness 9μm)

3.5 Adherence to the substratum

The layer adherence to the substratum – beside the microhardness – represents one of most important feature, which determinates the efficient coatings used for wear resistant growth.

Code	Growth, (μm)	$F_{NC,}$ (N)	Observation
P1	9	50	Detach marginally
P2	15	55	Detach marginally
P3	21	57	-

Tab. 3. Scratch test values

By the "scratch test" was realized the adherence determination, one of the most adequate methods in order to estimate adherence of hard layers applied on the cutting tools.

The results of these analyses are indicated in Table 3.

The method is accomplished by pressing a pyramidal or conic penetrator, made from diamond or ruby, with a progressive load from 0 to 100 N, on the surface area of a cutting tool having a displacement rate of 10 mm·min^{-1}.

The normally critical force is then determined; this one represents the pressing force which produces the layer damage by apparition of some breaches to the layer-substrate interface.

The method presented above utilizes a ruby conic penetrator with a radius of 0.2 mm.

With a microscope, in order to establish the values of the normally critical force (FNC), were examined the scratch stamps.

4. CONCLUSIONS

The results obtained have shown a superiority of the procedure presented in the paper in comparison with the conventional techniques, having in view the following points of view:

- The hardness of some TiN layers obtained by this new technique is about 3000 $HV_{0.04}$, in comparison with the usual hardness (of about 1000 – 1500 $HV_{0.04}$) obtained by the classical methods;
- It is possible to obtain layer thicknesses up to 15 μm, with a good adherence to the substratum. Usually, by application of classical methods there are obtained layers with a 2 – 5 μm thickness;
- The interface between the layer and the substratum could be of about 5 – 6 μm, in comparison with the standard thickness of 1 μm;
- The investigated coatings exhibited a high microhardness having good adhesion on the substrate (> 50 N).

5. REFERENCES

Kelly, P.J., Arnell, R.D., Ahmed, W., (1993), Some recent applications of materials deposited by unbalanced magnetron sputtering, *Surface Engineering*, Volume 9, Number 4, pp 287 – 291.

Brookes, K.J.A., (2001), Annapolis shows the cutting edge of hard metals technology, *Metal Powder Report*, Volume 56, Number 4, pp. 8-14.

Mateescu, G., (1998), *Advanced technologies – Vacuum deposition thin films*, Ed. Dorotea, Bucharest, Romania.

Wissmann, P., (1972), The effect of gas adsorption on the conductivity of thin metal films, *Thin Solid Films*, Volume 13, pp189.

Holmberg, K., Matthews, A., (1994), Coatings Tribology: Properties, Techniques and Applications in Surface Engineering, *Tribology Series*, Volume 28, Elsevier, Amsterdam, The Netherlands.

Ruset, C., Grigore, E., Maier, H., Neu, R., Li, X., Dong, H., Mitteau, R., and Courtois, X., (2007), Tungsten coatings deposited on CFC tiles by the combined magnetron sputtering and ion implantation technique, *Physica Scripta*, Volume T128, pp. 171-174.

Annals of DAAAM for 2008 & Proceedings of the 19th International DAAAM Symposium, ISSN 1726-9679
Editor B. Katalinic, Published by DAAAM International, Vienna, Austria 2008
Make Harmony Between Technology and Nature, and Your Mind will Fly Free as a Bird

EVALUATION OF THE CUTTING EDGE WEAR AND OF THE TECHNOLOGICAL SPEED FOR THE 13H11N2V2MFS STAINLESS STEEL TURNING

VLASE, A[uremia]; GHIONEA, I[onut]; MUNTEANU, G[abriel] & GHIORGHE, A[drian]

Abstract: *This paper presents some experimentally data from the turning process for the stainless steel 13H11N2V2MFS and a new methodology to determine the relations of regression for the tool wear and for the cutting speed. Some determinations are made, considering different process parameters. The measurable data defining the cutting process is processed mathematically and they were transposed under the shape of a relation in order to determine the wear of the cutting edge and speed. Putting the condition of limitation for VB=0.7 mm. Also, it is presented the form of the regression functions; as well the variation of the technological cutting speed in dependence of the cutting depth, the tool's feed.*

Key words: *turning, wear, cutting, regression, processing.*

1. INTRODUCTION

The stainless steel properties, like a very good resistance to corrosion factor actions and to high temperature, made these materials suitable to be used for a wide range of application from chemical or energetic to the alimentary or health industries. Even though, some characteristics like the high toughness, low thermal conductibility and cold-hardening during the cutting processing, are reducing the workability ratio. Therefore, a disadvantage we shall mention is the low workability of stainless steels have a (Barlier, C., 1999) but to obtain an increase for this parameter, some alloyed materials like S and Se are used (Shaw, 1999). The above mentioned research continues the author concern regarding the workability of stainless steel (Ghionea, A., Vlase, A., Ghionea, I., 2007).

2. CUTTING CONDITIONS

In Table 1 are presented some chemical characteristics of the steel 13H11N2V2MFS and the most important mechanical characteristics are shown in Table 2. Other information about the elaboration by casting, thermal treatments, mechanical characteristics, metallographic study is available in standards (STAS 3583-87, DIN EN 10988, NF EN 10027). The determinations and measurements were made according to some experimental conditions. Thus, the following elements of the technological system have been used:

- the semi-products used for experimental determinations are obtained by casting under the shape of bar;
- the high-speed steel cutter is suitable for stainless steel that have high plasticity and low hardness properties;
- the cooling and lubricating fluid: P 20% emulsion environment improves the cutting processing;
- machine tool: turning machine, SNA 500 type, 24 speed steps (D_{nc}: 16…1600 rpm), 21 feed steps (d_{fL}: 0.028…5 mm/rot), motor drive power PME = 7.5 kW;
- turning tool STAS 350-82 from Rp5 STAS 7382-88, 6-5-2;
- for cutting: UAS-200 machine with special cutting device.

C	Mo	Ni	Cr	Mn	Si	S	P	V	N
0.12	2.0	2.5	11.5	0.7	0.9	0.02	0.03	0.3	0.02

Tab. 1. Chemical characteristics

Tensile strength R_m [N/mm^2]	Flowing Limit R_{02} [N/mm^2]	Elongation δ [%]	Hardness HB
1080	788	9.2	185

Tab. 2. Mechanical characteristics

3. METHOD AND EXPERIMENTAL RESULTS

The method of parametric analysis was applied and the experimental data established were processed mathematically (Parakkal, G. & al., 2002), (Spur, G. & Stöferle T., 1979).
In order to establish a suitable relation of assessment for the tool cutting edge wear on the side face, it is taken into consideration the relation from (Vlase, A., 1973), under the form:

$$VB = C_{VB} \cdot d_L^{\,x} \cdot f_L^{\,y} \cdot v_c^{\,z} \cdot \tau^{w} \cdot [\text{mm}] \quad (1)$$

where: d_L is the depth of cut, in mm, VB – wear dimension, in mm; f_L- feed, mm/rev, v_c – technological cutting speed in m/min; τ - working time, in min.; x, y, z and w – polytrophic exponents, C_{VB} - the constant which depends on the piece's material and of the cutting tool.
The relation (1) becomes linear if we apply the logarithm function, as follows:

$$\lg VB = \lg C_{VB} + x \lg d_L + y \lg f_L + z \lg v_c + w \lg \tau \quad (2)$$

In equation (2) they were introduced five sets of values chosen among a greater number of data experimentally established and presented in Table 3.
The following restrictions were imposed: $d_L = 0.25…0.50$ mm and $f_L \leq 0.2$ mm/rev.

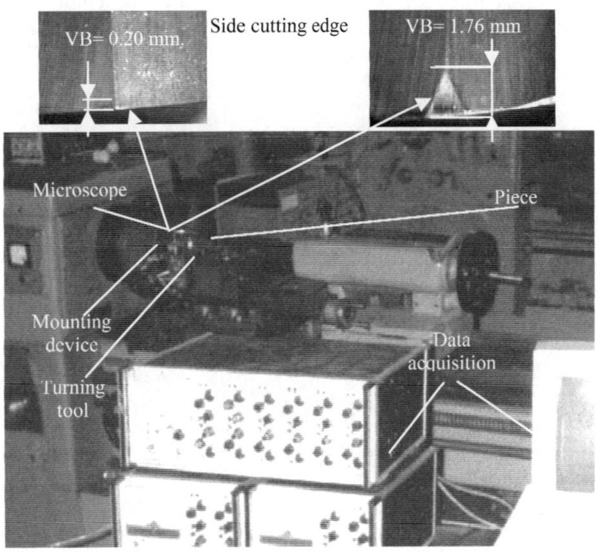

Fig. 1. Experimental equipments for measurement.

No. of det.	Depth, d_L, [mm]	Feed, f_l, [mm/rot]	Cutting speed, v_c, [m/min]	Working time, τ, [min]	Wear VB, [mm]
1	0.50	0.1	108.5	8	0.354
2	0.50	0.2	105.5	8	0.729
3	0.50	0.1	163.3	8	1.759
4	0.25	0.1	101.0	8	0.142
5	0.25	0.1	99.5	12	0.193

Tab. 3. Experimental results

In the Figure 1 is presented the equipment and the testing instruments used for measure the parameters (Vlase, I., 2001). The wear is measured using a microscope and the dimension is shown at different moments of wears expanding.
Some other processes have been made by the preliminary determinations with $f_{ax} > 0.25$ mm/rot.
In this last case the percent of the cutting edges' breaking was high.
Having the relation (2) and the data from the table 3 it can be formed a five equation system, where C_{VB}, x, y, z and w are the unknowns.

$$\begin{cases} \lg C_{VB} + x\lg 0.50 + y\lg 0.1 + z\lg 108.5 + w\lg 8 = \lg 0.354 \\ \lg C_{VB} + x\lg 0.50 + y\lg 0.2 + z\lg 105.5 + w\lg 8 = \lg 0.729 \\ \lg C_{VB} + x\lg 0.50 + y\lg 0.1 + z\lg 163.3 + w\lg 8 = \lg 1.759 \\ \lg C_{VB} + x\lg 0.25 + y\lg 0.1 + z\lg 101.0 + w\lg 8 = \lg 0.142 \\ \lg C_{VB} + x\lg 0.25 + y\lg 0.1 + z\lg 99.5 + w\lg 12 = \lg 0.193 \end{cases} \quad (3)$$

Then, we have determined the numerical values for the five unknowns: $C_{VB} = 1.74 \cdot 10^{-8}$, $x = 0.91$, $y = 1.2$ and $z = 3.92$, $w = 0.89$.
From the relation (2) results the formula of the wear for the 13H11N2V2MFS steel as a regression function, as shown:

$$VB = 1.74 \cdot 10^{-8} \cdot d_L^{0.91} \cdot f_L^{1.2} \cdot v_c^{3.92} \cdot \tau^{0.89} . [\text{mm}] \quad (4)$$

In the stainless steel turning process it is recommended the admissible cutting edge tool wear in a range of 0.5 to 0.7 mm. At the limit condition of VB=0.7, the cutting speed formula is:

$$v_c = \frac{87.08}{T^{0.227} \cdot d_L^{0.232} \cdot f_L^{0.306}} \cdot [\text{m/min}] \quad (5)$$

Relations (4) and (5) are used to determinate the calculus relations of the technological cutting speeds for different values of the admissible wear VB, in mm. The parameter τ is replaced with the tool life, in minutes.

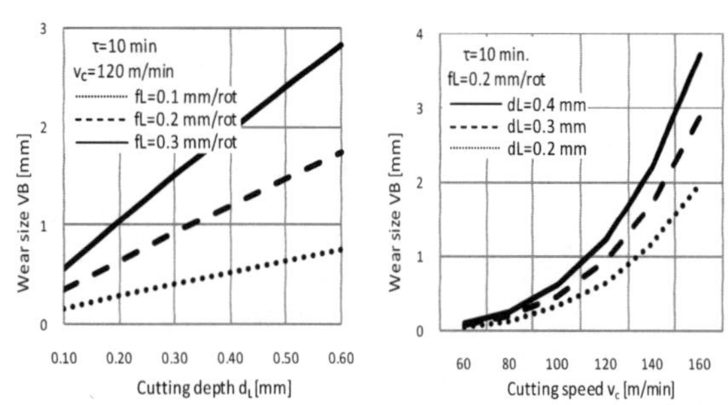

Fig. 2. Wear size dependency from d_L and v_C.

Based on the relations (4) and (5) there can be represented the variation curves of VB and v_c parameters considering D_t, f_L or τ. In Figure 2 are shown the graphical representations of the wear VB variation diagrams with the most important parameters D_t and v_c. The resulting chips are discontinuous. Also, a superior value of the cutting depth or speed causes an increase of the wear dimension, so the range shall be avoid.

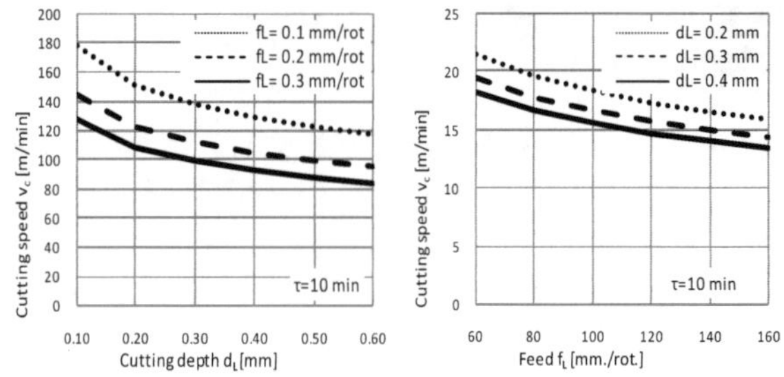

Fig. 3. Cutting speed dependency from d_L and f_L.

Based on the proposed equation (5) we represent in the figure 3 the graphical representation of the speed v_c variation diagrams with the parameters d_L and f_L.
From these graphics we can observe that as the cutting depth d_L or feed f_L is increasing, the cutting speed will decrease; therefore, a lower speed is helpful to improve the quality of cutting.

4. CONCLUSIONS

The objective of this paper consist in the establishing of two regression functions for the analytical evaluation with experimental data of two specific parameters for the cutting process: edge wear and the technological speed, applied for wide use.
The model consists of two equations, (4) and (5) suitable for 13H11N2V2MFS stainless steel cutting tool wear behavior. The established equations have the accuracy comparable with other methods that requires a larger amount of materials and tools.
Therefore, we consider that an economic gain is obtained by using the current model and the tool manufacturer as well as the final users may find it very helpful in order to reduce the time for experimental determination and to low the amount of materials or tools.
All the results obtained and above mentioned are a contribution in order to establish the index of workability for the stainless steels and confirm the classification of 13H11N2V2MFS stainless steel within the class of materials with a reduced workability through cutting.

5. REFERENCES

Barlier, C. & Girardin, L. (1999). *Memotech. Productique, matériaux et usinage (Memotech. Production, materials and machining)*, Editions Casteilla, ISBN 2.7135.2051.7, Paris
Ghionea, A., Vlase, A. & Ghionea, I. (2007). The wear evaluation of the cutting edge and the cutting speed in the drilling process of some manganese steels, *Proceedings of the 18th International DAAAM Symposium*, Katalinic, B., pp. 295-296, ISBN 3-901509-58-5, Vienna, October, 2007, Published by DAAAM International, Vienna
Parakkal, G. et al. (2002). Modeling of turning process cutting forces for grooved tools. *International Journal of Machine Tools & Manufacturing* 42 (2002), 179-191
Shaw, M. C. (1984) *Metal cutting principles,* Clarendon Press – Oxford
Spur, G. & Stöferle Th. (1979) *Handbuch der Fertigungstechnik*, Band 3/1 *Spanen*, Carl Hanser Verlag, München Wien
Vlase, A. (1973). Contributions regarding the genuine stainless steel workability, *Ph. D. Thesis*, University Politehnica of Bucharest
Vlase, I. (2001). Contribution regarding the determination of the assessment indicators concerning refractare steel workability, *Ph. D. Thesis*, University Politehnica of Bucharest

Annals of DAAAM for 2008 & Proceedings of the 19th International DAAAM Symposium, ISSN 1726-9679
Editor B. Katalinic, Published by DAAAM International, Vienna, Austria 2008
Make Harmony Between Technology and Nature, and Your Mind will Fly Free as a Bird

EXPERIMENTAL RESULTS CONCERNING THE GEOMETRIC, STRUCTURAL AND DIMENSIONAL DIFFERENCES IN FIBER REINFORCED COMPOSITES

VLASE, S[orin]; MODREA, A[rina]; BURCA, I[oan]; PURCAREA, R[amona]; MUNTEANU, M[ihaela] V[ioleta] & SCUTARU, M[aria] L[uminita]

Abstract: *During the manufacture process of a composite and after this, during the use of the material, there exists many ways in which the real material can have differences in comparison with the theoretical composite considered as a succession of a repeating cell. Sometime these differences can be great enough and they can have a major influence on the properties of the composite (Goia & Modrea, 1999). In the paper we try to present haw these differences can influence the elastic constants of such material. The part II of the paper is dedicated to present experimental results.*
Key words: *composite material, fibers, geometric differences, engineering constants*

1. INTRODUCTION

This part presents experimental results concerning composite materials reinforced with fibers. In the first part we have presented the influence of dimensional, structural or geometric differences on the values of the engineering constants of a composite. The theoretical considerations (Modrea, 2001, a,b,c,d) were proved by experimental procedures.

2. TESTA AND SPECIMENS

To verify the mentioned properties was tested specimens made respecting the existing standards. The tests was made for traction and bending.

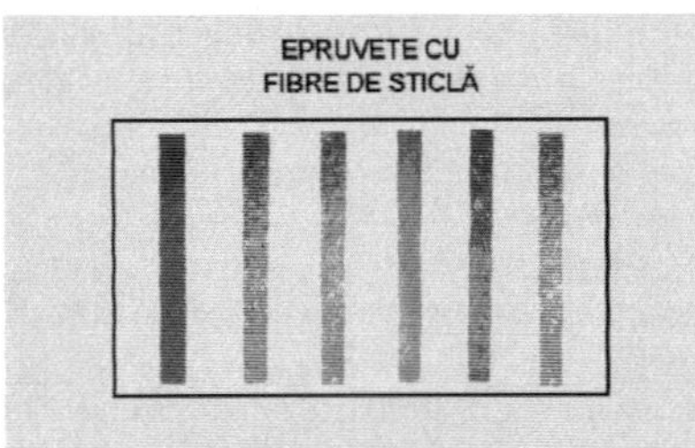

Fig. 1. Polyester PEÎD reinforced with glass fibers used for bending tests

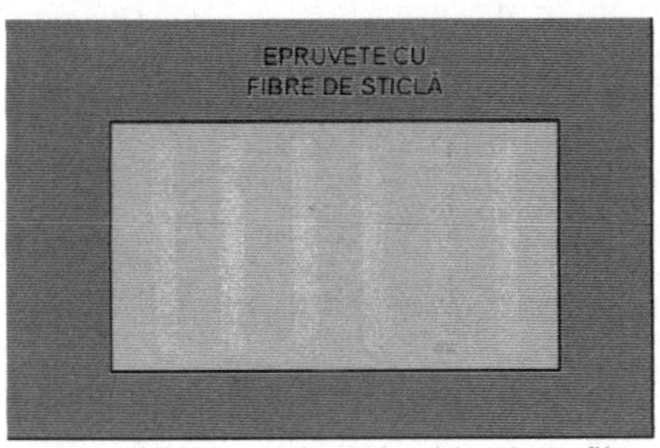

Fig. 2. Polyester PEJD reinforced with glass fibers used for bending tests

In fig.1 and 2 are presented some plastic specimens with the industrial name of PEJS (low density poly ethylene) and PEÎD (high density poly ethylene) used in the tests. These materials were made by Întreprinderea Mecanică Târgu-Mures. The specimens were organized after the type of the test, in many groups, every group having ten specimens and some reserved specimens. In the case when the specimens were destroying near the ends, they were changed by a reserve specimen.

In the same time were executed specimens for the bending test, respecting the existing standards.

Fig. 3. Specimen of composite made by resin reinforced with carbon fibers for bending tests

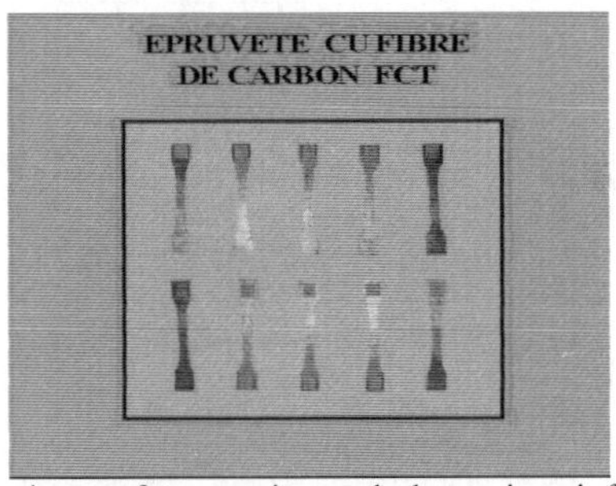

Fig. 4. Specimen of composite made by resin reinforced with carbon fibers for traction tests

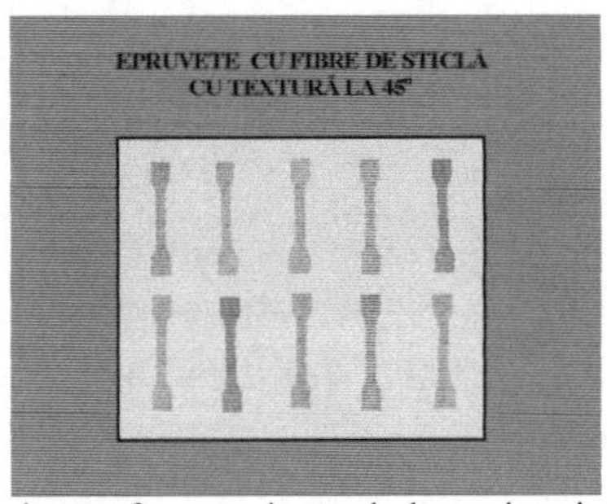

Fig. 5. Specimen of composite made by resin reinforced with glass fibers prevailed to 45° for traction tests

Other specimens for traction and bending tests for a composite material reinforced with carbon fibers. These were made to S.C.COMPOSITE srl Braşov, for traction (fig.3) and bending (fig.4) tests.

To the same enterprise was made specimens with resins matrix and glass fibers, oriented parallel, perpendicular and to an angle of 45° (fig.5).

3. EXPERIMENTAL RESULTS

For the traction tests was used the machine existing in the laboratory of TRANSILVANIA University. Tests for the materials PEÎD and PEJD with the series numbered T01 and T02 was made. In one series the specimen was numbered between one and ten.

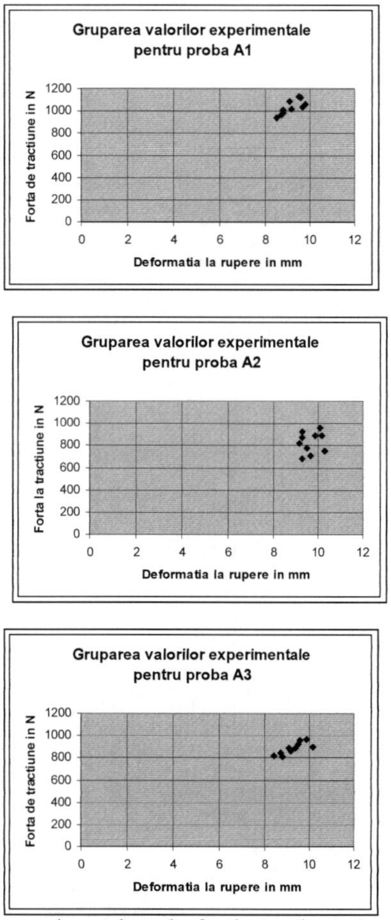

Fig. 6. The experimental results for the traction tests

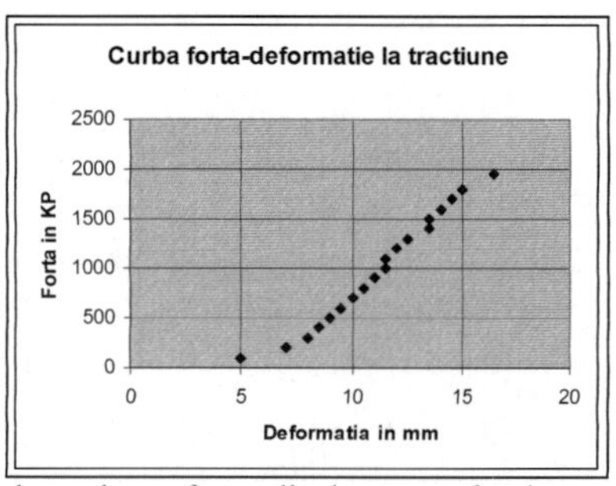

Fig.7. The dependence force-displacement for the traction tests

For the four series of tests was obtained the following values for the Young modulus: E_{T01}=271,145MPa; E_{T02}=203,951 MPa ; E_{T03} = 229,396 MPa ; E_{T04} = 196,108 MPa. The specimens made by resin reinforced with glass fibers, FSU and FSB, parallel with the specimen, were tested to traction. The dependence force- displacements was represented in fig.7. The series of these specimens was numbered T05 and T06.

The specimens numbered FSMB-1 to FSMB-10 were tested to bending and the dependence force displacement is presented in fig. 9.

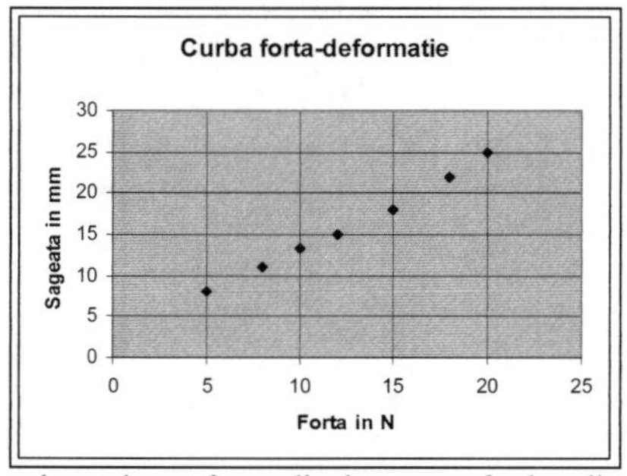

Fig. 8 The dependence force-displacement for bending testsFSMU-1

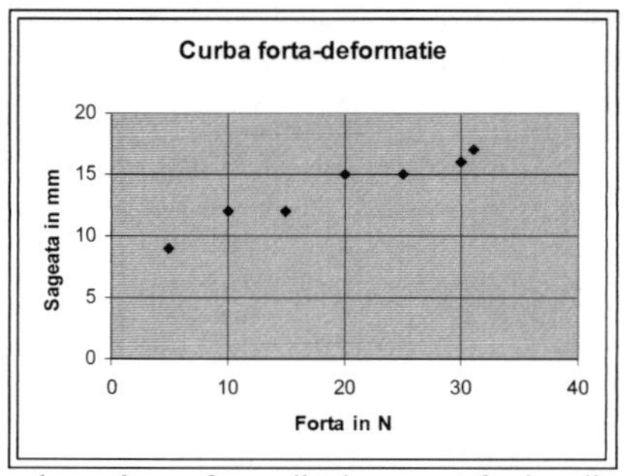

Fig. 9. The dependence force-displacement for bending tests FSMB-1and FSMB-2

4. CONCLUSIONS

The calculus made for the analyzed specimens compared with the experimental results shown that for a number of specimens the measured results are outside the calculated bounds of elastic moduli. This conclusions denote that the difference between the theoretic dimension, geometry and structure and the real composite conduce to difference, sometimes notable, between the calculated and real values of the mechanical properties. From these considerations results that is necessary to study how the differences mentioned can influence the properties of the composite material.

5. REFERENCES

Goia,I., Modrea,A., ş.a. (1999). Calculus of the Mechanical Properties for the Composite Materials. *A IX-a Conferinţă internaţională CONAT*, Braşov

Modrea, A., ş.a. (2001, a). Evaluation of the elastic parameter for a composite when the strain/stress field is obtain via finite element method. *A III-a Conferinţă de dinamica maşinilor*, Braşov, pp.365-370

Modrea,A., ş.a. (2001, b). Evaluation of homogenized coefficients for fiber reinforced plastic. *A III-a Conferinţă de dinamica maşinilor*, Braşov, pp.371-374

Modrea,A., ş.a. (2001, c). Mathematical Background for Evaluation of Homogenized Coefficients for Fiber Reinforced Plastic (FRP). *A III-a Conferinţă de dinamica maşinilor*, Braşov, pp.375-378

Modrea,A., ş.a. (2001, d). Mathematical Background for Evaluation of Homogenized Coefficients for Fiber Reinforced Plastic (FRP). *A III-a Conferinţă de dinamica maşinilor*, pp. 375-378, Braşov

Annals of DAAAM for 2008 & Proceedings of the 19th International DAAAM Symposium, ISSN 1726-9679
Editor B. Katalinic, Published by DAAAM International, Vienna, Austria 2008
Make Harmony Between Technology and Nature, and Your Mind will Fly Free as a Bird

ON THE TOPOLOGICAL DESCRIPTION OF THE MULTIBODY SYSTEMS

VLASE, S[orin]; MUNTEANU, M[ihaela] V[ioleta] & SCUTARU, M[aria] L[uminita]

Abstract: *The paper aims to provide a structural description of flat mechanics by using the theory of graphs (Berge,1958, Gross & Yellen,2003). Besides this description, fully defined by the incidence matrix (or by the equivalent basic cycle matrix), geometrical and kinematical description are also intended. The graph theory was used in the mechanical system analysis in two different ways: - in one way the constitutive elements of the system are considered nodes and the liaison between elements are considered edges. This analysis leads to the independent cycle method, developed by various authors in many variants; - another way is to consider the liaison between elements as nodes and the vector defined by the points of the liaison as edges. This description, less used in the literature permit to utilize the natural vector description of the system. The thus created formal framework allows representing the condition equations of speeds and accelerations suitable for numerical applications.*
Key words: *Graph Theory, Kinematics, Independent Cycle List.*

1. INTRODUCTION

The graph theory was used in the mechanical system analysis in two different ways:

- in one way the constitutive elements of the system are considered nodes and the liaison between elements are considered edges (Thoma, 1975), (Voinea & Atanasiu, 1964). This treatment is equivalent with the Kirchhoff equations for the electrical systems. This analysis leads to the independent cycle method, developed by various authors in many variants. Some descriptions (Shai & Pennock, 2006), (Tsai et al., 1998) continue this case.
- another way is to consider the liaison between elements as nodes and the vector defined by the points of the liaison as edges. This description, less used in the literature (Vlase, 2007) permit to utilize the natural vector description of the system. In the paper is analysed this treatment using a topological approach. The planar case, presented in the paper, can be extended to the spatial systems.

2. TOPOLOGICAL DESCRIPTION OF THE MECHANISM

2.1. Kinematical Equivalent Mechanism

In order to present the method we must to attach a graph to mechanism. The first step is to build a mechanical system made only by bars equivalent, from mechanical point of view, with the system done. We will analyze the planar mechanisms; the developed methods could be applied to other mechanical systems.

It is considered the mechanism presented in fig.1. The liaison between two elements it is accomplished by joints and slide bars.

We build a mechanism kinematical equivalent with the mechanism defined like:

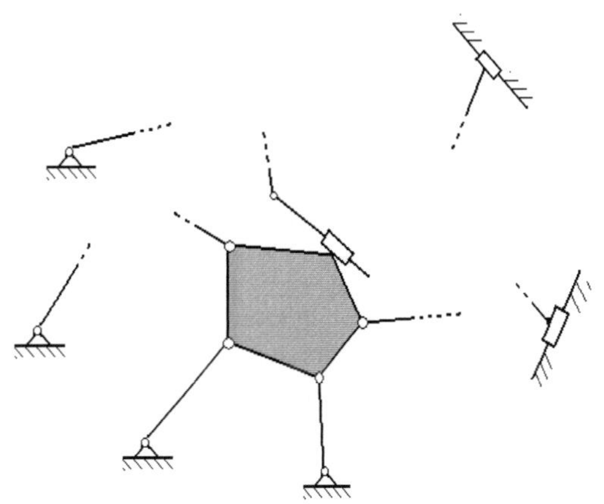

Fig. 1. A Planar Mechanism

I. We replace any mobile element type plate (fig.2a) with a rigid structure from fig. (2b), formed with triangles made by bars. It is introduced in this way the equations which describe the kinematical state of a fundamental cycles system (independent cycles) "false", but what they facilitate the equations representation.

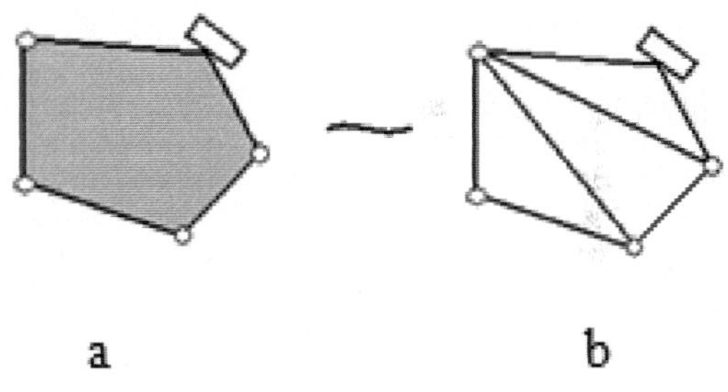

Fig. 2. Equivalence with a system made by bars

II. It bonds the points that accomplish the bonding with the fixed element through bars type elements that close the polygon (fig.3) and then it is eliminated one of these elements (it results from vectors sum with changed sign of all other elements). It is introduced like this the "false" elements which can modify the length, but not the position, so always angular speed and angular acceleration of such element are null.

By these two constructions it obtains an equivalent kinematical mechanism with the initial mechanism, but made it only with bars (fig.3).

In the moment of establishing the kinematical behavior of the mechanism elements, we consider that knowing all the bars length from the equivalent kinematical mechanism and the angles made of them with Ox axis of a xOy system.

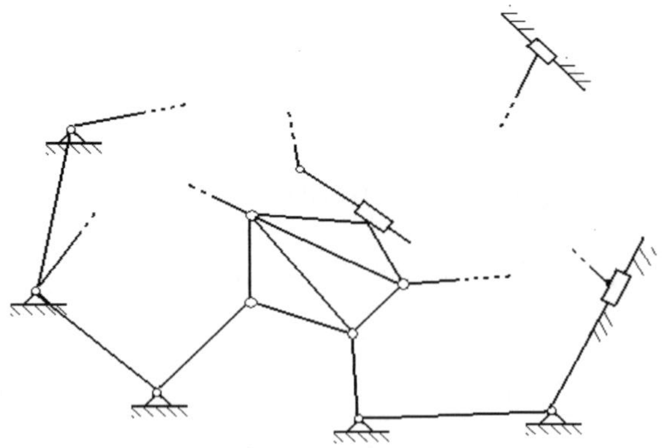

Fig. 3. Extended Mechanism

Putting to each bar element a vector with the initial point into a liaison and the end point into other liaison, we can build a graph which has, like nodes, the joints and slide of the mechanism, and like edges the mechanism bars, the sense of each line being determined by the direction of the vector attached to correspondent bar. (Fig.4).

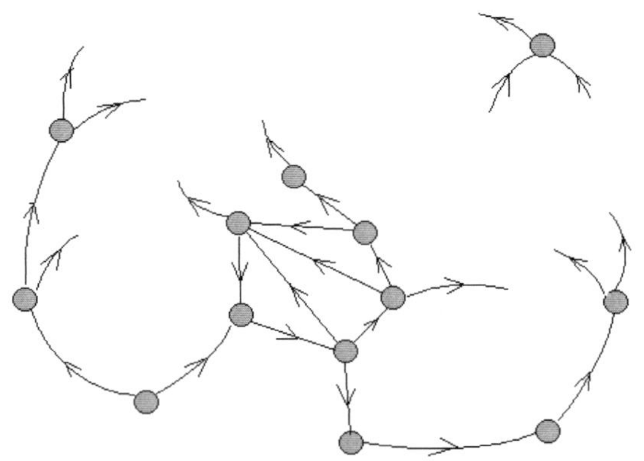

Fig. 4. Associated Graph

2.2. Independent Cycles

To understand the presentation we summarise some basic notions in graph theory[1],[2]. Being $r+1$ the number of the nodes and s the number of the edges of the obtained graph. The complete incidence matrix is the matrix $[A_a]=[a_{ij}]$ having the dimension $(r+1) \times s$ with:

- $a_{ij}=1$ if the edge j is incident to the node i and go out from node ;
- $a_{ij}=-1$ if the edge j is incident to the node i and go in to node ;
- $a_{ij}=0$ if the edge j is not incident to the node i.

The matrix $[A]$ obtained by eliminating a line in $[A_a]$ is the incidence matrix (or reduced incidence matrix) .

The selection of a tree into the graph allows a re-ordination and a partition of the $[A]$:

$$[A]=[A_t \vdots A_l] \tag{1}$$

following the edge of the tree and the branches (if we attach the branches to tree we reconstruct the initial graph).

The complete cycle's matrix is the matrix: $B_a=[b_{ij}]$ with:

- $b_{ij}=1$ if the edge j makes part from cycle i and its orientation coincide with the choice sense for passing the cycle;
- $b_{ij}=-1$ if the edge j makes part from cycle i and its orientation is in contrary sense with the choice sense for passing the cycle;
- $b_{ij}=0$ if the edge j makes not part from the cycle i.

The rank of the matrix B_a is s-r. The matrix B_f with rank s-r obtained from B_a by retaining of s-r independent lines is the cycles fundamental matrix. Between B_f and A there is the following relation:

$$B_f=[[-A_t^{-1}A_l]^T \vdots I] \tag{2}$$

3. KINEMATICAL DESCRIPTION

With some notations (Voinea & Atanasiu, 1964) the relations obtained for velocities and accelerations are:

$$\begin{aligned} \{\dot{q}\}_n &= -[P]_n^{-1}[P]_c\{\dot{q}\}_c \\ \{\ddot{q}\}_n &= -[P]_n^{-1}[P]_c\{\ddot{q}\}_c + [P]_n^{-1} \\ &([Q_2]\{q\}_2 + 2[Q_c]\{q\}_c) \end{aligned} \tag{3}$$

where index n refer to the unknown kinematical quantities and index c refer to the independent quantities.

4. CONCLUSIONS

The presented method offers the kinematical unknowns in the case of the planar mechanism, using the topological properties of the mechanical systems with the associated graph. This method, using the natural description of a mechanical system with the vector algebra leads to an automatic description of such system. It is possible to obtain the velocities and the accelerations. At the positions level the problem is more complicated due to the multiple solutions and the nonlinearity of the equations.

5. REFERENCES

Berge, C. (1958). *Théorie des graphes et ses applications. (Theory of graph)* Collection Universitaire de Mathématiques, II Dunod, Paris,

Gross, J.L.; Yellen, J. (2003). *Handbook of Graph Theory,* CRC

Shai, O.; Pennock, G.R. (2006). Extension of Graph Theory to the Duality Between Static Systems and Mechanisms. *Journal of Mechanical Design*, Volume 128, Issue 1, pp. 179-191

Thoma,U.J., (1975). *Introduction to Bond Graphs and Their Applications*, Pergamon Press, Oxford

Tsai, L.-W.; Chen, D.-Z.; Lin. T.W. (1998). Dynamic analysis of geared robotic mechanisms using graph theory. *Journal of mechanical design (J. mech. des.)*, vol. 120, No. 2, pp. 240-244

Vlase, S. (2007). *Mecanica. Cinematica (Mechanics. Kinematics.)* INFOMARKET Press, ISBN 978-973-8204-96-6, Braşov

Voinea, R.; Atanasiu, M. (1964). *New analytical methods in the theory of mechanisms*. Ed. Tehnica, Bucureşti

Annals of DAAAM for 2008 & Proceedings of the 19th International DAAAM Symposium, ISSN 1726-9679
Editor B. Katalinic, Published by DAAAM International, Vienna, Austria 2008
Make Harmony Between Technology and Nature, and Your Mind will Fly Free as a Bird

ON THE DYNAMIC ANALYSIS OF AN ELASTIC MULTI-BODIES SYSTEM

VLASE, S[orin]; PURCAREA, R[amona]; MUNTEANU, M[ihaela] V[ioleta] & SCUTARU, M[aria] L[uminita]

Abstract: *When is used a numerical method (like finite element method) is the analysis of a multi-bodies system there are two difficult and major problems when is used finite element method: one consist in the fact that the equation are more large and with more terms as in the classical procedures and the second is that the equations are only incremental valid, for a very short time interval; after this interval must generate new coefficient for the motions equations and the solutions previously obtained are the initial conditions for the new equations. In the paper are analyzed the problems involved by the use of finite element method in study of an elastic multi-bodies system.*

Key words: *Multibody, Elastic Elements, FEM*

1. INTRODUCTION

The first step is to establish the motion equations for an elastic finite element with a general three-dimensional motion together with an element of the system. The type of the shape function is determined by the type of the finite element. For this reason we will present the motion equations in three different situations: for a three-dimensional finite element with a general three-dimensional motion, for a two-dimensional finite element with a plane motion and for an one-dimensional element with a general three-dimensional motion. We will consider that the small deformations will not affect the general, rigid motion of the system. The major difficulty using FEM is the non-linearity of the motion equations. The coefficients that appears in equations are time-position dependent and, in some practical application (mechanisms with a periodical motion) they can be periodical. To solve this problem the motion must be considered "frozen" for a very short interval of time. In this case the obtained equations can be considered linear. Writing the principle of minimum energy is possible to obtain the motion equations for a finite element with a three-dimensional rigid motion. These equations have some important particularities: in the equations exists Coriolis terms (conservative) and the rigidity is modified by the some terms determined by the "rigid motion" of element. They depend on element distribution mass and on the field of velocities and accelerations. More, the force term of equations is modified by the effect of inertia forces and momentum due to the relative motion. Generally, the multi-bodies systems have a great complexity and a strong non-linearity.

2. MOTION EQUATIONS FOR ONE SINGLE ELEMENT

The equations describing the time dependent motion of an elastic mechanical system is obtained via Lagrange equations. We consider that, for the all elements of the system, we know the field of the velocities and of the accelerations. We refer the finite element to the local coordinate system Oxyz, mobile, and having a general motion with the part of system considered.

We note with $\vec{v}_o(\dot{X}_o, \dot{Y}_o, \dot{Z}_o)$ the velocity and with $\vec{a}_o(\ddot{X}_o, \ddot{Y}_o, \ddot{Z}_o)$ the acceleration of the origin of the local coordinate system. The motion of the whole system is refer to the general coordinate system O'XYZ. By [R] is denoted the rotation matrix. The velocity of point M' will be:

$$\begin{aligned} \{v_{M'}\} &= \{\dot{r}_o\} + [\dot{R}]\{r'\} + [\dot{R}]\{f\} + [R]\{\dot{f}\} = \\ &= \{\dot{r}_o\} + [\dot{R}]\{r'\} + [\dot{R}][N]\{\delta_e\} + [R][N]\{\dot{\delta}_e\} \end{aligned} \tag{1}$$

The kinetic energy of the finite element considered is:

$$E_c = \frac{1}{2}\int_V \rho v^2 dV = \frac{1}{2}\int_V \rho \{v_{M'}\}^T \{v_{M'}\} dV \tag{2}$$

where ρ is the mass density and the deformation energy is:

$$E_p = \frac{1}{2}\int_V \{\delta_e\}^T [k_e]\{\delta_e\} dV \tag{3}$$

where $[k_e]$ is the rigidity matrix for the e element. If we not with $\{p\} = \{p(x,y,z)\}$ the distributed forces vector, the external work of these is:

$$W = \int_V \{p\}^T \{f\} dV = \left(\int_V \{p\}^T [N] dV\right)\{\delta_e\} \tag{4}$$

and the nodal forces $\{q_e\}$ produce an external work:

$$W^c = \{q_e\}^T \{\delta_e\} \tag{5}$$

The Lagrangean for the considered element is obtained with the relation:

$$L = E_c - E_p + W + W^c \tag{6}$$

If we apply the Lagrange's equations:

$$\frac{d}{dt}\left\{\frac{\partial L}{\partial \dot{\delta}_e}\right\} - \left\{\frac{\partial L}{\partial \delta_e}\right\} = 0$$

with the notations explains in (Mocanu et al., 1990), (Vlase, 1985) and (Vlase, 1987), it result the motion equations for the finite element analyzed in a compact form:

$$\begin{aligned} &[m_e]\{\ddot{\delta}_e\} + 2[c_e]\{\dot{\delta}_e\} + ([k_e] + [k_e(E)] + [k_e(\Omega^2)])\{\delta_e\} = \\ &= \{q_e\} + \{\overset{*}{q}_e\} - \{q_e^i(E)\} - \{q_e^i(\Omega^2)\} - [m_{oe}^i][R]^T\{\ddot{r}_o\} \end{aligned} \tag{7}$$

where $\vec{\Omega}$ represent the angular velocity and $\vec{E}$ the angular acceleration with the components in the local coordinate system.

The notation:

$$\left\{\frac{\partial L}{\partial \delta_e}\right\}$$

we understand:

$$\left\{\frac{\partial L}{\partial X}\right\} = \begin{Bmatrix} \frac{\partial L}{\partial x_1} \\ \frac{\partial L}{\partial x_2} \\ \vdots \\ \frac{\partial L}{\partial x_n} \end{Bmatrix} \quad \text{where:} \quad \{X\} = \begin{Bmatrix} x_1 \\ x_2 \\ \vdots \\ x_n \end{Bmatrix}$$

For an one dimensional element with a three dimensional motion the motion equation are:

$$\begin{aligned} &[m_e]\{\ddot{\delta}_e\} + 2[c_e]\{\dot{\delta}_e\} + \\ &+ \left([k_e] + [k_e(E)] + [k_e(\Omega^2)] + [k_e^G]\right)\{\delta_e\} = \\ &= \{q_e\} + \{q_e^*\} - \{q_e^i(E)\} - \{q_e^i(\Omega^2)\} - \\ &- [m_{Ee}^i][I]\{E\} - [m_{oe}^i][R]^T\{\ddot{r}_o\} \end{aligned}$$

where $[k_e^G]$ take into account the second order effects.

3. ASSEMBLING PROCEDURES

The unknowns in the elasto-dynamic analysis of a mechanical system with liaisons are the nodal displacements and the liaison forces. By assembling the motion equations written for each finite element we try to eliminate the liaisons forces and the motion equations will contain only nodal displacements as unknowns. The liaisons between finite elements are realized by the nodes where the displacements can be equal or can be other type of functional relations between these. When two finite elements belong to two different elements (bodies) the liaison realized by node can imply relations more complicated between nodal displacement and their derivatives.

The motion equations for the whole structure, referred to the global coordinate system, are (Vlase, 1992), (Vlase, 1994), (Vlase, 1997):

$$\begin{aligned} &[M]\{\ddot{\Delta}\} + 2[C]\{\dot{\Delta}\} + \left([K] + [K(\varepsilon)] + [K(\omega^2)]\right)\{\Delta\} = \\ &\{Q\}^{ext} + \{Q^*\}^{ext} + \{Q\}^{leg} + \{Q\}^{inertie} \end{aligned} \tag{8}$$

4. THE INFLUENCE OF THE CORIOLIS TERMS

The matrix $[c]$ is skew-symmetric. If we want to obtain the energy balance by integration, we obtain that the variation of energy due to the term skew-symmetric is null. Consequently, the Coriolis term only transfer the energy between the independent coordinates of the system and had no role in the dissipation of the energy.

If we consider now a motion mode on the form:

$$\{q\} = \{A\}\sin(\omega t + \varphi)$$

and we introduce in the motion equations, where the forces are considered null, we obtain:

$$\begin{aligned} &-\omega^2[m]\{A\}\sin(\omega t+\varphi) + \omega[c]\{A\}\cos(\omega t+\varphi) + \\ &+([k]+[k_\varepsilon]+[k_{\omega^2}])\{A\}\sin(\omega t+\varphi) = \{0\} \end{aligned}$$

If we pre-multiply with $\{A\}^T$ we obtain:

$$\begin{aligned} &-\omega^2\{A\}^T[m]\{A\}\sin(\omega t+\varphi) + \\ &+\left(\{A\}^T[k]\{A\} + \{A\}^T[k_{\omega^2}]\{A\}\right)\sin(\omega t+\varphi) = \{0\} \end{aligned}$$

We have considered that:

$$\omega\{A\}^T[c]\{A\} = 0$$

and

$$\{A\}^T[k_\varepsilon]\{A\} = 0$$

because [c] and $[k_\varepsilon]$ are skew-symmetric. It results:

$$\omega^2 = \frac{\{A\}^T[k]\{A\} + \{A\}^T[k_{\omega^2}]\{A\}}{\{A\}^T[m]\{A\}}$$

This relation can not express, in a direct way, the influence of the matrix [c] in the eigen-values calculus, but this influence is present by the eigenvectors $\{A\}$.The term $[c]$ has an influence on the values of the eigen-values. Some of the eigen-values increase and the other decrease. This variation is presented, extended, in the paper. Between these values there exist some interesting relations.

5. CONCLUSIONS

The problems involved by use of finite element analysis of an elastic system are the followings:

- the strong geometric non-linearity of the motions equations and the additional term that appear in these;
- the motion equations are valid only for the "frozen" system, for a very short interval of time.

5. REFERRENCES

Mocanu, D.; Goia, I.; Vlase, S. and Vasu, O. (1990). Experimental Checking's in the Elasto-dynamic analysis of mechanism, by using finite elements. *International Congress On Experimental Mechanics*, Lyngby, Denmark, 1053-1058

Vlase, S. (1985). Elastodynamische Analyse der Mechanischen Systeme durch die Methode der Finiten Elemente. *Bul. Univ. Brasov*, pp.1-6, Brasov

Vlase, S. (1987). A Method of Eliminating Lagrangean Multipliers from the Equations of Motion of Interconnected Mechanical Systems. *Journal of Applied Mechanics, ASME* trans., vol.54, nr.1

Vlase, S. (1992). Finite Element Analysis of the Planar Mechanisms: Numerical Aspects. *Applied Mechanics* - 4. Elsevier, pp. 90-100

Vlase, S. (1994). Modeling of Multibody Systems with Elastic Elements. *Zwischenbericht. ZB-86*, Technische Universität, Sttutgart

Vlase, S. (1997). Elimination of Lagrangian Multipliers. Mechanics Research Communications, Vol. 14, pp. 17-22

Annals of DAAAM for 2008 & Proceedings of the 19th International DAAAM Symposium, ISSN 1726-9679
Editor B. Katalinic, Published by DAAAM International, Vienna, Austria 2008
Make Harmony Between Technology and Nature, and Your Mind will Fly Free as a Bird

EIGENVALUES AND EIGENVECTORS OF THE ELASTIC SYSTEMS WITH THREE IDENTICAL PARTS

VLASE, S[orin]; PURCAREA, R[amona]; SCUTARU, M[aria] L[uminita] & MUNTEANU, M[ihaela] V[ioleta]

Abstract: *There are many technical applications where the studied system can be considered composed by two or many identical parts or systems that present some symmetries. These symmetries can be used to simplify the analysis of su0ch systems in order to reduce the dimension of the equations that describe the motion. In the paper are identified, for the systems with three identical parts, some properties to vibrations. The use of the above mentioned properties permit to simplify the steady state analysis of the model*

Key words: *Eigenvalues, Eigenvectors, Vibrations, Multibody Systems, Symmetries*

1. INTRODUCTION

In the case of mechanical systems showing certain symmetries, the differential equations describing their evolution in time display a series of specific properties – a consequence of the very existence of these symmetries. The present work sets forth some of these properties, for a system with three identical parts, which may be used in the numerical resolution of these systems.

2. MOTION EQUATIONS

The equations describing the time-behavior, of an elastic mechanical system with constant concentrated parameters are the following (Tofan & Vlase, 1985), (Vlase, 2003), (Vlase, 2005), (Mangeron et al, 1991) :

$$[M]\{\ddot{X}\}+[C]\{\dot{X}\}+[K]\{X\}=\{F\} \quad (1)$$

where $[M],[C],[K]$ are symmetrical matrices positively defined, with constant elements, and $\{F\}$ generally depends on t, $\{X\}$ and $\{\dot{X}\}$. This type of equations results from many mechanical processes, particularly by analysis via finite element method.

3. SYSTEMS WITH THREE IDENTICAL SUBSYSTEMS

The engineering practice has recorded many cases in which a mechanical system (S) is composed by two or more identical sub-systems. We refer these identical sub-systems with (S_1). Be n the number of independent coordinates of the system (S) , n_1 the number of independent coordinates of system (S_1), $n_2 = n - pn_1$ where p is the number of identical sub-systems. The evolution of the system (S) containing two identical sub-systems (S_1) is described by the equations:

$$\begin{bmatrix} M_1 & 0 & 0 & M_{12} \\ 0 & M_1 & 0 & M_{12} \\ 0 & 0 & M_1 & M_{12} \\ M_{12}^T & M_{12}^T & M_{12}^T & M_2 \end{bmatrix}\begin{Bmatrix} \ddot{X}_1 \\ \ddot{X}_2 \\ \ddot{X}_3 \\ \ddot{X}_4 \end{Bmatrix}+ +\begin{bmatrix} C_1 & 0 & 0 & C_{12} \\ 0 & C_1 & 0 & C_{12} \\ 0 & 0 & C_1 & C_{12} \\ C_{12}^T & C_{12}^T & C_{12}^T & C_2 \end{bmatrix}\begin{Bmatrix} \dot{X}_1 \\ \dot{X}_2 \\ \dot{X}_3 \\ \dot{X}_4 \end{Bmatrix}+ +\begin{bmatrix} K_1 & 0 & 0 & K_{12} \\ 0 & K_1 & 0 & K_{12} \\ 0 & 0 & K_1 & K_{12} \\ K_{12}^T & K_{12}^T & K_{12}^T & K_2 \end{bmatrix}\begin{Bmatrix} X_1 \\ X_2 \\ X_3 \\ X_4 \end{Bmatrix}=\begin{Bmatrix} F_1 \\ F_2 \\ F_3 \\ F_4 \end{Bmatrix} \quad (2)$$

where: $[M_{12}],[C_{12}],[K_{12}] \in R^{n_1 x n_2}$ and $[0]$ is zero matrix of $n_1 x n_1$ dimension. The matrices indexed by 12 indicate the coupling of the identical sub-systems (S_1) with the rest of the system (S).

The existences of symmetries facilitate problem resolution for such system. In order to simplify the presentation of the properties of the structure we shall either consider the system as non-damped $([C]=0)$ or assume that $[C]$ meets the Caughey conditions.

4. PROPERTIES OF SYMMETRICAL BRANCHED SYSTEMS

Be $\{\Phi\}$ the solution of the eigenvector problem (Belmann, 1960):

$$[K]\{\Phi\}=p_i^2[M]\{\Phi\} \quad (3)$$

We have the following properties:

- **P1**. The eigenvalues of (S_1) are also eigenvalues for (S)
- **P2**. If $\{\Phi_1\}$ is an eigenvector of (S_1), then $[\,\alpha\{\Phi_1\}^T \;\beta\{\Phi_1\}^T \;\gamma\{\Phi_1\}^T \;\; 0]^T$
- **P3**. The other eigenvectors of (S) appear under the form of $[\,\{A\}^T \;\{A\}^T \;\{A\}^T \;\{B\}^T\,]^T$.

The first two properties can be verified via direct calculation. In order to demonstrate the third one, if we take the eigenvector under the form $[\{A\}^T \;\{A'\}^T \;\{B\}^T\,]^T$ we will necessarily have:

$$\begin{bmatrix} K_1 & 0 & 0 & K_{12} \\ 0 & K_1 & 0 & K_{12} \\ 0 & 0 & K_1 & K_{12} \\ K_{12}^T & K_{12}^T & K_{12}^T & K_2 \end{bmatrix} \begin{Bmatrix} A \\ A' \\ A'' \\ B \end{Bmatrix} = $$

$$= p_i^2 \begin{bmatrix} M_1 & 0 & 0 & M_{12} \\ 0 & M_1 & 0 & K_{12} \\ 0 & 0 & M_1 & K_{12} \\ K_{12}^T & K_{12}^T & K_{12}^T & K_2 \end{bmatrix} \begin{Bmatrix} A \\ A' \\ A'' \\ B \end{Bmatrix} \quad (4)$$

where p_i is not eigenvalue for the subsystem (S_1). Relation (4) can be written in an alternate form:

$$\begin{aligned} &\left([K_1] - p_i^2[M_1]\right)\{A\} + \left([K_{12}] - p_i^2[M_{12}]\right)\{B\} = \{0\} \\ &\left([K_1] - p_i^2[M_1]\right)\{A'\} + \left([K_{12}] - p_i^2[M_{12}]\right)\{B\} = \{0\} \\ &\left([K_1] - p_i^2[M_1]\right)\{A''\} + \left([K_{12}] - p_i^2[M_{12}]\right)\{B\} = \{0\} \end{aligned} \quad (5)$$

$$\begin{aligned} &\left([K_{12}]^T - p_i^2[M_{12}]^T\right)\{A\} + \left([K_{12}]^T - p_i^2[M_{12}]^T\right)\{A'\} + \\ &+ \left([K_{12}]^T - p_i^2[M_{12}]^T\right)\{A''\} + \left([K_2]^T - p_i^2[M_2]^T\right)\{B\} = \{0\} \end{aligned}$$

From the first three relations, because $\det\left([K_1] - p_i^2[M_1]\right) \neq 0$ it results $\{A\} = \{A'\} = \{A''\}$.

5. APPLICATION

We consider now a system with symmetries like in Fig.1. From this system we can identify three identical parts.

The computed eigenvalues for this system are: 3717, 2765, 1512, 702, 18, 284, 2884, 2000, 2884, 2000, 116, 116 [rad/s].

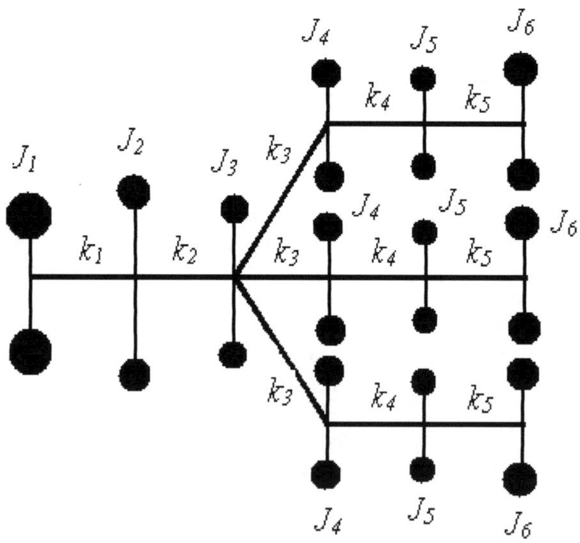

Fig. 1. System with symmetries

In Fig. 2 are presented the eigenvectors for the system. It is easily to see the proved properties.

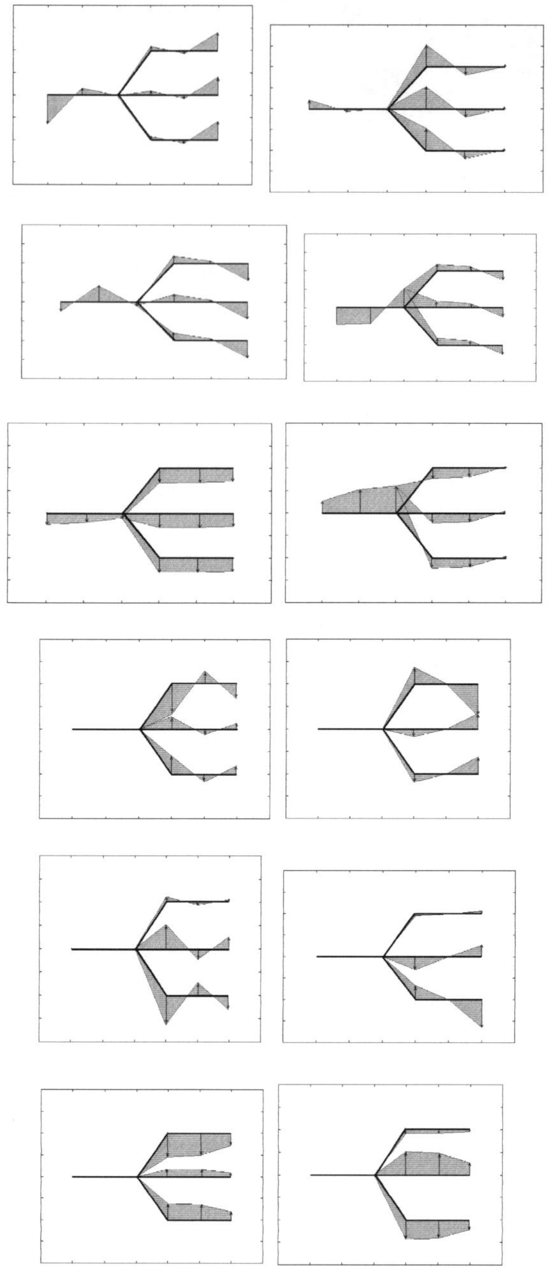

Fig.2. Eigenvectors for the system with symmetries

5. REFERENCES

Belmann, R. (1960). *Introduction to Matrix Analysis*. Mc Graw-Hill Book Company, Inc.

Mangeron, D., Goia, I., Vlase, S., (1991). Symmetrical Branched Systems Vibrations. *Memoriile secţiilor ştiinţifice ale Academiei (The memoirss of the scientific sections of the Academy)*, seria IV (4th series), Tomul XII, Nr.1, Bucuresti

Tofan, M., Vlase, S. (1985). *Vibraţiile sistemelor mecanice (The vibrations of the mechanical systems)*, Universitatea TRANSILVANIA din Braşov (Transilvania University of Brasov), Brasov

Vlase, S. (2003). *Mecanică III. Dinamica (Mechanics III. Dynamics)*. Universitatea TRANSILVANIA din Braşov (Transilvania University of Brasov), Brasov.

Vlase, S. (2005). *Computational Mechanics,* Ed. Infomarket, Brasov

Annals of DAAAM for 2008 & Proceedings of the 19th International DAAAM Symposium, ISSN 1726-9679
Editor B. Katalinic, Published by DAAAM International, Vienna, Austria 2008
Make Harmony Between Technology and Nature, and Your Mind will Fly Free as a Bird
*Young Researcher and Scientist Paper / * Supervisor, Mentor*

ROMANIA A BANKING NETWORK PUZZLE

VOICU, A[ndreea] R[aluca] & IONESCU, A[lexandru]

Abstract: *In the early of 2000's Romanian banking sector was dominated by the network branch activity and a high concentration of The Big 5's. Private capital has increased its market share, while the state banks have been on defensive positions. Additionally, the banking system was under the light of liquidity crises. While the state banks were overpopulated with employees and managers in order to balance the lack of IT solutions, the private ones started to look toward IT and small size branches as target strategy. Avoiding the human mistakes and focused on dedicated software solutions, the private banks showed a better performance for yield to employee and a better cross-sales force. The future gives no chance to the state-controlled banks, while the present is warning about this trend by clear signals.*
Key words: *Romania; banking; capital; branch; strategy.*

1. INTRODUCTION

The regular image of a champion is nothing else but a sportsman whose efforts allow him to go further or faster, or to sustain victorious campaigns.

Such principles of competition should stay at the base of economic relationships among the developed countries and less developed countries. Each competitor has weaknesses and strong characteristics, so the intellectual and leadership capacities of being aware about such aspects finally lead to or make the difference between success and non-success both on the international and domestic markets.

2. THE HISTORIC EVOLUTION OF FREE ECONOMY

There are some lucky countries those historic evolution toward free economy covers a time period of hundreds of years, but for the others that period is no longer than few tens of years or just few years.

Such an example of a non-continuous path on the free market principles could be considered Romanian economy that has kept going through economic and political reforms since the 1989's revolution.

The normal expected effects of the last years' reforms consisted in orientation toward a free market competition, political democracy, control of inflation, restructuring trends, free social strakes, and so on.

But there are still areas requiring work to be done because of the lack of performance or deregulations, all these being underwritten tasks according to European guide of economic integration.

It is no secret that the structure of Romanian economy is still far away from a targeted economic performance (e.g. of non-fulfilled tasks: a commercial surplus on foreign trade, inflation rates below 5-6 percents, total independence of justice).

3. THE BANKING SYSTEM

Under those circumstances, I will talk about a domain of Romanian economy: the banking system. The banking sector has known till year 2007 a high concentration as there were around 28 authorized banks covering demand and offer coming from a population around of 22 millions people. Under such financial conditions the investors used mainly as financing tools the banks loans.

That trend was sustained by the economic conjuncture on the Romanian market: higher and higher government expenses, adjusted commercial balance deficits, domestic falling down of interest rates but still far higher than European average. Thus it was no doubt that the banks' portfolio of assets and liabilities could do nothing but to flourish.

The essence of banks' activity is to make their network to pay back the investment having been done on each branch, agency, office or outlets (Ahmad & Hoffman, 2008).

Among the previous banks, there was one the most important: Romanian Commercial Bank (BCR). It was a state-own company being under the process of privatization toward a strategic investor as part of the I.M.F. funding conditions for Romania, but it has been delayed for several years. Privatization was likely to bring in a strategic partner or a strong consortium, of which the bank's development should benefit.

As a first step in this direction a share of 25% of the shareholder equity was sold toward I.F.C. and E.B.R.D. Its high level management consisted of one chairman and the four deputy-chairmen followed by the department managers and deputy-managers.

The bank acted both in the field of corporate business and the retail markets, offering a wide range of financial products and services of deposit and lending.

It had continued to be the largest bank in Romania with almost 31.9% market share of banking assets, 28.9% of total customers' loans, 33.9% of total customers' deposits (or available resources), but only 17% of all available Romanian customers, meaning that its portfolio had the characteristics of "golden eggs".

The territorial activity was based on a 280 offices network. The bank also had several 100% participation in abroad subsidiaries in countries like Germany, United Kingdom, France, Russia, so on.

The overall number of bank employees was around eleven thousands related to a market value estimated over USD 1 billion and an ascending rank for 2007 year, from 397 up to 387 among the most powerful world banks according to The Banker, being the 4-th among the Central and East European Banks.

Considered as the second important competitor according to the criteria of market share and aggressiveness, Raiffeisen Bank SA (former Banca Agricola SA) was a private operator (RFZ - Romania).

The bank had almost 3500 employees and a network consisting in 210 offices. As it can be noticed, this bank had only one third of BCR employees' number and a comparable network size.

4. HOW WAS THE BUSINESS CONDUCTED UNDER SUCH PRIVATE OWNERSHIP CONDITIONS

For example into a BCR branch it could be count up to 30 people, instead of 15-20 people for RFZ one.

Was it wrong? Yes vs. no - because there were advantages and disadvantages related to the number of services a branch could offer, the time of decision, the professionalism of human contact man-to-man, the IT system performance, customer preferences.

As an example it could be mentioned the card issues: BCR had almost 1,338,461 issued units while RFZ came with only 431,356 units, where the main influence factor had been the sale forces that according to the number criteria were in favor of BCR even though the card issuing time was favorable to RFZ (several days instead of a week).

What I wanted to pinpoint there it was the use of network potential as strategic tool on retail market: while the BCR preferred mainly the structure of big-size branches (167 branches) with manager + deputy manager + financial director + law adviser + heads of division + even more executive and control employees +/- security, its competitor RFZ had lower its wages cost by setting up mainly the agencies (132 agencies) having a less complex structure than a branch, but a strong IT communication channel between agencies and head office.

The decisional decentralization was in favor of RFZ. Another argues into that direction consisted in the presence of RFZ almost in every supermarket chains being built around the country capital – Bucharest, while BCR had only one or two agencies in such business areas.

But that strong points in favor of RFZ were balanced by the fees, commissions, and selling forces that were in favor of BCR. So that the clients used to chose their opportunity cost based on three criteria: ways of being contacted, timing and operating cost.

Under recession business periods their choice was definitely in favor of the lower operational banking fees (favorable to BCR), otherwise (in case of economic boom) the choice was oriented toward the operational timing (favorable to RFZ-Romania).

But the most important aspect it proved to remain the traditional human contact of doing business, and BCR by its superior number of employees remained the leader in doing that.

On Romanian market, there should be noticed the third important operator, by using the criteria of network structure, and that was Romanian Development Bank (BRD-GSG). Its branch network was smaller than that of previous two banks, but with a countrywide geographical spread, its strength was related to the shareholder reputation – Groupe Societté Générale (France).

In fact, on a Romanian financial market dominated by BCR with a market share higher than 30%, the competition among the RFZ-Romania and BRD Groupe Societté Générale was for being ranked "the second in rank" as operative importance, where BRD showed both a decisions' decentralization and lower fees.

Another banking operator being very active on Romanian retail market due to the network structure was Banc Post. Its top position in issuing debit and credit cards (1,059,150 units – second rank, behind BCR but in front of RFZ or BRD) was due to a focus on distribution by small outlets and a reliable IT channel in area of timing on card issue, although its capitalization was less important than BCR, RFZ or BRD. Thus that bank has succeeded to overwhelm the competition on the card market share.

The retail activity was intended to be the base of its activity (e.g. on December 31-th, this medium size bank had an average number of issued cards of 6,580 units per outlet/office compared to 4,712 units per office for BCR or 5,410 units per office for BRD:

	Aggregate network	**Issued Card**	**Card/network office**	**ATM**
Romanian Commercial Bank (BCR)	284	1,338,461	4,712	579
Raiffeisen Bank Romania (RFZ)	204	431,356	2,114	358
Romanian Development Bank (BRD)	177	957,251	5,410	368
Ion Tiriac Commercial Bank (BCIT)	50	137,161	2,743	125
Banc Post	161	1,059,150	6,580	451

Table 1. Card market share

5. CONCLUSION

All previous banks were offering full services for corporate and retail market, each of them with different structure on network management, but all having as common target the profit and shareholders wealth (Arkin, 2007).

On Romanian market although the banking sector was under an ongoing process of European integration and reform, the dominance of network size remained undeniably the factor that offered the premise of success in the case of a retail or corporate approach (Bussler, 2007). Until present days there has been a continuous growth of bank network, the Romanian banking sector not knowing the shutting down of working branches on large scale.

Sales forces are focused on human being participations, so the contacts with customers are done face-to-face. The implementation of automatic teller machine has still free market niches, e.g. invoices of utilities, account administrations.

Thus, in this particular perspective, the future impact of E.U. integration is interesting to be found out.

Especially when the technology and dedicated software platforms come with a better efficiency than the traditional employee (Andrews et al., 2007).

The reality shows that the sale forces have and can be allocated successfully from the back office activities.

6. REFERENCES

Activities of the EU (2008), SCADPlus: *A modern policy for SMEs*, http://europa.eu/scadplus *Accessed:* 2008-03-05.

Ahmad, N. & Hoffman, A. N. (2008), *A Framework for Addressing and Measuring Entrepreneurship, OECD Statistics Working Paper*. Washington D.C.: OECD, p. 10, *Available from:* http://www.oecd.org/std/research *Accessed:* 2008-01-24.

Andrews, T., F. Curbera, H. Dholakia, Y. Goland, J. Klein, F. Leymann, K. Liu, D. Roller, D. Smith, S. Thatte, I. Trickovic, S. Weerawarana (2007), *Business Process Execution Language for Web Services*, Version 1.1. OASIS.

Arkin, A.: *Business Process Modeling Language*, San Mateo, CA (2007), BPMI.org, Proposed Final Draft

Bussler, C. (2007), P2P in B2BI, *A paper delivered at the Proc. of the 35th Hawaii International Conf.* on System Sciences (HICSS 2002), Waikoloa (HI)

Annals of DAAAM for 2008 & Proceedings of the 19th International DAAAM Symposium, ISSN 1726-9679
Editor B. Katalinic, Published by DAAAM International, Vienna, Austria 2008
Make Harmony Between Technology and Nature, and Your Mind will Fly Free as a Bird

BUSINESS INTELLIGENCE - MAIN INSTRUMENT OF MODERN MANAGEMENT

VOICU, V[asilica]

Abstract: *The use of business intelligence technologies covers the need to manage more and more data, while there is allowed an adequate distribution of managerial competences into the knowledge-based economy. Business Intelligence is a concept widely used by successful entrepreneurs, being also a main instrument for the modern decision makers using IT support. The price of success is no more only hard work, but a mixture of odds where main player controls a very good IT infrastructure and applications. It is no use to reinvent the boil water, when it is easier to apply the existing "mechanical" forces of already existing steam, for the new criteria of efficiency. Romania is a case where the lack of strategy upon the business intelligence applications has led to a non-continuous implementation for this potential. So, it is the moment to promote a change in the knowledge-value concept in the new EU members even since the faculty stage.*
Key words: *Intelligence, Decision, OLAP, Warehouse, AI.*

1. INTRODUCTION

The informational modern society the people are building is defined as a knowledge society and an organizational one.

The knowledge-based organizations are the clever collective factors having a role of modeling the economic environment into the direction of informational competition. So these actors belong to modern times in their quality of managers and professionals. They sustain the scientific research in order to obtain the final result of joining both the ideas and the operational structure into a social and durable entity (Filip, 2005).

Using as brick and mortar the knowledge is the success formula against the limits of low-developed society as it was till few years ago. The function for the informational companies requires an enhanced activity in direction of innovation, learning and partnership interactions. So there is done the way for the new intelligent technologies.

"Business Intelligence" is a very and widely used business concept, and one of the most appealed modern management instruments. There are trends in management like Balance Core Card, Management Objectives, Key Performance Indicators, Empowerment, Organizational Outsourcing – that become a more and more important "power horse" for efficiency. But all such concepts are nothing in the situation of lack of IT powerful instruments for feed-back. So, it is a two ways conditional process: concept-IT and IT-concept.

Business Intelligence represents a set of methods and ideas having as target to enhancing decision making events by the use of IT technology. There are three level of integration into organization of BI – among them the first means the information consolidation into one single space, and then after that it is issued the diffusion of reports toward each manager according to their needs.

Business Intelligence is the process looking to forecast and influence a future performance for the company, starting with the momentum of understanding in depth about the present situation. Without BI it is consider that the companies are not capable to build and maintain a competitive advantage because of low capacity to understand the needs of customers for personalized services.

Another trigger-element sustaining the investment in BI is the capacity of the package of software applications to generate an aggregate picture upon the company performance and in the same time to offer a better transparency toward shareholders (Filip, 2005). The platform of BI allows the transformation of data into action decisions.

The dynamics of economy and the contemporaneous IT jump business make the BI application the indispensable instruments for every business. There are worldwide opinions sustaining a continuous development of IT applications in order to cover the information volume being into an increasing process estimated to 20 times more in 2012 than in 2007.

The threat in this area comes not only from the upward jump in the size of data bases, but also from the increasing complexity of such applications offering the possibility to the users to compete in the market.

For giving the solution to different situation, BI models are integrated by mathematical functions that are able to identify trends in companies' performance ratio. In the same the number of users without a background in IT has imposed a trend toward simplifying such IT solutions.

2. THE INTEGRATION OF BI APPLICATIONS INTO IT EXISTING SYSTEM

Having an activity history, the companies collect more and more data bases regarding the business flows and commercial transactions. The former classic IT data base administration solutions did not offer support for an in-deep analysis and structural evolution high-degree performance reports. Such old applications were especially focused on administration of receivables, inventories – having only the first level difficulty functions. The new applications come with enhance operative functions looking to cover all relations involved in the activity of a subject company, being in fact a package of services of Enterprise Relationship Planning topic. It can be said that the BI solutions are built in order to enhance data analysis. Under such premises the information are set on level of structures looking to offering answers for each kind of identified situation. Data deposits, multi-dimensions structures, and data mining are only few examples of technologies capable to offer a focused analysis.

Business intelligent applications can be considered as an intermediary operator between the data base and manager, offering both the possibility of analysis and also the chance of profitable decision issue.

3. INTEGRATION OF BUSINESS INTELLIGENCE WITHIN DATA BASES

The producers of data base administration software include high performance BI functions that efficiently communicate with traditional IT solutions. By this way, beside the support for the classical relational data bases and analysis there are offered advanced and performance techniques of information collect and administration.

BI applications are designated for collection, maintaining and use of available data in order to sustain the decisions makers, having in structure the following technologies:

-SSD – Decision Support Systems
-OLAP – Online Analytical Processing
-Data Warehouse
-Data Mining

These technologies allow a multi-dimensions analysis and a use of mathematical models for forecasting the economy trend based on the information existing in data bases. Additionally, there are added instruments (like Extract Transform Load) that allow a data consolidation inside the already created data base (Whalen et al., 2007).

So, all successful software applications include also functions of business intelligence. The share market of this kind of IT suppliers is more and more dynamic, existing more and more different IT programs.

The main companies very actively inside this domain, using such performance-oriented instruments, are:

-IBM (DB2 Universal Database): IBM DB2 Cube Views is a good quality accelerator of the Online Analytical Processing Family (OLAP), applied for multi-dimensions analyses; IBM DB2 Intelligent Miner is a good application using for its analysis the instruments of data mining.

-Microsoft (SQL Server) is equipped with complete intelligent business integrated services using a SQL data base. As example, Microsoft Reporting Services is a report solution that allows obtaining reports from BI application using WEB as link, and Microsoft Analysis Services is another integrated application being able to do OLAP analysis and data mining (Melomed et al., 2007).

- Oracle (Data Base Oracle) comes also with a set of BI solutions focused on different categories of users. The concept named CPM (Corporate Performance Management) includes methodologies, statistics, processes, and systems in order to fulfill the tasks regarding the performance measurement and management.

Additional to this international operators, on the Romanian market there are active several players offering pretty good solutions:

-Cognos products offer a complete set of BI functions, among them it can be noticed: OLAP, data base reviewing, reports, so on.

-SPSS and Business Object are two complete business applications allowing operations in area of statistics and data mining, reports and data analysis.

- ProClarity comes with an easier way for management analysis, and he WEB based application leads to an enhanced access of information.

- CorVu package is a business intelligence solution for management looking to cover needs in area of better performance for financial and operational divisions. This package is integrated with ERP solution produced by QAD.

The interest for business intelligence products has increased due to their possibility to come with flexibility and quick access to reports. Compared to traditional data bases, the new solutions prove to be better in generating analysis in real time or ad-hoc report leading to decreased expenses and time saving (Whalen et al., 2007). But the degree of implementing in economy for the new BI integrated software is still low because the companies prefer a development in only few sectors of business, considered as critic importance sectors, for the rest of them looking to cover by traditional solutions.

Romanian market knows a trend toward maturity, and so it is a general perception about the necessity to be ready for the challenges coming from united Europe. It is a sign about the increasing number of companies looking to create efficiency in their activity as a long term strategy by implementing business intelligence solutions.

4. ROMANIAN TIME: FEATURES AND PRIORITIES

For Romania the way for development will mean a society built on information as a condition of a long term success. In this context the idea according to which Romania knows a too high gap versus developed countries, and so the issue of information is only for the future, would be a dangerous one.

Being integrated into a worldwide economy, Romanian one was not able to create an exception related to knowledge orientation, but because of a lack of strategic policy there was a non-continuous implementation in all economic sectors for BI solutions that have lead to potential loses at all levels (execution, administration, intermediation, conception).

The big companies have an activity addicted to their hierarchic structure even in the case of a high degree of IT implementation. In return, the small companies but with foreign trade activity know already the managerial practices based on knowledge.

So there remains a national imperative way the creation of companies built on knowledge as a main growing source for sustaining a higher level of domestic competition and an international identity.

An interesting priority would and should be the implementation of modules or disciplines revealing the knowledge management in academic teaching programs, or the use of national-wide projects relating to Internet science of communication.

5. REFERENCES

Filip , F. G.(2005). *Computer-linked decision* , Technical Publisher, Romania

Melomed, E.; I. Gorbach; Al. Berger; PY Bateman (2007). *Microsoft SQL server 2005, Analysis Services*, Sams Publishing

Whalen, E.; M. Garcia; B. Patel; S. Misner; V. Isakov (2007). *Microsoft SQL Server 2005, Administrator's Companion*, Microsoft Press, 2007.

*** *Solid Quality Learning: Microsoft SQL Server 2005*, Database Essentials, Microsoft Press, 2007.

http://www.microsoft.com/sql/solutions/bi/default.mspx *Accessed:* 2008-01-06

http://www.siveco.ro *Accessed:* 2008-01-08

http://www.acigrup.com *Accessed:* 2008-01-07

http://www.microsoft.com/BusinessSolutions *Accessed:* 2008-01-06

Annals of DAAAM for 2008 & Proceedings of the 19th International DAAAM Symposium, ISSN 1726-9679
Editor B. Katalinic, Published by DAAAM International, Vienna, Austria 2008
Make Harmony Between Technology and Nature, and Your Mind will Fly Free as a Bird

NEW PROTOTYPE EQUIPMENT WITH UV RADIATION FOR WATER PURIFICATION PROCESSES

VOLF, I[rina]; BADEA, C[ristinel]; CIOBANU, B[ogdan] & CARAUSU, C[onstantin]

Abstract: *The aim of this paper is to establish the main steps of technologies to obtain an equipment which corresponds to the current requirements, compatible with the European and International norms, which wish to achieve the elimination of biological agents in water purification process. The research regarding the possibilities of water disinfection with ultraviolet radiations aims at integrating a big number of functions such as: LCD screen layout of the programming module, UV radiation level, number of functioning hours for every UV lamp; the automatic control of the UV radiation flux between 100%...150%; audio and visual alarm in case of decrease of the level of UV radiation between 100%...110%; external audio and visual alarm, GSM signal transmission in case of decrease of the level of UV radiation under 100%; the possibility of blocking the exit circuit by the automatic command of an electric valve. Numerical simulations, using FEM, are made in order to optimise hydraulic circuit.*

Key words: *UV radiation, prototype, water disinfection*

1. INTRODUCTION

Water represents one of the critical resources concerning the sustainable development, both by its position of a base resource and also for its contribution in generally development for life support. Water is essential for industrial, agricultural, services development and for energy production, for ecosystems conservation or to ensure a good population health. In September 2005, the UN Summit has launched an appeal for helping in course of development countries to prepare the Integrated Water Resources Management and also to establish efficient water resources utilization plans as well defined parts of national development strategies with the aim to fulfill *Millennium Development Goals.* Most of these objectives (extreme poverty and famine eradication, durability of environment assurance, global partnership development for sustainable development) are in strong connection with water supply recycle and also with control and risks minimization due to pollutants sloop in aquatic environment, */UN Millennium Project Task Force on Water and Sanitation, Final Report, 2005/.* Development strategies discussed at UN Summit has raised an important challenge: how to prepare the Millennium Development Objectives at national level in order to take into consideration water resources in such strategies. A general definition of Integrated Water Resources Management was given by Global Water Partnership through the Technical Committee: "Integrated Water Resources Management is a process which promote the coordination the development and the management of water supplies, durable utilization of land and other resources with the aim to maximize the economic results and social welfare in an equitable manner without compromising the durability of vital ecosystems." /GWP, 2004/. This definition implies aspects which aim: the soil, the water resources (underground and surface water), riverbank, coastal and maritime areas, upstream and downstream users, and also aspects which target human capacity of using and benefit in a durable way of physical resources.

Over the last decades the use of the ultraviolet radiation for the disinfection (bacteria, viruses and other water pathogens) has increased remarkably. World widely, the need of cleaning water is continually growing; the responsibility of the water companies is the supply of water of the highest quality possibly (according to the Guideline 98/83/CE).

International general tendencies are the elimination of chemical substances and boiling process in water purification and use of installations with ultraviolet radiations (Carona, 2007).

The main preoccupations and achievements on the world plane are directed to installations using as a purification procedure the exposure of water to ultraviolet radiations. This category of installation is already produced and sold by companies such as: Berson Milieutechnick BV - Holland, Wedeco AG, Atlantic Corporation, Delta UV Germany, Willand UV Systems Ltd Great Britain and so on.

There are not known systems with GSM signal that can transmit the main data regarding the functioning of the UV installation, nor procedures of automatic tracing of the weakest lamp in order to replace it.

The degree of novelty of the prototype proposed consists in elimination of the chemical substances used in water disinfection, quick and easy installing, possibility of a quick service, efficiency in the destruction of micro–organisms; GSM distance monitoring, good hydraulic behaviour.

2. TECHNOLOGY USE

The basic diagram of the installation is presented in figure 1.

The main body of the installation consists of a 120 mm diameter stainless steel pipe, 900 mm long. Inside it, there are mounted three quartz tubes at 120^0, each containing an UV lamp with a power of 55 W. Water is flooding in inside the pipe and exterior to the quartz tubes being submitted to UV discharges. Three UV sensors are mounted outside the stainless steel pipe for the control and diagnose of UV discharges.

In case of bigger water flows the UV reactors can be connected in batteries of three or more reactors placed in parallel (fig. 2).

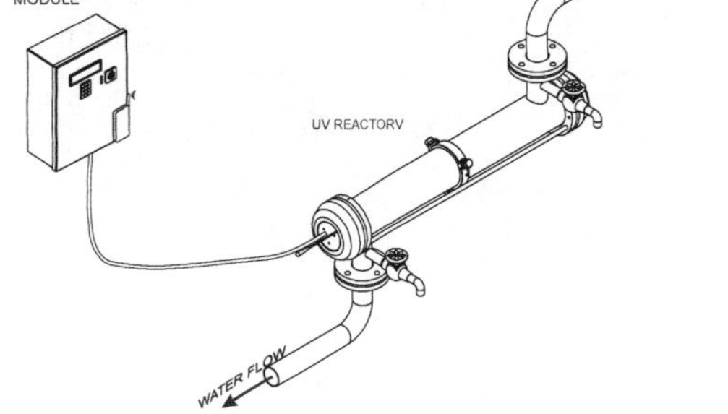

Fig.1. Basic diagram of the installation

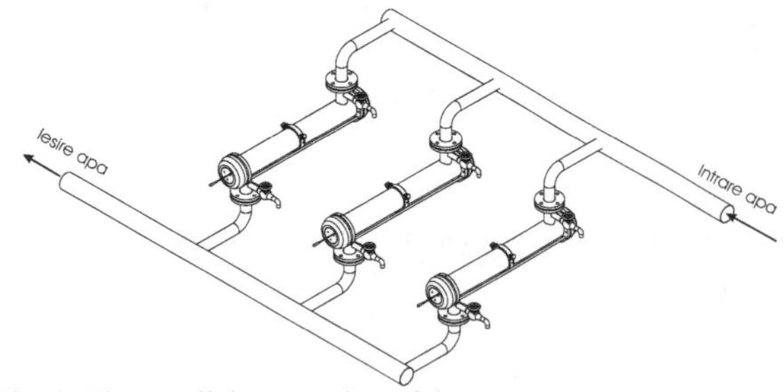

Fig. 2. The parallel connection of the reactors

Further on, there is presented an installation of the UV device in a system with collection and treatment of the drinking water (fig. 3).

The general view of equipment is presented on the figure 4.

Using this equipment the following main problems will be solved:

- comparative study of micro-organisms elimination through classic techniques of disinfection (in chlorine presence or his derivatives) and by UV treatment (Volf & Teodosiu, 2008);
- study concerning the factors that affect the disinfection process with UV radiation;
- testing the performances of proposed technological solution in water contaminated with different micro-organisms (levures and bacterium)
- demonstrate the possibilities and utilities of application of proposed solution by testing at an industrial level the proposed model.

For the proposed prototype we realize a flow simulation through hydraulic circuit using finite element method (Lewis, 2004) and (Dhaubhadel, 1994). The aim of this simulation was, at one hand, to estimate the total pressure loss when fluid pass through prototype and, at the other hand, to estimate the velocity of flow in UV radiation area. It's important to estimate this velocity because starting from this value we can calculate the pass-through time of a micro-volume of fluid through the UV radiation area, time that is directly connected to the efficiency of micro-organisms inactivation process.

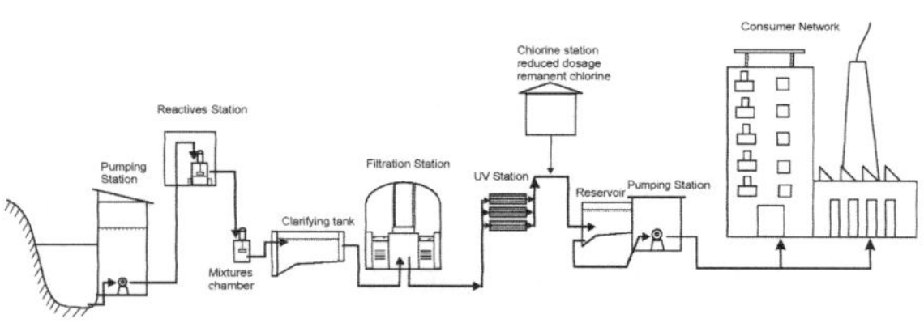

Fig. 3. Technologic Flux of Surface Water Treatment with UV Radiation

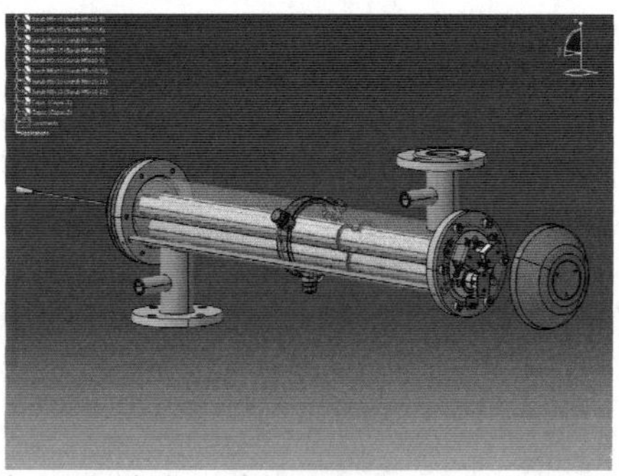

Fig. 4. The general view of the prototype

In order to realize the simulation, the model of hydraulic circuit and the finite element (wedge type) network, was created.

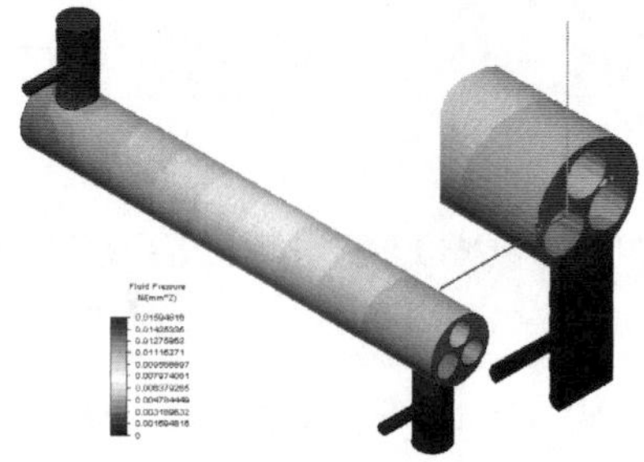

Fig. 5 – The pressure loss (detail at outlet)

The enforced boundary layer conditions was the average flow speed (v=1.4 m/s) on the inlet, corresponding to a flow rate Q=4 l/s, and zero flow speed (v=0 m/s) in vicinity of pipe boundary.

The FEA reveals a total pressure loss $\Delta p=0.16\cdot 10^5$ Pa. That means a total head loss $\Delta h=1.63$ mH_2O (fig. 5). Considering the whole circuit like a local hydraulic resistance we obtain a global hydraulic loss coefficient $\zeta=16.3$ for the proposed prototype.

The fluid flow speed in UV radiation area is, according to the analysis, about 0.9÷1.1 m/s.

The finite element analysis applied to the hydraulic circuit reveals two important conclusions:

- global hydraulic loss coefficient is similar like the one of a hydraulic valve with the same diameter (that means a very good value);
- pass-through time of a micro-volume of fluid through the UV radiation area is at least 1 sec.

3. CONCLUSIONS

The *technical impact* can be evaluated through the introduction of a new non pollutant technology which does not change the taste and smell of water, a technology which does not need chemical additives and uses a series of small and easy maintainable equipments with lower costs.

The *economic and social impact* can be evaluated as follows:

– risk of illness by consuming drinking water will be diminished
– necessary costs for assembly, maintenance and exploitation for this type of equipments will be reduced
– water producer productivity will be increased
– acquired experience will improve the ability of the producer of the equipment to develop a large range of such equipments.

4. REFERENCES

Carona, E., et al (2007). *Impact of microparticles on UV disinfection of indigenous aerobic spores,* Water Research 41, 4546 – 4556.

Dhaubhadel M.N., Habashi W.G., Engelman M.S., (1994). *Advances in Finite Element Analysis in Fluid Dynamics*, International Mechanical Engineering Congress and Exposition, Chicago, Illinois.

Lewis R.W., Nithiarasu P., Seetharam K., (2004). *Fundamentals of the Finite Element Method for Heat and Fluid Flow*, Wiley Publisher.

Volf, I., Teodosiu, C., (2008). *Inactivation of some bacteria and fungi using a UV radiation system for treatment of drinking water,* le Cinquieme Colloque Franco-Roumain de Chimie Appliquée, Bacau, 2008.

Annals of DAAAM for 2008 & Proceedings of the 19th International DAAAM Symposium, ISSN 1726-9679
Editor B. Katalinic, Published by DAAAM International, Vienna, Austria 2008
Make Harmony Between Technology and Nature, and Your Mind will Fly Free as a Bird
*Young Researcher and Scientist Paper / * Supervisor, Mentor*

DEVELOPMENT OF A PACKAGE OF THE APPLIED PROGRAMS IN THE FIELD OF POWER SAVINGS

VOLOGDIN, S[ergey]

Abstract: *The collective of scientists of the Izhevsk state technical university develops a package of applied programs for the decision of various problems by calculation and optimisation of parameters of systems of a heat supply, water supply and power savings. Program complex IASHPS 3.1 is intended for carrying out of large-scale mathematical modelling and optimisation of heating and hydraulic controls of heating networks with use of GIS-TECHNOLOGIES. Within 1997-2008 the developed package of applied programs was applied to performance more than 30 projects and programs.*

Key words: *Program complex, modelling, heat supply, power savings*

1. INTRODUCTION

The collective of scientists of the Izhevsk state technical university develops a package of applied programs for the decision of various problems by calculation and optimization of parameters of systems of a heat supply, water supply and power savings.

The package consists of following programs:

- GIS “It is information analytical system of a heat supply and power savings” 3.1 (IASHPS 3.1);
- the editor of schemes of network of a heat supply;
- the editor of schemes of system of a heat supply of buildings;
- calculation of heating losses of a building, designing of thermal protection of buildings;
- calculation of a standard cold and hot water consumption of buildings;
- calculation of tariffs for heating energy and water supply;
- heating and hydraulic calculation of system of a heat supply of buildings;
- hydraulic and heating calculation of network of a heat supply;
- optimization of parameters of network of a heat supply;
- calculation of temperature of premises in buildings on the basis of system of the equations of heating balance.

Computer programs were written with using the advanced mathematical approaches of calculation and optimization of parameters of various elements of network of a heat supply (Vologdin et al., 2001; Vologdin, 2007) taking into account operating standard documents were applied.

2. THE PROGRAM COMPLEX IASHPS 3.1

The Program complex IASHPS 3.1 is intended for carrying out of large-scale mathematical modeling and optimization of thermal and hydraulic controls of system of a heat supply with use of GIS-TECHNOLOGIES.

The information analytical system of a heat supply and power savings represents set of software for storage of the necessary information on elements of structure of systems of a heat supply, powerful toolkit of input and updating of the graphic and parametrical data, and also operative conducting hydraulic and thermal calculations of modes of operation and adjustment of heating networks. Thus, functioning IASHPS is based on the uniform information field which borders are, on the one hand, manufacturers, on the other hand, heating energy consumers.

The developed program complex has following possibilities:

- hydraulic calculation of network of a heat supply;
- processing of the graphic and digital information on a condition of objects of supervision and the account;
- a binding to elements of network of a heat supply corresponding attributive information;
- operative updating of the information;
- search of demanded object on an electronic card;
- possibility of escalating of a database and introduction in it of new elements;
- export of the initial information and the settlement data in Microsoft Excel;
- schedule construction hydraulic pressure head;
- working out of the block of generation of forms of account by results of calculations.

Experience on creation IASHPS shows that for effective operation of a program complex it is necessary for the developer and the user to carry out following stages of works:

1. Information Gathering on operating modes heat supply source, central heating station.
2. Gathering and specification of initial engineering specifications under characteristics of heating mains, schemes of their lining, designs of buildings and schemes of their heat supply.
3. Creation of a database of the digital information of various sites of network of a heat supply, characteristics of buildings and its premises.
4. Creation of a database of the standard, help information.
5. Working out of the functional block of mathematical models (Samarski, 1989; Gromov & Shubin, 1988; Merenkov, 1985; Uong, 1979):

- calculation of hydraulic and thermal balance of buildings, of network of a heat supply;
- parametric optimization network of a heat supply (diameters, ways of a lining and isolation of pipelines, an arrangement of pump stations).

8. User's guide working out.

The system consists of seven base subsystems.

The first subsystem – the scheme of transit pipeline.

The second subsystem – the scheme of municipal heat conductors with following characteristics: places of inserts and branching; geometrical parameters of linear sites of pipelines: length, difference of heights, diameter, a thickness of walls, steel mark, a parameter of multilayered isolation; heat-carrier parameters on an input and a building exit.

The third subsystem – the scheme of network of a heat supply in buildings with following characteristics: heat-carrier parameters; quantity of struts, quantity and type of heating devices on each strut; geometrical and thermophysical parameters of linear sites of pipelines: length, difference of heights, diameters, a thickness of walls, steel mark, isolation parameters.

The fourth subsystem – characteristics of premises: the name; a floor; the area building envelope; orientation on parts of the world; type of protecting surfaces; characteristics of protecting surfaces.

The fifth subsystem – a database of consumers of thermal energy: consumption on separate premises; consumption on separate buildings; consumption on all complex of buildings.

The sixth subsystem – the database of the help information consisting of following sections: user's inputs; armature, a thermal protection; building envelope; heating devices, ground.

The seventh subsystem – the functional block of mathematical models.

3. THE PROGRAM "CALCULATION OF TARIFFS FOR HEATING ENERGY AND WATER SUPPLY"

The Program "Calculation of tariffs for heating energy and water supply" is intended for definition of necessary expenses for manufacture of thermal energy for needs of systems of heating and hot water supply from heat supply source of various capacity working on any kinds of fuel. The given computer complex corresponding to requirements [4], is recommended to application for calculation and the analysis of structure of storage of the information necessary tariffs and has following possibilities:

- storage of the information necessary for calculation of tariffs;
- calculation of thermal loadings and thermal losses of the consumer;
- requirement calculation heat supply source in fuel, the electric power and water;
- calculation of expenses on water demineralization, auxiliary materials of a boiler-house;
- calculation of number of the personnel, fund of a payment of a boiler-house;
- calculation of expenses on amortization of the equipment of a boiler and of network of a heat supply, expenses for their maintenance and operation;
- calculation shop expenses;
- calculation of obligatory deductions, payments, taxes;
- calculation of necessary profit and the manufacture cost price;
- the generator of reports of the various information.

Heating energy development on boiler-house heating develops of thermal energy for own needs and thermal energy released in a network. Thermal energy which has been released in network of a heat supply, consists of useful thermal energy on heating, thermal energy of the boiler-house spent for needs and thermal energy losses in networks at transportation. The volume of useful thermal energy is defined on the basis of the contracts concluded with consumers.

4. THE PROGRAM "CALCULATION OF STANDARD WATER CONSUMPTION OF BUILDINGS"

The Program "Calculation of standard cold and hot water consumption of buildings" solves a problem of automation of process of calculation of water consumption of buildings of different function. In the program following calculations are realized:

- water consumption calculation by individual consumers, buildings as a whole, and also groups of buildings in average days and whole for a year;
- calculation of requirements for water in hour and days of the greatest water consumption;
- calculation of the maximum expense of water;
- calculation in requirement for water for fire-prevention needs;
- calculation of the average expense of water by one device for all building;
- calculation of cost of water consumption.

The technique of calculations corresponds to operating rules. Besides, the program contains a database on sanitary – to the technical devices corresponding to norms of consumption and gives possibility of generation of reports on water consumption of various groups of consumers for the various periods of time.

5. RESULTS

Within 1997-2008 the developed package of applied programs was applied to performance more than 30 projects and programs.

IASHPS 3.1 was used at carrying out of hydraulic and thermal calculation of network of a heat supply of Izhevsk.

Calculation of optimization of parameters of network of a heat supply of Izhevsk State Technical University has shown that almost on all sites pipelines with not optimum diameter are laid. If to lay a thermal network with optimum diameters in the underground way of a lining of pipelines its resulted expenses will be less expenses of an existing thermal network for 28 %.

Reduction of expenses is reached at the expense of reduction almost twice expenses for thermal losses, against insignificant increase of expenses for a construction and on an electricity. As a result of application of modern technologies of a thermal protection has shown that optimum diameters remained the same, as well as at use of "standard" isolation, and an optimum way of a lining of pipelines mainly elevated that has allowed reducing expenses for a heating main construction. In this case the economy already makes 54 %.

For definition of an actual situation on system of a heat supply of buildings of settlement complex power investigation in which result has been carried optimum parameters of a thermal network and heating systems of buildings have been defined, thermal loadings of subscribers are specified, and also recommendations about improvement of quality of system of a heat supply of settlement are developed.

Operation of a program complex and realization in practice of the offered actions only at the expense of elimination of a disbalance of system of a heat supply of micro district allows to save to 20 % of thermal energy.

6. CONCLUSION

The developed program complex can be used at carrying out of energy audit, design works. Introduction of a program complex allows raising energy efficiency of the enterprises of budgetary sphere, the industrial enterprises, systems of a heat supply of buildings and cities.

The plan of the further researches: modernization of a program complex, working out of new programs.

7. REFERENCES

Gromov, N. & Shubin, E. (1988). *Water thermal networks: the handbook on designing*, Moscow: Energoatomizdat

Merenkov, A. (1985). *The theory of hydraulic clains*, Moscow: Science

Samarski, A. (1989). *Numerical method*, Moscow: Science

Uong, H. (1979). *Basic formulas and data on heat exchange for engineers: the director*, Moscow: Atomizdat

Vologdin, S; Rusjak, I. & Gorokhov, M. (2001). Application of an information technology for the decision of problems of a heat supply and power savings. *Power efficiency: experience, problems, decisions*, №4, 2001, pp. 50–57

Vologdin, S. (2007). Working out of power saving up actions on objects of budgetary sphere and housing and communal services. *Vestnik ISTU*, №2, 2007, pp. 115–119, ISSN 1813-7903

Annals of DAAAM for 2008 & Proceedings of the 19th International DAAAM Symposium, ISSN 1726-9679
Editor B. Katalinic, Published by DAAAM International, Vienna, Austria 2008
Make Harmony Between Technology and Nature, and Your Mind will Fly Free as a Bird

PRELIMINARY TECHNICAL DESIGN METHODOLOGY OF MANUFACTURING SYSTEMS

VOSNIAKOS, G[eorge] C[hristopher]; GOGOUVITIS, X[enophon]; FIDLER, B[ranislav] & VALCUHA, S[tefan]

Abstract: *Preliminary design of a manufacturing system is focused on in this paper, concerning its technical rather than managerial aspects. The concept of the design process using the notion of generic and specific process plans is presented first, followed by a data model to support its implementation. Finally, an analysis of implementation requirements using UML according to the collaborative network paradigm is briefly outlined.*
Key words: *Manufacturing System, Preliminary Design, Collaborative Network, Process plans.*

1. INTRODUCTION

Manufacturing system (MS) design is mainly pursued by managers in order to improve productivity, integration of data being a major concern (Putnik, 2002 ; Paquet et al, 2004).

However, preliminary design of manufacturing systems involves technical issues, too, e.g. whether there is a suitable machine-tool for processing certain product families and their variations, whether there is enough space for robot movements during loading or unloading of another machine, whether an operator can work simultaneously at two machining centers, etc. This approach is common ground for different research studies either trying to model relevant knowledge (Feng et al, 2001) or employing 3D simulation (Bilalis et al, 2002).

In order to shape an initial technical solution for a given manufacturing system, there is need for scientific cooperation (Yang & Xue, 2003). Cooperation culture has been cultivated, so as to provide holistic solutions (Cunha et al, 2003) though virtual models utilization, especially though the internet (Jiang & Zhang, 2002 ; Westkämper & von Briel, 2001).

The work reported next is placed within the framework of this modern tendency and it stems from a bilateral cooperation project jointly funded by Greece and Slovakia. It is original in that it puts forward a model for the procedure of preliminary manufacturing systems design and formulates data models supporting it, including parametric geometry models to visualise the elements making up the system.

2. PRELIMINARY MS DESIGN MODEL

Product technical specifications and the required quantities as well as time frames for production are the drivers for alternative generic process plans, involving just manufacturing process types and no specific equipment or processing time and similar detail. These are created based on experience, in the first place, but it is foreseen that an expert system might be constructed to replace the human. The manufacturing system, however, needs to be configured in terms of machine tools and auxiliary material handling equipment. This is achieved by transforming generic process plans into specific process plans, including specific machine tools.

Transformation is achieved by a mapping process between requirements on manufacturing processes and characteristics of machine tools that can perform these processes. Thus, a relevant database is necessary.

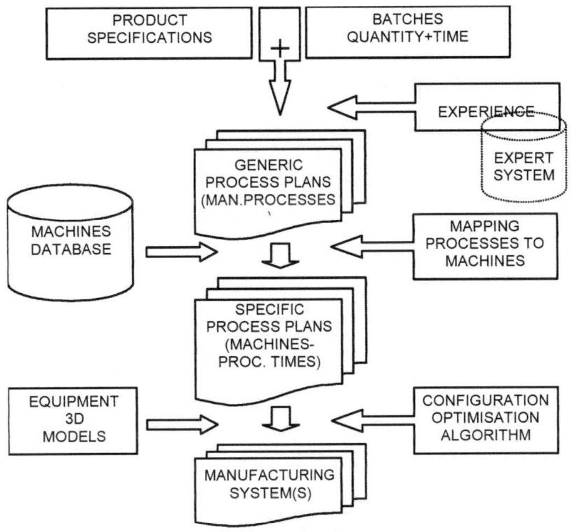

Fig. 1. Manufacturing System preliminary design model.

However, this is the way to potentially reach a huge number of alternative manufacturing systems, since any alternative generic process plan can be materialised by a combination of machine tools, still respecting any relevant constraints regarding part size, weight or machine capabilities, each combination yielding a specific process plan and accompanying manufacturing system. Therefore, it is necessary to find ways to prune the tree of alternatives. One way is to assess compatibility of machine tools within the same potential manufacturing system, e.g. by considering geometric constraints, i.e. distance, orientation etc. This can form the basis of subsequent detailed design of the manufacturing system, see Figure 1. Another way to prune the alternatives is to consider productivity of each one of them for different order scenarios, especially for virtual manufacturing systems, i.e. those that are geographically dispersed. This needs an optimisation algorithm with inbuilt means to assess productivity.

3. SUPPORTING DATA MODEL

In order to construct specific process plan alternatives available machine tools need to be considered in detail. Their characteristics are compared with the corresponding characteristics of the manufacturing processes contained in the generic process plan, e.g. for high-speed machining high rpm should be sought, and with the properties of the part to be manufactured, e.g. machine tool table needs to be larger than the part. The machine tools that do not match such specifications are rejected, whilst the rest are combined into potential manufacturing systems.

Emphasis was put on constructing a relational data model supporting the above logic, see Figure 2. The model can be implemented in stand alone as well as in web applications. So far, the assessment of the resulting manufacturing system configurations is left to the user.

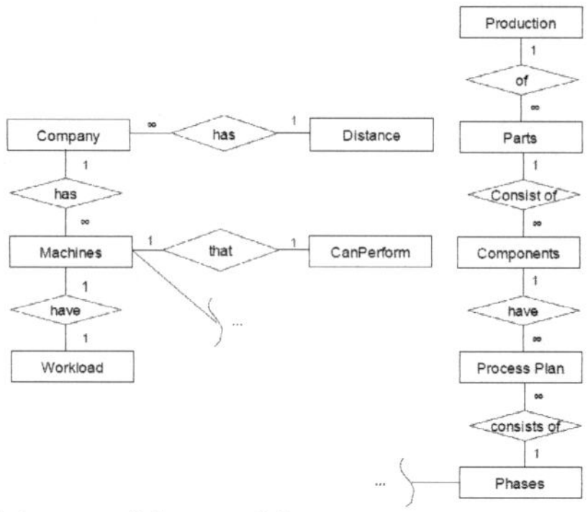

Fig. 2 Main part of data model

4. COLLABORATIVE MS DESIGN

The preliminary design methodology outlined above can easily be supported on the wed; indeed it benefits from being executed in a collaborative fashion, both when designing a system from scratch and when configuring a virtual manufacturing system. In the first case, it is necessary to collect machine tool data from manufacturers worldwide to populate the machine database, whereas in the second case it is necessary to register in the database machines that already work in other manufacturing systems. All data implied in Figure 1 can, thus, reside in different nodes of a local or wide area network.

The process of collaboratively configuring manufacturing systems was modeled using the UML-based GRAPLE methodology, see Figure 3. Manufacturability, time and cost determination form a central part of this model and involve both generic and specific process plan determination. The latter is tackled according to the methodology presented in section 2. Generic process plans may be either taken for granted, or they may formulated according to basic features of the product and rules of thumb, which is more difficult as the range of part families becomes wider and diverse.

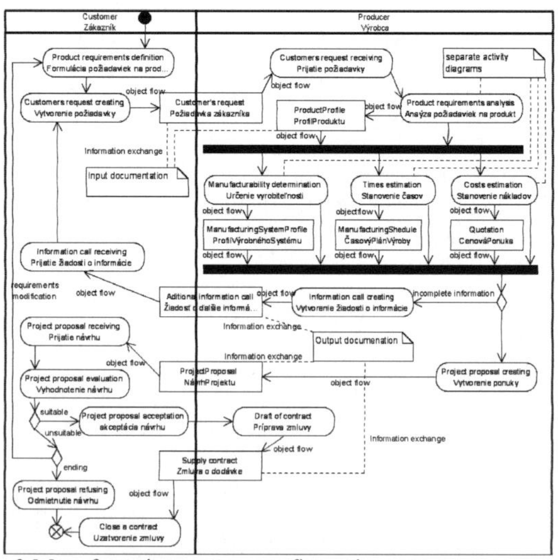

Fig. 3 Manufacturing system configuration process map.

The system is deployed as a web application, see Figure 4.

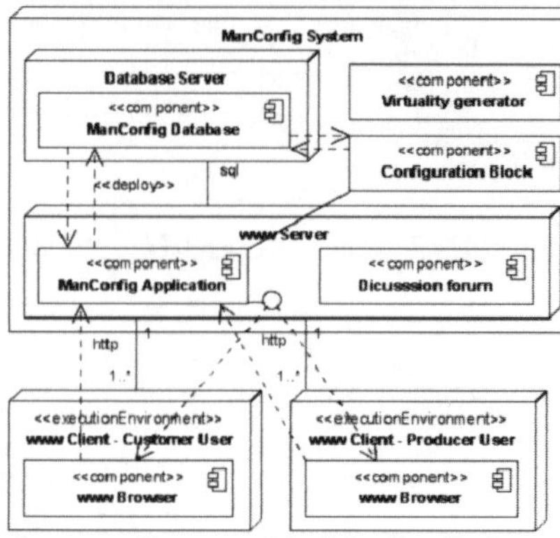

Fig. 4 Deployment diagram of the ManConfig System.

5. CONCLUSIONS

Preliminary design of manufacturing systems needs to start from generic process plans, which are transformed into specific process plans according to specific machines data models, relational in this case. This process needs to be executed collaboratively, the best tool being the internet, according to a suitable formal model, which has been designed.

Future developments concern addition of algorithms to help decision making at various sub-processes of preliminary design of manufacturing systems in a collaborative fashion.

6. ACKNOWLEDGMENTS

This work is supported by the General Secretariat of Research and Technology (GR) Joint Research and Technology Programs 2004-2006: Greece-Slovakia project 766 'Intelligent collaborative web supported configuration of manufacturing systems'.

7. REFERENCES

Bilalis, N.; Scroubelos, G.; Antoniadis, A.; Emiris, D. & Koulouriotis, D. (2002). Visual factory: basic principles and the zoning approach. *Int Jnl Production Research*, 40, 15, 3575-3588, ISSN 0020-7543.

Cunha, P.; Dionísio, J. & Henriques, E. (2003). An architecture to support the manufacturing system design and planning. *Int Jnl Computer Integrated Manufacturing*, 16, 7-8, 605-612, ISSN 0951-192X.

Feng, S.; Li, L.X. & Cen, L. (2001). An object-oriented intelligent design tool to aid the design of manufacturing systems. *Knowledge-Based Systems*, 14, 5-6, 225-232, ISSN 0950-7051.

Jiang, P. & Zhang, Y. (2002) Visualised part manufacturing via an online e-service platform on Web. *Concurrent Engineering: Research & Applications*, 10, 4, 267-277, ISSN 1063-293X.

Paquet, M.; Martel, A. & Desaulniers, G. (2004), Including technology selection decisions in manufacturing network design models. *Int Jnl Computer Integrated Manufacturing*, 17, 2, 117-125, ISSN 0951-192X.

Putnik, G.D. (2002) Part 1 - Towards an OPIM system: manufacturing SME organisation for the future. *Int Jnl Networking & Virtual Organisations*, 1, 2, 125-150, ISSN 1470-9503.

Westkämper, E. & von Briel, R. (2001). Continuous improvement and participative factory planning by computer systems. *Annals CIRP*, 50, 1, 347-352, ISSN 0007-8506.

Yang, H. & Xue, D. (2003). Recent research on developing Web-based manufacturing systems: a review. *Int Jnl Production Research*, 41, 15, 3601-3629, ISSN 0020-7543.

Annals of DAAAM for 2008 & Proceedings of the 19th International DAAAM Symposium, ISSN 1726-9679
Editor B. Katalinic, Published by DAAAM International, Vienna, Austria 2008
Make Harmony Between Technology and Nature, and Your Mind will Fly Free as a Bird

QUALITATIVE APPROACH OF VALUE CREATION MEASUREMENT IN REALISATION OF ORGANIZATIONAL EFFECTIVENESS

VRDOLJAK RAGUZ, I[vona] & JELENC, L[ara]

Abstract: *In last twenty years business literature has been fundamentally enriched by different studies that prove clearly the variety of value creation indicators. Traditional measurement has been largely criticized as being focused only on financial measures and functional level performance. The goal of every enterprise is the rational value creation and thus this paper offers some new tools for measuring the value creation with the accent on qualitative approach. A brief overview of the possible non-financial measures, methodologies, frameworks and models that could be employed as the secondary measures in the firms. The paper suggests several hypotheses for further research. The hypothesis would correlate the traditional-financial measurement and secondary-non-financial measurement in order to depict the nature of their relation.*

Key words*: value creation measurement, qualitative approach, organizational effectiveness*

1. INTRODUCTION

There is a growing importance of variety of performance measurement both the quantitative and qualitative measurements in enterprises in today's economy. The financial measures grow in complexity and are very specific in nature due to the characteristics of the industry, processes and goals. A broader conceptualization of the performance rather than just financial measures will offer more objective point of view of the level of performance in the enterprises.

The dominance of financial measures is accompanied with the new approaches, non-financial measures that are becoming more important to top management, investors and stakeholders. They offer the view on the performance as the signals of bad or poor performance before it will be reflected in the financial measures in the enterprises. Addition to that, they offer the qualitative approach to value creation that cannot be measured with numbers and represent the indicators of a stable and healthy core of the enterprises, no matter what the financial measures report.

Demand for changes in corporate reporting are also likely to force organizations to adopt a more balanced approach to performance measurement.

As a consequence organisation are forced to search beyond traditional financial measures and place greater emphasis on performance metrics related to softer issues emended in people and processes, whose strengths or weaknesses do not show up on a balance sheet (Bromwich & Bhimani, 1994.). Organizational theorists argue that overall organizational effectiveness is assessed only when it is viewed from multiple consistencies (Connoly, Conlon, Deutsch, 1980 cited in Karri, 2001, p.38; Boyd, 1991; Chakravarthy, 1986; Venkatraman & Ramanujam, 1986).

The aim of this paper is put more elaborated attention to the review of the literature about the possibility of qualitative measurement approaches of value creation apart from the financial ones. This article is directed toward the researchers as the theoretical background for the hypothesis generating and empirical research performed.

The paper consists of the general overview of; Balanced Scorecard, Total Quality Management, EFQM and Key Performance Indicators. This models, methodologies and frameworks are created in order to improve the quality of work, processes and indirectly indicators of performances.

2. QUALITATIVE APPROACH TO VALUE CREATION MEASUREMENT- LITERATURE REVIEW

Balanced Scorecard (BSC) is a synthesis of traditional financial measures and the development of elements of the competitive advantage. The BSC considers financial and non- financial measures such customer voice, internal business processes, learning and growth and financial perspective (Kaplan & Norton, 1993).

In 2001 Kaplan and Norton (Kaplan & Norton, 2001) introduced the five principles to keep strategy the focus of organizational management processes: translate the strategy into operational terms, align the organization to the strategy, make strategy everyone's everyday job, make strategy a continual process, and mobilize change through executive leadership.

The BSC is developed along the four well known perspectives of Financial, Customer, Internal Business Process, and Learning and Growth Performance. These perspectives measure the current status and future potential of organisations.

Balanced Scorecard represents the main indicator of controlling and has a central role in the control process in companies that exists in the companies where the planning is the central function of managing system (this especially refers to the large companies).

Another very popular approach is Total Quality Management. Total Quality Management (TQM) emerged as a philosophy in the UK in the late 1970s and early 1980s. It was born out of a much earlier quest for quality, which can be traced, back to the pioneering work of Juran, Deming and others with Japanese industry in the 1950s. The primary focus of TQM is not so much on quality as it is on the customer who defines what quality really is. Implementation of TQM necessitates adherence, discipline and constant employee effort. TQM includes everything, and is dependent on everyone. It is a system of enhancing and improving flexibility, effective and efficient business performance. Today more than ever TQM is being accepted and becoming a way of thinking and a way of life.

In the effective terms of extremely strong competitiveness and more demanding customers, product and service quality become one of the most significant competitive method and one of the fundamental key to business achievement.

The real benefits can be realized only when the companies that apply them truly understand both their capabilities and their limits (Gotzamani, K. D. & Tsiotras, G. D., 2001.). ISO 9000 certification is the main key to obtaining customer satisfaction. There are eights steps to this certification (Swalens, G. J., 1992.). The term Business Excellence has grown up from the TQM. In the world, enterprises have turned to the national and regional award policy rewarding the most successful enterprises whose main goal is the promotion of excellent quality systems and business excellence (Deming prize, The Malcolm Baldrige National Quality Award, ISO, European Quality Award etc.).

The Business Excellence (http://en.wikipedia.org/wiki/Business_Excellence) is based on these principles: orientation on the results, consumers satisfaction, leading the purpose, process management based on facts, employee development, team work, permanent education, innovation, partnership development and corporate social responsibility. A few models have been developed from these principles and they are giving the base for the development of Business Excellence.

In this paper the most famous European model of Business Excellence will be presented - European Foundation for Quality Management (EFQM). The EFQM has been founded as a non profit organisation in 1988 and represents an example of regional quality and business excellence award whose mission is to promote quality and business excellence in the European enterprises and which has given the preposition of the model for reaching the business excellence (http://www.efqm.org/). In January 2001 EFQM have had more than 800 members from almost each of the European countries and is the owner of EFQM model which is made of nine aspects. Five of them have the role to help enterprises to reach business excellence and remarkable business results, and the four of them are some kind of control system that show reached results.

For every of these nine aspects a set of questions has been developed that considers exact answers for reaching the model and its potentials in the implementation process.

The EFQM model predicts the five levels of excellence which helps enterprises continuous and objective process for reaching the business excellence and also the best solution and concept for an enterprise.

There is one more concept from the quantification aspect that is interesting and possible for the implementation. It reflects to the critical factors for the realisation of the organizational effectiveness, it is called Key Performance Indicators (http://en.wikipedia.org/wiki/Key_performance_indicator).

These indicators are different in every organization, primarily for the different organisation types that are characterised with different organisation units on which measurement depends.

It is useful to know the organisation goals that depend on the enterprises vision and mission but also on the stakeholders, before the implementation process. Their aim is to be critic to the success of the same enterprise.

This concept is best to use in a project management and in all fields of project management such as information technology, risk management, supply chain management, financial management, quality, production, sales etc. These indicators are best understood when they are presented as a different charts that are understandable to all employees and at all organisation levels. The most important in this concept, as in others, is the fact that all employees have to understand these concepts so that they can reach them.

3. INSTEAD OF CONCLUSION

The dominance of financial measures is accompanied with the new approaches as it can be seen from this paper. It is obvious that non-financial measures are becoming more important to top management, investors and stakeholders that is forced with the new business condition in the business environment. They all have their principles but also have the same goal and that is organizational effectiveness and the best business result possible.

The further research should be focused on the several aspects of this issue. First of all the brief literature review could be the theoretical background for the empirical research about financial and non-financial measurements. Hypothesis could be organized as such:

Hypothesis 1. The relation between traditional/primary/ financial measurements and contemporary/secondary/non-financial measurements could be presented in a correlation.

Hypothesis 1.1. The correlation between financial and non-financial measurements could reveal the contradiction of the subject they measure.

Hypothesis 1.2. The correlation between financial and non-financial measurements could reveal the interplay of the subject they measure, actually measuring the same subject from a different perspective.

Hypothesis 2. The introduction of the non-financial measurements in the enterprise is the signal of certain stage of growth in financial measurements. It is possible to point the stage of the enterprise that should welcome and introduce the non-financial measurements.

Researchers have suggested, regardless of the basis for assessment, the most important issue in performance is an preparedness of the enterprise for future action (Keats, Hitt, 1988; Thompson, 1967 cited in Harrington, 2001).

4. REFERENCES

Boyd, B.K. (1991). Strategic Planning and financial performance: A meta-analytic review, *Journal of Management Studies*, 28(4), 353-374, 1467-6486

Bromwich, M., Bhimani, A., (1994). Management Accounting-Pathways to Progress, CIMA Research, first edition, 1874784272, London

Chakravarthy, B.S. (1986). Measuring strategic performance, *Strategic Management Journal*, 7, 5, 437-458, 01432095

Gotzamani, K. D., Tsiotras, G. D. (2001). An empirical study of the ISO 9000 standards' contribution towards total quality management, *International Journal of Operations & Production Management*, 21, 10, 1326-1342, 0144-3577

Harrington, Robert, J. (2001). Dimensions in the strategy-making process, Washington State University, PhD thesis

Kaplan, Norton (1993) Putting the Balanced Scorecard to Work, *Harvard Business Review*, 71,5, 134-147, Harvard Business School Press, 0017-8012, Cambridge

Kaplan, R.S., Norton D.P. (2001). *The Strategy-Focused Organization: how Balanced Scorecard Companies Thrive in the New Business Environment*, Harvard Business School Press, 1578512506, Cambridge

Karri, Ranjan V.N. (2001). Strategic Flexibility and Firm Performance, Washington State University, PhD thesis

Swalens, G. J. (1992), ISO 9000 quality standards in 24 questions, *ISO 9000 News,* International Organization for Standardisation, Compuserve, 100112

Venkatraman, N., Ramanujam, V. (1986). Measurement of business performance in strategy research; A comparison of approaches, *Academy of Management Review*, 11, 4, 801-814, 0363-7425

http://en.wikipedia.org/wiki/Business_Excellence, Accessed: 2008-07-01

http://www.efqm.org/ Accessed: 2008-07-01

http://en.wikipedia.org/wiki/Key_performance_indicator Accessed: 2008-07-01

Annals of DAAAM for 2008 & Proceedings of the 19th International DAAAM Symposium, ISSN 1726-9679
Editor B. Katalinic, Published by DAAAM International, Vienna, Austria 2008
Make Harmony Between Technology and Nature, and Your Mind will Fly Free as a Bird

SURFACE MICROSTRUCTURING OF POLYESTER FABRICS BY UV EXCIMER LASER IRRADIATION

VRINCEANU, N[arcisa]; COMAN, D[iana]; SANDU, I[on] & GRIGORIU, A[urelia]

Abstract *In this study, laser was used to treat PES fabrics. The effects of laser treatment on the surface morphology of the fibres were characterized using scanning electron microscopy (SEM) and atomic force microscopy (AFM). The microscopic evaluation revealed the changes in the surface morphology of the fibres caused by laser treatment.. By surface microstructuring, the treatment achieved can be considered an intelligent solution for the improvement of surface properties of textile materials.*
Key words: microstructuring, UV excimer laser irradiation, PES fabrics, morphology.

1. INTRODUCTION

Today, pulsed UV laser is one of the most commonly used non-contact treatment techniques in modifying surface properties of polymers physically as well as chemically (Knittel et al, 1998). It is regarded as an environmentally friendly process since no chemicals are involved (Wong et al, 2000). The surfaces of the irradiated fibers develop ripple-like structures that result in an enhancement of fiber surface area. Surface structures are important for the physical and chemical properties of textile fibers since wetting, adhesion and optical properties are strongly dependent on them. These properties are also affected after laser irradiation (Bahners et al, 1993). Chemical surface treatment has been traditionally used to modify fibre materials but it has some disadvantages, such as influence on bulk properties and environmental pollution. Therefore, to improve the hydrophilicity of polyester without affecting its bulk properties, a lot of studies have been performed on the laser treatment for improving surface properties (Kan, 2007);
The laser irradiation of highly absorbing polymers such as polyester can generate characteristic modifications of the surface morphology (Bahners et al, 1990). Hence, it is reasonable to believe that such surface modification of a polymer may have an important impact on its textile properties (Wong et al, 2001).

2. EXPERIMENTAL

2.1. Materials and laser treatement

In this study, a 100% polyester fabric was used in all experiments.White plain weaves samples were conditioned at 20 ± 2 °C and 65 ± 2% relative humidity before laser irradiation. Irradiation was performed using a a LPX 200 Excimer 248 nm KrF. In high-fluence laser irradiation, samples were irradiated directly from the laser beam without using either special photomask or focusing lens. Laser energies like fluence and number of pulses vary from experiments in order to study their effects upon samples. The laser fluence was regulated in the range from 29 to 43 mJ/cm^2 and the number of pulses varied between 0 and 4, the pulse repetition was kept constant at 1Hz to avoid any possible heat accumulation. During the laser treatment, the control of treatment was very important, considering the low melting point (260°C) of the fibres.

2.2. Morphological study

The morphology of samples was investigated by:

Scanning Electron Microscopy (SEM) (TESCAN). All samples were gold coated prior to SEM examination.

- **Atomic Force Microscopy (AFM)**. AFM is the one of the effective tools to examine the microstructures of fibres. It is able to scan materials without any special preparation at normal temperature and pressure. The AFM used in this study was CSPM3300 produced by Benyuan Company. The vertical resolution of the machine is 0.1 nm, while the horizontal solution is 0.2 nm. The scanning mode used was contact mode in this study, and the scanning range was set at a size of 5.0 umX5.0 um.

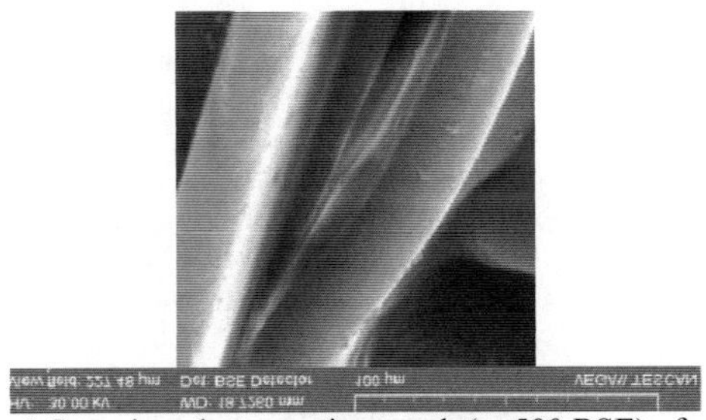

Fig. 1 Scanning electron micrograph (x 500 BSE) of control (no treatment) PES fabric sample

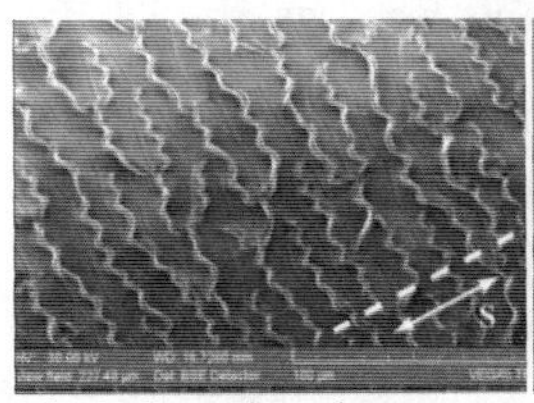

Fig.2. Scanning electron photo (x 1000 BSE) of laser treated polyester fabric: 248 nm, 2 pulses, 37 mJ/cm^2

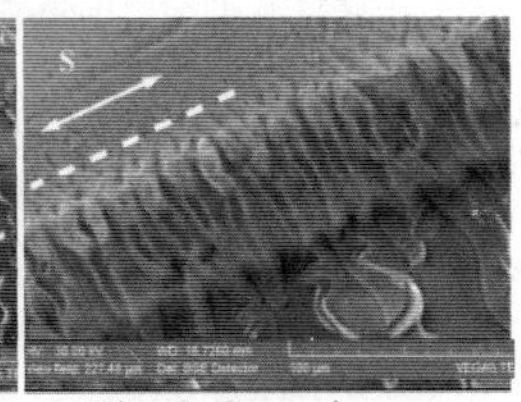

Fig. 3. Scanning electron photo (x 2000 BSE) of laser treated polyester fabric: 248 nm, 2 pulses, 37 mJ/cm^2

3. RESULTS AND DISCUSSION

Fig. 2, 3 show SEM micrographs of untreated and laser treated PES samples. Before the treatment, the sample had a smooth surface. After the laser treatment the surface developed a certain roughness or periodic roll, the so called ripple like or roll like structure, as shown in Fig. 3. The orientation of this type of structure is perpendicular to the fibrilar orientation of the fibres. Despite the fact that PES fibres without laser treatment appear to be smooth, in the AFM image, the lines and the roughness of the surface can be clearly shown.

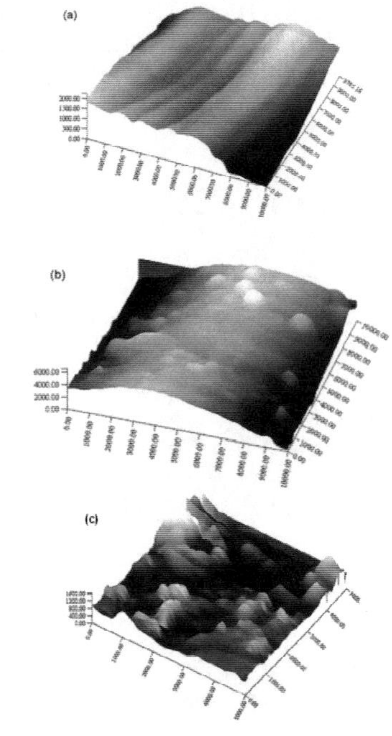

Fig. 4. AFM micrographs of fibres (a) untreated; (b) laser treated for 1 pulses, 248 nm, 37 mJ/cm^2; (c) laser treated for 4 pulses, 248 nm, 37 mJ/cm^2.

The height of the lines ranges from 30 to 353 nm (Fig. 4a). Laser treatment with 37 mJ/cm^2 and one laser pulse, etches the surface of PES fibres, forming aggregates on the surface as illustrated in Fig. 4(b). The surface topography has changed from an original smoothness to a fibrous one characterized by hills and groves. The AFM image also reveals the size of aggregates in the range between 10 and 314 nm. The longer exposure to the laser, the rougher the surface becomes. It can be seen from Fig. 4(c) that almost all the surface is etched and the etching effect makes the height of the surface fall by 10–716 nm. It can be noticed that the increasing the number of pulses, the surface becomes coarser and the height of the surface is now 716 nm. More fluence would give a larger distance between hils. The more number of pulses, the more the ripple will approach saturation. The surface roughness computed from AFM images are summarized in fig. 5 and 6.
In figures 5 and 6 both R_a and ripple spacing show almost linearity (linear relation) with laser number of pulses. R_a and ripple spacing increase with the number of laser pulses. For the roughness (R_a), the value increased dramatically after the second pulse. An increase in R_a represents an increase in ripple size.

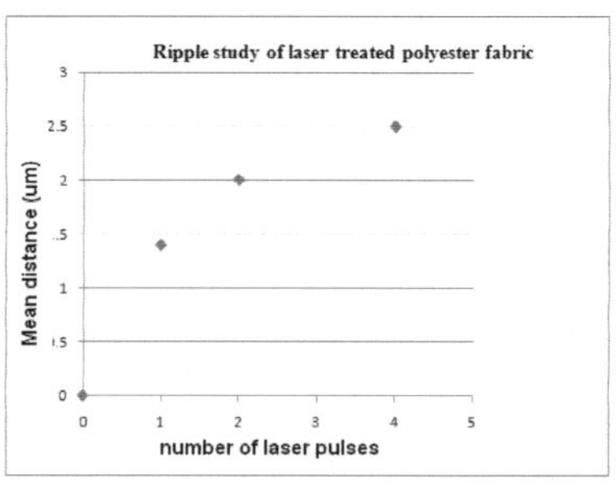

Fig. 5 Ripple modification of laser treated polyester fabric

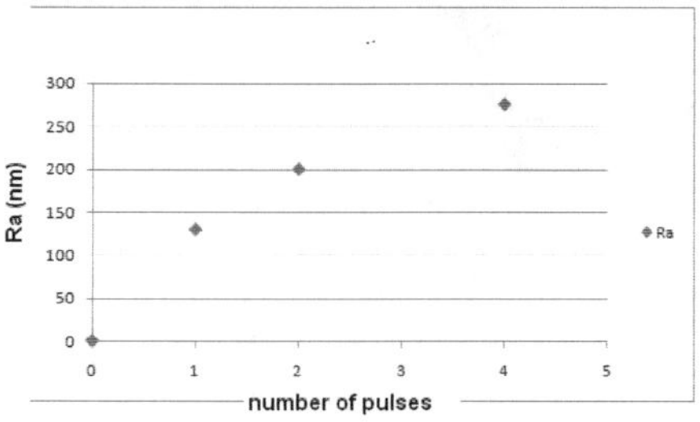

Fig. 6. Roughness evaluation of laser treated PES fabric

4. CONCLUSION

The microscopic observations have revealed that the use of UV excimer laser upon PES textile fabrics is a valuable tool to change the surface morpfology and consequently, the properties of the textile materials will be modified, even improved. The laser irradiation of highly absorbing polymers such as polyester can generate characteristic modifications of the surface morphology. In this case, some well-oriented structure of grooves or ripple structures with dimensions in the range of micrometre are developed on surface with irradiation fluence above the so-called ablation threshold. The study has proven the etching effect on the surface roughness of the polyester fabrics. SEM and AFM have been proven to be powerful tools in the examination of surface morphology. The surface roughness and the ripple spacing increased with laser pulses and fluences. It has been concluded that a specific modification of polyester fibers may be perfomed by UV excimer laser irradiation, resulting in a roughness at the surface textiles. The surface roughness depends on laser treatment parameters: wavelength, pulse duration, fluence and number of pulses. The ablation threshold values depend upon both the UV radiation wavelength and the polymer absorptivity. We have reason to believe that after this number of pulses the topography has not been changed too much.

Acknowledgement
The authors would like to thank to Prof. FOTAKIS Costas, the Director of Institute of Electronics Structure and Lasers (IESL), Foundation of Research Technology Hellas (FORTH), Heraklion, Greece, for infrastructure access.

5. REFERENCES

Bahners, T.; Kesting, W.; Schollmeyer, E. (1993) Designing surface properties of textile fibers by UV-laser irradiation. *Appl Surf Sci,*69(1-4):12-5.

Kan, C.W. (2008) Impact on textile properties of polyester with laser, *Optics & Laser Technology*, 40 ,113–119.

Knittel, D.; Schollmeyer, E. (1998) Surface Structuring of Synthetic Polymers by UV-Laser Irradiation. Part IV. Applications of Excimer Laser Induced Surface Modification of Textile Materials*, Polymer International* 45 , 103-117.

Wong, W.; Chan, K.; Yuen, KW.; Lau, KS (2000) Morphological modification of poly(ethyleneterephthalate)Surface after Excimer Laser Irradiation, *Journal of Textile Engineering,*Vol. 46 No. 2 pp.25-31.

Wong, W.; Chan, K.; Yeung, KW; Lau, KS. (2001) Chemical modification of poly(ethylene terephthalate) induced by laser treatment, *Textile Res J*, 71(2):117-20,

Annals of DAAAM for 2008 & Proceedings of the 19th International DAAAM Symposium, ISSN 1726-9679
Editor B. Katalinic, Published by DAAAM International, Vienna, Austria 2008
Make Harmony Between Technology and Nature, and Your Mind will Fly Free as a Bird

STUDIES ON THE ADHERENCE OF THE THERMOGLUED TEXTILE ENSEMBLES - METHOD AND INSTRUMENT USED

VRINCEANU, N[arcisa]; SANDU, I[on]; STEFANUTA, I[oan] & VLAD, D[orin]

Abstract: *The paper approaches a less treated topic by the specialists from textile clothing field: the adherence of the thermo-glued underensembles. This characteristic influences in a big extent the properties of wearing and easy-care of clothing items. In the first part, experimental work and instrument used are highlighted according to the international standards. The tests have been performed on 7 types of basic materials and 2 types of insertions considered more representative from point of view of utilisation level.*

Key words: *adherence, insertion, adhesive, thermogluing*

1. INTRODUCTION

Thermogluing is the operation by which two ore more textile materials, superposed in layers, are joined by means of adhesives. Nowadays, the thermogluing is the only method of obtaining undetachable joinings between both different materials and of the same nature materials, as well as joinings of textile materials with different destinations (Stefanuta, 2001)

The textile products subjected to thermogluing show improved functional characteristics: spatial forming of some underensembles, in order to the modelling of the items shape against the human body anatomy (or the shape of objects, in case of covering type guide marks), the increase of rigidity and strength (according to the type of solicitations the ensemble is subjected to, within the manufacturing, wearing or care processes), high dimensional stability, thanks to margins consolidation, detail doubling with auxiliary.

The adhesion or the adherence can be defined as the adhesive ability to provide a mechanical strength to the joining between two objects. From physical point of view, the adhesion can be considered as a resultant of the attraction, intermolecular forces, expressing at the contact surfaces level of different solids or liquids (Stefanuta, 2001).

2. THE PRINCIPLE OF THE ENSEMBLING

The basic of the ensembling consists of adhesion of the adhesive to those two types material surfaces, under a regime well-determined by some certain parameters. These parameters are: temperature– generated by heated metallic surfaces, pressure – exercised by calender cylinders onto heated layers, and time – of heating and pressure exposure, parameter expressed by movement speed of the subensemble through work area (Cioara et all, 2000).

The work regime is an important component of the technology design and, basically, it depends upon some factors as: structure and nature of the raw material from basic material (Stefanuta, 2001), the adhesive's nature and properties, structure and nature of the raw material from auxilliary (insertion), the type of adhesion.

3. THE PHYSICO-MECHANICAL CHARACTERISTICS OF BOTH BASIC MATERIAL AND ADHESIVE INSERTION

3.1. Basic Materials

The basic materials used are as follows:

Cotton type fabric

-cotton – 100%
-cotton – 98% + ellastan – 2%
-cotton – 80% + polyamide – 18% + ellastan – 2%
-cotton – 53% + polyester – 43% + ellastan – 4%

Woollen type fabric

-wool – 96% + ellastan – 4%
-wool – 60% + polyester – 38% + ellastan – 2%
-wool – 44% + polyester – 54% + ellastan – 2%

3.2. The Insertion

CHARACTERISTICS	Insertion A (EE 6035)	Insertion B (CY 1022)
Composition	70% polyamide +30% polyester, and has insertions on the wrap direction	100% polyester
Weight	9 g/m^2	12 g/m^2
Utilisation	For small and big items; for women and men items	For basic textile,especially for those subjected to some modern finishing treatments
THERMOGLUING by means of calender cyllinders		
temperature used	116 – 127 °C	138 – 149 °C
time	8 – 12 sec	12 – 15 sec
pressure	2 – 4 barr or 15 – 30 N/cm^2	3 – 5 barr or 30 – 45 N/cm^2
THERMOGLUING by means of press machine		
temperature used	116 – 127 °C	138 – 149 °C
time	8 – 12 sec	12 – 15 sec
pressure	200 – 350 mbarr	300 – 450 mbarr
ITEMS EASY-CARE		
Washing	60 °C	95 °C
Ironning	130 °C	130 °C
Spinning and drying	yes	yes
Chemical treatments	whitening	whitening

Tab. 1. Characteristics of the insertion

4. METHOD, INSTRUMENT USED AND EXPERIMENTAL WORK

4.1. Method Used

In order to test the adherence strength of textile thermoglued underensembles, the samples are subjected to the traction solicitation. Rectangular samples (160 mm X 50 mm), samples consisting of basic material layer and a layer of insertion were used, highlighting that the gluing of these two layers was made only on a distance of 120 mm.

The distance between jaws is 50 mm, and the speed of dynamometer jaw movement is 100 mm/min. After the testing, some parameters were displayed: effort – extention diagram, the minimum, maximum and the averaged tension, which were used for adherence calculus (DIN 53859 Part 2).

The adherence relation is as follows (Preda et all, 1996):

$$A_F = \frac{F_m}{l_E} \text{ [N/cm]}; \quad (1)$$

where: - F_m - the averaged tension
- l_E - the sample width, expressed in cm.

4.2. The Instrument

4.2.1.General considerations

Titan's software has been pre-programmed with British (BS), European (EN), International (ISO), American (ASTM) and Retailers' Standards (e.g. Next, M&S); the instrument's software is easy to follow. The tests are both quick and easy. The standards are grouped in three application modules: tensile, seam slippage and security of attachments; tear, peel bond and wing rip; stretch and recovery (cyclic) testing (Technical book of Titan^2).

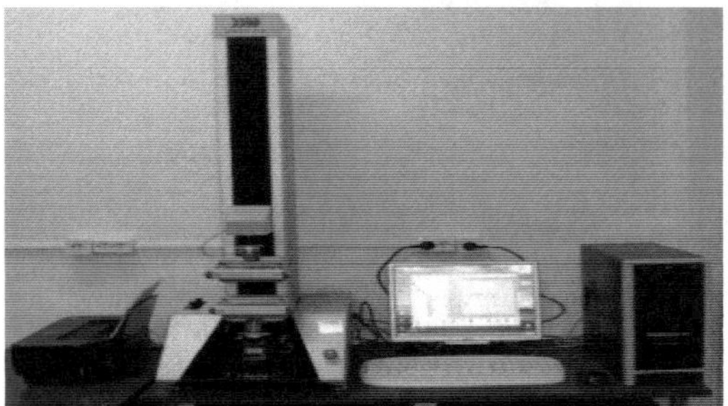

Fig.1. Titan^2 – Universal Strength Tester

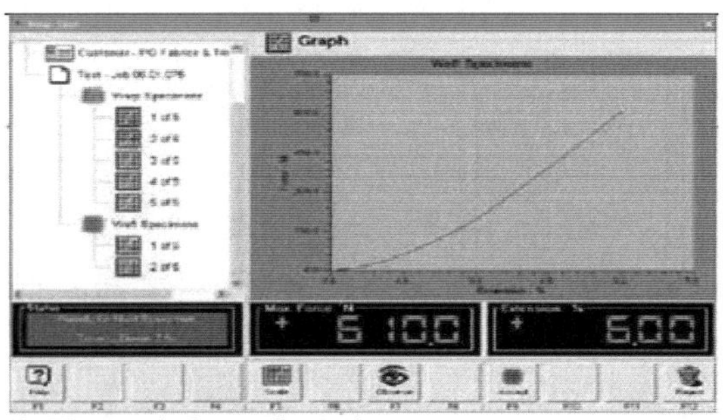

Fig.2. The graph force-elongation

4.2.2.Technical data

LOAD	
Capacity	3000 N
Range	1 – 100 %
Available load cells	3000/600/120 N
EXTENSION	
Maximum speed	4000 mm / min
Testing/return/jog speed	2 - 4000 mm / min
Maximum stroke	600 mm
Total vertical test space	840 mm
PNEUMATICS	
Input pressure	800 kPa / 8 bar / 110 PSI
Testing pressure	800 kPa / 8 bar / 110 PSI
Air preparation	Filtered to 5 microns (absolute) or better to remove excess particulates, oil and moisture
SAFETY	
Gripping	Low pressure during specimen loading
Jaw movement	Impact protection
Load cell	Force overload inhibit
Conformity	Complies with CE Directives
CONTROL SYSTEM	
Type	CRE
PC Control	Windows XP or VISTA running dedicated software
Drive control	Digital precision closed loop A.C. servo system via RS232 serial port
Power requirements	110 – 230V ± 10 %, 50 - 60 Hz, 500 W

Tab. 2. Technical data of Titan^2

4.3. Experimental Work

- 8 samples are made on longitudinal direction, from each type of basic material, so that 4 samples should be from each type of material with one of two types of insertions;
- for each sample, 40 mm will be left without being glued;
- the samples were conditioned ;
- the distance between the dynamometer jaws should be 50 mm, and the speed of lifting the jaw C_1, up to 100 mm/min, according to the DIN 53859 Part 2;
- the data necessary for the test, the test's name, the number of samples, the type of basic material and the type of insertion are introduced on the computer;
- the sample is laid between the dynamometer jaws and the instrument is running;
- the adherence is expressed on the basis of data provided by the computer.

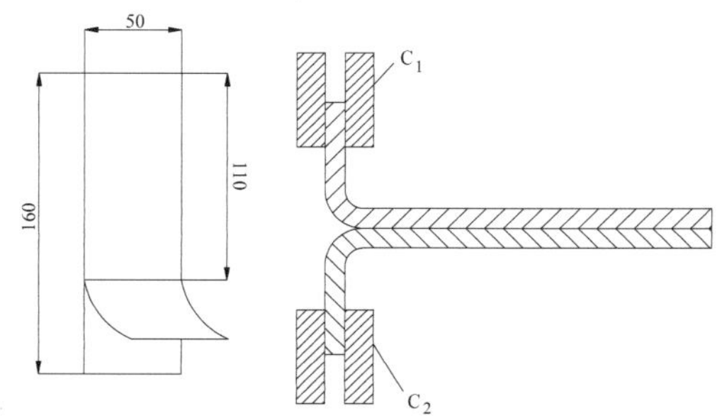

Fig.3. Size and type of the samples (Preda et all, 1996).

5. CONCLUSION

An adequate and high quality process of thermogluing influences positively the wearing and easy-care characterisitics of clothes.

The studied method and the instrument used in this paper are updated and according to the standards;

The cotton and woollen basic materials prevail in manufacturing processes;

Insertions used for tests provide strength and stability to the underensembles.

6. REFERENCES

Cioară, L., Cristian, I., Onofrei E., (2000) *Textile tissues structure characteristics and properties.*, Performantica, ISBN 973-730-029-7

DIN 53859 Part 2 – Tear growth testing of textile fabrics.

Preda, C.; Preda, Cr. (1996). *Methods and apparatus for Quality Control of textiles designed for wearing clothes manufacturing, BIT, ISBN 973-97907-2-0, Iasi;*

Stefanuta, I. (2001). *Processes and machines for textile clothing industry,* ULB, ISBN: 973-651-229-0, Sibiu;

Technical book of Titan^2 – *Universal Strength Tester*, Model 710

Annals of DAAAM for 2008 & Proceedings of the 19th International DAAAM Symposium, ISSN 1726-9679
Editor B. Katalinic, Published by DAAAM International, Vienna, Austria 2008
Make Harmony Between Technology and Nature, and Your Mind will Fly Free as a Bird

VALUE STREAM MAPPING FOR EFFECTIVE LEAN MANUFACTURING

VUJICA HERZOG, N[atasa]; POLAJNAR, A[ndrej] & KOSTANJEVEC, T[omaz]

Abstract: *Value stream mapping (VSM) has become up-to date implementation method for Lean Manufacturing (LM). Using VSM every process in material and information flow, along with key data can be presented. These abilities also distinguish VSM significantly from tools such as 'process mapping' or 'layout diagrams'. Despite of this obvious priority VSM is still too rarely used in manufacturing companies. The main goal of the paper is to expose this deficiency. Use of this applicable tool will be augmented, including manufacturing case study.*
Key words: *Lean manufacturing, Value stream mapping, Continuous improvement, Manufacturing case*

1. INTRODUCTION

Lean manufacturing (LM) was developed by Taiichi Ohno at Toyota Motor Company in the 1950s. The publication of the book The Machine that Changed the World (Womack et al, 1990) started the diffusion of lean manufacturing practices first of all in automotive industry and later in all other industries from agriculture to aerospace.

Womack et al. (1990) defined 'Lean' as the elimination of waste and for this reason the main goal of lean manufacturing is the elimination of everything that does not add value to the product or service.

Although lean manufacturing have been around for a long time its use is again increasing recently. Lean manufacturing is nowadays frequently mentioned together with another quality improvement tools such as Six-sigma or Theory of constrains. The new approach believes that for an organization to label itself as strictly a Lean or e.g. Six-Sigma is a mistake. Being a company that is dedicated to continuous improvement and employs the best tool available for a given application is better. Continuous improvement has been around forever and is a part of any industry. The performance of the changes may be slightly different depending on the industry but the objectives are the same – to do what we do better, cheaper and faster.

In the paper the use of a very simple but also very valuable tool called VSM will be presented. VSM can help us to understand how the work site currently operates, design a lean flow and plan how to get there.

2. LITERATURE REVIEW

2.1 Lean manufacturing and its tools

In their book Lean Thinking (1996), James P. Womack and Daniel T. Jones defined a set of five basic principles that characterize a lean enterprise:

- Define 'value' from the perspective of the customer.
- Identify the 'value streams' and eliminate 'waste' from them.
- Create 'Pull'.
- Introduce 'Pull'.
- Strive to 'Perfection'.

The brief description of the most common lean tools includes:

- Cellular manufacturing: Organizes the entire process for a particular product or similar products into a group or cell, including all the necessary machines, equipment and operators. Resources within cells are arranged to easily facilitate all operations.
- Just-in-time (JIT): A system where a customer initiates demand and demand is then transmitted backward from the final assembly all the way to raw material, thus 'pulling' all requirements just when they are required.
- Kanbans: A signalling system for implementing JIT production.
- Total preventive maintenance (TPM). Workers carry out regular equipment maintenance to detect any anomalies. The focus is changed from fixing breakdowns to preventing them. Since operators are the closest to the machines, they are included in maintenance and monitoring activities in order to prevent and provide warning of malfunctions.
- Setup time reduction: Continuously try to reduce the setup time on the machine.
- Total quality management (TQM): A system of continuous improvement employing participative management that is centred on the needs of customers. Key components are employee involvement and training, problem-solving teams, statistical methods, long-term goals, and recognition that inefficiencies are produced by the system, not people.
- 5S: Focuses on effective work place organization and standardized work procedures.

2.2 Overview of VSM

A value stream is a collection of all actions (value-added as well as non-value-added) that are required to bring a product through the main flows, starting with raw material and ending with the customer (Rother and Shook, 1999; Abdulmalek and Rajgopal, 2007). These actions consider the flow of both information and materials within the overall supply chain. The ultimate goal of VSM is to identify all types of waste in the value stream and to take steps to try and eliminate them. While researchers have developed a number of tools to optimize individual operations within a supply chain, most of these tools fall short in linking and visualizing the nature of the material and information flow throughout the company's entire supply chain. Taking the value stream viewpoint means working on the big picture and not individual processes. VSM creates a common basis for the production process, thus facilitation more thoughtful decisions to improve the value stream (McDonald et al., 2002).

VSM is a pencil and paper tool or computer added tool, which is created using a predefined set of standardized icons (Fig.1). The first step is to choose a particular product or product family as the target for improvement. The next step is to draw a current state map that is essentially a snapshot capturing how things are currently being done. This is accomplished while walking along the actual process, and provides one with a basis for analyzing the system and identifying its weaknesses. The third step in VSM is to create the future state map, which is a picture of how the system should look after the inefficiencies in it have been removed. Creating a future state map is done by answering a set of questions on issues related to efficiency, and on technical implementation related to the use of lean tools. This map then becomes the basis for making the necessary changes to the system.

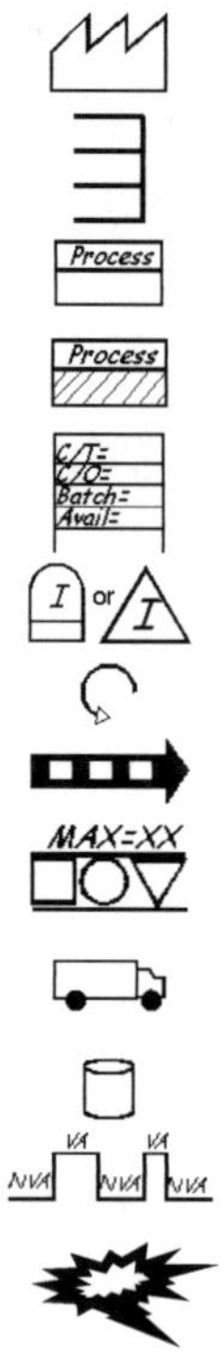

Customer or supplier

Supermarket/ small inventory

Process step

Process Step shared by other value stream

Data box

Inventory between processes, stores

Pull

Push

FIFO

Shipment

MRP or ERP system

Time line. VA- Value added, NVA – Non value add time

Important improvement opportunity

Fig. 1. Some symbols used for VSM

3. CURRENT AND FUTURE STATE MAP

VSM enables us to easily create, share and manage our processes. Current state mapping process can be summarized in the following items:

- Select a Product Family
- Form a Team
- Understand Customer Demand
- Map the Process Flow
- Map the Material Flow
- Map the Information Flow
- Understand Waste
- Identify and Eliminate Waste

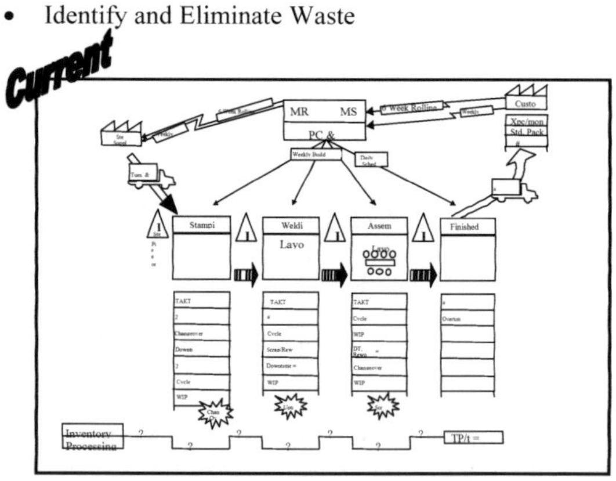

Fig. 2. Current state map

The process of defining and describing the future state map starts with developing the current state map, where target areas for improvement start to show up. When designing the future state map the following topics should be considered:

- Key Questions
- Future State Icons and Examples
- Map the Future State Process Flow
- Map the Future State Material Flow
- Map the Future State Information Flow
- Outline a Plan
- Metrics

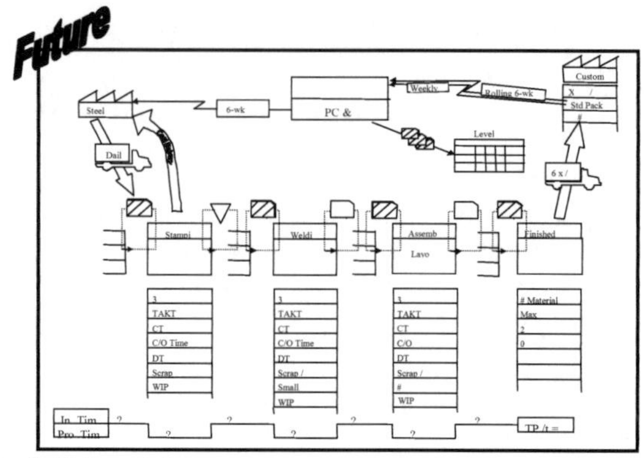

Fig. 3. Future state map

4. DISCUSSION

In creating the ideal future state map we try to identify lean manufacturing tools to drive the identified problems down, while looking at the schedule across the entire value stream. Inventory and lead time may be viewed as two related issues since the more inventories, the longer any item must for its turn and thus, the longer the lead time. We follow a systematic procedure where we try to answer a series of structured questions. In lean manufacturing understanding, identifying and elimination of waste is the key for success. Waste is anything that does not directly add value to the final product or contribute to the products transformation. There are seven basic types of waste known as COMMWIP what stands for Correction, Overproduction, Material Movement, Motion, Waiting, Inventory and Processing.

5. CONCLUSION

The presented VSM is one of the most effective and powerful tool that should be used at every improvement process in the company. It makes the flow of material and information visible and helps us to sort out what to keep and what eliminate and as such can be jointly used with any other quality improvement tool or methodology.

The future research could focus on simulation based VSM generators and thus make the decision processes about LM easier and more evident.

6. REFERENCES

Abdulmalek, F.A.; Rajgopal, J. (2007). Analyzing the benefits of lean manufacturing and value stream mapping via simulation: A process sector case study, *International journal of production economics*, 107, 223-236.

McDonald, T.; Van Aken, E.M. & Rentes, A.F. (2002). Utilizing simulation to enhance value stream mapping: a manufacturing case application. *International Journal of Logistics: Research and Applications* 5 (2), 213-232.

Rother, M.; Shook, J. (1999). *Learning to see: Value Stream Mapping to Add Value and Eliminate Muda,* The Lean Enterprise Institute, Inc., Brookline, MA.

Womack, J.P.; Jones, D.T. & Ross, D. (1990), *The Machine that changed the World.* Macmillan Publishing Company, Canada.

Womack, J.P.; Jones, D.T. (1996), *Lean thinking,* Simon & Schuster, New York.

Annals of DAAAM for 2008 & Proceedings of the 19th International DAAAM Symposium, ISSN 1726-9679
Editor B. Katalinic, Published by DAAAM International, Vienna, Austria 2008
Make Harmony Between Technology and Nature, and Your Mind will Fly Free as a Bird

IS OPEN SOURCE REALLY FREE OF CHARGE?

VUJICIC TOMIC, B[rankica]; POGARCIC, I[van] & NIZIC, B[erislav]

Abstract: *Paper is considering a problem of open software system. Since the open software is regularly identified with free software, problem was primarily analyzed from the financial aspect respectively is open software truly free of charge. Within frames of global e-Business open software can significantly influence on business costs. Even though open software's application in a concrete situation need not necessarily lower business costs it would be recommendable to examine possibilities of its usage. The authors also express their opinion regarding indicated problem from the aspects of their own business.*

Key words: *Open Source, Freeware, free of charge, Licence, Public Software*

1. INTRODUCTION

There are freeware and free software. In most cases these two categories overlap, but not always. Freeware is a kind of software which can be used free of charge. Free software is software which offers complete freedom in its usage. Many developers believe in releasing their software freely, but have hesitated to do so because they're concerned about losing control over their software. Licensing issues are complicated, and both the facts and fallacies you hear word-of-mouth can add to the confusion. (St. Laurent, 2004):

The most famous licence of free software is **GNU GPL** (General Public Licence) which declares that free software meets the following conditions:

- Freedom to execute a program, regardless of its purpose.
- Freedom to study how a program works and how to make adjustments to it to satisfy our needs. Access to source code is necessary condition for this freedom.
- Freedom to distribute program copies so we can help others with it.
- Freedom to improve program and to give improved program to public, which means allowing that all community benefits from it. Access to the source code is necessary condition for this freedom.

According to those criteria, software which is freeware, but has no free access to the source code isn't free software. However, **all free software is at the same time also free of charge**. When we know that, the answer to the subject of this paper is obvious. However, a fact that exists OS software and that it is completely free of charge, opens a few very interesting questions (Goldman&Gabriel, 2005) Namely, these are:

- Why is someone ready to spend so much time needed to develop quality software and after that give away all rights to economic benefits of that work?
- Is the development of OS community a threat to the software industry?
- How can we make profit on OS software?
- How reliable is OS software?

2. WHY WOULD SOMEBODY DEVELOP OS?

Many people wonder why somebody is ready to spend a lot of time developing the software and after completing it, give freedom to others - not just to use it free of charge, but also all other freedoms (Cusumano, 2005). Sometimes, to create quality software, time horizon is few months, in some cases even years. At first glance, that kind of behaviour is at least unusual.

The main reason why this is strange to us is that we live in a capitalist society where most of the things/services are viewed through the prism of economic gain and loss. However, the system of values in OS community is somehow different, which leads to strengthening of the OS community and quality of the OS software which is for the end user completely free of charge. After studying different texts about the question of this chapter, one can see that the answer to that question has two levels: an individual development of OS software and an organization which finances the development of OS software. It is very important to make the distinction between these two levels, because each one has different reasons for the development of OS software.

2.1. An individual starts with the development

Scenario: There are people with great software development skill. They decide for whatever reason to create quality and useful software. They put it under the OS license. Why? They did it for the personal satisfaction. Components of that satisfaction could be challenge, enjoying in programming, creation of something which will be their legacy, pleasure because they are participating is something very significant, potential indirect economic benefit: quality publicity on the job market. If the result of that work is also useful to others, that can be the initial spark of very positive energy. With participation of others their work achieves higher level of quality. They give away something which they would do anyway, so they do not lose anything. On the contrary, contribution of others creates something more valuable also for the original author. Because all of us need some income in our life to live normally, we can make assumption that people involved in OS activity have some sort of income in their life. For example, they have a job and in spare time they enjoy programming. In their spare time people do what they love. If somebody likes programming, spending time like that is not in any way different from spending time doing some hobby.

2.2. Big company is financing the development of OS

Scenario: Big organization decides to give source code of its software under the GNU LGPL licence. Why? The reason is much more different than for individuals, because in the case of companies there is economic motive which is easy to comprehend: according to that life of a company depends of their profit. (Goldman&Gabriel, 2005). Economic motives could be very positive publicity, creation of the market for some other non free products which are the great addition to OS software and creation of the market for giving support which becomes the source of income (www.opensource.org).

An example of the corporation like that is Sun Microsystems (www.sun.org). Their product OpenOffice.org is under LGPL licence. In addition to OpenOffice, they have also Java whose source code is available, but it is under the Sun licence. Partially, they give for free also Solaris operating system.

Authors of this paper deeply respect that kind of approach to the market because the end user benefits from that. This paper has been done with the OpenOffice.org Writer (in its first version).

3. IS OPEN SOURCE SOFTWARE THREAT TO THE COMMERCIAL SOFTWARE INDUSTRY?

This question rises after strengthening of OS community. The answer is: partially yes. But not so much that it can cause serious damage to it. Furthermore, in some areas it can even make it stronger. The beauty is that the end user can just benefit from the strong OS community. Commercial software industry is too strong so OS community cannot make some serious harm to it. But, in some cases, OS community has influence on the commercial software industry.

It is difficult to say in which ways exactly the development of OS community makes influence on software industry, and that is because of its dynamics, but it seems that it is making it better and it is giving a motive to the commercial companies to make an extra effort in the improvement of their products. This is important because only with extra features their product can motivate customers to buy it and not to use completely free OS alternative.

By using OS software, companies can reduce their costs. This statement is valid also for the software companies - e.g. companies which make custom made software for other companies. In the OS community there is not big interest for the development of that kind of software. This means, development of OS is a threat just to the part of software industry which produces popular software. Commercial companies which produce that kind of software must make an extra effort, but everybody else has better position in the development of custom made software.

Studying different sources, the author of this paper has found two ideas related to OS:

- Hardware manufactures can give OS drivers and with that give an opportunity to the market to improve it further on.
- the theory "what is valuable, will be paid" which says that if some software product has the value to the user, even if it is free of charge, the user will make donation for that software. This is very questionable theory because it seems too idealistic.

4. HOW RELIABLE IS OS SOFTWARE?

The answer to that question depends primarily on the level of the development and the **interest** of OS community for some tool. In case that there is great interest, the answer is definitely yes. Reasons for that are:

- great number of people are testing that kind of software and together they participate in the correction of found errors
- because of the openness of source code, it is great probability that certain security threats will be discovered and immediately resolved
- Great number of people participates in further improvements through suggestions, documentation, translation, addition of new functions etc.

If there is enough interest, sometimes it is just a question of time when will some tools become more quality and stableness than their commercial alternatives. Of course, if some software is the result of work of just one person, and OS community has no interest in that product, it is very possible that software isn't of high quality and stability.

5. USE CASE: AUTHORS OPINION OF OS

Professionally - The author and his company are not threatened by the OS projects. At the contrary, the company can even profit from OS by reducing its costs. There is no threat since the company develops custom made software adjusted to special needs of the particular client. Because of the nature of company's software, OS shows no interest in it.

Personally - The author desired to create two programs /games using C# programming language: **Sapere** - dictionary and program for studying foreign words **Abadel** – game Othello/Reversi.

When they will be finished, the following choices are optional:

Commercial version - try to sell it to someone: it is too much hassle for what it offers. Earning is too low and only enormous amounts of money could be a proper motive. Even if that would be commercial, because of the high rate of piracy, no matter of used protection, people who would like to use it without paying would find the way. Commercial version would help to reduce the program usage.

Freeware version - free of charge, but closed source code. Not bad, but there is a better option.

OS version - the best option. That significantly improves the chances for further improvements of programs. In the case the final products would have real quality there is not any reason why the program would be in some Linux distributions. This would initiate a great publicity, which could give more benefit than harm on the job market (e.g. one can write in CV that programs can be downloaded from the Internet and to mention in which Linux distributions programs are). This is the reason why the programs are being developed in C# in a way that they are compatible with Mono and .net 1.1 platforms. By working like that we will have software which has better chance to spread to both platforms: Linux and Windows.

We give advantage to the OS community. We also respect freeware versions if they have high quality. No matter why is someone active within the OS community, this is one very positive thing which brings benefits to many people.

6. INSTEAD OF CONCLUSION: NEW PARADIGM

Speaking of profit, OS approach put emphasis on service and not on the products. OS is response to monopoly. There is whole new paradigm: not just Windows or Linux platform, but Internet platform. (E.g. Google is using 100 000 Linux servers), and everybody is using Google, which indirectly means that we are also using the service of Linux platform. Amazon is also using OS. Internet platform - Web (Internet) has its foundation in OS: TCP/IP protocol, BIND, DNS, sendmail, etc., Apache Server and Linux as servers. Within the new paradigm, real money lies in service which is based on the free products and not any more in products themselves. For example: BIND is the protocol for DNS. Author of BIND does not make money from it, but Domain Name Services have.

7. REFERENCES AND RECOMMENDED LINKS

Cusumano, M. at al. (2005): *Perspectives on Free and Open Source Software* (Hardcover), the MIT Press

Goldman, R.; Gabriel, R.P. (2005): *Innovation Happens Elsewhere: Open Source as Business Strategy*, Morgan Kaufmann, 2005

St. Laurent, A.M. (2004): *Understanding Open Source and Free Software Licensing*, O'Reilly Media, Inc.

www.fsf.org: The GNU Project Free Software Foundation (Last *Accessed*: 2008-06-20)

www.opensource.org: Open Source Initiative OSI–Welcome *Accessed*: 2008-06-20

www.linux.org: The Linux Home Page, *Accessed*: 2008-06-05)

Annals of DAAAM for 2008 & Proceedings of the 19th International DAAAM Symposium, ISSN 1726-9679
Editor B. Katalinic, Published by DAAAM International, Vienna, Austria 2008
Make Harmony Between Technology and Nature, and Your Mind will Fly Free as a Bird

EXSTENSION OF IEC'S GENERIC DATA ACCESS WITH A LOCKING MECHANISM

VUKMIROVIC, S[rdan]; ERDELJAN, A[leksandar] & CAPKO, D[arko]

Abstract: *The aim of this paper is to propose an extension of IEC's Generic Data Access standard (GDA) that will allow locking of resources and/or resources associated with specified resource, which will allow safe multi-user work. The GDA defines a set of simple interfaces allowing access to data from any data model. The new set of functions proposed in this paper would allow clients to lock and unlock resources, get list of locked resources or ask if specific resource is locked. The paper describes a solution that is universal and configurable, so locking rules can be changed without changing the source code. The new interface is implemented and tested in industry environment.*

Key words: *generic data access, security, locking.*

1. INTRODUCTION

The International Electrotechnical Commission (IEC) uses the umbrella term Utility Management System (UMS) to refer to a class of systems which cover Water Quality and Energy Management Systems (WQEMS) in the water sector, and Energy Management Systems (EMS) or Distribution Management Systems (DMS) in the power sector. Utilities operate their water, gas, or power assets through control systems. The scope of control may include production facilities, bulk transmission networks, distribution networks, and supply points.

In order to make access to data from UMS easy and flexible IEC introduced the Data Access Facility – DAF (OMG, 2005) and its extension Generic Data Access - GDA (IEC, 2005) standards which describe simple interfaces for accessing data maintained by another system without any knowledge of the logical schema used for internal data storage. Although GDA has been available for a few years, it is seldom implemented and seldom written about in scientific papers.

This paper proposes an extension of the GDA standard that would allow safe multi-user work by resource locking. The proposal includes four functions which will lock/unlock resources and allow querying the lock state of the resources. In order to make locking rules as flexible as possible an XML configuration is proposed which defines locking rules. This way configuration of locking could be changed without need of changing the source code and building the executable files.

2. THE PROBLEM

The DAF specification defines functions for reading values of resource. Resource descriptions are obtained from operations on the resource query service. The interface provides a family of three operations intended to be easy to use: get_values(), get_extent_values(), and get_related_values(). A fourth operation, get_descendent_values(), is a generalization of the other three and is capable of greater optimization.

The GDA standard extends DAF in order to provide a way to change values of specified resources. The function apply_updates() has one input parameter, a DifferenceModel that describes the differences between the model before apply_updates() is invoked and the model after apply_updates() successfully completes. DifferenceModel contains four lists: headers, forwardDifferences, reverseDifferences and preconditions.

Problem with updating values with apply_updates function is that resource values can change between reading of resource and the call of update function. GDA standard defines optimistic approach to this problem. Each update operation is accompanied by precondition information that the server uses to detect conflicting concurrent updates. The precondition information represents the client's assumptions about the data, at the time the update was formulated. If these assumptions hold true later, at the time the server applies the update, then no conflicting updates have intervened. Conversely, if the assumptions no longer hold, the updates must be rejected.

Pessimistic way to control state of resource between reading and updating is proposed in this paper - a set of functions that will allow locking of the resources.

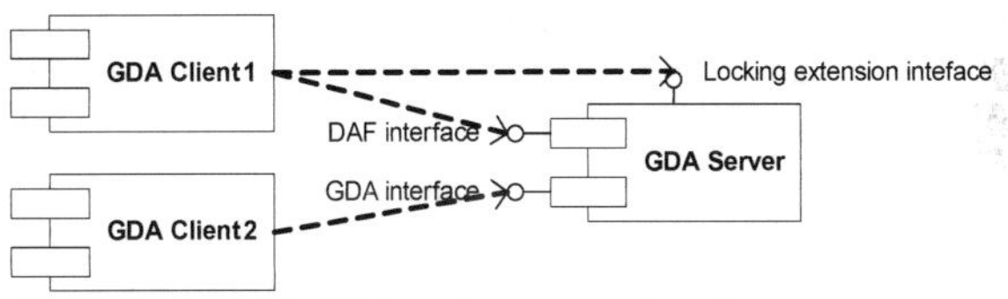

Fig. 1. Server with GDA and extension interfaces

Figure 1. shows proposed architecture with Server that implements GDA and DAF interfaces and a additional Locking interface.

3. THE SOLUTION

One possible solution for this problem is described in this paper. It proposes adding four new functions to the GDA interface which would allow users to lock specific resources by specifying the array of unique resource identifiers (ResourceID), unlock previously locked resources, get lock state of specified resource or of all locked resources.

Not all attributes of the object have same influence for the other parts of the system, for example changing Voltage level or number of input and output bays of a power substation will change topology of the system, whereas changing the name of a substation will not have such impact on the rest of the system. Therefore this solution proposes three modes of locking:

1. **Read only** - meaning that user can only view resource
2. **Restricted edit** – user can edit only attributes that do not impact functionality of other objects
3. **Full edit** – user can change any attribute of the object, or delete object

Formal Interface Description Languate (IDL) of new functions is shown in Listing 1.

```
interface gda_locking{
void Lock(in ResourceIDSequence ids, in long mode);
void Unlock(in ResourceIDSequence ids);
long IsLocked(in ResourceID rid);
void GetEditableElements(out ResourceIDSequence ids, out IntSequence modes);}
```

Listing 1. Extension functions definition (IDL)

Lock function is used to define list of resource identifiers which should be locked. It has two input parameters: ids – list of resource, and mode – list of modes in which resources should be locked. If user tries to lock the resource that is locked by another user an exception is thrown. Some types of resources could be permanently locked (e.g. system parameters that can be set only during the start-up of the system).

Unlock function is used to define list of resource identifiers which should be unlocked. It has input parameter, ids – list of resource identifiers. Only the user who locked the resource can unlock it. If another user tries to unlock that resource an exception is thrown.

IsLocked function is used to ask server for lock mode of the given resource. It has one input parameter, id – resource identifiers of the resource that is of interest to the user. Return value is mode in which resource is locked.

GetEditableElements function is used to retrieve lock state of all elements. It has two output parameters, ids – list of resource identifiers of resources which are locked and modes – list of modes in which resources are locked.

When a client application wants to change the state of the resource, it should first ask for the lock state of the resource with call of IsLocked() function. If resource is not locked it can lock it with call of function Lock(), make changes with apply_updates() function and unlock the resource by calling of Unlock() function.

With the use of these four functions a user can select whether it will use optimistic or pessimistic approach to changing the resource values. This means that a user can ignore these functions and try to update resources without locking (optimistic approach), but if a resource is locked an exception will be thrown. Another user can use pessimistic approach and lock resources before updating. This also means that existing clients can be used with server that support locking interface without changing.

4. EXTENSIBLE CONFIGURATION

In many cases locking rules could be introduced to lock resources that are logically connected. For example if a user locks a medium voltage substation, it is logical that all equipment inside this substation should be locked as well.

One possible XML configuration of locking rules is proposed in this paper. It contains one or more elements **rule**. A rule element has **name** as attribute, and up to three optional groups of elements: classes, associations and attributes. **Classes** elements contains unique identifiers of the types of resource that this rule applies to.

Associations element contain one or more **association** elements, which contain one attribute – id, the unique identifier of the association that exists in the system between elements. Element **attributes** contains one or more **attribute**. This element has two attributes, **class_id** that identifies type of object and **attrib_id** that identifies attribute of the object that should be locked

Example of XML configuration for universal locking is shown in Listing 2.

```
<locking_rules>
  <rule name="POWER OBJECT lock">
    <classes>
      <class id="Junction"/>
      <class id="Medium Voltage Substation"/>
      <class id="High Voltage Substation"/>
    </classes>
    <associations>
      <association id="Power Object Bays"/>
      <association id="Bay inner Elements"/>
      <association id="Bay connected elements"/>
    </associations>
    <attributes>
        <attribute class_id="Switchgear" attrib_id="Nominal
        voltage" />
        <attribute class_id="Fuse" attrib_id="Nominal
        current" />
        <attribute class_id="Medium Voltage Substation"
        attrib_id = "Medium voltage bays" />
    </attributes>
  </rule>
</locking_rules>
```

Listing 2. Example of XML configuration for locking

Configuration from the example illustrates the locking rule for DMS system objects: junctions, medium and high voltage substations. Rule defines that when one of this object is locked, its input elements, connected elements and its power object are locked with it in the same locking mode. Attributes that are locked with this rule are nominal voltage level for switchgear, nominal electrical current for fuses and medium voltage bays for substations.

5. IMLEMENTATIONS

Proposed solution is implemented in the GDA server for Power Distribution Management System software manufactured by DMS Group from Novi Sad (Ilic 2008). It is also used for a prototype system for software for Water distribution system (Lendak 2007, Erdeljan 2007) which was part of the scientific project sponsored by the Serbian government.

Server and clients are implemented using Microsoft Visual Studio 2005 in C++ and OmniORB CORBA implementation. Locking rules are configured in XML and successfully used for DMS projects.

6. CONCLUSION

The proposed extension of GDA interfaces allows control of resource locking in an UMS system. New functions allow locking and unlocking specified resources and querying lock state of resources. Using proposed functions, a user can ensure that no other user changes the resource state before he/she finishes working with the object.

Proposed architecture is accompanied with XML configuration which allows easy and universal definition of locking rules.

Solution is implemented in the Distribution Management System and it had shown its applicability in production systems.

7. REFERENCES

Aleksandar Erdeljan, Darko Čapko (2007), *Analysis of DAF and GDA specifications usability for UMS object model access, ,* ETRAN 2007, Herceg Novi, Montenegro

IEC - International Electrotechnical Commision (2005), *IEC 61970: Energy Management System Application Program Interface (EMS-API) – Part 403: Generic Data Access, draft Revision 7*

Ilić Slobodan, Aleksandar Erdeljan, Slobodan Zarić *(2008), Accessing O-DMS data stored in a relational database through DAF and GDA interfaces,* ETRAN 2008, Serbia

Lendak I., Erdeljan A. & Zarić S. (2007), *A generic DAF server architecture for water distribution systems*, 9th International Symposium Interdisciplinary Regional Research, May 2007, Novi Sad, Serbia

Object Management Group (2005), *Utility Management System (UMS) Data Access Facility Specification*, v2.0.1, July 2005

Annals of DAAAM for 2008 & Proceedings of the 19th International DAAAM Symposium, ISSN 1726-9679
Editor B. Katalinic, Published by DAAAM International, Vienna, Austria 2008
Make Harmony Between Technology and Nature, and Your Mind will Fly Free as a Bird
*Young Researcher and Scientist Paper / * Supervisor, Mentor*

NEW PRODUCT DEFINITION PROCESS WITH PRODUCT LIFECYCLE MANAGEMENT CONCEPT APPROACH

VUKOVIC, A[leksandar]; MIKAC, T[onci] & IKONIC, M[ilan]

Abstract: *The subject of this review is Product Lifecycle Management or PLM. It is very important to realize that PLM does not define a piece of technology, but it is a business concept that defines the approach for solving the problems in product and product information definition and management process. The main reasons for a review of this subject emerges from the need of today's enterprises for support in business decision process, collaboration and integration of tools, information's and skills across the entire enterprise. This paper presents the basic approach to PLM and its solutions today.*
Key words: *Product, Lifecycle, Management, Design, Manufacturing.*

1. INTRODUCTION

The only constant in the business world today is change. To survive and prosper in these surroundings, businesses must be able to quickly adapt to changes in the market place. They must identify and address problems, leverage technology and customer requirements faster than their competitors. Intense international competition and rapid technological change requires that companies be lean and dynamic. According to this the significance of Product Lifecycle Management is increasing, especially for companies in the high technology industries like automotive, telecommunications, and other industries where the new product development and production planning process need to be faster. In development departments widespread use of CAD technology leads to supplementation of the hardware models or prototypes by digital models. These models are tested in the similar way like prototypes.

Modern industry in a networked operational business environment almost invariably uses various information systems, like Enterprise Resource Planning systems, as aids in production planning, delivery and customer service, (Antti, S. & Anselmi, I., 2005). When it comes to implementation of new products, as well as reengineering already implemented ones, this kind of information systems are not sufficient, because in these processes enterprise becomes involved in product design generating the idea, developing concept, prototyping and manufacturing. Integration of tools like CAD, CAM, CAPP and CIM with ERP and other concepts generate PLM, which is not just about a software and technology, but also about business strategy, (Stackpole, B. 2003).

Conventional methods of planning and development are no longer sufficient to speed up planning and development processes, to reach ambitious budget targets and to manage the increasing complexity of new products. The solution lies in PLM which represents the creation of the central management system of all product and technology data and the technology used to access this information and knowledge, (Stark, J. 2006).

2. PLM DEFINITION

Product Lifecycle Management is defined as the set of activities for managing company's products across the complete lifecycle, from the early stages of product conception to the final disposal or recycling of the product (Stark, J. 2006).

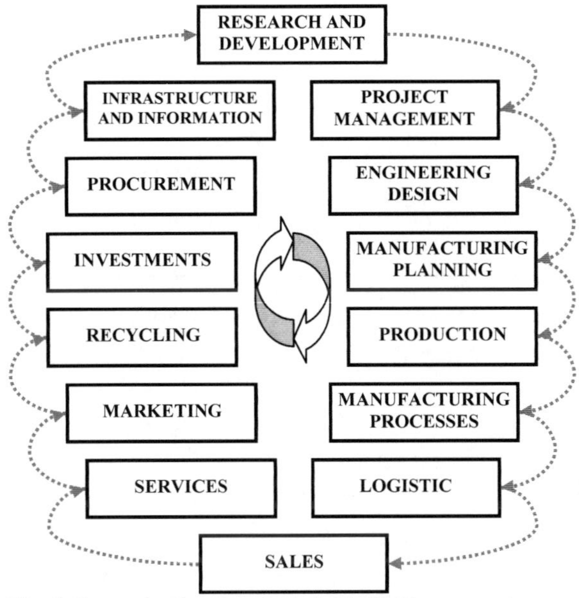

Fig. 1. Enterprise functions included in PLM

That means that PLM includes various activities from various functions of the enterprise, as shown in figure 1.
PLM, as we look it today, is relatively new concept for manufacturing companies, and it allows companies to manage their products across their lifecycles. PLM helps companies getting their products to market faster, provides better support for their use, and manage end of life better. In today's highly competitive global markets, companies must meet the increasing demands of customers to rapidly and continually improve their products and services, (Stark, J. 2006).
During product development processes, collaboration has become a common practice between different departments and companies that are involved in their activities. PLM tools can facilitate collaboration among distributed teams within the context of an extended enterprise, (Luo, Y. 2007).

3. PLM TOOLS AND METHODS

In the PLM concept great number of tools and methods are used for solving specific problems in the industrial enterprises which are confronted with new product development and innovation of products that are already in production and in the market. In this article, and the research that was carried out before it, we observed first stage only of the product lifecycle. First stage of product lifecycle encompasses research, development, design, analysis and planning.

The advantage of using PLM for first stage product development derives from the enterprise ability to effectively use knowledge's, skills and experiences from the entire network of employees, suppliers, vendors, buyers and customers. Similarly to ERP, PLM uses the same technology but with orientation to product in all stages.

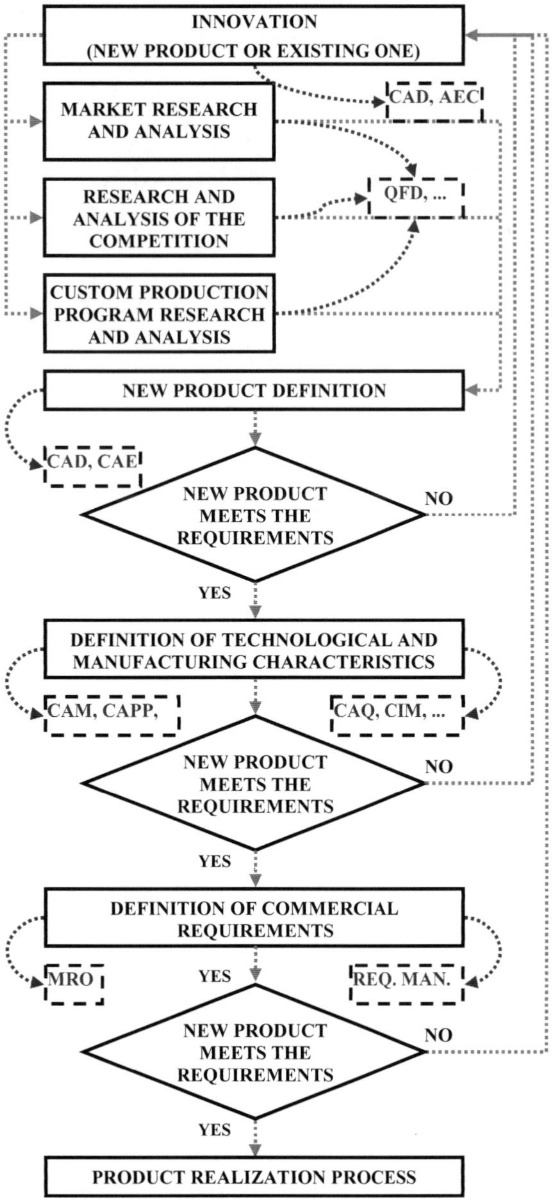

Fig 2. New product definition with PLM

The use of collaborative tools in the product development environment has been extensive in recent years. However, in other phases of product lifecycle there is no systematic methodology available for collaboration among knowledge owners, (Ameri, F., Dutta, D. 2005).

Figure 2 shows the procedure for new product definition process, from the basic idea or concept for the new product, or innovation of the existing one, to final realization of the product in the manufacturing facilities. Some of the tools the PLM concept uses in the first stage of product lifecycle are named in the dashed boxes.

PLM uses product design methodology that shapes products, having in mind future requirements for the basic design, technological and commercial requirements. These methodologies, among others, are the following:

- Design for manufacture and assembly, DFMA.
- Design for Six Sigma, DFSS.
- Engineering to order, ETO.
- New product development and introduction, NPDI.
- Concurrent engineering.
- Concurrent design.

DFMA represents the methodology for simplification of current and future product design and manufacturing process to achieve cost savings, especially in the manufacturing and assembly processes. DFSS is the methodology for developing products which will have needed characteristics for future efficient usage of six sigma tools. ETO concept is used for engineering customized products with specific requirements. Customers want products that fit their needs and product that are created on their timelines. NPDI encompasses all tools and methodologies mentioned before and it is an integration method. Concurrent Engineering is a systematic approach to integrated product development that emphasizes the response to customer expectations. It embodies team values of co-operation, trust and sharing in such a manner that decision making is by consensus, involving all perspectives in parallel, from the beginning of the product life cycle. That takes us to the Concurrent Design concept which uses concurrent engineering methodology to perform effective and fast product design studies and research using flexible teams with members from most enterprise functions included in PLM system, as shown in figure 1.

4. CONCLUSION

In the current business environment it is necessary for industrial enterprises to innovate, develop, plan and produce new products faster than the competition, and that means that the whole organization needs to rethink, reorganize, plan and perfect processes to satisfy the customers and market needs, always having in mind to keep the processes costs low.

The idea for new product is doomed if it is not objectively reviewed according to given requirements. That happens when new product ideas are selected according to subjective parameters, and also, if the process of new product definition is not followed.

In this article the process of new product development and PLM tools that follow it, is presented in its simplest form, but further research needs to be done in order to define the importance of each requirement for the product. According to overall product market success of a product every requirement has its own importance and that represent the basis for further research.

5. ACKNOWLEDGEMENTS

This paper is result of researches on scientific project supported by Ministry of Science, education and sport, Republic of Croatia, Modeling of advanced production structures at intelligent manufacture 069-0692976-1740.

6. REFERENCES

Ameri, F.; Dutta, D. (2005). Product Lifecycle Management: Closing the Knowledge Loops, *Available from:* http://www.cadanda.com/V2No5_01.pdf, *Acessed:* 2008-04-15

Antti, S.; Anselmi, I. (2005). *Product Lifecycle Management*, Springer, ISBN 3540257314, New York

Luo, Y.; (2007). *Cooperative Design, Visualization, and Engineering: 4th International Conference, CDVE 2007, Shanghai, China, September (Lecture Notes in Computer Science) (Paperback)*, Springer, ISBN:3540747796, New York

Stackpole, B.; (2003). There's a new App in Town, CIO Magazine, *Available from:* http://cio.com, *Accessed:* 2008-02-10

Stark, J.; (2006). *Product lifecycle management: 21st century paradigm for product realization*, Springer, ISBN:1852338105, London

Annals of DAAAM for 2008 & Proceedings of the 19th International DAAAM Symposium, ISSN 1726-9679
Editor B. Katalinic, Published by DAAAM International, Vienna, Austria 2008
Make Harmony Between Technology and Nature, and Your Mind will Fly Free as a Bird

AN OPEN-SOURCE IMPLEMENTATION FOR RFC3881

VULPE, A[nca] D[iana]; VULPE, M[arius] C[ristian]; SISAK, F[rancisc]; MORARU, S[orin] A[urel]; KRISTALY, D[ominic] M[ircea] & BUJDEI, C[atalin]

Abstract: *The legislation from many countries mandates that the healthcare application systems (HAS) must keep logs/audit records of all the actions they perform. For interoperability reasons, a few standards are available (like HL7 or the IETF's RFC3881) that instruct how such data has to be collected and formatted. This document proposes an overall approach for a system that manages the audit information collected from heterogeneous HASes. The format of the audit information is documented in RFC3881 and the system will be named generically AMS (Audit Management System).*
Key words: *Healthcare, Auditing, Java, Open-Source, RFC3881.*

1. MOTIVATION

The majority of the HASes have to keep records of the operations they perform preferably in a format that is standardized (like the XML schema definition format proposed in RFC3881 and in DICOM's Supplement 95). Additional requirements in this respect can also be found in IHE's "IT Infrastructure Technical Framework – Integration profiles".

However in most of the cases this information has a low-business value for the system that produces it and therefore is not in the focus of a particular HAS provider.

As a consequence, the people working for and HAS solution need a possibility to keep these records in a **secure** and **reliable** way with a **minimal effort** (the used API must be easy to use). Additionally, keeping this records must not affect dramatically the **performance** of the HAS. **Interoperability** is also in focus as systems from different vendors can be integrated at the same enterprise site. Here is a short analysis of the minimum features:

- **Secure**. Because of the privacy laws (e.g. HIPPA in United States), the audit messages' content in HAS applications are confidential. Therefore all the records must be end-to-end (from the generation to the storage) secured (either by restricting the access to them or by encrypting the transmitted information).
- **Reliable**. This implies 0% information loss (every audit message must be stored).
- **Fast**. The used API must not affect significantly the application's performance. By analogy with the logging systems, recording an audit event should have a minimal performance impact.
- **Easy to use**. The API's used for audit message generation must be easy to use by the HAS developers and integration/configuration of the API must be minimal.
- **Interoperable**. The best way to explain this feature is a quote from (http://www.ihe-europe.net/links/): "IHE in Europe promotes interoperability between IT systems in European health care. With over 280 autonomous regions, achieving a seamless flow of relevant information between IT systems of different healthcare systems presents unique challenges for cross-border interoperability."
- **Cheap**. Using the system must imply cost reductions for the customer. When referring to cost decreases two areas are in focus: maintenance costs and licensing costs.

2. AVAILABLE IMPLEMENTATIONS

The only (and best-known) publicly-available implementation for the mentioned RFC is dcm4che (http://www.dcm4che.org/). This is "a collection of open source applications and utilities for the healthcare enterprise".

Analyzing this implementation one may observe that the focus is the implementation of the DICOM standard and not the auditing part, so this is not a purely dedicated implementation for the generation, storage and management of healthcare audit information.

3. TARGET ARCHITECTURE

Figure 1 (AMS Architecture) presents the components available within the proposed Audit Management System. Essentially the system has two main development areas:
1) **Client API** that represents what typically an HAS developer would use.
2) **AMS server** representing the system that manages the client-produced audit messages.

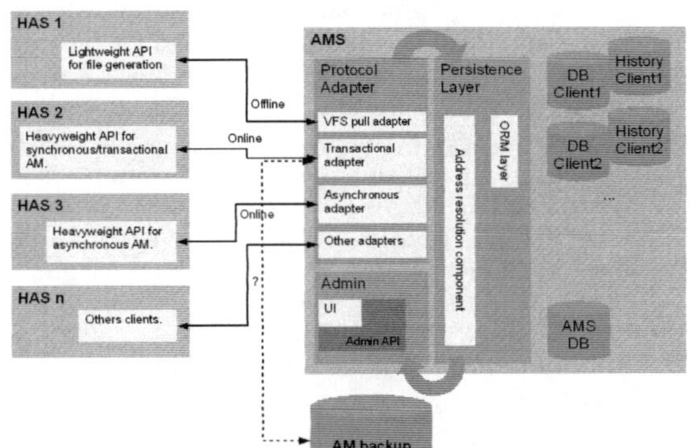

Fig. 1. AMS Architecture.

3.1 Client API

The HAS developer should have the possibility to easily integrate an AMS client API into his application. This API allows the user to:

- ✓ Create an audit message with all the parameters.
- ✓ Fire audit message events.
- ✓ Register/deregister listeners for audit messages occurrence.

The component that handles the audit message (AM) is the listener. Some possible behaviors for AM listeners are:

- ✓ Storing temporarily the audit message on the HAS application; this has the advantage that is very fast and the used libraries (dependencies) can be very small.

- ✓ Connecting to the AMS server and submitting the AM in a transactional manner; this is a heavyweight solution and can affect the performance of the HAS.
- ✓ Connecting to the AMS in an asynchronous and non-transactional manner; this is also a heavyweight solution as it might involve more dependencies for the AMS connection.

Note that other client API solutions can be imagined and the important aspect is that the user must use the same interface, but the application integrator has the freedom to switch the implementations based on the application's requirements.

Preferably, the client API will have more "flavors" (i.e. it can be composed out of Java API, C++ API, .Net API etc.). As an example, the Java API can use:

- JAXB for verification against XSD, marshaling and unmarshaling of XML-formatted audit messages. This ensures compatibility with the JEE standards documented in JSR 222 and described in the Java EE 5 Tutorial.
- Different audit message listeners (used for flat/compressed/encrypted file storage, EJB client, MDB client, Web service client), depending on the functionality the user needs.

The development of such client APIs will be done using a "per request" model and can vary a lot, depending on the platform where it will be used. As an example a tomograph can have an embedded library that generates RFC3881-compliant audit messages.

3.2 AMS Server

The idea of the AMS server is to have a centralized management solution, instead of disparate deployments (i.e. each HAS to have its own audit message system).

The main components of the AMS system are:

1. **Protocol adapter** is the part that helps gathering the audit messages from various/heterogeneous systems. Normally each type of client API has an associated protocol adapter. For example, one protocol adapter can use SCP protocol to collect the audit messages generated by several equipments while another protocol adapter can provide an end point for online EJB client.
2. **Persistence layer** facilitates keeping the audit messages in a format that allows easy handling of the information (e.g. generating reports, searching for certain audit messages, etc.). It also facilitates the discrimination between the different HAS applications/enterprise sites because it is not yet very well defined, from an architectural perspective, if the EIS persisting the audit messages will be a centralized one or different customer's data will be stored in different EISes.
3. **Enterprise information system** is the repository for audit messages. It might be composed of several sub-repositories (e.g. depending on the enterprise site ID). For easiness in data handling, in most of the cases the EIS is represented by a relational database.
4. **Administrative interface** is the interface that allows configuring the AMS server, viewing the audit information or generating reports. It is composed from two parts: an API used to connect to the persistence layer (that can also be used to develop applications) and a visual component in web format. One important scenario for the administrative interface is the capability of the user to track the evolution of a patient from the hospitalization moment until the exit from hospital (this might include both medical care operations and financial operations).

Technically, the plan for AMS server is to have it implemented as a mix between a standard Java application and a JEE application running on UNIX and using some open-source application server (like JBoss). However the main technical focus on the AMS will be the ease in maintenance and the portability (especially in the EIS area).

3.2 Backup System

Because all the audit messages have to be kept for archiving purposes, a backup system has to be provided. This system must keep the audit messages in the raw, unaltered format.

As an example, the backup system can be realized as a data storage machine equipped (as example) with Solaris and ZFS, as a huge amount of data can be stored (order of terabytes).

4. FURTHER STEPS

Currently the team working for this project has set up the architecture and started the first iterations regarding the client API.

The very next steps are to:

- ✓ create a basic end-to-end solution, as presented in the target architecture;
- ✓ find an adequate licensing model (as a business-friendly model is intended, probably the Apache license will be selected);
- ✓ choose the most convenient public hosting service (Sourceforge or Google are the considered options).

5. CONCLUSION

As the healthcare enterprises rely more and more on computing systems, the necessity of tracking the activities in such enterprises increases. There are many examples of hazard cases where patients suffered because of malfunctioning in some hardware or software, therefore some laws force the HASes vendors to provide the mentioned tracking capability.

Even if some parts of the legislation are clear, currently the healthcare IT industry is divided by various standards produced by many heterogeneous working groups (sometimes organized on geopolitical basis).

One alternative to get one of these standards the most used standard in industry is to provide publicly available reference implementations. Historically, some good examples are TCP/IP standards (versus OSI stack) or the APIs standardized using the Java Specification Request.

The current project has as a main goal the promotion of the RFC3881, by providing an open-source implementation that can be easily integrated into commercial products.

6. REFERENCES

ACC; HIMMS; RSNA, (2007). *IT Infrastructure Technical Framework*, vol 1., pp. 56-65 (http://www.ihe.net/Technical_Framework/upload/IHE_ITI_TF_4_0_Vol1_FT_2007_08_22.pdf)

DICOM Standards Committee, Working Group 14, (2004). *Digital Imaging and Communications in Medicine, Supplement 95: Audit Trail Messages*, (pp. 2-10) (ftp://medical.nema.org/medical/dicom/supps/sup95_fz.pdf)

DCM4CHE (http://www.dcm4che.org/)

Jendrock, E.; Ball, J.; Carson, D.; Evans, I.; Fordin, S. & Haase, K. (2007). *Java(TM) EE 5 Tutorial*, Prentice Hall PTR, 978-0321490292.

RFC3881, (2004). (http://www.faqs.org/rfcs/rfc3881.html)

Annals of DAAAM for 2008 & Proceedings of the 19th International DAAAM Symposium, ISSN 1726-9679
Editor B. Katalinic, Published by DAAAM International, Vienna, Austria 2008
Make Harmony Between Technology and Nature, and Your Mind will Fly Free as a Bird
2nd DAAAM-ICAT

RAPID PROTOTYPING AND VACUUM CASING TECHNOLOGY IN THE PRODUCTION OF INJECTION MOULDING TOOLS – CASE STUDY

WEISS, E[dmund] & WEISS, Z[enobia]

Abstract: *This paper concerns the use of an epoxy resin and aluminium composite in the production of forming inserts for injection moulding machines used for the production of parts for finished products and prototypes. Using the example of a knob for household equipment, the technology of the insert is presented. The problems meet by the forming of parts on the mould machine by the use of mould inserts from the composite are discussed and the recommended working methods are presented. Simulation possibilities and advantages are also described.*
Key words: *rapid tooling, mould insert, resin-composite tool, case study*

1. INTRODUCTION

Market competition requires enterprises to manufacture high quality products with a good design and to bring them on to the market as quickly as possible. All products will be judged by the investors, by the distributors, by the service support providers and finally by the end user and it is therefore necessary to limit imperfections in the new product. This can be achieved using RP, RT and VC techniques which allow for the production of the finished product, as well as prototypes and tools, quicker than using more traditional and more expensive technology such as machining, EDM etc.

2. AIM OF THE WORK

Plastics products are generally produced on moulding machines which often require complicated and expensive tools. A method is required to make the tools quickly and inexpensively and the tools should be able to produce a few dozen different models or prototypes of different shapes, colours etc. Making the moulding tool from a vacuum-caste resin-composite allows for such a possibility. These composites have distinctly different mechanical and thermal conductivity properties in comparison to steel and aluminium tools which require adjustments for individual pressing conditions. This article considers the approaches connected with the creation of mould inserts and presents a simulation of the injection moulding process. The results of experiments into the selection of special injection moulding conditions are detailed and the threats to customer acceptance and quality control. are analyzed.

3. THE MANUFACTURING OF THE MOULD INSERTS

The experiments conducted were based on a knob used for common household equipment (Fig. 1). The manufacturing process for such an element consists of a number of stages. At first a model of the product (using any technology) should be prepared and then the equipment is prepared including a ring insert, the dimensions of which are compatible with the dimensions of moulding machine's form. The following main stages of the mould insert manufacturing process can be distinguished:

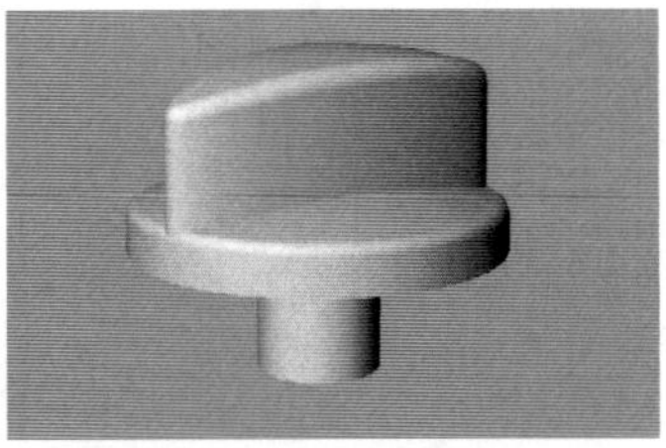

Fig. 1. Electronic model of the work.

- preparation of the prototype model (original product or its model using any technology), including plastic thermal contract,
- preparation of the equipment, especially ring insert, the dimensions of which are compatible with the dimensions of the moulding machine's form (Fig. 2 and 3),
- the determination of the model parting planes and their fixing in the tool,
- tool resin preparation, casting of the form and its vacuum degassing (EP 310 composite by MCP-HEK was used),
- solidification of composite insert at 25°C estimated time - 18-24 h.

A similar process should be applied to the second part of the form (Fig. 3). In order to strengthen the material both halves of the mould were heat treated at 40°C for about 4 hours. To further strengthen the properties of the material and thermal properties, the mould was stored at the following temperatures:

- 2h at 70°C,
- 1h at 100°C,
- 1h at 120°C,
- all night (12h) in 150°C.

The storage at the above temperatures should ensure an increase in the lifetime of the tools. The next operation involves removing the inserts and cleaning them with a machining processes. The aim of machining is to level off the contact surfaces of the moulds and the inserts. The drilling for the pusher follows.

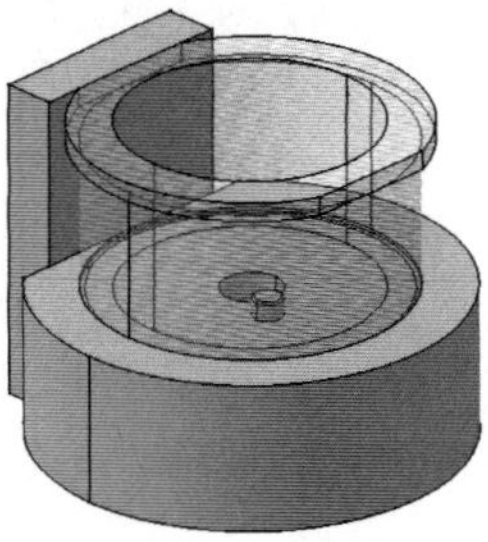

Fig. 2. Tools for preparing the first half of the mould inserts.

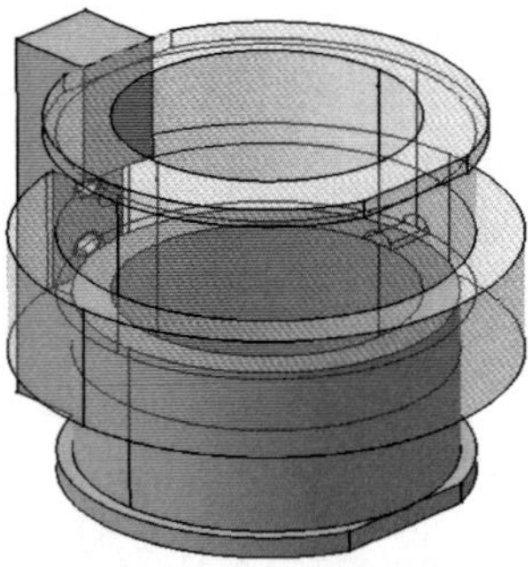

Fig. 3. Tools for preparing the second half of the mould inserts.

4. INJECTION PROCESS SIMULATION

The simulations of the flowing and cooling processes by injection enable for the early detection and assessment of behaviour patterns, and, as a consequence, can eliminate many future possible errors. It should be added that existing software supporting this area, does not deal with inserts produced by resin composites.

During the experiment the following simulation was carried out:

- injection point location analysis,
- time and probability of fulfilling of mould form,
- pressure analysis,
- changes in plastic temperatures after the injection,
- plastic bounding lines and existence of air bubbles in product,
- cooling quality and deformation possibility analysis,
- cooling quality of moulded piece using moulds produced by different materials (steel, aluminium, resin).

5. EXPERIMENTS WITH MOULDING OF KNOB IN RESIN COMPOSITE TYPE FORM

The inserts, produced using the described method (point 3), were assembled into a mould, a plate with pushers was added, the mould closing surface was checked and then the whole mould was mounted in an injection moulding machine. The next step was to carry out injection testing. The material used to made the moulded parts was PBT (Polybutylene terephthalate).The first attempt was made in similar conditions to those that would exist when using metal moulds. The experiments were conducted using the following parameters: injection pressure 70 MPa, cooling time 25 s, injection temperature 240°C, mould temperature 50°C, injection time 1.6 s.

Only 18 cycles of injections were carried out as the mandrel of the knob broke off and the plastic remained in one part of the mould. The reason for the break was an incomplete solidification process - the plastic did not solidify as the insert received too little heat from the part and the cause of this was the low thermal conductivity of the composite (1.45 W/mK). During the experiment, as a consequence of the compression of the insert's resin material under the high injection pressure (70 MPa) the material fashes at the borders of the die parting were observed, It was decided to carry out a set of tests with a reduction in the pressure. At a pressure of 50 Mpa, the moulded parts did not have such big fashes as in the first test. Further tests were carried out under the pressures of 40, 35 and 30 MPa. The moulded parts made at 50-35 MPa pressure had small fashes at the border of the die parting surfaces (about 0,05 mm). The making of moulded parts without fashes was possible at 30 MPa.

After the analysis of the results of the first experiment and after carrying out tests at a reduced pressure, the technological conditions for the second set of tests were planned. The following conditions were established: pressure: 30MPa, cooling time: 25 sec, injection temperature: 240°C, mould temperature: 40°C, injection time: 1,6 sec.

50 cycles of knob moulding were carried out and all the parts produced were of good quality. No problems of cooling were observed.

6. VALIDATION OF THE SIMULATION PROCESSES

The insert pressure established by technologists, based on experience, was, in-fact, 15 MPa higher then that estimated by the CAD/CAE system. Further experiments have shown that the best results can be achieved at a smaller pressure.

The temperature of the plastic was selected in accordance with calculations in simulation and, as in the simulation, 100% of form fulfilment was achieved.

The simulation of the cooling process for the moulded parts did show that some difficulties with the cooling of the knob mandrel are possible. The time for the cooling of the material in the form (25s) was calculated by the system. In the first set of experiments it was determined that this time was too little as the composite insert was too hot and this is why, after 17 cycles, the moulded parts solidified and remained in the form.

The analysis of the solidification shrinkage did demonstrate that there is the possibility to get a small deflection in form on the exterior of the product and this was observed in all parts.

7. CONCLUSIONS

The technology of Epoxy Tooling using Vacuum Casting for the making of the mould insert for the moulding machine from a resin composite is very useful for the production of a short-term product batch such as a prototype or information series in plastic (50-1000 pieces according to the composite used to made insert and according to the injected materials).

The strength and thermal conduction of the forms made from the resin composites are worse than metal moulding forms. That is why, in order to achieve a good quality product, there is the need to select special (lower) injections and cooling parameters.

Simulation programs are helpful in determining and eliminating problems connected with the forming of plastic materials.

8. REFERENCES

Chlebus E. (2003). *Innowacyjne technologie Rapid Prototyping – Rapid Tooling w rozwoju produktu (Innovation production technology Rapid Prototyping-Rapid Tooling in product development)*, Oficyna Wydawnicza Politechniki Wrocławskiej, ISBN 8370857477, Wrocław

Lokiec A. (2006). Symulacje procesu wtrysku w oparciu o program Cadmould (Simulation of the injection process on the basis of Cadmould programe), *Available from:* http://www.MESco.com.pl, 2006-04-20

MCP-HEK Tooling GmbH (April 2006). *Firm information*, Paderborn

Ratajczak P. (2006). *Wytwarzanie wkładek do form wtryskowych z materiałów kompozytowych w oparciu o technikę Vacuum Casting (Making of the inserts from resin composite with use of Vacuum Casting technique)*, MSc work, PUT Poznań

Weiss E. & Warczyński B. (2007). Using of Vacuum Casting Technique for Producing Functional Prototype (in Polish). In: *Zeszyty Naukowe Politechniki Poznańskiej, Budowa Maszyn i Zarządzanie Produkcją No. 5*, Wyd. Pol. Poznańskiej, pp. 89-96, ISSN 1733-1919

Annals of DAAAM for 2008 & Proceedings of the 19th International DAAAM Symposium, ISSN 1726-9679
Editor B. Katalinic, Published by DAAAM International, Vienna, Austria 2008
Make Harmony Between Technology and Nature, and Your Mind will Fly Free as a Bird

EFFICIENT MANAGEMENT IN PRODUCTION COMPANIES BY IMPLEMENTATION OF PLM

WESSELY, E[mil]; KRALIKOVA, R[uzena] & KRUPA, M[arek]

Abstract: *In present a company managed in a modern style must be market-oriented and must tend towards greater attractiveness even in the area of environmentally suitable performance. A significant step in the area of effective management of the product life cycle in production companies is also the implementation of integrated manufacturing policy, which is focused on products and services and their environmental aspects.*
Keywords: *Environment, management, product, manufacturing, innovation*

1. INTRODUCTION

Permanently changing more and more demanding requirements for the tasks related to the management and assurance of prosperity of companies as well as increasing pressure from competitors evoke constantly more urgent need of implementation of progressive methods and technologies which should assist to the companies by satisfaction of wants of their customers (Badida et al., 2001). The organisations do need a informational infrastructure for their survival, which enables to perform exact decisions in the real time, and which allows to focus fully on the customer's satisfaction. In the last years has been dynamically developing a strategy of product life cycle management (PLM), which significantly influences and conditions the success and competitive capacity of the companies.

2. LIFE CYCLE ASSESSMENT

The term product life cycle assessment (LCA) means all stages which each product goes through, i.e. the stage of raw material exploitation and material treatment for manufacturing of a particular product, production of the product itself, utilisation of a product connected eventually with the maintenance and product disposal as waste.

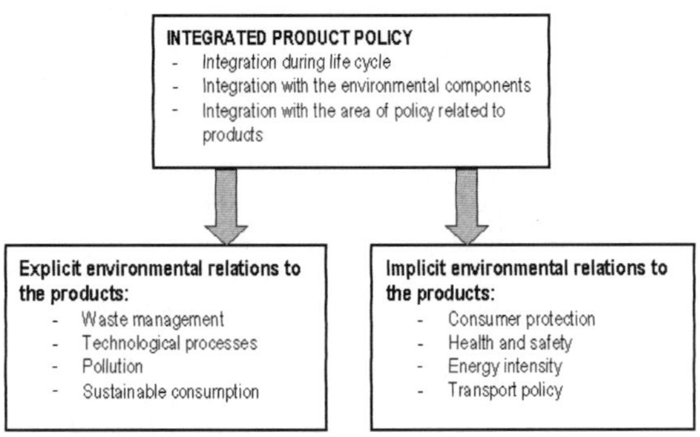

Fig. 1: Integrated product policy

An important step in the area of efficient management of the product life cycle in manufacturing companies is the implementation of an integrated product policy, which is focused on products and services and their environmental aspects. Integrated view is oriented on various aspects, to which belong:

- *Life-long cycle*
- *Environmental components*
- *Tools*
- *Areas* Fig.1.

Integrated product policy (IPP) has becoming a part of the European environmental policy and this strategy covers the measures for internalisation of environmental consciousness and responsibility into the individual decision-making processes. Integrated product policy is based on the market and stimulates the processes towards the sustainable development.

Acceptance of this strategy in the area of environmentally oriented quality management of products and services will bring to the mechanical-engineering companies significant economical and environmental benefits, which in the feedback directly influences the efficiency of functioning of every company (Katalinic et al., 2001). In fact, who approaches the market with a product – regardless of the size of company, if it is a small company or international company, follows the constantly cyclic order of the key processes:

1. **Planning of product** – covers the time period, when is considered a new product with regard on the market needs, when the management decides which product is appropriate to invest the resources to achieve the profit in the future (Product Innovation Charters - PIC).
2. **Product development** – is focused on the development of new products (Products in Development - PID), which will be immediately installed in the market including planned company sources.
3. **Product administration** – is focused on existing products in the market (Products in the Market - PIM), on their administration from their introduction in the market, when they need the support and service.

In the present is preferred predominantly the ***procedure oriented approach*** with the goal to eliminate the pollution from the industrial and other procedures. Integrated product policy is based on product orientation and approaches to the issue of pollution in integrated way from the aspect of the whole life cycle of the product and its functional parts and its utilisation (e.g. vehicle and its operation - manufacturing, fuel and lubricants consumption, maintenance, scrapping after end of its useful life). Integrated product policy also puts the accent on so called public procurement, which defines the principles of purchase of products by public institutions and proposes the instruments for the purchase with the goal to prefer the products, which are environmentally friendly during their whole life cycle.

Even though that product oriented environmental policy shall perform in the whole life cycle of a product and influence all in it participating subjects, its activity is focused in the recent practise predominantly on the producer, i.e. on the activities, which change the character of a product with the purpose to accomplish lower negative environmental impacts. Management, informational and educational instruments play the major role. At the same time it leads to the support of green markets (new instruments, for example the support of green procurement, and green orders).

3. UTILISATION OF INFORMATIONAL TECHNOLOGIES

The utilisation of information technologies and electronic communication enables to achieve the competition benefits, which is crucial factor for producers to utilise secure media for information share with partners, data from development, marketing etc. in real time. Production companies are the first, which understanding the outputs and the value „Chain of the digital product", shown below Figure 2. This chain shows like the developing teams cooperate by the information and change creation, which are essential for a new or innovated product (Ručinský et al., 2000).

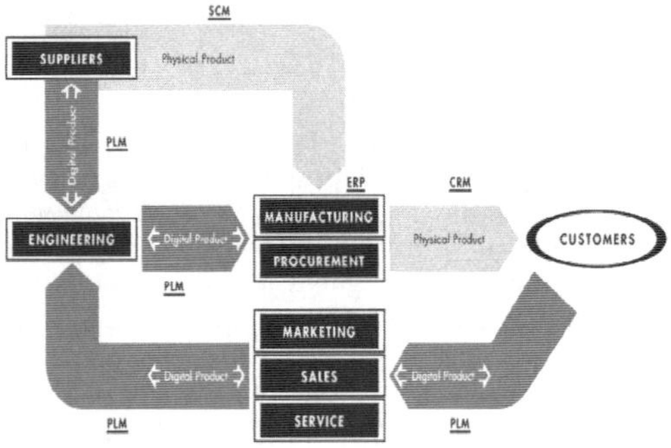

Figure 2: Chain of the digital product

4. PRODUCT LIFECYCLE MANAGEMENT

Product lifecycle management can be defined as the background created to control crucial information during life cycle including all key processes within the whole organisation and among cooperating organisations. It enables to share common processes and knowledge for the concerned product. PLM is the process of managing the entire lifecycle of a product from its conception, through design and manufacture, to service and disposal. It is one of the four cornerstones of a corporation's information technology structure. PLM is an essential element in effectively creating and utilizing global information networks, as it empowers businesses to make unified, information-driven decisions at every stage in the product lifecycle: *Plan – Define - Build –Support*.

PLM is company strategy, by means of which the organisation controls the life cycle of their product. Product Lifecycle Management or PLM is company strategy, by means of which the organisation controls the life cycle of their product. PLM is a strategic construction that covers wide scale of applications, such as Product Development System (PDM), Computer Aided Design (CAD), Enterprise Resource Planning (ERP) etc..

All companies need to manage communications and information with their customers (CRM-Customer Relationship Management) and their suppliers (SCM-Supply Chain Management) and the resources within the enterprise (ERP-Enterprise Resource Planning) (Ručinský et al., 2004). In addition, manufacturing engineering companies must also develop, describe, manage and communicate information about their products (PLM).

The companies can this way improve their infrastructure and cooperation in the value string and significantly decrease the time and costs necessary for the product development. It is a complex solution of all data administration concerning the product, or process changes. At the same time it is the basis, which supports the management and automation of product life cycle process from the first conceptual activities to its introduction in the market. It enables to communicate with the cooperating companies at the required level by using the Internet tools (Beca et al., 2004).

In present a company managed in a modern style must be market-oriented and must tend towards greater attractiveness in the area of environmentally suitable performance. This is possible to achieve by means two basic factors, which lead towards prosperity:

1. Accepting of qualitative requirements of a customer.
2. Respecting environmental standards, which guarantee sustainable development.

Proposed factors determine the trend to achieve strategic plans of the manufacturing companies in the area of environmentally oriented quality of products and services. The product quality cannot be combined only with the group of top technical parameters; it is important also operational reliability, regular supply, design, assembly, disassembly, undemanding disposal or recycling, efficient service and environmental standard.

4. CONCLUSION

In present time of global markets must the companies constantly innovate their products, processes and informational systems to remain on the smart. They must collect, store and utilise their intellectual capital. The need of efficient management of the product life cycle is given by competitive pressure and constantly growing demanding ness of customer requirements. From this view it is the strategy PLM, which offers to the producers the space for cooperation with business partners to speed up the introduction of a product on the market, decrease of manufacturing costs, increase of quality and last but not least improvement of customers satisfaction. It supports long-term environmental goals, contributes to the social and economical destructuralisation and focuses on fulfilment of social and environmental needs on the basis of sustainable development. This conclusion is just a rehash of the paper.

The paper was elaborated in connection with the projects VEGA: "Modelling of factors of working environment and their optimisation in specific conditions of the engineering companies", solved at the Department of Environmental Studies and Process Control Faculty -TU Košice.

5. REFERENCES

Badida, M., Králiková, R. & Pauliková, A. (2001): *Environmental impact assessment in mechanical engineering*, Acta Mechanica Slovaca, Vienala, č. 3, roč. 5, 1/2001, pp.57-66, ISSN 1335-2393

Katalinic, B., Pliana, S., Lioljic, B. & K Stuja (2001): *Modelling Flexible Manufacturing Systems With UML*, Proceedings of the 12th International DAAAM Symposium, ISBN 3-901509-19-4, Vienna, Austria., pp. 215-216,

Ručinský, R. & Mrázko, S. (2004): *Integration electronic markets and task information in the regulated supply of the chain. In: Almanac work from international workshop „ development small and central businesses lands of central and eastern Europe".* Praha:VŠE, 2004. ISBN 80-2450809-

Beca, M. – Ručinský, R. (2004): *E - business as strategic access to the businesses in the condition of globalization.* In: Sudzina, F. – Bašistová, A. – Mesároš, P. (eds.): Theoretic aspects of sectional economy II. 1. publ. Kosice : Royal Unicorn. ISBN 80-969181-1-7

Ručinský, R. (2002): *E - commerce and globalization and its hold on economy in Slovak republic.* In: Acta academica karviniensia. Karviná : OPF SU, 2002. ISSN 1212-415X

Annals of DAAAM for 2008 & Proceedings of the 19th International DAAAM Symposium, ISSN 1726-9679
Editor B. Katalinic, Published by DAAAM International, Vienna, Austria 2008
Make Harmony Between Technology and Nature, and Your Mind will Fly Free as a Bird

ANALYSIS OF MACHINE-TOOL SPINDLE ASSEMBLIES

WOLNY, R[yszard]

Abstract: *A computational method is presented, which enables the separation and optimization of the constructional variables of any variant of the spindle assembly. The computational analysis employed the statistical method of constructional variables separation and the optimization of significant variables by the Monte Carlo method. The separation of the constructional variables and the optimization of the significant variables of the spindle assembly of a medium-size turning lathe was made. For the real model of the turning lathe spindle assembly, a modified solution is proposed, which is characterized by better properties. In order to verify the obtained results, an attempt was made to analyse the spindle assembly using the finite element method (FEM) using an off-the-shelf computer program.*

Key words*: spindle, machine tool, variables, significance of variables, separation, optimization, FEM.*

1. INTRODUCTION

A spindle assembly constitutes a basic working system of rotary main movement. Increasingly high requirements are being set for spindle assemblies (*SA*), inducing thereby a necessity of improving their design and constructional process. The current state of computational techniques allows the determination of quantities describing the properties of an (*SA*) as early as at the initial phase of computational analysis. This enables the evaluation of these properties to be incorporated into the systems of computer-aided design. Depending on the evaluation result, a need for the modification of the (*SA*) might arise in order to improve its properties, or an attempt of its optimization might be made. In any of these cases, a need arises to separate a certain limited subset of variables from the set of all design variables describing the structure and state of the (*SA*), which influence most the evaluated properties (or the objective function), whereby they can constitute decision factors in the modification or optimization process.
A computational method has been shown, which enables the separation and optimization of design variables on the example of the real solution of a machine-tool (*SA*). In the computational analysis, a statistical method of design variables separation (Wolny, 1992) and a procedure for the optimization of significant variables by the Monte Carlo method (Findeisen et al., 1980; Polańscy, 1992) were used. The verification of the obtained computation results was performed by modelling the spindle assembly by the *FEM* method using an off-the-shelf computer program.

2. METHOD OF THE SEPARATION OF SPINDLE ASSEMBLY DESIGN VARIABLES

For the separation of the design variables of the spindle assembly, statistical methods were used, which, as combined with the computer aiding methods, form the applied theory of experiment. The experimental quantities were replaced by a set of the spindle assembly design variables (as generated using a generator of pseudo-random numbers with an equal probability distribution) and that of the results of calculations performed for the cases of matching the above variables (Wolny, 1992).

The result of using the method of the separation of (*SA*) design variables are regression equations containing all independent variables (not removed as a result of elimination) – the most important design variables of the spindle assembly. The separated significant design variables of the spindle assembly, while considering the adopted separation criterion, should represent decision variables in the successive stages of modification or optimization.

3. SEPARATION OF THE DESIGN VARIABLES OF SPINDLE ASSEMBLY

The separation of the design variables of a spindle assembly (*SA*) has been made based on a medium-size machine-tool. On the basis of an existing constructional solution (Fig. 1.a), a substitute model has been developed (Fig. 1.b). The spindle was divided into segments, each of a constant cross-section. These segments corresponded to the sections of a real spindle. As active loads, the forces and moments resulting from the machining process and the force originating from the drive were taken in the computational model.
The computational analysis was performed at five different points of application of cutting forces in the three areas A, B, C of the machining field . These areas were separated from the spindle end by a chuck length of l_u = 200 mm (Wolny 2002, Wolny 2003).
As a result of the performed analysis of the existing (*SA*) design solution, with a view to the possibility of making modifications, the initial set of ten variables was established and the intervals of variability were determined for the analyzed case of simplified model .
By using the computational capabilities of the separation method, the statistical verification of the significance of variables was made.

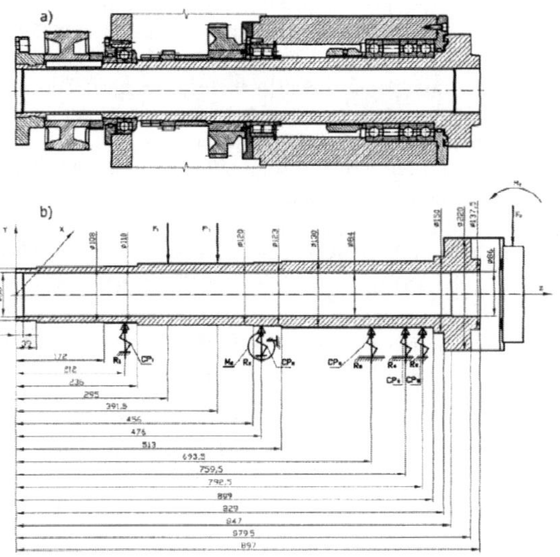

Fig. 1. Machine-tool spindle assembly: a) construction of a real system, b) substitute model.

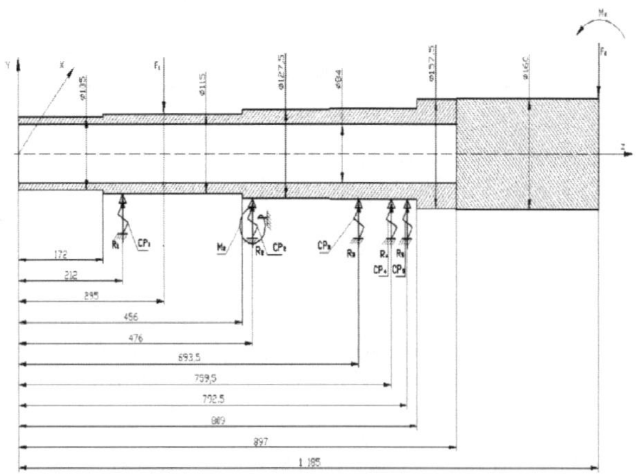

Fig. 2. Simplified model of lathe spindle assembly for MES analysis.

The result of the statistical procedure were regression equations containing significant (*SA*) design variables in the three areas A, B, C of the machining field.

The separation of the significant machine-tool (*SA*) variables made a basis for their optimization and then the modification of the existing design solution of the spindle assembly.

4. OPTIMIZATION OF THE SIGNIFICANT DESIGN VARIABLES

The application of the computational optimization methods required the decision variables, the limiting conditions and the optimization criterion to be defined, which were presented in a function form, i.e. as the objective function.

For the case of optimization of design (*SA*) variable values in the area B of the machining field, the separation of six significant variables out of the initial set of ten variables considerably reduced the size of the computational task. The significant variables (*SA*) were taken as decision variables, while considering the established limitations (the actual headstock size), whereas the regression equations formed simple mathematical models of the objective function (Wolny, 2002; Wolny, 2003).

The tasks of optimization of significant machine-tool (*SA*) variables were carried out by the Monte Carlo method available in the CADEX: PROFES program package for the number of iterations N=10000 (Polańscy, 1992).

After separating the significant variables and based on the optimization results an attempt was made to modify the actual machine-tool (*SA*) design solution.

Three solutions resulted, which were a trade-off between the computational analysis results and the design and constructional expertise and experience.

Ultimately, a modified (*SA*) solution was proposed, which was characterized by the best static properties.

5. THE ANALYSIS OF THE SPINDLE ASSEMBLY BY THE FINITE ELEMENT METHOD

Based on the equivalent system of the spindle assembly, a simplified model (*SA*), of the turning lathe was built (Fig. 2), which constituted a basis for the computational analysis by finite element method in the COMSOL computer system used.

The geometry of the spindle model was created using CAD tools available in the *Structural Mechanics* module. The available 3D primitives in the form of cylinders were used, which were connected into a single object.

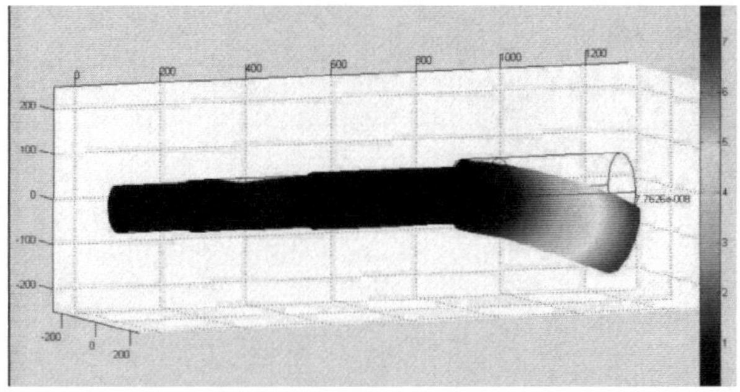

Fig. 3. Geometry deformation of lathe spindle assembly model.

The active loads in the form of generalized forces resulting from carrying out the cutting process and a force originating from the drive were modelled.

The passive loads were reactions in the bearing supports modelled as surfaces with a uniformly distributed load corresponding to the rigidity of the bearings.

Figure 3 shows deformations of the geometry of the turning lathe model (*SA*) analyzed. Locations which ate the most susceptible to static deformations under the active loads are marked by change in colour.

The performed (*SA*) computational analysis of the turning lathe has confirmed the rightness of use of the *FEM* as a means of verification of previously used computational methods.

6. CONCLUSION

The computational analysis of spindle assemblies presented in the paper, consisting in the use of the constructional variables separation method, analytical optimization methods and computational verification by the finite element method (*FEM*), should constitute an integral part of computational systems in the computer-aided process of designing machine tools and other technological machines.

7. REFERENCES

Findeisen, W.; Szymanowski, I.; Wierzbicki, A. (1980). *Teoria i metody obliczeniowe optymalizacji*, (The theory and the computational methods of optimiaztion), PWN, Warszawa

Polańscy, R. and Z. (1992). *Program CADEX: PROFES 5.0*, (CADEX: PROFES 5.0 Program), CERMET, Zakład Postępu Technicznego i Wdrożeń, Dział Zastosowań Informatyki, Kraków

Wolny, R. (1992). Metoda rozdziału parametrów konstrukcyjnych zespołu wrzecionowego obrabiarki w procesie projektowania, (The metod of separation of constructional parameters of machine tool spindle assembly in design process), *PhD thesis, ITMiA Politechniki Wrocławskiej,* Wrocław

Wolny, R. (2002). Optymalizacja zespołów wrzecionowych obrabiarek poprzedzona analizą statystyczną, (Optimization of the machine tool spindle units after statistical analysis), *Zeszyty Naukowe Wydziału Mechanicznego nr31*, Wydawnictwo Uczelniane Politechniki Koszalińskiej, Koszalin

Wolny, R. (2003). Analiza obliczeniowa zespołów wrzecionowych obrabiarek w komputerowo wspomaganym procesie projektowania, (Computational analysis of machine tool spindle units in computer aided design process), *Prace Naukowe ITMiA Pol. Wrocławskiej Nr 84, Seria: Konferencje Nr 41,* Oficyna Wydawnicza Politechniki Wrocławskiej, Wrocław

Annals of DAAAM for 2008 & Proceedings of the 19th International DAAAM Symposium, ISSN 1726-9679
Editor B. Katalinic, Published by DAAAM International, Vienna, Austria 2008
Make Harmony Between Technology and Nature, and Your Mind will Fly Free as a Bird

NEW SOLUTION OF THE TECHNOLOGICAL PROCESS OF PRESSURE CASTING DIE

WOLNY, R[yszard]

Abstract: *In this paper now solution of the technological process of movable and immovable pressure casting die insert of pump executed in one of motor industrial plant was presented. Machining processes of casting die elements were described in detail. Individual operations and the process of machining parameters selection were characterized. Machine applications were generated for the virtual 3D model of movable and immovable die inserts in Mastercam application operating in Windows environment. An attempt at modification of technological process for selected HSM-applied milling operations (high-speed machining) in Mastercam was made. As the result of modifications it turned out that machining time of rough milling operation has been significantly reduced, which means achieving considerable savings.*

Key words*:* *pressure casting die, technological process, machine tool, Mastercam, HSM, optimization.*

1. INTRODUCTION

The pressure casting die is a tool for making a pressure casting on a pressure die-casting machine. The basic components of the pressure casting die include: a movable and a fixed insert with a forming shape, a casing, slide blocks, servomotors and a cooling system. A technological process of making pump body inserts, as carried out in an automotive industry works (Walczak, 2004) is presented. The pump body profile (Fig. 1) was a complex profile, therefore the die inserts reproducing its shape had also a respectively complex construction. The repeatable production of such complicated shapes and, above all, the verification of the product as early as at the design stage, was only possible owing to the use of state-of-the-art CAD/CAM software applications. They enabled the design of virtual 3D shapes and automatic generation of the machine codes of working for numerically controlled (NC) machine tools. The reduction of manufacturing costs was possible through the application of new manufacturing strategies, in particular the *HSM* working (Zalewski et. al., 2005).

Fig. 1. Pump body – top view.

2. THE TECHNOLOGICAL PROCESS OF MAKING THE PRESSURE CASTING DIE

The technological process of making the pump body pressure casting die concerned the making of the inserts forming the casting shape on a pressure die-casting machine, i.e. the movable and the fixed insert.
The framework technological process of making the forming inserts was carried out in the following stages (Wolny & Rygałło, 2005):

- normalizing of the starting material,
- roughing on conventional machine tools (milling machines, drillers),
- stress relief annealing after the rough machining,
- pre-machining on NC machine tools,
- execution accuracy check no. 1
- heat treatment – hardening,
- grinding of the forming shape and holes,
- finishing by milling on NC machine tools,
- spark machining of complex shapes,
- grinding, retouching, polishing and fitting of the inserts,
- execution accuracy check no. 2,
- testing of the die on the pressure die-casting machine,
- approval and authorization of the die for production,
- covering of the forming shape with a protective coating.

The basic stages of the technological process of making the pump body insert pressure casting dies were machining operations.

3. ROUGING OF THE STARTING MATERIAL

The basic roughing operation was the milling of starting material bars of dimensions of 120 x 260 x 320 for the movable insert and 100 x 260 x 320 for the fixed insert, and the so called "skinning" of the profile, so that the least possible material allowance was left before the milling of the proper forming shape on the numerically controlled machine tools. The roughing was carried out on conventional machine tools.
The next operation was the making of holes in the starting material bars according to the insert working drawings. Drilling of holes was performed on a jig drilling machine to an accuracy of 0.01 mm.

4. MACHINING OF THE MOVABLE INSERT

The generation of working machine programs required building virtual 3D models of movable and fixed insert shapes. The insert models were created in the Ideas application (www.ideas), and then were exported in the Parasolid format to the Mastercam program (www.zalco). Milling machine programs and machining simulation were made in the Mastercam application running under the MS Windows environment.

The roughing on the NC milling machine was the pre-selection of the material from the movable insert forming shape region using a computer-generated program.

The rough milling program was generated for a Ø25/R5 head the following parameters:
-axial travel Z = 1 mm,
-tool rotations, n = 2200 rpm,
-working feed, v_f = 880 mm/min.,
-sinking feed, f_w = 100 mm/min.,
-allowance on the machined surface 0.3 mm.

The advantage of the CAD/CAM software application was the possibility of making machining simulation and verifying its correctness.

The finishing operation was the milling of the insert shape by the contouring methods, that is moving the mill along the contour by horizontal layers.

A program for the milling of the movable insert contour with a Ø16/R1 mill was generated for the fixed parameters:
-travel in the Z axis = 0.3 mm,
-tool rotations, n = 3000 rpm,
-working feed, v_f = 1400 mm/min.,
-sinking feed, f_w = 100 mm/min.

The Bacplot function used within the Mastercam enables the visualization and evaluation of mill paths for the contouring program. The verification of the contouring machining was carried out, as, irrespective of the correctness of the mathematical surface description, an incorrect tool pass might have occurred, leading to material damage.

The subsequent finishing milling operation was contouring the movable insert profile with a Ø12/R1 mill.

To obtain the final movable insert shape, a program for the linear finishing of the overflow basins of the pressure casting die was created. These are hollows of a specified shape serving for the collection of any access of liquid metal forced in under pressure in the casting process. As the basins had a radius-ended bottom, the most favourable in this case was to use a Ø10/R5 ball end mill.

The final finishing milling operation was the formation of the movable insert profile with the Ø4/R1 hob by the lining method. This made it possible to finally bring off the shape and obtain the proper surface quality. The contour milling provided little roughness of steep surface, while the linear finishing method was the most appropriate for horizontal surfaces or those with little inclination. After performing the linear finishing and making verification, a virtual shape of the movable insert of the pump body pressure casting die was obtained.

The obtained movable insert shape required additionally the removal of material from locations inaccessible to the mills. Therefore, the finishing of the shape on a high-speed milling machine (a Ø3 mill; n=20000 rpm) and spark machining of the rib shape, performed after heat treatment by hardening, was necessary.

5. MACHINING OF THE FIXED INSERT

The machining of the fixed insert was similar to that of the movable insert. The only differences resulted from a different forming shape (Wolny & Rygałło, 2006).

The roughing on an NC milling machine was pre-milling of the shape using a computer-generated contouring program for a Ø25/R5 mill.

The finishing working was made up of three operations of milling the insert shape by the contouring method using a Ø12/R1, Ø8/R1 and a Ø6/R1 mills. An additional finishing milling operation was the formation of the insert profile with the Ø6/R1 mill by the lining method.

After completion of the finishing milling operation and making final verification, a virtual model of the fixed insert shape was obtained.

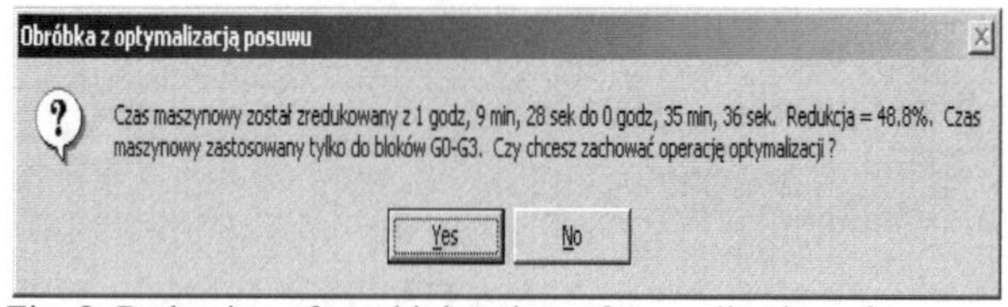

Fig. 2. Reduction of machining time after application of Mastercam HSM function.

The obtained insert shape required additionally the spark machining of three recesses on the conical surface, to be made after heat treatment by hardening.

6. HSM WORKING AND DIE INSERT FEED OPTIMIZATION

When modifying the technological process of making die inserts, the *HSM* milling strategy with variable feed values available within the Mastercam was used. The program offered a possibility of shortening the milling time from about ten to several tens percent through the automatic modification of machining parameters.

The high-speed feed (*Highfeed*) module dynamically varied (optimized) the feed during milling. The result was a reduction of the machining time without any deterioration in the surface quality and an extension of the tool working lifetime.

A modification to the technological process of machining roughing of pressure casting die inserts on NC machine tools was made. The result of this modification was a reduction of the machine time of machining operations by approx. 49% for the movable insert (Fig. 2) and 29% for the fixed insert.

7. CONCLUSION

Modern NC machine tools, properly selected cutting tools, implemented CAD/CAM systems and the *HSM* (high-speed machining) technology all enabled a considerable reduction of the machine time of selected machining operations. The savings were the greater, the larger were the series of parts machined and higher cutting speeds and feeds.

8. REFERENCES

Walczak, W. (2004). Proces technologiczny wybranej części formy ciśnieniowej wraz z programem na OSN. (Technological process of chosen element of the pressure casting die with the program on CNC machine tool). *Praca dyplomowa inżynierska, Politechnika Częstochowska, ITMiAP,* Częstochowa.

Wolny, R.; Rygałło, A. (2005). Proces technologiczny formy ciśnieniowej korpusu pompy z wykorzystaniem HSM. (Technological process of pressure casting die for pump body with HSM application). *Inżynieria Maszyn, R. 10, z. 4,* Oficyna Wydawnicza Pol. Wrocławskiej, ISSN-1426-708X, Wrocław.

Wolny, R.; Rygałło, A. (2006). Technologia HSM na przykładzie obróbki formy ciśnieniowej. (Technology of HSM on the exemple of pressure casting die machining). *International Conference „Production Engineering",* IBMiA, Pol. Wrocławskiej, Oficyna Wydawnicza Pol. Wrocławskiej, ISBN 83-7085-989-5, Wrocław.

Zalewski, A.; Osiak, A.; Sobieski, S.; Chłodnicki, Z. (2005). *Mastercam 9, część 2.(Mastercam 9, part 2).*TIZ IMPLEMENTS, ISBN 83-920967-1-1, Warszawa.

http://www.ideas.pl, Accessed: 2008-07-29

http://www.zalco.pl/mastercam.php, Accessed: 2008-07-29

Annals of DAAAM for 2008 & Proceedings of the 19th International DAAAM Symposium, ISSN 1726-9679
Editor B. Katalinic, Published by DAAAM International, Vienna, Austria 2008
Make Harmony Between Technology and Nature, and Your Mind will Fly Free as a Bird

MATHEMATICAL MODEL OF A CLOSED LOOP OSCILLATING HEAT PIPE IN PRESENCE OF MAGNETIC FIELD

WONGKASEM, K[an] & RITTIDECH, S[ampan]

Abstract: *A mathematical model for predicting the oscillating motion and heat transfer in a closed loop oscillating heat pipe in presence of magnetic field is presented. A theoretical of Magnetrohydrodynamic (MHD) is conducted to investigate the driving force due to oscillating motion. From the results, it can be concluded that magnetic field can increase the oscillating motion and raise heat transfer rate. Efficiency of oscillating heat pipe is enhanced more 8 percents. The results of this study will significantly help to interpret the activation of oscillating motion by magnetic field and offer a solution on how to enhance the heat transfer rate in an oscillating heat pipe*
Keywords: *Mathematical Model, Heat pipe ,Magnetic Field , Magnetrohydrodynamic*

1. INTRODUCTION

Nowadays, the world is in a constant energy crisis and energy seldom lacks. Energy recovery is a very important thing to consider. Heat pipe is the one equipment which referred to energy recovery. It is very useful for heat ventilation in the system. Many preceding research have been studied to enhance the efficiency of heat pipe by many procedures. Magnetic field is method that many researchers used. Magnetic field can produce force and energy on working fluid (Zuo et al, 2001). Elizbieta et al., 2005 studied the effect of magnetic field on vertical heat pipe. The effects of magnetic field on liquid- gas were investigated (Feng et al., 2005). From the past study, there are only a few studies used magnetic field to enhance oscillating heat pipe in spite of it is equipment with high efficiency and heat transfer rate (Zhang et al, 2004). Hence, the efficiency enhancement of oscillating heat pipe by magnetic field is to deserve mentioning. To reach full potential of a closed loop oscillating heat pipe in presence of magnetic field, developmental of mathematical modeling is necessary.

The goal of this study is to develop a mathematical modeling of oscillating motion and heat transfer of heat pipe in order to predict and investigate the effects of magnetic field on oscillating motion and heat transfer rate of oscillating heat pipe from the simulation results.

2. MATHEMATICAL MODEL OF OSCILLATING MOTION

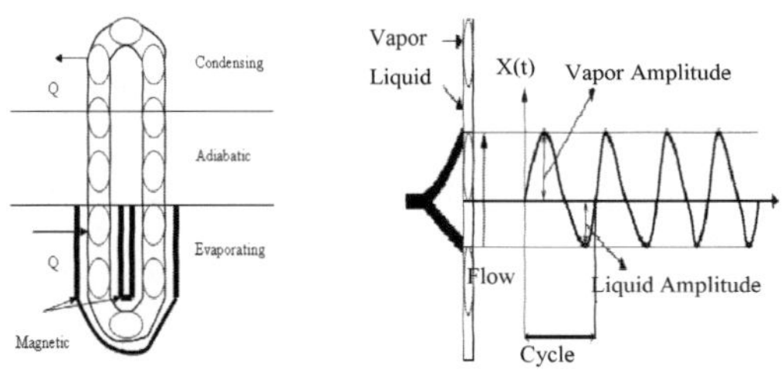

Fig. 1. Model of oscillating motion

A model of oscillating motion of closed loop heat pipe with magnetic field is shown in Fig.1. To obtain a reasonably accurate mathematical model of oscillating motion, Magnetrohydrodynamics (MHD) theoretical and conventional force in closed loop heat pipe were utilized. Model was developed under the following assumptions:

1. At steady state, liquid slug and vapor plug are of equal size and flow in the same direction.
2. Magnetic field is present only in evaporating section.
3. The length of evaporating, adiabatic and condensing section is equal.
4. Liquid is incompressible and vapor plug is assumed to be ideal gas.
5. Simplify the accumulation of force by Newton's second law.

From the second law of Newton

$$\sum F = m\left(\frac{d^2x}{dt^2}\right) \tag{1}$$

For oscillating heat pipe, there are hydrodynamics force and conventional force acted on liquid slug and vapor plug.

$$m\left(\frac{d^2x}{dt^2}\right) + F_f\frac{dx}{dT} = F_d + F_{mag} - F_v \tag{2}$$

where

$F_d = A\Delta P_d = A\left(\frac{h_{fg}\rho_{vc}}{Te}\right)\left(\frac{Te - Tc}{2}\right)(1+\cos\omega t)$ driving force (N)

$F_f = A\Delta P_f = A\left[(f_{l=}.Re_l)\left(\frac{\mu_l L_l}{2D_h^{\ 2}}\right) + (f_v.Re_v)\left(\frac{\mu_v L_v}{2D_h^{\ 2}}\right)\right]\frac{dx}{dT}$ frictional force (N)

$F_v = A\Delta P_v = \frac{\rho_v RTx}{L_v}$ vapor force (N)

$F_{mag} = A\Delta P_{mag} = \sigma\mu_r^{\ 2}\mu_o^{\ 2}H^2u$ magnetrohydrodynamics force (N)

Equation (2) became to

$$(L_l\rho_l + L_v\rho_v)A\frac{d^2x}{dT^2} + \left[\frac{8Pe}{D_h}(\mu_l L_l + \mu_v L_v)\right]A\frac{dx}{dT} + \frac{A\rho_v RTx}{L_v} = \left(\frac{Ah_{fg}\rho_{vc}}{Te}\right)\left(\frac{Te-Tc}{2}\right)(1+\cos\omega t) + \sigma\mu_r^{\ 2}\mu_o^{\ 2}H^2u \tag{3}$$

Equation (3) can be then written in the form of the equivalent ordinary differential equation of force damp vibration.

$$\frac{d^2x}{dT^2} + \frac{c}{m}\frac{dx}{dT} + \frac{k}{m}x = \frac{B}{m} + \frac{B}{m}\cos\omega t + \frac{F_{mag}}{m} \tag{4}$$

$$\frac{d^2x}{dT^2} + 2\varpi\xi\frac{dx}{dT} + \varpi^2 x = \frac{B}{m} + \frac{B}{m}\cos\omega t + \frac{F_{mag}}{m} \tag{5}$$

Hence, the Oscillating motion of closed loop heat pipe with magnetic field is

$$x(t) = \frac{B + F_{mag}}{m}\left[\frac{\sqrt{\zeta^2-1}\sin(\omega t) - e^{-\zeta\omega t}\sinh\left[\omega(\sqrt{\zeta^2-1})t\right]}{2\zeta\omega^2\sqrt{\zeta^2-1}} + \frac{1 - e^{-\zeta\omega t}\left[\cosh((\sqrt{\zeta^2-1})\omega t) + \frac{\zeta\sinh((\sqrt{\zeta^2-1})\omega t}{\sqrt{\zeta^2-1}}\right]t}{\omega^2}\right] \tag{6}$$

where $m = (L_l\rho_l + L_v\rho_v)A$, $c = A\frac{8P}{D_h}(\mu_l L_l + \mu_v L_v)$

$k = \frac{A\rho_v RT}{L_v}$, $B = \left(\frac{Ah_{fg}\rho_{vc}}{Te}\right)\left(\frac{\Delta T\max - \Delta T\min}{2}\right)$

$$F_{mag} = \sigma \mu_r^{\,2} \mu_o^{\,2} H^2 u \quad , \quad \omega = \sqrt{\frac{k}{m}} \quad , \quad \xi = \frac{c}{2m\omega}$$

A = cross-sectional area(m^2), C = specific heat (J/kg·K)
D = diameter(m) , f =friction factor, G = mass flux (kg/s·m^2)
h = heat transfer coefficient (W/cm^2·°C),
h_{fg} =latent heat(kJ/kg), L= length (m) , m = mass(kg)
p = pressure (N/m^2), P = perimeter(m), Pr = Prandtl Number
R = gas constant(J/kg-K), Re= Reynolds number
T = temperature(°C) , u = velocity (cm/s), x =coordinate(m)

3. MATHEMATICAL MODEL OF HEAT TRANSFER

Heat transfer of oscillating heat pipe can be modeled as a system of thermal resistance.
The total heat transfer can be defined in equation (7).

$$h_{TP} = \frac{Q_{hp}}{(Te - Tc)}\left(\frac{1}{Ae} + \frac{1}{Ac}\right) \quad (7)$$

Heat balance at evaporating section is

$$Q_{hp} = Q_{hw} - Q_w - Q_{ea} \quad (8)$$

And heat balance at condensing section is

$$Q_{hp} = Q_{cw} - Q_w + Q_{ca} \quad (9)$$

The total heat transfer coefficient in evaporating section is

$$htt = h_{NB} + h_C + h_{MHD} \quad (10)$$

where Q_{hp} is net heat transfer from evaporating to condensing
Q_{hw} is heat in put at evaporating section
Q_{cw} is heat out put at condensing section
Q_{ea} is heat loss to environmental at evaporating
Q_{ca} is heat loss to environmental at condensing
Q_w is heat transfer to wall by conduction
h_{NB} is Nucleate boiling heat transfer coefficien(w/cm^2.°c)
hc is force convection heat transfer coefficient(w/cm^2.°c)
h_{MHD} is magnetrohydrodynamic heat transfer coefficient (w/cm^2.°c)

4. RESULTS AND DISCUSSION

The results can be divided into 2 sections. All parameters used in the simulation were followed the previous study of Ma, (2006) to compare and confirm the results.

4.1 Results of oscillating motion

Model was simulated under the conditions: working fluids were water and salt solution; operating temperatures were varied among 40 °C, 60 °C, and 80 °C; pipe lengths were 1 m., 2 m., and 3 m.; and magnetic field strength was risen from 0-1.2 Tesla. The results of oscillating motion are presented in Fig.2. The results indicate that magnetic field increases the oscillating motion and amplifies the oscillating in the same phase. Magnetic fluid (salt solution) can receive more energy from magnetic field than water. Both of operating temperature and the different temperature effect for oscillation of working fluid. Higher operating temperature can increase more oscillation. In addition, higher different temperature also provides more oscillation. Total elongate length of tube gives more oscillation and can change phase oscillation.

4.2 Results of heat transfer

Model was simulated under the conditions: magnetic field strength rised from 0-1.2 Tesla, working fluid was salt solution, condensing temperature was 20°C, evaporating temperature was 80°C, pipe length was 1 m., and pipe diameter was 1.65 mm. The results of heat transfer are presented in Fig.3. Magnetic field can raise 50 Watts of heat output. It means that the efficiency is enhanced 8% more. Heat output and heat pipe efficiency are slightly increase, when magnetic field strength is increased.

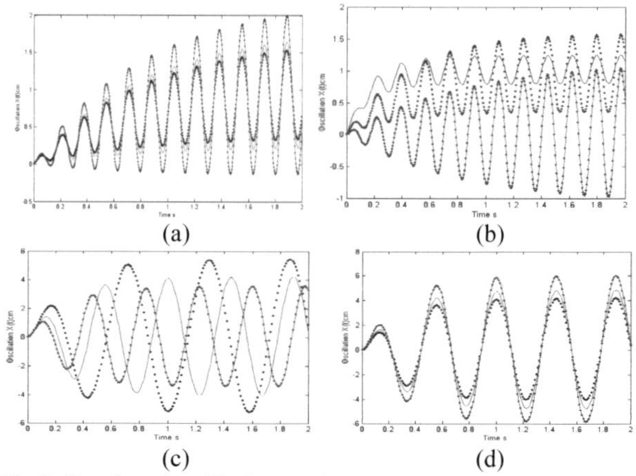

Fig.2. Results of oscillating motion
(a) Motions of working fluid when operating temperature 80 °C
(b) Motions of salt solution, operating temperature was varied
(c) Motions of salt solution when total length was varied
(d) Motion of salt solution with various magnetic field strength

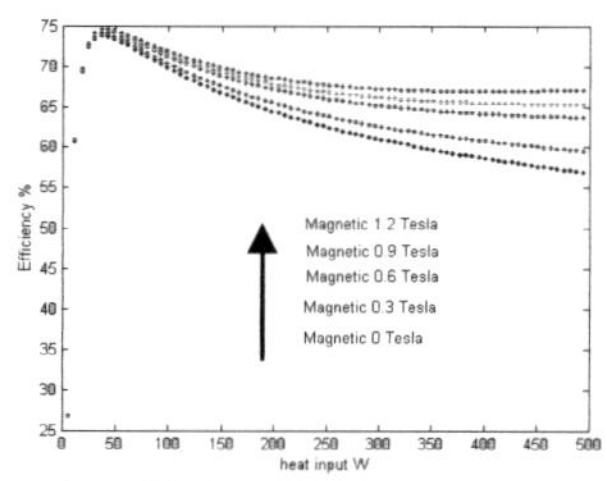

Fig.3. Heat transfer efficiency with magnetic field strength

5. CONCLUSIONS

Mathematical model of closed loop oscillating heat pipe is developed to study and describe the influence of magnetic field on oscillating motion and heat transfer. The accuracy of the model is conformed by others study. The results have a good agreement.This study will significantly help to interpret the activation of oscillating motion by magnetic field and offer a solution on how to enhance the heat transfer rate in an oscillating heat pipe.

6. REFERENCES

Elizbieta Fornalik, Piotr Filar, Toshio Tagawa, Hiroyuki Ozoe, and Janusz S.Szmyd (2005). *Experimental study on the magnetic convection in a vertical cylinderr,* J. Experimental Thermal and Fluid Science

Feng -Chen Li, Tomoki Kunugi, and Akimi Serizawa (2005). *MHD effect on flow structures and heat transfer characteristics of liquid metal–gas annular flow in a vertical pipe,* J. International Journal of Heat and Mass Transfer

Ma, H. B.,Hanlon, M.A., and Chen, C. L.(2006). *An Investigation of Oscillating Motionsin a Miniature Pulsating Heat Pipe,* Microfluidics and Nanofluidics

Zhang,X.M.,Xu,J.L.,and Zhou Z.Q.(2004). *Experimental Study of a Pulsating Heat PipeUsing FC-72, Ethanol, and Water as Working Fluids*, Experimental Heat Transfer

Zuo,J., North M. T., and Wert K. L.(2001). *High Heat Flux Heat Pipe Mechanism for Cooling of Electronics,* IEEE Transactions on Components and Packaging Technology.

Annals of DAAAM for 2008 & Proceedings of the 19th International DAAAM Symposium, ISSN 1726-9679
Editor B. Katalinic, Published by DAAAM International, Vienna, Austria 2008
Make Harmony Between Technology and Nature, and Your Mind will Fly Free as a Bird
2nd DAAAM-ICAT

SELECTIVE LASER MELTING FOR MEDICAL APPLICATIONS

YADROITSEV, I[gor]; BERTRAND, P[hilippe]; LAGET, B[ernard] & SMUROV, I[gor]

Abstract: *Laser additive technology of free-form manufacturing of objects from functionally graded materials represents promising direction to solve challenging medical problems: (a) medical applications require exclusive design and individual functionality for each single case; (b) biocompatibility and biointegration of a porous implant in the living tissue depend strongly on its surface structure. Selective Laser Melting makes it possible to create fully functional parts directly from metals without using any intermediate binders or any additional processing steps. Experiments were carried out on PHENIX PM100 machine with different metal powders: stainless steel grades 904L, 316L, CoCr, Inconel 625. The influence of different manufacturing strategies on the internal structure and mechanical properties of samples was analysed. The following features were studied: (a) specially structured surface, (b) miniaturized and (c) regular porous structures.*

Key words: *Selective laser melting; Metallic powders; Porous materials; Implants*

1. INTRODUCTION

Selective laser melting (SLM) is advantageous in comparison with the other methods for manufacturing parts with complex geometry and inner structure since SLM is a manufacturing process where parts are made directly from 3D CAD data without any special tooling (Wohlers, 2006; Bourell et al., 2002). Of the existing engineering processes for the manufacture of porous materials for biomedical and life science applications SLM technology is of particular interest (Wehmöller et al., 2005). Implants and endoprostheses for repair of damaged and lost bone tissue have been produced (Webb, 2000; Gibson, 2006). They are constructed as spatial lattices with ordered or random configuration and size of the cells. Such structures provide optimal conditions for the direct bone regeneration (invasion by tissue from adjacent bone).

The SLM technology allows to create a special rough and porous surface of the dental crowns permitting to cover them with resistant ceramic coatings. Fabricated by this technology, an implant or an artificial root for supporting the denture introduced into the jaw bone represent constructions similar to a natural tooth. It is possible to form a partly opened cavity in a body implant with the porous surface (Wohlers, 2006). Pores of the specific size and configuration allow the bone tissue to sprout freely through the implant and form a strong connection.

2. RESULTS AND DISCUSSION

2.1 Experimental procedure

Experiments were carried out on PHENIX PM100 machine with different gas-atomized spherical metal powders: stainless steel grades 904L (-16 µm), 316L (-25 µm), CoCr (-25 µm), Inconel 625 (-25 µm). PM 100 machine has the typical for SLM equipment design (Bourell et al., 1992; Levy et al., 2003) and earlier it was reported in details ([a, b]Yadroitsev et al., 2007). The source of radiation is YLR-50 cw fiber laser manufactured by IPG Photonics with a maximum power P = 50 W, the wavelength λ = 1075 nm and the laser spot size $d \sim$ 70 µm. The particle size distribution and morphology was analyzed by a granulo-morphometer ALPAGA 500NANO (OCCHIO s.a.).

2.2 Regular inner structures

The SLM technology allows producing samples with regular periodic structures representing the so-called opened porosity where the pores have the required size and a specially designed geometry.

A sample with micro-sized lattice structure is shown in figure 1. The size of the cell is about 150×150 µm, the walls are about 120 µm-thick. As it can be seen, the cells are practically regularly orthogonal.

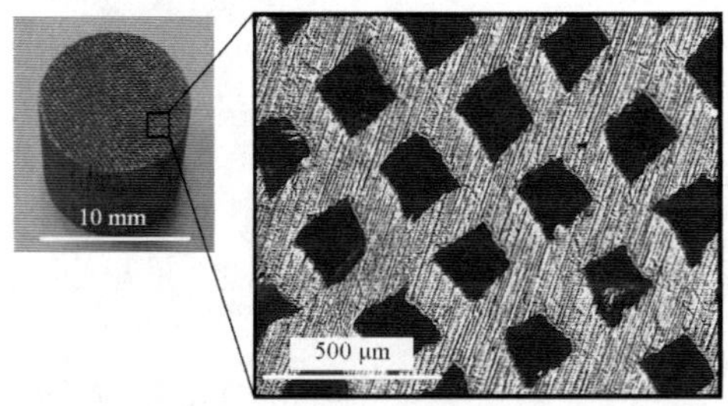

Fig. 1. Sample from stainless steel grade 316L (-25 µm) with micro-sized lattice structure. Size of the cell is 150×150 µm, the wall thickness is 120 µm.

A number of samples from stainless steel grade 316L characterized by a fine-walled porous structure were manufactured (Fig. 2).

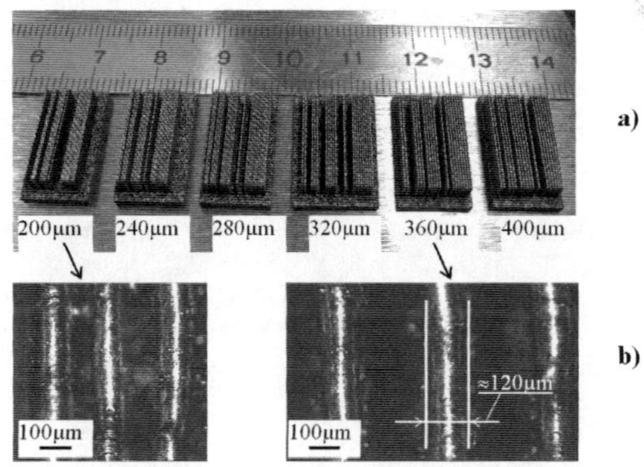

Fig. 2. Fine-walled model from stainless steel grade 316L powder. The laser power 50 W, spot diameter 70 µm, scan speed 0.14 m/s, layer thickness 50 µm, hatch distance 200÷400 µm with a step of 40 µm: (a) general view; (b) top view of the thin wall.

Each sample consisted of a set (1, 2, 3, 5 и 10) of consecutive fine walls with the minimum possible width corresponding to one laser pass. Hatch distance between vectors changed from 140 to 400 µm with a step of 20 µm. Samples with approximately 120 µm-wide walls were obtained. Fine structure of a sample relies on the possibility to achieve a narrow wall-pattern with an open pore structure. A regular gap between the walls was achieved at the hatch distance ~200 µm. At a smaller hatch distance the walls were often connected between them and thus a regular through-porosity was replaced by the closed one.

2.3 Models with specially structured surface

Using X-ray photographs and plotting the CAD-models of a joint, it is possible to manufacture an artificial limb taking into consideration the individual features of an organism for each given case. An artificial hip joint duplicates the natural hip joint elements. It includes a socket and a swirling head of femur thus allowing a normal range of movements (Fig. 3).

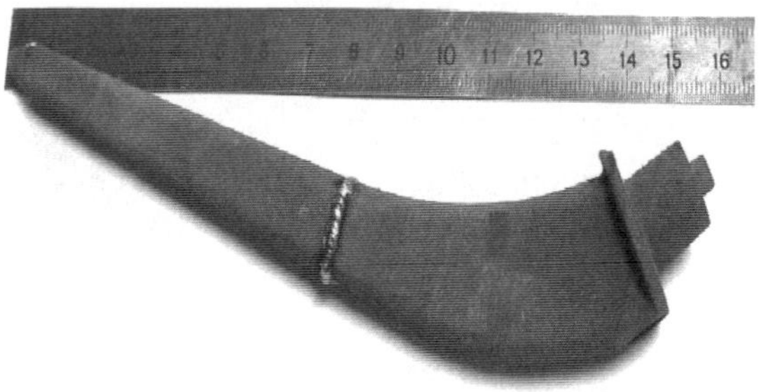

Fig. 3. General view of the fabricated hip joint model.

Figure 4 represents a model of the sacral vertebra with specially structured surface morphology consisting of ridges and hollows on which a biocompatible coating may be deposited.

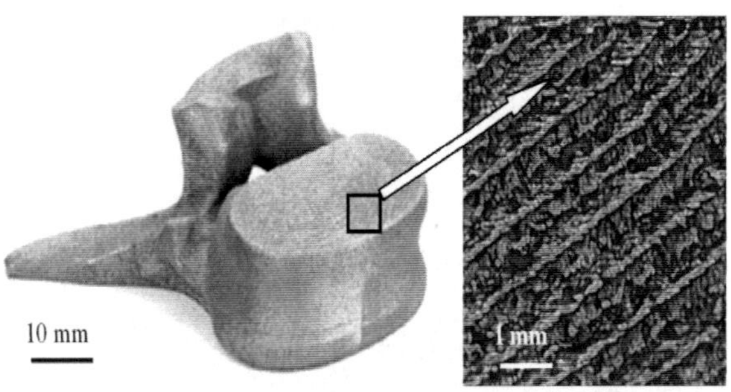

Fig. 4. Model of the sacral vertebra with specially structured surface.

2.4 Influence of process parameters on mechanical properties

Metal alloys continue to be used as the principal material for fixed prosthetic restorations and will likely be the principal material in the near future. No other material has the combination of strength, modulus, wear resistance, and biologic compatibility that a material must have to survive long term in the human body as a fixed prosthesis (Wataha & Messer, 2004).

SLM technology is designed to produce near-full-density objects with mechanical properties comparable to those of conventional materials. A number of experiments were carried out to study the influence of the different manufacturing strategies on the anisotropy of the samples inner structure and their mechanical properties such as the breaking strength. The existence of the anisotropy of the modulus of elasticity was found out while manufacturing samples by the SLM method. The average value of Young Modulus for the samples produced from Inconel 625 is 141±5 GPa for the laser scanning perpendicular to the samples axes and is 204±2 GPa for the parallel ones. The latter value is close to Young Modulus for the cast Inconel 625 (205.8 GPa) ([b]Yadroitsev et al., 2007).

As known, bones are porous media; their structure organized according to the direction of the physiological load and is highly variable and it depends on the anatomical site. When manufacturing an implant, it is necessary to take into account vectors of the loadings that occur during the functioning of the osteo-articular system and ligaments apparatus in order to develop the best mechanical properties in the direction of the maximal loading.

At the given laser parameters the optimum thickness of the powder layer is 20–60 µm. A further increase in layer thickness leads to the drops formation that means in a considerable deterioration of sample quality ([a]Yadroitsev et al., 2007).

3. CONCLUSION

Selective Laser Melting technology allows synthesizing complex-shaped objects with tailored properties for medical applications. Creating implants with non-porous internal structure that exhibits strength of a wrought material and porous coating that enhances bone ingrowth into the implant can broaden the range of their functional possibilities. Spatial resolution of the applied SLM technology is about 100 µm. These characteristics satisfy for conditions of bone tissue regeneration.

4. REFERENCES

Bourell, D., Marcus, H., Barlow, J., Beaman, J. (1992). Selective laser sintering of metals and ceramics. *Int. J. Powder Metall.*, Vol. 28, No. 4, 369-381, 0888-7462.

Bourell, D., Wohlert, M., Harlan, N., Das, S., Beaman, J. (2002). Powder Densification Maps in Selective Laser Sintering. *Adv. Eng. Mater.,* Vol. 4, No 9, 663-669, 1438-1656.

Gibson, I. (Ed.) (2006). *Advanced Manufacturing Technology for Medical Applications,* Wiley, 0-470-01688-4, 254 p.

Levy, G.N., Schindel, R., Kruth, J.-P. (2003). Rapid manufacturing and rapid tooling with layer manufacturing (LM) technologies, state of the art and future perspectives, *CIRP Ann. Manuf. Technol.* Vol. 52, No. 2, 589-609, 0007-8506.

Wataha, John C., Messer, Regina L. (2004). Casting alloys. *Dent. Clin. N. Am.,* Vol. 48, 499–512, 0011-8532.

Webb, P.A. (2000). A review of rapid prototyping (RP) techniques in the medical and biomedical sector. *J. Med. Eng. & Tech.,* Vol. 24, 149-153, 0309-1902.

Wehmöller, M., Warnke, P., Zilian, C., Eufinger, H. (2005). Implant design and production – a new approach by selective laser melting. *International Congress, CARS 2005*: Computer Assisted Radiology and Surgery, Vol. 1281, 690-695, Lemke, H.U., Inamura, K. , Doi, K., Vannier, M.W., Farman, A.G. (Ed.), June 2005, Elsevier, Berlin.

Wohlers, T. (2006). *Wohlers Report. Rapid Prototyping & Manufacturing State of the Industry, Annual Worldwide Progress Report*, 0-9754429-2-9, 250 p.

[a]Yadroitsev, I., Bertrand, Ph., Smurov, I. (2007). Parametric analysis of the selective laser melting process. *Appl. Surf. Sci.,* Vol. 253, Issue 19, 8064-8069, 0169-4332.

[b]Yadroitsev, I., Thivillon, L., Bertrand, Ph., Smurov, I. (2007). Strategy of manufacturing components with designed internal structure by selective laser melting of metallic powder. *Appl. Surf. Sci.,* Vol. 254, Issue 4, 980-983, 0169-4332.

Annals of DAAAM for 2008 & Proceedings of the 19th International DAAAM Symposium, ISSN 1726-9679
Editor B. Katalinic, Published by DAAAM International, Vienna, Austria 2008
Make Harmony Between Technology and Nature, and Your Mind will Fly Free as a Bird

PRIORITY ARRANGEMENT FOR PRODUCT DESIGN ASSESSMENT

YIANGKAMOLSING, C[hana] & ERIK, B[ohez L.J.]

Abstract: *Priority arrangement of customer requirements is indispensable in product design. Product designer can choose properly the important features from the priority of customer demand. However, there are many assessments in product design that product designers have to be considered carefully. The Universal design concept is a one of the design evaluation for everyone can use the product or service. In addition, many customer requirements or evaluations are not crisp. It is very difficult to capture uncertainty resulting from vagueness of linguistic terms in natural language. Fuzzy set can be defined by assigning customer needs from product design assessment. The objective of this research is to establish a priority of customer demand in product design assessment by using fuzzy logic inference. With this method, product designer can improve the product to respond to most customer needs in short time.*
Key words: *fuzzy logic, product design assessment*

1. INTRODUCTION

Product assessment is very necessary in the product development process which is the sequence of steps or activities which enterprise employs to conceive, design, and commercialize a product (Ulrich &, 2003). Customer requirements and assessment are important to product designer. Moreover, it is difficult to design the products which response to all customer requirements. Hence, priority arranging of customer requirements are needed to make up during design step.

Priority of customer demand was solved by fuzzy logic and the output variable is the singleton. The most priority demand is the total score of an important score multiply with a satisfaction score. (Chen et.al, 2004) Qualitative and qualitative data for evaluation of industrial robotic system were assessed in evaluation process. Fuzzy hierarchical structure was purpose to describe multi-criteria evaluation. (Kahraman et.al, 2007) Moreover, Competitor's information was purposed with new customer requirements ranking method. The success of a product depends not only on whether it met customer requirement, but also on how compares with competitor products. The purpose method was considered competitor position, current performance and customer's view point to product rating. (Lai, et.al, 2008) In contrast, Employee evaluation and selection system were presented in a hierarchical structure to carry out the business goals and strategy of organization. The process of matching an employee with certain job is performed through a competency-based fuzzy model. (Golec, & Kahya, 2007) Universal design assessment was evaluated in packaging design by elderly, disability, children, women, and men. Because the aging people are increasing and people more interested in convenience, the product need to be consider all requirements in different groups. (Yiangkamolsing et,al, 2008) Nevertheless, the set of customer requirements is not precise, and sharp. Proablity theory is not capable to capture uncertainty resulting from the vaguenessof linguistic terms in natural language. Product designer should know the priority requirement of the customer to improve their product as demand priority precisely. The objective of this research is to prioritize customer demand by fuzzy logic with universal design criteria. A case study in this research is a special size calculator in Thailand. The result shows that the product designer can use to make of the prioritized demand to redesign the product meeting most customer need in time. Because product design needs to answer the requirements of general customer, disability, and elderly, mass customization should be considered with product manufacturer. QFD will be applied in the future research.

2. FUZZY LOGIC

The process of demand priority in this research is starting with rating important and satisfaction scores in every criterion from customer. Priority demand is the important score multiply with a satisfaction score. The next step is the fuzzification converting the scores into fuzzy number. Fuzzy inference rules can be fired from the input score. Outputs from the rules are calculated from defuzzification.

3. EXPERIMENT

P1 Equitable use	1A This product is as usable for me as it is for anyone else.
	2A Using this product doesn't make me feel segregated or stigmatized.
	3A This product gives me needed privacy, security, and safety.
	4A The design of this product appeals to me.
P2 Flexibility in use	1B I can use this product inwhatever ways is effective for me.
	2B I can use this product with either left or right side (hand or foot) alone.
	3B I can use this product precisely or accurately.
	4B I can use this product whatever pace I want (slowly or quickly)
P3 Simple and intuitive use	1C This product is as simple and straight forward as it can be.
	2C This product works just like I expect it to work.
	3C I understand the language used in this product.
	4C The most important features of this product are most 1bvious.
	5C This product let me know I using the right way.
P4 Perception inormation	1D I can use this product without hearing.
	2D I can use this product without vision.
	3D I can easily identify the feature of this product in order to use instruction or telephone help line.
	4D I can use this product with the aids devices or techniques that I use.
P5 Tolerance for error	1E The product features I use most are the easiest to reach.
	2E This product protects me from potential hazard.
	3E If I make a mistake, It won't cause damage or hurt me.
	4E This product forces me to pay attention during critical tasks.
P6 Low physical effort	1F I can use this product comfortably without awkward movement or uncomfortable posture.
	2F I can use this product without exerting movement.
	3F I can use this product without having to repeat any motion enough to cause fatique or pain
	4F I don't have to rest after using this product.
P7 Size and space for approach & use	1G It's easy to me to see all important elements of this product from any position.
	2G It's easy to me to reach all important elements of this product from any position.
	3G This product fits my hand size
	4G There is enough space for me to use this product with the device or assistance that I need.

Tab. 1. Universal design principle

The assessment in this research is used the key concepts of product assessment for universal design to evaluate the calculator product. Its details are shown in table 1. (Story et.al,

1998) There are 10 disability evaluators who evaluted the calculator and provided satisfaction score and importance score.

3.1 Input and Output

The important scale and the satisfaction scale estimated by evaluator by a number in the interval [0, 10]. Fuzzy input express in *short*, *medium*, and *high*. Assume further that these linguistic terms are represented by the triangular fuzzy numbers as shown in fig.1. Demand priority from fuzzy logic has five distinctions characteristics expressed in natural language as *very low*, *short*, *medium*, *high*, and *very high* in the interval [0 1]. A reasonable definition of fuzzy numbers representing this linguistic expression is shown in fig 2. Metadata in this research are not only importance score and satisfaction score but also the users' information such as gender, age, type of disability.

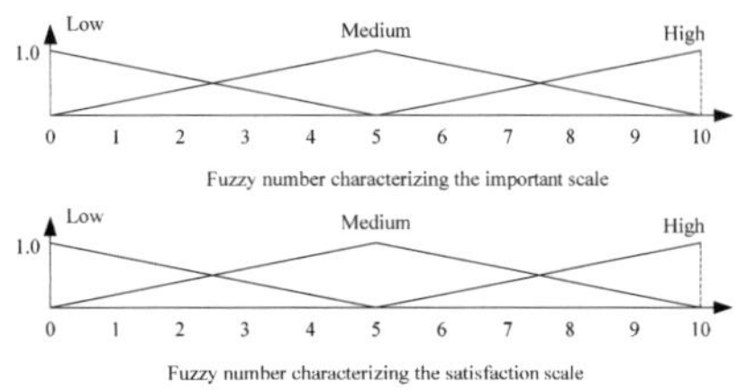

Fig 1. Fuzzy number representing three levels of important scale and satisfaction scale.

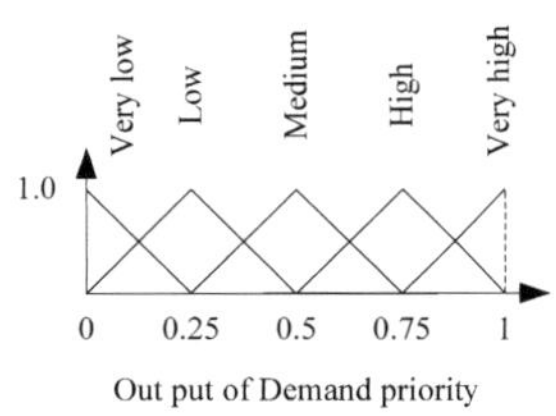

Fig 2. Fuzzy number representing five levels of demand priority.

3.2 Rule evaluation

Knowledge of experienced users can be expressed by conditional fuzzy preposition called fuzzy inference rules as following rule and table 2.

IF the **importance** is *low AND* the **satisfaction** is *low THEN* the **priority of demand** is *very low*.
IF the **importance** is *low AND* the **satisfaction** is *medium THEN* the **priority of demand** is *low*.
IF the **importance** is *low AND* the **satisfaction** is *high THEN* the **priority of demand** is *medium*.
IF the **importance** is *medium AND* the **satisfaction** is *low THEN* the **priority of demand** is *low*.
IF the **importance** is *medium AND* the **satisfaction** is *medium THEN* the **priority of demand** is *medium*.
IF the **importance** is *medium AND* the **satisfaction** is *high THEN* the **priority of demand** is *high*.
IF the **importance** is *high AND* the **satisfaction** is *low THEN* the **priority of demand** is *medium*.
IF the **importance** is *high AND* the **satisfaction** is *medium THEN* the **priority of demand** is *high*.
IF the **importance** is *high AND* the **satisfaction** is *high THEN* the **priority of demand** is *very high*.

	Satisfaction		
Importance	*Low*	*Medium*	*High*
Low	Very low	Low	Medium
Medium	Low	Medium	High
High	Medium	High	Very high

Tab. 2. Inference rule for demand priority

4. RESULT AND CONCLUSION

The priority score can be calculated from the important score and the satisfaction score and the result developed from MATLAB's fuzzy logic toolbox as shown in table 3. Percentage of priority rating of P. can be calculated from $\Sigma Wi/n$, n represents the number of criteria in the rule. The priority can be arranged from P3 (69.28%), P1 (69.275%), P2 (66.55%), P7 (58.975%), P6 (58.225%), P5 (56.1%)

This research use fuzzy logic to assess users' requirements of calculator product. It is very easy to capture uncertainty resulting from vagueness of linguistic terms in natural language. From this case study, it is found that users interest in P3 (simple and intuitive to use the product), P1 (every body can use it equitably), and P2 (flexibility in use) respectively.

No.	Importance	Score	Satisfaction	Score	W_i	P
1	I1A	10	S1A	9	0.84	2.77
2	I2A	8	S2A	6	0.652	
3	I3A	6	S3A	7	0.612	
4	I4A	7	S4A	8	0.667	
5	I1B	7	S1B	7	0.629	2.66
6	I2B	8	S2B	7	0.667	
7	I3B	10	S3B	8	0.797	
8	I4B	6	S4B	6	0.569	
9	I1C	8	S1C	9	0.72	3.46
10	I2C	7	S2C	8	0.667	
11	I3C	8	S3C	9	0.72	
12	I4C	8	S4C	10	0.797	
13	I5C	6	S5C	5	0.56	
14	I1D	4	S1D	6	0.489	2.31
15	I2D	9	S2D	10	0.84	
16	I3D	2	S3D	6	0.413	
17	I4D	6	S4D	6	0.569	
18	I1E	9	S1E	10	0.84	2.24
19	I2E	6	S2E	4	0.5	
20	I3E	7	S3E	4	0.542	
21	I4E	3	S4E	3	0.362	
22	I1F	7	S1F	5	0.605	2.33
23	I2F	6	S2F	7	0.612	
24	I3F	7	S3F	6	0.612	
25	I4F	5	S4F	5	0.5	
26	I1G	8	S1G	7	0.667	2.36
27	I2G	7	S2G	7	0.629	
28	I3G	7	S3G	5	0.605	
29	I4G	6	S4G	3	0.458	

Tab. 3. Important score, satisfaction score, and priority score

5. REFERENCES

Chen, C.C., Chen, L.C., & Lin, L.(2004), Methods for Processing and Prioritizing Customer Demands in Variant Product Design, *IEE Tsransaction*, Vol.36, pp.203-219

Golec, A., Kahya, E., (2007), A fuzzy model for competency-based employee evaluation and selection, *Computer & industrial engineering*, Vol.52, pp.146-161

Kahraman C., Cevik, S., Ates N.Y., & Gubay, M. (2007), Fuzzy Multi-criteria evaluation of industrial robot system, *Computer & industrial engineering*, Vol.52, pp.414-433

Lai, X., Xie, M., Tan, K.C., & Yang, B. (2008), Ranking of customer requirements in a competitive environment, *Computer & industrial engineering*, Vol.54, pp.202-214

Story M.F., Mueller J.L., & Mace R.L.,(1998) *universal design file*, Center of Universal Design, pp.33-45.

Ulrich, K.T., & Eppinger, S.D.(2003), *Product Design and Development*, McGraw-Hill, Singapore

Yiangkamolsing, C., Bohez, E.L.J., Jirapinyo, O., & Bueren, I., (2008), Universal design in plastic pouch packaging, *proceeding of 16th IAPRI world conference on packaging*, Thailand, 8-12 June 2008, Bangkok

Annals of DAAAM for 2008 & Proceedings of the 19th International DAAAM Symposium, ISSN 1726-9679
Editor B. Katalinic, Published by DAAAM International, Vienna, Austria 2008
Make Harmony Between Technology and Nature, and Your Mind will Fly Free as a Bird
*Young Researcher and Scientist Paper / * Supervisor, Mentor*

OPTIMIZING THE LIFE CYCLE OF AEROSPACE PRODUCTS USING ACCELERATED LIFE TESTING

ZAHARIA, S[ebastian] M[arian] & MARTINESCU, I[onel]*

Abstract: *Today reliability testing represents an important field in a product's development cycle. This paper presents the importance of implementation of accelerated life testing (ALT) in a product's life cycle, but also the major advantages yielded by it. The case study introduced the concept of accelerated life testing in aviation by testing 12 helicopter blades. The statistical analysis and data testing plan were realized using the ALTA 7 software.*

Key words: *Reliability, Lifecycle, ALT, ALT plans.*

1. INTRODUCTION

The reliability appeared as a result of the significant importance that the problems of operating safety of industrial equipment, devices and components have gained, representing today a top technique indispensable to engineers (Martinescu, 2005). Figure 1 presents a product's life cycle; accelerated life testing is part of the phase of internal verification (Ireson et al., 1995).

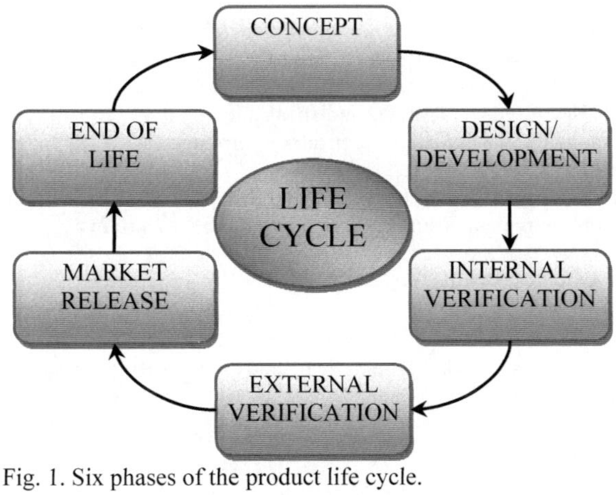

Fig. 1. Six phases of the product life cycle.

2. BACKGROUND THEORY

Increasing demand on product reliability and shortening of product development cycle have motivated research in accelerated life testing which aims alleviating such pressure (Nelson, 2004).

The accelerated tests have been introduced in the lifecycle of a product from the necessity of ascertaining its behavior in conditions of overstressing in the stages of qualification and production (Guo & Pann 2007). The principle of accelerated tests is to intensify the stress parameters with the purpose of obtaining data regarding the product's reliability in conditions of economic efficiency and to shorten the duration of the test (fig. 2). The accelerated tests don't need to be realized by increasing the restrain in relation to the conditions of exploitation, unless the acceleration factor is known and controlled. The accelerated life tests have as objective the determination of reliability indicators (usually the failure rate or

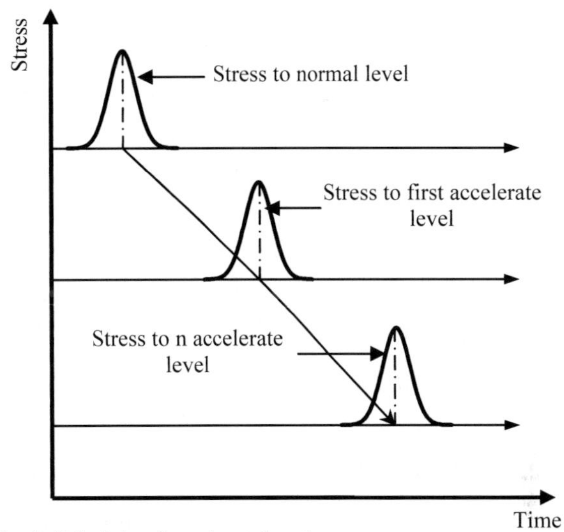

Fig. 2. Principle of accelerated testing.

the mean life), and are achieved by increasing the stress over the nominal margins, with the condition of maintaining the mechanisms responsible for the occurence of failures. The following can be used as stress parameters: temperature, tension, humidity, mechanical stress (shocks, vibrations).

Accelerated life testing (Nelson, 2004) is used in electronics, (resistors, lasers, liquid crystal displays, electronic bounds, switches, circuit breakers, relays, cells and batteries) in the study of metals and composite materials, but also for certain components and mechanical assemblies (automobile parts, hydraulic components, tools, bearings, gears). In aerospace industry, the accelerated life testing is used to test certain components: the engine, the oil pumps, fuel pumps, the landing gear, onboard electronic equipment and stiffening components (strips, spear).

3. CASE STUDY

This paper describes the accelerated life tests on the Puma IAR 330 helicopter blades with the purpose of determining the mean life in normal operating conditions. These components were chosen for testing because of the extended testing time for the blades that can reach up to 1.6 million cycles in normal operating conditions. The major advantages of introducing accelerated life testing in the aviation industry are as follows (Zaharia et al., 2007):

- It reduces the testing time, resulting in faster marketing time;
- Lower product development costs;
- Lower warranty costs;
- It improves product design and manufacturing.

To conduct the accelerated life tests a test plan has been made using the ALTA 7 software (it is the first, and still the only, commercially available software package designed expressly for quantitative accelerated life testing data analysis).

The parameter that intensifies in order to determine the mean life is the bending moment; in normal operating conditions the bending moment has the value of 200 Nm (Newton meter) with a maximum value of 300 Nm.

The mechanism of blade degradation is fatigue; the Weibull distribution and Inverse Power Law model are the best suited for mechanical stress. The test plan is described in figure 3.

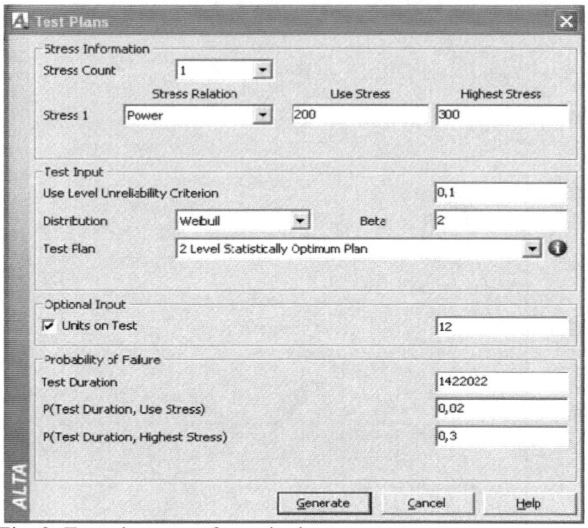

Fig. 3. Test plan setup for o single stress test.

Using this dialogue box two level of acceleration will result: the first level of acceleration will be 230 Nm and the second level 300 Nm. The tests have been made on a Brequet test bench at acceleration levels calculated with the software ALTA7 and the following results occurred (Table 1):

The data resulted from the tests were introduced in the ALTA7 software, with which the data analysis will be made and the mean life of the blades at 200 Nm in normal conditions will be determined. In figure 4 we can see the two distributions of failure time at the accelerated stress levels (230 and 300 Nm). The plot representation Life vs. Stress is frequently used for the Inverse Power Law model to ascertain the mean life. From figure 4 results that for the normal testing level the mean life for the blades is of 1.414.619 cycles. At the tests for the approval of the blades made by the producer in normal testing lavel a mean life of 1.422.022 cycles has resulted. We can see that the mean life obtained in accelerated level and presented above (Table 1) and that obtained by the producer is very similar (in accelerated testing level 1.414.691 cycles and in normal testing level 1.422.022 cycles). By using the accelerated tests in the life cycle of the tail rotor blades the testing time was reduced 4.34 times (the total testing time in accelerated testing level was of 3.929.035 cycles and the total testing time in normal level obtained by the producer was of 17.064.270 cycles).

Crt. No.	Time to failure [cycles]	Bending Moment [Nm]
1.	387.234	230
2.	456.723	230
3.	560.819	230
4.	589.090	230
5.	623.120	230
6.	732.109	230
7.	78.902	300
8.	80.132	300
9.	89.672	300
10.	98.098	300
11.	109.458	300
12.	123.678	300

Table 1. Results of accelerated tests.

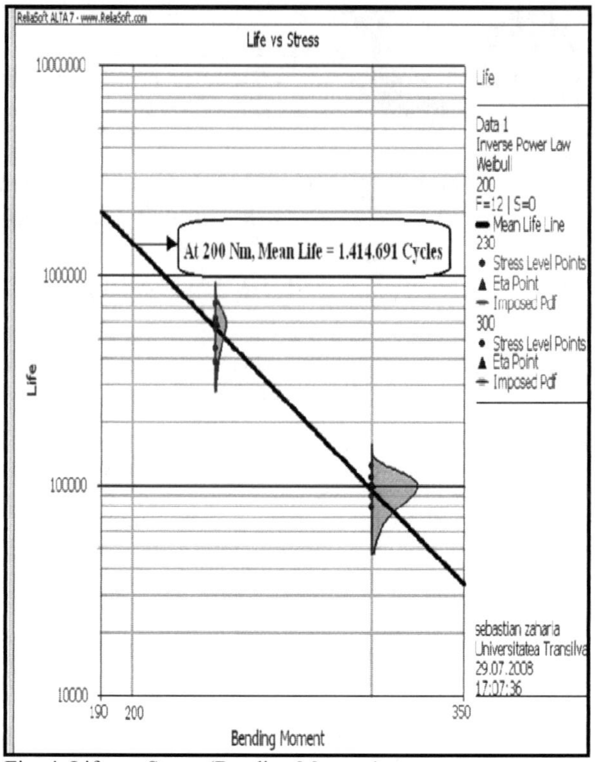

Fig. 4. Life vs. Stress (Bending Moment).

4. CONCLUSION

The testing method described in this paper presents significant interest in the lifecycle of the helicopter blades because of the decrease in testing time and of the development costs. In this field's fierce competition accelerated life tests will represent in the future indispensable method to all the companies that wish to introduce products with a higher reliability on the market.

As further research I will expand the accelerated tests on other components from aviation and engineering products.

5. ACKNOWLEDGEMENTS

This research has been supported by CNCSIS Project TD_189 with title: Theoretical and experimental research regarding reliability testing.

6. REFERENCES

Guo, H. & Pan, R. (2007). D – Optimal Reliability Test Design for Two - Stress Accelerated Life Tests, *Available from:* http://www.reliasoft.com/pubs/2007IEEEPID469965.pdf *Accessed*: 2008-07-19

Ireson, W. G., Clyde F. C. & Richard Y. M. (1996). *Handbook of Reliability Engineering and Management*, 2nd Edition, McGraw-Hill, ISBN 0-07-012750-6, New York, USA

Martinescu, I. (2005). *Fiabilitatea si Securitatea Sistemelor* (*The Reliability and Security Systems*), Editura Lux Libris, ISBN 673-9458-42-4, Brasov, Romania

Nelson, W. (2004). *Accelerated Testing; Statistical Methods, Test Plans, and Data Analyses"*, Wiley, ISBN 0-471-69736-2, New Jersey, USA

Zaharia, S.M., Buican, G. & Martinescu, I. (2007). A study about reliability tests with suspended elements, *Proceedings of the Fifth International Conference on Challenges in Higher Education and Research in the 21-st Century,* Kolev .N., Dimitrov, L., (Ed.), pp. 161-164, Heron Press, ISBN 978-954-580-227-0, Sofia, Bulgaria

Annals of DAAAM for 2008 & Proceedings of the 19th International DAAAM Symposium, ISSN 1726-9679
Editor B. Katalinic, Published by DAAAM International, Vienna, Austria 2008
Make Harmony Between Technology and Nature, and Your Mind will Fly Free as a Bird

TOWARDS AGENT-BASED MANUFACTURING SYSTEMS

ZAJAC, J[erzy] & CHWAJOL, G[rzegorz]

Abstract: *The paper presents a concept of multiagent system AIM which integrates process planning and manufacturing control in automated manufacturing systems. Types of agents are specified. Special attention is paid to the complexity of information integration in manufacturing environment. Web services, one of Internet technologies that offers new opportunities for integration of information in industrial systems, are used as an implementation platform.*
Key words: *control system, manufacturing, agent.*

1. INTRODUCTION

A modern discrete part manufacturing system can be successful only if it is able to cope with the complexity and frequent changes resulting from the dynamic nature of the manufacture-to-order environment. A modular and reconfigurable automated manufacturing system can fulfil such requirements and helps to achieve many other desirable features. It requires however novel functionalities provided by its control system. There are three most popular modern manufacturing control concepts i.e. multiagent (Shen & Nordie 1998), holonic (Van Brussel et al., 1998) and reconfigurable (Koren et al., 1999; Mehrabi et al., 2000). All the concepts take advantage of distributed approach to manufacturing control. Distributed manufacturing system consists of a collection of resources that must work together to achieve factory business goals. Increasing the throughput, quality and reconfigurability of manufacturing systems is the major issue to reach the goals. In recent years manufacturing control systems evolved from dedicated control system (for mass production) through tailor made control system (for flexible manufacturing systems) to reconfigurable control system (for mass customization). Reconfigurable control system should adapt manufacturing system to rapidly changing environment (demand and product changes) through a modular design that provides plug and play capability for its configuring and operating.

When planning to build reconfigurable manufacturing control system the essential steps to be performed are decentralization of decisions and simultaneous execution of tasks that are executed sequentially in typical manufacturing environment. These concern also an integration of process planning and manufacturing control.

Writing software for a complex, distributed, heterogeneous manufacturing environment requires an implementation of widely accepted communication standards. Contemporary Internet technologies offer the best approach for constructing a common architecture for manufacturing control system.

2. AIM SYSTEM

The AIM multiagent manufacturing control system consists of six types of agents. An essential role is played by resource agents which represent physical machines and equipment in manufacturing system. Each of them has knowledge about elementary activities performed by represented unit and is entitled to take part in decision making processes. In order to fulfil these tasks, resource agents have to exchange information with other agents by means of built-in communication mechanisms. Additionally, resource agents are involved in conditions checking processes which prevent from collisions as well as undesirable states like deadlocks or starvation.

Part agents constitute the second group of agents in the AIM system. They represent particular parts being produced during manufacturing processes. Depending on applied decision making algorithm, part agents, together with resource agents, may be actively involved in decision making processes. They are also responsible for monitoring and registration of any changes in parts state that occur during manufacturing process. Each of part agents has knowledge related to manufacturing route of represented part. Part agents are created while semi-finished products are passed into manufacturing system. Consequently, they are removed after finished products are ready.

Part agents remain in close relation with job agents which constitute the third group of agents in the AIM system. In contrary to part agents, job agents represent all parts of a particular type being produced in the system. Each of job agents has a full set of information concerning technical details of expected products, their manufacturing processes, due dates, costs limits etc. A job agent is created after a new order is submitted by a customer. It is removed from the system either when the order has been fulfilled or if the order is rejected (e.g. due to technical or economical reasons).

A significant role is played by the process planner agent constituting the fourth type of agents in the AIM system. Its main responsibility is to prepare and/or modify manufacturing processes for each incoming order. It is initialized at the control system startup and is active during the whole worktime of the system.

Another group is formed by driver agents which are responsible for information exchange between controllers of machines and manufacturing equipment and high level logic encapsulated within resource agents. Driver agents are dedicated to concrete drivers already at the stage of their design and implementation.

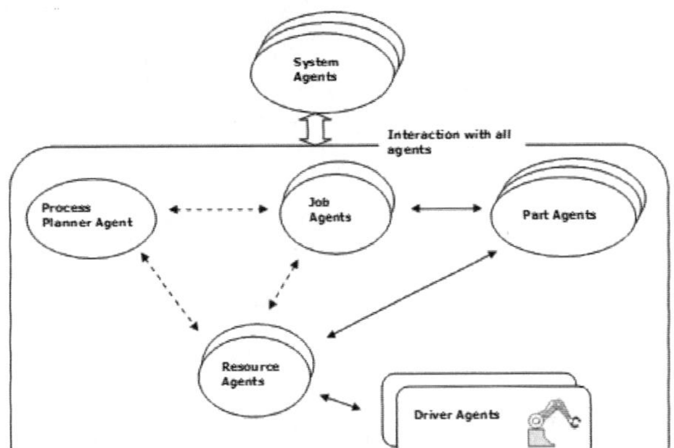

Fig. 1. The AIM manufacturing control system.

The last group comprises system agents which play administrative roles in the AIM system. They are responsible for agents registration, decision rights entitlement, system monitoring etc.

3. COMMUNICATION MECHANISMS

Diversity of solutions and heterogeneity of environments is a typical characteristic of contemporary production systems. Integration of information accomplished by manufacturing control system is a significant challenge in such circumstances, especially as far as distributed control system architectures are concerned. Distributed object technologies, like CORBA, DCOM and Java RMI used so far for implementation purposes apply dedicated communication mechanisms which are incompatible. Similarly, diversity of industrial networks (Profibus, Interbus, CAN etc.) where each of them uses its own standards of physical layers, data formats, data transmission rules etc. also generates lots of problems related to integration of devices supplied by different producers as well as communication between field level and control and planning subsystems. Lack of common standards which could be used in a process of information integration in manufacturing systems is therefore an important issue to be addressed.

During last few years there has been emerging a novel technology called Web services that gives a chance to overcome interoperability issues. Web services technology (Manes, 2003) is based on a textual data format defined by Extensible Markup Language (XML) specification. The essential advantage of using a text-based information format is easy processing of such data regardless of software platform or programming language. Web services may be implemented in various development environments, by means of different languages (Java, C++, C#, scripting languages as Perl or Python) and may be run on different operating systems (Windows, Unix). Web services solutions are based on TCP/IP and usually HTTP protocols, however, use of other protocols such as SMTP is also possible. Since these protocols typically use well-known TCP ports, it makes Web services insensitive to firewalls. Another significant advantage is possible integration of already existing distributed applications built upon middleware technologies like DCOM or CORBA. Last but not least, Web services technology is provided under an open license issued by World Wide Web Consortium and is actively developed and supported by leaders of IT market as well as wide groups of open source developers. It has been also noticed valuable for application in manufacturing control systems (Brandl, 2004).

Concerning field level of manufacturing systems, an industrial version of Ethernet standard, commonly known from local area networks (LANs), provides prerequisites for decreasing interoperability issues (Brenes, 2006). The use of Ethernet allows for easy integration of all levels of manufacturing planning and control processes. Ethernet is already a standard "de facto" at the level of long term and short term production planning, where servers and PC workstations hosting ERP/MRPII software are connected with each other by means of standardized Ethernet devices, links and connectors. As far as control level is concerned, there are a lot of solutions including PC based controllers, PLCs, HMIs etc. that already make use of Ethernet in order to transfer data. Finally, available parameters of contemporary Ethernet solutions allow for successful application of Ethernet also at the field level of manufacturing processes. Such an approach with use of Ethernet at every level would significantly decrease complexity flowing from diversity of network types designated to particular level. It would also facilitate flow and integration of information and enhance total efficiency and productivity. In consequence, total cost of maintenance would become lower as well.

Web services technology has been used for AIM manufacturing control system implementation [Fig. 2]. Production environment used for AIM system testing purposes is based on Industrial Ethernet solution.

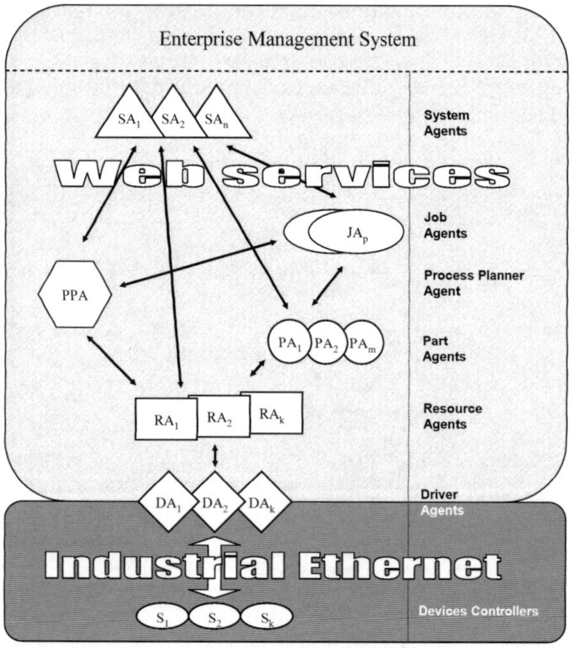

Fig. 2. The AIM system implementation platform.

4. CONCLUSION

The changes of customer requirements create the need for adaptation of a production environment. This requires usually an implementation of modern information technologies that are able to cope with complexity of contemporary manufacturing systems. Multiagent control systems offer new opportunities for building reconfigurable manufacturing systems. It requires however the application of commonly accepted communication standards. Web services, one of widely applied Internet technologies, together with Industrial Ethernet, a single network standard that can serve all purposes, seem to be a good base for such a task at the beginning of the 21st Century.

5. REFERENCES

Brandl D. (2004). Web services for manufacturing. *Control Engineering,* 51(2), February 2004, p. 58

Brenes M. (2002). Industrial Ethernet gaining factory floor strength. *World Bus Journal*, September 2002, pp. 6-9

Koren, Y.; Heisel, U.; Jovane, F.; Moriwaki, T.; Pritchow, G.; Van Brussel, H. & Ulsoy, A.G. (1999). Reconfigurable Manufacturing Systems, CIRP Annals, Vol. 48, No. 2, (keynote paper)

Manes A. T. (2003). *Web Services: A Manager's Guide*. Addison Wesley Professional, ISBN 0-321-18577-3

Mehrabi, M.G.; Ulsoy, A.G. & Koren, Y. (2000). Reconfigurable Manufacturing System and Their Enabling Technologies, International J. of Manufacturing Technology and Management, Vol. 1, No. 1, pp. 113-130

Shen, W. & Nordie. D.H. (1998). Agent-Based Systems for Intelligent Manufacturing: a State-of-the-Art Survey. *Knowledge and Information Systems*. pp. 129-156

Van Brussel, H.; Wyns, J.; Valckenaers, P.; Bongaerts, L. & Peeters, P. (1998). Reference Architecture for Holonic Manufacturing Systems: PROSA. *Computers In Industry, Special Issue on Intelligent Manufacturing Systems*, Vol. 37 No. 3, 1998, pp. 255-276

Annals of DAAAM for 2008 & Proceedings of the 19th International DAAAM Symposium, ISSN 1726-9679
Editor B. Katalinic, Published by DAAAM International, Vienna, Austria 2008
Make Harmony Between Technology and Nature, and Your Mind will Fly Free as a Bird

EXPERIMENTAL DYNAMIC CHARACTERIZATION OF MACHINE TOOL SPINDLE

ZAPCIU, M[iron]; K'NEVEZ, J[ean] Y[ves]; CAHUC, O[livier]; GERARD, A[lain] & BISU, C[laudiu]

Abstract: *Vibration analysis has long been used for the detection and identification of the machine tool dynamic condition. Predictive maintenance is directed towards recognizing the earliest significant changes in machinery condition. Contrasted with protective condition monitoring in which fast response is the primary requirement, predictive monitoring is not limited by time and may use a greater range of complex characteristics. Main focus is to identify a procedure to obtain Eigen values frequencies for machine tool spindle using tracking dynamic analysis.*
Key words*: spindle, vibration, dynamic, maintenance, method.*

1. INTRODUCTION

The modern approach on the dynamic behavior of machine tools is to measure specific parameters for its operation in order to have an intervention only when parameters indicate. This method is called conditional maintenance (predictive). For the conditional maintenance to be effective, we must make reliable and accurate measurements of machines. A number of variables machines can be used as indicators: temperature, oil pressure etc. However, experience has shown that the most reliable parameter that gives the earliest and the best way the deteriorated condition of a rotating machine is vibration. By measuring and monitoring the level of vibration produced by a machine, we obtain an ideal indicator on his condition. If the increased vibration of the machine can detect a defect, the analysis of vibration characteristics of the machine helps to identify the cause. Usually it's not possible to make an assumption about the causes and excitatory vibration. This information could be obtained by making a frequency analysis.

The vibration diagnosis is based on identification of mechanical phenomenon, the frequency of vibration it generates. To do this we must represent the vibrations in the space of frequencies, this operation is carried out by making a mathematically transform FFT using signal time vibration. Each component of the spectrum is a well-defined characteristic frequency (imbalance, resonance or misalignment). The frequency analysis is generally when the vibration level of the machine is considered higher than the permissible threshold.

The measurements systems used in this paper are Vibroport 41 (Schenck) and LMS Pimento module (Zapciu & Paraschiv, 2007), (Bisu, 2007). These devices are very efficient and help to develop a database, but do not allow the collection with an expert system. Recently, the MPPIV module uses a new measuring device built into a program CM400 expert, to establish ways of surveying and exploiting measurement results.

2. LEVELS OF PREDICTIVE MAINTENANCE

Predictive monitoring programs have three main objectives: warning earlier possible about the potential defects on equipment equipped with continuous monitoring systems; monitoring the dynamic condition of general purpose machinery; primary warning of defects that may remain hidden within complex characteristic, for example rolling-element bearing flaws.

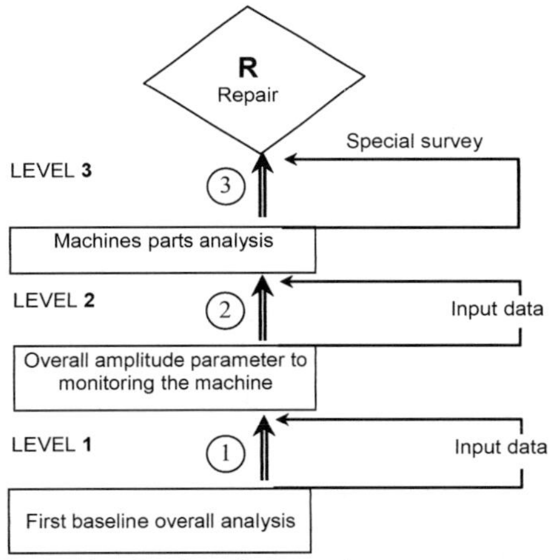

Fig. 1. Typical levels of predictive maintenance program.

Generally, most of effective predictive-monitoring programs will incorporate a wide variety of parameters to accurately characterize condition and provide earliest warning of significant changes. The process of typical predictive monitoring program consists in three levels starting with first baseline overall analysis and finishing with a detailed vibration signature on each piece of monitored equipment (Fig.1).

If the component at running frequency dominates the spectrum no greater than 20 ÷ 25 % of the amplitude at running frequency, then monitoring overall amplitude is two or three broad bands is a appropriate method for early detection on rolling element bearings and other mechanical rotating parts of machines and equipments in industrial mechanical engineering.

Experienced organizations report that a fully implemented of regularly predictive monitoring program will eliminate unexpected failures and reduce the number of machines in questionable marginal condition to less then 6 – 8 % of the total.

Fourier discreet function has a very important role in many applications of digital signal processing, such as linear filtering, spectral analysis and estimation. The main reason of its importance lies in the existence of efficient algorithms for calculation of the function using formula:

$$X_k = \sum_{n=0}^{N-1} x_n \exp\left(-\frac{2\pi i}{N} nk\right) \quad k=0,\dots,N-1 \qquad (1)$$

where x_0 and x_{N-1} numbers are complex.

Unlike discreet Fourier transformation, which requires N^2 operations for a set of N points, Fast Fourier transformation uses only N·log·2(N) operations, which constitutes a significant improvement in performance, especially for large sets of data.

Bearing machine tool spindle frequencies are best identified in the envelope spectrum. Studies of the envelope time history may be valuable in determining when and how many significant events take place.

Bearing frequencies (monitoring defect formulas) assuming pure-rolling motion are the following:

Outer race defect:

$$f_{OR} = \frac{n}{2} f_r \left(1 - \frac{BD}{PD}\cos\beta\right) \quad [Hz] \tag{2}$$

Inner race defect:

$$f_{IR} = \frac{n}{2} f_r \left(1 + \frac{BD}{PD}\cos\beta\right) \quad [Hz] \tag{3}$$

Ball defect:

$$f_B = \frac{PD}{BD} f_r \left(1 - (\frac{BD}{PD}\cos\beta)^2\right) \quad [Hz] \tag{4}$$

where:

n=number of balls or rollers; f_r=relative rev/s between inner and outer races; PD=Pitch Diameter; BD=Ball Diameter; β = contact angle.

The formulae apply to the shown deep groove bearing and only in the case of pure rolling operation. It is recommended to use Bearing Frequency Calculators supplied by bearing manufacturers (e.g. www.skf.com).

3. MEASUREMENT OF SPINDLE VIBRATIONS – APPLICATION FOR MILLING MACHINE

In order to know the proper speed domain for a High Speed Milling Machine tool the authors experiment the vibration of the HSM 600U spindle using *Tracking* analysis module of Vibroport 41 (Rigal et al., 2005), (Sutter & Molinari, 2005).

Considering Fast Fourier Time spectrum filtered out of operational domain of the machine tool, one of the most important frequencies was 780 Hz (46800 rpm) (Fig. 2).

Eigen frequencies of the assembly spindle-bearings (without cutting process) obtained experimentally (run-down spindle by 36000 rpm to 7000 rpm were: 29130 rpm (first harmonic of 242.5 Hz), 14550 rpm (242.5 Hz) and 12.600 rpm (210 Hz).

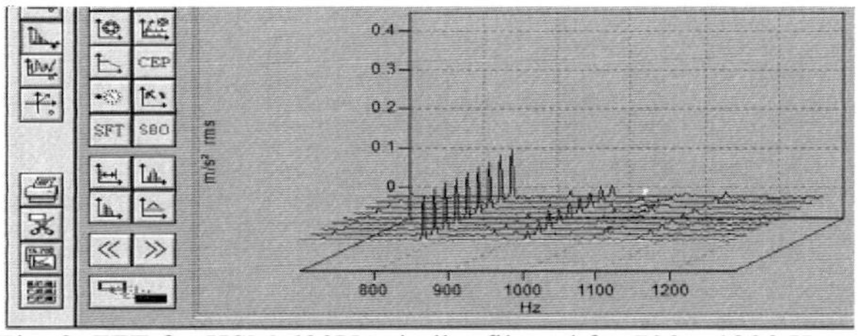

Fig. 2. FFT for HSM 600U spindle; filtered for 700 – 1200 Hz.

In the figures 3 and 4 are presented the tracking signals of speed vibration acquired for two speed domains (Zapciu et al., 2007). The maximum level of vibrations speed was 0,23 mm/s rms (level of vibrations is very acceptable).

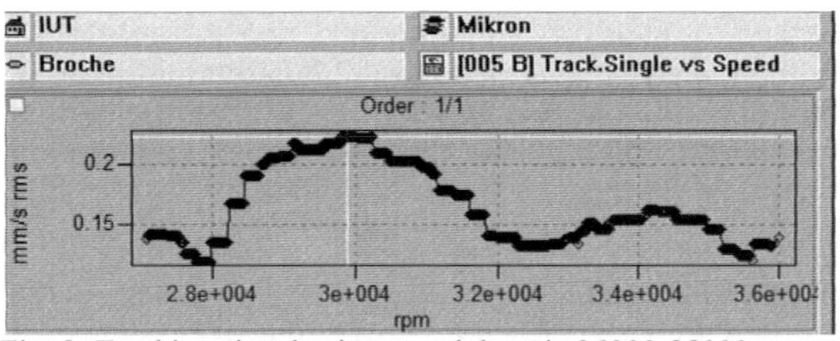

Fig. 3. Tracking signal using speed domain 36000-28000 rpm.

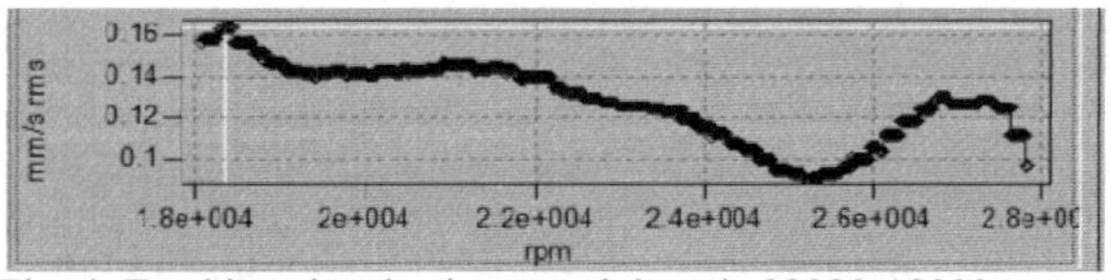

Fig. 4. Tracking signal using speed domain 28000-18000 rpm.

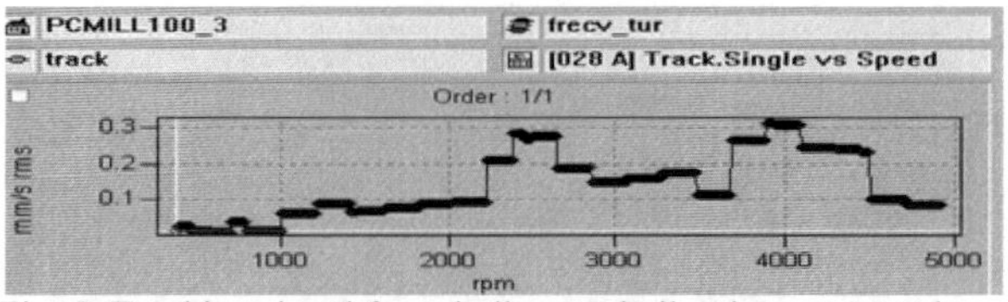

Fig. 5. Tracking signal for spindle speed vibration; case study of machine PC MILL 100 during 5000-200 rpm.

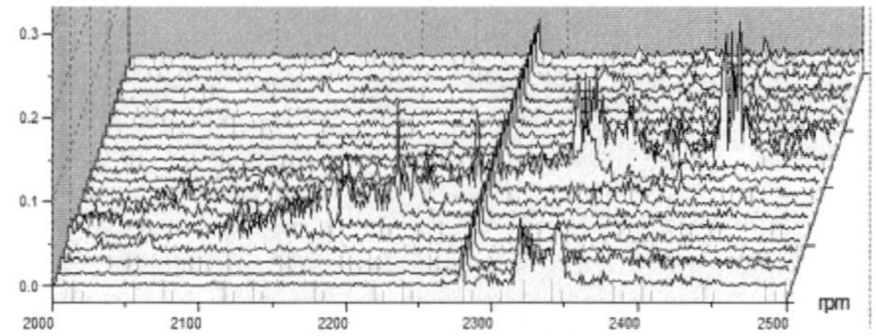

Fig.6. Acceleration of vibration for machine PC MILL 100 (zoom focus on Eigen frequency 2280 rpm).

Speed vibration of the spindle for another machine tool (PC MILL 100 by EMCO) is presented in the figure 5.

For this milling machine was obtained the following speed, recommended to be avoided in the work domain: 2280 rpm (Fig. 6) and 4000 rpm. The level of speed vibration was 0,3 mm/s; the threshold specific for future attention is about 1,2-1,4 mm/s rms for this machine tool.

4. CONCLUSIONS

This research paper proposes a maintenance survey for the machine tool spindle using vibration level (Tracking vs Speed). The second important objective was to separate the dynamics of the spindle of the machine tool in order to have a good control for the cutting processes. In this context, the actual subject is important and it can help to elaborate a proper model for predictive maintenance based on vibration.

This work was validated by the experimental results based on the measuring of the level of vibration of the spindle of two milling machines. Future work will include analyzing and filtering signal in order to better diagnose the bearing condition of the machine tool spindle. An effective predictive monitoring program will be designed to accurately recognize the earliest significant changes, and use the simplest and least costly means to separate the machine that have problems from the large number of machines in a good dynamic condition.

5. REFERENCES

Bisu, C. (2007). Study of Self-Maintained Vibration in 3D Cutting – A new modelling applied on turning, *PhD Thesis*, IMST faculty, University POLITEHNICA of Bucharest.

Rigal, J.F.; Zapciu, M.; Mabrouki, T.; Belhadi, S. (2006). Sawtooth chip formation in hard turning and the approach to separate process and machine vibration frequencies. *ICMaS*, Bucharest, ISSN 1842-3183, p.133-136.

Sutter, G.; Molinari, A. (2005). Analysis of the Cutting Force Components and Friction in High Speed Machining. *Journal of Manufacturing Science and Engineering*, Trans. of the ASME, Vol.127, pp. 245-250.

Zapciu, M.; K'nevez, J.Y.; Gerard, A. (2007).Using tracking analysis of predictive maintenance concept in order to obtain dynamics of machine tool spindles. *International Conference ICAMaT 2007*, Sibiu, Romania.

Zapciu, M.; Paraschiv, M. (2007). Predictive maintenance and use of tracking concept to analyze dynamics of machine tool spindle. *International Conference - TMCR'2007*, pp.512-516, May 31th – June 3th, 2007, Chisinau.

Annals of DAAAM for 2008 & Proceedings of the 19th International DAAAM Symposium, ISSN 1726-9679
Editor B. Katalinic, Published by DAAAM International, Vienna, Austria 2008
Make Harmony Between Technology and Nature, and Your Mind will Fly Free as a Bird

RISK MANAGEMENT FOR NEW PRODUCTS

ZAPCIU, M[iron]; TILINA, D[ragos] & TILINA, D[ana]

Abstract: *The fast rate of the products development generates not only benefits but also a series of events that can hinder the satisfaction of the client's needs. That's why, the launch of a product on the market suppose to assume some inevitable risks. The risk analysis and management as institutionalised process at the corporation level is not a systematic step yet. A systematic analysis of risk exposure, for example risk management for technical risks described in this paper and a risk classification doesn't generate important quantitative results but it can support actions of the manager. In addition the utilisation of the specific instruments of project management together with the risk management methods allows managing the three main elements that insure the success of a new launch product: costs, delays, performance.*
Key words: *risk management, project, new product.*

1. INTRODUCTION

A fundamental characteristic of the organizations is their capacity for an efficient and rapid adaptment to market requirements. The life cycle of large utilization products is of maximum 3 years, and the industrial ones of 5 years. In addition, the life durations of equipements are between 5 and 10 years, the exceeding of this duration going to moral wear and to the diminuation of the moral use. All these influenced the economic efficiency of the organizations. The organizations are obliged to a permanent renew of their products to be able to satisfy the market requires from qualitative and quantitative point of view. That's why, the launch of a product on the market suppose to assume some inevitable risks.

The risk management is often quoted in many works, but none clearly define the instruments that integrate in this. These works usually treat the analyses methods of the risks concentrated on a product or on the manufacturing methods of the product and do not permit the management of the three main elements wich assure the success of a product in the market: costs, delay and performances. That's why it is necessary the utilization of the specific management risk methods, together with the specific project management methods (Royer, 2000).

2. THE METHOD INTEGRATION PRINCIPLE

A study made on 27 searching works abouts the method integration in a organization showed that 77% of documents treate only the description of the methods and not the integration subject. Only 4.2% from these describe the possibilities of integration and the ancountered options (Perrin, 2005).

The studies lied to the conclusion that a risk evaluation made when the product was launched is not enough. A late identification of the risks may have only negative consequences. That's why, the evaluation of the risks is necessary when the product is conceived so to establish the

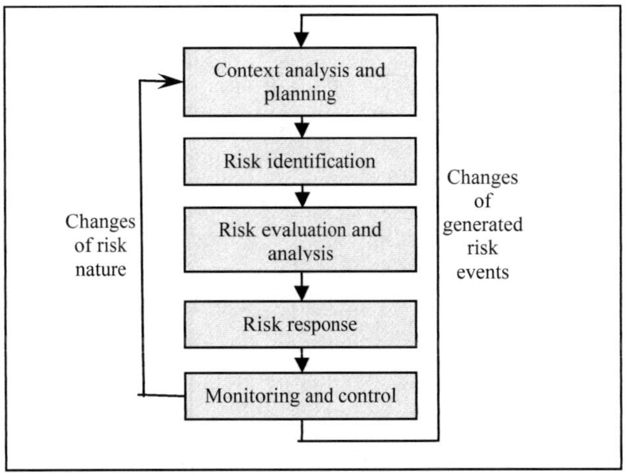

Fig 1. Project and risk management complementarity.

cause and the effects of some threads, and also the risks specific to the launched process in order to prevent its failure on the market, even in good conditions of planification and manufacturing (Ispas & Cotet, 1998).

In our approach of defining the stages of risks methodology we came up to the conclusion that any approach of risks mastering is based on some stages that facilitate a good development. In analogy with life cycle stages of a product so to analyse the phenomenon it is necessary identification and evaluation of the events that generate the risk, finally arriving to a risk analyses and to establish and implement risk prevention method (Tilina, 2008).

An important aspect is the interative character. This allows a continue unrolled of the development that starts in the some time with the initiation stage and finishes in the some time with the product materialization. More than that, the complementarity between the risk management and the project management lied us to the conclusion that risk analyses permits a better unrolled of the project.

3. RISK MANAGEMENT FOR TECHNICAL RISK

Many technical domains deal with the problems of incidents appearance, proposing attenuation methods for threads effects. The reliability put at the disposal, methods and technics for technical and functional viability achievement (Masala, 2004). The management proposes solutions to organise and lead an efficient process. The medium frequency of appearance may be associated as a common parameter for the incidents and threads categories. So a dependency between the frequencies of events appearance of the threads and the taking measures for assuring the stability and process security may be achieved (eq.1).

$$\bar{f} = -\frac{1}{T}\ln\left(1-\varepsilon\right);\ \varepsilon \in \left(0.1\right) \tag{1}$$

where:

- $\bar{f}$ - the medium frequency of an event appearence;
- T – the monitorisation time of the events appearance;
- ε – a positive subunit coefficient (the diminuation potential of an unwanted event appearance).

The medium frequency of technical events is the sum of medium frequencies divided at four.

$$\bar{f}_t = \frac{\bar{f}_{th} + \bar{f}_m + \bar{f}_i + \bar{f}_s}{4} \quad (2)$$

If we achieve surveillance and maintenance of the process, we can consider that the duration's distribution between two interventions belongs to the normal functional repartitions. The cutting of is asymmetric in accordance with the durations media between two normal interventions.

$$K_f(t) \in N(\mu,\sigma);\ t \in (0, T_i) \quad (3)$$

$$K = \int_0^{T_i} N(\mu,\sigma)\,dt \quad (4)$$

$$\mu = (\bar{f})^{-1} \quad (5)$$

and

$$\sigma^2 = \lim_{n\to\infty} \frac{1}{n}\sum_{i=1}^{n} (ti-\mu)^2 \quad (6)$$

The medium values of dispertion and media between two interventions along a cycle (T_c) are:

$$\bar{\mu} = \frac{1}{T_c}\int_0^{T_c} \mu(\delta)\,d\delta \quad (7)$$

and

$$\bar{\sigma} = \frac{1}{T_c}\int_0^{T_c} \sigma(\delta)\,d\delta \quad (8)$$

With these values the possibilities of some incidents appearance on the time of an unrolled normal cycle is calculated. For a better appreciation of a process evolution the possibilities of the unwanted events can be calculated for each thread category.

From a monitoring strategy of a process point of view, each definitory parameter x_i may be characterized by a critical limit XC_i in which the incident may appear, a value XE_i, over which measures of events appearance avoiding canot be taken and a alarm value XA_i which attention the incident appearance.

According to the characteristics of the process and the speed of a thread transformation in an incident, the intervention periods are defined ($t_e - t_a$) so to prevent the incident and the action, in order to annihilate the consequences of the incident ($t_c - t_e$) or for a preventive diminuation of this consequences.

The annihilation or attenuation methods have a short period of prevention, these obviously continuing also after the incident appearance.

If for the period ($t_e - t_a$) the intervention mode is established by the security strategy, the annihilation and attenuation actions of the consequences of an event preventive provoqued, represent a full security component and are achieved after specifical professional studies for each process.

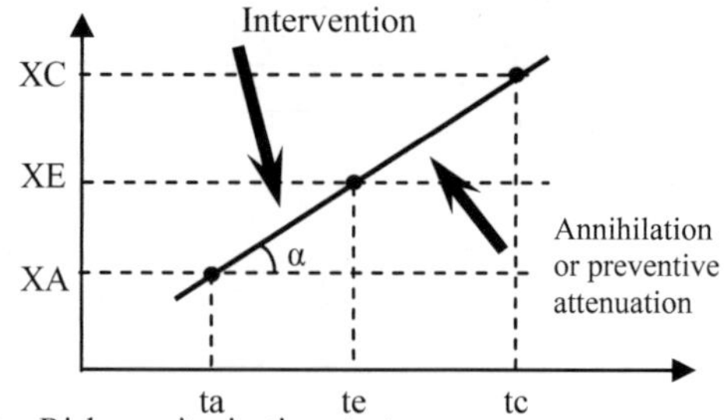

Fig 2. Risk monitorisation strategy

The intervention duration is determined with the following relation:

$$T_i = \frac{XE_i - XA_i}{tg\,\alpha} = (XE_i - XA_i)\cdot\left(\lim \frac{dx_i}{dt}\right)^{-1} \quad (9)$$

The medium probability of an unwell event appearance (e_i) during a normal cycle unroll with the medium probability of the delaying of the alarm signal detection after a time period longer than the critical time period (for intervention) of the process variable is:

$$P(e_i) = \frac{p[(t > Ti) \cap (0 \le t < Tc)]}{p(0 \le t < Tc)} \quad (10)$$

4. CONCLUSION

The study results allow parallel identifications of the project and risk management stages, considering a commune integration structure.

The choosing of the methods could be different from an organization to another. That is why we propose a risk management based on clearly defined steps, as we showed on the second chapter and detailed for technical risks category on the third chapter. Every step can be developed depending on risk type. For the technical risks, we can calculate the intervention duration T_i, as well as the medium probability of an unwell event appearance e_i. We also can choose the strategy based on risk analysis results.

In addition we propose, for a better adaptation to any type of organisation, the consideration of the following main aspects:

- The study of the methods applied in the organisation;
- The integration of the methods according to the defined structure;
- The validation of the integration in a general character study;

5. REFERENCES

Ispas, C.; Cotet, C. (1998). *Project Management concepts*, Bren, Bucharest

Masala, F. (2004) *Contributions regarding risk evaluation for investment projects,* Doctor's degree thesis, Bucharest

Perrin, F. (2005) *Improvement of design reflexes by integrating new methods,* Doctor's degree thesis, Bucharest

Royer, S. (2000). *Project Risk Management – A Proactive Aproach, Management Concepts,* Project Management Institute, Inc.

Tilina, D. (2008), *Contributions concerning risk assessment for new product launching,* Doctor's degree thesis, UPB

Annals of DAAAM for 2008 & Proceedings of the 19th International DAAAM Symposium, ISSN 1726-9679
Editor B. Katalinic, Published by DAAAM International, Vienna, Austria 2008
Make Harmony Between Technology and Nature, and Your Mind will Fly Free as a Bird

STATE ESTIMATION USING INTERVAL ANALYSIS ON SYNCHRONOUS MOTOR MODEL SIMULATION

ZBRANEK, P[avel] & VESELY, L[ibor]

Abstract: *The purpose of this paper is to use interval analysis on guaranteed nonlinear parameter computation and state estimation for synchronous motor – PMSM (Permanent Magnet Synchronous Motor).In this case, the mere knowledge of input and output parameters is not sufficient for control.However the state variables are technically or financially unnecessarily consumptive. The design of universal reliable estimator is not simple part. That is why we will try to estimate state of synchronous motors using the method of interval analysis. The results of this estimation are not point, but intervals. These intervals are guaranteed. It relies on the technician how he uses this interval.*

Key words: *Guaranteed state estimation, synchronous motor.*

1. INTRODUCTION

The purpose of this paper is to use interval analysis on guaranteed nonlinear parameter computation and state estimation for synchronous motor – PMSM (Permanent Magnet Synchronous Motor). Permanent magnet synchronous drives become very popular in many industrial applications. The main advantage of this drives is their more reliable mechanical construction comparing to DC drives. PMSM sensorless control is not so optimized, there are many things to research. That is why we selected this nonlinear dynamic system for our enquiry.

The method state estimation using interval analysis (Moore, 1979) assumes measurement noise; each noise component must belong to some known interval. The most often used reconstructors based mostly on Kalman filter principle estimate the value of state with the probability so it can be supposed, that it is not fully reliable. It is sufficient in many case studies, but in plenty of others this inaccuracy under specific conditions can lead to the malfunction of the system. So many scientists team search other possibilities. That is why we will try to estimate state of synchronous motors using the method of interval analysis where the result is guaranteed. It relies on the technician how he uses this interval.

2. INTERVAL ANALYSIS

Non-linear and time-varying system defined as follows:

$$\begin{cases} x_{k+1} = f_k(x_k, u_k, v_k), \\ y_k = h_k(x_k) + w_k, \end{cases} \quad k = 0,1,... \tag{1}$$

where $u_k \in \mathrm{R}^m, x_k \in \mathrm{R}^n$ and $y_k \in \mathrm{R}^p$ are defined as input, state and output vector. The beginning state x_0 is considered to be a part of some preceding compact set $X_0 \subset \mathrm{R}^n$. Concerning terms $\{v_k\}$ and $\{w_k\}$, they are unknown state and measurement noise sequences, presumed to be part of the known intervals $\{[v]_k\}$ and $\{[w]_k\}$.

The main issue to be dealt with is the recursive estimation of the minimal set X_l, securely incorporating all values of X_l compatible with the information accessible at an instant l, i.e., with $Z_l = \left\{X_0, \{u_k, y_k, [v]_k, [w]_k\}_{k=0}^{l}\right\}$. Let X_l be a set consisting of all possible values of x_l compatible with Z_l, and X_{l+} is a set of all values of a certain state.

$$X_{l+} = f_l(X_l, u_l, [v]_l) = \{f_l(x, u_l, v_l) \mid x \in X_l, v_l \in [v]_l\}$$

Let's consider Y_{l+1} a set of all acceptable values of the output, of course with measured value y_{l+1}

$$Y_{l+1} = y_{l+1} - [w_{l+1}] = \{y_{l+1} - w_{l+1} \mid w_{l+1} \in [w]_{l+1}$$

and let X^0_{l+1} be the set of all possible values of x. This would logically lead to a conclusion that $y \in Y_{l+1}$

$$X^0_{l+1} = h^{-1}_{l+1}(Y_{l+1}) = \{x \in \mathrm{R}^n \mid h_{l+1}(x) \in Y_{l+1}\}$$

A set including all values of x_l compatible with Z_{l+1} is therefore defined by $X_{l+1} = X_{l+} \cap X^0_{l+1}$. If X_0 comprises all possible values of the beginning state vector x_0, then a logical conclusion is to recursively evaluate a set X_l, which consists of all values of the state vector at each step l. This process can be summarized in following algorithm (Jaulin et al. 1993):

1. Prediction $X_{l+} = f_l(X_l, u_l, [v]_l)$.
2. Correction $X_{l+1} = h^{-1}_{l+1}(Y_{l+1}) \cap X_{l+}$.

If the estimation use interval analysis, the state space must be partioned into smaller subspaces – subpavings.

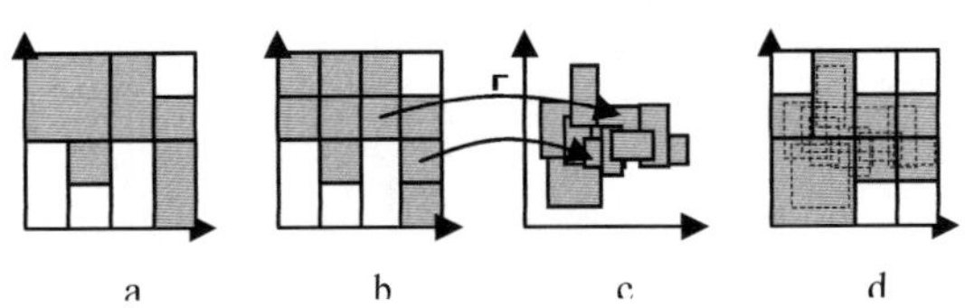

Fig. 1. Principle of interval computation – subpavings (Kieffer et al. 1998).

3. PERMANENT MAGNET SYNCHRONOUS MOTOR

Permanent magnet synchronous (PMSM) drives become very popular in many industrial applications. The main advantage of this drives is their more reliable mechanical construction comparing to DC drives. These drives can run for relatively long time without any service. On the other hand rotor position information is necessary even to be able to control drive speed. That is why in common applications a position sensor has to be used increasing significantly the drive cost (Vaclavek & Blaha 2007). The PMSM is highly nonlinear dynamic system. Its precisely state estimaton is not simple task. Therethrough we selected this system as the instance of our research.
The used simplified mathematical model of a PMSM is

$$\frac{d}{dt}\begin{bmatrix} i_\alpha \\ i_\beta \\ \omega_s \\ \theta \end{bmatrix} = \begin{bmatrix} -\frac{R}{L} & 0 & \frac{\psi_{af}\sin\theta}{L} & 0 \\ 0 & -\frac{R}{L} & -\frac{\psi_{af}\cos\theta}{L} & 0 \\ 0 & 0 & 0 & 0 \\ 0 & 0 & 1 & 0 \end{bmatrix} \cdot \begin{bmatrix} i_\alpha \\ i_\beta \\ \omega_s \\ \theta \end{bmatrix} + \begin{bmatrix} \frac{1}{L} & 0 \\ 0 & \frac{1}{L} \\ 0 & 0 \\ 0 & 0 \end{bmatrix} \cdot \begin{bmatrix} u_\alpha \\ u_\beta \end{bmatrix} \quad (2)$$

where

- i_α, i_β are stator current components
- u_α,ui_β are voltage current components
- R is stator resistance
- L is stator inductance
- ω is rotor speed
- θ is rotor position
- Ψ_{af} is EMF constant

4. SIMULATION

The simulation is imlepmented in programme Matlab Simulink. The first task is to implement the function after study the issue (Jaulin et al. 2001) about counting with intervals. The best solution was found in using Matlab environment where was toolbox INTLAB implemented – INTerval LABoratory, Version 5.3. This allows to define variables right in Matlab environment and at the same time leads the complier how to carry on with variables if we use common operators. There are many functions open to operate with intervals.

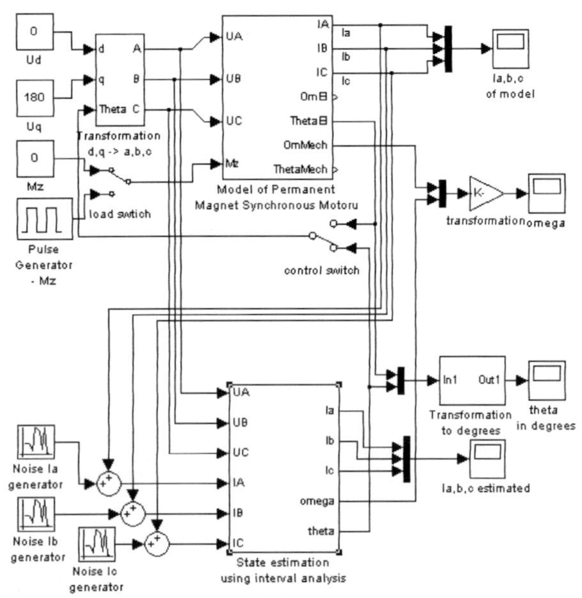

Fig. 2. Simulation scheme in Matlab Simulink.

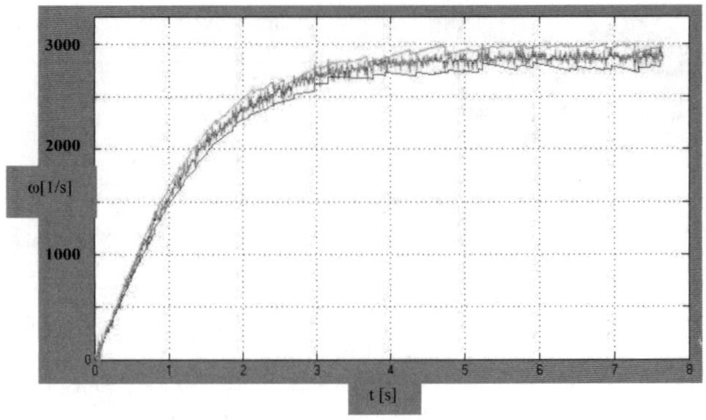

Fig. 3. Measuring data of speed estimation

5. CONCLUSION

In this paper, a nonlinear state estimation has been presented. Using interval analysis and set inversion is the main idea of approach of parameters identification. It is supposed to reach better results. In algorithm used there are the rates of states hand on between particular steps only in one interval for every state value. The area of this interval is maximum possible, the size of other state valuables are not take into account. If the algorithm hands on between the steps not only these intervals but whole particular sector of state space, the results of estimations will be narrowed. The future research will be algorithm optimization for wider application for more systems, in the best case for real systems. The biggest disadvantage consists in counting effort the reconstructor needs. That is meanwhile using in practice is not possible. Information technology is improving very quickly, so it can be perpective in the future.

6. ACKNOWLEDGEMENTS

The research has been supported by Czech Science Foundation under the project GA 102/06/0949 "Intelligent Control of Induction Machine and Synchronous Machine Electrical Drives", the Ministry of Education of the Czech Republic under the project 1M0567 "Centre for Applied Cybernetics".

7. REFERENCES

Di-Loreto, M., Dao, M., Jaulin, L., Lafay, J.F. & Loiseau, J.J. (2007). Applied interval computation: a new approach for time-delay system analysis. In *Application of Time-Delay Systems*, Lecture Notes in Control and Information Science, pp 175-197, Vol 352, ISBN: 978-3-540-49555-0

Jaulin, L. & Walter, E. (1993). Set inversion via interval analysis for nonlinear bounded-error estimation. *Automatica*, 29(4):1053–1064

Jaulin, L. & Walter, E. (1993). Guaranteed nonlinear parameter estimation from bounded error data via interval analysis. *Matematics and Computers in Simulation*, 5: 123-137

Jaulin, L.; Kiffer, M.; Didrit, O. & Walter, E. (2001). Applied Interval Analysis, Springer, ISBN 1-85233-219-0, London

Kieffer, M.; Jaulin, L. & Walter, E. (1998). Guaranteed recursive nonlinear state estimation using interval analysis. *International Journal of Adaptative Control and Signal Processing*, 6(3):193–218

Moore, R. E. (1979). *Methods and Applications of Interval Analysis*. SIAM, Philadelphia, PA

Vaclavek, P. & Blaha, P. (2007). Synchronous machine drive observability analysis for sensorless control design, *International Conference on Control Applications,* Part of IEEE Multi-conference on Systems and Control, pp 1113-1117, ISBN: 978-1-4244-0442-1, Singapore

Annals of DAAAM for 2008 & Proceedings of the 19th International DAAAM Symposium, ISSN 1726-9679
Editor B. Katalinic, Published by DAAAM International, Vienna, Austria 2008
Make Harmony Between Technology and Nature, and Your Mind will Fly Free as a Bird

RESEARCH OF THE ARITHMETIC AVERAGE OF PROCESSING BY LATHING OF HARDENED STEEL 42CrMo4

ZEQIRI, H[akif]; SALIHU, A[vdi]; BUNJAKU, A[vdyl]; OSMANI, H[ysni] & ZEQIRI, F[itim]

Abstract: *In this work are recorded the researched results of the processed surface roughness during the processing with turning of tempered steel 42CrMo4. The processing is implemented by changing various independent variables; cutting velocity (v), feed (s), cutting depth (a) and nose radius (r). For research is used the statistic method with five factors (2^5+6)*

Key words*: Cutting, roughness, hardening, chip, tool*

1. INTRODUCTION

The roughness of the processed surface is determined by the geometrical characteristics of the surface, accurateness of the proportions and physical characteristics. With the geometrical characteristics of the surface we understand (Aleksander, 1976):

1. Macro geometry
2. Micro geometry of the processed surface

With physical characteristics of the surface we understand the deflection of the physical-mechanical characteristics of the surface layer of the detail from the base material. The characteristics of the detail usage depend on the quality of the processed surface because exactly from the surface layer begins its destruction. Micro geometry of the surface depends on the form and geometry of the metal-cutting tool, pace and impact of other technological factors that greatly change the theoretical profile of the detail surface. The main factors of this are:

1. Plastic and elastic deformations
2. Friction of the back side of the tool
3. The phenomenon of increase in the cutting edge
4. Vibrations (dynamic phenomena in the system of machine-appliance- processed scrap-tool
5. Modification of the geometry of cutting edge of the tool as a result of its consumption.

As a function of the processed metal and cutting conditions, these factors change. During the processing of the plastic metals a great influence on the micro geometry of the processed surface have the magnitude of plastic deformation on the cutting area, the phenomenon of increase and the vibrations.

The cutting speed has a great influence on the roughness of the surface; with the increase of the cutting speed the value of the plastic deformation of processed material also increases and that is why the height of micro elevation decreases. The cutting speed during the clean processing should be relatively greater than during the rough processing. The height parameters Ra & Rz were taken for the research. Ra- stands for the average arithmetical value of all deflection points of the profile from the average line m-m; fig. 1 is applied for defining the roughness of the processed surface and it is calculated with this formula:

$$R_a = \frac{1}{L}\int_0^L |y(x)|dx \qquad (1)$$

Rz- the average height of uneven levels- is calculated by measuring five highest points and five lowest points along L length. Fig. 2 is calculated with formula (Bodinaku, 2004):

$$R_z = \frac{(y_{p1}+y_{p2}+y_{p3}+y_{p4}+y_{p5})-(y_{v1}+y_{v2}+y_{v3}+y_{v4}+y_{v5})}{5} \qquad (2)$$

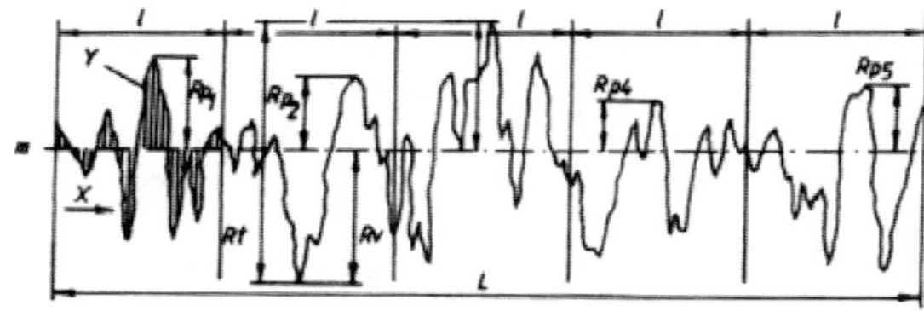

Fig. 1. Graphical interpretation of *Ra*

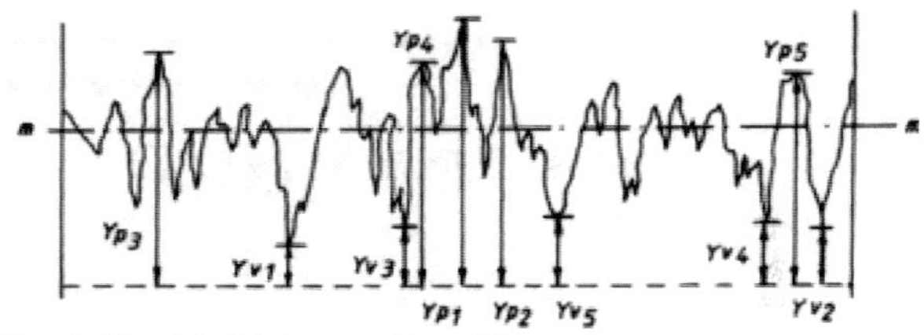

Fig. 2. Graphical interpretation of *Rz*

2. CONDITIONS DURING THE EXPERIMENT

2.1. Machine-Horizontal Lathe IK62, with these characteristics: P=10 kW, n=12.5÷2000 rpm, & feed s=0.035÷2.08mm/rev.

2.2. Tool-Cutting plate from the hard metal P30, SINTAL-ZAGREG, ISO SNMM120404, SNMM120408, SNMM120412, enforced in the body with sign ISO PSDNN2525P12, outcome 25 mm, with cutting geometry: γ=-60, α=60, λ=-60, r_ε=0.4mm, b_f=0.2mm, γ_f=-20^0.

2.3. Researching material–Steel 42CrMo4 in the cylinder shape with dimensions Φ68X750/Φ38.5X750mm.

2.4 Apparatus for roughness measurement–Duplex Microscope CARL SEISS type STOLLBERG/ERCGEBTYPS15A/GA6Og

2.5. Hardening of steel- electric automatic furnace of type JPG80/140 with these characteristics: P=130 kW, 3X380/220V, 50 Hz, tmax=950^{0}C, d=800 mm, h=1400 mm.

2.6. Experimental researching plan–based on the rating number of the rotations of the machine, consulted literature, professional experience, chemical composition, mechanic features of processed piece and Tool are defined cutting conditions: v, s, a, r & χ, Tab. 1 (Zeqiri, 2005 & Salihu 2001).

Characteristics of independent various sizes					
Nr	Note	Lev Code	Max 1	Aver 0	Min -1
1	v m/min]	X_1	67.000	53.000	42.000
2	s [mm/rev]	X_2	0.042	0.038	0.035
3	a [mm]	X_3	1.000	0.707	0.035
4	r [mm]	X_4	1.200	0.800	0.400
5	κ [^{0}]	X_5	60.000	51.961	45.000

Tab. 1. Cutting conditions

3. RESULTS AND DISCUSSION

Once defining the input and output parameters, selecting the appropriate method for measurement and measuring apparatus, analyzing the independent and changeable input parameters, selecting the function form for defining the area of experiment, executing the experiment, statistically processing the researched parameters and verifying the mathematical models, the impact of parameters (v, s, a, r & χ) acquired on the roughness of the processed surface is defined (Bunjaku et al., 2002). The surface form, acquired during the cutting process, is a result of the moving of the tool through the processed scrape and it also depends on the deformations caused by the removal of splinter from the processed scrape, by friction, by consumption of the metal-cutting instrument, by temperature and by the dynamic phenomena in the system: Machine-Appliance-Processed Scrape and tool. During the cutting it is impossible to acquire ideal and precise processed surface because of the big or small errors or deflections from the form depicted in the drawing. These deflections result in the disproportion between the real surface (real profile) and the geometrical form of given surface (geometrical profile). The researched parameters must be presented with adequate mathematical model through which the changes of physical phenomena and technological effects on the cutting-processed surface could be observed.
The results acquired during the experiment were given in table 2. The mathematical model (3&4) was acquired with the processing of the results acquired during the experiment. The graphical interpretation of the mathematical model (3&4) was given in fig. 3.

$$R_a = 71.099 \cdot v^{-0.088} \cdot s^{0.472} \cdot a^{-0.054} \cdot r^{-0.08} \cdot \chi^{-0.417} \quad (3)$$

$$R_z = 284.343 \cdot v^{-0.088} \cdot s^{0.448} \cdot a^{-0.054} \cdot r^{-0.08} \chi^{-0.417} \quad (4)$$

Nr	Real plan of metrics					Results	
	v m/min	s mm/rev	a mm	r mm	κ °	Ra μm	Rz μm
1	42	0.035	0.5	0.4	45	1.889	7.555
2	67	0.035	0.5	0.4	45	1.674	6.696
3	42	0.042	0.5	0.4	45	2.748	10.991
4	67	0.042	0.5	0.4	45	2.281	9.122
5	42	0.035	1.0	04	45	2.108	8.432
6	67	0.035	1.0	0.4	45	2.003	8.013
7	42	0.042	1.0	0.4	45	3.313	13.250
8	67	0.042	1.0	0.4	45	2.491	9.964
9	42	0.035	0.5	1.2	45	1.550	6.201
10	67	0.035	0.5	1.2	45	2.882	11.526
11	42	0.042	0.5	1.2	45	2.548	10.190
12	67	0.042	0.5	1.2	45	1.969	7.874
13	42	0.035	1.0	1.2	45	1.869	7.476
14	67	0.035	1.0	1.2	45	2.548	10.190
15	42	0.042	1.0	1.2	45	2.214	8.855
16	67	0.042	1.0	1.2	45	2.047	8.188
17	42	0.035	0.5	0.4	60	2.662	10.647
18	67	0.035	0.5	0.4	60	1.713	6.853
19	42	0.042	0.5	0.4	60	1.758	7.030
20	67	0.042	0.5	0.4	60	1.647	6.586
21	42	0.035	1.0	0.4	60	2.003	8.010
22	67	0.035	1.0	0.4	60	1.780	7.120
23	42	0.042	1.0	0.4	60	1.997	7.987
24	67	0.042	1.0	0.4	60	1.954	7.815
25	42	0.035	0.5	1.2	60	1.947	7.787
26	67	0.035	0.5	1.2	60	1.936	7.743
27	42	0.042	0.5	1.2	60	2.113	8.453
28	67	0.042	0.5	1.2	60	2.036	8.143
29	42	0.035	1.0	1.2	60	1.869	7.476
30	67	0.035	1.0	1.2	60	1.983	7.932
31	42	0.042	1.0	1.2	60	2.139	8.554
32	67	0.042	1.0	1.2	60	2.080	8.320
33	53	0.038	0.7	0.8	51	1.759	7.035
34	53	0.038	0.7	0.8	51	1.759	7.035
35	53	0.038	0.7	0.8	51	1.760	7.038
36	53	0.038	0.7	0.8	51	1.760	7.038
37	53	0.038	0.7	0.8	51	1.760	7.038
38	53	0.038	0.7	0.8	51	1.760	7.038

Tab. 2. Experimental results

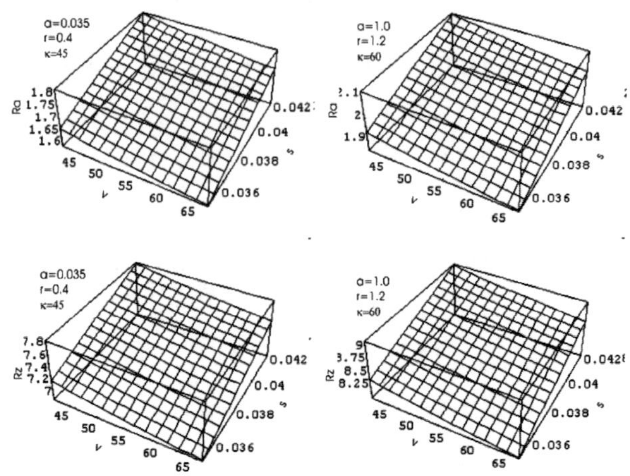

Fig.3. Graphical interpretation of the mathematical models 3&4

4. CONCLUSION

The statistic analysis of mathematical models, confirms the right solution of the function shape, where appears the physical researched manifestation of manufacturing process with the scrap removal, and technological effects in the processed surface.

From the Mathematical models (3&4) ad their graphic interpretation we conclude that:

1. The cutting speed (v) is a parameter which more or less depends from the quality of processed surface, increasing the cutting-speed, the roughness of the processed surface is reduced.
2. By researching is confirmed that increasing a cutting-step (s) the parameters of the roughness processed-surface is increased.
3. The cutting-deepness (a) has a small influence in the parameters of processed surface roughness, the influence of cutting-deepness in a proportion with other influenced factors is not important.
4. Increasing the ray of roundness (r) and decreasing the attack edge (χ) the roughness of processed surface is decreased.

5. REFERENCES

Aleksander, B. (1976), *Mechanical technology*, Volume 2, Faculty of Mechanical Engineering, Tiranë.

Bodinaku, A. (2004), *Mechanical technology*, Volume 2, Faculty of Mechanical Engineering, Tiranë.

Bunjaku, A., Bodinaku, A., Osmani, H. & Zeqiri, H. (2002), *The influence of cutting process on consuming of cutting plates during turning process of stele Ck 60,* Scientific Journal for the Theory and Application in Mechanical Engineering "Makieria" nr.1, p.15-19, Faculty of Mechanical Engineering, Prishtinë.

Salihu, A. (2001), *Research of machinability of cutting material with increased speed,* doctoral dissertation, Faculty of Mechanical Engineering, Prishtina.

Zeqiri, H. (2005), *Research of machinability by turning of 42CrMo4 steel,* doctoral dissertation, Faculty of Mechanical Engineering, Prishtina.

Annals of DAAAM for 2008 & Proceedings of the 19th International DAAAM Symposium, ISSN 1726-9679
Editor B. Katalinic, Published by DAAAM International, Vienna, Austria 2008
Make Harmony Between Technology and Nature, and Your Mind will Fly Free as a Bird

ADVANCED IMPROVEMENT OF QUALITY

ZGODAVOVA, K[ristina] & SLIMAK, I[van]

Abstract: *The goal of the paper is to present premises and also partially discuss issues of theory and methods of transition from the production quality management, that is focused on meeting demands of the customer on quality of products, to such a system of improving the production quality that will be aimed at global competitiveness management, primacy, and excellence. The paper is targeted at professionals employed in the research; teaching and practice in the field of production quality improvement in any sector and can help anybody who wants to improve something in everyday life. The paper was written with support of KEGA 3-4121-06 "Education Web Portal for Tools, Methods and Case Studies in Quality Management" project.*

Key words: *production quality, labor quality, quality of life, advanced improvement, proportionality, guarantee*

1. INTRODUCTION

Motto:

"*Methods already known do not exclude creation of new methods that meet better new conditions and circumstances*".

It has been felt that many large or small projects on the quality improvement as well as different systems of quality management, environmental management, safety and health management and social responsibility management in many institutions are not effective from the point of view of technology, economy, science, trade. Following issues are considered to be problems:

- What issues people in given time and place pay their attention to?
- How do they feel about those issues and how do they present their concerns?
- What do people consider as better and worse?
- Who is responsible for the process of improvement, how is it supposed to be implemented and what resources can be employed?
- Who should be affected by improvements and how should improvements be implemented and how should they affect people?
- What is the significance and contribution of improvements?

The basic agreement of the above mentioned issues deserves a unified platform based on approximately similar values and quality culture which takes into consideration predicted behavior and attitudes of people towards labor, production and life in a given community.

The aim of this paper is to initiate theoretical and practical activities which will lead to gradual but relatively fast transition from the quality management characterized by the system of meeting customers' demands (ISO 9000:2007 chapter 3.1.1) on products quality to more advanced improvements of the production quality, labor quality and quality of life by which the global competitiveness, primacy and excellence can be achieved.

2. PREMISES, METHODS AND PRACTICE OF IMPROVEMENT OF QUALITY

Starting points of our research are following premises:

- Quality of anything (any entity) is perceived as an object of study and not as a result of level determination by what the set of characteristics satisfies the needs according to ISO 9000:2005 chapter 3.1.1.
- Production is understood as a process and result of target-oriented activity, i.e. production processes and products (processed materials, products and services including artistic and spiritual work).
- Improvement is such measurement of function attributes, which is acting positively in certain environment and time, from the point of view of observed purpose.
- Advanced improvement of quality is concerned not only to process of improvement of product quality, but it also a complex and systematical improvement of production quality, labor quality and quality of life.

A lot of different methods can be used to improve the production quality. Among the best known methods are the Deming Cycle of improvement PDCA (Plan, Do, Check, Act), Marsh recommendations on improvement DACP (Define, Analyze, Correct, Prevent), and the complete procedures DMAIC (Define, Measure, Analyze, Improve, Control) of Six Sigma method and the ISO 9004:2000 as whole.

Specific standards ISO 14004:2004 "Environmental management systems – General guidelines on principles, systems" and supportive techniques and tools Life Cycle Management (LCM), Design for Environment (DfE), Life Cycle Assessment (LCA) and Environmental Effect Analysis (EEA) regulate the environment improvement (Tingström, 2005); OHSAS 18001:2007's Occupational Health and Safety Management Systems greatest strengths over other management systems is its requirement for continual improvement; SA 8000:2001 Social Accountability standard; and different theories of the quality of life.

Quality management and other areas of management are finally based on the systems theory, or theory of control and information theories.

According to (Golemanov, 2006) "All qualities create real and imaginary worlds". Quality is, in modern definition, inversely proportional to variability, we can expect – and actually we have seen already in practice – that the control and the improvement of quality can be also realized through appropriate applications of information technology (Pan, 2006).

Quality is multidimensional and interdisciplinary, and can be applied to any specific domain with principles independent to the specific domain. The organization of the whole defines a system, which itself as a phenomenon can be considered independent of the physical or conceptual manifestation of its elements. That is, the abstract organization of phenomena is independent of their substance, type, or spatial or temporal scale of existence (Heylighen, 1997). Based on those facts it is considered (Mlčoch&Slimák, 1987), that the total quality control and (Beckford, 1998) total quality management should consistently respect systems theory and cybernetics. (Schroeder et all., 2002), states that systems theory is one of the base principles of Six Sigma model and modern approach to the production quality improvement is without information and communication technology not conceivable today.

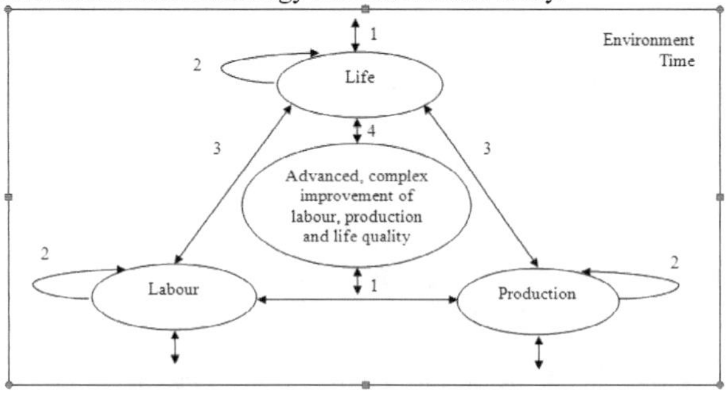

Fig. 1 Basic model for assessment of advanced, complex improvement of labor, production and life quality.

1 – out-bound management and other effects of environment and time
2 – in-bound management and learning
3 – mutual influence of labor, production and life
4 – activities of gradual, complex improvement of quality of labor, production and life

3. BRIEF DESCRIPTION OF ADVANCED IMPROVEMENT OF PRODUCTION QUALITY

Process of advanced improvement of production quality is characterized by two basic features:
1) It should exist as a subsystem of a complex process of improvement of labor, production and life.
2) It should be directed from the management oriented to meet customers' demands, requirements of supported standards, so that the state of competitiveness, primacy and excellence can be achieved. The excellence is according to (Loukas, 2002) „Harmony, creation of stability (but not stagnation) in a dynamic self improving circle, constant monitoring, team working, continuous learning and personal advancement through reflection and conjecture, passion in the acquisition of knowledge and the advancement of science are factors that explain and describe "excellence" but also they constitute the teaching of quality management".

The process of advanced quality improvement comprises these two sub processes or stages:
Stage 1: Defining the object of improvement as well as the tasks in a common procedure which should correspond with needs and requirements of quality improvement of labor, production and life of employees and other persons involved in the process.
Stage 2: Designing the project and documentation on implementation of improvement. The design should guarantee achievement of given requirements, mainly by keeping proportionality of interventions into the main factors of quality.
Stage 3: Implementation of suggested improvements, and launching them into operation.
Stage 4: Offering improved products, improved labor, improved life, improved services providing, maintenance and recycling.
Stage 5: Regular controlling and correction of implemented improvements, i.e. general controlling focused on the labor quality, quality of production and life as well as on environmental, safety and sanitary and social accountability issues in all parties involved.

At individual stages well known methods of engineering and quality management are employed according to current needs (Zgodavová at all., 2002) however at the first stage the most significant is an appropriate analysis of internal – in-bound environment and especially of the external – out-bound environment employing such methods as a PEST (Political, Economical, Social, Technological) or SLEPT (Social, Legal, Economical, Political, Technological) analysis. Except for these are used the tools of decision-making like Pareto analysis, Decision trees, Methods of prediction and Strategic Decision Making.

The entire process of advanced quality improvement should be in accordance with the principles of project management and should meet the requirements of a particular community in a given environment and time.

4. REVIEW OF FINDINGS

Theoretical basis of **advanced improvement of quality** understands the quality of entities as a system.

- Quality of production is a complex of its properties, characteristics and features which are expressed by typical functions in a given environment and time.
- Advanced improvement of production quality is related to production processes and products in a given in-bound environment and given out-bound environment and in time and it forms a subsystem of the complex advanced improvement of production quality, improvement of labor and life quality.
- The process of complex improvement of quality consists of three main sub processes:
  - Designing improvements in production quality
  - Production processes of higher quality
  - Providing production of higher quality
- Advanced improvement of production quality is defined by three characteristic features:
  - Improvements are designed systematically
  - Interventions into improvement factors are proportional
  - Consequences of improvements are guaranteed.

5. SUMMARY

In sum, viewing advanced improvement of quality improvement with system theory, cybernetics, ICT, and advanced improvement of quantity tools and methods are necessary. With information and communication technologies and of user-friendly software, it is always easy to use for professionals.

6. REFERENCES

Beckford, J. (1998). *Quality: A Critical Introduction, Routledge.* New York, NY. ISBN 041518164X

Golemanov, L. A. (2006). *Quality Control and Production Management, Industrial, Process, Production, Quality and Systems Engineering.* Textbook manuscript. Alexander Dubcek University of Trencin, Slovakia.

Heylighen, F. (1997). The economy as a distributed, learning control system. *Communication & Cognition – AI*, Vol. 13, pp. 207-24.

Loukas, N. A. (2007). The archetype of excellence in universities and TQM*: Journal of Management History*, 2007, Volume: 13, Issue: 4, Page: 307 – 321

Mlčoch, L. & Slimák, I. (1987): *Řízení kvality a strojírenská metrologie.* ALFA Praha

Schroeder, R., Linderman, K., Liedtke, C. & Choo, A. (2002). *Six Sigma: definition, theory and research propositions*, working paper, Curtis L. Carlson School of Management, University of Minnesota.

Pan, X. (2006). Computer-aided SPC and cybernetic view on quality control, *International Journal of Quality & Reliability Management*, 2006, Volume: 23, Issue: 9, Page: 1192 – 1203.

Annals of DAAAM for 2008 & Proceedings of the 19th International DAAAM Symposium, ISSN 1726-9679
Editor B. Katalinic, Published by DAAAM International, Vienna, Austria 2008
Make Harmony Between Technology and Nature, and Your Mind will Fly Free as a Bird
*Young Researcher and Scientist Paper / * Supervisor, Mentor*

RESEARCH OF MUTANT SCREEN ELEMENT STOCHASTIC APPLICATION

ZILJAK, I[vana]; MRCELIC, Z[eljka]; DUJIC, L[ucijana]; MARINCEL, P[etra] & ZILJAK VUJIC, J[ana]*

Abstract: *This paper presents research of implementation of mutant screen element in stochastic application. that applies an indefinite number of screening models and screening shapes in the same page. Example of mutating screen element is mutant R73 wich is a newly developed screen element. The goal is to achieve full individualization of reproduction with the intention of applying in security printing. Experiment results of testing mutant screen for different gray levels, different angles and screen rullings are presented. In this manner three stochastic parameters have been used in the application algorithm. Data on the input image move with the algorithm to RIP and the SEED parameter that individualizes the image, generates a chain of random numbers from the RIP random numbe generator.*

Key words: *Screening, Mutant, Stochastic, Security*

1. INTRODUCTION

This article proposes a new manner of creating areas of programmed screen shapes (Foley at al., 1997) in order to select shapes from this area by the stohastic selection, as well as to define angles from 0 to 90 degrees and screen rulings in the acceptable range of screen rulings (Adobe Systems I., 1999) of a printer. The creation of PostScript algorithms that, depending on the seed generator of random numbers implement pseudo-random N of individual screen shapes, angles and screen rulings, has resulted in a unique print by using one single number (SEED). Stochastic application of newly developed mutant screen elements (Pap at al., 2008) demands preliminary testing.

2. MUTANT SCREEN ELEMENT R73 IN GRAY LEVEL AND ANGLE TESTING

The software solution has been made with functions that include the transformation model that can be generated either continuously or stochastically. The screen R73 has been tested in detail in the definition area (Fig. 1). Low screen ruling are analyzed by print scanning. Some print scanning has been done with 900 dpi to study the dot gain in the environment of individual screen element (Fig. 2).

Experiment results of testing mutant screen R73 for different gray levels and different angles are presented in figures 3, 4, 5 and 6. A rough screen ruling of 3 lpi is used in all the examples in order to observe the parameter mutation development in the best possible way. Each row represents one mutant.

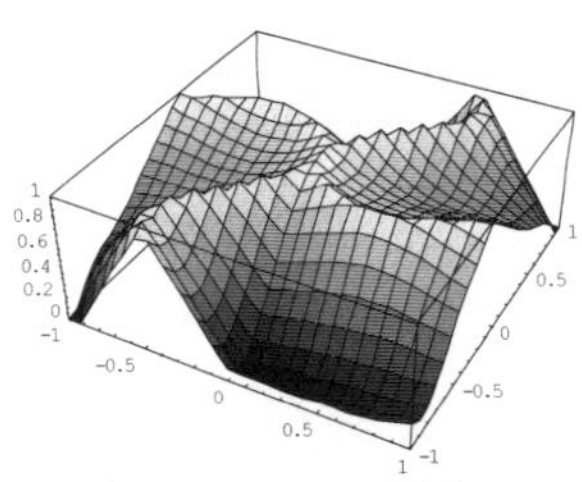

Fig. 1. The 3D model of screen mutant R73

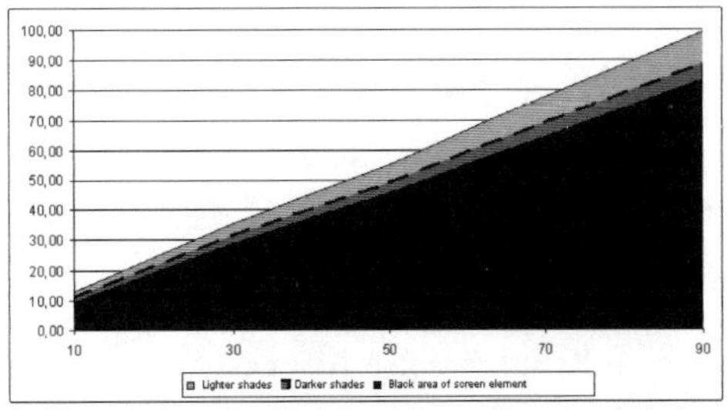

Fig. 2. Coverage and colored surface of mutant / 5lpi

The zero mutant is in the first row where the mutant parameter amounted to 0, and in the tenth row the mutant parameter is 1. On basis of analyzing such experimental results under low screen ruling and various angles, decision is made as to the allowed initial and final mutation shapes, and even as to discrete mutations, if necessary.

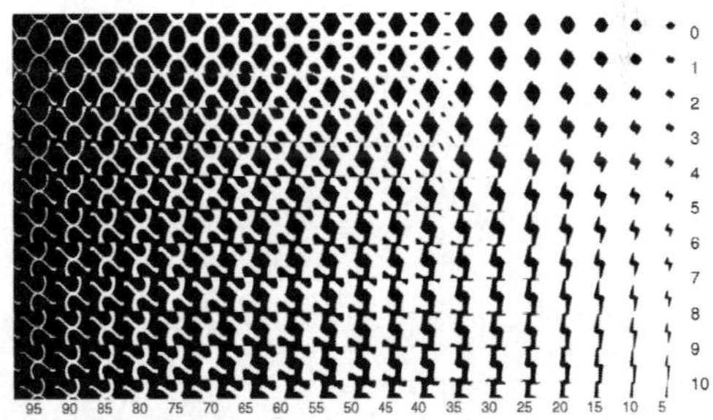

Fig. 3. Mutant R73 in gray level (95%-5%) and 0^0 angle

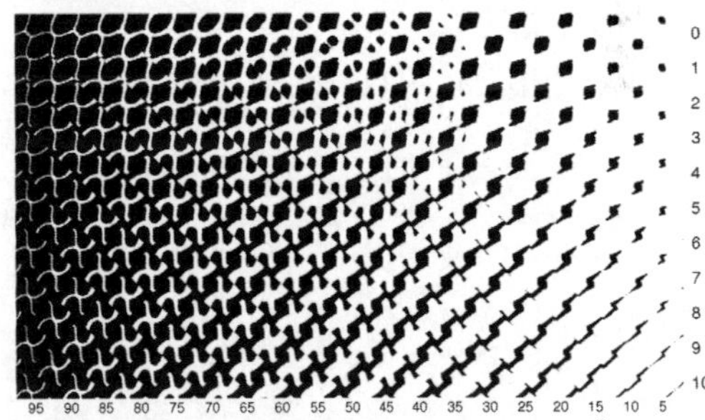

Fig. 4. Mutant R73 in gray level (95%-5%) and 45^0 angle

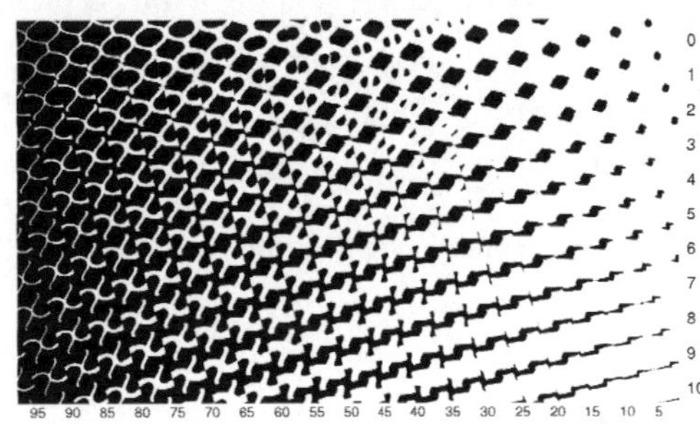

Fig. 5. Mutant R73 in gray level (95%-5%) and 75^0 angle

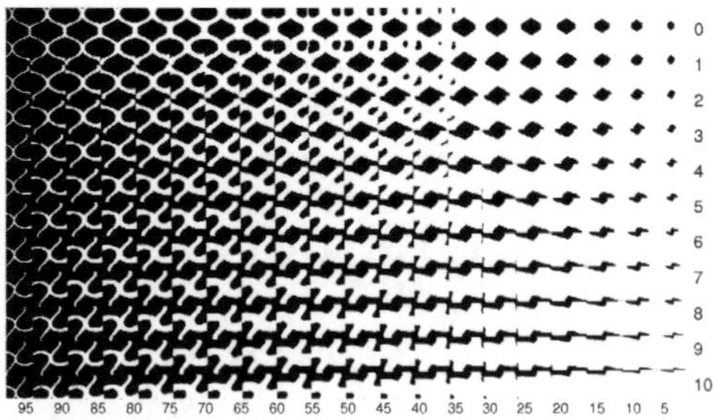

Fig. 6. Mutant R73 in gray level (95%-5%) and 90^0 angle

3. STOCHASTIC APPLICATION

The stochastic part of the discussion refers to application of an amplitude coverage model of each pixel separately (Žiljak & Pap, 2000). This applies to color printing as well where the usual procedure is to set various angles of screen. As in this case, a randomly chosen angle is joined with each pixel, there is no limiting in the number of colors: it is possible to mix process and spot inks.

Algorithms enable connecting of the color image with the angle and ruling screening alterations. We extend our research work with new screen element R73. The goal is to achieve full individualization of reproduction with the intention of applying in security printing. There is dot gain of each element's environment after printing, so the real border could not be determined any longer. Care should be taken only in respect to achieving the same coverage level: before printing and after printing.

Stochastic application of R73 halftone screen in 3 prepared images is presented in figures 7, 8 and 9. In this manner three stochastic parameters have been used in the Postscript program:

/kory 0.7 rn mul 0.2 add def % mutat. param. from 0.2 to 0.9,
/L {15 15 rn mul add } def % screen ruling from 15 to 30,
/K {0 90 rn mul add } def % screen angle from 0 to 90.

Fig. 7. Stochastic application of mutant R73 on 45% fill

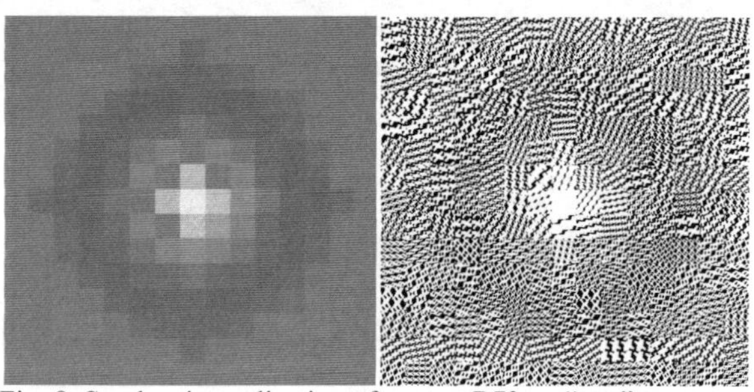
Fig. 8. Stochastic application of mutant R73 on "star"

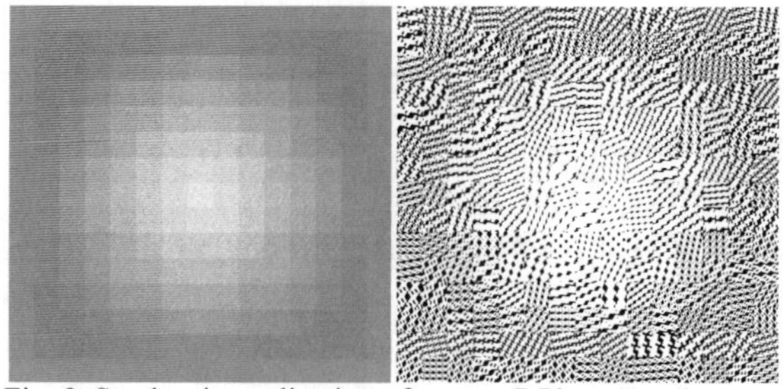
Fig. 9. Stochastic application of mutant R73 on "sun"

A novelty in this paper is the processing methodology in terms of the PostScript RIP printer. It implies that data on the input image move with the algorithm to RIP (Brett, 2000) and the SEED parameter that individualizes the image, generates a chain of random numbers from the RIP random number generator and not from the input PC. This is how the print becomes individualized not only by algorithms of new screen shapes, but also by a RIP random number. It is shown in our examples that the some SEED number is used.

4. CONCLUSION

A stohastic solution of print individualization is a border area of digital printing. Changes in PostScript commands enable a programming intervention during printing. Firstly, each printing sheet can have a new number, different image or text, prepared in a separate file. Secondly, individualization may be generated by an algorithm (commonly used in numeration), meaning that printed data are not in the memory, but created by logic defined in the PostScript program. This article proposes a method of randomly selected parameter mutant shape of a screen element for each pixel and presents the highest degree of individualization.

If various SEED values are defined for each input color chanel, it is possible to achieve multiple combinations of protection against counterfeiters. Program solution is created in a way which can make each researcher interested in it, continue the development of his own algorithms, procedures and finally apply it in specialised usages .

5. REFERENCES

Adobe Systems I., (1999). *PostScript Language Reference, third edition,* Adobe Systems Incorporated, pp. 482, ISBN 0-201-37922-8

Brett, G. (2000), *Digital Prepress Technologies*, Pira International Ltd, pp. 122, ISBN 1-85802-261-4

Foley, K.; van Dam; Feiner & Hughes (1997). *Computer Graphics Principles and Practice,* pp. 569, Addison-Wesley, ISBN 0-321-21056-5

Pap K.; Žiljak I. & Žiljak Vujić J. (2008). *Design of Digital Screening,* pp. 73, FotoSoft, ISBN 978-953-7064-10-5, Zagreb

Žiljak, V. & Pap, K. (2000). Mathematical Model of a Stochastic algorithm for Digital Printing, *Advances in Printing Science and Technology*, pp. 295-302, ISBN 1-85802-355-6, Leatherhead, 2000, Pira International Ltd

Annals of DAAAM for 2008 & Proceedings of the 19th International DAAAM Symposium, ISSN 1726-9679
Editor B. Katalinic, Published by DAAAM International, Vienna, Austria 2008
Make Harmony Between Technology and Nature, and Your Mind will Fly Free as a Bird

CORRELATIONS BETWEEN ENTREPRENEURSHIP FRAMEWORK AND ECONOMIC PERFORMANCES

ZIRRA, D[aniela]

Abstract: *The economic development depends essentially on the dynamic of business performances. In the same time, the characteristics of business environment have an important contribution on competitiveness, research and innovation capacity, more and better jobs creation, high quality of entrepreneurial activity, adaptability of workers and enterprises, increasing investments and development of knowledge-based economy. Considering these, in present day economy, it is very important to create a stable and performing entrepreneurship framework and also an extended and well functional network for SMEs, in order to stimulate business growth and to increase long-term company performance.*

Key words: *entrepreneurship, framework, SMEs, development, performances.*

¶

1. INTRODUCTION

¶

In actual circumstances of economic evolution, it is more and more difficult to realise economic performance and to evaluate how a public decision can stimulate entrepreneurial activity, new firms creation, the ambitions of entrepreneurs and managers to create wealth. Also, it is very important to have all the elements for measuring the effects of allocating funds to sustain economic development, to increase the dynamic of companies' activity and to improve the characteristics of business environment, with all the consequences involved.

For this reason, entrepreneurship (E-ship), as phenomena that create and exchange value (SCADPlus, 2008), need a specific and complex framework which is capable to stimulate and support the economic performances of firms in the new era. At international level there are various models of E-ship framework, but they need to be improved and updated such as the requirements and the dynamic of internal and international environment are asking. In these conditions, an E-ship framework must provide flexible and updated services, especially for the new born firms.

¶

2. AN EFFECTIVE E-SHIP FRAMEWORK

¶

Into a dynamic business environment, considering here the rapid technological, economic, social and political changing, SMEs development is a very difficult and complex process, but with many benefits, including human resources development, social cohesion, economic growth, increasing employment, a better flexibility and adaptability of individuals and organizations, competitiveness of national economies, etc.

Also, we have to mention the close correlations between a stable and performing entrepreneurship framework and long-term company performance. This is the reason of the central position in economic development strategies and the importance accorded to E-ship in present day economy, not only in developing countries, but in highest developed regions too. OECD has already elaborated a Demand and Supply Side Model of E-ship Framework (Ahmad & Hoffman, 2008). We appreciate that an adapted Model of this modern concept of the E-ship framework could be represented as in Figure-1.

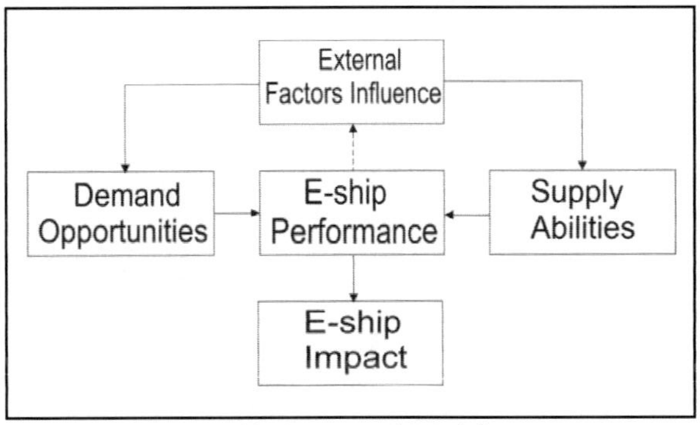

Fig. 1. Entrepreneurship Framework Model.

¶

E-ship meaning in modern economy is to offer the basis for the healthy circumstances of increasing and stimulating SMEs sustainable development. In actual stage, without an entrepreneurial education and culture, continuous learning abilities and skills it is very difficult to realise positive effects for business performances and SMEs environment.

The elements with a direct influence on E-ship performance, and as a propagate effect with a strong influence on business performances, are:

- External Factors Influence (EFI): individual incentives, culture and motivation, education system, infrastructure, framework conditions.
- Demand Opportunities (DO): technological transfer, R&D, communication, market access conditions, competition.
- Supply Abilities (SA): financial capital, business angel, venture capital, social capital, the quality of human capital.

All the three elements mentioned above represents the key factors which could be named the group of determinants for E-ship Performance (E-SP) and E-ship Impact (E-SI). In the same time, we cannot ignore that E-SP have an indirect influence on EFI, which is changing periodically because of economic and social conditions, in continuous changing and evolution.

As examples for the elements which represent criteria for appreciation the E-SP and E-SI, included also in the adapted Model, we can mention:

- E-SP: the rhythm of firm creation, business density, self-employment, new business survival, high growth firms.
- E-SI: poverty reduction, job creation ratio, productivity growth, economic growth, increasing of firm size.

But, the elements with a very strong influence on business performance are extremely diverse (OECD, 2007). For this reasons it is difficult to mark them out into distinguished categories.

According to the Statistical Business Registers (based on Administrative Registers), E-ship is the effort made to initiate an economic activity in the formal sector. The OECD/Eurostat Approach is to create a framework for Business Demography Statistics. This approach is very important because it gives us the opportunity to create more effective strategies to stimulate and support new firm creation and economic performance of existing firms.

3. WHY TO SUPPORT AND STIMULATE E-SHIP

¶

The general aim of business support policies in EU industrialized areas is to increase competitiveness through various kinds of interventions (European Committee, 2007), especially for SMEs development, such as:

- Providing training, financial, information, conciliation and business services.
- Granting funds for investments and high technology transfer.
- Improving communication and transport infrastructure.

We have to mention again that one of the key for a sustainable economic performance is to increase the competitiveness of SMEs and the dynamic of E-ship (Klapper et. al., 2007). For example, if the goal of government is to increase the flexibility and adaptability of workforce on management level, than the place of an E-ship stimulating policy is: INPUT - state subsidy for SMEs human resources development in business management; COMPONENT - Increasing E-ship; TARGET - Increasing workers quality (abilities and skills); GOAL - Increase the flexibility and adaptability of workforce on management level, with direct effects on economic performance.

In the new era, it can be notice that we are going towards an entrepreneurial society. In these conditions, if the goal of an organisation is to increase competitiveness or to maintain it at a high level, the modern management must be in principle entrepreneurial, for all kind of firms, including micro, small and medium too (SMEs). This requirement is more necessary in a very dynamic, aggressive and competitive environment. In this direction, if we are referring on the managerial team of a company, it is often used the concept of dynamic abilities, which could be classified in three categories (Teece, 2007), such as:

- Ability for identifying the new market opportunities, like technological, investments, informational, commercial, financial, etc. and the risks or the potential sources of risks from the economic and social environment.
- Capacity of a correct quantification or measuring the opportunities, risks and trends of a complex and continuous changing economy.
- Skills to catch on the opportunities and to transform them in sustainable business or in real factors for enhance company performances, through reengineering the assets of the organisation and the structure of its activities.

It must be noticed that the public strategies devoted to stimulate SMEs development must be based on correct, in time and relevant information.

All of these conditions are necessary to give the opportunity and to create a technical support for all interested decision makers for a better design, implement and evaluate the effectiveness of resources allocated and used to support SMEs. But, having the information could be a difficult process (Klapper et. al., 2007). We could outline here many influence factors with a strong impact on the quality of gathering and processing data (Table 2).

¶

No	Influence factors	Description
1.	Data availability	Not compiling data about new entry business or collecting data but not processing them
2.	Data "purity"	Legal or institutional reforms which change the rules for registration firms or the status of the firms
3.	Limitations regarding data on firm closures	Not in all countries exists requirements for reporting the closure of the businesses
4.	The existence of "shell" corporations	Created only for tax purpose, without being active businesses

Tab. 1. Influence factors on gathering and processing data.

4. E-SHIP DEVELOPMENT PROJECT

¶

The Human Resources Development Project Phare/2004/016-772.04.02.02.01.01.811, named „Today Managers, Entrepreneurs Tomorrow! Project for entrepreneurial and managerial skills development" itself is promoting the development of E-ship through increasing the quality of all level management for industrial sector. Based on the results of our research, we found that new firms creation depends essentially on policies, institutions and tax systems, specific for each country. Also, we can use 2 indicators for measuring the dynamic of entrepreneurial activity or new born firms (Klapper et. al., 2007): business entry (AER – Average Entry Rates) and density rate (ABD – Average Business Density). Based on WB Study, using data available from 84 countries, we can appreciate that the highest ABD per 1,000 active people, by region, in 2003-2005 is registered in Industrialized Countries (64.23%) and the lowest in Asia (1.57%). The highest AER, by region, in 2003-2005, using data from 82 countries is found also in Industrialized Countries (10.24%) and the lowest in Asia (6.7%). In Romania, ABD is evaluating ascendant in the last 2-3 years, but the indicator is still 2.5 less than EU media. So, the propagate goal of the project, is to create a network between business and high education, for a better evolution of the specialists supply for managerial activities in Romania, using the infrastructure and the experience of all the organisations involved in the project.

¶

5. CONCLUSION

¶

Yet, taking into account the evolution of science and technologies, the continuous changing of business environment, building an effective E-ship framework represents a permanent challenge. In these conditions, the main important requirements for a high performance of national economies are: the E-ship services must be updated based on the most relevant, complete and correct information; a continuous process of developing entrepreneurial abilities and skills; more collaborative projects between firms, universities and public sector.

¶

6. REFERENCES

¶

Activities of the EU (2008), SCADPlus: A modern policy for SMEs, http://europa.eu/scadplus *Accessed:* 2008-03-05

Ahmad, N. & Hoffman, A. N. (2008), A Framework for Addressing and Measuring Entrepreneurship, OECD Statistics Working Paper. Washington D.C.: OECD, p. 10, *Available from:* http://www.oecd.org/std/research *Accessed:* 2008-01-24

European Committee (2007), Evaluating Socio Economic Development, Sourcebook 1: Themes and Policy Areas Enterprise Policy, *Available from:* http://evalsed.info Accessed 2008-03-12

Klapper, L.; Amit, R.; Guillén, M. F. & Quesada, M. D. (2007), Entrepreneurship and Firm Formation Across Countries. World Bank Policy Research Working Paper 4313. Washington D.C.: World Bank, Development Research Group, Finance and Private Sector Team, pp.2-15, *Available from:* http://go.worldbank.org *Accessed:* 2008-14-03

OECD Entrepreneurship Indicators Project: Overview (2007), World Bank Seminar on Entrepreneurship and Development, Washington D.C., *Available from:* http://oecd.org/statistics/industry-services/entrepreneurship *Accessed:* 2008-03-09

Teece, D.J. (2007), The role of managers, entrepreneurs and the literati in enterprise performance and economic growth, *International Journal of Technological, Learning, Innovation and Development*, Vol. 1, No 1, August 2007, pp. 48-61, ISSN 1753-1950

Annals of DAAAM for 2008 & Proceedings of the 19th International DAAAM Symposium, ISSN 1726-9679
Editor B. Katalinic, Published by DAAAM International, Vienna, Austria 2008
Make Harmony Between Technology and Nature, and Your Mind will Fly Free as a Bird

MODELLING OF DISPERSION AND REFLECTION OF LIGHT ON PAPER SURFACE

ZITINSKI ELIAS, P[aula]; TOMASEGOVIC, T[amara] & MODRIC, D[amir]

Abstract: *Manufacturing significance of optical properties of paper motivates us to construct a theoretical approach which could be used to determine the reflection and dispersion of light from paper surface. The Monte Carlo method was used as a frame within we were able to describe the diffuse scattering of light which suffers multiple subsurface reflections. Monte Carlo simulation of photon propagation offers a flexible, but also rigorous approach to transport of photons in a medium. Our model, based on macroscopic optical properties of the paper components, describes this problem more physically plausible and gives possibility to model some desired optical properties of paper without making the actual experiment.*

Key words: *substrate surface, paper optics, light dispersion, Monte Carlo method.*

1. INTRODUCTION

Paper is today, despite the increasing of virtual mass-media, still one of the favorite media for information distribution. It is used in almost all areas of work and industry. In graphic technology its optical properties are very important, as they often identify types of paper according to its purpose.

The motivation of this simulation was to describe and anticipate the transport of light through a media such as paper and to see the impact the dispersion has had on the optical dot gain. The tone resolution and a reproduction characteristic for all graphics paper products is significantly conditioned by the way that the way the light scatters in a paper. The light that enters the paper between the raster elements can be scattered in the paper in all directions, including below raster element where it remains absorbed which causes the gray scale becoming darker than originally required. To manufacture a paper that successfully fulfils the needed optical properties, it is necessary to understand the physical principles of the structure and composition of a paper sheet. This theoretical work was performed within frame of Monte Carlo method which describes the transport of light. This is a numerical method for solving mathematical problems based on a random sampling from well defined probability distributions. Starting from real physical assumptions the subsurface light scattering in a substrate with a complex structure was modeled (Kubelka, 1931, Emmel, 1999, Mourad, 2002). For this kind of problem, where statistics approach offer the best insight and approximation for exact results, a method as Monte Carlo offers a more flexible approach to the transport of photons in a medium such as paper. Even though this model has pure stochastic nature it makes feasible quasi experimental approach in optical dot gain studying.

2. THE MONTE CARLO METHOD

Monte Carlo method was originally designed for needs of the studying of the interaction of elemental particles (neutrons, mesons…) with matter, but is also used to solve problems of radiation transport in high energetic physics, analysis of nuclear reactors, calculation of protection from harmful radiation, in treatment of cancer cells with radiation, etc. (Kalos (1986)).

For implementation of this method a stochastic model is constructed in which expected variable value (or combination of variables), is equivalent to the physical value that needs to be determined. The expected value is defined by medium value of multiple independent samples that represent this random variable. We use random generated numbers, which follow the prior selected natural distribution, to construct the desired array of independent samples.

The Monte Carlo method simulations of this type are based on the macroscopic optical properties for which it is assumed that they prevail over small parts of the volume of paper (e.g. cellulose fibers, fillers, adhesives, etc). (Veach (1997)).

We would like to present a typical trajectory of a single photon and the method describes the local rules of propagation of photons. According to well known procedure each step size between photon – substrate interaction positions is variable and equals

$$\frac{-\ln\xi}{(\mu_a+\mu_s)} \tag{1}$$

where ξ is a random number and μ_a and μ_s are the absorption and scattering coefficients, respectively. Every photon has assigned statistical weight which decrease from an initial value of 1 as it moves through the substrate, and equals a^n after n steps, where a is the albedo:

$$a=\frac{\mu_s}{(\mu_a+\mu_s)} \tag{2}$$

Once the photon packet has been moved the photon packet is ready to be scattered. There will be a deflection angle, $\theta \in [0, \pi)$, and an azimuthal angle, $\psi \in [0, 2\pi>$ to be sampled statistically. The probability distribution for the cosine of the deflection angle, $\cos\theta$, is described by the scattering function that Henyey and Greenstein (1941) originally proposed for galactic scattering:

$$p(\cos\theta)=\frac{1-g^2}{2\left(1+g^2-2g\cos\theta\right)^{\frac{3}{2}}} \tag{3}$$

where the anisotropy, g, equals $<\cos\theta>$ and has a value between –1 and 1(details in Modrić, 2007).

When the photon strikes the surface, a fraction of the photon weight escapes as reflectance and the remaining weight is internally reflected and continues to propagate. Eventually, the photon weight drops below a threshold level and the simulation for that photon is terminated. In this example, termination occured when the last significant fraction of remaining photon weight escaped at the surface at some position which differs from incoming point. To satisfy energy conservation law we used photon packet of hundred photons.

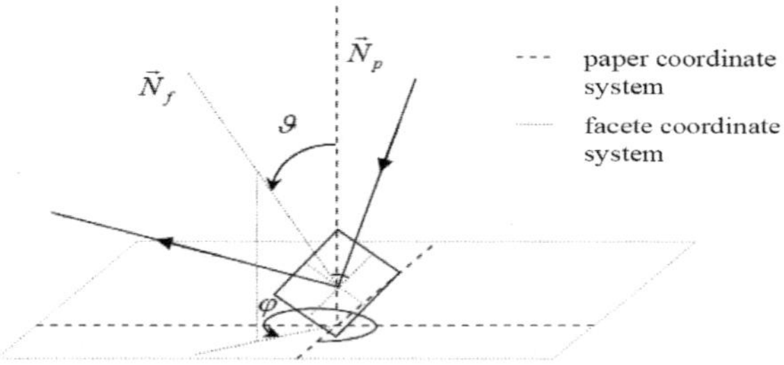

Fig. **1.** Review of basic values when describing paper surface with microfacets. ($\varphi \in [0, 2\pi]$ and $\vartheta \in [0, \vartheta_m]$)

We calculated numerous photon trajectories (10^4 to 10^6) to yield a plausible statistical description of photon distribution in the medium.
It is to emphasize that our simulation does not consider the wave nature of light, and that ignores values such as phase or light polarization. Inside the paper as an extremely complex media photons experience multiple scattering and the phase and polarization randomizes very quickly so that initially they don't affect much the energy transport.

3. MODELLING AND SOLVING THE PROBLEM

In this work we've approached the problem of light scattering in the paper in the way that describes the real situation in a more physical way. Our simulation offers a flexible and yet rigorous approach to the problem of transporting light in a media such as paper.
The complexity of the paper surface structure encouraged us to improve the models with the aim of approaching a more realistic description of the surface. In our first approximation we assumed that our paper surface was perfectly flat and that stretches infinitely in the X and Y plane. Above presented scattering profile is averaged overall scattering angle distribution realized with light beam which lighten whole paper sheet (Modrić, 2007).
Fig. 2. confirms the initial idea that increasing of paper roughness spreads distribution of scattered light due to the additional randomization of initial direction of incoming light.

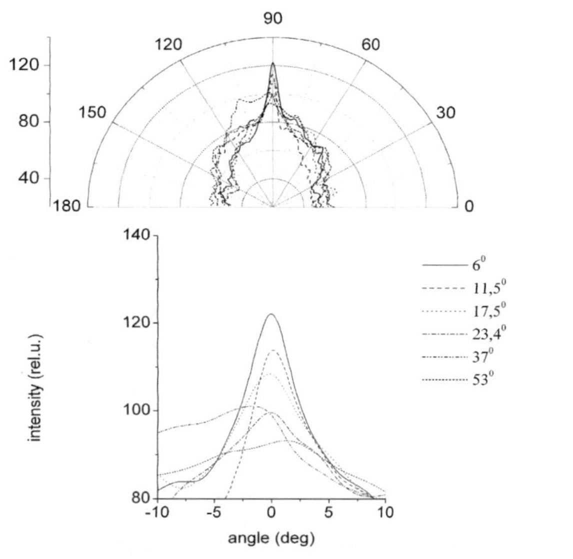

Fig. **2.** Calculated scattered light from paper for different paper roughness defined with correspondent angle (parameter) ϑ_m (specular reflection was not calculated). Incoming light penetrated vertically.

By varying parameter ϑ_m we can also predict light behavior for calendared papers or papers with some additional surface superstructure. As it could be seen the most spread distribution is for papers with parameter ϑ_m=53° which was presented only for illustration because paper with such surface doesn't have any commercial significance (Modrić, 2007).

4. CONCLUSION

The contribution of this work is clearly connected with the demands of an optimal reproduction and print quality. All paper components such as mechanical and/or chemical pulp, whiteners, fillings, adhesives etc., affect the way the light scatters in paper as well its surface properties. It is evident that paper appearance is not consequence generally of its surface topography but also of its subsurface optical properties.
The light that enters the paper between the raster elements can be scattered in the paper in all directions, including below, where the dot element remains absorbed which causes the gray scale becoming darker than originally required. Starting from real physical assumptions the undersurface light scattering in a substrate with a complex structure was modelled.
Our model offers the possibility of "experimenting" with various combinations of paper components to verify some ideas without the long-lasting and expensive realizing an actual paper. Given the combination and variation possibilities of the composition share of each component of a paper, this model can study the optical properties of any type of paper, including the recycled ones, where as one component appears the particle remains of the dyes and the treated components of previous paper.
We expect that improvement and optimization of our model will lead to its implementation in manufacturing process. Our future efforts will be directed to improve our model of paper (with generalization on all substrates) and its interactions with dye, caused with various printing techniques and dyes, and to examine potential relationships between optical and mechanical properties of substrate. Beside paper – dye interaction (mainly, cross-section profile of raster element) our interest will also be focused on investigation of influence of anisotropy factor in our model which could be important for optimization of paper components initial mixture.
It should be pointed that we haven't include wavelength dependence of scattered light because there are no data in literature for scattering and absorption coefficients of every component. However, implementation of such dependence in model is trivial.

5. REFERENCES

Emmel P. and Hersch R.D., *Towards a color prediction model for printed patches*, IEEE Comp. Graphics and Appl. 19 (1999), 54–60.

Kalos, M.H., Whitlock P.A., (1986). *Monte Carlo Methods, I: Basics*. John Wiley & Sons, Inc.

Kubelka P., Munk F. (1931), *Ein Beitrag zur Optik der Farbanstriche*, Z.Tech.Phys., 11a (1931), pp 593-601.

Modrić, D., Mikac Dadić, V., Džimbeg-Malčić V., Light Scattering Numerical Modeling Compared with Kubelka-Munk Method, Proceedings of the 11th International Conference on Printing, Design and Graphic Communications, Zadar, (2007) 107-111

Modrić, D. (2007), PhD Thesis, University of Zagreb

Mourad M.S., *Color predicting model for electrophotographic prints on common office paper*, (2002), Ph.D Thesis, École Polytechnique Fédérale de Lausnne.

Veach E. (1997), *Robust Monte Carlo Methods for Light Transport Simulation*. PhD thesis, Stanford University

Annals of DAAAM for 2008 & Proceedings of the 19th International DAAAM Symposium, ISSN 1726-9679
Editor B. Katalinic, Published by DAAAM International, Vienna, Austria 2008
Make Harmony Between Technology and Nature, and Your Mind will Fly Free as a Bird

ANALYSIS OF THE COMMERCIAL QUALITY FOR ELECTRICITY GRID USERS

ZIVIC DJUROVIC, M[arijana]; MOROVIC, M[ira] & DJUROVIC, G[ordan]

Abstract: *The introduction of electricity market and process of liberalization put uses interests into focus, such as ability of procuring cheaper electricity and ability of producing electrical energy of better quality. Apart from that, the privatization of electric power system segments attracts the interest of capital and profit making, which is in conflict with users interests. Therefore it is necessary to monitor the quality of electricity supply within requirements of quality management systems.*

Key words: *commercial quality, electricity supply, standard.*

1. INTRODUCTION

As a result of electric power market liberalization, electric power has become a freely traded open market product, meeting predetermined quality standards (Zeithami, V. et al. 1990.). New and more complex relationship between involved participants have resulted from deregulation in production, transmission and distribution systems. This type of approach is adopted in nearly all countries that are undergoing the process of liberalization. Still, a unique concept of making electric power systems to work optimally in new market conditions does not exist. The appropriate level of quality of electrical energy is requested from all parts of electric power system, but the most important influence on quality of electricity supply has distribution system (grid). Major repercussions regarding deregulating of electric power industry are requests for: reducing prices for electricity production; reducing prices of electrical energy for all customers; reducing compensations for all grid users. Request for reducing compensation for all grid users have direct impact on development and building transmission and distribution grids and its efficiency (Tahvanainen et al., 2004.). While doing these reductions in prices it's not allowed to reduce security i.e. quality of electrical energy. Security of electricity supply means that customers must get electricity when they need it, with standard quality and acceptable price. Commercial quality system in Croatia uses several coefficients, presented in this paper, which allows insight to quality of electricity supply, its analysis and development.

2. THE QUALITY OF ELECTRICITY SUPPLY

A customer service indicator measures direct contact with the customer. In setting the customer service standards, minimum standard guidelines are provided that are intended to maintain customer service quality while providing the utilities with flexibility to set service levels to the demands of their customers above the minimum guidelines. There are three main indicator groups for the quality of electricity supply: continuity of supply, voltage quality and commercial quality (Fumagalli et al., 2007.). Continuity of supply measure the grids ability to supply the customers with electricity. It is generally characterized as the number and duration of interruptions. Main parameters of voltage quality are frequency, voltage magnitude and its variation, voltage dips, temporary or transient overvoltages and harmonic distortion. There are several technical standards for voltage quality criteria, but in the end the quality is determined by the ability of customer equipment to perform properly. Therefore the definition of voltage quality should contain the impact of voltage disturbances on the customers (Blume et al., 1999.). Commercial quality relates to the nature and quality of customer service provided to the customers by the distribution companies. Commercial quality is generally monitored through standards. Standards can be Overall or Guaranteed standards according to CEER - Second benchmarking report on quality of electricity supply (2003.). Guaranteed standards set minimum service level which must be met in each individual case. If the company does not meet these standards, compensation at fixed rates must be paid. Guaranteed standards include:

- Service covered (e.g. estimated charges)
- Required performance level - usually whit a response time (e.g. certain number of working days)
- Penalty payment to be paid to a customer who fails to receive this level of service (e.g. amount of money).

Overall standards cover areas of service where it may not be possible to give individual guarantees but where companies are expected to deliver predetermined levels of service. Overall standards do not carry penalty payments but are used for monitoring purposes and for promoting quality of service. Overall standards are defined as followed:

- Service covered (e.g. connecting new customers to the grid)
- Minimum performance level to be achieved over a defined period of time.

2.1 Electricity supply services

General conditions of electricity supply defines public services which HEP Distribution System Operator (DSO) is obligated to provide to all electricity market participants in Croatia. Services are classified as: electricity distribution services and electricity supply services. In this paper electricity supply services and its coefficients will be analyzed. According to its nature and method of financing, electricity supply services defined in general conditions of electricity supply are classified as standard and nonstandard service. Standard services are defined as services in which customer, nature and performance circumstance are determined by acts of law. Service expenses are paid through determined tariff models. Non-standard services are services which HEP DSO must provide at customers demand or are unavoidable in their mutual cooperation, where DSO acts as a public service distributor. Non-standard service expenses are regulated within the Price List of Non-Standard Services. Distribution systems operator in Croatia had to establish commercial quality control system by July, 1st. 2006. The data from the year 2007. from 21 distribution centers are first analyzed using the coefficients for electricity supply services. The coefficient K_{sob} for commercial quality regarding billing and payment is defined as:

$$K_{sob} = \frac{\text{No. of justified complaints}}{\text{No. of receipts}} \cdot 100 \qquad (1)$$

Minimum service level which must be met are 0,5% for residential and 0,3% for commercial users.

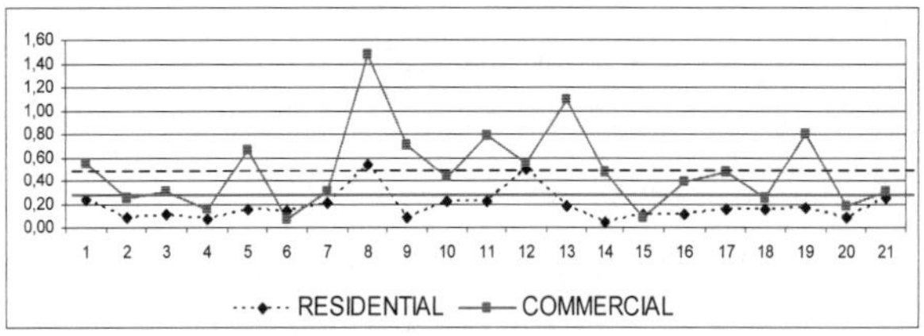

Fig. 1. The coefficient K_{sob}

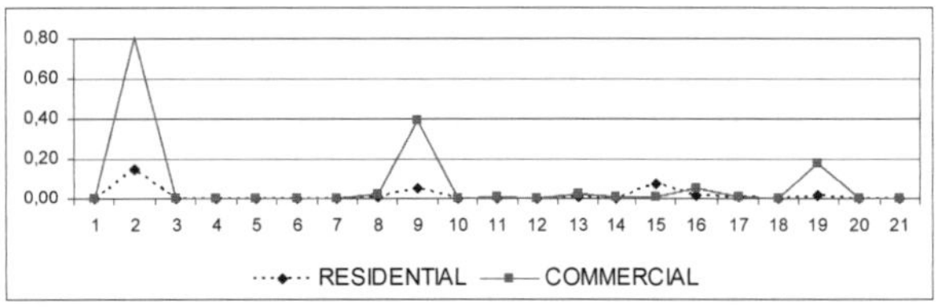

Fig. 2. The coefficient K_{snb}

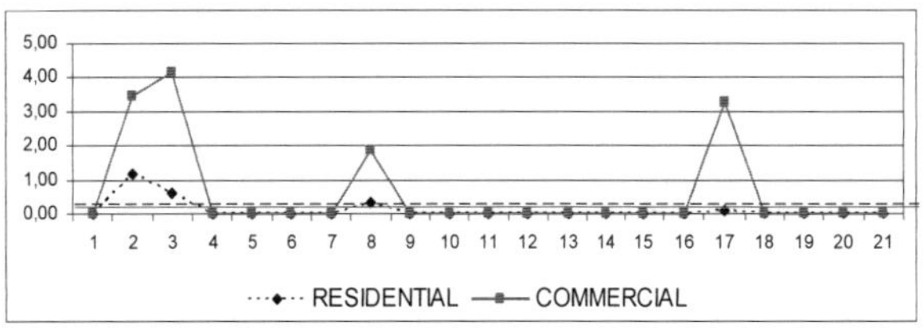

Fig. 3. The coefficient K_{nob}

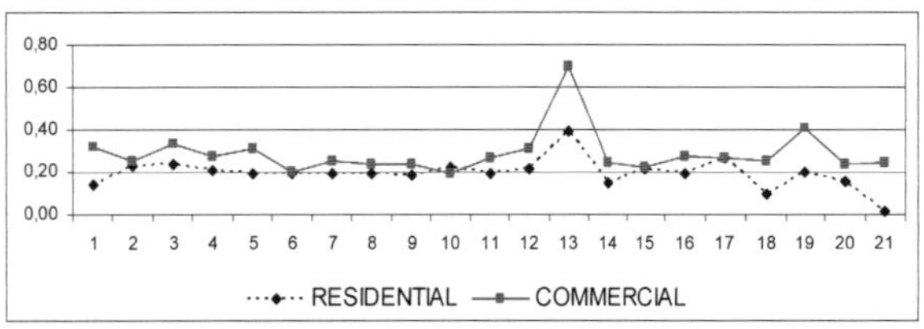

Fig. 4. The coefficient K_{nnp}

The coefficient K_{snp} for commercial quality regarding regular collection of receivables is defined as:

$$K_{snp} = \frac{\text{No. of justified complaints on billing procedure}}{\text{No. of paid bills without reminders}} \cdot 100 \qquad (2)$$

Minimum service level which must be met are 5% for residential and 3% for commercial users.

The coefficient K_{nob} for commercial quality regarding accounts settlement and issuance is defined as:

$$K_{nob} = \frac{\text{No. of justified complaints}}{\text{No. of non - standard accounting}} \cdot 100 \qquad (3)$$

Minimum service level which must be met are 0,3% for residential and 0,2% for commercial users.

The coefficient K_{nnp} for commercial quality regarding non-standard receivables collection is defined as:

$$K_{nnp} = \frac{\text{No. of justified complaints on billing procedure}}{\text{No. of reminders}} \cdot 100 \qquad (4)$$

Minimum service level which must be met are 3% for residential and 2% for commercial users.

Service level towards which must be met are 5% for residential and 3% for commercial users. The coefficients K_{snp} and K_{nnp} are within defined minimum service level, while coefficients K_{sob} and K_{nob} which are related to number of justified complaints for standard and non-standard services are, in some distribution centers, outside defined level. However, the data from the year 2007. are the first to observe, so the results are satisfactory, but there is plenty of room for improvements.

3. METHODS FOR COMMERCIAL QUALITY IMPROVEMENT

Main goal for commercial quality improvement are perfecting present values and obtaining values for mentioned coefficients according to Table 1. which defines not minimum, but excellent service level.

	Residential	**Commercial**
K_{sob}	0,1%	0,3%
K_{snp}	1%	3%
K_{nob}	0,1%	0,2%
K_{nnp}	1%	2%

Tab. 1. Excellent service level for electricity supply services

The customer information becomes a central aspect of commercial quality. Methods of provision include newspapers, Internet sites, Info desk and providing data via electricity bills, e-mail and SMS. Strategies to encourage customer participation can include standards associated with time of response to claim and requests for information, active participation of customers associations in the development of electricity sector regulation and diverse ways of contacting companies. It is also necessary to apply continuity with the full engagement of all employees (Zivic Djurovic et al., 2008.), using the latest quality management and quality assurance standards in order to build the structure of a modern and opened quality control system.

4. CONCLUSION

Grid operators are obliged to establish individual measurement system and system of continuous monitoring of the parameters of quality of electrical energy, with the purpose of determining the parameters condition and its improvements to standard levels. Conducted research and determining commercial quality conditions in distribution grid are groundwork for implementation of technical measures regarding improvement of quality of electrical energy parameters and overall planning distribution grid development.

5. REFERENCES

Blume, D.; Schlabbach, J. & Stephanblome, T. (1999.) *Spannungsqualitätin elektrischen Netzen (Vvoltage quality in electricity grids)*, Verlag, 3-8007-2265-8, Berlin

Fumagalli, E.; Lo Schiavo, L. & Delestre, F. (2007). *Service Quality Regulation in Electricity Distribution and Retail*, Springer-Verlag and Heidelberg GmbH & Co. K, 978-3-540-73442-0, Germany, UK

Tahvanainen, K.; Viljainen, S.; Honkapuro, S.; Lassila, J. & Partanen, J. (2004), Quality regulation in electricity distribution business, *Available from*: http://powersystems. tkk.fi/nordac2004/papers/nordac2004_tahvanainen_et_ al_paper.pdf *Accessed:* 2008-05-21

Zeithami, V.; Parasuraman, A. & Barry, L. (1990.) Delivering Quality Service, The Free Press, 978-0029357019, UK

Zivic Djurovic, M.; Morovic, M. & Varda, V. (2008). Approach to quality service improvement for electricity grid users, 1st International Conference on Electricity Distribution, HO CIRED, Šibenik 18.-21.May, 2008.

Annals of DAAAM for 2008 & Proceedings of the 19th International DAAAM Symposium, ISSN 1726-9679
Editor B. Katalinic, Published by DAAAM International, Vienna, Austria 2008
Make Harmony Between Technology and Nature, and Your Mind will Fly Free as a Bird

MACHINING PARAMETER OPTIMIZATION USING ANT COLONY SYSTEM

ZUPERL, U[ros]; CUS, F[ranc] & BALIC, J[oze]

Abstract: *This paper presents a new multi-objective optimization technique, based on ant colony optimization algorithm (ACO), to optimize the machining parameters in turning processes. Three conflicting objectives, production cost, operation time and cutting quality are simultaneously optimized. The proposed approach uses adaptive neuro fuzzy inference system (ANFIS) system to represent the manufacturer objective function and on ant colony optimization algorithm to obtain the optimal objective value.*
Key words: *Machining, turning, optimization, cutting conditions.*

1. INTRODUCTION

The selection of optimal cutting parameters is a very important issue for every machining process. In workshop practice, cutting parameters are selected from machining databases or specialized handbooks, but they don't consider economic aspects of machining.

Optimization of cutting parameters is a difficult work (Cus & Balic, 2000), where the following aspects are required: knowledge of machining; empirical equations relating the tool life, forces, power, surface finish, etc., to develop realistic constrains; specification of machine tool capabilities; development of an effective optimization criterion; and knowledge of mathematical and numerical optimization techniques.

Optimization of machining parameters is complicated when a lot of constraints are included, so it is difficult for the non-deterministic methods to solve this problem. Consequently, non-traditional techniques were used in the optimization problem (Liu & Wang, 1999). (Zuperl & Cus, 2003) have described the multi objective technique of optimization of cutting conditions for turning process by means of the neural networks. Further genetic algorithm and simulated annealing techniques have been applied to solve the continuous machining profile problem by (Milfelner et al., 2004).

In this paper, a multi-objective optimization method, based on combination of ANFIS and ACO evolutionary algorithms, is proposed to obtain the optimal parameters in turning processes.

2. MACHINING MODEL FORMULATION

The objective of this optimization machining model is to determine the optimal machining parameters including cutting speed, feed rate and depth of cut in order to minimize the operation cost and to maximize production rate (represented by manufacturing time (T_p) and cutting quality (R_a).

$$C_p = T_p \cdot \left(C_t / T + C_l + \cdot C_0 \right) \tag{1}$$

where C_t, C_l and C_0 are the tool cost, the labour cost and the overhead cost respectively; T is tool life. The objectives used in this work are determined according to (Zuperl & Cus, 2003). In order to ensure the evaluation of mutual influences and the effects between the objectives and to be able to obtain an overall survey of the manufacturer's value system the multi-attribute function of the manufacturer (y) is determined. The cutting parameter optimization problem is formulated as the following multi-objective optimization problem: min T_p (v, f, a), min C_p (v, f, a), min R_a (v, f, a)

$$y = 0{,}42 \cdot e^{(-0{,}22Tp)} + 0{,}17 \cdot e^{(-0{,}26Ra)} + \frac{0{,}05}{\left(1 + 1{,}22 \cdot T_p \cdot C_p \cdot R_a\right)} \tag{2}$$

A multiattribute value function is defined as a real-valued function that assigns a real value to each multiattribute alternative, such that more preferable alternative is associated with a larger value index than less preferable alternative.

The following limitations are taken into account: Permissible range of cutting conditions: $v_{min} \le v \le v_{max}$, $f_{min} \le f \le f_{max}$, $a_{min} \le a \le a_{max}$; Implied limitations issuing from the tool characteristics and the machine capacity; The limitations of the power and cutting force are equal to: $P(v, f, a) \le P_{max}$, $F(v, f, a) \le F_{max.}$

The proposed approach consists of two steps. First, experimental data are prepared to train and test ANFIS system to represent the objective functiony (y). Finally, an ACO algorithm is utilized to obtain the optimal objective value. Figure 1 shows the flowchart of the approach.

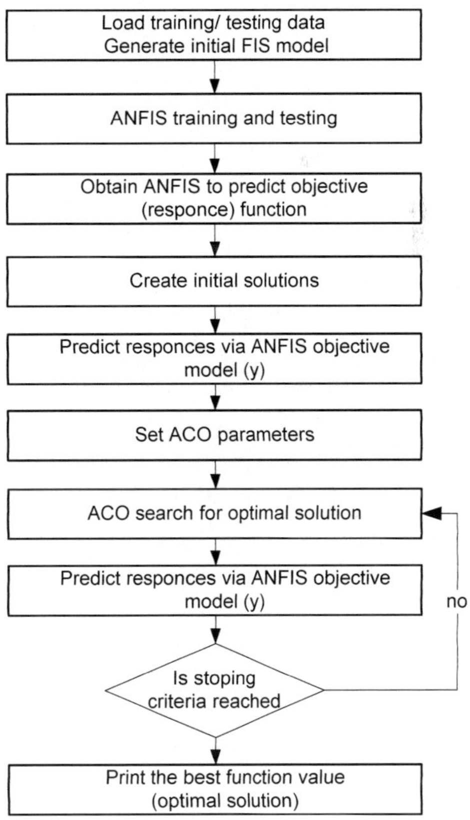

Fig. 1. Scheme of the proposed approach.

3. OBJECTIVE FUNCTION MODELLING

First step uses an adaptive neuro fuzzy inference system (ANFIS) to model the response (manufacturer's implicit multiattribute) function (y). The variables of this problem are velocity, feed rate and depth of cut, which can have any continuous value subject to the limits available. The ANFIS system needs three input neurons for three parameters: v, f and a. The output from the system is a real value (y). Figure 2 shows the fuzzy rule architecture of ANFIS when the triangular membership function is adopted, respectively. The architectures shown in Figure 2 consist of 32 fuzzy rules. During training in ANFIS, 140 sets of experimental data were used to conduct 400 cycles of training. ANFIS has proved to be an excellent universal approximator of non-linear functions. If it is capable to represent the manufacturer's implicit multiattribute function.

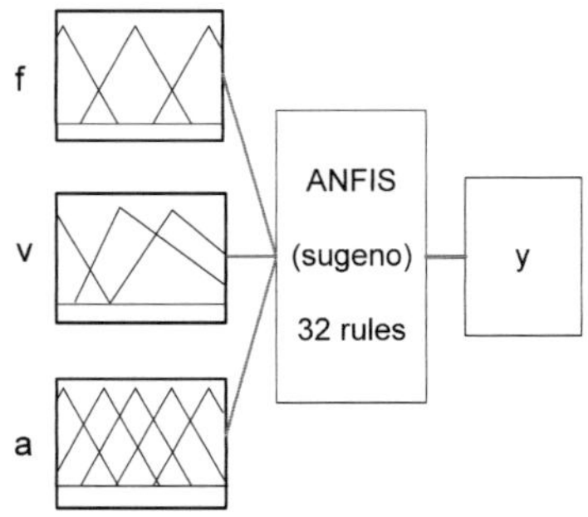

Fig. 2. Fuzzy rule architecture of the triangular and the membership function

4. ANT COLONY OPTIMIZATION (ACO)

Special insects like ants, termites, and bees that live in a colony are capable of solving their daily complex life problems. These behaviours which are seen in a special group of insects are called swarm intelligence. Swarm intelligence techniques focus on the group's behaviour and study the decartelized reactions of group agents with each other and with the environment. The swarm intelligence system includes a mixture of simple local behaviours for creating a complicated general behaviour and there is no central control in it. Ants have the ability to deposit pheromone on the ground and to follow, in probability, pheromone previously deposited by other ants. By depositing this chemical substance, the ants leave a trace on their paths. By detecting this trace, the other ants of the colony can follow the path discovered by other ants to find food. For finding the shortest way to get food, these ants can always follow the pheromone trails. The first ACO algorithm, called ant system (AS) has been applied to the travelling salesman problem (TSP). (Dorigo, 1996) proposed an ant colony optimization methodology for machining parameters optimization in a multi-pass turning model, which originally was developed by (*Vijayakumar* et al., 2002).

4.1 Ant colony algorithm

An ACO utilizes bi-level procedures which include local and global searches. Local search ants select a local trail I with a probability $P_i(t)=\tau_i(t)/\sum\tau_k(t)$, where i is the region index and $t_i(k)$ is the pheromone trail on region i at time t. After selecting the destination, the ant moves through a short distance ($\Delta(T,R)=R(1-r^{10(1-T)})$, where R is maximum search radius, r is a random number from [0,1], T is the total number of iterations of the algorithm. A global search is done sequentially by crossover and mutation operations. The subsequent values of the variables of the child are set to the corresponding value of a randomly chosen parent with a crossover probability (P_c). Mutation operation adds or subtracts a value to/from each variable with mutation probability (P_m). The mutation step size is the same as the above distance $\Delta(T,R)$. Performing an ACO, ants are repeatedly sent to trail solutions in order to optimize the objective value. The total number of ants (denoted by A) is set as half the total number of trail solutions (denoted by S). The number of global ants (denoted by G) and the number of local ants (denoted by L) are set as 80% and 20% of the total number of ants, respectively. The ACO algorithm:

Step 1. Set parameter values including: S, A, ρ, τ_0, P_c, P_m, T, R, and bounds of each control factor.
Step 2. Create S trail solutions (v, f, a). Estimate the objective value of the trail solutions through the ANFIS model (y).
Step 3. Set the initial pheromone value of all trails.
Step 4. Repeat steps 6—9 until the stopping criteria has reached.
Step 5. Send L ants to the selected trail solutions for local search.
Step 6. If the solution is improved, move the ants to the new solution and update the pheromone value.
Step 7. Send G ants to global trails and generate their offspring by crossover and mutation.
Step 8. Evaporate pheromone for all trails.

5. RESULTS AND DISCUSSION

The ant colony optimization method combined with ANFIS prediction system was tested. Proposed ACO approach was compared with Method using ANN routine, genetic algorithms and LP technique. The results revelled that the proposed method significantly outperforms the GA and LP approach. The proposed approach found an optimal solution of 0.30 for as low as 1-18 runs the genetic-based approach require as much as 900-1300 runs to find an solution of 0.4 This means that the proposed approach has 8.1% improvement over the solution found by GA approach and 17.3% over LP approach.

6. CONCLUSION

In this work, non-conventional optimization techniques ACO has been studied for the optimization of machining parameters in turning operations. The ACO algorithm is completely generalized and problem independent so that it can be easily modified to optimize this turning operation under various economic criteria. The algorithm can also be extended to other machining problems such as milling operations and threading operations.

7. REFERENCES

Cus, F. & Balic, J. (2000). Selection of cutting conditions and tool flow in flexible manufacturing system. *The international journal for manufacturing science & technology*, Vol. 2, pp. 101-106

Dorigo, E. (1996). The ant system: Optimization by a colony of cooperating agents. IEEE *Transaction on Systems, Man and Cybernetics*, Vol. 26, pp. 1-13.

Liu, Y. & Wang, C. (1999). Neural Network based Adaptive Control and Optimisation in the Milling Process. *International Journal of Advanced Manufacturing Technology*, Vol. 15, pp. 791-795

Milfelner, M.; Zuperl, U. & Cus, F. (2004). Optimisation of cutting parameters in high speed milling process by GA. *Int. j. simul. model.*, Vol. 3, pp. 121-131

Vijayakumar, K.; Prabhaharan, P.; Asokan, R. & Saravanan, M. (2002). Optimization of multi-pass turning operations using ant colony system. *International Journal of Machine Tools and Manufacture*, Vol. 3, pp. 633–639

Zuperl, U. & Cus, F. (2003). Optimization of cutting conditions during cutting by using neural networks. *Robot. comput.-integr. manuf.*, Vol. 19, pp. 189-199

AUTHOR INDEX

A

B

C

D

I

J

K

L

M

N

O

P

R

S

T

SUBJECT INDEX

A

B

D

E

H

I

J

K

L

M

Q

R

U

V

W

X

Z

MTF STU TRNAVA EXPRESSES THANKS TO SUPPORTERS AND SPONSORS OF 19TH DAAAM INTERNATIONAL SYMPOSIUM „INTELLIGENT MANUFACTURING & AUTOMATION" DAAAM-2008

SUPPORTERS:

Town of Trnava
Mesto Trnava

Self-governmental region of Trnava
Trnavský samosprávny kraj

SPONSORS:

Výskumný ústav jadrovej energetiky, a. s. Trnava

BEKAERT, a. s. Hlohovec

Jadrová a vyraďovacia spoločnosť, a. s. Jaslovské Bohunice

SLOVNAFT, a. s. Bratislava

PRVÁ ZVÁRAČSKÁ, a. s. Bratislava

DRILL, s. r. o. Bratislava

Nadácia Ján Korca Trnava

Železničné opravovne a strojárne, a. s. Trnava

INVEX, s. r. o. Trnava

Slovenské elektrárne, a. s., člen skupiny ENEL Bratislava

KESA Trnava

Nadácia ARX Jaslovské Bohunice

SEPS, a.s. Bratislava

IBOK

VUZ PI SR

TECHNODAT, CAE – systémy, spol. s r.o. Trenčín

Závody všeobecného strojárstva – holding, a. s. Dubnica n./Váhom

ZLIEVÁREŇ TRNAVA s.r.o.

RESEARCH IS WHAT I'M DOING WHEN I DON'T KNOW WHAT I'M DOING.

WERNHER VON BRAUN

- Acceptance without proof is the fundamental characteristic of Western religion, Rejection without proof is the fundamental characteristic of Western science. Gary Zukav, "The Dancing Wu Li Masters"
- Any sufficiently advanced technology is indistinguishable from magic. Arthur C. Clarke
- Basic research is like shooting an arrow into the air and, where it lands, painting a target. Homer Adkins
- Every revolutionary idea - in science, politics, art, or whatever - evokes three stages of reaction in a hearer: It is completely impossible - don't waste my time. It is possible, but it is not worth doing. I said it was a good idea all along.
- Every successful person has had failures but repeated failure is no guarantee of eventual success.
- I do not believe that it is possible to create living art out of anything but the direct visual experience of nature, combined with the heritage of a Tradition, unless it be by the practice of magic ritual. Michael Ayrton 1921-1975
- In theory, there is no difference between theory and practice. In practice, however, there is.
- Murphy's Law of Research: Enough research will tend to support your theory.
- Never worry about theory as long as the machinery does what it's supposed to do. R. A. Heinlein
- No committee could ever come up with anything as revolutionary as a camel anything as practical and as perfectly designed to perform effectively under such difficult conditions. Laurence J. Peter
- Not until a program has been in production for at least six months will the most harmful error then be discovered.
- Practical sciences proceed by building up; theoretical sciences by resolving into components.[Saint] Thomas Aquinas ca 1225-1274
- • Putt's Law: Technology is dominated by two types of people: Those who understand what they do not manage. Those who manage what they do not understand.
- • Rule of Creative Research: (1) Never draw what you can copy. (2) Never copy what you can trace. (3) Never trace what you can cut out and paste down.
- Science is the meeting place of two kinds of poetry: the poetry of thought and the poetry of action. George Agostinho da Silva
- Science is what happens when preconception meets verification.
- So, He was a great teacher. But what did he publish? From an Israeli academic
- The goal of Computer Science is to build something that will last at least until we've finished building it.
- The goal of science is to build better mousetraps. The goal of nature is to build better mice.
- The habit of apprehending a technology in its completeness-, this is the essence of technological humanism, and this is .what we should expect education in higher technology to achieve. I believe it could be achieved by making specialist studies the core around which are grouped liberal studies which are relevant to these specialist studies. But they must be relevant; the path to culture should be through a man's specialism, not by-passing it.... A student who can weave his technology into the fabric of society can claim to have a liberal education; a student who cannot weave bis technology into the fabric of society cannot claim even to be a good technologist [Lord] Eric Ashby 1904-
- Theft from a single author is plagiarism. Theft from two is comparative study. Theft from three or more is research.
- To get to know, to discover, to publish-this is the destiny of a scientist. Frangois Arago 1786-1853
- To know / To know what to do / To do it / To make it known. Savoir / Savoir faire / Faire Faire savoir. Department of Physics, Trinity College, Dublin
- Watson's Law: The reliability of machinery is inversely proportional to the number and significance of any persons watching it.
- What is a magician but a practicing theorist? Obi-Wan Kenobi
- Whenever anyone says, "theoretically," they really mean, "not really". Dave Parnas
- There is no national science just as there is no national multiplication tables; what is national is no longer science. Anton Chekhov (1860-1904)

AFTER ALL IS SAID AND DONE, A HELL LOT OF A LOT MORE IS SAID THAN DONE

DAAAM International Vienna
Karlsplatz 13/311, A-1040
Wien, Austria, European Union

ISBN 978-1-5108-4000-3

9 781510 840003

CURRAN ASSOCIATES INC.
proceedings.com